RAPHAEL'S 101-YEAR EPHEMERIS 1950–2050

RAPHAEL'S 101-YEAR EPHEMERIS 1950–2050

foulsham
LONDON • NEW YORK • TORONTO • SYDNEY

foulsham
The Publishing House, Bennetts Close, Cippenham, Slough, Berkshire, SL1 5AP, England

Foulsham books can be found in all good bookshops and direct from www.foulsham.com

ISBN: 978-0-572-03363-7

A CIP record for this book is available from the British Library

Black Moon and True Node calculations from Solar Fire v6, published by Astrolabe Inc at www.alabe.com

Printed in China through Colorcraft Ltd., Hong Kong

Contents

Raphael's Ephemeris 1950–2050

Raphael's Ephemeris has been a mainstay of the astrological community for over two hundred years, each year providing astronomical data to the best available precision recognised by the astronomers of that period. Throughout its illustrious history, Raphael's has maintained the standard of noon GMT so beloved of a large numbers of students, teachers and many schools of astrology. Various textbooks and schools' course notes have been written utilising *Raphael's Ephemeris* as their standard for teaching the basic calculations to the budding astrologer. This volume covers the comprehensive period of 1950 to 2050.

This is the most accurate ephemeris available at the time of going to press. Astronomical knowledge keeps moving on, not only in its understanding of the physical make-up of this solar system, but also in the precise calculation of celestial bodies. The traditional bodies up to Saturn have been reliably understood for a long while, although knowledge of the Moon's elements were significantly improved in the 1960s. The understanding of the calculations for the 'newer' planets – Uranus, Neptune and Pluto – have now been even further refined. Because these are slower-moving bodies, these refinements do have a significant effect in the timing of sign ingresses and slower aspects.

In the past, to calculate the planets' positions for the time of birth, students used *diurnal proportional logarithms* to simplify the calculation of the planets' positions for the exact time of birth. With logarithms, the written workings-out can appear simple and elegant but unless there is a regimen of careful double-checking of all the look-ups, there is a danger of making errors. The essential problem is that logarithms are not explicitly meaningful; they use 'magic numbers'. A more understandable tool – Raphael's Ready Reckoner – has now been provided in this book as an improved alternative for these calculations.

Technical Information

Basis

The positions of the Sun, Moon and planets have been derived from the standard Jet Propulsion Laboratory (JPL) ephemeris (DE406/LE406), with a standard epoch of 2000 January 1.5. The positions of the minor planets and comet Chiron have also been derived from the JPL data, using the Horizon System. The longitude of the Moon's true node and the Black Moon Lilith have been derived from figures obtained from Solar Fire.

All ephemerides were calculated at an interval of 0.5 days, and the times of all the aspects, parallels, stationary points and zodiacal sign entries were determined by an iterative procedure using inverse interpolation with sixth differences. All critical cases were examined in great detail to ensure that no aspects were omitted.

Time

In this publication, the time argument is Universal Time (UT), also referred to as Greenwich Mean Time (GMT). Ephemerides are calculated originally in Terrestrial Time (TT) and the difference TT-UT (*Delta T*) can only be ascertained by observation. Thus from 2000 onwards, an intelligent estimate has, of course, been used. In the immediate future any differences from our projections will be negligible. Even by 2050 it is unlikely that this difference will be more than about 30 seconds – for which the effect on the tabulated positions and upon the timing of aspects is insignificant.

Ephemeris

The apparent geocentric ecliptic longitudes and declinations of the major planets and the Sun are tabulated daily at 12h UT to a precision of 1 arcminute (except for the Sun's longitude which is to 1 arcsecond). The latitudes of the major planets are given to a precision of 1 arcminute at three-day intervals. (Our knowledge of the positions of these bodies is such that these ephemerides are accurate to a much higher degree.)

The longitude of the Moon is tabulated twice daily, at 12h and 24h UT, to a precision of 1 arcsecond. Its declination and latitude, together with the longitudes of the Moon's mean and true nodes, are tabulated daily to 1 arcminute.

The longitudes of the four earliest discovered asteroids (Ceres, Pallas, Juno and Vesta), Chiron and the Black Moon Lilith are tabulated at ten-day intervals from the first of each month, to 1 arcminute. For ease of interpolation and irrespective of the length of the calendar month, there is an entry for the '31st', which is ten days after the 21st (thus 31 February actually gives 3 March – or 2 March in a leap year).

The latitudes of the planets are given at three-day intervals for the 1st of each month to 1 arcminute. For ease of interpolation, there is an entry for the '31st' irrespective of the length of the calendar month – as described above.

The sidereal time is given in hours, minutes and seconds and is accurate to 1 second.

All the positions are apparent geocentric positions, i.e. they are referred to the apparent equinox, and as seen from the Earth's centre.

Aspectarian

The Aspectarian provides the times of all mutual aspects and planetary stations.

Aspects

The aspectarian gives the times of mutual aspects of the Sun, Moon and the planets in both longitude and declination. The list of aspects used also includes some aspects ignored by other publications. The full list is given below.

	Aspect	*Angle (in degrees)*
☌	conjunction	0
⚺	semi-sextile	30
⊥	semi-quintile	36
∠	semi-square	45
⚹	sextile	60
Q	quintile	72
□	square	90

	Aspect	*Angle (in degrees)*
△	trine	120
⚼	sesqui-quadrate	135
±	bi-quintile	144
⚻	quincunx	150
☍	opposition	180
∥	parallel	0 (dec)
⋕	contra-parallel	0 (dec)

Planetary stations

The letter in the last column (R or D) indicates whether the planet is at its first stationary point and beginning to retrograde (R), or at its second stationary point and resuming its direct motion (D).

During a planet's retrograde phase that column of figures are shaded for quick visual recognition.

Other Astrological Data

Moon's Phases and Longitudes

Times of the lunar phases are given, together with the Moon's longitude at that instant.

Eclipses

Eclipses are indicated as follows:

Annular	of the Sun
Ann-Total	Annular-Total of the Sun
Total	of the Sun (at New Moon)
total	of the Moon (at Full Moon)
Partial	of the Sun
partial	of the Moon

Moon's Apsides and Positions

Times of perigee and apogee, and also of the maximum and zero declinations, are tabulated to the nearest minute. These apsides are not usually given in other publications and some may find them worthy of study.

Black Moon Lilith

There are a variety of Liliths mentioned in astrological literature (including an asteroid and a hypothetical non-reflective earth satellite). The version used here is based on the apogee of the Moon's mean orbit.

Julian Date

The Julian Date is given for 12h on the first day of the month and is the astronomical / historical Julian Day Number (which is a count of all the days since BC 4713 January 1.5).

Delta T

The value of Delta T is given once each month. Values are estimated for the years after 2000 with the currently accepted projection for the future. In future years, any variance from that projected in this volume does not affect sidereal time; it only affects the 'planetary' co-ordinates (which will be insignificant).

Synetic Vernal Point

The Synetic Vernal Point is given for 12h on the first day of the month and has been calculated from 335.957955 – (precession + nutation from 1950.0) expressed in degrees.

Ayanamsa

The ayanamsa given is the Lahiri version which has been officially adopted by the Indian Government. This is given for 12h on the first day of the month and is calculated from 23.857072 + (precession + nutation from 2000.0) expressed in degrees.

True Obliquity

The true obliquity has been calculated from $84381.448 - 46.815T - 0.00059T^2 + 0.00183T^3 +$ nutation (expressed in arc-seconds) where T is measured in centuries from 2000 January 1.5 TT and is given for 12h on the first day of the month.

The Solar System – IAU 2006

In August 2006, the International Astronomical Union debated the definition of 'planet' at their conference in Prague.

IAU Resolution: Definition of a Planet in the Solar System

Contemporary observations are changing our understanding of planetary systems and it is important that our nomenclature for objects reflect our current understanding. This applies, in particular, to the designation 'planets'. The word 'planet' originally described 'wanderers' that were known only as moving lights in the sky. Recent discoveries lead us to create a new definition, which we can make using currently available scientific information.

The IAU therefore resolves that planets and other bodies in our Solar System be defined into three distinct categories in the following way:

1 *A **planet** is a celestial body that (a) is in orbit around the Sun, (b) has sufficient mass for its self-gravity to overcome rigid body forces so that it assumes a hydrostatic equilibrium (nearly round) shape, and (c) has cleared the neighbourhood around its orbit.*

2 *A **dwarf planet** is a celestial body that (a) is in orbit around the Sun, (b) has sufficient mass for its self-gravity to overcome rigid body forces so that it assumes a hydrostatic equilibrium (nearly round) shape, (c) has not cleared the neighbourhood around its orbit, and (d) is not a satellite.*

3 *All other objects orbiting the Sun shall be referred to collectively as '**Small Solar System Bodies**'.*

The IAU also resolved that:

Pluto is a dwarf planet by the above definition and is recognised as the prototype of a new category of trans-Neptunian objects.

The bottom line is that according to the IAU, there are now only eight planets (Mercury, Venus, Earth, Mars, Jupiter, Saturn, Uranus and Neptune). Pluto is **not** a planet (despite using the word 'planet') and is now deemed to be a 'dwarf planet'. The largest asteroid, Ceres, is also expected to be re-styled a dwarf planet in the future – and *many* other lumps of rock subsequently.

There remains controversy over the IAU's decision, about lack of thought and lack of democracy. Of the membership of nearly 9,000, only 30 per cent were able to contribute to the last-minute debate, and only 424 members remained at the conference to place their votes.

There may be panic in some quarters, but it is unlikely that Mickey Mouse's dog will be re-drawn as a chihuahua to match its new 'dwarf' status! Nor should astrologers be concerned about what this may mean to their astrological charts as there is no reason for it to have any practical impact. Astrology is an empirical subject, so that when experience suggests that Pluto had an impact on astrological charts, this experience cannot be discounted because of a change of name. (Astrologers may enjoy the synchronicity that on the afternoon of the vote, Mercury was closely applying to a trine of Pluto!)

Glossary of Symbols

♈	Aries	☉	Sun	⚷	Chiron	☌	conjunction (0°)
♉	Taurus	☽	Moon	⚳	Ceres	⚺	semi-sextile (30°)
♊	Gemini	☿	Mercury	⚴	Pallas	⊥	semi-quintile (36°)
♋	Cancer	♀	Venus	⚵	Juno	∠	semi-square (45°)
♌	Leo	♂	Mars	⚶	Vesta	⚹	sextile (60°)
♍	Virgo	♃	Jupiter	*Moon's phases*		Q	quintile (72°)
♎	Libra	♄	Saturn	●	New Moon	□	square (90°)
♏	Scorpio	♅	Uranus	☽	First quarter	△	trine (120°)
♐	Sagittarius	♆	Neptune	○	Full Moon	⚼	sesqui-quadrate (135°)
♑	Capricorn	♇	Pluto	☾	Last quarter	±	bi-quintile (144°)
♒	Aquarius	☊	Moon's node	॥	parallel	⚻	quincunx (150°)
♓	Pisces	⚸	Black Moon Lilith	⧺	contra-parallel	☍	opposition (180°)

Shading

In the ephemeris, all retrograde movement is highlighted by being shaded. This is a quick and helpful visual aid for the practising astrologer.

Additionally, to prevent misreading, the Moon's longitude at midnight is also shaded to distinguish it from the normal daily (noon) position.

Raphael's Ready Reckoner

Tables to assist in the interpolation of the motions of the Sun, Moon and planets are given on pages 21–30. Whilst the positions of all bodies are provided accurately, subsequent calculations by the astrologer to find the intermediate time of birth allow errors to accumulate. Raphael's Ready Reckoner (RRR) is based on the 24-hourly motions of the Sun and planets and, significantly, the **12-hourly motions of the Moon**. Previous generations have interpolated the Moon based on 24-hour motions; however, they unknowingly incurred errors of up to three minutes of arc – whether they used logarithms or calculators! This was because the Moon's motion could significantly accelerate or decelerate during a full day. When using half-daily positions this potential error is significantly reduced. Raphael's Ready Reckoner provides a decimal place higher precision for the Sun and planets to reduce accumulated rounding errors. However, the likely error for the Moon will now be reduced to a few seconds of arc – so no additional precision is provided in the intermediate positions. (If an astrologer wishes to calculate the Moon to the nearest second of arc, then they will have to resort to higher mathematics!) In essence, these tables give a greater sense of meaning inasmuch as you can see the proportions of the motion. Thus a quarter of a day (6 hours) finds the planet moving a quarter of the daily distance.

Instructions on their applications are provided on pages 16–20.

How to Use Raphael's Ready Reckoner

The RRR tables give, for each possible daily motion of the Sun and planets, a breakdown for how far it would move for each hour and number of minutes, so if the planet moved a single degree in 24 hours, then in half a day (12 hours) it will have moved half a degree (30'). However, most people have less convenient fractions of day upon which to do their calculations – the example birth data is typical in this respect.

Example data: **Queen Elizabeth II**
21 April 1926 at 2.40am BST
London (51N30 0W10)

The use of these tables is flexible and the few example calculations that follow should suffice to give an idea for their use. For greater clarity, extracts from the ephemeris are given to better illustrate their use.

APRIL 1926

Date	Sidereal time h m s	Sun ☉ ° ' "	Moon ☽ ° ' "	Moon ☽ 24.00 ° ' "	Mercury ☿ ° '	Venus ♀ ° '	Mars ♂ ° '	Jupiter ♃ ° '	Saturn ♄ ° '	Uranus ♅ ° '	Neptune ♆ ° '	Pluto ♇ ° '
19	01 47 20	28 40 27	23 ♋ 28 16	29 ♋ 26 42	03 52	12 25	19 44	22 16	24 33	27 17	22 03	12 41
20	01 51 17	29 ♈ 39 01	05 ♌ 23 05	11 ♌ 18 04	04 21	13 24	20 27	22 25	24 29	27 20	22 02	12 42
21	01 55 13	00 ♉ 37 33	17 ♌ 12 21	23 ♌ 06 36	04 55	14 23	21 11	22 35	24 25	27 23	22 02	12 43
22	01 59 10	01 36 03	29 ♌ 01 24	04 ♍ 57 23	05 32	15 23	21 54	22 44	24 21	27 26	22 02	12 43
23	02 03 06	02 34 31	10 ♍ 55 03	16 ♍ 54 55	06 13	16 23	22 37	22 53	24 17	27 29	22 01	12 44
24	02 07 03	03 32 56	22 ♍ 57 23	29 ♍ 02 49	06 58	17 23	23 21	23 01	24 13	27 32	22 01	12 44
25	02 10 59	04 31 20	05 ♎ 11 31	11 ♎ 23 41	07 46	18 24	24 04	23 10	24 09	27 34	22 01	12 45
26	02 14 56	05 29 42	17 ♎ 39 29	23 ♎ 58 58	08 38	19 25	24 48	23 19	24 05	27 37	22 00	12 46
27	02 18 53	06 28 01	00 ♏ 22 08	06 ♏ 48 57	09 33	20 27	25 31	23 27	24 01	27 40	22 00	12 47

Sun and Planetary Calculations

Given that the birth falls between noon on the 20th and noon on the 21st of April we need to first find the daily motion of each planet.

1 Find the daily motion

The *daily motion* is found by subtracting the smaller from the larger.

	Sun	*Mercury*	*Venus*	
21st	0♉37'33"	4♈55	14♓23	
20th	29♈39'01"	4♈21	13♓24	*subtract*
Motion	58'32"	0º 34'	0º 59'	

Note that the Sun has changed sign during the day and thus the calculation is:

30:37:33 – 29:39:01 = 58:32

2 Find the actual distance moved to the time of birth

Our ephemeris is in UT and therefore the given time of 2.40am BST is translated to 1.40am UT. This time is therefore 13 hours and 40 minutes from the previous noon.

Let us look at those daily motions for the Sun and Mercury respectively. We need to find how far each would have moved in 13 hours, and also 40 minutes.

	24h	*58'33"*	*0º 34'*	
	13h	31 42.9	0 18.4	
	40m	1 37.6	0 00.9	*add*
thus	13h 40m	33 20.5	0 19.3	

For Venus, the motion falls between 0º 58' and 1º 00" in the RRR. The solution is not rocket science – just take the values half way between the columns.

	24h	*0º 59'*	
	13h	0 31.9	
	40m	0 01.5	*add*
thus	13h 40m	0 33.4	

Thus the natal positions are:

☉	0♉12'21"	(29♈39' 01" + 33'20")
☿	4♈40	(4♈21 + 0º 19')
♀	13♓57	(13♓24 + 0º 33')

Retrograde planets

When a planet's motion is retrograde, you still take the smaller from the larger, and you still add the motions for 13h and for 40m together, but the resultant figure would now be **subtracted** from the previous noon's position.

'Awkward' birth times

If Queen Elizabeth had been born three minutes later (1.43am), then for each daily motion you would take values for 13h, 40m *and* 3m.

Moon Calculation

The calculations for the Moon are similar except that we are basing it on half-daily positions and motions instead of daily ones.

1 Find the half-daily motion

Find the two twelve-hour positions that *straddle* the birth time. Then subtract the earlier position from the later (remembering to take account of any sign change).

noon on 21st	17♌12' 21"	
midnight on 20th	11♌18' 04"	*subtract*
12h motion	5° 54' 17"	

2 Find the actual distance moved to the time of birth

In the tables, this motion falls between the columns headed 5:54 and 5:55 – it is enough to make rough adjustments to account for the 17" (three-tenths between).

	12h	*5° 54'(17")*	
	1h	0° 29' 32"	
	40m	0° 19' 41"	*add*
thus	1h 40m	0° 49' 13"	

Thus the natal position is

☽ 12♌07'17" (11♌18'04" + 49'13")

Ten-daily Calculations

The longitudes of the four primary asteroids (Ceres, Pallas, Juno and Vesta) as well as Chiron are given for ten-day intervals commencing from the 1st of each month (including an entry for the 31st which is ten days after the 21st – irrespective of the length of the month). An additional table has been added which aids interpolation for these too.

The ten-daily tables only give results to the whole day, but the first entry is for a single day, and this figure can be utilised with the more comprehensive daily planetary tables. The tables are quite flexible, so where figures are given every ***three days***, you can start looking at ten-daily tables using the line marked for three days and continue in a similar manner as for the ten-daily routine. However, the distances involved in the planetary latitudes are relatively small and most may find this unnecessary.

So – first find the ten-day motion and then, referring to the tables, you can find out how far it can move for each whole number of days. Irrespective of the number of whole days, extract the movement for a single day. You use that distance together with the normal daily planetary movement table, when you are then able to find the movement for the appropriate number of hours and minutes.

LONGITUDES

Date	Chiron ⚷ ° '	Ceres ⚳ ° '	Pallas ⚴ ° '	Juno ⚵ ° '	Vesta ⚶ ° '	Black Moon Lilith ⚸ ° '
01	26 ♈ 50	13 ♌ 00	01 ♋ 37	19 ♒ 42	28 ♊ 25	22 ♌ 15
11	27 ♈ 26	13 ♌ 41	06 ♋ 16	23 ♒ 19	01 ♋ 27	23 ♌ 22
21	28 ♈ 03	15 ♌ 01	11 ♋ 04	26 ♒ 49	04 ♋ 47	24 ♌ 29
31	28 ♈ 39	16 ♌ 53	15 ♋ 58	00 ♓ 09	08 ♋ 21	25 ♌ 35

An example may make this clearer. The position of Vesta in the example chart is calculated like this:

1 Find the ten-daily motion

The ten-daily motion is found by subtracting the smaller from the larger.

	Vesta	
21st	4♋47	
11th	1♋27	*subtract*
10-day motion	3° 20'	

2 Find the actual distance moved to the time of birth

In our example, the consecutive entries in the ephemeris preceding entry are on 11 and 21 April. The time difference is therefore 9 days, 13 hours and 40 minutes from the previous entry.

Take the entry for nine days from the ten-daily table. Note that a single day's movement is 20', so we use that figure in the normal daily planetary table to get the 13h and the 40 minutes.

	10 d	***3° 20'***	(and from that table ***1d is 20'***)
	9d	3 00.0	
	13h	0 10.8	
	40m	0 0.6	*add*
thus	9 d 13h 40m	3 11.4	

So the natal position of Vesta is 4♋38 (1♋27 + 3° 11.4')

An alternative route for the more confident

In our example, it may be recognised that the birth time is actually only a few hours before the second date and sometimes you may prefer to go *backwards*.

To show this – it is already known that the ten-daily motion is 3° 20' and that consequently the daily motion is 20'. Count from the following noon backwards to the birth time and you find a time difference of (0 days,) 10 hours 20 minutes.

Referring only to the daily table:

	24h	***0° 20'***	
	10h	0 08.3	
	20m	0 00.3	*add*
thus	10h 20m	0 08.6	

So Vesta is again found to be is 4♋38 (4♋47 – 0° 08.6' subtracting this time)

Accuracy

Positions provided in the ephemeris are correct to the provided accuracy. It is only when the astrologer interpolates between given positions that inaccuracies arise – whether using calculators, logarithms or the RRR. In practice, an astrologer can expect to achieve the Sun's position correct to within a second, planetary positions to within a minute, and the Moon to within a few seconds. (It is possible to get higher accuracy for the Moon – but not without extra tedious maths!) For the ten-daily positions, the error could be a minute or two (or three) of arc, but this is only when the body is noticeably accelerating/decelerating.

In using the RRR where the motions calculated do not have an explicit column in the tables, it may have seemed to be a cavalier attitude to take such 'rough approximations' as given in the examples above – but this is not pragmatically so. The resultant positions for the planets will usually not have any change in resultant value. If they do, then it is within the bounds of *rounding errors*.

To minimise errors in interpolation (which can be caused by adding together *rounded* intermediate results) the precision of the tables have been calculated for a further decimal position for the Sun and the planets. However, for the Moon, intermediate positions are merely given to the nearest second only as any further precision would give a false sense of security. The problem of the Moon's accelerating or decelerating movement is much reduced by basing calculations on the half-daily positions, but they nevertheless have errors amounting to a few seconds of arc. It is important to keep the achievable accuracy into perspective – there is very little practical astrology that requires this hyper-precision.

An example of when this precision may be of significance could be when a body changes sign – it takes a fraction of a moment to be in one sign and then be in the next. If you are in this circumstance, then look to the aspectarian to find the time of the sign change. However, if you were born at 5.55am and the Sun is stated to change signs at 5.55am then deciding which sign your Sun resides in is not an ephemeris problem, it is an astrological one.

Reversing the Process (or Solar and Lunar Returns)

These tables also allow you turn their use upside-down by being able to compute the time of a Solar or Lunar Return.

Queen Elizabeth II had her coronation on 2 June 1953.

By looking at the ephemeris, the natal Sun is found to return to its natal place between 20 and 21 April 1953.

APRIL 1953

Date	Sidereal time (h m s)	LONGITUDES Sun ☉ (° ' ")	Moon ☽ (° ' ")	Moon ☽ 24.00 (° ' ")	Mercury ☿ (° ')	Venus ♀ (° ')	Mars ♂ (° ')	Jupiter ♃ (° ')	Saturn ♄ (° ')	Uranus ♅ (° ')	Neptune ♆ (° ')	Pluto ♇ (° ')
20	01 53 06	00 ♉ 06 17	24 ♋ 11 08	00 ♌ 16 45	03 13	18 51	22 27	25 37	23 29	14 48	22 19	20 50
21	01 57 02	01 04 50	06 ♌ 18 43	12 ♌ 17 42	04 26	18 21	23 09	25 50	23 25	14 49	22 17	20 50

1 Find the distance from the previous noon to the natal position

Subtract the previous day's Sun position from the natal Sun

birth Sun	0♉12'21"	
Sun at previous noon	0♉06'17"	*subtract*
distance to solar return	0° 06' 04"	

2 Find the daily motion

Sun on 21st	1♉04'50"	
Sun on 20th	0♉06'17"	*subtract*
daily motion	58' 33"	

3 Deduce the time of the return

Refer to the RRR for the daily motion of 58'33" and gradually subtract from the 6'04" until there is nothing left. (Start with the highest number of hours which is less than the target amount and subtract it, then the highest number of minutes etc.)

Under the column for 58'33", the highest number of hours is two:

2h = 4' 52.8" thus 6'04" – 4'52.8" = 1'11.2"

Then continue with the minutes:

25m = 1'01.0" now 1'11.2" – 1'01.0" = 10.2"

Then the individual minutes:

4m = 9.8" and 10.2" – 9.8" = 0.4"

And the remaining 0.4" equates to about 10 seconds.

Thus the solar return can be calculated for **2h 29m10s am GMT on the 20th April 1953.**

Lunar returns follow the same principle – try it.

JUNE 1953

Date	Sidereal time h m s	LONGITUDES Sun ☉ ° ' "	Moon ☽ ° ' "	Moon ☽ 24.00 ° ' "	Mercury ☿ ° '	Venus ♀ ° '	Mars ♂ ° '	Jupiter ♃ ° '	Saturn ♄ ° '	Uranus ♅ ° '	Neptune ♆ ° '	Pluto ♇ ° '
15	05 33 53	24 04 25	10 ♌ 03 44	16 ♌ 09 31	15 52	08 34	00 54	08 37	20 36	17 14	21 11	21 14
16	05 37 49	25 01 44	22 ♌ 12 09	28 ♌ 12 10	17 23	09 28	01 34	08 50	20 36	17 17	21 11	21 15

The answer for the time of the first lunar return after her coronation is . . .

15 June 1953 at 4h 02m 30s pm GMT

which casts a chart with Pluto on the Midheaven.

⊙ ⊙

Table 1 – Solar Movements

24h	57'12"	57'15"	57'18"	57'21"	57'24"	57'27"	57'30"	57'33"	57'36"	57'39"	57'42"	57'45"	57'48"	57'51"	57'54"	24h
1m	0 02.4	0 02.4	0 02.4	0 02.4	0 02.4	0 02.4	0 02.4	0 02.4	0 02.4	0 02.4	0 02.4	0 02.4	0 02.4	0 02.4	0 02.4	1m
2m	0 04.8	0 04.8	0 04.8	0 04.8	0 04.8	0 04.8	0 04.8	0 04.8	0 04.8	0 04.8	0 04.8	0 04.8	0 04.8	0 04.8	0 04.8	2m
3m	0 07.2	0 07.2	0 07.2	0 07.2	0 07.2	0 07.2	0 07.2	0 07.2	0 07.2	0 07.2	0 07.2	0 07.2	0 07.2	0 07.2	0 07.2	3m
4m	0 09.5	0 09.5	0 09.6	0 09.6	0 09.6	0 09.6	0 09.6	0 09.6	0 09.6	0 09.6	0 09.6	0 09.6	0 09.6	0 09.6	0 09.7	4m
5m	0 11.9	0 11.9	0 11.9	0 11.9	0 12.0	0 12.0	0 12.0	0 12.0	0 12.0	0 12.0	0 12.0	0 12.0	0 12.0	0 12.1	0 12.1	5m
10m	0 23.8	0 23.9	0 23.9	0 23.9	0 23.9	0 23.9	0 24.0	0 24.0	0 24.0	0 24.0	0 24.0	0 24.1	0 24.1	0 24.1	0 24.1	10m
15m	0 35.8	0 35.8	0 35.8	0 35.8	0 35.9	0 35.9	0 35.9	0 36.0	0 36.0	0 36.0	0 36.1	0 36.1	0 36.1	0 36.2	0 36.2	15m
20m	0 47.7	0 47.7	0 47.8	0 47.8	0 47.8	0 47.9	0 47.9	0 48.0	0 48.0	0 48.0	0 48.1	0 48.1	0 48.2	0 48.2	0 48.3	20m
25m	0 59.6	0 59.6	0 59.7	0 59.7	0 59.8	0 59.8	0 59.9	0 59.9	1 00.0	1 00.1	1 00.1	1 00.2	1 00.2	1 00.3	1 00.3	25m
30m	1 11.5	1 11.6	1 11.6	1 11.7	1 11.8	1 11.8	1 11.9	1 11.9	1 12.0	1 12.1	1 12.1	1 12.2	1 12.3	1 12.3	1 12.4	30m
35m	1 23.4	1 23.5	1 23.6	1 23.6	1 23.7	1 23.8	1 23.9	1 23.9	1 24.0	1 24.1	1 24.1	1 24.2	1 24.3	1 24.4	1 24.4	35m
40m	1 35.3	1 35.4	1 35.5	1 35.6	1 35.7	1 35.8	1 35.8	1 35.9	1 36.0	1 36.1	1 36.2	1 36.3	1 36.3	1 36.4	1 36.5	40m
45m	1 47.3	1 47.3	1 47.4	1 47.5	1 47.6	1 47.7	1 47.8	1 47.9	1 48.0	1 48.1	1 48.2	1 48.3	1 48.4	1 48.5	1 48.6	45m
50m	1 59.2	1 59.3	1 59.4	1 59.5	1 59.6	1 59.7	1 59.8	1 59.9	2 00.0	2 00.1	2 00.2	2 00.3	2 00.4	2 00.5	2 00.6	50m
55m	2 11.1	2 11.2	2 11.3	2 11.4	2 11.5	2 11.7	2 11.8	2 11.9	2 12.0	2 12.1	2 12.2	2 12.3	2 12.5	2 12.6	2 12.7	55m
1h	2 23.0	2 23.1	2 23.3	2 23.4	2 23.5	2 23.6	2 23.8	2 23.9	2 24.0	2 24.1	2 24.3	2 24.4	2 24.5	2 24.6	2 24.8	1h
2h	4 46.0	4 46.3	4 46.5	4 46.8	4 47.0	4 47.3	4 47.5	4 47.8	4 48.0	4 48.3	4 48.5	4 48.8	4 49.0	4 49.3	4 49.5	2h
3h	7 09.0	7 09.4	7 09.8	7 10.1	7 10.5	7 10.9	7 11.3	7 11.6	7 12.0	7 12.4	7 12.8	7 13.1	7 13.5	7 13.9	7 14.3	3h
4h	9 32.0	9 32.5	9 33.0	9 33.5	9 34.0	9 34.5	9 35.0	9 35.5	9 36.0	9 36.5	9 37.0	9 37.5	9 38.0	9 38.5	9 39.0	4h
5h	11 55.0	11 55.6	11 56.3	11 56.9	11 57.5	11 58.1	11 58.8	11 59.4	12 00.0	12 00.6	12 01.3	12 01.9	12 02.5	12 03.1	12 03.8	5h
6h	14 18.0	14 18.8	14 19.5	14 20.3	14 21.0	14 21.8	14 22.5	14 23.3	14 24.0	14 24.8	14 25.5	14 26.3	14 27.0	14 27.8	14 28.5	6h
7h	16 41.0	16 41.9	16 42.8	16 43.6	16 44.5	16 45.4	16 46.3	16 47.1	16 48.0	16 48.9	16 49.8	16 50.6	16 51.5	16 52.4	16 53.3	7h
8h	19 04.0	19 05.0	19 06.0	19 07.0	19 08.0	19 09.0	19 10.0	19 11.0	19 12.0	19 13.0	19 14.0	19 15.0	19 16.0	19 17.0	19 18.0	8h
9h	21 27.0	21 28.1	21 29.3	21 30.4	21 31.5	21 32.6	21 33.8	21 34.9	21 36.0	21 37.1	21 38.3	21 39.4	21 40.5	21 41.6	21 42.8	9h
10h	23 50.0	23 51.3	23 52.5	23 53.8	23 55.0	23 56.3	23 57.5	23 58.8	24 00.0	24 01.3	24 02.5	24 03.8	24 05.0	24 06.3	24 07.5	10h
11h	26 13.0	26 14.4	26 15.8	26 17.1	26 18.5	26 19.9	26 21.3	26 22.6	26 24.0	26 25.4	26 26.8	26 28.1	26 29.5	26 30.9	26 32.3	11h
12h	28 36.0	28 37.5	28 39.0	28 40.5	28 42.0	28 43.5	28 45.0	28 46.5	28 48.0	28 49.5	28 51.0	28 52.5	28 54.0	28 55.5	28 57.0	12h
13h	30 59.0	31 00.6	31 02.3	31 03.9	31 05.5	31 07.1	31 08.8	31 10.4	31 12.0	31 13.6	31 15.3	31 16.9	31 18.5	31 20.1	31 21.8	13h
14h	33 22.0	33 23.8	33 25.5	33 27.3	33 29.0	33 30.8	33 32.5	33 34.3	33 36.0	33 37.8	33 39.5	33 41.3	33 43.0	33 44.8	33 46.5	14h
15h	35 45.0	35 46.9	35 48.8	35 50.6	35 52.5	35 54.4	35 56.3	35 58.1	36 00.0	36 01.9	36 03.8	36 05.6	36 07.5	36 09.4	36 11.3	15h
16h	38 08.0	38 10.0	38 12.0	38 14.0	38 16.0	38 18.0	38 20.0	38 22.0	38 24.0	38 26.0	38 28.0	38 30.0	38 32.0	38 34.0	38 36.0	16h
17h	40 31.0	40 33.1	40 35.3	40 37.4	40 39.5	40 41.6	40 43.8	40 45.9	40 48.0	40 50.1	40 52.3	40 54.4	40 56.5	40 58.6	41 00.8	17h
18h	42 54.0	42 56.3	42 58.5	43 00.8	43 03.0	43 05.3	43 07.5	43 09.8	43 12.0	43 14.3	43 16.5	43 18.8	43 21.0	43 23.3	43 25.5	18h
19h	45 17.0	45 19.4	45 21.8	45 24.1	45 26.5	45 28.9	45 31.3	45 33.6	45 36.0	45 38.4	45 40.8	45 43.1	45 45.5	45 47.9	45 50.3	19h
20h	47 40.0	47 42.5	47 45.0	47 47.5	47 50.0	47 52.5	47 55.0	47 57.5	48 00.0	48 02.5	48 05.0	48 07.5	48 10.0	48 12.5	48 15.0	20h
21h	50 03.0	50 05.6	50 08.3	50 10.9	50 13.5	50 16.1	50 18.8	50 21.4	50 24.0	50 26.6	50 29.3	50 31.9	50 34.5	50 37.1	50 39.8	21h
22h	52 26.0	52 28.8	52 31.5	52 34.3	52 37.0	52 39.8	52 42.5	52 45.3	52 48.0	52 50.8	52 53.5	52 56.3	52 59.0	53 01.8	53 04.5	22h
23h	54 49.0	54 51.9	54 54.8	54 57.6	55 00.5	55 03.4	55 06.3	55 09.1	55 12.0	55 14.9	55 17.8	55 20.6	55 23.5	55 26.4	55 29.3	23h
24h	57'12"	57'15"	57'18"	57'21"	57'24"	57'27"	57'30"	57'33"	57'36"	57'39"	57'42"	57'45"	57'48"	57'51"	57'54"	24h

24h	57'54"	57'57"	58'00"	58'03"	58'06"	58'09"	58'12"	58'15"	58'18"	58'21"	58'24"	58'27"	58'30"	58'33"	58'36"	24h
1m	0 02.4	0 02.4	0 02.4	0 02.4	0 02.4	0 02.4	0 02.4	0 02.4	0 02.4	0 02.4	0 02.4	0 02.4	0 02.4	0 02.4	0 02.4	1m
2m	0 04.8	0 04.8	0 04.8	0 04.8	0 04.8	0 04.8	0 04.9	0 04.9	0 04.9	0 04.9	0 04.9	0 04.9	0 04.9	0 04.9	0 04.9	2m
3m	0 07.2	0 07.2	0 07.3	0 07.3	0 07.3	0 07.3	0 07.3	0 07.3	0 07.3	0 07.3	0 07.3	0 07.3	0 07.3	0 07.3	0 07.3	3m
4m	0 09.7	0 09.7	0 09.7	0 09.7	0 09.7	0 09.7	0 09.7	0 09.7	0 09.7	0 09.7	0 09.7	0 09.7	0 09.8	0 09.8	0 09.8	4m
5m	0 12.1	0 12.1	0 12.1	0 12.1	0 12.1	0 12.1	0 12.1	0 12.1	0 12.1	0 12.2	0 12.2	0 12.2	0 12.2	0 12.2	0 12.2	5m
10m	0 24.1	0 24.1	0 24.2	0 24.2	0 24.2	0 24.2	0 24.3	0 24.3	0 24.3	0 24.3	0 24.3	0 24.4	0 24.4	0 24.4	0 24.4	10m
15m	0 36.2	0 36.2	0 36.3	0 36.3	0 36.3	0 36.3	0 36.4	0 36.4	0 36.4	0 36.5	0 36.5	0 36.5	0 36.6	0 36.6	0 36.6	15m
20m	0 48.3	0 48.3	0 48.3	0 48.4	0 48.4	0 48.5	0 48.5	0 48.5	0 48.6	0 48.6	0 48.7	0 48.7	0 48.8	0 48.8	0 48.8	20m
25m	1 00.3	1 00.4	1 00.4	1 00.5	1 00.5	1 00.6	1 00.6	1 00.7	1 00.7	1 00.8	1 00.8	1 00.9	1 00.9	1 01.0	1 01.0	25m
30m	1 12.4	1 12.4	1 12.5	1 12.6	1 12.6	1 12.7	1 12.8	1 12.8	1 12.9	1 12.9	1 13.0	1 13.1	1 13.1	1 13.2	1 13.3	30m
35m	1 24.4	1 24.5	1 24.6	1 24.7	1 24.7	1 24.8	1 24.9	1 24.9	1 25.0	1 25.1	1 25.2	1 25.2	1 25.3	1 25.4	1 25.5	35m
40m	1 36.5	1 36.6	1 36.7	1 36.8	1 36.8	1 36.9	1 37.0	1 37.1	1 37.2	1 37.3	1 37.3	1 37.4	1 37.5	1 37.6	1 37.7	40m
45m	1 48.6	1 48.7	1 48.8	1 48.8	1 48.9	1 49.0	1 49.1	1 49.2	1 49.3	1 49.4	1 49.5	1 49.6	1 49.7	1 49.8	1 49.9	45m
50m	2 00.6	2 00.7	2 00.8	2 00.9	2 01.0	2 01.1	2 01.3	2 01.4	2 01.5	2 01.6	2 01.7	2 01.8	2 01.9	2 02.0	2 02.1	50m
55m	2 12.7	2 12.8	2 12.9	2 13.0	2 13.1	2 13.3	2 13.4	2 13.5	2 13.6	2 13.7	2 13.8	2 13.9	2 14.1	2 14.2	2 14.3	55m
1h	2 24.8	2 24.9	2 25.0	2 25.1	2 25.3	2 25.4	2 25.5	2 25.6	2 25.8	2 25.9	2 26.0	2 26.1	2 26.3	2 26.4	2 26.5	1h
2h	4 49.5	4 49.8	4 50.0	4 50.3	4 50.5	4 50.8	4 51.0	4 51.3	4 51.5	4 51.8	4 52.0	4 52.3	4 52.5	4 52.8	4 53.0	2h
3h	7 14.3	7 14.6	7 15.0	7 15.4	7 15.8	7 16.1	7 16.5	7 16.9	7 17.3	7 17.6	7 18.0	7 18.4	7 18.8	7 19.1	7 19.5	3h
4h	9 39.0	9 39.5	9 40.0	9 40.5	9 41.0	9 41.5	9 42.0	9 42.5	9 43.0	9 43.5	9 44.0	9 44.5	9 45.0	9 45.5	9 46.0	4h
5h	12 03.8	12 04.4	12 05.0	12 05.6	12 06.3	12 06.9	12 07.5	12 08.1	12 08.8	12 09.4	12 10.0	12 10.6	12 11.3	12 11.9	12 12.5	5h
6h	14 28.5	14 29.3	14 30.0	14 30.8	14 31.5	14 32.3	14 33.0	14 33.8	14 34.5	14 35.3	14 36.0	14 36.8	14 37.5	14 38.3	14 39.0	6h
7h	16 53.3	16 54.1	16 55.0	16 55.9	16 56.8	16 57.6	16 58.5	16 59.4	17 00.3	17 01.1	17 02.0	17 02.9	17 03.8	17 04.6	17 05.5	7h
8h	19 18.0	19 19.0	19 20.0	19 21.0	19 22.0	19 23.0	19 24.0	19 25.0	19 26.0	19 27.0	19 28.0	19 29.0	19 30.0	19 31.0	19 32.0	8h
9h	21 42.8	21 43.9	21 45.0	21 46.1	21 47.3	21 48.4	21 49.5	21 50.6	21 51.8	21 52.9	21 54.0	21 55.1	21 56.3	21 57.4	21 58.5	9h
10h	24 07.5	24 08.8	24 10.0	24 11.3	24 12.5	24 13.8	24 15.0	24 16.3	24 17.5	24 18.8	24 20.0	24 21.3	24 22.5	24 23.8	24 25.0	10h
11h	26 32.3	26 33.6	26 35.0	26 36.4	26 37.8	26 39.1	26 40.5	26 41.9	26 43.3	26 44.6	26 46.0	26 47.4	26 48.8	26 50.1	26 51.5	11h
12h	28 57.0	28 58.5	29 00.0	29 01.5	29 03.0	29 04.5	29 06.0	29 07.5	29 09.0	29 10.5	29 12.0	29 13.5	29 15.0	29 16.5	29 18.0	12h
13h	31 21.8	31 23.4	31 25.0	31 26.6	31 28.3	31 29.9	31 31.5	31 33.1	31 34.8	31 36.4	31 38.0	31 39.6	31 41.3	31 42.9	31 44.5	13h
14h	33 46.5	33 48.3	33 50.0	33 51.8	33 53.5	33 55.3	33 57.0	33 58.8	34 00.5	34 02.3	34 04.0	34 05.8	34 07.5	34 09.3	34 11.0	14h
15h	36 11.3	36 13.1	36 15.0	36 16.9	36 18.8	36 20.6	36 22.5	36 24.4	36 26.3	36 28.1	36 30.0	36 31.9	36 33.8	36 35.6	36 37.5	15h
16h	38 36.0	38 38.0	38 40.0	38 42.0	38 44.0	38 46.0	38 48.0	38 50.0	38 52.0	38 54.0	38 56.0	38 58.0	39 00.0	39 02.0	39 04.0	16h
17h	41 00.8	41 02.9	41 05.0	41 07.1	41 09.3	41 11.4	41 13.5	41 15.6	41 17.8	41 19.9	41 22.0	41 24.1	41 26.3	41 28.4	41 30.5	17h
18h	43 25.5	43 27.8	43 30.0	43 32.3	43 34.5	43 36.8	43 39.0	43 41.3	43 43.5	43 45.8	43 48.0	43 50.3	43 52.5	43 54.8	43 57.0	18h
19h	45 50.3	45 52.6	45 55.0	45 57.4	45 59.8	46 02.1	46 04.5	46 06.9	46 09.3	46 11.6	46 14.0	46 16.4	46 18.8	46 21.1	46 23.5	19h
20h	48 15.0	48 17.5	48 20.0	48 22.5	48 25.0	48 27.5	48 30.0	48 32.5	48 35.0	48 37.5	48 40.0	48 42.5	48 45.0	48 47.5	48 50.0	20h
21h	50 39.8	50 42.4	50 45.0	50 47.6	50 50.3	50 52.9	50 55.5	50 58.1	51 00.8	51 03.4	51 06.0	51 08.6	51 11.3	51 13.9	51 16.5	21h
22h	53 04.5	53 07.3	53 10.0	53 12.8	53 15.5	53 18.3	53 21.0	53 23.8	53 26.5	53 29.3	53 32.0	53 34.8	53 37.5	53 40.3	53 43.0	22h
23h	55 29.3	55 32.1	55 35.0	55 37.9	55 40.8	55 43.6	55 46.5	55 49.4	55 52.3	55 55.1	55 58.0	56 00.9	56 03.8	56 06.6	56 09.5	23h
24h	57'54"	57'57"	58'00"	58'03"	58'06"	58'09"	58'12"	58'15"	58'18"	58'21"	58'24"	58'27"	58'30"	58'33"	58'36"	24h

☉ ☉

24h	58'36"	58'39"	58'42"	58'45"	58'48"	58'51"	58'54"	58'57"	59'00"	59'03"	59'06"	59'09"	59'12"	59'15"	59'18"	24h
1m	0 02.4	0 02.4	0 02.4	0 02.4	0 02.5	0 02.5	0 02.5	0 02.5	0 02.5	0 02.5	0 02.5	0 02.5	0 02.5	0 02.5	0 02.5	1m
2m	0 04.9	0 04.9	0 04.9	0 04.9	0 04.9	0 04.9	0 04.9	0 04.9	0 04.9	0 04.9	0 04.9	0 04.9	0 04.9	0 04.9	0 04.9	2m
3m	0 07.3	0 07.3	0 07.3	0 07.3	0 07.4	0 07.4	0 07.4	0 07.4	0 07.4	0 07.4	0 07.4	0 07.4	0 07.4	0 07.4	0 07.4	3m
4m	0 09.8	0 09.8	0 09.8	0 09.8	0 09.8	0 09.8	0 09.8	0 09.8	0 09.8	0 09.8	0 09.9	0 09.9	0 09.9	0 09.9	0 09.9	4m
5m	0 12.2	0 12.2	0 12.2	0 12.2	0 12.3	0 12.3	0 12.3	0 12.3	0 12.3	0 12.3	0 12.3	0 12.3	0 12.3	0 12.3	0 12.4	5m
10m	0 24.4	0 24.4	0 24.5	0 24.5	0 24.5	0 24.5	0 24.5	0 24.6	0 24.6	0 24.6	0 24.6	0 24.6	0 24.7	0 24.7	0 24.7	10m
15m	0 36.6	0 36.7	0 36.7	0 36.7	0 36.8	0 36.8	0 36.8	0 36.8	0 36.9	0 36.9	0 36.9	0 37.0	0 37.0	0 37.0	0 37.1	15m
20m	0 48.8	0 48.9	0 48.9	0 49.0	0 49.0	0 49.0	0 49.1	0 49.1	0 49.2	0 49.2	0 49.3	0 49.3	0 49.3	0 49.4	0 49.4	20m
25m	1 01.0	1 01.1	1 01.1	1 01.2	1 01.3	1 01.3	1 01.4	1 01.4	1 01.5	1 01.5	1 01.6	1 01.6	1 01.7	1 01.7	1 01.8	25m
30m	1 13.3	1 13.3	1 13.4	1 13.4	1 13.5	1 13.6	1 13.6	1 13.7	1 13.8	1 13.8	1 13.9	1 13.9	1 14.0	1 14.1	1 14.1	30m
35m	1 25.5	1 25.5	1 25.6	1 25.7	1 25.8	1 25.8	1 25.9	1 26.0	1 26.0	1 26.1	1 26.2	1 26.3	1 26.3	1 26.4	1 26.5	35m
40m	1 37.7	1 37.8	1 37.8	1 37.9	1 38.0	1 38.1	1 38.2	1 38.3	1 38.3	1 38.4	1 38.5	1 38.6	1 38.7	1 38.8	1 38.8	40m
45m	1 49.9	1 50.0	1 50.1	1 50.2	1 50.3	1 50.3	1 50.4	1 50.5	1 50.6	1 50.7	1 50.8	1 50.9	1 51.0	1 51.1	1 51.2	45m
50m	2 02.1	2 02.2	2 02.3	2 02.4	2 02.5	2 02.6	2 02.7	2 02.8	2 02.9	2 03.0	2 03.1	2 03.2	2 03.3	2 03.4	2 03.5	50m
55m	2 14.3	2 14.4	2 14.5	2 14.6	2 14.8	2 14.9	2 15.0	2 15.1	2 15.2	2 15.3	2 15.4	2 15.6	2 15.7	2 15.8	2 15.9	55m
1h	2 26.5	2 26.6	2 26.8	2 26.9	2 27.0	2 27.1	2 27.3	2 27.4	2 27.5	2 27.6	2 27.8	2 27.9	2 28.0	2 28.1	2 28.3	1h
2h	4 53.0	4 53.3	4 53.5	4 53.8	4 54.0	4 54.3	4 54.5	4 54.8	4 55.0	4 55.3	4 55.5	4 55.8	4 56.0	4 56.3	4 56.5	2h
3h	7 19.5	7 19.9	7 20.3	7 20.6	7 21.0	7 21.4	7 21.8	7 22.1	7 22.5	7 22.9	7 23.3	7 23.6	7 24.0	7 24.4	7 24.8	3h
4h	9 46.0	9 46.5	9 47.0	9 47.5	9 48.0	9 48.5	9 49.0	9 49.5	9 50.0	9 50.5	9 51.0	9 51.5	9 52.0	9 52.5	9 53.0	4h
5h	12 12.5	12 13.1	12 13.8	12 14.4	12 15.0	12 15.6	12 16.3	12 16.9	12 17.5	12 18.1	12 18.8	12 19.4	12 20.0	12 20.6	12 21.3	5h
6h	14 39.0	14 39.8	14 40.5	14 41.3	14 42.0	14 42.8	14 43.5	14 44.3	14 45.0	14 45.8	14 46.5	14 47.3	14 48.0	14 48.8	14 49.5	6h
7h	17 05.5	17 06.4	17 07.3	17 08.1	17 09.0	17 09.9	17 10.8	17 11.6	17 12.5	17 13.4	17 14.3	17 15.1	17 16.0	17 16.9	17 17.8	7h
8h	19 32.0	19 33.0	19 34.0	19 35.0	19 36.0	19 37.0	19 38.0	19 39.0	19 40.0	19 41.0	19 42.0	19 43.0	19 44.0	19 45.0	19 46.0	8h
9h	21 58.5	21 59.6	22 00.8	22 01.9	22 03.0	22 04.1	22 05.3	22 06.4	22 07.5	22 08.6	22 09.8	22 10.9	22 12.0	22 13.1	22 14.3	9h
10h	24 25.0	24 26.3	24 27.5	24 28.8	24 30.0	24 31.3	24 32.5	24 33.8	24 35.0	24 36.3	24 37.5	24 38.8	24 40.0	24 41.3	24 42.5	10h
11h	26 51.5	26 52.9	26 54.3	26 55.6	26 57.0	26 58.4	26 59.8	27 01.1	27 02.5	27 03.9	27 05.3	27 06.6	27 08.0	27 09.4	27 10.8	11h
12h	29 18.0	29 19.5	29 21.0	29 22.5	29 24.0	29 25.5	29 27.0	29 28.5	29 30.0	29 31.5	29 33.0	29 34.5	29 36.0	29 37.5	29 39.0	12h
13h	31 44.5	31 46.1	31 47.8	31 49.4	31 51.0	31 52.6	31 54.3	31 55.9	31 57.5	31 59.1	32 00.8	32 02.4	32 04.0	32 05.6	32 07.3	13h
14h	34 11.0	34 12.8	34 14.5	34 16.3	34 18.0	34 19.8	34 21.5	34 23.3	34 25.0	34 26.8	34 28.5	34 30.3	34 32.0	34 33.8	34 35.5	14h
15h	36 37.5	36 39.4	36 41.3	36 43.1	36 45.0	36 46.9	36 48.8	36 50.6	36 52.5	36 54.4	36 56.3	36 58.1	37 00.0	37 01.9	37 03.8	15h
16h	39 04.0	39 06.0	39 08.0	39 10.0	39 12.0	39 14.0	39 16.0	39 18.0	39 20.0	39 22.0	39 24.0	39 26.0	39 28.0	39 30.0	39 32.0	16h
17h	41 30.5	41 32.6	41 34.8	41 36.9	41 39.0	41 41.1	41 43.3	41 45.4	41 47.5	41 49.6	41 51.8	41 53.9	41 56.0	41 58.1	42 00.3	17h
18h	43 57.0	43 59.3	44 01.5	44 03.8	44 06.0	44 08.3	44 10.5	44 12.8	44 15.0	44 17.3	44 19.5	44 21.8	44 24.0	44 26.3	44 28.5	18h
19h	46 23.5	46 25.9	46 28.3	46 30.6	46 33.0	46 35.4	46 37.8	46 40.1	46 42.5	46 44.9	46 47.3	46 49.6	46 52.0	46 54.4	46 56.8	19h
20h	48 50.0	48 52.5	48 55.0	48 57.5	49 00.0	49 02.5	49 05.0	49 07.5	49 10.0	49 12.5	49 15.0	49 17.5	49 20.0	49 22.5	49 25.0	20h
21h	51 16.5	51 19.1	51 21.8	51 24.4	51 27.0	51 29.6	51 32.3	51 34.9	51 37.5	51 40.1	51 42.8	51 45.4	51 48.0	51 50.6	51 53.3	21h
22h	53 43.0	53 45.8	53 48.5	53 51.3	53 54.0	53 56.8	53 59.5	54 02.3	54 05.0	54 07.8	54 10.5	54 13.3	54 16.0	54 18.8	54 21.5	22h
23h	56 09.5	56 12.4	56 15.3	56 18.1	56 21.0	56 23.9	56 26.8	56 29.6	56 32.5	56 35.4	56 38.3	56 41.1	56 44.0	56 46.9	56 49.8	23h
24h	58'36"	58'39"	58'42"	58'45"	58'48"	58'51"	58'54"	58'57"	59'00"	59'03"	59'06"	59'09"	59'12"	59'15"	59'18"	24h

24h	59'18"	59'21"	59'24"	59'27"	59'30"	59'33"	59'36"	59'39"	59'42"	59'45"	59'48"	59'51"	59'54"	59'57"	60'00"	24h
1m	0 02.5	0 02.5	0 02.5	0 02.5	0 02.5	0 02.5	0 02.5	0 02.5	0 02.5	0 02.5	0 02.5	0 02.5	0 02.5	0 02.5	0 02.5	1m
2m	0 04.9	0 04.9	0 05.0	0 05.0	0 05.0	0 05.0	0 05.0	0 05.0	0 05.0	0 05.0	0 05.0	0 05.0	0 05.0	0 05.0	0 05.0	2m
3m	0 07.4	0 07.4	0 07.4	0 07.4	0 07.4	0 07.4	0 07.5	0 07.5	0 07.5	0 07.5	0 07.5	0 07.5	0 07.5	0 07.5	0 07.5	3m
4m	0 09.9	0 09.9	0 09.9	0 09.9	0 09.9	0 09.9	0 09.9	0 09.9	0 10.0	0 10.0	0 10.0	0 10.0	0 10.0	0 10.0	0 10.0	4m
5m	0 12.4	0 12.4	0 12.4	0 12.4	0 12.4	0 12.4	0 12.4	0 12.4	0 12.4	0 12.4	0 12.5	0 12.5	0 12.5	0 12.5	0 12.5	5m
10m	0 24.7	0 24.7	0 24.8	0 24.8	0 24.8	0 24.8	0 24.8	0 24.9	0 24.9	0 24.9	0 24.9	0 24.9	0 25.0	0 25.0	0 25.0	10m
15m	0 37.1	0 37.1	0 37.1	0 37.2	0 37.2	0 37.2	0 37.3	0 37.3	0 37.3	0 37.3	0 37.4	0 37.4	0 37.4	0 37.5	0 37.5	15m
20m	0 49.4	0 49.5	0 49.5	0 49.5	0 49.6	0 49.6	0 49.7	0 49.7	0 49.8	0 49.8	0 49.8	0 49.9	0 49.9	0 50.0	0 50.0	20m
25m	1 01.8	1 01.8	1 01.9	1 01.9	1 02.0	1 02.0	1 02.1	1 02.1	1 02.2	1 02.2	1 02.3	1 02.3	1 02.4	1 02.4	1 02.5	25m
30m	1 14.1	1 14.2	1 14.3	1 14.3	1 14.4	1 14.4	1 14.5	1 14.6	1 14.6	1 14.7	1 14.8	1 14.8	1 14.9	1 14.9	1 15.0	30m
35m	1 26.5	1 26.6	1 26.6	1 26.7	1 26.8	1 26.8	1 26.9	1 27.0	1 27.1	1 27.1	1 27.2	1 27.3	1 27.4	1 27.4	1 27.5	35m
40m	1 38.8	1 38.9	1 39.0	1 39.1	1 39.2	1 39.3	1 39.3	1 39.4	1 39.5	1 39.6	1 39.7	1 39.8	1 39.8	1 39.9	1 40.0	40m
45m	1 51.2	1 51.3	1 51.4	1 51.5	1 51.6	1 51.7	1 51.8	1 51.8	1 51.9	1 52.0	1 52.1	1 52.2	1 52.3	1 52.4	1 52.5	45m
50m	2 03.5	2 03.6	2 03.8	2 03.9	2 04.0	2 04.1	2 04.2	2 04.3	2 04.4	2 04.5	2 04.6	2 04.7	2 04.8	2 04.9	2 05.0	50m
55m	2 15.9	2 16.0	2 16.1	2 16.2	2 16.4	2 16.5	2 16.6	2 16.7	2 16.8	2 16.9	2 17.0	2 17.2	2 17.3	2 17.4	2 17.5	55m
1h	2 28.3	2 28.4	2 28.5	2 28.6	2 28.8	2 28.9	2 29.0	2 29.1	2 29.3	2 29.4	2 29.5	2 29.6	2 29.8	2 29.9	2 30.0	1h
2h	4 56.5	4 56.8	4 57.0	4 57.3	4 57.5	4 57.8	4 58.0	4 58.3	4 58.5	4 58.8	4 59.0	4 59.3	4 59.5	4 59.8	5 00.0	2h
3h	7 24.8	7 25.1	7 25.5	7 25.9	7 26.3	7 26.6	7 27.0	7 27.4	7 27.8	7 28.1	7 28.5	7 28.9	7 29.3	7 29.6	7 30.0	3h
4h	9 53.0	9 53.5	9 54.0	9 54.5	9 55.0	9 55.5	9 56.0	9 56.5	9 57.0	9 57.5	9 58.0	9 58.5	9 59.0	9 59.5	10 00.0	4h
5h	12 21.3	12 21.9	12 22.5	12 23.1	12 23.8	12 24.4	12 25.0	12 25.6	12 26.3	12 26.9	12 27.5	12 28.1	12 28.8	12 29.4	12 30.0	5h
6h	14 49.5	14 50.3	14 51.0	14 51.8	14 52.5	14 53.3	14 54.0	14 54.8	14 55.5	14 56.3	14 57.0	14 57.8	14 58.5	14 59.3	15 00.0	6h
7h	17 17.8	17 18.6	17 19.5	17 20.4	17 21.3	17 22.1	17 23.0	17 23.9	17 24.8	17 25.6	17 26.5	17 27.4	17 28.3	17 29.1	17 30.0	7h
8h	19 46.0	19 47.0	19 48.0	19 49.0	19 50.0	19 51.0	19 52.0	19 53.0	19 54.0	19 55.0	19 56.0	19 57.0	19 58.0	19 59.0	20 00.0	8h
9h	22 14.3	22 15.4	22 16.5	22 17.6	22 18.8	22 19.9	22 21.0	22 22.1	22 23.3	22 24.4	22 25.5	22 26.6	22 27.8	22 28.9	22 30.0	9h
10h	24 42.5	24 43.8	24 45.0	24 46.3	24 47.5	24 48.8	24 50.0	24 51.3	24 52.5	24 53.8	24 55.0	24 56.3	24 57.5	24 58.8	25 00.0	10h
11h	27 10.8	27 12.1	27 13.5	27 14.9	27 16.3	27 17.6	27 19.0	27 20.4	27 21.8	27 23.1	27 24.5	27 25.9	27 27.3	27 28.6	27 30.0	11h
12h	29 39.0	29 40.5	29 42.0	29 43.5	29 45.0	29 46.5	29 48.0	29 49.5	29 51.0	29 52.5	29 54.0	29 55.5	29 57.0	29 58.5	30 00.0	12h
13h	32 07.3	32 08.9	32 10.5	32 12.1	32 13.8	32 15.4	32 17.0	32 18.6	32 20.3	32 21.9	32 23.5	32 25.1	32 26.8	32 28.4	32 30.0	13h
14h	34 35.5	34 37.3	34 39.0	34 40.8	34 42.5	34 44.3	34 46.0	34 47.8	34 49.5	34 51.3	34 53.0	34 54.8	34 56.5	34 58.3	35 00.0	14h
15h	37 03.8	37 05.6	37 07.5	37 09.4	37 11.3	37 13.1	37 15.0	37 16.9	37 18.8	37 20.6	37 22.5	37 24.4	37 26.3	37 28.1	37 30.0	15h
16h	39 32.0	39 34.0	39 36.0	39 38.0	39 40.0	39 42.0	39 44.0	39 46.0	39 48.0	39 50.0	39 52.0	39 54.0	39 56.0	39 58.0	40 00.0	16h
17h	42 00.3	42 02.4	42 04.5	42 06.6	42 08.8	42 10.9	42 13.0	42 15.1	42 17.3	42 19.4	42 21.5	42 23.6	42 25.8	42 27.9	42 30.0	17h
18h	44 28.5	44 30.8	44 33.0	44 35.3	44 37.5	44 39.8	44 42.0	44 44.3	44 46.5	44 48.8	44 51.0	44 53.3	44 55.5	44 57.8	45 00.0	18h
19h	46 56.8	46 59.1	47 01.5	47 03.9	47 06.3	47 08.6	47 11.0	47 13.4	47 15.8	47 18.1	47 20.5	47 22.9	47 25.3	47 27.6	47 30.0	19h
20h	49 25.0	49 27.5	49 30.0	49 32.5	49 35.0	49 37.5	49 40.0	49 42.5	49 45.0	49 47.5	49 50.0	49 52.5	49 55.0	49 57.5	50 00.0	20h
21h	51 53.3	51 55.9	51 58.5	52 01.1	52 03.8	52 06.4	52 09.0	52 11.6	52 14.3	52 16.9	52 19.5	52 22.1	52 24.8	52 27.4	52 30.0	21h
22h	54 21.5	54 24.3	54 27.0	54 29.8	54 32.5	54 35.3	54 38.0	54 40.8	54 43.5	54 46.3	54 49.0	54 51.8	54 54.5	54 57.3	55 00.0	22h
23h	56 49.8	56 52.6	56 55.5	56 58.4	57 01.3	57 04.1	57 07.0	57 09.9	57 12.8	57 15.6	57 18.5	57 21.4	57 24.3	57 27.1	57 30.0	23h
24h	59'18"	59'21"	59'24"	59'27"	59'30"	59'33"	59'36"	59'39"	59'42"	59'45"	59'48"	59'51"	59'54"	59'57"	60'00"	24h

☉ ☉

24h	60'00"	60'03"	60'06"	60'09"	60'12"	60'15"	60'18"	60'21"	60'24"	60'27"	60'30"	60'33"	60'36"	24h
1m	0 02.5	0 02.5	0 02.5	0 02.5	0 02.5	0 02.5	0 02.5	0 02.5	0 02.5	0 02.5	0 02.5	0 02.5	0 02.5	1m
2m	0 05.0	0 05.0	0 05.0	0 05.0	0 05.0	0 05.0	0 05.0	0 05.0	0 05.0	0 05.0	0 05.0	0 05.0	0 05.1	2m
3m	0 07.5	0 07.5	0 07.5	0 07.5	0 07.5	0 07.5	0 07.5	0 07.5	0 07.6	0 07.6	0 07.6	0 07.6	0 07.6	3m
4m	0 10.0	0 10.0	0 10.0	0 10.0	0 10.0	0 10.0	0 10.1	0 10.1	0 10.1	0 10.1	0 10.1	0 10.1	0 10.1	4m
5m	0 12.5	0 12.5	0 12.5	0 12.5	0 12.5	0 12.6	0 12.6	0 12.6	0 12.6	0 12.6	0 12.6	0 12.6	0 12.6	5m
10m	0 25.0	0 25.0	0 25.0	0 25.1	0 25.1	0 25.1	0 25.1	0 25.1	0 25.2	0 25.2	0 25.2	0 25.2	0 25.3	10m
15m	0 37.5	0 37.5	0 37.6	0 37.6	0 37.6	0 37.7	0 37.7	0 37.7	0 37.8	0 37.8	0 37.8	0 37.8	0 37.9	15m
20m	0 50.0	0 50.0	0 50.1	0 50.1	0 50.2	0 50.2	0 50.3	0 50.3	0 50.3	0 50.4	0 50.4	0 50.5	0 50.5	20m
25m	1 02.5	1 02.6	1 02.6	1 02.7	1 02.7	1 02.8	1 02.8	1 02.9	1 02.9	1 03.0	1 03.0	1 03.1	1 03.1	25m
30m	1 15.0	1 15.1	1 15.1	1 15.2	1 15.3	1 15.3	1 15.4	1 15.4	1 15.5	1 15.6	1 15.6	1 15.7	1 15.8	30m
35m	1 27.5	1 27.6	1 27.6	1 27.7	1 27.8	1 27.9	1 27.9	1 28.0	1 28.1	1 28.2	1 28.2	1 28.3	1 28.4	35m
40m	1 40.0	1 40.1	1 40.2	1 40.3	1 40.3	1 40.4	1 40.5	1 40.6	1 40.7	1 40.8	1 40.8	1 40.9	1 41.0	40m
45m	1 52.5	1 52.6	1 52.7	1 52.8	1 52.9	1 53.0	1 53.1	1 53.2	1 53.3	1 53.3	1 53.4	1 53.5	1 53.6	45m
50m	2 05.0	2 05.1	2 05.2	2 05.3	2 05.4	2 05.5	2 05.6	2 05.7	2 05.8	2 05.9	2 06.0	2 06.1	2 06.3	50m
55m	2 17.5	2 17.6	2 17.7	2 17.8	2 18.0	2 18.1	2 18.2	2 18.3	2 18.4	2 18.5	2 18.6	2 18.8	2 18.9	55m
1h	2 30.0	2 30.1	2 30.3	2 30.4	2 30.5	2 30.6	2 30.8	2 30.9	2 31.0	2 31.1	2 31.3	2 31.4	2 31.5	1h
2h	5 00.0	5 00.3	5 00.5	5 00.8	5 01.0	5 01.3	5 01.5	5 01.8	5 02.0	5 02.3	5 02.5	5 02.8	5 03.0	2h
3h	7 30.0	7 30.4	7 30.8	7 31.1	7 31.5	7 31.9	7 32.3	7 32.6	7 33.0	7 33.4	7 33.8	7 34.1	7 34.5	3h
4h	10 00.0	10 00.5	10 01.0	10 01.5	10 02.0	10 02.5	10 03.0	10 03.5	10 04.0	10 04.5	10 05.0	10 05.5	10 06.0	4h
5h	12 30.0	12 30.6	12 31.3	12 31.9	12 32.5	12 33.1	12 33.8	12 34.4	12 35.0	12 35.6	12 36.3	12 36.9	12 37.5	5h
6h	15 00.0	15 00.8	15 01.5	15 02.3	15 03.0	15 03.8	15 04.5	15 05.3	15 06.0	15 06.8	15 07.5	15 08.3	15 09.0	6h
7h	17 30.0	17 30.9	17 31.8	17 32.6	17 33.5	17 34.4	17 35.3	17 36.1	17 37.0	17 37.9	17 38.8	17 39.6	17 40.5	7h
8h	20 00.0	20 01.0	20 02.0	20 03.0	20 04.0	20 05.0	20 06.0	20 07.0	20 08.0	20 09.0	20 10.0	20 11.0	20 12.0	8h
9h	22 30.0	22 31.1	22 32.3	22 33.4	22 34.5	22 35.6	22 36.8	22 37.9	22 39.0	22 40.1	22 41.3	22 42.4	22 43.5	9h
10h	25 00.0	25 01.3	25 02.5	25 03.8	25 05.0	25 06.3	25 07.5	25 08.8	25 10.0	25 11.3	25 12.5	25 13.8	25 15.0	10h
11h	27 30.0	27 31.4	27 32.8	27 34.1	27 35.5	27 36.9	27 38.3	27 39.6	27 41.0	27 42.4	27 43.8	27 45.1	27 46.5	11h
12h	30 00.0	30 01.5	30 03.0	30 04.5	30 06.0	30 07.5	30 09.0	30 10.5	30 12.0	30 13.5	30 15.0	30 16.5	30 18.0	12h
13h	32 30.0	32 31.6	32 33.3	32 34.9	32 36.5	32 38.1	32 39.8	32 41.4	32 43.0	32 44.6	32 46.3	32 47.9	32 49.5	13h
14h	35 00.0	35 01.8	35 03.5	35 05.3	35 07.0	35 08.8	35 10.5	35 12.3	35 14.0	35 15.8	35 17.5	35 19.3	35 21.0	14h
15h	37 30.0	37 31.9	37 33.8	37 35.6	37 37.5	37 39.4	37 41.3	37 43.1	37 45.0	37 46.9	37 48.8	37 50.6	37 52.5	15h
16h	40 00.0	40 02.0	40 04.0	40 06.0	40 08.0	40 10.0	40 12.0	40 14.0	40 16.0	40 18.0	40 20.0	40 22.0	40 24.0	16h
17h	42 30.0	42 32.1	42 34.3	42 36.4	42 38.5	42 40.6	42 42.8	42 44.9	42 47.0	42 49.1	42 51.3	42 53.4	42 55.5	17h
18h	45 00.0	45 02.3	45 04.5	45 06.8	45 09.0	45 11.3	45 13.5	45 15.8	45 18.0	45 20.3	45 22.5	45 24.8	45 27.0	18h
19h	47 30.0	47 32.4	47 34.8	47 37.1	47 39.5	47 41.9	47 44.3	47 46.6	47 49.0	47 51.4	47 53.8	47 56.1	47 58.5	19h
20h	50 00.0	50 02.5	50 05.0	50 07.5	50 10.0	50 12.5	50 15.0	50 17.5	50 20.0	50 22.5	50 25.0	50 27.5	50 30.0	20h
21h	52 30.0	52 32.6	52 35.3	52 37.9	52 40.5	52 43.1	52 45.8	52 48.4	52 51.0	52 53.6	52 56.3	52 58.9	53 01.5	21h
22h	55 00.0	55 02.8	55 05.5	55 08.3	55 11.0	55 13.8	55 16.5	55 19.3	55 22.0	55 24.8	55 27.5	55 30.3	55 33.0	22h
23h	57 30.0	57 32.9	57 35.8	57 38.6	57 41.5	57 44.4	57 47.3	57 50.1	57 53.0	57 55.9	57 58.8	58 01.6	58 04.5	23h
24h	60'00"	60'03"	60'06"	60'09"	60'12"	60'15"	60'18"	60'21"	60'24"	60'27"	60'30"	60'33"	60'36"	24h

24h	60'36"	60'39"	60'42"	60'45"	60'48"	60'51"	60'54"	60'57"	61'00"	61'03"	61'06"	61'09"	61'12"	24h
1m	0 02.5	0 02.5	0 02.5	0 02.5	0 02.5	0 02.5	0 02.5	0 02.5	0 02.5	0 02.5	0 02.5	0 02.5	0 02.6	1m
2m	0 05.1	0 05.1	0 05.1	0 05.1	0 05.1	0 05.1	0 05.1	0 05.1	0 05.1	0 05.1	0 05.1	0 05.1	0 05.1	2m
3m	0 07.6	0 07.6	0 07.6	0 07.6	0 07.6	0 07.6	0 07.6	0 07.6	0 07.6	0 07.6	0 07.6	0 07.6	0 07.7	3m
4m	0 10.1	0 10.1	0 10.1	0 10.1	0 10.1	0 10.1	0 10.2	0 10.2	0 10.2	0 10.2	0 10.2	0 10.2	0 10.2	4m
5m	0 12.6	0 12.6	0 12.6	0 12.7	0 12.7	0 12.7	0 12.7	0 12.7	0 12.7	0 12.7	0 12.7	0 12.7	0 12.8	5m
10m	0 25.3	0 25.3	0 25.3	0 25.3	0 25.3	0 25.4	0 25.4	0 25.4	0 25.4	0 25.4	0 25.5	0 25.5	0 25.5	10m
15m	0 37.9	0 37.9	0 37.9	0 38.0	0 38.0	0 38.0	0 38.1	0 38.1	0 38.1	0 38.2	0 38.2	0 38.2	0 38.3	15m
20m	0 50.5	0 50.5	0 50.6	0 50.6	0 50.7	0 50.7	0 50.8	0 50.8	0 50.8	0 50.9	0 50.9	0 51.0	0 51.0	20m
25m	1 03.1	1 03.2	1 03.2	1 03.3	1 03.3	1 03.4	1 03.4	1 03.5	1 03.5	1 03.6	1 03.6	1 03.7	1 03.8	25m
30m	1 15.8	1 15.8	1 15.9	1 15.9	1 16.0	1 16.1	1 16.1	1 16.2	1 16.3	1 16.3	1 16.4	1 16.4	1 16.5	30m
35m	1 28.4	1 28.4	1 28.5	1 28.6	1 28.7	1 28.7	1 28.8	1 28.9	1 29.0	1 29.0	1 29.1	1 29.2	1 29.3	35m
40m	1 41.0	1 41.1	1 41.2	1 41.3	1 41.3	1 41.4	1 41.5	1 41.6	1 41.7	1 41.8	1 41.8	1 41.9	1 42.0	40m
45m	1 53.6	1 53.7	1 53.8	1 53.9	1 54.0	1 54.1	1 54.2	1 54.3	1 54.4	1 54.5	1 54.6	1 54.7	1 54.8	45m
50m	2 06.3	2 06.4	2 06.5	2 06.6	2 06.7	2 06.8	2 06.9	2 07.0	2 07.1	2 07.2	2 07.3	2 07.4	2 07.5	50m
55m	2 18.9	2 19.0	2 19.1	2 19.2	2 19.3	2 19.4	2 19.6	2 19.7	2 19.8	2 19.9	2 20.0	2 20.1	2 20.3	55m
1h	2 31.5	2 31.6	2 31.8	2 31.9	2 32.0	2 32.1	2 32.3	2 32.4	2 32.5	2 32.6	2 32.8	2 32.9	2 33.0	1h
2h	5 03.0	5 03.3	5 03.5	5 03.8	5 04.0	5 04.3	5 04.5	5 04.8	5 05.0	5 05.3	5 05.5	5 05.8	5 06.0	2h
3h	7 34.5	7 34.9	7 35.3	7 35.6	7 36.0	7 36.4	7 36.8	7 37.1	7 37.5	7 37.9	7 38.3	7 38.6	7 39.0	3h
4h	10 06.0	10 06.5	10 07.0	10 07.5	10 08.0	10 08.5	10 09.0	10 09.5	10 10.0	10 10.5	10 11.0	10 11.5	10 12.0	4h
5h	12 37.5	12 38.1	12 38.8	12 39.4	12 40.0	12 40.6	12 41.3	12 41.9	12 42.5	12 43.1	12 43.8	12 44.4	12 45.0	5h
6h	15 09.0	15 09.8	15 10.5	15 11.3	15 12.0	15 12.8	15 13.5	15 14.3	15 15.0	15 15.8	15 16.5	15 17.3	15 18.0	6h
7h	17 40.5	17 41.4	17 42.3	17 43.1	17 44.0	17 44.9	17 45.8	17 46.6	17 47.5	17 48.4	17 49.3	17 50.1	17 51.0	7h
8h	20 12.0	20 13.0	20 14.0	20 15.0	20 16.0	20 17.0	20 18.0	20 19.0	20 20.0	20 21.0	20 22.0	20 23.0	20 24.0	8h
9h	22 43.5	22 44.6	22 45.8	22 46.9	22 48.0	22 49.1	22 50.3	22 51.4	22 52.5	22 53.6	22 54.8	22 55.9	22 57.0	9h
10h	25 15.0	25 16.3	25 17.5	25 18.8	25 20.0	25 21.3	25 22.5	25 23.8	25 25.0	25 26.3	25 27.5	25 28.8	25 30.0	10h
11h	27 46.5	27 47.9	27 49.3	27 50.6	27 52.0	27 53.4	27 54.8	27 56.1	27 57.5	27 58.9	28 00.3	28 01.6	28 03.0	11h
12h	30 18.0	30 19.5	30 21.0	30 22.5	30 24.0	30 25.5	30 27.0	30 28.5	30 30.0	30 31.5	30 33.0	30 34.5	30 36.0	12h
13h	32 49.5	32 51.1	32 52.8	32 54.4	32 56.0	32 57.6	32 59.3	33 00.9	33 02.5	33 04.1	33 05.8	33 07.4	33 09.0	13h
14h	35 21.0	35 22.8	35 24.5	35 26.3	35 28.0	35 29.8	35 31.5	35 33.3	35 35.0	35 36.8	35 38.5	35 40.3	35 42.0	14h
15h	37 52.5	37 54.4	37 56.3	37 58.1	38 00.0	38 01.9	38 03.8	38 05.6	38 07.5	38 09.4	38 11.3	38 13.1	38 15.0	15h
16h	40 24.0	40 26.0	40 28.0	40 30.0	40 32.0	40 34.0	40 36.0	40 38.0	40 40.0	40 42.0	40 44.0	40 46.0	40 48.0	16h
17h	42 55.5	42 57.6	42 59.8	43 01.9	43 04.0	43 06.1	43 08.3	43 10.4	43 12.5	43 14.6	43 16.8	43 18.9	43 21.0	17h
18h	45 27.0	45 29.3	45 31.5	45 33.8	45 36.0	45 38.3	45 40.5	45 42.8	45 45.0	45 47.3	45 49.5	45 51.8	45 54.0	18h
19h	47 58.5	48 00.9	48 03.3	48 05.6	48 08.0	48 10.4	48 12.8	48 15.1	48 17.5	48 19.9	48 22.3	48 24.6	48 27.0	19h
20h	50 30.0	50 32.5	50 35.0	50 37.5	50 40.0	50 42.5	50 45.0	50 47.5	50 50.0	50 52.5	50 55.0	50 57.5	51 00.0	20h
21h	53 01.5	53 04.1	53 06.8	53 09.4	53 12.0	53 14.6	53 17.3	53 19.9	53 22.5	53 25.1	53 27.8	53 30.4	53 33.0	21h
22h	55 33.0	55 35.8	55 38.5	55 41.3	55 44.0	55 46.8	55 49.5	55 52.3	55 55.0	55 57.8	56 00.5	56 03.3	56 06.0	22h
23h	58 04.5	58 07.4	58 10.3	58 13.1	58 16.0	58 18.9	58 21.8	58 24.6	58 27.5	58 30.4	58 33.3	58 36.1	58 39.0	23h
24h	60'36"	60'39"	60'42"	60'45"	60'48"	60'51"	60'54"	60'57"	61'00"	61'03"	61'06"	61'09"	61'12"	24h

Table 2 – Half-daily Lunar Movements

12h	5°52'	5°53'	5°54'	5°55'	5°56'	5°57'	5°58'	5°59'	6°00'	6°01'	6°02'	6°03'	6°04'	6°05'	6°06'	12h
	° ' "	° ' "	° ' "	° ' "	° ' "	° ' "	° ' "	° ' "	° ' "	° ' "	° ' "	° ' "	° ' "	° ' "	° ' "	
1m	0 00 29	0 00 29	0 00 30	0 00 30	0 00 30	0 00 30	0 00 30	0 00 30	0 00 30	0 00 30	0 00 30	0 00 30	0 00 30	0 00 30	0 00 31	1m
2m	0 00 59	0 00 59	0 00 59	0 00 59	0 00 59	0 01 00	0 01 00	0 01 00	0 01 00	0 01 00	0 01 00	0 01 01	0 01 01	0 01 01	0 01 01	2m
3m	0 01 28	0 01 28	0 01 29	0 01 29	0 01 29	0 01 29	0 01 30	0 01 30	0 01 30	0 01 30	0 01 31	0 01 31	0 01 31	0 01 31	0 01 32	3m
4m	0 01 57	0 01 58	0 01 58	0 01 58	0 01 59	0 01 59	0 01 59	0 02 00	0 02 00	0 02 00	0 02 01	0 02 01	0 02 01	0 02 02	0 02 02	4m
5m	0 02 27	0 02 27	0 02 28	0 02 28	0 02 28	0 02 29	0 02 29	0 02 30	0 02 30	0 02 30	0 02 31	0 02 31	0 02 32	0 02 32	0 02 33	5m
10m	0 04 53	0 04 54	0 04 55	0 04 56	0 04 57	0 04 58	0 04 58	0 04 59	0 05 00	0 05 01	0 05 02	0 05 03	0 05 03	0 05 04	0 05 05	10m
15m	0 07 20	0 07 21	0 07 23	0 07 24	0 07 25	0 07 26	0 07 28	0 07 29	0 07 30	0 07 31	0 07 33	0 07 34	0 07 35	0 07 36	0 07 38	15m
20m	0 09 47	0 09 48	0 09 50	0 09 52	0 09 53	0 09 55	0 09 57	0 09 58	0 10 00	0 10 02	0 10 03	0 10 05	0 10 07	0 10 08	0 10 10	20m
25m	0 12 13	0 12 15	0 12 18	0 12 20	0 12 22	0 12 24	0 12 26	0 12 28	0 12 30	0 12 32	0 12 34	0 12 36	0 12 38	0 12 40	0 12 43	25m
30m	0 14 40	0 14 43	0 14 45	0 14 48	0 14 50	0 14 53	0 14 55	0 14 58	0 15 00	0 15 03	0 15 05	0 15 08	0 15 10	0 15 13	0 15 15	30m
35m	0 17 07	0 17 10	0 17 13	0 17 15	0 17 18	0 17 21	0 17 24	0 17 27	0 17 30	0 17 33	0 17 36	0 17 39	0 17 42	0 17 45	0 17 48	35m
40m	0 19 33	0 19 37	0 19 40	0 19 43	0 19 47	0 19 50	0 19 53	0 19 57	0 20 00	0 20 03	0 20 07	0 20 10	0 20 13	0 20 17	0 20 20	40m
45m	0 22 00	0 22 04	0 22 08	0 22 11	0 22 15	0 22 19	0 22 23	0 22 26	0 22 30	0 22 34	0 22 38	0 22 41	0 22 45	0 22 49	0 22 53	45m
50m	0 24 27	0 24 31	0 24 35	0 24 39	0 24 43	0 24 48	0 24 52	0 24 56	0 25 00	0 25 04	0 25 08	0 25 13	0 25 17	0 25 21	0 25 25	50m
55m	0 26 53	0 26 58	0 27 03	0 27 07	0 27 12	0 27 16	0 27 21	0 27 25	0 27 30	0 27 35	0 27 39	0 27 44	0 27 48	0 27 53	0 27 58	55m
1h	0 29 20	0 29 25	0 29 30	0 29 35	0 29 40	0 29 45	0 29 50	0 29 55	0 30 00	0 30 05	0 30 10	0 30 15	0 30 20	0 30 25	0 30 30	1h
2h	0 58 40	0 58 50	0 59 00	0 59 10	0 59 20	0 59 30	0 59 40	0 59 50	1 00 00	1 00 10	1 00 20	1 00 30	1 00 40	1 00 50	1 01 00	2h
3h	1 28 00	1 28 15	1 28 30	1 28 45	1 29 00	1 29 15	1 29 30	1 29 45	1 30 00	1 30 15	1 30 30	1 30 45	1 31 00	1 31 15	1 31 30	3h
4h	1 57 20	1 57 40	1 58 00	1 58 20	1 58 40	1 59 00	1 59 20	1 59 40	2 00 00	2 00 20	2 00 40	2 01 00	2 01 20	2 01 40	2 02 00	4h
5h	2 26 40	2 27 05	2 27 30	2 27 55	2 28 20	2 28 45	2 29 10	2 29 35	2 30 00	2 30 25	2 30 50	2 31 15	2 31 40	2 32 05	2 32 30	5h
6h	2 56 00	2 56 30	2 57 00	2 57 30	2 58 00	2 58 30	2 59 00	2 59 30	3 00 00	3 00 30	3 01 00	3 01 30	3 02 00	3 02 30	3 03 00	6h
7h	3 25 20	3 25 55	3 26 30	3 27 05	3 27 40	3 28 15	3 28 50	3 29 25	3 30 00	3 30 35	3 31 10	3 31 45	3 32 20	3 32 55	3 33 30	7h
8h	3 54 40	3 55 20	3 56 00	3 56 40	3 57 20	3 58 00	3 58 40	3 59 20	4 00 00	4 00 40	4 01 20	4 02 00	4 02 40	4 03 20	4 04 00	8h
9h	4 24 00	4 24 45	4 25 30	4 26 15	4 27 00	4 27 45	4 28 30	4 29 15	4 30 00	4 30 45	4 31 30	4 32 15	4 33 00	4 33 45	4 34 30	9h
10h	4 53 20	4 54 10	4 55 00	4 55 50	4 56 40	4 57 30	4 58 20	4 59 10	5 00 00	5 00 50	5 01 40	5 02 30	5 03 20	5 04 10	5 05 00	10h
11h	5 22 40	5 23 35	5 24 30	5 25 25	5 26 20	5 27 15	5 28 10	5 29 05	5 30 00	5 30 55	5 31 50	5 32 45	5 33 40	5 34 35	5 35 30	11h
12h	5°52'	5°53'	5°54'	5°55'	5°56'	5°57'	5°58'	5°59'	6°00'	6°01'	6°02'	6°03'	6°04'	6°05'	6°06'	12h

12h	6 06 00	6 07 00	6 08 00	6 09 00	6 10 00	6 11 00	6 12 00	6 13 00	6 14 00	6 15 00	6 16 00	6 17 00	6 18 00	6 19 00	6 20 00	12h
	° ' "	° ' "	° ' "	° ' "	° ' "	° ' "	° ' "	° ' "	° ' "	° ' "	° ' "	° ' "	° ' "	° ' "	° ' "	
1m	0 00 31	0 00 31	0 00 31	0 00 31	0 00 31	0 00 31	0 00 31	0 00 31	0 00 31	0 00 31	0 00 31	0 00 31	0 00 32	0 00 32	0 00 32	1m
2m	0 01 01	0 01 01	0 01 01	0 01 02	0 01 02	0 01 02	0 01 02	0 01 02	0 01 02	0 01 03	0 01 03	0 01 03	0 01 03	0 01 03	0 01 03	2m
3m	0 01 32	0 01 32	0 01 32	0 01 32	0 01 33	0 01 33	0 01 33	0 01 33	0 01 34	0 01 34	0 01 34	0 01 34	0 01 35	0 01 35	0 01 35	3m
4m	0 02 02	0 02 02	0 02 03	0 02 03	0 02 03	0 02 04	0 02 04	0 02 04	0 02 05	0 02 05	0 02 05	0 02 06	0 02 06	0 02 06	0 02 07	4m
5m	0 02 33	0 02 33	0 02 33	0 02 34	0 02 34	0 02 35	0 02 35	0 02 35	0 02 36	0 02 36	0 02 37	0 02 37	0 02 38	0 02 38	0 02 38	5m
10m	0 05 05	0 05 06	0 05 07	0 05 08	0 05 08	0 05 09	0 05 10	0 05 11	0 05 12	0 05 13	0 05 13	0 05 14	0 05 15	0 05 16	0 05 17	10m
15m	0 07 38	0 07 39	0 07 40	0 07 41	0 07 43	0 07 44	0 07 45	0 07 46	0 07 48	0 07 49	0 07 50	0 07 51	0 07 53	0 07 54	0 07 55	15m
20m	0 10 10	0 10 12	0 10 13	0 10 15	0 10 17	0 10 18	0 10 20	0 10 22	0 10 23	0 10 25	0 10 27	0 10 28	0 10 30	0 10 32	0 10 33	20m
25m	0 12 43	0 12 45	0 12 47	0 12 49	0 12 51	0 12 53	0 12 55	0 12 57	0 12 59	0 13 01	0 13 03	0 13 05	0 13 08	0 13 10	0 13 12	25m
30m	0 15 15	0 15 18	0 15 20	0 15 23	0 15 25	0 15 28	0 15 30	0 15 33	0 15 35	0 15 38	0 15 40	0 15 43	0 15 45	0 15 48	0 15 50	30m
35m	0 17 48	0 17 50	0 17 53	0 17 56	0 17 59	0 18 02	0 18 05	0 18 08	0 18 11	0 18 14	0 18 17	0 18 20	0 18 23	0 18 25	0 18 28	35m
40m	0 20 20	0 20 23	0 20 27	0 20 30	0 20 33	0 20 37	0 20 40	0 20 43	0 20 47	0 20 50	0 20 53	0 20 57	0 21 00	0 21 03	0 21 07	40m
45m	0 22 53	0 22 56	0 23 00	0 23 04	0 23 08	0 23 11	0 23 15	0 23 19	0 23 23	0 23 26	0 23 30	0 23 34	0 23 38	0 23 41	0 23 45	45m
50m	0 25 25	0 25 29	0 25 33	0 25 38	0 25 42	0 25 46	0 25 50	0 25 54	0 25 58	0 26 03	0 26 07	0 26 11	0 26 15	0 26 19	0 26 23	50m
55m	0 27 58	0 28 02	0 28 07	0 28 11	0 28 16	0 28 20	0 28 25	0 28 30	0 28 34	0 28 39	0 28 43	0 28 48	0 28 53	0 28 57	0 29 02	55m
1h	0 30 30	0 30 35	0 30 40	0 30 45	0 30 50	0 30 55	0 31 00	0 31 05	0 31 10	0 31 15	0 31 20	0 31 25	0 31 30	0 31 35	0 31 40	1h
2h	1 01 00	1 01 10	1 01 20	1 01 30	1 01 40	1 01 50	1 02 00	1 02 10	1 02 20	1 02 30	1 02 40	1 02 50	1 03 00	1 03 10	1 03 20	2h
3h	1 31 30	1 31 45	1 32 00	1 32 15	1 32 30	1 32 45	1 33 00	1 33 15	1 33 30	1 33 45	1 34 00	1 34 15	1 34 30	1 34 45	1 35 00	3h
4h	2 02 00	2 02 20	2 02 40	2 03 00	2 03 20	2 03 40	2 04 00	2 04 20	2 04 40	2 05 00	2 05 20	2 05 40	2 06 00	2 06 20	2 06 40	4h
5h	2 32 30	2 32 55	2 33 20	2 33 45	2 34 10	2 34 35	2 35 00	2 35 25	2 35 50	2 36 15	2 36 40	2 37 05	2 37 30	2 37 55	2 38 20	5h
6h	3 03 00	3 03 30	3 04 00	3 04 30	3 05 00	3 05 30	3 06 00	3 06 30	3 07 00	3 07 30	3 08 00	3 08 30	3 09 00	3 09 30	3 10 00	6h
7h	3 33 30	3 34 05	3 34 40	3 35 15	3 35 50	3 36 25	3 37 00	3 37 35	3 38 10	3 38 45	3 39 20	3 39 55	3 40 30	3 41 05	3 41 40	7h
8h	4 04 00	4 04 40	4 05 20	4 06 00	4 06 40	4 07 20	4 08 00	4 08 40	4 09 20	4 10 00	4 10 40	4 11 20	4 12 00	4 12 40	4 13 20	8h
9h	4 34 30	4 35 15	4 36 00	4 36 45	4 37 30	4 38 15	4 39 00	4 39 45	4 40 30	4 41 15	4 42 00	4 42 45	4 43 30	4 44 15	4 45 00	9h
10h	5 05 00	5 05 50	5 06 40	5 07 30	5 08 20	5 09 10	5 10 00	5 10 50	5 11 40	5 12 30	5 13 20	5 14 10	5 15 00	5 15 50	5 16 40	10h
11h	5 35 30	5 36 25	5 37 20	5 38 15	5 39 10	5 40 05	5 41 00	5 41 55	5 42 50	5 43 45	5 44 40	5 45 35	5 46 30	5 47 25	5 48 20	11h
12h	6 06 00	6 07 00	6 08 00	6 09 00	6 10 00	6 11 00	6 12 00	6 13 00	6 14 00	6 15 00	6 16 00	6 17 00	6 18 00	6 19 00	6 20 00	12h

12h	6 20 00	6 21 00	6 22 00	6 23 00	6 24 00	6 25 00	6 26 00	6 27 00	6 28 00	6 29 00	6 30 00	6 31 00	6 32 00	6 33 00	6 34 00	12h
	° ' "	° ' "	° ' "	° ' "	° ' "	° ' "	° ' "	° ' "	° ' "	° ' "	° ' "	° ' "	° ' "	° ' "	° ' "	
1m	0 00 32	0 00 32	0 00 32	0 00 32	0 00 32	0 00 32	0 00 32	0 00 32	0 00 32	0 00 32	0 00 33	0 00 33	0 00 33	0 00 33	0 00 33	1m
2m	0 01 03	0 01 04	0 01 04	0 01 04	0 01 04	0 01 04	0 01 04	0 01 05	0 01 05	0 01 05	0 01 05	0 01 05	0 01 05	0 01 06	0 01 06	2m
3m	0 01 35	0 01 35	0 01 36	0 01 36	0 01 36	0 01 36	0 01 37	0 01 37	0 01 37	0 01 37	0 01 38	0 01 38	0 01 38	0 01 38	0 01 39	3m
4m	0 02 07	0 02 07	0 02 07	0 02 08	0 02 08	0 02 08	0 02 09	0 02 09	0 02 09	0 02 10	0 02 10	0 02 10	0 02 11	0 02 11	0 02 11	4m
5m	0 02 38	0 02 39	0 02 39	0 02 40	0 02 40	0 02 40	0 02 41	0 02 41	0 02 42	0 02 42	0 02 43	0 02 43	0 02 43	0 02 44	0 02 44	5m
10m	0 05 17	0 05 18	0 05 18	0 05 19	0 05 20	0 05 21	0 05 22	0 05 23	0 05 23	0 05 24	0 05 25	0 05 26	0 05 27	0 05 28	0 05 28	10m
15m	0 07 55	0 07 56	0 07 58	0 07 59	0 08 00	0 08 01	0 08 03	0 08 04	0 08 05	0 08 06	0 08 08	0 08 09	0 08 10	0 08 11	0 08 13	15m
20m	0 10 33	0 10 35	0 10 37	0 10 38	0 10 40	0 10 42	0 10 43	0 10 45	0 10 47	0 10 48	0 10 50	0 10 52	0 10 53	0 10 55	0 10 57	20m
25m	0 13 12	0 13 14	0 13 16	0 13 18	0 13 20	0 13 22	0 13 24	0 13 26	0 13 28	0 13 30	0 13 33	0 13 35	0 13 37	0 13 39	0 13 41	25m
30m	0 15 50	0 15 53	0 15 55	0 15 58	0 16 00	0 16 03	0 16 05	0 16 08	0 16 10	0 16 13	0 16 15	0 16 18	0 16 20	0 16 23	0 16 25	30m
35m	0 18 28	0 18 31	0 18 34	0 18 37	0 18 40	0 18 43	0 18 46	0 18 49	0 18 52	0 18 55	0 18 58	0 19 00	0 19 03	0 19 06	0 19 09	35m
40m	0 21 07	0 21 10	0 21 13	0 21 17	0 21 20	0 21 23	0 21 27	0 21 30	0 21 33	0 21 37	0 21 40	0 21 43	0 21 47	0 21 50	0 21 53	40m
45m	0 23 45	0 23 49	0 23 53	0 23 56	0 24 00	0 24 04	0 24 08	0 24 11	0 24 15	0 24 19	0 24 23	0 24 26	0 24 30	0 24 34	0 24 38	45m
50m	0 26 23	0 26 28	0 26 32	0 26 36	0 26 40	0 26 44	0 26 48	0 26 53	0 26 57	0 27 01	0 27 05	0 27 09	0 27 13	0 27 18	0 27 22	50m
55m	0 29 02	0 29 06	0 29 11	0 29 15	0 29 20	0 29 25	0 29 29	0 29 34	0 29 38	0 29 43	0 29 48	0 29 52	0 29 57	0 30 01	0 30 06	55m
1h	0 31 40	0 31 45	0 31 50	0 31 55	0 32 00	0 32 05	0 32 10	0 32 15	0 32 20	0 32 25	0 32 30	0 32 35	0 32 40	0 32 45	0 32 50	1h
2h	1 03 20	1 03 30	1 03 40	1 03 50	1 04 00	1 04 10	1 04 20	1 04 30	1 04 40	1 04 50	1 05 00	1 05 10	1 05 20	1 05 30	1 05 40	2h
3h	1 35 00	1 35 15	1 35 30	1 35 45	1 36 00	1 36 15	1 36 30	1 36 45	1 37 00	1 37 15	1 37 30	1 37 45	1 38 00	1 38 15	1 38 30	3h
4h	2 06 40	2 07 00	2 07 20	2 07 40	2 08 00	2 08 20	2 08 40	2 09 00	2 09 20	2 09 40	2 10 00	2 10 20	2 10 40	2 11 00	2 11 20	4h
5h	2 38 20	2 38 45	2 39 10	2 39 35	2 40 00	2 40 25	2 40 50	2 41 15	2 41 40	2 42 05	2 42 30	2 42 55	2 43 20	2 43 45	2 44 10	5h
6h	3 10 00	3 10 30	3 11 00	3 11 30	3 12 00	3 12 30	3 13 00	3 13 30	3 14 00	3 14 30	3 15 00	3 15 30	3 16 00	3 16 30	3 17 00	6h
7h	3 41 40	3 42 15	3 42 50	3 43 25	3 44 00	3 44 35	3 45 10	3 45 45	3 46 20	3 46 55	3 47 30	3 48 05	3 48 40	3 49 15	3 49 50	7h
8h	4 13 20	4 14 00	4 14 40	4 15 20	4 16 00	4 16 40	4 17 20	4 18 00	4 18 40	4 19 20	4 20 00	4 20 40	4 21 20	4 22 00	4 22 40	8h
9h	4 45 00	4 45 45	4 46 30	4 47 15	4 48 00	4 48 45	4 49 30	4 50 15	4 51 00	4 51 45	4 52 30	4 53 15	4 54 00	4 54 45	4 55 30	9h
10h	5 16 40	5 17 30	5 18 20	5 19 10	5 20 00	5 20 50	5 21 40	5 22 30	5 23 20	5 24 10	5 25 00	5 25 50	5 26 40	5 27 30	5 28 20	10h
11h	5 48 20	5 49 15	5 50 10	5 51 05	5 52 00	5 52 55	5 53 50	5 54 45	5 55 40	5 56 35	5 57 30	5 58 25	5 59 20	6 00 15	6 01 10	11h
12h	6 20 00	6 21 00	6 22 00	6 23 00	6 24 00	6 25 00	6 26 00	6 27 00	6 28 00	6 29 00	6 30 00	6 31 00	6 32 00	6 33 00	6 34 00	12h

☽ ☽

12h	6 34 00	6 35 00	6 36 00	6 37 00	6 38 00	6 39 00	6 40 00	6 41 00	6 42 00	6 43 00	6 44 00	6 45 00	6 46 00	6 47 00	6 48 00	12h
	° ' "	° ' "	° ' "	° ' "	° ' "	° ' "	° ' "	° ' "	° ' "	° ' "	° ' "	° ' "	° ' "	° ' "	° ' "	
1m	0 00 33	0 00 33	0 00 33	0 00 33	0 00 33	0 00 33	0 00 33	0 00 33	0 00 34	0 00 34	0 00 34	0 00 34	0 00 34	0 00 34	0 00 34	1m
2m	0 01 06	0 01 06	0 01 06	0 01 06	0 01 06	0 01 07	0 01 07	0 01 07	0 01 07	0 01 07	0 01 07	0 01 08	0 01 08	0 01 08	0 01 08	2m
3m	0 01 39	0 01 39	0 01 39	0 01 39	0 01 40	0 01 40	0 01 40	0 01 40	0 01 41	0 01 41	0 01 41	0 01 41	0 01 42	0 01 42	0 01 42	3m
4m	0 02 11	0 02 12	0 02 12	0 02 12	0 02 13	0 02 13	0 02 13	0 02 14	0 02 14	0 02 14	0 02 15	0 02 15	0 02 15	0 02 16	0 02 16	4m
5m	0 02 44	0 02 45	0 02 45	0 02 45	0 02 46	0 02 46	0 02 47	0 02 47	0 02 48	0 02 48	0 02 48	0 02 49	0 02 49	0 02 50	0 02 50	5m
10m	0 05 28	0 05 29	0 05 30	0 05 31	0 05 32	0 05 33	0 05 33	0 05 34	0 05 35	0 05 36	0 05 37	0 05 38	0 05 38	0 05 39	0 05 40	10m
15m	0 08 13	0 08 14	0 08 15	0 08 16	0 08 18	0 08 19	0 08 20	0 08 21	0 08 23	0 08 24	0 08 25	0 08 26	0 08 28	0 08 29	0 08 30	15m
20m	0 10 57	0 10 58	0 11 00	0 11 02	0 11 03	0 11 05	0 11 07	0 11 08	0 11 10	0 11 12	0 11 13	0 11 15	0 11 17	0 11 18	0 11 20	20m
25m	0 13 41	0 13 43	0 13 45	0 13 47	0 13 49	0 13 51	0 13 53	0 13 55	0 13 58	0 14 00	0 14 02	0 14 04	0 14 06	0 14 08	0 14 10	25m
30m	0 16 25	0 16 28	0 16 30	0 16 33	0 16 35	0 16 38	0 16 40	0 16 43	0 16 45	0 16 48	0 16 50	0 16 53	0 16 55	0 16 58	0 17 00	30m
35m	0 19 09	0 19 12	0 19 15	0 19 18	0 19 21	0 19 24	0 19 27	0 19 30	0 19 33	0 19 35	0 19 38	0 19 41	0 19 44	0 19 47	0 19 50	35m
40m	0 21 53	0 21 57	0 22 00	0 22 03	0 22 07	0 22 10	0 22 13	0 22 17	0 22 20	0 22 23	0 22 27	0 22 30	0 22 33	0 22 37	0 22 40	40m
45m	0 24 38	0 24 41	0 24 45	0 24 49	0 24 53	0 24 56	0 25 00	0 25 04	0 25 08	0 25 11	0 25 15	0 25 19	0 25 23	0 25 26	0 25 30	45m
50m	0 27 22	0 27 26	0 27 30	0 27 34	0 27 38	0 27 43	0 27 47	0 27 51	0 27 55	0 27 59	0 28 03	0 28 08	0 28 12	0 28 16	0 28 20	50m
55m	0 30 06	0 30 10	0 30 15	0 30 20	0 30 24	0 30 29	0 30 33	0 30 38	0 30 43	0 30 47	0 30 52	0 30 56	0 31 01	0 31 05	0 31 10	55m
1h	0 32 50	0 32 55	0 33 00	0 33 05	0 33 10	0 33 15	0 33 20	0 33 25	0 33 30	0 33 35	0 33 40	0 33 45	0 33 50	0 33 55	0 34 00	1h
2h	1 05 40	1 05 50	1 06 00	1 06 10	1 06 20	1 06 30	1 06 40	1 06 50	1 07 00	1 07 10	1 07 20	1 07 30	1 07 40	1 07 50	1 08 00	2h
3h	1 38 30	1 38 45	1 39 00	1 39 15	1 39 30	1 39 45	1 40 00	1 40 15	1 40 30	1 40 45	1 41 00	1 41 15	1 41 30	1 41 45	1 42 00	3h
4h	2 11 20	2 11 40	2 12 00	2 12 20	2 12 40	2 13 00	2 13 20	2 13 40	2 14 00	2 14 20	2 14 40	2 15 00	2 15 20	2 15 40	2 16 00	4h
5h	2 44 10	2 44 35	2 45 00	2 45 25	2 45 50	2 46 15	2 46 40	2 47 05	2 47 30	2 47 55	2 48 20	2 48 45	2 49 10	2 49 35	2 50 00	5h
6h	3 17 00	3 17 30	3 18 00	3 18 30	3 19 00	3 19 30	3 20 00	3 20 30	3 21 00	3 21 30	3 22 00	3 22 30	3 23 00	3 23 30	3 24 00	6h
7h	3 49 50	3 50 25	3 51 00	3 51 35	3 52 10	3 52 45	3 53 20	3 53 55	3 54 30	3 55 05	3 55 40	3 56 15	3 56 50	3 57 25	3 58 00	7h
8h	4 22 40	4 23 20	4 24 00	4 24 40	4 25 20	4 26 00	4 26 40	4 27 20	4 28 00	4 28 40	4 29 20	4 30 00	4 30 40	4 31 20	4 32 00	8h
9h	4 55 30	4 56 15	4 57 00	4 57 45	4 58 30	4 59 15	5 00 00	5 00 45	5 01 30	5 02 15	5 03 00	5 03 45	5 04 30	5 05 15	5 06 00	9h
10h	5 28 20	5 29 10	5 30 00	5 30 50	5 31 40	5 32 30	5 33 20	5 34 10	5 35 00	5 35 50	5 36 40	5 37 30	5 38 20	5 39 10	5 40 00	10h
11h	6 01 10	6 02 05	6 03 00	6 03 55	6 04 50	6 05 45	6 06 40	6 07 35	6 08 30	6 09 25	6 10 20	6 11 15	6 12 10	6 13 05	6 14 00	11h
12h	6 34 00	6 35 00	6 36 00	6 37 00	6 38 00	6 39 00	6 40 00	6 41 00	6 42 00	6 43 00	6 44 00	6 45 00	6 46 00	6 47 00	6 48 00	12h

12h	6 48 00	6 49 00	6 50 00	6 51 00	6 52 00	6 53 00	6 54 00	6 55 00	6 56 00	6 57 00	6 58 00	6 59 00	7 00 00	7 01 00	7 02 00	12h
	° ' "	° ' "	° ' "	° ' "	° ' "	° ' "	° ' "	° ' "	° ' "	° ' "	° ' "	° ' "	° ' "	° ' "	° ' "	
1m	0 00 34	0 00 34	0 00 34	0 00 34	0 00 34	0 00 34	0 00 35	0 00 35	0 00 35	0 00 35	0 00 35	0 00 35	0 00 35	0 00 35	0 00 35	1m
2m	0 01 08	0 01 08	0 01 08	0 01 09	0 01 09	0 01 09	0 01 09	0 01 09	0 01 09	0 01 10	0 01 10	0 01 10	0 01 10	0 01 10	0 01 10	2m
3m	0 01 42	0 01 42	0 01 43	0 01 43	0 01 43	0 01 43	0 01 44	0 01 44	0 01 44	0 01 44	0 01 45	0 01 45	0 01 45	0 01 45	0 01 46	3m
4m	0 02 16	0 02 16	0 02 17	0 02 17	0 02 17	0 02 18	0 02 18	0 02 18	0 02 19	0 02 19	0 02 19	0 02 20	0 02 20	0 02 20	0 02 21	4m
5m	0 02 50	0 02 50	0 02 51	0 02 51	0 02 52	0 02 52	0 02 53	0 02 53	0 02 53	0 02 54	0 02 54	0 02 55	0 02 55	0 02 55	0 02 56	5m
10m	0 05 40	0 05 41	0 05 42	0 05 43	0 05 43	0 05 44	0 05 45	0 05 46	0 05 47	0 05 48	0 05 48	0 05 49	0 05 50	0 05 51	0 05 52	10m
15m	0 08 30	0 08 31	0 08 33	0 08 34	0 08 35	0 08 36	0 08 38	0 08 39	0 08 40	0 08 41	0 08 43	0 08 44	0 08 45	0 08 46	0 08 48	15m
20m	0 11 20	0 11 22	0 11 23	0 11 25	0 11 27	0 11 28	0 11 30	0 11 32	0 11 33	0 11 35	0 11 37	0 11 38	0 11 40	0 11 42	0 11 43	20m
25m	0 14 10	0 14 12	0 14 14	0 14 16	0 14 18	0 14 20	0 14 23	0 14 25	0 14 27	0 14 29	0 14 31	0 14 33	0 14 35	0 14 37	0 14 39	25m
30m	0 17 00	0 17 03	0 17 05	0 17 08	0 17 10	0 17 13	0 17 15	0 17 18	0 17 20	0 17 23	0 17 25	0 17 28	0 17 30	0 17 33	0 17 35	30m
35m	0 19 50	0 19 53	0 19 56	0 19 59	0 20 02	0 20 05	0 20 08	0 20 10	0 20 13	0 20 16	0 20 19	0 20 22	0 20 25	0 20 28	0 20 31	35m
40m	0 22 40	0 22 43	0 22 47	0 22 50	0 22 53	0 22 57	0 23 00	0 23 03	0 23 07	0 23 10	0 23 13	0 23 17	0 23 20	0 23 23	0 23 27	40m
45m	0 25 30	0 25 34	0 25 38	0 25 41	0 25 45	0 25 49	0 25 53	0 25 56	0 26 00	0 26 04	0 26 08	0 26 11	0 26 15	0 26 19	0 26 23	45m
50m	0 28 20	0 28 24	0 28 28	0 28 33	0 28 37	0 28 41	0 28 45	0 28 49	0 28 53	0 28 58	0 29 02	0 29 06	0 29 10	0 29 14	0 29 18	50m
55m	0 31 10	0 31 15	0 31 19	0 31 24	0 31 28	0 31 33	0 31 38	0 31 42	0 31 47	0 31 51	0 31 56	0 32 00	0 32 05	0 32 10	0 32 14	55m
1h	0 34 00	0 34 05	0 34 10	0 34 15	0 34 20	0 34 25	0 34 30	0 34 35	0 34 40	0 34 45	0 34 50	0 34 55	0 35 00	0 35 05	0 35 10	1h
2h	1 08 00	1 08 10	1 08 20	1 08 30	1 08 40	1 08 50	1 09 00	1 09 10	1 09 20	1 09 30	1 09 40	1 09 50	1 10 00	1 10 10	1 10 20	2h
3h	1 42 00	1 42 15	1 42 30	1 42 45	1 43 00	1 43 15	1 43 30	1 43 45	1 44 00	1 44 15	1 44 30	1 44 45	1 45 00	1 45 15	1 45 30	3h
4h	2 16 00	2 16 20	2 16 40	2 17 00	2 17 20	2 17 40	2 18 00	2 18 20	2 18 40	2 19 00	2 19 20	2 19 40	2 20 00	2 20 20	2 20 40	4h
5h	2 50 00	2 50 25	2 50 50	2 51 15	2 51 40	2 52 05	2 52 30	2 52 55	2 53 20	2 53 45	2 54 10	2 54 35	2 55 00	2 55 25	2 55 50	5h
6h	3 24 00	3 24 30	3 25 00	3 25 30	3 26 00	3 26 30	3 27 00	3 27 30	3 28 00	3 28 30	3 29 00	3 29 30	3 30 00	3 30 30	3 31 00	6h
7h	3 58 00	3 58 35	3 59 10	3 59 45	4 00 20	4 00 55	4 01 30	4 02 05	4 02 40	4 03 15	4 03 50	4 04 25	4 05 00	4 05 35	4 06 10	7h
8h	4 32 00	4 32 40	4 33 20	4 34 00	4 34 40	4 35 20	4 36 00	4 36 40	4 37 20	4 38 00	4 38 40	4 39 20	4 40 00	4 40 40	4 41 20	8h
9h	5 06 00	5 06 45	5 07 30	5 08 15	5 09 00	5 09 45	5 10 30	5 11 15	5 12 00	5 12 45	5 13 30	5 14 15	5 15 00	5 15 45	5 16 30	9h
10h	5 40 00	5 40 50	5 41 40	5 42 30	5 43 20	5 44 10	5 45 00	5 45 50	5 46 40	5 47 30	5 48 20	5 49 10	5 50 00	5 50 50	5 51 40	10h
11h	6 14 00	6 14 55	6 15 50	6 16 45	6 17 40	6 18 35	6 19 30	6 20 25	6 21 20	6 22 15	6 23 10	6 24 05	6 25 00	6 25 55	6 26 50	11h
12h	6 48 00	6 49 00	6 50 00	6 51 00	6 52 00	6 53 00	6 54 00	6 55 00	6 56 00	6 57 00	6 58 00	6 59 00	7 00 00	7 01 00	7 02 00	12h

12h	7 02 00	7 03 00	7 04 00	7 05 00	7 06 00	7 07 00	7 08 00	7 09 00	7 10 00	7 11 00	7 12 00	7 13 00	7 14 00	7 15 00	7 16 00	12h
	° ' "	° ' "	° ' "	° ' "	° ' "	° ' "	° ' "	° ' "	° ' "	° ' "	° ' "	° ' "	° ' "	° ' "	° ' "	
1m	0 00 35	0 00 35	0 00 35	0 00 35	0 00 36	0 00 36	0 00 36	0 00 36	0 00 36	0 00 36	0 00 36	0 00 36	0 00 36	0 00 36	0 00 36	1m
2m	0 01 10	0 01 11	0 01 11	0 01 11	0 01 11	0 01 11	0 01 11	0 01 12	0 01 12	0 01 12	0 01 12	0 01 12	0 01 12	0 01 13	0 01 13	2m
3m	0 01 46	0 01 46	0 01 46	0 01 46	0 01 47	0 01 47	0 01 47	0 01 47	0 01 48	0 01 48	0 01 48	0 01 48	0 01 49	0 01 49	0 01 49	3m
4m	0 02 21	0 02 21	0 02 21	0 02 22	0 02 22	0 02 22	0 02 23	0 02 23	0 02 23	0 02 24	0 02 24	0 02 24	0 02 25	0 02 25	0 02 25	4m
5m	0 02 56	0 02 56	0 02 57	0 02 57	0 02 58	0 02 58	0 02 58	0 02 59	0 02 59	0 03 00	0 03 00	0 03 00	0 03 01	0 03 01	0 03 02	5m
10m	0 05 52	0 05 53	0 05 53	0 05 54	0 05 55	0 05 56	0 05 57	0 05 58	0 05 58	0 05 59	0 06 00	0 06 01	0 06 02	0 06 03	0 06 03	10m
15m	0 08 48	0 08 49	0 08 50	0 08 51	0 08 53	0 08 54	0 08 55	0 08 56	0 08 58	0 08 59	0 09 00	0 09 01	0 09 03	0 09 04	0 09 05	15m
20m	0 11 43	0 11 45	0 11 47	0 11 48	0 11 50	0 11 52	0 11 53	0 11 55	0 11 57	0 11 58	0 12 00	0 12 02	0 12 03	0 12 05	0 12 07	20m
25m	0 14 39	0 14 41	0 14 43	0 14 45	0 14 48	0 14 50	0 14 52	0 14 54	0 14 56	0 14 58	0 15 00	0 15 02	0 15 04	0 15 06	0 15 08	25m
30m	0 17 35	0 17 38	0 17 40	0 17 43	0 17 45	0 17 48	0 17 50	0 17 53	0 17 55	0 17 58	0 18 00	0 18 03	0 18 05	0 18 08	0 18 10	30m
35m	0 20 31	0 20 34	0 20 37	0 20 40	0 20 43	0 20 45	0 20 48	0 20 51	0 20 54	0 20 57	0 21 00	0 21 03	0 21 06	0 21 09	0 21 12	35m
40m	0 23 27	0 23 30	0 23 33	0 23 37	0 23 40	0 23 43	0 23 47	0 23 50	0 23 53	0 23 57	0 24 00	0 24 03	0 24 07	0 24 10	0 24 13	40m
45m	0 26 23	0 26 26	0 26 30	0 26 34	0 26 38	0 26 41	0 26 45	0 26 49	0 26 53	0 26 56	0 27 00	0 27 04	0 27 08	0 27 11	0 27 15	45m
50m	0 29 18	0 29 23	0 29 27	0 29 31	0 29 35	0 29 39	0 29 43	0 29 48	0 29 52	0 29 56	0 30 00	0 30 04	0 30 08	0 30 13	0 30 17	50m
55m	0 32 14	0 32 19	0 32 23	0 32 28	0 32 33	0 32 37	0 32 42	0 32 46	0 32 51	0 32 55	0 33 00	0 33 05	0 33 09	0 33 14	0 33 18	55m
1h	0 35 10	0 35 15	0 35 20	0 35 25	0 35 30	0 35 35	0 35 40	0 35 45	0 35 50	0 35 55	0 36 00	0 36 05	0 36 10	0 36 15	0 36 20	1h
2h	1 10 20	1 10 30	1 10 40	1 10 50	1 11 00	1 11 10	1 11 20	1 11 30	1 11 40	1 11 50	1 12 00	1 12 10	1 12 20	1 12 30	1 12 40	2h
3h	1 45 30	1 45 45	1 46 00	1 46 15	1 46 30	1 46 45	1 47 00	1 47 15	1 47 30	1 47 45	1 48 00	1 48 15	1 48 30	1 48 45	1 49 00	3h
4h	2 20 40	2 21 00	2 21 20	2 21 40	2 22 00	2 22 20	2 22 40	2 23 00	2 23 20	2 23 40	2 24 00	2 24 20	2 24 40	2 25 00	2 25 20	4h
5h	2 55 50	2 56 15	2 56 40	2 57 05	2 57 30	2 57 55	2 58 20	2 58 45	2 59 10	2 59 35	3 00 00	3 00 25	3 00 50	3 01 15	3 01 40	5h
6h	3 31 00	3 31 30	3 32 00	3 32 30	3 33 00	3 33 30	3 34 00	3 34 30	3 35 00	3 35 30	3 36 00	3 36 30	3 37 00	3 37 30	3 38 00	6h
7h	4 06 10	4 06 45	4 07 20	4 07 55	4 08 30	4 09 05	4 09 40	4 10 15	4 10 50	4 11 25	4 12 00	4 12 35	4 13 10	4 13 45	4 14 20	7h
8h	4 41 20	4 42 00	4 42 40	4 43 20	4 44 00	4 44 40	4 45 20	4 46 00	4 46 40	4 47 20	4 48 00	4 48 40	4 49 20	4 50 00	4 50 40	8h
9h	5 16 30	5 17 15	5 18 00	5 18 45	5 19 30	5 20 15	5 21 00	5 21 45	5 22 30	5 23 15	5 24 00	5 24 45	5 25 30	5 26 15	5 27 00	9h
10h	5 51 40	5 52 30	5 53 20	5 54 10	5 55 00	5 55 50	5 56 40	5 57 30	5 58 20	5 59 10	6 00 00	6 00 50	6 01 40	6 02 30	6 03 20	10h
11h	6 26 50	6 27 45	6 28 40	6 29 35	6 30 30	6 31 25	6 32 20	6 33 15	6 34 10	6 35 05	6 36 00	6 36 55	6 37 50	6 38 45	6 39 40	11h
12h	7 02 00	7 03 00	7 04 00	7 05 00	7 06 00	7 07 00	7 08 00	7 09 00	7 10 00	7 11 00	7 12 00	7 13 00	7 14 00	7 15 00	7 16 00	12h

☽ ☽

12h	7 16 00	7 17 00	7 18 00	7 19 00	7 20 00	7 21 00	7 22 00	7 23 00	7 24 00	7 25 00	7 26 00	7 27 00	7 28 00	7 29 00	7 30 00	12h
	° ' "	° ' "	° ' "	° ' "	° ' "	° ' "	° ' "	° ' "	° ' "	° ' "	° ' "	° ' "	° ' "	° ' "	° ' "	
1m	0 00 36	0 00 36	0 00 37	0 00 37	0 00 37	0 00 37	0 00 37	0 00 37	0 00 37	0 00 37	0 00 37	0 00 37	0 00 37	0 00 37	0 00 38	1m
2m	0 01 13	0 01 13	0 01 13	0 01 13	0 01 13	0 01 14	0 01 14	0 01 14	0 01 14	0 01 14	0 01 14	0 01 15	0 01 15	0 01 15	0 01 15	2m
3m	0 01 49	0 01 49	0 01 50	0 01 50	0 01 50	0 01 50	0 01 51	0 01 51	0 01 51	0 01 51	0 01 52	0 01 52	0 01 52	0 01 52	0 01 53	3m
4m	0 02 25	0 02 26	0 02 26	0 02 26	0 02 27	0 02 27	0 02 27	0 02 28	0 02 28	0 02 28	0 02 29	0 02 29	0 02 29	0 02 30	0 02 30	4m
5m	0 03 02	0 03 02	0 03 03	0 03 03	0 03 03	0 03 04	0 03 04	0 03 05	0 03 05	0 03 05	0 03 06	0 03 06	0 03 07	0 03 07	0 03 08	5m
10m	0 06 03	0 06 04	0 06 05	0 06 06	0 06 07	0 06 08	0 06 08	0 06 09	0 06 10	0 06 11	0 06 12	0 06 13	0 06 13	0 06 14	0 06 15	10m
15m	0 09 05	0 09 06	0 09 08	0 09 09	0 09 10	0 09 11	0 09 13	0 09 14	0 09 15	0 09 16	0 09 18	0 09 19	0 09 20	0 09 21	0 09 23	15m
20m	0 12 07	0 12 08	0 12 10	0 12 12	0 12 13	0 12 15	0 12 17	0 12 18	0 12 20	0 12 22	0 12 23	0 12 25	0 12 27	0 12 28	0 12 30	20m
25m	0 15 08	0 15 10	0 15 13	0 15 15	0 15 17	0 15 19	0 15 21	0 15 23	0 15 25	0 15 27	0 15 29	0 15 31	0 15 33	0 15 35	0 15 38	25m
30m	0 18 10	0 18 13	0 18 15	0 18 18	0 18 20	0 18 23	0 18 25	0 18 28	0 18 30	0 18 33	0 18 35	0 18 38	0 18 40	0 18 43	0 18 45	30m
35m	0 21 12	0 21 15	0 21 18	0 21 20	0 21 23	0 21 26	0 21 29	0 21 32	0 21 35	0 21 38	0 21 41	0 21 44	0 21 47	0 21 50	0 21 53	35m
40m	0 24 13	0 24 17	0 24 20	0 24 23	0 24 27	0 24 30	0 24 33	0 24 37	0 24 40	0 24 43	0 24 47	0 24 50	0 24 53	0 24 57	0 25 00	40m
45m	0 27 15	0 27 19	0 27 23	0 27 26	0 27 30	0 27 34	0 27 38	0 27 41	0 27 45	0 27 49	0 27 53	0 27 56	0 28 00	0 28 04	0 28 08	45m
50m	0 30 17	0 30 21	0 30 25	0 30 29	0 30 33	0 30 38	0 30 42	0 30 46	0 30 50	0 30 54	0 30 58	0 31 03	0 31 07	0 31 11	0 31 15	50m
55m	0 33 18	0 33 23	0 33 28	0 33 32	0 33 37	0 33 41	0 33 46	0 33 50	0 33 55	0 34 00	0 34 04	0 34 09	0 34 13	0 34 18	0 34 23	55m
1h	0 36 20	0 36 25	0 36 30	0 36 35	0 36 40	0 36 45	0 36 50	0 36 55	0 37 00	0 37 05	0 37 10	0 37 15	0 37 20	0 37 25	0 37 30	1h
2h	1 12 40	1 12 50	1 13 00	1 13 10	1 13 20	1 13 30	1 13 40	1 13 50	1 14 00	1 14 10	1 14 20	1 14 30	1 14 40	1 14 50	1 15 00	2h
3h	1 49 00	1 49 15	1 49 30	1 49 45	1 50 00	1 50 15	1 50 30	1 50 45	1 51 00	1 51 15	1 51 30	1 51 45	1 52 00	1 52 15	1 52 30	3h
4h	2 25 20	2 25 40	2 26 00	2 26 20	2 26 40	2 27 00	2 27 20	2 27 40	2 28 00	2 28 20	2 28 40	2 29 00	2 29 20	2 29 40	2 30 00	4h
5h	3 01 40	3 02 05	3 02 30	3 02 55	3 03 20	3 03 45	3 04 10	3 04 35	3 05 00	3 05 25	3 05 50	3 06 15	3 06 40	3 07 05	3 07 30	5h
6h	3 38 00	3 38 30	3 39 00	3 39 30	3 40 00	3 40 30	3 41 00	3 41 30	3 42 00	3 42 30	3 43 00	3 43 30	3 44 00	3 44 30	3 45 00	6h
7h	4 14 20	4 14 55	4 15 30	4 16 05	4 16 40	4 17 15	4 17 50	4 18 25	4 19 00	4 19 35	4 20 10	4 20 45	4 21 20	4 21 55	4 22 30	7h
8h	4 50 40	4 51 20	4 52 00	4 52 40	4 53 20	4 54 00	4 54 40	4 55 20	4 56 00	4 56 40	4 57 20	4 58 00	4 58 40	4 59 20	5 00 00	8h
9h	5 27 00	5 27 45	5 28 30	5 29 15	5 30 00	5 30 45	5 31 30	5 32 15	5 33 00	5 33 45	5 34 30	5 35 15	5 36 00	5 36 45	5 37 30	9h
10h	6 03 20	6 04 10	6 05 00	6 05 50	6 06 40	6 07 30	6 08 20	6 09 10	6 10 00	6 10 50	6 11 40	6 12 30	6 13 20	6 14 10	6 15 00	10h
11h	6 39 40	6 40 35	6 41 30	6 42 25	6 43 20	6 44 15	6 45 10	6 46 05	6 47 00	6 47 55	6 48 50	6 49 45	6 50 40	6 51 35	6 52 30	11h
12h	7 16 00	7 17 00	7 18 00	7 19 00	7 20 00	7 21 00	7 22 00	7 23 00	7 24 00	7 25 00	7 26 00	7 27 00	7 28 00	7 29 00	7 30 00	12h

12h	7 30 00	7 31 00	7 32 00	7 33 00	7 34 00	7 35 00	7 36 00	7 37 00	7 38 00	7 39 00	7 40 00	7 41 00	12h
	° ' "	° ' "	° ' "	° ' "	° ' "	° ' "	° ' "	° ' "	° ' "	° ' "	° ' "	° ' "	
1m	0 00 38	0 00 38	0 00 38	0 00 38	0 00 38	0 00 38	0 00 38	0 00 38	0 00 38	0 00 38	0 00 38	0 00 38	1m
2m	0 01 15	0 01 15	0 01 15	0 01 16	0 01 16	0 01 16	0 01 16	0 01 16	0 01 16	0 01 17	0 01 17	0 01 17	2m
3m	0 01 53	0 01 53	0 01 53	0 01 53	0 01 54	0 01 54	0 01 54	0 01 54	0 01 55	0 01 55	0 01 55	0 01 55	3m
4m	0 02 30	0 02 30	0 02 31	0 02 31	0 02 31	0 02 32	0 02 32	0 02 32	0 02 33	0 02 33	0 02 33	0 02 34	4m
5m	0 03 08	0 03 08	0 03 08	0 03 09	0 03 09	0 03 10	0 03 10	0 03 10	0 03 11	0 03 11	0 03 12	0 03 12	5m
10m	0 06 15	0 06 16	0 06 17	0 06 18	0 06 18	0 06 19	0 06 20	0 06 21	0 06 22	0 06 23	0 06 23	0 06 24	10m
15m	0 09 23	0 09 24	0 09 25	0 09 26	0 09 28	0 09 29	0 09 30	0 09 31	0 09 33	0 09 34	0 09 35	0 09 36	15m
20m	0 12 30	0 12 32	0 12 33	0 12 35	0 12 37	0 12 38	0 12 40	0 12 42	0 12 43	0 12 45	0 12 47	0 12 48	20m
25m	0 15 38	0 15 40	0 15 42	0 15 44	0 15 46	0 15 48	0 15 50	0 15 52	0 15 54	0 15 56	0 15 58	0 16 00	25m
30m	0 18 45	0 18 48	0 18 50	0 18 53	0 18 55	0 18 58	0 19 00	0 19 03	0 19 05	0 19 08	0 19 10	0 19 13	30m
35m	0 21 53	0 21 55	0 21 58	0 22 01	0 22 04	0 22 07	0 22 10	0 22 13	0 22 16	0 22 19	0 22 22	0 22 25	35m
40m	0 25 00	0 25 03	0 25 07	0 25 10	0 25 13	0 25 17	0 25 20	0 25 23	0 25 27	0 25 30	0 25 33	0 25 37	40m
45m	0 28 08	0 28 11	0 28 15	0 28 19	0 28 23	0 28 26	0 28 30	0 28 34	0 28 38	0 28 41	0 28 45	0 28 49	45m
50m	0 31 15	0 31 19	0 31 23	0 31 28	0 31 32	0 31 36	0 31 40	0 31 44	0 31 48	0 31 53	0 31 57	0 32 01	50m
55m	0 34 23	0 34 27	0 34 32	0 34 36	0 34 41	0 34 45	0 34 50	0 34 55	0 34 59	0 35 04	0 35 08	0 35 13	55m
1h	0 37 30	0 37 35	0 37 40	0 37 45	0 37 50	0 37 55	0 38 00	0 38 05	0 38 10	0 38 15	0 38 20	0 38 25	1h
2h	1 15 00	1 15 10	1 15 20	1 15 30	1 15 40	1 15 50	1 16 00	1 16 10	1 16 20	1 16 30	1 16 40	1 16 50	2h
3h	1 52 30	1 52 45	1 53 00	1 53 15	1 53 30	1 53 45	1 54 00	1 54 15	1 54 30	1 54 45	1 55 00	1 55 15	3h
4h	2 30 00	2 30 20	2 30 40	2 31 00	2 31 20	2 31 40	2 32 00	2 32 20	2 32 40	2 33 00	2 33 20	2 33 40	4h
5h	3 07 30	3 07 55	3 08 20	3 08 45	3 09 10	3 09 35	3 10 00	3 10 25	3 10 50	3 11 15	3 11 40	3 12 05	5h
6h	3 45 00	3 45 30	3 46 00	3 46 30	3 47 00	3 47 30	3 48 00	3 48 30	3 49 00	3 49 30	3 50 00	3 50 30	6h
7h	4 22 30	4 23 05	4 23 40	4 24 15	4 24 50	4 25 25	4 26 00	4 26 35	4 27 10	4 27 45	4 28 20	4 28 55	7h
8h	5 00 00	5 00 40	5 01 20	5 02 00	5 02 40	5 03 20	5 04 00	5 04 40	5 05 20	5 06 00	5 06 40	5 07 20	8h
9h	5 37 30	5 38 15	5 39 00	5 39 45	5 40 30	5 41 15	5 42 00	5 42 45	5 43 30	5 44 15	5 45 00	5 45 45	9h
10h	6 15 00	6 15 50	6 16 40	6 17 30	6 18 20	6 19 10	6 20 00	6 20 50	6 21 40	6 22 30	6 23 20	6 24 10	10h
11h	6 52 30	6 53 25	6 54 20	6 55 15	6 56 10	6 57 05	6 58 00	6 58 55	6 59 50	7 00 45	7 01 40	7 02 35	11h
12h	7 30 00	7 31 00	7 32 00	7 33 00	7 34 00	7 35 00	7 36 00	7 37 00	7 38 00	7 39 00	7 40 00	7 41 00	12h

Table 3 – Daily Planetary Movements

24h	0°02'	0°04'	0°06'	0°08'	0°10'	0°12'	0°14'	0°16'	0°18'	0°20'	0°22'	0°24'	0°26'	0°28'	0°30'	0°32'	0°34'	0°36'	24h
1m	0 00.0	0 00.0	0 00.0	0 00.0	0 00.0	0 00.0	0 00.0	0 00.0	0 00.0	0 00.0	0 00.0	0 00.0	0 00.0	0 00.0	0 00.0	0 00.0	0 00.0	0 00.0	*1m*
2m	0 00.0	0 00.0	0 00.0	0 00.0	0 00.0	0 00.0	0 00.0	0 00.0	0 00.0	0 00.0	0 00.0	0 00.0	0 00.0	0 00.0	0 00.0	0 00.0	0 00.0	0 00.1	*2m*
3m	0 00.0	0 00.0	0 00.0	0 00.0	0 00.0	0 00.0	0 00.0	0 00.0	0 00.0	0 00.0	0 00.0	0 00.1	0 00.1	0 00.1	0 00.1	0 00.1	0 00.1	0 00.1	*3m*
4m	0 00.0	0 00.0	0 00.0	0 00.0	0 00.0	0 00.0	0 00.0	0 00.0	0 00.1	0 00.1	0 00.1	0 00.1	0 00.1	0 00.1	0 00.1	0 00.1	0 00.1	0 00.1	*4m*
5m	0 00.0	0 00.0	0 00.0	0 00.0	0 00.0	0 00.0	0 00.0	0 00.1	0 00.1	0 00.1	0 00.1	0 00.1	0 00.1	0 00.1	0 00.1	0 00.1	0 00.1	0 00.1	*5m*
10m	0 00.0	0 00.0	0 00.0	0 00.1	0 00.1	0 00.1	0 00.1	0 00.1	0 00.1	0 00.1	0 00.2	0 00.2	0 00.2	0 00.2	0 00.2	0 00.2	0 00.2	0 00.3	*10m*
15m	0 00.0	0 00.0	0 00.1	0 00.1	0 00.1	0 00.1	0 00.1	0 00.2	0 00.2	0 00.2	0 00.2	0 00.3	0 00.3	0 00.3	0 00.3	0 00.3	0 00.4	0 00.4	*15m*
20m	0 00.0	0 00.1	0 00.1	0 00.1	0 00.1	0 00.2	0 00.2	0 00.2	0 00.3	0 00.3	0 00.3	0 00.3	0 00.4	0 00.4	0 00.4	0 00.4	0 00.5	0 00.5	*20m*
25m	0 00.0	0 00.1	0 00.1	0 00.1	0 00.2	0 00.2	0 00.2	0 00.3	0 00.3	0 00.3	0 00.4	0 00.4	0 00.5	0 00.5	0 00.5	0 00.6	0 00.6	0 00.6	*25m*
30m	0 00.0	0 00.1	0 00.1	0 00.2	0 00.2	0 00.3	0 00.3	0 00.3	0 00.4	0 00.4	0 00.5	0 00.5	0 00.5	0 00.6	0 00.6	0 00.7	0 00.7	0 00.8	*30m*
35m	0 00.0	0 00.1	0 00.1	0 00.2	0 00.2	0 00.3	0 00.3	0 00.4	0 00.4	0 00.5	0 00.5	0 00.6	0 00.6	0 00.7	0 00.7	0 00.8	0 00.8	0 00.9	*35m*
40m	0 00.1	0 00.1	0 00.2	0 00.2	0 00.3	0 00.3	0 00.4	0 00.4	0 00.5	0 00.6	0 00.6	0 00.7	0 00.7	0 00.8	0 00.8	0 00.9	0 00.9	0 01.0	*40m*
45m	0 00.1	0 00.1	0 00.2	0 00.3	0 00.3	0 00.4	0 00.4	0 00.5	0 00.6	0 00.6	0 00.7	0 00.8	0 00.8	0 00.9	0 00.9	0 01.0	0 01.1	0 01.1	*45m*
50m	0 00.1	0 00.1	0 00.2	0 00.3	0 00.3	0 00.4	0 00.5	0 00.6	0 00.6	0 00.7	0 00.8	0 00.8	0 00.9	0 01.0	0 01.0	0 01.1	0 01.2	0 01.3	*50m*
55m	0 00.1	0 00.2	0 00.2	0 00.3	0 00.4	0 00.5	0 00.5	0 00.6	0 00.7	0 00.8	0 00.8	0 00.9	0 01.0	0 01.1	0 01.1	0 01.2	0 01.3	0 01.4	*55m*
1h	0 00.1	0 00.2	0 00.3	0 00.3	0 00.4	0 00.5	0 00.6	0 00.7	0 00.8	0 00.8	0 00.9	0 01.0	0 01.1	0 01.2	0 01.3	0 01.3	0 01.4	0 01.5	*1h*
2h	0 00.2	0 00.3	0 00.5	0 00.7	0 00.8	0 01.0	0 01.2	0 01.3	0 01.5	0 01.7	0 01.8	0 02.0	0 02.2	0 02.3	0 02.5	0 02.7	0 02.8	0 03.0	*2h*
3h	0 00.3	0 00.5	0 00.8	0 01.0	0 01.3	0 01.5	0 01.8	0 02.0	0 02.3	0 02.5	0 02.8	0 03.0	0 03.3	0 03.5	0 03.8	0 04.0	0 04.3	0 04.5	*3h*
4h	0 00.3	0 00.7	0 01.0	0 01.3	0 01.7	0 02.0	0 02.3	0 02.7	0 03.0	0 03.3	0 03.7	0 04.0	0 04.3	0 04.7	0 05.0	0 05.3	0 05.7	0 06.0	*4h*
5h	0 00.4	0 00.8	0 01.3	0 01.7	0 02.1	0 02.5	0 02.9	0 03.3	0 03.8	0 04.2	0 04.6	0 05.0	0 05.4	0 05.8	0 06.3	0 06.7	0 07.1	0 07.5	*5h*
6h	0 00.5	0 01.0	0 01.5	0 02.0	0 02.5	0 03.0	0 03.5	0 04.0	0 04.5	0 05.0	0 05.5	0 06.0	0 06.5	0 07.0	0 07.5	0 08.0	0 08.5	0 09.0	*6h*
7h	0 00.6	0 01.2	0 01.8	0 02.3	0 02.9	0 03.5	0 04.1	0 04.7	0 05.3	0 05.8	0 06.4	0 07.0	0 07.6	0 08.2	0 08.8	0 09.3	0 09.9	0 10.5	*7h*
8h	0 00.7	0 01.3	0 02.0	0 02.7	0 03.3	0 04.0	0 04.7	0 05.3	0 06.0	0 06.7	0 07.3	0 08.0	0 08.7	0 09.3	0 10.0	0 10.7	0 11.3	0 12.0	*8h*
9h	0 00.8	0 01.5	0 02.3	0 03.0	0 03.8	0 04.5	0 05.3	0 06.0	0 06.8	0 07.5	0 08.3	0 09.0	0 09.8	0 10.5	0 11.3	0 12.0	0 12.8	0 13.5	*9h*
10h	0 00.8	0 01.7	0 02.5	0 03.3	0 04.2	0 05.0	0 05.8	0 06.7	0 07.5	0 08.3	0 09.2	0 10.0	0 10.8	0 11.7	0 12.5	0 13.3	0 14.2	0 15.0	*10h*
11h	0 00.9	0 01.8	0 02.8	0 03.7	0 04.6	0 05.5	0 06.4	0 07.3	0 08.3	0 09.2	0 10.1	0 11.0	0 11.9	0 12.8	0 13.8	0 14.7	0 15.6	0 16.5	*11h*
12h	0 01.0	0 02.0	0 03.0	0 04.0	0 05.0	0 06.0	0 07.0	0 08.0	0 09.0	0 10.0	0 11.0	0 12.0	0 13.0	0 14.0	0 15.0	0 16.0	0 17.0	0 18.0	*12h*
13h	0 01.1	0 02.2	0 03.3	0 04.3	0 05.4	0 06.5	0 07.6	0 08.7	0 09.8	0 10.8	0 11.9	0 13.0	0 14.1	0 15.2	0 16.3	0 17.3	0 18.4	0 19.5	*13h*
14h	0 01.2	0 02.3	0 03.5	0 04.7	0 05.8	0 07.0	0 08.2	0 09.3	0 10.5	0 11.7	0 12.8	0 14.0	0 15.2	0 16.3	0 17.5	0 18.7	0 19.8	0 21.0	*14h*
15h	0 01.3	0 02.5	0 03.8	0 05.0	0 06.3	0 07.5	0 08.8	0 10.0	0 11.3	0 12.5	0 13.8	0 15.0	0 16.3	0 17.5	0 18.8	0 20.0	0 21.3	0 22.5	*15h*
16h	0 01.3	0 02.7	0 04.0	0 05.3	0 06.7	0 08.0	0 09.3	0 10.7	0 12.0	0 13.3	0 14.7	0 16.0	0 17.3	0 18.7	0 20.0	0 21.3	0 22.7	0 24.0	*16h*
17h	0 01.4	0 02.8	0 04.3	0 05.7	0 07.1	0 08.5	0 09.9	0 11.3	0 12.8	0 14.2	0 15.6	0 17.0	0 18.4	0 19.8	0 21.3	0 22.7	0 24.1	0 25.5	*17h*
18h	0 01.5	0 03.0	0 04.5	0 06.0	0 07.5	0 09.0	0 10.5	0 12.0	0 13.5	0 15.0	0 16.5	0 18.0	0 19.5	0 21.0	0 22.5	0 24.0	0 25.5	0 27.0	*18h*
19h	0 01.6	0 03.2	0 04.8	0 06.3	0 07.9	0 09.5	0 11.1	0 12.7	0 14.3	0 15.8	0 17.4	0 19.0	0 20.6	0 22.2	0 23.8	0 25.3	0 26.9	0 28.5	*19h*
20h	0 01.7	0 03.3	0 05.0	0 06.7	0 08.3	0 10.0	0 11.7	0 13.3	0 15.0	0 16.7	0 18.3	0 20.0	0 21.7	0 23.3	0 25.0	0 26.7	0 28.3	0 30.0	*20h*
21h	0 01.8	0 03.5	0 05.3	0 07.0	0 08.8	0 10.5	0 12.3	0 14.0	0 15.8	0 17.5	0 19.3	0 21.0	0 22.8	0 24.5	0 26.3	0 28.0	0 29.8	0 31.5	*21h*
22h	0 01.8	0 03.7	0 05.5	0 07.3	0 09.2	0 11.0	0 12.8	0 14.7	0 16.5	0 18.3	0 20.2	0 22.0	0 23.8	0 25.7	0 27.5	0 29.3	0 31.2	0 33.0	*22h*
23h	0 01.9	0 03.8	0 05.8	0 07.7	0 09.6	0 11.5	0 13.4	0 15.3	0 17.3	0 19.2	0 21.1	0 23.0	0 24.9	0 26.8	0 28.8	0 30.7	0 32.6	0 34.5	*23h*
24h	0°02'	0°04'	0°06'	0°08'	0°10'	0°12'	0°14'	0°16'	0°18'	0°20'	0°22'	0°24'	0°26'	0°28'	0°30'	0°32'	0°34'	0°36'	24h

24h	0°36'	0°38'	0°40'	0°42'	0°44'	0°46'	0°48'	0°50'	0°52'	0°54'	0°56'	0°58'	1°00'	1°02'	1°04'	1°06'	1°08'	1°10'	24h
1m	0 00.0	0 00.0	0 00.0	0 00.0	0 00.0	0 00.0	0 00.0	0 00.0	0 00.0	0 00.0	0 00.0	0 00.0	0 00.0	0 00.0	0 00.0	0 00.0	0 00.0	0 00.0	*1m*
2m	0 00.1	0 00.1	0 00.1	0 00.1	0 00.1	0 00.1	0 00.1	0 00.1	0 00.1	0 00.1	0 00.1	0 00.1	0 00.1	0 00.1	0 00.1	0 00.1	0 00.1	0 00.1	*2m*
3m	0 00.1	0 00.1	0 00.1	0 00.1	0 00.1	0 00.1	0 00.1	0 00.1	0 00.1	0 00.1	0 00.1	0 00.1	0 00.1	0 00.1	0 00.1	0 00.1	0 00.1	0 00.1	*3m*
4m	0 00.1	0 00.1	0 00.1	0 00.1	0 00.1	0 00.1	0 00.1	0 00.1	0 00.1	0 00.2	0 00.2	0 00.2	0 00.2	0 00.2	0 00.2	0 00.2	0 00.2	0 00.2	*4m*
5m	0 00.1	0 00.1	0 00.1	0 00.1	0 00.2	0 00.2	0 00.2	0 00.2	0 00.2	0 00.2	0 00.2	0 00.2	0 00.2	0 00.2	0 00.2	0 00.2	0 00.2	0 00.2	*5m*
10m	0 00.3	0 00.3	0 00.3	0 00.3	0 00.3	0 00.3	0 00.3	0 00.3	0 00.4	0 00.4	0 00.4	0 00.4	0 00.4	0 00.4	0 00.4	0 00.5	0 00.5	0 00.5	*10m*
15m	0 00.4	0 00.4	0 00.4	0 00.4	0 00.5	0 00.5	0 00.5	0 00.5	0 00.5	0 00.6	0 00.6	0 00.6	0 00.6	0 00.6	0 00.7	0 00.7	0 00.7	0 00.7	*15m*
20m	0 00.5	0 00.5	0 00.6	0 00.6	0 00.6	0 00.6	0 00.7	0 00.7	0 00.7	0 00.8	0 00.8	0 00.8	0 00.8	0 00.9	0 00.9	0 00.9	0 00.9	0 01.0	*20m*
25m	0 00.6	0 00.7	0 00.7	0 00.7	0 00.8	0 00.8	0 00.8	0 00.9	0 00.9	0 00.9	0 01.0	0 01.0	0 01.0	0 01.1	0 01.1	0 01.1	0 01.2	0 01.2	*25m*
30m	0 00.8	0 00.8	0 00.8	0 00.9	0 00.9	0 01.0	0 01.0	0 01.0	0 01.1	0 01.1	0 01.2	0 01.2	0 01.3	0 01.3	0 01.3	0 01.4	0 01.4	0 01.5	*30m*
35m	0 00.9	0 00.9	0 01.0	0 01.0	0 01.1	0 01.1	0 01.2	0 01.2	0 01.3	0 01.3	0 01.4	0 01.4	0 01.5	0 01.5	0 01.6	0 01.6	0 01.7	0 01.7	*35m*
40m	0 01.0	0 01.1	0 01.1	0 01.2	0 01.2	0 01.3	0 01.3	0 01.4	0 01.4	0 01.5	0 01.6	0 01.6	0 01.7	0 01.7	0 01.8	0 01.8	0 01.9	0 01.9	*40m*
45m	0 01.1	0 01.2	0 01.3	0 01.3	0 01.4	0 01.4	0 01.5	0 01.6	0 01.6	0 01.7	0 01.8	0 01.8	0 01.9	0 01.9	0 02.0	0 02.1	0 02.1	0 02.2	*45m*
50m	0 01.3	0 01.3	0 01.4	0 01.5	0 01.5	0 01.6	0 01.7	0 01.7	0 01.8	0 01.9	0 01.9	0 02.0	0 02.1	0 02.2	0 02.2	0 02.3	0 02.4	0 02.4	*50m*
55m	0 01.4	0 01.5	0 01.5	0 01.6	0 01.7	0 01.8	0 01.8	0 01.9	0 02.0	0 02.1	0 02.1	0 02.2	0 02.3	0 02.4	0 02.4	0 02.5	0 02.6	0 02.7	*55m*
1h	0 01.5	0 01.6	0 01.7	0 01.8	0 01.8	0 01.9	0 02.0	0 02.1	0 02.2	0 02.3	0 02.3	0 02.4	0 02.5	0 02.6	0 02.7	0 02.8	0 02.8	0 02.9	*1h*
2h	0 03.0	0 03.2	0 03.3	0 03.5	0 03.7	0 03.8	0 04.0	0 04.2	0 04.3	0 04.5	0 04.7	0 04.8	0 05.0	0 05.2	0 05.3	0 05.5	0 05.7	0 05.8	*2h*
3h	0 04.5	0 04.8	0 05.0	0 05.3	0 05.5	0 05.8	0 06.0	0 06.3	0 06.5	0 06.8	0 07.0	0 07.3	0 07.5	0 07.8	0 08.0	0 08.3	0 08.5	0 08.8	*3h*
4h	0 06.0	0 06.3	0 06.7	0 07.0	0 07.3	0 07.7	0 08.0	0 08.3	0 08.7	0 09.0	0 09.3	0 09.7	0 10.0	0 10.3	0 10.7	0 11.0	0 11.3	0 11.7	*4h*
5h	0 07.5	0 07.9	0 08.3	0 08.8	0 09.2	0 09.6	0 10.0	0 10.4	0 10.8	0 11.3	0 11.7	0 12.1	0 12.5	0 12.9	0 13.3	0 13.8	0 14.2	0 14.6	*5h*
6h	0 09.0	0 09.5	0 10.0	0 10.5	0 11.0	0 11.5	0 12.0	0 12.5	0 13.0	0 13.5	0 14.0	0 14.5	0 15.0	0 15.5	0 16.0	0 16.5	0 17.0	0 17.5	*6h*
7h	0 10.5	0 11.1	0 11.7	0 12.3	0 12.8	0 13.4	0 14.0	0 14.6	0 15.2	0 15.8	0 16.3	0 16.9	0 17.5	0 18.1	0 18.7	0 19.3	0 19.8	0 20.4	*7h*
8h	0 12.0	0 12.7	0 13.3	0 14.0	0 14.7	0 15.3	0 16.0	0 16.7	0 17.3	0 18.0	0 18.7	0 19.3	0 20.0	0 20.7	0 21.3	0 22.0	0 22.7	0 23.3	*8h*
9h	0 13.5	0 14.3	0 15.0	0 15.8	0 16.5	0 17.3	0 18.0	0 18.8	0 19.5	0 20.3	0 21.0	0 21.8	0 22.5	0 23.3	0 24.0	0 24.8	0 25.5	0 26.3	*9h*
10h	0 15.0	0 15.8	0 16.7	0 17.5	0 18.3	0 19.2	0 20.0	0 20.8	0 21.7	0 22.5	0 23.3	0 24.2	0 25.0	0 25.8	0 26.7	0 27.5	0 28.3	0 29.2	*10h*
11h	0 16.5	0 17.4	0 18.3	0 19.3	0 20.2	0 21.1	0 22.0	0 22.9	0 23.8	0 24.8	0 25.7	0 26.6	0 27.5	0 28.4	0 29.3	0 30.3	0 31.2	0 32.1	*11h*
12h	0 18.0	0 19.0	0 20.0	0 21.0	0 22.0	0 23.0	0 24.0	0 25.0	0 26.0	0 27.0	0 28.0	0 29.0	0 30.0	0 31.0	0 32.0	0 33.0	0 34.0	0 35.0	*12h*
13h	0 19.5	0 20.6	0 21.7	0 22.8	0 23.8	0 24.9	0 26.0	0 27.1	0 28.2	0 29.3	0 30.3	0 31.4	0 32.5	0 33.6	0 34.7	0 35.8	0 36.8	0 37.9	*13h*
14h	0 21.0	0 22.2	0 23.3	0 24.5	0 25.7	0 26.8	0 28.0	0 29.2	0 30.3	0 31.5	0 32.7	0 33.8	0 35.0	0 36.2	0 37.3	0 38.5	0 39.7	0 40.8	*14h*
15h	0 22.5	0 23.8	0 25.0	0 26.3	0 27.5	0 28.8	0 30.0	0 31.3	0 32.5	0 33.8	0 35.0	0 36.3	0 37.5	0 38.8	0 40.0	0 41.3	0 42.5	0 43.8	*15h*
16h	0 24.0	0 25.3	0 26.7	0 28.0	0 29.3	0 30.7	0 32.0	0 33.3	0 34.7	0 36.0	0 37.3	0 38.7	0 40.0	0 41.3	0 42.7	0 44.0	0 45.3	0 46.7	*16h*
17h	0 25.5	0 26.9	0 28.3	0 29.8	0 31.2	0 32.6	0 34.0	0 35.4	0 36.8	0 38.3	0 39.7	0 41.1	0 42.5	0 43.9	0 45.3	0 46.8	0 48.2	0 49.6	*17h*
18h	0 27.0	0 28.5	0 30.0	0 31.5	0 33.0	0 34.5	0 36.0	0 37.5	0 39.0	0 40.5	0 42.0	0 43.5	0 45.0	0 46.5	0 48.0	0 49.5	0 51.0	0 52.5	*18h*
19h	0 28.5	0 30.1	0 31.7	0 33.3	0 34.8	0 36.4	0 38.0	0 39.6	0 41.2	0 42.8	0 44.3	0 45.9	0 47.5	0 49.1	0 50.7	0 52.3	0 53.8	0 55.4	*19h*
20h	0 30.0	0 31.7	0 33.3	0 35.0	0 36.7	0 38.3	0 40.0	0 41.7	0 43.3	0 45.0	0 46.7	0 48.3	0 50.0	0 51.7	0 53.3	0 55.0	0 56.7	0 58.3	*20h*
21h	0 31.5	0 33.3	0 35.0	0 36.8	0 38.5	0 40.3	0 42.0	0 43.8	0 45.5	0 47.3	0 49.0	0 50.8	0 52.5	0 54.3	0 56.0	0 57.8	0 59.5	1 01.3	*21h*
22h	0 33.0	0 34.8	0 36.7	0 38.5	0 40.3	0 42.2	0 44.0	0 45.8	0 47.7	0 49.5	0 51.3	0 53.2	0 55.0	0 56.8	0 58.7	1 00.5	1 02.3	1 04.2	*22h*
23h	0 34.5	0 36.4	0 38.3	0 40.3	0 42.2	0 44.1	0 46.0	0 47.9	0 49.8	0 51.8	0 53.7	0 55.6	0 57.5	0 59.4	1 01.3	1 03.3	1 05.2	1 07.1	*23h*
24h	0°36'	0°38'	0°40'	0°42'	0°44'	0°46'	0°48'	0°50'	0°52'	0°54'	0°56'	0°58'	1°00'	1°02'	1°04'	1°06'	1°08'	1°10'	24h

24h	1°10'	1°12'	1°14'	1°16'	1°18'	1°20'	1°22'	1°24'	1°26'	1°28'	1°30'	1°32'	1°34'	1°36'	1°38'	1°40'	1°42'	1°44'	24h
1m	0 00.0	0 00.1	0 00.1	0 00.1	0 00.1	0 00.1	0 00.1	0 00.1	0 00.1	0 00.1	0 00.1	0 00.1	0 00.1	0 00.1	0 00.1	0 00.1	0 00.1	0 00.1	1m
2m	0 00.1	0 00.1	0 00.1	0 00.1	0 00.1	0 00.1	0 00.1	0 00.1	0 00.1	0 00.1	0 00.1	0 00.1	0 00.1	0 00.1	0 00.1	0 00.1	0 00.1	0 00.1	2m
3m	0 00.1	0 00.2	0 00.2	0 00.2	0 00.2	0 00.2	0 00.2	0 00.2	0 00.2	0 00.2	0 00.2	0 00.2	0 00.2	0 00.2	0 00.2	0 00.2	0 00.2	0 00.2	3m
4m	0 00.2	0 00.2	0 00.2	0 00.2	0 00.2	0 00.2	0 00.2	0 00.2	0 00.2	0 00.2	0 00.3	0 00.3	0 00.3	0 00.3	0 00.3	0 00.3	0 00.3	0 00.3	4m
5m	0 00.2	0 00.3	0 00.3	0 00.3	0 00.3	0 00.3	0 00.3	0 00.3	0 00.3	0 00.3	0 00.3	0 00.3	0 00.3	0 00.3	0 00.3	0 00.3	0 00.4	0 00.4	5m
10m	0 00.5	0 00.5	0 00.5	0 00.5	0 00.5	0 00.6	0 00.6	0 00.6	0 00.6	0 00.6	0 00.6	0 00.6	0 00.7	0 00.7	0 00.7	0 00.7	0 00.7	0 00.7	10m
15m	0 00.7	0 00.8	0 00.8	0 00.8	0 00.8	0 00.8	0 00.9	0 00.9	0 00.9	0 00.9	0 00.9	0 01.0	0 01.0	0 01.0	0 01.0	0 01.0	0 01.1	0 01.1	15m
20m	0 01.0	0 01.0	0 01.0	0 01.1	0 01.1	0 01.1	0 01.1	0 01.2	0 01.2	0 01.2	0 01.3	0 01.3	0 01.3	0 01.3	0 01.4	0 01.4	0 01.4	0 01.4	20m
25m	0 01.2	0 01.3	0 01.3	0 01.3	0 01.4	0 01.4	0 01.4	0 01.5	0 01.5	0 01.5	0 01.6	0 01.6	0 01.6	0 01.7	0 01.7	0 01.7	0 01.8	0 01.8	25m
30m	0 01.5	0 01.5	0 01.5	0 01.6	0 01.6	0 01.7	0 01.7	0 01.8	0 01.8	0 01.8	0 01.9	0 01.9	0 02.0	0 02.0	0 02.0	0 02.1	0 02.1	0 02.2	30m
35m	0 01.7	0 01.8	0 01.8	0 01.8	0 01.9	0 01.9	0 02.0	0 02.0	0 02.1	0 02.1	0 02.2	0 02.2	0 02.3	0 02.3	0 02.4	0 02.4	0 02.5	0 02.5	35m
40m	0 01.9	0 02.0	0 02.1	0 02.1	0 02.2	0 02.2	0 02.3	0 02.3	0 02.4	0 02.4	0 02.5	0 02.6	0 02.6	0 02.7	0 02.7	0 02.8	0 02.8	0 02.9	40m
45m	0 02.2	0 02.3	0 02.3	0 02.4	0 02.4	0 02.5	0 02.6	0 02.6	0 02.7	0 02.8	0 02.8	0 02.9	0 02.9	0 03.0	0 03.1	0 03.1	0 03.2	0 03.3	45m
50m	0 02.4	0 02.5	0 02.6	0 02.6	0 02.7	0 02.8	0 02.8	0 02.9	0 03.0	0 03.1	0 03.1	0 03.2	0 03.3	0 03.3	0 03.4	0 03.5	0 03.5	0 03.6	50m
55m	0 02.7	0 02.8	0 02.8	0 02.9	0 03.0	0 03.1	0 03.1	0 03.2	0 03.3	0 03.4	0 03.4	0 03.5	0 03.6	0 03.7	0 03.7	0 03.8	0 03.9	0 04.0	55m
1h	0 02.9	0 03.0	0 03.1	0 03.2	0 03.3	0 03.3	0 03.4	0 03.5	0 03.6	0 03.7	0 03.8	0 03.8	0 03.9	0 04.0	0 04.1	0 04.2	0 04.3	0 04.3	1h
2h	0 05.8	0 06.0	0 06.2	0 06.3	0 06.5	0 06.7	0 06.8	0 07.0	0 07.2	0 07.3	0 07.5	0 07.7	0 07.8	0 08.0	0 08.2	0 08.3	0 08.5	0 08.7	2h
3h	0 08.8	0 09.0	0 09.3	0 09.5	0 09.8	0 10.0	0 10.3	0 10.5	0 10.8	0 11.0	0 11.3	0 11.5	0 11.8	0 12.0	0 12.3	0 12.5	0 12.8	0 13.0	3h
4h	0 11.7	0 12.0	0 12.3	0 12.7	0 13.0	0 13.3	0 13.7	0 14.0	0 14.3	0 14.7	0 15.0	0 15.3	0 15.7	0 16.0	0 16.3	0 16.7	0 17.0	0 17.3	4h
5h	0 14.6	0 15.0	0 15.4	0 15.8	0 16.3	0 16.7	0 17.1	0 17.5	0 17.9	0 18.3	0 18.8	0 19.2	0 19.6	0 20.0	0 20.4	0 20.8	0 21.3	0 21.7	5h
6h	0 17.5	0 18.0	0 18.5	0 19.0	0 19.5	0 20.0	0 20.5	0 21.0	0 21.5	0 22.0	0 22.5	0 23.0	0 23.5	0 24.0	0 24.5	0 25.0	0 25.5	0 26.0	6h
7h	0 20.4	0 21.0	0 21.6	0 22.2	0 22.8	0 23.3	0 23.9	0 24.5	0 25.1	0 25.7	0 26.3	0 26.8	0 27.4	0 28.0	0 28.6	0 29.2	0 29.8	0 30.3	7h
8h	0 23.3	0 24.0	0 24.7	0 25.3	0 26.0	0 26.7	0 27.3	0 28.0	0 28.7	0 29.3	0 30.0	0 30.7	0 31.3	0 32.0	0 32.7	0 33.3	0 34.0	0 34.7	8h
9h	0 26.3	0 27.0	0 27.8	0 28.5	0 29.3	0 30.0	0 30.8	0 31.5	0 32.3	0 33.0	0 33.8	0 34.5	0 35.3	0 36.0	0 36.8	0 37.5	0 38.3	0 39.0	9h
10h	0 29.2	0 30.0	0 30.8	0 31.7	0 32.5	0 33.3	0 34.2	0 35.0	0 35.8	0 36.7	0 37.5	0 38.3	0 39.2	0 40.0	0 40.8	0 41.7	0 42.5	0 43.3	10h
11h	0 32.1	0 33.0	0 33.9	0 34.8	0 35.8	0 36.7	0 37.6	0 38.5	0 39.4	0 40.3	0 41.3	0 42.2	0 43.1	0 44.0	0 44.9	0 45.8	0 46.8	0 47.7	11h
12h	0 35.0	0 36.0	0 37.0	0 38.0	0 39.0	0 40.0	0 41.0	0 42.0	0 43.0	0 44.0	0 45.0	0 46.0	0 47.0	0 48.0	0 49.0	0 50.0	0 51.0	0 52.0	12h
13h	0 37.9	0 39.0	0 40.1	0 41.2	0 42.3	0 43.3	0 44.4	0 45.5	0 46.6	0 47.7	0 48.8	0 49.8	0 50.9	0 52.0	0 53.1	0 54.2	0 55.3	0 56.3	13h
14h	0 40.8	0 42.0	0 43.2	0 44.3	0 45.5	0 46.7	0 47.8	0 49.0	0 50.2	0 51.3	0 52.5	0 53.7	0 54.8	0 56.0	0 57.2	0 58.3	0 59.5	1 00.7	14h
15h	0 43.8	0 45.0	0 46.3	0 47.5	0 48.8	0 50.0	0 51.3	0 52.5	0 53.8	0 55.0	0 56.3	0 57.5	0 58.8	1 00.0	1 01.3	1 02.5	1 03.8	1 05.0	15h
16h	0 46.7	0 48.0	0 49.3	0 50.7	0 52.0	0 53.3	0 54.7	0 56.0	0 57.3	0 58.7	1 00.0	1 01.3	1 02.7	1 04.0	1 05.3	1 06.7	1 08.0	1 09.3	16h
17h	0 49.6	0 51.0	0 52.4	0 53.8	0 55.3	0 56.7	0 58.1	0 59.5	1 00.9	1 02.3	1 03.8	1 05.2	1 06.6	1 08.0	1 09.4	1 10.8	1 12.3	1 13.7	17h
18h	0 52.5	0 54.0	0 55.5	0 57.0	0 58.5	1 00.0	1 01.5	1 03.0	1 04.5	1 06.0	1 07.5	1 09.0	1 10.5	1 12.0	1 13.5	1 15.0	1 16.5	1 18.0	18h
19h	0 55.4	0 57.0	0 58.6	1 00.2	1 01.8	1 03.3	1 04.9	1 06.5	1 08.1	1 09.7	1 11.3	1 12.8	1 14.4	1 16.0	1 17.6	1 19.2	1 20.8	1 22.3	19h
20h	0 58.3	1 00.0	1 01.7	1 03.3	1 05.0	1 06.7	1 08.3	1 10.0	1 11.7	1 13.3	1 15.0	1 16.7	1 18.3	1 20.0	1 21.7	1 23.3	1 25.0	1 26.7	20h
21h	1 01.3	1 03.0	1 04.8	1 06.5	1 08.3	1 10.0	1 11.8	1 13.5	1 15.3	1 17.0	1 18.8	1 20.5	1 22.3	1 24.0	1 25.8	1 27.5	1 29.3	1 31.0	21h
22h	1 04.2	1 06.0	1 07.8	1 09.7	1 11.5	1 13.3	1 15.2	1 17.0	1 18.8	1 20.7	1 22.5	1 24.3	1 26.2	1 28.0	1 29.8	1 31.7	1 33.5	1 35.3	22h
23h	1 07.1	1 09.0	1 10.9	1 12.8	1 14.8	1 16.7	1 18.6	1 20.5	1 22.4	1 24.3	1 26.3	1 28.2	1 30.1	1 32.0	1 33.9	1 35.8	1 37.8	1 39.7	23h
24h	1°10'	1°12'	1°14'	1°16'	1°18'	1°20'	1°22'	1°24'	1°26'	1°28'	1°30'	1°32'	1°34'	1°36'	1°38'	1°40'	1°42'	1°44'	24h

24h	1°44'	1°46'	1°48'	1°50'	1°52'	1°54'	1°56'	1°58'	2°00'	2°02'	2°04'	2°06'	2°08'	2°10'	2°12'	2°14'	2°16'	2°18'	24h
1m	0 00.1	0 00.1	0 00.1	0 00.1	0 00.1	0 00.1	0 00.1	0 00.1	0 00.1	0 00.1	0 00.1	0 00.1	0 00.1	0 00.1	0 00.1	0 00.1	0 00.1	0 00.1	1m
2m	0 00.1	0 00.1	0 00.2	0 00.2	0 00.2	0 00.2	0 00.2	0 00.2	0 00.2	0 00.2	0 00.2	0 00.2	0 00.2	0 00.2	0 00.2	0 00.2	0 00.2	0 00.2	2m
3m	0 00.2	0 00.2	0 00.2	0 00.2	0 00.2	0 00.2	0 00.2	0 00.2	0 00.3	0 00.3	0 00.3	0 00.3	0 00.3	0 00.3	0 00.3	0 00.3	0 00.3	0 00.3	3m
4m	0 00.3	0 00.3	0 00.3	0 00.3	0 00.3	0 00.3	0 00.3	0 00.3	0 00.3	0 00.3	0 00.3	0 00.4	0 00.4	0 00.4	0 00.4	0 00.4	0 00.4	0 00.4	4m
5m	0 00.4	0 00.4	0 00.4	0 00.4	0 00.4	0 00.4	0 00.4	0 00.4	0 00.4	0 00.4	0 00.4	0 00.4	0 00.4	0 00.5	0 00.5	0 00.5	0 00.5	0 00.5	5m
10m	0 00.7	0 00.7	0 00.8	0 00.8	0 00.8	0 00.8	0 00.8	0 00.8	0 00.8	0 00.8	0 00.9	0 00.9	0 00.9	0 00.9	0 00.9	0 00.9	0 00.9	0 01.0	10m
15m	0 01.1	0 01.1	0 01.1	0 01.1	0 01.2	0 01.2	0 01.2	0 01.2	0 01.3	0 01.3	0 01.3	0 01.3	0 01.3	0 01.4	0 01.4	0 01.4	0 01.4	0 01.4	15m
20m	0 01.4	0 01.5	0 01.5	0 01.5	0 01.6	0 01.6	0 01.6	0 01.6	0 01.7	0 01.7	0 01.7	0 01.8	0 01.8	0 01.8	0 01.8	0 01.9	0 01.9	0 01.9	20m
25m	0 01.8	0 01.8	0 01.9	0 01.9	0 01.9	0 02.0	0 02.0	0 02.0	0 02.1	0 02.1	0 02.2	0 02.2	0 02.2	0 02.3	0 02.3	0 02.3	0 02.4	0 02.4	25m
30m	0 02.2	0 02.2	0 02.3	0 02.3	0 02.3	0 02.4	0 02.4	0 02.5	0 02.5	0 02.5	0 02.6	0 02.6	0 02.7	0 02.7	0 02.8	0 02.8	0 02.8	0 02.9	30m
35m	0 02.5	0 02.6	0 02.6	0 02.7	0 02.7	0 02.8	0 02.8	0 02.9	0 02.9	0 03.0	0 03.0	0 03.1	0 03.1	0 03.2	0 03.2	0 03.3	0 03.3	0 03.4	35m
40m	0 02.9	0 02.9	0 03.0	0 03.1	0 03.1	0 03.2	0 03.2	0 03.3	0 03.3	0 03.4	0 03.4	0 03.5	0 03.6	0 03.6	0 03.7	0 03.7	0 03.8	0 03.8	40m
45m	0 03.3	0 03.3	0 03.4	0 03.4	0 03.5	0 03.6	0 03.6	0 03.7	0 03.8	0 03.8	0 03.9	0 03.9	0 04.0	0 04.1	0 04.1	0 04.2	0 04.3	0 04.3	45m
50m	0 03.6	0 03.7	0 03.8	0 03.8	0 03.9	0 04.0	0 04.0	0 04.1	0 04.2	0 04.2	0 04.3	0 04.4	0 04.4	0 04.5	0 04.6	0 04.7	0 04.7	0 04.8	50m
55m	0 04.0	0 04.0	0 04.1	0 04.2	0 04.3	0 04.4	0 04.4	0 04.5	0 04.6	0 04.7	0 04.7	0 04.8	0 04.9	0 05.0	0 05.0	0 05.1	0 05.2	0 05.3	55m
1h	0 04.3	0 04.4	0 04.5	0 04.6	0 04.7	0 04.8	0 04.8	0 04.9	0 05.0	0 05.1	0 05.2	0 05.3	0 05.3	0 05.4	0 05.5	0 05.6	0 05.7	0 05.8	1h
2h	0 08.7	0 08.8	0 09.0	0 09.2	0 09.3	0 09.5	0 09.7	0 09.8	0 10.0	0 10.2	0 10.3	0 10.5	0 10.7	0 10.8	0 11.0	0 11.2	0 11.3	0 11.5	2h
3h	0 13.0	0 13.3	0 13.5	0 13.8	0 14.0	0 14.3	0 14.5	0 14.8	0 15.0	0 15.3	0 15.5	0 15.8	0 16.0	0 16.3	0 16.5	0 16.8	0 17.0	0 17.3	3h
4h	0 17.3	0 17.7	0 18.0	0 18.3	0 18.7	0 19.0	0 19.3	0 19.7	0 20.0	0 20.3	0 20.7	0 21.0	0 21.3	0 21.7	0 22.0	0 22.3	0 22.7	0 23.0	4h
5h	0 21.7	0 22.1	0 22.5	0 22.9	0 23.3	0 23.8	0 24.2	0 24.6	0 25.0	0 25.4	0 25.8	0 26.3	0 26.7	0 27.1	0 27.5	0 27.9	0 28.3	0 28.8	5h
6h	0 26.0	0 26.5	0 27.0	0 27.5	0 28.0	0 28.5	0 29.0	0 29.5	0 30.0	0 30.5	0 31.0	0 31.5	0 32.0	0 32.5	0 33.0	0 33.5	0 34.0	0 34.5	6h
7h	0 30.3	0 30.9	0 31.5	0 32.1	0 32.7	0 33.3	0 33.8	0 34.4	0 35.0	0 35.6	0 36.2	0 36.8	0 37.3	0 37.9	0 38.5	0 39.1	0 39.7	0 40.3	7h
8h	0 34.7	0 35.3	0 36.0	0 36.7	0 37.3	0 38.0	0 38.7	0 39.3	0 40.0	0 40.7	0 41.3	0 42.0	0 42.7	0 43.3	0 44.0	0 44.7	0 45.3	0 46.0	8h
9h	0 39.0	0 39.8	0 40.5	0 41.3	0 42.0	0 42.8	0 43.5	0 44.3	0 45.0	0 45.8	0 46.5	0 47.3	0 48.0	0 48.8	0 49.5	0 50.3	0 51.0	0 51.8	9h
10h	0 43.3	0 44.2	0 45.0	0 45.8	0 46.7	0 47.5	0 48.3	0 49.2	0 50.0	0 50.8	0 51.7	0 52.5	0 53.3	0 54.2	0 55.0	0 55.8	0 56.7	0 57.5	10h
11h	0 47.7	0 48.6	0 49.5	0 50.4	0 51.3	0 52.3	0 53.2	0 54.1	0 55.0	0 55.9	0 56.8	0 57.8	0 58.7	0 59.6	1 00.5	1 01.4	1 02.3	1 03.3	11h
12h	0 52.0	0 53.0	0 54.0	0 55.0	0 56.0	0 57.0	0 58.0	0 59.0	1 00.0	1 01.0	1 02.0	1 03.0	1 04.0	1 05.0	1 06.0	1 07.0	1 08.0	1 09.0	12h
13h	0 56.3	0 57.4	0 58.5	0 59.6	1 00.7	1 01.8	1 02.8	1 03.9	1 05.0	1 06.1	1 07.2	1 08.3	1 09.3	1 10.4	1 11.5	1 12.6	1 13.7	1 14.8	13h
14h	1 00.7	1 01.8	1 03.0	1 04.2	1 05.3	1 06.5	1 07.7	1 08.8	1 10.0	1 11.2	1 12.3	1 13.5	1 14.7	1 15.8	1 17.0	1 18.2	1 19.3	1 20.5	14h
15h	1 05.0	1 06.3	1 07.5	1 08.8	1 10.0	1 11.3	1 12.5	1 13.8	1 15.0	1 16.3	1 17.5	1 18.8	1 20.0	1 21.3	1 22.5	1 23.8	1 25.0	1 26.3	15h
16h	1 09.3	1 10.7	1 12.0	1 13.3	1 14.7	1 16.0	1 17.3	1 18.7	1 20.0	1 21.3	1 22.7	1 24.0	1 25.3	1 26.7	1 28.0	1 29.3	1 30.7	1 32.0	16h
17h	1 13.7	1 15.1	1 16.5	1 17.9	1 19.3	1 20.8	1 22.2	1 23.6	1 25.0	1 26.4	1 27.8	1 29.3	1 30.7	1 32.1	1 33.5	1 34.9	1 36.3	1 37.8	17h
18h	1 18.0	1 19.5	1 21.0	1 22.5	1 24.0	1 25.5	1 27.0	1 28.5	1 30.0	1 31.5	1 33.0	1 34.5	1 36.0	1 37.5	1 39.0	1 40.5	1 42.0	1 43.5	18h
19h	1 22.3	1 23.9	1 25.5	1 27.1	1 28.7	1 30.3	1 31.8	1 33.4	1 35.0	1 36.6	1 38.2	1 39.8	1 41.3	1 42.9	1 44.5	1 46.1	1 47.7	1 49.3	19h
20h	1 26.7	1 28.3	1 30.0	1 31.7	1 33.3	1 35.0	1 36.7	1 38.3	1 40.0	1 41.7	1 43.3	1 45.0	1 46.7	1 48.3	1 50.0	1 51.7	1 53.3	1 55.0	20h
21h	1 31.0	1 32.8	1 34.5	1 36.3	1 38.0	1 39.8	1 41.5	1 43.3	1 45.0	1 46.8	1 48.5	1 50.3	1 52.0	1 53.8	1 55.5	1 57.3	1 59.0	2 00.8	21h
22h	1 35.3	1 37.2	1 39.0	1 40.8	1 42.7	1 44.5	1 46.3	1 48.2	1 50.0	1 51.8	1 53.7	1 55.5	1 57.3	1 59.2	2 01.0	2 02.8	2 04.7	2 06.5	22h
23h	1 39.7	1 41.6	1 43.5	1 45.4	1 47.3	1 49.3	1 51.2	1 53.1	1 55.0	1 56.9	1 58.8	2 00.8	2 02.7	2 04.6	2 06.5	2 08.4	2 10.3	2 12.3	23h
24h	1°44'	1°46'	1°48'	1°50'	1°52'	1°54'	1°56'	1°58'	2°00'	2°02'	2°04'	2°06'	2°08'	2°10'	2°12'	2°14'	2°16'	2°18'	24h

Table 4 – Ten-daily Planetary Movements

10d	0°02'	0°04'	0°06'	0°08'	0°10'	0°12'	0°14'	0°16'	0°18'	0°20'	0°22'	0°24'	0°26'	0°28'	0°30'	0°32'	0°34'	0°36'	10d
1d	0 00.2	0 00.4	0 00.6	0 00.8	0 01.0	0 01.2	0 01.4	0 01.6	0 01.8	0 02.0	0 02.2	0 02.4	0 02.6	0 02.8	0 03.0	0 03.2	0 03.4	0 03.6	1d
2d	0 00.4	0 00.8	0 01.2	0 01.6	0 02.0	0 02.4	0 02.8	0 03.2	0 03.6	0 04.0	0 04.4	0 04.8	0 05.2	0 05.6	0 06.0	0 06.4	0 06.8	0 07.2	2d
3d	0 00.6	0 01.2	0 01.8	0 02.4	0 03.0	0 03.6	0 04.2	0 04.8	0 05.4	0 06.0	0 06.6	0 07.2	0 07.8	0 08.4	0 09.0	0 09.6	0 10.2	0 10.8	3d
4d	0 00.8	0 01.6	0 02.4	0 03.2	0 04.0	0 04.8	0 05.6	0 06.4	0 07.2	0 08.0	0 08.8	0 09.6	0 10.4	0 11.2	0 12.0	0 12.8	0 13.6	0 14.4	4d
5d	0 01.0	0 02.0	0 03.0	0 04.0	0 05.0	0 06.0	0 07.0	0 08.0	0 09.0	0 10.0	0 11.0	0 12.0	0 13.0	0 14.0	0 15.0	0 16.0	0 17.0	0 18.0	5d
6d	0 01.2	0 02.4	0 03.6	0 04.8	0 06.0	0 07.2	0 08.4	0 09.6	0 10.8	0 12.0	0 13.2	0 14.4	0 15.6	0 16.8	0 18.0	0 19.2	0 20.4	0 21.6	6d
7d	0 01.4	0 02.8	0 04.2	0 05.6	0 07.0	0 08.4	0 09.8	0 11.2	0 12.6	0 14.0	0 15.4	0 16.8	0 18.2	0 19.6	0 21.0	0 22.4	0 23.8	0 25.2	7d
8d	0 01.6	0 03.2	0 04.8	0 06.4	0 08.0	0 09.6	0 11.2	0 12.8	0 14.4	0 16.0	0 17.6	0 19.2	0 20.8	0 22.4	0 24.0	0 25.6	0 27.2	0 28.8	8d
9d	0 01.8	0 03.6	0 05.4	0 07.2	0 09.0	0 10.8	0 12.6	0 14.4	0 16.2	0 18.0	0 19.8	0 21.6	0 23.4	0 25.2	0 27.0	0 28.8	0 30.6	0 32.4	9d
10d	0°02'	0°04'	0°06'	0°08'	0°10'	0°12'	0°14'	0°16'	0°18'	0°20'	0°22'	0°24'	0°26'	0°28'	0°30'	0°32'	0°34'	0°36'	10d

10d	0°36'	0°38'	0°40'	0°42'	0°44'	0°46'	0°48'	0°50'	0°52'	0°54'	0°56'	0°58'	1°00'	1°02'	1°04'	1°06'	1°08'	1°10'	10d
1d	0 03.6	0 03.8	0 04.0	0 04.2	0 04.4	0 04.6	0 04.8	0 05.0	0 05.2	0 05.4	0 05.6	0 05.8	0 06.0	0 06.2	0 06.4	0 06.6	0 06.8	0 07.0	1d
2d	0 07.2	0 07.6	0 08.0	0 08.4	0 08.8	0 09.2	0 09.6	0 10.0	0 10.4	0 10.8	0 11.2	0 11.6	0 12.0	0 12.4	0 12.8	0 13.2	0 13.6	0 14.0	2d
3d	0 10.8	0 11.4	0 12.0	0 12.6	0 13.2	0 13.8	0 14.4	0 15.0	0 15.6	0 16.2	0 16.8	0 17.4	0 18.0	0 18.6	0 19.2	0 19.8	0 20.4	0 21.0	3d
4d	0 14.4	0 15.2	0 16.0	0 16.8	0 17.6	0 18.4	0 19.2	0 20.0	0 20.8	0 21.6	0 22.4	0 23.2	0 24.0	0 24.8	0 25.6	0 26.4	0 27.2	0 28.0	4d
5d	0 18.0	0 19.0	0 20.0	0 21.0	0 22.0	0 23.0	0 24.0	0 25.0	0 26.0	0 27.0	0 28.0	0 29.0	0 30.0	0 31.0	0 32.0	0 33.0	0 34.0	0 35.0	5d
6d	0 21.6	0 22.8	0 24.0	0 25.2	0 26.4	0 27.6	0 28.8	0 30.0	0 31.2	0 32.4	0 33.6	0 34.8	0 36.0	0 37.2	0 38.4	0 39.6	0 40.8	0 42.0	6d
7d	0 25.2	0 26.6	0 28.0	0 29.4	0 30.8	0 32.2	0 33.6	0 35.0	0 36.4	0 37.8	0 39.2	0 40.6	0 42.0	0 43.4	0 44.8	0 46.2	0 47.6	0 49.0	7d
8d	0 28.8	0 30.4	0 32.0	0 33.6	0 35.2	0 36.8	0 38.4	0 40.0	0 41.6	0 43.2	0 44.8	0 46.4	0 48.0	0 49.6	0 51.2	0 52.8	0 54.4	0 56.0	8d
9d	0 32.4	0 34.2	0 36.0	0 37.8	0 39.6	0 41.4	0 43.2	0 45.0	0 46.8	0 48.6	0 50.4	0 52.2	0 54.0	0 55.8	0 57.6	0 59.4	1 01.2	1 03.0	9d
10d	0°36'	0°38'	0°40'	0°42'	0°44'	0°46'	0°48'	0°50'	0°52'	0°54'	0°56'	0°58'	1°00'	1°02'	1°04'	1°06'	1°08'	1°10'	10d

10d	1°10'	1°12'	1°14'	1°16'	1°18'	1°20'	1°22'	1°24'	1°26'	1°28'	1°30'	1°32'	1°34'	1°36'	1°38'	1°40'	1°42'	1°44'	10d
1d	0 07.0	0 07.2	0 07.4	0 07.6	0 07.8	0 08.0	0 08.2	0 08.4	0 08.6	0 08.8	0 09.0	0 09.2	0 09.4	0 09.6	0 09.8	0 10.0	0 10.2	0 10.4	1d
2d	0 14.0	0 14.4	0 14.8	0 15.2	0 15.6	0 16.0	0 16.4	0 16.8	0 17.2	0 17.6	0 18.0	0 18.4	0 18.8	0 19.2	0 19.6	0 20.0	0 20.4	0 20.8	2d
3d	0 21.0	0 21.6	0 22.2	0 22.8	0 23.4	0 24.0	0 24.6	0 25.2	0 25.8	0 26.4	0 27.0	0 27.6	0 28.2	0 28.8	0 29.4	0 30.0	0 30.6	0 31.2	3d
4d	0 28.0	0 28.8	0 29.6	0 30.4	0 31.2	0 32.0	0 32.8	0 33.6	0 34.4	0 35.2	0 36.0	0 36.8	0 37.6	0 38.4	0 39.2	0 40.0	0 40.8	0 41.6	4d
5d	0 35.0	0 36.0	0 37.0	0 38.0	0 39.0	0 40.0	0 41.0	0 42.0	0 43.0	0 44.0	0 45.0	0 46.0	0 47.0	0 48.0	0 49.0	0 50.0	0 51.0	0 52.0	5d
6d	0 42.0	0 43.2	0 44.4	0 45.6	0 46.8	0 48.0	0 49.2	0 50.4	0 51.6	0 52.8	0 54.0	0 55.2	0 56.4	0 57.6	0 58.8	1 00.0	1 01.2	1 02.4	6d
7d	0 49.0	0 50.4	0 51.8	0 53.2	0 54.6	0 56.0	0 57.4	0 58.8	1 00.2	1 01.6	1 03.0	1 04.4	1 05.8	1 07.2	1 08.6	1 10.0	1 11.4	1 12.8	7d
8d	0 56.0	0 57.6	0 59.2	1 00.8	1 02.4	1 04.0	1 05.6	1 07.2	1 08.8	1 10.4	1 12.0	1 13.6	1 15.2	1 16.8	1 18.4	1 20.0	1 21.6	1 23.2	8d
9d	1 03.0	1 04.8	1 06.6	1 08.4	1 10.2	1 12.0	1 13.8	1 15.6	1 17.4	1 19.2	1 21.0	1 22.8	1 24.6	1 26.4	1 28.2	1 30.0	1 31.8	1 33.6	9d
10d	1°10'	1°12'	1°14'	1°16'	1°18'	1°20'	1°22'	1°24'	1°26'	1°28'	1°30'	1°32'	1°34'	1°36'	1°38'	1°40'	1°42'	1°44'	10d

10d	1°44'	1°46'	1°48'	1°50'	1°52'	1°54'	1°56'	1°58'	2°00'	2°02'	2°04'	2°06'	2°08'	2°10'	2°12'	2°14'	2°16'	2°18'	10d
1d	0 10.4	0 10.6	0 10.8	0 11.0	0 11.2	0 11.4	0 11.6	0 11.8	0 12.0	0 12.2	0 12.4	0 12.6	0 12.8	0 13.0	0 13.2	0 13.4	0 13.6	0 13.8	1d
2d	0 20.8	0 21.2	0 21.6	0 22.0	0 22.4	0 22.8	0 23.2	0 23.6	0 24.0	0 24.4	0 24.8	0 25.2	0 25.6	0 26.0	0 26.4	0 26.8	0 27.2	0 27.6	2d
3d	0 31.2	0 31.8	0 32.4	0 33.0	0 33.6	0 34.2	0 34.8	0 35.4	0 36.0	0 36.6	0 37.2	0 37.8	0 38.4	0 39.0	0 39.6	0 40.2	0 40.8	0 41.4	3d
4d	0 41.6	0 42.4	0 43.2	0 44.0	0 44.8	0 45.6	0 46.4	0 47.2	0 48.0	0 48.8	0 49.6	0 50.4	0 51.2	0 52.0	0 52.8	0 53.6	0 54.4	0 55.2	4d
5d	0 52.0	0 53.0	0 54.0	0 55.0	0 56.0	0 57.0	0 58.0	0 59.0	1 00.0	1 01.0	1 02.0	1 03.0	1 04.0	1 05.0	1 06.0	1 07.0	1 08.0	1 09.0	5d
6d	1 02.4	1 03.6	1 04.8	1 06.0	1 07.2	1 08.4	1 09.6	1 10.8	1 12.0	1 13.2	1 14.4	1 15.6	1 16.8	1 18.0	1 19.2	1 20.4	1 21.6	1 22.8	6d
7d	1 12.8	1 14.2	1 15.6	1 17.0	1 18.4	1 19.8	1 21.2	1 22.6	1 24.0	1 25.4	1 26.8	1 28.2	1 29.6	1 31.0	1 32.4	1 33.8	1 35.2	1 36.6	7d
8d	1 23.2	1 24.8	1 26.4	1 28.0	1 29.6	1 31.2	1 32.8	1 34.4	1 36.0	1 37.6	1 39.2	1 40.8	1 42.4	1 44.0	1 45.6	1 47.2	1 48.8	1 50.4	8d
9d	1 33.6	1 35.4	1 37.2	1 39.0	1 40.8	1 42.6	1 44.4	1 46.2	1 48.0	1 49.8	1 51.6	1 53.4	1 55.2	1 57.0	1 58.8	2 00.6	2 02.4	2 04.2	9d
10d	1°44'	1°46'	1°48'	1°50'	1°52'	1°54'	1°56'	1°58'	2°00'	2°02'	2°04'	2°06'	2°08'	2°10'	2°12'	2°14'	2°16'	2°18'	10d

10d	2°18'	2°20'	2°22'	2°24'	2°26'	2°28'	2°30'	2°32'	2°34'	2°36'	2°38'	2°40'	2°42'	2°44'	2°46'	2°48'	2°50'	2°52'	10d
1d	0 13.8	0 14.0	0 14.2	0 14.4	0 14.6	0 14.8	0 15.0	0 15.2	0 15.4	0 15.6	0 15.8	0 16.0	0 16.2	0 16.4	0 16.6	0 16.8	0 17.0	0 17.2	1d
2d	0 27.6	0 28.0	0 28.4	0 28.8	0 29.2	0 29.6	0 30.0	0 30.4	0 30.8	0 31.2	0 31.6	0 32.0	0 32.4	0 32.8	0 33.2	0 33.6	0 34.0	0 34.4	2d
3d	0 41.4	0 42.0	0 42.6	0 43.2	0 43.8	0 44.4	0 45.0	0 45.6	0 46.2	0 46.8	0 47.4	0 48.0	0 48.6	0 49.2	0 49.8	0 50.4	0 51.0	0 51.6	3d
4d	0 55.2	0 56.0	0 56.8	0 57.6	0 58.4	0 59.2	1 00.0	1 00.8	1 01.6	1 02.4	1 03.2	1 04.0	1 04.8	1 05.6	1 06.4	1 07.2	1 08.0	1 08.8	4d
5d	1 09.0	1 10.0	1 11.0	1 12.0	1 13.0	1 14.0	1 15.0	1 16.0	1 17.0	1 18.0	1 19.0	1 20.0	1 21.0	1 22.0	1 23.0	1 24.0	1 25.0	1 26.0	5d
6d	1 22.8	1 24.0	1 25.2	1 26.4	1 27.6	1 28.8	1 30.0	1 31.2	1 32.4	1 33.6	1 34.8	1 36.0	1 37.2	1 38.4	1 39.6	1 40.8	1 42.0	1 43.2	6d
7d	1 36.6	1 38.0	1 39.4	1 40.8	1 42.2	1 43.6	1 45.0	1 46.4	1 47.8	1 49.2	1 50.6	1 52.0	1 53.4	1 54.8	1 56.2	1 57.6	1 59.0	2 00.4	7d
8d	1 50.4	1 52.0	1 53.6	1 55.2	1 56.8	1 58.4	2 00.0	2 01.6	2 03.2	2 04.8	2 06.4	2 08.0	2 09.6	2 11.2	2 12.8	2 14.4	2 16.0	2 17.6	8d
9d	2 04.2	2 06.0	2 07.8	2 09.6	2 11.4	2 13.2	2 15.0	2 16.8	2 18.6	2 20.4	2 22.2	2 24.0	2 25.8	2 27.6	2 29.4	2 31.2	2 33.0	2 34.8	9d
10d	2°18'	2°20'	2°22'	2°24'	2°26'	2°28'	2°30'	2°32'	2°34'	2°36'	2°38'	2°40'	2°42'	2°44'	2°46'	2°48'	2°50'	2°52'	10d

10d	2°52'	2°54'	2°56'	2°58'	3°00'	3°02'	3°04'	3°06'	3°08'	3°10'	3°12'	3°14'	3°16'	3°18'	3°20'	3°22'	3°24'	3°26'	10d
1d	0 17.2	0 17.4	0 17.6	0 17.8	0 18.0	0 18.2	0 18.4	0 18.6	0 18.8	0 19.0	0 19.2	0 19.4	0 19.6	0 19.8	0 20.0	0 20.2	0 20.4	0 20.6	1d
2d	0 34.4	0 34.8	0 35.2	0 35.6	0 36.0	0 36.4	0 36.8	0 37.2	0 37.6	0 38.0	0 38.4	0 38.8	0 39.2	0 39.6	0 40.0	0 40.4	0 40.8	0 41.2	2d
3d	0 51.6	0 52.2	0 52.8	0 53.4	0 54.0	0 54.6	0 55.2	0 55.8	0 56.4	0 57.0	0 57.6	0 58.2	0 58.8	0 59.4	1 00.0	1 00.6	1 01.2	1 01.8	3d
4d	1 08.8	1 09.6	1 10.4	1 11.2	1 12.0	1 12.8	1 13.6	1 14.4	1 15.2	1 16.0	1 16.8	1 17.6	1 18.4	1 19.2	1 20.0	1 20.8	1 21.6	1 22.4	4d
5d	1 26.0	1 27.0	1 28.0	1 29.0	1 30.0	1 31.0	1 32.0	1 33.0	1 34.0	1 35.0	1 36.0	1 37.0	1 38.0	1 39.0	1 40.0	1 41.0	1 42.0	1 43.0	5d
6d	1 43.2	1 44.4	1 45.6	1 46.8	1 48.0	1 49.2	1 50.4	1 51.6	1 52.8	1 54.0	1 55.2	1 56.4	1 57.6	1 58.8	2 00.0	2 01.2	2 02.4	2 03.6	6d
7d	2 00.4	2 01.8	2 03.2	2 04.6	2 06.0	2 07.4	2 08.8	2 10.2	2 11.6	2 13.0	2 14.4	2 15.8	2 17.2	2 18.6	2 20.0	2 21.4	2 22.8	2 24.2	7d
8d	2 17.6	2 19.2	2 20.8	2 22.4	2 24.0	2 25.6	2 27.2	2 28.8	2 30.4	2 32.0	2 33.6	2 35.2	2 36.8	2 38.4	2 40.0	2 41.6	2 43.2	2 44.8	8d
9d	2 34.8	2 36.6	2 38.4	2 40.2	2 42.0	2 43.8	2 45.6	2 47.4	2 49.2	2 51.0	2 52.8	2 54.6	2 56.4	2 58.2	3 00.0	3 01.8	3 03.6	3 05.4	9d
10d	2°52'	2°54'	2°56'	2°58'	3°00'	3°02'	3°04'	3°06'	3°08'	3°10'	3°12'	3°14'	3°16'	3°18'	3°20'	3°22'	3°24'	3°26'	10d

10d	3°26'	3°28'	3°30'	3°32'	3°34'	3°36'	3°38'	3°40'	3°42'	3°44'	3°46'	3°48'	3°50'	3°52'	3°54'	3°56'	3°58'	4°00'	10d
1d	0 20.6	0 20.8	0 21.0	0 21.2	0 21.4	0 21.6	0 21.8	0 22.0	0 22.2	0 22.4	0 22.6	0 22.8	0 23.0	0 23.2	0 23.4	0 23.6	0 23.8	0 24.0	1d
2d	0 41.2	0 41.6	0 42.0	0 42.4	0 42.8	0 43.2	0 43.6	0 44.0	0 44.4	0 44.8	0 45.2	0 45.6	0 46.0	0 46.4	0 46.8	0 47.2	0 47.6	0 48.0	2d
3d	1 01.8	1 02.4	1 03.0	1 03.6	1 04.2	1 04.8	1 05.4	1 06.0	1 06.6	1 07.2	1 07.8	1 08.4	1 09.0	1 09.6	1 10.2	1 10.8	1 11.4	1 12.0	3d
4d	1 22.4	1 23.2	1 24.0	1 24.8	1 25.6	1 26.4	1 27.2	1 28.0	1 28.8	1 29.6	1 30.4	1 31.2	1 32.0	1 32.8	1 33.6	1 34.4	1 35.2	1 36.0	4d
5d	1 43.0	1 44.0	1 45.0	1 46.0	1 47.0	1 48.0	1 49.0	1 50.0	1 51.0	1 52.0	1 53.0	1 54.0	1 55.0	1 56.0	1 57.0	1 58.0	1 59.0	2 00.0	5d
6d	2 03.6	2 04.8	2 06.0	2 07.2	2 08.4	2 09.6	2 10.8	2 12.0	2 13.2	2 14.4	2 15.6	2 16.8	2 18.0	2 19.2	2 20.4	2 21.6	2 22.8	2 24.0	6d
7d	2 24.2	2 25.6	2 27.0	2 28.4	2 29.8	2 31.2	2 32.6	2 34.0	2 35.4	2 36.8	2 38.2	2 39.6	2 41.0	2 42.4	2 43.8	2 45.2	2 46.6	2 48.0	7d
8d	2 44.8	2 46.4	2 48.0	2 49.6	2 51.2	2 52.8	2 54.4	2 56.0	2 57.6	2 59.2	3 00.8	3 02.4	3 04.0	3 05.6	3 07.2	3 08.8	3 10.4	3 12.0	8d
9d	3 05.4	3 07.2	3 09.0	3 10.8	3 12.6	3 14.4	3 16.2	3 18.0	3 19.8	3 21.6	3 23.4	3 25.2	3 27.0	3 28.8	3 30.6	3 32.4	3 34.2	3 36.0	9d
10d	3°26'	3°28'	3°30'	3°32'	3°34'	3°36'	3°38'	3°40'	3°42'	3°44'	3°46'	3°48'	3°50'	3°52'	3°54'	3°56'	3°58'	4°00'	10d

10d	4°00'	4°02'	4°04'	4°06'	4°08'	4°10'	4°12'	4°14'	4°16'	4°18'	4°20'	4°22'	4°24'	4°26'	4°28'	4°30'	4°32'	4°34'	10d
1d	0 24.0	0 24.2	0 24.4	0 24.6	0 24.8	0 25.0	0 25.2	0 25.4	0 25.6	0 25.8	0 26.0	0 26.2	0 26.4	0 26.6	0 26.8	0 27.0	0 27.2	0 27.4	1d
2d	0 48.0	0 48.4	0 48.8	0 49.2	0 49.6	0 50.0	0 50.4	0 50.8	0 51.2	0 51.6	0 52.0	0 52.4	0 52.8	0 53.2	0 53.6	0 54.0	0 54.4	0 54.8	2d
3d	1 12.0	1 12.6	1 13.2	1 13.8	1 14.4	1 15.0	1 15.6	1 16.2	1 16.8	1 17.4	1 18.0	1 18.6	1 19.2	1 19.8	1 20.4	1 21.0	1 21.6	1 22.2	3d
4d	1 36.0	1 36.8	1 37.6	1 38.4	1 39.2	1 40.0	1 40.8	1 41.6	1 42.4	1 43.2	1 44.0	1 44.8	1 45.6	1 46.4	1 47.2	1 48.0	1 48.8	1 49.6	4d
5d	2 00.0	2 01.0	2 02.0	2 03.0	2 04.0	2 05.0	2 06.0	2 07.0	2 08.0	2 09.0	2 10.0	2 11.0	2 12.0	2 13.0	2 14.0	2 15.0	2 16.0	2 17.0	5d
6d	2 24.0	2 25.2	2 26.4	2 27.6	2 28.8	2 30.0	2 31.2	2 32.4	2 33.6	2 34.8	2 36.0	2 37.2	2 38.4	2 39.6	2 40.8	2 42.0	2 43.2	2 44.4	6d
7d	2 48.0	2 49.4	2 50.8	2 52.2	2 53.6	2 55.0	2 56.4	2 57.8	2 59.2	3 00.6	3 02.0	3 03.4	3 04.8	3 06.2	3 07.6	3 09.0	3 10.4	3 11.8	7d
8d	3 12.0	3 13.6	3 15.2	3 16.8	3 18.4	3 20.0	3 21.6	3 23.2	3 24.8	3 26.4	3 28.0	3 29.6	3 31.2	3 32.8	3 34.4	3 36.0	3 37.6	3 39.2	8d
9d	3 36.0	3 37.8	3 39.6	3 41.4	3 43.2	3 45.0	3 46.8	3 48.6	3 50.4	3 52.2	3 54.0	3 55.8	3 57.6	3 59.4	4 01.2	4 03.0	4 04.8	4 06.6	9d
10d	4°00'	4°02'	4°04'	4°06'	4°08'	4°10'	4°12'	4°14'	4°16'	4°18'	4°20'	4°22'	4°24'	4°26'	4°28'	4°30'	4°32'	4°34'	10d

10d	4°34'	4°36'	4°38'	4°40'	0°42'	4°44'	4°46'	4°48'	4°50'	4°52'	4°54'	4°56'	4°58'	5°00'	5°02'	5°04'	5°06'	5°08'	10d
1d	0 27.4	0 27.6	0 27.8	0 28.0	0 28.2	0 28.4	0 28.6	0 28.8	0 29.0	0 29.2	0 29.4	0 29.6	0 29.8	0 30.0	0 30.2	0 30.4	0 30.6	0 30.8	1d
2d	0 54.8	0 55.2	0 55.6	0 56.0	0 56.4	0 56.8	0 57.2	0 57.6	0 58.0	0 58.4	0 58.8	0 59.2	0 59.6	1 00.0	1 00.4	1 00.8	1 01.2	1 01.6	2d
3d	1 22.2	1 22.8	1 23.4	1 24.0	1 24.6	1 25.2	1 25.8	1 26.4	1 27.0	1 27.6	1 28.2	1 28.8	1 29.4	1 30.0	1 30.6	1 31.2	1 31.8	1 32.4	3d
4d	1 49.6	1 50.4	1 51.2	1 52.0	1 52.8	1 53.6	1 54.4	1 55.2	1 56.0	1 56.8	1 57.6	1 58.4	1 59.2	2 00.0	2 00.8	2 01.6	2 02.4	2 03.2	4d
5d	2 17.0	2 18.0	2 19.0	2 20.0	2 21.0	2 22.0	2 23.0	2 24.0	2 25.0	2 26.0	2 27.0	2 28.0	2 29.0	2 30.0	2 31.0	2 32.0	2 33.0	2 34.0	5d
6d	2 44.4	2 45.6	2 46.8	2 48.0	2 49.2	2 50.4	2 51.6	2 52.8	2 54.0	2 55.2	2 56.4	2 57.6	2 58.8	3 00.0	3 01.2	3 02.4	3 03.6	3 04.8	6d
7d	3 11.8	3 13.2	3 14.6	3 16.0	3 17.4	3 18.8	3 20.2	3 21.6	3 23.0	3 24.4	3 25.8	3 27.2	3 28.6	3 30.0	3 31.4	3 32.8	3 34.2	3 35.6	7d
8d	3 39.2	3 40.8	3 42.4	3 44.0	3 45.6	3 47.2	3 48.8	3 50.4	3 52.0	3 53.6	3 55.2	3 56.8	3 58.4	4 00.0	4 01.6	4 03.2	4 04.8	4 06.4	8d
9d	4 06.6	4 08.4	4 10.2	4 12.0	4 13.8	4 15.6	4 17.4	4 19.2	4 21.0	4 22.8	4 24.6	4 26.4	4 28.2	4 30.0	4 31.8	4 33.6	4 35.4	4 37.2	9d
10d	4°34'	4°36'	4°38'	4°40'	0°42'	4°44'	4°46'	4°48'	4°50'	4°52'	4°54'	4°56'	4°58'	5°00'	5°02'	5°04'	5°06'	5°08'	10d

10d	5°08'	5°10'	5°12'	5°14'	5°16'	5°18'	5°20'	5°22'	5°24'	5°26'	5°28'	5°30'	5°32'	5°34'	5°36'	5°38'	5°40'	5°42'	10d
1d	0 30.8	0 31.0	0 31.2	0 31.4	0 31.6	0 31.8	0 32.0	0 32.2	0 32.4	0 32.6	0 32.8	0 33.0	0 33.2	0 33.4	0 33.6	0 33.8	0 34.0	0 34.2	1d
2d	1 01.6	1 02.0	1 02.4	1 02.8	1 03.2	1 03.6	1 04.0	1 04.4	1 04.8	1 05.2	1 05.6	1 06.0	1 06.4	1 06.8	1 07.2	1 07.6	1 08.0	1 08.4	2d
3d	1 32.4	1 33.0	1 33.6	1 34.2	1 34.8	1 35.4	1 36.0	1 36.6	1 37.2	1 37.8	1 38.4	1 39.0	1 39.6	1 40.2	1 40.8	1 41.4	1 42.0	1 42.6	3d
4d	2 03.2	2 04.0	2 04.8	2 05.6	2 06.4	2 07.2	2 08.0	2 08.8	2 09.6	2 10.4	2 11.2	2 12.0	2 12.8	2 13.6	2 14.4	2 15.2	2 16.0	2 16.8	4d
5d	2 34.0	2 35.0	2 36.0	2 37.0	2 38.0	2 39.0	2 40.0	2 41.0	2 42.0	2 43.0	2 44.0	2 45.0	2 46.0	2 47.0	2 48.0	2 49.0	2 50.0	2 51.0	5d
6d	3 04.8	3 06.0	3 07.2	3 08.4	3 09.6	3 10.8	3 12.0	3 13.2	3 14.4	3 15.6	3 16.8	3 18.0	3 19.2	3 20.4	3 21.6	3 22.8	3 24.0	3 25.2	6d
7d	3 35.6	3 37.0	3 38.4	3 39.8	3 41.2	3 42.6	3 44.0	3 45.4	3 46.8	3 48.2	3 49.6	3 51.0	3 52.4	3 53.8	3 55.2	3 56.6	3 58.0	3 59.4	7d
8d	4 06.4	4 08.0	4 09.6	4 11.2	4 12.8	4 14.4	4 16.0	4 17.6	4 19.2	4 20.8	4 22.4	4 24.0	4 25.6	4 27.2	4 28.8	4 30.4	4 32.0	4 33.6	8d
9d	4 37.2	4 39.0	4 40.8	4 42.6	4 44.4	4 46.2	4 48.0	4 49.8	4 51.6	4 53.4	4 55.2	4 57.0	4 58.8	5 00.6	5 02.4	5 04.2	5 06.0	5 07.8	9d
10d	5°08'	5°10'	5°12'	5°14'	5°16'	5°18'	5°20'	5°22'	5°24'	5°26'	5°28'	5°30'	5°32'	5°34'	5°36'	5°38'	5°40'	5°42'	10d

10d	5°42'	5°44'	5°46'	5°48'	5°50'	5°52'	5°54'	5°56'	5°58'	6°00'	6°02'	6°04'	6°06'	6°08'	6°10'	6°12'	6°14'	6°16'	10d
1d	0 34.2	0 34.4	0 34.6	0 34.8	0 35.0	0 35.2	0 35.4	0 35.6	0 35.8	0 36.0	0 36.2	0 36.4	0 36.6	0 36.8	0 37.0	0 37.2	0 37.4	0 37.6	1d
2d	1 08.4	1 08.8	1 09.2	1 09.6	1 10.0	1 10.4	1 10.8	1 11.2	1 11.6	1 12.0	1 12.4	1 12.8	1 13.2	1 13.6	1 14.0	1 14.4	1 14.8	1 15.2	2d
3d	1 42.6	1 43.2	1 43.8	1 44.4	1 45.0	1 45.6	1 46.2	1 46.8	1 47.4	1 48.0	1 48.6	1 49.2	1 49.8	1 50.4	1 51.0	1 51.6	1 52.2	1 52.8	3d
4d	2 16.8	2 17.6	2 18.4	2 19.2	2 20.0	2 20.8	2 21.6	2 22.4	2 23.2	2 24.0	2 24.8	2 25.6	2 26.4	2 27.2	2 28.0	2 28.8	2 29.6	2 30.4	4d
5d	2 51.0	2 52.0	2 53.0	2 54.0	2 55.0	2 56.0	2 57.0	2 58.0	2 59.0	3 00.0	3 01.0	3 02.0	3 03.0	3 04.0	3 05.0	3 06.0	3 07.0	3 08.0	5d
6d	3 25.2	3 26.4	3 27.6	3 28.8	3 30.0	3 31.2	3 32.4	3 33.6	3 34.8	3 36.0	3 37.2	3 38.4	3 39.6	3 40.8	3 42.0	3 43.2	3 44.4	3 45.6	6d
7d	3 59.4	4 00.8	4 02.2	4 03.6	4 05.0	4 06.4	4 07.8	4 09.2	4 10.6	4 12.0	4 13.4	4 14.8	4 16.2	4 17.6	4 19.0	4 20.4	4 21.8	4 23.2	7d
8d	4 33.6	4 35.2	4 36.8	4 38.4	4 40.0	4 41.6	4 43.2	4 44.8	4 46.4	4 48.0	4 49.6	4 51.2	4 52.8	4 54.4	4 56.0	4 57.6	4 59.2	5 00.8	8d
9d	5 07.8	5 09.6	5 11.4	5 13.2	5 15.0	5 16.8	5 18.6	5 20.4	5 22.2	5 24.0	5 25.8	5 27.6	5 29.4	5 31.2	5 33.0	5 34.8	5 36.6	5 38.4	9d
10d	5°42'	5°44'	5°46'	5°48'	5°50'	5°52'	5°54'	5°56'	5°58'	6°00'	6°02'	6°04'	6°06'	6°08'	6°10'	6°12'	6°14'	6°16'	10d

RAPHAEL'S 101-YEAR EPHEMERIS 1950–2050

JANUARY 1950

Date	Sidereal time h m s	Sun ☉ ° ' "	Moon ☽ ° ' "	Moon ☽ 24.00 ° ' "	Mercury ☿ ° '	Venus ♀ ° '	Mars ♂ ° '	Jupiter ♃ ° '	Saturn ♄ ° '	Uranus ♅ ° '	Neptune ♆ ° '	Pluto ♇ ° '
01	18 42 16	10 ♑ 30 52	07 ♊ 32 40	13 ♊ 44 08	29 ♑ 58	17 ♒ 09	02 ♎ 24	06 ♒ 37	19 ♍ 26	02 ♋ 40	17 ♎ 16	17 ♌ 47
02	18 46 13	11 32 00	19 ♊ 59 30	26 ♊ 18 54	00 ♒ 57	17 28	02 45	06 51	19 R 26	02 R 37	17 17	17 R 46
03	18 50 10	12 33 09	02 ♋ 42 20	09 ♋ 09 47	01 50	17 46	03 07	07 04	19 25	02 35	17 17	17 45
04	18 54 06	13 34 17	15 ♋ 41 06	22 ♋ 16 05	02 35	18 01	03 28	07 18	19 25	02 32	17 18	17 44
05	18 58 03	14 35 26	28 ♋ 54 32	05 ♌ 36 08	03 12	18 14	03 49	07 32	19 24	02 30	17 18	17 43
06	19 01 59	15 36 34	12 ♌ 20 36	19 ♌ 07 38	03 40	18 25	04 09	07 45	19 23	02 27	17 19	17 42
07	19 05 56	16 37 42	25 ♌ 56 58	02 ♍ 48 18	03 59	18 33	04 29	07 59	19 23	02 25	17 19	17 41
08	19 09 52	17 38 50	09 ♍ 41 27	16 ♍ 36 13	04 07	18 39	04 49	08 13	19 22	02 22	17 20	17 39
09	19 13 49	18 39 59	23 ♍ 32 27	00 ♎ 30 04	04 R 03	18 43	05 08	08 27	19 21	02 20	17 20	17 38
10	19 17 45	19 41 07	07 ♎ 29 00	14 ♎ 29 11	03 48	18 44	05 27	08 41	19 19	02 17	17 20	17 37
11	19 21 42	20 42 15	21 ♎ 30 36	28 ♎ 33 11	03 21	18 R 43	05 45	08 55	19 18	02 15	17 20	17 36
12	19 25 39	21 43 24	05 ♏ 36 51	12 ♏ 41 31	02 43	18 40	06 03	09 08	19 17	02 12	17 21	17 34
13	19 29 35	22 44 32	19 ♏ 46 57	26 ♏ 52 56	01 53	18 34	06 21	09 22	19 15	02 10	17 21	17 33
14	19 33 32	23 45 40	03 ♐ 59 08	11 ♐ 05 08	00 ♒ 54	18 25	06 38	09 36	19 14	02 08	17 21	17 32
15	19 37 28	24 46 48	18 ♐ 10 27	25 ♐ 14 33	29 ♑ 47	18 14	06 55	09 50	19 12	02 05	17 21	17 31
16	19 41 25	25 47 56	02 ♑ 16 51	09 ♑ 16 45	28 34	18 01	07 11	10 05	19 10	02 03	17 21	17 29
17	19 45 21	26 49 04	16 ♑ 13 38	23 ♑ 06 57	27 17	17 45	07 27	10 19	19 08	02 01	17 21	17 28
18	19 49 18	27 50 11	29 ♑ 56 11	06 ♒ 40 54	25 59	17 26	07 42	10 33	19 06	01 58	17 21	17 27
19	19 53 14	28 51 17	13 ♒ 20 47	19 ♒ 55 35	24 41	17 06	07 57	10 47	19 04	01 56	17 R 21	17 25
20	19 57 11	29 ♑ 52 23	26 ♒ 25 12	02 ♓ 49 40	23 27	16 43	08 11	11 01	19 02	01 54	17 21	17 24
21	20 01 08	00 ♒ 53 28	09 ♓ 09 05	15 ♓ 23 42	22 19	16 17	08 25	11 15	18 59	01 52	17 21	17 23
22	20 05 04	01 54 31	21 ♓ 33 50	27 ♓ 39 54	21 17	15 50	08 38	11 29	18 57	01 50	17 21	17 21
23	20 09 01	02 55 34	03 ♈ 42 25	09 ♈ 41 56	20 23	15 21	08 51	11 44	18 55	01 48	17 21	17 20
24	20 12 57	03 56 36	15 ♈ 39 02	21 ♈ 34 24	19 37	14 50	09 04	11 58	18 52	01 45	17 21	17 18
25	20 16 54	04 57 37	27 ♈ 28 40	03 ♉ 22 33	19 01	14 17	09 15	12 12	18 49	01 43	17 21	17 17
26	20 20 50	05 58 37	09 ♉ 16 44	15 ♉ 11 55	18 34	13 44	09 27	12 26	18 46	01 41	17 20	17 16
27	20 24 47	06 59 35	21 ♉ 08 46	27 ♉ 07 56	18 15	13 08	09 37	12 41	18 44	01 39	17 20	17 14
28	20 28 43	08 00 33	03 ♊ 10 01	09 ♊ 15 37	18 05	12 33	09 47	12 55	18 41	01 37	17 20	17 13
29	20 32 40	09 01 29	15 ♊ 25 12	21 ♊ 39 14	18 D 04	11 56	09 57	13 09	18 38	01 36	17 19	17 11
30	20 36 37	10 02 25	27 ♊ 58 02	04 ♋ 21 52	18 10	11 19	10 06	13 24	18 34	01 34	17 19	17 10
31	20 40 33	11 ♒ 03 19	10 ♋ 50 53	17 ♋ 25 06	18 ♑ 23	10 ♒ 42	10 ♎ 14	13 ♒ 38	18 ♍ 31	01 ♋ 32	17 ♎ 19	17 ♌ 09

Date	Moon True ☊ ° '	Moon Mean ☊ ° '	Moon ☽ Latitude ° '
01	12 ♈ 29	12 ♈ 05	04 N 07
02	12 R 17	12 02	04 38
03	12 04	11 59	04 56
04	11 50	11 56	04 59
05	11 37	11 52	04 46
06	11 25	11 49	04 18
07	11 16	11 46	03 34
08	11 11	11 43	02 38
09	11 08	11 40	01 32
10	11 D 07	11 37	00 N 19
11	11 R 07	11 33	00 S 55
12	11 06	11 30	02 06
13	11 04	11 27	03 09
14	10 59	11 24	04 01
15	10 51	11 21	04 38
16	10 40	11 18	04 58
17	10 28	11 14	05 00
18	10 16	11 11	04 44
19	10 04	11 08	04 13
20	09 55	11 05	03 29
21	09 48	11 02	02 35
22	09 44	10 58	01 35
23	09 42	10 55	00 S 32
24	09 D 42	10 52	00 N 32
25	09 43	10 49	01 33
26	09 R 43	10 46	02 31
27	09 42	10 43	03 23
28	09 38	10 39	04 06
29	09 32	10 36	04 38
30	09 24	10 33	04 59
31	09 ♈ 15	10 ♈ 30	05 N 04

DECLINATIONS

Date	Sun ☉ ° '	Moon ☽ ° '	Mercury ☿ ° '	Venus ♀ ° '	Mars ♂ ° '	Jupiter ♃ ° '	Saturn ♄ ° '	Uranus ♅ ° '	Neptune ♆ ° '	Pluto ♇ ° '
01	23 S 02	25 N 38	21 S 16	15 S 00	01 N 22	19 S 11	06 N 02	23 N 41	05 S 19	23 N 18
02	22 57	27 41	20 53	14 43	01 14	19 08	06 02	23 41	05 19	23 18
03	22 51	28 21	20 29	14 26	01 07	19 05	06 03	23 42	05 19	23 19
04	22 45	27 29	20 05	14 10	00 59	19 01	06 03	23 42	05 19	23 20
05	22 39	25 03	19 42	13 53	00 52	18 58	06 04	23 42	05 19	23 20
06	22 32	21 14	19 20	13 38	00 45	18 54	06 04	23 42	05 19	23 21
07	22 25	16 14	19 00	13 22	00 38	18 51	06 05	23 42	05 20	23 21
08	22 17	10 22	18 41	13 07	00 31	18 47	06 05	23 42	05 20	23 22
09	22 09	03 N 58	18 24	12 53	00 25	18 44	06 06	23 42	05 20	23 22
10	22 00	02 S 41	18 09	12 38	00 18	18 40	06 07	23 42	05 20	23 23
11	21 51	09 14	17 57	12 25	00 12	18 37	06 07	23 42	05 20	23 23
12	21 42	15 22	17 47	12 12	00 N 06	18 33	06 08	23 42	05 20	23 24
13	21 32	20 44	17 41	12 00	00 00	18 29	06 09	23 42	05 20	23 25
14	21 21	24 54	17 37	11 48	00 S 06	18 26	06 10	23 42	05 20	23 25
15	21 11	27 32	17 35	11 37	00 11	18 22	06 11	23 42	05 20	23 26
16	21 00	28 24	17 36	11 27	00 17	18 18	06 12	23 42	05 20	23 26
17	20 48	27 25	17 39	11 17	00 22	18 14	06 13	23 42	05 20	23 27
18	20 36	24 47	17 45	11 08	00 27	18 11	06 14	23 42	05 20	23 27
19	20 24	20 51	17 51	11 00	00 31	18 07	06 15	23 42	05 20	23 28
20	20 11	15 59	17 59	10 53	00 36	18 03	06 16	23 42	05 20	23 29
21	19 58	10 32	18 08	10 46	00 40	17 59	06 17	23 42	05 20	23 29
22	19 45	04 S 48	18 18	10 40	00 44	17 55	06 18	23 42	05 20	23 30
23	19 31	00 N 59	18 28	10 35	00 48	17 51	06 19	23 42	05 19	23 30
24	19 17	06 39	18 39	10 31	00 52	17 48	06 21	23 42	05 19	23 31
25	19 02	12 02	18 50	10 28	00 56	17 44	06 22	23 42	05 19	23 31
26	18 47	16 59	19 01	10 26	00 59	17 40	06 23	23 42	05 19	23 32
27	18 32	21 18	19 11	10 24	01 02	17 36	06 25	23 43	05 19	23 33
28	18 16	24 49	19 22	10 23	01 05	17 32	06 26	23 43	05 19	23 33
29	18 00	27 16	19 32	10 23	01 07	17 28	06 27	23 43	05 18	23 34
30	17 44	28 25	19 41	10 24	01 10	17 24	06 29	23 43	05 18	23 34
31	17 S 28	28 N 04	19 S 50	10 S 25	01 S 12	17 S 20	06 N 30	23 N 43	05 S 18	23 N 35

ZODIAC SIGN ENTRIES

Date	h	m	Planets
01	12	39	☿ ♒
03	06	56	☽ ♋
05	13	58	☽ ♌
07	19	06	☽ ♍
09	23	08	☽ ♎
12	02	28	☽ ♏
14	05	16	☽ ♐
15	07	35	☿ ♑
16	08	06	☽ ♑
18	12	07	☽ ♒
20	15	00	☉ ♒
20	18	41	☽ ♓
23	04	37	☽ ♈
25	17	08	☽ ♉
28	05	43	☽ ♊
30	15	50	☽ ♋

LATITUDES

Date	Mercury ☿ ° '	Venus ♀ ° '	Mars ♂ ° '	Jupiter ♃ ° '	Saturn ♄ ° '	Uranus ♅ ° '	Neptune ♆ ° '	Pluto ♇ ° '
01	01 S 08	00 N 44	02 N 31	00 S 35	02 N 01	00 N 16	01 N 36	08 N 12
04	00 S 31	01 20	02 35	00 35	02 01	00 16	01 36	08 13
07	00 N 17	02 00	02 38	00 35	02 02	00 16	01 36	08 13
10	01 11	02 42	02 41	00 36	02 03	00 16	01 36	08 14
13	02 07	03 26	02 45	00 36	02 04	00 16	01 36	08 15
16	02 55	04 12	02 48	00 36	02 04	00 16	01 36	08 15
19	03 24	04 57	02 52	00 36	02 05	00 16	01 37	08 16
22	03 30	05 40	02 55	00 36	02 06	00 16	01 37	08 16
25	03 18	06 21	02 59	00 36	02 07	00 16	01 37	08 16
28	02 53	06 56	03 03	00 37	02 07	00 16	01 37	08 17
31	02 N 22	07 N 25	03 N 06	00 S 37	02 N 08	00 N 16	01 N 37	08 N 17

DATA

Julian Date	2433283
Delta T	+29 seconds
Ayanamsa	23° 09' 28"
Synetic vernal point	05° ♓ 57' 32"
True obliquity of ecliptic	23° 26' 53"

MOON'S PHASES, APSIDES AND POSITIONS ☽

Date	h	m	Phase	Longitude ° '	Eclipse Indicator
04	07	48	○	13 ♋ 24	
11	10	31	☾	20 ♎ 38	
18	08	00	●	27 ♑ 40	
26	04	39	☽	05 ♉ 40	

Day	h	m		
13	05	51	Perigee	
25	21	33	Apogee	
03	10	34	Max dec	28° N 21'
10	02	22	0S	
16	11	04	Max dec	28° S 24'
23	07	54	0N	
30	18	41	Max dec	28° N 28'

LONGITUDES

Date	Chiron ⚷ ° '	Ceres ⚳ ° '	Pallas ⚴ ° '	Juno ⚵ ° '	Vesta ⚶ ° '	Black Moon Lilith ⚸ ° '
01	15 ♐ 50	24 ♏ 53	06 ♏ 03	02 ♎ 43	11 ♒ 00	28 ♈ 50
11	16 ♐ 56	28 ♏ 43	09 ♏ 24	03 ♎ 51	15 ♒ 58	29 ♈ 57
21	17 ♐ 58	02 ♐ 22	12 ♏ 25	04 ♎ 21	20 ♒ 58	01 ♉ 03
31	18 ♐ 54	05 ♐ 48	15 ♏ 05	04 ♎ 10	25 ♒ 59	02 ♉ 10

All ephemeris data is given at 12.00 UT and the Moon's longitude is additionally given for 24.00 UT

ASPECTARIAN

01 Sunday
01 37 ☽ △ ♂
01 40 ☽ ⚼ ♆
02 29 ☽ ⚺ ♅
05 33 ☽ ± ☉
08 35 ☽ Q ♇
10 10 ☽ △ ♃
18 17 ☽ ⚻ ☉
20 50 ♀ △ ♆

02 Monday
03 38 ☽ ⚼ ☿
03 47 ♂ □ ♅
06 49 ☽ △ ♆
07 04 ☽ △ ♀
07 46 ☽ ⚹ ♇
10 00 ☉ ⊥ ♀
10 56 ☽ □ ♄
12 56 ♂ ∠ ♇
15 36 ☽ ⚼ ♃
17 30 ☉ ± ♇
22 09 ☽ ± ☿
22 16 ♀ ⚼ ♅

03 Tuesday
08 53 ☽ ± ♃
10 15 ☽ ⚻ ☿
11 19 ♀ ☍ ♇
11 46 ☽ ☌ ♅
12 05 ☽ ∠ ♇
12 06 ☽ ⚼ ♀
12 47 ☽ □ ♂
20 16 ☽ ⚻ ♃
20 46 ☽ Q ♄

04 Wednesday
04 45 ☽ ⊥ ♇
05 08 ☽ ± ♀
07 48 ☽ ☍ ☉
10 29 ☿ ⚻ ♅
14 57 ☽ □ ♆
15 45 ☽ ⚺ ♇
16 20 ☽ ⚻ ♀
18 48 ☽ ⚹ ♄
22 50 ☽ Q ♂

05 Thursday
18 25 ☽ ⚺ ♅
20 01 ☽ ☍ ☿
21 02 ☽ ⚹ ♂
21 30 ☽ ∥ ♅
21 51 ☽ ∠ ♄
23 29 ☽ Q ♆
23 46 ☽ ∥ ♇

06 Friday
03 42 ☽ ☍ ♃
04 26 ☽ # ☉
05 06 ☽ ⊥ ♅
13 51 ☽ ⊥ ♄
18 15 ☽ ⚻ ☉
20 48 ☽ ⚹ ♆
21 01 ☽ ∠ ♅
21 28 ☽ ☌ ♇
22 23 ☽ # ☿
22 51 ☽ ☍ ♀
23 54 ☽ # ♃

07 Saturday
00 21 ☽ ∠ ♂
00 27 ☽ ⚺ ♄
05 42 ☽ ± ☉
16 33 ☽ ⊥ ♂
22 44 ☽ ⚼ ☉
23 09 ☽ ∠ ♆
23 17 ☽ ⚹ ♅

08 Sunday
00 38 ☽ # ♀
01 41 ☿ ∥ ♃
02 14 ☽ ⚻ ☿
03 18 ☽ ⚺ ♂
04 24 ☉ □ ♆
09 23 ☽ ⚻ ♃
12 12 ☉ ⚻ ♇
12 44 ☽ ± ☿
14 51 ☽ ⊥ ♆
16 53 ☿ St R
19 59 ☽ ± ♃
20 06 ☽ Q ♅

09 Monday
01 15 ☽ ⚺ ♆
01 35 ♃ ± ♅
01 48 ☽ ⚺ ♇
02 54 ☽ △ ☉
03 38 ☽ ⚻ ♀
04 10 ☽ ∥ ♄
04 19 ☽ ⚼ ☿
04 45 ☽ ☌ ♄
07 00 ☽ # ♆
11 50 ☽ ⚼ ♃
12 10 ☽ ⊥ ♇
13 15 ☉ ⚺ ♀
14 02 ☽ ± ♀

10 Tuesday
01 05 ☽ ∥ ♂
03 06 ☽ □ ♅
03 35 ☽ # ♂
03 39 ☉ △ ♄
03 39 ☽ ∠ ♇
05 34 ☽ ⚼ ♀
05 50 ☽ △ ☿
08 25 ☽ ☌ ♂
13 35 ♀ St R
14 05 ☽ △ ♃
21 36 ☽ ∥ ♆

11 Wednesday
00 29 ☽ # ♄
04 53 ☽ ☌ ♆
05 20 ☽ ⚹ ♇
07 15 ☽ △ ♀
08 14 ☽ ⚺ ♄
10 31 ☽ □ ☉
18 27 ☽ ⊥ ♄
23 44 ☽ ∥ ♀

12 Thursday
01 45 ☽ Q ♇
06 14 ☽ △ ♅
07 19 ☽ □ ☿
09 44 ☽ ∠ ♄
12 46 ☽ ⚺ ♂
18 05 ☽ □ ♃
19 31 ☽ Q ☉
22 04 ☽ ∥ ☿
23 09 ☽ ⊥ ♂

13 Friday
01 32 ☽ ∥ ♃
04 07 ☿ ⚻ ♅
07 35 ☽ ⚼ ♅
07 53 ☽ ⚺ ♆
08 14 ☽ □ ♇
09 57 ☽ □ ♀
11 06 ☽ ⚹ ♄
12 10 ☽ Q ☿
14 42 ☽ ∠ ♂
15 57 ☽ ∥ ☉
17 23 ☽ ⚹ ☉
18 02 ☽ ⊥ ♆
22 46 ☽ ± ♅

14 Saturday
01 03 ☽ Q ♃
02 31 ☽ # ♇
04 14 ☽ # ♅
07 09 ☽ ⚹ ☿
07 21 ☽ Q ♄
08 52 ☽ ⚻ ♅
09 14 ☽ ∠ ♆
16 04 ☽ Q ♀
16 34 ☽ ⚹ ♂
20 42 ☽ ∠ ☉
21 40 ☽ ⚹ ♃

15 Sunday
06 42 ☽ ∠ ☿
10 37 ☽ ⚹ ♆
10 53 ☽ △ ♇
12 06 ☽ ⚹ ♀
13 06 ☽ ⊥ ☉
13 16 ☽ Q ♂
13 44 ☽ □ ♄
20 47 ☽ ⊥ ☿
23 31 ☽ ∠ ♃

16 Monday
00 05 ☽ ⚺ ☉
06 10 ☽ ⚺ ☿
07 00 ☽ Q ♆
11 36 ☽ ☍ ♅
12 21 ☽ ⚼ ♇
13 14 ☽ ∠ ♀
15 07 ☽ ⊥ ♃
20 33 ☽ □ ♂

17 Tuesday
01 36 ☽ ⚺ ♃
03 47 ☽ ± ♇
04 24 ☽ ⊥ ♀
13 58 ☽ □ ♆
14 09 ☽ ⚻ ♇
14 35 ☽ ⚺ ♀
16 48 ☉ ☌ ☿
17 03 ☽ △ ♄

18 Wednesday
05 38 ☽ ☌ ☿
08 00 ☽ ☌ ☉
11 32 ♀ ☍ ♇
15 36 ☽ ⚻ ♅
17 53 ♀ △ ♆
19 18 ♆ St R
19 22 ☽ # ♅
19 23 ☽ ⚼ ♄
20 54 ☽ # ♇

19 Thursday
02 05 ☽ △ ♂
02 17 ☽ ± ♅
07 17 ☽ ☌ ♃
11 30 ☽ ± ♄
14 32 ☽ ∥ ☉
18 31 ☽ ⚼ ♅
18 38 ☽ ☌ ♀
19 18 ☽ △ ♆
19 24 ☽ ☍ ♇
22 24 ☽ ⚻ ♄
23 04 ♀ ⚼ ♅

20 Friday
02 06 ☽ ∥ ♃
02 48 ☽ ∥ ☿
05 54 ☽ ⚼ ♂
06 58 ☽ ⚺ ☿
17 11 ☽ ⊥ ☿
19 00 ☽ ⚺ ☉
19 08 ☿ ∥ ♃
22 13 ☽ △ ♅
23 00 ☽ ± ♂
23 07 ☽ ⚼ ♆

21 Saturday
07 19 ☽ ⊥ ☉
08 47 ☽ ∠ ☿
10 35 ☽ ⚻ ♂
10 59 ☽ ∥ ♀
16 06 ☽ ⚺ ♃
16 13 ☽ ± ♆

22 Sunday
01 15 ☽ ⚺ ♀
02 07 ☽ ∠ ☉
03 48 ☽ ⚻ ♆
03 49 ☽ ⚻ ♇
03 54 ☽ ⊥ ♃
05 47 ☽ # ♄
06 55 ☽ ☍ ♄
09 50 ☽ ∥ ♆
10 10 ☉ ⚻ ♅
11 29 ☽ ⚹ ☿
12 31 ☽ ⊥ ♀
12 53 ♆ ⚹ ♇
15 30 ☽ ± ♇
21 52 ☽ ∠ ♃

23 Monday
04 38 ☽ ∥ ♂
05 35 ☽ ∠ ♀
08 12 ☽ □ ♅
09 16 ☽ ⚼ ♇
09 32 ☽ Q ☿
10 18 ☽ ⚹ ☉
11 14 ☽ # ♂
22 29 ☽ ☍ ♂

24 Tuesday
04 25 ☽ ⚹ ♃
06 17 ☽ # ♆
10 15 ☉ ⚼ ♄
10 25 ☽ ⚹ ♀
10 41 ☽ ∥ ♄
12 39 ☽ Q ☉
15 21 ☽ △ ♇
15 26 ☽ ☍ ♆
18 29 ☽ ⚻ ♄
19 37 ☽ □ ☿
20 18 ☽ Q ♅

25 Wednesday
04 55 ☽ # ♀
05 12 ☽ Q ♃
06 37 ☽ ± ♄
09 42 ☽ Q ♀
20 37 ☽ ⚹ ♅
22 25 ☿ △ ♄
23 25 ☉ ∥ ☿

26 Thursday
00 51 ☽ ⚼ ♄
04 39 ☽ □ ☉
12 20 ☽ ⚻ ♂
15 30 ☽ # ♃
18 33 ☽ □ ♃
20 36 ☽ □ ♀
21 02 ☽ # ☉
23 15 ☽ # ☿

27 Friday
00 41 ☽ ± ♂
02 58 ☽ ∠ ♅
04 08 ☽ □ ♇
04 19 ☽ ⚻ ♆
06 18 ☽ △ ☿
07 09 ☽ △ ♄
16 02 ♃ ± ♄
16 24 ☽ ± ♆
19 05 ☽ ⚼ ♂
21 02 ☽ ⊥ ♅

28 Saturday
01 19 ♀ ☌ ♃
02 34 ☽ ∥ ♇
03 12 ☉ ± ♅
03 42 ☽ ∥ ♅
06 09 ♀ ± ♄
08 57 ☽ ⚺ ♅
10 21 ☽ ⚼ ♆
11 51 ☽ ⚼ ☿
16 02 ☽ Q ♇
22 25 ☽ △ ☉

29 Sunday
01 12 ☽ △ ♂
05 04 ☿ St D
05 28 ☽ ± ☿
05 33 ☽ △ ♀
07 31 ☽ △ ♃
15 25 ☽ ⚹ ♇
15 41 ☽ △ ♆
17 08 ☽ ⚻ ☿
18 10 ☽ □ ♄

30 Monday
05 58 ☽ ⚼ ☉
09 01 ☽ ⚼ ♀
12 49 ☽ ⚼ ♃
13 29 ☉ △ ♂
18 45 ☽ ☌ ♅
19 53 ☽ ∠ ♇

31 Tuesday
00 22 ☽ ± ☉
01 08 ☽ ± ♀
04 02 ☽ Q ♄
05 57 ☽ ± ♃
06 40 ☉ ☌ ♀
10 51 ☽ □ ♂
11 44 ☽ ⚻ ♀
12 25 ☽ ⚻ ☉
12 32 ☽ ⊥ ♇
17 12 ☽ ⚻ ♃
21 32 ☿ △ ♄
23 29 ☽ ⚺ ♇
23 48 ☽ □ ♆

FEBRUARY 1950

LONGITUDES

Date	Sidereal time h m s	Sun ☉ ° ' "	Moon ☽ ° ' "	Moon ☽ 24.00 ° ' "	Mercury ☿ ° '	Venus ♀ ° '	Mars ♂ ° '	Jupiter ♃ ° '	Saturn ♄ ° '	Uranus ♅ ° '	Neptune ♆ ° '	Pluto ♇ ° '
01	20 44 30	12 ♒ 04 12	24 ♋ 04 28	00 ♌ 48 47	18 ♑ 43	10 ♒ 04	10 ♎ 22	13 ♒ 52	18 ♍ 28	01 ♋ 30	17 ♎ 18	17 ♌ 07
02	20 48 26	13 05 03	07 ♌ 37 44	14 ♌ 30 58	19 08	09 R 28	10 29	14 06	18 R 25	01 R 28	17 R 18	17 R 06
03	20 52 23	14 05 54	21 ♌ 27 58	28 ♌ 28 15	19 40	08 51	10 35	14 21	18 21	01 27	17 17	17 04
04	20 56 19	15 06 43	05 ♍ 31 12	12 ♍ 36 16	20 16	08 16	10 41	14 35	18 18	01 25	17 17	17 03
05	21 00 16	16 07 32	19 ♍ 42 51	26 ♍ 50 27	20 57	07 42	10 46	14 49	18 14	01 23	17 16	17 01
06	21 04 12	17 08 19	03 ♎ 58 31	11 ♎ 06 39	21 42	07 09	10 50	15 04	18 10	01 22	17 16	17 00
07	21 08 09	18 09 05	18 ♎ 14 27	25 ♎ 21 37	22 31	06 38	10 54	15 18	18 06	01 20	17 15	16 59
08	21 12 06	19 09 50	02 ♏ 27 52	09 ♏ 33 00	23 24	06 08	10 57	15 32	18 03	01 19	17 14	16 57
09	21 16 02	20 10 35	16 ♏ 36 52	23 ♏ 39 19	24 20	05 40	11 00	15 47	17 59	01 17	17 14	16 56
10	21 19 59	21 11 18	00 ♐ 40 13	07 ♐ 39 27	25 19	05 15	11 01	16 01	17 55	01 16	17 13	16 54
11	21 23 55	22 12 01	14 ♐ 36 54	21 ♐ 32 25	26 20	04 51	11 02	16 15	17 51	01 14	17 12	16 53
12	21 27 52	23 12 42	28 ♐ 25 49	05 ♑ 16 57	27 24	04 30	11 R 02	16 30	17 47	01 13	17 11	16 51
13	21 31 48	24 13 22	12 ♑ 05 35	18 ♑ 51 31	28 31	04 11	11 02	16 44	17 43	01 12	17 10	16 50
14	21 35 45	25 14 02	25 ♑ 34 32	02 ♒ 14 23	29 ♑ 39	03 54	11 00	16 58	17 38	01 11	17 10	16 48
15	21 39 41	26 14 39	08 ♒ 50 52	15 ♒ 23 48	00 ♒ 50	03 40	10 58	17 12	17 34	01 09	17 09	16 47
16	21 43 38	27 15 15	21 ♒ 53 03	28 ♒ 18 30	02 02	03 28	10 56	17 27	17 30	01 08	17 08	16 46
17	21 47 35	28 15 50	04 ♓ 40 06	10 ♓ 57 52	03 16	03 19	10 52	17 41	17 25	01 07	17 07	16 44
18	21 51 31	29 ♒ 16 23	17 ♓ 11 53	23 ♓ 22 16	04 32	03 12	10 48	17 55	17 21	01 06	17 06	16 43
19	21 55 28	00 ♓ 16 55	29 ♓ 29 15	05 ♈ 33 06	05 49	03 08	10 42	18 09	17 17	01 05	17 05	16 41
20	21 59 24	01 17 24	11 ♈ 34 09	17 ♈ 32 48	07 08	03 06	10 36	18 23	17 12	01 04	17 04	16 40
21	22 03 21	02 17 52	23 ♈ 29 29	29 ♈ 24 44	08 28	03 D 07	10 30	18 37	17 08	01 03	17 03	16 39
22	22 07 17	03 18 19	05 ♉ 19 03	11 ♉ 13 01	09 50	03 10	10 22	18 52	17 03	01 02	17 02	16 37
23	22 11 14	04 18 43	17 ♉ 07 16	23 ♉ 02 23	11 13	03 15	10 14	19 06	16 58	01 02	17 01	16 36
24	22 15 10	05 19 05	28 ♉ 59 01	04 ♊ 57 50	12 37	03 23	10 05	19 20	16 54	01 01	17 00	16 35
25	22 19 07	06 19 26	10 ♊ 59 26	17 ♊ 04 27	14 02	03 33	09 55	19 34	16 49	01 00	16 58	16 33
26	22 23 04	07 19 45	23 ♊ 13 29	29 ♊ 27 04	15 29	03 45	09 44	19 48	16 44	01 00	16 57	16 32
27	22 27 00	08 20 01	05 ♋ 45 42	12 ♋ 09 49	16 57	03 59	09 33	20 02	16 40	00 59	16 56	16 31
28	22 30 57	09 ♓ 20 16	18 ♋ 39 44	25 ♋ 15 42	18 ♒ 26	04 ♒ 15	09 ♎ 21	20 ♒ 16	16 ♍ 35	00 ♋ 59	16 ♎ 55	16 ♌ 29

MOON

Date	Moon True ☊ ° '	Moon Mean ☊ ° '	Moon ☽ Latitude ° '
01	09 ♈ 04	10 ♈ 27	04 N 54
02	08 R 54	10 24	04 28
03	08 45	10 20	03 45
04	08 39	10 17	02 48
05	08 35	10 14	01 40
06	08 33	10 11	00 N 25
07	08 D 34	10 08	00 S 52
08	08 35	10 04	02 05
09	08 35	10 01	03 10
10	08 R 35	09 58	04 04
11	08 33	09 55	04 42
12	08 29	09 52	05 04
13	08 22	09 49	05 08
14	08 15	09 45	04 55
15	08 07	09 42	04 26
16	08 00	09 39	03 43
17	07 54	09 36	02 50
18	07 50	09 33	01 50
19	07 48	09 30	00 S 45
20	07 D 48	09 26	00 N 21
21	07 49	09 23	01 25
22	07 51	09 20	02 25
23	07 52	09 17	03 19
24	07 54	09 14	04 04
25	07 R 54	09 10	04 40
26	07 52	09 07	05 03
27	07 50	09 04	05 13
28	07 ♈ 46	09 ♈ 01	05 N 08

DECLINATIONS

Date	Sun ☉ ° '	Moon ☽ ° '	Mercury ☿ ° '	Venus ♀ ° '	Mars ♂ ° '	Jupiter ♃ ° '	Saturn ♄ ° '	Uranus ♅ ° '	Neptune ♆ ° '	Pluto ♇ ° '
01	17 S 11	26 N 08	19 S 58	10 S 27	01 S 14	17 S 16	06 N 32	23 N 43	05 S 18	23 N 35
02	16 54	22 41	20 06	10 30	01 15	17 12	06 33	23 43	05 18	23 36
03	16 36	17 54	20 13	10 33	01 17	17 07	06 35	23 43	05 17	23 36
04	16 19	12 05	20 18	10 37	01 18	17 03	06 36	23 43	05 17	23 37
05	16 01	05 N 36	20 23	10 41	01 19	16 59	06 38	23 43	05 17	23 37
06	15 42	01 S 12	20 27	10 46	01 20	16 55	06 40	23 43	05 17	23 38
07	15 24	07 57	20 30	10 51	01 20	16 51	06 41	23 43	05 16	23 39
08	15 05	14 18	20 32	10 57	01 20	16 47	06 43	23 43	05 16	23 39
09	14 46	19 51	20 33	11 02	01 20	16 43	06 45	23 43	05 16	23 40
10	14 27	24 16	20 33	11 08	01 20	16 39	06 46	23 43	05 15	23 40
11	14 07	27 14	20 31	11 15	01 19	16 34	06 48	23 43	05 15	23 41
12	13 47	28 30	20 29	11 21	01 18	16 30	06 50	23 43	05 15	23 41
13	13 27	28 00	20 25	11 28	01 17	16 26	06 52	23 43	05 14	23 42
14	13 07	25 51	20 20	11 35	01 15	16 22	06 54	23 43	05 14	23 42
15	12 46	22 19	20 14	11 41	01 14	16 17	06 55	23 43	05 14	23 43
16	12 26	17 44	20 06	11 48	01 12	16 13	06 57	23 43	05 13	23 43
17	12 05	12 27	19 57	11 55	01 09	16 09	06 59	23 43	05 13	23 44
18	11 44	06 45	19 48	12 01	01 07	16 05	07 01	23 43	05 12	23 44
19	11 23	00 S 54	19 36	12 08	01 04	16 00	07 03	23 43	05 12	23 44
20	11 01	04 N 54	19 24	12 14	01 01	15 56	07 05	23 43	05 11	23 45
21	10 40	10 26	19 10	12 21	00 57	15 52	07 06	23 43	05 11	23 45
22	10 18	15 34	18 55	12 27	00 54	15 48	07 08	23 43	05 11	23 46
23	09 56	20 07	18 39	12 32	00 50	15 43	07 10	23 43	05 10	23 46
24	09 34	23 54	18 21	12 38	00 45	15 39	07 12	23 43	05 10	23 47
25	09 12	26 43	18 02	12 43	00 41	15 35	07 14	23 43	05 09	23 47
26	08 49	28 19	17 42	12 48	00 36	15 30	07 16	23 43	05 09	23 47
27	08 27	28 32	17 21	12 53	00 31	15 26	07 18	23 43	05 08	23 48
28	08 S 04	27 N 14	16 S 58	12 S 58	00 S 26	15 S 22	07 N 20	23 N 43	05 S 08	23 N 48

ZODIAC SIGN ENTRIES

Date	h	m	Planets
01	22	34	☽ ♌
04	02	37	☽ ♍
06	05	19	☽ ♎
08	07	50	☽ ♏
10	10	51	☽ ♐
12	14	45	☽ ♑
14	19	12	☿ ♒
14	19	57	☽ ♒
17	03	11	☽ ♓
19	05	18	☉ ♓
19	13	01	☽ ♈
22	01	12	☽ ♉
24	14	03	☽ ♊
27	01	03	☽ ♋

LATITUDES

Date	Mercury ☿ ° '	Venus ♀ ° '	Mars ♂ ° '	Jupiter ♃ ° '	Saturn ♄ ° '	Uranus ♅ ° '	Neptune ♆ ° '	Pluto ♇ ° '
01	02 N 11	07 N 33	03 N 07	00 S 37	02 N 08	00 N 16	01 N 37	08 N 17
04	01 38	07 51	03 11	00 37	02 09	00 16	01 38	08 17
07	01 04	08 01	03 14	00 37	02 09	00 16	01 38	08 18
10	00 33	08 03	03 18	00 38	02 10	00 16	01 38	08 18
13	00 N 03	07 58	03 21	00 38	02 11	00 16	01 38	08 18
16	00 S 24	07 47	03 24	00 38	02 11	00 16	01 38	08 18
19	00 48	07 32	03 27	00 39	02 11	00 16	01 38	08 18
22	01 10	07 12	03 30	00 39	02 12	00 16	01 38	08 18
25	01 29	06 50	03 32	00 39	02 12	00 16	01 39	08 18
28	01 44	06 25	03 33	00 40	02 13	00 16	01 39	08 18
31	01 S 57	06 N 00	03 N 35	00 S 40	02 N 13	00 N 16	01 N 39	08 N 18

DATA

Julian Date	2433314
Delta T	+29 seconds
Ayanamsa	23° 09' 34"
Synetic vernal point	05° ♓ 57' 26"
True obliquity of ecliptic	23° 26' 54"

MOON'S PHASES, APSIDES AND POSITIONS ☽

Date	h	m	Phase	Longitude ° '	Eclipse Indicator
02	22	16	○	13 ♌ 31	
09	18	32	☾	20 ♏ 27	
16	22	53	●	27 ♒ 43	
25	01	52	☽	05 ♊ 54	

Day	h	m	
07	00	22	Perigee
22	18	12	Apogee

Day	h	m		
06	07	47	0S	
12	17	11	Max dec	28° S 33'
19	15	40	0N	
27	03	32	Max dec	28° N 38'

LONGITUDES

Date	Chiron ⚷ ° '	Ceres ⚳ ° '	Pallas ⚴ ° '	Juno ⚵ ° '	Vesta ⚶ ° '	Black Moon Lilith ⚸ ° '
01	18 ♐ 59	06 ♐ 08	15 ♏ 19	04 ♎ 07	26 ♒ 29	02 ♉ 16
11	19 ♐ 48	09 ♐ 17	17 ♏ 29	03 ♎ 10	01 ♓ 05	03 ♉ 23
21	20 ♐ 29	12 ♐ 07	19 ♏ 07	01 ♎ 35	06 ♓ 28	04 ♉ 30
31	21 ♐ 02	14 ♐ 36	20 ♏ 08	29 ♍ 29	11 ♓ 26	05 ♉ 36

All ephemeris data is given at 12.00 UT and the Moon's longitude is additionally given for 24.00 UT

ASPECTARIAN

h m	Aspects
01 Wednesday	
01 56	☽ ✶ ♄
02 04	☽ ☍ ☿
02 42	♀ △ ♂
03 05	☉ ∥ ♃
19 43	☽ Q ♂
20 52	☉ ± ♄
02 Thursday	
01 11	☽ ⊻ ♅
04 37	☽ ∠ ♄
05 47	☽ ∥ ♅
06 30	☽ ∥ ♇
07 55	☽ Q ♆
11 44	☽ ⊥ ♅
15 04	☽ ☍ ♀
17 01	☽ ✶ ♂
20 18	☽ ⊥ ♄
22 16	☽ ☍ ☉
23 29	☽ ☍ ♃
03 Friday	
01 22	☽ # ☿
03 21	☽ ∠ ♅
04 26	☽ ☌ ♇
04 48	☽ ✶ ♆
06 39	☽ ⊻ ♄
08 46	☽ ⊼ ☿
15 28	☽ # ♃
18 00	☽ # ☉
19 07	☽ ∠ ♂
19 30	☽ ± ☿
19 41	☉ ☌ ♃
04 Saturday	
05 02	☽ ✶ ♅
06 30	☽ ∠ ♆
10 34	☽ ⊥ ♂
11 33	☽ ⚼ ☿
16 29	☽ ⊼ ♀
17 35	☽ # ♀
20 48	☽ ⊻ ♂
21 45	☽ ⊥ ♆
05 Sunday	
01 21	☽ Q ♅
02 14	☽ ± ♀
03 37	☽ ⊼ ♃
05 29	☽ ⊼ ☉
07 28	☽ ⊻ ♇
07 53	☽ ⊻ ♆
08 18	☽ ∥ ♄
09 31	☽ ☌ ♄
13 08	☽ # ♆
13 54	☽ ± ♃
14 11	☽ △ ☿
16 22	☽ ± ☉
16 50	☽ ⚼ ♀
17 34	☽ ⊥ ♇
18 04	☉ ⚼ ♅
06 Monday	
02 06	♀ ± ♅
03 08	☽ # ♂
05 18	☽ ⚼ ♃
07 37	☽ □ ♅
08 41	☽ ⚼ ☉
08 41	☽ ∠ ♇
08 47	☉ ☍ ♇
12 26	☽ ∥ ♂
14 49	☉ △ ♆
17 09	☽ △ ♀
23 36	☽ ☌ ♂
07 Tuesday	
02 22	☽ ∥ ♆
06 58	☽ △ ♃
07 24	☽ # ♄
09 52	☽ ✶ ♇
10 20	☽ ☌ ♆
11 01	☉ ⊼ ♄
11 47	☽ ⊻ ♄
11 50	☽ △ ☉
19 40	☽ □ ☿
21 51	☽ ⊥ ♄
22 50	☽ ∥ ♀
08 Wednesday	
06 04	☽ Q ♇
10 03	☽ △ ♅
12 59	☽ ∠ ♄
15 01	☽ ∥ ☉
18 00	☽ □ ♀
22 07	☽ ∥ ♃
09 Thursday	
02 26	☽ ⊻ ♂
04 11	☽ Q ☿
10 33	☽ □ ♃
11 27	☽ ⚼ ♅
12 32	☽ □ ♇
12 39	☽ ⊥ ♂
13 02	☽ ⊻ ♆
14 19	☽ ✶ ♄
15 24	☽ ∥ ☿
18 32	☽ □ ☉
23 15	☽ ⊥ ♆
23 40	☽ Q ♀
10 Friday	
02 08	☽ ✶ ☿
02 46	☽ ± ♅
04 02	☽ ∠ ♂
08 14	☽ # ♇
08 31	☽ # ♅
10 43	☽ Q ♄
13 01	☽ ⊼ ♅
14 39	☽ ∠ ♆
17 50	☽ Q ♃
19 38	☽ ✶ ♀
11 Saturday	
03 47	☽ Q ☉
05 49	☽ ✶ ♂
05 53	☽ ∠ ☿
10 43	♃ ⚼ ♅
14 53	☽ ✶ ♃
15 55	☽ △ ♇
16 28	☽ ✶ ♆
17 34	☽ □ ♄
20 50	☽ ∠ ♀
22 44	☽ ⊥ ☿
12 Sunday	
02 11	☽ ✶ ☉
02 36	☽ Q ♂
05 48	♂ St R
10 03	☽ ⊻ ☿
12 07	☽ ⊥ ♀
13 19	☽ Q ♆
16 52	☽ ☍ ♅
17 27	☽ ∠ ♃
17 59	☽ ⚼ ♇
22 22	☽ ⊻ ♀
13 Monday	
06 32	☽ ∠ ☉
09 33	☽ ⊥ ♃
09 47	☽ ± ♇
10 07	☽ □ ♂
20 22	☽ ⊻ ♃
20 23	☽ ⊼ ♇
21 00	☽ □ ♆
21 20	♃ ☍ ♇
21 54	☽ △ ♄
23 45	☽ ⊥ ☉
14 Tuesday	
11 20	☽ ⊻ ☉
20 02	☽ ☌ ☿
22 04	☽ ⊼ ♅
15 Wednesday	
00 39	☽ ⚼ ♄
02 44	☽ ☌ ♀
03 30	☽ # ♅
03 32	☽ # ♇
05 49	☉ ⚼ ♂
06 16	♃ △ ♆
08 55	☽ ± ♅
15 52	☽ △ ♂
16 57	☽ ± ♄
18 32	☿ ⊼ ♅
23 50	☽ ∥ ☿
16 Thursday	
01 23	☽ ⚼ ♅
01 29	♅ ∥ ♇
02 32	☽ ☍ ♇
03 12	☽ △ ♆
03 37	☽ ☌ ♃
03 55	☽ ⊼ ♄
16 09	♃ ⊼ ♄
19 15	☽ ∥ ♃
19 30	☽ ⚼ ♂
20 34	☿ ⚼ ♄
22 53	☽ ☌ ☉
17 Friday	
05 18	☽ △ ♅
07 10	☽ ⚼ ♆
09 04	☽ ⊻ ☿
09 28	☽ ⊻ ♀
12 22	☽ ± ♂
12 52	☿ ☌ ♀
13 40	☽ ∥ ☉
14 14	☽ ∥ ♀
20 46	☽ ⊥ ♀
20 46	☉ ∥ ♀
21 44	☽ ⊥ ☿
23 45	☽ ⊼ ♂
18 Saturday	
00 16	☽ ± ♆
10 53	☽ # ♄
11 04	☽ ⊼ ♇
11 49	☽ ⊼ ♆
12 18	☽ ☍ ♄
13 25	☽ ⊻ ♃
13 56	☽ ∠ ♀
17 03	☽ ∠ ☿
18 20	☽ ∥ ♆
22 42	☽ ± ♇
19 Sunday	
01 19	☽ ⊥ ♃
11 18	☽ ∥ ♂
13 42	☽ ⊻ ☉
15 09	☽ □ ♅
16 20	☽ ⚼ ♇
19 11	☽ ✶ ♀
19 23	☽ ∠ ♃
19 59	☽ # ♂
20 Monday	
02 04	☽ ✶ ☿
02 41	☽ ⊥ ☉
06 51	☉ △ ♅
10 05	☽ ☍ ♂
10 50	☿ ± ♅
13 15	☽ # ♆
18 04	♀ St D
19 05	☽ Q ♀
21 20	☽ ∥ ♄
22 13	☽ △ ♇
22 21	☽ ∠ ☉
23 01	☽ ☍ ♆
23 14	☽ ⊼ ♄
21 Tuesday	
01 58	☽ ✶ ♃
03 03	☽ Q ♅
05 07	☽ Q ☿
06 09	☉ ⚼ ♆
11 16	☽ ± ♄
12 56	☽ # ☉
20 49	☽ # ♀
22 Wednesday	
02 45	☽ Q ♃
03 19	☽ ✶ ♅
05 24	☽ ⚼ ♄
07 32	☽ ✶ ☉
07 36	☽ □ ♀
08 25	☉ ⊻ ♀
13 04	☽ # ♃
20 01	♄ ⊻ ♆
20 30	☿ △ ♂
22 09	☽ ⊼ ♂
22 24	☽ □ ☿
23 Thursday	
04 17	☽ # ☿
08 03	☿ ± ♄
09 47	☽ ∠ ♅
10 12	☽ Q ☉
10 13	☽ ± ♂
10 16	☉ ± ♂
10 56	☽ □ ♇
11 42	☽ △ ♄
11 47	☽ ⊼ ♆
16 05	☽ □ ♃
23 56	☽ ± ♆
24 Friday	
04 00	☽ ⊥ ♅
04 13	☽ ⚼ ♂
10 37	☽ ∥ ♅
11 04	☽ ∥ ♇
16 05	☽ ⊻ ♅
18 02	☽ ⚼ ♆
20 57	☽ △ ♀
23 12	☽ Q ♇
25 Saturday	
01 52	☽ □ ☉
09 54	☽ △ ♂
18 50	☽ △ ☿
22 57	☽ ✶ ♇
23 26	☽ □ ♄
23 47	☽ △ ♆
26 Sunday	
03 07	☽ ⚼ ♀
05 12	☽ △ ♃
20 22	☿ ⚼ ♅
20 53	☽ ± ♀
27 Monday	
02 56	☽ ☌ ♅
03 49	☽ ⚼ ☿
03 57	☽ ∠ ♇
04 58	☿ ☍ ♇
07 36	☿ ⊼ ♄
08 33	☽ ⊼ ♀
09 56	☽ Q ♄
10 35	☽ ⚼ ♃
11 50	☿ △ ♆
17 15	☽ △ ☉
19 00	☽ □ ♂
20 54	☽ ⊥ ♇
23 00	☽ ± ☿
28 Tuesday	
03 45	☽ ± ♃
08 01	☽ ⊻ ♇
08 12	☽ ✶ ♄
08 48	☽ □ ♆
11 31	☽ ⊼ ☿
12 07	☉ ⊼ ♂
14 59	☽ ⊼ ♃
23 11	☽ ⚼ ☉

MARCH 1950

LONGITUDES

Date	Sidereal time h m s	Sun ☉ ° ′ ″	Moon ☽ ° ′ ″	Moon ☽ 24.00 ° ′ ″	Mercury ☿ ° ′	Venus ♀ ° ′	Mars ♂ ° ′	Jupiter ♃ ° ′	Saturn ♄ ° ′	Uranus ♅ ° ′	Neptune ♆ ° ′	Pluto ♇ ° ′
01	22 34 53	10 ♓ 20 29	01 ♌ 57 48	08 ♌ 46 04	19 ♒ 56	04 ♒ 33	09 ♎ 08	20 ♒ 30	16 ♍ 30	00 ♋ 58	16 ♎ 54	16 ♌ 28
02	22 38 50	11 20 39	15 ♌ 40 19	22 ♌ 40 16	21 27	04 53	08 R 54	20 44	16 R 26	00 R 58	16 R 52	16 R 27
03	22 42 46	12 20 48	29 ♌ 45 30	06 ♍ 55 27	22 59	05 15	08 40	20 57	16 21	00 57	16 51	16 25
04	22 46 43	13 20 54	14 ♍ 09 27	21 ♍ 26 45	24 32	05 39	08 25	21 11	16 16	00 57	16 50	16 24
05	22 50 39	14 20 59	28 ♍ 46 30	06 ♎ 07 50	26 07	06 04	08 09	21 25	16 11	00 57	16 49	16 23
06	22 54 36	15 21 02	13 ♎ 29 55	20 ♎ 51 52	27 42	06 31	07 53	21 39	16 06	00 57	16 47	16 22
07	22 58 33	16 21 03	28 ♎ 12 55	05 ♏ 32 20	29 ♒ 19	07 00	07 35	21 52	16 02	00 56	16 46	16 20
08	23 02 29	17 21 03	12 ♏ 49 29	20 ♏ 03 49	00 ♓ 57	07 30	07 18	22 06	15 57	00 56	16 45	16 19
09	23 06 26	18 21 01	27 ♏ 14 54	04 ♐ 22 26	02 36	08 02	06 59	22 20	15 52	00 56	16 43	16 18
10	23 10 22	19 20 58	11 ♐ 26 07	18 ♐ 25 48	04 16	08 35	06 41	22 33	15 47	00 D 56	16 42	16 17
11	23 14 19	20 20 53	25 ♐ 21 25	02 ♑ 12 55	05 57	09 10	06 21	22 47	15 43	00 56	16 40	16 16
12	23 18 15	21 20 46	09 ♑ 00 20	15 ♑ 43 43	07 39	09 46	06 01	23 00	15 38	00 56	16 39	16 15
13	23 22 12	22 20 38	22 ♑ 23 09	28 ♑ 58 44	09 23	10 23	05 40	23 14	15 33	00 57	16 37	16 13
14	23 26 08	23 20 28	05 ♒ 30 35	11 ♒ 58 50	11 08	11 02	05 19	23 27	15 28	00 57	16 36	16 12
15	23 30 05	24 20 16	18 ♒ 23 36	24 ♒ 45 00	12 53	11 42	04 58	23 40	15 24	00 57	16 34	16 11
16	23 34 02	25 20 03	01 ♓ 03 11	07 ♓ 18 15	14 40	12 22	04 36	23 54	15 19	00 57	16 33	16 10
17	23 37 58	26 19 47	13 ♓ 30 23	19 ♓ 39 42	16 29	13 04	04 14	24 07	15 14	00 58	16 31	16 09
18	23 41 55	27 19 29	25 ♓ 46 22	01 ♈ 50 34	18 18	13 47	03 52	24 20	15 10	00 58	16 30	16 08
19	23 45 51	28 19 10	07 ♈ 52 31	13 ♈ 52 25	20 09	14 31	03 29	24 33	15 05	00 59	16 28	16 07
20	23 49 48	29 ♓ 18 48	19 ♈ 50 32	25 ♈ 47 09	22 01	15 16	03 06	24 46	15 00	00 59	16 27	16 06
21	23 53 44	00 ♈ 18 25	01 ♉ 42 34	07 ♉ 37 09	23 54	16 02	02 43	24 59	14 56	01 00	16 25	16 05
22	23 57 41	01 17 59	13 ♉ 31 16	19 ♉ 25 21	25 49	16 48	02 19	25 12	14 51	01 01	16 24	16 04
23	00 01 37	02 17 31	25 ♉ 19 50	01 ♊ 15 12	27 44	17 36	01 56	25 25	14 47	01 01	16 22	16 03
24	00 05 34	03 17 01	07 ♊ 11 57	13 ♊ 10 37	29 ♓ 41	18 24	01 32	25 38	14 42	01 02	16 21	16 02
25	00 09 31	04 16 28	19 ♊ 11 45	25 ♊ 15 56	01 ♈ 39	19 13	01 09	25 51	14 38	01 03	16 19	16 01
26	00 13 27	05 15 53	01 ♋ 23 42	07 ♋ 35 37	03 38	20 03	00 46	26 04	14 34	01 04	16 17	16 01
27	00 17 24	06 15 16	13 ♋ 52 14	20 ♋ 14 03	05 38	20 53	00 ♎ 22	26 16	14 29	01 05	16 16	16 00
28	00 21 20	07 14 37	26 ♋ 41 32	03 ♌ 15 06	07 39	21 44	29 ♍ 59	26 29	14 25	01 06	16 14	15 59
29	00 25 17	08 13 55	09 ♌ 55 03	16 ♌ 41 36	09 41	22 36	29 36	26 41	14 21	01 07	16 12	15 58
30	00 29 13	09 13 11	23 ♌ 34 53	00 ♍ 34 49	11 44	23 29	29 13	26 54	14 17	01 08	16 11	15 57
31	00 33 10	10 ♈ 12 25	07 ♍ 41 13	14 ♍ 53 45	13 ♈ 47	24 ♒ 22	28 ♍ 51	27 ♒ 06	14 ♍ 13	01 ♋ 09	16 ♎ 09	15 ♌ 57

Date	Moon True ☊ ° ′	Moon Mean ☊ ° ′	Moon ☽ Latitude ° ′
01	07 ♈ 42	08 ♈ 58	04 N 46
02	07 R 38	08 55	04 07
03	07 34	08 51	03 13
04	07 31	08 48	02 05
05	07 30	08 45	00 N 48
06	07 D 30	08 42	00 S 33
07	07 31	08 39	01 52
08	07 32	08 36	03 03
09	07 33	08 32	04 01
10	07 34	08 29	04 44
11	07 R 34	08 26	05 09
12	07 33	08 23	05 16
13	07 32	08 20	05 06
14	07 30	08 16	04 39
15	07 28	08 13	03 59
16	07 26	08 10	03 08
17	07 25	08 07	02 09
18	07 24	08 04	01 S 04
19	07 D 24	08 01	00 N 03
20	07 24	07 57	01 09
21	07 25	07 54	02 11
22	07 26	07 51	03 07
23	07 26	07 48	03 56
24	07 27	07 45	04 35
25	07 27	07 41	05 02
26	07 27	07 38	05 16
27	07 R 27	07 35	05 16
28	07 27	07 32	05 00
29	07 D 27	07 29	04 28
30	07 27	07 26	03 40
31	07 ♈ 27	07 ♈ 22	02 N 38

DECLINATIONS

Date	Sun ☉ ° ′	Moon ☽ ° ′	Mercury ☿ ° ′	Venus ♀ ° ′	Mars ♂ ° ′	Jupiter ♃ ° ′	Saturn ♄ ° ′	Uranus ♅ ° ′	Neptune ♆ ° ′	Pluto ♇ ° ′
01	07 S 42	24 N 22	16 S 34	13 S 02	00 S 21	15 S 17	07 N 22	23 N 43	05 S 07	23 N 49
02	07 19	20 05	16 05	13 05	00 15	15 13	07 24	23 43	05 07	23 49
03	06 56	14 34	15 42	13 08	00 09	15 09	07 26	23 43	05 06	23 49
04	06 33	08 09	15 14	13 11	00 S 03	15 04	07 28	23 43	05 06	23 50
05	06 10	01 N 13	14 45	13 14	00 N 04	15 00	07 30	23 43	05 05	23 50
06	05 47	05 S 50	14 14	13 16	00 10	14 56	07 32	23 43	05 05	23 50
07	05 23	12 35	13 43	13 17	00 17	14 52	07 33	23 43	05 04	23 51
08	05 00	18 36	13 10	13 19	00 24	14 47	07 35	23 43	05 03	23 51
09	04 37	23 28	12 35	13 19	00 31	14 43	07 37	23 43	05 03	23 51
10	04 13	26 51	12 00	13 20	00 39	14 39	07 39	23 43	05 02	23 52
11	03 50	28 31	11 23	13 19	00 46	14 34	07 41	23 43	05 02	23 52
12	03 26	28 24	10 45	13 19	00 54	14 30	07 43	23 43	05 01	23 52
13	03 02	26 37	10 06	13 17	01 01	14 26	07 45	23 43	05 01	23 53
14	02 39	23 25	09 25	13 16	01 09	14 21	07 47	23 43	05 00	23 53
15	02 15	19 07	08 44	13 13	01 17	14 17	07 49	23 43	04 59	23 53
16	01 51	14 02	08 01	13 11	01 25	14 13	07 50	23 43	04 59	23 53
17	01 28	08 28	07 17	13 07	01 33	14 09	07 52	23 43	04 58	23 54
18	01 04	02 S 40	06 31	13 04	01 42	14 04	07 54	23 43	04 58	23 54
19	00 40	03 N 10	05 45	12 59	01 50	14 00	07 56	23 43	04 57	23 54
20	00 S 16	08 49	04 57	12 55	01 58	13 56	07 58	23 43	04 56	23 54
21	00 N 07	14 07	04 08	12 49	02 06	13 52	07 59	23 43	04 56	23 55
22	00 31	18 53	03 19	12 44	02 14	13 47	08 01	23 43	04 55	23 55
23	00 55	22 55	02 28	12 37	02 23	13 43	08 03	23 43	04 55	23 55
24	01 18	26 02	01 36	12 31	02 31	13 39	08 05	23 43	04 54	23 55
25	01 42	28 01	00 S 43	12 23	02 39	13 35	08 06	23 43	04 53	23 55
26	02 06	28 42	00 N 10	12 16	02 46	13 31	08 08	23 43	04 53	23 56
27	02 29	27 57	01 05	12 07	02 54	13 27	08 10	23 43	04 52	23 56
28	02 53	25 44	02 00	11 59	03 02	13 22	08 11	23 43	04 51	23 56
29	03 16	22 04	02 56	11 49	03 10	13 18	08 13	23 43	04 51	23 56
30	03 39	17 08	03 52	11 39	03 17	13 13	08 14	23 43	04 50	23 56
31	04 N 03	11 N 08	04 N 48	11 S 29	03 N 24	13 S 10	08 N 16	23 N 43	04 S 49	23 N 56

ZODIAC SIGN ENTRIES

Date	h m	Planets
01	08 30	☽ ♌
03	12 24	☽ ♍
05	14 00	☽ ♎
07	14 55	☽ ♏
07	22 04	☿ ♓
09	16 37	☽ ♐
11	20 07	☽ ♑
14	01 52	☽ ♒
16	09 59	☽ ♓
18	20 21	☽ ♈
21	04 35	☉ ♈
21	08 32	☽ ♉
23	21 28	☽ ♊
24	15 52	☿ ♈
26	09 17	☽ ♋
28	11 05	♂ ♍
28	18 05	☽ ♌
30	23 01	☽ ♍

LATITUDES

Date	Mercury ☿ ° ′	Venus ♀ ° ′	Mars ♂ ° ′	Jupiter ♃ ° ′	Saturn ♄ ° ′	Uranus ♅ ° ′	Neptune ♆ ° ′	Pluto ♇ ° ′
01	01 S 49	06 N 17	03 N 34	00 S 40	02 N 13	00 N 16	01 N 39	08 N 18
04	02 00	05 51	03 35	00 40	02 13	00 16	01 39	08 18
07	02 08	05 25	03 35	00 40	02 13	00 16	01 39	08 18
10	02 08	04 58	03 35	00 41	02 13	00 16	01 39	08 18
13	02 12	04 31	03 34	00 41	02 13	00 16	01 39	08 18
16	02 08	04 05	03 33	00 42	02 13	00 16	01 39	08 18
19	02 00	03 39	03 30	00 42	02 13	00 16	01 39	08 17
22	01 47	03 14	03 27	00 42	02 14	00 16	01 39	08 17
25	01 30	02 49	03 23	00 43	02 13	00 16	01 39	08 17
28	01 08	02 25	03 18	00 43	02 13	00 16	01 40	08 16
31	00 S 41	02 N 02	03 N 12	00 S 44	02 N 13	00 N 16	01 N 40	08 N 16

DATA

Julian Date	2433342
Delta T	+29 seconds
Ayanamsa	23° 09′ 38″
Synetic vernal point	05° ♓ 57′ 22″
True obliquity of ecliptic	23° 26′ 54″

MOON'S PHASES, APSIDES AND POSITIONS ☽

Date	h m	Phase	Longitude ° ′	Eclipse Indicator
04	10 34	○	13 ♍ 17	
11	02 38	☾	19 ♐ 58	
18	15 20	●	27 ♓ 28	Annular
26	20 10	☽	05 ♋ 36	

Day	h m	
06	13 32	Perigee
22	10 57	Apogee

05	16 08	0S	
11	22 21	Max dec	28° S 41′
18	22 55	0N	
26	11 43	Max dec	28° N 42′

LONGITUDES

Date	Chiron ⚷ ° ′	Ceres ⚳ ° ′	Pallas ⚴ ° ′	Juno ⚵ ° ′	Vesta ⚶ ° ′	Black Moon Lilith ⚸ ° ′
01	20 ♐ 56	14 ♐ 08	19 ♏ 59	29 ♍ 56	10 ♓ 26	05 ♉ 23
11	21 ♐ 22	16 ♐ 18	20 ♏ 26	27 ♍ 35	15 ♓ 22	06 ♉ 29
21	21 ♐ 38	17 ♐ 59	20 ♏ 05	25 ♍ 07	20 ♓ 15	07 ♉ 36
31	21 ♐ 44	19 ♐ 09	18 ♏ 56	22 ♍ 49	25 ♓ 05	08 ♉ 43

All ephemeris data is given at 12.00 UT and the Moon's longitude is additionally given for 24.00 UT

ASPECTARIAN

01 Wednesday
03 30 ☽ Q ♂; 10 14 ☽ ⊻ ♅; 11 11 ☽ ∠ ♄; 15 37 ☽ ∥ ♇; 16 13 ☽ ∥ ♅; 16 33 ☽ ± ☉; 16 41 ☽ ☍ ♀; 17 11 ☽ Q ♆; 19 18 ☉ ⊥ ♀; 20 50 ☽ ⊥ ♅; 22 36 ☿ ☌ ♃

02 Thursday
00 26 ☽ ✶ ♂; 00 59 ☉ ± ♆; 02 57 ☽ ⊥ ♄; 03 55 ☽ ⊼ ☉; 04 17 ♄ ⊻ ♇; 07 08 ☉ # ♄; 12 30 ☽ ∠ ♅; 13 18 ☽ ⊻ ♄; 13 20 ☽ ☌ ♇; 14 04 ☽ ✶ ♆; 20 50 ☽ ☍ ♃; 23 07 ☽ ☍ ☿

03 Friday
01 52 ☽ ∠ ♂; 07 03 ☽ # ☿; 09 39 ☽ # ♃; 14 01 ☽ ✶ ♅; 15 31 ☽ ∠ ♆; 16 48 ☽ ⊥ ♂; 17 33 ☽ # ♀; 21 04 ☿ bQ ♂; 21 27 ☽ ⊼ ♀

04 Saturday
02 38 ☽ ⊻ ♂; 06 30 ☽ ⊥ ♆; 07 43 ☽ ± ♀; 10 00 ☽ Q ♅; 10 34 ☽ ☍ ☉; 14 28 ☽ ∥ ♄; 15 28 ☽ ☌ ♄; 15 42 ☽ ⊻ ♇; 16 24 ☽ ⊻ ♆; 18 01 ☽ # ☉; 21 22 ☿ ∥ ♃; 22 45 ☽ # ♆; 23 00 ☽ bQ ♀; 23 46 ☽ ⊼ ♃

05 Sunday
01 33 ☽ ⊥ ♇; 07 07 ☽ ⊼ ☿; 09 45 ☽ ± ♃; 15 33 ☽ □ ♅; 15 52 ☽ ∥ ♂; 16 15 ☽ ∠ ♇; 16 25 ☽ # ♂; 18 07 ☽ ± ☿

06 Monday
00 16 ☽ △ ♀; 00 40 ☽ bQ ♃; 03 01 ☽ ☌ ♂; 09 23 ☽ ∥ ♆; 10 33 ☽ bQ ☿; 11 48 ☽ ∥ ☉; 15 14 ☽ ⊼ ☉; 16 14 ☽ ⊻ ♄; 16 39 ☽ ✶ ♇; 17 21 ☽ ☌ ♆; 17 52 ☽ # ♄

07 Tuesday
01 29 ☽ △ ♃; 01 44 ☽ ± ☉; 01 57 ☽ ⊥ ♄; 04 49 ☉ ☍ ♄; 09 04 ♀ ± ♅; 11 45 ☉ ⊼ ♇; 12 12 ☽ Q ♇; 14 02 ☽ △ ☿; 14 39 ☽ ∥ ♀; 15 54 ☽ ∥ ☿; 16 27 ☽ △ ♅; 16 35 ☽ ∠ ♄; 17 30 ☽ bQ ☉; 20 33 ☽ ∥ ♃; 21 43 ☉ ⊼ ♆

08 Wednesday
02 55 ☽ □ ♀; 03 04 ☽ ⊻ ♂; 05 46 ☿ ∥ ♀; 05 49 ♀ △ ♂; 08 17 ☉ ∥ ♆; 11 52 ☿ △ ♅; 12 46 ☽ ⊥ ♂; 16 18 ☿ ± ♂; 17 08 ☽ ✶ ♄; 17 09 ☽ bQ ♅; 17 47 ☽ □ ♇; 18 28 ☽ ⊻ ♆; 20 03 ☽ △ ☉; 20 50 ♂ bQ ♃; 23 26 ☿ bQ ♆

09 Thursday
03 24 ☽ ∠ ♂; 03 38 ☽ □ ♃; 04 26 ☽ ⊥ ♆; 08 08 ☽ ± ♅; 09 53 ☽ Q ♀; 13 02 ☽ Q ♄; 13 27 ☽ # ♅; 14 19 ☽ # ♇; 18 12 ☽ ⊼ ♅; 19 23 ♅ St D; 19 30 ☽ ∠ ♆; 22 11 ☽ □ ☿

10 Friday
04 05 ☽ ✶ ♂; 06 57 ☽ ✶ ♀; 10 28 ☽ Q ♃; 19 25 ☽ □ ♄; 20 17 ☽ △ ♇; 21 00 ☽ ✶ ♆

11 Saturday
00 08 ☽ Q ♂; 02 38 ☽ □ ☉; 07 27 ☽ ✶ ♃; 09 13 ☽ Q ☿; 09 51 ☽ ∠ ♀; 16 44 ☿ ⊼ ♂; 17 47 ☽ Q ♆; 21 46 ☽ ☍ ♅; 22 19 ☽ bQ ♇

12 Sunday
02 18 ☽ ⊥ ♀; 06 50 ☽ □ ♂; 07 06 ♀ ± ♄; 09 16 ☽ ✶ ☿; 10 12 ☽ ∠ ♃; 12 39 ☽ Q ☉; 13 25 ☽ ⊻ ♀; 14 12 ☽ ± ♇; 23 45 ☽ △ ♄

13 Monday
00 54 ☽ ⊼ ♇; 01 38 ☽ □ ♆; 02 32 ☽ ⊥ ♃; 09 10 ☉ ± ♇; 11 55 ☽ ✶ ☉; 13 33 ☽ ⊻ ♃; 16 10 ☽ ∠ ☿

14 Tuesday
02 47 ☽ bQ ♄; 03 36 ☽ ⊼ ♅; 04 53 ☿ ± ♆; 08 58 ☽ # ♇; 09 58 ☿ ⊻ ♀; 10 03 ☽ # ♅; 11 11 ☽ ⊥ ☿; 11 40 ☽ △ ♂; 14 39 ☽ ± ♅; 15 27 ☉ ⊻ ♃; 17 40 ☽ ∠ ☉; 19 17 ☽ ± ♄; 22 47 ☽ ☌ ♀

15 Wednesday
00 03 ☽ ⊻ ☿; 06 24 ☽ ⊼ ♄; 07 25 ☽ bQ ♅; 07 52 ☽ ☍ ♇; 08 36 ☽ △ ♆; 11 53 ☽ ⊥ ☉; 14 53 ☽ bQ ♂; 22 08 ☽ ☌ ♃

16 Thursday
00 11 ☽ ⊻ ☉; 07 27 ☽ ± ♂; 11 10 ☽ ∥ ♃; 11 49 ☽ △ ♅; 12 57 ☽ bQ ♆; 15 50 ☽ ∥ ♀; 17 24 ☿ # ♄; 18 37 ☽ ⊼ ♂; 20 11 ☿ ☍ ♄

17 Friday
06 14 ☽ ± ♆; 07 37 ☉ # ♂; 07 42 ☿ ⊼ ♇; 11 06 ☽ ⊻ ♀; 12 35 ☿ ⊼ ♆; 14 28 ☽ # ♄; 15 21 ☽ ☍ ♄; 17 08 ☽ ⊼ ♇; 17 43 ☽ ∥ ☿; 17 52 ☽ ⊼ ♆; 18 47 ☽ ☌ ☿; 23 31 ☽ ⊥ ♀

18 Saturday
02 32 ☽ ∥ ♆; 04 51 ☽ ± ♇; 09 07 ☽ ⊻ ♃; 15 20 ☽ ☌ ☉; 15 53 ☽ # ♂; 18 20 ☽ ∠ ♀; 19 01 ☽ ∥ ☉; 21 11 ☽ ⊥ ♃; 22 17 ☽ □ ♅; 22 35 ☽ bQ ♇

19 Sunday
02 20 ☽ # ☉; 03 31 ☽ ☍ ♂; 06 20 ☽ ∥ ♂; 15 25 ☽ ∠ ♃; 19 27 ☽ # ♆; 19 53 ☿ ⊥ ♀; 21 30 ☽ # ☿

20 Monday
02 10 ☽ ✶ ♀; 02 20 ☽ ⊼ ♄; 04 29 ☽ △ ♇; 04 31 ♀ ⊼ ♄; 05 11 ☽ ☍ ♆; 08 15 ☽ ∥ ♄; 10 17 ☽ Q ♅; 12 22 ☿ ∥ ♆; 13 04 ☿ ± ♇; 14 20 ☽ ± ♄; 17 12 ☽ ⊻ ☿; 22 08 ☽ ✶ ♃

21 Tuesday
04 01 ☽ Q ♀; 06 03 ☽ # ♀; 07 38 ☽ ⊥ ☿; 08 25 ☽ bQ ♄; 08 54 ☽ ⊻ ☉; 10 33 ☽ ✶ ♅; 10 48 ☽ # ♃; 11 06 ♀ bQ ♅; 13 46 ♀ ☍ ♇; 13 58 ☽ ⊼ ♂; 22 11 ☽ ⊥ ☉; 22 55 ☽ Q ♃; 23 49 ♀ △ ♆

22 Wednesday
01 46 ☽ ± ♂; 03 27 ☿ ⊻ ♃; 04 54 ☉ □ ♅; 05 25 ☽ ∠ ☿; 06 31 ☉ bQ ♇; 09 06 ☉ ⊥ ♃; 14 42 ☽ △ ♄; 17 04 ☽ ∠ ♅; 17 10 ☽ □ ♇; 17 50 ☽ ⊼ ♆; 18 10 ☽ ∠ ☉; 19 09 ☽ □ ♀; 19 29 ☽ bQ ♂; 22 33 ♀ bQ ♂

23 Thursday
05 44 ☉ ☍ ♂; 06 00 ☽ ± ♆; 11 22 ☽ ⊥ ♅; 12 11 ☽ □ ♃; 14 07 ☿ # ♂; 17 26 ☽ ∥ ♅; 17 50 ☽ ✶ ☿; 18 55 ☽ ∥ ♇; 23 33 ☽ ⊻ ♅

24 Friday
00 12 ☽ bQ ♆; 00 57 ☽ △ ♂; 03 23 ☽ ✶ ☉; 05 38 ☽ Q ♇; 08 16 ♂ ± ♃; 15 52 ♅ ∠ ♇; 17 33 ☉ # ☿; 22 47 ☽ Q ☿

25 Saturday
02 58 ☽ □ ♄; 03 46 ☉ ∠ ♀; 04 26 ☿ bQ ♇; 04 36 ☿ □ ♅; 05 40 ☽ Q ☉; 05 42 ☽ ✶ ♇; 06 17 ☽ △ ♆; 06 55 ☿ ☍ ♂; 12 03 ☽ △ ♀; 14 40 ☿ ⊥ ♃; 18 03 ♂ □ ♅; 20 02 ♂ ∠ ♇

26 Sunday
01 23 ☽ △ ♃; 10 48 ☽ □ ♂; 11 15 ☽ ∠ ♇; 11 21 ☽ ☌ ♅; 14 15 ☽ Q ♄; 17 11 ☽ □ ☿; 19 36 ☽ bQ ♀; 20 10 ☽ □ ☉

27 Monday
04 37 ☽ ⊥ ♇; 06 58 ☽ bQ ♃; 13 10 ☽ ✶ ♄; 14 04 ☽ ± ♀; 16 01 ☽ ⊻ ♇; 16 31 ☽ □ ♆; 17 11 ☿ ∠ ♀; 20 15 ☽ Q ♂

28 Tuesday
00 16 ☽ ± ♃; 02 10 ☽ ⊼ ♀; 02 28 ☉ ☌ ☿; 11 36 ☽ ⊼ ♃; 16 59 ☽ ∠ ♄; 17 53 ☽ ✶ ♂; 20 05 ☽ ⊻ ♅

29 Wednesday
00 54 ☽ ∥ ♇; 01 45 ☽ Q ♆; 02 14 ☉ ∥ ♂; 02 20 ☽ ∥ ♅; 06 58 ☽ ⊥ ♅; 08 45 ☽ △ ☉; 09 13 ☽ ⊥ ♄; 11 31 ☽ △ ☿; 18 50 ☿ ∥ ♂; 19 50 ☽ ⊻ ♄; 20 05 ☽ ∠ ♂; 22 43 ☽ ☌ ♇; 22 59 ☽ ∠ ♅; 23 07 ☽ ✶ ♆

30 Thursday
02 53 ☉ ∥ ☿; 07 06 ♀ ± ♂; 11 24 ☽ ⊥ ♂; 11 49 ☽ ☍ ♀; 13 11 ☽ bQ ☉; 14 09 ☿ ∠ ♃; 17 47 ☽ ☍ ♃; 18 21 ☽ bQ ☿; 21 26 ☽ ⊻ ♂

31 Friday
00 57 ☽ ✶ ♅; 01 00 ☽ ∠ ♆; 04 10 ☽ # ♃; 05 43 ☽ ± ☉; 10 38 ☽ # ♀; 12 11 ☽ ± ☿; 12 26 ☿ # ♆; 16 07 ☽ ⊥ ♆; 16 31 ☽ ⊼ ☉; 16 54 ☿ ⊼ ♄; 21 07 ☽ Q ♅; 22 25 ☽ ∥ ♄; 22 49 ☽ ☌ ♄; 23 50 ☽ ⊼ ☿

APRIL 1950

LONGITUDES

Date	Sidereal time h m s	Sun ☉ ° ' "	Moon ☽ ° ' "	Moon ☽ 24.00 ° ' "	Mercury ☿ ° '	Venus ♀ ° '	Mars ♂ ° '	Jupiter ♃ ° '	Saturn ♄ ° '	Uranus ♅ ° '	Neptune ♆ ° '	Pluto ♇ ° '
01	00 37 06	11 ♈ 11 36	22 ♍ 11 53	29 ♍ 34 56	15 ♈ 50	25 ♒ 16	28 ♍ 29	27 ♒ 18	14 ♍ 09	01 ♋ 10	16 ♎ 08	15 ♌ 56
02	00 41 03	12 10 45	07 ♎ 02 03	14 ♎ 32 17	17 54	26 10	28 R 07	27 30	14 R 05	01 11	16 R 06	15 R 55
03	00 45 00	13 09 52	22 ♎ 04 33	29 ♎ 37 44	19 58	27 05	27 45	27 42	14 01	01 12	16 04	15 55
04	00 48 56	14 08 58	07 ♏ 10 42	14 ♏ 42 17	22 01	28 00	27 24	27 55	13 57	01 14	16 03	15 54
05	00 52 53	15 08 01	22 ♏ 11 28	29 ♏ 37 17	24 03	28 56	27 04	28 06	13 53	01 15	16 01	15 53
06	00 56 49	16 07 03	06 ♐ 58 53	14 ♐ 15 36	26 05	29 ♒ 52	26 44	28 18	13 50	01 17	15 59	15 53
07	01 00 46	17 06 02	21 ♐ 26 55	28 ♐ 32 29	28 ♈ 05	00 ♓ 49	26 24	28 30	13 46	01 18	15 58	15 52
08	01 04 42	18 05 00	05 ♑ 32 04	12 ♑ 25 37	00 ♉ 04	01 47	26 05	28 42	13 43	01 20	15 56	15 52
09	01 08 39	19 03 57	19 ♑ 13 11	25 ♑ 54 55	02 00	02 44	25 46	28 54	13 39	01 21	15 54	15 51
10	01 12 35	20 02 51	02 ♒ 31 04	09 ♒ 01 57	03 55	03 43	25 29	29 05	13 36	01 23	15 53	15 50
11	01 16 32	21 01 44	15 ♒ 27 55	21 ♒ 49 19	05 46	04 41	25 11	29 17	13 32	01 24	15 51	15 50
12	01 20 29	22 00 35	28 ♒ 06 34	04 ♓ 20 03	07 34	05 40	24 55	29 28	13 29	01 26	15 49	15 50
13	01 24 25	22 59 25	10 ♓ 30 10	16 ♓ 37 18	09 19	06 40	24 39	29 39	13 26	01 28	15 48	15 49
14	01 28 22	23 58 12	22 ♓ 41 48	28 ♓ 44 01	11 00	07 39	24 24	29 ♒ 50	13 23	01 30	15 46	15 49
15	01 32 18	24 56 57	04 ♈ 44 16	10 ♈ 42 52	12 37	08 40	24 09	00 ♓ 01	13 20	01 32	15 45	15 48
16	01 36 15	25 55 41	16 ♈ 40 05	22 ♈ 36 13	14 10	09 40	23 55	00 12	13 17	01 33	15 43	15 48
17	01 40 11	26 54 23	28 ♈ 31 30	04 ♉ 26 13	15 39	10 41	23 42	00 23	13 14	01 35	15 41	15 48
18	01 44 08	27 53 03	10 ♉ 20 36	16 ♉ 14 55	17 03	11 42	23 30	00 34	13 12	01 37	15 40	15 47
19	01 48 04	28 51 40	22 ♉ 09 26	28 ♉ 04 25	18 22	12 43	23 18	00 45	13 09	01 39	15 38	15 47
20	01 52 01	29 ♈ 50 16	04 ♊ 00 12	09 ♊ 57 04	19 36	13 45	23 08	00 55	13 06	01 41	15 36	15 47
21	01 55 58	00 ♉ 48 50	15 ♊ 55 22	21 ♊ 55 28	20 45	14 47	22 58	01 06	13 04	01 44	15 35	15 47
22	01 59 54	01 47 22	27 ♊ 57 46	04 ♋ 02 42	21 49	15 49	22 49	01 16	13 02	01 46	15 33	15 47
23	02 03 51	02 45 52	10 ♋ 10 41	16 ♋ 22 11	22 47	16 52	22 40	01 26	12 59	01 48	15 32	15 46
24	02 07 47	03 44 19	22 ♋ 37 40	28 ♋ 57 38	23 41	17 55	22 33	01 37	12 57	01 50	15 30	15 46
25	02 11 44	04 42 44	05 ♌ 22 32	11 ♌ 52 50	24 28	18 58	22 26	01 47	12 55	01 52	15 29	15 46
26	02 15 40	05 41 08	18 ♌ 28 56	25 ♌ 11 12	25 11	20 01	22 20	01 57	12 53	01 55	15 27	15 46
27	02 19 37	06 39 29	01 ♍ 59 56	08 ♍ 55 19	25 48	21 04	22 15	02 06	12 51	01 57	15 26	15 46
28	02 23 33	07 37 48	15 ♍ 57 24	23 ♍ 06 08	26 19	22 08	22 10	02 16	12 50	02 00	15 24	15 D 46
29	02 27 30	08 36 04	00 ♎ 21 15	07 ♎ 42 20	26 44	23 12	22 07	02 26	12 48	02 02	15 23	15 46
30	02 31 27	09 ♉ 34 19	15 ♎ 08 47	22 ♎ 39 47	27 ♉ 04	24 ♓ 16	22 ♍ 04	02 ♓ 35	12 ♍ 46	02 ♋ 04	15 ♎ 21	15 ♌ 46

Date	Moon True ☊ ° '	Moon Mean ☊ ° '	Moon ☽ Latitude ° '
01	07 ♈ 28	07 ♈ 19	01 N 24
02	07 D 28	07 16	00 N 02
03	07 R 28	07 13	01 S 20
04	07 27	07 10	02 38
05	07 26	07 07	03 44
06	07 25	07 03	04 34
07	07 24	07 00	05 05
08	07 24	06 57	05 17
09	07 D 24	06 54	05 10
10	07 24	06 51	04 47
11	07 25	06 47	04 10
12	07 26	06 44	03 21
13	07 27	06 41	02 24
14	07 28	06 38	01 21
15	07 29	06 35	00 S 15
16	07 R 29	06 32	00 N 51
17	07 27	06 28	01 54
18	07 25	06 25	02 51
19	07 22	06 22	03 42
20	07 18	06 19	04 23
21	07 14	06 16	04 53
22	07 11	06 13	05 10
23	07 08	06 09	05 14
24	07 06	06 06	05 03
25	07 D 06	06 03	04 37
26	07 06	06 00	03 56
27	07 08	05 57	03 01
28	07 09	05 53	01 54
29	07 10	05 50	00 37
30	07 ♈ 10	05 ♈ 47	00 S 44

DECLINATIONS

Date	Sun ☉ ° '	Moon ☽ ° '	Mercury ☿ ° '	Venus ♀ ° '	Mars ♂ ° '	Jupiter ♃ ° '	Saturn ♄ ° '	Uranus ♅ ° '	Neptune ♆ ° '	Pluto ♇ ° '
01	04 N 26	04 N 23	05 N 45	11 S 18	03 N 31	13 S 06	08 N 17	23 N 43	04 S 49	23 N 56
02	04 49	02 S 45	06 42	11 07	03 38	13 02	08 19	23 43	04 48	23 57
03	05 12	09 51	07 39	10 56	03 45	12 58	08 20	23 43	04 48	23 57
04	05 35	16 24	08 35	10 43	03 51	12 54	08 22	23 43	04 47	23 57
05	05 58	21 55	09 31	10 31	03 57	12 50	08 23	23 43	04 46	23 57
06	06 21	25 59	10 26	10 18	04 03	12 46	08 24	23 43	04 46	23 57
07	06 43	28 15	11 20	10 04	04 09	12 42	08 26	23 42	04 45	23 57
08	07 06	28 37	12 12	09 50	04 14	12 38	08 27	23 42	04 44	23 57
09	07 28	27 11	13 04	09 36	04 19	12 34	08 28	23 42	04 44	23 57
10	07 50	24 16	13 53	09 21	04 24	12 30	08 29	23 42	04 43	23 57
11	08 13	20 11	14 41	09 06	04 29	12 27	08 31	23 42	04 42	23 57
12	08 35	15 17	15 27	08 50	04 33	12 23	08 32	23 42	04 42	23 57
13	08 56	09 51	16 11	08 34	04 37	12 19	08 33	23 42	04 41	23 57
14	09 18	04 S 08	16 52	08 17	04 41	12 15	08 34	23 42	04 41	23 57
15	09 40	01 N 39	17 31	08 00	04 44	12 12	08 35	23 42	04 40	23 57
16	10 01	07 20	18 08	07 43	04 47	12 08	08 36	23 42	04 39	23 57
17	10 22	12 43	18 42	07 25	04 50	12 04	08 37	23 42	04 39	23 57
18	10 44	17 38	19 13	07 07	04 52	12 01	08 38	23 42	04 38	23 57
19	11 04	21 53	19 42	06 49	04 54	11 57	08 39	23 42	04 37	23 57
20	11 25	25 16	20 08	06 30	04 56	11 53	08 40	23 42	04 37	23 57
21	11 46	27 33	20 31	06 11	04 58	11 50	08 41	23 42	04 36	23 57
22	12 06	28 36	20 51	05 52	04 59	11 46	08 41	23 42	04 36	23 57
23	12 26	28 16	21 09	05 32	05 00	11 43	08 42	23 42	04 35	23 56
24	12 46	26 31	21 25	05 13	05 00	11 39	08 43	23 42	04 34	23 56
25	13 06	23 24	21 37	04 52	05 01	11 36	08 43	23 42	04 34	23 56
26	13 25	19 02	21 47	04 32	05 00	11 33	08 44	23 42	04 33	23 56
27	13 45	13 35	21 55	04 11	05 00	11 29	08 45	23 42	04 33	23 56
28	14 04	07 17	21 59	03 50	04 59	11 26	08 45	23 42	04 32	23 56
29	14 22	00 26	22 01	03 29	04 58	11 23	08 46	23 42	04 32	23 56
30	14 N 41	06 S 38	22 N 01	03 S 07	04 N 57	11 S 20	08 N 46	23 N 42	04 S 31	23 N 55

ZODIAC SIGN ENTRIES

Date	h	m	Planets
02	00	41	☽ ♎
04	00	35	☽ ♏
06	00	37	☽ ♐
06	15	13	♀ ♓
08	02	29	☽ ♑
08	11	13	☿ ♉
10	07	24	☽ ♒
12	15	38	☽ ♓
15	02	32	☽ ♈
15	08	58	♃ ♓
17	15	00	☽ ♉
20	03	54	☽ ♊
20	15	59	☉ ♉
22	16	02	☽ ♋
25	01	57	☽ ♌
27	08	30	☽ ♍
29	11	25	☽ ♎

LATITUDES

Date	Mercury ☿ ° '	Venus ♀ ° '	Mars ♂ ° '	Jupiter ♃ ° '	Saturn ♄ ° '	Uranus ♅ ° '	Neptune ♆ ° '	Pluto ♇ ° '
01	00 S 31	01 N 54	03 N 11	00 S 44	02 N 13	00 N 16	01 N 40	08 N 16
04	00 00	01 33	03 04	00 44	02 13	00 16	01 40	08 16
07	00 N 34	01 12	02 57	00 45	02 13	00 16	01 40	08 15
10	01 08	00 52	02 50	00 46	02 12	00 16	01 40	08 15
13	01 40	00 33	02 43	00 46	02 12	00 16	01 40	08 14
16	02 08	00 N 15	02 35	00 47	02 12	00 16	01 40	08 14
19	02 30	00 S 02	02 27	00 47	02 12	00 16	01 40	08 14
22	02 44	00 18	02 19	00 48	02 11	00 16	01 40	08 13
25	02 49	00 33	02 11	00 48	02 11	00 16	01 40	08 13
28	02 44	00 46	02 03	00 49	02 10	00 16	01 39	08 12
31	02 N 28	00 S 59	01 N 55	00 S 50	02 N 10	00 N 16	01 N 39	08 N 12

DATA

Julian Date	2433373
Delta T	+29 seconds
Ayanamsa	23° 09' 41"
Synetic vernal point	05° ♓ 57' 19"
True obliquity of ecliptic	23° 26' 54"

MOON'S PHASES, APSIDES AND POSITIONS ☽

Date	h	m	Phase	Longitude ° '	Eclipse Indicator
02	20	49	○	12 ♎ 32	total
09	11	42	☾	19 ♑ 03	
17	08	25	●	26 ♈ 46	
25	10	40	☽	04 ♌ 39	

Day	h	m		
03	20	06	Perigee	
18	19	40	Apogee	
02	02	48	0S	
08	04	39	Max dec	28° S 42'
15	05	08	0N	
22	18	23	Max dec	28° N 39'
29	13	28	0S	

LONGITUDES

Date	Chiron ⚷ ° '	Ceres ⚳ ° '	Pallas ⚴ ° '	Juno ⚵ ° '	Vesta ⚶ ° '	Black Moon Lilith ⚸ ° '
01	21 ♐ 44	19 ♐ 14	18 ♏ 46	22 ♍ 36	25 ♓ 34	08 ♉ 49
11	21 ♐ 40	19 ♐ 45	16 ♏ 45	20 ♍ 44	00 ♈ 20	09 ♉ 56
21	21 ♐ 26	19 ♐ 38	14 ♏ 07	19 ♍ 25	05 ♈ 01	11 ♉ 03
31	21 ♐ 04	18 ♐ 52	11 ♏ 06	18 ♍ 42	09 ♈ 37	12 ♉ 09

ASPECTARIAN

h m Aspects

01 Saturday
01 43 ☽ ⊻ ♇
02 03 ☽ ⊻ ♆
07 50 ☽ ∥ ☿
10 31 ☽ # ♆
11 34 ☽ ⊥ ♇
11 50 ☽ ∥ ☉
13 04 ☿ △ ♇
14 53 ☽ ∥ ♂
15 18 ☿ ☍ ♆
17 19 ☽ ⊼ ♀
20 26 ☽ ⊼ ♃
21 58 ☽ ☌ ♂

02 Sunday
02 10 ☽ ∠ ♇
02 35 ☽ □ ♅
03 40 ☽ ± ♀
06 15 ☽ ± ♃
11 13 ☉ # ♆
14 58 ☽ # ♂
18 50 ☽ ∥ ♆
19 02 ☽ bQ ♀
19 17 ☽ # ☉
20 49 ☽ ☍ ☉
20 53 ☽ bQ ♃
22 01 ☉ ∠ ♃
23 13 ☽ ⊻ ♄

03 Monday
02 12 ☽ ⚹ ♇
02 28 ☽ ☌ ♆
03 09 ☿ Q ♅
03 16 ☽ # ☿
06 47 ☽ # ♄
08 06 ☽ ☍ ☿
08 44 ☽ ⊥ ♄
12 39 ☿ ± ♄
14 05 ♂ ⊼ ♃
15 41 ☽ ∥ ♀
20 28 ☽ △ ♀
20 49 ☽ ⊻ ♂
21 04 ☽ △ ♃
21 16 ☽ Q ♇
22 59 ☽ ∥ ♃
22 59 ☽ ∠ ♄

04 Tuesday
00 46 ♀ ⊼ ♂
02 32 ☽ △ ♅
06 08 ☽ ⊥ ♂
06 11 ☿ ∥ ♄
07 30 ☉ ⊼ ♄
08 56 ♀ ☌ ♃
20 08 ☽ ∠ ♂
22 45 ☽ ⚹ ♄
23 54 ☽ ⊼ ☉

05 Wednesday
01 54 ☽ □ ♇
02 07 ☽ ⊻ ♆
02 28 ☽ bQ ♅
10 11 ☽ ± ☉
11 43 ☽ ⊥ ♆
15 29 ☽ ⊼ ☿
16 56 ☽ ± ♅
17 56 ☽ Q ♄
19 41 ☽ ⚹ ♂
21 24 ☽ # ♅
21 41 ☽ □ ♃
22 46 ☽ # ♇
23 37 ☽ □ ♀

06 Thursday
01 45 ☽ bQ ☉
02 14 ☽ ∠ ♆
02 40 ☽ ⊼ ♅
02 44 ☽ ± ☿
06 12 ☉ △ ♇
08 56 ☉ ☍ ♆
09 13 ☿ # ♀
14 48 ☽ Q ♂
18 35 ☿ ⊼ ♂
19 50 ☽ bQ ☿
23 14 ☽ □ ♄

07 Friday
02 41 ☽ △ ♇
02 51 ☽ ⚹ ♆
03 37 ☽ Q ♃
04 11 ☽ △ ☉
07 17 ☽ Q ♀
15 25 ♀ bQ ♆
17 33 ☿ ⚹ ♃
19 59 ☿ bQ ♄
20 10 ☽ □ ♂
22 59 ☽ Q ♆

08 Saturday
00 06 ☽ ⚹ ♃
00 25 ♀ △ ♅
01 04 ☽ △ ☿
03 58 ☽ bQ ♇
04 45 ☽ ☍ ♅
05 03 ☽ ⚹ ♀
19 30 ☽ ± ♇
23 09 ☿ # ♃

09 Sunday
02 12 ☽ △ ♄
02 26 ☽ ∠ ♃
03 45 ☿ ⚹ ♅
06 02 ☽ ⊼ ♇
06 08 ☽ □ ♆
09 10 ☽ ∠ ♀
09 30 ☿ ± ♂
11 42 ☽ □ ☉
18 39 ☽ ⊥ ♃
19 13 ☉ Q ♅
23 29 ☽ △ ♂

10 Monday
01 34 ☉ ± ♄
02 33 ☽ ⊥ ♀
04 53 ☽ bQ ♄
05 39 ☽ ⊻ ♃
06 46 ☿ ⚹ ♀
09 55 ☽ ⊼ ♅
14 05 ☽ # ♇
14 22 ☽ ⊻ ♀
14 59 ☽ □ ☿
15 39 ☽ # ♅
20 58 ☽ ± ♅
21 18 ☽ ± ♄
23 01 ☽ Q ☉

11 Tuesday
02 21 ☽ bQ ♂
08 25 ☽ ⊼ ♄
12 41 ☽ ☍ ♇
12 43 ☽ △ ♆
13 46 ☽ bQ ♅
18 52 ☽ ± ♂
23 23 ☽ ⚹ ☉

12 Wednesday
06 01 ☽ ⊼ ♂
06 20 ☽ Q ☿
08 45 ☉ ∥ ♄
09 11 ♆ ⚹ ♇
11 18 ☽ # ☿
14 39 ☽ ☌ ♃
17 12 ☽ bQ ♆
18 25 ☽ △ ♅
21 41 ☉ # ♀

13 Thursday
01 11 ☽ ∥ ♃
03 04 ♄ Q ♅
03 52 ☽ ☌ ♀
06 41 ☽ ∠ ☉
09 19 ☽ ⚹ ☿
10 37 ☽ ± ♆
13 13 ♀ # ♄
15 39 ☽ # ☉
16 00 ☿ bQ ♂
17 32 ☽ # ♄
17 43 ☽ ☍ ♄
17 45 ☽ ∥ ♀
22 21 ☽ ⊼ ♆
22 25 ☽ ⊼ ♇

14 Friday
01 50 ☽ ⊥ ☉
09 46 ☽ ∥ ♆
09 47 ☽ # ♂
10 15 ☽ ± ♇
11 27 ♂ # ♆
14 45 ☽ ⊻ ☉
15 18 ☽ ☍ ♂
19 36 ☽ ∠ ☿
20 16 ☉ ⊼ ♂

15 Saturday
01 50 ☿ Q ♃
02 26 ☽ ⊻ ♃
04 08 ☽ bQ ♇
05 33 ☽ □ ♅
14 37 ☽ ⊥ ♃
16 21 ☽ ⊥ ☿
20 36 ☽ ⊻ ♀
22 33 ☿ △ ♄

16 Sunday
00 36 ☽ # ♆
01 02 ☽ ∥ ♂
05 12 ☽ ⊼ ♄
06 13 ☽ ⊻ ☿
09 00 ☽ ∠ ♃
09 48 ☽ ⊥ ♀
10 05 ☽ ☍ ♆
10 15 ☽ △ ♇
13 07 ♀ ± ♆
13 35 ☽ # ♀
17 16 ☽ ± ♄
17 31 ☽ ∥ ♄
17 51 ☽ Q ♅

17 Monday
00 35 ☽ ∥ ☉
02 24 ☽ ⊼ ♂
05 41 ☽ ∠ ♀
08 25 ☽ ☌ ☉
09 02 ☽ # ♃
11 25 ☽ bQ ♄
12 41 ☿ ⊼ ♆
14 21 ☽ ± ♂
14 30 ☿ □ ♇
15 50 ☽ ⚹ ♃
18 14 ☽ ⚹ ♅

18 Tuesday
04 27 ☿ ∠ ♅
08 19 ☽ bQ ♂
15 01 ☽ ⚹ ♀
16 35 ☽ Q ♃
17 46 ☽ △ ♄
19 16 ☉ bQ ♄
21 30 ☽ ∥ ☿
22 47 ☽ ⊼ ♆
22 59 ♃ bQ ♆
23 04 ☽ □ ♇

19 Wednesday
00 48 ☽ ∠ ♅
03 21 ☽ ☌ ☿
10 56 ☽ ± ♆
14 18 ☽ △ ♂
17 42 ☽ Q ♀
19 07 ☽ ⊥ ♅
21 13 ☉ ± ♂
21 36 ♀ ☍ ♄

20 Thursday
00 04 ☽ ∥ ♅
01 48 ☽ ∥ ♇
02 49 ☽ ⊻ ☉
05 09 ☽ bQ ♆
05 40 ☽ □ ♃
07 19 ☽ ⊻ ♅
11 33 ☽ Q ♇
16 02 ☽ ⊥ ☉

21 Friday
06 17 ☽ □ ♄
09 30 ☽ □ ♀
11 19 ☽ △ ♆
11 43 ☽ ⚹ ♇
11 46 ☽ ∠ ☉
16 08 ☉ # ♃
20 27 ☉ ⚹ ♃
22 37 ☽ ⊻ ☿

22 Saturday
01 53 ☽ □ ♂
06 01 ♀ ⊼ ♆
06 09 ☿ ± ♆
10 58 ♀ ⊼ ♇
11 18 ☉ ⚹ ♅
11 41 ☽ ⊥ ☿
17 34 ☽ ∠ ♇
18 02 ☽ Q ♄
18 38 ☽ △ ♃
19 32 ☽ ☌ ♅
20 13 ☽ ⚹ ☉

23 Sunday
06 57 ☽ ∠ ☿
09 19 ☿ △ ♂
11 13 ☽ ⊥ ♇
12 57 ☽ Q ♂
17 27 ☽ ⚹ ♄
21 40 ☽ Q ☉
22 21 ☽ □ ♆
22 51 ☽ ⊻ ♇

24 Monday
00 18 ☽ bQ ♃
02 08 ☽ △ ♀
11 51 ☽ ⚹ ♂
14 09 ☽ ⚹ ☿
17 45 ☽ ± ♃
22 04 ☽ ∠ ♄

25 Tuesday
02 19 ♀ # ♂
05 12 ☽ ⊼ ♃
05 27 ☽ ⊻ ♅
08 28 ☽ Q ♆
08 30 ☽ ∥ ♇
09 08 ☽ bQ ♀
10 04 ☽ ∥ ♅
10 40 ☽ □ ☉
14 10 ☽ Q ☿
14 51 ☽ ⊥ ♄
15 47 ☽ ∠ ♂
16 39 ☽ ⊥ ♅
22 09 ☽ ∥ ☿

26 Wednesday
01 52 ☽ ⊻ ♄
03 12 ☽ ± ♀
06 23 ♃ △ ♅
06 32 ☽ ⚹ ♆
07 05 ☽ ☌ ♇
08 08 ☽ ⊥ ♂
09 09 ☽ ∠ ♅
10 08 ♀ ∥ ♆
15 00 ☽ ⊼ ♀
18 52 ☽ ⊻ ♂

27 Thursday
00 35 ☽ □ ☿
09 15 ☽ ∠ ♆
11 22 ☽ ∥ ☉
11 55 ☽ ⚹ ♅
12 11 ☽ ☍ ♃
19 28 ☿ ⊥ ♅
20 22 ☽ # ♃
20 42 ☽ △ ☉

28 Friday
00 51 ☽ ⊥ ♆
01 31 ☉ bQ ♂
03 44 ♀ ± ♇
06 38 ☽ ∥ ♄
06 42 ☽ ☌ ♄
08 39 ☽ Q ♅
08 40 ♇ St D
11 04 ☽ ⊻ ♆
11 41 ☽ ⊻ ♇
12 51 ♀ ☍ ♂
20 13 ☽ ∥ ♂
21 46 ☽ ⊥ ♇
21 49 ☽ # ♆
22 24 ☽ ☌ ♂
23 13 ☽ ☍ ♀

29 Saturday
00 01 ☽ bQ ☉
00 56 ☽ # ♀
05 52 ☽ △ ☿
12 41 ☽ ∠ ♇
14 46 ☽ □ ♅
15 26 ☽ ⊼ ♃
15 57 ☽ ± ☉

30 Sunday
00 38 ☽ ∥ ♀
01 19 ☽ ± ♃
02 24 ☽ ⊼ ☉
04 48 ☽ ∥ ♆
06 17 ☽ # ♂
06 57 ☽ bQ ☿
08 11 ☽ ⊻ ♄
12 20 ☽ ☌ ♆
13 00 ☽ ⚹ ♇
15 57 ☽ bQ ♃
17 47 ☽ ⊥ ♄
19 20 ☽ # ♄
21 38 ☽ ± ☿
23 02 ☽ ⊻ ♂

All ephemeris data is given at 12.00 UT and the Moon's longitude is additionally given for 24.00 UT

MAY 1950

Date	Sidereal time h m s	Sun ☉ ° ' "	Moon ☽ ° ' "	Moon ☽ 24.00 ° ' "	Mercury ☿ ° '	Venus ♀ ° '	Mars ♂ ° '	Jupiter ♃ ° '	Saturn ♄ ° '	Uranus ♅ ° '	Neptune ♆ ° '	Pluto ♇ ° '
01	02 35 23	10 ♉ 32 32	00 ♏ 14 20	07 ♏ 51 18	27 ♉ 19	25 ♓ 21	22 ♍ 02	02 ♓ 44	12 ♍ 45	02 ♋ 07	15 ♎ 20	15 ♌ 46
02	02 39 20	11 30 43	15 ♏ 29 28	23 ♏ 07 28	27 28	26 25	22 R 01	02 54	12 R 43	02 09	15 R 18	15 46
03	02 43 16	12 28 53	00 ♐ 44 00	08 ♐ 17 48	27 31	27 30	22 00	03 03	12 42	02 12	15 17	15 46
04	02 47 13	13 27 01	15 ♐ 47 39	23 ♐ 12 33	27 R 30	28 35	22 D 01	03 12	12 41	02 15	15 15	15 47
05	02 51 09	14 25 07	00 ♑ 31 37	07 ♑ 44 14	27 23	29 ♓ 40	22 02	03 21	12 40	02 17	15 14	15 47
06	02 55 06	15 23 12	14 ♑ 49 55	21 ♑ 48 28	27 11	00 ♈ 45	22 03	03 29	12 39	02 20	15 13	15 47
07	02 59 02	16 21 16	28 ♑ 39 49	05 ♒ 24 04	26 55	01 51	22 06	03 38	12 38	02 23	15 11	15 47
08	03 02 59	17 19 18	12 ♒ 01 28	18 ♒ 32 25	26 35	02 57	22 09	03 46	12 37	02 25	15 10	15 48
09	03 06 56	18 17 19	25 ♒ 57 19	01 ♓ 16 44	26 11	04 03	22 13	03 55	12 37	02 28	15 09	15 48
10	03 10 52	19 15 18	07 ♓ 31 11	13 ♓ 41 14	25 43	05 09	22 17	04 03	12 36	02 31	15 07	15 48
11	03 14 49	20 13 16	19 ♓ 47 30	25 ♓ 50 30	25 13	06 15	22 23	04 11	12 36	02 34	15 06	15 49
12	03 18 45	21 11 13	01 ♈ 50 50	07 ♈ 48 59	24 41	07 21	22 29	04 19	12 35	02 37	15 05	15 49
13	03 22 42	22 09 08	13 ♈ 45 27	19 ♈ 40 42	24 07	08 28	22 35	04 27	12 35	02 40	15 03	15 49
14	03 26 38	23 07 02	25 ♈ 35 07	01 ♉ 29 07	23 32	09 34	22 42	04 34	12 35	02 43	15 02	15 50
15	03 30 35	24 04 55	07 ♉ 23 01	13 ♉ 17 07	22 56	10 41	22 50	04 42	12 D 35	02 46	15 01	15 50
16	03 34 31	25 02 46	19 ♉ 11 42	25 ♉ 07 01	22 21	11 48	22 59	04 49	12 35	02 49	15 00	15 51
17	03 38 28	26 00 36	01 ♊ 03 15	07 ♊ 00 39	21 46	12 55	23 08	04 56	12 35	02 52	14 59	15 51
18	03 42 25	26 58 25	12 ♊ 59 24	18 ♊ 59 40	21 13	14 02	23 18	05 03	12 35	02 55	14 58	15 52
19	03 46 21	27 56 12	25 ♊ 01 40	01 ♋ 05 37	20 42	15 09	23 28	05 10	12 36	02 58	14 56	15 52
20	03 50 18	28 53 58	07 ♋ 11 43	13 ♋ 20 14	20 13	16 17	23 40	05 17	12 36	03 01	14 55	15 53
21	03 54 14	29 ♉ 51 42	19 ♋ 31 25	25 ♋ 45 33	19 47	17 24	23 51	05 24	12 37	03 04	14 54	15 54
22	03 58 11	00 ♊ 49 24	02 ♌ 02 59	08 ♌ 24 03	19 25	18 32	24 03	05 30	12 37	03 07	14 53	15 54
23	04 02 07	01 47 05	14 ♌ 49 07	21 ♌ 18 33	19 06	19 40	24 16	05 36	12 38	03 10	14 52	15 55
24	04 06 04	02 44 45	27 ♌ 52 44	04 ♍ 32 02	18 51	20 47	24 29	05 42	12 39	03 14	14 51	15 56
25	04 10 00	03 42 22	11 ♍ 16 47	18 ♍ 07 16	18 40	21 55	24 43	05 48	12 40	03 17	14 50	15 56
26	04 13 57	04 39 59	25 ♍ 03 42	02 ♎ 06 11	18 33	23 03	24 58	05 54	12 41	03 20	14 49	15 57
27	04 17 54	05 37 34	09 ♎ 14 42	16 ♎ 29 05	18 31	24 11	25 13	06 00	12 42	03 23	14 48	15 58
28	04 21 50	06 35 07	23 ♎ 49 01	01 ♏ 13 56	18 D 33	25 20	25 28	06 05	12 44	03 27	14 48	15 59
29	04 25 47	07 32 39	08 ♏ 43 09	16 ♏ 15 43	18 39	26 28	25 44	06 11	12 45	03 30	14 47	15 59
30	04 29 43	08 30 10	23 ♏ 50 34	01 ♐ 26 30	18 50	27 36	26 01	06 16	12 47	03 33	14 46	16 00
31	04 33 40	09 ♊ 27 39	09 ♐ 02 12	16 ♐ 36 20	19 ♉ 06	28 ♈ 45	26 ♍ 17	06 ♓ 21	12 ♍ 48	03 ♋ 37	14 ♎ 45	16 ♌ 01

Date	Moon True ☊ ° '	Moon Mean ☊ ° '	Moon ☽ Latitude ° '
01	07 ♈ 09	05 ♈ 44	02 S 03
02	07 R 05	05 41	03 14
03	07 01	05 38	04 11
04	06 56	05 34	04 50
05	06 50	05 31	05 09
06	06 46	05 28	05 08
07	06 43	05 25	04 49
08	06 42	05 22	04 14
09	06 D 42	05 19	03 27
10	06 43	05 15	02 32
11	06 44	05 12	01 31
12	06 45	05 09	00 S 27
13	06 R 45	05 06	00 N 38
14	06 43	05 03	01 40
15	06 39	04 59	02 38
16	06 33	04 56	03 28
17	06 25	04 53	04 10
18	06 15	04 50	04 41
19	06 06	04 47	05 00
20	05 56	04 44	05 06
21	05 49	04 40	04 57
22	05 43	04 37	04 34
23	05 39	04 34	03 58
24	05 38	04 31	03 08
25	05 D 38	04 28	02 07
26	05 38	04 25	00 N 56
27	05 R 39	04 21	00 S 19
28	05 37	04 18	01 36
29	05 34	04 15	02 47
30	05 28	04 12	03 47
31	05 ♈ 20	04 ♈ 09	04 S 32

DECLINATIONS

Date	Sun ☉ ° '	Moon ☽ ° '	Mercury ☿ ° '	Venus ♀ ° '	Mars ♂ ° '	Jupiter ♃ ° '	Saturn ♄ ° '	Uranus ♅ ° '	Neptune ♆ ° '	Pluto ♇ ° '
01	14 N 59	13 S 29	21 N 58	02 S 45	04 N 55	11 S 16	08 N 47	23 N 42	04 S 31	23 N 55
02	15 17	19 35	21 53	02 24	04 54	11 13	08 47	23 42	04 30	23 55
03	15 35	24 24	21 45	02 01	04 51	11 10	08 47	23 42	04 29	23 55
04	15 53	27 30	21 35	01 39	04 49	11 07	08 48	23 42	04 29	23 55
05	16 10	28 36	21 22	01 16	04 46	11 04	08 48	23 42	04 28	23 55
06	16 27	27 43	21 08	00 53	04 43	11 01	08 48	23 42	04 28	23 54
07	16 44	25 09	20 51	00 30	04 40	10 58	08 48	23 41	04 27	23 54
08	17 01	21 15	20 32	00 S 07	04 37	10 55	08 49	23 41	04 27	23 54
09	17 17	16 28	20 12	00 N 16	04 33	10 53	08 49	23 41	04 26	23 54
10	17 33	11 06	19 50	00 40	04 29	10 50	08 49	23 41	04 26	23 53
11	17 48	05 S 26	19 26	01 03	04 25	10 47	08 49	23 41	04 25	23 53
12	18 04	00 N 20	19 02	01 27	04 20	10 44	08 49	23 41	04 25	23 53
13	18 19	06 01	18 37	01 51	04 15	10 42	08 49	23 41	04 25	23 53
14	18 34	11 27	18 11	02 14	04 10	10 39	08 49	23 41	04 24	23 52
15	18 48	16 28	17 45	02 38	04 05	10 37	08 48	23 41	04 24	23 52
16	19 02	20 52	17 19	03 02	04 00	10 34	08 48	23 41	04 23	23 52
17	19 16	24 27	16 54	03 26	03 54	10 32	08 48	23 41	04 23	23 51
18	19 29	27 01	16 30	03 49	03 48	10 29	08 48	23 41	04 22	23 51
19	19 42	28 21	16 06	04 15	03 42	10 27	08 48	23 41	04 22	23 51
20	19 55	28 20	15 44	04 39	03 35	10 25	08 47	23 41	04 22	23 50
21	20 08	26 56	15 23	05 03	03 29	10 23	08 47	23 41	04 21	23 50
22	20 20	24 10	15 04	05 27	03 22	10 21	08 46	23 41	04 21	23 50
23	20 32	20 11	14 47	05 51	03 15	10 19	08 46	23 40	04 21	23 49
24	20 43	15 09	14 32	06 15	03 08	10 17	08 45	23 40	04 20	23 49
25	20 54	09 17	14 19	06 39	03 01	10 15	08 45	23 40	04 20	23 49
26	21 05	02 N 50	14 09	07 04	02 53	10 13	08 44	23 40	04 20	23 48
27	21 15	03 S 58	14 01	07 28	02 45	10 11	08 44	23 40	04 19	23 48
28	21 25	10 44	13 55	07 51	02 37	10 09	08 43	23 40	04 19	23 48
29	21 35	17 03	13 51	08 15	02 29	10 07	08 42	23 40	04 19	23 47
30	21 44	22 25	13 50	08 39	02 21	10 06	08 42	23 40	04 18	23 47
31	21 N 53	26 S 18	13 N 50	09 N 03	02 N 13	10 S 04	08 N 41	23 N 40	04 S 18	23 N 46

ZODIAC SIGN ENTRIES

Date	h	m	Planets
01	11	37	☽ ♏
03	10	50	☽ ♐
05	11	08	☽ ♑
05	19	19	♀ ♈
07	14	22	☽ ♒
09	21	34	☽ ♓
12	08	18	☽ ♈
14	20	59	☽ ♉
17	09	52	☽ ♊
19	21	50	☽ ♋
21	15	27	☉ ♊
22	08	06	☽ ♌
24	15	51	☽ ♍
26	20	26	☽ ♎
28	22	01	☽ ♏
30	21	43	☽ ♐

LATITUDES

Date	Mercury ☿ ° '	Venus ♀ ° '	Mars ♂ ° '	Jupiter ♃ ° '	Saturn ♄ ° '	Uranus ♅ ° '	Neptune ♆ ° '	Pluto ♇ ° '
01	02 N 28	00 S 59	01 N 55	00 S 50	02 N 10	00 N 16	01 N 39	08 N 12
04	02 01	01 11	01 48	00 50	02 09	00 16	01 39	08 11
07	01 24	01 21	01 40	00 51	02 09	00 16	01 39	08 11
10	00 N 39	01 31	01 33	00 52	02 09	00 16	01 39	08 10
13	00 S 12	01 39	01 26	00 52	02 08	00 16	01 39	08 10
16	01 04	01 46	01 19	00 53	02 08	00 16	01 39	08 09
19	01 54	01 52	01 12	00 54	02 07	00 16	01 39	08 09
22	02 37	01 58	01 06	00 55	02 07	00 16	01 39	08 08
25	03 11	02 02	01 00	00 55	02 06	00 16	01 39	08 08
28	03 35	02 05	00 54	00 56	02 06	00 16	01 39	08 08
31	03 S 48	02 S 07	00 N 48	00 S 57	02 N 05	00 N 16	01 N 39	08 N 07

DATA

Julian Date	2433403
Delta T	+29 seconds
Ayanamsa	23° 09' 45"
Synetic vernal point	05° ♓ 57' 15"
True obliquity of ecliptic	23° 26' 54"

MOON'S PHASES, APSIDES AND POSITIONS ☽

Date	h	m	Phase	Longitude ° '	Eclipse Indicator
02	05	19	○	11 ♏ 15	
08	22	32	☾	17 ♒ 45	
17	00	54	●	25 ♉ 34	
24	21	28	☽	03 ♍ 07	
31	12	43	○	09 ♐ 29	

Day	h	m		
02	06	39	Perigee	
15	22	02	Apogee	
30	16	28	Perigee	
05	13	02	Max dec	28° S 36'
12	10	38	0N	
19	23	47	Max dec	28° N 31'
26	22	04	0S	

LONGITUDES

Date	Chiron ⚷ ° '	Ceres ⚳ ° '	Pallas ⚴ ° '	Juno ⚵ ° '	Vesta ⚶ ° '	Black Moon Lilith ⚸ ° '
01	21 ♐ 04	18 ♐ 52	11 ♏ 06	18 ♍ 42	09 ♈ 37	12 ♉ 09
11	20 ♐ 34	17 ♐ 31	08 ♏ 06	18 ♍ 36	14 ♈ 06	13 ♉ 16
21	19 ♐ 59	15 ♐ 40	05 ♏ 25	19 ♍ 05	18 ♈ 30	14 ♉ 23
31	19 ♐ 19	13 ♐ 32	03 ♏ 21	20 ♍ 04	22 ♈ 45	15 ♉ 30

All ephemeris data is given at 12.00 UT and the Moon's longitude is additionally given for 24.00 UT

ASPECTARIAN

h m Aspects

01 Monday
03 40 ☽ ⚻ ♀
04 08 ☽ ∥ ♃
07 19 ☽ ⚻ ☿
08 04 ☽ ∠ ♄
08 06 ☽ Q ♇
08 31 ☽ ⊥ ♂
13 53 ☽ ± ♀
14 58 ☽ △ ♅
15 59 ☽ △ ♃
17 54 ☽ # ☉
22 42 ☽ ∠ ♂

02 Tuesday
05 07 ☽ bQ ♀
05 19 ☽ ☍ ☉
07 39 ☽ ⚹ ♄
11 42 ☽ ⚺ ♆
12 27 ☽ □ ♇
14 38 ☽ bQ ♅
21 07 ☽ ⊥ ♆
22 15 ☽ ⚹ ♂
22 16 ☽ # ☿

03 Wednesday
02 30 ☽ Q ♄
04 49 ☽ ± ♅
06 30 ☽ △ ♀
06 55 ☽ ☍ ☿
07 55 ☽ # ♅
08 14 ☉ ∠ ♀
09 09 ☽ # ♇
11 17 ☽ ∠ ♆
12 29 ☿ ⚹ ♀
14 20 ☽ ⚻ ♅
15 42 ☽ □ ♃
15 51 ♂ St D
16 07 ☿ St R
17 11 ☽ Q ♂
17 21 ☉ △ ♄

04 Thursday
07 01 ☽ □ ♄
07 58 ☽ ⚻ ☉
11 08 ☽ ⚹ ♆
11 58 ☽ △ ♇
18 19 ☽ ± ☉
20 49 ☽ Q ♃
22 04 ☽ □ ♂

05 Friday
06 35 ☽ Q ♆
06 52 ☽ ⚻ ☿
10 02 ☽ bQ ☉
10 28 ☽ □ ♀
12 25 ☽ bQ ♇
14 55 ☽ ☍ ♅
16 40 ☽ ± ☿
16 43 ☽ ⚹ ♃

06 Saturday
03 26 ☽ ± ♇
07 34 ☽ bQ ☿
07 43 ☉ ⚻ ♆
08 17 ☽ △ ♄
12 36 ♀ bQ ♇
12 39 ☽ □ ♆
13 01 ☽ △ ☉
13 38 ☽ ⚻ ♇
14 57 ☉ Q ♃
18 20 ☽ ∠ ♃
19 18 ☽ Q ♀
21 54 ☉ □ ♇

07 Sunday
00 28 ☽ △ ♂
08 59 ☽ △ ☿
10 10 ☽ ⊥ ♃
10 11 ☽ bQ ♄
18 09 ☽ ⚹ ♀
18 37 ☽ ⚻ ♅
20 27 ☽ # ♇
20 55 ☽ ⚺ ♃
21 47 ☽ # ♅

08 Monday
00 07 ♀ □ ♅
02 12 ☽ ± ♄
03 07 ☽ bQ ♂
05 26 ☽ ± ♅
13 05 ☽ ⚻ ♄
14 41 ☉ ∠ ♅
16 10 ☽ # ☿
17 45 ☽ △ ♆
18 55 ☽ ☍ ♇
19 37 ☽ ± ♂
20 40 ☿ ⊥ ♅
21 58 ☽ bQ ♅
22 32 ☽ □ ☉
23 54 ☽ ∠ ♀

09 Tuesday
06 49 ☽ ⚻ ♂
08 20 ☽ # ☉
08 43 ♀ ⚺ ♃
14 14 ☽ □ ☿
18 24 ☽ ⊥ ♀
21 49 ☽ bQ ♆

10 Wednesday
02 20 ☽ △ ♅
05 14 ☽ ☌ ♃
06 58 ☽ ⚺ ♀
11 27 ☽ Q ☉
13 11 ☽ ∥ ♃
15 06 ☽ ± ♆
21 48 ☽ # ♄
21 52 ☽ ☍ ♄
23 36 ☽ Q ☿

11 Thursday
02 47 ☽ ⚻ ♆
04 09 ☽ ⚻ ♇
06 25 ♂ # ♆
12 55 ☽ ⚹ ☉
16 00 ☽ ± ♇
16 13 ☽ ∥ ♆
16 20 ☽ # ♂
17 09 ☽ ☍ ♂
22 18 ☽ ⚹ ☿

12 Friday
05 05 ☽ # ♀
09 22 ☉ ± ♆
09 56 ☽ bQ ♇
13 33 ☽ □ ♅
17 00 ☽ ⚺ ♃
17 01 ☽ ∥ ♀
21 29 ☽ ∠ ☉

13 Saturday
00 12 ☽ ☌ ♀
03 03 ☽ ∠ ☿
04 37 ☽ ∥ ♂
05 10 ☽ # ♆
05 14 ☽ ⊥ ♃
09 37 ☽ ⚻ ♄
14 38 ☽ ☍ ♆
16 11 ☽ △ ♇
17 17 ☽ ⊥ ☉
20 24 ☽ ⊥ ☿
21 14 ☿ ∠ ♀
21 46 ☽ ± ♄
22 37 ☉ ∥ ☿
23 39 ☽ ∠ ♃

14 Sunday
00 11 ☽ ∥ ♄
00 19 ☉ △ ♂
02 03 ☽ Q ♅
06 05 ☽ ⚻ ♂
06 32 ☽ ⚺ ☉
08 01 ☽ ⚺ ☿
08 26 ☽ # ♃
16 03 ☽ bQ ♄
18 17 ☉ ☌ ☿
18 25 ☽ ± ♂

15 Monday
02 33 ☽ ⚹ ♅
06 28 ☽ ⚹ ♃
09 22 ♄ St D
12 18 ♀ ⊥ ♃
12 56 ☽ bQ ♂
15 02 ☿ △ ♂
18 04 ☽ ∥ ☿
19 25 ☽ ⚺ ♀
22 34 ☽ △ ♄

16 Tuesday
00 56 ☽ ∥ ☉
03 30 ☽ ⚻ ♆
05 12 ☽ □ ♇
07 08 ☽ Q ♃
08 52 ☽ ⊥ ♀
09 11 ☽ ∠ ♅
15 39 ☽ ± ♆
18 05 ☽ ☌ ☿
19 47 ☽ △ ♂

17 Wednesday
00 54 ☽ ☌ ☉
03 30 ☽ ⊥ ♅
04 51 ♀ ⚻ ♄
05 00 ☽ ∠ ♀
06 14 ☽ ∥ ♅
07 30 ☽ ∥ ♇
09 50 ☽ bQ ♆
15 40 ☽ ⚺ ♅
17 39 ☽ Q ♇
19 54 ☽ □ ♃

18 Thursday
10 02 ♀ ∥ ♂
10 23 ☉ ⊥ ♅
11 12 ☽ □ ♄
14 18 ☽ ⚹ ♀
15 56 ☽ △ ♆
17 45 ☽ ⚹ ♇

19 Friday
00 04 ☿ ± ♆
03 44 ☽ ⚺ ☿
05 14 ☿ ⊥ ♀
07 29 ♀ ☍ ♆
08 52 ☽ □ ♂
15 11 ☽ ⊥ ☿
16 39 ☽ Q ♀
18 16 ☽ ⚺ ☉
19 17 ♀ # ♆
23 01 ☽ Q ♄
23 34 ☽ ∠ ♇

20 Saturday
03 27 ♀ △ ♇
03 45 ☽ ☌ ♅
07 06 ☽ ⊥ ☉
08 13 ☽ △ ♃
08 15 ☽ ∠ ☿
17 16 ☽ ⊥ ♇
20 52 ☽ Q ♂
22 35 ☽ ⚹ ♄

21 Sunday
02 12 ☽ ∠ ☉
03 04 ☽ □ ♆
04 58 ☽ ⚺ ♇
07 29 ☽ □ ♀
12 29 ☽ ⚹ ☿
13 03 ☉ bQ ♆
13 42 ☽ bQ ♃
20 29 ☽ ⚹ ♂

22 Monday
03 34 ☽ ∠ ♄
07 07 ☽ ± ♃
09 29 ☽ ⚹ ☉
10 49 ☽ Q ☿
13 35 ☽ Q ♆
14 01 ♀ ± ♄
14 02 ☽ ⚺ ♅
14 20 ☽ ∥ ♇
15 22 ☽ ∥ ♅
18 35 ☽ ⚻ ♃
20 40 ☽ ⊥ ♄

23 Tuesday
01 24 ☽ ⊥ ♅
01 27 ☽ ∠ ♂
02 29 ☿ ⚺ ♀
07 56 ☽ ⚺ ♄
09 55 ☽ Q ☉
10 14 ☽ ∥ ☉
12 06 ☽ ⚹ ♆
14 02 ☽ ☌ ♇
18 15 ☽ ∠ ♅
18 30 ☽ ⊥ ♂
19 45 ☽ □ ☿
21 49 ☽ △ ♀

24 Wednesday
05 43 ☽ ⚺ ♂
10 06 ♀ ∠ ♃
14 47 ☽ ∥ ☿
15 34 ☽ ∠ ♆
21 28 ☽ □ ☉
21 42 ☽ ⚹ ♅
21 46 ♀ Q ♅

25 Thursday
00 44 ☉ ⚺ ♅
02 12 ☽ ☍ ♃
03 33 ☽ bQ ♀
07 41 ☽ ⊥ ♆
08 15 ☽ # ♃
14 05 ☽ ∥ ♄
14 27 ☽ ☌ ♄
17 52 ☉ Q ♇
18 15 ☽ ⚺ ♆
19 04 ☽ Q ♅
20 12 ☽ ⚺ ♇
20 54 ☽ ± ♀
21 24 ☽ ∥ ♀

26 Friday
00 49 ☽ △ ☿
06 34 ☽ # ♆
06 38 ☽ ⊥ ♇
08 14 ☽ ⚻ ♀
11 47 ☽ ∥ ♂
11 50 ☽ ☌ ♂
22 03 ☽ ∠ ♇

27 Saturday
02 08 ☽ □ ♅
02 23 ☽ bQ ☿
05 30 ☽ △ ☉
06 32 ☽ ⚻ ♃
07 51 ☽ # ♂
12 28 ☿ St D
13 16 ☽ ∥ ♆
16 37 ☽ ± ♃
17 26 ☽ ± ☿
17 46 ☽ ⚺ ♄
21 13 ☽ ☌ ♆
22 16 ☉ □ ♃
23 09 ☽ ⚹ ♇

28 Sunday
01 05 ☽ # ♀
03 21 ☽ ⚻ ☿
03 41 ☽ ⊥ ♄
04 48 ☽ # ♄
07 31 ☽ bQ ♃
08 07 ☽ bQ ☉
09 56 ☽ ∥ ♃
14 39 ☽ ☍ ♀
14 44 ☽ ⚺ ♂
15 55 ♀ ⚻ ♂
18 21 ☽ ∠ ♄
18 45 ☽ Q ♇
23 38 ☽ # ☿
23 43 ☽ ± ☉

29 Monday
00 36 ☽ ⊥ ♂
03 37 ☽ △ ♅
07 55 ☽ △ ♃
10 00 ☽ ⚻ ☉
15 16 ☽ ∠ ♂
18 26 ☽ ⚹ ♄
21 38 ☽ ⚺ ♆
23 35 ☽ □ ♇

30 Tuesday
03 36 ☽ bQ ♅
03 59 ☽ ☍ ☿
07 08 ☽ ⊥ ♆
08 30 ☽ # ☉
13 29 ☽ Q ♄
14 24 ♀ ∥ ♄
15 29 ☽ ⚹ ♂
15 44 ♀ bQ ♄
17 53 ☽ ± ♅
18 25 ☽ ⚻ ♀
18 42 ☽ # ♅
19 20 ☽ # ♇
21 21 ☽ ∠ ♆

31 Wednesday
03 24 ☽ ⚻ ♅
04 40 ☽ ± ♀
07 43 ☽ □ ♃
10 48 ☽ Q ♂
12 43 ☽ ☍ ☉
17 59 ☽ □ ♄
20 04 ☽ bQ ♀
21 03 ☽ ⚹ ♆
23 05 ☽ △ ♇

JUNE 1950

LONGITUDES

Date	Sidereal time h m s	Sun ☉	Moon ☽	Moon ☽ 24.00	Mercury ☿	Venus ♀	Mars ♂	Jupiter ♃	Saturn ♄	Uranus ♅	Neptune ♆	Pluto ♇
01	04 37 36	10 ♊ 25 08	24 ♐ 07 37	01 ♑ 34 49	19 ♉ 26	29 ♈ 53	26 ♍ 35	06 ♓ 26	12 ♍ 50	03 ♋ 40	14 ♎ 44	16 ♌ 02
02	04 41 33	11 22 36	08 ♑ 56 54	16 ♑ 12 57	19 51	01 ♉ 02	26 53	06 30	12 52	03 43	14 R 44	16 03
03	04 45 29	12 20 03	23 ♑ 22 19	00 ♒ 24 31	20 19	02 11	27 11	06 35	12 54	03 47	14 43	16 04
04	04 49 26	13 17 29	07 ♒ 19 20	14 ♒ 06 42	20 52	03 20	27 30	06 39	12 56	03 50	14 42	16 05
05	04 53 23	14 14 54	20 ♒ 46 46	27 ♒ 19 50	21 29	04 29	27 49	06 43	12 58	03 54	14 41	16 06
06	04 57 19	15 12 19	03 ♓ 46 17	10 ♓ 06 39	22 10	05 38	28 08	06 47	13 00	03 57	14 41	16 07
07	05 01 16	16 09 43	16 ♓ 21 31	22 ♓ 31 30	22 55	06 47	28 28	06 51	13 02	04 01	14 40	16 08
08	05 05 12	17 07 06	28 ♓ 37 16	04 ♈ 39 28	23 44	07 56	28 49	06 54	13 05	04 04	14 40	16 09
09	05 09 09	18 04 29	10 ♈ 38 46	16 ♈ 35 48	24 36	09 05	29 10	06 58	13 07	04 08	14 39	16 10
10	05 13 05	19 01 51	22 ♈ 31 12	28 ♈ 25 33	25 32	10 15	29 31	07 01	13 10	04 11	14 39	16 11
11	05 17 02	19 59 13	04 ♉ 19 21	10 ♉ 13 09	26 32	11 24	29 ♍ 52	07 04	13 13	04 15	14 38	16 12
12	05 20 58	20 56 35	16 ♉ 07 21	22 ♉ 02 21	27 36	12 34	00 ♎ 14	07 07	13 15	04 18	14 38	16 13
13	05 24 55	21 53 55	27 ♉ 58 32	03 ♊ 56 09	28 42	13 43	00 37	07 10	13 18	04 22	14 37	16 15
14	05 28 52	22 51 15	09 ♊ 55 28	15 ♊ 56 40	29 ♉ 52	14 53	00 59	07 12	13 21	04 25	14 37	16 16
15	05 32 48	23 48 35	21 ♊ 59 57	28 ♊ 05 24	01 ♊ 06	16 02	01 22	07 14	13 24	04 29	14 36	16 17
16	05 36 45	24 45 54	04 ♋ 13 10	10 ♋ 23 18	02 23	17 12	01 46	07 17	13 27	04 32	14 36	16 18
17	05 40 41	25 43 13	16 ♋ 35 54	22 ♋ 51 02	03 43	18 22	02 10	07 18	13 31	04 36	14 36	16 20
18	05 44 38	26 40 30	29 ♋ 08 47	05 ♌ 29 16	05 06	19 32	02 34	07 20	13 34	04 39	14 36	16 21
19	05 48 34	27 37 48	11 ♌ 52 37	18 ♌ 18 58	06 32	20 42	02 58	07 22	13 37	04 43	14 35	16 22
20	05 52 31	28 35 04	24 ♌ 48 30	01 ♍ 21 26	08 02	21 52	03 23	07 23	13 41	04 47	14 35	16 23
21	05 56 27	29 ♊ 32 20	07 ♍ 57 59	14 ♍ 38 23	09 34	23 02	03 48	07 24	13 44	04 50	14 35	16 25
22	06 00 24	00 ♋ 29 34	21 ♍ 22 54	28 ♍ 11 44	11 10	24 12	04 14	07 25	13 48	04 54	14 35	16 26
23	06 04 21	01 26 49	05 ♎ 05 05	12 ♎ 03 06	12 49	25 22	04 39	07 26	13 52	04 57	14 35	16 28
24	06 08 17	02 24 02	19 ♎ 05 51	26 ♎ 13 17	14 31	26 33	05 05	07 27	13 56	05 01	14 35	16 29
25	06 12 14	03 21 15	03 ♏ 25 15	10 ♏ 41 27	16 16	27 43	05 32	07 27	14 00	05 05	14 35	16 30
26	06 16 10	04 18 28	18 ♏ 01 24	25 ♏ 24 29	18 03	28 ♉ 53	05 58	07 27	14 04	05 08	14 D 35	16 32
27	06 20 07	05 15 40	02 ♐ 49 55	10 ♐ 16 47	19 54	00 ♊ 04	06 25	07 R 27	14 08	05 12	14 35	16 33
28	06 24 03	06 12 51	17 ♐ 44 03	25 ♐ 10 35	21 47	01 14	06 52	07 27	14 12	05 15	14 35	16 35
29	06 28 00	07 10 03	02 ♑ 35 17	09 ♑ 57 02	23 43	02 25	07 20	07 27	14 16	05 19	14 35	16 36
30	06 31 56	08 ♋ 07 14	17 ♑ 14 49	24 ♑ 27 41	25 ♊ 41	03 ♊ 35	07 ♎ 48	07 ♓ 26	14 ♍ 21	05 ♋ 23	14 ♎ 35	16 ♌ 38

Date	Moon True ☊	Moon Mean ☊	Moon ☽ Latitude
01	05 ♈ 11	04 ♈ 05	04 S 58
02	05 R 01	04 02	05 02
03	04 53	03 59	05 48
04	04 46	03 56	04 16
05	04 41	03 53	03 31
06	04 39	03 50	02 36
07	04 D 39	03 46	01 36
08	04 39	03 43	00 S 32
09	04 R 39	03 40	00 N 32
10	04 37	03 37	01 33
11	04 34	03 34	02 30
12	04 28	03 30	03 21
13	04 19	03 27	04 03
14	04 08	03 24	04 34
15	03 55	03 21	04 54
16	03 42	03 18	05 00
17	03 29	03 15	04 52
18	03 18	03 11	04 31
19	03 09	03 08	03 55
20	03 03	03 05	03 07
21	03 00	03 02	02 08
22	02 D 59	02 59	01 N 01
23	02 R 59	02 56	00 S 11
24	02 59	02 52	01 24
25	02 57	02 49	02 33
26	02 52	02 46	03 34
27	02 46	02 43	04 21
28	02 36	02 40	04 50
29	02 26	02 36	05 00
30	02 ♈ 15	02 ♈ 33	04 S 50

DECLINATIONS

Date	Sun ☉	Moon ☽	Mercury ☿	Venus ♀	Mars ♂	Jupiter ♃	Saturn ♄	Uranus ♅	Neptune ♆	Pluto ♇
01	22 N 01	28 S 16	13 N 53	09 N 26	02 N 04	10 S 03	08 N 40	23 N 40	04 S 18	23 N 46
02	22 09	28 10	13 58	09 50	01 55	10 01	08 39	23 40	04 18	23 46
03	22 17	26 09	14 05	10 13	01 46	10 00	08 38	23 40	04 17	23 45
04	22 24	22 34	14 14	10 36	01 37	09 58	08 37	23 39	04 17	23 45
05	22 31	17 54	14 25	10 59	01 28	09 57	08 36	23 39	04 17	23 44
06	22 38	12 33	14 38	11 22	01 18	09 56	08 35	23 39	04 17	23 44
07	22 44	06 51	14 52	11 45	01 09	09 55	08 34	23 39	04 16	23 43
08	22 49	01 S 02	15 08	12 07	00 59	09 54	08 33	23 39	04 16	23 43
09	22 55	04 N 42	15 25	12 30	00 49	09 53	08 32	23 39	04 16	23 43
10	23 00	10 13	15 43	12 52	00 40	09 52	08 31	23 39	04 16	23 42
11	23 04	15 19	16 03	13 13	00 29	09 51	08 30	23 39	04 16	23 42
12	23 08	19 52	16 24	13 35	00 19	09 50	08 29	23 39	04 16	23 41
13	23 12	23 39	16 45	13 56	00 N 09	09 49	08 27	23 39	04 16	23 41
14	23 15	26 28	17 08	14 17	00 S 02	09 49	08 26	23 38	04 15	23 40
15	23 18	28 06	17 31	14 38	00 12	09 48	08 25	23 38	04 15	23 40
16	23 21	28 23	17 55	14 59	00 23	09 48	08 23	23 38	04 15	23 39
17	23 23	27 15	18 19	15 19	00 34	09 47	08 22	23 38	04 15	23 39
18	23 24	24 45	18 44	15 39	00 45	09 47	08 21	23 38	04 15	23 38
19	23 26	21 00	19 09	15 59	00 56	09 46	08 19	23 38	04 15	23 38
20	23 26	16 11	19 34	16 18	01 07	09 46	08 18	23 38	04 15	23 37
21	23 27	10 34	19 59	16 37	01 18	09 46	08 16	23 38	04 15	23 37
22	23 27	04 N 21	20 24	16 55	01 30	09 46	08 14	23 37	04 15	23 36
23	23 26	02 S 11	20 48	17 13	01 41	09 46	08 13	23 37	04 15	23 36
24	23 26	08 47	21 12	17 31	01 53	09 46	08 11	23 37	04 15	23 35
25	23 24	15 04	21 35	17 49	02 05	09 46	08 10	23 37	04 15	23 35
26	23 23	20 38	21 57	18 06	02 16	09 46	08 08	23 37	04 15	23 34
27	23 21	25 00	22 18	18 22	02 28	09 47	08 06	23 37	04 15	23 34
28	23 18	27 42	22 37	18 38	02 40	09 47	08 04	23 37	04 15	23 33
29	23 15	28 26	22 55	18 54	02 52	09 47	08 03	23 37	04 15	23 33
30	23 N 12	27 S 08	23 N 12	19 N 09	03 S 05	09 S 48	08 N 01	23 N 36	04 S 15	23 N 32

ZODIAC SIGN ENTRIES

Date	h m	Planets
01	14 19	♀ ♉
01	21 27	☽ ♑
03	23 18	☽ ♒
06	04 57	☽ ♓
08	14 44	☽ ♈
11	03 12	☽ ♉
11	20 27	♂ ♎
13	16 05	☽ ♊
14	14 33	☿ ♊
16	03 45	☽ ♋
18	13 37	☽ ♌
20	21 31	☽ ♍
21	23 36	☉ ♋
23	03 09	☽ ♎
25	06 19	☽ ♏
27	07 26	☽ ♐
27	10 45	♀ ♊
29	07 48	☽ ♑

LATITUDES

Date	Mercury ☿	Venus ♀	Mars ♂	Jupiter ♃	Saturn ♄	Uranus ♅	Neptune ♆	Pluto ♇
01	03 S 51	02 S 08	00 N 46	00 S 57	02 N 05	00 N 16	01 N 38	08 N 07
04	03 52	02 09	00 41	00 58	02 04	00 16	01 38	08 07
07	03 46	02 09	00 35	00 59	02 04	00 16	01 38	08 06
10	03 32	02 09	00 30	01 00	02 03	00 16	01 38	08 06
13	03 12	02 07	00 26	01 01	02 03	00 16	01 38	08 06
16	02 47	02 05	00 21	01 01	02 02	00 16	01 38	08 05
19	02 17	02 02	00 16	01 02	02 02	00 16	01 38	08 05
22	01 45	01 58	00 12	01 03	02 01	00 16	01 38	08 05
25	01 10	01 54	00 08	01 04	02 01	00 16	01 37	08 04
28	00 S 34	01 49	00 04	01 05	02 01	00 16	01 37	08 04
31	00 N 01	01 S 43	00 00	01 S 06	02 N 00	00 N 16	01 N 37	08 N 04

DATA

Julian Date	2433434
Delta T	+29 seconds
Ayanamsa	23° 09' 50"
Synetic vernal point	05° ♓ 57' 10"
True obliquity of ecliptic	23° 26' 53"

MOON'S PHASES, APSIDES AND POSITIONS ☽

Date	h m	Phase	Longitude	Eclipse Indicator
07	11 35	☾	16 ♓ 09	
15	15 53	●	23 ♊ 58	
23	05 13	☽	01 ♎ 11	
29	19 58	○	07 ♑ 29	

Day	h m	
12	06 19	Apogee
27	21 37	Perigee

Day	h m		
01	22 48	Max dec	28° S 29'
08	16 18	ON	
16	04 56	Max dec	28° N 26'
23	04 02	OS	
29	08 27	Max dec	28° S 27'

LONGITUDES

Date	Chiron ⚷	Ceres ⚳	Pallas ⚴	Juno ⚵	Vesta ⚶	Black Moon Lilith ⚸
01	19 ♐ 15	13 ♐ 18	03 ♏ 11	20 ♍ 11	23 ♈ 10	15 ♉ 36
11	18 ♐ 34	11 ♐ 06	01 ♏ 56	21 ♍ 39	27 ♈ 16	16 ♉ 43
21	17 ♐ 53	09 ♐ 06	01 ♏ 29	23 ♍ 29	01 ♉ 11	17 ♉ 50
31	17 ♐ 14	07 ♐ 29	01 ♏ 47	25 ♍ 39	04 ♉ 54	18 ♉ 57

ASPECTARIAN

01 Thursday
04 19 ☽ ⊼ ☿
12 29 ☽ Q ♃
14 09 ☽ ± ☿
16 01 ☽ □ ♂
16 11 ☽ Q ♆
22 02 ☽ △ ♀
23 08 ☽ Ᵽ ♇

02 Friday
03 27 ☽ ☍ ♅
05 05 ☽ Ᵽ ☿
07 59 ☽ ⚹ ♃
13 49 ☽ ± ♇
16 16 ☽ ⊼ ☉
18 27 ☽ △ ♄
21 31 ☽ □ ♆
23 00 ♀ ⋕ ♃
23 44 ☽ ⊼ ♇

03 Saturday
02 56 ☽ ± ☉
06 40 ☽ △ ☿
08 57 ☽ ∠ ♃
18 37 ☽ △ ♂
19 13 ☽ Ᵽ ☉
19 43 ☽ Ᵽ ♄

04 Sunday
00 21 ☽ ⊥ ♃
02 35 ☉ □ ♄
04 25 ☽ □ ♀
05 00 ☽ ⋕ ♇
05 09 ☿ ± ♆
05 33 ☽ ⋕ ♅
05 54 ☽ ⊼ ♅
10 49 ☽ ⊻ ♃
11 19 ☽ ± ♄
12 56 ☽ ⋕ ☉
16 26 ☽ ± ♅
16 46 ♀ ± ♂
21 20 ☽ Ᵽ ♂
21 55 ☽ ⊼ ♄
23 11 ♀ ⚹ ♅
23 21 ☽ △ ☉

05 Monday
01 03 ☽ △ ♆
03 33 ☽ ☍ ♇
08 34 ☽ Ᵽ ♅
13 21 ☽ □ ☿
13 56 ☽ ± ♂
15 23 ☽ Q ♀
22 58 ☉ △ ♆

06 Tuesday
01 13 ☽ ⊼ ♂
03 18 ☽ ⋕ ☿
04 22 ☽ Ᵽ ♆
12 20 ☽ △ ♅
15 51 ☽ ⚹ ♀
16 46 ☽ ⋕ ♀
17 43 ☽ ☌ ♃
21 16 ☽ ± ♆
23 12 ☽ ∥ ♃

07 Wednesday
00 52 ☽ Q ☿
04 50 ☽ ⋕ ♄
05 35 ☽ ☍ ♄
08 45 ☽ ⊼ ♆
11 15 ☉ ⚹ ♇
11 34 ☽ ⊼ ♇
11 35 ☽ □ ☉
13 29 ♀ ⚹ ♃
22 40 ☽ ∥ ♆
23 15 ☽ ± ♇
23 38 ☽ ∠ ♀

08 Thursday
01 40 ☽ ⚹ ☿
12 13 ☽ ⋕ ♂
12 23 ☽ ☍ ♂
17 01 ☽ Ᵽ ♇
19 16 ☽ ⊥ ♀
20 10 ☽ ∥ ♂
22 52 ☽ □ ♅

09 Friday
02 02 ☽ Q ☉
04 35 ☽ ⊻ ♃
08 32 ☽ ⊻ ♀
09 44 ☽ ∠ ☿
10 09 ☽ ⋕ ♆
16 41 ☽ ⊥ ♃
17 00 ☽ ⊼ ♄
20 04 ☽ ☍ ♆
23 09 ☽ △ ♇

10 Saturday
04 18 ☽ ⚹ ☉
04 29 ☽ ∥ ♄
05 10 ☽ ± ♄
05 26 ☽ ⊥ ☿
10 27 ☽ ⋕ ♃
10 59 ☽ ∠ ♃
11 19 ☽ Q ♅
18 41 ☽ ⊻ ☿
23 31 ☽ Ᵽ ♄

11 Sunday
01 05 ☽ ∥ ♀
02 39 ☽ ⊼ ♂
11 50 ☽ ⚹ ♅
13 28 ☽ ∠ ☉
15 15 ☽ ± ♂
15 53 ☽ ∥ ☿
17 37 ☽ ⚹ ♃

12 Monday
03 58 ☽ ☌ ♀
06 09 ☽ △ ♄
08 58 ☽ ⊼ ♆
09 24 ☽ ⊥ ☉
10 09 ☽ Ᵽ ♂
12 12 ☽ □ ♇
18 06 ☽ Q ♃
18 29 ☽ ∠ ♅
21 08 ☽ ± ♆
22 38 ☽ ⊻ ☉

13 Tuesday
03 02 ♀ △ ♄
04 18 ☿ ⊥ ♅
08 43 ☽ ∥ ☉
11 54 ☽ ∥ ♅
12 09 ☽ ∥ ♇
12 47 ☽ ⊥ ♅
13 38 ☽ ☌ ☿
15 19 ☽ Ᵽ ♆
17 29 ☽ △ ♂

14 Wednesday
00 38 ☽ Q ♇
00 55 ☽ ⊻ ♅
06 32 ☽ □ ♃
06 32 ♀ ⊼ ♆
06 48 ☿ Ᵽ ♆
18 52 ☽ □ ♄
21 21 ☽ △ ♆
22 56 ☽ ⊻ ♀

15 Thursday
00 39 ☽ ⚹ ♇
06 08 ♂ ∠ ♇
12 05 ☽ ⊥ ♀
15 53 ☽ ☌ ☉
17 07 ♀ □ ♇
19 37 ☿ △ ♂
22 20 ♀ Ᵽ ♂

16 Friday
06 18 ☽ ∠ ♇
06 35 ☽ Q ♄
07 03 ☽ □ ♂
07 39 ☽ ∠ ♀
07 58 ☽ ⊻ ☿
12 37 ☽ ☌ ♅
17 58 ☽ △ ♃
21 04 ☽ ⊥ ☿
23 52 ☽ ⊥ ♇

17 Saturday
06 01 ☽ ⚹ ♄
08 09 ☽ □ ♆
11 28 ☽ ⊻ ♇
15 45 ☽ ⚹ ♀
16 34 ☽ ∠ ☿
19 04 ☽ Q ♂
22 57 ☿ Q ♇
22 59 ☽ Ᵽ ♃

18 Sunday
04 10 ☿ ⊻ ♅
06 55 ☽ ⊻ ☉
07 51 ♀ Q ♃
10 53 ☽ ∠ ♄
14 42 ♀ ∠ ♅
16 10 ☽ ± ♃
16 59 ☽ Q ♀
18 32 ☽ Q ♆
18 41 ☽ ⚹ ♂
19 14 ☽ ⊥ ☉
19 56 ☽ ∥ ♇
19 57 ☽ ∥ ♅
21 24 ☽ ∥ ☉
22 29 ☽ ⊻ ♅

19 Monday
00 41 ☽ ⚹ ☿
03 31 ☽ ⊼ ♃
03 59 ☽ ⊥ ♄
06 32 ♅ ∥ ♇
09 45 ♀ ± ♆
09 49 ☽ ⊥ ♅
13 31 ☽ ∠ ☉
15 17 ☽ ⊻ ♄
17 04 ☽ ⚹ ♆
20 24 ☽ ☌ ♇
20 59 ☽ ∥ ☿
23 44 ☽ ∠ ♂

20 Tuesday
01 35 ☿ □ ♃
02 01 ☽ Q ☿
02 40 ☽ ∠ ♅
06 02 ☽ □ ♀
11 33 ☽ ∥ ♀
16 53 ☽ ⊥ ♂
19 29 ☽ ⚹ ☉
20 46 ☽ ∠ ♆

21 Wednesday
04 12 ☽ ⊻ ♂
06 18 ☽ ⚹ ♅
10 59 ☽ ☍ ♃
13 07 ☽ ⊥ ♆
15 11 ☽ ⋕ ♃
15 17 ☽ □ ☿
18 56 ☽ Q ☉
21 08 ☽ ∥ ♄
22 26 ☽ ☌ ♄
23 54 ☽ ⊻ ♆

22 Thursday
03 12 ☽ ⊻ ♇
04 00 ☽ Q ♅
12 23 ☽ ⋕ ♆
13 52 ☽ ⊥ ♇
17 27 ☽ △ ♀
22 16 ☽ ⋕ ♂

23 Friday
05 13 ☽ □ ☉
05 41 ☽ ∠ ♇
10 07 ☽ ∥ ♂
11 14 ☽ ☌ ♂
11 47 ☽ □ ♅
12 18 ☉ ∠ ♇
16 04 ☽ ⊼ ♃
19 28 ☽ ∥ ♆
20 16 ☉ ⊥ ♀
21 57 ☽ Ᵽ ♀
23 16 ☉ Q ♄

24 Saturday
02 22 ☽ ± ♃
03 08 ☽ △ ☿
03 10 ☽ ⊻ ♄
03 29 ☿ □ ♄
04 19 ☽ ☌ ♆
07 21 ♂ □ ♅
07 33 ☽ ⚹ ♇
09 50 ☽ ⋕ ♄
12 52 ☿ △ ♆
13 25 ☽ ⊥ ♄
14 40 ☽ ± ♀
15 40 ☽ ∥ ♃
17 39 ☽ Ᵽ ♃

25 Sunday
01 39 ☽ ⊼ ♀
03 48 ☽ Q ♇
04 36 ☽ ∠ ♄
07 55 ☽ Ᵽ ☿
11 53 ☽ △ ☉
14 45 ☽ △ ♅
15 20 ☿ ⚹ ♇
15 36 ☽ ⊻ ♂
18 40 ☽ △ ♃
23 57 ☽ ⋕ ♀

26 Monday
00 52 ☽ ± ☿
01 48 ☽ ⊥ ♂
05 30 ☽ ⚹ ♄
06 22 ☽ ⊻ ♆
08 04 ♆ St D
09 33 ☽ □ ♇
12 04 ☽ ⊼ ☿
14 14 ☽ Ᵽ ☉
15 27 ☽ Ᵽ ♅
16 09 ☽ ⊥ ♆
16 57 ☽ ∠ ♂
17 23 ♀ ⊥ ♅
19 02 ☽ ⋕ ☿

27 Tuesday
00 14 ♃ St R
01 07 ☽ Q ♄
02 02 ☽ ⋕ ☉
02 05 ♀ Ᵽ ♆
03 12 ☽ ⋕ ♇
03 30 ☽ ⋕ ♅
05 50 ☽ ± ☉
06 06 ☽ ± ♅
06 45 ☽ ∠ ♆
07 09 ☽ ☍ ♀
10 16 ☉ ☌ ♅
15 50 ☽ ⊼ ♅
16 11 ☽ ⊼ ☉
17 58 ☽ ⚹ ♂
19 27 ☽ □ ♃

28 Wednesday
06 17 ☽ □ ♄
06 55 ☽ ⚹ ♆
10 08 ☽ △ ♇
13 54 ☽ Q ♂
19 30 ☽ ☍ ☿

29 Thursday
00 26 ☽ Q ♃
02 16 ☽ Q ♆
10 24 ☽ Ᵽ ♇
11 41 ☽ ⊼ ♀
16 27 ☽ ☍ ♅
17 39 ♂ ⊼ ♃
18 51 ☉ △ ♃
19 54 ☽ ⚹ ♃
19 58 ☽ ☍ ☉
19 58 ☽ □ ♂
20 00 ☉ □ ♂
22 19 ☽ ± ♀

30 Friday
01 05 ☽ ± ♇
07 11 ☽ △ ♄
07 36 ☽ □ ♆
10 58 ☽ ⊼ ♇
12 10 ☉ ∥ ☿
14 25 ☽ Ᵽ ♀
20 36 ☽ ∠ ♃

All ephemeris data is given at 12.00 UT and the Moon's longitude is additionally given for 24.00 UT

JULY 1950

LONGITUDES

Date	Sidereal time h m s	Sun ☉ ° ' "	Moon ☽ ° ' "	Moon ☽ 24.00 ° ' "	Mercury ☿ ° '	Venus ♀ ° '	Mars ♂ ° '	Jupiter ♃ ° '	Saturn ♄ ° '	Uranus ♅ ° '	Neptune ♆ ° '	Pluto ♇ ° '
01	06 35 53	09 ♋ 04 25	01 ♒ 34 56	08 ♒ 35 57	27 ♊ 42	04 ♊ 46	08 ♎ 16	07 ♓ 25	14 ♍ 25	05 ♋ 26	14 ♎ 35	16 ♌ 39
02	06 39 50	10 01 36	15 ♒ 30 21	22 ♒ 17 56	29 ♊ 45	05 57	08 44	07 R 24	14 29	05 30	14 35	16 41
03	06 43 46	10 58 47	28 ♒ 58 41	05 ♓ 32 44	01 ♋ 49	07 07	09 12	07 23	14 34	05 33	14 35	16 42
04	06 47 43	11 55 58	12 ♓ 00 21	18 ♓ 21 57	03 55	08 18	09 41	07 22	14 39	05 37	14 36	16 44
05	06 51 39	12 53 09	24 ♓ 38 01	00 ♈ 49 06	06 03	09 29	10 10	07 20	14 43	05 41	14 36	16 45
06	06 55 36	13 50 21	06 ♈ 55 51	12 ♈ 58 54	08 11	10 40	10 39	07 18	14 48	05 44	14 36	16 47
07	06 59 32	14 47 33	18 ♈ 58 55	24 ♈ 56 36	10 21	11 51	11 09	07 17	14 53	05 48	14 37	16 48
08	07 03 29	15 44 46	00 ♉ 52 35	06 ♉ 47 33	12 30	13 02	11 39	07 14	14 58	05 51	14 37	16 50
09	07 07 25	16 41 58	12 ♉ 42 06	18 ♉ 36 50	14 40	14 13	12 09	07 12	15 03	05 55	14 37	16 52
10	07 11 22	17 39 12	24 ♉ 32 18	00 ♊ 28 59	16 50	15 25	12 39	07 10	15 08	05 58	14 38	16 53
11	07 15 19	18 36 25	06 ♊ 27 21	12 ♊ 27 46	19 00	16 36	13 09	07 07	15 13	06 02	14 38	16 55
12	07 19 15	19 33 39	18 ♊ 30 35	24 ♊ 36 03	21 08	17 47	13 40	07 04	15 18	06 06	14 39	16 57
13	07 23 12	20 30 54	00 ♋ 44 23	06 ♋ 55 44	23 16	18 58	14 11	07 01	15 24	06 09	14 39	16 58
14	07 27 08	21 28 09	13 ♋ 10 11	19 ♋ 27 47	25 23	20 10	14 42	06 58	15 29	06 13	14 40	17 00
15	07 31 05	22 25 24	25 ♋ 48 31	02 ♌ 12 22	27 29	21 21	15 13	06 54	15 35	06 16	14 40	17 02
16	07 35 01	23 22 40	08 ♌ 39 17	15 ♌ 09 11	29 ♋ 34	22 33	15 45	06 51	15 40	06 20	14 41	17 03
17	07 38 58	24 19 55	21 ♌ 42 00	28 ♌ 17 40	01 ♌ 37	23 44	16 16	06 47	15 46	06 23	14 42	17 05
18	07 42 54	25 17 11	04 ♍ 56 07	11 ♍ 37 21	03 38	24 56	16 48	06 43	15 51	06 27	14 42	17 07
19	07 46 51	26 14 28	18 ♍ 21 20	25 ♍ 08 05	05 38	26 07	17 20	06 39	15 57	06 30	14 43	17 09
20	07 50 48	27 11 44	01 ♎ 57 37	08 ♎ 50 00	07 36	27 19	17 53	06 35	16 03	06 33	14 44	17 10
21	07 54 44	28 09 01	15 ♎ 45 14	22 ♎ 43 22	09 33	28 31	18 25	06 30	16 08	06 37	14 45	17 12
22	07 58 41	29 ♋ 06 18	29 ♎ 44 21	06 ♏ 48 10	11 28	29 ♊ 43	18 58	06 25	16 14	06 40	14 45	17 14
23	08 02 37	00 ♌ 03 35	13 ♏ 54 40	21 ♏ 03 38	13 21	00 ♋ 54	19 31	06 21	16 20	06 44	14 46	17 16
24	08 06 34	01 00 53	28 ♏ 14 47	05 ♐ 27 42	15 12	02 06	20 04	06 16	16 26	06 47	14 47	17 17
25	08 10 30	01 58 11	12 ♐ 41 53	19 ♐ 56 44	17 02	03 18	20 37	06 11	16 32	06 50	14 48	17 19
26	08 14 27	02 55 29	27 ♐ 11 34	04 ♑ 25 38	18 50	04 30	21 11	06 05	16 38	06 54	14 49	17 21
27	08 18 23	03 52 48	11 ♑ 38 09	18 ♑ 48 19	20 36	05 42	21 44	06 00	16 44	06 57	14 50	17 23
28	08 22 20	04 50 08	25 ♑ 55 24	02 ♒ 58 40	22 20	06 54	22 18	05 54	16 50	07 00	14 51	17 25
29	08 26 17	05 47 28	09 ♒ 57 31	16 ♒ 51 26	24 03	08 06	22 52	05 49	16 57	07 04	14 52	17 26
30	08 30 13	06 44 49	23 ♒ 40 01	00 ♓ 23 02	25 44	09 18	23 26	05 43	17 03	07 07	14 53	17 28
31	08 34 10	07 ♌ 42 11	07 ♓ 00 22	13 ♓ 32 00	27 ♌ 23	10 ♋ 31	24 ♎ 01	05 ♓ 37	17 ♍ 09	07 ♋ 10	14 ♎ 54	17 ♌ 30

Date	Moon True ☊ ° '	Moon Mean ☊ ° '	Moon ☽ Latitude ° '
01	02 ♈ 04	02 ♈ 30	04 S 22
02	01 R 56	02 27	03 39
03	01 50	02 24	02 44
04	01 47	02 21	01 43
05	01 46	02 17	00 S 38
06	01 D 46	02 14	00 N 27
07	01 R 46	02 11	01 30
08	01 45	02 08	02 28
09	01 42	02 05	03 19
10	01 37	02 02	04 01
11	01 29	01 58	04 34
12	01 19	01 55	04 54
13	01 08	01 52	05 02
14	00 55	01 49	04 55
15	00 44	01 46	04 34
16	00 34	01 42	03 58
17	00 26	01 39	03 10
18	00 21	01 36	02 11
19	00 18	01 33	01 N 03
20	00 D 18	01 30	00 S 09
21	00 18	01 27	01 22
22	00 R 19	01 23	02 30
23	00 18	01 20	03 31
24	00 15	01 17	04 19
25	00 10	01 14	04 51
26	00 03	01 11	05 04
27	29 ♓ 55	01 08	04 59
28	29 46	01 04	04 34
29	29 38	01 01	03 53
30	29 31	00 58	03 00
31	29 ♓ 27	00 ♈ 55	01 S 57

DECLINATIONS

Date	Sun ☉ ° '	Moon ☽ ° '	Mercury ☿ ° '	Venus ♀ ° '	Mars ♂ ° '	Jupiter ♃ ° '	Saturn ♄ ° '	Uranus ♅ ° '	Neptune ♆ ° '	Pluto ♇ ° '
01	23 N 08	24 S 04	23 N 26	19 N 24	03 S 17	09 S 48	07 N 59	23 N 36	04 S 15	23 N 32
02	23 04	19 40	23 39	19 39	03 29	09 49	07 57	23 36	04 16	23 31
03	23 00	14 24	23 49	19 53	03 41	09 50	07 55	23 36	04 16	23 30
04	22 55	08 38	23 56	20 06	03 54	09 51	07 53	23 36	04 16	23 30
05	22 49	02 S 43	24 01	20 19	04 06	09 51	07 51	23 36	04 16	23 29
06	22 44	03 N 10	24 04	20 31	04 19	09 52	07 49	23 36	04 16	23 29
07	22 38	08 49	24 03	20 43	04 32	09 53	07 47	23 35	04 16	23 28
08	22 31	14 05	24 00	20 55	04 44	09 54	07 45	23 35	04 17	23 28
09	22 24	18 49	23 54	21 06	04 57	09 55	07 43	23 35	04 17	23 27
10	22 17	22 49	23 45	21 16	05 10	09 57	07 41	23 35	04 17	23 27
11	22 09	25 53	23 34	21 26	05 23	09 58	07 39	23 35	04 17	23 26
12	22 01	27 50	23 20	21 35	05 36	09 59	07 37	23 35	04 17	23 26
13	21 53	28 28	23 03	21 43	05 49	10 01	07 35	23 35	04 18	23 25
14	21 44	27 41	22 44	21 51	06 02	10 02	07 33	23 34	04 18	23 24
15	21 35	25 28	22 23	21 59	06 15	10 04	07 31	23 34	04 18	23 24
16	21 25	21 56	21 59	22 06	06 28	10 05	07 28	23 34	04 19	23 23
17	21 16	17 16	21 34	22 12	06 41	10 07	07 26	23 34	04 19	23 23
18	21 05	11 44	21 07	22 18	06 55	10 09	07 24	23 34	04 19	23 22
19	20 55	05 N 35	20 38	22 23	07 08	10 10	07 21	23 34	04 19	23 22
20	20 44	00 S 55	20 07	22 27	07 21	10 12	07 19	23 33	04 20	23 21
21	20 32	07 27	19 35	22 31	07 35	10 14	07 17	23 33	04 20	23 21
22	20 21	13 44	19 02	22 34	07 48	10 16	07 14	23 33	04 20	23 20
23	20 09	19 22	18 27	22 37	08 01	10 18	07 12	23 33	04 21	23 19
24	19 56	23 59	17 52	22 39	08 15	10 20	07 10	23 33	04 21	23 19
25	19 44	27 08	17 15	22 40	08 28	10 22	07 07	23 33	04 22	23 18
26	19 31	28 30	16 38	22 41	08 42	10 24	07 05	23 32	04 22	23 18
27	19 17	27 54	16 00	22 41	08 55	10 27	07 02	23 32	04 22	23 17
28	19 04	25 27	15 22	22 40	09 09	10 29	07 00	23 32	04 23	23 17
29	18 50	21 30	14 43	22 39	09 22	10 31	06 57	23 32	04 23	23 16
30	18 36	16 28	14 03	22 37	09 36	10 34	06 55	23 32	04 24	23 16
31	18 N 21	10 S 46	13 N 23	22 N 35	09 S 49	10 S 36	06 N 52	23 N 32	04 S 24	23 N 15

ZODIAC SIGN ENTRIES

Date	h	m	Planets
01	09	19	☽ ♒
02	14	57	☿ ♋
03	13	51	☽ ♓
05	22	24	☽ ♈
08	10	13	☽ ♉
10	23	02	☽ ♊
13	10	34	☽ ♋
15	19	52	☽ ♌
16	17	08	☿ ♌
18	03	05	☽ ♍
20	08	34	☽ ♎
22	12	27	☽ ♏
22	17	50	♀ ♋
23	10	30	☉ ♌
24	14	55	☽ ♐
26	16	39	☽ ♑
28	18	55	☽ ♒
30	23	19	☽ ♓

LATITUDES

Date	Mercury ☿ ° '	Venus ♀ ° '	Mars ♂ ° '	Jupiter ♃ ° '	Saturn ♄ ° '	Uranus ♅ ° '	Neptune ♆ ° '	Pluto ♇ ° '
01	00 N 01	01 S 43	00 00	01 S 06	02 N 00	00 N 16	01 N 37	08 N 04
04	00 33	01 37	00 S 04	01 07	02 00	00 16	01 37	08 04
07	01 01	01 31	00 08	01 08	01 59	00 16	01 37	08 04
10	01 23	01 24	00 11	01 08	01 59	00 16	01 37	08 03
13	01 38	01 16	00 14	01 09	01 59	00 16	01 36	08 03
16	01 47	01 09	00 18	01 10	01 58	00 16	01 36	08 03
19	01 49	01 01	00 21	01 11	01 58	00 16	01 36	08 03
22	01 45	00 53	00 24	01 12	01 58	00 16	01 36	08 03
25	01 36	00 44	00 27	01 12	01 57	00 16	01 36	08 03
28	01 22	00 36	00 30	01 13	01 57	00 16	01 36	08 03
31	01 N 04	00 S 27	00 S 32	01 S 14	01 N 57	00 N 16	01 N 36	08 N 03

DATA

Julian Date	2433464
Delta T	+29 seconds
Ayanamsa	23° 09' 56"
Synetic vernal point	05° ♓ 57' 04"
True obliquity of ecliptic	23° 26' 53"

MOON'S PHASES, APSIDES AND POSITIONS ☽

Date	h	m	Phase	Longitude ° '	Eclipse Indicator
07	02	53	☾	14 ♈ 26	
15	05	05	●	22 ♋ 09	
22	10	50	☽	29 ♎ 04	
29	04	18	○	05 ♒ 29	

Day	h	m		
09	21	06	Apogee	
25	13	29	Perigee	
05	22	59	0N	
13	10	54	Max dec	28° N 28'
20	08	39	0S	
26	16	40	Max dec	28° S 32'

LONGITUDES

Date	Chiron ⚷ ° '	Ceres ⚳ ° '	Pallas ⚴ ° '	Juno ⚵ ° '	Vesta ⚶ ° '	Black Moon Lilith ⚸ ° '
01	17 ♐ 14	07 ♐ 29	01 ♏ 47	25 ♍ 39	04 ♉ 54	18 ♉ 57
11	16 ♐ 40	06 ♐ 25	02 ♏ 45	28 ♍ 04	08 ♉ 24	20 ♉ 04
21	16 ♐ 12	05 ♐ 57	04 ♏ 18	00 ♎ 43	11 ♉ 37	21 ♉ 11
31	15 ♐ 52	06 ♐ 05	06 ♏ 21	03 ♎ 33	14 ♉ 31	22 ♉ 18

All ephemeris data is given at 12.00 UT and the Moon's longitude is additionally given for 24.00 UT

ASPECTARIAN

h m	Aspects
01 Saturday	
04 21	☽ ⊼ ☿
08 19	☽ ⚼ ♄
09 38	♀ Q ♇
11 43	☽ ⊥ ♃
14 52	☽ # ♅
15 21	☽ # ♇
15 40	☽ # ☿
16 13	☽ ± ☿
17 44	☽ # ☉
17 55	☽ △ ♀
18 36	☽ ⊼ ♅
21 23	☿ ∥ ♇
21 58	☽ ⊻ ♃
23 45	☽ ± ♄
23 49	☽ △ ♂
02 Sunday	
01 46	☽ ⊼ ☉
02 24	♀ ⊻ ♅
04 59	☽ ± ♅
06 59	☿ ∥ ♅
10 13	☽ ⊼ ♄
10 23	☽ △ ♆
10 26	☽ ⚼ ☿
12 07	☽ # ♀
12 59	☽ ± ☉
14 04	☽ ☍ ♇
20 50	☽ ⚼ ♅
03 Monday	
03 05	☽ ⚼ ♂
04 51	☉ ⊥ ♇
06 10	☽ ⚼ ☉
10 37	☿ ∠ ♇
13 07	☽ ⚼ ♆
17 12	♀ □ ♃
18 09	☽ △ ☿
19 20	♄ ⊻ ♆
20 00	☽ ± ♂
20 52	☿ Q ♄
04 Tuesday	
00 05	☽ △ ♅
03 22	☽ ☌ ♃
04 25	☽ □ ♀
05 38	☽ ± ♆
07 06	☽ ∥ ♃
07 30	☽ ⊼ ♂
11 51	☽ △ ☉
15 05	☽ # ♄
16 52	☽ ⊼ ♆
16 59	☽ ☍ ♄
20 55	☽ ⊼ ♇
05 Wednesday	
05 43	☽ ∥ ♆
06 33	☽ ∥ ♂
07 41	☿ ☌ ♅
08 23	☽ ± ♇
18 06	☽ Q ♀
06 Thursday	
01 52	☽ ⚼ ♇
02 15	☿ △ ♃
06 30	♂ ∥ ♆
09 38	☽ □ ♅
11 33	♀ △ ♂
12 44	☽ ⊻ ♃
15 02	☽ □ ☿
16 34	☽ # ♆
16 57	☽ # ♂
19 41	☽ ☍ ♂
20 12	☽ ✶ ♀
07 Friday	
00 37	☽ ⊥ ♃
02 53	☽ □ ☉
03 15	☽ ☍ ♆
03 44	☽ ⊼ ♄
07 22	☉ □ ♆
07 32	☽ ∥ ♄
07 38	☽ △ ♇
14 32	☉ ✶ ♄
15 51	☽ ± ♄
16 42	☽ # ♃
17 12	☿ ⊥ ♇
18 36	☽ ∠ ♃
21 44	☽ Q ♅
23 33	☿ □ ♂
08 Saturday	
05 37	☽ ∠ ♀
10 09	☽ ⚼ ♄
11 05	☽ Q ☿
18 20	☽ Q ☉
22 09	☽ ✶ ♅
09 Sunday	
00 52	☽ ✶ ♃
00 57	☿ ⊻ ♀
01 53	☽ ⊥ ♀
10 49	☽ ⊼ ♂
11 26	☿ □ ♆
15 26	☽ ⊻ ♀
15 54	☽ ⊼ ♆
16 13	☉ ⊻ ♇
16 21	☿ ✶ ♄
16 48	☽ △ ♄
16 54	☽ ✶ ☿
20 09	♀ △ ♆
20 28	☽ □ ♇
20 50	☽ ✶ ☉
23 32	☽ ± ♂
10 Monday	
01 09	☽ Q ♃
01 40	☽ ∥ ♀
04 05	☽ ± ♆
04 45	☽ ∠ ♅
06 03	♀ □ ♄
08 38	☽ ∥ ☉
12 36	☿ ⊻ ♇
16 21	☽ ∥ ♇
17 20	☽ ∥ ♅
18 15	☽ ∥ ☿
18 33	☽ ⚼ ♂
22 17	☽ ⚼ ♆
23 02	☽ ⊥ ♅
11 Tuesday	
04 17	☉ ☌ ☿
05 47	☽ ∠ ☉
05 58	☽ ∠ ☿
08 54	☽ Q ♇
10 02	☿ ∥ ♅
10 23	♂ ± ♃
11 09	☽ ⊻ ♅
13 19	☽ □ ♃
18 39	♀ ✶ ♇
12 Wednesday	
01 21	☽ ⊥ ☉
01 58	☽ △ ♂
02 16	☿ ∥ ♇
03 53	☽ ⊥ ☿
04 20	☽ △ ♆
05 37	☽ □ ♄
08 54	☽ ✶ ♇
10 24	☽ ☌ ♀
14 15	☽ ⊻ ☉
18 18	☽ ⊻ ☿
22 10	☿ ⚼ ♃
13 Thursday	
14 24	☽ ∠ ♇
17 12	☽ Q ♄
22 33	☽ ☌ ♅
14 Friday	
00 07	☽ △ ♃
01 29	☉ ∥ ♀
07 50	☽ ⊥ ♇
10 31	♂ ☌ ♆
14 52	☽ □ ♆
15 03	☽ □ ♂
16 28	☽ ✶ ♄
19 20	☽ ⊻ ♇
23 43	☉ ⚼ ♃
15 Saturday	
02 43	☽ ⊻ ♀
04 40	☽ ⚼ ♃
05 05	☽ ☌ ☉
08 28	☿ ⊥ ♀
15 12	☽ ⊥ ♀
15 46	☽ ☌ ☿
21 01	☽ ∠ ♄
21 31	☽ ± ♃
16 Sunday	
00 53	☽ Q ♆
01 55	☽ ∥ ♅
02 29	☽ Q ♂
03 05	☽ ∥ ♇
07 00	☿ ∥ ♀
07 40	☽ ⊻ ♅
07 49	♂ ⊻ ♄
08 39	☽ ⊼ ♃
09 44	☽ ∠ ♀
11 05	☽ ∥ ♀
11 39	☽ ∥ ☿
13 53	☽ ⊥ ♄
15 02	☽ ∥ ☉
18 49	☽ ⊥ ♅
23 09	☽ ✶ ♆
17 Monday	
01 02	☽ ⊻ ♄
01 33	☿ ∠ ♄
01 38	☽ ✶ ♂
02 35	☿ ± ♃
03 32	☽ ☌ ♇
11 25	☽ ∠ ♅
16 05	☽ ✶ ♀
17 10	☽ ⊻ ☉
18 Tuesday	
00 54	☿ Q ♆
02 33	☽ ∠ ♆
04 54	☽ ⊥ ☉
06 07	☽ ∠ ♂
09 14	☽ ⊻ ☿
13 49	☉ ∥ ☿
14 43	☽ ✶ ♅
15 11	☽ ☍ ♃
15 56	☽ Q ♀
18 21	☽ # ♃
18 47	☽ ⊥ ♆
21 56	☽ ⊥ ☿
22 21	☽ ∠ ☉
22 58	☽ ⊥ ♂
19 Wednesday	
02 45	♂ ✶ ♇
05 12	☽ ∥ ♄
05 31	☽ ⊻ ♆
06 18	☽ # ♂
07 09	☿ Q ♂
07 41	☽ ☌ ♄
09 51	☽ ⊻ ♇
10 07	☽ ⊻ ♂
12 15	☽ Q ♅
16 41	☽ # ♆
16 44	☽ ∠ ☿
20 30	☽ ⊥ ♇
22 50	☿ ⊻ ♅
23 46	☉ ⊻ ♀
23 53	☿ ⊼ ♃
20 Thursday	
03 00	☽ ✶ ☉
03 04	☽ □ ♀
08 44	♂ # ♄
12 22	☽ ∠ ♇
15 19	♃ △ ♅
20 01	☽ ⊼ ♃
20 04	☽ □ ♅
23 29	☽ ✶ ☿
21 Friday	
00 29	☽ ∥ ♆
01 34	☽ Q ☉
06 24	☽ ± ♃
10 15	☽ ☌ ♆
11 21	☽ # ♄
12 28	☽ ∥ ♂
12 40	☽ ⊻ ♄
14 30	☽ ✶ ♇
16 47	☽ ☌ ♂
19 48	☿ ⊥ ♄
21 51	☽ ⚼ ♃
22 29	☽ ∥ ♃
23 05	☽ ⊥ ♄
23 34	☽ Q ☿
22 Saturday	
10 50	☽ □ ☉
11 08	☽ Q ♇
11 57	☽ △ ♀
14 34	☽ ∠ ♄
23 18	☽ △ ♃
23 49	☽ △ ♅
23 Sunday	
03 51	☿ ⊥ ♅
08 15	☽ # ☿
10 54	☽ □ ☿
13 27	☽ ⊻ ♆
15 29	☽ # ☉
15 40	☽ ⚼ ♀
16 06	☽ ✶ ♄
17 39	☽ □ ♇
18 34	☉ ± ♃
21 47	☽ ⊻ ♂
23 32	☽ ⊥ ♆
24 Monday	
01 10	☽ ⚼ ♅
04 17	☽ # ♀
06 33	☿ ✶ ♆
08 04	☽ # ♇
08 06	☽ ± ♀
08 13	☽ ⊥ ♂
09 25	☽ # ♅
12 19	☽ Q ♄
14 34	☽ ∠ ♆
15 50	♀ ∠ ♇
16 15	☽ ± ♅
16 56	☽ △ ☉
19 00	☽ ⊼ ♀
23 48	☽ ∠ ♂
23 48	☉ ∠ ♄
25 Tuesday	
01 15	☽ □ ♃
02 15	☽ ⊼ ♅
05 05	☿ ⊻ ♄
15 29	☽ ✶ ♆
15 55	☿ ☌ ♇
18 24	☽ □ ♄
19 34	☽ ⚼ ☉
19 40	☽ △ ♇
20 12	☽ △ ☿
26 Wednesday	
01 39	☽ ✶ ♂
06 53	☽ Q ♃
08 37	♂ ⚼ ♃
09 16	☉ Q ♆
11 23	☽ Q ♆
11 31	☽ ± ☉
14 56	♀ Q ♄
20 34	☽ ⚼ ♇
22 11	☽ ⊼ ☉
22 20	☽ Q ♂
27 Thursday	
00 33	☽ ⚼ ☿
01 13	☽ ☍ ♀
02 40	☽ ✶ ♃
04 10	☽ ☍ ♅
11 34	☽ ± ♇
16 34	☿ ∠ ♀
17 21	☽ □ ♆
17 29	♀ △ ♃
17 38	☽ ± ☿
20 36	☽ △ ♄
21 38	☽ ⊼ ♇
28 Friday	
03 35	☽ ∠ ♃
05 06	☽ ⊼ ☿
05 38	☽ □ ♂
07 14	☿ ∠ ♅
11 17	☿ ✶ ♂
14 09	♀ ☌ ♅
18 43	☽ ⊥ ♃
22 08	☽ ⚼ ♄
29 Saturday	
00 38	☽ # ♅
02 12	☽ # ♇
04 18	☽ ☍ ☉
04 54	☽ ⊻ ♃
05 44	☽ # ♀
06 59	☽ ⊼ ♅
08 30	☽ ⊼ ♀
12 26	☉ ⊼ ♃
13 43	☽ ± ♄
15 54	♂ ⊥ ♄
17 24	☽ ± ♅
19 53	☽ ± ♀
20 32	☽ △ ♆
30 Sunday	
00 15	☽ ⊼ ♄
01 03	☽ ☍ ♇
01 52	☽ # ☉
09 14	☽ ⚼ ♅
11 35	☽ △ ♂
13 15	☽ ⚼ ♀
16 12	☽ ☍ ☿
21 47	☉ ⊻ ♅
23 07	☽ ⚼ ♆
23 45	☽ # ☿
31 Monday	
09 29	☽ ☌ ♃
12 18	☽ △ ♅
12 39	☽ ∥ ♃
13 22	☽ ⊼ ☉
15 28	☽ ± ♆
15 39	☽ ∥ ♂
15 50	☽ ⚼ ♂
19 05	☽ △ ♀

AUGUST 1950

LONGITUDES

Date	Sidereal time h m s	Sun ☉ ° ' "	Moon ☽ ° ' "	Moon ☽ 24.00 ° ' "	Mercury ☿ ° '	Venus ♀ ° '	Mars ♂ ° '	Jupiter ♃ ° '	Saturn ♄ ° '	Uranus ♅ ° '	Neptune ♆ ° '	Pluto ♇ ° '
01	08 38 06	08 ♌ 39 34	19 ♓ 58 05	26 ♓ 18 52	29 ♌ 01	11 ♋ 43	24 ♎ 35	05 ♓ 31	17 ♍ 16	07 ♋ 13	14 ♎ 55	17 ♌ 32
02	08 42 03	09 36 57	02 ♈ 34 40	08 ♈ 45 57	00 ♍ 37	12 55	25 10	05 R 24	17 22	07 17	14 56	17 34
03	08 45 59	10 34 22	14 ♈ 53 10	20 ♈ 56 54	02 11	14 08	25 45	05 18	17 29	07 20	14 58	17 36
04	08 49 56	11 31 48	26 ♈ 57 42	02 ♉ 56 13	03 44	15 20	26 20	05 11	17 35	07 23	14 59	17 37
05	08 53 52	12 29 16	08 ♉ 53 05	14 ♉ 48 55	05 14	16 33	26 55	05 05	17 42	07 26	15 00	17 39
06	08 57 49	13 26 44	20 ♉ 44 22	26 ♉ 40 05	06 44	17 45	27 30	04 58	17 48	07 29	15 01	17 41
07	09 01 46	14 24 14	02 ♊ 36 38	08 ♊ 34 37	08 11	18 58	28 05	04 51	17 55	07 32	15 03	17 43
08	09 05 42	15 21 46	14 ♊ 34 35	20 ♊ 37 00	09 37	20 10	28 41	04 44	18 02	07 35	15 04	17 45
09	09 09 39	16 19 18	26 ♊ 42 20	02 ♋ 50 58	11 00	21 23	29 17	04 37	18 08	07 38	15 05	17 47
10	09 13 35	17 16 52	09 ♋ 03 13	15 ♋ 19 19	12 22	22 36	29 ♎ 53	04 30	18 15	07 41	15 07	17 48
11	09 17 32	18 14 28	21 ♋ 39 28	28 ♋ 03 46	13 43	23 48	00 ♏ 29	04 23	18 22	07 44	15 08	17 50
12	09 21 28	19 12 04	04 ♌ 32 13	11 ♌ 04 47	15 01	25 01	01 05	04 16	18 29	07 47	15 09	17 52
13	09 25 25	20 09 42	17 ♌ 41 22	24 ♌ 21 47	16 18	26 14	01 41	04 08	18 36	07 50	15 11	17 54
14	09 29 21	21 07 21	01 ♍ 05 50	07 ♍ 53 13	17 32	27 27	02 18	04 01	18 43	07 53	15 12	17 56
15	09 33 18	22 05 01	14 ♍ 43 42	21 ♍ 36 57	18 44	28 40	02 55	03 53	18 50	07 56	15 14	17 58
16	09 37 15	23 02 42	28 ♍ 32 42	05 ♎ 30 36	19 55	29 ♋ 53	03 32	03 46	18 56	07 58	15 15	18 00
17	09 41 11	24 00 24	12 ♎ 30 25	19 ♎ 31 51	21 03	01 ♌ 06	04 08	03 38	19 03	08 01	15 17	18 01
18	09 45 08	24 58 08	26 ♎ 34 39	03 ♏ 38 35	22 09	02 19	04 46	03 30	19 11	08 04	15 18	18 03
19	09 49 04	25 55 52	10 ♏ 43 24	17 ♏ 48 54	23 12	03 32	05 23	03 22	19 18	08 07	15 20	18 05
20	09 53 01	26 53 38	24 ♏ 54 50	02 ♐ 00 57	24 13	04 45	06 00	03 15	19 25	08 09	15 21	18 07
21	09 56 57	27 51 24	09 ♐ 07 00	16 ♐ 12 43	25 11	05 59	06 38	03 07	19 32	08 12	15 23	18 09
22	10 00 54	28 49 12	23 ♐ 17 46	00 ♑ 21 49	26 07	07 12	07 15	02 59	19 39	08 15	15 25	18 11
23	10 04 50	29 ♌ 47 01	07 ♑ 24 30	14 ♑ 25 27	26 59	08 25	07 53	02 51	19 46	08 17	15 26	18 12
24	10 08 47	00 ♍ 44 51	21 ♑ 24 15	28 ♑ 20 32	27 49	09 39	08 31	02 43	19 53	08 20	15 28	18 14
25	10 12 44	01 42 43	05 ♒ 13 53	12 ♒ 03 57	28 35	10 52	09 09	02 35	20 01	08 22	15 30	18 16
26	10 16 40	02 40 36	18 ♒ 50 24	25 ♒ 32 58	29 18	12 05	09 47	02 27	20 08	08 25	15 32	18 18
27	10 20 37	03 38 30	02 ♓ 11 25	08 ♓ 45 36	29 ♍ 57	13 19	10 26	02 20	20 15	08 27	15 33	18 20
28	10 24 33	04 36 25	15 ♓ 15 25	21 ♓ 40 53	00 ♎ 32	14 32	11 04	02 12	20 22	08 29	15 35	18 21
29	10 28 30	05 34 23	28 ♓ 02 01	04 ♈ 19 00	01 02	15 46	11 42	02 04	20 30	08 32	15 37	18 23
30	10 32 26	06 32 22	10 ♈ 32 02	16 ♈ 41 22	01 29	17 00	12 21	01 56	20 37	08 34	15 39	18 25
31	10 36 23	07 ♍ 30 22	22 ♈ 47 22	28 ♈ 50 25	01 ♎ 50	18 ♌ 13	13 ♏ 00	01 ♓ 48	20 ♍ 44	08 ♋ 36	15 ♎ 40	18 ♌ 27

MOON NODES AND LATITUDE; DECLINATIONS

Date	Moon True ☊ ° '	Moon Mean ☊ ° '	Moon ☽ Latitude ° '	Sun ☉ ° '	Moon ☽ ° '	Mercury ☿ ° '	Venus ♀ ° '	Mars ♂ ° '	Jupiter ♃ ° '	Saturn ♄ ° '	Uranus ♅ ° '	Neptune ♆ ° '	Pluto ♇ ° '
01	29 ♓ 25	00 ♈ 52	00 S 51	18 N 06	04 S 45	12 N 43	22 N 31	10 S 03	10 S 38	06 N 50	23 N 32	04 S 25	23 N 14
02	29 D 24	00 48	00 N 17	17 51	01 N 17	12 03	22 28	10 16	10 41	06 47	23 31	04 25	23 14
03	29 25	00 45	01 23	17 36	07 08	11 23	22 23	10 30	10 44	06 45	23 31	04 26	23 13
04	29 27	00 42	02 23	17 20	12 37	10 42	22 18	10 43	10 46	06 42	23 31	04 26	23 13
05	29 27	00 39	03 17	17 04	17 34	10 02	22 12	10 57	10 49	06 39	23 31	04 27	23 12
06	29 R 27	00 36	04 01	16 48	21 49	09 21	22 06	11 10	10 51	06 37	23 31	04 27	23 12
07	29 25	00 33	04 36	16 31	25 12	08 41	21 59	11 24	10 54	06 34	23 31	04 28	23 11
08	29 22	00 29	04 59	16 14	27 30	08 00	21 51	11 37	10 57	06 31	23 30	04 28	23 11
09	29 16	00 26	05 09	15 57	28 33	07 21	21 43	11 51	11 00	06 29	23 30	04 29	23 10
10	29 10	00 23	05 04	15 40	28 12	06 41	21 34	12 04	11 02	06 26	23 30	04 29	23 10
11	29 03	00 20	04 45	15 22	26 24	06 01	21 24	12 18	11 05	06 23	23 30	04 30	23 09
12	28 56	00 17	04 12	15 04	23 12	05 23	21 14	12 31	11 08	06 21	23 30	04 30	23 09
13	28 50	00 14	03 24	14 46	18 46	04 44	21 03	12 45	11 11	06 18	23 30	04 31	23 08
14	28 45	00 10	02 24	14 28	13 20	04 06	20 52	12 58	11 14	06 15	23 30	04 32	23 08
15	28 43	00 07	01 15	14 09	07 10	03 29	20 40	13 11	11 17	06 12	23 29	04 32	23 07
16	28 42	00 04	00 N 01	13 50	00 N 36	02 52	20 27	13 25	11 20	06 09	23 29	04 33	23 07
17	28 D 43	00 ♈ 01	01 S 15	13 31	06 S 05	02 16	20 14	13 38	11 22	06 07	23 29	04 34	23 06
18	28 44	29 ♓ 58	02 26	13 12	12 32	01 41	20 00	13 51	11 25	06 04	23 29	04 34	23 06
19	28 45	29 54	03 29	12 53	18 22	01 06	19 46	14 04	11 28	06 01	23 29	04 35	23 05
20	28 46	29 51	04 20	12 33	23 12	00 33	19 31	14 17	11 31	05 58	23 29	04 35	23 05
21	28 R 46	29 48	04 55	12 13	26 40	00 N 01	19 16	14 31	11 34	05 55	23 28	04 36	23 04
22	28 44	29 45	05 11	11 53	28 28	00 S 31	19 00	14 44	11 37	05 53	23 28	04 37	23 04
23	28 41	29 42	05 09	11 33	28 23	01 01	18 43	14 57	11 40	05 50	23 28	04 37	23 03
24	28 37	29 39	04 48	11 13	26 29	01 30	18 26	15 09	11 43	05 47	23 28	04 38	23 03
25	28 33	29 35	04 11	10 52	23 02	01 57	18 08	15 22	11 46	05 44	23 28	04 39	23 02
26	28 29	29 32	03 20	10 31	18 21	02 23	17 50	15 35	11 49	05 41	23 28	04 39	23 02
27	28 26	29 29	02 19	10 11	12 52	02 47	17 31	15 48	11 52	05 38	23 28	04 40	23 01
28	28 24	29 26	01 12	09 49	06 55	03 10	17 12	16 01	11 55	05 35	23 28	04 41	23 01
29	28 24	29 23	00 S 02	09 28	00 S 49	03 30	16 53	16 13	11 58	05 33	23 27	04 42	23 00
30	28 D 24	29 19	01 N 06	09 07	05 N 11	03 49	16 33	16 26	12 01	05 30	23 27	04 42	23 00
31	28 ♓ 26	29 ♓ 16	02 N 10	08 N 45	10 N 53	04 S 05	16 N 12	16 S 38	12 S 03	05 N 27	23 N 27	04 S 43	23 N 00

ZODIAC SIGN ENTRIES

Date	h	m	Planets
02	02	44	☿ ♍
02	07	03	☽ ♈
04	18	06	☽ ♉
07	06	44	☽ ♊
09	18	27	☽ ♋
10	16	48	♂ ♏
12	03	36	☽ ♌
14	10	03	☽ ♍
16	14	18	♀ ♌
16	14	31	☽ ♎
18	17	49	☽ ♏
20	20	36	☽ ♐
22	23	23	☽ ♑
23	17	23	☉ ♍
25	02	53	☽ ♒
27	08	02	☽ ♓
27	14	17	☿ ♎
29	15	44	☽ ♈

LATITUDES

Date	Mercury ☿ ° '	Venus ♀ ° '	Mars ♂ ° '	Jupiter ♃ ° '	Saturn ♄ ° '	Uranus ♅ ° '	Neptune ♆ ° '	Pluto ♇ ° '
01	00 N 58	00 S 25	00 S 33	01 S 14	01 N 57	00 N 16	01 N 36	08 N 03
04	00 36	00 16	00 36	01 15	01 57	00 17	01 35	08 04
07	00 N 11	00 S 08	00 39	01 15	01 57	00 17	01 35	08 04
10	00 S 16	00 N 01	00 41	01 16	01 56	00 17	01 35	08 04
13	00 44	00 09	00 43	01 16	01 56	00 17	01 35	08 04
16	01 14	00 17	00 46	01 17	01 56	00 17	01 35	08 04
19	01 44	00 24	00 48	01 17	01 56	00 17	01 35	08 05
22	02 15	00 32	00 50	01 18	01 56	00 17	01 35	08 05
25	02 45	00 39	00 52	01 18	01 56	00 17	01 35	08 05
28	03 13	00 46	00 54	01 18	01 56	00 17	01 35	08 06
31	03 S 39	00 N 52	00 S 56	01 S 18	01 N 56	00 N 17	01 N 35	08 N 06

DATA

Julian Date	2433495
Delta T	+29 seconds
Ayanamsa	23° 10' 02"
Synetic vernal point	05° ♓ 56' 58"
True obliquity of ecliptic	23° 26' 54"

LONGITUDES

Date	Chiron ⚷ ° '	Ceres ⚳ ° '	Pallas ⚴ ° '	Juno ⚵ ° '	Vesta ⚶ ° '	Black Moon Lilith ⚸ ° '
01	15 ♐ 50	06 ♐ 08	06 ♏ 34	03 ♎ 51	14 ♉ 47	22 ♉ 24
11	15 ♐ 39	06 ♐ 53	09 ♏ 04	06 ♎ 51	17 ♉ 17	23 ♉ 31
21	15 ♐ 37	08 ♐ 10	11 ♏ 54	09 ♎ 58	19 ♉ 19	24 ♉ 38
31	15 ♐ 44	09 ♐ 54	15 ♏ 00	13 ♎ 12	20 ♉ 50	25 ♉ 45

MOON'S PHASES, APSIDES AND POSITIONS ☽

Date	h	m	Phase	Longitude ° '	Eclipse Indicator
05	19	56	☾	12 ♉ 48	
13	16	48	●	20 ♌ 21	
20	15	35	☽	27 ♏ 02	
27	14	51	○	03 ♓ 45	

Day	h	m		
06	14	51	Apogee	
20	04	31	Perigee	
02	06	51	0N	
09	18	11	Max dec	28° N 36'
16	14	07	0S	
22	23	01	Max dec	28° S 39'
29	15	13	0N	

ASPECTARIAN

01 Tuesday
01 18 ☽ ± ☉
02 34 ☽ ⊼ ♆
03 42 ☽ # ♄
06 53 ☽ ☍ ♄
07 26 ☽ ⊼ ♇
08 14 ♀ ⊥ ♇
09 17 ☽ ± ♂
13 21 ☽ ∥ ♆
18 44 ☽ ± ♇
19 32 ☽ bQ ☉
21 08 ☽ ⊼ ♂

02 Wednesday
01 42 ☿ ∠ ♆
07 40 ☽ ⊼ ☿
11 58 ☽ bQ ♇
17 25 ☽ ⊻ ♃
20 57 ☽ ± ☿
21 08 ☽ □ ♅

03 Thursday
00 43 ☽ # ♆
02 49 ☽ △ ☉
05 01 ☽ ⊥ ♃
10 21 ☽ □ ♀
10 21 ☽ ∥ ♄
12 09 ☽ ☍ ♆
17 10 ☽ ⊼ ♄
17 12 ☽ bQ ☿
17 22 ☽ △ ♇
22 37 ☽ ∠ ♃

04 Friday
03 07 ☽ # ♂
03 37 ☽ # ♃
04 20 ☽ ∥ ☿
04 49 ♀ □ ♆
05 11 ☽ ± ♄
08 49 ☽ Q ♅
09 44 ☿ # ♃
10 40 ☽ ☍ ♂
11 23 ☿ # ♂
13 33 ☉ ⊥ ♄
17 53 ♂ ∥ ♃
23 24 ☽ bQ ♄
23 48 ♄ ⊻ ♇

05 Saturday
02 15 ☽ Q ♀
03 35 ☽ △ ☿
04 23 ☽ ✶ ♃
09 03 ☽ ✶ ♅
09 33 ☽ ∥ ☉
09 38 ☿ ☍ ♃
19 56 ☽ □ ☉

06 Sunday
00 24 ☽ ⊼ ♆
04 26 ☽ Q ♃
05 16 ☽ ✶ ♀
05 48 ☽ □ ♇
06 00 ☽ △ ♄
10 39 ♀ ⊻ ♇
12 34 ☽ ± ♆
13 01 ☉ ⊥ ♅
13 09 ♀ ✶ ♄
13 41 ☽ ∥ ♀
15 33 ☽ ∠ ♅
20 56 ☽ ∥ ♇
23 09 ☽ ∥ ♅

07 Monday
00 54 ☿ ✶ ♅
02 24 ☽ ⊼ ♂
06 49 ☽ bQ ♆
09 49 ☽ ⊥ ♅
11 33 ☽ Q ☉
15 02 ☽ ∠ ♀
15 08 ☽ ± ♂
16 29 ☽ □ ♃
18 16 ☽ Q ♇
21 57 ☽ ⊻ ♅

08 Tuesday
00 44 ☽ □ ☿
02 38 ☿ ⊥ ♆
04 11 ♀ bQ ♃
04 21 ☉ ✶ ♆
10 08 ☽ bQ ♂
11 06 ☽ ⊥ ♀
12 58 ☽ △ ♆
13 42 ☽ ✶ ☉
18 20 ☽ ✶ ♇
18 56 ☽ □ ♄

09 Wednesday
00 21 ☽ ⊻ ♀
17 05 ☽ Q ☿
17 18 ☽ △ ♂
21 47 ☽ ∠ ☉
23 53 ☽ ∠ ♇

10 Thursday
03 18 ☽ △ ♃
06 33 ☽ Q ♄
08 58 ♂ Q ♇
09 21 ☽ ☌ ♅
16 38 ☽ ⊥ ☉
17 18 ☽ ⊥ ♇
19 08 ☽ ✶ ☿
21 43 ☿ ∥ ♄
23 37 ☽ □ ♆

11 Friday
01 37 ☉ ☌ ♇
04 46 ☽ ⊻ ♇
05 01 ☽ ⊻ ☉
05 43 ☽ ✶ ♄
07 45 ☽ bQ ♃
15 31 ☉ ⊻ ♄
16 28 ☽ ☌ ♀

12 Saturday
00 29 ☽ ± ♃
02 42 ☽ ∠ ☿
04 11 ☉ Q ♂
05 18 ☽ □ ♂
09 27 ☽ Q ♆
10 02 ☽ ∠ ♄
10 10 ☽ ∥ ♅
11 30 ☽ ⊼ ♃
12 24 ☽ ∥ ♇
14 35 ☿ ⊻ ♆
18 00 ☽ ⊻ ♅
21 08 ☽ ⊥ ☿
23 55 ☽ ∥ ♀

13 Sunday
02 41 ☽ ⊥ ♄
04 59 ☽ ⊥ ♅
07 27 ☽ ✶ ♆
09 13 ☽ ⊻ ☿
12 23 ☽ ☌ ♇
13 39 ☽ ⊻ ♄
15 47 ☽ Q ♂
16 48 ☽ ☌ ☉
20 04 ☿ # ♆
21 17 ☽ ∠ ♅

14 Monday
02 59 ☿ ∠ ♂
04 52 ☽ ⊻ ♀
07 03 ☽ ∥ ☉
10 25 ☽ ∠ ♆
13 27 ☽ # ♂
14 14 ☽ ✶ ♂
16 35 ☽ ⊥ ♀
17 07 ☽ ☍ ♃
20 02 ☿ ⊻ ♇
20 25 ☽ # ♃
22 03 ♀ ± ♃

15 Tuesday
00 02 ☽ ✶ ♅
02 20 ☽ ⊥ ♆
09 58 ☽ ∠ ♀
12 52 ☽ ⊻ ♆
13 53 ☿ ☌ ♄
15 38 ☽ ∥ ♄
17 39 ☽ ⊻ ♇
17 49 ☽ ∠ ♂
19 12 ☽ ☌ ♄
19 39 ☽ ☌ ☿
21 06 ☽ Q ♅
21 43 ☽ # ♆

16 Wednesday
01 46 ☽ ⊻ ☉
02 59 ☽ ∥ ☿
04 06 ☽ ⊥ ♇
10 08 ☉ ∠ ♅
10 09 ☽ ⊥ ♂
12 56 ☽ ⊥ ☉
13 20 ☿ Q ♅
14 32 ☽ ✶ ♀
19 35 ♂ △ ♃
19 41 ☽ ∠ ♇
20 55 ☽ ⊼ ♃
20 59 ☽ ⊻ ♂
23 22 ☽ # ☿

17 Thursday
04 17 ☽ □ ♅
05 34 ☽ ∠ ☉
06 28 ☽ ∥ ♆
07 07 ☽ ± ♃
07 10 ☉ # ♂
08 01 ♂ ∠ ♄
12 05 ☽ # ♄
13 07 ☽ Q ♀
16 45 ☽ ☌ ♆
21 27 ☽ ✶ ♇
22 22 ☽ bQ ♃
23 17 ☽ ⊻ ♄

18 Friday
03 50 ☽ ⊻ ☿
07 44 ☽ ∥ ♃
09 04 ☽ ✶ ☉
09 36 ☽ ⊥ ♄
14 30 ☽ # ☉
14 53 ☽ ⊥ ☿
17 23 ☽ ∥ ♂
17 55 ☽ Q ♇
22 40 ☽ □ ♀
23 39 ☽ △ ♃

19 Saturday
01 01 ☽ ∠ ♄
02 32 ☽ ☌ ♂
06 56 ☽ Q ☉
07 23 ☽ ∠ ☿
07 34 ☽ △ ♅
07 50 ♀ Q ♆
09 05 ♀ ⊼ ♃
18 07 ☽ # ♀
19 49 ☽ ⊻ ♆

20 Sunday
00 29 ☽ □ ♇
02 37 ☽ ✶ ♄
04 28 ♀ ∠ ♄
05 59 ☽ ⊥ ♆
09 01 ☽ bQ ♅
09 29 ☿ ⊥ ♇
10 44 ☽ ✶ ☿
11 18 ☽ # ♇
13 35 ☽ # ♅
15 35 ☽ □ ☉
21 13 ☽ ∠ ♆
23 04 ☽ Q ♄

21 Monday
00 16 ☽ ± ♅
01 57 ☽ □ ♃
06 12 ☽ △ ♀
07 36 ☽ ⊻ ♂
08 30 ☽ Q ☿
10 27 ☽ ⊼ ♅
18 13 ☽ ⊥ ♂
22 37 ☽ ✶ ♆

22 Tuesday
03 18 ☽ △ ♇
05 46 ☽ □ ♄
08 07 ☽ Q ♃
09 58 ☽ bQ ♀
10 09 ☽ ∠ ♂
14 19 ♀ □ ♂
17 06 ☽ □ ☿
19 00 ☽ Q ♆
22 04 ☽ △ ☉

23 Wednesday
02 42 ☽ ± ♀
04 18 ☽ ✶ ♃
04 48 ☉ # ♃
04 49 ☽ bQ ♇
09 14 ♀ ⊻ ♅
12 51 ☽ ✶ ♂
13 30 ☽ ☍ ♅
13 54 ☽ ⊼ ♀
20 13 ☽ ± ♇

24 Thursday
01 33 ☽ bQ ☉
01 46 ☽ □ ♆
04 15 ♂ △ ♅
04 50 ☉ ∠ ♆
05 43 ☽ ∠ ♃
06 32 ☽ ⊼ ♇
09 22 ☽ △ ♄
10 24 ☽ Q ♂
18 12 ☽ ± ☉
21 06 ☽ ⊥ ♃
23 45 ☽ △ ☿

25 Friday
05 24 ☽ ⊼ ☉
07 26 ☽ ⊻ ♃
09 24 ☽ # ♅
11 37 ☽ bQ ♄
11 56 ☽ # ♇
17 31 ☽ ⊼ ♅
19 12 ☽ □ ♂
22 52 ☽ ☍ ♀

26 Saturday
03 30 ☽ bQ ☿
03 34 ☽ ± ♄
04 07 ☽ ± ♅
06 06 ☽ △ ♆
07 13 ☉ ☍ ♃
11 02 ☽ ☍ ♇
14 19 ☽ ⊼ ♄
14 32 ☽ # ♀
20 11 ☽ bQ ♅
20 22 ☽ ± ☿
23 58 ☽ ∥ ♂

27 Sunday
07 43 ☽ ⊼ ☿
09 02 ☽ bQ ♆
12 15 ☽ ☌ ♃
14 51 ☽ ☍ ☉
16 05 ☽ ∥ ♃
23 28 ☽ △ ♅
23 40 ☽ # ☉

28 Monday
01 29 ☽ ± ♆
03 50 ☽ △ ♂
08 25 ♀ ⊥ ♄
10 32 ☽ ⊼ ♀
10 58 ♀ ⊥ ♅
12 37 ☽ ⊼ ♆
17 16 ☽ # ♄
17 47 ☽ ⊼ ♇
20 47 ☽ ∥ ♆
21 38 ☽ ☍ ♄
22 54 ☽ ± ♀

29 Tuesday
01 58 ☽ ∥ ☿
05 05 ☽ ± ♇
08 58 ♀ ✶ ♆
09 21 ☽ bQ ♂
17 46 ☽ bQ ♀
17 57 ☽ ☍ ☿
19 36 ☽ ⊻ ♃
20 38 ☿ ∠ ♀
20 54 ♄ Q ♅
22 14 ☽ bQ ♇

30 Wednesday
03 28 ☽ ± ♂
03 38 ☽ ⊼ ☉
06 07 ☽ # ☿
07 01 ☽ ⊥ ♃
08 11 ☽ □ ♅
10 02 ☽ # ♆
13 14 ☽ ∥ ♄
15 44 ☽ ⊼ ♂
16 14 ☽ ± ☉
17 04 ♀ # ♂
21 59 ☽ ☍ ♆

31 Thursday
00 21 ☽ ∠ ♃
02 00 ☽ △ ♀
03 23 ☽ ∥ ☉
03 25 ☽ △ ♇
07 55 ☽ ⊼ ♄
09 54 ☿ ⊼ ♃
11 24 ☽ bQ ☉
16 31 ♀ ☌ ♇
17 14 ☽ # ♃
19 35 ☽ Q ♅
19 54 ☽ ± ♄

All ephemeris data is given at 12.00 UT and the Moon's longitude is additionally given for 24.00 UT

SEPTEMBER 1950

LONGITUDES

Date	Sidereal time h m s	Sun ☉ ° ' "	Moon ☽ ° ' "	Moon ☽ 24.00 ° ' "	Mercury ☿ ° '	Venus ♀ ° '	Mars ♂ ° '	Jupiter ♃ ° '	Saturn ♄ ° '	Uranus ♅ ° '	Neptune ♆ ° '	Pluto ♇ ° '
01	10 40 19	08 ♍ 28 25	04 ♉ 50 57	10 ♉ 49 29	02 ♎ 07	19 ♌ 27	13 ♏ 39	01 ♓ 40	20 ♍ 52	08 ♋ 38	15 ♎ 42	18 ♌ 29
02	10 44 16	09 26 30	16 ♉ 46 31	22 ♉ 42 37	02 19	20 41	14 18	01 R 33	20 59	08 41	15 44	18 30
03	10 48 13	10 24 36	28 ♉ 38 20	04 ♊ 34 17	02 25	21 54	14 57	01 25	21 07	08 43	15 46	18 32
04	10 52 09	11 22 44	10 ♊ 31 02	16 ♊ 29 13	02 R 25	23 08	15 36	01 17	21 14	08 45	15 48	18 34
05	10 56 06	12 20 55	22 ♊ 29 23	28 ♊ 32 07	02 19	24 22	16 15	01 09	21 21	08 47	15 50	18 35
06	11 00 02	13 19 07	04 ♋ 37 57	10 ♋ 47 25	02 06	25 36	16 55	01 02	21 29	08 49	15 52	18 37
07	11 03 59	14 17 22	17 ♋ 00 57	23 ♋ 18 58	01 48	26 50	17 35	00 54	21 36	08 51	15 54	18 39
08	11 07 55	15 15 38	29 ♋ 41 48	06 ♌ 09 42	01 22	28 04	18 14	00 47	21 44	08 53	15 56	18 41
09	11 11 52	16 13 56	12 ♌ 42 50	19 ♌ 21 17	00 51	29 ♌ 18	18 54	00 40	21 51	08 55	15 58	18 42
10	11 15 48	17 12 17	26 ♌ 05 02	02 ♍ 53 56	00 ♎ 13	00 ♍ 32	19 34	00 32	21 59	08 56	16 00	18 44
11	11 19 45	18 10 39	09 ♍ 47 45	16 ♍ 46 08	29 ♍ 29	01 46	20 14	00 25	22 06	08 58	16 02	18 46
12	11 23 42	19 09 03	23 ♍ 48 41	00 ♎ 54 52	28 39	03 00	20 54	00 18	22 14	09 00	16 04	18 47
13	11 27 38	20 07 29	08 ♎ 04 07	15 ♎ 15 48	27 45	04 14	21 35	00 11	22 21	09 02	16 06	18 49
14	11 31 35	21 05 56	22 ♎ 29 16	29 ♎ 43 51	26 47	05 29	22 15	00 ♓ 04	22 29	09 03	16 08	18 50
15	11 35 31	22 04 25	06 ♏ 58 54	14 ♏ 13 46	25 46	06 43	22 56	29 ♒ 57	22 36	09 05	16 10	18 52
16	11 39 28	23 02 56	21 ♏ 27 55	28 ♏ 40 46	24 43	07 57	23 36	29 51	22 44	09 06	16 12	18 54
17	11 43 24	24 01 29	05 ♐ 51 52	13 ♐ 00 49	23 40	09 11	24 17	29 44	22 51	09 08	16 14	18 55
18	11 47 21	25 00 04	20 ♐ 07 17	27 ♐ 10 58	22 39	10 26	24 58	29 38	22 59	09 09	16 16	18 57
19	11 51 17	25 58 40	04 ♑ 11 40	11 ♑ 09 12	21 40	11 40	25 39	29 31	23 06	09 11	16 18	18 58
20	11 55 14	26 57 17	18 ♑ 03 28	24 ♑ 54 22	20 45	12 55	26 20	29 25	23 14	09 12	16 21	19 00
21	11 59 11	27 55 56	01 ♒ 41 50	08 ♒ 25 52	19 57	14 09	27 01	29 19	23 21	09 13	16 23	19 01
22	12 03 07	28 54 37	15 ♒ 06 25	21 ♒ 43 31	19 15	15 23	27 42	29 13	23 29	09 15	16 25	19 03
23	12 07 04	29 ♍ 53 20	28 ♒ 17 09	04 ♓ 47 21	18 41	16 38	28 24	29 07	23 36	09 16	16 27	19 04
24	12 11 00	00 ♎ 52 04	11 ♓ 14 10	17 ♓ 37 38	18 17	17 52	29 05	29 02	23 44	09 17	16 29	19 06
25	12 14 57	01 50 50	23 ♓ 57 49	00 ♈ 14 48	18 02	19 07	29 ♏ 46	28 56	23 51	09 18	16 31	19 07
26	12 18 53	02 49 38	06 ♈ 28 41	12 ♈ 39 37	17 D 57	20 22	00 ♐ 28	28 51	23 58	09 19	16 33	19 09
27	12 22 50	03 48 29	18 ♈ 47 44	24 ♈ 53 13	18 02	21 36	01 10	28 46	24 06	09 20	16 36	19 10
28	12 26 46	04 47 21	00 ♉ 56 19	06 ♉ 57 17	18 17	22 51	01 52	28 41	24 13	09 21	16 38	19 12
29	12 30 43	05 46 15	12 ♉ 56 24	18 ♉ 54 00	18 42	24 05	02 34	28 36	24 21	09 22	16 40	19 13
30	12 34 40	06 ♎ 45 12	24 ♉ 50 28	00 ♊ 46 12	19 ♍ 17	25 ♍ 20	03 ♐ 16	28 ♒ 31	24 ♍ 28	09 ♋ 23	16 ♎ 42	19 ♌ 14

Date	Moon True ☊ ° '	Moon Mean ☊ ° '	Moon ☽ Latitude ° '
01	28 ♓ 27	29 ♓ 13	03 N 08
02	28 D 29	29 10	03 56
03	28 30	29 07	04 34
04	28 30	29 04	05 00
05	28 R 30	29 00	05 14
06	28 29	28 57	05 14
07	28 27	28 54	04 59
08	28 25	28 51	04 30
09	28 24	28 48	03 46
10	28 22	28 45	02 49
11	28 21	28 41	01 41
12	28 21	28 38	00 N 25
13	28 D 21	28 35	00 S 54
14	28 21	28 32	02 10
15	28 22	28 29	03 18
16	28 22	28 25	04 14
17	28 23	28 22	04 53
18	28 23	28 19	05 14
19	28 R 23	28 16	05 15
20	28 23	28 13	04 59
21	28 23	28 10	04 25
22	28 D 23	28 06	03 38
23	28 23	28 03	02 39
24	28 23	28 00	01 34
25	28 23	27 57	00 S 25
26	28 R 23	27 54	00 N 45
27	28 23	27 51	01 51
28	28 22	27 47	02 51
29	28 21	27 44	03 42
30	28 ♓ 20	27 ♓ 41	04 N 24

DECLINATIONS

Date	Sun ☉ ° '	Moon ☽ ° '	Mercury ☿ ° '	Venus ♀ ° '	Mars ♂ ° '	Jupiter ♃ ° '	Saturn ♄ ° '	Uranus ♅ ° '	Neptune ♆ ° '	Pluto ♇ ° '
01	08 N 24	16 N 05	04 S 18	15 N 51	16 S 50	12 S 06	05 N 24	23 N 27	04 S 44	22 N 59
02	08 02	20 37	04 29	15 29	17 03	12 09	05 21	23 27	04 45	22 59
03	07 40	24 19	04 37	15 07	17 15	12 12	05 18	23 27	04 45	22 58
04	07 18	26 59	04 42	14 45	17 27	12 15	05 15	23 27	04 46	22 58
05	06 56	28 28	04 44	14 22	17 39	12 17	05 12	23 27	04 47	22 58
06	06 34	28 36	04 43	13 59	17 51	12 20	05 09	23 26	04 48	22 57
07	06 11	27 19	04 37	13 35	18 02	12 23	05 06	23 26	04 48	22 57
08	05 49	24 37	04 28	13 12	18 14	12 26	05 03	23 26	04 49	22 56
09	05 26	20 37	04 15	12 47	18 26	12 28	05 00	23 26	04 50	22 56
10	05 03	15 29	03 58	12 22	18 37	12 31	04 58	23 26	04 51	22 56
11	04 41	09 27	03 37	11 57	18 49	12 33	04 55	23 26	04 51	22 55
12	04 18	02 N 51	03 12	11 32	19 00	12 36	04 52	23 26	04 52	22 55
13	03 55	04 S 01	02 44	11 06	19 11	12 38	04 49	23 26	04 53	22 55
14	03 32	10 46	02 12	10 40	19 22	12 41	04 46	23 26	04 54	22 54
15	03 09	16 58	01 37	10 14	19 33	12 43	04 43	23 26	04 55	22 54
16	02 46	22 13	01 00	09 47	19 43	12 45	04 40	23 26	04 55	22 54
17	02 22	26 06	00 S 21	09 20	19 54	12 48	04 37	23 25	04 56	22 53
18	01 59	28 18	00 N 18	08 53	20 05	12 50	04 34	23 25	04 57	22 53
19	01 36	28 38	00 58	08 26	20 15	12 52	04 31	23 25	04 58	22 53
20	01 13	27 10	01 37	07 58	20 25	12 54	04 28	23 25	04 59	22 52
21	00 49	24 06	02 14	07 30	20 35	12 56	04 25	23 25	05 00	22 52
22	00 26	19 46	02 48	07 02	20 45	12 58	04 22	23 25	05 00	22 52
23	00 N 03	14 34	03 20	06 34	20 55	13 00	04 20	23 25	05 01	22 51
24	00 S 21	08 48	03 48	06 05	21 04	13 02	04 17	23 25	05 02	22 51
25	00 44	02 S 46	04 12	05 37	21 14	13 04	04 14	23 25	05 03	22 51
26	01 07	03 N 15	04 31	05 08	21 23	13 06	04 11	23 25	05 04	22 51
27	01 31	09 04	04 45	04 39	21 32	13 07	04 08	23 25	05 05	22 50
28	01 54	14 28	04 55	04 10	21 41	13 09	04 05	23 25	05 05	22 50
29	02 18	19 16	04 59	03 40	21 50	13 11	04 02	23 25	05 06	22 50
30	02 S 41	23 N 15	04 N 59	03 N 11	21 S 58	13 S 12	03 N 59	23 N 25	05 S 07	22 N 50

ZODIAC SIGN ENTRIES

Date	h	m	Planets
01	02	19	☽ ♉
03	14	45	☽ ♊
06	02	54	☽ ♋
08	12	34	☽ ♌
10	01	37	♀ ♍
10	18	55	☽ ♍
10	19	16	☿ ♍
12	22	28	☽ ♎
15	00	27	☽ ♏
15	02	23	♃ ♒
17	02	12	☽ ♐
19	04	49	☽ ♑
21	08	59	☽ ♒
23	14	44	☉ ♎
23	15	09	☽ ♓
25	19	48	♂ ♐
25	23	32	☽ ♈
28	10	08	☽ ♉
30	22	26	☽ ♊

LATITUDES

Date	Mercury ☿ ° '	Venus ♀ ° '	Mars ♂ ° '	Jupiter ♃ ° '	Saturn ♄ ° '	Uranus ♅ ° '	Neptune ♆ ° '	Pluto ♇ ° '
01	03 S 46	00 N 54	00 S 56	01 S 18	01 N 56	00 N 17	01 N 35	08 N 06
04	04 05	01 00	00 58	01 19	01 56	00 17	01 34	08 07
07	04 16	01 05	01 00	01 19	01 56	00 17	01 34	08 07
10	04 14	01 10	01 02	01 19	01 56	00 17	01 34	08 08
13	03 57	01 14	01 03	01 19	01 56	00 17	01 34	08 08
16	03 22	01 18	01 05	01 19	01 56	00 17	01 34	08 09
19	02 33	01 21	01 06	01 18	01 56	00 17	01 34	08 09
22	01 34	01 23	01 07	01 18	01 57	00 18	01 34	08 10
25	00 S 35	01 25	01 09	01 18	01 57	00 18	01 34	08 11
28	00 N 18	01 26	01 10	01 18	01 57	00 18	01 34	08 11
31	01 N 01	01 N 27	01 S 11	01 S 17	01 N 57	00 N 18	01 N 34	08 N 12

DATA

Julian Date	2433526
Delta T	+29 seconds
Ayanamsa	23° 10' 05"
Synetic vernal point	05° ♓ 56' 54"
True obliquity of ecliptic	23° 26' 54"

MOON'S PHASES, APSIDES AND POSITIONS ☽

Date	h	m	Phase	Longitude ° '	Eclipse Indicator
04	13	53	☾	11 ♊ 27	
12	03	29	●	18 ♍ 48	Total
18	20	54	☽	25 ♐ 22	
26	04	21	○	02 ♈ 31	total

Day	h	m		
03	09	48	Apogee	
15	07	16	Perigee	
06	02	23	Max dec	28° N 43'
12	21	58	0S	
19	04	21	Max dec	28° S 44'
25	22	59	0N	

LONGITUDES

Date	Chiron ⚷ ° '	Ceres ⚳ ° '	Pallas ⚴ ° '	Juno ⚵ ° '	Vesta ⚶ ° '	Black Moon Lilith ⚸ ° '
01	15 ♐ 45	10 ♐ 06	15 ♏ 20	13 ♎ 32	20 ♉ 57	25 ♉ 52
11	16 ♐ 02	12 ♐ 15	18 ♏ 41	16 ♎ 51	21 ♉ 48	26 ♉ 59
21	16 ♐ 28	14 ♐ 45	22 ♏ 14	20 ♎ 14	21 ♉ 58	28 ♉ 06
31	17 ♐ 03	17 ♐ 32	25 ♏ 57	23 ♎ 39	21 ♉ 23	29 ♉ 13

All ephemeris data is given at 12.00 UT and the Moon's longitude is additionally given for 24.00 UT

ASPECTARIAN

h m Aspects

01 Friday
05 43 ☽ ⚹ ♃
06 26 ☽ ⚻ ☿
10 55 ☽ ∥ ♀
14 03 ☽ ⚼ ♄
15 54 ☽ # ♂
16 18 ☉ ⚹ ♅
18 41 ☽ ± ☿
19 38 ☽ ⚹ ♅
19 55 ☽ △ ☉
02 Saturday
05 33 ☽ Q ♃
06 42 ☽ ☍ ♂
09 54 ☽ ⚻ ♆
13 06 ☽ ⚼ ☿
15 30 ☽ □ ♇
18 44 ♀ ⊻ ♄
19 34 ☉ ⊥ ♆
20 36 ☽ △ ♄
20 48 ☽ □ ♀
22 03 ☽ ± ♆
03 Sunday
02 00 ☽ ∠ ♅
02 33 ☽ ∥ ♇
05 46 ☽ ∥ ♅
16 19 ☽ ⚼ ♆
17 33 ☽ □ ♃
19 39 ☽ △ ☿
20 16 ☽ ⊥ ♅
04 Monday
00 13 ☿ St R
04 00 ☽ Q ♇
08 25 ☽ ⊻ ♅
13 24 ☽ Q ♀
13 53 ☽ □ ☉
19 37 ♂ ⊻ ♆
22 39 ☽ △ ♆
22 49 ☽ ⚻ ♂
05 Tuesday
00 14 ♀ ∠ ♅
04 12 ☽ ⚹ ♇
09 43 ☽ □ ♄
11 31 ☽ ± ♂
16 10 ☽ ⚹ ♀
06 Wednesday
04 56 ☽ Q ☉
05 00 ☽ △ ♃
06 22 ☽ ⚼ ♂
07 08 ☽ □ ☿
10 01 ☽ ∠ ♇
16 51 ☿ ∠ ♂
20 01 ☿ ⊥ ♀
20 11 ☽ ☌ ♅
21 33 ☽ Q ♄
07 Thursday
00 55 ☽ ∠ ♀
03 35 ☽ ⊥ ♇
06 19 ☽ ⚹ ☉
09 51 ☽ □ ♆
09 54 ☽ ⚼ ♃
13 08 ☽ △ ♂
13 29 ♃ ⚼ ♆
15 08 ☽ ⊻ ♇
17 09 ☽ Q ☿
20 05 ☽ ⊥ ♀
20 51 ☽ ⚹ ♄
08 Friday
02 52 ☽ ± ♃
08 37 ☽ ⊻ ♀
13 08 ☽ ∠ ☉
14 00 ☽ ⚻ ♃
15 01 ☽ ⚹ ☿
19 54 ☽ Q ♆
19 56 ☽ ∥ ♅
23 01 ☽ ∥ ♇
09 Saturday
01 10 ☽ ∠ ♄
04 32 ♂ □ ♇
05 02 ☽ ⊻ ♅
05 06 ☉ ⊻ ♆
07 07 ☽ ⊥ ☉
16 00 ☽ ⊥ ♅
17 26 ☽ ∠ ☿
17 45 ☽ ⊥ ♄
17 54 ☽ ⚹ ♆
18 53 ☽ ⊻ ☉
21 14 ☿ ⚻ ♃
22 25 ☽ # ♂
22 51 ☽ ☌ ♇
23 46 ☽ □ ♂
10 Sunday
04 38 ☽ ⊻ ♄
04 46 ♀ # ♃
07 57 ☿ ⊻ ♀
08 11 ☽ ∠ ♅
08 50 ☽ ⊥ ☿
12 05 ♀ ☍ ♃
18 56 ☽ ⊻ ☿
19 02 ☉ ∥ ♄
19 48 ☽ ☍ ♃
20 38 ☽ ☌ ♀
20 41 ☽ ∠ ♆
21 14 ♀ ∠ ♆
11 Monday
00 07 ☽ # ♃
00 56 ☉ # ♆
01 41 ☽ ∥ ♀
09 10 ☽ Q ♂
10 34 ☽ ⚹ ♅
12 24 ☽ ⊥ ♆
22 46 ☽ ⊻ ♆
12 Tuesday
02 47 ☉ ⊻ ♇
03 26 ☽ ⊻ ♇
03 29 ☽ ☌ ☉
04 47 ☽ ∥ ♄
04 49 ☽ # ♆
06 32 ☽ ∥ ☉
06 49 ☽ ⚹ ♂
07 13 ☽ Q ♅
08 20 ♄ # ♆
09 17 ☽ ☌ ♄
10 39 ☽ # ☿
13 39 ☽ ⊥ ♇
19 43 ☽ ☌ ☿
22 53 ☽ ⚻ ♃
13 Wednesday
04 52 ☽ ∠ ♇
04 59 ☽ ⊻ ♀
07 47 ☽ ∥ ☿
08 52 ☽ ± ♃
09 23 ☽ ∠ ♂
11 38 ☽ # ☉
13 36 ☽ □ ♅
14 45 ☽ # ♄
15 01 ☽ ∥ ♆
15 58 ☽ ⊥ ♀
23 46 ☽ ⚼ ♃
14 Thursday
01 08 ☽ ⊥ ♂
01 25 ☽ ☌ ♆
05 56 ☽ ⚹ ♇
07 31 ♂ ⊥ ♆
08 21 ☽ ∠ ♀
09 32 ☽ ⊻ ☉
10 54 ☉ Q ♅
11 35 ☽ ⊻ ♂
11 41 ☽ # ♀
11 59 ☽ ⊻ ♄
18 39 ☽ ⊻ ☿
19 10 ☽ ∥ ♃
20 12 ☽ ⊥ ☉
21 57 ♂ ⚹ ♄
22 01 ☽ ⊥ ♄
15 Friday
00 28 ☽ △ ♃
01 52 ☽ Q ♇
03 56 ☽ ⊥ ☿
11 31 ☽ ⚹ ♀
12 10 ☽ ∠ ☉
13 02 ☽ ∠ ♄
15 29 ☽ △ ♅
17 50 ☽ ∠ ☿
23 31 ☽ ∥ ♂
16 Saturday
02 58 ☉ ☌ ♄
03 15 ☽ ⊻ ♆
07 44 ☽ □ ♇
08 04 ☿ ⊥ ♇
09 15 ☽ Q ♀
13 13 ☽ ⊥ ♆
14 07 ☽ ⚹ ♄
14 49 ☽ ⚹ ☉
15 36 ☽ # ♇
15 44 ☽ ☌ ♂
16 24 ☽ ⚼ ♅
17 02 ☽ ⚹ ☿
18 33 ☽ # ♅
17 Sunday
01 50 ☽ □ ♃
03 29 ☿ ⚹ ♂
04 15 ☽ ∠ ♆
06 31 ♂ ⚼ ♅
07 25 ☽ ± ♅
07 47 ☉ ☌ ☿
10 18 ☽ Q ♄
10 50 ♀ ⚹ ♅
11 42 ☽ Q ☿
12 17 ☽ Q ☉
17 29 ☽ ⚻ ♅
18 06 ☽ □ ♀
18 Monday
04 57 ☿ ☌ ♄
05 28 ☽ ⚹ ♆
07 49 ☽ Q ♃
08 49 ♀ ⊥ ♆
08 55 ☉ ⚹ ♂
10 01 ☽ △ ♇
10 39 ☉ ⊥ ♇
16 00 ☽ □ ☿
16 53 ☽ □ ♄
20 38 ☽ ⊻ ♂
20 54 ☽ □ ☉
19 Tuesday
01 53 ☽ Q ♆
04 03 ☽ ⚹ ♃
07 24 ☽ ⊥ ♂
11 37 ☽ ⚼ ♇
20 36 ☽ ☍ ♅
23 42 ☽ ∠ ♂
20 Wednesday
00 13 ☿ Q ♅
02 10 ☽ △ ♀
02 42 ☉ ∥ ☿
03 11 ☽ ± ♇
05 43 ☽ ∠ ♃
09 00 ☽ □ ♆
13 39 ☽ ⚻ ♇
16 26 ☽ △ ☿
21 08 ☽ △ ♄
21 19 ☽ ⊥ ♃
21 Thursday
03 17 ☽ ⚹ ♂
04 49 ☽ △ ☉
07 02 ☽ ⚼ ♀
07 49 ☽ ⊻ ♃
16 12 ☽ # ♅
17 28 ☽ ⚼ ☿
19 29 ☽ # ♇
23 58 ☽ ⚼ ♄
22 Friday
00 40 ☽ ± ♀
01 27 ☽ ⚻ ♅
01 45 ☽ Q ♂
07 13 ☽ ∥ ♂
08 48 ☽ ± ☿
09 40 ☽ ⚼ ☉
12 15 ☽ ± ♅
12 34 ☽ ⚻ ♀
14 22 ☽ △ ♆
16 20 ☽ ± ♄
18 54 ☉ ⚻ ♃
19 09 ☽ ☍ ♇
19 10 ☽ ⚻ ☿
19 40 ☿ ⊻ ♇
23 Saturday
01 32 ♀ Q ♂
03 18 ☽ ± ☉
03 20 ☽ ⚻ ♄
04 37 ☽ ⚼ ♅
08 21 ♀ ⊻ ♆
12 12 ☽ □ ♂
13 32 ☽ ☌ ♃
15 11 ☽ ⚻ ☉
17 50 ☽ ⚼ ♆
18 38 ☽ ∥ ♃
24 Sunday
08 21 ☽ △ ♅
10 23 ♂ □ ♃
10 36 ☽ ± ♆
18 17 ☿ ☌ ♀
21 53 ☽ ⚻ ♆
23 47 ☽ # ♀
25 Monday
00 56 ☽ ☍ ☿
01 49 ☽ ☍ ♀
02 48 ☽ ⚻ ♇
03 01 ☽ ∥ ♆
06 11 ☽ # ♄
06 41 ☽ # ☿
11 47 ☽ ☍ ♄
12 06 ♀ ⊻ ♇
13 59 ☿ ∥ ♄
14 13 ☽ ± ♇
19 21 ☿ Q ♂
19 35 ☽ ∥ ☉
21 26 ☽ ⊻ ♃
23 45 ☽ △ ♂
26 Tuesday
02 52 ☽ # ☉
04 21 ☽ ☍ ☉
07 29 ☽ ⚼ ♇
08 53 ☽ ⊥ ♃
11 59 ☿ St D
15 19 ♀ # ♆
15 42 ☽ ∥ ♄
17 20 ☽ ∥ ☿
17 31 ☽ □ ♅
19 01 ☽ ∥ ♀
19 20 ☽ # ♆
27 Wednesday
02 13 ☽ ∠ ♃
06 32 ☽ ⚼ ♂
06 49 ♀ Q ♅
07 40 ☽ ☍ ♆
08 16 ☿ ∥ ♀
10 29 ☽ ⚻ ☿
12 44 ☽ △ ♇
18 09 ☽ ⚻ ♀
21 04 ☉ ∠ ♇
22 29 ☽ ± ☿
22 33 ☽ ⚻ ♄
28 Thursday
01 19 ☽ ± ♂
03 38 ♂ ∠ ♆
04 52 ☽ Q ♅
05 53 ☽ # ♃
07 22 ☽ ± ♀
07 33 ☽ ⚹ ♃
09 33 ☉ ± ♃
10 34 ☽ ± ♄
13 57 ☽ ⚻ ♂
16 10 ♀ ∥ ♄
16 49 ☽ ⚼ ☿
20 21 ☽ ⚻ ☉
29 Friday
03 23 ☽ ⚼ ♀
04 43 ☽ ⚼ ♄
04 49 ☽ ⚹ ♅
07 20 ☽ Q ♃
09 26 ☽ ± ☉
17 27 ♀ ☌ ♄
19 31 ☽ ⚻ ♆
30 Saturday
00 09 ☽ △ ☿
00 40 ☽ □ ♇
03 22 ☽ # ♂
05 12 ☽ ⚼ ☉
07 40 ☽ ± ♆
09 11 ☽ ∥ ♇
10 05 ♀ ⊥ ♇
10 32 ☿ ⊻ ♇
11 04 ☽ ∠ ♅
11 14 ☽ △ ♄
13 06 ☽ ∥ ♅
13 07 ☽ △ ♀
16 18 ♂ ± ♅
19 24 ☽ □ ♃

OCTOBER 1950

Date	Sidereal time h m s	Sun ☉ ° ' "	Moon ☽ ° ' "	Moon ☽ 24.00 ° ' "	Mercury ☿ ° '	Venus ♀ ° '	Mars ♂ ° '	Jupiter ♃ ° '	Saturn ♄ ° '	Uranus ♅ ° '	Neptune ♆ ° '	Pluto ⯓ ° '
01	12 38 36	07 ♎ 44 11	06 ♊ 41 39	12 ♊ 37 17	20 ♍ 00	26 ♍ 35	03 ♐ 58	28 ♒ 27	24 ♍ 35	09 ♋ 24	16 ♎ 44	19 ♌ 16
02	12 42 33	08 43 12	18 ♊ 33 37	24 ♊ 31 10	20 52	27 50	04 40	28 R 23	24 43	09 25	16 47	19 17
03	12 46 29	09 42 15	00 ♋ 30 30	06 ♋ 32 11	21 51	29 ♍ 04	05 22	28 18	24 50	09 25	16 49	19 18
04	12 50 26	10 41 21	12 ♋ 36 46	18 ♋ 44 50	22 58	00 ♎ 19	06 05	28 14	24 58	09 26	16 51	19 20
05	12 54 22	11 40 29	24 ♋ 56 58	01 ♌ 13 40	24 11	01 34	06 47	28 11	25 05	09 26	16 53	19 21
06	12 58 19	12 39 39	07 ♌ 35 26	14 ♌ 02 44	25 29	02 49	07 30	28 07	25 12	09 27	16 56	19 22
07	13 02 15	13 38 52	20 ♌ 35 55	27 ♌ 15 18	26 52	04 04	08 12	28 04	25 19	09 27	16 58	19 23
08	13 06 12	14 38 07	04 ♍ 01 03	10 ♍ 53 13	28 19	05 19	08 55	28 00	25 27	09 28	17 00	19 24
09	13 10 09	15 37 24	17 ♍ 51 45	24 ♍ 56 24	29 ♍ 50	06 34	09 38	27 57	25 34	09 28	17 02	19 26
10	13 14 05	16 36 43	02 ♎ 06 48	09 ♎ 22 23	01 ♎ 23	07 49	10 21	27 54	25 41	09 29	17 04	19 27
11	13 18 02	17 36 04	16 ♎ 42 27	24 ♎ 06 10	02 59	09 03	11 04	27 52	25 48	09 29	17 07	19 28
12	13 21 58	18 35 28	01 ♏ 32 36	09 ♏ 00 41	04 37	10 18	11 47	27 49	25 55	09 29	17 09	19 29
13	13 25 55	19 34 53	16 ♏ 29 23	23 ♏ 57 36	06 17	11 33	12 30	27 47	26 02	09 29	17 11	19 30
14	13 29 51	20 34 21	01 ♐ 24 18	08 ♐ 48 33	07 58	12 48	13 13	27 45	26 09	09 29	17 13	19 31
15	13 33 48	21 33 50	16 ♐ 09 30	23 ♐ 26 27	09 39	14 04	13 57	27 43	26 17	09 29	17 16	19 32
16	13 37 44	22 33 21	00 ♑ 38 51	07 ♑ 46 17	11 22	15 19	14 40	27 41	26 24	09 R 29	17 18	19 33
17	13 41 41	23 32 54	14 ♑ 48 30	21 ♑ 45 22	13 05	16 34	15 24	27 40	26 30	09 29	17 20	19 34
18	13 45 37	24 32 28	28 ♑ 36 54	05 ♒ 23 11	14 47	17 49	16 07	27 39	26 37	09 29	17 22	19 35
19	13 49 34	25 32 04	12 ♒ 04 24	18 ♒ 40 49	16 30	19 04	16 51	27 38	26 44	09 29	17 25	19 36
20	13 53 31	26 31 42	25 ♒ 12 40	01 ♓ 40 18	18 13	20 19	17 35	27 37	26 51	09 29	17 27	19 37
21	13 57 27	27 31 22	08 ♓ 04 01	14 ♓ 24 09	19 56	21 34	18 19	27 36	26 58	09 29	17 29	19 38
22	14 01 24	28 31 03	20 ♓ 40 59	26 ♓ 54 49	21 39	22 49	19 03	27 36	27 05	09 28	17 31	19 39
23	14 05 20	29 ♎ 30 46	03 ♈ 05 57	09 ♈ 14 37	23 21	24 04	19 47	27 35	27 12	09 28	17 33	19 39
24	14 09 17	00 ♏ 30 31	15 ♈ 21 03	21 ♈ 25 28	25 03	25 19	20 31	27 D 35	27 18	09 27	17 36	19 40
25	14 13 13	01 30 17	27 ♈ 28 04	03 ♉ 29 04	26 44	26 35	21 15	27 35	27 25	09 27	17 38	19 41
26	14 17 10	02 30 06	09 ♉ 28 38	15 ♉ 26 57	28 ♎ 25	27 50	21 59	27 36	27 31	09 26	17 40	19 42
27	14 21 06	03 29 57	21 ♉ 24 14	27 ♉ 20 42	00 ♏ 06	29 ♎ 05	22 43	27 36	27 38	09 26	17 42	19 43
28	14 25 03	04 29 50	03 ♊ 16 34	09 ♊ 12 07	01 46	00 ♏ 20	23 28	27 37	27 45	09 25	17 44	19 43
29	14 29 00	05 29 45	15 ♊ 07 37	21 ♊ 03 25	03 26	01 35	24 12	27 38	27 51	09 25	17 46	19 44
30	14 32 56	06 29 42	26 ♊ 59 51	02 ♋ 57 19	05 05	02 51	24 57	27 39	27 58	09 24	17 49	19 45
31	14 36 53	07 ♏ 29 41	08 ♋ 56 16	14 ♋ 57 11	06 ♏ 43	04 ♏ 06	25 ♐ 41	27 ♒ 41	28 ♍ 04	09 ♋ 23	17 ♎ 51	19 ♌ 45

Date	Moon True ☊ ° '	Moon Mean ☊ ° '	Moon ☽ Latitude ° '
01	28 ♓ 19	27 ♓ 38	04 N 54
02	28 R 18	27 35	05 12
03	28 17	27 31	05 16
04	28 D 17	27 28	05 07
05	28 17	27 25	04 43
06	28 18	27 22	04 05
07	28 20	27 19	03 14
08	28 21	27 16	02 10
09	28 22	27 12	00 N 58
10	28 R 22	27 09	00 S 21
11	28 21	27 06	01 39
12	28 19	27 03	02 53
13	28 17	27 00	03 55
14	28 14	26 57	04 41
15	28 11	26 53	05 07
16	28 09	26 50	05 14
17	28 07	26 47	05 01
18	28 D 07	26 44	04 31
19	28 08	26 41	03 46
20	28 09	26 37	02 51
21	28 11	26 34	01 48
22	28 12	26 31	00 S 41
23	28 R 12	26 28	00 N 27
24	28 11	26 25	01 32
25	28 08	26 22	02 33
26	28 03	26 18	03 26
27	27 57	26 15	04 09
28	27 50	26 12	04 42
29	27 43	26 09	05 02
30	27 37	26 06	05 10
31	27 ♓ 31	26 ♓ 03	05 N 04

DECLINATIONS

Date	Sun ☉ ° '	Moon ☽ ° '	Mercury ☿ ° '	Venus ♀ ° '	Mars ♂ ° '	Jupiter ♃ ° '	Saturn ♄ ° '	Uranus ♅ ° '	Neptune ♆ ° '	Pluto ⯓ ° '
01	03 S 04	26 N 16	04 N 53	02 N 41	22 S 07	13 S 14	03 N 57	23 N 25	05 S 08	22 N 50
02	03 28	28 08	04 43	02 12	22 15	13 15	03 54	23 25	05 09	22 49
03	03 51	28 43	04 29	01 42	22 23	13 16	03 51	23 25	05 10	22 49
04	04 14	27 56	04 11	01 12	22 31	13 18	03 48	23 25	05 11	22 49
05	04 37	25 47	03 49	00 42	22 38	13 19	03 45	23 25	05 11	22 49
06	05 00	22 19	03 23	00 N 13	22 46	13 20	03 42	23 25	05 12	22 49
07	05 23	17 41	02 55	00 S 17	22 53	13 21	03 40	23 25	05 13	22 49
08	05 46	12 04	02 24	00 47	23 00	13 22	03 37	23 25	05 14	22 48
09	06 09	05 N 41	01 50	01 17	23 07	13 23	03 34	23 25	05 15	22 48
10	06 32	01 S 09	01 14	01 47	23 14	13 24	03 31	23 25	05 16	22 48
11	06 55	08 06	00 N 37	02 17	23 20	13 25	03 29	23 25	05 17	22 48
12	07 17	14 43	00 S 02	02 47	23 26	13 26	03 26	23 25	05 17	22 48
13	07 40	20 31	00 42	03 17	23 32	13 26	03 23	23 25	05 18	22 48
14	08 02	25 02	01 24	03 47	23 38	13 27	03 21	23 25	05 19	22 48
15	08 25	27 49	02 06	04 17	23 44	13 27	03 18	23 25	05 20	22 48
16	08 47	28 41	02 48	04 46	23 49	13 28	03 15	23 25	05 21	22 48
17	09 09	27 37	03 31	05 16	23 54	13 28	03 12	23 25	05 22	22 48
18	09 31	24 52	04 15	05 46	23 59	13 28	03 10	23 25	05 22	22 47
19	09 53	20 48	04 58	06 15	24 04	13 29	03 07	23 25	05 23	22 47
20	10 14	15 48	05 42	06 44	24 08	13 29	03 05	23 25	05 24	22 47
21	10 36	10 13	06 25	07 13	24 12	13 29	03 02	23 25	05 25	22 47
22	10 57	04 S 19	07 08	07 42	24 16	13 29	02 59	23 25	05 26	22 47
23	11 18	01 N 38	07 51	08 11	24 20	13 29	02 57	23 25	05 27	22 47
24	11 39	07 28	08 34	08 40	24 23	13 29	02 54	23 25	05 27	22 47
25	12 00	12 57	09 16	09 08	24 27	13 28	02 52	23 25	05 28	22 47
26	12 21	17 54	09 58	09 37	24 30	13 28	02 49	23 25	05 29	22 47
27	12 41	22 08	10 39	10 05	24 32	13 28	02 47	23 25	05 30	22 48
28	13 01	25 26	11 20	10 32	24 35	13 27	02 44	23 25	05 31	22 48
29	13 22	27 37	12 00	11 00	24 37	13 27	02 42	23 25	05 31	22 48
30	13 41	28 34	12 40	11 27	24 39	13 26	02 40	23 25	05 32	22 48
31	14 S 01	28 N 12	13 S 19	11 S 54	24 S 41	13 S 26	02 N 37	23 N 25	05 S 33	22 N 48

ZODIAC SIGN ENTRIES

Date	h	m	Planets
03	10	59	☽ ♋
04	05	51	♀ ♎
05	21	40	☽ ♌
08	04	54	☽ ♍
09	14	40	☿ ♎
10	08	29	☽ ♎
12	09	31	☽ ♏
14	09	44	☽ ♐
16	10	55	☽ ♑
18	14	27	☽ ♒
20	20	53	☽ ♓
23	05	59	☽ ♈
23	23	45	☉ ♏
25	17	03	☽ ♉
27	10	36	☿ ♏
28	05	22	☽ ♊
28	05	33	♀ ♏
30	18	03	☽ ♋

LATITUDES

Date	Mercury ☿ ° '	Venus ♀ ° '	Mars ♂ ° '	Jupiter ♃ ° '	Saturn ♄ ° '	Uranus ♅ ° '	Neptune ♆ ° '	Pluto ⯓ ° '
01	01 N 01	01 N 27	01 S 11	01 S 17	01 N 57	00 N 18	01 N 34	08 N 12
04	01 31	01 27	01 12	01 17	01 58	00 18	01 34	08 13
07	01 49	01 27	01 13	01 17	01 58	00 18	01 34	08 14
10	01 57	01 26	01 14	01 16	01 58	00 18	01 34	08 14
13	01 57	01 24	01 15	01 16	01 58	00 18	01 34	08 15
16	01 50	01 22	01 15	01 16	01 59	00 18	01 34	08 16
19	01 39	01 19	01 16	01 15	01 59	00 18	01 34	08 17
22	01 24	01 16	01 17	01 15	02 00	00 18	01 34	08 18
25	01 07	01 12	01 17	01 14	02 00	00 18	01 34	08 19
28	00 48	01 07	01 18	01 14	02 01	00 18	01 34	08 20
31	00 N 29	01 N 03	01 S 18	01 S 13	02 N 01	00 N 19	01 N 34	08 N 20

DATA

Julian Date	2433556
Delta T	+30 seconds
Ayanamsa	23° 10' 09"
Synetic vernal point	05° ♓ 56' 51"
True obliquity of ecliptic	23° 26' 54"

MOON'S PHASES, APSIDES AND POSITIONS ☽

Date	h	m	Phase	Longitude ° '	Eclipse Indicator
04	07	53	☾	10 ♋ 31	
11	13	34	●	17 ♎ 40	
18	04	18	☽	24 ♑ 13	
25	20	46	○	01 ♉ 52	

Day	h	m		
01	04	25	Apogee	
13	04	05	Perigee	
28	19	23	Apogee	
03	10	24	Max dec	28° N 43'
10	08	00	0S	
16	10	25	Max dec	28° S 41'
23	05	22	0N	
30	17	15	Max dec	28° N 36'

LONGITUDES

Date	Chiron ⚷ ° '	Ceres ⚳ ° '	Pallas ⚴ ° '	Juno ⚵ ° '	Vesta ⚶ ° '	Black Moon Lilith ⚸ ° '
01	17 ♐ 03	17 ♐ 32	25 ♏ 57	23 ♎ 39	21 ♉ 23	29 ♉ 13
11	17 ♐ 44	20 ♐ 34	29 ♏ 46	27 ♎ 07	20 ♉ 04	00 ♊ 20
21	18 ♐ 33	23 ♐ 48	03 ♐ 42	00 ♏ 35	18 ♉ 07	01 ♊ 27
31	19 ♐ 27	27 ♐ 12	07 ♐ 42	04 ♏ 03	15 ♉ 41	02 ♊ 34

ASPECTARIAN

h m Aspects

01 Sunday
01 41 ☉ # ♀
01 56 ☽ ♇ ♆
05 19 ☽ ⊥ ♅
06 07 ☽ ☍ ♂
13 09 ☽ Q ⯓
14 18 ☽ △ ☉
17 29 ☽ ⊻ ♅

02 Monday
08 23 ☽ △ ♆
13 28 ☽ ⚹ ⯓
17 03 ☽ □ ☿
22 00 ♀ ⊼ ♃

03 Tuesday
00 31 ☽ □ ♄
01 41 ☿ Q ♅
05 00 ☉ □ ♅
07 37 ☽ △ ♃
08 48 ☽ □ ♀
12 09 ☉ # ♄
19 35 ☽ ∠ ⯓
22 17 ☽ ⊼ ♂

04 Wednesday
05 43 ☽ ☌ ♅
07 53 ☽ □ ☉
08 25 ☽ Q ☿
10 20 ☉ # ☿
10 53 ☽ ± ♂
12 41 ☽ Q ♄
13 14 ☽ ♇ ♃
13 24 ☽ ⊥ ⯓
20 20 ☽ □ ♆

05 Thursday
00 25 ☽ Q ♀
01 09 ☽ ⊻ ⯓
05 31 ☽ ♇ ♂
06 41 ☽ ± ♃
10 21 ☽ ⚹ ☿
12 15 ☽ ⚹ ♄
16 05 ☿ ∥ ♄
18 09 ☽ ⊼ ♃
21 49 ☽ Q ☉

06 Friday
00 07 ♂ Q ♄
02 02 ☽ ⚹ ♀
05 20 ☽ ∥ ♅
06 30 ☿ ☌ ♄
06 59 ☽ Q ♆
09 05 ☽ ∥ ⯓
09 27 ☽ # ♂
09 56 ☿ ⊥ ⯓
11 48 ☽ △ ♂
12 29 ☿ Q ♂
15 29 ☽ ⊻ ♅
16 55 ☽ ∠ ♄
18 02 ☽ ∠ ☿
21 17 ♂ # ⯓
22 13 ☽ ⚹ ☉
22 28 ☉ ♇ ♃

07 Saturday
01 03 ☉ ∥ ♆
02 36 ☽ ⊥ ♅
05 21 ☽ ⚹ ♆
08 55 ☽ ∠ ♀
09 39 ☽ ⊥ ♄
09 48 ☽ ☌ ⯓
11 57 ♀ ± ♃
12 32 ☽ ⊥ ☿
18 21 ♀ ∠ ⯓
18 59 ☽ ∠ ♅
20 37 ☽ ⊻ ♄

08 Sunday
00 39 ☽ ⊻ ☿
01 23 ☽ ☍ ♃
02 49 ☽ ⊥ ♀
03 20 ♄ ⊥ ⯓
03 38 ☽ ∠ ☉
06 46 ☽ # ♃
07 07 ☿ ⊼ ♃
08 26 ☽ ∠ ♆
14 30 ☽ ⊻ ♀
20 43 ☽ ⊥ ☉
21 03 ☽ □ ♂
21 32 ☽ ⚹ ♅

09 Monday
00 14 ☽ ⊥ ♆
06 35 ♂ ⊼ ♅
07 53 ☽ ⊻ ☉
10 24 ☽ # ☉
10 35 ☽ ⊻ ♆
13 34 ☽ # ♆
14 40 ☽ ⊻ ⯓
18 08 ☽ Q ♅
19 35 ☽ ∥ ♄

10 Tuesday
00 03 ☿ # ♀
00 50 ☽ ⊥ ⯓
01 10 ☽ ☌ ♄
02 29 ☽ # ♀
02 54 ☽ ∥ ☿
05 00 ☽ ⊼ ♃
05 23 ☽ Q ♂
10 39 ☽ ☌ ☿
12 16 ☽ # ☿
14 21 ☽ ∥ ♀
14 58 ☽ ± ♃
15 52 ☽ ∠ ⯓
20 06 ☽ # ♄
22 19 ☽ ☌ ♀
23 40 ☉ ☌ ♆

11 Wednesday
00 10 ☽ □ ♅
02 10 ☽ ∥ ♆
02 18 ☽ ⚹ ♂
05 45 ☽ ♇ ♃
07 37 ☽ ∥ ☉
12 40 ☽ ☌ ♆
13 34 ☽ ☌ ☉
16 29 ☽ ⚹ ⯓
20 09 ♀ □ ♅

12 Thursday
00 33 ☿ ± ♃
02 52 ☽ ⊻ ♄
03 56 ☽ ∠ ♂
05 25 ♂ # ♅
06 01 ☽ △ ♃
07 08 ☽ ∥ ♃
09 55 ☿ ∠ ⯓
11 54 ☽ Q ⯓
12 37 ☽ ⊥ ♄
17 34 ☽ ⊻ ☿
19 10 ☽ ⊥ ♂

13 Friday
00 46 ☽ △ ♅
03 11 ☽ ∠ ♄
03 22 ☽ ⊻ ♀
04 24 ☽ ⊥ ☿
05 17 ☽ ⊻ ♂
10 00 ☉ ⚹ ⯓
13 07 ☽ ⊻ ♆
13 52 ☽ ⊥ ♀
16 29 ♀ # ♄
16 50 ☽ □ ⯓
17 19 ☽ ⊻ ☉
20 40 ☽ ∠ ☿
22 47 ☽ ⊥ ♆
23 07 ☽ # ⯓

14 Saturday
00 51 ☽ ♇ ♅
02 25 ☽ # ♅
03 28 ☽ ∥ ♂
03 28 ☽ ⚹ ♄
03 39 ☽ ⊥ ☉
05 40 ☽ ∠ ♀
06 07 ☽ □ ♃
10 54 ♀ ♇ ♃
13 20 ☽ ∠ ♆
15 22 ☽ ± ♅
19 14 ☽ ∠ ☉
23 02 ☽ Q ♄

15 Sunday
00 00 ☽ ⚹ ☿
01 06 ☽ ⊼ ♅
06 54 ♀ ⚹ ♂
08 11 ☽ ☌ ♂
08 14 ☽ ⚹ ♀
09 37 ☿ □ ♅
11 17 ☽ Q ♃
13 49 ☽ ⚹ ♆
17 33 ☽ △ ⯓
21 33 ☽ ⚹ ☉
22 15 ☽ Q ☿

16 Monday
00 18 ♅ St R
04 50 ☽ □ ♄
05 54 ☽ Q ♀
07 04 ☽ ⚹ ♃
09 44 ☽ Q ♆
18 34 ☽ ♇ ⯓
19 03 ☽ Q ☉

17 Tuesday
02 10 ☿ # ♄
02 55 ☽ ☍ ♅
06 21 ☿ ♇ ♃
08 20 ☽ ∠ ♃
08 37 ☽ □ ☿
09 52 ☽ ± ⯓
13 04 ☽ ⊻ ♂
15 19 ☽ □ ♀
16 22 ☽ □ ♆
16 35 ♀ ∥ ♆
20 13 ☽ ⊼ ⯓
20 38 ♂ Q ♃
23 49 ☽ ⊥ ♃

18 Wednesday
00 00 ☽ ⊥ ♂
03 19 ♀ ☌ ♆
04 18 ☽ □ ☉
08 28 ☽ △ ♄
10 18 ☽ ⊻ ♃
16 41 ☽ ∠ ♂
17 42 ☽ ∥ ♂
21 22 ☽ # ♅

19 Thursday
01 03 ☽ # ⯓
07 20 ☽ ⊼ ♅
11 23 ☽ ♇ ♄
18 11 ☽ ± ♅
20 21 ☿ ⚹ ♂
21 10 ☽ ⚹ ♂
21 15 ☽ △ ☿
21 43 ☽ △ ♆
22 25 ♀ ⚹ ⯓

20 Friday
00 54 ☿ ☌ ♆
01 42 ☽ ☍ ⯓
02 02 ☽ △ ♀
02 09 ☿ ∥ ♆
03 54 ☽ ± ♄
07 22 ♂ ⚹ ♆
10 39 ☽ ♇ ♅
14 38 ☽ △ ☉
15 04 ☽ ⊼ ♄
16 26 ☽ ☌ ♃
20 35 ☽ Q ♂
20 51 ☉ ⊻ ♄
22 14 ☽ ∥ ♃

21 Saturday
01 29 ☽ ♇ ♆
05 12 ☽ ♇ ☿
07 40 ☿ ⚹ ⯓
08 52 ☽ ♇ ♀
10 31 ☽ ∥ ☉
13 53 ☉ △ ♃
14 40 ☽ △ ♅
18 29 ☽ ± ♆
21 09 ☽ ♇ ☉
23 21 ☽ ∥ ♀

22 Sunday
00 51 ☽ ± ☿
01 52 ☽ ∥ ☿
03 47 ☽ ± ♀
05 56 ☽ ⊼ ♆
07 33 ☽ ∥ ♆
08 40 ☽ □ ♂
10 00 ☽ ⊼ ⯓
14 08 ☽ ⊼ ☿
15 50 ☽ ± ☉
16 34 ☽ ⊼ ♀
17 23 ☽ # ♄
21 34 ☽ ± ⯓

23 Monday
00 26 ☽ ☍ ♄
01 19 ☽ ⊻ ♃
04 25 ☽ ⊼ ☉
08 03 ♂ △ ⯓
12 57 ☽ ⊥ ♃
15 02 ☽ ♇ ⯓
17 16 ☽ ∥ ♄

24 Tuesday
00 26 ☽ □ ♅
03 35 ☽ # ♆
06 31 ♃ St D
06 34 ☽ ∠ ♃
16 26 ☽ ☍ ♆
17 21 ☽ # ☿
17 34 ☽ # ♀
20 32 ☽ △ ⯓
22 16 ☿ ∥ ♀
22 51 ☽ △ ♂

25 Wednesday
03 10 ☿ ☌ ♀
07 24 ☽ # ☉
10 01 ☽ ☍ ♀
10 19 ☽ ☍ ☿
11 54 ☽ ⊼ ♄
11 58 ☽ Q ♅
12 15 ☽ ⚹ ♃
14 24 ☽ # ♃
16 23 ☉ Q ⯓
20 46 ☽ ☍ ☉
22 19 ☿ ⊻ ♄
23 58 ☽ ± ♄

26 Thursday
00 11 ☿ △ ♃
05 35 ♀ ⊻ ♄
06 40 ☽ ♇ ♂
07 30 ♀ △ ♃
11 56 ☽ ⚹ ♅
12 14 ☽ Q ♃
18 11 ☽ ♇ ♄

27 Friday
01 56 ☽ ± ♂
04 31 ☽ ⊼ ♆
05 05 ♃ ⊼ ♄
08 35 ☽ □ ⯓
14 50 ☽ ⊼ ♂
15 39 ☉ ⊥ ♄
16 17 ☽ ∥ ⯓
16 39 ☽ ± ♆
18 06 ☽ ∠ ♅
20 33 ☽ ∥ ♅

28 Saturday
00 32 ☽ □ ♃
00 42 ☽ △ ♄
04 58 ☽ # ♂
05 21 ☽ ⊼ ♀
08 27 ☽ ⊼ ☿
10 55 ☽ ♇ ♆
11 21 ☿ Q ⯓
12 17 ☽ ⊥ ♅
14 42 ☽ ⊼ ☉
18 56 ☽ ± ♀
21 01 ☽ Q ⯓
22 34 ☽ ± ☿

29 Sunday
00 26 ☽ ⊻ ♅
03 58 ☽ ± ☉
14 44 ♀ Q ⯓
15 19 ☽ ♇ ♀
17 23 ☽ △ ♆
18 19 ☉ ∥ ♃
18 36 ☿ ⊥ ♄
19 46 ☽ ♇ ☿
21 20 ☽ ⚹ ⯓
23 52 ☽ ♇ ☉

30 Monday
07 35 ☽ ☍ ♂
13 20 ☽ △ ♃
13 57 ☽ □ ♄

31 Tuesday
01 10 ☽ △ ♀
03 36 ☽ ∠ ⯓
06 51 ☽ △ ☿
08 51 ☽ △ ☉
11 17 ♀ ⊥ ♄
12 53 ☽ ☌ ♅
16 13 ☿ ∥ ♃
19 29 ☽ ♇ ♃
21 37 ☽ ⊥ ⯓

All ephemeris data is given at 12.00 UT and the Moon's longitude is additionally given for 24.00 UT

NOVEMBER 1950

LONGITUDES

Date	Sidereal time h m s	Sun ☉ ° ' "	Moon ☽ ° ' "	Moon ☽ 24.00 ° ' "	Mercury ☿ ° '	Venus ♀ ° '	Mars ♂ ° '	Jupiter ♃ ° '	Saturn ♄ ° '	Uranus ♅ ° '	Neptune ♆ ° '	Pluto ♇ ° '
01	14 40 49	08 ♏ 29 42	21 ♋ 00 32	27 ♋ 06 53	08 ♏ 22	05 ♏ 21	26 ♐ 26	27 ♒ 42	28 ♍ 10	09 ♋ 22	17 ♎ 53	19 ♌ 46
02	14 44 46	09 29 45	03 ♌ 16 46	09 ♌ 30 45	09 59	06 36	27 10	27 44	28 16	09 R 21	17 55	19 46
03	14 48 42	10 29 51	15 ♌ 49 26	22 ♌ 13 20	11 37	07 52	27 55	27 46	28 23	09 20	17 57	19 47
04	14 52 39	11 29 58	28 ♌ 43 01	05 ♍ 18 56	13 13	09 07	28 40	27 48	28 29	09 19	17 59	19 47
05	14 56 35	12 30 08	12 ♍ 01 30	18 ♍ 51 02	14 50	10 22	29 ♐ 25	27 50	28 35	09 18	18 01	19 48
06	15 00 32	13 30 20	25 ♍ 47 42	02 ♎ 51 31	16 26	11 38	00 ♑ 10	27 53	28 41	09 17	18 03	19 48
07	15 04 29	14 30 33	10 ♎ 02 21	17 ♎ 19 50	18 01	12 53	00 55	27 56	28 47	09 16	18 05	19 49
08	15 08 25	15 30 49	24 ♎ 43 25	02 ♏ 12 18	19 37	14 08	01 40	27 59	28 53	09 15	18 07	19 49
09	15 12 22	16 31 06	09 ♏ 45 28	17 ♏ 21 47	21 12	15 24	02 25	28 02	28 59	09 14	18 10	19 50
10	15 16 18	17 31 26	24 ♏ 59 56	02 ♐ 38 30	22 46	16 39	03 11	28 05	29 05	09 12	18 12	19 50
11	15 20 15	18 31 47	10 ♐ 16 06	17 ♐ 51 21	24 20	17 54	03 56	28 09	29 11	09 11	18 14	19 50
12	15 24 11	19 32 10	25 ♐ 23 00	02 ♑ 49 59	25 54	19 10	04 41	28 12	29 16	09 10	18 16	19 51
13	15 28 08	20 32 34	10 ♑ 11 22	17 ♑ 26 30	27 28	20 25	05 27	28 16	29 22	09 08	18 17	19 51
14	15 32 04	21 33 00	24 ♑ 34 54	01 ♒ 36 21	29 ♏ 01	21 40	06 12	28 20	29 27	09 07	18 19	19 51
15	15 36 01	22 33 26	08 ♒ 30 47	15 ♒ 18 20	00 ♐ 34	22 56	06 58	28 25	29 33	09 05	18 21	19 51
16	15 39 58	23 33 55	21 ♒ 59 16	28 ♒ 33 57	02 07	24 11	07 43	28 29	29 38	09 04	18 23	19 51
17	15 43 54	24 34 24	05 ♓ 02 49	11 ♓ 26 23	03 40	25 27	08 29	28 34	29 44	09 02	18 25	19 51
18	15 47 51	25 34 55	17 ♓ 45 11	23 ♓ 59 44	05 12	26 42	09 15	28 39	29 49	09 00	18 27	19 52
19	15 51 47	26 35 26	00 ♈ 10 35	06 ♈ 18 15	06 44	27 57	10 00	28 44	29 54	08 59	18 29	19 52
20	15 55 44	27 36 00	12 ♈ 23 13	18 ♈ 25 54	08 16	29 ♏ 13	10 46	28 49	29 ♍ 59	08 57	18 31	19 52
21	15 59 40	28 36 34	24 ♈ 26 45	00 ♉ 26 07	09 47	00 ♐ 28	11 32	28 54	00 ♎ 04	08 55	18 33	19 52
22	16 03 37	29 ♏ 37 10	06 ♉ 24 19	12 ♉ 21 38	11 19	01 43	12 18	29 00	00 09	08 54	18 34	19 R 52
23	16 07 33	00 ♐ 37 47	18 ♉ 18 19	24 ♉ 14 36	12 50	02 59	13 04	29 06	00 14	08 52	18 36	19 52
24	16 11 30	01 38 26	00 ♊ 10 39	06 ♊ 06 39	14 21	04 14	13 50	29 12	00 19	08 50	18 38	19 52
25	16 15 27	02 39 06	12 ♊ 02 46	17 ♊ 59 10	15 51	05 30	14 36	29 18	00 24	08 48	18 40	19 51
26	16 19 23	03 39 47	23 ♊ 56 01	29 ♊ 53 30	17 22	06 45	15 22	29 24	00 28	08 46	18 41	19 51
27	16 23 20	04 40 30	05 ♋ 51 51	11 ♋ 51 16	18 52	08 00	16 08	29 30	00 33	08 44	18 43	19 51
28	16 27 16	05 41 14	17 ♋ 52 03	23 ♋ 54 31	20 21	09 16	16 54	29 37	00 38	08 42	18 45	19 51
29	16 31 13	06 42 00	29 ♋ 59 01	06 ♌ 05 58	21 51	10 31	17 41	29 44	00 42	08 40	18 46	19 51
30	16 35 09	07 ♐ 42 47	12 ♌ 15 46	18 ♌ 28 56	23 ♐ 20	11 ♐ 47	18 ♑ 27	29 ♒ 51	00 ♎ 46	08 ♋ 38	18 ♎ 48	19 ♌ 51

Moon

Date	Moon True ☊ ° '	Moon Mean ☊ ° '	Moon ☽ Latitude ° '
01	27 ♓ 28	25 ♓ 59	04 N 44
02	27 R 26	25 56	04 11
03	27 D 26	25 53	03 26
04	27 27	25 50	02 29
05	27 28	25 47	01 23
06	27 29	25 43	00 N 09
07	27 R 28	25 40	01 S 07
08	27 26	25 37	02 21
09	27 21	25 34	03 27
10	27 14	25 31	04 19
11	27 06	25 28	04 53
12	26 58	25 24	05 06
13	26 50	25 21	04 58
14	26 45	25 18	04 31
15	26 42	25 15	03 49
16	26 41	25 12	02 55
17	26 D 42	25 08	01 53
18	26 42	25 05	00 S 48
19	26 R 43	25 02	00 N 19
20	26 41	24 59	01 23
21	26 37	24 56	02 22
22	26 31	24 53	03 15
23	26 22	24 49	03 58
24	26 10	24 46	04 31
25	25 57	24 43	04 53
26	25 44	24 40	05 01
27	25 31	24 37	04 56
28	25 20	24 34	04 39
29	25 12	24 30	04 08
30	25 ♓ 06	24 ♓ 27	03 N 26

DECLINATIONS

Date	Sun ☉ ° '	Moon ☽ ° '	Mercury ☿ ° '	Venus ♀ ° '	Mars ♂ ° '	Jupiter ♃ ° '	Saturn ♄ ° '	Uranus ♅ ° '	Neptune ♆ ° '	Pluto ♇ ° '
01	14 S 20	26 N 29	13 S 57	12 S 21	24 S 42	13 S 25	02 N 35	23 N 26	05 S 34	22 N 48
02	14 40	23 30	14 35	12 48	24 43	13 24	02 33	23 26	05 35	22 48
03	14 59	19 22	15 11	13 14	24 44	13 24	02 30	23 26	05 35	22 48
04	15 17	14 15	15 47	13 40	24 45	13 23	02 28	23 26	05 36	22 48
05	15 36	08 20	16 22	14 05	24 45	13 22	02 26	23 26	05 37	22 48
06	15 54	01 N 49	16 57	14 30	24 45	13 21	02 23	23 26	05 38	22 49
07	16 12	05 S 00	17 30	14 55	24 45	13 19	02 21	23 26	05 38	22 49
08	16 30	11 46	18 03	15 19	24 45	13 18	02 19	23 26	05 39	22 49
09	16 47	18 01	18 34	15 43	24 44	13 17	02 17	23 26	05 40	22 49
10	17 04	23 13	19 05	16 07	24 43	13 16	02 15	23 26	05 41	22 49
11	17 21	26 50	19 35	16 30	24 43	13 14	02 13	23 27	05 41	22 49
12	17 37	28 25	20 04	16 53	24 41	13 13	02 11	23 27	05 42	22 50
13	17 54	28 00	20 32	17 15	24 39	13 11	02 09	23 27	05 43	22 50
14	18 09	25 39	20 58	17 37	24 37	13 10	02 06	23 27	05 44	22 50
15	18 25	21 49	21 24	17 58	24 35	13 08	02 04	23 27	05 44	22 50
16	18 40	16 56	21 49	18 19	24 32	13 06	02 03	23 27	05 45	22 50
17	18 55	11 22	22 13	18 40	24 29	13 05	02 01	23 27	05 46	22 51
18	19 10	05 S 35	22 35	19 00	24 26	13 03	01 59	23 27	05 46	22 51
19	19 24	00 N 21	22 57	19 19	24 23	13 01	01 57	23 28	05 47	22 51
20	19 38	06 10	23 17	19 38	24 19	12 59	01 55	23 28	05 48	22 52
21	19 51	11 41	23 36	19 56	24 15	12 57	01 53	23 28	05 48	22 52
22	20 05	16 43	23 54	20 14	24 11	12 55	01 51	23 28	05 49	22 52
23	20 17	21 06	24 11	20 31	24 07	12 53	01 50	23 28	05 50	22 52
24	20 30	24 37	24 26	20 47	24 02	12 50	01 48	23 28	05 50	22 53
25	20 42	27 05	24 41	21 03	23 57	12 48	01 46	23 28	05 51	22 53
26	20 54	28 19	24 54	21 19	23 51	12 46	01 45	23 29	05 51	22 53
27	21 05	28 15	25 05	21 34	23 46	12 43	01 43	23 29	05 52	22 54
28	21 16	26 51	25 15	21 48	23 40	12 41	01 41	23 29	05 53	22 54
29	21 26	24 12	25 24	22 01	23 34	12 38	01 40	23 29	05 53	22 54
30	21 S 36	20 N 25	25 S 32	22 S 14	23 S 28	12 S 36	01 N 38	23 N 29	05 S 54	22 N 55

ZODIAC SIGN ENTRIES

Date	h	m	Planets
02	05	38	☽ ♌
04	14	21	☽ ♍
06	06	40	♂ ♑
06	19	10	☽ ♎
08	20	29	☽ ♏
10	19	51	☽ ♐
12	19	25	☽ ♑
14	21	10	☽ ♒
15	03	10	☿ ♐
17	02	38	☽ ♓
19	11	39	☽ ♈
20	15	50	♄ ♎
21	03	03	♀ ♐
21	23	08	☽ ♉
22	21	03	☉ ♐
24	11	38	☽ ♊
27	00	13	☽ ♋
29	12	02	☽ ♌

LATITUDES

Date	Mercury ☿ ° '	Venus ♀ ° '	Mars ♂ ° '	Jupiter ♃ ° '	Saturn ♄ ° '	Uranus ♅ ° '	Neptune ♆ ° '	Pluto ♇ ° '
01	00 N 22	01 N 01	01 S 18	01 S 13	02 N 01	00 N 19	01 N 34	08 N 21
04	00 N 02	00 56	01 18	01 12	02 02	00 19	01 34	08 22
07	00 S 18	00 50	01 18	01 12	02 03	00 19	01 35	08 23
10	00 38	00 44	01 19	01 11	02 03	00 19	01 35	08 24
13	00 57	00 38	01 19	01 11	02 03	00 19	01 35	08 25
16	01 15	00 31	01 19	01 10	02 04	00 19	01 35	08 25
19	01 32	00 24	01 19	01 10	02 05	00 19	01 35	08 26
22	01 47	00 17	01 19	01 09	02 05	00 19	01 35	08 27
25	01 59	00 10	01 18	01 09	02 06	00 19	01 35	08 28
28	02 10	00 N 03	01 18	01 09	02 07	00 19	01 35	08 29
31	02 S 18	00 S 04	01 S 18	01 S 08	02 N 08	00 N 19	01 N 35	08 N 30

DATA

Julian Date	2433587
Delta T	+30 seconds
Ayanamsa	23° 10' 13"
Synetic vernal point	05° ♓ 56' 47"
True obliquity of ecliptic	23° 26' 54"

MOON'S PHASES, APSIDES AND POSITIONS ☽

Date	h	m	Phase	Longitude ° '	Eclipse Indicator
03	01	00	☾	10 ♌ 02	
09	23	25	●	17 ♏ 00	
16	15	06	☽	23 ♒ 42	
24	15	14	○	01 ♊ 47	

Day	h	m		
10	12	49	Perigee	
25	00	11	Apogee	
06	18	26	0S	
12	18	37	Max dec	28° S 33'
19	10	34	0N	
26	22	47	Max dec	28° N 27'

LONGITUDES

Date	Chiron ⚷ ° '	Ceres ⚳ ° '	Pallas ⚴ ° '	Juno ⚵ ° '	Vesta ⚶ ° '	Black Moon Lilith ⚸ ° '
01	19 ♐ 33	27 ♐ 33	08 ♐ 07	04 ♏ 23	15 ♉ 26	02 ♊ 41
11	20 ♐ 32	01 ♑ 06	12 ♐ 11	07 ♏ 50	12 ♉ 49	03 ♊ 48
21	21 ♐ 35	04 ♑ 47	16 ♐ 17	11 ♏ 14	10 ♉ 23	04 ♊ 55
31	22 ♐ 41	08 ♑ 33	20 ♐ 25	14 ♏ 35	08 ♉ 25	06 ♊ 02

All ephemeris data is given at 12.00 UT and the Moon's longitude is additionally given for 24.00 UT

ASPECTARIAN

h m Aspects

01 Wednesday
02 20 ☽ Q ♄
05 48 ☽ □ ♆
09 32 ☽ ⊻ ♇
13 22 ☽ ± ♃
17 12 ☉ ☌ ☿
23 21 ☽ ⊼ ♂

02 Thursday
01 11 ☽ ⊼ ♃
02 11 ☽ ✶ ♄
02 45 ☿ △ ♅
03 21 ☽ ⋕ ♂
08 40 ☉ △ ♅
11 47 ☽ ± ♂
12 30 ☽ ∥ ♅
16 34 ☽ ∥ ♇
17 07 ☽ Q ♆
18 46 ☉ ∥ ☿
19 09 ☽ □ ♀
23 41 ☽ ⊻ ♅

03 Friday
01 00 ☽ □ ☉
02 49 ☽ □ ☿
06 09 ☽ ꝗ ♂
06 44 ♂ ✶ ♃
07 20 ☽ ∠ ♄
11 05 ☽ ⊥ ♅
16 01 ☽ ✶ ♆
19 27 ☽ ☌ ♇
20 41 ♀ ∥ ♃

04 Saturday
00 23 ☽ ⊥ ♄
03 55 ☽ ∠ ♅
05 03 ♂ □ ♄
05 55 ☽ ⋕ ☿
07 43 ☽ ⋕ ☉
08 45 ☽ Q ♀
10 19 ☽ ☍ ♃
11 34 ☽ ⊻ ♄
11 54 ☽ △ ♂
13 33 ☽ Q ☉
14 22 ☽ ⋕ ♀
15 46 ☽ ⋕ ♃
15 52 ♀ △ ♅
16 07 ☿ ∠ ♄
17 13 ☽ Q ☿
19 49 ☽ ∠ ♆

05 Sunday
00 26 ☿ ∠ ♂
07 10 ☽ ✶ ♅
08 46 ☽ ✶ ♀
12 00 ☽ ⊥ ♆
12 55 ☽ ✶ ☉
17 37 ☽ ✶ ☿
22 10 ☽ ⋕ ♆
22 35 ☽ ⊻ ♆

06 Monday
01 39 ☽ ⊻ ♇
04 15 ☽ Q ♅
08 19 ♂ Q ♆
09 55 ☽ ∥ ♄
12 01 ☽ ⊥ ♇
13 34 ☽ ∠ ♀
15 35 ☽ ⊼ ♃
16 45 ☉ ∠ ♄
16 58 ☽ ☌ ♄
16 59 ☽ ∠ ☉
19 52 ☽ □ ♂
22 48 ☽ ∠ ☿

07 Tuesday
01 46 ☽ ± ♃
02 46 ☽ ⋕ ♄
03 17 ☽ ∠ ♇
06 15 ☽ ⊥ ♀
09 16 ☽ ⊥ ☉
10 43 ☽ □ ♅
13 02 ☿ ⊻ ♆
14 13 ☽ ∥ ♆
15 41 ☽ ⊥ ☿
16 48 ☽ ꝗ ♃
17 09 ☽ ⊻ ♀
19 55 ☽ ⊻ ☉

08 Wednesday
01 16 ☽ ☌ ♆
02 43 ☽ ⊻ ☿
03 22 ☽ Q ♂
04 03 ☽ ✶ ♇
06 42 ♀ ∠ ♄
15 10 ☿ □ ♇
17 15 ☽ △ ♃
17 37 ☽ ∥ ♃
18 44 ☽ ⊻ ♄
23 23 ☽ Q ♇
23 44 ☽ ✶ ♂

09 Thursday
02 11 ☽ ∥ ♀
04 22 ☽ ⊥ ♄
06 44 ☽ ∥ ☉
11 10 ☽ △ ♅
14 32 ☽ ∥ ☿
18 43 ☽ ∠ ♄
21 42 ☽ ☌ ♀
23 25 ☽ ☌ ☉

10 Friday
00 43 ☽ ∠ ♂
01 17 ☽ ⊻ ♆
03 53 ☽ □ ♇
08 06 ☽ ☌ ☿
09 55 ☽ ⋕ ♇
10 44 ☽ ⊥ ♆
10 45 ☽ ꝗ ♅
13 13 ☽ ⋕ ♅
15 36 ☽ ⊥ ♂
16 52 ☽ □ ♃
18 27 ☽ ✶ ♄
20 36 ☽ ∥ ♂

11 Saturday
00 52 ☽ ± ♅
00 54 ☽ ∠ ♆
01 30 ☽ ⊻ ♂
04 30 ☉ ⊻ ♆
09 40 ☿ ꝗ ♅
10 14 ☿ ⊥ ♆
10 18 ☽ ⊼ ♅
13 26 ☽ Q ♄
18 16 ♀ ⊻ ♆
21 19 ☽ Q ♃

12 Sunday
00 37 ☽ ✶ ♆
01 10 ☽ ⊻ ♀
02 00 ☽ ⊻ ☉
03 09 ☽ △ ♇
11 37 ☽ ⊥ ♀
12 16 ☽ ⊥ ☉
12 56 ☽ ⊻ ☿
16 33 ☽ ✶ ♃
16 53 ♂ ꝗ ♇
18 17 ☽ □ ♄
19 18 ☉ □ ♇
19 51 ☽ Q ♆
23 44 ☽ ⊥ ☿

13 Monday
01 01 ♀ □ ♇
02 35 ☉ ∠ ♂
03 16 ☽ ꝗ ♇
03 28 ☽ ∠ ♀
03 49 ☽ ☌ ♂
03 51 ☽ ∠ ☉
10 17 ☽ ☍ ♅
13 17 ♀ ∠ ♂
16 12 ☽ ∠ ☿
17 06 ☽ ∠ ♃
18 02 ☽ ± ♇
23 59 ☉ ☌ ♀

14 Tuesday
01 02 ☿ □ ♃
01 27 ☽ □ ♆
04 01 ☽ ⊼ ♇
06 30 ☽ ✶ ☉
06 37 ☽ ✶ ♀
08 12 ☽ ⊥ ♃
18 26 ☽ ⊻ ♃
19 10 ☿ ✶ ♄
19 26 ☽ ∥ ♂
20 22 ☽ △ ♄
20 30 ☽ ✶ ☿

15 Wednesday
02 43 ☽ ⋕ ♅
04 34 ☽ Q ☉
05 08 ☽ Q ♀
06 19 ☽ ⋕ ♇
09 08 ☽ ⊻ ♂
13 00 ☽ ⊼ ♅
14 01 ☽ ∥ ☿
20 04 ☽ Q ☿
20 18 ☽ ⊥ ♂
22 43 ☽ ꝗ ♄
23 35 ☽ ± ♅
23 57 ☿ ⊥ ♂

16 Thursday
04 18 ☽ ∥ ☉
05 30 ☽ △ ♆
05 59 ☽ ∥ ♀
08 09 ☽ ☍ ♇
09 40 ♀ ꝗ ♅
13 25 ☽ ∠ ♂
15 01 ☽ ± ♄
15 06 ☽ □ ☉
15 45 ☽ ꝗ ♅
15 57 ♀ ⊥ ♆
16 25 ☽ □ ♀
23 31 ☉ ꝗ ♅
23 56 ☽ ☌ ♃

17 Friday
02 04 ☽ ⊼ ♄
02 28 ☿ ± ♅
04 58 ☽ ∥ ♃
08 12 ☿ ∠ ♆
08 14 ☉ ⊥ ♆
08 58 ☽ ꝗ ♆
09 04 ☽ □ ☿
18 50 ☽ ✶ ♂
19 27 ☽ △ ♅

18 Saturday
01 53 ☽ ± ♆
04 48 ♂ ☍ ♅
11 12 ☽ ∥ ♆
13 20 ☽ ⊼ ♆
16 02 ☽ ⊼ ♇
19 08 ☽ Q ♂

19 Sunday
02 37 ☽ ⋕ ♄
03 36 ☽ ± ♇
04 24 ☽ △ ☉
05 40 ☿ ⋕ ♇
07 11 ☽ △ ♀
09 10 ☽ ⊻ ♃
11 28 ☽ ☍ ♄
18 28 ☽ ∥ ♄
20 58 ☽ ⊥ ♃
21 10 ☽ ꝗ ♇

20 Monday
02 41 ☽ △ ☿
03 52 ♀ □ ♃
05 14 ☽ □ ♅
08 35 ☽ □ ♂
10 26 ☽ ⋕ ♆
12 28 ☽ ꝗ ☉
12 38 ☉ ∥ ♀
14 51 ☽ ∠ ♃
16 02 ☽ ꝗ ♀
22 38 ☿ ⊼ ♅

21 Tuesday
00 12 ☽ ☍ ♆
01 16 ☿ ⋕ ♅
02 51 ☽ △ ♇
03 51 ♀ ✶ ♄
08 00 ☽ ± ☉
12 03 ☽ ± ♀
12 47 ☽ ꝗ ☿
15 57 ♇ St R
16 57 ☽ Q ♅
17 45 ☽ ⋕ ♃
19 45 ☉ □ ♃
21 00 ☽ ✶ ♃
21 06 ☽ ⊼ ☉
23 21 ☽ ⊼ ♄

22 Wednesday
01 29 ☽ ⊼ ♀
09 28 ☽ ± ☿
11 29 ☽ ± ♄
17 00 ☽ ✶ ♅
21 20 ☽ Q ♃
23 20 ☽ ⊼ ☿

23 Thursday
00 42 ☽ △ ♂
01 49 ☉ ✶ ♄
02 05 ☿ Q ♄
05 46 ☽ ꝗ ♄
06 58 ☽ ⋕ ☉
06 59 ☿ ∥ ♂
08 17 ☽ ⋕ ♀
09 46 ♀ ± ♅
12 36 ☽ ⊼ ♆
15 08 ☽ □ ♇
19 35 ☿ ⊻ ♂
23 12 ☽ ∠ ♅
23 19 ☽ ∥ ♇

24 Friday
00 10 ♀ ∠ ♆
00 46 ☽ ± ♆
03 26 ☽ ∥ ♅
07 37 ☽ ⋕ ♂
09 06 ☽ ꝗ ♂
09 59 ☽ □ ♃
10 30 ☽ ⋕ ☿
12 17 ☽ △ ♄
12 50 ♂ ± ♇
15 14 ☽ ☍ ☉
17 21 ☽ ⊥ ♅
19 00 ☽ ꝗ ♆
21 11 ☽ ☍ ♀

25 Saturday
00 58 ♂ ∠ ♃
03 32 ☽ Q ♇
04 33 ☽ ± ♂
05 27 ☽ ⊻ ♅
15 22 ☉ ± ♅
17 31 ☽ ⊼ ♂
20 49 ☽ ☍ ☿

26 Sunday
01 24 ☽ △ ♆
03 47 ☽ ✶ ♇
12 39 ☿ Q ♃
12 40 ☉ ∠ ♆
23 06 ☽ △ ♃

27 Monday
01 15 ☽ □ ♄
09 24 ☽ ⊼ ☉
09 40 ☿ ✶ ♆
09 58 ☽ ∠ ♇
16 48 ☽ ⊼ ♀
17 44 ☽ ☌ ♅
22 32 ☽ ± ☉

28 Tuesday
01 29 ♀ ⊼ ♅
03 53 ☿ △ ♇
03 59 ☽ ⊥ ♇
05 27 ☽ ꝗ ♃
06 12 ☽ ± ♀
09 57 ☽ ☍ ♂
13 31 ☽ Q ♄
13 45 ☽ □ ♆
15 57 ☽ ⊻ ♇
17 39 ☽ ⊼ ☿
18 07 ☽ ꝗ ☉
23 31 ☽ ± ♃

29 Wednesday
02 10 ☽ ꝗ ♀
02 52 ☽ ⋕ ☿
07 12 ☽ ± ☿
11 29 ☽ ⊼ ♃
13 25 ☽ ✶ ♄
16 41 ☽ ⋕ ♂
17 07 ☽ ∥ ♅
20 59 ☽ ∥ ♇

30 Thursday
01 21 ☽ Q ♆
01 51 ☽ ⋕ ♀
02 21 ☽ △ ☉
03 18 ☽ ꝗ ☿
04 58 ☽ ⊻ ♅
05 25 ☽ ⋕ ☉
06 39 ♂ ⋕ ♅
10 57 ☽ △ ♀
16 34 ☽ ⊥ ♅
18 49 ☽ ∠ ♄
23 20 ♂ □ ♆

DECEMBER 1950

LONGITUDES

Date	Sidereal time h m s	Sun ☉	Moon ☽	Moon ☽ 24.00	Mercury ☿	Venus ♀	Mars ♂	Jupiter ♃	Saturn ♄	Uranus ♅	Neptune ♆	Pluto ♇
01	16 39 06	08 ♐ 43 35	24 ♌ 45 57	01 ♍ 07 22	24 ♐ 48	13 ♐ 02	19 ♑ 13	29 ♒ 58	00 ♎ 51	08 ♋ 36	18 ♎ 50	19 ♌ 50
02	16 43 02	09 44 25	07 ♍ 33 42	14 ♍ 05 29	26 16	14 17	20 00	00 ♓ 05	00 55	08 R 34	18 51	19 R 50
03	16 46 59	10 45 16	20 ♍ 43 13	27 ♍ 27 20	27 43	15 33	20 46	00 12	00 59	08 31	18 53	19 50
04	16 50 56	11 46 09	04 ♎ 18 13	11 ♎ 16 06	29 ♐ 10	16 48	21 32	00 20	01 03	08 29	18 54	19 49
05	16 54 52	12 47 03	18 ♎ 21 05	25 ♎ 33 05	00 ♑ 36	18 04	22 19	00 27	01 07	08 27	18 56	19 49
06	16 58 49	13 47 59	02 ♏ 51 51	10 ♏ 16 50	02 01	19 19	23 05	00 35	01 11	08 25	18 57	19 49
07	17 02 45	14 48 55	17 ♏ 47 17	25 ♏ 22 15	03 24	20 35	23 52	00 43	01 14	08 22	18 59	19 48
08	17 06 42	15 49 53	03 ♐ 00 30	10 ♐ 40 41	04 46	21 50	24 39	00 52	01 18	08 20	19 00	19 48
09	17 10 38	16 50 52	18 ♐ 21 20	26 ♐ 00 55	06 07	23 05	25 25	01 00	01 22	08 18	19 02	19 47
10	17 14 35	17 51 52	03 ♑ 37 57	11 ♑ 11 06	07 26	24 21	26 12	01 08	01 25	08 15	19 03	19 47
11	17 18 31	18 52 53	18 ♑ 39 07	26 ♑ 01 02	08 43	25 36	26 59	01 17	01 28	08 13	19 04	19 46
12	17 22 28	19 53 54	03 ♒ 16 05	10 ♒ 23 47	09 57	26 52	27 45	01 26	01 32	08 11	19 06	19 46
13	17 26 25	20 54 56	17 ♒ 23 51	24 ♒ 16 15	11 09	28 07	28 32	01 34	01 35	08 08	19 07	19 45
14	17 30 21	21 55 59	01 ♓ 01 08	07 ♓ 38 48	12 17	29 ♐ 23	29 ♑ 19	01 43	01 38	08 06	19 08	19 44
15	17 34 18	22 57 01	14 ♓ 09 42	20 ♓ 34 21	13 21	00 ♑ 38	00 ♒ 06	01 53	01 41	08 03	19 09	19 44
16	17 38 14	23 58 05	26 ♓ 53 21	03 ♈ 07 19	14 21	01 53	00 53	02 02	01 44	08 01	19 11	19 43
17	17 42 11	24 59 08	09 ♈ 16 54	15 ♈ 22 46	15 15	03 09	01 40	02 11	01 46	07 58	19 12	19 42
18	17 46 07	26 00 12	21 ♈ 25 30	27 ♈ 25 45	16 04	04 24	02 27	02 21	01 49	07 56	19 13	19 42
19	17 50 04	27 01 16	03 ♉ 24 03	09 ♉ 20 56	16 46	05 40	03 14	02 31	01 52	07 53	19 14	19 41
20	17 54 00	28 02 21	15 ♉ 16 53	21 ♉ 12 18	17 21	06 55	04 00	02 40	01 54	07 51	19 15	19 40
21	17 57 57	29 ♐ 03 26	27 ♉ 07 35	03 ♊ 03 02	17 47	08 10	04 47	02 50	01 57	07 48	19 16	19 39
22	18 01 54	00 ♑ 04 31	08 ♊ 58 56	14 ♊ 55 32	18 04	09 26	05 34	03 00	01 59	07 46	19 17	19 38
23	18 05 50	01 05 37	20 ♊ 53 07	26 ♊ 51 30	18 11	10 41	06 21	03 10	02 01	07 43	19 18	19 38
24	18 09 47	02 06 44	02 ♋ 51 11	08 ♋ 52 10	18 R 06	11 57	07 09	03 21	02 03	07 40	19 19	19 37
25	18 13 43	03 07 50	14 ♋ 54 34	20 ♋ 58 30	17 50	13 12	07 56	03 31	02 05	07 38	19 20	19 36
26	18 17 40	04 08 57	27 ♋ 04 07	03 ♌ 11 33	17 23	14 27	08 43	03 42	02 07	07 35	19 21	19 35
27	18 21 36	05 10 04	09 ♌ 20 59	15 ♌ 32 39	16 43	15 43	09 30	03 52	02 09	07 33	19 22	19 34
28	18 25 33	06 11 12	21 ♌ 46 47	28 ♌ 03 40	15 53	16 58	10 17	04 03	02 10	07 30	19 22	19 33
29	18 29 29	07 12 20	04 ♍ 23 40	10 ♍ 47 07	14 52	18 13	11 04	04 14	02 12	07 28	19 23	19 32
30	18 33 26	08 13 29	17 ♍ 14 24	23 ♍ 45 57	13 43	19 29	11 51	04 25	02 13	07 25	19 24	19 31
31	18 37 23	09 ♑ 14 38	00 ♎ 22 09	07 ♎ 03 23	12 ♑ 27	20 ♑ 44	12 ♒ 38	04 ♓ 36	02 ♎ 15	07 ♋ 22	19 ♎ 25	19 ♌ 30

Date	Moon True ☊	Moon Mean ☊	Moon ☽ Latitude
01	25 ♓ 03	24 ♓ 24	02 N 33
02	25 R 03	24 21	01 31
03	25 D 03	24 18	00 N 23
04	25 R 02	24 14	00 S 49
05	25 01	24 11	02 00
06	24 56	24 08	03 06
07	24 49	24 05	04 00
08	24 40	24 02	04 39
09	24 28	23 59	04 58
10	24 17	23 55	04 56
11	24 06	23 52	04 33
12	23 57	23 49	03 53
13	23 52	23 46	02 59
14	23 49	23 43	01 57
15	23 D 47	23 40	00 S 51
16	23 R 47	23 36	00 N 16
17	23 47	23 33	01 21
18	23 45	23 30	02 20
19	23 41	23 27	03 13
20	23 33	23 24	03 56
21	23 23	23 20	04 29
22	23 10	23 17	04 50
23	22 56	23 14	04 59
24	22 41	23 11	04 55
25	22 27	23 08	04 37
26	22 14	23 05	04 07
27	22 05	23 01	03 24
28	21 58	22 58	02 32
29	21 54	22 55	01 31
30	21 53	22 52	00 25
31	21 ♓ 53	22 ♓ 49	00 S 45

DECLINATIONS

Date	Sun ☉	Moon ☽	Mercury ☿	Venus ♀	Mars ♂	Jupiter ♃	Saturn ♄	Uranus ♅	Neptune ♆	Pluto ♇
01	21 S 46	15 N 40	25 S 38	22 S 26	23 S 21	12 S 33	01 N 37	23 N 29	05 S 54	22 N 55
02	21 55	10 09	25 43	22 38	23 14	12 30	01 35	23 29	05 55	22 56
03	22 04	04 N 02	25 46	22 49	23 07	12 28	01 34	23 30	05 55	22 56
04	22 12	02 S 27	25 48	22 59	23 00	12 25	01 33	23 30	05 56	22 56
05	22 20	09 03	25 49	23 08	22 52	12 22	01 31	23 30	05 57	22 57
06	22 28	15 22	25 48	23 17	22 44	12 19	01 30	23 30	05 57	22 57
07	22 35	20 59	25 45	23 25	22 36	12 16	01 29	23 30	05 58	22 58
08	22 42	25 20	25 42	23 32	22 27	12 13	01 28	23 30	05 58	22 58
09	22 48	27 53	25 36	23 39	22 19	12 10	01 26	23 31	05 59	22 58
10	22 54	28 20	25 30	23 45	22 10	12 07	01 25	23 31	05 59	22 59
11	22 59	26 40	25 22	23 50	22 01	12 04	01 24	23 31	05 59	22 59
12	23 04	23 12	25 13	23 55	21 51	12 00	01 23	23 31	06 00	23 00
13	23 08	18 28	25 02	23 58	21 42	11 57	01 22	23 31	06 00	23 00
14	23 12	12 56	24 50	24 01	21 32	11 54	01 21	23 31	06 01	23 01
15	23 16	07 01	24 37	24 04	21 22	11 50	01 20	23 32	06 01	23 01
16	23 19	00 S 59	24 23	24 05	21 11	11 47	01 19	23 32	06 02	23 02
17	23 21	04 N 55	24 08	24 06	21 01	11 43	01 18	23 32	06 02	23 02
18	23 23	10 32	23 52	24 06	20 50	11 40	01 18	23 32	06 02	23 03
19	23 25	15 40	23 36	24 05	20 39	11 36	01 17	23 32	06 03	23 03
20	23 26	20 11	23 19	24 04	20 27	11 33	01 16	23 32	06 03	23 04
21	23 27	23 53	23 02	24 01	20 16	11 29	01 15	23 33	06 03	23 04
22	23 27	26 35	22 45	23 58	20 04	11 25	01 15	23 33	06 04	23 05
23	23 27	28 06	22 28	23 55	19 52	11 21	01 14	23 33	06 04	23 05
24	23 26	28 22	22 11	23 50	19 40	11 17	01 14	23 33	06 04	23 06
25	23 25	27 12	21 55	23 45	19 27	11 14	01 13	23 33	06 05	23 06
26	23 23	24 47	21 40	23 39	19 15	11 10	01 12	23 33	06 05	23 07
27	23 21	21 12	21 26	23 32	19 02	11 06	01 12	23 34	06 05	23 07
28	23 18	16 39	21 12	23 25	18 49	11 02	01 12	23 34	06 05	23 08
29	23 15	11 19	21 00	23 17	18 36	10 58	01 11	23 34	06 06	23 09
30	23 12	05 25	20 48	23 08	18 22	10 54	01 11	23 34	06 06	23 09
31	23 S 08	00 S 50	20 S 38	22 S 58	18 S 09	10 S 49	01 N 11	23 N 34	06 S 06	23 N 10

ZODIAC SIGN ENTRIES

Date	h	m	Planets
01	19	57	♃ ♓
01	21	53	☽ ♍
04	04	29	☽ ♎
05	01	57	☿ ♑
06	07	19	☽ ♏
08	07	17	☽ ♐
10	06	16	☽ ♑
12	06	34	☽ ♒
14	10	10	☽ ♓
14	23	54	♀ ♑
15	08	59	♂ ♒
16	17	58	☽ ♈
19	05	10	☽ ♉
21	17	49	☽ ♊
22	10	13	☉ ♑
24	06	18	☽ ♋
26	17	45	☽ ♌
29	03	41	☽ ♍
31	11	20	☽ ♎

LATITUDES

Date	Mercury ☿	Venus ♀	Mars ♂	Jupiter ♃	Saturn ♄	Uranus ♅	Neptune ♆	Pluto ♇
01	02 S 18	00 S 04	01 S 18	01 S 08	02 N 08	00 N 19	01 N 35	08 N 30
04	02 22	00 11	01 17	01 08	02 08	00 19	01 36	08 31
07	02 21	00 18	01 17	01 07	02 09	00 19	01 36	08 32
10	02 15	00 25	01 16	01 07	02 10	00 19	01 36	08 33
13	02 03	00 32	01 16	01 06	02 11	00 19	01 36	08 34
16	01 43	00 39	01 15	01 06	02 11	00 19	01 36	08 35
19	01 13	00 46	01 14	01 06	02 12	00 20	01 36	08 35
22	00 S 31	00 52	01 13	01 05	02 13	00 20	01 36	08 36
25	00 N 20	00 58	01 12	01 05	02 14	00 20	01 37	08 37
28	01 19	01 03	01 11	01 04	02 15	00 20	01 37	08 38
31	02 N 14	01 S 08	01 S 10	01 S 04	02 N 16	00 N 20	01 N 37	08 N 38

DATA

Julian Date	2433617
Delta T	+30 seconds
Ayanamsa	23° 10' 18"
Synetic vernal point	05° ♓ 56' 42"
True obliquity of ecliptic	23° 26' 53"

MOON'S PHASES, APSIDES AND POSITIONS ☽

Date	h	m	Phase	Longitude	Eclipse Indicator
02	16	22	☾	09 ♍ 55	
09	09	29	●	16 ♐ 44	
16	05	56	☽	23 ♓ 43	
24	10	23	○	02 ♋ 03	

Day	h	m		
09	01	23	Perigee	
22	01	23	Apogee	
04	03	01	0S	
10	04	50	Max dec	28° S 26'
16	14	57	0N	
24	03	57	Max dec	28° N 24'
31	08	51	0S	

LONGITUDES

Date	Chiron ⚷	Ceres ⚳	Pallas ⚴	Juno ⚵	Vesta ⚶	Black Moon Lilith ☾
01	22 ♐ 41	08 ♑ 33	20 ♐ 25	14 ♏ 35	08 ♉ 25	06 ♊ 02
11	23 ♐ 48	12 ♑ 24	24 ♐ 32	17 ♏ 51	07 ♉ 05	07 ♊ 10
21	24 ♐ 56	16 ♑ 28	28 ♐ 39	21 ♏ 01	06 ♉ 29	08 ♊ 17
31	26 ♐ 03	20 ♑ 15	02 ♑ 44	24 ♏ 10	06 ♉ 38	09 ♊ 24

ASPECTARIAN

01 Friday
00 38 ☽ ⚹ ♆
00 43 ☽ ⊼ ♂
02 36 ☽ ☌ ♇
08 07 ♀ Q ♄
09 00 ☉ ⊼ ♅
09 47 ☽ ∠ ♅
12 05 ☽ △ ☿
12 09 ☽ ⊥ ♄
12 55 ☽ ± ♂
21 14 ♀ ⊥ ♂
21 55 ☽ ☍ ♃
23 32 ☽ ⊻ ♄

02 Saturday
02 02 ☽ # ♃
05 06 ☽ ∠ ♆
06 56 ☽ ⚼ ♂
07 04 ♂ ⊼ ♇
13 50 ☽ ⚹ ♅
16 22 ☽ □ ☉
21 46 ☽ ⊥ ♆

03 Sunday
01 40 ☽ □ ♀
04 46 ☽ # ♆
08 41 ☽ ⊻ ♆
10 24 ☽ ⊻ ♇
11 39 ☽ Q ♅
12 05 ☽ △ ♂
21 07 ☽ ⊥ ♇
21 15 ☽ ∥ ♄

04 Monday
01 57 ☽ □ ☿
03 26 ☽ Q ☉
05 00 ☽ ⊼ ♃
05 56 ♀ # ♇
06 17 ☽ ☌ ♄
08 41 ☽ # ♄
12 54 ☽ ∠ ♇
12 58 ☽ Q ♀
13 15 ♀ ∥ ♂
15 33 ☽ ± ♃
19 13 ☽ □ ♅
21 55 ♂ # ♇

05 Tuesday
00 39 ☽ ∥ ♆
01 51 ☽ ⚹ ☉
07 05 ☽ ⚼ ♃
09 26 ☿ ⚹ ♃
11 28 ☽ ⚹ ♀
12 27 ☽ Q ☿
12 58 ☽ ☌ ♆
14 27 ☽ ⚹ ♇
17 45 ☿ Q ♆
19 00 ☽ □ ♂
20 17 ☉ Q ♄
20 25 ♀ Q ♃
21 09 ☿ □ ♄

06 Wednesday
00 17 ☽ ∥ ♃
04 51 ☽ ∠ ☉
04 54 ♀ ⚹ ♆
08 15 ☽ △ ♃
09 14 ☽ ⊻ ♄
10 17 ☽ Q ♇
10 28 ☽ ⚹ ☿
14 35 ☽ ∠ ♀
19 02 ☽ ⊥ ♄
20 35 ☽ ⊥ ☉
20 58 ☽ △ ♅
21 17 ♀ △ ♇

07 Thursday
02 02 ☽ Q ♂
06 25 ☽ ⊥ ♀
06 55 ☽ ⊻ ☉
09 31 ☽ ∠ ♄
13 05 ☽ ∠ ☿
13 28 ☉ ∥ ♂
13 54 ☽ ⊻ ♆
15 12 ☽ □ ♇
16 50 ☽ ⊻ ♀
19 46 ☽ ∥ ♂
20 07 ☽ ∥ ☉
20 50 ☽ ⚼ ♅
21 55 ☽ # ♇
22 09 ☽ ⚹ ♂
23 24 ☽ ⊥ ♆

08 Friday
00 45 ☽ ∥ ♀
00 52 ☽ # ♅
04 42 ☽ ⊥ ☿
05 02 ♀ # ♅
08 36 ☽ □ ♃
09 19 ☽ ⚹ ♄
10 57 ☽ ± ♅
12 20 ☿ ⚼ ♇
13 34 ☽ ∠ ♆
14 28 ☽ ∥ ☿
15 02 ☽ ⊻ ☿
20 02 ♂ ⊥ ♃
20 19 ☽ ⊼ ♅
22 56 ☽ ∠ ♂

09 Saturday
04 10 ☽ Q ♄
09 29 ☽ ☌ ☉
13 01 ☽ Q ♃
13 03 ☽ ⚹ ♆
13 45 ☽ ⊥ ♂
14 14 ☽ △ ♇
20 05 ☽ ☌ ♀
23 40 ☽ ⊻ ♂

10 Sunday
07 55 ☽ Q ♆
08 01 ☽ ⚹ ♃
08 29 ☽ □ ♄
13 49 ☽ ⚼ ♇
18 36 ☽ ☌ ☿
19 19 ☽ ☍ ♅

11 Monday
02 50 ☿ ☍ ♅
04 08 ☽ ± ♇
08 08 ☽ ∠ ♃
12 24 ☽ ⊻ ☉
12 41 ☽ □ ♆
13 48 ☽ ⊼ ♇
14 17 ☉ # ♇
16 35 ☉ ⚹ ♆
22 51 ☽ ∥ ☿
22 54 ☽ ⊥ ☉
22 54 ☽ ⊥ ♃
23 00 ☉ Q ♃

12 Tuesday
00 23 ☽ ⊻ ♀
02 21 ☽ ☌ ♂
07 55 ☽ ∥ ♀
08 44 ☉ △ ♇
08 54 ☽ ⊻ ♃
09 05 ☽ △ ♄
10 12 ☽ # ♅
11 16 ☽ ⊥ ♀
12 49 ☽ ∥ ☉
13 12 ☽ # ♇
14 56 ☽ ∠ ☉
19 42 ☽ ∥ ♂
20 13 ☽ ⊼ ♅

13 Wednesday
00 18 ☽ ⊻ ☿
03 55 ☽ ∠ ♀
06 24 ☽ ± ♅
10 35 ☽ ⚼ ♄
11 31 ☽ ⊥ ☿
13 09 ♃ ⊼ ♄
14 59 ☽ △ ♆
16 05 ☽ ☍ ♇
18 36 ☽ ⚹ ☉
21 58 ☽ ⚼ ♅

14 Thursday
02 22 ☽ ± ♄
04 43 ☽ ∠ ☿
08 45 ☽ ⚹ ♀
08 46 ☽ ⊻ ♂
09 01 ♀ ⊻ ♂
13 06 ☽ ⊼ ♄
13 17 ☽ ☌ ♃
16 21 ☽ ∥ ♃
17 37 ☽ ⚼ ♆
17 41 ☽ Q ☉
20 15 ☽ ⊥ ♂

15 Friday
00 47 ☽ △ ♅
08 52 ☽ Q ♀
10 08 ☽ ± ♆
10 22 ☽ ⚹ ☿
13 51 ☽ ∠ ♂
15 57 ☽ ∥ ♆
20 46 ☿ ± ♇
21 21 ☽ ⊼ ♆
22 08 ♀ Q ♆
22 24 ☽ ⊼ ♇

16 Saturday
05 56 ☽ □ ☉
08 46 ♀ □ ♄
09 46 ☽ ± ♇
10 39 ☽ # ♄
10 53 ☽ Q ☿
15 05 ♀ ⚹ ♃
20 11 ☽ ⚹ ♂
21 16 ☽ ∥ ♄
21 20 ☽ ☍ ♄
22 01 ☽ ⊻ ♃
22 42 ☽ □ ♀

17 Sunday
03 05 ☽ ⚼ ♇
09 27 ☽ □ ♅
09 50 ☽ ⊥ ♃
14 50 ☿ ∥ ♀
15 41 ♂ △ ♄
16 39 ☽ # ♆
21 12 ☽ Q ♂

18 Monday
00 38 ☽ □ ☿
03 47 ☽ ∠ ♃
07 36 ☽ ☍ ♆
08 21 ♂ ⊻ ♃
08 33 ☽ △ ♇
17 03 ☽ # ♃
17 28 ♀ ⚼ ♇
20 58 ☽ Q ♅
22 00 ☽ △ ☉

19 Tuesday
08 53 ☽ ⊼ ♄
10 11 ☽ ⚹ ♃
11 37 ☽ □ ♂
16 55 ☿ # ♅
17 05 ☽ △ ♀
21 01 ☽ ⚹ ♅
21 02 ☽ ± ♄

20 Wednesday
02 31 ☉ ∥ ☿
07 02 ☽ ⚼ ☉
08 48 ☉ ⊥ ♂
10 45 ☽ Q ♃
13 31 ☽ # ♂
15 18 ☽ ⚼ ♄
16 22 ☽ △ ☿
20 03 ☽ ⊼ ♆
20 52 ☽ □ ♇

21 Thursday
02 59 ☽ ± ☉
03 02 ☽ ⚼ ♀
03 16 ☽ ∠ ♅
05 09 ♀ ☍ ♅
06 10 ☽ ∥ ♇
06 22 ☽ # ☿
08 14 ☽ ± ♆
08 48 ☽ # ☉
08 49 ☿ # ♇
09 30 ☽ ∥ ♅
13 01 ☽ # ♀
16 17 ☽ ⊼ ☉
19 40 ☿ ∠ ♃
21 26 ☽ ⊥ ♅
21 47 ☽ △ ♄
23 26 ☽ ± ♀
23 44 ☽ □ ♃
23 47 ☽ ⚼ ☿

22 Friday
02 29 ☽ ⚼ ♆
04 37 ☽ △ ♂
09 17 ☽ Q ♇
09 32 ☽ ⊻ ♅
13 01 ☽ ⊼ ♀
18 19 ☽ ± ☿

23 Saturday
06 32 ☽ ⊼ ☿
08 49 ☽ △ ♆
09 28 ☽ ⚹ ♇
12 02 ☿ ∠ ♃
13 01 ☽ ⚼ ♂
14 47 ☿ St R
16 58 ☉ Q ♆

24 Sunday
08 20 ☽ ± ♂
10 23 ☽ ☍ ☉
10 24 ☽ □ ♄
10 30 ☉ □ ♄
13 00 ☽ △ ♃
15 31 ☽ ∠ ♇
21 09 ☽ ⊼ ♂
21 35 ☽ ☌ ♅

25 Monday
03 26 ♂ ⊼ ♅
08 13 ☽ ☍ ♀
09 24 ☽ ⊥ ♇
17 38 ☽ ☍ ☿
19 15 ☽ ⚼ ♃
19 32 ♀ ± ♇
20 46 ☽ □ ♆
21 16 ☽ ⊻ ♇
22 16 ☽ Q ♄
23 04 ☉ ⚹ ♃

26 Tuesday
13 15 ☽ ± ♃
20 42 ☽ # ♀
21 05 ☽ ∥ ♅
21 55 ☽ ⚹ ♄
22 03 ☉ ⚼ ♇
22 23 ☽ # ☉

27 Wednesday
00 04 ☽ ∥ ♇
01 10 ☽ ⊼ ♃
03 07 ☽ ⊼ ☉
07 58 ♀ # ♅
08 08 ☽ Q ♆
08 30 ☽ ⊻ ♅
10 37 ☽ # ☿
12 18 ☽ ☍ ♂
15 51 ☽ ± ☉
20 06 ☽ ⊥ ♅
23 51 ☿ ☌ ♀

28 Thursday
00 37 ☽ # ♂
01 25 ☽ ⊼ ☿
01 42 ☽ ⊼ ♀
03 07 ☽ ∠ ♄
07 23 ☽ ⚹ ♆
07 44 ☽ ☌ ♇
10 46 ☽ ⚼ ☉
12 11 ☽ ± ☿
13 23 ☽ ∠ ♅
14 32 ☽ ± ♀
20 25 ☽ ⊥ ♄

29 Friday
04 06 ☽ ⚼ ☿
07 50 ☽ ⊻ ♄
09 33 ☽ ⚼ ♀
11 41 ☽ ☍ ♃
11 59 ☽ ∠ ♆
13 32 ☽ # ♃
17 44 ☉ ☍ ♅
17 45 ☽ ⚹ ♅
17 45 ☽ △ ☉
19 58 ☉ ∥ ♀

30 Saturday
01 21 ☽ ⊼ ♂
04 52 ☽ ⊥ ♆
06 01 ☽ △ ☿
09 17 ♀ # ♇
09 19 ☽ # ♆
10 04 ♃ ⚼ ♆
10 31 ♀ □ ♆
10 35 ♀ ∠ ♃
12 45 ♀ ⊼ ♇
13 12 ☽ ± ♂
15 51 ☿ ± ♇
16 00 ☽ ⊻ ♆
16 00 ☽ Q ♅
16 12 ☽ ⊻ ♇
16 34 ☽ △ ♀

31 Sunday
01 11 ☉ # ♇
03 11 ☽ ⊥ ♇
04 22 ☽ ∥ ♄
06 45 ☽ ⚼ ♂
09 47 ☿ ⊻ ♂
13 19 ☽ # ♄
15 23 ☽ ☌ ♄
19 25 ☽ ∠ ♇
19 43 ☽ ⊼ ♃

All ephemeris data is given at 12.00 UT and the Moon's longitude is additionally given for 24.00 UT

JANUARY 1951

Date	Sidereal time h m s	Sun ☉	Moon ☽	Moon ☽ 24.00	Mercury ☿	Venus ♀	Mars ♂	Jupiter ♃	Saturn ♄	Uranus ♅	Neptune ♆	Pluto ♇
01	18 41 19	10 ♑ 15 47	13 ♎ 50 01	20 ♎ 42 21	11 ♑ 07	21 ♑ 59	13 ♒ 26	04 ♓ 47	02 ♎ 16	07 ♋ 20	19 ♎ 25	19 ♌ 29
02	18 45 16	11 16 57	27 ♎ 40 33	04 ♏ 44 42	09 R 45	23 15	14 13	04 58	02 17	07 R 17	19 26	19 R 28
03	18 49 12	12 18 07	11 ♏ 54 45	19 ♏ 10 25	08 25	24 30	15 00	05 10	02 18	07 15	19 26	19 27
04	18 53 09	13 19 17	26 ♏ 31 17	03 ♐ 56 43	07 08	25 45	15 47	05 21	02 19	07 12	19 27	19 26
05	18 57 05	14 20 28	11 ♐ 25 50	18 ♐ 57 39	05 57	27 01	16 34	05 33	02 20	07 09	19 28	19 25
06	19 01 02	15 21 39	26 ♐ 30 59	04 ♑ 04 33	04 54	28 16	17 22	05 44	02 20	07 07	19 28	19 23
07	19 04 58	16 22 50	11 ♑ 37 02	19 ♑ 07 09	04 00	29 ♑ 31	18 09	05 56	02 21	07 04	19 29	19 22
08	19 08 55	17 24 01	26 ♑ 33 39	03 ♒ 55 26	03 15	00 ♒ 47	18 56	06 08	02 21	07 02	19 29	19 21
09	19 12 52	18 25 11	11 ♒ 11 37	18 ♒ 21 28	02 41	02 02	19 43	06 20	02 22	06 59	19 30	19 20
10	19 16 48	19 26 21	25 ♒ 24 28	02 ♓ 20 20	02 16	03 17	20 31	06 32	02 22	06 57	19 30	19 19
11	19 20 45	20 27 31	09 ♓ 08 59	15 ♓ 50 29	02 01	04 32	21 18	06 44	02 22	06 54	19 31	19 18
12	19 24 41	21 28 40	22 ♓ 25 06	28 ♓ 53 12	01 55	05 48	22 05	06 56	02 R 22	06 52	19 31	19 16
13	19 28 38	22 29 48	05 ♈ 15 16	11 ♈ 31 49	01 D 58	07 03	22 52	07 09	02 22	06 49	19 31	19 15
14	19 32 34	23 30 55	17 ♈ 43 30	23 ♈ 50 56	02 09	08 18	23 40	07 21	02 22	06 47	19 31	19 14
15	19 36 31	24 32 02	29 ♈ 54 46	05 ♉ 55 40	02 28	09 33	24 27	07 34	02 21	06 44	19 32	19 13
16	19 40 27	25 33 09	11 ♉ 54 16	17 ♉ 51 11	02 53	10 48	25 14	07 46	02 21	06 42	19 32	19 11
17	19 44 24	26 34 14	23 ♉ 47 01	29 ♉ 42 19	03 24	12 04	26 02	07 59	02 20	06 40	19 32	19 10
18	19 48 21	27 35 19	05 ♊ 37 35	11 ♊ 33 18	04 01	13 19	26 49	08 11	02 20	06 37	19 32	19 09
19	19 52 17	28 36 23	17 ♊ 29 52	23 ♊ 27 40	04 43	14 34	27 36	08 24	02 19	06 35	19 32	19 07
20	19 56 14	29 ♑ 37 27	29 ♊ 26 58	05 ♋ 28 05	05 30	15 49	28 23	08 37	02 18	06 33	19 32	19 06
21	20 00 10	00 ♒ 38 29	11 ♋ 31 11	17 ♋ 36 28	06 20	17 04	29 11	08 50	02 17	06 30	19 R 32	19 05
22	20 04 07	01 39 31	23 ♋ 44 03	29 ♋ 54 02	07 15	18 19	29 ♒ 58	09 03	02 16	06 28	19 32	19 03
23	20 08 03	02 40 32	06 ♌ 06 30	12 ♌ 21 30	08 12	19 34	00 ♓ 45	09 16	02 15	06 26	19 32	19 02
24	20 12 00	03 41 32	18 ♌ 39 05	24 ♌ 59 18	09 13	20 49	01 32	09 29	02 14	06 23	19 32	19 00
25	20 15 56	04 42 32	01 ♍ 22 13	07 ♍ 47 53	10 16	22 04	02 20	09 42	02 12	06 21	19 32	18 59
26	20 19 53	05 43 30	14 ♍ 16 24	20 ♍ 47 53	11 22	23 19	03 07	09 55	02 11	06 19	19 32	18 58
27	20 23 50	06 44 28	27 ♍ 22 27	04 ♎ 00 15	12 30	24 34	03 54	10 09	02 09	06 17	19 31	18 56
28	20 27 46	07 45 26	10 ♎ 41 26	17 ♎ 26 10	13 40	25 49	04 41	10 22	02 07	06 15	19 31	18 55
29	20 31 43	08 46 23	24 ♎ 14 36	01 ♏ 06 50	14 52	27 04	05 28	10 35	02 06	06 13	19 31	18 53
30	20 35 39	09 47 19	08 ♏ 02 58	15 ♏ 03 00	16 06	28 19	06 15	10 49	02 04	06 11	19 31	18 52
31	20 39 36	10 ♒ 48 14	22 ♏ 06 54	29 ♏ 14 29	17 ♑ 22	29 ♒ 34	07 ♓ 03	11 ♓ 02	02 ♎ 02	06 ♋ 09	19 ♎ 30	18 ♌ 51

Date	Moon True ☊	Moon Mean ☊	Moon ☽ Latitude
01	21 ♓ 53	22 ♓ 46	01 S 54
02	21 R 52	22 42	02 58
03	21 49	22 39	03 53
04	21 43	22 36	04 34
05	21 35	22 33	04 58
06	21 25	22 30	05 01
07	21 14	22 26	04 44
08	21 05	22 23	04 07
09	20 57	22 20	03 14
10	20 51	22 17	02 11
11	20 48	22 14	01 S 02
12	20 D 48	22 11	00 N 09
13	20 48	22 07	01 17
14	20 49	22 04	02 19
15	20 R 49	22 01	03 13
16	20 48	21 58	03 59
17	20 44	21 55	04 33
18	20 37	21 52	04 55
19	20 29	21 48	05 05
20	20 19	21 45	05 02
21	20 09	21 42	04 45
22	20 00	21 39	04 15
23	19 51	21 36	03 33
24	19 45	21 32	02 39
25	19 41	21 29	01 37
26	19 39	21 26	00 N 29
27	19 D 39	21 23	00 S 42
28	19 41	21 20	01 52
29	19 42	21 17	02 56
30	19 43	21 13	03 52
31	19 ♓ 42	21 ♓ 10	04 S 35

DECLINATIONS

Date	Sun ☉	Moon ☽	Mercury ☿	Venus ♀	Mars ♂	Jupiter ♃	Saturn ♄	Uranus ♅	Neptune ♆	Pluto ♇
01	23 S 03	07 S 12	20 S 29	22 S 48	17 S 55	10 S 45	01 N 11	23 N 34	06 S 06	23 N 10
02	22 58	13 25	20 22	22 37	17 41	10 41	01 11	23 34	06 07	23 11
03	22 53	19 06	20 16	22 26	17 27	10 37	01 10	23 35	06 07	23 11
04	22 47	23 49	20 11	22 13	17 12	10 32	01 10	23 35	06 07	23 12
05	22 41	27 04	20 08	22 00	16 58	10 28	01 10	23 35	06 07	23 13
06	22 34	28 25	20 06	21 47	16 43	10 24	01 10	23 35	06 07	23 13
07	22 27	27 39	20 06	21 33	16 28	10 19	01 10	23 35	06 07	23 14
08	22 19	24 53	20 08	21 18	16 13	10 15	01 10	23 35	06 07	23 14
09	22 11	20 32	20 11	21 02	15 58	10 10	01 10	23 35	06 08	23 15
10	22 02	15 07	20 15	20 46	15 42	10 06	01 11	23 36	06 08	23 15
11	21 53	09 06	20 20	20 29	15 27	10 01	01 11	23 36	06 08	23 16
12	21 44	02 S 53	20 27	20 12	15 11	09 57	01 11	23 36	06 08	23 17
13	21 34	03 N 16	20 34	19 54	14 55	09 52	01 11	23 36	06 08	23 17
14	21 24	09 06	20 42	19 35	14 39	09 47	01 12	23 36	06 08	23 18
15	21 13	14 28	20 50	19 16	14 23	09 42	01 12	23 36	06 08	23 18
16	21 02	19 00	20 59	18 56	14 07	09 38	01 13	23 36	06 08	23 19
17	20 51	23 05	21 08	18 36	13 50	09 33	01 13	23 37	06 08	23 19
18	20 39	26 06	21 16	18 16	13 34	09 28	01 13	23 37	06 08	23 20
19	20 27	27 55	21 25	17 54	13 17	09 23	01 14	23 37	06 08	23 21
20	20 14	28 29	21 33	17 33	13 00	09 18	01 15	23 37	06 08	23 21
21	20 01	27 41	21 41	17 10	12 44	09 14	01 15	23 37	06 08	23 22
22	19 48	25 33	21 48	16 48	12 26	09 09	01 16	23 37	06 08	23 22
23	19 34	22 11	21 55	16 24	12 09	09 04	01 17	23 37	06 08	23 23
24	19 20	17 46	22 01	16 01	11 52	08 59	01 18	23 37	06 08	23 24
25	19 06	12 30	22 06	15 37	11 35	08 54	01 18	23 37	06 07	23 24
26	18 51	06 38	22 10	15 12	11 17	08 49	01 19	23 38	06 07	23 25
27	18 36	00 N 24	22 14	14 47	11 00	08 43	01 20	23 38	06 07	23 25
28	18 20	05 S 57	22 16	14 22	10 42	08 38	01 21	23 38	06 07	23 26
29	18 04	12 08	22 18	13 56	10 24	08 33	01 22	23 38	06 07	23 27
30	17 48	17 51	22 18	13 30	10 06	08 28	01 23	23 38	06 07	23 27
31	17 S 32	22 S 44	22 S 17	13 S 04	09 S 48	08 S 23	01 N 24	23 N 38	06 S 07	23 N 28

ZODIAC SIGN ENTRIES

Date	h	m	Planets
02	15	58	☽ ♏
04	17	38	☽ ♐
06	17	32	☽ ♑
07	21	10	♀ ♒
08	17	35	☽ ♒
10	19	56	☽ ♓
13	02	05	☽ ♈
15	12	10	☽ ♉
18	00	36	☽ ♊
20	13	06	☽ ♋
20	20	52	☉ ♒
22	13	05	♂ ♓
23	00	12	☽ ♌
25	09	26	☽ ♍
27	16	46	☽ ♎
29	22	04	☽ ♏
31	20	14	♀ ♓

LATITUDES

Date	Mercury ☿	Venus ♀	Mars ♂	Jupiter ♃	Saturn ♄	Uranus ♅	Neptune ♆	Pluto ♇
01	02 N 30	01 S 10	01 S 10	01 S 04	02 N 16	00 N 20	01 N 37	08 N 39
04	03 05	01 15	01 09	01 04	02 17	00 20	01 37	08 39
07	03 17	01 19	01 08	01 03	02 18	00 20	01 38	08 40
10	03 11	01 22	01 06	01 03	02 18	00 20	01 38	08 41
13	02 52	01 25	01 05	01 03	02 19	00 20	01 38	08 41
16	02 26	01 28	01 04	01 03	02 20	00 20	01 38	08 42
19	01 57	01 30	01 02	01 02	02 20	00 20	01 38	08 42
22	01 27	01 31	01 01	01 02	02 22	00 20	01 38	08 43
25	00 57	01 32	01 00	01 02	02 23	00 20	01 39	08 43
28	00 29	01 32	00 58	01 02	02 23	00 20	01 39	08 44
31	00 N 02	01 S 32	00 S 57	01 S 02	02 N 24	00 N 20	01 N 39	08 N 44

DATA

Julian Date	2433648
Delta T	+30 seconds
Ayanamsa	23° 10' 24"
Synetic vernal point	05° ♓ 56' 36"
True obliquity of ecliptic	23° 26' 53"

MOON'S PHASES, APSIDES AND POSITIONS ☽

Date	h	m	Phase	Longitude	Eclipse Indicator
01	05	11	☾	09 ♎ 58	
07	20	10	●	16 ♑ 44	
15	00	23	☽	24 ♈ 02	
23	04	47	○	02 ♌ 22	
30	15	14	☾	09 ♏ 56	

Day	h	m		
06	12	42	Perigee	
18	14	12	Apogee	
06	15	19	Max dec	28° S 26'
12	23	09	0N	
20	09	56	Max dec	28° N 29'
27	13	32	0S	

LONGITUDES

Date	Chiron ⚷	Ceres ⚳	Pallas ⚴	Juno ⚵	Vesta ⚶	Black Moon Lilith ⚸
01	26 ♐ 10	20 ♑ 38	03 ♑ 08	24 ♏ 21	06 ♉ 41	09 ♊ 31
11	27 ♐ 15	24 ♑ 36	07 ♑ 10	27 ♏ 13	07 ♉ 35	10 ♊ 38
21	28 ♐ 17	28 ♑ 34	11 ♑ 27	29 ♏ 53	09 ♉ 05	11 ♊ 45
31	29 ♐ 15	02 ♒ 31	14 ♑ 59	02 ♐ 20	11 ♉ 06	12 ♊ 52

ASPECTARIAN

01 Monday
00 31 ☽ □ ♅
05 11 ☽ □ ☉
06 33 ☽ ± ♃
07 38 ☽ □ ☿
07 51 ☽ ∥ ♆
09 14 ♂ ± ♅
11 14 ☽ △ ♂
20 34 ☉ ☌ ☿
21 47 ☽ ☌ ♆
21 52 ☽ ⚹ ♇
22 33 ☽ ꝑ ♃

02 Tuesday
01 24 ☽ ∥ ♃
03 38 ☽ □ ♀
12 07 ☽ Q ☿
14 57 ☽ Q ☉
18 27 ☽ Q ♇
19 51 ☽ ⊻ ♄

03 Wednesday
00 33 ☽ △ ♃
04 13 ☽ △ ♅
04 59 ☽ ∥ ♂
05 20 ☿ ⊥ ♂
05 58 ☽ ⊥ ♄
06 39 ☽ ⚹ ☿
12 42 ☽ ⚹ ☉
13 04 ☽ Q ♀
13 30 ♆ ⚹ ♇
17 17 ☽ ∥ ☿
17 25 ☽ □ ♂
20 55 ☽ ∠ ♄

04 Thursday
00 26 ☽ □ ♇
00 27 ☽ ⊻ ♆
03 32 ☽ ∥ ♀
04 59 ☽ ꝑ ♅
05 25 ☽ ∠ ☿
06 15 ☽ ∥ ☉
08 25 ☽ # ♇
10 16 ☽ ⊥ ♆
10 36 ☽ # ♅
10 38 ☽ ⚹ ♀
10 42 ☿ ☍ ♅
14 29 ☉ ± ♇
15 08 ☽ ∠ ☉
18 54 ☽ ⊥ ☿
19 33 ☽ ± ♅
21 23 ☽ ⚹ ♄

05 Friday
00 24 ☽ Q ♂
00 50 ☽ ∠ ♆
02 27 ☽ □ ♃
03 51 ☽ ⊻ ☿
05 11 ☽ ⊼ ♅
06 42 ☽ ⊥ ☉
13 01 ☽ ∠ ♀
16 37 ☽ Q ♄
16 59 ☽ ⊻ ☉
19 37 ☿ ⚹ ♃
20 39 ☽ ⚹ ♂

06 Saturday
00 42 ☽ △ ♇
00 49 ☽ ⚹ ♆
04 38 ☽ ⊥ ♀
07 32 ☽ Q ♃
11 16 ♂ ꝑ ♄
15 02 ☽ ⊻ ♀
19 52 ☽ Q ♆
21 15 ☽ □ ♄
21 47 ☽ ∠ ♂

07 Sunday
00 29 ☽ ꝑ ♇
00 32 ☽ ☌ ☿
01 23 ☿ ꝑ ♇
02 50 ☽ ⚹ ♃
04 47 ☽ ☍ ♅
12 54 ☽ ⊥ ♂
14 48 ☽ ± ♇
20 10 ☽ ☌ ☉
21 27 ♀ ⊥ ♃
23 01 ☽ ⊻ ♂

08 Monday
00 23 ☽ ⊼ ♇
00 35 ☽ □ ♆
01 01 ☿ ∠ ♂
03 07 ☽ ∠ ♃
17 53 ☽ ⊥ ♃
19 29 ☽ ☌ ♀
20 00 ☽ # ♅
21 26 ☽ △ ♄
21 59 ☽ # ♇
22 27 ☽ ⊻ ☿

09 Tuesday
00 24 ♂ ☍ ♇
03 29 ☽ ∥ ☉
03 51 ☽ ⊻ ♃
05 01 ♂ △ ♆
05 03 ☽ ⊼ ♅
07 58 ☽ ⊥ ☿
09 26 ☽ ∥ ♀
13 43 ☽ ∥ ☿
14 59 ☽ ± ♅
18 18 ♀ △ ♄
21 02 ☿ ⊻ ♀
22 19 ☽ ꝑ ♄
22 31 ☽ ∠ ☿

10 Wednesday
01 02 ☽ ⊻ ☉
01 38 ☽ ☍ ♇
01 56 ☽ △ ♆
03 09 ☽ ☌ ♂
05 25 ☿ □ ♄
06 06 ☽ ꝑ ♅
09 05 ☉ ⊼ ♇
09 24 ☽ ∥ ♂
12 03 ☽ ⊥ ☉
13 30 ☉ □ ♆
13 38 ☽ ± ♄
23 37 ☽ ⚹ ☿

11 Thursday
00 03 ☽ ⊼ ♄
03 02 ☽ ⊻ ♀
03 48 ☽ ꝑ ♆
04 57 ☽ ∠ ☉
07 40 ☽ ☌ ♃
08 02 ☽ △ ♅
08 22 ☽ ∥ ♃
14 44 ☽ ⊥ ♀
19 48 ☽ ± ♆
20 37 ☽ Q ☿
21 03 ☿ ∥ ♀
23 28 ☽ ∥ ♆

12 Friday
01 17 ♄ St R
04 27 ♃ △ ♅
05 31 ♂ ꝑ ♅
06 15 ☽ ⊼ ♇
06 40 ☽ ⊼ ♆
08 42 ☽ ∠ ♀
10 07 ☽ ⚹ ☉
11 21 ☽ ⊻ ♂
15 35 ☿ St D
17 16 ☽ ± ♇
18 32 ☽ # ♄
23 12 ☽ ⊥ ♂

13 Saturday
01 37 ☉ ∠ ♃
03 48 ☽ ∥ ♄
05 44 ☽ □ ☿
06 32 ☽ ☍ ♄
07 49 ♀ ⊼ ♅
10 06 ☽ ꝑ ♇
10 26 ☽ Q ☉
14 14 ♀ ⊻ ♃
14 58 ☽ □ ♅
15 39 ☽ ⊻ ♃
15 47 ☽ ⚹ ♀
17 19 ☽ ∠ ♂
23 36 ☽ # ♆

14 Sunday
03 22 ☽ ⊥ ♃
08 30 ☿ ⊥ ♀
14 55 ☽ # ♃
14 56 ☽ △ ♇
15 31 ☽ ☍ ♆
17 36 ☽ Q ♀
21 13 ☽ ∠ ♃

15 Monday
00 23 ☽ □ ☉
00 26 ☽ ⚹ ♂
01 47 ☽ Q ♅
03 21 ☉ ⊻ ♂
04 48 ☿ □ ♄
11 40 ☽ # ♂
16 51 ☽ ⊼ ♄
17 14 ☽ △ ☿

16 Tuesday
01 35 ☽ ⚹ ♅
01 58 ☽ Q ♂
03 32 ☽ ⚹ ♃
04 51 ☽ ± ♄
09 32 ☽ □ ♀
10 43 ☽ # ♀
16 02 ☉ ∥ ☿
22 08 ☽ # ☉
22 39 ☽ # ☿
22 58 ☽ ꝑ ♄

17 Wednesday
00 35 ☽ ꝑ ☿
02 40 ☽ □ ♇
03 23 ☽ ⊼ ♆
04 09 ☽ Q ♃
07 43 ☽ ∠ ♅
13 23 ☽ ∥ ♇
15 24 ☽ ∥ ♅
15 32 ☽ ± ♆
16 52 ☽ □ ♂
18 11 ☽ △ ☉
19 43 ☽ ± ☿
21 22 ♂ ± ♄
23 07 ♀ ± ♅

18 Thursday
01 53 ☽ ⊥ ♅
05 19 ☽ △ ♄
08 33 ☽ ⊼ ☿
09 47 ☽ ꝑ ♆
14 00 ☽ ⊻ ♅
15 04 ☽ Q ♇
16 18 ☿ ꝑ ♇
17 17 ☽ □ ♃

19 Friday
03 25 ☽ ꝑ ☉
05 23 ☽ △ ♀
15 16 ☽ ⚹ ♇
16 06 ☽ △ ♆
23 14 ☽ ± ☉

20 Saturday
09 44 ☽ △ ♂
12 23 ☽ ⊼ ☉
15 03 ☽ ꝑ ♀
17 41 ☽ □ ♄
21 16 ☽ ∠ ♇

21 Sunday
00 57 ☽ ☍ ☿
02 05 ☽ ☌ ♅
05 30 ♆ St R
06 35 ☽ △ ♃
11 00 ☽ ± ♀
15 04 ☽ ⊥ ♇
16 04 ♀ ꝑ ♄
16 19 ☿ ☍ ♅
17 37 ☽ ꝑ ♂

22 Monday
00 12 ☽ ⊼ ♀
02 51 ☽ ⊻ ♇
03 47 ☽ □ ♆
05 14 ☽ Q ♄
12 29 ☽ ± ♂
12 37 ☽ ꝑ ♃

23 Tuesday
00 57 ☽ ⊼ ♂
01 48 ♀ ☍ ♇
02 05 ☉ △ ♄
02 43 ☽ ∥ ♅
04 20 ☽ ∥ ♇
04 33 ☽ ⚹ ♄
04 47 ☽ ☍ ☉
06 25 ☽ ± ♃
11 16 ♀ △ ♆
12 37 ☽ ⊻ ♅
13 35 ☽ # ☿
14 45 ☽ Q ♆
16 23 ☽ ⊼ ☿
18 11 ☽ ⊼ ♃

24 Wednesday
00 06 ☽ ⊥ ♅
03 36 ☽ # ☉
04 52 ☽ ± ☿
05 40 ☉ ⊥ ♃
09 18 ☽ ∠ ♄
12 41 ☽ ☌ ♇
13 40 ☽ ⚹ ♆
16 34 ☽ ☍ ♀
17 11 ☽ ∠ ♅
19 48 ☿ ⚹ ♃
21 04 ☽ # ♀
22 35 ♀ ꝑ ♅
23 29 ☽ ꝑ ☿

25 Thursday
02 18 ☽ ⊥ ♄
08 21 ♂ ⊼ ♄
13 33 ☽ ⊻ ♄
13 54 ☽ ☍ ♂
16 09 ☽ # ♂
17 55 ☽ ∠ ♆
18 47 ☽ ⊼ ☉
21 17 ☽ ⚹ ♅

26 Friday
03 13 ☽ # ♃
03 48 ☽ ☍ ♃
06 07 ☽ △ ☿
06 53 ☽ ± ☉
10 37 ☽ ⊥ ♆
14 02 ☽ # ♆
19 26 ☽ Q ♅
20 37 ☽ ⊻ ♇
21 40 ☽ ⊻ ♆

27 Saturday
00 52 ☽ ꝑ ☉
01 32 ☉ ⊼ ♅
06 22 ☽ ⊼ ♀
07 34 ☽ ⊥ ♇
08 29 ☽ ∥ ♄
18 24 ☽ ± ♀
18 36 ☽ # ♄
20 38 ☽ ☌ ♄
20 50 ☿ ± ♇
23 52 ☽ ∠ ♇

28 Sunday
00 33 ☽ ⊼ ♂
04 03 ☽ □ ♅
06 19 ☽ △ ☉
07 00 ♂ ꝑ ♆
11 25 ☽ ⊼ ♃
11 59 ☽ ± ♂
12 16 ☽ ꝑ ♀
12 40 ☽ ∥ ♆
17 36 ♀ ± ♄
17 50 ☽ □ ☿
22 10 ☽ ∥ ♃
22 16 ☽ ± ♃

29 Monday
02 35 ☽ ⚹ ♇
03 41 ☽ ☌ ♆
04 57 ☽ ꝑ ♂
05 28 ☽ ∥ ♂
14 24 ☽ ꝑ ♃
17 27 ☽ △ ♀
18 47 ☽ ∥ ♀
23 36 ☽ Q ♇

30 Tuesday
01 40 ☽ ⊻ ♄
04 31 ☽ Q ☿
08 43 ☽ △ ♂
08 47 ☽ △ ♅
09 42 ♂ △ ♅
11 47 ☽ ∥ ☉
12 01 ☽ ⊥ ♄
15 14 ☽ □ ☉
16 50 ☽ △ ♃

31 Wednesday
03 09 ☽ ⚹ ☿
03 23 ☽ ∠ ♄
06 28 ☽ □ ♇
07 35 ☽ ⊻ ♆
09 35 ☽ ∥ ☿
10 22 ☽ ꝑ ♅
16 09 ☽ # ♇
17 10 ☽ # ♅
17 43 ☽ ⊥ ♆
19 07 ☉ ⊻ ♃

All ephemeris data is given at 12.00 UT and the Moon's longitude is additionally given for 24.00 UT

FEBRUARY 1951

LONGITUDES

Date	Sidereal time h m s	Sun ☉ ° ' "	Moon ☽ ° ' "	Moon ☽ 24.00 ° ' "	Mercury ☿ ° '	Venus ♀ ° '	Mars ♂ ° '	Jupiter ♃ ° '	Saturn ♄ ° '	Uranus ♅ ° '	Neptune ♆ ° '	Pluto ♇ ° '
01	20 43 32	11 ♒ 49 09	06 ♐ 25 31	13 ♐ 39 35	18 ♑ 39	00 ♓ 49	07 ♓ 50	11 ♓ 16	02 ♎ 00	06 ♋ 07	19 ♎ 30	18 ♌ 49
02	20 47 29	12 50 03	20 ♐ 56 12	28 ♐ 14 45	19 57	02 04	08 37	11 29	01 R 57	06 R 05	19 R 30	18 R 48
03	20 51 25	13 50 57	05 ♑ 34 28	12 ♑ 54 32	21 17	03 19	09 24	11 43	01 55	06 03	19 29	18 46
04	20 55 22	14 51 49	20 ♑ 14 06	27 ♑ 32 13	22 38	04 34	10 11	11 57	01 53	06 01	19 29	18 45
05	20 59 19	15 52 41	04 ♒ 48 00	12 ♒ 00 36	24 00	05 49	10 58	12 11	01 50	05 59	19 28	18 43
06	21 03 15	16 53 31	19 ♒ 09 15	26 ♒ 13 15	25 23	07 03	11 45	12 24	01 48	05 57	19 28	18 42
07	21 07 12	17 54 20	03 ♓ 12 04	10 ♓ 05 18	26 47	08 18	12 32	12 38	01 45	05 56	19 27	18 41
08	21 11 08	18 55 07	16 ♓ 52 40	23 ♓ 34 04	28 13	09 33	13 19	12 52	01 42	05 54	19 26	18 39
09	21 15 05	19 55 54	00 ♈ 09 31	06 ♈ 39 08	29 ♑ 39	10 48	14 06	13 06	01 39	05 52	19 26	18 38
10	21 19 01	20 56 38	13 ♈ 03 12	19 ♈ 22 02	01 ♒ 06	12 02	14 53	13 20	01 36	05 51	19 25	18 36
11	21 22 58	21 57 21	25 ♈ 36 05	01 ♉ 45 50	02 34	13 17	15 40	13 34	01 33	05 49	19 25	18 35
12	21 26 54	22 58 03	07 ♉ 51 48	13 ♉ 54 36	04 04	14 32	16 27	13 48	01 30	05 48	19 24	18 33
13	21 30 51	23 58 43	19 ♉ 54 48	25 ♉ 53 02	05 34	15 46	17 14	14 02	01 27	05 46	19 23	18 32
14	21 34 48	24 59 21	01 ♊ 49 53	07 ♊ 45 59	07 05	17 01	18 01	14 16	01 24	05 45	19 22	18 30
15	21 38 44	25 59 58	13 ♊ 41 54	19 ♊ 38 14	08 36	18 15	18 48	14 31	01 20	05 43	19 21	18 29
16	21 42 41	27 00 32	25 ♊ 35 29	01 ♋ 34 12	10 09	19 30	19 35	14 45	01 17	05 42	19 21	18 28
17	21 46 37	28 01 06	07 ♋ 34 48	13 ♋ 37 44	11 43	20 44	20 21	14 59	01 14	05 41	19 20	18 26
18	21 50 34	29 ♒ 01 37	19 ♋ 43 22	25 ♋ 52 00	13 18	21 59	21 08	15 13	01 10	05 39	19 19	18 25
19	21 54 30	00 ♓ 02 07	02 ♌ 03 54	08 ♌ 19 16	14 53	23 13	21 55	15 28	01 06	05 38	19 18	18 23
20	21 58 27	01 02 35	14 ♌ 38 15	21 ♌ 00 56	16 30	24 28	22 42	15 42	01 03	05 37	19 17	18 22
21	22 02 23	02 03 01	27 ♌ 27 20	03 ♍ 57 27	18 07	25 42	23 28	15 56	00 59	05 36	19 16	18 20
22	22 06 20	03 03 26	10 ♍ 31 13	17 ♍ 08 31	19 45	26 56	24 15	16 10	00 55	05 35	19 15	18 19
23	22 10 17	04 03 49	23 ♍ 49 15	00 ♎ 33 13	21 24	28 10	25 02	16 25	00 51	05 34	19 14	18 18
24	22 14 13	05 04 10	07 ♎ 19 55	14 ♎ 10 13	23 05	29 ♓ 25	25 48	16 39	00 47	05 33	19 13	18 16
25	22 18 10	06 04 30	21 ♎ 02 50	27 ♎ 57 56	24 46	00 ♈ 39	26 35	16 54	00 43	05 32	19 12	18 15
26	22 22 06	07 04 48	04 ♏ 55 18	11 ♏ 54 45	26 28	01 53	27 21	17 08	00 39	05 31	19 11	18 14
27	22 26 03	08 05 05	18 ♏ 56 02	25 ♏ 58 58	28 11	03 07	28 08	17 22	00 35	05 30	19 10	18 12
28	22 29 59	09 ♓ 05 21	03 ♐ 03 17	10 ♐ 08 46	29 ♒ 55	04 ♈ 21	28 ♓ 54	17 ♓ 37	00 ♎ 31	05 ♋ 29	19 ♎ 09	18 ♌ 11

Date	Moon True ☊ ° '	Moon Mean ☊ ° '	Moon ☽ Latitude ° '
01	19 ♓ 40	21 ♓ 07	05 S 02
02	19 R 36	21 04	05 10
03	19 31	21 01	04 58
04	19 25	20 57	04 27
05	19 20	20 54	03 38
06	19 16	20 51	02 37
07	19 13	20 48	01 26
08	19 12	20 45	00 S 13
09	19 D 13	20 42	01 N 00
10	19 14	20 38	02 07
11	19 16	20 35	03 06
12	19 17	20 32	03 56
13	19 18	20 29	04 34
14	19 R 18	20 26	05 00
15	19 17	20 23	05 13
16	19 15	20 19	05 12
17	19 12	20 16	04 58
18	19 08	20 13	04 31
19	19 05	20 10	03 50
20	19 02	20 07	02 58
21	19 00	20 03	01 56
22	18 59	20 00	00 N 47
23	18 D 59	19 57	00 S 27
24	19 00	19 54	01 40
25	19 01	19 51	02 48
26	19 02	19 48	03 47
27	19 03	19 44	04 34
28	19 ♓ 03	19 ♓ 41	05 S 04

DECLINATIONS

Date	Sun ☉ ° '	Moon ☽ ° '	Mercury ☿ ° '	Venus ♀ ° '	Mars ♂ ° '	Jupiter ♃ ° '	Saturn ♄ ° '	Uranus ♅ ° '	Neptune ♆ ° '	Pluto ♇ ° '
01	17 S 15	26 S 21	22 S 16	12 S 37	09 S 30	08 S 18	01 N 25	23 N 38	06 S 06	23 N 28
02	16 58	28 18	22 13	12 10	09 12	08 12	01 26	23 38	06 06	23 29
03	16 41	28 18	22 09	11 42	08 54	08 07	01 27	23 38	06 06	23 29
04	16 23	26 19	22 03	11 15	08 36	08 02	01 28	23 38	06 06	23 30
05	16 05	22 36	21 57	10 47	08 17	07 57	01 30	23 38	06 05	23 30
06	15 47	17 34	21 49	10 18	07 59	07 51	01 31	23 39	06 05	23 31
07	15 28	11 41	21 40	09 50	07 40	07 46	01 32	23 39	06 05	23 32
08	15 09	05 S 23	21 30	09 21	07 22	07 41	01 33	23 39	06 05	23 32
09	14 50	00 N 58	21 19	08 52	07 03	07 35	01 35	23 39	06 04	23 33
10	14 31	07 06	21 06	08 22	06 44	07 30	01 36	23 39	06 04	23 33
11	14 12	12 47	20 52	07 53	06 26	07 24	01 38	23 39	06 04	23 34
12	13 52	17 51	20 36	07 23	06 07	07 19	01 39	23 39	06 03	23 34
13	13 32	22 07	20 20	06 53	05 48	07 14	01 41	23 39	06 03	23 35
14	13 12	25 26	20 02	06 23	05 29	07 08	01 42	23 39	06 03	23 35
15	12 51	27 37	19 42	05 53	05 10	07 03	01 44	23 39	06 02	23 36
16	12 31	28 34	19 22	05 22	04 52	06 57	01 45	23 39	06 02	23 36
17	12 10	28 12	19 00	04 52	04 33	06 52	01 47	23 39	06 02	23 37
18	11 49	26 28	18 37	04 21	04 14	06 46	01 48	23 39	06 01	23 37
19	11 28	23 27	18 12	03 50	03 55	06 41	01 50	23 39	06 01	23 38
20	11 06	19 17	17 46	03 19	03 36	06 35	01 52	23 39	06 01	23 38
21	10 45	14 11	17 19	02 48	03 17	06 29	01 53	23 39	06 00	23 39
22	10 23	08 21	16 50	02 17	02 58	06 24	01 55	23 39	06 00	23 39
23	10 01	02 N 03	16 20	01 46	02 39	06 18	01 57	23 40	05 59	23 40
24	09 39	04 S 26	15 49	01 15	02 19	06 13	01 58	23 40	05 59	23 40
25	09 17	10 49	15 16	00 44	02 00	06 07	02 00	23 40	05 58	23 40
26	08 55	16 44	14 42	00 S 13	01 41	06 01	02 02	23 40	05 58	23 41
27	08 32	21 51	14 07	00 N 19	01 22	05 56	02 04	23 40	05 57	23 41
28	08 S 10	25 S 45	13 S 30	00 N 50	01 S 03	05 S 50	02 N 06	23 N 40	05 S 57	23 N 42

ZODIAC SIGN ENTRIES

Date	h	m	Planets	
01	01	16	☽	♐
03	02	52	☽	♑
05	04	04	☽	♒
07	06	29	☽	♓
09	11	43	☽	♈
09	17	50	☿	♒
11	20	33	☽	♉
14	08	18	☽	♊
16	20	51	☽	♋
19	08	01	☽	♌
19	11	10	☉	♓
21	16	43	☽	♍
23	23	01	☽	♎
24	23	26	♀	♈
26	03	31	☽	♏
28	06	49	☽	♐
28	13	04	☿	♓

LATITUDES

Date	Mercury ☿ ° '	Venus ♀ ° '	Mars ♂ ° '	Jupiter ♃ ° '	Saturn ♄ ° '	Uranus ♅ ° '	Neptune ♆ ° '	Pluto ♇ ° '
01	00 S 07	01 S 32	00 S 56	01 S 02	02 N 25	00 N 20	01 N 39	08 N 44
04	00 31	01 31	00 54	01 02	02 25	00 20	01 39	08 44
07	00 53	01 29	00 53	01 02	02 26	00 20	01 39	08 45
10	01 12	01 26	00 51	01 02	02 27	00 20	01 39	08 45
13	01 29	01 23	00 49	01 02	02 27	00 20	01 40	08 45
16	01 43	01 19	00 48	01 01	02 28	00 20	01 40	08 45
19	01 55	01 15	00 46	01 01	02 29	00 20	01 40	08 45
22	02 03	01 10	00 44	01 01	02 29	00 20	01 40	08 45
25	02 07	01 05	00 42	01 01	02 30	00 20	01 40	08 45
28	02 08	00 59	00 40	01 01	02 30	00 20	01 40	08 45
31	02 S 05	00 S 52	00 S 39	01 S 01	02 N 31	00 N 20	01 N 40	08 N 45

DATA

Julian Date	2433679
Delta T	+30 seconds
Ayanamsa	23° 10' 30"
Synetic vernal point	05° ♓ 56' 30"
True obliquity of ecliptic	23° 26' 53"

MOON'S PHASES, APSIDES AND POSITIONS ☽

Date	h	m	Phase	Longitude ° '	Eclipse Indicator
06	07	54	●	16 ♒ 43	
13	20	55	☽	24 ♉ 21	
21	21	12	○	02 ♍ 26	
28	22	59	☾	09 ♐ 33	

Day	h	m		
03	15	22	Perigee	
15	09	35	Apogee	
03	00	02	Max dec	28° S 33'
09	08	17	ON	
16	17	14	Max dec	28° N 36'
23	19	36	OS	

LONGITUDES

Date	Chiron ⚷ ° '	Ceres ⚳ ° '	Pallas ⚴ ° '	Juno ⚵ ° '	Vesta ⚶ ° '	Black Moon Lilith ⚸ ° '
01	29 ♐ 20	02 ♒ 55	15 ♑ 21	02 ♐ 34	11 ♉ 20	12 ♊ 59
11	00 ♑ 12	06 ♒ 50	19 ♑ 06	04 ♐ 43	13 ♉ 50	14 ♊ 06
21	00 ♑ 58	10 ♒ 43	22 ♑ 43	06 ♐ 32	16 ♉ 41	15 ♊ 14
31	01 ♑ 37	14 ♒ 32	26 ♑ 09	07 ♐ 59	16 ♉ 50	16 ♊ 21

ASPECTARIAN

h m Aspects

01 Thursday
00 08 ☽ Q ☉
01 29 ☽ ± ♅
01 45 ☽ □ ♀
04 37 ☽ ✱ ♄
06 55 ☽ ∠ ☿
08 48 ☽ ∠ ♆
11 29 ☽ ⊼ ♅
14 28 ☽ □ ♂
15 11 ☿ ⊼ ♇
18 43 ☉ ± ♅
20 10 ☽ □ ♃
21 38 ☽ ✱ ☉
23 20 ☽ ⊥ ☿

02 Friday
00 31 ☽ Q ♄
03 40 ☿ □ ♆
08 29 ☽ △ ♇
09 37 ☽ ✱ ♆
09 55 ♀ ⊼ ♄
10 13 ☽ ⊻ ☿
10 26 ☽ Q ♀
21 51 ☽ Q ♂

03 Saturday
00 10 ☽ ∠ ☉
02 16 ☽ Q ♃
05 19 ☽ Q ♆
06 02 ☽ □ ♄
07 58 ☽ ✱ ♀
09 03 ☽ bQ ♇
12 46 ☽ ☍ ♅
16 00 ☽ ⊥ ☉
18 37 ☽ ✱ ♂
22 13 ☽ ✱ ♃
23 45 ☽ ± ♇

04 Sunday
02 33 ☽ ⊻ ☉
09 34 ☽ ⊼ ♇
10 22 ♀ bQ ♆
10 46 ☽ □ ♆
10 48 ☽ ∠ ♀
16 20 ☽ ☌ ☿
20 35 ☽ ∠ ♂
23 12 ☽ ∠ ♃

05 Monday
02 58 ☽ ⊥ ♀
06 12 ☽ # ♅
06 59 ☽ # ♇
07 07 ☽ △ ♄
12 18 ☽ ⊥ ♂
13 50 ☽ ⊻ ♀
13 58 ☽ ⊼ ♅
14 19 ☽ ⊥ ♃
15 19 ♀ △ ♅
15 33 ☽ ∥ ☿
22 51 ☽ ⊻ ♂
23 56 ☽ ± ♅

06 Tuesday
00 29 ☽ ⊻ ♃
07 54 ☽ ☌ ☉
08 02 ☽ bQ ♄
09 47 ☉ bQ ♄
11 14 ☽ ☍ ♇
12 31 ☽ △ ♆
15 03 ☽ bQ ♅
19 59 ☽ ∥ ☉
23 14 ☽ ± ♄
23 44 ☽ ⊻ ☿

07 Wednesday
01 27 ♂ ∥ ♃
09 30 ☽ ⊼ ♄
11 12 ☽ ⊥ ☿
14 10 ☽ bQ ♆
16 24 ♂ ☌ ♃
16 43 ☽ △ ♅
19 44 ☽ ∥ ♀
21 46 ☽ ☌ ♀

08 Thursday
03 12 ☽ ∥ ♃
04 09 ☽ ∥ ♂
04 44 ☽ ∠ ☿
04 46 ☽ ☌ ♃
05 12 ☿ ∠ ♃
05 18 ☽ ☌ ♂
05 49 ☉ ☍ ♇
05 55 ☽ ± ♆
09 22 ☽ ∥ ♆
15 09 ☽ ⊼ ♇
15 39 ♂ ± ♆
15 56 ☽ ⊻ ☉
16 09 ☿ ∠ ♂
16 34 ☽ ⊼ ♆

09 Friday
00 16 ☉ △ ♆
01 56 ☽ ± ♇
02 20 ☽ # ♄
03 39 ☽ ⊥ ☉
10 57 ☽ ✱ ☿
14 20 ☽ ∥ ♄
14 44 ☽ ☍ ♄
18 23 ☽ bQ ♇
21 33 ☽ ∠ ☉
22 32 ☽ □ ♅

10 Saturday
07 51 ☽ # ♆
09 42 ☉ bQ ♅
09 53 ☽ ⊻ ♀
10 36 ☽ # ♂
12 06 ☽ Q ☿
12 33 ☽ ⊻ ♃
13 34 ☽ # ♃
15 42 ☽ ⊻ ♂
16 46 ☽ # ♀
20 00 ☿ △ ♄
20 18 ♃ ± ♆
22 30 ☽ ⊥ ♀
22 31 ☽ △ ♇

11 Sunday
00 05 ☽ ☍ ♆
00 10 ☽ ⊥ ♃
03 54 ☽ ⊥ ♂
04 21 ☽ ✱ ☉
08 34 ☽ Q ♅
14 22 ♀ ± ♆
17 48 ☽ ∠ ♀
17 52 ☽ ∠ ♃
17 58 ☽ # ☉
18 46 ♀ ☌ ♃
22 32 ☽ ∠ ♂
23 33 ☽ ⊼ ♄

12 Monday
03 28 ☽ □ ☿
05 46 ☽ Q ☉
07 55 ☽ ✱ ♅
11 18 ☽ ± ♄
16 06 ♀ ∥ ♃
16 28 ♂ ∥ ♆

13 Tuesday
00 01 ☽ ✱ ♃
02 00 ☽ # ☿
02 45 ☽ ✱ ♀
05 06 ☽ bQ ♄
06 16 ☽ ✱ ♂
09 14 ☽ □ ♇
10 56 ☽ ⊼ ♆
13 43 ☽ ∠ ♅
15 15 ☿ ⊼ ♅
20 55 ☽ □ ☉
21 44 ☽ ∥ ♇
22 13 ☽ ∥ ♅
22 59 ☽ ± ♆

14 Wednesday
00 34 ☽ Q ♃
05 39 ☽ Q ♀
07 48 ☽ ⊥ ♅
08 04 ☽ Q ♂
11 08 ☽ △ ♄
17 08 ☽ bQ ♆
19 54 ☽ ⊻ ♅
21 11 ☉ ± ♄
21 26 ☽ Q ♇

15 Thursday
00 10 ☽ △ ☿
02 39 ♂ ⊼ ♇
04 19 ♀ ∥ ♆
10 11 ☿ ⊥ ♃
13 40 ☽ □ ♃
16 17 ♀ ⊼ ♇
21 39 ☽ ✱ ♇
22 17 ☽ □ ♀
23 02 ☽ □ ♂
23 25 ☽ △ ♆

16 Friday
04 59 ♂ ⊼ ♆
09 03 ♀ ⊼ ♆
10 59 ☽ bQ ☿
15 07 ☽ △ ☉
16 02 ♀ ☌ ♂
23 22 ☽ □ ♄

17 Saturday
03 45 ☽ ∠ ♇
07 44 ☽ ± ☿
08 13 ☽ ☌ ♅
11 24 ☿ ± ♅
21 26 ☽ ⊼ ☿
21 37 ☽ ⊥ ♇
23 46 ☽ bQ ☉

18 Sunday
02 58 ☽ △ ♃
09 26 ☽ ⊻ ♇
10 55 ☽ Q ♄
11 12 ☽ □ ♆
14 57 ☽ △ ♂
16 55 ☽ △ ♀
19 03 ☽ ± ☉

19 Monday
03 23 ♀ ∥ ♂
07 44 ☽ ⊼ ☉
08 50 ☽ bQ ♃
10 10 ☽ ✱ ♄
10 36 ☽ ∥ ♅
10 47 ☽ ∥ ♇
18 51 ☽ ⊻ ♅
21 56 ☽ bQ ♂
22 02 ☽ Q ♆
22 06 ☿ ⊻ ♃

20 Tuesday
01 06 ☽ bQ ♀
02 27 ☽ ± ♃
05 35 ☿ bQ ♄
06 17 ☽ ⊥ ♅
10 11 ♀ ± ♇
12 01 ☉ ⊼ ♄
14 02 ☽ ⊼ ♃
14 39 ☽ ∠ ♄
16 01 ☽ ☍ ☿
16 08 ☽ ± ♂
17 46 ☿ ⊥ ♂
19 01 ☽ ☌ ♇
19 59 ☽ ± ♀
20 18 ☽ # ☿
20 45 ☽ ✱ ♆
23 14 ☽ ∠ ♅

21 Wednesday
04 07 ☽ ⊼ ♂
07 26 ☽ ⊥ ♄
08 23 ☽ ⊼ ♀
15 17 ☿ ☍ ♇
18 29 ☽ ⊻ ♄
21 12 ☽ ☍ ☉

22 Thursday
00 33 ☽ ∠ ♆
02 59 ☽ ✱ ♅
03 20 ☽ # ☉
04 45 ☿ △ ♆
14 04 ♂ ± ♇
16 57 ☽ ⊥ ♆
19 40 ☽ # ♃
21 07 ☽ # ♆
22 26 ☽ ☍ ♃
23 54 ☿ bQ ♅

23 Friday
00 46 ☽ Q ♅
02 06 ☽ ⊻ ♇
03 47 ☽ ⊻ ♆
04 23 ♀ # ♄
07 04 ☽ ⊼ ☿
08 07 ♅ ∥ ♇
09 40 ☽ # ♂
12 23 ☽ ∥ ♄
12 51 ☽ ⊥ ♇
13 07 ☽ # ♀
14 17 ☽ ☍ ♂
16 00 ☉ bQ ♆
19 18 ☽ ± ☿
20 33 ☽ ☍ ♀

24 Saturday
00 28 ☽ ☌ ♄
01 05 ☽ ∥ ♀
02 51 ☽ # ♄
04 33 ☽ ∥ ♂
04 50 ☽ ∠ ♇
07 41 ☽ ⊼ ☉
08 51 ☽ □ ♅
13 29 ☽ bQ ☿
17 43 ☽ ∥ ♆
18 29 ☽ ∥ ♃
19 05 ☽ ± ☉
23 11 ☉ △ ♅

25 Sunday
04 38 ☽ ⊼ ♃
05 54 ☿ ⊥ ♀
06 29 ☽ ∥ ☉
07 08 ☽ ✱ ♇
08 47 ☽ ☌ ♆
11 24 ☿ ± ♄
12 03 ☽ bQ ☉
12 18 ♂ # ♄
13 19 ♀ ☍ ♄
15 16 ☽ ± ♃
19 21 ☽ △ ☿
22 10 ☽ ⊼ ♂

26 Monday
03 55 ☽ Q ♇
04 14 ☽ ∥ ☿
04 41 ☽ ⊻ ♄
06 15 ☽ ⊼ ♀
07 07 ☽ bQ ♃
09 08 ☽ ± ♂
13 01 ☽ △ ♅
14 57 ☽ ⊥ ♄
16 00 ☽ △ ☉
17 35 ☽ ± ♀

27 Tuesday
01 30 ☽ bQ ♂
04 15 ♃ ∥ ♆
06 18 ☽ ∠ ♄
09 18 ☽ △ ♃
10 28 ☽ bQ ♀
10 33 ☿ ⊻ ♂
10 45 ☽ □ ♇
12 23 ☽ ⊻ ♆
13 38 ♀ bQ ♇
14 40 ☽ bQ ♅
22 07 ☽ # ♅
22 19 ☽ # ♇
22 35 ☽ ⊥ ♆

28 Wednesday
04 33 ☽ △ ♂
05 57 ☽ □ ☿
05 58 ☽ ± ♅
07 43 ☽ ✱ ♄
13 50 ☽ ∠ ♆
14 24 ☽ △ ♀
16 07 ☽ ⊼ ♅
18 18 ♀ # ♂
19 47 ☿ ⊼ ♄
22 59 ☽ □ ☉

All ephemeris data is given at 12.00 UT and the Moon's longitude is additionally given for 24.00 UT

MARCH 1951

Date	Sidereal time h m s	Sun ☉	Moon ☽	Moon ☽ 24.00	Mercury ☿	Venus ♀	Mars ♂	Jupiter ♃	Saturn ♄	Uranus ♅	Neptune ♆	Pluto ⯓
01	22 33 56	10 ♓ 05 35	17 ♐ 15 08	24 ♐ 22 07	01 ♓ 41	05 ♈ 35	29 ♓ 41	17 ♓ 51	00 ♎ 26	05 ♋ 29	19 ♎ 07	18 ♌ 10
02	22 37 52	11 05 48	01 ♑ 29 21	08 ♑ 36 32	03 27	06 49	00 ♈ 27	18 06	00 R 22	05 R 28	19 R 06	18 R 08
03	22 41 49	12 06 00	15 ♑ 43 16	22 ♑ 49 09	05 14	08 03	01 13	18 20	00 18	05 27	19 05	18 07
04	22 45 46	13 06 09	29 ♑ 53 46	06 ♒ 56 40	07 03	09 17	02 00	18 35	00 13	05 27	19 04	18 06
05	22 49 42	14 06 17	13 ♒ 57 24	20 ♒ 55 34	08 52	10 31	02 46	18 49	00 09	05 26	19 02	18 04
06	22 53 39	15 06 24	27 ♒ 50 43	04 ♓ 42 28	10 43	11 45	03 32	19 04	00 05	05 26	19 01	18 03
07	22 57 35	16 06 28	11 ♓ 30 30	18 ♓ 14 31	12 34	12 58	04 18	19 19	00 ♎ 00	05 25	19 00	18 02
08	23 01 32	17 06 31	24 ♓ 54 18	01 ♈ 29 41	14 27	14 12	05 05	19 33	29 ♍ 56	05 25	18 59	18 01
09	23 05 28	18 06 32	08 ♈ 00 36	14 ♈ 27 03	16 20	15 26	05 51	19 48	29 51	05 25	18 57	17 59
10	23 09 25	19 06 30	20 ♈ 49 06	27 ♈ 06 53	18 15	16 39	06 37	20 02	29 46	05 25	18 56	17 58
11	23 13 21	20 06 27	03 ♉ 20 39	09 ♉ 30 41	20 11	17 53	07 23	20 17	29 42	05 24	18 54	17 57
12	23 17 18	21 06 22	15 ♉ 37 20	21 ♉ 41 00	22 07	19 07	08 09	20 31	29 37	05 24	18 53	17 56
13	23 21 15	22 06 14	27 ♉ 42 08	03 ♊ 41 15	24 05	20 20	08 55	20 46	29 32	05 24	18 52	17 55
14	23 25 11	23 06 04	09 ♊ 38 51	15 ♊ 35 31	26 03	21 33	09 41	21 00	29 28	05 D 24	18 50	17 53
15	23 29 08	24 05 52	21 ♊ 31 48	27 ♊ 28 18	28 ♓ 01	22 47	10 27	21 15	29 23	05 24	18 49	17 52
16	23 33 04	25 05 38	03 ♋ 25 37	09 ♋ 24 18	00 ♈ 01	24 00	11 13	21 29	29 18	05 24	18 47	17 51
17	23 37 01	26 05 22	15 ♋ 24 58	21 ♋ 28 09	02 00	25 13	11 58	21 44	29 14	05 24	18 46	17 50
18	23 40 57	27 05 03	27 ♋ 34 22	03 ♌ 44 07	04 00	26 27	12 44	21 58	29 09	05 25	18 44	17 49
19	23 44 54	28 04 42	09 ♌ 57 51	16 ♌ 15 56	05 59	27 40	13 30	22 13	29 04	05 25	18 43	17 48
20	23 48 50	29 ♓ 04 19	22 ♌ 38 41	29 ♌ 06 21	07 58	28 ♈ 53	14 16	22 27	28 59	05 25	18 41	17 47
21	23 52 47	00 ♈ 03 54	05 ♍ 39 06	12 ♍ 16 58	09 57	00 ♉ 06	15 01	22 42	28 55	05 25	18 40	17 46
22	23 56 44	01 03 26	18 ♍ 59 56	25 ♍ 47 51	11 54	01 19	15 47	22 56	28 50	05 26	18 38	17 45
23	00 00 40	02 02 57	02 ♎ 40 29	09 ♎ 37 31	13 50	02 32	16 32	23 11	28 45	05 26	18 37	17 44
24	00 04 37	03 02 25	16 ♎ 38 29	23 ♎ 42 55	15 45	03 45	17 18	23 25	28 41	05 27	18 35	17 43
25	00 08 33	04 01 51	00 ♏ 50 14	07 ♏ 59 49	17 37	04 57	18 03	23 40	28 36	05 27	18 34	17 42
26	00 12 30	05 01 16	15 ♏ 11 02	22 ♏ 23 14	19 27	06 10	18 49	23 54	28 31	05 28	18 32	17 41
27	00 16 26	06 00 39	29 ♏ 35 47	06 ♐ 48 07	21 13	07 23	19 34	24 08	28 26	05 29	18 30	17 40
28	00 20 23	07 00 00	13 ♐ 59 40	21 ♐ 09 58	22 57	08 35	20 19	24 23	28 22	05 29	18 29	17 39
29	00 24 19	07 59 19	28 ♐ 18 35	05 ♑ 25 10	24 36	09 48	21 05	24 37	28 17	05 30	18 27	17 39
30	00 28 16	08 58 37	12 ♑ 29 27	19 ♑ 31 13	26 11	11 00	21 50	24 51	28 13	05 31	18 26	17 38
31	00 32 13	09 ♈ 57 53	26 ♑ 30 18	03 ♒ 26 35	27 ♈ 42	12 ♉ 13	22 ♈ 35	25 ♓ 06	28 ♍ 08	05 ♋ 32	18 ♎ 24	17 ♌ 37

Date	Moon True ☊	Moon Mean ☊	Moon ☽ Latitude	Decl. Sun ☉	Moon ☽	Mercury ☿	Venus ♀	Mars ♂	Jupiter ♃	Saturn ♄	Uranus ♅	Neptune ♆	Pluto ⯓
01	19 ♓ 03	19 ♓ 38	05 S 16	07 S 47	28 S 05	12 S 52	01 N 21	00 S 44	05 S 45	02 N 07	23 N 40	05 S 57	23 N 42
02	19 R 02	19 35	05 09	07 24	28 35	12 13	01 53	00 25	05 39	02 09	23 40	05 56	23 43
03	19 02	19 32	04 43	07 02	27 14	11 32	02 24	00 S 06	05 33	02 11	23 40	05 56	23 43
04	19 01	19 29	04 00	06 39	24 05	10 50	02 55	00 N 13	05 28	02 13	23 40	05 55	23 43
05	19 01	19 25	03 02	06 15	19 33	10 07	03 26	00 32	05 22	02 15	23 40	05 55	23 44
06	19 00	19 22	01 55	05 52	14 01	09 22	03 57	00 51	05 16	02 17	23 40	05 54	23 44
07	19 00	19 19	00 S 41	05 29	07 53	08 37	04 28	01 10	05 11	02 19	23 40	05 53	23 44
08	19 D 00	19 16	00 N 33	05 06	01 S 31	07 49	04 59	01 29	05 05	02 21	23 40	05 53	23 45
09	19 00	19 13	01 43	04 42	04 N 46	07 01	05 30	01 47	04 59	02 22	23 40	05 52	23 45
10	19 R 00	19 09	02 47	04 19	10 43	06 12	06 01	02 06	04 53	02 24	23 40	05 52	23 45
11	19 00	19 06	03 42	03 55	16 06	05 21	06 31	02 25	04 48	02 26	23 40	05 51	23 46
12	19 00	19 03	04 25	03 32	20 45	04 30	07 01	02 44	04 42	02 28	23 40	05 51	23 46
13	18 59	19 00	04 55	03 08	24 27	03 37	07 32	03 03	04 36	02 30	23 40	05 50	23 46
14	18 59	18 57	05 13	02 44	27 03	02 43	08 02	03 21	04 31	02 32	23 40	05 50	23 47
15	18 59	18 54	05 16	02 21	28 26	01 49	08 32	03 40	04 25	02 34	23 40	05 49	23 47
16	18 D 59	18 50	05 07	01 57	28 31	00 S 54	09 01	03 58	04 19	02 36	23 40	05 48	23 47
17	18 59	18 47	04 44	01 33	27 15	00 N 02	09 31	04 17	04 14	02 38	23 40	05 48	23 48
18	19 00	18 44	04 08	01 10	24 42	00 58	10 00	04 35	04 08	02 40	23 40	05 47	23 48
19	19 01	18 41	03 20	00 46	20 58	01 54	10 29	04 54	04 02	02 42	23 40	05 47	23 48
20	19 02	18 38	02 21	00 S 22	16 12	02 51	10 58	05 12	03 57	02 44	23 40	05 46	23 48
21	19 03	18 35	01 N 13	00 N 02	10 35	03 48	11 27	05 30	03 51	02 46	23 40	05 45	23 49
22	19 03	18 31	00 00	00 25	04 N 22	04 44	11 55	05 48	03 45	02 48	23 40	05 45	23 49
23	19 R 03	18 28	01 S 15	00 49	02 S 12	05 40	12 23	06 07	03 40	02 49	23 40	05 44	23 49
24	19 02	18 25	02 27	01 13	08 48	06 35	12 51	06 25	03 34	02 51	23 40	05 44	23 49
25	19 00	18 22	03 31	01 36	15 03	07 29	13 18	06 43	03 28	02 53	23 40	05 43	23 49
26	18 58	18 19	04 22	02 00	20 34	08 22	13 45	07 00	03 23	02 55	23 40	05 42	23 50
27	18 56	18 15	04 57	02 23	24 55	09 14	14 12	07 18	03 17	02 57	23 39	05 42	23 50
28	18 54	18 12	05 14	02 47	27 41	10 03	14 38	07 36	03 11	02 59	23 39	05 41	23 50
29	18 53	18 09	05 11	03 10	28 37	10 51	15 04	07 54	03 06	03 01	23 39	05 41	23 50
30	18 D 53	18 06	04 49	03 34	27 39	11 37	15 30	08 11	03 00	03 02	23 39	05 40	23 50
31	18 ♓ 53	18 ♓ 03	04 S 10	03 N 57	24 S 57	12 N 21	15 N 55	08 N 28	02 S 54	03 N 04	23 N 39	05 S 39	23 N 50

ZODIAC SIGN ENTRIES

Date	h m	Planets
01	22 03	♂ ♈
02	09 29	☽ ♑
04	12 11	☽ ♒
06	15 45	☽ ♓
07	12 12	♄ ♍
08	21 16	☽ ♈
11	05 33	☽ ♉
13	16 36	☽ ♊
16	05 06	☽ ♋
16	11 53	☿ ♈
18	16 44	☽ ♌
21	01 39	☽ ♍
21	10 05	♀ ♉
21	10 26	☉ ♈
23	07 21	☽ ♎
25	10 36	☽ ♏
27	12 40	☽ ♐
29	14 51	☽ ♑
31	18 02	☽ ♒

LATITUDES

Date	Mercury ☿	Venus ♀	Mars ♂	Jupiter ♃	Saturn ♄	Uranus ♅	Neptune ♆	Pluto ⯓
01	02 S 08	00 S 56	00 S 40	01 S 01	02 N 30	00 N 20	01 N 40	08 N 45
04	02 03	00 50	00 38	01 01	02 31	00 20	01 40	08 45
07	01 54	00 43	00 36	01 01	02 31	00 20	01 41	08 45
10	01 41	00 35	00 34	01 02	02 31	00 20	01 41	08 45
13	01 23	00 27	00 32	01 02	02 32	00 20	01 41	08 45
16	00 59	00 19	00 30	01 02	02 32	00 19	01 41	08 45
19	00 S 31	00 10	00 28	01 02	02 32	00 19	01 41	08 44
22	00 N 02	00 S 01	00 26	01 02	02 32	00 19	01 41	08 44
25	00 37	00 N 08	00 24	01 02	02 32	00 19	01 41	08 44
28	01 13	00 17	00 22	01 02	02 32	00 19	01 41	08 43
31	01 N 48	00 N 26	00 S 20	01 S 03	02 N 32	00 N 19	01 N 41	08 N 43

DATA

Julian Date	2433707
Delta T	+30 seconds
Ayanamsa	23° 10' 34"
Synetic vernal point	05° ♓ 56' 26"
True obliquity of ecliptic	23° 26' 54"

MOON'S PHASES, APSIDES AND POSITIONS ☽

Date	h m	Phase	Longitude	Eclipse Indicator
07	20 51	●	16 ♓ 29	Annular
15	17 40	☽	24 ♊ 20	
23	10 50	○	02 ♎ 09	
30	05 35	☾	08 ♑ 43	

Day	h m		
02	06 56	Perigee	
15	06 19	Apogee	
27	08 33	Perigee	
02	06 23	Max dec	28° S 39'
08	17 45	ON	
16	01 22	Max dec	28° N 39'
23	04 01	OS	
29	11 43	Max dec	28° S 37'

LONGITUDES

Date	Chiron ⚷	Ceres ⚳	Pallas ⚴	Juno ⚵	Vesta ⚶	Black Moon Lilith ⚸
01	01 ♑ 30	13 ♒ 46	25 ♑ 29	07 ♐ 43	19 ♉ 11	16 ♊ 07
11	02 ♑ 02	17 ♒ 31	28 ♑ 47	08 ♐ 50	22 ♉ 32	17 ♊ 15
21	02 ♑ 26	21 ♒ 11	01 ♒ 52	09 ♐ 28	26 ♉ 06	18 ♊ 22
31	02 ♑ 41	24 ♒ 43	04 ♒ 42	09 ♐ 36	29 ♉ 49	19 ♊ 29

ASPECTARIAN

01 Thursday
03 55 ☽ Q ♄
09 55 ♀ □ ♅
13 02 ☽ □ ♃
13 32 ☽ △ ⯓
15 09 ☽ ⚹ ♆
16 40 ☽ Q ☿
02 Friday
07 40 ☽ Q ☉
09 43 ♂ ☍ ♄
10 07 ☽ □ ♄
10 09 ☽ □ ♂
11 21 ☽ Q ♆
14 46 ☽ ♇ ⯓
15 27 ♃ ⊼ ⯓
15 46 ☽ ⚹ ☿
18 42 ☽ ☍ ♅
19 54 ☽ Q ♃
20 43 ☿ ♇ ♆
21 50 ☽ □ ♀
03 Saturday
01 35 ♀ ∥ ♄
05 25 ☽ ⚹ ☉
05 55 ☽ ± ⯓
14 55 ☿ △ ♅
16 02 ☽ ⊼ ⯓
16 30 ☽ ⚹ ♃
17 40 ☽ □ ♆
18 15 ☽ Q ♂
20 44 ☽ ∠ ☿
04 Sunday
07 08 ☽ Q ♀
08 43 ☽ ∠ ☉
11 03 ☉ ± ♆
12 33 ☽ △ ♄
14 10 ☽ # ⯓
14 14 ☽ ⊥ ☿
14 32 ☽ # ♅
15 46 ☽ ⚹ ♂
18 23 ☽ ∠ ♃
21 26 ☽ ⊼ ♅
05 Monday
01 13 ☽ ⊥ ☉
01 59 ☽ ⊻ ☿
05 32 ☽ ⚹ ♀
07 41 ☽ ± ♅
10 01 ☽ ⊥ ♃
12 16 ☽ ⊻ ☉
14 02 ☽ ♇ ♄
18 56 ☽ ∠ ♂
19 04 ☽ ☍ ⯓
20 31 ☽ ⊻ ♃
20 44 ☽ △ ♆
21 16 ♂ ♇ ⯓
23 09 ☽ ♇ ♅
06 Tuesday
05 29 ☽ ± ♄
07 45 ♃ ⊼ ♆
09 54 ☽ ∠ ♀
10 07 ☉ ∥ ♆
11 26 ☽ ⊥ ♂
15 52 ☽ ⊼ ♄
22 32 ☽ ⊻ ♂
22 46 ☽ ♇ ♆
07 Wednesday
01 16 ☽ △ ♅
03 11 ☽ ⊥ ♀
08 52 ☽ ∥ ☿
14 11 ☽ ☌ ☿
14 38 ☽ ± ♆
14 52 ☽ ⊻ ♀
17 26 ☿ ± ♆
19 35 ☽ ∥ ♆
20 51 ☽ ☌ ☉
21 42 ☽ ∥ ☉
22 25 ☽ ∥ ♃
23 36 ☽ ⊼ ⯓
23 57 ☽ # ♀
08 Thursday
01 20 ☽ ⊼ ♆
02 10 ☽ ☌ ♃
03 01 ☿ ⊻ ♀
08 56 ☽ # ♄
10 23 ☽ ± ⯓
12 10 ☽ # ♂
13 03 ☉ ∥ ♃
14 53 ☉ # ♀
15 47 ♀ # ♃
21 05 ☽ ☍ ♄
22 34 ♂ □ ♅
23 57 ☽ ∥ ♂
09 Friday
02 44 ☽ ∥ ♄
02 45 ☽ ♇ ⯓
07 12 ☽ □ ♅
07 45 ☽ ☌ ♂
09 10 ☉ ⊼ ⯓
11 47 ☽ # ☉
12 51 ☽ # ♃
15 08 ☽ ∥ ♀
16 21 ☽ # ♆
19 50 ☽ # ☿
10 Saturday
03 18 ☽ ☌ ♀
05 19 ♀ # ♆
06 17 ☽ ⊻ ☿
06 37 ☽ △ ⯓
07 49 ☉ ⊼ ♆
08 26 ☽ ☍ ♆
08 29 ☿ ⊼ ⯓
08 29 ☽ ⊻ ☉
10 29 ☽ ⊻ ♃
15 21 ☿ # ♀
16 55 ☽ Q ♅
19 42 ☽ ⊥ ☿
20 22 ☿ ⊼ ♆
20 52 ☽ ⊥ ☉
21 37 ☿ ∥ ♆
22 08 ☽ ⊥ ♃
11 Sunday
05 00 ☽ ⊼ ♄
10 10 ☉ ☌ ☿
13 15 ♀ △ ⯓
13 24 ☿ ☌ ♃
13 40 ♂ ∥ ♄
15 43 ☽ ∠ ☉
15 49 ☽ ∠ ♃
16 00 ☽ ⚹ ♅
16 13 ☽ ∠ ☿
16 32 ☽ ± ♄
17 24 ☉ ☌ ♃
20 22 ☽ ⊻ ♂
12 Monday
05 32 ☿ ∥ ♃
07 41 ♀ ☍ ♆
08 54 ☽ ⊥ ♂
10 02 ☽ ♇ ♄
16 33 ☽ □ ⯓
18 26 ☽ ⊼ ♆
19 40 ☽ ⊻ ♀
21 27 ☽ ∠ ♅
21 53 ☽ ⚹ ♃
23 50 ☽ ⚹ ☉
13 Tuesday
03 21 ☽ ⚹ ☿
03 56 ☽ ∠ ♂
06 20 ☽ ∥ ♅
06 20 ☽ ± ♆
07 06 ☽ ∥ ⯓
08 57 ☽ ⊥ ♀
09 58 ☿ ± ⯓
15 05 ☉ # ♂
15 24 ☽ ⊥ ♅
15 39 ☽ △ ♄
22 21 ☽ Q ♃
22 30 ♀ ⊻ ♃
23 28 ☿ # ♂
14 Wednesday
00 19 ☽ ♇ ♆
02 00 ☽ Q ☉
03 27 ☽ ⊻ ♅
04 26 ☽ Q ⯓
05 04 ☽ ∠ ♀
08 08 ☽ Q ☿
10 40 ♅ St D
11 17 ☉ ∥ ☿
12 04 ☽ ⚹ ♂
16 53 ☿ # ♄
23 31 ☉ # ♄
15 Thursday
04 37 ☽ ⚹ ⯓
06 31 ☽ △ ♆
06 38 ☉ ± ⯓
11 25 ☽ □ ♃
13 59 ☽ Q ♂
14 49 ☽ ⚹ ♀
17 40 ☽ □ ☉
16 Friday
00 14 ♀ Q ♅
03 44 ☽ □ ☿
03 45 ☽ □ ♄
03 50 ☿ ☍ ♄
10 51 ☽ ∠ ⯓
15 59 ☽ ☌ ♅
17 46 ☽ Q ♀
17 Saturday
04 40 ☽ □ ♂
04 52 ☽ ⊥ ⯓
08 42 ♂ # ♃
15 35 ☽ Q ♄
16 48 ☽ ⊻ ⯓
18 38 ☽ □ ♆
21 56 ☿ ♇ ⯓
18 Sunday
00 46 ☽ △ ♃
09 33 ☽ □ ♀
10 58 ☽ △ ☉
15 04 ☽ ⚹ ♄
15 30 ☉ # ☿
18 31 ☽ ∥ ⯓
19 27 ☽ ∥ ♅
19 Monday
02 54 ☽ △ ☿
03 15 ☽ ⊻ ♅
05 04 ☿ □ ♅
05 46 ☽ Q ♆
06 37 ☽ ♇ ♃
14 46 ☽ ⊥ ♅
18 27 ☽ ♇ ☉
19 11 ☽ △ ♂
19 47 ☽ ∠ ♄
20 Tuesday
00 08 ☽ ± ♃
01 35 ♀ ⊥ ♃
02 53 ☽ ☌ ⯓
04 35 ☽ ⚹ ♆
07 50 ☽ ∠ ♅
08 46 ☿ ∥ ♄
10 10 ☉ ☍ ♄
11 38 ☽ ⊼ ♃
12 39 ☽ ⊥ ♄
12 43 ☽ ♇ ☿
12 52 ☽ ± ☉
14 03 ♀ ⊼ ♄
23 43 ☽ ⊻ ♄
21 Wednesday
00 47 ☽ △ ♀
00 55 ☽ ⊼ ☉
01 03 ☽ ♇ ♂
08 20 ☽ ± ☿
08 23 ☽ ∠ ♆
08 35 ☉ ⊻ ♀
08 46 ☽ ∥ ♀
11 35 ☽ ⚹ ♅
13 14 ☿ # ♃
18 29 ☽ ± ♂
21 08 ☽ ⊼ ☿
22 Thursday
00 39 ☽ ⊥ ♆
05 56 ☽ ⊼ ♂
06 45 ☽ ♇ ♀
06 47 ☽ # ♆
06 49 ☽ ∥ ♂
07 23 ♂ # ♆
09 13 ☽ Q ♅
09 47 ☽ ⊻ ⯓
10 47 ☽ ∥ ☿
11 21 ☽ ⊻ ♆
14 17 ☽ # ♃
17 47 ☽ ∥ ♄
19 06 ☽ ☍ ♃
20 23 ☽ ⊥ ⯓
23 Friday
00 14 ☽ ± ♀
01 40 ☽ ∥ ☉
05 13 ☽ ☌ ♄
06 39 ☽ # ☉
10 50 ☽ ☍ ☉
11 43 ☽ ⊼ ♀
12 06 ☽ ∠ ⯓
13 53 ☿ # ♆
14 15 ☽ # ♄
16 47 ☽ □ ♅
17 11 ☽ ∥ ♃
24 Saturday
00 46 ☽ ∥ ♆
02 33 ☽ # ☿
02 49 ☽ # ♂
04 16 ☉ ♇ ⯓
05 14 ☿ ∥ ♂
10 14 ☽ ☍ ☿
13 11 ☽ ☍ ♂
13 50 ☽ ⚹ ⯓
15 18 ☽ ☌ ♆
23 42 ☽ ⊼ ♃
25 Sunday
00 58 ♂ △ ⯓
04 29 ☽ # ♀
05 21 ♀ ± ♄
08 15 ☽ ⊻ ♄
09 59 ☽ ± ♃
10 06 ☽ Q ⯓
13 05 ☿ △ ⯓
15 47 ♃ ± ⯓
17 45 ☽ ⊼ ☉
18 16 ☽ ⊥ ♄
19 33 ☽ ☍ ♀
19 45 ☽ △ ♅
21 43 ☿ ☌ ♂
22 01 ♀ ⚹ ♅
26 Monday
00 07 ☿ ☍ ♆
01 20 ☽ ♇ ♃
03 24 ♂ ☍ ♆
04 32 ☽ ± ☉
09 14 ☽ ∠ ♄
16 10 ☽ □ ⯓
17 34 ☽ ⊻ ♆
18 23 ☽ ⊼ ♂
20 07 ☽ ⊼ ☿
20 39 ☽ ♇ ☉
20 49 ☽ ♇ ♅
22 56 ☉ □ ♅
27 Tuesday
02 46 ☽ △ ♃
03 33 ☽ ⊥ ♆
04 13 ☽ # ♅
04 56 ☽ ± ♂
05 14 ☽ # ⯓
07 30 ☽ ± ☿
10 05 ☽ ⚹ ♄
11 48 ☽ ± ♅
18 30 ☽ ∠ ♆
20 44 ☽ ♇ ♂
21 48 ☽ ⊼ ♅
23 28 ☽ △ ☉
28 Wednesday
00 33 ☽ ♇ ☿
02 09 ☽ ⊼ ♀
05 58 ☽ Q ♄
13 05 ☽ ± ♀
18 07 ☽ △ ⯓
19 29 ☽ ⚹ ♆
19 52 ☿ Q ♅
23 11 ☽ △ ♂
29 Thursday
01 24 ☉ ∥ ♄
04 58 ☽ △ ☿
05 33 ☽ ♇ ♀
05 41 ☽ □ ♃
07 37 ♀ ∠ ♃
08 13 ☉ # ♃
11 58 ☽ □ ♄
12 19 ☿ ⊻ ♃
15 36 ☽ Q ♆
19 18 ☽ ♇ ⯓
30 Friday
00 09 ☽ ☍ ♅
03 58 ♃ # ♄
05 35 ☽ □ ☉
09 14 ☽ △ ♀
10 32 ☽ ± ⯓
12 38 ☽ Q ♃
20 45 ☽ ⊼ ⯓
22 06 ☽ □ ♆
31 Saturday
04 52 ☽ □ ♂
09 32 ☽ ⚹ ♃
14 18 ☽ □ ☿
14 43 ☽ Q ☉
14 48 ☽ △ ♄
18 48 ☿ ⊼ ♄
19 11 ☽ # ⯓
20 20 ☽ # ♅

All ephemeris data is given at 12.00 UT and the Moon's longitude is additionally given for 24.00 UT

APRIL 1951

Date	Sidereal time h m s	LONGITUDES Sun ☉ ° ' "	Moon ☽ ° ' "	Moon ☽ 24.00 ° ' "	Mercury ☿ ° '	Venus ♀ ° '	Mars ♂ ° '	Jupiter ♃ ° '	Saturn ♄ ° '	Uranus ♅ ° '	Neptune ♆ ° '	Pluto ⯓ ° '
01	00 36 09	10 ♈ 57 07	10 ♒ 19 58	17 ♒ 10 24	29 ♈ 08	13 ♉ 25	23 ♈ 20	25 ♓ 20	28 ♍ 03	05 ♋ 33	18 ♎ 22	17 ♌ 36
02	00 40 06	11 56 19	23 ♒ 57 51	00 ♓ 42 16	00 ♉ 28	14 37	24 05	25 34	27 R 59	05 34	18 R 21	17 R 35
03	00 44 02	12 55 29	07 ♓ 23 38	14 ♓ 01 55	01 43	15 49	24 50	25 49	27 54	05 35	18 19	17 35
04	00 47 59	13 54 38	20 ♓ 37 06	27 ♓ 09 08	02 52	17 02	25 35	26 03	27 50	05 36	18 17	17 34
05	00 51 55	14 53 44	03 ♈ 38 00	10 ♈ 03 41	03 55	18 14	26 20	26 17	27 45	05 37	18 16	17 33
06	00 55 52	15 52 49	16 ♈ 26 12	22 ♈ 45 31	04 52	19 26	27 05	26 31	27 41	05 38	18 14	17 33
07	00 59 48	16 51 51	29 ♈ 01 43	05 ♉ 14 52	05 43	20 38	27 50	26 45	27 37	05 40	18 12	17 32
08	01 03 45	17 50 51	11 ♉ 25 03	17 ♉ 32 26	06 27	21 49	28 35	26 59	27 32	05 41	18 11	17 31
09	01 07 42	18 49 50	23 ♉ 37 14	29 ♉ 39 39	07 04	23 01	29 ♈ 20	27 13	27 28	05 42	18 09	17 31
10	01 11 38	19 48 46	05 ♊ 40 00	11 ♊ 38 38	07 35	24 13	00 ♉ 04	27 27	27 24	05 44	18 07	17 30
11	01 15 35	20 47 40	17 ♊ 35 55	23 ♊ 32 19	08 00	25 25	00 49	27 41	27 20	05 45	18 06	17 30
12	01 19 31	21 46 32	29 ♊ 28 16	05 ♋ 24 19	08 17	26 36	01 34	27 55	27 16	05 46	18 04	17 29
13	01 23 28	22 45 22	11 ♋ 21 00	17 ♋ 18 53	08 29	27 48	02 18	28 09	27 12	05 48	18 03	17 29
14	01 27 24	23 44 09	23 ♋ 18 35	29 ♋ 20 41	08 33	28 ♉ 59	03 03	28 23	27 08	05 50	18 01	17 28
15	01 31 21	24 42 54	05 ♌ 25 49	11 ♌ 34 35	08 R 32	00 ♊ 10	03 47	28 37	27 04	05 51	17 59	17 28
16	01 35 17	25 41 37	17 ♌ 47 34	24 ♌ 05 20	08 24	01 21	04 32	28 50	27 00	05 53	17 58	17 28
17	01 39 14	26 40 17	00 ♍ 28 21	06 ♍ 57 06	08 11	02 33	05 16	29 04	26 56	05 55	17 56	17 27
18	01 43 11	27 38 55	13 ♍ 31 54	20 ♍ 13 02	07 52	03 44	06 00	29 18	26 52	05 56	17 54	17 27
19	01 47 07	28 37 32	27 ♍ 00 36	03 ♎ 54 37	07 28	04 55	06 45	29 31	26 49	05 58	17 53	17 27
20	01 51 04	29 ♈ 36 05	10 ♎ 54 53	18 ♎ 01 05	07 00	06 05	07 29	29 45	26 45	06 00	17 51	17 26
21	01 55 00	00 ♉ 34 37	25 ♎ 12 43	02 ♏ 29 05	06 28	07 16	08 13	29 ♓ 58	26 42	06 02	17 50	17 26
22	01 58 57	01 33 08	09 ♏ 49 25	17 ♏ 12 44	05 53	08 27	08 57	00 ♈ 12	26 38	06 04	17 48	17 26
23	02 02 53	02 31 36	24 ♏ 38 02	02 ♐ 04 15	05 15	09 37	09 41	00 25	26 35	06 06	17 46	17 26
24	02 06 50	03 30 03	09 ♐ 30 19	16 ♐ 55 12	04 36	10 48	10 25	00 39	26 31	06 08	17 45	17 25
25	02 10 46	04 28 28	24 ♐ 17 57	01 ♑ 37 46	03 56	11 58	11 09	00 52	26 28	06 10	17 43	17 25
26	02 14 43	05 26 51	08 ♑ 53 57	16 ♑ 05 59	03 15	13 08	11 53	01 05	26 25	06 12	17 42	17 25
27	02 18 40	06 25 13	23 ♑ 13 29	00 ♒ 16 14	02 35	14 19	12 37	01 18	26 22	06 14	17 40	17 25
28	02 22 36	07 23 33	07 ♒ 14 06	14 ♒ 07 08	01 56	15 29	13 21	01 31	26 19	06 17	17 39	17 25
29	02 26 33	08 21 52	20 ♒ 55 26	27 ♒ 39 10	01 18	16 39	14 04	01 44	26 16	06 19	17 37	17 25
30	02 30 29	09 ♉ 20 09	04 ♓ 18 34	10 ♓ 53 54	00 ♉ 43	17 ♊ 48	14 ♉ 48	01 ♈ 57	26 ♍ 13	06 ♋ 21	17 ♎ 36	17 ♌ 25

Date	Moon True ☊ ° '	Moon Mean ☊ ° '	Moon ☽ Latitude ° '
01	18 ♓ 55	18 ♓ 00	03 S 17
02	18 D 56	17 56	02 13
03	18 57	17 53	01 S 03
04	18 R 58	17 50	00 N 09
05	18 57	17 47	01 20
06	18 55	17 44	02 25
07	18 51	17 40	03 22
08	18 47	17 37	04 08
09	18 41	17 34	04 43
10	18 36	17 31	05 04
11	18 31	17 28	05 12
12	18 28	17 25	05 06
13	18 25	17 21	04 47
14	18 25	17 18	04 16
15	18 D 25	17 15	03 33
16	18 27	17 12	02 39
17	18 28	17 09	01 36
18	18 29	17 06	00 N 27
19	18 R 29	17 02	00 S 46
20	18 27	16 59	01 59
21	18 23	16 56	03 05
22	18 18	16 53	04 02
23	18 11	16 50	04 42
24	18 04	16 46	05 04
25	17 58	16 43	05 06
26	17 54	16 40	04 48
27	17 51	16 37	04 11
28	17 50	16 34	03 21
29	17 D 51	16 31	02 20
30	17 ♓ 52	16 ♓ 27	01 S 12

DECLINATIONS

Date	Sun ☉ ° '	Moon ☽ ° '	Mercury ☿ ° '	Venus ♀ ° '	Mars ♂ ° '	Jupiter ♃ ° '	Saturn ♄ ° '	Uranus ♅ ° '	Neptune ♆ ° '	Pluto ⯓ ° '
01	04 N 20	20 S 49	13 N 01	16 N 20	08 N 46	02 S 49	03 N 06	23 N 39	05 S 39	23 N 51
02	04 43	15 38	13 40	16 45	09 03	02 43	03 08	23 39	05 38	23 51
03	05 06	09 47	14 15	17 09	09 20	02 38	03 10	23 39	05 37	23 51
04	05 29	03 S 35	14 48	17 33	09 37	02 32	03 11	23 39	05 37	23 51
05	05 52	02 N 40	15 17	17 56	09 54	02 26	03 13	23 39	05 36	23 51
06	06 15	08 42	15 43	18 19	10 11	02 21	03 15	23 39	05 35	23 51
07	06 38	14 17	16 07	18 41	10 28	02 15	03 17	23 39	05 35	23 51
08	07 00	19 12	16 27	19 03	10 44	02 10	03 18	23 39	05 34	23 51
09	07 23	23 14	16 43	19 25	11 01	02 04	03 20	23 39	05 34	23 51
10	07 45	26 14	16 56	19 46	11 17	01 59	03 21	23 39	05 33	23 51
11	08 07	28 02	17 06	20 06	11 33	01 53	03 23	23 39	05 32	23 51
12	08 29	28 33	17 13	20 26	11 50	01 48	03 25	23 39	05 32	23 51
13	08 51	27 44	17 16	20 46	12 06	01 43	03 26	23 39	05 31	23 51
14	09 13	25 38	17 15	21 05	12 21	01 37	03 28	23 38	05 30	23 51
15	09 35	22 21	17 11	21 23	12 37	01 32	03 29	23 38	05 30	23 51
16	09 56	18 02	17 04	21 41	12 53	01 26	03 31	23 38	05 29	23 51
17	10 17	12 49	16 54	21 58	13 08	01 21	03 32	23 38	05 29	23 51
18	10 38	06 53	16 40	22 15	13 24	01 16	03 33	23 38	05 28	23 51
19	10 59	00 N 29	16 24	22 31	13 39	01 10	03 35	23 38	05 27	23 51
20	11 20	06 S 08	16 04	22 46	13 54	01 05	03 36	23 38	05 27	23 51
21	11 41	12 38	15 43	23 01	14 09	01 00	03 37	23 38	05 26	23 51
22	12 01	18 35	15 18	23 15	14 24	00 55	03 39	23 38	05 26	23 51
23	12 21	23 30	14 53	23 29	14 38	00 49	03 40	23 38	05 25	23 51
24	12 41	26 54	14 25	23 42	14 53	00 44	03 41	23 38	05 24	23 51
25	13 01	28 25	13 56	23 54	15 07	00 39	03 42	23 37	05 24	23 51
26	13 21	27 56	13 27	24 06	15 21	00 34	03 43	23 37	05 23	23 50
27	13 40	25 35	12 57	24 17	15 35	00 29	03 44	23 37	05 23	23 50
28	13 59	21 42	12 28	24 28	15 49	00 24	03 45	23 37	05 22	23 50
29	14 18	16 44	11 59	24 38	16 03	00 19	03 46	23 37	05 21	23 50
30	14 N 37	11 S 03	11 N 31	24 N 47	16 N 16	00 S 13	03 N 47	23 N 37	05 S 21	23 N 50

ZODIAC SIGN ENTRIES

Date	h	m	Planets
02	03	27	☿ ♉
02	22	44	☽ ♓
05	05	16	☽ ♈
07	13	52	☽ ♉
10	00	41	☽ ♊
10	09	37	♂ ♉
12	13	04	☽ ♋
15	01	18	☽ ♌
15	08	33	♀ ♊
17	11	07	☽ ♍
19	17	13	☽ ♎
20	21	48	☉ ♉
21	14	57	♃ ♈
21	19	55	☽ ♏
23	20	40	☽ ♐
25	21	19	☽ ♑
27	23	32	☽ ♒
30	04	13	☽ ♓

LATITUDES

Date	Mercury ☿ ° '	Venus ♀ ° '	Mars ♂ ° '	Jupiter ♃ ° '	Saturn ♄ ° '	Uranus ♅ ° '	Neptune ♆ ° '	Pluto ⯓ ° '
01	01 N 59	00 N 29	00 S 20	01 S 03	02 N 32	00 N 19	01 N 41	08 N 43
04	02 28	00 39	00 18	01 03	02 32	00 19	01 41	08 43
07	02 51	00 48	00 16	01 03	02 32	00 19	01 41	08 42
10	03 04	00 58	00 14	01 03	02 32	00 19	01 41	08 42
13	03 05	01 07	00 12	01 04	02 32	00 19	01 41	08 41
16	02 55	01 16	00 10	01 04	02 31	00 19	01 41	08 41
19	02 31	01 25	00 08	01 04	02 31	00 19	01 41	08 40
22	01 56	01 33	00 06	01 05	02 31	00 19	01 41	08 40
25	01 11	01 41	00 04	01 05	02 30	00 19	01 41	08 39
28	00 N 20	01 49	00 02	01 05	02 30	00 19	01 41	08 39
31	00 S 31	01 N 56	00 00	01 S 06	02 N 29	00 N 19	01 N 41	08 N 38

DATA

Julian Date	2433738
Delta T	+30 seconds
Ayanamsa	23° 10' 37"
Synetic vernal point	05° ♓ 56' 22"
True obliquity of ecliptic	23° 26' 54"

LONGITUDES

Date	Chiron ⚷ ° '	Ceres ⚳ ° '	Pallas ⚴ ° '	Juno ⚵ ° '	Vesta ⚶ ° '	Black Moon Lilith ⚸ ° '
01	02 ♑ 43	25 ♒ 04	04 ♒ 58	09 ♐ 35	00 ♊ 12	19 ♊ 36
11	02 ♑ 48	28 ♒ 27	07 ♒ 29	09 ♐ 07	04 ♊ 05	20 ♊ 43
21	02 ♑ 44	01 ♓ 40	09 ♒ 40	08 ♐ 05	08 ♊ 05	21 ♊ 50
31	02 ♑ 31	04 ♓ 41	11 ♒ 27	06 ♐ 34	12 ♊ 11	22 ♊ 57

MOON'S PHASES, APSIDES AND POSITIONS ☽

Date	h	m	Phase	Longitude ° '	Eclipse Indicator
06	10	52	●	15 ♈ 50	
14	12	56	☽	23 ♋ 46	
21	21	30	○	00 ♏ 58	
28	12	18	☾	07 ♒ 24	

Day	h	m		
12	01	34	Apogee	
23	22	57	Perigee	
05	01	43	0N	
12	09	12	Max dec	28° N 33'
19	13	46	0S	
25	18	05	Max dec	28° S 29'

ASPECTARIAN

h m Aspects

01 Sunday
03 39 ☽ ⊼ ♅
05 16 ♀ ♇ ♄
12 00 ☽ ∠ ♃
13 10 ☽ ✶ ☉
13 52 ☽ Q ♂
14 08 ☽ ± ♅
16 44 ☽ ♇ ♄
17 55 ☽ □ ♀
18 49 ♂ Q ♅

02 Monday
00 45 ☽ ☍ ⯓
01 14 ☽ Q ☿
02 05 ☽ △ ♆
04 05 ☽ ⊥ ♃
05 58 ☽ ♇ ♅
07 28 ☽ # ♀
08 30 ☽ ± ♄
12 14 ☽ ✶ ♂
14 54 ☽ ⊻ ♃
17 42 ☽ ∠ ☉
19 06 ☽ ⊼ ♄
19 31 ☽ # ☿

03 Tuesday
00 47 ☽ ✶ ☿
04 41 ☽ ♇ ♆
04 57 ☽ Q ♀
08 44 ☽ △ ♅
11 05 ☽ ⊥ ☉
13 39 ☽ # ♂
14 19 ☿ ⊥ ♃
16 41 ☽ ∠ ♂
20 52 ☽ ± ♆
22 48 ☽ ⊻ ☉

04 Wednesday
04 11 ☽ ∥ ♆
04 47 ☽ ✶ ♀
05 06 ☽ # ☉
06 26 ☽ ⊼ ⯓
06 31 ☽ ∠ ☿
07 45 ☽ ⊼ ♆
10 00 ☽ ⊥ ♂
13 29 ☽ # ♄
16 04 ☽ ∥ ♃
17 24 ☽ ± ⯓
19 31 ☉ # ♆
21 41 ☽ ⊻ ♂
22 09 ☽ ☌ ♃
22 41 ♀ □ ⯓

05 Thursday
00 31 ☽ ⊥ ☿
01 11 ☽ ☍ ♄
08 24 ☿ ± ♄
09 15 ♂ ⊻ ♃
10 00 ☽ ♇ ⯓
11 09 ☽ # ♃
11 10 ☽ ∠ ♀
12 35 ☽ ⊻ ☿
12 40 ♀ ⊼ ♆
14 10 ☽ ∥ ♄
15 42 ☽ □ ♅
23 30 ☽ # ♆

06 Friday
01 27 ☽ ∥ ☉
05 44 ☽ ⊥ ♀
10 52 ☽ ☌ ☉
14 06 ☽ △ ⯓
15 24 ☽ ☍ ♆
18 15 ☽ ⊻ ♀
18 30 ☽ ∥ ♂

07 Saturday
01 42 ☽ Q ♅
05 27 ♂ ⊼ ♄
07 33 ☽ ⊻ ♃
09 18 ☽ ⊼ ♄
09 34 ☽ ☌ ♂
10 20 ☿ ✶ ♅
12 39 ♀ ∠ ♅
19 19 ☽ ⊥ ♃
20 47 ☽ ± ♄
21 10 ☽ ∥ ☿

08 Sunday
00 49 ☽ ✶ ♅
01 45 ☽ ☌ ☿
04 10 ☉ △ ⯓
11 13 ☽ ∥ ♀
13 08 ☽ ∠ ♃
14 11 ☽ ♇ ♄
19 52 ☉ ☍ ♆
23 57 ☽ □ ⯓

09 Monday
01 14 ☽ ⊼ ♆
01 43 ☽ ⊻ ☉
06 13 ☽ ∠ ♅
10 41 ☽ ☌ ♀
13 03 ☽ ± ♆
14 37 ☽ ⊥ ☉
14 47 ☽ ∥ ♅
16 15 ☽ ∥ ⯓
19 17 ☽ ✶ ♃
19 36 ☽ △ ♄

10 Tuesday
00 05 ☽ ⊻ ♂
00 06 ☽ ⊥ ♅
06 55 ☽ ♇ ♆
07 44 ♃ ☍ ♄
10 08 ☽ ∠ ☉
10 13 ♀ ± ♆
11 41 ☽ Q ⯓
12 07 ☽ ⊻ ♅
12 52 ☽ ⊥ ♂
16 00 ☽ ⊻ ☿
19 45 ☽ Q ♃

11 Wednesday
04 30 ☽ ⊥ ☿
08 10 ☽ ∠ ♂
11 48 ☽ ✶ ⯓
13 00 ☽ △ ♆
19 02 ☽ ✶ ☉
23 12 ☽ ∠ ☿

12 Thursday
05 33 ☽ ⊻ ♀
07 33 ☽ □ ♄
08 48 ☽ □ ♃
16 31 ☽ ✶ ♂
18 06 ☽ ∠ ⯓
19 02 ☽ ⊥ ♀
21 29 ☽ Q ☉

13 Friday
00 35 ♀ △ ♄
00 46 ☽ ☌ ♅
06 08 ☽ ✶ ☿
12 16 ☽ ⊥ ⯓
15 14 ☽ ∠ ♀
18 21 ☽ Q ♂
19 42 ☽ Q ♄
20 55 ♀ ✶ ♃

14 Saturday
00 19 ☽ ⊻ ⯓
01 26 ☽ □ ♆
06 28 ☽ Q ☿
12 56 ☽ □ ☉
14 17 ☉ Q ♅
14 26 ♂ ± ♄
17 50 ☿ St R
19 34 ☽ ✶ ♄
22 17 ☽ △ ♃

15 Sunday
00 31 ☽ ✶ ♀
02 02 ☽ ∥ ⯓
03 32 ☽ ∥ ♅
05 27 ♀ ⊥ ♅
08 34 ☽ □ ♂
12 50 ☽ ⊻ ♅
13 05 ☽ Q ♆
17 30 ☽ ∥ ♀
18 01 ☽ □ ☿

16 Monday
00 34 ☽ ⊥ ♅
00 53 ☽ ∠ ♄
02 33 ☽ Q ♀
04 14 ☽ ♇ ♃
11 22 ☽ ☌ ⯓
12 19 ☽ ✶ ♆
16 52 ☽ ∥ ☿
17 55 ☽ ∠ ♅
18 06 ☽ ⊥ ♄
21 48 ☽ ± ♃

17 Tuesday
02 37 ♂ ⊥ ♃
04 17 ☽ △ ☉
05 24 ☽ ⊻ ♄
09 20 ☽ ⊼ ♃
10 40 ☽ ∥ ♂
16 15 ☽ □ ♀
16 34 ☽ ∠ ♆
18 05 ☉ ⊼ ♄
19 44 ♀ ♇ ♆
21 26 ☽ △ ♂
21 56 ☽ ∥ ☉
22 06 ☽ ✶ ♅

18 Wednesday
01 56 ☽ △ ☿
02 58 ☉ ⊥ ♀
09 04 ☽ ⊥ ♆
09 48 ♂ ✶ ♅
10 16 ☽ ♇ ☉
17 28 ☽ # ♆
19 03 ☽ ⊻ ⯓
19 51 ☽ ⊻ ♆
19 57 ☽ Q ♅

19 Thursday
00 37 ☽ ∥ ♄
02 11 ☽ ♇ ♂
03 41 ☽ ± ☉
04 15 ☽ ♇ ☿
05 43 ☽ ⊥ ⯓
09 26 ☽ # ♃
11 39 ☽ ☌ ♄
15 03 ☽ ⊼ ☉
16 28 ☽ ☍ ♃
17 57 ☽ ∥ ♃
18 53 ☽ ± ♂
19 32 ☽ ± ☿
21 28 ☽ ∠ ⯓
22 48 ♀ Q ⯓

20 Friday
02 39 ☿ ☌ ♂
02 47 ☽ # ♄
02 59 ☽ △ ♀
03 35 ☽ □ ♅
05 33 ☽ ⊼ ☿
05 49 ☽ ⊼ ♂
09 30 ☽ ∥ ♆
10 10 ♀ ⊻ ♅
16 40 ☉ ⊻ ♃
23 01 ☽ ✶ ⯓
23 42 ☽ ☌ ♆

21 Saturday
00 54 ☿ ⊻ ♀
06 40 ☽ ♇ ♀
08 11 ☽ # ☉
14 27 ☽ ⊻ ♄
18 04 ☽ # ♂
18 59 ☽ Q ⯓
19 59 ☽ ⊼ ♃
21 30 ☽ ☍ ☉
22 53 ☽ ± ♀
23 16 ☽ # ☿

22 Sunday
00 18 ☽ ⊥ ♄
02 54 ☿ ⊥ ♃
05 08 ☿ ✶ ♅
05 49 ☽ ☍ ☿
05 51 ☽ △ ♅
05 59 ☽ ± ♃
09 34 ☽ ⊼ ♀
10 30 ☽ ☍ ♂
14 56 ☽ ∠ ♄
20 52 ☽ ♇ ♃

23 Monday
00 21 ☽ □ ⯓
00 56 ☽ ⊻ ♆
06 17 ☽ ♇ ♅
10 37 ☽ ⊥ ♆
11 55 ☽ # ♀
12 44 ☽ # ♅
13 14 ☉ ± ♄
14 01 ☽ # ⯓
15 08 ☽ ✶ ♄
15 23 ♀ ⊻ ♂
20 19 ☿ ∥ ♂
20 50 ☽ ± ♅
21 29 ☽ △ ♃

24 Tuesday
01 06 ☽ ∠ ♆
01 38 ☽ ⊼ ☉
03 42 ♀ ∥ ♅
04 25 ☽ ⊼ ☿
06 33 ☽ ⊼ ♅
09 27 ☿ ⊥ ♀
10 25 ☽ Q ♄
12 00 ☽ ± ☉
13 33 ☽ ⊼ ♂
13 42 ☽ ± ☿
14 16 ☽ ☍ ♀
23 46 ☽ ± ♂

25 Wednesday
00 49 ☽ △ ⯓
01 19 ☽ ✶ ♆
03 36 ☽ ♇ ☉
03 38 ☽ ♇ ☿
04 02 ☉ ☌ ☿
04 32 ♀ ∥ ⯓
15 11 ☽ ♇ ♂
15 32 ☽ □ ♄
20 51 ☽ Q ♆
21 49 ♂ ♇ ♄
22 54 ☽ □ ♃

26 Thursday
01 18 ☽ ♇ ⯓
03 05 ☽ △ ☿
05 53 ☽ △ ☉
07 32 ☽ ☍ ♅
10 37 ♀ Q ♃
15 14 ☉ ∥ ☿
16 11 ☽ ± ⯓
17 13 ☽ △ ♂
19 41 ☽ ⊼ ♀

27 Friday
02 13 ☽ ⊼ ⯓
02 39 ☽ □ ♆
05 17 ☽ Q ♃
06 38 ☽ ± ♀
07 22 ☉ ✶ ♅
17 19 ☽ △ ♄
20 17 ☿ ± ♄
20 33 ☽ # ♀
23 18 ☽ ♇ ♀
23 48 ☽ # ⯓

28 Saturday
01 08 ☽ # ♅
02 00 ☽ ✶ ♃
03 15 ☽ □ ☿
10 20 ☽ ⊼ ♅
12 18 ☽ □ ☉
16 10 ☉ ⊥ ♃
19 04 ☽ ♇ ♄
20 48 ☽ ± ♅
23 14 ☽ □ ♂
23 23 ☿ ⊻ ♃

29 Sunday
03 44 ☽ △ ♀
04 29 ☽ ∠ ♃
05 48 ☽ ☍ ⯓
06 10 ☽ △ ♆
07 25 ☿ ∠ ♀
09 15 ☽ Q ☿
10 51 ☽ ± ♄
12 42 ☽ ♇ ♅
14 55 ☽ # ♂
20 43 ☽ ⊥ ♃
21 29 ☽ ⊼ ♄
22 01 ☽ # ☉
22 27 ☽ Q ☉

30 Monday
03 56 ♀ ✶ ⯓
05 16 ⯓ St D
05 47 ☽ ✶ ☿
07 40 ♀ △ ♆
07 41 ☽ ⊻ ♃
08 54 ☽ ♇ ♆
09 06 ☽ Q ♂
09 59 ☽ # ☿
15 43 ☽ △ ♅
21 52 ☽ ✶ ☉

All ephemeris data is given at 12.00 UT and the Moon's longitude is additionally given for 24.00 UT

MAY 1951

Date	Sidereal time h m s	Sun ☉ ° ' "	Moon ☽ ° ' "	Moon ☽ 24.00 ° ' "	Mercury ☿ ° '	Venus ♀ ° '	Mars ♂ ° '	Jupiter ♃ ° '	Saturn ♄ ° '	Uranus ♅ ° '	Neptune ♆ ° '	Pluto ⯓ ° '
01	02 34 26	10 ♉ 18 24	17 ♓ 25 25	23 ♓ 53 23	00 ♉ 12	18 ♊ 58	15 ♉ 32	02 ♈ 10	26 ♍ 11	06 ♋ 23	17 ♎ 34	17 ♌ 25
02	02 38 22	11 16 38	00 ♈ 18 04	06 ♈ 39 42	29 ♈ 43	20 08	16 15	02 23	26 R 08	06 26	17 R 33	17 D 25
03	02 42 19	12 14 51	12 ♈ 58 29	19 ♈ 14 38	29 R 18	21 17	16 59	02 36	26 05	06 28	17 31	17 25
04	02 46 15	13 13 02	25 ♈ 28 16	01 ♉ 39 33	28 58	22 27	17 42	02 49	26 03	06 31	17 30	17 25
05	02 50 12	14 11 11	07 ♉ 48 37	13 ♉ 55 34	28 42	23 36	18 26	03 02	26 00	06 33	17 28	17 25
06	02 54 09	15 09 19	20 ♉ 00 31	26 ♉ 03 35	28 30	24 45	19 09	03 14	25 58	06 36	17 27	17 26
07	02 58 05	16 07 25	02 ♊ 04 55	08 ♊ 04 40	28 23	25 54	19 52	03 27	25 56	06 38	17 25	17 26
08	03 02 02	17 05 29	14 ♊ 03 02	20 ♊ 00 12	28 D 21	27 03	20 36	03 39	25 54	06 41	17 24	17 26
09	03 05 58	18 03 32	25 ♊ 56 28	01 ♋ 52 07	28 23	28 12	21 19	03 52	25 52	06 44	17 23	17 26
10	03 09 55	19 01 32	07 ♋ 47 31	13 ♋ 43 04	28 30	29 ♊ 21	22 02	04 04	25 50	06 46	17 21	17 26
11	03 13 51	19 59 31	19 ♋ 39 11	25 ♋ 36 24	28 42	00 ♋ 29	22 45	04 16	25 48	06 49	17 20	17 27
12	03 17 48	20 57 29	01 ♌ 35 13	07 ♌ 36 13	28 58	01 38	23 28	04 28	25 46	06 52	17 19	17 27
13	03 21 44	21 55 24	13 ♌ 40 00	19 ♌ 47 11	29 19	02 46	24 11	04 40	25 45	06 54	17 17	17 27
14	03 25 41	22 53 18	25 ♌ 58 24	02 ♍ 14 17	29 ♈ 44	03 54	24 54	04 52	25 43	06 57	17 16	17 28
15	03 29 38	23 51 09	08 ♍ 35 26	15 ♍ 02 26	00 ♉ 13	05 02	25 37	05 04	25 42	07 00	17 15	17 28
16	03 33 34	24 48 59	21 ♍ 35 49	28 ♍ 16 00	00 46	06 10	26 20	05 16	25 41	07 03	17 14	17 29
17	03 37 31	25 46 48	05 ♎ 03 18	11 ♎ 57 55	01 24	07 18	27 03	05 28	25 39	07 06	17 12	17 29
18	03 41 27	26 44 34	18 ♎ 59 50	26 ♎ 08 55	02 05	08 25	27 46	05 39	25 38	07 09	17 11	17 30
19	03 45 24	27 42 19	03 ♏ 24 43	10 ♏ 46 40	02 50	09 33	28 28	05 51	25 37	07 12	17 10	17 30
20	03 49 20	28 40 03	18 ♏ 13 55	25 ♏ 45 25	03 38	10 40	29 11	06 02	25 36	07 15	17 09	17 31
21	03 53 17	29 ♉ 37 45	03 ♐ 19 58	10 ♐ 56 16	04 30	11 47	29 ♉ 54	06 14	25 35	07 18	17 08	17 31
22	03 57 13	00 ♊ 35 26	18 ♐ 32 56	26 ♐ 08 35	05 25	12 54	00 ♊ 36	06 25	25 35	07 21	17 07	17 32
23	04 01 10	01 33 06	03 ♑ 41 55	11 ♑ 11 46	06 23	14 01	01 19	06 36	25 34	07 24	17 05	17 33
24	04 05 07	02 30 44	18 ♑ 37 08	25 ♑ 57 13	07 25	15 07	02 01	06 47	25 33	07 27	17 04	17 33
25	04 09 03	03 28 22	03 ♒ 11 26	10 ♒ 19 24	08 30	16 14	02 44	06 58	25 33	07 30	17 03	17 34
26	04 13 00	04 25 59	17 ♒ 20 56	24 ♒ 16 03	09 38	17 20	03 26	07 09	25 33	07 33	17 02	17 35
27	04 16 56	05 23 34	01 ♓ 04 53	07 ♓ 47 41	10 48	18 26	04 08	07 20	25 33	07 36	17 01	17 35
28	04 20 53	06 21 09	14 ♓ 24 48	20 ♓ 56 39	12 02	19 32	04 51	07 31	25 32	07 40	17 00	17 36
29	04 24 49	07 18 43	27 ♓ 23 40	03 ♈ 46 18	13 18	20 37	05 33	07 41	25 D 32	07 43	17 00	17 37
30	04 28 46	08 16 16	10 ♈ 05 00	16 ♈ 20 11	14 37	21 43	06 15	07 52	25 32	07 46	16 59	17 38
31	04 32 42	09 ♊ 13 48	22 ♈ 32 20	28 ♈ 41 45	15 ♉ 59	22 ♋ 48	06 ♊ 57	08 ♈ 02	25 ♍ 33	07 ♋ 49	16 ♎ 58	17 ♌ 39

Date	Moon True ☊ ° '	Moon Mean ☊ ° '	Moon ☽ Latitude ° '	DECLINATIONS Sun ☉ ° '	Moon ☽ ° '	Mercury ☿ ° '	Venus ♀ ° '	Mars ♂ ° '	Jupiter ♃ ° '	Saturn ♄ ° '	Uranus ♅ ° '	Neptune ♆ ° '	Pluto ⯓ ° '
01	17 ♓ 52	16 ♓ 24	00 S 02	14 N 55	05 S 00	11 N 04	24 N 55	16 N 30	00 S 08	03 N 48	23 N 37	05 S 20	23 N 50
02	17 R 52	16 21	01 N 06	15 13	01 N 08	10 38	25 03	16 43	00 S 03	03 49	23 37	05 20	23 50
03	17 49	16 18	02 10	15 31	07 08	10 15	25 10	16 56	00 N 02	03 50	23 37	05 19	23 49
04	17 43	16 15	03 07	15 49	12 45	09 53	25 17	17 09	00 06	03 51	23 36	05 19	23 49
05	17 35	16 12	03 54	16 06	17 48	09 34	25 23	17 22	00 11	03 52	23 36	05 18	23 49
06	17 26	16 08	04 30	16 23	22 05	09 17	25 28	17 34	00 16	03 53	23 36	05 18	23 49
07	17 15	16 05	04 53	16 40	25 22	09 02	25 32	17 46	00 21	03 53	23 36	05 17	23 49
08	17 03	16 02	05 03	16 57	27 30	08 50	25 36	17 59	00 26	03 54	23 36	05 17	23 48
09	16 53	15 59	05 00	17 13	28 23	08 40	25 39	18 11	00 31	03 55	23 36	05 16	23 48
10	16 44	15 56	04 43	17 29	27 56	08 33	25 41	18 22	00 35	03 55	23 36	05 16	23 48
11	16 37	15 52	04 15	17 45	26 13	08 28	25 43	18 34	00 40	03 56	23 35	05 16	23 48
12	16 33	15 49	03 35	18 00	23 19	08 26	25 44	18 46	00 45	03 56	23 35	05 15	23 47
13	16 31	15 46	02 45	18 15	19 22	08 26	25 44	18 57	00 49	03 57	23 35	05 14	23 47
14	16 D 31	15 43	01 47	18 30	14 33	08 29	25 44	19 08	00 54	03 57	23 35	05 14	23 47
15	16 31	15 40	00 N 42	18 45	09 00	08 34	25 43	19 19	00 59	03 58	23 35	05 13	23 46
16	16 R 32	15 37	00 S 27	18 59	02 N 55	08 41	25 41	19 30	01 03	03 58	23 35	05 13	23 46
17	16 30	15 33	01 37	19 13	03 S 29	08 50	25 39	19 40	01 08	03 58	23 35	05 12	23 46
18	16 27	15 30	02 43	19 26	09 57	09 02	25 36	19 51	01 12	03 58	23 34	05 12	23 46
19	16 21	15 27	03 41	19 39	16 07	09 15	25 32	20 01	01 17	03 59	23 34	05 12	23 45
20	16 12	15 24	04 26	19 52	21 32	09 30	25 28	20 11	01 21	03 59	23 34	05 11	23 45
21	16 02	15 21	04 54	20 05	25 38	09 47	25 23	20 20	01 25	03 59	23 34	05 11	23 44
22	15 52	15 18	05 01	20 17	27 57	10 05	25 17	20 30	01 30	03 59	23 34	05 10	23 44
23	15 43	15 14	04 47	20 29	28 10	10 25	25 11	20 39	01 34	03 59	23 34	05 10	23 44
24	15 35	15 11	04 13	20 40	26 20	10 47	25 04	20 48	01 38	03 59	23 33	05 10	23 43
25	15 30	15 08	03 24	20 51	22 45	11 09	24 56	20 57	01 42	03 59	23 33	05 09	23 43
26	15 27	15 05	02 23	21 02	17 54	11 33	24 48	21 06	01 46	03 59	23 33	05 09	23 43
27	15 26	15 02	01 15	21 13	12 16	11 59	24 39	21 15	01 50	03 59	23 33	05 09	23 42
28	15 D 26	14 58	00 S 05	21 23	06 13	12 26	24 30	21 23	01 54	03 59	23 33	05 08	23 42
29	15 R 26	14 55	01 N 03	21 32	00 S 05	12 53	24 20	21 31	01 58	03 59	23 32	05 08	23 41
30	15 24	14 52	02 06	21 42	05 N 56	13 22	24 09	21 39	02 02	03 58	23 32	05 08	23 41
31	15 ♓ 20	14 ♓ 49	03 N 02	21 N 51	11 N 36	13 N 51	23 N 58	21 N 47	02 N 06	03 N 58	23 N 32	05 S 07	23 N 41

ZODIAC SIGN ENTRIES

Date	h	m	Planets
01	21	25	☿ ♈
02	11	26	☽ ♈
04	20	47	☽ ♉
07	07	51	☽ ♊
09	20	13	☽ ♋
11	01	41	♀ ♋
12	08	49	☽ ♌
14	19	44	☽ ♍
15	01	40	☿ ♉
17	03	05	☽ ♎
19	06	23	☽ ♏
21	06	44	☽ ♐
21	15	32	♂ ♊
21	21	15	☉ ♊
23	06	07	☽ ♑
25	06	41	☽ ♒
27	10	05	☽ ♓
29	16	53	☽ ♈

LATITUDES

Date	Mercury ☿ ° '	Venus ♀ ° '	Mars ♂ ° '	Jupiter ♃ ° '	Saturn ♄ ° '	Uranus ♅ ° '	Neptune ♆ ° '	Pluto ⯓ ° '
01	00 S 31	01 N 56	00 00	01 S 06	02 N 29	00 N 19	01 N 41	08 N 38
04	01 19	02 03	00 N 02	01 06	02 29	00 19	01 41	08 38
07	02 00	02 09	00 04	01 07	02 28	00 19	01 41	08 37
10	02 34	02 14	00 06	01 07	02 28	00 19	01 41	08 37
13	02 59	02 19	00 08	01 08	02 27	00 19	01 41	08 36
16	03 16	02 23	00 09	01 08	02 27	00 19	01 41	08 36
19	03 25	02 26	00 11	01 09	02 26	00 19	01 41	08 35
22	03 27	02 28	00 13	01 09	02 26	00 19	01 41	08 35
25	03 22	02 30	00 15	01 10	02 25	00 19	01 41	08 34
28	03 10	02 30	00 17	01 10	02 24	00 19	01 40	08 34
31	02 S 54	02 N 29	00 N 19	01 S 11	02 N 24	00 N 19	01 N 40	08 N 33

DATA

Julian Date	2433768
Delta T	+30 seconds
Ayanamsa	23° 10' 41"
Synetic vernal point	05° ♓ 56' 19"
True obliquity of ecliptic	23° 26' 53"

MOON'S PHASES, APSIDES AND POSITIONS ☽

Date	h	m	Phase	Longitude ° '	Eclipse Indicator
06	01	36	●	14 ♉ 44	
14	05	32	☽	22 ♌ 38	
21	05	45	○	29 ♏ 23	
27	20	17	☾	05 ♓ 43	

Day	h	m		
09	16	54	Apogee	
22	04	19	Perigee	
02	07	33	0N	
09	15	54	Max dec	28° N 24'
16	23	02	0S	
23	02	29	Max dec	28° S 20'
29	12	18	0N	

LONGITUDES

Date	Chiron ⚷ ° '	Ceres ⚳ ° '	Pallas ⚴ ° '	Juno ⚵ ° '	Vesta ⚶ ° '	Black Moon Lilith ⚸ ° '
01	02 ♑ 31	04 ♓ 41	11 ♒ 27	06 ♐ 34	12 ♊ 11	22 ♊ 57
11	02 ♑ 10	07 ♓ 29	12 ♒ 47	04 ♐ 38	16 ♊ 21	24 ♊ 05
21	01 ♑ 43	10 ♓ 00	13 ♒ 35	02 ♐ 27	20 ♊ 35	25 ♊ 38
31	01 ♑ 09	12 ♓ 12	13 ♒ 50	00 ♐ 12	24 ♊ 51	26 ♊ 19

All ephemeris data is given at 12.00 UT and the Moon's longitude is additionally given for 24.00 UT

ASPECTARIAN

h m	Aspects
01 Tuesday	
01 15	☽ ± ♆
08 02	☽ ∠ ☿
08 18	☽ ⚹ ♂
10 42	☽ ‖ ♆
11 59	☽ ⊼ ⯓
12 16	☽ ⊼ ♆
15 08	☽ □ ♀
16 41	☽ # ♄
23 07	☽ ± ⯓
02 Wednesday	
00 06	☽ ⊥ ☿
03 50	☽ ∠ ☉
04 13	☽ ☍ ♄
07 16	☽ ‖ ♃
07 50	☽ # ♃
08 33	☉ ♇ ♄
10 57	☽ ⊻ ☿
13 54	☽ ∠ ♂
15 16	♃ ♇ ⯓
15 59	☽ ♇ ⯓
16 00	☽ ☌ ♃
22 09	☽ ⊥ ☉
22 40	☽ ‖ ♄
23 36	☽ □ ♅
03 Thursday	
04 17	☽ Q ♀
04 40	☽ # ♆
07 58	☽ ⊥ ♂
10 30	☽ ⊻ ☉
20 08	☽ ⊻ ♂
20 30	☽ △ ⯓
20 40	☽ ☍ ♆
04 Friday	
00 15	☽ ‖ ☿
02 35	♂ □ ⯓
05 15	♂ ⊼ ♆
05 34	☽ ⚹ ♀
10 08	☽ Q ♅
13 07	☽ ⊼ ♄
17 10	♂ ∠ ♃
18 37	☽ ☌ ☿
05 Saturday	
00 43	☽ ± ♄
02 30	☽ ⊻ ♃
03 00	☽ ‖ ☉
09 32	☽ ⚹ ♅
09 37	☽ ‖ ♂
13 43	☽ ∠ ♀
14 25	☽ ⊥ ♃
18 15	☽ ♇ ♄
06 Sunday	
01 36	☽ ☌ ☉
06 54	☽ □ ⯓
06 57	☽ ⊼ ♆
08 26	☽ ∠ ♃
09 16	☽ ⊥ ♀
10 12	☽ ☌ ♂
15 09	☽ ∠ ♅
18 48	☽ ± ♆
22 10	☽ ‖ ♅
22 24	☽ ⊻ ♀
23 40	☽ ‖ ⯓
23 47	☽ △ ♄
07 Monday	
04 41	☽ ⊻ ☿
06 22	♆ ⚹ ⯓
09 06	☽ ⊥ ♅
10 11	♀ ⊥ ♂
12 33	♀ □ ♄
12 41	☽ ♇ ♆
13 31	☽ ‖ ♀
14 46	☽ ⚹ ♃
16 35	☽ ⊥ ☿
18 42	☽ Q ⯓
21 09	☽ ⊻ ♅
08 Tuesday	
10 35	☽ ∠ ☿
11 50	☿ St D
15 17	☽ Q ♃
18 40	☽ ⊻ ☉
18 44	☽ △ ♆
18 49	☽ ⚹ ⯓
19 27	☉ ⊼ ♆
20 29	☉ □ ⯓
09 Wednesday	
02 03	☽ ⊻ ♂
07 51	☽ ⊥ ☉
11 51	☽ □ ♄
14 58	☽ ⊥ ♂
16 10	☿ ⚹ ♀
16 59	☽ ⚹ ☿
17 04	☽ ☌ ♀
10 Thursday	
01 09	☽ ∠ ⯓
02 32	♂ ∠ ♅
03 41	☽ ∠ ☉
04 19	☽ □ ♃
09 55	☽ ☌ ♅
10 22	☽ ∠ ♂
13 12	☉ ∠ ♃
17 34	☽ Q ☿
19 24	☽ ⊥ ⯓
11 Friday	
00 12	☽ Q ♄
07 19	☽ □ ♆
07 32	☽ ⊻ ⯓
12 45	☽ ⚹ ☉
16 49	☽ ‖ ♀
18 40	☽ ⚹ ♂
12 Saturday	
00 22	☽ ⚹ ♄
06 37	☽ □ ☿
06 42	♂ ± ♆
08 36	☽ ‖ ⯓
10 02	☽ ‖ ♅
12 06	☽ ⊻ ♀
14 59	☽ Q ☉
17 52	☽ △ ♃
19 25	☽ Q ♆
20 15	☽ Q ♂
22 34	☽ ⊻ ♅
13 Sunday	
01 19	☽ ⊥ ♀
05 23	♀ ∠ ⯓
06 15	☽ ∠ ♄
10 30	☽ ⊥ ♅
11 34	☉ ∠ ♅
14 11	☽ ‖ ♂
17 37	☽ ‖ ☉
19 06	☽ ⚹ ♆
19 27	☽ ☌ ⯓
20 53	☽ ∠ ♀
23 54	☽ ⊥ ♄
23 58	☽ ♇ ♃
14 Monday	
04 11	☽ ∠ ♅
05 32	☽ □ ☉
09 49	☽ □ ♂
11 31	☽ ⊻ ♄
17 40	☽ ± ♃
19 30	☽ △ ☿
21 13	☉ ± ♆
15 Tuesday	
00 02	☽ ∠ ♆
04 39	☽ ⚹ ♀
05 16	☽ ⊼ ♃
09 00	☽ ⚹ ♅
12 48	♀ □ ♃
13 45	☽ ‖ ☿
14 28	♂ △ ♄
16 57	☽ ⊥ ♆
16 Wednesday	
00 52	☽ ♇ ☿
03 08	☽ # ♆
04 02	☽ ⊻ ♆
04 29	☽ ⊻ ⯓
05 10	☽ Q ♀
07 21	☽ Q ♅
08 00	☽ ‖ ♄
15 24	☽ ⊥ ⯓
18 01	☽ ± ☿
18 16	☽ △ ☉
19 00	☽ ‖ ♃
19 21	☽ ☌ ♄
21 02	☽ △ ♂
17 Thursday	
03 08	☽ # ♃
05 14	☽ ⊼ ☿
07 29	☽ ∠ ⯓
07 31	♀ ☌ ♅
08 56	☉ △ ♄
12 43	☽ ☍ ♃
13 47	☽ # ♄
15 35	☽ □ ♅
16 16	☽ □ ♀
18 20	☽ ‖ ♆
21 08	☿ ± ♄
22 42	☽ ♇ ☉
18 Friday	
00 48	☽ ♇ ♂
08 24	☽ # ☿
08 56	☽ ☌ ♆
09 27	☽ ⚹ ⯓
15 10	☽ ± ☉
16 54	☽ ± ♂
23 08	☽ ⊻ ♄
19 Saturday	
01 55	☽ ⊼ ☉
03 26	☽ ⊼ ♂
05 34	☽ Q ⯓
09 03	☽ ⊥ ♄
10 59	☽ ☍ ☿
16 02	☽ ⊼ ♃
18 12	☽ △ ♅
22 50	☽ △ ♀
23 44	☽ ∠ ♄
20 Sunday	
01 55	☽ ± ♃
03 47	☽ # ☉
05 20	☽ # ♂
10 16	☽ ⊻ ♆
10 51	☽ □ ⯓
16 33	☽ ♇ ♃
18 26	☽ ♇ ♅
19 50	☽ ⊥ ♆
22 47	☽ # ♅
23 45	☽ ⚹ ♄
23 49	☽ # ⯓
21 Monday	
00 48	☽ ♇ ♀
05 45	☽ ☍ ☉
06 17	♀ ⊥ ⯓
06 18	☽ ☍ ♂
08 46	☽ ± ♅
10 06	☽ ∠ ♆
10 13	☽ # ♀
13 57	☽ ⊼ ☿
16 11	☽ ± ♀
16 38	☽ △ ♃
18 17	☽ ⊼ ♅
18 43	☽ Q ♄
22 Tuesday	
00 01	☽ ± ☿
02 23	☽ ⊼ ♀
09 44	☽ ⚹ ♆
10 24	☽ △ ⯓
13 22	☉ ☌ ♂
15 09	☽ ♇ ☿
23 06	☽ □ ♄
23 Wednesday	
02 29	♀ Q ♄
04 41	☽ Q ♆
07 55	☉ ⊥ ♅
08 01	☽ ⊼ ♂
08 21	☽ ⊼ ☉
10 09	☽ ♇ ⯓
15 03	♂ ⊥ ♅
16 36	☽ △ ☿
16 42	☽ □ ♃
17 56	☽ ☍ ♅
18 03	☽ ± ♂
18 14	☿ ⊻ ♃
18 34	☽ ± ☉
24 Thursday	
00 34	☽ ± ⯓
01 15	☉ ♇ ♆
05 52	☽ ☍ ♀
09 16	☽ ♇ ♂
09 30	☽ □ ♆
10 05	☽ ♇ ☉
10 16	☽ ⊼ ⯓
12 43	☿ ⚹ ♅
13 45	♂ ♇ ♆
22 05	☽ # ♀
22 13	☽ Q ♃
23 21	☽ △ ♄
25 Friday	
06 23	☽ # ⯓
07 22	☽ # ♅
11 11	☽ △ ♂
12 30	☽ △ ☉
18 25	☽ ⚹ ♃
19 15	☽ ⊼ ♅
21 12	☽ # ♂
21 38	☽ # ☉
21 40	☽ □ ☿
26 Saturday	
00 23	☽ ♇ ♄
05 28	☽ ± ♅
05 47	♀ □ ♆
11 28	☽ △ ♆
11 58	☽ ⊼ ♀
12 24	☽ ☍ ⯓
15 47	☽ ± ♄
17 26	♀ ⊻ ⯓
20 26	☽ ∠ ♃
21 03	☽ ♇ ♅
23 16	☽ ± ♀
27 Sunday	
02 14	☽ ⊼ ♄
06 46	☿ ♇ ♄
07 35	☽ Q ☿
12 27	☽ ⊥ ♃
13 04	☽ # ☿
13 40	☽ ♇ ♆
16 33	☽ ♇ ♀
16 58	☉ Q ⯓
17 45	☽ □ ♂
20 17	☽ □ ☉
23 19	☽ ⊻ ♃
23 42	☽ △ ♅
28 Monday	
05 49	☽ ± ♆
07 12	☽ ⚹ ☿
16 14	☽ ‖ ♆
16 45	☽ ⊼ ♆
17 51	☽ ⊼ ⯓
20 26	☉ ‖ ♂
20 45	☽ # ♄
22 15	☽ △ ♀
29 Tuesday	
03 35	♄ St D
04 25	☽ Q ♂
04 39	☽ # ♃
04 57	☽ ± ⯓
07 32	♀ ∠ ♂
07 48	☽ Q ☉
08 32	☽ ☍ ♄
13 54	☽ ∠ ☿
14 20	♂ Q ⯓
16 24	♃ □ ♅
20 11	☽ ‖ ♃
20 15	☿ ⊥ ♃
21 49	☽ ♇ ⯓
22 37	☉ ⊻ ♅
23 35	☉ ⚹ ♃
30 Wednesday	
04 05	☽ ‖ ♄
04 16	☽ ⚹ ♂
07 34	☽ □ ♅
07 43	☽ ☌ ♃
08 15	☽ ⚹ ☉
08 44	☽ # ♆
08 53	☽ ⊥ ☿
21 45	☽ ⊻ ☿
31 Thursday	
01 13	☽ ☍ ♆
02 30	☽ △ ⯓
10 48	☽ ∠ ♂
12 33	☽ □ ♀
15 34	☽ ∠ ☉
17 51	☽ ⊼ ♄
18 25	☽ Q ♅
23 19	☽ ‖ ☿

JUNE 1951

Date	Sidereal time h m s	Sun ☉ ° ' "	Moon ☽ ° ' "	Moon ☽ 24.00 ° ' "	Mercury ☿ ° '	Venus ♀ ° '	Mars ♂ ° '	Jupiter ♃ ° '	Saturn ♄ ° '	Uranus ♅ ° '	Neptune ♆ ° '	Pluto ♇ ° '
01	04 36 39	10 ♊ 11 19	04 ♉ 48 48	10 ♉ 53 46	17 ♉ 24	23 ♋ 53	07 ♊ 39	08 ♈ 13	25 ♍ 33	07 ♋ 53	16 ♎ 57	17 ♌ 39
02	04 40 36	11 08 49	16 ♉ 56 56	22 ♉ 58 32	18 51	24 58	08 21	08 23	25 33	07 56	16 R 56	17 40
03	04 44 32	12 06 19	28 ♉ 58 46	04 ♊ 57 49	20 21	26 02	09 03	08 33	25 34	07 59	16 55	17 41
04	04 48 29	13 03 47	10 ♊ 55 50	16 ♊ 52 59	21 53	27 06	09 45	08 43	25 34	08 03	16 55	17 42
05	04 52 25	14 01 15	22 ♊ 49 26	28 ♊ 45 22	23 28	28 11	10 27	08 53	25 35	08 06	16 54	17 43
06	04 56 22	14 58 42	04 ♋ 40 58	10 ♋ 36 26	25 06	29 ♋ 14	11 09	09 02	25 36	08 09	16 53	17 44
07	05 00 18	15 56 08	16 ♋ 32 03	22 ♋ 28 04	26 46	00 ♌ 18	11 51	09 12	25 37	08 13	16 52	17 45
08	05 04 15	16 53 33	28 ♋ 24 51	04 ♌ 22 45	28 ♉ 29	01 21	12 32	09 21	25 38	08 16	16 52	17 46
09	05 08 11	17 50 56	10 ♌ 22 12	16 ♌ 23 39	00 ♊ 15	02 25	13 14	09 31	25 39	08 20	16 51	17 47
10	05 12 08	18 48 19	22 ♌ 27 36	28 ♌ 34 38	02 02	03 27	13 56	09 40	25 40	08 23	16 51	17 48
11	05 16 05	19 45 41	04 ♍ 45 17	11 ♍ 00 09	03 53	04 30	14 37	09 49	25 42	08 26	16 50	17 49
12	05 20 01	20 43 01	17 ♍ 19 50	23 ♍ 44 56	05 46	05 32	15 19	09 58	25 43	08 30	16 50	17 50
13	05 23 58	21 40 21	00 ♎ 16 00	06 ♎ 53 33	07 41	06 34	16 00	10 07	25 44	08 33	16 49	17 52
14	05 27 54	22 37 39	13 ♎ 38 00	20 ♎ 29 40	09 39	07 36	16 42	10 16	25 46	08 37	16 49	17 53
15	05 31 51	23 34 57	27 ♎ 28 45	04 ♏ 35 13	11 38	08 37	17 23	10 24	25 48	08 40	16 48	17 54
16	05 35 47	24 32 14	11 ♏ 48 54	19 ♏ 09 21	13 40	09 38	18 04	10 33	25 50	08 44	16 48	17 55
17	05 39 44	25 29 30	26 ♏ 35 56	04 ♐ 07 43	15 44	10 39	18 46	10 41	25 52	08 48	16 47	17 56
18	05 43 40	26 26 45	11 ♐ 43 37	19 ♐ 22 21	17 50	11 39	19 27	10 49	25 54	08 51	16 47	17 58
19	05 47 37	27 24 00	27 ♐ 02 29	04 ♑ 42 34	19 57	12 39	20 08	10 57	25 56	08 55	16 47	17 59
20	05 51 34	28 21 15	12 ♑ 21 08	19 ♑ 56 49	22 05	13 39	20 49	11 05	25 58	08 58	16 46	18 00
21	05 55 30	29 ♊ 18 29	27 ♑ 28 22	04 ♒ 54 44	24 15	14 38	21 30	11 13	26 00	09 02	16 46	18 01
22	05 59 27	00 ♋ 15 42	12 ♒ 15 07	19 ♒ 28 53	26 26	15 37	22 11	11 20	26 03	09 05	16 46	18 03
23	06 03 23	01 12 56	26 ♒ 35 42	03 ♓ 35 22	28 ♊ 37	16 35	22 53	11 28	26 05	09 09	16 46	18 04
24	06 07 20	02 10 09	10 ♓ 27 56	17 ♓ 13 34	00 ♋ 48	17 33	23 33	11 35	26 08	09 13	16 46	18 06
25	06 11 16	03 07 22	23 ♓ 52 36	00 ♈ 25 24	03 00	18 31	24 14	11 42	26 10	09 16	16 45	18 07
26	06 15 13	04 04 35	06 ♈ 52 29	13 ♈ 14 21	05 11	19 28	24 55	11 50	26 13	09 20	16 45	18 08
27	06 19 09	05 01 48	19 ♈ 31 31	25 ♈ 44 33	07 21	20 25	25 36	11 56	26 16	09 23	16 45	18 10
28	06 23 06	05 59 01	01 ♉ 53 59	08 ♉ 00 18	09 31	21 21	26 17	12 03	26 19	09 27	16 45	18 11
29	06 27 02	06 56 14	14 ♉ 04 01	20 ♉ 05 33	11 40	22 17	26 58	12 10	26 22	09 31	16 D 45	18 12
30	06 30 59	07 ♋ 53 28	26 ♉ 05 20	02 ♊ 03 42	13 ♋ 48	23 ♌ 12	27 ♊ 38	12 ♈ 16	26 ♍ 25	09 ♋ 34	16 ♎ 45	18 ♌ 14

Date	Moon True ☊ ° '	Moon Mean ☊ ° '	Moon ☽ Latitude ° '
01	15 ♓ 13	14 ♓ 46	03 N 49
02	15 R 03	14 43	04 25
03	14 51	14 39	04 48
04	14 37	14 36	04 59
05	14 23	14 33	04 56
06	14 10	14 30	04 41
07	13 58	14 27	04 13
08	13 49	14 24	03 34
09	13 43	14 20	02 46
10	13 40	14 17	01 49
11	13 39	14 14	00 N 47
12	13 D 38	14 11	00 S 20
13	13 R 38	14 08	01 27
14	13 37	14 04	02 31
15	13 34	14 01	03 30
16	13 28	13 58	04 17
17	13 20	13 55	04 48
18	13 10	13 52	05 01
19	13 00	13 49	04 52
20	12 50	13 45	04 22
21	12 42	13 42	03 35
22	12 37	13 39	02 33
23	12 34	13 36	01 24
24	12 33	13 33	00 S 11
25	12 D 33	13 29	01 N 00
26	12 R 34	13 26	02 05
27	12 33	13 23	03 00
28	12 30	13 20	03 51
29	12 24	13 17	04 27
30	12 ♓ 16	13 ♓ 14	04 N 51

DECLINATIONS

Date	Sun ☉	Moon ☽	Mercury ☿	Venus ♀	Mars ♂	Jupiter ♃	Saturn ♄	Uranus ♅	Neptune ♆	Pluto ♇
01	21 N 59	16 N 43	14 N 21	23 N 47	21 N 55	02 N 10	03 N 58	23 N 32	05 S 07	23 N 40
02	22 07	21 08	14 52	23 34	22 02	02 14	03 58	23 32	05 07	23 40
03	22 15	24 37	15 24	23 22	22 09	02 18	03 57	23 31	05 07	23 39
04	22 22	27 01	15 56	23 08	22 16	02 22	03 57	23 31	05 06	23 39
05	22 29	28 11	16 28	22 54	22 23	02 25	03 56	23 31	05 06	23 39
06	22 36	28 02	17 00	22 40	22 29	02 29	03 56	23 31	05 06	23 38
07	22 42	26 37	17 33	22 25	22 36	02 32	03 55	23 31	05 06	23 38
08	22 48	23 59	18 05	22 10	22 42	02 36	03 55	23 30	05 05	23 37
09	22 54	20 19	18 38	21 54	22 47	02 39	03 54	23 30	05 05	23 37
10	22 59	15 45	19 10	21 38	22 53	02 43	03 53	23 30	05 05	23 36
11	23 03	10 30	19 42	21 21	22 59	02 46	03 53	23 30	05 05	23 36
12	23 07	04 N 42	20 13	21 04	23 04	02 49	03 52	23 30	05 05	23 35
13	23 11	01 S 26	20 43	20 46	23 09	02 53	03 51	23 29	05 05	23 35
14	23 15	07 42	21 12	20 28	23 14	02 56	03 50	23 29	05 04	23 34
15	23 18	13 50	21 40	20 10	23 18	02 59	03 49	23 29	05 04	23 34
16	23 20	19 27	22 07	19 51	23 23	03 02	03 48	23 29	05 04	23 33
17	23 22	24 04	22 32	19 32	23 27	03 05	03 47	23 28	05 04	23 33
18	23 24	27 10	22 55	19 13	23 31	03 08	03 46	23 28	05 04	23 32
19	23 25	28 17	23 17	18 53	23 34	03 11	03 45	23 28	05 04	23 32
20	23 26	27 14	23 36	18 33	23 38	03 14	03 44	23 28	05 04	23 31
21	23 27	24 11	23 53	18 12	23 41	03 17	03 43	23 28	05 04	23 31
22	23 27	19 35	24 07	17 51	23 44	03 20	03 42	23 27	05 04	23 30
23	23 27	13 58	24 19	17 30	23 47	03 22	03 41	23 27	05 04	23 30
24	23 26	07 49	24 28	17 09	23 50	03 25	03 40	23 27	05 04	23 29
25	23 25	01 S 31	24 34	16 47	23 52	03 28	03 38	23 27	05 04	23 29
26	23 23	04 N 39	24 38	16 26	23 54	03 30	03 37	23 26	05 04	23 28
27	23 21	10 28	24 38	16 03	23 57	03 32	03 36	23 26	05 04	23 28
28	23 19	15 45	24 36	15 41	23 58	03 35	03 34	23 26	05 04	23 27
29	23 16	20 19	24 31	15 19	24 00	03 37	03 33	23 26	05 04	23 26
30	23 N 13	23 N 59	24 N 23	14 N 56	24 N 01	03 N 39	03 N 32	23 N 25	05 S 04	23 N 26

ZODIAC SIGN ENTRIES

Date	h m	Planets
01	02 33	☽ ♉
03	14 03	☽ ♊
06	02 31	☽ ♋
07	05 10	♀ ♌
08	15 12	☽ ♌
09	08 43	☿ ♊
11	02 47	☽ ♍
13	11 31	☽ ♎
15	16 17	☽ ♏
17	17 26	☽ ♐
19	16 38	☽ ♑
21	16 04	☽ ♒
22	05 25	☉ ♋
23	17 49	☽ ♓
24	03 13	☿ ♋
25	23 13	☽ ♈
28	08 17	☽ ♉
30	19 51	☽ ♊

LATITUDES

Date	Mercury ☿	Venus ♀	Mars ♂	Jupiter ♃	Saturn ♄	Uranus ♅	Neptune ♆	Pluto ♇
01	02 S 47	02 N 29	00 N 19	01 S 11	02 N 24	00 N 19	01 N 40	08 N 33
04	02 24	02 26	00 21	01 12	02 23	00 19	01 40	08 33
07	01 57	02 23	00 23	01 12	02 22	00 19	01 40	08 32
10	01 27	02 18	00 24	01 13	02 22	00 19	01 40	08 32
13	00 54	02 12	00 26	01 14	02 21	00 19	01 40	08 32
16	00 S 20	02 05	00 28	01 14	02 20	00 19	01 40	08 32
19	00 N 13	01 57	00 30	01 15	02 20	00 19	01 39	08 31
22	00 43	01 47	00 31	01 16	02 19	00 19	01 39	08 31
25	01 09	01 35	00 33	01 16	02 18	00 19	01 39	08 31
28	01 30	01 22	00 35	01 17	02 18	00 19	01 39	08 30
31	01 N 44	01 N 08	00 N 36	01 S 18	02 N 17	00 N 19	01 N 39	08 N 30

DATA

Julian Date	2433799
Delta T	+30 seconds
Ayanamsa	23° 10' 46"
Synetic vernal point	05° ♓ 56' 14"
True obliquity of ecliptic	23° 26' 53"

MOON'S PHASES, APSIDES AND POSITIONS ☽

Date	h m	Phase	Longitude ° '	Eclipse Indicator
04	16 40	●	13 ♊ 15	
12	18 52	☽	20 ♍ 59	
19	12 36	○	27 ♐ 25	
26	06 21	☾	03 ♈ 51	

Day	h m		
06	01 16	Apogee	
19	13 36	Perigee	
05	21 25	Max dec	28° N 17'
13	06 29	0S	
19	12 21	Max dec	28° S 17'
25	17 49	0N	

LONGITUDES

Date	Chiron ⚷ ° '	Ceres ⚳ ° '	Pallas ⚴ ° '	Juno ⚵ ° '	Vesta ⚶ ° '	Black Moon Lilith ⚸ ° '
01	01 ♑ 06	12 ♓ 24	13 ♒ 49	29 ♏ 59	25 ♊ 17	26 ♊ 26
11	00 ♑ 28	14 ♓ 13	13 ♒ 22	27 ♏ 54	27 ♊ 36	27 ♊ 33
21	29 ♐ 48	15 ♓ 35	12 ♒ 17	26 ♏ 10	03 ♋ 56	28 ♊ 40
31	29 ♐ 09	16 ♓ 29	10 ♒ 35	24 ♏ 52	08 ♋ 18	29 ♊ 47

All ephemeris data is given at 12.00 UT and the Moon's longitude is additionally given for 24.00 UT

ASPECTARIAN

h m Aspects

01 Friday
04 34 ☿ ⊼ ♆
05 25 ☽ ⊥ ♂
05 35 ☽ ± ♄
10 40 ☽ ⊥ ☉
16 25 ☿ □ ♇
17 56 ☽ ⊻ ♂
18 04 ☽ ✱ ♅
18 47 ☽ ⊻ ♃
20 16 ♂ ⊻ ♅
23 19 ☽ bQ ♄
23 31 ☽ ⊻ ☉

02 Saturday
01 01 ♀ ∥ ♇
03 19 ☽ Q ♀
06 50 ☽ ⊥ ♃
11 58 ☽ ⊼ ♆
13 08 ♂ ✱ ♃
13 26 ☽ □ ♇
16 18 ☽ ☌ ☿
17 17 ♀ ∥ ♅
17 47 ☽ ∥ ♂
18 23 ☽ ∥ ☉
23 54 ☽ ± ♆
23 58 ☽ ∠ ♅

03 Sunday
00 59 ☽ ∠ ♃
01 24 ♀ ✱ ♄
03 04 ☽ ∥ ♀
03 41 ☽ ∥ ♅
04 39 ☽ ∥ ♇
05 10 ☽ △ ♄
05 32 ☽ ✱ ♀
17 53 ☽ bQ ♆
18 03 ☽ ⊥ ♅

04 Monday
01 28 ☽ Q ♇
06 10 ☽ ⊻ ♅
07 28 ☽ ✱ ♃
09 29 ☽ ☌ ♂
14 36 ☽ ∠ ♀
16 40 ☽ ☌ ☉

05 Tuesday
00 02 ☽ △ ♆
01 40 ☽ ✱ ♇
03 26 ☿ ± ♆
06 11 ☿ ∠ ♅
08 01 ☽ Q ♃
10 34 ☽ ⊥ ♀
13 31 ☽ ⊻ ☿
17 36 ☽ □ ♄
18 41 ☿ ∠ ♃
23 54 ☽ ⊻ ♀

06 Wednesday
03 35 ☽ ⊥ ☿
08 03 ☽ ∠ ♇
16 35 ☉ ∥ ♀
19 04 ☽ ☌ ♅
19 18 ☿ △ ♄
20 57 ☽ □ ♃

07 Thursday
00 24 ♀ ∥ ♂
00 46 ☽ ∠ ☿
01 55 ☽ ⊻ ♂
02 18 ☽ ⊥ ♇
06 05 ☽ Q ♄
10 41 ☽ ⊻ ☉
12 41 ☽ □ ♆
14 28 ☽ ⊻ ♇
14 49 ☽ ⊥ ♂
23 53 ☽ ⊥ ☉

08 Friday
06 23 ☽ ✱ ♄
10 08 ☽ ∠ ♂
11 17 ☉ △ ♆
12 10 ☽ ✱ ☿
14 42 ☽ ∥ ♇
15 31 ☽ ∥ ♅
18 30 ☽ ☌ ♀
19 37 ☽ ∠ ☉
20 14 ☽ ∥ ☉
20 56 ☽ ∥ ♂

09 Saturday
00 58 ☽ Q ♆
01 37 ☽ ∥ ♀
07 54 ☽ ⊻ ♅
10 16 ☽ △ ♃
10 23 ☉ ✱ ♇
12 34 ☽ ∠ ♄
16 23 ☽ Q ☿
18 04 ☽ ✱ ♂
19 56 ☽ ⊥ ♅
20 21 ☽ ∥ ☿

10 Sunday
00 54 ☽ ✱ ♆
02 47 ☽ ☌ ♇
04 10 ☽ ✱ ☉
06 29 ☽ ⊥ ♄
09 24 ☿ bQ ♆
13 50 ☽ ∠ ♅
16 24 ☽ bQ ♃
16 39 ☿ ⊥ ♅
18 19 ☽ ⊻ ♄
19 14 ☽ Q ♂

11 Monday
05 43 ☽ Q ☉
06 21 ☽ ∠ ♆
10 01 ☽ □ ☿
10 10 ☽ ± ♃
11 28 ☽ ⊻ ♀
19 08 ☽ ✱ ♅
19 39 ♀ Q ♆
21 51. ☽ ⊼ ♃
23 40 ☽ ⊥ ♆

12 Tuesday
00 02 ☽ ⊥ ♀
05 44 ☿ ✱ ♀
07 58 ☽ □ ♂
10 31 ☽ # ♆
11 03 ☽ ⊻ ♆
12 58 ☽ ⊻ ♇
13 00 ☿ Q ♇
15 22 ☽ ∥ ♄
17 58 ☽ Q ♅
18 33 ☽ ∠ ♀
18 52 ☽ □ ☉
19 24 ☽ ∥ ♃

13 Wednesday
00 11 ☽ ⊥ ♇
03 40 ☽ ☌ ♄
13 50 ☿ ∥ ♀
16 44 ☽ ∠ ♇
17 36 ☽ # ♃
21 14 ☽ # ♄
23 06 ☿ ⊻ ♅

14 Thursday
00 22 ☽ ✱ ♀
01 06 ☉ Q ♃
01 56 ☽ ∥ ♆
03 03 ☽ □ ♅
03 42 ☽ △ ☿
05 57 ☽ ☍ ♃
15 54 ♂ △ ♆
17 35 ☽ ☌ ♆
17 40 ☽ △ ♂
19 28 ☽ ✱ ♇
20 03 ☿ ✱ ♃
22 11 ☉ ∠ ♀
23 17 ☽ Q ♀

15 Friday
02 28 ☉ ∥ ♂
04 50 ☽ △ ☉
09 07 ☽ ⊻ ♄
10 20 ☽ bQ ☿
13 20 ♀ ⊻ ♅
16 07 ☽ Q ♇
19 20 ☽ ⊥ ♄
20 44 ☽ bQ ♂

16 Saturday
04 01 ☽ ± ☿
06 25 ♂ ✱ ♇
06 53 ☽ △ ♅
07 58 ☽ bQ ☉
08 08 ☽ □ ♀
09 53 ☽ ⊼ ♃
10 22 ☽ ∠ ♄
12 27 ☽ ± ♂
13 46 ☽ # ♀
15 33 ☽ ⊼ ☿
19 49 ☽ ± ♃
20 09 ☽ ⊻ ♆
22 00 ☽ □ ♇
22 45 ☽ ⊼ ♂
23 45 ☽ ± ☉

17 Sunday
02 21 ☽ # ☿
05 53 ☽ ⊥ ♆
07 28 ☽ bQ ♅
07 50 ☽ # ☉
08 14 ☽ # ♂
08 27 ☽ # ♅
08 53 ☽ # ♇
10 06 ☽ ⊼ ☉
10 31 ☽ bQ ♃
10 49 ☽ ✱ ♄
12 55 ♀ △ ♃
17 10 ♀ ∠ ♄
20 17 ☽ ∠ ♆
21 34 ☉ □ ♄
21 44 ♂ ∥ ♅
21 55 ☽ ± ♅

18 Monday
00 06 ☿ △ ♆
05 57 ☽ Q ♄
07 27 ☽ ⊼ ♅
10 33 ☽ △ ♃
11 53 ☽ △ ♀
13 31 ☿ ✱ ♇
19 56 ☽ ✱ ♆
21 15 ♂ ∥ ♇
21 48 ☽ △ ♇
23 07 ☽ ☍ ☿

19 Tuesday
00 41 ☽ ☍ ♂
10 15 ☽ □ ♄
12 36 ☽ ☍ ☉
13 02 ☽ bQ ♀
14 43 ☽ Q ♆
15 08 ☿ ☌ ♂
21 18 ☽ bQ ♇
23 06 ☉ ∥ ☿

20 Wednesday
01 39 ☿ ∥ ♅
04 06 ☽ ± ♀
06 13 ☿ ∥ ♇
06 40 ☽ ☍ ♅
09 59 ☽ □ ♃
11 27 ☽ ± ♇
14 11 ☽ ⊼ ♀
15 25 ☿ ∥ ♂
18 58 ☽ □ ♆
20 56 ☽ ⊼ ♇
23 47 ☿ Q ♃

21 Thursday
02 02 ☽ ⊼ ♂
05 59 ☽ ⊼ ☿
09 38 ☽ △ ♄
12 03 ☽ ± ♂
13 43 ☽ # ☿
14 49 ☽ Q ♃
14 55 ☽ # ♂
15 09 ☽ ⊼ ☉
15 58 ☽ # ♇
16 17 ☽ # ♅
16 20 ☽ # ☉
17 14 ☽ ± ☿
22 15 ♀ ⊥ ♅

22 Friday
01 31 ☽ ± ☉
03 18 ☽ bQ ♂
06 47 ☽ ⊼ ♅
07 41 ☿ □ ♄
10 00 ☽ bQ ♄
10 24 ☽ bQ ☿
10 29 ☽ ✱ ♃
16 42 ☽ ± ♅
17 19 ☽ bQ ☉
17 58 ☽ ☍ ♀
19 28 ☽ △ ♆
20 15 ☽ # ♀
21 37 ☽ ☍ ♇

23 Saturday
00 59 ☽ ± ♄
05 23 ☽ △ ♂
07 50 ☽ bQ ♅
11 08 ☽ ⊼ ♄
11 47 ☽ ∠ ♃
16 04 ☽ △ ☿
16 14 ♀ ✱ ♆
20 29 ☽ △ ☉
20 51 ☽ bQ ♆

24 Sunday
03 24 ☽ ⊥ ♃
09 47 ☽ △ ♅
12 31 ☽ ± ♆
13 18 ♂ Q ♃
14 00 ☽ ⊻ ♃
22 30 ☽ ∥ ♆
23 10 ☽ ⊼ ♆

25 Monday
01 34 ☽ ⊼ ♀
01 34 ☽ ⊼ ♇
01 39 ♀ ☌ ♇
03 53 ☽ # ♄
04 39 ☽ # ♃
11 48 ☉ ∠ ♇
12 26 ☽ ± ♇
12 42 ☽ □ ♂
13 16 ☽ ± ♀
13 22 ☿ ∠ ♇
14 33 ☉ ☌ ☿
16 12 ☽ ☍ ♄
22 15 ☿ ∠ ♀

26 Tuesday
05 01 ☽ bQ ♇
06 21 ☽ □ ☉
07 09 ☽ bQ ♀
07 25 ☽ ∥ ♃
07 56 ☽ ∥ ♄
08 11 ☽ □ ☿
13 39 ☽ # ♆
16 38 ☽ □ ♅
21 24 ☽ ☌ ♃

27 Wednesday
00 03 ☽ Q ♂
06 42 ☽ ☍ ♆
07 55 ♀ ⊥ ♄
09 23 ☽ △ ♇
13 51 ☽ △ ♀
19 19 ☽ Q ☉

28 Thursday
00 25 ☽ ✱ ♂
01 04 ☽ ⊼ ♄
01 38 ☽ Q ☿
03 16 ☽ Q ♅
09 26 ♃ ∥ ♄
11 12 ☿ ☌ ♅
11 45 ☽ ∥ ♀
12 49 ☽ ± ♄
13 13 ♂ □ ♄
20 42 ☽ ✱ ☉
21 52 ♆ St D

29 Friday
02 56 ☽ ✱ ♅
06 13 ☽ ✱ ☿
06 37 ☽ bQ ♄
07 34 ☽ ∠ ♂
08 11 ☽ ⊻ ♃
17 21 ☽ ⊼ ♆
17 51 ☿ □ ♃
18 10 ☿ ⊥ ♇
20 13 ☽ ⊥ ♃
20 16 ☽ □ ♇

30 Saturday
02 33 ☽ ⊥ ♂
05 02 ☽ ∠ ☉
05 19 ☽ ± ♆
05 45 ☽ □ ♀
06 26 ☽ ∥ ☉
07 50 ☽ ∥ ♅
07 54 ☽ ∥ ♇
08 56 ☽ ∠ ♅
12 14 ☽ ∥ ♂
12 40 ☽ △ ♄
14 23 ☽ ∠ ♃
14 55 ☽ ∥ ☿
15 18 ☽ ⊻ ♂
18 36 ☽ ∠ ☿
19 12 ☿ Q ♄
23 23 ☽ bQ ♆

JULY 1951

Date	Sidereal time h m s	Sun ☉ ° ' "	Moon ☽ ° ' "	Moon ☽ 24.00 ° ' "	Mercury ☿ ° '	Venus ♀ ° '	Mars ♂ ° '	Jupiter ♃ ° '	Saturn ♄ ° '	Uranus ♅ ° '	Neptune ♆ ° '	Pluto ♇ ° '
01	06 34 56	08 ♋ 50 41	08 ♊ 01 01	13 ♊ 57 34	15 ♋ 54	24 ♌ 07	28 ♊ 19	12 ♈ 22	26 ♍ 28	09 ♋ 38	16 ♎ 45	18 ♌ 16
02	06 38 52	09 47 54	19 ♊ 53 36	25 ♊ 49 22	17 59	25 01	29 00	12 29	26 31	09 41	16 45	18 17
03	06 42 49	10 45 08	01 ♋ 45 04	07 ♋ 40 54	20 02	25 55	29 ♊ 40	12 34	26 35	09 45	16 46	18 19
04	06 46 45	11 42 21	13 ♋ 37 03	19 ♋ 33 44	22 04	26 48	00 ♋ 21	12 40	26 38	09 49	16 46	18 20
05	06 50 42	12 39 35	25 ♋ 31 08	01 ♌ 29 28	24 04	27 41	01 01	12 46	26 42	09 52	16 46	18 22
06	06 54 38	13 36 48	07 ♌ 28 58	13 ♌ 29 54	26 02	28 33	01 42	12 51	26 46	09 56	16 46	18 23
07	06 58 35	14 34 01	19 ♌ 32 33	25 ♌ 37 16	27 58	29 ♌ 24	02 22	12 57	26 49	10 00	16 46	18 25
08	07 02 32	15 31 14	01 ♍ 44 23	07 ♍ 54 20	29 ♋ 52	00 ♍ 15	03 02	13 02	26 53	10 03	16 47	18 26
09	07 06 28	16 28 27	14 ♍ 07 31	20 ♍ 24 25	01 ♌ 45	01 05	03 43	13 07	26 57	10 07	16 47	18 28
10	07 10 25	17 25 40	26 ♍ 45 28	03 ♎ 11 11	03 35	01 54	04 23	13 11	27 01	10 10	16 47	18 30
11	07 14 21	18 22 53	09 ♎ 42 01	16 ♎ 18 24	05 23	02 43	05 03	13 16	27 05	10 14	16 48	18 31
12	07 18 18	19 20 05	23 ♎ 00 44	29 ♎ 49 20	07 10	03 31	05 43	13 20	27 09	10 18	16 48	18 33
13	07 22 14	20 17 18	06 ♏ 44 25	13 ♏ 46 04	08 54	04 18	06 24	13 25	27 13	10 21	16 49	18 35
14	07 26 11	21 14 31	20 ♏ 54 15	28 ♏ 08 41	10 37	05 04	07 04	13 29	27 18	10 25	16 49	18 36
15	07 30 07	22 11 44	05 ♐ 28 58	12 ♐ 54 25	12 17	05 49	07 44	13 33	27 22	10 28	16 50	18 38
16	07 34 04	23 08 57	20 ♐ 24 14	27 ♐ 57 23	13 56	06 33	08 24	13 36	27 27	10 32	16 50	18 40
17	07 38 01	24 06 10	05 ♑ 32 41	13 ♑ 08 53	15 33	07 17	09 04	13 40	27 31	10 35	16 51	18 41
18	07 41 57	25 03 23	20 ♑ 44 40	28 ♑ 18 43	17 07	07 59	09 44	13 43	27 36	10 39	16 51	18 43
19	07 45 54	26 00 37	05 ♒ 49 48	13 ♒ 16 47	18 40	08 41	10 23	13 46	27 40	10 42	16 52	18 45
20	07 49 50	26 57 51	20 ♒ 38 44	27 ♒ 54 52	20 11	09 21	11 03	13 49	27 45	10 46	16 53	18 47
21	07 53 47	27 55 06	05 ♓ 04 37	12 ♓ 07 35	21 40	10 01	11 43	13 52	27 50	10 50	16 53	18 48
22	07 57 43	28 52 22	19 ♓ 03 37	25 ♓ 52 42	23 06	10 39	12 23	13 55	27 55	10 53	16 54	18 50
23	08 01 40	29 ♋ 49 38	02 ♈ 34 58	09 ♈ 10 41	24 31	11 16	13 03	13 57	28 00	10 56	16 55	18 52
24	08 05 36	00 ♌ 46 55	15 ♈ 40 14	22 ♈ 04 02	25 54	11 52	13 42	13 59	28 05	11 00	16 56	18 54
25	08 09 33	01 44 13	28 ♈ 22 37	04 ♉ 36 31	27 15	12 27	14 22	14 01	28 10	11 03	16 57	18 56
26	08 13 30	02 41 32	10 ♉ 46 16	16 ♉ 52 27	28 33	13 00	15 02	14 03	28 15	11 07	16 57	18 57
27	08 17 26	03 38 52	22 ♉ 55 37	28 ♉ 56 20	29 ♌ 49	13 32	15 41	14 05	28 20	11 10	16 58	18 59
28	08 21 23	04 36 13	04 ♊ 55 05	10 ♊ 52 23	01 ♍ 03	14 03	16 21	14 06	28 26	11 14	16 59	19 01
29	08 25 19	05 33 35	16 ♊ 48 41	22 ♊ 44 24	02 15	14 32	17 00	14 07	28 31	11 17	17 00	19 03
30	08 29 16	06 30 58	28 ♊ 39 57	04 ♋ 35 40	03 25	15 00	17 40	14 08	28 37	11 20	17 01	19 05
31	08 33 12	07 ♌ 28 22	10 ♋ 31 52	16 ♋ 28 51	04 ♍ 32	15 ♍ 27	18 ♋ 19	14 ♈ 09	28 ♍ 42	11 ♋ 24	17 ♎ 02	19 ♌ 06

Date	Moon True ☊ ° '	Moon Mean ☊ ° '	Moon ☽ Latitude ° '
01	12 ♓ 07	13 ♓ 10	05 N 02
02	11 R 55	13 07	05 00
03	11 44	13 04	04 45
04	11 33	13 01	04 17
05	11 23	12 58	03 39
06	11 16	12 55	02 50
07	11 11	12 51	01 53
08	11 09	12 48	00 N 50
09	11 D 09	12 45	00 S 16
10	11 09	12 42	01 23
11	11 10	12 39	02 27
12	11 R 11	12 35	03 26
13	11 09	12 32	04 14
14	11 06	12 29	04 49
15	11 01	12 26	05 06
16	10 55	12 23	05 03
17	10 48	12 20	04 39
18	10 41	12 16	03 56
19	10 36	12 13	02 56
20	10 33	12 10	01 45
21	10 31	12 07	00 S 29
22	10 D 31	12 04	00 N 46
23	10 33	12 01	01 57
24	10 34	11 57	02 59
25	10 35	11 54	03 51
26	10 R 35	11 51	04 30
27	10 33	11 48	04 56
28	10 29	11 45	05 09
29	10 25	11 41	05 09
30	10 19	11 38	04 55
31	10 ♓ 13	11 ♓ 35	04 N 29

DECLINATIONS

Date	Sun ☉ ° '	Moon ☽ ° '	Mercury ☿ ° '	Venus ♀ ° '	Mars ♂ ° '	Jupiter ♃ ° '	Saturn ♄ ° '	Uranus ♅ ° '	Neptune ♆ ° '	Pluto ♇ ° '
01	23 N 09	26 N 37	24 N 13	14 N 33	24 N 02	03 N 42	03 N 30	23 N 25	05 S 04	23 N 25
02	23 05	28 03	24 00	14 10	24 03	03 44	03 29	23 25	05 04	23 25
03	23 01	28 11	23 45	13 47	24 04	03 46	03 27	23 25	05 04	23 24
04	22 56	27 01	23 28	13 23	24 05	03 48	03 26	23 24	05 04	23 24
05	22 51	24 37	23 09	13 00	24 05	03 50	03 24	23 24	05 04	23 23
06	22 45	21 09	22 47	12 36	24 05	03 52	03 22	23 24	05 05	23 23
07	22 39	16 45	22 24	12 13	24 05	03 53	03 21	23 23	05 05	23 22
08	22 33	11 38	21 59	11 49	24 05	03 55	03 19	23 23	05 05	23 21
09	22 26	06 00	21 33	11 25	24 04	03 57	03 17	23 23	05 05	23 21
10	22 19	00 N 01	21 05	11 01	24 03	03 58	03 16	23 22	05 05	23 20
11	22 11	06 S 06	20 36	10 37	24 02	04 00	03 14	23 22	05 05	23 20
12	22 03	12 08	20 06	10 14	24 01	04 01	03 12	23 22	05 06	23 19
13	21 55	17 46	19 35	09 50	24 00	04 03	03 10	23 22	05 06	23 19
14	21 46	22 37	19 03	09 26	23 58	04 04	03 08	23 22	05 06	23 18
15	21 37	26 14	18 30	09 02	23 57	04 05	03 06	23 21	05 06	23 17
16	21 28	28 08	17 56	08 38	23 55	04 07	03 04	23 21	05 07	23 17
17	21 18	27 59	17 22	08 15	23 53	04 08	03 02	23 21	05 07	23 16
18	21 08	25 44	16 47	07 51	23 50	04 09	03 00	23 21	05 07	23 16
19	20 57	21 40	16 12	07 28	23 48	04 10	02 58	23 20	05 07	23 15
20	20 46	16 17	15 37	07 05	23 45	04 11	02 56	23 20	05 08	23 15
21	20 35	10 07	15 01	06 41	23 42	04 11	02 54	23 20	05 08	23 14
22	20 24	03 S 37	14 25	06 19	23 39	04 12	02 52	23 19	05 08	23 13
23	20 12	02 N 49	13 49	05 56	23 36	04 13	02 50	23 19	05 09	23 13
24	19 59	08 55	13 13	05 33	23 32	04 13	02 48	23 19	05 09	23 12
25	19 47	14 29	12 36	05 11	23 29	04 14	02 46	23 19	05 09	23 12
26	19 34	19 20	12 00	04 49	23 25	04 14	02 44	23 18	05 10	23 11
27	19 21	23 17	11 25	04 27	23 21	04 15	02 41	23 18	05 10	23 11
28	19 07	26 11	10 49	04 06	23 16	04 15	02 39	23 18	05 11	23 10
29	18 53	27 55	10 14	03 45	23 12	04 15	02 37	23 18	05 11	23 09
30	18 39	28 22	09 39	03 24	23 07	04 15	02 35	23 17	05 11	23 09
31	18 N 25	27 N 30	09 N 05	03 N 04	23 N 03	04 N 15	02 N 32	23 N 17	05 S 12	23 N 08

ZODIAC SIGN ENTRIES

Date	h	m	Planets
03	08	27	☽ ♋
03	23	42	♂ ♋
05	21	00	☽ ♌
08	04	54	♀ ♍
08	08	36	☽ ♍
08	13	39	☿ ♌
10	18	04	☽ ♎
13	00	19	☽ ♏
15	03	03	☽ ♐
17	03	14	☽ ♑
19	02	41	☽ ♒
21	03	29	☽ ♓
23	07	21	☽ ♈
23	16	21	☉ ♌
25	15	07	☽ ♉
27	15	24	☿ ♍
28	02	08	☽ ♊
30	14	42	☽ ♋

LATITUDES

Date	Mercury ☿ ° '	Venus ♀ ° '	Mars ♂ ° '	Jupiter ♃ ° '	Saturn ♄ ° '	Uranus ♅ ° '	Neptune ♆ ° '	Pluto ♇ ° '
01	01 N 44	01 N 08	00 N 36	01 S 18	02 N 17	00 N 19	01 N 39	08 N 30
04	01 51	00 52	00 38	01 19	02 17	00 19	01 39	08 29
07	01 52	00 34	00 39	01 20	02 16	00 19	01 39	08 29
10	01 47	00 N 14	00 40	01 20	02 16	00 19	01 38	08 29
13	01 36	00 S 07	00 42	01 21	02 15	00 19	01 38	08 29
16	01 21	00 30	00 44	01 22	02 15	00 19	01 38	08 29
19	01 01	00 55	00 45	01 23	02 14	00 19	01 38	08 29
22	00 38	01 22	00 47	01 24	02 14	00 19	01 38	08 29
25	00 N 11	01 51	00 48	01 25	02 13	00 19	01 38	08 29
28	00 S 18	02 21	00 50	01 26	02 13	00 20	01 37	08 29
31	00 S 50	02 S 54	00 N 51	01 S 26	02 N 12	00 N 20	01 N 37	08 N 29

DATA

Julian Date	2433829
Delta T	+30 seconds
Ayanamsa	23° 10' 52"
Synetic vernal point	05° ♓ 56' 08"
True obliquity of ecliptic	23° 26' 52"

MOON'S PHASES, APSIDES AND POSITIONS ☽

Date	h	m	Phase	Longitude ° '	Eclipse Indicator
04	07	48	●	11 ♋ 32	
12	04	56	☽	19 ♎ 03	
18	19	17	○	25 ♑ 21	
25	18	59	☾	02 ♉ 01	

Day	h	m		
03	04	12	Apogee	
17	22	50	Perigee	
30	12	12	Apogee	
03	02	31	Max dec	28° N 17'
10	12	05	0S	
16	22	17	Max dec	28° S 20'
23	01	25	0N	
30	08	09	Max dec	28° N 23'

LONGITUDES

Date	Chiron ⚷ ° '	Ceres ⚳ ° '	Pallas ⚴ ° '	Juno ⚵ ° '	Vesta ⚶ ° '	Black Moon Lilith ⚸ ° '
01	29 ♐ 09	16 ♓ 29	10 ♒ 35	24 ♏ 52	08 ♋ 18	29 ♊ 47
11	28 ♐ 31	16 ♓ 50	08 ♒ 23	24 ♏ 07	12 ♋ 40	00 ♋ 55
21	27 ♐ 57	16 ♓ 36	05 ♒ 50	23 ♏ 54	17 ♋ 02	02 ♋ 02
31	27 ♐ 29	15 ♓ 47	03 ♒ 09	24 ♏ 13	21 ♋ 24	03 ♋ 09

All ephemeris data is given at 12.00 UT and the Moon's longitude is additionally given for 24.00 UT

ASPECTARIAN

h m	Aspects
01 Sunday	
00 40	☽ ⊥ ☉
03 07	☽ ⊥ ♅
08 27	☽ Q ♇
13 49	☽ ⊻ ☉
15 16	☽ ⊻ ♅
16 37	☽ ⊥ ☿
20 52	☽ ✶ ♃
20 58	☽ Q ♀
21 48	☿ □ ♆
02 Monday	
02 29	♀ ∠ ♅
05 39	☽ △ ♆
07 00	☿ ∥ ♂
07 19	☽ ⊻ ☿
08 44	☽ ✶ ♇
09 07	☉ ☌ ♅
13 35	♅ ∥ ♇
15 31	☿ ⊻ ♇
21 21	☽ Q ♃
23 14	☽ ✶ ♀
03 Tuesday	
01 29	☽ □ ♄
07 32	☽ ☌ ♂
09 29	☿ ⊥ ♀
15 10	☽ ∠ ♇
04 Wednesday	
04 16	☽ ☌ ♅
07 08	♀ ⊻ ♄
07 48	☽ ☌ ☉
08 03	☽ ∠ ♀
09 24	☽ ⊥ ♇
10 04	☽ □ ♃
14 04	☽ Q ♄
16 54	☿ ∥ ♅
17 44	☿ ∥ ♇
18 21	☽ □ ♆
21 33	☽ ⊻ ♇
05 Thursday	
03 40	☽ ⊥ ♀
04 18	☉ ⊥ ♇
08 29	☽ ☌ ☿
14 23	☽ ✶ ♄
14 32	♀ ⚼ ♃
14 55	☉ □ ♃
16 14	☽ ∥ ♂
16 42	☽ ⊻ ♀
18 34	☉ ∠ ♀
21 16	☽ ∥ ♅
21 23	☽ ∥ ♇
23 43	☽ ⊻ ♂
06 Friday	
00 18	☽ ∥ ☿
01 28	☽ ∥ ☉
06 34	☽ Q ♆
12 27	☽ ⊥ ♂
15 14	☉ ∥ ☿
16 55	☽ ⊻ ♅
20 35	☽ ∠ ♄
21 17	☿ ✶ ♄
22 48	☽ △ ♃
07 Saturday	
01 17	☽ ⊻ ☉
04 55	☽ ⊥ ♅
06 31	☽ ✶ ♆
07 26	☽ ∠ ♂
09 46	☽ ☌ ♇
14 12	☽ ⊥ ☉
14 33	☽ ⊥ ♄
18 53	☉ Q ♄
22 49	☽ ∠ ♅
08 Sunday	
02 26	☽ ⊻ ♄
04 41	☽ ⚼ ♃
07 41	☽ ⊻ ☿
08 52	☽ ☌ ♀
09 25	☽ ∠ ☉
11 10	☽ ∥ ♀
12 05	☽ ∠ ♆
14 41	☽ ✶ ♂
20 41	☿ ⊻ ♀
21 30	☽ ⊥ ☿
22 22	☽ ± ♃
09 Monday	
02 54	♂ ∠ ♇
04 14	☽ ✶ ♅
05 34	☽ ⊥ ♆
10 02	☽ ⊼ ♃
15 13	☽ Q ♂
15 45	☽ # ♆
16 52	☽ ✶ ☉
17 06	☽ ⊻ ♆
17 53	☽ ∠ ☿
19 52	☉ □ ♆
20 19	☽ ∥ ♃
20 20	☽ ⊻ ♇
23 04	☽ ∥ ♄
10 Tuesday	
03 19	☽ Q ♅
07 44	☽ ⊥ ♇
08 39	♀ ∠ ♆
12 29	☽ ☌ ♄
17 24	☽ Q ☉
22 16	☽ ⊻ ♀
11 Wednesday	
00 36	☽ ∠ ♇
00 49	☽ # ♄
02 48	☽ ✶ ☿
03 00	☽ □ ♂
03 44	☽ # ♃
04 05	☿ Q ♆
04 54	☿ ⊻ ♂
08 02	☽ ∥ ♆
10 04	☽ ⊥ ♀
12 59	☽ □ ♅
15 40	☉ ⊻ ♇
18 32	☽ ☍ ♃
12 Thursday	
00 53	☽ ☌ ♆
03 27	☽ ∠ ♀
04 01	☽ ✶ ♇
04 05	☽ Q ☿
04 48	☽ # ♀
04 56	☽ □ ☉
19 22	☽ ⊻ ♄
13 Friday	
01 18	☽ Q ♇
05 54	☽ ⊥ ♄
07 31	☽ ✶ ♀
11 22	☽ △ ♂
16 14	☽ □ ☿
18 13	☽ △ ♅
19 39	☽ # ☿
21 26	☽ ∠ ♄
23 27	☽ ⊼ ♃
14 Saturday	
05 09	☽ ⊻ ♆
05 12	☽ Q ♀
07 32	☽ # ☉
08 09	☽ □ ♇
09 05	☿ ⊻ ♅
09 36	☽ ± ♃
12 36	☽ △ ☉
14 01	☽ ⚼ ♂
15 12	☽ ⊥ ♆
15 50	☽ # ♇
16 12	☽ # ♅
19 31	☽ ⚼ ♅
19 50	☽ # ♂
22 39	☽ ✶ ♄
15 Sunday	
00 36	☽ ⚼ ♃
05 35	☽ ± ♂
06 02	☽ ∠ ♆
10 21	☽ ± ♅
12 34	☽ □ ♀
13 14	☿ ∠ ♄
14 58	☽ ⚼ ☉
15 49	☽ ⊼ ♂
18 20	☽ Q ♄
20 07	☽ ⊼ ♅
16 Monday	
00 22	☽ △ ☿
01 05	☽ △ ♃
06 18	☽ ✶ ♆
06 27	☽ ± ☉
07 01	☿ △ ♃
09 13	☽ △ ♇
16 40	☽ ⊼ ☉
23 15	☽ □ ♄
23 41	☿ ⊥ ♂
17 Tuesday	
01 24	☽ Q ♆
03 10	☽ ⚼ ☿
09 04	☽ ⚼ ♇
14 53	☽ △ ♀
17 48	☽ ☍ ♂
19 03	☽ ± ☿
20 00	☽ ☍ ♅
23 18	☽ ± ♇
18 Wednesday	
00 52	☽ □ ♃
01 58	♀ ± ♃
04 30	☿ ⊥ ♅
05 37	☽ ⊼ ☿
05 51	☽ □ ♆
07 57	☿ ✶ ♆
08 48	☽ ⊼ ♇
15 44	☽ ⚼ ♀
19 17	☽ ☍ ☉
22 55	☽ △ ♄
19 Thursday	
00 28	☽ # ♂
03 12	☽ # ♅
03 40	☽ # ♇
05 29	☽ Q ♃
06 43	☽ ± ♀
13 19	☿ ☌ ♇
15 37	☽ # ☉
16 48	☽ ⊼ ♀
19 41	☽ ⊼ ♂
19 52	☽ ⊼ ♅
23 05	☽ ⚼ ♄
20 Friday	
00 34	♂ ☌ ♅
00 50	☽ ✶ ♃
05 38	☽ ± ♅
05 50	☽ △ ♆
05 51	☽ ± ♂
08 56	☽ ☍ ♇
11 09	☽ ☍ ☿
13 50	☽ ± ♄
15 01	☽ # ☿
20 28	☽ ⚼ ♅
21 21	☽ ⚼ ♂
23 09	☽ ⊼ ☉
23 48	☽ ⊼ ♄
21 Saturday	
01 33	☽ ∠ ♃
06 38	☽ ⚼ ♆
09 40	☉ ✶ ♄
09 54	☽ ± ☉
15 01	☿ ⊥ ♄
16 45	☽ ⊥ ♃
20 47	☽ ☍ ♀
21 49	☽ △ ♅
21 54	☽ ± ♆
23 52	☽ △ ♂
22 Sunday	
01 30	☽ # ♀
02 20	☽ ⚼ ☉
03 02	☽ ⊻ ♃
06 25	☽ ∥ ♆
08 15	☽ ⊼ ♆
09 52	☽ # ♃
11 37	☽ ⊼ ♇
14 47	☽ # ♄
19 56	☽ ⊼ ☿
21 47	♀ ✶ ♅
21 51	♀ ⊥ ♆
22 11	☽ ± ♇
23 Monday	
03 43	☽ ☍ ♄
05 17	♂ ⊥ ♇
06 40	☽ △ ☉
07 51	☽ ± ☿
12 05	☽ ∥ ♄
14 20	☽ ⚼ ♇
17 22	☽ ∥ ♃
20 58	☽ # ♆
23 20	☽ ∥ ♀
24 Tuesday	
02 07	☽ ⚼ ☿
03 19	☽ □ ♅
04 37	☽ ⊼ ♀
08 09	☽ □ ♂
08 52	☽ ☌ ♃
13 52	☿ ∠ ♅
14 21	☽ ☍ ♆
16 18	☽ ± ♀
18 03	☽ △ ♇
22 51	♂ □ ♃
25 Wednesday	
04 24	☽ ∥ ☿
09 34	☽ △ ☿
10 09	☽ ⚼ ♀
11 36	☽ ⊼ ♄
13 18	☽ Q ♅
13 52	♀ # ♆
18 59	☽ □ ☉
20 06	☽ Q ♂
23 13	☽ ± ♄
26 Thursday	
06 05	☿ ⊻ ♄
12 40	☽ ✶ ♅
13 13	☽ ∥ ☉
16 35	☽ △ ♀
16 54	☽ ⚼ ♄
18 27	☽ ⊻ ♃
20 50	☽ ✶ ♂
21 35	☿ ⚼ ♃
27 Friday	
00 11	☽ ⊼ ♆
04 09	☽ □ ♇
06 20	☽ ⊥ ♃
09 14	☽ Q ☉
11 15	☽ ∥ ♇
12 05	☽ ± ♆
12 08	☽ ∥ ♅
12 25	☽ ∥ ♂
18 30	☽ ∠ ♅
22 53	☽ △ ♄
28 Saturday	
00 18	☽ ∠ ♃
02 06	♀ ∥ ♃
03 22	☽ □ ☿
03 24	♂ ∥ ♅
04 24	☽ ∠ ♂
06 06	☽ ⚼ ♆
11 19	☽ ✶ ☉
12 38	☽ ⊥ ♅
14 30	♀ ⊼ ♃
15 35	♂ Q ♄
16 14	☽ Q ♇
21 49	☉ Q ♆
23 34	☽ ⊥ ♂
29 Sunday	
00 30	☿ ∠ ♂
00 47	☽ ⊻ ♅
06 33	☽ ✶ ♃
06 50	☿ ∠ ♆
07 13	☽ □ ♀
12 05	♂ □ ♆
12 23	☽ △ ♆
12 24	☽ ⊻ ♂
16 32	☽ ✶ ♇
19 44	☽ Q ☿
20 15	☽ ∠ ☉
30 Monday	
03 45	♂ ∥ ♇
06 53	☽ Q ♃
11 53	☽ □ ♄
16 05	☽ ⊥ ☉
21 08	☽ Q ♀
22 37	☽ ✶ ☿
22 59	☽ ∠ ♇
31 Tuesday	
05 17	☽ ⊻ ☉
13 45	☽ ☌ ♅
17 13	☽ ⊥ ♇
19 19	☽ □ ♃
22 16	☽ ✶ ♀

AUGUST 1951

Date	Sidereal time h m s	LONGITUDES Sun ☉ ° ′ ″	Moon ☽ ° ′ ″	Moon ☽ 24.00 ° ′ ″	Mercury ☿ ° ′	Venus ♀ ° ′	Mars ♂ ° ′	Jupiter ♃ ° ′	Saturn ♄ ° ′	Uranus ♅ ° ′	Neptune ♆ ° ′	Pluto ♇ ° ′
01	08 37 09	08 ♌ 25 47	22 ♋ 26 53	28 ♋ 26 11	05 ♍ 36	15 ♍ 51	18 ♋ 58	14 ♈ 10	28 ♍ 48	11 ♋ 27	17 ♎ 03	19 ♌ 08
02	08 41 05	09 23 13	04 ♌ 26 59	10 ♌ 29 29	06 38	16 14	19 38	14 10	28 53	11 30	17 04	19 10
03	08 45 02	10 20 39	16 ♌ 33 51	22 ♌ 40 18	07 37	16 35	20 17	14 11	28 59	11 34	17 05	19 12
04	08 48 59	11 18 07	28 ♌ 49 01	05 ♍ 00 11	08 33	16 55	20 56	14 R 11	29 05	11 37	17 07	19 14
05	08 52 55	12 15 35	11 ♍ 14 01	17 ♍ 30 45	09 27	17 13	21 35	14 11	29 10	11 40	17 08	19 16
06	08 56 52	13 13 05	23 ♍ 50 36	00 ♎ 13 48	10 17	17 28	22 15	14 10	29 16	11 43	17 09	19 18
07	09 00 48	14 10 35	06 ♎ 40 37	13 ♎ 11 19	11 04	17 42	22 54	14 10	29 22	11 47	17 10	19 19
08	09 04 45	15 08 05	19 ♎ 46 09	26 ♎ 25 21	11 47	17 54	23 33	14 09	29 28	11 50	17 11	19 21
09	09 08 41	16 05 37	03 ♏ 09 08	09 ♏ 57 41	12 27	18 03	24 12	14 08	29 34	11 53	17 13	19 23
10	09 12 38	17 03 10	16 ♏ 51 06	23 ♏ 49 26	13 03	18 11	24 51	14 07	29 40	11 56	17 14	19 25
11	09 16 34	18 00 43	00 ♐ 52 37	08 ♐ 00 31	13 35	18 16	25 30	14 06	29 46	11 59	17 15	19 27
12	09 20 31	18 58 17	15 ♐ 12 49	22 ♐ 29 07	14 02	18 19	26 09	14 04	29 53	12 02	17 17	19 29
13	09 24 28	19 55 53	29 ♐ 48 52	07 ♑ 11 23	14 25	18 R 20	26 48	14 02	29 ♍ 59	12 05	17 18	19 31
14	09 28 24	20 53 29	14 ♑ 35 53	22 ♑ 01 29	14 44	18 18	27 27	14 00	00 ♎ 05	12 08	17 19	19 32
15	09 32 21	21 51 06	29 ♑ 27 14	06 ♒ 52 08	14 57	18 15	28 05	13 58	00 11	12 11	17 21	19 34
16	09 36 17	22 48 44	14 ♒ 15 13	21 ♒ 35 33	15 06	18 08	28 44	13 56	00 18	12 14	17 22	19 36
17	09 40 14	23 46 24	28 ♒ 52 15	06 ♓ 04 34	15 09	17 59	29 ♋ 23	13 54	00 24	12 17	17 24	19 38
18	09 44 10	24 44 04	13 ♓ 11 51	20 ♓ 13 37	15 R 07	17 48	00 ♌ 02	13 51	00 31	12 20	17 25	19 40
19	09 48 07	25 41 46	27 ♓ 09 30	03 ♈ 59 17	14 59	17 35	00 40	13 48	00 37	12 23	17 27	19 42
20	09 52 03	26 39 30	10 ♈ 42 54	17 ♈ 20 24	14 45	17 19	01 19	13 45	00 44	12 26	17 28	19 44
21	09 56 00	27 37 15	23 ♈ 51 58	00 ♉ 17 52	14 26	17 01	01 58	13 42	00 50	12 29	17 30	19 45
22	09 59 57	28 35 02	06 ♉ 38 28	12 ♉ 54 10	14 01	16 40	02 36	13 38	00 57	12 31	17 31	19 47
23	10 03 53	29 ♌ 32 51	19 ♉ 05 28	25 ♉ 12 54	13 30	16 18	03 15	13 35	01 04	12 34	17 33	19 49
24	10 07 50	00 ♍ 30 41	01 ♊ 16 58	07 ♊ 18 17	12 54	15 53	03 53	13 31	01 10	12 37	17 35	19 51
25	10 11 46	01 28 33	13 ♊ 17 23	19 ♊ 14 51	12 13	15 27	04 32	13 27	01 17	12 40	17 36	19 53
26	10 15 43	02 26 27	25 ♊ 11 15	01 ♋ 07 05	11 27	14 58	05 10	13 23	01 24	12 42	17 38	19 55
27	10 19 39	03 24 23	07 ♋ 02 55	12 ♋ 59 13	10 38	14 28	05 49	13 18	01 31	12 45	17 40	19 56
28	10 23 36	04 22 20	18 ♋ 56 26	24 ♋ 55 01	09 45	13 56	06 27	13 14	01 37	12 47	17 41	19 58
29	10 27 32	05 20 20	00 ♌ 55 20	06 ♌ 57 44	08 50	13 23	07 05	13 09	01 44	12 50	17 43	20 00
30	10 31 29	06 18 21	13 ♌ 02 32	19 ♌ 10 00	07 54	12 48	07 44	13 04	01 51	12 52	17 45	20 02
31	10 35 26	07 ♍ 16 23	25 ♌ 20 21	01 ♍ 33 46	06 ♍ 58	12 ♍ 13	08 ♌ 22	12 ♈ 59	01 ♎ 58	12 ♋ 55	17 ♎ 47	20 ♌ 04

Date	Moon True ☊ ° ′	Moon Mean ☊ ° ′	Moon ☽ Latitude ° ′	DECLINATIONS Sun ☉ ° ′	Moon ☽ ° ′	Mercury ☿ ° ′	Venus ♀ ° ′	Mars ♂ ° ′	Jupiter ♃ ° ′	Saturn ♄ ° ′	Uranus ♅ ° ′	Neptune ♆ ° ′	Pluto ♇ ° ′
01	10 ♓ 07	11 ♓ 32	03 N 51	18 N 10	25 N 22	08 N 31	02 N 44	22 N 58	04 N 15	02 N 30	23 N 17	05 S 12	23 N 08
02	10 R 03	11 29	03 02	17 55	22 06	07 58	02 25	22 52	04 15	02 28	23 16	05 13	23 07
03	09 59	11 26	02 05	17 39	17 51	07 25	02 06	22 47	04 15	02 25	23 16	05 13	23 06
04	09 57	11 22	01 N 01	17 24	12 50	06 54	01 48	22 42	04 15	02 23	23 16	05 14	23 06
05	09 D 57	11 19	00 S 07	17 08	07 15	06 23	01 30	22 36	04 15	02 20	23 16	05 14	23 05
06	09 58	11 16	01 16	16 51	01 N 17	05 54	01 13	22 30	04 14	02 18	23 15	05 15	23 05
07	09 59	11 13	02 22	16 35	04 S 49	05 25	00 56	22 24	04 14	02 15	23 15	05 15	23 04
08	10 00	11 10	03 22	16 18	10 51	04 58	00 41	22 18	04 13	02 13	23 15	05 16	23 04
09	10 02	11 07	04 12	16 01	16 31	04 32	00 25	22 12	04 13	02 11	23 15	05 16	23 03
10	10 02	11 03	04 49	15 44	21 30	04 08	00 N 11	22 05	04 12	02 08	23 14	05 17	23 03
11	10 R 02	11 00	05 11	15 26	25 24	03 45	00 S 03	21 58	04 11	02 05	23 14	05 17	23 02
12	10 00	10 57	05 13	15 08	27 49	03 24	00 15	21 52	04 10	02 03	23 14	05 18	23 01
13	09 58	10 54	04 56	14 50	28 23	03 05	00 27	21 45	04 09	02 00	23 14	05 18	23 01
14	09 56	10 51	04 19	14 32	26 56	02 48	00 38	21 37	04 08	01 58	23 13	05 19	23 00
15	09 53	10 47	03 25	14 14	23 36	02 33	00 48	21 30	04 07	01 55	23 13	05 19	23 00
16	09 52	10 44	02 17	13 55	18 44	02 21	00 57	21 23	04 06	01 52	23 13	05 20	22 59
17	09 51	10 41	01 S 00	13 36	12 49	02 11	01 05	21 15	04 05	01 50	23 13	05 21	22 59
18	09 D 50	10 38	00 N 19	13 17	06 S 19	02 04	01 11	21 07	04 04	01 47	23 12	05 21	22 58
19	09 51	10 35	01 34	12 58	00 N 19	02 00	01 17	20 59	04 02	01 45	23 12	05 22	22 58
20	09 52	10 32	02 43	12 38	06 44	01 59	01 21	20 51	04 01	01 42	23 12	05 22	22 57
21	09 53	10 28	03 40	12 18	12 40	02 01	01 25	20 43	03 59	01 39	23 12	05 23	22 57
22	09 54	10 25	04 25	11 58	17 54	02 06	01 27	20 35	03 58	01 36	23 11	05 24	22 56
23	09 54	10 22	04 56	11 38	22 14	02 15	01 28	20 26	03 56	01 34	23 11	05 24	22 56
24	09 R 55	10 19	05 13	11 18	25 32	02 27	01 27	20 18	03 54	01 31	23 11	05 25	22 55
25	09 54	10 16	05 16	10 57	27 38	02 42	01 26	20 09	03 53	01 28	23 11	05 26	22 55
26	09 54	10 13	05 05	10 37	28 27	03 01	01 23	20 00	03 51	01 25	23 10	05 26	22 54
27	09 53	10 09	04 42	10 16	27 57	03 23	01 19	19 51	03 49	01 23	23 10	05 27	22 54
28	09 52	10 06	04 06	09 55	26 10	03 47	01 14	19 42	03 47	01 20	23 10	05 28	22 53
29	09 52	10 03	03 19	09 33	23 12	04 15	01 07	19 33	03 45	01 17	23 10	05 28	22 53
30	09 51	10 00	02 23	09 12	19 23	04 44	01 00	19 23	03 43	01 14	23 10	05 29	22 52
31	09 ♓ 51	09 ♓ 57	01 N 20	08 N 51	14 N 20	05 N 15	00 S 51	19 N 14	03 N 41	01 N 11	23 N 09	05 S 30	22 N 52

ZODIAC SIGN ENTRIES

Date	h	m	Planets
02	03	08	☽ ♌
04	14	18	☽ ♍
06	23	34	☽ ♎
09	06	24	☽ ♏
11	10	31	☽ ♐
13	12	18	☽ ♑
13	16	44	♄ ♎
15	12	53	☽ ♒
17	13	52	☽ ♓
18	10	55	♂ ♌
19	16	58	☽ ♈
21	23	26	☽ ♉
23	23	16	☉ ♍
24	09	27	☽ ♊
26	21	44	☽ ♋
29	10	10	☽ ♌
31	21	00	☽ ♍

LATITUDES

Date	Mercury ☿ ° ′	Venus ♀ ° ′	Mars ♂ ° ′	Jupiter ♃ ° ′	Saturn ♄ ° ′	Uranus ♅ ° ′	Neptune ♆ ° ′	Pluto ♇ ° ′
01	01 S 01	03 S 05	00 N 52	01 S 27	02 N 12	00 N 20	01 N 37	08 N 29
04	01 35	03 40	00 53	01 28	02 12	00 20	01 37	08 29
07	02 10	04 16	00 55	01 28	02 11	00 20	01 37	08 29
10	02 45	04 53	00 56	01 29	02 11	00 20	01 37	08 29
13	03 18	05 30	00 57	01 30	02 11	00 20	01 37	08 29
16	03 49	06 07	00 59	01 31	02 10	00 20	01 37	08 30
19	04 15	06 43	01 00	01 32	02 10	00 20	01 37	08 30
22	04 32	07 17	01 01	01 32	02 10	00 20	01 36	08 30
25	04 37	07 46	01 03	01 33	02 10	00 20	01 36	08 31
28	04 27	08 10	01 04	01 34	02 09	00 20	01 36	08 31
31	04 S 00	08 S 28	01 N 05	01 S 35	02 N 09	00 N 20	01 N 36	08 N 31

DATA

Julian Date	2433860
Delta T	+30 seconds
Ayanamsa	23° 10′ 57″
Synetic vernal point	05° ♓ 56′ 02″
True obliquity of ecliptic	23° 26′ 53″

MOON'S PHASES, APSIDES AND POSITIONS ☽

Date	h	m	Phase	Longitude ° ′	Eclipse Indicator
02	22	39	●	09 ♌ 49	
10	12	22	☽	17 ♏ 04	
17	02	59	○	23 ♒ 25	
24	10	20	☾	00 ♊ 27	

Day	h	m		
15	04	13	Perigee	
27	02	51	Apogee	
06	17	05	0S	
13	06	51	Max dec	28° S 26′
19	10	52	0N	
26	14	57	Max dec	28° N 27′

LONGITUDES

Date	Chiron ⚷ ° ′	Ceres ⚳ ° ′	Pallas ⚴ ° ′	Juno ⚵ ° ′	Vesta ⚶ ° ′	Black Moon Lilith ⚸ ° ′
01	27 ♐ 26	15 ♓ 40	02 ♒ 54	24 ♏ 17	21 ♋ 50	03 ♋ 16
11	27 ♐ 06	14 ♓ 14	00 ♒ 23	25 ♏ 09	26 ♋ 10	04 ♋ 23
21	26 ♐ 52	12 ♓ 22	28 ♑ 14	26 ♏ 27	00 ♌ 29	05 ♋ 30
31	26 ♐ 48	10 ♓ 13	26 ♑ 36	28 ♏ 08	04 ♌ 45	06 ♋ 37

All ephemeris data is given at 12.00 UT and the Moon's longitude is additionally given for 24.00 UT

ASPECTARIAN

h m	Aspects
01 Wednesday	
00 32	☽ Q ♄
01 08	☽ □ ♆
04 37	☽ ☌ ♂
05 20	☽ ⊻ ♇
07 56	☽ ∠ ☿
18 20	♂ ⊻ ♇
02 Thursday	
00 49	☽ ⚹ ♄
03 42	☽ ⊥ ☿
04 12	☽ ‖ ♅
05 16	☽ ‖ ♇
05 23	☽ ∠ ♀
06 49	☽ ‖ ♂
08 16	♀ ‖ ♄
13 14	☽ Q ♆
16 45	☽ ⊻ ☿
22 39	☽ ☌ ☉
23 51	☽ ⊥ ♀
03 Friday	
02 05	☽ ⊻ ♅
06 52	☽ ∠ ♄
07 18	☽ △ ♃
12 03	☽ ⊻ ♀
13 02	☽ ⚹ ♆
13 05	☽ ‖ ☉
13 58	☽ ⊥ ♅
17 12	☽ ☌ ♇
19 44	☽ ⊻ ♂
21 28	☉ ⊥ ♀
04 Saturday	
00 42	☽ ⊥ ♄
02 09	☿ ± ♃
06 51	♃ St R
07 42	☽ ∠ ♅
08 08	☽ ⊥ ♂
12 31	☽ ⊻ ♄
12 42	☽ bQ ♃
18 24	☽ ∠ ♆
20 21	☉ ⊻ ♅
05 Sunday	
02 34	☽ ∠ ♂
04 40	♀ ⊻ ♆
06 08	☽ ± ♃
08 19	☽ ☌ ☿
11 48	☽ ⊥ ♆
12 50	☽ ⚹ ♅
14 08	☽ ⊻ ☉
15 52	☽ ‖ ☿
17 38	☽ ⊼ ♃
20 13	☽ # ♆
23 17	☽ ⊻ ♆
23 40	☽ ☌ ♀
06 Monday	
00 15	☽ ‖ ♃
02 31	☽ ⊥ ☉
03 22	☽ ⊻ ♇
07 58	☽ ‖ ♄
08 49	☽ ⚹ ♂
11 46	☽ Q ♅
12 19	☽ ‖ ♀
14 44	☽ ⊥ ♇
20 54	☽ ∠ ☉
21 26	☽ # ♀
22 17	☽ ☌ ♄
07 Tuesday	
02 01	☽ # ♄
07 37	☽ ∠ ♇
08 31	☽ Q ♂
09 41	☽ # ♃
11 39	☉ △ ♃
13 42	☽ ‖ ♆
14 12	☽ # ☿
15 31	☿ ⊥ ♆
17 24	☉ ∠ ♄
20 35	☽ ⊻ ☿
20 35	☿ # ♆
21 27	☽ □ ♅
08 Wednesday	
01 46	☽ ☍ ♃
02 54	☽ ⚹ ☉
07 18	☽ ☌ ♆
08 12	☽ ⊥ ☿
08 33	☽ ⊻ ♀
11 15	☽ ⚹ ♇
13 44	☿ ⚹ ♅
19 11	☽ □ ♂
19 33	☽ ⊥ ♀
09 Thursday	
01 19	☽ ∠ ☿
02 18	☽ Q ☉
05 35	☽ ⊻ ♄
08 51	☽ Q ♇
09 56	☽ # ☉
11 50	☽ ∠ ♀
16 18	☽ ⊥ ♄
10 Friday	
03 25	☽ △ ♅
05 06	☽ ⚹ ☿
07 16	☽ ⊼ ♃
07 43	☿ ‖ ♃
08 11	☽ ∠ ♄
12 22	☽ □ ☉
12 39	☽ ⊻ ♆
14 19	☽ ⚹ ♀
15 06	☽ # ♂
16 26	☽ □ ♇
16 34	☉ ⚹ ♆
17 38	☽ ± ♃
20 35	☽ # ♇
21 45	☽ # ♅
23 00	☽ ⊥ ♆
11 Saturday	
02 25	☽ △ ♂
02 39	☽ Q ☿
05 22	☽ bQ ♅
08 59	☽ bQ ♃
10 07	☽ ⚹ ♄
10 58	☽ Q ♀
11 20	☉ ⊥ ♅
14 20	☽ ∠ ♆
18 56	☉ ⊻ ♀
20 39	☽ ± ♅
12 Sunday	
04 55	☽ bQ ♂
06 25	☽ Q ♄
06 42	☽ ⊼ ♅
09 59	☽ □ ☿
10 06	☽ △ ♃
13 43	☿ ⊼ ♃
15 25	☽ ⚹ ♆
17 09	☽ □ ♀
18 39	☽ △ ☉
19 04	☽ △ ♇
20 32	☽ ± ♂
13 Monday	
01 05	☉ ☌ ♇
06 50	☽ ⊼ ♂
07 51	♀ St R
11 09	☽ Q ♆
12 16	☽ □ ♄
19 40	☽ bQ ♇
20 55	☽ bQ ☉
14 Tuesday	
08 00	☽ ☍ ♅
10 17	☽ ± ♇
11 03	☽ □ ♃
12 13	☽ △ ☿
12 30	☽ ± ☉
16 25	☽ □ ♆
17 59	☽ △ ♀
20 00	☽ ⊼ ♇
22 52	☽ ⊼ ☉
15 Wednesday	
09 42	☽ ☍ ♂
12 49	☽ bQ ☿
13 12	☽ △ ♄
14 12	☽ # ♅
15 25	☽ # ♇
16 04	☽ Q ♃
18 05	☽ bQ ♀
23 28	☽ # ♂
16 Thursday	
03 33	☽ ± ☿
08 35	☽ ± ♀
08 42	☽ ⊼ ♅
11 29	☽ ⚹ ♃
13 23	☽ ⊼ ☿
13 43	☽ bQ ♄
17 06	☽ △ ♆
18 17	☽ ⊼ ♀
18 32	☽ ± ♅
20 45	☽ ☍ ♇
17 Friday	
02 59	☽ ☍ ☉
04 34	☽ ± ♄
08 48	☽ # ☉
09 22	☽ bQ ♅
12 02	☽ ∠ ♃
12 53	☽ ⊼ ♂
14 04	☿ St R
14 34	☽ ⊼ ♄
17 52	☽ bQ ♆
23 21	☽ ± ♂
18 Saturday	
03 00	☽ ⊥ ♃
05 41	☉ ⊥ ♄
08 59	☽ ± ♆
10 32	☽ △ ♅
13 06	☽ ⊻ ♃
14 44	☿ ∠ ♂
15 14	☽ ☍ ☿
15 16	☽ bQ ♂
15 29	☽ ‖ ♆
19 12	☽ ⊼ ♆
19 44	☽ ☍ ♀
20 12	☽ # ♃
23 03	☽ ⊼ ♇
19 Sunday	
03 32	☽ # ☿
04 29	☽ # ♄
06 17	☽ ‖ ♀
09 16	☽ ⊼ ☉
09 27	☽ ± ♇
09 31	♂ ⚹ ♄
15 36	☽ # ♀
17 12	☽ ‖ ♄
18 06	☽ ☍ ♄
18 09	☽ ‖ ☿
18 28	☽ △ ♂
20 33	☽ ± ☉
23 44	♀ ⊻ ♆
20 Monday	
01 17	☽ bQ ♇
01 42	☽ ‖ ♃
06 47	☽ # ♆
13 50	☽ bQ ☉
15 06	☽ □ ♅
17 27	☽ ☌ ♃
19 08	☽ ⊼ ☿
23 42	☽ ⊼ ♀
21 Tuesday	
00 16	☽ ☍ ♆
04 24	☽ △ ♇
05 50	☽ ± ☿
08 15	☉ ∠ ♅
10 28	☽ ± ♀
10 31	☽ ‖ ☉
13 19	♀ ∠ ♂
19 33	☽ △ ☉
22 04	☽ bQ ☿
22 Wednesday	
00 23	☽ Q ♅
01 08	☽ ⊼ ♄
02 51	☽ bQ ♀
03 56	☽ □ ♂
12 35	☽ ± ♄
13 14	☉ bQ ♃
23 19	☽ ⚹ ♅
23 Thursday	
01 21	☽ ⊻ ♃
01 36	☽ △ ☿
01 42	☽ ‖ ♂
06 03	☽ bQ ♄
06 44	☽ △ ♀
08 16	☿ ⊼ ♃
08 59	☽ ⊼ ♆
12 56	☽ ⊥ ♃
13 25	☽ □ ♇
16 25	☽ ‖ ♇
16 27	☽ Q ♂
18 07	☽ ‖ ♅
20 45	☽ ± ♆
24 Friday	
04 42	☽ ∠ ♅
06 32	☽ ∠ ♃
10 20	☽ □ ☉
11 46	☽ △ ♄
14 35	☽ bQ ♆
17 28	☽ ⚹ ♂
21 38	☿ ⚹ ♅
22 40	☽ ⊥ ♅
25 Saturday	
01 07	☽ Q ♇
06 34	☉ ⊻ ♄
09 58	☽ □ ☿
10 44	☽ ⊻ ♅
12 19	☽ ⚹ ♃
16 10	☽ □ ♀
20 42	☽ △ ♆
26 Sunday	
01 17	☽ ∠ ♂
01 19	☽ ⚹ ♇
01 34	☽ Q ☉
06 43	☿ ⊥ ♆
12 23	☽ Q ♃
16 42	☿ ⊥ ♂
16 54	☉ ∠ ♆
20 05	☽ Q ☿
20 31	☽ ⊥ ♂
27 Monday	
00 41	☽ □ ♄
03 06	☽ Q ♀
03 59	☽ ⚹ ☉
06 07	♂ Q ♆
07 44	☽ ∠ ♇
09 21	☽ ⊻ ♂
18 45	☽ ⚹ ☿
23 33	☽ ☌ ♅
28 Tuesday	
00 34	☽ □ ♃
01 58	☽ ⊥ ♇
02 21	☽ ⚹ ♀
09 29	☽ □ ♆
11 35	☿ ‖ ♃
12 57	☽ ∠ ☉
13 23	☽ Q ♄
14 05	☽ ⊻ ♇
22 51	☽ ∠ ☿
29 Wednesday	
07 09	☽ ∠ ♀
08 34	☽ ⊥ ☉
12 16	☽ ‖ ♅
13 39	☽ ⚹ ♄
14 10	☽ ‖ ♇
15 32	☽ ⊥ ☿
17 47	♀ ⊥ ♂
21 33	☽ ⊻ ☉
21 34	☽ Q ♆
23 18	♀ ⊼ ♃
30 Thursday	
00 15	☽ ⊥ ♀
00 56	☽ ☌ ♂
02 35	☽ ⊻ ☿
09 23	♀ ⚹ ♅
10 55	☽ ‖ ♂
11 33	☽ ⊻ ♀
11 40	☽ ⊻ ♅
12 03	☽ △ ♃
14 40	☿ ⊻ ♂
19 33	☽ ∠ ♄
21 15	☽ ⚹ ♆
23 28	☽ ⊥ ♅
31 Friday	
01 43	☽ ☌ ♇
05 25	☉ ± ♃
08 09	☉ ☌ ☿
11 33	☿ ± ♃
13 14	☽ ⊥ ♄
17 00	☽ ∠ ♅
17 05	☽ bQ ♃
23 49	☿ # ♆

SEPTEMBER 1951

LONGITUDES

Date	Sidereal time h m s	Sun ☉ ° ' "	Moon ☽ ° ' "	Moon ☽ 24.00 ° ' "	Mercury ☿ ° '	Venus ♀ ° '	Mars ♂ ° '	Jupiter ♃ ° '	Saturn ♄ ° '	Uranus ♅ ° '	Neptune ♆ ° '	Pluto ♇ ° '
01	10 39 22	08 ♍ 14 27	07 ♍ 50 24	14 ♍ 10 22	06 ♍ 03	11 ♍ 37	09 ♌ 00	12 ♈ 54	02 ♎ 05	12 ♋ 57	17 ♎ 48	20 ♌ 06
02	10 43 19	09 12 33	20 ♍ 33 44	27 ♍ 00 33	05 R 11	11 R 00	09 38	12 R 48	02 12	13 00	17 50	20 07
03	10 47 15	10 10 41	03 ♎ 30 50	10 ♎ 04 35	04 22	10 23	10 16	12 43	02 19	13 02	17 52	20 09
04	10 51 12	11 08 50	16 ♎ 41 48	23 ♎ 22 24	03 37	09 46	10 55	12 37	02 26	13 04	17 54	20 11
05	10 55 08	12 07 00	00 ♏ 06 22	06 ♏ 53 36	02 59	09 09	11 33	12 31	02 33	13 07	17 56	20 13
06	10 59 05	13 05 12	13 ♏ 44 03	20 ♏ 37 35	02 27	08 33	12 11	12 25	02 41	13 09	17 58	20 14
07	11 03 01	14 03 26	27 ♏ 34 06	04 ♐ 33 27	02 03	07 57	12 49	12 19	02 48	13 11	18 00	20 16
08	11 06 58	15 01 41	11 ♐ 35 27	18 ♐ 39 55	01 47	07 21	13 27	12 13	02 55	13 13	18 02	20 18
09	11 10 55	15 59 58	25 ♐ 46 36	02 ♑ 55 12	01 39	06 47	14 05	12 07	03 02	13 15	18 04	20 20
10	11 14 51	16 58 16	10 ♑ 05 22	17 ♑ 16 45	01 D 40	06 15	14 42	12 00	03 09	13 17	18 06	20 21
11	11 18 48	17 56 36	24 ♑ 28 53	01 ♒ 41 18	01 51	05 43	15 20	11 54	03 16	13 19	18 08	20 23
12	11 22 44	18 54 57	08 ♒ 53 28	16 ♒ 04 50	02 11	05 14	15 58	11 47	03 24	13 21	18 09	20 25
13	11 26 41	19 53 20	23 ♒ 14 49	00 ♓ 22 50	02 39	04 46	16 36	11 40	03 31	13 23	18 12	20 26
14	11 30 37	20 51 45	07 ♓ 28 19	14 ♓ 30 44	03 16	04 20	17 14	11 33	03 38	13 25	18 14	20 28
15	11 34 34	21 50 11	21 ♓ 29 32	28 ♓ 24 19	04 02	03 56	17 51	11 26	03 46	13 27	18 16	20 30
16	11 38 30	22 48 39	05 ♈ 14 41	12 ♈ 00 20	04 56	03 34	18 29	11 19	03 53	13 29	18 18	20 31
17	11 42 27	23 47 09	18 ♈ 41 03	25 ♈ 16 45	05 57	03 15	19 07	11 11	04 00	13 30	18 20	20 33
18	11 46 24	24 45 41	01 ♉ 47 24	08 ♉ 13 04	07 05	02 58	19 44	11 04	04 08	13 32	18 22	20 34
19	11 50 20	25 44 16	14 ♉ 33 56	20 ♉ 50 13	08 20	02 43	20 22	10 57	04 15	13 34	18 24	20 36
20	11 54 17	26 42 52	27 ♉ 02 16	03 ♊ 10 28	09 40	02 31	20 59	10 49	04 22	13 35	18 26	20 38
21	11 58 13	27 41 31	09 ♊ 15 17	15 ♊ 17 10	11 06	02 21	21 37	10 41	04 30	13 37	18 28	20 39
22	12 02 10	28 40 12	21 ♊ 16 42	27 ♊ 14 25	12 36	02 14	22 14	10 34	04 37	13 38	18 30	20 41
23	12 06 06	29 ♍ 38 55	03 ♋ 10 56	09 ♋ 06 49	14 10	02 09	22 52	10 26	04 44	13 40	18 32	20 42
24	12 10 03	00 ♎ 37 40	15 ♋ 02 42	20 ♋ 59 09	15 47	02 07	23 29	10 18	04 52	13 41	18 34	20 44
25	12 13 59	01 36 28	26 ♋ 56 47	02 ♌ 56 09	17 27	02 D 06	24 06	10 10	04 59	13 43	18 37	20 45
26	12 17 56	02 35 18	08 ♌ 57 48	15 ♌ 02 14	19 10	02 09	24 44	10 02	05 07	13 44	18 39	20 47
27	12 21 53	03 34 10	21 ♌ 09 55	27 ♌ 21 16	20 54	02 13	25 21	09 54	05 14	13 45	18 41	20 48
28	12 25 49	04 33 04	03 ♍ 36 36	09 ♍ 56 13	22 39	02 20	25 58	09 46	05 21	13 46	18 43	20 50
29	12 29 46	05 32 00	16 ♍ 20 20	22 ♍ 49 03	24 26	02 29	26 36	09 38	05 29	13 48	18 45	20 51
30	12 33 42	06 ♎ 30 59	29 ♍ 22 24	06 ♎ 00 21	26 ♍ 13	02 ♍ 40	27 ♌ 13	09 ♈ 30	05 ♎ 36	13 ♋ 49	18 ♎ 47	20 ♌ 53

MOON NODES, LATITUDE AND DECLINATIONS

Date	Moon True ☊ ° '	Moon Mean ☊ ° '	Moon ☽ Latitude ° '	Sun ☉ ° '	Moon ☽ ° '	Mercury ☿ ° '	Venus ♀ ° '	Mars ♂ ° '	Jupiter ♃ ° '	Saturn ♄ ° '	Uranus ♅ ° '	Neptune ♆ ° '	Pluto ♇ ° '
01	09 ♓ 51	09 ♓ 53	00 N 11	08 N 29	08 N 48	05 N 46	00 S 41	19 N 04	03 N 38	01 N 09	23 N 09	05 S 31	22 N 51
02	09 D 51	09 50	00 S 59	08 07	02 N 50	06 19	00 31	18 54	03 36	01 06	23 09	05 31	22 51
03	09 R 51	09 47	02 08	07 45	03 S 21	06 51	00 19	18 44	03 34	01 03	23 09	05 32	22 51
04	09 51	09 44	03 10	07 23	09 30	07 23	00 S 07	18 34	03 31	01 00	23 08	05 33	22 50
05	09 51	09 41	04 04	07 01	15 19	07 54	00 N 07	18 24	03 29	00 57	23 08	05 33	22 50
06	09 50	09 38	04 45	06 39	20 29	08 22	00 20	18 14	03 26	00 54	23 08	05 34	22 49
07	09 50	09 34	05 10	06 16	24 39	08 49	00 35	18 04	03 24	00 52	23 08	05 35	22 49
08	09 50	09 31	05 17	05 54	27 25	09 13	00 50	17 53	03 21	00 49	23 08	05 36	22 48
09	09 D 50	09 28	05 05	05 31	28 27	09 33	01 05	17 43	03 19	00 46	23 08	05 36	22 48
10	09 50	09 25	04 34	05 09	27 37	09 51	01 21	17 32	03 16	00 43	23 07	05 37	22 48
11	09 51	09 22	03 46	04 46	24 56	10 04	01 37	17 21	03 13	00 40	23 07	05 38	22 47
12	09 52	09 18	02 43	04 23	20 40	10 14	01 52	17 10	03 11	00 37	23 07	05 39	22 47
13	09 52	09 15	01 31	04 00	15 12	10 20	02 08	16 59	03 08	00 34	23 07	05 40	22 47
14	09 53	09 12	00 S 13	03 37	08 59	10 22	02 24	16 48	03 05	00 31	23 07	05 40	22 46
15	09 R 52	09 09	01 N 04	03 14	02 S 24	10 19	02 40	16 37	03 02	00 28	23 07	05 41	22 46
16	09 52	09 06	02 16	02 51	04 N 10	10 13	02 55	16 26	02 59	00 25	23 06	05 42	22 45
17	09 50	09 03	03 18	02 28	10 23	10 02	03 10	16 14	02 56	00 23	23 06	05 43	22 45
18	09 48	08 59	04 09	02 05	15 59	09 48	03 25	16 03	02 53	00 20	23 06	05 44	22 45
19	09 46	08 56	04 46	01 42	20 45	09 30	03 39	15 51	02 50	00 17	23 06	05 44	22 44
20	09 44	08 53	05 08	01 18	24 29	09 08	03 53	15 40	02 47	00 14	23 06	05 45	22 44
21	09 42	08 50	05 15	00 55	27 02	08 44	04 06	15 28	02 44	00 11	23 06	05 46	22 44
22	09 41	08 47	05 09	00 32	28 18	08 16	04 18	15 16	02 41	00 08	23 06	05 47	22 44
23	09 D 41	08 44	04 49	00 N 08	28 14	07 45	04 30	15 04	02 38	00 05	23 05	05 48	22 43
24	09 41	08 40	04 17	00 S 15	26 52	07 12	04 42	14 52	02 35	00 N 02	23 05	05 48	22 43
25	09 43	08 37	03 34	00 38	24 17	06 36	04 52	14 40	02 32	00 S 01	23 05	05 49	22 43
26	09 45	08 34	02 41	01 02	20 37	05 59	05 02	14 28	02 28	00 04	23 05	05 50	22 42
27	09 46	08 31	01 41	01 25	16 02	05 19	05 11	14 16	02 25	00 07	23 05	05 51	22 42
28	09 47	08 28	00 N 34	01 49	10 43	04 38	05 20	14 04	02 22	00 10	23 05	05 52	22 42
29	09 R 47	08 24	00 S 36	02 12	04 50	03 56	05 27	13 51	02 19	00 13	23 05	05 53	22 42
30	09 ♓ 46	08 ♓ 21	01 S 45	02 S 35	01 S 22	03 N 13	05 N 34	13 N 39	02 N 16	00 S 15	23 N 05	05 S 53	22 N 41

ZODIAC SIGN ENTRIES

Date	h	m	Planets
03	05	32	☽ ♎
05	11	49	☽ ♏
07	16	11	☽ ♐
09	19	06	☽ ♑
11	21	11	☽ ♒
13	23	21	☽ ♓
16	02	47	☽ ♈
18	08	41	☽ ♉
20	17	47	☽ ♊
23	05	34	☽ ♋
23	20	37	☉ ♎
25	18	08	☽ ♌
28	05	05	☽ ♍
30	13	08	☽ ♎

LATITUDES

Date	Mercury ☿ ° '	Venus ♀ ° '	Mars ♂ ° '	Jupiter ♃ ° '	Saturn ♄ ° '	Uranus ♅ ° '	Neptune ♆ ° '	Pluto ♇ ° '
01	03 S 47	08 S 32	01 N 06	01 S 35	02 N 09	00 N 20	01 N 36	08 N 31
04	03 00	08 39	01 07	01 35	02 09	00 20	01 36	08 32
07	02 04	08 38	01 08	01 36	02 09	00 20	01 36	08 32
10	01 07	08 28	01 09	01 36	02 09	00 20	01 36	08 33
13	00 S 13	08 11	01 10	01 37	02 09	00 20	01 36	08 33
16	00 N 33	07 48	01 12	01 37	02 09	00 21	01 36	08 34
19	01 08	07 20	01 13	01 38	02 09	00 21	01 36	08 35
22	01 33	06 49	01 14	01 38	02 09	00 21	01 36	08 35
25	01 47	06 16	01 15	01 38	02 09	00 21	01 36	08 36
28	01 53	05 42	01 16	01 38	02 09	00 21	01 36	08 37
31	01 N 51	05 S 07	01 N 17	01 S 38	02 N 09	00 N 21	01 N 36	08 N 37

DATA

Julian Date	2433891
Delta T	+30 seconds
Ayanamsa	23° 11' 02"
Synetic vernal point	05° ♓ 55' 58"
True obliquity of ecliptic	23° 26' 53"

MOON'S PHASES, APSIDES AND POSITIONS ☽

Date	h	m	Phase	Longitude ° '	Eclipse Indicator
01	12	50	●	08 ♍ 16	Annular
08	18	16	☽	15 ♐ 17	
15	12	38	○	21 ♓ 52	
23	04	13	☾	29 ♊ 20	

Day	h	m	
11	20	34	Perigee
23	21	11	Apogee

02	23	03	OS	
09	13	26	Max dec	28° S 28'
15	20	42	ON	
22	22	45	Max dec	28° N 26'
30	06	47	OS	

LONGITUDES

Date	Chiron ⚷ ° '	Ceres ⚳ ° '	Pallas ⚴ ° '	Juno ⚵ ° '	Vesta ⚶ ° '	Black Moon Lilith ⚸ ° '
01	26 ♐ 48	09 ♓ 59	26 ♑ 29	28 ♏ 20	05 ♌ 11	06 ♋ 44
11	26 ♐ 53	07 ♓ 47	25 ♑ 31	00 ♐ 24	09 ♌ 24	07 ♋ 51
21	27 ♐ 06	05 ♓ 47	25 ♑ 10	02 ♐ 45	13 ♌ 32	08 ♋ 58
31	27 ♐ 28	04 ♓ 10	25 ♑ 23	05 ♐ 22	17 ♌ 35	10 ♋ 05

All ephemeris data is given at 12.00 UT and the Moon's longitude is additionally given for 24.00 UT

ASPECTARIAN

h m	Aspects
01 Saturday	
00 54	☽ ⚺ ♄
01 09	♃ □ ♅
02 22	☽ ∠ ♆
04 42	♀ ⊥ ♆
08 49	☽ ☌ ☿
10 13	☽ ± ♃
12 50	☽ ☌ ☉
13 25	☽ ∥ ☉
14 20	☽ ⚺ ♂
18 50	☽ ☌ ♀
19 33	☽ ⊥ ♆
21 32	☽ ⚻ ♃
21 44	☽ ✶ ♅
23 20	☽ ∥ ☿
02 Sunday	
01 23	☽ ♯ ♆
02 17	☽ ⊥ ♂
06 53	☽ ⚺ ♆
08 58	☽ ∥ ♃
11 10	☽ ⚺ ♇
18 51	☽ ∥ ♄
20 00	☽ ∠ ♂
20 17	☽ Q ♅
21 22	☽ ♯ ♀
22 23	☽ ⊥ ♇
03 Monday	
00 39	☽ ∥ ♀
03 11	☽ ♯ ♄
09 47	☽ ☌ ♄
12 50	☽ ♯ ♃
13 28	☽ ⚺ ☿
14 08	♀ ⚺ ♂
15 01	☽ ∠ ♇
15 08	☉ ☌ ♀
18 56	☉ ⚺ ♂
20 29	☽ ∥ ♆
23 50	☽ ⊥ ☿
04 Tuesday	
00 00	☽ ⚺ ♀
00 59	☽ ✶ ♂
01 09	☽ ⚺ ☉
02 55	☽ ♯ ☿
04 10	☽ ♯ ☉
04 41	☽ ☍ ♃
05 25	☽ □ ♅
10 24	☽ ⊥ ♀
12 03	☉ ∥ ☿
12 53	☽ ⊥ ☉
14 11	☽ ☌ ♆
15 18	☽ ∠ ☿
18 17	☽ ✶ ♇
23 43	☽ Q ♂
05 Wednesday	
01 51	☽ ∠ ♀
06 16	☽ ∠ ☉
07 15	☉ ⊥ ♆
13 59	☿ ∠ ♆
15 44	☽ Q ♇
16 23	☽ ⚺ ♄
16 53	☽ ✶ ☿
21 09	☉ ⚻ ♃
06 Thursday	
01 24	☽ ♯ ♂
03 03	☽ ⊥ ♄
03 14	☿ ⚺ ♄
03 17	☽ ✶ ♀
09 09	☽ □ ♂
09 44	☽ ⚻ ♃
10 47	☽ ✶ ☉
10 58	☽ △ ♅
13 13	☽ Q ☿
13 33	☉ ✶ ♅
18 56	☽ ∠ ♄
19 23	☽ ⚺ ♆
20 03	♂ △ ♃
20 07	☽ ± ♃
23 21	☽ □ ♇
23 22	☽ Q ♀
07 Friday	
00 35	☽ ♯ ♇
02 25	☽ ♯ ♅
05 49	☽ ⊥ ♆
09 12	☽ Q ☉
11 35	☽ ⚼ ♃
13 04	☽ ⚼ ♅
19 32	☽ □ ☿
21 03	☽ ✶ ♄
21 21	☽ ∠ ♆
08 Saturday	
02 57	♂ ⚺ ♅
04 32	☽ ± ♅
05 04	☽ □ ♀
10 24	♀ ∥ ♄
13 04	☽ △ ♃
14 46	☽ ⚻ ♅
15 18	☽ △ ♂
17 42	☽ Q ♄
18 16	☽ □ ☉
22 57	☽ ✶ ♆
09 Sunday	
02 47	☽ △ ♇
06 52	☉ ♯ ♆
17 48	☽ ⚼ ♂
19 13	☽ Q ♆
20 22	☿ St D
21 52	☽ △ ☿
10 Monday	
00 18	☽ □ ♄
04 04	☽ ⚼ ♇
05 48	☽ △ ♀
09 35	☽ ± ♂
15 10	☽ □ ♃
17 21	☽ ☍ ♅
19 08	☽ ± ♇
20 04	☽ ⚻ ♂
23 06	☽ ⚼ ☿
11 Tuesday	
00 19	☽ △ ☉
01 23	☽ □ ♆
01 54	♀ ± ♃
05 10	☽ ⚻ ♇
05 57	☽ ⚼ ♀
14 19	☽ ± ☿
16 39	☉ ⚺ ♆
20 26	☽ ± ♀
20 56	☽ Q ♃
23 12	☽ ♯ ♅
12 Wednesday	
00 31	☽ ⚻ ☿
01 04	☽ ♯ ♇
02 46	☽ △ ♄
03 07	☽ ⚼ ☉
06 06	☽ ⚻ ♀
16 47	☽ ✶ ♃
19 12	☽ ± ☉
19 28	☽ ⚻ ♅
13 Thursday	
00 21	☽ ☍ ♂
03 31	☽ △ ♆
04 00	☽ ⚼ ♄
04 22	☽ ♯ ♂
05 31	☽ ± ♅
05 58	☽ ⚻ ☉
07 17	☽ ☍ ♇
17 42	☽ ∠ ♃
19 14	☽ ± ♄
20 39	☽ ⚼ ♅
14 Friday	
01 57	☉ ⚺ ♇
04 32	☽ ☍ ☿
04 47	☽ ⚼ ♆
05 27	☽ ⚻ ♄
06 50	☽ ☍ ♀
06 50	☽ ♯ ☿
08 46	☽ ⊥ ♃
10 15	☿ ∠ ♆
18 53	☽ ⚺ ♃
20 07	☽ ± ♆
22 09	☽ △ ♅
15 Saturday	
00 06	☽ ∥ ♆
02 16	☿ ⚺ ♄
05 26	☽ ⚻ ♂
06 25	☽ ⚻ ♆
08 46	☽ ♯ ☉
09 41	☽ ♯ ♃
10 01	☿ ☌ ♀
10 16	☽ ⚻ ♇
11 05	☽ ♯ ♀
12 38	☽ ☍ ☉
16 17	☽ ± ♂
19 02	☽ ♯ ♄
20 33	♀ ⚺ ♄
20 41	☽ ± ♇
22 21	☽ ∥ ♄
16 Sunday	
02 45	☉ ∥ ♃
04 19	♂ ✶ ♆
07 12	☽ ∥ ♀
07 25	☽ ∥ ☉
07 40	☽ ∥ ♃
08 45	☽ ⚼ ♂
09 08	☽ ⚻ ♀
09 35	☽ ☍ ♄
09 37	☉ ∥ ♀
11 24	☽ ⚻ ☿
12 29	☽ ⚼ ♇
17 15	♀ ∥ ♃
17 47	☽ ♯ ♆
19 29	☽ ± ♀
20 15	☿ ± ♃
22 40	☽ ☌ ♃
22 53	☽ ± ☿
17 Monday	
02 40	☽ □ ♅
06 40	♀ ∠ ♆
06 54	♂ ∠ ♄
10 40	☽ ∥ ☿
11 14	☽ ⚼ ♀
11 21	☽ ☍ ♆
12 49	☽ △ ♂
15 23	☽ △ ♇
16 29	☽ ⚼ ☿
22 01	☽ ⚻ ☉
18 Tuesday	
03 52	♂ ⊥ ♅
09 56	☽ ± ☉
11 32	☽ Q ♅
12 16	☽ ∥ ♂
14 09	☽ △ ♀
16 23	☽ ⚻ ♄
22 55	☽ △ ☿
19 Wednesday	
03 45	☽ ± ♄
04 09	☽ ⚼ ☉
05 12	☽ ⚺ ♃
07 34	☉ Q ♅
10 05	☽ ✶ ♅
16 29	☽ ⊥ ♃
19 20	☽ ⚻ ♆
21 02	☽ ⚼ ♄
21 32	♂ ☌ ♇
23 34	☽ □ ♇
23 40	☽ □ ♂
23 51	☽ ∥ ♇
20 Thursday	
02 11	☽ ∥ ♅
06 56	☽ ± ♆
09 39	☽ ∠ ♃
09 49	☉ ⊥ ♇
11 19	☽ △ ☉
15 02	☽ ∠ ♅
22 34	☽ □ ♀
21 Friday	
00 33	☽ ⚼ ♆
02 30	☽ △ ♄
05 46	☿ ⚻ ♃
06 43	☉ ⊥ ♂
08 45	☽ ⊥ ♅
10 48	☽ Q ♇
12 45	☽ Q ♂
14 49	☽ ✶ ♃
16 10	☽ □ ☿
20 41	☽ ⚺ ♅
22 Saturday	
06 25	☽ △ ♆
09 55	☽ Q ♀
10 26	☿ ⊥ ♆
10 48	☽ ✶ ♇
14 02	☽ ✶ ♂
14 33	☽ Q ♃
23 Sunday	
04 13	☽ □ ☉
04 17	☿ ✶ ♅
09 38	☽ Q ☿
09 56	☽ ✶ ♀
15 11	☽ □ ♄
16 01	☉ ∥ ♄
17 07	☽ ∠ ♇
22 00	☽ ∠ ♂
24 Monday	
00 13	☉ ♯ ♄
02 30	☽ □ ♃
09 15	☽ ☌ ♅
11 22	☽ ⊥ ♇
13 45	☽ ✶ ☿
16 10	☽ ∠ ♀
17 12	☽ ⊥ ♂
19 09	☽ □ ♆
19 53	☽ Q ☉
23 31	☽ ⚺ ♇
25 Tuesday	
00 58	♀ St D
03 57	☽ Q ♄
05 59	☽ ⚺ ♂
10 19	☽ ⊥ ♀
20 35	☽ ∥ ♅
22 11	☽ ✶ ☉
22 22	☽ ⚺ ♀
23 07	☽ ∥ ♇
26 Wednesday	
00 37	☉ ⚺ ♀
00 51	☽ ∠ ☿
04 16	☽ ✶ ♄
04 40	☿ ⚺ ♆
07 23	☽ Q ♆
14 06	☽ △ ♃
17 08	☿ ♯ ♆
21 27	☽ ⚺ ♅
21 41	☽ ⊥ ☿
21 48	♂ ⚼ ♃
27 Thursday	
06 29	☽ ∠ ☉
07 08	☽ ✶ ♆
09 14	☽ ⊥ ♅
10 10	☽ ∠ ♄
10 44	☿ ⚺ ♇
11 18	☽ ☌ ♇
11 23	☽ ⚺ ☿
15 56	☿ ∥ ♀
19 11	☽ ⚼ ♃
20 34	☽ ☌ ♂
20 40	☽ ∥ ♂
28 Friday	
01 29	☽ ⊥ ☉
02 43	☽ ∠ ♅
03 47	☽ ⊥ ♄
09 33	☽ ☌ ♀
12 12	☽ ∠ ♆
12 18	☽ ± ♃
13 57	☽ ⚺ ☉
15 22	☽ ⚺ ♄
23 34	☽ ⚻ ♃
29 Saturday	
05 17	☽ ⊥ ♆
07 14	☽ ✶ ♅
07 54	☽ ♯ ♆
09 37	☽ ∥ ♀
10 33	☉ ☌ ♄
16 01	☽ ∥ ☿
16 30	☽ ⚺ ♆
18 17	☉ ♯ ♃
19 59	☉ ∠ ♇
20 23	☽ ⚺ ♇
21 43	☽ ♯ ☉
21 58	☽ ∥ ♃
30 Sunday	
05 20	☽ ☌ ☿
05 29	☽ Q ♅
05 51	☽ ♯ ♄
06 27	☿ Q ♅
07 26	☽ ⊥ ♇
07 45	☽ ∥ ♄
07 52	☽ ⚺ ♂
15 24	☽ ♯ ♃
16 59	☽ ∥ ☉
18 05	☽ ⚺ ♀
18 21	☽ ♯ ☿
19 18	☽ ⊥ ♂
20 52	☿ ⊥ ♇
23 23	☽ ☌ ♄
23 47	☽ ∠ ♇

OCTOBER 1951

LONGITUDES

Date	Sidereal time h m s	Sun ☉ ° ' "	Moon ☽ ° ' "	Moon ☽ 24.00 ° ' "	Mercury ☿ ° '	Venus ♀ ° '	Mars ♂ ° '	Jupiter ♃ ° '	Saturn ♄ ° '	Uranus ♅ ° '	Neptune ♆ ° '	Pluto ♇ ° '
01	12 37 39	07 ♎ 29 59	12 ♎ 42 46	19 ♎ 29 25	28 ♍ 01	02 ♍ 54	27 ♌ 50	09 ♈ 22	05 ♎ 44	13 ♋ 50	18 ♎ 50	20 ♌ 54
02	12 41 35	08 29 01	26 ♎ 20 00	03 ♏ 14 10	29 ♍ 49	03 09	28 27	09 R 14	05 51	13 51	18 52	20 55
03	12 45 32	09 28 06	10 ♏ 11 29	17 ♏ 11 31	01 ♎ 37	03 27	29 04	09 06	05 59	13 52	18 54	20 57
04	12 49 28	10 27 12	24 ♏ 13 46	01 ♐ 17 47	03 25	03 46	29 ♌ 41	08 58	06 06	13 53	18 56	20 58
05	12 53 25	11 26 21	08 ♐ 23 04	15 ♐ 29 10	05 13	04 07	00 ♍ 18	08 50	06 13	13 53	18 59	20 59
06	12 57 22	12 25 31	22 ♐ 35 42	29 ♐ 42 16	07 00	04 30	00 55	08 42	06 21	13 54	19 01	21 01
07	13 01 18	13 24 43	06 ♑ 48 33	13 ♑ 54 17	08 47	04 55	01 32	08 34	06 28	13 55	19 03	21 02
08	13 05 15	14 23 56	20 ♑ 59 11	28 ♑ 03 05	10 33	05 21	02 09	08 26	06 36	13 56	19 05	21 03
09	13 09 11	15 23 12	05 ♒ 05 46	12 ♒ 07 05	12 19	05 49	02 45	08 18	06 43	13 56	19 07	21 04
10	13 13 08	16 22 29	19 ♒ 06 51	26 ♒ 04 54	14 04	06 19	03 22	08 10	06 50	13 57	19 10	21 06
11	13 17 04	17 21 47	03 ♓ 01 05	09 ♓ 55 10	15 49	06 50	03 59	08 02	06 58	13 57	19 12	21 07
12	13 21 01	18 21 08	16 ♓ 46 59	23 ♓ 36 17	17 32	07 23	04 36	07 54	07 05	13 58	19 14	21 08
13	13 24 57	19 20 30	00 ♈ 22 51	07 ♈ 06 27	19 15	07 57	05 12	07 47	07 12	13 58	19 16	21 09
14	13 28 54	20 19 54	13 ♈ 46 50	20 ♈ 23 48	20 58	08 32	05 49	07 39	07 20	13 59	19 19	21 10
15	13 32 51	21 19 21	26 ♈ 57 11	03 ♉ 26 48	22 39	09 09	06 25	07 31	07 27	13 59	19 21	21 11
16	13 36 47	22 18 49	09 ♉ 52 34	16 ♉ 14 27	24 20	09 47	07 02	07 24	07 34	13 59	19 23	21 12
17	13 40 44	23 18 20	22 ♉ 32 27	28 ♉ 46 40	26 01	10 26	07 38	07 16	07 41	13 59	19 25	21 13
18	13 44 40	24 17 53	04 ♊ 57 16	11 ♊ 04 28	27 40	11 07	08 15	07 09	07 49	14 00	19 28	21 14
19	13 48 37	25 17 28	17 ♊ 08 35	23 ♊ 09 58	29 ♎ 19	11 48	08 51	07 02	07 56	14 00	19 30	21 15
20	13 52 33	26 17 05	29 ♊ 09 03	05 ♋ 06 19	00 ♏ 58	12 31	09 28	06 54	08 03	14 00	19 32	21 16
21	13 56 30	27 16 45	11 ♋ 02 18	16 ♋ 57 33	02 36	13 15	10 04	06 47	08 10	14 R 00	19 34	21 17
22	14 00 26	28 16 26	22 ♋ 52 43	28 ♋ 48 24	04 13	14 00	10 40	06 40	08 17	14 00	19 36	21 18
23	14 04 23	29 ♎ 16 10	04 ♌ 45 15	10 ♌ 43 57	05 49	14 45	11 16	06 33	08 24	13 59	19 39	21 19
24	14 08 20	00 ♏ 15 56	16 ♌ 45 08	22 ♌ 49 27	07 25	15 32	11 53	06 27	08 31	13 59	19 41	21 20
25	14 12 16	01 15 45	28 ♌ 57 32	05 ♍ 09 57	09 01	16 20	12 29	06 20	08 38	13 59	19 43	21 21
26	14 16 13	02 15 35	11 ♍ 27 13	17 ♍ 49 49	10 36	17 08	13 05	06 14	08 46	13 59	19 45	21 22
27	14 20 09	03 15 28	24 ♍ 18 05	00 ♎ 52 18	12 10	17 58	13 41	06 07	08 53	13 58	19 47	21 22
28	14 24 06	04 15 23	07 ♎ 32 35	14 ♎ 18 58	13 44	18 48	14 17	06 01	08 59	13 58	19 50	21 23
29	14 28 02	05 15 20	21 ♎ 11 16	28 ♎ 09 13	15 17	19 39	14 53	05 55	09 06	13 58	19 52	21 24
30	14 31 59	06 15 19	05 ♏ 12 21	12 ♏ 20 03	16 50	20 31	15 29	05 49	09 13	13 57	19 54	21 25
31	14 35 55	07 ♏ 15 20	19 ♏ 31 38	26 ♏ 46 16	18 ♏ 23	21 ♍ 23	16 ♍ 05	05 ♈ 43	09 ♎ 20	13 ♋ 57	19 ♎ 56	21 ♌ 25

Moon

Date	Moon True ☊ ° '	Moon Mean ☊ ° '	Moon ☽ Latitude ° '
01	09 ♓ 43	08 ♓ 18	02 S 51
02	09 R 39	08 15	03 48
03	09 35	08 12	04 32
04	09 30	08 09	05 01
05	09 26	08 05	05 12
06	09 23	08 02	05 04
07	09 22	07 59	04 37
08	09 D 22	07 56	03 53
09	09 23	07 53	02 56
10	09 24	07 50	01 48
11	09 25	07 46	00 S 35
12	09 R 25	07 43	00 N 40
13	09 23	07 40	01 51
14	09 20	07 37	02 55
15	09 14	07 34	03 49
16	09 06	07 30	04 29
17	08 58	07 27	04 55
18	08 50	07 24	05 07
19	08 43	07 21	05 04
20	08 38	07 18	04 48
21	08 34	07 15	04 20
22	08 32	07 11	03 41
23	08 D 32	07 08	02 52
24	08 33	07 05	01 55
25	08 34	07 02	00 N 51
26	08 R 35	06 59	00 S 15
27	08 33	06 56	01 23
28	08 30	06 52	02 29
29	08 24	06 49	03 27
30	08 16	06 46	04 15
31	08 ♓ 06	06 ♓ 43	04 S 48

DECLINATIONS

Date	Sun ☉ ° '	Moon ☽ ° '	Mercury ☿ ° '	Venus ♀ ° '	Mars ♂ ° '	Jupiter ♃ ° '	Saturn ♄ ° '	Uranus ♅ ° '	Neptune ♆ ° '	Pluto ♇ ° '
01	02 S 59	07 S 39	02 N 29	05 N 40	13 N 27	02 N 13	00 S 18	23 N 05	05 S 54	22 N 41
02	03 22	13 42	01 44	05 46	13 14	02 09	00 21	23 05	05 55	22 41
03	03 45	19 11	00 59	05 50	13 01	02 06	00 24	23 04	05 56	22 41
04	04 08	23 42	00 N 13	05 54	12 49	02 03	00 27	23 04	05 57	22 41
05	04 32	26 50	00 S 32	05 57	12 36	02 00	00 30	23 04	05 58	22 40
06	04 55	28 18	01 19	05 59	12 23	01 57	00 33	23 04	05 58	22 40
07	05 18	27 53	02 05	06 01	12 10	01 54	00 36	23 04	05 59	22 40
08	05 41	25 39	02 51	06 01	11 58	01 50	00 39	23 04	06 00	22 40
09	06 04	21 51	03 37	06 01	11 45	01 47	00 41	23 04	06 01	22 40
10	06 26	16 49	04 22	06 01	11 32	01 44	00 44	23 04	06 02	22 40
11	06 49	10 57	05 08	05 59	11 19	01 41	00 47	23 04	06 03	22 39
12	07 12	04 S 36	05 53	05 57	11 06	01 38	00 50	23 04	06 03	22 39
13	07 34	01 N 51	06 37	05 54	10 52	01 35	00 53	23 04	06 04	22 39
14	07 57	08 08	07 21	05 50	10 39	01 32	00 56	23 04	06 05	22 39
15	08 19	13 56	08 05	05 46	10 26	01 29	00 58	23 04	06 06	22 39
16	08 41	19 02	08 48	05 41	10 13	01 27	01 01	23 04	06 07	22 39
17	09 03	23 10	09 31	05 35	10 00	01 24	01 04	23 04	06 08	22 39
18	09 25	26 10	10 13	05 29	09 46	01 21	01 07	23 04	06 09	22 39
19	09 47	27 53	10 54	05 22	09 33	01 18	01 09	23 04	06 09	22 39
20	10 09	28 15	11 35	05 14	09 20	01 15	01 12	23 04	06 10	22 39
21	10 30	27 18	12 15	05 06	09 06	01 13	01 15	23 04	06 11	22 39
22	10 52	25 08	12 55	04 57	08 53	01 10	01 18	23 04	06 12	22 39
23	11 13	21 52	13 34	04 47	08 39	01 08	01 20	23 04	06 13	22 39
24	11 34	17 39	14 12	04 37	08 26	01 05	01 23	23 04	06 13	22 39
25	11 55	12 39	14 49	04 27	08 12	01 03	01 26	23 04	06 14	22 39
26	12 16	07 02	15 25	04 15	07 59	01 00	01 28	23 04	06 15	22 39
27	12 36	00 N 59	16 01	04 04	07 45	00 58	01 31	23 04	06 16	22 39
28	12 57	05 S 16	16 36	03 51	07 32	00 56	01 34	23 04	06 17	22 39
29	13 17	11 28	17 10	03 38	07 18	00 53	01 36	23 05	06 18	22 39
30	13 37	17 16	17 43	03 25	07 04	00 51	01 39	23 05	06 18	22 39
31	13 S 56	22 S 14	18 S 15	03 N 11	06 N 51	00 N 49	01 S 41	23 N 05	06 S 19	22 N 39

ZODIAC SIGN ENTRIES

Date	h	m	Planets	
02	14	25	☿	♎
02	18	23	☽	♏
04	21	48	☽	♐
05	00	20	♂	♍
07	00	30	☽	♑
09	03	19	☽	♒
11	06	46	☽	♓
13	11	19	☽	♈
15	17	37	☽	♉
18	02	22	☽	♊
19	21	52	☿	♏
20	13	42	☽	♋
23	02	25	☽	♌
24	05	36	☉	♏
25	14	01	☽	♍
27	22	25	☽	♎
30	03	09	☽	♏

LATITUDES

Date	Mercury ☿ ° '	Venus ♀ ° '	Mars ♂ ° '	Jupiter ♃ ° '	Saturn ♄ ° '	Uranus ♅ ° '	Neptune ♆ ° '	Pluto ♇ ° '
01	01 N 51	05 S 07	01 N 17	01 S 38	02 N 09	00 N 21	01 N 36	08 N 37
04	01 44	04 32	01 19	01 38	02 09	00 21	01 36	08 38
07	01 32	03 58	01 20	01 38	02 09	00 21	01 36	08 39
10	01 17	03 26	01 21	01 38	02 09	00 21	01 36	08 40
13	01 00	02 54	01 22	01 38	02 09	00 21	01 36	08 41
16	00 41	02 24	01 23	01 38	02 10	00 21	01 36	08 41
19	00 21	01 55	01 24	01 37	02 10	00 21	01 36	08 42
22	00 N 01	01 28	01 25	01 37	02 10	00 22	01 36	08 43
25	00 S 20	01 02	01 26	01 36	02 10	00 22	01 36	08 44
28	00 40	00 38	01 27	01 36	02 11	00 22	01 36	08 45
31	00 S 59	00 S 15	01 N 28	01 S 35	02 N 11	00 N 22	01 N 36	08 N 46

DATA

Julian Date	2433921
Delta T	+30 seconds
Ayanamsa	23° 11' 05"
Synetic vernal point	05° ♓ 55' 55"
True obliquity of ecliptic	23° 26' 53"

MOON'S PHASES, APSIDES AND POSITIONS ☽

Date	h	m	Phase	Longitude ° '	Eclipse Indicator
01	01	57	●	07 ♎ 05	
08	00	00	☽	13 ♑ 54	
15	00	51	○	20 ♈ 52	
22	23	55	☾	28 ♋ 46	
30	13	55	●	06 ♏ 20	

Day	h	m		
07	06	39	Perigee	
21	17	09	Apogee	
06	18	46	Max dec	28° S 22'
13	05	06	0N	
20	06	43	Max dec	28° N 17'
27	15	49	0S	

LONGITUDES

Date	Chiron ⚷ ° '	Ceres ⚳ ° '	Pallas ⚴ ° '	Juno ⚵ ° '	Vesta ⚶ ° '	Black Moon Lilith ⚸ ° '
01	27 ♐ 28	04 ♓ 10	25 ♑ 23	05 ♐ 22	17 ♌ 35	10 ♋ 05
11	27 ♐ 59	03 ♓ 06	26 ♑ 09	08 ♐ 12	21 ♌ 31	11 ♋ 12
21	28 ♐ 36	02 ♓ 38	27 ♑ 23	11 ♐ 13	25 ♌ 30	12 ♋ 19
31	29 ♐ 20	02 ♓ 46	29 ♑ 01	14 ♐ 24	28 ♌ 56	13 ♋ 26

All ephemeris data is given at 12.00 UT and the Moon's longitude is additionally given for 24.00 UT

ASPECTARIAN

h m	Aspects
01 Monday	
01 30	☉ ⧣ ☿
01 57	☽ ☌ ☉
04 19	☽ ⧣ ♀
05 04	☽ ⊥ ♀
05 19	☽ ∥ ♆
06 06	☽ ☍ ♃
08 12	☿ ⊻ ♂
12 13	☽ ∠ ♂
13 59	☽ □ ♅
21 22	☽ ∠ ♀
21 36	☿ ∥ ♃
22 52	☽ ☌ ♆
02 Tuesday	
02 30	☽ ⚹ ♇
10 10	☽ ⧣ ♂
15 52	☽ ⚹ ♂
16 46	♀ ± ♃
18 59	☽ ⊻ ☿
23 28	☽ Q ♇
03 Wednesday	
00 06	☽ ⚹ ♀
03 47	♂ ∠ ♅
04 07	☉ ☍ ♃
04 27	♄ ∠ ♇
04 40	☽ ⊻ ♄
06 55	☽ ⊥ ☿
10 09	☽ ⊼ ♃
10 40	☽ ⊻ ☉
11 08	☉ ⊥ ♀
13 34	☽ Q ♂
15 06	☽ ⊥ ♄
18 18	☽ △ ♅
20 21	☽ ± ♃
21 13	☽ Q ♀
21 44	☽ ⊥ ☉
04 Thursday	
00 39	☽ ∠ ☿
02 58	☽ ⊻ ♆
05 17	☿ ⧣ ♄
06 00	☽ ⧣ ♇
06 26	☽ □ ♇
06 37	☽ ∠ ♄
08 17	☽ ⧣ ♅
11 33	☽ ⚼ ♃
13 13	☽ ⊥ ♆
14 14	☽ ∠ ☉
17 43	☿ ⊻ ♀
19 54	☽ ⚼ ♅
21 41	☽ □ ♂
05 Friday	
01 25	♀ ∠ ♆
04 31	☽ ∠ ♆
04 35	☽ □ ♀
05 52	☽ ⚹ ☿
08 16	♂ ⊥ ♄
08 19	☽ ⚹ ♄
10 33	☿ ∥ ♄
11 10	☽ ± ♅
12 45	☽ △ ♃
17 33	☽ ⚹ ☉
21 19	☽ ⊼ ♅
22 21	♀ ⧣ ♆
22 31	☿ ∠ ♇
06 Saturday	
02 32	☿ ☌ ♄
04 46	☽ Q ♄
05 04	☽ Q ☿
05 56	☽ ⚹ ♆
09 19	☽ △ ♇
10 13	☿ ⊥ ♂
15 19	☽ Q ☉
07 Sunday	
02 15	☽ Q ♆
02 41	☽ △ ♂
06 37	☿ ⧣ ♃
08 42	☽ △ ♀
09 16	☿ ☍ ♃
10 41	☽ ⚼ ♇
11 25	☽ □ ♄
14 56	☽ □ ♃
15 49	☽ □ ☿
08 Monday	
00 00	☽ □ ☉
00 02	☽ ☍ ♅
00 23	☉ □ ♅
01 56	☽ ± ♇
05 11	☽ ⚼ ♂
08 46	☽ □ ♆
10 53	☽ ⚼ ♀
12 07	☽ ⊼ ♇
21 09	☽ ± ♂
21 10	☽ Q ♃
21 14	♂ ± ♃
09 Tuesday	
02 42	☽ ± ♀
02 45	☿ ⊥ ♀
05 10	☽ ⧣ ♅
07 30	☽ ⧣ ♇
07 50	☽ ⊼ ♂
09 06	☉ ∥ ♆
09 42	☉ ⧣ ♀
13 17	☽ ⊼ ♀
14 47	☽ △ ♄
17 25	☽ ⚹ ♃
19 45	♀ ⧣ ♆
10 Wednesday	
02 06	☽ △ ☿
03 08	☽ ⊼ ♅
06 56	☽ △ ☉
10 20	☿ □ ♅
12 05	☽ △ ♆
13 26	☽ ± ♅
15 24	☽ ☍ ♇
16 44	☽ ⚼ ♄
18 55	☽ ∠ ♃
11 Thursday	
04 58	☽ ⚼ ♅
07 37	☽ ⚼ ☿
08 24	☽ ± ♄
10 19	☽ ⊥ ♃
10 30	☽ ⧣ ♂
10 47	☽ ⚼ ☉
13 45	☽ ☍ ♂
14 03	☽ ⚼ ♆
18 54	☽ ☍ ♀
18 54	☽ ⊼ ♄
19 17	♀ ⊻ ♄
20 38	☽ ⊻ ♃
21 03	♂ ∠ ♆
12 Friday	
01 30	☽ ± ☿
02 51	☽ ∥ ☉
03 39	☽ ± ☉
05 46	☽ ± ♆
06 36	☽ ∥ ♆
06 58	☽ ⧣ ♀
07 04	☽ △ ♅
07 46	☽ ∥ ☿
13 31	☽ ⊼ ☿
14 14	☿ ⧣ ♀
14 58	☽ ⊼ ☉
16 19	☽ ⊼ ♆
17 58	☿ ∥ ♆
19 39	☽ ⊼ ♇
23 06	☽ ⧣ ♃
13 Saturday	
01 54	☽ ∥ ♄
06 14	♀ ⊼ ♃
06 16	☽ ± ♇
08 20	☽ ⧣ ♄
10 16	☉ ☌ ♆
11 01	☽ ∥ ♃
12 15	☿ ☌ ♆
14 52	☉ ☌ ☿
21 00	☽ ⊼ ♂
22 18	☽ ⚼ ♇
14 Sunday	
00 17	☽ ☍ ♄
01 04	☽ ☌ ♃
02 07	☽ ⊼ ♀
03 11	☽ ∥ ♀
04 01	☽ ⧣ ♆
08 17	☽ ± ♂
08 33	☽ ⧣ ☿
08 44	☿ ∠ ♂
11 14	☽ ⧣ ☉
12 21	☽ □ ♅
13 26	☽ ± ♀
14 57	☿ ⚹ ♇
21 46	☽ ∥ ♂
22 03	☽ ☍ ♆
15 Monday	
00 51	☽ ☍ ☉
01 23	☽ ⚼ ♂
01 25	☽ △ ♇
02 58	☽ ☍ ☿
06 36	☽ ⚼ ♀
08 39	☉ ⚹ ♇
18 19	☉ ∠ ♂
19 12	♃ ☍ ♄
21 17	☽ Q ♅
16 Tuesday	
03 58	☉ ∥ ☿
06 25	☽ △ ♂
07 24	☽ ⊻ ♃
07 38	☽ ⊼ ♄
11 49	☽ △ ♀
18 33	☽ ⊥ ♃
19 01	☽ ± ♄
19 44	☽ ⚹ ♅
22 21	☿ ∠ ♀
23 55	♂ ⊼ ♃
17 Wednesday	
06 02	☽ ⊼ ♆
08 36	☽ ∥ ♇
09 28	☽ □ ♇
11 20	☽ ∥ ♅
11 29	☽ ∠ ♃
12 17	☽ ⚼ ♄
13 35	☽ ⊼ ☉
14 25	♂ ⊻ ♄
17 33	☽ ± ♆
19 42	☽ ⊼ ☿
18 Thursday	
00 24	☿ ⧣ ♂
00 25	☽ ∠ ♅
02 09	☽ ± ☉
09 07	☽ ± ☿
11 02	☽ ⚼ ♆
16 15	☽ ⚹ ♃
17 38	☽ △ ♄
17 57	☽ ⊥ ♅
18 47	☽ □ ♂
20 24	☽ Q ♇
21 16	☽ ⚼ ☉
19 Friday	
00 48	☽ □ ♀
02 16	☉ ⧣ ♂
05 32	☽ ⚼ ☿
05 46	☽ ⊻ ♅
15 42	☽ Q ♃
16 42	☽ △ ♆
20 12	☽ ⚹ ♇
20 Saturday	
05 43	☽ △ ☉
08 25	☽ Q ♂
14 56	☽ Q ♀
16 14	☽ △ ☿
20 56	♅ St R
21 Sunday	
02 22	♃ ⧣ ♄
02 22	☽ ∠ ♇
03 29	☽ □ ♃
06 08	☽ □ ♄
09 55	☽ ⚹ ♂
16 46	☽ ⚹ ♀
17 59	☽ ☌ ♅
20 37	☽ ⊥ ♇
22 22	☿ Q ♇
23 01	♀ ⊥ ♆
22 Monday	
05 21	☽ □ ♆
08 48	☽ ⊻ ♇
11 59	♀ ⚹ ♅
17 57	☽ ∠ ♂
18 58	☽ Q ♄
23 55	☽ □ ☉
23 Tuesday	
01 13	☽ ∠ ♀
03 57	☽ ∥ ♅
06 52	☽ ∥ ♇
13 06	☽ ⊥ ♂
14 29	☽ □ ☿
15 36	☽ △ ♃
17 50	☽ Q ♆
19 25	☽ ⚹ ♄
20 36	☽ ⊥ ♀
22 18	☿ ⊼ ♃
24 Wednesday	
01 46	☽ ⊻ ♂
06 30	☽ ⊻ ♅
09 25	☽ ⊻ ♀
15 16	☽ Q ☉
17 49	☽ ⚹ ♆
18 25	☽ ⊥ ♅
21 04	☽ ☌ ♇
21 12	☽ ⚼ ♃
25 Thursday	
01 30	☽ ∠ ♄
03 04	☽ ⧣ ☿
05 57	☿ ⊻ ♄
07 39	☽ Q ☿
10 01	♃ ⚼ ♇
12 03	☽ ∠ ♅
14 39	☽ ± ♃
15 03	☽ ⧣ ☉
16 52	☽ ⚹ ☉
19 12	☽ ⊥ ♄
20 23	☉ ∠ ♀
23 10	☽ ∠ ♆
26 Friday	
02 07	☽ ⊼ ♃
06 50	☽ ⊻ ♄
07 56	☽ ∥ ♂
10 08	☽ ⚹ ☿
15 11	☽ ⧣ ♆
15 14	☽ ☌ ♂
16 22	☽ ⊥ ♆
16 46	☽ ⚹ ♅
23 26	☽ ☌ ♀
23 34	☽ ∥ ♀
23 51	☽ ∠ ☉
27 Saturday	
03 38	☽ ⊻ ♆
06 35	☽ ⊻ ♇
09 58	☽ ⧣ ♄
11 20	☿ ± ♃
12 06	☽ ∥ ♃
14 47	☉ Q ♇
15 04	☽ Q ♅
16 43	♂ ⊥ ♆
17 38	☽ ⊥ ♇
17 52	☽ ⊥ ☉
17 58	☽ ∠ ☿
19 29	☽ ⧣ ♃
21 43	☽ ∥ ♄
23 39	♂ ⚹ ♅
28 Sunday	
05 38	☽ ⊻ ☉
06 46	☽ ⧣ ♀
09 18	☽ ☍ ♃
09 56	☽ ∠ ♇
12 23	☽ ⊥ ☿
14 36	☽ ☌ ♄
15 37	☿ △ ♅
15 52	☽ ∥ ♆
20 21	☽ ⧣ ♂
23 23	☽ □ ♅
29 Monday	
00 23	☽ ⊻ ☿
00 29	☽ ⊻ ♂
01 44	☿ ⚹ ♂
08 58	☿ ⊥ ♄
09 09	☽ ⊻ ♀
09 42	☽ ☌ ♆
11 26	☽ ⊥ ♂
12 22	☽ ⚹ ♇
18 21	♀ ⊻ ♆
19 40	☽ ∥ ☉
20 12	☽ ⊥ ♀
30 Tuesday	
02 25	☉ ⊼ ♃
03 37	☽ ∠ ♂
08 57	☽ Q ♇
12 33	☽ ∠ ♀
13 02	☽ ⊼ ♃
13 55	☽ ☌ ☉
14 13	☽ ∥ ☿
18 50	☽ ⊻ ♄
23 04	☽ ± ♃
31 Wednesday	
02 42	☽ △ ♅
04 58	☽ ⊥ ♄
06 00	☽ ⚹ ♂
09 52	☽ ☌ ☿
12 41	☽ ⊻ ♆
13 02	♀ ⊻ ♇
13 58	☽ ⚼ ♃
14 16	☽ ⧣ ♇
15 09	☽ □ ♇
15 17	☽ ⚹ ♀
16 44	☽ ⧣ ♅
20 02	☽ ∠ ♄
22 39	☽ ⊥ ♆

NOVEMBER 1951

LONGITUDES

Date	Sidereal time h m s	Sun ☉ ° ' "	Moon ☽ ° ' "	Moon ☽ 24.00 ° ' "	Mercury ☿ ° '	Venus ♀ ° '	Mars ♂ ° '	Jupiter ♃ ° '	Saturn ♄ ° '	Uranus ♅ ° '	Neptune ♆ ° '	Pluto ♇ ° '
01	14 39 52	08 ♏ 15 22	04 ♐ 03 03	11 ♐ 21 06	19 ♏ 55	22 ♍ 16	16 ♍ 40	05 ♈ 38	09 ♎ 27	13 ♋ 56	19 ♎ 58	21 ♌ 26
02	14 43 49	09 15 27	18 ♐ 39 29	25 ♐ 57 23	21 26	23 10	17 16	05 R 32	09 34	13 R 55	20 00	21 26
03	14 47 45	10 15 33	03 ♑ 14 01	10 ♑ 28 43	22 57	24 04	17 52	05 27	09 41	13 55	20 03	21 27
04	14 51 42	11 15 41	17 ♑ 40 57	24 ♑ 50 17	24 28	24 59	18 27	05 22	09 47	13 54	20 05	21 28
05	14 55 38	12 15 51	01 ♒ 56 25	08 ♒ 59 12	25 58	25 55	19 03	05 17	09 54	13 53	20 07	21 28
06	14 59 35	13 16 01	15 ♒ 58 32	22 ♒ 54 25	27 28	26 51	19 39	05 12	10 01	13 52	20 09	21 29
07	15 03 31	14 16 14	29 ♒ 46 54	06 ♓ 36 06	28 ♏ 57	27 48	20 14	05 07	10 07	13 51	20 11	21 29
08	15 07 28	15 16 27	13 ♓ 22 08	20 ♓ 05 07	00 ♐ 26	28 46	20 49	05 03	10 14	13 50	20 13	21 30
09	15 11 24	16 16 42	26 ♓ 45 10	03 ♈ 22 24	01 54	29 ♍ 44	21 25	04 59	10 20	13 49	20 15	21 30
10	15 15 21	17 16 59	09 ♈ 56 52	16 ♈ 28 37	03 22	00 ♎ 42	22 00	04 55	10 27	13 48	20 17	21 30
11	15 19 18	18 17 17	22 ♈ 57 39	29 ♈ 24 00	04 49	01 41	22 35	04 51	10 33	13 47	20 19	21 31
12	15 23 14	19 17 37	05 ♉ 47 36	12 ♉ 08 26	06 16	02 40	23 11	04 47	10 39	13 46	20 21	21 31
13	15 27 11	20 17 58	18 ♉ 26 28	24 ♉ 41 39	07 42	03 40	23 46	04 44	10 46	13 45	20 23	21 32
14	15 31 07	21 18 21	00 ♊ 54 02	07 ♊ 03 35	09 08	04 40	24 21	04 40	10 52	13 44	20 25	21 32
15	15 35 04	22 18 46	13 ♊ 10 25	19 ♊ 14 38	10 33	05 41	24 56	04 37	10 58	13 42	20 27	21 32
16	15 39 00	23 19 13	25 ♊ 16 26	01 ♋ 16 00	11 57	06 42	25 31	04 34	11 04	13 41	20 29	21 32
17	15 42 57	24 19 41	07 ♋ 13 40	13 ♋ 09 47	13 20	07 44	26 06	04 32	11 10	13 40	20 31	21 32
18	15 46 53	25 20 11	19 ♋ 04 44	24 ♋ 59 01	14 43	08 46	26 41	04 29	11 16	13 38	20 33	21 33
19	15 50 50	26 20 43	00 ♌ 53 09	06 ♌ 47 41	16 04	09 49	27 15	04 27	11 22	13 37	20 35	21 33
20	15 54 47	27 21 16	12 ♌ 43 05	18 ♌ 40 30	17 24	10 51	27 50	04 25	11 28	13 35	20 37	21 33
21	15 58 43	28 21 52	24 ♌ 40 06	00 ♍ 42 45	18 43	11 54	28 25	04 23	11 34	13 34	20 39	21 33
22	16 02 40	29 ♏ 22 29	06 ♍ 49 08	12 ♍ 59 56	20 00	12 58	28 59	04 21	11 40	13 32	20 41	21 33
23	16 06 36	00 ♐ 23 07	19 ♍ 15 50	25 ♍ 37 26	21 16	14 02	29 ♍ 34	04 19	11 46	13 30	20 42	21 33
24	16 10 33	01 23 47	02 ♎ 05 17	08 ♎ 39 51	22 30	15 06	00 ♎ 08	04 18	11 51	13 29	20 44	21 R 33
25	16 14 29	02 24 29	15 ♎ 21 27	22 ♎ 10 17	23 41	16 10	00 43	04 17	11 57	13 27	20 46	21 33
26	16 18 26	03 25 13	29 ♎ 06 21	06 ♏ 09 31	24 50	17 15	01 17	04 16	12 03	13 25	20 48	21 33
27	16 22 22	04 25 58	13 ♏ 19 23	20 ♏ 35 21	25 56	18 20	01 51	04 16	12 08	13 23	20 50	21 33
28	16 26 19	05 26 44	27 ♏ 56 37	05 ♐ 22 13	26 59	19 26	02 26	04 15	12 13	13 21	20 51	21 33
29	16 30 16	06 27 32	12 ♐ 51 01	20 ♐ 21 47	27 58	20 31	03 00	04 15	12 19	13 20	20 53	21 33
30	16 34 12	07 ♐ 28 22	27 ♐ 53 16	05 ♑ 24 12	28 ♐ 53	21 ♎ 37	03 ♎ 34	04 ♈ 15	12 ♎ 24	13 ♋ 18	20 ♎ 55	21 ♌ 32

Date	Moon True ☊ ° '	Moon Mean ☊ ° '	Moon ☽ Latitude ° '
01	07 ♓ 56	06 ♓ 40	05 S 03
02	07 R 49	06 36	04 58
03	07 40	06 33	04 34
04	07 35	06 30	03 53
05	07 33	06 27	02 58
06	07 D 33	06 24	01 52
07	07 33	06 21	00 S 41
08	07 R 33	06 17	00 N 31
09	07 32	06 14	01 40
10	07 28	06 11	02 43
11	07 21	06 08	03 36
12	07 11	06 05	04 17
13	06 59	06 02	04 45
14	06 45	05 58	04 59
15	06 32	05 55	04 58
16	06 19	05 52	04 44
17	06 08	05 49	04 18
18	06 00	05 46	03 41
19	05 55	05 42	02 54
20	05 53	05 39	01 59
21	05 D 52	05 36	00 N 59
22	05 R 52	05 33	00 S 05
23	05 52	05 30	01 10
24	05 50	05 27	02 14
25	05 45	05 23	03 12
26	05 38	05 20	04 02
27	05 28	05 17	04 38
28	05 16	05 14	04 58
29	05 04	05 11	04 57
30	04 ♓ 53	05 ♓ 07	04 S 36

DECLINATIONS

Date	Sun ☉	Moon ☽	Mercury ☿	Venus ♀	Mars ♂	Jupiter ♃	Saturn ♄	Uranus ♅	Neptune ♆	Pluto ♇
01	14 S 16	25 S 55	18 S 47	02 N 57	06 N 37	00 N 47	01 S 44	23 N 05	06 S 20	22 N 39
02	14 35	27 55	19 17	02 42	06 24	00 45	01 47	23 05	06 21	22 39
03	14 54	27 59	19 46	02 27	06 10	00 43	01 49	23 05	06 21	22 39
04	15 13	26 08	20 15	02 11	05 56	00 41	01 52	23 05	06 22	22 39
05	15 31	22 37	20 42	01 55	05 43	00 40	01 54	23 05	06 23	22 39
06	15 50	17 50	21 09	01 38	05 29	00 38	01 57	23 05	06 24	22 39
07	16 08	12 12	21 34	01 21	05 15	00 36	01 59	23 05	06 25	22 40
08	16 25	06 S 04	21 59	01 04	05 02	00 35	02 01	23 06	06 25	22 40
09	16 43	00 N 14	22 22	00 46	04 48	00 33	02 04	23 06	06 26	22 40
10	17 00	06 26	22 44	00 28	04 34	00 32	02 06	23 06	06 27	22 40
11	17 17	12 16	23 05	00 N 10	04 21	00 31	02 09	23 06	06 28	22 40
12	17 33	17 30	23 25	00 S 09	04 07	00 29	02 11	23 06	06 28	22 40
13	17 50	21 53	23 44	00 28	03 54	00 28	02 13	23 06	06 29	22 41
14	18 06	25 13	24 01	00 47	03 40	00 27	02 16	23 06	06 30	22 41
15	18 21	27 19	24 18	01 07	03 26	00 26	02 18	23 07	06 30	22 41
16	18 37	28 06	24 33	01 27	03 13	00 25	02 20	23 07	06 31	22 41
17	18 52	27 33	24 47	01 47	02 59	00 24	02 22	23 07	06 32	22 42
18	19 06	25 44	24 59	02 08	02 46	00 24	02 24	23 07	06 33	22 42
19	19 21	22 48	25 10	02 28	02 32	00 23	02 27	23 07	06 33	22 42
20	19 35	18 54	25 20	02 49	02 19	00 22	02 29	23 07	06 34	22 42
21	19 48	14 14	25 28	03 10	02 05	00 22	02 31	23 08	06 35	22 43
22	20 01	08 56	25 35	03 32	01 52	00 21	02 33	23 08	06 35	22 43
23	20 14	03 N 10	25 41	03 53	01 38	00 21	02 35	23 08	06 36	22 43
24	20 27	02 S 53	25 45	04 15	01 25	00 21	02 37	23 08	06 37	22 44
25	20 39	09 00	25 48	04 36	01 11	00 21	02 39	23 08	06 37	22 44
26	20 51	14 56	25 49	04 58	00 58	00 21	02 41	23 09	06 38	22 44
27	21 02	20 16	25 49	05 20	00 45	00 21	02 43	23 09	06 38	22 45
28	21 13	24 32	25 47	05 42	00 31	00 21	02 45	23 09	06 39	22 45
29	21 24	27 16	25 44	06 05	00 18	00 21	02 47	23 09	06 40	22 45
30	21 S 34	28 S 02	25 S 39	06 S 27	00 N 05	00 N 21	02 S 49	23 N 09	06 S 40	22 N 46

ZODIAC SIGN ENTRIES

Date	h	m	Planets
01	05	20	☽ ♐
03	06	40	☽ ♑
05	08	43	☽ ♒
07	12	23	☽ ♓
08	04	59	☿ ♐
09	17	53	☽ ♈
09	18	48	♀ ♎
12	01	07	☽ ♉
14	10	15	☽ ♊
16	21	27	☽ ♋
19	10	12	☽ ♌
21	22	35	☽ ♍
23	02	51	☉ ♐
24	06	11	♂ ♎
24	08	09	☽ ♎
26	13	32	☽ ♏
28	15	20	☽ ♐
30	15	22	☽ ♑

LATITUDES

Date	Mercury ☿	Venus ♀	Mars ♂	Jupiter ♃	Saturn ♄	Uranus ♅	Neptune ♆	Pluto ♇
01	01 S 06	00 S 08	01 N 28	01 S 35	02 N 11	00 N 22	01 N 36	08 N 46
04	01 24	00 N 12	01 29	01 34	02 12	00 22	01 36	08 47
07	01 41	00 31	01 30	01 34	02 12	00 22	01 36	08 48
10	01 56	00 49	01 31	01 33	02 13	00 22	01 36	08 49
13	02 10	01 05	01 32	01 32	02 13	00 22	01 36	08 50
16	02 20	01 19	01 33	01 31	02 14	00 22	01 36	08 51
19	02 28	01 32	01 34	01 31	02 14	00 22	01 36	08 52
22	02 31	01 44	01 35	01 30	02 15	00 22	01 36	08 53
25	02 30	01 54	01 36	01 29	02 15	00 22	01 37	08 54
28	02 22	02 03	01 37	01 28	02 16	00 22	01 37	08 55
31	02 S 07	02 N 11	01 N 38	01 S 27	02 N 16	00 N 23	01 N 37	08 N 56

DATA

Julian Date	2433952
Delta T	+30 seconds
Ayanamsa	23° 11' 08"
Synetic vernal point	05° ♓ 55' 51"
True obliquity of ecliptic	23° 26' 52"

MOON'S PHASES, APSIDES AND POSITIONS ☽

Date	h	m	Phase	Longitude ° '	Eclipse Indicator
06	06	59	☽	13 ♒ 03	
13	15	52	○	20 ♉ 28	
21	20	01	☾	28 ♌ 42	
29	01	00	●	06 ♐ 00	

Day	h	m		
02	12	37	Perigee	
18	12	30	Apogee	
30	12	20	Perigee	
03	00	45	Max dec	28° S 12'
09	11	05	0N	
16	13	54	Max dec	28° N 06'
24	00	41	0S	
30	09	00	Max dec	28° S 03'

LONGITUDES

Date	Chiron ⚷	Ceres ⚳	Pallas ⚴	Juno ⚵	Vesta ⚶	Black Moon Lilith ⚸
01	29 ♐ 24	02 ♓ 49	29 ♑ 12	14 ♐ 43	29 ♌ 17	13 ♋ 32
11	00 ♑ 15	03 ♓ 37	01 ♒ 12	18 ♐ 03	02 ♍ 40	14 ♋ 39
21	01 ♑ 09	04 ♓ 56	03 ♒ 30	21 ♐ 30	05 ♍ 46	15 ♋ 46
31	02 ♑ 08	06 ♓ 43	06 ♒ 03	25 ♐ 02	08 ♍ 32	16 ♋ 53

All ephemeris data is given at 12.00 UT and the Moon's longitude is additionally given for 24.00 UT

ASPECTARIAN

h m	Aspects
01 Thursday	
02 46	☽ Q ♂
03 35	☽ ⚼ ♅
12 23	☽ Q ♀
13 00	☿ ⊻ ♆
13 31	☽ ∠ ♆
14 35	☽ △ ♃
18 23	☽ ± ♅
19 25	☽ ⊻ ☉
20 57	☽ ✶ ♄
22 37	☿ ⚼ ♃
02 Friday	
04 14	☽ ⊼ ♅
06 00	☽ ⊥ ☉
09 37	☽ □ ♂
12 05	☿ □ ♇
14 13	☽ ✶ ♆
16 35	☽ △ ♇
16 49	☽ Q ♄
16 52	♂ # ♆
17 06	☽ ⊻ ☿
19 54	☽ □ ♀
20 15	☉ ⊻ ♄
21 53	☽ ∠ ☉
03 Saturday	
04 07	☽ ⊥ ☿
10 02	☽ Q ♆
15 38	☽ □ ♃
17 20	☽ ⚼ ♇
20 43	☽ ∠ ☿
22 45	☽ □ ♄
04 Sunday	
00 30	☽ ✶ ☉
05 42	☽ ☍ ♅
08 17	☽ ± ♇
13 21	☽ △ ♂
14 14	☉ ± ♃
16 01	☽ □ ♆
17 34	☿ ∠ ♄
18 20	☽ ⊼ ♇
21 28	☽ Q ♃
22 03	☽ Q ☉
05 Monday	
00 43	☽ ✶ ☿
01 07	☽ △ ♀
09 17	☽ # ♅
09 56	☿ ✶ ♀
11 09	♀ Q ♅
11 47	☽ # ♇
12 47	♀ # ♄
14 24	☿ ⊥ ♆
15 44	☽ ⚼ ♂
17 39	☽ ✶ ♃
21 22	☽ ∥ ☿
23 29	☽ Q ☿
06 Tuesday	
01 40	☽ △ ♄
04 24	☽ ⚼ ♀
06 59	☽ □ ☉
07 48	☽ ± ♂
08 23	☽ ⊼ ♅
18 37	☽ ⊼ ♂
18 44	☽ ± ♅
19 14	☽ △ ♆
19 16	☽ ∠ ♃
20 26	☽ ∥ ☉
21 03	☽ ± ♀
21 32	☽ ☍ ♇
07 Wednesday	
02 14	☉ △ ♅
03 47	☽ ⚼ ♄
03 56	♀ ⊥ ♇
08 17	☽ ⊼ ♀
09 53	♂ ⊻ ♆
10 22	☽ □ ☿
10 23	☽ ⚼ ♅
10 28	☿ ⚼ ♅
10 51	☽ ⊥ ♃
19 41	☽ ± ♄
21 20	☽ ⊻ ♃
21 31	☽ ⚼ ♆
08 Thursday	
06 22	☽ ⊼ ♄
10 38	☽ ∥ ♆
12 50	☽ △ ♅
13 31	☽ ± ♆
15 40	☽ △ ☉
16 06	☽ # ♂
09 Friday	
00 16	☽ ⊼ ♆
01 56	☽ ☍ ♂
02 32	☽ ⊼ ♇
03 17	☽ ∥ ♄
07 58	☽ # ♀
08 58	☽ # ♃
13 12	☽ ∥ ♃
13 21	☽ ± ♇
13 33	☉ ⊥ ♄
13 56	☽ ∥ ♀
15 39	♂ ⊻ ♇
17 48	☽ ☍ ♀
19 01	☽ # ♄
20 52	☽ ⚼ ☉
22 30	☽ △ ☿
10 Saturday	
02 51	☽ ☌ ♃
04 57	☽ ∥ ♂
05 43	☽ ⚼ ♇
06 45	♀ ∥ ♃
07 19	☿ # ♇
12 03	☽ # ♆
12 55	☽ ☍ ♄
14 39	☽ ± ☉
19 04	☽ □ ♅
11 Sunday	
02 37	☽ ⊼ ☉
05 27	☽ ⚼ ☿
05 27	☿ Q ♂
07 06	☽ ☍ ♆
09 19	☽ △ ♇
11 17	☽ ⊼ ♂
12 20	☿ △ ♃
12 47	☿ # ♅
20 25	☿ ∠ ♆
22 59	☽ ± ♂
12 Monday	
00 19	☽ ± ☿
04 27	☽ Q ♅
05 38	☽ ⊼ ♀
10 07	☽ ⊻ ♃
12 18	☽ # ☉
13 01	☽ ⊼ ☿
16 43	☽ ⚼ ♂
17 53	☽ ± ♀
21 16	☽ ⊼ ♄
21 23	☽ ⊥ ♃
23 04	☉ ⚼ ♃
13 Tuesday	
03 04	☽ ✶ ♅
08 46	☽ ± ♄
12 07	♀ # ♃
12 28	☽ ⚼ ♀
12 40	☿ ± ♅
14 11	☉ ⊻ ♆
14 27	☽ ∠ ♃
15 44	☽ ⊼ ♆
15 52	☽ ☍ ☉
17 02	☽ ∥ ♇
17 55	☽ □ ♇
19 53	☽ ∥ ♅
22 42	☽ △ ♂
14 Wednesday	
01 30	☽ # ☿
02 11	☽ ⚼ ♄
03 18	☽ ± ♆
07 48	☽ ∠ ♅
11 55	♀ ☍ ♃
17 22	☉ □ ♇
19 18	☽ ✶ ♃
20 00	☽ △ ♀
20 49	☽ ⚼ ♆
15 Thursday	
01 17	☽ ⊥ ♅
04 51	☽ Q ♇
06 10	☽ ☍ ☿
07 38	☽ △ ♄
13 03	☽ ⊻ ♅
18 46	☽ Q ♃
19 44	☿ ✶ ♄
16 Friday	
02 27	☽ △ ♆
04 33	☽ ✶ ♇
07 45	☽ ⊼ ☉
08 00	♀ ∠ ♇
12 30	☽ □ ♂
18 46	♂ Q ♅
20 50	☽ ± ☉
17 Saturday	
06 34	☽ □ ♃
10 37	☽ ∠ ♇
13 07	☽ □ ♀
16 38	☽ ⚼ ☉
17 30	☿ ⊼ ♅
20 02	☽ □ ♄
18 Sunday	
00 59	☽ ☌ ♅
01 59	☽ ⊼ ☿
02 36	☽ Q ♂
04 49	☽ ⊥ ♇
15 00	☽ □ ♆
15 45	☽ ± ☿
17 01	☽ ⊻ ♇
18 31	☽ # ☿
19 Monday	
01 54	☽ △ ☉
04 14	☽ ✶ ♂
05 08	☽ Q ♀
08 54	☽ Q ♄
09 41	☽ ∥ ♅
09 53	♀ ∥ ♄
12 25	☽ ⚼ ☿
12 38	☽ ∥ ♇
12 44	☉ ∠ ♄
14 43	♀ # ♂
17 51	☉ ⊥ ♆
19 13	☽ △ ♃
20 29	♂ # ♄
20 Tuesday	
00 06	♂ ⊥ ♇
03 40	☽ Q ♆
07 52	☽ ✶ ♀
08 24	☽ # ☉
09 27	☽ ✶ ♄
12 14	☽ ∠ ♂
13 45	☽ ⊻ ♅
22 37	☽ △ ☿
21 Wednesday	
01 26	☽ ⚼ ♃
01 48	☽ ⊥ ♅
03 31	♀ ☌ ♄
03 56	☽ ✶ ♆
05 46	☽ ☌ ♇
07 16	☽ ⊥ ♂
14 39	☉ ✶ ♂
15 49	☽ ∠ ♄
16 32	☉ ⚼ ♅
16 53	☽ ∠ ♀
19 22	☽ ± ♃
19 43	☽ ∠ ♅
19 49	☽ ⊻ ♂
20 01	☽ □ ☉
22 Thursday	
07 10	☽ ⊼ ♃
09 44	☽ ⊥ ♄
09 46	☽ ∠ ♆
12 19	☽ ⊥ ♀
21 30	☽ ⊻ ♄
21 57	☽ # ♆
23 Friday	
00 30	♀ □ ♅
01 00	☽ ✶ ♅
01 01	☿ ✶ ♆
01 03	☽ ⊻ ♀
03 16	☽ ⊥ ♆
09 17	☽ # ♀
10 11	☽ Q ☉
14 22	☽ # ♄
14 45	☽ ⊻ ♆
16 13	☽ □ ☿
16 20	☽ ⊻ ♇
16 32	♇ St R
17 29	☿ △ ♇
18 25	☽ ∥ ♂
23 18	☽ ∥ ♃
23 45	☽ Q ♅
24 Saturday	
02 04	☽ # ♃
03 36	☽ ⊥ ♇
06 28	☽ # ♂
08 14	☽ ☌ ♂
10 37	☽ ✶ ☉
10 59	☽ ∥ ♄
16 04	☽ ☍ ♃
17 41	☽ ∥ ♀
20 10	☽ ∠ ♇
25 Sunday	
02 37	☽ ∥ ♆
04 49	☽ Q ☿
05 53	☽ ☌ ♄
08 36	☽ □ ♅
13 34	☽ ☌ ♀
15 56	☽ ∠ ☉
17 52	☿ Q ♄
21 34	☽ ☌ ♆
22 55	☽ ✶ ♇
26 Monday	
03 59	☽ ✶ ☿
08 52	☽ ⊥ ☉
15 53	☽ ⊻ ♂
19 35	☽ Q ♇
19 56	☽ ⊻ ☉
20 48	☽ ⊼ ♃
27 Tuesday	
02 29	☽ ⊥ ♂
06 54	☽ ± ♃
07 42	☽ ∠ ☿
07 55	☉ △ ♃
10 00	☽ ⊻ ♄
12 07	☽ △ ♅
16 02	☽ ∥ ☉
18 06	☽ ∠ ♂
20 01	☽ ⊥ ♄
20 58	☽ ⊻ ♀
21 49	☽ ⚼ ♃
23 48	☽ ⊥ ☿
28 Wednesday	
00 25	☽ ⊻ ♆
01 01	☽ # ♇
01 34	☽ □ ♇
03 21	☽ # ♅
07 35	☽ ⊥ ♀
10 14	☽ ⊥ ♆
10 20	☽ ⊻ ☿
10 49	☽ ∠ ♄
12 40	☽ ⚼ ♅
19 33	☽ ✶ ♂
21 00	☽ ∥ ☿
21 59	☉ ∠ ♆
22 12	☽ △ ♃
23 19	☽ ∠ ♀
29 Thursday	
00 48	☽ ∠ ♆
01 00	☽ ☌ ☉
03 10	☽ ± ♅
07 17	♂ ∥ ♃
11 08	☽ ✶ ♄
12 46	☽ ⊼ ♅
15 34	☽ Q ♂
20 07	♀ ☌ ♆
30 Friday	
00 51	☽ ✶ ♆
01 13	☽ ✶ ♀
01 53	☽ △ ♇
04 06	♃ St D
06 24	☽ Q ♄
07 55	☉ ± ♅
10 15	♀ ✶ ♇
13 42	☽ ☌ ☿
20 02	☽ Q ♆
21 25	☽ □ ♂
21 52	☽ Q ♀
22 09	☽ □ ♃

DECEMBER 1951

LONGITUDES

Date	Sidereal time h m s	Sun ☉	Moon ☽	Moon ☽ 24.00	Mercury ☿	Venus ♀	Mars ♂	Jupiter ♃	Saturn ♄	Uranus ♅	Neptune ♆	Pluto ♇
01	16 38 09	08 ♐ 29 12	12 ♑ 53 25	20 ♑ 19 52	29 ♐ 43	22 ♎ 43	04 ♎ 08	04 ♈ 15	12 ♎ 29	13 ♋ 16	20 ♎ 56	21 ♌ 32
02	16 42 05	09 30 03	27 ♑ 42 41	05 ♒ 01 07	00 ♑ 28	23 50	04 42	04 D 15	12 34	13 R 14	20 58	21 R 32
03	16 46 02	10 30 55	12 ♒ 14 41	19 ♒ 23 01	01 06	24 56	05 15	04 16	12 39	13 12	21 00	21 32
04	16 49 58	11 31 48	26 ♒ 25 59	03 ♓ 23 34	01 37	26 03	05 49	04 17	12 44	13 10	21 01	21 31
05	16 53 55	12 32 42	10 ♓ 15 50	17 ♓ 03 01	02 00	27 10	06 23	04 18	12 49	13 08	21 03	21 31
06	16 57 51	13 33 36	23 ♓ 45 22	00 ♈ 23 12	02 15	28 18	06 56	04 19	12 54	13 05	21 04	21 31
07	17 01 48	14 34 31	06 ♈ 56 49	13 ♈ 26 34	02 R 20	29 ♎ 25	07 30	04 20	12 59	13 03	21 06	21 30
08	17 05 45	15 35 27	19 ♈ 52 47	26 ♈ 15 43	02 15	00 ♏ 33	08 03	04 22	13 04	13 01	21 07	21 30
09	17 09 41	16 36 23	02 ♉ 35 40	08 ♉ 52 51	01 58	01 41	08 36	04 24	13 08	12 59	21 09	21 29
10	17 13 38	17 37 20	15 ♉ 07 28	21 ♉ 19 41	01 31	02 49	09 10	04 26	13 13	12 57	21 10	21 29
11	17 17 34	18 38 18	27 ♉ 29 39	03 ♊ 37 27	00 52	03 57	09 43	04 28	13 17	12 54	21 12	21 28
12	17 21 31	19 39 17	09 ♊ 43 13	15 ♊ 47 01	00 ♑ 02	05 06	10 16	04 31	13 22	12 52	21 13	21 28
13	17 25 27	20 40 17	21 ♊ 48 59	27 ♊ 49 13	29 ♐ 01	06 15	10 49	04 33	13 26	12 50	21 14	21 27
14	17 29 24	21 41 17	03 ♋ 47 51	09 ♋ 45 02	27 52	07 24	11 22	04 36	13 30	12 47	21 16	21 27
15	17 33 20	22 42 18	15 ♋ 41 00	21 ♋ 35 59	26 35	08 33	11 54	04 39	13 34	12 45	21 17	21 26
16	17 37 17	23 43 20	27 ♋ 30 15	03 ♌ 24 10	25 14	09 42	12 27	04 42	13 38	12 43	21 18	21 26
17	17 41 14	24 44 23	09 ♌ 18 07	15 ♌ 12 32	23 51	10 51	13 00	04 46	13 42	12 40	21 19	21 25
18	17 45 10	25 45 27	21 ♌ 07 55	27 ♌ 04 48	22 29	12 01	13 32	04 49	13 46	12 38	21 21	21 24
19	17 49 07	26 46 31	03 ♍ 03 45	09 ♍ 05 24	21 11	13 11	14 04	04 53	13 50	12 35	21 22	21 24
20	17 53 03	27 47 36	15 ♍ 10 23	21 ♍ 19 22	19 58	14 20	14 37	04 57	13 53	12 33	21 23	21 23
21	17 57 00	28 48 42	27 ♍ 33 01	03 ♎ 51 58	18 54	15 30	15 09	05 01	13 57	12 30	21 24	21 22
22	18 00 56	29 ♐ 49 49	10 ♎ 16 50	16 ♎ 48 12	17 59	16 41	15 41	05 06	14 00	12 28	21 25	21 21
23	18 04 53	00 ♑ 50 56	23 ♎ 26 32	00 ♏ 12 13	17 14	17 51	16 13	05 10	14 04	12 25	21 26	21 21
24	18 08 49	01 52 04	07 ♏ 05 29	14 ♏ 06 23	16 40	19 01	16 45	05 15	14 07	12 23	21 27	21 20
25	18 12 46	02 53 12	21 ♏ 14 46	28 ♏ 30 17	16 17	20 12	17 16	05 20	14 10	12 20	21 28	21 19
26	18 16 43	03 54 23	05 ♐ 52 21	13 ♐ 20 08	16 04	21 22	17 48	05 25	14 13	12 18	21 29	21 18
27	18 20 39	04 55 33	20 ♐ 47 30	28 ♐ 28 33	16 D 02	22 33	18 19	05 31	14 16	12 15	21 30	21 17
28	18 24 36	05 56 43	06 ♑ 06 38	13 ♑ 45 26	16 08	23 44	18 51	05 36	14 19	12 12	21 31	21 16
29	18 28 32	06 57 54	21 ♑ 23 32	28 ♑ 59 36	16 23	24 55	19 22	05 42	14 22	12 10	21 32	21 15
30	18 32 29	07 59 05	06 ♒ 32 24	14 ♒ 00 53	16 47	26 06	19 53	05 48	14 25	12 07	21 33	21 14
31	18 36 25	09 ♑ 00 15	21 ♒ 24 10	28 ♒ 41 37	17 ♐ 17	27 ♏ 17	20 ♎ 24	05 ♈ 54	14 ♎ 27	12 ♋ 05	21 ♎ 34	21 ♌ 13

DECLINATIONS

Date	Moon True ☊	Moon Mean ☊	Moon ☽ Latitude	Sun ☉	Moon ☽	Mercury ☿	Venus ♀	Mars ♂	Jupiter ♃	Saturn ♄	Uranus ♅	Neptune ♆	Pluto ♇
01	04 ♓ 43	05 ♓ 04	03 S 57	21 S 44	26 S 45	25 S 33	06 S 49	00 S 08	00 N 21	02 S 51	23 N 10	06 S 41	22 N 46
02	04 R 37	05 01	03 01	21 53	23 35	25 26	07 12	00 21	00 22	02 52	23 10	06 41	22 46
03	04 33	04 58	01 55	22 02	18 58	25 17	07 34	00 34	00 22	02 54	23 10	06 42	22 47
04	04 32	04 55	00 S 43	22 10	13 23	25 07	07 56	00 48	00 23	02 56	23 10	06 42	22 47
05	04 D 32	04 52	00 N 30	22 18	07 15	24 56	08 19	01 01	00 24	02 58	23 11	06 43	22 48
06	04 R 32	04 48	01 40	22 26	00 S 57	24 43	08 41	01 14	00 24	02 59	23 11	06 44	22 48
07	04 30	04 45	02 42	22 33	05 N 15	24 29	09 04	01 26	00 25	03 01	23 11	06 44	22 48
08	04 26	04 42	03 35	22 40	11 06	24 14	09 26	01 39	00 26	03 02	23 11	06 45	22 49
09	04 19	04 39	04 16	22 46	16 23	23 58	09 48	01 52	00 27	03 04	23 11	06 45	22 49
10	04 09	04 36	04 44	22 52	20 54	23 41	10 10	02 05	00 28	03 06	23 12	06 46	22 50
11	03 56	04 33	04 59	22 58	24 27	23 22	10 33	02 18	00 29	03 07	23 12	06 46	22 50
12	03 43	04 29	04 59	23 03	26 50	23 03	10 55	02 30	00 31	03 08	23 12	06 46	22 51
13	03 28	04 26	04 45	23 07	27 57	22 42	11 17	02 43	00 32	03 10	23 12	06 47	22 51
14	03 15	04 23	04 20	23 11	27 43	22 21	11 38	02 56	00 33	03 11	23 13	06 47	22 52
15	03 04	04 20	03 42	23 15	26 12	22 00	12 00	03 08	00 35	03 13	23 13	06 48	22 52
16	02 56	04 17	02 56	23 18	23 32	21 39	12 21	03 20	00 36	03 14	23 13	06 48	22 52
17	02 50	04 13	02 02	23 21	19 53	21 18	12 43	03 33	00 38	03 15	23 13	06 49	22 53
18	02 48	04 10	01 N 02	23 23	15 26	20 58	13 04	03 45	00 40	03 17	23 14	06 49	22 53
19	02 D 47	04 07	00 S 02	23 24	10 22	20 40	13 25	03 57	00 41	03 18	23 14	06 49	22 54
20	02 47	04 04	01 06	23 26	04 N 50	20 24	13 46	04 10	00 43	03 19	23 14	06 50	22 55
21	02 48	04 01	02 08	23 27	00 S 59	20 10	14 06	04 22	00 45	03 20	23 14	06 50	22 55
22	02 R 47	03 58	03 06	23 27	06 56	19 59	14 27	04 34	00 47	03 21	23 15	06 51	22 56
23	02 45	03 54	03 56	23 27	12 46	19 51	14 47	04 46	00 49	03 22	23 15	06 51	22 56
24	02 40	03 51	04 35	23 26	18 13	19 46	15 06	04 58	00 51	03 23	23 15	06 51	22 57
25	02 33	03 48	04 59	23 25	22 53	19 43	15 26	05 10	00 54	03 24	23 15	06 52	22 57
26	02 24	03 45	05 04	23 23	26 17	19 44	15 45	05 21	00 56	03 25	23 16	06 52	22 58
27	02 15	03 42	04 49	23 21	27 56	19 46	16 04	05 33	00 58	03 26	23 16	06 52	22 58
28	02 06	03 39	04 13	23 19	27 31	19 51	16 22	05 45	01 01	03 27	23 16	06 52	22 59
29	01 58	03 35	03 19	23 16	25 02	19 58	16 41	05 56	01 03	03 28	23 16	06 53	22 59
30	01 53	03 32	02 12	23 12	20 46	20 07	16 58	06 08	01 06	03 29	23 17	06 53	23 00
31	01 ♓ 51	03 ♓ 29	00 S 56	23 S 09	15 S 15	20 S 16	17 S 16	06 S 19	01 N 08	03 S 29	23 N 17	06 S 53	23 N 01

ZODIAC SIGN ENTRIES

Date	h m	Planets
01	20 41	☿ ♑
02	15 45	☽ ♒
04	18 08	☽ ♓
06	23 18	☽ ♈
08	00 19	♀ ♏
09	07 04	☽ ♉
11	16 54	☽ ♊
12	12 40	☿ ♐
14	04 22	☽ ♋
16	17 05	☽ ♌
19	05 52	☽ ♍
21	16 41	☽ ♎
22	16 00	☉ ♑
23	23 38	☽ ♏
26	02 27	☽ ♐
28	02 24	☽ ♑
30	01 36	☽ ♒

LATITUDES

Date	Mercury ☿	Venus ♀	Mars ♂	Jupiter ♃	Saturn ♄	Uranus ♅	Neptune ♆	Pluto ♇
01	02 S 07	02 N 11	01 N 38	01 S 27	02 N 16	00 N 23	01 N 37	08 N 56
04	01 41	02 17	01 39	01 26	02 17	00 23	01 37	08 57
07	01 04	02 22	01 40	01 25	02 18	00 23	01 37	08 58
10	00 S 14	02 26	01 41	01 24	02 18	00 23	01 37	08 59
13	00 N 44	02 28	01 42	01 24	02 19	00 23	01 37	09 00
16	01 43	02 30	01 43	01 23	02 20	00 23	01 38	09 01
19	02 30	02 30	01 44	01 22	02 20	00 23	01 38	09 01
22	02 56	02 29	01 45	01 21	02 21	00 23	01 38	09 02
25	03 02	02 28	01 46	01 20	02 22	00 23	01 38	09 03
28	02 53	02 25	01 46	01 19	02 23	00 23	01 38	09 04
31	02 N 35	02 N 21	01 N 47	01 S 18	02 N 24	00 N 23	01 N 38	09 N 05

DATA

Julian Date	2433982
Delta T	+30 seconds
Ayanamsa	23° 11' 14"
Synetic vernal point	05° ♓ 55' 46"
True obliquity of ecliptic	23° 26' 52"

MOON'S PHASES, APSIDES AND POSITIONS ☽

Date	h m	Phase	Longitude	Eclipse Indicator
05	16 20	☽	12 ♓ 44	
13	09 30	○	20 ♊ 34	
21	14 37	☾	28 ♍ 55	
28	11 43	●	05 ♑ 56	

Day	h m		
16	02 48	Apogee	
28	22 50	Perigee	
06	15 38	0N	
13	19 54	Max dec	28° N 01'
21	07 59	0S	
27	19 21	Max dec	28° S 02'

LONGITUDES

Date	Chiron ⚷	Ceres ⚳	Pallas ⚴	Juno ⚵	Vesta ⚶	Black Moon Lilith ⚸
01	02 ♑ 08	06 ♓ 43	06 ♒ 03	25 ♐ 02	08 ♍ 32	16 ♋ 53
11	03 ♑ 09	08 ♓ 55	08 ♒ 47	28 ♐ 38	10 ♍ 54	18 ♋ 00
21	04 ♑ 11	11 ♓ 27	11 ♒ 41	02 ♑ 18	12 ♍ 46	19 ♋ 07
31	05 ♑ 14	14 ♓ 17	14 ♒ 43	06 ♑ 00	12 ♍ 04	20 ♋ 14

ASPECTARIAN

h m	Aspects
01 Saturday	
01 49	☽ ⚼ ♇
02 40	♀ ∥ ♆
04 25	☽ ⚺ ☉
11 21	☽ □ ♄
12 36	☽ ☍ ♅
14 45	☽ ⊥ ☉
16 15	☽ ± ♇
17 13	♂ ☍ ♃
23 03	☽ ∥ ☿
02 Sunday	
01 01	☽ □ ♆
01 57	☽ ⚻ ♇
03 07	☽ Q ♃
05 10	☽ □ ♀
06 23	☽ ∠ ☉
12 49	♂ # ♃
14 30	☽ # ♅
16 43	☽ ⚺ ☿
16 45	☽ # ♇
21 18	☽ ∥ ☉
22 45	☽ ✶ ♃
23 55	☽ △ ♂
03 Monday	
03 04	☽ ⊥ ☿
08 54	☽ ✶ ☉
12 42	☽ △ ♄
13 35	☽ ⚻ ♅
18 43	☽ ∠ ☿
23 39	☽ ± ♅
23 49	☽ ∠ ♃
04 Tuesday	
02 02	☽ ⚼ ♂
02 45	☽ △ ♆
03 38	☽ ☍ ♇
06 39	☽ Q ☉
11 18	☽ △ ♀
14 15	☽ ⚼ ♄
14 57	☽ ⚼ ♅
15 10	☽ ⊥ ♃
18 04	☽ ± ♂
21 13	☽ ✶ ☿
05 Wednesday	
01 33	☽ ⚺ ♃
04 36	☽ ⚼ ♆
04 54	☽ ⚻ ♂
05 56	☽ ± ♄
08 10	☽ ∥ ♀
14 04	☽ ∥ ♆
15 40	☽ ⚼ ♀
16 20	☽ □ ☉
16 32	☽ ⚻ ♄
17 02	☽ △ ♅
17 55	♂ ∠ ♇
18 45	☽ Q ☿
19 09	☉ ✶ ♄
20 27	☽ ± ♆
06 Thursday	
01 15	☉ ⚻ ♅
04 18	☽ ∥ ♄
07 10	☽ ⚻ ♆
07 58	☽ ⚻ ♇
09 08	☽ ± ♀
11 00	☽ ∥ ♂
14 05	☽ # ♃
17 12	☽ ∥ ♃
18 47	☽ ± ♇
20 39	☽ # ♂
20 58	☽ ⚻ ♀
07 Friday	
03 13	☽ # ♄
03 32	☽ □ ☿
07 12	☽ ☌ ♃
11 11	☽ ⚼ ♇
11 56	☿ St R
13 03	☽ ☍ ♂
17 57	☽ # ♆
23 13	☽ ☍ ♄
23 15	☽ □ ♅
08 Saturday	
02 48	♄ □ ♅
03 18	☽ △ ☉
04 30	☽ # ♀
14 20	☽ ☍ ♆
15 02	☽ △ ♇
20 22	☉ ∠ ♀
09 Sunday	
08 56	☽ Q ♅
09 57	☽ ⚼ ☉
10 06	☽ ☍ ♀
10 51	☽ △ ☿
15 26	☽ ⚺ ♃
16 35	☿ ✶ ♀
10 Monday	
00 00	☽ ⚻ ♂
00 14	☉ # ♇
02 57	☽ ⊥ ♃
04 40	☽ ± ☉
07 49	☽ ✶ ♅
08 18	☽ ⚻ ♄
12 04	☽ ± ♂
14 34	☽ ⚼ ☿
17 15	☽ ⚻ ☉
19 57	☽ ± ♄
20 21	☽ ∠ ♃
23 43	☽ ⚻ ♆
11 Tuesday	
00 08	☽ ∥ ♇
00 18	☽ □ ♇
00 45	☽ # ☉
01 54	♀ Q ♇
02 40	☽ ∥ ♅
04 34	☽ # ☿
06 19	☽ ⚼ ♂
07 09	☽ ± ☿
11 25	☽ ± ♆
12 48	☽ ∠ ♅
13 33	☽ ⚼ ♄
18 12	☽ ⚻ ☿
23 08	♀ ⚻ ♃
12 Wednesday	
00 31	☿ # ♅
01 42	☽ ✶ ♃
01 58	☽ ⚻ ♀
05 05	☽ ⚼ ♆
06 23	☽ ⊥ ♅
11 30	☽ Q ♇
12 00	☉ ∥ ☿
13 07	☽ △ ♂
15 00	☽ ± ♀
18 12	☽ ⚺ ♅
19 14	☽ △ ♄
13 Thursday	
01 29	☽ Q ♃
02 01	☿ # ♇
09 30	☽ ☍ ☉
10 44	☽ ⚼ ♀
10 51	☽ △ ♆
11 17	☽ ✶ ♇
14 Friday	
01 10	☽ ☍ ☿
01 41	☉ ✶ ♆
06 23	☉ △ ♇
13 37	☽ □ ♃
17 20	☽ ∠ ♇
20 01	☽ △ ♀
22 10	☉ # ♅
15 Saturday	
03 59	☽ □ ♂
06 05	☽ ☌ ♅
07 42	☽ □ ♄
11 30	☽ ⊥ ♇
22 04	♂ ∥ ♄
23 23	☽ □ ♆
23 40	☽ ⚺ ♇
16 Sunday	
03 35	☽ ⚻ ☉
05 20	☿ Q ♄
07 52	☽ ⚻ ☿
13 46	☽ # ☉
14 22	☽ ∥ ♅
16 50	☽ ∥ ♇
16 56	☽ ± ☉
17 06	☿ ∠ ♀
18 17	☽ Q ♂
18 48	☽ ± ☿
20 27	☽ Q ♄
21 49	☿ Q ♂
22 40	♂ □ ♅
17 Monday	
02 32	☽ # ☿
02 43	☽ △ ♃
03 09	☉ ☌ ☿
10 00	♀ ± ♃
11 11	☽ ⚼ ☿
12 03	☽ Q ♆
12 58	☽ ⚼ ☉
15 30	☽ □ ♀
18 49	☽ ⚺ ♅
19 52	☽ ✶ ♂
20 59	☽ ✶ ♄
18 Tuesday	
00 44	☉ Q ♂
06 57	☽ ⊥ ♅
09 20	☽ ⚼ ♃
12 11	☉ Q ♄
12 26	☽ ✶ ♆
12 33	☽ ☌ ♇
14 27	☽ △ ☿
22 13	☽ △ ☉
22 46	☽ # ♀
23 38	♂ ☌ ♄
19 Wednesday	
00 15	♀ △ ♅
01 03	☽ ∠ ♅
03 28	☽ ∠ ♄
03 35	☽ ± ♃
03 38	☽ ∠ ♂
07 49	☽ Q ♀
07 53	☿ △ ♇
08 32	☿ ✶ ♆
15 40	☽ ⚻ ♃
18 36	☽ ∠ ♆
21 33	☽ ⊥ ♄
22 27	☽ ⊥ ♂
20 Thursday	
02 11	♀ ⚺ ♄
03 32	☽ # ♆
06 51	☽ ✶ ♅
08 10	☿ ⊥ ♀
09 28	☽ ⚺ ♄
10 11	☽ ✶ ♀
10 37	♆ ✶ ♇
10 51	☽ ⚺ ♂
12 25	☽ ⊥ ♆
14 44	☽ # ♂
18 20	☽ # ♄
20 35	☽ □ ☿
22 19	♀ ⚺ ♂
21 Friday	
00 06	☽ ⚺ ♇
00 08	☽ ⚺ ♆
04 57	☽ ∥ ♃
06 10	☽ Q ♅
11 03	☽ # ♃
11 39	☽ ⊥ ♇
14 37	☽ □ ☉
18 13	☽ ∠ ♀
21 32	☽ ∥ ♄
22 Saturday	
02 08	☽ ∥ ♂
02 16	☽ ☍ ♃
04 29	☽ Q ☿
04 41	☽ ∠ ♇
11 40	☽ ∥ ♆
12 48	☽ ⊥ ♀
16 01	☽ □ ♅
18 55	☽ ☌ ♄
22 22	☽ ☌ ♂
23 Sunday	
00 55	☽ ⚺ ♀
01 21	☽ ✶ ☿
03 02	☽ Q ☉
04 04	☿ ⚺ ♀
08 14	☽ ✶ ♇
08 24	☽ ☌ ♆
21 07	☽ ∥ ♀
24 Monday	
02 11	☽ ✶ ☉
02 54	☽ ∠ ☿
05 29	☽ Q ♇
08 48	☽ ⚻ ♃
10 12	☿ ✶ ♂
17 57	☽ ⊥ ☿
19 11	☽ ± ♃
19 20	☽ ∥ ☿
21 02	☽ △ ♅
25 Tuesday	
00 04	☽ ⚺ ♄
03 52	☽ ⚺ ☿
05 05	☽ ⚺ ♂
05 57	☽ ∠ ☉
10 06	☽ ☌ ♀
10 12	☽ ⊥ ♄
10 28	☽ ⚼ ♃
11 27	♀ ⊥ ♄
12 07	☽ □ ♇
12 22	☽ ⚺ ♆
12 26	☽ # ♇
14 14	☽ # ♅
15 04	♀ ⚼ ♃
15 11	☽ ∥ ☉
15 30	☽ ⊥ ♂
22 03	☽ ⊥ ☉
22 03	☽ ⚼ ♅
22 19	☽ ⊥ ♆
26 Wednesday	
01 08	☽ ∠ ♄
01 58	☉ Q ♆
06 50	☽ ∠ ♂
08 35	☽ ⚺ ☉
10 30	♀ □ ♇
11 16	☽ △ ♃
12 41	☽ ± ♅
13 00	☽ ∠ ♆
14 20	♀ ⚺ ♆
22 18	☽ ⚻ ♅
27 Thursday	
01 28	☽ ✶ ♄
04 17	☽ ☌ ☿
06 39	☿ St D
07 48	☽ ✶ ♂
12 39	☽ △ ♇
13 00	☽ ✶ ♆
14 53	☽ ⚺ ♀
20 33	☽ Q ♄
28 Friday	
01 08	☽ ⊥ ♀
03 07	☉ □ ♃
03 26	☽ Q ♂
07 55	☽ Q ♆
11 12	☽ □ ♃
11 43	☽ ☌ ☉
12 15	☽ ⚼ ♇
16 28	☽ ∠ ♀
19 31	☉ ⚼ ♇
21 32	☽ ☍ ♅
29 Saturday	
00 55	☽ □ ♄
02 22	☽ ± ♇
03 59	☽ ⚺ ☿
08 24	☉ # ♅
08 42	☽ □ ♂
11 47	☽ ⚻ ♇
12 13	☽ □ ♆
13 36	☽ ⊥ ☿
15 39	☽ Q ♃
18 02	☽ ✶ ♀
22 56	☽ # ♅
23 08	☽ ∥ ☉
30 Sunday	
00 31	☽ # ♇
04 12	☽ ∠ ☿
04 12	♀ ⊥ ♂
10 48	☽ ✶ ♃
14 29	☽ ⚺ ☉
14 43	☽ Q ♀
15 02	☽ ∥ ☿
20 55	☽ ⚻ ♅
31 Monday	
00 41	☽ △ ♄
00 50	☽ ⊥ ☉
04 07	☽ ∥ ♀
05 01	☽ ✶ ☿
06 36	☽ ± ♅
07 54	♀ ⚼ ♅
10 18	☽ △ ♂
11 10	☽ ∠ ♃
11 42	☽ ☍ ♇
12 15	☽ △ ♆
16 35	☽ ∠ ☉
17 32	♀ ⊥ ♆
21 18	☽ ⚼ ♅
22 32	☽ □ ♀

All ephemeris data is given at 12.00 UT and the Moon's longitude is additionally given for 24.00 UT

JANUARY 1952

LONGITUDES

Date	Sidereal time h m s	Sun ☉	Moon ☽	Moon ☽ 24.00	Mercury ☿	Venus ♀	Mars ♂	Jupiter ♃	Saturn ♄	Uranus ♅	Neptune ♆	Pluto ♇
01	18 40 22	10 ♑ 01 26	05 ♓ 52 46	12 ♓ 57 24	17 ♐ 53	28 ♏ 29	20 ♎ 55	06 ♈ 00	14 ♎ 30	12 ♋ 02	21 ♎ 34	21 ♌ 12
02	18 44 18	11 02 36	19 ♓ 55 27	26 ♓ 46 58	18 35	29 ♏ 40	21 26	06 06	14 32	12 R 00	21 35	21 R 11
03	18 48 15	12 03 46	03 ♈ 32 12	10 ♈ 11 27	19 22	00 ♐ 51	21 56	06 13	14 35	11 57	21 36	21 10
04	18 52 12	13 04 55	16 ♈ 45 04	23 ♈ 13 29	20 14	02 03	22 27	06 20	14 37	11 54	21 36	21 09
05	18 56 08	14 06 05	29 ♈ 37 09	05 ♉ 56 30	21 10	03 15	22 57	06 26	14 39	11 52	21 37	21 08
06	19 00 05	15 07 14	12 ♉ 12 00	18 ♉ 24 02	22 09	04 26	23 27	06 34	14 41	11 49	21 38	21 07
07	19 04 01	16 08 22	24 ♉ 33 06	00 ♊ 39 29	23 12	05 38	23 57	06 41	14 43	11 47	21 38	21 06
08	19 07 58	17 09 31	06 ♊ 43 34	12 ♊ 45 39	24 17	06 50	24 27	06 48	14 45	11 44	21 39	21 05
09	19 11 54	18 10 39	18 ♊ 46 02	24 ♊ 44 57	25 25	08 02	24 56	06 56	14 46	11 41	21 39	21 03
10	19 15 51	19 11 47	00 ♋ 42 39	06 ♋ 39 21	26 35	09 14	25 26	07 03	14 48	11 39	21 40	21 02
11	19 19 47	20 12 54	12 ♋ 35 14	18 ♋ 30 32	27 47	10 26	25 55	07 11	14 49	11 36	21 40	21 01
12	19 23 44	21 14 01	24 ♋ 25 27	00 ♌ 20 10	29 ♐ 01	11 38	26 25	07 19	14 51	11 34	21 41	21 00
13	19 27 41	22 15 07	06 ♌ 14 56	12 ♌ 10 00	00 ♑ 17	12 50	26 54	07 27	14 52	11 31	21 41	20 59
14	19 31 37	23 16 14	18 ♌ 05 38	24 ♌ 02 09	01 34	14 03	27 23	07 36	14 53	11 29	21 41	20 57
15	19 35 34	24 17 20	29 ♌ 59 52	05 ♍ 59 11	02 52	15 15	27 51	07 44	14 54	11 26	21 42	20 56
16	19 39 30	25 18 25	12 ♍ 00 30	18 ♍ 04 16	04 12	16 28	28 20	07 53	14 55	11 24	21 42	20 55
17	19 43 27	26 19 31	24 ♍ 10 57	00 ♎ 21 04	05 33	17 40	28 48	08 01	14 56	11 21	21 42	20 54
18	19 47 23	27 20 36	06 ♎ 35 07	12 ♎ 53 38	06 55	18 53	29 16	08 10	14 57	11 19	21 42	20 52
19	19 51 20	28 21 40	19 ♎ 17 09	25 ♎ 46 09	08 19	20 05	29 ♎ 44	08 19	14 57	11 16	21 42	20 51
20	19 55 16	29 ♑ 22 45	02 ♏ 21 07	09 ♏ 02 26	09 43	21 18	00 ♏ 12	08 28	14 58	11 14	21 43	20 50
21	19 59 13	00 ♒ 23 49	15 ♏ 50 24	22 ♏ 45 15	11 08	22 31	00 40	08 37	14 58	11 11	21 43	20 48
22	20 03 10	01 24 53	29 ♏ 47 00	06 ♐ 55 34	12 33	23 43	01 07	08 47	14 58	11 09	21 43	20 47
23	20 07 06	02 25 56	14 ♐ 10 40	21 ♐ 31 49	14 00	24 56	01 34	08 56	14 59	11 07	21 43	20 46
24	20 11 03	03 26 59	28 ♐ 58 18	06 ♑ 29 15	15 27	26 09	02 01	09 06	14 59	11 04	21 R 43	20 44
25	20 14 59	04 28 02	14 ♑ 03 38	21 ♑ 40 15	16 55	27 22	02 28	09 16	14 R 59	11 02	21 43	20 43
26	20 18 56	05 29 03	29 ♑ 17 49	06 ♒ 55 02	18 24	28 35	02 55	09 26	14 59	11 00	21 43	20 41
27	20 22 52	06 30 04	14 ♒ 30 35	22 ♒ 03 16	19 54	29 ♐ 48	03 21	09 36	14 58	10 57	21 42	20 40
28	20 26 49	07 31 04	29 ♒ 31 59	06 ♓ 55 46	21 24	01 ♑ 01	03 47	09 46	14 58	10 55	21 42	20 39
29	20 30 45	08 32 03	14 ♓ 13 51	21 ♓ 25 40	22 55	02 14	04 13	09 56	14 57	10 53	21 42	20 37
30	20 34 42	09 33 00	28 ♓ 30 50	05 ♈ 29 08	24 27	03 27	04 38	10 07	14 57	10 51	21 42	20 36
31	20 38 39	10 ♒ 33 57	12 ♈ 20 32	19 ♈ 05 09	25 ♑ 59	04 ♑ 40	05 ♏ 04	10 ♈ 17	14 ♎ 56	10 ♋ 48	21 ♎ 42	20 ♌ 34

MOON NODES AND LATITUDE

Date	Moon True ☊	Moon Mean ☊	Moon ☽ Latitude
01	01 ♓ 50	03 ♓ 26	00 N 22
02	01 D 51	03 23	01 36
03	01 52	03 19	02 42
04	01 R 53	03 16	03 38
05	01 51	03 13	04 21
06	01 48	03 10	04 50
07	01 42	03 07	05 06
08	01 35	03 04	05 07
09	01 26	03 00	04 54
10	01 17	02 57	04 29
11	01 09	02 54	03 52
12	01 03	02 51	03 05
13	00 58	02 48	02 10
14	00 55	02 45	01 09
15	00 54	02 41	00 N 05
16	00 D 54	02 38	01 S 00
17	00 56	02 35	02 04
18	00 58	02 32	03 03
19	00 59	02 29	03 54
20	00 R 59	02 25	04 35
21	00 58	02 22	05 03
22	00 56	02 19	05 13
23	00 53	02 16	05 05
24	00 48	02 13	04 37
25	00 45	02 10	03 49
26	00 41	02 06	02 44
27	00 39	02 03	01 28
28	00 39	02 00	00 S 06
29	00 D 39	01 57	01 N 14
30	00 40	01 54	02 30
31	00 ♓ 42	01 ♓ 51	03 N 30

DECLINATIONS

Date	Sun ☉	Moon ☽	Mercury ☿	Venus ♀	Mars ♂	Jupiter ♃	Saturn ♄	Uranus ♅	Neptune ♆	Pluto ♇
01	23 S 04	09 S 01	20 S 27	17 S 33	06 S 30	01 N 11	03 S 30	23 N 17	06 S 53	23 N 01
02	22 59	02 S 31	20 38	17 50	06 41	01 14	03 31	23 17	06 54	23 02
03	22 54	03 N 53	20 50	18 06	06 53	01 17	03 31	23 18	06 54	23 02
04	22 48	09 56	21 03	18 22	07 04	01 20	03 32	23 18	06 54	23 03
05	22 42	15 24	21 15	18 37	07 15	01 23	03 33	23 18	06 54	23 03
06	22 35	20 06	21 28	18 52	07 25	01 26	03 33	23 18	06 54	23 04
07	22 28	23 51	21 40	19 07	07 36	01 29	03 34	23 19	06 55	23 05
08	22 21	26 29	21 52	19 21	07 47	01 32	03 34	23 19	06 55	23 05
09	22 13	27 52	22 04	19 35	07 57	01 35	03 34	23 19	06 55	23 06
10	22 04	27 56	22 15	19 48	08 08	01 38	03 35	23 19	06 55	23 06
11	21 56	26 42	22 25	20 00	08 18	01 42	03 35	23 19	06 55	23 07
12	21 46	24 17	22 35	20 13	08 29	01 45	03 35	23 20	06 55	23 08
13	21 37	20 49	22 44	20 24	08 39	01 49	03 36	23 20	06 55	23 08
14	21 26	16 31	22 52	20 35	08 49	01 52	03 36	23 20	06 55	23 09
15	21 16	11 33	23 00	20 46	08 59	01 56	03 36	23 20	06 55	23 09
16	21 05	06 08	23 06	20 56	09 09	01 59	03 36	23 21	06 55	23 10
17	20 54	00 N 25	23 11	21 05	09 19	02 03	03 36	23 21	06 55	23 11
18	20 42	05 S 25	23 15	21 14	09 28	02 07	03 36	23 21	06 55	23 11
19	20 30	11 10	23 18	21 22	09 38	02 10	03 36	23 21	06 55	23 12
20	20 17	16 36	23 20	21 30	09 47	02 14	03 36	23 21	06 55	23 12
21	20 04	21 24	23 21	21 37	09 57	02 18	03 36	23 22	06 55	23 13
22	19 51	25 13	23 21	21 44	10 06	02 22	03 36	23 22	06 55	23 14
23	19 37	27 33	23 19	21 50	10 15	02 26	03 35	23 22	06 55	23 14
24	19 23	28 03	23 17	21 55	10 24	02 30	03 35	23 22	06 55	23 15
25	19 09	26 30	23 13	21 59	10 33	02 34	03 35	23 22	06 55	23 15
26	18 54	22 59	23 07	22 03	10 42	02 38	03 35	23 23	06 55	23 16
27	18 39	17 53	23 01	22 07	10 51	02 42	03 34	23 23	06 55	23 17
28	18 24	11 44	22 53	22 10	11 00	02 46	03 34	23 23	06 55	23 17
29	18 08	05 S 04	22 44	22 12	11 08	02 50	03 33	23 23	06 55	23 18
30	17 52	01 N 40	22 33	22 13	11 17	02 54	03 33	23 23	06 55	23 18
31	17 S 36	08 N 06	22 S 21	22 S 14	11 S 25	02 N 59	03 S 32	23 N 24	06 S 54	23 N 19

ZODIAC SIGN ENTRIES

Date	h	m	Planets
01	02	10	☽ ♓
02	18	44	♀ ♐
03	05	42	☽ ♈
05	12	43	☽ ♉
07	22	42	☽ ♊
10	10	34	☽ ♋
12	23	19	☽ ♌
13	06	44	☿ ♑
15	12	00	☽ ♍
17	23	19	☽ ♎
20	01	33	♂ ♏
20	07	44	☽ ♏
21	02	38	☉ ♒
22	12	57	☽ ♐
24	13	39	☽ ♑
26	13	06	☽ ♒
27	15	58	♀ ♑
28	12	45	☽ ♓
30	14	32	☽ ♈

LATITUDES

Date	Mercury ☿	Venus ♀	Mars ♂	Jupiter ♃	Saturn ♄	Uranus ♅	Neptune ♆	Pluto ♇
01	02 N 27	02 N 20	01 N 48	01 S 18	02 N 24	00 N 23	01 N 38	09 N 05
04	02 03	02 15	01 48	01 17	02 25	00 23	01 39	09 06
07	01 36	02 10	01 49	01 17	02 26	00 23	01 39	09 06
10	01 09	02 04	01 50	01 16	02 27	00 23	01 39	09 07
13	00 43	01 57	01 51	01 15	02 27	00 23	01 39	09 08
16	00 N 17	01 50	01 51	01 14	02 28	00 23	01 39	09 08
19	00 S 07	01 43	01 52	01 14	02 29	00 23	01 40	09 09
22	00 30	01 34	01 53	01 13	02 30	00 23	01 40	09 09
25	00 50	01 26	01 54	01 12	02 31	00 23	01 40	09 10
28	01 09	01 17	01 54	01 12	02 32	00 23	01 40	09 10
31	01 S 26	01 N 08	01 N 55	01 S 11	02 N 33	00 N 23	01 N 40	09 N 11

DATA

Julian Date	2434013
Delta T	+30 seconds
Ayanamsa	23° 11' 20"
Synetic vernal point	05° ♓ 55' 40"
True obliquity of ecliptic	23° 26' 52"

MOON'S PHASES, APSIDES AND POSITIONS ☽

Date	h	m	Phase	Longitude	Eclipse Indicator
04	04	42	☽	12 ♈ 46	
12	04	55	○	20 ♋ 56	
20	06	09	☾	29 ♎ 08	
26	22	26	●	05 ♒ 56	

Day	h	m		
12	05	21	Apogee	
26	11	59	Perigee	
02	21	21	0N	
10	01	14	Max dec	28° N 04'
17	13	43	0S	
24	05	57	Max dec	28° S 07'
30	06	00	0N	

LONGITUDES

Date	Chiron ⚷	Ceres ⚳	Pallas ⚴	Juno ⚵	Vesta ⚶	Black Moon Lilith ⚸
01	05 ♑ 21	14 ♓ 35	15 ♒ 02	06 ♑ 23	14 ♍ 09	20 ♋ 21
11	06 ♑ 23	17 ♓ 42	18 ♒ 11	10 ♑ 06	14 ♍ 41	21 ♋ 27
21	07 ♑ 21	21 ♓ 00	21 ♒ 13	13 ♑ 50	14 ♍ 28	22 ♋ 34
31	08 ♑ 21	24 ♓ 29	24 ♒ 41	13 ♑ 33	13 ♍ 27	23 ♋ 41

ASPECTARIAN

h m	Aspects
01 Tuesday	
01 18	☽ bQ ♄
01 31	☽ Q ☿
02 06	☽ ⊥ ♃
12 04	☽ bQ ♂
12 12	☽ ⊻ ♃
13 10	☽ bQ ♆
16 26	☽ ± ♄
19 33	☽ ⚹ ☉
19 54	☽ ∥ ♆
21 04	☽ ∥ ♂
22 24	☽ △ ♅
02 Wednesday	
01 06	♂ ⚹ ♇
01 43	☉ # ♇
02 41	☽ ⊼ ♄
03 56	☽ ± ♂
04 30	☽ ± ♆
08 21	☽ ∥ ♄
09 19	♀ ∠ ♄
09 33	☽ □ ☿
14 12	☽ ⊼ ♇
14 43	☽ ⊼ ♂
14 53	☽ ⊼ ♆
16 44	☽ # ♃
17 52	☽ Q ☉
19 36	♂ ☌ ♆
03 Thursday	
00 42	☽ ± ♇
02 03	☽ ∥ ♃
03 32	♃ bQ ♇
06 45	☽ △ ♀
09 25	☉ ☍ ♅
10 37	☽ # ♄
14 49	♂ ∥ ♆
16 43	☽ bQ ♇
16 51	☽ ☌ ♃
23 43	☽ # ♆
23 59	☽ # ♂
04 Friday	
03 09	☽ □ ♅
04 42	☽ □ ☉
08 04	☽ ☍ ♄
12 36	☽ bQ ♀
18 55	☽ △ ☿
20 08	☽ △ ♇
21 00	☽ ☍ ♆
22 58	☽ ☍ ♂
05 Saturday	
07 04	☽ ± ♀
11 18	☿ △ ♇
12 27	☽ Q ♅
19 35	☽ ⊼ ♀
23 21	☿ ⚹ ♆
06 Sunday	
01 05	☽ ⊻ ♃
01 18	☉ □ ♄
01 28	☽ bQ ☿
04 52	☽ # ♀
11 16	☽ ⚹ ♅
11 52	☉ ± ♇
12 42	☽ ⊥ ♃
16 48	☽ ⊼ ♄
18 09	☽ △ ☉
20 19	☽ ± ☿
20 26	☽ # ☿
07 Monday	
02 40	☽ # ☉
04 29	☽ ± ♄
05 15	☽ □ ♇
06 18	☽ ⊼ ♆
06 20	☽ ∠ ♃
06 26	☽ ∥ ♇
08 04	☽ ∥ ♅
09 05	☽ ⊼ ☿
10 46	☽ ⊼ ♂
14 44	♀ ± ♅
16 21	☽ ∠ ♅
18 03	☽ ± ♆
22 10	☽ bQ ♄
23 03	☽ ± ♂
08 Tuesday	
02 08	☽ bQ ☉
08 15	♀ ∠ ♆
10 02	☽ ⊥ ♅
11 19	♀ △ ♃
11 51	☽ bQ ♆
12 09	☽ ⚹ ♃
12 14	☽ ☍ ♀
16 40	☽ Q ♇
17 38	☽ bQ ♂
18 25	☿ ⚹ ♂
21 37	☽ ± ☉
21 55	☽ ⊻ ♅
09 Wednesday	
03 59	☽ △ ♄
10 43	☽ ⊼ ☉
12 19	☽ Q ♃
16 35	☽ ⚹ ♇
17 47	☽ △ ♆
22 54	☉ ∥ ☿
10 Thursday	
00 55	☽ △ ♂
02 46	☽ ☍ ☿
16 27	☿ Q ♄
22 44	☽ ∠ ♇
11 Friday	
00 57	☽ □ ♃
07 09	☽ ⊼ ♀
10 01	☽ ☌ ♅
16 32	☽ □ ♄
16 55	☽ ⊥ ♇
20 40	☽ ± ♀
12 Saturday	
04 25	♀ ∠ ♂
04 55	☽ ☍ ☉
05 04	☽ ⊻ ♇
06 25	☽ □ ♆
06 34	☉ ⊼ ♇
10 34	♀ ⊼ ♅
16 12	☽ □ ♂
17 00	☽ bQ ♀
19 16	☽ ∥ ♅
20 43	☽ ∥ ♇
22 26	☽ ⊼ ☿
22 30	☉ □ ♆
13 Sunday	
00 01	☽ # ☿
05 07	☽ Q ♄
06 48	☽ # ☉
12 04	☽ ± ☿
14 23	☽ # ♀
14 28	☽ △ ♃
18 58	☽ Q ♆
22 39	☽ ⊻ ♅
14 Monday	
02 53	☽ △ ♀
05 30	☽ ⚹ ♄
06 16	☽ Q ♂
08 32	☽ bQ ☿
10 45	☽ ⊥ ♅
17 47	☽ ☌ ♇
19 16	☽ ⚹ ♆
21 12	☽ bQ ♃
23 26	☽ ⊼ ☉
15 Tuesday	
04 52	☽ ∠ ♅
04 57	♀ ⚹ ♄
07 31	☽ ⚹ ♂
11 48	☽ ∠ ♄
12 38	☽ ± ☉
15 31	☽ ± ♃
18 29	☽ △ ☿
23 15	☽ # ♂
16 Wednesday	
01 25	☽ ∠ ♆
02 51	☿ Q ♆
03 40	☽ ⊼ ♃
05 51	☽ ⊥ ♄
08 18	☽ bQ ☉
08 35	☽ # ♆
10 47	☽ ⚹ ♅
14 44	☽ ∠ ♂
17 47	☽ ⊻ ♄
19 19	☽ ⊥ ♆
21 47	☽ □ ♀
22 35	☉ ∥ ♀
22 44	☽ # ♄
17 Thursday	
03 38	☉ Q ♃
05 18	☽ ∥ ♃
05 34	☽ ⊻ ♇
07 08	☽ ⊻ ♆
09 11	☽ ⊥ ♂
09 35	☿ # ♇
10 23	☽ Q ♅
16 34	☽ △ ☉
17 17	☽ ⊥ ♇
17 51	☿ bQ ♇
21 21	☽ ⊻ ♂
22 16	☽ # ♃
18 Friday	
04 32	☽ ∥ ♄
10 38	☽ ∠ ♇
12 37	☽ Q ♀
12 44	☽ □ ☿
15 04	☽ ☍ ♃
18 14	☽ ∥ ♆
20 58	☽ □ ♅
19 Saturday	
03 53	☽ ☌ ♄
05 21	☽ ∥ ♂
12 11	☿ □ ♃
13 39	☽ ⚹ ♀
14 54	☽ ⚹ ♇
16 30	☽ ☌ ♆
20 Sunday	
02 33	☽ Q ☿
02 51	♀ △ ♇
06 09	☽ □ ☉
07 58	☽ ☌ ♂
12 51	☽ Q ♇
19 48	☽ ∠ ♀
20 10	♀ ⚹ ♆
23 07	☽ ⊼ ♃
21 Monday	
02 44	☽ ⚹ ☿
03 50	☽ △ ♅
05 14	☽ ∥ ☉
09 51	☽ ± ♃
10 28	☽ ⊻ ♄
13 02	☿ ☍ ♅
13 12	☽ ∥ ♀
13 17	☽ ⊥ ♀
16 49	☽ Q ☉
20 37	☽ □ ♇
20 55	☽ ⊥ ♄
22 12	☽ ⊻ ♆
22 30	☽ # ♇
23 19	☽ ∥ ☿
23 20	☉ □ ♂
23 23	☽ # ♅
22 Tuesday	
00 41	☽ ⊻ ♀
01 39	☽ bQ ♃
05 50	☽ bQ ♅
07 47	☽ ∠ ☿
08 29	☽ ⊥ ♆
12 19	☽ ∠ ♄
14 20	☽ ⊻ ♂
14 58	☽ ⚹ ☉
21 00	☽ ± ♅
23 39	☽ ∠ ♆
23 Wednesday	
00 39	☽ ⊥ ☿
00 43	☽ ⊥ ♂
01 41	☿ Q ♂
03 15	☽ △ ♃
06 57	☽ ⊼ ♅
11 40	☽ ⊻ ☿
13 19	☽ ⚹ ♄
16 03	☽ ∠ ♂
16 29	♆ St R
17 44	☽ ∠ ☉
22 44	☽ △ ♇
24 Thursday	
00 18	☽ ⚹ ♆
00 22	☿ ± ♇
04 09	☿ □ ♄
07 04	☽ ☌ ♀
08 48	☽ Q ♄
09 23	☽ ⊥ ☉
17 02	☽ ⚹ ♂
17 54	♄ St R
19 35	☽ Q ♆
19 41	☽ ⊻ ☉
22 03	☿ # ♇
22 47	☽ bQ ♇
25 Friday	
04 20	☽ □ ♃
04 20	♀ Q ♄
07 13	☽ ☍ ♅
12 40	☽ Q ♂
13 02	☽ ± ♇
13 27	☽ □ ♄
17 00	☽ ☌ ☿
22 29	☽ ⊼ ♇
26 Saturday	
00 04	☽ □ ♆
00 42	♂ Q ♇
09 02	☽ Q ♃
09 46	☽ # ♅
10 24	☽ # ♇
10 47	☽ ⊻ ♀
11 11	☽ ∥ ☿
16 48	☽ ∥ ♀
17 51	☽ □ ♂
21 03	☽ ⊥ ♀
22 26	☽ ☌ ☉
27 Sunday	
04 08	☽ ⚹ ♃
06 23	☽ ⊼ ♅
08 34	☽ ∥ ☉
12 30	☽ ∠ ♀
12 44	☽ △ ♄
15 52	☽ ± ♅
21 31	☽ ⊻ ☿
21 46	☽ ☍ ♇
23 27	☽ △ ♆
28 Monday	
00 05	☿ ⊼ ♇
04 15	☽ ∠ ♃
06 12	☽ bQ ♅
08 11	☽ ⊥ ☿
12 42	☽ bQ ♄
14 37	☽ ⚹ ♀
14 41	☽ ∥ ♂
16 45	☿ □ ♆
18 56	☽ ⊥ ♃
19 05	☽ △ ♂
23 38	☽ bQ ♆
29 Tuesday	
00 25	☽ ∠ ☿
01 56	☽ ⊻ ☉
03 19	☽ ± ♄
04 50	☽ ⊻ ♃
05 26	☽ ∥ ♆
06 29	☽ △ ♅
12 00	☽ Q ♀
12 32	☽ ⊥ ☉
13 12	☽ ⊼ ♄
14 26	☽ ± ♆
17 21	☽ ∥ ♄
19 48	☽ # ♃
20 33	☽ bQ ♂
22 38	☽ ⊼ ♇
30 Wednesday	
00 27	☽ ⊼ ♆
04 16	☽ ⚹ ☿
04 45	☽ ∠ ☉
08 44	☽ ± ♇
12 13	☽ ± ♂
16 34	☽ ∥ ♃
16 49	♀ Q ♆
18 51	☽ # ♄
21 18	☽ □ ♀
22 52	☽ ⊼ ♂
23 32	☉ ⊥ ♀
31 Thursday	
00 10	☽ bQ ♇
03 24	☽ Q ☿
03 58	☉ ⚹ ♃
07 24	☽ # ♆
08 20	☽ ☌ ♃
08 37	☽ ⚹ ☉
09 18	☽ □ ♅
16 35	☽ ☍ ♄
17 31	☉ ⊼ ♅
23 45	♀ ⚹ ♂

All ephemeris data is given at 12.00 UT and the Moon's longitude is additionally given for 24.00 UT

FEBRUARY 1952

LONGITUDES

Date	Sidereal time h m s	Sun ☉ ° ' "	Moon ☽ ° ' "	Moon ☽ 24.00 ° ' "	Mercury ☿ ° '	Venus ♀ ° '	Mars ♂ ° '	Jupiter ♃ ° '	Saturn ♄ ° '	Uranus ♅ ° '	Neptune ♆ ° '	Pluto ♇ ° '
01	20 42 35	11 ♒ 34 52	25 ♈ 43 13	02 ♉ 15 03	27 ♑ 33	05 ♑ 53	05 ♏ 29	10 ♈ 28	14 ♎ 55	10 ♋ 46	21 ♎ 41	20 ♌ 33
02	20 46 32	12 35 46	08 ♉ 41 03	15 ♉ 01 42	29 ♑ 06	07 07	05 54	10 38	14 R 55	10 R 44	21 R 41	20 R 32
03	20 50 28	13 36 38	21 ♉ 17 30	27 ♉ 28 57	00 ♒ 41	08 20	06 18	10 49	14 54	10 42	21 41	20 30
04	20 54 25	14 37 29	03 ♊ 36 36	09 ♊ 40 58	02 16	09 33	06 43	11 00	14 52	10 40	21 40	20 29
05	20 58 21	15 38 19	15 ♊ 42 33	21 ♊ 41 53	03 52	10 46	07 07	11 11	14 51	10 38	21 40	20 27
06	21 02 18	16 39 07	27 ♊ 39 24	03 ♋ 35 34	05 29	12 00	07 31	11 22	14 50	10 36	21 39	20 26
07	21 06 14	17 39 54	09 ♋ 30 46	15 ♋ 25 26	07 06	13 13	07 54	11 34	14 49	10 34	21 39	20 24
08	21 10 11	18 40 40	21 ♋ 19 52	27 ♋ 14 26	08 45	14 26	08 17	11 45	14 47	10 32	21 39	20 23
09	21 14 08	19 41 24	03 ♌ 09 26	09 ♌ 05 07	10 24	15 40	08 40	11 56	14 45	10 30	21 38	20 21
10	21 18 04	20 42 06	15 ♌ 01 46	20 ♌ 59 37	12 04	16 53	09 03	12 08	14 44	10 29	21 37	20 20
11	21 22 01	21 42 48	26 ♌ 58 54	02 ♍ 59 52	13 44	18 07	09 25	12 20	14 42	10 27	21 37	20 18
12	21 25 57	22 43 28	09 ♍ 02 43	15 ♍ 07 42	15 26	19 20	09 47	12 31	14 40	10 25	21 36	20 17
13	21 29 54	23 44 06	21 ♍ 15 03	27 ♍ 25 00	17 08	20 34	10 09	12 43	14 38	10 23	21 35	20 16
14	21 33 50	24 44 43	03 ♎ 37 49	09 ♎ 53 47	18 51	21 47	10 30	12 55	14 36	10 22	21 35	20 14
15	21 37 47	25 45 20	16 ♎ 13 09	22 ♎ 36 13	20 35	23 01	10 51	13 07	14 34	10 20	21 34	20 13
16	21 41 43	26 45 54	29 ♎ 03 17	05 ♏ 34 37	22 20	24 14	11 12	13 19	14 31	10 19	21 33	20 11
17	21 45 40	27 46 28	12 ♏ 10 29	18 ♏ 51 09	24 06	25 28	11 32	13 31	14 29	10 17	21 32	20 10
18	21 49 37	28 47 01	25 ♏ 36 48	02 ♐ 27 35	25 53	26 41	11 52	13 44	14 27	10 16	21 32	20 08
19	21 53 33	29 ♒ 47 32	09 ♐ 23 37	16 ♐ 24 52	27 40	27 55	12 12	13 56	14 24	10 14	21 31	20 07
20	21 57 30	00 ♓ 48 02	23 ♐ 31 13	00 ♑ 42 28	29 ♒ 29	29 ♑ 09	12 31	14 08	14 21	10 13	21 30	20 05
21	22 01 26	01 48 31	07 ♑ 58 15	15 ♑ 18 06	01 ♓ 18	00 ♒ 22	12 50	14 21	14 19	10 11	21 29	20 04
22	22 05 23	02 48 58	22 ♑ 41 21	00 ♒ 07 17	03 08	01 36	13 08	14 33	14 16	10 10	21 28	20 03
23	22 09 19	03 49 24	07 ♒ 35 02	15 ♒ 03 39	04 59	02 50	13 26	14 46	14 13	10 09	21 27	20 01
24	22 13 16	04 49 49	22 ♒ 32 07	29 ♒ 59 24	06 51	04 03	13 44	14 59	14 10	10 08	21 26	20 00
25	22 17 12	05 50 12	07 ♓ 24 29	14 ♓ 46 24	08 44	05 17	14 01	15 11	14 07	10 07	21 25	19 58
26	22 21 09	06 50 33	22 ♓ 04 16	29 ♓ 17 18	10 37	06 31	14 18	15 24	14 04	10 05	21 24	19 57
27	22 25 06	07 50 52	06 ♈ 24 53	13 ♈ 26 30	12 31	07 45	14 34	15 37	14 00	10 04	21 23	19 56
28	22 29 02	08 51 09	20 ♈ 21 50	27 ♈ 10 42	14 25	08 58	14 50	15 50	13 57	10 03	21 22	19 54
29	22 32 59	09 ♓ 51 24	03 ♉ 53 03	10 ♉ 29 00	16 ♓ 20	10 ♒ 12	15 ♏ 05	16 ♈ 03	13 ♎ 54	10 ♋ 02	21 ♎ 21	19 ♌ 53

Moon

Date	Moon True ☊ ° '	Moon Mean ☊ ° '	Moon ☽ Latitude ° '
01	00 ♓ 43	01 ♓ 47	04 N 19
02	00 D 44	01 44	04 53
03	00 R 44	01 41	05 12
04	00 43	01 38	05 16
05	00 41	01 35	05 05
06	00 39	01 31	04 42
07	00 37	01 28	04 06
08	00 35	01 25	03 21
09	00 33	01 22	02 26
10	00 32	01 19	01 25
11	00 32	01 16	00 N 20
12	00 D 32	01 12	00 S 47
13	00 32	01 09	01 53
14	00 33	01 06	02 54
15	00 34	01 03	03 47
16	00 34	01 00	04 31
17	00 34	00 56	05 02
18	00 34	00 53	05 17
19	00 R 34	00 50	05 14
20	00 D 34	00 47	04 53
21	00 34	00 44	04 13
22	00 35	00 41	03 15
23	00 35	00 37	02 04
24	00 35	00 34	00 S 45
25	00 R 35	00 31	00 N 38
26	00 35	00 28	01 57
27	00 34	00 25	03 06
28	00 33	00 22	04 03
29	00 ♓ 32	00 ♓ 18	04 N 44

DECLINATIONS

Date	Sun ☉ ° '	Moon ☽ ° '	Mercury ☿ ° '	Venus ♀ ° '	Mars ♂ ° '	Jupiter ♃ ° '	Saturn ♄ ° '	Uranus ♅ ° '	Neptune ♆ ° '	Pluto ♇ ° '
01	17 S 19	13 N 58	22 S 08	22 S 14	11 S 33	03 N 03	03 S 32	23 N 24	06 S 54	23 N 20
02	17 02	19 01	21 54	22 14	11 41	03 08	03 31	23 24	06 54	23 20
03	16 45	23 06	21 38	22 13	11 49	03 12	03 31	23 24	06 54	23 21
04	16 27	26 03	21 21	22 11	11 57	03 17	03 30	23 24	06 54	23 21
05	16 09	27 44	21 02	22 08	12 04	03 21	03 29	23 24	06 54	23 22
06	15 51	28 07	20 42	22 05	12 12	03 26	03 29	23 25	06 54	23 22
07	15 33	27 12	20 20	22 02	12 19	03 30	03 28	23 25	06 53	23 23
08	15 14	25 03	19 57	21 57	12 27	03 35	03 27	23 25	06 53	23 23
09	14 55	21 50	19 33	21 52	12 34	03 39	03 26	23 25	06 53	23 24
10	14 36	17 41	19 07	21 47	12 41	03 44	03 25	23 25	06 52	23 24
11	14 16	12 50	18 40	21 40	12 48	03 49	03 24	23 25	06 52	23 25
12	13 57	07 27	18 12	21 33	12 55	03 54	03 23	23 25	06 52	23 26
13	13 37	01 N 45	17 42	21 26	13 01	03 58	03 22	23 26	06 51	23 26
14	13 17	04 S 06	17 10	21 17	13 08	04 03	03 21	23 26	06 51	23 27
15	12 56	09 53	16 37	21 09	13 14	04 08	03 20	23 26	06 51	23 27
16	12 36	15 22	16 03	20 59	13 21	04 13	03 19	23 26	06 50	23 28
17	12 15	20 16	15 27	20 49	13 27	04 18	03 18	23 26	06 50	23 28
18	11 54	24 17	14 50	20 38	13 33	04 23	03 17	23 26	06 50	23 29
19	11 33	27 02	14 12	20 27	13 39	04 27	03 16	23 26	06 49	23 30
20	11 12	28 10	13 32	20 15	13 45	04 32	03 14	23 26	06 49	23 30
21	10 50	27 25	12 52	20 03	13 50	04 37	03 13	23 26	06 49	23 30
22	10 28	24 45	12 08	19 49	13 56	04 42	03 12	23 26	06 48	23 31
23	10 07	20 23	11 24	19 36	14 01	04 47	03 10	23 27	06 48	23 31
24	09 45	14 43	10 38	19 22	14 06	04 52	03 09	23 27	06 47	23 32
25	09 22	08 12	09 52	19 07	14 11	04 57	03 08	23 27	06 47	23 32
26	09 00	01 S 22	09 04	18 51	14 16	05 01	03 06	23 27	06 47	23 33
27	08 38	05 N 24	08 15	18 35	14 21	05 05	03 05	23 27	06 46	23 33
28	08 15	11 42	07 25	18 19	14 26	05 13	03 03	23 27	06 46	23 34
29	07 S 53	17 N 15	06 S 33	18 S 02	14 S 30	05 N 18	03 S 02	23 N 27	06 S 45	23 N 34

ZODIAC SIGN ENTRIES

Date	h	m	Planets
01	19	51	☽ ♉
03	01	38	☿ ♒
04	04	55	☽ ♊
06	16	44	☽ ♋
09	05	36	☽ ♌
11	18	02	☽ ♍
14	05	00	☽ ♎
16	13	45	☽ ♏
18	19	42	☽ ♐
19	16	57	☉ ♓
20	18	55	☿ ♓
20	22	49	☽ ♑
21	04	48	♀ ♒
22	23	48	☽ ♒
25	00	01	☽ ♓
27	01	11	☽ ♈
29	05	02	☽ ♉

LATITUDES

Date	Mercury ☿ ° '	Venus ♀ ° '	Mars ♂ ° '	Jupiter ♃ ° '	Saturn ♄ ° '	Uranus ♅ ° '	Neptune ♆ ° '	Pluto ♇ ° '
01	01 S 31	01 N 05	01 N 55	01 S 11	02 N 33	00 N 23	01 N 40	09 N 11
04	01 44	00 55	01 55	01 10	02 34	00 23	01 41	09 11
07	01 54	00 46	01 56	01 10	02 34	00 23	01 41	09 11
10	02 01	00 37	01 56	01 09	02 35	00 23	01 41	09 12
13	02 05	00 27	01 57	01 09	02 36	00 23	01 41	09 12
16	02 06	00 18	01 57	01 08	02 37	00 23	01 41	09 12
19	02 02	00 N 08	01 57	01 08	02 37	00 23	01 41	09 12
22	01 54	00 S 01	01 57	01 07	02 38	00 23	01 41	09 12
25	01 41	00 10	01 57	01 07	02 39	00 23	01 42	09 12
28	01 23	00 18	01 57	01 06	02 39	00 23	01 42	09 12
31	00 S 59	00 S 27	01 N 56	01 S 06	02 N 40	00 N 23	01 N 42	09 N 12

DATA

Julian Date	2434044
Delta T	+30 seconds
Ayanamsa	23° 11' 25"
Synetic vernal point	05° ♓ 55' 35"
True obliquity of ecliptic	23° 26' 52"

MOON'S PHASES, APSIDES AND POSITIONS ☽

Date	h	m	Phase	Longitude ° '	Eclipse Indicator
02	20	01	☽	12 ♉ 56	
11	00	28	○	21 ♌ 14	partial
18	18	01	☾	29 ♏ 02	
25	09	16	●	05 ♓ 43	Total

Day	h	m		
08	08	16	Apogee	
23	22	24	Perigee	
06	06	58	Max dec	28° N 09'
13	19	11	0S	
20	14	41	Max dec	28° S 10'
26	16	46	0N	

LONGITUDES

Date	Chiron ⚷ ° '	Ceres ⚳ ° '	Pallas ⚴ ° '	Juno ⚵ ° '	Vesta ⚶ ° '	Black Moon Lilith ⚸ ° '
01	08 ♑ 27	24 ♓ 51	25 ♒ 01	17 ♑ 55	13 ♍ 18	23 ♋ 48
11	09 ♑ 20	28 ♓ 29	28 ♒ 20	21 ♑ 36	11 ♍ 30	24 ♋ 54
21	10 ♑ 08	01 ♈ 13	01 ♓ 39	25 ♑ 09	09 ♍ 08	26 ♋ 01
31	10 ♑ 50	06 ♈ 03	04 ♓ 59	28 ♑ 47	06 ♍ 31	27 ♋ 08

All ephemeris data is given at 12.00 UT and the Moon's longitude is additionally given for 24.00 UT

ASPECTARIAN

h m Aspects

01 Friday
01 30 ☿ ∥ ♀
01 31 ☽ # ♂
02 39 ☽ △ ♇
04 41 ☽ ☍ ♆
05 25 ♀ ⚼ ♇
07 47 ☽ Q ☉
15 47 ☽ □ ☿
17 34 ☽ Q ♅

02 Saturday
02 35 ☽ # ☉
03 56 ☿ Q ♃
06 36 ☽ ☍ ♂
08 44 ☽ △ ♀
15 44 ☽ ⊻ ♃
15 51 ☽ ✱ ♅
20 01 ☽ □ ☉
22 47 ♃ □ ♅
23 45 ☽ ⊼ ♄

03 Sunday
03 13 ☽ # ☿
03 18 ☽ ⊥ ♃
06 14 ☽ # ♀
10 29 ☽ □ ♇
11 14 ☽ ± ♄
12 45 ☽ ⊼ ♆
13 42 ☽ ∥ ♇
14 05 ☽ ∥ ♅
16 22 ☽ ⚼ ♀
20 31 ☽ ∠ ♅
20 54 ☽ ∠ ♃

04 Monday
00 23 ☽ ± ♆
04 41 ☽ ⚼ ♄
08 58 ☽ △ ☿
11 52 ☽ ± ♀
14 05 ☽ ⊥ ♅
17 48 ☉ △ ♄
18 02 ☽ ⚼ ♆
18 20 ☽ ⊼ ♂
21 35 ☽ Q ♇

05 Tuesday
01 04 ☽ ⊼ ♀
01 55 ☽ ⊻ ♅
02 51 ☽ ✱ ♃
06 38 ☽ ± ♂
09 21 ♀ ☍ ♅
10 18 ☽ △ ♄
11 51 ☽ △ ☉
19 18 ☽ ⚼ ☿
21 29 ☽ ✱ ♇
21 35 ♀ □ ♃
23 56 ☽ △ ♆

06 Wednesday
01 16 ☽ ⚼ ♂
03 14 ☽ Q ♃
10 52 ☉ ± ♅
16 16 ☽ ± ☿
20 50 ☽ ⚼ ☉

07 Thursday
00 53 ♃ # ♄
03 41 ☽ ∠ ♇
06 20 ☽ ⊼ ☿
08 37 ☽ △ ♂
14 08 ☽ ☌ ♅
16 13 ☽ □ ♃
16 47 ☽ ± ☉
20 23 ☽ ☍ ♀
21 55 ☽ ⊥ ♇
22 44 ☽ □ ♄

08 Friday
03 16 ☿ □ ♂
06 06 ☽ ⊼ ☉
10 04 ☽ ⊻ ♇
10 51 ♀ ± ♇
12 38 ☽ □ ♆
18 36 ♀ □ ♄

09 Saturday
01 07 ☽ ∥ ♅
01 14 ☽ ∥ ♇
11 11 ☽ Q ♄
11 42 ☽ # ♀
13 36 ☿ ⊼ ♅
23 32 ☽ □ ♂

10 Sunday
01 06 ☽ Q ♆
02 50 ☽ ⊻ ♅
03 18 ☽ # ☿
03 26 ☉ ☍ ♇
05 02 ☽ ☍ ☿
06 04 ☽ △ ♃
11 24 ☽ ✱ ♄
13 12 ☿ ✱ ♃
14 55 ☽ ⊥ ♅
16 10 ☽ ⊼ ♀
22 39 ☽ ☌ ♇

11 Monday
00 28 ☽ ☍ ☉
01 15 ☽ ✱ ♆
04 42 ☽ # ☉
05 36 ☽ ± ♀
08 49 ♅ ∥ ♇
08 56 ☽ ∠ ♅
09 37 ☉ △ ♆
12 10 ☽ # ♂
12 42 ☽ ⚼ ♃
12 54 ☽ Q ♂
17 25 ☽ ∠ ♄

12 Tuesday
01 26 ☿ △ ♄
01 36 ☽ ⚼ ♀
06 55 ☽ ± ♃
07 10 ☽ ∠ ♆
11 15 ☽ ⊥ ♄
13 31 ☽ ✱ ♂
14 32 ☽ # ♆
14 43 ☽ ✱ ♅
18 59 ☽ ⊼ ♃
23 04 ☽ ⊻ ♄

13 Wednesday
00 55 ☽ ⊥ ♆
01 45 ☿ ± ♅
02 38 ☽ ⊼ ☿
02 53 ☽ ∥ ♃
05 14 ☽ # ♄
06 12 ♀ ⊼ ♇
10 04 ☽ ⊻ ♇
10 30 ☽ △ ♀
12 40 ☽ ⊻ ♆
14 13 ☽ Q ♅
16 16 ☽ ± ☿
17 17 ☽ ⊼ ☉
19 49 ☽ ∠ ♂
21 44 ☽ ⊥ ♇

14 Thursday
03 13 ♂ △ ♅
05 57 ☽ ± ☉
07 58 ♀ □ ♆
08 57 ☽ ∥ ♄
11 48 ☽ # ♃
12 30 ☽ ⚼ ☿
13 44 ☽ ⊥ ♂
15 05 ☽ ∠ ♇
19 57 ☉ ∥ ♂
23 20 ☽ ∥ ♆

15 Friday
00 44 ☽ ⚼ ☉
00 52 ☽ □ ♅
01 32 ☽ ⊻ ♂
02 18 ☉ ⚼ ♅
06 02 ☽ ☍ ♃
06 54 ☿ ☍ ♇
07 40 ♀ Q ♂
08 52 ☽ ☌ ♄
19 30 ☽ ✱ ♇
21 31 ☽ △ ☿
22 03 ☽ ☌ ♆

16 Saturday
00 24 ☽ ∥ ☉
01 23 ☿ △ ♆
02 07 ☽ □ ♀
02 46 ☽ ∥ ♂
07 24 ☽ △ ☉
14 51 ☽ ∥ ☿
17 46 ☽ Q ♇

17 Sunday
08 35 ☽ △ ♅
10 49 ☽ ☌ ♂
14 28 ☽ ⊼ ♃
14 34 ☽ Q ♀
14 49 ☽ ∥ ♀
16 09 ☽ ⊻ ♄

18 Monday
01 24 ☽ ± ♃
02 19 ☽ □ ♇
02 52 ☽ ⊥ ♄
03 49 ☿ ⚼ ♅
04 46 ☽ ⊻ ♆
06 18 ☽ # ♅
06 36 ☽ # ♇
10 16 ☉ ∠ ♃
11 23 ☽ ⚼ ♅
12 32 ☽ □ ☿
14 05 ☽ ✱ ♀
15 22 ☽ ⊥ ♆
17 33 ☽ ⚼ ♃
18 01 ☽ □ ☉
18 43 ☽ ∠ ♄

19 Tuesday
03 03 ☉ ⚼ ♄
03 06 ☽ ± ♅
07 02 ☽ ∠ ♆
13 27 ☽ ⊼ ♅
16 55 ☽ ⊻ ♂
18 37 ☽ ∠ ♀
19 53 ☽ △ ♃
20 33 ☽ ✱ ♄
22 22 ☿ ⊻ ♀

20 Wednesday
00 18 ☽ Q ☿
03 22 ☽ ⊥ ♂
03 26 ☽ Q ☉
05 17 ☿ ∥ ♂
06 14 ☽ △ ♇
06 56 ☿ ∠ ♃
08 36 ☽ ✱ ♆
10 26 ☿ ⚼ ♄
11 19 ☽ ⊥ ♀
16 44 ☽ Q ♄
18 50 ☽ ∠ ♂
22 17 ☽ ⊻ ♀
23 23 ☽ ✱ ☿

21 Thursday
01 04 ☽ ✱ ☉
04 36 ☽ Q ♆
07 13 ☽ ⚼ ♇
08 36 ♃ ☍ ♄
15 38 ☽ ☍ ♅
20 08 ☽ ✱ ♂
21 58 ☽ ± ♇
22 21 ☽ □ ♄
22 35 ☽ □ ♃

22 Friday
02 47 ☉ ☌ ☿
03 31 ☽ ∠ ☉
03 34 ☽ ∠ ☿
07 43 ☽ ⊼ ♇
10 01 ☽ □ ♆
16 02 ☽ Q ♂
19 09 ☽ ⊥ ☉
19 42 ☽ # ♇
20 09 ☽ # ♅
20 12 ☽ ⊥ ☿

23 Saturday
03 40 ☽ ☌ ♀
04 09 ☽ Q ♃
05 31 ☽ ⊻ ☉
07 14 ☽ ⊻ ☿
10 30 ♀ Q ♃
15 49 ☽ ∥ ♀
16 07 ☽ ⊼ ♅
21 35 ☽ □ ♂
22 36 ☽ △ ♄
23 42 ☽ ✱ ♃

24 Sunday
01 44 ☽ ± ♅
06 42 ☿ ⚼ ♆
07 56 ☽ ☍ ♇
10 14 ☽ △ ♆
14 18 ☽ ∥ ♂
16 10 ☽ ⚼ ♅
22 38 ☽ ⚼ ♄

25 Monday
00 09 ☽ ∠ ♃
04 21 ☿ ± ♄
05 18 ☽ ∥ ☿
07 35 ☽ ∥ ☉
08 15 ☽ ⊻ ♀
09 16 ☽ ☌ ☉
10 24 ☽ ⚼ ♆
13 08 ☽ ± ♄
14 27 ☽ ☌ ☿
14 56 ☽ ⊥ ♃
16 23 ☽ △ ♅
17 03 ☽ ∥ ♆
18 39 ☿ ⊥ ♃
18 53 ☽ ⊥ ♀
18 54 ♂ ⊻ ♄
22 53 ☽ ⊼ ♄
22 58 ☽ △ ♂
23 18 ☽ # ♃

26 Tuesday
00 52 ☽ ⊻ ♃
01 02 ☽ ± ♆
01 38 ☉ ⚼ ♆
05 23 ☿ △ ♅
05 54 ☽ ∥ ♄
08 30 ☽ ⊼ ♇
10 54 ☽ ⊼ ♆
11 00 ☽ ∠ ♀
15 12 ☉ ∥ ☿
18 25 ☽ ± ♇

27 Wednesday
00 15 ☽ ⚼ ♂
03 41 ☽ # ♄
09 29 ☽ ⚼ ♇
11 01 ☽ ∥ ♃
14 28 ☽ ✱ ♀
14 37 ☽ ⊻ ☉
15 34 ☉ ± ♄
15 44 ☽ ± ♂
17 02 ☽ # ♆
18 13 ☽ □ ♅
21 20 ☽ # ☿
23 06 ☉ ⊻ ♀
23 21 ☽ # ☉

28 Thursday
00 03 ☽ ⊻ ☿
00 55 ☽ ☍ ♄
01 41 ☽ ⊥ ☉
02 12 ☽ ⊼ ♂
04 00 ☽ ☌ ♃
06 15 ☿ ⊼ ♄
11 12 ☽ △ ♇
12 07 ☽ ⊥ ☿
13 10 ☽ Q ♀
13 45 ☽ ☍ ♆
17 57 ☿ △ ♂
18 36 ☽ ∠ ☉
23 29 ☽ # ♂
23 45 ☿ ± ♆

29 Friday
01 33 ☽ Q ♅
06 25 ☿ ∥ ♆
06 39 ☽ ∠ ☿
07 26 ☿ ⊥ ♀
08 03 ☿ ⊻ ♃
08 52 ♀ ⊼ ♅
15 30 ☽ # ♀
16 18 ☉ △ ♅
18 01 ☉ ⊥ ♃
23 10 ☽ ✱ ♅
23 45 ☽ ✱ ☉

MARCH 1952

LONGITUDES

Date	Sidereal time h m s	Sun ☉ ° ' "	Moon ☽ ° ' "	Moon ☽ 24.00 ° ' "	Mercury ☿ ° '	Venus ♀ ° '	Mars ♂ ° '	Jupiter ♃ ° '	Saturn ♄ ° '	Uranus ♅ ° '	Neptune ♆ ° '	Pluto ♇ ° '
01	22 36 55	10 ♓ 51 38	16 ♉ 58 47	23 ♉ 22 42	18 ♓ 15	11 ♒ 26	15 ♏ 20	16 ♈ 16	13 ♎ 50	10 ♋ 01	21 ♎ 20	19 ♌ 52
02	22 40 52	11 51 49	29 ♉ 41 09	05 ♊ 54 38	20 10	12 40	15 34	16 29	13 R 47	10 R 01	21 R 18	19 R 50
03	22 44 48	12 51 59	12 ♊ 03 40	18 ♊ 08 48	22 05	13 53	15 48	16 43	13 43	10 00	21 17	19 49
04	22 48 45	13 52 06	24 ♊ 10 38	00 ♋ 09 46	23 59	15 07	16 02	16 56	13 39	09 59	21 16	19 48
05	22 52 41	14 52 11	06 ♋ 06 47	12 ♋ 02 16	25 53	16 21	16 15	17 09	13 36	09 58	21 15	19 46
06	22 56 38	15 52 14	17 ♋ 56 48	23 ♋ 50 56	27 45	17 35	16 27	17 23	13 32	09 58	21 14	19 45
07	23 00 35	16 52 15	29 ♋ 45 11	05 ♌ 40 03	29 ♓ 36	18 49	16 39	17 36	13 28	09 57	21 12	19 44
08	23 04 31	17 52 14	11 ♌ 35 58	17 ♌ 33 23	01 ♈ 26	20 02	16 50	17 50	13 24	09 57	21 11	19 42
09	23 08 28	18 52 11	23 ♌ 32 39	29 ♌ 34 07	03 13	21 16	17 01	18 03	13 20	09 56	21 10	19 41
10	23 12 24	19 52 05	05 ♍ 38 04	11 ♍ 44 44	04 57	22 30	17 11	18 17	13 16	09 56	21 08	19 40
11	23 16 21	20 51 58	17 ♍ 54 21	24 ♍ 07 03	06 38	23 44	17 21	18 30	13 12	09 55	21 07	19 39
12	23 20 17	21 51 49	00 ♎ 22 58	06 ♎ 42 11	08 15	24 58	17 30	18 44	13 08	09 55	21 06	19 37
13	23 24 14	22 51 37	13 ♎ 04 46	19 ♎ 30 45	09 48	26 11	17 38	18 58	13 03	09 55	21 04	19 36
14	23 28 10	23 51 24	26 ♎ 00 08	02 ♏ 32 55	11 17	27 25	17 46	19 12	12 59	09 54	21 03	19 35
15	23 32 07	24 51 09	09 ♏ 09 04	15 ♏ 48 33	12 40	28 39	17 53	19 25	12 55	09 54	21 01	19 34
16	23 36 04	25 50 53	22 ♏ 31 22	29 ♏ 17 27	13 57	29 ♒ 53	18 00	19 39	12 51	09 54	21 00	19 33
17	23 40 00	26 50 35	06 ♐ 06 45	12 ♐ 59 14	15 09	01 ♓ 07	18 06	19 53	12 46	09 54	20 59	19 32
18	23 43 57	27 50 15	19 ♐ 54 50	26 ♐ 53 28	16 13	02 21	18 11	20 07	12 42	09 D 54	20 57	19 31
19	23 47 53	28 49 53	03 ♑ 55 02	10 ♑ 59 24	17 11	03 34	18 16	20 21	12 37	09 54	20 56	19 30
20	23 51 50	29 ♓ 49 30	18 ♑ 06 23	25 ♑ 15 45	18 02	04 48	18 20	20 35	12 33	09 54	20 54	19 29
21	23 55 46	00 ♈ 49 05	02 ♒ 27 12	09 ♒ 40 22	18 45	06 02	18 23	20 49	12 28	09 54	20 53	19 27
22	23 59 43	01 48 38	16 ♒ 54 49	24 ♒ 10 02	19 21	07 16	18 25	21 03	12 24	09 55	20 51	19 26
23	00 03 39	02 48 09	01 ♓ 25 25	08 ♓ 40 22	19 48	08 30	18 27	21 17	12 19	09 55	20 49	19 25
24	00 07 36	03 47 39	15 ♓ 54 10	23 ♓ 06 07	20 08	09 44	18 28	21 31	12 15	09 55	20 48	19 24
25	00 11 33	04 47 06	00 ♈ 15 30	07 ♈ 21 39	20 21	10 58	18 R 29	21 46	12 10	09 55	20 46	19 24
26	00 15 29	05 46 31	14 ♈ 23 55	21 ♈ 21 44	20 25	12 12	18 28	22 00	12 05	09 56	20 45	19 23
27	00 19 26	06 45 55	28 ♈ 14 38	05 ♉ 02 15	20 R 22	13 25	18 27	22 14	12 01	09 56	20 43	19 22
28	00 23 22	07 45 16	11 ♉ 44 20	18 ♉ 20 47	20 11	14 39	18 25	22 28	11 56	09 57	20 42	19 21
29	00 27 19	08 44 35	24 ♉ 51 34	01 ♊ 16 50	19 54	15 53	18 22	22 43	11 51	09 57	20 40	19 20
30	00 31 15	09 43 52	07 ♊ 36 47	13 ♊ 51 46	19 30	17 07	18 19	22 57	11 47	09 58	20 38	19 19
31	00 35 12	10 ♈ 43 06	20 ♊ 02 10	26 ♊ 08 29	19 ♈ 01	18 ♓ 21	18 ♏ 15	23 ♈ 11	11 ♎ 42	09 ♋ 59	20 ♎ 37	19 ♌ 18

Moon Nodes and Latitude

Date	Moon True ☊ ° '	Moon Mean ☊ ° '	Moon ☽ Latitude ° '
01	00 ♓ 31	00 ♓ 15	05 N 08
02	00 R 30	00 12	05 17
03	00 30	00 09	05 11
04	00 D 30	00 06	04 50
05	00 31	00 ♓ 02	04 18
06	00 32	29 ♒ 59	03 35
07	00 34	29 56	02 43
08	00 35	29 53	01 43
09	00 36	29 50	00 N 39
10	00 R 36	29 47	00 S 28
11	00 35	29 43	01 34
12	00 33	29 40	02 37
13	00 30	29 37	03 33
14	00 26	29 34	04 19
15	00 23	29 31	04 53
16	00 19	29 28	05 11
17	00 17	29 24	05 12
18	00 15	29 21	04 56
19	00 D 15	29 18	04 22
20	00 16	29 15	03 31
21	00 17	29 12	02 27
22	00 19	29 08	01 S 13
23	00 R 19	29 05	00 N 06
24	00 18	29 02	01 24
25	00 16	28 59	02 36
26	00 11	28 56	03 37
27	00 06	28 53	04 24
28	00 00	28 49	04 55
29	29 ♒ 53	28 46	05 09
30	29 48	28 43	05 08
31	29 ♒ 44	28 ♒ 40	04 N 51

DECLINATIONS

Date	Sun ☉	Moon ☽	Mercury ☿	Venus ♀	Mars ♂	Jupiter ♃	Saturn ♄	Uranus ♅	Neptune ♆	Pluto ♇
01	07 S 30	21 N 50	05 S 41	17 S 45	14 S 35	05 N 23	03 S 00	23 N 27	06 S 45	23 N 35
02	07 07	25 15	04 48	17 27	14 39	05 28	02 59	23 27	06 44	23 35
03	06 44	27 22	03 55	17 08	14 43	05 33	02 57	23 27	06 44	23 35
04	06 21	28 09	03 01	16 49	14 47	05 39	02 56	23 27	06 43	23 36
05	05 58	27 36	02 06	16 30	14 51	05 44	02 54	23 27	06 43	23 36
06	05 34	25 47	01 12	16 10	14 55	05 49	02 52	23 27	06 42	23 37
07	05 11	22 51	00 S 17	15 49	14 58	05 54	02 51	23 27	06 42	23 37
08	04 48	18 58	00 N 37	15 29	15 02	06 00	02 49	23 27	06 41	23 37
09	04 24	14 17	01 31	15 07	15 05	06 05	02 47	23 27	06 41	23 38
10	04 01	09 01	02 24	14 46	15 08	06 10	02 46	23 27	06 40	23 38
11	03 37	03 N 20	03 15	14 24	15 11	06 15	02 44	23 27	06 40	23 38
12	03 14	02 S 33	04 06	14 01	15 14	06 21	02 42	23 27	06 39	23 39
13	02 50	08 24	04 55	13 38	15 17	06 26	02 40	23 27	06 39	23 39
14	02 26	14 04	05 41	13 15	15 19	06 31	02 39	23 27	06 38	23 39
15	02 03	19 10	06 26	12 51	15 22	06 37	02 37	23 27	06 38	23 40
16	01 39	23 25	07 08	12 27	15 24	06 42	02 35	23 27	06 37	23 40
17	01 15	26 28	07 47	12 03	15 26	06 47	02 33	23 27	06 36	23 40
18	00 52	27 59	08 23	11 38	15 28	06 53	02 31	23 27	06 36	23 41
19	00 28	27 45	08 56	11 14	15 30	06 58	02 30	23 27	06 35	23 41
20	00 S 04	25 42	09 26	10 48	15 31	07 03	02 28	23 27	06 35	23 41
21	00 N 20	22 00	09 52	10 23	15 33	07 09	02 26	23 27	06 34	23 41
22	00 43	16 56	10 14	09 57	15 34	07 14	02 24	23 27	06 33	23 42
23	01 07	10 53	10 32	09 31	15 35	07 19	02 22	23 27	06 33	23 42
24	01 31	04 S 16	10 46	09 04	15 36	07 25	02 20	23 27	06 32	23 42
25	01 54	02 N 29	10 56	08 38	15 37	07 30	02 19	23 27	06 32	23 42
26	02 18	09 01	11 02	08 11	15 38	07 35	02 17	23 27	06 31	23 42
27	02 41	14 58	11 04	07 44	15 38	07 41	02 15	23 27	06 30	23 43
28	03 05	20 02	11 01	07 17	15 39	07 46	02 13	23 27	06 30	23 43
29	03 28	23 59	10 55	06 49	15 39	07 51	02 11	23 27	06 29	23 43
30	03 51	26 39	10 44	06 21	15 39	07 57	02 09	23 27	06 29	23 43
31	04 N 15	27 N 55	10 N 30	05 S 54	15 S 39	08 N 02	02 S 07	23 N 27	06 S 28	23 N 43

ZODIAC SIGN ENTRIES

Date	h	m	Planets
02	12	36	☽ ♊
04	23	40	☽ ♋
07	12	30	☽ ♌
07	17	10	☿ ♈
10	00	51	☽ ♍
12	11	16	☽ ♎
14	19	20	☽ ♏
16	14	18	♀ ♓
17	01	15	☽ ♐
19	05	19	☽ ♑
20	16	14	☉ ♈
21	07	55	☽ ♒
23	09	39	☽ ♓
25	11	34	☽ ♈
27	15	05	☽ ♉
29	21	36	☽ ♊

LATITUDES

Date	Mercury ☿	Venus ♀	Mars ♂	Jupiter ♃	Saturn ♄	Uranus ♅	Neptune ♆	Pluto ♇
01	01 S 08	00 S 24	01 N 57	01 S 06	02 N 40	00 N 23	01 N 42	09 N 12
04	00 41	00 32	01 56	01 06	02 40	00 23	01 42	09 12
07	00 S 09	00 40	01 55	01 05	02 41	00 23	01 42	09 12
10	00 N 28	00 47	01 55	01 05	02 41	00 23	01 42	09 12
13	01 07	00 54	01 54	01 05	02 42	00 23	01 42	09 12
16	01 46	01 00	01 52	01 04	02 42	00 23	01 42	09 11
19	02 22	01 06	01 51	01 04	02 42	00 23	01 42	09 11
22	02 53	01 11	01 49	01 04	02 43	00 23	01 43	09 11
25	03 14	01 16	01 47	01 04	02 43	00 23	01 43	09 11
28	03 23	01 20	01 44	01 03	02 43	00 23	01 43	09 10
31	03 N 17	01 S 24	01 N 42	01 S 03	02 N 43	00 N 23	01 N 43	09 N 10

DATA

Julian Date	2434073
Delta T	+30 seconds
Ayanamsa	23° 11' 29"
Synetic vernal point	05° ♓ 55' 31"
True obliquity of ecliptic	23° 26' 52"

MOON'S PHASES, APSIDES AND POSITIONS ☽

Date	h	m	Phase	Longitude ° '	Eclipse Indicator
03	13	43	☽	12 ♊ 56	
11	18	14	○	21 ♍ 08	
19	02	40	☾	28 ♐ 27	
25	20	13	●	05 ♈ 07	

Day	h	m		
06	22	48	Apogee	
22	22	19	Perigee	
04	13	54	Max dec	28° N 10'
12	01	39	0S	
18	20	57	Max dec	28° S 06'
25	03	08	0N	
31	21	55	Max dec	28° N 02'

LONGITUDES

Date	Chiron ⚷ ° '	Ceres ⚳ ° '	Pallas ⚴ ° '	Juno ⚵ ° '	Vesta ⚶ ° '	Black Moon Lilith ⚸ ° '
01	10 ♑ 46	05 ♈ 40	04 ♓ 39	28 ♑ 26	06 ♍ 47	27 ♋ 01
11	10 ♑ 23	09 ♈ 34	07 ♓ 57	01 ♒ 54	04 ♍ 15	28 ♋ 08
21	11 ♑ 52	13 ♈ 30	11 ♓ 01	05 ♒ 16	02 ♍ 06	29 ♋ 14
31	12 ♑ 13	17 ♈ 29	14 ♓ 26	08 ♒ 29	00 ♍ 37	00 ♌ 21

All ephemeris data is given at 12.00 UT and the Moon's longitude is additionally given for 24.00 UT

ASPECTARIAN

01 Saturday
00 40 ☽ □ ♀
06 12 ☽ ⊼ ♄
08 53 ☽ ☍ ♂
10 40 ☽ ⊻ ♃
14 47 ☽ ⚹ ☿
17 19 ☽ ± ♄
17 22 ☽ □ ♇
19 35 ☿ # ♃
20 07 ☽ ⊼ ♆
22 05 ☽ ⊥ ♃
22 19 ☽ ∥ ♅
23 12 ☽ ∥ ♇
23 58 ☽ Q ☉

02 Sunday
03 06 ☽ ∠ ♅
07 28 ☽ ± ♆
07 56 ☿ ⊼ ♇
10 16 ☽ ⚼ ♄
15 32 ☽ ∠ ♃
17 37 ☽ Q ☿
20 19 ☽ ⊥ ♅

03 Monday
00 45 ☽ ⚼ ♆
02 11 ☿ ⊼ ♆
03 43 ☽ Q ♇
07 58 ☽ ⊻ ♅
08 45 ♀ △ ♄
12 05 ☉ ∥ ♆
13 43 ☽ □ ☉
15 14 ☽ △ ♄
16 00 ☽ △ ♀
19 31 ☽ ⊼ ♂
21 20 ☽ ⚹ ♃

04 Tuesday
03 17 ☽ ⚹ ♇
06 12 ☽ △ ♆
07 11 ☉ ⊼ ♄
07 38 ☽ ± ♂
11 32 ☽ □ ☿
14 24 ☿ ∥ ♄
21 43 ☽ Q ♃

05 Wednesday
01 17 ☽ ⚼ ♀
02 00 ☽ ⚼ ♂
04 43 ♀ ± ♅
09 18 ☽ ∠ ♇
09 31 ♀ □ ♂
10 39 ☿ ± ♇
19 48 ☽ ☌ ♅
20 52 ☉ ± ♆
21 34 ☽ ± ♀
23 41 ☉ # ♃

06 Thursday
03 04 ☽ □ ♄
03 29 ☽ ⊥ ♇
07 13 ♀ ⚹ ♃
07 23 ☽ △ ☉
08 54 ☽ △ ♂
10 49 ☽ □ ♃
11 10 ☽ ⊼ ♀
15 39 ☽ ⊻ ♇
18 39 ☽ □ ♆

07 Friday
05 22 ☉ △ ♂
06 32 ☽ ∥ ♇
07 43 ☽ ∥ ♅
11 39 ☽ △ ☿
15 27 ☽ Q ♄
16 42 ☽ ⚼ ☉

08 Saturday
05 37 ♀ ☍ ♇
07 08 ☽ Q ♆
08 39 ☽ ⊻ ♅
10 41 ☉ ⊻ ♃
12 36 ☽ ± ☉
15 37 ☽ ⚹ ♄
18 05 ☿ ⚼ ♂
20 45 ☽ ⊥ ♅
22 43 ☽ □ ♂
23 27 ☽ ⚼ ☿

09 Sunday
00 47 ☽ △ ♃
01 47 ☽ ⊼ ☉
04 17 ☽ ☌ ♇
06 56 ☽ ☍ ♀
07 15 ☽ ⚹ ♆
07 39 ☽ # ♀
08 11 ☽ # ♂
09 56 ♀ △ ♆
14 04 ♀ ∥ ♂
14 46 ☽ ∠ ♅
20 33 ☽ ± ☿
21 29 ☽ ∠ ♄

10 Monday
07 13 ☉ ⊼ ♇
07 16 ☽ ⚼ ♃
08 04 ☿ ⚼ ♇
10 26 ☽ ⊼ ☿
11 06 ☽ Q ♂
13 00 ☽ ∠ ♆
15 12 ☽ ⊥ ♄
20 26 ☽ ⚹ ♅
21 49 ☿ # ♄
22 06 ☽ # ♆

11 Tuesday
00 01 ☽ ∥ ♃
01 17 ☽ ± ♃
02 53 ☽ ⊻ ♄
06 35 ☽ ⊥ ♆
10 45 ☽ # ☉
10 54 ☽ ⚹ ♂
12 18 ☽ ∥ ☿
13 11 ☽ ⊼ ♃
14 31 ☽ # ♄
15 22 ☽ ⊻ ♇
17 54 ☉ ⊼ ♆
18 12 ☽ ⊻ ♆
18 14 ☽ ☍ ☉
19 04 ☉ # ☿
19 46 ☽ Q ♅

12 Wednesday
00 29 ☽ ⊼ ♀
02 55 ☽ ⊥ ♇
11 08 ♀ ⚼ ♅
12 36 ☽ ∥ ♄
13 13 ☽ ± ♀
14 34 ☽ ∥ ☉
16 04 ☽ ∠ ♂
19 16 ☽ # ☿
20 03 ☽ ∠ ♇

13 Thursday
03 38 ☽ # ♃
04 38 ☽ ∥ ♆
05 01 ☽ ☍ ☿
06 03 ☽ □ ♅
08 05 ☽ ⚼ ♀
09 16 ☽ ⊥ ♂
11 58 ☽ ☌ ♄
13 38 ☿ □ ♅
20 36 ☽ ⊻ ♂
22 38 ☉ ∥ ♄
23 11 ☽ ☍ ♃

14 Friday
00 09 ☽ ⚹ ♇
02 52 ☽ ☌ ♆
07 43 ☽ ⊼ ☉
08 38 ☽ ∥ ♀
14 53 ☽ △ ♀
17 40 ☽ ∥ ♂
19 40 ☽ ± ☉
21 03 ☿ ± ♂
22 14 ☽ Q ♇
22 26 ♀ ⚼ ♄

15 Saturday
13 22 ☽ △ ♅
13 22 ☽ ⚼ ☉
15 45 ♃ # ♆
16 15 ☿ ☍ ♄
18 24 ☿ # ♆
18 46 ☽ ⊻ ♄
18 49 ☿ ∥ ♃
19 03 ☽ ⊼ ☿

16 Sunday
01 38 ♃ △ ♇
03 51 ☽ ☌ ♂
04 53 ☉ ± ♇
05 28 ☽ ⊥ ♄
06 42 ☽ □ ♇
06 48 ☽ ⊼ ♃
06 57 ☽ ± ☿
09 17 ☽ ⊻ ♆
12 17 ☽ # ♅
13 39 ☽ # ♇
16 14 ☽ ⚼ ♅
17 40 ☽ ± ♃
18 23 ☽ △ ☉
19 56 ☽ ⊥ ♆
21 23 ☽ ∠ ♄

17 Monday
00 32 ☽ ⚼ ☿
02 20 ☽ □ ♀
08 07 ☽ ± ♅
09 49 ☽ ⚼ ♃
11 46 ☽ ∠ ♆
18 37 ☽ ⊼ ♅
23 34 ☽ ⚹ ♄

18 Tuesday
04 12 ♅ St D
05 06 ☽ △ ☿
09 00 ☽ ⊻ ♂
11 18 ☽ △ ♇
12 22 ☽ △ ♃
12 49 ☽ Q ♀
13 47 ☽ ⚹ ♆
19 24 ☽ ⊥ ♂
20 11 ☽ Q ♄

19 Wednesday
02 40 ☽ □ ☉
10 19 ☽ Q ♆
10 53 ☽ ∠ ♂
11 22 ☽ ⚹ ♀
12 59 ☽ ⚼ ♇
22 10 ☽ ☍ ♅

20 Thursday
02 41 ☽ □ ♄
04 12 ☽ ± ♇
11 29 ☽ Q ☉
11 52 ☽ □ ☿
12 22 ☽ ⚹ ♂
14 18 ☽ ⊼ ♇
15 07 ☽ ∠ ♀
16 14 ☽ □ ♃
16 41 ☽ □ ♆
22 05 ☿ ⊼ ♂

21 Friday
02 17 ☽ # ♇
03 42 ☽ # ♅
06 46 ♀ ∠ ♃
07 36 ☽ ⊥ ♀
08 32 ☽ Q ♂
08 55 ♀ ⚼ ♆
09 04 ☽ ⚹ ☉
17 12 ♃ ☍ ♆
18 31 ☽ ⊻ ♀
19 29 ☽ Q ☿
20 00 ♀ ± ♄
22 46 ☽ Q ♃

22 Saturday
00 23 ☽ ⊼ ♅
03 13 ☿ # ♀
04 33 ☽ △ ♄
06 48 ♂ ⊥ ♄
10 20 ☽ ± ♅
11 49 ☽ ∠ ☉
14 30 ☽ □ ♂
16 10 ☽ ⚹ ☿
16 11 ☽ ☍ ♇
16 19 ☿ △ ♇
17 39 ☽ ∥ ♂
18 30 ☽ △ ♆
18 58 ☽ ⚹ ♃

23 Sunday
01 14 ☽ ⚼ ♅
03 48 ☽ ⊥ ☉
05 15 ☽ ⚼ ♄
13 14 ☽ # ☿
14 27 ☽ ⊻ ☉
17 27 ☽ ∥ ♀
17 45 ☽ ∠ ☿
19 16 ☽ ⚼ ♆
20 04 ☽ ± ♄
20 11 ☽ ∠ ♃

24 Monday
00 48 ☽ ☌ ♀
00 54 ☽ # ♃
02 04 ☽ △ ♅
03 53 ☽ ∥ ♆
04 02 ☉ ⚼ ♂
05 57 ☽ ⊼ ♄
09 01 ☽ ⊥ ☿
10 10 ☽ ± ♆
11 22 ☽ ⊥ ♃
15 41 ♀ △ ♅
16 16 ☽ △ ♂
17 50 ☽ ⊼ ♇
18 52 ☽ ∥ ♄
19 11 ☽ ⊻ ☿
20 08 ☽ ⊼ ♆
21 15 ☽ # ☉
21 31 ☽ ⊻ ♃

25 Tuesday
02 38 ☉ ⚼ ♇
03 50 ☽ ± ♇
09 46 ☽ ∥ ☉
11 07 ♂ St R
11 21 ☽ # ♄
17 25 ☽ ⚼ ♂
18 58 ☽ ⚼ ♇
20 13 ☽ ☌ ☉

26 Wednesday
02 38 ☽ # ♆
04 22 ☽ □ ♅
06 32 ☽ ∥ ♃
07 52 ☽ ⊻ ♀
08 04 ☽ ☍ ♄
08 42 ☽ ± ♂
09 03 ☽ # ♀
10 08 ♀ ⊼ ♄
11 05 ☉ # ♄
13 55 ☿ St R
18 59 ☽ ⊼ ♂
19 09 ☽ ⊥ ♀
19 54 ☽ ∥ ☿
20 33 ☽ △ ♇
22 21 ☽ ☌ ☿
22 55 ☽ ☍ ♆

27 Thursday
01 20 ☽ ☌ ♃
11 28 ☽ Q ♅
12 21 ☽ ∠ ♀
14 17 ♀ # ♃
14 57 ☽ # ♂

28 Friday
04 17 ☽ ⊻ ☉
04 18 ☿ ⊥ ♀
08 47 ☽ ⚹ ♅
12 21 ☽ ⊼ ♄
12 45 ♀ ± ♆
15 56 ☽ ⊥ ☉
17 49 ☽ ⚹ ♀
23 11 ☽ ± ♄

29 Saturday
00 06 ☽ ☍ ♂
01 49 ☽ □ ♇
03 04 ☽ ⊻ ☿
04 16 ☽ ⊼ ♆
07 57 ☽ ⊻ ♃
08 16 ☽ ∥ ♅
09 46 ☽ ∠ ☉
10 05 ☽ ∥ ♇
12 11 ☽ ∠ ♅
13 53 ☽ ⊥ ☿
15 21 ☽ ± ♆
15 42 ☽ ⚼ ♄
18 14 ☽ Q ♀
19 19 ☽ ⊥ ♃

30 Sunday
05 04 ☽ ⊥ ♅
05 39 ♀ ∥ ♆
06 17 ☽ ∠ ☿
07 54 ♀ ⊥ ♃
08 15 ☽ ⚼ ♆
11 26 ☽ Q ♇
12 39 ☽ ∠ ♃
16 24 ☽ ⚹ ☉
16 31 ☽ ⊻ ♅
17 51 ☉ □ ♅
19 56 ☽ △ ♄
22 05 ☿ △ ♇

31 Monday
08 20 ☽ □ ♀
08 32 ☽ ⊼ ♂
10 05 ☽ ⚹ ☿
10 10 ♀ △ ♂
10 34 ☽ ⚹ ♇
13 08 ☽ △ ♆
17 43 ☽ Q ☉
18 18 ☽ ⚹ ♃
20 13 ☽ ± ♂
20 59 ☿ ⊻ ♀

APRIL 1952

LONGITUDES

Date	Sidereal time h m s	Sun ☉ ° ' "	Moon ☽ ° ' "	Moon ☽ 24.00 ° ' "	Mercury ☿ ° '	Venus ♀ ° '	Mars ♂ ° '	Jupiter ♃ ° '	Saturn ♄ ° '	Uranus ♅ ° '	Neptune ♆ ° '	Pluto ♇ ° '
01	00 39 08	11 ♈ 42 19	02 ♋ 11 16	08 ♋ 11 04	18 ♈ 26	19 ♓ 35	18 ♏ 10	23 ♈ 25	11 ♎ 37	10 ♋ 00	20 ♎ 35	19 ♌ 17
02	00 43 05	12 41 29	14 ♋ 08 32	20 ♋ 04 18	17 R 47	20 49	18 R 04	23 40	11 R 33	10 00	20 R 34	19 R 17
03	00 47 02	13 40 36	25 ♋ 59 00	01 ♌ 53 19	17 04	22 02	17 58	23 54	11 28	10 01	20 32	19 16
04	00 50 58	14 39 42	07 ♌ 47 52	13 ♌ 43 17	16 19	23 16	17 51	24 09	11 23	10 02	20 30	19 15
05	00 54 55	15 38 45	19 ♌ 40 11	25 ♌ 39 08	15 32	24 30	17 43	24 23	11 19	10 03	20 29	19 15
06	00 58 51	16 37 45	01 ♍ 40 39	07 ♍ 45 12	14 45	25 44	17 34	24 37	11 14	10 04	20 27	19 14
07	01 02 48	17 36 44	13 ♍ 53 14	20 ♍ 05 04	13 57	26 58	17 24	24 52	11 10	10 05	20 25	19 13
08	01 06 44	18 35 40	26 ♍ 21 00	02 ♎ 41 13	13 11	28 11	17 14	25 06	11 05	10 06	20 24	19 13
09	01 10 41	19 34 34	09 ♎ 05 50	15 ♎ 34 53	12 26	29 ♓ 25	17 03	25 21	11 00	10 07	20 22	19 12
10	01 14 37	20 33 26	22 ♎ 08 19	28 ♎ 46 00	11 44	00 ♈ 39	16 51	25 35	10 56	10 09	20 20	19 11
11	01 18 34	21 32 16	05 ♏ 27 42	12 ♏ 13 09	11 06	01 53	16 39	25 49	10 51	10 10	20 19	19 11
12	01 22 31	22 31 04	19 ♏ 02 02	25 ♏ 53 58	10 31	03 07	16 26	26 04	10 47	10 11	20 17	19 10
13	01 26 27	23 29 51	02 ♐ 48 34	09 ♐ 45 27	10 00	04 20	16 12	26 18	10 42	10 13	20 16	19 10
14	01 30 24	24 28 36	16 ♐ 44 13	23 ♐ 44 32	09 35	05 34	15 57	26 33	10 38	10 14	20 14	19 09
15	01 34 20	25 27 19	00 ♑ 46 04	07 ♑ 48 32	09 14	06 48	15 42	26 47	10 33	10 15	20 12	19 09
16	01 38 17	26 26 00	14 ♑ 51 41	21 ♑ 55 19	08 58	08 02	15 26	27 02	10 29	10 17	20 11	19 08
17	01 42 13	27 24 40	28 ♑ 59 16	06 ♒ 03 22	08 47	09 16	15 10	27 16	10 24	10 18	20 09	19 08
18	01 46 10	28 23 17	13 ♒ 07 30	20 ♒ 11 30	08 42	10 29	14 52	27 31	10 20	10 20	20 07	19 08
19	01 50 06	29 ♈ 21 54	27 ♒ 15 13	04 ♓ 18 28	08 D 41	11 43	14 35	27 45	10 16	10 22	20 06	19 07
20	01 54 03	00 ♉ 20 28	11 ♓ 21 01	18 ♓ 22 36	08 46	12 57	14 16	27 59	10 12	10 23	20 04	19 07
21	01 58 00	01 19 01	25 ♓ 22 54	02 ♈ 21 33	08 56	14 11	13 57	28 14	10 07	10 25	20 02	19 07
22	02 01 56	02 17 32	09 ♈ 18 11	16 ♈ 12 22	09 10	15 25	13 38	28 28	10 03	10 27	20 01	19 07
23	02 05 53	03 16 02	23 ♈ 03 42	29 ♈ 51 44	09 30	16 38	13 18	28 43	09 59	10 29	19 59	19 06
24	02 09 49	04 14 29	06 ♉ 36 07	13 ♉ 16 31	09 53	17 52	12 58	28 57	09 55	10 31	19 58	19 06
25	02 13 46	05 12 55	19 ♉ 52 37	26 ♉ 24 16	10 21	19 06	12 37	29 12	09 51	10 33	19 56	19 06
26	02 17 42	06 11 19	02 ♊ 51 19	09 ♊ 13 47	10 54	20 20	12 16	29 26	09 47	10 35	19 55	19 06
27	02 21 39	07 09 41	15 ♊ 31 43	21 ♊ 45 19	11 30	21 33	11 55	29 40	09 43	10 37	19 53	19 06
28	02 25 35	08 08 01	27 ♊ 54 49	04 ♋ 00 35	12 10	22 47	11 34	29 ♈ 55	09 39	10 39	19 51	19 06
29	02 29 32	09 06 19	10 ♋ 03 59	16 ♋ 02 40	12 54	24 01	11 12	00 ♉ 09	09 36	10 41	19 50	19 06
30	02 33 29	10 ♉ 04 35	22 ♋ 00 01	27 ♋ 55 42	13 ♈ 41	25 ♈ 15	10 ♏ 50	00 ♉ 23	09 ♎ 32	10 ♋ 43	19 ♎ 48	19 ♌ 06

Date	Moon True ☊ ° '	Moon Mean ☊ ° '	Moon ☽ Latitude ° '
01	29 ♒ 42	28 ♒ 37	04 N 22
02	29 D 41	28 34	03 42
03	29 42	28 30	02 53
04	29 43	28 27	01 56
05	29 45	28 24	00 N 54
06	29 R 45	28 21	00 S 10
07	29 44	28 18	01 16
08	29 41	28 14	02 18
09	29 36	28 11	03 16
10	29 28	28 08	04 04
11	29 20	28 05	04 40
12	29 11	28 02	05 01
13	29 02	27 59	05 05
14	28 55	27 55	04 51
15	28 50	27 52	04 20
16	28 48	27 49	03 33
17	28 D 47	27 46	02 33
18	28 48	27 43	01 23
19	28 48	27 40	00 S 08
20	28 R 48	27 36	01 N 07
21	28 45	27 33	02 17
22	28 40	27 30	03 18
23	28 32	27 27	04 07
24	28 22	27 24	04 41
25	28 10	27 20	04 59
26	27 59	27 17	05 01
27	27 48	27 14	04 48
28	27 40	27 11	04 22
29	27 34	27 08	03 44
30	27 ♒ 30	27 ♒ 05	02 N 57

DECLINATIONS

Date	Sun ☉ ° '	Moon ☽ ° '	Mercury ☿ ° '	Venus ♀ ° '	Mars ♂ ° '	Jupiter ♃ ° '	Saturn ♄ ° '	Uranus ♅ ° '	Neptune ♆ ° '	Pluto ♇ ° '
01	04 N 38	27 N 48	10 N 11	05 S 26	15 S 38	08 N 08	02 S 06	23 N 27	06 S 27	23 N 43
02	05 01	26 23	09 50	04 57	15 38	08 13	02 04	23 27	06 27	23 43
03	05 24	23 47	09 26	04 29	15 37	08 18	02 02	23 27	06 26	23 44
04	05 47	20 12	08 59	04 00	15 36	08 24	02 00	23 27	06 25	23 44
05	06 10	15 47	08 31	03 32	15 35	08 29	01 58	23 27	06 25	23 44
06	06 32	10 43	08 03	03 03	15 34	08 34	01 56	23 26	06 24	23 44
07	06 55	05 N 11	07 29	02 35	15 33	08 39	01 55	23 26	06 24	23 44
08	07 17	00 S 40	06 58	02 06	15 31	08 45	01 53	23 26	06 23	23 44
09	07 40	06 36	06 26	01 37	15 29	08 50	01 51	23 26	06 22	23 44
10	08 02	12 24	05 55	01 08	15 28	08 55	01 49	23 26	06 22	23 44
11	08 24	17 45	05 25	00 39	15 26	09 01	01 48	23 26	06 21	23 44
12	08 46	22 18	04 56	00 S 10	15 23	09 06	01 46	23 26	06 20	23 44
13	09 08	25 43	04 29	00 N 19	15 21	09 11	01 44	23 26	06 20	23 44
14	09 29	27 37	04 03	00 48	15 18	09 16	01 42	23 26	06 19	23 44
15	09 51	27 47	03 40	01 18	15 16	09 22	01 41	23 25	06 19	23 44
16	10 12	26 09	03 20	01 47	15 13	09 27	01 39	23 25	06 18	23 44
17	10 33	22 51	03 01	02 16	15 10	09 32	01 37	23 25	06 17	23 44
18	10 54	18 12	02 46	02 45	15 07	09 37	01 36	23 25	06 17	23 44
19	11 15	12 34	02 33	03 14	15 04	09 43	01 34	23 25	06 16	23 44
20	11 36	06 S 17	02 22	03 43	15 00	09 48	01 33	23 25	06 16	23 44
21	11 56	00 N 15	02 14	04 12	14 57	09 53	01 31	23 25	06 15	23 44
22	12 16	06 43	02 09	04 41	14 53	09 58	01 29	23 25	06 14	23 44
23	12 36	12 47	02 07	05 09	14 49	10 03	01 28	23 24	06 14	23 44
24	12 56	18 09	02 06	05 38	14 45	10 08	01 26	23 24	06 13	23 44
25	13 16	22 31	02 09	06 07	14 41	10 13	01 25	23 24	06 13	23 43
26	13 35	25 40	02 13	06 35	14 37	10 19	01 24	23 24	06 12	23 43
27	13 54	27 26	02 20	07 03	14 33	10 24	01 22	23 24	06 11	23 43
28	14 13	27 48	02 29	07 31	14 28	10 29	01 21	23 24	06 11	23 43
29	14 32	26 47	02 40	07 59	14 24	10 34	01 19	23 23	06 10	23 43
30	14 N 51	24 N 33	02 N 53	08 N 27	14 S 20	10 N 39	01 S 18	23 N 23	06 S 10	23 N 43

ZODIAC SIGN ENTRIES

Date	h	m	Planets
01	07	39	☽ ♋
03	20	10	☽ ♌
06	08	40	☽ ♍
08	18	56	☽ ♎
09	23	17	♀ ♈
11	02	13	☽ ♏
13	07	08	☽ ♐
15	10	41	☽ ♑
17	13	43	☽ ♒
19	16	40	☽ ♓
20	03	37	☉ ♉
21	19	56	☽ ♈
24	00	15	☽ ♉
26	06	40	☽ ♊
28	16	06	☽ ♋
28	20	50	♃ ♉

LATITUDES

Date	Mercury ☿ ° '	Venus ♀ ° '	Mars ♂ ° '	Jupiter ♃ ° '	Saturn ♄ ° '	Uranus ♅ ° '	Neptune ♆ ° '	Pluto ♇ ° '
01	03 N 12	01 S 25	01 N 40	01 S 03	02 N 43	00 N 23	01 N 43	09 N 10
04	02 47	01 27	01 37	01 03	02 43	00 23	01 43	09 10
07	02 09	01 29	01 33	01 03	02 43	00 23	01 43	09 09
10	01 23	01 31	01 29	01 03	02 43	00 23	01 43	09 08
13	00 N 33	01 32	01 24	01 02	02 43	00 22	01 43	09 08
16	00 S 15	01 32	01 19	01 02	02 43	00 22	01 43	09 07
19	00 59	01 32	01 13	01 02	02 43	00 22	01 43	09 07
22	01 37	01 31	01 06	01 02	02 43	00 22	01 43	09 06
25	02 08	01 29	01 00	01 02	02 42	00 22	01 43	09 06
28	02 32	01 27	00 53	01 02	02 42	00 22	01 43	09 05
31	02 S 49	01 S 24	00 N 45	01 S 02	02 N 41	00 N 22	01 N 43	09 N 05

DATA

Julian Date	2434104
Delta T	+30 seconds
Ayanamsa	23° 11' 33"
Synetic vernal point	05° ♓ 55' 27"
True obliquity of ecliptic	23° 26' 52"

MOON'S PHASES, APSIDES AND POSITIONS ☽

Date	h	m	Phase	Longitude ° '	Eclipse Indicator
02	08	48	☽	12 ♋ 34	
10	08	53	○	20 ♎ 26	
17	09	07	☾	27 ♑ 18	
24	07	27	●	04 ♉ 03	

Day	h	m		
03	18	03	Apogee	
18	08	22	Perigee	
08	09	18	0S	
15	02	10	Max dec	27° S 56'
21	11	03	0N	
28	06	05	Max dec	27° N 50'

LONGITUDES

Date	Chiron ⚷ ° '	Ceres ⚳ ° '	Pallas ⚴ ° '	Juno ⚵ ° '	Vesta ⚶ ° '	Black Moon Lilith ⚸ ° '
01	12 ♑ 14	17 ♈ 53	14 ♓ 45	08 ♒ 48	00 ♍ 31	00 ♌ 28
11	12 ♑ 26	21 ♈ 53	17 ♓ 52	11 ♒ 49	29 ♌ 54	01 ♌ 34
21	12 ♑ 29	25 ♈ 53	20 ♓ 54	14 ♒ 38	00 ♍ 06	02 ♌ 41
31	12 ♑ 24	29 ♈ 53	23 ♓ 49	17 ♒ 12	01 ♍ 02	03 ♌ 47

All ephemeris data is given at 12.00 UT and the Moon's longitude is additionally given for 24.00 UT

ASPECTARIAN

h m	Aspects
01 Tuesday	
06 27	♀ ⊼ ♇
08 41	☽ Q ☿
10 11	☉ ☍ ♄
13 56	☽ ⚼ ♂
16 12	☽ ∠ ♇
18 36	☽ Q ♃
22 15	☉ ± ♂
23 59	☿ ⊼ ♂
02 Wednesday	
03 39	☽ ☌ ♅
06 48	☽ □ ♄
07 14	♀ ⊼ ♆
08 48	☽ □ ☉
10 15	☽ ⊥ ♇
10 20	☉ # ♀
18 58	☽ □ ☿
19 53	☽ △ ♂
22 23	☽ ⊻ ♇
03 Thursday	
00 58	☽ □ ♆
03 04	☽ △ ♀
07 41	☽ □ ♃
12 28	☽ ‖ ♇
14 35	☽ ‖ ♅
19 02	☽ Q ♄
04 Friday	
13 04	☽ ⚼ ♀
13 26	☽ Q ♆
16 33	☽ ⊻ ♅
19 14	☽ ⚹ ♄
05 Saturday	
03 09	☽ △ ☉
04 11	☽ △ ☿
04 42	☽ ⊥ ♅
08 06	☽ □ ♂
09 07	♀ ⊻ ♃
09 22	☽ ± ♀
10 33	☉ ☌ ☿
11 08	☽ ☌ ♇
13 00	☽ # ♂
13 12	☿ ‖ ♃
13 37	☽ ⚹ ♆
21 39	☽ △ ♃
22 48	☽ ⊼ ♀
22 49	☽ ∠ ♅
06 Sunday	
01 14	☽ ∠ ♄
02 19	♀ ± ♇
03 39	☉ # ♆
08 24	☽ ⚼ ☿
11 54	☽ ⚼ ☉
19 00	☽ ⊥ ♄
19 27	☽ ∠ ♆
19 36	☽ Q ♂
21 22	☽ ‖ ♃
07 Monday	
01 06	☽ ± ☿
01 11	☽ ‖ ☿
03 59	☽ ⚼ ♃
04 34	☽ ⚹ ♅
05 06	☽ ‖ ☉
06 43	☽ ⊻ ♄
06 52	☽ # ♆
07 11	☽ ± ☉
07 40	☉ ⊼ ♂
12 07	☽ ⊼ ☿
13 02	☽ ⊥ ♆
18 44	☽ ⚹ ♂
19 51	☽ ⊼ ☉
21 50	☽ ± ♃
22 19	☽ ⊻ ♇
23 45	☽ # ♀
08 Tuesday	
00 37	☽ ⊻ ♆
01 35	☽ # ♄
03 14	☉ ‖ ☿
03 52	☽ Q ♅
09 34	☽ ⊼ ♃
09 50	☽ ⊥ ♇
15 53	☽ ☍ ♀
16 53	☽ ‖ ♄
17 21	☽ ‖ ♀
23 00	☽ ∠ ♂
23 23	♀ ‖ ♄
09 Wednesday	
02 51	☽ ∠ ♇
02 52	☉ △ ♇
11 04	☽ ‖ ♆
11 22	☽ # ☿
13 55	☽ □ ♅
14 49	☿ # ♆
15 31	☽ ☌ ♄
15 35	☽ ⊥ ♂
16 36	☽ # ☉
17 52	☽ ☍ ☿
21 15	☽ # ♃
21 39	♂ ⊥ ♄
10 Thursday	
02 29	☽ ⊻ ♂
06 37	☽ ⚹ ♇
06 51	☉ ☍ ♆
08 44	☽ ☌ ♆
08 53	☽ ☍ ☉
18 22	☽ ☍ ♃
11 Friday	
01 20	☽ ‖ ♂
04 21	☽ Q ♇
04 57	☽ ⊼ ♀
08 07	♀ ⚼ ♂
16 44	☽ ± ♀
20 22	☽ △ ♅
21 32	☽ ⊻ ♄
21 35	☽ ⊼ ☿
23 01	☿ ☍ ♄
12 Saturday	
07 30	☽ ☌ ♂
07 45	☽ ± ☿
08 03	☽ ⊥ ♄
10 13	☽ ⚼ ♀
12 14	☽ □ ♇
14 11	☽ ⊻ ♆
18 34	☽ ⊼ ☉
18 36	☿ ± ♂
18 58	☽ # ♅
20 58	☽ # ♇
22 46	☽ ⚼ ♅
22 55	☽ ⚼ ☿
23 43	☽ ∠ ♄
13 Sunday	
00 30	☽ ⊼ ♃
00 39	☽ ⊥ ♆
02 25	☿ □ ♅
05 49	☽ ± ☉
08 33	♀ ⚼ ♇
11 07	☽ ± ♃
14 26	☽ ± ♅
14 55	☽ △ ♀
16 14	☽ ∠ ♆
17 05	☉ ‖ ♃
22 34	☽ ⚼ ☉
14 Monday	
00 02	☽ △ ☿
00 48	☽ ⊼ ♅
01 33	☽ ⚹ ♄
02 55	☽ ⚼ ♃
10 41	☽ ⊻ ♂
16 09	☽ △ ♇
17 59	☽ ⚹ ♆
20 47	☽ ⊥ ♂
22 02	☽ Q ♄
15 Tuesday	
02 15	☽ △ ☉
05 05	☽ △ ♃
11 53	☽ ∠ ♂
14 27	☽ Q ♆
17 46	☽ ⚼ ♇
23 16	☽ □ ♀
16 Wednesday	
02 08	☽ □ ☿
04 12	☽ ☍ ♅
04 35	☽ □ ♄
06 03	♀ # ♄
09 05	☽ ± ♇
12 57	☽ ⚹ ♂
19 16	☽ ⊼ ♇
21 01	☽ □ ♆
17 Thursday	
03 43	☿ ☌ ♀
06 33	☽ # ♇
07 21	☉ ☌ ♃
08 18	☽ Q ☿
08 33	☽ # ♅
08 47	☽ Q ♀
08 57	☽ Q ♂
09 02	☽ □ ♃
09 07	☽ □ ☉
10 23	♀ ± ♂
18 Friday	
04 31	☽ ⚹ ☿
07 06	☽ ⚹ ♀
07 15	☽ ⊼ ♅
07 17	☽ △ ♄
08 52	♀ □ ♅
09 07	♀ ☍ ♄
10 38	☉ Q ♅
12 13	♄ □ ♅
12 29	☿ ‖ ♀
14 55	☽ □ ♂
16 07	☽ Q ♃
17 28	☽ ± ♅
17 57	☽ Q ☉
22 11	☽ ☍ ♇
23 52	☽ △ ♆
19 Saturday	
01 33	☿ St D
01 42	☽ ‖ ♂
03 43	☿ ± ♂
05 56	☽ ∠ ☿
08 38	☽ ⚼ ♄
08 47	☽ ⚼ ♅
11 00	☽ ∠ ♀
12 51	☽ ⚹ ♃
15 51	☽ ⚹ ☉
16 53	☽ # ☉
21 17	☽ ⊥ ☿
22 58	☽ # ♃
23 52	☽ ± ♄
20 Sunday	
01 19	☽ ⚼ ♆
03 47	☽ ⊥ ♀
07 34	☽ ⊻ ☿
10 02	☽ ⊼ ♄
10 21	☽ △ ♅
12 06	☽ ‖ ♆
14 51	☽ ∠ ♃
15 00	☽ ⊻ ♀
16 38	☽ ± ♆
16 53	☽ △ ♂
19 19	☽ ∠ ☉
20 49	☽ # ♀
21 Monday	
01 16	☽ ⊼ ♇
02 42	☽ # ☿
02 52	☽ ⊼ ♆
05 29	☽ ‖ ♄
06 30	☽ ⊥ ♃
08 33	♀ ⊼ ♂
11 32	☽ ± ♇
11 53	☽ ⊥ ☉
16 36	☽ # ♄
16 59	☽ ⊻ ♃
18 00	☽ ⚼ ♂
19 10	☽ ‖ ☿
22 58	☽ ⊻ ☉
22 Tuesday	
03 01	☽ ⚼ ♇
03 40	☽ ‖ ♀
09 11	☽ ± ♂
09 19	♃ Q ♅
10 10	☽ # ♆
11 46	☽ ☌ ☿
13 18	☽ ☍ ♄
13 59	☽ □ ♅
19 21	☽ ⊼ ♂
23 39	☽ ☌ ♀
23 Wednesday	
00 44	☽ ‖ ♃
05 04	☽ △ ♇
06 37	☽ ☍ ♆
11 11	☽ ‖ ☉
20 32	☽ # ♂
21 34	☽ Q ♅
22 09	☽ ☌ ♃
24 Thursday	
07 27	☽ ☌ ☉
13 18	☿ ☍ ♄
17 55	☽ ⊼ ♄
18 06	☽ ⊻ ☿
19 02	☽ ⚹ ♅
23 09	☽ ☍ ♂
25 Friday	
04 42	☽ ± ♄
05 20	☽ ⊥ ☿
10 26	☽ ⊻ ♀
10 35	☽ □ ♇
12 01	♀ △ ♇
12 06	☽ ⊼ ♆
17 01	♀ # ♆
17 50	☽ ‖ ♅
20 06	☽ ‖ ♇
21 05	☽ ⚼ ♄
21 07	☿ □ ♅
22 26	☽ ∠ ♅
22 29	☽ ∠ ☿
22 35	☽ ⊥ ♀
23 07	☽ ± ♆
26 Saturday	
03 59	♀ ☍ ♆
05 30	☽ ⊻ ♃
15 14	☽ ⊥ ♅
15 51	☽ ⚼ ♆
16 56	☽ ⊥ ♃
17 08	☽ ∠ ♀
18 46	☽ ⊻ ☉
19 58	☽ Q ♇
27 Sunday	
00 59	☽ △ ♄
02 35	☽ ⊻ ♅
03 55	☽ ⚹ ☿
05 18	☽ ⊼ ♂
07 06	☽ ⊥ ☉
10 20	☽ ∠ ♃
16 28	☽ ± ♂
18 51	☽ ⚹ ♇
20 22	☽ △ ♆
21 55	☿ ⊼ ♂
28 Monday	
00 54	☽ ⚹ ♀
01 53	☽ ∠ ☉
04 16	☽ Q ☿
09 26	☽ ⚼ ♂
16 00	☽ ⚹ ♃
29 Tuesday	
00 10	☽ ∠ ♇
03 04	☽ Q ♀
03 32	☉ # ♂
09 57	☽ ⚹ ☉
11 05	☽ □ ♄
13 15	☽ ☌ ♅
14 13	☽ △ ♂
16 17	☽ Q ♃
18 05	☽ ⊥ ♇
18 05	☽ □ ☿
23 19	☉ ⊼ ♄
30 Wednesday	
06 08	☽ ⊻ ♇
07 35	☽ □ ♆
12 10	☽ Q ☉
18 55	☽ ‖ ♇
18 56	♂ △ ♅
19 19	☽ □ ♀
21 25	☽ ‖ ♅
23 08	☽ Q ♄
23 19	♇ St D

MAY 1952

LONGITUDES

Date	Sidereal time h m s	Sun ☉	Moon ☽	Moon ☽ 24.00	Mercury ☿	Venus ♀	Mars ♂	Jupiter ♃	Saturn ♄	Uranus ♅	Neptune ♆	Pluto ♇
01	02 37 25	11 ♉ 02 49	03 ♌ 50 20	09 ♌ 44 34	14 ♈ 32	26 ♈ 28	10 ♏ 28	00 ♉ 38	09 ♎ 28	10 ♋ 45	19 ♎ 47	19 ♌ 06
02	02 41 22	12 01 00	15 ♌ 39 07	21 ♌ 34 38	15 26	27 42	10 R 06	00 52	09 R 25	10 47	19 R 45	19 D 06
03	02 45 18	12 59 10	27 ♌ 31 49	03 ♍ 31 21	16 23	28 ♈ 56	09 43	01 06	09 21	10 50	19 44	19 06
04	02 49 15	13 57 18	09 ♍ 33 51	15 ♍ 39 57	17 23	00 ♉ 09	09 21	01 21	09 18	10 52	19 42	19 06
05	02 53 11	14 55 24	21 ♍ 50 11	28 ♍ 05 05	18 26	01 23	08 59	01 35	09 14	10 54	19 41	19 06
06	02 57 08	15 53 28	04 ♎ 25 01	10 ♎ 50 19	19 32	02 37	08 37	01 49	09 11	10 57	19 39	19 06
07	03 01 04	16 51 30	17 ♎ 21 13	23 ♎ 57 47	20 41	03 51	08 15	02 03	09 08	10 59	19 38	19 06
08	03 05 01	17 49 30	00 ♏ 39 58	07 ♏ 27 37	21 52	05 04	07 53	02 17	09 05	11 01	19 37	19 06
09	03 08 58	18 47 29	14 ♏ 20 26	21 ♏ 17 57	23 06	06 18	07 32	02 31	09 02	11 04	19 35	19 07
10	03 12 54	19 45 26	28 ♏ 22 35	05 ♐ 24 52	24 22	07 32	07 11	02 46	08 59	11 06	19 34	19 07
11	03 16 51	20 43 22	12 ♐ 32 56	19 ♐ 43 06	25 40	08 45	06 50	03 00	08 56	11 09	19 32	19 07
12	03 20 47	21 41 16	26 ♐ 54 38	04 ♑ 06 49	27 01	09 59	06 29	03 14	08 53	11 12	19 31	19 07
13	03 24 44	22 39 09	11 ♑ 19 01	18 ♑ 30 37	28 25	11 13	06 09	03 28	08 50	11 14	19 30	19 08
14	03 28 40	23 37 01	25 ♑ 41 08	02 ♒ 50 11	29 ♈ 50	12 26	05 49	03 42	08 48	11 17	19 29	19 08
15	03 32 37	24 34 51	09 ♒ 57 28	17 ♒ 02 45	01 ♉ 18	13 40	05 30	03 56	08 45	11 20	19 27	19 08
16	03 36 33	25 32 41	24 ♒ 05 53	01 ♓ 06 48	02 48	14 54	05 11	04 10	08 43	11 22	19 26	19 09
17	03 40 30	26 30 29	08 ♓ 05 27	15 ♓ 01 49	04 20	16 07	04 53	04 24	08 40	11 25	19 25	19 09
18	03 44 27	27 28 15	21 ♓ 55 54	28 ♓ 47 40	05 55	17 21	04 35	04 37	08 38	11 28	19 23	19 10
19	03 48 23	28 26 01	05 ♈ 37 05	12 ♈ 24 07	07 32	18 35	04 18	04 51	08 36	11 31	19 22	19 10
20	03 52 20	29 ♉ 23 46	19 ♈ 08 39	25 ♈ 50 35	09 11	19 48	04 01	05 05	08 34	11 34	19 21	19 11
21	03 56 16	00 ♊ 21 29	02 ♉ 29 47	09 ♉ 06 06	10 52	21 02	03 46	05 19	08 32	11 37	19 20	19 11
22	04 00 13	01 19 12	15 ♉ 39 21	22 ♉ 09 23	12 35	22 16	03 30	05 32	08 30	11 40	19 19	19 12
23	04 04 09	02 16 53	28 ♉ 36 04	04 ♊ 59 18	14 21	23 29	03 16	05 46	08 28	11 43	19 18	19 13
24	04 08 06	03 14 33	11 ♊ 18 59	17 ♊ 35 07	16 09	24 43	03 02	06 00	08 26	11 46	19 17	19 13
25	04 12 02	04 12 12	23 ♊ 47 45	29 ♊ 56 58	17 59	25 57	02 49	06 13	08 24	11 49	19 16	19 14
26	04 15 59	05 09 49	06 ♋ 02 57	12 ♋ 05 56	19 51	27 10	02 37	06 27	08 23	11 52	19 15	19 15
27	04 19 55	06 07 25	18 ♋ 06 15	24 ♋ 04 14	21 45	28 24	02 25	06 40	08 21	11 55	19 13	19 15
28	04 23 52	07 05 00	00 ♌ 00 22	05 ♌ 55 07	23 41	29 ♉ 38	02 15	06 54	08 20	11 58	19 12	19 16
29	04 27 49	08 02 34	11 ♌ 49 02	17 ♌ 42 42	25 40	00 ♊ 51	02 05	07 07	08 19	12 01	19 12	19 17
30	04 31 45	09 00 06	23 ♌ 36 45	29 ♌ 31 51	27 40	02 05	01 56	07 20	08 18	12 04	19 11	19 18
31	04 35 42	09 ♊ 57 36	05 ♍ 28 40	11 ♍ 27 53	29 ♉ 42	03 ♊ 19	01 ♏ 47	07 ♉ 33	08 ♎ 17	12 ♋ 07	19 ♎ 10	19 ♌ 18

DECLINATIONS

Date	Moon True ☊	Moon Mean ☊	Moon ☽ Latitude	Sun ☉	Moon ☽	Mercury ☿	Venus ♀	Mars ♂	Jupiter ♃	Saturn ♄	Uranus ♅	Neptune ♆	Pluto ♇
01	27 ♒ 29	27 ♒ 01	02 N 02	15 N 09	21 N 17	03 N 08	08 N 55	14 S 15	10 N 44	01 S 17	23 N 23	06 S 09	23 N 42
02	27 D 28	26 58	01 N 02	15 27	17 09	03 25	09 22	14 11	10 49	01 15	23 23	06 09	23 42
03	27 R 29	26 55	00 00	15 45	12 20	03 43	09 49	14 06	10 54	01 14	23 23	06 08	23 42
04	27 28	26 52	01 S 04	16 02	07 00	04 03	10 16	14 02	10 59	01 13	23 22	06 08	23 42
05	27 26	26 49	02 05	16 19	01 N 19	04 25	10 43	13 57	11 03	01 12	23 22	06 07	23 42
06	27 21	26 45	03 02	16 36	04 S 33	04 49	11 09	13 53	11 08	01 11	23 22	06 07	23 41
07	27 14	26 42	03 52	16 53	10 23	05 14	11 36	13 48	11 13	01 10	23 22	06 06	23 41
08	27 04	26 39	04 30	17 09	15 55	05 40	12 02	13 44	11 18	01 09	23 22	06 05	23 41
09	26 53	26 36	04 54	17 25	20 49	06 07	12 27	13 39	11 23	01 07	23 21	06 05	23 41
10	26 41	26 33	05 00	17 41	24 40	06 36	12 53	13 35	11 28	01 06	23 21	06 04	23 41
11	26 30	26 30	04 49	17 56	27 05	07 06	13 18	13 31	11 33	01 06	23 21	06 04	23 40
12	26 21	26 26	04 19	18 12	27 44	07 38	13 43	13 26	11 37	01 05	23 21	06 03	23 40
13	26 14	26 23	03 33	18 26	26 30	08 10	14 07	13 22	11 42	01 04	23 20	06 03	23 40
14	26 10	26 20	02 33	18 41	23 32	08 43	14 31	13 18	11 47	01 03	23 20	06 03	23 39
15	26 08	26 17	01 24	18 55	19 07	09 17	14 55	13 14	11 52	01 02	23 20	06 02	23 39
16	26 D 08	26 14	00 S 11	19 09	13 40	09 53	15 18	13 11	11 56	01 01	23 20	06 02	23 39
17	26 R 08	26 11	01 N 03	19 23	07 34	10 28	15 41	13 07	12 01	01 00	23 20	06 01	23 38
18	26 06	26 07	02 12	19 36	01 S 11	11 05	16 04	13 04	12 05	01 00	23 19	06 01	23 38
19	26 03	26 04	03 12	19 49	05 N 11	11 42	16 26	13 00	12 10	00 59	23 19	06 00	23 38
20	25 57	26 01	04 01	20 02	11 13	12 20	16 48	12 57	12 15	00 58	23 19	06 00	23 37
21	25 48	25 58	04 36	20 14	16 40	12 59	17 09	12 54	12 19	00 58	23 19	06 00	23 37
22	25 36	25 55	04 56	20 26	21 15	13 38	17 30	12 52	12 24	00 57	23 18	05 59	23 37
23	25 23	25 51	05 00	20 37	24 43	14 17	17 51	12 49	12 28	00 57	23 18	05 59	23 36
24	25 11	25 48	04 49	20 49	26 54	14 56	18 11	12 47	12 33	00 56	23 18	05 58	23 36
25	24 59	25 45	04 24	21 00	27 42	15 35	18 31	12 45	12 37	00 56	23 17	05 58	23 36
26	24 49	25 42	03 47	21 10	27 05	16 15	18 50	12 43	12 41	00 55	23 17	05 58	23 35
27	24 41	25 39	03 00	21 20	25 12	16 54	19 08	12 42	12 46	00 55	23 17	05 57	23 35
28	24 36	25 36	02 06	21 30	22 13	17 33	19 26	12 40	12 50	00 55	23 17	05 57	23 34
29	24 34	25 32	01 07	21 39	18 20	18 12	19 44	12 39	12 54	00 54	23 16	05 57	23 34
30	24 34	25 29	00 05	21 48	13 44	18 50	20 01	12 38	12 59	00 54	23 16	05 56	23 34
31	24 ♒ 34	25 ♒ 26	00 S 58	21 N 57	08 N 37	19 N 27	20 N 18	12 S 37	13 N 03	00 S 54	23 N 16	05 S 56	23 N 33

ZODIAC SIGN ENTRIES

Date	h	m	Planets
01	04	12	☽ ♌
03	16	57	☽ ♍
04	08	55	♀ ♉
06	03	39	☽ ♎
08	10	49	☽ ♏
10	14	50	☽ ♐
12	17	09	☽ ♑
14	14	43	☿ ♉
14	19	14	☽ ♒
16	22	05	☽ ♓
19	02	07	☽ ♈
21	03	04	☉ ♊
21	07	29	☽ ♉
23	14	37	☽ ♊
26	00	06	☽ ♋
28	11	59	☽ ♌
28	19	19	♀ ♊
31	00	57	☽ ♍
31	15	26	☿ ♊

LATITUDES

Date	Mercury ☿	Venus ♀	Mars ♂	Jupiter ♃	Saturn ♄	Uranus ♅	Neptune ♆	Pluto ♇
01	02 S 49	01 S 24	00 N 45	01 S 02	02 N 41	00 N 22	01 N 43	09 N 05
04	03 00	01 21	00 37	01 02	02 41	00 22	01 43	09 04
07	03 05	01 17	00 29	01 02	02 41	00 22	01 43	09 04
10	03 03	01 13	00 21	01 02	02 40	00 22	01 43	09 03
13	02 56	01 08	00 13	01 02	02 39	00 22	01 42	09 03
16	02 44	01 03	00 N 05	01 02	02 39	00 22	01 42	09 02
19	02 27	00 58	00 S 03	01 02	02 38	00 22	01 42	09 01
22	02 06	00 52	00 11	01 02	02 38	00 22	01 42	09 01
25	01 40	00 46	00 19	01 02	02 37	00 22	01 42	09 00
28	01 11	00 39	00 26	01 02	02 36	00 22	01 42	09 00
31	00 S 40	00 S 32	00 S 33	01 S 03	02 N 36	00 N 22	01 N 42	08 N 59

DATA

Julian Date	2434134
Delta T	+30 seconds
Ayanamsa	23° 11' 37"
Synetic vernal point	05° ♓ 55' 23"
True obliquity of ecliptic	23° 26' 51"

MOON'S PHASES, APSIDES AND POSITIONS ☽

Date	h	m	Phase	Longitude	Eclipse Indicator
02	03	58	☽	11 ♌ 42	
09	20	16	○	19 ♏ 07	
16	14	39	☾	25 ♒ 39	
23	19	28	●	02 ♊ 35	
31	21	46	☽	10 ♍ 21	

Day	h	m		
01	13	47	Apogee	
13	16	27	Perigee	
29	08	06	Apogee	
05	17	26	OS	
12	08	19	Max dec	27° S 45'
18	16	25	ON	
25	13	23	Max dec	27° N 42'

LONGITUDES

Date	Chiron ⚷	Ceres ⚳	Pallas ⚴	Juno ⚵	Vesta ⚶	Black Moon Lilith ⚸
01	12 ♑ 24	29 ♈ 53	23 ♓ 49	17 ♒ 12	01 ♍ 02	03 ♌ 47
11	12 ♑ 10	03 ♉ 51	26 ♓ 35	19 ♒ 28	02 ♍ 40	04 ♌ 54
21	11 ♑ 49	07 ♉ 46	29 ♓ 11	21 ♒ 22	04 ♍ 52	06 ♌ 01
31	11 ♑ 22	11 ♉ 41	01 ♈ 35	22 ♒ 50	07 ♍ 35	07 ♌ 07

ASPECTARIAN

h m	Aspects
01 Thursday	
01 31	☉ ☍ ♂
04 23	☉ ⚹ ♅
05 21	☽ □ ♃
20 00	☽ Q ♆
23 23	☽ ⚹ ♄
02 Friday	
01 03	☽ □ ♂
02 05	☽ ⚺ ♅
03 58	☽ □ ☉
11 32	☽ △ ☿
14 19	☽ ⊥ ♅
18 58	☽ ☌ ♇
20 17	☽ ∥ ☉
20 18	☽ ⚹ ♆
03 Saturday	
03 22	☽ # ♂
05 38	☽ ∠ ♄
08 34	☽ ∠ ♅
09 53	♀ Q ♅
12 23	☽ Q ♂
15 08	☽ △ ♀
18 34	☽ ∥ ♃
19 19	☽ △ ♃
20 26	☽ ⚼ ☿
22 39	☽ ∥ ♀
23 36	☽ ⊥ ♄
04 Sunday	
02 23	☽ ∠ ♆
11 28	☽ ⚺ ♄
11 36	☽ ⚹ ♂
14 34	☽ ⚹ ♅
15 47	☽ # ♆
15 56	☽ ± ☿
16 26	♂ ⚺ ♄
20 09	☽ ⊥ ♆
21 24	☽ △ ☉
23 50	☽ ∥ ☿
05 Monday	
00 14	☽ ⚼ ♀
01 35	☽ ⚼ ♃
04 47	☽ ⚻ ☿
06 41	☽ ⚺ ♇
07 50	☽ ⚺ ♆
12 30	☽ # ♄
14 04	☽ Q ♅
16 01	☽ ∠ ♂
16 41	♀ ☌ ♃
18 17	☽ ⊥ ♇
19 21	☽ ± ♃
19 25	☉ ± ♄
19 35	☽ ± ♀
22 18	☽ ∥ ♄
06 Tuesday	
02 29	☿ △ ♇
04 47	☽ ⚼ ☉
07 00	☽ ⚻ ♃
08 14	☽ ⚻ ♀
08 42	☽ ⊥ ♂
10 51	♀ ∥ ♃
11 24	☽ ∠ ♇
13 09	☽ # ☿
14 30	☿ ☍ ♆
18 22	☽ ∥ ♆
19 39	☽ ⚺ ♂
20 53	☽ ☌ ♄
23 04	☽ ± ☉
07 Wednesday	
00 14	☽ □ ♅
11 01	☽ ⚻ ☉
15 11	☽ ⚹ ♇
15 35	☽ # ♃
16 09	☽ ☌ ♆
17 32	☽ # ♀
18 39	☽ ☍ ☿
08 Thursday	
02 24	☽ ∥ ♂
12 47	☽ Q ♇
14 56	☽ ☍ ♃
18 01	☽ # ☉
20 34	☽ ☍ ♀
09 Friday	
00 26	☽ ☌ ♂
02 47	☽ ⚺ ♄
06 17	☽ △ ♅
09 58	☿ # ♆
13 11	☽ ⊥ ♄
19 55	☉ □ ♇
20 14	☽ □ ♇
20 16	☽ ☍ ☉
21 03	☽ ⚺ ♆
10 Saturday	
02 52	☽ # ♅
04 34	☽ ⚻ ☿
04 37	☽ ∠ ♄
04 59	☽ # ♇
06 42	♀ ☍ ♂
07 18	☽ ⊥ ♆
07 18	☉ ⚻ ♆
08 13	☽ ⚼ ♅
15 48	☽ ± ☿
19 39	☽ ⚻ ♃
22 33	☽ ∠ ♆
23 31	☽ ± ♅
11 Sunday	
02 37	☽ ⚺ ♂
05 02	☽ ⚻ ♀
05 56	☽ ± ♃
05 57	☽ ⚹ ♄
08 32	☽ ⚼ ☿
09 39	☽ ⚻ ♅
12 28	☽ ⊥ ♂
15 19	♀ ⚻ ♄
16 03	☽ ± ♀
21 16	☽ ⚼ ♃
22 31	♀ # ♂
23 00	☽ △ ♇
23 41	☽ ⚹ ♆
12 Monday	
01 59	☽ Q ♄
02 40	☽ ⚻ ☉
03 10	☽ ∠ ♂
08 29	☽ ⚼ ♀
12 12	☽ △ ☿
13 23	☽ ± ☉
19 40	☽ Q ♆
22 42	☽ △ ♃
13 Tuesday	
00 01	☽ ⚼ ♇
03 35	☽ ⚹ ♂
05 27	☽ ⚼ ☉
07 53	☽ □ ♄
11 48	☽ △ ♀
11 52	☽ ☍ ♅
12 35	♀ ⚹ ♅
15 01	☽ ± ♇
23 08	☽ Q ♂
14 Wednesday	
01 02	☽ ⚻ ♇
01 38	☽ □ ♆
02 29	☿ Q ♅
08 17	☽ △ ☉
11 10	☽ # ♇
13 12	☽ # ♅
16 13	☉ ⚼ ♄
19 45	☽ □ ☿
15 Thursday	
01 40	☽ □ ♃
04 39	☽ □ ♂
09 58	☽ △ ♄
12 53	☽ # ☉
14 19	☽ ⚻ ♅
18 52	☽ □ ♀
16 Friday	
00 31	☽ ± ♅
03 34	☽ ☍ ♇
04 04	☽ △ ♆
05 34	☽ # ♀
05 42	☽ Q ☿
08 32	♀ ± ♄
08 38	☽ Q ♃
09 16	☉ ± ♆
11 20	☽ ⚼ ♄
14 00	☽ ∥ ♂
14 39	☽ □ ☉
15 54	☽ ⚼ ♅
18 54	☽ # ♃
17 Saturday	
01 47	☽ # ☿
02 42	☽ ± ♄
04 30	☽ Q ♀
04 44	☽ ⚹ ☿
05 31	☽ ⚹ ♃
05 40	☽ ⚼ ♆
06 35	☽ △ ♂
09 43	☉ ∠ ♅
12 55	☿ ☌ ♃
13 00	☽ ⚻ ♄
17 46	☽ △ ♅
17 52	☽ ∥ ♆
18 57	☿ ☍ ♂
21 11	☽ ± ♆
23 55	☽ Q ☉
18 Sunday	
03 15	☽ ⚹ ♀
07 11	☽ ⚻ ♇
07 35	☽ ⚻ ♆
07 55	☽ ∠ ♃
08 00	☽ ⚼ ♂
10 00	☽ ∠ ☿
10 15	♂ ☍ ♃
12 42	☽ ∥ ♄
17 39	☽ ± ♇
20 08	☽ # ♄
22 25	☽ ⚹ ☉
23 24	☽ ± ♂
23 54	☽ ⊥ ♃
19 Monday	
03 50	☽ ⊥ ☿
08 03	☽ ∠ ♀
09 27	☽ ⚼ ♇
09 43	☽ ⚻ ♂
10 38	☽ ⚺ ♃
15 12	☽ # ♆
15 50	☽ ⚺ ☿
17 15	☽ ☍ ♄
22 28	☽ □ ♅
23 43	♀ □ ♇
20 Tuesday	
01 32	☽ ⊥ ♀
02 54	☽ ∠ ☉
03 14	☿ ⚻ ♄
03 17	♀ ⚻ ♆
07 53	☿ ∥ ♃
12 04	☽ △ ♇
12 22	☽ ☍ ♆
13 18	☽ ⚺ ♀
16 21	☽ ∥ ♃
17 18	☽ ∥ ☿
19 16	☽ # ♂
20 12	☽ ⊥ ☉
21 Wednesday	
06 46	☽ Q ♅
07 50	☽ ⚺ ☉
09 29	☿ # ♂
14 15	☽ ☍ ♂
14 33	☽ ∥ ♀
17 12	☽ ☌ ♃
22 45	☿ ⚹ ♅
22 55	☽ ⚻ ♄
22 Thursday	
04 39	☽ ⚹ ♅
05 31	☽ ☌ ☿
07 06	☽ ∥ ☉
09 52	☽ ± ♄
18 32	☽ □ ♇
18 44	☽ ⚻ ♆
23 Friday	
01 09	☽ ∥ ♅
01 28	☽ ☌ ♀
02 27	☽ ⚼ ♄
03 20	☽ ∥ ♇
05 50	☽ ± ♆
08 27	☽ ∠ ♅
11 33	♀ ⚼ ♄
13 32	☿ ± ♄
19 28	☽ ☌ ☉
20 36	☽ ⚻ ♂
22 41	☽ ⚼ ♆
24 Saturday	
01 25	☽ ⊥ ♅
01 43	☽ ⚺ ♃
04 13	☽ Q ♇
06 32	☽ △ ♄
07 44	☽ ± ♂
07 48	☉ ⚻ ♂
12 51	☽ ⚺ ♅
13 19	☽ ⊥ ♃
22 48	☽ ⚺ ☿
22 49	♀ ± ♆
25 Sunday	
00 39	☽ ⚼ ♂
03 10	☽ ⚹ ♇
03 14	☽ △ ♆
06 55	☽ ∠ ♃
12 25	☽ ⊥ ☿
13 22	☉ ⚼ ♆
16 38	☽ ⚺ ♀
26 Monday	
04 17	☿ □ ♇
04 23	☿ ⚻ ♆
05 20	☽ △ ♂
05 41	☽ ⊥ ♀
05 42	♀ ∠ ♅
08 26	☽ ∠ ♇
09 11	☽ ∠ ☿
10 06	☽ ⚺ ☉
10 46	♆ ⚹ ♇
12 48	☽ ⚹ ♃
16 36	☽ □ ♄
17 44	♂ # ♃
23 01	☽ ⊥ ☉
23 35	☽ ☌ ♅
27 Tuesday	
01 31	☽ ∠ ♀
02 18	☽ ⊥ ♇
06 27	☉ ⊥ ♅
13 09	☽ Q ♃
14 15	☽ □ ♆
14 19	☽ ⚺ ♇
18 36	☽ ∠ ☉
20 44	☽ ⚹ ☿
28 Wednesday	
02 01	☽ ∥ ♇
04 19	☽ ∥ ♅
04 35	☽ Q ♄
05 47	☉ ⚺ ♃
07 43	☿ ⚼ ♄
11 08	☽ ⚹ ♀
16 28	☽ □ ♂
16 36	☽ ∥ ☉
16 40	☉ Q ♇
29 Thursday	
01 50	☽ Q ☿
02 15	☽ □ ♃
02 36	☽ Q ♆
03 38	☽ ⚹ ☉
04 24	☽ ∥ ♀
04 53	☽ ⚹ ♄
06 23	☿ ± ♆
12 25	☽ ⚺ ♅
12 38	☽ ∥ ☿
12 48	☉ ± ♂
14 21	☽ Q ♀
18 39	☉ △ ♄
30 Friday	
00 41	☽ ⊥ ♅
03 00	☽ ⚹ ♆
03 13	☽ ☌ ♇
04 36	☽ Q ♂
04 43	☿ ∠ ♅
06 13	☽ Q ☉
09 20	♀ ⚻ ♂
11 21	☽ ∠ ♄
15 37	☽ ∥ ♃
17 22	☽ # ♂
19 03	☽ ∠ ♅
21 56	☽ □ ☿
31 Saturday	
04 39	☽ ⚹ ♂
05 34	☽ ⊥ ♄
07 08	☽ □ ♀
09 21	☽ ∠ ♆
16 15	☽ △ ♃
17 37	☽ ⚺ ♄
21 46	☽ □ ☉
23 52	☽ # ♆

All ephemeris data is given at 12.00 UT and the Moon's longitude is additionally given for 24.00 UT

JUNE 1952

Date	Sidereal time h m s	Sun ☉ ° ' "	Moon ☽ ° ' "	Moon ☽ 24.00 ° ' "	Mercury ☿ ° '	Venus ♀ ° '	Mars ♂ ° '	Jupiter ♃ ° '	Saturn ♄ ° '	Uranus ♅ ° '	Neptune ♆ ° '	Pluto ♇ ° '
01	04 39 38	10 ♊ 55 06	17 ♍ 30 11	23 ♍ 36 14	01 ♊ 47	04 ♊ 32	01 ♏ 40	07 ♉ 47	08 ♎ 16	12 ♋ 11	19 ♎ 09	19 ♌ 19
02	04 43 35	11 52 34	29 ♍ 46 41	06 ♎ 02 08	03 52	05 46	01 R 33	08 00	08 R 15	12 14	19 R 08	19 20
03	04 47 31	12 50 00	12 ♎ 23 07	18 ♎ 50 05	06 00	06 59	01 28	08 13	08 14	12 17	19 07	19 21
04	04 51 28	13 47 26	25 ♎ 23 23	02 ♏ 03 16	08 08	08 13	01 23	08 26	08 13	12 20	19 06	19 22
05	04 55 24	14 44 50	08 ♏ 49 47	15 ♏ 42 54	10 18	09 27	01 18	08 39	08 13	12 24	19 06	19 23
06	04 59 21	15 42 14	22 ♏ 42 21	29 ♏ 47 44	12 29	10 40	01 15	08 52	08 12	12 27	19 05	19 24
07	05 03 18	16 39 36	06 ♐ 58 27	14 ♐ 13 47	14 41	11 54	01 13	09 05	08 12	12 30	19 04	19 25
08	05 07 14	17 36 58	21 ♐ 32 52	28 ♐ 54 44	16 53	13 08	01 11	09 17	08 12	12 34	19 03	19 26
09	05 11 11	18 34 19	06 ♑ 18 23	13 ♑ 42 48	19 05	14 21	01 10	09 30	08 12	12 37	19 03	19 27
10	05 15 07	19 31 39	21 ♑ 07 00	28 ♑ 30 04	21 17	15 35	01 D 10	09 43	08 12	12 40	19 02	19 28
11	05 19 04	20 28 59	05 ♒ 51 11	13 ♒ 09 41	23 28	16 49	01 11	09 55	08 D 12	12 44	19 01	19 29
12	05 23 00	21 26 18	20 ♒ 24 59	27 ♒ 36 41	25 39	18 02	01 12	10 08	08 12	12 47	19 01	19 30
13	05 26 57	22 23 36	04 ♓ 44 29	11 ♓ 48 12	27 49	19 16	01 14	10 20	08 12	12 51	19 00	19 31
14	05 30 53	23 20 54	18 ♓ 47 46	25 ♓ 43 11	29 ♊ 58	20 30	01 17	10 33	08 12	12 54	19 00	19 32
15	05 34 50	24 18 12	02 ♈ 34 31	09 ♈ 21 52	02 ♋ 06	21 43	01 21	10 45	08 13	12 58	18 59	19 34
16	05 38 47	25 15 30	16 ♈ 05 21	22 ♈ 45 08	04 12	22 57	01 26	10 57	08 13	13 01	18 59	19 35
17	05 42 43	26 12 47	29 ♈ 21 21	05 ♉ 54 07	06 16	24 11	01 31	11 09	08 14	13 05	18 58	19 36
18	05 46 40	27 10 04	12 ♉ 23 34	18 ♉ 49 49	08 19	25 24	01 37	11 22	08 15	13 08	18 58	19 37
19	05 50 36	28 07 21	25 ♉ 12 56	01 ♊ 33 00	10 19	26 38	01 44	11 34	08 16	13 12	18 58	19 39
20	05 54 33	29 ♊ 04 37	07 ♊ 50 05	14 ♊ 04 16	12 18	27 52	01 51	11 45	08 16	13 15	18 57	19 40
21	05 58 29	00 ♋ 01 53	20 ♊ 15 38	26 ♊ 24 15	14 14	29 ♊ 05	02 00	11 57	08 18	13 19	18 57	19 41
22	06 02 26	00 59 09	02 ♋ 30 15	08 ♋ 33 45	16 09	00 ♋ 19	02 09	12 09	08 19	13 22	18 57	19 42
23	06 06 22	01 56 24	14 ♋ 34 56	20 ♋ 34 00	18 01	01 33	02 18	12 21	08 20	13 26	18 57	19 44
24	06 10 19	02 53 39	26 ♋ 31 13	02 ♌ 26 51	19 52	02 47	02 29	12 32	08 21	13 30	18 56	19 45
25	06 14 16	03 50 54	08 ♌ 21 15	14 ♌ 14 50	21 40	04 00	02 40	12 44	08 23	13 33	18 56	19 47
26	06 18 12	04 48 08	20 ♌ 07 59	26 ♌ 01 12	23 25	05 14	02 51	12 55	08 24	13 37	18 56	19 48
27	06 22 09	05 45 21	01 ♍ 55 01	07 ♍ 49 57	25 09	06 28	03 04	13 07	08 26	13 40	18 56	19 49
28	06 26 05	06 42 34	13 ♍ 46 37	19 ♍ 45 37	26 50	07 42	03 17	13 18	08 28	13 44	18 56	19 51
29	06 30 02	07 39 47	25 ♍ 47 34	01 ♎ 53 07	28 ♋ 29	08 55	03 30	13 29	08 30	13 48	18 56	19 52
30	06 33 58	08 ♋ 36 59	08 ♎ 02 54	14 ♎ 17 32	00 ♌ 06	10 ♋ 09	03 ♏ 44	13 ♉ 40	08 ♎ 31	13 ♋ 51	18 ♎ 56	19 ♌ 54

Date	Moon True ☊ ° '	Moon Mean ☊ ° '	Moon Latitude ° '
01	24 ♒ 33	25 ♒ 23	01 S 59
02	24 R 32	25 20	02 56
03	24 29	25 17	03 46
04	24 23	25 13	04 25
05	24 15	25 10	04 52
06	24 05	25 07	05 03
07	23 55	25 04	04 55
08	23 45	25 01	04 28
09	23 37	24 57	03 44
10	23 31	24 54	02 43
11	23 28	24 51	01 32
12	23 27	24 48	00 S 16
13	23 D 27	24 45	01 N 00
14	23 28	24 42	02 11
15	23 R 27	24 38	03 13
16	23 25	24 35	04 03
17	23 21	24 32	04 39
18	23 15	24 29	05 00
19	23 07	24 26	05 05
20	22 57	24 23	04 55
21	22 47	24 19	04 32
22	22 38	24 16	03 55
23	22 31	24 13	03 09
24	22 26	24 10	02 15
25	22 22	24 07	01 15
26	22 21	24 03	00 N 12
27	22 D 22	24 00	00 S 52
28	22 23	23 57	01 53
29	22 24	23 54	02 51
30	22 ♒ 25	23 ♒ 51	03 S 42

DECLINATIONS

Date	Sun ☉ ° '	Moon ☽ ° '	Mercury ☿ ° '	Venus ♀ ° '	Mars ♂ ° '	Jupiter ♃ ° '	Saturn ♄ ° '	Uranus ♅ ° '	Neptune ♆ ° '	Pluto ♇ ° '
01	22 N 05	03 N 07	20 N 03	20 N 34	12 S 37	13 N 07	00 S 54	23 N 15	05 S 56	23 N 33
02	22 13	02 S 36	20 38	20 49	12 36	13 11	00 54	23 15	05 55	23 32
03	22 21	08 21	21 11	21 04	12 36	13 16	00 54	23 15	05 55	23 32
04	22 28	13 56	21 43	21 18	12 37	13 20	00 54	23 15	05 55	23 31
05	22 34	19 03	22 14	21 32	12 37	13 24	00 54	23 14	05 55	23 31
06	22 41	23 20	22 42	21 45	12 38	13 28	00 54	23 14	05 54	23 31
07	22 47	26 20	23 08	21 57	12 39	13 32	00 54	23 14	05 54	23 30
08	22 52	27 39	23 32	22 09	12 41	13 36	00 54	23 13	05 54	23 30
09	22 57	27 01	23 53	22 20	12 42	13 40	00 54	23 13	05 54	23 29
10	23 02	24 29	24 12	22 31	12 44	13 44	00 54	23 13	05 54	23 29
11	23 06	20 19	24 28	22 41	12 46	13 48	00 54	23 12	05 53	23 28
12	23 10	14 57	24 41	22 50	12 48	13 52	00 55	23 12	05 53	23 28
13	23 14	08 51	24 51	22 59	12 51	13 55	00 55	23 12	05 53	23 27
14	23 17	02 S 25	24 59	23 07	12 53	13 59	00 55	23 11	05 53	23 27
15	23 19	03 N 59	25 04	23 14	12 56	14 03	00 56	23 11	05 53	23 26
16	23 22	10 04	25 06	23 21	13 00	14 07	00 56	23 11	05 53	23 26
17	23 24	15 36	25 05	23 27	13 03	14 10	00 57	23 10	05 52	23 25
18	23 25	20 19	25 02	23 32	13 07	14 14	00 57	23 10	05 52	23 25
19	23 26	24 00	24 56	23 36	13 10	14 18	00 58	23 10	05 52	23 24
20	23 27	26 29	24 48	23 40	13 15	14 21	00 58	23 09	05 52	23 23
21	23 27	27 36	24 37	23 44	13 19	14 25	00 59	23 09	05 52	23 23
22	23 27	27 21	24 25	23 46	13 23	14 28	01 00	23 09	05 52	23 22
23	23 26	25 47	24 10	23 48	13 28	14 32	01 00	23 08	05 52	23 22
24	23 25	23 04	23 54	23 49	13 33	14 35	01 01	23 08	05 52	23 21
25	23 23	19 23	23 36	23 49	13 38	14 38	01 02	23 08	05 52	23 21
26	23 22	14 58	23 16	23 49	13 43	14 42	01 03	23 07	05 52	23 20
27	23 19	10 00	22 55	23 48	13 49	14 45	01 04	23 07	05 52	23 20
28	23 17	04 N 38	22 32	23 46	13 55	14 48	01 04	23 06	05 52	23 19
29	23 14	00 S 57	22 08	23 44	14 00	14 52	01 05	23 06	05 52	23 18
30	23 N 10	06 S 36	21 N 43	23 N 41	14 S 06	14 N 55	01 S 06	23 N 06	05 S 52	23 N 18

ZODIAC SIGN ENTRIES

Date	h	m	Planets
02	12	26	☽ ♎
04	20	19	☽ ♏
07	00	21	☽ ♐
09	01	46	☽ ♑
11	02	26	☽ ♒
13	04	00	☽ ♓
14	12	22	☿ ♋
15	07	29	☽ ♈
17	13	11	☽ ♉
19	21	03	☽ ♊
21	11	13	☉ ♋
22	05	46	♀ ♋
22	07	04	☽ ♋
24	19	02	☽ ♌
27	08	06	☽ ♍
29	20	18	☽ ♎
30	10	27	☿ ♌

LATITUDES

Date	Mercury ☿ ° '	Venus ♀ ° '	Mars ♂ ° '	Jupiter ♃ ° '	Saturn ♄ ° '	Uranus ♅ ° '	Neptune ♆ ° '	Pluto ♇ ° '
01	00 S 29	00 S 30	00 S 35	01 S 03	02 N 35	00 N 22	01 N 42	08 N 59
04	00 N 03	00 23	00 42	01 03	02 35	00 22	01 42	08 59
07	00 34	00 16	00 48	01 03	02 34	00 22	01 42	08 58
10	01 02	00 09	00 54	01 03	02 33	00 22	01 41	08 58
13	01 25	00 S 02	01 00	01 03	02 33	00 22	01 41	08 57
16	01 43	00 N 05	01 05	01 04	02 32	00 22	01 41	08 57
19	01 53	00 12	01 10	01 04	02 31	00 22	01 41	08 57
22	01 57	00 19	01 15	01 04	02 30	00 22	01 41	08 56
25	01 55	00 26	01 19	01 04	02 30	00 22	01 41	08 56
28	01 46	00 33	01 23	01 05	02 29	00 22	01 41	08 56
31	01 N 32	00 N 39	01 S 27	01 S 05	02 N 28	00 N 22	01 N 40	08 N 55

DATA

Julian Date	2434165
Delta T	+30 seconds
Ayanamsa	23° 11' 42"
Synetic vernal point	05° ♓ 55' 18"
True obliquity of ecliptic	23° 26' 51"

MOON'S PHASES, APSIDES AND POSITIONS ☽

Date	h	m	Phase	Longitude ° '	Eclipse Indicator
08	05	07	○	17 ♐ 21	
14	20	28	☾	23 ♓ 41	
22	08	45	●	00 ♋ 51	
30	13	11	☽	08 ♎ 40	

Day	h	m		
10	06	39	Perigee	
25	23	18	Apogee	
02	01	09	0S	
08	16	27	Max dec	27° S 41'
14	21	00	0N	
21	19	29	Max dec	27° N 40'
29	07	58	0S	

LONGITUDES

Date	Chiron ⚷ ° '	Ceres ⚳ ° '	Pallas ⚴ ° '	Juno ⚵ ° '	Vesta ⚶ ° '	Black Moon Lilith ⚸ ° '
01	11 ♑ 19	12 ♉ 04	01 ♈ 48	22 ♒ 57	07 ♍ 53	07 ♌ 14
11	10 ♑ 46	15 ♉ 54	03 ♈ 56	23 ♒ 53	11 ♍ 03	08 ♌ 20
21	10 ♑ 10	19 ♉ 38	05 ♈ 47	24 ♒ 14	14 ♍ 34	09 ♌ 27
31	09 ♑ 32	23 ♉ 17	07 ♈ 17	23 ♒ 58	18 ♍ 23	10 ♌ 33

ASPECTARIAN

h m Aspects

01 Sunday
01 22 ☽ ✶ ♅
03 22 ☽ ⊥ ♆
04 29 ♀ ⚼ ♆
10 22 ☽ ∠ ♂
10 49 ☿ ⊼ ♂
15 15 ☽ ⊻ ♆
15 36 ☽ ⊻ ♇
21 24 ☽ # ♄
22 34 ☽ ⚼ ♃

02 Monday
01 11 ☽ Q ♅
03 22 ☽ ⊥ ♇
03 53 ☽ ⊥ ♂
04 54 ☽ ∥ ♄
14 56 ☿ ⚼ ♆
15 24 ☽ ⊻ ♂
16 21 ☽ ± ♃
20 46 ☽ ∠ ♇
21 26 ☉ ⊻ ♅
21 28 ☽ △ ☿
21 34 ♀ ⊥ ♅

03 Tuesday
00 43 ☽ △ ♀
01 49 ☽ ∥ ♆
02 04 ☿ ∥ ♀
04 00 ☽ ⊼ ♃
04 11 ☽ ☌ ♄
11 49 ☽ □ ♅
12 54 ☽ △ ☉
14 00 ♃ ⊼ ♄
15 21 ☿ ⊥ ♅
19 08 ♀ Q ♇
20 34 ♀ ± ♂

04 Wednesday
00 31 ☽ ☌ ♆
00 58 ☽ ✶ ♇
03 18 ☿ Q ♇
03 47 ☿ ± ♂
06 10 ☽ ∥ ♂
07 06 ☽ ⚼ ☿
07 38 ☽ ⚼ ♀
09 16 ☽ # ♃
12 06 ♀ △ ♄
12 55 ☿ △ ♄
13 59 ☿ ☌ ♀
14 52 ♂ Q ♇
15 36 ☿ ⊻ ♃
17 05 ♀ ⊻ ♃
18 37 ☽ ⚼ ☉
22 44 ☽ ☌ ♂
22 47 ☽ Q ♇

05 Thursday
01 32 ☽ ± ♀
02 29 ☽ ± ☿
08 47 ☉ ⊥ ♃
10 55 ☽ ⊻ ♄
11 41 ☽ ☍ ♃
11 51 ☽ ± ☉
13 11 ☽ ⊼ ♀
15 05 ☽ ⊼ ☿
18 16 ☽ △ ♅
21 24 ☽ ⊥ ♄
23 05 ☽ ⊼ ☉

06 Friday
01 53 ☽ # ♀
05 48 ☽ ⊻ ♆
06 20 ☽ □ ♇
07 31 ☽ # ☿
07 49 ☽ # ☉
10 41 ☉ ∥ ☿
11 23 ☽ # ♅
11 36 ☿ ⊻ ♅
12 51 ☽ ∠ ♄
13 10 ☽ # ♇
16 02 ☽ ⊥ ♆
20 05 ☽ ⚼ ♅

07 Saturday
01 09 ☉ ⚼ ♂
02 24 ☽ ⊻ ♂
07 10 ☽ ∠ ♆
11 13 ☽ ± ♅
12 24 ☽ ⊥ ♂
14 02 ☽ ✶ ♄
15 33 ☽ ⊼ ♃
16 50 ☿ ⊥ ♃
17 20 ☿ ∥ ♅
20 55 ☽ ☍ ♀
21 12 ☽ ⊼ ♅

08 Sunday
00 25 ♀ ⊻ ♅
01 36 ☽ ± ♃
03 00 ☽ ☍ ☿
03 14 ☽ ∠ ♂
04 30 ☿ ⚼ ♂
05 07 ☽ ☍ ☉
07 56 ☽ ✶ ♆
08 32 ☽ △ ♇
09 40 ☿ ∥ ♇
09 47 ☽ Q ♄
16 33 ☽ ⚼ ♃

09 Monday
02 14 ☉ ☌ ☿
03 28 ☽ Q ♆
03 40 ☽ ✶ ♂
08 59 ☽ ⚼ ♇
11 38 ☿ △ ♆
15 03 ☽ □ ♄
16 04 ☿ ✶ ♇
17 15 ☽ △ ♃
22 16 ☽ ☍ ♅
23 07 ☽ Q ♂
23 35 ☽ ± ♇
23 46 ☉ △ ♆

10 Tuesday
02 13 ☽ ⊼ ♀
02 45 ♂ St D
08 38 ☽ □ ♆
09 15 ☽ ⊼ ☉
09 19 ☽ ⊼ ♇
10 27 ☉ ✶ ♇
12 18 ☽ ⊼ ☿
12 49 ☽ ± ♀
13 20 ♄ St D
13 48 ☽ # ☿
15 06 ♀ ⊥ ♃
18 37 ☽ # ♇
19 40 ☽ ± ☉
20 15 ☽ # ♅
21 08 ☽ # ☉
23 30 ♀ ⚼ ♂
23 46 ☽ ± ☿
23 51 ☽ # ♀

11 Wednesday
04 21 ☽ □ ♂
04 47 ☽ ⚼ ♀
11 21 ☽ ⚼ ☉
15 50 ☽ △ ♄
17 03 ☽ ⚼ ☿
18 46 ☽ □ ♃
23 20 ☽ ⊼ ♅

12 Thursday
05 39 ☿ ∠ ♃
07 41 ☽ △ ♀
09 17 ☽ ± ♅
09 41 ☽ △ ♆
10 29 ☽ ☍ ♇
13 49 ☽ △ ☉
16 23 ☽ # ♃
16 37 ☽ ⚼ ♄
20 37 ☽ ∥ ♂
22 17 ☽ △ ☿
22 42 ☉ ∥ ♅

13 Friday
00 21 ☽ ⚼ ♅
01 04 ☽ Q ♃
06 05 ☽ △ ♂
06 59 ♀ △ ♆
07 42 ☽ ± ♄
10 46 ☽ ⚼ ♆
17 06 ♀ ✶ ♇
17 52 ☽ ⊼ ♄
21 39 ☽ ✶ ♃
23 08 ☽ ∥ ♆

14 Saturday
01 50 ☽ △ ♅
02 03 ☽ ± ♆
07 40 ☽ ⚼ ♂
12 21 ☽ ⊼ ♆
13 17 ☽ ⊼ ♇
15 13 ☽ □ ♀
17 34 ☽ ∥ ♄
20 28 ☽ □ ☉
23 18 ☽ ± ♂
23 42 ☽ ± ♇
23 52 ☽ ∠ ♃

15 Sunday
00 28 ☽ # ♄
02 15 ♀ ∥ ♅
03 21 ☿ △ ♂
09 51 ☽ ⊼ ♂
11 00 ☽ □ ☿
15 30 ☽ ⚼ ♇
15 53 ☽ ⊥ ♃
19 18 ☽ # ♆
21 58 ☽ ☍ ♄

16 Monday
01 54 ☽ Q ♀
02 41 ☽ ⊻ ♃
06 29 ☽ □ ♅
06 33 ☽ Q ☉
16 30 ☿ ∠ ♇
17 12 ☽ ☍ ♆
18 13 ☉ ∥ ♀
18 17 ☽ △ ♇

17 Tuesday
00 26 ☽ # ♂
01 02 ☽ Q ☿
01 37 ☽ ✶ ♀
05 25 ☽ ∥ ♃
05 50 ☽ ✶ ☉
05 56 ♀ ∥ ♇
10 13 ☉ ∠ ♃
15 10 ☽ Q ♅
15 59 ☽ ☍ ♂

18 Wednesday
03 02 ☽ ✶ ☿
04 19 ☽ ⊼ ♄
05 11 ☉ ∥ ♇
07 56 ☽ ∠ ♀
10 03 ☽ ☌ ♃
11 14 ☿ □ ♄
11 33 ☽ ∠ ☉
12 44 ♂ Q ♇
13 23 ☽ ✶ ♅
15 27 ☽ ± ♄

19 Thursday
00 15 ☽ ⊼ ♆
01 30 ☽ □ ♇
02 28 ☽ ⊥ ♀
05 42 ☽ ⊥ ☉
05 47 ☽ ∥ ♅
07 29 ☽ ∥ ♇
07 42 ☽ ∥ ☉
08 19 ☽ ⚼ ♄
08 55 ☽ ∥ ♀
10 15 ♀ ∠ ♃
11 31 ☽ ± ♆
12 14 ☽ ∠ ☿
14 58 ☽ ⊻ ♀
17 40 ☽ ∠ ♅
17 57 ☽ ⊻ ☉
19 17 ☽ ∥ ☿

20 Friday
00 28 ☽ ⊼ ♂
04 35 ☽ ⚼ ♆
04 39 ☿ ✶ ♃
08 31 ☽ ⊥ ☿
10 53 ☽ ⊥ ♅
11 40 ☽ Q ♇
12 03 ☽ ± ♂
12 51 ☽ △ ♄
19 39 ☽ ⊻ ♃
22 11 ☽ ⊻ ☿
22 28 ☽ ⊻ ♅

21 Saturday
00 08 ☿ ☌ ♅
05 01 ☿ ⊥ ♇
05 35 ☽ ⚼ ♂
07 27 ☽ ⊥ ♃
09 27 ☽ △ ♆
10 53 ☽ ✶ ♇

22 Sunday
01 18 ☽ ∠ ♃
07 13 ☽ ☌ ♀
08 45 ☽ ☌ ☉
11 17 ☽ △ ♂
16 22 ☽ ∠ ♇
23 31 ☽ □ ♄

23 Monday
07 28 ☽ ✶ ♃
09 42 ☽ ☌ ♅
10 18 ☽ ⊥ ♇
20 09 ☽ ☌ ☿
20 44 ☽ □ ♆
22 20 ☽ ⊻ ♇
23 08 ☉ △ ♂
23 57 ☿ □ ♆

24 Tuesday
04 51 ☽ ∥ ☿
05 09 ♀ △ ♂
06 14 ☽ ∥ ♀
07 56 ☽ Q ♃
09 19 ☽ ∥ ☉
09 48 ☽ ∥ ♇
10 35 ☿ ⊻ ♇
11 29 ☽ ∥ ♅
11 40 ☽ Q ♄
18 31 ☿ ∥ ♀
22 17 ☉ ☌ ♀

25 Wednesday
00 15 ☽ □ ♂
02 02 ☽ ⊻ ☉
02 08 ☽ ⊻ ♀
09 07 ☽ Q ♆
12 03 ☽ ✶ ♄
15 18 ☽ ⊥ ☉
15 45 ☽ ⊥ ♀
21 03 ☽ □ ♃
22 38 ☽ ⊻ ♅

26 Thursday
03 21 ♀ ∠ ♇
04 34 ☉ ∥ ☿
06 56 ☿ ∥ ♇
09 33 ☽ ✶ ♆
10 56 ☽ ⊥ ♅
11 16 ☽ ∠ ☉
11 19 ☽ ☌ ♇
11 56 ☉ ∠ ♇
12 14 ☽ ∠ ♀
13 21 ☽ ∥ ♃
13 30 ☽ Q ♂
18 07 ☽ # ♂
18 41 ☽ ∠ ♄
19 52 ☽ ⊻ ☿
22 20 ☿ ∥ ♅

27 Friday
05 22 ☽ ∠ ♅
08 49 ☉ ∥ ♇
10 10 ☽ ⊥ ☿
11 23 ☿ Q ♃
13 03 ☽ ⊥ ♄
14 22 ☽ ✶ ♂
16 06 ☽ ∠ ♆
20 29 ☽ ✶ ☉
22 18 ☽ ✶ ♀

28 Saturday
01 15 ☽ ⊻ ♄
06 31 ☿ Q ♄
06 36 ☽ # ♆
07 27 ☽ ∠ ☿
10 18 ☽ ⊥ ♆
11 01 ☽ △ ♃
11 55 ☽ ✶ ♅
21 12 ☽ ∠ ♂
22 21 ☽ ⊻ ♆
22 45 ☽ Q ☉

29 Sunday
00 12 ☽ ⊻ ♇
01 13 ☽ Q ♀
03 20 ☽ # ♄
03 24 ♀ □ ♄
12 00 ☽ Q ♅
12 09 ☽ ⊥ ♇
12 37 ☽ ∥ ♄
15 27 ☽ ⊥ ♂
17 24 ☽ ⚼ ♃
18 09 ☽ ✶ ☿

30 Monday
03 27 ☽ ⊻ ♂
05 52 ☽ ∠ ♇
08 53 ☽ ∥ ♆
09 23 ♆ St D
09 35 ☉ □ ♄
11 15 ☽ ± ♃
12 55 ☽ ☌ ♄
13 11 ☽ □ ☉
16 30 ☽ □ ♀
20 57 ☽ Q ☿
22 58 ☽ ⊼ ♃
23 13 ☽ □ ♅

All ephemeris data is given at 12.00 UT and the Moon's longitude is additionally given for 24.00 UT

JULY 1952

Date	Sidereal time h m s	Sun ☉ ° ' "	Moon ☽ ° ' "	Moon ☽ 24.00 ° ' "	Mercury ☿ ° '	Venus ♀ ° '	Mars ♂ ° '	Jupiter ♃ ° '	Saturn ♄ ° '	Uranus ♅ ° '	Neptune ♆ ° '	Pluto ♇ ° '
01	06 37 55	09 ♋ 34 11	20 ♎ 37 34	27 ♎ 03 34	01 ♌ 41	11 ♋ 23	03 ♏ 59	13 ♉ 51	08 ♎ 33	13 ♋ 55	18 ♎ 56	19 ♌ 55
02	06 41 51	10 31 22	03 ♏ 35 58	10 ♏ 15 08	03 13	12 37	04 14	14 02	08 36	13 58	18 D 56	19 57
03	06 45 48	11 28 34	17 ♏ 01 18	23 ♏ 54 33	04 43	13 50	04 30	14 12	08 38	14 02	18 56	19 58
04	06 49 45	12 25 45	00 ♐ 54 51	08 ♐ 01 55	06 11	15 04	04 47	14 23	08 40	14 06	18 56	20 00
05	06 53 41	13 22 56	15 ♐ 15 20	22 ♐ 34 29	07 37	16 18	05 04	14 34	08 43	14 09	18 56	20 01
06	06 57 38	14 20 06	29 ♐ 58 34	07 ♑ 26 38	09 00	17 32	05 22	14 44	08 45	14 13	18 56	20 03
07	07 01 34	15 17 17	14 ♑ 57 36	22 ♑ 30 21	10 20	18 45	05 40	14 54	08 48	14 17	18 57	20 05
08	07 05 31	16 14 28	00 ♒ 03 40	07 ♒ 36 24	11 39	19 59	05 59	15 05	08 50	14 20	18 57	20 06
09	07 09 27	17 11 39	15 ♒ 07 27	22 ♒ 35 48	12 55	21 13	06 18	15 15	08 53	14 24	18 57	20 08
10	07 13 24	18 08 50	00 ♓ 00 36	07 ♓ 21 08	14 08	22 27	06 38	15 25	08 56	14 28	18 58	20 09
11	07 17 20	19 06 02	14 ♓ 36 48	21 ♓ 47 13	15 19	23 40	06 58	15 35	08 59	14 31	18 58	20 11
12	07 21 17	20 03 14	28 ♓ 52 07	05 ♈ 51 23	16 27	24 54	07 18	15 44	09 02	14 35	18 58	20 13
13	07 25 14	21 00 26	12 ♈ 45 00	19 ♈ 33 04	17 32	26 08	07 40	15 54	09 05	14 38	18 59	20 14
14	07 29 10	21 57 40	26 ♈ 15 45	02 ♉ 53 16	18 35	27 22	08 01	16 04	09 08	14 42	18 59	20 16
15	07 33 07	22 54 54	09 ♉ 25 54	15 ♉ 53 56	19 34	28 36	08 23	16 13	09 12	14 46	19 00	20 18
16	07 37 03	23 52 08	22 ♉ 17 40	28 ♉ 37 27	20 31	29 ♋ 50	08 46	16 22	09 15	14 49	19 00	20 19
17	07 41 00	24 49 24	04 ♊ 53 34	11 ♊ 06 20	21 25	01 ♌ 03	09 09	16 32	09 18	14 53	19 01	20 21
18	07 44 56	25 46 40	17 ♊ 16 03	23 ♊ 22 58	22 15	02 17	09 32	16 41	09 22	14 56	19 01	20 23
19	07 48 53	26 43 56	29 ♊ 27 22	05 ♋ 29 32	23 02	03 31	09 56	16 50	09 26	15 00	19 02	20 25
20	07 52 49	27 41 13	11 ♋ 29 40	17 ♋ 28 03	23 45	04 45	10 20	16 58	09 29	15 04	19 02	20 26
21	07 56 46	28 38 31	23 ♋ 24 55	29 ♋ 20 31	24 25	05 59	10 45	17 07	09 33	15 07	19 03	20 28
22	08 00 43	29 ♋ 35 49	05 ♌ 15 06	11 ♌ 08 57	25 00	07 13	11 10	17 16	09 37	15 11	19 04	20 30
23	08 04 39	00 ♌ 33 08	17 ♌ 02 20	22 ♌ 55 35	25 32	08 27	11 36	17 24	09 41	15 14	19 04	20 32
24	08 08 36	01 30 27	28 ♌ 49 01	04 ♍ 43 00	26 00	09 41	12 02	17 32	09 45	15 18	19 05	20 34
25	08 12 32	02 27 47	10 ♍ 37 55	16 ♍ 34 11	26 23	10 54	12 28	17 41	09 49	15 21	19 06	20 35
26	08 16 29	03 25 08	22 ♍ 32 15	28 ♍ 32 35	26 42	12 08	12 55	17 49	09 54	15 25	19 07	20 37
27	08 20 25	04 22 28	04 ♎ 35 41	10 ♎ 42 03	26 56	13 22	13 22	17 56	09 58	15 28	19 08	20 39
28	08 24 22	05 19 50	16 ♎ 52 12	23 ♎ 06 40	27 05	14 36	13 49	18 04	10 02	15 32	19 09	20 41
29	08 28 18	06 17 12	29 ♎ 25 57	05 ♏ 50 34	27 10	15 50	14 17	18 12	10 07	15 35	19 09	20 43
30	08 32 15	07 14 34	12 ♏ 20 58	18 ♏ 57 32	27 R 09	17 04	14 45	18 19	10 11	15 39	19 10	20 44
31	08 36 12	08 ♌ 11 57	25 ♏ 40 35	02 ♐ 30 22	27 ♌ 03	18 ♌ 18	15 ♏ 13	18 ♉ 26	10 ♎ 16	15 ♋ 42	19 ♎ 11	20 ♌ 46

Date	Moon True ☊ ° '	Moon Mean ☊ ° '	Moon ☽ Latitude ° '
01	22 ♒ 24	23 ♒ 48	04 S 24
02	22 R 22	23 45	04 54
03	22 18	23 41	05 09
04	22 13	23 38	05 07
05	22 07	23 35	04 46
06	22 02	23 32	04 05
07	21 57	23 29	03 08
08	21 54	23 25	01 56
09	21 52	23 22	00 S 37
10	21 D 52	23 19	00 N 44
11	21 54	23 16	02 01
12	21 55	23 13	03 09
13	21 56	23 09	04 03
14	21 R 56	23 06	04 43
15	21 55	23 03	05 06
16	21 53	23 00	05 13
17	21 49	22 57	05 05
18	21 45	22 54	04 43
19	21 41	22 50	04 08
20	21 37	22 47	03 23
21	21 34	22 44	02 29
22	21 32	22 41	01 29
23	21 31	22 38	00 N 25
24	21 D 31	22 34	00 S 40
25	21 32	22 31	01 44
26	21 34	22 28	02 43
27	21 35	22 25	03 36
28	21 36	22 22	04 21
29	21 37	22 19	04 54
30	21 R 37	22 15	05 13
31	21 ♒ 36	22 ♒ 12	05 S 16

DECLINATIONS

Date	Sun ☉ ° '	Moon ☽ ° '	Mercury ☿ ° '	Venus ♀ ° '	Mars ♂ ° '	Jupiter ♃ ° '	Saturn ♄ ° '	Uranus ♅ ° '	Neptune ♆ ° '	Pluto ♇ ° '
01	23 N 06	12 S 08	21 N 17	23 N 37	14 S 13	14 N 58	01 S 07	23 N 05	05 S 52	23 N 17
02	23 02	17 19	20 51	23 32	14 19	15 01	01 08	23 05	05 52	23 17
03	22 57	21 52	20 23	23 27	14 25	15 04	01 10	23 05	05 52	23 16
04	22 52	25 21	19 55	23 21	14 32	15 07	01 11	23 04	05 52	23 16
05	22 46	27 22	19 26	23 14	14 39	15 10	01 12	23 04	05 52	23 15
06	22 41	27 32	18 56	23 07	14 46	15 13	01 13	23 03	05 53	23 14
07	22 34	25 43	18 27	22 59	14 53	15 16	01 14	23 03	05 53	23 14
08	22 28	22 02	17 57	22 50	15 00	15 19	01 16	23 03	05 53	23 13
09	22 20	16 53	17 26	22 41	15 07	15 22	01 17	23 02	05 53	23 13
10	22 13	10 47	16 56	22 31	15 15	15 25	01 18	23 02	05 53	23 12
11	22 05	04 S 12	16 25	22 20	15 22	15 27	01 20	23 02	05 53	23 11
12	21 57	02 N 26	15 55	22 09	15 30	15 30	01 21	23 01	05 54	23 11
13	21 48	08 46	15 25	21 57	15 38	15 33	01 22	23 01	05 54	23 10
14	21 39	14 32	14 55	21 44	15 46	15 35	01 24	23 00	05 54	23 10
15	21 30	19 28	14 25	21 31	15 54	15 38	01 25	23 00	05 54	23 09
16	21 20	23 24	13 56	21 17	16 02	15 40	01 27	23 00	05 54	23 08
17	21 10	26 07	13 28	21 03	16 10	15 43	01 29	22 59	05 55	23 08
18	21 00	27 32	13 00	20 48	16 18	15 45	01 30	22 59	05 55	23 07
19	20 49	27 35	12 32	20 32	16 27	15 48	01 32	22 58	05 55	23 07
20	20 38	26 19	12 06	20 16	16 35	15 50	01 33	22 58	05 56	23 06
21	20 26	23 52	11 41	19 59	16 44	15 52	01 35	22 58	05 56	23 05
22	20 15	20 24	11 16	19 41	16 52	15 55	01 37	22 57	05 56	23 05
23	20 02	16 08	10 53	19 23	17 01	15 57	01 39	22 57	05 56	23 04
24	19 50	11 16	10 31	19 05	17 09	15 59	01 40	22 57	05 57	23 04
25	19 37	05 59	10 11	18 46	17 18	16 01	01 42	22 56	05 57	23 03
26	19 24	00 N 28	09 52	18 26	17 27	16 03	01 44	22 56	05 57	23 02
27	19 10	05 S 08	09 35	18 06	17 36	16 05	01 46	22 55	05 58	23 02
28	18 57	10 38	09 20	17 45	17 45	16 07	01 48	22 55	05 58	23 01
29	18 42	15 51	09 07	17 24	17 53	16 09	01 50	22 55	05 59	23 01
30	18 28	20 30	08 56	17 03	18 02	16 11	01 52	22 54	05 59	23 00
31	18 N 13	24 S 18	08 N 47	16 N 41	18 S 11	16 N 13	01 S 54	22 N 54	05 S 59	22 N 59

ZODIAC SIGN ENTRIES

Date	h	m	Planets
02	05	25	☽ ♏
04	10	27	☽ ♐
06	12	02	☽ ♑
08	11	54	☽ ♒
10	11	59	☽ ♓
12	13	56	☽ ♈
14	18	45	☽ ♉
16	15	23	♀ ♌
17	02	37	☽ ♊
19	13	05	☽ ♋
22	01	20	☽ ♌
22	22	07	☉ ♌
24	14	24	☽ ♍
27	02	54	☽ ♎
29	13	04	☽ ♏
31	19	37	☽ ♐

LATITUDES

Date	Mercury ☿ ° '	Venus ♀ ° '	Mars ♂ ° '	Jupiter ♃ ° '	Saturn ♄ ° '	Uranus ♅ ° '	Neptune ♆ ° '	Pluto ♇ ° '
01	01 N 32	00 N 39	01 S 27	01 S 05	02 N 28	00 N 22	01 N 40	08 N 55
04	01 13	00 45	01 30	01 05	02 28	00 22	01 40	08 55
07	00 49	00 51	01 33	01 06	02 27	00 22	01 40	08 55
10	00 N 21	00 57	01 36	01 06	02 26	00 22	01 40	08 55
13	00 S 11	01 02	01 39	01 06	02 25	00 22	01 40	08 54
16	00 46	01 07	01 42	01 07	02 25	00 22	01 40	08 54
19	01 23	01 11	01 44	01 07	02 24	00 22	01 39	08 54
22	02 02	01 15	01 46	01 08	02 23	00 22	01 39	08 54
25	02 42	01 19	01 48	01 08	02 23	00 22	01 39	08 54
28	03 21	01 22	01 50	01 08	02 22	00 23	01 39	08 54
31	03 S 57	01 N 24	01 S 52	01 S 09	02 N 22	00 N 23	01 N 39	08 N 54

DATA

Julian Date	2434195
Delta T	+30 seconds
Ayanamsa	23° 11' 47"
Synetic vernal point	05° ♓ 55' 13"
True obliquity of ecliptic	23° 26' 51"

MOON'S PHASES, APSIDES AND POSITIONS ☽

Date	h	m	Phase	Longitude ° '	Eclipse Indicator
07	12	33	○	15 ♑ 19	
14	03	42	☾	21 ♈ 38	
21	23	31	●	29 ♋ 06	
30	01	51	☽	06 ♏ 50	

Day	h	m		
08	11	18	Perigee	
23	08	21	Apogee	
06	02	10	Max dec	27° S 42'
12	03	07	0N	
19	00	52	Max dec	27° N 44'
26	13	59	0S	

LONGITUDES

Date	Chiron ⚷ ° '	Ceres ⚳ ° '	Pallas ⚴ ° '	Juno ⚵ ° '	Vesta ⚶ ° '	Black Moon Lilith ⚸ ° '
01	09 ♑ 32	23 ♉ 17	07 ♈ 17	23 ♒ 58	18 ♍ 23	10 ♌ 33
11	08 ♑ 54	26 ♉ 49	08 ♈ 23	23 ♒ 02	22 ♍ 27	11 ♌ 40
21	08 ♑ 18	00 ♊ 13	09 ♈ 03	21 ♒ 29	26 ♍ 44	12 ♌ 46
31	07 ♑ 45	03 ♊ 26	09 ♈ 09	19 ♒ 25	01 ♎ 12	13 ♌ 53

All ephemeris data is given at 12.00 UT and the Moon's longitude is additionally given for 24.00 UT

ASPECTARIAN

h m	Aspects
01 Tuesday	
08 48	☽ ☌ ♆
10 40	☽ ✶ ♇
16 30	♅ ⊥ ♇
17 01	☉ ∥ ♅
21 32	☽ ∥ ♂
02 Wednesday	
00 57	☽ # ♃
01 19	♃ ✶ ♅
08 59	☽ Q ♇
11 13	☽ □ ☿
13 11	☽ ☌ ♂
21 03	☽ ⊻ ♄
03 Thursday	
01 26	☽ △ ☉
04 28	☽ # ☿
05 49	☽ △ ♀
06 43	☽ △ ♅
06 58	☽ ☍ ♃
07 46	☽ ⊥ ♄
07 48	☿ □ ♂
14 40	♀ ⊥ ♇
15 21	☽ ⊻ ♆
16 03	♀ ☌ ♅
17 10	☽ □ ♇
18 31	☽ # ☉
19 27	☽ # ♅
20 26	♀ ✶ ♃
20 41	☽ # ♇
21 38	☽ # ♀
23 33	☽ ∠ ♄
04 Friday	
01 46	☽ ⊥ ♆
05 37	☽ ⚼ ☉
08 53	☽ ⚼ ♅
10 25	☽ ⚼ ♀
17 07	☽ ∠ ♆
18 41	☽ ⊻ ♂
21 54	☽ △ ☿
21 59	☽ ± ☉
05 Saturday	
00 10	☽ ± ♅
00 36	☿ Q ♆
01 06	☽ ✶ ♄
03 01	☽ ± ♀
04 55	☽ ⊥ ♂
08 41	☽ ⊼ ☉
09 29	♀ ∥ ♇
10 11	☽ ⊼ ♅
10 50	☽ ⊼ ♃
13 52	☽ ⊼ ♀
18 03	☽ ✶ ♆
19 51	☽ △ ♇
20 04	☽ ∠ ♂
20 49	☽ ± ♃
20 59	☽ Q ♄
06 Sunday	
01 19	☽ ⚼ ☿
04 36	☉ ⊥ ♇
07 35	☿ ✶ ♄
08 50	☉ ☌ ♅
11 36	☽ ⚼ ♃
13 33	☽ Q ♆
15 03	♃ ± ♄
17 21	☽ ± ☿
20 11	☽ ⚼ ♇
20 51	☽ ✶ ♂
23 21	♀ ∥ ♅
07 Monday	
00 16	☉ ✶ ♃
02 08	☽ □ ♄
03 55	☽ ⊼ ☿
10 35	☽ ± ♇
10 55	☽ ☍ ♅
11 55	☽ △ ♃
12 33	☽ ☍ ☉
15 43	♀ □ ♆
16 24	☽ Q ♂
18 21	☽ □ ♆
18 35	☽ ☍ ♀
20 09	☽ ⊼ ♇
08 Tuesday	
05 21	☽ # ♇
06 23	☽ # ♅
07 25	☽ # ♀
09 39	☽ # ☉
14 21	♀ ⊻ ♇
21 37	☽ □ ♂
09 Wednesday	
02 00	☽ △ ♄
08 08	☽ ☍ ☿
09 28	☽ # ☿
10 50	☽ ⊼ ♅
12 12	☽ □ ♃
15 32	☽ ⊼ ☉
18 08	☽ △ ♆
18 14	☽ # ♃
19 06	☽ ∥ ♂
20 03	☽ ☍ ♇
20 30	☽ ± ♅
22 39	☽ ⊼ ♀
10 Thursday	
01 51	☽ ± ☉
02 07	☽ ⚼ ♄
09 14	☽ ± ♀
11 06	☽ ⚼ ♅
16 47	☽ ± ♄
17 28	☽ ⚼ ☉
17 37	☽ Q ♃
18 26	☽ ⚼ ♆
18 54	☿ ⊻ ♅
23 04	☽ △ ♂
11 Friday	
01 16	☽ ⚼ ♀
02 39	☽ ⊼ ♄
05 54	☽ ∥ ♆
08 33	☉ □ ♆
09 16	☽ ± ♆
11 51	☽ △ ♅
13 16	☽ ⊼ ☿
13 37	☽ ✶ ♃
18 26	☿ □ ♃
19 16	☽ ⊼ ♆
20 01	☽ △ ☉
21 19	☽ ⊼ ♇
22 17	☽ ∥ ♄
12 Saturday	
00 11	☽ ± ☿
00 36	☽ ⚼ ♂
04 37	☽ △ ♀
07 28	☽ ± ♇
08 01	☽ # ♄
11 22	♂ # ♃
15 14	☽ ∠ ♃
16 05	☉ ⊻ ♇
16 17	☽ ± ♂
16 47	☽ ⚼ ☿
22 54	☽ ⚼ ♇
13 Sunday	
00 54	☽ # ♆
02 53	☽ ⊼ ♂
03 44	☿ # ♂
05 34	☽ ☍ ♄
06 16	☿ ∥ ♃
06 58	☽ ⊥ ♃
15 20	☽ □ ♅
17 36	☽ ⊻ ♃
21 09	☽ △ ☿
22 59	☽ ☍ ♆
14 Monday	
01 15	☽ △ ♇
03 42	☽ □ ☉
07 22	♀ Q ♄
13 35	☽ ∥ ☿
14 11	☽ □ ♀
16 51	☽ ∥ ♃
17 45	☽ # ♂
21 43	☿ ✶ ♆
23 43	☽ Q ♅
15 Tuesday	
03 33	♀ Q ♃
10 01	☽ ☍ ♂
11 34	☽ ⊼ ♄
14 57	☽ Q ☉
17 54	☉ ∥ ♀
21 56	☽ ✶ ♅
22 44	☽ ± ♄
23 02	☽ ∥ ♀
23 07	☽ ∥ ☉
16 Wednesday	
00 45	☽ ☌ ♃
02 43	☽ Q ♀
05 48	☽ ⊼ ♆
06 50	☿ ☌ ♇
08 17	☽ □ ♇
08 23	☽ □ ☿
09 12	☽ ∥ ♅
10 13	☽ ∥ ♇
15 13	☽ ✶ ☉
15 43	☽ ⚼ ♄
17 07	☽ ± ♆
20 36	☿ ⊥ ♅
17 Thursday	
02 21	☽ ∠ ♅
03 51	☽ ✶ ♀
10 18	☽ ⚼ ♆
18 41	☽ Q ♇
19 44	☽ ⊥ ♅
20 28	☽ ⊼ ♂
20 33	☽ △ ♄
21 21	☽ Q ☿
22 19	☽ ∠ ☉
23 37	♂ ⊻ ♄
18 Friday	
07 26	☽ ⊻ ♅
08 31	☽ ± ♂
10 50	☽ ⊻ ♃
12 03	☽ ∠ ♀
15 26	☽ △ ♆
17 20	☽ ⊥ ☉
18 07	☽ ✶ ♇
22 27	☽ ✶ ☿
22 45	☽ ⊥ ♃
19 Saturday	
02 46	☽ ⚼ ♂
06 09	☽ ⊻ ☉
07 44	☽ ⊥ ♀
16 46	☽ ∠ ♃
20 59	☽ ⊻ ♀
23 52	☽ ∠ ♇
20 Sunday	
06 10	☽ ∠ ☿
06 43	☉ Q ♄
07 58	☽ □ ♄
09 37	☽ △ ♂
17 56	☽ ⊥ ♇
19 12	☽ ☌ ♅
23 08	☽ ✶ ♃
21 Monday	
01 19	☽ ⊥ ☿
03 11	☽ □ ♆
06 02	☽ ⊻ ♇
14 07	☽ ⊻ ☿
17 56	☽ ∥ ♇
18 14	☿ ∠ ♄
18 53	☽ ∥ ♅
20 25	☽ Q ♄
23 31	☽ ☌ ☉
23 41	☽ Q ♃
22 Tuesday	
02 05	☉ Q ♃
09 02	♀ Q ♆
12 59	☽ ∥ ☉
15 41	☽ Q ♆
16 27	☽ ☌ ♀
16 36	☽ ∥ ♀
20 56	☽ ✶ ♄
23 Wednesday	
00 30	☽ □ ♂
07 28	☽ # ♂
08 19	☽ ⊻ ♅
12 45	☽ □ ♃
12 55	☽ ∥ ♃
16 09	☽ ✶ ♆
19 08	☽ ☌ ♇
20 36	☽ ⊥ ♅
24 Thursday	
03 40	☽ ∠ ♄
06 02	☽ ☌ ☿
13 35	♀ ✶ ♄
14 34	☽ Q ♂
15 02	☽ ∠ ♅
15 42	☽ ∥ ☿
17 58	☽ ⊻ ☉
22 06	☽ ⊥ ♄
22 44	☽ ∠ ♆
25 Friday	
07 13	☽ ⊥ ☉
10 21	☽ ⊻ ♄
12 08	☽ # ♆
12 37	☽ ⊻ ♀
15 51	☽ ✶ ♂
17 00	☽ ⊥ ♆
21 36	☽ ✶ ♅
26 Saturday	
02 09	☽ ⊥ ♀
02 23	☽ △ ♃
03 00	☽ ∠ ☉
05 07	☽ ⊻ ♆
06 34	☽ # ♄
08 08	☽ ⊻ ♇
20 11	☽ ⊥ ♇
20 30	☽ ⊻ ☿
21 28	☽ ∥ ♄
21 48	☽ Q ♅
22 15	☽ ∠ ♀
23 09	☽ ∠ ♂
27 Sunday	
08 40	☽ ⊥ ☿
08 42	☽ ⚼ ♃
11 32	☽ ✶ ☉
11 43	♀ □ ♂
14 05	☽ ∠ ♇
15 34	☽ ∥ ♆
17 40	☽ ⊥ ♂
22 37	☽ ☌ ♄
28 Monday	
02 34	☽ ± ♃
02 37	☽ ∠ ☿
05 51	☽ ⊻ ♂
06 29	☽ # ☿
07 07	☽ ✶ ♀
09 24	☽ □ ♅
12 45	♀ # ♂
12 58	☽ Q ☉
14 20	☽ ⊼ ♃
16 23	☽ ☌ ♆
19 22	☽ ✶ ♇
29 Tuesday	
06 57	♀ ⊻ ♅
07 41	☽ ✶ ☿
08 39	☽ Q ♀
13 30	☽ # ♃
18 10	☽ Q ♇
19 08	☽ # ♀
20 31	☿ St R
22 25	☽ ∥ ♂
30 Wednesday	
01 36	☽ # ☉
01 51	☽ □ ☉
06 08	☽ Q ☿
08 00	☽ ⊻ ♄
10 13	☉ Q ♆
16 33	☽ ☌ ♂
18 02	☽ △ ♅
19 02	☽ ⊥ ♄
21 28	☽ □ ♀
22 57	☽ ☍ ♃
31 Thursday	
00 24	☽ ⊻ ♆
02 21	☽ # ♅
02 58	☽ # ♇
03 14	☽ □ ♇
11 08	☽ ⊥ ♆
11 16	☽ ∠ ♄
14 05	☉ # ♂
14 24	☽ □ ☿
15 04	♀ □ ♃
20 53	☽ ⚼ ♅

AUGUST 1952

LONGITUDES

Date	Sidereal time h m s	Sun ☉ ° ' "	Moon ☽ ° ' "	Moon ☽ 24.00 ° ' "	Mercury ☿ ° '	Venus ♀ ° '	Mars ♂ ° '	Jupiter ♃ ° '	Saturn ♄ ° '	Uranus ♅ ° '	Neptune ♆ ° '	Pluto ♇ ° '
01	08 40 08	09 ♌ 09 21	09 ♐ 26 58	16 ♐ 30 21	26 ♌ 52	19 ♌ 32	15 ♏ 42	18 ♉ 34	10 ♎ 20	15 ♋ 46	19 ♎ 12	20 ♌ 48
02	08 44 05	10 06 45	23 ♐ 40 18	00 ♑ 56 29	26 R 36	20 46	16 11	18 41	10 25	15 49	19 13	20 50
03	08 48 01	11 04 10	08 ♑ 18 20	15 ♑ 45 08	26 14	22 00	16 41	18 47	10 30	15 52	19 14	20 52
04	08 51 58	12 01 36	23 ♑ 16 00	00 ♒ 49 55	25 48	23 14	17 10	18 54	10 35	15 56	19 16	20 54
05	08 55 54	12 59 02	08 ♒ 25 44	16 ♒ 02 16	25 17	24 27	17 41	19 01	10 40	15 59	19 17	20 56
06	08 59 51	13 56 30	23 ♒ 38 18	01 ♓ 12 38	24 41	25 41	18 11	19 07	10 45	16 02	19 18	20 57
07	09 03 47	14 53 58	08 ♓ 44 09	16 ♓ 11 51	24 02	26 55	18 41	19 13	10 50	16 06	19 19	20 59
08	09 07 44	15 51 28	23 ♓ 34 51	00 ♈ 52 26	23 19	28 09	19 12	19 19	10 55	16 09	19 20	21 01
09	09 11 41	16 48 59	08 ♈ 04 02	15 ♈ 09 19	22 34	29 ♌ 23	19 44	19 25	11 00	16 12	19 21	21 03
10	09 15 37	17 46 31	22 ♈ 08 02	29 ♈ 00 09	21 46	00 ♍ 37	20 15	19 31	11 06	16 15	19 23	21 05
11	09 19 34	18 44 05	05 ♉ 45 43	12 ♉ 24 56	20 57	01 51	20 47	19 37	11 11	16 19	19 24	21 07
12	09 23 30	19 41 40	18 ♉ 58 05	25 ♉ 25 30	20 08	03 05	21 19	19 42	11 16	16 22	19 25	21 09
13	09 27 27	20 39 17	01 ♊ 47 37	08 ♊ 04 52	19 19	04 19	21 51	19 47	11 22	16 25	19 26	21 11
14	09 31 23	21 36 55	14 ♊ 17 44	20 ♊ 26 40	18 32	05 33	22 24	19 52	11 27	16 28	19 28	21 12
15	09 35 20	22 34 35	26 ♊ 32 11	02 ♋ 34 44	17 47	06 47	22 57	19 57	11 33	16 31	19 29	21 14
16	09 39 16	23 32 17	08 ♋ 34 46	14 ♋ 32 44	17 06	08 01	23 30	20 02	11 39	16 34	19 31	21 16
17	09 43 13	24 29 59	20 ♋ 29 03	26 ♋ 24 06	16 29	09 15	24 03	20 06	11 44	16 37	19 32	21 18
18	09 47 10	25 27 44	02 ♌ 18 16	08 ♌ 11 54	15 57	10 29	24 37	20 11	11 50	16 40	19 34	21 20
19	09 51 06	26 25 29	14 ♌ 05 19	19 ♌ 58 50	15 31	11 43	25 11	20 15	11 56	16 43	19 35	21 22
20	09 55 03	27 23 16	25 ♌ 52 44	01 ♍ 47 19	15 11	12 57	25 45	20 19	12 02	16 46	19 37	21 24
21	09 58 59	28 21 05	07 ♍ 42 51	13 ♍ 39 37	14 59	14 10	26 19	20 23	12 08	16 49	19 38	21 25
22	10 02 56	29 ♌ 18 55	19 ♍ 37 53	25 ♍ 37 56	14 53	15 24	26 53	20 26	12 14	16 52	19 40	21 27
23	10 06 52	00 ♍ 16 46	01 ♎ 40 02	07 ♎ 44 30	14 D 56	16 38	27 28	20 30	12 20	16 55	19 41	21 29
24	10 10 49	01 14 38	13 ♎ 51 39	20 ♎ 01 46	15 06	17 52	28 03	20 33	12 26	16 58	19 43	21 31
25	10 14 45	02 12 32	26 ♎ 15 13	02 ♏ 32 19	15 24	19 06	28 39	20 36	12 32	17 01	19 44	21 33
26	10 18 42	03 10 27	08 ♏ 53 26	15 ♏ 18 54	15 50	20 20	29 14	20 39	12 38	17 04	19 46	21 35
27	10 22 39	04 08 24	21 ♏ 49 04	28 ♏ 24 14	16 24	21 34	29 ♏ 50	20 42	12 44	17 06	19 48	21 37
28	10 26 35	05 06 22	05 ♐ 04 41	11 ♐ 50 41	17 06	22 48	00 ♐ 26	20 44	12 51	17 09	19 49	21 39
29	10 30 32	06 04 21	18 ♐ 42 22	25 ♐ 39 52	17 55	24 02	01 03	20 47	12 57	17 12	19 51	21 40
30	10 34 28	07 02 21	02 ♑ 43 09	09 ♑ 52 06	18 52	25 16	01 38	20 49	13 03	17 14	19 53	21 42
31	10 38 25	08 ♍ 00 23	17 ♑ 06 28	24 ♑ 25 51	19 ♌ 55	26 ♍ 30	02 ♐ 15	20 ♉ 51	13 ♎ 10	17 ♋ 17	19 ♎ 55	21 ♌ 44

MOON

Date	Moon True ☊ ° '	Moon Mean ☊ ° '	Moon ☽ Latitude ° '
01	21 ♒ 35	22 ♒ 09	05 S 02
02	21 R 34	22 06	04 29
03	21 33	22 03	03 38
04	21 32	22 00	02 31
05	21 31	21 56	01 S 12
06	21 D 31	21 53	00 N 12
07	21 31	21 50	01 34
08	21 32	21 47	02 49
09	21 32	21 44	03 51
10	21 33	21 40	04 37
11	21 33	21 37	05 06
12	21 33	21 34	05 17
13	21 R 33	21 31	05 13
14	21 D 33	21 28	04 53
15	21 33	21 25	04 21
16	21 33	21 21	03 37
17	21 33	21 18	02 44
18	21 34	21 15	01 45
19	21 34	21 12	00 N 41
20	21 R 34	21 09	00 S 25
21	21 34	21 06	01 29
22	21 33	21 02	02 30
23	21 32	20 59	03 25
24	21 31	20 56	04 11
25	21 29	20 53	04 47
26	21 28	20 50	05 09
27	21 27	20 46	05 17
28	21 D 27	20 43	05 08
29	21 27	20 40	04 42
30	21 28	20 37	03 58
31	21 ♒ 29	20 ♒ 34	02 S 59

DECLINATIONS

Date	Sun ☉ ° '	Moon ☽ ° '	Mercury ☿ ° '	Venus ♀ ° '	Mars ♂ ° '	Jupiter ♃ ° '	Saturn ♄ ° '	Uranus ♅ ° '	Neptune ♆ ° '	Pluto ♇ ° '
01	17 N 58	26 S 51	08 N 41	16 N 18	18 S 20	16 N 15	01 S 56	22 N 53	06 S 00	22 N 59
02	17 43	27 46	08 37	15 55	18 29	16 17	01 58	22 53	06 00	22 58
03	17 27	26 48	08 36	15 32	18 38	16 18	02 00	22 53	06 01	22 58
04	17 12	23 55	08 37	15 08	18 47	16 20	02 02	22 52	06 01	22 57
05	16 55	19 19	08 41	14 44	18 56	16 22	02 04	22 52	06 02	22 56
06	16 39	13 28	08 48	14 19	19 05	16 23	02 06	22 52	06 02	22 56
07	16 22	06 S 50	08 57	13 54	19 14	16 25	02 08	22 51	06 03	22 55
08	16 05	00 N 02	09 09	13 29	19 23	16 26	02 10	22 51	06 03	22 55
09	15 48	06 44	09 23	13 03	19 32	16 28	02 13	22 50	06 04	22 55
10	15 31	12 54	09 40	12 37	19 41	16 29	02 15	22 50	06 04	22 54
11	15 13	18 15	09 58	12 11	19 50	16 31	02 17	22 50	06 05	22 53
12	14 55	22 33	10 18	11 44	19 59	16 32	02 19	22 49	06 05	22 53
13	14 37	25 38	10 40	11 17	20 07	16 33	02 22	22 49	06 06	22 52
14	14 18	27 22	11 03	10 50	20 16	16 34	02 24	22 49	06 06	22 51
15	14 00	27 45	11 26	10 22	20 25	16 35	02 26	22 48	06 07	22 51
16	13 41	26 47	11 50	09 54	20 34	16 37	02 28	22 48	06 07	22 50
17	13 22	24 36	12 14	09 26	20 42	16 38	02 31	22 48	06 08	22 50
18	13 02	21 21	12 37	08 58	20 51	16 39	02 33	22 47	06 08	22 49
19	12 43	17 16	13 00	08 29	20 59	16 40	02 36	22 47	06 09	22 49
20	12 23	12 31	13 21	08 00	21 08	16 40	02 38	22 46	06 10	22 48
21	12 03	07 19	13 42	07 31	21 16	16 41	02 41	22 46	06 10	22 48
22	11 43	01 N 49	14 00	07 02	21 24	16 42	02 43	22 46	06 11	22 47
23	11 23	03 S 48	14 17	06 32	21 33	16 43	02 45	22 45	06 12	22 46
24	11 02	09 20	14 31	06 03	21 41	16 44	02 48	22 45	06 12	22 46
25	10 41	14 35	14 43	05 33	21 49	16 44	02 50	22 45	06 13	22 45
26	10 21	19 21	14 53	05 03	21 57	16 45	02 53	22 44	06 13	22 45
27	10 00	23 19	14 59	04 33	22 05	16 45	02 56	22 44	06 14	22 44
28	09 39	26 12	15 03	04 03	22 12	16 46	02 58	22 44	06 15	22 44
29	09 17	27 39	15 04	03 32	22 20	16 46	03 01	22 43	06 15	22 43
30	08 56	27 24	15 01	03 02	22 28	16 47	03 03	22 43	06 16	22 43
31	08 N 34	25 S 19	14 N 55	02 N 31	22 S 35	16 N 47	03 S 06	22 N 43	06 S 17	22 N 42

ZODIAC SIGN ENTRIES

Date	h	m	Planets
02	22	27	☽ ♑
04	22	41	☽ ♒
06	22	05	☽ ♓
08	22	33	☽ ♈
09	23	58	♀ ♍
11	01	46	☽ ♉
13	08	36	☽ ♊
15	18	52	☽ ♋
18	07	19	☽ ♌
20	20	22	☽ ♍
23	05	03	☉ ♍
23	08	42	☽ ♎
25	19	10	☽ ♏
27	18	53	♂ ♐
28	02	53	☽ ♐
30	07	24	☽ ♑

LATITUDES

Date	Mercury ☿ ° '	Venus ♀ ° '	Mars ♂ ° '	Jupiter ♃ ° '	Saturn ♄ ° '	Uranus ♅ ° '	Neptune ♆ ° '	Pluto ♇ ° '
01	04 S 07	01 N 25	01 S 52	01 S 09	02 N 21	00 N 23	01 N 39	08 N 54
04	04 34	01 26	01 54	01 09	02 21	00 23	01 39	08 54
07	04 50	01 27	01 55	01 10	02 20	00 23	01 39	08 54
10	04 51	01 28	01 56	01 10	02 20	00 23	01 38	08 54
13	04 36	01 27	01 57	01 11	02 19	00 23	01 38	08 55
16	04 04	01 27	01 58	01 11	02 19	00 23	01 38	08 55
19	03 20	01 25	01 59	01 12	02 18	00 23	01 38	08 55
22	02 29	01 23	02 00	01 12	02 18	00 23	01 38	08 55
25	01 34	01 21	02 00	01 12	02 18	00 23	01 38	08 56
28	00 S 42	01 17	02 01	01 13	02 17	00 23	01 38	08 56
31	00 N 05	01 N 14	02 S 01	01 S 14	02 N 17	00 N 23	01 N 38	08 N 57

DATA

Julian Date	2434226
Delta T	+30 seconds
Ayanamsa	23° 11' 52"
Synetic vernal point	05° ♓ 55' 07"
True obliquity of ecliptic	23° 26' 51"

MOON'S PHASES, APSIDES AND POSITIONS ☽

Date	h	m	Phase	Longitude ° '	Eclipse Indicator
05	19	40	○	13 ♒ 17	partial
12	13	27	☾	19 ♉ 45	
20	15	20	●	27 ♌ 31	Annular
28	12	03	☽	05 ♐ 06	

Day	h	m	
05	20	34	Perigee
19	11	16	Apogee

Day	h	m		
02	12	06	Max dec	27° S 46'
08	11	52	0N	
15	06	28	Max dec	27° N 47'
22	19	46	0S	
29	20	39	Max dec	27° S 46'

LONGITUDES

Date	Chiron ⚷ ° '	Ceres ⚳ ° '	Pallas ⚴ ° '	Juno ⚵ ° '	Vesta ⚶ ° '	Black Moon Lilith ⚸ ° '
01	07 ♑ 42	03 ♊ 44	09 ♈ 07	19 ♒ 11	01 ♎ 40	13 ♌ 59
11	07 ♑ 15	06 ♊ 44	08 ♈ 36	16 ♒ 47	06 ♎ 18	15 ♌ 02
21	06 ♑ 55	09 ♊ 28	07 ♈ 28	14 ♒ 21	11 ♎ 05	16 ♌ 16
31	06 ♑ 43	11 ♊ 55	05 ♈ 43	12 ♒ 11	15 ♎ 59	17 ♌ 19

All ephemeris data is given at 12.00 UT and the Moon's longitude is additionally given for 24.00 UT

ASPECTARIAN

h m Aspects

01 Friday
02 57 ☽ ∠ ♆
05 35 ♀ ✶ ♆
11 28 ☽ △ ☉
12 32 ☽ ± ♅
13 32 ☽ ✶ ♄
15 10 ♂ △ ♅
15 16 ♀ ∥ ♃
22 47 ☽ ⚻ ♅
23 01 ☽ ⚺ ♂

02 Saturday
03 35 ☽ ⚻ ♃
04 34 ☽ ✶ ♆
06 42 ☽ △ ♀
07 15 ☽ △ ♇
09 27 ☽ ⊥ ♂
09 54 ☽ Q ♄
13 24 ♀ ☌ ♇
13 41 ☽ ± ♃
14 34 ☽ bQ ☉
16 44 ☽ △ ☿
20 24 ☉ ✶ ♄

03 Sunday
00 29 ☽ Q ♆
00 50 ☽ ∠ ♂
01 32 ♂ ⊥ ♄
04 36 ☽ bQ ♃
06 23 ☽ ± ☉
08 02 ☽ bQ ♇
09 31 ♀ ⊥ ♅
09 41 ☽ bQ ♀
15 34 ☽ □ ♄
16 37 ☽ bQ ☿
16 47 ☽ ⚻ ☉
22 36 ☽ ± ♇

04 Monday
00 14 ☽ ☍ ♅
01 30 ☽ ± ♀
01 57 ☽ ✶ ♂
04 59 ☽ △ ♃
05 36 ☽ □ ♆
06 38 ☽ ± ☿
08 13 ☽ ⚻ ♇
11 56 ☽ ⚻ ♀
15 54 ☽ ⚻ ☿
17 47 ☽ # ♇
18 14 ☽ # ♅
21 02 ☿ ∠ ♄
21 42 ☽ Q ♂

05 Tuesday
13 41 ☽ ∥ ♂
15 33 ☽ △ ♄
19 40 ☽ ☍ ☉
22 51 ☽ # ☉
22 57 ☿ ☌ ♀
23 58 ☽ ⚻ ♅

06 Wednesday
00 35 ☽ # ♃
03 05 ☽ □ ♂
04 49 ☽ □ ♃
05 08 ☽ △ ♆
07 45 ☽ ☍ ♇
08 28 ☽ # ♀
09 28 ☽ ± ♅
13 12 ♀ ∠ ♄
13 36 ☽ ☍ ☿
15 21 ☽ bQ ♄
15 32 ☽ ☍ ♀
23 46 ☽ bQ ♅

07 Thursday
04 42 ☽ # ☿
04 56 ☽ bQ ♆
05 44 ☽ ± ♄
08 36 ☉ ∥ ♃
09 34 ☽ Q ♃
14 48 ☽ ∥ ♆
15 23 ☽ ⚻ ♄
19 22 ☽ ± ♆
22 35 ☽ ⚻ ☉
23 53 ☽ △ ♅

08 Friday
04 20 ☽ ∥ ♄
04 37 ☽ △ ♂
05 01 ☽ ✶ ♃
05 04 ☽ ⚻ ♆
07 49 ☽ ⚻ ♇
08 59 ☽ ± ☉
11 36 ☽ ⚻ ☿
16 01 ♃ ⚻ ♆
17 39 ☽ ± ♇
18 11 ♂ ⚺ ♆
18 35 ♂ ☍ ♃
19 33 ☽ # ♄
19 45 ☉ ⚺ ♅
20 12 ☽ ⚻ ♀
20 58 ☽ ± ☿

09 Saturday
00 49 ☽ bQ ☉
05 51 ☽ ∠ ♃
06 12 ☽ bQ ♂
07 05 ☽ ± ♀
08 37 ☽ bQ ♇
09 30 ☽ # ♆
11 11 ☽ bQ ☿
16 59 ☽ ☍ ♄
21 06 ☽ ⊥ ♃
21 56 ☽ ± ♂
22 08 ☿ ⊥ ♅
22 26 ☽ ∥ ☿
23 43 ☽ bQ ♀

10 Sunday
01 51 ☽ □ ♅
03 56 ☽ △ ☉
07 14 ☽ ☍ ♆
07 27 ☽ ⚺ ♃
08 37 ☽ ⚻ ♂
10 10 ☽ △ ♇
10 56 ☽ ∥ ♀
11 24 ☽ △ ☿
22 36 ☽ ∥ ☉

11 Monday
01 02 ♀ ∠ ♅
03 40 ☽ ∥ ♃
04 20 ☽ △ ♀
07 22 ☿ ☌ ♇
09 24 ☽ Q ♅
14 59 ☿ □ ♂
20 23 ☽ # ♂
21 50 ☽ ⚻ ♄

12 Tuesday
03 53 ♂ □ ♇
04 58 ☉ ✶ ♆
07 11 ☽ ✶ ♅
08 51 ☽ ± ♄
12 08 ☉ □ ♃
12 50 ☽ ⚻ ♆
13 22 ☽ ☌ ♃
13 27 ☽ □ ☉
13 46 ☽ ∥ ♅
14 01 ☽ □ ☿
14 07 ☽ ∥ ♇
16 02 ☽ □ ♇
16 32 ☽ ☍ ♂
17 51 ☉ ☌ ☿
23 22 ☿ □ ♃

13 Wednesday
00 01 ☽ ± ♆
01 41 ☽ bQ ♄
08 21 ☿ ✶ ♆
11 17 ☽ ∠ ♅
14 31 ♀ ∠ ♆
17 03 ☽ bQ ♆
17 19 ☽ □ ♀
21 54 ☽ Q ☿

14 Thursday
01 28 ☉ ☌ ♇
02 08 ☽ Q ♇
02 12 ☽ Q ☉
04 34 ☽ ⊥ ♅
05 57 ☿ ∥ ♀
06 27 ☽ △ ♄
10 04 ♀ ⊥ ♄
16 15 ☽ ⚺ ♅
19 46 ☽ ✶ ☿
22 06 ☽ △ ♆
22 57 ☽ ⚺ ♃

15 Friday
01 32 ☽ ✶ ♇
03 31 ☽ ✶ ☉
04 35 ☽ ⚻ ♂
08 08 ☽ Q ♀
10 33 ☉ ⊥ ♅
10 50 ☽ ⊥ ♃
17 00 ☽ ± ♂
23 44 ☽ ∠ ☿

16 Saturday
04 51 ☽ ∠ ♃
07 22 ☽ ∠ ♇
09 29 ☉ □ ♂
10 44 ☽ ✶ ♀
11 49 ☽ bQ ♂
11 55 ☽ ∠ ☉
16 48 ☽ ⊥ ☿
18 12 ☽ □ ♄

17 Sunday
01 30 ☽ ⊥ ♇
04 10 ☽ ☌ ♅
04 17 ☽ ⚺ ☿
06 39 ☿ ⚺ ♅
07 38 ☽ ⊥ ☉
10 04 ☽ □ ♆
11 14 ☽ ✶ ♃
13 40 ☽ ⚺ ♇
19 35 ☽ △ ♂
20 30 ☽ ∠ ♀
20 52 ☽ ⚺ ☉

18 Monday
01 59 ☽ ∥ ♇
02 14 ☽ ∥ ♅
06 56 ☽ Q ♄
11 45 ☽ Q ♃
15 09 ☽ # ♂
16 56 ☽ ⊥ ♀
22 43 ☽ Q ♆

19 Tuesday
02 18 ☉ ∥ ☿
06 35 ☽ ⚺ ♀
07 34 ☽ ✶ ♄
14 49 ☽ ☌ ☿
15 15 ☽ ∥ ♃
16 41 ♀ ⚺ ♄
17 24 ☽ ⚺ ♅
23 13 ☽ ✶ ♆

20 Wednesday
00 37 ☽ □ ♃
02 03 ☉ ∠ ♄
02 51 ☽ ☌ ♇
05 40 ☽ ⊥ ♅
06 04 ♂ ⊥ ♆
08 14 ☽ ∥ ☿
11 43 ☽ □ ♂
12 42 ☽ ∥ ☉
14 21 ☽ ∠ ♄
15 20 ☽ ☌ ☉

21 Thursday
00 01 ☽ ∠ ♅
01 15 ♀ ⊥ ♆
05 45 ☽ ∠ ♆
08 46 ☽ ⊥ ♄
10 58 ☽ ∥ ♀
17 02 ☽ # ♆
17 09 ♀ Q ♂
20 59 ☽ ⚺ ♄
23 58 ☽ ⊥ ♆

22 Friday
02 00 ☽ Q ♂
02 20 ☿ ⚺ ♀
02 31 ☽ ⚺ ☿
02 32 ☽ ☌ ♀
06 26 ☽ ✶ ♅
08 08 ☽ # ♄
12 03 ☽ ⚺ ♆
13 38 ☽ △ ♃
14 31 ☽ ⊥ ☿
15 40 ☽ ⚺ ♇
16 54 ☿ St D

23 Saturday
03 15 ☽ ✶ ♂
03 41 ☽ ⊥ ♇
04 52 ♂ ∠ ♄
06 32 ☽ Q ♅
07 32 ☽ ∥ ♄
08 31 ☽ ∠ ☿
09 00 ☽ ⚺ ☉
17 41 ♀ ✶ ♅
19 37 ☽ bQ ♃
21 33 ☽ ∠ ♇
21 54 ☽ ⊥ ☉
22 21 ☽ ∥ ♆
22 52 ☽ # ♀

24 Sunday
04 32 ♀ # ♆
09 11 ☽ ☌ ♄
10 21 ☽ ∠ ♂
13 21 ☽ ± ♃
14 28 ☽ ✶ ☿
17 03 ☽ ∠ ☉
18 05 ☽ □ ♅
19 10 ☽ # ☉
20 41 ☽ ⚺ ♀
23 25 ☽ ☌ ♆

25 Monday
01 04 ☽ ⚻ ♃
02 55 ☽ ✶ ♇
04 42 ☽ ⊥ ♂
06 56 ☉ ∠ ♅
09 33 ☽ ⊥ ♀
12 40 ☽ # ☿
14 16 ☽ Q ☿
16 48 ☽ ⚺ ♂
22 29 ☽ # ♃

26 Tuesday
00 19 ☽ ✶ ☉
00 40 ♀ ⚺ ♆
01 57 ☽ Q ♇
04 35 ☽ ∠ ♀
18 25 ♀ △ ♃
19 04 ☽ ⚺ ♄

27 Wednesday
00 41 ☽ Q ☉
01 31 ☽ □ ☿
03 17 ☽ △ ♅
03 36 ☽ ∥ ♂
06 18 ☽ ⊥ ♄
08 04 ☽ # ♅
08 07 ☽ # ♇
08 17 ☽ ⚺ ♆
09 56 ☽ ☍ ♃
11 30 ☽ ✶ ♀
11 37 ☽ □ ♇
12 53 ♀ ⚺ ♇
19 17 ☽ ⊥ ♆
22 53 ☽ ∠ ♄

28 Thursday
03 16 ☽ ☌ ♂
04 47 ☉ ∠ ♆
06 45 ☽ bQ ♅
11 27 ☽ Q ♀
11 33 ☽ ∠ ♆
12 03 ☽ □ ☉
13 55 ☿ ⚺ ♅
22 49 ☽ ± ♅

29 Friday
01 52 ☽ ✶ ♄
04 27 ☿ ⊥ ♀
09 22 ☽ ⚻ ♅
10 32 ☽ △ ☿
13 57 ♅ ∥ ♇
14 00 ☽ ✶ ♆
15 36 ☽ ⚻ ♃
17 09 ☽ △ ♇
22 05 ☽ □ ♀
22 51 ☽ Q ♄

30 Saturday
01 57 ☽ ± ♃
10 05 ☽ ⚺ ♂
10 35 ☽ Q ♆
10 55 ♀ # ♄
12 29 ☉ ⊥ ♄
14 04 ☽ bQ ☿
17 14 ☽ bQ ♃
18 43 ☽ bQ ♇
19 48 ☽ △ ☉
20 38 ☽ ⊥ ♂

31 Sunday
05 26 ☽ □ ♄
06 18 ☽ ± ☿
09 44 ☽ ± ♇
11 46 ☿ ✶ ♆
12 14 ☽ ∠ ♂
12 18 ☽ ☍ ♅
13 48 ♂ bQ ♅
16 37 ☽ □ ♆
17 01 ☽ ⚻ ☿
18 09 ☽ △ ♃
19 37 ☽ ⚻ ♇
22 21 ☽ bQ ☉

SEPTEMBER 1952

LONGITUDES

Date	Sidereal time h m s	Sun ☉ ° ' "	Moon ☽ ° ' "	Moon ☽ 24.00 ° ' "	Mercury ☿ ° '	Venus ♀ ° '	Mars ♂ ° '	Jupiter ♃ ° '	Saturn ♄ ° '	Uranus ♅ ° '	Neptune ♆ ° '	Pluto ♇ ° '
01	10 42 21	08 ♍ 58 26	01 ♒ 49 41	09 ♒ 17 16	21 ♌ 05	27 ♍ 44	02 ♐ 51	20 ♉ 53	13 ♎ 16	17 ♋ 20	19 ♎ 56	21 ♌ 46
02	10 46 18	09 56 31	16 ♒ 47 46	24 ♒ 20 12	22 22	28 ♍ 58	03 28	20 54	13 23	17 22	19 58	21 48
03	10 50 14	10 54 37	01 ♓ 53 30	09 ♓ 26 33	23 44	00 ♎ 11	04 05	20 55	13 29	17 25	20 00	21 50
04	10 54 11	11 52 45	16 ♓ 58 12	24 ♓ 27 18	25 11	01 25	04 42	20 57	13 36	17 27	20 02	21 51
05	10 58 08	12 50 54	01 ♈ 52 50	09 ♈ 13 50	26 44	02 39	05 20	20 58	13 43	17 30	20 04	21 53
06	11 02 04	13 49 05	16 ♈ 29 28	23 ♈ 39 06	28 ♌ 20	03 53	05 57	20 58	13 49	17 32	20 06	21 55
07	11 06 01	14 47 19	00 ♉ 42 14	07 ♉ 38 35	00 ♍ 00	05 07	06 35	20 59	13 56	17 34	20 07	21 57
08	11 09 57	15 45 34	14 ♉ 28 01	21 ♉ 10 32	01 43	06 21	07 13	20 59	14 03	17 37	20 09	21 58
09	11 13 54	16 43 51	27 ♉ 46 19	04 ♊ 15 39	03 29	07 35	07 51	20 59	14 09	17 39	20 11	22 00
10	11 17 50	17 42 11	10 ♊ 38 56	16 ♊ 56 38	05 17	08 48	08 30	20 R 59	14 16	17 41	20 13	22 02
11	11 21 47	18 40 33	23 ♊ 09 16	29 ♊ 17 25	07 07	10 02	09 08	20 59	14 23	17 43	20 15	22 04
12	11 25 43	19 38 56	05 ♋ 21 39	11 ♋ 22 36	08 58	11 16	09 47	20 59	14 30	17 45	20 17	22 05
13	11 29 40	20 37 22	17 ♋ 20 52	23 ♋ 17 03	10 50	12 30	10 26	20 58	14 37	17 48	20 19	22 07
14	11 33 37	21 35 50	29 ♋ 11 42	05 ♌ 05 23	12 42	13 44	11 05	20 57	14 44	17 50	20 21	22 09
15	11 37 33	22 34 20	10 ♌ 58 37	16 ♌ 51 53	14 35	14 58	11 44	20 56	14 51	17 52	20 23	22 10
16	11 41 30	23 32 52	22 ♌ 45 39	28 ♌ 40 17	16 28	16 11	12 23	20 55	14 58	17 54	20 25	22 12
17	11 45 26	24 31 26	04 ♍ 36 11	10 ♍ 33 39	18 21	17 25	13 02	20 53	15 05	17 56	20 27	22 14
18	11 49 23	25 30 02	16 ♍ 33 00	22 ♍ 34 26	20 14	18 39	13 42	20 52	15 12	17 57	20 29	22 15
19	11 53 19	26 28 40	28 ♍ 38 13	04 ♎ 44 29	22 06	19 53	14 22	20 50	15 19	17 59	20 31	22 17
20	11 57 16	27 27 20	10 ♎ 53 25	17 ♎ 05 07	23 58	21 07	15 02	20 48	15 26	18 01	20 33	22 19
21	12 01 12	28 26 02	23 ♎ 19 44	29 ♎ 37 20	25 49	22 20	15 42	20 46	15 33	18 03	20 36	22 20
22	12 05 09	29 ♍ 24 46	05 ♏ 58 02	12 ♏ 21 56	27 40	23 34	16 22	20 43	15 40	18 05	20 38	22 22
23	12 09 06	00 ♎ 23 31	18 ♏ 49 07	25 ♏ 19 43	29 ♍ 29	24 48	17 02	20 41	15 47	18 06	20 40	22 23
24	12 13 02	01 22 19	01 ♐ 53 50	08 ♐ 31 35	01 ♎ 18	26 02	17 43	20 38	15 55	18 08	20 42	22 25
25	12 16 59	02 21 08	15 ♐ 13 06	21 ♐ 58 31	03 06	27 15	18 24	20 35	16 02	18 09	20 44	22 26
26	12 20 55	03 19 59	28 ♐ 47 57	05 ♑ 41 28	04 53	28 29	19 04	20 31	16 09	18 11	20 46	22 28
27	12 24 52	04 18 51	12 ♑ 39 08	19 ♑ 40 59	06 39	29 ♎ 43	19 45	20 28	16 16	18 12	20 48	22 29
28	12 28 48	05 17 46	26 ♑ 46 54	03 ♒ 56 47	08 24	00 ♏ 57	20 26	20 24	16 23	18 14	20 51	22 31
29	12 32 45	06 16 42	11 ♒ 10 21	18 ♒ 27 16	10 09	02 10	21 07	20 21	16 31	18 15	20 53	22 32
30	12 36 41	07 ♎ 15 40	25 ♒ 47 02	03 ♓ 09 03	11 ♎ 52	03 ♏ 24	21 ♐ 49	20 ♉ 17	16 ♎ 38	18 ♋ 16	20 ♎ 55	22 ♌ 34

Date	Moon True ☊ ° '	Moon Mean ☊ ° '	Moon ☽ Latitude ° '
01	21 ♒ 30	20 ♒ 31	01 S 47
02	21 D 31	20 27	00 S 26
03	21 R 31	20 24	00 N 57
04	21 30	20 21	02 16
05	21 27	20 18	03 25
06	21 25	20 15	04 18
07	21 21	20 12	04 55
08	21 18	20 08	05 12
09	21 16	20 05	05 13
10	21 15	20 02	04 57
11	21 D 15	19 59	04 27
12	21 16	19 56	03 46
13	21 17	19 52	02 56
14	21 19	19 49	01 59
15	21 20	19 46	00 N 57
16	21 R 21	19 43	00 S 08
17	21 20	19 40	01 12
18	21 17	19 37	02 13
19	21 13	19 33	03 09
20	21 07	19 30	03 57
21	21 01	19 27	04 35
22	20 54	19 24	05 00
23	20 48	19 21	05 10
24	20 43	19 18	05 04
25	20 40	19 14	04 42
26	20 38	19 11	04 04
27	20 D 39	19 08	03 11
28	20 40	19 05	02 06
29	20 41	19 02	00 51
30	20 ♒ 41	18 ♒ 58	00 N 28

DECLINATIONS

Date	Sun ☉ ° '	Moon ☽ ° '	Mercury ☿ ° '	Venus ♀ ° '	Mars ♂ ° '	Jupiter ♃ ° '	Saturn ♄ ° '	Uranus ♅ ° '	Neptune ♆ ° '	Pluto ♇ ° '
01	08 N 12	21 S 30	14 N 46	02 N 00	22 S 43	16 N 47	03 S 08	22 N 42	06 S 18	22 N 42
02	07 51	16 13	14 34	01 30	22 50	16 48	03 11	22 42	06 18	22 42
03	07 29	09 55	14 18	00 59	22 57	16 48	03 14	22 42	06 19	22 41
04	07 07	03 S 04	13 59	00 N 28	23 04	16 48	03 16	22 42	06 20	22 41
05	06 44	03 N 53	13 38	00 S 03	23 11	16 48	03 19	22 41	06 20	22 40
06	06 22	10 28	13 13	00 34	23 18	16 48	03 22	22 41	06 21	22 40
07	06 00	16 19	12 45	01 04	23 24	16 48	03 24	22 41	06 22	22 39
08	05 37	21 09	12 15	01 35	23 31	16 48	03 27	22 40	06 23	22 39
09	05 14	24 45	11 42	02 06	23 37	16 48	03 30	22 40	06 23	22 38
10	04 52	26 57	11 07	02 37	23 43	16 48	03 32	22 40	06 24	22 38
11	04 29	27 43	10 30	03 08	23 49	16 47	03 35	22 40	06 25	22 38
12	04 06	27 06	09 51	03 39	23 55	16 47	03 38	22 39	06 26	22 37
13	03 43	25 14	09 10	04 09	24 01	16 47	03 41	22 39	06 26	22 37
14	03 20	22 16	08 28	04 40	24 06	16 46	03 43	22 39	06 27	22 36
15	02 57	18 23	07 45	05 11	24 12	16 46	03 46	22 39	06 28	22 36
16	02 34	13 49	07 01	05 41	24 17	16 46	03 49	22 38	06 29	22 36
17	02 11	08 43	06 16	06 11	24 22	16 45	03 52	22 38	06 30	22 35
18	01 47	03 N 16	05 30	06 42	24 27	16 44	03 54	22 38	06 30	22 35
19	01 24	02 S 21	04 43	07 12	24 31	16 44	03 57	22 38	06 31	22 35
20	01 01	07 57	03 57	07 42	24 36	16 43	04 00	22 38	06 32	22 34
21	00 37	13 19	03 10	08 11	24 40	16 42	04 03	22 37	06 33	22 34
22	00 N 14	18 13	02 22	08 41	24 44	16 42	04 05	22 37	06 34	22 34
23	00 S 09	22 23	01 35	09 10	24 48	16 41	04 08	22 37	06 34	22 33
24	00 33	25 31	00 N 48	09 40	24 52	16 40	04 11	22 37	06 35	22 33
25	00 56	27 18	00 00	10 09	24 55	16 39	04 14	22 37	06 36	22 33
26	01 20	27 31	00 S 47	10 37	24 58	16 38	04 17	22 36	06 37	22 32
27	01 43	26 01	01 34	11 06	25 01	16 37	04 19	22 36	06 38	22 32
28	02 06	22 52	02 20	11 34	25 04	16 36	04 22	22 36	06 38	22 32
29	02 30	18 15	03 07	12 02	25 07	16 35	04 25	22 36	06 39	22 32
30	02 S 53	12 S 30	03 S 53	12 S 30	25 S 09	16 N 34	04 S 28	22 N 36	06 S 40	22 N 31

ZODIAC SIGN ENTRIES

Date	h	m	Planets
01	09	03	☽ ♒
03	08	17	♀ ♎
03	09	00	☽ ♓
05	08	57	☽ ♈
07	10	48	☽ ♉
07	12	02	☿ ♍
09	16	06	☽ ♊
12	01	24	☽ ♋
14	13	38	☽ ♌
17	02	42	☽ ♍
19	14	41	☽ ♎
22	00	43	☽ ♏
23	02	24	☉ ♎
23	18	45	☿ ♎
24	08	33	☽ ♐
26	14	06	☽ ♑
27	17	36	♀ ♏
28	17	24	☽ ♒
30	18	52	☽ ♓

LATITUDES

Date	Mercury ☿ ° '	Venus ♀ ° '	Mars ♂ ° '	Jupiter ♃ ° '	Saturn ♄ ° '	Uranus ♅ ° '	Neptune ♆ ° '	Pluto ♇ ° '
01	00 N 19	01 N 12	02 S 01	01 S 14	02 N 17	00 N 23	01 N 38	08 N 57
04	00 55	01 08	02 01	01 15	02 16	00 23	01 37	08 57
07	01 22	01 03	02 01	01 15	02 16	00 23	01 37	08 58
10	01 39	00 57	02 01	01 16	02 16	00 23	01 37	08 58
13	01 48	00 51	02 01	01 16	02 16	00 24	01 37	08 59
16	01 49	00 45	02 01	01 17	02 15	00 24	01 37	08 59
19	01 44	00 38	02 00	01 17	02 15	00 24	01 37	09 00
22	01 34	00 30	02 00	01 17	02 15	00 24	01 37	09 01
25	01 21	00 23	01 59	01 18	02 15	00 24	01 37	09 01
28	01 05	00 15	01 59	01 18	02 15	00 24	01 37	09 02
31	00 N 47	00 N 07	01 S 58	01 S 19	02 N 15	00 N 24	01 N 37	09 N 03

DATA

Julian Date	2434257
Delta T	+30 seconds
Ayanamsa	23° 11' 57"
Synetic vernal point	05° ♓ 55' 03"
True obliquity of ecliptic	23° 26' 51"

MOON'S PHASES, APSIDES AND POSITIONS ☽

Date	h	m	Phase	Longitude ° '	Eclipse Indicator
04	03	19	○	11 ♓ 32	
11	02	36	☾	18 ♊ 18	
19	07	22	●	26 ♍ 17	
26	20	31	☽	03 ♑ 41	

Day	h	m		
03	06	32	Perigee	
15	18	54	Apogee	
04	22	31	0N	
11	13	08	Max dec	27° N 43'
19	01	58	0S	
26	03	03	Max dec	27° S 38'

LONGITUDES

Date	Chiron ⚷ ° '	Ceres ⚳ ° '	Pallas ⚴ ° '	Juno ⚵ ° '	Vesta ⚶ ° '	Black Moon Lilith ⚸ ° '
01	06 ♑ 42	12 ♊ 08	05 ♈ 31	12 ♒ 00	16 ♎ 29	17 ♌ 25
11	06 ♑ 38	14 ♊ 11	03 ♈ 13	10 ♒ 23	21 ♎ 30	18 ♌ 32
21	06 ♑ 42	15 ♊ 48	00 ♈ 34	09 ♒ 27	26 ♎ 36	19 ♌ 38
31	06 ♑ 55	16 ♊ 54	27 ♓ 49	09 ♒ 16	01 ♏ 47	20 ♌ 45

ASPECTARIAN

h m	Aspects
01 Monday	
04 46	☽ △ ♀
05 22	☽ # ♅
05 25	☽ # ♇
05 32	☽ ∥ ♂
07 38	☿ □ ♃
09 29	♂ # ♇
11 00	♂ # ♅
12 46	♀ ⊥ ♇
13 44	☽ ✶ ♂
13 59	☽ ± ☉
02 Tuesday	
00 17	☽ ⊼ ☉
01 15	☿ ☌ ♇
06 30	☽ △ ♄
07 04	☽ ꝗ ♀
09 39	☽ # ♃
09 48	☽ Q ♂
12 55	☽ ⊼ ♅
17 04	☽ △ ♆
18 33	☽ □ ♃
18 54	☽ # ☿
19 59	☽ ☍ ♇
20 18	♀ Q ♅
21 44	☽ ☍ ☿
22 30	☽ ± ♅
22 41	☽ ± ♀
03 Wednesday	
06 20	☿ ⊥ ♅
06 33	☽ ꝗ ♄
09 04	☽ ⊼ ♀
12 50	☽ ꝗ ♅
15 38	☽ □ ♂
16 57	☽ ꝗ ♆
20 57	☽ ± ♄
21 10	☽ # ☉
23 11	☽ Q ♃
04 Thursday	
00 43	☽ ∥ ♆
03 19	☽ ☍ ☉
06 35	☽ ⊼ ♄
07 18	☽ ± ♆
11 16	☽ ∥ ♄
12 46	☽ △ ♅
16 54	☽ ⊼ ♆
18 22	☽ ✶ ♃
19 50	☽ ⊼ ♇
21 37	☽ # ♀
23 18	☽ ∥ ♀
05 Friday	
01 06	♂ ∠ ♆
02 41	☽ ⊼ ☿
05 31	☽ ± ♇
06 45	☿ ⊥ ♀
10 01	☽ # ♄
13 22	☽ ☍ ♀
13 32	☽ ± ☿
17 52	☽ △ ♂
18 39	☽ ∠ ♃
20 10	☽ ꝗ ♇
20 46	☽ # ♆
21 39	☽ ∥ ☉
06 Saturday	
06 06	☽ ꝗ ☿
07 15	☽ ⊼ ☉
07 32	☽ ☍ ♄
09 28	☽ ⊥ ♃
12 03	☉ ⊻ ♄
12 54	☉ # ♆
13 44	☽ □ ♅
17 57	☽ ± ☉
18 02	☽ ☍ ♆
19 01	☉ ⊥ ♆
19 30	☽ ⊻ ♃
19 39	☿ ∠ ♄
19 49	☽ ꝗ ♂
21 06	☽ △ ♇
22 04	☽ ∥ ☿
07 Sunday	
10 19	☽ ꝗ ☉
10 37	☽ △ ☿
11 48	☽ ± ♂
14 10	☽ ∥ ♃
20 21	☽ ⊼ ♀
20 25	☽ Q ♅
22 39	☽ ⊼ ♂
08 Monday	
04 59	♀ ꝗ ♃
07 52	☽ ± ♀
11 15	☽ ⊼ ♄
14 28	☽ △ ☉
17 37	☽ ✶ ♅
20 56	☽ ∥ ♇
21 07	☽ ∥ ♅
22 03	☽ ± ♄
22 11	☽ ⊼ ♆
23 40	☽ ☌ ♃
09 Tuesday	
00 31	☿ ∠ ♅
00 33	♀ ∠ ♇
01 28	☽ □ ♇
01 34	☽ ꝗ ♀
03 11	☽ # ♂
09 06	☽ ± ♆
14 34	☽ ꝗ ♄
19 44	♃ St R
21 02	☽ ∠ ♅
23 19	♀ ✶ ♂
10 Wednesday	
00 15	☽ □ ☿
01 46	☽ ꝗ ♆
07 43	☽ ☍ ♂
08 09	☽ △ ♀
10 50	☽ Q ♇
11 12	☿ ∠ ♆
11 34	☉ ✶ ♅
13 58	☽ ⊥ ♅
18 57	☽ △ ♄
11 Thursday	
01 28	☽ ⊻ ♅
02 36	☽ □ ☉
06 22	☽ △ ♆
07 48	☽ ⊻ ♃
09 52	☽ ✶ ♇
16 29	☽ Q ☿
19 28	☽ ⊥ ♃
12 Friday	
05 39	☿ ⊥ ♄
11 19	♀ ∥ ♄
13 14	☽ ∠ ♃
15 27	☽ ∠ ♇
16 57	☽ Q ☉
20 29	☽ ✶ ☿
21 18	☽ ⊼ ♂
13 Saturday	
00 14	☉ # ♀
01 08	☽ □ ♀
04 10	☿ □ ♂
04 15	☉ ⊻ ♆
06 27	☽ □ ♄
09 31	☽ ⊥ ♇
10 02	☽ ± ♂
12 54	☽ ☌ ♅
14 18	☉ # ♄
18 01	☽ □ ♆
19 12	☽ ✶ ☉
19 18	☽ ✶ ♃
20 23	☉ △ ♃
21 40	☽ ⊻ ♇
22 30	☽ # ♂
14 Sunday	
05 18	☽ ꝗ ♂
08 24	☽ ∠ ☿
09 15	☽ ∥ ♅
09 33	☽ ∥ ♇
17 45	☽ Q ♀
19 16	☽ Q ♄
19 38	☽ Q ♃
15 Monday	
01 55	☉ ⊻ ♇
04 26	☽ ∠ ☉
06 12	☽ ⊥ ☿
06 42	☽ Q ♆
09 23	☿ ⊥ ♆
09 32	♀ ☌ ♄
11 33	♀ ± ♃
13 37	☽ △ ♂
15 30	☿ ⊻ ♄
17 07	♂ ± ♅
19 58	☽ ✶ ♄
20 46	☽ ⊻ ☿
20 54	☽ ∥ ♃
21 04	☽ ✶ ♀
16 Tuesday	
00 26	☽ ⊥ ☉
01 38	☿ ⊻ ♀
02 04	☽ ⊻ ♅
03 53	♃ ± ♄
07 14	☽ ✶ ♆
08 15	☽ □ ♃
10 52	☽ ☌ ♇
13 45	☽ ⊻ ☉
14 19	☽ ⊥ ♅
17 Wednesday	
02 46	☽ ∠ ♄
04 45	☿ # ♆
06 25	☿ ✶ ♅
07 05	☽ ∠ ♀
08 36	☽ ∠ ♅
13 20	☿ # ♀
13 43	☽ ∠ ♆
21 07	☽ ⊥ ♄
21 53	☽ # ♆
22 08	♀ □ ♅
22 17	☽ # ♀
18 Thursday	
00 42	☽ ∥ ☿
02 48	♀ ∥ ♆
03 18	☽ ⊥ ♀
05 58	☽ □ ♂
07 52	☽ ⊥ ♆
09 14	☽ # ♄
09 16	☽ ⊻ ♄
14 49	☽ ✶ ♅
15 18	☿ ⊻ ♆
16 40	☽ ⊻ ♀
18 48	☽ ∥ ☉
19 53	☽ ⊻ ♆
19 55	☿ △ ♃
20 35	☽ △ ♃
20 42	☽ ☌ ☿
23 24	☽ ⊻ ♇
19 Friday	
07 22	☽ ☌ ☉
08 12	☽ # ☉
11 18	☽ ⊥ ♇
14 18	☿ ⊻ ♇
14 40	☽ Q ♅
18 52	☽ ∥ ♄
19 46	☽ Q ♂
20 52	☽ # ☿
20 Saturday	
00 55	♀ ☌ ♆
02 06	☽ ꝗ ♃
05 01	☽ ∠ ♇
05 51	☽ ∥ ♆
06 05	♀ ⊼ ♃
10 27	☿ # ♄
10 46	☽ ∥ ♀
19 33	☽ ± ♃
20 29	☽ ✶ ♂
20 54	☽ ☌ ♄
21 Sunday	
01 50	☽ □ ♅
05 37	♂ ✶ ♄
06 44	☽ ☌ ♆
07 05	☽ ⊼ ♃
09 34	☉ ⊥ ♇
09 54	☽ ☌ ♀
10 06	☽ ✶ ♇
11 57	♀ ✶ ♇
17 35	☽ ⊻ ☿
22 34	☽ ⊻ ☉
22 Monday	
02 49	☽ ∠ ♂
04 13	☽ # ♃
06 55	☽ ⊥ ☿
08 58	☽ Q ♇
10 52	☽ ⊥ ☉
20 43	☽ ⊥ ♂
21 20	☿ ⊥ ♇
23 Tuesday	
02 38	☽ ∠ ☿
04 42	☉ Q ♅
05 07	☽ ∠ ☉
06 20	☽ ⊻ ♄
08 31	☽ ⊻ ♂
10 40	☽ △ ♅
13 06	☽ # ♇
13 31	☽ # ♅
15 25	☽ ⊻ ♆
15 25	☽ ☍ ♃
15 41	♃ ⊼ ♆
17 32	☽ ⊥ ♄
18 37	☽ □ ♇
20 14	☿ Q ♅
24 Wednesday	
00 10	☽ ⊻ ♀
02 29	☽ ⊥ ♆
06 01	☽ ∥ ♂
10 11	☽ ∠ ♄
10 45	☽ ✶ ☿
10 58	☽ ✶ ☉
12 16	☽ ⊥ ♀
14 03	☉ ☌ ☿
14 15	☽ ꝗ ♅
17 03	☉ # ☿
18 55	☽ ∠ ♆
25 Thursday	
03 17	♂ ⊼ ♅
06 10	☽ ∠ ♀
06 31	☽ ± ♅
10 20	☽ Q ☉
11 45	☽ Q ☿
13 28	☽ ✶ ♄
17 14	☽ ⊼ ♅
17 57	☽ ☌ ♂
21 29	☽ ⊼ ♃
21 50	☽ ✶ ♆
26 Friday	
00 51	☽ △ ♇
08 02	☽ ± ♃
10 51	☽ Q ♄
11 24	☽ ✶ ♀
18 57	☽ Q ♆
20 23	☿ ꝗ ♃
20 31	☽ □ ☉
23 40	☽ ꝗ ♃
27 Saturday	
00 10	☽ □ ☿
03 06	☽ ꝗ ♇
10 14	☽ Q ♀
18 15	☽ □ ♄
18 35	☽ ± ♇
20 50	☽ ∥ ♂
21 22	☉ ∥ ☿
21 30	☽ ☍ ♅
23 35	☿ ∠ ♇
28 Sunday	
00 44	☽ ⊻ ♂
01 16	☽ △ ♃
01 56	☽ □ ♆
04 47	☽ ⊼ ♇
11 01	♂ ⊼ ♃
11 23	☽ ⊥ ♂
12 39	☿ Q ♂
13 35	☽ # ♅
14 01	☽ # ♇
14 32	☉ ꝗ ♃
19 38	☽ □ ♀
29 Monday	
02 58	♂ ✶ ♆
03 12	☽ ∠ ♂
03 17	☽ △ ☉
10 04	☽ △ ☿
19 26	☽ # ♃
20 53	☽ △ ♄
23 41	☽ ⊼ ♅
30 Tuesday	
03 02	☽ □ ♃
04 01	☽ △ ♆
05 11	☽ ✶ ♂
05 50	☽ ꝗ ☉
06 44	☽ ☍ ♇
09 32	☽ ± ♅
11 59	☽ ∥ ♀
14 01	☽ ꝗ ☿
19 35	☉ ∠ ♇
21 34	☽ ± ☉
21 37	☽ ꝗ ♄

All ephemeris data is given at 12.00 UT and the Moon's longitude is additionally given for 24.00 UT

OCTOBER 1952

LONGITUDES

Date	Sidereal time h m s	Sun ☉ ° ' "	Moon ☽ ° ' "	Moon ☽ 24.00 ° ' "	Mercury ☿ ° '	Venus ♀ ° '	Mars ♂ ° '	Jupiter ♃ ° '	Saturn ♄ ° '	Uranus ♅ ° '	Neptune ♆ ° '	Pluto ♇ ° '
01	12 40 38	08 ♎ 14 39	10 ♓ 32 34	17 ♓ 56 46	13 ♎ 35	04 ♏ 38	22 ♐ 30	20 ♉ 12	16 ♎ 45	18 ♋ 18	20 ♎ 57	22 ♌ 35
02	12 44 35	09 13 40	25 ♓ 20 44	02 ♈ 43 30	15 17	05 51	23 12	20 R 08	16 53	18 19	20 59	22 37
03	12 48 31	10 12 44	10 ♈ 04 05	17 ♈ 21 31	16 58	07 05	23 53	20 04	17 00	18 20	21 01	22 38
04	12 52 28	11 11 49	24 ♈ 34 55	01 ♉ 43 29	18 38	08 18	24 35	19 59	17 07	18 21	21 04	22 39
05	12 56 24	12 10 56	08 ♉ 46 33	15 ♉ 43 36	20 17	09 32	25 17	19 54	17 14	18 22	21 06	22 41
06	13 00 21	13 10 06	22 ♉ 34 17	29 ♉ 18 25	21 55	10 46	25 59	19 49	17 22	18 23	21 08	22 42
07	13 04 17	14 09 18	05 ♊ 55 58	12 ♊ 27 04	23 33	11 59	26 41	19 44	17 29	18 24	21 10	22 43
08	13 08 14	15 08 32	18 ♊ 51 58	25 ♊ 11 03	25 10	13 13	27 23	19 38	17 36	18 25	21 13	22 45
09	13 12 10	16 07 49	01 ♋ 24 46	07 ♋ 33 41	26 46	14 26	28 05	19 33	17 44	18 26	21 15	22 46
10	13 16 07	17 07 08	13 ♋ 38 23	19 ♋ 39 31	28 21	15 40	28 48	19 27	17 51	18 27	21 17	22 47
11	13 20 04	18 06 29	25 ♋ 37 45	01 ♌ 33 46	29 ♎ 56	16 53	29 ♐ 30	19 21	17 58	18 27	21 19	22 48
12	13 24 00	19 05 53	07 ♌ 28 14	13 ♌ 21 51	01 ♏ 30	18 07	00 ♑ 13	19 15	18 06	18 28	21 21	22 50
13	13 27 57	20 05 19	19 ♌ 15 14	25 ♌ 09 01	03 03	19 20	00 56	19 09	18 13	18 29	21 24	22 51
14	13 31 53	21 04 47	01 ♍ 03 47	07 ♍ 00 05	04 36	20 34	01 38	19 03	18 20	18 29	21 26	22 52
15	13 35 50	22 04 17	12 ♍ 58 23	18 ♍ 59 08	06 07	21 47	02 21	18 56	18 28	18 30	21 28	22 53
16	13 39 46	23 03 49	25 ♍ 02 41	01 ♎ 09 22	07 39	23 01	03 04	18 50	18 35	18 30	21 30	22 54
17	13 43 43	24 03 24	07 ♎ 19 24	13 ♎ 32 56	09 09	24 14	03 48	18 43	18 42	18 31	21 33	22 55
18	13 47 39	25 03 01	19 ♎ 50 06	26 ♎ 10 54	10 39	25 28	04 31	18 36	18 50	18 31	21 35	22 56
19	13 51 36	26 02 39	02 ♏ 35 19	09 ♏ 03 16	12 08	26 41	05 14	18 29	18 57	18 31	21 37	22 57
20	13 55 33	27 02 20	15 ♏ 34 37	22 ♏ 09 12	13 37	27 55	05 58	18 22	19 04	18 32	21 39	22 58
21	13 59 29	28 02 03	28 ♏ 46 52	05 ♐ 27 25	15 05	29 ♏ 08	06 41	18 15	19 12	18 32	21 42	22 59
22	14 03 26	29 ♎ 01 48	12 ♐ 10 39	18 ♐ 56 27	16 32	00 ♐ 21	07 25	18 08	19 19	18 32	21 44	23 00
23	14 07 22	00 ♏ 01 34	25 ♐ 47 48	02 ♑ 35 08	17 59	01 35	08 08	18 01	19 26	18 32	21 46	23 01
24	14 11 19	01 01 22	09 ♑ 27 49	16 ♑ 22 41	19 25	02 48	08 52	17 53	19 33	18 32	21 48	23 02
25	14 15 15	02 01 12	23 ♑ 19 40	00 ♒ 18 45	20 50	04 01	09 36	17 45	19 41	18 R 32	21 50	23 03
26	14 19 12	03 01 04	07 ♒ 19 55	14 ♒ 23 06	22 14	05 15	10 20	17 38	19 48	18 32	21 53	23 04
27	14 23 08	04 00 57	21 ♒ 28 14	28 ♒ 35 11	23 38	06 28	11 04	17 30	19 55	18 32	21 55	23 05
28	14 27 05	05 00 51	05 ♓ 43 44	12 ♓ 53 36	25 00	07 41	11 48	17 22	20 02	18 32	21 57	23 05
29	14 31 02	06 00 48	20 ♓ 04 24	27 ♓ 15 40	26 22	08 54	12 32	17 15	20 09	18 31	21 59	23 06
30	14 34 58	07 00 46	04 ♈ 26 51	11 ♈ 37 18	27 43	10 08	13 16	17 07	20 17	18 31	22 01	23 07
31	14 38 55	08 ♏ 00 45	18 ♈ 46 22	25 ♈ 53 18	29 ♏ 02	11 ♐ 21	14 ♑ 01	16 ♉ 59	20 ♎ 24	18 ♋ 31	22 ♎ 04	23 ♌ 08

Date	Moon True ☊ ° '	Moon Mean ☊ ° '	Moon ☽ Latitude ° '
01	20 ♒ 39	18 ♒ 55	01 N 45
02	20 R 36	18 52	02 56
03	20 30	18 49	03 54
04	20 22	18 46	04 36
05	20 14	18 43	05 01
06	20 05	18 39	05 06
07	19 58	18 36	04 55
08	19 53	18 33	04 28
09	19 49	18 30	03 50
10	19 48	18 27	03 02
11	19 D 48	18 23	02 06
12	19 49	18 20	01 06
13	19 50	18 17	00 N 03
14	19 R 49	18 14	01 S 00
15	19 46	18 11	02 00
16	19 41	18 08	02 56
17	19 33	18 04	03 45
18	19 23	18 01	04 24
19	19 11	17 58	04 50
20	18 59	17 55	05 02
21	18 47	17 52	04 58
22	18 38	17 49	04 37
23	18 31	17 45	04 01
24	18 27	17 42	03 11
25	18 25	17 39	02 09
26	18 D 25	17 36	00 S 58
27	18 R 25	17 33	00 N 16
28	18 24	17 29	01 30
29	18 21	17 26	02 39
30	18 15	17 23	03 38
31	18 ♒ 06	17 ♒ 20	04 N 23

DECLINATIONS

Date	Sun ☉ ° '	Moon ☽ ° '	Mercury ☿ ° '	Venus ♀ ° '	Mars ♂ ° '	Jupiter ♃ ° '	Saturn ♄ ° '	Uranus ♅ ° '	Neptune ♆ ° '	Pluto ♇ ° '
01	03 S 16	06 S 00	04 S 39	12 S 58	25 S 12	16 N 32	04 S 31	22 N 36	06 S 41	22 N 31
02	03 39	00 N 50	05 24	13 25	25 14	16 31	04 33	22 36	06 42	22 31
03	04 03	07 35	06 09	13 52	25 15	16 30	04 36	22 35	06 43	22 31
04	04 26	13 49	06 53	14 18	25 17	16 28	04 39	22 35	06 44	22 30
05	04 49	19 10	07 37	14 45	25 18	16 27	04 42	22 35	06 44	22 30
06	05 12	23 21	08 20	15 10	25 19	16 26	04 45	22 35	06 45	22 30
07	05 35	26 08	09 02	15 36	25 20	16 24	04 47	22 35	06 46	22 30
08	05 58	27 26	09 44	16 01	25 21	16 23	04 50	22 35	06 47	22 30
09	06 21	27 16	10 26	16 26	25 21	16 21	04 53	22 35	06 48	22 29
10	06 44	25 46	11 07	16 50	25 21	16 20	04 56	22 35	06 48	22 29
11	07 06	23 05	11 47	17 14	25 21	16 18	04 59	22 35	06 49	22 29
12	07 29	19 28	12 26	17 38	25 21	16 16	05 01	22 35	06 50	22 29
13	07 51	15 06	13 05	18 01	25 20	16 14	05 04	22 35	06 51	22 29
14	08 14	10 13	13 43	18 23	25 20	16 13	05 07	22 34	06 52	22 29
15	08 36	04 N 50	14 20	18 45	25 19	16 11	05 10	22 34	06 52	22 29
16	08 58	00 S 44	14 56	19 07	25 17	16 09	05 12	22 34	06 53	22 29
17	09 20	06 21	15 32	19 28	25 16	16 07	05 15	22 34	06 54	22 29
18	09 42	11 49	16 07	19 49	25 14	16 05	05 18	22 34	06 55	22 28
19	10 04	16 54	16 41	20 09	25 12	16 03	05 21	22 34	06 56	22 28
20	10 25	21 19	17 14	20 29	25 10	16 01	05 23	22 34	06 57	22 28
21	10 47	24 44	17 46	20 48	25 07	15 59	05 26	22 34	06 57	22 28
22	11 08	26 50	18 18	21 06	25 04	15 57	05 29	22 34	06 58	22 28
23	11 29	27 24	18 48	21 24	25 01	15 55	05 31	22 34	06 59	22 28
24	11 50	26 17	19 18	21 42	24 58	15 53	05 34	22 34	07 00	22 28
25	12 11	23 33	19 46	21 58	24 54	15 51	05 37	22 34	07 01	22 28
26	12 31	19 23	20 14	22 15	24 50	15 49	05 39	22 34	07 02	22 28
27	12 52	14 06	20 40	22 30	24 46	15 47	05 42	22 35	07 02	22 28
28	13 12	08 01	21 06	22 45	24 42	15 45	05 45	22 35	07 03	22 28
29	13 32	01 S 30	21 30	23 00	24 37	15 43	05 47	22 35	07 04	22 28
30	13 52	05 N 05	21 54	23 13	24 33	15 40	05 50	22 35	07 05	22 28
31	14 S 11	11 N 24	22 S 16	23 S 26	24 S 28	15 N 38	05 S 53	22 N 35	07 S 06	22 N 28

ZODIAC SIGN ENTRIES

Date	h	m	Planets
02	19	34	☽ ♈
04	21	05	☽ ♉
07	01	15	☽ ♊
09	09	16	☽ ♋
11	13	05	☿ ♏
11	20	50	☽ ♌
12	04	45	♂ ♑
14	09	51	☽ ♍
16	21	44	☽ ♎
19	07	10	☽ ♏
21	14	12	☽ ♐
22	05	02	♀ ♐
23	11	22	☉ ♏
23	19	28	☽ ♑
25	23	28	☽ ♒
28	02	23	☽ ♓
30	04	34	☽ ♈

LATITUDES

Date	Mercury ☿ ° '	Venus ♀ ° '	Mars ♂ ° '	Jupiter ♃ ° '	Saturn ♄ ° '	Uranus ♅ ° '	Neptune ♆ ° '	Pluto ♇ ° '
01	00 N 47	00 N 07	01 S 58	01 S 19	02 N 15	00 N 24	01 N 37	09 N 03
04	00 27	00 S 02	01 57	01 19	02 15	00 24	01 37	09 03
07	00 N 07	00 10	01 56	01 19	02 15	00 24	01 37	09 04
10	00 S 14	00 19	01 55	01 19	02 15	00 24	01 37	09 04
13	00 35	00 28	01 54	01 20	02 15	00 24	01 37	09 06
16	00 55	00 36	01 53	01 20	02 15	00 24	01 37	09 07
19	01 15	00 45	01 51	01 20	02 15	00 25	01 37	09 08
22	01 35	00 54	01 50	01 20	02 15	00 25	01 37	09 09
25	01 52	01 02	01 49	01 20	02 15	00 25	01 37	09 10
28	02 08	01 10	01 47	01 20	02 15	00 25	01 37	09 11
31	02 S 22	01 S 18	01 S 45	01 S 20	02 N 16	00 N 25	01 N 37	09 N 12

DATA

Julian Date	2434287
Delta T	+30 seconds
Ayanamsa	23° 12' 00"
Synetic vernal point	05° ♓ 55' 00"
True obliquity of ecliptic	23° 26' 51"

MOON'S PHASES, APSIDES AND POSITIONS ☽

Date	h	m	Phase	Longitude ° '	Eclipse Indicator
03	12	15	○	10 ♈ 13	
10	19	33	☾	17 ♋ 26	
18	22	42	●	25 ♎ 30	
26	04	04	☽	02 ♒ 41	

Day	h	m		
01	13	08	Perigee	
13	10	10	Apogee	
29	05	51	Perigee	
02	09	04	0N	
08	21	06	Max dec	27° N 33'
16	08	53	0S	
23	08	06	Max dec	27° S 25'
29	17	25	0N	

LONGITUDES

Date	Chiron ⚷ ° '	Ceres ⚳ ° '	Pallas ⚴ ° '	Juno ⚵ ° '	Vesta ⚶ ° '	Black Moon Lilith ⚸ ° '
01	06 ♑ 55	16 ♊ 54	27 ♓ 49	09 ♒ 16	01 ♏ 47	20 ♌ 45
11	07 ♑ 15	17 ♊ 27	25 ♓ 14	09 ♒ 49	07 ♏ 02	21 ♌ 51
21	07 ♑ 43	17 ♊ 21	23 ♓ 01	11 ♒ 03	12 ♏ 20	22 ♌ 58
31	08 ♑ 18	16 ♊ 36	21 ♓ 24	12 ♒ 55	17 ♏ 41	24 ♌ 04

All ephemeris data is given at 12.00 UT and the Moon's longitude is additionally given for 24.00 UT

ASPECTARIAN

01 Wednesday
00 13 ☽ ⚼ ♅
01 32 ☽ △ ♀
01 43 ☽ Q ♂
04 32 ☽ ⚼ ♆
06 34 ☽ ± ☿
07 32 ☿ ∥ ♄
08 00 ☽ ⚻ ☉
08 14 ☽ Q ♃
09 33 ☽ ∥ ♆
11 15 ♀ Q ♇
12 21 ☽ ± ♄
15 07 ♂ △ ♇
16 18 ☽ ∥ ☿
17 13 ☽ ∥ ♄
17 34 ☽ ⚻ ☿
19 10 ☽ ± ♆
20 28 ☿ ± ♃
21 06 ☽ ∥ ☉
22 09 ☽ ⚻ ♄

02 Thursday
00 35 ☽ △ ♅
03 35 ☽ ✶ ♃
04 03 ☽ ⚼ ♀
04 55 ☽ ⚻ ♆
07 33 ☽ ⚻ ♇
08 20 ☽ □ ♂
17 19 ☽ ± ♇
19 59 ☽ ± ♀
22 31 ☽ # ☉

03 Friday
01 12 ☽ # ♄
03 51 ☽ ∠ ♃
06 08 ☽ # ☿
06 40 ☽ ⚻ ♀
08 00 ☽ ⚼ ♇
08 50 ☽ # ♆
12 15 ☽ ☍ ☉
12 36 ☿ ☌ ♄
18 31 ☽ ⊥ ♃
23 30 ☽ ☍ ♄

04 Saturday
00 49 ☽ ☍ ☿
01 38 ☽ □ ♅
04 23 ☽ ⚺ ♃
06 07 ☽ ☍ ♆
06 45 ☿ ∥ ♆
07 59 ☿ □ ♅
08 47 ☽ △ ♇
12 00 ☽ △ ♂
14 13 ☽ # ♀
23 18 ☽ ∥ ♃

05 Sunday
03 32 ☉ ∥ ♄
06 45 ☿ ⚻ ♃
07 53 ☽ Q ♅
13 25 ☽ ☍ ♀
14 43 ☽ ⚼ ♂
18 18 ☽ ⚻ ☉

06 Monday
00 14 ☿ ☌ ♆
02 46 ☽ ⚻ ♄
04 38 ☽ ✶ ♅
05 33 ☽ ± ☉
06 27 ☽ ∥ ♇
06 58 ☽ ∥ ♅
07 01 ♂ ± ♃
07 11 ☽ ☌ ♃
07 11 ☽ ± ♂
09 28 ☽ ⚻ ♆
10 41 ☽ ⚻ ☿
12 14 ☽ □ ♇
13 25 ☽ ± ♄
18 23 ☽ ⚻ ♂
20 08 ☽ ± ♆
22 03 ♀ ∠ ♂
22 45 ☽ ⚼ ☉
22 50 ☽ ± ☿
23 40 ☿ ✶ ♇

07 Tuesday
02 29 ☉ ± ♃
03 37 ☽ # ♂
05 40 ☽ ⚼ ♄
07 23 ☽ ∠ ♅
12 26 ☽ ⚼ ♆
17 28 ☽ ⚼ ☿
20 49 ☽ Q ♇
23 55 ☽ ⊥ ♅

08 Wednesday
00 18 ☽ ⚻ ♀
04 26 ☽ △ ☉
09 36 ☽ △ ♄
11 09 ☽ ⚺ ♅
12 43 ☽ ± ♀
13 27 ☽ ⚺ ♃
16 27 ☽ △ ♆
19 21 ☽ ✶ ♇

09 Thursday
00 47 ☽ ⊥ ♃
01 43 ☽ △ ☿
05 11 ☽ ☍ ♂
07 42 ♀ # ♃
07 45 ☽ ⚼ ♀
08 29 ☉ Q ♂
18 03 ☽ ∠ ♃

10 Friday
00 25 ☽ ∠ ♇
16 19 ☽ # ♂
16 29 ☽ △ ♀
17 14 ☉ ∥ ♆
18 16 ☽ ⊥ ♇
19 33 ☽ □ ☉
20 28 ☽ □ ♄
21 35 ☽ ☌ ♅
23 30 ☽ ✶ ♃

11 Saturday
00 14 ☿ ✶ ♂
03 18 ☽ □ ♆
06 18 ☽ ⚺ ♇
08 17 ☉ ☌ ♄
15 47 ☽ ∥ ♅
16 26 ☽ ∥ ♇
20 20 ☽ ⚻ ♂
20 31 ☉ □ ♅
22 01 ☽ □ ☿
23 29 ☽ Q ♃

12 Sunday
07 10 ♂ Q ♄
09 11 ☽ Q ♄
09 17 ☽ ± ♂
11 10 ☽ Q ☉
11 37 ♀ ⚺ ♄
15 27 ☉ ⚻ ♃
15 51 ☽ Q ♆
18 59 ♀ △ ♅
21 43 ☽ # ♀

13 Monday
04 47 ☽ ⚼ ♂
06 02 ☽ ∥ ♃
08 38 ♀ ☍ ♃
09 52 ☽ ✶ ♄
10 25 ☽ ⚺ ♅
11 48 ☽ □ ♃
12 12 ☽ □ ♀
13 51 ☽ ✶ ☉
16 12 ☽ Q ☿
16 22 ☽ ✶ ♆
19 19 ☽ ☌ ♇
20 58 ☽ # ☿
22 38 ☽ ⊥ ♅

14 Tuesday
13 15 ☽ △ ♂
16 19 ☿ Q ♇
16 39 ☽ ∠ ♄
16 55 ☽ ∠ ♅
20 12 ☽ ✶ ☿
20 19 ☽ # ☉
20 53 ☉ ☌ ♆
22 53 ☽ ∠ ♆
23 04 ☽ ∠ ☉

15 Wednesday
03 01 ☽ # ♆
04 53 ☽ Q ♀
05 34 ♀ ⚺ ♆
10 35 ☽ # ♄
10 58 ☽ ⊥ ♄
17 01 ☽ ⊥ ♆
18 45 ☽ ⊥ ☉
18 51 ♄ □ ♅
23 02 ☽ ✶ ♅
23 04 ☽ ⚺ ♄
23 48 ☽ △ ♃

16 Thursday
04 59 ☽ ⚺ ♆
06 35 ☽ ∠ ☿
07 32 ☽ ✶ ♀
07 44 ☽ ⚺ ☉
07 46 ☽ ⚺ ♇
08 00 ☉ ✶ ♇
09 47 ♀ □ ♇
17 16 ☉ ⚺ ♀
19 36 ☽ ⊥ ♇
22 44 ☽ Q ♅

17 Friday
02 46 ☽ ⊥ ☿
03 16 ♂ Q ♆
04 43 ☽ □ ♂
05 04 ☽ ⚼ ♃
07 15 ☽ ∥ ♄
09 54 ♂ ⚼ ♃
13 09 ☽ ∠ ♇
13 14 ♃ ⚻ ♄
14 23 ☽ ∥ ♆
16 02 ☽ ⚺ ☿
16 06 ☽ ∠ ♀
22 14 ♀ ⊥ ♄
22 19 ☽ ± ♃

18 Saturday
01 52 ☽ ∥ ☉
09 29 ☽ □ ♅
09 41 ☽ ⚻ ♃
10 04 ☽ ☌ ♄
10 56 ☿ # ♃
11 13 ☽ ⊥ ♀
15 19 ☽ ☌ ♆
17 23 ☽ Q ♂
17 53 ☽ ✶ ♇
22 42 ☽ ☌ ☉
23 46 ☽ ⚺ ♀

19 Sunday
05 44 ♃ ✶ ♅
07 49 ☽ # ♃
10 44 ☽ ∥ ☿
16 24 ☽ Q ♇
17 13 ☽ ✶ ♂

20 Monday
06 39 ☽ ∥ ♀
06 54 ♀ ⊥ ♆
07 57 ☽ ☌ ☿
17 04 ☽ ☍ ♃
17 24 ☽ △ ♅
18 27 ☽ ⚺ ♄
19 18 ☽ # ♇
19 57 ☽ # ♅
22 24 ☽ ∠ ♂
23 08 ☽ ⚺ ♆

21 Tuesday
01 30 ☽ □ ♇
05 27 ☽ ⊥ ♄
10 02 ☽ ⊥ ♆
10 33 ☽ ⚺ ☉
12 42 ☽ ☌ ♀
15 22 ☽ ∥ ♂
15 38 ☽ ⊥ ♂
20 33 ☽ ⚼ ♅
21 50 ☽ ∠ ♄
22 13 ☽ ⊥ ☉

22 Wednesday
02 15 ☽ ∠ ♆
03 01 ☽ ⚺ ♂
12 38 ☽ ± ♅
15 33 ☽ ∠ ☉
20 40 ☽ ⚺ ☿
22 28 ☽ ⚻ ♃
23 17 ☽ ⚻ ♅

23 Thursday
00 47 ☽ ✶ ♄
04 59 ☽ ✶ ♆
07 12 ☽ △ ♇
07 57 ♂ ⚼ ♇
08 32 ☽ ⊥ ☿
08 58 ☽ ± ♃
12 26 ☿ ☍ ♃
20 07 ☽ ✶ ☉
21 15 ☿ △ ♅
22 05 ☽ Q ♄
23 14 ☽ ⚺ ♀

24 Friday
00 38 ☽ ⚼ ♃
02 06 ☽ Q ♆
02 10 ☽ ∠ ☿
09 30 ☽ ⚼ ♇
10 44 ☽ ⊥ ♀
10 54 ☽ ☌ ♂
14 42 ☿ ⚺ ♄
15 19 ♀ ⊥ ♂
16 47 ♅ St R
18 40 ☽ Q ☉

25 Saturday
01 09 ☽ ± ♇
01 32 ☽ ∥ ♂
02 24 ♀ ⚼ ♅
02 28 ☽ △ ♃
03 44 ☽ ☍ ♅
03 51 ☽ ∠ ♀
05 39 ☽ □ ♄
07 12 ☽ ✶ ☿
09 26 ☽ □ ♆
11 31 ☽ ⚻ ♇
18 22 ☽ # ♅
19 00 ☽ # ♇
21 22 ☽ ∥ ♀

26 Sunday
02 18 ♀ ∠ ♄
04 04 ☽ □ ☉
05 43 ☿ ⚺ ♆
06 08 ☽ Q ☿
08 03 ☽ ∥ ☿
08 06 ☽ ✶ ♀
17 23 ☽ ⚺ ♂

27 Monday
02 22 ☿ □ ♇
04 08 ☽ ⊥ ♂
04 45 ☽ # ♃
05 21 ☽ □ ♃
06 26 ☽ Q ♀
07 02 ☽ ⚻ ♅
08 54 ♀ # ♇
09 21 ☽ △ ♄
12 45 ☽ △ ♆
14 43 ☽ ☍ ♇
16 02 ☽ □ ☿
16 48 ☽ ∥ ☉
17 10 ☽ ± ♅
18 49 ♀ # ♅
20 11 ☽ ∠ ♂
21 06 ♀ ∠ ♆

28 Tuesday
08 18 ☽ ⚼ ♅
10 43 ☽ △ ☉
10 50 ☽ ⚼ ♄
11 24 ☽ Q ♃
13 49 ☉ Q ♇
14 03 ☽ ⚼ ♆
15 35 ☽ □ ♀
15 36 ☽ ∥ ♆
20 24 ☽ ∥ ♄
22 43 ☽ ✶ ♂

29 Wednesday
02 02 ☽ ± ♄
05 09 ☽ ± ♆
07 19 ☽ ✶ ♃
07 57 ☿ ⊥ ♄
09 25 ☽ △ ♅
12 09 ☽ ⚻ ♄
13 41 ☽ ⚼ ☉
15 12 ☽ ⚻ ♆
17 04 ☽ ⚻ ♇
19 51 ☽ Q ♂
23 36 ☽ △ ☿

30 Thursday
03 05 ☽ ± ♇
05 50 ☽ ± ☉
08 08 ☽ ∠ ♃
14 43 ☽ # ♄
16 37 ☽ ⚻ ☉
17 46 ☿ ⊥ ♆
18 08 ☽ ⚼ ♇
19 22 ☽ # ♆
22 23 ☽ △ ♀
23 02 ☽ ⊥ ♃

31 Friday
03 15 ☽ ⚼ ☿
03 34 ☽ □ ♂
09 01 ☽ ⚺ ♃
10 49 ☿ ∠ ♂
11 34 ☽ □ ♅
14 45 ☽ ☍ ♄
17 33 ☽ ☍ ♆
19 20 ☽ △ ♇
19 55 ☽ ± ☿

NOVEMBER 1952

LONGITUDES

Date	Sidereal time h m s	Sun ☉ ° ' "	Moon ☽ ° ' "	Moon ☽ 24.00 ° ' "	Mercury ☿ ° '	Venus ♀ ° '	Mars ♂ ° '	Jupiter ♃ ° '	Saturn ♄ ° '	Uranus ♅ ° '	Neptune ♆ ° '	Pluto ♇ ° '
01	14 42 51	09 ♏ 00 47	02 ♉ 57 23	09 ♉ 57 57	00 ♐ 21	12 ♐ 34	14 ♑ 45	16 ♉ 51	20 ♎ 31	18 ♋ 30	22 ♎ 06	23 ♌ 08
02	14 46 48	10 00 50	16 ♉ 54 21	23 ♉ 46 04	01 38	13 47	15 30	16 R 43	20 38	18 R 30	22 08	23 09
03	14 50 44	11 00 55	00 ♊ 32 38	07 ♊ 13 46	02 54	15 00	16 14	16 34	20 45	18 29	22 10	23 10
04	14 54 41	12 01 03	13 ♊ 49 17	20 ♊ 19 09	04 08	16 13	16 59	16 26	20 52	18 29	22 12	23 10
05	14 58 37	13 01 12	26 ♊ 43 26	03 ♋ 02 21	05 21	17 26	17 43	16 18	20 59	18 28	22 14	23 11
06	15 02 34	14 01 23	09 ♋ 16 13	15 ♋ 25 28	06 32	18 39	18 28	16 10	21 06	18 28	22 17	23 11
07	15 06 31	15 01 36	21 ♋ 30 35	27 ♋ 32 09	07 41	19 52	19 13	16 02	21 13	18 27	22 19	23 12
08	15 10 27	16 01 52	03 ♌ 30 46	09 ♌ 27 07	08 47	21 05	19 58	15 54	21 20	18 26	22 21	23 12
09	15 14 24	17 02 09	15 ♌ 21 52	21 ♌ 15 44	09 51	22 18	20 42	15 46	21 27	18 25	22 23	23 13
10	15 18 20	18 02 28	27 ♌ 09 25	03 ♍ 03 37	10 52	23 31	21 27	15 37	21 34	18 24	22 25	23 13
11	15 22 17	19 02 49	08 ♍ 59 01	14 ♍ 56 17	11 49	24 44	22 12	15 29	21 40	18 24	22 27	23 14
12	15 26 13	20 03 12	20 ♍ 56 01	26 ♍ 58 47	12 43	25 57	22 57	15 21	21 47	18 23	22 29	23 14
13	15 30 10	21 03 37	03 ♎ 05 07	09 ♎ 15 25	13 32	27 10	23 43	15 13	21 54	18 22	22 31	23 14
14	15 34 06	22 04 04	15 ♎ 30 04	21 ♎ 49 18	14 17	28 23	24 28	15 05	22 01	18 20	22 33	23 15
15	15 38 03	23 04 32	28 ♎ 13 18	04 ♏ 42 08	14 57	29 ♐ 36	25 13	14 57	22 07	18 19	22 35	23 15
16	15 42 00	24 05 02	11 ♏ 15 44	17 ♏ 53 58	15 30	00 ♑ 48	25 58	14 49	22 14	18 18	22 37	23 15
17	15 45 56	25 05 34	24 ♏ 36 35	01 ♐ 23 15	15 57	02 01	26 44	14 41	22 21	18 17	22 39	23 15
18	15 49 53	26 06 08	08 ♐ 13 33	15 ♐ 07 02	16 17	03 14	27 29	14 33	22 27	18 16	22 41	23 16
19	15 53 49	27 06 43	22 ♐ 03 14	29 ♐ 01 40	16 28	04 26	28 14	14 25	22 34	18 14	22 43	23 16
20	15 57 46	28 07 20	06 ♑ 01 49	13 ♑ 03 17	16 R 31	05 39	29 00	14 18	22 40	18 13	22 45	23 16
21	16 01 42	29 ♏ 07 57	20 ♑ 05 39	27 ♑ 08 35	16 24	06 52	29 ♑ 45	14 10	22 46	18 12	22 47	23 16
22	16 05 39	00 ♐ 08 36	04 ♒ 11 48	11 ♒ 15 06	16 07	08 04	00 ♒ 31	14 02	22 53	18 10	22 49	23 16
23	16 09 35	01 09 16	18 ♒ 18 18	25 ♒ 21 17	15 39	09 17	01 17	13 55	22 59	18 09	22 50	23 16
24	16 13 32	02 09 57	02 ♓ 23 58	09 ♓ 26 14	15 00	10 29	02 02	13 48	23 05	18 07	22 52	23 16
25	16 17 29	03 10 39	16 ♓ 28 02	23 ♓ 29 13	14 11	11 42	02 48	13 40	23 12	18 06	22 54	23 R 16
26	16 21 25	04 11 22	00 ♈ 29 41	07 ♈ 29 14	13 12	12 54	03 34	13 33	23 18	18 04	22 56	23 16
27	16 25 22	05 12 07	14 ♈ 27 39	21 ♈ 24 39	12 04	14 06	04 19	13 26	23 24	18 02	22 58	23 16
28	16 29 18	06 12 52	28 ♈ 19 55	05 ♉ 13 05	10 49	15 19	05 05	13 19	23 30	18 01	22 59	23 16
29	16 33 15	07 13 38	12 ♉ 03 47	18 ♉ 51 36	09 29	16 31	05 51	13 12	23 36	17 59	23 01	23 16
30	16 37 11	08 ♐ 14 25	25 ♉ 36 12	02 ♊ 17 12	08 ♐ 06	17 ♑ 43	06 ♒ 37	13 ♉ 06	23 ♎ 42	17 ♋ 57	23 ♎ 03	23 ♌ 16

DECLINATIONS

Date	Moon True ☊ ° '	Moon Mean ☊ ° '	Moon ☽ Latitude ° '	Sun ☉ ° '	Moon ☽ ° '	Mercury ☿ ° '	Venus ♀ ° '	Mars ♂ ° '	Jupiter ♃ ° '	Saturn ♄ ° '	Uranus ♅ ° '	Neptune ♆ ° '	Pluto ♇ ° '
01	17 ♒ 55	17 ♒ 17	04 N 50	14 S 30	17 N 03	22 S 37	23 S 39	24 S 22	15 N 36	05 S 55	22 N 35	07 S 06	22 N 28
02	17 R 43	17 14	05 00	14 49	21 41	22 57	23 50	24 17	15 34	05 58	22 35	07 07	22 29
03	17 30	17 10	04 53	15 08	25 02	23 15	24 01	24 11	15 31	06 00	22 35	07 08	22 29
04	17 19	17 07	04 29	15 27	26 55	23 32	24 12	24 05	15 29	06 03	22 35	07 09	22 29
05	17 10	17 04	03 53	15 45	27 17	23 48	24 21	23 58	15 27	06 05	22 35	07 09	22 29
06	17 03	17 01	03 05	16 03	26 12	24 03	24 30	23 52	15 25	06 08	22 35	07 10	22 29
07	17 00	16 58	02 10	16 21	23 52	24 16	24 39	23 45	15 23	06 10	22 35	07 11	22 29
08	16 58	16 55	01 11	16 38	20 31	24 28	24 46	23 38	15 20	06 13	22 36	07 12	22 29
09	16 D 58	16 51	00 N 08	16 56	16 22	24 38	24 53	23 30	15 18	06 15	22 36	07 12	22 29
10	16 R 58	16 48	00 S 54	17 13	11 37	24 47	24 59	23 23	15 16	06 18	22 36	07 13	22 30
11	16 56	16 45	01 54	17 29	06 27	24 54	25 05	23 15	15 13	06 20	22 36	07 14	22 30
12	16 53	16 42	02 49	17 46	01 N 00	24 59	25 09	23 07	15 11	06 23	22 36	07 15	22 30
13	16 48	16 39	03 38	18 02	04 S 34	25 03	25 13	22 58	15 09	06 25	22 36	07 15	22 30
14	16 39	16 35	04 18	18 17	10 00	25 05	25 16	22 50	15 07	06 27	22 37	07 16	22 30
15	16 28	16 32	04 46	18 33	15 17	25 05	25 19	22 41	15 05	06 30	22 37	07 17	22 31
16	16 15	16 29	04 59	18 48	19 57	25 03	25 20	22 32	15 02	06 32	22 37	07 17	22 31
17	16 02	16 26	04 57	19 03	23 44	24 59	25 21	22 23	15 00	06 34	22 37	07 18	22 31
18	15 49	16 23	04 38	19 17	26 16	24 53	25 22	22 13	14 58	06 37	22 37	07 19	22 31
19	15 39	16 20	04 03	19 31	27 15	24 45	25 21	22 03	14 56	06 39	22 38	07 20	22 32
20	15 31	16 16	03 12	19 45	26 31	24 34	25 20	21 53	14 54	06 41	22 38	07 20	22 32
21	15 26	16 13	02 10	19 58	24 05	24 21	25 18	21 43	14 52	06 44	22 38	07 21	22 32
22	15 24	16 10	00 S 59	20 11	20 10	24 06	25 15	21 33	14 50	06 46	22 38	07 22	22 32
23	15 D 24	16 07	00 N 15	20 24	15 06	23 47	25 11	21 22	14 48	06 48	22 38	07 22	22 33
24	15 R 24	16 04	01 29	20 36	09 14	23 26	25 07	21 11	14 46	06 50	22 39	07 23	22 33
25	15 23	16 01	02 37	20 48	02 S 56	23 02	25 02	21 00	14 44	06 52	22 39	07 24	22 33
26	15 21	15 57	03 35	20 59	03 N 29	22 36	24 57	20 48	14 42	06 54	22 39	07 24	22 34
27	15 15	15 54	04 20	21 10	09 42	22 07	24 50	20 37	14 40	06 56	22 39	07 25	22 34
28	15 07	15 51	04 50	21 21	15 23	21 37	24 43	20 25	14 38	06 59	22 40	07 25	22 34
29	14 57	15 48	05 02	21 31	20 15	21 05	24 35	20 13	14 36	07 01	22 40	07 26	22 35
30	14 ♒ 45	15 ♒ 45	04 N 57	21 S 41	23 N 58	20 S 33	24 S 26	20 S 01	14 N 34	07 S 03	22 N 40	07 S 27	22 N 35

ZODIAC SIGN ENTRIES

Date	h	m	Planets
01	05	34	☿ ♐
01	06	58	☽ ♉
03	11	02	☽ ♊
05	18	12	☽ ♋
08	04	56	☽ ♌
10	17	47	☽ ♍
13	05	57	☽ ♎
15	15	18	☽ ♏
15	20	03	♀ ♑
17	21	33	☽ ♐
20	01	40	☽ ♑
21	19	40	♂ ♒
22	04	52	☽ ♒
22	08	36	☉ ♐
24	07	55	☽ ♓
26	11	09	☽ ♈
28	14	54	☽ ♉
30	19	53	☽ ♊

LATITUDES

Date	Mercury ☿ ° '	Venus ♀ ° '	Mars ♂ ° '	Jupiter ♃ ° '	Saturn ♄ ° '	Uranus ♅ ° '	Neptune ♆ ° '	Pluto ♇ ° '
01	02 S 26	01 S 21	01 S 45	01 S 20	02 N 16	00 N 25	01 N 37	09 N 12
04	02 36	01 28	01 43	01 20	02 16	00 25	01 37	09 13
07	02 42	01 35	01 42	01 19	02 16	00 25	01 37	09 14
10	02 44	01 42	01 40	01 19	02 17	00 25	01 37	09 15
13	02 38	01 48	01 38	01 19	02 17	00 25	01 37	09 16
16	02 24	01 54	01 36	01 18	02 17	00 25	01 38	09 17
19	02 00	01 59	01 34	01 18	02 18	00 25	01 38	09 18
22	01 23	02 03	01 32	01 17	02 18	00 26	01 38	09 19
25	00 S 32	02 07	01 30	01 17	02 18	00 26	01 38	09 20
28	00 N 28	02 10	01 28	01 16	02 19	00 26	01 38	09 21
31	01 N 27	02 S 12	01 S 26	01 S 15	02 N 19	00 N 26	01 N 38	09 N 22

DATA

Julian Date	2434318
Delta T	+30 seconds
Ayanamsa	23° 12' 04"
Synetic vernal point	05° ♓ 54' 56"
True obliquity of ecliptic	23° 26' 50"

MOON'S PHASES, APSIDES AND POSITIONS ☽

Date	h	m	Phase	Longitude ° '	Eclipse Indicator
01	23	10	○	09 ♉ 29	
09	15	43	☾	17 ♌ 11	
17	12	56	●	25 ♏ 08	
24	11	34	☽	02 ♓ 09	

Day	h	m		
10	05	43	Apogee	
23	08	15	Perigee	
05	05	44	Max dec	27° N 20'
12	16	19	0S	
19	13	54	Max dec	27° S 15'
25	22	56	0N	

LONGITUDES

Date	Chiron ⚷ ° '	Ceres ⚳ ° '	Pallas ⚴ ° '	Juno ⚵ ° '	Vesta ⚶ ° '	Black Moon Lilith ⚸ ° '
01	08 ♑ 22	16 ♊ 29	21 ♓ 16	13 ♒ 08	18 ♏ 13	24 ♌ 11
11	09 ♑ 04	15 ♊ 02	20 ♓ 22	15 ♒ 37	23 ♏ 36	25 ♌ 17
21	09 ♑ 51	13 ♊ 04	20 ♓ 10	18 ♒ 34	29 ♏ 00	26 ♌ 24
31	10 ♑ 43	10 ♊ 48	20 ♓ 38	21 ♒ 56	04 ♐ 25	27 ♌ 30

All ephemeris data is given at 12.00 UT and the Moon's longitude is additionally given for 24.00 UT

ASPECTARIAN

h m	Aspects
01 Saturday	
00 05	☽ # ☉
01 59	☽ Ꝑ ♀
02 15	☿ # ♇
05 32	☽ ∥ ♃
07 07	☽ ⊼ ☿
09 39	☿ # ♅
10 49	♀ ± ♅
18 04	☽ Q ♅
18 45	☽ ± ♀
23 10	☽ ☍ ☉
02 Sunday	
06 04	☽ ⊼ ♀
09 24	☽ △ ♂
11 40	☽ ☌ ♃
14 46	☽ ⚹ ♅
16 51	☽ ∥ ♇
17 32	☽ ∥ ♅
18 33	☽ ⊼ ♄
20 39	☽ # ☿
21 09	☽ ⊼ ♆
22 55	☽ □ ♇
03 Monday	
03 03	☽ # ♀
04 59	☽ # ♂
05 12	☽ ± ♄
07 46	☽ ± ♆
13 18	☽ Ꝑ ♂
16 39	☽ ☍ ☿
17 16	☽ ∠ ♅
21 19	♂ △ ♃
21 25	☽ Ꝑ ♄
23 16	☿ Ꝑ ♅
23 55	☽ Ꝑ ♆
04 Tuesday	
01 26	♀ ∥ ♂
06 29	☽ ± ♂
07 09	☽ Q ♇
08 26	☽ ⊼ ☉
09 33	☽ ⊥ ♅
14 52	☉ # ♃
15 51	♀ ⊼ ♃
16 46	☽ ⊻ ♃
16 52	☽ ☍ ♀
18 09	☽ ⊼ ♂
18 18	♂ ± ♇
20 23	☽ ± ☉
20 35	☽ ⊻ ♅
05 Wednesday	
01 08	☽ △ ♄
03 33	☽ △ ♆
03 47	☽ ⊥ ♃
05 20	☽ ⚹ ♇
14 40	☽ Ꝑ ☉
20 36	☽ ∠ ♃
23 02	☿ ∥ ♂
06 Thursday	
02 10	☿ ∠ ♄
02 16	♀ ⊻ ♂
06 10	☽ ⊼ ☿
08 09	♀ ⊼ ♅
09 54	☽ ∠ ♇
11 49	♂ ☍ ♅
19 00	☽ ± ☿
22 05	☽ △ ☉
07 Friday	
01 19	☽ ⚹ ♃
03 29	☽ ⊥ ♇
03 59	☿ ∠ ♆
05 34	☽ # ♀
05 57	☽ ☌ ♅
07 09	☽ ☍ ♂
08 24	☽ ⊼ ♀
08 53	☽ # ☿
11 25	☽ □ ♄
13 07	☽ # ♂
13 36	☽ □ ♆
14 33	☽ Ꝑ ☿
15 21	☽ ⊻ ♇
21 39	☽ ± ♀
21 58	☽ ∥ ♅
22 44	☽ ∥ ♇
08 Saturday	
00 51	☽ Q ♃
09 09	☉ ☍ ♃
17 14	♀ ⚹ ♄
17 48	☽ Ꝑ ♀
23 42	☽ △ ☿
23 52	☽ Q ♄
09 Sunday	
01 51	☽ Q ♆
02 17	♀ ± ♃
09 10	☽ # ☉
12 48	☽ □ ♃
13 29	♀ ⚹ ♆
15 43	☽ □ ☉
17 41	☽ ∥ ♃
18 13	☽ ⊻ ♅
23 36	☽ ⊼ ♂
10 Monday	
00 30	☽ ⚹ ♄
02 19	☽ ⚹ ♆
03 45	☽ △ ♀
03 59	☽ ☌ ♇
06 00	♀ △ ♇
06 25	☽ ⊥ ♅
12 39	☽ ± ♂
15 54	♂ □ ♄
20 37	☉ △ ♅
11 Tuesday	
00 41	☽ ∠ ♅
07 17	☽ ∠ ♄
07 43	☽ Q ☉
08 10	☽ Ꝑ ♂
08 27	☽ # ♆
08 53	☽ ∠ ♆
12 29	☽ # ♄
18 12	☽ □ ☿
20 07	♂ □ ♆
12 Wednesday	
00 57	☽ △ ♃
01 36	☽ ⊥ ♄
02 57	☿ ± ♅
03 05	☽ ⊥ ♆
06 54	☽ ⚹ ♅
10 05	☽ ⚹ ☉
13 43	☽ ⊻ ♄
15 06	☽ ⊻ ♆
16 18	☽ △ ♂
16 35	☽ ⊻ ♇
20 52	♂ ⊼ ♇
23 05	☽ □ ♀
23 42	☉ ⊥ ♀
13 Thursday	
04 27	☽ ⊥ ♇
06 26	☽ Ꝑ ♃
06 40	☽ Q ♅
08 46	☽ Q ☿
18 19	☽ ∠ ☉
20 04	☽ ∥ ♄
22 02	☽ ∠ ♇
23 41	☽ ∥ ♆
23 48	☽ ± ♃
14 Friday	
09 32	☽ ⚹ ☿
10 28	☉ ⊻ ♄
11 12	☽ ⊼ ♃
13 11	☽ ⊥ ☉
13 51	☽ Q ♀
17 24	☽ □ ♅
23 53	☉ ⊻ ♆
15 Saturday	
00 28	☽ ☌ ♄
01 25	☽ ☌ ♆
01 32	☽ ⊻ ☉
02 41	☽ ⚹ ♇
06 02	☽ □ ♂
11 00	☽ # ♃
12 06	☿ ⊼ ♃
14 49	☽ ⚹ ♀
15 22	☽ ∠ ☿
16 08	☉ □ ♇
18 20	♀ Ꝑ ♃
22 58	♂ # ♅
16 Sunday	
01 01	☽ Q ♇
05 21	☽ ∥ ☉
08 40	☽ ⊥ ☿
14 34	♂ # ♇
17 13	☽ Q ♂
18 23	☽ ☍ ♃
19 59	☽ ⊻ ☿
21 03	☽ ∠ ♀
17 Monday	
00 42	☽ △ ♅
03 00	☽ ∥ ♂
03 32	☽ # ♇
04 12	☽ # ♅
07 55	☽ ⊻ ♄
08 30	☽ ⊻ ♆
09 35	☽ □ ♇
12 56	☽ ☌ ☉
14 45	☽ ⊥ ♀
15 59	☽ ⚹ ♂
18 41	☽ ⊥ ♄
19 11	☽ ⊥ ♆
21 59	☽ ∥ ☿
18 Tuesday	
01 50	☽ ∥ ♀
02 23	☽ ⊻ ♀
03 19	☽ Ꝑ ♅
10 38	☽ ∠ ♄
11 03	☽ ∠ ♆
19 02	☽ ± ♅
19 51	☽ ∠ ♂
22 55	☽ ⊼ ♃
19 Wednesday	
02 15	☽ ☌ ☿
05 25	☽ ⊼ ♅
09 13	☽ ± ♃
12 20	☽ ⊥ ♂
12 53	☽ ⚹ ♄
13 08	☽ ⚹ ♆
14 05	☽ △ ♇
14 34	♀ Q ♄
17 33	♀ Q ♆
21 23	☽ ⊻ ☉
23 16	☽ ⊻ ♂
20 Thursday	
00 34	☽ Ꝑ ♃
06 42	☿ St R
08 29	☽ ⊥ ☉
09 39	☽ Q ♄
09 48	☽ Q ♆
11 17	☽ ☌ ♀
15 49	☽ Ꝑ ♇
21 Friday	
01 03	☽ ∠ ☉
01 48	☽ ∥ ♀
01 59	☽ △ ♃
02 28	☉ ⊥ ♄
03 18	☉ ⊥ ♆
05 47	☽ ⊻ ☿
07 11	☽ ± ♇
08 46	☽ ☍ ♅
09 45	☽ ∥ ☿
13 17	♄ ☌ ♆
15 52	☽ ⊥ ☿
16 35	☽ □ ♆
16 36	☽ □ ♄
17 24	☽ ⊼ ♇
21 53	☽ # ♅
22 29	☽ # ♇
22 Saturday	
04 07	☽ ∥ ♂
04 34	☽ ⚹ ☉
05 23	☽ ☌ ♂
06 53	☽ ∠ ☿
11 55	☽ ∥ ☉
15 56	♀ Ꝑ ♇
19 12	☽ ⊻ ♀
23 Sunday	
00 06	☿ ∠ ♂
02 34	☽ Q ☉
04 36	☽ □ ♃
06 22	☽ ⊥ ♀
07 39	☽ ⚹ ☿
11 44	☽ ⊼ ♅
13 21	☽ # ♃
19 44	☽ △ ♆
20 01	☽ △ ♄
20 27	☽ ☍ ♇
21 55	☽ ± ♅
23 07	☽ ∠ ♀
23 47	☉ ⚹ ♂
24 Monday	
03 15	☽ Q ☿
10 58	☽ Q ♃
11 21	☽ ⊻ ♂
11 34	☽ □ ☉
13 14	☽ Ꝑ ♅
17 08	♇ St R
19 10	☽ ∥ ♆
21 13	☽ ∥ ♄
21 21	☽ Ꝑ ♆
21 46	☽ Ꝑ ♄
22 10	☽ ⊥ ♂
25 Tuesday	
03 05	☽ ⚹ ♀
07 16	☽ ⚹ ♃
08 20	☽ □ ☿
10 03	☉ Ꝑ ♅
12 45	☽ ± ♆
13 15	☽ ± ♄
14 24	☽ ∠ ♂
14 46	☽ △ ♅
23 01	☽ ⊼ ♆
23 35	☽ ⊼ ♄
23 38	☽ ⊼ ♇
26 Wednesday	
00 29	☉ ∥ ♂
01 31	☽ Q ♀
02 46	☿ ⊼ ♃
05 53	♄ ⚹ ♇
08 42	☽ ∠ ♃
09 23	☿ # ♅
09 54	☽ ± ♇
14 01	☿ # ♇
15 09	☿ ⊻ ♀
17 34	☽ ⚹ ♂
18 50	☽ △ ☉
23 49	♀ △ ♃
27 Thursday	
00 00	☽ ⊥ ♃
01 06	☽ # ♄
01 21	☽ Ꝑ ♇
02 59	☽ # ♆
08 12	☽ △ ☿
10 15	☽ ⊻ ♃
11 20	☽ □ ♀
12 32	☿ ± ♅
15 24	☽ Q ♂
18 10	☽ □ ♅
22 41	☽ Ꝑ ☉
28 Friday	
02 43	☽ ☍ ♆
03 13	☽ △ ♇
03 33	☽ ☍ ♄
08 00	☽ Ꝑ ☿
08 38	☽ ∥ ♃
15 32	☽ ± ☉
21 01	☉ ∥ ☿
22 18	☽ ± ☿
29 Saturday	
00 28	☽ □ ♂
01 21	☽ Q ♅
02 17	☿ ⊥ ♀
02 50	☽ ⊼ ☉
07 52	☽ ⊼ ☿
11 52	☽ # ♂
13 59	☽ ☌ ♃
16 14	☽ # ☿
19 42	☽ # ☉
20 36	☽ △ ♀
22 25	☽ ⚹ ♅
30 Sunday	
02 04	☽ ∥ ♇
02 21	☿ ∠ ♄
02 37	☽ ∥ ♅
03 00	♀ ± ♇
07 18	☉ ∠ ♆
07 26	☽ ⊼ ♆
07 49	☽ □ ♇
08 34	☽ ⊼ ♄
10 37	☉ ☌ ☿
12 56	☿ ∠ ♆
15 37	☽ # ♀
16 38	♀ ☍ ♅
18 11	☽ ± ♆
19 23	☽ ± ♄
23 59	☉ ∠ ♄

DECEMBER 1952

LONGITUDES

Date	Sidereal time h m s	Sun ☉ ° ' "	Moon ☽ ° ' "	Moon ☽ 24.00 ° ' "	Mercury ☿ ° '	Venus ♀ ° '	Mars ♂ ° '	Jupiter ♃ ° '	Saturn ♄ ° '	Uranus ♅ ° '	Neptune ♆ ° '	Pluto ♇ ° '
01	16 41 08	09 ♐ 15 14	08 ♊ 54 17	15 ♊ 27 15	06 ♐ 44	18 ♑ 55	07 ♒ 23	12 ♉ 59	23 ♎ 48	17 ♋ 55	23 ♎ 05	23 ♌ 16
02	16 45 04	10 16 03	21 ♊ 55 53	28 ♊ 20 07	05 R 25	20 07	08 09	12 R 53	23 54	17 R 53	23 06	23 R 15
03	16 49 01	11 16 54	04 ♋ 39 57	10 ♋ 55 29	04 12	21 19	08 55	12 46	23 59	17 51	23 08	23 15
04	16 52 58	12 17 46	17 ♋ 06 52	23 ♋ 14 24	03 06	22 31	09 41	12 40	24 05	17 50	23 09	23 15
05	16 56 54	13 18 40	29 ♋ 18 24	05 ♌ 19 18	02 11	23 43	10 27	12 34	24 11	17 48	23 11	23 15
06	17 00 51	14 19 34	11 ♌ 17 35	17 ♌ 13 48	01 25	24 54	11 13	12 28	24 16	17 46	23 13	23 14
07	17 04 47	15 20 30	23 ♌ 08 31	29 ♌ 02 22	00 52	26 06	11 59	12 23	24 22	17 44	23 14	23 14
08	17 08 44	16 21 27	04 ♍ 56 01	10 ♍ 50 08	00 29	27 18	12 45	12 17	24 27	17 41	23 16	23 13
09	17 12 40	17 22 25	16 ♍ 45 24	22 ♍ 42 30	00 18	28 29	13 31	12 12	24 33	17 39	23 17	23 13
10	17 16 37	18 23 24	28 ♍ 42 08	04 ♎ 44 56	00 D 17	29 ♑ 41	14 17	12 06	24 38	17 37	23 19	23 13
11	17 20 33	19 24 24	10 ♎ 51 31	17 ♎ 02 27	00 26	00 ♒ 52	15 03	12 01	24 43	17 35	23 20	23 12
12	17 24 30	20 25 25	23 ♎ 18 16	29 ♎ 39 22	00 45	02 03	15 49	11 57	24 48	17 33	23 22	23 12
13	17 28 27	21 26 28	06 ♏ 06 05	12 ♏ 38 39	01 12	03 15	16 35	11 52	24 53	17 31	23 23	23 11
14	17 32 23	22 27 31	19 ♏ 17 09	26 ♏ 01 34	01 46	04 26	17 21	11 47	24 58	17 28	23 24	23 11
15	17 36 20	23 28 35	02 ♐ 51 42	09 ♐ 47 16	02 27	05 37	18 08	11 43	25 03	17 26	23 26	23 10
16	17 40 16	24 29 41	16 ♐ 47 49	23 ♐ 52 47	03 13	06 48	18 54	11 39	25 08	17 24	23 27	23 09
17	17 44 13	25 30 47	01 ♑ 01 29	08 ♑ 13 13	04 06	07 59	19 40	11 35	25 13	17 21	23 28	23 09
18	17 48 09	26 31 53	15 ♑ 27 12	22 ♑ 42 39	05 02	09 10	20 26	11 31	25 18	17 19	23 29	23 08
19	17 52 06	27 33 00	29 ♑ 58 47	07 ♒ 14 53	06 03	10 20	21 13	11 28	25 23	17 17	23 31	23 07
20	17 56 02	28 34 07	14 ♒ 30 17	21 ♒ 44 27	07 07	11 31	21 59	11 24	25 27	17 14	23 32	23 07
21	17 59 59	29 ♐ 35 15	28 ♒ 56 51	06 ♓ 07 08	08 14	12 41	22 45	11 21	25 32	17 12	23 33	23 06
22	18 03 56	00 ♑ 36 22	13 ♓ 14 58	20 ♓ 20 08	09 24	13 52	23 31	11 18	25 36	17 09	23 34	23 05
23	18 07 52	01 37 30	27 ♓ 22 30	04 ♈ 21 56	10 37	15 02	24 18	11 16	25 40	17 07	23 35	23 04
24	18 11 49	02 38 38	11 ♈ 18 24	18 ♈ 11 51	11 51	16 12	25 04	11 13	25 45	17 04	23 36	23 04
25	18 15 45	03 39 45	25 ♈ 02 16	01 ♉ 49 39	13 08	17 22	25 50	11 11	25 49	17 02	23 37	23 03
26	18 19 42	04 40 53	08 ♉ 33 58	15 ♉ 15 13	14 26	18 32	26 36	11 09	25 53	16 59	23 38	23 02
27	18 23 38	05 42 01	21 ♉ 53 22	28 ♉ 28 22	15 46	19 42	27 23	11 07	25 57	16 57	23 39	23 01
28	18 27 35	06 43 09	05 ♊ 00 11	11 ♊ 28 47	17 06	20 51	28 09	11 05	26 01	16 54	23 40	23 00
29	18 31 31	07 44 17	17 ♊ 54 08	24 ♊ 16 12	18 28	22 01	28 55	11 03	26 05	16 52	23 41	22 59
30	18 35 28	08 45 25	00 ♋ 35 00	06 ♋ 50 32	19 51	23 10	29 ♒ 42	11 02	26 08	16 49	23 42	22 58
31	18 39 25	09 ♑ 46 34	13 ♋ 02 54	19 ♋ 12 10	21 ♐ 15	24 ♒ 19	00 ♓ 28	11 ♉ 01	26 ♎ 12	16 ♋ 47	23 ♎ 43	22 ♌ 57

Moon Nodes and Latitude / DECLINATIONS

Date	Moon True ☊ ° '	Moon Mean ☊ ° '	Moon ☽ Latitude ° '	Sun ☉ ° '	Moon ☽ ° '	Mercury ☿ ° '	Venus ♀ ° '	Mars ♂ ° '	Jupiter ♃ ° '	Saturn ♄ ° '	Uranus ♅ ° '	Neptune ♆ ° '	Pluto ♇ ° '
01	14 ♒ 34	15 ♒ 41	04 N 36	21 S 51	26 N 20	20 S 01	24 S 17	19 S 49	14 N 33	07 S 05	22 N 40	07 S 27	22 N 36
02	14 R 23	15 38	04 01	22 00	27 12	19 30	24 07	19 36	14 31	07 07	22 41	07 28	22 36
03	14 14	15 35	03 14	22 08	26 36	19 02	23 57	19 23	14 29	07 09	22 41	07 28	22 36
04	14 08	15 32	02 19	22 17	24 39	18 37	23 45	19 10	14 27	07 10	22 41	07 29	22 37
05	14 04	15 29	01 18	22 24	21 35	18 15	23 33	18 57	14 26	07 12	22 41	07 29	22 37
06	14 03	15 26	00 N 15	22 32	17 38	17 58	23 21	18 44	14 24	07 14	22 42	07 30	22 38
07	14 D 03	15 22	00 S 49	22 38	13 03	17 45	23 07	18 30	14 23	07 16	22 42	07 30	22 38
08	14 04	15 19	01 50	22 45	08 00	17 36	22 53	18 16	14 21	07 18	22 42	07 31	22 38
09	14 05	15 16	02 46	22 51	02 N 41	17 31	22 39	18 02	14 20	07 20	22 43	07 31	22 39
10	14 R 04	15 13	03 36	22 56	02 S 47	17 30	22 24	17 48	14 19	07 21	22 43	07 32	22 39
11	14 02	15 10	04 17	23 01	08 15	17 32	22 08	17 34	14 17	07 23	22 43	07 32	22 40
12	13 57	15 07	04 47	23 06	13 30	17 38	21 51	17 20	14 16	07 25	22 44	07 33	22 40
13	13 50	15 03	05 04	23 10	18 20	17 47	21 34	17 05	14 15	07 27	22 44	07 33	22 41
14	13 41	15 00	05 05	23 14	22 27	17 57	21 17	16 50	14 14	07 28	22 44	07 34	22 41
15	13 32	14 57	04 50	23 17	25 28	18 10	20 59	16 35	14 13	07 30	22 45	07 34	22 42
16	13 24	14 54	04 17	23 20	27 03	18 25	20 40	16 20	14 12	07 31	22 45	07 35	22 42
17	13 16	14 51	03 27	23 22	26 54	18 40	20 21	16 05	14 11	07 33	22 45	07 35	22 43
18	13 11	14 47	02 24	23 24	24 56	18 57	20 01	15 50	14 10	07 34	22 45	07 36	22 43
19	13 08	14 44	01 S 11	23 25	21 19	19 15	19 40	15 34	14 09	07 36	22 46	07 36	22 44
20	13 D 08	14 41	00 N 07	23 26	16 22	19 33	19 20	15 18	14 08	07 37	22 46	07 36	22 44
21	13 08	14 38	01 25	23 27	10 31	19 51	18 58	15 02	14 08	07 39	22 46	07 37	22 45
22	13 10	14 35	02 36	23 27	04 S 11	20 10	18 37	14 46	14 07	07 40	22 47	07 37	22 45
23	13 11	14 32	03 37	23 26	02 N 16	20 28	18 14	14 30	14 07	07 42	22 47	07 37	22 46
24	13 R 10	14 28	04 24	23 25	08 31	20 46	17 52	14 14	14 06	07 43	22 47	07 38	22 46
25	13 08	14 25	04 55	23 24	14 16	21 04	17 29	13 58	14 06	07 44	22 48	07 38	22 47
26	13 04	14 22	05 10	23 22	19 15	21 21	17 05	13 41	14 05	07 45	22 48	07 38	22 48
27	12 59	14 19	05 07	23 19	23 11	21 38	16 41	13 24	14 05	07 47	22 48	07 39	22 48
28	12 53	14 16	04 48	23 17	25 51	21 55	16 17	13 08	14 05	07 48	22 49	07 39	22 49
29	12 46	14 12	04 15	23 13	27 07	22 10	15 52	12 51	14 04	07 49	22 49	07 39	22 49
30	12 40	14 09	03 29	23 09	26 56	22 25	15 27	12 34	14 04	07 50	22 49	07 40	22 50
31	12 ♒ 36	14 ♒ 06	02 N 34	23 S 05	25 N 22	22 S 39	15 S 01	12 S 17	14 N 04	07 S 51	22 N 50	07 S 40	22 N 50

ZODIAC SIGN ENTRIES

Date	h	m	Planets
03	03	09	☽ ♋
05	13	23	☽ ♌
08	01	57	☽ ♍
10	14	35	☽ ♎
10	18	30	♀ ♒
13	00	39	☽ ♏
15	07	00	☽ ♐
17	10	17	☽ ♑
19	12	02	☽ ♒
21	13	45	☽ ♓
21	21	43	☉ ♑
23	16	30	☽ ♈
25	20	46	☽ ♉
28	02	48	☽ ♊
30	10	53	☽ ♋
30	21	35	♂ ♓

LATITUDES

Date	Mercury ☿ ° '	Venus ♀ ° '	Mars ♂ ° '	Jupiter ♃ ° '	Saturn ♄ ° '	Uranus ♅ ° '	Neptune ♆ ° '	Pluto ♇ ° '
01	01 N 27	02 S 12	01 S 26	01 S 15	02 N 19	00 N 26	01 N 38	09 N 22
04	02 13	02 13	01 23	01 15	02 20	00 26	01 38	09 23
07	02 39	02 14	01 21	01 14	02 20	00 26	01 38	09 24
10	02 47	02 13	01 19	01 13	02 21	00 26	01 39	09 25
13	02 41	02 12	01 16	01 12	02 22	00 26	01 39	09 26
16	02 26	02 09	01 14	01 12	02 22	00 26	01 39	09 27
19	02 07	02 06	01 12	01 11	02 23	00 26	01 39	09 28
22	01 44	02 01	01 09	01 10	02 24	00 26	01 39	09 29
25	01 20	01 56	01 07	01 09	02 24	00 26	01 39	09 30
28	00 55	01 50	01 04	01 08	02 25	00 26	01 40	09 30
31	00 N 31	01 S 42	01 S 02	01 S 07	02 N 26	00 N 26	01 N 40	09 N 31

DATA

Julian Date	2434348
Delta T	+30 seconds
Ayanamsa	23° 12' 09"
Synetic vernal point	05° ♓ 54' 51"
True obliquity of ecliptic	23° 26' 50"

MOON'S PHASES, APSIDES AND POSITIONS ☽

Date	h	m	Phase	Longitude ° '	Eclipse Indicator
01	12	41	○	09 ♊ 17	
09	13	22	☾	17 ♍ 26	
17	02	02	●	25 ♐ 05	
23	19	52	☽	01 ♈ 58	
31	05	06	○	09 ♋ 29	

Day	h	m		
08	02	49	Apogee	
19	20	25	Perigee	
02	13	54	Max dec	27° N 13'
09	23	47	0S	
16	22	02	Max dec	27° S 12'
23	03	31	0N	
29	20	47	Max dec	27° N 13'

LONGITUDES

Date	Chiron ⚷ ° '	Ceres ⚳ ° '	Pallas ⚴ ° '	Juno ⚵ ° '	Vesta ⚶ ° '	Black Moon Lilith ⚸ ° '
01	10 ♑ 43	10 ♊ 48	20 ♓ 38	21 ♒ 56	04 ♐ 25	27 ♌ 30
11	11 ♑ 37	08 ♊ 28	21 ♓ 43	25 ♒ 39	09 ♐ 50	28 ♌ 37
21	12 ♑ 35	06 ♊ 22	23 ♓ 20	29 ♒ 42	15 ♐ 15	29 ♌ 43
31	13 ♑ 33	04 ♊ 44	25 ♓ 27	04 ♓ 00	20 ♐ 38	00 ♍ 50

ASPECTARIAN

h m Aspects

01 Monday
- 01 10 ☽ ∠ ♅
- 02 02 ☽ ⚼ ♀
- 04 41 ☿ ⚹ ♂
- 08 25 ☽ ☍ ☿
- 09 03 ☽ △ ♂
- 10 29 ☽ ⚼ ♆
- 11 48 ☽ ⚼ ♄
- 12 41 ☽ ☍ ☉
- 16 18 ☽ Q ♇
- 17 30 ☽ ⊥ ♅
- 19 24 ☽ ⊻ ♃
- 20 04 ☽ ± ♀

02 Tuesday
- 04 17 ☿ ∥ ♂
- 04 31 ☽ ⊻ ♅
- 06 22 ☽ ⊥ ♃
- 08 17 ☽ ⊼ ♀
- 14 11 ☽ △ ♆
- 14 24 ☽ ⚼ ♂
- 14 28 ☽ ⚹ ♇
- 14 56 ☿ ∠ ♀
- 15 41 ☽ △ ♄
- 23 03 ☽ ∠ ♃

03 Wednesday
- 08 27 ☽ ± ♂
- 11 11 ☽ ⊼ ☿
- 18 51 ☽ ∠ ♇
- 20 39 ☽ ⊼ ♂
- 21 43 ☽ ± ☿

04 Thursday
- 01 13 ☉ ± ♅
- 01 49 ☽ ⊼ ☉
- 03 26 ☽ ⚹ ♃
- 12 15 ☽ ⊥ ♇
- 13 23 ☽ ☌ ♅
- 13 47 ☽ ⚼ ☿
- 14 31 ☽ ± ☉
- 19 01 ☿ ⚼ ♅
- 20 01 ☉ ⊼ ♃
- 20 24 ☽ ⋕ ♀
- 23 43 ☽ ☍ ♀
- 23 52 ☽ □ ♆

05 Friday
- 00 01 ☽ ⊻ ♇
- 01 14 ♀ □ ♆
- 01 46 ☽ □ ♄
- 02 40 ♀ ⊼ ♇
- 02 42 ☽ Q ♃
- 04 08 ☽ ∥ ♅
- 04 41 ☽ ∥ ♇
- 06 26 ☽ ⋕ ☉
- 09 51 ☽ ⚼ ☉
- 17 21 ☽ △ ☿
- 22 11 ♀ □ ♄

06 Saturday
- 05 24 ☽ ⋕ ♂
- 10 00 ☽ ⋕ ☿
- 11 49 ☽ ☍ ♂
- 11 50 ☽ Q ♆
- 13 59 ☽ Q ♄
- 14 21 ☽ □ ♃
- 18 42 ☽ △ ☉

07 Sunday
- 01 02 ☽ ⊻ ♅
- 05 15 ☽ ∥ ♃
- 07 24 ♆ ⚹ ♇
- 10 10 ☉ ⋕ ♇
- 12 11 ☽ ☌ ♇
- 12 12 ☽ ⚹ ♆
- 13 11 ☽ ⊥ ♅
- 14 30 ☽ ⚹ ♄
- 18 42 ☽ ⊼ ♀
- 23 08 ♂ □ ♃

08 Monday
- 01 59 ☉ ⋕ ♅
- 03 11 ☽ □ ☿
- 06 10 ☿ Q ♂
- 07 27 ☽ ∠ ♅
- 08 17 ☽ ± ♀
- 14 05 ☿ ⊥ ♄
- 14 14 ☽ ⋕ ♆
- 15 13 ☽ ⋕ ♄
- 18 47 ☽ ∠ ♆
- 21 16 ☽ ∠ ♄
- 21 54 ☉ ∥ ♀

09 Tuesday
- 02 50 ☽ △ ♃
- 04 38 ☽ ⚼ ♀
- 04 59 ☽ ⊼ ♂
- 05 47 ♀ ⋕ ♅
- 11 51 ♀ ⋕ ♇
- 13 04 ☽ ⊥ ♆
- 13 22 ☽ □ ☉
- 13 49 ☽ ⚹ ♅
- 15 05 ☽ Q ☿
- 15 38 ☽ ⊥ ♄
- 17 57 ☽ ± ♂
- 18 26 ☉ ⊼ ♅

10 Wednesday
- 01 01 ☽ ⊻ ♇
- 01 11 ☽ ⊻ ♆
- 01 29 ☿ St D
- 03 48 ☽ ⊻ ♄
- 05 52 ☉ ± ♃
- 08 50 ☽ ⚼ ♃
- 13 01 ☽ ⊥ ♇
- 13 14 ☽ ⚼ ♂
- 13 49 ☽ Q ♅
- 14 09 ☽ △ ♀
- 15 11 ☽ ⚹ ☿

11 Thursday
- 01 39 ☿ ⚹ ♀
- 02 35 ☽ ± ♃
- 04 37 ☽ Q ☉
- 06 48 ☽ ∠ ♇
- 08 11 ☽ ∥ ♄
- 08 53 ☽ ∥ ♆
- 14 08 ☿ ∥ ♂
- 14 16 ☽ ⊼ ♃
- 20 41 ☽ △ ♂
- 21 06 ☽ ∠ ☿

12 Friday
- 01 00 ☽ □ ♅
- 06 01 ☽ ⚹ ☉
- 11 47 ☽ ⚹ ♇
- 12 06 ☽ ☌ ♆
- 14 50 ☽ ⊥ ☿
- 14 52 ☽ ☌ ♄
- 15 38 ☽ ⋕ ♃
- 16 33 ☿ ⊥ ♄

13 Saturday
- 02 31 ☽ ⊻ ☿
- 05 50 ☽ ∥ ♂
- 06 10 ☽ □ ♀
- 08 58 ☽ ∥ ☿
- 10 18 ☽ Q ♇
- 12 41 ☽ ∠ ☉
- 22 31 ☽ ☍ ♃

14 Sunday
- 05 08 ☽ ∥ ♀
- 06 30 ☽ ⊥ ☉
- 08 19 ☽ □ ♂
- 08 45 ☽ △ ♅
- 13 37 ☽ ⋕ ♇
- 13 57 ☽ ⋕ ♅
- 15 23 ♂ ⊼ ♅
- 17 29 ☽ ∥ ☉
- 18 08 ☽ ⊻ ☉
- 18 10 ☽ Q ♀
- 18 57 ☽ □ ♇
- 19 22 ☽ ⊻ ♆
- 22 12 ☽ ⊻ ♄

15 Monday
- 04 45 ☉ △ ♇
- 05 59 ☽ ⊥ ♆
- 08 50 ☽ ⊥ ♄
- 10 49 ☉ ⚹ ♆
- 11 14 ☽ ☌ ☿
- 11 15 ☽ ⚼ ♅
- 11 39 ☿ ⚼ ♅
- 17 14 ☽ ⚹ ♀
- 18 01 ☽ Q ♂
- 21 40 ☽ ∠ ♆

16 Tuesday
- 00 32 ☽ ∠ ♄
- 02 47 ☽ ± ♅
- 03 15 ☽ ⊼ ♃
- 13 01 ☽ ⊼ ♅
- 13 27 ☽ ± ♃
- 15 47 ☽ ⚹ ♂
- 21 15 ☽ ∠ ♀
- 22 46 ☽ △ ♇
- 23 17 ☽ ⚹ ♆

17 Wednesday
- 02 02 ☽ ☌ ☉
- 02 12 ☽ ⚹ ♄
- 04 29 ☉ ⚹ ♄
- 04 35 ☽ ⚼ ♃
- 13 44 ☽ ⊥ ♀
- 17 28 ☽ ⊻ ☿
- 18 26 ☽ ∠ ♂
- 19 26 ☽ Q ♆
- 22 23 ☽ Q ♄
- 23 52 ☽ ⚼ ♇

18 Thursday
- 00 38 ☽ ⊻ ♀
- 04 09 ☽ ⊥ ☿
- 05 31 ☽ △ ♃
- 10 13 ☽ ⊥ ♂
- 11 48 ☉ ⚼ ♃
- 14 47 ☽ ± ♇
- 15 04 ☽ ☍ ♅
- 20 08 ☽ ∠ ☿
- 20 42 ☽ ⊻ ♂
- 23 20 ☽ ∥ ☉

19 Friday
- 00 41 ☽ ⊼ ♇
- 01 18 ☽ □ ♆
- 03 31 ☽ ⋕ ♅
- 03 45 ☽ ⋕ ♇
- 04 22 ☽ □ ♄
- 07 41 ☽ ⊻ ☉
- 11 55 ♄ ∥ ♆
- 18 20 ☽ ⊥ ☉
- 21 14 ☽ ∥ ♀
- 22 05 ☽ ∥ ☿
- 22 48 ☽ ⚹ ☿

20 Saturday
- 03 57 ☿ ∥ ♀
- 06 37 ☽ ☌ ♀
- 06 54 ☽ □ ♃
- 09 56 ♀ □ ♃
- 10 20 ☽ ∠ ☉
- 16 31 ☽ ⊼ ♅
- 16 49 ☽ ∥ ♂
- 20 17 ☽ Q ☿
- 21 30 ☽ ⋕ ♃

21 Sunday
- 01 06 ☽ ☌ ♂
- 02 16 ☽ ☍ ♇
- 02 27 ☽ ± ♅
- 03 00 ☽ △ ♆
- 06 16 ☽ △ ♄
- 12 41 ☽ Q ♃
- 13 09 ☽ ⚹ ☉
- 17 25 ☽ ⚼ ♅
- 18 36 ☿ ∠ ♆
- 22 40 ♂ ☍ ♇
- 22 59 ☽ ∥ ♄
- 23 08 ☽ ∥ ♆

22 Monday
- 01 09 ♂ ± ♅
- 04 06 ☽ ⚼ ♆
- 04 56 ☽ □ ☿
- 07 31 ☽ ⚼ ♄
- 08 44 ☽ ⚹ ♃
- 10 50 ☽ Q ☉
- 13 08 ☽ ⊻ ♀
- 13 31 ♂ △ ♆
- 18 35 ☽ △ ♅
- 19 19 ☽ ± ♆
- 22 48 ☽ ± ♄

23 Tuesday
- 00 12 ☽ ⊥ ♀
- 04 40 ☽ ⊼ ♇
- 05 32 ☽ ⊼ ♆
- 06 26 ☽ ⊻ ♂
- 09 05 ☽ ⊼ ♄
- 10 06 ☽ ∠ ♃
- 13 12 ☿ ∠ ♄
- 14 54 ☽ ± ♇
- 16 58 ☽ ∠ ♀
- 17 17 ☽ ⊥ ♂
- 19 52 ☽ □ ☉
- 21 24 ☿ ± ♅

24 Wednesday
- 00 07 ☿ ⊼ ♃
- 01 30 ☽ ⊥ ♃
- 06 23 ☽ ⚼ ♇
- 08 28 ☽ ⋕ ♆
- 08 48 ☽ ⋕ ♄
- 09 43 ☽ ∠ ♂
- 11 51 ☽ ⊻ ♃
- 13 03 ☽ △ ☿
- 21 18 ☽ ⚹ ♀
- 22 00 ☽ □ ♅
- 22 47 ♅ ⊥ ♇
- 23 54 ♂ ⋕ ♃

25 Thursday
- 05 19 ♀ ⊼ ♅
- 08 30 ☽ △ ♇
- 09 30 ☽ ☍ ♆
- 10 41 ☽ ⋕ ♂
- 11 13 ☽ ∥ ♃
- 11 13 ♂ △ ♄
- 13 22 ☽ ☍ ♄
- 13 29 ☽ ⚹ ♂
- 18 02 ☽ ⚼ ☿
- 20 22 ☽ Q ♀

26 Friday
- 01 50 ☽ ⋕ ♀
- 04 30 ☽ △ ☉
- 05 38 ☽ Q ♅
- 11 44 ☽ ± ☿
- 12 05 ☽ Q ♂
- 16 36 ☽ ☌ ♃
- 19 40 ☿ Q ♂
- 23 41 ☽ ⊼ ☿

27 Saturday
- 00 56 ☽ ⋕ ☿
- 03 05 ☽ ⚹ ♅
- 07 38 ☽ □ ♀
- 09 20 ☽ ∥ ♇
- 09 22 ☽ ∥ ♅
- 09 40 ☽ ⚼ ☉
- 10 57 ☉ Q ♆
- 13 02 ☽ ⋕ ☉
- 14 03 ☽ □ ♇
- 15 13 ☽ ⊼ ♆
- 19 25 ☽ ⊼ ♄
- 22 37 ☽ □ ♂

28 Sunday
- 02 11 ☽ ± ♆
- 03 27 ☽ ± ☉
- 06 19 ☽ ∠ ♅
- 06 28 ☽ ± ♄
- 08 33 ☿ ⊼ ♅
- 11 35 ☿ ± ♃
- 15 26 ☽ ⊼ ☉
- 18 48 ☽ ⚼ ♆
- 21 57 ♅ ∥ ♇
- 22 54 ☽ ⊥ ♅
- 23 06 ☽ Q ♇
- 23 11 ☽ ⚼ ♄
- 23 14 ☽ ⊻ ♃

29 Monday
- 10 03 ☽ ⊻ ♅
- 10 25 ☽ ⊥ ♃
- 13 12 ☽ ☍ ☿
- 16 06 ♂ Q ♃
- 17 46 ☉ ⚼ ♇
- 20 30 ☉ Q ♄
- 20 30 ☽ △ ♀
- 21 34 ☽ ⚹ ♇
- 22 55 ☽ △ ♆

30 Tuesday
- 03 21 ☽ ∠ ♃
- 03 30 ☽ △ ♄
- 05 03 ♀ ± ♅
- 08 00 ♀ ☍ ♇
- 10 11 ☽ △ ♂
- 23 18 ♀ △ ♆

31 Wednesday
- 02 09 ☽ ∠ ♇
- 04 02 ☽ ⚼ ♀
- 05 06 ☽ ☍ ☉
- 08 04 ☽ ⚹ ♃
- 17 01 ☽ ⚼ ♂
- 19 14 ☽ ☌ ♅
- 19 36 ☽ ⊥ ♇
- 23 20 ☽ ± ♀

All ephemeris data is given at 12.00 UT and the Moon's longitude is additionally given for 24.00 UT

JANUARY 1953

Date	Sidereal time h m s	LONGITUDES Sun ☉ ° ' "	Moon ☽ ° ' "	Moon ☽ 24.00 ° ' "	Mercury ☿ ° '	Venus ♀ ° '	Mars ♂ ° '	Jupiter ♃ ° '	Saturn ♄ ° '	Uranus ♅ ° '	Neptune ♆ ° '	Pluto ♇ ° '
01	18 43 21	10 ♑ 47 42	25 ♋ 18 28	01 ♌ 22 02	22 ♐ 40	25 ♒ 28	01 ♓ 14	11 ♉ 00	26 ♎ 16	16 ♋ 44	23 ♎ 44	22 ♌ 56
02	18 47 18	11 48 50	07 ♌ 23 03	13 ♌ 21 51	24 06	26 37	02 00	10 R 59	26 19	16 R 41	23 45	22 R 55
03	18 51 14	12 49 59	19 ♌ 18 45	25 ♌ 14 08	25 32	27 46	02 47	10 59	26 22	16 39	23 45	22 54
04	18 55 11	13 51 07	01 ♍ 08 25	07 ♍ 02 06	26 59	28 ♒ 54	03 33	10 59	26 26	16 36	23 46	22 53
05	18 59 07	14 52 16	12 ♍ 55 42	18 ♍ 49 44	28 27	00 ♓ 02	04 19	10 D 59	26 29	16 34	23 47	22 52
06	19 03 04	15 53 25	24 ♍ 44 48	00 ♎ 41 30	29 ♐ 55	01 10	05 05	10 59	26 32	16 31	23 47	22 51
07	19 07 00	16 54 34	06 ♎ 40 27	12 ♎ 42 17	01 ♑ 24	02 18	05 51	10 59	26 35	16 28	23 48	22 50
08	19 10 57	17 55 43	18 ♎ 47 36	24 ♎ 57 02	02 53	03 26	06 38	11 00	26 38	16 26	23 49	22 49
09	19 14 54	18 56 52	01 ♏ 11 10	07 ♏ 30 32	04 23	04 34	07 24	11 00	26 41	16 23	23 49	22 48
10	19 18 50	19 58 01	13 ♏ 55 37	20 ♏ 26 50	05 53	05 41	08 10	11 01	26 43	16 21	23 50	22 46
11	19 22 47	20 59 10	27 ♏ 04 28	03 ♐ 48 44	07 24	06 48	08 56	11 03	26 46	16 18	23 50	22 45
12	19 26 43	22 00 19	10 ♐ 39 42	17 ♐ 37 15	08 55	07 55	09 42	11 04	26 49	16 15	23 51	22 44
13	19 30 40	23 01 28	24 ♐ 41 10	01 ♑ 51 02	10 27	09 02	10 28	11 05	26 51	16 13	23 51	22 43
14	19 34 36	24 02 37	09 ♑ 06 16	16 ♑ 26 08	12 00	10 08	11 15	11 07	26 53	16 10	23 51	22 42
15	19 38 33	25 03 45	23 ♑ 49 49	01 ♒ 16 21	13 32	11 15	12 01	11 09	26 55	16 08	23 52	22 40
16	19 42 29	26 04 53	08 ♒ 44 41	16 ♒ 13 48	15 06	12 21	12 47	11 11	26 58	16 05	23 52	22 39
17	19 46 26	27 06 00	23 ♒ 42 38	01 ♓ 10 12	16 40	13 26	13 33	11 14	27 00	16 03	23 52	22 38
18	19 50 23	28 07 06	08 ♓ 35 34	15 ♓ 57 57	18 14	14 32	14 19	11 17	27 02	16 00	23 53	22 36
19	19 54 19	29 ♑ 08 12	23 ♓ 16 38	00 ♈ 31 04	19 49	15 37	15 05	11 19	27 03	15 58	23 53	22 35
20	19 58 16	00 ♒ 09 16	07 ♈ 40 50	14 ♈ 45 40	21 24	16 42	15 51	11 22	27 05	15 55	23 53	22 34
21	20 02 12	01 10 20	21 ♈ 45 22	28 ♈ 39 53	23 00	17 47	16 37	11 26	27 07	15 53	23 53	22 33
22	20 06 09	02 11 23	05 ♉ 29 16	12 ♉ 13 37	24 37	18 51	17 23	11 29	27 08	15 50	23 53	22 31
23	20 10 05	03 12 24	18 ♉ 53 06	25 ♉ 27 55	26 14	19 55	18 09	11 32	27 09	15 48	23 53	22 30
24	20 14 02	04 13 25	01 ♊ 56 44	08 ♊ 24 35	27 52	20 59	18 55	11 36	27 11	15 45	23 53	22 28
25	20 17 58	05 14 25	14 ♊ 46 57	21 ♊ 05 42	29 ♑ 30	22 02	19 41	11 40	27 12	15 43	23 R 53	22 27
26	20 21 55	06 15 23	27 ♊ 21 07	03 ♋ 33 26	01 ♒ 09	23 05	20 27	11 44	27 13	15 40	23 53	22 26
27	20 25 52	07 16 21	09 ♋ 42 54	15 ♋ 49 47	02 49	24 08	21 13	11 49	27 14	15 38	23 53	22 24
28	20 29 48	08 17 17	21 ♋ 54 17	27 ♋ 56 39	04 30	25 10	21 58	11 53	27 15	15 36	23 53	22 23
29	20 33 45	09 18 13	03 ♌ 57 06	09 ♌ 55 52	06 11	26 12	22 44	11 58	27 16	15 33	23 53	22 21
30	20 37 41	10 19 07	15 ♌ 53 11	21 ♌ 49 17	07 52	27 13	23 30	12 03	27 16	15 31	23 53	22 20
31	20 41 38	11 ♒ 20 00	27 ♌ 44 25	03 ♍ 38 53	09 ♒ 35	28 ♓ 14	24 ♓ 16	12 ♉ 08	27 ♎ 17	15 ♋ 29	23 ♎ 53	22 ♌ 19

Date	Moon True ☊ ° '	Moon Mean ☊ ° '	Moon ☽ Latitude ° '
01	12 ♒ 32	14 ♒ 03	01 N 33
02	12 R 31	14 00	00 N 28
03	12 D 31	13 57	00 S 37
04	12 32	13 53	01 41
05	12 34	13 50	02 40
06	12 36	13 47	03 32
07	12 37	13 44	04 15
08	12 R 38	13 41	04 49
09	12 37	13 38	05 09
10	12 35	13 34	05 15
11	12 33	13 31	05 05
12	12 30	13 28	04 38
13	12 27	13 25	03 54
14	12 24	13 22	02 54
15	12 23	13 18	01 41
16	12 22	13 15	00 S 20
17	12 D 22	13 12	01 N 01
18	12 23	13 09	02 20
19	12 24	13 06	03 28
20	12 25	13 03	04 21
21	12 26	12 59	04 57
22	12 R 26	12 56	05 15
23	12 26	12 53	05 15
24	12 25	12 50	04 59
25	12 24	12 47	04 28
26	12 23	12 44	03 45
27	12 22	12 40	02 52
28	12 22	12 37	01 51
29	12 21	12 34	00 N 47
30	12 D 21	12 31	00 S 20
31	12 ♒ 22	12 ♒ 28	01 S 25

Date	DECLINATIONS Sun ☉ ° '	Moon ☽ ° '	Mercury ☿ ° '	Venus ♀ ° '	Mars ♂ ° '	Jupiter ♃ ° '	Saturn ♄ ° '	Uranus ♅ ° '	Neptune ♆ ° '	Pluto ♇ ° '
01	23 S 00	22 N 36	22 S 52	14 S 36	11 S 59	14 N 04	07 S 52	22 N 50	07 S 40	22 N 51
02	22 55	18 53	23 04	14 09	11 42	14 04	07 53	22 50	07 40	22 52
03	22 50	14 27	23 15	13 43	11 25	14 04	07 54	22 51	07 41	22 52
04	22 44	09 30	23 25	13 16	11 07	14 05	07 55	22 51	07 41	22 53
05	22 37	04 N 15	23 34	12 49	10 50	14 05	07 56	22 51	07 41	22 53
06	22 30	01 S 09	23 42	12 22	10 32	14 05	07 57	22 52	07 41	22 54
07	22 23	06 34	23 49	11 54	10 14	14 06	07 58	22 52	07 41	22 55
08	22 15	11 49	23 54	11 26	09 56	14 06	07 59	22 52	07 41	22 55
09	22 06	16 43	23 58	10 58	09 38	14 06	07 59	22 53	07 42	22 56
10	21 58	21 02	24 01	10 30	09 20	14 07	08 00	22 53	07 42	22 56
11	21 49	24 28	24 03	10 01	09 02	14 08	08 01	22 53	07 42	22 57
12	21 39	26 38	24 04	09 33	08 44	14 09	08 02	22 54	07 42	22 58
13	21 29	27 14	24 03	09 04	08 26	14 09	08 02	22 54	07 42	22 58
14	21 18	26 01	24 01	08 35	08 07	14 10	08 03	22 54	07 42	22 59
15	21 08	23 08	23 58	08 06	07 49	14 11	08 03	22 55	07 42	23 00
16	20 56	18 24	23 53	07 36	07 31	14 12	08 04	22 55	07 42	23 00
17	20 45	12 38	23 47	07 07	07 12	14 13	08 04	22 55	07 42	23 01
18	20 33	06 S 15	23 40	06 37	06 54	14 14	08 05	22 56	07 42	23 01
19	20 20	00 N 31	23 31	06 07	06 35	14 15	08 05	22 56	07 42	23 02
20	20 07	07 07	23 20	05 37	06 16	14 17	08 05	22 56	07 42	23 03
21	19 54	13 04	23 09	05 06	05 58	14 18	08 06	22 56	07 42	23 03
22	19 41	18 18	22 56	04 38	05 39	14 19	08 06	22 57	07 42	23 04
23	19 27	22 29	22 41	04 08	05 20	14 20	08 06	22 57	07 42	23 04
24	19 12	25 27	22 25	03 37	05 01	14 22	08 07	22 57	07 42	23 05
25	18 58	27 01	22 08	03 07	04 43	14 23	08 07	22 58	07 42	23 06
26	18 43	27 10	21 49	02 37	04 24	14 25	08 07	22 58	07 42	23 06
27	18 28	25 57	21 28	02 07	04 05	14 27	08 07	22 58	07 42	23 07
28	18 12	23 30	21 06	01 37	03 46	14 28	08 07	22 58	07 42	23 08
29	17 56	20 02	20 43	01 07	03 28	14 30	08 08	22 59	07 42	23 08
30	17 40	15 46	20 18	00 37	03 09	14 32	08 08	22 59	07 42	23 09
31	17 S 23	10 N 56	19 S 52	00 S 07	02 S 50	14 N 33	08 S 07	22 N 59	07 S 42	23 N 09

ZODIAC SIGN ENTRIES

Date	h	m	Planets
01	21	17	☽ ♌
04	09	41	☽ ♍
05	11	10	♀ ♓
06	13	24	☿ ♑
06	22	36	☽ ♎
09	09	44	☽ ♏
11	17	14	☽ ♐
13	20	55	☽ ♑
15	21	57	☽ ♒
17	22	07	☽ ♓
19	23	08	☽ ♈
20	08	21	☉ ♒
22	02	20	☽ ♉
24	08	21	☽ ♊
25	19	10	☿ ♒
26	17	07	☽ ♋
29	04	06	☽ ♌
31	16	35	☽ ♍

Date	LATITUDES Mercury ☿ ° '	Venus ♀ ° '	Mars ♂ ° '	Jupiter ♃ ° '	Saturn ♄ ° '	Uranus ♅ ° '	Neptune ♆ ° '	Pluto ♇ ° '
01	00 N 23	01 S 39	01 S 01	01 S 07	02 N 26	00 N 26	01 N 40	09 N 32
04	00 00	01 13	00 59	01 06	02 27	00 26	01 40	09 32
07	00 S 22	01 20	00 56	01 05	02 27	00 26	01 40	09 33
10	00 43	01 09	00 54	01 04	02 28	00 26	01 40	09 33
13	01 01	00 57	00 51	01 03	02 29	00 26	01 40	09 34
16	01 18	00 44	00 48	01 02	02 30	00 26	01 41	09 35
19	01 33	00 29	00 46	01 02	02 31	00 26	01 41	09 35
22	01 45	00 S 14	00 43	01 02	02 31	00 27	01 41	09 36
25	01 55	00 N 02	00 41	01 02	02 32	00 27	01 41	09 37
28	02 01	00 20	00 38	00 59	02 33	00 27	01 41	09 37
31	02 S 05	00 N 38	00 S 36	00 S 58	02 N 34	00 N 26	01 N 42	09 N 38

DATA

Julian Date	2434379
Delta T	+30 seconds
Ayanamsa	23° 12' 15"
Synetic vernal point	05° ♓ 54' 45"
True obliquity of ecliptic	23° 26' 49"

MOON'S PHASES, APSIDES AND POSITIONS ☽

Date	h	m	Phase	Longitude ° '	Eclipse Indicator
08	10	09	☾	17 ♎ 51	
15	14	08	●	25 ♑ 09	
22	05	43	☽	01 ♉ 55	
29	23	44	○	09 ♌ 48	total

Day	h	m		
04	22	27	Apogee	
16	22	53	Perigee	
06	06	54	0S	
13	08	11	Max dec	27° S 16'
19	10	09	0N	
26	02	30	Max dec	27° N 17'

LONGITUDES

Date	Chiron ⚷ ° '	Ceres ⚳ ° '	Pallas ⚴ ° '	Juno ⚵ ° '	Vesta ⚶ ° '	Black Moon Lilith ⚸ ° '
01	13 ♑ 39	04 ♊ 36	25 ♓ 41	04 ♓ 27	21 ♐ 10	00 ♍ 56
11	14 ♑ 38	03 ♊ 39	28 ♓ 15	09 ♓ 02	26 ♐ 30	02 ♍ 03
21	15 ♑ 36	03 ♊ 23	01 ♈ 10	13 ♓ 49	01 ♑ 48	03 ♍ 09
31	16 ♑ 32	03 ♊ 49	04 ♈ 25	18 ♓ 47	07 ♑ 02	04 ♍ 16

All ephemeris data is given at 12.00 UT and the Moon's longitude is additionally given for 24.00 UT

ASPECTARIAN

h m	Aspects
01 Thursday	
06 07	☽ ⊼ ☿
07 20	☽ ⊻ ♇
07 28	☽ Q ♃
08 41	☿ # ♅
08 53	☽ □ ♆
08 55	☽ # ☉
10 12	☽ ∥ ♇
10 12	☽ # ☿
10 18	☽ ∥ ♅
10 23	☿ # ♇
11 51	☽ ± ♂
12 21	☽ ⊼ ♀
13 53	☽ □ ♄
16 30	☿ △ ♇
16 48	☉ △ ♃
19 32	☽ ± ☿
02 Friday	
00 00	☉ ∥ ☿
00 32	☽ ⊼ ♂
02 44	♂ ⚼ ♅
05 26	♀ △ ♄
06 02	☿ ✶ ♆
15 54	☽ ⚼ ☿
16 34	♀ # ♃
19 13	☽ □ ♃
20 45	☽ Q ♆
21 43	☽ ⊼ ☉
03 Saturday	
01 59	☽ Q ♄
02 35	☉ # ♇
06 38	☽ ⊻ ♅
07 45	☉ # ♅
10 56	☽ ± ☉
13 52	☽ ∥ ♃
16 03	☽ # ♀
18 44	☽ ⊥ ♅
19 16	☽ ☌ ♇
19 24	☿ ⚼ ♃
21 00	☽ ✶ ♆
04 Sunday	
02 22	☽ △ ☿
02 22	☉ ∠ ♀
02 23	☽ ✶ ♄
02 28	☿ ✶ ♄
03 53	☽ # ♂
06 54	☽ ⚼ ☉
06 58	☽ ☍ ♀
12 56	☽ ∠ ♅
13 36	♀ Q ♃
17 14	☽ ☍ ♂
19 20	☽ # ♄
20 28	☽ # ♆
05 Monday	
03 33	☽ ∠ ♆
07 53	♃ St D
08 02	☽ △ ♃
09 02	☽ ∠ ♄
16 20	☽ △ ☉
19 22	☽ ✶ ♅
21 53	☽ ⊥ ♆
06 Tuesday	
03 25	☽ ⊥ ♄
08 10	☽ ⊻ ♇
10 04	☽ ⊻ ♆
14 30	☽ ⚼ ♃
15 38	☽ ⊻ ♄
19 00	♀ ⚼ ♅
19 35	☽ Q ♅
20 17	☽ ⊥ ♇
23 55	☽ □ ☿
07 Wednesday	
02 10	☉ ☍ ♅
02 20	☽ ⊼ ♀
08 37	☽ ± ♃
10 12	☉ ± ♇
10 15	☽ ⊼ ♂
14 18	☽ ∠ ♇
15 36	☽ ± ♀
17 05	☽ ∥ ♆
18 20	☽ ∥ ♄
20 36	☽ ⊼ ♃
23 01	☽ ± ♂
08 Thursday	
03 47	☽ ∥ ♂
07 22	☽ □ ♅
10 09	☽ □ ☉
10 25	☽ ∥ ♀
11 14	☽ ⚼ ♀
16 39	☽ Q ☿
17 55	☽ ⚼ ♂
19 50	☽ ✶ ♇
21 48	☽ ☌ ♆
22 58	☽ # ♃
09 Friday	
03 18	☽ ☌ ♄
18 51	☽ Q ♇
18 54	☽ ✶ ☿
19 03	☽ △ ♀
23 31	☿ ✶ ♀
23 54	☽ Q ☉
10 Saturday	
00 33	☽ △ ♂
06 35	☽ ☍ ♃
11 04	☿ Q ♆
16 27	☽ △ ♅
17 37	☽ ∥ ☉
11 Sunday	
00 03	☽ ✶ ☉
00 04	☽ # ♅
00 28	☽ # ♇
02 28	☽ ∠ ☿
04 13	☽ □ ♇
06 09	☽ ⊻ ♆
08 42	☽ ∥ ☿
08 50	♂ ⚼ ♆
11 27	☽ ⊻ ♄
16 57	☽ ⊥ ♆
17 32	☿ ⚼ ♇
19 32	☽ ⚼ ♅
20 42	☽ ⊥ ☿
22 11	☽ ⊥ ♄
12 Monday	
05 06	☽ ∠ ☉
06 48	☽ □ ♀
08 36	☽ ⊻ ☿
08 50	☽ ∠ ♆
10 10	☿ Q ♄
10 14	☽ □ ♂
11 18	☽ ± ♅
12 42	☽ ⊼ ♃
14 00	☽ ∠ ♄
21 38	☽ ⊼ ♅
21 57	☽ ⊥ ☉
23 04	☽ ± ♃
13 Tuesday	
04 50	☉ ⊼ ♇
08 03	♀ ⚼ ♆
08 40	☽ △ ♇
08 59	☽ ⊻ ☉
10 35	☽ ✶ ♆
12 39	☿ ✶ ♂
14 22	☽ ⚼ ♃
15 39	☽ ✶ ♄
16 17	☽ Q ♀
18 44	☽ Q ♂
22 09	☿ △ ♃
14 Wednesday	
06 38	☽ Q ♆
07 33	☉ □ ♆
08 02	♂ ✶ ♃
09 41	☽ ⚼ ♇
11 39	☽ Q ♄
13 50	☽ ✶ ♀
15 19	☽ △ ♃
15 42	☽ ✶ ♂
17 18	☽ ☌ ☿
17 58	♂ ∥ ♄
23 32	☽ ☍ ♅
15 Thursday	
00 24	☽ ± ♇
05 33	☽ ∥ ☿
09 08	♂ ⚼ ♄
10 01	♀ ✶ ♃
10 08	☽ ⊼ ♇
12 03	☽ # ♇
12 03	☽ □ ♆
12 33	☽ # ♅
13 49	♀ ∥ ♄
14 08	☽ ☌ ☉
16 13	☽ ∠ ♀
17 01	☽ □ ♄
17 25	☽ ∠ ♂
20 57	♂ ∥ ♆
23 08	☽ ∥ ☉
16 Friday	
03 21	♀ ⚼ ♄
07 03	♀ ∥ ♆
07 51	☽ ⊥ ♀
08 41	☽ ⊥ ♂
15 56	☽ □ ♃
18 14	☽ ⊻ ♀
18 49	☽ ⊻ ♂
23 22	☽ ⊻ ☿
23 44	☽ ⊼ ♅
23 58	♀ ∥ ♂
17 Saturday	
02 48	☿ ☍ ♅
05 48	☽ # ♃
09 20	☽ ± ♅
09 24	☉ □ ♄
10 07	☽ ⊥ ☿
10 16	☽ ☍ ♇
11 32	☿ ± ♇
12 16	☽ △ ♆
17 17	☽ △ ♄
17 51	☽ ⊻ ☉
20 05	♀ ☌ ♂
20 54	☽ Q ♃
23 46	☽ ⚼ ♅
18 Sunday	
02 18	☽ ∠ ☿
04 14	☽ ⊥ ☉
05 07	☽ ∥ ♄
06 28	☽ ∥ ♆
09 18	☽ ∥ ♂
10 18	☽ ∥ ♀
12 28	☽ ⚼ ♆
16 22	☽ ✶ ♃
17 35	☽ ⚼ ♄
19 54	☽ ∠ ☉
21 49	☽ ☌ ♂
22 26	☽ ☌ ♀
19 Monday	
00 01	☽ △ ♅
03 08	☽ ± ♆
05 37	☽ ✶ ☿
08 20	☽ ± ♄
10 52	☽ ⊼ ♇
13 00	☽ ⊼ ♆
17 03	☽ ∠ ♃
18 16	☽ ⊼ ♄
19 17	♀ △ ♅
20 46	☽ ± ♇
22 26	☽ ✶ ☉
20 Tuesday	
03 56	☽ Q ☿
07 03	☽ # ♀
08 06	☽ ⊥ ♃
09 14	☽ # ♂
11 48	☽ ⚼ ♇
13 58	♂ △ ♅
14 32	☽ # ♆
16 01	☽ # ♄
18 16	☽ ⊻ ♃
20 09	☽ Q ☉
21 Wednesday	
01 56	☽ □ ♅
02 40	☽ ⊻ ♂
04 35	☽ ⊻ ♀
05 07	☿ ⊼ ♇
13 21	☽ △ ♇
13 35	☽ ⊥ ♂
14 23	♀ ± ♆
14 27	☽ □ ☿
15 41	☽ ☍ ♆
15 47	☽ ⊥ ♀
17 18	☽ ∥ ♃
21 18	☽ ☍ ♄
21 56	☿ # ♇
22 Thursday	
01 05	☿ □ ♆
05 43	☽ □ ☉
06 12	☽ ∠ ♂
08 51	☽ ∠ ♀
09 05	☽ Q ♅
10 07	☿ # ♅
18 52	☽ # ☉
22 43	☽ ☌ ♃
23 Friday	
03 43	♂ ± ♆
06 26	☽ ✶ ♅
06 39	☉ ∠ ♂
10 35	☽ ✶ ♂
13 13	☽ # ☿
14 02	☽ ✶ ♀
15 08	☽ ∥ ♅
16 01	☽ ∥ ♇
18 34	☽ □ ♇
21 07	☽ ⊼ ♆
24 Saturday	
01 45	☿ □ ♄
03 08	☽ ⊼ ♄
03 20	☽ △ ☿
08 08	☽ ± ♆
09 45	☽ ∠ ♅
09 55	☽ Q ♂
14 02	☽ Q ♀
14 15	☽ ± ♄
16 32	☽ △ ☉
16 39	♀ ± ♄
25 Sunday	
00 54	☽ ⚼ ♆
00 55	♆ St R
02 28	☽ ⊥ ♅
03 51	☽ Q ♇
06 06	☽ ⊻ ♃
07 07	☽ ⚼ ♄
11 24	☽ ⚼ ☿
13 45	☽ ⊻ ♅
17 30	☽ ⊥ ♃
21 19	♀ ⊼ ♇
21 54	☽ □ ♂
23 17	☽ ⚼ ☉
26 Monday	
02 34	☽ ✶ ♇
03 03	☽ □ ♀
05 21	☽ △ ♆
07 08	☽ ± ☿
10 49	☽ ∠ ♃
11 44	☽ △ ♄
18 06	☽ ± ☉
20 29	☽ ⊼ ☿
27 Tuesday	
06 26	♀ ⊼ ♆
06 48	☽ ⊼ ☉
07 30	☽ ∠ ♇
12 43	♂ ± ♄
16 08	☽ ✶ ♃
23 35	☽ ☌ ♅
28 Wednesday	
01 06	☽ ⊥ ♇
12 09	☽ △ ♂
12 57	☽ ⊻ ♇
14 53	☽ ∥ ♇
15 55	☽ □ ♆
15 58	☽ Q ♃
16 03	☽ ∥ ♅
19 05	☽ △ ♀
22 37	☽ □ ♄
29 Thursday	
00 24	♂ ⊼ ♇
07 10	☽ # ☿
17 11	☽ ☍ ☿
20 07	☽ ⚼ ♂
23 44	☽ ☍ ☉
30 Friday	
01 06	☽ # ☉
03 55	☽ ⚼ ♀
03 55	☽ Q ♆
04 12	☽ □ ♃
07 33	♀ ∠ ♃
10 45	☽ Q ♄
11 15	☽ ⊻ ♅
13 09	♀ ⊼ ♄
15 29	☽ ± ♂
18 21	☽ ∥ ♃
23 21	☽ ⊥ ♅
23 48	☽ ± ♀
23 52	♂ ⊼ ♆
31 Saturday	
01 01	☽ ☌ ♇
04 03	☿ ∠ ♂
04 10	☽ ✶ ♆
04 28	☽ ⊼ ♂
11 04	☽ ✶ ♄
13 07	☽ ⊼ ♀
13 37	♀ ± ♇
17 33	☽ ∠ ♅

FEBRUARY 1953

Date	Sidereal time h m s	Sun ☉ ° ' "	Moon ☽ ° ' "	Moon ☽ 24.00 ° ' "	Mercury ☿ ° '	Venus ♀ ° '	Mars ♂ ° '	Jupiter ♃ ° '	Saturn ♄ ° '	Uranus ♅ ° '	Neptune ♆ ° '	Pluto ⯓ ° '
01	20 45 34	12 ♒ 20 53	09 ♍ 32 58	15 ♍ 27 00	11 ♒ 18	29 ♓ 15	25 ♓ 01	12 ♉ 13	27 ♎ 17	15 ♋ 27	23 ♎ 52	22 ♌ 17
02	20 49 31	13 21 44	21 ♍ 21 19	27 ♍ 16 18	13 01	00 ♈ 15	25 47	12 19	27 18	15 R 24	23 R 52	22 R 16
03	20 53 27	14 22 35	03 ♎ 12 21	09 ♎ 09 54	14 46	01 15	26 33	12 24	27 18	15 22	23 52	22 14
04	20 57 24	15 23 24	15 ♎ 09 26	21 ♎ 11 26	16 31	02 14	27 18	12 30	27 18	15 20	23 51	22 13
05	21 01 21	16 24 13	27 ♎ 16 23	03 ♏ 24 50	18 17	03 13	28 04	12 36	27 R 18	15 18	23 51	22 11
06	21 05 17	17 25 00	09 ♏ 37 18	15 ♏ 54 18	20 04	04 11	28 50	12 42	27 18	15 16	23 51	22 10
07	21 09 14	18 25 47	22 ♏ 16 20	28 ♏ 43 55	21 51	05 09	29 ♓ 35	12 48	27 18	15 14	23 50	22 08
08	21 13 10	19 26 33	05 ♐ 17 28	11 ♐ 57 20	23 39	06 07	00 ♈ 21	12 55	27 17	15 12	23 50	22 07
09	21 17 07	20 27 18	18 ♐ 43 49	25 ♐ 37 05	25 27	07 03	01 06	13 01	27 17	15 10	23 49	22 06
10	21 21 03	21 28 02	02 ♑ 37 12	09 ♑ 44 02	27 16	07 59	01 52	13 08	27 16	15 08	23 49	22 04
11	21 25 00	22 28 44	16 ♑ 57 21	24 ♑ 16 39	29 ♒ 05	08 55	02 37	13 15	27 16	15 06	23 48	22 03
12	21 28 56	23 29 26	01 ♒ 41 21	09 ♒ 10 37	00 ♓ 55	09 50	03 22	13 22	27 15	15 04	23 48	22 01
13	21 32 53	24 30 06	16 ♒ 43 30	24 ♒ 18 52	02 45	10 44	04 08	13 29	27 14	15 02	23 47	22 00
14	21 36 50	25 30 45	01 ♓ 55 32	09 ♓ 32 15	04 35	11 38	04 53	13 37	27 13	15 00	23 46	21 58
15	21 40 46	26 31 22	17 ♓ 07 46	24 ♓ 40 52	06 25	12 31	05 38	13 44	27 12	14 58	23 46	21 57
16	21 44 43	27 31 58	02 ♈ 10 26	09 ♈ 35 30	08 14	13 23	06 23	13 52	27 11	14 57	23 45	21 55
17	21 48 39	28 32 31	16 ♈ 55 14	24 ♈ 09 01	10 03	14 15	07 09	14 00	27 10	14 55	23 44	21 54
18	21 52 36	29 ♒ 33 03	01 ♉ 16 24	08 ♉ 17 07	11 51	15 06	07 54	14 08	27 08	14 53	23 43	21 52
19	21 56 32	00 ♓ 33 34	15 ♉ 11 03	21 ♉ 58 18	13 38	15 56	08 39	14 16	27 07	14 52	23 42	21 51
20	22 00 29	01 34 02	28 ♉ 39 01	05 ♊ 13 31	15 24	16 45	09 24	14 24	27 05	14 50	23 42	21 49
21	22 04 25	02 34 29	11 ♊ 42 09	18 ♊ 05 22	17 07	17 33	10 09	14 33	27 04	14 49	23 41	21 48
22	22 08 22	03 34 54	24 ♊ 23 38	00 ♋ 37 27	18 48	18 20	10 54	14 41	27 02	14 47	23 40	21 47
23	22 12 19	04 35 17	06 ♋ 47 19	12 ♋ 53 44	20 26	19 07	11 39	14 50	27 00	14 46	23 39	21 45
24	22 16 15	05 35 37	18 ♋ 57 11	24 ♋ 58 09	22 01	19 52	12 24	14 59	26 58	14 44	23 38	21 44
25	22 20 12	06 35 56	00 ♌ 57 03	06 ♌ 54 18	23 32	20 37	13 09	15 08	26 56	14 43	23 37	21 42
26	22 24 08	07 36 13	12 ♌ 50 17	18 ♌ 45 21	24 58	21 20	13 54	15 17	26 54	14 42	23 36	21 41
27	22 28 05	08 36 29	24 ♌ 39 49	00 ♍ 34 00	26 18	22 03	14 39	15 26	26 52	14 41	23 35	21 40
28	22 32 01	09 ♓ 36 42	06 ♍ 28 10	12 ♍ 22 33	27 ♓ 33	22 ♈ 44	15 ♈ 23	15 ♉ 35	26 ♎ 50	14 ♋ 39	23 ♎ 34	21 ♌ 38

Date	Moon True ☊ ° '	Moon Mean ☊ ° '	Moon ☽ Latitude ° '
01	12 ♒ 22	12 ♒ 24	02 S 26
02	12 R 21	12 21	03 20
03	12 21	12 18	04 07
04	12 21	12 15	04 43
05	12 21	12 12	05 07
06	12 20	12 09	05 17
07	12 D 20	12 05	05 13
08	12 21	12 02	04 53
09	12 21	11 59	04 16
10	12 22	11 56	03 23
11	12 23	11 53	02 17
12	12 23	11 50	00 S 59
13	12 R 24	11 46	00 N 24
14	12 23	11 43	01 46
15	12 22	11 40	03 01
16	12 20	11 37	04 02
17	12 18	11 34	04 46
18	12 16	11 30	05 11
19	12 14	11 27	05 16
20	12 14	11 24	05 03
21	12 D 14	11 21	04 36
22	12 15	11 18	03 55
23	12 17	11 15	03 04
24	12 18	11 11	02 05
25	12 20	11 08	01 N 02
26	12 R 20	11 05	00 S 03
27	12 19	11 02	01 07
28	12 ♒ 17	10 ♒ 59	02 S 09

DECLINATIONS

Date	Sun ☉ ° '	Moon ☽ ° '	Mercury ☿ ° '	Venus ♀ ° '	Mars ♂ ° '	Jupiter ♃ ° '	Saturn ♄ ° '	Uranus ♅ ° '	Neptune ♆ ° '	Pluto ⯓ ° '
01	17 S 06	05 N 45	19 S 24	00 N 23	02 S 31	14 N 35	08 S 07	23 N 00	07 S 41	23 N 10
02	16 49	00 N 22	18 54	00 53	02 12	14 37	08 07	23 00	07 41	23 11
03	16 31	05 S 03	18 23	01 23	01 53	14 39	08 06	23 00	07 41	23 11
04	16 14	10 19	17 51	01 52	01 34	14 41	08 06	23 00	07 41	23 12
05	15 56	15 16	17 17	02 22	01 15	14 43	08 06	23 00	07 41	23 12
06	15 37	19 43	16 42	02 51	00 56	14 45	08 06	23 01	07 41	23 13
07	15 19	23 23	16 05	03 21	00 38	14 48	08 05	23 01	07 40	23 14
08	15 00	25 59	15 26	03 50	00 S 19	14 50	08 05	23 01	07 40	23 14
09	14 41	27 13	14 47	04 19	00 00	14 52	08 05	23 01	07 40	23 15
10	14 21	26 49	14 06	04 48	00 N 19	14 54	08 04	23 02	07 40	23 15
11	14 01	24 38	13 23	05 16	00 38	14 57	08 04	23 02	07 39	23 16
12	13 42	20 45	12 39	05 45	00 56	14 59	08 03	23 02	07 39	23 16
13	13 22	15 27	11 54	06 13	01 15	15 01	08 03	23 02	07 39	23 17
14	13 01	09 08	11 08	06 41	01 34	15 04	08 02	23 02	07 39	23 18
15	12 41	02 S 18	10 21	07 09	01 53	15 06	08 02	23 03	07 38	23 18
16	12 20	04 N 34	09 33	07 36	02 11	15 09	08 01	23 03	07 38	23 19
17	11 59	11 03	08 44	08 04	02 30	15 11	08 00	23 03	07 37	23 19
18	11 38	16 46	07 54	08 30	02 48	15 14	07 59	23 03	07 37	23 20
19	11 17	21 25	07 04	08 57	03 07	15 17	07 59	23 03	07 37	23 20
20	10 55	24 48	06 14	09 24	03 25	15 19	07 58	23 03	07 37	23 21
21	10 34	26 45	05 23	09 50	03 44	15 22	07 57	23 04	07 36	23 21
22	10 12	27 14	04 33	10 15	04 02	15 25	07 56	23 04	07 36	23 22
23	09 50	26 20	03 43	10 41	04 21	15 28	07 55	23 04	07 35	23 22
24	09 28	24 11	02 54	11 06	04 39	15 30	07 54	23 04	07 35	23 23
25	09 06	20 58	02 06	11 30	04 57	15 33	07 54	23 04	07 35	23 23
26	08 43	16 55	01 19	11 54	05 15	15 36	07 53	23 04	07 34	23 24
27	08 21	12 15	00 34	12 18	05 33	15 39	07 52	23 04	07 34	23 24
28	07 S 58	07 N 09	00 N 09	12 N 41	05 N 51	15 N 42	07 S 51	23 N 05	07 S 33	23 N 25

ZODIAC SIGN ENTRIES

Date	h	m	Planets
02	05	54	♀ ♈
03	05	31	☽ ♎
05	17	21	☽ ♏
08	01	07	♂ ♈
08	02	20	☽ ♐
10	07	32	☽ ♑
11	23	57	☿ ♓
12	09	17	☽ ♒
14	08	58	☽ ♓
16	08	30	☽ ♈
18	09	50	☽ ♉
18	22	41	☉ ♓
20	14	27	☽ ♊
22	22	48	☽ ♋
25	10	05	☽ ♌
27	22	51	☽ ♍

LATITUDES

Date	Mercury ☿ ° '	Venus ♀ ° '	Mars ♂ ° '	Jupiter ♃ ° '	Saturn ♄ ° '	Uranus ♅ ° '	Neptune ♆ ° '	Pluto ⯓ ° '
01	02 S 05	00 N 45	00 S 35	00 S 58	02 N 34	00 N 26	01 N 42	09 N 38
04	02 03	01 04	00 33	00 57	02 35	00 26	01 42	09 38
07	01 57	01 25	00 30	00 56	02 36	00 26	01 42	09 38
10	01 47	01 46	00 28	00 56	02 36	00 26	01 42	09 39
13	01 31	02 08	00 25	00 55	02 37	00 26	01 42	09 39
16	01 09	02 31	00 23	00 54	02 38	00 26	01 43	09 39
19	00 41	02 55	00 21	00 53	02 39	00 26	01 43	09 39
22	00 S 08	03 19	00 18	00 53	02 40	00 26	01 43	09 39
25	00 N 31	03 43	00 16	00 52	02 40	00 26	01 43	09 39
28	01 13	04 09	00 13	00 51	02 41	00 26	01 43	09 39
31	01 N 56	04 N 34	00 S 11	00 S 51	02 N 42	00 N 26	01 N 43	09 N 39

DATA

Julian Date	2434410
Delta T	+30 seconds
Ayanamsa	23° 12' 20"
Synetic vernal point	05° ♓ 54' 39"
True obliquity of ecliptic	23° 26' 50"

MOON'S PHASES, APSIDES AND POSITIONS ☽

Date	h	m	Phase	Longitude ° '	Eclipse Indicator
07	04	09	☾	18 ♏ 06	
14	01	10	●	25 ♒ 03	Partial
20	17	44	☽	01 ♊ 48	
28	18	59	○	09 ♍ 54	

Day	h	m		
01	11	31	Apogee	
14	10	06	Perigee	
28	13	16	Apogee	
02	13	36	0S	
09	18	22	Max dec	27° S 17'
15	19	59	0N	
22	08	12	Max dec	27° N 15'

LONGITUDES

Date	Chiron ⚷ ° '	Ceres ⚳ ° '	Pallas ⚴ ° '	Juno ⚵ ° '	Vesta ⚶ ° '	Black Moon Lilith ⚸ ° '
01	16 ♑ 37	03 ♊ 54	04 ♈ 45	19 ♓ 17	07 ♑ 33	04 ♍ 23
11	17 ♑ 30	05 ♊ 00	08 ♈ 18	24 ♓ 25	12 ♑ 42	05 ♍ 29
21	18 ♑ 19	06 ♊ 41	12 ♈ 05	29 ♓ 45	17 ♑ 42	06 ♍ 36
31	19 ♑ 02	08 ♊ 50	16 ♈ 05	03 ♈ 06	22 ♑ 40	07 ♍ 42

All ephemeris data is given at 12.00 UT and the Moon's longitude is additionally given for 24.00 UT

ASPECTARIAN

h m	Aspects
01 Sunday	
01 12	☽ # ♄
03 09	☽ # ♆
08 41	☉ □ ♃
10 37	☽ ∠ ♆
16 09	☽ ⊼ ☿
17 28	☽ △ ♃
17 34	☽ ∠ ♄
18 14	☽ ⊼ ☉
23 57	☽ ⚹ ♅
02 Monday	
01 33	☿ □ ♃
03 20	☽ # ♂
04 55	☽ ⊥ ♆
06 27	☽ ± ☿
07 34	☽ ± ☉
09 53	☽ ∥ ♀
11 52	☽ ⊥ ♄
13 50	☽ ⊻ ⯓
17 06	☽ ⊻ ♆
18 02	☽ # ♀
21 36	☽ ☍ ♂
22 43	☽ ∥ ♂
23 11	☉ ☌ ☿
03 Tuesday	
00 03	☽ ⊻ ♄
00 10	☽ ♇ ♃
00 14	☽ Q ♅
01 59	☽ ⊥ ⯓
03 32	☽ ♇ ☉
03 51	☽ ♇ ☿
07 41	☽ ☍ ♀
18 30	☽ ± ♃
20 05	☿ ⊼ ♅
20 07	☽ ∠ ⯓
23 53	☽ ∥ ♆
04 Wednesday	
01 49	☽ ∥ ♄
03 05	♀ # ♂
06 39	☽ ⊼ ♃
10 41	☉ ⊼ ♅
11 42	♂ ⊼ ♄
12 21	☽ □ ♅
12 30	☽ △ ☉
15 11	☽ △ ☿
18 59	♂ ∠ ♃
05 Thursday	
02 00	☽ ⚹ ⯓
02 30	♄ St R
05 16	☽ ☌ ♆
09 13	☽ # ♃
10 00	☿ ∠ ♀
12 03	☽ ☌ ♄
13 40	☽ ⊼ ♂
15 08	☽ ∥ ☉
15 45	♂ ± ⯓
21 16	☽ ∥ ☿
06 Friday	
00 37	☽ ⊼ ♀
01 29	☽ Q ⯓
02 08	☽ ± ♂
13 11	☽ ± ♀
17 57	☽ ☍ ♃
20 34	☽ ♇ ♂
22 45	☽ △ ♅
07 Saturday	
03 50	☿ ± ♅
04 09	☽ □ ☉
07 42	☽ ♇ ♀
09 19	☽ # ♅
10 51	☽ # ⯓
11 05	☽ □ ☿
11 45	☽ □ ⯓
14 55	☽ ⊻ ♆
15 52	☿ ☍ ⯓
21 20	☽ ⊻ ♄
08 Sunday	
02 02	☽ ⊥ ♆
02 25	☽ △ ♂
02 43	☽ ♇ ♅
08 21	☽ ⊥ ♄
13 36	☽ △ ♀
14 25	☿ △ ♆
16 13	☽ Q ☉
18 23	☽ ∠ ♆
19 02	☽ ± ♅
23 11	☉ # ♃
09 Monday	
00 35	☽ ∠ ♄
01 13	☽ Q ☿
01 49	☽ ⊼ ♃
04 00	☿ ⊥ ♂
05 43	☽ ⊼ ♅
09 02	☿ # ♃
11 06	♀ ⊥ ♃
12 31	☽ ± ♃
12 56	♀ ♇ ⯓
15 16	☽ ⚹ ☉
17 52	☽ △ ⯓
18 59	☉ ∥ ☿
20 53	☽ ⚹ ♆
10 Tuesday	
01 28	☽ ⚹ ☿
02 52	☽ ⚹ ♄
04 12	☉ ± ♅
04 16	☽ ♇ ♃
10 38	☽ □ ♂
12 04	☿ △ ♄
17 24	☽ Q ♆
19 01	☽ ∠ ☉
19 31	☽ ♇ ⯓
21 42	☽ □ ♀
23 13	☽ Q ♄
11 Wednesday	
01 55	☉ ☍ ⯓
05 49	☽ △ ♃
06 35	☽ ∠ ☿
08 56	☽ ☍ ♅
10 30	☽ ± ⯓
11 09	☽ ⊥ ☉
18 21	☽ Q ♂
20 20	☽ ⊼ ⯓
21 37	☽ # ⯓
21 44	☽ ⊻ ☉
23 07	☽ # ♅
23 13	☽ □ ♆
23 29	☽ ⊥ ☿
12 Thursday	
00 59	☿ ♇ ♅
04 50	☽ □ ♄
05 22	☽ Q ♀
10 35	☽ ⊻ ☿
14 51	☽ ⚹ ♂
18 20	☿ Q ♃
19 05	☉ △ ♆
13 Friday	
01 53	☽ ⚹ ♀
06 50	☽ □ ♃
09 19	☽ ⊼ ♅
13 42	☽ # ♃
16 00	☽ ∠ ♂
18 48	☽ ± ♅
20 19	☽ ☍ ⯓
20 46	☽ ∥ ☉
23 09	☽ △ ♆
14 Saturday	
01 10	☽ ☌ ☉
03 08	☽ ∠ ♀
03 42	☽ ∥ ☿
04 36	☽ △ ♄
06 57	☽ ⊥ ♂
08 59	☽ ♇ ♅
11 30	☽ Q ♃
15 58	☽ ∥ ♄
16 46	☽ ☌ ☿
16 54	☽ ⊻ ♂
17 21	☽ ∥ ♆
18 12	☽ ⊥ ♀
18 41	☿ ⊻ ♂
20 11	☽ # ♀
22 47	☽ ♇ ♆
15 Sunday	
04 13	☽ ♇ ♄
04 15	☽ ⊻ ♀
06 35	☽ ⚹ ♃
08 36	☽ △ ♅
13 00	☽ ± ♆
13 25	☽ # ♂
14 36	☿ ⊥ ♀
18 27	☽ ± ♄
19 38	☽ ⊼ ⯓
22 31	☽ ⊼ ♆
16 Monday	
03 13	☽ ∥ ♂
03 53	☉ △ ♄
04 00	☽ ⊼ ♄
04 01	☽ ⊻ ☉
05 11	☽ ± ⯓
06 39	☽ ∠ ♃
13 27	♀ # ♆
14 21	☽ ⊥ ☉
18 39	☿ ♇ ♆
19 10	☽ ☌ ♂
19 39	☽ ♇ ⯓
21 17	☽ ⊥ ♃
23 04	☽ # ♆
23 11	☽ ⊻ ☿
23 48	☽ ∥ ♀
17 Tuesday	
00 28	☽ # ♄
03 46	♀ ⊻ ♃
04 10	☽ # ☿
04 20	♂ ♇ ⯓
06 02	☽ ∠ ☉
07 09	☽ ⊻ ♃
07 20	☽ ☌ ♀
08 43	☽ □ ♅
09 09	♀ # ♄
10 22	☽ ⊥ ☿
15 29	☽ # ☉
20 14	☽ △ ⯓
23 18	☽ ☍ ♆
18 Wednesday	
00 40	☿ # ♀
03 28	☽ ∠ ☿
05 02	☽ ☍ ♄
05 06	☽ ∥ ♃
06 22	♀ □ ♅
08 51	☽ ⚹ ☉
09 27	☿ ∥ ♄
14 45	☽ Q ♅
15 45	☿ ♇ ♄
19 51	☉ ♇ ♅
20 12	☿ ∥ ♆
21 07	♂ ⊥ ♃
23 59	☽ ⊻ ♂
19 Thursday	
07 03	☽ Q ☉
08 54	☽ ⚹ ☿
10 23	☽ ☌ ♃
11 01	☽ ⊥ ♂
11 26	☽ ⚹ ♅
13 23	☽ ⊻ ♀
21 15	☿ ⚹ ♃
22 26	☽ ∥ ♅
23 46	☽ □ ⯓
20 Friday	
00 26	☽ ∥ ⯓
00 42	☽ ⊥ ♀
03 05	☽ ⊼ ♆
03 53	☽ ∠ ♂
04 25	☿ △ ♅
09 11	☽ ⊼ ♄
09 24	☽ Q ☿
13 53	☽ ± ♆
14 09	☽ ∠ ♅
17 44	☽ □ ☉
18 00	☽ ∠ ♀
20 04	☽ ± ♄
21 Saturday	
06 23	☽ ♇ ♆
06 38	☽ ⊥ ♅
08 28	☽ Q ⯓
08 56	☽ ⚹ ♂
09 16	☉ ∠ ♀
11 13	☉ Q ♃
12 40	☽ ♇ ♄
17 23	☽ ⊻ ♃
17 49	☽ ⊻ ♅
19 49	☿ ± ♆
23 25	☿ ⊻ ♀
23 43	☽ ⚹ ♀
23 44	☽ □ ☿
22 Sunday	
04 51	☽ ⊥ ♃
07 01	☽ ⚹ ⯓
08 58	☽ Q ♂
10 16	☉ # ♀
10 36	☽ △ ♆
17 03	☽ △ ♄
22 19	☽ ∠ ♃
22 41	☿ # ♂
23 Monday	
00 13	☽ Q ♀
01 57	♃ ⚹ ♅
07 19	☽ △ ☉
11 56	☽ ∠ ⯓
20 17	☿ ± ♄
22 10	☽ □ ♂
24 Tuesday	
03 40	☽ ☌ ♅
04 01	☽ ⚹ ♃
05 37	☽ ⊥ ⯓
07 35	☿ ⊼ ⯓
13 57	☽ □ ♀
15 34	☽ ♇ ☉
17 31	☽ ⊻ ⯓
18 40	☽ ∥ ⯓
19 00	☽ △ ☿
21 08	☽ ∥ ♅
21 19	☽ □ ♆
25 Wednesday	
03 58	☽ □ ♄
04 14	☽ Q ♃
11 14	☽ ± ☉
13 27	☿ ⊼ ♆
26 Thursday	
00 26	☽ ⊼ ☉
05 24	☽ ♇ ☿
09 30	☽ Q ♆
14 17	☽ △ ♂
15 46	☽ ⊻ ♅
16 10	☽ Q ♄
17 01	☽ □ ♃
18 59	☽ ∥ ♃
23 15	♀ △ ⯓
27 Friday	
02 02	☽ ± ☿
03 55	☽ ⊥ ♅
05 54	☽ ☌ ⯓
06 20	☽ △ ♀
09 49	☽ ⚹ ♆
11 26	☉ ♇ ♆
11 46	☽ ∥ ♀
13 04	♂ □ ♅
15 10	☉ ⊥ ♂
15 45	☽ ⊼ ☿
16 28	☽ ⚹ ♄
22 10	☽ ∠ ♅
22 13	☿ ⊼ ♄
22 48	☽ ♇ ♂
28 Saturday	
07 56	☽ # ☉
08 48	☽ # ♄
10 07	☽ # ♆
13 39	☿ ± ⯓
14 43	☽ ♇ ♀
16 15	☽ ∠ ♆
17 33	☽ ∥ ♂
18 20	☽ ± ♂
18 59	☽ ☍ ☉
20 19	♂ ⊻ ♃
20 24	☉ ∥ ♄
22 51	☽ ∠ ♄

MARCH 1953

LONGITUDES

Date	Sidereal time h m s	Sun ☉ ° ' "	Moon ☽ ° ' "	Moon ☽ 24.00 ° ' "	Mercury ☿ ° '	Venus ♀ ° '	Mars ♂ ° '	Jupiter ♃ ° '	Saturn ♄ ° '	Uranus ♅ ° '	Neptune ♆ ° '	Pluto ♇ ° '
01	22 35 58	10 ♓ 36 54	18 ♍ 17 25	24 ♍ 13 00	28 ♓ 41	23 ♈ 24	16 ♈ 08	15 ♉ 45	26 ♎ 47	14 ♋ 38	23 ♎ 33	21 ♌ 37
02	22 39 54	11 37 03	00 ♎ 09 31	06 ♎ 07 13	29 ♓ 43	24 03	16 53	15 55	26 R 45	14 R 37	23 R 32	21 R 35
03	22 43 51	12 37 11	12 ♎ 06 19	18 ♎ 07 07	00 ♈ 36	24 40	17 37	16 04	26 42	14 36	23 31	21 34
04	22 47 48	13 37 18	24 ♎ 09 50	00 ♏ 14 49	01 22	25 16	18 22	16 14	26 39	14 35	23 30	21 33
05	22 51 44	14 37 22	06 ♏ 22 21	12 ♏ 32 47	01 59	25 51	19 06	16 24	26 37	14 34	23 28	21 31
06	22 55 41	15 37 26	18 ♏ 46 28	25 ♏ 03 49	02 28	26 25	19 51	16 34	26 34	14 33	23 27	21 30
07	22 59 37	16 37 27	01 ♐ 25 12	07 ♐ 51 03	02 47	26 57	20 35	16 44	26 31	14 32	23 26	21 29
08	23 03 34	17 37 27	14 ♐ 21 45	20 ♐ 57 41	02 58	27 27	21 20	16 55	26 28	14 32	23 25	21 27
09	23 07 30	18 37 26	27 ♐ 39 13	04 ♑ 26 40	02 R 59	27 56	22 04	17 05	26 25	14 31	23 23	21 26
10	23 11 27	19 37 22	11 ♑ 20 14	18 ♑ 20 05	02 52	28 23	22 48	17 15	26 22	14 30	23 22	21 25
11	23 15 23	20 37 18	25 ♑ 26 14	02 ♒ 38 31	02 36	28 49	23 33	17 26	26 18	14 30	23 21	21 24
12	23 19 20	21 37 11	09 ♒ 56 41	17 ♒ 20 14	02 12	29 12	24 17	17 37	26 15	14 29	23 20	21 22
13	23 23 17	22 37 03	24 ♒ 48 30	02 ♓ 20 38	01 40	29 34	25 01	17 48	26 12	14 28	23 18	21 21
14	23 27 13	23 36 52	09 ♓ 55 37	17 ♓ 32 15	01 02	29 ♈ 55	25 45	17 58	26 08	14 28	23 17	21 20
15	23 31 10	24 36 40	25 ♓ 09 18	02 ♈ 45 26	00 ♈ 18	00 ♉ 13	26 29	18 09	26 05	14 28	23 15	21 19
16	23 35 06	25 36 26	10 ♈ 19 19	17 ♈ 49 43	29 ♓ 30	00 29	27 13	18 21	26 01	14 27	23 14	21 18
17	23 39 03	26 36 10	25 ♈ 15 29	02 ♉ 35 41	28 38	00 43	27 58	18 32	25 57	14 27	23 13	21 16
18	23 42 59	27 35 52	09 ♉ 49 31	16 ♉ 56 27	27 44	00 55	28 42	18 43	25 54	14 27	23 11	21 15
19	23 46 56	28 35 31	23 ♉ 56 05	00 ♊ 48 19	26 48	01 05	29 ♈ 25	18 54	25 50	14 26	23 10	21 14
20	23 50 52	29 ♓ 35 09	07 ♊ 33 10	14 ♊ 10 52	25 53	01 12	00 ♉ 09	19 06	25 46	14 26	23 08	21 13
21	23 54 49	00 ♈ 34 44	20 ♊ 41 44	27 ♊ 06 13	24 59	01 17	00 53	19 17	25 42	14 26	23 07	21 12
22	23 58 46	01 34 17	03 ♋ 24 52	09 ♋ 38 15	24 07	01 20	01 37	19 29	25 38	14 26	23 05	21 11
23	00 02 42	02 33 47	15 ♋ 47 00	21 ♋ 51 45	23 18	01 R 20	02 21	19 41	25 34	14 D 26	23 04	21 10
24	00 06 39	03 33 15	27 ♋ 53 08	03 ♌ 51 48	22 34	01 18	03 04	19 53	25 30	14 26	23 02	21 09
25	00 10 35	04 32 41	09 ♌ 48 21	15 ♌ 43 22	21 54	01 14	03 48	20 05	25 26	14 26	23 01	21 08
26	00 14 32	05 32 05	21 ♌ 37 22	27 ♌ 30 53	21 19	01 07	04 32	20 16	25 22	14 26	22 59	21 07
27	00 18 28	06 31 26	03 ♍ 24 21	09 ♍ 18 11	20 49	00 57	05 15	20 29	25 18	14 26	22 58	21 06
28	00 22 25	07 30 45	15 ♍ 12 44	21 ♍ 08 21	20 25	00 45	05 59	20 41	25 13	14 27	22 56	21 05
29	00 26 21	08 30 02	27 ♍ 05 17	03 ♎ 03 46	20 07	00 31	06 42	20 53	25 09	14 27	22 54	21 04
30	00 30 18	09 29 17	09 ♎ 04 01	15 ♎ 06 10	19 55	00 ♉ 14	07 26	21 05	25 05	14 28	22 53	21 03
31	00 34 15	10 ♈ 28 29	21 ♎ 10 23	27 ♎ 16 47	19 ♓ 48	29 ♈ 54	08 ♉ 09	21 ♉ 17	25 ♎ 00	14 ♋ 28	22 ♎ 51	21 ♌ 02

DECLINATIONS

Date	Moon True ☊ ° '	Moon Mean ☊ ° '	Moon ☽ Latitude ° '	Sun ☉ ° '	Moon ☽ ° '	Mercury ☿ ° '	Venus ♀ ° '	Mars ♂ ° '	Jupiter ♃ ° '	Saturn ♄ ° '	Uranus ♅ ° '	Neptune ♆ ° '	Pluto ♇ ° '
01	12 ♒ 14	10 ♒ 56	03 S 05	07 S 35	01 N 48	00 N 49	13 N 04	06 N 09	15 N 45	07 S 49	23 N 05	07 S 33	23 N 25
02	12 R 09	10 52	03 53	07 13	03 S 37	01 27	13 26	06 27	15 48	07 48	23 05	07 32	23 26
03	12 04	10 49	04 31	06 50	08 56	02 01	13 48	06 45	15 51	07 47	23 05	07 32	23 26
04	11 58	10 46	04 57	06 27	13 59	02 32	14 09	07 02	15 54	07 46	23 05	07 32	23 26
05	11 53	10 43	05 11	06 03	18 32	02 59	14 30	07 20	15 57	07 45	23 05	07 31	23 27
06	11 48	10 40	05 10	05 40	22 22	03 22	14 50	07 38	16 00	07 44	23 05	07 31	23 27
07	11 46	10 36	04 54	05 17	25 15	03 41	15 10	07 55	16 03	07 42	23 05	07 30	23 27
08	11 44	10 33	04 23	04 54	26 53	03 55	15 29	08 12	16 06	07 41	23 05	07 30	23 28
09	11 D 45	10 30	03 37	04 30	27 03	04 05	15 47	08 30	16 09	07 40	23 05	07 29	23 28
10	11 46	10 27	02 38	04 07	25 36	04 10	16 05	08 47	16 12	07 39	23 05	07 29	23 29
11	11 47	10 24	01 28	03 43	22 30	04 10	16 22	09 04	16 16	07 37	23 06	07 29	23 29
12	11 48	10 21	00 S 10	03 19	17 55	04 05	16 38	09 21	16 19	07 36	23 06	07 28	23 29
13	11 R 47	10 17	01 N 10	02 56	12 09	03 56	16 53	09 38	16 22	07 35	23 06	07 28	23 30
14	11 45	10 14	02 27	02 32	05 S 35	03 43	17 08	09 55	16 25	07 33	23 06	07 27	23 30
15	11 40	10 11	03 33	02 08	01 N 20	03 25	17 22	10 11	16 28	07 32	23 06	07 26	23 30
16	11 34	10 08	04 24	01 45	08 08	03 04	17 35	10 28	16 32	07 30	23 06	07 25	23 31
17	11 27	10 05	04 57	01 21	14 22	02 39	17 47	10 44	16 35	07 29	23 06	07 25	23 31
18	11 20	10 01	05 09	00 57	19 38	02 12	17 58	11 01	16 38	07 27	23 06	07 24	23 31
19	11 14	09 58	05 01	00 34	23 37	01 43	18 08	11 17	16 41	07 26	23 06	07 24	23 31
20	11 10	09 55	04 37	00 S 10	26 08	01 13	18 17	11 33	16 45	07 24	23 06	07 23	23 32
21	11 07	09 52	03 58	00 N 14	27 05	00 41	18 25	11 49	16 48	07 23	23 06	07 23	23 32
22	11 D 07	09 49	03 09	00 38	26 33	00 N 10	18 32	12 05	16 51	07 21	23 06	07 22	23 32
23	11 08	09 46	02 12	01 01	24 42	00 S 22	18 38	12 21	16 55	07 20	23 06	07 21	23 33
24	11 09	09 42	01 11	01 25	21 45	00 52	18 42	12 36	16 58	07 18	23 06	07 21	23 33
25	11 10	09 39	00 N 07	01 48	17 55	01 22	18 45	12 52	17 01	07 16	23 06	07 20	23 33
26	11 R 09	09 36	00 S 56	02 12	13 25	01 50	18 47	13 07	17 05	07 15	23 06	07 20	23 33
27	11 06	09 33	01 57	02 35	08 27	02 15	18 48	13 22	17 08	07 13	23 06	07 19	23 34
28	11 01	09 30	02 52	02 59	03 N 11	02 39	18 47	13 38	17 12	07 11	23 06	07 19	23 34
29	10 53	09 27	03 40	03 22	02 S 13	03 01	18 45	13 53	17 15	07 10	23 06	07 18	23 34
30	10 44	09 23	04 19	03 46	07 34	03 20	18 41	14 07	17 18	07 08	23 05	07 17	23 34
31	10 ♒ 33	09 ♒ 20	04 S 47	04 N 09	12 S 41	03 S 36	18 N 36	14 N 22	17 N 22	07 S 07	23 N 05	07 S 17	23 N 34

ZODIAC SIGN ENTRIES

Date	h	m	Planets
02	11	41	☽ ♎
02	19	21	☿ ♈
04	23	31	☽ ♏
07	09	20	☽ ♐
09	16	10	☽ ♑
11	19	37	☽ ♒
13	20	17	☽ ♓
14	18	58	♀ ♉
15	19	39	☽ ♈
15	21	16	☿ ♓
17	19	44	☽ ♉
19	22	35	☽ ♊
20	06	54	♂ ♉
20	22	01	☉ ♈
22	05	29	☽ ♋
24	16	14	☽ ♌
27	05	04	☽ ♍
29	17	51	☽ ♎
31	05	17	♀ ♈

LATITUDES

Date	Mercury ☿ ° '	Venus ♀ ° '	Mars ♂ ° '	Jupiter ♃ ° '	Saturn ♄ ° '	Uranus ♅ ° '	Neptune ♆ ° '	Pluto ♇ ° '
01	01 N 28	04 N 17	00 S 13	00 S 51	02 N 41	00 N 26	01 N 43	09 N 39
04	02 10	04 42	00 10	00 50	02 42	00 26	01 43	09 39
07	02 49	05 08	00 08	00 50	02 42	00 26	01 43	09 39
10	03 18	05 33	00 06	00 49	02 43	00 26	01 44	09 39
13	03 34	05 57	00 04	00 48	02 44	00 26	01 44	09 39
16	03 34	06 21	00 S 02	00 48	02 44	00 26	01 44	09 38
19	03 16	06 43	00 N 01	00 47	02 45	00 26	01 44	09 38
22	02 43	07 03	00 03	00 47	02 45	00 26	01 44	09 38
25	02 01	07 19	00 05	00 46	02 45	00 26	01 44	09 38
28	01 15	07 32	00 07	00 46	02 46	00 26	01 44	09 37
31	00 N 28	07 N 40	00 N 09	00 S 45	02 N 46	00 N 26	01 N 44	09 N 37

DATA

Julian Date	2434438
Delta T	+30 seconds
Ayanamsa	23° 12' 24"
Synetic vernal point	05° ♓ 54' 36"
True obliquity of ecliptic	23° 26' 50"

MOON'S PHASES, APSIDES AND POSITIONS ☽

Date	h	m	Phase	Longitude ° '	Eclipse Indicator
08	18	26	☾	17 ♐ 54	
15	11	05	●	24 ♓ 34	
22	08	11	☽	01 ♋ 25	
30	12	55	○	09 ♎ 32	

Day	h	m		
14	22	25	Perigee	
27	17	49	Apogee	
01	19	58	0S	
09	02	35	Max dec	27° S 10'
15	07	23	0N	
21	15	11	Max dec	27° N 06'
29	02	11	0S	

LONGITUDES

Date	Chiron ⚷ ° '	Ceres ⚳ ° '	Pallas ⚴ ° '	Juno ⚵ ° '	Vesta ⚶ ° '	Black Moon Lilith ⚸ ° '
01	18 ♑ 54	08 ♊ 22	15 ♈ 16	04 ♈ 01	21 ♑ 41	07 ♍ 29
11	19 ♑ 33	10 ♊ 51	19 ♈ 25	09 ♈ 30	26 ♑ 30	08 ♍ 36
21	20 ♑ 06	13 ♊ 41	23 ♈ 45	15 ♈ 06	01 ♒ 10	09 ♍ 42
31	20 ♑ 32	16 ♊ 50	28 ♈ 15	20 ♈ 46	05 ♒ 39	10 ♍ 49

All ephemeris data is given at 12.00 UT and the Moon's longitude is additionally given for 24.00 UT

ASPECTARIAN

h m	Aspects
01 Sunday	
04 36	☽ ✱ ♅
06 47	☽ △ ♃
07 20	☽ ⊼ ♂
10 05	☽ ± ♀
10 30	☽ ⊥ ♆
14 36	☉ ∥ ♆
15 52	☽ ∥ ☿
17 02	☽ ⊥ ♄
17 24	♀ ☍ ♆
18 43	☽ ⊻ ♇
22 38	☽ ⊻ ♆
22 57	☽ ⊼ ♀
02 Monday	
01 09	☽ # ☿
04 52	☽ Q ♅
05 08	☽ ⊻ ♄
06 50	☽ ⊥ ♇
11 01	☽ ☍ ☿
13 32	☽ bQ ♃
14 56	☉ bQ ♄
03 Tuesday	
00 55	☽ ∠ ♇
01 24	☽ # ♂
03 01	☽ ∥ ☉
05 35	☽ ∥ ♆
06 45	☽ ∥ ♄
07 52	☽ ± ♃
13 07	☽ ⊼ ☉
14 50	☉ # ♂
16 59	☽ □ ♅
20 02	☽ ⊼ ♃
23 44	☽ ☍ ♂
04 Wednesday	
02 10	☽ ± ☉
06 06	☿ ∠ ♃
06 49	☽ ✱ ♇
10 40	☽ ☌ ♆
12 58	☽ # ♀
14 18	☽ ☍ ♀
16 55	☽ ☌ ♄
21 35	☽ bQ ☉
21 52	☽ # ♃
05 Thursday	
03 00	☽ ⊼ ☿
06 26	☽ Q ♇
10 44	☉ △ ♅
15 18	☽ ± ☿
06 Friday	
02 41	♂ # ♆
03 53	☽ △ ♅
05 25	☽ △ ☉
07 42	☽ ☍ ♃
09 24	☽ bQ ☿
14 11	☽ ⊼ ♂
17 12	☽ □ ♇
17 13	☽ # ♅
18 16	♀ ☍ ♄
19 48	♂ # ♄
20 03	☽ # ♇
20 55	☽ ⊻ ♆
07 Saturday	
02 20	☽ ± ♂
02 47	☽ ⊻ ♄
03 12	☽ ⊼ ♀
08 16	☽ ⊥ ♆
08 28	☽ bQ ♅
14 03	☽ ⊥ ♄
14 37	☽ △ ☿
14 59	☽ ± ♀
15 16	☉ ✱ ♃
20 16	☽ bQ ♂
08 Sunday	
01 03	☽ ∠ ♆
01 16	☽ ± ♅
06 42	☽ ∠ ♄
07 00	☉ ± ♆
08 21	☽ bQ ♀
12 18	☽ ⊼ ♅
16 02	♂ △ ♇
16 43	☽ ⊼ ♃
18 26	☽ □ ☉
09 Monday	
00 52	☽ △ ♇
01 25	☽ △ ♂
03 43	☽ ± ♃
03 45	☿ St R
04 23	☽ ✱ ♆
09 48	☽ ✱ ♄
12 31	☽ △ ♀
19 57	☽ bQ ♃
21 23	☽ □ ☿
10 Tuesday	
01 38	☽ Q ♆
03 28	☽ bQ ♇
05 03	☽ Q ☉
06 52	☽ Q ♄
09 01	☉ # ☿
17 27	☽ ☍ ♅
19 00	☽ ± ♇
22 18	☽ △ ♃
11 Wednesday	
01 10	♀ ∥ ♃
03 16	☽ ✱ ☉
04 01	☽ Q ☿
04 51	☉ ± ♄
05 12	☽ ⊼ ♇
05 34	☽ # ♇
05 45	♂ ☍ ♆
08 13	☽ # ♅
08 30	☽ □ ♆
08 39	☽ □ ♂
13 27	☽ □ ♄
17 48	☽ □ ♀
19 15	☿ ∠ ♃
23 38	☽ ✱ ☿
12 Thursday	
06 09	☽ ∠ ☉
06 11	☉ ⊼ ♇
16 01	☽ Q ♂
17 31	☽ # ♀
19 05	☽ # ♃
19 23	☽ ⊼ ♅
21 53	☽ ⊥ ☉
23 23	☽ ∠ ☿
13 Friday	
00 06	☽ Q ♀
00 36	☽ □ ♃
05 03	☽ ± ♅
06 28	☽ ☍ ♇
08 15	☽ ⊻ ☉
09 36	☽ △ ♆
12 21	☽ ✱ ♂
13 19	☽ ⊥ ☿
14 13	☽ △ ♄
19 26	☽ bQ ♅
19 47	☽ ✱ ♀
21 05	☽ # ♂
22 30	☽ ⊻ ☿
14 Saturday	
04 08	☉ ⊼ ♆
05 00	☽ ∥ ♄
05 24	☽ ∥ ♆
05 41	☽ Q ♃
09 24	☽ bQ ♆
13 22	☽ ∠ ♂
13 54	☽ bQ ♄
18 49	☽ # ☿
19 09	☽ △ ♅
20 01	☽ ∠ ♀
22 19	☉ ⊥ ♀
23 16	☽ ∥ ☉
23 35	☽ ± ♆
23 35	♂ ☍ ♄
15 Sunday	
00 50	☽ ✱ ♃
04 02	☽ ± ♄
04 17	☽ ⊥ ♂
05 57	☽ ⊼ ♇
09 01	☽ ⊼ ♆
10 29	☽ ⊥ ♀
11 05	☽ ☌ ☉
13 27	☽ ⊼ ♄
14 05	☿ ⊻ ♀
14 13	☽ ⊻ ♂
14 38	☽ # ☉
15 24	☽ ± ♇
18 53	☽ ∥ ☿
19 43	☽ ☌ ☿
20 08	☽ ⊻ ♀
16 Monday	
00 47	☽ ∠ ♃
05 36	☽ bQ ♇
09 25	☽ # ♆
09 42	☽ # ♄
15 16	☽ ⊥ ♃
18 35	☽ □ ♅
21 01	☽ ∥ ♂
21 22	☉ ⊼ ♄
17 Tuesday	
00 59	☽ ⊻ ♃
05 33	☽ △ ♇
08 41	☽ ☍ ♆
13 08	☽ ☍ ♄
14 21	☽ ⊻ ☉
16 38	☽ ☌ ♂
17 11	☽ ⊻ ☿
21 02	☽ ☌ ♀
21 34	☽ ∥ ♃
21 53	☿ ⊻ ♂
23 45	☽ Q ♅
18 Wednesday	
00 54	☽ ⊥ ☉
02 28	☽ ⊥ ☿
03 30	☽ ∥ ♀
03 52	☉ ± ♇
13 36	☉ ☌ ☿
16 34	☽ ∠ ☿
17 00	☽ ∠ ☉
19 46	☽ ✱ ♅
19 Thursday	
00 32	☿ ± ♇
03 14	☽ ☌ ♃
07 21	☽ □ ♇
08 17	☽ ∥ ♅
10 40	☽ ⊼ ♆
11 19	☽ ∥ ♇
15 17	☽ ⊼ ♄
16 40	☽ ✱ ☿
20 45	☽ ✱ ☉
21 06	☽ ± ♆
21 36	☽ ∠ ♅
22 07	☽ ⊻ ♂
20 Friday	
00 36	☽ ⊻ ♀
01 45	☽ ± ♄
09 21	☽ ⊥ ♂
11 22	☽ ⊥ ♀
12 33	☽ Q ☿
13 03	☽ bQ ♆
13 35	☽ ⊥ ♅
14 59	☽ Q ♇
15 10	☿ ⊼ ♄
17 46	☽ bQ ♄
19 52	☽ Q ☉
21 Saturday	
00 28	☽ ⊻ ♅
02 36	☽ ∠ ♂
03 49	☽ ∠ ♀
04 20	☿ ⊥ ♀
09 21	☽ ⊻ ♃
12 56	☽ ✱ ♇
13 14	♄ ∥ ♆
13 24	☿ ⊥ ♂
16 30	☽ △ ♆
19 29	☽ □ ☿
20 43	☽ ⊥ ♃
21 19	☽ △ ♄
23 57	☉ ∥ ☿
22 Sunday	
02 16	♀ ☌ ♂
06 05	☉ ⊻ ♀
08 01	☽ ✱ ♀
08 11	☽ □ ☉
08 21	☽ ✱ ♂
14 05	☽ ∠ ♃
16 06	☉ ⊻ ♂
17 18	☽ ∠ ♇
21 22	♅ St D
23 Monday	
03 52	♀ St R
07 13	☽ Q ♀
09 00	☽ Q ♂
09 21	☽ ☌ ♅
10 47	☽ ⊥ ♇
14 56	♂ Q ♅
19 48	☽ ✱ ♃
19 50	☿ ⊼ ♆
22 24	☽ ∥ ♇
22 36	☽ ⊻ ♇
24 Tuesday	
01 59	☽ △ ☿
02 03	☽ ∥ ♅
02 21	☽ □ ♆
07 16	☽ □ ♄
18 49	☽ □ ♀
20 08	☽ Q ♃
23 05	☽ □ ♂
25 Wednesday	
00 25	☽ △ ☉
06 25	☽ bQ ☿
07 11	☽ ∥ ♀
14 26	☽ Q ♆
16 57	☽ ∥ ♃
19 19	☽ Q ♄
21 23	☽ ⊻ ♅
23 45	☽ ± ☿
26 Thursday	
04 06	☉ ∠ ♃
09 13	☽ □ ♃
09 35	☽ bQ ☉
09 35	☽ ⊥ ♅
10 58	☽ ☌ ♇
11 23	☽ ⊼ ☿
13 25	☽ ∥ ♂
14 46	☽ ✱ ♆
19 35	☽ ✱ ♄
21 16	☿ ⊼ ♇
27 Friday	
01 50	☉ bQ ♇
03 56	☽ ∠ ♅
05 36	☽ ± ☉
07 05	☽ △ ♀
16 01	☽ △ ♂
17 15	☽ # ♆
17 44	☽ # ♄
18 56	☽ ⊼ ☉
21 15	☽ ∠ ♆
23 45	☿ ∠ ♂
28 Saturday	
01 08	☿ ✱ ♃
01 56	☽ ∠ ♄
10 27	☽ ✱ ♅
12 50	☽ ∥ ☉
13 05	☽ bQ ♀
14 13	☽ # ☿
15 29	☽ ⊥ ♆
20 05	☽ ⊥ ♄
22 15	☽ ☍ ☿
23 16	☽ △ ♃
23 52	☽ ⊻ ♇
29 Sunday	
00 26	☽ bQ ♂
03 36	☽ ⊻ ♆
06 55	☽ ± ♀
08 07	☽ ⊻ ♄
10 43	☽ Q ♅
11 57	☽ ⊥ ♇
15 48	☽ ∥ ☿
17 34	☽ # ☉
18 44	☽ ⊼ ♀
19 44	☽ ± ♂
30 Monday	
05 57	☽ bQ ♃
05 59	☽ ∠ ♇
08 26	♃ □ ♇
08 31	☽ ⊼ ♂
10 04	☽ ∥ ♄
10 45	☽ ∥ ♆
12 55	☽ ☍ ☉
22 44	☽ □ ♅
31 Tuesday	
00 10	☽ ± ♃
09 18	☽ ⊼ ☿
11 44	☽ ✱ ♇
12 14	☽ ⊼ ♃
15 18	☽ ☌ ♆
19 30	☽ ☌ ♄
20 45	☽ # ♂
21 04	☽ ± ☿

APRIL 1953

Date	Sidereal time h m s	Sun ☉ ° ' "	Moon ☽ ° ' "	Moon ☽ 24.00 ° ' "	Mercury ☿ ° '	Venus ♀ ° '	Mars ♂ ° '	Jupiter ♃ ° '	Saturn ♄ ° '	Uranus ♅ ° '	Neptune ♆ ° '	Pluto ♇ ° '
01	00 38 11	11 ♈ 27 40	03 ♏ 25 29	09 ♏ 36 35	19 ♓ 47	29 ♈ 33	08 ♉ 52	21 ♉ 30	24 ♎ 56	14 ♋ 29	22 ♎ 50	21 ♌ 01
02	00 42 08	12 26 49	15 ♏ 50 13	22 ♏ 06 31	19 D 52	29 R 09	09 36	21 42	24 R 52	14 29	22 R 48	21 R 01
03	00 46 04	13 25 56	28 ♏ 25 38	04 ♐ 47 44	20 02	28 43	10 19	21 55	24 47	14 30	22 46	21 00
04	00 50 01	14 25 02	11 ♐ 13 01	17 ♐ 41 42	20 17	28 14	11 02	22 07	24 43	14 30	22 45	20 59
05	00 53 57	15 24 05	24 ♐ 14 04	00 ♑ 50 20	20 37	27 45	11 45	22 20	24 38	14 31	22 43	20 58
06	00 57 54	16 23 07	07 ♑ 30 47	14 ♑ 15 41	21 02	27 13	12 28	22 33	24 34	14 32	22 42	20 58
07	01 01 50	17 22 07	21 ♑ 05 17	27 ♑ 59 46	21 31	26 40	13 11	22 46	24 29	14 33	22 40	20 57
08	01 05 47	18 21 05	04 ♒ 59 16	12 ♒ 03 50	22 05	26 05	13 54	22 58	24 25	14 33	22 38	20 56
09	01 09 44	19 20 02	19 ♒ 13 24	26 ♒ 27 46	22 43	25 29	14 37	23 11	24 20	14 34	22 37	20 55
10	01 13 40	20 18 56	03 ♓ 46 34	11 ♓ 09 16	23 24	24 53	15 20	23 24	24 15	14 35	22 35	20 55
11	01 17 37	21 17 49	18 ♓ 35 11	26 ♓ 03 27	24 09	24 16	16 03	23 37	24 11	14 36	22 33	20 54
12	01 21 33	22 16 40	03 ♈ 33 04	11 ♈ 02 55	24 58	23 38	16 46	23 50	24 06	14 37	22 32	20 54
13	01 25 30	23 15 30	18 ♈ 31 49	25 ♈ 58 35	25 50	23 00	17 28	24 03	24 01	14 38	22 30	20 53
14	01 29 26	24 14 17	03 ♉ 22 04	10 ♉ 41 12	26 45	22 23	18 11	24 17	23 57	14 40	22 28	20 53
15	01 33 23	25 13 02	17 ♉ 55 04	25 ♉ 02 54	27 43	21 45	18 54	24 30	23 52	14 41	22 27	20 52
16	01 37 19	26 11 46	02 ♊ 04 09	08 ♊ 58 28	28 43	21 09	19 36	24 43	23 48	14 42	22 25	20 52
17	01 41 16	27 10 27	15 ♊ 45 41	22 ♊ 25 48	29 ♓ 47	20 32	20 19	24 56	23 43	14 43	22 24	20 51
18	01 45 13	28 09 06	28 ♊ 59 02	05 ♋ 25 42	00 ♈ 53	19 57	21 02	25 10	23 38	14 45	22 22	20 51
19	01 49 09	29 ♈ 07 43	11 ♋ 46 14	18 ♋ 01 11	02 02	19 24	21 44	25 23	23 34	14 46	22 20	20 50
20	01 53 06	00 ♉ 06 17	24 ♋ 11 08	00 ♌ 16 45	03 13	18 51	22 27	25 37	23 29	14 48	22 19	20 50
21	01 57 02	01 04 50	06 ♌ 18 24	12 ♌ 17 42	04 26	18 21	23 09	25 50	23 25	14 49	22 17	20 50
22	02 00 59	02 03 20	18 ♌ 14 24	24 ♌ 09 30	05 42	17 52	23 51	26 04	23 20	14 51	22 15	20 49
23	02 04 55	03 01 48	00 ♍ 03 38	05 ♍ 57 25	07 00	17 25	24 34	26 17	23 16	14 52	22 14	20 49
24	02 08 52	04 00 14	11 ♍ 51 26	17 ♍ 46 13	08 20	17 00	25 16	26 31	23 11	14 54	22 12	20 49
25	02 12 48	04 58 38	23 ♍ 42 15	29 ♍ 39 57	09 42	16 37	25 58	26 44	23 07	14 56	22 11	20 49
26	02 16 45	05 56 59	05 ♎ 39 41	11 ♎ 41 46	11 06	16 16	26 40	26 58	23 02	14 58	22 09	20 48
27	02 20 42	06 55 19	17 ♎ 46 26	23 ♎ 53 52	12 32	15 58	27 22	27 12	22 58	14 59	22 07	20 48
28	02 24 38	07 53 37	00 ♏ 04 13	06 ♏ 17 32	14 00	15 42	28 04	27 25	22 53	15 01	22 06	20 48
29	02 28 35	08 51 53	12 ♏ 33 52	18 ♏ 53 12	15 30	15 29	28 46	27 39	22 49	15 03	22 04	20 48
30	02 32 31	09 ♉ 50 08	25 ♏ 15 32	01 ♐ 40 47	17 ♈ 02	15 ♈ 18	29 ♉ 28	27 ♉ 53	22 ♎ 45	15 ♋ 05	22 ♎ 03	20 ♌ 48

Date	Moon True ☊ ° '	Moon Mean ☊ ° '	Moon ☽ Latitude ° '
01	10 ♒ 21	09 ♒ 17	05 S 01
02	10 R 10	09 14	05 02
03	10 01	09 11	04 48
04	09 54	09 07	04 20
05	09 49	09 04	03 38
06	09 47	09 01	02 43
07	09 D 47	08 58	01 38
08	09 48	08 55	00 S 26
09	09 R 47	08 52	00 N 50
10	09 45	08 48	02 04
11	09 41	08 45	03 10
12	09 34	08 42	04 05
13	09 24	08 39	04 42
14	09 13	08 36	05 00
15	09 02	08 33	04 58
16	08 52	08 29	04 37
17	08 44	08 26	04 01
18	08 39	08 23	03 13
19	08 36	08 20	02 17
20	08 35	08 17	01 16
21	08 D 35	08 13	00 N 12
22	08 R 35	08 10	00 S 51
23	08 33	08 07	01 51
24	08 30	08 04	02 46
25	08 23	08 01	03 34
26	08 14	07 58	04 14
27	08 03	07 54	04 42
28	07 50	07 51	04 57
29	07 36	07 48	04 59
30	07 ♒ 24	07 ♒ 45	04 S 46

DECLINATIONS

Date	Sun ☉ ° '	Moon ☽ ° '	Mercury ☿ ° '	Venus ♀ ° '	Mars ♂ ° '	Jupiter ♃ ° '	Saturn ♄ ° '	Uranus ♅ ° '	Neptune ♆ ° '	Pluto ♇ ° '
01	04 N 32	17 S 23	03 S 50	18 N 29	14 N 37	17 N 25	07 S 05	23 N 05	07 S 16	23 N 34
02	04 55	21 24	04 02	18 21	14 51	17 28	07 03	23 05	07 15	23 35
03	05 18	24 30	04 11	18 11	15 05	17 32	07 01	23 05	07 15	23 35
04	05 41	26 25	04 17	18 00	15 19	17 35	07 00	23 05	07 14	23 35
05	06 04	26 57	04 21	17 47	15 33	17 39	06 58	23 05	07 14	23 35
06	06 27	25 57	04 22	17 33	15 47	17 42	06 56	23 05	07 13	23 35
07	06 49	23 24	04 21	17 18	16 01	17 45	06 55	23 05	07 12	23 35
08	07 12	19 26	04 18	17 01	16 14	17 49	06 53	23 05	07 12	23 35
09	07 34	14 16	04 13	16 43	16 28	17 52	06 51	23 05	07 11	23 35
10	07 56	08 12	04 05	16 24	16 41	17 56	06 50	23 04	07 11	23 35
11	08 19	01 S 36	03 56	16 04	16 54	17 59	06 48	23 04	07 10	23 35
12	08 41	05 N 09	03 44	15 42	17 07	18 02	06 46	23 04	07 09	23 35
13	09 02	11 36	03 31	15 20	17 20	18 06	06 44	23 04	07 09	23 35
14	09 24	17 20	03 16	14 58	17 32	18 09	06 43	23 04	07 08	23 35
15	09 46	21 56	02 59	14 34	17 45	18 12	06 41	23 04	07 08	23 35
16	10 07	25 07	02 40	14 10	17 57	18 16	06 39	23 04	07 07	23 35
17	10 28	26 41	02 19	13 46	18 09	18 19	06 38	23 03	07 06	23 35
18	10 49	26 40	01 57	13 22	18 21	18 22	06 36	23 03	07 06	23 35
19	11 10	25 12	01 34	12 58	18 32	18 26	06 34	23 03	07 05	23 35
20	11 31	22 31	01 09	12 33	18 44	18 29	06 33	23 03	07 04	23 35
21	11 51	18 54	00 42	12 10	18 55	18 32	06 31	23 03	07 04	23 35
22	12 11	14 34	00 S 15	11 46	19 07	18 36	06 29	23 03	07 03	23 35
23	12 32	09 43	00 N 15	11 23	19 18	18 39	06 28	23 02	07 03	23 35
24	12 51	04 N 33	00 45	11 00	19 29	18 42	06 26	23 02	07 02	23 35
25	13 11	00 S 47	01 17	10 38	19 39	18 45	06 25	23 02	07 02	23 35
26	13 31	06 08	01 50	10 17	19 49	18 49	06 23	23 02	07 01	23 35
27	13 50	11 19	02 24	09 57	20 00	18 52	06 22	23 02	07 00	23 35
28	14 09	16 08	02 59	09 38	20 10	18 55	06 20	23 01	07 00	23 34
29	14 27	20 21	03 35	09 19	20 20	18 58	06 18	23 01	06 59	23 34
30	14 N 46	23 S 42	04 N 12	09 N 02	20 N 29	19 N 02	06 S 17	23 N 01	06 S 59	23 N 34

ZODIAC SIGN ENTRIES

Date	h	m	Planets
01	05	19	☽ ♏
03	14	58	☽ ♐
05	22	29	☽ ♑
08	03	27	☽ ♒
10	05	49	☽ ♓
12	06	19	☽ ♈
14	06	31	☽ ♉
16	08	27	☽ ♊
17	16	48	☿ ♈
18	13	53	☽ ♋
20	09	25	☉ ♉
20	23	27	☽ ♌
23	11	53	☽ ♍
26	00	40	☽ ♎
28	11	52	☽ ♏
30	20	52	☽ ♐

LATITUDES

Date	Mercury ☿ ° '	Venus ♀ ° '	Mars ♂ ° '	Jupiter ♃ ° '	Saturn ♄ ° '	Uranus ♅ ° '	Neptune ♆ ° '	Pluto ♇ ° '
01	00 N 13	07 N 41	00 N 10	00 S 45	02 N 46	00 N 26	01 N 44	09 N 37
04	00 S 28	07 40	00 12	00 45	02 46	00 26	01 44	09 36
07	01 05	07 32	00 13	00 44	02 46	00 26	01 44	09 36
10	01 36	07 17	00 15	00 44	02 46	00 26	01 44	09 35
13	02 01	06 54	00 17	00 43	02 47	00 26	01 44	09 35
16	02 21	06 24	00 19	00 43	02 47	00 26	01 44	09 34
19	02 35	05 49	00 21	00 42	02 46	00 26	01 44	09 34
22	02 44	05 09	00 23	00 42	02 46	00 26	01 44	09 33
25	02 47	04 27	00 24	00 42	02 46	00 26	01 44	09 33
28	02 46	03 44	00 26	00 41	02 46	00 25	01 44	09 32
31	02 S 39	03 N 02	00 N 28	00 S 41	02 N 46	00 N 25	01 N 44	09 N 31

DATA

Julian Date	2434469
Delta T	+31 seconds
Ayanamsa	23° 12' 27"
Synetic vernal point	05° ♓ 54' 33"
True obliquity of ecliptic	23° 26' 50"

MOON'S PHASES, APSIDES AND POSITIONS ☽

Date	h	m	Phase	Longitude ° '	Eclipse Indicator
07	04	58	☾	17 ♑ 05	
13	20	09	●	23 ♈ 35	
21	00	41	☽	00 ♌ 37	
29	04	20	○	08 ♏ 33	

Day	h	m		
12	06	59	Perigee	
24	08	10	Apogee	
05	08	29	Max dec	26° S 58'
11	17	40	0N	
17	23	37	Max dec	26° N 52'
25	08	30	0S	

LONGITUDES

Date	Chiron ⚷ ° '	Ceres ⚳ ° '	Pallas ⚴ ° '	Juno ⚵ ° '	Vesta ⚶ ° '	Black Moon Lilith ⚸ ° '
01	20 ♑ 34	17 ♊ 09	28 ♈ 43	21 ♈ 20	06 ♒ 05	10 ♍ 55
11	20 ♑ 51	20 ♊ 34	03 ♉ 22	27 ♈ 05	10 ♒ 20	12 ♍ 02
21	21 ♑ 01	24 ♊ 11	08 ♉ 10	02 ♉ 54	14 ♒ 19	13 ♍ 09
31	21 ♑ 02	27 ♊ 59	13 ♉ 06	08 ♉ 45	18 ♒ 00	14 ♍ 16

All ephemeris data is given at 12.00 UT and the Moon's longitude is additionally given for 24.00 UT

ASPECTARIAN

01 Wednesday
03 36 ☿ St D
04 39 ☽ ☍ ♀
11 13 ☽ Q ♇
12 14 ☽ # ♃
14 39 ☽ Ꝑ ☿
17 59 ☽ # ♀
23 13 ☽ ☍ ♂

02 Thursday
04 55 ☽ ⊼ ☉
09 24 ☽ △ ♅
17 26 ☽ ± ☉
19 48 ☽ △ ☿
21 54 ☽ □ ♇
23 25 ☽ ☍ ♃

03 Friday
00 04 ☽ # ♅
01 18 ☽ ⊻ ♆
03 58 ☽ # ♇
05 08 ☽ ⊻ ♄
12 01 ☽ Ꝑ ☉
12 31 ☽ ⊼ ♀
12 39 ☽ ⊥ ♆
14 01 ☽ Ꝑ ♅
16 26 ☽ ⊥ ♄
23 26 ☽ ± ♀

04 Saturday
05 33 ☽ ∠ ♆
06 56 ☽ ± ♅
09 13 ☽ ∠ ♄
11 38 ☽ ⊼ ♂
14 10 ☉ □ ♅
15 38 ☽ Ꝑ ♀
18 07 ☽ ⊼ ♅
18 26 ☽ △ ☉
23 24 ☽ ± ♂

05 Sunday
05 12 ☽ □ ☿
06 02 ☽ △ ♇
08 28 ☽ ⊼ ♃
09 14 ☽ ✶ ♆
12 44 ☽ ✶ ♄
16 51 ☽ Ꝑ ♂
18 09 ☽ △ ♀
19 35 ☽ ± ♃

06 Monday
00 21 ♀ ∥ ♃
06 57 ☽ Q ♆
08 01 ☿ ⊼ ♇
09 13 ☽ Ꝑ ♇
10 18 ☽ Q ♄
12 04 ☽ Ꝑ ♃
14 49 ☽ Q ☿
16 18 ☿ ⊥ ♀
17 01 ☉ ⊥ ♃
21 19 ☽ △ ♂

07 Tuesday
00 29 ☽ ☍ ♅
01 13 ☽ ± ♇
02 43 ♃ ⊼ ♆
04 58 ☽ □ ☉
10 40 ☽ # ♇
11 45 ☽ ⊼ ♇
12 47 ☽ ✶ ☿
14 19 ☽ # ♅
14 45 ☽ □ ♆
14 58 ☽ △ ♃
17 21 ☉ # ♄
17 53 ☽ □ ♄
21 19 ☽ □ ♀

08 Wednesday
12 01 ☉ # ♆
14 30 ☽ Q ☉
15 44 ☽ ∠ ☿
20 02 ☽ # ♃

09 Thursday
00 35 ☽ # ♀
02 47 ☽ Q ♀
02 47 ☽ # ♂
03 53 ☽ □ ♂
04 13 ☽ ⊼ ♅
07 36 ☽ ⊥ ☿
08 33 ☿ ⊼ ♆
10 25 ♂ ✶ ♅
12 12 ☽ ✶ ☉
14 15 ☽ ± ♅
14 50 ☽ ☍ ♇
17 37 ☽ △ ♆
18 05 ☽ ⊻ ☿
18 41 ☽ □ ♃
20 26 ☽ △ ♄
21 59 ☽ ✶ ♀
23 23 ♀ ∥ ♂

10 Friday
05 08 ☽ Ꝑ ♅
11 14 ☽ Q ♂
12 10 ☿ ✶ ♃
12 56 ☽ # ☉
14 42 ☽ ∠ ☉
15 50 ☽ ∥ ♆
17 08 ☽ ∥ ♄
18 12 ☽ Ꝑ ♆
20 52 ☽ Ꝑ ♄
21 32 ☽ ∠ ♀

11 Saturday
00 35 ☽ Q ♃
02 29 ☉ △ ♇
03 26 ☽ ∥ ☿
05 34 ☽ △ ♅
06 19 ☽ ⊥ ☉
07 42 ☽ ✶ ♂
08 44 ☽ ± ♆
11 21 ☽ ± ♄
11 30 ☽ ⊥ ♀
12 43 ☿ ⊼ ♄
13 52 ☿ ⊻ ♀
15 41 ♀ ☍ ♄
15 44 ☽ ⊼ ♇
16 40 ☽ ⊻ ☉
18 22 ☽ ⊼ ♆
20 13 ☽ ✶ ♃
20 45 ☽ ⊻ ♀
20 56 ☽ ⊼ ♄
21 27 ☽ ☌ ☿

12 Sunday
01 21 ☽ ± ♇
06 14 ♀ ⊻ ♃
07 05 ☽ # ☿
09 00 ☽ ∠ ♂
15 45 ☽ Ꝑ ♇
17 49 ☽ # ♄
18 00 ☉ ☍ ♆
19 15 ☽ # ♆
20 35 ☽ ∠ ♃

13 Monday
00 07 ☽ ⊥ ♂
01 38 ☽ ∥ ☉
05 45 ☽ □ ♅
08 15 ☉ ☌ ♀
09 20 ♃ ⊼ ♄
10 13 ☽ ⊻ ♂
11 14 ☽ ⊥ ♃
15 47 ☽ △ ♇
18 23 ☽ ☍ ♆
18 55 ☽ ☌ ♀
20 09 ☽ ☌ ☉
20 48 ☽ ☍ ♄
21 02 ☽ ⊻ ♃

14 Tuesday
00 31 ☽ ⊻ ☿
02 13 ☽ ∥ ♀
05 25 ☉ ☍ ♄
08 08 ♀ ☍ ♆
10 51 ☽ Q ♅
10 55 ☽ ⊥ ☿
12 58 ☽ ∥ ♂
13 14 ☉ ⊻ ♃
15 20 ☿ ± ♇
15 51 ☽ ∥ ♃

15 Wednesday
02 43 ☽ ∠ ☿
06 36 ☽ ✶ ♅
13 43 ☽ ☌ ♂
16 56 ☽ □ ♇
18 10 ☽ ⊻ ♀
19 17 ☽ ∥ ♅
19 35 ☽ ⊼ ♆
21 57 ☽ ⊼ ♄
23 03 ☽ ∥ ♇
23 15 ☽ ☌ ♃

16 Thursday
01 12 ☽ ⊻ ☉
03 54 ☽ ⊥ ♀
05 45 ☽ ± ♆
05 48 ☽ ✶ ☿
07 56 ☽ ∠ ♅
08 07 ☽ ± ♄
12 14 ☽ ⊥ ☉
18 45 ☽ ∠ ♀
21 16 ☽ Ꝑ ♆
23 18 ♀ △ ♇
23 33 ☽ ⊥ ♅
23 37 ☽ Ꝑ ♄
23 48 ☽ Q ♇

17 Friday
04 20 ☽ Q ☿
05 08 ☽ ∠ ☉
10 09 ☽ ⊻ ♅
16 08 ♀ ⊻ ♂
20 13 ☽ ✶ ♀
20 38 ☽ ⊻ ♂
21 08 ☽ ✶ ♇
23 54 ☽ △ ♆

18 Saturday
02 15 ☽ △ ♄
04 51 ☽ ⊻ ♃
05 57 ♂ □ ♇
08 11 ☽ ⊥ ♂
10 20 ☽ ✶ ☉
15 51 ☽ □ ☿
16 06 ☽ ⊥ ♃
16 29 ♂ ∥ ♃
17 17 ☽ Q ♀

19 Sunday
00 47 ☽ ∠ ♇
01 54 ☽ ∠ ♂
09 19 ☽ ∠ ♃
10 40 ☽ Q ☉
12 15 ♀ ⊥ ♃
17 45 ☽ ☌ ♅
17 52 ☽ ⊥ ♇

20 Monday
02 03 ☽ □ ♀
03 33 ☽ ∥ ♇
05 28 ☽ ⊻ ♇
07 42 ♂ ⊼ ♆
07 56 ☽ ∥ ♅
08 21 ☽ □ ♆
08 23 ☽ ✶ ♂
10 39 ☽ □ ♄
14 51 ☽ ✶ ♃

21 Tuesday
00 41 ☽ □ ☉
07 50 ☽ △ ☿
09 32 ☽ Q ♂
11 51 ☽ ∥ ♂
14 05 ☽ ∥ ♃
15 06 ☽ Q ♃
17 33 ♅ ⊥ ♇
19 56 ☽ Q ♆
20 04 ♂ ⊼ ♄
21 58 ☉ ∥ ♀
22 10 ☽ Q ♄

22 Wednesday
05 08 ☽ ⊻ ♅
11 16 ☽ △ ♀
12 07 ♀ ⊥ ♂
14 21 ☿ Ꝑ ♇
17 14 ☽ ☌ ♇
17 18 ☽ ⊥ ♅
17 35 ☽ Ꝑ ☿
20 07 ☽ ✶ ♆
22 16 ☽ ✶ ♄
23 13 ☽ ∥ ☉

23 Thursday
00 06 ☽ □ ♂
03 19 ☽ ∥ ♀
04 10 ☽ □ ♃
08 03 ☉ Q ♅
11 37 ☽ ∠ ♅
14 08 ☽ ± ☿
16 37 ☽ Ꝑ ♀
18 35 ☽ △ ☉

24 Friday
00 35 ☽ # ♆
02 33 ☽ ∠ ♆
03 20 ☽ # ♄
03 54 ☽ ⊼ ☿
04 35 ☽ ∠ ♄
10 18 ☽ ± ♀
18 12 ☽ ✶ ♅
20 48 ☽ ⊥ ♆
22 05 ☽ ⊼ ♀
22 45 ☽ ⊥ ♄

25 Saturday
03 37 ☽ ∥ ☿
03 48 ☽ Ꝑ ☉
06 09 ☽ ⊻ ♇
08 55 ☽ ⊻ ♆
10 49 ☽ ⊻ ♄
14 28 ☽ # ☿
16 51 ☽ △ ♂
18 14 ☽ △ ♃
18 16 ☽ ⊥ ♇
18 31 ☽ Q ♅
23 33 ☽ ± ☉

26 Sunday
12 17 ☽ ∠ ♇
12 38 ☽ ⊼ ☉
13 10 ☽ ∥ ♄
16 02 ☽ ∥ ♆

27 Monday
00 15 ☽ ☍ ☿
00 41 ☽ Ꝑ ♂
00 47 ☽ Ꝑ ♃
03 03 ♂ ☌ ♃
05 24 ☿ ∠ ♃
05 58 ☽ # ♀
06 30 ☽ □ ♅
06 55 ☿ ∠ ♂
08 31 ☽ ☍ ♀
17 57 ☽ ✶ ♇
18 51 ☽ ± ♃
19 29 ☽ ± ♂
20 31 ☽ ☌ ♆
22 07 ☽ ☌ ♄

28 Tuesday
01 07 ☽ # ☉
06 46 ☽ ⊼ ♃
07 54 ☽ ⊼ ♂
12 49 ♂ ± ♆
17 17 ☽ Q ♇

29 Wednesday
03 35 ☽ # ♃
04 20 ☽ ☍ ☉
04 48 ☿ □ ♅
11 44 ☿ ☌ ♀
11 51 ☽ # ♂
13 24 ♂ ± ♄
16 45 ☽ △ ♅
17 27 ☽ ⊼ ♀
18 20 ☽ ⊼ ☿

30 Thursday
03 37 ☽ □ ♇
04 38 ☽ ± ♀
05 58 ☽ ⊻ ♆
06 26 ☽ # ♅
07 14 ☽ ± ☿
07 18 ☽ ⊻ ♄
10 53 ☽ # ♇
17 00 ☽ ☍ ♃
17 12 ☽ ⊥ ♆
18 30 ☽ ⊥ ♄
20 20 ☽ ☍ ♂
21 03 ☽ Ꝑ ♅
21 18 ☽ Ꝑ ♀

MAY 1953

LONGITUDES

Date	Sidereal time h m s	Sun ☉ ° ' "	Moon ☽ ° ' "	Moon ☽ 24.00 ° ' "	Mercury ☿ ° '	Venus ♀ ° '	Mars ♂ ° '	Jupiter ♃ ° '	Saturn ♄ ° '	Uranus ♅ ° '	Neptune ♆ ° '	Pluto ⯓ ° '
01	02 36 28	10 ♉ 48 21	08 ♐ 08 55	14 ♐ 39 53	18 ♈ 35	15 ♈ 09	00 ♊ 10	28 ♉ 07	22 ♎ 40	15 ♋ 07	22 ♎ 01	20 ♌ 48
02	02 40 24	11 46 32	21 ♐ 13 39	27 ♐ 50 12	20 11	15 R 03	00 52	28 21	22 R 36	15 09	22 R 00	20 R 48
03	02 44 21	12 44 42	04 ♑ 29 32	11 ♑ 11 43	21 49	14 59	01 34	28 34	22 32	15 11	21 58	20 D 48
04	02 48 17	13 42 50	17 ♑ 56 48	24 ♑ 44 51	23 28	14 58	02 16	28 48	22 28	15 13	21 57	20 48
05	02 52 14	14 40 57	01 ♒ 35 59	08 ♒ 30 16	25 10	14 D 59	02 58	29 02	22 24	15 15	21 55	20 48
06	02 56 11	15 39 02	15 ♒ 27 48	22 ♒ 28 35	26 53	15 03	03 39	29 16	22 20	15 18	21 54	20 48
07	03 00 07	16 37 06	29 ♒ 32 37	06 ♓ 39 48	28 ♈ 38	15 08	04 21	29 30	22 16	15 20	21 52	20 48
08	03 04 04	17 35 08	13 ♓ 49 55	21 ♓ 02 41	00 ♉ 25	15 16	05 03	29 44	22 12	15 22	21 51	20 48
09	03 08 00	18 33 10	28 ♓ 17 41	05 ♈ 34 20	02 14	15 26	05 44	29 ♉ 58	22 08	15 24	21 49	20 48
10	03 11 57	19 31 10	12 ♈ 52 01	20 ♈ 09 56	04 05	15 39	06 26	00 ♊ 12	22 04	15 27	21 48	20 49
11	03 15 53	20 29 08	27 ♈ 27 16	04 ♉ 43 08	05 58	15 53	07 07	00 26	22 00	15 29	21 46	20 49
12	03 19 50	21 27 05	11 ♉ 56 37	19 ♉ 06 52	07 53	16 09	07 49	00 40	21 57	15 32	21 45	20 49
13	03 23 46	22 25 01	26 ♉ 13 06	03 ♊ 14 36	09 49	16 27	08 30	00 54	21 53	15 34	21 44	20 49
14	03 27 43	23 22 55	10 ♊ 10 48	17 ♊ 01 16	11 48	16 48	09 11	01 08	21 49	15 37	21 42	20 50
15	03 31 40	24 20 48	23 ♊ 45 46	00 ♋ 24 08	13 48	17 09	09 53	01 22	21 46	15 39	21 41	20 50
16	03 35 36	25 18 39	06 ♋ 56 26	13 ♋ 22 48	15 50	17 33	10 34	01 36	21 42	15 42	21 40	20 50
17	03 39 33	26 16 29	19 ♋ 43 33	25 ♋ 59 04	17 53	17 58	11 15	01 50	21 39	15 44	21 38	20 51
18	03 43 29	27 14 17	02 ♌ 09 49	08 ♌ 16 22	19 59	18 25	11 56	02 04	21 36	15 47	21 37	20 51
19	03 47 26	28 12 03	14 ♌ 19 18	20 ♌ 19 16	22 05	18 53	12 38	02 18	21 32	15 50	21 36	20 52
20	03 51 22	29 ♉ 09 47	26 ♌ 16 56	02 ♍ 12 57	24 13	19 23	13 19	02 32	21 29	15 52	21 35	20 52
21	03 55 19	00 ♊ 07 30	08 ♍ 08 01	14 ♍ 02 46	26 23	19 54	14 00	02 47	21 26	15 55	21 33	20 53
22	03 59 15	01 05 11	19 ♍ 57 52	25 ♍ 53 54	28 ♉ 33	20 27	14 41	03 01	21 23	15 58	21 32	20 53
23	04 03 12	02 02 51	01 ♎ 51 29	07 ♎ 51 07	00 ♊ 44	21 01	15 22	03 15	21 20	16 01	21 31	20 54
24	04 07 09	03 00 29	13 ♎ 53 16	19 ♎ 58 23	02 55	21 36	16 03	03 29	21 17	16 04	21 30	20 54
25	04 11 05	03 58 06	26 ♎ 06 48	02 ♏ 18 48	05 07	22 13	16 44	03 43	21 14	16 07	21 29	20 55
26	04 15 02	04 55 41	08 ♏ 34 34	14 ♏ 54 16	07 19	22 50	17 25	03 57	21 12	16 10	21 28	20 56
27	04 18 58	05 53 16	21 ♏ 17 57	27 ♏ 45 34	09 31	23 29	18 05	04 11	21 09	16 13	21 27	20 56
28	04 22 55	06 50 49	04 ♐ 17 04	10 ♐ 52 18	11 42	24 09	18 46	04 25	21 07	16 15	21 26	20 57
29	04 26 51	07 48 20	17 ♐ 31 04	24 ♐ 13 09	13 52	24 50	19 27	04 39	21 04	16 19	21 25	20 58
30	04 30 48	08 45 51	00 ♑ 58 19	07 ♑ 46 17	16 01	25 32	20 08	04 53	21 02	16 22	21 24	20 59
31	04 34 44	09 ♊ 43 21	14 ♑ 36 49	21 ♑ 29 40	18 ♊ 09	26 ♈ 15	20 ♊ 48	05 ♊ 07	21 ♎ 00	16 ♋ 25	21 ♎ 23	20 ♌ 59

Date	Moon True ☊ ° '	Moon Mean ☊ ° '	Moon ☽ Latitude ° '
01	07 ♒ 12	07 ♒ 42	04 S 18
02	07 R 07	07 39	03 36
03	06 58	07 35	02 42
04	06 55	07 32	01 38
05	06 54	07 29	00 S 28
06	06 D 54	07 26	00 N 45
07	06 R 54	07 23	01 57
08	06 52	07 19	03 02
09	06 48	07 16	03 56
10	06 40	07 13	04 36
11	06 31	07 10	04 58
12	06 20	07 07	05 00
13	06 09	07 04	04 43
14	05 59	07 00	04 10
15	05 51	06 57	03 23
16	05 45	06 54	02 26
17	05 42	06 51	01 24
18	05 41	06 48	00 N 19
19	05 D 41	06 45	00 S 46
20	05 41	06 41	01 48
21	05 R 41	06 38	02 44
22	05 39	06 35	03 33
23	05 35	06 32	04 13
24	05 29	06 29	04 43
25	05 20	06 25	05 00
26	05 10	06 22	05 03
27	05 00	06 19	04 52
28	04 50	06 16	04 25
29	04 41	06 13	03 44
30	04 35	06 10	02 50
31	04 ♒ 31	06 ♒ 06	01 S 45

DECLINATIONS

Date	Sun ☉ ° '	Moon ☽ ° '	Mercury ☿ ° '	Venus ♀ ° '	Mars ♂ ° '	Jupiter ♃ ° '	Saturn ♄ ° '	Uranus ♅ ° '	Neptune ♆ ° '	Pluto ⯓ ° '
01	15 N 04	25 S 55	04 N 50	08 N 46	20 N 39	19 N 05	06 S 15	23 N 01	06 S 58	23 N 34
02	15 22	26 45	05 29	08 31	20 48	19 08	06 14	23 00	06 58	23 34
03	15 40	26 04	06 09	08 17	20 57	19 11	06 13	23 00	06 57	23 34
04	15 58	23 52	06 50	08 04	21 06	19 14	06 11	23 00	06 56	23 33
05	16 15	20 16	07 32	07 52	21 15	19 17	06 10	23 00	06 56	23 33
06	16 32	15 29	08 14	07 41	21 23	19 21	06 08	22 59	06 55	23 33
07	16 49	09 49	08 57	07 32	21 32	19 24	06 07	22 59	06 55	23 33
08	17 05	03 S 34	09 40	07 23	21 40	19 27	06 06	22 59	06 54	23 33
09	17 21	02 N 56	10 25	07 16	21 48	19 30	06 04	22 59	06 54	23 32
10	17 37	09 19	11 09	07 10	21 55	19 33	06 03	22 58	06 53	23 32
11	17 53	15 12	11 54	07 05	22 03	19 36	06 02	22 58	06 53	23 32
12	18 08	20 10	12 39	07 01	22 10	19 39	06 00	22 58	06 52	23 31
13	18 23	23 54	13 24	06 58	22 17	19 42	05 59	22 57	06 52	23 31
14	18 37	26 06	14 09	06 56	22 24	19 45	05 58	22 57	06 51	23 31
15	18 52	26 41	14 55	06 55	22 31	19 48	05 57	22 57	06 51	23 31
16	19 06	25 42	15 40	06 54	22 38	19 51	05 56	22 56	06 50	23 30
17	19 20	23 23	16 24	06 55	22 44	19 54	05 55	22 56	06 50	23 30
18	19 33	19 59	17 08	06 57	22 50	19 56	05 54	22 56	06 50	23 30
19	19 46	15 48	17 51	06 59	22 56	19 59	05 53	22 56	06 49	23 29
20	19 59	11 04	18 34	07 02	23 02	20 02	05 52	22 55	06 49	23 29
21	20 11	05 59	19 15	07 06	23 07	20 05	05 51	22 55	06 48	23 29
22	20 23	00 N 42	19 55	07 11	23 12	20 08	05 50	22 55	06 48	23 28
23	20 35	04 S 37	20 33	07 17	23 17	20 11	05 49	22 54	06 47	23 28
24	20 46	09 49	21 10	07 23	23 22	20 13	05 48	22 54	06 47	23 28
25	20 57	14 44	21 44	07 30	23 27	20 16	05 47	22 53	06 47	23 27
26	21 07	19 09	22 17	07 37	23 31	20 19	05 46	22 53	06 46	23 27
27	21 18	22 47	22 47	07 46	23 36	20 21	05 45	22 53	06 46	23 26
28	21 28	25 21	23 15	07 54	23 40	20 24	05 45	22 52	06 46	23 26
29	21 37	26 35	23 40	08 04	23 43	20 27	05 44	22 52	06 45	23 26
30	21 46	26 16	24 03	08 13	23 47	20 29	05 43	22 52	06 45	23 25
31	21 N 55	24 S 23	24 N 23	08 N 24	23 N 50	20 N 32	05 S 43	22 N 51	06 S 45	23 N 25

ZODIAC SIGN ENTRIES

Date	h m	Planets
01	06 08	♂ ♊
03	03 55	☽ ♑
05	09 12	☽ ♒
07	12 46	☽ ♓
08	06 24	☿ ♉
09	14 49	☽ ♈
09	15 33	♃ ♊
11	16 12	☽ ♉
13	18 27	☽ ♊
15	23 16	☽ ♋
18	07 47	☽ ♌
20	19 31	☽ ♍
21	08 53	☉ ♊
23	03 58	☿ ♊
23	08 16	☽ ♎
25	19 32	☽ ♏
28	04 08	☽ ♐
30	10 17	☽ ♑

LATITUDES

Date	Mercury ☿ ° '	Venus ♀ ° '	Mars ♂ ° '	Jupiter ♃ ° '	Saturn ♄ ° '	Uranus ♅ ° '	Neptune ♆ ° '	Pluto ⯓ ° '
01	02 S 39	03 N 02	00 N 28	00 S 41	02 N 46	00 N 25	01 N 44	09 N 31
04	02 27	02 21	00 29	00 41	02 45	00 25	01 44	09 31
07	02 11	01 42	00 31	00 40	02 45	00 25	01 44	09 30
10	01 51	01 05	00 33	00 40	02 45	00 25	01 44	09 30
13	01 26	00 32	00 34	00 40	02 44	00 25	01 44	09 29
16	00 58	00 N 01	00 36	00 39	02 44	00 25	01 44	09 29
19	00 S 27	00 S 27	00 37	00 39	02 43	00 25	01 44	09 28
22	00 N 04	00 52	00 39	00 39	02 43	00 25	01 44	09 28
25	00 35	01 14	00 40	00 39	02 42	00 25	01 44	09 27
28	01 04	01 34	00 41	00 39	02 42	00 25	01 43	09 26
31	01 N 28	01 S 52	00 N 43	00 S 38	02 N 41	00 N 25	01 N 43	09 N 26

DATA

Julian Date	2434499
Delta T	+31 seconds
Ayanamsa	23° 12' 31"
Synetic vernal point	05° ♓ 54' 29"
True obliquity of ecliptic	23° 26' 49"

MOON'S PHASES, APSIDES AND POSITIONS ☽

Date	h m	Phase	Longitude ° '	Eclipse Indicator
06	12 21	☾	15 ♒ 40	
13	05 06	●	22 ♉ 08	
20	18 20	☽	29 ♌ 25	
28	17 03	○	07 ♐ 03	

Day	h m		
10	04 39	Perigee	
22	02 17	Apogee	
02	13 27	Max dec	26° S 45'
09	01 11	0N	
15	08 35	Max dec	26° N 42'
22	15 11	0S	
29	19 21	Max dec	26° S 39'

LONGITUDES

Date	Chiron ⚷ ° '	Ceres ⚳ ° '	Pallas ⚴ ° '	Juno ⚵ ° '	Vesta ⚶ ° '	Black Moon Lilith ⚸ ° '
01	21 ♑ 02	27 ♊ 59	13 ♉ 06	08 ♉ 45	18 ♒ 00	14 ♍ 16
11	20 ♑ 55	01 ♋ 55	18 ♉ 09	14 ♉ 39	21 ♒ 20	15 ♍ 22
21	20 ♑ 41	05 ♋ 59	23 ♉ 03	20 ♉ 34	24 ♒ 16	16 ♍ 29
31	20 ♑ 20	10 ♋ 09	28 ♉ 37	26 ♉ 29	26 ♒ 46	17 ♍ 36

ASPECTARIAN

h m	Aspects
01 Friday	
02 23	☽ ♇ ☿
03 21	♃ ± ♆
09 55	☽ ∠ ♆
10 04	♂ ∠ ♅
11 08	☽ ∠ ♄
11 25	♀ ∠ ♂
13 48	☽ ± ♅
17 18	☽ ⊼ ☉
17 23	♀ □ ♅
02 Saturday	
00 47	☽ △ ♀
00 52	☽ ⊼ ♅
05 12	☽ ± ☉
09 50	☽ △ ☿
11 13	☽ △ ⯓
13 24	☽ ✶ ♆
14 30	☽ ✶ ♄
20 20	⯓ St D
21 05	☿ △ ⯓
22 52	☽ ♇ ☉
03 Sunday	
01 09	☽ ⊼ ♃
06 27	☽ ⊼ ♂
08 53	♃ ± ♄
11 04	☽ Q ♆
12 04	☽ Q ♄
12 09	☽ ± ♃
13 52	☿ # ♄
14 15	☿ ☍ ♆
14 21	☽ ♇ ⯓
17 49	☽ ± ♂
22 04	☿ ☍ ♄
04 Monday	
00 51	☿ ⊥ ♃
03 54	☽ △ ☉
04 31	☽ ♇ ♃
06 25	☽ ± ⯓
06 43	☽ □ ♀
07 09	☽ ☍ ♅
10 43	☽ ♇ ♂
12 34	♀ St D
14 30	☽ # ⯓
15 38	☿ # ♆
17 03	☽ ⊼ ⯓
18 42	☽ # ♅
19 03	☽ □ ♆
19 57	☽ □ ♄
23 08	☽ □ ☿
05 Tuesday	
06 27	☽ # ♂
07 27	☽ △ ♃
14 25	☽ Q ♀
14 30	☽ △ ♂
17 18	☽ # ♃
19 54	☉ ⊻ ♀
21 08	☿ ∥ ♀
06 Wednesday	
02 47	☉ ✶ ♅
07 25	☽ # ☉
10 52	☽ Q ☿
11 17	☽ ✶ ♀
11 42	☽ ⊼ ♅
12 21	☽ □ ☉
21 09	☽ ☍ ⯓
22 00	☽ ± ♅
22 59	☽ △ ♆
23 42	☽ △ ♄
07 Thursday	
05 34	☿ ⊥ ♂
10 15	☽ ✶ ☿
11 55	☽ □ ♃
13 01	☽ ∠ ♀
13 20	☽ ♇ ♅
15 04	☽ # ☿
20 31	☽ □ ♂
21 10	☽ # ♀
21 11	☽ Q ☉
23 20	☽ ∥ ♆
08 Friday	
00 20	☽ ♇ ♆
00 57	☽ ♇ ♄
01 26	☿ ⊻ ♃
02 26	☽ ∥ ♄
04 18	☽ ⊥ ♀
14 26	☽ ⊻ ♀
14 34	☽ △ ♅
15 02	☽ ∠ ☿
15 21	☽ ± ♆
15 55	☽ ± ♄
18 36	☽ Q ♃
18 42	☽ ✶ ☉
23 36	☽ ⊼ ⯓
09 Saturday	
01 18	☽ ⊼ ♆
01 51	☽ ⊼ ♄
04 05	☽ Q ♂
06 15	♀ □ ♅
08 06	☽ ⊥ ☿
09 33	☽ ± ⯓
14 48	☽ ✶ ♃
19 27	☽ ⊻ ☿
21 17	☽ ∠ ☉
23 37	☽ # ♄
10 Sunday	
00 23	☽ ♇ ⯓
00 53	☽ ✶ ♂
02 44	☽ # ♆
03 35	☿ Q ♅
03 54	☽ ∥ ♀
13 09	☽ ⊥ ☉
15 54	☽ ∠ ♃
16 15	☽ □ ♅
16 38	☽ ☌ ♀
20 13	☽ ∥ ☿
23 43	☽ ⊻ ☉
11 Monday	
00 26	♂ ♇ ♆
01 04	☽ △ ⯓
02 40	☽ ☍ ♆
02 47	☽ ∠ ♂
03 04	☽ ☍ ♄
06 56	☽ ⊥ ♃
08 21	♂ ♇ ♄
17 00	☽ ⊻ ♃
18 21	☽ ⊥ ♂
18 44	♃ ∠ ♅
20 13	☉ □ ⯓
21 59	☽ Q ♅
12 Tuesday	
00 57	☽ ∥ ☉
04 12	☽ ☌ ☿
04 47	☽ ⊻ ♂
09 09	☽ ∥ ♃
10 43	☿ ⊻ ♂
18 00	☽ ✶ ♅
19 11	☽ ⊻ ♀
19 17	☉ ⊼ ♆
23 28	☉ ⊼ ♄
13 Wednesday	
00 08	☽ ∥ ♂
02 52	☽ □ ⯓
04 25	☽ ⊼ ♆
04 41	☽ ⊼ ♄
05 00	☽ ∥ ♅
05 06	☽ ☌ ☉
05 29	☽ ⊥ ♀
09 05	☽ ∥ ⯓
13 32	☉ ⊥ ♀
14 34	☽ ± ♆
14 49	☽ ± ♄
19 26	☽ ∠ ♅
20 07	☽ ☌ ♃
21 09	☽ ∠ ♀
23 21	♂ Q ⯓
14 Thursday	
05 59	☽ ♇ ♆
06 12	☽ ♇ ♄
09 39	☽ Q ⯓
10 11	☽ ☌ ♂
11 00	☽ ⊥ ♅
15 17	☽ ⊻ ☿
21 32	☽ ⊻ ♅
23 54	☽ ✶ ♀
15 Friday	
03 37	♂ ⊥ ♅
03 41	☽ ⊥ ☿
06 46	☽ ✶ ⯓
08 17	☽ △ ♆
08 26	☽ △ ♄
13 08	☽ ⊻ ☉
22 01	☽ Q ♀
22 43	☽ ∠ ☿
16 Saturday	
00 50	☽ ⊥ ☉
02 01	☽ ⊻ ♃
09 58	☽ ∠ ⯓
10 25	☿ ✶ ♅
13 15	☽ ⊥ ♃
18 46	☽ ∠ ☉
19 07	☽ ⊻ ♂
17 Sunday	
02 45	☽ ⊥ ⯓
04 25	☽ ☌ ♅
06 25	☽ ∠ ♃
07 02	☽ ⊥ ♂
07 50	☽ ✶ ☿
08 32	☽ □ ♀
10 59	☽ ∥ ⯓
13 09	☿ ⊻ ♀
14 08	☽ ⊻ ⯓
15 34	☽ ∥ ♅
15 39	☽ □ ♆
15 39	☽ □ ♄
16 59	☽ ∥ ♂
17 28	♄ ☌ ♆
18 Monday	
01 15	☽ ∠ ♂
01 37	☽ ✶ ☉
11 34	☽ Q ☿
11 49	☽ ✶ ♃
12 17	☽ ∥ ♃
14 35	☽ ∥ ☉
20 22	☉ ± ♄
21 18	☉ ± ♆
22 03	☿ □ ⯓
19 Tuesday	
02 20	☽ ∥ ☿
02 33	☽ Q ♄
02 38	☽ Q ♆
03 06	☽ Q ☉
05 55	☿ ⊼ ♄
06 30	☿ ⊼ ♆
08 25	☽ ✶ ♂
10 31	♂ ∥ ♅
11 58	☽ Q ♃
15 01	☽ ⊻ ♅
21 31	☽ △ ♀
20 Wednesday	
01 06	☽ ☌ ⯓
02 23	☽ ✶ ♄
02 32	☽ ✶ ♆
03 05	☽ ⊥ ♅
06 56	☽ □ ☿
09 55	☽ Q ♂
18 20	☽ □ ☉
20 51	☉ ∥ ♃
21 19	☽ ∠ ♅
21 Thursday	
00 55	☽ □ ♃
05 02	☿ ⊥ ♀
05 09	☽ ♇ ♀
06 53	☽ ∥ ♀
08 13	☽ # ♆
08 34	☽ ∠ ♄
08 48	☽ ∠ ♆
12 40	☽ # ♄
23 26	☿ ± ♄
22 Friday	
00 16	☽ ± ♀
00 38	☽ □ ♂
00 57	☿ ± ♆
02 45	☽ ⊥ ♄
03 02	☽ ⊥ ♆
03 52	☽ ✶ ♅
08 53	☉ ∠ ♅
13 02	☽ ⊼ ♀
13 52	☽ ⊻ ⯓
14 52	☽ ⊻ ♄
15 11	☽ ⊻ ♆
20 36	☿ ∥ ♃
23 Saturday	
02 00	☽ ⊥ ⯓
04 14	☽ Q ♅
07 03	♀ △ ⯓
09 14	☽ △ ☿
12 25	☽ △ ☉
13 19	☉ ∥ ☿
14 50	☽ △ ♃
15 10	☿ ∠ ♅
17 26	☽ ∥ ♄
20 06	☽ ∠ ⯓
21 54	☽ ∥ ♆
24 Sunday	
00 13	♀ ☍ ♄
00 25	☽ # ♀
08 00	♀ ☍ ♆
12 37	♂ ⊻ ♅
13 39	☉ ☌ ☿
16 19	☽ □ ♅
16 31	☽ △ ♂
18 49	☿ ☌ ♃
20 50	☽ ♇ ☉
21 15	☽ ♇ ♃
21 43	☽ ♇ ☿
25 Monday	
01 51	☽ ✶ ⯓
02 31	☽ ☌ ♄
02 58	☽ ☌ ♆
03 38	☉ ☌ ♃
03 59	☽ ☍ ♀
13 14	♂ ∥ ⯓
15 10	☽ ± ♃
15 54	☽ ± ☉
19 05	☽ ± ☿
23 30	☽ ♇ ♂
23 59	☿ ♇ ♄
26 Tuesday	
01 10	☽ Q ⯓
02 44	☿ ♇ ♆
02 59	☽ ⊼ ♃
04 27	☽ ⊼ ☉
09 05	☽ ⊼ ☿
17 42	☽ ± ♂
19 10	☽ # ♃
20 01	☿ ∠ ♀
27 Wednesday	
00 55	☽ # ☉
02 25	☽ △ ♅
05 39	☽ ⊼ ♂
05 43	☿ Q ⯓
11 20	☽ □ ⯓
11 44	☽ ⊻ ♄
12 04	☽ # ☿
12 16	☽ ⊻ ♆
12 45	☽ # ♅
16 17	☽ ⊼ ♀
16 28	☿ ∥ ♅
17 11	☽ # ⯓
18 20	☉ ♇ ♄
18 38	☽ # ♂
19 51	☿ ⊥ ♅
22 50	☽ ⊥ ♄
23 24	☽ ⊥ ♆
28 Thursday	
01 40	☉ ♇ ♆
04 00	☽ ± ♀
06 26	☽ ♇ ♅
12 15	☽ ☍ ♃
15 20	☽ ∠ ♄
15 55	☽ ∠ ♆
17 03	☽ ☍ ☉
21 21	☽ ♇ ♀
21 45	☿ ∥ ⯓
22 56	☽ ± ♅
29 Friday	
04 08	☽ ☍ ☿
09 49	☽ ⊼ ♅
15 25	☿ ∥ ♂
15 39	☽ ☍ ♂
18 11	☽ △ ⯓
18 21	☽ ✶ ♄
18 58	☽ ✶ ♆
30 Saturday	
01 48	☽ △ ♀
15 38	☽ Q ♄
15 58	☿ ⊻ ♅
16 17	☽ Q ♆
17 22	☉ Q ⯓
19 03	☽ ⊼ ♃
20 51	☽ ♇ ⯓
31 Sunday	
02 47	☽ ⊼ ☉
05 47	☽ ± ♃
11 57	☽ # ☿
12 39	☽ ± ⯓
13 29	♄ ✶ ⯓
14 05	☽ ± ☉
15 09	☽ ☍ ♅
16 33	☽ # ♂
18 19	♂ △ ♄
18 41	♂ ✶ ⯓
19 17	☽ ⊼ ☿
20 01	☽ # ⯓
21 47	☽ ♇ ♃
23 06	☽ □ ♄
23 08	☽ ⊼ ⯓
23 21	☽ ⊼ ♂
23 47	☽ □ ♆

All ephemeris data is given at 12.00 UT and the Moon's longitude is additionally given for 24.00 UT

JUNE 1953

LONGITUDES

Date	Sidereal time h m s	Sun ☉ ° ' "	Moon ☽ ° ' "	Moon ☽ 24.00 ° ' "	Mercury ☿ ° '	Venus ♀ ° '	Mars ♂ ° '	Jupiter ♃ ° '	Saturn ♄ ° '	Uranus ♅ ° '	Neptune ♆ ° '	Pluto ♇ ° '
01	04 38 41	10 ♊ 40 50	28 ♑ 24 37	05 ♒ 21 28	20 ♊ 15	26 ♈ 59	21 ♊ 29	05 ♊ 21	20 ♎ 57	16 ♋ 28	21 ♎ 22	21 ♌ 00
02	04 42 38	11 38 18	12 ♒ 20 02	19 ♒ 20 10	22 19	27 43	22 09	05 35	20 R 55	16 31	21 R 21	21 01
03	04 46 34	12 35 46	26 ♒ 21 45	03 ♓ 24 37	24 22	28 29	22 50	05 49	20 53	16 34	21 20	21 02
04	04 50 31	13 33 13	10 ♓ 28 39	17 ♓ 33 41	26 23	29 ♈ 16	23 31	06 03	20 51	16 37	21 19	21 03
05	04 54 27	14 30 39	24 ♓ 39 32	01 ♈ 45 58	28 ♊ 21	00 ♉ 03	24 11	06 17	20 49	16 40	21 18	21 04
06	04 58 24	15 28 04	08 ♈ 52 42	15 ♈ 59 25	00 ♋ 17	00 51	24 51	06 31	20 48	16 44	21 17	21 05
07	05 02 20	16 25 29	23 ♈ 05 43	00 ♉ 11 11	02 11	01 40	25 32	06 45	20 46	16 47	21 17	21 05
08	05 06 17	17 22 53	07 ♉ 15 19	14 ♉ 17 38	04 03	02 29	26 12	06 59	20 45	16 50	21 16	21 06
09	05 10 13	18 20 17	21 ♉ 17 36	28 ♉ 14 42	05 52	03 19	26 53	07 13	20 43	16 53	21 15	21 07
10	05 14 10	19 17 40	05 ♊ 08 27	11 ♊ 58 24	07 38	04 10	27 33	07 27	20 42	16 57	21 14	21 09
11	05 18 07	20 15 03	18 ♊ 44 12	25 ♊ 25 32	09 22	05 02	28 13	07 41	20 41	17 00	21 14	21 10
12	05 22 03	21 12 25	02 ♋ 02 10	08 ♋ 34 01	11 03	05 54	28 53	07 55	20 39	17 03	21 13	21 11
13	05 26 00	22 09 46	15 ♋ 01 02	21 ♋ 23 19	12 42	06 47	29 ♊ 34	08 09	20 38	17 07	21 12	21 12
14	05 29 56	23 07 06	27 ♋ 41 00	03 ♌ 54 22	14 19	07 40	00 ♋ 14	08 23	20 37	17 10	21 12	21 13
15	05 33 53	24 04 25	10 ♌ 03 44	16 ♌ 09 31	15 52	08 34	00 54	08 37	20 36	17 14	21 11	21 14
16	05 37 49	25 01 44	22 ♌ 12 09	28 ♌ 12 10	17 23	09 28	01 34	08 50	20 36	17 17	21 11	21 15
17	05 41 46	25 59 01	04 ♍ 10 07	10 ♍ 06 34	18 52	10 23	02 14	09 04	20 35	17 20	21 10	21 16
18	05 45 42	26 56 18	16 ♍ 02 09	21 ♍ 57 29	20 18	11 18	02 54	09 18	20 34	17 24	21 10	21 18
19	05 49 39	27 53 35	27 ♍ 53 11	03 ♎ 49 52	21 41	12 14	03 34	09 31	20 34	17 27	21 09	21 19
20	05 53 36	28 50 50	09 ♎ 48 10	15 ♎ 48 39	23 01	13 10	04 14	09 45	20 34	17 31	21 09	21 20
21	05 57 32	29 ♊ 48 05	21 ♎ 51 53	27 ♎ 58 24	24 18	14 07	04 54	09 59	20 33	17 34	21 09	21 21
22	06 01 29	00 ♋ 45 19	04 ♏ 08 39	10 ♏ 23 03	25 33	15 04	05 34	10 12	20 33	17 38	21 08	21 23
23	06 05 25	01 42 32	16 ♏ 41 56	23 ♏ 05 34	26 45	16 01	06 13	10 26	20 33	17 41	21 08	21 24
24	06 09 22	02 39 45	29 ♏ 34 08	06 ♐ 07 42	27 54	16 59	06 53	10 39	20 D 33	17 45	21 08	21 25
25	06 13 18	03 36 58	12 ♐ 46 16	19 ♐ 29 44	29 ♋ 00	17 58	07 33	10 53	20 33	17 49	21 07	21 27
26	06 17 15	04 34 10	26 ♐ 17 52	03 ♑ 10 24	00 ♌ 03	18 56	08 13	11 06	20 33	17 52	21 07	21 28
27	06 21 11	05 31 22	10 ♑ 06 58	17 ♑ 07 08	01 02	19 55	08 53	11 20	20 34	17 56	21 07	21 29
28	06 25 08	06 28 34	24 ♑ 10 24	01 ♒ 16 16	01 59	20 55	09 32	11 33	20 34	17 59	21 07	21 31
29	06 29 05	07 25 45	08 ♒ 24 11	15 ♒ 33 37	02 52	21 54	10 12	11 46	20 35	18 03	21 07	21 32
30	06 33 01	08 ♋ 22 57	22 ♒ 44 03	29 ♒ 54 58	03 ♌ 41	22 ♉ 55	10 ♋ 51	12 ♊ 00	20 ♎ 35	18 ♋ 06	21 ♎ 07	21 ♌ 34

Moon / DECLINATIONS

Date	Moon True ☊ ° '	Moon Mean ☊ ° '	Moon ☽ Latitude ° '	Sun ☉ ° '	Moon ☽ ° '	Mercury ☿ ° '	Venus ♀ ° '	Mars ♂ ° '	Jupiter ♃ ° '	Saturn ♄ ° '	Uranus ♅ ° '	Neptune ♆ ° '	Pluto ♇ ° '
01	04 ♒ 29	06 ♒ 03	00 S 33	22 N 03	21 S 01	24 N 40	08 N 35	23 N 53	20 N 35	05 S 42	22 N 51	06 S 44	23 N 24
02	04 D 29	06 00	00 N 42	22 11	16 26	24 55	08 46	23 56	20 37	05 41	22 50	06 44	23 24
03	04 30	05 57	01 55	22 19	10 56	25 06	08 58	23 59	20 40	05 41	22 50	06 44	23 23
04	04 31	05 54	03 01	22 26	04 S 51	25 15	09 10	24 02	20 42	05 40	22 50	06 43	23 23
05	04 R 30	05 50	03 57	22 33	01 N 30	25 22	09 22	24 04	20 45	05 40	22 49	06 43	23 22
06	04 28	05 47	04 38	22 39	07 46	25 26	09 35	24 06	20 47	05 39	22 49	06 43	23 22
07	04 24	05 44	05 02	22 45	13 39	25 27	09 48	24 08	20 50	05 39	22 48	06 43	23 21
08	04 18	05 41	05 07	22 51	18 46	25 26	10 02	24 10	20 52	05 39	22 48	06 42	23 21
09	04 11	05 38	04 54	22 56	22 49	25 23	10 15	24 11	20 54	05 38	22 48	06 42	23 20
10	04 04	05 35	04 24	23 01	25 29	25 18	10 29	24 12	20 57	05 38	22 47	06 42	23 20
11	03 57	05 31	03 39	23 05	26 37	25 10	10 44	24 13	20 59	05 38	22 47	06 42	23 19
12	03 52	05 28	02 43	23 09	26 09	25 01	10 58	24 14	21 01	05 38	22 46	06 41	23 19
13	03 48	05 25	01 40	23 13	24 15	24 50	11 13	24 15	21 04	05 38	22 46	06 41	23 18
14	03 47	05 22	00 N 33	23 16	21 10	24 38	11 27	24 15	21 06	05 37	22 46	06 41	23 18
15	03 D 47	05 19	00 S 34	23 19	17 11	24 24	11 42	24 16	21 08	05 37	22 45	06 41	23 17
16	03 48	05 16	01 39	23 21	12 34	24 08	11 58	24 16	21 10	05 37	22 45	06 41	23 17
17	03 49	05 12	02 38	23 23	07 32	23 51	12 13	24 15	21 12	05 37	22 44	06 41	23 16
18	03 51	05 09	03 30	23 25	02 N 17	23 33	12 28	24 15	21 14	05 37	22 44	06 40	23 16
19	03 R 51	05 06	04 13	23 26	03 S 01	23 14	12 44	24 14	21 17	05 37	22 43	06 40	23 15
20	03 51	05 03	04 45	23 26	08 15	22 55	12 59	24 14	21 19	05 37	22 43	06 40	23 15
21	03 49	05 00	05 05	23 27	13 14	22 34	13 15	24 13	21 21	05 38	22 42	06 40	23 14
22	03 45	04 56	05 12	23 27	17 47	22 12	13 30	24 11	21 23	05 38	22 42	06 40	23 14
23	03 41	04 53	05 04	23 26	21 41	21 50	13 46	24 10	21 25	05 38	22 41	06 40	23 13
24	03 36	04 50	04 41	23 25	24 38	21 28	14 01	24 08	21 27	05 38	22 41	06 40	23 13
25	03 31	04 47	04 02	23 24	26 20	21 05	14 17	24 06	21 29	05 39	22 41	06 40	23 12
26	03 27	04 44	03 09	23 22	26 33	20 41	14 32	24 04	21 31	05 39	22 40	06 40	23 12
27	03 24	04 41	02 04	23 20	25 07	20 18	14 48	24 02	21 32	05 39	22 40	06 40	23 11
28	03 23	04 37	00 S 50	23 17	22 07	19 54	15 03	24 00	21 34	05 40	22 39	06 40	23 11
29	03 D 22	04 34	00 N 28	23 14	17 43	19 30	15 19	23 57	21 36	05 40	22 39	06 40	23 10
30	03 ♒ 23	04 ♒ 31	01 N 44	23 N 11	12 S 18	19 N 07	15 N 34	23 N 54	21 N 38	05 S 41	22 N 38	06 S 40	23 N 09

ZODIAC SIGN ENTRIES

Date	h m	Planets
01	14 45	☽ ♒
03	18 12	☽ ♓
05	10 34	♀ ♉
05	21 01	☽ ♈
06	08 23	☿ ♋
07	23 41	☽ ♉
10	03 03	☽ ♊
12	08 17	☽ ♋
14	03 49	♂ ♋
14	16 27	☽ ♌
17	03 37	☽ ♍
19	16 16	☽ ♎
21	17 00	☉ ♋
22	03 57	☽ ♏
24	12 48	☽ ♐
26	11 01	☿ ♌
26	18 29	☽ ♑
28	21 51	☽ ♒

LATITUDES

Date	Mercury ☿ ° '	Venus ♀ ° '	Mars ♂ ° '	Jupiter ♃ ° '	Saturn ♄ ° '	Uranus ♅ ° '	Neptune ♆ ° '	Pluto ♇ ° '
01	01 N 35	01 S 57	00 N 43	00 S 38	02 N 41	00 N 25	01 N 43	09 N 25
04	01 52	02 12	00 44	00 38	02 40	00 25	01 43	09 25
07	02 01	02 24	00 46	00 38	02 39	00 25	01 43	09 24
10	02 04	02 35	00 47	00 38	02 38	00 25	01 43	09 24
13	02 00	02 43	00 48	00 37	02 38	00 25	01 43	09 23
16	01 50	02 50	00 49	00 37	02 37	00 25	01 43	09 23
19	01 34	02 55	00 50	00 37	02 36	00 25	01 42	09 22
22	01 11	02 59	00 52	00 37	02 35	00 25	01 42	09 22
25	00 44	03 01	00 53	00 37	02 35	00 25	01 42	09 22
28	00 N 11	03 02	00 54	00 37	02 34	00 25	01 42	09 21
31	00 S 26	03 S 02	00 N 55	00 S 36	02 N 33	00 N 25	01 N 42	09 N 21

DATA

Julian Date	2434530
Delta T	+31 seconds
Ayanamsa	23° 12' 36"
Synetic vernal point	05° ♓ 54' 24"
True obliquity of ecliptic	23° 26' 48"

MOON'S PHASES, APSIDES AND POSITIONS ☽

Date	h m	Phase	Longitude ° '	Eclipse Indicator
04	17 35	☾	13 ♓ 47	
11	14 55	●	20 ♊ 22	
19	12 01	☽	27 ♍ 54	
27	03 29	○	05 ♑ 11	

Day	h m		
05	13 39	Perigee	
18	21 06	Apogee	
05	06 21	ON	
11	16 53	Max dec	26° N 39'
18	22 19	OS	
26	03 11	Max dec	26° S 39'

LONGITUDES

Date	Chiron ⚷ ° '	Ceres ⚳ ° '	Pallas ⚴ ° '	Juno ⚵ ° '	Vesta ⚶ ° '	Black Moon Lilith ⚸ ° '
01	20 ♑ 18	10 ♋ 34	29 ♉ 09	27 ♉ 05	26 ♒ 55	17 ♍ 43
11	19 ♑ 50	14 ♋ 49	04 ♊ 33	03 ♊ 00	28 ♒ 52	18 ♍ 49
21	19 ♑ 18	19 ♋ 09	10 ♊ 05	08 ♊ 55	29 ♒ 52	19 ♍ 56
31	18 ♑ 43	23 ♋ 31	15 ♊ 37	14 ♊ 47	00 ♓ 16	21 ♍ 03

ASPECTARIAN

h m	Aspects
01 Monday	
00 07	☽ ⌗ ♅
05 46	☽ ⌗ ☉
06 13	☉ ⊥ ♅
06 55	☽ ⚼ ☉
07 35	☽ ± ☿
07 50	♂ △ ♆
09 23	☽ □ ♀
10 18	☽ ± ♂
14 34	☽ ⌗ ♃
20 00	☿ △ ♄
20 44	☿ ⚹ ♇
02 Tuesday	
00 12	☽ △ ♃
00 43	☿ △ ♆
01 53	☽ ⚼ ☿
02 39	☽ ⚼ ♂
09 06	☿ ☌ ♂
10 43	☽ △ ☉
18 09	☽ Q ♀
19 12	☽ ⚻ ♅
03 Wednesday	
02 40	☽ △ ♄
02 53	☽ ☍ ♇
03 25	☽ △ ♆
05 30	☽ ± ♅
05 40	☽ △ ♂
08 02	☽ △ ☿
15 50	☽ ⚹ ♀
17 41	♃ ⚼ ♄
19 43	☽ ⌗ ♀
20 54	☽ ⚼ ♅
04 Thursday	
04 10	☽ ⚼ ♄
04 22	☽ □ ♃
04 46	☽ ∥ ♆
04 57	☽ ⚼ ♆
08 49	☽ ∥ ♄
17 35	☽ □ ☉
18 47	☽ ∠ ♀
19 24	☽ ± ♄
20 11	☽ ± ♆
22 27	☽ △ ♅
05 Friday	
05 32	☽ ⚻ ♄
05 55	☽ ⚻ ♇
06 20	☽ ⚻ ♆
10 54	☽ ⊥ ♀
11 09	☽ □ ♂
11 22	☽ Q ♃
13 03	♃ ⚼ ♆
16 04	☽ ± ♇
19 14	☽ □ ☿
21 38	☽ ⊻ ♀
22 22	♀ ⊥ ♃
06 Saturday	
02 13	☽ Q ☉
03 49	☽ ⌗ ♄
07 16	☽ ⚼ ♇
07 52	☽ ⌗ ♆
07 58	☽ ⚹ ♃
19 03	☽ Q ♂
19 26	☽ ∥ ♀
23 55	☽ ⚹ ☉
07 Sunday	
00 15	☿ ⚹ ♀
01 18	☽ □ ♅
06 20	☽ Q ☿
08 05	☽ ☍ ♄
08 37	☽ △ ♇
08 56	☽ ☍ ♆
09 42	☽ ∠ ♃
16 19	☽ ⚹ ♂
21 28	☉ ⊻ ♅
08 Monday	
00 43	☉ ∥ ♅
01 11	☽ ⊥ ♃
03 07	☽ ∠ ☉
03 24	☽ ☌ ♀
05 43	☽ ⚹ ☿
07 52	☽ Q ♅
11 32	☽ ⊻ ♃
19 04	☽ ∠ ♂
19 32	☽ ⊥ ☉
23 37	☽ ∥ ♃
09 Tuesday	
04 25	☽ ⚹ ♅
06 33	☽ ⊻ ☉
09 12	☉ ∠ ♀
11 01	☽ ⚻ ♄
11 09	☽ ∠ ☿
11 15	☽ ⊥ ♂
11 43	☽ □ ♇
11 51	☽ ∥ ♅
11 56	☽ ⚻ ♆
12 52	☽ ∥ ☉
15 35	☿ ∠ ♇
15 49	☽ ∥ ♇
21 21	☽ ± ♄
22 07	☽ ⊻ ♂
22 16	☽ ± ♆
22 44	☽ ∥ ♂
10 Wednesday	
05 00	☽ ⊥ ☿
06 24	☽ ∠ ♅
09 09	☿ ⊻ ♃
09 46	☽ ∥ ☿
10 12	☽ ⊻ ♀
12 58	☽ ⚼ ♄
13 55	☽ ⚼ ♆
16 07	☽ ☌ ♃
17 00	☽ ⊻ ☿
19 01	☽ Q ♇
21 25	☽ ⊥ ♀
22 14	☽ ⊥ ♅
11 Thursday	
08 54	☽ ⊻ ♅
11 07	♀ Q ♅
14 28	☽ ∠ ♀
14 55	☽ ☌ ☉
15 28	☽ △ ♄
16 20	☽ ⚹ ♇
16 27	☽ △ ♆
22 26	☉ △ ♄
12 Friday	
05 58	☽ ☌ ♂
11 15	☉ ⚹ ♇
12 15	☉ △ ♆
19 35	☽ ⚹ ♀
19 36	☽ ∠ ♇
23 00	☽ ⊻ ♃
13 Saturday	
05 37	☽ ∥ ☿
07 03	☽ ☌ ☿
10 21	☽ ⊥ ♃
12 05	☽ ∥ ♂
12 20	☽ ⊥ ♇
15 57	☽ ☌ ♅
19 35	☽ Q ♀
20 21	☽ ∥ ♇
20 55	☽ ∥ ☉
21 03	♆ ⚹ ♇
22 34	☽ □ ♄
23 39	☽ □ ♆
23 39	☽ ⊻ ♇
14 Sunday	
00 38	☽ ∥ ♅
02 34	☽ ⊻ ☉
03 38	☽ ∠ ♃
12 30	☽ ∥ ♃
13 12	☿ ⊥ ♃
14 59	☽ ⊥ ☉
17 10	☽ ⊻ ♂
15 Monday	
01 05	☉ ∥ ♇
01 59	☿ ⊥ ♇
05 28	☽ ⊥ ♂
08 50	☽ □ ♀
09 06	☽ ⚹ ♃
09 09	☽ Q ♄
09 54	☽ ∠ ☉
10 17	☽ Q ♆
13 42	♀ ⊻ ♃
16 Tuesday	
00 09	☽ ∠ ♂
00 41	☿ ∥ ♂
01 05	☽ ⊻ ☿
02 11	☽ ⊻ ♅
08 48	☽ ⚹ ♄
09 14	☽ Q ♃
09 58	☽ ⚹ ♆
10 06	☽ ☌ ♇
10 14	☿ ☌ ♅
14 10	☽ ⊥ ♅
14 42	☽ ⊥ ☿
14 48	☽ ∥ ♀
18 08	☽ ⚹ ☉
17 Wednesday	
07 52	☽ ⚹ ♂
08 18	☽ ∠ ♅
11 18	☽ ∠ ☿
14 51	☽ ∠ ♄
15 58	☽ ⌗ ♆
16 02	☽ ∠ ♆
20 23	☽ Q ☉
20 49	☽ ⌗ ♄
22 05	☽ □ ♃
18 Thursday	
01 36	☽ △ ♀
09 02	☽ ⊥ ♄
09 34	☽ Q ♂
10 14	☽ ⊥ ♆
11 39	♃ Q ♇
14 46	☽ ⚹ ♅
16 48	☿ □ ♄
21 11	☽ ⊻ ♄
21 47	☽ ⚹ ☿
22 23	☽ ⊻ ♆
22 33	☉ ∥ ☿
22 40	☽ ⊻ ♇
19 Friday	
02 55	☿ □ ♆
05 32	☿ ⊻ ♇
10 34	☽ ⚼ ♀
10 50	☽ ⊥ ♇
11 00	☿ ∥ ♇
12 01	☽ □ ☉
15 11	☽ Q ♅
23 51	☽ ∥ ♄
20 Saturday	
00 09	☽ □ ♂
01 11	☽ Q ☿
04 40	☽ ∥ ♆
05 02	☽ ∠ ♇
06 16	☽ ± ♀
11 54	☽ △ ♃
19 18	☽ ⚻ ♀
21 Sunday	
01 52	☿ ∥ ♅
03 28	☽ □ ♅
09 25	☽ ☌ ♄
10 34	☽ ☌ ♆
11 00	☽ ⚹ ♇
12 03	☽ ⌗ ♀
17 22	☽ □ ☿
18 15	☽ ⚼ ♃
22 Monday	
03 42	☿ ∠ ♃
04 52	☽ △ ☉
10 31	☽ Q ♇
12 07	☽ ± ♃
14 53	☽ △ ♂
23 52	☽ ⚻ ♃
23 Tuesday	
10 08	☽ ⌗ ♃
10 37	☽ ☍ ♀
12 01	☽ ⚼ ☉
12 58	☽ ⌗ ☿
13 53	☽ △ ♅
17 26	♄ St D
18 31	♂ ∠ ♇
19 15	☽ ⊻ ♄
19 16	☽ ⌗ ♅
20 20	☽ ⊻ ♆
20 51	☽ □ ♇
20 58	☽ ⚼ ♂
23 21	☽ ⌗ ♇
24 Wednesday	
01 06	☽ ⌗ ☉
06 12	☽ ± ☉
06 26	☽ ⊥ ♄
07 19	☽ ⌗ ♂
07 30	☽ ⊥ ♆
08 37	☽ △ ☿
12 49	☿ ∥ ♃
14 33	☽ ± ♂
17 52	☽ ⚼ ♅
18 07	☽ ⚻ ☉
22 57	☽ ∠ ♄
25 Thursday	
00 00	☽ ∠ ♆
02 05	☽ ⚻ ♂
08 03	♀ ⚹ ♅
08 33	☽ ☍ ♃
10 16	☽ ± ♅
14 23	☽ ⚼ ☿
21 03	☽ ⚻ ♅
22 00	☽ ⚻ ♀
26 Friday	
01 53	☽ ⚹ ♄
02 53	☽ ⚹ ♆
03 29	☽ △ ♇
07 43	☽ ± ☿
09 26	☽ ± ♀
19 04	☽ ⚻ ☿
22 56	☽ Q ♄
23 54	☽ Q ♆
27 Saturday	
02 21	☽ ⚼ ♀
03 29	☽ ☍ ☉
05 44	☽ ⚼ ♇
09 45	☽ ☍ ♂
14 07	☽ ⚻ ♃
21 14	☽ ± ♇
22 16	☽ ⌗ ♂
28 Sunday	
00 33	☽ ± ♃
01 26	☽ ☍ ♅
03 40	♀ ⚻ ♄
03 50	☽ ⌗ ☉
04 48	☽ ⌗ ♇
05 53	☽ □ ♄
06 03	☽ △ ♀
06 48	☽ □ ♆
07 29	☽ ⚻ ♇
08 25	☽ ⌗ ♅
12 59	☉ ∠ ♇
15 20	☽ ⌗ ♃
16 06	☽ ⚼ ♃
16 51	♀ ⚻ ♆
29 Monday	
02 04	☽ ☍ ☿
02 10	☽ ⌗ ☿
02 52	♀ □ ♇
10 15	☽ ⚻ ☉
15 09	☽ ⚻ ♂
17 45	☽ △ ♃
21 02	☽ ± ☉
22 38	☽ ⌗ ♀
30 Tuesday	
01 42	☽ ± ♂
04 14	☽ ⚻ ♅
08 25	☽ △ ♄
09 17	☽ △ ♆
10 02	☽ ☍ ♇
12 19	☽ □ ♀
13 10	☽ ⚼ ☉
14 18	☽ ± ♅
17 28	☽ ⚼ ♂

All ephemeris data is given at 12.00 UT and the Moon's longitude is additionally given for 24.00 UT

JULY 1953

Date	Sidereal time h m s	Sun ☉ ° ' "	Moon ☽ ° ' "	Moon ☽ 24.00 ° ' "	Mercury ☿ ° '	Venus ♀ ° '	Mars ♂ ° '	Jupiter ♃ ° '	Saturn ♄ ° '	Uranus ♅ ° '	Neptune ♆ ° '	Pluto ♇ ° '
01	06 36 58	09 ♋ 20 08	07 ♓ 05 55	14 ♓ 16 27	04 ♌ 27	23 ♉ 55	11 ♋ 31	12 ♊ 13	20 ♎ 36	18 ♋ 10	21 ♎ 07	21 ♌ 35
02	06 40 54	10 17 20	21 ♓ 26 11	28 ♓ 34 46	05 09	24 56	12 11	12 26	20 37	18 14	21 R 06	21 37
03	06 44 51	11 14 32	05 ♈ 41 53	12 ♈ 47 16	05 48	25 57	12 50	12 39	20 38	18 17	21 D 06	21 38
04	06 48 47	12 11 44	19 ♈ 50 39	26 ♈ 51 49	06 22	26 58	13 30	12 52	20 39	18 21	21 07	21 40
05	06 52 44	13 08 57	03 ♉ 50 34	10 ♉ 46 42	06 52	28 00	14 09	13 05	20 40	18 25	21 07	21 41
06	06 56 40	14 06 09	17 ♉ 40 02	24 ♉ 30 25	07 19	29 ♉ 02	14 48	13 18	20 41	18 28	21 07	21 43
07	07 00 37	15 03 23	01 ♊ 17 40	08 ♊ 01 40	07 41	00 ♊ 04	15 28	13 31	20 42	18 32	21 07	21 44
08	07 04 34	16 00 36	14 ♊ 42 15	21 ♊ 19 21	07 58	01 06	16 07	13 44	20 44	18 36	21 07	21 46
09	07 08 30	16 57 50	27 ♊ 52 51	04 ♋ 22 42	08 11	02 09	16 47	13 57	20 45	18 39	21 07	21 48
10	07 12 27	17 55 04	10 ♋ 48 51	17 ♋ 11 21	08 19	03 12	17 26	14 10	20 47	18 43	21 07	21 49
11	07 16 23	18 52 19	23 ♋ 30 13	29 ♋ 45 33	08 22	04 15	18 05	14 23	20 49	18 46	21 08	21 51
12	07 20 20	19 49 33	05 ♌ 57 30	12 ♌ 06 15	08 R 21	05 19	18 45	14 35	20 50	18 50	21 08	21 52
13	07 24 16	20 46 48	18 ♌ 12 01	24 ♌ 15 06	08 15	06 22	19 24	14 48	20 52	18 54	21 08	21 54
14	07 28 13	21 44 03	00 ♍ 15 49	06 ♍ 14 32	08 04	07 26	20 03	15 01	20 54	18 57	21 09	21 56
15	07 32 09	22 41 18	12 ♍ 11 41	18 ♍ 07 41	07 49	08 30	20 42	15 13	20 56	19 01	21 09	21 57
16	07 36 06	23 38 33	24 ♍ 03 02	29 ♍ 58 14	07 29	09 35	21 21	15 25	20 59	19 05	21 09	21 59
17	07 40 03	24 35 48	05 ♎ 53 51	11 ♎ 50 25	07 04	10 39	22 01	15 38	21 01	19 08	21 10	22 01
18	07 43 59	25 33 03	17 ♎ 48 30	23 ♎ 48 41	06 36	11 44	22 40	15 50	21 03	19 12	21 10	22 03
19	07 47 56	26 30 19	29 ♎ 51 34	05 ♏ 57 41	06 04	12 49	23 19	16 02	21 06	19 16	21 11	22 04
20	07 51 52	27 27 35	12 ♏ 07 35	18 ♏ 21 47	05 29	13 54	23 58	16 14	21 08	19 19	21 11	22 06
21	07 55 49	28 24 51	24 ♏ 40 46	01 ♐ 04 57	04 51	14 59	24 37	16 27	21 11	19 23	21 12	22 08
22	07 59 45	29 ♋ 22 07	07 ♐ 34 39	14 ♐ 10 10	04 10	16 04	25 16	16 39	21 14	19 26	21 13	22 10
23	08 03 42	00 ♌ 19 24	20 ♐ 51 38	27 ♐ 39 06	03 29	17 10	25 55	16 51	21 16	19 30	21 13	22 11
24	08 07 38	01 16 42	04 ♑ 32 31	11 ♑ 31 40	02 46	18 16	26 34	17 02	21 19	19 34	21 14	22 13
25	08 11 35	02 13 59	18 ♑ 36 15	25 ♑ 45 46	02 03	19 22	27 13	17 14	21 22	19 37	21 15	22 15
26	08 15 32	03 11 18	02 ♒ 59 39	10 ♒ 17 13	01 21	20 28	27 51	17 26	21 25	19 41	21 15	22 17
27	08 19 28	04 08 37	17 ♒ 37 40	25 ♒ 00 09	00 41	21 34	28 30	17 38	21 29	19 44	21 16	22 19
28	08 23 25	05 05 57	02 ♓ 23 48	09 ♓ 47 41	00 ♌ 03	22 41	29 09	17 49	21 32	19 48	21 17	22 20
29	08 27 21	06 03 17	17 ♓ 10 57	24 ♓ 32 47	29 ♋ 27	23 47	29 ♋ 48	18 01	21 35	19 52	21 18	22 22
30	08 31 18	07 00 39	01 ♈ 52 24	09 ♈ 09 09	28 55	24 54	00 ♌ 27	18 12	21 39	19 55	21 19	22 24
31	08 35 14	07 ♌ 58 01	16 ♈ 22 29	23 ♈ 31 58	28 ♋ 28	26 ♊ 01	01 ♌ 06	18 ♊ 23	21 ♎ 42	19 ♋ 59	21 ♎ 20	22 ♌ 26

Date	Moon True ☊ ° '	Moon Mean ☊ ° '	Moon ☽ Latitude ° '
01	03 ♒ 25	04 ♒ 28	02 N 55
02	03 D 26	04 25	03 54
03	03 27	04 22	04 39
04	03 R 27	04 18	05 06
05	03 26	04 15	05 15
06	03 24	04 12	05 05
07	03 21	04 09	04 38
08	03 19	04 06	03 57
09	03 17	04 02	03 03
10	03 15	03 59	02 01
11	03 14	03 56	00 N 54
12	03 D 14	03 53	00 S 15
13	03 14	03 50	01 22
14	03 15	03 47	02 24
15	03 16	03 43	03 19
16	03 17	03 40	04 06
17	03 18	03 37	04 42
18	03 18	03 34	05 06
19	03 R 19	03 31	05 16
20	03 18	03 28	05 13
21	03 18	03 24	04 55
22	03 17	03 21	04 22
23	03 17	03 18	03 34
24	03 17	03 15	02 33
25	03 17	03 12	01 21
26	03 D 17	03 08	00 S 02
27	03 R 17	03 05	01 N 19
28	03 17	03 02	02 35
29	03 16	02 59	03 40
30	03 16	02 56	04 31
31	03 ♒ 16	02 ♒ 53	05 N 03

DECLINATIONS

Date	Sun ☉ ° '	Moon ☽ ° '	Mercury ☿ ° '	Venus ♀ ° '	Mars ♂ ° '	Jupiter ♃ ° '	Saturn ♄ ° '	Uranus ♅ ° '	Neptune ♆ ° '	Pluto ♇ ° '
01	23 N 07	06 S 12	18 N 44	15 N 49	23 N 51	21 N 40	05 S 41	22 N 38	06 S 40	23 N 08
02	23 03	00 N 12	18 21	16 04	23 48	21 42	05 42	22 37	06 40	23 08
03	22 58	06 32	17 58	16 19	23 45	21 43	05 42	22 37	06 40	23 07
04	22 53	12 29	17 36	16 33	23 41	21 45	05 43	22 36	06 40	23 07
05	22 48	17 44	17 15	16 48	23 37	21 47	05 43	22 36	06 40	23 06
06	22 42	21 59	16 54	17 02	23 33	21 48	05 44	22 35	06 40	23 05
07	22 36	24 58	16 34	17 16	23 29	21 50	05 45	22 35	06 40	23 05
08	22 29	26 29	16 15	17 30	23 25	21 52	05 46	22 34	06 40	23 04
09	22 22	26 29	15 58	17 44	23 20	21 53	05 46	22 34	06 41	23 04
10	22 15	25 01	15 41	17 57	23 16	21 55	05 47	22 33	06 41	23 03
11	22 07	22 17	15 26	18 10	23 11	21 56	05 48	22 33	06 41	23 02
12	21 59	18 33	15 12	18 23	23 06	21 58	05 49	22 32	06 41	23 02
13	21 50	14 05	15 00	18 35	23 01	21 59	05 50	22 32	06 41	23 01
14	21 41	09 08	14 49	18 48	22 55	22 01	05 51	22 31	06 41	23 00
15	21 32	03 N 55	14 40	19 00	22 50	22 02	05 52	22 31	06 42	23 00
16	21 23	01 S 24	14 33	19 11	22 44	22 03	05 53	22 30	06 42	22 59
17	21 13	06 39	14 28	19 22	22 38	22 05	05 54	22 30	06 42	22 59
18	21 02	11 41	14 25	19 33	22 32	22 06	05 55	22 29	06 42	22 58
19	20 52	16 21	14 23	19 44	22 25	22 07	05 56	22 29	06 42	22 57
20	20 40	20 27	14 23	19 54	22 19	22 09	05 58	22 28	06 43	22 57
21	20 29	23 43	14 26	20 04	22 12	22 10	05 59	22 28	06 43	22 56
22	20 17	25 53	14 30	20 13	22 06	22 11	06 00	22 27	06 43	22 55
23	20 05	26 42	14 35	20 22	21 59	22 12	06 01	22 27	06 43	22 55
24	19 53	25 55	14 43	20 30	21 51	22 13	06 03	22 26	06 44	22 54
25	19 40	23 29	14 52	20 38	21 44	22 15	06 04	22 26	06 44	22 54
26	19 27	19 31	15 02	20 46	21 37	22 16	06 05	22 25	06 44	22 53
27	19 14	14 18	15 14	20 53	21 29	22 17	06 07	22 25	06 45	22 52
28	19 00	08 13	15 27	21 00	21 21	22 18	06 08	22 24	06 45	22 52
29	18 46	01 S 41	15 40	21 06	21 14	22 19	06 10	22 24	06 45	22 51
30	18 32	04 N 53	15 55	21 12	21 05	22 20	06 11	22 23	06 46	22 50
31	18 N 17	11 N 06	16 N 10	21 N 17	20 N 57	22 N 21	06 S 13	22 N 23	06 S 46	22 N 50

ZODIAC SIGN ENTRIES

Date	h	m	Planets
01	00	08	☽ ♓
03	02	23	☽ ♈
05	05	23	☽ ♉
07	09	42	☽ ♊
07	10	30	♀ ♊
09	15	54	☽ ♋
12	00	28	☽ ♌
14	11	28	☽ ♍
17	00	04	☽ ♎
19	12	17	☽ ♏
21	21	59	☽ ♐
23	03	52	☉ ♌
24	04	07	☽ ♑
26	07	03	☽ ♒
28	08	07	☽ ♓
28	13	40	☿ ♋
29	19	25	♂ ♌
30	08	56	☽ ♈

LATITUDES

Date	Mercury ☿ ° '	Venus ♀ ° '	Mars ♂ ° '	Jupiter ♃ ° '	Saturn ♄ ° '	Uranus ♅ ° '	Neptune ♆ ° '	Pluto ♇ ° '
01	00 S 26	03 S 02	00 N 55	00 S 36	02 N 33	00 N 25	01 N 42	09 N 21
04	01 07	03 00	00 56	00 36	02 32	00 25	01 42	09 21
07	01 51	02 58	00 57	00 36	02 31	00 25	01 42	09 20
10	02 36	02 54	00 58	00 36	02 31	00 25	01 41	09 20
13	03 19	02 50	00 59	00 36	02 30	00 25	01 41	09 20
16	03 59	02 44	00 59	00 36	02 29	00 25	01 41	09 20
19	04 31	02 38	01 00	00 36	02 28	00 25	01 41	09 20
22	04 52	02 31	01 01	00 36	02 28	00 25	01 41	09 19
25	04 58	02 23	01 02	00 36	02 27	00 25	01 41	09 19
28	04 49	02 15	01 03	00 36	02 26	00 25	01 40	09 19
31	04 S 24	02 S 06	01 N 04	00 S 35	02 N 25	00 N 25	01 N 40	09 N 19

DATA

Julian Date	2434560
Delta T	+31 seconds
Ayanamsa	23° 12' 42"
Synetic vernal point	05° ♓ 54' 18"
True obliquity of ecliptic	23° 26' 48"

MOON'S PHASES, APSIDES AND POSITIONS ☽

Date	h	m	Phase	Longitude ° '	Eclipse Indicator
03	22	03	☾	11 ♈ 38	
11	02	28	●	18 ♋ 30	Partial
19	04	47	☽	26 ♎ 13	
26	12	21	○	03 ♒ 12	total

Day	h	m		
01	00	24	Perigee	
16	15	15	Apogee	
28	13	52	Perigee	
02	11	17	ON	
08	23	49	Max dec	26° N 40'
16	05	41	OS	
23	12	32	Max dec	26° S 42'
29	18	06	ON	

LONGITUDES

Date	Chiron ⚷ ° '	Ceres ⚳ ° '	Pallas ⚴ ° '	Juno ⚵ ° '	Vesta ⚶ ° '	Black Moon Lilith ⚸ ° '
01	18 ♑ 43	23 ♋ 31	15 ♊ 37	14 ♊ 47	00 ♓ 16	21 ♍ 03
11	18 ♑ 06	27 ♋ 57	21 ♊ 16	20 ♊ 37	29 ♒ 53	22 ♍ 10
21	17 ♑ 30	02 ♌ 24	26 ♊ 58	26 ♊ 24	28 ♒ 43	23 ♍ 17
31	16 ♑ 55	06 ♌ 53	02 ♋ 43	02 ♋ 05	26 ♒ 51	24 ♍ 24

All ephemeris data is given at 12.00 UT and the Moon's longitude is additionally given for 24.00 UT

ASPECTARIAN

01 Wednesday
02 51 ☉ ∥ ♇
04 55 ♃ ⊥ ♅
05 24 ☽ ⚼ ♅
07 20 ☽ ⚻ ☿
09 30 ☽ ⚼ ♄
10 13 ☽ ∥ ♆
10 21 ☽ ⚼ ♆
13 57 ☽ ∥ ♄
16 00 ☽ △ ☉
17 54 ☽ ± ☿
19 45 ☽ △ ♂
20 40 ☽ Q ♀
20 41 ☽ □ ♃

02 Thursday
00 34 ☽ ± ♄
01 24 ☽ ± ♆
06 36 ☽ △ ♅
09 45 ☽ ⚼ ☿
10 37 ☽ ⚻ ♄
11 27 ☽ ⚻ ♆
12 18 ☽ ⚻ ♇
18 19 ☽ ✶ ♀
22 14 ♆ St D
22 23 ☽ ± ♇

03 Friday
02 09 ♂ ⊻ ♃
03 22 ☽ Q ♃
08 48 ☽ # ♄
12 10 ☽ △ ☿
12 32 ☽ # ♆
13 35 ☽ ⚼ ♇
21 34 ☽ ∠ ♀
22 03 ☽ □ ☉
23 58 ☽ ✶ ♃

04 Saturday
00 40 ☽ □ ♂
04 19 ♀ ± ♄
09 27 ☽ □ ♅
13 22 ☽ ☍ ♄
14 04 ☽ ⊥ ♀
14 09 ☽ ☍ ♆
15 06 ☽ △ ♇
15 16 ♀ ± ♆

05 Sunday
01 09 ☽ ⊻ ♀
01 57 ☽ ∠ ♃
07 01 ☽ Q ☉
07 11 ☽ ∥ ♀
08 56 ☽ Q ♂
09 44 ☽ ∥ ☿
10 05 ☉ ⊻ ♃
16 27 ☽ Q ♅
17 25 ☽ □ ☿
17 42 ☽ ⊥ ♃

06 Monday
04 16 ☽ ⊻ ♃
05 19 ☽ ✶ ☉
06 17 ☿ ∥ ♀
06 46 ☽ ✶ ♂
10 50 ☽ ∥ ♃
13 25 ☽ ✶ ♅
16 05 ☽ ∥ ♅
16 44 ☽ ∥ ☉
17 18 ☽ ⚻ ♄
18 02 ☽ ⚻ ♆
19 06 ☽ □ ♇
19 40 ☽ ∥ ♇
23 02 ☽ ∥ ♂

07 Tuesday
01 48 ☽ Q ☿
03 52 ☽ ± ♄
04 36 ☽ ± ♆
09 28 ☉ ∠ ♀
09 38 ☽ ∠ ☉
09 38 ☽ ☌ ♀
10 27 ☽ ∠ ♂
15 57 ☉ ∥ ♅
16 00 ☽ ∠ ♅
19 52 ☽ ⚼ ♄
20 35 ☽ ⚼ ♆
22 26 ♂ ⊥ ♇
23 38 ☽ ✶ ☿

08 Wednesday
02 54 ☽ ⊥ ☉
03 06 ☽ Q ♇
03 20 ☽ ⊥ ♂
05 41 ☉ ⊥ ♇
08 11 ☽ ⊥ ♅
10 14 ☽ ☌ ♃
12 58 ♀ ∠ ♂
14 33 ☽ ⊻ ☉
14 42 ☽ ⊻ ♂
19 04 ☽ ⊻ ♅
21 00 ☉ ☌ ♂
22 57 ☽ △ ♄
23 38 ☽ △ ♆

09 Thursday
00 50 ☽ ✶ ♇
03 16 ☽ ∠ ☿
20 02 ☽ ⊥ ☿
20 34 ☽ ⊻ ♀

10 Friday
04 31 ☽ ∠ ♇
07 17 ☽ ⊻ ☿
08 43 ☽ ⊥ ♀
18 24 ☽ ⊻ ♃
21 26 ☽ ⊥ ♇

11 Saturday
00 26 ♀ ∠ ♅
01 08 ☽ ☌ ♂
02 28 ☽ ☌ ☉
02 58 ☽ ☌ ♅
03 11 ☽ ∠ ♀
04 53 ☽ ∥ ♂
05 57 ☽ ⊥ ♃
06 13 ☽ ∥ ♇
06 52 ☽ □ ♄
07 28 ☽ □ ♆
08 50 ☽ ⊻ ♇
09 23 ☉ ☌ ♅
10 01 ☽ ∥ ♅
13 14 ☽ ∥ ☉
14 28 ☽ ∥ ♃
17 25 ☿ St R
23 28 ☽ ∠ ♃

12 Sunday
10 37 ☽ ✶ ♀
12 53 ☽ ∥ ♀
15 06 ☉ ∥ ♃
15 45 ♂ ☌ ♅
16 38 ☽ ☌ ☿
17 38 ☽ Q ♄
18 11 ☽ Q ♆

13 Monday
00 23 ♀ ⚼ ♄
05 11 ☽ ✶ ♃
06 42 ♀ ⚼ ♆
07 06 ☽ ∥ ☿
09 47 ♂ ∥ ♇
12 22 ☽ Q ♀
12 37 ☉ ⊥ ♃
13 23 ☽ ⊻ ♅
14 26 ☉ □ ♄
14 30 ☽ ⊻ ♂
17 18 ☽ ✶ ♄
17 33 ☽ ⊻ ☉
17 49 ☽ ✶ ♆
19 21 ☽ ☌ ♇
21 04 ☉ □ ♆

14 Tuesday
01 21 ☽ ⊥ ♅
03 06 ☽ ⊥ ♂
05 23 ☽ Q ♃
06 30 ☽ ⊥ ☉
17 04 ☉ ⊻ ♇
19 27 ☽ ∠ ♅
22 10 ☽ ∠ ♂
23 20 ☽ # ♆
23 21 ☽ ∠ ♄
23 39 ☿ ✶ ♀
23 49 ☽ ∠ ♆

15 Wednesday
02 07 ☽ ∠ ☉
03 08 ☽ # ♄
03 22 ☽ ⊻ ☿
03 49 ☽ □ ♀
15 11 ☽ ⊥ ☿
17 34 ☽ ⊥ ♄
17 59 ☽ ⊥ ♆
18 13 ☽ □ ♃
21 14 ♂ □ ♄

16 Thursday
01 52 ☽ ✶ ♅
04 38 ♂ □ ♆
05 45 ☽ ⊻ ♄
06 08 ☽ ⊻ ♆
06 13 ☽ ✶ ♂
07 48 ☽ ⊻ ♇
08 54 ☽ ∠ ☿
11 06 ☽ ✶ ☉
15 39 ♂ ⊥ ♃
20 00 ☽ ⊥ ♇
21 26 ♀ Q ♇

17 Friday
02 19 ☽ Q ♅
07 57 ☽ Q ♂
08 32 ☽ ∥ ♄
12 14 ☽ ∥ ♆
12 15 ♂ ⊻ ♇
13 32 ☽ Q ☉
14 16 ☽ ∠ ♇
14 17 ☽ ✶ ☿
22 33 ☽ △ ♀

18 Saturday
07 58 ☽ △ ♃
13 31 ☽ Q ☿
14 48 ☽ □ ♅
18 31 ☽ ☌ ♄
18 45 ☽ ☌ ♆
20 30 ☽ ✶ ♇
22 10 ♂ ∥ ♅
22 16 ☽ □ ♂

19 Sunday
01 35 ☽ # ☿
04 47 ☽ □ ☉
07 33 ☽ ⚼ ♀
14 22 ☽ ⚼ ♃
20 19 ☽ Q ♇
22 34 ♀ ⊥ ♅
23 39 ☽ □ ☿

20 Monday
02 59 ☽ ± ♀
08 17 ☽ ± ♃
08 23 ☽ # ♀
13 26 ☽ # ☉
15 45 ☽ ⚻ ♀
20 04 ☽ ⚻ ♃
23 41 ☽ # ♃

21 Tuesday
00 27 ☽ # ♂
01 54 ☽ △ ♅
01 59 ☽ # ♅
05 21 ☽ ⊻ ♄
05 24 ☽ ⊻ ♆
05 36 ☽ # ♇
07 10 ☽ □ ♇
11 52 ☽ △ ♂
16 44 ☽ ⊥ ♄
16 45 ☽ ⊥ ♆
19 35 ☽ △ ☉
19 36 ♂ ∥ ♃

22 Wednesday
01 22 ♄ ☌ ♆
06 03 ☽ △ ☿
06 12 ☽ ⚼ ♅
09 29 ☽ ∠ ♆
09 31 ☽ ∠ ♄
17 08 ☉ ∥ ♀
17 10 ☽ ⚼ ♂
22 44 ☽ ± ♅

23 Thursday
01 19 ☽ ⚼ ☉
03 19 ♀ ☌ ♃
04 42 ☽ ☍ ♃
04 48 ☽ ☍ ♀
07 57 ☽ ⚼ ☿
09 34 ☽ ⚻ ♅
10 13 ☽ ± ♂
12 39 ☽ ✶ ♆
12 44 ☽ ✶ ♄
14 22 ☽ △ ♇
18 36 ☽ ± ☉
21 23 ☽ ⚻ ♂
23 07 ☽ ± ☿

24 Friday
05 25 ☿ ∠ ♀
05 55 ☽ ⚻ ☉
09 05 ☽ ⚻ ☿
09 44 ☽ Q ♆
09 53 ☽ Q ♄
16 38 ☽ ⚼ ♇

25 Saturday
07 13 ☿ ∠ ♃
08 01 ☽ ± ♇
09 27 ☉ ☌ ☿
09 40 ☽ ⚻ ♃
12 07 ☉ ∠ ♃
13 23 ☽ ⚻ ♀
13 43 ☽ ☍ ♅
16 14 ☽ # ♇
16 27 ☽ □ ♆
16 40 ☽ □ ♄
17 57 ♀ ⊻ ♅
18 08 ☽ ⚻ ♇
19 20 ☽ # ♅
19 53 ☽ ± ♃
20 28 ☽ # ♃

26 Sunday
00 02 ☽ # ♂
00 16 ☽ ± ♀
03 05 ☽ ☍ ♂
05 30 ☽ # ♀
09 25 ☽ ☍ ☿
11 03 ☽ ⚼ ♃
12 21 ☽ ☍ ☉
12 23 ☽ # ☉
16 25 ☽ ⚼ ♀

27 Monday
05 25 ♀ △ ♆
08 12 ☽ # ☿
09 52 ♀ △ ♄
12 00 ☽ △ ♃
15 27 ☽ ⚻ ♅
17 56 ☽ △ ♆
18 17 ☽ △ ♄
18 57 ☽ △ ♀
19 38 ☽ ☍ ♇

28 Tuesday
01 15 ☽ ± ♅
04 28 ♀ ✶ ♇
06 30 ☽ ⚻ ♂
08 20 ☽ ⚻ ☿
15 55 ☽ ⚼ ♅
16 41 ☽ ± ♂
16 41 ☽ ⚻ ☉
17 28 ☽ ∥ ♆
17 41 ☽ ± ☿
18 19 ☽ ⚼ ♆
18 44 ☽ ⚼ ♄
19 43 ☽ ∥ ♄

29 Wednesday
03 06 ☽ ± ☉
05 09 ☿ ☌ ♂
07 11 ☿ ⊥ ♀
07 44 ☽ ⚼ ☿
07 57 ☽ ⚼ ♂
08 56 ☽ ± ♆
09 24 ☽ ± ♄
12 32 ♀ ⊥ ♂
13 22 ☽ □ ♃
16 22 ☽ △ ♅
18 42 ☽ ⚻ ♆
18 44 ☽ ⚼ ☉
19 12 ☽ ⚻ ♄
20 28 ☽ ⚻ ♇
23 39 ☽ □ ♀

30 Thursday
01 13 ♀ ∥ ♂
06 18 ☽ ± ♇
07 19 ☽ △ ☿
09 33 ☽ △ ♂
16 53 ☽ # ♄
19 03 ☽ # ♆
19 13 ☽ Q ♃
21 03 ☽ △ ☉
21 07 ☽ ⚼ ♇

31 Friday
07 45 ☽ Q ♀
15 25 ☽ ✶ ♃
18 03 ☽ □ ♅
20 18 ☽ ☍ ♆
20 57 ☽ ☍ ♄
22 10 ☽ △ ♇

AUGUST 1953

LONGITUDES

Date	Sidereal time h m s	Sun ☉ ° ′ ″	Moon ☽ ° ′ ″	Moon ☽ 24.00 ° ′ ″	Mercury ☿ ° ′	Venus ♀ ° ′	Mars ♂ ° ′	Jupiter ♃ ° ′	Saturn ♄ ° ′	Uranus ♅ ° ′	Neptune ♆ ° ′	Pluto ♇ ° ′
01	08 39 11	08 ♌ 55 25	00 ♉ 37 14	07 ♉ 38 05	28 ♋ 05	27 ♊ 08	01 ♌ 44	18 ♊ 35	21 ♎ 46	20 ♋ 02	21 ♎ 21	22 ♌ 28
02	08 43 07	09 52 50	14 ♉ 34 22	21 ♉ 26 03	27 R 48	28 16	02 23	18 46	21 49	20 06	21 21	22 30
03	08 47 04	10 50 17	28 ♉ 13 09	04 ♊ 55 46	27 36	29 ♊ 23	03 02	18 57	21 53	20 09	21 22	22 32
04	08 51 01	11 47 44	11 ♊ 34 01	18 ♊ 08 05	27 31	00 ♋ 31	03 40	19 08	21 57	20 13	21 23	22 33
05	08 54 57	12 45 13	24 ♊ 38 09	01 ♋ 04 24	27 D 31	01 38	04 19	19 19	22 01	20 16	21 25	22 35
06	08 58 54	13 42 43	07 ♋ 27 03	13 ♋ 46 20	27 38	02 46	04 58	19 29	22 05	20 19	21 26	22 37
07	09 02 50	14 40 14	20 ♋ 02 25	26 ♋ 15 32	27 52	03 54	05 36	19 40	22 09	20 23	21 27	22 39
08	09 06 47	15 37 47	02 ♌ 25 52	08 ♌ 33 38	28 13	05 02	06 15	19 51	22 13	20 26	21 28	22 41
09	09 10 43	16 35 20	14 ♌ 39 01	20 ♌ 42 13	28 41	06 11	06 54	20 01	22 17	20 30	21 29	22 43
10	09 14 40	17 32 54	26 ♌ 43 29	02 ♍ 43 00	29 15	07 19	07 32	20 11	22 22	20 33	21 30	22 45
11	09 18 36	18 30 30	08 ♍ 41 03	14 ♍ 37 51	29 ♋ 56	08 27	08 11	20 22	22 26	20 36	21 31	22 47
12	09 22 33	19 28 07	20 ♍ 33 43	26 ♍ 28 58	00 ♌ 44	09 36	08 49	20 32	22 31	20 40	21 33	22 48
13	09 26 30	20 25 44	02 ♎ 23 55	08 ♎ 18 57	01 38	10 45	09 28	20 42	22 35	20 43	21 34	22 50
14	09 30 26	21 23 23	14 ♎ 14 29	20 ♎ 10 55	02 39	11 54	10 06	20 52	22 40	20 46	21 35	22 52
15	09 34 23	22 21 03	26 ♎ 08 45	02 ♏ 08 27	03 46	13 03	10 45	21 02	22 44	20 49	21 36	22 54
16	09 38 19	23 18 43	08 ♏ 10 32	14 ♏ 15 32	04 59	14 12	11 23	21 12	22 49	20 53	21 38	22 56
17	09 42 16	24 16 25	20 ♏ 24 00	26 ♏ 36 29	06 17	15 21	12 02	21 21	22 54	20 56	21 39	22 58
18	09 46 12	25 14 08	02 ♐ 53 29	09 ♐ 15 33	07 41	16 30	12 40	21 31	22 59	20 59	21 41	23 00
19	09 50 09	26 11 52	15 ♐ 43 10	22 ♐ 16 45	09 10	17 40	13 18	21 40	23 04	21 02	21 42	23 02
20	09 54 05	27 09 38	28 ♐ 56 39	05 ♑ 43 10	10 44	18 49	13 57	21 49	23 09	21 05	21 43	23 04
21	09 58 02	28 07 24	12 ♑ 36 25	19 ♑ 36 26	12 22	19 59	14 35	21 59	23 14	21 08	21 45	23 06
22	10 01 59	29 ♌ 05 11	26 ♑ 43 05	03 ♒ 56 04	14 04	21 09	15 13	22 08	23 19	21 11	21 46	23 08
23	10 05 55	00 ♍ 03 00	11 ♒ 14 53	18 ♒ 38 52	15 49	22 19	15 52	22 17	23 24	21 15	21 48	23 09
24	10 09 52	01 00 50	26 ♒ 07 10	03 ♓ 38 48	17 37	23 29	16 30	22 25	23 30	21 18	21 49	23 11
25	10 13 48	01 58 42	11 ♓ 12 38	18 ♓ 47 28	19 28	24 39	17 08	22 34	23 35	21 21	21 51	23 13
26	10 17 45	02 56 35	26 ♓ 22 04	03 ♈ 55 11	21 21	25 49	17 47	22 43	23 40	21 23	21 53	23 15
27	10 21 41	03 54 29	11 ♈ 25 41	18 ♈ 52 31	23 16	26 59	18 25	22 51	23 46	21 26	21 54	23 17
28	10 25 38	04 52 26	26 ♈ 14 47	03 ♉ 31 46	25 12	28 10	19 03	22 59	23 51	21 29	21 56	23 19
29	10 29 34	05 50 24	10 ♉ 42 55	17 ♉ 47 54	27 09	29 ♋ 20	19 41	23 07	23 57	21 32	21 58	23 21
30	10 33 31	06 48 24	24 ♉ 46 30	01 ♊ 38 44	29 ♌ 06	00 ♌ 31	20 19	23 15	24 03	21 35	21 59	23 23
31	10 37 28	07 ♍ 46 26	08 ♊ 24 43	15 ♊ 04 38	01 ♍ 04	01 ♌ 41	20 ♌ 58	23 ♊ 23	24 ♎ 08	21 ♋ 38	22 ♎ 01	23 ♌ 24

Date	Moon True ☊ ° ′	Moon Mean ☊ ° ′	Moon ☽ Latitude ° ′
01	03 ♒ 15	02 ♒ 49	05 N 17
02	03 D 15	02 46	05 11
03	03 16	02 43	04 48
04	03 16	02 40	04 09
05	03 17	02 37	03 18
06	03 18	02 33	02 18
07	03 19	02 30	01 13
08	03 19	02 27	00 N 05
09	03 R 19	02 24	01 S 02
10	03 18	02 21	02 06
11	03 16	02 18	03 03
12	03 14	02 14	03 52
13	03 11	02 11	04 31
14	03 08	02 08	04 58
15	03 06	02 05	05 13
16	03 04	02 02	05 14
17	03 04	01 59	05 01
18	03 D 04	01 55	04 33
19	03 05	01 52	03 52
20	03 06	01 49	02 57
21	03 08	01 46	01 51
22	03 09	01 43	00 S 35
23	03 R 08	01 39	00 N 44
24	03 07	01 36	02 03
25	03 04	01 33	03 14
26	03 01	01 30	04 11
27	02 56	01 27	04 51
28	02 52	01 24	05 10
29	02 49	01 20	05 09
30	02 47	01 17	04 50
31	02 ♒ 46	01 ♒ 14	04 N 14

DECLINATIONS

Date	Sun ☉ ° ′	Moon ☽ ° ′	Mercury ☿ ° ′	Venus ♀ ° ′	Mars ♂ ° ′	Jupiter ♃ ° ′	Saturn ♄ ° ′	Uranus ♅ ° ′	Neptune ♆ ° ′	Pluto ♇ ° ′
01	18 N 02	16 N 38	16 N 25	21 N 22	20 N 49	22 N 22	06 S 14	22 N 22	06 S 47	22 N 49
02	17 47	21 10	16 40	21 26	20 40	22 23	06 16	22 22	06 47	22 49
03	17 31	24 26	16 55	21 30	20 32	22 24	06 17	22 21	06 47	22 48
04	17 15	26 17	17 10	21 33	20 23	22 25	06 19	22 21	06 48	22 47
05	16 59	26 38	17 25	21 35	20 14	22 26	06 21	22 20	06 48	22 47
06	16 43	25 32	17 39	21 37	20 05	22 27	06 22	22 20	06 49	22 46
07	16 26	23 09	17 52	21 39	19 56	22 27	06 24	22 19	06 49	22 46
08	16 09	19 42	18 03	21 40	19 46	22 28	06 26	22 19	06 50	22 45
09	15 52	15 27	18 14	21 41	19 37	22 29	06 28	22 18	06 50	22 44
10	15 35	10 38	18 24	21 41	19 27	22 30	06 29	22 18	06 51	22 44
11	15 17	05 29	18 31	21 40	19 17	22 31	06 31	22 17	06 51	22 43
12	14 59	00 N 11	18 37	21 39	19 08	22 31	06 33	22 17	06 52	22 43
13	14 41	05 S 06	18 41	21 37	18 58	22 32	06 35	22 16	06 52	22 42
14	14 23	10 12	18 43	21 35	18 47	22 33	06 37	22 16	06 53	22 41
15	14 04	14 57	18 43	21 32	18 37	22 33	06 39	22 15	06 53	22 41
16	13 45	19 11	18 40	21 28	18 27	22 34	06 41	22 15	06 54	22 40
17	13 26	22 41	18 35	21 24	18 16	22 35	06 43	22 14	06 54	22 40
18	13 07	25 13	18 27	21 19	18 06	22 35	06 45	22 14	06 55	22 39
19	12 47	26 31	18 16	21 14	17 55	22 36	06 47	22 13	06 55	22 38
20	12 28	26 24	18 03	21 08	17 44	22 36	06 49	22 13	06 56	22 38
21	12 08	24 41	17 47	21 02	17 33	22 37	06 51	22 13	06 57	22 37
22	11 48	21 24	17 28	20 55	17 22	22 38	06 53	22 12	06 57	22 37
23	11 27	16 42	17 06	20 48	17 11	22 38	06 55	22 12	06 58	22 36
24	11 07	10 53	16 41	20 40	16 59	22 39	06 57	22 11	06 58	22 36
25	10 46	04 S 23	16 14	20 31	16 48	22 39	06 59	22 11	06 59	22 35
26	10 26	02 N 24	15 44	20 22	16 36	22 40	07 01	22 10	07 00	22 35
27	10 05	08 59	15 12	20 12	16 25	22 40	07 04	22 10	07 00	22 34
28	09 44	14 57	14 38	20 01	16 13	22 41	07 06	22 09	07 01	22 34
29	09 22	19 56	14 02	19 50	16 01	22 41	07 08	22 09	07 02	22 33
30	09 01	23 39	13 24	19 39	15 49	22 41	07 10	22 09	07 02	22 33
31	08 N 39	25 N 54	12 N 44	19 N 27	15 N 37	22 N 42	07 S 12	22 N 08	07 S 03	22 N 32

ZODIAC SIGN ENTRIES

Date	h	m	Planets
01	10	57	☽ ♉
03	15	10	☽ ♊
04	01	08	♀ ♋
05	21	59	☽ ♋
08	07	16	☽ ♌
10	18	33	☽ ♍
11	14	04	☿ ♌
13	07	08	☽ ♎
15	19	43	☽ ♏
18	06	30	☽ ♐
20	13	53	☽ ♑
22	17	29	☽ ♒
23	10	45	☉ ♍
24	18	12	☽ ♓
26	17	46	☽ ♈
28	18	10	☽ ♉
30	01	35	♀ ♌
30	21	07	☽ ♊
30	22	59	☿ ♍

LATITUDES

Date	Mercury ☿ ° ′	Venus ♀ ° ′	Mars ♂ ° ′	Jupiter ♃ ° ′	Saturn ♄ ° ′	Uranus ♅ ° ′	Neptune ♆ ° ′	Pluto ♇ ° ′
01	04 S 13	02 S 03	01 N 04	00 S 35	02 N 25	00 N 25	01 N 40	09 N 19
04	03 34	01 54	01 05	00 35	02 25	00 25	01 40	09 19
07	02 47	01 44	01 05	00 35	02 24	00 26	01 40	09 19
10	01 58	01 34	01 06	00 35	02 23	00 26	01 40	09 20
13	01 08	01 24	01 07	00 35	02 23	00 26	01 40	09 20
16	00 S 22	01 14	01 07	00 35	02 22	00 26	01 40	09 20
19	00 N 19	01 03	01 08	00 35	02 21	00 26	01 39	09 20
22	00 53	00 53	01 08	00 35	02 21	00 26	01 39	09 20
25	01 19	00 42	01 09	00 35	02 20	00 26	01 39	09 21
28	01 36	00 31	01 09	00 35	02 20	00 26	01 39	09 21
31	01 N 45	00 S 21	01 N 10	00 S 35	02 N 19	00 N 26	01 N 39	09 N 21

DATA

Julian Date	2434591
Delta T	+31 seconds
Ayanamsa	23° 12′ 47″
Synetic vernal point	05° ♓ 54′ 13″
True obliquity of ecliptic	23° 26′ 48″

MOON'S PHASES, APSIDES AND POSITIONS ☽

Date	h	m	Phase	Longitude ° ′	Eclipse Indicator
02	03	16	☾	09 ♉ 32	
09	16	10	●	16 ♌ 45	Partial
17	20	08	☽	24 ♏ 36	
24	20	21	○	01 ♓ 21	
31	10	46	☾	07 ♊ 43	

Day	h	m		
13	07	02	Apogee	
25	18	32	Perigee	
05	05	37	Max dec	26° N 42′
12	12	50	0S	
19	22	01	Max dec	26° S 39′
26	03	31	0N	

LONGITUDES

Date	Chiron ⚷ ° ′	Ceres ⚳ ° ′	Pallas ⚴ ° ′	Juno ⚵ ° ′	Vesta ⚶ ° ′	Black Moon Lilith ⚸ ° ′
01	16 ♑ 52	07 ♌ 20	03 ♋ 18	02 ♋ 39	26 ♒ 38	24 ♍ 30
11	16 ♑ 22	11 ♌ 50	09 ♋ 05	08 ♋ 15	24 ♒ 17	25 ♍ 37
21	15 ♑ 56	16 ♌ 20	14 ♋ 51	13 ♋ 43	21 ♒ 48	26 ♍ 44
31	15 ♑ 37	20 ♌ 49	20 ♋ 35	19 ♋ 03	19 ♒ 33	27 ♍ 51

All ephemeris data is given at 12.00 UT and the Moon's longitude is additionally given for 24.00 UT

ASPECTARIAN

h m Aspects

01 Saturday
05 35 ☽ ⚹ ♀
07 48 ☽ □ ☿
10 55 ☽ ∥ ☿
14 00 ☽ □ ♂
16 16 ♃ ∥ ♅
17 06 ☽ ∠ ♃
18 29 ☽ ∥ ☉
22 40 ☉ Q ♆

02 Sunday
00 45 ☽ Q ♅
03 16 ☽ □ ☉
03 57 ☿ ⊻ ♀
08 49 ☽ ⊥ ♃
09 11 ☽ ∥ ♂
09 31 ☽ ∠ ♀
10 28 ☉ Q ♄
13 42 ☽ ∥ ♀
14 06 ☽ Q ☿
19 25 ☽ ⊻ ♃
19 43 ☽ ∥ ♅
19 54 ☽ ∥ ♃
21 41 ☽ ⚹ ♅
22 40 ☽ Q ♂
22 49 ☽ ∥ ♇
23 53 ☽ ⊼ ♆

03 Monday
00 45 ☽ ⊼ ♄
01 54 ☽ □ ♇
02 40 ☽ ⊥ ♀
10 30 ☽ ± ♆
10 55 ☽ ⚹ ☿
11 24 ☽ ± ♄
13 11 ☽ Q ☉
14 16 ☽ ⊻ ♀
21 01 ☽ ⚹ ♂

04 Tuesday
00 27 ☽ ∠ ♅
02 37 ☽ ⚼ ♆
03 36 ☽ ⚼ ♄
10 10 ☽ Q ♇
12 27 ☽ ⚹ ☉
13 42 ☽ ∠ ☿
16 03 ☉ ∥ ☿
16 50 ☽ ⊥ ♅
21 20 ☿ St D

05 Wednesday
01 40 ☽ ∠ ♂
02 01 ☽ ☌ ♃
03 53 ☽ ⊻ ♅
06 01 ☽ △ ♆
06 13 ☽ ⊥ ☿
07 07 ☽ △ ♄
08 12 ☽ ⚹ ♇
11 28 ♂ ∠ ♃
17 24 ☽ ⊻ ☿
18 16 ☽ ∠ ☉
19 13 ☽ ⊥ ♂

06 Thursday
02 19 ☽ ☌ ♀
07 03 ☽ ⊻ ♂
12 19 ☽ ∠ ♇
12 32 ☽ ⊥ ☉

07 Friday
00 52 ☽ ⊻ ☉
05 29 ☽ ⊥ ♇
11 16 ☽ ⊻ ♃
12 39 ☽ ☌ ♅
14 42 ☽ □ ♆
15 08 ☽ ∥ ♇
16 05 ☽ □ ♄
17 02 ☽ ⊻ ♇
17 24 ☽ ∥ ♃
18 27 ☽ ∥ ♅
23 01 ☽ ⊥ ♃
23 11 ☽ ∥ ♀

08 Saturday
03 32 ☽ ☌ ☿
11 33 ☽ ∥ ♂
16 47 ☽ ∠ ♃
17 37 ☽ ⊻ ♀
19 53 ☽ ☌ ♂
21 20 ☽ ∥ ☿

09 Sunday
01 48 ☽ Q ♆
03 21 ☽ Q ♄
06 37 ☽ ⊥ ♀
09 38 ☽ ∥ ☉
16 10 ☽ ☌ ☉
22 48 ☽ ⚹ ♃
23 38 ☽ ⊻ ♅

10 Monday
01 34 ☽ ⚹ ♆
02 17 ☽ ∠ ♀
03 15 ☽ ⚹ ♄
04 02 ☽ ☌ ♇
11 39 ☽ ⊥ ♅
17 20 ☽ ⊻ ☿
21 18 ♀ ∠ ♇
22 41 ♀ ⊻ ♂
23 06 ☽ Q ♃

11 Tuesday
05 45 ☽ # ♆
05 46 ☽ ∠ ♅
06 07 ☽ ⊥ ☿
07 16 ☽ # ♄
07 38 ☽ ∠ ♆
09 28 ☽ ∠ ♄
10 56 ☽ ⊻ ♂
11 30 ☽ ⚹ ♀
23 43 ☽ ⊥ ♂

12 Wednesday
01 30 ☽ ∠ ☿
01 50 ☽ ⊥ ♆
03 45 ☽ ⊥ ♄
09 35 ☽ ⊻ ☉
11 56 ☽ □ ♃
12 12 ☽ ⚹ ♅
13 59 ☽ ⊻ ♆
14 20 ☽ Q ♀
15 58 ☽ ⊻ ♄
16 34 ☽ ⊻ ♇
18 59 ☽ ∠ ♂
22 49 ☽ ⊥ ☉

13 Thursday
04 46 ☽ ⊥ ♇
10 19 ☽ ⚹ ☿
12 39 ☽ Q ♅
15 15 ♃ ⊻ ♅
15 52 ♂ Q ♆
18 41 ☽ ∠ ☉
18 56 ☽ ∥ ♄
19 35 ☉ ⊻ ♅
20 11 ☉ ⚹ ♃
20 13 ☽ ∥ ♆
23 04 ☽ ∠ ♇

14 Friday
03 09 ☽ ⚹ ♂
06 44 ☽ □ ♀
12 54 ☽ Q ☿
17 00 ☉ ⚹ ♆
20 58 ☿ ∥ ♂

15 Saturday
01 15 ☽ □ ♅
01 34 ☽ △ ♃
02 51 ☽ ☌ ♆
03 42 ☽ ⚹ ☉
04 47 ☽ Q ♂
05 07 ☽ ☌ ♄
05 28 ☽ ⚹ ♇
07 40 ☽ # ☉
11 44 ♂ Q ♄
22 36 ☉ ⚹ ♄

16 Sunday
02 16 ☉ ☌ ♇
04 55 ☽ □ ☿
05 33 ☽ Q ♇
05 50 ☽ Q ☉
07 45 ☽ # ♂
08 01 ☽ ⚼ ♃
08 58 ☽ # ☿
18 42 ☽ □ ♂

17 Monday
01 07 ☽ △ ♀
02 01 ☽ ± ♃
02 44 ☽ # ♀
08 38 ☽ # ♅
11 13 ☽ # ♃
11 51 ☽ # ♇
13 02 ☽ △ ♅
13 19 ☿ ∠ ♃
13 52 ☽ ⊼ ♃
14 26 ☽ ⊻ ♆
16 53 ☽ ⊻ ♄
16 59 ☽ □ ♇
20 08 ☽ □ ☉

18 Tuesday
02 02 ☽ ⊥ ♆
04 30 ☽ ⊥ ♄
09 06 ☽ ⚼ ♀
17 52 ☽ ⚼ ♅
19 10 ☽ ∠ ♆
19 52 ♄ ⚹ ♇
21 40 ☽ ∠ ♄
22 13 ☽ △ ☿
22 32 ♀ ⊥ ♇

19 Wednesday
03 45 ☽ ± ♀
07 19 ☽ △ ♂
10 44 ☽ ± ♅
15 55 ☽ ⊼ ♀
17 47 ♃ △ ♆
20 24 ☿ Q ♆
21 47 ☽ ⊼ ♅
22 58 ☽ ⚹ ♆
23 01 ☽ ☍ ♃

20 Thursday
01 24 ☽ △ ♇
01 31 ☽ ⚹ ♄
05 28 ☽ ⚼ ☿
08 34 ☽ △ ☉
10 06 ☉ ⊥ ♅
12 00 ☽ ⚼ ♂
18 34 ☿ Q ♄
20 30 ☽ Q ♆
23 04 ☽ Q ♄
23 38 ☽ ± ☿

21 Friday
04 08 ☽ ⚼ ♇
04 41 ☽ ± ♂
11 31 ☽ ⊼ ☿
12 57 ☽ ⚼ ☉
15 35 ☽ ⊼ ♂
18 01 ♃ ∥ ♇
19 44 ☽ ± ♇

22 Saturday
01 46 ☽ ☍ ♀
02 39 ☽ ☍ ♅
03 40 ☽ □ ♆
04 11 ☽ ⊼ ♃
04 14 ☽ # ♃
04 18 ☽ # ♇
05 27 ☽ ± ☉
05 57 ☽ ⊼ ♇
06 15 ☽ □ ♄
07 00 ☽ # ♅
13 00 ♀ ☌ ♅
14 23 ☽ ± ♃
14 52 ☽ # ♀
16 14 ☽ ⊼ ☉

23 Sunday
01 15 ♀ □ ♆
01 31 ☿ ∥ ♂
05 26 ☽ ⚼ ♃
09 44 ☽ # ♂
10 03 ☽ # ☿
11 11 ♀ ⊻ ♃
12 58 ☿ ☌ ♂
19 50 ☽ ☍ ♂
20 27 ☽ ☍ ☿

24 Monday
04 14 ☽ ⊼ ♅
05 06 ☽ △ ♆
05 55 ♀ ⊻ ♇
06 01 ☽ △ ♃
07 18 ☽ ☍ ♇
07 24 ☽ ⊼ ♀
07 46 ☽ △ ♄
11 04 ☽ # ☉
12 25 ♀ □ ♄
13 53 ☽ ± ♅
17 49 ☽ ± ♀
20 21 ☽ ☍ ☉

25 Tuesday
02 37 ☽ ∥ ♆
02 38 ☽ ∥ ♄
04 15 ☽ ⚼ ♅
05 05 ☽ ⚼ ♆
07 05 ♄ ∥ ♆
07 49 ☽ ⚼ ♄
09 19 ☽ ⚼ ♀
19 22 ☽ ± ♆
21 48 ☽ ⊼ ♂
22 09 ☽ ± ♄

26 Wednesday
02 55 ☽ ⊼ ☿
04 05 ☽ △ ♅
04 52 ☽ ⊼ ♆
06 09 ☽ □ ♃
07 03 ☽ ⊼ ♇
07 42 ☽ ⊼ ♄
07 43 ☽ ± ♂
11 03 ☽ △ ♀
12 31 ☿ ⊻ ♅
13 47 ☽ ± ☿
16 35 ☽ ± ♇
18 44 ☿ ⚹ ♆
22 38 ☽ ⚼ ♂
23 10 ☽ ⊼ ☉

27 Thursday
04 37 ☽ # ♆
04 48 ☽ # ♄
06 11 ☽ ⚼ ☿
06 25 ☿ ⚹ ♃
06 57 ☽ ⚼ ♇
09 24 ☽ ± ☉
11 04 ☽ Q ♃
12 15 ☿ ☌ ♇
15 59 ☽ ∥ ☉
18 35 ☿ ⚹ ♄
23 46 ☽ △ ♂

28 Friday
00 54 ☽ ⚼ ☉
04 13 ☽ □ ♅
04 57 ☽ ☍ ♆
06 38 ☽ ⚹ ♃
07 12 ☽ △ ♇
08 04 ☽ ☍ ♄
10 01 ☽ △ ☿
10 46 ☽ ∥ ☿
15 17 ☉ Q ♃
15 25 ☽ □ ♀
17 26 ☽ ∥ ♂

29 Saturday
03 15 ☽ △ ☉
07 08 ♀ ⊥ ♃
07 37 ☽ ∠ ♃
10 01 ☽ Q ♅
11 31 ☽ ∥ ♀
16 56 ☿ ⊥ ♅
22 57 ☽ ⊥ ♃

30 Sunday
00 14 ☽ Q ♀
01 11 ☽ ∥ ♅
03 52 ☽ ∥ ♇
03 57 ☽ □ ♂
04 51 ☽ ∥ ♃
06 10 ☉ ∠ ♅
06 28 ☽ ⚹ ♅
07 10 ☽ ⊼ ♆
09 21 ☽ ⊻ ♃
09 34 ☽ □ ♇
10 44 ☽ ⊼ ♄
16 37 ☉ ∠ ♆
17 36 ☽ ± ♆
20 48 ☽ □ ☿
21 15 ☽ ± ♄
22 57 ☽ ⚹ ♀

31 Monday
08 49 ☽ ∠ ♅
09 30 ☽ ⚼ ♆
10 46 ☽ □ ☉
13 02 ☽ Q ♂
13 19 ☽ ⚼ ♄
16 25 ♃ ⚹ ♇
17 23 ☽ Q ♇
21 30 ☉ ⊥ ♀

SEPTEMBER 1953

LONGITUDES

Date	Sidereal time h m s	Sun ☉	Moon ☽	Moon ☽ 24.00	Mercury ☿	Venus ♀	Mars ♂	Jupiter ♃	Saturn ♄	Uranus ♅	Neptune ♆	Pluto ♇
01	10 41 24	08 ♍ 44 30	21 ♊ 38 51	28 ♊ 07 42	03 ♍ 02	02 ♌ 52	21 ♌ 36	23 ♊ 31	24 ♎ 14	21 ♋ 41	22 ♎ 03	23 ♌ 26
02	10 45 21	09 42 36	04 ♋ 31 39	10 ♋ 51 06	04 59	04 03	22 14	23 39	24 20	21 43	22 04	23 28
03	10 49 17	10 40 44	17 ♋ 06 32	23 ♋ 18 23	06 57	05 14	22 52	23 46	24 26	21 46	22 06	23 30
04	10 53 14	11 38 54	29 ♋ 27 05	05 ♌ 33 02	08 53	06 25	23 30	23 53	24 32	21 49	22 08	23 32
05	10 57 10	12 37 06	11 ♌ 36 38	17 ♌ 38 13	10 50	07 36	24 08	24 01	24 38	21 51	22 10	23 34
06	11 01 07	13 35 20	23 ♌ 38 07	29 ♌ 36 37	12 45	08 48	24 46	24 08	24 44	21 54	22 12	23 35
07	11 05 03	14 33 35	05 ♍ 34 00	11 ♍ 30 30	14 39	09 59	25 24	24 15	24 50	21 56	22 13	23 37
08	11 09 00	15 31 52	17 ♍ 26 21	23 ♍ 21 47	16 33	11 11	26 03	24 21	24 56	21 59	22 15	23 39
09	11 12 57	16 30 11	29 ♍ 16 58	05 ♎ 12 10	18 25	12 22	26 41	24 28	25 02	22 01	22 17	23 41
10	11 16 53	17 28 32	11 ♎ 07 35	17 ♎ 03 27	20 16	13 34	27 19	24 34	25 08	22 04	22 19	23 43
11	11 20 50	18 26 54	23 ♎ 00 03	28 ♎ 57 38	22 07	14 45	27 57	24 40	25 15	22 06	22 21	23 44
12	11 24 46	19 25 18	04 ♏ 56 33	10 ♏ 57 08	23 56	15 57	28 35	24 47	25 21	22 09	22 23	23 46
13	11 28 43	20 23 44	16 ♏ 59 46	23 ♏ 04 52	25 44	17 09	29 13	24 53	25 27	22 11	22 25	23 48
14	11 32 39	21 22 11	29 ♏ 12 54	05 ♐ 24 19	27 31	18 21	29 ♌ 51	24 58	25 34	22 13	22 27	23 50
15	11 36 36	22 20 40	11 ♐ 39 39	17 ♐ 59 23	29 ♍ 17	19 33	00 ♍ 28	25 04	25 40	22 15	22 29	23 51
16	11 40 32	23 19 11	24 ♐ 24 03	00 ♑ 54 09	01 ♎ 02	20 45	01 06	25 09	25 47	22 18	22 31	23 53
17	11 44 29	24 17 44	07 ♑ 30 10	14 ♑ 12 31	02 46	21 57	01 44	25 15	25 53	22 20	22 33	23 55
18	11 48 26	25 16 18	21 ♑ 01 31	27 ♑ 57 27	04 29	23 10	02 22	25 20	26 00	22 22	22 35	23 56
19	11 52 22	26 14 53	05 ♒ 00 23	12 ♒ 10 17	06 11	24 22	03 00	25 25	26 06	22 24	22 37	23 58
20	11 56 19	27 13 31	19 ♒ 26 54	26 ♒ 49 47	07 51	25 34	03 38	25 29	26 13	22 26	22 39	24 00
21	12 00 15	28 12 09	04 ♓ 18 15	11 ♓ 51 25	09 31	26 47	04 16	25 34	26 20	22 28	22 41	24 01
22	12 04 12	29 ♍ 10 50	19 ♓ 28 12	27 ♓ 07 19	11 10	28 00	04 54	25 39	26 26	22 30	22 43	24 03
23	12 08 08	00 ♎ 09 33	04 ♈ 47 23	12 ♈ 26 59	12 48	29 ♌ 12	05 31	25 43	26 33	22 32	22 45	24 05
24	12 12 05	01 08 17	20 ♈ 04 39	27 ♈ 39 03	14 24	00 ♍ 25	06 09	25 47	26 40	22 34	22 47	24 06
25	12 16 01	02 07 04	05 ♉ 08 56	12 ♉ 33 16	16 00	01 38	06 47	25 51	26 47	22 35	22 49	24 08
26	12 19 58	03 05 53	19 ♉ 51 14	27 ♉ 02 15	17 35	02 51	07 25	25 54	26 54	22 37	22 52	24 09
27	12 23 55	04 04 45	04 ♊ 05 56	11 ♊ 02 09	19 09	04 03	08 03	25 58	27 01	22 39	22 54	24 11
28	12 27 51	05 03 38	17 ♊ 50 57	24 ♊ 32 34	20 42	05 16	08 40	26 01	27 07	22 40	22 56	24 12
29	12 31 48	06 02 34	01 ♋ 07 20	07 ♋ 35 43	22 15	06 30	09 18	26 04	27 14	22 42	22 58	24 14
30	12 35 44	07 ♎ 01 32	13 ♋ 58 15	20 ♋ 15 30	23 ♎ 46	07 ♍ 43	09 ♍ 56	26 ♊ 07	27 ♎ 21	22 ♋ 44	23 ♎ 00	24 ♌ 15

Moon nodes, latitude and DECLINATIONS

Date	Moon True ☊	Moon Mean ☊	Moon ☽ Latitude	Sun ☉	Moon ☽	Mercury ☿	Venus ♀	Mars ♂	Jupiter ♃	Saturn ♄	Uranus ♅	Neptune ♆	Pluto ♇
01	02 ♒ 47	01 ♒ 11	03 N 26	08 N 18	26 N 36	12 N 03	19 N 14	15 N 25	22 N 42	07 S 15	22 N 08	07 S 03	22 N 31
02	02 D 48	01 08	02 28	07 56	25 50	11 20	19 01	15 13	22 42	07 17	22 07	07 04	22 31
03	02 50	01 05	01 25	07 34	23 45	10 37	18 47	15 00	22 43	07 19	22 07	07 05	22 31
04	02 51	01 01	00 N 18	07 12	20 34	09 53	18 33	14 48	22 43	07 22	22 07	07 06	22 30
05	02 R 51	00 58	00 S 47	06 50	16 33	09 09	18 18	14 36	22 44	07 24	22 06	07 06	22 30
06	02 48	00 55	01 50	06 27	11 54	08 21	18 03	14 23	22 44	07 26	22 06	07 07	22 29
07	02 44	00 52	02 48	06 05	06 52	07 35	17 47	14 10	22 44	07 29	22 05	07 08	22 29
08	02 38	00 49	03 38	05 42	01 N 37	06 48	17 31	13 57	22 44	07 31	22 05	07 08	22 28
09	02 31	00 45	04 18	05 20	03 S 39	06 01	17 14	13 45	22 45	07 33	22 05	07 09	22 28
10	02 22	00 42	04 47	04 57	08 48	05 14	16 57	13 32	22 45	07 36	22 04	07 10	22 27
11	02 13	00 39	05 03	04 34	13 38	04 26	16 39	13 19	22 45	07 38	22 04	07 11	22 27
12	02 05	00 36	05 07	04 11	17 59	03 39	16 21	13 06	22 45	07 41	22 04	07 11	22 26
13	01 59	00 33	04 57	03 48	21 39	02 51	16 02	12 53	22 46	07 43	22 03	07 12	22 26
14	01 54	00 30	04 33	03 25	24 26	02 04	15 43	12 39	22 46	07 45	22 03	07 13	22 26
15	01 52	00 26	03 57	03 02	26 06	01 17	15 23	12 26	22 46	07 48	22 03	07 14	22 25
16	01 D 51	00 23	03 07	02 39	26 27	00 N 30	15 03	12 13	22 46	07 50	22 02	07 14	22 25
17	01 52	00 20	02 07	02 16	25 21	00 S 17	14 43	11 59	22 46	07 53	22 02	07 15	22 24
18	01 53	00 17	00 S 58	01 53	22 45	01 04	14 22	11 46	22 47	07 55	22 02	07 16	22 24
19	01 R 53	00 14	00 N 17	01 30	18 45	01 50	14 01	11 32	22 47	07 58	22 01	07 17	22 24
20	01 52	00 11	01 33	01 07	13 31	02 36	13 39	11 19	22 47	08 00	22 01	07 17	22 23
21	01 48	00 07	02 44	00 43	07 23	03 21	13 17	11 05	22 47	08 03	22 01	07 18	22 23
22	01 42	00 04	03 46	00 N 20	00 S 42	04 06	12 54	10 51	22 47	08 05	22 00	07 19	22 22
23	01 34	00 ♒ 01	04 32	00 S 04	06 N 04	04 51	12 31	10 37	22 47	08 08	22 00	07 20	22 22
24	01 26	29 ♑ 58	04 58	00 27	12 27	05 35	12 08	10 23	22 48	08 10	22 00	07 21	22 22
25	01 17	29 55	05 03	00 51	18 00	06 19	11 44	10 09	22 48	08 13	22 00	07 21	22 21
26	01 09	29 51	04 48	01 14	22 20	07 02	11 20	09 56	22 48	08 15	21 59	07 22	22 21
27	01 03	29 48	04 15	01 37	25 09	07 44	10 56	09 42	22 48	08 18	21 59	07 23	22 21
28	01 00	29 45	03 29	02 01	26 21	08 26	10 31	09 28	22 48	08 20	21 59	07 24	22 21
29	00 58	29 42	02 29	02 24	25 59	09 08	10 06	09 13	22 48	08 23	21 59	07 24	22 20
30	00 ♒ 58	29 ♑ 39	01 N 29	02 S 47	24 N 12	09 S 48	09 N 41	08 N 59	22 N 48	08 S 26	21 N 58	07 S 25	22 N 20

ZODIAC SIGN ENTRIES

Date	h	m	Planets
02	03	30	☽ ♋
04	13	05	☽ ♌
07	00	47	☽ ♍
09	13	27	☽ ♎
12	02	05	☽ ♏
14	13	32	☽ ♐
14	17	59	♂ ♍
15	21	45	☿ ♎
16	22	21	☽ ♑
19	03	30	☽ ♒
21	05	06	☽ ♓
23	04	30	☽ ♈
23	08	06	☉ ♎
24	03	48	♀ ♍
25	03	45	☽ ♉
27	05	01	☽ ♊
29	09	56	☽ ♋

LATITUDES

Date	Mercury ☿	Venus ♀	Mars ♂	Jupiter ♃	Saturn ♄	Uranus ♅	Neptune ♆	Pluto ♇
01	01 N 46	00 S 18	01 N 10	00 S 35	02 N 19	00 N 26	01 N 39	09 N 22
04	01 46	00 S 07	01 11	00 35	02 18	00 26	01 39	09 22
07	01 40	00 N 02	01 11	00 35	02 18	00 26	01 39	09 23
10	01 30	00 12	01 11	00 35	02 18	00 26	01 39	09 23
13	01 16	00 21	01 12	00 35	02 17	00 26	01 39	09 24
16	00 59	00 30	01 12	00 35	02 17	00 26	01 39	09 24
19	00 40	00 38	01 13	00 35	02 17	00 27	01 38	09 25
22	00 N 20	00 46	01 13	00 35	02 16	00 27	01 38	09 25
25	00 S 01	00 54	01 13	00 35	02 16	00 27	01 38	09 26
28	00 23	01 01	01 14	00 35	02 16	00 27	01 38	09 27
31	00 S 44	01 N 07	01 N 14	00 S 35	02 N 15	00 N 27	01 N 38	09 N 28

DATA

Julian Date	2434622
Delta T	+31 seconds
Ayanamsa	23° 12' 51"
Synetic vernal point	05° ♓ 54' 09"
True obliquity of ecliptic	23° 26' 48"

LONGITUDES

Date	Chiron ⚷	Ceres ⚳	Pallas ⚴	Juno ⚵	Vesta ⚶	Black Moon Lilith ⚸
01	15 ♑ 35	21 ♌ 16	21 ♋ 10	19 ♋ 34	19 ♒ 20	27 ♍ 58
11	15 ♑ 24	25 ♌ 44	26 ♋ 50	24 ♋ 43	17 ♒ 39	29 ♍ 05
21	15 ♑ 20	00 ♍ 10	02 ♌ 24	29 ♋ 40	16 ♒ 40	00 ♎ 12
31	15 ♑ 24	04 ♍ 33	07 ♌ 50	04 ♌ 23	16 ♒ 28	01 ♎ 19

MOON'S PHASES, APSIDES AND POSITIONS ☽

Date	h	m	Phase	Longitude	Eclipse Indicator
08	07	48	●	15 ♍ 22	
16	09	49	☽	23 ♐ 14	
23	04	16	○	29 ♓ 51	
29	21	51	☾	06 ♋ 27	

Day	h	m		
09	16	37	Apogee	
23	04	33	Perigee	
01	11	15	Max dec	26° N 36'
08	19	22	0S	
16	06	03	Max dec	26° S 30'
22	14	29	0N	
28	17	59	Max dec	26° N 24'

ASPECTARIAN

01 Tuesday
01 03 ☽ ⊥ ♅
04 24 ☽ ∠ ♀
07 07 ☿ ⊻ ♀
10 40 ☽ Q ☿
11 54 ☽ ✶ ♂
12 03 ☽ ⊻ ♅
12 44 ☽ △ ♆
15 13 ♂ ⊻ ♅
15 18 ☽ ✶ ♇
15 29 ☽ ☌ ♃
16 49 ☽ △ ♄
22 11 ☽ Q ☉
22 38 ☽ ⊥ ♀

02 Wednesday
01 37 ☉ ∠ ♄
05 42 ♂ ✶ ♆
11 01 ☽ ⊻ ♀
13 02 ☽ ✶ ☿
17 23 ☽ ∠ ♂
19 29 ☽ ∠ ♇
20 34 ☿ Q ♃
22 39 ☽ ✶ ☉

03 Thursday
09 45 ☿ ∠ ♅
11 31 ☽ ⊥ ♂
12 45 ☽ ⊥ ♇
13 58 ☿ ∠ ♆
20 40 ☽ ∥ ♃
21 02 ☽ ☌ ♅
21 41 ☽ □ ♆
22 19 ☽ ∥ ♇
23 06 ☽ ∠ ☿
23 45 ☽ ⊻ ♂

04 Friday
00 24 ☽ ⊻ ♇
01 02 ☽ ⊻ ♃
01 17 ☽ ∥ ♅
02 18 ☽ □ ♄
02 24 ☉ # ♄
06 03 ☽ ∠ ☉
12 52 ☽ ⊥ ♃
13 02 ♂ ☌ ♇
18 33 ☉ # ♆
20 02 ☽ ⊥ ☿
20 20 ☿ ∠ ♄

05 Saturday
01 15 ☽ ⊥ ☉
01 23 ☽ ∥ ♀
03 12 ☽ ☌ ♀
06 02 ♂ ✶ ♃
06 48 ☽ ∠ ♃
09 07 ☽ Q ♆
10 09 ☽ ⊻ ☿
14 03 ☽ Q ♄
14 11 ☽ ⊻ ☉
22 56 ☽ ∥ ♂

06 Sunday
08 30 ☽ ⊻ ♅
09 06 ☽ ✶ ♆
10 03 ♂ ✶ ♄
11 55 ☽ ☌ ♇
13 00 ☽ ✶ ♃
14 13 ☽ ✶ ♄
14 25 ☽ ☌ ♂
19 25 ♀ ∠ ♃
20 35 ☽ ⊥ ♅

07 Monday
08 05 ☽ ∥ ☿
09 12 ☽ # ♄
09 35 ☉ ☌ ☿
10 48 ☽ # ♆
13 23 ☽ Q ♃
14 47 ☽ ∠ ♅
15 09 ☿ # ♄
15 21 ☽ ∠ ♆
15 56 ☽ ∥ ☉
16 57 ♀ Q ♆
20 41 ☽ ∠ ♄
21 55 ☽ ⊻ ♀

08 Tuesday
01 50 ☿ # ♆
07 48 ☽ ☌ ☉
08 16 ☿ ⊥ ♆
09 36 ☽ ⊥ ♆
09 51 ☽ ☌ ☿
11 24 ☽ ⊥ ♀
15 03 ☽ ⊥ ♄
21 14 ☽ ✶ ♅
21 47 ☽ ⊻ ♆

09 Wednesday
00 37 ☽ ⊻ ♇
02 08 ☽ □ ♃
03 19 ☽ ⊻ ♄
06 25 ☽ ⊻ ♂
06 28 ☉ ⊥ ♆
07 41 ☽ ∠ ♀
10 16 ☿ ⊥ ♀
12 48 ☽ ⊥ ♇
19 11 ☽ # ☉
19 16 ☽ ⊥ ♂
20 27 ☿ ⊥ ♄
21 28 ☽ # ☿
21 39 ☽ Q ♅

10 Thursday
02 42 ♀ Q ♄
04 15 ☽ ∥ ♆
06 15 ☽ ∥ ♄
07 06 ☽ ∠ ♇
14 32 ☽ ∠ ♂
17 29 ☽ ✶ ♀

11 Friday
01 59 ☽ ⊻ ☉
04 18 ☉ ∥ ☿
09 53 ☽ ⊻ ☿
10 11 ☽ □ ♅
10 25 ☽ # ♂
10 41 ☽ ☌ ♆
11 53 ☿ ✶ ♅
13 30 ☽ ✶ ♇
15 10 ☿ ⊻ ♆
15 11 ☽ ⊥ ☉
15 24 ☽ △ ♃
16 34 ☽ ☌ ♄
18 29 ♂ ⊥ ♅
20 25 ☽ Q ♀
22 31 ☽ ✶ ♂

12 Saturday
00 09 ☽ ⊥ ☿
03 13 ☽ # ♀
09 46 ☿ ⊻ ♇
10 02 ☉ ⊥ ♄
10 52 ☽ ∠ ☉
13 39 ☽ Q ♇
21 23 ☽ ∠ ☿
21 44 ☽ ⚼ ♃
23 50 ☿ □ ♃
23 53 ☽ Q ♂

13 Sunday
08 01 ☿ ⊻ ♄
12 21 ☽ □ ♀
14 59 ☽ # ♅
15 45 ☽ ± ♃
17 55 ☽ # ♇
19 18 ☽ ✶ ☉
20 35 ☽ # ♃
22 16 ☽ △ ♅
22 43 ☽ ⊻ ♆

14 Monday
01 26 ☽ □ ♇
03 39 ☽ ⊼ ♃
04 48 ☽ ⊻ ♄
08 08 ☽ ✶ ☿
10 30 ☽ ⊥ ♆
13 17 ☽ □ ♂
16 36 ☽ ⊥ ♄
20 45 ☽ Q ☉

15 Tuesday
03 32 ☽ ⚼ ♅
03 58 ☽ ∠ ♆
09 46 ☉ ✶ ♅
10 06 ☽ ∠ ♄
11 10 ☽ Q ☿
15 28 ☉ ⊻ ♆
19 53 ☿ ⊥ ♇
20 45 ☽ ± ♅

16 Wednesday
04 29 ☽ △ ♀
08 04 ☽ ⊼ ♅
08 29 ☽ ✶ ♆
09 49 ☽ □ ☉
11 02 ☽ △ ♇
13 25 ☽ ☍ ♃
13 31 ☿ ⊻ ♂
14 35 ☽ ✶ ♄

17 Thursday
01 00 ☽ △ ♂
02 07 ☽ □ ☿
02 17 ☉ ⊻ ♇
06 38 ☽ Q ♆
10 55 ☽ ⚼ ♀
12 42 ☽ Q ♄
14 33 ☽ ⚼ ♇
19 39 ♀ ⊻ ♅

18 Friday
00 06 ♀ ✶ ♆
04 34 ☽ ± ♀
05 17 ☽ ⚼ ♂
06 35 ☽ ± ♇
10 20 ☿ Q ♅
11 51 ☽ # ♃
13 34 ☉ □ ♃
14 20 ☽ ☍ ♅
14 32 ☽ # ♇
14 43 ☽ □ ♆
16 05 ☽ ⊼ ♀
17 04 ☽ # ♅
17 05 ☽ ⊼ ♇
19 31 ☽ ⊼ ♃
19 56 ☽ △ ☉
20 42 ☽ □ ♄
21 43 ☽ ± ♂

19 Saturday
03 51 ♀ ☌ ♇
04 53 ☉ # ☿
05 52 ☽ ± ♃
08 06 ☉ ⊻ ♄
08 27 ☽ ⊼ ♂
14 14 ☽ △ ☿
21 07 ☽ ⚼ ♃
23 13 ☽ ⚼ ☉

20 Sunday
10 14 ♀ ✶ ♃
11 27 ☽ # ♀
15 07 ☽ ± ☉
16 53 ☽ ⊼ ♅
17 14 ☽ △ ♆
18 16 ☽ ⚼ ☿
19 25 ☽ ☍ ♇
21 21 ☽ # ♂
21 53 ☽ △ ♃
22 51 ☽ ☍ ♀
23 06 ☽ △ ♄

21 Monday
01 32 ☽ ⊼ ☉
02 05 ♀ ✶ ♄
02 37 ☽ ± ♅
04 42 ☿ ∠ ♇
09 32 ☽ ∥ ♄
10 35 ☽ ± ☿
11 56 ☽ ☍ ♂
12 17 ☽ ∥ ♆
17 03 ☽ ⚼ ♅
17 24 ☽ ⚼ ♆
21 19 ☽ ⊼ ☿
23 15 ☽ ⚼ ♄

22 Tuesday
01 07 ☽ ∥ ☿
05 36 ☿ ⊥ ♂
07 40 ☽ ± ♆
13 25 ☽ # ☉
13 32 ☽ ± ♄
15 26 ☽ ∥ ☉
16 46 ☽ △ ♅
17 07 ☽ ⊼ ♆
19 12 ☽ ⊼ ♇
21 44 ☽ □ ♃
22 18 ♀ ⊥ ♅
23 01 ☽ ⊼ ♄

23 Wednesday
02 30 ☽ ⊼ ♀
04 16 ☽ ☍ ☉
04 37 ☽ ± ♇
07 07 ☽ # ☿
09 54 ☉ ⊥ ♇
12 42 ☽ ± ♀
13 12 ☽ ⊼ ♂
16 36 ☽ # ♆
18 43 ☽ ⚼ ♇
19 35 ☽ # ♄
23 00 ☽ ± ♂

24 Thursday
02 01 ☽ ☍ ☿
02 03 ☽ Q ♃
04 01 ☽ ⚼ ♀
04 17 ☽ ∥ ♂
10 49 ☽ ∥ ♀
13 46 ☽ ⚼ ♂
15 56 ☽ □ ♅
16 18 ☽ ☍ ♆
18 23 ☽ △ ♇
21 04 ☽ ✶ ♃
22 31 ☽ ☍ ♄

25 Friday
05 51 ☽ △ ♀
06 48 ☽ ⊼ ☉
14 45 ☽ △ ♂
17 08 ☽ ± ☉
20 49 ☽ Q ♅
21 15 ☽ ∠ ♃

26 Saturday
07 48 ☽ ⊼ ☿
08 53 ☽ ⚼ ☉
09 47 ☽ ∥ ♅
12 05 ☽ ⊥ ♃
12 10 ☽ ∥ ♇
15 12 ☽ ∥ ♃
16 36 ☽ ✶ ♅
17 00 ☽ ⊼ ♆
18 59 ☽ ± ☿
19 10 ☽ □ ♇
20 10 ♂ ∠ ♅
22 08 ☽ ⊻ ♃
23 41 ☿ ∥ ♆
23 51 ☽ ⊼ ♄

27 Sunday
03 07 ☽ ± ♆
05 13 ☿ ∠ ♀
05 57 ♂ ∠ ♆
08 43 ♂ Q ♃
10 07 ☽ ± ♄
11 55 ☽ □ ♀
11 58 ☽ △ ☉
12 07 ☽ ⚼ ☿
14 12 ☉ ⊻ ♀
18 07 ☽ ∠ ♅
18 33 ☽ ⚼ ♆
19 07 ☽ □ ♂

28 Monday
01 49 ☽ ⚼ ♄
02 02 ☽ Q ♇
02 18 ☉ Q ♅
08 28 ☿ ∥ ♄
09 55 ☽ ⊥ ♅
17 46 ☽ △ ☿
20 39 ☽ ⊻ ♅
21 07 ☽ △ ♆
22 41 ☽ Q ♀
23 25 ☽ ✶ ♇

29 Tuesday
02 44 ☽ ☌ ♃
04 40 ☽ Q ♂
04 50 ☽ △ ♄
14 32 ☿ # ♂
19 17 ☿ □ ♅
21 51 ☽ □ ☉
22 59 ☽ ✶ ♀
23 35 ☿ ☌ ♆

30 Wednesday
03 06 ☽ ∠ ♇
03 59 ☽ ✶ ♂
09 24 ☿ # ♀
12 20 ♀ ∠ ♅
17 54 ♀ ∠ ♆
19 51 ☿ ✶ ♇
20 11 ☽ ⊥ ♇
20 27 ♀ Q ♃

All ephemeris data is given at 12.00 UT and the Moon's longitude is additionally given for 24.00 UT

OCTOBER 1953

LONGITUDES

Date	Sidereal time h m s	Sun ☉ ° ' "	Moon ☽ ° ' "	Moon ☽ 24.00 ° ' "	Mercury ☿ ° '	Venus ♀ ° '	Mars ♂ ° '	Jupiter ♃ ° '	Saturn ♄ ° '	Uranus ♅ ° '	Neptune ♆ ° '	Pluto ♇ ° '
01	12 39 41	08 ♎ 00 33	26 ♋ 28 04	02 ♌ 36 35	25 ♎ 17	08 ♍ 56	10 ♍ 34	26 ♊ 10	27 ♎ 28	22 ♋ 45	23 ♎ 02	24 ♌ 17
02	12 43 37	08 59 36	08 ♌ 41 37	14 ♌ 43 47	26 46	10 09	11 11	26 13	27 35	22 47	23 04	24 18
03	12 47 34	09 58 40	20 ♌ 43 36	26 ♌ 41 36	28 15	11 23	11 49	26 15	27 42	22 48	23 07	24 20
04	12 51 30	10 57 48	02 ♍ 38 15	08 ♍ 33 57	29 ♎ 43	12 36	12 27	26 17	27 49	22 49	23 09	24 21
05	12 55 27	11 56 57	14 ♍ 29 06	20 ♍ 24 01	01 ♏ 10	13 49	13 04	26 19	27 56	22 51	23 11	24 23
06	12 59 24	12 56 08	26 ♍ 18 59	02 ♎ 14 16	02 36	15 03	13 42	26 21	28 04	22 52	23 13	24 24
07	13 03 20	13 55 22	08 ♎ 10 04	14 ♎ 06 35	04 01	16 16	14 20	26 23	28 11	22 53	23 15	24 25
08	13 07 17	14 54 38	20 ♎ 03 58	26 ♎ 02 24	05 26	17 30	14 57	26 24	28 18	22 54	23 18	24 27
09	13 11 13	15 53 56	02 ♏ 02 00	08 ♏ 02 56	06 49	18 44	15 35	26 25	28 25	22 55	23 20	24 28
10	13 15 10	16 53 15	14 ♏ 05 24	20 ♏ 09 33	08 11	19 58	16 13	26 27	28 32	22 56	23 22	24 29
11	13 19 06	17 52 37	26 ♏ 15 38	02 ♐ 23 53	09 32	21 11	16 50	26 27	28 39	22 57	23 24	24 31
12	13 23 03	18 52 01	08 ♐ 34 36	14 ♐ 48 07	10 52	22 25	17 28	26 28	28 46	22 58	23 27	24 32
13	13 26 59	19 51 26	21 ♐ 04 47	27 ♐ 25 01	12 11	23 39	18 05	26 28	28 54	22 59	23 29	24 33
14	13 30 56	20 50 54	03 ♑ 49 13	10 ♑ 17 50	13 29	24 53	18 43	26 29	29 01	23 00	23 31	24 34
15	13 34 53	21 50 23	16 ♑ 51 19	23 ♑ 30 06	14 46	26 07	19 20	26 R 29	29 08	23 01	23 33	24 35
16	13 38 49	22 49 53	00 ♒ 14 09	07 ♒ 05 04	16 01	27 21	19 58	26 28	29 15	23 02	23 36	24 37
17	13 42 46	23 49 26	14 ♒ 01 50	21 ♒ 05 01	17 14	28 35	20 35	26 28	29 23	23 02	23 38	24 38
18	13 46 42	24 49 00	28 ♒ 14 34	05 ♓ 30 19	18 26	29 ♍ 49	21 13	26 28	29 30	23 03	23 40	24 39
19	13 50 39	25 48 36	12 ♓ 51 52	20 ♓ 18 36	19 36	01 ♎ 04	21 50	26 27	29 37	23 04	23 42	24 40
20	13 54 35	26 48 13	27 ♓ 49 41	05 ♈ 24 06	20 44	02 18	22 28	26 26	29 44	23 04	23 45	24 41
21	13 58 32	27 47 53	13 ♈ 00 37	20 ♈ 37 56	21 51	03 32	23 05	26 25	29 52	23 05	23 47	24 42
22	14 02 28	28 47 34	28 ♈ 14 37	05 ♉ 49 16	22 54	04 46	23 43	26 23	29 ♎ 59	23 05	23 49	24 43
23	14 06 25	29 ♎ 47 18	13 ♉ 20 33	20 ♉ 47 14	23 56	06 01	24 20	26 22	00 ♏ 06	23 05	23 51	24 44
24	14 10 22	00 ♏ 47 03	28 ♉ 06 18	05 ♊ 22 55	24 54	07 15	24 58	26 20	00 13	23 06	23 53	24 45
25	14 14 18	01 46 51	12 ♊ 30 28	19 ♊ 30 37	25 50	08 30	25 35	26 18	00 21	23 06	23 56	24 46
26	14 18 15	02 46 41	26 ♊ 23 13	03 ♋ 08 18	26 42	09 44	26 12	26 16	00 28	23 06	23 58	24 47
27	14 22 11	03 46 33	09 ♋ 46 07	16 ♋ 17 02	27 30	10 59	26 50	26 13	00 35	23 06	24 00	24 48
28	14 26 08	04 46 27	22 ♋ 41 32	29 ♋ 00 09	28 14	12 13	27 27	26 11	00 43	23 06	24 02	24 49
29	14 30 04	05 46 23	05 ♌ 13 33	11 ♌ 22 22	28 53	13 28	28 04	26 08	00 50	23 06	24 05	24 49
30	14 34 01	06 46 22	17 ♌ 27 17	23 ♌ 28 57	29 28	14 42	28 42	26 05	00 57	23 R 06	24 07	24 50
31	14 37 57	07 ♏ 46 23	29 ♌ 28 01	05 ♍ 25 09	29 ♏ 56	15 ♎ 57	29 ♍ 19	26 ♊ 02	01 ♏ 04	23 ♋ 06	24 ♎ 09	24 ♌ 51

DECLINATIONS

Date	Moon True ☊ ° '	Moon Mean ☊ ° '	Moon ☽ Latitude ° '	Sun ☉ ° '	Moon ☽ ° '	Mercury ☿ ° '	Venus ♀ ° '	Mars ♂ ° '	Jupiter ♃ ° '	Saturn ♄ ° '	Uranus ♅ ° '	Neptune ♆ ° '	Pluto ♇ ° '
01	00 ♒ 59	29 ♑ 36	00 N 24	03 S 11	21 N 16	10 S 28	09 N 16	08 N 45	22 N 48	08 S 28	21 N 58	07 S 26	22 N 20
02	00 R 59	29 32	00 S 41	03 34	17 26	11 08	08 50	08 31	22 48	08 31	21 58	07 27	22 20
03	00 57	29 29	01 43	03 57	12 58	11 46	08 24	08 16	22 48	08 33	21 58	07 28	22 19
04	00 52	29 26	02 40	04 20	08 03	12 24	07 57	08 02	22 49	08 36	21 58	07 29	22 19
05	00 45	29 23	03 29	04 44	02 N 54	13 02	07 31	07 48	22 49	08 38	21 57	07 29	22 19
06	00 35	29 20	04 09	05 07	02 S 21	13 38	07 04	07 33	22 49	08 41	21 57	07 30	22 19
07	00 23	29 17	04 38	05 30	07 30	14 14	06 37	07 19	22 49	08 44	21 57	07 31	22 18
08	00 09	29 13	04 56	05 53	12 24	14 49	06 10	07 04	22 49	08 46	21 57	07 32	22 18
09	29 ♑ 55	29 10	05 00	06 15	16 52	15 23	05 42	06 50	22 49	08 49	21 57	07 33	22 18
10	29 42	29 07	04 51	06 38	20 42	15 56	05 14	06 35	22 49	08 51	21 57	07 34	22 18
11	29 31	29 04	04 29	07 01	23 40	16 28	04 47	06 21	22 49	08 54	21 56	07 34	22 18
12	29 23	29 01	03 54	07 24	25 35	17 00	04 19	06 06	22 49	08 56	21 56	07 35	22 17
13	29 17	28 57	03 07	07 46	26 15	17 30	03 50	05 52	22 49	08 59	21 56	07 36	22 17
14	29 14	28 54	02 10	08 08	25 33	18 00	03 22	05 37	22 49	09 02	21 56	07 37	22 17
15	29 D 14	28 51	01 S 05	08 31	23 27	18 28	02 54	05 22	22 49	09 04	21 56	07 38	22 17
16	29 R 14	28 48	00 N 05	08 53	20 01	18 56	02 25	05 07	22 49	09 07	21 56	07 38	22 17
17	29 13	28 45	01 17	09 15	15 23	19 22	01 57	04 53	22 49	09 09	21 56	07 39	22 17
18	29 11	28 42	02 27	09 37	09 48	19 47	01 28	04 38	22 49	09 12	21 56	07 40	22 17
19	29 06	28 38	03 28	09 59	03 S 31	20 11	00 59	04 23	22 49	09 14	21 56	07 41	22 17
20	28 59	28 35	04 17	10 20	03 N 04	20 33	00 30	04 09	22 49	09 17	21 56	07 42	22 17
21	28 49	28 32	04 49	10 42	09 34	20 55	00 N 01	03 54	22 49	09 20	21 56	07 43	22 16
22	28 38	28 29	05 00	11 03	15 31	21 15	00 S 28	03 39	22 49	09 22	21 55	07 43	22 16
23	28 26	28 26	04 50	11 24	20 27	21 33	00 57	03 24	22 49	09 25	21 55	07 44	22 16
24	28 16	28 22	04 21	11 45	23 59	21 50	01 26	03 09	22 49	09 27	21 55	07 45	22 16
25	28 08	28 19	03 35	12 06	25 52	22 05	01 55	02 55	22 49	09 30	21 55	07 46	22 16
26	28 02	28 16	02 39	12 26	26 02	22 19	02 24	02 40	22 49	09 32	21 55	07 47	22 16
27	28 00	28 13	01 35	12 47	24 40	22 31	02 53	02 25	22 49	09 35	21 55	07 47	22 16
28	27 59	28 10	00 N 28	13 07	22 00	22 41	03 22	02 10	22 49	09 37	21 55	07 48	22 16
29	27 D 59	28 07	00 S 38	13 27	18 21	22 49	03 51	01 55	22 49	09 40	21 55	07 49	22 16
30	27 R 58	28 03	01 40	13 47	14 00	22 55	04 20	01 40	22 49	09 42	21 55	07 50	22 16
31	27 ♑ 58	28 ♑ 00	02 S 38	14 S 06	09 N 11	22 S 58	04 S 49	01 N 26	22 N 49	09 S 45	21 N 56	07 S 51	22 N 16

ZODIAC SIGN ENTRIES

Date	h	m	Planets
01	18	53	☽ ♌
04	06	40	☽ ♍
04	16	40	☿ ♏
06	19	28	☽ ♎
09	07	56	☽ ♏
11	19	19	☽ ♐
14	04	51	☽ ♑
16	11	34	☽ ♒
18	14	55	☽ ♓
18	15	27	♀ ♎
20	15	27	☽ ♈
22	14	47	☽ ♉
22	15	36	♄ ♏
23	17	06	☉ ♏
24	15	04	☽ ♊
26	18	24	☽ ♋
29	01	55	☽ ♌
31	13	04	☽ ♍
31	15	49	☿ ♐

LATITUDES

Date	Mercury ☿ ° '	Venus ♀ ° '	Mars ♂ ° '	Jupiter ♃ ° '	Saturn ♄ ° '	Uranus ♅ ° '	Neptune ♆ ° '	Pluto ♇ ° '
01	00 S 44	01 N 07	01 N 14	00 S 35	02 N 15	00 N 27	01 N 38	09 N 28
04	01 06	01 13	01 14	00 35	02 15	00 27	01 38	09 28
07	01 27	01 18	01 14	00 35	02 15	00 27	01 38	09 29
10	01 48	01 22	01 15	00 35	02 15	00 27	01 38	09 30
13	02 06	01 26	01 15	00 35	02 15	00 27	01 38	09 31
16	02 24	01 29	01 15	00 35	02 15	00 27	01 38	09 32
19	02 38	01 32	01 15	00 35	02 15	00 27	01 38	09 33
22	02 50	01 34	01 15	00 35	02 15	00 28	01 38	09 34
25	02 57	01 35	01 15	00 35	02 15	00 28	01 38	09 35
28	02 59	01 35	01 15	00 35	02 15	00 28	01 38	09 36
31	02 S 54	01 N 35	01 N 15	00 S 35	02 N 15	00 N 28	01 N 38	09 N 37

DATA

Julian Date	2434652
Delta T	+31 seconds
Ayanamsa	23° 12' 54"
Synetic vernal point	05° ♓ 54' 06"
True obliquity of ecliptic	23° 26' 48"

MOON'S PHASES, APSIDES AND POSITIONS ☽

Date	h	m	Phase	Longitude ° '	Eclipse Indicator
08	00	40	●	14 ♎ 27	
15	21	44	☽	22 ♑ 15	
22	12	56	○	28 ♈ 50	
29	13	09	☾	05 ♌ 49	

Day	h	m	
06	18	36	Apogee
21	15	38	Perigee

06	01	15	0S	
13	11	58	Max dec	26° S 15'
20	00	52	0N	
26	02	30	Max dec	26° N 10'

LONGITUDES

Date	Chiron ⚷ ° '	Ceres ⚳ ° '	Pallas ⚴ ° '	Juno ⚵ ° '	Vesta ⚶ ° '	Black Moon Lilith ⚸ ° '
01	15 ♑ 24	04 ♍ 33	07 ♌ 50	04 ♌ 23	16 ♒ 28	01 ♎ 19
11	15 ♑ 35	08 ♍ 52	13 ♌ 05	08 ♌ 02	17 ♒ 01	02 ♎ 26
21	15 ♑ 55	13 ♍ 06	18 ♌ 04	12 ♌ 57	18 ♒ 16	03 ♎ 33
31	16 ♑ 21	17 ♍ 13	22 ♌ 47	16 ♌ 42	20 ♒ 07	04 ♎ 40

All ephemeris data is given at 12.00 UT and the Moon's longitude is additionally given for 24.00 UT

ASPECTARIAN

h m	Aspects
01 Thursday	
00 25	☽ ∥ ♃
04 09	☽ ∥ ♇
04 47	☽ ☌ ♅
05 20	☽ □ ♆
06 33	☽ ∠ ♀
06 53	☽ ∥ ♅
07 45	☽ ⊻ ♇
09 22	☽ □ ☿
10 09	☽ ∠ ♂
11 02	☽ Q ☉
11 25	☽ ⊻ ♃
13 58	☽ □ ♄
19 47	☿ ∠ ♂
23 11	☽ ⊥ ♃
02 Friday	
02 02	☽ ⊥ ♀
02 44	☿ △ ♃
04 42	☽ ⊥ ♂
12 03	♂ # ♄
12 39	☽ ⚹ ☉
15 13	☽ ⊻ ♀
16 44	☽ Q ♆
17 01	☽ ∠ ♃
17 13	☽ ⊻ ♂
19 48	☉ ∠ ♇
03 Saturday	
01 47	☽ Q ☿
01 51	☽ Q ♄
02 21	☿ ☌ ♄
03 59	♀ # ♄
16 10	☽ ⊻ ♅
16 48	☽ ⚹ ♆
17 16	☽ # ☿
19 15	☽ ☌ ♇
21 19	☽ ∠ ☉
23 09	☽ ⚹ ♃
04 Sunday	
02 11	☽ ⚹ ♄
02 30	♀ ∥ ♂
04 17	☽ ⊥ ♅
05 16	☽ ⚹ ☿
05 50	♀ ☌ ♂
09 26	☽ # ♄
12 05	☽ ∥ ♂
12 30	☽ ∥ ♀
14 43	☽ # ♆
16 52	♀ ∠ ♄
17 08	☽ ⊥ ☉
22 31	☽ ∠ ♅
23 11	☽ ∠ ♆
23 28	☽ Q ♃
05 Monday	
04 09	☽ # ☉
05 45	♂ ∠ ♄
06 23	☽ ⊻ ☉
08 50	☽ ∠ ♄
08 59	☽ ☌ ♂
10 30	☽ ☌ ♀
13 06	♀ # ♆
15 53	☽ ∠ ☿
17 30	☽ ⊥ ♆
06 Tuesday	
03 17	☽ ⊥ ♄
04 59	☽ ⚹ ♅
05 42	☽ ⊻ ♆
08 06	☽ ⊻ ♇
12 05	☽ □ ♃
12 39	☽ ⊥ ☿
15 34	☽ ⊻ ♄
16 47	♂ # ♆
20 17	☽ ⊥ ♇
07 Wednesday	
01 48	☽ ∥ ☉
02 28	☽ ⊻ ☿
05 21	☽ Q ♅
08 09	☽ # ♀
11 09	☽ # ♂
12 05	☽ ∥ ♆
14 32	☽ ∠ ♇
17 54	☽ ∥ ♄
08 Thursday	
00 40	☽ ☌ ☉
01 08	☽ ⊻ ♂
06 15	☽ ⊻ ♀
07 49	♀ ⊥ ♆
13 53	☽ ⊥ ♂
15 03	☉ ⊻ ♂
17 43	☽ □ ♅
18 31	☽ ☌ ♆
19 42	☽ ⊥ ♀
20 06	☉ # ♀
20 49	☽ ⚹ ♇
09 Friday	
00 45	☽ △ ♃
02 27	☽ ∥ ☿
04 41	☽ ☌ ♄
05 52	☿ Q ♇
08 56	☽ ∠ ♂
15 47	☽ ∠ ♀
20 52	☽ Q ♇
22 46	☽ ☌ ☿
10 Saturday	
06 44	☽ ⚼ ♃
10 07	☉ # ♂
16 26	☽ ⚹ ♂
18 02	☽ ⊻ ☉
21 08	☽ # ♅
23 55	☽ # ♇
11 Sunday	
00 34	☽ ± ♃
00 55	☽ ⚹ ♀
04 13	☽ # ♃
05 30	☽ △ ♅
06 23	☽ ⊻ ♆
06 54	☽ ⊥ ☉
08 33	☽ □ ♇
12 23	☽ ⚻ ♃
16 44	☽ ⊻ ♄
17 19	☽ Q ♂
18 11	☽ ⊥ ♆
12 Monday	
02 04	☽ ∠ ☉
03 03	☽ Q ♀
04 33	☽ ⊥ ♄
10 50	☽ ⚼ ♅
11 12	♂ ⊥ ♆
11 44	☽ ∠ ♆
16 58	☽ ⊻ ☿
19 39	♀ ⊥ ♄
22 07	☽ ∠ ♄
22 50	☿ ⚼ ♃
22 55	♀ ⚹ ♅
13 Tuesday	
00 53	☉ ∥ ♆
04 11	☽ ± ♅
05 51	☽ ⊥ ☿
06 00	☽ □ ♂
08 33	♀ ⊻ ♆
09 28	☽ ⚹ ☉
15 38	☽ ⚻ ♅
16 34	☽ ⚹ ♆
17 25	☽ □ ♀
18 36	☽ △ ♇
22 13	☽ ☍ ♃
14 Wednesday	
00 53	☽ ∠ ☿
02 55	☽ ⚹ ♄
05 47	♀ ⊻ ♇
10 02	☽ Q ☉
15 10	☽ Q ♆
22 41	☽ ⚼ ♇
15 Thursday	
01 27	☽ Q ♄
02 55	♃ St R
07 46	☽ ⚹ ☿
15 09	☽ ± ♇
16 44	☽ △ ♂
17 15	☽ # ♃
18 59	♀ □ ♃
21 15	☽ # ♇
21 44	☽ □ ☉
23 08	☽ ☍ ♅
23 46	☽ # ♅
16 Friday	
00 08	☽ □ ♆
01 58	☽ ⚻ ♇
05 19	☽ ⚻ ♃
06 22	☽ △ ♀
07 39	☽ Q ☿
10 14	☽ □ ♄
15 56	☽ ± ♃
16 48	☉ □ ♅
17 38	☽ ∥ ☿
20 42	☽ ⚼ ♂
17 Saturday	
05 06	☉ ∥ ♄
07 08	☉ ☌ ♆
07 36	☽ ⚼ ♃
11 10	☽ ⚼ ♀
13 00	☽ ± ♂
17 59	☽ □ ☿
23 41	☽ ⚻ ♂
18 Sunday	
03 18	☽ ⚻ ♅
03 55	☽ ± ♀
04 20	☽ △ ♆
05 01	♀ ⊻ ♄
05 51	☽ △ ☉
05 59	☽ ☍ ♇
07 49	☉ ⚹ ♇
09 02	☽ △ ♃
12 40	☽ ∥ ☉
13 21	☽ ± ♅
14 06	☽ △ ♄
14 21	☽ ∥ ♄
14 52	☽ ⚻ ♀
20 20	☽ ∥ ♆
19 Monday	
04 11	☽ ⚼ ♅
04 14	♀ ⊥ ♇
05 13	☽ ⚼ ♆
08 25	☽ ⚼ ☉
08 39	☽ # ♂
14 52	☽ ⚼ ♄
19 50	☽ ± ♆
22 02	☽ # ♀
23 46	☽ △ ☿
20 Tuesday	
00 00	☽ ± ☉
03 05	☽ ☍ ♂
03 07	☉ △ ♃
03 20	☽ ∥ ♀
04 25	☽ △ ♅
05 26	☽ ± ♄
05 27	☿ ± ♃
05 28	☽ ⚻ ♆
06 59	☽ ⚻ ♇
09 47	☽ □ ♃
10 15	☽ ⚻ ☉
15 04	☽ ⚻ ♄
15 45	☽ ∥ ♂
16 32	☽ ± ♇
19 43	☽ ☍ ♀
21 Wednesday	
01 31	☽ ⚼ ☿
04 59	☽ # ♆
06 47	☽ ⚼ ♇
11 04	☽ # ♄
11 30	♂ ⚹ ♅
14 12	☽ Q ♃
16 35	☽ # ☉
16 48	☽ ± ☿
22 Thursday	
02 58	☽ ⚻ ☿
03 51	☽ □ ♅
04 33	☽ ⚻ ♂
05 00	☽ ☍ ♆
06 26	☽ △ ♇
09 04	☽ ⚹ ♃
12 56	☽ ☍ ☉
14 25	☽ ± ♂
14 46	☽ ☍ ♄
16 03	☿ △ ♅
16 15	♂ ⊻ ♆
18 03	♀ Q ♅
23 15	☽ ⚻ ♀
23 Friday	
00 52	♂ ⊥ ♄
05 19	☽ ⚼ ♂
08 23	☽ Q ♅
08 50	☽ ∠ ♃
09 41	☽ ± ♀
10 06	☿ ⊻ ♆
18 58	☽ # ☿
20 38	☉ ☌ ♄
20 47	☽ ∥ ♅
23 03	☽ ∥ ♇
23 17	☽ ⊥ ♃
24 Saturday	
01 30	☽ ⚼ ♀
02 47	☽ ∥ ♃
03 41	♂ ⊻ ♇
03 45	☽ ⚹ ♅
05 02	☽ ⚻ ♆
06 20	☽ ☍ ☿
06 26	☽ □ ♇
06 34	☽ △ ♂
08 00	☿ □ ♇
09 02	☽ ⊻ ♃
14 54	☽ ± ♆
15 28	☽ ⚻ ♄
16 01	☿ ⚹ ♂
16 41	☽ ⚻ ☉
20 05	☿ # ♅
23 17	☉ ⊥ ♂
25 Sunday	
01 31	☽ ± ♄
03 25	☽ ± ☉
04 33	☽ ∠ ♅
04 34	☽ △ ♀
05 56	☽ ⚼ ♆
12 26	☽ Q ♇
16 53	☽ ⚼ ♄
19 51	☽ ⊥ ♅
19 51	☽ ⚼ ☉
26 Monday	
00 14	☿ ⚻ ♃
06 14	☽ ⊻ ♅
06 50	☿ # ♇
07 44	☽ △ ♆
09 10	☽ ⚹ ♇
11 40	☽ □ ♂
11 47	☽ ☌ ♃
12 35	☽ ⚻ ☿
12 55	♀ ∠ ♇
14 00	♂ □ ♃
19 17	☽ △ ♄
20 37	♀ # ♂
23 56	☽ ± ☿
27 Tuesday	
00 16	☽ △ ☉
12 03	☽ ∠ ♇
14 26	☽ □ ♀
17 19	☽ ⚼ ☿
21 46	☽ Q ♂
28 Wednesday	
04 42	☽ ⊥ ♇
05 38	☽ ∥ ♃
06 56	☽ # ☿
09 54	☽ ∥ ♇
12 32	☽ ∥ ♅
12 47	☽ ☌ ♅
14 33	☽ □ ♆
16 01	☽ ⊻ ♇
18 35	☽ ⊻ ♃
21 30	☽ ⚹ ♂
23 08	☽ △ ☿
29 Thursday	
03 25	☽ □ ♄
03 55	☽ Q ♀
06 02	☽ ⊥ ♃
11 08	☿ # ♃
13 09	☽ □ ☉
14 19	♅ St R
23 29	☽ ∠ ♃
30 Friday	
01 25	☽ Q ♆
03 38	☿ ∠ ♀
04 10	☽ ∠ ♂
05 57	☽ ⚹ ♀
13 04	☽ # ☉
13 33	☉ Q ♇
15 00	☽ Q ♄
23 00	☽ ⊥ ♂
23 15	☽ ⊻ ♅
31 Saturday	
01 18	☽ ⚹ ♆
02 43	☽ ☌ ♇
03 54	☽ Q ☉
05 08	☽ ⚹ ♃
09 20	☽ # ♄
11 16	☽ ⊥ ♅
11 41	☽ ⊻ ♂
12 58	☽ □ ☿
15 16	☽ ⚹ ♄
15 20	☽ ∠ ♀
18 26	☽ # ♆

NOVEMBER 1953

LONGITUDES

Date	Sidereal time h m s	Sun ☉ ° ' "	Moon ☽ ° ' "	Moon ☽ 24.00 ° ' "	Mercury ☿ ° '	Venus ♀ ° '	Mars ♂ ° '	Jupiter ♃ ° '	Saturn ♄ ° '	Uranus ♅ ° '	Neptune ♆ ° '	Pluto ♇ ° '
01	14 41 54	08 ♏ 46 25	11 ♍ 20 55	17 ♍ 15 52	00 ♐ 18	17 ♎ 12	29 ♍ 56	25 ♊ 58	01 ♏ 12	23 ♋ 06	24 ♎ 11	24 ♌ 52
02	14 45 51	09 46 30	23 ♍ 10 30	29 ♍ 05 18	00 33	18 27	00 ♎ 34	25 R 55	01 19	23 R 06	24 13	24 52
03	14 49 47	10 46 37	05 ♎ 00 39	10 ♎ 56 54	00 41	19 41	01 11	25 51	01 26	23 06	24 15	24 53
04	14 53 44	11 46 46	16 ♎ 54 21	22 ♎ 53 15	00 R 40	20 56	01 48	25 47	01 33	23 05	24 18	24 54
05	14 57 40	12 46 57	28 ♎ 53 48	04 ♏ 56 08	00 30	22 11	02 25	25 43	01 40	23 05	24 20	24 54
06	15 01 37	13 47 10	11 ♏ 00 24	17 ♏ 06 42	00 ♐ 11	23 26	03 03	25 39	01 48	23 05	24 22	24 55
07	15 05 33	14 47 24	23 ♏ 15 06	29 ♏ 25 40	29 ♏ 42	24 41	03 40	25 34	01 55	23 04	24 24	24 56
08	15 09 30	15 47 41	05 ♐ 38 29	11 ♐ 53 38	29 04	25 56	04 17	25 30	02 02	23 04	24 26	24 56
09	15 13 26	16 47 59	18 ♐ 11 13	24 ♐ 31 21	28 15	27 11	04 54	25 25	02 09	23 03	24 28	24 57
10	15 17 23	17 48 18	00 ♑ 54 12	07 ♑ 19 57	27 18	28 26	05 31	25 20	02 16	23 02	24 30	24 57
11	15 21 20	18 48 40	13 ♑ 48 35	20 ♑ 21 03	26 12	29 ♎ 41	06 09	25 15	02 23	23 02	24 32	24 58
12	15 25 16	19 49 02	26 ♑ 56 55	03 ♒ 36 42	24 59	00 ♏ 56	06 46	25 09	02 30	23 01	24 35	24 58
13	15 29 13	20 49 26	10 ♒ 20 41	17 ♒ 09 05	23 41	02 11	07 23	25 04	02 37	23 00	24 37	24 59
14	15 33 09	21 49 51	24 ♒ 02 09	01 ♓ 00 00	22 21	03 26	08 00	24 58	02 44	22 59	24 39	24 59
15	15 37 06	22 50 18	08 ♓ 02 43	15 ♓ 10 12	21 00	04 41	08 37	24 52	02 51	22 59	24 41	24 59
16	15 41 02	23 50 46	22 ♓ 22 16	29 ♓ 38 35	19 43	05 56	09 14	24 46	02 58	22 58	24 43	25 00
17	15 44 59	24 51 15	06 ♈ 58 38	14 ♈ 21 42	18 30	07 11	09 51	24 40	03 05	22 57	24 45	25 00
18	15 48 55	25 51 45	21 ♈ 47 00	29 ♈ 13 32	17 25	08 26	10 28	24 34	03 12	22 56	24 47	25 00
19	15 52 52	26 52 17	06 ♉ 40 14	14 ♉ 06 01	16 29	09 41	11 05	24 28	03 19	22 55	24 49	25 00
20	15 56 49	27 52 50	21 ♉ 29 44	28 ♉ 50 18	15 43	10 57	11 42	24 21	03 26	22 53	24 51	25 01
21	16 00 45	28 53 25	06 ♊ 11 06	13 ♊ 18 11	15 09	12 12	12 19	24 15	03 33	22 52	24 53	25 01
22	16 04 42	29 ♏ 54 01	20 ♊ 23 57	27 ♊ 23 31	14 47	13 27	12 56	24 08	03 39	22 51	24 55	25 01
23	16 08 38	00 ♐ 54 39	04 ♋ 16 32	11 ♋ 02 51	14 36	14 42	13 33	24 01	03 46	22 50	24 56	25 01
24	16 12 35	01 55 18	17 ♋ 42 29	24 ♋ 15 35	14 D 37	15 57	14 10	23 54	03 53	22 48	24 58	25 01
25	16 16 31	02 55 59	00 ♌ 42 28	07 ♌ 03 31	14 48	17 13	14 47	23 47	04 00	22 47	25 00	25 01
26	16 20 28	03 56 42	13 ♌ 19 14	19 ♌ 30 11	15 09	18 28	15 23	23 40	04 06	22 46	25 02	25 01
27	16 24 24	04 57 26	25 ♌ 36 57	01 ♍ 40 11	15 38	19 43	16 00	23 32	04 13	22 44	25 04	25 R 01
28	16 28 21	05 58 11	07 ♍ 40 31	13 ♍ 38 38	16 16	20 59	16 37	23 25	04 19	22 43	25 06	25 01
29	16 32 18	06 58 58	19 ♍ 35 10	25 ♍ 30 45	17 00	22 14	17 14	23 17	04 26	22 41	25 07	25 01
30	16 36 14	07 ♐ 59 46	01 ♎ 25 59	07 ♎ 21 26	17 ♏ 51	23 ♏ 29	17 ♎ 51	23 ♊ 10	04 ♏ 32	22 ♋ 40	25 ♎ 09	25 ♌ 01

Date	Moon True ☊ ° '	Moon Mean ☊ ° '	Moon ☽ Latitude ° '
01	27 ♑ 52	27 ♑ 57	03 S 28
02	27 R 45	27 54	04 08
03	27 36	27 51	04 38
04	27 24	27 48	04 55
05	27 10	27 44	05 00
06	26 56	27 41	04 52
07	26 43	27 38	04 29
08	26 32	27 35	03 54
09	26 23	27 32	03 07
10	26 17	27 28	02 10
11	26 15	27 25	01 S 06
12	26 D 14	27 22	00 N 04
13	26 14	27 19	01 14
14	26 R 15	27 16	02 22
15	26 14	27 13	03 23
16	26 10	27 09	04 13
17	26 05	27 06	04 47
18	25 57	27 03	05 03
19	25 48	27 00	04 59
20	25 38	26 57	04 34
21	25 30	26 54	03 52
22	25 25	26 50	02 55
23	25 18	26 47	01 50
24	25 16	26 44	00 N 41
25	25 D 16	26 41	00 S 29
26	25 17	26 38	01 36
27	25 18	26 34	02 36
28	25 R 19	26 31	03 28
29	25 18	26 28	04 10
30	25 ♑ 15	26 ♑ 25	04 S 42

DECLINATIONS

Date	Sun ☉ ° '	Moon ☽ ° '	Mercury ☿ ° '	Venus ♀ ° '	Mars ♂ ° '	Jupiter ♃ ° '	Saturn ♄ ° '	Uranus ♅ ° '	Neptune ♆ ° '	Pluto ♇ ° '
01	14 S 26	04 N 06	22 S 59	05 S 18	01 N 11	22 N 48	09 S 47	21 N 56	07 S 51	22 N 16
02	14 45	01 S 05	22 57	05 46	00 56	22 48	09 50	21 56	07 52	22 17
03	15 04	06 16	22 52	06 15	00 41	22 48	09 52	21 56	07 53	22 17
04	15 22	11 11	22 44	06 43	00 26	22 48	09 55	21 56	07 54	22 17
05	15 41	15 45	22 33	07 12	00 N 11	22 48	09 57	21 56	07 54	22 17
06	15 59	19 45	22 19	07 40	00 S 03	22 48	10 00	21 56	07 55	22 17
07	16 17	22 56	22 00	08 08	00 18	22 48	10 02	21 56	07 56	22 17
08	16 34	25 06	21 38	08 36	00 33	22 48	10 04	21 56	07 57	22 17
09	16 52	26 02	21 11	09 04	00 48	22 48	10 07	21 56	07 58	22 17
10	17 09	25 37	20 41	09 31	01 02	22 48	10 09	21 57	07 58	22 17
11	17 25	23 49	20 08	09 58	01 17	22 47	10 11	21 57	07 59	22 18
12	17 42	20 43	19 31	10 25	01 32	22 47	10 14	21 57	08 00	22 18
13	17 58	16 28	18 52	10 52	01 47	22 47	10 16	21 57	08 00	22 18
14	18 14	11 16	18 12	11 19	02 01	22 47	10 19	21 57	08 01	22 18
15	18 29	05 S 24	17 31	11 45	02 16	22 47	10 21	21 57	08 02	22 18
16	18 44	00 N 51	16 52	12 11	02 30	22 47	10 23	21 58	08 03	22 19
17	18 59	07 10	16 14	12 37	02 45	22 47	10 25	21 58	08 03	22 19
18	19 14	13 10	15 40	13 02	02 59	22 46	10 28	21 58	08 04	22 19
19	19 28	18 26	15 09	13 28	03 14	22 46	10 30	21 58	08 05	22 19
20	19 42	22 33	14 44	13 52	03 28	22 46	10 32	21 58	08 05	22 20
21	19 55	25 08	14 23	14 17	03 43	22 46	10 34	21 59	08 06	22 20
22	20 08	26 01	14 08	14 41	03 57	22 46	10 37	21 59	08 07	22 20
23	20 21	25 13	13 58	15 05	04 12	22 45	10 39	21 59	08 07	22 20
24	20 33	22 57	13 53	15 28	04 26	22 45	10 41	21 59	08 08	22 21
25	20 45	19 32	13 53	15 51	04 40	22 45	10 43	22 00	08 09	22 21
26	20 57	15 18	13 58	16 14	04 55	22 45	10 45	22 00	08 09	22 21
27	21 08	10 32	14 06	16 36	05 09	22 44	10 47	22 00	08 10	22 22
28	21 19	05 28	14 17	16 58	05 23	22 44	10 50	22 00	08 11	22 22
29	21 29	00 17	14 31	17 19	05 37	22 44	10 52	22 01	08 11	22 22
30	21 S 39	04 S 53	14 S 48	17 S 40	05 S 51	22 N 44	10 S 54	22 N 01	08 S 12	22 N 23

ZODIAC SIGN ENTRIES

Date	h	m	Planets
01	14	19	♂ ♎
03	01	51	☽ ♎
05	14	12	☽ ♏
06	22	19	☿ ♏
08	01	06	☽ ♐
10	10	18	☽ ♑
11	18	12	♀ ♏
12	17	31	☽ ♒
14	22	17	☽ ♓
17	00	35	☽ ♈
19	01	15	☽ ♉
21	01	54	☽ ♊
22	14	22	☉ ♐
23	04	31	☽ ♋
25	10	40	☽ ♌
27	20	41	☽ ♍
30	09	06	☽ ♎

LATITUDES

Date	Mercury ☿ ° '	Venus ♀ ° '	Mars ♂ ° '	Jupiter ♃ ° '	Saturn ♄ ° '	Uranus ♅ ° '	Neptune ♆ ° '	Pluto ♇ ° '
01	02 S 50	01 N 35	01 N 16	00 S 35	02 N 15	00 N 28	01 N 38	09 N 37
04	02 30	01 34	01 16	00 35	02 15	00 28	01 38	09 38
07	01 57	01 32	01 16	00 34	02 15	00 28	01 39	09 39
10	01 10	01 30	01 15	00 34	02 15	00 28	01 39	09 40
13	00 S 11	01 27	01 15	00 34	02 15	00 28	01 39	09 41
16	00 N 50	01 24	01 15	00 34	02 16	00 29	01 39	09 42
19	01 41	01 20	01 15	00 34	02 16	00 29	01 39	09 43
22	02 14	01 15	01 15	00 33	02 16	00 29	01 39	09 44
25	02 30	01 10	01 15	00 33	02 16	00 29	01 39	09 45
28	02 32	01 05	01 15	00 33	02 16	00 29	01 39	09 46
31	02 N 23	00 N 59	01 N 14	00 S 33	02 N 17	00 N 29	01 N 39	09 N 48

DATA

Julian Date	2434683
Delta T	+31 seconds
Ayanamsa	23° 12' 58"
Synetic vernal point	05° ♓ 54' 02"
True obliquity of ecliptic	23° 26' 48"

MOON'S PHASES, APSIDES AND POSITIONS ☽

Date	h	m	Phase	Longitude ° '	Eclipse Indicator
06	17	58	●	14 ♏ 02	
14	07	52	☽	21 ♒ 39	
20	23	12	○	28 ♉ 21	
28	08	16	☾	05 ♍ 49	

Day	h	m		
03	02	08	Apogee	
18	23	02	Perigee	
30	18	30	Apogee	
02	06	58	0S	
09	16	48	Max dec	26° S 03'
16	08	47	0N	
22	12	15	Max dec	26° N 01'
29	13	18	0S	

LONGITUDES

Date	Chiron ⚷ ° '	Ceres ⚳ ° '	Pallas ⚴ ° '	Juno ⚵ ° '	Vesta ⚶ ° '	Black Moon Lilith ⚸ ° '
01	16 ♑ 24	17 ♍ 37	23 ♌ 14	17 ♌ 03	20 ♒ 20	04 ♎ 47
11	16 ♑ 58	21 ♍ 36	27 ♌ 31	20 ♌ 19	22 ♒ 45	05 ♎ 54
21	17 ♑ 38	25 ♍ 23	01 ♍ 20	23 ♌ 03	25 ♒ 36	07 ♎ 01
31	18 ♑ 22	28 ♍ 58	04 ♍ 36	25 ♌ 11	28 ♒ 48	08 ♎ 09

All ephemeris data is given at 12.00 UT and the Moon's longitude is additionally given for 24.00 UT

ASPECTARIAN

h m	Aspects
01 Sunday	
02 35	☿ ⊥ ♆
05 12	☽ Q ♃
05 26	☽ ∠ ♅
06 18	☽ ⚹ ☉
06 56	☽ # ♀
07 36	☽ ∠ ♆
11 39	☽ ⊥ ♀
21 55	☽ ∠ ♄
02 Monday	
01 15	☽ ⊻ ♀
01 55	☽ ⊥ ♆
02 13	☽ ∥ ♂
02 27	☽ Q ☿
11 18	☽ # ♂
11 33	☿ ⚹ ♂
11 51	☽ ⚹ ♅
14 08	☽ ⊻ ♆
15 27	☽ ⊻ ♇
15 33	☽ ∠ ☉
16 23	☽ ⊥ ♄
17 32	☽ □ ♃
03 Tuesday	
00 18	♂ ⊥ ♇
03 10	☽ ⚹ ☿
03 38	☽ ⊥ ♇
03 49	☽ ☌ ♂
04 41	☽ ⊻ ♄
11 29	☽ ⊥ ☉
12 02	☽ ∥ ♀
12 10	☽ Q ♅
13 43	☉ ⚼ ♃
19 50	☽ ∥ ♆
21 49	☿ St R
21 52	☽ ∠ ♇
23 57	♂ ⊻ ♄
04 Wednesday	
00 44	☽ ⊻ ☉
01 42	☿ # ♃
05 37	☽ ∥ ♄
09 31	☽ ∠ ☿
21 02	☽ ☌ ♀
05 Thursday	
00 24	☽ □ ♅
02 51	☽ ☌ ♆
03 23	☽ ⊥ ☿
04 02	☽ ⚹ ♇
05 42	☽ △ ♃
11 32	☽ ∥ ☉
15 08	☽ ⊻ ☿
17 35	☽ ☌ ♄
19 24	☽ ⊻ ♂
06 Friday	
01 10	☿ ⊥ ♆
03 55	☽ Q ♇
05 14	♀ □ ♅
07 55	☽ ⊥ ♂
11 18	☽ ⚼ ♃
14 13	☿ # ♇
17 58	☽ ☌ ☉
22 46	☿ ⊥ ♀
07 Saturday	
01 30	♀ ∥ ♆
02 34	☽ ∠ ♂
03 42	☽ # ♅
04 52	☽ ± ♃
04 59	☽ ∥ ☿
06 29	☽ # ♇
06 30	♀ ☌ ♆
10 49	☽ # ♃
11 39	☽ △ ♅
14 15	☽ ⊻ ♆
15 06	☽ ⊻ ♀
15 16	☽ □ ♇
16 25	☿ # ♅
16 29	☽ ⊼ ♃
16 48	♀ ⚹ ♇
23 57	☽ ☌ ☿
08 Sunday	
01 55	☽ ⊥ ♆
04 02	☽ ⊥ ♀
04 11	♀ △ ♃
04 58	☽ ⊻ ♄
09 15	☽ ⚹ ♂
16 38	☽ ⊥ ♄
16 39	☽ ⚼ ♅
19 19	☽ ∠ ♆
23 16	☽ ∠ ♀
09 Monday	
09 08	☽ ⊻ ☉
09 26	☽ Q ♂
09 51	☽ ± ♅
10 00	☽ ∠ ♄
17 35	♂ Q ♅
21 13	☽ ⊼ ♅
21 30	☽ ⊥ ☉
23 53	☿ ⊻ ♀
23 56	☽ ⚹ ♆
10 Tuesday	
00 48	☽ △ ♇
01 36	☽ ☍ ♃
05 43	☽ ⊻ ☿
06 51	☽ ⚹ ♀
14 35	☽ ⚹ ♄
15 52	☽ ∠ ☉
16 08	☽ ⊥ ☿
21 04	☽ □ ♂
22 30	☽ Q ♆
11 Wednesday	
04 53	☽ ⚼ ♇
07 33	☽ ∠ ☿
07 38	☽ Q ♀
13 04	☽ Q ♄
21 04	☽ # ♃
21 28	☽ ± ♇
21 32	☉ ± ♃
21 57	☽ ⚹ ☉
12 Thursday	
00 47	♀ ∥ ♄
00 57	☽ # ♇
03 33	☽ # ♅
04 52	☽ ☍ ♅
07 41	☽ □ ♆
08 20	☿ ⊼ ♃
08 25	☽ ⊼ ♇
08 44	☽ ⚹ ☿
08 46	☽ ⊼ ♃
12 12	☿ □ ♇
19 24	☿ ⊻ ♆
19 33	☽ ± ♃
19 55	☽ □ ♀
20 44	☽ ∥ ☿
21 30	☽ Q ☉
22 06	☽ □ ♄
13 Friday	
04 27	☽ Q ☿
04 35	☽ ∥ ☉
06 29	☽ △ ♂
11 30	☽ ⚼ ♃
21 22	♀ ☌ ♄
14 Saturday	
00 21	☿ △ ♅
04 01	☿ ∠ ♂
07 52	☽ □ ☉
09 20	☽ □ ☿
09 25	♃ ⚹ ♇
10 07	☽ ⚼ ♂
10 11	☽ ⊼ ♅
11 16	☉ ∥ ☿
11 50	☽ ∥ ♀
13 03	☽ △ ♆
13 37	☽ △ ♃
13 38	☽ ☍ ♇
16 05	☽ ∥ ♄
17 15	☉ ☌ ☿
20 33	☽ ± ♅
15 Sunday	
01 33	☽ ∥ ♆
02 20	☽ ± ♂
03 06	☽ △ ♄
05 44	☽ △ ♀
11 53	☽ ⚼ ♅
13 01	☽ ⊼ ♂
14 46	☽ ⚼ ♆
15 15	☉ △ ♅
23 43	☽ ∥ ♂
16 Monday	
04 37	☽ ⚼ ♄
05 54	☽ ± ♆
07 56	☽ △ ☿
09 23	☽ ⚼ ♀
12 59	☽ △ ♅
14 38	☽ △ ☉
15 53	☽ ⊼ ♆
15 57	☽ □ ♃
16 20	☽ ⊼ ♇
18 31	☽ # ♂
19 40	☽ ± ♄
23 11	♃ △ ♆
17 Tuesday	
01 39	☽ ± ♀
02 13	☽ ± ♇
05 36	☽ ⊼ ♄
06 44	☽ ⚼ ☿
08 05	☿ ± ♃
08 05	☉ ⊼ ♃
08 25	♀ Q ♇
09 20	☉ ⊻ ♆
11 48	☉ ∠ ♂
12 22	☽ ⊼ ♀
15 26	☽ # ♆
15 27	☉ □ ♇
16 53	☽ ☍ ♂
16 55	☽ ⚼ ♇
17 02	☽ ⚼ ☉
17 46	♂ ∠ ♇
20 21	☽ ± ☿
21 12	☽ Q ♃
18 Wednesday	
00 50	☽ # ♄
05 24	☽ ⊼ ☿
08 40	☽ ± ☉
11 24	☽ # ♀
13 51	☽ □ ♅
16 28	☽ ⚹ ♃
16 51	☽ ☍ ♆
17 12	☽ △ ♇
19 03	☽ ⊼ ☉
21 51	☽ # ☿
19 Thursday	
02 21	☿ ⊥ ♂
06 33	☽ ☍ ♄
08 00	♀ ⚼ ♃
16 28	☽ ∠ ♃
17 19	☽ ☍ ♀
17 36	☽ # ☉
18 50	☽ Q ♅
19 26	☽ ⊼ ♂
20 Friday	
03 03	☽ ☍ ☿
05 34	☽ ± ♂
06 56	☽ ⊥ ♃
08 05	☽ ∥ ♅
10 28	☽ ∥ ♇
13 35	☽ ∥ ♃
14 16	☽ ⚹ ♅
16 38	☽ ⊻ ♃
17 28	☽ ⊼ ♆
17 44	☽ □ ♇
20 52	☽ ⚼ ♂
23 12	☽ ☍ ☉
21 Saturday	
03 20	☽ ± ♆
07 43	☽ ⊼ ♄
14 55	☽ ∠ ♅
15 45	☿ ∥ ♀
16 32	♀ ⊻ ♂
17 45	☽ ± ♄
18 17	☽ ⚼ ♆
22 49	☽ △ ♂
23 07	☽ ⊼ ♀
23 31	☽ Q ♇
22 Sunday	
02 42	☽ ⊼ ☿
05 59	☽ ⊥ ♅
09 01	☽ ⚼ ♄
10 14	☽ ± ♀
12 39	☽ ± ☿
16 11	☽ ⊻ ♅
18 20	☽ ☌ ♃
19 44	☽ △ ♆
19 54	☽ ⚹ ♇
23 Monday	
03 13	☽ ⚼ ♀
03 55	☽ ⚼ ☿
05 39	☽ ⊼ ☉
10 18	☿ ☌ ♀
11 06	☽ △ ♄
12 44	☉ ⊥ ♆
17 01	☽ ± ☉
22 10	☽ ∠ ♇
22 56	☿ St D
24 Tuesday	
05 17	☽ □ ♂
06 22	☽ △ ☿
08 30	☽ △ ♀
10 27	☽ ⚼ ☉
13 34	☽ ∥ ♃
14 23	☽ ⊥ ♇
16 45	☽ ∥ ♇
19 27	☽ ∥ ♅
21 18	☽ ☌ ♅
23 14	☽ ⊻ ♃
25 Wednesday	
01 21	☽ □ ♆
01 24	☽ ⊻ ♇
04 33	☽ # ☉
10 17	☽ ⊥ ♃
13 33	☿ ⊻ ♂
16 05	☽ Q ♂
16 33	☽ △ ☉
18 14	☽ □ ♄
21 55	♀ ± ♃
26 Thursday	
00 50	♆ ⚹ ♇
03 08	☽ ∠ ♃
07 24	☽ # ♀
11 27	☽ Q ♆
15 39	☽ □ ☿
16 13	☽ ⚹ ♂
16 15	☉ ⊻ ♄
18 49	☽ # ☿
20 44	♇ St R
23 07	☽ □ ♀
27 Friday	
05 15	☽ Q ♄
06 21	☽ ⊻ ♅
07 57	☽ ⚹ ♃
10 47	☽ # ♄
10 49	☽ ☌ ♇
10 55	☽ ⚹ ♆
18 10	☽ ⊥ ♅
23 15	☽ ∠ ♂
23 20	☽ # ♆
28 Saturday	
04 46	☽ Q ☿
05 14	☽ ⚹ ♄
07 31	☽ Q ♃
08 16	☽ □ ☉
12 04	☽ ∠ ♅
12 23	☽ # ♂
14 55	☽ Q ♀
16 52	☽ ∠ ♆
18 14	☽ ⊥ ♂
29 Sunday	
06 25	☽ ⚹ ☿
06 59	☽ ⊻ ♂
11 04	☽ ⊥ ♆
11 41	☽ ∠ ♄
17 59	☽ ⚹ ♀
18 16	☽ ⚹ ♅
19 09	☿ ± ♃
19 25	☽ □ ♃
20 30	♀ △ ♅
23 00	☽ ⊻ ♇
23 14	☽ ⊻ ♆
23 57	☽ Q ☉
30 Monday	
04 14	☉ ⚼ ♅
06 05	☽ ⊥ ♄
06 19	♀ ⊼ ♃
11 08	☿ ⊻ ♂
11 09	☽ ⊥ ♇
15 07	☽ ∠ ☿
16 50	☽ ∥ ♂
18 21	☽ ⊻ ♄
18 31	☽ Q ♅

DECEMBER 1953

LONGITUDES

Date	Sidereal time h m s	Sun ☉ ° ' "	Moon ☽ ° ' "	Moon ☽ 24.00 ° ' "	Mercury ☿ ° '	Venus ♀ ° '	Mars ♂ ° '	Jupiter ♃ ° '	Saturn ♄ ° '	Uranus ♅ ° '	Neptune ♆ ° '	Pluto ♇ ° '
01	16 40 11	09 ♐ 00 36	13 ♎ 17 38	19 ♎ 15 03	18 ♏ 48	24 ♏ 45	18 ♎ 27	23 ♊ 02	04 ♏ 39	22 ♋ 38	25 ♎ 11	25 ♌ 01
02	16 44 07	10 01 28	25 ♎ 14 09	01 ♏ 15 17	19 49	26 00	19 04	22 R 54	04 45	22 R 36	25 13	25 R 01
03	16 48 04	11 02 20	07 ♏ 18 47	13 ♏ 24 54	20 54	27 15	19 41	22 46	04 52	22 35	25 14	25 01
04	16 52 00	12 03 14	19 ♏ 33 52	25 ♏ 45 49	22 03	28 31	20 18	22 38	04 58	22 33	25 16	25 00
05	16 55 57	13 04 10	02 ♐ 00 51	08 ♐ 19 02	23 16	29 ♏ 46	20 54	22 30	05 04	22 31	25 18	25 00
06	16 59 53	14 05 06	14 ♐ 40 21	21 ♐ 04 48	24 31	01 ♐ 02	21 31	22 22	05 10	22 29	25 19	25 00
07	17 03 50	15 06 03	27 ♐ 32 21	04 ♑ 02 54	25 48	02 17	22 08	22 14	05 17	22 27	25 21	25 00
08	17 07 47	16 07 01	10 ♑ 36 26	17 ♑ 12 50	27 07	03 32	22 44	22 06	05 23	22 25	25 23	24 59
09	17 11 43	17 08 00	23 ♑ 52 06	00 ♒ 34 08	28 28	04 48	23 21	21 58	05 29	22 23	25 24	24 59
10	17 15 40	18 09 00	07 ♒ 18 56	14 ♒ 06 29	29 ♏ 50	06 03	23 57	21 50	05 35	22 21	25 26	24 58
11	17 19 36	19 10 00	20 ♒ 56 46	27 ♒ 49 46	01 ♐ 14	07 19	24 34	21 42	05 41	22 19	25 27	24 58
12	17 23 33	20 11 01	04 ♓ 45 28	11 ♓ 43 51	02 39	08 34	25 10	21 34	05 47	22 17	25 29	24 58
13	17 27 29	21 12 03	18 ♓ 44 50	25 ♓ 48 19	04 05	09 50	25 47	21 26	05 52	22 15	25 30	24 57
14	17 31 26	22 13 04	02 ♈ 54 07	10 ♈ 02 00	05 31	11 05	26 23	21 17	05 58	22 13	25 32	24 57
15	17 35 22	23 14 06	17 ♈ 11 40	24 ♈ 22 42	06 58	12 21	26 59	21 09	06 04	22 11	25 33	24 56
16	17 39 19	24 15 09	01 ♉ 34 38	08 ♉ 46 55	08 26	13 36	27 36	21 01	06 10	22 09	25 34	24 56
17	17 43 16	25 16 12	15 ♉ 58 56	23 ♉ 10 02	09 55	14 51	28 12	20 53	06 15	22 07	25 36	24 55
18	17 47 12	26 17 16	00 ♊ 19 31	07 ♊ 26 42	11 24	16 07	28 48	20 45	06 21	22 04	25 37	24 54
19	17 51 09	27 18 20	14 ♊ 30 57	21 ♊ 31 38	12 53	17 22	29 ♎ 25	20 37	06 26	22 02	25 38	24 54
20	17 55 05	28 19 24	28 ♊ 28 13	05 ♋ 20 15	14 23	18 38	00 ♏ 01	20 29	06 32	22 00	25 40	24 53
21	17 59 02	29 ♐ 20 29	12 ♋ 07 24	18 ♋ 49 24	15 54	19 53	00 37	20 21	06 37	21 57	25 41	24 52
22	18 02 58	00 ♑ 21 35	25 ♋ 26 09	01 ♌ 57 36	17 24	21 09	01 13	20 13	06 42	21 55	25 42	24 52
23	18 06 55	01 22 41	08 ♌ 23 53	14 ♌ 45 09	18 55	22 24	01 49	20 05	06 47	21 53	25 43	24 51
24	18 10 51	02 23 47	21 ♌ 01 42	27 ♌ 13 53	20 26	23 40	02 26	19 57	06 53	21 50	25 44	24 50
25	18 14 48	03 24 54	03 ♍ 22 08	09 ♍ 26 55	21 58	24 55	03 02	19 49	06 58	21 48	25 45	24 49
26	18 18 44	04 26 02	15 ♍ 28 46	21 ♍ 28 16	23 30	26 11	03 38	19 42	07 03	21 45	25 46	24 49
27	18 22 41	05 27 10	27 ♍ 25 58	03 ♎ 22 30	25 02	27 26	04 14	19 34	07 08	21 43	25 48	24 48
28	18 26 38	06 28 18	09 ♎ 18 28	15 ♎ 14 28	26 34	28 42	04 50	19 27	07 12	21 41	25 49	24 47
29	18 30 34	07 29 28	21 ♎ 11 07	27 ♎ 08 58	28 07	29 ♐ 57	05 26	19 19	07 17	21 38	25 50	24 46
30	18 34 31	08 30 37	03 ♏ 08 37	09 ♏ 10 34	29 ♐ 40	01 ♑ 13	06 02	19 12	07 22	21 35	25 51	24 45
31	18 38 27	09 ♑ 31 47	15 ♏ 15 17	21 ♏ 23 14	01 ♑ 13	02 ♑ 28	06 ♏ 37	19 ♊ 05	07 ♏ 26	21 ♋ 33	25 ♎ 51	24 ♌ 44

Date	Moon True ☊ ° '	Moon Mean ☊ ° '	Moon ☽ Latitude ° '
01	25 ♑ 10	26 ♑ 22	05 S 01
02	25 R 03	26 19	05 08
03	24 54	26 15	05 01
04	24 46	26 12	04 40
05	24 38	26 09	04 05
06	24 31	26 06	03 18
07	24 26	26 03	02 20
08	24 23	26 00	01 14
09	24 22	25 56	00 S 03
10	24 D 23	25 53	01 N 10
11	24 24	25 50	02 20
12	24 26	25 47	03 22
13	24 27	25 44	04 14
14	24 R 26	25 40	04 51
15	24 24	25 37	05 10
16	24 21	25 34	05 10
17	24 17	25 31	04 50
18	24 12	25 28	04 13
19	24 08	25 25	03 20
20	24 05	25 21	02 16
21	24 03	25 18	01 N 05
22	24 D 03	25 15	00 S 08
23	24 03	25 12	01 18
24	24 05	25 09	02 23
25	24 07	25 06	03 20
26	24 08	25 02	04 07
27	24 09	24 59	04 42
28	24 R 09	24 56	05 05
29	24 09	24 53	05 15
30	24 07	24 50	05 12
31	24 ♑ 05	24 ♑ 46	04 S 55

DECLINATIONS

Date	Sun ☉ ° '	Moon ☽ ° '	Mercury ☿ ° '	Venus ♀ ° '	Mars ♂ ° '	Jupiter ♃ ° '	Saturn ♄ ° '	Uranus ♅ ° '	Neptune ♆ ° '	Pluto ♇ ° '
01	21 S 48	09 S 52	15 S 07	18 S 00	06 S 05	22 N 43	10 S 56	22 N 01	08 S 12	22 N 23
02	21 58	14 32	15 28	18 20	06 19	22 43	10 58	22 01	08 13	22 23
03	22 06	18 41	15 50	18 39	06 33	22 43	11 00	22 02	08 14	22 24
04	22 15	22 07	16 13	18 58	06 47	22 42	11 02	22 02	08 14	22 24
05	22 24	24 35	16 37	19 17	07 01	22 42	11 04	22 02	08 15	22 25
06	22 30	25 51	17 01	19 34	07 15	22 42	11 06	22 03	08 15	22 25
07	22 37	25 46	17 26	19 52	07 29	22 41	11 08	22 03	08 16	22 26
08	22 43	24 15	17 51	20 08	07 42	22 41	11 09	22 03	08 16	22 26
09	22 49	21 23	18 16	20 24	07 56	22 41	11 11	22 04	08 17	22 26
10	22 55	17 19	18 41	20 40	08 09	22 40	11 13	22 04	08 17	22 27
11	23 00	12 19	19 05	20 55	08 23	22 40	11 15	22 05	08 18	22 27
12	23 05	06 38	19 30	21 09	08 36	22 40	11 17	22 05	08 18	22 28
13	23 09	00 S 33	19 53	21 23	08 50	22 39	11 19	22 05	08 19	22 28
14	23 13	05 N 36	20 16	21 36	09 03	22 39	11 20	22 06	08 19	22 29
15	23 16	11 31	20 39	21 49	09 16	22 38	11 22	22 06	08 20	22 29
16	23 19	16 52	21 00	22 00	09 29	22 38	11 24	22 06	08 20	22 30
17	23 21	21 15	21 21	22 12	09 43	22 38	11 25	22 07	08 21	22 30
18	23 24	24 21	21 41	22 22	09 56	22 37	11 27	22 07	08 21	22 31
19	23 25	25 51	22 00	22 32	10 09	22 37	11 29	22 08	08 22	22 31
20	23 26	25 42	22 19	22 41	10 21	22 36	11 30	22 08	08 22	22 32
21	23 27	23 58	22 36	22 50	10 34	22 36	11 32	22 08	08 22	22 32
22	23 27	20 56	22 52	22 57	10 47	22 35	11 33	22 09	08 22	22 33
23	23 26	16 55	23 07	23 04	11 00	22 35	11 35	22 09	08 23	22 34
24	23 25	12 14	23 21	23 11	11 12	22 35	11 36	22 09	08 23	22 34
25	23 24	07 10	23 34	23 16	11 25	22 34	11 38	22 10	08 24	22 35
26	23 22	01 N 56	23 46	23 21	11 37	22 34	11 39	22 10	08 24	22 35
27	23 20	03 S 18	23 57	23 26	11 49	22 33	11 41	22 11	08 24	22 36
28	23 17	08 22	24 06	23 29	12 02	22 33	11 42	22 11	08 25	22 37
29	23 14	13 08	24 14	23 32	12 14	22 33	11 44	22 12	08 25	22 37
30	23 10	17 27	24 21	23 34	12 26	22 32	11 45	22 12	08 25	22 38
31	23 S 06	21 S 06	24 S 27	23 S 35	12 S 38	22 N 32	11 S 46	22 N 12	08 S 26	22 N 38

ZODIAC SIGN ENTRIES

Date	h	m	Planets
02	21	30	☽ ♏
05	08	09	☽ ♐
05	16	24	♀ ♐
07	16	33	☽ ♑
09	22	59	☽ ♒
10	14	48	☿ ♐
12	03	46	☽ ♓
14	07	06	☽ ♈
16	09	22	☽ ♉
18	11	27	☽ ♊
20	11	22	♂ ♏
20	14	40	☽ ♋
22	03	31	☉ ♑
22	20	23	☽ ♌
25	05	24	☽ ♍
27	17	11	☽ ♎
29	12	53	♀ ♑
30	05	43	☽ ♏
30	17	14	☿ ♑

LATITUDES

Date	Mercury ☿ ° '	Venus ♀ ° '	Mars ♂ ° '	Jupiter ♃ ° '	Saturn ♄ ° '	Uranus ♅ ° '	Neptune ♆ ° '	Pluto ♇ ° '
01	02 N 23	00 N 59	01 N 14	00 S 33	02 N 17	00 N 29	01 N 39	09 N 48
04	02 08	00 53	01 14	00 32	02 17	00 29	01 39	09 49
07	01 50	00 47	01 14	00 32	02 17	00 29	01 40	09 50
10	01 28	00 40	01 13	00 31	02 18	00 29	01 40	09 51
13	01 06	00 33	01 13	00 31	02 18	00 29	01 40	09 52
16	00 43	00 26	01 13	00 31	02 19	00 29	01 40	09 53
19	00 N 21	00 19	01 12	00 30	02 19	00 29	01 40	09 54
22	00 S 01	00 12	01 12	00 30	02 20	00 29	01 40	09 55
25	00 22	00 N 04	01 11	00 29	02 20	00 29	01 40	09 56
28	00 42	00 S 03	01 11	00 29	02 20	00 29	01 41	09 56
31	01 S 00	00 S 10	01 N 10	00 S 28	02 N 21	00 N 29	01 N 41	09 N 57

DATA

Julian Date	2434713
Delta T	+31 seconds
Ayanamsa	23° 13' 02"
Synetic vernal point	05° ♓ 53' 58"
True obliquity of ecliptic	23° 26' 47"

MOON'S PHASES, APSIDES AND POSITIONS ☽

Date	h	m	Phase	Longitude ° '	Eclipse Indicator
06	10	48	●	14 ♐ 02	
13	16	30	☽	21 ♓ 23	
20	11	43	○	28 ♊ 19	
28	05	43	☾	06 ♎ 12	

Day	h	m		
16	13	58	Perigee	
28	15	06	Apogee	
06	22	34	Max dec	25° S 59'
13	14	10	0N	
19	21	42	Max dec	26° N 00'
26	20	50	0S	

LONGITUDES

Date	Chiron ⚷ ° '	Ceres ⚳ ° '	Pallas ⚴ ° '	Juno ⚵ ° '	Vesta ⚶ ° '	Black Moon Lilith ⚸ ° '
01	18 ♑ 22	28 ♍ 58	04 ♍ 36	25 ♌ 11	28 ♒ 48	08 ♎ 09
11	19 ♑ 11	02 ♎ 18	07 ♍ 12	26 ♌ 35	02 ♓ 19	09 ♎ 16
21	20 ♑ 03	05 ♎ 19	08 ♍ 59	27 ♌ 13	06 ♓ 04	10 ♎ 23
31	20 ♑ 57	07 ♎ 57	09 ♍ 51	26 ♌ 59	10 ♓ 00	11 ♎ 30

ASPECTARIAN

h m Aspects

01 Tuesday
01 21 ♀ ⊥ ♂
02 32 ☽ ✶ ☉
03 50 ☽ ∥ ♆
03 59 ☽ ∠ ♀
05 23 ☽ ∠ ♇
10 54 ☽ ⊥ ☿
17 10 ♀ □ ♇
17 18 ☽ ∥ ♄
20 35 ♀ ⊻ ♆
22 58 ☽ ☌ ♂

02 Wednesday
00 06 ☽ ⊻ ☿
00 16 ☽ ⊥ ♀
06 45 ☽ □ ♅
07 23 ☽ △ ♃
11 32 ☽ ∠ ☉
11 33 ☽ ✶ ♇
11 57 ☽ ☌ ♆
13 42 ☽ ⊻ ♀
16 34 ☉ ∠ ♆
17 33 ☽ ∥ ☿
23 07 ☉ # ♅

03 Thursday
07 06 ☽ ⊥ ☉
07 07 ☽ ☌ ♄
07 19 ☉ ⊥ ♄
11 24 ☽ Q ♇
11 46 ☽ ∥ ♀
12 54 ☽ bQ ♃
20 00 ☽ ⊻ ☉

04 Friday
06 22 ☽ ± ♃
11 22 ☽ # ♅
13 05 ☽ ∥ ☉
13 29 ☽ ⊻ ♂
14 23 ☽ # ♇
16 55 ☽ # ♃
17 21 ☽ ☌ ☿
17 46 ☽ △ ♅
17 54 ☽ ⊼ ♃
21 38 ☿ △ ♅
22 32 ☽ □ ♇
22 35 ☿ ⊼ ♃
23 04 ☽ ⊻ ♆

05 Saturday
01 42 ☽ ⊥ ♂
07 13 ☽ ☌ ♀
09 58 ♃ ⊻ ♅
10 37 ☽ ⊥ ♆
17 53 ☽ ⊻ ♄
19 48 ☽ ∠ ♂
19 52 ☉ # ♇
22 27 ☽ bQ ♅

06 Sunday
03 47 ☽ ∠ ♆
05 21 ☽ ⊥ ♄
10 48 ☽ ☌ ☉
15 24 ☽ ± ♅
17 47 ♀ ⊥ ♆
21 09 ☿ □ ♇
22 24 ☽ ∠ ♄

07 Monday
01 27 ☽ ✶ ♂
02 16 ☽ ☍ ♃
02 35 ☽ ⊼ ♅
03 35 ☿ ⊻ ♆
07 17 ☽ △ ♇
07 56 ☽ ✶ ♆
08 24 ☽ ⊻ ☿
15 41 ♂ △ ♃
20 44 ☽ ⊥ ☿
21 41 ☽ ⊻ ♀

08 Tuesday
00 17 ♂ □ ♅
00 44 ☽ Q ♂
02 22 ☽ ✶ ♄
03 41 ☉ # ♃
06 05 ☽ Q ♆
09 51 ☽ ⊥ ♀
10 52 ☽ bQ ♇
15 03 ☽ ∠ ☿
19 00 ☉ ± ♅
22 51 ☽ ⊻ ☉

09 Wednesday
00 24 ☽ Q ♄
01 38 ☽ ∥ ☉
02 27 ☽ # ♃
03 12 ☽ ± ♇
03 54 ☽ ∠ ♀
04 20 ☽ # ♇
07 10 ☽ # ♅
08 37 ☽ ⊼ ♃
09 21 ☽ ☍ ♅
10 34 ☽ ⊥ ☉
11 01 ☽ □ ♂
14 00 ☽ ⊼ ♇
14 46 ☽ □ ♆
17 59 ☽ ∥ ♀
19 17 ☽ ± ♃
21 10 ☽ ✶ ☿

10 Thursday
02 09 ♀ ⊻ ♄
04 00 ☽ ∠ ☉
05 16 ☽ ∥ ☿
08 54 ☽ □ ♄
09 32 ☽ ✶ ♀
11 09 ☽ bQ ♃
15 35 ☿ ⊥ ♂
20 54 ☽ Q ☿

11 Friday
02 32 ♂ ∥ ♆
08 38 ☽ ✶ ☉
08 51 ☽ Q ♀
12 12 ♀ bQ ♅
13 18 ☽ △ ♃
14 24 ☽ ⊼ ♅
15 49 ☿ ⊥ ♆
16 38 ☽ ∥ ♄
18 36 ☽ △ ♂
19 01 ☽ ☍ ♇
19 53 ☽ △ ♆

12 Saturday
00 50 ☽ ± ♅
03 47 ♂ ✶ ♇
04 11 ☽ ∥ ♂
05 08 ☽ ∥ ♆
07 12 ☽ Q ☉
07 56 ☽ □ ☿
13 46 ☽ △ ♄
16 21 ☽ bQ ♅
19 13 ☽ □ ♀
21 45 ☽ bQ ♂
21 52 ☽ bQ ♆

13 Sunday
00 38 ♂ ☌ ♆
03 30 ☉ ∠ ♄
13 17 ☽ ± ♆
13 50 ☽ ± ♂
15 39 ☽ bQ ♄
16 30 ☽ □ ☉
16 31 ☽ □ ♃
16 41 ☉ ☍ ♃
17 57 ☽ △ ♅
22 33 ☽ ⊼ ♇
23 30 ☽ ⊼ ♆

14 Monday
00 29 ☽ ⊼ ♂
01 07 ♀ ∠ ♆
07 01 ☽ ± ♄
08 42 ☽ ± ♇
12 01 ☉ ⊼ ♅
16 55 ☽ △ ☿
17 12 ☽ ⊼ ♄
20 00 ☿ ⊻ ♄
22 39 ☽ Q ♃
22 51 ☽ # ♆
23 03 ♀ ∠ ♂
23 51 ☽ bQ ♇

15 Tuesday
02 19 ☽ # ♂
03 05 ☽ △ ♀
06 18 ♀ ⊥ ♄
11 21 ☽ # ♄
15 20 ☿ bQ ♅
18 33 ☽ ✶ ♃
20 19 ☽ □ ♅
20 54 ☽ bQ ☿
21 06 ♃ bQ ♄
22 52 ☽ △ ☉

16 Wednesday
00 55 ☽ △ ♇
01 58 ☽ ☍ ♆
05 05 ☽ ☍ ♂
06 34 ☽ bQ ♀
13 36 ☽ ± ☿
19 20 ☽ ∠ ♃
19 41 ☽ ☍ ♄
22 59 ☽ ± ♀

17 Thursday
00 44 ☽ ⊼ ☿
01 11 ♀ # ♅
01 45 ☽ bQ ☉
02 14 ☽ Q ♅
03 44 ☉ △ ♇
09 57 ☽ ⊼ ♀
10 11 ☽ ⊥ ♃
12 41 ☽ # ☿
17 40 ☽ ∥ ♅
17 54 ☽ ± ☉
18 33 ☽ # ♀
19 49 ☉ ✶ ♆
20 06 ☽ ⊻ ♃
20 27 ☽ ∥ ♇
21 18 ☽ ∥ ♃
22 12 ☽ ✶ ♅
23 09 ☿ ∠ ♆

18 Friday
02 55 ☽ □ ♇
03 13 ☽ # ☉
04 05 ☽ ⊼ ♆
04 42 ☽ ⊼ ☉
09 20 ☽ ⊼ ♂
11 11 ♀ ± ♅
14 10 ☽ ± ♆
19 53 ☽ ± ♂
22 13 ☽ ⊼ ♄
23 20 ☽ ∠ ♅

19 Saturday
04 15 ☿ ⊥ ♄
05 24 ☽ bQ ♆
08 26 ☽ ± ♄
08 54 ☽ ☍ ☿
09 15 ☽ Q ♇
10 38 ♀ # ♇
11 49 ☽ bQ ♂
14 35 ☽ ⊥ ♅
17 21 ☽ ☍ ♀
21 25 ☿ # ♅
22 20 ☽ ☌ ♃
23 55 ☽ bQ ♄

20 Sunday
00 00 ♀ # ♃
00 50 ☽ ⊻ ♅
05 48 ☽ ✶ ♇
07 07 ☽ △ ♆
11 43 ☽ ☍ ☉
14 49 ☽ △ ♂

21 Monday
02 11 ☽ △ ♄
04 45 ☿ ∠ ♂
06 58 ☿ # ♇
08 00 ☽ ∠ ♇
12 06 ☿ # ♃
13 00 ☿ ± ♅
16 56 ☽ # ☉
19 35 ☽ ⊼ ☿
19 53 ♀ ☍ ♃
21 45 ☽ # ♀
23 00 ☽ # ☿

22 Tuesday
00 01 ☽ ∥ ♃
00 05 ☽ ⊥ ♇
00 25 ☽ ∥ ♇
02 36 ☽ ⊻ ♃
03 23 ☽ ⊼ ♀
03 29 ☽ ∥ ♅
05 37 ☽ ☌ ♅
07 49 ☽ ± ☿
10 57 ☽ ⊻ ♇
12 29 ☽ □ ♆
13 24 ☽ ⊥ ♃
15 28 ☽ ± ♀
21 49 ☽ ⊼ ☉
23 09 ☽ □ ♂
23 26 ♀ ∠ ♄

23 Wednesday
02 16 ♀ ⊼ ♅
02 31 ☽ bQ ☿
03 35 ☿ ∥ ♀
05 52 ☽ ∠ ♃
08 58 ☽ □ ♄
09 55 ☽ ± ☉
09 56 ☽ bQ ♀
22 03 ☽ Q ♆

24 Thursday
02 09 ♄ Q ♇
04 25 ☽ bQ ☉
04 57 ☿ ☍ ♃
09 57 ☽ ✶ ♃
10 42 ☽ △ ☿
10 47 ☽ Q ♂
13 33 ☽ ⊻ ♅
13 44 ☉ ✶ ♂
15 01 ☽ # ♄
16 47 ☽ # ♂
17 39 ☽ △ ♀
18 59 ☉ ∥ ☿
19 21 ☽ ☌ ♇
19 29 ☽ Q ♄
21 07 ☽ ✶ ♆

25 Friday
01 08 ☽ ⊥ ♅
02 21 ♃ ∥ ♇
06 17 ☽ # ♆
09 00 ☽ Q ♃
09 29 ☿ ⊼ ♅
10 09 ♀ △ ♇
11 18 ☽ ✶ ♂
11 58 ☿ ∠ ♄
12 06 ☽ △ ☉
18 44 ☽ ∠ ♅
19 08 ☽ ✶ ♄

26 Saturday
02 37 ☽ ∠ ♆
04 11 ♀ ✶ ♆
15 03 ☉ ∥ ♀
17 05 ♂ ∥ ♄
18 38 ☽ ∠ ♂
20 21 ☽ □ ♃
20 37 ☽ ⊥ ♆

27 Sunday
00 32 ☽ ✶ ♅
01 15 ☽ ∠ ♄
06 26 ☽ □ ☿
06 41 ☽ ⊻ ♇
08 25 ☿ △ ♇
08 41 ☽ ⊻ ♆
12 01 ☽ □ ♀
13 41 ☽ ⊥ ♂
18 47 ☽ ⊥ ♇
19 30 ☽ ⊥ ♄
23 13 ♂ bQ ♃

28 Monday
00 04 ☿ ✶ ♆
00 39 ☽ Q ♅
02 27 ☽ ⊻ ♂
05 43 ☽ □ ☉
07 43 ☽ ⊻ ♄
12 15 ☽ ∥ ♆
12 57 ☽ ∠ ♇

29 Tuesday
00 14 ☽ Q ☿
04 41 ☽ ∥ ♄
04 42 ☽ Q ♀
06 46 ☉ ✶ ♄
07 06 ☽ ∥ ♂
08 16 ☽ △ ♃
12 54 ☽ □ ♅
19 12 ☽ ✶ ♇
20 01 ☉ Q ♆
21 21 ☽ ☌ ♆
21 29 ☽ Q ☉

30 Wednesday
04 00 ☽ ✶ ☿
07 41 ☽ ✶ ♀
14 05 ☽ bQ ♃
18 03 ☽ ☌ ♂
19 11 ☽ Q ♇
20 28 ☽ ☌ ♄
23 40 ☽ ✶ ☉

31 Thursday
07 45 ☽ ± ♃
14 10 ☽ ∠ ☿
16 22 ♂ Q ♇
16 46 ☉ bQ ♇
16 51 ☽ ∠ ♀
19 25 ☽ ⊼ ♃
20 31 ☽ # ♅
23 09 ☽ # ♃

All ephemeris data is given at 12.00 UT and the Moon's longitude is additionally given for 24.00 UT

JANUARY 1954

LONGITUDES

Date	Sidereal time h m s	Sun ☉ ° ' "	Moon ☽ ° ' "	Moon ☽ 24.00 ° ' "	Mercury ☿ ° '	Venus ♀ ° '	Mars ♂ ° '	Jupiter ♃ ° '	Saturn ♄ ° '	Uranus ♅ ° '	Neptune ♆ ° '	Pluto ♇ ° '
01	18 42 24	10 ♑ 32 57	27 ♏ 34 46	03 ♐ 50 14	02 ♑ 47	03 ♑ 44	07 ♏ 13	18 ♊ 57	07 ♏ 31	21 ♋ 30	25 ♎ 52	24 ♌ 43
02	18 46 20	11 34 08	10 ♐ 09 50	16 ♐ 33 48	04 21	04 59	07 49	18 R 50	07 35	21 R 28	25 53	24 R 42
03	18 50 17	12 35 19	23 ♐ 02 11	29 ♐ 35 02	05 55	06 15	08 25	18 44	07 40	21 25	25 54	24 41
04	18 54 13	13 36 30	06 ♑ 12 17	12 ♑ 53 48	07 29	07 30	09 01	18 37	07 44	21 23	25 55	24 40
05	18 58 10	14 37 41	19 ♑ 39 23	26 ♑ 28 48	09 04	08 46	09 36	18 30	07 48	21 20	25 56	24 39
06	19 02 07	15 38 52	03 ♒ 21 41	10 ♒ 17 44	10 40	10 01	10 12	18 24	07 53	21 18	25 56	24 38
07	19 06 03	16 40 02	17 ♒ 16 31	24 ♒ 17 39	12 15	11 17	10 47	18 17	07 57	21 15	25 57	24 37
08	19 10 00	17 41 12	01 ♓ 20 42	08 ♓ 25 16	13 52	12 32	11 23	18 11	08 01	21 12	25 58	24 36
09	19 13 56	18 42 22	15 ♓ 30 57	22 ♓ 37 21	15 28	13 48	11 58	18 05	08 04	21 10	25 58	24 35
10	19 17 53	19 43 31	29 ♓ 44 06	06 ♈ 50 53	17 05	15 03	12 34	17 59	08 08	21 07	25 59	24 34
11	19 21 49	20 44 39	13 ♈ 57 21	21 ♈ 03 13	18 43	16 19	13 09	17 53	08 12	21 05	25 59	24 32
12	19 25 46	21 45 47	28 ♈ 08 13	05 ♉ 12 05	20 20	17 34	13 44	17 48	08 16	21 02	26 00	24 31
13	19 29 42	22 46 54	12 ♉ 14 35	19 ♉ 15 26	21 59	18 50	14 20	17 42	08 19	20 59	26 00	24 30
14	19 33 39	23 48 01	26 ♉ 14 26	03 ♊ 11 20	23 38	20 05	14 55	17 37	08 23	20 57	26 01	24 29
15	19 37 36	24 49 07	10 ♊ 05 53	16 ♊ 57 54	25 17	21 20	15 30	17 32	08 26	20 54	26 01	24 28
16	19 41 32	25 50 12	23 ♊ 47 07	00 ♋ 33 20	26 57	22 36	16 05	17 27	08 29	20 52	26 02	24 26
17	19 45 29	26 51 17	07 ♋ 16 21	13 ♋ 56 00	28 ♑ 37	23 51	16 40	17 22	08 32	20 49	26 02	24 25
18	19 49 25	27 52 21	20 ♋ 32 07	27 ♋ 04 34	00 ♒ 18	25 07	17 15	17 18	08 35	20 46	26 02	24 24
19	19 53 22	28 53 24	03 ♌ 33 17	09 ♌ 58 15	01 59	26 22	17 50	17 13	08 38	20 44	26 03	24 23
20	19 57 18	29 ♑ 54 27	16 ♌ 19 27	22 ♌ 36 57	03 41	27 38	18 25	17 09	08 41	20 41	26 03	24 21
21	20 01 15	00 ♒ 55 29	28 ♌ 50 54	05 ♍ 01 26	05 24	28 ♑ 53	19 00	17 05	08 44	20 39	26 03	24 20
22	20 05 11	01 56 30	11 ♍ 08 50	17 ♍ 13 21	07 06	00 ♒ 08	19 35	17 01	08 47	20 36	26 03	24 19
23	20 09 08	02 57 31	23 ♍ 15 20	29 ♍ 15 11	08 50	01 24	20 10	16 57	08 49	20 34	26 04	24 17
24	20 13 05	03 58 31	05 ♎ 13 19	11 ♎ 10 14	10 33	02 39	20 44	16 54	08 52	20 31	26 04	24 16
25	20 17 01	04 59 31	17 ♎ 06 26	23 ♎ 02 27	12 18	03 55	21 19	16 50	08 54	20 29	26 04	24 15
26	20 20 58	06 00 30	28 ♎ 58 52	04 ♏ 56 16	14 02	05 10	21 53	16 47	08 56	20 26	26 04	24 13
27	20 24 54	07 01 29	10 ♏ 55 14	16 ♏ 56 21	15 47	06 25	22 28	16 44	08 59	20 24	26 R 04	24 12
28	20 28 51	08 02 27	23 ♏ 00 15	29 ♏ 07 28	17 31	07 41	23 02	16 42	09 01	20 21	26 04	24 10
29	20 32 47	09 03 24	05 ♐ 18 34	11 ♐ 34 04	19 16	08 56	23 37	16 39	09 03	20 19	26 04	24 09
30	20 36 44	10 04 21	17 ♐ 54 24	24 ♐ 19 58	21 01	10 11	24 11	16 37	09 05	20 16	26 04	24 08
31	20 40 40	11 ♒ 05 16	00 ♑ 51 05	07 ♑ 27 57	22 ♒ 45	11 ♒ 27	24 ♏ 45	16 ♊ 35	09 ♏ 06	20 ♋ 14	26 ♎ 04	24 ♌ 06

DECLINATIONS

Date	Moon True ☊ ° '	Moon Mean ☊ ° '	Moon ☽ Latitude ° '	Sun ☉ ° '	Moon ☽ ° '	Mercury ☿ ° '	Venus ♀ ° '	Mars ♂ ° '	Jupiter ♃ ° '	Saturn ♄ ° '	Uranus ♅ ° '	Neptune ♆ ° '	Pluto ♇ ° '
01	24 ♑ 03	24 ♑ 43	04 S 24	23 S 02	23 S 54	24 S 31	23 S 36	12 S 50	22 N 31	11 S 47	22 N 13	08 S 26	22 N 39
02	24 R 01	24 40	03 39	22 57	25 36	24 34	23 36	13 02	22 31	11 49	22 13	08 26	22 39
03	24 00	24 37	02 43	22 51	25 59	24 36	23 35	13 13	22 31	11 50	22 14	08 26	22 40
04	23 59	24 34	01 37	22 45	24 55	24 36	23 34	13 25	22 30	11 51	22 14	08 27	22 41
05	23 58	24 31	00 S 24	22 39	22 24	24 35	23 31	13 36	22 30	11 52	22 14	08 27	22 41
06	23 D 58	24 27	00 N 52	22 32	18 34	24 32	23 28	13 48	22 29	11 53	22 15	08 27	22 42
07	23 59	24 24	02 06	22 24	13 40	24 28	23 24	13 59	22 29	11 55	22 15	08 27	22 42
08	23 59	24 21	03 13	22 17	08 00	24 23	23 20	14 10	22 29	11 56	22 16	08 27	22 43
09	24 00	24 18	04 08	22 08	01 S 53	24 16	23 15	14 22	22 28	11 57	22 16	08 28	22 44
10	24 00	24 15	04 49	22 00	04 N 19	24 08	23 09	14 33	22 28	11 58	22 16	08 28	22 44
11	24 00	24 11	05 12	21 51	10 18	23 58	23 02	14 44	22 28	11 59	22 17	08 28	22 45
12	24 R 01	24 08	05 16	21 41	15 44	23 47	22 55	14 54	22 27	12 00	22 17	08 28	22 45
13	24 00	24 05	05 01	21 31	20 17	23 34	22 47	15 05	22 27	12 00	22 18	08 28	22 46
14	24 00	24 02	04 28	21 21	23 40	23 20	22 38	15 16	22 27	12 01	22 18	08 28	22 47
15	24 D 00	23 59	03 40	21 10	25 36	23 04	22 29	15 26	22 26	12 02	22 19	08 28	22 47
16	24 01	23 56	02 40	20 59	25 58	22 47	22 18	15 37	22 26	12 03	22 19	08 28	22 48
17	24 01	23 52	01 32	20 48	24 46	22 28	22 08	15 47	22 26	12 04	22 19	08 29	22 49
18	24 01	23 49	00 N 19	20 36	22 12	22 08	21 56	15 57	22 26	12 05	22 20	08 29	22 49
19	24 R 01	23 46	00 S 53	20 23	18 31	21 46	21 44	16 07	22 25	12 05	22 20	08 29	22 50
20	24 00	23 43	02 01	20 11	14 02	21 22	21 31	16 17	22 25	12 06	22 21	08 29	22 51
21	24 00	23 40	03 01	19 57	09 03	20 57	21 18	16 27	22 25	12 07	22 21	08 29	22 51
22	23 59	23 37	03 52	19 44	03 N 48	20 30	21 04	16 37	22 25	12 07	22 21	08 29	22 52
23	23 57	23 33	04 32	19 30	01 S 29	20 02	20 49	16 47	22 25	12 08	22 22	08 29	22 53
24	23 56	23 30	05 00	19 16	06 40	19 33	20 34	16 56	22 25	12 08	22 22	08 29	22 53
25	23 55	23 27	05 14	19 01	11 33	19 01	20 18	17 06	22 24	12 09	22 22	08 29	22 54
26	23 54	23 24	05 15	18 47	16 01	18 29	20 01	17 15	22 24	12 10	22 23	08 29	22 54
27	23 D 54	23 21	05 03	18 31	19 54	17 54	19 44	17 24	22 24	12 10	22 23	08 29	22 55
28	23 55	23 17	04 37	18 16	22 59	17 19	19 26	17 33	22 24	12 10	22 24	08 29	22 56
29	23 56	23 14	03 58	18 00	25 05	16 42	19 08	17 42	22 24	12 11	22 24	08 28	22 56
30	23 57	23 11	03 06	17 44	25 59	16 04	18 49	17 51	22 24	12 11	22 24	08 28	22 57
31	23 ♑ 59	23 ♑ 08	02 S 04	17 S 27	25 S 31	15 S 24	18 S 30	18 S 00	22 N 24	12 S 12	22 N 25	08 S 28	22 N 58

ZODIAC SIGN ENTRIES

Date	h	m	Planets
01	16	39	☽ ♐
04	00	45	☽ ♑
06	06	09	☽ ♒
08	09	43	☽ ♓
10	12	27	☽ ♈
12	15	10	☽ ♉
14	18	29	☽ ♊
16	23	01	☽ ♋
18	07	43	☿ ♒
19	05	24	☽ ♌
20	14	11	☉ ♒
21	14	14	☽ ♍
22	09	20	♀ ♒
24	01	30	☽ ♎
26	14	03	☽ ♏
29	01	42	☽ ♐
31	10	27	☽ ♑

LATITUDES

Date	Mercury ☿ ° '	Venus ♀ ° '	Mars ♂ ° '	Jupiter ♃ ° '	Saturn ♄ ° '	Uranus ♅ ° '	Neptune ♆ ° '	Pluto ♇ ° '
01	01 S 06	00 S 13	01 N 10	00 S 28	02 N 22	00 N 30	01 N 41	09 N 58
04	01 22	00 20	01 09	00 27	02 22	00 30	01 41	09 58
07	01 36	00 27	01 08	00 27	02 23	00 30	01 41	09 59
10	01 47	00 33	01 07	00 26	02 23	00 30	01 41	10 00
13	01 57	00 40	01 06	00 26	02 24	00 30	01 42	10 01
16	02 03	00 46	01 05	00 25	02 25	00 30	01 42	10 01
19	02 05	00 52	01 04	00 25	02 25	00 30	01 42	10 02
22	02 04	00 57	01 03	00 24	02 26	00 30	01 42	10 03
25	01 59	01 03	01 02	00 23	02 27	00 30	01 42	10 03
28	01 49	01 07	01 01	00 23	02 28	00 30	01 43	10 04
31	01 S 33	01 S 11	01 N 00	00 S 22	02 N 28	00 N 30	01 N 43	10 N 04

DATA

Julian Date	2434744
Delta T	+31 seconds
Ayanamsa	23° 13' 08"
Synetic vernal point	05° ♓ 53' 52"
True obliquity of ecliptic	23° 26' 46"

MOON'S PHASES, APSIDES AND POSITIONS ☽

Date	h	m	Phase	Longitude ° '	Eclipse Indicator
05	02	21	●	14 ♑ 13	Annular
12	00	22	☽	21 ♈ 16	
19	02	37	○	28 ♋ 30	total
27	03	28	☾	06 ♏ 40	

Day	h	m		
10	09	31	Perigee	
25	12	16	Apogee	
03	06	32	Max dec	26° S 01'
09	19	18	ON	
16	05	27	Max dec	26° N 01'
23	05	13	OS	
30	16	04	Max dec	26° S 01'

LONGITUDES

Date	Chiron ⚷ ° '	Ceres ⚳ ° '	Pallas ⚴ ° '	Juno ⚵ ° '	Vesta ⚶ ° '	Black Moon Lilith ⚸ ° '
01	21 ♑ 02	08 ♎ 11	09 ♍ 53	26 ♌ 54	10 ♓ 25	11 ♎ 37
11	21 ♑ 57	10 ♎ 20	09 ♍ 35	25 ♌ 44	14 ♓ 32	12 ♎ 44
21	22 ♑ 52	11 ♎ 56	08 ♍ 12	23 ♌ 50	18 ♓ 47	13 ♎ 51
31	23 ♑ 46	12 ♎ 56	05 ♍ 50	21 ♌ 25	23 ♓ 08	14 ♎ 59

All ephemeris data is given at 12.00 UT and the Moon's longitude is additionally given for 24.00 UT

ASPECTARIAN

h m Aspects

01 Friday
00 08 ☽ # ♇
00 16 ☽ △ ♅
03 47 ☽ ∥ ☉
06 29 ☽ □ ♇
07 44 ☽ ∠ ☉
08 42 ☽ ⊻ ♆
08 57 ☽ ∥ ♀
10 14 ☽ ⊥ ☿
12 19 ☽ ⊥ ♀
19 10 ☽ ∥ ☿
20 15 ☽ ⊥ ♆
23 24 ☽ ⊻ ☿

02 Saturday
01 06 ☽ ⊻ ♀
01 37 ♂ ☌ ♄
02 32 ☽ ⊥ ☉
05 01 ☽ ᑫ ♅
07 07 ☽ ⊻ ♄
07 21 ☽ ⊻ ♂
13 22 ☽ ∠ ♆
14 53 ☽ ⊻ ☉
18 29 ☽ ⊥ ♄
19 12 ☽ ⊥ ♂
21 55 ☽ ± ♅

03 Sunday
04 06 ☽ ☍ ♃
09 02 ☽ ⊼ ♅
11 19 ☽ ∠ ♄
12 44 ☽ ∠ ♂
15 02 ☽ △ ♇
17 16 ☽ ⚹ ♆

04 Monday
13 01 ☿ ☌ ♀
14 35 ☽ ☌ ♀
14 38 ☽ ☌ ☿
14 46 ☽ ⚹ ♄
15 05 ☽ Q ♆
15 55 ☿ ⚹ ♄
15 58 ☽ ∥ ☿
16 43 ♀ ⚹ ♄
17 17 ☽ ⚹ ♂
18 13 ☽ ᑫ ♇
18 29 ☿ Q ♆
19 54 ♀ Q ♆

05 Tuesday
02 21 ☽ ☌ ☉
02 43 ☽ ∥ ♀
03 32 ☉ # ♇
09 50 ☽ # ♇
09 59 ☽ ⊼ ♃
10 05 ☽ ∥ ☉
10 14 ☽ ± ♇
11 16 ☽ # ♃
12 16 ☽ Q ♄
13 11 ☽ # ♅
14 57 ☽ ☍ ♅
15 35 ☽ Q ♂
20 28 ☽ ± ♃
20 40 ☿ ᑫ ♇
20 47 ☽ ⊼ ♇
23 02 ☽ □ ♆

06 Wednesday
00 48 ☿ ⚹ ♂
04 45 ♀ ᑫ ♇
12 03 ☽ ᑫ ♃
18 22 ♀ ⚹ ♂
19 52 ☽ □ ♄
20 13 ☉ # ♃

07 Thursday
00 21 ☽ □ ♂
00 40 ☽ ⊻ ♀
02 16 ☽ ⊻ ☿
10 36 ☽ ∥ ♂
10 53 ☽ ⊻ ☉
12 00 ☽ ⊥ ♀
13 43 ☽ △ ♃
13 54 ☽ ⊥ ☿
18 47 ☽ ⊼ ♅
19 42 ☽ ∥ ♄
21 57 ☽ ⊥ ☉

08 Friday
00 32 ☽ ☍ ♇
02 50 ☽ △ ♆
04 53 ☽ ∠ ♀
04 59 ☽ ± ♅
07 14 ☽ ∠ ☿
10 08 ☽ ∥ ♆
14 27 ☽ ∠ ☉
14 39 ☉ # ♅
20 13 ☽ ᑫ ♅
22 40 ☉ ⊼ ♃
23 21 ☽ △ ♄

09 Saturday
04 19 ☽ ᑫ ♆
05 45 ☽ △ ♂
08 48 ☽ ⚹ ♀
09 04 ☉ ± ♇
11 55 ☽ ⚹ ☿
15 52 ♂ ± ♃
16 18 ☽ □ ♃
17 48 ☽ ⚹ ☉
19 32 ☽ ± ♆
21 30 ☽ △ ♅

10 Sunday
00 49 ☽ ᑫ ♄
03 17 ☽ ⊼ ♇
05 40 ☽ ⊼ ♆
07 02 ☽ Q ♀
08 11 ☽ ᑫ ♂
10 46 ☽ Q ☿
13 23 ☽ ± ♇
15 37 ☽ Q ☉
16 04 ☽ ± ♄
22 20 ☉ Q ♄
22 28 ☽ Q ♃

11 Monday
00 01 ☽ ± ♂
00 34 ☿ ⊼ ♃
02 14 ☽ ⊼ ♄
04 28 ☽ # ♆
04 33 ☽ ᑫ ♇
09 33 ☿ ± ♇
10 35 ☽ ⊼ ♂
16 22 ☽ □ ♀
18 36 ☽ ⚹ ♃
19 07 ☽ # ♄
19 30 ☉ ☍ ♅
21 05 ☽ □ ☿

12 Tuesday
00 00 ☽ □ ♅
00 22 ☽ □ ☉
05 53 ☽ △ ♇
07 59 ☽ # ♂
08 22 ☽ ☍ ♆
10 46 ☿ Q ♄
16 02 ♀ ⊼ ♃
19 51 ☽ ∠ ♃
21 53 ☿ ☍ ♅

13 Wednesday
05 17 ☽ ☍ ♄
05 55 ♀ ± ♇
06 28 ☽ Q ♅
11 05 ☽ ⊥ ♃
13 42 ♀ # ♇
15 43 ☽ ☍ ♂
19 24 ☽ # ☉
21 17 ☽ ⊻ ♃

14 Thursday
00 22 ☽ △ ♀
01 08 ☽ ∥ ♅
02 12 ☽ ∥ ♃
02 56 ☽ ⚹ ♅
03 58 ☽ # ♀
04 41 ☽ ∥ ♇
06 54 ☽ △ ☿
07 28 ☽ △ ☉
08 00 ☉ ± ♃
08 59 ☽ □ ♇
09 22 ☽ # ☿
11 37 ☽ ⊼ ♆
11 51 ☿ ± ♃
17 51 ♀ Q ♄
18 33 ☉ ☌ ☿
21 58 ☽ ± ♆

15 Friday
00 15 ☿ ⊼ ♇
03 44 ☉ ⊼ ♇
03 55 ♀ ☍ ♅
04 44 ☽ ∠ ♅
04 49 ☽ ᑫ ♀
09 05 ☽ ⊼ ♄
11 28 ☽ ᑫ ☉
12 22 ☽ ᑫ ☿
13 37 ☽ ᑫ ♆
16 07 ☽ Q ♇
17 25 ♀ # ♃
19 36 ☽ ± ♄
20 22 ☽ ⊥ ♅
21 51 ☽ ⊼ ♂
22 05 ☽ ± ♀
22 44 ☿ □ ♆

16 Saturday
00 55 ☽ ☌ ♃
04 29 ☽ ± ☉
06 18 ☽ ± ☿
06 51 ☽ ⊻ ♅
08 52 ☽ ± ♂
09 41 ☽ ⊼ ♀
10 23 ☿ # ♇
10 53 ♀ # ♅
11 28 ☽ ᑫ ♄
13 09 ☽ ⚹ ♇
15 55 ☽ ⊼ ☉
15 58 ☽ △ ♆
16 34 ☉ □ ♆
18 23 ☽ ⊼ ☿

17 Sunday
01 32 ☽ ᑫ ♂
03 16 ♀ ± ♃
13 09 ☿ Q ♂
14 17 ☽ △ ♄
14 31 ☿ # ♃
15 51 ☽ ∠ ♇
22 07 ☿ # ♅
22 36 ♀ ⊼ ♇

18 Monday
05 45 ☽ △ ♂
06 08 ☽ ⊻ ♃
07 06 ☽ ∥ ♇
08 07 ☽ ⊥ ♇
10 13 ☽ ∥ ♃
10 59 ☽ ∥ ♅
12 26 ☽ ☌ ♅
12 35 ☽ # ☿
13 20 ♂ ⊼ ♃
14 03 ☽ # ♀
17 01 ☽ ⊥ ♃
19 04 ☽ ⊻ ♇
21 17 ☽ ☍ ♀
22 06 ☽ □ ♆
23 52 ☽ # ☉

19 Tuesday
02 37 ☽ ☍ ☉
05 48 ♀ □ ♆
08 39 ☽ ☍ ☿
09 32 ☽ ∠ ♃
15 06 ☿ ᑫ ♃
15 39 ☿ ∥ ♀
21 32 ☽ □ ♄

20 Wednesday
00 45 ☽ # ♂
07 41 ☽ Q ♆
13 33 ☽ ⚹ ♃
16 11 ☽ □ ♂
20 17 ☽ ⊻ ♅
21 29 ☽ # ♄

21 Thursday
03 19 ☽ ☌ ♇
06 36 ☽ ⚹ ♆
07 46 ☽ ⊥ ♅
07 54 ☽ Q ♄
12 04 ☽ ⊼ ♀
12 27 ☽ Q ♃
14 37 ☽ # ♆
16 12 ☉ Q ♂
16 23 ☽ ⊼ ☉

22 Friday
01 04 ☽ ± ♀
01 10 ☽ ∠ ♅
02 47 ☽ ⊼ ☿
04 39 ☽ Q ♂
05 08 ☽ ± ☉
07 20 ☽ ⚹ ♄
11 49 ☽ ∠ ♆
13 39 ☉ ᑫ ♃
16 30 ☽ ± ☿
20 47 ☽ ᑫ ♀
23 32 ☽ □ ♃

23 Saturday
00 30 ☽ ᑫ ☉
05 31 ☽ ⚹ ♂
05 38 ☽ ⊥ ♆
06 39 ☽ ⚹ ♅
11 53 ☿ □ ♄
13 08 ☽ ∠ ♄
13 20 ☽ ᑫ ☿
14 03 ☽ ⊻ ♇
17 36 ☽ ⊻ ♆
22 12 ♀ ᑫ ♃

24 Sunday
02 03 ☽ ⊥ ♇
03 29 ♂ △ ♅
06 13 ☽ △ ♀
06 35 ☽ Q ♅
07 14 ☽ ⊥ ♄
09 15 ☽ △ ☉
13 06 ☽ ∠ ♂
15 02 ♀ Q ♂
19 22 ☽ ⊻ ♄
20 08 ☽ ∠ ♇
20 44 ☽ ∥ ♆

25 Monday
00 36 ☽ △ ☿
08 11 ☽ ⊥ ♂
11 28 ☽ △ ♃
11 48 ☉ ∥ ☿
15 04 ☽ ∥ ♄
18 47 ☽ □ ♅
20 57 ☽ ⊻ ♂

26 Tuesday
02 24 ☽ ⚹ ♇
06 07 ☽ ☌ ♆
17 38 ☽ ᑫ ♃
19 29 ☽ ∥ ♂

27 Wednesday
00 50 ☽ ∥ ☿
01 55 ☽ □ ♀
02 33 ☽ Q ♇
03 28 ☽ □ ☉
03 35 ☽ ∥ ☉
08 06 ☽ ☌ ♄
10 50 ♆ St R
11 00 ☽ ∥ ♀
11 39 ☽ ± ♃
23 19 ☽ □ ☿
23 34 ☽ ⊼ ♃

28 Thursday
00 56 ☿ △ ♃
04 17 ☿ ∥ ♂
06 47 ☽ △ ♅
06 48 ☽ # ♅
06 53 ☽ # ♃
11 29 ☽ # ♇
12 04 ☽ ☌ ♂
14 18 ☽ □ ♇
17 51 ☽ Q ♀
18 01 ☽ ⊻ ♆
18 31 ☽ Q ☉

29 Friday
05 43 ☽ ⊥ ♆
11 45 ☉ □ ♄
12 00 ☽ ᑫ ♅
14 13 ♀ □ ♄
16 23 ☽ Q ☿
18 20 ♃ ∥ ♅
19 12 ☽ ⊻ ♄
19 45 ☽ ⚹ ♀
19 50 ☽ ⚹ ☉
23 02 ☽ ∠ ♆

30 Saturday
00 17 ☉ ☌ ♀
02 03 ☿ ⊼ ♅
04 47 ☉ ∥ ♂
05 10 ☽ ± ♅
06 39 ☽ ⊥ ♄
09 35 ☽ ☍ ♃
09 39 ♂ □ ♇
16 25 ☽ ⊼ ♅
18 44 ☽ ⚹ ☿
23 33 ☽ ∠ ♄
23 36 ☽ △ ♇

31 Sunday
00 16 ☽ ⊻ ♂
02 30 ☽ ∠ ☉
03 03 ☽ ∠ ♀
03 12 ☽ ⚹ ♆
11 49 ☽ ⊥ ♂
20 21 ☽ ⊥ ☉
21 14 ☽ ⊥ ♀

FEBRUARY 1954

LONGITUDES

Date	Sidereal time h	m	s	Sun ☉ ° ' "	Moon ☽ ° ' "	Moon ☽ 24.00 ° ' "	Mercury ☿ ° '	Venus ♀ ° '	Mars ♂ ° '	Jupiter ♃ ° '	Saturn ♄ ° '	Uranus ♅ ° '	Neptune ♆ ° '	Pluto ♇ ° '
01	20	44	37	12 ♒ 06 12	14 ♑ 10 41	20 ♑ 59 17	24 ♒ 29	12 ♒ 42	25 ♏ 19	16 ♊ 33	09 ♏ 08	20 ♋ 12	26 ♎ 03	24 ♌ 05
02	20	48	34	13 07 06	27 ♑ 53 35	04 ♒ 53 19	26 12	13 57	25 54	16 R 31	09 10	20 R 09	26 R 03	24 R 03
03	20	52	30	14 07 59	11 ♒ 58 04	19 ♒ 07 18	27 54	15 13	26 28	16 30	09 11	20 07	26 03	24 02
04	20	56	27	15 08 51	26 ♒ 20 21	03 ♓ 36 27	29 ♒ 35	16 28	27 01	16 28	09 13	20 05	26 03	24 00
05	21	00	23	16 09 41	10 ♓ 54 48	18 ♓ 14 30	01 ♓ 14	17 43	27 35	16 27	09 14	20 02	26 02	23 59
06	21	04	20	17 10 30	25 ♓ 34 41	02 ♈ 54 30	02 51	18 59	28 09	16 26	09 15	20 00	26 02	23 57
07	21	08	16	18 11 18	10 ♈ 13 09	17 ♈ 29 54	04 25	20 14	28 43	16 26	09 16	19 58	26 02	23 56
08	21	12	13	19 12 05	24 ♈ 44 08	01 ♉ 55 19	05 56	21 29	29 16	16 25	09 17	19 56	26 01	23 55
09	21	16	09	20 12 49	09 ♉ 03 05	16 ♉ 07 07	07 24	22 44	29 ♏ 50	16 25	09 18	19 53	26 01	23 53
10	21	20	06	21 13 33	23 ♉ 07 15	00 ♊ 03 23	08 47	24 00	00 ♐ 23	16 D 25	09 19	19 51	26 00	23 52
11	21	24	03	22 14 14	06 ♊ 55 31	13 ♊ 43 43	10 05	25 15	00 57	16 25	09 19	19 49	26 00	23 50
12	21	27	59	23 14 54	20 ♊ 28 03	27 ♊ 08 40	11 17	26 30	01 30	16 25	09 20	19 47	25 59	23 49
13	21	31	56	24 15 33	03 ♋ 45 43	10 ♋ 19 21	12 23	27 45	02 03	16 26	09 20	19 45	25 59	23 47
14	21	35	52	25 16 09	16 ♋ 49 44	23 ♋ 16 59	13 22	29 ♒ 00	02 36	16 26	09 21	19 43	25 58	23 46
15	21	39	49	26 16 44	29 ♋ 41 15	06 ♌ 02 39	14 12	00 ♓ 16	03 09	16 27	09 21	19 41	25 58	23 44
16	21	43	45	27 17 17	12 ♌ 21 17	18 ♌ 37 15	14 55	01 31	03 42	16 29	09 21	19 39	25 57	23 43
17	21	47	42	28 17 49	24 ♌ 50 38	01 ♍ 01 32	15 27	02 46	04 15	16 30	09 R 21	19 37	25 56	23 41
18	21	51	38	29 ♒ 18 19	07 ♍ 10 04	13 ♍ 16 19	15 51	04 01	04 47	16 31	09 21	19 36	25 56	23 40
19	21	55	35	00 ♓ 18 48	19 ♍ 20 25	25 ♍ 22 34	16 04	05 16	05 20	16 33	09 21	19 34	25 55	23 38
20	21	59	32	01 19 15	01 ♎ 22 55	07 ♎ 21 44	16 R 07	06 31	05 52	16 35	09 21	19 32	25 54	23 37
21	22	03	28	02 19 41	13 ♎ 19 15	19 ♎ 15 49	16 00	07 46	06 25	16 37	09 20	19 30	25 53	23 35
22	22	07	25	03 20 05	25 ♎ 11 47	01 ♏ 07 33	15 44	09 01	06 57	16 39	09 20	19 29	25 53	23 34
23	22	11	21	04 20 28	07 ♏ 03 34	13 ♏ 00 20	15 17	10 16	07 29	16 42	09 19	19 27	25 52	23 32
24	22	15	18	05 20 49	18 ♏ 58 22	24 ♏ 58 15	14 42	11 31	08 01	16 45	09 18	19 25	25 51	23 31
25	22	19	14	06 21 09	01 ♐ 00 32	07 ♐ 05 52	13 59	12 46	08 33	16 48	09 18	19 24	25 50	23 30
26	22	23	11	07 21 28	13 ♐ 14 50	19 ♐ 28 04	13 09	14 01	09 05	16 51	09 17	19 22	25 49	23 28
27	22	27	07	08 21 45	25 ♐ 46 10	02 ♑ 09 42	12 13	15 16	09 37	16 54	09 16	19 21	25 48	23 27
28	22	31	04	09 ♓ 22 01	08 ♑ 39 10	15 ♑ 15 01	11 ♓ 13	16 ♓ 31	10 ♐ 08	16 ♊ 57	09 ♏ 15	19 ♋ 19	25 ♎ 47	23 ♌ 25

DECLINATIONS

Date	Moon True ☊ ° '	Moon Mean ☊ ° '	Moon ☽ Latitude ° '	Sun ☉ ° '	Moon ☽ ° '	Mercury ☿ ° '	Venus ♀ ° '	Mars ♂ ° '	Jupiter ♃ ° '	Saturn ♄ ° '	Uranus ♅ ° '	Neptune ♆ ° '	Pluto ♇ ° '
01	24 ♑ 00	23 ♑ 05	00 S 54	17 S 10	23 S 35	14 S 43	18 S 10	18 S 09	22 N 24	12 S 12	22 N 25	08 S 28	22 N 58
02	24 R 00	23 02	00 N 21	16 53	20 14	14 02	17 50	18 17	22 24	12 12	22 25	08 28	22 59
03	23 59	22 58	01 37	16 36	15 39	13 19	17 29	18 25	22 24	12 12	22 26	08 28	22 59
04	23 57	22 55	02 49	16 18	10 06	12 35	17 07	18 34	22 24	12 13	22 26	08 28	23 00
05	23 54	22 52	03 50	16 00	03 S 56	11 51	16 45	18 42	22 24	12 13	22 26	08 28	23 01
06	23 51	22 49	04 37	15 42	02 N 28	11 07	16 23	18 50	22 24	12 13	22 27	08 27	23 01
07	23 48	22 46	05 05	15 23	08 43	10 22	16 00	18 58	22 24	12 13	22 27	08 27	23 02
08	23 45	22 43	05 14	15 04	14 27	09 38	15 37	19 06	22 25	12 13	22 27	08 27	23 03
09	23 43	22 39	05 03	14 45	19 18	08 54	15 13	19 13	22 25	12 13	22 28	08 27	23 03
10	23 D 43	22 36	04 34	14 26	22 58	08 10	14 49	19 21	22 25	12 13	22 28	08 27	23 04
11	23 44	22 33	03 49	14 06	25 14	07 28	14 25	19 28	22 25	12 13	22 28	08 26	23 04
12	23 45	22 30	02 52	13 46	25 58	06 47	14 00	19 36	22 25	12 13	22 29	08 26	23 05
13	23 47	22 27	01 48	13 26	25 11	06 08	13 34	19 43	22 26	12 13	22 29	08 26	23 05
14	23 48	22 23	00 N 38	13 06	23 01	05 32	13 09	19 50	22 26	12 13	22 29	08 26	23 06
15	23 R 48	22 20	00 S 32	12 46	19 42	04 58	12 43	19 57	22 26	12 13	22 29	08 25	23 07
16	23 46	22 17	01 40	12 25	15 30	04 27	12 16	20 04	22 26	12 13	22 30	08 25	23 07
17	23 43	22 14	02 41	12 04	10 43	04 00	11 50	20 11	22 27	12 12	22 30	08 25	23 08
18	23 38	22 11	03 34	11 43	05 34	03 36	11 23	20 17	22 27	12 12	22 30	08 24	23 08
19	23 31	22 08	04 17	11 22	00 N 17	03 17	10 55	20 24	22 27	12 12	22 30	08 24	23 09
20	23 24	22 04	04 47	11 00	04 S 56	03 03	10 28	20 30	22 28	12 11	22 31	08 24	23 09
21	23 17	22 01	05 05	10 39	09 56	02 53	10 00	20 36	22 28	12 11	22 31	08 23	23 10
22	23 10	21 58	05 09	10 17	14 33	02 48	09 32	20 42	22 29	12 11	22 31	08 23	23 10
23	23 05	21 55	05 00	09 55	18 36	02 48	09 04	20 48	22 29	12 10	22 31	08 23	23 11
24	23 01	21 52	04 38	09 33	21 56	02 53	08 35	20 54	22 29	12 10	22 32	08 22	23 11
25	23 00	21 49	04 04	09 11	24 21	03 02	08 06	21 00	22 30	12 09	22 32	08 22	23 12
26	23 D 00	21 45	03 18	08 49	25 40	03 16	07 37	21 06	22 30	12 09	22 32	08 22	23 12
27	23 01	21 42	02 22	08 26	25 45	03 34	07 08	21 11	22 31	12 08	22 32	08 22	23 13
28	23 ♑ 02	21 ♑ 39	01 S 17	08 S 04	24 S 27	03 S 55	06 S 38	21 S 17	22 N 31	12 S 08	22 N 32	08 S 21	23 N 13

ZODIAC SIGN ENTRIES

Date	h	m	Planets
02	15	38	☽ ♒
04	18	03	☿ ♓
04	18	03	☽ ♓
06	19	14	☽ ♈
08	20	47	☽ ♉
09	19	18	♂ ♐
10	23	54	☽ ♊
13	05	10	☽ ♋
15	07	01	♀ ♓
15	12	35	☽ ♌
17	22	00	☽ ♍
19	04	32	☉ ♓
20	09	14	☽ ♎
22	21	43	☽ ♏
25	10	00	☽ ♐
27	19	58	☽ ♑

LATITUDES

Date	Mercury ☿ ° '	Venus ♀ ° '	Mars ♂ ° '	Jupiter ♃ ° '	Saturn ♄ ° '	Uranus ♅ ° '	Neptune ♆ ° '	Pluto ♇ ° '
01	01 S 26	01 S 13	00 N 59	00 S 22	02 N 29	00 N 30	01 N 43	10 N 04
04	01 02	01 16	00 58	00 21	02 29	00 30	01 43	10 05
07	00 S 31	01 20	00 56	00 21	02 30	00 30	01 43	10 05
10	00 N 07	01 22	00 55	00 20	02 31	00 30	01 43	10 05
13	00 50	01 24	00 53	00 20	02 31	00 30	01 43	10 06
16	01 37	01 26	00 51	00 19	02 32	00 30	01 44	10 06
19	02 24	01 27	00 49	00 19	02 33	00 30	01 44	10 06
22	03 04	01 27	00 47	00 18	02 34	00 30	01 44	10 06
25	03 32	01 27	00 45	00 18	02 34	00 30	01 44	10 06
28	03 43	01 26	00 42	00 17	02 35	00 29	01 44	10 06
31	03 N 34	01 S 24	00 N 40	00 S 17	02 N 36	00 N 29	01 N 44	10 N 06

DATA

Julian Date	2434775
Delta T	+31 seconds
Ayanamsa	23° 13' 14"
Synetic vernal point	05° ♓ 53' 46"
True obliquity of ecliptic	23° 26' 46"

MOON'S PHASES, APSIDES AND POSITIONS ☽

Date	h	m	Phase	Longitude ° '	Eclipse Indicator
03	15	55	●	14 ♒ 18	
10	08	29	☽	21 ♉ 05	
17	19	17	○	28 ♌ 36	
25	23	29	☾	06 ♐ 50	

Day	h	m		
06	05	55	Perigee	
22	06	38	Apogee	
06	02	44	ON	
12	11	24	Max dec	25° N 58'
19	13	18	0S	
27	01	19	Max dec	25° S 53'

LONGITUDES

Date	Chiron ⚷ ° '	Ceres ⚳ ° '	Pallas ⚴ ° '	Juno ⚵ ° '	Vesta ⚶ ° '	Black Moon Lilith ⚸ ° '
01	23 ♑ 52	13 ♎ 00	05 ♍ 33	21 ♌ 10	23 ♓ 35	15 ♎ 05
11	24 ♑ 44	13 ♎ 14	02 ♍ 24	18 ♌ 34	28 ♓ 01	16 ♎ 12
21	25 ♑ 32	12 ♎ 46	29 ♌ 02	16 ♌ 10	02 ♈ 37	17 ♎ 20
31	26 ♑ 17	11 ♎ 35	25 ♌ 56	14 ♌ 15	07 ♈ 03	18 ♎ 27

ASPECTARIAN

h m Aspects

01 Monday
- 01 04 ☽ Q ♆
- 02 23 ☽ ∠ ☿
- 02 55 ☽ ⚼ ♇
- 02 59 ☽ ✶ ♄
- 04 50 ☽ ∠ ♂
- 06 29 ☿ ☍ ♇
- 08 00 ☽ ⊻ ☉
- 09 06 ☽ ⊻ ♀
- 13 10 ♀ ∥ ♂
- 16 11 ☽ ⚻ ♃
- 17 14 ☽ ⫲ ♇
- 18 53 ☽ ± ♇
- 20 42 ☽ ⊥ ☿
- 21 32 ☽ ⫲ ♅
- 21 40 ☽ ⫲ ♃
- 22 35 ☽ ☍ ♅

02 Tuesday
- 00 17 ☽ Q ♄
- 02 42 ☽ ± ♃
- 05 22 ☽ ⚻ ♇
- 05 34 ☿ □ ♂
- 08 24 ☽ ✶ ♂
- 08 40 ☽ ⊻ ☿
- 08 49 ☽ □ ♆
- 09 58 ☿ △ ♆
- 11 22 ☿ ± ♅
- 18 14 ☽ ⚼ ♃
- 18 47 ♂ ⊻ ♆
- 22 37 ☽ ∥ ♂

03 Wednesday
- 02 24 ☽ ∥ ♀
- 05 49 ☽ Q ♂
- 07 12 ☽ ∥ ☉
- 07 18 ☽ □ ♄
- 15 55 ☽ ☌ ☉
- 17 59 ☽ ☌ ♀
- 19 36 ☽ △ ♃

04 Thursday
- 00 05 ☽ ∥ ☿
- 01 37 ☽ ⚻ ♅
- 03 15 ☽ ∥ ♄
- 08 08 ☽ ☍ ♇
- 11 31 ☽ △ ♆
- 11 34 ☽ ± ♅
- 12 05 ♀ △ ♃
- 13 11 ☽ □ ♂
- 18 03 ☽ ☌ ☿
- 18 31 ☽ ∥ ♆

05 Friday
- 00 26 ☿ ∥ ♄
- 02 23 ☽ ⚼ ♅
- 09 14 ☽ △ ♄
- 12 13 ☽ ⚼ ♆
- 18 49 ☉ △ ♃
- 21 04 ☽ □ ♃
- 21 14 ☽ ⊻ ☉

06 Saturday
- 00 12 ☽ ⊻ ♀
- 02 54 ☽ △ ♅
- 02 56 ☽ ± ♆
- 07 47 ☽ ⊥ ☉
- 09 21 ☽ ⚻ ♇
- 09 50 ☽ ⚼ ♄
- 10 56 ☽ ⊥ ♀
- 12 45 ☽ ⚻ ♆
- 16 23 ☽ △ ♂
- 19 09 ☽ ± ♇
- 23 36 ☽ ∠ ☉

07 Sunday
- 00 35 ☽ ± ♄
- 01 21 ☽ ⊻ ☿
- 02 30 ☽ Q ♃
- 03 02 ☽ ∠ ♀
- 07 01 ♀ ⚻ ♅
- 09 53 ☽ ⚼ ♇
- 10 26 ☽ ⚻ ♄
- 10 56 ☽ ⫲ ♆
- 12 22 ☽ ⊥ ☿
- 17 54 ☽ ⫲ ☿
- 17 59 ☽ ⚼ ♂
- 20 17 ☿ ⚼ ♅
- 22 13 ☽ ✶ ♃

08 Monday
- 02 08 ☽ ✶ ☉
- 02 18 ☽ ⫲ ♄
- 04 02 ☽ □ ♅
- 04 58 ☽ ∠ ☿
- 06 05 ☽ ✶ ♀
- 09 28 ☽ ± ♂
- 10 38 ☽ △ ♇
- 14 09 ☽ ☍ ♆
- 14 41 ☽ ⫲ ☉
- 16 59 ☽ ⫲ ♀
- 19 52 ☽ ⚻ ♂
- 23 09 ☽ ∠ ♃
- 23 37 ☽ Q ☉

09 Tuesday
- 04 02 ☽ Q ♀
- 04 37 ☉ ⚻ ♅
- 08 54 ☽ ✶ ☿
- 10 03 ☽ Q ♅
- 11 36 ☽ ⫲ ♂
- 12 25 ☽ ☍ ♄
- 14 18 ☽ ⊥ ♃

10 Wednesday
- 00 30 ☽ ⊻ ♃
- 02 54 ☿ ∥ ♆
- 06 24 ☽ ✶ ♅
- 07 33 ☽ Q ☿
- 07 46 ☽ ∥ ♃
- 08 08 ☽ ∥ ♅
- 08 29 ☽ □ ☉
- 09 22 ♃ St D
- 09 28 ♀ ☍ ♇
- 12 46 ☽ ∥ ♇
- 13 16 ☽ □ ♇
- 13 39 ☽ □ ♀
- 16 59 ☽ ⚻ ♆
- 21 41 ☿ △ ♄

11 Thursday
- 01 06 ☽ ☍ ♂
- 03 23 ☽ ± ♆
- 08 19 ☽ ∠ ♅
- 16 13 ☽ ⚻ ♄
- 18 07 ☽ □ ☿
- 19 10 ☽ ⚼ ♆
- 20 38 ☽ Q ♇
- 22 40 ♀ ± ♅

12 Friday
- 00 08 ☽ ⊥ ♅
- 02 18 ♀ △ ♆
- 02 50 ☽ ± ♄
- 04 47 ☽ ☌ ♃
- 05 57 ☿ ⚼ ♆
- 10 47 ☽ ⊻ ♅
- 17 24 ☽ △ ☉
- 17 59 ☽ ✶ ♇
- 18 56 ☽ ⚼ ♄
- 21 55 ☽ △ ♆
- 23 58 ☽ △ ♀

13 Saturday
- 01 02 ☉ ☍ ♇
- 08 45 ☽ ⚻ ♂
- 09 07 ♂ ⊥ ♆
- 20 11 ☽ ± ♂
- 21 10 ☽ ∠ ♇
- 22 12 ☽ △ ♄
- 22 53 ☽ ⚼ ☉

14 Sunday
- 05 06 ☽ △ ☿
- 06 14 ☽ ⚼ ♀
- 11 16 ☽ ∥ ♇
- 11 17 ☽ ⊻ ♃
- 13 30 ☽ ⚼ ♂
- 13 43 ☽ ⊥ ♇
- 16 24 ☽ ∥ ♅
- 16 50 ☽ ∥ ♃
- 16 55 ☽ ± ☉
- 17 21 ☽ ☌ ♅
- 22 22 ☉ ± ♅
- 22 27 ☽ ⊥ ♃
- 23 22 ☉ ∥ ♀

15 Monday
- 00 44 ☽ ± ♀
- 00 52 ☽ ⊻ ♇
- 04 33 ☉ △ ♆
- 05 01 ☽ □ ♆
- 05 03 ☽ ⚻ ☉
- 10 27 ☽ ⫲ ♂
- 11 02 ☽ ⚼ ☿
- 13 12 ☽ ⚻ ♀
- 15 20 ☽ ∠ ♃
- 18 49 ☽ △ ♂

16 Tuesday
- 05 05 ☽ ± ☿
- 06 17 ☽ □ ♄
- 15 03 ☽ Q ♆
- 15 37 ♀ ∥ ♄
- 17 08 ☽ ⚻ ☿
- 19 54 ☽ ✶ ♃

17 Wednesday
- 01 57 ☽ ⊻ ♅
- 02 41 ☉ ∥ ♄
- 04 46 ☽ ⫲ ♄
- 04 56 ☽ ⫲ ☉
- 06 04 ☽ ⫲ ♀
- 06 15 ♄ St R
- 09 46 ☽ ☌ ♇
- 13 30 ☽ ⊥ ♅
- 14 07 ☽ ✶ ♆
- 16 52 ☽ Q ♄
- 19 06 ☽ Q ♃
- 19 17 ☽ ☍ ☉
- 22 52 ☽ ⫲ ♆

18 Thursday
- 03 49 ♂ ⚼ ♅
- 05 08 ☽ ☍ ♀
- 06 58 ☽ ∠ ♅
- 07 08 ☽ □ ♂
- 16 17 ☽ ✶ ♄
- 19 23 ☽ ∠ ♆
- 21 33 ☽ ⫲ ☿
- 22 48 ♀ ⚼ ♅

19 Friday
- 05 26 ☽ ☍ ☿
- 06 28 ☽ □ ♃
- 12 26 ☽ ✶ ♅
- 13 08 ☽ ⊥ ♆
- 14 11 ♀ □ ♂
- 20 18 ☽ Q ♂
- 20 31 ☽ ⊻ ♇
- 21 57 ☽ ∠ ♄

20 Saturday
- 01 04 ☽ ⊻ ♆
- 03 34 ☽ ∥ ☿
- 07 33 ☿ St R
- 08 28 ☽ ⊥ ♇
- 11 52 ☽ ⚻ ☉
- 12 18 ☽ Q ♅
- 15 56 ☽ ⊥ ♄
- 21 26 ☽ ✶ ♂
- 23 31 ☽ ⚻ ♀

21 Sunday
- 01 01 ☽ ± ☉
- 02 29 ☽ ∠ ♇
- 03 59 ☽ ⊻ ♄
- 04 25 ☽ ∥ ♆
- 12 18 ☽ ∥ ♀
- 13 01 ☽ ± ♀
- 15 19 ☽ ∥ ☉
- 17 20 ☽ ⚻ ☿
- 18 41 ☽ △ ♃
- 20 50 ☽ ⚼ ☉
- 23 23 ☽ ∥ ♄

22 Monday
- 00 28 ☽ □ ♅
- 05 07 ☽ ∠ ♂
- 05 10 ☽ ± ☿
- 08 42 ☽ ✶ ♇
- 09 20 ☽ ⚼ ♀
- 13 23 ☽ ☌ ♆
- 17 53 ♀ △ ♄
- 22 50 ☽ ⚼ ☿

23 Tuesday
- 00 12 ☽ ⊥ ♂
- 01 07 ☽ ⚼ ♃
- 06 00 ☽ △ ☉
- 08 56 ☽ Q ♇
- 12 54 ☽ ⊻ ♂
- 14 32 ☉ ⚼ ♅
- 16 34 ☽ ☌ ♄
- 19 15 ☽ △ ♀
- 19 23 ☽ ± ♃
- 23 13 ♀ ⚼ ♆

24 Wednesday
- 03 43 ☽ ∥ ♂
- 03 50 ☽ △ ☿
- 07 30 ☽ ⚻ ♃
- 12 54 ☽ △ ♅
- 16 50 ☽ ⫲ ♃
- 17 08 ☽ ⫲ ♅
- 21 05 ☽ □ ♇
- 22 33 ♀ ∥ ♆
- 23 20 ☽ ⫲ ♇

25 Thursday
- 01 44 ☽ ⊻ ♆
- 13 38 ☽ ⊥ ♆
- 18 41 ☽ ⚼ ♅
- 23 29 ☽ □ ☉

26 Friday
- 02 01 ☿ ☌ ♀
- 03 31 ☽ ☌ ♂
- 04 17 ☽ ⊻ ♄
- 07 17 ☽ ∠ ♆
- 11 49 ☽ □ ☿
- 12 15 ☽ ± ♅
- 13 40 ☽ □ ♀
- 15 56 ☽ ⊥ ♄
- 18 59 ☽ ☍ ♃
- 20 37 ♂ ⊻ ♄
- 23 48 ☽ ⚻ ♅

27 Saturday
- 07 36 ☽ △ ♇
- 09 09 ☽ ∠ ♄
- 12 04 ☽ ✶ ♆
- 13 13 ☽ Q ☉
- 17 21 ☉ ∥ ♆
- 19 46 ☽ Q ☿

28 Sunday
- 03 34 ☽ Q ♀
- 09 07 ☉ △ ♄
- 10 25 ☽ Q ♆
- 11 35 ☽ ⚼ ♇
- 13 05 ☽ ✶ ♄
- 13 25 ☽ ✶ ☉
- 14 50 ☽ ⊻ ♂
- 16 22 ☽ ✶ ☿
- 20 46 ♀ □ ♃
- 22 23 ☿ ⚼ ♆

All ephemeris data is given at 12.00 UT and the Moon's longitude is additionally given for 24.00 UT

MARCH 1954

LONGITUDES

Date	Sidereal time h m s	Sun ☉ ° ' "	Moon ☽ ° ' "	Moon ☽ 24.00 ° ' "	Mercury ☿	Venus ♀	Mars ♂	Jupiter ♃	Saturn ♄	Uranus ♅	Neptune ♆	Pluto ♇
01	22 35 01	10 ♓ 22 15	21 ♑ 57 35	28 ♑ 47 07	10 ♓ 11	17 ♓ 46	10 ♐ 40	17 ♊ 01	09 ♏ 13	19 ♋ 18	25 ♎ 46	23 ♌ 24
02	22 38 57	11 22 27	05 ♒ 43 40	12 ♒ 47 09	09 R 08	19 01	11 11	17 05	09 R 12	19 R 17	25 R 45	23 R 23
03	22 42 54	12 22 38	19 ♒ 57 16	27 ♒ 13 33	08 05	20 16	11 42	17 09	09 11	19 15	25 44	23 21
04	22 46 50	13 22 48	04 ♓ 35 18	12 ♓ 01 39	07 05	21 31	12 13	17 13	09 09	19 14	25 43	23 20
05	22 50 47	14 22 55	19 ♓ 31 31	27 ♓ 03 46	06 07	22 46	12 44	17 17	09 08	19 13	25 42	23 18
06	22 54 43	15 23 00	04 ♈ 37 06	12 ♈ 10 06	05 14	24 01	13 15	17 22	09 06	19 12	25 41	23 17
07	22 58 40	16 23 04	19 ♈ 41 58	27 ♈ 11 05	04 26	25 15	13 45	17 26	09 04	19 11	25 39	23 16
08	23 02 36	17 23 05	04 ♉ 36 34	11 ♉ 57 33	03 43	26 30	14 16	17 31	09 02	19 10	25 38	23 14
09	23 06 33	18 23 05	19 ♉ 13 23	26 ♉ 23 34	03 07	27 45	14 46	17 36	09 00	19 09	25 37	23 13
10	23 10 30	19 23 02	03 ♊ 27 51	10 ♊ 26 06	02 37	29 ♓ 00	15 16	17 42	08 58	19 08	25 36	23 12
11	23 14 26	20 22 57	17 ♊ 18 22	24 ♊ 04 50	02 13	00 ♈ 14	15 46	17 47	08 56	19 07	25 35	23 10
12	23 18 23	21 22 50	00 ♋ 45 45	07 ♋ 21 28	01 56	01 29	16 16	17 53	08 54	19 06	25 33	23 09
13	23 22 19	22 22 41	13 ♋ 52 21	20 ♋ 18 50	01 46	02 44	16 46	17 58	08 51	19 05	25 32	23 08
14	23 26 16	23 22 29	26 ♋ 41 18	03 ♌ 00 11	01 42	03 58	17 15	18 04	08 49	19 05	25 31	23 07
15	23 30 12	24 22 15	09 ♌ 15 52	15 ♌ 28 41	01 D 45	05 13	17 44	18 10	08 46	19 04	25 29	23 05
16	23 34 09	25 21 59	21 ♌ 38 59	27 ♌ 47 02	01 53	06 28	18 14	18 17	08 44	19 03	25 28	23 04
17	23 38 05	26 21 41	03 ♍ 53 07	09 ♍ 57 25	02 06	07 42	18 43	18 23	08 41	19 03	25 27	23 03
18	23 42 02	27 21 21	16 ♍ 00 09	22 ♍ 01 29	02 25	08 57	19 11	18 29	08 38	19 02	25 25	23 02
19	23 45 59	28 20 58	28 ♍ 01 33	04 ♎ 00 31	02 49	10 11	19 40	18 36	08 35	19 02	25 24	23 01
20	23 49 55	29 ♓ 20 34	09 ♎ 58 32	15 ♎ 55 45	03 18	11 26	20 08	18 43	08 33	19 01	25 22	23 00
21	23 53 52	00 ♈ 20 07	21 ♎ 52 20	27 ♎ 48 29	03 52	12 40	20 37	18 50	08 29	19 01	25 21	22 58
22	23 57 48	01 19 39	03 ♏ 44 27	09 ♏ 40 28	04 29	13 55	21 05	18 57	08 26	19 01	25 20	22 57
23	00 01 45	02 19 09	15 ♏ 36 53	21 ♏ 34 01	05 10	15 09	21 32	19 04	08 23	19 00	25 18	22 56
24	00 05 41	03 18 37	27 ♏ 32 18	03 ♐ 32 09	05 56	16 24	22 00	19 12	08 20	19 00	25 17	22 55
25	00 09 38	04 18 03	09 ♐ 34 05	15 ♐ 38 37	06 44	17 38	22 27	19 19	08 17	19 00	25 15	22 54
26	00 13 34	05 17 28	21 ♐ 46 20	27 ♐ 57 49	07 36	18 52	22 54	19 27	08 13	19 00	25 14	22 53
27	00 17 31	06 16 50	04 ♑ 13 40	10 ♑ 34 31	08 31	20 07	23 21	19 35	08 10	19 00	25 12	22 52
28	00 21 28	07 16 11	17 ♑ 00 56	23 ♑ 33 29	09 29	21 21	23 48	19 43	08 06	19 D 00	25 10	22 51
29	00 25 24	08 15 31	00 ♒ 12 41	06 ♒ 58 54	10 30	22 35	24 14	19 51	08 03	19 00	25 09	22 50
30	00 29 21	09 14 48	13 ♒ 52 28	20 ♒ 53 28	11 33	23 49	24 41	19 59	07 59	19 00	25 07	22 49
31	00 33 17	10 ♈ 14 04	28 ♒ 01 53	05 ♓ 17 25	12 ♓ 39	25 ♈ 04	25 ♐ 07	20 ♊ 07	07 ♏ 55	19 ♋ 00	25 ♎ 06	22 ♌ 48

Date	Moon True ☊ ° '	Moon Mean ☊ ° '	Moon ☽ Latitude ° '
01	23 ♑ 03	21 ♑ 36	00 S 06
02	23 R 02	21 33	01 N 08
03	22 59	21 30	02 20
04	22 54	21 26	03 24
05	22 47	21 23	04 16
06	22 39	21 20	04 51
07	22 31	21 17	05 05
08	22 23	21 14	04 59
09	22 18	21 10	04 33
10	22 14	21 07	03 50
11	22 13	21 04	02 56
12	22 D 13	21 01	01 52
13	22 14	20 58	00 N 45
14	22 R 14	20 55	00 S 24
15	22 13	20 51	01 30
16	22 09	20 48	02 30
17	22 02	20 45	03 22
18	21 53	20 42	04 05
19	21 41	20 39	04 36
20	21 29	20 35	04 55
21	21 15	20 32	05 01
22	21 03	20 29	04 54
23	20 53	20 26	04 34
24	20 45	20 23	04 02
25	20 40	20 20	03 19
26	20 37	20 16	02 26
27	20 36	20 13	01 26
28	20 D 36	20 10	00 S 19
29	20 R 36	20 07	00 N 51
30	20 34	20 04	01 59
31	20 ♑ 30	20 ♑ 00	03 N 04

DECLINATIONS

Date	Sun ☉ ° '	Moon ☽ ° '	Mercury ☿ ° '	Venus ♀ ° '	Mars ♂ ° '	Jupiter ♃ ° '	Saturn ♄ ° '	Uranus ♅ ° '	Neptune ♆ ° '	Pluto ♇ ° '
01	07 S 41	21 S 45	04 S 19	06 S 09	21 S 22	22 N 32	12 S 07	22 N 33	08 S 20	23 N 14
02	07 18	17 45	04 46	05 39	21 27	22 32	12 06	22 33	08 20	23 14
03	06 55	12 37	05 14	05 09	21 32	22 33	12 06	22 33	08 20	23 15
04	06 32	06 40	05 43	04 39	21 37	22 34	12 05	22 33	08 19	23 15
05	06 09	00 S 13	06 12	04 09	21 42	22 34	12 04	22 33	08 19	23 16
06	05 46	06 N 17	06 41	03 38	21 47	22 35	12 04	22 33	08 18	23 16
07	05 22	12 25	07 09	03 08	21 52	22 35	12 03	22 34	08 18	23 17
08	04 59	17 45	07 36	02 38	21 56	22 36	12 02	22 34	08 17	23 17
09	04 36	21 54	08 01	02 07	22 00	22 37	12 01	22 34	08 17	23 17
10	04 12	24 37	08 25	01 36	22 05	22 37	12 00	22 34	08 16	23 18
11	03 49	25 45	08 46	01 06	22 09	22 38	12 00	22 34	08 16	23 18
12	03 25	25 19	09 06	00 35	22 13	22 39	11 59	22 34	08 15	23 19
13	03 02	23 28	09 23	00 S 04	22 17	22 39	11 58	22 34	08 15	23 19
14	02 38	20 26	09 38	00 N 27	22 21	22 40	11 57	22 34	08 14	23 19
15	02 14	16 30	09 51	00 57	22 25	22 41	11 56	22 34	08 14	23 20
16	01 51	11 56	10 01	01 28	22 29	22 41	11 55	22 34	08 13	23 20
17	01 27	06 56	10 09	01 59	22 32	22 42	11 54	22 34	08 13	23 20
18	01 03	01 N 45	10 15	02 30	22 36	22 43	11 53	22 35	08 12	23 21
19	00 39	03 S 27	10 19	03 00	22 39	22 44	11 52	22 35	08 12	23 21
20	00 S 16	08 29	10 20	03 31	22 42	22 44	11 50	22 35	08 11	23 21
21	00 N 08	13 11	10 19	04 01	22 46	22 45	11 49	22 35	08 11	23 22
22	00 32	17 22	10 17	04 32	22 49	22 46	11 48	22 35	08 10	23 22
23	00 55	20 53	10 12	05 02	22 52	22 47	11 47	22 35	08 09	23 22
24	01 19	23 33	10 05	05 32	22 55	22 47	11 46	22 35	08 09	23 23
25	01 43	25 10	09 57	06 02	22 58	22 48	11 45	22 35	08 08	23 23
26	02 06	25 38	09 46	06 32	23 01	22 49	11 44	22 35	08 08	23 23
27	02 30	24 48	09 34	07 02	23 03	22 50	11 42	22 35	08 07	23 23
28	02 53	22 41	09 21	07 32	23 06	22 50	11 41	22 35	08 07	23 23
29	03 17	19 17	09 05	08 01	23 09	22 51	11 40	22 35	08 06	23 23
30	03 40	14 45	08 48	08 31	23 11	22 52	11 38	22 35	08 06	23 24
31	04 N 03	09 S 17	08 S 29	09 N 00	23 S 14	22 N 53	11 S 37	22 N 35	08 S 05	23 N 24

ZODIAC SIGN ENTRIES

Date	h	m	Planets
02	02	07	☽ ♒
04	04	32	☽ ♓
06	04	40	☽ ♈
08	04	32	☽ ♉
10	06	06	☽ ♊
11	07	22	♀ ♈
12	10	37	☽ ♋
14	18	17	☽ ♌
17	04	21	☽ ♍
19	15	57	☽ ♎
21	03	53	☉ ♈
22	04	26	☽ ♏
24	16	56	☽ ♐
27	03	55	☽ ♑
29	11	37	☽ ♒
31	15	16	☽ ♓

LATITUDES

Date	Mercury ☿ ° '	Venus ♀ ° '	Mars ♂ ° '	Jupiter ♃ ° '	Saturn ♄ ° '	Uranus ♅ ° '	Neptune ♆ ° '	Pluto ♇ ° '
01	03 N 42	01 S 25	00 N 41	00 S 17	02 N 35	00 N 29	01 N 44	10 N 06
04	03 27	01 24	00 39	00 16	02 36	00 29	01 44	10 06
07	02 57	01 22	00 36	00 16	02 37	00 29	01 45	10 06
10	02 17	01 19	00 33	00 15	02 37	00 29	01 45	10 06
13	01 34	01 16	00 30	00 15	02 38	00 29	01 45	10 06
16	00 51	01 12	00 27	00 14	02 38	00 29	01 45	10 05
19	00 N 10	01 07	00 24	00 14	02 39	00 29	01 45	10 05
22	00 S 26	01 03	00 20	00 14	02 39	00 29	01 45	10 05
25	00 58	00 57	00 16	00 13	02 40	00 29	01 45	10 04
28	01 26	00 52	00 12	00 13	02 40	00 29	01 45	10 04
31	01 S 49	00 S 46	00 N 08	00 S 12	02 N 41	00 N 29	01 N 45	10 N 04

DATA

Julian Date	2434803
Delta T	+31 seconds
Ayanamsa	23° 13' 17"
Synetic vernal point	05° ♓ 53' 42"
True obliquity of ecliptic	23° 26' 47"

MOON'S PHASES, APSIDES AND POSITIONS ☽

Date	h	m	Phase	Longitude ° '	Eclipse Indicator
05	03	11	●	14 ♓ 01	
11	17	52	☽	20 ♊ 38	
19	12	42	○	28 ♍ 23	
27	16	14	☾	06 ♑ 27	

Day	h	m		
06	09	29	Perigee	
21	17	41	Apogee	
05	12	48	ON	
11	17	05	Max dec	25° N 47'
18	20	04	OS	
26	08	45	Max dec	25° S 38'

LONGITUDES

Date	Chiron ⚷ ° '	Ceres ⚳ ° '	Pallas ⚴ ° '	Juno ⚵ ° '	Vesta ⚶ ° '	Black Moon Lilith ⚸ ° '
01	26 ♑ 08	11 ♎ 52	26 ♌ 31	14 ♌ 35	06 ♈ 08	18 ♎ 13
11	26 ♑ 49	10 ♎ 13	23 ♌ 55	13 ♌ 13	10 ♈ 42	19 ♎ 21
21	27 ♑ 24	08 ♎ 06	22 ♌ 11	12 ♌ 34	15 ♈ 16	20 ♎ 28
31	27 ♑ 54	05 ♎ 49	21 ♌ 22	12 ♌ 38	19 ♈ 50	21 ♎ 35

All ephemeris data is given at 12.00 UT and the Moon's longitude is additionally given for 24.00 UT

ASPECTARIAN

h m Aspects

01 Monday
00 22 ☽ # ♇
02 09 ☽ ⊥ ♂
03 08 ☽ ⊼ ♃
03 46 ☽ ✶ ♀
03 53 ☽ ± ♇
04 43 ☿ □ ♂
06 05 ☽ # ♅
06 11 ☽ # ♃
07 17 ☽ ☍ ♅
09 51 ☉ ☌ ☿
10 42 ☽ Q ♄
13 53 ☽ ± ♃
14 33 ☽ ⊼ ♇
14 37 ☽ ∥ ♂
16 43 ♂ ∠ ♆
17 17 ☽ ∠ ☿
18 30 ☽ ∠ ☉
18 43 ☽ □ ♆
18 48 ☽ ∠ ♂
21 20 ☉ ⚼ ♆

02 Tuesday
02 31 ☉ □ ♂
05 41 ☽ ⚼ ♃
07 51 ☽ ⊥ ☿
08 46 ☽ ∠ ♀
10 25 ☿ △ ♄
11 21 ☽ ⊥ ☉
16 56 ♀ △ ♅
17 25 ☽ ⊻ ☿
17 55 ☽ □ ♄
21 39 ☽ ✶ ♂
22 21 ☽ ⊻ ☉

03 Wednesday
01 35 ☽ ⊥ ♀
01 53 ♀ ± ♆
04 41 ♃ ∥ ♅
07 18 ☽ △ ♃
10 03 ☿ ∥ ♀
10 51 ☽ ⊼ ♅
12 34 ☽ ⊻ ♀
14 15 ☽ ∥ ♄
17 37 ☽ ☍ ♇
18 26 ☽ Q ♂
20 45 ☽ ± ♅
21 32 ☽ △ ♆

04 Thursday
05 34 ☽ ∥ ♆
11 26 ☽ ⚼ ♅
12 31 ☽ ∥ ☉
15 22 ☽ ∥ ☿
15 47 ☽ ☌ ☿
19 22 ☽ △ ♄
20 16 ☽ ∥ ♀
21 53 ☽ ⚼ ♆

05 Friday
00 45 ☽ □ ♂
03 11 ☽ ☌ ☉
08 25 ☽ □ ♃
10 43 ☉ ∥ ☿
11 31 ☽ △ ♅
12 16 ☽ ± ♆
17 37 ☽ ☌ ♀
18 01 ☽ ⊼ ♇
19 19 ☽ ⚼ ♄
21 49 ☽ ⊼ ♆
22 17 ♀ ⊼ ♇

06 Saturday
02 53 ☽ # ♀
03 33 ☽ ± ♇
09 35 ☽ ± ♄
09 48 ♂ ± ♅
10 10 ☽ # ☉
12 55 ☽ ⊻ ☿
13 11 ☽ Q ♃
13 36 ☽ # ☿
13 41 ♀ ⚼ ♄
17 49 ☽ ⚼ ♇
19 06 ☽ ⊼ ♄
19 40 ☽ # ♆
21 57 ☽ ⊥ ☿

07 Sunday
02 12 ☽ △ ♂
06 20 ☽ ⊻ ☉
08 23 ☽ ✶ ♃
10 30 ☽ # ♄
11 10 ☽ □ ♅
11 35 ☽ ∠ ☿
16 36 ☽ ⊥ ☉
17 41 ☽ △ ♇
19 36 ♀ ⊼ ♆
20 08 ☿ ⚼ ♅
21 32 ☽ ☍ ♆
21 43 ☽ ⊻ ♀

08 Monday
03 03 ☽ ⚼ ♂
08 08 ☽ ∠ ☉
08 16 ☽ ⊥ ♀
08 36 ☽ ∠ ♃
10 37 ☽ ✶ ☿
15 37 ☉ □ ♃
16 09 ☽ Q ♅
18 10 ☽ ± ♂
19 12 ☽ ☍ ♄
23 21 ☽ ⊥ ♃

09 Tuesday
00 18 ☽ ∠ ♀
04 22 ☽ ⊼ ♂
05 27 ☽ Q ☿
09 18 ☽ ⊻ ♃
10 30 ☽ ✶ ☉
11 52 ☽ ✶ ♅
12 43 ☽ # ♂
16 44 ☽ ∥ ♅
17 07 ☽ ∥ ♃
18 39 ☽ □ ♇
22 29 ♂ ⊥ ♄
22 35 ☽ ∥ ♇
22 41 ☽ ⊼ ♆

10 Wednesday
03 23 ☿ ∥ ♆
03 40 ☽ ✶ ♀
06 01 ☉ △ ♅
08 11 ☽ Q ☉
08 49 ☽ ± ♆
10 35 ☽ □ ☿
13 08 ☽ ∠ ♅
15 48 ♀ ± ♇
17 01 ☉ ± ♆
21 26 ☽ ⊼ ♄

11 Thursday
00 16 ☽ ⚼ ♆
01 18 ☽ Q ♇
02 15 ☽ Q ♀
04 40 ☽ ⊥ ♅
07 51 ☽ ± ♄
09 12 ☽ ☍ ♂
12 51 ☽ ☌ ♃
15 11 ☽ ⊻ ♅
17 52 ☽ □ ☉
22 22 ☽ ✶ ♇
23 42 ☽ ⚼ ♄

12 Friday
02 39 ☽ △ ♆
13 27 ☽ □ ♀
14 06 ☽ △ ☿
19 32 ☿ ⊻ ♀

13 Saturday
01 26 ☽ ∠ ♇
02 47 ☽ △ ♄
13 24 ☽ ∥ ♇
14 21 ♀ ± ♄
17 20 ☽ ⚼ ☿
17 35 ☽ ⊼ ♂
18 03 ☽ ⊥ ♇
19 16 ☽ ∥ ♃
19 41 ☽ ⊻ ♃
20 00 ☽ ∥ ♅
21 42 ☽ ☌ ♅
22 08 ☽ # ♂

14 Sunday
05 13 ☽ △ ☉
05 16 ☽ ± ♂
05 16 ☽ ⊻ ♇
05 47 ☉ ⊼ ♇
07 01 ☽ ⊥ ♃
09 47 ☽ □ ♆
10 09 ☽ ± ☿
15 07 ☿ St D
21 32 ☽ ⊼ ☿
22 11 ☉ ⚼ ♄
23 00 ☽ ⚼ ♂

15 Monday
00 14 ☽ ∠ ♃
03 23 ☽ △ ♀
11 04 ☽ □ ♄
12 13 ☽ ⚼ ☉
20 08 ☽ Q ♆

16 Tuesday
05 04 ☽ △ ♂
05 22 ☽ ✶ ♃
06 57 ☽ ⊻ ♅
07 10 ☽ ± ☉
08 04 ♀ Q ♃
11 36 ☽ ⚼ ♀
12 04 ☽ # ♄
14 22 ☉ ⊼ ♆
14 46 ☽ ☌ ♇
15 04 ♂ ☍ ♃
18 39 ☽ ⊥ ♅
19 27 ☽ ✶ ♆
19 54 ☽ ⊼ ☉
21 05 ☽ # ☿
21 53 ☉ # ♀
21 54 ☽ Q ♄

17 Wednesday
05 02 ☽ Q ♃
05 59 ☽ # ♆
07 13 ☽ ± ♀
08 25 ☽ ☍ ☿
12 19 ☽ ∠ ♅
18 34 ♀ ⚼ ♇
20 24 ☽ ⊼ ♀
21 27 ☽ ✶ ♄
22 09 ☿ ⊥ ♀

18 Thursday
00 56 ☽ ∠ ♆
04 09 ♂ # ♅
04 33 ♂ ⊼ ♅
06 16 ♀ ⊼ ♄
08 55 ☽ ∥ ♀
15 30 ☽ # ☉
17 00 ☽ □ ♃
18 02 ☽ ✶ ♅
18 36 ☽ □ ♂
18 48 ☽ ⊥ ♆

19 Friday
00 00 ☽ ∥ ☉
01 59 ☽ ⊻ ♇
03 10 ☽ ∠ ♄
06 45 ☽ ⊻ ♆
09 44 ☽ # ♀
12 42 ☽ ☍ ☉
13 58 ☽ ⊥ ♇
18 01 ☽ Q ♅
21 07 ☽ ⊥ ♄
22 00 ☽ ⊼ ☿

20 Saturday
03 41 ☉ ± ♇
08 01 ☽ ∠ ♇
08 09 ☽ Q ♂
09 08 ☽ ⊻ ♄
10 35 ☽ ∥ ♆
10 35 ☽ ± ☿
15 16 ☽ ☍ ♀
21 14 ☽ ∥ ☿

21 Sunday
04 53 ☽ ∥ ♄
05 26 ♂ # ♃
05 36 ☽ ⚼ ☿
05 48 ☽ △ ♃
06 14 ☽ □ ♅
09 21 ☽ ✶ ♂
14 13 ☽ ✶ ♇
18 15 ☿ ⚼ ♅
19 01 ☽ ☌ ♆

22 Monday
06 40 ☽ ⊼ ☉
12 26 ☽ ⚼ ♃
13 35 ☽ △ ☿
14 27 ☽ Q ♇
16 55 ☽ ∠ ♂
19 55 ☽ ± ☉
21 28 ☽ ☌ ♄
23 57 ♃ ⊻ ♅

23 Tuesday
06 49 ☽ ± ♃
10 58 ☽ ⊼ ♀
11 50 ☽ ⊥ ♂
13 34 ☉ ± ♄
15 45 ☽ ⚼ ☉
18 50 ☽ △ ♅
19 03 ☽ ⊼ ♃

24 Wednesday
00 25 ☽ ⊻ ♂
00 28 ☽ ± ♀
02 14 ☽ # ♅
02 44 ☽ □ ♇
04 11 ☽ # ♃
05 19 ☽ ∥ ♂
07 28 ☽ ⊻ ♆
10 07 ☽ # ♇
19 28 ☽ ⊥ ♆
20 36 ☽ ⚼ ♀

25 Thursday
00 35 ☽ △ ☉
00 56 ☽ ⚼ ♅
05 58 ☽ □ ☿
09 27 ☽ ⊻ ♄
13 21 ☽ ∠ ♆
18 48 ☽ ± ♅
21 16 ☽ ⊥ ♄

26 Friday
05 42 ☽ △ ♀
06 35 ☽ ⊼ ♅
07 25 ☽ ☍ ♃
10 50 ♂ △ ♇
14 10 ☽ △ ♇
14 18 ☽ ☌ ♂
14 29 ♀ □ ♅
14 48 ☽ ∠ ♄
18 42 ☽ ✶ ♆
20 01 ☽ Q ☿

27 Saturday
00 29 ♀ ✶ ♃
02 52 ♂ ∠ ♄
03 26 ☿ △ ♄
16 14 ☽ □ ☉
17 31 ♅ St D
17 38 ☽ Q ♆
18 54 ☽ ⚼ ♇
19 26 ☽ ✶ ♄
20 47 ☽ ✶ ☿

28 Sunday
05 27 ☽ # ♇
08 15 ☽ ∥ ♂
10 36 ☽ # ♃
11 42 ☽ ± ♇
12 51 ☽ # ♅
15 40 ☽ ☍ ♅
17 01 ☽ ⊼ ♃
17 40 ☽ Q ♄
20 48 ☽ □ ♀
22 42 ☽ ⊼ ♇

29 Monday
00 24 ☉ Q ♃
00 52 ☽ ⊻ ♂
01 53 ☉ ⚼ ♇
02 48 ☽ ∠ ☿
02 54 ☽ □ ♆
04 04 ☽ ± ♃
04 08 ☿ ⚼ ♆
04 19 ☽ Q ☉
07 07 ☉ ⊼ ♄
12 03 ☽ ⊥ ♂
15 46 ♀ # ♆
16 46 ♀ △ ♇
20 15 ☽ ⊥ ☿
20 19 ☽ ⚼ ♃

30 Tuesday
01 48 ☽ □ ♄
03 21 ☽ ✶ ☉
04 29 ☽ ∠ ♂
07 38 ☽ ⊻ ☿
08 06 ☽ Q ♀
20 48 ☽ ⊼ ♅
20 49 ☿ # ♀
22 34 ☽ △ ♃

31 Wednesday
02 09 ☽ ∥ ♄
03 15 ☽ ☍ ♇
06 34 ☽ ✶ ♀
06 56 ☽ ± ♅
06 58 ☽ ✶ ♂
06 59 ☽ ∠ ☉
07 06 ☽ △ ♆
11 16 ♂ ✶ ♆
12 41 ♀ ☍ ♆
13 05 ☽ # ♀
13 30 ♀ △ ♂
15 24 ☽ ∥ ☿
16 52 ☽ ∥ ♆
21 54 ☽ ⚼ ♅
23 01 ☽ ⊥ ☉

APRIL 1954

LONGITUDES

Date	Sidereal time h m s	Sun ☉	Moon ☽	Moon ☽ 24.00	Mercury ☿	Venus ♀	Mars ♂	Jupiter ♃	Saturn ♄	Uranus ♅	Neptune ♆	Pluto ♇
01	00 37 14	11 ♈ 13 17	12 ♓ 39 37	20 ♓ 07 43	13 ♓ 47	26 ♈ 18	25 ♐ 32	20 ♊ 16	07 ♏ 52	19 ♋ 01	25 ♎ 04	22 ♌ 47
02	00 41 10	12 12 29	27 ♓ 40 45	05 ♈ 17 32	14 57	27 32	25 58	20 24	07 R 48	19 01	25 R 03	22 R 46
03	00 45 07	13 11 39	12 ♈ 56 44	20 ♈ 36 53	16 10	28 ♈ 46	26 23	20 33	07 44	19 01	25 01	22 46
04	00 49 03	14 10 47	28 ♈ 16 29	05 ♉ 54 06	17 25	00 ♉ 00	26 48	20 42	07 40	19 02	24 59	22 45
05	00 53 00	15 09 53	13 ♉ 28 21	20 ♉ 58 04	18 42	01 14	27 12	20 51	07 36	19 02	24 58	22 44
06	00 56 57	16 08 56	28 ♉ 22 16	05 ♊ 40 10	20 00	02 28	27 37	21 00	07 32	19 03	24 56	22 43
07	01 00 53	17 07 58	12 ♊ 51 17	19 ♊ 55 18	21 21	03 42	28 01	21 09	07 28	19 03	24 55	22 42
08	01 04 50	18 06 57	26 ♊ 52 09	03 ♋ 41 57	22 43	04 56	28 24	21 19	07 24	19 04	24 53	22 42
09	01 08 46	19 05 54	10 ♋ 24 57	17 ♋ 01 30	24 08	06 10	28 48	21 28	07 19	19 04	24 51	22 41
10	01 12 43	20 04 49	23 ♋ 32 04	29 ♋ 57 09	25 33	07 24	29 11	21 38	07 15	19 05	24 50	22 40
11	01 16 39	21 03 41	06 ♌ 17 18	12 ♌ 33 04	27 01	08 38	29 33	21 47	07 11	19 06	24 48	22 40
12	01 20 36	22 02 31	18 ♌ 44 59	24 ♌ 53 36	28 ♓ 31	09 52	29 ♐ 56	21 57	07 07	19 07	24 46	22 39
13	01 24 32	23 01 18	00 ♍ 59 24	07 ♍ 02 51	00 ♈ 02	11 06	00 ♑ 18	22 07	07 02	19 08	24 45	22 38
14	01 28 29	24 00 04	13 ♍ 04 21	19 ♍ 04 18	01 34	12 20	00 40	22 17	06 58	19 08	24 43	22 38
15	01 32 26	24 58 47	25 ♍ 03 01	01 ♎ 00 48	03 09	13 33	01 01	22 27	06 53	19 09	24 41	22 37
16	01 36 22	25 57 28	06 ♎ 57 54	12 ♎ 54 32	04 45	14 47	01 22	22 37	06 49	19 10	24 40	22 37
17	01 40 19	26 56 07	18 ♎ 50 54	24 ♎ 47 11	06 22	16 01	01 43	22 48	06 45	19 12	24 38	22 36
18	01 44 15	27 54 44	00 ♏ 43 32	06 ♏ 40 07	08 02	17 14	02 03	22 58	06 40	19 13	24 37	22 36
19	01 48 12	28 53 20	12 ♏ 37 07	18 ♏ 34 43	09 43	18 28	02 23	23 09	06 36	19 14	24 35	22 35
20	01 52 08	29 ♈ 51 53	24 ♏ 33 06	00 ♐ 32 32	11 25	19 42	02 42	23 19	06 31	19 15	24 33	22 35
21	01 56 05	00 ♉ 50 25	06 ♐ 33 16	12 ♐ 35 37	13 09	20 55	03 01	23 30	06 27	19 16	24 32	22 35
22	02 00 01	01 48 55	18 ♐ 39 56	24 ♐ 46 36	14 55	22 09	03 20	23 41	06 22	19 18	24 30	22 34
23	02 03 58	02 47 23	00 ♑ 56 04	07 ♑ 08 47	16 43	23 22	03 38	23 52	06 18	19 19	24 28	22 34
24	02 07 55	03 45 50	13 ♑ 25 16	19 ♑ 46 01	18 32	24 36	03 56	24 02	06 13	19 20	24 27	22 34
25	02 11 51	04 44 15	26 ♑ 11 34	02 ♒ 42 25	20 23	25 49	04 13	24 13	06 08	19 22	24 25	22 33
26	02 15 48	05 42 38	09 ♒ 19 05	16 ♒ 01 58	22 16	27 03	04 30	24 25	06 04	19 24	24 24	22 33
27	02 19 44	06 41 00	22 ♒ 51 27	29 ♒ 47 44	24 10	28 16	04 47	24 36	05 59	19 25	24 22	22 33
28	02 23 41	07 39 21	06 ♓ 50 58	14 ♓ 01 02	26 06	29 ♉ 29	05 03	24 47	05 55	19 27	24 20	22 33
29	02 27 37	08 37 39	21 ♓ 17 40	28 ♓ 40 24	28 ♈ 04	00 ♊ 43	05 18	24 58	05 50	19 28	24 19	22 32
30	02 31 34	09 ♉ 35 57	06 ♈ 08 31	13 ♈ 41 03	00 ♉ 03	01 ♊ 56	05 ♑ 33	25 ♊ 10	05 ♏ 46	19 ♋ 30	24 ♎ 17	22 ♌ 32

Date	Moon True ☊	Moon Mean ☊	Moon ☽ Latitude
01	20 ♑ 24	19 ♑ 57	03 N 58
02	20 R 14	19 54	04 38
03	20 03	19 51	04 58
04	19 52	19 48	04 57
05	19 42	19 45	04 35
06	19 34	19 41	03 55
07	19 28	19 38	03 00
08	19 25	19 35	01 56
09	19 D 24	19 32	00 N 47
10	19 R 24	19 29	00 S 22
11	19 24	19 26	01 28
12	19 22	19 22	02 28
13	19 18	19 19	03 21
14	19 10	19 16	04 03
15	19 00	19 13	04 34
16	18 48	19 10	04 53
17	18 34	19 06	05 00
18	18 20	19 03	04 53
19	18 07	19 00	04 33
20	17 55	18 57	04 01
21	17 46	18 54	03 19
22	17 40	18 51	02 27
23	17 37	18 47	01 27
24	17 36	18 44	00 S 22
25	17 D 36	18 41	00 N 45
26	17 R 36	18 38	01 53
27	17 35	18 35	02 56
28	17 32	18 32	03 50
29	17 27	18 28	04 32
30	17 ♑ 19	18 ♑ 25	04 N 57

DECLINATIONS

Date	Sun ☉	Moon ☽	Mercury ☿	Venus ♀	Mars ♂	Jupiter ♃	Saturn ♄	Uranus ♅	Neptune ♆	Pluto ♇
01	04 N 26	03 S 08	08 S 09	09 N 29	23 S 16	22 N 53	11 S 36	22 N 35	08 S 04	23 N 24
02	04 50	03 N 19	07 48	09 57	23 19	22 54	11 34	22 34	08 04	23 24
03	05 13	09 41	07 24	10 26	23 21	22 55	11 33	22 34	08 03	23 24
04	05 36	15 29	07 00	10 54	23 23	22 56	11 32	22 34	08 02	23 24
05	05 58	20 15	06 34	11 22	23 25	22 56	11 30	22 34	08 02	23 25
06	06 21	23 37	06 07	11 50	23 27	22 57	11 29	22 34	08 01	23 25
07	06 44	25 19	05 38	12 17	23 30	22 58	11 27	22 34	08 01	23 25
08	07 06	25 21	05 08	12 44	23 32	22 59	11 26	22 34	08 00	23 25
09	07 29	23 49	04 37	13 11	23 34	22 59	11 25	22 34	07 59	23 25
10	07 51	21 02	04 05	13 38	23 36	23 00	11 23	22 34	07 59	23 25
11	08 13	17 17	03 31	14 04	23 38	23 01	11 22	22 34	07 58	23 25
12	08 35	12 52	02 56	14 30	23 40	23 02	11 20	22 33	07 58	23 25
13	08 57	08 00	02 20	14 55	23 41	23 02	11 19	22 33	07 57	23 25
14	09 19	02 N 55	01 43	15 20	23 43	23 03	11 17	22 33	07 56	23 25
15	09 40	02 S 14	01 05	15 45	23 45	23 04	11 16	22 33	07 56	23 25
16	10 02	07 15	00 S 26	16 09	23 47	23 04	11 14	22 33	07 55	23 25
17	10 23	12 00	00 N 14	16 33	23 49	23 05	11 13	22 33	07 55	23 25
18	10 44	16 18	00 55	16 57	23 51	23 06	11 11	22 33	07 54	23 25
19	11 05	19 58	01 37	17 20	23 53	23 07	11 10	22 32	07 53	23 25
20	11 26	22 49	02 20	17 42	23 54	23 07	11 08	22 32	07 53	23 25
21	11 46	24 40	03 04	18 05	23 56	23 08	11 07	22 32	07 52	23 25
22	12 06	25 24	03 49	18 26	23 58	23 09	11 06	22 32	07 52	23 25
23	12 27	24 54	04 35	18 48	24 00	23 09	11 04	22 32	07 51	23 25
24	12 47	23 08	05 21	19 08	24 02	23 10	11 03	22 31	07 50	23 25
25	13 06	20 10	06 08	19 29	24 04	23 10	11 01	22 31	07 50	23 25
26	13 26	16 07	06 56	19 48	24 06	23 11	11 00	22 31	07 49	23 25
27	13 45	11 08	07 45	20 07	24 07	23 12	10 58	22 31	07 49	23 24
28	14 04	05 S 26	08 34	20 26	24 09	23 12	10 57	22 30	07 48	23 24
29	14 23	00 N 43	09 23	20 44	24 11	23 13	10 55	22 30	07 48	23 24
30	14 N 42	06 N 59	10 N 13	21 N 02	24 S 13	23 N 13	10 S 54	22 N 30	07 S 47	23 N 24

ZODIAC SIGN ENTRIES

Date	h	m	Planets
02	15	40	☽ ♈
04	11	55	♀ ♉
04	14	43	☽ ♉
06	14	40	☽ ♊
08	17	29	☽ ♋
11	00	05	☽ ♌
12	16	28	♂ ♑
13	10	03	☽ ♍
13	11	34	☿ ♈
15	21	58	☽ ♎
18	10	32	☽ ♏
20	15	20	☉ ♉
20	22	55	☽ ♐
23	10	11	☽ ♑
25	19	02	☽ ♒
28	00	21	☽ ♓
28	22	03	♀ ♊
30	02	08	☽ ♈
30	11	26	☿ ♉

LATITUDES

Date	Mercury ☿	Venus ♀	Mars ♂	Jupiter ♃	Saturn ♄	Uranus ♅	Neptune ♆	Pluto ♇
01	01 S 55	00 S 43	00 N 06	00 S 12	02 N 41	00 N 29	01 N 45	10 N 03
04	02 12	00 37	00 N 01	00 12	02 41	00 29	01 45	10 03
07	02 24	00 30	00 S 04	00 11	02 42	00 29	01 45	10 03
10	02 31	00 23	00 09	00 11	02 42	00 29	01 45	10 02
13	02 34	00 15	00 15	00 10	02 42	00 29	01 46	10 01
16	02 32	00 S 08	00 21	00 10	02 42	00 29	01 46	10 01
19	02 25	00 00	00 27	00 10	02 42	00 29	01 46	10 00
22	02 14	00 N 08	00 34	00 09	02 42	00 29	01 46	10 00
25	01 58	00 16	00 41	00 09	02 42	00 29	01 46	09 59
28	01 38	00 24	00 48	00 08	02 42	00 28	01 45	09 59
31	01 S 14	00 N 32	00 S 56	00 S 08	02 N 42	00 N 28	01 N 45	09 N 58

DATA

Julian Date	2434834
Delta T	+31 seconds
Ayanamsa	23° 13' 21"
Synetic vernal point	05° ♓ 53' 39"
True obliquity of ecliptic	23° 26' 47"

MOON'S PHASES, APSIDES AND POSITIONS ☽

Date	h	m	Phase	Longitude	Eclipse Indicator
03	12	25	●	13 ♈ 13	
10	05	05	☽	19 ♋ 48	
18	05	48	○	27 ♎ 40	
26	04	57	☾	05 ♒ 26	

Day	h	m		
03	19	46	Perigee	
17	19	33	Apogee	
01	23	43	0N	
08	00	13	Max dec	25° N 32'
15	01	34	0S	
22	14	19	Max dec	25° S 24'
29	09	16	0N	

LONGITUDES

Date	Chiron ⚷	Ceres ⚳	Pallas ⚴	Juno ⚵	Vesta ⚶	Black Moon Lilith ⚸
01	27 ♑ 56	05 ♎ 35	21 ♌ 20	12 ♌ 41	20 ♈ 18	21 ♎ 42
11	28 ♑ 18	03 ♎ 27	21 ♌ 29	13 ♌ 28	24 ♈ 52	22 ♎ 49
21	28 ♑ 32	01 ♎ 42	22 ♌ 25	14 ♌ 50	29 ♈ 24	23 ♎ 56
31	28 ♑ 39	00 ♎ 32	24 ♌ 00	16 ♌ 41	03 ♉ 55	25 ♎ 04

ASPECTARIAN

01 Thursday
03 26 ☽ Q ♂
04 14 ☽ △ ♄
07 21 ☽ # ☉
07 49 ☽ ⚼ ♆
09 30 ☽ ⊻ ☉
09 36 ☽ ∠ ♀
13 58 ☽ ☌ ☿
17 47 ☿ ∥ ♆
22 13 ☽ △ ♅
22 17 ☽ ± ♆

02 Friday
00 20 ☽ □ ♃
01 22 ☽ ⊥ ♀
04 14 ☽ ⊼ ♇
04 17 ☽ ⚼ ♄
07 50 ☽ ⊼ ♆
09 12 ☽ □ ♂
11 45 ☽ ⊻ ♀
13 44 ☽ ± ♇
17 56 ☽ ∥ ☉
18 28 ☽ ± ♄

03 Saturday
03 46 ☽ # ☿
03 52 ☽ ⊼ ♄
03 53 ☽ ⚼ ♇
05 03 ☽ Q ♃
05 43 ☽ # ♆
12 25 ☽ ☌ ☉
15 11 ☽ ∥ ♀
17 29 ☽ ⊻ ☿
19 21 ☽ # ♄
21 31 ☽ □ ♅

04 Sunday
00 01 ☽ ⚹ ♃
03 21 ☽ △ ♇
03 42 ☽ ⊥ ☿
06 52 ☽ ☍ ♆
09 37 ☽ △ ♂
14 57 ☽ ☌ ♀
19 06 ☽ ∠ ☿
23 48 ☽ ∠ ♃

05 Monday
01 47 ☽ Q ♅
02 43 ☽ ☍ ♄
03 05 ♂ # ♇
09 56 ☽ ⚼ ♂
14 13 ☽ ⊥ ♃
14 53 ☽ ⊻ ☉
16 54 ☿ ± ♆
18 19 ☿ △ ♅
18 42 ♀ # ♄
20 54 ☽ ⚹ ♅
21 08 ☽ ⚹ ☿
23 56 ☽ ⊻ ♃

06 Tuesday
00 44 ☽ ± ♂
01 11 ☽ ⊥ ☉
02 50 ☽ □ ♇
03 14 ☽ ∥ ♅
05 04 ☉ # ☿
06 13 ☽ ∥ ♃
06 25 ☽ ⊼ ♆
10 08 ☽ ∥ ♇
10 32 ☽ # ♂
10 43 ☽ ⊼ ♂
16 11 ☽ ± ♆
16 52 ☽ ∠ ☉
18 33 ☽ Q ☿
19 21 ☽ ⊻ ♀
21 19 ☽ ∠ ♅

07 Wednesday
03 01 ☽ ⊼ ♄
06 13 ☽ ⊥ ♀
07 04 ☽ ⚼ ♆
08 11 ☿ □ ♃
08 24 ☽ Q ♇
12 20 ☽ ⊥ ♅
13 01 ☽ ± ♄
19 47 ☽ ⚹ ☉
22 31 ☽ ⊻ ♅
22 53 ☽ ∠ ♀

08 Thursday
02 17 ☽ ☌ ♃
04 01 ☽ □ ☿
04 17 ☽ ⚼ ♄
04 46 ☽ ⚹ ♇
06 34 ☿ ⚼ ♄
08 33 ☽ △ ♆
11 32 ☿ ⊼ ♇
14 46 ☽ ☍ ♂
18 07 ☽ Q ☉

09 Friday
03 38 ☽ ⚹ ♀
06 29 ☽ △ ♄
07 06 ☽ ∠ ♇
11 22 ☉ □ ♅
14 44 ☽ # ♂
16 14 ☽ ∥ ♇
18 39 ♀ ∠ ♃
20 14 ☽ ∥ ♃
23 22 ☽ ⊥ ♇

10 Saturday
00 03 ☽ ∥ ♅
00 04 ☿ ⊼ ♆
03 34 ☽ Q ♀
03 46 ☽ ☌ ♅
05 05 ☽ □ ☉
05 41 ♀ Q ♅
08 26 ☽ ⊻ ♃
09 11 ♀ ☍ ♄
10 24 ☽ ⊻ ♇
14 24 ☽ □ ♆
16 15 ☽ △ ☿
19 44 ☽ ⊥ ♃
20 11 ☉ # ♆
22 52 ☽ ⊼ ♂

11 Sunday
10 34 ☽ ± ♂
12 58 ☽ ∠ ♃
13 42 ☽ □ ♄
16 58 ☽ □ ♀

12 Monday
00 27 ☽ Q ♆
00 28 ☽ ⚼ ☿
04 12 ☽ ∥ ♀
04 22 ☽ ⚼ ♂
09 25 ☉ ⚹ ♃
12 42 ☽ ⊻ ♅
14 14 ☿ ± ♇
18 20 ☽ ⚹ ♃
18 58 ☽ △ ☉
19 36 ☽ ☌ ♇
19 44 ☽ # ♄
20 22 ☽ ± ☿
23 44 ☽ ⚹ ♆

13 Tuesday
00 21 ☽ Q ♄
00 26 ☽ ⊥ ♅
02 45 ☉ △ ♇
03 46 ♃ ⚼ ♄
07 43 ☽ ∥ ☉
09 50 ☽ ⊼ ☿
10 36 ☽ △ ♂
12 14 ☽ # ♆
17 34 ☿ □ ♂
18 13 ☽ ∠ ♅
18 17 ☽ Q ♃
23 55 ☽ ⚹ ♄

14 Wednesday
03 02 ☿ ± ♄
03 10 ☽ ⚼ ☉
05 20 ☽ ∠ ♆
10 21 ☽ △ ♀
18 18 ☽ # ☿
22 44 ☽ ± ☉
23 16 ☽ ⊥ ♆

15 Thursday
00 09 ☽ ⚹ ♅
05 07 ☉ ☍ ♆
05 42 ☽ ∠ ♄
06 43 ☽ □ ♃
07 07 ☽ ⊻ ♇
07 15 ☽ ∥ ☿
11 17 ☽ ⊻ ♆
11 51 ☽ ⊼ ☉
19 11 ☽ ⊥ ♇
19 52 ☽ ⚼ ♀
23 41 ☽ ⊥ ♄

16 Friday
00 18 ☽ Q ♅
00 22 ☽ □ ♂
06 49 ☽ ☍ ☿
10 30 ♃ ⚹ ♇
11 42 ☽ ⊻ ♄
13 18 ☽ ∠ ♇
15 16 ☽ ∥ ♆
16 06 ☽ ± ♀

17 Saturday
02 56 ☽ # ☉
05 37 ☽ ⊼ ♀
07 56 ☽ ∥ ♄
12 42 ☽ □ ♅
13 48 ☽ Q ♂
17 10 ☿ ⊼ ♄
19 35 ☽ ⚹ ♇
20 06 ☽ △ ♃
23 40 ☽ ☌ ♆

18 Sunday
05 47 ♀ ⊥ ♃
05 48 ☿ ⚼ ♇
05 48 ☽ ☍ ☉
06 50 ♀ ⚼ ♂
14 45 ☽ ⚹ ♂
16 23 ☽ # ♀
19 49 ☽ Q ♇
23 56 ☽ ☌ ♄

19 Monday
02 50 ☽ ⚼ ♃
05 10 ☽ ⊼ ☿
17 25 ☉ # ♄
19 16 ☽ ± ☿
21 15 ☽ ± ♃
21 52 ☽ ∠ ♂

20 Tuesday
01 08 ☽ ☍ ♀
01 20 ☽ △ ♅
03 10 ♀ ⚹ ♅
08 03 ☽ □ ♇
09 17 ☽ # ♅
09 29 ☽ ⊼ ♃
10 26 ☿ Q ♃
12 00 ☽ ⊻ ♆
15 12 ☽ # ♃
16 23 ☽ ⚼ ☿
16 26 ☽ ⊥ ♂
18 28 ☽ # ♇
23 35 ☽ ⊼ ☉

21 Wednesday
00 00 ☽ ⊥ ♆
00 35 ☽ ∥ ♂
04 46 ☽ ⊻ ♂
07 27 ☽ ⚼ ♅
11 47 ☽ ⊻ ♄
12 37 ☽ ± ☉
17 54 ☽ ∠ ♆
23 38 ☽ ⊥ ♄

22 Thursday
01 22 ☽ ± ♅
03 21 ☽ △ ☿
08 02 ☽ ⚼ ☉
13 14 ☽ ⊼ ♅
17 17 ☽ ∠ ♄
19 36 ☽ ⊼ ♀
19 40 ☽ △ ♇
20 15 ♀ □ ♇
22 00 ☽ ☍ ♃
23 26 ☽ ⚹ ♆

23 Friday
08 38 ☽ ± ♀
15 54 ☽ △ ☉
17 22 ☽ ☌ ♂
22 18 ☽ ⚹ ♄
22 41 ☽ Q ♆
23 12 ♀ ⊻ ♃

24 Saturday
00 48 ☽ ⚼ ♇
02 03 ☽ ∥ ♂
03 55 ☽ ⚼ ♀
09 06 ☽ # ♇
09 08 ♀ ⊼ ♆
11 44 ☽ # ♃
14 21 ☿ ⊥ ♀
17 52 ☽ # ♅
17 57 ☽ ± ♇
18 02 ☉ △ ♂
21 02 ☽ Q ♄
22 38 ☿ □ ♅
23 13 ☽ ☍ ♅
23 19 ☽ □ ☿

25 Sunday
05 14 ☽ ⊼ ♇
08 17 ☽ ⊼ ♃
08 43 ☽ □ ♆
11 14 ☽ △ ♀
16 14 ☽ # ♀
19 33 ☽ ± ♃

26 Monday
03 06 ☽ ⊻ ♂
04 57 ☽ □ ☉
06 09 ☽ □ ♄
10 03 ♃ △ ♆
12 10 ☽ ⚼ ♃
13 59 ☽ Q ☿
14 11 ☽ ⊥ ♂
15 39 ☿ △ ♇
20 06 ☉ ☍ ♄

27 Tuesday
00 38 ☽ # ☉
05 58 ☽ ⊼ ♅
06 30 ☽ ∠ ♂
11 27 ☽ ☍ ♇
12 45 ☽ ∥ ♄
13 59 ☿ # ♆
14 28 ☿ ☍ ♆
14 37 ☽ △ ♆
14 39 ☽ ⚹ ☿
15 04 ☽ △ ♃
15 25 ☽ Q ☉
16 28 ☽ ± ♅
17 57 ☿ ⚹ ♃
22 16 ☽ □ ♀

28 Wednesday
00 46 ☽ # ☿
00 55 ♀ ± ♂
02 19 ☽ ∥ ♆
06 39 ☉ Q ♅
07 55 ☽ ⚼ ♅
08 54 ☽ ⚹ ♂
10 26 ☽ △ ♄
13 28 ☽ ⚹ ☉
16 11 ☽ ⚼ ♆
20 16 ☽ ∠ ☿

29 Thursday
04 22 ♀ ± ♆
05 19 ☽ Q ♂
07 07 ☽ ± ♆
07 22 ☽ Q ♀
09 00 ☽ △ ♅
11 15 ☽ ⚼ ♄
13 27 ☽ ⊥ ☿
14 02 ☽ ⊼ ♇
16 05 ☽ ∠ ☉
16 55 ☽ ⊼ ♆
18 05 ☽ □ ♃
23 47 ☽ ± ♇

30 Friday
00 42 ☽ ⊻ ☿
01 48 ☽ ± ♄
04 39 ☽ ⚹ ♀
07 39 ☽ ⊥ ☉
11 02 ☽ □ ♂
11 24 ☽ ⊼ ♄
14 14 ☽ ⚼ ♇
15 07 ☽ # ♆
17 54 ☽ ⊻ ☉
23 19 ☽ Q ♃

All ephemeris data is given at 12.00 UT and the Moon's longitude is additionally given for 24.00 UT

MAY 1954

Date	Sidereal time h m s	Sun ☉ ° ' "	Moon ☽ ° ' "	Moon ☽ 24.00 ° ' "	Mercury ☿ ° '	Venus ♀ ° '	Mars ♂ ° '	Jupiter ♃ ° '	Saturn ♄ ° '	Uranus ♅ ° '	Neptune ♆ ° '	Pluto ♇ ° '
01	02 35 30	10 ♉ 34 12	21 ♈ 16 55	28 ♈ 54 47	02 ♉ 04	03 ♊ 09	05 ♑ 47	25 ♊ 21	05 ♏ 41	19 ♋ 32	24 ♎ 16	22 ♌ 32
02	02 39 27	11 32 26	06 ♉ 33 18	14 ♉ 11 00	04 06	04 22	06 01	25 33	05 R 37	19 34	24 R 14	22 R 32
03	02 43 24	12 30 39	21 ♉ 46 32	29 ♉ 18 33	06 10	05 35	06 15	25 45	05 32	19 36	24 13	22 32
04	02 47 20	13 28 49	06 ♊ 45 56	14 ♊ 07 42	08 15	06 49	06 27	25 56	05 28	19 37	24 11	22 32
05	02 51 17	14 26 58	21 ♊ 23 07	28 ♊ 31 40	10 22	08 02	06 40	26 08	05 23	19 39	24 09	22 D 32
06	02 55 13	15 25 06	05 ♋ 33 03	12 ♋ 27 10	12 30	09 15	06 51	26 20	05 19	19 41	24 08	22 32
07	02 59 10	16 23 11	19 ♋ 14 07	25 ♋ 54 07	14 39	10 28	07 02	26 32	05 14	19 43	24 06	22 32
08	03 03 06	17 21 14	02 ♌ 27 32	08 ♌ 54 49	16 48	11 41	07 13	26 44	05 10	19 46	24 05	22 32
09	03 07 03	18 19 15	15 ♌ 16 30	21 ♌ 33 08	18 58	12 54	07 23	26 56	05 05	19 48	24 03	22 32
10	03 10 59	19 17 14	27 ♌ 45 17	03 ♍ 53 34	21 09	14 07	07 32	27 08	05 01	19 50	24 02	22 32
11	03 14 56	20 15 12	09 ♍ 58 33	16 ♍ 00 48	23 20	15 20	07 40	27 20	04 57	19 52	24 01	22 33
12	03 18 53	21 13 08	22 ♍ 00 51	27 ♍ 59 13	25 30	16 32	07 48	27 33	04 52	19 54	23 59	22 33
13	03 22 49	22 11 01	03 ♎ 56 21	09 ♎ 52 41	27 40	17 45	07 56	27 45	04 48	19 57	23 58	22 33
14	03 26 46	23 08 54	15 ♎ 48 36	21 ♎ 44 26	29 ♉ 50	18 58	08 03	27 57	04 44	19 59	23 56	22 33
15	03 30 42	24 06 44	27 ♎ 40 30	03 ♏ 37 04	01 ♊ 58	20 11	08 09	28 10	04 40	20 01	23 55	22 34
16	03 34 39	25 04 33	09 ♏ 34 22	15 ♏ 32 37	04 05	21 23	08 14	28 22	04 36	20 04	23 54	22 34
17	03 38 35	26 02 21	21 ♏ 31 59	27 ♏ 32 40	06 11	22 36	08 19	28 35	04 31	20 06	23 52	22 34
18	03 42 32	27 00 07	03 ♐ 34 50	09 ♐ 38 39	08 14	23 48	08 23	28 47	04 27	20 09	23 51	22 35
19	03 46 28	27 57 52	15 ♐ 44 18	21 ♐ 51 59	10 16	25 01	08 26	29 00	04 23	20 11	23 50	22 35
20	03 50 25	28 55 35	28 ♐ 01 54	04 ♑ 14 17	12 16	26 14	08 29	29 12	04 20	20 14	23 48	22 36
21	03 54 22	29 ♉ 53 18	10 ♑ 29 24	16 ♑ 47 32	14 13	27 26	08 30	29 25	04 16	20 16	23 47	22 36
22	03 58 18	00 ♊ 50 59	23 ♑ 09 00	29 ♑ 34 06	16 08	28 38	08 32	29 38	04 12	20 19	23 46	22 36
23	04 02 15	01 48 39	06 ♒ 03 12	12 ♒ 36 38	18 00	29 ♊ 51	08 32	29 ♊ 51	04 08	20 22	23 45	22 37
24	04 06 11	02 46 18	19 ♒ 14 43	25 ♒ 57 45	19 49	01 ♋ 03	08 R 32	00 ♋ 04	04 04	20 24	23 43	22 38
25	04 10 08	03 43 56	02 ♓ 45 59	09 ♓ 39 36	21 36	02 15	08 30	00 17	04 00	20 27	23 42	22 38
26	04 14 04	04 41 33	16 ♓ 38 41	23 ♓ 43 13	23 19	03 28	08 29	00 30	03 57	20 30	23 41	22 39
27	04 18 01	05 39 09	00 ♈ 53 02	08 ♈ 07 51	25 00	04 40	08 26	00 43	03 53	20 33	23 40	22 39
28	04 21 57	06 36 44	15 ♈ 27 09	22 ♈ 50 19	26 38	05 52	08 23	00 56	03 50	20 35	23 39	22 40
29	04 25 54	07 34 19	00 ♉ 16 33	07 ♉ 44 55	28 13	07 04	08 18	01 09	03 47	20 38	23 38	22 41
30	04 29 51	08 31 53	15 ♉ 14 23	22 ♉ 43 50	29 ♊ 44	08 16	08 14	01 22	03 43	20 41	23 37	22 41
31	04 33 47	09 ♊ 29 25	00 ♊ 12 07	07 ♊ 38 09	01 ♋ 13	09 ♋ 28	08 ♑ 08	01 ♋ 35	03 ♏ 40	20 ♋ 44	23 ♎ 36	22 ♌ 42

Date	Moon True ☊ ° '	Moon Mean ☊ ° '	Moon ☽ Latitude ° '	DECLINATIONS Sun ☉ ° '	Moon ☽ ° '	Mercury ☿ ° '	Venus ♀ ° '	Mars ♂ ° '	Jupiter ♃ ° '	Saturn ♄ ° '	Uranus ♅ ° '	Neptune ♆ ° '	Pluto ♇ ° '
01	17 ♑ 10	18 ♑ 22	05 N 02	15 N 00	12 N 58	11 N 03	21 N 19	24 S 16	23 N 14	10 S 52	22 N 30	07 S 46	23 N 24
02	17 R 00	18 19	04 45	15 18	18 12	11 53	21 35	24 18	23 14	10 51	22 29	07 46	23 24
03	16 51	18 16	04 09	15 36	22 13	12 43	21 51	24 20	23 15	10 49	22 29	07 45	23 24
04	16 44	18 12	03 15	15 53	24 39	13 32	22 06	24 22	23 15	10 48	22 29	07 45	23 23
05	16 39	18 09	02 10	16 11	25 20	14 22	22 21	24 24	23 16	10 46	22 28	07 44	23 23
06	16 36	18 06	00 N 59	16 28	24 18	15 11	22 35	24 27	23 16	10 45	22 28	07 44	23 23
07	16 D 36	18 03	00 S 14	16 45	21 50	15 59	22 48	24 29	23 17	10 44	22 28	07 43	23 23
08	16 37	18 00	01 24	17 01	18 15	16 47	23 01	24 32	23 17	10 42	22 27	07 43	23 23
09	16 37	17 57	02 27	17 17	13 55	17 34	23 13	24 34	23 17	10 41	22 27	07 42	23 22
10	16 R 37	17 53	03 22	17 33	09 06	18 19	23 24	24 37	23 18	10 40	22 27	07 42	23 22
11	16 35	17 50	04 06	17 49	04 N 02	19 02	23 35	24 40	23 18	10 38	22 26	07 41	23 22
12	16 30	17 47	04 38	18 04	01 S 06	19 45	23 45	24 42	23 19	10 37	22 26	07 41	23 21
13	16 24	17 44	04 58	18 19	06 08	20 25	23 55	24 45	23 19	10 36	22 26	07 40	23 21
14	16 16	17 41	05 05	18 34	10 55	21 03	24 03	24 48	23 19	10 34	22 25	07 40	23 21
15	16 06	17 38	04 59	18 48	15 08	21 39	24 11	24 51	23 19	10 33	22 25	07 39	23 21
16	15 56	17 34	04 40	19 02	19 06	22 12	24 18	24 54	23 20	10 32	22 25	07 39	23 20
17	15 47	17 31	04 09	19 16	22 09	22 44	24 25	24 58	23 20	10 30	22 24	07 38	23 20
18	15 39	17 28	03 26	19 30	24 15	23 12	24 31	25 01	23 20	10 29	22 24	07 38	23 20
19	15 33	17 25	02 33	19 43	25 13	23 38	24 37	25 04	23 20	10 28	22 24	07 37	23 19
20	15 30	17 22	01 33	19 56	24 59	24 02	24 41	25 08	23 21	10 27	22 23	07 37	23 19
21	15 28	17 18	00 S 27	20 08	23 29	24 22	24 45	25 11	23 21	10 26	22 23	07 36	23 19
22	15 D 28	17 15	00 N 41	20 20	20 47	24 40	24 48	25 15	23 21	10 25	22 22	07 36	23 18
23	15 29	17 12	01 49	20 32	17 00	24 56	24 51	25 19	23 21	10 23	22 22	07 36	23 18
24	15 31	17 09	02 53	20 43	12 19	25 09	24 52	25 23	23 21	10 22	22 22	07 35	23 18
25	15 31	17 06	03 48	20 54	06 56	25 20	24 53	25 27	23 21	10 21	22 21	07 35	23 17
26	15 R 31	17 03	04 32	21 05	01 S 06	25 28	24 54	25 31	23 21	10 20	22 21	07 34	23 17
27	15 29	16 59	05 00	21 15	04 N 56	25 34	24 54	25 35	23 21	10 19	22 20	07 34	23 16
28	15 25	16 56	05 10	21 25	10 51	25 37	24 52	25 39	23 21	10 18	22 20	07 34	23 16
29	15 20	16 53	05 00	21 35	16 15	25 39	24 50	25 43	23 21	10 17	22 19	07 33	23 15
30	15 14	16 50	04 29	21 44	20 42	25 39	24 47	25 48	23 21	10 16	22 19	07 33	23 15
31	15 ♑ 09	16 ♑ 47	03 N 40	21 N 53	23 N 47	25 N 37	24 N 44	25 S 52	23 N 21	10 S 15	22 N 18	07 S 33	23 N 15

ZODIAC SIGN ENTRIES

Date	h	m	Planets	
02	01	42	☽	♉
04	01	06	☽	♊
06	02	30	☽	♋
08	07	29	☽	♌
10	16	23	☽	♍
13	04	03	☽	♎
14	13	57	☿	♊
15	16	42	☽	♏
18	04	53	☽	♐
20	15	49	☽	♑
21	14	47	☉	♊
23	00	48	☽	♒
23	15	04	♀	♋
24	04	43	♃	♋
25	07	08	☽	♓
27	10	32	☽	♈
29	11	33	☽	♉
30	16	13	☿	♋
31	11	40	☽	♊

LATITUDES

Date	Mercury ☿ ° '	Venus ♀ ° '	Mars ♂ ° '	Jupiter ♃ ° '	Saturn ♄ ° '	Uranus ♅ ° '	Neptune ♆ ° '	Pluto ♇ ° '
01	01 S 14	00 N 32	00 S 56	00 S 08	02 N 42	00 N 28	01 N 45	09 N 58
04	00 46	00 40	01 05	00 08	02 42	00 28	01 45	09 57
07	00 S 15	00 47	01 14	00 07	02 42	00 28	01 45	09 57
10	00 N 16	00 55	01 23	00 07	02 42	00 28	01 45	09 56
13	00 47	01 02	01 33	00 07	02 41	00 28	01 45	09 55
16	01 16	01 09	01 43	00 06	02 41	00 28	01 45	09 55
19	01 40	01 16	01 54	00 06	02 40	00 28	01 45	09 54
22	01 58	01 22	02 05	00 06	02 40	00 28	01 45	09 53
25	02 09	01 28	02 17	00 05	02 40	00 28	01 45	09 53
28	02 13	01 33	02 29	00 05	02 39	00 28	01 45	09 52
31	02 N 10	01 N 38	02 S 41	00 S 05	02 N 39	00 N 28	01 N 45	09 N 52

DATA

Julian Date	2434864
Delta T	+31 seconds
Ayanamsa	23° 13' 24"
Synetic vernal point	05° ♓ 53' 36"
True obliquity of ecliptic	23° 26' 46"

LONGITUDES

Date	Chiron ⚷ ° '	Ceres ⚳ ° '	Pallas ⚴ ° '	Juno ⚵ ° '	Vesta ⚶ ° '	Black Moon Lilith ⚸ ° '
01	28 ♑ 39	00 ♎ 32	24 ♌ 00	16 ♌ 41	03 ♉ 55	25 ♎ 04
11	28 ♑ 38	00 ♎ 03	26 ♌ 07	18 ♌ 55	08 ♉ 24	26 ♎ 11
21	28 ♑ 30	00 ♎ 14	28 ♌ 40	21 ♌ 29	12 ♉ 50	27 ♎ 18
31	28 ♑ 15	01 ♎ 03	01 ♍ 34	24 ♌ 19	17 ♉ 12	28 ♎ 25

MOON'S PHASES, APSIDES AND POSITIONS ☽

Date	h	m	Phase	Longitude ° '	Eclipse Indicator
02	20	22	●	11 ♉ 53	
09	18	17	☽	18 ♌ 34	
17	21	47	○	26 ♏ 26	
25	13	49	☾	03 ♓ 48	

Day	h	m		
02	06	25	Perigee	
15	01	14	Apogee	
30	13	25	Perigee	
05	09	12	Max dec	25° N 21'
12	06	52	0S	
19	19	20	Max dec	25° S 17'
26	16	23	0N	

ASPECTARIAN

h m Aspects

01 Saturday
02 46 ☽ ∥ ☿
03 24 ☽ ⋕ ♄
03 53 ♂ ⚹ ♄
05 25 ☉ ∠ ♃
06 38 ☽ ∠ ♀
07 12 ☿ ⋕ ♄
09 14 ☽ □ ♅
13 58 ☽ △ ♇
16 41 ☽ ☍ ♆
18 30 ☽ ⚹ ♃
21 21 ☽ ∥ ☉
22 02 ☽ ⊥ ♀

02 Sunday
07 33 ☽ ☌ ☿
08 16 ☽ ⊻ ♀
10 31 ☽ ☍ ♄
11 09 ☽ △ ♂
13 35 ☽ Q ♅
15 49 ♀ ∠ ♅
18 22 ☽ ∠ ♃
19 44 ☿ ⊻ ♀
20 22 ☽ ☌ ☉

03 Monday
04 56 ☿ ☍ ♄
08 31 ♂ Q ♆
08 32 ☽ ⚹ ♅
08 44 ☽ ⊥ ♃
09 11 ☽ ∥ ♀
10 56 ♀ ⊼ ♄
11 09 ☽ bQ ♂
12 59 ☿ △ ♂
13 12 ☽ □ ♇
14 00 ☽ ∥ ♅
15 51 ☽ ⊼ ♆
18 23 ☽ ⊻ ♃
20 17 ☽ ∥ ♃
21 34 ☽ ∥ ♇

04 Tuesday
01 25 ☽ ± ♆
01 42 ☽ ± ♂
03 35 ♀ ⊼ ♂
04 39 ☿ Q ♅
07 56 ☽ ⋕ ♂
08 32 ☽ ∠ ♅
09 54 ☽ ⊼ ♄
11 30 ☽ ⊼ ♂
12 05 ☽ ☌ ♀
14 49 ☽ ⊻ ☿
15 55 ☽ bQ ♆
17 42 ♇ St D
18 07 ☽ Q ♇
19 36 ☽ ± ♄
23 12 ☽ ⊥ ♅
23 43 ☽ ⊻ ☉

05 Wednesday
02 16 ☽ ⊥ ☿
09 07 ☽ ⊻ ♅
10 20 ☽ ⊥ ☉
10 21 ☽ bQ ♄
13 55 ☽ ⚹ ♇
16 38 ☽ △ ♆
19 50 ☽ ∠ ☿
20 04 ☽ ☌ ♃
21 36 ☿ ∠ ♃

06 Thursday
00 38 ♀ ∥ ♅
02 34 ☽ ∠ ☉
09 47 ♀ bQ ♆
10 07 ☽ ⋕ ♂
11 35 ☽ △ ♄
14 17 ☽ ☍ ♂
15 26 ☽ ∠ ♇
19 01 ☽ ⊻ ♀
22 31 ☽ ∥ ♇
23 35 ☽ ∥ ♃

07 Friday
02 19 ☽ ⚹ ☿
04 34 ☽ ∥ ♀
06 33 ☽ ⚹ ☉
06 35 ☽ ⊥ ♀
06 54 ☽ ∥ ♅
07 12 ☽ ⊥ ♇
12 52 ☽ ☌ ♅
13 24 ♀ Q ♇
17 55 ☽ ⊻ ♇
20 44 ☽ □ ♆

08 Saturday
00 20 ☽ ∠ ♀
01 21 ☽ ⊻ ♃
02 21 ♀ ± ♄
03 57 ☽ Q ☿
05 50 ☽ Q ☉
12 31 ☽ ⊥ ♃
16 59 ☽ □ ♄
18 47 ☽ ∥ ☉
19 13 ☽ ∥ ☿
20 56 ☽ ⊼ ♂
23 02 ☉ ∥ ☿
23 03 ☉ ☌ ☿

09 Sunday
05 35 ☽ ∠ ♃
05 55 ☽ Q ♆
07 01 ☽ ⚹ ♀
08 21 ☽ ± ♂
18 17 ☽ □ ☉
20 32 ☽ □ ☿
20 39 ☽ ⊻ ♅
21 14 ☿ ⚹ ♅
21 34 ♀ ∥ ♃

10 Monday
01 46 ☽ bQ ♂
01 54 ☽ ☌ ♇
02 52 ☽ Q ♄
04 23 ☽ ⋕ ♄
04 48 ☽ ⚹ ♆
06 16 ♀ ⊥ ♅
07 03 ♀ ∥ ♇
08 15 ☽ ⊥ ♅
08 28 ☽ Q ♀
10 47 ☽ ⚹ ♃
11 53 ☿ ⊥ ♃
13 43 ♂ bQ ♇
18 45 ☽ ⋕ ♆

11 Tuesday
01 53 ☽ ∠ ♅
02 01 ☉ ⚹ ♅
02 08 ☽ ⚹ ♄
03 23 ☿ □ ♇
04 20 ☿ bQ ♂
07 24 ☽ △ ♂
10 06 ☽ ∠ ♆
10 43 ☽ Q ♃
19 28 ☿ ⊼ ♆
23 49 ☽ □ ♀

12 Wednesday
03 57 ☽ ⊥ ♆
07 44 ☽ ∠ ♄
07 46 ☽ ⚹ ♅
10 16 ☽ △ ☉
13 04 ☽ ⊻ ♇
15 57 ☽ ⊻ ♆
20 33 ☽ △ ☿
22 18 ☉ ⊥ ♃
23 18 ☽ □ ♃

13 Thursday
01 08 ☽ ⊥ ♇
01 42 ☽ ⊥ ♄
07 57 ☽ Q ♅
13 00 ☿ ⊻ ♃
13 44 ☽ ⊻ ♄
19 08 ☽ bQ ☉
19 18 ☽ ∠ ♇
19 33 ☽ ∥ ♆
20 08 ☽ □ ♂
21 12 ☉ □ ♇

14 Friday
09 02 ☉ bQ ♂
09 34 ☽ bQ ☿
10 14 ☽ ∥ ♄
13 16 ☿ ± ♆
14 57 ☽ ± ☉
19 07 ☽ △ ♀
20 28 ☽ □ ♅

15 Saturday
01 39 ☽ ⚹ ♇
02 21 ♀ bQ ♄
04 10 ☽ ⊼ ☉
04 25 ☽ ☌ ♆
07 14 ☉ ⊼ ♆
07 47 ☽ ± ☿
08 48 ♀ ⊻ ♅
08 53 ☽ Q ♂
13 00 ☽ △ ♃
14 06 ☿ ± ♂
22 33 ☽ ⊼ ☿

16 Sunday
01 54 ☽ Q ♇
02 01 ☽ ☌ ♄
04 52 ☽ bQ ♀
09 17 ☽ ⚹ ♂
11 30 ☽ ⋕ ☉
17 38 ☿ ⊼ ♄
19 46 ☽ bQ ♃
20 43 ♄ Q ♇
21 06 ☿ ∥ ♅
23 24 ☿ ∠ ♅

17 Monday
01 00 ☽ ± ♀
09 08 ☽ △ ♅
11 28 ♀ ⚹ ♇
12 12 ♃ ∥ ♇
14 05 ☽ □ ♇
14 08 ☽ ± ♃
14 22 ☽ ⊼ ♀
14 24 ☽ ⋕ ♅
15 34 ☽ ∠ ♂
16 40 ☽ ⊻ ♆
19 03 ☽ ⋕ ☿
21 47 ☽ ☍ ☉

18 Tuesday
00 03 ☽ ⋕ ♇
00 06 ☽ ⋕ ♃
02 19 ☽ ⊼ ♃
04 36 ☽ ⊥ ♆
09 36 ☽ ⊥ ♂
12 48 ♀ △ ♆
13 40 ☿ ⊼ ♂
13 44 ☽ ⊻ ♄
15 07 ☽ bQ ♅
17 00 ☽ ⋕ ♀
18 34 ☿ ∥ ♇
19 06 ☿ bQ ♆
19 13 ☿ ∥ ♃
21 33 ☽ ⊻ ♂
22 25 ☽ ∠ ♆
23 05 ☽ ☍ ☿

19 Wednesday
01 32 ☽ ⊥ ♄
04 56 ☽ ∥ ♂
08 56 ☽ ± ♅
13 25 ☿ ± ♄
15 48 ☿ Q ♇
19 07 ☽ ∠ ♄
20 45 ☽ ⊼ ♅

20 Thursday
01 25 ☽ △ ♇
03 48 ☽ ⚹ ♆
07 25 ☽ ∥ ♂
08 07 ☽ ☍ ♀
13 53 ☽ ⊼ ☉
14 20 ☽ ☍ ♃
18 16 ☽ ⋕ ♀
21 08 ☉ ⊻ ♃

21 Friday
00 06 ☽ ⚹ ♄
02 15 ☽ ⋕ ☿
02 26 ☽ ± ☉
03 00 ☽ Q ♆
06 28 ☽ bQ ♇
08 12 ☽ ☌ ♂
09 29 ☉ ± ♆
12 43 ☿ ⊥ ♅
13 28 ☽ ⋕ ♃
13 52 ☽ ⋕ ♇
20 23 ☽ ⊼ ☿
21 05 ☽ bQ ☉
22 56 ☽ Q ♄
23 04 ☽ ⋕ ♅
23 39 ☽ ± ♇

22 Saturday
06 39 ☽ ☍ ♅
09 44 ☽ ± ☿
10 59 ☽ ⊼ ♇
13 09 ☽ □ ♆
15 01 ☽ ⋕ ☉
23 20 ☽ ⊼ ♀

23 Sunday
00 20 ☽ ⊼ ♃
01 50 ☿ ∥ ♀
03 32 ☽ △ ☉
05 25 ☽ bQ ☿
08 29 ☽ □ ♄
11 35 ☽ ± ♀
11 37 ☽ ± ♃
12 07 ♀ ☌ ♃
12 47 ♂ St R
16 33 ☽ ⊻ ♂

24 Monday
02 26 ☿ bQ ♄
03 29 ☽ ⊥ ♂
04 20 ☽ bQ ♃
05 40 ☽ bQ ♀
05 56 ☉ ± ♂
13 11 ☽ △ ☿
14 05 ☽ ⊼ ♅
18 04 ☽ ☍ ♇
19 39 ☽ ∠ ♂
20 00 ☽ △ ♆
20 05 ☿ ⊻ ♅
21 02 ☽ ∥ ♄

25 Tuesday
00 50 ☽ ± ♅
07 34 ☽ △ ♃
09 17 ☽ ∥ ♆
11 01 ☽ △ ♀
13 49 ☽ □ ☉
14 10 ☽ △ ♄
16 42 ☽ bQ ♅
18 32 ☉ ⊼ ♄
21 59 ☽ ⚹ ♂
22 20 ☽ bQ ♆

26 Wednesday
02 29 ☿ ⚹ ♇
13 46 ☽ ± ♆
15 55 ☽ bQ ♄
17 06 ☿ △ ♆
18 30 ☽ Q ♂
18 34 ☽ △ ♅
21 18 ♀ △ ♄
22 12 ☽ ⊼ ♇
23 01 ☽ Q ☉
23 56 ☽ ⊼ ♆

27 Thursday
00 51 ☽ □ ☿
07 01 ☽ ± ♄
08 17 ☽ ± ♇
09 08 ☉ ∠ ♅
11 43 ☽ □ ♃
16 59 ☽ ⊼ ♄
18 51 ☽ □ ♀
20 28 ☽ ⚹ ☉
22 30 ☽ ⋕ ♆
23 14 ☽ bQ ♇

28 Friday
00 27 ☽ □ ♂
09 44 ☽ ⋕ ♄
10 29 ☽ Q ☿
17 45 ☽ Q ♃
20 23 ☽ □ ♅
22 42 ☽ ∠ ☉
23 44 ☽ △ ♇

29 Saturday
01 18 ☽ ☍ ♆
02 52 ☽ Q ♀
08 17 ☽ ⚹ ☿
13 25 ☽ ⚹ ♃
14 14 ☽ ⊥ ☉
17 36 ☽ ☍ ♄
23 52 ☽ ⚹ ♀

30 Sunday
00 17 ♀ ∠ ♇
00 31 ☽ ⊻ ☉
00 50 ☽ △ ♂
01 28 ☽ Q ♅
04 58 ☉ ⊼ ♂
11 06 ☽ ∠ ☿
11 09 ♀ ☍ ♂
13 50 ☽ ∠ ♃
14 00 ☉ bQ ♆
19 11 ☽ ∥ ☉
20 45 ☽ ⚹ ♅
23 06 ☽ ∥ ♅
23 57 ☽ □ ♇

31 Monday
00 43 ☽ bQ ♂
01 24 ☽ ⊼ ♆
01 59 ☽ ∠ ♀
03 07 ☽ ⊥ ☿
04 28 ☽ ⊥ ♃
06 49 ☽ ∥ ♇
07 50 ☽ ∥ ♃
11 01 ☽ ± ♆
13 48 ☽ ⊻ ☿
13 56 ☉ ⊻ ♀
14 16 ☽ ⊻ ♃
15 05 ☽ ± ♂
16 10 ☉ ± ♄
17 34 ☽ ⊼ ♄
17 44 ☽ ⊥ ♀
19 16 ☿ ☌ ♃
20 57 ☽ ∠ ♅
23 58 ☽ ∥ ♀

All ephemeris data is given at 12.00 UT and the Moon's longitude is additionally given for 24.00 UT

JUNE 1954

LONGITUDES

Date	Sidereal time h m s	Sun ☉ ° ' "	Moon ☽ ° ' "	Moon ☽ 24.00 ° ' "	Mercury ☿ ° '	Venus ♀ ° '	Mars ♂ ° '	Jupiter ♃ ° '	Saturn ♄ ° '	Uranus ♅ ° '	Neptune ♆ ° '	Pluto ♇ ° '
01	04 37 44	10 ♊ 26 57	15 ♊ 00 52	22 ♊ 19 21	02 ♋ 38	10 ♋ 40	08 ♑ 01	01 ♋ 48	03 ♏ 37	20 ♋ 47	23 ♎ 35	22 ♌ 43
02	04 41 40	11 24 28	29 ♊ 32 47	06 ♋ 40 34	04 01	11 52	07 R 54	02 01	03 R 34	20 50	23 R 34	22 44
03	04 45 37	12 21 57	13 ♋ 42 13	20 ♋ 37 27	05 20	13 04	07 46	02 15	03 31	20 53	23 33	22 44
04	04 49 33	13 19 26	27 ♋ 26 09	04 ♌ 08 20	06 36	14 16	07 38	02 28	03 28	20 56	23 32	22 45
05	04 53 30	14 16 53	10 ♌ 44 10	17 ♌ 13 56	07 49	15 28	07 28	02 41	03 25	20 59	23 31	22 46
06	04 57 26	15 14 19	23 ♌ 37 59	29 ♌ 56 47	08 58	16 39	07 18	02 55	03 22	21 02	23 30	22 47
07	05 01 23	16 11 44	06 ♍ 10 50	12 ♍ 20 39	10 04	17 51	07 07	03 08	03 19	21 06	23 29	22 48
08	05 05 20	17 09 08	18 ♍ 26 49	24 ♍ 29 55	11 06	19 03	06 56	03 21	03 17	21 09	23 28	22 49
09	05 09 16	18 06 31	00 ♎ 30 30	06 ♎ 29 09	12 05	20 14	06 44	03 35	03 14	21 12	23 28	22 50
10	05 13 13	19 03 52	12 ♎ 26 25	18 ♎ 22 50	13 01	21 26	06 31	03 48	03 11	21 15	23 27	22 51
11	05 17 09	20 01 13	24 ♎ 18 54	00 ♏ 15 04	13 52	22 37	06 18	04 02	03 09	21 18	23 26	22 52
12	05 21 06	20 58 33	06 ♏ 11 48	12 ♏ 09 28	14 40	23 49	06 04	04 15	03 07	21 22	23 25	22 53
13	05 25 02	21 55 51	18 ♏ 08 26	24 ♏ 09 02	15 24	25 00	05 49	04 28	03 04	21 25	23 25	22 54
14	05 28 59	22 53 09	00 ♐ 11 31	06 ♐ 16 10	16 04	26 11	05 34	04 42	03 02	21 28	23 24	22 55
15	05 32 55	23 50 27	12 ♐ 23 11	18 ♐ 32 45	16 40	27 23	05 19	04 55	03 00	21 32	23 23	22 56
16	05 36 52	24 47 44	24 ♐ 45 00	01 ♑ 00 06	17 12	28 34	05 03	05 09	02 58	21 35	23 23	22 57
17	05 40 49	25 45 00	07 ♑ 18 09	13 ♑ 39 15	17 39	29 ♋ 45	04 47	05 23	02 56	21 38	23 22	22 58
18	05 44 45	26 42 15	20 ♑ 03 29	26 ♑ 30 57	18 02	00 ♌ 56	04 30	05 36	02 55	21 42	23 22	23 00
19	05 48 42	27 39 31	03 ♒ 01 43	09 ♒ 35 52	18 21	02 07	04 13	05 50	02 53	21 45	23 21	23 01
20	05 52 38	28 36 46	16 ♒ 13 29	22 ♒ 54 36	18 35	03 18	03 55	06 03	02 51	21 48	23 21	23 02
21	05 56 35	29 ♊ 34 00	29 ♒ 39 17	06 ♓ 27 35	18 44	04 29	03 37	06 17	02 50	21 52	23 20	23 03
22	06 00 31	00 ♋ 31 14	13 ♓ 19 30	20 ♓ 15 00	18 49	05 40	03 19	06 30	02 48	21 55	23 20	23 05
23	06 04 28	01 28 29	27 ♓ 14 03	04 ♈ 16 30	18 R 50	06 50	03 01	06 44	02 47	21 59	23 19	23 06
24	06 08 24	02 25 43	11 ♈ 22 10	18 ♈ 30 48	18 46	08 01	02 43	06 58	02 46	22 02	23 19	23 07
25	06 12 21	03 22 57	25 ♈ 42 04	02 ♉ 55 32	18 37	09 12	02 24	07 11	02 45	22 06	23 19	23 09
26	06 16 18	04 20 11	10 ♉ 10 42	17 ♉ 27 00	18 24	10 22	02 05	07 25	02 44	22 09	23 18	23 10
27	06 20 14	05 17 26	24 ♉ 43 47	02 ♊ 00 22	18 07	11 33	01 47	07 39	02 43	22 13	23 18	23 11
28	06 24 11	06 14 40	09 ♊ 16 01	16 ♊ 30 01	17 46	12 43	01 28	07 52	02 42	22 16	23 18	23 13
29	06 28 07	07 11 54	23 ♊ 41 38	00 ♋ 50 12	17 21	13 53	01 09	08 06	02 41	22 20	23 18	23 14
30	06 32 04	08 ♋ 09 09	07 ♋ 55 05	14 ♋ 55 46	16 ♋ 54	15 ♌ 04	00 ♑ 51	08 ♋ 19	02 ♏ 40	22 ♋ 24	23 ♎ 18	23 ♌ 15

Date	Moon True ☊ ° '	Moon Mean ☊ ° '	Moon ☽ Latitude ° '
01	15 ♑ 05	16 ♑ 44	02 N 36
02	15 R 03	16 40	01 24
03	15 02	16 37	00 N 07
04	15 D 03	16 34	01 S 07
05	15 04	16 31	02 16
06	15 05	16 28	03 16
07	15 07	16 24	04 04
08	15 R 07	16 21	04 40
09	15 06	16 18	05 03
10	15 04	16 15	05 13
11	15 02	16 12	05 09
12	14 58	16 09	04 52
13	14 54	16 05	04 23
14	14 51	16 02	03 41
15	14 48	15 59	02 49
16	14 46	15 56	01 48
17	14 45	15 53	00 S 41
18	14 D 45	15 49	00 N 29
19	14 45	15 46	01 39
20	14 46	15 43	02 45
21	14 48	15 40	03 43
22	14 48	15 37	04 30
23	14 49	15 34	05 01
24	14 R 49	15 30	05 16
25	14 48	15 27	05 10
26	14 47	15 24	04 46
27	14 46	15 21	04 03
28	14 45	15 18	03 04
29	14 44	15 15	01 54
30	14 ♑ 44	15 ♑ 11	00 N 38

DECLINATIONS

Date	Sun ☉ ° '	Moon ☽ ° '	Mercury ☿ ° '	Venus ♀ ° '	Mars ♂ ° '	Jupiter ♃ ° '	Saturn ♄ ° '	Uranus ♅ ° '	Neptune ♆ ° '	Pluto ♇ ° '
01	22 N 01	25 N 11	25 N 33	24 N 40	25 S 57	23 N 21	10 S 14	22 N 18	07 S 32	23 N 14
02	22 09	24 50	25 27	24 35	26 01	23 21	10 14	22 18	07 32	23 14
03	22 17	22 52	25 20	24 30	26 06	23 21	10 13	22 17	07 32	23 13
04	22 24	19 35	25 12	24 24	26 11	23 21	10 12	22 17	07 31	23 13
05	22 31	15 22	25 02	24 17	26 15	23 21	10 11	22 16	07 31	23 12
06	22 38	10 34	24 51	24 09	26 20	23 21	10 10	22 16	07 31	23 12
07	22 44	05 28	24 39	24 01	26 25	23 20	10 10	22 15	07 30	23 11
08	22 49	00 N 16	24 26	23 52	26 30	23 20	10 09	22 15	07 30	23 11
09	22 55	04 S 50	24 12	23 43	26 34	23 20	10 08	22 14	07 30	23 10
10	23 00	09 43	23 57	23 33	26 39	23 20	10 08	22 14	07 30	23 10
11	23 04	14 13	23 42	23 22	26 44	23 19	10 07	22 13	07 29	23 09
12	23 08	18 11	23 26	23 10	26 49	23 19	10 06	22 13	07 29	23 09
13	23 12	21 26	23 09	22 58	26 53	23 19	10 06	22 12	07 29	23 08
14	23 15	23 48	22 52	22 46	26 58	23 18	10 05	22 12	07 29	23 08
15	23 18	25 04	22 35	22 32	27 03	23 18	10 05	22 11	07 29	23 07
16	23 21	25 08	22 17	22 18	27 07	23 17	10 05	22 10	07 28	23 07
17	23 23	23 56	22 00	22 04	27 12	23 17	10 04	22 10	07 28	23 06
18	23 24	21 28	21 42	21 48	27 16	23 17	10 04	22 09	07 28	23 06
19	23 26	17 53	21 25	21 33	27 20	23 16	10 03	22 09	07 28	23 05
20	23 26	13 21	21 08	21 16	27 24	23 16	10 03	22 08	07 28	23 04
21	23 27	08 07	20 51	20 59	27 29	23 15	10 03	22 08	07 28	23 04
22	23 27	02 S 24	20 34	20 42	27 33	23 14	10 03	22 07	07 28	23 03
23	23 26	03 N 31	20 18	20 24	27 36	23 14	10 02	22 07	07 28	23 03
24	23 25	09 20	20 02	20 06	27 40	23 13	10 02	22 06	07 27	23 02
25	23 24	14 45	19 47	19 47	27 44	23 13	10 02	22 06	07 27	23 02
26	23 22	19 33	19 33	19 27	27 47	23 12	10 02	22 05	07 27	23 01
27	23 20	22 53	19 20	19 07	27 50	23 11	10 02	22 04	07 27	23 00
28	23 18	24 53	19 07	18 47	27 53	23 11	10 02	22 04	07 27	23 00
29	23 15	25 12	18 56	18 26	27 56	23 10	10 02	22 03	07 27	22 59
30	23 N 12	23 N 50	18 N 45	18 N 04	27 S 59	23 N 09	10 S 02	22 N 03	07 S 27	22 N 59

ZODIAC SIGN ENTRIES

Date	h	m	Planets
02	12	46	☽ ♋
04	16	34	☽ ♌
07	00	06	☽ ♍
09	10	59	☽ ♎
11	23	30	☽ ♏
14	11	37	☽ ♐
16	22	05	☽ ♑
17	17	04	♀ ♌
19	06	26	☽ ♒
21	12	37	☽ ♓
21	12	54	☉ ♋
23	16	44	☽ ♈
25	19	09	☽ ♉
27	20	41	☽ ♊
29	22	35	☽ ♋

LATITUDES

Date	Mercury ☿ ° '	Venus ♀ ° '	Mars ♂ ° '	Jupiter ♃ ° '	Saturn ♄ ° '	Uranus ♅ ° '	Neptune ♆ ° '	Pluto ♇ ° '
01	02 N 08	01 N 39	02 S 45	00 S 05	02 N 38	00 N 28	01 N 45	09 N 51
04	01 55	01 43	02 57	00 04	02 38	00 28	01 44	09 51
07	01 35	01 47	03 10	00 04	02 37	00 28	01 44	09 50
10	01 09	01 50	03 22	00 04	02 36	00 28	01 44	09 50
13	00 N 36	01 52	03 34	00 04	02 36	00 28	01 44	09 49
16	00 S 03	01 53	03 46	00 03	02 35	00 28	01 44	09 49
19	00 47	01 54	03 58	00 03	02 34	00 28	01 44	09 48
22	01 34	01 54	04 08	00 03	02 34	00 28	01 44	09 48
25	02 23	01 53	04 18	00 02	02 33	00 28	01 43	09 47
28	03 10	01 51	04 27	00 02	02 32	00 28	01 43	09 47
31	03 S 52	01 N 49	04 S 35	00 S 02	02 N 31	00 N 28	01 N 43	09 N 46

DATA

Julian Date	2434895
Delta T	+31 seconds
Ayanamsa	23° 13' 29"
Synetic vernal point	05° ♓ 53' 31"
True obliquity of ecliptic	23° 26' 45"

MOON'S PHASES, APSIDES AND POSITIONS ☽

Date	h	m	Phase	Longitude ° '	Eclipse Indicator
01	04	03	●	10 ♊ 08	
08	09	14	☽	17 ♍ 03	
16	12	06	○	24 ♐ 48	
23	19	46	☾	01 ♈ 47	
30	12	26	●	08 ♋ 10	Total

Day	h	m		
11	15	01	Apogee	
27	10	03	Perigee	
01	19	05	Max dec	25° N 16'
08	13	15	0S	
16	01	16	Max dec	25° S 16'
22	21	47	0N	
29	04	24	Max dec	25° N 17'

LONGITUDES

Date	Chiron ⚷ ° '	Ceres ⚳ ° '	Pallas ⚴ ° '	Juno ⚵ ° '	Vesta ⚶ ° '	Black Moon Lilith ⚸ ° '
01	28 ♑ 13	01 ♎ 10	01 ♍ 52	24 ♌ 37	17 ♉ 38	28 ♎ 32
11	27 ♑ 50	02 ♎ 36	05 ♍ 05	27 ♌ 40	21 ♉ 56	29 ♎ 39
21	27 ♑ 23	04 ♎ 32	08 ♍ 32	00 ♍ 54	26 ♉ 09	00 ♏ 46
31	26 ♑ 51	06 ♎ 53	12 ♍ 10	04 ♍ 17	00 ♊ 17	01 ♏ 54

All ephemeris data is given at 12.00 UT and the Moon's longitude is additionally given for 24.00 UT

ASPECTARIAN

h m	Aspects
01 Tuesday	
00 43	☽ ⊼ ♂
01 32	☽ bQ ♆
03 14	☽ ± ♄
04 03	☽ ☌ ☉
04 18	☽ ⊻ ♀
04 59	☽ Q ♇
11 38	☽ ⊥ ♅
17 52	☽ bQ ♄
18 44	☉ Q ♇
21 30	☽ ⊻ ♅
02 Wednesday	
00 40	☽ ⚹ ♇
02 04	☽ △ ♆
04 17	☿ △ ♄
16 13	☽ ☌ ♃
16 39	☽ ∥ ♀
18 43	☽ △ ♄
20 17	☽ ☌ ☿
03 Thursday	
01 48	☽ ∠ ♇
01 57	☽ ☍ ♂
07 24	☽ ∥ ♃
08 40	☽ ∥ ♇
09 32	☽ ⊻ ☉
10 48	☽ ☌ ♀
12 12	☉ ∥ ♅
16 30	♂ bQ ♇
16 43	☽ ∥ ☉
16 55	☽ ∥ ♅
17 15	☽ ⊥ ♇
20 40	☽ ⊥ ☉
04 Friday	
00 30	☽ ☌ ♅
03 43	☽ ⊻ ♇
05 06	☽ □ ♆
13 42	☽ ∠ ☉
21 08	☽ ⊻ ♃
22 44	☽ □ ♄
05 Saturday	
05 56	☿ ☍ ♂
06 06	☽ ⊼ ♂
06 07	☽ ⊻ ☿
08 11	☽ ⊥ ♃
11 10	☿ ∠ ♇
13 26	☽ Q ♆
16 58	☽ ± ♂
18 13	☽ ⊥ ☿
19 03	☽ ⚹ ☉
21 36	☽ ⊻ ♀
06 Sunday	
01 04	☽ ∠ ♃
06 45	☉ ⊥ ♅
07 06	☽ ⊻ ♅
07 45	☽ Q ♄
09 31	☽ bQ ♂
09 58	☽ ⊥ ♀
10 24	☽ ☌ ♇
11 45	☽ ⚹ ♆
12 41	☽ ∠ ☿
13 54	☽ # ♄
14 34	♀ ⊥ ♇
18 29	☽ ⊥ ♅
19 24	☽ Q ☉
07 Monday	
02 29	☽ # ♆
04 54	☽ ∠ ♀
06 00	☽ ⚹ ♃
06 30	☽ ⚹ ♄
11 50	☽ ∠ ♅
13 48	☽ △ ♂
16 28	☽ ∠ ♆
20 16	☽ ⚹ ☿
08 Tuesday	
04 48	♃ △ ♄
05 48	☽ Q ♃
09 14	☽ □ ☉
10 05	☽ ⊥ ♆
11 40	☽ ∠ ♄
13 19	☽ ⚹ ♀
17 22	☽ ⚹ ♅
20 40	☽ ⊻ ♇
21 57	☽ ⊻ ♆
22 04	☽ Q ☿
09 Wednesday	
05 28	☽ ⊥ ♄
08 38	☽ ⊥ ♇
14 59	☉ bQ ♄
15 51	☽ Q ♀
17 25	☽ Q ♅
17 27	☽ ⊻ ♄
18 16	☽ □ ♃
10 Thursday	
00 16	☽ □ ♂
00 53	☽ ∥ ♆
02 44	☽ ∠ ♇
08 12	♀ ☌ ♅
13 15	☽ □ ☿
14 06	☽ ∥ ♄
11 Friday	
02 33	☽ △ ☉
05 53	☽ □ ♅
08 12	☽ □ ♀
09 04	☽ ⚹ ♇
10 13	☽ ☌ ♆
11 58	☽ Q ♂
16 56	♀ ⊻ ♇
17 33	♀ ∥ ♃
12 Saturday	
04 10	♀ □ ♆
05 48	☽ ☌ ♄
08 00	☽ △ ♃
09 21	☽ Q ♇
11 31	☽ bQ ☉
11 44	☽ ⚹ ♂
14 47	☉ ∥ ♇
15 10	☉ ∥ ♀
15 19	♀ ∥ ♇
21 44	☿ ∥ ♃
22 15	☉ ⊻ ♅
13 Sunday	
06 10	☽ △ ☿
07 11	☽ ± ☉
08 26	☉ ∥ ☿
13 02	☿ ∥ ♇
14 43	☽ bQ ♃
17 15	☽ ∠ ♂
18 35	☽ △ ♅
18 45	☽ # ♅
20 14	☽ ⊼ ☉
21 31	☽ □ ♇
22 31	☽ ⊻ ♆
14 Monday	
01 12	☽ # ♀
02 30	☽ # ☿
03 12	☽ △ ♀
04 08	☽ # ♇
05 20	☽ # ☉
06 05	☽ # ♃
08 59	☽ ± ♃
10 26	☽ ⊥ ♆
10 48	☽ ⊥ ♂
12 49	☉ ⚹ ♇
13 50	☽ bQ ☿
17 37	☽ ⊻ ♄
21 05	☽ ⊼ ♃
22 25	☽ ⊻ ♂
15 Tuesday	
00 27	☽ bQ ♅
00 46	☉ △ ♆
04 10	☽ ∠ ♆
05 01	♂ Q ♆
05 24	☽ ⊥ ♄
08 28	☽ ± ☿
10 10	☉ ∥ ♃
11 59	☽ bQ ♀
18 09	☽ ± ♅
20 44	☽ ⊼ ☿
22 55	☽ ∠ ♄
16 Wednesday	
00 25	☿ ⊥ ♇
05 51	☽ ⊼ ♅
06 25	☿ ∥ ♀
07 08	♂ ☍ ♃
07 20	☽ ± ♀
08 32	☽ △ ♇
09 21	☽ ⚹ ♆
12 06	☽ ☍ ☉
20 06	☽ ⊼ ♀
21 46	☿ ∥ ♅
17 Thursday	
01 15	♀ ∥ ♅
03 43	☽ ⚹ ♄
07 18	☽ ☌ ♂
08 16	☽ ☍ ♃
08 20	☽ Q ♆
13 17	☽ bQ ♇
18 25	☽ # ☉
19 33	☽ # ♃
21 28	☽ # ♇
18 Friday	
02 23	☽ Q ♄
06 16	☽ ± ♇
06 17	☽ # ♅
08 07	☽ ☍ ☿
08 58	☽ # ♀
09 51	☽ # ☿
15 04	☽ ☍ ♅
17 29	☽ ⊼ ♇
18 09	☽ □ ♆
19 Saturday	
01 20	☽ ⊼ ☉
10 10	☽ ☍ ♀
11 44	☽ □ ♄
13 15	☽ ± ☉
14 07	☽ ⊻ ♂
17 13	☽ ⊼ ♃
20 Sunday	
00 50	☽ ⊥ ♂
03 12	♀ □ ♄
04 20	☽ ± ♃
06 55	☽ bQ ☉
16 18	☽ ⊼ ☿
16 45	☽ ∠ ♂
20 50	☽ bQ ♃
22 04	☽ ⊼ ♅
22 05	♀ ⊼ ♂
21 Monday	
00 14	☽ ☍ ♇
00 46	☽ △ ♆
03 10	☽ ± ☿
03 25	☽ ∥ ♄
08 49	☽ ± ♅
11 50	☽ △ ☉
14 48	☽ ∥ ♆
17 36	☽ △ ♄
18 52	☽ ⚹ ♂
19 16	☽ bQ ☿
21 20	☽ ⊼ ♀
23 53	☽ △ ♃
22 Tuesday	
00 46	☽ bQ ♅
03 17	☽ bQ ♆
08 50	☽ ± ♀
15 24	☽ Q ♂
18 57	☽ ± ♆
19 46	☽ bQ ♄
21 33	☽ △ ☿
23 Wednesday	
01 53	☽ bQ ♀
02 09	☿ St R
02 57	☽ △ ♅
04 54	☽ ⊼ ♇
05 18	☽ ⊼ ♆
09 21	♀ ⊻ ♃
11 14	☽ ± ♄
15 11	☽ ± ♇
19 46	☽ □ ☉
21 27	☽ ⊼ ♄
21 39	☽ □ ♂
24 Thursday	
04 09	☽ # ♆
04 26	☽ □ ♃
05 50	☽ △ ♀
06 30	☽ bQ ♇
07 30	♂ ⚹ ♄
14 59	☽ # ♄
17 21	☉ ☍ ♂
20 16	☉ △ ♄
23 12	♀ ± ♂
25 Friday	
00 18	☽ □ ☿
04 17	☽ Q ☉
05 58	☽ □ ♅
07 44	☽ △ ♇
08 00	☿ ∥ ♀
08 01	☽ ☍ ♆
11 08	☽ Q ♃
22 54	☽ △ ♂
23 41	☽ ☍ ♄
26 Saturday	
01 39	☽ ⚹ ☉
05 51	☽ Q ☿
07 22	☽ ⚹ ♃
08 31	☉ ⊥ ♀
11 58	☽ Q ♅
12 20	☽ ∥ ♀
12 21	☽ □ ♀
12 54	☽ ∥ ☿
23 10	☽ bQ ♂
27 Sunday	
01 19	☽ ⚹ ☿
04 10	☽ ∠ ☉
05 34	☽ ∥ ♅
07 05	♀ Q ♆
07 50	☽ ⚹ ♅
08 30	☽ ∠ ♃
09 27	☽ □ ♇
09 39	☽ ⊼ ♆
13 08	☽ ∥ ♇
13 41	☽ ± ♂
14 46	☽ ∥ ♃
16 07	☽ ∥ ☉
19 32	☽ ± ♆
20 03	☽ ⊥ ☉
20 38	☽ Q ♀
23 23	☽ ⊼ ♂
23 35	☽ ⊥ ♃
28 Monday	
01 09	☽ ⊼ ♄
01 31	☽ ∠ ☿
06 39	☽ ⊻ ☉
08 41	☽ ∠ ♅
09 39	☽ ⊻ ♃
10 24	☽ bQ ♆
11 03	☽ ± ♄
15 13	☽ Q ♇
16 02	☽ ⊥ ☿
18 13	☽ ⚹ ♀
23 40	☽ ⊥ ♅
29 Tuesday	
01 44	☽ ⊻ ☿
01 59	☽ bQ ♄
09 43	☽ ⊻ ♅
11 14	☽ ⚹ ♇
11 20	☽ △ ♆
17 16	♀ ⊥ ♃
18 23	☿ ⊥ ♇
21 30	☽ ∠ ♀
30 Wednesday	
00 16	☽ ☍ ♂
03 07	☽ △ ♄
04 14	♃ ∠ ♇
12 26	☽ ☌ ☉
12 35	☽ ∠ ♇
12 42	☽ ☌ ♃
14 07	☽ ⊥ ♀
14 43	☉ ∠ ♇
17 39	☉ ☌ ♃
18 45	☽ ∥ ☉
19 02	☽ ∥ ♃
20 40	☽ ∥ ♇

JULY 1954

LONGITUDES

Date	Sidereal time h m s	Sun ☉ ° ' "	Moon ☽ ° ' "	Moon ☽ 24.00 ° ' "	Mercury ☿ ° '	Venus ♀ ° '	Mars ♂ ° '	Jupiter ♃ ° '	Saturn ♄ ° '	Uranus ♅ ° '	Neptune ♆ ° '	Pluto ⯓ ° '
01	06 36 00	09 ♋ 06 23	21 ♋ 51 46	28 ♋ 42 45	16 ♋ 23	16 ♌ 14	00 ♑ 33	08 ♋ 33	02 ♏ 40	22 ♋ 27	23 ♎ 17	23 ♌ 17
02	06 39 57	10 03 37	05 ♌ 28 30	12 ♌ 08 51	15 R 50	17 24	00 ♑ 14	08 47	02 R 39	22 31	23 R 17	23 18
03	06 43 53	11 00 50	18 ♌ 43 48	25 ♌ 13 26	15 18	18 34	29 ♐ 57	09 00	02 39	22 34	23 17	23 20
04	06 47 50	11 58 04	01 ♍ 37 55	07 ♍ 57 32	14 38	19 44	29 R 39	09 14	02 39	22 38	23 17	23 21
05	06 51 47	12 55 17	14 ♍ 12 37	20 ♍ 23 33	14 01	20 54	29 22	09 27	02 39	22 42	23 D 17	23 23
06	06 55 43	13 52 29	26 ♍ 30 48	02 ♎ 34 53	13 24	22 04	29 05	09 41	02 39	22 45	23 17	23 24
07	06 59 40	14 49 42	08 ♎ 36 18	14 ♎ 35 37	12 48	23 13	28 49	09 55	02 D 39	22 49	23 17	23 26
08	07 03 36	15 46 55	20 ♎ 33 25	26 ♎ 30 15	12 12	24 23	28 33	10 08	02 39	22 52	23 17	23 28
09	07 07 33	16 44 07	02 ♏ 26 41	08 ♏ 23 17	11 39	25 32	28 17	10 22	02 39	22 56	23 17	23 29
10	07 11 29	17 41 19	14 ♏ 20 37	20 ♏ 19 11	11 07	26 42	28 03	10 35	02 39	23 00	23 18	23 31
11	07 15 26	18 38 32	26 ♏ 19 29	02 ♐ 21 59	10 39	27 51	27 48	10 49	02 40	23 03	23 18	23 32
12	07 19 22	19 35 44	08 ♐ 27 06	14 ♐ 35 13	10 14	29 ♌ 00	27 35	11 02	02 40	23 07	23 18	23 34
13	07 23 19	20 32 56	20 ♐ 46 39	27 ♐ 01 41	09 54	00 ♍ 09	27 22	11 16	02 41	23 11	23 18	23 36
14	07 27 16	21 30 09	03 ♑ 20 33	09 ♑ 43 22	09 37	01 19	27 09	11 29	02 42	23 14	23 19	23 37
15	07 31 12	22 27 22	16 ♑ 10 17	22 ♑ 41 17	09 25	02 27	26 58	11 43	02 42	23 18	23 19	23 39
16	07 35 09	23 24 35	29 ♑ 16 22	05 ♒ 55 27	09 19	03 36	26 47	11 56	02 43	23 22	23 19	23 41
17	07 39 05	24 21 49	12 ♒ 38 23	19 ♒ 24 59	09 D 17	04 45	26 37	12 10	02 44	23 25	23 20	23 42
18	07 43 02	25 19 02	26 ♒ 15 01	03 ♓ 08 12	09 21	05 54	26 27	12 23	02 45	23 29	23 20	23 44
19	07 46 58	26 16 17	10 ♓ 04 14	17 ♓ 02 50	09 31	07 02	26 18	12 37	02 47	23 33	23 20	23 46
20	07 50 55	27 13 32	24 ♓ 03 38	01 ♈ 06 18	09 46	08 10	26 10	12 50	02 48	23 36	23 21	23 48
21	07 54 51	28 10 48	08 ♈ 10 30	15 ♈ 15 55	10 06	09 19	26 03	13 03	02 49	23 40	23 21	23 49
22	07 58 48	29 ♋ 08 05	22 ♈ 22 12	29 ♈ 29 02	10 33	10 27	25 57	13 17	02 51	23 44	23 22	23 51
23	08 02 45	00 ♌ 05 22	06 ♉ 36 08	13 ♉ 43 09	11 05	11 35	25 51	13 30	02 52	23 47	23 22	23 53
24	08 06 41	01 02 41	20 ♉ 49 49	27 ♉ 55 49	11 43	12 43	25 47	13 43	02 54	23 51	23 23	23 55
25	08 10 38	02 00 01	05 ♊ 00 51	12 ♊ 04 35	12 27	13 51	25 43	13 57	02 56	23 55	23 24	23 56
26	08 14 34	02 57 21	19 ♊ 06 43	26 ♊ 06 56	13 17	14 58	25 40	14 10	02 58	23 58	23 24	23 58
27	08 18 31	03 54 43	03 ♋ 04 53	10 ♋ 00 16	14 12	16 06	25 37	14 23	03 00	24 02	23 25	24 00
28	08 22 27	04 52 05	16 ♋ 52 45	23 ♋ 42 02	15 12	17 13	25 36	14 36	03 02	24 05	23 26	24 02
29	08 26 24	05 49 28	00 ♌ 27 51	07 ♌ 09 57	16 18	18 21	25 35	14 49	03 04	24 09	23 27	24 04
30	08 30 20	06 46 52	13 ♌ 48 08	20 ♌ 22 16	17 29	19 28	25 D 36	15 02	03 06	24 13	23 27	24 06
31	08 34 17	07 ♌ 44 17	26 ♌ 52 13	03 ♍ 18 00	18 ♋ 45	20 ♍ 35	25 ♐ 37	15 ♋ 16	03 ♏ 09	24 ♋ 16	23 ♎ 28	24 ♌ 07

Date	Moon True ☊ ° '	Moon Mean ☊ ° '	Moon ☽ Latitude ° '
01	14 ♑ 44	15 ♑ 08	00 S 39
02	14 D 44	15 05	01 52
03	14 45	15 02	02 57
04	14 45	14 59	03 52
05	14 45	14 55	04 33
06	14 45	14 52	05 01
07	14 R 45	14 49	05 15
08	14 D 45	14 46	05 15
09	14 45	14 43	05 02
10	14 46	14 40	04 36
11	14 46	14 36	03 57
12	14 47	14 33	03 08
13	14 47	14 30	02 09
14	14 47	14 27	01 S 03
15	14 R 48	14 24	00 N 08
16	14 47	14 21	01 19
17	14 47	14 17	02 28
18	14 46	14 14	03 30
19	14 44	14 11	04 20
20	14 43	14 08	04 56
21	14 42	14 05	05 14
22	14 41	14 01	05 13
23	14 D 41	13 58	04 53
24	14 42	13 55	04 15
25	14 43	13 52	03 22
26	14 44	13 49	02 17
27	14 45	13 46	01 N 04
28	14 R 45	13 42	00 S 12
29	14 45	13 39	01 25
30	14 43	13 36	02 33
31	14 ♑ 40	13 ♑ 33	03 S 31

DECLINATIONS

Date	Sun ☉ ° '	Moon ☽ ° '	Mercury ☿ ° '	Venus ♀ ° '	Mars ♂ ° '	Jupiter ♃ ° '	Saturn ♄ ° '	Uranus ♅ ° '	Neptune ♆ ° '	Pluto ⯓ ° '
01	23 N 08	21 N 01	18 N 36	17 N 42	28 S 02	23 N 08	10 S 02	22 N 01	07 S 27	22 N 58
02	23 04	17 05	18 28	17 20	28 04	23 08	10 02	22 01	07 27	22 57
03	22 59	12 24	18 21	16 57	28 07	23 07	10 02	22 01	07 27	22 57
04	22 54	07 17	18 15	16 34	28 09	23 06	10 02	22 00	07 27	22 56
05	22 49	02 N 01	18 10	16 11	28 11	23 05	10 02	22 00	07 27	22 55
06	22 43	03 S 13	18 07	15 47	28 12	23 04	10 03	21 59	07 27	22 55
07	22 37	08 14	18 05	15 22	28 14	23 03	10 03	21 59	07 27	22 54
08	22 31	12 54	18 05	14 58	28 16	23 02	10 03	21 58	07 27	22 54
09	22 24	17 03	18 06	14 33	28 17	23 01	10 04	21 57	07 28	22 53
10	22 17	20 32	18 08	14 07	28 18	23 00	10 04	21 57	07 28	22 53
11	22 09	23 11	18 11	13 42	28 19	22 59	10 04	21 56	07 28	22 52
12	22 01	24 49	18 16	13 16	28 20	22 58	10 05	21 55	07 28	22 51
13	21 52	25 16	18 22	12 49	28 20	22 57	10 05	21 55	07 28	22 50
14	21 44	24 27	18 28	12 23	28 21	22 56	10 06	21 54	07 28	22 50
15	21 34	22 20	18 36	11 56	28 22	22 55	10 06	21 53	07 28	22 49
16	21 25	19 01	18 45	11 29	28 22	22 54	10 07	21 53	07 28	22 48
17	21 15	14 39	18 54	11 01	28 22	22 53	10 07	21 52	07 29	22 48
18	21 05	09 29	19 04	10 34	28 22	22 52	10 08	21 52	07 29	22 47
19	20 54	03 S 47	19 15	10 06	28 22	22 51	10 09	21 51	07 29	22 47
20	20 43	02 N 10	19 26	09 38	28 22	22 50	10 09	21 51	07 29	22 46
21	20 32	08 03	19 37	09 09	28 22	22 48	10 10	21 50	07 30	22 45
22	20 20	13 33	19 48	08 41	28 22	22 47	10 11	21 49	07 30	22 45
23	20 08	18 20	20 00	08 12	28 22	22 46	10 12	21 49	07 30	22 44
24	19 56	22 04	20 11	07 43	28 22	22 45	10 13	21 48	07 30	22 44
25	19 43	24 27	20 21	07 14	28 22	22 43	10 13	21 47	07 31	22 43
26	19 30	25 16	20 32	06 45	28 20	22 42	10 14	21 47	07 31	22 42
27	19 17	24 28	20 41	06 16	28 19	22 41	10 15	21 46	07 31	22 41
28	19 03	22 11	20 49	05 46	28 19	22 40	10 16	21 46	07 32	22 41
29	18 49	18 40	20 57	05 16	28 18	22 38	10 17	21 45	07 32	22 40
30	18 35	14 15	21 03	04 47	28 17	22 37	10 18	21 44	07 32	22 40
31	18 N 20	09 N 15	21 N 07	04 N 17	28 S 17	22 N 35	10 S 19	21 N 44	07 S 33	22 N 39

ZODIAC SIGN ENTRIES

Date	h	m	Planets
02	02	16	☽ ♌
03	07	23	♂ ♐
04	08	56	☽ ♍
06	18	53	☽ ♎
09	07	04	☽ ♏
11	19	19	☽ ♐
13	08	43	♀ ♍
14	05	40	☽ ♑
16	13	19	☽ ♒
18	18	33	☽ ♓
20	22	07	☽ ♈
23	00	52	☽ ♉
23	09	45	☉ ♌
25	03	30	☽ ♊
27	06	41	☽ ♋
29	11	10	☽ ♌
31	17	49	☽ ♍

LATITUDES

Date	Mercury ☿ ° '	Venus ♀ ° '	Mars ♂ ° '	Jupiter ♃ ° '	Saturn ♄ ° '	Uranus ♅ ° '	Neptune ♆ ° '	Pluto ⯓ ° '
01	03 S 52	01 N 49	04 S 35	00 S 02	02 N 31	00 N 28	01 N 43	09 N 46
04	04 25	01 45	04 42	00 02	02 30	00 28	01 43	09 46
07	04 46	01 41	04 48	00 01	02 30	00 28	01 43	09 46
10	04 52	01 36	04 52	00 01	02 29	00 28	01 43	09 45
13	04 44	01 30	04 55	00 S 01	02 28	00 28	01 43	09 45
16	04 23	01 23	04 58	00 00	02 27	00 28	01 42	09 45
19	03 52	01 15	04 59	00 00	02 26	00 28	01 42	09 45
22	03 14	01 07	04 59	00 00	02 26	00 28	01 42	09 45
25	02 31	00 57	04 58	00 N 01	02 25	00 28	01 42	09 45
28	01 46	00 47	04 57	00 01	02 24	00 28	01 42	09 44
31	01 S 01	00 N 36	04 S 54	00 N 01	02 N 23	00 N 28	01 N 42	09 N 44

DATA

Julian Date	2434925
Delta T	+31 seconds
Ayanamsa	23° 13' 34"
Synetic vernal point	05° ♓ 53' 25"
True obliquity of ecliptic	23° 26' 45"

MOON'S PHASES, APSIDES AND POSITIONS ☽

Date	h	m	Phase	Longitude ° '	Eclipse Indicator
08	01	33	☽	15 ♎ 22	
16	00	29	○	22 ♑ 57	partial
23	00	14	☾	29 ♈ 37	
29	22	20	●	06 ♌ 14	

Day	h	m		
09	08	25	Apogee	
23	18	27	Perigee	
05	21	10	0S	
13	08	49	Max dec	25° S 17'
20	03	18	0N	
26	12	06	Max dec	25° N 16'

LONGITUDES

Date	Chiron ⚷ ° '	Ceres ⚳ ° '	Pallas ⚴ ° '	Juno ⚵ ° '	Vesta ⚶ ° '	Black Moon Lilith ⚸ ° '
01	26 ♑ 51	06 ♎ 53	12 ♍ 10	04 ♍ 17	00 ♊ 17	01 ♏ 54
11	26 ♑ 17	09 ♎ 35	15 ♍ 58	07 ♍ 46	04 ♊ 18	03 ♏ 01
21	25 ♑ 42	12 ♎ 35	19 ♍ 54	11 ♍ 20	08 ♊ 11	04 ♏ 08
31	25 ♑ 07	15 ♎ 50	23 ♍ 56	14 ♍ 58	11 ♊ 55	05 ♏ 15

All ephemeris data is given at 12.00 UT and the Moon's longitude is additionally given for 24.00 UT

ASPECTARIAN

h m Aspects

01 Thursday
00 49 ♀ ♇ ♂
01 21 ☽ ⊻ ♀
02 51 ☽ ☌ ☿
04 03 ☽ ⊥ ⯓
04 31 ☽ ∥ ♅
08 40 ☉ ∥ ♃
13 02 ☽ ☌ ♅
14 09 ☿ ⊻ ♀
14 29 ☽ ⊻ ⯓
14 29 ☽ □ ♆
19 09 ♆ ⚹ ⯓

02 Friday
02 54 ☽ ⊼ ♂
03 59 ☽ ∥ ☿
06 59 ☽ □ ♄
10 31 ☽ ∥ ♀
13 20 ☽ ± ♂
18 01 ☽ ⊻ ♃
20 52 ☽ ⊻ ☉
22 27 ☽ Q ♆

03 Saturday
05 04 ☽ ⊥ ♃
05 14 ☽ ♇ ♂
05 54 ☽ ⊻ ☿
08 37 ☽ ⊥ ☉
11 40 ☽ ☌ ♀
15 32 ☽ Q ♄
16 26 ☽ ⊥ ☿
19 07 ☽ ⊻ ♅
20 24 ☽ ⚹ ♆
20 30 ☽ ☌ ⯓
21 55 ☽ ∠ ♃
23 16 ☽ ⋕ ♄

04 Sunday
02 33 ☽ ∠ ☉
02 48 ☉ ∥ ⯓
06 20 ☽ ⊥ ♅
08 21 ☽ △ ♂
08 25 ☽ ∠ ☿
11 14 ☽ ⋕ ♆
13 55 ☽ ⚹ ♄
23 26 ☽ ∠ ♅

05 Monday
00 15 ☿ ⊥ ♀
00 37 ☽ ∠ ♆
02 42 ☽ ⚹ ♃
06 49 ♀ Q ♄
08 31 ♆ St D
09 19 ☽ ⚹ ☉
11 39 ☽ ⚹ ☿
17 57 ☽ ⊥ ♆
18 39 ☽ ∠ ♄

06 Tuesday
02 20 ☽ ⊻ ♀
02 21 ☽ Q ♃
04 35 ☽ ⚹ ♅
04 48 ☉ ☌ ☿
05 40 ☽ ⊻ ♆
05 53 ☽ ⊻ ⯓
09 55 ☽ Q ☿
10 38 ☽ Q ☉
12 15 ☽ ⊥ ♄
15 22 ☽ ⊥ ♀
15 53 ♄ St D
16 58 ☽ □ ♂
17 43 ☽ ⊥ ⯓

07 Wednesday
00 07 ☽ ⊻ ♄
03 07 ♀ ⊻ ♅
04 24 ☽ Q ♅
08 10 ☽ ∥ ♆
11 09 ☽ ∠ ♀
11 39 ☽ ∠ ⯓
13 22 ♀ ⚹ ♆
14 40 ☽ □ ♃
16 29 ♀ ☌ ⯓
19 59 ☽ □ ☿
21 05 ☽ ∥ ♄

08 Thursday
01 33 ☽ □ ☉
04 06 ☽ Q ♂
16 42 ☽ □ ♅
17 31 ☽ ☌ ♆
17 52 ☽ ⚹ ⯓
20 33 ☽ ⚹ ♀
22 29 ☽ ⋕ ♀

09 Friday
03 47 ☽ ⚹ ♂
07 29 ♀ ∠ ♃
12 25 ☽ ☌ ♄
18 09 ☽ Q ⯓
18 47 ☽ ⋕ ☿
23 24 ☽ Q ♀

10 Saturday
03 43 ☿ ∠ ♀
04 18 ☽ △ ♃
05 47 ☽ △ ☿
07 25 ☉ ⊥ ⯓
09 26 ☽ ∠ ♂
19 18 ☽ △ ☉
23 42 ☽ ⋕ ♅

11 Sunday
02 06 ☽ ⋕ ☉
03 09 ☽ ⊥ ♂
05 27 ☽ △ ♅
05 57 ☽ ⊻ ♆
06 12 ☿ ☌ ♃
06 26 ☽ □ ⯓
08 36 ☽ ⋕ ⯓
09 58 ☽ ⋕ ♃
10 42 ☽ ♇ ☿
10 58 ☽ ♇ ♃
11 13 ♀ △ ♂
14 54 ☽ ⊻ ♂
15 22 ☽ □ ♀
17 55 ☽ ⊥ ♆

12 Monday
00 36 ☽ ⊻ ♄
03 46 ☽ ♇ ☉
03 58 ☽ ± ☿
05 10 ☽ ± ♃
11 20 ☽ ♇ ♅
11 42 ☽ ∠ ♆
12 26 ☽ ⊥ ♄
14 27 ♀ ⊥ ♅
15 24 ☽ ⊼ ☿
17 10 ☽ ⊼ ♃
22 55 ☽ ± ☉

13 Tuesday
04 37 ☉ ∥ ♅
05 00 ☽ ± ♅
06 00 ☽ ∠ ♄
11 31 ☽ ⊼ ☉
16 39 ☽ ⊼ ♅
16 52 ☽ ⚹ ♆
17 26 ☽ △ ⯓

14 Wednesday
00 26 ☽ ☌ ♂
07 46 ☽ △ ♀
10 46 ☽ ⚹ ♄
15 43 ☽ Q ♆
21 58 ☽ ♇ ⯓
23 37 ☽ ☍ ☿

15 Thursday
03 35 ☽ ☍ ♃
06 36 ☽ ⋕ ♃
07 37 ☽ ⋕ ⯓
09 17 ☽ Q ♄
14 36 ☽ ♇ ♀
14 44 ☽ ± ⯓
15 47 ☽ ⋕ ♅
17 18 ♀ ⚹ ♄
17 36 ♅ □ ♆
18 40 ☽ ⋕ ☉

16 Friday
00 29 ☽ ☍ ☉
01 09 ☽ □ ♆
01 11 ☽ ☍ ♅
01 47 ☽ ⊼ ⯓
07 32 ☽ ⊻ ♂
08 41 ☽ ± ♀
09 43 ☉ □ ♆
10 42 ☉ ☌ ♅
13 35 ☽ ⋕ ☿
18 15 ☽ □ ♄
18 16 ☽ ⊥ ♂
18 57 ☉ ⊻ ⯓
20 34 ☽ ⊼ ♀

17 Saturday
06 01 ☽ ⊼ ☿
06 50 ☿ St D
10 11 ☽ ∠ ♂
11 08 ☽ ⊼ ♃
16 43 ☽ ± ☿
21 57 ☽ ± ♃

18 Sunday
06 46 ☽ ⋕ ♀
06 53 ☽ △ ♆
07 08 ☽ ⊼ ♅
07 35 ☽ ☍ ⯓
08 39 ☽ ♇ ☿
09 08 ☽ ∥ ♄
10 15 ☽ ⊼ ☉
12 21 ☽ ⚹ ♂
14 01 ☽ ♇ ♃
17 40 ☽ ± ♅
20 36 ☽ ∥ ♆
21 30 ☽ ± ☉
23 22 ☽ △ ♄

19 Monday
06 17 ☽ ☍ ♀
08 59 ☽ Q ♂
09 01 ☽ ♇ ♆
09 21 ☽ ♇ ♅
09 32 ♀ ⋕ ♄
11 01 ☽ △ ☿
12 46 ☉ ⊼ ♂
14 13 ☽ ♇ ☉
16 27 ☽ △ ♃

20 Tuesday
00 31 ☽ ± ♆
01 16 ☽ ♇ ♄
10 47 ☽ ⊼ ♆
11 13 ☽ △ ♅
11 32 ☽ ⊼ ⯓
15 34 ☽ □ ♂
15 41 ♀ ∠ ♆
16 41 ☽ ± ♄
17 47 ☽ △ ☉
21 37 ♀ ∠ ♅
21 47 ☽ ± ⯓

21 Wednesday
02 54 ☽ ⊼ ♄
09 42 ☽ ⋕ ♆
13 06 ☽ ♇ ⯓
14 06 ☽ ⊼ ♀
15 22 ☽ □ ☿
16 18 ☽ ∥ ♀
20 24 ☽ □ ♃
21 01 ☽ ⋕ ♄

22 Thursday
01 08 ☽ ± ♀
13 41 ☽ ☍ ♆
14 18 ☽ □ ♅
14 30 ☽ △ ⯓
15 52 ☿ ⚹ ♀
17 39 ☽ ♇ ♀
18 00 ☽ △ ♂
22 49 ☽ Q ☿

23 Friday
00 14 ☽ □ ☉
03 16 ☽ Q ♃
05 42 ☽ ☍ ♄
19 08 ☽ ♇ ♂
19 54 ☽ ⚹ ☿
20 42 ☉ ∥ ☿
20 47 ☽ Q ♅
21 07 ☽ △ ♀
22 06 ☽ ∥ ☉
22 16 ☽ ∥ ☿
23 49 ☽ ⚹ ♃

24 Saturday
08 46 ☽ Q ☉
10 00 ☽ ∥ ♅
10 14 ☽ ± ♂
16 19 ☽ ⊼ ♆
17 07 ☽ ⚹ ♅
17 13 ☽ □ ⯓
17 18 ☽ ∥ ⯓
17 29 ☽ ∥ ♃
20 19 ☽ ⊼ ♂
22 29 ☽ ∠ ☿
22 31 ♀ ⋕ ♆

25 Sunday
01 33 ☽ ∠ ♃
02 28 ☽ ± ♆
05 09 ☉ ± ♂
06 31 ☽ ⚹ ☉
08 28 ☽ ⊼ ♄
14 35 ☽ ⊥ ☿
14 38 ♀ ⚹ ♃
17 03 ☽ ⊥ ♃
17 45 ☽ ♇ ♆
18 39 ☽ ∠ ♅
18 40 ☽ ± ♄
23 48 ☽ Q ⯓

26 Monday
01 24 ☽ ⊻ ☿
03 25 ☽ ⊻ ♃
04 19 ☽ □ ♀
09 53 ☽ ∠ ☉
10 02 ☽ ♇ ♄
10 02 ☽ ⊥ ♅
11 58 ♅ ⊻ ⯓
12 13 ☉ □ ♄
18 30 ♃ ∥ ⯓
19 21 ☽ △ ♆
20 20 ☽ ⚹ ⯓
20 21 ☽ ⊻ ♅
23 11 ☽ ☍ ♂

27 Tuesday
02 26 ☽ ⊥ ☉
11 51 ☽ △ ♄
13 32 ☽ ⊻ ☉
13 55 ☽ Q ♀
18 06 ☿ ☌ ♃
22 17 ☽ ∠ ⯓

28 Wednesday
07 50 ☽ ∥ ⯓
07 57 ☽ ☌ ♃
07 59 ☽ ∥ ♃
08 49 ☽ ☌ ☿
12 39 ☽ ⚹ ♀
14 01 ☽ ⊥ ⯓
15 22 ☽ ∥ ♅
16 28 ♀ ⊥ ♆
21 49 ☽ ∥ ☿
23 32 ☽ □ ♆

29 Thursday
00 37 ☽ ⊻ ⯓
00 45 ☽ ☌ ♅
03 21 ☽ ⊼ ♂
05 53 ♀ ∠ ♄
11 01 ☽ ∥ ☉
14 00 ☽ ± ♂
15 19 ♂ St D
16 40 ☽ □ ♄
17 37 ☽ ∠ ♀
22 20 ☽ ☌ ☉

30 Friday
06 11 ☽ ♇ ♂
07 44 ☽ Q ♆
11 20 ☽ ⊥ ♀
14 17 ☽ ⊻ ♃
19 25 ☽ ⊻ ☿
23 18 ☽ ⊻ ♀
23 57 ☿ ⊥ ⯓

31 Saturday
01 23 ☽ Q ♄
01 27 ☽ ⊥ ♃
05 42 ☽ ⚹ ♆
06 54 ☽ ☌ ⯓
07 02 ☽ ⋕ ♄
07 10 ☽ ⊻ ♅
07 39 ☽ ⊥ ☿
09 40 ☽ △ ♂
18 22 ☽ ⊥ ♅
18 25 ☽ ∠ ♃
19 50 ☽ ⋕ ♆
23 45 ☽ ⚹ ♄

AUGUST 1954

LONGITUDES

Date	Sidereal time h m s	Sun ☉ ° ' "	Moon ☽ ° ' "	Moon ☽ 24.00 ° ' "	Mercury ☿ ° '	Venus ♀ ° '	Mars ♂ ° '	Jupiter ♃ ° '	Saturn ♄ ° '	Uranus ♅ ° '	Neptune ♆ ° '	Pluto ♇ ° '
01	08 38 14	08 ♌ 41 42	09 ♍ 39 37	15 ♍ 57 11	20 ♋ 07	21 ♍ 42	25 ♐ 39	15 ♋ 29	03 ♏ 11	24 ♋ 20	23 ♎ 29	24 ♌ 09
02	08 42 10	09 39 08	22 ♍ 10 53	28 ♍ 20 57	21 33	22 49	25 42	15 42	03 14	24 23	23 30	24 11
03	08 46 07	10 36 35	04 ♎ 27 41	10 ♎ 31 28	23 04	23 55	25 45	15 54	03 16	24 27	23 31	24 13
04	08 50 03	11 34 02	16 ♎ 32 43	22 ♎ 31 53	24 39	25 02	25 50	16 07	03 19	24 31	23 32	24 15
05	08 54 00	12 31 31	28 ♎ 29 30	04 ♏ 26 06	26 18	26 08	25 55	16 20	03 22	24 34	23 33	24 17
06	08 57 56	13 29 00	10 ♏ 22 15	16 ♏ 18 32	28 01	27 14	26 01	16 33	03 25	24 38	23 34	24 19
07	09 01 53	14 26 29	22 ♏ 15 35	28 ♏ 13 59	29 ♋ 48	28 20	26 08	16 46	03 28	24 41	23 35	24 21
08	09 05 49	15 24 00	04 ♐ 14 21	10 ♐ 17 15	01 ♌ 37	29 ♍ 26	26 16	16 59	03 31	24 45	23 36	24 22
09	09 09 46	16 21 31	16 ♐ 23 17	22 ♐ 32 57	03 30	00 ♎ 31	26 25	17 11	03 34	24 48	23 37	24 24
10	09 13 43	17 19 04	28 ♐ 46 46	05 ♑ 05 08	05 25	01 37	26 34	17 24	03 37	24 52	23 38	24 26
11	09 17 39	18 16 37	11 ♑ 28 26	17 ♑ 56 57	07 22	02 42	26 44	17 36	03 40	24 55	23 39	24 28
12	09 21 36	19 14 11	24 ♑ 30 50	01 ♒ 10 12	09 21	03 47	26 55	17 49	03 44	24 58	23 40	24 30
13	09 25 32	20 11 46	07 ♒ 55 00	14 ♒ 45 03	11 21	04 52	27 07	18 01	03 47	25 02	23 41	24 32
14	09 29 29	21 09 23	21 ♒ 40 07	28 ♒ 39 47	13 22	05 56	27 19	18 14	03 51	25 05	23 43	24 34
15	09 33 25	22 07 00	05 ♓ 43 32	12 ♓ 50 47	15 24	07 01	27 32	18 26	03 55	25 09	23 44	24 36
16	09 37 22	23 04 39	20 ♓ 00 51	27 ♓ 13 02	17 26	08 05	27 46	18 38	03 58	25 12	23 45	24 38
17	09 41 18	24 02 19	04 ♈ 26 35	11 ♈ 40 45	19 28	09 09	28 00	18 51	04 02	25 15	23 46	24 40
18	09 45 15	25 00 01	18 ♈ 54 52	26 ♈ 08 16	21 30	10 13	28 15	19 03	04 06	25 19	23 48	24 42
19	09 49 12	25 57 44	03 ♉ 20 24	10 ♉ 30 46	23 32	11 16	28 31	19 15	04 10	25 22	23 49	24 43
20	09 53 08	26 55 29	17 ♉ 38 59	24 ♉ 44 45	25 33	12 19	28 47	19 27	04 14	25 25	23 51	24 45
21	09 57 05	27 53 16	01 ♊ 47 52	08 ♊ 48 10	27 33	13 23	29 04	19 39	04 18	25 28	23 52	24 47
22	10 01 01	28 51 05	15 ♊ 45 35	22 ♊ 40 05	29 ♌ 32	14 25	29 22	19 51	04 22	25 32	23 53	24 49
23	10 04 58	29 ♌ 48 55	29 ♊ 31 39	06 ♋ 20 18	01 ♍ 30	15 28	29 40	20 03	04 27	25 35	23 55	24 51
24	10 08 54	00 ♍ 46 47	13 ♋ 06 03	19 ♋ 48 55	03 27	16 30	29 ♐ 59	20 15	04 31	25 38	23 56	24 53
25	10 12 51	01 44 41	26 ♋ 28 55	03 ♌ 06 00	05 23	17 32	00 ♑ 18	20 26	04 35	25 41	23 58	24 55
26	10 16 47	02 42 36	09 ♌ 40 12	16 ♌ 11 25	07 18	18 34	00 38	20 38	04 40	25 44	23 59	24 57
27	10 20 44	03 40 33	22 ♌ 39 40	29 ♌ 04 51	09 11	19 35	00 59	20 50	04 45	25 47	24 01	24 59
28	10 24 41	04 38 31	05 ♍ 26 58	11 ♍ 45 59	11 04	20 37	01 20	21 01	04 49	25 50	24 03	25 01
29	10 28 37	05 36 31	18 ♍ 01 53	24 ♍ 14 44	12 55	21 38	01 42	21 13	04 54	25 53	24 04	25 03
30	10 32 34	06 34 33	00 ♎ 24 36	06 ♎ 31 37	14 44	22 38	02 04	21 24	04 59	25 56	24 06	25 04
31	10 36 30	07 ♍ 32 36	12 ♎ 35 56	18 ♎ 37 49	16 ♍ 33	23 ♎ 38	02 ♑ 27	21 ♋ 35	05 ♏ 04	25 ♋ 59	24 ♎ 07	25 ♌ 06

Date	Moon True ☊ ° '	Moon Mean ☊ ° '	Moon ☽ Latitude ° '
01	14 ♑ 36	13 ♑ 30	04 S 17
02	14 R 33	13 27	04 50
03	14 29	13 23	05 08
04	14 26	13 20	05 13
05	14 23	13 17	05 04
06	14 23	13 14	04 42
07	14 D 23	13 11	04 07
08	14 24	13 07	03 22
09	14 26	13 04	02 27
10	14 27	13 01	01 25
11	14 28	12 58	00 S 16
12	14 R 28	12 55	00 N 55
13	14 26	12 52	02 04
14	14 22	12 48	03 08
15	14 17	12 45	04 03
16	14 11	12 42	04 43
17	14 05	12 39	05 05
18	14 00	12 36	05 08
19	13 57	12 33	04 52
20	13 55	12 29	04 18
21	13 D 54	12 26	03 28
22	13 55	12 23	02 26
23	13 56	12 20	01 17
24	13 57	12 17	00 N 05
25	13 R 56	12 13	01 S 07
26	13 54	12 10	02 14
27	13 48	12 07	03 13
28	13 41	12 04	04 01
29	13 32	12 01	04 36
30	13 23	11 58	04 58
31	13 ♑ 13	11 ♑ 54	05 S 05

DECLINATIONS

Date	Sun ☉ ° '	Moon ☽ ° '	Mercury ☿ ° '	Venus ♀ ° '	Mars ♂ ° '	Jupiter ♃ ° '	Saturn ♄ ° '	Uranus ♅ ° '	Neptune ♆ ° '	Pluto ♇ ° '
01	18 N 05	03 N 59	21 N 10	03 N 47	28 S 16	22 N 34	10 S 20	21 N 43	07 S 33	22 N 38
02	17 50	01 S 20	21 11	03 17	28 15	22 33	10 21	21 42	07 33	22 37
03	17 35	06 29	21 10	02 47	28 14	22 31	10 22	21 42	07 34	22 37
04	17 19	11 19	21 07	02 17	28 13	22 30	10 24	21 41	07 34	22 36
05	17 03	15 40	21 01	01 47	28 12	22 28	10 25	21 41	07 35	22 36
06	16 47	19 23	20 53	01 16	28 11	22 27	10 26	21 40	07 35	22 35
07	16 30	22 19	20 42	00 46	28 10	22 25	10 27	21 39	07 35	22 34
08	16 13	24 19	20 29	00 N 16	28 09	22 24	10 29	21 39	07 36	22 34
09	15 56	25 12	20 13	00 S 14	28 08	22 22	10 30	21 38	07 36	22 33
10	15 39	24 51	19 54	00 45	28 07	22 21	10 31	21 38	07 37	22 32
11	15 21	23 13	19 33	01 15	28 06	22 19	10 32	21 37	07 37	22 32
12	15 03	20 20	19 09	01 45	28 05	22 18	10 34	21 36	07 38	22 31
13	14 45	16 18	18 42	02 15	28 03	22 16	10 35	21 36	07 38	22 31
14	14 27	11 19	18 14	02 46	28 02	22 15	10 37	21 35	07 39	22 30
15	14 09	05 S 39	17 43	03 16	28 01	22 13	10 38	21 35	07 39	22 29
16	13 50	00 N 23	17 10	03 46	27 59	22 12	10 40	21 34	07 40	22 29
17	13 31	06 26	16 35	04 16	27 58	22 10	10 41	21 33	07 40	22 28
18	13 12	12 10	15 58	04 46	27 57	22 08	10 43	21 33	07 41	22 27
19	12 52	17 12	15 20	05 15	27 55	22 07	10 44	21 32	07 41	22 27
20	12 32	21 13	14 40	05 45	27 54	22 05	10 46	21 32	07 42	22 26
21	12 13	23 55	13 59	06 15	27 52	22 04	10 47	21 31	07 42	22 26
22	11 53	25 07	13 17	06 44	27 50	22 02	10 49	21 31	07 43	22 25
23	11 32	24 44	12 34	07 13	27 49	22 00	10 51	21 30	07 43	22 25
24	11 12	22 53	11 50	07 42	27 47	21 59	10 52	21 29	07 44	22 24
25	10 51	19 46	11 05	08 11	27 45	21 57	10 54	21 29	07 45	22 23
26	10 31	15 41	10 20	08 40	27 43	21 55	10 56	21 28	07 45	22 23
27	10 10	10 56	09 35	09 09	27 41	21 54	10 57	21 28	07 46	22 22
28	09 49	05 47	08 49	09 38	27 39	21 52	10 59	21 27	07 46	22 22
29	09 27	00 N 30	08 02	10 06	27 37	21 50	11 01	21 27	07 47	22 21
30	09 06	04 S 43	07 16	10 34	27 35	21 49	11 03	21 26	07 48	22 21
31	08 N 45	09 S 40	06 N 29	11 S 02	27 S 33	21 N 47	11 S 05	21 N 26	07 S 48	22 N 20

ZODIAC SIGN ENTRIES

Date	h	m	Planets
03	03	14	☽ ♎
05	15	03	☽ ♏
07	14	44	☿ ♌
08	03	32	☽ ♐
09	00	34	♀ ♎
10	14	20	☽ ♑
12	21	54	☽ ♒
15	02	17	☽ ♓
17	04	37	☽ ♈
19	06	26	☽ ♉
21	08	56	☽ ♊
22	17	42	☿ ♍
23	12	50	☽ ♋
23	16	36	☉ ♍
24	13	22	♂ ♑
25	18	22	☽ ♌
28	01	44	☽ ♍
30	11	12	☽ ♎

LATITUDES

Date	Mercury ☿ ° '	Venus ♀ ° '	Mars ♂ ° '	Jupiter ♃ ° '	Saturn ♄ ° '	Uranus ♅ ° '	Neptune ♆ ° '	Pluto ♇ ° '
01	00 S 47	00 N 32	04 S 54	00 N 01	02 N 23	00 N 28	01 N 42	09 N 44
04	00 S 06	00 20	04 51	00 01	02 22	00 28	01 41	09 44
07	00 N 31	00 N 07	04 47	00 02	02 21	00 28	01 41	09 44
10	01 01	00 S 07	04 43	00 02	02 21	00 28	01 41	09 45
13	01 23	00 21	04 39	00 02	02 20	00 28	01 41	09 45
16	01 38	00 36	04 34	00 03	02 19	00 28	01 41	09 45
19	01 45	00 52	04 29	00 03	02 19	00 28	01 41	09 45
22	01 45	01 08	04 24	00 03	02 18	00 29	01 41	09 45
25	01 40	01 25	04 18	00 04	02 17	00 29	01 40	09 46
28	01 30	01 42	04 13	00 04	02 17	00 29	01 40	09 46
31	01 N 17	02 S 00	04 S 07	00 N 04	02 N 16	00 N 29	01 N 40	09 N 46

DATA

Julian Date	2434956
Delta T	+31 seconds
Ayanamsa	23° 13' 40"
Synetic vernal point	05° ♓ 53' 20"
True obliquity of ecliptic	23° 26' 45"

MOON'S PHASES, APSIDES AND POSITIONS ☽

Date	h	m	Phase	Longitude ° '	Eclipse Indicator
06	18	51	☽	13 ♏ 45	
14	11	03	○	21 ♒ 07	
21	04	51	☾	27 ♉ 36	
28	10	21	●	04 ♍ 35	

Day	h	m		
06	03	04	Apogee	
18	05	42	Perigee	
02	05	55	0S	
09	17	32	Max dec	25° S 14'
16	10	31	0N	
22	18	08	Max dec	25° N 10'
29	14	16	0S	

LONGITUDES

Date	Chiron ⚷ ° '	Ceres ⚳ ° '	Pallas ⚴ ° '	Juno ⚵ ° '	Vesta ⚶ ° '	Black Moon Lilith ⚸ ° '
01	25 ♑ 03	16 ♎ 10	24 ♍ 21	15 ♍ 20	12 ♊ 17	05 ♏ 22
11	24 ♑ 31	19 ♎ 38	28 ♍ 29	19 ♍ 01	15 ♊ 50	06 ♏ 29
21	24 ♑ 02	23 ♎ 17	02 ♎ 41	22 ♍ 45	19 ♊ 09	07 ♏ 36
31	23 ♑ 38	27 ♎ 04	06 ♎ 58	26 ♍ 30	22 ♊ 14	08 ♏ 43

All ephemeris data is given at 12.00 UT and the Moon's longitude is additionally given for 24.00 UT

ASPECTARIAN

h m	Aspects
01 Sunday	
02 22	☽ ∠ ☿
09 46	☽ ∠ ♆
10 01	☽ ⊻ ☉
11 22	☽ ∠ ♅
12 58	☽ ∥ ♀
22 23	☽ ⊥ ☉
23 17	☽ ✶ ♃
02 Monday	
02 57	☽ ⊥ ♆
04 21	☽ ∠ ♄
10 36	☽ ✶ ☿
13 20	☽ ☌ ♀
14 33	☽ ⊻ ♆
15 54	☽ ⊻ ♇
16 18	☽ ✶ ♅
17 12	☽ ∠ ☉
18 51	☽ □ ♂
20 09	☽ # ♀
21 51	☽ ⊥ ♄
22 54	☽ Q ♃
03 Tuesday	
03 06	♀ ⊻ ♆
03 38	☽ ⊥ ♇
09 39	☽ ⊻ ♄
13 21	☽ Q ☿
15 57	☽ Q ♅
15 58	☉ ⚼ ♂
17 10	☽ ∥ ♆
18 38	♀ ⊻ ♇
19 02	☿ □ ♆
21 26	☽ ∠ ♇
04 Wednesday	
00 10	♀ ✶ ♅
01 13	☽ ✶ ☉
05 58	☿ ⊻ ♇
06 33	☽ Q ♂
07 13	☽ ∥ ♄
09 55	☿ ☌ ♅
11 00	☉ Q ♆
11 08	☽ □ ♃
05 Thursday	
02 01	☽ ☌ ♆
03 19	☽ Q ☉
03 29	☽ ✶ ♇
04 03	☽ □ ♅
04 57	☿ ✶ ♀
06 17	☿ ⊼ ♂
06 45	☽ ⊻ ♀
06 47	☽ ✶ ♂
06 51	☽ □ ☿
07 02	♀ □ ♂
19 52	☽ # ☉
20 06	☽ ⊥ ♀
21 52	☽ ☌ ♄
06 Friday	
03 46	☽ Q ♇
13 20	☽ ∠ ♂
16 05	♀ ⊥ ♄
16 09	☽ ∠ ♀
18 51	☽ □ ☉
22 44	☽ # ☿
07 Saturday	
00 43	☽ △ ♃
05 54	☽ # ♅
07 41	☽ ⊥ ♂
12 58	☽ # ♃
14 26	☽ # ♇
14 39	☽ ⊻ ♆
16 12	☽ □ ♇
16 54	☽ △ ♅
19 53	☽ ⊻ ♂
23 43	♀ Q ♃
08 Sunday	
01 26	☽ ✶ ♀
02 43	☽ ⊥ ♆
05 50	☽ △ ☿
07 24	☽ ⚼ ♃
10 33	☽ ⊻ ♄
20 40	☽ ∠ ♆
20 59	☿ ± ♂
22 31	☽ ⊥ ♄
22 59	☽ ⚼ ♅
09 Monday	
01 36	☽ ± ♃
03 39	☽ Q ♀
09 23	♀ ⊥ ♇
11 56	☽ △ ☉
12 49	☿ □ ♄
13 35	☽ ⊼ ♃
16 16	☽ ∠ ♄
16 44	☽ ± ♅
16 53	☽ ⚼ ☿
10 Tuesday	
02 05	☽ ✶ ♆
03 38	☽ △ ♇
04 26	☽ ⊼ ♅
07 42	☽ ☌ ♂
13 27	☽ ± ☿
14 32	☉ ⊻ ♃
17 55	☽ □ ♀
19 18	☽ ⚼ ☉
21 16	☽ ✶ ♄
11 Wednesday	
01 03	☽ Q ♆
02 55	☽ ⊼ ☿
08 14	☽ ⚼ ♇
13 37	☽ ± ☉
18 48	☽ # ♇
19 50	☽ Q ♄
20 42	☽ # ♃
23 33	☽ ☍ ♃
12 Thursday	
00 59	☽ ± ♇
01 36	☽ ⊼ ☉
02 34	☽ # ♅
10 28	☽ □ ♆
10 49	♀ ⊻ ♄
11 59	☽ ⊼ ♇
12 50	☽ ☍ ♅
16 25	☽ ⊻ ♂
20 41	☽ # ☿
13 Friday	
03 20	☽ ⊥ ♂
04 39	☽ □ ♄
06 07	☽ △ ♀
16 02	☿ Q ♆
19 06	☽ ☍ ☿
19 29	☽ ∠ ♂
20 22	☽ # ☉
21 59	☿ ⚼ ♂
14 Saturday	
05 58	☽ ⊼ ♃
10 38	☽ ⚼ ♀
11 03	☽ ☍ ☉
15 05	☽ ∥ ♄
15 31	☽ △ ♆
16 28	☽ ± ♃
17 00	☽ ☍ ♇
17 54	☽ ⊼ ♅
21 51	☽ ✶ ♂
15 Sunday	
03 21	☽ ± ♀
03 47	☽ ∥ ♆
04 12	☽ ± ♅
06 28	☉ Q ♄
08 04	☽ ⚼ ♃
08 55	☽ △ ♄
14 21	☽ ⊼ ♀
15 05	♀ Q ♅
17 05	☽ ⚼ ♆
18 32	☽ Q ♂
19 29	☽ ⚼ ♅
20 53	☽ ∥ ♀
16 Monday	
06 59	☽ ⊼ ☿
08 13	☽ ± ♆
09 40	☽ △ ♃
10 15	☽ ⚼ ♄
10 16	♃ ⊥ ♇
10 56	☉ ∠ ♀
17 28	☽ ⊼ ☉
18 15	☽ ⊼ ♆
18 39	☽ ± ☿
19 43	☽ ⊼ ♇
20 41	☽ △ ♅
17 Tuesday	
01 07	☽ □ ♂
01 19	☽ ± ♄
02 32	☽ # ♀
03 46	☿ ⊻ ♃
04 10	☽ ± ☉
05 16	☉ ✶ ♆
05 42	☽ ± ♇
11 19	☽ ⊼ ♄
12 03	☽ ⚼ ☿
17 02	☽ # ♆
20 10	☽ ⚼ ☉
20 25	☽ ☍ ♀
20 40	☽ ⚼ ♇
23 55	♀ ∠ ♇
18 Wednesday	
04 03	☉ ☌ ♇
05 41	☽ # ♄
12 13	☽ □ ♃
13 31	☉ ⊥ ♃
16 21	☽ ∥ ☉
17 00	☽ △ ☿
19 17	☿ Q ♄
20 07	☽ ☍ ♆
20 11	☉ ⊻ ♅
21 37	☽ △ ♇
22 40	☽ □ ♅
22 50	☽ △ ☉
19 Thursday	
03 41	☽ ∥ ☿
03 48	☽ △ ♂
13 23	☽ ☍ ♄
15 30	☿ ✶ ♆
18 38	☽ Q ♃
20 Friday	
02 20	☽ ⊼ ♀
02 27	☿ ☌ ♇
04 51	☽ Q ♅
05 22	☽ ⚼ ♂
10 28	☿ ⊻ ♅
10 47	☿ ⊥ ♃
13 14	☽ ± ♀
14 13	☽ ∥ ♅
15 05	☽ ✶ ♃
18 25	☽ ∥ ♃
20 51	☽ ± ♂
21 18	☽ ∥ ♇
22 29	☽ ⊼ ♆
21 Saturday	
00 03	☽ □ ♇
01 12	☽ ✶ ♅
03 34	☽ □ ☿
04 51	☽ □ ☉
05 42	☽ ⚼ ♀
07 15	☽ ⊼ ♂
08 42	☽ ± ♆
16 18	☽ ⊼ ♄
16 57	☽ ∠ ♃
20 01	☉ ☌ ☿
22 Sunday	
00 08	☽ ⚼ ♆
02 39	☽ ± ♄
02 56	☽ ∠ ♅
06 54	☽ Q ♇
08 39	☽ ⊥ ♃
09 12	☿ ∠ ♀
09 30	☽ △ ♀
09 36	☿ △ ♂
14 02	☽ Q ☉
15 35	☽ Q ☿
18 18	☽ ⚼ ♄
18 33	☽ ⊥ ♅
19 12	☽ ⊻ ♃
23 Monday	
02 09	☽ △ ♆
03 48	☽ ✶ ♇
05 03	☽ ⊻ ♅
06 35	☉ △ ♂
12 15	☽ ☍ ♂
12 33	☽ ✶ ☉
12 59	☿ ⊥ ♅
16 03	☽ ✶ ☿
20 42	☽ △ ♄
24 Tuesday	
06 16	☽ ∠ ♇
13 20	♀ ∥ ♆
16 21	☽ ∥ ♇
17 09	☽ ∠ ☉
18 35	☽ □ ♀
19 59	☽ ∥ ♃
22 21	☽ ⊥ ♇
23 10	☽ ∠ ☿
23 49	☽ ∥ ♅
25 Wednesday	
00 58	☽ ☌ ♃
01 43	☿ ✶ ♄
07 27	☽ □ ♆
09 10	☽ ⊻ ♇
09 16	☉ # ♄
10 25	☉ ⊥ ♅
10 33	☽ ☌ ♅
10 34	☽ ⊥ ☉
12 46	☿ ∠ ♃
17 54	☿ # ♄
18 09	☽ ⊥ ☿
19 06	☽ ⊼ ♂
22 17	☽ ⊻ ☉
26 Thursday	
01 50	☉ ∥ ☿
02 48	☽ □ ♄
05 50	☽ Q ♀
06 19	☽ ± ♂
06 55	☽ ⊻ ☿
14 34	♀ Q ♂
16 16	☽ Q ♆
23 17	☽ ⚼ ♂
27 Friday	
05 48	☽ ✶ ♀
08 32	☽ ⊻ ♃
09 45	☿ ∠ ♆
11 51	☽ # ♄
12 09	☽ Q ♄
14 32	☽ ✶ ♆
15 55	☽ ∥ ☉
16 20	☽ ☌ ♇
17 51	☽ ⊻ ♅
19 33	☽ ∥ ☿
19 43	☽ # ♀
19 54	☽ ⊥ ♃
20 16	☿ # ♀
28 Saturday	
02 52	☽ # ♆
04 01	☽ △ ♂
05 09	☽ ⊥ ♅
09 03	☿ ∠ ♅
10 21	☽ ☌ ☉
10 48	☽ ✶ ♄
12 20	☽ ∠ ♀
13 06	☽ ∠ ♃
16 50	☉ ✶ ♄
17 24	☉ # ♀
18 50	☽ ∠ ♆
22 16	☽ ∠ ♅
23 55	♀ □ ♃
29 Sunday	
00 30	☽ ☌ ☿
06 58	☽ ⊥ ♀
12 04	☽ ⊥ ♆
15 37	☽ ∠ ♄
18 13	☽ ✶ ♃
19 33	☽ ⊻ ♀
19 47	☿ # ♆
23 41	☽ ⊻ ♆
30 Monday	
01 35	☽ ⊻ ♇
03 15	☽ ✶ ♅
06 37	☉ ∠ ♃
09 11	☽ ⊥ ♄
13 18	☽ ⊥ ♇
15 21	☽ □ ♂
17 57	☽ Q ♃
21 01	☽ ⊻ ♄
22 31	☽ # ☿
31 Tuesday	
01 08	☽ ⊻ ☉
02 47	☽ ∥ ♆
02 51	☽ Q ♅
07 03	☽ ∠ ♇
07 44	☽ # ☉
14 02	☽ ⊥ ☉
14 28	♀ ∥ ♄
19 18	☽ ∥ ♄
19 48	☽ ∥ ♀
21 13	☽ ⊻ ☿
23 57	♀ ☌ ♆

SEPTEMBER 1954

Date	Sidereal time h m s	Sun ☉ ° ' "	Moon ☽ ° ' "	Moon ☽ 24.00 ° ' "	Mercury ☿ ° '	Venus ♀ ° '	Mars ♂ ° '	Jupiter ♃ ° '	Saturn ♄ ° '	Uranus ♅ ° '	Neptune ♆ ° '	Pluto ♇ ° '
01	10 40 27	08 ♍ 30 40	24 ♎ 37 33	00 ♏ 35 28	18 ♍ 20	24 ♎ 38	02 ♑ 51	21 ♋ 47	05 ♏ 09	26 ♋ 02	24 ♎ 09	25 ♌ 08
02	10 44 23	09 28 46	06 ♏ 31 59	12 ♏ 27 33	20 06	25 38	03 14	21 58	05 14	26 05	24 11	25 10
03	10 48 20	10 26 54	18 ♏ 22 40	24 ♏ 17 53	21 51	26 37	03 38	22 09	05 19	26 08	24 13	25 12
04	10 52 16	11 25 02	00 ♐ 13 47	06 ♐ 10 59	23 34	27 36	04 04	22 20	05 24	26 11	24 14	25 14
05	10 56 13	12 23 13	12 ♐ 10 08	18 ♐ 11 53	25 16	28 34	04 29	22 31	05 29	26 14	24 16	25 16
06	11 00 10	13 21 25	24 ♐ 16 53	00 ♑ 25 47	26 57	29 ♎ 32	04 55	22 41	05 34	26 16	24 18	25 18
07	11 04 06	14 19 38	06 ♑ 39 13	12 ♑ 57 46	28 ♍ 37	00 ♏ 30	05 21	22 52	05 40	26 19	24 20	25 19
08	11 08 03	15 17 53	19 ♑ 21 58	25 ♑ 52 17	00 ♎ 16	01 27	05 48	23 03	05 45	26 22	24 21	25 21
09	11 11 59	16 16 10	02 ♒ 29 02	09 ♒ 12 29	01 54	02 24	06 15	23 13	05 50	26 25	24 23	25 23
10	11 15 56	17 14 27	16 ♒ 02 43	22 ♒ 59 38	03 30	03 20	06 42	23 24	05 56	26 27	24 25	25 25
11	11 19 52	18 12 47	00 ♓ 02 59	07 ♓ 12 21	05 05	04 16	07 10	23 34	06 01	26 30	24 27	25 27
12	11 23 49	19 11 08	14 ♓ 27 05	21 ♓ 46 24	06 39	05 11	07 39	23 44	06 07	26 32	24 29	25 28
13	11 27 45	20 09 31	29 ♓ 09 20	06 ♈ 34 51	08 12	06 06	08 08	23 54	06 13	26 35	24 31	25 30
14	11 31 42	21 07 56	14 ♈ 01 49	21 ♈ 29 06	09 44	07 00	08 37	24 04	06 18	26 37	24 33	25 32
15	11 35 39	22 06 23	28 ♈ 55 36	06 ♉ 20 18	11 15	07 54	09 06	24 14	06 24	26 40	24 35	25 34
16	11 39 35	23 04 52	13 ♉ 42 17	21 ♉ 00 49	12 45	08 47	09 36	24 24	06 30	26 42	24 37	25 36
17	11 43 32	24 03 23	28 ♉ 15 19	05 ♊ 25 20	14 14	09 40	10 06	24 34	06 36	26 45	24 39	25 37
18	11 47 28	25 01 57	12 ♊ 30 38	19 ♊ 31 05	15 41	10 32	10 37	24 43	06 42	26 47	24 41	25 39
19	11 51 25	26 00 33	26 ♊ 26 42	03 ♋ 17 35	17 07	11 23	11 08	24 53	06 48	26 49	24 43	25 41
20	11 55 21	26 59 11	10 ♋ 03 54	16 ♋ 45 54	18 33	12 14	11 39	25 02	06 54	26 51	24 45	25 42
21	11 59 18	27 57 51	23 ♋ 23 50	29 ♋ 57 59	19 57	13 04	12 11	25 12	07 00	26 54	24 47	25 44
22	12 03 14	28 56 33	06 ♌ 28 36	12 ♌ 55 56	21 20	13 53	12 43	25 21	07 06	26 56	24 49	25 46
23	12 07 11	29 ♍ 55 18	19 ♌ 20 13	25 ♌ 41 39	22 42	14 42	13 15	25 30	07 12	26 58	24 51	25 47
24	12 11 08	00 ♎ 54 04	02 ♍ 00 23	08 ♍ 16 34	24 02	15 30	13 47	25 39	07 19	27 00	24 53	25 49
25	12 15 04	01 52 53	14 ♍ 30 17	20 ♍ 41 40	25 21	16 17	14 20	25 48	07 25	27 02	24 55	25 51
26	12 19 01	02 51 44	26 ♍ 50 45	02 ♎ 57 39	26 39	17 03	14 53	25 56	07 31	27 04	24 57	25 52
27	12 22 57	03 50 37	09 ♎ 02 25	15 ♎ 05 09	27 56	17 48	15 27	26 05	07 38	27 06	24 59	25 54
28	12 26 54	04 49 31	21 ♎ 05 59	27 ♎ 05 04	29 ♎ 11	18 33	16 00	26 13	07 44	27 08	25 01	25 56
29	12 30 50	05 48 28	03 ♏ 02 36	08 ♏ 58 48	00 ♏ 24	19 16	16 34	26 22	07 51	27 10	25 03	25 57
30	12 34 47	06 ♎ 47 27	14 ♏ 53 59	20 ♏ 48 28	01 ♏ 36	19 ♏ 59	17 ♑ 09	26 ♋ 30	07 ♏ 57	27 ♋ 12	25 ♎ 05	25 ♌ 59

Date	Moon True ☊ ° '	Moon Mean ☊ ° '	Moon ☽ Latitude ° '
01	13 ♑ 04	11 ♑ 51	04 S 59
02	12 R 57	11 48	04 40
03	12 52	11 45	04 09
04	12 50	11 42	03 27
05	12 49	11 38	02 36
06	12 D 49	11 35	01 37
07	12 50	11 32	00 S 33
08	12 R 50	11 29	00 N 35
09	12 48	11 26	01 43
10	12 44	11 23	02 47
11	12 38	11 19	03 43
12	12 29	11 16	04 27
13	12 19	11 13	04 54
14	12 09	11 10	05 02
15	12 00	11 07	04 49
16	11 53	11 04	04 17
17	11 48	11 00	03 29
18	11 46	10 57	02 28
19	11 D 45	10 54	01 20
20	11 45	10 51	00 N 09
21	11 R 45	10 48	01 S 01
22	11 43	10 44	02 07
23	11 38	10 41	03 05
24	11 30	10 38	03 52
25	11 20	10 35	04 28
26	11 07	10 32	04 51
27	10 53	10 29	04 59
28	10 39	10 25	04 55
29	10 26	10 22	04 37
30	10 ♑ 15	10 ♑ 19	04 S 07

DECLINATIONS

Date	Sun ☉ ° '	Moon ☽ ° '	Mercury ☿ ° '	Venus ♀ ° '	Mars ♂ ° '	Jupiter ♃ ° '	Saturn ♄ ° '	Uranus ♅ ° '	Neptune ♆ ° '	Pluto ♇ ° '
01	08 N 23	14 S 10	05 N 42	11 S 30	27 S 30	21 N 45	11 S 06	21 N 25	07 S 49	22 N 20
02	08 01	18 06	04 56	11 57	27 28	21 44	11 08	21 25	07 50	22 19
03	07 39	21 17	04 09	12 24	27 25	21 42	11 10	21 24	07 50	22 19
04	07 17	23 35	03 23	12 51	27 23	21 40	11 12	21 24	07 51	22 18
05	06 55	24 50	02 36	13 18	27 20	21 39	11 14	21 23	07 52	22 18
06	06 33	24 57	01 50	13 44	27 17	21 37	11 16	21 23	07 52	22 17
07	06 10	23 50	01 04	14 10	27 14	21 35	11 18	21 22	07 53	22 17
08	05 48	21 28	00 N 19	14 36	27 11	21 34	11 20	21 22	07 54	22 16
09	05 25	17 57	00 S 27	15 02	27 08	21 32	11 22	21 21	07 54	22 16
10	05 02	13 22	01 12	15 27	27 04	21 30	11 24	21 21	07 55	22 16
11	04 40	07 58	01 56	15 52	27 01	21 29	11 26	21 20	07 56	22 15
12	04 17	02 S 01	02 40	16 17	26 57	21 27	11 28	21 20	07 57	22 14
13	03 54	04 N 10	03 24	16 41	26 54	21 26	11 30	21 19	07 57	22 14
14	03 31	10 10	04 07	17 05	26 50	21 24	11 32	21 19	07 58	22 13
15	03 08	15 36	04 50	17 28	26 46	21 22	11 34	21 18	07 59	22 13
16	02 45	20 35	05 32	17 51	26 42	21 21	11 36	21 18	07 59	22 12
17	02 22	23 10	06 14	18 14	26 38	21 19	11 38	21 17	08 00	22 12
18	01 58	24 45	06 55	18 37	26 33	21 17	11 40	21 17	08 01	22 12
19	01 35	24 44	07 35	18 59	26 29	21 16	11 42	21 17	08 02	22 11
20	01 12	23 13	08 15	19 20	26 24	21 14	11 44	21 16	08 02	22 11
21	00 49	20 25	08 54	19 42	26 19	21 13	11 46	21 16	08 03	22 10
22	00 25	16 37	09 33	20 03	26 14	21 11	11 48	21 15	08 04	22 10
23	00 N 02	12 06	10 10	20 23	26 09	21 10	11 51	21 15	08 05	22 10
24	00 S 22	07 09	10 47	20 43	26 04	21 08	11 53	21 15	08 06	22 09
25	00 45	01 N 59	11 23	21 02	25 58	21 07	11 55	21 14	08 06	22 09
26	01 08	03 S 11	11 59	21 21	25 53	21 05	11 57	21 14	08 07	22 09
27	01 32	08 10	12 33	21 40	25 47	21 04	11 59	21 14	08 08	22 08
28	01 55	12 47	13 07	21 58	25 41	21 02	12 01	21 13	08 09	22 08
29	02 18	16 52	13 39	22 16	25 35	21 01	12 04	21 13	08 09	22 08
30	02 S 42	20 S 15	14 S 11	22 S 33	25 S 28	20 N 59	12 S 06	21 N 13	08 S 10	22 N 07

ZODIAC SIGN ENTRIES

Date	h	m	Planets
01	22	49	☽ ♏
04	11	32	☽ ♐
06	23	10	☽ ♑
06	23	29	♀ ♏
08	08	05	☿ ♎
09	07	31	☽ ♒
11	11	55	☽ ♓
13	13	22	☽ ♈
15	13	44	☽ ♉
17	14	55	☽ ♊
19	18	13	☽ ♋
22	00	04	☽ ♌
23	13	55	☉ ♎
24	08	11	☽ ♍
26	18	11	☽ ♎
29	04	06	☿ ♏
29	05	52	☽ ♏

LATITUDES

Date	Mercury ☿ ° '	Venus ♀ ° '	Mars ♂ ° '	Jupiter ♃ ° '	Saturn ♄ ° '	Uranus ♅ ° '	Neptune ♆ ° '	Pluto ♇ ° '
01	01 N 11	02 S 05	04 S 05	00 N 04	02 N 16	00 N 29	01 N 40	09 N 46
04	00 54	02 23	04 00	00 05	02 15	00 29	01 40	09 47
07	00 34	02 42	03 54	00 05	02 15	00 29	01 40	09 47
10	00 N 13	03 00	03 48	00 05	02 14	00 29	01 40	09 48
13	00 S 09	03 19	03 42	00 06	02 14	00 29	01 40	09 48
16	00 32	03 37	03 36	00 06	02 13	00 29	01 40	09 49
19	00 56	03 55	03 31	00 06	02 13	00 29	01 40	09 50
22	01 19	04 14	03 25	00 07	02 12	00 29	01 40	09 50
25	01 42	04 31	03 19	00 07	02 12	00 29	01 40	09 51
28	02 04	04 49	03 13	00 08	02 12	00 30	01 39	09 52
31	02 S 24	05 S 05	03 S 07	00 N 08	02 N 11	00 N 30	01 N 39	09 N 52

DATA

Julian Date	2434987
Delta T	+31 seconds
Ayanamsa	23° 13' 43"
Synetic vernal point	05° ♓ 53' 17"
True obliquity of ecliptic	23° 26' 45"

MOON'S PHASES, APSIDES AND POSITIONS ☽

Date	h	m	Phase	Longitude ° '	Eclipse Indicator
05	12	28	☽	12 ♐ 24	
12	20	19	○	19 ♓ 31	
19	11	11	☾	25 ♊ 59	
27	00	50	●	03 ♎ 23	

Day	h	m		
02	21	40	Apogee	
14	19	55	Perigee	
30	14	02	Apogee	
06	02	11	Max dec	25° S 03'
12	19	50	0N	
18	23	39	Max dec	24° N 57'
25	21	08	0S	

LONGITUDES

Date	Chiron ⚷ ° '	Ceres ⚳ ° '	Pallas ⚴ ° '	Juno ⚵ ° '	Vesta ⚶ ° '	Black Moon Lilith ⚸ ° '
01	23 ♑ 36	27 ♎ 27	07 ♎ 24	26 ♍ 52	22 ♊ 32	08 ♏ 50
11	23 ♑ 19	01 ♏ 22	11 ♎ 44	00 ♎ 38	25 ♊ 16	09 ♏ 57
21	23 ♑ 08	05 ♏ 22	16 ♎ 06	04 ♎ 25	27 ♊ 39	11 ♏ 04
31	23 ♑ 05	09 ♏ 27	20 ♎ 31	08 ♎ 07	29 ♊ 36	12 ♏ 11

All ephemeris data is given at 12.00 UT and the Moon's longitude is additionally given for 24.00 UT

ASPECTARIAN

h m	Aspects
01 Wednesday	
04 10	☽ Q ♂
06 12	☽ □ ♃
09 29	☿ ⊥ ♆
09 34	☽ ∠ ☉
11 03	☽ ☌ ♆
11 19	☽ ⊥ ☿
12 02	☽ ☌ ♀
13 02	☽ ⚹ ♇
14 51	☽ □ ♅
21 22	☿ ⊥ ♀
02 Thursday	
00 27	♀ ⚹ ♇
04 21	☉ ∠ ♆
05 07	☽ ⚹ ♂
08 36	☽ ∠ ☿
09 20	☽ ☌ ♄
13 17	☽ Q ♇
13 49	☿ ∠ ♄
18 30	☽ ⚹ ☉
23 37	♀ □ ♅
03 Friday	
00 08	☉ # ♆
12 34	☽ ∠ ♂
12 59	☽ # ♅
15 38	☽ # ♃
16 39	☿ ⚹ ♃
19 46	☽ △ ♃
20 14	☽ ⚹ ☿
20 59	☽ Q ☉
21 27	☽ # ♇
23 51	☽ ⊻ ♆
04 Saturday	
01 52	☽ □ ♇
03 47	☽ △ ♅
05 52	☉ ∠ ♅
06 12	☽ ⊻ ♀
07 27	☽ ⊥ ♂
12 01	☽ ⊥ ♆
19 24	☽ ⊥ ♀
20 01	☽ ⊻ ♂
21 32	☿ ⊻ ♆
22 30	☽ ⊻ ♄
05 Sunday	
00 34	☽ Q ☿
02 31	☽ Ꝑ ♃
06 11	☽ ∠ ♆
10 07	☽ Ꝑ ♅
10 37	☽ ⊥ ♄
11 50	☿ ⊻ ♇
12 28	☽ □ ☉
15 03	☽ ∠ ♀
20 47	☽ ± ♃
06 Monday	
01 57	☿ ⚹ ♅
04 05	☽ ± ♅
04 38	☽ ∠ ♄
08 49	☽ ⊼ ♃
12 02	☽ ⚹ ♆
13 59	☽ △ ♇
15 55	☽ ⊼ ♅
18 04	☽ □ ☿
23 08	☽ ⚹ ♀
07 Tuesday	
09 25	☽ ☌ ♂
10 05	☽ ⚹ ♄
11 22	☽ Q ♆
19 01	☽ Ꝑ ♇
08 Wednesday	
00 02	☽ Q ♀
03 46	☽ △ ☉
03 58	☿ ⊥ ♄
05 09	☽ # ♇
08 50	♂ ⚹ ♄
08 58	☽ Q ♄
11 16	☽ # ♃
11 59	☽ ± ♇
12 55	☽ # ♅
18 54	☽ ☍ ♃
21 15	☽ □ ♆
23 05	☽ ⊼ ♇
09 Thursday	
00 57	☽ ☍ ♅
04 19	☿ ⊥ ♇
09 38	☽ Ꝑ ☉
10 47	☽ △ ☿
11 50	☽ □ ♀
18 03	☽ □ ♄
18 59	☽ ⊻ ♂
19 48	♂ Q ♆
10 Friday	
02 32	☽ ∥ ♀
02 57	☽ ± ☉
05 58	☽ ⊥ ♂
06 10	☿ ⊻ ♀
14 14	☽ ⊼ ☉
16 49	☽ Ꝑ ☿
21 07	☽ ∥ ♄
22 08	☽ ∠ ♂
11 Saturday	
00 51	☽ ⊼ ♃
02 28	☽ △ ♆
04 10	☽ ☍ ♇
05 58	☽ ⊼ ♅
10 10	☽ ± ☿
11 10	☽ ± ♃
12 10	☽ ∥ ♆
16 08	☽ ± ♅
18 02	☉ ⊥ ♆
19 35	☽ △ ♀
20 10	☿ Q ♃
21 30	☽ ⊼ ☿
22 06	☽ △ ♄
12 Sunday	
00 21	☽ ⚹ ♂
02 26	☽ # ☉
02 26	☽ Ꝑ ♃
03 13	☿ ⊻ ♄
03 46	☽ Ꝑ ♆
07 11	☽ Ꝑ ♅
09 41	☽ ∥ ☿
18 38	☽ ± ♆
20 19	☽ ☍ ☉
20 49	☽ Q ♂
22 02	☽ Ꝑ ♀
23 00	☽ Ꝑ ♄
13 Monday	
03 22	☽ △ ♃
04 27	☽ ⊼ ♆
06 04	☽ ⊼ ♇
07 49	☽ △ ♅
08 38	☽ # ☿
10 11	☿ □ ♂
11 02	☽ ∥ ☉
13 38	☽ ± ♀
13 43	☽ ± ♄
15 14	♀ ☌ ♄
15 49	☽ ± ♇
18 01	☿ Q ♅
22 51	☉ # ☿
23 29	☽ ⊼ ♄
23 57	☽ ⊼ ♀
14 Tuesday	
02 59	☽ □ ♂
03 00	☽ # ♆
04 18	☽ ☍ ☿
06 22	☽ Ꝑ ♇
16 49	☉ ∠ ♄
17 44	☽ # ♄
15 Wednesday	
00 14	☽ ⊼ ☉
00 49	☿ ∠ ♇
02 33	♀ Q ♇
04 21	☽ □ ♃
04 58	☽ ☍ ♆
06 34	☽ △ ♇
08 20	☽ □ ♅
10 35	☽ ± ☉
22 15	☽ # ♀
16 Thursday	
00 11	☽ ☍ ♄
02 11	☽ Ꝑ ☉
03 28	☽ ☍ ♀
05 04	☽ △ ♂
09 51	☽ Q ♃
10 16	☽ ⊼ ☿
13 38	☽ Q ♅
20 21	☽ ∥ ♅
20 38	☽ ∥ ♃
21 12	☽ ± ☿
17 Friday	
03 16	☽ ∥ ♇
04 31	☽ △ ☉
05 48	☽ ⚹ ♃
05 59	☽ ⊼ ♆
06 35	☽ Ꝑ ♂
07 37	☽ □ ♇
09 29	☽ ⚹ ♅
13 48	☽ Ꝑ ☿
16 00	☽ ± ♆
22 09	☽ ± ♂
18 Saturday	
02 05	☽ ⊼ ♄
02 54	☉ ⚹ ♃
02 55	☉ ⊻ ♆
03 03	♃ □ ♆
07 10	☽ Ꝑ ♆
07 13	☽ ∠ ♃
08 25	☽ ⊼ ♀
08 39	☽ ⊼ ♂
10 45	☽ ∠ ♅
12 19	☽ ± ♄
13 47	♂ Ꝑ ♇
13 57	☽ Q ♇
17 47	♀ ⚹ ♂
18 02	☽ △ ☿
19 19	☽ ± ♀
21 35	♃ ∥ ♅
22 45	☽ ⊥ ♃
19 Sunday	
02 13	☽ ⊥ ♅
03 39	☉ ⊻ ♇
03 53	☽ Ꝑ ♄
08 58	☽ △ ♆
09 15	☽ ⊻ ♃
10 40	☽ ⚹ ♇
11 11	☽ □ ☉
11 53	☽ Ꝑ ♀
12 39	☽ ⊻ ♅
20 Monday	
04 16	☿ ∥ ♆
06 20	☽ △ ♄
08 43	☉ ⚹ ♅
13 09	☽ ∠ ♇
14 57	☽ ☍ ♂
16 07	☽ △ ♀
21 30	☽ Q ☉
22 05	☽ ∥ ♇
21 Tuesday	
05 01	☽ □ ☿
05 21	☽ ⊥ ♇
05 37	☽ ∥ ♅
05 59	☽ ∥ ♃
14 31	☽ □ ♆
15 18	☽ ☌ ♃
16 16	☽ ⊻ ♇
16 32	☽ # ♀
18 24	☽ ☌ ♅
21 00	☽ ⚹ ☉
22 Wednesday	
03 40	☉ ∠ ♀
13 10	☽ □ ♄
17 55	☽ Q ☿
23 48	☽ Q ♆
23 Thursday	
00 05	☽ ⊼ ♂
02 43	☽ □ ♀
03 02	☽ ∠ ☉
11 49	☽ ± ♂
13 17	☽ # ♄
19 04	☽ ⚹ ☿
20 28	☽ # ☿
21 22	☽ ⊥ ☉
22 25	☽ ⚹ ♆
23 10	☽ Q ♄
23 46	☽ ⊻ ♃
24 Friday	
00 13	☽ ☌ ♇
02 27	☽ ⊻ ♅
05 36	☽ Ꝑ ♂
07 32	☽ # ♆
09 43	☽ ⊻ ☉
11 18	☽ ⊥ ♃
13 54	☽ ⊥ ♅
15 02	☽ Q ♀
22 14	☽ ⚹ ♄
23 15	☉ ⊥ ♄
25 Saturday	
03 03	☽ ∠ ☿
03 07	☽ ∠ ♆
03 45	☿ ☌ ♆
04 45	☽ ∠ ♃
07 13	☽ ∠ ♅
11 06	☉ ⊥ ♇
11 39	☽ △ ♂
15 40	☽ ⚹ ♀
16 57	♀ # ♃
17 17	☽ # ☉
20 34	☽ ⊥ ♆
21 04	☿ □ ♃
21 14	☿ ⚹ ♇
22 30	☽ ⊥ ☿
22 47	♃ ⊻ ♇
26 Sunday	
01 38	☽ ∥ ☉
02 41	♀ # ♅
03 29	☽ ∠ ♄
08 17	☽ ⊻ ♆
10 05	☽ ⊻ ♇
10 12	☽ ⚹ ♃
10 45	☿ ∥ ♄
11 35	☽ ⊻ ☿
12 26	☽ ⚹ ♅
19 59	☿ □ ♅
21 15	☽ ⊥ ♄
21 53	☽ ⊥ ♇
22 53	☽ ∠ ♀
27 Monday	
00 50	☽ ☌ ☉
09 11	☽ ⊻ ♄
10 05	☽ Q ♃
11 49	☽ ∥ ♆
12 07	☽ Q ♅
15 41	☽ ∠ ♇
17 50	☽ ⊥ ♀
28 Tuesday	
01 20	☽ □ ♂
06 34	☽ ⊻ ♀
07 50	☽ ∥ ♄
14 04	☽ ∥ ☿
19 53	☽ ☌ ♆
21 42	☽ ⚹ ♇
22 23	☽ □ ♃
29 Wednesday	
00 07	☽ □ ♅
01 00	♀ # ♇
06 04	☽ ☌ ☿
15 15	☽ Q ♂
18 05	☽ ⊻ ☉
21 47	☽ ☌ ♄
21 56	☽ Q ♇
30 Thursday	
07 20	☽ ⊥ ☉
16 47	☽ ⚹ ♂
18 08	☽ # ♃
19 40	♄ Q ♇
20 07	☽ # ♅
22 58	☽ ☌ ♀

OCTOBER 1954

LONGITUDES

Date	Sidereal time h m s	Sun ☉ ° ' "	Moon ☽ ° ' "	Moon ☽ 24.00 ° ' "	Mercury ☿ ° '	Venus ♀ ° '	Mars ♂ ° '	Jupiter ♃ ° '	Saturn ♄ ° '	Uranus ♅ ° '	Neptune ♆ ° '	Pluto ♇ ° '
01	12 38 43	07 ♎ 46 27	26 ♏ 42 39	02 ♐ 36 59	02 ♏ 46	20 ♏ 40	17 ♑ 43	26 ♋ 38	08 ♏ 04	27 ♋ 13	25 ♎ 08	26 ♌ 00
02	12 42 40	08 45 30	08 ♐ 31 57	14 ♐ 28 05	03 54	21 21	18 18	26 46	08 10	27 15	25 10	26 02
03	12 46 37	09 44 34	20 ♐ 26 00	26 ♐ 26 19	05 00	22 00	18 53	26 53	08 17	27 17	25 12	26 03
04	12 50 33	10 43 40	02 ♑ 29 40	08 ♑ 36 43	06 04	22 39	19 28	27 01	08 23	27 18	25 14	26 05
05	12 54 30	11 42 48	14 ♑ 48 10	21 ♑ 04 39	07 06	23 16	20 04	27 09	08 30	27 20	25 16	26 06
06	12 58 26	12 41 58	27 ♑ 26 50	03 ♒ 55 17	08 05	23 52	20 40	27 16	08 37	27 21	25 19	26 07
07	13 02 23	13 41 09	10 ♒ 30 31	17 ♒ 12 57	09 01	24 26	21 16	27 23	08 43	27 23	25 21	26 09
08	13 06 19	14 40 22	24 ♒ 02 51	01 ♓ 00 20	09 55	25 00	21 52	27 30	08 50	27 24	25 23	26 10
09	13 10 16	15 39 37	08 ♓ 05 19	15 ♓ 17 31	10 45	25 31	22 28	27 37	08 57	27 26	25 25	26 12
10	13 14 12	16 38 53	22 ♓ 36 24	00 ♈ 01 13	11 31	26 02	23 05	27 44	09 04	27 27	25 27	26 13
11	13 18 09	17 38 12	07 ♈ 31 01	15 ♈ 04 38	12 14	26 31	23 42	27 51	09 11	27 28	25 30	26 14
12	13 22 06	18 37 32	22 ♈ 40 45	00 ♉ 17 59	12 52	26 58	24 19	27 57	09 18	27 29	25 32	26 16
13	13 26 02	19 36 55	07 ♉ 54 57	15 ♉ 30 17	13 25	27 24	24 56	28 04	09 25	27 30	25 34	26 17
14	13 29 59	20 36 20	23 ♉ 02 44	00 ♊ 31 13	13 54	27 48	25 33	28 10	09 31	27 32	25 36	26 18
15	13 33 55	21 35 47	07 ♊ 54 51	15 ♊ 12 56	14 17	28 10	26 11	28 16	09 38	27 33	25 38	26 19
16	13 37 52	22 35 16	22 ♊ 25 02	29 ♊ 30 50	14 33	28 30	26 49	28 22	09 45	27 34	25 41	26 21
17	13 41 48	23 34 48	06 ♋ 30 18	13 ♋ 23 29	14 43	28 49	27 27	28 27	09 52	27 35	25 43	26 22
18	13 45 45	24 34 22	20 ♋ 10 35	26 ♋ 51 53	14 R 46	29 06	28 05	28 33	09 59	27 35	25 45	26 23
19	13 49 41	25 33 58	03 ♌ 27 47	09 ♌ 58 39	14 40	29 20	28 43	28 38	10 07	27 36	25 47	26 24
20	13 53 38	26 33 36	16 ♌ 24 55	22 ♌ 47 02	14 27	29 33	29 ♑ 21	28 44	10 14	27 37	25 50	26 25
21	13 57 35	27 33 17	29 ♌ 05 25	05 ♍ 20 27	14 05	29 43	00 ♒ 00	28 49	10 21	27 38	25 52	26 26
22	14 01 31	28 33 00	11 ♍ 32 30	17 ♍ 41 55	13 35	29 52	00 39	28 53	10 28	27 39	25 54	26 28
23	14 05 28	29 ♎ 32 45	23 ♍ 48 59	29 ♍ 53 57	12 55	29 ♏ 58	01 18	28 58	10 35	27 39	25 56	26 29
24	14 09 24	00 ♏ 32 32	05 ♎ 57 05	11 ♎ 58 33	12 06	00 ♐ 02	01 57	29 03	10 42	27 40	25 59	26 30
25	14 13 21	01 32 22	17 ♎ 58 32	23 ♎ 57 13	11 10	00 04	02 36	29 07	10 49	27 40	26 01	26 31
26	14 17 17	02 32 13	29 ♎ 54 44	05 ♏ 51 16	10 06	00 R 03	03 15	29 11	10 56	27 41	26 03	26 32
27	14 21 14	03 32 06	11 ♏ 46 58	17 ♏ 42 02	08 56	00 ♐ 00	03 55	29 15	11 03	27 41	26 05	26 33
28	14 25 10	04 32 02	23 ♏ 36 41	29 ♏ 31 08	07 42	29 ♏ 54	04 34	29 19	11 11	27 41	26 08	26 33
29	14 29 07	05 31 59	05 ♐ 25 42	11 ♐ 20 40	06 25	29 46	05 14	29 23	11 18	27 42	26 10	26 34
30	14 33 04	06 31 58	17 ♐ 16 26	23 ♐ 13 23	05 08	29 36	05 54	29 26	11 25	27 42	26 12	26 35
31	14 37 00	07 ♏ 31 59	29 ♐ 12 00	05 ♑ 12 45	03 ♏ 53	29 ♏ 23	06 ♒ 34	29 ♋ 30	11 ♏ 32	27 ♋ 42	26 ♎ 14	26 ♌ 36

Date	Moon True ☊ ° '	Moon Mean ☊ ° '	Moon ☽ Latitude ° '
01	10 ♑ 07	10 ♑ 16	03 S 27
02	10 R 02	10 13	02 38
03	09 59	10 10	01 41
04	09 D 58	10 06	00 S 39
05	09 R 58	10 03	00 N 26
06	09 58	10 00	01 31
07	09 56	09 57	02 34
08	09 52	09 54	03 30
09	09 45	09 50	04 16
10	09 35	09 47	04 48
11	09 24	09 44	05 00
12	09 13	09 41	04 52
13	09 03	09 38	04 24
14	08 55	09 35	03 37
15	08 49	09 31	02 35
16	08 46	09 28	01 25
17	08 45	09 25	00 N 12
18	08 D 45	09 22	01 S 00
19	08 R 45	09 19	02 07
20	08 44	09 16	03 06
21	08 40	09 12	03 54
22	08 33	09 09	04 29
23	08 23	09 06	04 52
24	08 12	09 03	05 02
25	07 59	09 00	04 57
26	07 46	08 56	04 40
27	07 34	08 53	04 11
28	07 24	08 50	03 30
29	07 16	08 47	02 41
30	07 12	08 44	01 44
31	07 ♑ 09	08 ♑ 41	00 S 42

DECLINATIONS

Date	Sun ☉ ° '	Moon ☽ ° '	Mercury ☿ ° '	Venus ♀ ° '	Mars ♂ ° '	Jupiter ♃ ° '	Saturn ♄ ° '	Uranus ♅ ° '	Neptune ♆ ° '	Pluto ♇ ° '
01	03 S 05	22 S 47	14 S 42	22 S 50	25 S 22	20 N 58	12 S 08	21 N 12	08 S 11	22 N 07
02	03 28	24 20	15 11	23 06	25 15	20 57	12 10	21 12	08 12	22 07
03	03 52	24 47	15 39	23 21	25 08	20 55	12 12	21 12	08 13	22 06
04	04 15	24 05	16 06	23 36	25 01	20 54	12 15	21 11	08 13	22 06
05	04 38	22 12	16 32	23 51	24 54	20 53	12 17	21 11	08 14	22 06
06	05 01	19 11	16 57	24 05	24 47	20 51	12 19	21 11	08 15	22 06
07	05 24	15 08	17 20	24 18	24 39	20 50	12 21	21 11	08 16	22 05
08	05 47	10 12	17 41	24 31	24 32	20 49	12 24	21 10	08 17	22 05
09	06 10	04 S 34	18 01	24 43	24 24	20 47	12 26	21 10	08 17	22 05
10	06 33	01 N 28	18 19	24 54	24 16	20 46	12 28	21 10	08 18	22 05
11	06 55	07 35	18 35	25 05	24 07	20 45	12 30	21 10	08 19	22 05
12	07 18	13 21	18 49	25 16	23 59	20 44	12 33	21 10	08 20	22 04
13	07 41	18 18	19 01	25 25	23 50	20 43	12 35	21 09	08 21	22 04
14	08 03	22 02	19 10	25 34	23 42	20 42	12 37	21 09	08 22	22 04
15	08 25	24 11	19 17	25 42	23 33	20 41	12 39	21 09	08 22	22 04
16	08 47	24 39	19 20	25 50	23 23	20 40	12 42	21 09	08 23	22 04
17	09 09	23 29	19 21	25 56	23 14	20 39	12 44	21 09	08 24	22 04
18	09 31	20 56	19 19	26 02	23 05	20 38	12 46	21 09	08 25	22 04
19	09 53	17 20	19 12	26 07	22 55	20 37	12 48	21 08	08 26	22 03
20	10 15	12 58	19 02	26 12	22 45	20 36	12 51	21 08	08 26	22 03
21	10 36	08 09	18 48	26 15	22 35	20 35	12 53	21 08	08 27	22 03
22	10 58	03 N 05	18 30	26 18	22 25	20 34	12 55	21 08	08 28	22 03
23	11 19	02 S 01	18 07	26 19	22 14	20 33	12 58	21 08	08 29	22 03
24	11 40	06 59	17 40	26 20	22 04	20 32	13 00	21 08	08 30	22 03
25	12 01	11 38	17 08	26 20	21 53	20 32	13 02	21 08	08 30	22 03
26	12 21	15 48	16 32	26 18	21 42	20 31	13 04	21 08	08 31	22 03
27	12 42	19 21	15 53	26 16	21 31	20 30	13 07	21 08	08 32	22 03
28	13 02	22 04	15 11	26 12	21 20	20 29	13 09	21 08	08 33	22 03
29	13 22	23 51	14 26	26 08	21 08	20 29	13 11	21 08	08 34	22 03
30	13 42	24 34	13 41	26 02	20 57	20 28	13 13	21 08	08 34	22 03
31	14 S 02	24 S 09	12 S 57	25 S 55	20 S 45	20 N 28	13 S 16	21 N 08	08 S 35	22 N 03

ZODIAC SIGN ENTRIES

Date	h	m	Planets
01	18	41	☽ ♐
04	07	04	☽ ♑
06	16	45	☽ ♒
08	22	17	☽ ♓
10	23	58	☽ ♈
12	23	32	☽ ♉
14	23	10	☽ ♊
17	00	50	☽ ♋
19	05	41	☽ ♌
21	12	03	♂ ♒
21	13	44	☽ ♍
23	22	07	♀ ♐
23	22	56	☉ ♏
24	00	12	☽ ♎
26	12	11	☽ ♏
27	10	42	♀ ♏
29	00	59	☽ ♐
31	13	36	☽ ♑

LATITUDES

Date	Mercury ☿ ° '	Venus ♀ ° '	Mars ♂ ° '	Jupiter ♃ ° '	Saturn ♄ ° '	Uranus ♅ ° '	Neptune ♆ ° '	Pluto ♇ ° '
01	02 S 24	05 S 05	03 S 07	00 N 08	02 N 11	00 N 30	01 N 39	09 N 52
04	02 43	05 21	03 01	00 08	02 11	00 30	01 39	09 53
07	02 59	05 35	02 56	00 09	02 10	00 30	01 39	09 54
10	03 11	05 49	02 50	00 09	02 10	00 30	01 39	09 55
13	03 17	06 00	02 44	00 10	02 10	00 30	01 39	09 56
16	03 17	06 09	02 38	00 10	02 09	00 30	01 39	09 57
19	03 06	06 16	02 33	00 10	02 09	00 30	01 39	09 58
22	02 42	06 19	02 27	00 11	02 09	00 30	01 39	09 59
25	02 03	06 18	02 22	00 11	02 09	00 30	01 39	10 00
28	01 10	06 13	02 16	00 12	02 09	00 31	01 39	10 01
31	00 S 08	06 S 03	02 S 11	00 N 12	02 N 09	00 N 31	01 N 39	10 N 02

DATA

Julian Date	2435017
Delta T	+31 seconds
Ayanamsa	23° 13' 46"
Synetic vernal point	05° ♓ 53' 14"
True obliquity of ecliptic	23° 26' 45"

LONGITUDES

Date	Chiron ⚷ ° '	Ceres ⚳ ° '	Pallas ⚴ ° '	Juno ⚵ ° '	Vesta ⚶ ° '	Black Moon Lilith ⚸ ° '
01	23 ♑ 05	09 ♏ 27	20 ♎ 31	08 ♎ 07	29 ♊ 36	12 ♏ 11
11	23 ♑ 10	13 ♏ 36	24 ♎ 56	11 ♎ 49	01 ♋ 01	13 ♏ 18
21	23 ♑ 22	17 ♏ 48	29 ♎ 22	15 ♎ 29	01 ♋ 52	14 ♏ 25
31	23 ♑ 41	22 ♏ 01	03 ♏ 48	19 ♎ 05	02 ♋ 01	15 ♏ 32

MOON'S PHASES, APSIDES AND POSITIONS ☽

Date	h	m	Phase	Longitude ° '	Eclipse Indicator
05	05	31	☽	11 ♑ 27	
12	05	10	○	18 ♈ 21	
18	20	30	☾	24 ♋ 55	
26	17	47	●	02 ♏ 47	

Day	h	m		
13	01	51	Perigee	
27	22	58	Apogee	
03	09	32	Max dec	24° S 48'
10	06	14	0N	
16	06	29	Max dec	24° N 42'
23	02	28	0S	
30	15	17	Max dec	24° S 35'

ASPECTARIAN

h m	Aspects
01 Friday	
03 16	☽ ∠ ☉
04 46	☽ # ♇
08 46	☽ ⊻ ♆
10 33	☽ □ ♇
11 50	☽ △ ♃
12 33	☽ ǁ ♀
13 02	☽ △ ♅
19 49	☉ ⊻ ♄
21 00	☽ ⊥ ♆
02 Saturday	
00 50	☽ ∠ ♂
01 37	☽ ⊻ ☿
11 15	☽ ⊻ ♄
12 05	☉ Q ♃
12 30	☽ ✱ ☉
15 03	☽ ⊥ ☿
15 19	☽ ∠ ♆
18 36	☽ bQ ♃
19 32	☽ bQ ♅
20 01	☽ ⊥ ♂
23 30	☽ ⊥ ♄
03 Sunday	
00 19	☉ Q ♅
08 44	☽ ⊻ ♂
11 03	☽ ∠ ☿
12 56	☽ ± ♃
13 42	☽ ± ♅
14 52	☽ Q ☉
15 20	☽ ⊻ ♀
17 45	☽ ∠ ♄
21 34	☽ ✱ ♆
23 15	☽ △ ♇
04 Monday	
01 02	☽ ⊼ ♃
01 42	☽ ⊼ ♅
03 58	☽ ⊥ ♀
18 42	☽ ǁ ♀
19 41	☽ ✱ ☿
20 43	☉ ∠ ♇
21 20	☽ Q ♆
22 39	☽ ∠ ♀
23 40	☽ ✱ ♄
05 Tuesday	
04 50	☽ bQ ♇
05 31	☽ □ ☉
12 59	☽ # ♇
13 33	♂ ± ♇
20 56	☽ Q ☿
21 07	☽ # ♅
22 10	☽ ± ♇
22 34	☽ ☌ ♂
23 00	☽ Q ♄
23 44	☽ # ♃
06 Wednesday	
04 57	☽ ✱ ♀
07 59	☽ □ ♆
09 31	☽ ⊼ ♇
09 39	♂ Q ♄
11 40	☽ ☍ ♃
11 50	☽ ☍ ♅
13 07	☿ Q ♇
07 Thursday	
00 47	☽ ǁ ☿
03 15	☿ ☌ ♄
04 17	☽ Q ♀
08 45	☽ □ ♄
09 06	☽ □ ☿
10 01	♃ ☌ ♅
18 10	☽ △ ☉
08 Friday	
01 53	☽ ǁ ♄
07 45	☿ Q ♂
08 01	☽ ⊻ ♂
13 00	♀ ǁ ♂
13 43	☽ □ ♀
14 20	☽ △ ♆
15 42	☽ ☍ ♇
17 49	☽ ⊼ ♅
18 02	☽ ⊼ ♃
18 55	☽ ⊥ ♂
20 27	☽ ǁ ♆
22 27	☽ bQ ☉
09 Saturday	
04 07	☽ ± ♅
04 24	☽ ± ♃
05 50	☽ ǁ ☉
06 50	♀ ⊻ ♆
10 55	☽ ∠ ♂
13 28	☽ △ ♄
14 50	☽ ± ☉
15 55	☽ bQ ♆
16 43	☽ △ ☿
19 16	☽ bQ ♅
19 38	☽ bQ ♃
10 Sunday	
01 32	☽ ⊼ ☉
06 50	☽ ± ♆
12 48	☽ ✱ ♂
14 23	☽ bQ ♄
16 38	☽ ⊼ ♆
17 45	☽ △ ♀
17 52	☽ ⊼ ♇
18 41	☽ bQ ☿
19 52	☽ △ ♅
20 23	☽ △ ♃
21 38	♀ □ ♇
11 Monday	
03 33	☽ ± ♇
05 01	☽ ± ♄
08 58	☽ Q ♂
09 13	☽ # ☉
09 51	☽ ± ☿
14 40	☽ ⊼ ♄
14 58	☽ # ♆
17 56	☽ bQ ♇
18 33	☽ bQ ♀
19 50	☽ ⊼ ☿
12 Tuesday	
05 10	☽ ☍ ☉
08 29	☽ # ♄
09 13	☽ ± ♀
14 41	☽ □ ♂
16 30	☽ ☍ ♆
17 39	☽ △ ♇
18 57	☽ ⊼ ♀
19 35	☽ □ ♅
20 22	☽ □ ♃
13 Wednesday	
14 22	☽ ☍ ♄
16 05	☽ # ☿
18 58	♀ △ ♅
21 00	☽ ☍ ☿
14 Thursday	
00 01	☽ Q ♅
00 58	☽ Q ♃
02 19	☽ ǁ ♃
05 25	☽ ǁ ♅
07 50	☽ ⊼ ☉
12 18	☽ ǁ ♇
14 01	♂ □ ♆
16 06	☽ ⊼ ♆
16 11	☽ △ ♂
17 13	☽ □ ♇
18 06	☽ ± ☉
19 11	☽ ✱ ♅
19 48	☽ ☍ ♀
20 15	☽ ✱ ♃
15 Friday	
01 47	☽ ± ♆
03 15	☽ # ♂
08 43	☉ ǁ ♆
09 42	☽ bQ ☉
14 51	☽ ⊼ ♄
16 29	☽ bQ ♆
17 35	☽ bQ ♂
17 39	♂ ⊼ ♇
19 10	☿ Q ♂
19 36	☽ ∠ ♅
20 50	☽ ∠ ♃
21 19	♀ △ ♃
22 32	☽ Q ♇
22 40	☽ ⊼ ☿
16 Saturday	
00 48	☽ ± ♄
08 49	☽ ± ☿
09 02	☉ ⊥ ♀
09 11	☽ ± ♂
10 34	☽ ⊥ ♅
11 54	☽ ⊥ ♃
12 18	☽ △ ☉
15 58	☽ bQ ♄
17 30	☽ △ ♆
18 38	☽ ✱ ♇
19 45	☽ ⊼ ♂
20 41	☽ ⊻ ♅
22 06	☽ ⊻ ♃
22 31	☽ ⊼ ♀
17 Sunday	
00 14	☽ bQ ☿
09 01	☽ ± ♀
15 14	☽ # ♂
17 11	♂ ☍ ♅
17 54	☽ △ ♄
20 28	☽ ∠ ♇
18 Monday	
01 01	☽ bQ ♀
02 24	☽ △ ☿
02 47	☽ ǁ ♇
08 24	☿ St R
10 25	☽ ǁ ♅
12 22	☽ ⊥ ♇
14 22	☽ ǁ ♃
20 30	☽ □ ☉
22 01	☽ □ ♆
23 09	☽ ⊻ ♇
23 51	☽ # ☿
19 Tuesday	
01 20	☽ ☌ ♅
02 55	☽ ☍ ♂
03 09	☽ ☌ ♃
04 21	☽ △ ♀
08 39	♂ ☍ ♃
17 38	☉ ☌ ♆
20 Wednesday	
00 21	☽ □ ♄
07 09	☽ Q ♆
08 14	☽ Q ☉
08 24	☽ □ ☿
08 36	☉ ✱ ♇
12 38	☽ # ♄
22 10	♀ ✱ ♂
21 Thursday	
00 47	☽ # ☉
05 50	☽ ✱ ♆
06 56	☽ ☌ ♇
08 49	☽ ✱ ☉
09 13	☽ ⊻ ♅
10 30	☽ # ♆
10 34	☽ Q ♄
11 28	☽ ⊻ ♃
13 14	☽ □ ♀
13 50	☽ ⊼ ♂
13 51	☉ □ ♅
17 32	☽ Q ☿
20 43	☽ ⊥ ♅
23 03	☽ ⊥ ♃
22 Friday	
02 00	☽ ± ♂
09 53	☽ ✱ ♄
10 45	☽ ∠ ♆
14 08	☽ ∠ ♅
15 46	☽ ✱ ☿
16 15	☽ ∠ ☉
16 36	☽ ∠ ♃
20 26	☽ bQ ♂
20 56	☉ □ ♃
23 Saturday	
00 26	☽ Q ♀
04 22	☽ ⊥ ♆
11 25	☽ ⊥ ☉
15 31	☽ ∠ ♄
16 12	☽ ⊻ ♆
17 15	☽ ⊻ ♇
19 34	☽ ✱ ♅
19 36	☽ ∠ ☿
22 14	☽ ✱ ♃
23 00	☉ ⊻ ♀
24 Sunday	
00 13	☽ ✱ ♀
00 19	☽ ⊻ ☉
03 36	☽ △ ♂
05 08	☽ ⊥ ♇
09 29	☽ ⊥ ♄
12 17	☽ ⊥ ☿
13 56	♂ # ♇
19 23	☽ Q ♅
19 38	☽ ǁ ♆
21 33	☽ ⊻ ♄
22 12	☽ Q ♃
23 03	☽ ∠ ♇
23 24	☽ ⊻ ☿
25 Monday	
06 09	☽ ∠ ♀
14 15	☽ ǁ ☉
16 36	♀ St R
19 16	☿ ☌ ♄
19 48	☽ ǁ ♄
26 Tuesday	
00 13	☽ ⊥ ♀
04 12	☽ ☌ ♆
05 10	☽ ✱ ♇
07 30	☽ □ ♅
10 32	☽ □ ♃
12 16	☽ ⊻ ♀
15 55	☽ ǁ ☿
17 47	☽ ☌ ☉
19 08	☽ □ ♂
27 Wednesday	
05 26	☽ Q ♇
06 46	☽ ☌ ☿
10 31	☽ ☌ ♄
19 37	☿ Q ♇
21 14	☽ # ♃
28 Thursday	
02 46	☽ # ♅
05 08	☽ ǁ ♂
09 46	☽ Q ♂
11 44	☽ # ♇
14 44	☉ □ ♂
17 07	☽ ⊻ ♆
17 59	☽ □ ♇
20 17	☽ △ ♅
20 58	☉ ǁ ♄
23 39	☽ △ ♃
29 Friday	
00 39	☽ ☌ ♀
05 21	☽ ⊥ ♆
11 35	☽ ✱ ♂
12 14	☽ ⊻ ☉
13 26	♂ # ♅
13 48	☽ ⊻ ☿
21 15	☉ ☌ ☿
23 40	☽ ∠ ♆
30 Saturday	
00 02	☽ ⊻ ♄
00 47	☽ ⊥ ☿
01 31	☽ ⊥ ☉
02 30	☿ □ ♂
02 45	☽ bQ ♅
06 14	☽ bQ ♃
11 43	☉ ǁ ☿
12 18	☽ ⊥ ♄
17 13	☽ ∠ ☿
19 45	☽ ∠ ♂
20 56	☽ ± ♅
21 23	☽ ∠ ☉
31 Sunday	
00 30	☽ ± ♃
02 15	☿ ǁ ♄
02 45	♀ △ ♃
06 02	☽ ✱ ♆
06 36	☽ ∠ ♄
06 47	☽ △ ♇
09 00	☽ ⊼ ♅
12 22	☽ ⊻ ♀
12 35	☽ ⊼ ♃
14 54	☽ ⊥ ♂
20 31	☽ ✱ ☿

All ephemeris data is given at 12.00 UT and the Moon's longitude is additionally given for 24.00 UT

NOVEMBER 1954

LONGITUDES

Date	Sidereal time h m s	Sun ☉ ° ' "	Moon ☽ ° ' "	Moon ☽ 24.00 ° ' "	Mercury ☿ ° '	Venus ♀ ° '	Mars ♂ ° '	Jupiter ♃ ° '	Saturn ♄ ° '	Uranus ♅ ° '	Neptune ♆ ° '	Pluto ♇ ° '
01	14 40 57	08 ♏ 32 01	11 ♑ 16 12	17 ♑ 22 54	02 ♏ 43	29 ♏ 08	07 ♒ 14	29 ♋ 33	11 ♏ 39	27 ♋ 42	26 ♎ 16	26 ♌ 37
02	14 44 53	09 32 05	23 ♑ 33 27	29 ♑ 48 28	01 R 40	28 R 50	07 54	29 36	11 47	27 42	26 19	26 38
03	14 48 50	10 32 11	06 ♒ 08 33	12 ♒ 34 17	00 45	28 30	08 35	29 38	11 54	27 R 42	26 21	26 38
04	14 52 46	11 32 18	19 ♒ 06 13	25 ♒ 44 51	00 ♏ 01	28 08	09 15	29 41	12 01	27 42	26 23	26 39
05	14 56 43	12 32 26	02 ♓ 30 33	09 ♓ 23 37	29 ♎ 28	27 44	09 56	29 43	12 08	27 42	26 25	26 40
06	15 00 39	13 32 36	16 ♓ 24 09	23 ♓ 32 06	29 06	27 18	10 37	29 45	12 15	27 42	26 27	26 41
07	15 04 36	14 32 48	00 ♈ 47 10	08 ♈ 08 53	28 55	26 49	11 17	29 47	12 23	27 42	26 29	26 41
08	15 08 33	15 33 01	15 ♈ 36 31	23 ♈ 09 07	28 D 56	26 19	11 58	29 49	12 30	27 42	26 32	26 42
09	15 12 29	16 33 16	00 ♉ 45 32	08 ♉ 24 28	29 09	25 48	12 39	29 51	12 37	27 41	26 34	26 42
10	15 16 26	17 33 32	16 ♉ 04 31	23 ♉ 44 15	29 ♎ 31	25 15	13 20	29 52	12 44	27 41	26 36	26 43
11	15 20 22	18 33 50	01 ♊ 22 16	08 ♊ 57 15	00 ♏ 02	24 41	14 02	29 53	12 51	27 41	26 38	26 43
12	15 24 19	19 34 10	16 ♊ 28 02	23 ♊ 53 38	00 42	24 06	14 43	29 54	12 59	27 40	26 40	26 44
13	15 28 15	20 34 32	01 ♋ 13 17	08 ♋ 26 26	01 30	23 30	15 24	29 55	13 06	27 40	26 42	26 44
14	15 32 12	21 34 56	15 ♋ 32 44	22 ♋ 32 02	02 25	22 53	16 06	29 56	13 13	27 39	26 44	26 45
15	15 36 08	22 35 22	29 ♋ 24 21	06 ♌ 09 53	03 25	22 17	16 47	29 56	13 20	27 39	26 46	26 45
16	15 40 05	23 35 49	12 ♌ 48 53	19 ♌ 21 45	04 30	21 40	17 29	29 57	13 27	27 38	26 48	26 46
17	15 44 02	24 36 18	25 ♌ 48 55	02 ♍ 10 53	05 40	21 04	18 10	29 R 57	13 34	27 37	26 50	26 46
18	15 47 58	25 36 49	08 ♍ 28 09	14 ♍ 41 13	06 54	20 29	18 52	29 56	13 42	27 36	26 53	26 46
19	15 51 55	26 37 22	20 ♍ 50 38	26 ♍ 56 51	08 11	19 54	19 34	29 56	13 49	27 35	26 55	26 47
20	15 55 51	27 37 57	03 ♎ 00 22	09 ♎ 01 36	09 30	19 20	20 16	29 55	13 56	27 35	26 57	26 47
21	15 59 48	28 38 33	15 ♎ 00 58	20 ♎ 58 52	10 52	18 47	20 57	29 55	14 03	27 34	26 58	26 47
22	16 03 44	29 ♏ 39 11	26 ♎ 55 36	02 ♏ 51 30	12 16	18 16	21 39	29 54	14 10	27 33	27 00	26 47
23	16 07 41	00 ♐ 39 51	08 ♏ 46 50	14 ♏ 41 53	13 42	17 46	22 21	29 53	14 17	27 32	27 02	26 48
24	16 11 37	01 40 32	20 ♏ 36 52	26 ♏ 32 00	15 09	17 19	23 04	29 51	14 24	27 31	27 04	26 48
25	16 15 34	02 41 14	02 ♐ 27 31	08 ♐ 23 37	16 37	16 53	23 46	29 50	14 31	27 29	27 06	26 48
26	16 19 31	03 41 58	14 ♐ 20 32	20 ♐ 18 30	18 07	16 29	24 28	29 48	14 38	27 28	27 08	26 48
27	16 23 27	04 42 44	26 ♐ 17 45	02 ♑ 18 34	19 37	16 07	25 10	29 46	14 45	27 27	27 10	26 48
28	16 27 24	05 43 30	08 ♑ 21 15	14 ♑ 26 06	21 07	15 48	25 53	29 44	14 52	27 26	27 12	26 48
29	16 31 20	06 44 18	20 ♑ 33 29	26 ♑ 43 46	22 38	15 31	26 35	29 41	14 59	27 24	27 14	26 R 48
30	16 35 17	07 ♐ 45 06	02 ♒ 57 22	09 ♒ 14 42	24 ♏ 10	15 ♏ 16	27 ♒ 17	29 ♋ 39	15 ♏ 05	27 ♋ 23	27 ♎ 16	26 ♌ 48

Moon Nodes and Latitude / DECLINATIONS

Date	Moon True ☊ ° '	Moon Mean ☊ ° '	Moon ☽ Latitude ° '	Sun ☉ ° '	Moon ☽ ° '	Mercury ☿ ° '	Venus ♀ ° '	Mars ♂ ° '	Jupiter ♃ ° '	Saturn ♄ ° '	Uranus ♅ ° '	Neptune ♆ ° '	Pluto ♇ ° '
01	07 ♑ 09	08 ♑ 37	00 N 22	14 S 21	22 S 36	12 S 14	25 S 47	20 S 33	20 N 27	13 S 18	21 N 08	08 S 36	22 N 03
02	07 D 10	08 34	01 27	14 40	19 58	11 34	25 38	20 21	20 27	13 20	21 08	08 37	22 03
03	07 11	08 31	02 29	14 59	16 20	10 57	25 28	20 08	20 26	13 22	21 08	08 37	22 03
04	07 R 10	08 28	03 26	15 18	11 50	10 26	25 16	19 56	20 26	13 24	21 08	08 38	22 03
05	07 09	08 25	04 13	15 36	06 39	10 00	25 03	19 43	20 26	13 27	21 08	08 39	22 03
06	07 04	08 22	04 47	15 54	00 S 57	09 40	24 49	19 31	20 25	13 29	21 08	08 40	22 03
07	06 58	08 18	05 05	16 12	04 N 58	09 26	24 34	19 18	20 25	13 31	21 08	08 41	22 03
08	06 51	08 15	05 03	16 30	10 48	09 17	24 17	19 05	20 25	13 33	21 08	08 41	22 04
09	06 43	08 12	04 40	16 47	16 07	09 14	23 59	18 51	20 25	13 35	21 08	08 42	22 04
10	06 35	08 09	03 57	17 04	20 26	09 16	23 41	18 38	20 25	13 38	21 08	08 43	22 04
11	06 29	08 06	02 57	17 21	23 20	09 23	23 21	18 24	20 24	13 40	21 08	08 44	22 04
12	06 26	08 02	01 46	17 38	24 30	09 34	23 00	18 11	20 24	13 42	21 08	08 44	22 04
13	06 24	07 59	00 N 28	17 54	23 54	09 49	22 39	17 57	20 24	13 44	21 09	08 45	22 04
14	06 D 24	07 56	00 S 49	18 10	21 43	10 07	22 16	17 43	20 24	13 46	21 09	08 46	22 04
15	06 25	07 53	02 01	18 25	18 18	10 28	21 53	17 29	20 24	13 48	21 09	08 46	22 05
16	06 27	07 50	03 04	18 41	14 01	10 52	21 30	17 14	20 25	13 50	21 09	08 47	22 05
17	06 R 27	07 47	03 56	18 56	09 13	11 17	21 06	17 00	20 25	13 52	21 09	08 48	22 05
18	06 26	07 43	04 34	19 10	04 N 09	11 44	20 42	16 45	20 25	13 55	21 09	08 49	22 05
19	06 23	07 40	04 59	19 24	00 S 57	12 13	20 18	16 31	20 25	13 57	21 10	08 49	22 06
20	06 19	07 37	05 10	19 38	05 56	12 42	19 54	16 16	20 26	13 59	21 10	08 50	22 06
21	06 13	07 34	05 07	19 52	10 38	13 12	19 30	16 01	20 26	14 01	21 10	08 51	22 06
22	06 06	07 31	04 50	20 05	14 53	13 43	19 06	15 46	20 26	14 03	21 10	08 51	22 06
23	05 59	07 27	04 21	20 18	18 33	14 14	18 43	15 31	20 27	14 05	21 10	08 52	22 07
24	05 52	07 24	03 41	20 30	21 28	14 45	18 21	15 15	20 27	14 07	21 11	08 53	22 07
25	05 47	07 21	02 52	20 42	23 28	15 17	17 58	15 00	20 28	14 09	21 11	08 53	22 07
26	05 43	07 18	01 54	20 54	24 25	15 48	17 37	14 44	20 28	14 11	21 11	08 54	22 08
27	05 41	07 15	00 S 51	21 05	24 15	16 19	17 17	14 29	20 29	14 13	21 12	08 55	22 08
28	05 D 40	07 12	00 N 15	21 16	22 56	16 49	16 57	14 13	20 29	14 15	21 12	08 55	22 08
29	05 41	07 08	01 21	21 26	20 32	17 19	16 38	13 57	20 30	14 17	21 12	08 56	22 09
30	05 ♑ 43	07 ♑ 05	02 N 24	21 S 36	17 S 10	17 S 49	16 S 21	13 S 41	20 N 31	14 S 19	21 N 12	08 S 56	22 N 09

ZODIAC SIGN ENTRIES

Date	h	m	Planets
03	00	22	☽ ♒
04	12	36	☿ ♎
05	07	34	☽ ♓
07	10	42	☽ ♈
09	10	48	☽ ♉
11	09	50	☽ ♊
11	10	25	☿ ♏
13	09	59	☽ ♋
15	13	03	☽ ♌
17	19	52	☽ ♍
20	06	02	☽ ♎
22	18	13	☽ ♏
22	20	14	☉ ♐
25	07	01	☽ ♐
27	19	24	☽ ♑
30	06	19	☽ ♒

LATITUDES

Date	Mercury ☿ ° '	Venus ♀ ° '	Mars ♂ ° '	Jupiter ♃ ° '	Saturn ♄ ° '	Uranus ♅ ° '	Neptune ♆ ° '	Pluto ♇ ° '
01	00 N 12	05 S 58	02 S 09	00 N 12	02 N 09	00 N 31	01 N 39	10 N 02
04	01 07	05 40	02 04	00 13	02 08	00 31	01 40	10 03
07	01 47	05 15	01 59	00 13	02 08	00 31	01 40	10 04
10	02 10	04 44	01 53	00 14	02 08	00 31	01 40	10 05
13	02 20	04 08	01 48	00 14	02 08	00 31	01 40	10 06
16	02 18	03 26	01 43	00 15	02 08	00 31	01 40	10 07
19	02 08	02 41	01 38	00 15	02 08	00 31	01 40	10 08
22	01 54	01 55	01 33	00 16	02 09	00 31	01 40	10 10
25	01 36	01 08	01 29	00 16	02 09	00 32	01 40	10 11
28	01 16	00 S 24	01 24	00 17	02 09	00 32	01 40	10 12
31	00 N 55	00 N 18	01 S 19	00 N 17	02 N 09	00 N 32	01 N 40	10 N 13

DATA

Julian Date	2435048
Delta T	+31 seconds
Ayanamsa	23° 13' 50"
Synetic vernal point	05° ♓ 53' 10"
True obliquity of ecliptic	23° 26' 44"

MOON'S PHASES, APSIDES AND POSITIONS ☽

Date	h	m	Phase	Longitude ° '	Eclipse Indicator
03	20	55	☽	10 ♒ 55	
10	14	29	○	17 ♉ 40	
17	09	33	☾	24 ♌ 30	
25	12	30	●	02 ♐ 43	

Day	h	m		
10	13	25	Perigee	
23	23	45	Apogee	
06	15	54	0N	
12	15	40	Max dec	24° N 32'
19	07	30	0S	
26	20	28	Max dec	24° S 29'

LONGITUDES

Date	Chiron ⚷ ° '	Ceres ⚳ ° '	Pallas ⚴ ° '	Juno ⚵ ° '	Vesta ⚶ ° '	Black Moon Lilith ⚸ ° '
01	23 ♑ 43	22 ♏ 27	04 ♏ 15	19 ♎ 26	02 ♋ 00	15 ♏ 39
11	24 ♑ 10	26 ♏ 41	08 ♏ 40	22 ♎ 57	01 ♋ 22	16 ♏ 46
21	24 ♑ 43	00 ♐ 56	13 ♏ 04	26 ♎ 22	00 ♋ 00	17 ♏ 52
31	25 ♑ 21	05 ♐ 10	17 ♏ 25	29 ♎ 39	27 ♊ 59	18 ♏ 59

All ephemeris data is given at 12.00 UT and the Moon's longitude is additionally given for 24.00 UT

ASPECTARIAN

01 Monday
00 06 ☽ ⊥ ♀
03 33 ☽ ⊻ ♂
06 03 ☽ Q ♆
06 06 ☽ ⚹ ☉
12 41 ☽ bQ ♇
12 46 ☽ ⚹ ♄
13 58 ☉ Q ♇
17 31 ☽ ∠ ♀
17 54 ☽ # ♇
18 14 ☽ Q ☿
23 44 ♂ # ♃

02 Tuesday
02 32 ☽ # ♅
06 19 ☽ ± ♇
07 44 ☽ Q ☉
08 13 ☽ # ♃
08 49 ☽ ‖ ♂
12 26 ☽ Q ♄
17 19 ☽ □ ♆
17 55 ☽ ⊼ ♇
19 59 ☽ ☍ ♅
21 54 ☽ ⚹ ♀
23 38 ☽ ☍ ♃

03 Wednesday
02 28 ☽ □ ☿
10 57 ♅ St R
16 50 ☽ ☌ ♂
19 09 ☽ ‖ ☉
19 57 ☽ Q ♀
20 55 ☽ □ ☉
22 51 ☽ □ ♄

04 Thursday
04 09 ☽ ‖ ♄
19 28 ☽ ‖ ☿

05 Friday
00 28 ☿ □ ♃
01 02 ☉ ☌ ♄
01 10 ☽ △ ♆
01 38 ☽ ☍ ♇
03 05 ☽ ‖ ♆
03 30 ☽ ⊼ ♅
03 48 ☽ □ ♀
06 48 ☽ △ ☿
07 04 ☽ ⊼ ♃
07 39 ♀ Q ♂
13 38 ♀ △ ♅
14 06 ☽ ± ♅
17 38 ☽ ± ♃

06 Saturday
01 36 ☽ ⊻ ♂
03 32 ☽ bQ ♆
04 52 ☽ △ ♄
05 42 ☽ bQ ♅
06 45 ☽ △ ☉
08 09 ☽ bQ ☿
09 12 ☽ bQ ♃
12 22 ☽ ⊥ ♂
18 52 ☽ ± ♆
23 05 ☽ ± ☿

07 Sunday
04 13 ☽ ∠ ♂
04 54 ☽ ⊼ ♆
05 14 ☽ ⊼ ♇
05 41 ☽ △ ♀
06 20 ☽ bQ ♄
06 55 ☽ △ ♅
08 57 ☽ ⊼ ☿
09 49 ☽ bQ ☉
10 22 ☽ △ ♃
15 07 ☽ ± ♇
18 37 ♀ □ ♇
21 12 ☽ ± ♄
21 31 ☿ St D

08 Monday
01 34 ☽ ± ☉
03 02 ♀ ⊻ ♆
03 07 ☽ # ♆
05 21 ☽ bQ ♀
05 43 ☽ bQ ♇
05 44 ☽ # ☿
05 53 ☽ ⚹ ♂
06 58 ☽ ⊼ ♄
11 54 ☽ ⊼ ☉
19 16 ☽ ± ♀

09 Tuesday
00 05 ☽ # ♄
01 55 ☽ Q ♂
04 27 ☽ ⊼ ♀
05 23 ☽ ☍ ♆
05 37 ☽ △ ♇
07 10 ☽ □ ♅
09 25 ☽ ☍ ☿
10 24 ♂ □ ♄
10 34 ☽ □ ♃
15 34 ☽ # ☉

10 Wednesday
01 40 ☽ # ♂
06 44 ☽ ☍ ♄
07 31 ☽ □ ♂
11 23 ☽ Q ♅
11 48 ☽ ‖ ♃
14 29 ☽ ☍ ☉
14 49 ☽ Q ♃
16 45 ☽ ‖ ♅
23 48 ☽ ‖ ♇

11 Thursday
01 51 ☽ ☍ ♀
04 32 ☽ ⊼ ♆
04 41 ☽ □ ♇
05 39 ☿ □ ♃
06 11 ☽ ⚹ ♅
09 40 ☽ ⚹ ♃
09 49 ☽ ⊼ ☿
12 07 ☽ # ♀
14 00 ☽ ± ♆
19 42 ☽ ± ☿

12 Friday
04 18 ☽ bQ ♆
05 55 ☽ ∠ ♅
06 22 ☽ ⊼ ♄
09 03 ☽ △ ♂
09 13 ☽ Q ♇
09 30 ☽ ∠ ♃
10 43 ☽ bQ ☿
16 04 ☽ ± ♄
17 21 ☽ ⊼ ☉
20 23 ☽ ⊥ ♅
23 51 ☽ ⊼ ♀

13 Saturday
00 02 ☽ ⊥ ♃
03 49 ☽ ± ☉
04 34 ☽ △ ♆
04 38 ☽ ⚹ ♇
06 09 ☽ ⊻ ♅
06 49 ☽ bQ ♄
09 16 ☽ ± ♀
09 51 ☽ ⊻ ♃
10 35 ☽ bQ ♂
12 30 ☽ △ ☿
14 15 ☉ ‖ ♂
19 45 ☽ bQ ☉
23 36 ☽ bQ ♀

14 Sunday
02 18 ☽ ± ♂
05 34 ☽ ∠ ♇
06 22 ☽ # ♀
08 01 ☽ △ ♄
08 58 ☽ ‖ ♇
12 59 ☽ ⊼ ♂
16 38 ☽ ‖ ♅
20 48 ♆ ⚹ ♇
20 55 ☽ ⊥ ♇
22 08 ☽ ‖ ♃
23 10 ☽ △ ☉

15 Monday
00 05 ☽ △ ♀
00 21 ♀ # ♇
07 21 ☽ ⊻ ♇
07 22 ☽ □ ♆
07 26 ☉ ☌ ♀
08 54 ☽ ☌ ♅
11 18 ☽ # ☉
12 56 ☽ ☌ ♃
17 15 ☽ # ♂
19 42 ☽ □ ☿

16 Tuesday
12 56 ☽ # ♄
13 10 ☽ □ ♄
15 39 ☽ Q ♆
21 00 ☽ ☍ ♂

17 Wednesday
02 43 ☽ # ☿
03 05 ♃ St R
03 33 ☽ □ ♀
07 35 ☽ Q ☿
09 03 ♀ # ♅
09 33 ☽ □ ☉
13 47 ☽ ☌ ♇
13 56 ☽ ⚹ ♆
14 00 ☽ # ♆
15 23 ☽ ⊻ ♅
19 46 ☽ ⊻ ♃
22 57 ☽ Q ♄

18 Thursday
02 43 ☽ ⊥ ♅
07 09 ☽ ⊥ ♃
08 39 ☽ ⚹ ☿
12 01 ☽ Q ♀
18 35 ☽ ∠ ♆
19 58 ☽ ∠ ♅
22 10 ☽ ⚹ ♄
22 48 ☽ Q ☉

19 Friday
00 29 ☽ ∠ ♃
05 03 ♀ # ♃
09 20 ☽ ⊼ ♂
10 14 ☽ ⚹ ♀
12 08 ☽ ⊥ ♆
15 43 ☉ □ ♇
17 08 ☽ ∠ ☿
18 19 ♀ □ ♂
19 02 ☉ ⊻ ♆
21 50 ☽ ± ♂
22 59 ☿ Q ♇
23 40 ☽ ⊻ ♇
23 57 ☽ ⊻ ♆

20 Saturday
00 23 ☽ ⚹ ☉
01 15 ☽ ⚹ ♅
03 50 ☽ ∠ ♄
05 53 ☽ ⚹ ♃
10 42 ☉ △ ♅
11 33 ☽ ⊥ ♇
13 07 ☽ ⊥ ☿
14 31 ☽ ∠ ♀
16 45 ☽ bQ ♂
21 54 ☽ ⊥ ♄
22 06 ☉ ‖ ♀

21 Sunday
01 05 ☽ Q ♅
02 37 ☽ ⊻ ☿
02 39 ☽ ‖ ♆
05 31 ☽ ∠ ♇
05 47 ☽ Q ♃
07 43 ☽ ⊥ ♀
08 59 ☽ ∠ ☉
10 02 ☽ ⊻ ♄
19 16 ☽ ⊻ ♀

22 Monday
00 42 ☽ △ ♂
04 09 ☽ ‖ ☿
04 46 ☽ ⊥ ☉
07 00 ☽ ‖ ♄
11 43 ☽ ⚹ ♇
12 10 ☽ ☌ ♆
13 15 ☽ □ ♅
17 03 ☽ ‖ ♂
17 38 ☉ △ ♃
18 00 ☽ □ ♃
18 02 ☽ ⊻ ☉

23 Tuesday
04 23 ☿ ‖ ♄
12 02 ☽ Q ♇
13 06 ☽ ‖ ♀
22 27 ☿ ☌ ♄
23 16 ☽ ☌ ♄
23 22 ☽ ☌ ☿

24 Wednesday
02 35 ☽ ‖ ☉
02 49 ☽ # ♃
05 33 ☽ ☌ ♀
05 49 ☉ # ♃
09 18 ☽ # ♅
17 16 ☽ □ ♂
18 41 ☽ # ♇

25 Thursday
00 32 ☽ □ ♇
01 08 ☽ ⊻ ♆
01 57 ☽ △ ♅
03 21 ☿ ‖ ♂
06 41 ☽ △ ♃
12 30 ☽ ☌ ☉
13 19 ☽ ⊥ ♆
15 16 ☿ ☌ ♀
22 12 ☉ ⊥ ♆

26 Friday
07 32 ☽ ∠ ♆
07 59 ☽ Q ♂
08 14 ☽ bQ ♅
12 35 ☽ ⊻ ♄
12 55 ☽ bQ ♃
16 11 ☽ ⊻ ♀
20 40 ☽ ⊻ ☿

27 Saturday
00 46 ☽ ⊥ ♄
02 18 ☽ ± ♅
03 53 ☽ ⊥ ♀
06 57 ☽ ± ♃
09 37 ☽ ⚹ ♂
10 26 ☽ ⊥ ☿
13 01 ☽ △ ♇
13 45 ☽ ⚹ ♆
14 18 ☽ ⊼ ♅
18 54 ☽ ⊼ ♃
18 57 ☽ ∠ ♄
21 23 ☽ ∠ ♀

28 Sunday
02 31 ☉ # ♅
06 19 ☽ ⊻ ☉
06 56 ☽ ∠ ☿
09 09 ♂ ‖ ♄
13 40 ☽ Q ♆
15 44 ☿ ‖ ♀
17 18 ☽ ∠ ♂
18 49 ☽ bQ ♇
19 16 ☽ ⊥ ☉
21 14 ☽ # ♇
23 43 ♇ St R

29 Monday
00 58 ☽ ⚹ ♄
02 21 ☽ ⚹ ♀
04 37 ☽ ‖ ☉
06 19 ☽ # ♅
12 03 ☽ ⊥ ♂
12 20 ☽ # ♃
12 28 ☽ ± ♇
14 31 ☽ ∠ ☉
16 38 ☽ ⚹ ☿
19 22 ♂ ☍ ♇

30 Tuesday
00 08 ☽ ⊼ ♇
00 26 ☽ ⊻ ♂
00 36 ☽ Q ♄
01 00 ☽ □ ♆
01 15 ☽ Q ♀
01 17 ☽ ☍ ♅
05 39 ☽ ☍ ♃
08 18 ☽ ‖ ☿
10 53 ♂ △ ♆
15 02 ♂ ⊼ ♅
17 24 ☽ ‖ ♀
18 59 ☽ Q ☿
21 58 ☽ ⚹ ☉

DECEMBER 1954

LONGITUDES

Date	Sidereal time h m s	Sun ☉ ° ' "	Moon ☽ ° ' "	Moon ☽ 24.00 ° ' "	Mercury ☿ ° '	Venus ♀ ° '	Mars ♂ ° '	Jupiter ♃ ° '	Saturn ♄ ° '	Uranus ♅ ° '	Neptune ♆ ° '	Pluto ♇ ° '
01	16 39 13	08 ♐ 45 56	15 ♒ 36 11	22 ♒ 02 16	25 ♏ 42	15 ♏ 04	28 ♒ 00	29 ♋ 36	15 ♏ 12	27 ♋ 22	27 ♎ 17	26 ♌ 48
02	16 43 10	09 46 46	28 ♒ 33 22	05 ♓ 09 52	27 14	14 R 55	28 42	29 R 33	15 19	27 R 20	27 19	26 R 48
03	16 47 06	10 47 38	11 ♓ 52 07	18 ♓ 40 22	28 ♏ 47	14 47	29 ♒ 25	29 30	15 26	27 19	27 21	26 48
04	16 51 03	11 48 30	25 ♓ 34 50	02 ♈ 35 34	00 ♐ 19	14 43	00 ♓ 08	29 27	15 33	27 17	27 23	26 48
05	16 55 00	12 49 23	09 ♈ 42 30	16 ♈ 55 23	01 52	14 40	00 50	29 23	15 39	27 15	27 24	26 47
06	16 58 56	13 50 16	24 ♈ 13 50	01 ♉ 37 16	03 25	14 D 40	01 33	29 19	15 46	27 14	27 26	26 47
07	17 02 53	14 51 11	09 ♉ 04 55	16 ♉ 35 53	04 58	14 43	02 16	29 15	15 53	27 12	27 28	26 47
08	17 06 49	15 52 06	24 ♉ 09 05	01 ♊ 43 23	06 31	14 48	02 58	29 11	15 59	27 10	27 29	26 47
09	17 10 46	16 53 02	09 ♊ 17 34	16 ♊ 50 25	08 04	14 55	03 41	29 07	16 06	27 09	27 31	26 46
10	17 14 42	17 53 59	24 ♊ 20 46	01 ♋ 47 33	09 37	15 05	04 24	29 03	16 12	27 07	27 33	26 46
11	17 18 39	18 54 57	09 ♋ 09 46	16 ♋ 26 39	11 11	15 17	05 07	28 58	16 19	27 05	27 34	26 46
12	17 22 35	19 55 56	23 ♋ 37 32	00 ♌ 42 00	12 44	15 31	05 50	28 53	16 25	27 03	27 36	26 45
13	17 26 32	20 56 56	07 ♌ 39 44	14 ♌ 30 40	14 18	15 47	06 33	28 48	16 31	27 01	27 37	26 45
14	17 30 29	21 57 57	21 ♌ 14 49	27 ♌ 52 21	15 51	16 05	07 15	28 43	16 38	26 59	27 39	26 44
15	17 34 25	22 58 59	04 ♍ 23 35	10 ♍ 48 52	17 25	16 25	07 58	28 38	16 44	26 57	27 40	26 44
16	17 38 22	24 00 01	17 ♍ 08 40	23 ♍ 23 29	18 59	16 47	08 41	28 33	16 50	26 55	27 42	26 44
17	17 42 18	25 01 05	29 ♍ 33 51	05 ♎ 40 20	20 33	17 11	09 24	28 27	16 57	26 53	27 43	26 43
18	17 46 15	26 02 10	11 ♎ 43 30	17 ♎ 43 54	22 07	17 37	10 07	28 21	17 03	26 51	27 44	26 42
19	17 50 11	27 03 15	23 ♎ 42 06	29 ♎ 38 38	23 41	18 04	10 50	28 16	17 09	26 49	27 46	26 42
20	17 54 08	28 04 22	05 ♏ 34 01	11 ♏ 28 43	25 15	18 34	11 33	28 09	17 15	26 47	27 47	26 41
21	17 58 04	29 ♐ 05 29	17 ♏ 23 12	23 ♏ 17 53	26 50	19 04	12 16	28 03	17 21	26 44	27 48	26 41
22	18 02 01	00 ♑ 06 36	29 ♏ 13 09	05 ♐ 09 20	28 24	19 37	12 59	27 57	17 27	26 42	27 50	26 40
23	18 05 58	01 07 45	11 ♐ 06 47	17 ♐ 05 45	29 ♐ 59	20 11	13 42	27 51	17 33	26 40	27 51	26 39
24	18 09 54	02 08 54	23 ♐ 06 31	29 ♐ 09 17	01 ♑ 35	20 46	14 25	27 44	17 39	26 38	27 52	26 39
25	18 13 51	03 10 03	05 ♑ 14 16	11 ♑ 21 39	03 10	21 22	15 08	27 37	17 44	26 35	27 53	26 38
26	18 17 47	04 11 13	17 ♑ 31 37	23 ♑ 44 18	04 46	22 00	15 52	27 30	17 50	26 33	27 54	26 37
27	18 21 44	05 12 23	29 ♑ 59 52	06 ♒ 18 28	06 21	22 40	16 35	27 24	17 56	26 31	27 56	26 36
28	18 25 40	06 13 33	12 ♒ 40 15	19 ♒ 05 21	07 58	23 20	17 18	27 16	18 01	26 28	27 57	26 35
29	18 29 37	07 14 43	25 ♒ 33 56	02 ♓ 06 08	09 34	24 02	18 01	27 09	18 07	26 26	27 58	26 35
30	18 33 33	08 15 53	08 ♓ 42 06	15 ♓ 21 58	11 11	24 44	18 44	27 02	18 12	26 23	27 59	26 34
31	18 37 30	09 ♑ 17 03	22 ♓ 05 51	28 ♓ 53 52	12 ♑ 48	25 ♏ 28	19 ♓ 27	26 ♋ 55	18 ♏ 18	26 ♋ 21	28 ♎ 00	26 ♌ 33

Date	Moon True ☊ ° '	Moon Mean ☊ ° '	Moon ☽ Latitude ° '
01	05 ♑ 45	07 ♑ 02	03 N 22
02	05 D 46	06 59	04 11
03	05 R 46	06 56	04 48
04	05 46	06 53	05 10
05	05 44	06 49	05 14
06	05 41	06 46	04 59
07	05 39	06 43	04 23
08	05 36	06 40	03 29
09	05 34	06 37	02 20
10	05 32	06 33	01 N 02
11	05 D 32	06 30	00 S 20
12	05 33	06 27	01 39
13	05 34	06 24	02 49
14	05 35	06 21	03 47
15	05 36	06 18	04 31
16	05 37	06 14	05 01
17	05 R 37	06 11	05 15
18	05 36	06 08	05 15
19	05 36	06 05	05 01
20	05 34	06 02	04 35
21	05 33	05 59	03 57
22	05 33	05 55	03 08
23	05 32	05 52	02 12
24	05 32	05 49	01 08
25	05 D 32	05 46	00 S 02
26	05 32	05 43	01 N 06
27	05 32	05 39	02 12
28	05 R 32	05 36	03 12
29	05 32	05 33	04 04
30	05 31	05 30	04 44
31	05 ♑ 31	05 ♑ 27	05 N 09

DECLINATIONS

Date	Sun ☉ ° '	Moon ☽ ° '	Mercury ☿ ° '	Venus ♀ ° '	Mars ♂ ° '	Jupiter ♃ ° '	Saturn ♄ ° '	Uranus ♅ ° '	Neptune ♆ ° '	Pluto ♇ ° '
01	21 S 46	12 S 57	18 S 18	16 S 04	13 S 25	20 N 31	14 S 21	21 N 13	08 S 57	22 N 09
02	21 55	08 03	18 46	15 48	13 08	20 32	14 23	21 13	08 58	22 10
03	22 04	02 S 40	19 14	15 34	12 52	20 33	14 24	21 13	08 58	22 10
04	22 13	02 N 59	19 41	15 21	12 36	20 34	14 26	21 14	08 59	22 10
05	22 20	08 40	20 07	15 09	12 19	20 35	14 28	21 14	08 59	22 11
06	22 28	14 01	20 32	14 58	12 02	20 36	14 30	21 14	09 00	22 11
07	22 35	18 41	20 56	14 48	11 46	20 37	14 32	21 15	09 00	22 12
08	22 42	22 11	21 19	14 39	11 29	20 38	14 34	21 15	09 01	22 12
09	22 48	24 09	21 41	14 32	11 12	20 39	14 35	21 15	09 02	22 13
10	22 54	24 21	22 03	14 25	10 55	20 40	14 37	21 16	09 02	22 13
11	22 59	22 48	22 23	14 20	10 38	20 41	14 39	21 16	09 03	22 13
12	23 04	19 46	22 42	14 15	10 21	20 42	14 41	21 17	09 03	22 14
13	23 08	15 38	23 00	14 12	10 04	20 43	14 42	21 17	09 04	22 14
14	23 12	10 50	23 17	14 09	09 47	20 45	14 44	21 17	09 04	22 15
15	23 16	05 41	23 32	14 08	09 29	20 46	14 46	21 18	09 05	22 15
16	23 19	00 N 28	23 47	14 07	09 12	20 47	14 48	21 18	09 05	22 16
17	23 21	04 S 39	24 00	14 08	08 54	20 48	14 49	21 19	09 05	22 16
18	23 23	09 28	24 12	14 09	08 37	20 50	14 51	21 19	09 06	22 17
19	23 25	13 52	24 23	14 11	08 19	20 51	14 52	21 20	09 06	22 17
20	23 26	17 42	24 33	14 13	08 02	20 52	14 54	21 20	09 07	22 18
21	23 27	20 48	24 41	14 16	07 44	20 54	14 56	21 20	09 07	22 18
22	23 27	23 03	24 48	14 20	07 26	20 55	14 57	21 21	09 08	22 19
23	23 26	24 17	24 53	14 25	07 09	20 57	14 59	21 21	09 08	22 20
24	23 26	24 24	24 57	14 30	06 51	20 58	15 00	21 22	09 08	22 20
25	23 24	23 22	25 00	14 36	06 33	21 00	15 02	21 22	09 09	22 21
26	23 23	21 12	25 02	14 42	06 15	21 01	15 03	21 23	09 09	22 21
27	23 21	18 00	25 01	14 49	05 57	21 03	15 04	21 23	09 09	22 22
28	23 18	13 56	25 00	14 56	05 39	21 04	15 06	21 24	09 10	22 22
29	23 15	09 10	24 57	15 04	05 21	21 06	15 07	21 24	09 10	22 23
30	23 11	03 55	24 53	15 12	05 03	21 07	15 09	21 25	09 10	22 24
31	23 S 07	01 N 36	24 S 47	15 S 20	04 S 45	21 N 09	15 S 10	21 N 25	09 S 11	22 N 24

ZODIAC SIGN ENTRIES

Date	h	m	Planets
02	14	38	☽ ♓
04	07	02	☿ ♐
04	07	41	♂ ♓
04	19	35	☽ ♈
06	21	23	☽ ♉
08	21	16	☽ ♊
10	21	06	☽ ♋
12	22	48	☽ ♌
15	03	54	☽ ♍
17	12	51	☽ ♎
20	00	43	☽ ♏
22	09	24	☉ ♑
22	13	35	☽ ♐
23	12	10	☿ ♑
25	01	40	☽ ♑
27	12	00	☽ ♒
29	20	09	☽ ♓

LATITUDES

Date	Mercury ☿ ° '	Venus ♀ ° '	Mars ♂ ° '	Jupiter ♃ ° '	Saturn ♄ ° '	Uranus ♅ ° '	Neptune ♆ ° '	Pluto ♇ ° '
01	00 N 55	00 N 18	01 S 19	00 N 17	02 N 09	00 N 32	01 N 40	10 N 13
04	00 33	00 57	01 15	00 18	02 09	00 32	01 40	10 14
07	00 N 12	01 32	01 10	00 19	02 09	00 32	01 41	10 15
10	00 S 09	02 02	01 06	00 19	02 10	00 32	01 41	10 16
13	00 29	02 29	01 01	00 20	02 10	00 32	01 41	10 17
16	00 48	02 51	00 57	00 20	02 10	00 32	01 41	10 18
19	01 06	03 10	00 53	00 21	02 10	00 32	01 41	10 19
22	01 22	03 26	00 49	00 21	02 11	00 32	01 41	10 20
25	01 36	03 38	00 45	00 22	02 11	00 32	01 41	10 21
28	01 48	03 48	00 41	00 22	02 11	00 32	01 42	10 22
31	01 S 57	03 N 55	00 S 37	00 N 23	02 N 12	00 N 32	01 N 42	10 N 23

DATA

Julian Date	2435078
Delta T	+31 seconds
Ayanamsa	23° 13' 55"
Synetic vernal point	05° ♓ 53' 05"
True obliquity of ecliptic	23° 26' 43"

MOON'S PHASES, APSIDES AND POSITIONS ☽

Date	h	m	Phase	Longitude ° '	Eclipse Indicator
03	09	56	☽	10 ♓ 42	
10	00	57	○	17 ♊ 26	
17	02	21	☾	24 ♍ 37	
25	07	33	●	02 ♑ 59	Annular

Day	h	m		
09	01	44	Perigee	
21	09	02	Apogee	
03	23	25	0N	
10	02	33	Max dec	24° N 29'
16	14	07	0S	
24	02	33	Max dec	24° S 30'
31	05	07	0N	

LONGITUDES

Date	Chiron ⚷ ° '	Ceres ⚳ ° '	Pallas ⚴ ° '	Juno ⚵ ° '	Vesta ⚶ ° '	Black Moon Lilith ⚸ ° '
01	25 ♑ 21	05 ♐ 10	17 ♏ 25	29 ♎ 39	27 ♊ 59	18 ♏ 59
11	26 ♑ 04	09 ♐ 23	21 ♏ 43	02 ♏ 48	25 ♊ 31	20 ♏ 06
21	26 ♑ 50	13 ♐ 33	25 ♏ 56	05 ♏ 47	22 ♊ 54	21 ♏ 13
31	27 ♑ 40	17 ♐ 40	00 ♐ 03	08 ♏ 33	20 ♊ 25	22 ♏ 20

All ephemeris data is given at 12.00 UT and the Moon's longitude is additionally given for 24.00 UT

ASPECTARIAN

h m	Aspects
01 Wednesday	
01 36	♀ ☌ ♄
04 32	☽ ∥ ♄
09 22	☽ ∥ ♂
11 01	☽ □ ♀
11 15	☽ □ ♄
22 28	☽ Q ☉
02 Thursday	
05 12	☿ □ ♇
07 45	☽ ∥ ♆
08 47	☽ ☍ ♇
09 15	☽ □ ☿
09 44	☽ △ ♆
09 46	☽ ⊼ ♅
12 18	☽ ☌ ♂
13 20	☿ ⊻ ♆
13 32	☿ △ ♅
13 49	☽ ⊼ ♃
19 06	♅ □ ♆
20 41	☽ ± ♅
03 Friday	
00 39	☽ ± ♃
09 56	☽ □ ☉
12 47	☽ ⚼ ♅
12 51	☽ ⚼ ♆
14 32	♂ ⊼ ♃
16 38	☽ ⚼ ♃
17 08	☽ △ ♀
18 21	☽ △ ♄
22 51	☿ △ ♃
04 Saturday	
04 42	☽ ± ♆
05 32	☉ # ♇
06 30	☿ □ ♂
14 05	☽ ⊼ ♇
14 56	☽ △ ♅
15 06	☽ ⊼ ♆
18 36	☽ △ ♃
19 04	☽ ⚼ ♀
20 13	☽ ⊻ ♂
20 35	☽ ⚼ ♄
21 08	☽ △ ☿
22 57	☉ ⚼ ♅
05 Sunday	
00 20	☽ ± ♇
01 50	☉ ∠ ♆
06 56	☽ ⊥ ♂
10 16	☽ ± ♀
11 54	☽ ± ♄
13 25	☽ # ♆
15 29	☽ ⚼ ♇
17 36	☽ △ ☉
20 16	☽ ⊼ ♀
21 59	☽ ⊼ ♄
22 39	♀ St D
22 44	☽ ∠ ♂
06 Monday	
01 19	☽ ⚼ ☿
03 19	☽ # ♂
12 19	☿ ⊥ ♆
14 17	☽ # ♄
16 06	☿ # ♃
16 10	☽ △ ♇
16 20	☽ # ♀
16 53	☽ □ ♅
17 14	☽ ☍ ♆
17 47	☽ ± ☿
20 03	☽ ⚼ ☉
20 15	☽ □ ♃
22 47	☉ ⚼ ♃
07 Tuesday	
00 29	☽ ⚹ ♂
04 37	☽ ⊼ ☿
08 34	☉ ⊻ ♀
11 36	☽ ± ☉
20 41	☽ Q ♂
21 03	☽ ☍ ♀
21 45	☽ Q ♅
21 53	☽ ⊼ ☉
22 56	☽ ☍ ♄
08 Wednesday	
00 07	☽ ∥ ♃
01 00	☽ Q ♃
04 09	☽ # ☿
04 34	☽ ∥ ♅
07 40	☿ # ♅
12 06	☽ ∥ ♇
15 06	☉ ⊻ ♄
16 10	☽ □ ♇
16 47	☽ ⚹ ♅
16 48	☽ # ☉
17 18	☽ ⊼ ♆
18 25	♂ ± ♅
19 57	☽ ⚹ ♃
09 Thursday	
01 36	♀ ∥ ♄
02 40	☽ □ ♂
02 49	☽ ± ♆
09 50	☽ ☍ ☿
16 31	☽ ∠ ♅
17 08	☽ ⚼ ♆
19 38	☽ ∠ ♃
20 42	☽ Q ♇
21 02	☽ ⊼ ♀
22 53	☽ ⊼ ♄
10 Friday	
00 57	☽ ☍ ☉
06 42	☽ ± ♀
06 50	☽ ⊥ ♅
08 32	☽ ± ♄
09 56	☽ ⊥ ♃
15 53	☽ ⚹ ♇
16 26	☽ ⊻ ♅
17 09	☽ △ ♆
19 31	☽ ⊻ ♃
21 21	☽ ⚼ ♀
23 08	☽ ⚼ ♄
11 Saturday	
00 24	☿ # ♇
05 03	☽ △ ♂
07 37	♂ ± ♃
10 07	☽ # ☉
15 34	☽ # ☿
15 42	☽ ⊼ ☿
16 16	☽ ∠ ♇
17 25	☽ ∥ ♇
22 13	☽ △ ♀
23 52	☽ △ ♄
12 Sunday	
01 17	☽ ∥ ♅
01 39	☿ ⚼ ♅
02 49	☽ ± ☿
05 20	☽ ⊼ ☉
05 34	☽ ∥ ♃
07 04	☽ ⚼ ♂
07 11	☽ ⊥ ♇
09 47	☿ ∠ ♆
16 12	☽ ± ☉
17 17	☽ ⊻ ♇
17 46	☽ ☌ ♅
18 43	☽ □ ♆
19 49	☽ ⚼ ☿
20 52	☽ ☌ ♃
23 04	☽ ± ♂
13 Monday	
04 54	☿ ⚼ ♃
08 48	☽ ⚼ ☉
09 57	☽ ⊼ ♂
13 35	☉ ± ♅
16 51	☽ # ♄
19 34	☽ # ♀
14 Tuesday	
01 07	☽ △ ☿
01 59	☽ Q ♆
02 34	☽ □ ♀
03 11	☉ ∥ ☿
03 41	☽ □ ♄
13 24	☽ △ ☉
15 59	☉ ⊥ ♀
16 25	☿ ⊻ ♀
17 19	☽ # ♂
20 20	☽ # ♆
21 56	☽ ☌ ♇
22 21	☽ ⊻ ♅
23 36	☽ ⚹ ♆
15 Wednesday	
00 48	☿ ⊻ ♄
01 28	☽ ⊻ ♃
04 27	☉ ± ♃
05 29	☉ ⊥ ♄
09 20	☽ ⊥ ♅
12 03	☽ Q ♀
12 27	☽ ⊥ ♃
12 38	☽ Q ♄
19 04	☽ ☍ ♂
16 Thursday	
02 06	☽ ∠ ♅
03 32	☽ ∠ ♆
05 12	☽ ∠ ♃
11 18	☽ ⚹ ♀
11 25	☽ ⚹ ♄
16 00	☽ □ ☿
16 30	♀ ☌ ♄
20 44	☽ ⊥ ♆
21 12	♂ ∥ ♆
17 Friday	
02 21	☽ □ ☉
06 27	☽ ⊻ ♇
06 47	☽ ⚹ ♅
08 23	☽ ⊻ ♆
09 51	☽ ⚹ ♃
16 42	☽ ∠ ♄
17 07	☿ ± ♅
17 19	☽ ∠ ♀
18 10	☽ ⊥ ♇
18 Saturday	
06 18	☽ Q ♅
07 52	☽ ∥ ♂
08 19	☽ Q ☿
08 36	☽ ⊼ ♂
09 18	☽ Q ♃
10 05	☽ ∥ ♆
10 38	☽ ⊥ ♄
11 46	☽ ⊥ ♀
11 58	☽ ∠ ♇
15 33	☿ ± ♃
17 02	☽ Q ☉
21 20	☽ ± ♂
22 43	☽ ⊻ ♄
19 Sunday	
00 14	☽ ⊻ ♀
03 17	☿ ⊥ ♄
03 40	☉ △ ♇
06 31	☉ ⊼ ♅
11 57	☽ ⚹ ☿
13 49	☽ ∥ ♀
16 35	☽ ⚼ ♂
17 59	☽ ∥ ♄
18 02	☽ ⚹ ♇
18 16	☽ □ ♅
19 24	☽ ⚹ ☉
20 13	☽ ☌ ♆
20 38	☿ ⊥ ♀
21 07	☽ □ ♃
20 Monday	
05 03	☉ ⚹ ♆
13 50	☉ ⊼ ♃
18 20	☽ Q ♇
19 08	♂ ⚼ ♅
22 58	☽ ∠ ☿
21 Tuesday	
00 56	☽ △ ♂
04 41	☽ ∠ ☉
09 42	☿ △ ♇
10 41	☿ ⊼ ♅
11 55	☽ ☌ ♄
12 49	☽ # ♃
15 35	☽ ☌ ♀
16 55	☽ # ♅
20 04	☽ ⊥ ☿
22 Wednesday	
00 40	☽ ⊥ ☉
02 56	☽ # ♇
03 04	☿ ⚹ ♆
05 29	☿ ⊼ ♃
06 29	♂ ⚼ ♆
06 50	☽ □ ♇
06 55	☽ △ ♅
09 11	☽ ⊻ ♆
09 27	☽ △ ♃
10 06	☽ ⊻ ☿
10 54	♂ ⚼ ♃
13 58	☽ ⊻ ☉
17 47	☽ ∥ ☉
21 20	☽ ⊥ ♆
23 Thursday	
10 57	♃ □ ♆
13 06	☽ ⚼ ♅
15 27	☽ ⚼ ♃
15 30	☽ ∠ ♆
17 32	☽ □ ♂
22 40	♅ ⊻ ♇
24 Friday	
01 00	☽ ⊻ ♄
07 04	☽ ± ♅
07 05	☽ ⊻ ♀
09 17	☽ ± ♃
13 04	☽ ⊥ ♄
18 58	☽ ⊼ ♅
19 01	☽ △ ♇
19 39	☽ ⊥ ♀
21 06	☽ ⊼ ♃
21 28	☽ ⚹ ♆
25 Saturday	
00 52	☉ ∠ ♄
05 10	☿ ∠ ♄
07 02	☽ ∠ ♄
07 18	☽ ☌ ☿
07 33	☽ ☌ ☉
07 37	☽ Q ♂
09 52	☉ Q ♂
11 20	☿ Q ♂
11 24	☽ ∥ ☉
12 07	☉ ☌ ☿
14 21	☽ ∠ ♀
21 08	☽ Q ♆
26 Sunday	
00 31	☽ ⚼ ♇
00 47	☽ # ♇
08 34	☽ ⚹ ♂
10 25	☽ # ♅
12 36	☽ ⚹ ♄
13 35	☽ # ♃
17 58	☽ ± ♇
21 08	☽ ⚹ ♀
27 Monday	
05 21	☽ ☍ ♅
05 31	☽ ⊼ ♇
07 04	☽ ☍ ♃
08 02	☽ □ ♆
11 52	☽ Q ♄
15 12	☽ ∠ ♂
21 22	☽ Q ♀
22 47	☽ ⊻ ☉
28 Tuesday	
01 50	☽ ⊻ ☿
05 38	☽ ∥ ♄
06 39	☽ ∥ ♀
09 16	☽ ⊥ ♂
11 05	☽ ⊥ ☉
14 46	☽ ⊥ ☿
21 10	☽ ⊻ ♂
21 45	☿ ∠ ♀
22 05	☽ □ ♄
29 Wednesday	
05 20	☽ ∠ ☉
09 00	☽ □ ♀
09 54	☽ ∠ ☿
12 01	☽ ∥ ♆
13 35	☽ ⊼ ♅
13 52	☽ ☍ ♇
14 54	☽ ⊼ ♃
15 50	♂ △ ♄
16 25	☽ △ ♆
17 57	☿ Q ♆
30 Thursday	
00 34	☽ ± ♅
00 56	♀ ∥ ♄
01 48	☽ ± ♃
06 42	☽ ∥ ♂
11 09	☽ ⚹ ☉
16 50	☽ ⚼ ♅
17 06	☽ ⚹ ☿
17 38	☿ ⚼ ♇
17 58	☽ ⚼ ♃
19 44	☽ ⚼ ♆
31 Friday	
05 12	☽ △ ♄
07 02	☽ ☌ ♂
10 26	☽ Q ☉
11 49	☽ ± ♆
17 25	☽ Q ☿
18 18	☽ △ ♀
19 30	☽ △ ♅
19 51	☽ ⊼ ♇
20 26	☽ △ ♃
22 26	☽ ⊼ ♆

JANUARY 1955

LONGITUDES

Date	Sidereal time h m s	Sun ☉ ° ′ ″	Moon ☽ ° ′ ″	Moon ☽ 24.00 ° ′ ″	Mercury ☿ ° ′	Venus ♀ ° ′	Mars ♂ ° ′	Jupiter ♃ ° ′	Saturn ♄ ° ′	Uranus ♅ ° ′	Neptune ♆ ° ′	Pluto ♇ ° ′
01	18 41 27	10 ♑ 18 12	05 ♈ 46 04	12 ♈ 42 28	14 ♑ 25	26 ♏ 13	20 ♓ 10	26 ♋ 47	18 ♏ 23	26 ♋ 19	28 ♎ 01	26 ♌ 32
02	18 45 23	11 19 22	19 ♈ 43 01	26 ♈ 47 37	16 03	26 58	20 53	26 R 40	18 28	26 R 16	28 02	26 R 31
03	18 49 20	12 20 31	03 ♉ 56 03	11 ♉ 08 03	17 41	27 45	21 36	26 32	18 34	26 14	28 03	26 30
04	18 53 16	13 21 40	18 ♉ 23 12	25 ♉ 41 02	19 19	28 33	22 20	26 25	18 39	26 11	28 03	26 29
05	18 57 13	14 22 49	03 ♊ 00 56	10 ♊ 22 15	20 58	29 ♏ 21	23 03	26 17	18 44	26 08	28 04	26 28
06	19 01 09	15 23 57	17 ♊ 44 11	25 ♊ 05 56	22 36	00 ♐ 11	23 46	26 09	18 49	26 06	28 05	26 27
07	19 05 06	16 25 05	02 ♋ 26 38	09 ♋ 45 26	24 15	01 01	24 29	26 01	18 54	26 03	28 06	26 26
08	19 09 02	17 26 13	17 ♋ 01 30	24 ♋ 14 01	25 55	01 52	25 12	25 53	18 59	26 01	28 07	26 25
09	19 12 59	18 27 21	01 ♌ 22 18	08 ♌ 25 43	27 34	02 44	25 55	25 45	19 03	25 58	28 07	26 24
10	19 16 56	19 28 28	15 ♌ 23 48	22 ♌ 16 09	29 ♑ 14	03 36	26 38	25 37	19 08	25 56	28 08	26 23
11	19 20 52	20 29 36	29 ♌ 02 33	05 ♍ 42 55	00 ♒ 54	04 29	27 21	25 29	19 13	25 53	28 09	26 22
12	19 24 49	21 30 43	12 ♍ 17 15	18 ♍ 45 44	02 33	05 23	28 04	25 21	19 17	25 50	28 09	26 20
13	19 28 45	22 31 50	25 ♍ 08 35	01 ♎ 26 11	04 13	06 18	28 47	25 13	19 22	25 48	28 10	26 19
14	19 32 42	23 32 57	07 ♎ 38 57	13 ♎ 47 21	05 52	07 13	29 ♓ 30	25 05	19 26	25 45	28 10	26 18
15	19 36 38	24 34 03	19 ♎ 51 58	25 ♎ 53 21	07 32	08 09	00 ♈ 13	24 57	19 31	25 43	28 11	26 17
16	19 40 35	25 35 10	01 ♏ 52 07	07 ♏ 48 52	09 10	09 05	00 56	24 49	19 35	25 40	28 11	26 16
17	19 44 31	26 36 16	13 ♏ 44 15	19 ♏ 38 51	10 48	10 02	01 39	24 41	19 39	25 37	28 12	26 14
18	19 48 28	27 37 22	25 ♏ 33 17	01 ♐ 28 09	12 25	10 59	02 22	24 33	19 43	25 35	28 12	26 13
19	19 52 25	28 38 28	07 ♐ 23 58	13 ♐ 21 18	14 00	11 57	03 05	24 25	19 47	25 32	28 13	26 12
20	19 56 21	29 ♑ 39 33	19 ♐ 20 35	25 ♐ 22 17	15 35	12 56	03 48	24 17	19 51	25 30	28 13	26 11
21	20 00 18	00 ♒ 40 38	01 ♑ 26 47	07 ♑ 34 23	17 07	13 55	04 31	24 09	19 55	25 27	28 13	26 09
22	20 04 14	01 41 43	13 ♑ 45 24	20 ♑ 00 01	18 37	14 54	05 14	24 01	19 59	25 24	28 14	26 08
23	20 08 11	02 42 47	26 ♑ 18 22	02 ♒ 40 35	20 04	15 54	05 57	23 53	20 02	25 22	28 14	26 07
24	20 12 07	03 43 50	09 ♒ 06 38	15 ♒ 36 32	21 27	16 54	06 40	23 45	20 06	25 19	28 14	26 05
25	20 16 04	04 44 52	22 ♒ 10 09	28 ♒ 47 24	22 47	17 55	07 23	23 37	20 09	25 17	28 14	26 04
26	20 20 00	05 45 53	05 ♓ 28 04	12 ♓ 11 59	24 02	18 56	08 06	23 30	20 13	25 14	28 14	26 03
27	20 23 57	06 46 53	18 ♓ 58 56	25 ♓ 48 40	25 11	19 57	08 49	23 22	20 16	25 12	28 14	26 01
28	20 27 54	07 47 52	02 ♈ 40 59	09 ♈ 35 38	26 14	20 59	09 31	23 14	20 19	25 09	28 14	26 00
29	20 31 50	08 48 50	16 ♈ 32 25	23 ♈ 31 08	27 10	22 01	10 14	23 07	20 22	25 06	28 14	25 58
30	20 35 47	09 49 47	00 ♉ 31 37	07 ♉ 33 40	27 59	23 03	10 57	22 59	20 25	25 04	28 R 14	25 57
31	20 39 43	10 ♒ 50 42	14 ♉ 37 07	21 ♉ 41 48	28 ♒ 39	24 ♐ 06	11 ♈ 40	22 ♋ 52	20 ♏ 28	25 ♋ 01	28 ♎ 14	25 ♌ 56

Moon / DECLINATIONS

Date	Moon True ☊ ° ′	Moon Mean ☊ ° ′	Moon ☽ Latitude ° ′	Sun ☉ ° ′	Moon ☽ ° ′	Mercury ☿ ° ′	Venus ♀ ° ′	Mars ♂ ° ′	Jupiter ♃ ° ′	Saturn ♄ ° ′	Uranus ♅ ° ′	Neptune ♆ ° ′	Pluto ♇ ° ′
01	05 ♑ 31	05 ♑ 24	05 N 18	23 S 03	07 N 09	24 S 39	15 S 29	04 S 27	21 N 11	15 S 11	21 N 26	09 S 11	22 N 25
02	05 D 31	05 20	05 08	22 58	12 28	24 30	15 38	04 09	21 12	15 13	21 26	09 11	22 25
03	05 31	05 17	04 39	22 52	17 13	24 20	15 47	03 51	21 14	15 14	21 27	09 12	22 26
04	05 32	05 14	03 53	22 46	21 02	24 08	15 56	03 32	21 15	15 15	21 27	09 12	22 27
05	05 33	05 11	02 51	22 40	23 34	23 54	16 06	03 14	21 17	15 16	21 28	09 12	22 27
06	05 33	05 08	01 37	22 33	24 30	23 39	16 15	02 56	21 19	15 18	21 28	09 12	22 28
07	05 34	05 05	00 N 17	22 26	23 43	23 22	16 25	02 38	21 20	15 19	21 29	09 13	22 29
08	05 R 33	05 01	01 S 03	22 18	21 19	23 04	16 35	02 20	21 22	15 20	21 29	09 13	22 29
09	05 33	04 58	02 18	22 10	17 37	22 44	16 45	02 02	21 23	15 21	21 30	09 13	22 30
10	05 31	04 55	03 23	22 02	12 59	22 22	16 55	01 43	21 25	15 22	21 30	09 13	22 30
11	05 29	04 52	04 15	21 53	07 50	21 59	17 05	01 25	21 27	15 23	21 31	09 13	22 31
12	05 27	04 49	04 51	21 44	02 N 28	21 35	17 15	01 07	21 28	15 24	21 31	09 13	22 32
13	05 25	04 45	05 11	21 34	02 S 50	21 09	17 26	00 49	21 30	15 26	21 32	09 13	22 32
14	05 23	04 42	05 16	21 23	07 52	20 41	17 36	00 31	21 31	15 27	21 32	09 14	22 33
15	05 22	04 39	05 06	21 13	12 29	20 12	17 45	00 S 13	21 33	15 28	21 33	09 14	22 34
16	05 D 22	04 36	04 43	21 02	16 33	19 42	17 55	00 N 06	21 35	15 28	21 33	09 14	22 34
17	05 23	04 33	04 08	20 50	19 54	19 10	18 05	00 24	21 36	15 29	21 34	09 14	22 35
18	05 25	04 30	03 23	20 39	22 26	18 37	18 15	00 42	21 38	15 30	21 34	09 14	22 36
19	05 27	04 26	02 29	20 26	24 00	18 03	18 24	01 00	21 39	15 31	21 34	09 14	22 36
20	05 28	04 23	01 28	20 14	24 29	17 29	18 33	01 18	21 41	15 32	21 35	09 14	22 37
21	05 29	04 20	00 S 22	20 01	23 48	16 53	18 43	01 36	21 42	15 33	21 35	09 14	22 38
22	05 R 29	04 17	00 N 45	19 47	21 59	16 16	18 51	01 54	21 44	15 34	21 36	09 14	22 38
23	05 27	04 14	01 52	19 33	19 04	15 39	19 00	02 12	21 45	15 35	21 36	09 14	22 39
24	05 24	04 11	02 54	19 19	15 11	15 02	19 08	02 30	21 47	15 35	21 37	09 14	22 40
25	05 20	04 07	03 48	19 05	10 31	14 25	19 16	02 48	21 48	15 36	21 37	09 14	22 40
26	05 15	04 04	04 31	18 50	05 S 18	13 49	19 24	03 06	21 50	15 37	21 38	09 14	22 41
27	05 10	04 01	05 00	18 35	00 N 14	13 13	19 32	03 23	21 51	15 37	21 38	09 14	22 42
28	05 05	03 58	05 12	18 19	05 50	12 38	19 39	03 41	21 53	15 38	21 39	09 14	22 42
29	05 01	03 55	05 06	18 04	11 12	12 04	19 46	03 59	21 54	15 39	21 39	09 14	22 43
30	05 00	03 51	04 42	17 47	16 03	11 33	19 52	04 17	21 55	15 39	21 40	09 14	22 43
31	04 ♑ 59	03 ♑ 48	04 N 01	17 S 31	20 N 03	11 S 04	19 S 58	04 N 34	21 N 57	15 S 40	21 N 40	09 S 14	22 N 44

ZODIAC SIGN ENTRIES

Date	h m	Planets
01	01 56	☽ ♈
03	05 24	☽ ♉
05	07 04	☽ ♊
06	06 48	♀ ♐
07	08 00	☽ ♋
09	09 41	☽ ♌
10	23 05	☿ ♒
11	13 43	☽ ♍
13	21 15	☽ ♎
15	04 33	♂ ♈
16	08 15	☽ ♏
18	21 01	☽ ♐
20	20 02	☉ ♒
21	09 09	☽ ♑
23	18 58	☽ ♒
26	02 11	☽ ♓
28	07 19	☽ ♈
30	11 06	☽ ♉

LATITUDES

Date	Mercury ☿ ° ′	Venus ♀ ° ′	Mars ♂ ° ′	Jupiter ♃ ° ′	Saturn ♄ ° ′	Uranus ♅ ° ′	Neptune ♆ ° ′	Pluto ♇ ° ′
01	02 S 00	03 N 56	00 S 36	00 N 23	02 N 12	00 N 33	01 N 42	10 N 23
04	02 06	04 00	00 32	00 23	02 12	00 33	01 42	10 24
07	02 08	04 02	00 29	00 24	02 13	00 33	01 42	10 25
10	02 06	04 00	00 25	00 24	02 13	00 33	01 42	10 26
13	02 00	04 00	00 22	00 24	02 14	00 33	01 43	10 26
16	01 48	03 56	00 18	00 25	02 14	00 33	01 43	10 27
19	01 30	03 52	00 15	00 25	02 15	00 33	01 43	10 28
22	01 04	03 45	00 12	00 26	02 16	00 33	01 43	10 28
25	00 S 31	03 38	00 09	00 26	02 16	00 33	01 43	10 29
28	00 N 09	03 30	00 06	00 26	02 17	00 33	01 43	10 30
31	00 N 57	03 N 21	00 S 03	00 N 27	02 N 17	00 N 33	01 N 44	10 N 30

DATA

Julian Date	2435109
Delta T	+31 seconds
Ayanamsa	23° 14′ 00″
Synetic vernal point	05° ♓ 53′ 00″
True obliquity of ecliptic	23° 26′ 43″

MOON'S PHASES, APSIDES AND POSITIONS ☽

Date	h m	Phase	Longitude ° ′	Eclipse Indicator
01	20 29	☽	10 ♈ 40	
08	12 44	○	17 ♋ 28	
15	22 14	☾	25 ♎ 00	
24	01 07	●	03 ♒ 16	
31	05 06	☽	10 ♉ 33	

Day	h m		
06	08 59	Perigee	
18	03 10	Apogee	
06	13 03	Max dec	24° N 30′
12	23 06	0S	
20	10 12	Max dec	24° S 29′
27	10 59	0N	

LONGITUDES

Date	Chiron ⚷ ° ′	Ceres ⚳ ° ′	Pallas ⚴ ° ′	Juno ⚵ ° ′	Vesta ⚶ ° ′	Black Moon Lilith ⚸ ° ′
01	27 ♑ 45	18 ♐ 05	00 ♐ 28	08 ♏ 49	20 ♊ 12	22 ♏ 27
11	28 ♑ 36	22 ♐ 07	04 ♐ 27	11 ♏ 18	18 ♊ 13	23 ♏ 33
21	29 ♑ 28	26 ♐ 03	08 ♐ 17	13 ♏ 30	16 ♊ 54	24 ♏ 40
31	00 ♒ 20	29 ♐ 53	11 ♐ 55	15 ♏ 22	16 ♊ 20	25 ♏ 47

All ephemeris data is given at 12.00 UT and the Moon's longitude is additionally given for 24.00 UT

ASPECTARIAN

h m	Aspects
01 Saturday	
00 52	☽ # ♂
05 02	☉ Q ♆
06 22	☽ ± ♇
07 50	☽ ⚼ ♄
14 55	♀ △ ♅
20 29	☽ □ ☉
21 00	☽ # ♆
21 55	♀ □ ♇
21 58	☽ ⚼ ♇
21 58	☽ ⚼ ♀
23 31	☽ ± ♄
02 Sunday	
03 38	♀ △ ♃
04 54	☽ □ ☿
09 52	☽ ⊼ ♄
14 06	☽ ⊻ ♂
14 16	☽ ± ♀
16 29	☉ ⚼ ♇
23 05	☽ □ ♅
23 31	☽ △ ♇
23 41	☽ □ ♃
03 Monday	
00 49	☽ ⊥ ♂
01 01	☽ ⊼ ♀
01 30	☽ # ♄
02 06	☽ ☍ ♆
04 05	☽ # ♀
16 42	☽ ∠ ♂
19 56	♃ ⊻ ♇
20 58	♀ ⊻ ♆
04 Tuesday	
01 38	☿ ⚹ ♄
02 50	♂ ± ♆
03 04	☽ △ ☉
05 05	☽ Q ♅
05 29	☽ Q ♃
12 26	☽ ☍ ♄
13 41	☽ ∥ ♃
13 44	☽ △ ☿
15 11	☽ ∥ ♅
18 49	☽ ⚹ ♂
23 42	☽ ∥ ♇
05 Wednesday	
00 47	☽ ⚹ ♅
01 04	☽ ⚹ ♃
01 18	☽ □ ♇
02 14	☽ # ☉
03 54	☽ ⊼ ♆
04 53	☿ ± ♇
05 37	☽ ⚼ ☉
05 40	☽ ☍ ♀
09 12	☉ ∠ ♀
13 44	☽ ± ♆
15 29	☽ Q ♂
16 20	☽ # ☿
17 25	☽ ⚼ ☿
21 24	☽ ± ☉
06 Thursday	
01 13	☽ ∠ ♅
01 22	☽ ∠ ♃
04 25	☽ ⚼ ♆
06 39	☽ Q ♇
07 55	☽ ⊼ ☉
09 56	☽ ± ☿
13 46	☽ ⊼ ♄
15 50	☽ ⊥ ♅
15 54	☽ ⊥ ♃
20 56	☽ ⊼ ☿
22 20	☽ □ ♂
23 36	☽ ± ♄
07 Friday	
01 36	☽ ⊻ ♅
01 36	☽ ⊻ ♃
02 00	♃ ☌ ♅
02 11	☽ ⚹ ♇
04 53	☉ # ♇
04 53	☽ △ ♆
09 31	☽ ⊼ ♀
14 23	☽ ⚼ ♄
17 29	☽ # ☿
17 44	☿ ⚹ ♂
19 57	☽ ± ♀
08 Saturday	
02 07	☽ ∥ ♇
02 44	☽ ∠ ♇
03 18	☽ # ☉
10 42	☽ ∥ ♅
11 38	☽ ∥ ♃
11 40	☿ ☍ ♃
11 43	☽ ⚼ ♀
12 44	☽ ☍ ☉
13 25	☿ ☍ ♅
15 15	☽ △ ♄
17 37	☽ ⊥ ♇
19 11	☿ ⊼ ♇
09 Sunday	
02 20	☽ △ ♂
02 38	☽ ☌ ♃
02 56	☽ ☌ ♅
03 38	☽ ⊻ ♇
04 45	☽ ☍ ☿
06 31	☽ □ ♆
07 25	♂ △ ♃
13 40	♂ △ ♅
14 27	☽ △ ♀
16 35	☽ # ♀
20 01	☿ □ ♆
10 Monday	
00 08	☽ # ♄
03 17	☿ # ♇
03 22	☉ ⚹ ♄
03 35	♂ ⊼ ♇
05 09	☽ ⚼ ♂
13 17	☽ Q ♆
18 33	☽ □ ♄
19 40	☽ ⊼ ☉
21 38	☽ ± ♂
11 Tuesday	
02 34	♀ ⊥ ♆
05 40	☽ # ♆
05 45	☽ ⊻ ♃
06 24	☽ ⊻ ♅
07 06	☽ ± ☉
07 14	☽ ☌ ♇
08 49	☽ ⊼ ♂
08 54	☉ ± ♇
10 24	☽ ⚹ ♆
15 47	☽ ⊼ ☿
16 20	☽ ⊥ ♃
16 49	☿ Q ♄
17 04	☽ ⊥ ♅
22 11	☉ ∥ ☿
22 29	☽ □ ♀
12 Wednesday	
00 34	☽ ⚼ ☉
02 48	☽ Q ♄
04 11	☽ ± ☿
08 29	☽ ∠ ♃
09 21	☽ ∠ ♅
13 36	☽ ∠ ♆
14 51	♂ ⊼ ♆
15 23	☿ # ♅
17 40	☿ # ♃
18 26	☽ # ♂
23 12	☽ ⚼ ☿
13 Thursday	
01 04	☽ ⚹ ♄
03 17	☽ ∥ ♂
06 22	☽ ⊥ ♆
06 38	☽ △ ☉
10 17	☽ Q ♀
12 09	☽ ⚹ ♃
13 14	☽ ⚹ ♅
14 14	☽ ⊻ ♇
17 00	☉ # ♅
17 45	☽ ⊻ ♆
19 21	☽ ☍ ♂
19 53	☉ # ♃
14 Friday	
01 41	☽ ⊥ ♇
05 45	☽ ∠ ♄
08 02	☽ △ ☿
10 55	☽ Q ♃
11 06	☽ ⚹ ♀
12 12	☽ Q ♅
18 49	☽ ∥ ♆
19 07	☽ ∠ ♇
23 23	☽ ⊥ ♄
23 44	♃ ∥ ♅
15 Saturday	
11 17	☽ ⊻ ♄
19 04	☽ ∠ ♀
20 01	☉ ☍ ♃
22 01	☽ □ ♃
22 14	☽ □ ☉
23 36	☽ □ ♅
16 Sunday	
00 46	☽ ⚹ ♇
04 36	☽ ☌ ♆
05 15	☽ ∥ ♄
09 07	☿ ⚹ ♀
10 01	☽ ⊼ ♂
13 50	☉ ☍ ♅
14 40	☽ ⊥ ♀
21 38	☽ ∥ ♀
22 53	☽ ± ♂
17 Monday	
00 53	☽ Q ♇
03 34	☉ ⊼ ♇
03 50	☽ ⊻ ♀
04 16	♀ ⚼ ♃
05 06	☽ □ ☿
07 04	☽ ∥ ☿
13 56	☽ Q ☉
18 18	☽ ⚼ ♂
19 23	☽ ∥ ☉
18 Tuesday	
00 05	☽ ☌ ♄
02 11	♀ ⚼ ♅
02 42	☽ # ♅
03 16	☽ # ♃
07 00	♂ ± ♇
09 59	☽ △ ♃
12 03	☽ △ ♅
13 21	☽ □ ♇
13 54	☽ # ♇
16 36	☽ ⚹ ☉
17 23	☽ ⊻ ♆
23 24	☽ Q ☿
19 Wednesday	
00 37	☿ ∥ ♀
01 47	☉ □ ♆
02 43	☽ △ ♂
05 33	☽ ⊥ ♆
16 01	☽ ⚼ ♃
18 18	☽ ⚼ ♅
22 00	☽ ☌ ♀
23 43	☽ ∠ ♆
20 Thursday	
01 45	☽ ∠ ☉
03 20	☽ ⚹ ☿
09 54	☽ ± ♃
12 18	☽ ± ♅
13 01	☽ ⊻ ♄
19 02	♀ ∠ ♆
21 24	☽ ⊥ ☉
21 44	☽ ⊼ ♃
21 Friday	
00 12	☽ ⊼ ♅
01 01	☽ ⊥ ♄
01 34	☽ △ ♇
05 38	☽ ⚹ ♆
10 21	☽ ⊻ ☉
13 30	☽ ∠ ☿
18 25	☽ □ ♂
18 50	☽ ∠ ♄
22 Saturday	
02 30	♂ ⚼ ♄
04 58	☽ # ♇
05 10	☽ Q ♆
06 56	☽ ⚼ ♇
09 29	☽ ⊥ ☿
14 24	☽ ⊻ ♀
14 26	☽ # ♃
15 43	☽ # ♅
19 05	☉ Q ♄
22 34	☽ ⊻ ☿
23 Sunday	
00 01	☽ ⚹ ♄
00 14	☽ ± ♇
02 54	☽ ⊥ ♀
07 16	☽ Q ♂
07 27	☽ ☍ ♃
08 12	☽ ∥ ☉
10 13	☽ ☍ ♅
11 38	☽ ⊼ ♇
11 38	☿ □ ♄
12 26	☽ ∥ ♀
15 06	☿ ∥ ♄
15 38	☽ □ ♆
21 24	☽ ∠ ♀
22 51	☽ Q ♄
24 Monday	
01 07	☽ ☌ ☉
07 11	☽ ⚹ ♂
09 43	☽ ∥ ♄
12 55	☽ ∥ ☿
20 06	☿ ∠ ♂
23 44	☉ ∥ ♀
25 Tuesday	
03 35	☽ ⚹ ♀
05 54	♀ ± ♃
08 19	☽ □ ♄
12 24	☽ ∠ ♂
13 14	☽ ☌ ☿
14 37	☽ ⊼ ♃
17 38	☽ ⊼ ♅
18 05	☽ ∥ ♆
19 04	☽ ☍ ♇
23 00	☽ △ ♆
26 Wednesday	
01 22	☽ ± ♃
02 31	☿ ⊼ ♃
03 11	☽ Q ♀
04 26	☽ ± ♅
05 37	☽ ⊥ ♂
12 34	☽ ⊻ ☉
16 58	☽ ⊻ ♂
17 21	☽ ⚼ ♃
18 53	♀ ± ♅
20 29	☽ ⚼ ♅
21 11	☽ # ♂
27 Thursday	
00 08	☽ ⊥ ☉
01 51	☽ ⚼ ♆
12 10	☿ ⊼ ♅
13 51	☽ □ ♀
14 16	☽ △ ♄
17 19	☽ ∠ ☉
17 44	☽ ± ♆
19 38	☽ △ ♃
19 47	♀ ⊻ ♄
22 53	☽ △ ♅
23 50	☽ ⊻ ☿
28 Friday	
00 21	☽ ⊼ ♇
02 13	☽ ∥ ♂
04 15	☽ ⊼ ♆
06 25	☿ ☍ ♇
10 48	☽ ± ♇
11 10	☽ ⊥ ☿
16 36	☽ ⚼ ♄
21 36	☽ ⚹ ☉
29 Saturday	
00 31	☽ ☌ ♂
02 24	☽ ⚼ ♇
03 01	☽ # ♆
03 57	☽ ∠ ☿
08 15	☽ ± ♄
15 40	☽ # ☿
18 37	☽ ⊼ ♄
19 17	♆ St R
19 56	☽ Q ☉
22 11	☽ △ ♀
23 12	☽ □ ♃
30 Sunday	
02 41	☽ □ ♅
04 11	☽ △ ♇
07 24	☽ ⚹ ☿
08 05	☽ ☍ ♆
09 53	☽ # ♄
10 34	♀ ⊼ ♃
11 57	♂ ⚼ ♇
20 48	☿ △ ♆
21 06	☽ # ☉
31 Monday	
01 52	☽ ⚼ ♀
04 56	☽ Q ☿
05 06	☽ □ ☉
05 41	☽ Q ♃
06 43	☽ ⊻ ♂
09 18	☽ Q ♅
11 25	☽ # ♀
17 27	☽ ⊥ ♂
18 23	☽ ± ♀
19 49	☿ ± ♃
21 58	☽ ☍ ♄

FEBRUARY 1955

LONGITUDES

Date	Sidereal time h m s	Sun ☉ ° ' "	Moon ☽ ° ' "	Moon ☽ 24.00 ° ' "	Mercury ☿ ° '	Venus ♀ ° '	Mars ♂ ° '	Jupiter ♃ ° '	Saturn ♄ ° '	Uranus ♅ ° '	Neptune ♆ ° '	Pluto ♇ ° '
01	20 43 40	11 ♒ 51 37	28 ♉ 47 31	05 ♊ 54 04	29 ♒ 09	25 ♐ 09	12 ♈ 23	22 ♋ 45	20 ♏ 31	24 ♋ 59	28 ♎ 14	25 ♌ 54
02	20 47 36	12 52 29	13 ♊ 01 13	20 ♊ 08 41	29 29	26 13	13 05	22 R 37	20 34	24 R 57	28 R 14	25 R 53
03	20 51 33	13 53 21	27 ♊ 16 08	04 ♋ 23 12	29 39	27 16	13 48	22 30	20 37	24 54	28 14	25 51
04	20 55 29	14 54 11	11 ♋ 26 46	18 ♋ 34 26	29 R 37	28 20	14 31	22 23	20 39	24 52	28 14	25 50
05	20 59 26	15 55 00	25 ♋ 37 40	02 ♌ 38 36	29 25	29 ♐ 24	15 13	22 16	20 42	24 49	28 14	25 48
06	21 03 23	16 55 47	09 ♌ 36 46	16 ♌ 31 37	29 01	00 ♑ 29	15 56	22 10	20 44	24 47	28 13	25 47
07	21 07 19	17 56 34	23 ♌ 22 43	00 ♍ 09 39	28 27	01 34	16 39	22 03	20 46	24 45	28 13	25 45
08	21 11 16	18 57 18	06 ♍ 52 05	13 ♍ 29 45	27 44	02 39	17 21	21 57	20 48	24 42	28 13	25 44
09	21 15 12	19 58 02	20 ♍ 02 30	26 ♍ 30 17	26 51	03 44	18 04	21 50	20 51	24 40	28 13	25 43
10	21 19 09	20 58 44	02 ♎ 53 09	09 ♎ 11 14	25 52	04 49	18 46	21 44	20 52	24 38	28 12	25 41
11	21 23 05	21 59 26	15 ♎ 24 46	21 ♎ 34 06	24 47	05 55	19 29	21 38	20 54	24 35	28 12	25 40
12	21 27 02	23 00 06	27 ♎ 39 38	03 ♏ 41 51	23 39	07 01	20 11	21 32	20 56	24 33	28 11	25 38
13	21 30 58	24 00 45	09 ♏ 41 17	15 ♏ 38 30	22 29	08 07	20 54	21 26	20 58	24 31	28 11	25 37
14	21 34 55	25 01 23	21 ♏ 34 09	27 ♏ 28 52	21 20	09 14	21 36	21 21	20 59	24 29	28 10	25 35
15	21 38 52	26 02 00	03 ♐ 23 19	09 ♐ 18 11	20 12	10 20	22 18	21 15	21 01	24 27	28 10	25 34
16	21 42 48	27 02 36	15 ♐ 14 07	21 ♐ 11 47	19 08	11 27	23 01	21 10	21 02	24 25	28 09	25 32
17	21 46 45	28 03 10	27 ♐ 11 50	03 ♑ 14 50	18 08	12 34	23 43	21 05	21 03	24 23	28 09	25 31
18	21 50 41	29 ♒ 03 43	09 ♑ 21 22	15 ♑ 31 54	17 15	13 41	24 25	20 59	21 05	24 21	28 08	25 29
19	21 54 38	00 ♓ 04 15	21 ♑ 46 54	28 ♑ 06 41	16 28	14 48	25 08	20 55	21 06	24 19	28 07	25 28
20	21 58 34	01 04 46	04 ♒ 31 31	11 ♒ 01 33	15 48	15 56	25 50	20 50	21 07	24 17	28 07	25 26
21	22 02 31	02 05 15	17 ♒ 36 49	24 ♒ 17 14	15 16	17 03	26 32	20 45	21 08	24 15	28 06	25 25
22	22 06 27	03 05 42	01 ♓ 02 37	07 ♓ 52 38	14 51	18 11	27 14	20 41	21 08	24 13	28 05	25 23
23	22 10 24	04 06 07	14 ♓ 46 53	21 ♓ 44 52	14 34	19 19	27 57	20 37	21 09	24 11	28 04	25 22
24	22 14 21	05 06 31	28 ♓ 46 01	05 ♈ 49 44	14 24	20 27	28 39	20 33	21 09	24 09	28 04	25 20
25	22 18 17	06 06 53	12 ♈ 55 21	20 ♈ 02 17	14 D 21	21 35	29 ♈ 21	20 29	21 10	24 07	28 03	25 19
26	22 22 14	07 07 14	27 ♈ 09 55	04 ♉ 17 45	14 25	22 43	00 ♉ 03	20 25	21 10	24 06	28 02	25 17
27	22 26 10	08 07 32	11 ♉ 25 17	18 ♉ 32 09	14 35	23 52	00 45	20 22	21 10	24 04	28 01	25 16
28	22 30 07	09 ♓ 07 48	25 ♉ 38 02	02 ♊ 42 43	14 ♒ 51	25 ♑ 01	01 ♉ 27	20 ♋ 19	21 ♏ 11	24 ♋ 02	28 ♎ 00	25 ♌ 14

MOON

Date	Moon True ☊ ° '	Moon Mean ☊ ° '	Moon ☽ Latitude ° '
01	05 ♑ 00	03 ♑ 45	03 N 05
02	05 D 02	03 42	01 57
03	05 03	03 39	00 N 42
04	05 R 03	03 36	00 S 35
05	05 01	03 32	01 49
06	04 57	03 29	02 56
07	04 51	03 26	03 52
08	04 44	03 23	04 33
09	04 36	03 20	04 58
10	04 28	03 16	05 08
11	04 21	03 13	05 02
12	04 15	03 10	04 43
13	04 12	03 07	04 11
14	04 10	03 04	03 29
15	04 D 11	03 01	02 38
16	04 12	02 57	01 40
17	04 13	02 54	00 S 38
18	04 R 13	02 51	00 N 28
19	04 11	02 48	01 33
20	04 07	02 45	02 35
21	04 00	02 42	03 31
22	03 52	02 38	04 16
23	03 41	02 35	04 48
24	03 31	02 32	05 03
25	03 21	02 29	05 00
26	03 13	02 26	04 38
27	03 08	02 22	03 59
28	03 ♑ 05	02 ♑ 19	03 N 05

DECLINATIONS

Date	Sun ☉ ° '	Moon ☽ ° '	Mercury ☿ ° '	Venus ♀ ° '	Mars ♂ ° '	Jupiter ♃ ° '	Saturn ♄ ° '	Uranus ♅ ° '	Neptune ♆ ° '	Pluto ♇ ° '
01	17 S 14	22 N 54	10 S 37	20 S 04	04 N 52	21 N 58	15 S 41	21 N 41	09 S 14	22 N 45
02	16 57	24 18	10 14	20 10	05 09	21 59	15 41	21 41	09 14	22 45
03	16 40	24 07	09 55	20 15	05 27	22 01	15 42	21 41	09 14	22 46
04	16 22	22 22	09 39	20 19	05 44	22 02	15 42	21 42	09 14	22 47
05	16 04	19 14	09 28	20 23	06 01	22 03	15 42	21 42	09 14	22 47
06	15 46	15 01	09 22	20 27	06 19	22 04	15 43	21 43	09 14	22 48
07	15 27	10 05	09 19	20 30	06 36	22 05	15 43	21 43	09 13	22 49
08	15 09	04 N 46	09 22	20 33	06 53	22 06	15 44	21 43	09 13	22 49
09	14 50	00 S 38	09 28	20 35	07 10	22 08	15 44	21 44	09 13	22 50
10	14 30	05 51	09 39	20 37	07 27	22 09	15 44	21 44	09 13	22 50
11	14 11	10 43	09 54	20 38	07 44	22 10	15 45	21 45	09 13	22 51
12	13 51	15 03	10 11	20 39	08 01	22 11	15 45	21 45	09 12	22 52
13	13 31	18 41	10 31	20 40	08 18	22 12	15 45	21 46	09 12	22 52
14	13 11	21 31	10 53	20 39	08 34	22 13	15 45	21 46	09 12	22 53
15	12 51	23 26	11 16	20 39	08 51	22 14	15 45	21 46	09 12	22 53
16	12 30	24 17	11 39	20 38	09 08	22 14	15 46	21 47	09 11	22 54
17	12 09	24 03	12 03	20 36	09 24	22 15	15 46	21 47	09 11	22 55
18	11 48	22 39	12 26	20 34	09 40	22 16	15 46	21 47	09 11	22 55
19	11 27	20 09	12 49	20 31	09 57	22 17	15 46	21 48	09 11	22 56
20	11 06	16 37	13 11	20 27	10 13	22 18	15 46	21 48	09 10	22 56
21	10 44	12 13	13 31	20 24	10 29	22 18	15 46	21 48	09 10	22 57
22	10 22	07 07	13 49	20 19	10 45	22 19	15 46	21 49	09 10	22 57
23	10 00	01 S 34	14 06	20 14	11 01	22 20	15 46	21 49	09 09	22 58
24	09 38	04 N 08	14 22	20 09	11 16	22 21	15 46	21 49	09 09	22 59
25	09 16	09 42	14 35	20 03	11 32	22 21	15 46	21 50	09 09	22 59
26	08 54	14 47	14 47	19 56	11 48	22 22	15 46	21 50	09 08	23 00
27	08 31	19 03	14 57	19 49	12 03	22 22	15 46	21 50	09 08	23 00
28	08 S 09	22 N 10	15 S 05	19 S 41	12 N 18	22 N 23	15 S 45	21 N 50	09 S 08	23 N 01

ZODIAC SIGN ENTRIES

Date	h	m	Planets
01	14	02	☽ ♊
03	16	36	☽ ♋
05	19	28	☽ ♌
06	01	15	♀ ♑
07	23	43	☽ ♍
10	06	33	☽ ♎
12	16	38	☽ ♏
15	05	07	☽ ♐
17	17	34	☽ ♑
19	10	19	☉ ♓
20	03	33	☽ ♒
22	10	09	☽ ♓
24	14	06	☽ ♈
26	10	22	♂ ♉
26	16	46	☽ ♉
28	19	24	☽ ♊

LATITUDES

Date	Mercury ☿ ° '	Venus ♀ ° '	Mars ♂ ° '	Jupiter ♃ ° '	Saturn ♄ ° '	Uranus ♅ ° '	Neptune ♆ ° '	Pluto ♇ ° '
01	01 N 14	03 N 17	00 S 02	00 N 27	02 N 17	00 N 33	01 N 44	10 N 31
04	02 05	03 07	00 N 01	00 27	02 18	00 33	01 44	10 31
07	02 52	02 56	00 04	00 27	02 19	00 33	01 44	10 31
10	03 27	02 45	00 06	00 28	02 19	00 33	01 44	10 32
13	03 42	02 32	00 09	00 28	02 20	00 33	01 44	10 32
16	03 37	02 20	00 11	00 28	02 21	00 33	01 45	10 32
19	03 14	02 07	00 14	00 28	02 21	00 33	01 45	10 32
22	02 40	01 55	00 16	00 28	02 22	00 33	01 45	10 33
25	02 02	01 41	00 19	00 29	02 22	00 33	01 45	10 33
28	01 22	01 28	00 21	00 29	02 23	00 33	01 45	10 33
31	00 N 43	01 N 15	00 N 23	00 N 29	02 N 24	00 N 32	01 N 45	10 N 33

DATA

Julian Date	2435140
Delta T	+31 seconds
Ayanamsa	23° 14' 05"
Synetic vernal point	05° ♓ 52' 55"
True obliquity of ecliptic	23° 26' 43"

MOON'S PHASES, APSIDES AND POSITIONS ☽

Date	h	m	Phase	Longitude ° '	Eclipse Indicator
07	01	43	○	17 ♌ 31	
14	19	40	☾	25 ♏ 21	
22	15	54	●	03 ♓ 16	

Day	h	m		
02	19	38	Perigee	
15	00	12	Apogee	
27	13	12	Perigee	
02	21	18	Max dec	24° N 25'
09	09	11	0S	
16	18	47	Max dec	24° S 20'
23	18	37	0N	

LONGITUDES

Date	Chiron ⚷ ° '	Ceres ⚳ ° '	Pallas ⚴ ° '	Juno ⚵ ° '	Vesta ⚶ ° '	Black Moon Lilith ⚸ ° '
01	00 ♒ 25	00 ♑ 16	12 ♐ 17	15 ♏ 31	16 ♊ 19	25 ♏ 54
11	01 ♒ 15	03 ♑ 56	15 ♐ 40	16 ♏ 56	16 ♊ 33	27 ♏ 00
21	02 ♒ 03	07 ♑ 27	18 ♐ 48	17 ♏ 54	17 ♊ 28	28 ♏ 07
31	02 ♒ 48	10 ♑ 47	21 ♐ 35	18 ♏ 21	19 ♊ 00	29 ♏ 14

All ephemeris data is given at 12.00 UT and the Moon's longitude is additionally given for 24.00 UT

ASPECTARIAN

01 Tuesday
00 16 ☽ ∥ ♅
01 51 ☽ ⚹ ♃
02 45 ☽ ∥ ♃
05 21 ☽ ⊼ ♀
05 35 ☽ ⚹ ♅
07 07 ☽ □ ♇
08 16 ♀ ⊼ ♅
09 29 ☽ ∠ ♂
10 22 ☽ ∥ ♇
11 04 ☽ ⊼ ♆
12 37 ☽ □ ☿
21 12 ☽ ± ♆

02 Wednesday
02 59 ☽ ∠ ♃
04 39 ♀ △ ♇
06 50 ☽ ∠ ♅
11 44 ☽ △ ☉
12 07 ☽ ⚹ ♂
12 22 ☽ Ᵽ ♆
13 27 ☽ Q ♇
18 01 ☽ ⊥ ♃
20 26 ♀ ⊥ ♄
21 57 ☽ ⊥ ♅

03 Thursday
00 45 ☽ ⊼ ♄
04 02 ☽ ⊻ ♃
05 00 ☉ ⚹ ♂
08 01 ☽ ⊻ ♅
09 24 ☽ Q ♂
09 37 ☽ ⚹ ♇
10 53 ☽ ± ♄
12 00 ☽ ☍ ♀
13 38 ☽ △ ♆
14 56 ☽ Ᵽ ☉
16 01 ☽ △ ☿
20 54 ☿ St R

04 Friday
02 06 ☽ Ᵽ ♄
07 17 ☽ ± ☉
07 51 ☽ ∥ ♇
09 40 ♀ ⚹ ♆
10 53 ☽ ∠ ♇
15 02 ☿ ∠ ♂
15 07 ☽ ∥ ♃
17 03 ♂ ± ♄
17 15 ☽ Ᵽ ☿
17 23 ☽ □ ♂
18 01 ☽ ∥ ♅
18 13 ☽ ⊼ ☉

05 Saturday
02 07 ☽ ⊥ ♇
03 35 ☽ △ ♄
04 20 ☽ # ♀
06 20 ☽ ☌ ♃
08 18 ☽ ± ☿
10 38 ☽ ☌ ♅
12 03 ☿ ⚹ ♀
12 18 ☽ ⊻ ♇
16 26 ☽ □ ♆
18 19 ☽ ⊼ ☿
18 59 ☽ ⊼ ♀

06 Sunday
06 09 ☽ ± ♀
07 46 ☽ # ☉
08 20 ☽ # ♄
15 55 ☉ ∥ ♄
23 03 ☽ Ᵽ ♀
23 28 ☽ Q ♆
23 34 ☽ △ ♂

07 Monday
01 43 ☽ ☍ ☉
07 24 ☽ □ ♄
09 41 ☽ ⊻ ♃
14 24 ☽ ⊻ ♅
15 29 ☽ # ☿
15 57 ☽ # ♆
16 11 ☽ ☌ ♇
20 11 ☽ ⊥ ♃
20 19 ☿ △ ♆
20 32 ☽ ☍ ☿
20 33 ☽ ⚹ ♆

08 Tuesday
01 00 ☽ ⊥ ♅
03 01 ☽ ∥ ♂
03 27 ☽ Ᵽ ♂
03 46 ☽ △ ♀
04 11 ☿ ± ♃
12 08 ☽ ∠ ♃
15 30 ☽ Q ♄
17 06 ☽ ∠ ♅
20 34 ☽ ± ♂
23 29 ☽ ∠ ♆

09 Wednesday
08 09 ☽ ⊼ ♂
11 51 ☽ ⊼ ☉
13 29 ☽ ⚹ ♄
15 18 ☽ ⚹ ♃
16 00 ☽ ⊥ ♆
20 33 ☽ ⚹ ♅
22 30 ☽ ⊻ ♇
23 47 ☽ ⊼ ☿
23 56 ☽ ± ☉

10 Thursday
03 11 ☽ ⊻ ♆
09 27 ☉ □ ♄
09 44 ☽ ⊥ ♇
10 13 ☽ ± ☿
13 36 ☽ Q ♃
16 01 ☽ □ ♀
16 21 ☿ ☍ ♇
17 41 ☽ ∠ ♄
18 23 ☽ Ᵽ ☉
19 05 ☽ Q ♅
20 07 ☽ # ♂

11 Friday
02 02 ☽ Ᵽ ☿
02 51 ☽ ∠ ♇
04 17 ☉ ⊼ ♃
04 19 ☽ ∥ ♆
07 31 ☽ ∥ ☿
11 01 ☽ ⊥ ♄
11 43 ♀ ∠ ♄
16 27 ☿ ⊼ ♅
20 24 ☽ ☍ ♂
22 44 ☽ ⊻ ♄

12 Saturday
00 02 ☽ □ ♃
01 59 ☽ △ ☉
04 47 ☽ △ ☿
05 34 ☽ ∥ ☉
05 53 ☽ □ ♅
06 16 ☽ Q ♀
08 00 ☽ ⚹ ♇
13 03 ☽ ☌ ♆
16 18 ☽ ∥ ♄
19 12 ☉ ☌ ☿

13 Sunday
05 19 ☿ ∠ ♀
07 50 ☽ Q ♇
08 32 ☽ ⚹ ♀
14 28 ♂ ⊼ ♄
23 34 ☉ ⊼ ♅

14 Monday
03 50 ☽ ∥ ♀
04 16 ♂ □ ♃
08 27 ☿ ⚹ ♂
10 49 ☽ ☌ ♄
11 33 ☽ △ ♃
11 33 ☽ □ ☿
11 36 ☿ ⊼ ♃
12 04 ☽ ⊼ ♂
14 29 ☽ # ♅
17 53 ☽ △ ♅
17 57 ☽ ∠ ♀
18 53 ☿ □ ♄
19 24 ☽ # ♃
19 40 ☽ □ ☉
20 08 ☽ □ ♇

15 Tuesday
01 01 ☽ ± ♂
01 02 ☉ ☍ ♇
01 24 ☽ ⊻ ♆
03 48 ☽ # ♇
08 19 ♀ Q ♆
13 34 ☽ ⊥ ♆
14 07 ☽ ⊥ ♀
16 45 ♀ Ᵽ ♇
17 46 ☽ Ᵽ ♃
20 27 ☽ Ᵽ ♂
20 56 ☽ Q ☿

16 Wednesday
00 15 ☽ Ᵽ ♅
03 33 ☽ ⊻ ♀
07 48 ☽ ∠ ♆
11 35 ☽ Q ☉
11 51 ☽ ± ♃
14 35 ☉ ± ♃
17 40 ♂ # ♆
18 23 ☽ ± ♅
19 13 ☽ ⚹ ☿
23 42 ☽ ⊻ ♄
23 51 ☽ ⊼ ♃

17 Thursday
04 37 ☽ △ ♂
06 23 ☽ ⊼ ♅
07 02 ☿ ⊥ ♀
08 39 ☽ △ ♇
11 43 ☽ ⊥ ♄
13 52 ☽ ⚹ ☉
13 53 ☽ ⚹ ♆
14 09 ☉ △ ♆
15 20 ☉ ∥ ☿
15 55 ♃ △ ♄
22 57 ☽ ∠ ☿

18 Friday
05 34 ☽ ∠ ♄
08 38 ☽ # ♇
09 23 ♂ □ ♅
13 31 ☽ Q ♆
14 12 ☽ Ᵽ ♇
15 27 ☽ ⊥ ☿
16 29 ☽ # ♃
21 16 ☽ ☌ ♀
21 32 ☽ # ♅
21 58 ☽ ∠ ☉

19 Saturday
02 23 ☽ ⊻ ☿
07 34 ☽ ± ♇
09 03 ☽ ∥ ♀
10 21 ☽ ☍ ♃
10 41 ☽ ⚹ ♄
16 44 ☽ ⊥ ☉
16 48 ☽ ☍ ♅
17 31 ☉ ± ♅
18 44 ☽ □ ♂
18 59 ☽ ⊼ ♇
22 58 ♂ △ ♇

20 Sunday
00 01 ☽ □ ♆
05 01 ☽ ⊻ ☉
09 22 ☽ Q ♄
10 17 ☿ ⊻ ♀
17 02 ☽ ∥ ♄

21 Monday
05 48 ☽ ∥ ☿
06 06 ☽ Q ♂
07 54 ☽ ☌ ☿
10 54 ☽ ⊻ ♀
17 38 ☽ ⊼ ♃
18 20 ☽ □ ♄
18 29 ☉ ∠ ♀
19 49 ☽ ∥ ☉
20 03 ☽ # ♂
21 43 ☉ # ♂
22 42 ☽ ⊥ ♀
23 54 ☽ ⊼ ♅

22 Tuesday
01 59 ☽ ☍ ♇
02 40 ☽ ∥ ♆
03 28 ☿ Q ♂
04 19 ☽ ± ♃
04 53 ☽ ⚹ ♂
06 46 ☽ △ ♆
10 32 ☽ ± ♅
15 54 ☽ ☌ ☉
16 07 ☽ ∠ ♀
20 07 ☽ Ᵽ ♃

23 Wednesday
02 18 ☽ Ᵽ ♅
08 39 ☽ ∠ ♂
09 03 ☽ Ᵽ ♆
11 39 ☽ ⊻ ☿
12 58 ♀ ± ♇
16 25 ♂ ☍ ♆
20 31 ☽ ⚹ ♀
21 50 ☽ ⊥ ☿
22 00 ☽ △ ♃
22 59 ☽ △ ♄

24 Thursday
00 33 ☽ ± ♆
00 59 ☽ ⊥ ♂
04 08 ☽ △ ♅
06 10 ☽ ⊼ ♇
10 48 ☽ ⊼ ♆
11 47 ☽ ⊻ ♂
13 05 ☽ ∠ ☿
13 58 ♀ ☍ ♃
16 22 ☽ ± ♇
18 49 ☽ Q ♀
21 52 ☉ Ᵽ ♃
23 36 ☽ ⊻ ☉

25 Friday
00 34 ☽ Ᵽ ♄
03 03 ♀ ⚹ ♄
07 36 ☽ Ᵽ ♇
09 33 ☽ # ♆
10 13 ☽ # ☉
10 18 ☿ St D
10 32 ☽ ⊥ ☉
14 25 ☽ ⚹ ☿
15 47 ☽ ± ♄
20 13 ☉ ∥ ♆
20 46 ☽ ∥ ♂

26 Saturday
00 42 ☽ □ ♃
01 54 ☽ ⊼ ♄
02 52 ☽ ∠ ☉
03 53 ☽ □ ♀
06 51 ☽ □ ♅
08 51 ☽ △ ♇
10 44 ☽ Q ☿
12 00 ☽ # ☿
13 27 ☽ ☍ ♆
17 04 ☽ # ♄
17 06 ☽ ☌ ♂

27 Sunday
06 02 ☽ ⚹ ☉
06 53 ☽ Q ♃
13 05 ☽ Q ♅
16 09 ♀ ☍ ♅
16 55 ☽ # ♀
17 25 ☽ □ ☿

28 Monday
03 02 ☽ ⚹ ♃
03 48 ☽ Q ☉
04 28 ☽ ☍ ♄
08 52 ☽ ∥ ♅
09 19 ☽ ⚹ ♅
09 55 ☉ Ᵽ ♅
10 51 ☽ △ ♀
11 20 ☽ □ ♇
14 05 ☽ ∥ ♃
16 00 ☽ ⊼ ♆
16 47 ♀ ⊼ ♇
21 11 ☽ ∥ ♇
22 22 ☽ ⊻ ♂

MARCH 1955

Date	Sidereal time h m s	Sun ☉ ° ' "	Moon ☽ ° ' "	Moon ☽ 24.00 ° ' "	Mercury ☿ ° '	Venus ♀ ° '	Mars ♂ ° '	Jupiter ♃ ° '	Saturn ♄ ° '	Uranus ♅ ° '	Neptune ♆ ° '	Pluto ⯓ ° '
01	22 34 03	10 ♓ 08 03	09 ♊ 46 00	16 ♊ 47 48	15 ♒ 13	26 ♑ 09	02 ♉ 09	20 ♋ 16	21 ♏ 11	24 ♋ 01	27 ♎ 59	25 ♌ 13
02	22 38 00	11 08 15	23 ♊ 48 02	00 ♋ 46 39	15 40	27 18	02 51	20 R 13	21 R 11	23 R 59	27 R 58	25 R 12
03	22 41 56	12 08 25	07 ♋ 43 37	14 ♋ 38 53	16 11	28 27	03 33	20 10	21 10	23 58	27 57	25 10
04	22 45 53	13 08 33	21 ♋ 32 22	28 ♋ 23 58	16 47	29 ♑ 36	04 15	20 08	21 10	23 56	27 56	25 09
05	22 49 50	14 08 39	05 ♌ 13 34	12 ♌ 00 59	17 28	00 ♒ 45	04 57	20 05	21 10	23 55	27 55	25 07
06	22 53 46	15 08 43	18 ♌ 46 02	25 ♌ 28 29	18 12	01 54	05 38	20 03	21 09	23 54	27 54	25 06
07	22 57 43	16 08 45	02 ♍ 08 05	08 ♍ 44 37	19 00	03 04	06 20	20 01	21 09	23 52	27 53	25 05
08	23 01 39	17 08 44	15 ♍ 17 50	21 ♍ 47 33	19 51	04 13	07 02	20 00	21 08	23 51	27 52	25 03
09	23 05 36	18 08 42	28 ♍ 13 35	04 ♎ 35 52	20 45	05 23	07 44	19 58	21 07	23 50	27 51	25 02
10	23 09 32	19 08 38	10 ♎ 54 19	17 ♎ 08 59	21 42	06 32	08 25	19 57	21 06	23 49	27 49	25 01
11	23 13 29	20 08 33	23 ♎ 19 58	29 ♎ 27 27	22 42	07 42	09 07	19 56	21 05	23 47	27 48	24 59
12	23 17 25	21 08 25	05 ♏ 31 42	11 ♏ 33 02	23 45	08 52	09 49	19 55	21 04	23 46	27 47	24 58
13	23 21 22	22 08 16	17 ♏ 31 53	23 ♏ 28 42	24 50	10 02	10 30	19 54	21 03	23 45	27 46	24 57
14	23 25 19	23 08 05	29 ♏ 24 00	05 ♐ 18 24	25 58	11 12	11 12	19 53	21 02	23 44	27 45	24 55
15	23 29 15	24 07 52	11 ♐ 12 29	17 ♐ 06 56	27 07	12 22	11 53	19 53	21 00	23 43	27 43	24 54
16	23 33 12	25 07 38	23 ♐ 02 24	28 ♐ 59 37	28 19	13 32	12 35	19 53	20 59	23 43	27 42	24 53
17	23 37 08	26 07 22	04 ♑ 59 15	11 ♑ 02 00	29 ♒ 32	14 42	13 16	19 D 53	20 57	23 42	27 41	24 52
18	23 41 05	27 07 04	17 ♑ 08 32	23 ♑ 19 30	00 ♓ 48	15 53	13 58	19 53	20 56	23 41	27 39	24 50
19	23 45 01	28 06 44	29 ♑ 35 29	05 ♒ 56 58	02 05	17 03	14 39	19 54	20 54	23 40	27 38	24 49
20	23 48 58	29 ♓ 06 23	12 ♒ 24 24	18 ♒ 58 05	03 24	18 14	15 20	19 54	20 52	23 40	27 37	24 48
21	23 52 54	00 ♈ 06 00	25 ♒ 38 12	02 ♓ 24 46	04 45	19 24	16 02	19 55	20 50	23 39	27 35	24 47
22	23 56 51	01 05 35	09 ♓ 17 40	16 ♓ 16 36	06 08	20 35	16 43	19 56	20 48	23 38	27 34	24 46
23	00 00 48	02 05 08	23 ♓ 21 06	00 ♈ 30 32	07 32	21 45	17 24	19 57	20 46	23 38	27 32	24 45
24	00 04 44	03 04 39	07 ♈ 44 08	15 ♈ 01 01	08 57	22 56	18 05	19 58	20 44	23 37	27 31	24 43
25	00 08 41	04 04 08	22 ♈ 20 15	29 ♈ 40 51	10 24	24 07	18 47	20 00	20 41	23 37	27 29	24 42
26	00 12 37	05 03 34	07 ♉ 01 51	14 ♉ 22 19	11 53	25 18	19 28	20 02	20 39	23 37	27 28	24 41
27	00 16 34	06 02 59	21 ♉ 41 26	28 ♉ 58 29	13 23	26 29	20 09	20 04	20 36	23 36	27 26	24 40
28	00 20 30	07 02 22	06 ♊ 12 53	13 ♊ 24 12	14 54	27 39	20 50	20 06	20 34	23 36	27 25	24 39
29	00 24 27	08 01 42	20 ♊ 32 07	27 ♊ 36 26	16 27	28 ♒ 50	21 31	20 08	20 31	23 36	27 23	24 38
30	00 28 23	09 01 00	04 ♋ 37 04	11 ♋ 34 01	18 01	00 ♓ 01	22 12	20 11	20 28	23 36	27 22	24 37
31	00 32 20	10 ♈ 00 15	18 ♋ 27 20	25 ♋ 17 09	19 ♓ 37	01 ♓ 13	22 ♉ 53	20 ♋ 13	20 ♏ 26	23 ♋ 36	27 ♎ 20	24 ♌ 36

Date	Moon True ☊ ° '	Moon Mean ☊ ° '	Moon ☽ Latitude ° '
01	03 ♑ 04	02 ♑ 16	02 N 01
02	03 D 05	02 13	00 N 49
03	03 R 05	02 10	00 S 25
04	03 04	02 07	01 36
05	03 00	02 03	02 42
06	02 53	02 00	03 37
07	02 44	01 57	04 20
08	02 32	01 54	04 48
09	02 19	01 51	05 00
10	02 06	01 48	04 57
11	01 54	01 44	04 40
12	01 44	01 41	04 11
13	01 37	01 38	03 30
14	01 33	01 35	02 41
15	01 30	01 32	01 45
16	01 D 30	01 28	00 S 45
17	01 R 30	01 25	00 N 18
18	01 29	01 22	01 22
19	01 27	01 19	02 23
20	01 22	01 16	03 18
21	01 15	01 13	04 05
22	01 05	01 09	04 39
23	00 53	01 06	04 58
24	00 41	01 03	04 58
25	00 29	01 00	04 39
26	00 20	00 57	04 02
27	00 13	00 54	03 08
28	00 09	00 50	02 03
29	00 08	00 47	00 N 51
30	00 D 08	00 44	00 S 24
31	00 ♑ 07	00 ♑ 41	01 S 35

DECLINATIONS

Date	Sun ☉ ° '	Moon ☽ ° '	Mercury ☿ ° '	Venus ♀ ° '	Mars ♂ ° '	Jupiter ♃ ° '	Saturn ♄ ° '	Uranus ♅ ° '	Neptune ♆ ° '	Pluto ⯓ ° '
01	07 S 46	23 N 55	15 S 11	19 S 33	12 N 34	22 N 23	15 S 45	21 N 51	09 S 07	23 N 01
02	07 23	24 07	15 15	19 24	12 49	22 24	15 45	21 51	09 07	23 02
03	07 01	22 49	15 18	19 15	13 04	22 24	15 45	21 51	09 07	23 02
04	06 38	20 08	15 19	19 05	13 19	22 25	15 44	21 51	09 06	23 03
05	06 14	16 21	15 18	18 55	13 33	22 25	15 44	21 52	09 06	23 03
06	05 51	11 46	15 15	18 44	13 48	22 25	15 44	21 52	09 05	23 03
07	05 28	06 40	15 11	18 32	14 02	22 26	15 43	21 52	09 05	23 04
08	05 05	01 N 22	15 05	18 20	14 17	22 26	15 43	21 52	09 04	23 04
09	04 41	03 S 53	14 58	18 08	14 31	22 26	15 43	21 52	09 04	23 05
10	04 18	08 52	14 49	17 55	14 45	22 27	15 42	21 53	09 03	23 05
11	03 54	13 24	14 39	17 41	14 59	22 27	15 42	21 53	09 03	23 06
12	03 31	17 18	14 27	17 27	15 13	22 27	15 41	21 53	09 02	23 06
13	03 07	20 26	14 13	17 13	15 27	22 27	15 41	21 53	09 02	23 06
14	02 44	22 39	13 58	16 58	15 40	22 27	15 40	21 53	09 02	23 07
15	02 20	23 52	13 42	16 42	15 54	22 27	15 40	21 53	09 01	23 07
16	01 56	24 00	13 24	16 26	16 07	22 27	15 39	21 54	09 01	23 08
17	01 32	23 03	13 05	16 10	16 20	22 27	15 39	21 54	09 00	23 08
18	01 09	21 00	12 44	15 53	16 33	22 27	15 38	21 54	08 59	23 08
19	00 45	17 55	12 22	15 36	16 46	22 27	15 37	21 54	08 59	23 09
20	00 S 21	13 55	11 59	15 18	16 59	22 27	15 37	21 54	08 58	23 09
21	00 N 02	09 08	11 34	15 00	17 12	22 27	15 36	21 54	08 58	23 09
22	00 26	03 S 46	11 08	14 41	17 24	22 27	15 35	21 54	08 57	23 10
23	00 50	01 N 55	10 41	14 22	17 36	22 27	15 35	21 54	08 57	23 10
24	01 13	07 38	10 12	14 02	17 48	22 27	15 34	21 54	08 56	23 10
25	01 37	13 01	09 43	13 42	18 00	22 26	15 33	21 54	08 56	23 10
26	02 01	17 40	09 13	13 22	18 12	22 26	15 32	21 54	08 55	23 11
27	02 24	21 13	08 39	13 01	18 24	22 26	15 32	21 54	08 55	23 11
28	02 48	23 22	08 06	12 40	18 35	22 26	15 31	21 54	08 54	23 11
29	03 11	23 57	07 31	12 19	18 47	22 25	15 30	21 54	08 54	23 11
30	03 35	22 58	06 56	11 57	18 58	22 25	15 29	21 54	08 53	23 12
31	03 N 58	20 N 36	06 S 19	11 S 35	19 N 09	22 N 25	15 S 28	21 N 54	08 S 52	23 N 12

ZODIAC SIGN ENTRIES

Date	h	m	Planets
02	22	40	☽ ♋
04	20	22	♀ ♒
05	02	48	☽ ♌
07	08	09	☽ ♍
09	15	20	☽ ♎
12	01	04	☽ ♏
14	13	13	☽ ♐
17	02	01	☽ ♑
17	20	49	☿ ♓
19	12	47	☽ ♒
21	09	35	☉ ♈
21	19	45	☽ ♓
23	23	09	☽ ♈
26	00	31	☽ ♉
28	01	42	☽ ♊
30	04	05	☽ ♋
30	11	30	♀ ♓

LATITUDES

Date	Mercury ☿ ° '	Venus ♀ ° '	Mars ♂ ° '	Jupiter ♃ ° '	Saturn ♄ ° '	Uranus ♅ ° '	Neptune ♆ ° '	Pluto ⯓ ° '
01	01 N 09	01 N 24	00 N 22	00 N 29	02 N 23	00 N 32	01 N 45	10 N 33
04	00 31	01 11	00 24	00 29	02 24	00 32	01 45	10 33
07	00 S 03	00 58	00 26	00 29	02 25	00 32	01 46	10 33
10	00 35	00 45	00 28	00 29	02 25	00 32	01 46	10 32
13	01 02	00 33	00 30	00 29	02 26	00 32	01 46	10 32
16	01 25	00 20	00 32	00 29	02 26	00 32	01 46	10 32
19	01 45	00 N 09	00 33	00 29	02 27	00 32	01 46	10 31
22	02 01	00 S 03	00 35	00 29	02 27	00 32	01 46	10 31
25	02 12	00 14	00 37	00 29	02 28	00 32	01 46	10 31
28	02 20	00 24	00 39	00 30	02 28	00 32	01 46	10 31
31	02 S 23	00 S 34	00 N 40	00 N 30	02 N 29	00 N 32	01 N 46	10 N 30

DATA

Julian Date	2435168
Delta T	+31 seconds
Ayanamsa	23° 14' 09"
Synetic vernal point	05° ♓ 52' 51"
True obliquity of ecliptic	23° 26' 43"

MOON'S PHASES, APSIDES AND POSITIONS ☽

Date	h	m	Phase	Longitude ° '	Eclipse Indicator
01	12	40	☽	10 ♊ 10	
08	15	41	○	17 ♍ 18	
16	16	36	☾	25 ♐ 19	
24	03	42	●	02 ♈ 44	
30	20	10	☽	09 ♋ 21	

Day	h	m		
14	20	33	Apogee	
26	15	22	Perigee	
02	03	13	Max dec	24° N 13'
08	18	12	0S	
16	03	05	Max dec	24° S 05'
23	03	59	0N	
29	08	44	Max dec	23° N 58'

LONGITUDES

Date	Chiron ⚷ ° '	Ceres ⚳ ° '	Pallas ⚴ ° '	Juno ⚵ ° '	Vesta ⚶ ° '	Black Moon Lilith ⚸ ° '
01	02 ♒ 39	10 ♑ 08	21 ♐ 04	18 ♏ 18	18 ♊ 39	29 ♏ 00
11	03 ♒ 21	13 ♑ 16	23 ♐ 33	18 ♏ 19	20 ♊ 35	00 ♐ 07
21	03 ♒ 58	16 ♑ 09	25 ♐ 35	17 ♏ 46	22 ♊ 59	01 ♐ 14
31	04 ♒ 30	18 ♑ 44	27 ♐ 06	16 ♏ 38	25 ♊ 46	02 ♐ 20

ASPECTARIAN

h m	Aspects
01 Tuesday	
02 10	☽ ± ♆
04 22	☽ ∠ ♃
06 19	♄ St R
09 06	☽ ⊥ ♂
10 43	☽ ∠ ♅
12 40	☽ □ ☉
14 34	☽ ♇ ♀
17 29	☽ ♇ ♆
17 53	☽ Q ⯓
19 39	☽ ⊥ ♃
21 35	☽ △ ☿
02 Wednesday	
01 16	☽ ∠ ♂
02 03	☽ ⊥ ♅
05 52	☽ ⊻ ♃
07 20	☽ ± ♀
07 30	☽ ⊼ ♄
12 19	☽ ⊻ ♅
14 23	☽ ⚹ ⯓
17 48	☽ ± ♄
18 33	☽ ⊼ ♀
19 09	☽ △ ♆
03 Thursday	
00 15	☽ ♇ ☿
01 47	♀ □ ♆
04 24	☽ ⚹ ♂
09 14	☽ ∥ ⯓
09 19	☽ ♇ ♄
16 13	☽ ∠ ⯓
16 26	☽ ± ☿
16 29	☽ ∥ ♃
20 15	☽ △ ☉
21 56	☽ ∥ ♅
04 Friday	
02 17	☽ Q ♂
03 20	☽ ⊼ ☿
07 07	☉ ♇ ♆
07 50	☽ ⊥ ⯓
09 33	☽ ☌ ♃
11 21	☽ △ ♄
16 11	☽ ☌ ♅
18 18	☽ ⊻ ⯓
19 41	☽ # ♀
23 10	☽ □ ♆
05 Saturday	
00 28	☽ ♇ ☉
03 24	☽ ☍ ♀
11 28	☽ □ ♂
15 26	☽ # ♄
17 33	☽ ± ☉
17 51	☽ # ☿
06 Sunday	
02 15	☽ ∥ ♂
05 02	☽ ⊼ ☉
06 54	☽ Q ♆
10 55	☽ ☍ ☿
14 17	☽ ⊻ ♃
16 15	☽ □ ♄
21 09	☽ ⊻ ♅
23 19	☽ ☌ ⯓
07 Monday	
00 50	☽ # ♆
01 01	☽ ⊥ ♃
04 20	☽ ⚹ ♆
07 55	☽ ⊥ ♅
13 42	♀ Q ♄
13 50	☽ ⊼ ♀
17 13	☽ ∠ ♃
17 56	☽ # ☉
20 02	☽ △ ♂
08 Tuesday	
00 13	☽ ∠ ♅
00 43	☽ Q ♄
01 48	☽ ± ♀
07 32	☽ ∠ ♆
15 41	☽ ☍ ☉
15 55	☿ ⊼ ♃
19 56	☽ ♇ ♀
20 39	☽ ⚹ ♃
21 01	☽ ⊼ ☿
22 46	☽ ⚹ ♄
09 Wednesday	
00 07	☽ ⊥ ♆
01 09	☽ ♇ ♂
03 48	☽ ⚹ ♅
06 02	☽ ⊻ ⯓
09 01	☽ ± ☿
11 17	☽ ⊻ ♆
15 30	☽ ∥ ☉
17 16	☽ ⊥ ⯓
18 57	☽ ± ♂
19 01	☽ Q ♃
20 05	♂ Q ♃
21 16	☿ □ ♄
10 Thursday	
02 19	☽ Q ♅
02 50	☽ △ ♀
02 52	☽ ∠ ♄
03 20	☽ ♇ ☿
06 59	☽ ⊼ ♂
10 18	☽ ∠ ⯓
12 55	☽ ∥ ♆
15 54	☿ # ♂
20 03	☽ ⊥ ♄
11 Friday	
05 15	☽ ⊼ ☉
05 23	☽ □ ♃
06 55	☉ △ ♃
07 38	☽ ⊻ ♄
10 40	☽ △ ☿
12 54	☽ □ ♅
15 14	☽ ⚹ ⯓
17 59	☽ ± ☉
18 51	☽ ∥ ☿
20 44	☽ ☌ ♆
21 48	☽ # ♂
12 Saturday	
01 32	☽ ∥ ♄
10 20	☉ △ ♄
12 29	☿ ⊼ ♅
12 57	☽ ∥ ♀
13 19	☽ ♇ ☉
14 51	☽ Q ⯓
19 21	☽ □ ♀
21 03	☽ ☍ ♂
13 Sunday	
03 10	☉ ± ♆
14 19	☿ ☍ ⯓
16 46	☽ △ ♃
19 05	☽ ☌ ♄
22 08	☽ △ ☉
14 Monday	
00 33	☽ △ ♅
02 30	☽ # ♅
02 56	☽ □ ⯓
04 17	☽ □ ☿
08 38	☽ ⊻ ♆
09 20	☽ # ♃
10 33	☿ ± ♃
11 32	☽ Q ♀
11 52	♂ # ♄
11 54	♀ □ ♂
19 02	☽ # ⯓
20 48	☽ ⊥ ♆
23 09	☽ ♇ ♃
15 Tuesday	
02 21	☉ △ ♅
06 29	♂ Q ♅
06 57	☽ ♇ ♅
13 28	☽ ⊼ ♂
14 37	☽ ⚹ ♀
15 04	☽ ∠ ♆
17 26	☽ ± ♃
20 50	☽ Q ☿
23 57	☿ △ ♆
16 Wednesday	
01 13	☽ ± ♅
02 24	☽ ± ♂
05 36	☽ ⊼ ♃
06 11	☉ ⊼ ⯓
07 51	☽ ⊻ ♄
13 21	☽ ⊼ ♅
15 43	☽ △ ⯓
16 36	☽ □ ☉
19 56	☽ ⊥ ♄
20 34	♃ St D
21 23	☽ ⚹ ♆
21 43	☽ ♇ ♂
23 51	☽ ⚹ ☿
17 Thursday	
00 17	☽ ∠ ♀
03 34	♀ # ♂
10 34	☽ # ⯓
13 55	☽ ∠ ♄
14 57	☿ ± ♅
20 11	☽ ⊥ ♀
20 18	☽ # ♃
21 18	☽ Q ♆
21 40	☽ ♇ ⯓
18 Friday	
02 53	☽ # ♅
05 23	☽ △ ♂
07 41	☽ Q ☉
09 04	☽ ∠ ☿
09 16	☽ ⊻ ♀
15 18	☽ ± ⯓
17 21	☽ ☍ ♃
19 21	☽ ⚹ ♄
19 Saturday	
00 39	☉ ⊼ ♆
00 41	☽ ☍ ♅
02 54	☽ ⊼ ⯓
04 32	☽ ⊥ ☿
08 16	☽ □ ♆
08 56	☽ ⚹ ☉
09 31	♀ ∥ ♄
17 17	☽ ⊻ ☿
18 15	☽ Q ♄
19 01	☽ # ♂
20 Sunday	
02 23	☽ ∥ ♄
03 38	☽ ∥ ♀
09 24	☿ Q ♂
15 23	☽ ∠ ☉
17 41	☽ □ ♂
23 04	☽ ∥ ☿
23 41	☽ ☌ ♀
21 Monday	
01 42	☽ ⊼ ♃
03 24	☽ □ ♄
08 27	☽ ⊼ ♅
09 02	☽ ⊥ ☉
10 28	☽ ☍ ⯓
12 30	☽ ± ♃
12 46	☽ ∥ ♆
14 53	☿ ♇ ♃
15 28	☽ △ ♆
19 07	☽ ± ♅
20 33	☽ ⊻ ☉
22 38	♀ ⊼ ♃
22 Tuesday	
03 37	☽ Q ♂
04 08	☉ ± ⯓
04 24	☽ ♇ ♃
05 53	☽ ☌ ☿
10 52	☽ ♇ ♅
16 24	♀ □ ♄
17 38	☽ ♇ ♆
23 Wednesday	
01 15	☽ # ☉
01 24	☽ ⚹ ♂
06 15	☽ △ ♃
07 07	☽ ∥ ☉
07 38	☽ △ ♄
08 57	☽ ± ♆
09 04	☽ ⊻ ♀
12 28	☽ △ ♅
14 20	☽ ⊼ ⯓
19 01	☽ ⊼ ♆
20 04	☽ ⊥ ♀
24 Thursday	
00 22	☽ ± ⯓
03 42	☽ ☌ ☉
03 55	☽ ∠ ♂
06 31	☿ ♇ ♅
08 41	☽ ♇ ♄
12 21	☽ ∠ ♀
12 59	☉ ∠ ♂
14 14	☽ ⊻ ☿
15 17	☽ ♇ ⯓
17 37	☽ # ♆
19 32	☽ ⊥ ♂
22 16	☽ # ☿
23 30	☽ ± ♄
25 Friday	
01 12	☽ ⊥ ☿
01 57	♀ ⊼ ♅
05 53	☽ ⊻ ♂
08 10	☽ □ ♃
09 18	☽ ⊼ ♄
14 05	☽ □ ♅
15 07	☽ # ♀
15 10	☽ ⚹ ♀
15 52	☽ △ ⯓
17 34	☽ ∠ ☿
20 24	☽ ☍ ♆
23 51	♀ ☍ ⯓
26 Saturday	
00 29	☽ # ♄
08 33	☽ ⊻ ☉
12 28	☽ Q ♀
13 38	☽ Q ♃
15 19	☽ ∥ ♂
19 04	☽ ⊥ ☉
19 29	☽ Q ♅
20 49	☽ ⚹ ☿
21 15	☿ ♇ ♆
27 Sunday	
00 33	☿ ∥ ♆
01 42	☉ ♇ ♄
03 22	♀ ± ♃
08 50	♂ ⚹ ♃
09 19	☽ ⚹ ♃
09 21	☽ ☌ ♂
10 14	☽ ☍ ♄
10 52	☽ ∠ ☉
15 09	☽ ⚹ ♅
16 54	☽ □ ⯓
18 06	☽ ∥ ♅
18 46	☽ Q ☿
20 34	☽ □ ♀
21 27	☽ ⊼ ♆
23 25	☽ ∥ ♃
28 Monday	
03 07	♂ ☍ ♄
07 10	♀ △ ♆
07 21	☽ ± ♆
09 01	☽ ∥ ⯓
10 08	☽ ∠ ♃
13 28	☽ ⚹ ☉
15 58	☽ ∠ ♅
22 20	☽ ♇ ♆
22 44	☽ Q ⯓
29 Tuesday	
01 12	☽ ⊥ ♃
04 17	☽ □ ☿
07 03	☽ ⊥ ♅
11 05	☽ Q ☉
11 19	☽ ⊻ ♃
11 58	☽ ⊼ ♄
13 45	☽ ⊻ ♂
17 11	☽ ⊻ ♅
18 56	☽ ⚹ ⯓
22 07	☽ ± ♄
23 36	☽ △ ♆
30 Wednesday	
00 27	☽ ⊥ ♂
03 20	♀ ± ♅
03 24	☽ △ ♀
08 42	☽ ∥ ⯓
13 28	☽ ♇ ♄
16 41	☽ ∠ ♂
18 57	☽ ∥ ♃
20 10	☽ □ ☉
20 37	☽ ∠ ⯓
31 Thursday	
00 20	☽ ∥ ♅
02 24	☉ ♇ ⯓
07 42	☽ ♇ ♀
12 15	☽ ⊥ ⯓
14 19	☽ △ ☿
15 06	☽ ☌ ♃
15 26	☽ △ ♄
20 11	☽ ⚹ ♂
21 01	☽ ☌ ♅
21 15	☿ △ ♃
22 15	☽ ∥ ♂
22 47	☽ ⊻ ⯓
23 40	☿ △ ♄

All ephemeris data is given at 12.00 UT and the Moon's longitude is additionally given for 24.00 UT

APRIL 1955

Date	Sidereal time h m s	Sun ☉ ° ' "	Moon ☽ ° ' "	Moon ☽ 24.00 ° ' "	Mercury ☿ ° '	Venus ♀ ° '	Mars ♂ ° '	Jupiter ♃ ° '	Saturn ♄ ° '	Uranus ♅ ° '	Neptune ♆ ° '	Pluto ⯓ ° '
01	00 36 17	10 ♈ 59 28	02 ♌ 03 33	08 ♌ 46 42	21 ♓ 14	02 ♓ 24	23 ♉ 34	20 ♋ 16	20 ♏ 23	23 ♋ 36	27 ♎ 19	24 ♌ 35
02	00 40 13	11 58 39	15 ♌ 26 44	22 ♌ 03 44	22 53	03 35	24 15	20 19	20 R 20	23 D 36	27 R 17	24 R 34
03	00 44 10	12 57 47	28 ♌ 37 48	05 ♍ 09 00	24 33	04 46	24 56	20 23	20 17	23 36	27 16	24 33
04	00 48 06	13 56 53	11 ♍ 37 22	18 ♍ 02 54	26 14	05 57	25 37	20 26	20 14	23 36	27 14	24 33
05	00 52 03	14 55 57	24 ♍ 25 37	00 ♎ 45 29	27 57	07 09	26 18	20 30	20 10	23 36	27 12	24 32
06	00 55 59	15 54 59	07 ♎ 02 30	13 ♎ 16 40	29 ♓ 41	08 20	26 58	20 33	20 07	23 36	27 11	24 31
07	00 59 56	16 53 58	19 ♎ 28 01	25 ♎ 36 35	01 ♈ 27	09 31	27 39	20 37	20 04	23 37	27 09	24 30
08	01 03 52	17 52 56	01 ♏ 42 27	07 ♏ 45 46	03 14	10 43	28 20	20 41	20 00	23 37	27 08	24 29
09	01 07 49	18 51 52	13 ♏ 46 43	19 ♏ 45 32	05 03	11 54	29 00	20 46	19 57	23 37	27 06	24 28
10	01 11 46	19 50 46	25 ♏ 42 30	01 ♐ 37 59	06 53	13 06	29 ♉ 41	20 50	19 53	23 38	27 04	24 28
11	01 15 42	20 49 38	07 ♐ 32 22	13 ♐ 26 08	08 45	14 17	00 ♊ 22	20 55	19 50	23 38	27 03	24 27
12	01 19 39	21 48 28	19 ♐ 19 47	25 ♐ 13 52	10 38	15 29	01 02	20 59	19 46	23 39	27 01	24 26
13	01 23 35	22 47 16	01 ♑ 08 59	07 ♑ 05 46	12 33	16 40	01 43	21 04	19 42	23 40	26 59	24 26
14	01 27 32	23 46 03	13 ♑ 04 51	19 ♑ 06 56	14 29	17 52	02 23	21 09	19 39	23 40	26 58	24 25
15	01 31 28	24 44 48	25 ♑ 12 41	01 ♒ 22 45	16 27	19 04	03 04	21 15	19 35	23 41	26 56	24 24
16	01 35 25	25 43 31	07 ♒ 37 48	13 ♒ 58 25	18 26	20 16	03 44	21 20	19 31	23 42	26 55	24 24
17	01 39 21	26 42 13	20 ♒ 25 09	26 ♒ 58 27	20 26	21 27	04 25	21 26	19 27	23 42	26 53	24 23
18	01 43 18	27 40 53	03 ♓ 38 39	10 ♓ 25 58	22 28	22 39	05 05	21 31	19 23	23 43	26 51	24 23
19	01 47 15	28 39 31	17 ♓ 20 26	24 ♓ 21 56	24 31	23 51	05 46	21 37	19 19	23 44	26 50	24 22
20	01 51 11	29 ♈ 38 07	01 ♈ 30 08	08 ♈ 44 31	26 36	25 03	06 26	21 43	19 15	23 45	26 48	24 22
21	01 55 08	00 ♉ 36 42	16 ♈ 04 20	23 ♈ 28 44	28 ♈ 41	26 15	07 06	21 49	19 11	23 46	26 46	24 21
22	01 59 04	01 35 14	00 ♉ 56 38	08 ♉ 26 55	00 ♉ 48	27 26	07 46	21 56	19 06	23 47	26 45	24 21
23	02 03 01	02 33 45	15 ♉ 58 24	23 ♉ 29 51	02 55	28 38	08 27	22 02	19 02	23 49	26 43	24 21
24	02 06 57	03 32 14	01 ♊ 00 09	08 ♊ 28 15	05 03	29 ♓ 50	09 07	22 09	18 58	23 50	26 41	24 20
25	02 10 54	04 30 42	15 ♊ 53 14	23 ♊ 14 20	07 12	01 ♈ 02	09 47	22 16	18 54	23 51	26 40	24 20
26	02 14 50	05 29 07	00 ♋ 30 57	07 ♋ 42 40	09 20	02 14	10 27	22 23	18 49	23 52	26 38	24 20
27	02 18 47	06 27 29	14 ♋ 49 12	21 ♋ 50 26	11 29	03 26	11 07	22 30	18 45	23 54	26 37	24 19
28	02 22 44	07 25 50	28 ♋ 46 21	05 ♌ 37 02	13 37	04 38	11 48	22 37	18 41	23 55	26 35	24 19
29	02 26 40	08 24 09	12 ♌ 22 39	19 ♌ 03 25	15 45	05 50	12 28	22 44	18 36	23 56	26 33	24 19
30	02 30 37	09 ♉ 22 25	25 ♌ 39 37	02 ♍ 11 30	17 ♉ 51	07 ♈ 02	13 ♊ 08	22 ♋ 52	18 ♏ 32	23 ♋ 58	26 ♎ 32	24 ♌ 19

Date	Moon True ☊ ° '	Moon Mean ☊ ° '	Moon ☽ Latitude ° '
01	00 ♑ 06	00 ♑ 38	02 S 40
02	00 R 02	00 34	03 35
03	29 ♐ 56	00 31	04 18
04	29 47	00 28	04 46
05	29 35	00 25	05 00
06	29 22	00 22	04 58
07	29 09	00 19	04 43
08	28 58	00 15	04 14
09	28 48	00 12	03 34
10	28 40	00 09	02 45
11	28 36	00 06	01 50
12	28 34	00 03	00 S 49
13	28 D 33	00 ♑ 00	00 N 14
14	28 34	29 ♐ 56	01 17
15	28 R 34	29 53	02 17
16	28 33	29 50	03 13
17	28 31	29 47	04 01
18	28 26	29 44	04 37
19	28 19	29 40	05 00
20	28 10	29 37	05 05
21	28 01	29 34	04 50
22	27 52	29 31	04 16
23	27 45	29 28	03 24
24	27 40	29 25	02 18
25	27 38	29 21	01 N 03
26	27 D 37	29 18	00 S 16
27	27 38	29 15	01 31
28	27 39	29 12	02 40
29	27 R 39	29 09	03 37
30	27 ♐ 38	29 ♐ 05	04 S 22

DECLINATIONS

Date	Sun ☉ ° '	Moon ☽ ° '	Mercury ☿ ° '	Venus ♀ ° '	Mars ♂ ° '	Jupiter ♃ ° '	Saturn ♄ ° '	Uranus ♅ ° '	Neptune ♆ ° '	Pluto ⯓ ° '
01	04 N 21	17 N 06	05 S 40	11 S 13	19 N 20	22 N 24	15 S 27	21 N 54	08 S 52	23 N 12
02	04 44	12 47	05 01	10 50	19 30	22 24	15 26	21 54	08 51	23 12
03	05 07	07 55	04 21	10 27	19 41	22 23	15 25	21 54	08 51	23 12
04	05 30	02 N 47	03 39	10 03	19 51	22 23	15 24	21 54	08 50	23 12
05	05 53	02 S 22	02 57	09 40	20 01	22 22	15 23	21 54	08 49	23 13
06	06 16	07 22	02 13	09 16	20 11	22 22	15 22	21 54	08 49	23 13
07	06 38	11 59	01 29	08 52	20 21	22 21	15 21	21 54	08 48	23 13
08	07 01	16 02	00 S 43	08 27	20 31	22 20	15 20	21 54	08 48	23 13
09	07 23	19 23	00 N 03	08 02	20 40	22 20	15 19	21 54	08 47	23 13
10	07 46	21 52	00 51	07 37	20 49	22 19	15 18	21 54	08 47	23 13
11	08 08	23 22	01 39	07 12	20 58	22 19	15 17	21 54	08 46	23 13
12	08 30	23 50	02 28	06 47	21 07	22 18	15 16	21 54	08 45	23 13
13	08 52	23 13	03 19	06 21	21 16	22 17	15 15	21 54	08 45	23 13
14	09 14	21 32	04 09	05 55	21 25	22 16	15 14	21 53	08 44	23 13
15	09 35	18 51	05 01	05 29	21 33	22 16	15 13	21 53	08 44	23 13
16	09 57	15 15	05 53	05 03	21 41	22 15	15 12	21 53	08 43	23 13
17	10 18	10 53	06 46	04 36	21 49	22 14	15 11	21 53	08 42	23 13
18	10 39	05 52	07 39	04 10	21 57	22 13	15 10	21 53	08 42	23 13
19	11 00	00 S 24	08 32	03 43	22 05	22 12	15 08	21 53	08 41	23 13
20	11 21	05 N 15	09 26	03 16	22 12	22 11	15 07	21 52	08 41	23 13
21	11 41	10 47	10 19	02 49	22 19	22 10	15 06	21 52	08 40	23 13
22	12 02	15 48	11 13	02 22	22 26	22 09	15 05	21 52	08 39	23 13
23	12 22	19 53	12 06	01 55	22 33	22 08	15 04	21 52	08 39	23 13
24	12 42	22 37	12 59	01 28	22 40	22 07	15 03	21 52	08 38	23 13
25	13 02	23 44	13 52	01 00	22 46	22 06	15 01	21 51	08 38	23 13
26	13 21	23 11	14 43	00 33	22 52	22 05	15 00	21 51	08 37	23 13
27	13 41	21 06	15 34	00 S 05	22 59	22 04	14 59	21 51	08 36	23 13
28	14 00	17 48	16 23	00 N 22	23 04	22 02	14 58	21 51	08 36	23 13
29	14 19	13 37	17 11	00 50	23 10	22 01	14 56	21 50	08 35	23 13
30	14 N 37	08 N 51	17 N 57	01 N 17	23 N 16	22 N 00	14 S 55	21 N 50	08 S 35	23 N 12

ZODIAC SIGN ENTRIES

Date	h	m	Planets
01	08	20	☽ ♌
03	14	31	☽ ♍
05	22	34	☽ ♎
06	16	14	☿ ♈
08	08	38	☽ ♏
10	20	41	☽ ♐
10	23	09	♂ ♊
13	09	40	☽ ♑
15	21	20	☽ ♒
18	05	28	☽ ♓
20	09	29	☽ ♈
20	20	58	☉ ♉
22	02	57	☿ ♉
22	10	29	☽ ♉
24	10	24	☽ ♊
24	15	13	♀ ♈
26	11	09	☽ ♋
28	14	08	☽ ♌
30	19	58	☽ ♍

LATITUDES

Date	Mercury ☿ ° '	Venus ♀ ° '	Mars ♂ ° '	Jupiter ♃ ° '	Saturn ♄ ° '	Uranus ♅ ° '	Neptune ♆ ° '	Pluto ⯓ ° '
01	02 S 24	00 S 38	00 N 41	00 N 30	02 N 29	00 N 32	01 N 46	10 N 30
04	02 21	00 47	00 42	00 30	02 29	00 32	01 46	10 30
07	02 15	00 56	00 44	00 30	02 30	00 32	01 47	10 29
10	02 03	01 03	00 45	00 30	02 30	00 32	01 47	10 29
13	01 47	01 11	00 46	00 30	02 30	00 32	01 47	10 28
16	01 27	01 17	00 48	00 30	02 31	00 32	01 47	10 28
19	01 03	01 23	00 49	00 30	02 31	00 32	01 47	10 27
22	00 34	01 28	00 50	00 30	02 31	00 31	01 47	10 26
25	00 S 04	01 33	00 51	00 30	02 31	00 31	01 47	10 26
28	00 N 29	01 36	00 53	00 30	02 31	00 31	01 47	10 25
31	01 N 00	01 S 39	00 N 54	00 N 30	02 N 31	00 N 31	01 N 47	10 N 24

DATA

Julian Date	2435199
Delta T	+31 seconds
Ayanamsa	23° 14' 12"
Synetic vernal point	05° ♓ 52' 48"
True obliquity of ecliptic	23° 26' 43"

MOON'S PHASES, APSIDES AND POSITIONS ☽

Date	h	m	Phase	Longitude ° '	Eclipse Indicator
07	06	35	○	16 ♎ 41	
15	11	01	☾	24 ♑ 42	
22	13	06	●	01 ♉ 38	
29	04	23	☽	08 ♌ 06	

Day	h	m	
11	13	32	Apogee
23	18	39	Perigee

05	00	55	0S	
12	10	11	Max dec	23° S 50'
19	13	43	0N	
25	15	56	Max dec	23° N 46'

LONGITUDES

Date	Chiron ⚷ ° '	Ceres ⚳ ° '	Pallas ⚴ ° '	Juno ⚵ ° '	Vesta ⚶ ° '	Black Moon Lilith ⚸ ° '
01	04 ♒ 32	18 ♑ 59	27 ♐ 13	16 ♏ 30	26 ♊ 04	02 ♐ 27
11	04 ♒ 57	21 ♑ 11	28 ♐ 03	14 ♏ 50	29 ♊ 12	03 ♐ 34
21	05 ♒ 16	22 ♑ 58	28 ♐ 11	12 ♏ 48	02 ♋ 35	04 ♐ 40
31	05 ♒ 27	24 ♑ 18	27 ♐ 34	10 ♏ 33	06 ♋ 12	05 ♐ 47

All ephemeris data is given at 12.00 UT and the Moon's longitude is additionally given for 24.00 UT

ASPECTARIAN

h m	Aspects
01 Friday	
01 00	☽ ± ♀
03 36	☽ □ ♆
12 39	☽ ⊼ ♀
12 51	♅ St D
12 56	♂ ⚹ ♅
13 04	☿ ± ♆
18 35	☽ Q ♂
20 29	☽ ♇ ☿
21 38	☽ # ♄
02 Saturday	
05 15	☽ △ ☉
11 43	☽ Q ♆
13 41	♃ △ ♄
14 58	☽ ± ☿
18 29	☉ # ☿
20 49	☽ □ ♄
20 52	☽ ⊻ ♃
22 19	☿ △ ♅
22 42	☽ # ♀
23 05	♂ □ ⯓
03 Sunday	
02 48	☽ ⊻ ♅
03 27	☽ ⊼ ☿
04 33	☽ ☌ ⯓
04 52	☽ □ ♂
07 33	☽ # ♆
07 51	☽ ⊥ ♃
09 30	☽ ⚹ ♆
10 41	☽ ♇ ☉
12 08	☿ ⊼ ⯓
13 46	☽ ⊥ ♅
21 11	☿ ⚹ ♂
04 Monday	
00 16	☽ ∥ ☉
00 26	☽ ☍ ♀
00 28	☽ ∠ ♃
00 55	♀ ♇ ♃
04 37	☽ ± ☉
05 43	☽ Q ♄
06 23	☽ ∠ ♅
07 21	☽ # ☿
13 08	☽ ∠ ♆
16 42	☽ ⊼ ☉
18 26	☉ ± ♄
05 Tuesday	
01 46	☿ ⊼ ♆
04 01	☽ ⚹ ♄
04 33	☽ ⚹ ♃
05 57	☽ ⊥ ♆
10 26	☽ ⚹ ♅
12 11	☽ ⊻ ⯓
14 22	☽ ∥ ☿
15 44	☽ △ ♂
17 15	☽ ⊻ ♆
19 44	☽ ☍ ☿
23 33	☽ ⊥ ⯓
06 Wednesday	
03 23	☽ Q ♃
06 08	☽ # ☉
08 20	☽ ∠ ♄
09 15	☽ Q ♅
14 44	☽ ⊼ ♀
16 44	☽ ∠ ⯓
17 32	♀ ♇ ♅
19 05	♂ ⊼ ♆
19 17	☽ ∥ ♆
20 50	☽ ∥ ♀
22 02	☽ ♇ ♂
23 10	☿ ± ⯓
07 Thursday	
01 34	☽ ⊥ ♄
03 32	☽ ± ♀
06 35	☽ ☍ ☉
13 09	☽ ⊻ ♄
14 16	☽ □ ♃
15 15	♀ ∥ ♆
16 30	☽ ± ♂
20 05	☽ □ ♅
21 49	☽ ⚹ ⯓
22 56	☽ ♇ ♀
08 Friday	
03 00	☽ ☌ ♆
04 57	☽ ⊼ ♂
07 36	☽ ∥ ♄
15 33	☽ ⊼ ☿
21 27	☽ Q ⯓
09 Saturday	
05 35	☽ ± ☿
07 50	☽ △ ♀
10 41	☿ ♇ ♄
15 52	♀ ♇ ♆
23 07	☽ ⊼ ☉
10 Sunday	
00 03	☽ # ♂
00 19	☽ ☌ ♄
02 06	☽ △ ♃
02 52	☽ ♇ ☿
07 40	☉ # ♀
07 48	☽ △ ♅
09 29	☽ □ ⯓
12 18	☽ ± ☉
12 22	☽ # ♅
13 00	☉ ⊼ ♄
14 45	☽ ⊻ ♆
17 42	☽ # ♃
20 32	☽ ☍ ♂
11 Monday	
02 53	☽ ⊥ ♆
08 12	☽ ♇ ☉
08 31	☽ # ⯓
08 40	☽ ♇ ♃
14 13	☉ □ ♃
14 14	☽ ♇ ♅
14 56	☽ △ ☿
20 54	☿ ♇ ⯓
21 09	☽ ∠ ♆
12 Tuesday	
03 06	☽ ± ♃
03 17	☽ □ ♀
08 34	☽ ± ♅
12 53	☽ ⊻ ♄
15 24	☽ ⊼ ♃
17 30	☽ △ ☉
20 47	☽ ⊼ ♅
22 23	☽ △ ⯓
13 Wednesday	
01 01	☽ ⊥ ♄
03 36	☽ ⚹ ♆
04 20	☉ # ♆
11 44	☽ # ⯓
13 13	☽ ⊼ ♂
19 09	☽ ∠ ♄
19 55	☽ Q ♀
14 Thursday	
00 44	☉ ⊥ ♀
01 58	☿ ± ♄
02 02	☽ ± ♂
03 03	☽ # ♃
03 46	☽ Q ♆
04 40	☽ ♇ ⯓
07 52	☽ # ♅
09 34	☉ □ ♅
13 11	☽ # ♂
15 20	☽ □ ☿
21 05	☽ ♇ ♂
22 34	☽ ⚹ ♀
22 37	☽ ± ⯓
15 Friday	
00 59	☽ ⚹ ♄
03 46	☉ △ ⯓
04 09	☽ ☍ ♃
07 37	♂ ± ♆
09 00	☽ ☍ ♅
10 26	☽ ⊼ ⯓
11 01	☽ □ ☉
15 22	☽ □ ♆
20 38	☿ # ♀
21 50	♀ △ ♄
16 Saturday	
00 19	☽ Q ♄
04 08	☽ △ ♂
06 59	☽ ∠ ♀
09 16	☽ Q ☿
12 21	☽ ∥ ♄
17 38	☿ ∠ ♂
17 Sunday	
00 29	☽ Q ☉
00 36	☿ ⊼ ♄
00 46	♀ ± ♆
01 50	☽ ⊥ ♀
10 13	☽ □ ♄
11 24	♀ △ ♃
12 02	☽ ⚹ ☿
13 52	☽ ⊼ ♃
14 06	☽ ⊻ ♀
14 43	☽ # ☉
16 16	☉ ☍ ♆
18 03	☽ ⊼ ♅
19 17	☽ ☍ ⯓
22 44	☽ ∥ ♆
23 20	♂ ∥ ♅
23 49	☽ △ ♆
18 Monday	
00 18	☿ □ ♃
00 26	☽ ⚹ ☉
00 55	☽ ± ♃
04 58	☽ ± ♅
04 59	☽ # ☿
14 42	☽ □ ♂
17 09	☽ ♇ ♃
17 10	☿ ⊻ ♀
19 59	☽ ∠ ☿
20 19	☽ ∥ ♀
21 00	☽ ♇ ♅
19 Tuesday	
02 28	☽ ♇ ♆
02 48	☿ □ ♅
05 09	☽ ∠ ☉
07 56	♀ Q ♂
09 46	♀ △ ♅
10 15	☿ △ ⯓
14 23	☽ ⊥ ☿
15 23	☽ △ ♄
16 03	☿ # ♆
17 58	☽ ± ♆
19 23	☽ △ ♃
21 47	☽ ⊥ ☉
22 25	♀ ⊼ ⯓
22 57	☽ △ ♅
23 31	☽ Q ♂
20 Wednesday	
00 00	☽ ⊼ ⯓
00 09	☽ ☌ ♀
02 22	☽ ⊻ ☿
04 08	☽ ⊼ ♆
04 13	☽ # ♀
08 39	☽ ⊻ ☉
09 10	♂ ∥ ♃
10 06	☽ ± ⯓
14 19	☿ ☍ ♆
16 33	☽ ♇ ♄
20 35	☽ ⚹ ♂
21 Thursday	
00 12	♂ ∠ ♃
01 01	☽ ♇ ⯓
02 40	☽ # ♆
07 18	☽ ± ♄
09 31	☽ ∥ ☿
16 23	☽ ∥ ☉
17 01	☽ ⊼ ♄
21 24	☽ □ ♃
22 15	☽ ∠ ♂
22 22	♀ ⊼ ♆
22 Friday	
00 29	☽ □ ♅
01 24	☽ △ ⯓
05 16	☽ ☍ ♆
05 53	☽ ⊻ ♀
08 19	☽ # ♄
11 43	☽ ☌ ☿
13 06	☽ ☌ ☉
13 24	☽ ⊥ ♂
16 21	☽ ⊥ ♀
23 26	☽ ⊻ ♂
23 Saturday	
02 28	☽ Q ♃
02 46	☉ ⊥ ♂
04 12	☿ ⊥ ♂
04 35	☉ ☌ ☿
05 21	☽ Q ♅
07 58	☽ ∠ ♀
16 52	☽ ☍ ♄
21 44	☽ ⚹ ♃
23 19	☉ ∥ ☿
24 Sunday	
00 31	☽ ⚹ ♅
01 21	☽ □ ⯓
01 25	♂ ∠ ♅
03 53	☽ ∥ ♅
05 07	☽ ⊼ ♆
06 28	☽ ∥ ♃
09 58	☽ ⚹ ♀
12 38	☽ ∥ ♂
14 42	☽ ± ♆
16 21	☽ ⊻ ☉
19 35	☽ ⊻ ☿
20 45	☽ ∥ ⯓
21 54	♀ ± ⯓
21 57	☽ ∠ ♃
25 Monday	
00 36	☽ ∠ ♅
01 39	☽ ☌ ♂
02 41	☽ ⊥ ☉
05 10	☽ ♇ ♆
06 14	☽ Q ⯓
06 53	☽ ⊥ ☿
06 58	☽ Q ♀
08 05	☿ ⊥ ♀
12 37	☽ ⊥ ♃
15 12	☽ ⊥ ♅
16 52	☽ ⊼ ♄
18 19	☽ ∠ ☉
22 29	☽ ⊻ ♃
26 Tuesday	
00 03	☽ ∠ ☿
01 01	☽ ⊻ ♅
01 47	☽ ⚹ ⯓
02 39	☽ ± ♄
05 36	☽ △ ♆
11 26	☽ ∥ ⯓
15 07	☽ □ ♀
16 38	☽ ∥ ♂
17 28	☽ ♇ ♄
19 48	☿ # ♄
20 52	☽ ⚹ ☉
27 Wednesday	
00 19	☿ Q ♃
02 42	☽ ∠ ⯓
02 48	☽ ∥ ♃
05 04	☽ ∥ ♅
05 20	☽ ⚹ ☿
05 26	☽ ⊻ ♂
06 13	☿ ⊻ ♂
16 07	☽ ⊥ ♂
16 42	☿ Q ♅
17 54	♀ ♇ ♄
17 58	☽ ⊥ ⯓
18 40	☽ Q ☉
18 40	☽ △ ♄
28 Thursday	
01 14	☽ ☌ ♃
03 34	☽ ☌ ♅
04 16	☽ ⊻ ⯓
04 47	♂ ♇ ♆
05 31	☽ Q ☿
08 12	☽ □ ♆
08 23	☽ ∠ ♂
19 11	☽ ∥ ☿
23 16	☽ △ ♀
29 Friday	
04 23	☽ □ ☉
04 46	☽ # ♄
06 45	♂ Q ⯓
08 31	☽ ∥ ☉
12 09	☽ ⚹ ♂
15 53	☽ Q ♆
19 10	☽ □ ☿
22 24	♂ ∥ ⯓
23 07	☽ □ ♄
30 Saturday	
04 45	☽ ♇ ♀
06 51	☽ ⊻ ♃
08 54	☽ ⊻ ♅
09 32	☽ ☌ ⯓
10 58	☽ Q ♂
13 21	☽ # ♆
13 35	☽ ⚹ ♆
17 55	☽ ⊥ ♃
19 30	☿ ☍ ♄
19 55	☽ ⊥ ♅
22 53	☽ ± ♀

MAY 1955

LONGITUDES

Date	Sidereal time h m s	Sun ☉ ° ' "	Moon ☽ ° ' "	Moon ☽ 24.00 ° ' "	Mercury ☿ ° '	Venus ♀ ° '	Mars ♂ ° '	Jupiter ♃ ° '	Saturn ♄ ° '	Uranus ♅ ° '	Neptune ♆ ° '	Pluto ♇ ° '
01	02 34 33	10 ♉ 20 40	08 ♍ 39 21	15 ♍ 03 27	19 ♉ 57	08 ♈ 15	13 ♊ 48	22 ♋ 59	18 ♏ 27	23 ♋ 59	26 ♎ 30	24 ♌ 18
02	02 38 30	11 18 52	21 ♍ 24 05	27 ♍ 41 28	22 00	09 27	14 28	23 07	18 R 23	24 01	26 R 29	24 R 18
03	02 42 26	12 17 02	03 ♎ 55 51	10 ♎ 07 26	24 02	10 39	15 08	23 15	18 19	24 03	26 27	24 18
04	02 46 23	13 15 11	16 ♎ 16 26	22 ♎ 23 02	26 02	11 51	15 47	23 23	18 14	24 04	26 25	24 18
05	02 50 19	14 13 17	28 ♎ 27 23	04 ♏ 29 41	28 00	13 03	16 27	23 31	18 10	24 06	26 24	24 18
06	02 54 16	15 11 22	10 ♏ 30 07	16 ♏ 28 50	29 ♉ 55	14 15	17 07	23 39	18 05	24 08	26 22	24 18
07	02 58 13	16 09 26	22 ♏ 26 04	28 ♏ 22 02	01 ♊ 47	15 27	17 47	23 47	18 01	24 10	26 21	24 D 18
08	03 02 09	17 07 27	04 ♐ 16 59	10 ♐ 11 11	03 36	16 40	18 27	23 56	17 56	24 12	26 19	24 18
09	03 06 06	18 05 28	16 ♐ 04 58	21 ♐ 58 41	05 22	17 52	19 06	24 04	17 52	24 13	26 18	24 18
10	03 10 02	19 03 26	27 ♐ 52 42	03 ♑ 47 28	07 05	19 04	19 46	24 13	17 47	24 15	26 16	24 18
11	03 13 59	20 01 24	09 ♑ 43 26	15 ♑ 41 06	08 44	20 16	20 26	24 22	17 43	24 17	26 15	24 18
12	03 17 55	20 59 19	21 ♑ 41 00	27 ♑ 43 40	10 20	21 29	21 06	24 31	17 38	24 20	26 13	24 18
13	03 21 52	21 57 14	03 ♒ 49 42	09 ♒ 59 41	11 53	22 41	21 45	24 40	17 34	24 22	26 12	24 19
14	03 25 48	22 55 07	16 ♒ 14 11	22 ♒ 33 46	13 22	23 53	22 25	24 49	17 29	24 24	26 11	24 19
15	03 29 45	23 52 59	28 ♒ 58 59	05 ♓ 30 18	14 47	25 06	23 04	24 58	17 25	24 26	26 09	24 19
16	03 33 42	24 50 50	12 ♓ 08 08	18 ♓ 52 49	16 08	26 18	23 44	25 07	17 20	24 28	26 08	24 19
17	03 37 38	25 48 40	25 ♓ 44 32	02 ♈ 43 20	17 26	27 30	24 24	25 17	17 16	24 30	26 06	24 20
18	03 41 35	26 46 28	09 ♈ 49 06	17 ♈ 01 33	18 40	28 43	25 03	25 26	17 11	24 33	26 05	24 20
19	03 45 31	27 44 15	24 ♈ 20 10	01 ♉ 44 18	19 50	29 ♈ 55	25 43	25 36	17 07	24 35	26 04	24 20
20	03 49 28	28 42 02	09 ♉ 13 03	16 ♉ 45 26	20 56	01 ♉ 08	26 22	25 46	17 02	24 37	26 02	24 21
21	03 53 24	29 ♉ 39 47	24 ♉ 20 17	01 ♊ 56 23	21 58	02 20	27 01	25 56	16 58	24 40	26 01	24 21
22	03 57 21	00 ♊ 37 30	09 ♊ 32 29	17 ♊ 07 22	22 56	03 33	27 41	26 06	16 54	24 42	26 00	24 22
23	04 01 17	01 35 13	24 ♊ 39 51	02 ♋ 08 55	23 50	04 45	28 20	26 16	16 49	24 45	25 58	24 22
24	04 05 14	02 32 54	09 ♋ 33 39	16 ♋ 53 19	24 39	05 58	29 00	26 26	16 45	24 47	25 57	24 23
25	04 09 11	03 30 33	24 ♋ 07 21	01 ♌ 15 22	25 25	07 10	29 ♊ 39	26 36	16 41	24 50	25 56	24 23
26	04 13 07	04 28 11	08 ♌ 17 07	15 ♌ 12 33	26 06	08 23	00 ♋ 18	26 46	16 37	24 53	25 55	24 24
27	04 17 04	05 25 48	22 ♌ 01 41	28 ♌ 44 43	26 42	09 35	00 58	26 57	16 33	24 55	25 54	24 24
28	04 21 00	06 23 23	05 ♍ 21 51	11 ♍ 53 26	27 14	10 48	01 37	27 07	16 28	24 58	25 52	24 25
29	04 24 57	07 20 56	18 ♍ 19 50	24 ♍ 41 25	27 42	12 00	02 16	27 18	16 24	25 01	25 51	24 25
30	04 28 53	08 18 28	00 ♎ 58 37	07 ♎ 11 51	28 05	13 13	02 55	27 29	16 20	25 03	25 50	24 26
31	04 32 50	09 ♊ 15 59	13 ♎ 21 32	19 ♎ 28 05	28 ♊ 23	14 ♉ 25	03 ♋ 34	27 ♋ 39	16 ♏ 16	25 ♋ 06	25 ♎ 49	24 ♌ 27

Date	Moon True ☊ ° '	Moon Mean ☊ ° '	Moon ☽ Latitude ° '
01	27 ♐ 35	29 ♐ 02	04 S 52
02	27 R 29	28 59	05 06
03	27 22	28 56	05 06
04	27 14	28 53	04 52
05	27 06	28 50	04 24
06	26 59	28 46	03 45
07	26 53	28 43	02 56
08	26 49	28 40	01 59
09	26 46	28 37	00 S 58
10	26 D 46	28 34	00 N 06
11	26 46	28 31	01 10
12	26 48	28 27	02 12
13	26 50	28 24	03 09
14	26 51	28 21	03 58
15	26 R 51	28 18	04 37
16	26 50	28 15	05 03
17	26 47	28 11	05 13
18	26 43	28 08	05 05
19	26 39	28 05	04 38
20	26 35	28 02	03 51
21	26 32	27 59	02 48
22	26 30	27 56	01 32
23	26 29	27 52	00 N 10
24	26 D 30	27 49	01 S 11
25	26 31	27 46	02 27
26	26 32	27 43	03 31
27	26 33	27 40	04 20
28	26 34	27 37	04 55
29	26 R 34	27 33	05 13
30	26 32	27 30	05 15
31	26 ♐ 30	27 ♐ 27	05 S 02

DECLINATIONS

Date	Sun ☉ ° '	Moon ☽ ° '	Mercury ☿ ° '	Venus ♀ ° '	Mars ♂ ° '	Jupiter ♃ ° '	Saturn ♄ ° '	Uranus ♅ ° '	Neptune ♆ ° '	Pluto ♇ ° '
01	14 N 56	03 N 49	18 N 42	01 N 45	23 N 21	21 N 59	14 S 54	21 N 50	08 S 34	23 N 12
02	15 14	01 S 17	19 24	02 12	23 26	21 58	14 53	21 49	08 34	23 12
03	15 32	06 15	20 04	02 40	23 31	21 56	14 52	21 49	08 33	23 12
04	15 49	10 53	20 42	03 08	23 36	21 55	14 50	21 49	08 33	23 12
05	16 07	15 02	21 18	03 35	23 40	21 53	14 49	21 48	08 32	23 12
06	16 24	18 32	21 51	04 03	23 44	21 52	14 48	21 48	08 31	23 11
07	16 41	21 13	22 22	04 30	23 48	21 51	14 47	21 48	08 31	23 11
08	16 57	22 58	22 50	04 57	23 52	21 49	14 46	21 47	08 30	23 11
09	17 13	23 41	23 16	05 25	23 56	21 48	14 44	21 47	08 30	23 11
10	17 29	23 20	23 39	05 52	23 59	21 46	14 43	21 47	08 29	23 10
11	17 45	21 55	23 59	06 19	24 03	21 45	14 42	21 46	08 29	23 10
12	18 00	19 32	24 18	06 46	24 06	21 43	14 41	21 46	08 28	23 10
13	18 16	16 14	24 33	07 13	24 09	21 42	14 39	21 46	08 28	23 10
14	18 30	12 11	24 47	07 40	24 11	21 40	14 38	21 45	08 27	23 09
15	18 45	07 30	24 58	08 06	24 14	21 38	14 37	21 45	08 27	23 09
16	18 59	02 S 20	25 08	08 33	24 16	21 37	14 36	21 44	08 26	23 09
17	19 13	03 N 06	25 15	08 59	24 18	21 35	14 35	21 44	08 26	23 08
18	19 26	08 34	25 20	09 25	24 20	21 33	14 34	21 44	08 25	23 08
19	19 40	13 44	25 23	09 51	24 22	21 31	14 32	21 43	08 25	23 08
20	19 52	18 13	25 24	10 17	24 23	21 30	14 31	21 43	08 25	23 07
21	20 05	21 34	25 24	10 42	24 24	21 28	14 30	21 42	08 24	23 07
22	20 17	23 24	25 22	11 08	24 25	21 26	14 29	21 42	08 24	23 07
23	20 29	23 30	25 19	11 33	24 26	21 24	14 28	21 41	08 23	23 06
24	20 40	21 55	25 14	11 58	24 27	21 22	14 27	21 41	08 23	23 06
25	20 52	18 53	25 07	12 22	24 27	21 20	14 26	21 40	08 22	23 06
26	21 02	14 48	25 00	12 46	24 28	21 18	14 25	21 40	08 22	23 05
27	21 13	10 04	24 51	13 10	24 28	21 16	14 24	21 39	08 22	23 05
28	21 23	04 N 59	24 41	13 34	24 27	21 14	14 23	21 39	08 21	23 04
29	21 32	00 S 11	24 30	13 58	24 27	21 12	14 22	21 38	08 21	23 04
30	21 42	05 12	24 18	14 21	24 27	21 10	14 21	21 38	08 20	23 04
31	21 N 51	09 S 55	24 N 05	14 N 43	24 N 26	21 N 08	14 S 20	21 N 37	08 S 20	23 N 03

ZODIAC SIGN ENTRIES

Date	h	m	Planets
03	04	26	☽ ♎
05	15	04	☽ ♏
06	13	05	☿ ♊
08	03	19	☽ ♐
10	16	19	☽ ♑
13	04	29	☽ ♒
15	13	53	☽ ♓
17	19	21	☽ ♈
19	13	35	♀ ♉
19	21	12	☽ ♉
21	20	24	☉ ♊
21	20	56	☽ ♊
23	20	33	☽ ♋
25	21	52	☽ ♌
26	00	50	♂ ♋
28	02	16	☽ ♍
30	10	08	☽ ♎

LATITUDES

Date	Mercury ☿ ° '	Venus ♀ ° '	Mars ♂ ° '	Jupiter ♃ ° '	Saturn ♄ ° '	Uranus ♅ ° '	Neptune ♆ ° '	Pluto ♇ ° '
01	01 N 00	01 S 39	00 N 54	00 N 30	02 N 31	00 N 31	01 N 47	10 N 24
04	01 29	01 42	00 55	00 30	02 31	00 31	01 47	10 24
07	01 53	01 43	00 56	00 30	02 31	00 31	01 46	10 23
10	02 11	01 44	00 57	00 30	02 31	00 31	01 46	10 22
13	02 22	01 44	00 57	00 30	02 31	00 31	01 46	10 22
16	02 25	01 43	00 58	00 30	02 31	00 31	01 46	10 21
19	02 20	01 42	00 59	00 30	02 31	00 31	01 46	10 20
22	02 07	01 40	01 00	00 30	02 30	00 31	01 46	10 20
25	01 46	01 38	01 01	00 30	02 30	00 31	01 46	10 19
28	01 16	01 35	01 01	00 31	02 30	00 31	01 46	10 18
31	00 N 39	01 S 31	01 N 02	00 N 31	02 N 29	00 N 31	01 N 46	10 N 18

DATA

Julian Date	2435229
Delta T	+31 seconds
Ayanamsa	23° 14' 15"
Synetic vernal point	05° ♓ 52' 44"
True obliquity of ecliptic	23° 26' 42"

MOON'S PHASES, APSIDES AND POSITIONS ☽

Date	h	m	Phase	Longitude ° '	Eclipse Indicator
06	22	14	○	15 ♏ 36	
15	01	42	☾	23 ♒ 28	
21	20	59	●	00 ♊ 01	
28	14	01	☽	06 ♍ 28	

Day	h	m		
08	23	48	Apogee	
22	03	48	Perigee	
02	05	55	OS	
09	16	09	Max dec	23° S 42'
16	22	25	ON	
23	01	23	Max dec	23° N 41'
29	11	09	OS	

LONGITUDES

Date	Chiron ⚷ ° '	Ceres ⚳ ° '	Pallas ⚴ ° '	Juno ⚵ ° '	Vesta ⚶ ° '	Black Moon Lilith ⚸ ° '
01	05 ♒ 27	24 ♑ 18	27 ♐ 34	10 ♏ 33	06 ♋ 12	05 ♐ 47
11	05 ♒ 31	25 ♑ 06	26 ♐ 10	08 ♏ 19	10 ♋ 00	06 ♐ 53
21	05 ♒ 27	25 ♑ 20	24 ♐ 04	06 ♏ 16	13 ♋ 57	08 ♐ 00
31	05 ♒ 17	24 ♑ 58	21 ♐ 25	04 ♏ 36	18 ♋ 02	09 ♐ 06

All ephemeris data is given at 12.00 UT and the Moon's longitude is additionally given for 24.00 UT

ASPECTARIAN

h m	Aspects
01 Sunday	
07 56	☽ Q ♄
10 09	☉ # ♄
10 44	☽ ∠ ♃
11 09	☽ ⊼ ♀
12 38	☽ ∠ ♅
15 25	☽ △ ☉
17 19	☽ ∠ ♆
20 54	☽ ∥ ♀
22 09	☽ □ ♂
02 Monday	
06 17	☉ Q ♃
06 19	☽ ⚹ ♄
09 14	♀ ᕋ ♇
10 15	☽ ⊥ ♆
13 23	☽ △ ☿
15 18	☽ ⚹ ♃
16 49	☽ # ♀
16 59	☽ ⚹ ♅
17 32	☽ ⊻ ♇
21 39	☽ ⊻ ♆
22 09	☽ ᕋ ☉
03 Tuesday	
01 54	☿ ⚹ ♃
05 01	☽ ⊥ ♇
05 55	☉ Q ♅
10 48	☽ ∠ ♄
12 03	☿ ⚹ ♅
14 34	☽ Q ♃
15 07	☿ □ ♇
16 06	☽ Q ♅
16 56	☽ ± ☉
22 24	☽ ∠ ♇
23 39	☽ ∥ ♆
23 49	☽ ᕋ ☿
04 Wednesday	
02 25	☽ ☍ ♀
04 09	☽ ⊥ ♄
05 35	☽ ⊼ ☉
11 00	☽ △ ♂
15 49	☽ ⊻ ♄
16 37	☿ ⊼ ♆
19 17	♀ ± ♄
20 49	☽ ± ☿
05 Thursday	
02 07	☽ □ ♃
03 22	☽ □ ♅
03 47	☽ ⚹ ♇
07 56	☽ ☌ ♆
10 41	☽ ∥ ♄
10 55	☽ ⊼ ☿
13 40	☿ ∠ ♀
18 18	☽ ᕋ ♂
19 29	☽ # ☉
06 Friday	
03 36	☽ Q ♇
09 49	☿ ∥ ♅
12 43	☿ ∥ ♃
13 18	☽ ± ♂
18 24	♇ St D
20 22	☽ ⊼ ♀
22 14	☽ ☍ ☉
07 Saturday	
02 04	☽ ⊼ ♂
03 08	☽ ☌ ♄
09 48	☽ ± ♀
12 17	♂ ⊥ ♃
14 46	☽ △ ♃
15 30	☽ △ ♅
15 46	☽ □ ♇
18 40	☽ # ♅
19 11	☽ # ♃
19 16	☿ ± ♆
19 24	♂ ⊼ ♄
19 54	☽ ⊻ ♆
08 Sunday	
02 25	♂ ⊥ ♅
06 04	☽ ᕋ ♀
08 02	☽ ⊥ ♆
08 31	☽ # ☿
10 22	☽ ☍ ☿
16 41	☽ # ♇
21 33	☽ ᕋ ♃
22 00	☽ ᕋ ♅
09 Monday	
02 17	☽ ∠ ♆
06 40	☉ ☍ ♄
07 15	☿ ∥ ♇
11 55	♀ ⊼ ♄
15 36	☽ ⊻ ♄
16 02	☽ △ ♀
16 06	☽ ± ♃
16 22	☽ ± ♅
16 27	☽ ⊼ ☉
18 31	☽ ☍ ♂
10 Tuesday	
01 24	♃ ∥ ♅
03 44	☽ ⊥ ♄
04 28	☽ ⊼ ♃
04 37	☽ ⊼ ♅
04 44	☽ △ ♇
05 02	☽ # ☿
05 45	☽ ± ☉
08 45	☽ ⚹ ♆
10 56	☉ ⊻ ♀
15 47	☽ # ♇
20 38	♃ ☌ ♅
21 54	☽ ∠ ♄
11 Wednesday	
01 39	☽ ᕋ ☉
02 26	♃ ⊻ ♇
09 02	☽ Q ♆
09 42	☽ ⊼ ☿
11 09	☽ ᕋ ♇
13 50	☽ # ♅
14 11	☽ # ♃
16 56	☿ ∥ ♂
19 02	♀ ⚹ ♂
20 19	☿ ∠ ♅
22 10	☿ ∠ ♃
23 19	♅ ⊻ ♇
23 41	☽ ± ☿
12 Thursday	
03 58	☽ ⚹ ♄
05 15	☽ ± ♇
10 30	☽ △ ☉
10 45	☽ ⊼ ♂
11 33	☽ □ ♀
17 14	☽ ⊼ ♇
17 16	☽ ☍ ♅
17 42	☽ ☍ ♃
20 12	☉ ⊻ ♂
20 21	☽ ᕋ ☿
21 00	☽ □ ♆
23 00	☽ # ☉
23 22	☽ ± ♂
13 Friday	
01 25	☿ ᕋ ♆
03 40	☽ Q ♄
18 02	☽ ᕋ ♂
18 52	☿ Q ♇
21 55	☽ ∥ ♄
14 Saturday	
02 46	☽ Q ♀
05 46	☽ △ ☿
14 22	☽ □ ♄
20 31	♀ △ ♇
22 25	♀ □ ♅
15 Sunday	
00 22	☽ △ ♂
01 42	☽ □ ☉
03 18	☽ ☍ ♇
03 29	☽ ⊼ ♅
04 00	☽ ⚹ ♀
04 26	☽ ⊼ ♃
06 44	☽ △ ♆
07 20	☽ ∥ ♆
09 09	♀ □ ♃
09 17	☽ # ♀
14 41	☽ ± ♅
15 43	☽ ± ♃
22 53	☉ □ ♇
22 59	♂ ± ♄
16 Monday	
02 12	☉ ⚹ ♅
06 18	♀ # ♆
07 11	☽ ᕋ ♅
08 20	☽ ᕋ ♃
08 39	♀ ☍ ♆
10 12	☽ ᕋ ♆
10 21	☽ ∠ ♀
13 22	☽ Q ☉
19 55	☽ □ ☿
20 15	☉ ⚹ ♃
21 13	☽ △ ♄
17 Tuesday	
02 10	☽ ± ♆
03 54	☽ ⊥ ♀
08 54	☿ ⊼ ♄
09 32	☽ □ ♂
09 33	☽ ⊼ ♇
09 39	♂ ⚹ ♇
09 51	☽ △ ♅
11 12	☽ △ ♃
12 08	☽ ⚹ ☉
12 38	☽ ⊼ ♆
15 21	☽ ⊻ ♀
16 25	♂ ⊻ ♅
19 10	☉ ⊼ ♆
19 55	☽ ± ♇
23 09	☽ ᕋ ♄
18 Wednesday	
06 12	☽ Q ☿
09 31	☿ ⊥ ♅
11 11	☽ ᕋ ♇
11 23	☽ # ♆
14 17	☽ ± ♄
15 31	☽ ∠ ☉
16 11	☽ ∥ ♀
17 40	☽ Q ♂
19 Thursday	
00 12	☽ ⊼ ♄
04 00	☽ ⚹ ☿
06 23	☿ ⊥ ♃
06 51	♂ ⊻ ♃
07 27	☽ ⊥ ☉
12 00	☽ △ ♇
12 24	☽ □ ♅
14 05	☽ □ ♃
14 20	☽ ⚹ ♂
14 48	☽ ☍ ♆
15 58	☽ # ♄
17 55	☽ ⊻ ☉
21 52	☽ ☌ ♀
20 Friday	
00 24	♂ △ ♆
06 20	☽ ∠ ☿
15 35	☽ ∠ ♂
17 27	☽ Q ♅
19 20	☽ Q ♃
21 47	☽ ⊥ ☿
23 24	☽ ∥ ☉
21 Saturday	
00 23	☽ ☍ ♄
06 31	☽ ⊥ ♂
07 59	☽ ⊻ ☿
11 05	☽ ∥ ♃
12 01	☽ □ ♇
12 31	☽ ⚹ ♅
13 15	☽ ∥ ♅
14 32	☽ ⚹ ♃
14 39	☽ ⊼ ♆
16 26	☽ ⊻ ♂
20 59	☽ ☌ ☉
23 11	♃ □ ♆
22 Sunday	
00 06	☽ ± ♆
01 43	☽ ⊻ ♀
06 13	☽ ∥ ♇
11 08	☿ ± ♄
12 00	☽ ⊥ ♀
12 16	☽ ∠ ♅
14 18	☽ ᕋ ♆
14 29	☽ ∠ ♃
16 27	☽ Q ♇
23 35	☽ ⊼ ♄
23 Monday	
02 33	☽ ⊥ ♅
03 30	☽ ∠ ♀
04 54	☽ ⊥ ♃
09 05	☽ ± ♄
10 35	☽ ☌ ☿
11 32	☽ ⚹ ♇
12 08	☽ ⊻ ♅
14 05	☽ △ ♆
14 35	☽ ⊻ ♃
18 09	☽ ☌ ♂
20 52	☽ ∥ ♇
21 27	☉ ± ♆
23 25	☽ ᕋ ♄
23 52	☽ ⊻ ☉
24 Tuesday	
03 33	☿ ⚹ ♇
05 38	☽ ⚹ ♀
10 14	☽ ⊥ ☉
11 42	☽ ∠ ♇
14 17	☽ ∥ ♅
16 17	☿ ⊻ ♅
17 14	☽ ∥ ♃
22 24	☽ ∥ ☉
23 43	☽ △ ♄
25 Wednesday	
02 01	☽ ∠ ☉
02 28	☽ ⊥ ♇
03 01	☽ Q ♀
12 26	☽ ⊻ ♇
13 11	☽ ☌ ♅
14 16	☽ ⊻ ☿
15 01	☽ □ ♆
16 12	☽ ☌ ♃
21 44	☽ ⊻ ♂
26 Thursday	
00 54	☽ ⊥ ☿
04 59	☽ ⚹ ☉
05 31	☿ △ ♆
08 26	☽ ⊥ ♂
12 10	☽ □ ♀
14 07	☽ # ♄
17 05	☽ ∠ ☿
21 43	☽ Q ♆
21 46	☽ ∥ ♀
27 Friday	
00 47	☽ ∠ ♂
02 23	☽ □ ♄
03 17	☽ Q ☉
16 14	☽ ☌ ♇
17 10	☽ ⊻ ♅
18 52	☽ ⚹ ♆
19 00	☉ ∥ ♃
20 09	☽ # ♆
20 42	☽ ⚹ ☿
20 53	☽ ⊻ ♃
28 Saturday	
03 37	☿ ⊻ ♃
03 59	☽ ⊥ ♅
04 50	☽ ⚹ ♂
07 21	♂ ᕋ ♄
07 52	☽ ⊥ ♃
10 23	☽ Q ♄
14 01	☽ □ ☉
19 23	☽ Q ☿
20 28	☽ ∠ ♅
22 06	☽ ∠ ♆
23 00	☽ △ ♀
29 Sunday	
00 36	☽ ∠ ♃
04 00	☽ Q ♂
08 25	☽ ⚹ ♄
14 51	☽ ⊥ ♆
17 51	☿ ∥ ♂
23 30	☽ ⊻ ♇
30 Monday	
00 39	☽ ⚹ ♅
02 11	☽ ⊻ ♆
02 22	☉ ∥ ♅
05 12	☽ ⚹ ♃
06 09	☽ ᕋ ♀
06 18	☽ □ ☿
08 21	☿ ∠ ♀
08 45	♀ Q ♅
10 58	☽ ⊥ ♇
11 57	♀ # ♄
12 42	☽ ∠ ♄
15 57	☽ □ ♂
23 46	☽ Q ♅
31 Tuesday	
01 20	☽ ± ♀
03 21	☽ △ ☉
03 45	☽ ∥ ♆
04 22	☽ ∠ ♇
04 40	☽ Q ♃
06 00	☽ ⊥ ♄
14 19	☽ ⊼ ♀
17 41	☽ ⊻ ♄

JUNE 1955

LONGITUDES

Date	Sidereal time h m s	Sun ☉ ° ' "	Moon ☽ ° ' "	Moon ☽ 24.00 ° ' "	Mercury ☿ ° '	Venus ♀ ° '	Mars ♂ ° '	Jupiter ♃ ° '	Saturn ♄ ° '	Uranus ♅ ° '	Neptune ♆ ° '	Pluto ♇ ° '
01	04 36 46	10 ♊ 13 29	25 ♎ 31 53	01 ♏ 33 20	28 ♊ 37	15 ♉ 38	04 ♋ 14	27 ♋ 50	16 ♏ 12	25 ♋ 09	25 ♎ 48	24 ♌ 27
02	04 40 43	11 10 57	07 ♏ 32 46	13 ♏ 30 33	28 46	16 50	04 53	28 01	16 R 09	25 12	25 R 47	24 28
03	04 44 40	12 08 24	19 ♏ 26 59	25 ♏ 22 23	28 50	18 03	05 32	28 12	16 05	25 15	25 46	24 29
04	04 48 36	13 05 50	01 ♐ 17 02	07 ♐ 11 14	28 R 50	19 16	06 11	28 23	16 01	25 18	25 45	24 30
05	04 52 33	14 03 15	13 ♐ 05 15	18 ♐ 59 21	28 45	20 28	06 50	28 34	15 57	25 20	25 44	24 31
06	04 56 29	15 00 40	24 ♐ 53 51	00 ♑ 49 00	28 36	21 41	07 29	28 45	15 54	25 23	25 43	24 31
07	05 00 26	15 58 03	06 ♑ 45 07	12 ♑ 42 30	28 23	22 54	08 08	28 57	15 50	25 26	25 42	24 32
08	05 04 22	16 55 26	18 ♑ 41 29	24 ♑ 42 26	28 06	24 06	08 47	29 08	15 46	25 30	25 41	24 33
09	05 08 19	17 52 47	00 ♒ 45 41	06 ♒ 51 37	27 45	25 19	09 26	29 19	15 43	25 33	25 40	24 34
10	05 12 15	18 50 09	13 ♒ 00 40	19 ♒ 13 13	27 21	26 32	10 05	29 31	15 40	25 36	25 39	24 35
11	05 16 12	19 47 29	25 ♒ 29 42	01 ♓ 50 31	26 54	27 45	10 44	29 43	15 36	25 39	25 39	24 36
12	05 20 09	20 44 49	08 ♓ 16 07	14 ♓ 46 51	26 25	28 ♉ 57	11 23	29 ♋ 54	15 33	25 42	25 38	24 37
13	05 24 05	21 42 09	21 ♓ 23 06	28 ♓ 05 09	25 53	00 ♊ 10	12 02	00 ♌ 06	15 30	25 45	25 37	24 38
14	05 28 02	22 39 28	04 ♈ 53 14	11 ♈ 47 30	25 21	01 23	12 41	00 18	15 27	25 48	25 36	24 39
15	05 31 58	23 36 47	18 ♈ 47 57	25 ♈ 54 32	24 47	02 36	13 20	00 29	15 24	25 52	25 36	24 40
16	05 35 55	24 34 06	03 ♉ 06 59	10 ♉ 24 54	24 13	03 49	13 58	00 41	15 21	25 55	25 35	24 41
17	05 39 51	25 31 24	17 ♉ 47 44	25 ♉ 14 46	23 39	05 02	14 37	00 53	15 18	25 58	25 34	24 42
18	05 43 48	26 28 42	02 ♊ 45 09	10 ♊ 17 53	23 06	06 14	15 16	01 05	15 15	26 01	25 34	24 44
19	05 47 44	27 26 00	17 ♊ 51 54	25 ♊ 26 03	22 35	07 27	15 55	01 17	15 12	26 05	25 33	24 45
20	05 51 41	28 23 17	02 ♋ 59 11	10 ♋ 30 10	22 05	08 40	16 34	01 29	15 09	26 08	25 33	24 46
21	05 55 38	29 ♊ 20 34	17 ♋ 57 56	25 ♋ 21 32	21 38	09 53	17 12	01 41	15 07	26 11	25 32	24 47
22	05 59 34	00 ♋ 17 50	02 ♌ 40 08	09 ♌ 53 04	21 14	11 06	17 51	01 54	15 04	26 15	25 32	24 48
23	06 03 31	01 15 06	16 ♌ 59 50	24 ♌ 00 05	20 53	12 19	18 30	02 06	15 02	26 18	25 31	24 50
24	06 07 27	02 12 21	00 ♍ 53 40	07 ♍ 40 32	20 36	13 32	19 09	02 18	14 59	26 22	25 31	24 51
25	06 11 24	03 09 35	14 ♍ 20 48	20 ♍ 54 43	20 23	14 45	19 47	02 30	14 57	26 25	25 30	24 52
26	06 15 20	04 06 49	27 ♍ 22 35	03 ♎ 44 50	20 14	15 58	20 26	02 43	14 55	26 28	25 30	24 54
27	06 19 17	05 04 02	10 ♎ 01 53	16 ♎ 14 16	20 10	17 11	21 05	02 55	14 53	26 32	25 30	24 55
28	06 23 13	06 01 15	22 ♎ 22 30	28 ♎ 27 07	20 D 10	18 24	21 43	03 08	14 51	26 35	25 29	24 56
29	06 27 10	06 58 27	04 ♏ 28 41	10 ♏ 27 43	20 15	19 37	22 22	03 20	14 49	26 39	25 29	24 58
30	06 31 07	07 ♋ 55 39	16 ♏ 24 45	22 ♏ 20 16	20 ♊ 25	20 ♊ 50	23 ♋ 01	03 ♌ 33	14 ♏ 47	26 ♋ 42	25 ♎ 29	24 ♌ 59

Moon / DECLINATIONS

Date	Moon True ☊ ° '	Moon Mean ☊ ° '	Moon ☽ Latitude ° '	Sun ☉ ° '	Moon ☽ ° '	Mercury ☿ ° '	Venus ♀ ° '	Mars ♂ ° '	Jupiter ♃ ° '	Saturn ♄ ° '	Uranus ♅ ° '	Neptune ♆ ° '	Pluto ♇ ° '
01	26 ♐ 28	27 ♐ 24	04 S 36	21 N 59	14 S 09	23 N 51	15 N 06	24 N 25	21 N 06	14 S 19	21 N 37	08 S 20	23 N 03
02	26 R 26	27 21	03 58	22 07	17 47	23 36	15 28	24 24	21 04	14 18	21 36	08 19	23 02
03	26 24	27 17	03 10	22 15	20 39	23 21	15 50	24 22	21 02	14 17	21 36	08 19	23 02
04	26 22	27 14	02 14	22 23	22 37	23 05	16 11	24 21	20 59	14 16	21 35	08 19	23 01
05	26 21	27 11	01 13	22 30	23 35	22 49	16 32	24 19	20 57	14 15	21 35	08 18	23 01
06	26 21	27 08	00 S 08	22 36	23 29	22 32	16 52	24 17	20 55	14 14	21 34	08 18	23 00
07	26 D 21	27 05	00 N 57	22 42	22 19	22 15	17 12	24 15	20 53	14 13	21 34	08 18	23 00
08	26 21	27 02	02 01	22 48	20 09	21 57	17 32	24 13	20 50	14 12	21 33	08 18	22 59
09	26 22	26 58	03 00	22 54	17 04	21 40	17 51	24 10	20 48	14 12	21 32	08 17	22 59
10	26 23	26 55	03 51	22 59	13 13	21 22	18 10	24 08	20 46	14 11	21 32	08 17	22 58
11	26 24	26 52	04 33	23 03	08 44	21 05	18 29	24 05	20 43	14 10	21 31	08 17	22 58
12	26 24	26 49	05 02	23 07	03 S 48	20 48	18 47	24 02	20 41	14 09	21 31	08 16	22 57
13	26 24	26 46	05 17	23 11	01 N 26	20 31	19 04	23 58	20 38	14 08	21 30	08 16	22 57
14	26 R 24	26 43	05 14	23 14	06 45	20 15	19 21	23 55	20 36	14 08	21 29	08 16	22 56
15	26 24	26 39	04 54	23 17	11 54	19 59	19 37	23 51	20 33	14 07	21 29	08 16	22 56
16	26 24	26 36	04 15	23 20	16 33	19 45	19 53	23 47	20 31	14 06	21 28	08 16	22 55
17	26 24	26 33	03 19	23 22	20 19	19 31	20 09	23 43	20 28	14 06	21 28	08 15	22 54
18	26 D 23	26 30	02 08	23 24	22 48	19 19	20 24	23 39	20 26	14 05	21 27	08 15	22 54
19	26 24	26 27	00 N 47	23 25	23 40	19 08	20 38	23 35	20 23	14 04	21 26	08 15	22 53
20	26 R 24	26 23	00 S 37	23 26	22 48	18 58	20 52	23 30	20 21	14 04	21 26	08 15	22 53
21	26 23	26 20	01 57	23 27	20 18	18 50	21 05	23 26	20 18	14 03	21 25	08 15	22 52
22	26 23	26 17	03 08	23 27	16 31	18 43	21 17	23 21	20 15	14 03	21 24	08 15	22 52
23	26 23	26 14	04 06	23 26	11 50	18 38	21 30	23 15	20 13	14 02	21 24	08 15	22 51
24	26 22	26 11	04 47	23 26	06 41	18 34	21 41	23 10	20 10	14 02	21 23	08 14	22 50
25	26 22	26 08	05 11	23 24	01 N 23	18 32	21 52	23 05	20 07	14 01	21 23	08 14	22 50
26	26 21	26 04	05 18	23 23	03 S 49	18 32	22 02	22 59	20 04	14 01	21 22	08 14	22 49
27	26 D 21	26 01	05 09	23 21	08 42	18 34	22 12	22 53	20 01	14 01	21 21	08 14	22 49
28	26 22	25 58	04 46	23 18	13 07	18 36	22 21	22 47	19 59	14 00	21 21	08 14	22 48
29	26 22	25 55	04 10	23 16	16 56	18 41	22 29	22 41	19 56	14 00	21 20	08 14	22 47
30	26 ♐ 23	25 ♐ 52	03 S 24	23 N 12	20 S 00	18 N 47	22 N 37	22 N 35	19 N 53	14 S 00	21 N 19	08 S 14	22 N 47

ZODIAC SIGN ENTRIES

Date	h	m	Planets
01	20	54	☽ ♏
04	09	24	☽ ♐
06	22	21	☽ ♑
09	10	30	☽ ♒
11	20	32	☽ ♓
13	00	07	♃ ♌
13	08	38	♀ ♊
14	03	24	☽ ♈
16	06	50	☽ ♉
18	07	37	☽ ♊
20	07	15	☽ ♋
22	04	31	☉ ♋
22	07	36	☽ ♌
24	10	26	☽ ♍
26	16	55	☽ ♎
29	03	04	☽ ♏

LATITUDES

Date	Mercury ☿ ° '	Venus ♀ ° '	Mars ♂ ° '	Jupiter ♃ ° '	Saturn ♄ ° '	Uranus ♅ ° '	Neptune ♆ ° '	Pluto ♇ ° '
01	00 N 25	01 S 29	01 N 02	00 N 31	02 N 29	00 N 31	01 N 46	10 N 17
04	00 S 21	01 25	01 03	00 31	02 29	00 31	01 46	10 17
07	01 11	01 20	01 04	00 31	02 28	00 31	01 46	10 16
10	02 03	01 15	01 04	00 31	02 28	00 31	01 45	10 16
13	02 52	01 09	01 05	00 31	02 27	00 31	01 45	10 15
16	03 34	01 03	01 05	00 31	02 26	00 31	01 45	10 15
19	04 07	00 56	01 06	00 31	02 26	00 31	01 45	10 14
22	04 27	00 50	01 06	00 31	02 25	00 31	01 45	10 13
25	04 34	00 43	01 06	00 32	02 24	00 31	01 45	10 13
28	04 29	00 36	01 07	00 32	02 24	00 31	01 45	10 12
31	04 S 13	00 S 28	01 N 07	00 N 32	02 N 23	00 N 31	01 N 44	10 N 12

DATA

Julian Date	2435260
Delta T	+31 seconds
Ayanamsa	23° 14' 20"
Synetic vernal point	05° ♓ 52' 40"
True obliquity of ecliptic	23° 26' 42"

MOON'S PHASES, APSIDES AND POSITIONS ☽

Date	h	m	Phase	Longitude ° '	Eclipse Indicator
05	14	08	○	14 ♐ 08	
13	12	37	☾	21 ♓ 44	
20	04	12	●	28 ♊ 05	Total
27	01	44	☽	04 ♎ 40	

Day	h	m		
05	02	31	Apogee	
19	13	35	Perigee	
05	21	47	Max dec	23° S 40'
13	05	30	0N	
19	12	00	Max dec	23° N 40'
25	18	18	0S	

LONGITUDES

Date	Chiron ⚷ ° '	Ceres ⚳ ° '	Pallas ⚴ ° '	Juno ⚵ ° '	Vesta ⚶ ° '	Black Moon Lilith ⚸ ° '
01	05 ♒ 16	24 ♑ 54	21 ♐ 08	04 ♏ 27	18 ♋ 27	09 ♐ 13
11	04 ♒ 58	23 ♑ 52	18 ♐ 13	03 ♏ 19	22 ♋ 40	10 ♐ 20
21	04 ♒ 35	22 ♑ 19	15 ♐ 22	02 ♏ 43	26 ♋ 58	11 ♐ 26
31	04 ♒ 07	20 ♑ 21	12 ♐ 54	02 ♏ 39	01 ♌ 22	12 ♐ 33

All ephemeris data is given at 12.00 UT and the Moon's longitude is additionally given for 24.00 UT

ASPECTARIAN

01 Wednesday
09 52 ☽ ⚹ ♇
09 59 ☉ ∠ ♅
11 14 ☽ □ ♅
11 20 ☽ ⚼ ☉
12 32 ☽ ☌ ♆
12 55 ☽ ∥ ♄
16 39 ☽ □ ♃
16 47 ♀ Q ♃
18 13 ☽ △ ☿
18 26 ☽ # ♀
22 50 ♀ ☍ ♄

02 Thursday
02 08 ☉ ⚼ ♆
06 21 ☽ △ ♂
06 51 ☽ ± ☉
09 50 ☽ Q ♇
19 57 ☽ ⊼ ☉

03 Friday
00 37 ☽ ⚼ ☿
05 13 ☽ ☌ ♄
08 51 ☽ ☍ ♀
14 19 ☽ ⚼ ♂
15 47 ☽ # ♃
18 52 ☽ ± ☿
20 43 ☉ Q ♇
22 06 ☽ # ♅
22 12 ☽ □ ♇
22 46 ☿ St R
23 47 ☽ △ ♅

04 Saturday
00 47 ☽ ⊻ ♆
06 01 ☽ △ ♃
07 02 ☽ ⊼ ☿
08 00 ☽ # ☉
09 38 ☽ ± ♂
12 57 ☽ ⊥ ♆
17 57 ☿ ∥ ♇
19 03 ☽ # ☿
19 17 ☽ # ♇
20 58 ☉ ∠ ♃
22 32 ☽ ⊼ ♂

05 Sunday
06 24 ☽ ⚼ ♅
07 13 ☽ ∠ ♆
13 00 ☽ ⚼ ♃
14 08 ☽ ☍ ☉
17 48 ☽ ⊻ ♄

06 Monday
00 46 ☽ ± ♅
01 36 ☿ ⊻ ♃
04 44 ☽ ⊼ ♀
05 56 ☽ ⊥ ♄
07 35 ☽ ± ♃
07 36 ☉ ∥ ☿
11 15 ☽ △ ♇
13 00 ☽ ⊼ ♅
13 40 ☽ ⚹ ♆
18 18 ☽ ± ♀
19 24 ☽ ☍ ☿
19 57 ☽ ⊼ ♃

07 Tuesday
00 06 ☽ ∠ ♄
00 35 ☽ # ♇
03 31 ☿ ⊥ ♀
06 23 ☽ # ☉
08 49 ☉ ⊼ ♄
13 13 ☽ # ☿
13 55 ☽ Q ♆
14 34 ☽ ⚼ ♀
14 57 ☽ ☍ ♂
17 38 ☽ ⚼ ♇
21 41 ☽ # ♅
22 15 ♀ ∠ ♂

08 Wednesday
05 16 ☽ # ♃
06 11 ☽ ⚹ ♄
08 09 ☽ ⊼ ☉
11 43 ☽ ± ♇
20 57 ♀ □ ♇
21 11 ☽ ± ☉
23 43 ☽ ⊼ ♇

09 Thursday
00 01 ☽ △ ♀
01 37 ☽ ☍ ♅
01 56 ☽ □ ♆
06 00 ☽ Q ♄
06 13 ☽ ⊼ ☿
06 54 ☽ # ♀
09 07 ☽ ☍ ♃
16 32 ☽ ⚼ ☉
16 37 ♀ ⚹ ♅
17 04 ♂ ∠ ♇
17 43 ☽ ± ☿
18 54 ♀ ⊼ ♆
22 15 ☿ ∥ ♅

10 Friday
05 59 ☽ ⊼ ♂
06 22 ☽ ∥ ♄
10 39 ☉ ∥ ♇
10 46 ☽ ⚼ ☿
17 07 ☽ □ ♄
18 17 ☽ ± ♂
23 55 ☿ ⊻ ♀

11 Saturday
00 12 ☽ △ ☉
08 09 ☉ ⊥ ♅
10 18 ☽ ☍ ♇
11 08 ♅ □ ♆
12 17 ☽ △ ♆
12 17 ☽ ⊼ ♅
12 29 ☽ ⚼ ♂
14 19 ☽ ∥ ♆
14 35 ☽ △ ☿
16 43 ☽ □ ♀
20 06 ☽ ⊼ ♃
23 41 ☽ ± ♅

12 Sunday
07 32 ☽ ± ♃
16 22 ☽ ⚼ ♆
16 31 ☽ ⚼ ♅
18 04 ☽ △ ♂
23 21 ☿ ∥ ♃

13 Monday
00 24 ☽ ⚼ ♃
01 21 ☽ △ ♄
05 35 ☽ Q ♀
07 04 ☉ ± ♄
08 49 ☽ ± ♆
10 15 ♀ ⚹ ♃
12 37 ☽ □ ☉
17 36 ☿ ⊻ ♅
17 51 ☽ ⊼ ♇
19 36 ☽ ⊼ ♆
19 46 ☽ □ ☿
19 52 ☽ △ ♅

14 Tuesday
00 18 ☿ △ ♆
03 48 ☽ △ ♃
04 13 ☽ ⚼ ♄
04 32 ☽ ± ♇
05 14 ☽ ⚹ ♀
16 21 ♀ ± ♆
18 56 ☽ # ♆
19 55 ☽ ± ♄
20 19 ☽ ⚼ ♇
22 47 ☽ Q ☉

15 Wednesday
02 06 ☽ Q ☿
02 11 ☽ □ ♂
06 12 ☽ ⊼ ♄
09 46 ☽ ∠ ♀
16 36 ☿ ⚹ ♇
20 44 ☽ ⚹ ☉
21 14 ☿ ⊥ ♃
21 43 ☽ ⚹ ☿
21 56 ☽ △ ♇
23 01 ☽ # ♄
23 28 ☽ ☍ ♆
23 58 ☽ □ ♅

16 Thursday
02 22 ☽ ⊥ ♀
05 19 ☿ ∥ ♀
06 26 ☉ ☌ ☿
07 55 ☽ □ ♃
10 01 ☽ Q ♂
13 15 ☽ ⊻ ♀
15 05 ☉ ⚹ ♇
15 44 ☉ ⊥ ♃
21 40 ☽ ∠ ☿
23 21 ☽ ∠ ☉

17 Friday
05 46 ☽ Q ♅
06 37 ☽ ⚹ ♂
06 39 ☽ ∥ ☿
07 58 ☽ ☍ ♄
10 37 ☽ ∥ ♀
11 47 ☽ ⊥ ☿
13 10 ☽ ∥ ♃
13 13 ☉ △ ♆
13 47 ☽ Q ♃
14 59 ☽ ⊥ ☉
21 06 ☽ ⊻ ☿
21 20 ☽ ∥ ♅
23 09 ☽ □ ♇
23 50 ☉ ⊻ ♅

18 Saturday
00 31 ☽ ⊼ ♆
01 12 ☽ ⚹ ♅
01 18 ☽ ⊻ ☉
07 52 ☽ ∠ ♂
09 18 ☽ ⚹ ♃
10 06 ☽ ± ♆
11 13 ♂ △ ♄
13 22 ☽ ∥ ♇
15 12 ♀ ∥ ♃
18 02 ☽ ☌ ♀
22 49 ☽ ∥ ☉
22 50 ☽ ⊥ ♂

19 Sunday
00 25 ☽ ⚼ ♆
01 12 ☽ ∠ ♅
03 52 ☽ Q ♇
05 06 ☽ ∥ ♂
07 47 ☽ ⊼ ♄
08 46 ☽ ⊻ ♂
09 28 ☽ ∠ ♃
15 31 ☽ ⊥ ♅
17 16 ☽ ± ♄
19 14 ☽ ☌ ☿
20 57 ☽ ∥ ♂
22 55 ☽ ⚹ ♇
23 55 ☽ ⊥ ♃

20 Monday
00 11 ☽ △ ♆
00 41 ☽ ∥ ☉
01 04 ☽ ⊻ ♅
04 12 ☽ ☌ ☉
07 31 ☽ ⚼ ♄
09 35 ☽ ⊻ ♃
10 53 ☽ ∥ ♇
21 52 ☽ ⊻ ♀
22 50 ☽ ∠ ♇

21 Tuesday
02 55 ☽ ∥ ♅
06 18 ☽ ∥ ♀
07 02 ☉ ∥ ♂
07 25 ☽ △ ♄
08 21 ☽ ⊥ ♀
10 43 ☽ ☌ ♂
12 02 ☽ ∥ ♃
13 20 ☽ ⊥ ♇
17 47 ☽ ⊻ ☿
22 31 ☽ ∥ ☿
23 05 ☽ ⊻ ♇

22 Wednesday
00 15 ☽ ∠ ♀
00 17 ☽ □ ♆
00 41 ♀ ⚼ ♆
01 24 ☽ ☌ ♅
03 18 ☽ ⊥ ☿
06 31 ☉ ⚼ ♄
07 49 ☽ ⊻ ☉
10 42 ☽ ☌ ♃
14 56 ♀ ∠ ♅
17 46 ☽ ∠ ☿
18 26 ☽ ⊥ ☉

23 Thursday
00 35 ☿ ± ♄
01 05 ♀ ∥ ♅
01 06 ☽ # ♄
03 21 ☽ ⚹ ♀
06 07 ☽ Q ♆
08 40 ☽ □ ♄
10 39 ☽ ∠ ☉
14 41 ☽ ⊻ ♂
18 30 ☽ ⚹ ☿
19 32 ♀ ⊥ ♂
22 11 ♀ Q ♇

24 Friday
00 37 ♂ ⊥ ♇
01 27 ☽ ☌ ♇
01 29 ☽ ⊥ ♂
01 45 ☽ Q ♀
02 37 ☽ ⚹ ♆
04 03 ☽ ⊻ ♅
04 53 ☽ # ♆
14 29 ☽ ⚹ ☉
14 31 ☽ ⊻ ♃
14 35 ☽ ⊥ ♅
14 57 ☽ Q ☿
15 02 ☉ ⊻ ♃
15 41 ☽ Q ♄
18 01 ☽ ∠ ♂

25 Saturday
01 19 ☽ ⊥ ♃
05 04 ☽ ∠ ♆
06 41 ☽ ∠ ♅
08 57 ☿ ⊥ ♅
12 49 ☽ □ ♀
13 06 ☽ ⚹ ♄
13 35 ☽ Q ☉
15 49 ♀ ⊼ ♄
17 51 ☽ ∠ ♃
21 24 ☽ ⊥ ♆
22 27 ☽ ⚹ ♂
22 53 ☽ □ ☿

26 Sunday
05 49 ☿ ⊻ ♂
07 22 ☽ ⊻ ♇
08 30 ☽ ⊻ ♆
10 18 ☽ ⚹ ♅
16 45 ☽ ∠ ♄
18 37 ☽ ⊥ ♇
22 01 ☽ Q ♂
22 12 ☽ ⚹ ♃

27 Monday
01 44 ☽ □ ☉
09 07 ☽ Q ♅
09 38 ☽ ∥ ♆
09 48 ☽ ⊥ ♄
11 47 ☽ ∠ ♇
21 20 ☽ ⊻ ♄
21 36 ☽ Q ♃
23 10 ☿ St D

28 Tuesday
03 22 ☽ △ ♀
05 24 ♀ ∠ ♃
07 39 ☽ △ ☿
09 34 ♂ ∥ ♇
10 39 ☽ □ ♂
17 03 ☽ ⚹ ♇
17 11 ☽ ∥ ♄
18 08 ☽ ☌ ♆
20 21 ☽ □ ♅

29 Wednesday
09 41 ☽ □ ♃
12 19 ☽ ⚼ ♀
13 33 ☽ ⚼ ☿
16 59 ☽ Q ♇
17 26 ☽ △ ☉

30 Thursday
01 16 ☽ # ☿
01 58 ☿ ☌ ♀
07 53 ☽ ± ☿
08 23 ♀ ∥ ♂
08 28 ☽ ± ♀
08 43 ☽ ☌ ♄
09 14 ♀ ⊥ ♅
10 53 ☽ # ♃
10 56 ♀ ± ♄
20 14 ☽ ⊼ ☿
22 00 ☽ ⊼ ♀

JULY 1955

LONGITUDES

Date	Sidereal time h m s	Sun ☉ ° ' "	Moon ☽ ° ' "	Moon ☽ 24.00 ° ' "	Mercury ☿ ° '	Venus ♀ ° '	Mars ♂ ° '	Jupiter ♃ ° '	Saturn ♄ ° '	Uranus ♅ ° '	Neptune ♆ ° '	Pluto ♇ ° '
01	06 35 03	08 ♋ 52 50	28 ♏ 14 46	04 ♐ 08 42	20 ♊ 39	22 ♊ 04	23 ♋ 39	03 ♌ 45	14 ♏ 45	26 ♋ 46	25 ♎ 29	25 ♌ 00
02	06 39 00	09 50 01	10 ♐ 02 29	15 ♐ 56 31	20 59	23 17	24 18	03 58	14 R 44	26 49	25 R 28	25 02
03	06 42 56	10 47 13	21 ♐ 51 09	27 ♐ 46 43	21 23	24 30	24 56	04 11	14 42	26 53	25 28	25 03
04	06 46 53	11 44 24	03 ♑ 43 33	09 ♑ 41 55	21 53	25 43	25 35	04 23	14 41	26 57	25 28	25 05
05	06 50 49	12 41 34	15 ♑ 42 05	21 ♑ 44 16	22 27	26 56	26 13	04 36	14 39	27 00	25 28	25 06
06	06 54 46	13 38 45	27 ♑ 48 43	03 ♒ 55 39	23 06	28 09	26 52	04 49	14 38	27 04	25 28	25 08
07	06 58 42	14 35 56	10 ♒ 05 15	16 ♒ 17 42	23 50	29 ♊ 23	27 30	05 01	14 37	27 07	25 28	25 09
08	07 02 39	15 33 07	22 ♒ 33 14	28 ♒ 52 02	24 39	00 ♋ 36	28 09	05 14	14 36	27 11	25 D 28	25 11
09	07 06 36	16 30 19	05 ♓ 14 16	11 ♓ 40 09	25 32	01 49	28 47	05 27	14 35	27 15	25 28	25 12
10	07 10 32	17 27 30	18 ♓ 09 53	24 ♓ 43 39	26 30	03 02	29 ♋ 26	05 40	14 34	27 18	25 28	25 14
11	07 14 29	18 24 43	01 ♈ 21 37	08 ♈ 03 58	27 33	04 16	00 ♌ 04	05 53	14 33	27 22	25 28	25 16
12	07 18 25	19 21 55	14 ♈ 50 50	21 ♈ 42 20	28 40	05 29	00 43	06 06	14 32	27 26	25 28	25 17
13	07 22 22	20 19 08	28 ♈ 38 29	05 ♉ 39 19	29 ♊ 52	06 42	01 21	06 19	14 32	27 29	25 28	25 19
14	07 26 18	21 16 22	12 ♉ 44 43	19 ♉ 54 31	01 ♋ 08	07 56	01 59	06 32	14 31	27 33	25 29	25 21
15	07 30 15	22 13 37	27 ♉ 08 27	04 ♊ 26 06	02 28	09 09	02 38	06 45	14 31	27 36	25 29	25 22
16	07 34 11	23 10 52	11 ♊ 46 58	19 ♊ 10 25	03 52	10 23	03 16	06 58	14 30	27 40	25 29	25 24
17	07 38 08	24 08 08	26 ♊ 35 41	04 ♋ 01 57	05 21	11 36	03 55	07 11	14 30	27 44	25 29	25 26
18	07 42 05	25 05 24	11 ♋ 28 18	18 ♋ 53 45	06 54	12 50	04 33	07 24	14 30	27 48	25 30	25 27
19	07 46 01	26 02 41	26 ♋ 17 19	03 ♌ 38 03	08 31	14 03	05 11	07 37	14 D 30	27 51	25 30	25 29
20	07 49 58	26 59 58	10 ♌ 55 04	18 ♌ 07 32	10 11	15 17	05 50	07 50	14 30	27 55	25 30	25 31
21	07 53 54	27 57 15	25 ♌ 14 46	02 ♍ 16 13	11 55	16 30	06 28	08 03	14 30	27 59	25 31	25 33
22	07 57 51	28 54 33	09 ♍ 11 29	16 ♍ 00 21	13 42	17 44	07 07	08 17	14 31	28 02	25 31	25 34
23	08 01 47	29 ♋ 51 52	22 ♍ 42 42	29 ♍ 18 36	15 33	18 58	07 45	08 30	14 31	28 06	25 32	25 36
24	08 05 44	00 ♌ 49 10	05 ♎ 48 15	12 ♎ 11 56	17 27	20 11	08 23	08 43	14 31	28 10	25 32	25 38
25	08 09 40	01 46 29	18 ♎ 30 04	24 ♎ 43 07	19 23	21 25	09 02	08 56	14 32	28 13	25 33	25 40
26	08 13 37	02 43 48	00 ♏ 51 39	06 ♏ 56 12	21 21	22 38	09 40	09 09	14 33	28 17	25 33	25 41
27	08 17 34	03 41 08	12 ♏ 57 25	18 ♏ 55 55	23 22	23 52	10 18	09 22	14 33	28 21	25 34	25 43
28	08 21 30	04 38 28	24 ♏ 52 20	00 ♐ 47 17	25 24	25 06	10 56	09 36	14 34	28 24	25 35	25 45
29	08 25 27	05 35 49	06 ♐ 41 23	12 ♐ 35 14	27 27	26 20	11 35	09 49	14 35	28 28	25 35	25 47
30	08 29 23	06 33 11	18 ♐ 29 22	24 ♐ 24 19	29 ♋ 32	27 33	12 13	10 02	14 36	28 32	25 36	25 49
31	08 33 20	07 ♌ 30 33	00 ♑ 20 35	06 ♑ 18 35	01 ♌ 37	28 ♋ 47	12 ♌ 51	10 ♌ 15	14 ♏ 37	28 ♋ 35	25 ♎ 37	25 ♌ 51

Date	Moon True ☊ ° '	Moon Mean ☊ ° '	Moon ☽ Latitude ° '
01	26 ♐ 24	25 ♐ 49	02 S 30
02	26 D 25	25 45	01 30
03	26 26	25 42	00 S 25
04	26 R 26	25 39	00 N 40
05	26 25	25 36	01 45
06	26 23	25 33	02 45
07	26 21	25 29	03 38
08	26 18	25 26	04 22
09	26 15	25 23	04 54
10	26 13	25 20	05 12
11	26 11	25 17	05 14
12	26 10	25 14	04 59
13	26 D 10	25 10	04 26
14	26 11	25 07	03 37
15	26 13	25 04	02 33
16	26 14	25 01	01 N 19
17	26 R 14	24 58	00 S 02
18	26 14	24 54	01 23
19	26 12	24 51	02 37
20	26 08	24 48	03 41
21	26 04	24 45	04 29
22	25 59	24 42	04 59
23	25 54	24 39	05 12
24	25 51	24 35	05 08
25	25 48	24 32	04 48
26	25 47	24 29	04 16
27	25 D 48	24 26	03 33
28	25 49	24 23	02 41
29	25 51	24 20	01 43
30	25 52	24 16	00 40
31	25 ♐ 52	24 ♐ 13	00 N 24

DECLINATIONS

Date	Sun ☉ ° '	Moon ☽ ° '	Mercury ☿ ° '	Venus ♀ ° '	Mars ♂ ° '	Jupiter ♃ ° '	Saturn ♄ ° '	Uranus ♅ ° '	Neptune ♆ ° '	Pluto ♇ ° '
01	23 N 09	22 S 13	18 N 54	22 N 44	22 N 28	19 N 50	13 S 59	21 N 18	08 S 14	22 N 46
02	23 05	23 26	19 03	22 51	22 22	19 47	13 59	21 18	08 14	22 45
03	23 00	23 37	19 12	22 57	22 15	19 44	13 59	21 17	08 14	22 45
04	22 56	22 43	19 23	23 02	22 08	19 41	13 59	21 16	08 14	22 44
05	22 50	20 47	19 35	23 06	22 01	19 38	13 59	21 16	08 14	22 44
06	22 45	17 54	19 47	23 10	21 54	19 35	13 58	21 15	08 14	22 43
07	22 39	14 13	20 00	23 13	21 46	19 32	13 58	21 14	08 14	22 42
08	22 32	09 52	20 14	23 16	21 39	19 29	13 58	21 14	08 14	22 42
09	22 25	05 S 02	20 28	23 18	21 31	19 26	13 58	21 13	08 14	22 41
10	22 18	00 N 06	20 42	23 19	21 23	19 23	13 58	21 12	08 14	22 40
11	22 11	05 20	20 56	23 19	21 15	19 20	13 58	21 12	08 14	22 40
12	22 03	10 26	21 10	23 19	21 07	19 17	13 58	21 11	08 14	22 39
13	21 54	15 08	21 24	23 18	20 58	19 13	13 58	21 10	08 15	22 38
14	21 46	19 07	21 37	23 16	20 50	19 10	13 58	21 09	08 15	22 38
15	21 37	22 01	21 49	23 14	20 41	19 07	13 58	21 09	08 15	22 37
16	21 27	23 30	22 01	23 11	20 32	19 04	13 59	21 08	08 15	22 36
17	21 17	23 22	22 11	23 08	20 23	19 00	13 59	21 07	08 15	22 36
18	21 07	21 35	22 21	23 03	20 14	18 57	13 59	21 06	08 15	22 35
19	20 57	18 19	22 28	22 58	20 05	18 54	13 59	21 06	08 15	22 34
20	20 46	13 57	22 34	22 53	19 56	18 51	14 00	21 05	08 16	22 34
21	20 35	08 53	22 39	22 46	19 46	18 47	14 00	21 04	08 16	22 33
22	20 23	03 N 30	22 41	22 39	19 36	18 44	14 00	21 04	08 16	22 32
23	20 11	01 S 53	22 41	22 32	19 27	18 41	14 01	21 03	08 16	22 32
24	19 59	07 01	22 38	22 23	19 17	18 37	14 01	21 02	08 17	22 31
25	19 46	11 42	22 33	22 14	19 07	18 34	14 01	21 01	08 17	22 30
26	19 33	15 46	22 26	22 05	18 56	18 30	14 02	21 01	08 17	22 30
27	19 20	19 06	22 15	21 54	18 46	18 27	14 02	21 00	08 17	22 29
28	19 06	21 35	22 02	21 43	18 36	18 23	14 03	20 59	08 18	22 28
29	18 53	23 07	21 47	21 32	18 25	18 20	14 03	20 58	08 18	22 28
30	18 38	23 37	21 29	21 20	18 14	18 17	14 04	20 58	08 18	22 27
31	18 N 24	23 S 02	21 N 08	21 N 07	18 N 04	18 N 13	14 S 04	20 N 57	08 S 19	22 N 26

ZODIAC SIGN ENTRIES

Date	h	m	Planets
01	15	34	☽ ♐
04	04	29	☽ ♑
06	16	18	☽ ♒
08	00	15	♀ ♋
09	02	09	☽ ♓
11	09	22	♂ ♌
11	09	33	☽ ♈
13	14	20	☽ ♉
13	14	44	☿ ♋
15	16	43	☽ ♊
17	17	30	☽ ♋
19	18	03	☽ ♌
21	20	06	☽ ♍
23	15	25	☉ ♌
24	01	16	☽ ♎
26	10	19	☽ ♏
28	22	24	☽ ♐
30	17	22	☿ ♌
31	11	18	☽ ♑

LATITUDES

Date	Mercury ☿ ° '	Venus ♀ ° '	Mars ♂ ° '	Jupiter ♃ ° '	Saturn ♄ ° '	Uranus ♅ ° '	Neptune ♆ ° '	Pluto ♇ ° '
01	04 S 13	00 S 28	01 N 07	00 N 32	02 N 23	00 N 31	01 N 44	10 N 12
04	03 49	00 21	01 07	00 32	02 22	00 31	01 44	10 11
07	03 18	00 13	01 08	00 32	02 22	00 31	01 44	10 11
10	02 42	00 S 06	01 08	00 32	02 21	00 31	01 44	10 11
13	02 03	00 N 02	01 08	00 32	02 20	00 31	01 44	10 10
16	01 22	00 09	01 08	00 33	02 19	00 31	01 44	10 10
19	00 42	00 16	01 08	00 33	02 18	00 31	01 43	10 10
22	00 S 04	00 23	01 09	00 33	02 18	00 31	01 43	10 10
25	00 N 31	00 30	01 09	00 33	02 17	00 31	01 43	10 10
28	01 00	00 37	01 09	00 33	02 16	00 31	01 43	10 09
31	01 N 22	00 N 43	01 N 09	00 N 34	02 N 15	00 N 31	01 N 43	10 N 09

DATA

Julian Date	2435290
Delta T	+31 seconds
Ayanamsa	23° 14' 25"
Synetic vernal point	05° ♓ 52' 35"
True obliquity of ecliptic	23° 26' 41"

MOON'S PHASES, APSIDES AND POSITIONS ☽

Date	h	m	Phase	Longitude ° '	Eclipse Indicator
05	05	29	○	12 ♑ 26	
12	20	31	☾	19 ♈ 42	
19	11	35	●	26 ♋ 02	
26	16	00	☽	02 ♏ 53	

Day	h	m		
02	08	25	Apogee	
17	20	20	Perigee	
29	21	45	Apogee	
03	03	59	Max dec	23° S 41'
10	11	30	0N	
16	22	04	Max dec	23° N 39'
23	03	32	0S	
30	11	11	Max dec	23° S 37'

LONGITUDES

Date	Chiron ⚷ ° '	Ceres ⚳ ° '	Pallas ⚴ ° '	Juno ⚵ ° '	Vesta ⚶ ° '	Black Moon Lilith ⚸ ° '
01	04 ♒ 07	20 ♑ 21	12 ♐ 54	02 ♏ 39	01 ♌ 22	12 ♐ 33
11	03 ♒ 36	18 ♑ 11	11 ♐ 02	03 ♏ 05	05 ♌ 49	13 ♐ 39
21	03 ♒ 03	16 ♑ 02	09 ♐ 52	04 ♏ 00	10 ♌ 21	14 ♐ 46
31	02 ♒ 28	14 ♑ 06	09 ♐ 26	05 ♏ 20	14 ♌ 55	15 ♐ 52

All ephemeris data is given at 12.00 UT and the Moon's longitude is additionally given for 24.00 UT

ASPECTARIAN

h m	Aspects
01 Friday	
00 58	☽ # ♅
02 08	☽ △ ♂
02 21	☽ ⅊ ☉
05 24	☽ □ ♇
06 22	☽ ⊻ ♆
08 59	☽ △ ♅
15 34	☽ # ♂
18 10	♀ ∥ ♇
18 34	☽ ⊥ ♆
19 32	☿ ± ♄
20 32	☽ # ♇
20 45	☽ # ♀
22 15	☽ ± ☉
22 59	☿ ⊥ ♅
23 24	☽ △ ♃
02 Saturday	
03 10	☽ # ☉
10 24	☽ ⅊ ♂
11 32	☽ ⊼ ☉
12 53	☽ ∠ ♆
15 39	☽ ⅊ ♅
17 06	☉ ∠ ♇
21 31	☽ ⊻ ♄
03 Sunday	
05 45	☽ ± ♂
06 28	☽ ⅊ ♃
09 40	☽ ⊥ ♄
10 02	☽ ± ♅
11 01	☽ ☍ ☿
16 35	♂ ⊻ ♇
17 58	☽ ☍ ♀
18 30	☽ △ ♇
18 37	☽ ⊼ ♂
19 20	☽ ✶ ♆
21 14	☉ ∥ ♀
22 15	☽ ⊼ ♅
23 13	♀ ✶ ♇
04 Monday	
01 02	☽ ± ♃
03 51	☽ ∠ ♄
06 20	♀ ⊻ ♂
06 42	☽ # ♀
07 06	♀ △ ♆
07 48	♂ □ ♆
08 05	☽ # ☉
11 45	☽ # ♇
13 21	☽ ⊼ ♃
19 31	☽ Q ♆
21 20	☽ # ♂
05 Tuesday	
00 47	☽ ⅊ ♇
05 29	☽ ☍ ☉
07 05	☽ # ♅
09 55	☽ ✶ ♄
13 23	♀ ⊻ ♅
17 16	☿ ∥ ♃
18 47	☽ ± ♇
22 15	☽ # ☿
22 43	☽ # ♃
06 Wednesday	
02 09	☽ ⊼ ☿
06 42	☽ ⊼ ♇
07 22	☽ □ ♆
09 41	☽ Q ♄
10 02	☽ ☍ ♂
10 31	☽ ☍ ♅
12 45	☽ ⊼ ♀
14 41	☽ ± ☿
20 12	♂ ☌ ♅
20 24	☉ ∥ ♇
07 Thursday	
01 49	☽ ± ♀
01 58	☽ ☍ ♃
03 37	♀ ⊥ ♃
09 24	☽ ⅊ ☿
12 22	☉ △ ♄
13 25	☽ ∥ ♄
16 36	♀ ⅊ ♄
19 40	♆ St D
20 45	☽ □ ♄
21 12	☽ ⅊ ♀
21 27	☽ ⊼ ☉
08 Friday	
09 56	☽ ± ☉
16 17	☽ △ ☿
17 01	☽ ☍ ♇
17 33	☽ △ ♆
20 17	☽ ∥ ♆
20 51	☽ ⊼ ♅
23 12	☽ ⊼ ♂
09 Saturday	
03 09	☿ ✶ ♇
04 25	☽ ⅊ ☉
04 54	☽ △ ♀
08 14	☽ ± ♅
10 11	☿ △ ♆
11 07	☽ ± ♂
12 25	☽ ⊼ ♃
21 46	☽ ⅊ ♆
23 48	☽ ± ♃
10 Sunday	
01 08	☽ ⅊ ♅
04 45	☽ ⅊ ♂
05 22	☽ △ ♄
10 36	☽ △ ☉
14 23	☽ ± ♆
16 40	☽ ⅊ ♃
11 Monday	
00 57	☽ ⊼ ♇
01 21	☽ ⊼ ♆
04 31	☽ □ ☿
04 45	☽ △ ♅
07 42	☿ ⊻ ♅
08 44	☽ ⅊ ♄
09 34	☽ △ ♂
11 49	☽ ± ♇
17 44	☽ □ ♀
20 14	☽ △ ♃
22 40	♂ ∥ ♅
12 Tuesday	
00 51	☽ ± ♄
01 30	☽ # ♆
03 56	☽ ⅊ ♇
08 15	☿ ∥ ♂
10 00	☉ ⊥ ♇
11 27	☽ ⊼ ♄
13 12	☿ ∥ ♅
15 29	☽ Q ☿
20 31	☽ □ ☉
13 Wednesday	
02 37	♀ ⊻ ♃
04 33	☽ Q ♀
05 33	☿ ⅊ ♄
05 46	☽ # ♄
06 15	☽ △ ♇
06 32	☽ ☍ ♆
10 00	☽ □ ♅
14 18	☽ ✶ ☿
16 52	☽ □ ♂
22 36	☿ ⊥ ♃
14 Thursday	
01 20	☽ □ ♃
03 06	☽ ✶ ♀
05 43	☽ Q ☉
12 25	☽ ∥ ♃
14 59	☽ ☍ ♄
16 44	☽ Q ♅
18 15	☽ ∠ ☿
21 46	☉ ∥ ☿
15 Friday	
00 19	☽ ∥ ♂
00 42	☽ Q ♂
03 17	☽ ✶ ☉
03 41	☽ ∥ ♅
06 36	☽ ∠ ♀
07 59	☽ Q ♃
08 11	☽ ∥ ☉
09 04	☽ □ ♇
09 15	☽ ⊼ ♆
09 52	☽ ∥ ☿
10 46	☽ ⊥ ☿
12 46	☽ ✶ ♅
17 24	☿ ⊻ ♂
19 07	☽ ∥ ♇
19 09	☽ ± ♆
21 27	☽ ✶ ♂
21 41	☽ ⊻ ☿
22 48	☽ ⊥ ♀
16 Saturday	
04 01	☽ ✶ ♃
04 29	☽ ∥ ♀
05 43	☽ ∠ ☉
09 30	☽ ⊻ ♀
09 53	☽ ⅊ ♆
12 23	♀ ∠ ♇
13 27	☽ ∠ ♅
14 38	☽ Q ♇
16 26	☽ ⊼ ♄
21 23	☽ ⊥ ☉
23 01	☽ ∠ ♂
17 Sunday	
02 09	☽ ± ♄
04 06	☽ ⊥ ♅
04 46	☽ ∠ ♃
07 45	☽ ⊻ ☉
10 07	☽ ✶ ♇
10 13	☽ △ ♆
13 50	☽ ⊻ ♅
14 13	☽ ⊥ ♂
16 42	☽ ⅊ ♄
17 17	☽ ∥ ♀
19 31	☽ ⊥ ♃
18 Monday	
00 20	☽ ⊻ ♂
01 06	☽ ∥ ♇
03 45	☽ ☌ ☿
04 35	☽ ∥ ☿
05 20	☽ ⊻ ♃
10 21	☽ ∠ ♇
13 46	☉ ∥ ♅
14 23	☽ ☌ ♀
16 11	☽ ∥ ♅
16 20	☽ ∥ ☉
16 54	☽ △ ♄
20 47	☿ ⊻ ♃
21 28	☉ ⊻ ♇
22 15	☉ □ ♆
23 38	☽ ∥ ♂
19 Tuesday	
00 56	☽ ⊥ ♇
07 29	♄ St D
08 17	☽ ∥ ♃
10 41	☽ ⊻ ♇
10 43	☽ □ ♆
11 35	☽ ☌ ☉
14 34	☽ ☌ ♅
20 45	♀ △ ♄
20 Wednesday	
03 13	☽ ☌ ♂
06 43	♆ ✶ ♇
06 50	☽ ☌ ♃
08 55	☿ ∥ ♇
10 37	☽ ⊻ ☿
11 49	☽ # ♄
16 18	☽ Q ♆
16 42	☿ ∠ ♇
17 57	☽ □ ♄
19 55	☽ ⊻ ♀
21 56	☽ ⊥ ☿
21 Thursday	
06 56	☽ ⊥ ♀
12 27	☽ ✶ ♆
12 30	☽ ☌ ♇
12 34	☉ ☌ ♅
14 49	☽ # ♆
15 15	☽ ∠ ☿
16 40	☽ ⊻ ♅
16 57	☽ ⊻ ☉
23 43	☽ ∠ ♀
22 Friday	
00 24	☽ Q ♄
03 00	☽ ⊥ ♅
04 00	☽ ⊥ ☉
08 06	☿ ∥ ♀
08 12	☽ ⊻ ♂
10 23	☽ ⊻ ♃
14 20	☽ ∠ ♆
18 47	☽ ∠ ♅
19 13	☽ ⊥ ♂
20 55	☽ ∠ ☉
21 05	☽ ⊥ ♃
21 10	☽ ✶ ☿
21 21	☽ ✶ ♄
22 33	☿ △ ♄
23 Saturday	
04 35	☽ ✶ ♀
06 17	☽ ⊥ ♆
11 54	♀ ∥ ♇
12 04	☽ ∠ ♂
13 26	☽ ∠ ♃
17 06	☽ ⊻ ♆
17 15	☽ ⊻ ♇
21 50	☽ ✶ ♅
22 15	☽ Q ☿
24 Sunday	
00 23	☽ ∠ ♄
00 52	♀ ⊥ ♇
02 03	☽ ✶ ☉
04 15	☽ ⊥ ♇
04 36	☽ Q ♀
17 05	☽ ✶ ♂
17 05	☽ ⊥ ♄
17 32	☽ ✶ ♃
18 14	☽ ∥ ♆
20 12	☽ Q ♅
21 04	☽ ∠ ♇
25 Monday	
02 15	☽ Q ☉
04 25	☽ ⊻ ♄
06 45	♂ ☌ ♃
14 00	☽ □ ☿
15 31	☿ ⊥ ♇
16 46	☽ Q ♃
17 07	☽ Q ♂
18 13	☽ □ ♀
23 02	☿ ∥ ♇
26 Tuesday	
01 15	☽ ∥ ♄
01 37	☽ ☌ ♆
01 52	☽ ✶ ♇
06 55	☽ □ ♅
16 00	☽ □ ☉
27 Wednesday	
01 32	☽ Q ♇
04 43	☽ □ ♃
06 24	☽ □ ♂
06 52	☽ # ♃
09 25	☽ # ♂
13 44	☽ # ☉
15 12	☽ ☌ ♄
28 Thursday	
03 10	☿ ☌ ♀
05 19	☽ # ♅
12 31	☽ △ ♀
13 17	☽ △ ☿
13 26	☽ ⊻ ♆
13 27	☽ # ♀
13 47	☽ □ ♇
14 08	☿ □ ♆
16 11	☿ ⊻ ♇
17 01	☽ # ☿
19 12	☽ △ ♅
21 30	♀ □ ♆
23 48	☽ # ♇
29 Friday	
01 05	♀ ⊻ ♇
01 37	☽ ⊥ ♆
09 35	☽ △ ☉
18 29	☽ △ ♃
19 57	☽ ∠ ♆
22 31	☽ △ ♂
22 32	☽ ⅊ ♀
30 Saturday	
00 01	☿ ☌ ♅
01 51	☽ ⅊ ♅
02 14	☽ ⅊ ☿
04 05	☽ ⊻ ♄
04 53	♂ ∥ ♃
16 18	☽ ⊥ ♄
18 46	☽ ⅊ ☉
18 57	☽ ± ♀
20 14	☽ ± ♅
31 Sunday	
00 25	☽ ± ☿
01 31	☽ ⅊ ♃
02 26	☽ ✶ ♆
02 53	☽ △ ♇
06 41	☽ ⅊ ♂
07 56	♀ ☌ ♅
08 26	☽ ⊼ ♅
08 30	☽ ⊼ ♀
10 33	☽ ∠ ♄
14 33	☽ ± ☉
15 07	☽ ⊼ ☿
15 31	☿ ∥ ♀
20 01	☽ ± ♃
22 50	☽ # ♇

LONGITUDES

Date	Sidereal time h m s	Sun ☉ ° ' "	Moon ☽ ° ' "	Moon ☽ 24.00 ° ' "	Mercury ☿ ° '	Venus ♀ ° '	Mars ♂ ° '	Jupiter ♃ ° '	Saturn ♄ ° '	Uranus ♅ ° '	Neptune ♆ ° '	Pluto ♇ ° '
01	08 37 16	08 ♌ 27 55	12 ♑ 18 44	18 ♑ 21 21	03 ♌ 43	00 ♌ 01	13 ♌ 29	10 ♌ 29	14 ♏ 39	28 ♋ 39	25 ♎ 38	25 ♌ 52
02	08 41 13	09 25 19	24 ♑ 26 44	00 ♒ 35 07	05 49	01 15	14 08	10 42	14 40	28 42	25 38	25 54
03	08 45 09	10 22 43	06 ♒ 46 41	13 ♒ 01 34	07 54	02 29	14 46	10 55	14 41	28 46	25 39	25 56
04	08 49 06	11 20 08	19 ♒ 19 49	25 ♒ 41 30	09 59	03 42	15 24	11 08	14 43	28 50	25 40	25 58
05	08 53 02	12 17 34	02 ♓ 06 36	08 ♓ 35 06	12 04	04 56	16 02	11 21	14 44	28 53	25 41	26 00
06	08 56 59	13 15 01	15 ♓ 06 55	21 ♓ 41 59	14 08	06 10	16 41	11 35	14 46	28 57	25 42	26 02
07	09 00 56	14 12 29	28 ♓ 20 13	05 ♈ 01 33	16 10	07 24	17 19	11 48	14 48	29 00	25 43	26 04
08	09 04 52	15 09 59	11 ♈ 45 53	18 ♈ 33 10	18 12	08 38	17 57	12 01	14 50	29 04	25 44	26 06
09	09 08 49	16 07 30	25 ♈ 23 20	02 ♉ 16 21	20 12	09 52	18 35	12 14	14 52	29 08	25 45	26 08
10	09 12 45	17 05 02	09 ♉ 12 09	16 ♉ 10 41	22 12	11 06	19 14	12 27	14 54	29 11	25 46	26 09
11	09 16 42	18 02 35	23 ♉ 11 54	00 ♊ 15 44	24 09	12 20	19 52	12 41	14 56	29 15	25 47	26 11
12	09 20 38	19 00 11	07 ♊ 22 02	14 ♊ 30 37	26 06	13 34	20 30	12 54	14 58	29 18	25 48	26 13
13	09 24 35	19 57 47	21 ♊ 41 16	28 ♊ 53 39	28 01	14 48	21 08	13 07	15 01	29 22	25 49	26 15
14	09 28 31	20 55 25	06 ♋ 07 21	13 ♋ 21 53	29 ♌ 55	16 02	21 46	13 20	15 03	29 25	25 50	26 17
15	09 32 28	21 53 05	20 ♋ 36 41	27 ♋ 51 05	01 ♍ 47	17 16	22 25	13 33	15 06	29 29	25 52	26 19
16	09 36 25	22 50 46	05 ♌ 04 23	12 ♌ 15 51	03 38	18 31	23 03	13 47	15 08	29 32	25 53	26 21
17	09 40 21	23 48 28	19 ♌ 24 43	26 ♌ 30 15	05 27	19 45	23 41	14 00	15 11	29 36	25 54	26 23
18	09 44 18	24 46 12	03 ♍ 31 48	10 ♍ 28 45	07 15	20 59	24 19	14 13	15 14	29 39	25 55	26 25
19	09 48 14	25 43 57	17 ♍ 20 37	24 ♍ 07 02	09 01	22 13	24 57	14 26	15 17	29 42	25 57	26 27
20	09 52 11	26 41 43	00 ♎ 47 46	07 ♎ 22 41	10 46	23 27	25 36	14 39	15 20	29 46	25 58	26 29
21	09 56 07	27 39 30	13 ♎ 51 50	20 ♎ 15 22	12 30	24 42	26 14	14 52	15 23	29 49	25 59	26 31
22	10 00 04	28 37 19	26 ♎ 33 33	02 ♏ 46 46	14 12	25 56	26 52	15 05	15 26	29 53	26 01	26 33
23	10 04 00	29 ♌ 35 08	08 ♏ 55 27	15 ♏ 00 09	15 53	27 10	27 30	15 18	15 29	29 56	26 02	26 35
24	10 07 57	00 ♍ 32 59	21 ♏ 01 26	26 ♏ 59 57	17 32	28 24	28 08	15 31	15 33	29 ♋ 59	26 03	26 37
25	10 11 54	01 30 51	02 ♐ 56 21	08 ♐ 51 18	19 10	29 ♌ 39	28 46	15 44	15 36	00 ♌ 02	26 05	26 38
26	10 15 50	02 28 45	14 ♐ 45 29	20 ♐ 39 35	20 47	00 ♍ 53	29 ♌ 25	15 57	15 40	00 06	26 06	26 40
27	10 19 47	03 26 40	26 ♐ 34 16	02 ♑ 30 10	22 23	02 07	00 ♍ 03	16 10	15 43	00 09	26 08	26 42
28	10 23 43	04 24 36	08 ♑ 27 53	14 ♑ 28 00	23 57	03 22	00 41	16 23	15 47	00 12	26 09	26 44
29	10 27 40	05 22 33	20 ♑ 31 01	26 ♑ 37 24	25 29	04 36	01 19	16 36	15 51	00 15	26 11	26 46
30	10 31 36	06 20 32	02 ♒ 47 30	09 ♒ 01 40	27 01	05 51	01 57	16 48	15 54	00 18	26 13	26 48
31	10 35 33	07 ♍ 18 32	15 ♒ 20 05	21 ♒ 42 54	28 ♍ 31	07 ♍ 05	02 ♍ 36	17 ♌ 01	15 ♏ 58	00 ♌ 22	26 ♎ 14	26 ♌ 50

DECLINATIONS

Date	Moon True ☊ ° '	Moon Mean ☊ ° '	Moon ☽ Latitude ° '	Sun ☉	Moon ☽	Mercury ☿	Venus ♀	Mars ♂	Jupiter ♃	Saturn ♄	Uranus ♅	Neptune ♆	Pluto ♇
01	25 ♐ 50	24 ♐ 10	01 N 28	18 N 09	21 S 25	20 N 45	20 N 53	17 N 53	18 N 10	14 S 05	20 N 56	08 S 19	22 N 26
02	25 R 47	24 07	02 29	17 54	18 48	20 19	20 39	17 42	18 06	14 06	20 56	08 19	22 25
03	25 42	24 04	03 23	17 39	15 18	19 51	20 25	17 30	18 02	14 06	20 55	08 20	22 24
04	25 35	24 00	04 09	17 23	11 05	19 21	20 09	17 19	17 59	14 07	20 54	08 20	22 24
05	25 27	23 57	04 43	17 07	06 19	18 49	19 54	17 08	17 55	14 08	20 53	08 20	22 23
06	25 19	23 54	05 03	16 51	01 S 13	18 16	19 37	16 56	17 52	14 09	20 53	08 21	22 22
07	25 11	23 51	05 07	16 34	04 N 02	17 40	19 20	16 45	17 48	14 09	20 52	08 21	22 22
08	25 06	23 48	04 55	16 17	09 10	17 04	19 03	16 33	17 45	14 10	20 51	08 21	22 21
09	25 02	23 45	04 26	16 00	13 56	16 25	18 45	16 21	17 41	14 11	20 50	08 22	22 20
10	25 00	23 41	03 41	15 43	18 03	15 46	18 26	16 09	17 37	14 12	20 50	08 22	22 20
11	25 D 00	23 38	02 43	15 26	21 12	15 06	18 07	15 57	17 34	14 13	20 49	08 22	22 19
12	25 00	23 35	01 33	15 08	23 05	14 24	17 47	15 45	17 30	14 14	20 48	08 23	22 18
13	25 01	23 32	00 N 18	14 50	23 29	13 42	17 27	15 33	17 26	14 14	20 48	08 23	22 18
14	25 R 01	23 29	00 S 59	14 31	22 19	13 00	17 07	15 20	17 23	14 15	20 47	08 24	22 17
15	24 58	23 26	02 13	14 13	19 41	12 16	16 46	15 08	17 19	14 16	20 46	08 25	22 16
16	24 53	23 22	03 17	13 54	15 49	11 32	16 24	14 55	17 15	14 17	20 45	08 25	22 16
17	24 46	23 19	04 09	13 35	11 04	10 48	16 02	14 43	17 12	14 18	20 45	08 26	22 15
18	24 37	23 16	04 44	13 16	05 48	10 04	15 39	14 30	17 08	14 19	20 44	08 26	22 14
19	24 27	23 13	05 02	12 57	00 N 22	09 19	15 17	14 17	17 04	14 21	20 43	08 27	22 14
20	24 17	23 10	05 02	12 37	04 S 56	08 34	14 53	14 04	17 00	14 22	20 43	08 27	22 13
21	24 09	23 06	04 47	12 17	09 52	07 49	14 30	13 51	16 57	14 23	20 42	08 28	22 13
22	24 02	23 03	04 17	11 57	14 14	07 04	14 05	13 38	16 53	14 24	20 41	08 28	22 12
23	23 58	23 00	03 36	11 37	17 53	06 19	13 41	13 25	16 49	14 25	20 41	08 29	22 11
24	23 56	22 57	02 46	11 17	20 41	05 34	13 16	13 12	16 45	14 26	20 40	08 29	22 11
25	23 D 55	22 54	01 50	10 56	22 33	04 49	12 51	12 58	16 42	14 27	20 39	08 30	22 10
26	23 56	22 51	00 S 49	10 36	23 23	04 05	12 25	12 45	16 38	14 29	20 39	08 30	22 10
27	23 R 56	22 47	00 N 14	10 15	23 10	03 20	11 59	12 31	16 34	14 30	20 38	08 31	22 09
28	23 55	22 44	01 17	09 54	21 54	02 36	11 33	12 18	16 30	14 31	20 37	08 32	22 08
29	23 52	22 41	02 16	09 33	19 38	01 52	11 06	12 04	16 27	14 33	20 37	08 32	22 08
30	23 46	22 38	03 11	09 11	16 26	01 09	10 40	11 50	16 23	14 34	20 36	08 33	22 07
31	23 ♐ 38	22 ♐ 35	03 N 57	08 N 50	12 S 28	00 N 25	10 N 12	11 N 37	16 N 19	14 S 35	20 N 35	08 S 33	22 N 07

ZODIAC SIGN ENTRIES

Date	h	m	Planets
01	11	43	♀ ♌
02	22	52	☽ ♒
05	08	04	☽ ♓
07	15	00	☽ ♈
09	20	03	☽ ♉
11	23	33	☽ ♊
14	01	50	☽ ♋
14	13	08	☿ ♍
16	03	34	☽ ♌
18	05	57	☽ ♍
20	10	34	☽ ♎
22	18	37	☽ ♏
23	22	19	☉ ♍
24	18	04	♅ ♌
25	06	03	☽ ♐
25	18	52	♀ ♍
27	10	13	♂ ♍
27	18	57	☽ ♑
30	06	35	☽ ♒

LATITUDES

Date	Mercury ☿ ° '	Venus ♀ ° '	Mars ♂ ° '	Jupiter ♃ ° '	Saturn ♄ ° '	Uranus ♅ ° '	Neptune ♆ ° '	Pluto ♇ ° '
01	01 N 28	00 N 45	01 N 09	00 N 34	02 N 15	00 N 31	01 N 43	10 N 09
04	01 40	00 51	01 09	00 34	02 14	00 31	01 43	10 09
07	01 46	00 57	01 09	00 34	02 13	00 31	01 42	10 09
10	01 45	01 02	01 09	00 34	02 13	00 31	01 42	10 10
13	01 39	01 06	01 09	00 35	02 12	00 31	01 42	10 10
16	01 28	01 10	01 09	00 35	02 11	00 31	01 42	10 10
19	01 13	01 14	01 09	00 35	02 10	00 31	01 42	10 10
22	00 55	01 17	01 09	00 36	02 10	00 31	01 42	10 10
25	00 35	01 20	01 08	00 36	02 09	00 31	01 42	10 10
28	00 N 13	01 22	01 08	00 36	02 08	00 31	01 41	10 10
31	00 S 11	01 N 24	01 N 08	00 N 37	02 N 08	00 N 31	01 N 41	10 N 11

DATA

Julian Date	2435321
Delta T	+31 seconds
Ayanamsa	23° 14' 30"
Synetic vernal point	05° ♓ 52' 30"
True obliquity of ecliptic	23° 26' 41"

MOON'S PHASES, APSIDES AND POSITIONS ☽

Date	h	m	Phase	Longitude ° '	Eclipse Indicator
03	19	30	○	10 ♒ 41	
11	02	33	☾	17 ♉ 40	
17	19	58	●	24 ♌ 08	
25	08	52	☽	01 ♐ 23	

Day	h	m		
14	17	30	Perigee	
26	15	11	Apogee	
06	17	33	0N	
13	06	11	Max dec	23° N 32'
19	13	37	0S	
26	19	08	Max dec	23° S 26'

LONGITUDES

Date	Chiron ⚷ ° '	Ceres ⚳ ° '	Pallas ⚴ ° '	Juno ⚵ ° '	Vesta ⚶ ° '	Black Moon Lilith ⚸ ° '
01	02 ♒ 25	13 ♑ 56	09 ♐ 26	05 ♏ 29	15 ♌ 23	15 ♐ 59
11	01 ♒ 52	12 ♑ 29	09 ♐ 45	07 ♏ 15	20 ♌ 00	17 ♐ 05
21	01 ♒ 21	11 ♑ 35	10 ♐ 41	09 ♏ 13	24 ♌ 40	18 ♐ 12
31	00 ♒ 54	11 ♑ 16	12 ♐ 09	11 ♏ 34	29 ♌ 21	19 ♐ 18

All ephemeris data is given at 12.00 UT and the Moon's longitude is additionally given for 24.00 UT

ASPECTARIAN

h m Aspects

01 Monday
00 04 ☿ ∥ ♅
01 50 ☽ ± ♂
02 38 ☽ Q ♆
03 39 ☽ ⊼ ☉
06 25 ♀ ∥ ♅
08 16 ☽ ⊼ ♃
09 07 ☽ bQ ♇
11 04 ☉ ∥ ♃
14 29 ☽ ⊼ ♂
16 39 ☽ ✶ ♄
17 06 ☽ # ♅
17 14 ♂ Q ♆
18 11 ☽ # ♀
20 26 ☽ # ☿

02 Tuesday
03 02 ☽ ± ♇
14 21 ☽ □ ♆
14 52 ☽ ⊼ ♇
16 21 ☽ Q ♄
17 22 ☽ # ♃
19 17 ☽ # ☉
20 23 ☽ ☍ ♅
20 41 ☽ # ♂

03 Wednesday
02 45 ☽ ☍ ♀
08 57 ♂ □ ♄
14 36 ☽ ☍ ☿
19 11 ☽ ∥ ♄
19 30 ☽ ☍ ☉
20 06 ☽ ☍ ♃

04 Thursday
03 12 ☽ □ ♄
04 08 ☽ ☍ ♂
05 31 ☉ ☌ ♃
23 58 ☽ △ ♆

05 Friday
00 33 ☽ ☍ ♇
02 08 ☽ ∥ ♆
02 50 ☿ ☌ ♃
05 58 ☽ ⊼ ♅
08 00 ☉ ∥ ♂
16 57 ☉ ☌ ☿
17 11 ☽ ± ♅
17 48 ☽ ⊼ ♀

06 Saturday
03 53 ☽ bQ ♆
05 24 ☽ ⊼ ♃
06 02 ☽ ± ♀
06 59 ☿ Q ♆
08 19 ☽ ⊼ ☉
09 51 ☽ ⊼ ☿
09 51 ☽ bQ ♅
11 22 ☽ △ ♄
15 00 ☽ ⊼ ♂
16 34 ☽ ± ♃
19 39 ☿ □ ♄
20 08 ☽ ± ☉
20 22 ☽ ± ♆
22 49 ☽ ± ☿
23 26 ☉ Q ♆

07 Sunday
00 10 ☽ bQ ♀
02 28 ☽ ± ♂
06 14 ☿ ∥ ♃
07 16 ☽ ⊼ ♆
07 53 ☽ ⊼ ♇
09 11 ☽ bQ ♃
13 13 ☽ △ ♅
13 41 ☽ bQ ☉
14 38 ☽ bQ ♄
18 01 ☽ bQ ☿
18 43 ☽ ± ♇
19 30 ☽ bQ ♂

08 Monday
03 19 ☉ □ ♄
05 53 ☽ △ ♀
06 46 ☽ ± ♄
07 43 ☿ ☌ ♂
08 07 ☽ # ♆
10 48 ☽ bQ ♇
12 27 ☽ △ ♃
17 27 ☽ ⊼ ♄
18 29 ☽ △ ☉
23 29 ☽ △ ♂

09 Tuesday
01 21 ☽ △ ☿
12 38 ☽ ☍ ♆
13 17 ☽ △ ♇
13 18 ☽ # ♄
15 54 ☿ ∥ ♂
18 33 ☽ □ ♅
22 44 ☽ ∥ ☉

10 Wednesday
00 00 ☽ ∥ ☿
00 51 ☽ ∥ ♂
09 16 ☽ ∥ ♃
14 19 ☽ ∥ ♀
15 10 ☉ ∥ ☿
15 35 ☽ □ ♀
17 42 ☽ □ ♃
21 50 ☽ ☍ ♄

11 Thursday
01 47 ☽ Q ♅
02 33 ☽ □ ☉
06 02 ☽ □ ♂
08 34 ☽ ∥ ♅
13 54 ☽ □ ☿
16 24 ☽ ⊼ ♆
17 06 ☽ □ ♇
20 08 ♀ ☌ ♃
22 19 ☽ ✶ ♅
23 58 ☽ ∥ ♇

12 Friday
00 54 ☽ Q ♃
01 17 ☽ Q ♀
02 36 ☽ ± ♆
08 16 ☿ ✶ ♆
11 21 ☽ Q ☉
13 33 ☿ ☌ ♇
14 00 ☽ Q ♂
16 36 ♀ Q ♆
17 47 ☽ bQ ♆
18 07 ☿ # ♄
21 26 ☽ ✶ ♃
23 24 ☽ ✶ ♀
23 33 ☽ Q ♇
23 42 ☽ ∠ ♅

13 Saturday
00 49 ☽ ⊼ ♄
01 04 ☽ Q ☿
08 55 ☽ ✶ ☉
10 52 ☽ ± ♄
11 02 ☽ ✶ ♂
13 19 ♀ ∥ ♃
14 48 ☽ ⊥ ♅
16 12 ♀ □ ♄
18 54 ☽ △ ♆
19 38 ☽ ✶ ♇
22 52 ☽ ∠ ♃

14 Sunday
00 08 ☽ ✶ ☿
00 50 ☽ ⊻ ♅
01 54 ☽ bQ ♄
02 46 ☽ ∠ ♀
05 34 ☿ ⊻ ♅
11 39 ☽ ∠ ☉
12 24 ☽ ∥ ♇
13 08 ☽ ∠ ♂
14 03 ☽ ⊥ ♃
19 06 ☽ ⊥ ♀
20 35 ☽ ∠ ♇

15 Monday
00 08 ☽ ⊻ ♃
02 50 ☽ △ ♄
03 27 ☽ ∥ ♅
03 37 ☽ ⊥ ☉
04 43 ☽ ⊥ ♂
04 43 ☽ ∠ ☿
05 57 ☽ ⊻ ♀
07 47 ☉ # ♄
11 31 ☽ ⊥ ♇
14 16 ☽ ⊻ ☉
15 07 ☽ ⊻ ♂
20 43 ☽ □ ♆
21 29 ☽ ⊻ ♇
21 49 ☽ ⊥ ☿

16 Tuesday
02 46 ☽ ☌ ♅
03 38 ☽ ∥ ♃
05 29 ☿ Q ♄
08 24 ☽ ∥ ♀
09 15 ☽ ⊻ ☿
17 00 ☽ ∥ ♂
20 03 ☽ # ♄
22 47 ☽ ∥ ☉

17 Wednesday
00 36 ♃ Q ♆
02 44 ☽ Q ♆
02 45 ☽ ☌ ♃
02 46 ☉ ☌ ♂
04 52 ☽ □ ♄
12 37 ☽ ☌ ♀
13 25 ☽ ∥ ☿
13 59 ☿ ⊥ ♅
19 33 ☽ ☌ ♂
19 58 ☽ ☌ ☉
22 59 ☽ ✶ ♆
23 49 ☽ ☌ ♇

18 Thursday
00 10 ☽ # ♆
05 20 ☽ ⊻ ♅
11 29 ☽ Q ♄
15 40 ☽ ⊥ ♅
19 21 ☽ ☌ ☿

19 Friday
00 47 ☽ ∠ ♆
06 04 ♂ # ♄
06 48 ☽ ⊻ ♃
07 21 ☽ ∠ ♅
08 22 ☽ ✶ ♄
16 36 ☽ ⊥ ♆
17 23 ☉ ✶ ♆
17 32 ☽ ⊥ ♃
21 29 ☽ ⊻ ♀

20 Saturday
02 10 ☽ ⊻ ♂
03 17 ☽ ⊻ ♆
04 02 ☽ ⊻ ☉
04 12 ☽ ⊻ ♇
06 27 ☉ ☌ ♇
09 20 ☽ ⊥ ♀
09 54 ☽ ∠ ♃
10 08 ☽ ✶ ♅
11 09 ☽ ∠ ♄
13 31 ☽ ⊥ ♂
14 43 ☿ ∠ ♆
15 04 ☽ ⊥ ♇
15 40 ☿ # ♆
15 43 ☽ ⊥ ☉

21 Sunday
02 34 ♂ ✶ ♆
03 05 ☽ # ☿
03 27 ☽ ∠ ♀
03 39 ☽ ⊥ ♄
04 53 ☽ ∥ ♆
06 51 ☽ ∠ ♂
07 37 ☽ ∠ ♇
08 11 ☽ Q ♅
09 04 ☽ ⊻ ☿
09 35 ☽ ∠ ☉
13 54 ☽ ✶ ♃
14 51 ☽ ⊻ ♄
18 29 ♀ # ♄
22 02 ☽ ⊥ ☿
23 15 ♂ ☌ ♇
23 56 ☽ # ☉

22 Monday
08 38 ☽ # ♂
10 40 ☽ ✶ ♀
10 57 ☽ ☌ ♆
11 12 ☽ # ♀
11 58 ☽ ✶ ♇
12 37 ☽ ✶ ♂
12 58 ☽ ∥ ♄
13 01 ☽ Q ♃
13 34 ♀ ✶ ♆
16 18 ☽ ✶ ☉
17 53 ☽ ∠ ☿
18 24 ☽ □ ♅
21 55 ☿ ∠ ♅

23 Tuesday
00 13 ♀ ☌ ♇
02 26 ☿ ⊻ ♃
04 33 ☽ # ♃
06 10 ☿ ✶ ♄
11 19 ☽ Q ♇
12 32 ☽ Q ♀
13 12 ☽ Q ♂
17 41 ☽ Q ☉
21 07 ☉ ⊻ ♅

24 Wednesday
00 49 ☽ □ ♃
01 01 ☽ ☌ ♄
01 17 ♀ ☌ ♂
03 56 ☽ ✶ ☿
11 46 ☽ # ♅
16 02 ♃ □ ♄
20 52 ♀ ∥ ♂
22 07 ☽ ⊻ ♆
23 15 ☽ □ ♇

25 Thursday
03 06 ☽ □ ♂
04 34 ☽ □ ♀
06 01 ☽ # ♇
06 07 ☽ △ ♅
07 52 ☽ Q ☿
08 52 ☽ □ ☉
10 16 ☽ ⊥ ♆
20 01 ♀ ⊻ ♅

26 Friday
01 40 ☿ ⊥ ♆
04 34 ☽ ∠ ♆
12 41 ☽ bQ ♅
13 51 ☽ ⊻ ♄
14 28 ☽ △ ♃

27 Saturday
02 06 ☽ ⊥ ♄
02 10 ☽ □ ☿
07 04 ☽ ± ♅
08 16 ☿ ⊥ ♃
11 06 ☽ ✶ ♆
12 16 ☽ △ ♇
16 11 ♂ ⊻ ♅
19 17 ☽ ⊼ ♅
19 18 ☉ Q ♄
19 26 ☽ △ ♂
20 26 ☽ ∠ ♄
21 28 ☽ bQ ♃

28 Sunday
00 32 ☽ △ ♀
03 08 ☽ △ ☉
08 35 ☽ # ♇
11 23 ☽ Q ♆
15 54 ☽ ± ♃
18 34 ☽ bQ ♇
20 32 ♀ Q ♄

29 Monday
02 41 ☽ ✶ ♄
02 48 ☽ # ♅
03 13 ☽ bQ ♂
04 05 ☽ ⊼ ♃
09 59 ☽ bQ ♀
11 42 ☽ bQ ☉
12 30 ☽ ± ♇
21 58 ☽ ± ♂
23 03 ☿ ⊻ ♆
23 10 ☽ □ ♆
23 11 ☽ △ ☿

30 Tuesday
00 19 ☽ ⊼ ♇
02 27 ☽ Q ♄
05 38 ☽ ± ♀
06 51 ☽ ± ☉
07 10 ☽ ☍ ♅
08 34 ☿ ⊻ ♇
10 18 ☽ ⊼ ♂
11 06 ☉ ⊥ ♅
12 24 ☽ # ♃
18 32 ☽ ⊼ ♀
19 25 ☽ ⊼ ☉
21 25 ♀ ⊥ ♅
23 47 ☽ ∥ ♄

31 Wednesday
08 05 ☽ bQ ☿
13 13 ☽ □ ♄
15 14 ☽ ☍ ♃
16 53 ☽ # ♂

SEPTEMBER 1955

LONGITUDES

Date	Sidereal time h m s	Sun ☉ ° ' "	Moon ☽ ° ' "	Moon ☽ 24.00 ° ' "	Mercury ☿ ° '	Venus ♀ ° '	Mars ♂ ° '	Jupiter ♃ ° '	Saturn ♄ ° '	Uranus ♅ ° '	Neptune ♆ ° '	Pluto ♇ ° '
01	10 39 29	08 ♍ 16 34	28 ♒ 10 10	04 ♓ 41 50	00 ♎ 00	08 ♍ 19	03 ♍ 14	17 ♌ 14	16 ♏ 02	00 ♌ 25	26 ♎ 16	26 ♌ 52
02	10 43 26	09 14 37	11 ♓ 17 45	17 ♓ 57 43	01 27	09 34	03 52	17 27	16 06	00 28	26 17	26 54
03	10 47 23	10 12 42	24 ♓ 41 27	01 ♈ 28 37	02 53	10 48	04 30	17 39	16 11	00 31	26 19	26 56
04	10 51 19	11 10 48	08 ♈ 18 51	15 ♈ 11 45	04 18	12 03	05 08	17 52	16 15	00 34	26 21	26 58
05	10 55 16	12 08 57	22 ♈ 06 56	29 ♈ 04 02	05 41	13 17	05 46	18 05	16 19	00 37	26 22	27 00
06	10 59 12	13 07 07	06 ♉ 02 41	13 ♉ 02 36	07 03	14 32	06 25	18 17	16 23	00 40	26 24	27 01
07	11 03 09	14 05 20	20 ♉ 03 31	27 ♉ 05 14	08 23	15 46	07 03	18 30	16 28	00 43	26 26	27 03
08	11 07 05	15 03 34	04 ♊ 07 34	11 ♊ 10 24	09 42	17 01	07 41	18 42	16 32	00 46	26 28	27 05
09	11 11 02	16 01 51	18 ♊ 13 37	25 ♊ 17 07	10 59	18 15	08 19	18 55	16 37	00 48	26 29	27 07
10	11 14 58	17 00 10	02 ♋ 20 47	09 ♋ 24 30	12 15	19 30	08 57	19 07	16 42	00 51	26 31	27 09
11	11 18 55	17 58 30	16 ♋ 28 07	23 ♋ 31 24	13 29	20 44	09 36	19 19	16 46	00 54	26 33	27 11
12	11 22 52	18 56 53	00 ♌ 34 06	07 ♌ 35 53	14 42	21 59	10 14	19 32	16 51	00 57	26 35	27 13
13	11 26 48	19 55 18	14 ♌ 36 23	21 ♌ 35 10	15 52	23 14	10 52	19 44	16 56	00 59	26 37	27 14
14	11 30 45	20 53 45	28 ♌ 31 47	05 ♍ 25 46	17 01	24 28	11 30	19 56	17 01	01 02	26 39	27 16
15	11 34 41	21 52 14	12 ♍ 16 38	19 ♍ 03 57	18 07	25 43	12 09	20 08	17 06	01 05	26 41	27 18
16	11 38 38	22 50 45	25 ♍ 47 18	02 ♎ 26 22	19 12	26 57	12 47	20 20	17 11	01 07	26 43	27 20
17	11 42 34	23 49 18	09 ♎ 00 52	15 ♎ 30 41	20 14	28 12	13 25	20 33	17 16	01 10	26 44	27 22
18	11 46 31	24 47 52	21 ♎ 55 42	28 ♎ 16 00	21 13	29 ♍ 27	14 03	20 45	17 21	01 13	26 46	27 23
19	11 50 27	25 46 29	04 ♏ 31 42	10 ♏ 43 03	22 11	00 ♎ 41	14 41	20 56	17 27	01 15	26 48	27 25
20	11 54 24	26 45 07	16 ♏ 50 21	22 ♏ 54 03	23 05	01 56	15 20	21 08	17 32	01 17	26 50	27 27
21	11 58 21	27 43 47	28 ♏ 54 37	04 ♐ 52 36	23 56	03 11	15 58	21 20	17 37	01 20	26 52	27 29
22	12 02 17	28 42 29	10 ♐ 48 34	16 ♐ 43 12	24 44	04 26	16 36	21 32	17 43	01 22	26 54	27 30
23	12 06 14	29 ♍ 41 12	22 ♐ 37 07	28 ♐ 31 02	25 29	05 40	17 15	21 44	17 48	01 25	26 56	27 32
24	12 10 10	00 ♎ 39 57	04 ♑ 25 39	10 ♑ 21 38	26 09	06 55	17 53	21 55	17 54	01 27	26 58	27 34
25	12 14 07	01 38 44	16 ♑ 19 42	22 ♑ 20 28	26 46	08 10	18 31	22 07	17 59	01 29	27 01	27 35
26	12 18 03	02 37 32	28 ♑ 24 36	04 ♒ 32 40	27 18	09 24	19 09	22 18	18 05	01 31	27 03	27 37
27	12 22 00	03 36 23	10 ♒ 45 10	17 ♒ 02 34	27 45	10 39	19 48	22 30	18 11	01 34	27 05	27 39
28	12 25 56	04 35 15	23 ♒ 25 10	29 ♒ 53 16	28 07	11 54	20 26	22 41	18 17	01 36	27 07	27 40
29	12 29 53	05 34 08	06 ♓ 26 57	13 ♓ 06 14	28 24	13 09	21 04	22 52	18 22	01 38	27 09	27 42
30	12 33 50	06 ♎ 33 04	19 ♓ 51 00	26 ♓ 40 59	28 ♎ 34	14 ♎ 23	21 ♍ 42	23 ♌ 04	18 ♏ 28	01 ♌ 40	27 ♎ 11	27 ♌ 44

Date	Moon True ☊ ° '	Moon Mean ☊ ° '	Moon ☽ Latitude ° '
01	23 ♐ 28	22 ♐ 32	04 N 33
02	23 R 16	22 28	04 55
03	23 04	22 25	05 01
04	22 53	22 22	04 50
05	22 44	22 19	04 23
06	22 37	22 16	03 40
07	22 33	22 12	02 43
08	22 31	22 09	01 36
09	22 D 31	22 06	00 N 23
10	22 R 31	22 03	00 S 52
11	22 30	22 00	02 03
12	22 26	21 57	03 07
13	22 20	21 53	03 58
14	22 10	21 50	04 36
15	21 59	21 47	04 56
16	21 46	21 44	05 00
17	21 33	21 41	04 47
18	21 22	21 38	04 19
19	21 13	21 34	03 40
20	21 06	21 31	02 50
21	21 03	21 28	01 54
22	21 01	21 25	00 S 54
23	21 D 01	21 22	00 N 09
24	21 R 01	21 18	01 11
25	21 00	21 15	02 10
26	20 57	21 12	03 05
27	20 52	21 09	03 52
28	20 45	21 06	04 29
29	20 35	21 03	04 53
30	20 ♐ 24	20 ♐ 59	05 N 02

DECLINATIONS

Date	Sun ☉ ° '	Moon ☽ ° '	Mercury ☿ ° '	Venus ♀ ° '	Mars ♂ ° '	Jupiter ♃ ° '	Saturn ♄ ° '	Uranus ♅ ° '	Neptune ♆ ° '	Pluto ♇ ° '
01	08 N 28	07 S 51	00 S 17	09 N 45	11 N 23	16 N 15	14 S 37	20 N 35	08 S 34	22 N 06
02	08 06	02 S 47	01 00	09 17	11 09	16 12	14 38	20 34	08 35	22 06
03	07 44	02 N 30	01 42	08 49	10 55	16 08	14 39	20 33	08 35	22 05
04	07 22	07 45	02 23	08 21	10 41	16 04	14 41	20 33	08 36	22 05
05	07 00	12 41	03 04	07 53	10 27	16 00	14 42	20 32	08 37	22 04
06	06 38	16 59	03 44	07 24	10 13	15 57	14 44	20 31	08 37	22 03
07	06 16	20 22	04 23	06 55	09 58	15 53	14 45	20 31	08 38	22 03
08	05 53	22 33	05 02	06 26	09 44	15 49	14 47	20 30	08 39	22 02
09	05 31	23 18	05 41	05 57	09 30	15 45	14 48	20 30	08 39	22 02
10	05 08	22 34	06 18	05 27	09 15	15 41	14 50	20 29	08 40	22 01
11	04 45	20 24	06 55	04 58	09 01	15 38	14 51	20 29	08 41	22 01
12	04 22	17 00	07 30	04 28	08 46	15 34	14 53	20 28	08 41	22 00
13	04 00	12 39	08 05	03 58	08 32	15 30	14 54	20 27	08 42	22 00
14	03 37	07 41	08 39	03 28	08 17	15 27	14 56	20 27	08 43	21 59
15	03 14	02 N 23	09 12	02 58	08 02	15 23	14 57	20 26	08 43	21 59
16	02 50	02 S 55	09 44	02 28	07 47	15 19	14 59	20 26	08 44	21 59
17	02 27	07 58	10 15	01 57	07 33	15 15	15 01	20 25	08 45	21 58
18	02 04	12 33	10 45	01 27	07 18	15 12	15 02	20 25	08 46	21 58
19	01 41	16 29	11 13	00 57	07 03	15 08	15 04	20 24	08 46	21 57
20	01 17	19 36	11 40	00 N 26	06 48	15 04	15 06	20 24	08 47	21 57
21	00 54	21 47	12 05	00 S 04	06 33	15 01	15 07	20 23	08 48	21 56
22	00 31	22 58	12 29	00 35	06 18	14 57	15 09	20 23	08 48	21 56
23	00 N 07	23 06	12 51	01 05	06 03	14 53	15 11	20 22	08 49	21 56
24	00 S 16	22 12	13 11	01 36	05 48	14 50	15 12	20 22	08 50	21 55
25	00 39	20 18	13 29	02 06	05 33	14 46	15 14	20 21	08 51	21 55
26	01 03	17 28	13 45	02 37	05 18	14 43	15 16	20 21	08 52	21 54
27	01 26	13 49	13 59	03 07	05 03	14 39	15 17	20 20	08 52	21 54
28	01 49	09 29	14 10	03 38	04 47	14 35	15 19	20 20	08 53	21 54
29	02 13	04 36	14 18	04 08	04 32	14 32	15 21	20 19	08 54	21 53
30	02 S 36	00 N 36	14 S 23	04 S 38	04 N 17	14 N 28	15 S 23	20 N 19	08 S 55	21 N 53

ZODIAC SIGN ENTRIES

Date	h	m	Planets
01	12	06	☿ ♎
01	15	23	☽ ♓
03	21	24	☽ ♈
06	01	36	☽ ♉
08	04	58	☽ ♊
10	08	01	☽ ♋
12	11	02	☽ ♌
14	14	33	☽ ♍
16	19	35	☽ ♎
18	22	41	♀ ♎
19	03	18	☽ ♏
21	14	11	☽ ♐
23	19	41	☉ ♎
24	03	01	☽ ♑
26	15	07	☽ ♒
29	00	12	☽ ♓

LATITUDES

Date	Mercury ☿ ° '	Venus ♀ ° '	Mars ♂ ° '	Jupiter ♃ ° '	Saturn ♄ ° '	Uranus ♅ ° '	Neptune ♆ ° '	Pluto ♇ ° '
01	00 S 19	01 N 24	01 N 08	00 N 37	02 N 07	00 N 31	01 N 41	10 N 11
04	00 44	01 25	01 08	00 37	02 07	00 31	01 41	10 11
07	01 09	01 25	01 08	00 37	02 06	00 31	01 41	10 12
10	01 35	01 24	01 07	00 38	02 05	00 32	01 41	10 12
13	02 00	01 23	01 07	00 38	02 05	00 32	01 41	10 13
16	02 24	01 22	01 07	00 38	02 04	00 32	01 41	10 14
19	02 47	01 20	01 06	00 39	02 04	00 32	01 41	10 14
22	03 07	01 17	01 06	00 39	02 03	00 32	01 41	10 15
25	03 24	01 14	01 06	00 40	02 02	00 32	01 41	10 15
28	03 36	01 10	01 05	00 40	02 02	00 32	01 40	10 16
31	03 S 40	01 N 06	01 N 05	00 N 41	02 N 02	00 N 32	01 N 40	10 N 17

DATA

Julian Date	2435352
Delta T	+31 seconds
Ayanamsa	23° 14' 34"
Synetic vernal point	05° ♓ 52' 26"
True obliquity of ecliptic	23° 26' 42"

MOON'S PHASES, APSIDES AND POSITIONS ☽

Date	h	m	Phase	Longitude ° '	Eclipse Indicator
02	07	59	○	09 ♓ 05	
09	07	59	☾	15 ♊ 52	
16	06	19	●	22 ♍ 37	
24	03	41	☽	00 ♑ 20	

Day	h	m		
10	01	03	Perigee	
23	10	35	Apogee	
03	00	43	0N	
09	12	09	Max dec	23° N 18'
15	22	46	0S	
23	03	07	Max dec	23° S 10'
30	09	15	0N	

LONGITUDES

Date	Chiron ⚷ ° '	Ceres ⚳ ° '	Pallas ⚴ ° '	Juno ⚵ ° '	Vesta ⚶ ° '	Black Moon Lilith ⚸ ° '
01	00 ♒ 52	11 ♑ 16	12 ♐ 19	11 ♏ 49	29 ♌ 49	19 ♐ 25
11	00 ♒ 30	11 ♑ 37	14 ♐ 16	14 ♏ 23	04 ♍ 31	20 ♐ 31
21	00 ♒ 15	12 ♑ 30	16 ♐ 36	17 ♏ 09	09 ♍ 14	21 ♐ 38
31	00 ♒ 06	13 ♑ 54	19 ♐ 14	20 ♏ 05	13 ♍ 58	22 ♐ 44

All ephemeris data is given at 12.00 UT and the Moon's longitude is additionally given for 24.00 UT

ASPECTARIAN

h m	Aspects
01 Thursday	
01 23	☽ # ♀
03 15	☽ ± ☿
05 40	☉ # ♆
07 57	☉ ☌ ♀
08 25	☽ ∥ ♆
08 28	☽ △ ♆
08 39	☽ # ☉
09 35	☽ ☍ ♇
15 48	☽ ⊼ ☿
16 09	☽ ⊼ ♅
19 05	☿ ⚹ ♅
21 47	☽ ☍ ♂
02 Friday	
03 11	☽ ± ♅
06 01	☿ ∠ ♄
07 59	☽ ☍ ☉
08 32	☽ ☍ ♀
11 59	☽ bQ ♆
19 14	☽ ∥ ☿
19 33	☽ bQ ♅
20 43	☽ △ ♄
22 13	♂ Q ♄
23 15	☽ ⊼ ♃
03 Saturday	
04 12	☽ ± ♆
07 29	☿ ∠ ♃
07 49	☽ # ☿
10 08	☽ ± ♃
12 46	☿ ⊥ ♇
14 53	☽ ⊼ ♆
15 59	☽ ⊼ ♇
22 10	♀ ∠ ♆
22 20	☽ △ ♅
23 32	☽ bQ ♄
23 38	♀ # ♆
04 Sunday	
02 18	☽ bQ ♃
02 35	☽ ± ♇
04 09	☽ ☍ ☿
06 10	☽ ⊼ ♂
10 23	☽ ∥ ☉
14 36	☽ ∥ ♀
15 24	☽ ± ♄
16 01	☽ # ♆
16 13	☉ ∠ ♆
17 10	☽ ± ♂
17 23	☽ ⊼ ☉
18 23	☽ bQ ♇
19 10	☽ ⊼ ♀
05 Monday	
01 22	☽ ∥ ♂
01 54	☽ ⊼ ♄
04 37	☽ ± ☉
04 54	☽ △ ♃
06 37	☽ ± ♀
09 34	☽ bQ ♂
14 57	☿ ⊻ ♂
19 22	☽ ☍ ♆
20 27	☽ △ ♇
21 21	☽ bQ ☉
22 48	☽ # ♄
23 42	☽ bQ ♀
06 Tuesday	
02 43	☽ □ ♅
05 50	☽ ∥ ♃
12 39	☽ △ ♂
13 54	☽ ⊼ ☿
22 17	♂ ⊥ ♅
07 Wednesday	
01 02	☽ △ ☉
01 17	☽ ± ☿
03 57	☽ △ ♀
05 49	☽ ☍ ♄
09 17	☽ □ ♃
09 41	☽ Q ♅
10 50	♀ ∠ ♅
13 13	☽ ∥ ♅
18 17	☽ bQ ☿
22 54	☽ ⊼ ♆
23 58	☽ □ ♇
08 Thursday	
02 17	♀ ⚹ ♄
04 55	☽ ∥ ♇
06 15	☽ ⚹ ♅
09 09	☽ ± ♆
16 27	☽ Q ♃
18 21	☽ □ ♂
22 27	☽ △ ☿
09 Friday	
00 31	☽ bQ ♆
04 31	☿ ⊥ ♄
06 11	☉ ∠ ♅
06 42	☽ Q ♇
07 52	☽ ∠ ♅
07 59	☽ □ ☉
08 06	☉ # ☿
09 15	☽ ⊼ ♄
12 03	☽ □ ♀
13 11	☽ ⚹ ♃
17 46	☿ # ♀
19 30	☽ ± ♄
23 13	☽ ⊥ ♅
10 Saturday	
02 05	☽ △ ♆
02 25	☽ Q ♂
03 09	☽ ⚹ ♇
03 14	♀ ⊻ ♃
03 41	☉ ⚹ ♄
09 27	☽ ⊻ ♅
09 56	☿ ∠ ♇
10 53	☽ bQ ♄
15 03	☽ ∠ ♃
16 51	☽ Q ☉
19 36	☽ ∥ ♇
21 36	☽ Q ♀
23 46	☽ ⚹ ♂
11 Sunday	
00 04	☿ Q ♅
04 42	☽ ∠ ♇
06 27	☽ □ ☿
06 35	☽ ⊥ ♃
08 17	♀ ⊥ ♆
11 18	☽ ∥ ♅
12 31	☽ △ ♄
14 45	☽ ⚹ ☉
16 56	☽ ⊻ ♃
19 58	☽ ⚹ ♀
20 02	☽ ⊥ ♇
12 Monday	
02 29	☽ ∠ ♂
05 12	☽ □ ♆
06 16	☽ ⊻ ♇
12 39	☽ ☌ ♅
15 57	☽ Q ☿
18 12	☽ ∠ ☉
18 33	☽ ⊥ ♂
19 36	♂ # ♆
20 31	☽ ∥ ♃
13 Tuesday	
00 01	☽ ∠ ♀
00 10	☽ # ♄
05 17	☽ ⊻ ♂
06 08	☉ ⊻ ♃
06 42	☉ ∥ ♀
10 44	☽ ⊥ ☉
12 01	☽ Q ♆
14 22	☽ ⚹ ☿
16 01	☽ □ ♄
16 56	☽ ⊥ ♀
20 56	☽ ☌ ♃
21 49	☽ ⊻ ☉
14 Wednesday	
00 49	☿ # ♂
04 17	☽ ⊻ ♀
05 37	☉ ⊥ ♆
07 11	☽ # ♆
07 53	☽ # ☿
08 44	☽ ⚹ ♆
09 03	☽ ∥ ♂
09 49	☽ ☌ ♇
12 05	☿ ⊻ ♄
14 23	☿ ∥ ♆
16 22	☽ ⊻ ♅
17 33	♂ ∠ ♆
18 35	☽ ∠ ☿
23 21	☽ Q ♄
15 Thursday	
02 51	☽ ⊥ ♅
07 58	☽ ∥ ☉
09 09	☽ ∥ ♀
10 56	☽ ∠ ♆
11 42	☽ ⊥ ☿
11 45	☽ ☌ ♂
13 08	☿ ⊥ ♂
18 44	☽ ∠ ♅
20 34	☽ ⚹ ♄
21 49	♀ ⊥ ♃
23 13	☽ ⊻ ☿
16 Friday	
02 07	☽ ⊻ ♃
02 54	☽ ⊥ ♆
06 19	☽ ☌ ☉
07 04	♀ ⊻ ♆
10 08	☽ # ♀
11 42	☽ # ☉
13 01	☽ ⊥ ♃
13 39	☽ ⊻ ♆
14 19	☽ ☌ ♀
14 47	☽ ⊻ ♇
19 21	♀ ⊻ ♇
21 39	☽ ⚹ ♅
23 37	☽ ∠ ♄
17 Saturday	
01 39	☽ ⊥ ♇
05 33	☽ ∠ ♃
10 01	☽ # ♂
15 54	☽ ∥ ♆
16 10	☽ ⊥ ♄
18 10	☽ ∠ ♇
19 41	☽ Q ♅
20 32	☽ ⊻ ♂
21 18	☿ ⚹ ♃
18 Sunday	
01 00	☽ ∥ ☿
03 22	☽ ⊻ ♄
08 18	☽ ⊥ ♂
09 44	☽ ⚹ ♃
10 34	☽ ☌ ☿
17 52	☽ ⊻ ☉
21 11	☽ ☌ ♆
22 21	☽ ⚹ ♇
19 Monday	
02 13	☽ ∠ ♂
02 45	☽ ∥ ♄
03 23	☽ # ♃
03 49	☽ ⊻ ♀
05 41	☽ □ ♅
06 15	☽ ⊥ ☉
08 54	☽ Q ♃
16 39	☽ ⊥ ♀
21 30	☽ Q ♇
23 09	♀ ⚹ ♅
20 Tuesday	
01 09	☽ ∠ ☉
06 28	♃ # ♄
08 52	☽ ⚹ ♂
12 13	☽ ∠ ♀
13 22	☽ ☌ ♄
14 13	☉ ⊻ ♆
19 36	☽ # ♅
20 39	☽ □ ♃
23 55	☉ ⊥ ♃
21 Wednesday	
00 23	♀ ∠ ♄
01 19	☽ ⊻ ☿
05 36	☉ ⊻ ♇
07 55	☽ ⊻ ♆
09 07	☽ □ ♇
09 25	☽ ⚹ ☉
10 00	☽ Q ♂
14 12	☽ ⊥ ☿
14 20	☽ # ♇
16 53	☽ △ ♅
17 51	♀ ⊥ ♇
19 59	☽ ⊥ ♆
21 35	☽ ⚹ ♀
22 Thursday	
02 39	♂ ∠ ♅
09 40	☽ ∠ ☿
10 12	☉ # ♀
11 47	☽ Q ☉
14 14	☽ ∠ ♆
23 20	☽ bQ ♅
23 Friday	
00 26	☽ □ ♂
00 45	☽ Q ♀
02 08	☽ ⊻ ♄
10 09	☽ △ ♃
14 26	☽ ⊥ ♄
17 42	☽ ± ♅
18 11	☽ ⚹ ☿
20 49	☽ ⚹ ♆
22 01	☽ △ ♇
24 Saturday	
03 41	☽ □ ☉
05 56	☽ ⊼ ♅
08 52	☽ ∠ ♄
12 08	♀ ∠ ♃
12 45	♂ ⚹ ♄
16 23	☽ # ♇
17 08	☽ bQ ♃
17 38	☽ □ ♀
19 58	☽ Q ☿
21 14	☽ Q ♆
25 Sunday	
04 28	☽ bQ ♇
07 56	☉ ⚹ ♅
11 24	☽ # ♅
11 34	☽ ± ♃
15 21	☽ ⚹ ♄
16 38	☽ △ ♂
22 32	☽ ± ♇
23 15	☿ ☌ ♆
23 44	☽ ⊼ ♃
26 Monday	
09 18	☽ □ ♆
09 43	☽ □ ☿
10 26	☽ ⊼ ♇
15 19	☽ Q ♄
18 07	☽ ☍ ♅
20 59	☽ △ ☉
23 51	☽ bQ ♂
27 Tuesday	
00 27	☉ ∠ ♄
02 59	☽ ∥ ♄
05 23	☿ ⚹ ♇
06 54	☽ # ♃
11 06	☽ ∥ ☿
11 47	☽ △ ♀
12 58	☉ ⊥ ♇
18 08	☽ ± ♂
28 Wednesday	
02 16	☽ □ ♄
04 13	☽ bQ ☉
06 06	☽ ⊼ ♂
10 36	☽ ☍ ♃
15 04	☽ ∥ ♆
18 53	☽ △ ♆
19 10	☽ bQ ♀
19 55	☽ ☍ ♇
19 56	♀ ⊥ ♄
20 57	☽ △ ☿
22 23	☽ ± ☉
29 Thursday	
03 11	☽ ⊼ ♅
03 16	♀ ∠ ♇
10 16	☽ ⊼ ☉
12 22	☽ # ♂
13 23	☽ ± ♀
14 01	☽ ∥ ♀
14 09	☽ ± ♅
15 06	♂ ⊥ ♆
21 38	♀ Q ♅
22 19	☽ bQ ♆
22 23	☽ ∥ ☉
30 Friday	
00 38	♀ # ♂
00 43	☽ bQ ☿
01 18	☽ ⊼ ♀
06 20	☽ bQ ♅
09 33	☽ △ ♄
14 22	☽ ± ♆
15 26	☽ ☍ ♂
16 50	☽ ± ☿
17 44	☽ ⊼ ♃
21 42	☽ # ☉

OCTOBER 1955

LONGITUDES

Date	Sidereal time h m s	Sun ☉ ° ' "	Moon ☽ ° ' "	Moon ☽ 24.00 ° ' "	Mercury ☿ ° '	Venus ♀ ° '	Mars ♂ ° '	Jupiter ♃ ° '	Saturn ♄ ° '	Uranus ♅ ° '	Neptune ♆ ° '	Pluto ♇ ° '
01	12 37 46	07 ♎ 32 02	03 ♈ 35 47	10 ♈ 34 57	28 ♎ 38	15 ♎ 38	22 ♍ 21	23 ♌ 15	18 ♏ 34	01 ♌ 42	27 ♎ 13	27 ♌ 45
02	12 41 43	08 31 01	17 ♈ 37 51	24 ♈ 43 52	28 R 35	16 53	22 59	23 26	18 40	01 44	27 15	27 47
03	12 45 39	09 30 03	01 ♉ 52 18	09 ♉ 02 26	28 25	18 08	23 37	23 37	18 46	01 46	27 17	27 48
04	12 49 36	10 29 07	16 ♉ 13 35	23 ♉ 25 08	28 07	19 22	24 16	23 48	18 52	01 47	27 20	27 50
05	12 53 32	11 28 13	00 ♊ 36 29	07 ♊ 47 08	27 42	20 37	24 54	23 58	18 58	01 49	27 22	27 51
06	12 57 29	12 27 21	14 ♊ 56 41	22 ♊ 04 48	27 08	21 52	25 32	24 09	19 05	01 51	27 24	27 53
07	13 01 25	13 26 32	29 ♊ 11 14	06 ♋ 15 48	26 27	23 07	26 11	24 20	19 11	01 53	27 26	27 54
08	13 05 22	14 25 46	13 ♋ 18 22	20 ♋ 18 52	25 38	24 21	26 49	24 30	19 17	01 54	27 28	27 56
09	13 09 19	15 25 01	27 ♋ 17 15	04 ♌ 13 26	24 42	25 36	27 27	24 41	19 23	01 56	27 30	27 57
10	13 13 15	16 24 19	11 ♌ 07 24	17 ♌ 59 04	23 40	26 51	28 06	24 51	19 30	01 58	27 33	27 58
11	13 17 12	17 23 39	24 ♌ 48 21	01 ♍ 35 08	22 33	28 06	28 44	25 02	19 36	01 59	27 35	28 00
12	13 21 08	18 23 01	08 ♍ 19 18	15 ♍ 00 40	21 22	29 ♎ 21	29 ♍ 23	25 12	19 43	02 01	27 37	28 01
13	13 25 05	19 22 26	21 ♍ 35 09	28 ♍ 14 24	20 10	00 ♏ 35	00 ♎ 01	25 22	19 49	02 02	27 39	28 03
14	13 29 01	20 21 53	04 ♎ 46 24	11 ♎ 14 59	18 58	01 50	00 39	25 32	19 56	02 03	27 42	28 04
15	13 32 58	21 21 21	17 ♎ 40 01	24 ♎ 01 25	17 48	03 05	01 18	25 42	20 02	02 05	27 44	28 05
16	13 36 54	22 20 52	00 ♏ 19 10	06 ♏ 33 17	16 43	04 20	01 56	25 52	20 09	02 06	27 46	28 06
17	13 40 51	23 20 25	12 ♏ 43 52	18 ♏ 51 04	15 44	05 35	02 35	26 01	20 15	02 07	27 48	28 08
18	13 44 48	24 20 00	24 ♏ 55 07	00 ♐ 56 18	14 53	06 50	03 13	26 11	20 22	02 08	27 50	28 09
19	13 48 44	25 19 37	06 ♐ 54 58	12 ♐ 51 32	14 11	08 04	03 52	26 20	20 29	02 09	27 53	28 10
20	13 52 41	26 19 16	18 ♐ 46 29	24 ♐ 40 19	13 40	09 19	04 30	26 30	20 36	02 10	27 55	28 11
21	13 56 37	27 18 56	00 ♑ 33 36	06 ♑ 26 56	13 19	10 34	05 09	26 39	20 42	02 11	27 57	28 13
22	14 00 34	28 18 39	12 ♑ 20 57	18 ♑ 16 19	13 10	11 49	05 47	26 48	20 49	02 12	27 59	28 14
23	14 04 30	29 ♎ 18 23	24 ♑ 13 42	00 ♒ 13 45	13 D 12	13 04	06 26	26 57	20 56	02 13	28 02	28 15
24	14 08 27	00 ♏ 18 08	06 ♒ 17 08	12 ♒ 24 31	13 25	14 18	07 04	27 06	21 03	02 14	28 04	28 16
25	14 12 23	01 17 56	18 ♒ 36 30	24 ♒ 53 38	13 49	15 33	07 43	27 15	21 10	02 15	28 06	28 17
26	14 16 20	02 17 44	01 ♓ 16 26	07 ♓ 45 16	14 22	16 48	08 21	27 24	21 16	02 16	28 08	28 18
27	14 20 17	03 17 35	14 ♓ 20 27	21 ♓ 02 11	15 04	18 03	09 00	27 32	21 23	02 16	28 11	28 19
28	14 24 13	04 17 28	27 ♓ 50 28	04 ♈ 45 12	15 54	19 18	09 38	27 41	21 30	02 17	28 13	28 20
29	14 28 10	05 17 22	11 ♈ 46 06	18 ♈ 52 43	16 51	20 33	10 17	27 49	21 37	02 17	28 15	28 21
30	14 32 06	06 17 18	26 ♈ 04 28	03 ♉ 20 38	17 54	21 47	10 55	27 57	21 44	02 18	28 17	28 22
31	14 36 03	07 ♏ 17 15	10 ♉ 40 20	18 ♉ 02 40	19 ♎ 04	23 ♏ 02	11 ♎ 34	28 ♌ 05	21 ♏ 51	02 ♌ 18	28 ♎ 20	28 ♌ 23

Moon

Date	Moon True ☊ ° '	Moon Mean ☊ ° '	Moon ☽ Latitude ° '
01	20 ♐ 12	20 ♐ 56	04 N 54
02	20 R 01	20 53	04 28
03	19 52	20 50	03 45
04	19 45	20 47	02 48
05	19 42	20 43	01 40
06	19 40	20 40	00 N 25
07	19 D 40	20 37	00 S 51
08	19 41	20 34	02 03
09	19 R 40	20 31	03 07
10	19 38	20 28	03 59
11	19 33	20 24	04 37
12	19 25	20 21	04 59
13	19 16	20 18	05 04
14	19 05	20 15	04 53
15	18 55	20 12	04 27
16	18 45	20 09	03 49
17	18 37	20 05	03 00
18	18 32	20 02	02 03
19	18 29	19 59	01 S 02
20	18 D 28	19 56	00 N 02
21	18 29	19 53	01 05
22	18 30	19 49	02 05
23	18 31	19 46	03 01
24	18 R 31	19 43	03 50
25	18 29	19 40	04 29
26	18 26	19 37	04 56
27	18 20	19 34	05 09
28	18 13	19 30	05 05
29	18 06	19 27	04 44
30	17 59	19 24	04 04
31	17 ♐ 54	19 ♐ 21	03 N 08

DECLINATIONS

Date	Sun ☉ ° '	Moon ☽ ° '	Mercury ☿ ° '	Venus ♀ ° '	Mars ♂ ° '	Jupiter ♃ ° '	Saturn ♄ ° '	Uranus ♅ ° '	Neptune ♆ ° '	Pluto ♇ ° '
01	02 S 59	05 N 55	14 S 25	05 S 09	04 N 02	14 N 25	15 S 25	20 N 19	08 S 55	21 N 53
02	03 23	11 03	14 23	05 39	03 46	14 21	15 26	20 18	08 56	21 52
03	03 46	15 39	14 18	06 09	03 31	14 18	15 28	20 18	08 57	21 52
04	04 09	19 22	14 08	06 39	03 16	14 14	15 30	20 17	08 58	21 52
05	04 32	21 54	13 54	07 08	03 00	14 11	15 32	20 17	08 58	21 52
06	04 55	23 01	13 36	07 38	02 45	14 07	15 34	20 17	08 59	21 51
07	05 18	22 36	13 13	08 07	02 30	14 04	15 35	20 16	09 00	21 51
08	05 41	20 45	12 46	08 36	02 14	14 01	15 37	20 16	09 01	21 51
09	06 04	17 39	12 14	09 05	01 59	13 57	15 39	20 16	09 02	21 51
10	06 27	13 37	11 38	09 34	01 43	13 54	15 41	20 15	09 02	21 50
11	06 50	08 54	10 59	10 03	01 28	13 51	15 43	20 15	09 03	21 50
12	07 13	03 N 50	10 16	10 31	01 12	13 48	15 45	20 15	09 04	21 50
13	07 35	01 S 21	09 32	10 59	00 57	13 44	15 46	20 14	09 05	21 50
14	07 58	06 23	08 46	11 27	00 42	13 41	15 48	20 14	09 05	21 49
15	08 20	11 03	08 01	11 55	00 26	13 38	15 50	20 14	09 06	21 49
16	08 42	15 09	07 17	12 22	00 N 11	13 35	15 52	20 14	09 07	21 49
17	09 04	18 31	06 35	12 49	00 S 05	13 32	15 54	20 13	09 08	21 49
18	09 26	21 00	05 57	13 16	00 20	13 29	15 56	20 13	09 09	21 49
19	09 48	22 29	05 23	13 43	00 36	13 25	15 58	20 13	09 10	21 49
20	10 10	22 57	04 54	14 09	00 51	13 22	15 59	20 13	09 10	21 49
21	10 31	22 22	04 31	14 34	01 07	13 19	16 01	20 13	09 11	21 48
22	10 53	20 47	04 13	15 00	01 22	13 16	16 03	20 13	09 12	21 48
23	11 14	18 18	04 01	15 25	01 38	13 14	16 05	20 12	09 13	21 48
24	11 35	15 00	03 56	15 49	01 53	13 11	16 07	20 12	09 14	21 48
25	11 56	11 00	03 55	16 14	02 08	13 08	16 09	20 12	09 14	21 48
26	12 16	06 25	04 00	16 38	02 24	13 05	16 11	20 12	09 15	21 48
27	12 37	01 S 25	04 10	17 01	02 39	13 02	16 13	20 12	09 16	21 48
28	12 57	03 N 48	04 24	17 24	02 55	13 00	16 14	20 12	09 17	21 48
29	13 17	09 00	04 42	17 46	03 10	12 57	16 16	20 12	09 18	21 48
30	13 37	13 52	05 04	18 08	03 25	12 54	16 18	20 11	09 18	21 48
31	13 S 57	18 N 00	05 S 28	18 S 30	03 S 40	12 N 52	16 S 20	20 N 11	09 S 19	21 N 48

ZODIAC SIGN ENTRIES

Date	h	m	Planets
01	05	46	☽ ♈
03	08	52	☽ ♉
05	10	59	☽ ♊
07	13	23	☽ ♋
09	16	41	☽ ♌
11	21	11	☽ ♍
13	00	39	♀ ♏
13	11	20	♂ ♎
14	03	13	☽ ♎
16	11	23	☽ ♏
18	22	07	☽ ♐
21	10	52	☽ ♑
23	23	33	☽ ♒
24	04	43	☉ ♏
26	09	37	☽ ♓
28	15	46	☽ ♈
30	18	30	☽ ♉

LATITUDES

Date	Mercury ☿ ° '	Venus ♀ ° '	Mars ♂ ° '	Jupiter ♃ ° '	Saturn ♄ ° '	Uranus ♅ ° '	Neptune ♆ ° '	Pluto ♇ ° '
01	03 S 40	01 N 06	01 N 05	00 N 41	02 N 02	00 N 32	01 N 40	10 N 17
04	03 34	01 01	01 04	00 41	02 01	00 32	01 40	10 18
07	03 14	00 56	01 04	00 42	02 01	00 32	01 40	10 19
10	02 38	00 50	01 03	00 42	02 00	00 32	01 40	10 19
13	01 47	00 44	01 03	00 43	02 00	00 33	01 40	10 20
16	00 S 46	00 38	01 02	00 43	01 59	00 33	01 40	10 21
19	00 N 14	00 31	01 01	00 44	01 59	00 33	01 40	10 22
22	01 04	00 24	01 01	00 44	01 59	00 33	01 40	10 23
25	01 40	00 17	01 00	00 45	01 58	00 33	01 40	10 24
28	02 01	00 10	00 59	00 45	01 58	00 33	01 40	10 25
31	02 N 09	00 02	00 N 59	00 N 46	01 N 58	00 N 33	01 N 40	10 N 26

DATA

Julian Date	2435382
Delta T	+31 seconds
Ayanamsa	23° 14' 37"
Synetic vernal point	05° ♓ 52' 23"
True obliquity of ecliptic	23° 26' 41"

LONGITUDES

Date	Chiron ⚷ ° '	Ceres ⚳ ° '	Pallas ⚴ ° '	Juno ⚵ ° '	Vesta ⚶ ° '	Black Moon Lilith ⚸ ° '
01	00 ♒ 06	13 ♑ 54	19 ♐ 14	20 ♏ 05	13 ♍ 58	22 ♐ 44
11	00 ♒ 05	15 ♑ 43	22 ♐ 07	23 ♏ 09	18 ♍ 40	23 ♐ 50
21	00 ♒ 10	17 ♑ 56	25 ♐ 12	26 ♏ 20	23 ♍ 36	24 ♐ 57
31	00 ♒ 23	20 ♑ 28	28 ♐ 28	29 ♏ 37	28 ♍ 00	26 ♐ 03

MOON'S PHASES, APSIDES AND POSITIONS ☽

Date	h	m	Phase	Longitude ° '	Eclipse Indicator
01	19	17	○	07 ♈ 50	
08	14	04	☾	14 ♋ 31	
15	19	32	●	21 ♎ 40	
23	23	05	☽	29 ♑ 46	
31	06	04	○	07 ♉ 02	

Day	h	m		
05	10	53	Perigee	
21	06	24	Apogee	
06	17	27	Max dec	23° N 03'
13	05	44	0S	
20	10	29	Max dec	22° S 57'
27	18	34	0N	

ASPECTARIAN

h m	Aspects
01 Saturday	
00 54	☽ ⊼ ♆
01 51	☽ ⊼ ♇
03 23	☽ ⊼ ☿
03 49	☽ ∥ ♂
04 22	☽ ± ♃
08 06	☽ # ♀
08 42	☽ △ ♅
11 12	☿ # ♃
11 57	☽ ⚼ ♄
12 16	☽ ± ♇
13 58	☿ St R
19 17	☽ ☍ ☉
20 06	☽ ⚼ ♃
22 36	☿ ⊥ ♂
02 Sunday	
01 53	☽ # ♆
03 30	☽ ± ♄
03 44	☽ ⚼ ♇
09 21	☉ ∠ ♃
10 36	☽ ☍ ♀
13 46	☽ ⊼ ♄
21 29	☽ ⊼ ♂
21 56	☽ △ ♃
03 Monday	
02 44	☉ # ♂
04 17	☽ ☍ ♆
04 39	☽ ∥ ♃
04 44	☽ # ☿
05 09	☽ △ ♇
06 17	☽ ☍ ☿
08 03	☽ ± ♂
10 59	☽ # ♄
11 13	☿ # ♃
11 24	♂ ⊻ ♃
11 49	☽ □ ♅
23 50	☽ ⚼ ♂
04 Tuesday	
01 33	♀ ⊻ ♄
01 43	☽ ⊼ ☉
12 28	☽ ± ☉
16 27	☽ ☍ ♄
17 45	☽ ⊼ ♀
17 58	☽ Q ♅
19 20	☽ ∥ ♅
05 Wednesday	
00 47	☽ □ ♃
02 02	☽ △ ♂
04 15	☿ ⚹ ♇
04 35	☽ ⚼ ☉
04 42	☽ ± ♀
06 34	☽ ⊼ ♆
07 17	☽ ⊼ ☿
07 24	☽ □ ♇
11 23	☽ ∥ ♇
14 02	☽ ⚹ ♅
16 37	☽ ± ♆
16 59	☽ ± ☿
21 10	☽ ⚼ ♀
06 Thursday	
02 02	☿ ☌ ♆
07 15	☽ Q ♃
07 29	☽ ⚼ ☿
07 31	☽ △ ☉
07 43	☽ ⚼ ♆
13 34	☽ Q ♇
15 12	☽ ∠ ♅
19 00	☽ ⊼ ♄
22 33	☉ ∠ ♇
07 Friday	
00 45	☽ △ ♀
03 41	☽ ⚹ ♃
04 55	☉ ⊥ ♄
05 11	☽ ± ♄
06 24	☽ ⊥ ♅
06 41	☽ □ ♂
07 36	☽ △ ☿
09 02	☽ △ ♆
09 49	☽ ⚹ ♇
16 33	☿ ⊻ ♂
16 34	☽ ⊻ ♅
20 32	☽ ⚼ ♄
22 55	☉ Q ♅
23 56	☽ ∥ ♇
08 Saturday	
05 26	☽ ∠ ♃
11 21	☽ ∠ ♇
14 04	☽ □ ☉
14 42	☽ Q ♂
15 22	♀ ⚹ ♃
16 24	☽ ∥ ♅
21 01	☽ ⊥ ♃
22 19	☽ △ ♄
09 Sunday	
02 11	☿ ☌ ♀
02 48	☽ ⊥ ♇
07 27	☽ ⊻ ♃
07 50	☽ □ ☿
08 44	♀ ∥ ♆
08 49	☽ □ ♀
12 19	☽ ⚹ ♂
12 22	☿ ⚹ ♃
12 23	☽ □ ♆
13 09	☽ ⊻ ♇
13 58	♂ ⊻ ♆
20 03	☽ ☌ ♅
23 25	☽ Q ☉
10 Monday	
00 24	☽ # ♄
07 12	♂ ⊻ ♇
10 24	☽ ∥ ♃
12 53	☽ Q ☿
15 37	☽ ∠ ♂
19 10	☽ Q ♀
19 45	☽ Q ♆
21 57	☽ ⚹ ☉
11 Tuesday	
00 02	☽ # ☿
01 47	♀ ☌ ♆
02 46	☽ □ ♄
06 53	☽ # ♀
08 10	☽ ⊥ ♂
08 20	☽ ⚹ ☿
10 04	♀ ⚹ ♇
11 16	☽ # ♆
12 24	☽ ☌ ♃
16 55	☽ ⚹ ♆
17 39	☽ ☌ ♇
18 24	☽ ⚹ ♀
19 18	☽ ⊻ ♂
21 15	☽ # ☉
12 Wednesday	
00 44	☽ ⊻ ♅
02 30	☽ ∠ ☉
06 57	☿ ∥ ♀
08 48	☽ ∠ ☿
10 54	☽ Q ♄
11 26	☽ ⊥ ♅
13 23	♀ ⊻ ♂
19 43	☽ ∠ ♆
19 52	☽ ⊥ ☉
23 36	☽ ⊥ ☿
23 55	☽ ∠ ♀
13 Thursday	
00 47	☽ ∥ ♂
03 38	☽ ∠ ♅
07 32	☽ ⊻ ☉
08 39	☽ ⚹ ♄
09 32	☽ ⊻ ☿
10 14	☽ # ♂
12 00	☽ ⊥ ♆
17 54	☽ ⊥ ♀
18 22	☿ ⊻ ♄
18 50	☽ ⊻ ♃
20 41	☉ ☌ ☿
22 58	☽ ⊻ ♆
23 40	☽ ⊻ ♇
14 Friday	
00 07	☉ ⊻ ♄
01 59	☿ ∥ ♆
04 02	☽ ☌ ♂
05 57	☽ ⊥ ♃
06 01	☽ ⊻ ♀
06 59	☽ ⚹ ♅
10 41	☽ ⊥ ♇
12 17	☽ ∠ ♄
16 17	♀ □ ♅
20 31	☽ ∥ ☉
22 19	☽ ∥ ☿
22 48	☽ ∠ ♃
15 Saturday	
01 41	☽ ∥ ♆
03 25	☽ ∠ ♇
05 08	☽ ⊥ ♄
05 14	☉ ∥ ☿
05 16	☽ Q ♅
12 14	☽ ☌ ☿
16 30	☽ ⊻ ♄
17 20	☽ ∥ ♀
19 32	☽ ☌ ☉
16 Sunday	
02 27	☽ # ♃
03 23	☽ ⚹ ♃
07 06	☽ ☌ ♆
07 46	☽ ⚹ ♇
07 59	♂ ⊥ ♃
15 16	☽ ⊻ ♂
15 25	☽ □ ♅
16 41	☽ ∥ ♄
18 09	♂ ⚹ ♅
20 34	☽ ☌ ♀
17 Monday	
02 43	☽ Q ♃
03 29	☽ ⊥ ♂
06 55	☽ Q ♇
16 21	☉ ∥ ♆
17 28	☽ ⊻ ☿
22 02	☽ ∠ ♂
18 Tuesday	
02 54	☽ ☌ ♄
03 28	☽ # ♅
04 30	☽ ⊥ ☿
10 44	☽ ⊻ ☉
14 33	☽ □ ♃
17 50	☽ ⊻ ♆
18 26	☽ □ ♇
21 18	☽ ∠ ☿
22 00	♀ # ♃
23 08	☽ # ♇
23 46	☽ ⊥ ☉
19 Wednesday	
02 25	☽ △ ♅
02 38	☿ ⊥ ♄
05 31	☽ ⚹ ♂
05 52	☽ ⊥ ♆
13 05	☿ Q ♅
14 36	☽ ⊻ ♀
17 53	♀ Q ♃
19 30	☽ ∠ ☉
23 53	♂ ⊥ ♇
20 Thursday	
00 05	☽ ∠ ♆
02 01	☽ ⚹ ☿
04 10	☽ ⊥ ♀
07 07	☽ Q ♂
08 45	☽ ⚼ ♅
15 44	☽ ⊻ ♄
17 00	☉ ⚹ ♃
21 Friday	
00 37	☽ ∠ ♀
01 35	☽ Q ☿
03 05	☽ ± ♅
03 56	☽ △ ♃
04 04	☽ ⊥ ♄
04 47	☽ ⚹ ☉
05 00	♀ Q ♇
06 40	☽ ⚹ ♆
07 12	☽ △ ♇
15 20	☽ ⊼ ♅
21 53	☽ □ ♂
22 17	☽ # ♇
22 35	☽ ∠ ♄
22 Saturday	
00 33	☿ ∠ ♇
03 58	☉ ☌ ♆
07 11	☽ Q ♆
07 29	☽ Q ☉
09 58	☉ ⚹ ♇
10 47	☽ ⚹ ♀
10 53	☽ ⚼ ♃
10 53	♀ ⊥ ♂
13 26	♂ ∠ ♄
13 40	☽ □ ☿
13 47	☽ ⚼ ♇
18 28	☽ # ♅
19 23	☿ St D
23 Sunday	
05 18	☽ ⚹ ♄
05 20	☽ ± ♃
08 01	☽ ± ♇
13 52	☽ Q ♀
15 08	☿ ⊻ ♀
17 32	☽ ⊼ ♃
19 10	☿ ∠ ♇
19 38	☽ □ ♆
20 04	☽ ⊼ ♇
23 05	☽ □ ☉
24 Monday	
03 59	☽ ☍ ♅
04 32	☽ ∥ ♄
05 32	☽ Q ♄
07 03	☽ ∥ ♀
13 38	☽ △ ♂
23 31	☽ # ♃
25 Tuesday	
02 23	☽ △ ☿
05 27	☽ □ ♀
06 43	♀ ∥ ♄
07 07	☽ ∥ ☉
16 56	☽ □ ♄
20 16	☽ ⚼ ♂
21 28	☽ ∥ ♆
26 Wednesday	
04 39	☽ ☍ ♃
06 07	☽ △ ♆
06 25	☽ ☍ ♇
07 51	☿ Q ♅
08 15	☽ ⚼ ☿
11 08	☉ □ ♅
13 50	☽ ⊼ ♅
14 04	☽ △ ☉
14 07	☽ ± ♂
23 27	☽ ∥ ☿
27 Thursday	
00 56	☽ ± ♅
01 46	☽ ⊼ ♂
01 50	☽ ± ☿
06 29	☽ ∥ ♂
09 53	☽ ⚼ ♆
13 23	☽ ⊼ ☿
17 17	☽ ⚼ ♅
19 21	☽ △ ♀
19 41	☽ ⚼ ☉
23 25	☿ ⊥ ♄
28 Friday	
00 44	☽ △ ♄
02 04	☽ ± ♆
07 42	☽ # ♂
11 43	☽ ⊼ ♃
12 39	☽ ⊼ ♆
12 51	☽ ± ☉
12 52	☽ ⊼ ♇
14 26	☉ # ♃
14 52	☽ # ☿
19 44	☽ △ ♅
22 15	☽ ± ♃
23 17	☽ ± ♇
29 Saturday	
00 04	☽ ⊼ ☉
00 19	☽ ⚼ ♀
03 08	☽ ⚼ ♄
09 21	☽ ☍ ♂
13 22	☽ # ♆
13 48	☽ ⚼ ♃
14 41	☽ ⚼ ♇
17 09	☽ ± ♀
18 34	☽ ± ♄
21 15	☽ ☍ ☿
30 Sunday	
04 12	☽ ⊼ ♀
04 43	☽ ⊼ ♄
07 06	☽ ∥ ♃
10 39	☽ # ☉
10 52	♀ ☌ ♄
15 09	☽ △ ♃
15 41	☽ ☍ ♆
15 48	☽ △ ♇
22 17	☽ □ ♅
31 Monday	
01 36	☽ # ♄
06 04	☽ ☍ ☉
13 31	☽ ⊼ ♂
15 41	☽ # ♀
23 44	☽ ± ♂

All ephemeris data is given at 12.00 UT and the Moon's longitude is additionally given for 24.00 UT

NOVEMBER 1955

LONGITUDES

Date	Sidereal time h m s	Sun ☉ ° ' "	Moon ☽ ° ' "	Moon ☽ 24.00 ° ' "	Mercury ☿ ° '	Venus ♀ ° '	Mars ♂ ° '	Jupiter ♃ ° '	Saturn ♄ ° '	Uranus ♅ ° '	Neptune ♆ ° '	Pluto ♇ ° '
01	14 39 59	08 ♏ 17 15	25 ♉ 26 40	02 ♊ 51 21	20 ♎ 17	24 ♏ 17	12 ♎ 12	28 ♌ 13	21 ♏ 58	02 ♌ 19	28 ♎ 22	28 ♌ 24
02	14 43 56	09 17 17	10 ♊ 15 46	17 ♊ 39 03	21 35	25 32	12 51	28 21	22 05	02 20	28 24	28 25
03	14 47 52	10 17 21	25 ♊ 00 24	02 ♋ 19 10	22 57	26 47	13 30	28 29	22 12	02 20	28 26	28 25
04	14 51 49	11 17 27	09 ♋ 34 46	16 ♋ 46 46	24 21	28 01	14 08	28 37	22 19	02 20	28 28	28 26
05	14 55 46	12 17 35	23 ♋ 54 51	00 ♌ 58 49	25 48	29 ♏ 16	14 47	28 44	22 26	02 20	28 31	28 27
06	14 59 42	13 17 45	07 ♌ 58 32	14 ♌ 53 58	27 17	00 ♐ 31	15 25	28 51	22 34	02 20	28 33	28 28
07	15 03 39	14 17 57	21 ♌ 45 08	28 ♌ 32 08	28 ♎ 48	01 46	16 04	28 58	22 41	02 20	28 35	28 28
08	15 07 35	15 18 11	05 ♍ 15 02	11 ♍ 54 00	00 ♏ 19	03 01	16 43	29 05	22 48	02 R 20	28 37	28 29
09	15 11 32	16 18 28	18 ♍ 29 08	25 ♍ 00 35	01 52	04 15	17 21	29 12	22 55	02 20	28 39	28 30
10	15 15 28	17 18 46	01 ♎ 28 31	07 ♎ 53 01	03 26	05 30	18 00	29 19	23 02	02 20	28 41	28 30
11	15 19 25	18 19 06	14 ♎ 14 15	20 ♎ 32 19	05 01	06 45	18 39	29 26	23 09	02 20	28 44	28 31
12	15 23 21	19 19 28	26 ♎ 47 21	02 ♏ 59 28	06 36	08 00	19 18	29 32	23 16	02 20	28 46	28 32
13	15 27 18	20 19 52	09 ♏ 08 47	15 ♏ 15 26	08 11	09 15	19 56	29 38	23 23	02 19	28 48	28 32
14	15 31 15	21 20 17	21 ♏ 19 36	27 ♏ 21 26	09 47	10 30	20 35	29 44	23 31	02 19	28 50	28 33
15	15 35 11	22 20 45	03 ♐ 21 09	09 ♐ 18 58	11 23	11 44	21 14	29 50	23 38	02 19	28 52	28 33
16	15 39 08	23 21 14	15 ♐ 15 09	21 ♐ 10 00	12 59	12 59	21 52	29 ♌ 56	23 45	02 18	28 54	28 34
17	15 43 04	24 21 44	27 ♐ 03 51	02 ♑ 57 06	14 35	14 14	22 31	00 ♍ 02	23 52	02 18	28 56	28 34
18	15 47 01	25 22 16	08 ♑ 50 08	14 ♑ 43 25	16 11	15 29	23 10	00 07	23 59	02 17	28 58	28 34
19	15 50 57	26 22 49	20 ♑ 37 26	26 ♑ 32 43	17 47	16 44	23 49	00 13	24 06	02 17	29 00	28 35
20	15 54 54	27 23 23	02 ♒ 29 48	08 ♒ 29 15	19 23	17 58	24 28	00 18	24 14	02 16	29 02	28 35
21	15 58 50	28 23 59	14 ♒ 31 41	20 ♒ 37 40	20 58	19 13	25 06	00 23	24 21	02 16	29 04	28 35
22	16 02 47	29 ♏ 24 36	26 ♒ 47 49	03 ♓ 02 43	22 34	20 28	25 45	00 28	24 28	02 15	29 06	28 36
23	16 06 44	00 ♐ 25 14	09 ♓ 22 54	15 ♓ 48 53	24 09	21 43	26 24	00 32	24 35	02 14	29 08	28 36
24	16 10 40	01 25 53	22 ♓ 21 07	28 ♓ 59 56	25 44	22 57	27 03	00 37	24 42	02 13	29 10	28 36
25	16 14 37	02 26 33	05 ♈ 45 37	12 ♈ 38 17	27 20	24 12	27 42	00 41	24 49	02 12	29 12	28 36
26	16 18 33	03 27 14	19 ♈ 37 55	26 ♈ 44 19	28 ♏ 54	25 27	28 20	00 45	24 56	02 11	29 14	28 36
27	16 22 30	04 27 57	03 ♉ 57 08	11 ♉ 15 50	00 ♐ 29	26 42	28 59	00 49	25 03	02 10	29 16	28 37
28	16 26 26	05 28 40	18 ♉ 39 41	26 ♉ 07 50	02 04	27 56	29 ♎ 38	00 53	25 11	02 09	29 18	28 37
29	16 30 23	06 29 25	03 ♊ 39 16	11 ♊ 12 52	03 39	29 ♐ 11	00 ♏ 17	00 57	25 18	02 08	29 20	28 37
30	16 34 19	07 ♐ 30 11	18 ♊ 47 28	26 ♊ 21 54	05 ♐ 13	00 ♑ 26	00 ♏ 56	01 ♍ 00	25 ♏ 25	02 ♌ 07	29 ♎ 22	28 ♌ 37

Moon Nodes and Latitude

Date	Moon True ☊ ° '	Moon Mean ☊ ° '	Moon ☽ Latitude ° '
01	17 ♐ 50	19 ♐ 18	01 N 59
02	17 R 48	19 15	00 N 41
03	17 D 48	19 11	00 S 39
04	17 49	19 08	01 56
05	17 51	19 05	03 04
06	17 52	19 02	04 00
07	17 R 52	18 59	04 41
08	17 50	18 55	05 05
09	17 47	18 52	05 13
10	17 43	18 49	05 04
11	17 38	18 46	04 40
12	17 32	18 43	04 03
13	17 28	18 40	03 15
14	17 24	18 36	02 18
15	17 22	18 33	01 16
16	17 21	18 30	00 S 12
17	17 D 22	18 27	00 N 53
18	17 23	18 24	01 56
19	17 25	18 21	02 54
20	17 26	18 17	03 45
21	17 28	18 14	04 26
22	17 28	18 11	04 57
23	17 R 28	18 08	05 14
24	17 27	18 05	05 16
25	17 25	18 01	05 01
26	17 23	17 58	04 28
27	17 22	17 55	03 38
28	17 20	17 52	02 33
29	17 19	17 49	01 15
30	17 ♐ 19	17 ♐ 46	00 S 08

DECLINATIONS

Date	Sun ☉ ° '	Moon ☽ ° '	Mercury ☿ ° '	Venus ♀ ° '	Mars ♂ ° '	Jupiter ♃ ° '	Saturn ♄ ° '	Uranus ♅ ° '	Neptune ♆ ° '	Pluto ♇ ° '
01	14 S 16	21 N 03	05 S 55	18 S 51	03 S 56	12 N 49	16 S 22	20 N 11	09 S 20	21 N 48
02	14 36	22 40	06 25	19 12	04 11	12 46	16 24	20 11	09 21	21 48
03	14 55	22 42	06 56	19 32	04 26	12 44	16 26	20 11	09 21	21 48
04	15 13	21 10	07 29	19 51	04 41	12 42	16 28	20 11	09 22	21 48
05	15 32	18 18	08 03	20 10	04 57	12 39	16 29	20 11	09 23	21 48
06	15 50	14 24	08 38	20 28	05 12	12 37	16 31	20 11	09 24	21 48
07	16 08	09 49	09 14	20 46	05 27	12 34	16 33	20 11	09 24	21 48
08	16 26	04 N 51	09 50	21 04	05 42	12 32	16 35	20 11	09 25	21 48
09	16 43	00 S 15	10 27	21 20	05 57	12 30	16 37	20 11	09 26	21 49
10	17 00	05 14	11 04	21 36	06 12	12 28	16 39	20 11	09 27	21 49
11	17 17	09 55	11 41	21 52	06 27	12 26	16 40	20 11	09 27	21 49
12	17 34	14 06	12 18	22 07	06 42	12 24	16 42	20 12	09 28	21 49
13	17 50	17 37	12 54	22 21	06 57	12 22	16 44	20 12	09 29	21 49
14	18 06	20 19	13 31	22 34	07 12	12 20	16 46	20 12	09 30	21 49
15	18 22	22 05	14 06	22 47	07 26	12 18	16 48	20 12	09 30	21 49
16	18 37	22 49	14 42	22 59	07 41	12 16	16 49	20 12	09 31	21 50
17	18 52	22 31	15 17	23 11	07 56	12 14	16 51	20 12	09 32	21 50
18	19 07	21 13	15 51	23 22	08 10	12 13	16 53	20 12	09 32	21 50
19	19 21	19 00	16 25	23 32	08 25	12 11	16 55	20 12	09 33	21 50
20	19 35	15 58	16 58	23 42	08 40	12 09	16 57	20 13	09 34	21 50
21	19 48	12 14	17 30	23 50	08 54	12 08	16 58	20 13	09 35	21 51
22	20 02	07 56	18 01	23 58	09 08	12 06	17 00	20 13	09 35	21 51
23	20 15	03 S 12	18 32	24 06	09 23	12 05	17 02	20 13	09 36	21 51
24	20 27	01 N 48	19 01	24 12	09 37	12 04	17 04	20 13	09 37	21 52
25	20 39	06 53	19 30	24 18	09 51	12 02	17 05	20 14	09 37	21 52
26	20 51	11 49	19 58	24 23	10 05	12 01	17 07	20 14	09 38	21 52
27	21 02	16 15	20 25	24 28	10 20	12 00	17 09	20 14	09 38	21 52
28	21 13	19 49	20 51	24 31	10 34	11 59	17 10	20 14	09 39	21 53
29	21 24	22 07	21 15	24 34	10 48	11 58	17 12	20 15	09 40	21 53
30	21 S 34	22 N 50	21 S 39	24 S 37	11 S 01	11 N 57	17 S 14	20 N 15	09 S 40	21 N 53

ZODIAC SIGN ENTRIES

Date	h	m	Planets
01	19	23	☽ ♊
03	20	11	☽ ♋
05	22	20	☽ ♌
06	02	02	♀ ♐
08	02	36	☽ ♍
08	06	57	☿ ♏
10	09	15	☽ ♎
12	18	12	☽ ♏
15	05	17	☽ ♐
17	03	59	♃ ♍
17	17	59	☽ ♑
20	06	58	☽ ♒
22	18	10	☽ ♓
23	02	01	☉ ♐
25	01	47	☽ ♈
27	04	34	☽ ♉
27	05	27	☿ ♐
29	01	33	♂ ♏
29	06	11	☽ ♊
30	03	42	♀ ♑

LATITUDES

Date	Mercury ☿ ° '	Venus ♀ ° '	Mars ♂ ° '	Jupiter ♃ ° '	Saturn ♄ ° '	Uranus ♅ ° '	Neptune ♆ ° '	Pluto ♇ ° '
01	02 N 10	00 00	00 N 58	00 N 46	01 N 58	00 N 33	01 N 40	10 N 27
04	02 06	00 S 08	00 58	00 47	01 58	00 33	01 40	10 28
07	01 56	00 15	00 57	00 47	01 57	00 34	01 40	10 29
10	01 42	00 23	00 56	00 48	01 57	00 34	01 41	10 30
13	01 25	00 31	00 55	00 49	01 57	00 34	01 41	10 31
16	01 06	00 38	00 54	00 49	01 57	00 34	01 41	10 32
19	00 45	00 45	00 53	00 50	01 57	00 34	01 41	10 33
22	00 25	00 52	00 52	00 51	01 57	00 34	01 41	10 34
25	00 N 04	00 59	00 51	00 52	01 57	00 34	01 41	10 36
28	00 S 16	01 06	00 50	00 52	01 57	00 34	01 41	10 37
31	00 S 36	01 S 12	00 N 49	00 N 53	01 N 57	00 N 34	01 N 41	10 N 38

DATA

Julian Date	2435413
Delta T	+31 seconds
Ayanamsa	23° 14' 40"
Synetic vernal point	05° ♓ 52' 20"
True obliquity of ecliptic	23° 26' 41"

MOON'S PHASES, APSIDES AND POSITIONS ☽

Date	h	m	Phase	Longitude ° '	Eclipse Indicator
06	21	56	☾	13 ♌ 43	
14	12	01	●	21 ♏ 20	
22	17	29	☽	29 ♒ 38	
29	16	50	○	06 ♊ 42	partial

Day	h	m		
02	03	26	Perigee	
17	23	34	Apogee	
30	11	32	Perigee	
03	00	24	Max dec	22° N 53'
09	10	51	0S	
16	17	04	Max dec	22° S 51'
24	03	28	0N	
30	10	12	Max dec	22° N 50'

LONGITUDES

Date	Chiron ⚷ ° '	Ceres ⚳ ° '	Pallas ⚴ ° '	Juno ⚵ ° '	Vesta ⚶ ° '	Black Moon Lilith ⚸ ° '
01	00 ♒ 25	20 ♑ 45	28 ♐ 48	29 ♏ 57	28 ♍ 28	26 ♐ 10
11	00 ♒ 45	23 ♑ 35	02 ♑ 13	03 ♐ 19	03 ♎ 04	27 ♐ 16
21	01 ♒ 12	26 ♑ 40	05 ♑ 44	06 ♐ 44	07 ♎ 35	28 ♐ 23
31	01 ♒ 44	29 ♑ 57	09 ♑ 20	10 ♐ 12	12 ♎ 00	29 ♐ 29

ASPECTARIAN

h m Aspects

01 Tuesday
02 52 ☽ ⊼ ☿
03 41 ☽ Q ♅
04 05 ☽ ∥ ♅
06 20 ☽ ☍ ♄
09 57 ☽ ☍ ♀
13 30 ☽ ± ☿
14 59 ☽ bQ ♂
16 32 ☽ □ ♃
16 44 ☽ ⊼ ♆
16 47 ☽ □ ♇
20 32 ☽ ∥ ♇
23 08 ☽ ✶ ♅

02 Wednesday
02 29 ☽ ± ♆
05 28 ☽ bQ ☿
10 18 ☽ ⊼ ☉
16 23 ☽ △ ♂
17 06 ☽ bQ ♆
20 45 ☽ ± ☉
21 47 ☿ ⊻ ♄
21 59 ☽ Q ♃
21 59 ☽ Q ♇
22 02 ♆ ✶ ♇
23 26 ♃ ☌ ♇
23 28 ☽ ∠ ♅
23 47 ♃ ✶ ♆

03 Thursday
07 23 ☽ ⊼ ♄
08 17 ☽ △ ☿
09 18 ♂ ∠ ♇
11 29 ♂ ∠ ♃
12 30 ☽ bQ ☉
14 10 ☽ ⊥ ♅
15 10 ☽ ⊼ ♀
15 15 ☉ Q ♇
17 17 ☽ ± ♄
17 18 ☉ Q ♃
17 36 ☽ ✶ ♇
17 38 ☽ △ ♆
17 45 ☽ ✶ ♃

04 Friday
00 01 ☽ ⊻ ♅
01 57 ☽ ± ♀
04 31 ☽ ∥ ♇
08 14 ☽ bQ ♄
15 03 ☽ △ ☉
18 16 ☽ bQ ♀
18 25 ☽ ∠ ♇
18 46 ☽ ∠ ♃
19 13 ♂ Q ♅
19 56 ☽ □ ♂
20 01 ♀ □ ♇
20 54 ♀ ⊻ ♆
21 25 ☽ ∥ ♅
23 02 ☽ ⋕ ♀

05 Saturday
00 30 ♀ □ ♃
09 29 ☽ △ ♄
09 31 ☽ ⊥ ♇
09 59 ☽ ⊥ ♃
13 32 ♀ ⋕ ♅
15 34 ☽ □ ☿
19 42 ☽ ⊻ ♇
19 49 ☽ □ ♆
20 14 ☽ ⊻ ♃
21 58 ☽ △ ♀
23 43 ☽ ⋕ ♄

06 Sunday
02 19 ☽ ☌ ♅
03 48 ☽ Q ♂
04 21 ☽ ⋕ ☉
08 18 ♀ ∠ ♂
21 48 ☽ ∥ ♃
21 56 ☽ □ ☉

07 Monday
01 33 ☽ ✶ ♂
02 13 ☽ Q ☿
02 55 ☽ Q ♆
06 56 ☿ ✶ ♇
08 37 ☿ ☌ ♆
13 39 ☽ □ ♄
14 02 ☽ ⋕ ♆
14 34 ☽ ⋕ ☿
15 06 ☿ ✶ ♃
18 58 ☿ ∥ ♆
23 01 ♀ △ ♅
23 54 ☽ ☌ ♇

08 Tuesday
00 07 ☽ ✶ ♆
00 53 ☽ ☌ ♃
02 03 ☽ ✶ ☿
05 21 ☽ ∠ ♂
06 47 ☽ ⊻ ♅
07 34 ☽ □ ♀
08 10 ☽ ⋕ ♂
08 13 ☽ Q ☉
09 30 ♅ St R
15 46 ♂ ⊥ ♄
17 33 ☽ ⊥ ♅
22 05 ☽ Q ♄
22 21 ☽ ⊥ ♂

09 Wednesday
02 08 ☉ ∥ ♄
03 10 ☽ ∠ ♆
07 41 ☽ ✶ ☉
08 40 ☽ ∠ ☿
09 50 ☽ ⊻ ♂
09 54 ☽ ∠ ♅
19 05 ☿ □ ♅
19 39 ☽ Q ♀
19 41 ☽ ⊥ ♆
19 51 ♀ ⊥ ♆
20 13 ☽ ✶ ♄

10 Thursday
03 28 ☽ ⊥ ☿
06 28 ☽ ⊻ ♇
06 48 ☽ ⊻ ♆
07 57 ☽ ⊻ ♃
13 36 ☽ ✶ ♅
13 42 ☽ ∠ ☉
16 11 ☽ ⊻ ☿
17 04 ☽ ∥ ♂
17 40 ☽ ⊥ ♇
19 15 ☽ ⊥ ♃
20 21 ☽ ✶ ♀

11 Friday
00 24 ☽ ∠ ♄
07 18 ♀ ⋕ ♇
08 03 ☽ ⊥ ☉
09 33 ☽ ∥ ♆
10 38 ☽ ∠ ♇
12 11 ☽ Q ♅
12 22 ☽ ∠ ♃
17 35 ☽ ⊥ ♄
20 26 ☽ ⊻ ☉
20 50 ☽ ☌ ♂
23 23 ☽ ∥ ☿

12 Saturday
01 54 ☽ ⋕ ♃
03 54 ☽ ∠ ♀
05 10 ☽ ⊻ ♄
09 51 ☉ ⊻ ♂
15 21 ☽ ✶ ♇
15 45 ☿ ⋕ ♃
15 49 ☽ ☌ ♆
17 21 ☽ ✶ ♃
22 43 ☽ □ ♅
23 12 ☽ ⊥ ♀

13 Sunday
05 22 ☽ ∥ ♄
09 51 ☽ ☌ ☿
12 13 ☽ ⊻ ♀
13 47 ☽ ∥ ☉
14 43 ☽ Q ♇
16 56 ☽ Q ♃

14 Monday
10 27 ☽ ⊻ ♂
10 39 ☽ ⋕ ♅
12 01 ☽ ☌ ☉
16 23 ☽ ☌ ♄
23 03 ☽ ⊥ ♂
23 27 ☿ Q ♇

15 Tuesday
02 23 ☽ □ ♇
02 59 ☽ ⊻ ♆
04 54 ☽ □ ♃
07 26 ☽ ⋕ ♇
09 55 ☽ △ ♅
15 03 ☽ ⊥ ♆
18 07 ☽ ∠ ♂
19 18 ☿ Q ♃

16 Wednesday
06 41 ☽ ⊻ ☿
06 52 ☽ ☌ ♀
09 15 ☽ ∠ ♆
12 17 ☿ ⊻ ♀
16 10 ☽ bQ ♅
20 45 ☽ ⊥ ☿
22 40 ☉ ☌ ♄

17 Thursday
02 13 ☽ ✶ ♂
05 26 ☽ ⊻ ♄
05 59 ☽ ⊻ ☉
06 08 ♀ ∠ ♆
10 27 ☽ ± ♅
15 04 ☽ △ ♇
15 50 ☽ ✶ ♆
17 46 ☽ ⊥ ♄
17 56 ☽ ∠ ☿
18 06 ☽ △ ♃
19 21 ☽ ⊥ ☉
22 40 ☽ ⊼ ♅

18 Friday
02 50 ☽ ⋕ ♇
04 05 ☽ Q ♂
12 19 ☽ ∠ ♄
15 25 ☽ ∠ ☉
16 22 ☽ Q ♆
21 40 ☽ bQ ♇

19 Saturday
00 09 ☽ ⋕ ♅
00 55 ☽ bQ ♃
03 08 ☽ ⊻ ♀
05 19 ☽ ✶ ☿
09 02 ☽ ∥ ☉
15 58 ☽ ± ♇
16 46 ☽ ⊥ ♀
18 50 ☽ □ ♂
19 08 ☽ ✶ ♄
19 20 ☽ ± ♃
22 35 ♀ bQ ♅

20 Sunday
00 45 ☽ ✶ ☉
01 25 ♂ ⊻ ♄
04 07 ☽ ⊼ ♇
04 55 ☽ ∥ ♄
05 01 ☽ □ ♆
05 48 ☽ ∥ ☿
07 33 ☽ ⊼ ♃
09 24 ☽ Q ☿
11 17 ☿ ∥ ♄
11 33 ☽ ☍ ♅
13 04 ☽ ∠ ♀
19 33 ☽ Q ♄

21 Monday
03 04 ☽ Q ☉
12 36 ☽ ⋕ ♃
16 32 ☉ □ ♇
22 17 ☽ ✶ ♀

22 Tuesday
02 33 ☽ □ ☿
03 06 ☽ ∥ ♆
04 33 ☉ ⊻ ♆
05 52 ☽ ∥ ♂
07 26 ☽ □ ♄
09 52 ☽ △ ♂
15 28 ☽ ☍ ♇
16 28 ☽ △ ♆
17 29 ☽ □ ☉
19 06 ☽ ☍ ♃
22 28 ☽ ⊼ ♅

23 Wednesday
00 06 ☽ Q ♀
09 21 ☉ ⋕ ♅
09 50 ☽ ± ♅
15 03 ☉ □ ♃
15 59 ☽ bQ ♂
19 03 ☿ ☌ ♄
20 55 ☽ bQ ♆

24 Thursday
02 36 ☽ bQ ♅
09 30 ☽ ± ♂
11 04 ♂ ∥ ♆
13 13 ☽ □ ♀
13 30 ☽ ± ♆
16 18 ☽ △ ♄
18 58 ☽ △ ☿
20 56 ☽ ⊼ ♂
23 17 ☽ ⊼ ♇

25 Friday
00 20 ☽ ⊼ ♆
02 58 ☽ ⊼ ♃
05 40 ☽ △ ☉
05 43 ☽ △ ♅
06 25 ☉ △ ♅
09 58 ☽ ± ♇
13 38 ☽ ± ♃
19 10 ☽ bQ ♄
21 24 ☿ ⊻ ♂

26 Saturday
00 56 ☽ bQ ☿
01 08 ☽ ⋕ ♆
01 09 ♀ ⊻ ♄
01 40 ☽ bQ ♇
02 58 ☽ ⋕ ♂
05 20 ☽ bQ ♃
07 26 ☿ □ ♇
09 50 ☽ bQ ☉
10 49 ☽ ± ♄
13 00 ☽ ∥ ♃
17 06 ☿ ⊻ ♆
18 15 ☽ ± ☿
21 03 ☽ ⊼ ♄
21 58 ♂ ✶ ♇
22 47 ☽ △ ♀

27 Sunday
02 03 ♀ ± ♅
02 12 ☽ ± ☉
02 24 ☿ ⋕ ♅
03 08 ☽ △ ♇
03 22 ☽ ☍ ♂
04 13 ☽ ☍ ♆
05 33 ☽ ⊼ ☿
06 47 ☽ △ ♃
09 03 ☽ □ ♅
12 55 ☽ ⊼ ☉
17 17 ☿ □ ♃
17 24 ☽ ⋕ ♄
23 01 ♂ ☌ ♆

28 Monday
01 53 ☽ bQ ♀
07 40 ☉ ⊥ ♆
13 18 ☿ △ ♅
14 24 ☽ Q ♅
15 27 ☽ ∥ ♅
17 46 ☽ ± ♀
22 18 ☉ ⊥ ♂
22 33 ☽ ☍ ♄
22 45 ☽ ⋕ ☿

29 Tuesday
00 57 ♀ △ ♇
01 49 ☽ ⋕ ☉
03 58 ☽ □ ♇
04 14 ☽ ⊼ ♀
05 06 ☽ ⊼ ♆
06 23 ☽ ⊼ ♂
07 40 ☽ □ ♃
08 36 ☽ ∥ ♇
09 35 ☽ ✶ ♅
11 59 ☽ ☍ ☿
14 40 ☽ ± ♆
14 55 ♀ ✶ ♆
16 22 ☽ ± ♂
16 50 ☽ ☍ ☉

30 Wednesday
02 31 ☉ ∥ ☿
04 58 ☽ bQ ♆
07 16 ☽ bQ ♂
08 33 ☽ Q ♇
09 21 ☽ ∠ ♅
12 20 ☽ Q ♃
14 14 ☿ ⊥ ♆
14 59 ♂ ✶ ♃
22 34 ☽ ⊼ ♄
23 33 ♀ △ ♃
23 35 ☽ ⊥ ♅

All ephemeris data is given at 12.00 UT and the Moon's longitude is additionally given for 24.00 UT

DECEMBER 1955

Date	Sidereal time h m s	Sun ☉ ° ' "	Moon ☽ ° ' "	Moon ☽ 24.00 ° ' "	Mercury ☿ ° '	Venus ♀ ° '	Mars ♂ ° '	Jupiter ♃ ° '	Saturn ♄ ° '	Uranus ♅ ° '	Neptune ♆ ° '	Pluto ♇ ° '
01	16 38 16	08 ♐ 30 59	03 ♋ 54 59	11 ♋ 25 39	06 ♐ 48	01 ♑ 41	01 ♏ 35	01 ♍ 03	25 ♏ 32	02 ♌ 06	29 ♎ 24	28 ♌ 37
02	16 42 13	09 31 47	18 ♋ 52 55	26 ♋ 15 56	08 22	02 55	02 14	01 07	25 39	02 R 05	29 25	28 R 37
03	16 46 09	10 32 37	03 ♌ 34 03	10 ♌ 46 42	09 56	04 10	02 53	01 09	25 46	02 03	29 27	28 37
04	16 50 06	11 33 29	17 ♌ 53 32	24 ♌ 54 20	11 30	05 25	03 31	01 12	25 53	02 02	29 29	28 37
05	16 54 02	12 34 22	01 ♍ 49 03	08 ♍ 37 42	13 05	06 39	04 10	01 15	26 00	02 01	29 31	28 36
06	16 57 59	13 35 16	15 ♍ 20 28	21 ♍ 57 34	14 39	07 54	04 49	01 17	26 07	01 59	29 33	28 36
07	17 01 55	14 36 11	28 ♍ 29 18	04 ♎ 56 01	16 13	09 09	05 28	01 19	26 14	01 58	29 34	28 36
08	17 05 52	15 37 08	11 ♎ 18 05	17 ♎ 35 54	17 47	10 23	06 07	01 21	26 21	01 56	29 36	28 36
09	17 09 48	16 38 05	23 ♎ 49 51	00 ♏ 00 19	19 22	11 38	06 46	01 23	26 27	01 55	29 38	28 36
10	17 13 45	17 39 04	06 ♏ 07 41	12 ♏ 12 19	20 56	12 52	07 25	01 25	26 34	01 53	29 39	28 35
11	17 17 42	18 40 05	18 ♏ 14 33	24 ♏ 14 43	22 30	14 07	08 04	01 26	26 41	01 51	29 41	28 35
12	17 21 38	19 41 06	00 ♐ 13 08	06 ♐ 10 04	24 05	15 22	08 43	01 27	26 48	01 50	29 43	28 35
13	17 25 35	20 42 08	12 ♐ 05 49	18 ♐ 00 38	25 39	16 36	09 22	01 28	26 55	01 48	29 44	28 34
14	17 29 31	21 43 11	23 ♐ 54 48	29 ♐ 48 33	27 14	17 51	10 01	01 29	27 01	01 46	29 46	28 34
15	17 33 28	22 44 15	05 ♑ 42 09	11 ♑ 35 54	28 ♐ 48	19 05	10 40	01 30	27 08	01 44	29 47	28 34
16	17 37 24	23 45 19	17 ♑ 30 02	23 ♑ 24 53	00 ♑ 23	20 20	11 19	01 30	27 15	01 42	29 49	28 33
17	17 41 21	24 46 24	29 ♑ 20 46	05 ♒ 18 01	01 58	21 35	11 58	01 30	27 21	01 40	29 50	28 33
18	17 45 17	25 47 29	11 ♒ 16 59	17 ♒ 18 05	03 33	22 49	12 38	01 R 30	27 28	01 39	29 52	28 32
19	17 49 14	26 48 35	23 ♒ 21 43	29 ♒ 28 20	05 08	24 04	13 17	01 30	27 35	01 37	29 53	28 32
20	17 53 11	27 49 41	05 ♓ 38 22	11 ♓ 52 19	06 44	25 18	13 56	01 30	27 41	01 35	29 55	28 31
21	17 57 07	28 50 47	18 ♓ 10 37	24 ♓ 33 46	08 19	26 33	14 35	01 29	27 48	01 32	29 56	28 31
22	18 01 04	29 ♐ 51 54	01 ♈ 02 13	07 ♈ 36 22	09 54	27 47	15 14	01 29	27 54	01 30	29 57	28 30
23	18 05 00	00 ♑ 53 01	14 ♈ 16 37	21 ♈ 03 14	11 29	29 ♑ 01	15 53	01 28	28 01	01 28	29 ♎ 59	28 29
24	18 08 57	01 54 07	27 ♈ 56 27	04 ♉ 56 21	13 05	00 ♒ 16	16 32	01 26	28 07	01 26	00 ♏ 00	28 29
25	18 12 53	02 55 14	12 ♉ 02 54	19 ♉ 15 54	14 40	01 30	17 11	01 25	28 13	01 24	00 01	28 28
26	18 16 50	03 56 22	26 ♉ 34 59	03 ♊ 59 36	16 15	02 45	17 50	01 24	28 20	01 22	00 02	28 27
27	18 20 46	04 57 29	11 ♊ 29 00	19 ♊ 02 18	17 50	03 59	18 30	01 22	28 26	01 20	00 04	28 27
28	18 24 43	05 58 37	26 ♊ 38 25	04 ♋ 16 11	19 25	05 13	19 09	01 20	28 32	01 17	00 05	28 26
29	18 28 40	06 59 44	11 ♋ 54 18	19 ♋ 31 30	21 00	06 28	19 48	01 18	28 38	01 15	00 06	28 25
30	18 32 36	08 00 52	27 ♋ 06 28	04 ♌ 38 03	22 33	07 42	20 27	01 16	28 44	01 13	00 07	28 24
31	18 36 33	09 ♑ 02 00	12 ♌ 05 07	19 ♌ 26 46	24 ♑ 07	08 ♒ 56	21 ♏ 06	01 ♍ 13	28 ♏ 50	01 ♌ 10	00 ♏ 08	28 ♌ 23

Date	Moon True ☊ ° '	Moon Mean ☊ ° '	Moon ☽ Latitude ° '
01	17 ♐ 19	17 ♐ 42	01 S 31
02	17 D 20	17 39	02 47
03	17 21	17 36	03 50
04	17 21	17 33	04 37
05	17 22	17 30	05 07
06	17 R 22	17 27	05 18
07	17 22	17 23	05 12
08	17 21	17 20	04 51
09	17 21	17 17	04 16
10	17 D 21	17 14	03 30
11	17 21	17 11	02 35
12	17 22	17 07	01 34
13	17 22	17 04	00 S 29
14	17 R 22	17 01	00 N 36
15	17 21	16 58	01 40
16	17 21	16 55	02 40
17	17 20	16 52	03 33
18	17 19	16 48	04 17
19	17 17	16 45	04 50
20	17 16	16 42	05 11
21	17 15	16 39	05 17
22	17 D 15	16 36	05 08
23	17 15	16 32	04 43
24	17 16	16 29	04 01
25	17 18	16 26	03 03
26	17 19	16 23	01 52
27	17 20	16 20	00 N 32
28	17 R 19	16 17	00 S 51
29	17 18	16 13	02 12
30	17 16	16 10	03 23
31	17 ♐ 13	16 ♐ 07	04 S 18

DECLINATIONS

Date	Sun ☉ ° '	Moon ☽ ° '	Mercury ☿ ° '	Venus ♀ ° '	Mars ♂ ° '	Jupiter ♃ ° '	Saturn ♄ ° '	Uranus ♅ ° '	Neptune ♆ ° '	Pluto ♇ ° '
01	21 S 44	21 N 52	22 S 02	24 S 38	11 S 15	11 N 56	17 S 15	20 N 15	09 S 41	21 N 54
02	21 53	19 22	22 24	24 39	11 29	11 55	17 17	20 16	09 42	21 54
03	22 02	15 38	22 44	24 39	11 43	11 54	17 19	20 16	09 42	21 55
04	22 10	11 04	23 04	24 38	11 56	11 53	17 20	20 16	09 43	21 55
05	22 19	06 03	23 22	24 36	12 10	11 53	17 22	20 17	09 43	21 55
06	22 26	00 N 53	23 39	24 34	12 23	11 52	17 24	20 17	09 44	21 56
07	22 33	04 S 10	23 55	24 31	12 37	11 51	17 25	20 17	09 45	21 56
08	22 40	08 56	24 10	24 27	12 50	11 51	17 27	20 18	09 45	21 57
09	22 46	13 13	24 23	24 22	13 03	11 51	17 28	20 18	09 46	21 57
10	22 52	16 52	24 35	24 17	13 16	11 50	17 30	20 18	09 46	21 57
11	22 58	19 45	24 46	24 11	13 29	11 50	17 31	20 19	09 47	21 58
12	23 03	21 44	24 55	24 04	13 42	11 50	17 33	20 19	09 47	21 58
13	23 07	22 44	25 04	23 56	13 55	11 50	17 34	20 20	09 48	21 59
14	23 11	22 42	25 10	23 48	14 08	11 50	17 36	20 20	09 48	21 59
15	23 15	21 39	25 16	23 39	14 20	11 50	17 37	20 21	09 49	22 00
16	23 18	19 39	25 20	23 29	14 33	11 50	17 39	20 21	09 49	22 00
17	23 20	16 49	25 22	23 19	14 45	11 50	17 40	20 21	09 50	22 01
18	23 23	13 16	25 23	23 08	14 57	11 50	17 42	20 22	09 50	22 01
19	23 24	09 10	25 23	22 56	15 10	11 50	17 43	20 22	09 51	22 02
20	23 26	04 S 37	25 21	22 44	15 22	11 51	17 45	20 23	09 51	22 02
21	23 26	00 N 12	25 18	22 30	15 34	11 51	17 46	20 23	09 51	22 03
22	23 27	05 07	25 13	22 17	15 46	11 52	17 47	20 24	09 52	22 03
23	23 26	09 58	25 06	22 02	15 57	11 52	17 49	20 24	09 52	22 04
24	23 26	14 29	24 59	21 47	16 09	11 53	17 50	20 25	09 53	22 05
25	23 25	18 22	24 49	21 31	16 21	11 54	17 51	20 25	09 53	22 05
26	23 23	21 13	24 38	21 15	16 32	11 54	17 53	20 26	09 53	22 06
27	23 21	22 42	24 26	20 58	16 43	11 55	17 54	20 26	09 54	22 06
28	23 19	22 33	24 11	20 40	16 54	11 56	17 55	20 27	09 54	22 07
29	23 16	20 43	23 56	20 22	17 06	11 57	17 57	20 27	09 55	22 07
30	23 12	17 26	23 39	20 03	17 17	11 58	17 58	20 28	09 55	22 08
31	23 S 08	13 N 02	23 S 20	19 S 44	17 S 27	11 N 59	17 S 59	20 N 28	09 S 55	22 N 09

ZODIAC SIGN ENTRIES

Date	h	m	Planets
01	05	46	☽ ♋
03	06	07	☽ ♌
05	08	50	☽ ♍
07	14	48	☽ ♎
09	23	59	☽ ♏
12	11	34	☽ ♐
15	00	23	☽ ♑
16	06	06	☿ ♑
17	13	19	☽ ♒
20	01	02	☽ ♓
22	10	05	☽ ♈
22	15	11	☉ ♑
24	06	52	♀ ♒
24	15	22	♆ ♏
24	15	33	☽ ♉
26	17	33	☽ ♊
28	17	17	☽ ♋
30	16	36	☽ ♌

LATITUDES

Date	Mercury ☿ ° '	Venus ♀ ° '	Mars ♂ ° '	Jupiter ♃ ° '	Saturn ♄ ° '	Uranus ♅ ° '	Neptune ♆ ° '	Pluto ♇ ° '
01	00 S 36	01 S 12	00 N 49	00 N 53	01 N 57	00 N 34	01 N 41	10 N 38
04	00 54	01 18	00 48	00 54	01 57	00 34	01 41	10 39
07	01 12	01 23	00 47	00 55	01 57	00 35	01 41	10 40
10	01 27	01 28	00 46	00 55	01 57	00 35	01 42	10 41
13	01 41	01 32	00 45	00 56	01 57	00 35	01 42	10 42
16	01 53	01 36	00 43	00 57	01 57	00 35	01 42	10 44
19	02 02	01 40	00 42	00 58	01 57	00 35	01 42	10 45
22	02 09	01 42	00 41	00 58	01 57	00 35	01 42	10 46
25	02 12	01 44	00 39	00 59	01 58	00 35	01 42	10 47
28	02 10	01 46	00 38	01 00	01 58	00 35	01 42	10 48
31	02 S 04	01 S 46	00 N 36	01 N 01	01 N 58	00 N 35	01 N 43	10 N 49

DATA

Julian Date	2435443
Delta T	+31 seconds
Ayanamsa	23° 14' 45"
Synetic vernal point	05° ♓ 52' 15"
True obliquity of ecliptic	23° 26' 40"

MOON'S PHASES, APSIDES AND POSITIONS ☽

Date	h	m	Phase	Longitude ° '	Eclipse Indicator
06	08	35	☾	13 ♍ 27	
14	07	07	●	21 ♐ 31	Annular
22	09	39	☽	29 ♓ 46	
29	03	44	○	06 ♋ 39	

Day	h	m		
15	07	29	Apogee	
29	00	33	Perigee	
06	16	09	0S	
13	23	16	Max dec	22° S 51'
21	11	03	0N	
27	21	53	Max dec	22° N 50'

LONGITUDES

Date	Chiron ⚷ ° '	Ceres ⚳ ° '	Pallas ⚴ ° '	Juno ⚵ ° '	Vesta ⚶ ° '	Black Moon Lilith ⚸ ° '
01	01 ♒ 44	29 ♑ 57	09 ♑ 20	10 ♐ 12	12 ♎ 00	29 ♐ 29
11	02 ♒ 22	03 ♒ 24	12 ♑ 59	13 ♐ 41	16 ♎ 18	00 ♑ 36
21	03 ♒ 03	06 ♒ 58	16 ♑ 41	17 ♐ 09	20 ♎ 27	01 ♑ 42
31	03 ♒ 48	10 ♒ 40	20 ♑ 23	20 ♐ 37	24 ♎ 23	02 ♑ 49

ASPECTARIAN

h m Aspects

01 Thursday
02 58 ☿ ⋕ ♇
03 34 ☽ ✶ ♇
04 46 ♇ St R
04 48 ☽ △ ♆
07 26 ☽ ✶ ♃
08 05 ♀ ✶ ♂
08 07 ☽ △ ♂
08 07 ☽ ☍ ♀
08 10 ☽ ± ♄
08 53 ♀ ⊥ ♄
09 06 ☽ ⊻ ♅
10 06 ☽ ⋕ ☿
11 39 ☽ ∥ ♇
13 42 ☽ ⋕ ☉
17 07 ☽ ⊼ ☿
19 52 ☽ ⊼ ☉
20 00 ♀ ⊼ ♅
22 39 ☽ bQ ♄

02 Friday
03 30 ☽ ∠ ♇
03 52 ☽ ± ☿
04 48 ☽ ∥ ♅
06 12 ☽ ± ☉
06 37 ♂ □ ♅
07 31 ☽ ∠ ♃
08 23 ☿ ⊥ ♂
14 58 ☉ ⋕ ♇
18 03 ☽ ⊥ ♇
20 08 ☽ bQ ☿
21 51 ☽ bQ ☉
22 09 ☽ ⊥ ♃
23 05 ☽ △ ♄

03 Saturday
02 03 ☽ ⋕ ♄
03 50 ☽ ⊻ ♇
05 13 ☽ □ ♆
08 01 ☽ ⊻ ♃
09 30 ☽ ☌ ♅
10 48 ☽ □ ♂
13 05 ☽ ⊼ ♀
23 54 ☽ △ ☿

04 Sunday
00 01 ☽ ± ♀
00 29 ☽ △ ☉
06 53 ♂ ⋕ ♃
07 52 ☽ ⋕ ♂
07 55 ☽ ∥ ♃
11 18 ☽ Q ♆
14 09 ☉ ☌ ☿
16 42 ☽ bQ ♀
18 30 ☽ Q ♂
18 38 ☽ ⋕ ♆

05 Monday
01 48 ☽ □ ♄
06 24 ☽ ☌ ♇
07 58 ☽ ✶ ♆
11 00 ☽ ☌ ♃
12 20 ☽ ⊻ ♅
16 20 ☽ ✶ ♂
21 22 ☽ △ ♀
22 53 ☽ ⊥ ♅

06 Tuesday
08 35 ☽ □ ☉
09 46 ☽ Q ♄
10 22 ☿ ∠ ♆
10 34 ☽ ∠ ♆
10 35 ☽ □ ☿
14 58 ☽ ∠ ♅
20 32 ☽ ∠ ♂

07 Wednesday
02 56 ☽ ⊥ ♆
07 47 ☽ ✶ ♄
11 15 ☉ ∠ ♆
12 13 ☽ ⊻ ♇
13 55 ☽ ⊥ ♂
14 01 ☽ ⊻ ♆
17 16 ☽ ⊻ ♃
18 26 ☽ ✶ ♅
20 18 ☽ Q ☉
23 10 ☿ bQ ♅
23 23 ☽ ⊥ ♇

08 Thursday
00 09 ☽ Q ☿
01 42 ☽ ⊻ ♂
04 32 ☽ ⊥ ♃
10 05 ☽ □ ♀
12 05 ☽ ∠ ♄
16 22 ☽ ∠ ♇
16 23 ☽ ∥ ♆
17 00 ☽ Q ♅
20 56 ☽ ✶ ☉
21 39 ☽ ∠ ♃

09 Friday
02 08 ☽ ✶ ☿
03 59 ☽ ⋕ ♃
05 26 ☽ ⊥ ♄
08 19 ♀ ∠ ♄
10 54 ☿ ∥ ♀
10 58 ☽ ∥ ♂
11 57 ♀ Q ♆
17 08 ☽ ⊻ ♄
18 19 ☉ bQ ♅
21 15 ☽ ✶ ♇
23 17 ☽ ☌ ♆

10 Saturday
00 32 ☽ Q ♀
02 44 ☽ ✶ ♃
03 41 ☽ □ ♅
04 33 ☽ ∠ ☉
11 33 ☽ ∠ ☿
14 41 ☽ ☌ ♂
16 45 ☽ ∥ ♄
20 48 ☽ Q ♇
23 54 ☽ ⊥ ☉

11 Sunday
01 46 ♀ bQ ♇
02 25 ☽ Q ♃
02 51 ☽ ✶ ♀
08 01 ☽ ⊥ ☿
12 56 ☽ ⊻ ☉
17 50 ☽ ⋕ ♅
21 48 ☽ ⊻ ☿

12 Monday
02 44 ☿ ∠ ♂
05 03 ☽ ☌ ♄
08 42 ☽ □ ♇
10 58 ☽ ⊻ ♆
12 19 ☽ ∠ ♀
14 29 ☽ □ ♃
15 14 ☽ △ ♅
16 06 ☽ ⋕ ♇
23 06 ☽ ⊥ ♆

13 Tuesday
06 10 ☽ ⊻ ♂
08 37 ☽ ⊥ ♀
09 22 ♀ bQ ♃
14 09 ☿ ± ♅
17 22 ☽ ∠ ♆
19 02 ☽ ⊥ ♂
21 31 ☽ bQ ♅
22 13 ☽ ⊻ ♀

14 Wednesday
07 07 ☽ ☌ ☉
08 38 ☿ ⊻ ♄
14 23 ☽ ∠ ♂
15 46 ☽ ± ♅
18 23 ☽ ⊻ ♄
19 47 ☽ ☌ ☿
21 28 ☽ △ ♇
23 56 ☽ ✶ ♆

15 Thursday
03 26 ☽ △ ♃
03 57 ☽ ⊼ ♅
06 12 ☽ ⋕ ♇
06 43 ☽ ⊥ ♄
07 55 ♂ Q ♇
08 16 ☿ △ ♇
22 43 ☽ ✶ ♂

16 Friday
00 25 ☽ Q ♆
01 13 ☽ ∠ ♄
03 07 ☿ ✶ ♆
03 59 ☽ bQ ♇
04 48 ☽ ⋕ ♅
09 58 ☽ bQ ♃
18 26 ☽ ☌ ♀
22 15 ☽ ± ♇

17 Saturday
00 30 ☽ Q ♂
01 53 ☽ ⊻ ☉
04 14 ☽ ± ♃
04 57 ☿ △ ♃
05 29 ☽ ∥ ♄
07 37 ☿ ⊼ ♅
07 57 ☽ ✶ ♄
09 24 ☉ ∥ ♀
10 23 ☽ ⊼ ♇
13 00 ☽ □ ♆
15 09 ☽ ⊥ ☉
16 22 ☽ ⊼ ♃
16 41 ☽ ☍ ♅
18 06 ☽ ⊻ ☿

18 Sunday
01 44 ☽ ∥ ♂
04 29 ♃ St R
06 36 ♀ ± ♇
08 01 ☽ ⊥ ☿
08 20 ☽ Q ♄
08 35 ☉ ± ♅
10 37 ☿ ⊥ ♄
10 56 ☽ ∠ ☉
14 50 ☽ □ ♂
20 44 ☽ ⋕ ♃

19 Monday
04 40 ☽ ∠ ☿
08 13 ☽ ∥ ♆
13 32 ☽ ⊻ ♀
19 24 ☽ ✶ ☉
20 17 ♂ Q ♃
20 22 ☽ □ ♄
22 09 ☽ ☍ ♇

20 Tuesday
00 50 ☽ △ ♆
02 38 ☽ ⊥ ♀
03 58 ☽ ☍ ♃
04 08 ☽ ⊼ ♅
08 17 ☉ ⊻ ♄
14 24 ☽ ✶ ☿
15 44 ☽ ± ♅
15 45 ♀ ± ♃
20 48 ☽ Q ☉
21 59 ☽ ∠ ♀

21 Wednesday
04 09 ☉ △ ♇
04 48 ☽ △ ♂
05 50 ☽ bQ ♆
08 55 ☽ bQ ♅
13 29 ♀ Q ♂
16 36 ☽ Q ☿
22 50 ☽ ± ♆

22 Thursday
05 21 ☽ ✶ ♀
06 10 ☽ △ ♄
07 19 ☽ ⊼ ♇
09 39 ☽ □ ☉
10 00 ☽ ⊼ ♆
10 26 ☽ bQ ♂
12 48 ☽ ⊼ ♃
12 52 ☽ △ ♅
14 08 ☉ ✶ ♆
14 30 ♀ ✶ ♄
18 21 ☽ ± ♇
23 45 ☽ ± ♃

23 Friday
01 43 ♀ ⊼ ♇
03 43 ☽ ± ♂
05 35 ☽ Q ♀
06 20 ☽ □ ☿
09 08 ♀ ⋕ ♇
09 43 ☽ bQ ♄
10 36 ☽ bQ ♇
11 29 ☽ ⋕ ♆
12 00 ☉ ∠ ♂
15 00 ☽ ⊼ ♂
15 53 ☽ bQ ♃
19 25 ☿ Q ♆
21 49 ☽ ∥ ♃

24 Saturday
01 21 ☉ △ ♃
01 23 ☉ ⊼ ♅
01 47 ☽ ± ♄
03 20 ♃ ⊻ ♅
06 44 ♀ □ ♆
12 18 ☽ ⊼ ♄
12 35 ☿ ∠ ♄
12 56 ☽ △ ♇
15 33 ☽ ☍ ♆
16 24 ☽ □ ♀
17 58 ☿ bQ ♇
18 00 ☽ □ ♅
18 01 ☽ △ ♃
19 21 ☽ △ ☉
22 11 ☽ ⋕ ♂

25 Sunday
08 31 ☽ ⋕ ♄
10 02 ♀ ☍ ♅
10 23 ♀ ⊼ ♃
16 55 ☽ △ ☿
20 58 ☽ ☍ ♂
22 31 ☽ bQ ☉

26 Monday
00 11 ☽ Q ♅
04 14 ☽ ∥ ♅
12 18 ☽ ⋕ ♀
14 04 ☿ bQ ♃
14 22 ☽ ± ☉
14 51 ☽ ☍ ♄
15 02 ☽ □ ♇
17 37 ☽ ⊼ ♆
19 44 ☽ ✶ ♅
19 48 ☽ □ ♃
20 29 ☽ bQ ☿
22 09 ☉ ⊥ ♄
22 54 ☽ △ ♀
23 12 ☽ ∥ ♇

27 Tuesday
00 47 ☽ ⊼ ☉
03 19 ☽ ± ♆
12 38 ☽ ± ☿
14 29 ♄ □ ♇
17 42 ☽ bQ ♆
19 41 ☽ ∠ ♅
19 53 ☽ Q ♇
23 17 ☽ ⊼ ☿
23 38 ☽ ⊼ ♂

28 Wednesday
00 30 ☽ Q ♃
00 58 ☽ bQ ♀
04 54 ☿ ✶ ♂
09 32 ☽ ± ♂
09 52 ☽ ⊥ ♅
14 49 ☽ ✶ ♇
15 00 ☽ ⊼ ♄
16 25 ☽ ± ♀
17 25 ☽ △ ♆
19 18 ☽ ⊻ ♅
19 22 ☽ ✶ ♃
20 01 ☽ ∥ ♇

29 Thursday
00 20 ☽ bQ ♂
00 30 ☽ ± ♄
02 41 ☽ ⊼ ♀
03 44 ☽ ☍ ☉
05 22 ♀ ⋕ ♅
14 22 ☽ ∥ ♅
14 22 ☽ ∠ ♇
14 44 ☽ bQ ♄
15 31 ☽ ⋕ ♀
18 54 ☽ ∠ ♃

30 Friday
00 59 ☽ △ ♂
03 58 ☽ ☍ ☿
04 33 ☽ ⊥ ♇
08 40 ☽ ⋕ ♄
09 05 ☽ ⊥ ♃
09 38 ☿ ± ♇
12 54 ☽ ⋕ ♂
14 03 ☽ ⊻ ♇
14 36 ☽ △ ♄
16 47 ☽ □ ♆
18 31 ☽ ☌ ♅
18 35 ☽ ⊻ ♃

31 Saturday
06 27 ☽ ☍ ♀
06 42 ☽ ⊼ ☉
17 09 ☽ ± ☉
17 09 ☽ ∥ ♃
21 52 ☽ Q ♆
23 04 ☉ ⊻ ♀

All ephemeris data is given at 12.00 UT and the Moon's longitude is additionally given for 24.00 UT

JANUARY 1956

LONGITUDES

Date	Sidereal time h m s	Sun ☉ ° ' "	Moon ☽ ° ' "	Moon ☽ 24.00 ° ' "	Mercury ☿ ° '	Venus ♀ ° '	Mars ♂ ° '	Jupiter ♃ ° '	Saturn ♄ ° '	Uranus ♅ ° '	Neptune ♆ ° '	Pluto ♇ ° '
01	18 40 29	10 ♑ 03 08	26 ♌ 42 16	03 ♍ 51 04	25 ♑ 39	10 ♒ 10	21 ♏ 45	01 ♍ 10	28 ♏ 56	01 ♌ 08	00 ♏ 09	28 ♌ 22
02	18 44 26	11 04 17	10 ♍ 52 50	17 ♍ 47 23	27 11	11 24	22 25	01 R 08	29 02	01 R 06	00 10	28 R 22
03	18 48 22	12 05 26	24 ♍ 34 46	01 ♎ 15 08	28 ♑ 41	12 38	23 04	01 05	29 08	01 03	00 11	28 21
04	18 52 19	13 06 35	07 ♎ 48 48	14 ♎ 16 08	00 ♒ 10	13 53	23 43	01 01	29 14	01 01	00 12	28 20
05	18 56 15	14 07 45	20 ♎ 37 39	26 ♎ 53 51	01 37	15 07	24 22	00 58	29 20	00 58	00 13	28 19
06	19 00 12	15 08 54	03 ♏ 05 18	09 ♏ 12 36	03 02	16 21	25 01	00 54	29 25	00 56	00 14	28 18
07	19 04 09	16 10 04	15 ♏ 16 18	21 ♏ 17 00	04 24	17 35	25 41	00 51	29 31	00 53	00 15	28 17
08	19 08 05	17 11 14	27 ♏ 15 15	03 ♐ 11 34	05 43	18 49	26 20	00 47	29 37	00 51	00 15	28 16
09	19 12 02	18 12 24	09 ♐ 06 27	15 ♐ 00 21	06 58	20 03	26 59	00 42	29 42	00 48	00 16	28 15
10	19 15 58	19 13 34	20 ♐ 53 43	26 ♐ 46 55	08 09	21 17	27 38	00 38	29 48	00 46	00 17	28 14
11	19 19 55	20 14 44	02 ♑ 40 18	08 ♑ 34 11	09 15	22 30	28 18	00 34	29 53	00 43	00 18	28 12
12	19 23 51	21 15 54	14 ♑ 28 51	20 ♑ 24 33	10 15	23 44	28 57	00 29	29 ♏ 59	00 41	00 18	28 11
13	19 27 48	22 17 03	26 ♑ 21 30	02 ♒ 19 54	11 09	24 58	29 ♏ 36	00 24	00 ♐ 04	00 38	00 19	28 10
14	19 31 44	23 18 12	08 ♒ 19 58	14 ♒ 21 51	11 55	26 12	00 ♐ 16	00 19	00 09	00 35	00 20	28 09
15	19 35 41	24 19 21	20 ♒ 25 46	26 ♒ 31 54	12 33	27 25	00 55	00 14	00 14	00 33	00 20	28 08
16	19 39 38	25 20 28	02 ♓ 40 26	08 ♓ 51 36	13 01	28 39	01 34	00 09	00 19	00 30	00 21	28 07
17	19 43 34	26 21 36	15 ♓ 05 37	21 ♓ 22 46	13 19	29 ♒ 53	02 14	00 ♍ 03	00 24	00 28	00 21	28 05
18	19 47 31	27 22 42	27 ♓ 43 17	04 ♈ 07 30	13 27	01 ♓ 06	02 53	29 ♌ 58	00 29	00 25	00 22	28 04
19	19 51 27	28 23 48	10 ♈ 35 43	17 ♈ 08 14	13 R 23	02 20	03 32	29 52	00 34	00 22	00 22	28 03
20	19 55 24	29 ♑ 24 52	23 ♈ 45 24	00 ♉ 27 30	13 07	03 33	04 11	29 46	00 39	00 20	00 23	28 02
21	19 59 20	00 ♒ 25 56	07 ♉ 14 48	14 ♉ 07 33	12 40	04 47	04 51	29 40	00 44	00 17	00 23	28 00
22	20 03 17	01 26 59	21 ♉ 05 52	28 ♉ 09 50	12 02	06 00	05 30	29 34	00 48	00 15	00 24	27 59
23	20 07 13	02 28 01	05 ♊ 19 22	12 ♊ 34 18	11 13	07 13	06 09	29 28	00 53	00 12	00 24	27 58
24	20 11 10	03 29 02	19 ♊ 54 15	27 ♊ 18 42	10 15	08 27	06 49	29 21	00 57	00 09	00 24	27 57
25	20 15 07	04 30 02	04 ♋ 46 55	12 ♋ 18 04	09 09	09 40	07 28	29 15	01 02	00 07	00 24	27 55
26	20 19 03	05 31 01	19 ♋ 51 05	27 ♋ 24 49	07 58	10 53	08 07	29 08	01 06	00 04	00 25	27 54
27	20 23 00	06 31 59	04 ♌ 58 04	12 ♌ 29 33	06 43	12 06	08 47	29 01	01 10	00 ♌ 02	00 25	27 52
28	20 26 56	07 32 56	19 ♌ 58 04	27 ♌ 22 27	05 27	13 19	09 26	28 54	01 15	29 ♋ 59	00 25	27 51
29	20 30 53	08 33 52	04 ♍ 41 42	11 ♍ 54 58	04 13	14 32	10 05	28 47	01 19	29 56	00 25	27 50
30	20 34 49	09 34 48	19 ♍ 01 36	26 ♍ 01 10	03 01	15 45	10 45	28 40	01 23	29 54	00 25	27 48
31	20 38 46	10 ♒ 35 42	02 ♎ 53 25	09 ♎ 38 18	01 ♒ 54	16 ♓ 57	11 ♐ 24	28 ♌ 33	01 ♐ 27	29 ♋ 51	00 ♏ 25	27 ♌ 47

Date	Moon True ☊ ° '	Moon Mean ☊ ° '	Moon ☽ Latitude ° '
01	17 ♐ 10	16 ♐ 04	04 S 55
02	17 R 07	16 01	05 13
03	17 05	15 58	05 13
04	17 04	15 54	04 55
05	17 D 04	15 51	04 23
06	17 05	15 48	03 39
07	17 07	15 45	02 46
08	17 08	15 42	01 47
09	17 10	15 38	00 S 44
10	17 R 10	15 35	00 N 20
11	17 09	15 32	01 24
12	17 06	15 29	02 24
13	17 01	15 26	03 18
14	16 54	15 23	04 03
15	16 47	15 19	04 38
16	16 40	15 16	05 01
17	16 33	15 13	05 10
18	16 28	15 10	05 04
19	16 24	15 07	04 43
20	16 22	15 04	04 07
21	16 D 22	15 00	03 16
22	16 23	14 57	02 13
23	16 25	14 54	01 N 00
24	16 R 25	14 51	00 S 19
25	16 24	14 48	01 38
26	16 20	14 44	02 50
27	16 14	14 41	03 51
28	16 06	14 38	04 36
29	15 58	14 35	05 01
30	15 49	14 32	05 06
31	15 ♐ 42	14 ♐ 29	04 S 53

DECLINATIONS

Date	Sun ☉ ° '	Moon ☽ ° '	Mercury ☿ ° '	Venus ♀ ° '	Mars ♂ ° '	Jupiter ♃ ° '	Saturn ♄ ° '	Uranus ♅ ° '	Neptune ♆ ° '	Pluto ♇ ° '
01	23 S 04	07 N 59	23 S 00	19 S 24	17 S 38	12 N 01	18 S 00	20 N 29	09 S 55	22 N 09
02	22 59	02 N 39	22 39	19 04	17 49	12 02	18 01	20 30	09 56	22 10
03	22 54	02 S 38	22 16	18 43	17 59	12 03	18 03	20 30	09 56	22 11
04	22 48	07 37	21 52	18 21	18 09	12 05	18 04	20 31	09 56	22 11
05	22 42	12 07	21 27	17 59	18 20	12 06	18 05	20 31	09 57	22 12
06	22 35	15 59	21 01	17 37	18 30	12 08	18 06	20 32	09 57	22 12
07	22 28	19 04	20 34	17 14	18 39	12 09	18 07	20 32	09 57	22 13
08	22 20	21 17	20 06	16 51	18 49	12 11	18 08	20 33	09 57	22 14
09	22 12	22 33	19 37	16 27	18 59	12 12	18 09	20 33	09 58	22 14
10	22 04	22 47	19 09	16 02	19 08	12 14	18 10	20 34	09 58	22 15
11	21 55	22 01	18 40	15 38	19 18	12 16	18 11	20 35	09 58	22 16
12	21 46	20 16	18 12	15 13	19 27	12 18	18 12	20 35	09 58	22 16
13	21 36	17 39	17 44	14 47	19 36	12 20	18 13	20 36	09 58	22 17
14	21 26	14 16	17 17	14 21	19 45	12 22	18 14	20 36	09 58	22 18
15	21 15	10 17	16 51	13 55	19 54	12 24	18 15	20 37	09 59	22 18
16	21 04	05 50	16 27	13 28	20 02	12 26	18 16	20 37	09 59	22 19
17	20 53	01 S 06	16 05	13 01	20 11	12 28	18 17	20 38	09 59	22 20
18	20 41	03 N 45	15 46	12 34	20 19	12 30	18 18	20 39	09 59	22 20
19	20 29	08 32	15 30	12 06	20 27	12 32	18 19	20 39	09 59	22 21
20	20 17	13 03	15 16	11 38	20 35	12 35	18 20	20 40	09 59	22 22
21	20 04	17 01	15 07	11 10	20 43	12 37	18 21	20 40	09 59	22 22
22	19 50	20 10	15 00	10 41	20 51	12 39	18 22	20 41	09 59	22 23
23	19 37	22 10	14 58	10 12	20 58	12 42	18 22	20 41	09 59	22 24
24	19 23	22 45	14 58	09 43	21 05	12 44	18 23	20 42	09 59	22 24
25	19 08	21 44	15 02	09 14	21 13	12 47	18 24	20 43	09 59	22 25
26	18 54	19 10	15 09	08 44	21 20	12 49	18 25	20 43	09 59	22 26
27	18 39	15 17	15 19	08 15	21 27	12 52	18 25	20 44	09 59	22 26
28	18 23	10 28	15 31	07 45	21 33	12 54	18 26	20 44	09 59	22 27
29	18 07	05 N 07	15 44	07 15	21 40	12 57	18 27	20 45	09 59	22 28
30	17 51	00 S 21	15 58	06 44	21 46	13 00	18 27	20 45	09 59	22 28
31	17 S 35	05 S 38	16 S 14	06 S 14	21 S 52	13 N 02	18 S 28	20 N 46	09 S 59	22 N 29

ZODIAC SIGN ENTRIES

Date	h m	Planets
01	17 31	☽ ♍
03	21 44	☽ ♎
04	09 16	☿ ♒
06	06 00	☽ ♏
08	17 32	☽ ♐
11	06 33	☽ ♑
12	18 46	♄ ♐
13	19 19	☽ ♒
14	02 28	♂ ♐
16	06 47	☽ ♓
17	14 22	♀ ♓
18	02 04	♃ ♌
18	16 17	☽ ♈
20	23 11	☽ ♉
21	01 48	☉ ♒
23	03 06	☽ ♊
25	04 20	☽ ♋
27	04 06	☽ ♌
28	01 57	♅ ♋
29	04 17	☽ ♍
31	06 56	☽ ♎

LATITUDES

Date	Mercury ☿ ° '	Venus ♀ ° '	Mars ♂ ° '	Jupiter ♃ ° '	Saturn ♄ ° '	Uranus ♅ ° '	Neptune ♆ ° '	Pluto ♇ ° '
01	02 S 01	01 S 46	00 N 36	01 N 01	01 N 58	00 N 35	01 N 43	10 N 49
04	01 47	01 46	00 34	01 02	01 58	00 35	01 43	10 50
07	01 26	01 45	00 32	01 03	01 59	00 35	01 43	10 51
10	00 57	01 43	00 31	01 03	01 59	00 35	01 43	10 52
13	00 S 18	01 41	00 29	01 04	01 59	00 35	01 43	10 53
16	00 N 29	01 37	00 27	01 05	02 00	00 35	01 44	10 53
19	01 22	01 33	00 25	01 05	02 00	00 35	01 44	10 54
22	02 16	01 29	00 23	01 06	02 00	00 36	01 44	10 55
25	03 02	01 23	00 21	01 07	02 01	00 36	01 44	10 55
28	03 30	01 17	00 19	01 07	02 01	00 36	01 44	10 56
31	03 N 36	01 S 10	00 N 17	01 N 08	02 N 02	00 N 36	01 N 44	10 N 57

DATA

Julian Date	2435474
Delta T	+31 seconds
Ayanamsa	23° 14' 50"
Synetic vernal point	05° ♓ 52' 09"
True obliquity of ecliptic	23° 26' 39"

MOON'S PHASES, APSIDES AND POSITIONS ☽

Date	h m	Phase	Longitude ° '	Eclipse Indicator
04	22 41	☾	13 ♎ 34	
13	03 01	●	21 ♑ 54	
20	22 58	☽	29 ♈ 53	
27	14 40	○	06 ♌ 39	

Day	h m		
11	07 50	Apogee	
26	12 58	Perigee	
02	23 57	0S	
10	05 44	Max dec	22° S 50'
17	17 30	0N	
24	08 51	Max dec	22° N 46'
30	10 26	0S	

LONGITUDES

Date	Chiron ⚷ ° '	Ceres ⚳ ° '	Pallas ⚴ ° '	Juno ⚵ ° '	Vesta ⚶ ° '	Black Moon Lilith ⚸ ° '
01	03 ♒ 53	11 ♒ 02	20 ♑ 45	20 ♐ 58	24 ♎ 46	02 ♑ 56
11	04 ♒ 40	14 ♒ 49	24 ♑ 28	24 ♐ 23	28 ♎ 26	04 ♑ 02
21	05 ♒ 29	18 ♒ 40	28 ♑ 09	27 ♐ 45	01 ♏ 48	05 ♑ 09
31	06 ♒ 19	22 ♒ 34	01 ♒ 48	01 ♑ 02	04 ♏ 48	06 ♑ 15

ASPECTARIAN

h m	Aspects
01 Sunday	
03 02	☽ ⋕ ♆
03 25	☽ □ ♂
04 43	☿ ± ♃
06 26	☉ ∥ ☿
09 03	☽ ⚼ ☉
10 03	☽ ⊼ ☿
14 47	☽ ☌ ♇
15 45	☽ □ ♄
17 46	☽ ⚹ ♆
19 23	☽ ⊻ ♅
19 28	☽ ☌ ♃
21 17	☽ ± ☿
02 Monday	
04 13	♀ Q ♄
05 32	☽ ⊥ ♅
11 09	☽ Q ♂
12 21	☽ △ ☉
13 00	☽ ⊼ ♀
14 31	☽ ⚼ ☿
19 26	☽ ∠ ♆
21 00	☽ ∠ ♅
22 46	☽ Q ♄
03 Tuesday	
00 27	☽ ± ♀
06 33	☿ ⊼ ♇
09 10	☽ ⚹ ♂
11 18	☽ ⊥ ♆
14 14	☉ Q ♆
17 25	☿ ⋕ ♇
18 02	☽ ⚼ ♀
18 44	☽ ⊻ ♇
19 44	☿ ⚹ ♄
20 14	☽ ⚹ ♄
20 17	☽ △ ☿
21 19	♂ ∥ ♄
22 05	☽ ⊻ ♆
23 36	☽ ⚹ ♅
23 38	☽ ⊻ ♃
04 Wednesday	
05 36	☽ ⊥ ♇
10 33	☽ ⊥ ♃
12 33	☿ □ ♆
13 45	☽ ∠ ♂
17 03	☉ ⚼ ♇
21 02	♀ ∥ ♂
21 37	☽ Q ♅
22 14	☽ ∠ ♇
22 41	☽ □ ☉
05 Thursday	
00 01	☽ ∠ ♄
00 02	☽ ∥ ♆
00 28	☽ △ ♀
01 32	☿ ☍ ♅
01 33	☿ ⊼ ♃
03 13	☽ ∠ ♃
03 58	♃ ⊻ ♅
06 21	♀ ∥ ♄
07 29	☽ ⊥ ♂
11 56	☽ ⋕ ♃
17 11	☉ ∠ ♄
17 11	☽ ⊥ ♄
19 32	☽ ⊻ ♂
06 Friday	
02 43	☽ ⚹ ♇
04 49	☽ ⊻ ♄
06 26	☽ ☌ ♆
07 46	☽ ⚹ ♃
07 49	☽ □ ♅
11 52	☽ □ ☿
12 08	☽ Q ☉
22 38	☽ ∥ ♀
07 Saturday	
02 07	☽ Q ♇
03 51	☽ ∥ ♄
04 47	☉ ⚼ ♃
07 12	☽ Q ♃
08 09	☽ ∥ ♂
13 04	☿ ⋕ ♅
13 57	☽ ⚹ ☉
17 07	☽ □ ♀
08 Sunday	
00 18	☽ ∥ ☿
02 45	☽ ⋕ ♅
04 01	☽ Q ☿
10 02	☽ ☌ ♂
14 02	☽ □ ♇
16 48	☽ ☌ ♄
18 04	☽ ⊻ ♆
19 04	☽ □ ♃
19 14	☽ △ ♅
22 54	☽ ∠ ☉
09 Monday	
03 55	☽ ⋕ ♇
04 15	☽ ∥ ☉
06 14	☽ ⊥ ♆
06 39	☉ ⋕ ♇
07 10	☽ ⚹ ☿
09 35	☽ Q ♀
18 54	☽ ⊥ ☉
10 Tuesday	
00 33	☽ ∠ ♆
01 35	☽ ⚼ ♅
08 17	☽ ⊻ ☉
12 16	☿ ∥ ♂
12 52	☽ ⚹ ♀
17 06	☽ ∠ ☿
19 51	☽ ± ♅
11 Wednesday	
02 34	☽ ⊻ ♂
02 55	☽ △ ♇
06 17	☽ ⊻ ♄
07 05	☽ ⋕ ♇
07 09	☽ ⚹ ♆
07 44	☽ △ ♃
08 02	☽ ⊼ ♅
08 51	♂ □ ♇
13 18	☽ ⊥ ☿
14 05	☽ ∥ ☉
15 30	☽ ⊥ ♂
18 36	☽ ⊥ ♄
22 59	☽ ∠ ♀
12 Thursday	
02 40	☽ ⊻ ☿
07 35	☽ Q ♆
08 30	☽ ⋕ ♅
09 23	☽ ⚼ ♇
10 52	☽ ∠ ♂
11 20	☿ ∥ ♄
13 01	☽ ∠ ♄
14 01	☽ ⚼ ♃
19 22	☽ ⊥ ♀
19 55	☽ ∥ ♂
13 Friday	
03 01	☽ ☌ ☉
03 34	☽ ± ♇
07 21	☽ ∥ ♄
08 05	☽ ± ♃
08 52	☽ ⊻ ♀
09 21	☉ ± ♇
11 14	☽ ∥ ☿
15 38	☽ ⊼ ♇
18 54	☽ ⚹ ♂
19 30	☽ ⚹ ♄
19 58	☽ □ ♆
20 04	☽ ⊼ ♃
20 34	☽ ☍ ♅
14 Saturday	
07 21	♂ ☌ ♄
09 31	♃ ⚹ ♆
11 23	☽ ∥ ♀
13 55	♂ □ ♃
14 34	♂ ⊻ ♆
19 34	☽ ☌ ☿
19 39	☽ Q ♄
20 16	☽ Q ♂
21 33	☿ Q ♄
23 21	♂ △ ♅
23 45	☽ ⋕ ♃
15 Sunday	
10 05	☉ ± ♃
11 39	♃ □ ♄
13 42	☽ ∥ ♆
20 22	☽ ⊻ ☉
16 Monday	
01 35	♀ ☍ ♇
03 06	☽ ☍ ♇
03 17	☽ ☌ ♀
07 06	☽ ☍ ♃
07 23	☽ □ ♄
07 28	☽ △ ♆
07 47	☽ ⊼ ♅
09 10	☽ ⊥ ☉
09 44	☽ □ ♂
19 25	☽ ± ♅
20 43	♄ ⊻ ♆
17 Tuesday	
04 11	☽ ∠ ☉
08 32	☽ ⊻ ☿
12 30	☽ ⚼ ♆
12 42	☽ ⚼ ♅
15 11	♀ ☍ ♃
20 12	☽ ⊥ ☿
21 25	♀ △ ♆
22 34	♄ △ ♅
22 59	♀ ⊼ ♅
23 02	♀ □ ♄
18 Wednesday	
05 40	☽ ± ♆
11 18	☽ ⚹ ☉
12 39	☽ ⊼ ♇
13 22	☽ ∠ ☿
14 46	♀ ⋕ ♃
15 51	☿ St R
16 11	☽ ⊼ ♃
16 58	☽ ⊼ ♆
17 03	☽ △ ♅
17 12	☉ ⋕ ♅
17 14	☽ △ ♄
19 02	☽ ⊻ ♀
22 12	☽ △ ♂
23 53	☽ ± ♇
19 Thursday	
03 18	☽ ± ♃
03 59	☉ ⊼ ♇
07 23	☽ ⊥ ♀
11 36	☽ Q ☉
12 12	♅ □ ♆
14 27	☉ ∥ ♂
16 31	☽ ⚼ ♇
17 03	☽ ⚹ ☿
19 31	☽ ⋕ ♆
19 47	☽ ⚼ ♃
21 11	☽ ⚼ ♄
20 Friday	
01 37	☽ ∠ ♀
03 18	☽ ⚼ ♂
05 00	☽ ⋕ ♀
09 24	☽ ∥ ♃
13 37	☽ ± ♄
14 23	☽ Q ☿
19 34	☉ ⊼ ♃
19 39	☽ △ ♇
20 22	☽ ± ♂
22 41	☽ △ ♃
22 58	☽ □ ☉
23 44	☽ □ ♅
23 52	☽ ☍ ♆
21 Saturday	
00 24	☽ ⋕ ☿
00 25	☽ ⊼ ♄
03 21	♂ ⋕ ♅
07 13	☽ ⚹ ♀
07 34	☽ ⊼ ♂
08 41	☉ ☍ ♅
09 41	☿ Q ♄
10 54	☉ □ ♆
14 52	♀ □ ♂
19 32	☉ ⚹ ♄
21 05	☽ □ ☿
21 15	☽ ⋕ ♄
22 Sunday	
06 11	☽ Q ♀
07 07	☽ Q ♅
09 17	☽ ⋕ ☉
16 35	♀ ± ♅
16 54	☽ ∥ ♅
18 55	☽ ⋕ ♂
23 41	☽ □ ♇
23 Monday	
02 15	☽ □ ♃
03 27	☽ ⚹ ♅
03 45	☽ ⊼ ♆
04 32	☽ ☍ ♄
06 52	☽ △ ☉
13 27	☽ ☍ ♂
13 47	☽ ± ♆
15 27	☽ □ ♀
16 29	☽ ∥ ♇
20 53	♂ ⊥ ♆
21 11	☽ △ ☿
22 41	♀ ∥ ♆
24 Tuesday	
04 16	☽ ∠ ♅
04 39	☽ ⚼ ♆
05 33	☽ Q ♇
07 52	☽ Q ♃
09 31	☽ ⚼ ☉
18 53	☽ ⊥ ♅
20 05	☽ ⚼ ☿
25 Wednesday	
00 41	☽ ∥ ♇
01 00	☽ ⚹ ♇
01 11	☽ ± ☉
03 11	☽ ⚹ ♃
04 32	☽ ⊻ ♅
04 59	☽ △ ♆
05 57	☽ ⊼ ♄
06 48	☿ ⊻ ♀
09 34	☽ ± ☿
11 31	☽ ⊼ ☉
15 37	☽ ± ♄
16 29	☽ ⊼ ♂
17 54	☽ ⋕ ♂
18 29	☽ ⊼ ☿
20 29	☽ △ ♀
23 14	☽ ∥ ♅
26 Thursday	
00 58	☽ ∠ ♇
02 29	☽ ± ♂
02 59	☽ ∠ ♃
06 01	☽ ⚼ ♄
09 59	☿ ⚹ ♂
14 06	☽ ⋕ ☉
15 15	☽ ⊥ ♇
17 10	☽ ⊥ ♃
17 17	☽ ⋕ ♄
17 26	☽ ⚼ ♂
22 24	☽ ⚼ ♀
27 Friday	
00 45	☽ ⊻ ♇
02 37	☽ ⊻ ♃
04 10	☽ ☌ ♅
04 46	☽ □ ♆
05 57	☽ △ ♄
11 50	☽ ⋕ ☿
13 57	☽ ± ♀
13 59	☉ ☌ ☿
14 35	☽ ☍ ☿
14 40	☽ ☍ ☉
18 02	☿ ⊥ ♀
18 21	☽ △ ♂
28 Saturday	
00 22	☽ ⊼ ♀
00 24	☽ ∥ ♃
07 45	☉ ∥ ♄
09 30	☽ Q ♆
14 11	☽ ⋕ ♆
29 Sunday	
00 46	☽ ☌ ♇
01 40	☽ ⋕ ♀
02 23	☽ ☌ ♃
04 13	☽ ⊻ ♅
04 58	☽ ⚹ ♆
06 25	☽ □ ♄
11 16	☽ ⊼ ☿
14 03	☽ ⊥ ♅
16 22	☉ ⊥ ♀
18 54	☽ ⊼ ☉
19 50	♀ ⚼ ♅
20 26	☽ ± ☿
21 23	☽ □ ♂
30 Monday	
05 02	☽ ∠ ♅
05 36	♀ ⚼ ♆
05 43	☽ ± ☉
05 53	☽ ∠ ♆
05 55	☽ ☍ ♀
10 25	☽ ⚼ ☿
12 36	☽ Q ♄
21 14	☽ ⊥ ♆
22 16	☽ ⚼ ☉
31 Tuesday	
03 04	☽ ⊻ ♇
04 28	☽ ⊻ ♃
05 35	☽ Q ♂
06 41	☽ ⚹ ♅
07 40	☽ ⊻ ♆
09 27	☽ ⚹ ♄
10 24	☽ △ ☿
11 29	☿ ∠ ♀
13 34	☽ ⊥ ♇
14 34	☽ ∥ ♀
14 54	☽ ⊥ ♃
21 57	☿ ⚹ ♄

All ephemeris data is given at 12.00 UT and the Moon's longitude is additionally given for 24.00 UT

FEBRUARY 1956

LONGITUDES

Date	Sidereal time h m s	Sun ☉ ° ' "	Moon ☽ ° ' "	Moon ☽ 24.00 ° ' "	Mercury ☿ ° '	Venus ♀ ° '	Mars ♂ ° '	Jupiter ♃ ° '	Saturn ♄ ° '	Uranus ♅ ° '	Neptune ♆ ° '	Pluto ♇ ° '
01	20 42 42	11 ♒ 36 36	16 ♎ 15 57	22 ♎ 46 40	00 ♒ 54	18 ♓ 10	12 ♐ 04	28 ♌ 26	01 ♐ 31	29 ♋ 49	00 ♏ 25	27 ♌ 46
02	20 46 39	12 37 29	29 ♎ 10 51	05 ♏ 28 59	00 ♒ 01	19 23	12 43	28 R 18	01 34	29 R 46	00 R 25	27 R 44
03	20 50 36	13 38 21	11 ♏ 41 42	17 ♏ 49 35	29 ♑ 15	20 35	13 22	28 11	01 38	29 44	00 25	27 43
04	20 54 32	14 39 13	23 ♏ 53 20	29 ♏ 53 38	28 R 39	21 48	14 02	28 04	01 42	29 41	00 25	27 41
05	20 58 29	15 40 03	05 ♐ 51 09	11 ♐ 46 34	28 10	23 00	14 41	27 56	01 45	29 39	00 25	27 40
06	21 02 25	16 40 53	17 ♐ 40 32	23 ♐ 33 40	27 51	24 12	15 21	27 48	01 49	29 36	00 25	27 38
07	21 06 22	17 41 42	29 ♐ 26 33	05 ♑ 19 43	27 39	25 25	16 00	27 41	01 52	29 34	00 25	27 37
08	21 10 18	18 42 29	11 ♑ 13 38	17 ♑ 08 46	27 35	26 37	16 39	27 33	01 55	29 31	00 24	27 35
09	21 14 15	19 43 16	23 ♑ 05 28	29 ♑ 04 05	27 D 39	27 49	17 19	27 25	01 58	29 29	00 24	27 34
10	21 18 11	20 44 01	05 ♒ 04 52	11 ♒ 08 02	27 49	29 ♓ 01	17 58	27 17	02 02	29 26	00 24	27 32
11	21 22 08	21 44 45	17 ♒ 13 45	23 ♒ 22 08	28 06	00 ♈ 13	18 37	27 09	02 05	29 24	00 23	27 31
12	21 26 05	22 45 28	29 ♒ 33 15	05 ♓ 47 08	28 29	01 24	19 17	27 02	02 07	29 22	00 23	27 29
13	21 30 01	23 46 09	12 ♓ 03 50	18 ♓ 23 20	28 57	02 36	19 56	26 54	02 10	29 19	00 23	27 28
14	21 33 58	24 46 49	24 ♓ 45 39	01 ♈ 10 46	29 ♑ 31	03 48	20 36	26 46	02 13	29 17	00 22	27 26
15	21 37 54	25 47 27	07 ♈ 38 44	14 ♈ 09 34	00 ♒ 09	04 59	21 15	26 38	02 16	29 15	00 22	27 25
16	21 41 51	26 48 04	20 ♈ 43 21	27 ♈ 20 11	00 51	06 11	21 54	26 30	02 18	29 13	00 21	27 23
17	21 45 47	27 48 39	04 ♉ 00 11	10 ♉ 43 29	01 38	07 22	22 34	26 22	02 20	29 10	00 21	27 22
18	21 49 44	28 49 12	17 ♉ 30 15	24 ♉ 20 38	02 28	08 33	23 13	26 14	02 23	29 08	00 20	27 20
19	21 53 40	29 ♒ 49 43	01 ♊ 14 47	08 ♊ 12 48	03 21	09 44	23 53	26 06	02 25	29 06	00 20	27 19
20	21 57 37	00 ♓ 50 13	15 ♊ 14 45	22 ♊ 20 36	04 17	10 55	24 32	25 58	02 27	29 04	00 19	27 17
21	22 01 34	01 50 40	29 ♊ 30 15	06 ♋ 43 26	05 17	12 06	25 11	25 50	02 29	29 02	00 18	27 16
22	22 05 30	02 51 06	13 ♋ 59 49	21 ♋ 18 53	06 19	13 17	25 51	25 43	02 31	29 00	00 18	27 14
23	22 09 27	03 51 30	28 ♋ 39 57	06 ♌ 02 17	07 23	14 27	26 30	25 35	02 33	28 58	00 17	27 13
24	22 13 23	04 51 52	13 ♌ 24 57	20 ♌ 47 00	08 29	15 38	27 09	25 27	02 35	28 56	00 16	27 12
25	22 17 20	05 52 12	28 ♌ 07 23	05 ♍ 25 07	09 38	16 48	27 49	25 19	02 36	28 54	00 15	27 10
26	22 21 16	06 52 30	12 ♍ 39 12	19 ♍ 48 47	10 49	17 58	28 28	25 12	02 38	28 52	00 15	27 09
27	22 25 13	07 52 46	26 ♍ 53 06	03 ♎ 51 32	12 02	19 09	29 07	25 04	02 39	28 50	00 14	27 07
28	22 29 09	08 53 01	10 ♎ 43 41	17 ♎ 29 18	13 16	20 18	29 ♐ 47	24 57	02 41	28 48	00 13	27 06
29	22 33 06	09 ♓ 53 15	24 ♎ 08 16	00 ♏ 40 42	14 ♒ 32	21 ♈ 28	00 ♑ 26	24 ♌ 49	02 ♐ 42	28 ♋ 46	00 ♏ 12	27 ♌ 04

Moon

Date	Moon True ☊ ° '	Moon Mean ☊ ° '	Moon ☽ Latitude ° '
01	15 ♐ 36	14 ♐ 25	04 S 24
02	15 R 33	14 22	03 43
03	15 32	14 19	02 51
04	15 D 33	14 16	01 54
05	15 33	14 13	00 S 52
06	15 R 34	14 10	00 N 11
07	15 33	14 06	01 14
08	15 29	14 03	02 13
09	15 23	14 00	03 06
10	15 14	13 57	03 52
11	15 03	13 54	04 28
12	14 50	13 50	04 51
13	14 37	13 47	05 02
14	14 25	13 44	04 57
15	14 15	13 41	04 37
16	14 07	13 38	04 03
17	14 02	13 35	03 15
18	14 00	13 31	02 16
19	14 D 00	13 28	01 N 07
20	14 R 00	13 25	00 S 07
21	13 59	13 22	01 21
22	13 56	13 19	02 32
23	13 51	13 15	03 33
24	13 43	13 12	04 20
25	13 32	13 09	04 50
26	13 20	13 06	05 01
27	13 08	13 03	04 52
28	12 57	13 00	04 26
29	12 ♐ 48	12 ♐ 56	03 S 47

DECLINATIONS

Date	Sun ☉ ° '	Moon ☽ ° '	Mercury ☿ ° '	Venus ♀ ° '	Mars ♂ ° '	Jupiter ♃ ° '	Saturn ♄ ° '	Uranus ♅ ° '	Neptune ♆ ° '	Pluto ♇ ° '
01	17 S 18	10 S 28	16 S 29	05 S 43	21 S 58	13 N 05	18 S 29	20 N 46	09 S 59	22 N 30
02	17 01	14 39	16 45	05 12	22 04	13 08	18 29	20 47	09 59	22 30
03	16 44	18 04	17 00	04 41	22 10	13 10	18 30	20 48	09 59	22 31
04	16 26	20 35	17 15	04 10	22 15	13 13	18 31	20 48	09 59	22 32
05	16 09	22 08	17 30	03 39	22 21	13 16	18 31	20 49	09 59	22 32
06	15 50	22 41	17 43	03 08	22 26	13 19	18 32	20 49	09 59	22 33
07	15 32	22 13	17 56	02 37	22 31	13 21	18 32	20 50	09 59	22 34
08	15 13	20 46	18 08	02 05	22 36	13 24	18 33	20 50	09 59	22 34
09	14 54	18 24	18 18	01 34	22 40	13 27	18 33	20 51	09 58	22 35
10	14 35	15 15	18 28	01 02	22 45	13 30	18 34	20 51	09 58	22 36
11	14 16	11 25	18 36	00 S 31	22 49	13 33	18 34	20 52	09 58	22 36
12	13 56	07 05	18 43	00 N 01	22 53	13 35	18 34	20 52	09 58	22 37
13	13 36	02 S 23	18 49	00 32	22 57	13 38	18 35	20 53	09 58	22 37
14	13 16	02 N 28	18 54	01 04	23 01	13 41	18 35	20 53	09 58	22 38
15	12 56	07 17	18 58	01 35	23 04	13 44	18 35	20 53	09 57	22 39
16	12 35	11 51	19 00	02 07	23 08	13 47	18 36	20 54	09 57	22 39
17	12 14	15 55	19 01	02 38	23 11	13 49	18 36	20 54	09 57	22 40
18	11 53	19 14	19 01	03 10	23 14	13 52	18 36	20 55	09 57	22 41
19	11 32	21 31	18 59	03 41	23 17	13 55	18 37	20 55	09 56	22 41
20	11 11	22 31	18 56	04 12	23 19	13 58	18 37	20 56	09 56	22 42
21	10 49	22 05	18 52	04 43	23 22	14 01	18 37	20 56	09 56	22 42
22	10 28	20 12	18 46	05 14	23 24	14 03	18 37	20 57	09 56	22 43
23	10 06	16 57	18 39	05 45	23 26	14 06	18 38	20 57	09 55	22 43
24	09 44	12 38	18 31	06 16	23 28	14 09	18 38	20 57	09 55	22 44
25	09 22	07 35	18 22	06 47	23 30	14 11	18 38	20 58	09 55	22 45
26	08 59	02 N 11	18 11	07 17	23 31	14 14	18 38	20 58	09 54	22 45
27	08 37	03 S 14	17 59	07 48	23 32	14 16	18 38	20 58	09 54	22 46
28	08 14	08 20	17 45	08 18	23 34	14 19	18 38	20 59	09 54	22 46
29	07 S 52	12 S 52	17 S 30	08 N 48	23 S 35	14 N 22	18 S 38	20 N 59	09 S 53	22 N 47

ZODIAC SIGN ENTRIES

Date	h	m	Planets
02	12	18	☿ ♑
02	13	33	☽ ♏
05	00	13	☽ ♐
07	13	08	☽ ♑
10	01	52	☽ ♒
11	07	46	♀ ♈
12	12	52	☽ ♓
14	21	48	☽ ♈
15	06	34	☿ ♒
17	04	48	☽ ♉
19	09	50	☽ ♊
19	16	05	☉ ♓
21	12	50	☽ ♋
23	14	10	☽ ♌
25	15	05	☽ ♍
27	17	20	☽ ♎
28	20	05	♂ ♑
29	22	45	☽ ♏

LATITUDES

Date	Mercury ☿ ° '	Venus ♀ ° '	Mars ♂ ° '	Jupiter ♃ ° '	Saturn ♄ ° '	Uranus ♅ ° '	Neptune ♆ ° '	Pluto ♇ ° '
01	03 N 33	01 S 08	00 N 16	01 N 08	02 N 02	00 N 36	01 N 45	10 N 57
04	03 15	01 00	00 14	01 08	02 02	00 36	01 45	10 57
07	02 45	00 51	00 12	01 09	02 03	00 36	01 45	10 58
10	02 11	00 42	00 09	01 09	02 03	00 36	01 45	10 58
13	01 35	00 32	00 07	01 09	02 04	00 36	01 45	10 58
16	01 00	00 22	00 04	01 10	02 04	00 35	01 45	10 59
19	00 N 26	00 S 11	00 N 02	01 10	02 05	00 35	01 46	10 59
22	00 S 04	00 00	00 S 01	01 10	02 05	00 35	01 46	10 59
25	00 32	00 N 11	00 04	01 10	02 06	00 35	01 46	10 59
28	00 57	00 24	00 07	01 10	02 06	00 35	01 46	10 59
31	01 S 19	00 N 36	00 S 10	01 N 11	02 N 07	00 N 35	01 N 46	10 N 59

DATA

Julian Date	2435505
Delta T	+31 seconds
Ayanamsa	23° 14' 55"
Synetic vernal point	05° ♓ 52' 05"
True obliquity of ecliptic	23° 26' 40"

MOON'S PHASES, APSIDES AND POSITIONS ☽

Date	h	m	Phase	Longitude ° '	Eclipse Indicator
03	16	08	☾	13 ♏ 49	
11	21	38	●	22 ♒ 09	
19	09	21	☽	29 ♉ 43	
26	01	42	○	06 ♍ 27	

Day	h	m		
07	18	49	Apogee	
23	18	31	Perigee	
06	12	52	Max dec	22° S 41'
13	23	51	0N	
20	17	03	Max dec	22° N 33'
26	21	36	0S	

LONGITUDES

Date	Chiron ⚷ ° '	Ceres ⚳ ° '	Pallas ⚴ ° '	Juno ⚵ ° '	Vesta ⚶ ° '	Black Moon Lilith ⚸ ° '
01	06 ♒ 24	22 ♒ 58	02 ♒ 09	01 ♑ 21	05 ♏ 04	06 ♑ 22
11	07 ♒ 12	26 ♒ 53	05 ♒ 44	04 ♑ 31	07 ♏ 33	07 ♑ 28
21	07 ♒ 59	00 ♓ 50	09 ♒ 14	07 ♑ 33	09 ♏ 29	08 ♑ 35
31	08 ♒ 44	04 ♓ 46	12 ♒ 39	10 ♑ 25	10 ♏ 46	09 ♑ 42

All ephemeris data is given at 12.00 UT and the Moon's longitude is additionally given for 24.00 UT

ASPECTARIAN

h m	Aspects
01 Wednesday	
02 51	☽ △ ☉
03 56	☽ Q ♅
03 58	☽ ⚹ ♂
05 38	☽ ∠ ♇
06 31	♆ St R
06 53	☽ ∠ ♃
09 32	☽ ∥ ♆
12 27	☽ ∠ ♄
15 50	☽ ⊼ ♀
02 Thursday	
00 29	☿ □ ♆
02 41	☽ ⋕ ♃
04 06	☽ ± ♀
05 11	☽ ⊥ ♄
09 05	☽ ∠ ♂
09 17	☽ ⚹ ♇
10 22	☽ ⚹ ♃
13 06	☽ □ ♅
13 28	☽ □ ☿
14 21	☽ ☌ ♆
16 33	☽ ⊻ ♄
18 06	☉ ⚹ ♂
19 40	☿ ☍ ♅
22 56	☽ ⚼ ♀
03 Friday	
00 00	☉ ∥ ☿
02 44	☽ ∥ ☉
03 10	☽ ⊥ ♂
03 14	☽ ∥ ☿
08 09	☽ Q ♇
09 06	☽ Q ♃
11 54	☉ Q ♄
15 27	☽ ⊻ ♂
15 37	☽ ∥ ♄
16 08	☽ □ ☉
22 20	☽ Q ☿
04 Saturday	
04 27	☿ ∠ ♂
07 23	☽ △ ♀
14 38	☽ ⋕ ♅
19 34	☽ □ ♇
20 14	☽ □ ♃
21 06	☽ ⚹ ☿
23 32	☽ △ ♅
05 Sunday	
01 03	☽ ⊻ ♆
03 42	☽ ☌ ♄
07 11	☽ Q ☉
10 33	♂ ⚼ ♅
13 08	☽ ⊥ ♆
17 52	☽ ∥ ♂
06 Monday	
00 13	☽ ⋕ ♇
02 24	☽ ∠ ☿
05 46	☽ ⚼ ♅
06 58	☽ ☌ ♂
07 24	☽ ∠ ♆
09 47	☽ ⚹ ☉
14 33	♂ ∠ ♆
16 07	♀ ± ♆
20 19	☿ ⊼ ♃
20 20	☽ ⊥ ☿
07 Tuesday	
00 02	☽ ± ♅
01 02	☽ ⋕ ♇
02 50	☽ □ ♀
03 56	☽ ∥ ♂
08 17	☽ △ ♇
08 23	☽ ⊻ ☿
08 26	☽ △ ♃
12 14	☽ ⊼ ♅
13 58	☽ ⚹ ♆
16 58	☽ ⊻ ♄
19 15	☽ ∠ ☉
08 Wednesday	
00 16	☿ ⊼ ♇
02 38	♃ ☌ ♇
03 33	☿ ⊼ ♃
04 06	♂ ⋕ ♇
05 15	☽ ⊥ ♄
11 08	☽ ⋕ ♅
12 11	☿ St D
14 23	☽ Q ♆
14 36	☿ ⊼ ♇
14 39	☽ ⚼ ♃
14 46	☽ ⚼ ♇
15 17	☽ ⊥ ☉
19 39	☽ Q ♀
23 36	☽ ∠ ♄
23 39	☽ ⊻ ♂
09 Thursday	
04 34	☽ ⊻ ☉
04 54	♀ ⊼ ♃
07 08	♀ ⊼ ♇
08 19	☿ ⚹ ♀
08 40	☽ ± ♃
08 56	☽ ± ♇
10 45	☽ ∥ ♄
12 28	☽ ⊥ ♂
12 50	☽ ∥ ☿
20 36	☽ ⊼ ♃
20 58	☽ ⊼ ♇
21 16	☽ ☌ ☿
22 33	☽ ⚹ ♀
10 Friday	
00 47	☽ ☍ ♅
02 40	☽ □ ♆
05 53	☽ ⚹ ♄
07 33	☽ ∠ ♂
16 53	☽ ∥ ☉
20 17	♀ △ ♅
23 18	☽ ⋕ ♃
11 Saturday	
05 00	☿ ∥ ♄
05 47	☽ Q ♄
07 36	☽ ∠ ♀
14 53	☽ ⚹ ♂
15 35	♀ ⊼ ♆
20 18	☽ ∥ ♆
21 38	☽ ☌ ☉
12 Sunday	
03 06	☽ ⊥ ♀
07 09	☽ ☍ ♃
08 01	☽ ☍ ♇
09 51	☽ ⊻ ☿
11 38	☽ ⊼ ♅
13 36	☽ △ ♆
15 31	☽ Q ♂
15 58	☽ ⊻ ♀
16 59	☽ □ ♄
21 51	☽ ⊥ ☿
23 09	☽ ± ♅
13 Monday	
02 58	♀ △ ♄
09 37	☉ ⋕ ♃
15 45	☽ ∠ ☿
16 17	☽ ⚼ ♅
17 16	♀ ± ♃
18 18	☽ ⚼ ♆
20 18	☽ ⋕ ♀
14 Tuesday	
03 04	☿ ☍ ♅
03 44	☽ □ ♂
04 16	☽ ∥ ♀
04 59	♀ ± ♇
11 16	☽ ± ♆
12 02	☽ ⊻ ☉
15 43	☽ ⊼ ♃
17 01	☽ ⊼ ♇
20 26	☽ △ ♅
21 20	☽ ⚹ ☿
22 29	☽ ⊼ ♆
15 Wednesday	
00 13	☽ ⊥ ☉
01 59	☽ △ ♄
02 48	☽ ± ♃
04 11	☽ ± ♇
06 35	☽ ☌ ♀
18 18	☽ ∠ ☉
19 17	☽ ⚼ ♃
19 23	☿ □ ♆
20 46	☽ Q ☿
20 47	☽ ⚼ ♇
16 Thursday	
01 49	☽ ⋕ ♆
05 38	☉ ☍ ♃
05 44	☽ ⚼ ♄
14 16	☽ △ ♂
15 48	☽ ⋕ ☉
22 10	☽ ± ♄
22 23	☽ △ ♃
23 05	☽ ∥ ♃
23 56	☽ ⚹ ☉
17 Friday	
00 05	☽ △ ♇
01 41	☉ ☍ ♇
03 20	☽ □ ♅
05 26	☽ ☍ ♆
07 28	☽ □ ☿
09 01	☽ ⊼ ♄
18 36	☽ ⊻ ♀
18 42	☽ ⚼ ♂
23 13	☽ Q ☉
18 Saturday	
06 17	☽ ⊥ ♀
06 57	☽ ⋕ ♄
09 04	♂ ± ♅
09 36	☿ ⚹ ♄
10 12	☽ ⋕ ☿
11 21	☽ Q ♅
11 28	☽ ± ♂
19 14	☉ ⊼ ♅
22 32	☽ ⊼ ♂
23 37	☽ ∠ ♀
19 Sunday	
03 09	☽ □ ♃
04 28	☽ ∥ ♅
05 12	☽ □ ♇
08 17	☽ ⚹ ♅
09 21	☽ □ ☉
10 25	☽ ⊼ ♆
14 02	☽ ☍ ♄
15 54	☽ △ ☿
20 45	☽ ± ♆
23 45	☉ △ ♆
20 Monday	
03 57	☽ ⚹ ♀
09 51	☽ Q ♃
10 00	☽ ∠ ♅
12 05	☽ Q ♇
12 07	☽ ⚼ ♆
12 54	♀ ⚼ ♃
19 21	☽ ⚼ ☿
21 Tuesday	
01 11	☽ ⊥ ♅
02 09	☽ Q ♀
04 26	☽ ☍ ♂
05 56	☽ ⚹ ♃
08 16	☽ ⚹ ♇
11 13	☽ ⊻ ♅
11 36	☽ ± ☿
13 20	☽ △ ♆
15 16	♀ ⚼ ♇
16 12	☽ △ ☉
16 59	☽ ⊼ ♄
22 20	☽ ⊼ ☿
22 Wednesday	
02 57	☽ ± ♄
03 48	☉ □ ♄
04 30	☽ ∥ ♅
06 38	☽ ∠ ♃
07 52	♂ △ ♃
09 07	☽ ∠ ♇
10 43	☽ □ ♀
17 48	☽ ⚼ ♄
18 48	☽ ⚼ ☉
21 17	☽ ⊥ ♃
23 52	☽ ⊥ ♇
23 Thursday	
00 07	☽ ⋕ ☿
00 46	☽ ⋕ ♄
07 01	☽ ⊻ ♃
08 18	☽ ⊼ ♂
09 38	☽ ⊻ ♇
10 35	☽ ± ☉
12 29	☽ ☌ ♅
14 38	☽ □ ♆
17 07	☿ ∥ ♄
18 20	☽ △ ♄
18 32	☽ ± ♂
21 04	☽ ⊼ ☉
23 37	☉ ∥ ♆
24 Friday	
03 20	☽ ☍ ☿
04 16	☽ ∥ ♃
09 51	☽ ⚼ ♂
13 16	♂ △ ♇
13 29	☉ ± ♅
15 55	☽ △ ♀
19 54	☽ Q ♆
25 Saturday	
01 16	☽ ⋕ ♆
03 14	☽ ⋕ ☉
07 27	☽ ☌ ♃
10 26	☽ ☌ ♇
11 28	☽ △ ♂
13 16	☽ ⊻ ♅
15 20	☽ ∥ ♀
15 30	☽ ⚹ ♆
18 34	☽ ⚼ ♀
19 23	☽ □ ♄
23 07	☽ ⊥ ♅
26 Sunday	
01 42	☽ ☍ ☉
04 48	♀ ⚼ ♄
08 40	☽ ⊼ ☿
10 46	☽ ± ♀
14 01	☽ ∠ ♅
16 19	☽ ∠ ♆
19 36	☽ ± ☿
21 42	☽ ⊼ ♀
27 Monday	
01 24	☽ Q ♄
01 51	♂ ⊼ ♅
07 29	☽ ⊥ ♆
08 56	☽ ⊻ ♃
12 16	☽ ⚼ ☿
12 24	☽ ⊻ ♇
15 20	☽ ⚹ ♅
16 01	☽ □ ♂
17 44	☽ ⊻ ♆
19 07	☽ ⊥ ♃
21 56	☽ ⚹ ♄
22 42	☽ ⊥ ♇
28 Tuesday	
08 30	☽ ⊼ ☉
10 23	☉ ⋕ ♀
10 38	☽ ∠ ♃
11 35	☽ ∥ ☉
11 50	☽ ⋕ ♀
12 08	☽ Q ♅
14 24	☽ ∠ ♇
16 57	☽ △ ☿
19 52	☽ ∥ ♆
19 57	☽ ± ☉
29 Wednesday	
00 21	☽ ∠ ♄
01 10	☽ Q ♂
03 40	♂ ⚹ ♆
06 42	☽ ☍ ♀
13 14	☽ ⚹ ♃
13 29	☽ ⚼ ☉
15 01	☿ Q ♄
16 41	☽ ⊥ ♄
17 21	☽ ⚹ ♇
20 28	☽ □ ♅
20 54	☽ ⋕ ♃
23 06	☽ ☌ ♆

MARCH 1956

LONGITUDES

Date	Sidereal time h m s	Sun ☉ ° ' "	Moon ☽ ° ' "	Moon ☽ 24.00 ° ' "	Mercury ☿ ° '	Venus ♀ ° '	Mars ♂ ° '	Jupiter ♃ ° '	Saturn ♄ ° '	Uranus ♅ ° '	Neptune ♆ ° '	Pluto ♇ ° '
01	22 37 03	10 ♓ 53 26	07 ♏ 06 48	13 ♏ 26 57	15 ♒ 50	22 ♈ 38	01 ♑ 05	24 ♌ 42	02 ♐ 43	28 ♋ 45	00 ♏ 11	27 ♌ 03
02	22 40 59	11 53 37	19 ♏ 41 36	25 ♏ 51 18	17 09	23 47	01 45	24 R 34	02 44	28 R 43	00 R 10	27 R 01
03	22 44 56	12 53 45	01 ♐ 56 39	07 ♐ 58 19	18 30	24 57	02 24	24 27	02 45	28 41	00 09	27 00
04	22 48 52	13 53 53	13 ♐ 56 59	19 ♐ 53 20	19 53	26 06	03 03	24 20	02 46	28 40	00 08	26 58
05	22 52 49	14 53 58	25 ♐ 48 07	01 ♑ 41 59	21 16	27 15	03 43	24 13	02 47	28 38	00 07	26 57
06	22 56 45	15 54 02	07 ♑ 35 38	13 ♑ 29 43	22 41	28 24	04 22	24 06	02 47	28 36	00 06	26 56
07	23 00 42	16 54 05	19 ♑ 24 50	25 ♑ 21 32	24 08	29 ♈ 33	05 01	23 59	02 48	28 35	00 05	26 54
08	23 04 38	17 54 05	01 ♒ 20 21	07 ♒ 21 44	25 35	00 ♉ 41	05 41	23 52	02 48	28 34	00 04	26 53
09	23 08 35	18 54 04	13 ♒ 26 03	19 ♒ 33 37	27 04	01 50	06 20	23 46	02 48	28 32	00 03	26 51
10	23 12 32	19 54 02	25 ♒ 44 40	01 ♓ 59 22	28 ♒ 35	02 58	06 59	23 39	02 49	28 31	00 02	26 50
11	23 16 28	20 53 57	08 ♓ 17 47	14 ♓ 39 58	00 ♓ 06	04 06	07 38	23 33	02 49	28 29	00 ♏ 01	26 49
12	23 20 25	21 53 50	21 ♓ 05 49	27 ♓ 35 15	01 39	05 14	08 18	23 27	02 R 49	28 28	29 ♎ 59	26 47
13	23 24 21	22 53 42	04 ♈ 08 06	10 ♈ 44 11	03 12	06 22	08 57	23 21	02 49	28 27	29 58	26 46
14	23 28 18	23 53 31	17 ♈ 23 16	24 ♈ 05 08	04 47	07 29	09 36	23 15	02 49	28 26	29 57	26 45
15	23 32 14	24 53 19	00 ♉ 49 35	07 ♉ 36 26	06 24	08 37	10 15	23 09	02 48	28 25	29 56	26 43
16	23 36 11	25 53 04	14 ♉ 25 29	21 ♉ 16 39	08 01	09 44	10 55	23 03	02 48	28 24	29 55	26 42
17	23 40 07	26 52 47	28 ♉ 09 47	05 ♊ 04 52	09 40	10 51	11 34	22 57	02 47	28 23	29 53	26 41
18	23 44 04	27 52 28	12 ♊ 01 50	19 ♊ 00 39	11 20	11 57	12 13	22 52	02 47	28 22	29 52	26 40
19	23 48 01	28 52 06	26 ♊ 01 18	03 ♋ 03 44	13 01	13 04	12 52	22 47	02 46	28 21	29 51	26 38
20	23 51 57	29 ♓ 51 43	10 ♋ 07 51	17 ♋ 13 33	14 44	14 10	13 31	22 42	02 45	28 20	29 49	26 37
21	23 55 54	00 ♈ 51 17	24 ♋ 20 38	01 ♌ 28 49	16 27	15 16	14 10	22 37	02 44	28 19	29 48	26 36
22	23 59 50	01 50 48	08 ♌ 37 45	15 ♌ 46 59	18 12	16 22	14 49	22 32	02 43	28 18	29 46	26 35
23	00 03 47	02 50 17	22 ♌ 56 00	00 ♍ 04 14	19 58	17 27	15 28	22 27	02 42	28 17	29 45	26 34
24	00 07 43	03 49 44	07 ♍ 11 02	14 ♍ 15 43	21 46	18 33	16 07	22 23	02 41	28 17	29 44	26 33
25	00 11 40	04 49 09	21 ♍ 17 39	28 ♍ 16 11	23 35	19 38	16 46	22 19	02 40	28 16	29 42	26 31
26	00 15 36	05 48 31	05 ♎ 10 45	12 ♎ 00 50	25 25	20 42	17 25	22 14	02 39	28 16	29 41	26 30
27	00 19 33	06 47 52	18 ♎ 46 04	25 ♎ 26 08	27 17	21 47	18 04	22 10	02 37	28 15	29 39	26 29
28	00 23 30	07 47 11	02 ♏ 00 54	08 ♏ 30 18	29 ♓ 09	22 51	18 43	22 07	02 35	28 15	29 38	26 28
29	00 27 26	08 46 27	14 ♏ 54 26	21 ♏ 13 28	01 ♈ 03	23 55	19 22	22 03	02 34	28 14	29 36	26 27
30	00 31 23	09 45 42	27 ♏ 27 42	03 ♐ 37 31	02 59	24 59	20 01	22 00	02 32	28 14	29 35	26 26
31	00 35 19	10 ♈ 44 55	09 ♐ 43 23	15 ♐ 45 50	04 ♈ 56	26 ♉ 02	20 ♑ 40	21 ♌ 56	02 ♐ 30	28 ♋ 14	29 ♎ 33	26 ♌ 25

Date	Moon True ☊ ° '	Moon Mean ☊ ° '	Moon ☽ Latitude ° '
01	12 ♐ 42	12 ♐ 53	02 S 56
02	12 R 39	12 50	01 58
03	12 38	12 47	00 S 56
04	12 D 38	12 44	00 N 07
05	12 R 37	12 41	01 09
06	12 36	12 37	02 08
07	12 32	12 34	03 02
08	12 26	12 31	03 47
09	12 17	12 28	04 24
10	12 06	12 25	04 48
11	11 52	12 21	05 00
12	11 39	12 18	04 56
13	11 26	12 15	04 38
14	11 15	12 12	04 04
15	11 07	12 09	03 16
16	11 02	12 06	02 16
17	10 59	12 02	01 N 07
18	10 D 59	11 59	00 S 06
19	10 59	11 56	01 19
20	10 R 59	11 53	02 28
21	10 56	11 50	03 29
22	10 52	11 47	04 17
23	10 45	11 43	04 48
24	10 36	11 40	05 02
25	10 25	11 37	04 57
26	10 14	11 34	04 35
27	10 04	11 31	03 57
28	09 57	11 27	03 07
29	09 51	11 24	02 09
30	09 48	11 21	01 06
31	09 ♐ 48	11 ♐ 18	00 00

DECLINATIONS

Date	Sun ☉ ° '	Moon ☽ ° '	Mercury ☿ ° '	Venus ♀ ° '	Mars ♂ ° '	Jupiter ♃ ° '	Saturn ♄ ° '	Uranus ♅ ° '	Neptune ♆ ° '	Pluto ♇ ° '
01	07 S 29	16 S 40	17 S 14	09 N 18	23 S 35	14 N 24	18 S 38	21 N 00	09 S 53	22 N 47
02	07 06	19 34	16 57	09 47	23 36	14 26	18 38	21 00	09 53	22 48
03	06 43	21 28	16 38	10 17	23 36	14 29	18 38	21 00	09 52	22 48
04	06 20	22 22	16 18	10 46	23 37	14 31	18 38	21 01	09 52	22 49
05	05 57	22 13	15 57	11 15	23 37	14 34	18 38	21 01	09 51	22 49
06	05 34	21 06	15 34	11 44	23 36	14 36	18 38	21 01	09 51	22 50
07	05 10	19 03	15 10	12 12	23 36	14 38	18 38	21 01	09 51	22 50
08	04 47	16 10	14 45	12 40	23 36	14 40	18 38	21 02	09 50	22 51
09	04 24	12 35	14 19	13 08	23 35	14 43	18 38	21 02	09 50	22 51
10	04 00	08 25	13 51	13 36	23 34	14 45	18 38	21 02	09 49	22 51
11	03 36	03 S 49	13 22	14 03	23 33	14 47	18 38	21 02	09 49	22 52
12	03 13	01 N 01	12 52	14 30	23 32	14 49	18 38	21 03	09 48	22 52
13	02 49	05 53	12 20	14 57	23 30	14 51	18 37	21 03	09 48	22 53
14	02 26	10 35	11 48	15 23	23 29	14 53	18 37	21 03	09 47	22 53
15	02 02	14 49	11 14	15 49	23 27	14 55	18 37	21 03	09 47	22 53
16	01 38	18 20	10 39	16 15	23 25	14 57	18 37	21 04	09 46	22 54
17	01 14	20 51	10 02	16 40	23 23	14 58	18 37	21 04	09 46	22 54
18	00 51	22 09	09 25	17 05	23 21	15 00	18 36	21 04	09 45	22 55
19	00 27	22 04	08 46	17 30	23 18	15 02	18 36	21 04	09 45	22 55
20	00 S 03	20 36	08 06	17 54	23 15	15 03	18 36	21 04	09 44	22 55
21	00 N 20	17 50	07 25	18 18	23 13	15 05	18 35	21 04	09 44	22 56
22	00 44	13 59	06 43	18 41	23 10	15 06	18 35	21 05	09 43	22 56
23	01 08	09 20	05 59	19 04	23 07	15 08	18 35	21 05	09 43	22 56
24	01 31	04 N 12	05 14	19 26	23 03	15 09	18 34	21 05	09 42	22 56
25	01 55	01 S 06	04 29	19 48	23 00	15 11	18 34	21 05	09 42	22 57
26	02 18	06 16	03 42	20 10	22 56	15 12	18 33	21 05	09 41	22 57
27	02 42	11 01	02 54	20 31	22 52	15 13	18 33	21 05	09 41	22 57
28	03 05	15 06	02 05	20 52	22 48	15 14	18 33	21 05	09 40	22 58
29	03 29	18 22	01 15	21 12	22 44	15 15	18 32	21 05	09 40	22 58
30	03 52	20 40	00 25	21 32	22 40	15 16	18 32	21 05	09 39	22 58
31	04 N 15	21 S 55	00 N 27	21 N 51	22 S 35	15 N 17	18 S 31	21 N 05	09 S 39	22 N 58

ZODIAC SIGN ENTRIES

Date	h	m	Planets
03	08	09	☽ ♐
05	20	32	☽ ♑
07	21	31	♀ ♉
08	09	19	☽ ♒
10	20	11	☽ ♓
11	10	27	☿ ♓
12	01	53	♆ ♎
13	04	26	☽ ♈
15	10	32	☽ ♉
17	15	11	☽ ♊
19	18	47	☽ ♋
20	15	20	☉ ♈
21	21	31	☽ ♌
23	23	53	☽ ♍
26	03	00	☽ ♎
28	08	18	☽ ♏
28	22	41	☿ ♈
30	16	56	☽ ♐

LATITUDES

Date	Mercury ☿ ° '	Venus ♀ ° '	Mars ♂ ° '	Jupiter ♃ ° '	Saturn ♄ ° '	Uranus ♅ ° '	Neptune ♆ ° '	Pluto ♇ ° '
01	01 S 12	00 N 32	00 S 09	01 N 10	02 N 06	00 N 35	01 N 46	10 N 59
04	01 31	00 44	00 12	01 11	02 07	00 35	01 46	10 59
07	01 48	00 57	00 15	01 11	02 08	00 35	01 46	10 59
10	02 00	01 10	00 18	01 10	02 08	00 35	01 47	10 59
13	02 09	01 23	00 22	01 10	02 09	00 35	01 47	10 59
16	02 15	01 36	00 25	01 10	02 09	00 35	01 47	10 58
19	02 16	01 49	00 29	01 10	02 10	00 35	01 47	10 58
22	02 13	02 02	00 33	01 10	02 10	00 35	01 47	10 58
25	02 06	02 15	00 37	01 10	02 10	00 35	01 47	10 58
28	01 55	02 27	00 40	01 09	02 11	00 35	01 47	10 57
31	01 S 38	02 N 39	00 S 45	01 N 09	02 N 11	00 N 35	01 N 47	10 N 57

DATA

Julian Date	2435534
Delta T	+31 seconds
Ayanamsa	23° 14' 58"
Synetic vernal point	05° ♓ 52' 01"
True obliquity of ecliptic	23° 26' 40"

MOON'S PHASES, APSIDES AND POSITIONS ☽

Date	h	m	Phase	Longitude ° '	Eclipse Indicator
04	11	53	☾	13 ♐ 54	
12	13	37	●	21 ♓ 58	
19	17	14	☽	29 ♊ 05	
26	13	11	○	05 ♎ 51	

Day	h	m		
06	13	19	Apogee	
22	00	29	Perigee	
04	20	42	Max dec	22° S 26'
12	07	02	0N	
18	22	42	Max dec	22° N 17'
25	07	00	0S	

LONGITUDES

Date	Chiron ⚷ ° '	Ceres ⚳ ° '	Pallas ⚴ ° '	Juno ⚵ ° '	Vesta ⚶ ° '	Black Moon Lilith ⚸ ° '
01	08 ♒ 40	04 ♓ 22	12 ♒ 19	10 ♑ 08	10 ♏ 40	09 ♑ 35
11	09 ♒ 21	08 ♓ 17	15 ♒ 37	12 ♑ 50	11 ♏ 17	10 ♑ 41
21	09 ♒ 59	12 ♓ 10	18 ♒ 46	15 ♑ 17	11 ♏ 06	11 ♑ 48
31	10 ♒ 32	16 ♓ 00	21 ♒ 46	17 ♑ 28	10 ♏ 05	12 ♑ 55

All ephemeris data is given at 12.00 UT and the Moon's longitude is additionally given for 24.00 UT

ASPECTARIAN

h m	Aspects
01 Thursday	
00 10	☽ ⚹ ♂
03 46	☽ ⊻ ♄
11 13	☽ Q ♃
15 38	☽ Q ♇
15 51	☽ ∥ ☿
19 45	☽ △ ☉
21 17	☿ ∠ ♂
02 Friday	
03 29	☽ ∥ ♄
06 00	☽ ∠ ♂
06 31	☽ □ ☿
16 04	♀ # ♆
20 47	☽ ⊼ ♀
21 24	☽ □ ♃
03 Saturday	
00 27	☽ ⊥ ♂
02 16	☽ □ ♇
02 41	♀ △ ♃
04 44	☽ # ♅
05 35	☽ △ ♅
08 28	☽ ⊻ ♆
09 49	☽ ± ♀
12 58	☽ ⊻ ♂
13 36	☽ ☌ ♄
20 22	☽ ⊥ ♆
22 14	☽ Q ☿
04 Sunday	
01 02	♂ ⊻ ♄
05 40	☽ ⚼ ♀
06 27	☉ ⚼ ♅
11 25	☽ ⚼ ♅
11 53	☽ □ ☉
14 24	☽ ∠ ♆
05 Monday	
01 34	☽ ⚹ ☿
01 58	♀ ± ♄
05 35	☽ ± ♅
05 50	♀ △ ♇
08 49	☽ △ ♃
14 20	☽ △ ♇
15 16	☽ △ ♀
17 14	☉ ⚼ ♆
17 45	☽ ⊼ ♅
20 47	☽ ⚹ ♆
06 Tuesday	
02 12	☽ ⊻ ♄
03 47	☽ Q ☉
05 03	☽ ☌ ♂
12 13	☽ ∠ ☿
13 07	☽ # ♅
14 26	☽ ⊥ ♄
15 02	☽ ⚼ ♃
16 15	♀ □ ♅
20 48	☽ ⚼ ♇
21 10	☽ Q ♆
07 Wednesday	
06 26	☽ ⚹ ☉
08 43	☽ ∠ ♄
09 02	☽ ⊥ ☿
09 08	☽ ± ♃
09 49	☿ ☍ ♃
15 00	☽ ± ♇
15 49	☽ ∥ ♄
20 28	☉ Q ♂
21 09	☽ ⊼ ♃
22 51	☽ ⊻ ☿
23 10	♀ ☍ ♆
08 Thursday	
03 05	☽ ⊼ ♇
06 26	☽ ☍ ♅
09 28	☽ □ ♆
10 34	☽ □ ♀
14 56	☽ ⚹ ♄
15 24	☽ ∠ ☉
15 58	☿ # ♃
21 09	☽ ⊻ ♂
22 24	☽ # ♃
23 22	☽ ∥ ☿
09 Friday	
08 36	☿ ☍ ♇
08 53	☽ # ♀
09 42	☽ ⊥ ♂
10 51	☽ ⊥ ☉
14 42	☽ Q ♄
23 40	☽ ⊻ ☉
10 Saturday	
01 48	☽ Q ♀
04 10	☽ ∥ ♆
04 19	☽ ∠ ♂
08 00	☽ ☍ ♃
08 43	♀ ⊼ ♄
11 01	☿ ⊼ ♅
14 06	☽ ☍ ♇
17 19	☽ ⊼ ♅
18 12	☽ ☌ ☿
18 27	☿ # ♀
20 14	☽ △ ♆
11 Sunday	
01 34	☽ □ ♄
03 15	☽ ⚹ ♀
04 47	☽ ± ♅
10 39	☿ △ ♆
10 42	☽ ⚹ ♂
13 11	☽ ∥ ☉
21 47	☽ ⚼ ♅
12 Monday	
00 38	☽ ⚼ ♆
03 28	♄ St R
10 14	☽ ∠ ♀
10 26	☽ Q ♂
13 37	☽ ☌ ☉
16 19	☽ ⊼ ♃
16 45	♂ ⚼ ♃
17 21	☽ ± ♆
22 00	☽ # ☉
22 31	☽ ⊼ ♇
13 Tuesday	
01 36	☽ △ ♅
03 18	☽ ± ♃
04 24	☽ ⊼ ♆
04 27	☽ ⊥ ♀
05 59	☿ □ ♄
06 05	♀ ∥ ♃
06 59	♂ ⊥ ♄
09 30	☽ ± ♇
09 35	☽ △ ♄
10 04	☽ ⊻ ☿
16 26	☽ ⊻ ♀
19 36	☽ ⚼ ♃
21 13	☽ □ ♂
21 48	☉ ⊼ ♃
22 29	☽ ⊥ ☿
14 Wednesday	
01 51	☽ ⚼ ♇
06 37	☿ ± ♅
07 52	☽ # ♆
12 45	☽ ⚼ ♄
13 23	☉ ± ♆
16 54	☽ ∠ ☿
17 51	☽ # ☿
22 25	☽ △ ♃
15 Thursday	
00 35	☽ ⊻ ☉
04 43	☽ △ ♇
04 51	☽ ± ♄
07 43	☽ □ ♅
10 25	☽ ☍ ♆
12 07	☽ ⊥ ☉
12 36	☽ ∥ ♃
15 30	☽ ⊼ ♄
19 09	☽ ∥ ♀
23 12	☽ ⚹ ☿
16 Friday	
03 00	☽ ☌ ♀
05 17	☽ ∠ ☉
05 31	☽ △ ♂
14 15	☽ # ♄
15 27	☽ Q ♅
23 08	☽ Q ☿
17 Saturday	
03 00	☽ □ ♃
07 20	☉ ⊼ ♇
09 05	☽ ⚼ ♂
09 26	☽ □ ♇
09 36	☽ ⚹ ☉
12 22	☽ ⚹ ♅
14 43	☽ ∥ ♅
14 59	☽ ⊼ ♆
16 17	♂ ⚼ ♇
20 02	☽ ☍ ♄
22 38	☿ ∥ ♆
23 35	♂ Q ♆
18 Sunday	
01 22	☽ ± ♆
01 28	☽ ± ♂
08 00	☽ Q ☉
10 01	☽ Q ♃
10 38	☽ □ ☿
11 51	☽ ⊻ ♀
12 20	☽ ⊼ ♂
14 17	☽ ∠ ♅
16 31	☽ Q ♇
16 52	☽ ⚼ ♆
23 04	☽ ⊥ ♀
23 32	☉ △ ♅
19 Monday	
01 39	♀ △ ♂
05 43	☽ ⊥ ♅
06 30	☽ ⚹ ♃
08 29	☿ ⚹ ♂
10 02	☉ ± ♃
13 03	☽ ⚹ ♇
13 46	☿ ⚹ ♀
15 47	☽ ∠ ♀
15 58	☽ ⊻ ♅
16 34	☿ ⚼ ♅
17 14	☽ □ ☉
18 31	☽ △ ♆
23 29	☽ ⊼ ♄
20 Tuesday	
06 15	☽ ∥ ♅
07 54	☽ ∠ ♃
09 40	☽ ± ♄
11 01	☉ ⊼ ♆
13 18	☿ ⚼ ♆
14 31	☽ ∠ ♇
18 01	☽ ☍ ♂
19 24	☽ ⚹ ♀
20 51	☽ △ ☿
23 02	☽ ⊥ ♃
21 Wednesday	
00 53	☽ ⚼ ♄
05 42	☽ ⊥ ♇
06 18	☽ # ♄
08 57	☽ ∥ ♀
09 06	☽ ⊻ ♃
15 48	☽ ⊻ ♇
17 20	☽ Q ♀
18 41	☽ ☌ ♅
21 09	☽ □ ♆
23 46	☽ △ ☉
22 Thursday	
01 37	☽ ⚼ ☿
02 06	☽ △ ♄
05 34	☽ ∥ ♃
05 59	♀ # ♄
10 43	♀ Q ♅
18 50	☽ ± ☿
22 53	☽ ⊼ ♂
23 Friday	
02 03	☽ □ ♀
02 49	☽ ⚼ ☉
03 19	☽ Q ♆
05 26	☉ ± ♇
06 20	☽ ⊼ ☿
08 51	☉ △ ♄
09 26	☽ ± ♂
10 06	☽ # ♆
11 12	☽ ☌ ♃
18 05	☽ ☌ ♇
19 03	☽ ± ☉
21 00	☽ ⊻ ♅
23 27	☽ ⚹ ♆
24 Saturday	
01 17	☽ ⚼ ♂
04 25	☽ □ ♄
05 55	☽ ⊼ ☉
06 25	☽ # ☿
07 06	☽ ⊥ ♅
19 51	☿ ⊼ ♃
20 36	♂ ± ♃
22 19	☽ ∠ ♅
23 16	☽ ∥ ☉
25 Sunday	
00 46	☽ ∠ ♆
03 54	☽ △ ♂
08 54	☽ △ ♀
10 55	☽ Q ♄
13 35	☿ ± ♆
13 44	☽ ⊻ ♃
16 00	☽ # ☉
16 08	☽ ⊥ ♆
16 31	☽ ☍ ☿
20 58	☽ ⊻ ♇
26 Monday	
00 00	☽ ⚹ ♅
00 01	☽ ⊥ ♃
01 30	☽ ∥ ☿
02 27	☽ ⊻ ♆
06 14	♂ # ♇
07 21	☽ ⊥ ♇
07 35	☽ ⚹ ♄
13 00	☽ ⚼ ♀
13 11	☽ ☍ ☉
15 35	☽ ∠ ♃
19 49	♂ ∠ ♄
20 54	☽ Q ♅
23 05	☽ ∠ ♇
27 Tuesday	
01 57	☿ ⊼ ♇
05 00	☽ ∥ ♆
06 13	☽ ± ♀
09 57	☽ ∠ ♄
10 42	☽ □ ♂
16 05	☉ # ☿
17 04	☉ ∠ ♀
17 52	☽ ⊼ ♀
18 05	☽ ⚹ ♃
20 21	♀ □ ♃
20 36	☉ ⚼ ♃
23 08	☿ ± ♃
28 Wednesday	
00 27	☿ △ ♅
01 53	☽ ⚹ ♇
02 07	☽ ⊥ ♄
05 06	☽ □ ♅
05 53	☽ ⊼ ☿
07 38	☽ ☌ ♆
12 52	☽ # ♃
13 03	☽ ⊻ ♄
15 50	☽ Q ♃
17 56	☿ ⊼ ♆
18 47	☽ ± ☿
21 09	☽ Q ♂
23 33	☽ ⊼ ☉
23 55	☽ Q ♇
29 Thursday	
03 59	♀ ∥ ♅
11 44	☽ ± ☉
13 28	☽ ∥ ♄
14 34	☽ ⚼ ☿
17 55	☿ Q ♂
20 55	☽ ⚹ ♂
30 Friday	
01 31	☽ □ ♃
05 15	☿ ± ♇
06 20	☽ ⚼ ☉
06 31	☿ △ ♄
06 45	☽ ☍ ♀
10 01	☽ □ ♇
13 30	☽ △ ♅
16 06	☽ ⊻ ♆
18 12	☽ # ♅
21 51	☽ ☌ ♄
31 Saturday	
00 46	☽ △ ☿
03 00	♂ ± ♇
03 33	☽ ∠ ♂
03 48	☽ ⊥ ♆
08 25	☽ # ♀
14 13	☽ △ ☉
18 57	☽ ⚼ ♅
20 43	♀ □ ♇
21 34	☽ ∠ ♆
22 22	☽ ⊥ ♂

APRIL 1956

LONGITUDES

Date	Sidereal time h m s	Sun ☉ ° ' "	Moon ☽ ° ' "	Moon ☽ 24.00 ° ' "	Mercury ☿ ° '	Venus ♀ ° '	Mars ♂ ° '	Jupiter ♃ ° '	Saturn ♄ ° '	Uranus ♅ ° '	Neptune ♆ ° '	Pluto ♇ ° '
01	00 39 16	11 ♈ 44 06	21 ♐ 45 25	27 ♐ 42 46	06 ♈ 54	27 ♉ 05	21 ♑ 19	21 ♌ 53	02 ♐ 28	28 ♋ 13	29 ♎ 32	26 ♌ 24
02	00 43 12	12 43 16	03 ♑ 38 33	09 ♑ 33 24	08 53	28 08	21 58	21 R 50	02 R 26	28 R 13	29 R 30	26 R 23
03	00 47 09	13 42 24	15 ♑ 27 59	21 ♑ 23 00	10 54	29 ♉ 10	22 37	21 48	02 24	28 13	29 29	26 22
04	00 51 05	14 41 29	27 ♑ 19 05	03 ♒ 16 52	12 55	00 ♊ 12	23 15	21 45	02 22	28 13	29 27	26 21
05	00 55 02	15 40 33	09 ♒ 16 58	15 ♒ 19 56	14 58	01 14	23 54	21 43	02 20	28 D 13	29 25	26 20
06	00 58 59	16 39 36	21 ♒ 26 16	27 ♒ 36 25	17 01	02 15	24 33	21 41	02 17	28 13	29 24	26 20
07	01 02 55	17 38 36	03 ♓ 50 46	10 ♓ 09 35	19 06	03 16	25 12	21 39	02 15	28 13	29 22	26 19
08	01 06 52	18 37 34	16 ♓ 33 05	23 ♓ 01 22	21 11	04 17	25 50	21 37	02 12	28 13	29 21	26 18
09	01 10 48	19 36 31	29 ♓ 34 27	06 ♈ 12 14	23 16	05 17	26 29	21 35	02 10	28 13	29 19	26 17
10	01 14 45	20 35 26	12 ♈ 54 34	19 ♈ 41 09	25 22	06 17	27 07	21 34	02 07	28 14	29 17	26 16
11	01 18 41	21 34 19	26 ♈ 31 40	03 ♉ 25 42	27 28	07 16	27 46	21 33	02 04	28 14	29 16	26 16
12	01 22 38	22 33 09	10 ♉ 22 49	17 ♉ 22 34	29 ♈ 33	08 15	28 24	21 32	02 01	28 14	29 14	26 15
13	01 26 34	23 31 58	24 ♉ 24 29	01 ♊ 28 05	01 ♉ 38	09 13	29 03	21 31	01 58	28 15	29 13	26 14
14	01 30 31	24 30 45	08 ♊ 32 57	15 ♊ 38 41	03 42	10 12	29 ♑ 41	21 30	01 55	28 15	29 11	26 14
15	01 34 28	25 29 29	22 ♊ 44 55	29 ♊ 51 20	05 45	11 09	00 ♒ 20	21 30	01 52	28 16	29 09	26 13
16	01 38 24	26 28 11	06 ♋ 57 40	14 ♋ 03 39	07 46	12 06	00 58	21 29	01 49	28 16	29 08	26 12
17	01 42 21	27 26 51	21 ♋ 09 06	28 ♋ 13 47	09 45	13 03	01 36	21 29	01 46	28 17	29 06	26 12
18	01 46 17	28 25 29	05 ♌ 17 32	12 ♌ 20 08	11 42	13 59	02 15	21 D 29	01 43	28 18	29 04	26 11
19	01 50 14	29 ♈ 24 04	19 ♌ 21 24	26 ♌ 21 07	13 37	14 55	02 53	21 29	01 39	28 18	29 03	26 11
20	01 54 10	00 ♉ 22 37	03 ♍ 19 01	10 ♍ 14 52	15 28	15 50	03 31	21 30	01 36	28 19	29 01	26 10
21	01 58 07	01 21 08	17 ♍ 08 23	23 ♍ 59 18	17 17	16 44	04 09	21 31	01 32	28 20	29 00	26 10
22	02 02 03	02 19 37	00 ♎ 47 20	07 ♎ 32 13	19 02	17 38	04 47	21 31	01 29	28 21	28 58	26 09
23	02 06 00	03 18 04	14 ♎ 13 40	20 ♎ 51 30	20 43	18 31	05 25	21 32	01 25	28 22	28 56	26 09
24	02 09 57	04 16 29	27 ♎ 25 31	03 ♏ 55 36	22 21	19 23	06 03	21 34	01 21	28 23	28 55	26 08
25	02 13 53	05 14 52	10 ♏ 21 39	16 ♏ 43 41	23 55	20 15	06 41	21 35	01 18	28 24	28 53	26 08
26	02 17 50	06 13 13	23 ♏ 01 44	29 ♏ 15 56	25 24	21 06	07 18	21 37	01 14	28 25	28 51	26 08
27	02 21 46	07 11 32	05 ♐ 26 28	11 ♐ 33 37	26 50	21 57	07 56	21 38	01 10	28 26	28 50	26 07
28	02 25 43	08 09 50	17 ♐ 37 40	23 ♐ 39 00	28 11	22 47	08 34	21 40	01 06	28 27	28 48	26 07
29	02 29 39	09 08 06	29 ♐ 38 03	05 ♑ 35 17	29 ♉ 27	23 35	09 11	21 42	01 02	28 28	28 47	26 07
30	02 33 36	10 ♉ 06 21	11 ♑ 31 13	17 ♑ 26 24	00 ♊ 39	24 ♊ 24	09 ♒ 49	21 ♌ 45	00 ♐ 58	28 ♋ 30	28 ♎ 45	26 ♌ 07

Date	Moon True ☊ ° '	Moon Mean ☊ ° '	Moon ☽ Latitude ° '
01	09 ♐ 48	11 ♐ 15	01 N 04
02	09 D 49	11 12	02 04
03	09 R 49	11 08	02 59
04	09 48	11 05	03 47
05	09 45	11 02	04 25
06	09 40	10 59	04 51
07	09 33	10 56	05 05
08	09 25	10 53	05 04
09	09 16	10 49	04 48
10	09 07	10 46	04 16
11	09 00	10 43	03 28
12	08 55	10 40	02 28
13	08 52	10 37	01 17
14	08 51	10 33	00 N 02
15	08 D 52	10 30	01 S 15
16	08 53	10 27	02 26
17	08 54	10 24	03 29
18	08 R 54	10 21	04 19
19	08 52	10 18	04 53
20	08 49	10 14	05 09
21	08 44	10 11	05 07
22	08 38	10 08	04 48
23	08 32	10 05	04 13
24	08 27	10 02	03 25
25	08 23	09 59	02 27
26	08 20	09 55	01 23
27	08 19	09 52	00 S 16
28	08 D 20	09 49	00 N 51
29	08 21	09 46	01 54
30	08 ♐ 23	09 ♐ 43	02 N 52

DECLINATIONS

Date	Sun ☉ ° '	Moon ☽ ° '	Mercury ☿ ° '	Venus ♀ ° '	Mars ♂ ° '	Jupiter ♃ ° '	Saturn ♄ ° '	Uranus ♅ ° '	Neptune ♆ ° '	Pluto ♇ ° '
01	04 N 38	22 S 08	01 N 20	22 N 09	22 S 31	15 N 18	18 S 31	21 N 05	09 S 38	22 N 58
02	05 02	21 19	02 13	22 28	22 26	15 19	18 30	21 05	09 37	22 59
03	05 25	19 35	03 07	22 45	22 21	15 20	18 30	21 05	09 37	22 59
04	05 47	17 00	04 02	23 03	22 16	15 21	18 29	21 05	09 36	22 59
05	06 10	13 41	04 57	23 19	22 11	15 21	18 28	21 05	09 36	22 59
06	06 33	09 45	05 53	23 35	22 05	15 22	18 28	21 05	09 35	22 59
07	06 56	05 21	06 49	23 51	22 00	15 22	18 27	21 05	09 35	22 59
08	07 18	00 S 38	07 45	24 06	21 54	15 23	18 27	21 05	09 34	22 59
09	07 40	04 N 14	08 41	24 21	21 48	15 24	18 26	21 05	09 33	23 00
10	08 03	09 01	09 37	24 35	21 42	15 24	18 25	21 05	09 33	23 00
11	08 25	13 28	10 32	24 48	21 36	15 24	18 25	21 05	09 32	23 00
12	08 47	17 16	11 27	25 01	21 30	15 24	18 24	21 05	09 32	23 00
13	09 08	20 08	12 21	25 14	21 24	15 24	18 23	21 05	09 31	23 00
14	09 30	21 46	13 14	25 25	21 17	15 24	18 23	21 05	09 31	23 00
15	09 52	22 00	14 06	25 37	21 10	15 24	18 22	21 04	09 30	23 00
16	10 13	20 49	14 56	25 47	21 04	15 24	18 21	21 04	09 29	23 00
17	10 34	18 20	15 45	25 58	20 57	15 24	18 21	21 04	09 29	23 00
18	10 55	14 46	16 31	26 07	20 50	15 24	18 20	21 04	09 28	23 00
19	11 16	10 23	17 16	26 16	20 43	15 24	18 19	21 04	09 28	23 00
20	11 36	05 29	17 59	26 25	20 35	15 24	18 18	21 04	09 27	23 00
21	11 57	00 N 22	18 39	26 33	20 28	15 23	18 18	21 03	09 26	23 00
22	12 17	04 S 43	19 17	26 40	20 21	15 23	18 17	21 03	09 26	23 00
23	12 37	09 30	19 53	26 47	20 13	15 23	18 16	21 03	09 25	23 00
24	12 57	13 44	20 26	26 53	20 05	15 22	18 15	21 03	09 25	23 00
25	13 16	17 15	20 56	26 59	19 57	15 21	18 14	21 03	09 24	23 00
26	13 36	19 52	21 24	27 04	19 50	15 21	18 13	21 02	09 24	23 00
27	13 55	21 28	21 50	27 09	19 42	15 20	18 13	21 02	09 23	22 59
28	14 14	22 01	22 13	27 13	19 33	15 19	18 12	21 02	09 22	22 59
29	14 33	21 32	22 33	27 17	19 25	15 19	18 11	21 02	09 22	22 59
30	14 N 51	20 S 05	22 N 51	27 N 20	19 S 17	15 N 18	18 S 10	21 N 01	09 S 21	22 N 59

ZODIAC SIGN ENTRIES

Date	h	m	Planets	
02	04	37	☽	♑
04	07	23	♀	♊
04	17	24	☽	♒
07	04	37	☽	♓
09	12	47	☽	♈
11	18	03	☽	♉
12	17	10	☿	♉
13	21	30	☽	♊
14	23	40	♂	♒
16	00	15	☽	♋
18	03	00	☽	♌
20	02	43	☉	♉
20	06	17	☽	♍
22	10	36	☽	♎
24	16	44	☽	♏
27	01	25	☽	♐
29	12	44	☽	♑
29	22	41	☿	♊

LATITUDES

Date	Mercury ☿ ° '	Venus ♀ ° '	Mars ♂ ° '	Jupiter ♃ ° '	Saturn ♄ ° '	Uranus ♅ ° '	Neptune ♆ ° '	Pluto ♇ ° '
01	01 S 32	02 N 43	00 S 46	01 N 09	02 N 11	00 N 35	01 N 47	10 N 57
04	01 10	02 55	00 50	01 09	02 12	00 35	01 47	10 56
07	00 43	03 06	00 55	01 08	02 12	00 35	01 47	10 56
10	00 S 13	03 16	00 59	01 08	02 13	00 34	01 48	10 55
13	00 N 20	03 26	01 04	01 08	02 13	00 34	01 48	10 54
16	00 53	03 34	01 09	01 07	02 13	00 34	01 48	10 54
19	01 24	03 42	01 13	01 07	02 13	00 34	01 48	10 53
22	01 53	03 49	01 18	01 07	02 14	00 34	01 48	10 52
25	02 15	03 55	01 24	01 06	02 14	00 34	01 48	10 52
28	02 31	03 59	01 29	01 06	02 14	00 34	01 48	10 51
31	02 N 39	04 N 02	01 S 35	01 N 05	02 N 14	00 N 34	01 N 48	10 N 50

DATA

Julian Date	2435565
Delta T	+32 seconds
Ayanamsa	23° 15' 01"
Synetic vernal point	05° ♓ 51' 58"
True obliquity of ecliptic	23° 26' 39"

MOON'S PHASES, APSIDES AND POSITIONS ☽

Date	h	m	Phase	Longitude ° '	Eclipse Indicator
03	08	06	☾	13 ♑ 33	
11	02	39	●	21 ♈ 11	
17	23	28	◐	27 ♋ 55	
25	01	41	○	04 ♏ 50	

Day	h	m		
03	09	32	Apogee	
15	21	20	Perigee	
01	04	47	Max dec	22° S 10'
08	15	09	0N	
15	04	04	Max dec	22° N 05'
21	13	41	0S	
28	12	35	Max dec	22° S 02'

LONGITUDES

Date	Chiron ⚷ ° '	Ceres ⚳ ° '	Pallas ⚴ ° '	Juno ⚵ ° '	Vesta ⚶ ° '	Black Moon Lilith ⚸ ° '
01	10 ♒ 35	16 ♓ 23	22 ♒ 03	17 ♑ 40	09 ♏ 57	13 ♑ 01
11	11 ♒ 01	20 ♓ 08	24 ♒ 50	19 ♑ 28	08 ♏ 08	14 ♑ 08
21	11 ♒ 22	23 ♓ 49	27 ♒ 24	20 ♑ 54	05 ♏ 50	15 ♑ 15
31	11 ♒ 36	27 ♓ 23	29 ♒ 41	21 ♑ 53	03 ♏ 21	16 ♑ 22

All ephemeris data is given at 12.00 UT and the Moon's longitude is additionally given for 24.00 UT

ASPECTARIAN

h m	Aspects
01 Sunday	
04 02	☉ Ᵽ ♇
10 46	☽ # ♀
11 04	☽ ⊻ ♂
11 54	☿ Ᵽ ♃
12 16	☽ △ ♃
12 56	☽ ± ♅
21 20	☽ △ ♇
23 45	☽ ⊼ ♀
02 Monday	
01 02	☽ ⊼ ♅
03 38	☽ ⚹ ♆
07 46	♂ ⊼ ♃
09 34	☽ ⊻ ♄
10 08	♀ # ♂
13 04	☽ ± ♀
14 11	♀ ⚹ ♅
16 06	☽ # ♅
18 28	☽ Ᵽ ♃
21 42	☽ ⊥ ♄
03 Tuesday	
00 48	☽ □ ☿
03 42	☽ Ᵽ ♇
03 55	☽ Q ♆
08 06	☽ □ ☉
09 06	☽ Ᵽ ♀
12 40	☽ ± ♃
15 55	☽ ∠ ♄
17 38	☿ Ᵽ ♇
19 04	♀ ⊼ ♆
21 56	☽ ± ♇
23 01	☽ ∥ ♄
04 Wednesday	
00 47	☽ ⊼ ♃
03 19	☽ ☌ ♂
06 47	♀ ∥ ♇
10 04	☽ ⊼ ♇
13 49	☽ ☍ ♅
16 17	☽ □ ♆
18 21	☽ △ ♀
20 45	☽ Q ☿
22 08	☽ ⚹ ♄
23 47	☽ Q ☉
05 Thursday	
00 30	☽ # ♃
11 21	♅ St D
21 59	☽ Q ♄
06 Friday	
01 34	☽ ⚹ ☿
01 48	☽ ⚹ ☉
03 55	☉ ☌ ☿
12 28	☽ ☍ ♃
12 54	♀ ☍ ♄
12 58	☽ ∥ ♆
15 01	☿ Ᵽ ♄
17 09	☿ ∠ ♀
18 24	☽ ⊻ ♂
21 31	☽ ☍ ♇
07 Saturday	
01 11	☽ ⊼ ♅
02 44	☉ Ᵽ ♄
03 26	☽ △ ♆
04 19	☽ # ☉
05 37	☽ # ☿
06 38	☽ ⊥ ♂
08 57	☽ □ ♄
09 30	☽ ∠ ☉
10 47	☽ □ ♀
12 34	☽ ∠ ☿
12 43	☽ ± ♅
16 48	☉ ∥ ☿
08 Sunday	
00 42	☽ ∠ ♂
04 02	☽ ⊥ ☉
05 46	☽ Ᵽ ♅
07 53	☽ Ᵽ ♆
08 56	☽ ⊥ ☿
16 11	☽ ⊻ ☉
16 55	☿ △ ♃
21 23	☽ ⊼ ♃
22 15	☽ ⊻ ☿
23 31	☽ Q ♀
09 Monday	
00 34	☽ ± ♆
04 51	♂ ⊼ ♇
06 00	☽ ⊼ ♇
06 04	☽ ⚹ ♂
08 23	☽ ± ♃
09 32	☽ △ ♅
11 32	☽ ⊼ ♆
12 56	♀ ± ♆
16 41	☽ △ ♄
16 55	☽ ± ♇
23 10	☽ ⚹ ♀
10 Tuesday	
00 40	☽ Ᵽ ♃
04 54	☽ Q ♂
06 35	☽ ∥ ☉
09 05	☽ Ᵽ ♇
10 20	☿ # ♆
14 43	☽ # ♆
15 50	☽ ∥ ☿
19 26	☽ Ᵽ ♄
20 24	☿ ± ♄
22 19	☿ △ ♇
11 Wednesday	
02 39	☽ ☌ ☉
03 17	☽ △ ♃
03 57	☽ ∠ ♀
11 12	☽ ± ♄
11 19	☉ △ ♃
11 32	☽ △ ♇
13 55	☽ ☌ ☿
14 16	☽ □ ♂
14 59	☽ □ ♅
16 46	☽ ☍ ♆
17 03	☿ □ ♂
20 52	☿ □ ♅
20 53	☽ ⊥ ♀
21 37	☽ ⊼ ♄
23 34	☽ ∥ ♃
12 Thursday	
05 36	♂ ☍ ♅
08 03	☽ ⊻ ♀
08 26	☿ ☍ ♆
20 25	☽ # ♄
22 03	☽ Q ♅
13 Friday	
07 04	☽ □ ♃
10 24	☽ ⊻ ☉
15 07	☽ □ ♇
15 51	☿ ⊼ ♄
17 48	♂ □ ♆
18 32	☽ ⚹ ♅
19 01	♀ Q ♃
20 09	☽ ⊼ ♆
20 16	☽ △ ♂
21 22	☽ ⊥ ☉
23 32	☽ ∥ ♅
14 Saturday	
00 49	☽ ☍ ♄
02 23	☽ ⊻ ☿
03 19	☽ # ♂
06 19	☽ ± ♆
12 31	☉ # ♆
13 37	☽ Q ♃
13 45	☽ ∠ ☉
14 16	☽ ⊥ ☿
14 59	☽ ☌ ♀
19 58	☽ ∠ ♅
21 31	☽ Ᵽ ♆
21 36	☽ Q ♇
22 53	☽ Ᵽ ♂
22 53	☿ ⊥ ♀
15 Sunday	
08 03	☽ ∠ ☿
09 53	☽ ⚹ ♃
11 11	☽ ⊥ ♅
14 48	☽ ± ♂
16 58	☽ ⚹ ☉
17 51	☽ ⚹ ♇
20 50	☉ ± ♄
21 19	☽ ⊻ ♅
22 48	☽ △ ♆
16 Monday	
01 24	☽ ⊼ ♂
03 21	☽ ⊼ ♄
05 34	☉ △ ♇
08 33	☽ # ♂
08 38	☽ ∥ ♅
09 36	♂ # ♅
11 12	☽ ∠ ♃
13 27	☽ ± ♄
13 35	☽ ⚹ ☿
14 44	☽ Q ☉
19 10	☽ ∠ ♇
21 19	☽ ⊻ ♀
17 Tuesday	
01 53	☿ ∥ ♃
02 25	☽ ⊥ ♃
04 36	☽ Ᵽ ♄
08 11	☽ ⊥ ♀
10 23	☽ ⊥ ♇
11 57	☽ # ♄
12 34	☽ ⊻ ♃
13 04	♃ St D
13 11	☽ Q ☿
17 31	♂ ⚹ ♄
18 03	♀ ∠ ♅
20 33	☽ ⊻ ♇
23 28	☽ □ ☉
18 Wednesday	
00 06	☽ ☌ ♅
00 31	☽ ∠ ♀
01 27	☽ □ ♆
02 52	☽ ∥ ☿
05 56	☽ △ ♄
06 34	☽ ☍ ♂
08 06	☽ ∥ ♃
08 44	☉ □ ♅
14 17	♀ Ᵽ ♆
17 11	♀ Q ♇
19 Thursday	
00 39	☽ □ ☿
03 31	☉ ☍ ♆
03 52	☽ ⚹ ♀
07 44	☽ ∥ ☉
08 03	☽ Q ♆
15 40	☽ ☌ ♃
16 40	☽ # ♆
23 42	☽ ☌ ♇
20 Friday	
01 52	☽ Q ♀
03 22	☽ ⊻ ♅
04 36	☽ ⚹ ♆
06 33	☽ △ ☉
09 02	☽ □ ♄
12 21	☽ ⊼ ♂
13 44	☽ ⊥ ♅
21 18	☿ ⊻ ♀
23 10	☿ # ♄
23 14	☽ ± ♂
23 15	☿ Q ♅
21 Saturday	
05 21	☽ ∠ ♅
06 31	☽ ∠ ♆
10 31	☽ Ᵽ ☉
11 14	☽ □ ♀
12 17	☽ △ ☿
15 41	☽ Ᵽ ♂
16 10	☽ Q ♄
16 17	☉ ⊼ ♄
19 39	☽ ⊻ ♃
22 14	☽ ⊥ ♆
22 Sunday	
03 31	☽ ± ☉
03 49	☽ ⊻ ♇
06 13	☽ ⊥ ♃
07 41	☽ ⚹ ♅
08 47	☽ ⊻ ♆
13 13	☽ ⚹ ♄
14 25	☽ ⊥ ♇
14 56	☽ ⊼ ☉
18 35	☽ Ᵽ ☿
19 26	☽ △ ♂
22 12	☽ ∠ ♃
23 Monday	
05 03	☽ Q ♅
06 28	☽ ∠ ♇
11 38	☽ ∥ ♆
13 01	☽ ± ☿
15 56	☽ ∠ ♄
20 18	☽ △ ♀
23 43	☿ # ♂
24 Tuesday	
00 04	☿ □ ♃
01 16	☽ ⚹ ♃
01 26	☽ ⊼ ☿
06 48	☽ # ☉
08 13	☽ ⊥ ♄
09 39	☽ ⚹ ♇
13 45	☽ □ ♅
14 43	☽ ☌ ♆
19 12	☽ ⊻ ♄
22 24	☽ # ♃
23 20	☽ Q ♃
25 Wednesday	
01 41	☽ ☍ ☉
01 47	☽ Ᵽ ♀
04 46	☽ □ ♂
07 50	☽ Q ♇
13 16	☉ ∠ ♀
17 08	☿ ∥ ♅
19 51	☽ ± ♀
20 01	☽ ∥ ♄
26 Thursday	
08 04	☽ ⊼ ♀
09 17	☽ □ ♃
11 33	☽ ∥ ♂
16 36	☽ Q ♂
17 10	☽ ☍ ☿
17 57	☽ □ ♇
22 22	☽ △ ♅
23 11	☽ ⊻ ♆
23 57	☿ □ ♇
27 Friday	
02 49	♀ ⚹ ♃
03 42	☽ # ♅
03 43	☽ ☌ ♄
10 48	☽ ⊥ ♆
15 43	☽ ⊼ ☉
17 09	☽ ⚹ ♂
28 Saturday	
02 25	♀ ⊥ ♅
03 43	☽ Ᵽ ♅
04 26	☽ ∠ ♆
04 32	☽ ± ☉
17 06	☿ ⚹ ♅
20 04	☽ △ ♃
21 38	☽ ± ♅
23 00	☽ ☍ ♀
23 16	☿ ⊼ ♆
29 Sunday	
00 00	☽ Ᵽ ☉
00 29	☽ ∠ ♂
04 56	☽ △ ♇
09 40	☽ ⊼ ♅
10 17	☽ ⚹ ♆
11 36	☽ ⊼ ☿
14 48	☽ ⊻ ♄
15 56	☉ □ ♂
19 34	☽ ⊥ ♂
22 22	☽ # ♅
30 Monday	
01 04	☽ ± ☿
02 18	☽ Ᵽ ♃
02 51	☽ ⊥ ♄
08 22	☽ ⊻ ♂
08 53	☽ △ ☉
10 26	☽ Q ♆
11 10	☽ Ᵽ ♇
18 11	☿ ☍ ♄
20 35	☽ ± ♃
20 58	☽ ∠ ♄
21 17	☽ Ᵽ ☿
21 56	☽ ∥ ♂
23 47	☿ ∥ ♇

MAY 1956

LONGITUDES

Date	Sidereal time h m s	Sun ☉ ° ' "	Moon ☽ ° ' "	Moon ☽ 24.00 ° ' "	Mercury ☿ ° '	Venus ♀ ° '	Mars ♂ ° '	Jupiter ♃ ° '	Saturn ♄ ° '	Uranus ♅ ° '	Neptune ♆ ° '	Pluto ♇ ° '
01	02 37 32	11 ♉ 04 34	23 ♑ 21 24	29 ♑ 16 49	01 ♊ 47	25 ♊ 11	10 ♒ 27	21 ♌ 47	00 ♐ 54	28 ♋ 31	28 ♎ 43	26 ♌ 06
02	02 41 29	12 02 45	05 ♒ 13 15	11 ♒ 11 20	02 49	25 57	11 04	21 50	00 R 50	28 33	28 R 42	26 R 06
03	02 45 26	13 00 55	17 ♒ 11 39	23 ♒ 14 50	03 47	26 43	11 41	21 52	00 46	28 34	28 40	26 06
04	02 49 22	13 59 03	29 ♒ 21 24	05 ♓ 31 57	04 40	27 28	12 19	21 55	00 42	28 35	28 39	26 06
05	02 53 19	14 57 11	11 ♓ 46 56	18 ♓ 06 47	05 28	28 11	12 56	21 59	00 38	28 37	28 37	26 06
06	02 57 15	15 55 16	24 ♓ 31 53	01 ♈ 02 30	06 11	28 54	13 33	22 02	00 33	28 39	28 35	26 06
07	03 01 12	16 53 20	07 ♈ 38 48	14 ♈ 20 52	06 49	29 ♊ 36	14 10	22 05	00 29	28 40	28 34	26 06
08	03 05 08	17 51 23	21 ♈ 08 38	28 ♈ 01 56	07 22	00 ♋ 16	14 47	22 09	00 25	28 42	28 32	26 D 06
09	03 09 05	18 49 24	05 ♉ 00 29	12 ♉ 03 52	07 50	00 56	15 24	22 13	00 20	28 44	28 31	26 06
10	03 13 01	19 47 24	19 ♉ 11 34	26 ♉ 22 58	08 13	01 34	16 00	22 17	00 16	28 45	28 29	26 06
11	03 16 58	20 45 22	03 ♊ 37 23	10 ♊ 54 05	08 30	02 11	16 37	22 21	00 12	28 47	28 28	26 06
12	03 20 55	21 43 19	18 ♊ 12 18	25 ♊ 31 15	08 43	02 47	17 14	22 25	00 07	28 49	28 26	26 06
13	03 24 51	22 41 14	02 ♋ 50 12	10 ♋ 08 27	08 50	03 22	17 50	22 30	00 ♐ 03	28 51	28 25	26 06
14	03 28 48	23 39 08	17 ♋ 25 21	24 ♋ 40 19	08 53	03 55	18 26	22 34	29 ♏ 58	28 53	28 23	26 07
15	03 32 44	24 36 59	01 ♌ 52 51	09 ♌ 02 34	08 R 50	04 27	19 02	22 39	29 54	28 55	28 22	26 07
16	03 36 41	25 34 49	16 ♌ 09 07	23 ♌ 12 14	08 43	04 58	19 39	22 44	29 50	28 57	28 21	26 07
17	03 40 37	26 32 37	00 ♍ 11 44	07 ♍ 07 31	08 32	05 27	20 15	22 49	29 45	28 59	28 19	26 07
18	03 44 34	27 30 23	13 ♍ 59 29	20 ♍ 47 38	08 16	05 54	20 50	22 55	29 41	29 01	28 18	26 08
19	03 48 30	28 28 08	27 ♍ 31 57	04 ♎ 12 29	07 56	06 20	21 26	23 00	29 36	29 04	28 16	26 08
20	03 52 27	29 ♉ 25 50	10 ♎ 49 16	17 ♎ 22 22	07 33	06 44	22 02	23 06	29 32	29 06	28 15	26 08
21	03 56 24	00 ♊ 23 32	23 ♎ 51 54	00 ♏ 17 55	07 07	07 06	22 37	23 11	29 27	29 08	28 14	26 09
22	04 00 20	01 21 12	06 ♏ 40 32	12 ♏ 59 52	06 38	07 27	23 13	23 17	29 23	29 10	28 12	26 09
23	04 04 17	02 18 50	19 ♏ 16 01	25 ♏ 29 08	06 07	07 46	23 48	23 23	29 18	29 13	28 11	26 09
24	04 08 13	03 16 27	01 ♐ 39 22	07 ♐ 46 52	05 34	08 02	24 23	23 29	29 14	29 15	28 10	26 10
25	04 12 10	04 14 04	13 ♐ 51 51	19 ♐ 54 31	05 01	08 17	24 58	23 36	29 09	29 17	28 09	26 10
26	04 16 06	05 11 38	25 ♐ 55 08	01 ♑ 53 57	04 27	08 30	25 33	23 42	29 05	29 20	28 07	26 11
27	04 20 03	06 09 12	07 ♑ 51 16	13 ♑ 47 27	03 53	08 40	26 07	23 48	29 00	29 22	28 06	26 11
28	04 23 59	07 06 45	19 ♑ 42 51	25 ♑ 37 52	03 20	08 49	26 42	23 55	28 56	29 25	28 05	26 12
29	04 27 56	08 04 17	01 ♒ 32 56	07 ♒ 28 32	02 48	08 55	27 16	24 02	28 52	29 28	28 04	26 13
30	04 31 53	09 01 48	13 ♒ 25 08	19 ♒ 23 16	02 18	08 59	27 51	24 09	28 47	29 30	28 03	26 13
31	04 35 49	09 ♊ 59 18	25 ♒ 23 27	01 ♓ 26 14	01 ♊ 50	09 ♋ 01	28 ♒ 25	24 ♌ 16	28 ♏ 43	29 ♋ 33	28 ♎ 01	26 ♌ 14

Date	Moon True ☊ ° '	Moon Mean ☊ ° '	Moon ☽ Latitude ° '
01	08 ♐ 24	09 ♐ 39	03 N 43
02	08 D 25	09 36	04 24
03	08 R 26	09 33	04 53
04	08 25	09 30	05 11
05	08 23	09 27	05 14
06	08 20	09 24	05 02
07	08 17	09 20	04 35
08	08 14	09 17	03 51
09	08 11	09 14	02 53
10	08 10	09 11	01 43
11	08 09	09 08	00 N 25
12	08 D 09	09 04	00 S 55
13	08 10	09 01	02 12
14	08 11	08 58	03 21
15	08 12	08 55	04 15
16	08 13	08 52	04 54
17	08 R 13	08 49	05 14
18	08 12	08 45	05 15
19	08 11	08 42	04 59
20	08 10	08 39	04 27
21	08 09	08 36	03 42
22	08 08	08 33	02 46
23	08 08	08 30	01 43
24	08 08	08 26	00 S 36
25	08 D 08	08 23	00 N 32
26	08 08	08 20	01 37
27	08 08	08 17	02 38
28	08 08	08 14	03 31
29	08 08	08 10	04 15
30	08 R 08	08 07	04 49
31	08 ♐ 08	08 ♐ 04	05 N 10

DECLINATIONS

Date	Sun ☉ ° '	Moon ☽ ° '	Mercury ☿ ° '	Venus ♀ ° '	Mars ♂ ° '	Jupiter ♃ ° '	Saturn ♄ ° '	Uranus ♅ ° '	Neptune ♆ ° '	Pluto ♇ ° '
01	15 N 09	17 S 46	23 N 07	27 N 23	19 S 08	15 N 17	18 S 09	21 N 01	09 S 21	22 N 59
02	15 27	14 42	23 20	27 25	19 00	15 16	18 08	21 01	09 20	22 59
03	15 45	11 01	23 31	27 27	18 51	15 15	18 08	21 00	09 20	22 59
04	16 02	06 51	23 39	27 28	18 43	15 14	18 07	21 00	09 19	22 58
05	16 20	02 S 18	23 46	27 28	18 34	15 13	18 06	21 00	09 19	22 58
06	16 36	02 N 27	23 50	27 29	18 25	15 11	18 05	20 59	09 18	22 58
07	16 53	07 14	23 52	27 28	18 16	15 10	18 04	20 59	09 18	22 58
08	17 09	11 49	23 52	27 28	18 07	15 09	18 03	20 59	09 17	22 58
09	17 26	15 54	23 49	27 27	17 58	15 07	18 02	20 58	09 16	22 57
10	17 41	19 10	23 45	27 25	17 49	15 06	18 01	20 58	09 16	22 57
11	17 57	21 17	23 39	27 23	17 40	15 05	18 00	20 57	09 15	22 57
12	18 12	22 00	23 31	27 21	17 31	15 03	18 00	20 57	09 15	22 57
13	18 27	21 12	23 21	27 18	17 22	15 01	17 59	20 57	09 14	22 56
14	18 41	19 00	23 09	27 15	17 13	15 00	17 58	20 56	09 14	22 56
15	18 56	15 36	22 56	27 11	17 03	14 58	17 57	20 56	09 13	22 56
16	19 10	11 19	22 41	27 08	16 54	14 56	17 56	20 55	09 13	22 55
17	19 23	06 30	22 24	27 03	16 44	14 55	17 55	20 55	09 13	22 55
18	19 36	01 N 27	22 06	26 59	16 35	14 53	17 54	20 54	09 12	22 55
19	19 49	03 S 35	21 47	26 54	16 25	14 51	17 53	20 54	09 12	22 54
20	20 02	08 22	21 27	26 48	16 16	14 49	17 52	20 53	09 11	22 54
21	20 14	12 42	21 05	26 43	16 06	14 47	17 51	20 53	09 11	22 54
22	20 26	16 21	20 43	26 37	15 57	14 45	17 50	20 52	09 10	22 53
23	20 38	19 12	20 21	26 31	15 47	14 43	17 50	20 52	09 10	22 53
24	20 49	21 05	19 58	26 24	15 38	14 41	17 49	20 51	09 09	22 53
25	21 00	21 57	19 35	26 17	15 28	14 39	17 48	20 51	09 09	22 52
26	21 10	21 46	19 12	26 10	15 18	14 36	17 47	20 50	09 09	22 52
27	21 20	20 35	18 50	26 03	15 09	14 34	17 46	20 50	09 08	22 51
28	21 30	18 31	18 28	25 55	14 59	14 32	17 45	20 49	09 08	22 51
29	21 39	15 40	18 08	25 47	14 49	14 29	17 44	20 49	09 07	22 50
30	21 49	12 11	17 48	25 39	14 40	14 27	17 43	20 48	09 07	22 50
31	21 N 57	08 S 12	17 N 30	25 N 31	14 S 30	14 N 24	17 S 43	20 N 48	09 S 07	22 N 50

ZODIAC SIGN ENTRIES

Date	h	m	Planets
02	01	27	☽ ♒
04	13	15	☽ ♓
06	22	05	☽ ♈
08	02	17	♀ ♋
09	03	24	☽ ♉
11	06	00	☽ ♊
13	07	21	☽ ♋
14	03	45	♄ ♏
15	08	52	☽ ♌
17	11	40	☽ ♍
19	16	25	☽ ♎
21	02	13	☉ ♊
21	23	26	☽ ♏
24	08	46	☽ ♐
26	20	11	☽ ♑
29	08	52	☽ ♒
31	21	09	☽ ♓

LATITUDES

Date	Mercury ☿ ° '	Venus ♀ ° '	Mars ♂ ° '	Jupiter ♃ ° '	Saturn ♄ ° '	Uranus ♅ ° '	Neptune ♆ ° '	Pluto ♇ ° '
01	02 N 39	04 N 02	01 S 35	01 N 05	02 N 14	00 N 34	01 N 48	10 N 50
04	02 37	04 03	01 40	01 05	02 14	00 34	01 48	10 50
07	02 27	04 02	01 46	01 05	02 14	00 34	01 47	10 49
10	02 06	03 59	01 52	01 04	02 14	00 34	01 47	10 48
13	01 36	03 54	01 58	01 04	02 14	00 34	01 47	10 48
16	00 56	03 47	02 04	01 04	02 14	00 34	01 47	10 47
19	00 N 09	03 36	02 11	01 03	02 14	00 34	01 47	10 46
22	00 S 43	03 23	02 17	01 03	02 14	00 34	01 47	10 45
25	01 35	03 07	02 24	01 03	02 14	00 33	01 47	10 45
28	02 24	02 47	02 31	01 02	02 14	00 33	01 47	10 44
31	03 S 06	02 N 23	02 S 38	01 N 02	02 N 13	00 N 33	01 N 47	10 N 43

DATA

Julian Date	2435595
Delta T	+32 seconds
Ayanamsa	23° 15' 05"
Synetic vernal point	05° ♓ 51' 55"
True obliquity of ecliptic	23° 26' 39"

MOON'S PHASES, APSIDES AND POSITIONS ☽

Date	h	m	Phase	Longitude ° '	Eclipse Indicator
03	02	55	☾	12 ♒ 39	
10	13	04	●	19 ♉ 50	
17	05	15	☽	26 ♌ 16	
24	15	26	○	03 ♐ 25	partial

Day	h	m		
01	04	44	Apogee	
13	00	52	Perigee	
28	21	00	Apogee	
05	23	42	ON	
12	11	22	Max dec	22° N 00'
18	18	50	OS	
25	19	46	Max dec	22° S 00'

LONGITUDES

Date	Chiron ⚷ ° '	Ceres ⚳ ° '	Pallas ⚴ ° '	Juno ⚵ ° '	Vesta ⚶ ° '	Black Moon Lilith ⚸ ° '
01	11 ♒ 36	27 ♓ 23	29 ♒ 41	21 ♑ 53	03 ♏ 21	16 ♑ 22
11	11 ♒ 43	00 ♈ 50	01 ♓ 40	22 ♑ 21	01 ♏ 05	17 ♑ 28
21	11 ♒ 43	04 ♈ 08	03 ♓ 18	22 ♑ 16	29 ♎ 20	18 ♑ 35
31	11 ♒ 37	07 ♈ 16	04 ♓ 32	21 ♑ 37	28 ♎ 19	19 ♑ 42

All ephemeris data is given at 12.00 UT and the Moon's longitude is additionally given for 24.00 UT

ASPECTARIAN

h m Aspects

01 Tuesday
05 25 ☽ ± ♇
08 28 ☽ || ♄
08 48 ☽ ⊼ ♃
15 58 ☽ ⊼ ♀
17 34 ☽ ⊼ ♇
21 37 ☉ || ♃
22 29 ☽ ☍ ♅
22 51 ☽ □ ♆

02 Wednesday
03 12 ☽ ⚹ ♄
04 57 ☽ ± ♀
06 42 ☽ △ ☿
07 02 ☽ # ☉
07 57 ☽ # ♃
16 39 ♀ ⚹ ♇

03 Thursday
00 19 ☽ bQ ♀
00 24 ☽ ☌ ♂
02 55 ☽ □ ☉
03 13 ☽ Q ♄
07 08 ♀ bQ ♂
21 19 ☽ ☍ ♃
22 03 ☽ || ♆

04 Friday
05 37 ☽ ☍ ♇
08 03 ☽ △ ♀
10 30 ☽ ⊼ ♅
10 36 ☽ △ ♆
11 22 ☿ ± ♆
14 36 ☽ □ ♄
17 34 ☽ Q ☉
22 12 ☽ ± ♅
23 03 ☽ □ ☿

05 Saturday
01 27 ♂ Q ♄
11 52 ♅ □ ♆
14 18 ☽ ⊻ ♂
15 29 ☽ bQ ♆
15 30 ☽ bQ ♅
18 32 ☽ ⚹ ☉

06 Sunday
01 53 ♀ △ ♆
02 13 ☽ ⊥ ♂
02 57 ♀ ⊻ ♅
07 20 ☽ ⊼ ♃
08 24 ☽ ± ♆
11 19 ☽ Q ☿
14 54 ☽ ⊼ ♇
18 30 ☽ ± ♃
19 29 ☽ ⊼ ♆
19 37 ☽ △ ♅
19 47 ☽ ∠ ♂
20 32 ☽ □ ♀
23 03 ☽ △ ♄

07 Monday
00 43 ☽ ∠ ☉
01 56 ☽ ± ♇
06 26 ☉ Q ♅
10 26 ☽ ⚹ ☿
10 59 ☽ bQ ♃
18 12 ☽ bQ ♇
18 17 ☽ ⊥ ☉
20 02 ♇ St D
22 32 ☽ # ♆

08 Tuesday
00 14 ☽ ⚹ ♂
01 57 ☽ bQ ♄
05 46 ☽ ⊻ ☉
06 41 ☽ Q ♀
13 46 ☽ △ ♃
14 13 ☽ ∠ ☿
16 34 ♀ ⊼ ♄
17 41 ☽ ± ♄
20 39 ☽ △ ♇
22 17 ☽ Q ♂

09 Wednesday
00 48 ♂ || ♄
00 51 ☽ ☍ ♆
01 11 ☽ □ ♅
04 02 ☽ ⊼ ♄
04 40 ☽ ⚹ ♀
06 23 ☽ ⊥ ☿
07 07 ☽ || ♃
16 58 ☽ ⊻ ☿
23 09 ☽ || ☉

10 Thursday
01 39 ☽ # ♂
02 42 ☽ # ♄
06 25 ☽ □ ♂
07 24 ☽ ∠ ♀
07 54 ☽ Q ♅
13 04 ☽ ☌ ☉
17 11 ☽ □ ♃
19 50 ☉ # ♂
23 08 ☽ ⊥ ♀
23 32 ☽ □ ♇

11 Friday
03 28 ☽ ⊼ ♆
03 59 ☽ ⚹ ♅
06 21 ☽ ☍ ♄
07 00 ☽ || ♅
09 32 ☽ ⊻ ♀
13 23 ☽ ± ♆
17 24 ☉ # ♄
20 11 ☽ ☌ ☿
23 09 ☽ Q ♃

12 Saturday
04 11 ☽ bQ ♆
04 47 ☽ ∠ ♅
05 16 ☽ Q ♇
10 19 ☽ △ ♂
18 11 ☽ ⊻ ☉
18 57 ☽ ⚹ ♃
19 35 ☽ ⊥ ♅

13 Sunday
00 57 ☽ ⚹ ♇
04 43 ☽ ⊥ ☉
04 45 ☽ △ ♆
05 27 ☽ ⊻ ♅
06 51 ☉ □ ♃
07 27 ☽ ⊼ ♄
12 00 ☽ bQ ♂
12 54 ☽ ☌ ♀
15 51 ☽ || ♅
17 15 ☽ ± ♄
19 42 ☽ ∠ ♃
20 32 ☽ ∠ ☉
21 54 ☽ ⊻ ☿

14 Monday
01 35 ☽ ∠ ♇
03 25 ☽ ± ♂
07 48 ☽ ⊥ ☿
07 59 ☽ bQ ♄
10 35 ☽ ⊥ ♃
12 13 ☿ St R
13 45 ☽ ⊼ ♂
14 18 ☽ || ☉
16 26 ☽ ⊥ ♇
20 09 ☽ # ♄
20 34 ☽ ⊻ ♃
22 40 ☽ ∠ ☿
23 03 ☽ ⚹ ☉

15 Tuesday
02 07 ☽ # ♂
02 23 ☽ ⊻ ♇
06 09 ☽ □ ♆
07 03 ☽ ☌ ♅
08 43 ☽ △ ♄
12 25 ☿ || ♇
15 48 ☽ || ♃
16 28 ☽ ⊻ ♀
20 30 ☽ Q ☉
23 34 ☽ ⚹ ☿

16 Wednesday
02 55 ☽ ⊥ ♀
12 19 ☽ Q ♆
18 11 ☽ ☍ ♂
18 42 ☽ ∠ ♀
19 40 ☽ Q ☿
22 45 ☽ # ♆
23 16 ☽ ☌ ♃

17 Thursday
01 24 ☉ □ ♇
04 59 ☽ ☌ ♇
05 15 ☽ □ ☉
08 47 ☽ ⚹ ♆
09 55 ☽ ⊻ ♅
11 14 ☽ □ ♄
20 18 ☽ ⊥ ♅
21 24 ☽ ⚹ ♀

18 Friday
01 39 ♀ ± ♄
02 11 ☽ □ ☿
10 47 ☽ ∠ ♆
12 03 ☽ ∠ ♅
18 27 ☽ Q ♄
19 07 ☽ Q ♀
21 38 ♀ bQ ♂

19 Saturday
00 38 ☽ ⊼ ♂
02 38 ☽ ⊥ ♆
03 52 ☽ ⊻ ♃
07 14 ☉ ⊼ ♆
09 30 ☽ ⊻ ♇
11 49 ☽ ± ♂
13 19 ☽ ⊻ ♆
13 48 ☽ △ ☉
14 39 ☽ ⊥ ♃
14 44 ☽ ⚹ ♅
15 41 ☽ ⚹ ♄
20 15 ☽ ⊥ ♇

20 Sunday
03 19 ☉ ⚹ ♅
04 21 ☽ □ ♀
04 47 ☽ bQ ♂
06 14 ☽ △ ☿
07 00 ☽ ∠ ♃
12 30 ☽ Q ♅
12 35 ☽ ∠ ♇
14 16 ☉ ☍ ♄
16 17 ☽ || ♆
18 44 ☽ ∠ ♄
19 07 ☽ bQ ☉

21 Monday
08 52 ☽ bQ ☿
09 35 ☽ △ ♂
10 44 ☽ ⚹ ♃
11 14 ☽ ⊥ ♄
12 12 ☿ ⊻ ♀
13 03 ☽ ± ☉
16 14 ☽ ⚹ ♇
20 07 ☽ ☌ ♆
21 51 ☽ □ ♅
22 21 ☽ ⊻ ♄

22 Tuesday
00 58 ☽ # ♃
01 03 ☽ ± ☿
01 10 ☽ ⊼ ☉
01 47 ☿ || ♅
09 11 ☽ || ♂
09 21 ☽ Q ♃
11 55 ☽ ⊼ ☿
13 30 ☽ △ ♀
14 47 ☽ Q ♇
15 36 ☉ ⊥ ♀
15 39 ♂ ☍ ♃
23 34 ☽ || ♄

23 Wednesday
00 00 ☉ || ☿
18 53 ☽ bQ ♀
20 00 ☽ □ ♃
21 10 ☽ □ ♂
22 51 ☽ # ☿

24 Thursday
01 19 ☽ □ ♇
05 13 ☽ ⊻ ♆
07 05 ☽ # ☉
07 18 ☽ △ ♅
07 18 ☽ ☌ ♄
07 37 ♄ △ ♅
08 20 ☽ # ♅
12 46 ☽ ± ♀
15 26 ☽ ☍ ☉
16 54 ☽ ⊥ ♆
17 19 ☉ || ♅
19 20 ☽ ☍ ☿

25 Friday
00 46 ☽ ⊼ ♀
09 46 ☉ ± ♆
10 08 ☽ Q ♂
10 35 ☽ ∠ ♆
12 51 ☽ bQ ♅

26 Saturday
00 11 ☉ ☌ ☿
06 49 ☽ ± ♅
07 31 ☽ △ ♃
11 13 ☽ ⚹ ♂
12 32 ☽ △ ♇
16 24 ☽ ⚹ ♆
18 18 ☽ ⊻ ♄
18 52 ☽ ⊼ ♅

27 Sunday
00 40 ☽ # ☉
02 04 ☿ ± ♆
04 21 ☽ ⊼ ☿
06 18 ☽ ⊥ ♄
08 15 ☽ # ♅
08 16 ☽ ⊼ ☉
13 41 ☽ ☍ ♀
13 57 ☽ bQ ♃
14 50 ♂ ☍ ♇
15 54 ☽ ± ☿
16 32 ☽ Q ♆
18 45 ☽ bQ ♇
18 57 ☽ ∠ ♂
21 27 ☽ ± ☉

28 Monday
00 22 ☽ ∠ ♄
08 20 ☽ ± ♃
09 19 ☽ bQ ☿
12 31 ☽ # ☿
12 59 ☽ ± ♇
14 06 ☽ ⊥ ♂
17 18 ☽ bQ ☉
19 05 ☽ || ♄
20 37 ☽ ⊼ ♃

29 Tuesday
01 10 ☽ ⊼ ♇
02 53 ☽ ⊻ ♂
04 57 ☽ □ ♆
06 35 ☽ ⚹ ♄
07 06 ☿ ⊥ ♀
07 45 ☽ ☍ ♅
14 25 ☽ △ ☿
18 32 ☽ || ♂
20 41 ☽ # ♃

30 Wednesday
02 22 ☽ △ ☉
03 01 ☽ ⊼ ♀
06 44 ☽ Q ♄
10 53 ☉ ⊻ ♀
15 10 ☽ ± ♀
18 22 ☿ # ♄
20 09 ♂ △ ♆

31 Thursday
06 42 ☽ || ♆
09 15 ☽ bQ ♀
09 44 ☽ ☍ ♃
13 41 ☽ ☍ ♇
17 14 ☽ △ ♆
18 04 ♀ St R
18 18 ☽ ☌ ♂
18 34 ☽ □ ♄
20 17 ☽ ⊼ ♅
23 28 ♂ □ ♄

JUNE 1956

Date	Sidereal time h m s	Sun ☉	Moon ☽	Moon ☽ 24.00	Mercury ☿	Venus ♀	Mars ♂	Jupiter ♃	Saturn ♄	Uranus ♅	Neptune ♆	Pluto ♇
01	04 39 46	10 ♊ 56 47	07 ♓ 32 11	13 ♓ 41 49	01 ♊ 26	09 ♋ 00	28 ♒ 59	24 ♌ 23	28 ♏ 39	29 ♋ 36	28 ♎ 00	26 ♌ 15
02	04 43 42	11 54 15	19 ♓ 55 42	26 ♓ 14 20	01 R 04	08 R 58	29 ♒ 32	24 31	28 R 34	29 38	27 R 59	26 15
03	04 47 39	12 51 43	02 ♈ 38 12	09 ♈ 07 43	00 46	08 52	00 ♓ 06	24 38	28 30	29 41	27 58	26 16
04	04 51 35	13 49 10	15 ♈ 43 13	22 ♈ 24 59	00 32	08 44	00 39	24 46	28 26	29 44	27 57	26 17
05	04 55 32	14 46 36	29 ♈ 13 11	06 ♉ 07 51	00 22	08 34	01 12	24 53	28 22	29 47	27 56	26 18
06	04 59 28	15 44 02	13 ♉ 08 52	20 ♉ 16 00	00 16	08 22	01 45	25 01	28 17	29 50	27 55	26 19
07	05 03 25	16 41 27	27 ♉ 28 51	04 ♊ 46 50	00 D 14	08 07	02 18	25 09	28 13	29 52	27 54	26 19
08	05 07 22	17 38 52	12 ♊ 09 15	19 ♊ 35 15	00 17	07 50	02 50	25 17	28 09	29 55	27 53	26 20
09	05 11 18	18 36 15	27 ♊ 03 53	04 ♋ 34 05	00 24	07 30	03 23	25 26	28 05	29 ♋ 58	27 52	26 21
10	05 15 15	19 33 38	12 ♋ 04 47	19 ♋ 34 54	00 36	07 09	03 55	25 34	28 01	00 ♌ 01	27 52	26 22
11	05 19 11	20 31 00	27 ♋ 03 23	04 ♌ 29 16	00 53	06 45	04 27	25 42	27 57	00 04	27 51	26 23
12	05 23 08	21 28 21	11 ♌ 51 40	19 ♌ 09 52	01 14	06 19	04 58	25 51	27 53	00 07	27 50	26 24
13	05 27 04	22 25 41	26 ♌ 23 17	03 ♍ 31 28	01 39	05 51	05 29	26 00	27 49	00 10	27 49	26 25
14	05 31 01	23 23 00	10 ♍ 34 08	17 ♍ 31 07	02 09	05 21	06 00	26 08	27 46	00 14	27 48	26 26
15	05 34 57	24 20 18	24 ♍ 22 25	01 ♎ 08 07	02 43	04 49	06 31	26 17	27 42	00 17	27 48	26 27
16	05 38 54	25 17 35	07 ♎ 48 22	14 ♎ 23 26	03 21	04 16	07 02	26 26	27 38	00 20	27 47	26 28
17	05 42 51	26 14 51	20 ♎ 53 36	27 ♎ 19 11	04 04	03 42	07 32	26 35	27 34	00 23	27 46	26 29
18	05 46 47	27 12 07	03 ♏ 40 33	09 ♏ 58 02	04 50	03 06	08 02	26 45	27 31	00 26	27 45	26 30
19	05 50 44	28 09 22	16 ♏ 12 02	22 ♏ 22 51	05 41	02 30	08 32	26 54	27 27	00 29	27 45	26 32
20	05 54 40	29 ♊ 06 36	28 ♏ 30 51	04 ♐ 36 21	06 36	01 53	09 01	27 03	27 24	00 33	27 44	26 33
21	05 58 37	00 ♋ 03 49	10 ♐ 39 38	16 ♐ 41 00	07 34	01 16	09 31	27 13	27 20	00 36	27 44	26 34
22	06 02 33	01 01 03	22 ♐ 40 44	28 ♐ 39 04	08 37	00 38	09 59	27 22	27 17	00 39	27 43	26 35
23	06 06 30	01 58 15	04 ♑ 36 16	10 ♑ 32 34	09 43	00 ♋ 00	10 28	27 32	27 14	00 43	27 43	26 36
24	06 10 26	02 55 28	16 ♑ 28 14	22 ♑ 23 29	10 53	29 ♊ 23	10 56	27 42	27 11	00 46	27 42	26 38
25	06 14 23	03 52 40	28 ♑ 18 37	04 ♒ 13 54	12 06	28 46	11 24	27 52	27 07	00 49	27 42	26 39
26	06 18 20	04 49 52	10 ♒ 09 38	16 ♒ 06 07	13 24	28 10	11 52	28 02	27 04	00 53	27 41	26 40
27	06 22 16	05 47 04	22 ♒ 03 42	28 ♒ 02 46	14 45	27 35	12 19	28 12	27 01	00 56	27 41	26 42
28	06 26 13	06 44 16	04 ♓ 03 43	10 ♓ 06 58	16 09	27 01	12 46	28 22	26 58	01 00	27 41	26 43
29	06 30 09	07 41 27	16 ♓ 12 58	22 ♓ 22 12	17 37	26 29	13 12	28 32	26 56	01 03	27 40	26 44
30	06 34 06	08 ♋ 38 39	28 ♓ 35 09	04 ♈ 52 18	19 ♊ 09	25 ♊ 58	13 ♓ 38	28 ♌ 42	26 ♏ 53	01 ♌ 06	27 ♎ 40	26 ♌ 46

Date	Moon True ☊	Moon Mean ☊	Moon ☽ Latitude
01	08 ♐ 08	08 ♐ 01	05 N 17
02	08 D 08	07 58	05 11
03	08 08	07 55	04 49
04	08 09	07 51	04 12
05	08 09	07 48	03 20
06	08 10	07 45	02 14
07	08 10	07 42	00 N 59
08	08 R 10	07 39	00 S 22
09	08 10	07 36	01 43
10	08 09	07 32	02 57
11	08 08	07 29	03 59
12	08 06	07 26	04 44
13	08 05	07 23	05 10
14	08 04	07 20	05 16
15	08 D 04	07 16	05 04
16	08 04	07 13	04 35
17	08 05	07 10	03 52
18	08 06	07 07	02 59
19	08 07	07 04	01 58
20	08 09	07 01	00 S 53
21	08 R 10	06 57	00 N 14
22	08 08	06 54	01 19
23	08 06	06 51	02 21
24	08 03	06 48	03 15
25	07 59	06 45	04 01
26	07 55	06 42	04 37
27	07 50	06 38	05 01
28	07 46	06 35	05 12
29	07 43	06 32	05 10
30	07 ♐ 41	06 ♐ 29	04 N 52

DECLINATIONS

Date	Sun ☉	Moon ☽	Mercury ☿	Venus ♀	Mars ♂	Jupiter ♃	Saturn ♄	Uranus ♅	Neptune ♆	Pluto ♇
01	22 N 05	03 S 50	17 N 14	25 N 22	14 S 20	14 N 22	17 S 42	20 N 47	09 S 06	22 N 49
02	22 13	00 N 47	16 59	25 13	14 11	14 19	17 41	20 46	09 06	22 49
03	22 21	05 28	16 46	25 03	14 01	14 17	17 40	20 46	09 06	22 48
04	22 28	10 04	16 35	24 54	13 52	14 14	17 39	20 45	09 05	22 48
05	22 34	14 19	16 26	24 44	13 42	14 11	17 38	20 45	09 05	22 47
06	22 41	17 55	16 20	24 33	13 33	14 09	17 38	20 44	09 05	22 47
07	22 47	20 34	16 15	24 23	13 23	14 06	17 37	20 43	09 04	22 46
08	22 52	21 53	16 13	24 12	13 14	14 03	17 36	20 43	09 04	22 46
09	22 57	21 42	16 12	24 01	13 04	14 00	17 35	20 42	09 04	22 45
10	23 02	19 57	16 14	23 49	12 55	13 57	17 35	20 41	09 03	22 45
11	23 06	16 51	16 17	23 38	12 46	13 54	17 34	20 41	09 03	22 44
12	23 10	12 41	16 22	23 26	12 37	13 51	17 33	20 40	09 03	22 43
13	23 14	07 52	16 30	23 14	12 27	13 48	17 32	20 39	09 03	22 43
14	23 17	02 N 44	16 39	23 01	12 18	13 45	17 32	20 39	09 02	22 42
15	23 19	02 S 25	16 49	22 48	12 09	13 42	17 31	20 38	09 02	22 42
16	23 22	07 18	17 01	22 35	12 01	13 39	17 30	20 37	09 02	22 41
17	23 23	11 45	17 15	22 22	11 52	13 36	17 30	20 37	09 02	22 41
18	23 25	15 33	17 29	22 09	11 43	13 33	17 29	20 36	09 02	22 40
19	23 26	18 34	17 45	21 56	11 34	13 29	17 28	20 35	09 01	22 40
20	23 26	20 42	18 02	21 42	11 26	13 26	17 28	20 35	09 01	22 39
21	23 27	21 49	18 20	21 29	11 17	13 23	17 27	20 34	09 01	22 38
22	23 26	21 55	18 39	21 15	11 09	13 19	17 27	20 33	09 01	22 38
23	23 26	21 01	18 58	21 02	11 00	13 16	17 26	20 32	09 01	22 37
24	23 25	19 12	19 18	20 48	10 52	13 12	17 26	20 32	09 01	22 36
25	23 23	16 34	19 39	20 35	10 44	13 09	17 25	20 31	09 01	22 36
26	23 21	13 15	19 59	20 22	10 36	13 05	17 25	20 30	09 01	22 35
27	23 19	09 24	20 20	20 10	10 28	13 02	17 24	20 29	09 00	22 35
28	23 16	05 10	20 40	19 58	10 21	12 58	17 24	20 29	09 00	22 34
29	23 13	00 41	21 01	19 46	10 13	12 55	17 23	20 28	09 00	22 33
30	23 N 10	03 N 55	21 N 20	19 N 34	10 S 06	12 N 51	17 S 23	20 N 27	09 S 00	22 N 33

ZODIAC SIGN ENTRIES

Date	h m	Planets
03	07 05	☽ ♈
03	07 51	♂ ♓
05	13 22	☽ ♉
07	16 09	☽ ♊
09	16 42	☽ ♋
10	01 48	♅ ♌
11	16 45	☽ ♌
13	18 03	☽ ♍
15	21 58	☽ ♎
18	05 03	☽ ♏
20	14 55	☽ ♐
21	10 24	☉ ♋
23	02 43	☽ ♑
23	12 10	♀ ♊
25	15 25	☽ ♒
28	03 54	☽ ♓
30	14 43	☽ ♈

LATITUDES

Date	Mercury ☿	Venus ♀	Mars ♂	Jupiter ♃	Saturn ♄	Uranus ♅	Neptune ♆	Pluto ♇
01	03 S 17	02 N 14	02 S 41	01 N 02	02 N 13	00 N 33	01 N 47	10 N 43
04	03 45	01 44	02 48	01 01	02 13	00 33	01 47	10 42
07	04 02	01 11	02 55	01 01	02 12	00 33	01 47	10 42
10	04 09	00 N 34	03 03	01 01	02 12	00 33	01 46	10 41
13	04 05	00 S 05	03 11	01 01	02 12	00 33	01 46	10 40
16	03 53	00 47	03 19	01 00	02 11	00 33	01 46	10 40
19	03 33	01 30	03 27	01 00	02 11	00 33	01 46	10 39
22	03 08	02 11	03 35	01 00	02 10	00 33	01 46	10 39
25	02 38	02 51	03 44	01 00	02 10	00 33	01 46	10 38
28	02 04	03 27	03 52	00 59	02 09	00 33	01 46	10 38
31	01 S 28	03 S 58	04 S 01	00 N 59	02 N 08	00 N 33	01 N 45	10 N 37

DATA

Julian Date	2435626
Delta T	+32 seconds
Ayanamsa	23° 15' 09"
Synetic vernal point	05° ♓ 51' 51"
True obliquity of ecliptic	23° 26' 38"

MOON'S PHASES, APSIDES AND POSITIONS ☽

Date	h m	Phase	Longitude	Eclipse Indicator
01	19 13	☾	11 ♓ 14	
08	21 29	●	18 ♊ 02	Total
15	11 56	☽	24 ♍ 20	
23	06 14	○	01 ♑ 44	

Day	h m	
10	03 02	Perigee
25	07 37	Apogee

Day	h m		
02	08 01	0N	
08	21 06	Max dec	22° N 00'
15	00 39	0S	
22	02 23	Max dec	22° S 00'
29	15 35	0N	

LONGITUDES

Date	Chiron ⚷	Ceres ⚳	Pallas ⚴	Juno ⚵	Vesta ⚶	Black Moon Lilith ⚸
01	11 ♒ 36	07 ♈ 34	04 ♓ 38	21 ♑ 31	28 ♎ 15	19 ♑ 48
11	11 ♒ 22	10 ♈ 28	05 ♓ 21	20 ♑ 13	28 ♎ 08	20 ♑ 55
21	11 ♒ 03	13 ♈ 07	05 ♓ 33	18 ♑ 25	28 ♎ 50	22 ♑ 02
31	10 ♒ 38	15 ♈ 29	05 ♓ 11	16 ♑ 14	00 ♏ 15	23 ♑ 09

All ephemeris data is given at 12.00 UT and the Moon's longitude is additionally given for 24.00 UT

ASPECTARIAN

h m Aspects

01 Friday
00 22 ☽ □ ☿
06 54 ♂ # ♃
08 10 ☽ ± ♅
14 52 ☽ △ ♀
19 13 ☽ □ ☉
22 39 ☽ ⚼ ♆

02 Saturday
01 47 ☽ ⚼ ♅
10 23 ☽ Q ☿
15 56 ☽ ± ♆
16 41 ♂ ⊼ ♅
20 49 ☽ ⊼ ♃

03 Sunday
00 03 ☽ ⊼ ♇
03 17 ☽ ⊼ ♆
04 18 ☽ △ ♄
05 29 ☉ Q ♃
06 28 ☽ △ ♅
07 02 ☽ ⊻ ♂
08 14 ☽ ± ♃
08 25 ☽ Q ☉
08 35 ☽ ⚹ ☿
11 19 ☽ ± ♇
14 40 ☉ ⚼ ♆
18 42 ☽ ⊥ ♂
23 25 ☽ □ ♀

04 Monday
01 03 ☽ ⚼ ♃
03 56 ☽ ⚼ ♇
06 50 ☽ # ♆
07 52 ☽ ⚼ ♄
08 08 ☿ □ ♂
08 17 ☽ ⚹ ☉
11 40 ☽ ∠ ☿
11 52 ☽ ∠ ♂
23 45 ☉ Q ♇
23 58 ☽ ± ♄

05 Tuesday
03 32 ☽ ⊥ ☿
04 19 ☽ △ ♃
06 52 ☽ △ ♇
07 25 ☽ Q ♀
08 31 ☽ # ♂
09 45 ☽ ☍ ♆
10 30 ☽ ⊼ ♄
11 17 ☽ ∥ ♃
12 01 ☉ ∠ ♅
12 59 ☽ □ ♅
13 03 ☽ ∠ ☉
13 58 ☽ ⊻ ☿
15 36 ☽ ⚹ ♂

06 Wednesday
01 03 ☽ ∥ ☿
03 58 ☽ ⚹ ♀
05 46 ☽ ⊥ ☉
09 49 ☽ # ♄
13 04 ☽ Q ♂
16 42 ☽ ⊻ ☉
19 56 ☽ Q ♅

07 Thursday
04 53 ☽ ∠ ♀
08 07 ☽ □ ♃
08 33 ☿ St D
09 53 ☉ ∥ ♇
10 05 ☽ □ ♇
12 42 ☽ ⊼ ♆
13 13 ☽ ☍ ♄
13 59 ☽ ∥ ♅
15 58 ☽ ⚹ ♅
16 33 ☽ ☌ ☿
19 30 ☽ ⊥ ♀
20 14 ☽ □ ♂
22 33 ☽ ± ♆

08 Friday
05 07 ☽ ⊻ ♀
13 11 ☽ ⚼ ♆
13 51 ☽ Q ♃
15 32 ☽ Q ♇
16 30 ☽ ∠ ♅
21 29 ☽ ☌ ☉

09 Saturday
07 02 ☽ ⊥ ♅
09 21 ☽ ⚹ ♃
10 52 ☽ ⚹ ♇
13 18 ☽ △ ♆
13 38 ☽ ⊼ ♄
16 40 ☽ ⊻ ♅
17 24 ☽ ⊻ ☿
22 28 ☽ △ ♂
23 11 ☽ ± ♄

10 Sunday
03 07 ☽ ⊥ ☿
04 05 ☽ ∥ ♅
04 18 ☽ ☌ ♀
09 34 ☽ ∠ ♃
10 52 ☽ ∠ ♇
13 30 ☽ ⚼ ♄
17 44 ☽ ∠ ☿
23 20 ☽ ⚼ ♂

11 Monday
00 05 ☽ ⊥ ♃
00 47 ☽ ⊻ ☉
01 16 ☽ ⊥ ♇
07 09 ☽ # ♄
07 31 ☿ ⊥ ♀
09 49 ☽ ⊻ ♃
10 55 ☽ ⊻ ♇
11 04 ☽ ⊥ ☉
13 16 ☽ □ ♆
13 26 ☽ △ ♄
14 19 ☽ ± ♂
15 29 ☽ ∥ ☿
16 53 ☽ ☌ ♅
18 18 ☽ ⚹ ☿

12 Tuesday
00 22 ☽ ⊼ ♂
02 37 ☽ ∠ ☉
03 13 ☽ ⊻ ♀
05 38 ☽ ∥ ♃
12 25 ☽ # ♂
12 43 ☽ ⊥ ♀
14 18 ☽ Q ☿
18 30 ☽ Q ♆

13 Wednesday
03 04 ☽ ∠ ♀
04 56 ☽ ⚹ ☉
06 18 ☽ # ♆
11 20 ☽ ☌ ♃
11 52 ☉ ∥ ♀
12 03 ☽ ☌ ♇
14 23 ♄ ⊻ ♆
14 23 ☽ □ ♄
14 23 ☽ ⚹ ♆
18 22 ☽ ⊻ ♅
20 25 ♀ △ ♂
21 08 ☽ □ ☿

14 Thursday
02 31 ☽ Q ☉
03 24 ☽ ⚹ ♀
03 55 ☽ ☍ ♂
04 33 ☽ ⊥ ♅
15 50 ☽ ∠ ♆
20 03 ☽ ∠ ♅
20 54 ☽ Q ♄
23 16 ♂ ± ♅
23 17 ☽ Q ♀

15 Friday
07 28 ☽ ⊥ ♆
10 23 ☉ ⊥ ♅
11 56 ☽ □ ☉
15 25 ☽ ⊻ ♃
15 41 ☽ ⊻ ♇
17 51 ☽ ⚹ ♄
18 02 ☽ ⊻ ♆
22 31 ☽ ⚹ ♅

16 Saturday
00 37 ♀ ∥ ♇
02 13 ☽ ⊥ ♃
02 23 ☽ ⊥ ♇
03 33 ☽ △ ☿
05 53 ☽ □ ♀
10 32 ☽ ⊼ ♂
17 45 ♃ ☌ ♇
18 40 ☽ ∠ ♇
18 41 ☽ ∠ ♃
20 16 ☽ Q ♅
20 45 ☽ ∠ ♄
21 00 ☽ ∥ ♆
21 53 ☽ ± ♂

17 Sunday
02 36 ☿ ± ♆
05 25 ☿ ⊻ ♀
08 24 ☽ ⚼ ☿
12 40 ☽ ∥ ♂
13 15 ☽ ⊥ ♄
15 11 ☽ ⚼ ♂
17 45 ♀ ± ♄
18 11 ☉ ⚹ ♇
22 16 ☉ ⚹ ♃
22 27 ☽ ⚹ ♇
22 46 ☽ ⚹ ♃
22 48 ☽ △ ☉
22 59 ☽ # ♃

18 Monday
00 25 ☽ ⊻ ♄
00 50 ☽ ☌ ♆
02 14 ☽ ± ☿
05 51 ☽ □ ♅
10 58 ☽ △ ♀
11 25 ☿ # ♄
14 21 ☽ ⊼ ☿
19 22 ☉ ⊼ ♄
20 39 ☽ △ ♂
21 13 ☽ Q ♇
21 47 ☽ Q ♃

19 Tuesday
01 50 ☉ △ ♆
02 29 ☽ ∥ ♄
04 03 ☽ # ☿
05 38 ☽ ⚼ ☉
14 24 ☽ ⚼ ♀

20 Wednesday
00 32 ☽ ± ☉
07 06 ☽ ± ♀
08 08 ☽ □ ♇
09 06 ☽ □ ♃
09 49 ☽ ☌ ♄
10 21 ☽ # ♅
10 29 ☽ ⊻ ♆
13 16 ☽ ⊼ ☉
16 01 ☽ △ ♅
18 18 ☽ ⊼ ♀
22 17 ☽ ⊥ ♆

21 Thursday
03 51 ☽ # ♀
05 19 ☽ ☍ ☿
09 37 ☽ □ ♂
16 07 ☽ ∠ ♆
21 53 ☽ ⚼ ♅

22 Friday
02 03 ♃ □ ♄
02 18 ☉ ⊻ ♅
06 09 ☉ ☌ ♀
11 12 ♀ ⊻ ♅
15 59 ☽ ± ♅
19 52 ☽ △ ♇
21 12 ☽ ⊻ ♄
21 33 ☽ △ ♃
22 07 ☽ ⚹ ♆
23 07 ☽ Q ♂

23 Saturday
03 11 ☽ ☍ ♀
04 07 ☽ ⊼ ♅
06 14 ☽ ☍ ☉
09 14 ☽ ⊥ ♄
11 55 ☽ # ♀
19 39 ☽ # ♅
22 19 ☽ Q ♆
23 26 ☽ ⊼ ☿

24 Sunday
00 20 ☽ ⚹ ♂
02 11 ☽ ⚼ ♇
03 21 ☽ ∠ ♄
04 15 ☽ ⚼ ♃
10 59 ☽ # ☿
12 39 ♃ ⚹ ♆
12 55 ☽ ± ☿
13 53 ☿ □ ♂
18 00 ☉ ± ♄
20 27 ☽ ± ♇
22 45 ☽ ± ♃

25 Monday
04 52 ☽ ∥ ♄
07 58 ☽ ∠ ♂
08 38 ☽ ⊼ ♇
09 16 ☽ ⚼ ☿
09 36 ☽ ⚹ ♄
10 45 ☽ □ ♆
11 05 ☽ ⊼ ♃
12 53 ☽ ⊼ ♀
17 07 ☽ ☍ ♅
20 23 ♀ ∥ ♅
23 02 ☿ ⚼ ♆

26 Tuesday
00 16 ☽ ⊼ ☉
00 27 ☽ ± ♀
02 57 ☽ ⊥ ♂
09 48 ☽ Q ♄
13 03 ☽ # ♃
13 28 ☽ ± ☉
15 35 ☽ ⊻ ♂
16 28 ♀ ⚹ ♃
17 48 ☽ ⚼ ♀
19 22 ☽ △ ☿

27 Wednesday
04 49 ☿ ∥ ♀
05 22 ☽ ∥ ♂
07 58 ♀ △ ♆
09 13 ☽ ⚼ ☉
11 07 ☿ Q ♇
14 19 ☽ ∥ ♆
21 19 ☽ ☍ ♇
21 55 ☽ □ ♄
22 35 ☽ △ ♀
22 59 ☿ ∥ ♅
23 16 ☽ △ ♆

28 Thursday
00 28 ☽ ☍ ♃
05 51 ☽ ⊼ ♅
07 21 ♂ ⚼ ♆
09 13 ☿ ∠ ♅
14 20 ♀ ⊼ ♄
16 01 ☿ Q ♃
17 46 ☽ △ ☉
17 51 ☽ ± ♅

29 Friday
00 54 ♀ ⚹ ♇
05 02 ☽ ⚼ ♆
05 52 ☽ ☌ ♂
11 40 ☽ ⚼ ♅
15 08 ☽ □ ☿
22 38 ☽ ± ♆

30 Saturday
07 10 ☽ □ ♀
08 29 ☽ ⊼ ♇
08 44 ☽ △ ♄
10 14 ☽ ⊼ ♆
12 14 ☽ ⊼ ♃
16 51 ☽ △ ♅
20 00 ☽ ± ♇
23 51 ☽ ± ♃

JULY 1956

LONGITUDES

Date	Sidereal time h m s	Sun ☉ ° ' "	Moon ☽ ° ' "	Moon ☽ 24.00 ° ' "	Mercury ☿ ° '	Venus ♀ ° '	Mars ♂ ° '	Jupiter ♃ ° '	Saturn ♄ ° '	Uranus ♅ ° '	Neptune ♆ ° '	Pluto ♇ ° '
01	06 38 02	09 ♋ 35 51	11 ♈ 14 11	17 ♈ 41 14	20 ♊ 44	25 ♊ 29	14 ♓ 04	28 ♌ 53	26 ♏ 50	01 ♌ 10	27 ♎ 40	26 ♌ 47
02	06 41 59	10 33 04	24 ♈ 13 55	00 ♉ 52 39	22 22	25 R 02	14 29	29 03	26 R 47	01 13	27 R 39	26 49
03	06 45 55	11 30 16	07 ♉ 37 44	14 ♉ 29 27	24 03	24 37	14 54	29 14	26 45	01 17	27 39	26 50
04	06 49 52	12 27 29	21 ♉ 27 54	28 ♉ 33 04	25 48	24 14	15 19	29 24	26 42	01 20	27 39	26 52
05	06 53 49	13 24 42	05 ♊ 44 48	13 ♊ 02 44	27 36	23 53	15 43	29 35	26 40	01 24	27 39	26 53
06	06 57 45	14 21 56	20 ♊ 26 19	27 ♊ 54 49	29 ♊ 27	23 34	16 06	29 46	26 38	01 28	27 39	26 55
07	07 01 42	15 19 09	05 ♋ 27 19	13 ♋ 02 42	01 ♋ 21	23 18	16 30	29 ♌ 57	26 36	01 31	27 39	26 56
08	07 05 38	16 16 23	20 ♋ 39 46	28 ♋ 17 12	03 17	23 04	16 52	00 ♍ 08	26 33	01 35	27 39	26 58
09	07 09 35	17 13 37	05 ♌ 53 42	13 ♌ 27 57	05 16	22 53	17 14	00 19	26 31	01 38	27 D 39	26 59
10	07 13 31	18 10 51	20 ♌ 58 45	28 ♌ 25 02	07 17	22 44	17 36	00 30	26 29	01 42	27 39	27 01
11	07 17 28	19 08 05	05 ♍ 45 53	13 ♍ 00 37	09 20	22 37	17 57	00 41	26 28	01 45	27 39	27 02
12	07 21 24	20 05 18	20 ♍ 08 44	27 ♍ 09 56	11 25	22 33	18 18	00 52	26 26	01 49	27 39	27 04
13	07 25 21	21 02 32	04 ♎ 04 07	10 ♎ 51 22	13 31	22 30	18 38	01 03	26 24	01 53	27 39	27 06
14	07 29 18	21 59 46	17 ♎ 31 52	24 ♎ 05 58	15 38	22 D 31	18 58	01 15	26 23	01 56	27 39	27 07
15	07 33 14	22 57 00	00 ♏ 34 04	06 ♏ 56 39	17 46	22 33	19 17	01 26	26 21	02 00	27 39	27 09
16	07 37 11	23 54 14	13 ♏ 14 15	19 ♏ 27 24	19 54	22 38	19 35	01 38	26 20	02 04	27 40	27 11
17	07 41 07	24 51 28	25 ♏ 36 38	01 ♐ 42 31	22 03	22 45	19 53	01 49	26 18	02 07	27 40	27 12
18	07 45 04	25 48 43	07 ♐ 45 33	13 ♐ 46 14	24 11	22 54	20 10	02 01	26 17	02 11	27 40	27 14
19	07 49 00	26 45 57	19 ♐ 45 02	25 ♐ 42 22	26 19	23 05	20 27	02 12	26 16	02 15	27 40	27 16
20	07 52 57	27 43 12	01 ♑ 38 38	07 ♑ 34 10	28 ♋ 27	23 19	20 43	02 24	26 15	02 18	27 41	27 17
21	07 56 53	28 40 28	13 ♑ 29 18	19 ♑ 24 19	00 ♌ 34	23 34	20 59	02 36	26 14	02 22	27 41	27 19
22	08 00 50	29 ♋ 37 44	25 ♑ 19 28	01 ♒ 14 58	02 40	23 51	21 14	02 48	26 13	02 26	27 41	27 21
23	08 04 47	00 ♌ 35 00	07 ♒ 11 03	13 ♒ 07 54	04 44	24 10	21 28	03 00	26 13	02 29	27 42	27 23
24	08 08 43	01 32 17	19 ♒ 05 44	25 ♒ 04 44	06 48	24 31	21 42	03 12	26 12	02 33	27 42	27 25
25	08 12 40	02 29 35	01 ♓ 05 07	07 ♓ 07 06	08 50	24 53	21 54	03 24	26 11	02 37	27 43	27 26
26	08 16 36	03 26 54	13 ♓ 10 56	19 ♓ 16 53	10 50	25 18	22 07	03 36	26 11	02 40	27 43	27 28
27	08 20 33	04 24 13	25 ♓ 25 15	01 ♈ 36 23	12 49	25 44	22 18	03 48	26 10	02 44	27 44	27 30
28	08 24 29	05 21 33	07 ♈ 50 37	14 ♈ 08 22	14 47	26 11	22 29	04 00	26 10	02 48	27 45	27 32
29	08 28 26	06 18 55	20 ♈ 30 02	26 ♈ 56 02	16 43	26 40	22 39	04 12	26 10	02 52	27 45	27 34
30	08 32 22	07 16 17	03 ♉ 26 48	10 ♉ 02 47	18 37	27 11	22 49	04 24	26 10	02 55	27 46	27 35
31	08 36 19	08 ♌ 13 41	16 ♉ 44 20	23 ♉ 31 49	20 ♌ 30	27 ♊ 43	22 ♓ 57	04 ♍ 36	26 ♏ 10	02 ♌ 59	27 ♎ 47	27 ♌ 37

Date	Moon True ☊ ° '	Moon Mean ☊ ° '	Moon ☽ Latitude ° '
01	07 ♐ 41	06 ♐ 26	04 N 21
02	07 D 42	06 22	03 36
03	07 43	06 19	02 37
04	07 44	06 16	01 28
05	07 45	06 13	00 N 11
06	07 R 45	06 10	01 S 09
07	07 42	06 07	02 25
08	07 39	06 03	03 32
09	07 33	06 00	04 24
10	07 28	05 57	04 57
11	07 22	05 54	05 10
12	07 17	05 51	05 02
13	07 14	05 48	04 37
14	07 13	05 44	03 57
15	07 D 13	05 41	03 05
16	07 14	05 38	02 06
17	07 15	05 35	01 S 03
18	07 R 16	05 32	00 N 03
19	07 15	05 28	01 07
20	07 12	05 25	02 07
21	07 07	05 22	03 02
22	07 00	05 19	03 49
23	06 51	05 16	04 26
24	06 41	05 13	04 51
25	06 30	05 09	05 03
26	06 21	05 06	05 03
27	06 13	05 03	04 48
28	06 07	05 00	04 20
29	06 04	04 57	03 38
30	06 03	04 53	02 45
31	06 ♐ 03	04 ♐ 50	01 N 41

DECLINATIONS

Date	Sun ☉ ° '	Moon ☽ ° '	Mercury ☿ ° '	Venus ♀ ° '	Mars ♂ ° '	Jupiter ♃ ° '	Saturn ♄ ° '	Uranus ♅ ° '	Neptune ♆ ° '	Pluto ♇ ° '
01	23 N 06	08 N 27	21 N 40	19 N 24	09 S 59	12 N 47	17 S 22	20 N 26	09 S 00	22 N 32
02	23 01	12 44	21 58	19 13	09 51	12 44	17 22	20 26	09 00	22 31
03	22 57	16 32	22 16	19 04	09 44	12 40	17 22	20 25	09 00	22 31
04	22 52	19 33	22 32	18 54	09 38	12 36	17 21	20 24	09 00	22 30
05	22 46	21 27	22 47	18 46	09 31	12 32	17 21	20 23	09 00	22 29
06	22 40	21 57	23 01	18 38	09 25	12 28	17 21	20 23	09 00	22 29
07	22 34	20 55	23 12	18 31	09 18	12 25	17 20	20 22	09 00	22 28
08	22 27	18 22	23 22	18 24	09 12	12 21	17 20	20 21	09 00	22 27
09	22 20	14 32	23 30	18 18	09 06	12 17	17 20	20 20	09 00	22 27
10	22 13	09 48	23 35	18 13	09 00	12 13	17 19	20 19	09 00	22 26
11	22 05	04 N 36	23 38	18 08	08 55	12 09	17 19	20 19	09 00	22 25
12	21 56	00 S 44	23 38	18 04	08 50	12 05	17 19	20 18	09 01	22 25
13	21 48	05 51	23 36	18 01	08 44	12 01	17 19	20 17	09 01	22 24
14	21 39	10 31	23 31	17 58	08 40	11 57	17 19	20 16	09 01	22 23
15	21 29	14 34	23 23	17 56	08 35	11 52	17 19	20 15	09 01	22 23
16	21 20	17 49	23 12	17 54	08 30	11 48	17 19	20 14	09 01	22 22
17	21 10	20 11	22 59	17 53	08 26	11 44	17 18	20 14	09 01	22 21
18	20 59	21 34	22 43	17 53	08 22	11 40	17 18	20 13	09 01	22 21
19	20 48	21 56	22 24	17 53	08 18	11 36	17 18	20 12	09 01	22 20
20	20 37	21 19	22 03	17 53	08 14	11 32	17 18	20 11	09 02	22 19
21	20 26	19 44	21 40	17 54	08 11	11 27	17 18	20 10	09 02	22 19
22	20 14	17 20	21 15	17 55	08 08	11 23	17 18	20 10	09 02	22 18
23	20 02	14 12	20 47	17 56	08 05	11 19	17 19	20 09	09 02	22 17
24	19 49	10 29	20 18	17 58	08 02	11 14	17 19	20 08	09 02	22 16
25	19 36	06 21	19 47	18 00	08 00	11 10	17 19	20 07	09 03	22 16
26	19 23	01 S 57	19 14	18 03	07 58	11 06	17 19	20 06	09 03	22 15
27	19 10	02 N 35	18 40	18 05	07 56	11 01	17 19	20 05	09 03	22 14
28	18 56	07 05	18 05	18 08	07 54	10 57	17 19	20 04	09 03	22 14
29	18 42	11 23	17 28	18 11	07 53	10 52	17 19	20 04	09 04	22 13
30	18 27	15 15	16 51	18 14	07 51	10 48	17 20	20 03	09 04	22 12
31	18 N 13	18 N 28	16 N 12	18 N 17	07 S 51	10 N 43	17 S 20	20 N 02	09 S 04	22 N 12

ZODIAC SIGN ENTRIES

Date	h	m	Planets
02	22	26	☽ ♉
05	02	26	☽ ♊
06	19	02	☿ ♋
07	03	20	☽ ♋
07	19	01	♃ ♍
09	02	42	☽ ♌
11	02	34	☽ ♍
13	04	54	☽ ♎
15	10	56	☽ ♏
17	20	38	☽ ♐
20	08	40	☽ ♑
21	05	35	☿ ♌
22	21	20	☉ ♌
22	21	28	☽ ♒
25	09	50	☽ ♓
27	20	54	☽ ♈
30	05	40	☽ ♉

LATITUDES

Date	Mercury ☿ ° '	Venus ♀ ° '	Mars ♂ ° '	Jupiter ♃ ° '	Saturn ♄ ° '	Uranus ♅ ° '	Neptune ♆ ° '	Pluto ♇ ° '
01	01 S 28	03 S 58	04 S 01	00 N 59	02 N 08	00 N 33	01 N 45	10 N 37
04	00 50	04 25	04 10	00 59	02 08	00 33	01 45	10 37
07	00 S 14	04 46	04 19	00 59	02 07	00 33	01 45	10 36
10	00 N 20	05 02	04 28	00 59	02 06	00 33	01 45	10 36
13	00 50	05 13	04 37	00 58	02 06	00 33	01 45	10 35
16	01 15	05 20	04 46	00 58	02 05	00 33	01 45	10 35
19	01 33	05 24	04 55	00 58	02 04	00 33	01 44	10 35
22	01 43	05 24	05 04	00 58	02 04	00 33	01 44	10 35
25	01 47	05 21	05 13	00 58	02 03	00 33	01 44	10 34
28	01 45	05 15	05 21	00 58	02 02	00 33	01 44	10 34
31	01 N 38	05 S 08	05 S 30	00 N 58	02 N 01	00 N 33	01 N 44	10 N 34

DATA

Julian Date	2435656
Delta T	+32 seconds
Ayanamsa	23° 15' 14"
Synetic vernal point	05° ♓ 51' 46"
True obliquity of ecliptic	23° 26' 38"

MOON'S PHASES, APSIDES AND POSITIONS ☽

Date	h	m	Phase	Longitude ° '	Eclipse Indicator
01	08	41	☾	09 ♈ 28	
08	04	38	●	15 ♋ 59	
14	20	47	☽	22 ♎ 21	
22	21	29	○	00 ♒ 00	
30	19	31	☾	07 ♉ 34	

Day	h	m		
08	11	20	Perigee	
22	10	55	Apogee	
06	08	04	Max dec	21° N 59'
12	08	42	0S	
19	08	50	Max dec	21° S 57'
26	22	20	0N	

LONGITUDES

Date	Chiron ⚷ ° '	Ceres ⚳ ° '	Pallas ⚴ ° '	Juno ⚵ ° '	Vesta ⚶ ° '	Black Moon Lilith ⚸ ° '
01	10 ♒ 38	15 ♈ 29	05 ♓ 11	16 ♑ 14	00 ♏ 15	23 ♑ 09
11	10 ♒ 10	17 ♈ 30	04 ♓ 14	13 ♑ 54	02 ♏ 20	24 ♑ 16
21	09 ♒ 38	19 ♈ 08	02 ♓ 43	11 ♑ 37	04 ♏ 57	25 ♑ 23
31	09 ♒ 05	20 ♈ 18	00 ♓ 42	09 ♑ 37	08 ♏ 02	26 ♑ 30

All ephemeris data is given at 12.00 UT and the Moon's longitude is additionally given for 24.00 UT

ASPECTARIAN

h m	Aspects
01 Sunday	
06 36	☽ Q ☿
08 41	☽ □ ☉
13 02	☽ bQ ♇
13 07	☽ bQ ♄
15 01	☽ # ♆
16 03	☽ Q ♀
17 00	☽ bQ ♃
17 28	☽ ⊻ ♂
20 06	☽ # ♂
02 Monday	
02 52	♀ ⊥ ♅
04 56	☽ ⊥ ♂
05 08	♄ □ ♇
05 44	☽ ± ♄
08 06	☽ ⚹ ☿
11 57	☽ ∥ ♃
13 25	☽ ⚹ ♀
16 38	☽ ⊼ ♄
16 41	☽ △ ♇
18 12	☽ ☍ ♆
20 25	☽ Q ☉
20 51	☽ △ ♃
21 49	☽ ∠ ♂
03 Tuesday	
00 40	☽ □ ♅
14 52	☽ ∠ ☿
15 23	☽ ∠ ♀
17 53	☽ # ♄
17 53	☉ bQ ♄
18 18	☿ ☌ ♀
19 19	☽ ⚹ ☉
20 31	☉ ∠ ♇
04 Wednesday	
01 06	☽ ⚹ ♂
05 30	☿ ⊥ ♅
06 31	☽ ∥ ♀
06 36	☽ ⊥ ♀
08 21	☽ Q ♅
08 45	☽ ⊥ ☿
08 54	☿ ∥ ♇
16 35	☽ ⊻ ♀
20 25	☽ ⊻ ☿
20 52	☽ ☍ ♄
20 55	☽ ∥ ♅
21 10	☽ □ ♇
22 12	☽ Q ♂
22 29	☽ ⊼ ♆
22 53	☽ ∠ ☉
23 54	☿ ⊼ ♄
05 Thursday	
01 36	☽ □ ♃
02 23	☿ ⚹ ♇
04 44	☽ ⚹ ♅
08 31	☽ ± ♆
10 37	☉ ∥ ☿
12 37	☿ △ ♆
14 57	☽ ⊥ ☉
23 21	☽ bQ ♆
06 Friday	
01 28	☽ ⊻ ☉
03 01	☽ Q ♇
04 48	☽ □ ♂
05 32	☽ ∠ ♅
07 37	☽ Q ♃
16 29	☿ ⚹ ♃
16 57	☽ ☌ ♀
20 06	☽ ⊥ ♅
21 55	☽ ⊼ ♄
22 25	☽ ⚹ ♇
23 34	☽ △ ♆
07 Saturday	
00 26	☉ ∠ ♃
03 08	☽ ⚹ ♃
04 31	☽ ☌ ☿
05 43	☽ ⊻ ♅
07 28	☽ ± ♄
13 51	♂ bQ ♅
14 15	☿ ⊻ ♅
18 34	☽ ∥ ♅
21 41	☽ bQ ♄
22 16	☽ ∠ ♇
08 Sunday	
03 11	☽ ∠ ♃
03 14	☿ ± ♄
04 38	☽ ☌ ☉
05 53	☽ △ ♂
10 48	☉ ∥ ♇
11 39	☽ ∥ ♀
12 28	☽ ⊥ ♇
15 44	☽ ⊻ ♀
17 31	☽ ⊥ ♃
19 11	☽ # ♄
21 15	☽ △ ♄
21 56	☽ ⊻ ♇
22 59	☽ □ ♆
09 Monday	
01 04	☽ ⊥ ♀
03 05	☽ ⊻ ♃
05 15	☽ ☌ ♅
06 05	☽ bQ ♂
06 11	♆ St D
10 51	☽ ⊻ ☿
12 33	☉ △ ♂
15 06	☽ ∠ ♀
20 40	☽ ± ♂
21 48	☽ ⊥ ☿
10 Tuesday	
00 03	☽ ∥ ♃
03 28	☽ Q ♆
06 28	☽ ⊼ ♂
07 13	☽ ⊻ ☉
12 34	♂ ∥ ♆
14 25	☽ ∠ ☿
14 47	☽ ⚹ ♀
15 47	☽ # ♆
15 51	☽ # ♂
17 30	☽ ⊥ ☉
20 52	☽ □ ♄
21 44	☽ ☌ ♇
22 45	☽ ⚹ ♆
11 Wednesday	
03 35	☽ ☌ ♃
05 25	☽ ⊻ ♅
09 08	☽ ∠ ☉
10 07	☽ Q ♀
15 18	☽ ⊥ ♅
18 52	☽ ⚹ ☿
23 24	☽ ∠ ♆
12 Thursday	
02 24	☽ Q ♄
06 21	☽ ∠ ♅
08 48	☽ ☍ ♂
11 54	☽ ⚹ ☉
12 15	☿ bQ ♄
14 33	☽ ⊥ ♆
16 04	☽ □ ♀
18 32	☽ Q ☿
19 38	☿ ∠ ♇
22 43	☽ ⚹ ♄
23 51	☽ ⊻ ♇
13 Friday	
00 50	☽ ⊻ ♆
06 40	☽ ⊻ ♃
08 09	☽ ⚹ ♅
10 04	☽ Q ☉
10 17	☽ ⊥ ♇
13 20	☉ ⊥ ♇
17 20	☽ ⊥ ♃
21 20	♀ St D
14 Saturday	
00 57	☽ ∠ ♄
02 14	☽ ∠ ♇
02 15	☽ ∥ ♂
03 55	☽ ∥ ♆
05 29	☽ Q ♅
07 55	☽ □ ☿
09 39	☽ ∠ ♃
14 40	☽ ⊼ ♂
17 10	☽ ⊥ ♄
19 38	☿ ∠ ♃
19 50	☽ # ♃
20 47	☽ □ ☉
21 06	☽ △ ♀
15 Sunday	
01 27	☉ ⊻ ♀
01 56	☽ ± ♂
04 10	☽ ⊻ ♄
05 37	☽ ⚹ ♇
06 34	☽ ☌ ♆
13 39	☽ ⚹ ♃
14 42	☽ □ ♅
19 08	☽ bQ ♂
16 Monday	
01 14	☽ bQ ♀
04 13	☽ Q ♇
07 46	☽ ∥ ♄
07 51	☿ △ ♂
12 44	☽ # ♀
12 46	☽ Q ♃
18 35	☽ ± ♀
17 Tuesday	
00 33	☽ △ ♂
02 27	☿ ⊥ ♇
03 34	☽ △ ☿
06 21	☽ ⊼ ♀
10 24	☽ △ ☉
12 36	☽ # ♅
13 22	☽ ☌ ♄
15 08	☽ □ ♇
16 02	☽ ⊻ ♆
20 25	☿ ⊻ ♀
18 Wednesday	
00 25	☽ □ ♃
00 53	☽ △ ♅
01 13	☽ # ☉
03 52	☽ ⊥ ♆
15 28	☽ bQ ☿
18 23	☉ ⊥ ♃
18 37	☽ bQ ☉
21 48	☽ ∠ ♆
23 42	☉ △ ♄
19 Thursday	
06 56	☽ bQ ♅
10 33	☿ ⊥ ♃
11 22	☿ △ ♄
13 24	☽ ± ☿
13 27	☽ □ ♂
14 13	☽ ± ☉
17 29	☿ ∥ ♇
18 41	♃ ⊻ ♅
18 50	☽ ☍ ♀
21 01	☉ ☌ ☿
22 43	☿ ⊻ ♇
20 Friday	
00 52	☉ ⊻ ♇
01 07	☽ ⊻ ♄
01 09	☽ ± ♅
03 11	☽ △ ♇
03 14	☿ □ ♆
03 22	☽ ⊼ ☉
03 59	☽ ⚹ ♆
04 08	☽ ⊼ ☿
10 57	☉ □ ♆
13 14	☽ ⊥ ♄
13 21	☽ ⊼ ♅
13 34	☽ △ ♃
23 00	☿ ⊥ ♀
21 Saturday	
01 52	☽ # ☉
02 39	☽ Q ♂
04 17	☽ Q ♆
06 31	☽ # ♅
07 26	☽ ∠ ♄
09 37	☽ bQ ♇
20 29	☽ bQ ♃
22 Sunday	
03 31	☽ ⚹ ♂
03 55	☽ ± ♇
06 50	☽ # ♀
08 56	☽ ⊼ ♀
09 16	☿ ☌ ♅
12 09	☽ ∥ ♄
13 43	☿ ⊻ ♃
13 49	☽ ⚹ ♄
15 02	☽ ± ♃
16 07	☽ ⊼ ♇
16 48	☽ □ ♆
20 10	☉ ⊥ ♀
21 25	☽ ± ♀
21 29	☽ ☍ ☉
21 29	☉ ∥ ♅
23 Monday	
02 28	☽ ☍ ♅
03 23	☽ ⊼ ♃
06 01	☽ ☍ ☿
10 31	☽ ∠ ♂
14 04	☽ Q ♄
16 07	☽ bQ ♀
24 Tuesday	
05 02	☽ ⊥ ♂
07 18	☽ # ♃
10 40	☿ bQ ♂
13 34	♂ ± ♆
17 19	☽ ⊻ ♂
20 11	☿ ∥ ♅
20 39	☽ ∥ ♆
23 13	☽ △ ♀
25 Wednesday	
02 14	☽ □ ♄
02 38	☽ ∥ ♂
04 42	☽ ☍ ♇
05 16	☽ △ ♆
15 03	☽ ⊼ ☉
15 04	☽ ⊼ ♅
15 14	☉ ☌ ♅
16 40	☽ ☍ ♃
26 Thursday	
01 05	☉ ∥ ☿
03 02	☽ ± ♅
03 47	☿ ∠ ♀
03 59	☽ ± ☉
06 27	☽ ⊼ ☿
11 06	☽ bQ ♆
16 35	☉ ⊻ ♃
20 37	☽ ± ☿
20 53	☽ bQ ♅
23 15	☽ bQ ☉
27 Friday	
04 48	☽ ± ♆
05 49	☽ ☌ ♂
12 37	☽ □ ♀
13 28	☽ △ ♄
16 03	☽ ⊼ ♇
16 30	☽ ⊼ ♆
17 34	☽ bQ ☿
28 Saturday	
02 15	☽ △ ♅
03 41	☽ ± ♇
04 29	☽ ⊼ ♃
06 50	☽ △ ☉
10 03	☿ ∥ ♀
11 10	♀ ⊼ ♄
16 11	☽ ± ♃
16 23	☽ # ♂
18 21	☽ bQ ♄
20 58	☽ bQ ♇
22 48	☽ # ♆
23 57	☿ Q ♆
29 Sunday	
00 34	☽ Q ♀
03 36	☽ △ ☿
09 07	☽ ∥ ♃
09 31	☽ bQ ♃
11 09	☿ ± ♂
11 22	☽ ± ♄
16 05	☽ ⊻ ♂
17 43	☿ # ♄
22 19	♀ ⊥ ♅
22 35	☽ ⊼ ♄
23 59	☽ ⚹ ♀
30 Monday	
01 12	☽ △ ♇
01 32	☽ ☍ ♆
03 22	☽ ⊥ ♂
11 02	☽ □ ♅
13 46	☽ △ ♃
18 35	♄ St D
19 31	☽ □ ☉
20 02	☽ ∠ ♂
21 23	☽ ∥ ☿
31 Tuesday	
02 51	☽ # ♄
03 54	☉ bQ ♂
04 30	☽ ∠ ♀
05 59	☉ ∥ ♀
07 50	♀ ⚹ ♇
10 04	☽ ∥ ☉
10 31	☽ ∥ ♀
14 58	♀ △ ♆
19 33	☽ Q ♅
19 44	☽ □ ☿
21 10	☽ ⊥ ♀
23 06	☽ ⚹ ♂

AUGUST 1956

LONGITUDES

Date	Sidereal time h m s	Sun ☉ ° ' "	Moon ☽ ° ' "	Moon ☽ 24.00 ° ' "	Mercury ☿ ° '	Venus ♀ ° '	Mars ♂ ° '	Jupiter ♃ ° '	Saturn ♄ ° '	Uranus ♅ ° '	Neptune ♆ ° '	Pluto ♇ ° '
01	08 40 16	09 ♌ 11 06	00 ♊ 25 29	07 ♊ 25 32	22 ♌ 21	28 ♊ 16	23 ♓ 05	04 ♍ 49	26 ♏ 10	03 ♌ 03	27 ♎ 47	27 ♌ 39
02	08 44 12	10 08 32	14 ♊ 31 58	21 ♊ 44 41	24 11	28 51	23 12	05 01	26 D 10	03 06	27 48	27 41
03	08 48 09	11 05 59	29 ♊ 03 24	06 ♋ 27 34	25 58	29 ♊ 26	23 18	05 13	26 11	03 10	27 49	27 43
04	08 52 05	12 03 28	13 ♋ 56 32	21 ♋ 29 21	27 45	00 ♋ 03	23 24	05 26	26 11	03 14	27 50	27 45
05	08 56 02	13 00 57	29 ♋ 04 56	06 ♌ 42 03	29 ♌ 29	00 42	23 28	05 38	26 12	03 17	27 51	27 47
06	08 59 58	13 58 28	14 ♌ 19 20	21 ♌ 55 25	01 ♍ 12	01 21	23 32	05 51	26 12	03 21	27 51	27 49
07	09 03 55	14 56 00	29 ♌ 28 56	06 ♍ 58 37	02 54	02 01	23 35	06 03	26 13	03 25	27 52	27 51
08	09 07 51	15 53 32	14 ♍ 23 21	21 ♍ 42 10	04 34	02 43	23 37	06 16	26 14	03 28	27 53	27 52
09	09 11 48	16 51 06	28 ♍ 54 22	05 ♎ 59 27	06 12	03 25	23 39	06 28	26 15	03 32	27 54	27 54
10	09 15 45	17 48 40	12 ♎ 57 08	19 ♎ 47 20	07 49	04 09	23 39	06 41	26 16	03 35	27 55	27 56
11	09 19 41	18 46 15	26 ♎ 30 12	03 ♏ 05 59	09 24	04 53	23 R 39	06 54	26 17	03 39	27 56	27 58
12	09 23 38	19 43 51	09 ♏ 35 06	15 ♏ 58 02	10 58	05 38	23 38	07 06	26 18	03 43	27 57	28 00
13	09 27 34	20 41 29	22 ♏ 15 23	28 ♏ 27 44	12 30	06 24	23 36	07 19	26 19	03 46	27 58	28 02
14	09 31 31	21 39 07	04 ♐ 35 45	10 ♐ 40 04	14 00	07 11	23 33	07 32	26 21	03 50	27 59	28 04
15	09 35 27	22 36 46	16 ♐ 41 20	22 ♐ 40 11	15 29	07 59	23 30	07 44	26 22	03 53	28 01	28 06
16	09 39 24	23 34 26	28 ♐ 37 11	04 ♑ 32 54	16 56	08 47	23 26	07 57	26 24	03 57	28 02	28 08
17	09 43 20	24 32 07	10 ♑ 27 51	16 ♑ 22 29	18 22	09 36	23 21	08 10	26 25	04 00	28 03	28 10
18	09 47 17	25 29 50	22 ♑ 17 15	28 ♑ 12 29	19 46	10 26	23 15	08 23	26 27	04 04	28 04	28 12
19	09 51 14	26 27 33	04 ♒ 08 32	10 ♒ 05 39	21 09	11 17	23 08	08 36	26 29	04 07	28 05	28 14
20	09 55 10	27 25 18	16 ♒ 04 04	22 ♒ 04 00	22 29	12 08	23 01	08 49	26 31	04 11	28 07	28 16
21	09 59 07	28 23 04	28 ♒ 05 34	04 ♓ 08 56	23 48	13 00	22 53	09 01	26 33	04 14	28 08	28 18
22	10 03 03	29 ♌ 20 52	10 ♓ 14 13	16 ♓ 21 29	25 06	13 53	22 44	09 14	26 35	04 18	28 09	28 20
23	10 07 00	00 ♍ 18 41	22 ♓ 30 53	28 ♓ 42 31	26 21	14 46	22 34	09 27	26 37	04 21	28 11	28 22
24	10 10 56	01 16 31	04 ♈ 56 31	11 ♈ 13 02	27 35	15 40	22 24	09 40	26 40	04 25	28 12	28 24
25	10 14 53	02 14 24	17 ♈ 32 16	23 ♈ 54 24	28 46	16 34	22 13	09 53	26 42	04 28	28 14	28 26
26	10 18 49	03 12 18	00 ♉ 19 42	06 ♉ 48 26	29 ♍ 56	17 29	22 02	10 06	26 44	04 31	28 15	28 28
27	10 22 46	04 10 13	13 ♉ 20 54	19 ♉ 57 24	01 ♎ 03	18 25	21 49	10 19	26 47	04 35	28 16	28 30
28	10 26 43	05 08 11	26 ♉ 38 16	03 ♊ 23 47	02 08	19 21	21 37	10 32	26 50	04 38	28 18	28 31
29	10 30 39	06 06 11	10 ♊ 14 13	17 ♊ 09 46	03 11	20 17	21 23	10 45	26 52	04 41	28 19	28 33
30	10 34 36	07 04 12	24 ♊ 10 33	01 ♋ 16 36	04 12	21 14	21 10	10 58	26 55	04 44	28 21	28 35
31	10 38 32	08 ♍ 02 16	08 ♋ 27 47	15 ♋ 43 50	05 ♎ 10	22 ♋ 12	20 ♓ 55	11 ♍ 11	26 ♏ 58	04 ♌ 48	28 ♎ 23	28 ♌ 37

Date	Moon True ☊ ° '	Moon Mean ☊ ° '	Moon ☽ Latitude ° '
01	06 ♐ 03	04 ♐ 47	00 N 30
02	06 R 03	04 44	00 S 45
03	06 01	04 41	01 59
04	05 57	04 38	03 07
05	05 50	04 34	04 03
06	05 42	04 31	04 42
07	05 32	04 28	05 01
08	05 22	04 25	04 59
09	05 13	04 22	04 37
10	05 07	04 19	04 00
11	05 03	04 15	03 09
12	05 01	04 12	02 11
13	05 D 01	04 09	01 07
14	05 01	04 06	00 S 02
15	05 R 00	04 03	01 N 02
16	04 58	03 59	02 02
17	04 54	03 56	02 56
18	04 46	03 53	03 42
19	04 36	03 50	04 19
20	04 24	03 47	04 45
21	04 11	03 44	04 58
22	03 57	03 40	04 58
23	03 44	03 37	04 44
24	03 34	03 34	04 17
25	03 25	03 31	03 36
26	03 20	03 28	02 45
27	03 18	03 25	01 43
28	03 D 17	03 21	00 N 35
29	03 R 17	03 18	00 S 37
30	03 16	03 15	01 48
31	03 ♐ 14	03 ♐ 12	02 S 54

DECLINATIONS

Date	Sun ☉ ° '	Moon ☽ ° '	Mercury ☿ ° '	Venus ♀ ° '	Mars ♂ ° '	Jupiter ♃ ° '	Saturn ♄ ° '	Uranus ♅ ° '	Neptune ♆ ° '	Pluto ♇ ° '
01	17 N 58	20 N 44	15 N 33	18 N 21	07 S 50	10 N 39	17 S 20	20 N 01	09 S 05	22 N 11
02	17 42	21 48	14 53	18 24	07 50	10 34	17 20	20 00	09 05	22 10
03	17 27	21 27	14 12	18 27	07 50	10 30	17 21	19 59	09 05	22 09
04	17 11	19 37	13 31	18 31	07 50	10 25	17 21	19 59	09 06	22 09
05	16 55	16 23	12 50	18 34	07 50	10 21	17 21	19 58	09 06	22 08
06	16 38	12 02	12 08	18 37	07 51	10 16	17 22	19 57	09 06	22 07
07	16 22	06 57	11 26	18 41	07 52	10 11	17 22	19 56	09 07	22 07
08	16 05	01 N 33	10 44	18 44	07 53	10 07	17 23	19 55	09 07	22 06
09	15 47	03 S 48	10 01	18 47	07 55	10 02	17 23	19 54	09 07	22 05
10	15 30	08 47	09 19	18 50	07 57	09 57	17 24	19 54	09 08	22 05
11	15 12	13 10	08 36	18 52	07 59	09 53	17 24	19 53	09 08	22 04
12	14 54	16 45	07 54	18 55	08 01	09 48	17 25	19 52	09 09	22 03
13	14 36	19 25	07 11	18 57	08 03	09 43	17 25	19 51	09 09	22 03
14	14 17	21 06	06 29	19 00	08 06	09 39	17 26	19 50	09 10	22 02
15	13 59	21 45	05 47	19 02	08 09	09 34	17 26	19 49	09 10	22 01
16	13 40	21 25	05 06	19 03	08 12	09 29	17 27	19 49	09 10	22 01
17	13 21	20 06	04 24	19 05	08 15	09 24	17 28	19 48	09 11	22 00
18	13 01	17 57	03 43	19 06	08 19	09 19	17 28	19 47	09 11	21 59
19	12 42	15 01	03 03	19 07	08 23	09 15	17 29	19 46	09 12	21 59
20	12 22	11 29	02 22	19 07	08 27	09 10	17 30	19 45	09 12	21 58
21	12 02	07 28	01 43	19 07	08 31	09 05	17 30	19 45	09 13	21 57
22	11 42	03 S 08	01 04	19 07	08 35	09 00	17 31	19 44	09 13	21 57
23	11 22	01 N 23	00 N 25	19 07	08 39	08 55	17 32	19 43	09 14	21 56
24	11 01	05 53	00 S 13	19 06	08 44	08 50	17 32	19 42	09 14	21 56
25	10 41	10 13	00 50	19 05	08 49	08 46	17 33	19 41	09 15	21 55
26	10 20	14 09	01 26	19 03	08 53	08 41	17 34	19 41	09 16	21 54
27	09 59	17 29	02 02	19 01	08 58	08 36	17 35	19 40	09 16	21 54
28	09 38	19 59	02 36	18 59	09 03	08 31	17 36	19 39	09 17	21 53
29	09 16	21 23	03 10	18 56	09 08	08 26	17 37	19 38	09 17	21 53
30	08 55	21 31	03 43	18 53	09 13	08 21	17 37	19 38	09 18	21 52
31	08 N 33	20 N 16	04 S 14	18 N 49	09 S 18	08 N 16	17 S 38	19 N 37	09 S 18	21 N 51

ZODIAC SIGN ENTRIES

Date	h	m	Planets
01	11	16	☽ ♊
03	13	32	☽ ♋
04	09	49	♀ ♋
05	13	27	☽ ♌
05	19	06	☿ ♍
07	12	50	☽ ♍
09	13	50	☽ ♎
11	18	20	☽ ♏
14	03	00	☽ ♐
16	14	47	☽ ♑
19	03	38	☽ ♒
21	15	47	☽ ♓
23	04	15	☉ ♍
24	02	30	☽ ♈
26	11	23	☽ ♉
26	13	30	☿ ♎
28	17	59	☽ ♊
30	21	51	☽ ♋

LATITUDES

Date	Mercury ☿ ° '	Venus ♀ ° '	Mars ♂ ° '	Jupiter ♃ ° '	Saturn ♄ ° '	Uranus ♅ ° '	Neptune ♆ ° '	Pluto ♇ ° '
01	01 N 34	05 S 05	05 S 32	00 N 58	02 N 01	00 N 33	01 N 44	10 N 34
04	01 21	04 56	05 40	00 58	02 00	00 33	01 43	10 34
07	01 04	04 45	05 48	00 58	02 00	00 33	01 43	10 34
10	00 43	04 33	05 55	00 58	01 59	00 33	01 43	10 34
13	00 N 21	04 20	06 00	00 58	01 58	00 33	01 43	10 34
16	00 S 04	04 06	06 06	00 58	01 57	00 33	01 43	10 34
19	00 31	03 52	06 10	00 58	01 57	00 33	01 43	10 34
22	00 58	03 37	06 12	00 58	01 56	00 33	01 43	10 35
25	01 26	03 22	06 14	00 58	01 55	00 34	01 42	10 35
28	01 55	03 06	06 14	00 58	01 54	00 34	01 42	10 35
31	02 S 23	02 S 50	06 S 13	00 N 58	01 N 54	00 N 34	01 N 42	10 N 36

DATA

Julian Date	2435687
Delta T	+32 seconds
Ayanamsa	23° 15' 19"
Synetic vernal point	05° ♓ 51' 41"
True obliquity of ecliptic	23° 26' 38"

MOON'S PHASES, APSIDES AND POSITIONS ☽

Date	h	m	Phase	Longitude ° '	Eclipse Indicator
06	11	25	●	13 ♌ 57	
13	08	45	☽	20 ♏ 34	
21	12	38	○	28 ♒ 25	
29	04	13	☾	05 ♊ 47	

Day	h	m		
05	20	54	Perigee	
18	16	01	Apogee	
02	18	22	Max dec	21° N 51'
08	18	51	0S	
15	15	35	Max dec	21° S 46'
23	04	41	0N	
30	02	26	Max dec	21° N 37'

LONGITUDES

Date	Chiron ⚷ ° '	Ceres ⚳ ° '	Pallas ⚴ ° '	Juno ⚵ ° '	Vesta ⚶ ° '	Black Moon Lilith ⚸ ° '
01	09 ♒ 02	20 ♈ 24	00 ♓ 29	09 ♑ 26	08 ♏ 22	26 ♑ 36
11	08 ♒ 29	20 ♈ 59	28 ♒ 04	07 ♑ 57	11 ♏ 53	27 ♑ 43
21	07 ♒ 58	21 ♈ 00	25 ♒ 31	07 ♑ 04	15 ♏ 43	28 ♑ 50
31	07 ♒ 30	20 ♈ 25	23 ♒ 01	06 ♑ 48	19 ♏ 49	29 ♑ 57

All ephemeris data is given at 12.00 UT and the Moon's longitude is additionally given for 24.00 UT

ASPECTARIAN

h m Aspects

01 Wednesday
03 11 ☽ ∥ ♅
04 37 ☽ ☍ ♄
05 58 ☽ Q ☉
07 11 ☽ □ ♇
07 26 ☽ ⊼ ♆
08 06 ☽ ⊻ ♀
16 32 ☽ ⚹ ♅
17 48 ☽ ± ♆
19 39 ☽ □ ♃
20 04 ☽ Q ♂
22 15 ☿ ⊼ ♂

02 Thursday
04 04 ☽ ⚹ ☉
07 28 ☽ Q ☿
09 05 ☽ ⚼ ♆
13 56 ☽ Q ♇
17 59 ☽ ∠ ♅

03 Friday
02 18 ☽ Q ♃
02 30 ☽ □ ♂
06 16 ☽ ⚹ ☿
06 50 ☽ ∠ ☉
07 17 ☽ ⊼ ♄
08 54 ☽ ⊥ ♅
09 48 ☽ ⚹ ♇
09 58 ☽ △ ♆
12 39 ☽ ☌ ♀
14 44 ☿ □ ♄
17 05 ☽ ± ♄
18 42 ☽ ⊻ ♅
20 55 ☉ # ♄
22 09 ☽ ⚹ ♃
22 29 ☽ ⊥ ☉

04 Saturday
07 35 ☽ ⚼ ♄
08 24 ☽ ∥ ♅
08 47 ☽ ⊻ ☉
09 50 ☽ ∠ ☿
10 05 ☽ ∠ ♇
12 02 ☿ ☌ ♇
13 09 ☿ ⚹ ♆
21 08 ☽ ∥ ♀
22 28 ☽ ∠ ♃

05 Sunday
00 26 ☽ ⊥ ♇
02 01 ☽ ⊥ ☿
03 06 ☽ △ ♂
05 39 ☽ # ♄
07 26 ☽ △ ♄
08 21 ☽ ∥ ☉
09 56 ☽ ⊻ ♇
10 03 ☽ □ ♆
12 43 ☽ ⊻ ☿
12 53 ☽ ⊥ ♃
14 39 ☽ ⊻ ♀
18 39 ☽ ☌ ♅
22 28 ☽ ⊻ ♃

06 Monday
00 31 ☽ ⊥ ♀
02 51 ☽ ⚼ ♂
11 25 ☽ ☌ ☉
11 25 ☽ ∥ ☿
14 25 ☽ Q ♆
15 19 ☿ ⚹ ♀
15 21 ☽ ∠ ♀
17 05 ☽ ± ♂
20 47 ☽ ∥ ♃

07 Tuesday
02 07 ☽ # ♆
02 36 ☽ ⊼ ♂
06 48 ☽ □ ♄
07 52 ☽ # ♂
09 23 ☽ ☌ ♇
09 26 ☽ ⚹ ♆
16 15 ☽ ⚹ ♀
18 08 ☽ ☌ ☿
18 18 ☽ ⊻ ♅
18 50 ♀ ± ♄
19 37 ☿ ⊻ ♅
22 40 ☽ ☌ ♃

08 Wednesday
03 59 ☽ ⊥ ♅
09 33 ☽ ∠ ♆
11 44 ☽ Q ♄
11 53 ☉ Q ♆
12 33 ☽ Q ♀
14 37 ☽ ⊻ ☉
18 42 ☽ ∠ ♅

09 Thursday
00 19 ☽ ⊥ ♆
01 11 ☽ ⊥ ☉
03 13 ☽ ☍ ♂
07 11 ♆ ⚹ ♇
07 32 ☽ ⚹ ♄
10 19 ☽ ⊻ ♆
10 19 ☽ ⊻ ♇
11 29 ☿ ∥ ♃
16 00 ♀ ⊻ ♅
16 37 ☿ ☌ ♃
17 19 ☽ ∠ ☉
19 51 ☽ ⚹ ♅
20 02 ☽ □ ♀
20 28 ☽ ⊥ ♇

10 Friday
01 01 ☽ ⊻ ♃
01 58 ☽ ⊻ ☿
07 42 ☽ ∥ ♂
08 07 ☉ ± ♂
09 04 ☽ ∠ ♄
11 31 ☽ ⊥ ♃
11 59 ☽ ∠ ♇
13 42 ☽ ⊥ ☿
13 44 ☽ ∥ ♆
14 19 ☽ # ☿
16 18 ♂ St R
16 38 ☽ Q ♅
17 56 ☽ # ♃
18 06 ☿ # ♆
21 09 ☽ ⚹ ☉

11 Saturday
00 51 ☽ ⊥ ♄
03 36 ☽ ∠ ♃
06 53 ☽ ⊼ ♂
07 43 ☽ ∠ ☿
11 36 ☽ ⊻ ♄
14 36 ☽ ☌ ♆
14 40 ☽ ⚹ ♇
16 00 ☿ ⊥ ♅
17 42 ☽ ± ♂
20 21 ☽ Q ☉
23 55 ☽ # ☉

12 Sunday
01 04 ☽ □ ♅
04 13 ☽ △ ♀
07 19 ☽ ⚹ ♃
08 16 ☿ # ♂
10 14 ☽ ⚼ ♂
12 47 ☽ Q ♇
14 55 ☽ ⚹ ☿
17 13 ☽ ∥ ♄

13 Monday
06 16 ☽ Q ♃
07 04 ☽ # ♀
08 45 ☽ □ ☉
10 15 ☽ ⚼ ♀
14 35 ☽ △ ♂
16 54 ☽ Q ☿
16 58 ☽ # ♅
19 41 ☿ ∠ ♆
19 51 ☽ ☌ ♄
23 04 ☽ ⊻ ♆
23 12 ☽ □ ♇

14 Tuesday
04 51 ☽ ± ♀
10 29 ☽ △ ♅
10 49 ☽ ⊥ ♆
17 27 ☽ ⊼ ♀
17 34 ☿ Q ♄
17 53 ☽ □ ♃

15 Wednesday
02 13 ♀ ⚹ ♃
04 39 ☽ ∠ ♆
09 15 ☽ □ ☿
16 26 ☽ ⚼ ♅

16 Thursday
00 56 ☽ △ ☉
01 36 ☽ □ ♂
07 30 ☽ ⊻ ♄
08 37 ☉ ⊼ ♂
10 38 ☽ ± ♅
10 48 ☽ ⚹ ♆
11 01 ☽ △ ♇
19 39 ☽ ⊥ ♄
22 50 ☽ ⊼ ♅

17 Friday
07 15 ☽ △ ♃
09 57 ☽ ⚼ ☉
10 08 ☽ ☍ ♀
11 10 ☽ Q ♆
13 46 ☽ Q ♂
13 57 ☽ ∠ ♄
16 07 ☽ # ♅
17 30 ☽ ⚼ ♇
23 21 ☿ ∠ ♅

18 Saturday
00 26 ☽ # ♀
01 00 ☉ ∠ ♀
05 50 ☽ ± ☉
06 13 ☽ △ ☿
11 49 ☽ ± ♇
13 56 ☽ ⚹ ♂
14 15 ☽ ⚼ ♃
16 18 ☽ ∥ ♄
19 05 ☽ ⊼ ☉
20 28 ☽ ⚹ ♄
23 45 ☽ □ ♆

19 Sunday
00 01 ☽ ⊼ ♇
08 49 ☽ ± ♃
11 58 ☽ ☍ ♅
12 34 ☉ □ ♄
16 34 ☽ ⚼ ☿
17 48 ♀ ⚼ ♄
19 59 ☽ ∠ ♂
20 47 ☽ Q ♄
21 09 ☽ ⊼ ♃

20 Monday
00 26 ♃ # ♆
03 30 ☽ ⊼ ♀
05 08 ☿ ⊥ ♆
05 46 ☽ # ☉
12 57 ☽ ± ☿
13 52 ☽ ⊥ ♂
16 28 ☽ ± ♀
20 37 ☿ ☍ ♂

21 Tuesday
01 44 ☽ ⊻ ♂
01 53 ☽ ∥ ♆
02 26 ☽ ⊼ ☿
02 27 ☽ # ♃
05 38 ☉ ⚹ ♆
06 05 ☽ ∥ ♂
08 55 ☽ □ ♄
09 44 ☉ ☌ ♇
11 49 ☽ ⚼ ♀
12 05 ☽ △ ♆
12 24 ☽ ☍ ♇
12 38 ☽ ☍ ☉
20 21 ♀ ∠ ♇

22 Wednesday
00 14 ☽ ⊼ ♅
10 00 ☽ ☍ ♃
12 07 ☽ ± ♅
17 45 ☽ ⚼ ♆
19 42 ☽ △ ♀

23 Thursday
00 54 ☽ # ☿
05 49 ☽ ⚼ ♅
07 32 ☽ ∥ ☿
11 21 ☽ ± ♆
12 06 ☽ ☌ ♂
17 23 ☿ ⚹ ♄
19 59 ☽ △ ♄
20 16 ☽ ☍ ☿
23 00 ☽ ⊼ ♆
23 22 ☽ ⊼ ♇

24 Friday
04 22 ☽ ⊼ ☉
10 57 ☽ ± ♇
10 58 ☽ △ ♅
16 51 ☽ ± ☉
18 58 ☿ Q ♀
21 12 ☽ ⊼ ♃

25 Saturday
00 46 ☿ ⊻ ♆
00 53 ☽ ⚼ ♄
03 52 ☽ # ♂
03 53 ☽ ∥ ♃
04 11 ☽ ⚼ ♇
04 25 ♂ # ♃
04 50 ☿ ⊻ ♇
06 29 ☽ # ♆
08 49 ☽ ± ♃
10 02 ☽ □ ♀
11 12 ♂ ± ♆
11 23 ☽ ⚼ ☉
14 28 ☽ ∥ ☉
17 59 ☽ ± ♄
20 42 ☽ ⊻ ♂

26 Sunday
02 04 ☽ ⚼ ♃
05 17 ☽ ⊼ ♄
07 47 ☽ ⊥ ♂
08 07 ☽ ☍ ♆
08 31 ☽ △ ♇
11 11 ☽ ⊼ ☿
17 46 ☽ △ ☉
19 49 ☽ □ ♅
22 18 ☽ Q ♀
23 22 ☽ ± ☿

27 Monday
00 13 ☽ ∠ ♂
06 21 ☽ △ ♃
12 45 ☽ # ♄
17 23 ☽ ⚼ ☿
21 54 ☽ ⚹ ♀
22 42 ☉ ⊻ ♅

28 Tuesday
01 24 ☽ ∥ ♀
03 08 ☽ ⚹ ♂
04 47 ☽ Q ♅
08 16 ☽ ∥ ♅
12 20 ☽ ☍ ♄
14 58 ☽ ⊼ ♆
15 22 ☽ □ ♇
22 36 ☽ △ ☿

29 Wednesday
00 11 ☽ Q ♂
01 37 ☽ ± ♆
02 14 ☽ ⚹ ♅
02 41 ☽ ∠ ♀
02 53 ♃ ⊥ ♅
04 13 ☽ □ ☉
11 08 ☉ # ♆
12 54 ☽ □ ♃
17 23 ☽ ⚼ ♆
19 30 ☉ # ♂
19 33 ☽ ⊥ ♀
22 59 ☽ Q ♇

30 Thursday
04 24 ☽ ∠ ♅
06 37 ☽ ⊻ ♀
06 57 ☽ □ ♂
10 29 ♀ △ ♂
13 38 ☽ Q ☉
16 40 ☽ ⊼ ♄
19 05 ☽ △ ♆
19 30 ☽ ⚹ ♇
19 46 ☽ ⊥ ♅
20 14 ☽ Q ♃
22 00 ☿ ⊥ ♇

31 Friday
02 15 ☿ ⚹ ♅
02 49 ☽ ± ♄
05 52 ☽ ⊻ ♅
06 07 ☽ □ ☿
11 15 ☽ ⚹ ☉
12 34 ♂ ∥ ♆
16 34 ☽ ⚹ ♃
17 50 ☽ ⚼ ♄
19 25 ☽ ∥ ♅
20 33 ☽ ∠ ♇
23 02 ♀ ⊥ ♇

SEPTEMBER 1956

LONGITUDES

Date	Sidereal time h m s	Sun ☉ ° ' "	Moon ☽ ° ' "	Moon ☽ 24.00 ° ' "	Mercury ☿ ° '	Venus ♀ ° '	Mars ♂ ° '	Jupiter ♃ ° '	Saturn ♄ ° '	Uranus ♅ ° '	Neptune ♆ ° '	Pluto ♇ ° '
01	10 42 29	09 ♍ 00 21	23 ♋ 04 17	00 ♌ 28 30	06 ♎ 25	23 ♋ 06	20 ♓ 41	11 ♍ 24	27 ♏ 01	04 ♌ 51	28 ♎ 24	28 ♌ 39
02	10 46 25	09 58 29	07 ♌ 55 41	15 ♌ 24 53	06 57	24 08	20 R 25	11 37	27 04	04 54	28 26	28 41
03	10 50 22	10 56 38	22 ♌ 54 58	00 ♍ 24 46	07 46	25 07	20 10	11 50	27 08	04 57	28 27	28 43
04	10 54 18	11 54 48	07 ♍ 53 05	15 ♍ 18 41	08 32	26 06	19 54	12 03	27 11	05 00	28 29	28 45
05	10 58 15	12 53 01	22 ♍ 40 28	29 ♍ 57 26	09 14	27 06	19 38	12 16	27 14	05 04	28 31	28 47
06	11 02 12	13 51 15	07 ♎ 08 45	14 ♎ 13 46	09 52	28 06	19 22	12 29	27 18	05 07	28 32	28 49
07	11 06 08	14 49 31	21 ♎ 12 03	28 ♎ 03 21	10 27	29 ♋ 06	19 05	12 42	27 21	05 10	28 34	28 51
08	11 10 05	15 47 49	04 ♏ 47 37	11 ♏ 24 58	10 57	00 ♌ 07	18 49	12 55	27 25	05 13	28 36	28 53
09	11 14 01	16 46 08	17 ♏ 55 39	24 ♏ 20 06	11 22	01 08	18 32	13 08	27 29	05 16	28 38	28 55
10	11 17 58	17 44 28	00 ♐ 38 46	06 ♐ 52 14	11 43	02 09	18 16	13 21	27 32	05 19	28 39	28 56
11	11 21 54	18 42 51	13 ♐ 01 07	19 ♐ 07 04	11 58	03 11	17 59	13 34	27 36	05 22	28 41	28 58
12	11 25 51	19 41 15	25 ♐ 07 45	01 ♑ 06 50	12 07	04 13	17 43	13 47	27 40	05 24	28 43	29 00
13	11 29 47	20 39 40	07 ♑ 03 59	12 ♑ 59 50	12 11	05 15	17 26	14 00	27 44	05 27	28 45	29 02
14	11 33 44	21 38 07	18 ♑ 55 00	24 ♑ 50 03	12 R 08	06 18	17 10	14 13	27 48	05 30	28 47	29 04
15	11 37 41	22 36 36	00 ♒ 45 30	06 ♒ 41 50	11 59	07 21	16 54	14 26	27 52	05 33	28 49	29 06
16	11 41 37	23 35 06	12 ♒ 39 30	18 ♒ 38 50	11 43	08 25	16 39	14 39	27 57	05 36	28 51	29 08
17	11 45 34	24 33 38	24 ♒ 40 11	00 ♓ 43 48	11 20	09 28	16 23	14 52	28 01	05 38	28 52	29 09
18	11 49 30	25 32 12	06 ♓ 49 52	12 ♓ 58 33	10 51	10 32	16 08	15 05	28 05	05 41	28 54	29 11
19	11 53 27	26 30 47	19 ♓ 09 56	25 ♓ 24 07	10 14	11 36	15 54	15 17	28 10	05 44	28 56	29 13
20	11 57 23	27 29 25	01 ♈ 41 05	08 ♈ 01 47	09 30	12 41	15 39	15 30	28 14	05 46	28 58	29 15
21	12 01 20	28 28 04	14 ♈ 23 28	20 ♈ 48 49	08 41	13 46	15 26	15 43	28 19	05 49	29 00	29 16
22	12 05 16	29 ♍ 26 46	27 ♈ 16 57	03 ♉ 47 51	07 45	14 51	15 13	15 56	28 24	05 52	29 02	29 18
23	12 09 13	00 ♎ 25 29	10 ♉ 21 32	16 ♉ 58 02	06 45	15 56	15 00	16 09	28 29	05 54	29 04	29 20
24	12 13 10	01 24 15	23 ♉ 37 26	00 ♊ 19 48	05 41	17 01	14 48	16 22	28 33	05 57	29 06	29 22
25	12 17 06	02 23 04	07 ♊ 05 14	13 ♊ 53 50	04 36	18 07	14 36	16 34	28 38	05 59	29 08	29 23
26	12 21 03	03 21 54	20 ♊ 45 44	27 ♊ 40 59	03 29	19 13	14 25	16 47	28 43	06 01	29 10	29 25
27	12 24 59	04 20 47	04 ♋ 39 38	11 ♋ 41 41	02 23	20 19	14 15	17 00	28 48	06 04	29 12	29 27
28	12 28 56	05 19 42	18 ♋ 47 03	25 ♋ 55 32	01 20	21 26	14 06	17 12	28 53	06 06	29 14	29 28
29	12 32 52	06 18 39	03 ♌ 06 53	10 ♌ 20 41	00 ♎ 21	22 32	13 57	17 25	28 59	06 08	29 17	29 30
30	12 36 49	07 ♎ 17 39	17 ♌ 36 26	24 ♌ 53 30	29 ♍ 29	23 ♌ 39	13 ♓ 49	17 ♍ 38	29 ♏ 04	06 ♌ 10	29 ♎ 19	29 ♌ 32

Date	Moon True ☊ ° '	Moon Mean ☊ ° '	Moon ☽ Latitude ° '
01	03 ♐ 09	03 ♐ 09	03 S 51
02	03 R 02	03 05	04 33
03	02 52	03 02	04 56
04	02 41	02 59	04 59
05	02 30	02 56	04 43
06	02 20	02 53	04 08
07	02 12	02 50	03 18
08	02 07	02 46	02 19
09	02 05	02 43	01 14
10	02 04	02 40	00 S 08
11	02 D 04	02 37	00 N 58
12	02 R 04	02 34	01 59
13	02 03	02 31	02 55
14	01 59	02 27	03 42
15	01 53	02 24	04 19
16	01 45	02 21	04 46
17	01 35	02 18	05 00
18	01 23	02 15	05 01
19	01 11	02 11	04 48
20	01 00	02 08	04 21
21	00 50	02 05	03 40
22	00 43	02 02	02 48
23	00 39	01 59	01 46
24	00 37	01 56	00 N 37
25	00 D 37	01 52	00 S 35
26	00 38	01 49	01 46
27	00 R 38	01 46	02 52
28	00 37	01 43	03 49
29	00 35	01 40	04 32
30	00 ♐ 30	01 ♐ 37	04 S 59

DECLINATIONS

Date	Sun ☉ ° '	Moon ☽ ° '	Mercury ☿ ° '	Venus ♀ ° '	Mars ♂ ° '	Jupiter ♃ ° '	Saturn ♄ ° '	Uranus ♅ ° '	Neptune ♆ ° '	Pluto ♇ ° '
01	08 N 12	17 N 41	04 S 45	18 N 45	09 S 23	08 N 11	17 S 39	19 N 36	09 S 19	21 N 51
02	07 50	13 54	05 14	18 40	09 28	08 06	17 40	19 35	09 20	21 50
03	07 28	09 13	05 41	18 35	09 34	08 02	17 41	19 35	09 20	21 50
04	07 06	03 N 59	06 07	18 30	09 39	07 57	17 42	19 34	09 21	21 49
05	06 43	01 S 25	06 32	18 24	09 43	07 52	17 43	19 33	09 22	21 48
06	06 21	06 38	06 54	18 18	09 48	07 47	17 44	19 32	09 22	21 48
07	05 59	11 20	07 15	18 11	09 53	07 42	17 45	19 32	09 23	21 47
08	05 36	15 18	07 33	18 03	09 58	07 37	17 46	19 31	09 23	21 47
09	05 13	18 22	07 49	17 56	10 02	07 32	17 47	19 30	09 24	21 46
10	04 51	20 25	08 03	17 47	10 06	07 27	17 48	19 30	09 25	21 46
11	04 28	21 24	08 14	17 39	10 10	07 22	17 49	19 29	09 25	21 45
12	04 05	21 22	08 23	17 29	10 14	07 17	17 50	19 28	09 26	21 45
13	03 42	20 21	08 28	17 20	10 18	07 12	17 51	19 28	09 27	21 44
14	03 19	18 27	08 29	17 09	10 22	07 07	17 53	19 27	09 27	21 44
15	02 56	15 46	08 28	16 59	10 25	07 02	17 54	19 26	09 28	21 43
16	02 33	12 26	08 22	16 48	10 28	06 57	17 55	19 26	09 29	21 43
17	02 10	08 35	08 12	16 36	10 31	06 52	17 56	19 25	09 30	21 42
18	01 46	04 S 20	07 59	16 24	10 33	06 47	17 57	19 24	09 30	21 42
19	01 23	00 N 08	07 41	16 11	10 35	06 43	17 58	19 24	09 31	21 41
20	01 00	04 39	07 19	15 58	10 37	06 38	18 00	19 23	09 32	21 41
21	00 37	09 04	06 52	15 45	10 39	06 33	18 01	19 22	09 32	21 41
22	00 N 13	13 07	06 22	15 31	10 40	06 28	18 02	19 22	09 33	21 40
23	00 S 10	16 36	05 48	15 16	10 41	06 23	18 03	19 21	09 34	21 40
24	00 34	19 17	05 10	15 01	10 42	06 18	18 04	19 21	09 35	21 39
25	00 57	20 56	04 30	14 46	10 42	06 13	18 06	19 20	09 35	21 39
26	01 20	21 22	03 49	14 30	10 42	06 08	18 07	19 19	09 36	21 38
27	01 44	20 30	03 06	14 14	10 42	06 03	18 08	19 19	09 37	21 38
28	02 07	18 21	02 24	13 57	10 42	05 58	18 10	19 18	09 38	21 38
29	02 30	15 03	01 42	13 40	10 41	05 54	18 11	19 18	09 38	21 37
30	02 S 54	10 N 49	01 S 02	13 N 23	10 S 40	05 N 49	18 S 12	19 N 17	09 S 39	21 N 37

ZODIAC SIGN ENTRIES

Date	h	m	Planets
01	23	14	☽ ♌
03	23	20	☽ ♍
06	00	04	☽ ♎
08	03	26	☽ ♏
08	09	23	♀ ♌
10	10	46	☽ ♐
12	21	46	☽ ♑
15	10	28	☽ ♒
17	22	34	☽ ♓
20	08	47	☽ ♈
22	17	01	☽ ♉
23	01	35	☉ ♎
24	23	25	☽ ♊
27	04	00	☽ ♋
29	06	49	☽ ♌
29	21	25	☿ ♍

LATITUDES

Date	Mercury ☿ ° '	Venus ♀ ° '	Mars ♂ ° '	Jupiter ♃ ° '	Saturn ♄ ° '	Uranus ♅ ° '	Neptune ♆ ° '	Pluto ♇ ° '
01	02 S 32	02 S 45	06 S 12	00 N 58	01 N 54	00 N 34	01 N 42	10 N 36
04	02 59	02 29	06 08	00 58	01 53	00 34	01 42	10 36
07	03 23	02 13	06 03	00 59	01 52	00 34	01 42	10 36
10	03 43	01 57	05 56	00 59	01 52	00 34	01 42	10 37
13	03 58	01 41	05 48	00 59	01 51	00 34	01 42	10 37
16	04 03	01 25	05 38	00 59	01 50	00 34	01 42	10 38
19	03 57	01 10	05 27	00 59	01 50	00 34	01 42	10 39
22	03 35	00 55	05 15	00 59	01 49	00 34	01 42	10 39
25	02 55	00 40	05 02	01 00	01 48	00 34	01 41	10 40
28	02 00	00 26	04 49	01 00	01 48	00 34	01 41	10 41
31	01 S 01	00 S 12	04 S 34	01 N 00	01 N 47	00 N 35	01 N 41	10 N 42

DATA

Julian Date	2435718
Delta T	+32 seconds
Ayanamsa	23° 15' 23"
Synetic vernal point	05° ♓ 51' 37"
True obliquity of ecliptic	23° 26' 38"

MOON'S PHASES, APSIDES AND POSITIONS ☽

Date	h	m	Phase	Longitude ° '	Eclipse Indicator
04	18	57	●	12 ♍ 12	
12	00	13	☽	19 ♐ 13	
20	03	19	○	27 ♓ 08	
27	11	25	☾	04 ♋ 19	

Day	h	m		
03	04	12	Perigee	
15	04	55	Apogee	
05	05	40	0S	
11	23	00	Max dec	21° S 31'
19	11	20	0N	
26	08	09	Max dec	21° N 23'

LONGITUDES

Date	Chiron ⚷ ° '	Ceres ⚳ ° '	Pallas ⚴ ° '	Juno ⚵ ° '	Vesta ⚶ ° '	Black Moon Lilith ⚸ ° '
01	07 ♒ 27	20 ♈ 19	22 ♒ 47	06 ♑ 48	20 ♏ 15	00 ♒ 04
11	07 ♒ 03	19 ♈ 04	20 ♒ 37	07 ♑ 14	24 ♏ 36	01 ♒ 11
21	06 ♒ 45	17 ♈ 19	18 ♒ 54	08 ♑ 14	29 ♏ 08	02 ♒ 18
31	06 ♒ 32	15 ♈ 11	17 ♒ 44	09 ♑ 46	03 ♐ 49	03 ♒ 25

All ephemeris data is given at 12.00 UT and the Moon's longitude is additionally given for 24.00 UT

ASPECTARIAN

01 Saturday
03 14 ☽ ∥ ♀
08 10 ☽ △ ♂
11 19 ☽ ⊥ ♇
12 09 ☽ ☌ ♀
12 10 ☽ # ♄
12 28 ☉ ∥ ♃
13 38 ☽ ∠ ☉
13 45 ☽ Q ☿
17 29 ☽ ∠ ♃
18 26 ☽ △ ♄
20 40 ☽ □ ♆
21 05 ☽ ⊻ ♇

02 Sunday
05 12 ☽ ⊥ ☉
07 07 ☽ ☌ ♅
08 03 ☽ ⚼ ♂
08 14 ☽ ⊥ ♃
10 20 ☽ ⚹ ☿
15 31 ☽ ⊻ ☉
18 00 ☽ ⊻ ♃
22 14 ☽ ± ♂

03 Monday
01 39 ☽ Q ♆
07 41 ☽ ⊼ ♂
10 21 ☽ # ♂
11 23 ☽ # ♆
11 45 ☽ ∠ ☿
12 18 ☉ ⊥ ♅
15 46 ☽ ⊻ ♀
17 41 ☽ ∥ ♃
18 46 ☽ □ ♄
20 50 ☽ ∥ ☉
20 53 ☽ ⚹ ♆
21 18 ☽ ☌ ♇

04 Tuesday
02 03 ☽ ⊥ ♀
02 56 ☽ ⊥ ☿
03 05 ☽ # ☿
04 05 ♂ ⚼ ♅
07 21 ☽ ⊻ ♅
13 05 ☽ ⊻ ☿
16 18 ☉ ☌ ♃
17 03 ☽ ⊥ ♅
17 33 ☽ ∠ ♀
18 49 ☽ ☌ ♃
18 57 ☽ ☌ ☉
21 03 ☽ ∠ ♆
23 50 ☽ Q ♄

05 Wednesday
07 07 ☽ ☍ ♂
07 43 ☽ ∠ ♅
11 44 ☽ ⊥ ♆
15 43 ♀ △ ♄
17 19 ♀ ∠ ♃
18 18 ☉ # ☿
19 32 ☽ ⚹ ♄
19 48 ☽ ⚹ ♀
21 38 ☽ ⊻ ♆
21 55 ☿ Q ♀
22 05 ☽ ⊻ ♇

06 Thursday
04 00 ☉ ∠ ♆
08 05 ☽ ⊥ ♇
08 34 ☽ ⚹ ♅
10 47 ☽ # ☉
13 26 ☽ ∥ ☿
16 48 ☽ ☌ ☿
17 21 ☽ Q ♀
17 30 ☽ # ♃
20 44 ☽ ∠ ♄
21 10 ☽ ⊻ ♃
23 01 ♀ □ ♆
23 19 ☽ ∠ ♇

07 Friday
00 12 ☽ ⊻ ☉
01 37 ☽ ∥ ♆
04 06 ☽ ∥ ♂
05 00 ☽ Q ♅
05 48 ♀ ⊻ ♇
07 36 ☽ ⊥ ♃
08 25 ☽ ⊼ ♂
11 18 ☽ ⊥ ☉
12 16 ☽ ⊥ ♄
18 39 ☽ ± ♂
22 49 ☽ ⊻ ♄
23 33 ☽ ∠ ♃

08 Saturday
00 56 ☽ ☌ ♆
01 26 ☽ ⚹ ♇
01 56 ☉ Q ♄
02 58 ☽ □ ♀
04 18 ☽ ∠ ☉
10 17 ☽ ⚼ ♂
12 45 ☽ □ ♅
15 52 ☿ # ♃
23 03 ☽ Q ♇
23 32 ☽ ⊻ ☿

09 Sunday
03 00 ☽ ⚹ ♃
06 46 ☽ ∥ ♄
08 12 ☽ # ♀
09 41 ☽ ⚹ ☉
10 56 ☽ ⊥ ☿
13 07 ☽ △ ♂
23 45 ☽ # ♅

10 Monday
01 44 ☽ Q ♃
04 18 ☽ ∠ ☿
06 02 ☽ ☌ ♄
08 11 ☽ ⊻ ♆
08 44 ☽ □ ♇
09 38 ♀ # ♄
10 07 ☽ Q ☉
15 09 ☽ △ ♀
19 44 ☽ ⊥ ♆
21 01 ☽ △ ♅
21 58 ☉ ☍ ♂

11 Tuesday
08 22 ♀ ⚼ ♂
09 54 ☽ ⚹ ☿
13 06 ☽ □ ♃
13 19 ☽ ∠ ♆
21 34 ☽ □ ♂
23 08 ☽ ⚼ ♀

12 Wednesday
00 13 ☽ □ ☉
02 33 ☽ ⚼ ♅
03 42 ♃ ∠ ♆
09 58 ☽ Q ☿
17 07 ☽ ⊻ ♄
18 46 ☽ ± ♀
19 12 ☽ ⚹ ♆
19 47 ☽ △ ♇
20 36 ☽ ± ♅

13 Thursday
05 15 ☽ ⊥ ♄
06 41 ☉ ∠ ♅
08 00 ☽ ⊼ ♀
08 44 ☽ ⊼ ♅
08 47 ☽ Q ♂
14 07 ☿ St R
16 47 ♀ ☌ ♅
19 28 ☽ Q ♆
22 21 ☽ □ ☿
23 32 ☽ ∠ ♄

14 Friday
00 37 ☽ # ♅
02 08 ☽ ⚼ ♇
02 17 ☽ △ ♃
08 32 ☽ ⚹ ♂
17 39 ☽ ∥ ♄
18 01 ☽ △ ☉
20 26 ☽ ± ♇

15 Saturday
01 13 ☽ # ♀
06 08 ☽ ⚹ ♄
08 03 ☽ □ ♆
08 37 ☽ ⊼ ♇
09 16 ☽ ⚼ ♃
14 16 ☽ ∠ ♂
17 06 ☉ ⊥ ♆
21 43 ☽ ☍ ♅

16 Sunday
02 37 ☽ ☍ ♀
03 04 ☽ ⚼ ☉
03 47 ☽ ± ♃
06 31 ☽ Q ♄
08 02 ☽ ⊥ ♂
10 10 ☽ △ ☿
16 04 ☽ ⊼ ♃
18 42 ♀ ⊥ ♃
19 49 ☽ ⊻ ♂
22 45 ☽ ± ☉

17 Monday
00 28 ☽ ∥ ♂
06 34 ☽ ∥ ♆
11 46 ☽ ⊼ ☉
14 18 ☽ ∥ ☿
15 12 ☽ ⚼ ☿
18 41 ☽ □ ♄
20 21 ☽ △ ♆
20 55 ☽ ☍ ♇
22 05 ☽ # ♃

18 Tuesday
04 43 ♀ ± ♂
08 16 ☽ ± ☿
09 45 ☽ ⊼ ♅
12 23 ☉ ∠ ♀
16 32 ☿ ⚹ ♀
19 30 ☽ ⊼ ☿
19 56 ☽ ⊼ ♀
21 31 ☽ ± ♅

19 Wednesday
01 51 ☽ ⚼ ♆
03 12 ☽ # ☉
04 22 ☽ ☍ ♃
05 47 ☽ ☌ ♂
08 42 ☽ ± ♀
15 02 ☽ ⚼ ♅
18 08 ☽ ∥ ☉
19 17 ☽ ± ♆

20 Thursday
03 19 ☽ ☍ ☉
03 39 ☽ ⚼ ♀
05 24 ☽ △ ♄
06 49 ☽ ⊼ ♆
07 20 ☽ ⊼ ♇
18 47 ☽ ± ♇
19 47 ☽ △ ♅
20 17 ♂ ☍ ♃
22 24 ☽ ∥ ♃

21 Friday
01 02 ☽ # ☿
01 56 ☽ ☍ ☿
08 01 ☉ ⚹ ♄
09 59 ☽ ⚼ ♄
10 42 ☽ △ ♀
11 47 ☽ ⚼ ♇
13 55 ☽ ⊻ ♂
14 32 ☽ ⊼ ♃
14 44 ☽ # ♆
21 08 ☽ # ♂

22 Saturday
00 55 ☽ ⊥ ♂
01 37 ☉ ⊻ ♆
01 55 ☽ ± ♃
02 53 ☽ ± ♄
06 34 ☿ # ♃
08 23 ☉ ⊻ ♇
14 04 ☽ ⊼ ♄
15 15 ☽ ☍ ♆
15 44 ☽ △ ♇
16 19 ☽ ⊼ ☉
17 19 ☽ ∠ ♂
18 48 ♀ ⊼ ♂
18 51 ☽ ⚼ ♃

23 Sunday
02 53 ☽ ∥ ♀
03 50 ☽ □ ♅
04 14 ☽ ± ☉
05 54 ☽ ⊼ ☿
16 02 ☽ ± ☿
17 54 ♀ ⊻ ♃
20 18 ☽ ⚹ ♂
21 57 ☽ ⚼ ☉
22 41 ☽ △ ♃
23 02 ☽ □ ♀

24 Monday
00 01 ☽ # ♄
06 38 ☿ ⚹ ♅
07 07 ☽ ⚼ ☿
12 34 ☽ Q ♅
12 39 ☽ ∥ ♅
13 52 ♀ Q ♆
17 37 ☽ Q ♂
19 06 ☿ ⊥ ♇
20 53 ☽ ☍ ♄
21 50 ☽ ⊼ ♆
22 18 ☽ □ ♇

25 Tuesday
03 00 ☽ △ ☉
07 55 ☽ △ ☿
08 32 ☽ ± ♆
10 02 ☽ ⚹ ♅
10 08 ☽ Q ♀

26 Wednesday
00 12 ♃ Q ♄
00 27 ☽ ⚼ ♆
01 04 ☽ □ ♂
04 00 ☿ ∠ ♀
04 57 ☽ □ ♃
06 09 ☽ Q ♇
09 04 ☽ ⚹ ♀
12 27 ☽ ∠ ♅
13 20 ☉ ☌ ☿

27 Thursday
01 52 ☽ ⊼ ♄
02 36 ☽ △ ♆
03 01 ☽ ⚹ ♇
04 05 ☽ ⊥ ♅
08 23 ☽ □ ☿
11 25 ☽ □ ☉
12 15 ☽ ± ♄
12 35 ☽ Q ♃
13 14 ☽ ∠ ♀
14 24 ☽ ⊻ ♅
17 41 ♂ ⚼ ♆

28 Friday
02 54 ☽ ∥ ♅
03 40 ☽ ⚼ ♄
04 09 ☽ △ ♂
04 42 ☽ ∠ ♇
05 51 ☽ ⊥ ♀
09 18 ☽ ⚹ ♃
12 52 ☽ Q ☿
13 38 ☽ # ♄
15 39 ☉ ⊥ ♇
16 50 ☽ ⊻ ♀
18 05 ☉ ∥ ☿
19 54 ☽ ⊥ ♇
20 13 ☽ Q ☉

29 Saturday
05 03 ☽ △ ♄
05 07 ☽ ⚼ ♂
05 35 ☽ □ ♆
05 58 ☽ ⊻ ♇
07 36 ☉ ⚹ ♅
07 41 ☽ ⚹ ☿
10 49 ☽ ∠ ♃
17 02 ☽ ☌ ♅
17 42 ☽ ⚹ ☉
19 57 ☽ ± ♂
20 59 ☽ ∥ ♀

30 Sunday
01 59 ☽ ⊥ ♃
05 47 ☽ ⊼ ♂
07 07 ☽ ∠ ☿
09 49 ☿ ⊥ ♀
10 38 ☿ ⊻ ♇
11 31 ☽ Q ♆
12 02 ☽ ⊻ ♃
12 48 ☽ # ♂
16 48 ☿ ⊻ ♆
17 56 ☽ # ♆
20 17 ☽ ∠ ☉
21 11 ☽ ⊥ ☿
22 48 ☽ ☌ ♀
23 26 ☿ ⚹ ♄

OCTOBER 1956

LONGITUDES

Date	Sidereal time h m s	Sun ☉ ° ' "	Moon ☽ ° ' "	Moon ☽ 24.00 ° ' "	Mercury ☿ ° '	Venus ♀ ° '	Mars ♂ ° '	Jupiter ♃ ° '	Saturn ♄ ° '	Uranus ♅ ° '	Neptune ♆ ° '	Pluto ♇ ° '
01	12 40 45	08 ♎ 16 41	02 ♍ 11 10	09 ♍ 28 36	28 ♍ 44	24 ♌ 46	13 ♓ 41	17 ♍ 50	29 ♏ 09	06 ♌ 13	29 ♎ 21	29 ♌ 33
02	12 44 42	09 15 45	16 ♍ 44 58	23 ♍ 59 24	28 R 07	25 54	13 R 34	18 03	29 14	06 15	29 23	29 35
03	12 48 39	10 14 51	01 ♎ 11 02	08 ♎ 19 04	27 40	27 01	13 28	18 15	29 20	06 17	29 25	29 36
04	12 52 35	11 13 59	15 ♎ 22 48	22 ♎ 21 37	27 23	28 09	13 23	18 28	29 25	06 19	29 27	29 38
05	12 56 32	12 13 10	29 ♎ 15 03	06 ♏ 02 48	27 17	29 ♌ 17	13 19	18 40	29 31	06 21	29 29	29 40
06	13 00 28	13 12 22	12 ♏ 44 39	19 ♏ 20 34	27 D 21	00 ♍ 25	13 15	18 53	29 36	06 23	29 32	29 41
07	13 04 25	14 11 36	25 ♏ 50 39	02 ♐ 15 08	27 36	01 33	13 13	19 05	29 42	06 25	29 34	29 43
08	13 08 21	15 10 52	08 ♐ 34 18	14 ♐ 48 34	28 00	02 42	13 11	19 17	29 48	06 27	29 36	29 44
09	13 12 18	16 10 10	20 ♐ 58 26	27 ♐ 04 26	28 35	03 50	13 09	19 30	29 53	06 28	29 38	29 46
10	13 16 14	17 09 30	03 ♑ 07 09	09 ♑ 07 11	29 ♍ 19	04 59	13 D 09	19 42	29 ♏ 59	06 30	29 40	29 47
11	13 20 11	18 08 52	15 ♑ 05 11	21 ♑ 01 47	00 ♎ 10	06 08	13 10	19 54	00 ♐ 05	06 32	29 42	29 48
12	13 24 08	19 08 15	26 ♑ 57 37	02 ♒ 53 18	01 10	07 17	13 11	20 06	00 11	06 34	29 45	29 50
13	13 28 04	20 07 40	08 ♒ 49 27	14 ♒ 46 39	02 16	08 26	13 13	20 18	00 17	06 35	29 47	29 51
14	13 32 01	21 07 07	20 ♒ 45 25	26 ♒ 46 15	03 28	09 36	13 16	20 31	00 23	06 37	29 49	29 53
15	13 35 57	22 06 35	02 ♓ 49 36	08 ♓ 55 51	04 45	10 45	13 19	20 43	00 29	06 38	29 51	29 54
16	13 39 54	23 06 06	15 ♓ 05 20	21 ♓ 18 19	06 07	11 55	13 24	20 54	00 35	06 40	29 54	29 55
17	13 43 50	24 05 38	27 ♓ 35 00	03 ♈ 55 30	07 32	13 05	13 29	21 06	00 41	06 41	29 56	29 57
18	13 47 47	25 05 12	10 ♈ 19 54	16 ♈ 48 10	09 01	14 15	13 35	21 18	00 47	06 43	29 ♎ 58	29 58
19	13 51 43	26 04 48	23 ♈ 20 17	29 ♈ 56 02	10 32	15 25	13 42	21 30	00 54	06 44	00 ♏ 00	29 ♌ 59
20	13 55 40	27 04 26	06 ♉ 35 23	13 ♉ 18 04	12 06	16 35	13 49	21 42	01 00	06 45	00 02	00 ♍ 00
21	13 59 37	28 04 06	20 ♉ 03 54	26 ♉ 52 37	13 41	17 46	13 57	21 53	01 06	06 46	00 05	00 01
22	14 03 33	29 ♎ 03 49	03 ♊ 44 00	10 ♊ 37 49	15 18	18 56	14 06	22 05	01 13	06 48	00 07	00 03
23	14 07 30	00 ♏ 03 33	17 ♊ 33 49	24 ♊ 31 47	16 56	20 07	14 16	22 16	01 19	06 49	00 09	00 04
24	14 11 26	01 03 20	01 ♋ 31 31	08 ♋ 32 49	18 35	21 18	14 26	22 28	01 25	06 50	00 11	00 05
25	14 15 23	02 03 09	15 ♋ 35 29	22 ♋ 39 19	20 15	22 29	14 37	22 39	01 32	06 51	00 14	00 06
26	14 19 19	03 03 00	29 ♋ 44 06	06 ♌ 49 37	21 55	23 40	14 48	22 51	01 38	06 52	00 16	00 07
27	14 23 16	04 02 54	13 ♌ 55 38	21 ♌ 01 51	23 35	24 51	15 01	23 02	01 45	06 53	00 18	00 08
28	14 27 12	05 02 49	28 ♌ 07 59	05 ♍ 13 40	25 15	26 02	15 14	23 13	01 52	06 53	00 20	00 09
29	14 31 09	06 02 47	12 ♍ 18 31	19 ♍ 22 09	26 55	27 14	15 27	23 24	01 58	06 54	00 23	00 10
30	14 35 06	07 02 47	26 ♍ 24 07	03 ♎ 23 58	28 ♎ 35	28 25	15 41	23 35	02 05	06 55	00 25	00 11
31	14 39 02	08 ♏ 02 49	10 ♎ 21 17	17 ♎ 15 38	00 ♏ 15	29 ♍ 37	15 ♓ 56	23 ♍ 46	02 ♐ 12	06 ♌ 56	00 ♏ 27	00 ♍ 12

MOON NODES, LATITUDE AND DECLINATIONS

Date	Moon True ☊ ° '	Moon Mean ☊ ° '	Moon ☽ Latitude ° '	Sun ☉ ° '	Moon ☽ ° '	Mercury ☿ ° '	Venus ♀ ° '	Mars ♂ ° '	Jupiter ♃ ° '	Saturn ♄ ° '	Uranus ♅ ° '	Neptune ♆ ° '	Pluto ♇ ° '
01	00 ♐ 23	01 ♐ 33	05 S 06	03 S 17	05 N 56	00 S 26	13 N 05	10 S 38	05 N 44	18 S 13	19 N 17	09 S 40	21 N 37
02	00 R 15	01 30	04 54	03 40	00 N 43	00 N 07	12 46	10 36	05 39	18 15	19 16	09 41	21 36
03	00 06	01 27	04 23	04 04	04 S 29	00 36	12 27	10 34	05 34	18 16	19 16	09 41	21 36
04	29 ♏ 59	01 24	03 36	04 27	09 23	01 00	12 08	10 32	05 29	18 17	19 16	09 42	21 36
05	29 53	01 21	02 37	04 50	13 40	01 19	11 49	10 29	05 25	18 19	19 15	09 43	21 35
06	29 50	01 17	01 31	05 13	17 07	01 32	11 29	10 26	05 20	18 20	19 14	09 44	21 35
07	29 48	01 14	00 S 21	05 36	19 34	01 40	11 08	10 22	05 15	18 21	19 14	09 44	21 35
08	29 D 49	01 11	00 N 47	05 59	20 57	01 43	10 47	10 19	05 10	18 23	19 14	09 45	21 35
09	29 50	01 08	01 52	06 22	21 16	01 41	10 26	10 14	05 06	18 24	19 13	09 46	21 34
10	29 51	01 05	02 51	06 44	20 34	01 33	10 05	10 10	05 01	18 26	19 13	09 47	21 34
11	29 52	01 02	03 41	07 07	18 56	01 21	09 43	10 05	04 56	18 27	19 12	09 48	21 34
12	29 R 52	00 58	04 21	07 30	16 30	01 05	09 21	10 01	04 52	18 28	19 12	09 48	21 34
13	29 50	00 55	04 50	07 52	13 23	00 44	08 59	09 55	04 47	18 30	19 12	09 49	21 33
14	29 46	00 52	05 07	08 14	09 44	00 N 21	08 36	09 50	04 42	18 31	19 11	09 50	21 33
15	29 41	00 49	05 10	08 37	05 39	00 S 06	08 13	09 44	04 38	18 33	19 11	09 51	21 33
16	29 35	00 46	04 59	08 59	01 S 16	00 36	07 50	09 38	04 33	18 34	19 11	09 51	21 33
17	29 29	00 42	04 34	09 21	03 N 14	01 08	07 26	09 32	04 29	18 35	19 10	09 52	21 33
18	29 23	00 39	03 55	09 43	07 42	01 42	07 02	09 25	04 24	18 37	19 10	09 53	21 32
19	29 18	00 36	03 03	10 04	11 54	02 18	06 38	09 18	04 20	18 38	19 10	09 54	21 32
20	29 14	00 33	02 00	10 26	15 37	02 56	06 13	09 11	04 15	18 40	19 09	09 55	21 32
21	29 12	00 30	00 N 50	10 47	18 34	03 34	05 49	09 04	04 11	18 41	19 09	09 55	21 32
22	29 D 12	00 27	00 S 25	11 09	20 30	04 14	05 24	08 56	04 06	18 42	19 09	09 56	21 32
23	29 13	00 23	01 39	11 30	21 13	04 54	04 59	08 49	04 02	18 44	19 09	09 57	21 32
24	29 14	00 20	02 48	11 51	20 38	05 35	04 33	08 41	03 57	18 45	19 08	09 58	21 32
25	29 16	00 17	03 48	12 11	18 46	06 16	04 08	08 32	03 53	18 47	19 08	09 59	21 32
26	29 17	00 14	04 34	12 32	15 45	06 58	03 42	08 24	03 48	18 48	19 08	09 59	21 32
27	29 R 16	00 11	05 03	12 52	11 49	07 39	03 16	08 15	03 44	18 50	19 08	10 00	21 32
28	29 15	00 08	05 14	13 12	07 08	08 20	02 50	08 06	03 40	18 51	19 08	10 01	21 31
29	29 12	00 04	05 06	13 32	02 N 14	09 02	02 24	07 57	03 36	18 52	19 07	10 02	21 31
30	29 09	00 ♐ 01	04 40	13 52	02 S 51	09 43	01 57	07 48	03 31	18 54	19 07	10 02	21 31
31	29 ♏ 06	29 ♏ 58	03 S 57	14 S 12	07 S 44	10 S 23	01 N 31	07 S 39	03 N 27	18 S 55	19 N 07	10 S 03	21 N 31

ZODIAC SIGN ENTRIES

Date	h	m	Planets
01	08	24	☽ ♍
03	10	01	☽ ♎
05	13	19	☽ ♏
06	03	12	♀ ♍
07	19	46	☽ ♐
10	05	48	☽ ♑
10	15	11	♄ ♐
11	07	30	☿ ♎
12	18	09	☽ ♒
15	06	25	☽ ♓
17	16	35	☽ ♈
19	09	28	♆ ♏
20	00	07	☽ ♉
20	06	12	♇ ♍
22	05	29	☽ ♊
23	10	34	☉ ♏
24	09	23	☽ ♋
26	12	27	☽ ♌
28	15	09	☽ ♍
30	18	10	☽ ♎
31	08	19	☿ ♏
31	19	40	♀ ♎

LATITUDES

Date	Mercury ☿ ° '	Venus ♀ ° '	Mars ♂ ° '	Jupiter ♃ ° '	Saturn ♄ ° '	Uranus ♅ ° '	Neptune ♆ ° '	Pluto ♇ ° '
01	01 S 01	00 S 12	04 S 34	01 N 00	01 N 47	00 N 35	01 N 41	10 N 42
04	00 S 03	00 N 01	04 20	01 01	01 47	00 35	01 41	10 42
07	00 N 47	00 14	04 05	01 01	01 46	00 35	01 41	10 43
10	01 24	00 26	03 51	01 01	01 46	00 35	01 41	10 44
13	01 47	00 37	03 36	01 02	01 45	00 35	01 41	10 45
16	01 59	00 48	03 22	01 02	01 45	00 35	01 41	10 46
19	02 02	00 57	03 08	01 02	01 45	00 35	01 41	10 47
22	01 57	01 06	02 54	01 03	01 44	00 35	01 41	10 48
25	01 46	01 15	02 41	01 03	01 44	00 35	01 41	10 49
28	01 32	01 22	02 29	01 03	01 43	00 35	01 41	10 50
31	01 N 15	01 N 29	02 S 17	01 N 04	01 N 43	00 N 36	01 N 41	10 N 51

DATA

Julian Date	2435748
Delta T	+32 seconds
Ayanamsa	23° 15' 26"
Synetic vernal point	05° ♓ 51' 34"
True obliquity of ecliptic	23° 26' 38"

MOON'S PHASES, APSIDES AND POSITIONS ☽

Date	h	m	Phase	Longitude ° '	Eclipse Indicator
04	04	25	●	10 ♎ 55	
11	18	44	☽	18 ♑ 26	
19	17	25	○	26 ♈ 18	
26	18	02	☾	03 ♌ 18	

Day	h	m		
01	02	14	Perigee	
12	22	55	Apogee	
27	05	56	Perigee	
02	15	17	OS	
09	07	07	Max dec	21° S 18'
16	18	49	ON	
23	13	22	Max dec	21° N 13'
29	22	30	OS	

LONGITUDES

Date	Chiron ⚷ ° '	Ceres ⚳ ° '	Pallas ⚴ ° '	Juno ⚵ ° '	Vesta ⚶ ° '	Black Moon Lilith ⚸ ° '
01	06 ♒ 32	15 ♈ 11	17 ♒ 44	09 ♑ 46	03 ♐ 49	03 ♒ 25
11	06 ♒ 26	12 ♈ 55	17 ♒ 09	11 ♑ 46	08 ♐ 39	04 ♒ 32
21	06 ♒ 27	10 ♈ 46	17 ♒ 10	14 ♑ 10	13 ♐ 36	05 ♒ 39
31	06 ♒ 35	08 ♈ 56	17 ♒ 42	16 ♑ 57	18 ♐ 37	06 ♒ 46

ASPECTARIAN

h m	Aspects
01 Monday	
06 34	☽ ⊻ ☿
06 59	☽ □ ♄
07 19	☽ ⚹ ♆
07 40	☽ ☌ ♇
12 10	☽ ⊥ ☉
12 57	☽ ∥ ♃
18 38	☽ ⊻ ♅
22 45	☽ ⊻ ☉
23 26	☽ # ☉
02 Tuesday	
04 33	☽ ⊥ ♅
06 48	☽ ☍ ♂
08 05	☽ ∠ ♆
12 49	☽ Q ♄
14 11	☽ ☌ ♃
14 29	☽ ∥ ☿
16 15	☽ # ☿
19 28	☽ ∠ ♅
23 01	☽ ⊥ ♆
03 Wednesday	
04 27	☽ ⊻ ♀
06 16	☽ ☌ ☿
08 53	☽ ⚹ ♄
09 02	☽ ⊻ ♆
09 22	☽ ⊻ ♇
09 49	☽ ∥ ☉
15 21	☽ ⊥ ♀
17 03	☽ # ♃
19 27	☽ ⊥ ♇
20 35	☽ ⚹ ♅
22 34	☿ ⊻ ♀
04 Thursday	
04 25	☽ ☌ ☉
07 52	☽ ∠ ♀
08 37	☽ ⊼ ♂
10 21	☽ ∠ ♄
10 43	☽ ∠ ♇
13 42	☽ ∥ ♆
17 02	☽ Q ♅
17 22	☽ ⊻ ♃
18 01	☽ ∥ ♂
18 50	☽ ± ♂
05 Friday	
01 54	☽ # ♀
01 56	☽ ⊥ ♄
02 03	♄ ⊻ ♆
03 53	☽ ⊥ ♃
08 33	☽ ⊻ ☿
10 22	☽ ⚼ ♂
12 04	☽ ⚹ ♀
12 25	☽ ☌ ♆
12 28	☽ ⊻ ♄
12 43	☽ ⚹ ♇
14 21	☿ St D
16 32	♀ ⚹ ♆
17 19	♀ □ ♄
19 05	☽ ⊥ ☿
19 55	☽ ∠ ♃
20 11	♀ ☌ ♇
06 Saturday	
00 34	☽ □ ♅
10 05	☽ Q ♇
11 17	☽ ∠ ☿
11 21	☽ Q ♀
12 54	☽ ⊻ ☉
12 55	☽ △ ♂
13 08	☉ ⊼ ♂
18 05	☉ # ♃
22 48	☽ ∥ ♄
23 20	☽ ⚹ ♃
07 Sunday	
00 42	☽ ⊥ ☉
08 02	☽ # ♅
15 21	☽ ⚹ ☿
15 42	♄ □ ♇
18 46	☽ ∠ ☉
18 58	☽ ⊻ ♆
19 14	☽ □ ♇
19 15	☽ ☌ ♄
21 58	☽ Q ♃
23 44	☽ □ ♀
08 Monday	
00 53	☉ ∠ ♇
01 36	☉ ∠ ♄
06 19	☽ ⊥ ♆
07 56	☽ △ ♅
14 52	☽ Q ☿
20 50	☽ □ ♂
23 37	☽ ∠ ♆
09 Tuesday	
01 49	☽ ⚹ ☉
09 04	☽ □ ♃
12 59	☽ ⚼ ♅
10 Wednesday	
03 26	☽ Q ☉
03 54	☽ □ ☿
04 47	♀ # ♂
05 07	☽ ⚹ ♆
05 21	☽ △ ♇
05 43	☽ ⊻ ♄
06 47	☽ ± ♅
08 05	☽ Q ♂
10 06	♂ St D
16 07	☽ △ ♀
17 46	☽ ⊥ ♄
18 47	☽ ⊼ ♅
23 00	☿ ⊻ ♆
11 Thursday	
02 03	☿ ⊻ ♇
05 11	☽ Q ♆
07 22	♀ # ♆
08 07	☽ ⚹ ♂
08 42	☽ # ♅
09 28	☿ ⚹ ♄
11 26	☽ ⚼ ♇
12 00	☽ ∠ ♄
17 22	☽ ∥ ♄
18 44	☽ □ ☉
20 32	♀ ⊻ ♅
21 36	☉ Q ♅
21 54	☽ △ ♃
12 Friday	
01 31	☽ ⚼ ♀
05 39	☽ ± ♇
13 04	☉ ± ♂
14 29	☽ ∠ ♂
17 39	☽ □ ♆
17 49	☽ ⊼ ♇
18 35	☽ ⚹ ♄
21 21	☽ △ ☿
21 42	☽ ± ♀
13 Saturday	
04 46	☽ ⚼ ♃
07 28	☽ ☍ ♅
08 44	☽ ⊥ ♂
11 08	☽ ⊼ ♀
17 28	☉ ⊻ ♃
19 02	☽ Q ♄
20 53	☽ ⊻ ♂
23 15	☽ ± ♃
14 Sunday	
06 52	☽ ⚼ ☿
11 21	☽ ∥ ♂
11 21	☽ ∥ ♆
11 22	♂ ∥ ♆
11 30	☽ ⊼ ♃
12 47	☽ △ ☉
19 34	☽ # ♀
20 14	☽ ∥ ☉
23 23	☉ # ♀
15 Monday	
02 57	☽ ± ☿
06 06	☽ △ ♆
06 12	☽ ☍ ♇
07 20	☽ □ ♄
16 15	☽ ⊼ ☿
17 47	☽ # ♃
19 31	☽ ⊼ ♅
21 10	☽ ⚼ ☉
16 Tuesday	
05 11	☽ ☍ ♀
07 17	☽ ± ♅
08 42	☽ ☌ ♂
08 43	☿ ⊥ ♇
11 37	☽ ⚼ ♆
15 14	☽ ∥ ☿
16 14	☽ ± ☉
21 37	☿ ⚹ ♅
23 21	☽ # ☿
23 25	☽ ☍ ♃
17 Wednesday	
00 43	☽ ⚼ ♅
03 42	♀ ⊥ ♅
04 46	☽ ⊼ ☉
05 01	☽ ± ♆
16 28	☽ ⊼ ♆
16 29	☽ ⊼ ♇
17 56	☽ △ ♄
18 30	☽ ∥ ♃
21 02	♀ ☍ ♂
21 07	☉ ∥ ♂
18 Thursday	
03 49	☽ ± ♇
04 01	☉ ⊥ ♄
05 13	☽ △ ♅
07 31	♆ ⚹ ♇
08 41	☽ ∥ ♀
09 13	☽ ☍ ☿
18 06	☽ ⊻ ♂
20 00	☽ ⊼ ♀
20 37	☽ ⚼ ♇
21 19	☽ # ♂
22 13	☽ ⚼ ♄
23 53	☉ ∥ ♆
19 Friday	
00 15	☽ # ♆
00 17	☽ # ☉
03 15	♀ ∠ ♆
05 16	☽ ⊥ ♂
08 08	☽ ± ♀
08 35	☽ ⊼ ♃
14 52	☽ ± ♄
17 25	☽ ☍ ☉
19 42	☽ ± ♃
21 51	☽ ∠ ♂
20 Saturday	
00 07	☽ △ ♇
00 10	☽ ☍ ♆
01 51	☽ ⊼ ♄
02 07	☽ ⚼ ♀
12 11	☽ ⚼ ♃
12 18	☽ □ ♅
19 55	♃ ∠ ♅
23 10	☽ ⊼ ☿
21 Sunday	
01 03	☽ ⚹ ♂
06 37	☉ ⊥ ♃
07 32	☽ △ ♀
11 15	☽ ± ☿
13 12	☽ # ♄
15 16	☽ △ ♃
16 20	☿ ⊼ ♂
18 01	☽ ∥ ♅
20 19	☽ Q ♅
22 29	☽ Q ♂
22 Monday	
03 12	☽ ⊼ ☉
05 12	☽ ⚼ ☿
05 33	☽ □ ♇
05 40	☽ ⊼ ♆
07 34	☽ ☍ ♄
07 42	☿ # ♃
08 06	☿ ∠ ♇
13 06	☉ ⚼ ♂
14 30	☽ ± ☉
16 10	☽ ± ♆
17 20	☽ ⚹ ♅
18 06	♀ Q ♄
23 Tuesday	
02 15	☿ ∠ ♄
06 13	☽ □ ♂
07 20	☽ ⚼ ☉
07 49	☽ ⚼ ♆
10 47	☽ △ ☿
12 07	☉ ⚹ ♇
12 52	☽ Q ♇
13 36	☿ # ♀
14 21	☉ ☌ ♆
16 49	☽ □ ♀
19 20	☽ ∠ ♅
20 14	☽ □ ♃
24 Wednesday	
09 32	☽ ⚹ ♇
09 42	☽ △ ♆
10 48	☽ ⊥ ♅
11 08	☽ △ ☉
11 50	☽ ⊼ ♄
15 32	☿ Q ♅
21 05	☽ ⊻ ♅
21 58	☉ ⊻ ♄
22 10	☽ ± ♄
22 56	♀ ∠ ♅
25 Thursday	
02 30	☽ Q ♀
03 29	☽ Q ♃
08 15	☽ ∥ ♅
10 19	☽ △ ♂
11 10	☽ ∠ ♇
11 52	☽ # ♄
13 37	☽ ⚼ ♄
16 15	♀ ☌ ♃
18 02	☿ ± ♂
20 58	☽ □ ☿
26 Friday	
00 10	☽ ⚹ ♃
00 46	☽ ⚹ ♀
02 28	☽ ⊥ ♇
04 44	♀ ∥ ♃
12 08	☽ ⚼ ♂
12 39	☽ ⊻ ♇
12 54	☽ □ ♆
15 15	☽ △ ♄
18 02	☽ □ ☉
27 Saturday	
00 04	☽ ☌ ♅
00 33	♀ ⊥ ♆
01 54	☽ ∠ ♃
03 11	☿ ⊻ ♃
03 34	☽ ± ♂
04 29	☽ ∠ ♀
06 26	☽ # ☉
07 30	☽ Q ☿
13 52	☽ ⊼ ♂
17 19	☽ ⊥ ♃
19 25	☽ Q ♆
21 05	☽ ⊥ ♀
21 44	☽ # ♆
28 Sunday	
02 45	☽ Q ☉
03 35	☽ ⊻ ♃
05 20	☿ ∥ ♂
06 29	☽ ⚹ ☿
07 02	☽ # ☿
07 21	☽ # ♂
08 08	☽ ⊻ ♀
15 25	☽ ☌ ♇
15 44	☽ ⚹ ♆
18 21	☽ □ ♄
21 25	☿ ⊥ ♄
29 Monday	
00 35	☽ ⚹ ☉
02 33	♂ ⚼ ♆
02 50	☽ ⊻ ♅
05 25	☽ ∥ ♃
11 08	☽ ∥ ♀
11 15	☽ ∠ ☿
13 01	☽ ⊥ ♅
17 13	☽ ∠ ♆
17 26	☽ ☍ ♂
30 Tuesday	
01 08	☽ Q ♄
03 41	☿ ⊻ ♀
04 00	☽ ∠ ☉
04 20	☽ ∠ ♅
04 37	☽ ⊥ ☿
07 08	☽ ☌ ♃
08 06	☽ # ♀
08 36	☽ ⊥ ♆
08 48	☉ □ ♅
15 12	☽ # ♃
15 47	☽ ☌ ♀
16 15	☽ ⊻ ☿
18 29	☽ ⊻ ♇
18 53	☽ ⊻ ♆
20 34	☽ ⊥ ☉
21 49	☽ ⚹ ♄
23 50	☿ ∥ ♆
31 Wednesday	
04 11	☿ ⊥ ♃
04 49	☽ ⊥ ♇
06 04	☽ ⚹ ♅
07 42	☽ ⊻ ☉
11 14	☿ ⚹ ♇
11 36	☽ ∥ ♂
14 54	☿ ☌ ♆
20 25	☽ ∠ ♇
21 52	☽ ⊼ ♂
23 34	☿ ⚼ ♂
23 53	♀ ⊻ ♇
23 59	☽ ∠ ♄

All ephemeris data is given at 12.00 UT and the Moon's longitude is additionally given for 24.00 UT

NOVEMBER 1956

LONGITUDES

Date	Sidereal time h m s	Sun ☉ ° ′ ″	Moon ☽ ° ′ ″	Moon ☽ 24.00 ° ′ ″	Mercury ☿ ° ′	Venus ♀ ° ′	Mars ♂ ° ′	Jupiter ♃ ° ′	Saturn ♄ ° ′	Uranus ♅ ° ′	Neptune ♆ ° ′	Pluto ♇ ° ′
01	14 42 59	09 ♏ 02 53	24 ♎ 06 37	00 ♏ 53 53	01 ♏ 55	00 ♎ 49	16 ♓ 12	23 ♍ 57	02 ♐ 18	06 ♌ 56	00 ♏ 29	00 ♍ 13
02	14 46 55	10 02 59	07 ♏ 37 08	14 ♏ 16 08	03 35	02 01	16 27	24 08	02 25	06 57	00 32	00 14
03	14 50 52	11 03 07	20 ♏ 50 43	27 ♏ 20 49	05 14	03 13	16 44	24 19	02 32	06 57	00 34	00 15
04	14 54 48	12 03 17	03 ♐ 46 25	10 ♐ 07 37	06 53	04 25	17 01	24 29	02 39	06 58	00 36	00 16
05	14 58 45	13 03 28	16 ♐ 24 33	22 ♐ 37 27	08 31	05 37	17 19	24 40	02 45	06 58	00 38	00 17
06	15 02 41	14 03 41	28 ♐ 46 37	04 ♑ 52 25	10 10	06 49	17 37	24 51	02 52	06 59	00 40	00 17
07	15 06 38	15 03 56	10 ♑ 55 17	16 ♑ 55 39	11 47	08 02	17 56	25 01	02 59	06 59	00 43	00 18
08	15 10 35	16 04 12	22 ♑ 54 03	28 ♑ 51 01	13 25	09 14	18 15	25 11	03 06	06 59	00 45	00 19
09	15 14 31	17 04 30	04 ♒ 47 07	10 ♒ 42 57	15 02	10 26	18 35	25 22	03 13	06 59	00 47	00 19
10	15 18 28	18 04 49	16 ♒ 39 05	22 ♒ 36 09	16 39	11 39	18 55	25 32	03 20	06 59	00 49	00 20
11	15 22 24	19 05 10	28 ♒ 34 43	04 ♓ 35 22	18 15	12 52	19 16	25 42	03 27	06 59	00 51	00 21
12	15 26 21	20 05 32	10 ♓ 38 41	16 ♓ 45 12	19 51	14 04	19 37	25 52	03 34	06 R 59	00 53	00 21
13	15 30 17	21 05 55	22 ♓ 55 22	29 ♓ 09 39	21 27	15 17	19 59	26 01	03 41	06 59	00 56	00 22
14	15 34 14	22 06 20	05 ♈ 28 26	11 ♈ 52 01	23 03	16 30	20 21	26 11	03 48	06 59	00 58	00 23
15	15 38 10	23 06 46	18 ♈ 20 37	24 ♈ 54 23	24 38	17 43	20 44	26 21	03 55	06 59	01 00	00 23
16	15 42 07	24 07 14	01 ♉ 33 21	08 ♉ 17 29	26 13	18 56	21 07	26 30	04 02	06 59	01 02	00 24
17	15 46 04	25 07 43	15 ♉ 06 36	22 ♉ 00 28	27 47	20 09	21 30	26 40	04 09	06 59	01 04	00 24
18	15 50 00	26 08 13	28 ♉ 58 42	06 ♊ 00 52	29 ♏ 22	21 23	21 54	26 49	04 16	06 58	01 06	00 24
19	15 53 57	27 08 45	13 ♊ 06 27	20 ♊ 14 54	00 ♐ 56	22 36	22 18	26 58	04 23	06 58	01 08	00 25
20	15 57 53	28 09 19	27 ♊ 25 33	04 ♋ 37 48	02 30	23 49	22 43	27 07	04 30	06 58	01 10	00 25
21	16 01 50	29 ♏ 09 55	11 ♋ 51 00	19 ♋ 04 30	04 04	25 03	23 08	27 16	04 37	06 57	01 12	00 26
22	16 05 46	00 ♐ 10 32	26 ♋ 17 44	03 ♌ 30 08	05 38	26 16	23 33	27 25	04 45	06 57	01 14	00 26
23	16 09 43	01 11 11	10 ♌ 41 13	17 ♌ 50 33	07 11	27 30	23 59	27 34	04 52	06 56	01 16	00 26
24	16 13 39	02 11 51	24 ♌ 57 46	02 ♍ 02 35	08 44	28 43	24 25	27 43	04 59	06 55	01 18	00 26
25	16 17 36	03 12 33	09 ♍ 04 45	16 ♍ 04 05	10 18	29 ♎ 57	24 51	27 51	05 06	06 55	01 20	00 27
26	16 21 33	04 13 17	23 ♍ 00 27	29 ♍ 53 45	11 51	01 ♏ 11	25 18	28 00	05 13	06 54	01 22	00 27
27	16 25 29	05 14 02	06 ♎ 43 54	13 ♎ 30 52	13 24	02 24	25 45	28 08	05 20	06 53	01 24	00 27
28	16 29 26	06 14 49	20 ♎ 14 37	26 ♎ 55 08	14 57	03 38	26 12	28 16	05 27	06 52	01 26	00 27
29	16 33 22	07 15 37	03 ♏ 32 23	10 ♏ 06 22	16 29	04 52	26 40	28 24	05 34	06 51	01 28	00 27
30	16 37 19	08 ♐ 16 26	16 ♏ 37 05	23 ♏ 04 33	18 ♐ 02	06 ♏ 06	27 ♓ 07	28 ♍ 32	05 ♐ 41	06 ♌ 50	01 ♏ 30	00 ♍ 27

Date	Moon True ☊ ° ′	Moon Mean ☊ ° ′	Moon ☽ Latitude ° ′
01	29 ♏ 03	29 ♏ 55	03 S 01
02	29 R 01	29 52	01 55
03	29 00	29 48	00 S 45
04	29 D 00	29 45	00 N 26
05	29 01	29 42	01 35
06	29 02	29 39	02 37
07	29 03	29 36	03 32
08	29 05	29 33	04 16
09	29 05	29 29	04 49
10	29 06	29 26	05 09
11	29 R 06	29 23	05 17
12	29 05	29 20	05 10
13	29 04	29 17	04 50
14	29 03	29 14	04 15
15	29 02	29 10	03 27
16	29 02	29 07	02 27
17	29 01	29 04	01 N 16
18	29 D 01	29 01	00 00
19	29 01	28 58	01 S 17
20	29 01	28 54	02 31
21	29 R 01	28 51	03 36
22	29 01	28 48	04 27
23	29 01	28 45	05 01
24	29 01	28 42	05 16
25	29 D 01	28 39	05 12
26	29 01	28 35	04 50
27	29 02	28 32	04 11
28	29 02	28 29	03 19
29	29 03	28 26	02 17
30	29 ♏ 04	28 ♏ 23	01 S 08

DECLINATIONS

Sun ☉ ° ′	Moon ☽ ° ′	Mercury ☿ ° ′	Venus ♀ ° ′	Mars ♂ ° ′	Jupiter ♃ ° ′	Saturn ♄ ° ′	Uranus ♅ ° ′	Neptune ♆ ° ′	Pluto ♇ ° ′
14 S 31	12 S 09	11 S 04	01 N 04	07 S 29	03 N 23	18 S 57	19 N 07	10 S 04	21 N 31
14 50	15 52	11 43	00 37	07 19	03 19	18 58	19 07	10 05	21 31
15 09	18 41	12 23	00 N 10	07 09	03 15	18 59	19 07	10 05	21 31
15 27	20 29	13 01	00 S 17	06 59	03 11	19 01	19 07	10 06	21 32
15 46	21 11	13 40	00 44	06 49	03 07	19 02	19 07	10 07	21 32
16 04	20 49	14 17	01 11	06 38	03 03	19 04	19 07	10 08	21 32
16 21	19 29	14 54	01 38	06 28	02 59	19 05	19 07	10 08	21 32
16 39	17 17	15 30	02 05	06 17	02 55	19 07	19 07	10 09	21 32
16 56	14 24	16 05	02 32	06 06	02 51	19 08	19 07	10 10	21 32
17 13	10 56	16 39	03 00	05 55	02 47	19 09	19 07	10 11	21 32
17 30	07 01	17 13	03 27	05 44	02 43	19 11	19 07	10 11	21 32
17 46	02 S 47	17 46	03 54	05 32	02 39	19 12	19 07	10 12	21 32
18 02	01 N 38	18 18	04 21	05 21	02 36	19 14	19 07	10 13	21 32
18 18	06 05	18 49	04 49	05 09	02 32	19 15	19 07	10 14	21 33
18 33	10 23	19 19	05 16	04 57	02 28	19 16	19 07	10 14	21 33
18 48	14 18	19 48	05 43	04 45	02 25	19 18	19 07	10 15	21 33
19 03	17 35	20 16	06 10	04 33	02 21	19 19	19 07	10 16	21 33
19 17	19 56	20 43	06 37	04 21	02 18	19 20	19 07	10 16	21 33
19 31	21 06	21 09	07 04	04 09	02 14	19 22	19 07	10 17	21 34
19 45	20 54	21 34	07 30	03 56	02 11	19 23	19 07	10 18	21 34
19 59	19 20	21 58	07 57	03 44	02 07	19 24	19 08	10 18	21 34
20 11	16 31	22 21	08 24	03 31	02 04	19 26	19 08	10 19	21 34
20 24	12 43	22 43	08 50	03 19	02 01	19 27	19 08	10 20	21 35
20 36	08 14	23 03	09 16	03 06	01 57	19 28	19 08	10 20	21 35
20 48	03 N 20	23 23	09 42	02 53	01 54	19 30	19 08	10 21	21 35
21 00	01 S 40	23 41	10 08	02 40	01 51	19 31	19 09	10 22	21 35
21 11	06 31	23 58	10 34	02 27	01 48	19 32	19 09	10 22	21 36
21 21	10 59	24 14	10 59	02 14	01 45	19 34	19 09	10 23	21 36
21 32	14 51	24 29	11 24	02 01	01 42	19 35	19 09	10 24	21 36
21 S 41	17 S 54	24 S 42	11 S 49	01 S 47	01 N 39	19 S 36	19 N 10	10 S 24	21 N 37

ZODIAC SIGN ENTRIES

Date	h	m	Planets
01	22	24	☽ ♏
04	04	56	☽ ♐
06	14	24	☽ ♑
09	02	19	☽ ♒
11	14	51	☽ ♓
14	01	36	☽ ♈
16	09	12	☽ ♉
18	13	45	☽ ♊
18	21	42	☿ ♐
20	16	18	☽ ♋
22	07	50	☉ ♐
22	18	10	☽ ♌
24	20	32	☽ ♍
25	13	01	♀ ♏
27	00	11	☽ ♎
29	05	34	☽ ♏

LATITUDES

Date	Mercury ☿ ° ′	Venus ♀ ° ′	Mars ♂ ° ′	Jupiter ♃ ° ′	Saturn ♄ ° ′	Uranus ♅ ° ′	Neptune ♆ ° ′	Pluto ♇ ° ′
01	01 N 09	01 N 31	02 S 13	01 N 04	01 N 43	00 N 36	01 N 41	10 N 51
04	00 50	01 37	02 01	01 05	01 43	00 36	01 41	10 53
07	00 30	01 41	01 50	01 05	01 42	00 36	01 41	10 54
10	00 N 10	01 45	01 40	01 06	01 42	00 36	01 41	10 55
13	00 S 10	01 48	01 30	01 06	01 42	00 36	01 41	10 56
16	00 30	01 51	01 20	01 07	01 42	00 36	01 41	10 57
19	00 49	01 52	01 11	01 07	01 41	00 36	01 42	10 58
22	01 07	01 53	01 03	01 08	01 41	00 37	01 42	11 00
25	01 24	01 52	00 55	01 09	01 41	00 37	01 42	11 01
28	01 39	01 51	00 47	01 09	01 41	00 37	01 42	11 02
31	01 S 52	01 N 50	00 S 40	01 N 10	01 N 41	00 N 37	01 N 42	11 N 03

DATA

Julian Date	2435779
Delta T	+32 seconds
Ayanamsa	23° 15′ 28″
Synetic vernal point	05° ♓ 51′ 31″
True obliquity of ecliptic	23° 26′ 37″

MOON'S PHASES, APSIDES AND POSITIONS ☽

Date	h	m	Phase	Longitude ° ′	Eclipse Indicator
02	16	44	●	10 ♏ 15	
10	15	09	☽	18 ♒ 13	
18	06	45	○	25 ♉ 55	total
25	01	13	☾	02 ♍ 45	

Day	h	m		
09	19	24	Apogee	
21	16	58	Perigee	
05	15	36	Max dec	21° S 11′
13	03	13	0N	
19	20	37	Max dec	21° N 11′
26	04	00	0S	

LONGITUDES

Date	Chiron ⚷ ° ′	Ceres ⚳ ° ′	Pallas ⚴ ° ′	Juno ⚵ ° ′	Vesta ⚶ ° ′	Black Moon Lilith ⚸ ° ′
01	06 ♒ 36	08 ♈ 47	17 ♒ 47	17 ♑ 15	19 ♐ 08	06 ♒ 53
11	06 ♒ 52	07 ♈ 31	18 ♒ 52	20 ♑ 22	24 ♐ 15	08 ♒ 00
21	07 ♒ 13	06 ♈ 53	20 ♒ 21	23 ♑ 45	29 ♐ 25	09 ♒ 07
31	07 ♒ 41	06 ♈ 54	22 ♒ 13	27 ♑ 23	04 ♑ 38	10 ♒ 14

All ephemeris data is given at 12.00 UT and the Moon's longitude is additionally given for 24.00 UT

ASPECTARIAN

h m	Aspects
01 Thursday	
00 15	☽ ∥ ♆
02 55	☽ Q ♅
04 37	☽ ∥ ☿
05 16	♀ ⊻ ♆
08 34	☽ ± ♂
09 16	☉ ∠ ♃
11 43	☽ ⊻ ♃
15 54	☽ ⊥ ♄
18 00	☿ ⊻ ♄
22 28	☽ ⊥ ♃
22 48	☽ ✶ ♇
23 18	☽ ☌ ♆
02 Friday	
00 46	☽ Ƥ ♂
01 00	☽ ⊻ ♀
02 37	☽ ⊻ ♄
03 45	☽ ☌ ☿
04 02	☽ ∥ ☉
10 48	☽ □ ♅
12 47	☽ ⊥ ♀
14 46	☽ ∠ ♃
16 44	☽ ☌ ☉
20 19	☽ Q ♇
20 58	♀ ✶ ♄
03 Saturday	
04 19	☽ △ ♂
06 42	☽ ∠ ♀
15 15	☽ ∥ ♄
16 34	☽ # ♅
18 28	☽ ✶ ♃
04 Sunday	
05 25	☽ □ ♇
06 02	☽ ⊻ ♆
09 52	☽ ☌ ♄
13 13	☿ □ ♅
13 20	☽ ✶ ♀
17 01	☉ Q ♇
17 11	☽ Q ♃
17 20	☽ ⊥ ♆
18 01	☽ △ ♅
18 43	☽ ⊻ ☿
05 Monday	
05 02	☽ ⊻ ☉
06 36	♃ ⊥ ♆
07 50	☽ ⊥ ☿
10 31	☽ ∠ ♆
13 47	☽ □ ♂
14 34	☽ Q ♀
17 33	☽ ⊥ ☉
22 44	☽ Ƥ ♅
06 Tuesday	
01 17	♀ ⊥ ♇
03 51	☽ ∠ ☿
04 12	☽ □ ♃
06 49	☿ ∠ ♃
12 36	☽ ∠ ☉
14 58	☽ △ ♇
15 06	♀ ✶ ♅
15 44	☽ ✶ ♆
16 19	☽ ± ♅
20 08	☽ ⊻ ♄
07 Wednesday	
01 50	☽ Q ♂
04 10	☽ ⊼ ♅
05 36	☽ □ ♀
08 07	☽ ⊥ ♄
14 00	☽ ✶ ☿
15 35	☽ Q ♆
16 44	☽ # ♅
16 58	☽ ∥ ♄
19 36	☿ Q ♇
20 45	☽ Ƥ ♇
21 02	☽ ✶ ☉
08 Thursday	
02 16	☽ ∠ ♄
02 24	☽ ✶ ♂
12 24	♄ # ♅
14 51	☽ ± ♇
16 41	☽ △ ♃
17 15	☽ ∥ ☉
17 52	☽ Q ☿
23 23	☽ Q ☉
09 Friday	
00 55	☽ ∥ ☿
02 58	☽ ⊼ ♇
03 53	☽ □ ♆
08 48	☽ ✶ ♄
09 30	☽ ∠ ♂
16 27	☽ ☍ ♅
23 26	☽ Ƥ ♃
10 Saturday	
00 45	☽ △ ♀
02 12	♀ # ♃
04 14	☽ ⊥ ♂
09 19	☽ Q ♄
11 59	☽ □ ☿
15 09	☽ □ ☉
16 43	☽ ⊻ ♂
16 46	☽ ∥ ♆
17 53	☽ ± ♃
11 Sunday	
06 08	☽ ⊼ ♃
10 24	☽ Ƥ ♀
15 32	☽ ☍ ♇
16 34	☽ △ ♆
18 42	☉ △ ♂
19 49	☽ ∥ ♂
21 50	☽ □ ♄
12 Monday	
04 46	☽ ⊼ ♅
06 21	☽ ± ♀
06 23	☽ ∥ ♀
06 50	♅ St R
07 31	☿ △ ♂
12 37	☉ ∥ ☿
12 43	☽ # ♃
14 00	☉ ⊥ ♀
16 37	☽ ± ♅
19 30	☽ ⊼ ♀
21 36	☉ ☌ ☿
22 21	☽ Ƥ ♆
13 Tuesday	
01 42	☿ ⊥ ♀
06 08	☽ ☌ ♂
08 09	☽ △ ☉
08 44	☽ △ ☿
10 12	☽ Ƥ ♅
13 33	♀ ∠ ♇
15 53	☽ ± ♆
17 07	☽ ∥ ♃
18 04	☽ ☍ ♃
14 Wednesday	
02 19	☽ ⊼ ♇
03 25	☽ ⊼ ♆
04 20	☽ # ♀
07 10	☽ # ♂
08 48	☽ △ ♄
13 42	☽ ± ♇
14 51	☽ △ ♅
15 20	☽ Ƥ ☉
17 32	☽ Ƥ ☿
15 Thursday	
00 37	♀ ∥ ♂
02 29	☿ # ♅
06 32	☽ Ƥ ♇
09 32	☽ ± ☉
10 04	☿ ∥ ♄
10 44	☽ ☍ ♀
11 10	☽ # ♆
12 36	☽ ± ☿
13 04	☽ Ƥ ♄
16 31	☽ ⊻ ♂
21 28	☽ ⊼ ☉
16 Friday	
01 04	☽ ⊼ ☿
02 47	☽ ⊼ ♃
03 46	☽ ⊥ ♂
05 36	☽ ± ♄
09 55	☽ △ ♇
11 03	☽ ☍ ♆
12 53	♀ Q ♅
13 43	☽ ± ♃
14 06	♀ ∠ ♄
16 28	☽ ⊼ ♄
16 57	☿ ✶ ♃
20 23	☽ ∠ ♂
21 41	☽ □ ♅
17 Saturday	
05 53	☽ Ƥ ♃
18 37	☉ # ♅
21 39	☽ ⊼ ♀
23 27	☽ ✶ ♂
18 Sunday	
02 17	☽ ∥ ♅
03 12	☽ # ☉
04 38	☽ # ♄
05 07	☽ Q ♅
06 45	☽ ☍ ☉
08 15	☽ △ ♃
08 59	☽ ± ♀
12 45	☽ ☍ ☿
14 27	☽ □ ♇
15 38	☽ ⊼ ♆
16 12	♂ Ƥ ♅
17 35	☉ ∥ ♄
20 39	☽ Q ♂
21 07	☽ ☍ ♄
19 Monday	
01 37	☽ ✶ ♅
01 48	☽ Ƥ ♀
01 53	☽ ± ♆
03 24	♀ ⊼ ♂
04 00	☿ □ ♇
07 07	☉ ✶ ♃
15 09	☿ ⊻ ♆
17 07	☽ Ƥ ♆
20 56	☽ Q ♇
20 Tuesday	
02 52	☽ ∠ ♅
03 54	☽ □ ♂
05 25	☽ △ ♀
11 29	☽ □ ♃
11 45	☿ # ♇
13 19	☽ ⊼ ☉
17 00	☽ ✶ ♇
17 53	☽ ⊥ ♅
18 15	☽ △ ♆
21 29	☽ ⊼ ☿
23 53	☽ ⊼ ♄
21 Wednesday	
00 03	☽ ± ☉
03 52	☽ ⊻ ♅
05 08	☽ # ☉
08 41	☽ ± ☿
09 57	☽ ± ♄
11 08	☽ # ♄
14 08	☽ ∥ ♅
16 08	☽ Ƥ ☉
17 45	☽ Q ♃
17 57	☽ ∠ ♇
21 17	☿ ☌ ♄
22 Thursday	
01 01	☽ Ƥ ♄
01 26	☽ Ƥ ☿
07 18	☽ △ ♂
08 54	☽ ⊥ ♇
11 57	☽ □ ♀
13 54	☽ ✶ ♃
18 06	☉ □ ♇
18 53	☽ ⊻ ♇
18 57	☽ △ ☉
20 15	☽ □ ♆
23 Friday	
02 11	☽ △ ♄
05 26	☽ △ ☿
05 44	☽ ☌ ♅
08 09	☿ △ ♅
09 03	☽ Ƥ ♂
13 23	☿ ⊥ ♆
13 39	♀ ⊻ ♃
14 06	☉ ⊻ ♆
15 11	☽ ∠ ♃
20 49	☽ Q ♀
24 Saturday	
00 37	☽ ± ♂
01 06	☽ # ♆
02 26	☽ Q ♆
06 28	☽ ⊥ ♃
07 08	☽ # ♀
11 03	☽ ⊼ ♂
16 42	☽ ⊻ ♃
17 36	♀ ⊥ ♄
18 58	☽ ✶ ♀
21 17	☽ ☌ ♇
22 46	☽ ✶ ♆
25 Sunday	
01 13	☽ □ ☉
04 32	☿ Q ♃
05 09	☽ □ ♄
08 18	☽ ⊻ ♅
14 16	☽ # ♂
14 20	☽ □ ☿
18 34	☽ ⊥ ♅
18 57	☽ ∥ ♃
21 43	♀ ✶ ♇
23 03	☽ ∠ ♀
26 Monday	
00 30	☽ ∠ ♆
10 05	☽ ∠ ♅
10 32	☽ Q ☉
12 22	☽ Q ♄
12 53	☽ # ♃
15 40	♀ ± ♂
15 54	♀ ☌ ♆
16 07	☽ ☍ ♂
16 07	☽ ⊥ ♆
16 08	☽ ⊥ ♀
16 20	♂ ± ♆
16 40	☽ ∥ ♂
20 46	☽ ☌ ♃
27 Tuesday	
00 58	☽ ⊻ ♇
01 12	♀ ∥ ♆
01 26	☽ Q ☿
02 37	☽ ⊻ ♆
03 39	☽ ⊻ ♀
09 09	☽ ✶ ☉
09 31	☽ ✶ ♄
11 30	☽ ⊥ ♇
12 16	☽ ✶ ♅
14 43	☉ ☌ ♄
28 Wednesday	
01 19	☽ ✶ ☿
03 27	☽ ∠ ♇
08 36	☽ ∥ ♆
09 33	☽ Q ♅
12 00	☽ ∥ ♀
12 23	☽ ∠ ♄
13 57	☽ ∠ ☉
23 05	☽ ⊼ ♂
29 Thursday	
01 51	♀ ⊥ ♃
02 35	☉ △ ♅
02 35	☽ ⊻ ♃
04 44	☽ ⊥ ♄
06 24	☽ ✶ ♇
07 31	☽ ⊥ ☉
07 47	☽ ∠ ☿
08 13	☽ ☌ ♆
10 21	☽ ± ♂
11 41	☿ ∠ ♆
13 35	☽ ⊥ ♃
14 40	☽ ☌ ♀
15 44	☽ ⊻ ♄
17 04	☉ ⊥ ♆
18 03	☽ □ ♅
19 21	☽ ⊻ ☉
30 Friday	
00 04	☉ # ♇
02 24	☽ ⊥ ☿
03 11	♀ ⊻ ♄
03 24	☽ Ƥ ♂
04 19	☽ Q ♇
06 15	☽ ∠ ♃
14 58	☽ ⊻ ☿

DECEMBER 1956

Date	Sidereal time h m s	Sun ☉ ° ' "	Moon ☽ ° ' "	Moon ☽ 24.00 ° ' "	Mercury ☿ ° '	Venus ♀ ° '	Mars ♂ ° '	Jupiter ♃ ° '	Saturn ♄ ° '	Uranus ♅ ° '	Neptune ♆ ° '	Pluto ♇ ° '
01	16 41 15	09 ♐ 17 18	29 ♏ 28 45	05 ♐ 49 45	19 ♐ 34	07 ♏ 20	27 ♓ 36	28 ♍ 40	05 ♐ 49	06 ♌ 49	01 ♏ 32	00 ♍ 27
02	16 45 12	10 18 10	12 ♐ 07 35	18 ♐ 22 19	21 07	08 34	28 04	28 48	05 56	06 R 48	01 34	00 R 27
03	16 49 08	11 19 03	24 ♐ 34 03	00 ♑ 42 55	22 39	09 48	28 33	28 55	06 03	06 47	01 36	00 27
04	16 53 05	12 19 58	06 ♑ 49 05	12 ♑ 52 46	24 11	11 02	29 02	29 02	06 10	06 46	01 37	00 27
05	16 57 02	13 20 53	18 ♑ 54 13	24 ♑ 53 42	25 43	12 16	29 ♓ 31	29 10	06 17	06 45	01 39	00 27
06	17 00 58	14 21 49	00 ♒ 51 34	06 ♒ 48 13	27 15	13 31	00 ♈ 01	29 17	06 24	06 44	01 41	00 27
07	17 04 55	15 22 46	12 ♒ 44 02	18 ♒ 39 30	28 ♐ 47	14 45	00 31	29 24	06 31	06 42	01 43	00 27
08	17 08 51	16 23 44	24 ♒ 35 07	00 ♓ 31 23	00 ♑ 18	15 59	01 01	29 31	06 38	06 41	01 44	00 27
09	17 12 48	17 24 42	06 ♓ 28 53	12 ♓ 28 10	01 49	17 13	01 31	29 37	06 45	06 40	01 46	00 27
10	17 16 44	18 25 41	18 ♓ 29 51	24 ♓ 34 30	03 20	18 28	02 01	29 44	06 52	06 38	01 48	00 26
11	17 20 41	19 26 40	00 ♈ 42 44	06 ♈ 55 07	04 50	19 42	02 32	29 50	06 59	06 37	01 49	00 26
12	17 24 37	20 27 40	13 ♈ 12 12	19 ♈ 34 29	06 20	20 56	03 03	29 ♍ 56	07 06	06 35	01 51	00 26
13	17 28 34	21 28 41	26 ♈ 02 26	02 ♉ 36 25	07 49	22 11	03 34	00 ♎ 02	07 13	06 34	01 53	00 25
14	17 32 31	22 29 42	09 ♉ 16 43	16 ♉ 03 29	09 18	23 25	04 05	00 08	07 20	06 32	01 54	00 25
15	17 36 27	23 30 44	22 ♉ 56 46	29 ♉ 56 27	10 45	24 40	04 37	00 14	07 27	06 30	01 56	00 25
16	17 40 24	24 31 46	07 ♊ 02 16	14 ♊ 13 45	12 11	25 54	05 09	00 20	07 34	06 29	01 57	00 24
17	17 44 20	25 32 49	21 ♊ 30 19	28 ♊ 51 11	13 37	27 09	05 41	00 25	07 41	06 27	01 59	00 24
18	17 48 17	26 33 52	06 ♋ 15 29	13 ♋ 42 11	15 00	28 23	06 13	00 31	07 48	06 25	02 00	00 23
19	17 52 13	27 34 57	21 ♋ 10 14	28 ♋ 38 31	16 22	29 ♏ 38	06 45	00 36	07 55	06 23	02 02	00 23
20	17 56 10	28 36 01	06 ♌ 05 57	13 ♌ 31 32	17 41	00 ♐ 53	07 17	00 41	08 02	06 21	02 03	00 22
21	18 00 06	29 ♐ 37 07	20 ♌ 54 19	28 ♌ 13 33	18 58	02 07	07 50	00 45	08 08	06 19	02 05	00 22
22	18 04 03	00 ♑ 38 13	05 ♍ 28 33	12 ♍ 38 51	20 12	03 22	08 23	00 50	08 15	06 17	02 06	00 21
23	18 08 00	01 39 19	19 ♍ 44 08	26 ♍ 44 11	21 23	04 37	08 56	00 55	08 22	06 15	02 07	00 21
24	18 11 56	02 40 27	03 ♎ 38 59	10 ♎ 28 35	22 29	05 51	09 29	00 59	08 29	06 13	02 09	00 20
25	18 15 53	03 41 35	17 ♎ 13 08	23 ♎ 52 52	23 31	07 06	10 02	01 03	08 35	06 11	02 10	00 19
26	18 19 49	04 42 44	00 ♏ 28 02	06 ♏ 58 58	24 28	08 21	10 35	01 07	08 42	06 09	02 11	00 19
27	18 23 46	05 43 53	13 ♏ 25 57	19 ♏ 49 19	25 18	09 36	11 09	01 11	08 49	06 07	02 13	00 18
28	18 27 42	06 45 03	26 ♏ 09 22	02 ♐ 26 22	26 01	10 50	11 42	01 14	08 55	06 05	02 14	00 17
29	18 31 39	07 46 13	08 ♐ 40 36	14 ♐ 52 18	26 36	12 05	12 16	01 18	09 02	06 03	02 15	00 16
30	18 35 35	08 47 23	21 ♐ 01 40	27 ♐ 08 53	27 03	13 20	12 50	01 21	09 08	06 01	02 16	00 16
31	18 39 32	09 ♑ 48 34	03 ♑ 14 09	09 ♑ 17 36	27 ♑ 20	14 ♐ 35	13 ♈ 24	01 ♎ 24	09 ♐ 15	05 ♌ 58	02 ♏ 17	00 ♍ 15

Date	Moon True ☊ ° '	Moon Mean ☊ ° '	Moon ☽ Latitude ° '
01	29 ♏ 04	28 ♏ 20	00 N 02
02	29 R 04	28 16	01 12
03	29 03	28 13	02 17
04	29 01	28 10	03 14
05	28 59	28 07	04 02
06	28 56	28 04	04 39
07	28 54	28 00	05 03
08	28 52	27 57	05 15
09	28 51	27 54	05 13
10	28 D 50	27 51	04 57
11	28 51	27 48	04 28
12	28 52	27 45	03 46
13	28 53	27 41	02 51
14	28 55	27 38	01 46
15	28 56	27 35	00 N 33
16	28 R 56	27 32	00 S 44
17	28 54	27 29	02 01
18	28 52	27 26	03 11
19	28 48	27 22	04 08
20	28 43	27 19	04 49
21	28 39	27 16	05 10
22	28 36	27 13	05 10
23	28 33	27 10	04 51
24	28 D 33	27 06	04 16
25	28 33	27 03	03 27
26	28 35	27 00	02 27
27	28 37	26 57	01 22
28	28 38	26 54	00 S 13
29	28 R 37	26 51	00 N 54
30	28 35	26 47	01 59
31	28 ♏ 30	26 ♏ 44	02 N 56

DECLINATIONS

Date	Sun ☉ ° '	Moon ☽ ° '	Mercury ☿ ° '	Venus ♀ ° '	Mars ♂ ° '	Jupiter ♃ ° '	Saturn ♄ ° '	Uranus ♅ ° '	Neptune ♆ ° '	Pluto ♇ ° '
01	21 S 51	20 00	24 S 54	12 S 14	01 S 34	01 N 36	19 S 38	19 N 10	10 S 25	21 N 37
02	22 00	21 04	25 05	12 38	01 20	01 33	19 39	19 10	10 25	21 37
03	22 08	21 03	25 14	13 03	01 07	01 30	19 40	19 10	10 26	21 38
04	22 17	20 02	25 22	13 26	00 53	01 28	19 41	19 11	10 27	21 38
05	22 24	18 07	25 29	13 50	00 40	01 25	19 43	19 11	10 27	21 39
06	22 32	15 26	25 34	14 13	00 26	01 22	19 44	19 11	10 28	21 39
07	22 38	12 08	25 38	14 36	00 S 12	01 20	19 45	19 12	10 28	21 39
08	22 45	08 23	25 40	14 59	00 N 02	01 17	19 46	19 12	10 29	21 40
09	22 51	04 S 17	25 41	15 21	00 16	01 15	19 47	19 13	10 29	21 40
10	22 56	00 N 01	25 41	15 43	00 30	01 13	19 49	19 13	10 30	21 41
11	23 01	04 23	25 38	16 04	00 44	01 10	19 50	19 13	10 31	21 41
12	23 06	08 41	25 35	16 25	00 58	01 08	19 51	19 14	10 31	21 42
13	23 10	12 43	25 30	16 46	01 12	01 06	19 52	19 14	10 32	21 42
14	23 14	16 16	25 23	17 06	01 26	01 04	19 53	19 15	10 32	21 43
15	23 17	19 03	25 16	17 26	01 40	01 02	19 54	19 15	10 33	21 43
16	23 20	20 46	25 06	17 45	01 55	01 00	19 55	19 16	10 33	21 44
17	23 22	21 10	24 55	18 04	02 09	00 58	19 57	19 16	10 34	21 44
18	23 24	20 07	24 43	18 23	02 23	00 56	19 58	19 17	10 34	21 45
19	23 25	17 42	24 30	18 40	02 38	00 54	19 59	19 17	10 34	21 45
20	23 26	14 05	24 15	18 58	02 52	00 52	20 00	19 18	10 35	21 46
21	23 27	09 38	23 59	19 15	03 06	00 51	20 01	19 18	10 35	21 46
22	23 27	04 N 41	23 42	19 31	03 21	00 49	20 02	19 19	10 36	21 47
23	23 26	00 S 24	23 23	19 47	03 35	00 47	20 03	19 19	10 36	21 47
24	23 25	05 22	23 04	20 02	03 50	00 46	20 04	19 20	10 37	21 48
25	23 24	09 57	22 45	20 17	04 04	00 45	20 05	19 20	10 37	21 49
26	23 22	13 56	22 24	20 31	04 19	00 43	20 06	19 21	10 37	21 49
27	23 19	17 10	22 04	20 45	04 33	00 42	20 07	19 21	10 38	21 50
28	23 16	19 31	21 43	20 58	04 48	00 41	20 08	19 22	10 38	21 50
29	23 13	20 51	21 23	21 10	05 02	00 40	20 09	19 22	10 38	21 51
30	23 09	21 10	21 03	21 22	05 17	00 39	20 10	19 23	10 39	21 51
31	23 S 05	20 S 28	20 S 44	21 S 33	05 N 31	00 N 38	20 S 11	19 N 24	10 S 39	21 N 52

ZODIAC SIGN ENTRIES

Date	h	m	Planets
01	12	59	☽ ♐
03	22	36	☽ ♑
06	10	16	☽ ♒
06	11	24	♂ ♈
08	07	11	☿ ♑
08	22	57	☽ ♓
11	10	37	☽ ♈
13	02	17	♃ ♎
13	19	15	☽ ♉
16	00	06	☽ ♊
18	01	52	☽ ♋
19	19	07	♀ ♐
20	02	11	☽ ♌
21	20	59	☉ ♑
22	02	56	☽ ♍
24	05	39	☽ ♎
26	11	09	☽ ♏
28	19	20	☽ ♐
31	05	37	☽ ♑

LATITUDES

Date	Mercury ☿ ° '	Venus ♀ ° '	Mars ♂ ° '	Jupiter ♃ ° '	Saturn ♄ ° '	Uranus ♅ ° '	Neptune ♆ ° '	Pluto ♇ ° '
01	01 S 52	01 N 50	00 S 40	01 N 10	01 N 41	00 N 37	01 N 42	11 N 03
04	02 03	01 47	00 33	01 11	01 41	00 37	01 42	11 04
07	02 12	01 44	00 26	01 11	01 41	00 37	01 42	11 06
10	02 16	01 41	00 20	01 12	01 40	00 37	01 42	11 07
13	02 17	01 36	00 14	01 13	01 40	00 37	01 42	11 08
16	02 14	01 31	00 09	01 13	01 40	00 37	01 43	11 09
19	02 04	01 26	00 S 03	01 14	01 40	00 37	01 43	11 10
22	01 47	01 20	00 N 02	01 15	01 40	00 38	01 43	11 11
25	01 22	01 14	00 06	01 16	01 40	00 38	01 43	11 12
28	00 47	01 07	00 11	01 17	01 41	00 38	01 43	11 13
31	00 S 02	01 N 00	00 N 15	01 N 17	01 N 41	00 N 38	01 N 43	11 N 14

DATA

Julian Date	2435809
Delta T	+32 seconds
Ayanamsa	23° 15' 33"
Synetic vernal point	05° ♓ 51' 27"
True obliquity of ecliptic	23° 26' 36"

MOON'S PHASES, APSIDES AND POSITIONS ☽

Date	h	m	Phase	Longitude ° '	Eclipse Indicator
02	08	13	●	10 ♐ 09	Partial
10	11	51	☽	18 ♓ 25	
17	19	06	○	25 ♊ 51	
24	10	10	☾	02 ♎ 36	

Day	h	m		
07	15	53	Apogee	
19	12	53	Perigee	
02	23	49	Max dec	21° S 12'
10	11	57	0N	
17	06	51	Max dec	21° N 12'
23	10	05	0S	
30	07	11	Max dec	21° S 11'

LONGITUDES

Date	Chiron ⚷ ° '	Ceres ⚳ ° '	Pallas ⚴ ° '	Juno ⚵ ° '	Vesta ⚶ ° '	Black Moon Lilith ⚸ ° '
01	07 ♒ 41	06 ♈ 54	22 ♒ 13	27 ♑ 23	04 ♑ 38	10 ♒ 14
11	08 ♒ 14	07 ♈ 31	24 ♒ 24	01 ♒ 13	09 ♑ 53	11 ♒ 22
21	08 ♒ 51	08 ♈ 43	26 ♒ 51	05 ♒ 14	15 ♑ 10	12 ♒ 29
31	09 ♒ 33	10 ♈ 25	29 ♒ 31	09 ♒ 24	20 ♑ 26	13 ♒ 36

ASPECTARIAN

h m Aspects

01 Saturday
00 57 ☽ ⋕ ♅
02 13 ♀ □ ♅
06 36 ☽ ∥ ♄
07 20 ♂ ⋕ ♃
08 19 ☽ △ ♂
10 27 ☽ ⚹ ♃
13 50 ☽ □ ♇
15 52 ☽ ⊻ ♆

02 Sunday
00 05 ☽ ☌ ♄
01 52 ☽ △ ♅
03 16 ☽ ⊥ ♆
04 28 ☽ ⊻ ♀
07 14 ♇ St R
08 13 ☽ ☌ ☉
09 25 ☽ Q ♃
17 11 ☽ ⊥ ♀
20 32 ☽ ∠ ♆
22 39 ☿ Ꝑ ♅

03 Monday
01 13 ☉ Q ♃
06 37 ☽ Ꝑ ♅
07 45 ☽ ☌ ☿
12 30 ☽ ∠ ♀
20 05 ☽ □ ♂
20 34 ☽ □ ♃
23 29 ☽ △ ♇

04 Tuesday
00 07 ☽ ± ♅
01 45 ☽ ⚹ ♆
10 42 ☽ ⊻ ♄
11 54 ☽ ⊼ ♅
12 35 ♂ ☍ ♃
17 13 ☽ ∥ ♄
21 17 ☽ ⚹ ♀
22 41 ☽ ⊥ ♄
23 55 ☽ ⊻ ☉
23 59 ☽ ⋕ ♅

05 Wednesday
01 31 ☽ Q ♆
05 07 ☽ Ꝑ ♇
09 07 ☽ Q ♂
12 58 ☽ ⊥ ☉
15 30 ♀ Q ♇
16 48 ☽ ∠ ♄
23 07 ☽ ± ♇
23 59 ☽ Q ♀

06 Thursday
03 40 ☽ ⊻ ☿
08 42 ☽ ∠ ☉
08 47 ☽ △ ♃
10 13 ☽ ⚹ ♂
11 11 ☽ ⊼ ♇
13 40 ☽ □ ♆
17 33 ☽ ⊥ ☿
20 19 ☽ ∥ ♀
23 18 ☽ ⚹ ♄
23 50 ☽ ☍ ♅

07 Friday
04 31 ♀ ∠ ♃
09 07 ♂ ⊼ ♇
14 26 ☽ ∠ ☿
15 24 ☽ Ꝑ ♃
16 33 ☽ □ ♀
17 52 ☽ ⚹ ☉
17 52 ☽ ∠ ♂
22 27 ☿ □ ♃
22 56 ☽ ∥ ♆
23 50 ☽ Q ♄

08 Saturday
09 48 ☽ ± ♃
12 51 ♀ Ꝑ ♂
12 54 ☽ ⊥ ♂
14 13 ☿ △ ♇
17 53 ☿ ± ♅
19 56 ♄ △ ♅
20 22 ☉ ∠ ♆
20 25 ☽ Q ☉
22 03 ☽ ⊼ ♃
23 50 ☽ ☍ ♇

09 Sunday
01 15 ☽ ⚹ ☿
01 33 ☽ ⊻ ♂
02 29 ☽ △ ♆
04 39 ☿ □ ♂
11 06 ☿ ⚹ ♆
12 22 ☽ ⊼ ♅
12 33 ☽ □ ♄

10 Monday
00 21 ☽ ± ♅
00 44 ♂ ⊼ ♆
04 49 ☽ Q ☿
05 13 ☽ ⋕ ♃
08 31 ☉ ⊻ ♀
08 37 ☽ Ꝑ ♆
09 21 ☽ ⋕ ♂
11 51 ☽ □ ☉
11 55 ☽ △ ♀
14 50 ☽ ∥ ♂
18 12 ☽ Ꝑ ♅
18 32 ☽ ∥ ♃
23 11 ☿ ∠ ♀

11 Tuesday
02 26 ☽ ± ♆
10 17 ☽ ☍ ♃
11 28 ☽ ⊼ ♇
14 10 ☽ ⊼ ♆
15 41 ☽ ☌ ♂
20 35 ☽ Ꝑ ♀
21 06 ☽ □ ☿
23 04 ☽ ± ♇
23 23 ☽ △ ♅

12 Wednesday
00 15 ☽ △ ♄
15 39 ☽ ± ♀
15 57 ☿ ⊼ ♅
16 13 ☽ Ꝑ ♇
22 41 ☽ ⋕ ♆

13 Thursday
01 28 ☿ ⊻ ♄
02 41 ♂ ∥ ♃
02 50 ☽ △ ☉
04 06 ☽ ⊼ ♀
04 52 ☽ Ꝑ ♄
13 52 ☉ Ꝑ ♅
19 23 ☽ ⊼ ♃
20 02 ☽ △ ♇
21 34 ☽ ± ♄
22 42 ☽ ☍ ♆

14 Friday
02 18 ☽ ⊻ ♂
06 20 ☽ ± ♃
07 06 ☽ □ ♅
08 30 ☽ ⊼ ♄
08 33 ☽ Ꝑ ☉
12 02 ☽ △ ☿
13 30 ☽ ⊥ ♂
19 18 ☽ ⋕ ♀
22 28 ☽ Ꝑ ♃

15 Saturday
01 47 ☽ ± ☉
06 00 ☽ ∠ ♂
13 03 ☽ ⊼ ☉
14 15 ☽ ∥ ♅
14 41 ☽ Q ♅
15 15 ☽ ☍ ♀
17 24 ☽ Ꝑ ☿
22 09 ☽ ⋕ ♄

16 Sunday
00 35 ☽ △ ♃
00 48 ☽ □ ♇
03 25 ☽ ⊼ ♆
08 42 ☽ ⚹ ♂
10 25 ☽ ± ☿
11 03 ☽ ⚹ ♅
12 54 ☽ ☍ ♄
13 33 ☽ ± ♆
21 33 ☽ ⊼ ☿

17 Monday
04 33 ☽ Ꝑ ♆
05 28 ☽ Q ♂
06 21 ♃ ⊻ ♇
06 54 ☽ Q ♇
11 54 ☽ ∠ ♅
13 25 ☿ ⊥ ♄
18 30 ☿ Q ♆
19 06 ☽ ☍ ☉
22 04 ☽ ⊼ ♀

18 Tuesday
02 30 ☽ ⚹ ♇
02 34 ☽ ⊥ ♅
02 38 ☽ □ ♃
05 07 ☽ △ ♆
08 42 ☽ ± ♀
11 55 ☽ □ ♂
12 15 ☽ ⊻ ♅
14 07 ☽ ⋕ ♄
14 30 ☽ ⊼ ♄
18 45 ☿ Ꝑ ♇
19 50 ♂ ± ♇
20 42 ♂ △ ♅
21 55 ☽ ∥ ♅

19 Wednesday
00 15 ☽ ± ♄
00 32 ☽ Ꝑ ♀
02 42 ☽ ∠ ♇
03 30 ☽ ☍ ☿
04 34 ☽ ⋕ ♀
07 50 ☽ Q ♃
14 49 ☽ Ꝑ ♄
17 09 ☽ ⊥ ♇
23 03 ☽ ⊼ ☉

20 Thursday
02 22 ♀ □ ♇
02 47 ☽ ⊻ ♇
02 50 ☽ △ ♀
03 13 ☽ ⚹ ♃
05 29 ☽ □ ♆
07 54 ♀ ⚹ ♃
09 24 ☽ ± ☉
12 25 ☽ ☌ ♅
14 00 ☽ △ ♂
15 08 ☽ △ ♄
19 20 ♄ ⊥ ♆

21 Friday
01 01 ☽ Ꝑ ☉
03 34 ☽ ∠ ♃
07 08 ☽ ⋕ ♆
08 33 ☽ ⊼ ☿
10 39 ☽ Q ♆
11 12 ♀ ⊻ ♆
15 16 ☽ Ꝑ ♂
16 50 ♀ ⋕ ♅
18 20 ☽ ⊥ ♃
19 16 ☽ ± ☿

22 Saturday
03 22 ☽ △ ☉
03 31 ☽ ☌ ♇
04 07 ☉ ± ♅
04 16 ☽ ⊻ ♃
05 08 ♂ △ ♄
05 23 ☉ △ ♇
06 23 ☽ ⚹ ♆
06 39 ☽ ± ♂
08 10 ☽ □ ♀
11 31 ☽ Ꝑ ☿
13 21 ☽ ⊻ ♅
16 40 ☽ □ ♄
17 02 ☉ □ ♃
17 02 ☽ ⊼ ♂
18 03 ☽ ∥ ♂
23 22 ☽ ⊥ ♅

23 Sunday
06 20 ☽ ∥ ♃
07 33 ☽ ∠ ♆
08 30 ☉ ∥ ☿
13 49 ☽ ⋕ ♃
14 35 ☽ ∠ ♅
15 03 ☽ △ ☿
17 23 ☽ Q ♀
22 58 ☽ ⊥ ♆
23 17 ☉ ⚹ ♆
23 27 ☽ Q ♄

24 Monday
04 02 ☽ ⋕ ♂
06 14 ☽ ⊻ ♇
07 20 ☽ ☌ ♃
09 22 ☽ ⊻ ♆
10 10 ☽ □ ☉
15 01 ♀ ∥ ♄
16 15 ☽ ⚹ ♀
16 30 ☽ ⚹ ♅
16 42 ☽ ⊥ ♇
18 57 ♀ △ ♅
20 33 ☽ ⚹ ♄
22 40 ☽ ☍ ♂

25 Tuesday
08 37 ☽ ∠ ♇
13 44 ☽ Q ♅
13 55 ☿ ∠ ♄
15 47 ☽ ∥ ♆
20 43 ☽ Q ☉
21 41 ☽ ∠ ♀
23 34 ☽ ∠ ♄

26 Wednesday
00 14 ☽ □ ☿
08 01 ☿ ± ♇
08 56 ♀ ⊥ ♆
11 43 ☽ ⚹ ♇
13 12 ☽ ⊻ ♃
15 10 ☽ ☌ ♆
15 49 ☽ ⊥ ♀
16 08 ☽ ⊥ ♄
19 31 ♀ ☌ ♄
20 28 ☽ ⚹ ☉
22 26 ☽ □ ♅

27 Thursday
00 18 ☽ ⊥ ♃
03 19 ☽ ⊻ ♄
04 05 ☽ ⊻ ♀
07 32 ☽ ⊼ ♂
09 53 ☽ Q ♇
11 44 ☽ Q ☿
17 10 ☽ ∠ ♃
19 17 ☽ ± ♂
20 51 ☉ ⊼ ♅

28 Friday
02 55 ☽ ∠ ☉
04 06 ☿ ⋕ ♇
10 06 ☽ ⋕ ♅
11 43 ☽ ⚹ ☿
13 06 ☽ Ꝑ ♂
19 04 ☿ ∠ ♀
19 52 ☽ □ ♇
21 04 ☽ ∥ ♄
21 32 ☽ ⊥ ☉
21 45 ☽ ⚹ ♃
23 37 ☽ ⊻ ♆

29 Saturday
06 57 ☽ △ ♅
10 06 ☽ ⊻ ☉
11 10 ☽ ⊥ ♆
12 41 ☽ ☌ ♄
17 54 ☽ ∠ ☿
18 23 ♀ △ ♂
19 17 ☽ △ ♂
19 20 ☽ ☌ ♀
20 59 ☽ Q ♃
21 26 ☿ ∥ ♀

30 Sunday
03 03 ☽ ∥ ☿
04 39 ☽ ∠ ♆
11 58 ☽ Ꝑ ♅
12 02 ☽ ⊥ ☿
12 21 ♀ Q ♃
21 13 ☉ ⊻ ♄

31 Monday
00 07 ☽ ⊻ ☿
02 42 ☽ ∥ ☿
05 35 ☽ ± ♅
06 06 ☽ △ ♇
08 22 ☽ □ ♃
10 07 ☽ ⚹ ♆
17 10 ☽ ∥ ♄
17 24 ☽ ⊼ ♅

All ephemeris data is given at 12.00 UT and the Moon's longitude is additionally given for 24.00 UT

JANUARY 1957

Date	Sidereal time h m s	LONGITUDES Sun ☉ ° ' "	Moon ☽ ° ' "	Moon ☽ 24.00 ° ' "	Mercury ☿ ° '	Venus ♀ ° '	Mars ♂ ° '	Jupiter ♃ ° '	Saturn ♄ ° '	Uranus ♅ ° '	Neptune ♆ ° '	Pluto ♇ ° '
01	18 43 29	10 ♑ 49 45	15 ♑ 19 24	21 ♑ 19 41	27 ♑ 26	15 ♐ 50	13 ♈ 58	01 ♎ 27	09 ♐ 21	05 ♌ 56	02 ♏ 18	00 ♍ 14
02	18 47 25	11 50 56	27 ♑ 18 37	03 ♒ 16 23	27 R 21	17 05	14 32	01 30	09 28	05 R 54	02 19	00 R 13
03	18 51 22	12 52 07	09 ♒ 13 11	15 ♒ 09 14	27 04	18 19	15 07	01 32	09 34	05 51	02 20	00 12
04	18 55 18	13 53 17	21 ♒ 04 47	27 ♒ 00 09	26 35	19 34	15 41	01 35	09 40	05 49	02 21	00 11
05	18 59 15	14 54 27	02 ♓ 55 41	08 ♓ 51 44	25 55	20 49	16 16	01 37	09 47	05 47	02 22	00 10
06	19 03 11	15 55 37	14 ♓ 48 45	20 ♓ 47 11	25 04	22 04	16 51	01 39	09 53	05 44	02 23	00 09
07	19 07 08	16 56 47	26 ♓ 47 34	02 ♈ 50 25	24 02	23 19	17 26	01 40	09 59	05 42	02 24	00 08
08	19 11 04	17 57 56	08 ♈ 56 20	15 ♈ 05 55	22 53	24 34	18 01	01 42	10 05	05 39	02 25	00 07
09	19 15 01	18 59 05	21 ♈ 19 45	27 ♈ 38 27	21 38	25 49	18 36	01 43	10 11	05 37	02 26	00 06
10	19 18 58	20 00 13	04 ♉ 02 37	10 ♉ 32 47	20 19	27 04	19 11	01 45	10 17	05 35	02 27	00 05
11	19 22 54	21 01 21	17 ♉ 09 28	23 ♉ 53 04	18 59	28 19	19 46	01 46	10 23	05 32	02 28	00 04
12	19 26 51	22 02 28	00 ♊ 43 53	07 ♊ 42 04	17 41	29 ♐ 34	20 21	01 47	10 29	05 30	02 28	00 03
13	19 30 47	23 03 35	14 ♊ 47 36	22 ♊ 00 16	16 26	00 ♑ 49	20 57	01 47	10 35	05 27	02 29	00 02
14	19 34 44	24 04 41	29 ♊ 19 39	06 ♋ 45 06	15 17	02 04	21 32	01 48	10 41	05 24	02 30	00 01
15	19 38 40	25 05 46	14 ♋ 15 42	21 ♋ 50 24	14 15	03 19	22 08	01 48	10 47	05 22	02 30	00 ♍ 00
16	19 42 37	26 06 51	29 ♋ 27 54	07 ♌ 06 50	13 22	04 34	22 43	01 R 48	10 52	05 19	02 31	29 ♌ 58
17	19 46 33	27 07 55	14 ♌ 45 46	22 ♌ 23 16	12 38	05 49	23 19	01 48	10 58	05 17	02 31	29 57
18	19 50 30	28 08 59	29 ♌ 57 57	07 ♍ 28 37	12 04	07 04	23 55	01 48	11 04	05 14	02 32	29 56
19	19 54 27	29 ♑ 10 02	14 ♍ 54 14	22 ♍ 13 59	11 39	08 19	24 31	01 47	11 09	05 12	02 32	29 55
20	19 58 23	00 ♒ 11 05	29 ♍ 27 16	06 ♎ 33 44	11 23	09 34	25 07	01 46	11 15	05 09	02 33	29 53
21	20 02 20	01 12 07	13 ♎ 33 13	20 ♎ 25 45	11 16	10 49	25 43	01 45	11 20	05 06	02 33	29 52
22	20 06 16	02 13 09	27 ♎ 11 32	03 ♏ 50 53	11 D 17	12 04	26 19	01 44	11 26	05 04	02 34	29 51
23	20 10 13	03 14 11	10 ♏ 24 12	16 ♏ 51 57	11 26	13 19	26 55	01 43	11 31	05 01	02 34	29 50
24	20 14 09	04 15 12	23 ♏ 14 41	29 ♏ 32 53	11 43	14 34	27 31	01 42	11 36	04 58	02 35	29 48
25	20 18 06	05 16 13	05 ♐ 47 07	11 ♐ 57 53	12 06	15 49	28 07	01 40	11 41	04 56	02 35	29 47
26	20 22 02	06 17 13	18 ♐ 05 40	24 ♐ 10 55	12 35	17 04	28 44	01 38	11 46	04 53	02 35	29 46
27	20 25 59	07 18 13	00 ♑ 14 02	06 ♑ 15 23	13 09	18 19	29 20	01 36	11 51	04 51	02 35	29 44
28	20 29 56	08 19 12	12 ♑ 15 19	18 ♑ 14 04	13 48	19 34	29 ♈ 56	01 34	11 56	04 48	02 36	29 43
29	20 33 52	09 20 10	24 ♑ 11 54	00 ♒ 09 02	14 32	20 49	00 ♉ 33	01 32	12 01	04 45	02 36	29 42
30	20 37 49	10 21 07	06 ♒ 05 37	12 ♒ 01 51	15 20	22 04	01 10	01 29	12 06	04 43	02 36	29 40
31	20 41 45	11 ♒ 22 03	17 ♒ 57 52	23 ♒ 53 50	16 ♑ 12	23 ♑ 19	01 ♉ 46	01 ♎ 26	12 ♐ 11	04 ♌ 40	02 ♏ 36	29 ♌ 39

Date	Moon True ☊ ° '	Moon Mean ☊ ° '	Moon ☽ Latitude ° '
01	28 ♏ 23	26 ♏ 41	03 N 46
02	28 R 15	26 38	04 24
03	28 06	26 35	04 51
04	27 57	26 31	05 05
05	27 48	26 28	05 06
06	27 42	26 25	04 54
07	27 37	26 22	04 29
08	27 35	26 19	03 51
09	27 D 34	26 16	03 02
10	27 35	26 12	02 03
11	27 36	26 09	00 N 56
12	27 R 37	26 06	00 S 17
13	27 35	26 03	01 31
14	27 31	26 00	02 41
15	27 25	25 57	03 43
16	27 17	25 53	04 29
17	27 08	25 50	04 57
18	26 58	25 47	05 04
19	26 50	25 44	04 49
20	26 44	25 41	04 16
21	26 40	25 37	03 28
22	26 39	25 34	02 30
23	26 D 39	25 31	01 25
24	26 39	25 28	00 S 18
25	26 R 39	25 25	00 N 48
26	26 37	25 22	01 51
27	26 32	25 18	02 48
28	26 24	25 15	03 37
29	26 13	25 12	04 15
30	26 00	25 09	04 42
31	25 ♏ 46	25 ♏ 06	04 N 57

Date	DECLINATIONS Sun ☉ ° '	Moon ☽ ° '	Mercury ☿ ° '	Venus ♀ ° '	Mars ♂ ° '	Jupiter ♃ ° '	Saturn ♄ ° '	Uranus ♅ ° '	Neptune ♆ ° '	Pluto ♇ ° '
01	23 S 00	18 S 50	20 S 26	21 S 44	05 N 46	00 N 37	20 S 12	19 N 24	10 S 39	21 N 53
02	22 55	16 23	20 09	21 54	06 06	00 36	20 13	19 25	10 40	21 53
03	22 49	13 16	19 54	22 03	06 15	00 35	20 14	19 25	10 40	21 54
04	22 43	09 39	19 40	22 12	06 29	00 34	20 15	19 26	10 40	21 55
05	22 37	05 40	19 29	22 20	06 44	00 34	20 16	19 26	10 41	21 55
06	22 30	01 S 28	19 19	22 27	06 58	00 33	20 16	19 27	10 41	21 56
07	22 22	02 N 50	19 12	22 34	07 13	00 33	20 17	19 28	10 41	21 57
08	22 14	07 05	19 07	22 40	07 27	00 32	20 18	19 28	10 41	21 57
09	22 06	11 08	19 04	22 45	07 42	00 32	20 19	19 29	10 42	21 58
10	21 57	14 48	19 02	22 50	07 56	00 32	20 20	19 30	10 42	21 59
11	21 48	17 51	19 02	22 53	08 11	00 32	20 21	19 30	10 42	21 59
12	21 38	20 02	19 04	22 57	08 25	00 32	20 21	19 31	10 42	22 00
13	21 28	21 04	19 07	22 59	08 39	00 32	20 22	19 31	10 42	22 01
14	21 18	20 45	19 11	23 01	08 54	00 32	20 23	19 32	10 43	22 01
15	21 07	18 59	19 17	23 02	09 08	00 32	20 24	19 33	10 43	22 02
16	20 56	15 52	19 23	23 03	09 22	00 32	20 24	19 33	10 43	22 03
17	20 44	11 40	19 30	23 03	09 36	00 32	20 25	19 34	10 43	22 03
18	20 32	06 45	19 37	23 01	09 51	00 33	20 26	19 35	10 43	22 04
19	20 20	01 N 30	19 46	23 00	10 05	00 33	20 26	19 35	10 43	22 05
20	20 07	03 S 42	19 54	22 57	10 19	00 34	20 27	19 36	10 43	22 05
21	19 54	08 33	20 03	22 54	10 33	00 34	20 28	19 37	10 43	22 06
22	19 40	12 48	20 12	22 50	10 47	00 35	20 28	19 37	10 44	22 07
23	19 26	16 18	20 20	22 46	11 01	00 36	20 29	19 38	10 44	22 07
24	19 12	18 53	20 29	22 41	11 15	00 37	20 30	19 39	10 44	22 08
25	18 57	20 29	20 37	22 35	11 28	00 38	20 30	19 39	10 44	22 09
26	18 42	21 04	20 45	22 28	11 42	00 39	20 31	19 40	10 44	22 09
27	18 27	20 38	20 52	22 21	11 56	00 40	20 32	19 40	10 44	22 10
28	18 11	19 17	20 59	22 13	12 09	00 41	20 32	19 41	10 44	22 11
29	17 55	17 05	21 05	22 04	12 23	00 42	20 33	19 42	10 44	22 12
30	17 39	14 11	21 10	21 55	12 37	00 44	20 33	19 42	10 44	22 12
31	17 S 22	10 S 44	21 S 15	21 S 45	12 N 50	00 N 45	20 S 34	19 N 43	10 S 44	22 N 13

ZODIAC SIGN ENTRIES

Date	h	m	Planets
02	17	25	☽ ♒
05	06	04	☽ ♓
07	18	23	☽ ♈
10	04	27	☽ ♉
12	10	44	☽ ♊
12	20	23	♀ ♑
14	13	06	☽ ♋
15	02	45	♇ ♌
16	12	50	☽ ♌
18	12	03	☽ ♍
20	07	39	☉ ♒
20	12	55	☽ ♎
22	17	02	☽ ♏
25	00	52	☽ ♐
27	11	32	☽ ♑
28	14	19	♂ ♉
29	23	42	☽ ♒

Date	LATITUDES Mercury ☿ ° '	Venus ♀ ° '	Mars ♂ ° '	Jupiter ♃ ° '	Saturn ♄ ° '	Uranus ♅ ° '	Neptune ♆ ° '	Pluto ♇ ° '
01	00 N 15	00 N 58	00 N 16	01 N 18	01 N 41	00 N 38	01 N 43	11 N 15
04	01 11	00 50	00 20	01 19	01 41	00 38	01 44	11 16
07	02 08	00 43	00 24	01 19	01 41	00 38	01 44	11 17
10	02 54	00 35	00 28	01 20	01 41	00 38	01 44	11 18
13	03 20	00 27	00 31	01 21	01 41	00 38	01 44	11 18
16	03 25	00 19	00 34	01 22	01 41	00 38	01 44	11 19
19	03 11	00 11	00 37	01 23	01 42	00 38	01 45	11 20
22	02 47	00 N 03	00 40	01 23	01 42	00 38	01 45	11 21
25	02 17	00 S 04	00 43	01 24	01 42	00 38	01 45	11 21
28	01 45	00 12	00 45	01 25	01 42	00 38	01 45	11 22
31	01 N 13	00 S 20	00 N 48	01 N 26	01 N 43	00 N 38	01 N 45	11 N 23

DATA

Julian Date	2435840
Delta T	+32 seconds
Ayanamsa	23° 15' 38"
Synetic vernal point	05° ♓ 51' 21"
True obliquity of ecliptic	23° 26' 36"

MOON'S PHASES, APSIDES AND POSITIONS ☽

Date	h	m	Phase	Longitude ° '	Eclipse Indicator
01	02	14	●	10 ♑ 25	
09	07	06	☽	18 ♈ 47	
16	06	21	○	25 ♋ 52	
22	21	48	☾	02 ♏ 38	
30	21	25	●	10 ♒ 45	

Day	h	m		
04	08	18	Apogee	
16	22	27	Perigee	
31	14	01	Apogee	
06	20	11	0N	
13	18	40	Max dec	21° N 08'
19	18	50	0S	
26	13	46	Max dec	21° S 04'

LONGITUDES

Date	Chiron ⚷ ° '	Ceres ⚳ ° '	Pallas ⚴ ° '	Juno ⚵ ° '	Vesta ⚶ ° '	Black Moon Lilith ⚸ ° '
01	09 ♒ 37	10 ♈ 36	29 ♒ 48	09 ♒ 50	20 ♑ 58	13 ♒ 43
11	10 ♒ 21	12 ♈ 47	02 ♓ 41	14 ♒ 09	26 ♑ 15	14 ♒ 50
21	11 ♒ 07	15 ♈ 20	05 ♓ 35	18 ♒ 35	01 ♒ 30	15 ♒ 57
31	11 ♒ 54	18 ♈ 11	08 ♓ 52	23 ♒ 08	06 ♒ 45	17 ♒ 04

ASPECTARIAN

h m Aspects

01 Tuesday
00 01 ☽ ⊻ ♄
02 14 ☽ ☌ ☉
04 56 ☽ # ♅
09 10 ☽ □ ♂
09 58 ☽ Q ♆
11 49 ☽ Ᵽ ♇
12 04 ☽ ⊥ ♄
13 07 ☽ ⊻ ♀
13 22 ☿ St R

02 Wednesday
02 31 ☽ ⊥ ♀
05 48 ☽ ± ♇
06 14 ☽ ∠ ♄
06 30 ☿ ∥ ♄
10 48 ♀ # ♇
12 04 ☽ ☌ ☿
16 49 ♀ ∠ ♆
17 50 ☽ ⊼ ♇
20 27 ☽ △ ♃
22 06 ☽ □ ♆
22 43 ☽ ∠ ♀
23 03 ☽ Q ♂

03 Thursday
05 14 ☽ ☍ ♅
12 43 ☽ ⚹ ♄
15 35 ♂ Ᵽ ♇
20 04 ☽ ⊻ ☉

04 Friday
00 32 ☽ ⚹ ♂
02 51 ☽ Ᵽ ♃
05 30 ☽ ∥ ♆
08 35 ☽ ⚹ ♀
09 22 ☽ ⊥ ☉
13 13 ☽ Q ♄
21 08 ☽ ± ♃
22 36 ☽ ⊻ ☿
23 13 ☉ Q ♆

05 Saturday
00 58 ☿ ⊥ ♀
05 19 ☽ ∠ ☉
06 06 ☽ # ♂
06 25 ☽ ☍ ♇
08 28 ☽ ∠ ♂
09 20 ☽ ⊼ ♃
10 04 ☽ ⊥ ☿
10 53 ☽ △ ♆
11 13 ♀ Ᵽ ♅
11 46 ☽ Q ♀
17 16 ☿ # ♅
17 45 ☽ ⊼ ♅
18 06 ☉ Ᵽ ♇

06 Sunday
01 58 ☽ □ ♄
03 06 ☽ ∠ ☿
03 36 ☽ ⊥ ♂
05 50 ☽ ± ♅
10 49 ☉ ⊥ ♄
14 27 ☽ ⚹ ☉
16 02 ☿ ∠ ♄
16 17 ☉ ∥ ♀
16 18 ☽ ⊻ ♂
17 05 ☽ # ♃
17 11 ☽ Ᵽ ♆
23 16 ☽ ∥ ♃
23 52 ☽ Ᵽ ♅

07 Monday
04 16 ☽ □ ♀
06 58 ☽ ⚹ ☿
09 49 ☿ ± ♇
11 13 ☽ ± ♆
16 41 ☽ Q ☉
18 38 ☽ ⊼ ♇
19 21 ☿ ⊻ ♀
21 43 ☽ ☍ ♃
23 09 ☽ ⊼ ♆

08 Tuesday
04 45 ☽ Q ☿
05 35 ☽ △ ♅
06 29 ☽ ± ♇
14 16 ☽ ∥ ♂
14 16 ☽ △ ♄
14 23 ☉ □ ♂

09 Wednesday
00 02 ☽ Ᵽ ♇
06 29 ☽ ☌ ♂
07 06 ☽ □ ☉
09 17 ☽ # ♆
12 32 ☽ □ ☿
19 25 ☽ Ᵽ ♄
21 29 ☽ △ ♀

10 Thursday
04 37 ☽ △ ♇
07 42 ☽ ⊼ ♃
08 33 ☉ # ♇
09 01 ☽ ☍ ♆
12 28 ☽ ± ♄
14 50 ☽ □ ♅
15 12 ☉ ☌ ☿
18 52 ☽ ± ♃
23 37 ☽ ⊼ ♄

11 Friday
02 14 ☿ □ ♂
04 19 ☽ Ᵽ ♀
11 17 ☽ Ᵽ ♃
14 59 ☽ △ ☿
16 54 ☽ ⊻ ♂
19 29 ☽ △ ☉
22 09 ☽ ± ♀
23 21 ☽ Q ♅
23 46 ☽ # ☿

12 Saturday
04 01 ☽ ⊥ ♂
05 05 ☽ ∥ ♅
09 46 ☽ ⊼ ♀
10 40 ♀ ± ♅
10 49 ☽ □ ♇
13 49 ☽ △ ♃
15 01 ☽ ⊼ ♆
15 05 ☽ Ᵽ ☿
17 05 ☽ # ♄
20 12 ☽ ⚹ ♅
20 20 ☽ ∠ ♂
21 12 ♀ △ ♇
23 43 ☽ Ᵽ ☉

13 Sunday
01 19 ☽ ± ♆
04 51 ☽ ☍ ♄
05 14 ☽ ± ☿
09 11 ☿ ⊥ ♄
14 32 ☽ ⊼ ☿
16 05 ☽ ± ☉
16 30 ☽ Ᵽ ♆
17 24 ☽ Q ♇
21 24 ☽ ∠ ♅
22 41 ☽ ⚹ ♂

14 Monday
02 46 ☽ ⊼ ☉
06 49 ♀ □ ♃
10 28 ☉ ± ♇
12 08 ☽ ⊥ ♅
13 07 ☽ ⚹ ♇
16 00 ☽ □ ♃
16 51 ☽ ☍ ♀
17 08 ☽ △ ♆
18 12 ☿ Ᵽ ♇
19 03 ☽ # ♄
19 06 ☽ Q ♂
20 21 ♀ ⚹ ♆
21 49 ☽ ⊻ ♅

15 Tuesday
06 00 ☿ Q ♆
06 14 ☽ ∥ ♅
06 25 ☽ ⊼ ♄
09 14 ☽ # ☿
12 00 ☽ ☍ ☿
13 10 ☽ ∠ ♇
16 01 ☽ ± ♄
20 47 ☽ Q ♃

16 Wednesday
00 58 ☽ □ ♂
03 22 ☽ ⊥ ♇
05 46 ☉ ∠ ♄
06 19 ☽ Ᵽ ♄
06 21 ☽ ☍ ☉
09 21 ♃ St R
12 48 ☽ ⊻ ♇
15 40 ☽ ⚹ ♃
16 47 ☽ □ ♆
20 43 ☽ ⊼ ♀
21 10 ☽ ☌ ♅

17 Thursday
02 07 ♀ ⊼ ♅
06 01 ☽ △ ♄
06 57 ☽ ± ♀
08 48 ☽ ⊼ ☿
15 12 ☽ ∠ ♃
16 52 ☽ # ♆
17 51 ☽ ± ☿
21 04 ☽ Q ♆
21 54 ☽ ∥ ♂
22 22 ☽ Ᵽ ♀

18 Friday
02 01 ☽ △ ♂
02 57 ☿ # ♅
05 23 ☽ ⊥ ♃
07 33 ☽ Ᵽ ☿
08 54 ☽ ⊼ ☉
11 57 ☽ ☌ ♇
14 54 ☽ ⊻ ♃
16 05 ☽ ⚹ ♆
19 09 ☽ ± ☉
20 23 ☽ ⊻ ♅
23 41 ☉ ∥ ♄

19 Saturday
00 22 ☽ △ ♀
02 55 ☽ Ᵽ ♂
05 53 ☽ □ ♄
06 00 ☽ ⊥ ♅
06 50 ☽ △ ☿
10 43 ☽ Ᵽ ☉
16 18 ☽ ∥ ♃
16 18 ☽ ∠ ♆
18 08 ☽ ± ♂
20 37 ☽ ∠ ♅
21 23 ☽ # ♃

20 Sunday
04 27 ☽ ⊼ ♂
05 13 ☉ ⊼ ♇
07 09 ☽ ⊥ ♆
11 39 ☽ Q ♄
12 44 ☽ ⊻ ♇
13 19 ☽ △ ☉
15 53 ☽ ☌ ♃
17 12 ☽ ⊻ ♆
21 34 ☽ ⚹ ♅
22 51 ☽ ⊥ ♇

21 Monday
01 45 ☿ ⊻ ♄
02 11 ☉ ∥ ☿
06 48 ☽ □ ♀
08 04 ☽ □ ☿
08 09 ☽ ⚹ ♄
14 17 ☽ ∠ ♇
18 09 ☽ Q ♅
19 57 ☿ St D
20 34 ☿ ☌ ♀
22 51 ♀ ⊻ ♄
23 23 ☽ # ♂
23 47 ☽ ∥ ♆

22 Tuesday
00 54 ☉ △ ♃
06 31 ♂ # ♆
10 21 ☽ ☍ ♂
10 37 ☽ ∠ ♄
15 47 ☽ Q ☿
16 46 ☽ ⚹ ♇
16 46 ☉ # ♅
17 16 ♂ Ᵽ ♄
17 41 ☽ Q ♀
20 10 ☽ ⊻ ♃
20 10 ☉ □ ♆
21 40 ☽ ☌ ♆
21 48 ☽ □ ☉

23 Wednesday
02 10 ☽ □ ♅
02 58 ☽ ⊥ ♄
07 04 ☽ ⊥ ♃
13 57 ☽ ⚹ ☿
14 04 ☽ ⊻ ♄
14 38 ☽ Q ♇
17 57 ☽ ⚹ ♀
23 28 ☿ ⊻ ♄
23 42 ☽ ∠ ♃

24 Thursday
09 58 ☽ Q ☉
12 14 ♀ Q ♆
15 00 ☿ ∥ ♄
15 21 ☽ ∥ ☉
16 34 ♀ Ᵽ ♇
18 46 ☽ ∠ ☿
20 32 ☽ ⊼ ♂
21 41 ☽ # ♅

25 Friday
00 28 ☽ □ ♇
01 22 ☽ ∠ ♀
04 05 ☽ ⚹ ♃
04 19 ☉ ☍ ♅
05 49 ☽ ⊻ ♆
08 38 ☽ ± ♂
10 21 ☽ △ ♅
10 55 ☽ ⚹ ☉
12 34 ☽ ∥ ♄
12 37 ☽ ⊥ ☿
15 44 ☽ ∥ ☿
17 25 ☽ ⊥ ♆
20 41 ☽ ⊥ ♀
23 32 ☽ ☌ ♄

26 Saturday
00 44 ☽ ⊻ ☿
03 00 ☽ Ᵽ ♂
03 17 ☽ Q ♃
09 45 ☽ ⊻ ♀
11 00 ☽ ∠ ♆
15 31 ☽ Ᵽ ♅
18 51 ☽ ∠ ☉

27 Sunday
02 35 ♀ ⊥ ♄
05 54 ☽ ∥ ☿
09 15 ☽ ± ♅
10 07 ☽ △ ♂
11 01 ☽ △ ♇
14 19 ☽ ⊥ ☉
14 43 ☽ □ ♃
14 51 ☽ ∥ ♄
16 41 ☽ ⚹ ♆
21 09 ☽ ⊼ ♅

28 Monday
03 23 ♂ △ ♇
03 24 ☽ ⊻ ☉
06 19 ☽ # ♅
11 22 ☽ ⊻ ♄
15 18 ☽ ☌ ☿
16 41 ☽ Q ♆
16 55 ☽ Ᵽ ♇
17 35 ♀ # ♇
23 29 ☽ ⊥ ♄

29 Tuesday
02 44 ☽ ∥ ☉
04 24 ☽ ☌ ♀
10 59 ☽ ± ♇
13 45 ☿ Q ♆
16 38 ☿ Ᵽ ♇
17 44 ☽ ∠ ♄
23 03 ☽ ⊼ ♇

30 Wednesday
01 30 ☽ □ ♂
02 43 ☽ △ ♃
04 56 ☽ □ ♆
09 13 ☽ ☍ ♅
21 25 ☽ ☌ ☉
22 40 ☽ # ♂
23 50 ♂ ⊼ ♃

31 Thursday
00 13 ☽ ⚹ ♄
08 09 ☽ ⊻ ☿
08 55 ☽ Ᵽ ♃
12 00 ☽ ∥ ♆
15 51 ☽ Q ♂
18 11 ♀ ± ♇
21 17 ☽ ⊥ ☿

All ephemeris data is given at 12.00 UT and the Moon's longitude is additionally given for 24.00 UT

FEBRUARY 1957

LONGITUDES

Date	Sidereal time h m s	Sun ☉ ° ' "	Moon ☽ ° ' "	Moon ☽ 24.00 ° ' "	Mercury ☿ ° '	Venus ♀ ° '	Mars ♂ ° '	Jupiter ♃ ° '	Saturn ♄ ° '	Uranus ♅ ° '	Neptune ♆ ° '	Pluto ♇ ° '
01	20 45 42	12 ♒ 22 57	29 ♒ 49 54	05 ♓ 46 15	17 ♑ 07	24 ♑ 07	02 ♉ 23	01 ♎ 23	12 ♐ 15	04 ♌ 38	02 ♏ 36	29 ♌ 37
02	20 49 38	13 23 51	11 ♓ 43 05	17 ♓ 40 38	18 06	25 49	02 59	01 R 20	12 20	04 R 35	02 36	29 R 36
03	20 53 35	14 24 44	23 ♓ 39 11	29 ♓ 39 01	19 07	27 04	03 36	01 17	12 25	04 32	02 R 36	29 34
04	20 57 31	15 25 35	05 ♈ 40 32	11 ♈ 44 06	20 10	28 19	04 13	01 13	12 29	04 30	02 36	29 33
05	21 01 28	16 26 24	17 ♈ 50 12	23 ♈ 59 18	21 16	29 ♑ 34	04 50	01 09	12 33	04 27	02 36	29 32
06	21 05 25	17 27 13	00 ♉ 11 56	06 ♉ 28 40	22 25	00 ♒ 49	05 27	01 06	12 38	04 25	02 36	29 30
07	21 09 21	18 28 00	12 ♉ 50 04	19 ♉ 16 42	23 35	02 04	06 04	01 02	12 42	04 22	02 36	29 29
08	21 13 18	19 28 45	25 ♉ 49 07	02 ♊ 27 49	24 47	03 19	06 41	00 57	12 46	04 20	02 35	29 27
09	21 17 14	20 29 29	09 ♊ 13 16	16 ♊ 05 48	26 01	04 34	07 18	00 53	12 50	04 17	02 35	29 26
10	21 21 11	21 30 11	23 ♊ 05 37	00 ♋ 12 47	27 16	05 49	07 55	00 49	12 54	04 15	02 35	29 24
11	21 25 07	22 30 52	07 ♋ 27 08	14 ♋ 48 17	28 33	07 04	08 32	00 44	12 58	04 12	02 35	29 23
12	21 29 04	23 31 31	22 ♋ 15 37	29 ♋ 48 17	29 ♑ 52	08 19	09 09	00 39	13 02	04 10	02 34	29 21
13	21 33 00	24 32 08	07 ♌ 25 11	15 ♌ 05 01	01 ♒ 12	09 34	09 46	00 34	13 06	04 07	02 34	29 20
14	21 36 57	25 32 44	22 ♌ 46 21	00 ♍ 27 39	02 33	10 49	10 23	00 29	13 09	04 05	02 34	29 18
15	21 40 54	26 33 18	08 ♍ 07 25	15 ♍ 44 09	03 55	12 04	11 00	00 24	13 13	04 03	02 33	29 17
16	21 44 50	27 33 51	23 ♍ 16 33	00 ♎ 43 29	05 19	13 19	11 37	00 18	13 16	04 00	02 33	29 15
17	21 48 47	28 34 23	08 ♎ 04 03	15 ♎ 17 35	06 43	14 34	12 15	00 12	13 20	03 58	02 32	29 14
18	21 52 43	29 ♒ 34 53	22 ♎ 23 42	29 ♎ 22 12	08 09	15 49	12 52	00 07	13 23	03 56	02 32	29 12
19	21 56 40	00 ♓ 35 22	06 ♏ 13 08	12 ♏ 56 42	09 36	17 04	13 29	00 ♎ 01	13 26	03 53	02 31	29 11
20	22 00 36	01 35 49	19 ♏ 33 14	26 ♏ 03 13	11 04	18 19	14 07	29 ♍ 55	13 29	03 51	02 31	29 09
21	22 04 33	02 36 16	02 ♐ 27 11	08 ♐ 45 44	12 34	19 34	14 44	29 49	13 32	03 49	02 30	29 08
22	22 08 29	03 36 41	14 ♐ 59 27	21 ♐ 09 00	14 04	20 49	15 21	29 43	13 35	03 47	02 29	29 06
23	22 12 26	04 37 05	27 ♐ 14 58	03 ♑ 17 58	15 35	22 04	15 59	29 36	13 38	03 45	02 29	29 05
24	22 16 23	05 37 27	09 ♑ 18 32	15 ♑ 17 13	17 07	23 18	16 36	29 30	13 41	03 43	02 28	29 03
25	22 20 19	06 37 48	21 ♑ 14 30	27 ♑ 10 47	18 40	24 33	17 13	29 23	13 43	03 40	02 27	29 02
26	22 24 16	07 38 07	03 ♒ 06 28	09 ♒ 01 53	20 15	25 48	17 51	29 16	13 46	03 38	02 27	29 00
27	22 28 12	08 38 24	14 ♒ 57 20	20 ♒ 53 04	21 50	27 03	18 28	29 09	13 49	03 36	02 26	28 59
28	22 32 09	09 ♓ 38 40	26 ♒ 49 17	02 ♓ 46 10	23 ♒ 26	28 ♒ 18	19 ♉ 06	29 ♍ 03	13 ♐ 51	03 ♌ 34	02 ♏ 25	28 ♌ 57

Date	Moon True ☊ ° '	Moon Mean ☊ ° '	Moon ☽ Latitude ° '
01	25 ♏ 31	25 ♏ 03	04 N 59
02	25 R 17	24 59	04 48
03	25 06	24 56	04 24
04	24 57	24 53	03 48
05	24 51	24 50	03 02
06	24 48	24 47	02 06
07	24 D 47	24 43	01 N 03
08	24 R 47	24 40	00 S 05
09	24 47	24 37	01 16
10	24 45	24 34	02 23
11	24 40	24 31	03 25
12	24 33	24 28	04 14
13	24 23	24 24	04 47
14	24 12	24 21	05 00
15	24 00	24 18	04 51
16	23 50	24 15	04 22
17	23 42	24 12	03 35
18	23 36	24 09	02 37
19	23 33	24 05	01 30
20	23 33	24 02	00 S 21
21	23 D 33	23 59	00 N 47
22	23 R 32	23 56	01 51
23	23 30	23 53	02 48
24	23 26	23 49	03 37
25	23 18	23 46	04 15
26	23 08	23 43	04 43
27	22 56	23 40	04 58
28	22 ♏ 42	23 ♏ 37	05 N 00

DECLINATIONS

Date	Sun ☉ ° '	Moon ☽ ° '	Mercury ☿ ° '	Venus ♀ ° '	Mars ♂ ° '	Jupiter ♃ ° '	Saturn ♄ ° '	Uranus ♅ ° '	Neptune ♆ ° '	Pluto ♇ ° '
01	17 S 05	06 S 52	21 S 19	21 S 34	13 N 03	00 N 46	20 S 34	19 N 44	10 S 44	22 N 14
02	16 48	02 S 44	21 21	21 23	13 17	00 48	20 35	19 44	10 44	22 14
03	16 31	01 N 31	21 23	21 11	13 30	00 49	20 35	19 45	10 44	22 15
04	16 13	05 45	21 24	20 59	13 43	00 51	20 36	19 45	10 44	22 16
05	15 55	09 48	21 23	20 46	13 56	00 52	20 36	19 46	10 43	22 16
06	15 36	13 31	21 22	20 32	14 09	00 54	20 37	19 47	10 43	22 17
07	15 18	16 41	21 19	20 17	14 22	00 56	20 37	19 47	10 43	22 18
08	14 59	19 08	21 15	20 02	14 35	00 58	20 38	19 48	10 43	22 18
09	14 40	20 36	21 10	19 47	14 48	01 00	20 38	19 48	10 43	22 19
10	14 20	20 52	21 04	19 31	15 01	01 02	20 38	19 49	10 43	22 20
11	14 01	19 50	20 57	19 14	15 13	01 04	20 39	19 50	10 43	22 20
12	13 41	17 25	20 48	18 57	15 26	01 06	20 39	19 50	10 43	22 21
13	13 21	13 47	20 39	18 39	15 38	01 08	20 39	19 51	10 42	22 22
14	13 00	09 12	20 28	18 21	15 50	01 10	20 40	19 51	10 42	22 22
15	12 40	04 N 01	20 15	18 02	16 03	01 13	20 40	19 52	10 42	22 23
16	12 19	01 S 20	20 02	17 42	16 15	01 15	20 40	19 52	10 42	22 23
17	11 58	06 30	19 47	17 22	16 27	01 18	20 41	19 53	10 42	22 24
18	11 37	11 08	19 31	17 02	16 39	01 20	20 41	19 54	10 41	22 25
19	11 16	15 01	19 13	16 41	16 50	01 23	20 41	19 54	10 41	22 25
20	10 55	17 58	18 55	16 20	17 02	01 25	20 42	19 55	10 41	22 26
21	10 33	19 53	18 35	15 58	17 14	01 28	20 42	19 55	10 41	22 27
22	10 11	20 46	18 14	15 36	17 25	01 30	20 42	19 56	10 40	22 27
23	09 49	20 37	17 51	15 13	17 37	01 33	20 42	19 56	10 40	22 28
24	09 27	19 31	17 27	14 50	17 48	01 36	20 42	19 57	10 40	22 28
25	09 05	17 34	17 02	14 26	17 59	01 39	20 43	19 57	10 39	22 29
26	08 42	14 53	16 35	14 02	18 10	01 41	20 43	19 58	10 39	22 29
27	08 20	11 36	16 08	13 38	18 21	01 44	20 43	19 58	10 39	22 30
28	07 S 57	07 S 53	15 S 38	13 S 13	18 N 32	01 N 47	20 S 43	19 N 58	10 S 39	22 N 31

ZODIAC SIGN ENTRIES

Date	h	m	Planets
01	12	20	☽ ♓
04	00	42	☽ ♈
05	20	16	♀ ♒
06	11	37	☽ ♉
08	19	34	☽ ♊
10	23	39	☽ ♋
12	14	30	☿ ♒
13	00	19	☽ ♌
14	23	17	☽ ♍
16	22	50	☽ ♎
18	21	58	☉ ♓
19	01	06	☽ ♏
19	15	37	♃ ♍
21	07	23	☽ ♐
23	17	27	☽ ♑
26	05	42	☽ ♒
28	18	25	☽ ♓

LATITUDES

Date	Mercury ☿ ° '	Venus ♀ ° '	Mars ♂ ° '	Jupiter ♃ ° '	Saturn ♄ ° '	Uranus ♅ ° '	Neptune ♆ ° '	Pluto ♇ ° '
01	01 N 03	00 S 22	00 N 48	01 N 26	01 N 43	00 N 38	01 N 45	11 N 23
04	00 32	00 29	00 51	01 27	01 43	00 38	01 45	11 23
07	00 N 04	00 36	00 53	01 28	01 43	00 38	01 46	11 24
10	00 S 22	00 43	00 55	01 28	01 43	00 38	01 46	11 24
13	00 46	00 49	00 56	01 29	01 44	00 38	01 46	11 25
16	01 07	00 55	00 58	01 30	01 44	00 38	01 46	11 25
19	01 26	01 00	01 00	01 30	01 44	00 38	01 46	11 25
22	01 41	01 05	01 01	01 31	01 45	00 38	01 47	11 25
25	01 54	01 10	01 03	01 31	01 45	00 38	01 47	11 25
28	02 03	01 14	01 04	01 32	01 46	00 38	01 47	11 25
31	02 S 09	01 S 17	01 N 05	01 N 32	01 N 46	00 N 38	01 N 47	11 N 25

DATA

Julian Date	2435871
Delta T	+32 seconds
Ayanamsa	23° 15' 43"
Synetic vernal point	05° ♓ 51' 16"
True obliquity of ecliptic	23° 26' 36"

MOON'S PHASES, APSIDES AND POSITIONS ☽

Date	h	m	Phase	Longitude ° '	Eclipse Indicator
07	23	23	☽	18 ♉ 57	
14	16	38	○	25 ♌ 44	
21	12	19	☾	02 ♐ 37	

Day	h	m		
14	11	19	Perigee	
27	15	27	Apogee	
03	03	26	0N	
10	05	19	Max dec	20° N 56'
16	05	59	0S	
22	20	23	Max dec	20° S 49'

LONGITUDES

Date	Chiron ⚷ ° '	Ceres ⚳ ° '	Pallas ⚴ ° '	Juno ⚵ ° '	Vesta ⚶ ° '	Black Moon Lilith ⚸ ° '
01	11 ♒ 59	18 ♈ 29	09 ♓ 11	23 ♒ 36	07 ♒ 16	17 ♒ 11
11	12 ♒ 46	21 ♈ 37	12 ♓ 28	28 ♒ 15	12 ♒ 28	18 ♒ 18
21	13 ♒ 31	24 ♈ 58	15 ♓ 49	02 ♓ 52	17 ♒ 37	19 ♒ 26
31	14 ♒ 15	28 ♈ 50	19 ♓ 13	07 ♓ 46	22 ♒ 42	20 ♒ 33

ASPECTARIAN

h m Aspects

01 Friday
00 06 ☽ ⊻ ♀
00 39 ☽ Q ♄
03 03 ☽ ± ♃
08 49 ☉ ⚹ ♄
11 35 ☽ ☍ ♇
13 40 ☽ ⊥ ♀
15 08 ☽ ⊼ ♃
17 02 ☽ ∠ ☿
17 26 ☽ ⚹ ♂
17 36 ☽ △ ♆
20 37 ♂ ☍ ♆
21 39 ☽ ⊼ ♅

02 Saturday
09 43 ☽ ± ♅
09 58 ☽ ∠ ♀
13 15 ☽ □ ♄
15 29 ☿ ∥ ♀
15 42 ☽ ⊻ ☉
15 51 ♆ St R
18 17 ☿ ⊥ ♄
22 54 ☽ # ♃
23 51 ☽ ⚼ ♆

03 Sunday
01 19 ☽ ∠ ♂
02 01 ☽ ⚹ ☿
03 46 ☽ ⚼ ♅
04 54 ☽ ⊥ ☉
08 00 ☽ ∥ ♃
17 54 ☽ ± ♆
18 59 ♀ ∠ ♄
19 38 ☽ ⚹ ♀
20 20 ☽ ⊥ ♂
23 49 ☽ ⊼ ♇

04 Monday
00 35 ☽ ∠ ☉
03 10 ☽ ☍ ♃
04 21 ☽ Q ☿
05 53 ☽ ⊼ ♆
08 57 ☽ ⊻ ♂
09 40 ☽ △ ♅
11 45 ☽ ± ♇
22 14 ♂ □ ♅
22 16 ☽ Q ♀

05 Tuesday
01 34 ☽ △ ♄
05 31 ☽ ⚼ ♇
05 43 ☉ ⚼ ♃
09 01 ☽ ⚹ ☉
11 11 ♀ ⊼ ♇
17 44 ☽ # ♆
19 23 ☽ □ ☿

06 Wednesday
03 54 ♀ ∥ ♄
07 01 ☽ ⚼ ♄
10 26 ☽ Q ☉
10 40 ☽ △ ♇
13 19 ☽ □ ♀
13 43 ☽ ⊼ ♃
16 36 ☽ ☍ ♆
16 50 ☽ ∥ ♂
17 00 ♀ △ ♃
20 02 ☽ □ ♅
22 33 ☽ ☌ ♂

07 Thursday
00 21 ☽ ± ♄
01 06 ☽ ± ♃
01 53 ☽ # ☉
09 57 ☿ ± ♇
11 45 ☽ ⊼ ♄
17 56 ☽ ⚼ ♃
22 01 ♀ □ ♆
23 23 ☽ □ ☉

08 Friday
05 38 ☽ Q ♅
09 55 ☽ △ ☿
15 58 ♂ ± ♄
18 34 ☽ □ ♇
20 59 ☽ ∥ ♅
21 15 ☽ △ ♃
21 45 ♂ ± ♃
22 51 ☽ # ♀

09 Saturday
00 13 ☽ ⊼ ♆
02 55 ☽ △ ♀
03 17 ☽ ⚹ ♅
05 51 ☉ # ♂
06 45 ♀ ☍ ♅
08 26 ☽ ⊻ ♂
09 35 ♀ # ♅
10 53 ☽ ± ♆
13 05 ☽ # ♄
15 28 ☽ ⚼ ☿
18 22 ☽ ☍ ♄
19 28 ☽ ⊥ ♂
20 14 ♃ Q ♄

10 Sunday
02 16 ☽ Q ♇
02 34 ☽ ⚼ ♆
05 27 ☽ ∠ ♅
07 45 ☽ ⚼ ♀
08 35 ☽ ± ☿
09 05 ☽ △ ☉
11 41 ☽ ∠ ♂
19 46 ☽ ⊼ ☿
20 41 ☽ ⊥ ♅
21 02 ☽ # ♄
22 38 ☽ ⚹ ♇

11 Monday
00 25 ☽ ± ♀
00 29 ☿ ∠ ♄
00 56 ☽ □ ♃
03 57 ☽ △ ♆
06 39 ☽ ⊻ ♅
11 19 ☽ ⊼ ♀
11 59 ☽ ∥ ♅
12 07 ☽ ⚼ ☉
13 51 ☽ ⚹ ♂
20 26 ☽ # ♀
21 03 ☽ ⊼ ♄
23 17 ☽ ∠ ♇

12 Tuesday
02 52 ☿ ⊼ ♇
03 51 ☽ ± ☉
06 14 ☽ Q ♃
06 48 ☽ ± ♄
10 08 ☽ Q ♂
13 45 ☽ ⊥ ♇
14 10 ☽ ⊼ ☉
21 14 ☽ ⚼ ♄
23 16 ☽ ⊻ ♇

13 Wednesday
01 15 ☽ ☍ ☿
01 16 ☽ ⚹ ♃
01 19 ☽ ∥ ♂
01 25 ☿ △ ♃
04 22 ☽ □ ♆
06 50 ☽ ☌ ♅
10 19 ☿ ∥ ♄
12 41 ☉ ± ♃
14 44 ☽ # ☉
15 40 ☽ ☍ ♀
15 50 ☽ □ ♂
19 31 ♀ □ ♂
20 55 ☽ △ ♄

14 Thursday
00 41 ☽ ∠ ♃
02 05 ☉ Q ♄
04 33 ☽ # ♆
08 33 ☽ Q ♆
12 15 ☿ □ ♆
14 39 ☽ ⊥ ♃
16 38 ☽ ☍ ☉
22 11 ☽ ☌ ♇
23 58 ☽ ⊻ ♃

15 Friday
03 17 ☽ ⚹ ♆
04 45 ☽ ⊼ ☿
05 37 ☽ ⊻ ♅
14 07 ☿ ☍ ♅
15 01 ☽ ⊥ ♅
15 06 ☽ ± ☿
16 43 ☽ △ ♂
18 46 ☽ ⊼ ♀
20 03 ☽ □ ♄

16 Saturday
00 28 ☽ ∥ ♃
02 52 ☽ ∠ ♆
05 07 ☽ ± ♀
05 12 ☽ ∠ ♅
06 47 ☽ ⚼ ☿
11 04 ♀ ⚹ ♄
11 36 ☽ # ♃
17 15 ☽ ⊥ ♆
17 36 ☽ ⚼ ♂
19 23 ☽ ⊼ ☉
20 51 ☽ ⚼ ♀
21 36 ☽ ⊻ ♇
23 15 ☽ ☌ ♃

17 Sunday
00 56 ☽ Q ♄
02 49 ☿ # ♅
02 57 ☽ ⊻ ♆
05 18 ☽ ⚹ ♅
05 51 ☽ ± ☉
07 21 ☽ ⊥ ♇
08 52 ☽ ± ♂
09 33 ☽ △ ☿
19 13 ☽ ⊼ ♂
20 45 ☽ ⚹ ♄
21 49 ☽ ⚼ ☉
22 12 ☽ ∠ ♇
23 29 ♀ ⚼ ♃
23 49 ☽ △ ♀

18 Monday
01 06 ☽ Q ♅
03 13 ☉ ☍ ♇
09 31 ☽ ∥ ♆
14 31 ☽ ∥ ☉
22 20 ☽ ∠ ♄
23 32 ☉ ⊼ ♃
23 41 ☽ ⚹ ♇

19 Tuesday
01 12 ☽ ⊻ ♃
01 20 ☽ △ ☉
05 07 ♀ # ♂
05 30 ☽ ☌ ♆
07 54 ☽ □ ♅
09 49 ♂ ⊼ ♄
11 39 ☽ ⊥ ♃
14 10 ☽ ⊥ ♄
18 45 ☽ □ ☿
20 49 ☽ Q ♇
23 14 ☽ ∥ ♀

20 Wednesday
00 56 ☽ ⊻ ♄
01 37 ☽ ☍ ♂
02 54 ☽ # ♂
03 37 ☽ ∠ ♃
09 30 ☽ □ ♀
20 50 ☽ ∥ ☿

21 Thursday
03 23 ☉ ∥ ♆
05 45 ☽ □ ♇
07 04 ☽ ⚹ ♃
07 57 ☽ Q ☿
09 34 ☉ △ ♆
12 05 ☽ ⊻ ♆
12 19 ☽ □ ☉
12 33 ☽ # ♅
14 34 ☽ △ ♅
14 43 ♂ ⚼ ♃
19 53 ☉ Q ♂
22 47 ☽ Q ♀
23 29 ☽ ⊥ ♆

22 Friday
04 11 ☿ ⚹ ♄
05 43 ☽ Q ♃
08 33 ☽ ∥ ♄
09 16 ☽ ☌ ♄
09 57 ☽ ⚹ ☿
12 44 ☽ ⊼ ♂
15 52 ☉ ⊼ ♅
16 51 ☽ ∠ ♆
19 21 ☽ ⚼ ♅
21 35 ☿ ⚼ ♃

23 Saturday
00 37 ☽ ⚹ ♀
01 04 ☽ ± ♂
02 04 ☽ Q ☉
08 07 ☽ ∥ ♄
12 58 ☽ ± ♅
15 37 ☽ △ ♇
16 37 ☽ □ ♃
19 33 ☽ ∠ ☿
19 47 ☽ ⚼ ♂
21 57 ☿ # ♂
22 22 ☽ ⚹ ♆
22 24 ☿ □ ♂

24 Sunday
00 51 ☽ ⊼ ♅
03 58 ☽ ⚹ ☉
04 43 ☽ # ♅
09 46 ☽ ∠ ♀
15 17 ♀ ± ♃
16 10 ☽ ⊥ ☿
20 48 ☽ ⊻ ♄
21 30 ☽ ⚼ ♇
22 21 ☽ Q ♆

25 Monday
03 27 ☽ △ ♂
05 57 ☽ ⊥ ♀
06 02 ☽ ⊻ ☿
07 53 ☽ # ♂
08 56 ☽ ⊥ ♄
12 51 ☽ ∠ ☉
15 36 ☽ ± ♇
18 20 ☽ ∥ ☿
19 29 ☽ ⊻ ♀

26 Tuesday
03 11 ☽ ∠ ♄
03 42 ☽ ⊼ ♇
04 18 ☽ △ ♃
08 44 ☽ ⊥ ☉
10 39 ☽ □ ♆
11 14 ♀ Q ♄
13 04 ☽ ☍ ♅
19 30 ☽ ∥ ♀
22 01 ☽ ⊻ ☉

27 Wednesday
09 40 ☽ ⚹ ♄
10 24 ☽ ⚼ ♃
18 25 ☽ ∥ ♆
19 31 ☽ □ ♂

28 Thursday
04 05 ☽ ☌ ☿
04 26 ☽ ± ♃
06 32 ☿ ± ♃
10 02 ☽ Q ♄
10 20 ☉ ± ♅
11 28 ☽ ∥ ☉
15 20 ☽ ☌ ♀
16 18 ☽ ☍ ♇
16 26 ☽ ⊼ ♃
23 17 ☽ △ ♆

All ephemeris data is given at 12.00 UT and the Moon's longitude is additionally given for 24.00 UT

MARCH 1957

LONGITUDES

Date	Sidereal time h m s	Sun ☉ ° ' "	Moon ☽ ° ' "	Moon ☽ 24.00 ° ' "	Mercury ☿ ° '	Venus ♀ ° '	Mars ♂ ° '	Jupiter ♃ ° '	Saturn ♄ ° '	Uranus ♅ ° '	Neptune ♆ ° '	Pluto ♇ ° '
01	22 36 05	10 ♓ 38 55	08 ♓ 43 53	14 ♓ 42 35	25 ♒ 04	29 ♒ 33	19 ♉ 43	28 ♍ 56	13 ♐ 53	03 ♌ 32	02 ♏ 24	28 ♌ 56
02	22 40 02	11 39 07	20 ♓ 42 25	26 ♓ 43 30	26 42	00 ♓ 48	20 21	28 R 48	13 55	03 R 31	02 R 23	28 R 54
03	22 43 58	12 39 17	02 ♈ 46 01	08 ♈ 50 09	28 ♒ 21	02 03	20 58	28 41	13 57	03 29	02 22	28 53
04	22 47 55	13 39 26	14 ♈ 56 06	21 ♈ 04 06	00 ♓ 02	03 18	21 36	28 34	13 59	03 27	02 21	28 51
05	22 51 52	14 39 33	27 ♈ 14 26	03 ♉ 27 24	01 43	04 32	22 13	28 27	14 01	03 25	02 21	28 50
06	22 55 48	15 39 37	09 ♉ 43 22	16 ♉ 02 42	03 26	05 47	22 51	28 19	14 03	03 23	02 20	28 48
07	22 59 45	16 39 40	22 ♉ 25 50	28 ♉ 53 11	05 10	07 02	23 29	28 12	14 05	03 22	02 19	28 47
08	23 03 41	17 39 41	05 ♊ 25 11	12 ♊ 02 16	06 55	08 17	24 06	28 04	14 06	03 20	02 18	28 46
09	23 07 38	18 39 39	18 ♊ 44 49	25 ♊ 33 09	08 41	09 32	24 44	27 57	14 08	03 18	02 16	28 44
10	23 11 34	19 39 35	02 ♋ 27 46	09 ♋ 28 11	10 28	10 46	25 21	27 49	14 09	03 17	02 15	28 43
11	23 15 31	20 39 29	16 ♋ 34 59	23 ♋ 47 49	12 16	12 01	25 59	27 41	14 11	03 15	02 14	28 41
12	23 19 27	21 39 21	01 ♌ 06 20	08 ♌ 29 59	14 05	13 16	26 37	27 34	14 12	03 14	02 12	28 40
13	23 23 24	22 39 10	15 ♌ 58 00	23 ♌ 29 26	15 56	14 31	27 14	27 26	14 13	03 12	02 12	28 39
14	23 27 20	23 38 57	01 ♍ 03 10	08 ♍ 37 59	17 48	15 45	27 52	27 18	14 14	03 11	02 11	28 37
15	23 31 17	24 38 42	16 ♍ 12 33	23 ♍ 45 32	19 40	17 00	28 30	27 11	14 15	03 09	02 10	28 36
16	23 35 14	25 38 25	01 ♎ 15 41	08 ♎ 41 49	21 34	18 15	29 07	27 03	14 16	03 08	02 08	28 35
17	23 39 10	26 38 06	16 ♎ 02 55	23 ♎ 18 09	23 30	19 29	29 ♉ 45	26 55	14 16	03 07	02 07	28 33
18	23 43 07	27 37 46	00 ♏ 26 53	07 ♏ 28 43	25 26	20 44	00 ♊ 23	26 47	14 17	03 06	02 06	28 32
19	23 47 03	28 37 23	14 ♏ 23 26	21 ♏ 11 00	27 23	21 59	01 00	26 39	14 17	03 04	02 05	28 31
20	23 51 00	29 ♓ 36 59	27 ♏ 51 34	04 ♐ 25 24	29 ♓ 21	23 13	01 38	26 32	14 18	03 03	02 03	28 30
21	23 54 56	00 ♈ 36 33	10 ♐ 52 54	17 ♐ 14 34	01 ♈ 21	24 28	02 16	26 24	14 18	03 02	02 02	28 28
22	23 58 53	01 36 05	23 ♐ 30 55	29 ♐ 42 33	03 21	25 43	02 53	26 16	14 18	03 01	02 01	28 27
23	00 02 49	02 35 35	05 ♑ 50 05	11 ♑ 54 08	05 21	26 57	03 31	26 08	14 18	03 00	01 59	28 26
24	00 06 46	03 35 04	17 ♑ 55 19	23 ♑ 54 15	07 23	28 12	04 09	26 01	14 R 18	02 59	01 58	28 25
25	00 10 43	04 34 31	29 ♑ 51 31	05 ♒ 47 38	09 25	29 ♓ 27	04 46	25 53	14 18	02 59	01 57	28 24
26	00 14 39	05 33 56	11 ♒ 43 08	17 ♒ 38 30	11 27	00 ♈ 41	05 24	25 46	14 18	02 58	01 55	28 22
27	00 18 36	06 33 19	23 ♒ 34 08	29 ♒ 30 26	13 29	01 56	06 02	25 38	14 18	02 57	01 54	28 21
28	00 22 32	07 32 40	05 ♓ 27 44	11 ♓ 26 21	15 30	03 10	06 40	25 30	14 17	02 56	01 52	28 20
29	00 26 29	08 32 00	17 ♓ 26 30	23 ♓ 28 25	17 32	04 25	07 17	25 23	14 17	02 56	01 51	28 19
30	00 30 25	09 31 17	29 ♓ 32 18	05 ♈ 38 16	19 32	05 39	07 55	25 16	14 16	02 55	01 49	28 18
31	00 34 22	10 ♈ 30 32	11 ♈ 46 29	17 ♈ 57 03	21 ♈ 31	06 ♈ 54	08 ♊ 33	25 ♍ 08	14 ♐ 16	02 ♌ 54	01 ♏ 48	28 ♌ 17

Moon Nodes and Latitude

Date	Moon True ☊ ° '	Moon Mean ☊ ° '	Moon ☽ Latitude ° '
01	22 ♏ 28	23 ♏ 34	04 N 49
02	22 R 15	23 30	04 25
03	22 04	23 27	03 49
04	21 55	23 24	03 02
05	21 50	23 21	02 07
06	21 47	23 18	01 N 04
07	21 D 46	23 15	00 S 04
08	21 47	23 11	01 12
09	21 R 47	23 08	02 19
10	21 46	23 05	03 19
11	21 44	23 02	04 09
12	21 38	22 59	04 45
13	21 31	22 55	05 03
14	21 23	22 52	05 00
15	21 14	22 49	04 36
16	21 05	22 46	03 53
17	20 59	22 43	02 55
18	20 55	22 40	01 47
19	20 53	22 36	00 S 35
20	20 D 53	22 33	00 N 37
21	20 54	22 30	01 46
22	20 55	22 27	02 46
23	20 R 56	22 24	03 38
24	20 54	22 20	04 19
25	20 51	22 17	04 48
26	20 46	22 14	05 04
27	20 39	22 11	05 08
28	20 31	22 08	04 58
29	20 23	22 05	04 35
30	20 15	22 01	03 59
31	20 ♏ 08	21 ♏ 58	03 N 12

DECLINATIONS

Date	Sun ☉ ° '	Moon ☽ ° '	Mercury ☿ ° '	Venus ♀ ° '	Mars ♂ ° '	Jupiter ♃ ° '	Saturn ♄ ° '	Uranus ♅ ° '	Neptune ♆ ° '	Pluto ♇ ° '
01	07 S 35	03 S 50	15 S 08	12 S 48	18 N 42	01 N 50	20 S 43	19 N 59	10 S 38	22 N 31
02	07 12	00 N 23	14 36	12 23	18 53	01 53	20 43	19 59	10 38	22 32
03	06 49	04 36	14 03	11 57	19 03	01 56	20 44	20 00	10 37	22 32
04	06 26	08 41	13 29	11 31	19 13	01 59	20 44	20 00	10 37	22 33
05	06 03	12 28	12 54	11 05	19 24	02 02	20 44	20 01	10 37	22 33
06	05 39	15 44	12 17	10 38	19 34	02 05	20 44	20 01	10 36	22 34
07	05 16	18 19	11 39	10 11	19 43	02 08	20 44	20 01	10 36	22 34
08	04 53	20 02	10 59	09 44	19 53	02 11	20 44	20 02	10 36	22 35
09	04 29	20 40	10 18	09 17	20 03	02 14	20 44	20 02	10 35	22 35
10	04 06	20 06	09 36	08 49	20 12	02 17	20 44	20 02	10 35	22 36
11	03 42	18 17	08 53	08 21	20 22	02 20	20 44	20 03	10 34	22 36
12	03 19	15 16	08 09	07 53	20 31	02 24	20 44	20 03	10 34	22 36
13	02 55	11 14	07 23	07 24	20 40	02 27	20 44	20 03	10 33	22 37
14	02 31	06 25	06 36	06 56	20 49	02 30	20 44	20 04	10 33	22 37
15	02 08	01 N 12	05 48	06 27	20 58	02 33	20 44	20 04	10 33	22 38
16	01 44	04 S 04	04 59	05 58	21 06	02 36	20 44	20 04	10 32	22 38
17	01 20	09 01	04 09	05 29	21 15	02 39	20 44	20 05	10 32	22 39
18	00 57	13 18	03 18	05 00	21 23	02 42	20 44	20 05	10 31	22 39
19	00 33	16 43	02 26	04 30	21 31	02 46	20 44	20 05	10 31	22 39
20	00 S 09	19 05	01 32	04 01	21 39	02 49	20 44	20 05	10 30	22 40
21	00 N 15	20 05	00 S 38	03 31	21 47	02 52	20 44	20 06	10 30	22 40
22	00 38	20 31	00 N 17	03 02	21 55	02 55	20 44	20 06	10 29	22 40
23	01 02	19 41	01 12	02 32	22 02	02 58	20 44	20 06	10 29	22 41
24	01 26	17 58	02 08	02 02	22 10	03 01	20 44	20 06	10 28	22 41
25	01 49	15 30	03 05	01 32	22 17	03 04	20 43	20 06	10 28	22 41
26	02 13	12 24	04 02	01 02	22 24	03 07	20 43	20 06	10 27	22 42
27	02 36	08 49	04 58	00 32	22 31	03 10	20 43	20 07	10 27	22 42
28	02 59	04 53	05 55	00 S 02	22 37	03 13	20 43	20 07	10 26	22 42
29	03 23	00 S 45	06 52	00 N 28	22 44	03 16	20 43	20 07	10 26	22 42
30	03 46	03 N 28	07 48	00 58	22 50	03 19	20 43	20 07	10 25	22 43
31	04 N 10	07 N 36	08 N 44	01 N 28	22 N 57	03 N 22	20 S 42	20 N 07	10 S 25	22 N 43

ZODIAC SIGN ENTRIES

Date	h	m	Planets
01	20	39	♀ ♓
03	06	31	☽ ♈
04	11	34	☿ ♓
05	17	20	☽ ♉
08	02	03	☽ ♊
10	07	45	☽ ♋
12	10	30	☽ ♌
14	10	20	☽ ♍
16	09	59	☽ ♎
17	21	34	♂ ♊
18	11	15	☽ ♏
20	15	53	☽ ♐
20	19	48	☿ ♈
20	21	16	☉ ♈
23	00	34	☽ ♑
25	12	17	☽ ♒
25	22	46	♀ ♈
28	01	00	☽ ♓
30	12	55	☽ ♈

LATITUDES

Date	Mercury ☿ ° '	Venus ♀ ° '	Mars ♂ ° '	Jupiter ♃ ° '	Saturn ♄ ° '	Uranus ♅ ° '	Neptune ♆ ° '	Pluto ♇ ° '
01	02 S 05	01 S 15	01 N 05	01 N 32	01 N 46	00 N 38	01 N 47	11 N 25
04	02 10	01 18	01 06	01 32	01 46	00 38	01 47	11 25
07	02 10	01 21	01 07	01 33	01 46	00 38	01 47	11 25
10	02 07	01 23	01 08	01 33	01 47	00 38	01 47	11 25
13	02 00	01 25	01 09	01 33	01 47	00 38	01 47	11 25
16	01 47	01 26	01 10	01 33	01 47	00 38	01 48	11 25
19	01 31	01 27	01 11	01 33	01 48	00 38	01 48	11 25
22	01 09	01 27	01 12	01 34	01 48	00 38	01 48	11 24
25	00 41	01 26	01 12	01 33	01 49	00 37	01 48	11 24
28	00 S 12	01 25	01 13	01 33	01 49	00 37	01 48	11 23
31	00 N 22	01 S 23	01 N 14	01 N 33	01 N 49	00 N 37	01 N 48	11 N 23

DATA

Julian Date	2435899
Delta T	+32 seconds
Ayanamsa	23° 15' 46"
Synetic vernal point	05° ♓ 51' 13"
True obliquity of ecliptic	23° 26' 37"

MOON'S PHASES, APSIDES AND POSITIONS ☽

Date	h	m	Phase	Longitude ° '	Eclipse Indicator
01	16	12	●	10 ♓ 49	
09	11	50	☽	18 ♊ 39	
16	02	22	○	25 ♍ 14	
23	05	04	☾	02 ♑ 18	
31	09	19	●	10 ♈ 24	

Day	h	m		
14	22	16	Perigee	
27	03	34	Apogee	
02	09	52	0N	
09	13	03	Max dec	20° N 40'
15	17	24	0S	
22	03	58	Max dec	20° S 34'
29	16	14	0N	

LONGITUDES

Date	Chiron ⚷ ° '	Ceres ⚳ ° '	Pallas ⚴ ° '	Juno ⚵ ° '	Vesta ⚶ ° '	Black Moon Lilith ⚸ ° '
01	14 ♒ 07	27 ♈ 47	18 ♓ 32	06 ♓ 48	21 ♒ 41	20 ♒ 19
11	14 ♒ 48	01 ♉ 25	21 ♓ 59	11 ♓ 39	26 ♒ 43	21 ♒ 27
21	15 ♒ 26	05 ♉ 12	25 ♓ 27	16 ♓ 32	01 ♓ 40	22 ♒ 34
31	16 ♒ 00	09 ♉ 04	28 ♓ 57	21 ♓ 28	06 ♓ 32	23 ♒ 41

ASPECTARIAN

01 Friday
00 17 ♀ ☍ ♇
01 02 ♀ ⚻ ♃
01 35 ☽ ⚻ ♅
09 51 ☽ Q ♂
11 16 ♃ ⊻ ♇
13 37 ☽ ± ♅
16 12 ☽ ☌ ☉
22 23 ☽ □ ♄
23 20 ☽ # ♃

02 Saturday
00 25 ☿ Q ♄
05 22 ☽ ⚼ ♆
07 37 ☽ ⚼ ♅
11 14 ☽ ⚹ ♂
20 37 ☽ ∥ ♃
23 19 ☽ ± ♆

03 Sunday
01 51 ☽ ⊻ ☿
03 59 ☽ ☍ ♃
04 18 ☽ ⚻ ♇
10 24 ☽ ⊻ ♀
11 13 ☽ ⚻ ♆
13 24 ☽ △ ♅
15 39 ☽ ⊥ ☿
16 11 ☽ ± ♇
16 27 ☿ ⚻ ♃
18 14 ♀ △ ♆
18 41 ☽ ∠ ♂
19 26 ☿ ☍ ♇
23 38 ☽ ⊥ ♀
23 43 ☽ # ☉

04 Monday
06 29 ♂ Q ♅
09 16 ☽ ⊻ ☉
09 53 ☽ ⚼ ♇
10 08 ☽ △ ♄
12 13 ☽ ∠ ☿
13 22 ☽ ⊥ ♂
14 53 ♀ ⚻ ♅
19 19 ☽ ∠ ♀
20 15 ☉ □ ♄
22 04 ☽ ⊥ ☉
23 57 ☽ # ♆

05 Tuesday
01 44 ☽ ⊻ ♂
03 53 ☽ # ♀
14 19 ☽ ⚻ ♃
14 31 ☽ # ☿
15 04 ☽ △ ♇
15 27 ☽ ⚼ ♄
17 05 ☽ ∠ ☉
20 38 ☿ △ ♆
21 51 ☽ ☍ ♆
22 02 ☽ ⚹ ☿
23 54 ☽ □ ♅

06 Wednesday
01 46 ☽ ± ♃
03 39 ☽ ⚹ ♀
08 48 ☽ ± ♄
11 22 ☿ ⚻ ♅
13 37 ♀ ∥ ♆
18 47 ☽ ⚼ ♃
20 15 ☽ ⚻ ♄

07 Thursday
00 14 ☽ ⚹ ☉
00 33 ☽ Q ☿
04 57 ☽ Q ♀
10 00 ☽ Q ♅
14 03 ☽ ☌ ♂
22 37 ☽ △ ♃
23 47 ☽ □ ♇

08 Friday
00 33 ☽ Q ☉
03 17 ☉ ⚼ ♆
06 17 ☽ ⚻ ♆
08 11 ☽ ⚹ ♅
08 58 ☽ ∥ ♂
12 02 ☽ ∥ ♅
15 08 ☽ □ ☿
17 13 ☽ ± ♆
17 45 ☽ □ ♀

09 Saturday
02 07 ☿ ∥ ♆
03 40 ☉ ⚼ ♅
03 45 ☽ ☍ ♄
07 47 ♀ ± ♅
08 26 ☽ Q ♇
09 23 ☽ ⚼ ♆
09 57 ♂ ∥ ♅
11 13 ☽ ∠ ♅
11 50 ☽ □ ☉
20 22 ☿ ± ♅
23 04 ☽ ⊻ ♂

10 Sunday
03 02 ☽ ⊥ ♅
04 02 ☽ □ ♃
05 31 ☽ ⚹ ♇
10 00 ☽ ⊥ ♂
10 04 ☽ ∥ ♂
11 39 ☽ △ ♆
13 12 ☽ ∥ ♅
13 24 ☽ ⊻ ♅

11 Monday
01 36 ☿ ☌ ♀
02 08 ☽ ∠ ♂
03 35 ☽ △ ♀
03 40 ☽ △ ☿
07 09 ☽ ∠ ♇
07 57 ☽ ⚻ ♄
10 31 ☽ Q ♃
18 00 ☽ ± ♄
19 18 ☽ △ ☉
22 09 ☽ ⊥ ♇

12 Tuesday
04 19 ☽ ⚹ ♂
06 15 ☽ ⚹ ♃
06 55 ☽ ⚼ ♀
08 01 ☽ ⊻ ♇
08 14 ☽ ⚼ ☿
08 53 ☽ ⚼ ♄
13 27 ☿ □ ♄
13 49 ☽ □ ♆
15 27 ☽ ☌ ♅
21 40 ☽ ⚼ ☉
22 24 ☿ Q ♂
22 55 ☽ ± ♀

13 Wednesday
00 43 ☽ Q ♂
00 56 ☽ ± ☿
06 13 ♀ □ ♄
06 23 ☽ ∠ ♃
09 12 ☽ △ ♄
09 27 ☽ ⚻ ♀
10 26 ☿ ∥ ♀
11 56 ☽ ⚻ ☿
13 11 ☽ ± ☉
15 33 ☽ # ♆
18 11 ♂ △ ♃
18 45 ☽ Q ♆
20 39 ☽ ⊥ ♃
23 25 ☽ ⚻ ☉
23 35 ♂ # ♄

14 Thursday
04 14 ☿ ⚼ ♆
06 07 ☽ ⊻ ♃
06 44 ☽ □ ♂
08 09 ☽ ☌ ♇
09 19 ☽ # ♀
10 57 ☽ # ☿
13 19 ☉ # ♃
13 47 ☽ ⚹ ♆
15 22 ☽ ⊻ ♅
16 17 ♀ Q ♂
16 55 ☿ ⚼ ♅

15 Friday
00 51 ☽ ⊥ ♅
05 55 ☽ ∥ ♃
07 26 ☽ # ☉
08 53 ☽ □ ♄
13 22 ☽ ☍ ♀
13 30 ☽ ∠ ♆
15 00 ♀ ⚼ ♆
15 05 ☽ ∠ ♅
15 55 ♂ □ ♇
18 17 ☽ ☍ ☿

16 Saturday
02 00 ☽ ∥ ☉
02 22 ☽ ☍ ☉
03 49 ☽ ⊥ ♆
05 09 ☽ # ♃
05 18 ☽ ☌ ♃
07 42 ☽ ⊻ ♇
08 25 ☽ △ ♂
09 54 ♀ ⚼ ♅
13 25 ☽ ⊻ ♆
13 36 ☽ Q ♄
15 01 ☽ ⚹ ♅
15 41 ☽ ∥ ☿
17 20 ☽ ⊥ ♇
20 10 ☽ ∥ ♀
23 48 ☉ ± ♆

17 Sunday
07 55 ☽ ∠ ♇
09 05 ☽ ⚹ ♄
09 46 ☽ ⚼ ♂
10 28 ☽ Q ♅
18 01 ☉ ☍ ♃
18 12 ☽ ⚻ ♀
19 59 ☽ ∥ ♆

18 Monday
01 20 ☽ ± ♂
02 14 ☽ ⚻ ☿
05 09 ☽ ± ♀
05 53 ☽ ⊻ ♃
06 53 ☽ ⚻ ☉
08 46 ☽ ⚹ ♇
10 02 ☽ ∠ ♄
11 53 ☽ ⚻ ♂
13 56 ☽ ± ☿
14 48 ☽ ☌ ♆
15 56 ☽ ⊥ ♃
16 29 ☽ □ ♅
17 49 ☽ ± ☉
20 09 ☿ ± ♆
21 53 ☽ ⚼ ♀

19 Tuesday
01 24 ☽ ⊥ ♄
03 24 ☿ # ♃
03 39 ☿ ☍ ♃
05 15 ☽ Q ♇
07 16 ☽ ∠ ♃
07 55 ☽ ⚼ ☿
09 25 ☉ ⚻ ♇
10 33 ☽ ⚼ ☉
11 50 ☽ ⊻ ♄

20 Wednesday
01 38 ☿ ⚻ ♇
02 48 ☽ △ ♀
09 37 ☽ ⚹ ♃
13 09 ☽ □ ♇
15 12 ☽ △ ☿
15 27 ☽ △ ☉
18 19 ☉ ☌ ☿
19 13 ☽ ☍ ♂
19 38 ☽ ⊻ ♆
21 28 ☽ △ ♅

21 Thursday
03 34 ♂ ⚻ ♆
05 08 ☽ # ♅
06 42 ☽ ⊥ ♆
07 25 ☽ Q ♃
18 26 ☽ ☌ ♄
19 16 ☉ # ☿
20 12 ☿ ⚻ ♆
23 35 ☽ ∠ ♆

22 Friday
01 30 ☽ ⚼ ♅
04 05 ☿ ⚹ ♂
08 10 ☿ △ ♅
16 43 ☽ □ ♀
16 55 ♂ ⚹ ♅
17 01 ♀ # ♃
17 16 ☽ □ ♃
17 40 ♀ ± ♆
18 46 ☽ ± ♅
21 32 ☽ △ ♇
21 41 ☉ ⚻ ♆
21 45 ♀ ☍ ♃

23 Saturday
01 06 ☿ ± ♇
03 14 ☽ # ♅
04 21 ☉ ∥ ☿
04 28 ☽ ⚹ ♆
05 04 ☽ □ ☉
06 27 ☽ ⚻ ♅
07 12 ☽ ⚻ ♂
10 52 ☽ □ ☿
19 40 ☽ ± ♂
21 50 ☉ △ ♅

24 Sunday
00 45 ♄ St R
03 01 ☽ ⚼ ♇
04 07 ☽ Q ♆
04 47 ☽ ⊻ ♄
08 10 ☽ Q ♀
10 16 ☿ # ♀
14 35 ☽ ⚼ ♂
16 03 ♀ ⚻ ♇
16 47 ☽ ⊥ ♄
20 00 ☽ Q ☉
20 59 ☽ ± ♇
22 58 ♃ ⊥ ♆

25 Monday
04 04 ☽ △ ♃
04 21 ☉ # ♀
06 03 ☽ Q ☿
07 40 ☉ ± ♇
09 03 ☽ ⚻ ♇
10 53 ☽ ∠ ♄
11 04 ☽ ⚹ ♀
11 38 ☿ ∥ ♃
16 12 ☽ □ ♆
18 17 ☽ ☍ ♅
22 24 ☽ ⚹ ☉
22 30 ☽ △ ♂

26 Tuesday
01 13 ☉ ⚹ ♂
10 04 ☽ ⚼ ♃
11 20 ☽ ⚹ ☿
17 14 ☽ ⚹ ♄
20 59 ☽ ∠ ♀

27 Wednesday
01 25 ☽ ∥ ♆
04 07 ☽ ± ♃
07 33 ☽ ∠ ☉
10 34 ☿ ⚼ ♇
11 22 ♀ ⚻ ♆
16 08 ☽ ⚻ ♃
17 20 ☽ ⊥ ♀
17 31 ☽ Q ♄
21 39 ☽ ☍ ♇
21 40 ☿ △ ♄
23 58 ☽ ∠ ☿

28 Thursday
03 24 ☽ ⊥ ☉
04 47 ☽ △ ♆
06 51 ☽ ⊻ ♀
06 55 ☽ ⚻ ♅
07 02 ☽ # ☿
07 32 ♀ △ ♅
14 33 ☽ □ ♂
16 34 ☽ ⊻ ☉
18 59 ☽ ± ♅
21 41 ☽ # ♃
21 47 ☽ ⊥ ☿
22 07 ☽ # ☉

29 Friday
03 29 ☉ ∥ ♃
05 35 ♂ ∥ ♇
05 42 ☽ □ ♄
10 11 ♀ ± ♇
10 49 ☽ ⚼ ♆
12 12 ☽ ⊻ ☿
12 58 ☽ ⚼ ♅
13 24 ☽ # ♀
19 50 ☽ ∥ ♀

30 Saturday
03 38 ☽ ☍ ♃
04 27 ☽ Q ♂
04 40 ☽ ± ♆
08 28 ♂ ± ♆
09 34 ☽ ⚻ ♇
11 03 ☽ ∥ ♃
13 53 ☽ ∥ ☉
16 29 ☽ ⚻ ♆
18 39 ☽ △ ♅
21 22 ☽ ± ♇

31 Sunday
01 24 ☽ ☌ ♀
05 21 ☽ ⚹ ♂
09 19 ☽ ☌ ☉
14 56 ☽ ⚼ ♇
16 50 ☽ △ ♄
20 45 ☽ ∥ ☿

All ephemeris data is given at 12.00 UT and the Moon's longitude is additionally given for 24.00 UT

APRIL 1957

LONGITUDES

Date	Sidereal time h m s	Sun ☉	Moon ☽	Moon ☽ 24.00	Mercury ☿	Venus ♀	Mars ♂	Jupiter ♃	Saturn ♄	Uranus ♅	Neptune ♆	Pluto ♇
01	00 38 18	11 ♈ 29 46	24 ♈ 10 05	00 ♉ 25 42	23 ♈ 28	08 ♈ 08	09 ♊ 11	25 ♍ 01	14 ♐ 15	02 ♌ 54	01 ♏ 46	28 ♌ 16
02	00 42 15	12 28 57	06 ♉ 43 59	13 ♉ 05 06	25 24	09 23	09 48	24 R 54	14 R 14	02 R 53	01 R 45	28 R 15
03	00 46 12	13 28 06	19 ♉ 29 09	25 ♉ 56 19	27 17	10 37	10 26	24 47	14 13	02 53	01 43	28 14
04	00 50 08	14 27 13	02 ♊ 26 45	09 ♊ 00 39	29 ♈ 08	11 52	11 04	24 40	14 12	02 53	01 42	28 13
05	00 54 05	15 26 18	15 ♊ 38 11	22 ♊ 19 33	00 ♉ 55	13 06	11 41	24 33	14 11	02 52	01 40	28 12
06	00 58 01	16 25 21	29 ♊ 04 55	05 ♋ 54 25	02 39	14 21	12 19	24 26	14 09	02 52	01 39	28 11
07	01 01 58	17 24 21	12 ♋ 48 11	19 ♋ 46 14	04 19	15 35	12 57	24 19	14 08	02 52	01 37	28 10
08	01 05 54	18 23 19	26 ♋ 48 32	03 ♌ 54 58	05 55	16 49	13 35	24 13	14 07	02 52	01 35	28 09
09	01 09 51	19 22 14	11 ♌ 05 18	18 ♌ 19 10	07 26	18 04	14 12	24 06	14 05	02 52	01 34	28 09
10	01 13 47	20 21 07	25 ♌ 36 05	02 ♍ 55 28	08 53	19 18	14 50	24 00	14 03	02 D 52	01 32	28 08
11	01 17 44	21 19 58	10 ♍ 16 34	17 ♍ 38 36	10 15	20 32	15 28	23 54	14 02	02 52	01 31	28 07
12	01 21 41	22 18 46	25 ♍ 00 40	02 ♎ 21 52	11 31	21 47	16 06	23 48	14 00	02 52	01 29	28 06
13	01 25 37	23 17 33	09 ♎ 41 15	16 ♎ 57 56	12 42	23 01	16 43	23 41	13 58	02 52	01 27	28 05
14	01 29 34	24 16 17	24 ♎ 11 05	01 ♏ 19 57	13 48	24 15	17 21	23 36	13 56	02 52	01 26	28 05
15	01 33 30	25 15 00	08 ♏ 23 57	15 ♏ 22 34	14 48	25 29	17 59	23 30	13 54	02 53	01 24	28 04
16	01 37 27	26 13 40	22 ♏ 15 27	29 ♏ 02 24	15 43	26 44	18 36	23 24	13 52	02 53	01 23	28 03
17	01 41 23	27 12 19	05 ♐ 43 22	12 ♐ 18 23	16 32	27 58	19 14	23 19	13 49	02 53	01 21	28 03
18	01 45 20	28 10 56	18 ♐ 47 38	25 ♐ 11 24	17 14	29 ♈ 12	19 52	23 13	13 47	02 54	01 19	28 02
19	01 49 16	29 ♈ 09 31	01 ♑ 30 02	07 ♑ 43 59	17 51	00 ♉ 26	20 29	23 08	13 45	02 54	01 18	28 01
20	01 53 13	00 ♉ 08 05	13 ♑ 53 43	19 ♑ 59 47	18 22	01 41	21 07	23 03	13 42	02 55	01 16	28 01
21	01 57 10	01 06 37	26 ♑ 02 44	02 ♒ 03 09	18 47	02 55	21 45	22 58	13 39	02 55	01 14	28 00
22	02 01 06	02 05 07	08 ♒ 01 37	13 ♒ 58 44	19 05	04 09	22 23	22 54	13 37	02 56	01 13	28 00
23	02 05 03	03 03 36	19 ♒ 55 03	25 ♒ 51 09	19 18	05 23	23 00	22 49	13 34	02 56	01 11	27 59
24	02 08 59	04 02 03	01 ♓ 47 34	07 ♓ 44 48	19 25	06 37	23 38	22 45	13 31	02 57	01 09	27 59
25	02 12 56	05 00 28	13 ♓ 43 20	19 ♓ 43 36	19 R 27	07 51	24 16	22 40	13 28	02 58	01 08	27 58
26	02 16 52	05 58 52	25 ♓ 46 00	01 ♈ 50 54	19 22	09 05	24 53	22 36	13 25	02 59	01 06	27 58
27	02 20 49	06 57 14	07 ♈ 58 35	14 ♈ 09 19	19 13	10 19	25 31	22 32	13 22	03 00	01 05	27 58
28	02 24 45	07 55 34	20 ♈ 23 18	26 ♈ 40 42	18 58	11 34	26 09	22 29	13 19	03 01	01 03	27 57
29	02 28 42	08 53 53	03 ♉ 01 37	09 ♉ 26 07	18 39	12 48	26 46	22 25	13 16	03 02	01 01	27 57
30	02 32 39	09 ♉ 52 10	15 ♉ 54 14	22 ♉ 25 56	18 ♉ 16	14 ♉ 02	27 ♊ 24	22 ♍ 21	13 ♐ 12	03 ♌ 03	01 ♏ 00	27 ♌ 57

Date	Moon True ☊	Moon Mean ☊	Moon ☽ Latitude
01	20 ♏ 04	21 ♏ 55	02 N 16
02	20 R 01	21 52	01 12
03	20 00	21 49	00 N 03
04	20 D 01	21 46	01 S 07
05	20 02	21 42	02 15
06	20 04	21 39	03 17
07	20 05	21 36	04 09
08	20 R 04	21 33	04 47
09	20 03	21 30	05 09
10	20 00	21 26	05 11
11	19 56	21 23	04 53
12	19 52	21 20	04 16
13	19 49	21 17	03 22
14	19 46	21 14	02 16
15	19 44	21 11	01 S 02
16	19 D 44	21 07	00 N 14
17	19 44	21 04	01 27
18	19 46	21 01	02 33
19	19 47	20 58	03 30
20	19 49	20 55	04 16
21	19 49	20 52	04 49
22	19 R 49	20 48	05 09
23	19 49	20 45	05 15
24	19 47	20 42	05 09
25	19 45	20 39	04 48
26	19 43	20 36	04 15
27	19 41	20 32	03 30
28	19 39	20 29	02 35
29	19 38	20 26	01 31
30	19 ♏ 38	20 ♏ 23	00 N 21

DECLINATIONS

Date	Sun ☉	Moon ☽	Mercury ☿	Venus ♀	Mars ♂	Jupiter ♃	Saturn ♄	Uranus ♅	Neptune ♆	Pluto ♇
01	04 N 33	11 N 29	09 N 38	01 N 58	23 N 03	03 N 24	20 S 42	20 N 07	10 S 24	22 N 43
02	04 56	14 54	10 31	02 28	23 09	03 27	20 42	20 07	10 23	22 43
03	05 19	17 39	11 23	02 58	23 14	03 30	20 42	20 07	10 23	22 43
04	05 42	19 33	12 14	03 28	23 20	03 33	20 41	20 07	10 22	22 44
05	06 05	20 26	13 02	03 58	23 25	03 35	20 41	20 07	10 22	22 44
06	06 27	20 09	13 49	04 28	23 31	03 38	20 41	20 07	10 21	22 44
07	06 50	18 42	14 33	04 58	23 36	03 40	20 41	20 07	10 21	22 44
08	07 13	16 06	15 15	05 27	23 40	03 43	20 40	20 07	10 20	22 44
09	07 35	12 30	15 55	05 57	23 45	03 45	20 40	20 07	10 20	22 44
10	07 57	08 06	16 32	06 26	23 50	03 48	20 40	20 07	10 19	22 45
11	08 19	03 N 11	17 06	06 55	23 54	03 50	20 40	20 07	10 18	22 45
12	08 41	01 S 56	17 38	07 24	23 58	03 53	20 39	20 07	10 18	22 45
13	09 03	06 56	18 06	07 53	24 02	03 55	20 39	20 07	10 17	22 45
14	09 25	11 29	18 32	08 22	24 06	03 57	20 39	20 07	10 17	22 45
15	09 46	15 17	18 55	08 50	24 10	03 59	20 38	20 07	10 16	22 45
16	10 08	18 07	19 15	09 18	24 13	04 01	20 38	20 07	10 16	22 45
17	10 29	19 50	19 33	09 47	24 16	04 03	20 38	20 07	10 15	22 45
18	10 50	20 26	19 47	10 14	24 19	04 05	20 37	20 07	10 14	22 45
19	11 11	19 56	19 58	10 42	24 22	04 07	20 37	20 07	10 14	22 45
20	11 31	18 29	20 07	11 09	24 25	04 09	20 36	20 07	10 13	22 45
21	11 52	16 13	20 12	11 37	24 28	04 11	20 36	20 06	10 13	22 45
22	12 12	13 17	20 14	12 03	24 30	04 13	20 36	20 06	10 12	22 45
23	12 32	09 51	20 14	12 30	24 32	04 14	20 35	20 06	10 12	22 45
24	12 52	06 02	20 11	12 56	24 34	04 16	20 35	20 06	10 11	22 45
25	13 12	01 S 58	20 05	13 22	24 36	04 17	20 34	20 06	10 11	22 45
26	13 31	02 N 14	19 56	13 48	24 38	04 19	20 34	20 05	10 10	22 45
27	13 50	06 23	19 44	14 13	24 39	04 20	20 34	20 05	10 09	22 45
28	14 09	10 21	19 30	14 38	24 40	04 21	20 33	20 05	10 09	22 45
29	14 28	13 57	19 13	15 03	24 41	04 23	20 33	20 05	10 08	22 44
30	14 N 47	16 N 56	18 N 54	15 N 27	24 N 42	04 N 24	20 S 32	20 N 04	10 S 08	22 N 44

ZODIAC SIGN ENTRIES

Date	h	m	Planets
01	23	11	☽ ♉
04	07	30	☽ ♊
04	23	37	☿ ♉
06	13	37	☽ ♋
08	17	24	☽ ♌
10	19	13	☽ ♍
12	20	08	☽ ♎
14	21	45	☽ ♏
17	01	43	☽ ♐
19	03	28	♀ ♉
19	09	08	☽ ♑
20	08	41	☉ ♉
21	19	53	☽ ♒
24	08	23	☽ ♓
26	20	22	☽ ♈
29	06	18	☽ ♉

LATITUDES

Date	Mercury ☿	Venus ♀	Mars ♂	Jupiter ♃	Saturn ♄	Uranus ♅	Neptune ♆	Pluto ♇
01	00 N 33	01 S 22	01 N 14	01 N 33	01 N 49	00 N 37	01 N 48	11 N 23
04	01 08	01 20	01 14	01 33	01 50	00 37	01 48	11 22
07	01 42	01 16	01 15	01 33	01 50	00 37	01 48	11 22
10	02 11	01 13	01 15	01 32	01 50	00 37	01 48	11 21
13	02 34	01 09	01 16	01 32	01 51	00 37	01 48	11 21
16	02 50	01 04	01 16	01 32	01 51	00 37	01 48	11 20
19	02 56	00 59	01 16	01 31	01 51	00 37	01 48	11 19
22	02 51	00 54	01 17	01 31	01 51	00 37	01 48	11 19
25	02 35	00 48	01 17	01 30	01 52	00 37	01 48	11 18
28	02 07	00 42	01 17	01 30	01 52	00 37	01 48	11 17
31	01 N 28	00 S 36	01 N 17	01 N 29	01 N 52	00 N 37	01 N 48	11 N 17

DATA

Julian Date	2435930
Delta T	+32 seconds
Ayanamsa	23° 15' 49"
Synetic vernal point	05° ♓ 51' 11"
True obliquity of ecliptic	23° 26' 36"

MOON'S PHASES, APSIDES AND POSITIONS ☽

Date	h	m	Phase	Longitude	Eclipse Indicator
07	20	33	☽	17 ♋ 45	
14	12	09	○	24 ♎ 17	
21	23	01	☾	01 ♒ 33	
29	23	54	●	09 ♉ 23	Annular

Day	h	m	
12	01	18	Perigee
23	21	27	Apogee

Day	h	m		
05	18	29	Max dec	20° N 28'
12	02	57	0S	
18	12	44	Max dec	20° S 26'
25	23	17	0N	

LONGITUDES

Date	Chiron ⚷	Ceres ⚳	Pallas ⚴	Juno ⚵	Vesta ⚶	Black Moon Lilith ⚸
01	16 ♒ 03	09 ♉ 28	29 ♓ 18	21 ♓ 58	07 ♓ 01	23 ♒ 48
11	16 ♒ 32	13 ♉ 25	02 ♈ 48	26 ♓ 57	11 ♓ 46	24 ♒ 55
21	16 ♒ 55	17 ♉ 27	06 ♈ 16	01 ♈ 57	16 ♓ 23	26 ♒ 02
31	17 ♒ 12	21 ♉ 31	09 ♈ 43	06 ♈ 59	20 ♓ 53	27 ♒ 09

ASPECTARIAN

h m Aspects

01 Monday
05 07 ☽ # ♆
10 25 ☽ ☌ ☿
12 01 ☽ ∠ ♂
13 37 ☽ ⚻ ♃
19 51 ☽ △ ♇
21 44 ☽ ⊡ ♄

02 Tuesday
00 53 ☿ ∠ ♂
01 00 ☽ ± ♃
02 32 ☽ ☍ ♆
04 42 ☽ □ ♅
06 04 ☿ ⚻ ♃
06 09 ☽ ⊥ ♂
08 27 ☿ # ♆
14 50 ☽ ± ♄
17 33 ☽ ⚺ ♀
17 56 ☽ ⊡ ♃
18 07 ☽ ⚺ ♂
23 47 ☽ ⚺ ☉

03 Wednesday
02 09 ☽ ⚻ ♄
04 38 ♀ ⚹ ♂
06 04 ☽ ⊥ ♀
06 20 ☉ ⊡ ♇
11 58 ☽ ⊥ ☉
14 36 ☽ Q ♅
21 46 ☽ △ ♃

04 Thursday
00 09 ☿ △ ♇
00 37 ☽ ∠ ♀
04 13 ☽ □ ♇
04 54 ☽ ⚺ ☿
05 54 ☉ △ ♄
06 03 ☽ ∠ ☉
10 37 ☽ ⚻ ♆
12 48 ☽ ⚹ ♅
12 56 ☿ ⊡ ♄
15 52 ♀ ∥ ♃
17 42 ☽ ⊥ ☿
21 35 ☽ ± ♆
23 53 ☽ ∥ ♅

05 Friday
04 31 ☽ ☌ ♂
06 57 ☽ ⚹ ♀
07 18 ☿ ± ♃
09 22 ☽ ☍ ♄
11 37 ☽ ⚹ ☉
12 35 ☽ ∠ ☿
13 01 ☽ Q ♇
13 51 ☽ ⊡ ♆
13 52 ♀ ⊡ ♇
16 02 ☽ ∠ ♅
22 12 ☿ ☍ ♆

06 Saturday
03 50 ☽ □ ♃
06 40 ☽ Q ♀
08 05 ☽ ⊥ ♅
08 29 ♀ △ ♄
10 25 ☽ ⚹ ♇
10 45 ☽ Q ☉
12 50 ☽ ∥ ♅
15 10 ☿ □ ♅
16 31 ☽ △ ♆
18 40 ☽ ⚺ ♅
19 10 ☽ ⚹ ☿

07 Sunday
11 10 ☽ Q ♃
12 16 ☽ ⚺ ♂
12 38 ☽ ∠ ♇
14 18 ☽ ⚻ ♄
17 16 ☽ □ ♀
18 52 ☽ Q ☿
20 33 ☽ □ ☉
23 05 ☽ ⊥ ♂

08 Monday
00 36 ☽ ± ♄
04 05 ☽ ⊥ ♇
07 37 ☽ ⚹ ♃
14 17 ☽ ⚺ ♇
15 08 ☽ ∠ ♂
15 53 ☽ ⊡ ♄
17 10 ☽ ∥ ☿
20 04 ☽ □ ♆
22 14 ☽ ☌ ♅

09 Tuesday
05 11 ☽ □ ☿
07 31 ♂ ☍ ♄
08 43 ☽ ∠ ♃
16 58 ☽ △ ♄
17 25 ☽ ⚹ ♂
22 26 ☿ ± ♄
23 34 ☽ ⊥ ♃

10 Wednesday
00 19 ☽ # ♆
00 39 ☽ △ ♀
02 02 ☽ Q ♆
02 44 ☽ △ ☉
08 20 ♅ St D
09 23 ☽ ⚺ ♃
10 39 ☿ ⊥ ♂
12 42 ☽ ∥ ☉
13 54 ☿ ⊡ ♃
14 07 ☽ Q ♂
16 09 ☽ ☌ ♇
19 36 ☽ ∥ ♀
21 43 ☽ ⚹ ♆
23 54 ☽ ⚺ ♅

11 Thursday
03 34 ☽ ⊡ ♀
05 06 ☽ ⊡ ☉
08 56 ☽ ∥ ♃
09 42 ☽ ⊥ ♅
11 56 ☽ △ ☿
18 06 ☽ □ ♄
19 35 ☽ ± ♀
20 49 ☽ ± ☉
20 50 ☽ □ ♂
22 08 ☽ ∠ ♆

12 Friday
00 22 ☽ ∠ ♅
06 15 ☽ ⚻ ♀
07 17 ☽ ⚻ ☉
10 02 ☽ ☌ ♃
12 20 ♂ Q ♇
12 46 ☽ ⊥ ♆
14 41 ☽ ⊡ ☿
17 02 ☽ ⚺ ♇
21 14 ☽ # ♃
22 33 ☽ ⚺ ♆
23 22 ☽ Q ♄

13 Saturday
00 49 ☽ ⚹ ♅
02 20 ♂ ⊡ ♆
02 50 ☽ ⊥ ♇
06 41 ☽ ± ☿
17 18 ☽ # ♀
17 23 ☽ ⚻ ☿
17 36 ☽ ∠ ♇
19 02 ☽ ⚹ ♄
20 32 ☽ Q ♅
20 53 ☉ ⚻ ♃
23 42 ☽ # ☉

14 Sunday
00 07 ☽ △ ♂
00 08 ♀ ⚻ ♃
05 20 ☽ ∥ ♆
11 01 ☽ ⚺ ♃
12 08 ☽ ☍ ♀
12 09 ☽ ☍ ☉
13 39 ☉ ☌ ♀
14 50 ☿ ⚻ ♄
18 31 ☽ ⚹ ♇
19 56 ☽ ∠ ♄
21 00 ☽ ⊥ ♃

15 Monday
00 09 ☽ ☌ ♆
02 21 ☽ ⊡ ♂
02 36 ☽ □ ♅
08 02 ♂ ∠ ♅
11 09 ☽ ⊥ ♄
12 10 ☽ ∠ ♃
14 51 ☽ Q ♇
18 26 ☽ ± ♂
21 25 ☽ ⚺ ♄
23 49 ☽ ☍ ☿

16 Tuesday
05 18 ☽ ⚻ ♂
14 00 ☽ ⚹ ♃
19 33 ☽ ⚻ ☉
20 41 ☽ ⚻ ♀
20 51 ☉ # ♆
22 14 ☽ □ ♇

17 Wednesday
04 09 ☽ ⚺ ♆
05 13 ☽ # ☿
06 53 ☽ △ ♅
07 06 ☽ ± ☉
08 30 ☽ ± ♀
11 16 ☽ Q ♃
13 30 ♀ △ ♇
14 56 ☽ ⊥ ♆
18 42 ☽ # ♅

18 Thursday
00 46 ☽ ⊡ ☉
02 36 ☽ ⊡ ♀
02 45 ☽ ☌ ♄
04 06 ♀ ⊡ ♄
07 25 ☽ ∠ ♆
08 23 ☉ △ ♇
08 57 ☽ ⚻ ☿
10 19 ☽ ⊡ ♅
12 04 ♀ # ♆
12 24 ♀ ± ♃
14 06 ☽ ☍ ♂
20 14 ☽ □ ♃
20 46 ☽ ± ☿

19 Friday
02 10 ☉ ⊡ ♄
03 14 ☽ ± ♅
05 22 ☽ △ ♇
07 09 ☽ △ ☉
07 14 ☽ # ♅
09 45 ☽ △ ♀
11 15 ☽ # ☿
11 32 ☉ ± ♃
11 36 ☽ ⚹ ♆
14 41 ☽ ⚻ ♅
14 42 ☽ ⊡ ☿

20 Saturday
04 14 ♀ ☍ ♆
10 17 ☽ ⊡ ♇
10 46 ☽ Q ♆
11 37 ☽ ⚺ ♄
12 01 ☿ ∥ ♅
21 07 ☽ △ ☿
23 22 ☽ ⊥ ♄

21 Sunday
03 00 ☽ ⚻ ♂
03 59 ☽ ± ♇
05 56 ☽ △ ♃
12 09 ♀ □ ♅
15 06 ☉ ☍ ♆
15 35 ☽ ± ♂
15 54 ☽ ⚻ ♇
17 11 ☽ ∠ ♄
22 21 ☽ □ ♆
23 01 ☽ □ ☉

22 Monday
01 45 ☽ ☍ ♅
03 18 ☽ □ ♀
10 37 ☽ ⊡ ♂
11 44 ☽ ⊡ ♃
19 12 ☽ # ☉
19 55 ☽ # ♀
23 13 ☽ ⚹ ♄

23 Tuesday
05 37 ♂ □ ♃
05 46 ☽ ± ♃
09 02 ☉ □ ♅
09 42 ☽ ∥ ♆
10 44 ☽ □ ☿
14 31 ☽ Q ☉
17 50 ☽ ⚻ ♃
18 35 ☽ △ ♂
19 49 ☽ Q ♀
20 04 ☉ ∥ ♀
23 22 ☽ Q ♄

24 Wednesday
04 18 ☽ ☍ ♇
10 43 ☽ △ ♆
14 21 ☽ ⚻ ♅
16 55 ☽ ⚹ ☉
22 29 ☽ # ♃
22 51 ☽ ⚹ ♀
23 23 ☽ Q ☿

25 Thursday
02 27 ☽ ± ♅
04 48 ♀ ± ♄
05 32 ☿ St R
08 39 ♀ ⊡ ♃
09 00 ☿ ∥ ♅
11 30 ☽ □ ♄
16 49 ☽ ⊡ ♆
20 30 ☽ ⊡ ♅
23 23 ☽ ⚹ ☿

26 Friday
01 40 ☽ ∠ ☉
05 46 ☽ ☍ ♃
08 18 ☽ ∠ ♀
10 10 ☽ □ ♂
10 41 ☽ ± ♆
16 21 ☽ ⚻ ♇
21 03 ☽ ⊥ ☉
22 31 ☽ ⚻ ♆

27 Saturday
00 02 ☽ ∥ ♃
02 14 ☽ △ ♅
03 03 ☿ ⊥ ♂
04 04 ☽ ⊥ ♀
04 09 ☽ ± ♇
04 45 ☽ ∠ ☿
09 50 ☽ ⚺ ☉
17 05 ☽ ⚺ ♀
19 37 ♀ ∠ ♂
21 41 ☽ ⊡ ♇
21 42 ☉ ± ♄
22 00 ☽ ⊥ ☿
22 26 ☽ △ ♄
23 20 ☽ Q ♂

28 Sunday
01 32 ☉ ⊡ ♃
09 20 ☽ ⚺ ☿
10 41 ☽ # ♆
15 58 ☽ ⚻ ♃
23 34 ☽ ⚹ ♂

29 Monday
02 25 ☽ △ ♇
03 03 ☽ ⊡ ♄
03 20 ☽ ± ♃
08 14 ☽ ☍ ♆
12 00 ☽ □ ♅
16 15 ☽ ∥ ☉
19 55 ☽ ± ♄
20 12 ☽ ⊡ ♃
20 45 ♀ ⚻ ♄
21 26 ☽ ∥ ♀
22 01 ♂ ⊥ ♅
23 54 ☽ ☌ ☉

30 Tuesday
05 11 ☽ ∠ ♂
07 02 ☽ ⚻ ♄
08 10 ☽ ☌ ♀
16 12 ☽ ☌ ☿
21 28 ☽ Q ♅
22 37 ☽ ⊥ ♂
23 49 ☽ △ ♃

All ephemeris data is given at 12.00 UT and the Moon's longitude is additionally given for 24.00 UT

MAY 1957

LONGITUDES

Date	Sidereal time h m s	Sun ☉ ° ' "	Moon ☽ ° ' "	Moon ☽ 24.00 ° ' "	Mercury ☿ ° '	Venus ♀ ° '	Mars ♂ ° '	Jupiter ♃ ° '	Saturn ♄ ° '	Uranus ♅ ° '	Neptune ♆ ° '	Pluto ♇ ° '
01	02 36 35	10 ♉ 50 25	29 ♉ 01 11	05 ♊ 39 54	17 ♉ 48	15 ♉ 16	28 ♊ 02	22 ♍ 18	13 ♐ 09	03 ♌ 04	00 ♏ 58	27 ♌ 56
02	02 40 32	11 48 38	12 ♊ 22 00	19 ♊ 07 20	17 R 18	16 30	28 39	22 R 15	13 R 06	03 05	00 R 56	27 R 56
03	02 44 28	12 46 50	25 ♊ 55 47	02 ♋ 47 11	16 45	17 44	29 17	22 12	13 02	03 06	00 55	27 56
04	02 48 25	13 44 59	09 ♋ 41 24	16 ♋ 38 14	16 10	18 58	29 ♊ 55	22 10	12 59	03 07	00 53	27 56
05	02 52 21	14 43 07	23 ♋ 37 30	00 ♌ 39 00	15 33	20 12	00 ♋ 32	22 07	12 55	03 09	00 52	27 56
06	02 56 18	15 41 13	07 ♌ 42 31	14 ♌ 47 47	14 56	21 26	01 10	22 05	12 51	03 10	00 50	27 56
07	03 00 14	16 39 16	21 ♌ 54 33	29 ♌ 02 30	14 18	22 39	01 48	22 02	12 48	03 11	00 49	27 56
08	03 04 11	17 37 18	06 ♍ 11 17	13 ♍ 20 33	13 41	23 53	02 25	22 00	12 44	03 13	00 47	27 56
09	03 08 08	18 35 17	20 ♍ 29 53	27 ♍ 38 51	13 06	25 07	03 03	21 59	12 40	03 14	00 45	27 55
10	03 12 04	19 33 15	04 ♎ 47 00	11 ♎ 53 49	12 32	26 21	03 41	21 57	12 36	03 16	00 44	27 D 55
11	03 16 01	20 31 11	18 ♎ 58 51	26 ♎ 01 37	12 01	27 35	04 18	21 55	12 32	03 18	00 42	27 56
12	03 19 57	21 29 06	03 ♏ 01 37	09 ♏ 58 27	11 32	28 ♉ 49	04 56	21 54	12 28	03 19	00 41	27 56
13	03 23 54	22 26 58	16 ♏ 51 41	23 ♏ 40 59	11 07	00 ♊ 03	05 34	21 53	12 24	03 21	00 39	27 56
14	03 27 50	23 24 50	00 ♐ 26 05	07 ♐ 06 45	10 45	01 16	06 11	21 52	12 20	03 23	00 38	27 56
15	03 31 47	24 22 40	13 ♐ 42 50	20 ♐ 14 18	10 28	02 30	06 49	21 51	12 16	03 24	00 36	27 56
16	03 35 43	25 20 28	26 ♐ 41 09	03 ♑ 03 30	10 14	03 44	07 26	21 51	12 12	03 26	00 35	27 56
17	03 39 40	26 18 15	09 ♑ 21 30	15 ♑ 35 25	10 05	04 58	08 04	21 50	12 08	03 28	00 34	27 56
18	03 43 37	27 16 01	21 ♑ 45 33	27 ♑ 52 18	10 00	06 12	08 42	21 50	12 04	03 30	00 32	27 57
19	03 47 33	28 13 46	03 ♒ 56 03	09 ♒ 57 18	10 D 00	07 25	09 19	21 D 50	12 00	03 32	00 31	27 57
20	03 51 30	29 ♉ 11 29	15 ♒ 56 33	21 ♒ 54 20	10 04	08 39	09 57	21 50	11 55	03 34	00 29	27 57
21	03 55 26	00 ♊ 09 12	27 ♒ 51 12	03 ♓ 47 44	10 13	09 53	10 34	21 51	11 51	03 36	00 28	27 57
22	03 59 23	01 06 53	09 ♓ 44 31	15 ♓ 42 08	10 27	11 06	11 12	21 51	11 47	03 38	00 27	27 58
23	04 03 19	02 04 34	21 ♓ 41 09	27 ♓ 42 07	10 44	12 20	11 50	21 52	11 42	03 40	00 25	27 58
24	04 07 16	03 02 13	03 ♈ 45 35	09 ♈ 52 03	11 07	13 34	12 27	21 53	11 38	03 43	00 24	27 59
25	04 11 12	03 59 51	16 ♈ 01 58	22 ♈ 15 46	11 33	14 48	13 05	21 54	11 34	03 45	00 23	27 59
26	04 15 09	04 57 29	28 ♈ 33 48	04 ♉ 56 21	12 04	16 01	13 42	21 55	11 29	03 47	00 21	28 00
27	04 19 06	05 55 05	11 ♉ 23 38	17 ♉ 55 46	12 39	17 15	14 20	21 57	11 25	03 49	00 20	28 00
28	04 23 02	06 52 40	24 ♉ 32 49	01 ♊ 14 43	13 18	18 29	14 58	21 58	11 20	03 52	00 19	28 01
29	04 26 59	07 50 15	08 ♊ 01 20	14 ♊ 52 25	14 00	19 42	15 35	22 00	11 16	03 54	00 18	28 01
30	04 30 55	08 47 48	21 ♊ 47 39	28 ♊ 46 37	14 47	20 56	16 13	22 02	11 12	03 57	00 16	28 02
31	04 34 52	09 ♊ 45 20	05 ♋ 48 53	12 ♋ 53 53	15 ♉ 37	22 ♊ 09	16 ♋ 50	22 ♍ 04	11 ♐ 07	03 ♌ 59	00 ♏ 15	28 ♌ 02

Date	Moon True ☊ ° '	Moon Mean ☊ ° '	Moon ☽ Latitude ° '
01	19 ♏ 38	20 ♏ 20	00 S 52
02	19 D 39	20 17	02 03
03	19 39	20 13	03 08
04	19 40	20 10	04 03
05	19 40	20 07	04 45
06	19 41	20 04	05 10
07	19 R 41	20 01	05 17
08	19 40	19 58	05 04
09	19 40	19 54	04 32
10	19 40	19 51	03 44
11	19 D 40	19 48	02 42
12	19 40	19 45	01 31
13	19 40	19 42	00 S 16
14	19 R 40	19 38	00 N 59
15	19 40	19 35	02 09
16	19 40	19 32	03 11
17	19 39	19 29	04 02
18	19 38	19 26	04 40
19	19 37	19 23	05 05
20	19 36	19 19	05 16
21	19 36	19 16	05 13
22	19 D 36	19 13	04 57
23	19 37	19 10	04 28
24	19 38	19 07	03 47
25	19 39	19 04	02 55
26	19 40	19 00	01 54
27	19 41	18 57	00 N 46
28	19 R 41	18 54	00 S 27
29	19 40	18 51	01 39
30	19 39	18 48	02 48
31	19 ♏ 36	18 ♏ 44	03 S 47

DECLINATIONS

Date	Sun ☉ ° '	Moon ☽ ° '	Mercury ☿ ° '	Venus ♀ ° '	Mars ♂ ° '	Jupiter ♃ ° '	Saturn ♄ ° '	Uranus ♅ ° '	Neptune ♆ ° '	Pluto ♇ ° '
01	15 N 05	19 N 06	18 N 33	15 N 51	24 N 43	04 N 25	20 S 32	20 N 04	10 S 07	22 N 44
02	15 23	20 15	18 10	16 14	24 43	04 26	20 31	20 04	10 07	22 44
03	15 41	20 15	17 46	16 37	24 44	04 27	20 31	20 04	10 06	22 44
04	15 58	19 03	17 20	17 00	24 44	04 28	20 30	20 03	10 06	22 44
05	16 15	16 42	16 53	17 22	24 44	04 29	20 30	20 03	10 05	22 43
06	16 32	13 21	16 26	17 43	24 44	04 29	20 29	20 03	10 04	22 43
07	16 49	09 12	15 58	18 05	24 43	04 30	20 29	20 02	10 04	22 43
08	17 05	04 N 32	15 31	18 25	24 43	04 31	20 28	20 02	10 03	22 43
09	17 22	00 S 25	15 03	18 46	24 42	04 31	20 28	20 01	10 03	22 43
10	17 37	05 20	14 37	19 05	24 41	04 32	20 27	20 01	10 02	22 42
11	17 53	09 56	14 12	19 25	24 40	04 32	20 27	20 01	10 02	22 42
12	18 08	13 57	13 48	19 43	24 38	04 32	20 26	20 00	10 01	22 42
13	18 23	17 08	13 26	20 02	24 37	04 33	20 26	20 00	10 01	22 42
14	18 38	19 17	13 05	20 19	24 35	04 33	20 25	19 59	10 00	22 41
15	18 52	20 19	12 47	20 36	24 33	04 33	20 25	19 59	10 00	22 41
16	19 06	20 13	12 31	20 53	24 31	04 33	20 24	19 59	09 59	22 41
17	19 20	19 06	12 17	21 09	24 29	04 33	20 24	19 58	09 59	22 40
18	19 33	17 04	12 06	21 24	24 27	04 33	20 23	19 58	09 58	22 40
19	19 46	14 20	11 56	21 39	24 24	04 33	20 23	19 57	09 58	22 40
20	19 59	11 02	11 50	21 53	24 21	04 32	20 22	19 57	09 57	22 39
21	20 11	07 19	11 45	22 07	24 18	04 32	20 22	19 56	09 57	22 39
22	20 23	03 S 20	11 44	22 20	24 15	04 31	20 21	19 56	09 57	22 39
23	20 35	00 N 49	11 44	22 32	24 12	04 31	20 20	19 55	09 56	22 38
24	20 46	04 58	11 47	22 44	24 08	04 30	20 20	19 55	09 56	22 38
25	20 57	09 00	11 52	22 55	24 05	04 30	20 19	19 54	09 55	22 37
26	21 08	12 45	11 59	23 06	24 01	04 29	20 19	19 53	09 55	22 37
27	21 18	15 59	12 08	23 15	23 57	04 28	20 18	19 53	09 54	22 37
28	21 28	18 29	12 19	23 24	23 53	04 27	20 18	19 52	09 54	22 36
29	21 37	20 01	12 31	23 33	23 48	04 27	20 17	19 52	09 54	22 36
30	21 46	20 24	12 46	23 40	23 44	04 26	20 17	19 51	09 53	22 35
31	21 N 55	19 N 32	13 N 02	23 N 47	23 N 39	04 N 25	20 S 16	19 N 50	09 S 53	22 N 35

ZODIAC SIGN ENTRIES

Date	h	m	Planets
01	13	47	☽ ♊
03	19	08	☽ ♋
04	15	22	♂ ♋
05	22	54	☽ ♌
08	01	37	☽ ♍
10	03	57	☽ ♎
12	06	48	☽ ♏
13	11	08	♀ ♊
14	11	13	☽ ♐
16	18	13	☽ ♑
19	04	12	☽ ♒
21	08	10	☉ ♊
21	16	20	☽ ♓
24	04	34	☽ ♈
26	14	43	☽ ♉
28	21	47	☽ ♊
31	02	05	☽ ♋

LATITUDES

Date	Mercury ☿ ° '	Venus ♀ ° '	Mars ♂ ° '	Jupiter ♃ ° '	Saturn ♄ ° '	Uranus ♅ ° '	Neptune ♆ ° '	Pluto ♇ ° '
01	01 N 28	00 S 36	01 N 17	01 N 29	01 N 52	00 N 37	01 N 48	11 N 17
04	00 N 41	00 29	01 17	01 28	01 52	00 36	01 48	11 16
07	00 S 10	00 22	01 17	01 28	01 52	00 36	01 48	11 15
10	01 02	00 15	01 17	01 27	01 52	00 36	01 48	11 14
13	01 50	00 08	01 17	01 26	01 52	00 36	01 48	11 13
16	02 30	00 S 01	01 17	01 26	01 52	00 36	01 48	11 13
19	03 02	00 N 06	01 17	01 25	01 52	00 36	01 48	11 12
22	03 24	00 13	01 17	01 24	01 52	00 36	01 48	11 11
25	03 37	00 21	01 17	01 24	01 52	00 36	01 48	11 10
28	03 41	00 28	01 17	01 23	01 52	00 36	01 48	11 10
31	03 S 38	00 N 35	01 N 17	01 N 22	01 N 52	00 N 36	01 N 48	11 N 09

DATA

Julian Date	2435960
Delta T	+32 seconds
Ayanamsa	23° 15' 52"
Synetic vernal point	05° ♓ 51' 08"
True obliquity of ecliptic	23° 26' 36"

MOON'S PHASES, APSIDES AND POSITIONS ☽

Date	h	m	Phase	Longitude ° '	Eclipse Indicator
07	02	29	☽	16 ♌ 16	
13	22	34	○	22 ♏ 52	total
21	17	03	☾	00 ♓ 21	
29	11	39	●	07 ♊ 49	

Day	h	m		
09	03	39	Perigee	
21	16	31	Apogee	
02	23	59	Max dec	20° N 24'
09	10	01	0S	
15	21	57	Max dec	20° S 24'
23	07	19	0N	
30	07	30	Max dec	20° N 25'

LONGITUDES

Date	Chiron ⚷ ° '	Ceres ⚳ ° '	Pallas ⚴ ° '	Juno ⚵ ° '	Vesta ⚶ ° '	Black Moon Lilith ⚸ ° '
01	17 ♒ 12	21 ♉ 31	09 ♈ 43	06 ♈ 59	20 ♓ 53	27 ♒ 09
11	17 ♒ 22	25 ♉ 37	13 ♈ 08	12 ♈ 01	25 ♓ 12	28 ♒ 17
21	17 ♒ 27	29 ♉ 44	16 ♈ 29	17 ♈ 04	29 ♓ 21	29 ♒ 24
31	17 ♒ 24	03 ♊ 52	19 ♈ 46	22 ♈ 07	03 ♈ 17	00 ♓ 31

All ephemeris data is given at 12.00 UT and the Moon's longitude is additionally given for 24.00 UT

ASPECTARIAN

h m	Aspects	h m	Aspects	h m	Aspects
01 Wednesday		**11 Saturday**		12 45	☽ Q ☿
05 59	☽ ∥ ☿	00 08	☽ ⚼ ♀	17 03	☽ □ ☉
08 42	♂ ⚹ ♇	00 36	☽ ⚻ ☿	17 16	☽ △ ♆
10 03	☽ □ ♇	01 08	☽ ⚹ ♄	19 36	☉ ⚻ ♆
10 07	☽ ⚺ ♂	01 44	☽ ∠ ♇	19 54	☿ ⚺ ♀
15 31	☽ ⚻ ♆	03 53	☽ ± ☉	23 39	☽ ⚻ ♅
19 19	☽ ⚹ ♅	05 44	☽ Q ♅	**22 Wednesday**	
02 Thursday		12 32	☽ ∥ ♆	04 54	☽ # ♃
02 19	☽ ± ♆	14 48	☽ ⚻ ☉	07 46	☉ # ♄
05 52	☽ ∥ ♅	16 51	☽ ± ♀	11 47	☽ ± ♅
10 56	☽ ⚺ ☉	17 00	☽ ⚺ ♃	13 27	☽ ⚹ ☿
13 18	☽ ☍ ♄	18 40	♀ □ ♇	15 04	☽ □ ♀
18 21	☽ ⚼ ♆	**12 Sunday**		15 06	☽ △ ♂
18 21	☽ Q ♇	02 31	☽ ∠ ♄	15 39	♀ ⚺ ♂
20 05	☽ ⚺ ♀	03 13	☽ ⊥ ♃	16 05	☽ □ ♄
20 26	☽ ⚺ ☿	03 15	☽ ⚹ ♇	23 27	☽ ⚼ ♆
22 10	☽ ∠ ♅	04 04	☽ ⚻ ♀	**23 Thursday**	
22 25	☽ ⊥ ☉	07 47	☉ Q ♅	00 22	♀ ☍ ♄
22 52	☿ ☌ ♀	07 59	☽ ☌ ♆	05 57	☽ ⚼ ♅
03 Friday		11 07	☽ # ☿	07 52	♂ ⚻ ♄
05 28	☽ □ ♃	12 30	☽ □ ♅	08 30	☽ Q ☉
06 37	☽ ⊥ ☿	15 26	☽ △ ♂	12 22	☽ ☍ ♃
07 45	☽ ⊥ ♀	16 41	♀ ⊥ ♂	17 27	☽ ± ♆
14 04	☽ ⊥ ♅	17 55	☽ ⊥ ♄	20 20	☽ ∠ ☿
15 30	☽ ∠ ☉	18 40	☽ ∠ ♃	23 34	♀ ∥ ♇
15 31	☽ ⚹ ♇	22 12	☉ △ ♃	**24 Friday**	
17 59	☉ ⚻ ♄	23 55	☽ Q ♇	00 32	☽ ⚻ ♇
18 09	☽ ∥ ♅	**13 Monday**		05 22	☽ ⚻ ♆
18 10	☽ ☌ ♂	02 16	☽ ☍ ☿	07 10	☽ Q ♀
20 43	☽ △ ♆	04 16	☽ ⚺ ♄	09 18	☽ ∥ ♃
21 46	☽ ∠ ☿	09 39	♀ ∥ ♅	10 27	☽ ⚹ ☉
04 Saturday		18 48	☽ ⚼ ♂	11 54	☽ △ ♅
00 34	☽ ⚺ ♅	20 49	☽ ⚹ ♃	12 26	☽ ± ♇
01 04	☽ ∠ ♀	22 34	☽ ☍ ☉	14 45	☽ ⊥ ☿
12 49	☽ Q ♃	23 42	♀ ⚻ ♆	**25 Saturday**	
17 36	☽ ∠ ♇	**14 Tuesday**		02 57	☽ ⚺ ☿
17 40	☽ ⚻ ♄	02 12	☽ # ☉	03 21	☽ △ ♄
19 33	☽ ⚹ ☉	07 32	☽ □ ♇	05 28	☉ ⚹ ♅
22 05	☿ ∥ ♀	11 32	☽ ± ♂	05 58	☽ □ ♂
22 43	☽ ⚹ ☿	12 21	☽ ⚺ ♆	06 05	☽ ⚼ ♇
05 Sunday		13 39	☽ ☍ ♀	08 20	♂ ∠ ♇
03 58	☽ ± ♄	17 17	☽ △ ♅	09 20	☽ ⚹ ♀
05 33	☽ ⚹ ♀	18 09	☽ Q ♃	12 16	☿ ⚻ ♄
06 55	☽ ∥ ♀	20 04	♀ # ♄	17 39	☽ # ♆
09 06	☽ ⊥ ♇	22 50	☽ ⚻ ♂	18 12	☽ ∠ ☉
09 25	☽ ⚹ ♃	23 07	☽ ⊥ ♆	23 13	♀ ⚼ ♆
10 04	☽ ∥ ☿	**15 Wednesday**		23 19	☽ ⚻ ♃
12 12	☿ ∠ ♂	01 12	☽ # ♅	**26 Sunday**	
15 16	☽ ∥ ☉	06 11	☽ ⚻ ☿	06 40	☽ ∥ ☿
17 41	☽ Q ☉	09 23	☽ ☌ ♄	08 05	☽ ⚼ ♄
18 26	☽ Q ☿	15 28	☽ ∠ ♆	10 47	☽ ± ♃
19 19	☽ ⚼ ♄	16 56	☽ ± ☿	10 55	☽ △ ♇
19 22	☽ ⚺ ♇	20 38	☽ ⚼ ♅	11 27	♀ Q ♇
23 49	♂ △ ♆	**16 Thursday**		12 48	☽ ⊥ ☉
06 Monday		02 59	☽ □ ♃	15 23	☽ ☍ ♆
00 20	☽ □ ♆	06 04	♀ ⚹ ♅	17 08	☽ ∠ ♀
00 22	☽ ⚺ ♂	09 17	☽ ⚻ ☉	18 14	☽ Q ♂
00 34	☉ ☌ ☿	09 19	☽ ⚼ ☿	21 52	☽ □ ♅
04 02	☽ Q ♀	13 25	☽ ± ♅	**27 Monday**	
04 16	☽ ☌ ♅	14 21	☽ △ ♇	00 57	☽ ± ♄
06 53	♀ Q ♅	19 18	☽ ⚹ ♆	01 00	☽ ⚺ ☉
08 30	☉ ∥ ☿	19 29	☽ # ♅	03 44	☽ ⚼ ♃
10 56	☽ ∠ ♃	21 28	☽ ± ☉	11 42	☽ ⊥ ♀
11 02	☽ ⊥ ♂	**17 Friday**		12 02	☽ ⚻ ♄
20 41	☽ △ ♄	00 45	☽ ⚻ ♅	14 26	☽ ☌ ☿
23 42	☽ □ ☿	02 43	☽ ⚻ ♀	17 41	☽ ⚹ ♂
07 Tuesday		08 46	☽ # ☉	22 10	☉ ± ♆
00 18	♀ △ ♃	09 24	☽ ☍ ♂	23 52	☽ ⚺ ♀
02 08	☽ ⊥ ♃	13 22	☽ △ ☿	**28 Tuesday**	
02 29	☽ □ ☉	15 25	☽ ± ♀	07 08	☽ Q ♅
02 59	☽ ∠ ♂	16 03	☽ ⚼ ☉	07 20	☽ △ ♃
06 47	☽ Q ♆	17 18	☽ ⚺ ♄	18 14	☽ □ ♇
07 18	☽ # ♆	18 08	☽ Q ♆	19 49	♀ ∠ ♅
12 13	☽ ⚺ ♃	18 53	☽ ⚼ ♇	22 11	☽ ∠ ♂
13 23	☽ □ ♀	**18 Saturday**		22 19	☽ ⚻ ♆
21 56	☉ ∠ ♂	04 50	☽ ⊥ ♄	**29 Wednesday**	
22 07	☽ ☌ ♇	10 46	☽ ⚼ ♀	04 42	☽ ⚹ ♅
08 Wednesday		12 09	☽ △ ♃	08 38	☽ ∥ ♅
02 57	☽ ⚹ ♆	12 22	☽ ± ♇	08 57	☽ ± ♆
05 23	☽ ⚹ ♂	18 33	♀ ± ♆	11 39	☽ ☌ ☉
07 00	☽ ⚺ ♅	22 21	☽ ∠ ♄	14 53	☽ ⊥ ♂
12 05	☽ ∥ ♃	23 44	☽ △ ☉	17 40	☽ ☍ ♄
17 05	☽ ⊥ ♅	**19 Sunday**		20 15	☽ # ♄
22 56	☽ □ ♄	00 09	☽ ⚻ ♇	23 06	☽ ⚺ ☿
09 Thursday		01 04	☿ St D	**30 Thursday**	
00 05	☽ △ ☿	02 22	♃ St D	00 43	☽ ⚼ ♆
02 27	☽ Q ♂	04 57	☉ □ ♇	01 52	☽ ⚺ ♂
04 04	☽ ∠ ♆	05 14	☽ □ ♆	02 00	☽ Q ♇
08 12	☽ ∠ ♅	11 12	☽ ☍ ♅	03 37	☿ ⊥ ♀
08 34	☽ △ ☉	17 47	☽ ⚼ ♃	07 03	☽ ∠ ♅
14 28	☽ ☌ ♃	19 44	☽ △ ♀	10 08	☽ ⊥ ☿
19 08	☽ ⊥ ♆	23 19	☽ ⚻ ♂	10 22	☽ ☌ ♀
19 36	♂ ⚺ ♅	**20 Monday**		12 25	☽ □ ♃
19 47	♇ St D	00 08	☽ □ ☿	18 57	♀ ∥ ♂
20 29	☽ △ ♀	03 59	☽ ⚹ ♄	19 02	☽ # ♄
10 Friday		06 20	☽ # ☿	22 36	☽ ⊥ ♅
00 16	☽ ⚼ ☿	07 54	☉ ∥ ♅	22 44	☽ ⚹ ♇
00 28	☽ ⚺ ♇	07 54	♂ Q ♃	**31 Friday**	
05 00	☽ Q ♄	11 48	☽ ± ♃	02 32	☽ △ ♆
05 12	☽ ⚺ ♆	12 01	☽ ± ♂	02 34	☽ ∠ ☿
08 02	☽ # ♃	17 51	☿ ⚹ ♂	06 24	☽ ∥ ♅
08 33	☿ ⚻ ♄	19 10	☽ ∥ ♆	08 53	☽ ⚺ ♅
09 26	☽ ⚹ ♅	23 52	☽ ⚻ ♃	10 13	♀ □ ♃
10 03	☽ □ ♂	**21 Tuesday**		19 10	☽ ⚺ ☉
10 33	☽ ⊥ ♇	03 58	☽ Q ♄	19 14	☽ Q ♃
11 35	☽ ⚼ ☉	07 08	☽ ⚼ ♂	20 57	☽ ⚻ ♄
14 51	☽ ± ☿	12 13	☽ ☍ ♇	21 33	♂ ± ♄

JUNE 1957

LONGITUDES

Date	Sidereal time h m s	Sun ☉ ° ' "	Moon ☽ ° ' "	Moon ☽ 24.00 ° ' "	Mercury ☿ ° '	Venus ♀ ° '	Mars ♂ ° '	Jupiter ♃ ° '	Saturn ♄ ° '	Uranus ♅ ° '	Neptune ♆ ° '	Pluto ♇ ° '
01	04 38 48	10 ♊ 42 51	20 ♋ 01 05	27 ♋ 09 54	16 ♉ 30	23 ♊ 23	17 ♋ 28	22 ♍ 06	11 ♐ 03	04 ♌ 02	00 ♏ 14	28 ♌ 03
02	04 42 45	11 40 21	04 ♌ 19 45	11 ♌ 30 07	17 27	24 37	18 06	22 09	10 R 58	04 04	00 R 13	28 04
03	04 46 41	12 37 49	18 ♌ 40 26	25 ♌ 50 14	18 28	25 50	18 43	22 11	10 54	04 07	00 12	28 04
04	04 50 38	13 35 16	02 ♍ 59 07	10 ♍ 06 41	19 32	27 04	19 21	22 14	10 49	04 09	00 11	28 05
05	04 54 35	14 32 42	17 ♍ 12 38	24 ♍ 16 42	20 39	28 17	19 58	22 17	10 45	04 12	00 10	28 06
06	04 58 31	15 30 06	01 ♎ 18 41	08 ♎ 18 24	21 49	29 ♊ 31	20 36	22 20	10 40	04 15	00 09	28 07
07	05 02 28	16 27 30	15 ♎ 15 42	22 ♎ 10 28	23 02	00 ♋ 44	21 13	22 24	10 36	04 17	00 07	28 07
08	05 06 24	17 24 52	29 ♎ 02 35	05 ♏ 51 58	24 18	01 58	21 51	22 27	10 32	04 20	00 07	28 08
09	05 10 21	18 22 13	12 ♏ 38 31	19 ♏ 22 07	25 38	03 11	22 29	22 31	10 27	04 23	00 06	28 09
10	05 14 17	19 19 34	26 ♏ 02 41	02 ♐ 40 07	27 00	04 25	23 06	22 35	10 23	04 26	00 05	28 10
11	05 18 14	20 16 53	09 ♐ 14 21	15 ♐ 45 17	28 26	05 38	23 44	22 39	10 18	04 29	00 04	28 11
12	05 22 10	21 14 12	22 ♐ 12 51	28 ♐ 37 02	29 ♉ 54	06 51	24 21	22 43	10 14	04 32	00 03	28 12
13	05 26 07	22 11 30	04 ♑ 57 49	11 ♑ 15 14	01 ♊ 25	08 05	24 59	22 47	10 10	04 34	00 02	28 13
14	05 30 04	23 08 48	17 ♑ 29 21	23 ♑ 40 18	02 59	09 18	25 36	22 52	10 06	04 37	00 01	28 14
15	05 34 00	24 06 05	29 ♑ 48 15	05 ♒ 53 25	04 36	10 32	26 14	22 56	10 01	04 40	00 ♏ 00	28 15
16	05 37 57	25 03 21	11 ♒ 56 06	17 ♒ 56 36	06 16	11 45	26 52	23 01	09 57	04 43	29 ♎ 59	28 16
17	05 41 53	26 00 37	23 ♒ 55 20	29 ♒ 52 43	07 58	12 58	27 29	23 06	09 53	04 47	29 59	28 17
18	05 45 50	26 57 53	05 ♓ 49 13	11 ♓ 45 22	09 44	14 12	28 07	23 11	09 49	04 50	29 58	28 18
19	05 49 46	27 55 08	17 ♓ 41 41	23 ♓ 38 47	11 32	15 25	28 44	23 16	09 45	04 53	29 57	28 19
20	05 53 43	28 52 24	29 ♓ 37 15	05 ♈ 37 42	13 23	16 38	29 ♋ 22	23 22	09 41	04 56	29 57	28 20
21	05 57 39	29 ♊ 49 39	11 ♈ 40 45	17 ♈ 47 00	15 16	17 52	00 ♌ 00	23 27	09 36	04 59	29 56	28 21
22	06 01 36	00 ♋ 46 53	23 ♈ 57 03	00 ♉ 11 28	17 12	19 05	00 37	23 33	09 32	05 02	29 55	28 23
23	06 05 33	01 44 08	06 ♉ 30 46	12 ♉ 55 25	19 11	20 18	01 15	23 38	09 29	05 05	29 55	28 24
24	06 09 29	02 41 23	19 ♉ 25 47	26 ♉ 02 10	21 11	21 31	01 52	23 44	09 25	05 09	29 54	28 25
25	06 13 26	03 38 38	02 ♊ 44 44	09 ♊ 33 31	23 14	22 45	02 30	23 51	09 21	05 12	29 54	28 26
26	06 17 22	04 35 52	16 ♊ 28 25	23 ♊ 29 10	25 19	23 58	03 08	23 57	09 17	05 15	29 53	28 28
27	06 21 19	05 33 06	00 ♋ 35 22	07 ♋ 46 25	27 25	25 11	03 45	24 03	09 13	05 19	29 53	28 29
28	06 25 15	06 30 21	15 ♋ 01 38	22 ♋ 20 08	29 ♊ 33	26 24	04 23	24 10	09 09	05 22	29 52	28 30
29	06 29 12	07 27 35	29 ♋ 41 03	07 ♌ 03 22	01 ♋ 41	27 38	05 00	24 16	09 06	05 25	29 52	28 32
30	06 33 08	08 ♋ 24 48	14 ♌ 26 06	21 ♌ 48 19	03 ♋ 51	28 ♋ 51	05 ♌ 38	24 ♍ 23	09 ♐ 02	05 ♌ 29	29 ♎ 52	28 ♌ 33

DECLINATIONS

Date	Moon True ☊ ° '	Moon Mean ☊ ° '	Moon ☽ Latitude ° '	Sun ☉ ° '	Moon ☽ ° '	Mercury ☿ ° '	Venus ♀ ° '	Mars ♂ ° '	Jupiter ♃ ° '	Saturn ♄ ° '	Uranus ♅ ° '	Neptune ♆ ° '	Pluto ♇ ° '
01	19 ♏ 34	18 ♏ 41	04 S 34	22 N 03	17 N 26	13 N 20	23 N 54	23 N 34	04 N 23	20 S 15	19 N 50	09 S 53	22 N 34
02	19 R 31	18 38	05 03	22 11	14 16	13 39	23 59	23 29	04 22	20 15	19 49	09 52	22 34
03	19 29	18 35	05 14	22 19	10 15	14 00	24 04	23 24	04 21	20 14	19 49	09 52	22 33
04	19 27	18 32	05 06	22 26	05 39	14 21	24 09	23 19	04 20	20 14	19 48	09 51	22 33
05	19 D 27	18 29	04 38	22 33	00 N 47	14 44	24 12	23 13	04 18	20 13	19 47	09 51	22 32
06	19 27	18 25	03 54	22 39	04 S 06	15 08	24 15	23 07	04 17	20 13	19 47	09 51	22 32
07	19 29	18 22	02 57	22 45	08 44	15 33	24 17	23 02	04 15	20 12	19 46	09 50	22 32
08	19 30	18 19	01 50	22 51	12 51	15 59	24 18	22 55	04 14	20 12	19 45	09 50	22 31
09	19 31	18 16	00 S 38	22 56	16 14	16 25	24 19	22 49	04 12	20 11	19 45	09 50	22 30
10	19 R 31	18 13	00 N 36	23 01	18 41	16 52	24 19	22 43	04 10	20 11	19 44	09 50	22 30
11	19 30	18 09	01 46	23 05	20 06	17 20	24 18	22 36	04 09	20 10	19 43	09 49	22 29
12	19 27	18 06	02 49	23 09	20 24	17 48	24 17	22 30	04 07	20 10	19 43	09 49	22 29
13	19 23	18 03	03 43	23 13	19 38	18 16	24 14	22 23	04 05	20 09	19 42	09 49	22 28
14	19 17	18 00	04 25	23 16	17 56	18 44	24 11	22 16	04 03	20 09	19 41	09 49	22 28
15	19 11	17 57	04 53	23 19	15 25	19 12	24 08	22 09	04 01	20 08	19 40	09 48	22 27
16	19 06	17 54	05 08	23 21	12 17	19 40	24 03	22 01	03 59	20 08	19 40	09 48	22 27
17	19 00	17 50	05 09	23 23	08 41	20 08	23 58	21 54	03 57	20 07	19 39	09 48	22 26
18	18 57	17 47	04 57	23 24	04 46	20 35	23 53	21 46	03 54	20 07	19 38	09 48	22 25
19	18 55	17 44	04 33	23 26	00 S 41	21 01	23 46	21 39	03 52	20 06	19 37	09 47	22 25
20	18 D 54	17 41	03 56	23 26	03 N 27	21 27	23 39	21 31	03 50	20 06	19 37	09 47	22 24
21	18 55	17 38	03 08	23 27	07 30	21 52	23 31	21 23	03 47	20 05	19 36	09 47	22 24
22	18 56	17 35	02 11	23 26	11 20	22 15	23 22	21 14	03 45	20 05	19 35	09 47	22 23
23	18 57	17 31	01 N 07	23 26	14 45	22 37	23 13	21 06	03 42	20 04	19 34	09 47	22 22
24	18 R 58	17 28	00 S 03	23 25	17 33	22 57	23 03	20 57	03 40	20 04	19 33	09 47	22 22
25	18 57	17 25	01 14	23 24	19 30	23 16	22 53	20 49	03 37	20 03	19 33	09 46	22 21
26	18 54	17 22	02 23	23 22	20 23	23 32	22 41	20 40	03 35	20 03	19 32	09 46	22 21
27	18 50	17 19	03 25	23 20	20 01	23 46	22 30	20 31	03 32	20 02	19 31	09 46	22 20
28	18 43	17 15	04 16	23 17	18 21	23 58	22 17	20 22	03 29	20 02	19 30	09 46	22 19
29	18 36	17 12	04 50	23 14	15 29	24 07	22 04	20 13	03 26	20 02	19 29	09 46	22 18
30	18 ♏ 29	17 ♏ 09	05 S 06	23 N 11	11 N 37	24 N 14	21 N 50	20 N 03	03 N 24	20 S 01	19 N 29	09 S 46	22 N 18

ZODIAC SIGN ENTRIES

Date	h	m	Planets
02	04	45	☽ ♌
04	06	59	☽ ♍
06	09	45	☽ ♎
06	21	35	♀ ♋
08	13	41	☽ ♏
10	19	09	☽ ♐
12	13	40	☿ ♊
13	02	36	☽ ♑
15	12	23	☽ ♒
15	20	07	♆ ♎
18	00	15	☽ ♓
20	12	46	☽ ♈
21	12	18	♂ ♌
21	16	21	☉ ♋
22	23	38	☽ ♉
25	07	07	☽ ♊
27	11	01	☽ ♋
28	17	08	☿ ♋
29	12	31	☽ ♌

LATITUDES

Date	Mercury ☿ ° '	Venus ♀ ° '	Mars ♂ ° '	Jupiter ♃ ° '	Saturn ♄ ° '	Uranus ♅ ° '	Neptune ♆ ° '	Pluto ♇ ° '
01	03 S 35	00 N 37	01 N 17	01 N 22	01 N 52	00 N 36	01 N 48	11 N 09
04	03 23	00 44	01 17	01 21	01 52	00 36	01 48	11 08
07	03 05	00 50	01 16	01 21	01 51	00 36	01 47	11 07
10	02 42	00 57	01 16	01 20	01 51	00 36	01 47	11 06
13	02 14	01 03	01 16	01 19	01 51	00 36	01 47	11 05
16	01 43	01 08	01 16	01 19	01 50	00 36	01 47	11 05
19	01 09	01 13	01 15	01 18	01 50	00 36	01 47	11 04
22	00 35	01 18	01 15	01 18	01 50	00 35	01 47	11 04
25	00 S 01	01 22	01 14	01 17	01 49	00 35	01 47	11 03
28	00 N 31	01 26	01 14	01 16	01 49	00 35	01 46	11 03
31	00 N 59	01 N 29	01 N 14	01 N 16	01 N 48	00 N 35	01 N 46	11 N 02

DATA

Julian Date	2435991
Delta T	+32 seconds
Ayanamsa	23° 15' 57"
Synetic vernal point	05° ♓ 51' 03"
True obliquity of ecliptic	23° 26' 35"

MOON'S PHASES, APSIDES AND POSITIONS ☽

Date	h	m	Phase	Longitude ° '	Eclipse Indicator
05	07	10	☽	14 ♍ 21	
12	10	02	○	21 ♐ 10	
20	10	22	☾	28 ♓ 49	
27	20	53	●	05 ♋ 54	

Day	h	m		
03	04	26	Perigee	
18	10	49	Apogee	
30	08	04	Perigee	
05	15	48	0S	
12	06	39	Max dec	20° S 26'
19	15	57	0N	
26	17	15	Max dec	20° N 25'

LONGITUDES

Date	Chiron ⚷ ° '	Ceres ⚳ ° '	Pallas ⚴ ° '	Juno ⚵ ° '	Vesta ⚶ ° '	Black Moon Lilith ⚸ ° '
01	17 ♒ 23	04 ♊ 17	20 ♈ 05	22 ♈ 37	03 ♈ 40	00 ♓ 38
11	17 ♒ 14	08 ♊ 25	23 ♈ 15	27 ♈ 39	07 ♈ 20	01 ♓ 45
21	16 ♒ 58	12 ♊ 32	26 ♈ 18	02 ♉ 39	10 ♈ 43	02 ♓ 52
31	16 ♒ 37	16 ♊ 37	29 ♈ 11	07 ♉ 36	13 ♈ 46	03 ♓ 59

All ephemeris data is given at 12.00 UT and the Moon's longitude is additionally given for 24.00 UT

ASPECTARIAN

h m Aspects

01 Saturday
00 15 ☽ ∠ ♇
05 41 ☽ ✶ ☿
06 02 ☽ ⊥ ☉
07 01 ☽ ± ♄
07 30 ☽ ☌ ♂
15 25 ☽ ⊥ ♇
15 31 ☽ ✶ ♃
18 11 ☽ ⊻ ♀
19 40 ☉ ☍ ♄
22 04 ☽ Ᵽ ♄
22 15 ☽ ∠ ☉

02 Sunday
01 30 ☽ ⊻ ♇
03 15 ☽ Q ☿
05 07 ☽ □ ♆
05 11 ☽ ⊥ ♀
11 34 ☽ ☌ ♅
15 39 ☽ ∥ ☿
16 44 ☽ ∠ ♃
21 40 ☽ ∠ ♀
23 03 ☽ △ ♄

03 Monday
01 10 ☽ ✶ ☉
07 50 ☽ ⊥ ♃
11 12 ☽ Q ♆
11 37 ☽ □ ☿
12 05 ☽ ⊻ ♂
14 06 ☽ # ♆
17 55 ☽ ⊻ ♃
18 25 ☉ ⊥ ♂
22 35 ☽ ⊥ ♂
22 41 ☽ Q ☉

04 Tuesday
01 07 ☽ ✶ ♀
02 20 ☿ ✶ ♂
03 46 ☽ ☌ ♇
07 17 ☽ ✶ ♆
13 58 ☽ ⊻ ♅
14 24 ☽ ∠ ♂
18 39 ☽ ∥ ♃
23 11 ☽ Q ♀

05 Wednesday
00 07 ☽ ⊥ ♅
01 08 ☽ □ ♄
07 10 ☽ □ ☉
08 16 ♀ ✶ ♇
08 32 ☽ ∠ ♆
10 14 ♀ ⊥ ♅
10 49 ☉ ∥ ♇
15 23 ☽ ∠ ♅
16 54 ☽ ✶ ♂
18 20 ☽ △ ☿
20 39 ☽ ☌ ♃
23 47 ☽ ⊥ ♆

06 Thursday
03 08 ☉ Ᵽ ♆
06 32 ☽ ⊻ ♇
07 31 ☽ Q ♄
08 38 ☽ □ ♀
10 00 ☽ ⊻ ♆
12 52 ☽ # ♃
14 18 ☽ Q ♂
16 48 ☽ ⊥ ♇
17 02 ☽ ✶ ♅
20 55 ☿ Q ♅
22 19 ☽ Ᵽ ☿
22 59 ☿ △ ♃

07 Friday
00 11 ♀ △ ♆
03 29 ☉ Q ♇
03 59 ☽ ✶ ♄
08 18 ☽ ∠ ♇
13 47 ☽ Q ♅
14 14 ☽ △ ☉
15 23 ☽ ± ☿
18 08 ☽ ∥ ♆
22 50 ☽ □ ♂

08 Saturday
00 26 ☽ ⊻ ♃
02 52 ☽ ⊼ ☿
05 53 ☽ ∠ ♄
10 25 ☽ ✶ ♇
10 58 ☽ ⊥ ♃
13 52 ☽ ☌ ♆
17 38 ☽ △ ♀
18 22 ☽ Ᵽ ☉
21 20 ☽ □ ♅
21 35 ☽ ⊥ ♄
21 42 ☉ ∥ ♂
23 15 ♂ ⊥ ♇

09 Sunday
02 52 ☽ ∠ ♃
07 35 ☽ Q ♇
08 08 ☽ ⊻ ♄
11 29 ☽ ± ☉
13 38 ♂ ✶ ♃
13 49 ☽ # ☿
22 52 ☽ Ᵽ ♀
23 00 ☽ ⊼ ☉

10 Monday
05 44 ☽ ✶ ♃
06 26 ☽ △ ♂
12 25 ♀ ⊻ ♅
13 56 ☽ ☍ ☿
14 45 ☉ ∠ ♅
15 50 ☽ □ ♇
16 42 ☽ ± ♀
19 17 ☽ ⊻ ♆

11 Tuesday
03 16 ☽ △ ♅
03 31 ☽ # ♅
03 33 ☽ Q ♃
04 43 ☽ ⊼ ♀
06 11 ☽ ⊥ ♆
07 55 ☿ □ ♇
11 01 ☽ Ᵽ ♂
13 57 ☽ ☌ ♄
14 06 ☽ ∥ ♄
22 42 ☽ ∠ ♆

12 Wednesday
04 27 ☽ ± ♂
06 58 ☽ Ᵽ ♅
10 02 ☽ ☍ ☉
12 56 ☽ □ ♃
14 23 ☿ ⊼ ♆
15 24 ♂ ∥ ♇
16 12 ☽ ⊼ ♂
23 13 ☽ △ ♇
23 47 ☽ ∥ ♄
23 52 ☽ ± ♅

13 Thursday
02 41 ☽ ✶ ♆
04 21 ☽ ⊼ ☿
10 53 ☽ # ♅
11 16 ☽ ⊼ ♅
17 19 ☽ ± ☿
18 16 ♂ Ᵽ ♄
18 34 ☽ ☍ ♀
21 51 ☽ ⊻ ♄

14 Friday
01 29 ☽ Q ♆
03 47 ☽ Ᵽ ♇
04 11 ☉ □ ♃
04 13 ☽ # ☿
09 19 ☽ ⊥ ♄
13 06 ☽ Ᵽ ☿
21 12 ☽ ± ♇
22 29 ☽ △ ♃
23 54 ☽ ⊼ ☉

15 Saturday
02 38 ♀ ⊼ ♄
02 41 ☽ ∠ ♄
04 38 ☽ ☍ ♂
07 42 ☿ ⊥ ♀
08 56 ☽ ⊼ ♇
12 24 ☽ □ ♆
12 38 ☽ ± ☉
13 07 ☿ ✶ ♅
20 36 ♀ Q ♃
21 38 ☽ ☍ ♅
22 56 ☽ △ ☿

16 Sunday
04 10 ☽ Ᵽ ♃
07 56 ☽ Ᵽ ☉
08 05 ☽ ✶ ♄
08 09 ☿ ± ♆
11 35 ☽ ⊼ ♀
11 40 ☿ ∥ ♅
22 13 ☽ ± ♃

17 Monday
00 56 ☽ ± ♀
04 47 ☽ ∥ ♆
07 55 ☽ Q ♄
10 20 ☽ ⊼ ♃
11 25 ☿ # ♄
16 34 ☽ △ ☉
18 11 ♀ ∠ ♇
19 35 ☽ ⊼ ♂
20 48 ☽ ☍ ♇
21 05 ☽ Ᵽ ♀

18 Tuesday
00 11 ☽ △ ♆
08 21 ☽ ± ♂
09 59 ☽ ⊼ ♅
13 03 ☿ ☍ ♄
17 09 ☽ # ♃
19 23 ♂ ⊻ ♇
20 01 ☽ □ ♄
21 19 ☽ □ ☿
22 10 ☽ ± ♅

19 Wednesday
03 34 ☽ Ᵽ ♂
06 28 ☽ Ᵽ ♆
06 52 ☽ △ ♀
16 26 ☽ Ᵽ ♅
18 06 ♀ ± ♄
22 15 ☉ ✶ ♇
23 20 ☽ ☍ ♃

20 Thursday
00 37 ☽ ± ♆
09 25 ☽ ⊼ ♇
10 22 ☽ □ ☉
11 28 ☽ △ ♂
12 39 ☽ ⊼ ♆
13 32 ☉ ⊥ ♅
14 10 ☽ ∥ ♃
14 36 ☿ ∥ ♂
16 11 ☽ Q ☿
21 27 ☽ ± ♇
22 39 ☽ △ ♅

21 Friday
06 45 ☿ ∠ ♂
07 45 ☿ Ᵽ ♆
07 56 ☽ △ ♄
09 46 ♂ □ ♆
14 38 ☉ △ ♆
15 19 ☽ Ᵽ ♇
20 24 ☽ ✶ ☿

22 Saturday
00 04 ☉ ⊻ ♂
00 44 ☉ ∥ ♀
01 06 ☽ Q ☉
01 30 ☽ □ ♀
01 40 ☿ Q ♇
02 03 ☽ # ♆
11 13 ☽ ⊼ ♃
13 08 ☽ Ᵽ ♄
20 19 ☿ ∥ ♇
20 32 ☽ △ ♇
22 51 ☽ ± ♃
23 29 ☽ ☍ ♆

23 Sunday
01 29 ☽ □ ♂
02 12 ☽ ✶ ☉
06 17 ☽ ± ♄
06 46 ☽ ∠ ☿
09 18 ☽ □ ♅
15 43 ☽ Q ♀
16 02 ☽ Ᵽ ♃
17 32 ☽ ⊼ ♄
23 15 ☿ ∠ ♅

24 Monday
02 45 ☽ ⊥ ☿
08 33 ☽ ∠ ☉
12 51 ☽ Q ♂
15 48 ☽ ⊻ ☿
16 13 ☽ ✶ ♀
17 01 ☿ ∥ ♀
18 48 ☽ Q ♅
19 55 ☽ △ ♃
21 54 ☿ ⊻ ♀

25 Tuesday
02 12 ☽ ⊥ ☉
04 18 ☽ □ ♇
05 52 ♀ ⊥ ♇
06 56 ☽ ⊼ ♆
11 32 ☽ ✶ ♂
12 40 ☽ ∥ ♅
13 43 ☽ ⊻ ☉
16 22 ☽ ✶ ♅
17 34 ☽ ± ♆
19 27 ☿ □ ♃
21 41 ☽ ∠ ♀
22 18 ☉ ∥ ☿
22 59 ☽ # ♄
23 21 ♃ ⊥ ♆
23 34 ☽ ☍ ♄

26 Wednesday
09 16 ☽ Ᵽ ♆
11 35 ♀ ✶ ♃
11 59 ☽ Q ♇
14 49 ☽ ⊥ ♀
14 58 ☽ ∠ ♂
17 55 ♀ Ᵽ ♄
18 31 ☽ ∠ ♅

27 Thursday
00 53 ☽ □ ♃
02 02 ☽ ⊻ ♀
05 30 ☉ ⊻ ♅
05 43 ☽ ☌ ☿
07 00 ☽ ⊥ ♂
08 27 ☽ ✶ ♇
09 50 ☽ ⊥ ♅
10 48 ☽ △ ♆
11 29 ☽ # ♄
17 26 ☿ ⊥ ♂
17 32 ☽ ⊻ ♂
19 56 ☽ ⊻ ♅
20 53 ☽ ☌ ☉
21 34 ☽ ∥ ♅

28 Friday
00 12 ☿ ✶ ♇
02 20 ☽ ⊼ ♄
07 14 ☽ Q ♃
07 54 ♀ ∥ ♇
09 29 ☽ ∠ ♇
09 57 ☿ ⊥ ♅
12 13 ☽ ± ♄
15 41 ☿ △ ♆

29 Saturday
00 18 ☽ ⊥ ♇
02 55 ☽ Ᵽ ♄
03 06 ☽ ✶ ♃
08 21 ☽ ☌ ♀
10 07 ☽ ⊻ ♇
12 18 ☽ □ ♆
15 50 ☽ ⊻ ☿
21 03 ☽ ☌ ♂
21 22 ☽ ☌ ♅

30 Sunday
01 32 ☽ ⊻ ☉
03 15 ☽ △ ♄
03 16 ☽ ⊥ ☿
03 43 ☽ ∠ ♃
05 23 ♂ ☌ ♅
06 02 ♀ ⊻ ♇
11 58 ☽ ⊥ ☉
17 30 ♂ # ♄
17 34 ☽ Q ♆
18 29 ☽ ⊥ ♃
20 26 ☽ ∠ ☿
22 06 ☽ # ♆

JULY 1957

LONGITUDES

Date	Sidereal time h m s	Sun ☉ ° ' "	Moon ☽ ° ' "	Moon ☽ 24.00 ° ' "	Mercury ☿ ° '	Venus ♀ ° '	Mars ♂ ° '	Jupiter ♃ ° '	Saturn ♄ ° '	Uranus ♅ ° '	Neptune ♆ ° '	Pluto ♇ ° '
01	06 37 05	09 ♋ 22 01	29 ♌ 09 06	06 ♍ 27 41	06 ♋ 01	00 ♌ 04	06 ♌ 16	24 ♍ 30	08 ♐ 59	05 ♌ 32	29 ♎ 51	28 ♌ 34
02	06 41 02	10 19 14	13 ♍ 43 23	20 ♍ 55 39	08 12	01 17	06 53	24 37	08 R 55	05 35	29 R 51	28 36
03	06 44 58	11 16 27	28 ♍ 04 07	05 ♎ 08 29	10 23	02 30	07 31	24 44	08 52	05 39	29 51	28 37
04	06 48 55	12 13 39	12 ♎ 08 39	19 ♎ 04 32	12 33	03 43	08 09	24 51	08 48	05 42	29 50	28 39
05	06 52 51	13 10 51	25 ♎ 56 13	02 ♏ 43 47	14 43	04 56	08 46	24 59	08 45	05 46	29 50	28 40
06	06 56 48	14 08 03	09 ♏ 27 24	16 ♏ 07 15	16 52	06 10	09 24	25 06	08 42	05 49	29 50	28 42
07	07 00 44	15 05 14	22 ♏ 43 32	29 ♏ 16 27	19 00	07 23	10 01	25 14	08 39	05 53	29 50	28 43
08	07 04 41	16 02 25	05 ♐ 46 10	12 ♐ 12 51	21 07	08 36	10 39	25 22	08 36	05 56	29 50	28 45
09	07 08 37	16 59 37	18 ♐ 36 38	24 ♐ 57 40	23 13	09 49	11 17	25 30	08 33	06 00	29 50	28 46
10	07 12 34	17 56 48	01 ♑ 16 02	07 ♑ 31 48	25 18	11 02	11 54	25 38	08 30	06 03	29 50	28 48
11	07 16 31	18 54 00	13 ♑ 45 05	19 ♑ 55 55	27 20	12 15	12 32	25 46	08 27	06 07	29 50	28 49
12	07 20 27	19 51 11	26 ♑ 04 23	02 ♒ 10 35	29 ♋ 22	13 28	13 10	25 54	08 24	06 11	29 D 50	28 51
13	07 24 24	20 48 23	08 ♒ 14 39	14 ♒ 16 42	01 ♌ 21	14 40	13 47	26 03	08 22	06 14	29 50	28 52
14	07 28 20	21 45 36	20 ♒ 16 55	26 ♒ 15 32	03 19	15 53	14 25	26 11	08 19	06 18	29 50	28 54
15	07 32 17	22 42 49	02 ♓ 12 48	08 ♓ 09 03	05 15	17 06	15 03	26 20	08 16	06 21	29 50	28 56
16	07 36 13	23 40 02	14 ♓ 04 38	19 ♓ 59 59	07 09	18 19	15 41	26 28	08 14	06 25	29 50	28 57
17	07 40 10	24 37 16	25 ♓ 55 32	01 ♈ 51 50	09 01	19 32	16 18	26 37	08 12	06 29	29 50	28 59
18	07 44 06	25 34 30	07 ♈ 49 24	13 ♈ 48 50	10 52	20 45	16 56	26 46	08 09	06 32	29 50	29 01
19	07 48 03	26 31 45	19 ♈ 50 45	25 ♈ 55 47	12 41	21 58	17 34	26 55	08 07	06 36	29 50	29 02
20	07 52 00	27 29 01	02 ♉ 04 35	08 ♉ 17 47	14 28	23 10	18 11	27 04	08 05	06 40	29 51	29 04
21	07 55 56	28 26 18	14 ♉ 36 02	20 ♉ 59 55	16 13	24 23	18 49	27 13	08 03	06 43	29 51	29 06
22	07 59 53	29 ♋ 23 36	27 ♉ 29 57	04 ♊ 06 35	17 56	25 36	19 27	27 23	08 01	06 47	29 51	29 08
23	08 03 49	00 ♌ 20 54	10 ♊ 50 11	17 ♊ 40 56	19 38	26 49	20 05	27 32	07 59	06 51	29 52	29 09
24	08 07 46	01 18 13	24 ♊ 38 54	01 ♋ 43 56	21 17	28 01	20 42	27 42	07 57	06 54	29 52	29 11
25	08 11 42	02 15 34	08 ♋ 55 41	16 ♋ 13 36	22 55	29 ♌ 14	21 20	27 51	07 56	06 58	29 53	29 13
26	08 15 39	03 12 55	23 ♋ 36 55	01 ♌ 04 39	24 31	00 ♍ 27	21 58	28 01	07 54	07 02	29 53	29 15
27	08 19 35	04 10 16	08 ♌ 35 43	16 ♌ 08 49	26 06	01 39	22 36	28 11	07 52	07 05	29 53	29 17
28	08 23 32	05 07 39	23 ♌ 42 42	01 ♍ 16 02	27 38	02 52	23 14	28 21	07 51	07 09	29 54	29 18
29	08 27 29	06 05 02	08 ♍ 47 34	16 ♍ 16 11	29 ♌ 09	04 05	23 51	28 31	07 50	07 13	29 55	29 20
30	08 31 25	07 02 25	23 ♍ 40 54	01 ♎ 00 55	00 ♍ 38	05 17	24 29	28 41	07 48	07 16	29 55	29 22
31	08 35 22	07 ♌ 59 49	08 ♎ 15 40	15 ♎ 24 45	02 ♍ 05	06 ♍ 30	25 ♌ 07	28 ♍ 51	07 ♐ 47	07 ♌ 20	29 ♎ 56	29 ♌ 24

DECLINATIONS

Date	Moon True ☊ ° '	Moon Mean ☊ ° '	Moon ☽ Latitude ° '	Sun ☉	Moon ☽	Mercury ☿	Venus ♀	Mars ♂	Jupiter ♃	Saturn ♄	Uranus ♅	Neptune ♆	Pluto ♇
01	18 ♏ 22	17 ♏ 06	05 S 01	23 N 07	07 N 03	24 N 18	21 N 36	19 N 54	03 N 21	20 S 01	19 N 28	09 S 46	22 N 17
02	18 R 17	17 03	04 37	23 02	02 N 08	24 19	21 21	19 44	03 18	20 01	19 27	09 46	22 16
03	18 15	17 00	03 56	22 58	02 S 51	24 17	21 05	19 35	03 15	20 00	19 26	09 46	22 16
04	18 14	16 56	03 01	22 53	07 35	24 13	20 49	19 25	03 12	20 00	19 25	09 46	22 15
05	18 D 15	16 53	01 57	22 47	11 50	24 05	20 32	19 15	03 08	20 00	19 24	09 46	22 14
06	18 15	16 50	00 S 47	22 42	15 23	23 55	20 15	19 04	03 05	19 59	19 23	09 46	22 14
07	18 R 16	16 47	00 N 24	22 35	18 04	23 43	19 57	18 54	03 02	19 59	19 23	09 46	22 13
08	18 14	16 44	01 32	22 29	19 45	23 28	19 38	18 44	02 59	19 59	19 22	09 46	22 12
09	18 10	16 41	02 35	22 22	20 23	23 10	19 19	18 33	02 56	19 58	19 21	09 46	22 12
10	18 04	16 37	03 29	22 14	19 58	22 50	19 00	18 23	02 52	19 58	19 20	09 46	22 11
11	17 56	16 34	04 11	22 07	18 34	22 29	18 40	18 12	02 49	19 58	19 19	09 46	22 10
12	17 45	16 31	04 42	21 58	16 19	22 05	18 19	18 01	02 45	19 58	19 18	09 46	22 10
13	17 34	16 28	04 59	21 50	13 24	21 39	17 58	17 50	02 42	19 57	19 17	09 46	22 09
14	17 23	16 25	05 02	21 41	09 56	21 12	17 37	17 39	02 38	19 57	19 16	09 46	22 08
15	17 12	16 21	04 53	21 32	06 08	20 43	17 15	17 28	02 35	19 57	19 15	09 46	22 08
16	17 04	16 18	04 30	21 22	02 S 06	20 13	16 52	17 16	02 32	19 57	19 15	09 46	22 07
17	16 58	16 15	03 56	21 12	02 N 00	19 41	16 29	17 05	02 27	19 57	19 14	09 46	22 06
18	16 54	16 12	03 12	21 02	06 02	19 08	16 06	16 53	02 24	19 56	19 13	09 46	22 05
19	16 53	16 09	02 18	20 51	09 54	18 34	15 42	16 42	02 20	19 56	19 12	09 47	22 05
20	16 D 52	16 06	01 18	20 40	13 25	18 00	15 18	16 30	02 16	19 56	19 11	09 47	22 04
21	16 53	16 02	00 N 12	20 29	16 25	17 24	14 53	16 18	02 12	19 56	19 10	09 47	22 03
22	16 R 52	15 59	00 S 56	20 17	18 42	16 48	14 28	16 06	02 09	19 56	19 09	09 47	22 03
23	16 50	15 56	02 03	20 05	20 02	16 11	14 03	15 54	02 05	19 56	19 08	09 47	22 02
24	16 46	15 53	03 06	19 52	20 14	15 34	13 37	15 41	02 01	19 56	19 07	09 47	22 01
25	16 39	15 50	03 59	19 40	19 10	14 56	13 11	15 29	01 57	19 56	19 06	09 48	22 00
26	16 30	15 47	04 38	19 26	16 49	14 18	12 45	15 17	01 53	19 56	19 05	09 48	22 00
27	16 20	15 43	04 58	19 13	13 19	13 39	12 18	15 04	01 49	19 56	19 04	09 48	21 59
28	16 09	15 40	04 58	18 59	08 55	13 00	11 51	14 52	01 45	19 56	19 03	09 48	21 58
29	15 59	15 37	04 38	18 45	03 N 59	12 22	11 24	14 39	01 41	19 56	19 02	09 49	21 58
30	15 52	15 34	03 58	18 31	01 S 08	11 43	10 56	14 26	01 37	19 56	19 01	09 49	21 57
31	15 ♏ 47	15 ♏ 31	03 S 04	18 N 16	06 S 06	11 N 04	10 N 29	14 N 13	01 N 32	19 S 56	19 N 01	09 S 49	21 N 56

ZODIAC SIGN ENTRIES

Date	h	m	Planets
01	10	42	♀ ♌
01	13	23	☽ ♍
03	15	16	☽ ♎
05	19	10	☽ ♏
08	01	20	☽ ♐
10	09	35	☽ ♑
12	19	41	☿ ♌
12	19	43	☽ ♒
15	07	32	☽ ♓
17	20	14	☽ ♈
20	07	58	☽ ♉
22	16	34	☽ ♊
23	03	15	☉ ♌
24	21	05	☽ ♋
26	03	10	♀ ♍
26	22	16	☽ ♌
28	21	59	☽ ♍
30	01	44	☿ ♍
30	22	20	☽ ♎

LATITUDES

Date	Mercury ☿ ° '	Venus ♀ ° '	Mars ♂ ° '	Jupiter ♃ ° '	Saturn ♄ ° '	Uranus ♅ ° '	Neptune ♆ ° '	Pluto ♇ ° '
01	00 N 59	01 N 29	01 N 14	01 N 16	01 N 48	00 N 35	01 N 46	11 N 02
04	00 21	01 32	01 13	01 15	01 48	00 35	01 46	11 02
07	01 38	01 34	01 13	01 15	01 47	00 35	01 46	11 01
10	01 47	01 35	01 12	01 14	01 47	00 35	01 46	11 01
13	01 50	01 36	01 12	01 14	01 46	00 35	01 46	11 00
16	01 47	01 36	01 11	01 13	01 46	00 35	01 45	11 00
19	01 38	01 36	01 10	01 13	01 45	00 35	01 45	11 00
22	01 25	01 35	01 10	01 12	01 44	00 35	01 45	10 59
25	01 07	01 33	01 09	01 12	01 44	00 35	01 45	10 59
28	00 46	01 30	01 09	01 11	01 43	00 35	01 45	10 59
31	00 N 21	01 N 27	01 N 08	01 N 11	01 N 42	00 N 35	01 N 45	10 N 59

DATA

Julian Date	2436021
Delta T	+32 seconds
Ayanamsa	23° 16' 02"
Synetic vernal point	05° ♓ 50' 58"
True obliquity of ecliptic	23° 26' 35"

MOON'S PHASES, APSIDES AND POSITIONS ☽

Date	h	m	Phase	Longitude ° '	Eclipse Indicator
04	12	09	☽	12 ♎ 14	
11	22	50	○	19 ♑ 20	
20	02	17	☾	27 ♈ 06	
27	04	28	●	03 ♌ 52	

Day	h	m		
16	02	51	Apogee	
28	09	37	Perigee	
02	22	14	0S	
09	14	11	Max dec	20° S 23'
17	00	20	0N	
24	04	02	Max dec	20° N 18'
30	06	38	0S	

LONGITUDES

Date	Chiron ⚷ ° '	Ceres ⚳ ° '	Pallas ⚴ ° '	Juno ⚵ ° '	Vesta ⚶ ° '	Black Moon Lilith ⚸ ° '
01	16 ♒ 37	16 ♊ 37	29 ♈ 11	07 ♉ 36	13 ♈ 46	03 ♓ 59
11	16 ♒ 12	20 ♊ 41	01 ♉ 53	12 ♉ 29	16 ♈ 25	05 ♓ 07
21	15 ♒ 43	24 ♊ 41	04 ♉ 21	17 ♉ 17	18 ♈ 36	06 ♓ 14
31	15 ♒ 12	28 ♊ 38	06 ♉ 32	21 ♉ 58	20 ♈ 15	07 ♓ 21

ASPECTARIAN

h m	Aspects
01 Monday	
02 47	☉ ⊼ ♄
03 38	☽ ∠ ☉
04 20	☽ ⊻ ♃
06 26	☿ ⊻ ♅
07 50	♀ □ ♆
11 03	☽ ☌ ♇
13 09	☽ ⚹ ♆
13 38	☽ ⊻ ♀
15 40	☿ ⊻ ♂
22 31	☽ ⊻ ♅
02 Tuesday	
00 12	☽ ⊻ ♂
00 23	☽ ⊥ ♀
01 16	☽ ⚹ ☿
04 05	☽ □ ♄
05 58	☽ ⚹ ☉
06 21	☽ ∥ ♃
08 27	☽ ⊥ ♅
10 33	☽ ⊥ ♂
13 52	☽ ∠ ♆
16 39	☽ ∠ ♀
19 44	☿ ⊼ ♄
23 29	☽ ∠ ♅
03 Wednesday	
00 43	☽ Q ☿
02 14	☽ ∠ ♂
03 21	☽ Q ☉
04 53	☽ ⊥ ♆
06 20	☽ ☌ ♃
09 59	☽ Q ♄
12 56	☽ ⊻ ♇
13 57	☽ # ♃
15 00	☽ ⊻ ♆
20 13	☽ ⚹ ♀
23 08	☽ ⊥ ♇
04 Thursday	
00 55	☽ ⚹ ♅
04 48	☽ ⚹ ♂
05 37	☉ ☌ ☿
06 17	☽ ⚹ ♄
10 38	♂ ∥ ♅
12 09	☽ □ ☉
12 50	☽ □ ☿
14 35	☽ ∠ ♇
15 36	☿ Q ♃
18 47	☽ Q ♀
21 39	☽ Q ♅
23 53	☽ ∥ ♆
05 Friday	
00 13	☿ ∠ ♇
02 31	☽ Q ♂
06 14	☉ Q ♃
08 11	☽ ∠ ♄
10 18	☽ ⊻ ♃
11 25	♂ △ ♄
12 24	☿ ± ♄
16 49	☽ ⚹ ♇
18 52	☽ ☌ ♆
20 59	☽ ⊥ ♃
06 Saturday	
00 00	☽ ⊥ ♄
00 34	☉ ∠ ♇
05 00	♀ ☌ ♅
05 28	☽ □ ♅
05 31	☽ □ ♀
10 39	☽ ⊻ ♄
11 53	☽ □ ♂
13 11	☽ ∠ ♃
14 13	☽ Q ♇
21 04	☽ △ ☉
07 Sunday	
01 30	☉ ± ♄
03 56	☽ △ ☿
09 04	♀ # ♄
16 38	☽ ⚹ ♃
21 10	☽ # ♂
22 09	♂ ∠ ♃
23 00	☽ □ ♇
08 Monday	
01 01	☽ ⊻ ♆
02 34	☽ bQ ☉
04 53	☽ # ♅
10 07	☽ # ♀
12 03	♀ △ ♄
12 07	☽ ⊥ ♆
12 19	☽ △ ♅
12 47	☽ bQ ☿
15 00	☽ Q ♃
17 01	☽ ∥ ♄
17 14	☽ ☌ ♄
17 48	☽ △ ♀
20 35	☽ ± ☉
21 33	☽ △ ♂
09 Tuesday	
04 53	☽ ∠ ♆
06 45	☿ ⊥ ♇
08 43	☽ ⊼ ☉
08 52	☽ ± ☿
10 00	♀ ∥ ♅
15 41	☿ bQ ♄
16 31	☽ bQ ♅
22 24	☽ ⊼ ☿
10 Wednesday	
00 58	☽ bQ ♀
01 09	☽ □ ♃
03 13	♀ ∠ ♃
03 16	☽ bQ ♂
07 16	☽ △ ♇
09 15	☽ ⚹ ♆
09 41	☽ ± ♅
11 50	☽ ∥ ♄
16 14	☿ ⚹ ♃
19 58	☽ ± ♀
21 13	☽ ⊼ ♅
21 21	☽ ± ♂
23 30	☉ ∥ ♇
11 Thursday	
00 55	☽ # ♅
01 48	☽ ⊻ ♄
08 17	☽ Q ♆
08 46	☽ ⊼ ♀
09 31	☽ ⊼ ♂
10 26	☽ # ♀
12 08	☽ bQ ♇
13 21	☽ ⊥ ♄
17 02	☽ # ♂
17 49	♆ St D
22 50	☽ ☍ ☉
23 55	♀ ☌ ♂
12 Friday	
05 40	☽ ± ♇
05 48	☿ ⊻ ♇
06 48	☽ ∠ ♄
07 07	☿ ∥ ♇
11 40	☽ △ ♃
17 27	☽ ⊼ ♇
17 36	☿ □ ♆
19 22	☽ □ ♆
19 43	☽ ☍ ☿
21 05	☉ ∥ ☿
13 Saturday	
08 00	☽ ☍ ♅
12 14	☽ ⚹ ♄
17 38	☽ bQ ♃
23 38	☽ ☍ ♂
14 Sunday	
02 14	☽ ☍ ♀
06 48	♀ ∥ ♂
11 48	☽ ± ♃
12 04	☽ Q ♄
13 09	☽ ∥ ♆
15 13	☽ ⊼ ☉
15 Monday	
00 00	☽ ⊼ ♃
04 20	☽ ± ☉
04 50	♃ Q ♄
05 22	☽ ☍ ♇
07 11	☽ △ ♆
17 34	☉ ⊥ ♇
19 18	☽ ⊼ ☿
20 25	☽ ⊼ ♅
16 Tuesday	
00 05	☽ bQ ☉
00 12	☽ □ ♄
01 30	☉ bQ ♄
02 21	♀ Q ♆
02 23	☿ ☌ ♅
08 37	☽ ± ♅
09 32	☽ # ♃
09 46	☽ ± ☿
13 32	☽ bQ ♆
15 25	☽ ⊼ ♂
21 35	☽ ⊼ ♀
17 Wednesday	
00 15	☿ # ♄
01 31	☿ △ ♄
02 57	☽ bQ ♅
04 15	☽ ± ♂
07 26	☽ bQ ☿
07 46	☽ ± ♆
09 08	☽ △ ☉
11 07	☽ ± ♀
13 25	☽ ☍ ♃
14 41	☽ ∥ ♃
18 12	☽ ⊼ ♇
19 54	☽ ⊼ ♆
23 29	☽ bQ ♂
18 Thursday	
06 20	☽ ± ♇
07 21	☽ bQ ♀
08 41	☿ ∥ ♅
09 24	☽ △ ♅
12 29	♀ ⊥ ♃
12 40	☽ △ ♄
19 12	☽ △ ☿
19 Friday	
00 26	☽ bQ ♇
00 57	☿ ∠ ♃
07 13	☽ △ ♂
11 13	☽ # ♆
16 39	☽ △ ♀
18 27	☽ bQ ♄
22 46	♂ Q ♆
23 39	☉ ⚹ ♃
20 Saturday	
02 07	☽ ⊼ ♃
02 17	☽ □ ☉
06 08	☽ △ ♇
07 40	☽ ☍ ♆
12 01	☽ ± ♄
13 57	☽ ± ♃
20 54	☽ □ ♅
23 33	☽ ⊼ ♄
21 Sunday	
00 44	☽ ∥ ♀
07 26	☽ bQ ♃
11 02	☽ ∥ ♂
15 31	☽ □ ☿
15 45	☽ Q ☉
19 31	☽ ∥ ☿
20 21	☽ □ ♂
22 Monday	
05 07	☉ ⊻ ♇
06 59	☽ Q ♅
08 09	☽ □ ♀
10 52	☿ Q ♆
11 47	☽ △ ♃
14 59	☽ □ ♇
15 44	☽ ⚹ ☉
16 18	☽ ⊼ ♆
18 18	☽ ∥ ♅
23 42	☉ □ ♆
23 Tuesday	
03 09	☽ ± ♆
04 52	☽ ⚹ ♅
05 29	☽ Q ☿
06 52	☽ Q ♂
06 57	☽ ☍ ♄
09 04	☽ # ♄
13 04	☽ ∥ ☉
19 05	☽ bQ ♆
19 40	☽ Q ♀
20 32	☽ ∠ ☉
22 22	☿ ☌ ♂
23 07	☽ Q ♇
24 Wednesday	
04 32	♀ ⊻ ♃
04 38	☿ ∥ ♂
04 55	☽ ⚹ ♂
05 28	☽ ⚹ ☿
05 33	☉ # ♄
07 17	☽ ∠ ♅
13 12	☽ ⊥ ☉
17 15	☽ □ ♃
18 17	☽ ⚹ ♀
18 34	☿ ⊥ ♃
19 44	☽ ⚹ ♇
20 52	☽ △ ♆
22 36	☽ # ♄
22 39	☽ ⊥ ♅
25 Thursday	
00 05	☽ ⊻ ☉
02 37	☽ ∥ ☉
07 30	☽ ∠ ♂
08 44	☽ ⊻ ♅
10 07	☽ ∠ ☿
10 20	☽ ⊼ ♄
11 38	♀ ☌ ♇
12 55	☽ ∥ ♅
20 13	☽ ± ♄
20 44	☽ ∠ ♇
21 32	☽ ∠ ♀
23 01	☽ ⊥ ♂
23 31	☽ Q ♃
26 Friday	
00 47	♀ ⚹ ♆
02 45	☽ ⊥ ☿
09 13	☽ ⊻ ♂
10 51	☽ bQ ♄
11 24	☽ ⊥ ♇
13 28	☽ ⊥ ♀
13 38	☽ ⊻ ☿
14 35	♂ ⊥ ♃
19 10	☽ ⚹ ♃
21 05	☽ ⊻ ♇
22 06	☽ □ ♆
23 57	☽ ⊻ ♀
27 Saturday	
00 09	☽ ∥ ♂
04 28	☽ ☌ ☉
09 32	☽ ∥ ☿
09 35	☽ ☌ ♅
10 51	☽ △ ♄
18 35	☽ ∥ ♀
19 22	☽ ∠ ♃
28 Sunday	
02 47	☽ Q ♆
04 22	☉ ∥ ♅
07 28	☽ # ♆
09 49	☽ ⊥ ♃
11 12	☽ ☌ ♂
18 56	☽ ☌ ☿
19 26	☽ ⊻ ♃
20 54	☽ ☌ ♇
21 50	☽ ⚹ ♆
29 Monday	
00 34	☿ ⊻ ♃
03 49	☽ ☌ ♀
07 23	☽ ⊻ ☉
09 28	☽ ⊻ ♅
10 28	☽ □ ♄
15 06	☿ ☌ ♇
17 38	☽ ⊥ ☉
19 07	☽ ⊥ ♅
21 49	☽ ∠ ♆
22 56	☽ ∥ ♃
30 Tuesday	
00 20	☿ ⚹ ♆
09 09	☽ ∠ ☉
09 42	☽ ∠ ♅
12 23	☽ ⊥ ♆
13 22	☽ ⊻ ♂
14 11	☽ # ♃
15 28	☽ Q ♄
18 16	☉ ☌ ♅
20 16	☽ ☌ ♃
21 19	☽ ⊻ ♇
22 12	☽ ⊻ ♆
23 38	☽ ⊥ ♂
31 Wednesday	
00 38	☽ ⊻ ☿
06 50	☉ △ ♄
07 14	☽ ⊥ ♇
08 48	☽ ⊻ ♀
10 27	☽ ⚹ ♅
11 13	☽ ⚹ ♄
11 32	☽ ⚹ ☉
11 40	☽ ⊥ ☿
15 15	☽ ∠ ♂
19 44	☽ ⊥ ♀
22 19	☽ ∠ ♇

All ephemeris data is given at 12.00 UT and the Moon's longitude is additionally given for 24.00 UT

AUGUST 1957

Date	Sidereal time h m s	Sun ☉ ° ' "	Moon ☽ ° ' "	Moon ☽ 24.00 ° ' "	Mercury ☿ ° '	Venus ♀ ° '	Mars ♂ ° '	Jupiter ♃ ° '	Saturn ♄ ° '	Uranus ♅ ° '	Neptune ♆ ° '	Pluto ♇ ° '
01	08 39 18	08 ♌ 57 14	22 ♎ 27 56	29 ♎ 25 12	03 ♍ 30	07 ♍ 42	25 ♌ 45	29 ♍ 01	07 ♐ 46	07 ♌ 24	29 ♎ 56	29 ♌ 26
02	08 43 15	09 54 40	06 ♏ 16 38	13 ♏ 02 27	04 53	08 55	26 23	29 12	07 R 45	07 28	29 57	29 28
03	08 47 11	10 52 06	19 ♏ 42 55	26 ♏ 18 24	06 15	10 07	27 01	29 22	07 44	07 31	29 58	29 30
04	08 51 08	11 49 32	02 ♐ 49 16	09 ♐ 15 55	07 34	11 20	27 39	29 32	07 44	07 35	29 59	29 31
05	08 55 04	12 47 00	15 ♐ 38 44	21 ♐ 58 05	08 51	12 32	28 17	29 43	07 43	07 39	29 ♎ 59	29 33
06	08 59 01	13 44 28	28 ♐ 14 20	04 ♑ 27 46	10 07	13 44	28 54	29 ♍ 54	07 42	07 42	00 ♏ 00	29 35
07	09 02 58	14 41 57	10 ♑ 38 41	16 ♑ 47 20	11 20	14 57	29 ♌ 32	00 ♎ 04	07 42	07 46	00 01	29 37
08	09 06 54	15 39 27	22 ♑ 53 55	28 ♑ 58 39	12 31	16 09	00 ♍ 10	00 15	07 42	07 50	00 02	29 39
09	09 10 51	16 36 58	05 ♒ 01 39	11 ♒ 03 07	13 40	17 21	00 48	00 26	07 41	07 53	00 03	29 41
10	09 14 47	17 34 30	17 ♒ 03 10	23 ♒ 01 56	14 46	18 33	01 26	00 37	07 41	07 57	00 04	29 43
11	09 18 44	18 32 03	28 ♒ 59 35	04 ♓ 56 18	15 50	19 46	02 04	00 48	07 41	08 01	00 05	29 45
12	09 22 40	19 29 37	10 ♓ 52 15	16 ♓ 47 41	16 51	20 58	02 42	00 59	07 D 41	08 04	00 06	29 47
13	09 26 37	20 27 12	22 ♓ 42 52	28 ♓ 38 05	17 50	22 10	03 20	01 10	07 41	08 08	00 07	29 49
14	09 30 33	21 24 49	04 ♈ 33 43	10 ♈ 30 10	18 46	23 22	03 58	01 21	07 41	08 12	00 08	29 51
15	09 34 30	22 22 27	16 ♈ 27 51	22 ♈ 27 17	19 39	24 34	04 36	01 33	07 42	08 15	00 09	29 53
16	09 38 27	23 20 07	28 ♈ 29 00	04 ♉ 33 34	20 29	25 46	05 14	01 44	07 42	08 19	00 10	29 55
17	09 42 23	24 17 48	10 ♉ 41 36	16 ♉ 53 41	21 15	26 58	05 52	01 56	07 42	08 22	00 11	29 57
18	09 46 20	25 15 31	23 ♉ 10 29	29 ♉ 32 35	21 58	28 10	06 30	02 07	07 43	08 26	00 12	29 ♌ 59
19	09 50 16	26 13 16	06 ♊ 00 35	12 ♊ 35 01	22 38	29 ♍ 22	07 09	02 19	07 44	08 30	00 13	00 ♍ 01
20	09 54 13	27 11 02	19 ♊ 16 21	26 ♊ 04 55	23 13	00 ♎ 34	07 47	02 30	07 44	08 33	00 14	00 03
21	09 58 09	28 08 50	03 ♋ 00 56	10 ♋ 04 26	23 45	01 46	08 25	02 42	07 45	08 37	00 16	00 05
22	10 02 06	29 ♌ 06 40	17 ♋ 15 15	24 ♋ 33 02	24 12	02 57	09 03	02 54	07 46	08 40	00 17	00 07
23	10 06 02	00 ♍ 04 31	01 ♌ 57 07	09 ♌ 26 41	24 35	04 09	09 41	03 05	07 47	08 44	00 18	00 09
24	10 09 59	01 02 24	17 ♌ 00 39	24 ♌ 37 48	24 53	05 21	10 19	03 17	07 48	08 47	00 20	00 11
25	10 13 56	02 00 18	02 ♍ 16 43	09 ♍ 56 00	25 06	06 33	10 57	03 29	07 50	08 51	00 21	00 13
26	10 17 52	02 58 14	17 ♍ 34 11	25 ♍ 09 55	25 13	07 44	11 36	03 41	07 51	08 54	00 22	00 14
27	10 21 49	03 56 11	02 ♎ 41 58	10 ♎ 09 15	25 R 15	08 56	12 14	03 53	07 53	08 58	00 24	00 16
28	10 25 45	04 54 10	17 ♎ 30 57	24 ♎ 46 25	25 11	10 07	12 52	04 05	07 54	09 01	00 25	00 18
29	10 29 42	05 52 10	01 ♏ 55 16	08 ♏ 57 17	25 01	11 19	13 30	04 17	07 56	09 05	00 27	00 20
30	10 33 38	06 50 12	15 ♏ 52 26	22 ♏ 40 54	24 45	12 31	14 09	04 29	07 57	09 08	00 28	00 22
31	10 37 35	07 ♍ 48 15	29 ♏ 22 55	05 ♐ 58 52	24 ♍ 23	13 ♎ 42	14 ♍ 47	04 ♎ 41	07 ♐ 59	09 ♌ 11	00 ♏ 30	00 ♍ 24

Date	Moon True ☊ ° '	Moon Mean ☊ ° '	Moon ☽ Latitude ° '
01	15 ♏ 45	15 ♏ 27	02 S 00
02	15 D 44	15 24	00 S 50
03	15 R 44	15 21	00 N 21
04	15 44	15 18	01 29
05	15 41	15 15	02 31
06	15 36	15 12	03 24
07	15 29	15 08	04 07
08	15 18	15 05	04 37
09	15 05	15 02	04 55
10	14 52	14 59	04 59
11	14 38	14 56	04 50
12	14 25	14 53	04 29
13	14 15	14 49	03 56
14	14 07	14 46	03 12
15	14 02	14 43	02 20
16	13 59	14 40	01 21
17	13 58	14 37	00 N 17
18	13 D 58	14 33	00 S 49
19	13 R 58	14 30	01 54
20	13 57	14 27	02 55
21	13 53	14 24	03 49
22	13 46	14 21	04 31
23	13 38	14 18	04 56
24	13 28	14 14	05 01
25	13 18	14 11	04 46
26	13 08	14 08	04 10
27	13 01	14 05	03 16
28	12 56	14 02	02 11
29	12 54	13 58	00 S 58
30	12 D 54	13 55	00 N 16
31	12 ♏ 54	13 ♏ 52	01 N 27

DECLINATIONS

Date	Sun ☉ ° '	Moon ☽ ° '	Mercury ☿ ° '	Venus ♀ ° '	Mars ♂ ° '	Jupiter ♃ ° '	Saturn ♄ ° '	Uranus ♅ ° '	Neptune ♆ ° '	Pluto ♇ ° '
01	18 N 01	10 S 36	10 N 25	10 00	14 N 00	01 N 28	19 S 56	19 N 00	09 S 49	21 N 55
02	17 46	14 24	09 46	09 32	13 47	01 24	19 56	18 59	09 50	21 55
03	17 31	17 20	09 08	09 03	13 34	01 20	19 56	18 58	09 50	21 54
04	17 15	19 16	08 30	08 35	13 21	01 15	19 56	18 57	09 50	21 53
05	16 59	20 10	07 52	08 05	13 07	01 11	19 56	18 56	09 51	21 52
06	16 42	20 02	07 15	07 36	12 54	01 07	19 56	18 55	09 51	21 52
07	16 26	18 55	06 38	07 07	12 40	01 02	19 56	18 54	09 51	21 51
08	16 09	16 56	06 01	06 37	12 27	00 58	19 56	18 53	09 52	21 50
09	15 52	14 14	05 25	06 07	12 13	00 53	19 57	18 52	09 52	21 50
10	15 34	10 58	04 50	05 37	11 59	00 49	19 57	18 51	09 52	21 49
11	15 16	07 17	04 16	05 07	11 46	00 45	19 57	18 50	09 53	21 48
12	14 59	03 S 20	03 42	04 37	11 32	00 40	19 57	18 49	09 53	21 48
13	14 40	00 N 43	03 09	04 06	11 18	00 35	19 57	18 48	09 53	21 47
14	14 22	04 45	02 37	03 36	11 04	00 31	19 58	18 47	09 54	21 46
15	14 03	08 38	02 07	03 05	10 50	00 26	19 58	18 47	09 54	21 45
16	13 45	12 12	01 37	02 35	10 35	00 22	19 58	18 46	09 55	21 45
17	13 26	15 19	01 09	02 04	10 21	00 17	19 59	18 45	09 55	21 44
18	13 06	17 47	00 43	01 33	10 07	00 12	19 59	18 44	09 56	21 43
19	12 47	19 26	00 N 17	01 02	09 52	00 08	19 59	18 43	09 56	21 43
20	12 27	20 06	00 S 06	00 N 31	09 38	00 N 03	20 00	18 42	09 57	21 42
21	12 07	19 36	00 28	00 00	09 23	00 S 02	20 00	18 41	09 57	21 41
22	11 47	17 51	00 48	00 S 31	09 09	00 06	20 01	18 40	09 58	21 41
23	11 27	14 55	01 05	01 02	08 54	00 11	20 01	18 39	09 58	21 40
24	11 06	10 57	01 21	01 33	08 39	00 16	20 01	18 38	09 59	21 39
25	10 46	06 13	01 33	02 04	08 25	00 21	20 01	18 37	09 59	21 39
26	10 25	01 N 05	01 44	02 35	08 10	00 26	20 02	18 37	10 00	21 38
27	10 04	04 S 05	01 51	03 06	07 55	00 30	20 02	18 36	10 00	21 38
28	09 43	08 53	01 55	03 37	07 40	00 35	20 03	18 35	10 01	21 37
29	09 22	13 03	01 57	04 08	07 25	00 40	20 03	18 34	10 01	21 36
30	09 00	16 20	01 54	04 38	07 10	00 45	20 04	18 33	10 02	21 36
31	08 N 39	18 S 37	01 S 48	05 S 09	06 N 55	00 S 50	20 S 04	18 N 32	10 S 02	21 N 35

ZODIAC SIGN ENTRIES

Date	h m	Planets
02	01 01	☽ ♏
04	06 47	☽ ♐
06	08 25	♆ ♏
06	15 23	☽ ♑
07	02 11	♃ ♎
08	05 27	♂ ♍
09	02 01	☽ ♒
11	14 02	☽ ♓
14	02 46	☽ ♈
16	15 00	☽ ♉
19	00 51	☽ ♊
19	04 23	♇ ♍
20	00 44	♀ ♎
21	06 48	☽ ♋
23	08 51	☽ ♌
23	10 08	☉ ♍
25	08 26	☽ ♍
27	07 41	☽ ♎
29	08 45	☽ ♏
31	13 07	☽ ♐

LATITUDES

Date	Mercury ☿ ° '	Venus ♀ ° '	Mars ♂ ° '	Jupiter ♃ ° '	Saturn ♄ ° '	Uranus ♅ ° '	Neptune ♆ ° '	Pluto ♇ ° '
01	00 N 12	01 N 26	01 N 08	01 N 11	01 N 42	00 N 35	01 N 44	10 N 59
04	00 S 15	01 22	01 07	01 10	01 42	00 35	01 44	10 59
07	00 45	01 17	01 06	01 10	01 41	00 35	01 44	10 59
10	01 16	01 11	01 06	01 09	01 40	00 35	01 44	10 59
13	01 48	01 05	01 05	01 09	01 40	00 35	01 44	10 59
16	02 20	00 58	01 04	01 09	01 39	00 36	01 44	10 59
19	02 52	00 51	01 04	01 09	01 38	00 36	01 44	10 59
22	03 22	00 43	01 03	01 08	01 38	00 36	01 43	10 59
25	03 49	00 35	01 02	01 08	01 37	00 36	01 43	10 59
28	04 11	00 26	01 01	01 08	01 36	00 36	01 43	10 59
31	04 S 24	00 N 17	01 N 00	01 N 07	01 N 36	00 N 36	01 N 43	11 N 00

DATA

Julian Date	2436052
Delta T	+32 seconds
Ayanamsa	23° 16' 06"
Synetic vernal point	05° ♓ 50' 54"
True obliquity of ecliptic	23° 26' 35"

MOON'S PHASES, APSIDES AND POSITIONS ☽

Date	h m	Phase	Longitude ° '	Eclipse Indicator
02	18 55	☽	10 ♏ 11	
10	13 08	○	17 ♒ 37	
18	16 16	☾	25 ♉ 26	
25	11 33	●	01 ♍ 59	

Day	h m		
12	13 43	Apogee	
25	18 00	Perigee	
05	20 39	Max dec	20° S 14'
13	07 46	ON	
20	13 57	Max dec	20° N 06'
26	16 58	OS	

LONGITUDES

Date	Chiron ⚷ ° '	Ceres ⚳ ° '	Pallas ⚴ ° '	Juno ⚵ ° '	Vesta ⚶ ° '	Black Moon Lilith ⚸ ° '
01	15 ♒ 08	29 ♊ 01	06 ♉ 44	22 ♉ 25	20 ♈ 23	07 ♓ 28
11	14 ♒ 36	02 ♋ 52	08 ♉ 30	26 ♉ 55	21 ♈ 21	08 ♓ 35
21	14 ♒ 05	06 ♋ 37	09 ♉ 51	01 ♊ 13	21 ♈ 36	09 ♓ 42
31	13 ♒ 35	10 ♋ 14	10 ♉ 38	05 ♊ 14	21 ♈ 06	10 ♓ 49

ASPECTARIAN

h m	Aspects
01 Thursday	
04 29	☽ ∠ ☿
05 35	♀ ⊻ ♅
06 44	☽ Q ♅
07 37	☽ ∥ ♆
08 59	☽ # ♀
09 13	☽ Q ☉
11 07	☽ # ☿
12 27	☽ ∠ ♀
12 31	☽ ∠ ♄
13 17	♀ □ ♄
17 55	☽ ⚹ ♂
21 19	♀ # ♆
23 27	☽ ⊻ ♃
02 Friday	
00 03	☽ ⚹ ♇
00 55	☽ ☌ ♆
04 04	☽ ⊥ ♄
08 00	☽ # ♂
09 17	☽ ⚹ ☿
10 04	☽ ⊥ ♃
10 04	☿ # ♆
14 06	☽ □ ♅
14 36	☽ ⊻ ♄
15 54	☽ Q ♂
17 07	☽ ⚹ ♀
18 55	☽ □ ☉
21 12	☽ Q ♇
03 Saturday	
02 15	☽ ∠ ♃
09 02	☽ Q ☿
13 36	☽ # ☉
16 48	☽ Q ♀
23 45	☿ ∥ ♀
04 Sunday	
01 58	☽ □ ♂
05 51	☽ ⚹ ♃
05 54	☽ □ ♇
06 44	☽ ⊻ ♆
06 59	☽ # ♅
09 18	♃ ⊻ ♇
12 19	☿ ⊻ ♅
14 56	☿ □ ♄
17 52	☽ ⊥ ♆
20 54	☽ △ ♅
21 07	☽ ☌ ♄
21 49	☽ □ ☿
05 Monday	
02 18	☽ ∥ ♄
04 30	☽ Q ♃
05 31	☽ □ ♀
06 10	☽ △ ☉
10 45	☽ ∠ ♆
06 Tuesday	
01 21	☽ bQ ♅
11 18	♀ ⊥ ♅
12 12	☉ ⊻ ♀
12 14	♄ △ ♅
13 03	☽ bQ ☉
13 21	☽ △ ♂
14 36	☽ △ ♇
15 09	☽ ∥ ♄
15 14	☽ □ ♃
15 24	☽ ⚹ ♆
18 42	☽ ± ♅
07 Wednesday	
03 40	♃ ⊻ ♆
06 16	☽ ⊻ ♄
06 23	☽ ⊼ ♅
07 54	☽ ± ☉
12 13	☽ # ♅
13 27	♀ ∠ ♆
13 29	☽ △ ☿
14 41	☽ Q ♆
15 12	♂ ☌ ♇
17 57	☽ ⊥ ♄
19 47	☽ bQ ♇
20 01	☽ bQ ♂
20 35	☽ ⊼ ☉
21 18	☽ △ ♀
23 32	☉ ∠ ♃
08 Thursday	
06 29	♂ ⚹ ♆
11 36	☽ ∠ ♄
13 29	☽ ± ♇
14 39	☽ ± ♂
16 17	♂ ⊻ ♃
20 37	☽ # ☉
22 04	☽ bQ ☿
09 Friday	
01 22	☽ ⊼ ♇
02 06	☽ □ ♆
02 45	☽ △ ♃
03 09	☽ ⊼ ♂
06 06	☽ bQ ♀
17 14	☿ ⊥ ♅
17 17	☽ ⚹ ♄
17 43	☽ ☍ ♅
17 46	☽ ± ☿
10 Saturday	
02 00	☽ ± ♀
04 20	☽ # ♂
06 58	☽ ⊼ ☿
09 05	☽ bQ ♃
13 08	☽ ☍ ☉
15 21	☽ ⊼ ♀
17 16	☽ Q ♄
18 40	☿ ∠ ♆
19 22	☽ ∥ ♆
11 Sunday	
00 22	☉ Q ♆
03 25	☽ ± ♃
13 32	☽ ☍ ♇
14 11	☽ △ ♆
15 42	☽ ⊼ ♃
18 33	☽ ☍ ♂
23 57	♄ St D
12 Monday	
03 16	☽ # ♀
05 33	☽ □ ♄
06 18	☽ ⊼ ♅
09 31	☽ # ☿
18 31	☽ ± ♅
20 34	☽ bQ ♆
13 Tuesday	
01 13	☽ ☍ ☿
04 08	☽ # ♃
07 01	☽ ⊼ ☉
10 46	☽ ☍ ♀
11 16	☽ ∥ ♃
12 51	☽ bQ ♅
14 50	☽ ± ♆
20 15	☽ ± ☉
14 Wednesday	
00 44	☽ ∥ ☿
02 26	☽ ⊼ ♇
03 01	☽ ⊼ ♆
05 25	☽ ☍ ♃
05 51	☽ ∥ ♀
08 22	♀ ∠ ♅
10 44	☽ ⊼ ♂
14 36	☽ ± ♇
16 04	☽ bQ ☉
18 19	☽ △ ♄
19 23	☽ △ ♅
23 32	☽ ± ♂
15 Thursday	
03 26	♀ ⊥ ♆
08 48	☽ bQ ♇
18 39	☽ bQ ♂
18 52	☽ ⊼ ☿
20 19	☽ # ♆
16 Friday	
00 29	☽ bQ ♄
00 52	☽ △ ☉
01 33	☽ ∥ ♂
06 00	☽ ⊼ ♀
07 44	☽ ± ☿
10 37	♀ Q ♄
14 50	☽ △ ♇
15 20	☽ ☍ ♆
18 22	☽ ± ♄
18 32	☽ ⊼ ♃
19 13	☽ ± ♀
22 21	☽ ∥ ☉
17 Saturday	
02 04	☽ △ ♂
02 45	☽ bQ ☿
06 10	☽ ⊼ ♄
06 31	☽ ± ♃
07 27	☽ □ ♅
14 44	☽ bQ ♀
18 Sunday	
00 15	☽ bQ ♃
09 35	☽ △ ☿
16 16	☽ □ ☉
18 12	☽ Q ♅
22 24	☽ △ ♀
23 55	☽ ∥ ♅
19 Monday	
00 51	☽ □ ♇
01 15	☽ ⊼ ♆
05 03	☽ △ ♃
06 00	♂ # ♆
12 23	☽ ± ♆
14 11	☽ □ ♂
14 45	☉ ⊥ ♃
15 09	☽ ☍ ♄
16 35	☽ ⚹ ♅
23 57	☿ ∥ ♃
20 Tuesday	
01 18	♀ ⊻ ♇
03 34	☽ # ♄
04 07	☽ Q ☉
04 47	☽ bQ ♆
05 27	♀ ⊻ ♆
09 17	☿ # ♃
09 48	☽ Q ♇
10 35	♂ □ ♄
19 17	☽ □ ☿
19 36	☽ ∠ ♅
23 15	☿ # ♀
23 58	☽ # ♄
21 Wednesday	
00 01	☽ Q ♂
02 58	☽ ⚹ ☉
04 33	☿ ∠ ♅
06 56	☽ ⚹ ♇
07 15	☽ △ ♆
09 38	☽ □ ♀
10 59	♀ # ♃
11 18	☽ ⊥ ♅
11 27	☽ □ ♃
13 38	♀ ∥ ♃
20 05	☽ ⊼ ♄
20 19	♂ ⊻ ♅
21 34	☽ ⊻ ♅
21 38	☽ ⚹ ♂
22 Thursday	
02 43	☽ ∥ ♅
03 19	☽ Q ☿
06 12	☽ ± ♄
06 24	☽ ∠ ☉
08 26	☽ ∠ ♇
10 27	♀ ☌ ♃
16 55	☿ ⊥ ♆
18 05	☽ Q ♃
18 39	☽ Q ♀
21 06	☽ bQ ♄
22 20	☽ ⊥ ☉
23 18	☽ ⊥ ♇
23 41	☽ ∠ ♂
23 45	☽ ⚹ ☿
23 Friday	
08 46	☽ ⊻ ☉
09 04	☽ ⊻ ♇
09 20	☽ □ ♆
13 44	☉ ☌ ♇
13 51	☽ ⚹ ♃
14 55	☽ ⊥ ♂
15 51	☽ ⚹ ♀
17 31	☿ ∥ ♀
17 50	☉ ⚹ ♆
21 22	☽ △ ♄
22 54	☽ ☌ ♅
24 Saturday	
00 29	☽ ∠ ☿
00 56	☽ ⊻ ♂
11 04	☽ ∥ ☉
14 02	☽ ∠ ♃
14 05	☽ Q ♆
15 00	☽ ⊥ ☿
17 12	☽ # ♆
17 43	☽ ∠ ♀
25 Sunday	
00 35	☽ ⊻ ☿
00 38	☽ ∥ ♂
04 23	☽ ⊥ ♃
05 04	♀ ⊥ ♇
08 45	☽ ☌ ♇
08 58	☽ ⚹ ♆
09 03	☽ ⊥ ♀
11 33	☽ ☌ ☉
13 55	☽ ⊻ ♃
19 15	☽ ⊻ ♀
20 43	☽ □ ♄
22 20	☽ ⊻ ♅
26 Monday	
02 12	☽ ☌ ♂
05 42	☽ # ♀
07 47	☽ ⊥ ♅
08 32	☽ ∠ ♆
09 05	☽ # ☿
14 20	♀ ⚹ ♄
14 58	☽ # ♃
19 03	☽ ∥ ♃
22 02	☽ ∠ ♅
22 46	☽ ⊥ ♆
27 Tuesday	
00 08	☽ ☌ ☿
01 07	☽ Q ♄
01 19	☽ ∥ ☿
06 51	☽ ∥ ♀
08 03	☿ St R
08 07	☽ ⊻ ♇
08 19	☽ ⊻ ♆
10 18	☉ ⊻ ♃
12 38	♀ ⚹ ♅
13 55	☽ ☌ ♃
14 07	☽ ⊻ ☉
16 22	☉ # ♆
17 45	☽ ⊥ ♇
20 20	☽ ⚹ ♄
22 07	☽ ⚹ ♅
22 54	☽ ☌ ♀
28 Wednesday	
00 28	☽ ⊥ ☉
04 04	☽ ⊻ ♂
05 59	☽ # ♂
08 22	☽ ∠ ♇
14 20	☽ ⊥ ♂
16 06	☽ # ☉
16 12	☽ ∠ ☉
17 48	☽ Q ♅
18 04	☽ ∥ ♆
20 54	☽ ∠ ♄
29 Thursday	
00 34	☽ ⊻ ☿
05 59	☽ ∠ ♂
09 19	☽ ⚹ ♇
09 30	☽ ☌ ♆
10 30	☽ ⊥ ☿
12 01	☽ ⊥ ♄
16 04	☽ ⊻ ♃
19 13	☽ ⚹ ☉
22 16	☽ ⊻ ♄
30 Friday	
00 16	☽ □ ♅
01 36	☽ ∠ ☿
02 30	☽ ⊥ ♃
05 35	☽ ⊻ ♀
05 53	☽ Q ♇
08 50	☽ ⚹ ♂
17 04	☽ ⊥ ♀
17 36	☽ Q ☉
18 26	☽ ∠ ♃
31 Saturday	
03 18	☽ ⚹ ☿
06 06	☿ ⊥ ♆
07 05	☽ Q ♂
10 39	☽ ∠ ♀
10 59	☽ # ♅
13 51	☽ □ ♇
14 01	☽ ⊻ ♆
16 44	☉ □ ♄
21 35	☿ ∠ ♅
21 47	☽ ⚹ ♃
23 19	☉ ⊥ ♀

All ephemeris data is given at 12.00 UT and the Moon's longitude is additionally given for 24.00 UT

SEPTEMBER 1957

Date	Sidereal time h m s	Sun ☉ ° ' "	Moon ☽ ° ' "	Moon ☽ 24.00 ° ' "	Mercury ☿ ° '	Venus ♀ ° '	Mars ♂ ° '	Jupiter ♃ ° '	Saturn ♄ ° '	Uranus ♅ ° '	Neptune ♆ ° '	Pluto ♇ ° '
01	10 41 31	08 ♍ 46 19	12 ♐ 29 09	18 ♐ 54 16	23 ♍ 55	14 ♎ 53	15 ♍ 25	04 ♎ 54	08 ♐ 01	09 ♌ 15	00 ♏ 31	00 ♍ 26
02	10 45 28	09 44 24	25 ♐ 14 42	01 ♑ 30 56	23 R 21	16 05	16 04	05 06	08 03	09 18	00 33	00 28
03	10 49 25	10 42 31	07 ♑ 43 29	13 ♑ 52 48	22 41	17 16	16 42	05 18	08 05	09 21	00 34	00 30
04	10 53 21	11 40 40	19 ♑ 59 20	26 ♑ 03 30	21 55	18 27	17 20	05 30	08 08	09 25	00 36	00 32
05	10 57 18	12 38 50	02 ♒ 05 39	08 ♒ 06 08	21 05	19 38	17 59	05 43	08 10	09 28	00 37	00 34
06	11 01 14	13 37 01	14 ♒ 05 14	20 ♒ 03 15	20 11	20 50	18 37	05 55	08 12	09 31	00 39	00 36
07	11 05 11	14 35 14	26 ♒ 00 23	01 ♓ 56 54	19 13	22 01	19 16	06 08	08 15	09 34	00 41	00 38
08	11 09 07	15 33 29	07 ♓ 52 57	13 ♓ 48 46	18 14	23 12	19 54	06 20	08 17	09 37	00 42	00 40
09	11 13 04	16 31 46	19 ♓ 44 31	25 ♓ 40 25	17 14	24 23	20 32	06 33	08 20	09 41	00 44	00 42
10	11 17 00	17 30 04	01 ♈ 36 40	07 ♈ 33 29	16 14	25 34	21 11	06 45	08 23	09 44	00 46	00 44
11	11 20 57	18 28 24	13 ♈ 31 09	19 ♈ 29 56	15 16	26 44	21 49	06 58	08 26	09 47	00 48	00 46
12	11 24 54	19 26 46	25 ♈ 30 10	01 ♉ 32 11	14 22	27 55	22 28	07 10	08 28	09 50	00 49	00 48
13	11 28 50	20 25 10	07 ♉ 36 23	13 ♉ 43 13	13 32	29 ♎ 06	23 07	07 23	08 31	09 53	00 51	00 49
14	11 32 47	21 23 36	19 ♉ 53 58	26 ♉ 06 37	12 48	00 ♏ 17	23 45	07 36	08 35	09 56	00 53	00 51
15	11 36 43	22 22 05	02 ♊ 24 10	08 ♊ 46 18	12 11	01 27	24 24	07 48	08 38	09 59	00 55	00 53
16	11 40 40	23 20 35	15 ♊ 13 33	21 ♊ 46 21	11 42	02 38	25 02	08 01	08 41	10 02	00 57	00 55
17	11 44 36	24 19 08	28 ♊ 25 11	05 ♋ 10 23	11 22	03 48	25 41	08 14	08 44	10 05	00 59	00 57
18	11 48 33	25 17 43	12 ♋ 02 14	19 ♋ 00 53	11 11	04 59	26 20	08 27	08 48	10 08	01 00	00 59
19	11 52 29	26 16 21	26 ♋ 06 19	03 ♌ 18 22	11 D 10	06 09	26 58	08 40	08 51	10 10	01 02	01 01
20	11 56 26	27 15 00	10 ♌ 36 39	18 ♌ 00 35	11 18	07 20	27 37	08 52	08 55	10 13	01 04	01 02
21	12 00 23	28 13 41	25 ♌ 29 22	02 ♍ 02 01	11 36	08 30	28 16	09 05	08 59	10 16	01 06	01 04
22	12 04 19	29 ♍ 12 25	10 ♍ 37 25	18 ♍ 14 17	12 03	09 40	28 55	09 18	09 02	10 19	01 08	01 06
23	12 08 16	00 ♎ 11 11	25 ♍ 51 17	03 ♎ 27 05	12 40	10 50	29 ♍ 33	09 31	09 06	10 22	01 10	01 08
24	12 12 12	01 09 58	11 ♎ 00 24	18 ♎ 30 05	13 25	12 00	00 ♎ 12	09 44	09 10	10 24	01 12	01 10
25	12 16 09	02 08 48	25 ♎ 55 04	03 ♏ 14 32	14 19	13 10	00 51	09 57	09 14	10 27	01 14	01 11
26	12 20 05	03 07 40	10 ♏ 27 49	17 ♏ 34 29	15 20	14 20	01 30	10 10	09 18	10 29	01 16	01 13
27	12 24 02	04 06 33	24 ♏ 34 17	01 ♐ 27 08	16 28	15 30	02 09	10 23	09 22	10 32	01 18	01 15
28	12 27 58	05 05 28	08 ♐ 13 07	14 ♐ 52 28	17 42	16 40	02 48	10 36	09 27	10 34	01 20	01 17
29	12 31 55	06 04 25	21 ♐ 25 31	27 ♐ 52 39	19 02	17 50	03 26	10 49	09 31	10 37	01 22	01 18
30	12 35 52	07 ♎ 03 24	04 ♑ 14 22	10 ♑ 31 10	20 ♍ 27	18 ♏ 59	04 ♎ 05	11 ♎ 02	09 ♐ 35	10 ♌ 39	01 ♏ 24	01 ♍ 20

Date	Moon True ☊ ° '	Moon Mean ☊ ° '	Moon ☽ Latitude ° '	DECLINATIONS Sun ☉ ° '	Moon ☽ ° '	Mercury ☿ ° '	Venus ♀ ° '	Mars ♂ ° '	Jupiter ♃ ° '	Saturn ♄ ° '	Uranus ♅ ° '	Neptune ♆ ° '	Pluto ♇ ° '
01	12 ♏ 54	13 ♏ 49	02 N 31	08 N 17	19 S 48	01 S 39	05 S 40	06 N 40	00 S 55	20 S 05	18 N 31	10 S 03	21 N 34
02	12 R 53	13 46	03 26	07 55	19 56	01 26	06 10	06 25	01 00	20 05	18 30	10 03	21 34
03	12 50	13 43	04 09	07 33	19 04	01 09	06 41	06 10	01 05	20 06	18 29	10 04	21 33
04	12 44	13 39	04 41	07 11	17 19	00 48	07 11	05 54	01 10	20 06	18 29	10 05	21 33
05	12 36	13 36	04 59	06 49	14 50	00 S 24	07 41	05 39	01 15	20 07	18 28	10 05	21 32
06	12 26	13 33	05 04	06 27	11 45	00 N 04	08 11	05 24	01 20	20 08	18 27	10 06	21 31
07	12 15	13 30	04 56	06 04	08 13	00 34	08 41	05 09	01 24	20 08	18 26	10 06	21 31
08	12 04	13 27	04 34	05 42	04 22	01 07	09 10	04 53	01 29	20 09	18 25	10 07	21 30
09	11 54	13 24	04 01	05 19	00 S 22	01 42	09 40	04 38	01 34	20 09	18 24	10 08	21 30
10	11 46	13 20	03 17	04 56	03 N 40	02 18	10 09	04 22	01 39	20 10	18 24	10 08	21 29
11	11 40	13 17	02 25	04 34	07 34	02 55	10 38	04 07	01 45	20 10	18 23	10 09	21 29
12	11 36	13 14	01 25	04 11	11 03	03 32	11 07	03 51	01 50	20 11	18 22	10 10	21 28
13	11 34	13 11	00 N 21	03 48	14 23	04 08	11 36	03 36	01 55	20 12	18 21	10 10	21 28
14	11 D 35	13 08	00 S 45	03 25	17 00	04 42	12 04	03 20	02 00	20 13	18 21	10 11	21 27
15	11 36	13 04	01 50	03 02	18 51	05 14	12 32	03 05	02 05	20 13	18 20	10 12	21 26
16	11 37	13 01	02 52	02 39	19 47	05 44	13 00	02 49	02 10	20 14	18 19	10 12	21 26
17	11 R 37	12 58	03 46	02 15	19 40	06 09	13 28	02 33	02 15	20 15	18 18	10 13	21 25
18	11 36	12 55	04 29	01 52	18 26	06 32	13 55	02 18	02 20	20 16	18 17	10 14	21 25
19	11 33	12 52	04 58	01 29	16 03	06 50	14 22	02 02	02 25	20 16	18 17	10 14	21 25
20	11 28	12 49	05 09	01 06	12 37	07 03	14 49	01 46	02 30	20 17	18 16	10 15	21 24
21	11 22	12 45	05 00	00 42	08 19	07 13	15 15	01 31	02 35	20 18	18 15	10 16	21 24
22	11 15	12 42	04 30	00 N 19	03 N 25	07 17	15 41	01 15	02 40	20 19	18 15	10 16	21 23
23	11 09	12 39	03 40	00 S 04	01 S 43	07 17	16 07	00 59	02 45	20 19	18 14	10 17	21 23
24	11 05	12 36	02 36	00 28	06 45	07 12	16 32	00 43	02 50	20 20	18 13	10 18	21 22
25	11 02	12 33	01 21	00 51	11 17	07 03	16 57	00 28	02 55	20 21	18 12	10 19	21 22
26	11 01	12 30	00 S 03	01 15	15 01	06 50	17 21	00 N 12	03 01	20 22	18 12	10 19	21 21
27	11 D 02	12 26	01 N 13	01 38	17 44	06 33	17 46	00 S 04	03 06	20 23	18 11	10 20	21 21
28	11 03	12 23	02 23	02 01	19 20	06 12	18 10	00 20	03 11	20 23	18 10	10 21	21 21
29	11 05	12 20	03 23	02 25	19 48	05 47	18 33	00 35	03 16	20 24	18 10	10 21	21 20
30	11 ♏ 05	12 ♏ 17	04 N 10	02 S 48	19 S 12	05 N 20	18 S 56	00 S 51	03 S 21	20 S 25	18 N 09	10 S 22	21 N 20

ZODIAC SIGN ENTRIES

Date	h	m	Planets
02	21	05	☽ ♑
05	07	50	☽ ♒
07	20	04	☽ ♓
10	08	45	☽ ♈
12	20	57	☽ ♉
14	06	20	♀ ♏
15	07	26	☽ ♊
17	14	50	☽ ♋
19	18	31	☽ ♌
21	19	11	☽ ♍
23	07	26	☉ ♎
23	18	33	☽ ♎
24	04	31	♂ ♎
25	18	40	☽ ♏
27	21	27	☽ ♐
30	03	59	☽ ♑

LATITUDES

Date	Mercury ☿ ° '	Venus ♀ ° '	Mars ♂ ° '	Jupiter ♃ ° '	Saturn ♄ ° '	Uranus ♅ ° '	Neptune ♆ ° '	Pluto ♇ ° '
01	04 S 26	00 N 13	01 N 00	01 N 07	01 N 35	00 N 36	01 N 43	11 N 00
04	04 22	00 N 03	00 59	01 07	01 35	00 36	01 43	11 00
07	04 01	00 S 07	00 58	01 07	01 34	00 36	01 43	11 01
10	03 24	00 17	00 57	01 07	01 33	00 36	01 43	11 01
13	02 32	00 28	00 56	01 07	01 33	00 36	01 42	11 02
16	01 34	00 39	00 55	01 07	01 32	00 36	01 42	11 02
19	00 S 36	00 50	00 54	01 06	01 32	00 36	01 42	11 03
22	00 N 16	01 02	00 53	01 06	01 31	00 36	01 42	11 03
25	00 57	01 13	00 52	01 06	01 30	00 36	01 42	11 04
28	01 27	01 24	00 51	01 06	01 30	00 37	01 42	11 05
31	01 N 46	01 S 35	00 N 50	01 N 06	01 N 29	00 N 37	01 N 42	11 N 06

DATA

Julian Date	2436083
Delta T	+32 seconds
Ayanamsa	23° 16' 10"
Synetic vernal point	05° ♓ 50' 50"
True obliquity of ecliptic	23° 26' 35"

MOON'S PHASES, APSIDES AND POSITIONS ☽

Date	h	m	Phase	Longitude ° '	Eclipse Indicator
01	04	35	☽	08 ♐ 28	
09	04	55	○	16 ♓ 15	
17	04	02	☾	24 ♊ 00	
23	19	18	●	00 ♎ 29	
30	17	49	☽	07 ♑ 18	

Day	h	m		
08	16	38	Apogee	
23	04	27	Perigee	
02	02	56	Max dec	20° S 00'
09	14	10	ON	
16	21	37	Max dec	19° N 52'
23	03	59	OS	
29	10	13	Max dec	19° S 48'

LONGITUDES

Date	Chiron ⚷ ° '	Ceres ⚳ ° '	Pallas ⚴ ° '	Juno ⚵ ° '	Vesta ⚶ ° '	Black Moon Lilith ⚸ ° '
01	13 ♒ 32	10 ♋ 35	10 ♉ 41	05 ♊ 37	21 ♈ 01	10 ♓ 56
11	13 ♒ 07	14 ♋ 01	10 ♉ 47	09 ♊ 15	19 ♈ 41	12 ♓ 03
21	12 ♒ 45	17 ♋ 16	10 ♉ 08	12 ♊ 26	17 ♈ 43	13 ♓ 10
31	12 ♒ 29	20 ♋ 16	08 ♉ 41	15 ♊ 04	15 ♈ 17	14 ♓ 17

All ephemeris data is given at 12.00 UT and the Moon's longitude is additionally given for 24.00 UT

ASPECTARIAN

h m	Aspects
01 Sunday	
00 19	☽ Q ☿
00 58	☽ ⊥ ♆
03 43	☽ ☌ ♄
04 35	☽ □ ☉
04 42	♂ ⊥ ♅
05 58	☽ △ ♅
15 47	♂ ∠ ♆
16 56	☽ ✶ ♀
17 40	☽ ∠ ♆
17 45	☽ □ ♂
20 21	☽ Q ♃
23 24	♀ ∠ ♇
02 Monday	
00 26	☉ ⊻ ♅
08 33	☽ □ ☿
10 12	☽ ⚼ ♅
11 14	♀ ⊻ ♂
17 58	☽ Q ♀
19 43	♀ ⋕ ♂
22 01	☽ △ ♇
22 09	☽ ✶ ♆
03 Tuesday	
03 30	☽ ± ♅
07 14	☽ □ ♃
12 43	☽ ⊻ ♄
15 11	☽ ⊼ ♅
16 05	☿ ∥ ♃
18 18	☽ △ ☉
21 21	☽ ⋕ ♅
21 28	☽ Q ♆
04 Wednesday	
00 27	☽ ⊥ ♄
03 13	☽ ⚼ ♇
06 30	☽ △ ♂
08 39	☽ □ ♀
12 05	☉ ⋕ ♀
15 34	☽ △ ☿
18 13	☽ ∠ ♄
21 01	☽ ± ♇
05 Thursday	
02 23	☽ ⚼ ☉
08 57	☽ ⊼ ♇
09 04	☽ □ ♆
13 52	☽ ⚼ ♂
19 21	☽ △ ♃
19 25	☽ ⚼ ☿
21 53	☽ ± ☉
06 Friday	
00 10	☽ ✶ ♄
02 48	☽ ☍ ♅
04 39	☿ ⊻ ♀
08 53	☽ ± ♂
10 58	☽ ⊼ ☉
12 10	☽ ± ☿
21 38	☽ ⊼ ♂
23 21	☽ ⊼ ☿
23 30	☽ ∥ ♆
07 Saturday	
00 21	☽ Q ♄
01 59	☽ ⚼ ♃
02 40	♀ Q ♅
03 03	☽ △ ♀
09 19	☽ ∥ ♀
11 28	☿ ☌ ♂
20 28	☽ ± ♃
21 22	☽ ☍ ♇
21 27	☽ △ ♆
08 Sunday	
03 01	☽ ⋕ ☉
08 38	☽ ⋕ ♂
08 49	☽ ⊼ ♃
12 42	☽ ⚼ ♀
12 49	☽ □ ♄
13 43	☉ ⊥ ♅
13 59	♀ ∠ ♄
15 32	☽ ⊼ ♅
15 48	☉ ∠ ♆
23 25	☿ ⊥ ♀
09 Monday	
03 44	☽ ± ♅
03 52	☽ ⚼ ♆
04 55	☽ ☍ ☉
04 57	☽ ∥ ♃
05 05	☽ ⋕ ☿
06 12	☿ ⋕ ♃
07 19	☽ ☍ ☿
08 56	☽ ± ♀
13 43	☽ ☍ ♂
20 32	☉ ☌ ☿
22 02	☽ ⚼ ♅
22 08	☽ ± ♆
22 25	☽ ⊼ ♀
23 46	☽ ⋕ ♃
10 Tuesday	
02 24	☽ ∥ ☿
08 37	♃ ⊥ ♇
10 13	☽ ⊼ ♇
10 17	☽ ⊼ ♆
11 14	♀ ∥ ♆
16 02	☽ ∥ ♂
19 04	☽ ∥ ☉
22 21	☽ ± ♇
22 34	☽ ☍ ♃
23 17	☿ ∠ ♆
23 53	☿ ⊥ ♅
11 Wednesday	
01 42	☽ △ ♄
04 27	☽ △ ♅
15 16	☽ ⊼ ☿
16 31	☽ ⚼ ♇
22 49	☽ ⊼ ☉
12 Thursday	
02 27	☽ ± ☿
04 56	☽ ⋕ ♆
05 36	☽ ⊼ ♂
07 56	☽ ⚼ ♄
11 27	☽ ⋕ ♀
11 53	☽ ± ☉
17 20	☽ ☍ ♀
18 14	☽ ± ♂
19 11	☽ ⚼ ☿
21 06	☿ ∥ ♂
22 33	☽ △ ♇
22 37	☽ ☍ ♆
13 Friday	
01 55	☽ ± ♄
03 51	☉ ∥ ☿
05 09	☿ ∠ ♀
07 19	☽ ⚼ ☉
11 33	☽ ⊼ ♃
12 24	♀ ⊥ ♂
13 03	☽ ⚼ ♂
13 49	☽ ⊼ ♄
16 30	☽ □ ♅
22 58	☽ △ ☿
23 33	☽ ± ♃
14 Saturday	
15 10	☽ △ ☉
17 20	☽ ⚼ ♃
19 52	☽ △ ♂
15 Sunday	
00 05	♀ ✶ ♇
00 40	♀ ☌ ♆
02 32	☉ ∥ ♂
03 33	☽ Q ♅
04 08	☽ ∥ ♅
09 07	☽ □ ♇
09 10	☽ ⊼ ♆
10 01	☽ ⊼ ♀
20 32	☽ ± ♆
22 22	☽ △ ♃
22 30	☽ ± ♀
23 47	☽ ☍ ♄
16 Monday	
02 19	☽ ✶ ♅
05 41	☽ □ ☿
08 20	♂ ⊥ ♆
11 41	♂ ∠ ♅
13 07	♀ ⊥ ♄
13 20	☽ ⚼ ♆
16 52	☽ ⚼ ♀
18 48	☽ Q ♇
17 Tuesday	
04 02	☽ □ ☉
05 58	☽ ∠ ♅
06 50	☽ □ ♂
12 32	☉ ⋕ ♃
13 40	☽ Q ☿
16 32	☽ ✶ ♇
16 35	☽ △ ♆
22 06	☽ ⊥ ♅
22 30	☽ △ ♀
18 Wednesday	
04 41	☉ ⊥ ♆
05 39	☽ □ ♃
06 20	☽ ⊼ ♄
07 39	☉ ∠ ♅
08 40	☽ ⊻ ♅
09 35	♂ ⋕ ♃
10 32	☽ ✶ ☿
13 48	☽ ∥ ♅
14 21	☽ Q ☉
16 09	☽ Q ♂
16 47	☽ ± ♄
18 49	☽ ∠ ♇
19 Thursday	
03 37	☿ St D
07 14	♂ Q ♄
08 12	☽ ⚼ ♄
10 09	☽ ⊥ ♇
12 06	☽ ∠ ☿
12 18	☽ ✶ ☉
12 57	☽ Q ♃
13 31	☽ ✶ ♂
20 13	☽ ⊻ ♇
20 16	☽ □ ♆
23 10	☽ ⋕ ♀
20 Friday	
03 10	☽ ⊥ ☿
03 15	☉ Q ♄
06 09	☽ □ ♀
09 07	☽ ✶ ♃
09 13	☽ △ ♄
11 22	☽ ☌ ♅
13 08	☽ ⊻ ☿
14 52	☽ ∠ ☉
15 25	☽ ∠ ♂
18 41	♃ ✶ ♄
21 Saturday	
01 41	☽ ⋕ ♆
01 44	☽ Q ♆
06 25	☽ ⊥ ☉
06 36	☽ ⊥ ♂
09 44	☽ ∠ ♃
13 45	☽ Q ♀
14 29	☉ ☌ ♂
16 37	☽ ⊻ ♂
16 40	☽ ⊻ ☉
17 31	☽ ∥ ☿
20 54	☽ ☌ ♇
20 57	☽ ✶ ♆
22 21	♀ ⊻ ♄
22 Sunday	
00 15	☽ ⊥ ♃
02 46	♀ ⊻ ♃
09 29	☽ □ ♄
09 53	☽ ⊻ ♃
10 22	☽ ✶ ♀
11 31	☽ ⊻ ♅
14 20	☽ ☌ ☿
15 29	☽ ⋕ ♃
20 42	☽ ∠ ♆
21 00	☽ ⊥ ♅
22 43	☽ ∥ ♂
23 Monday	
01 45	♀ □ ♅
03 42	☽ ∥ ☉
04 13	☽ ⋕ ☉
08 44	☽ ⋕ ♂
10 55	☽ ⊥ ♆
11 13	☽ ∠ ♅
11 58	☽ ∠ ♀
13 59	☽ Q ♄
16 54	☽ ∥ ♃
18 06	☽ ☌ ♂
19 18	☽ ☌ ☉
20 21	☽ ⊻ ♇
20 24	☽ ⊻ ♆
24 Tuesday	
03 23	☽ ⊥ ♀
05 52	☽ ⊥ ♇
09 04	☽ ✶ ♄
09 56	☽ ☌ ♃
11 02	☽ ✶ ♅
11 50	☉ ⊻ ♇
12 55	☉ ⊻ ♆
13 44	☽ ⊻ ♀
14 15	☽ ⋕ ☿
16 05	☽ ⊻ ☿
20 15	☽ ∠ ♇
21 34	☉ ⋕ ♂
25 Wednesday	
02 19	☽ ⊥ ☿
06 21	☽ Q ♅
06 33	☽ ∥ ♆
09 15	☽ ∠ ♄
12 19	♀ Q ♇
17 56	☽ ∠ ☿
20 26	☽ ⊻ ♂
20 38	☽ ✶ ♇
20 43	☽ ☌ ♆
22 56	☽ ⊻ ☉
26 Thursday	
00 03	☽ ⊥ ♄
01 13	♂ ⊻ ♇
03 07	♂ ⊻ ♆
06 49	☽ ⊥ ♂
09 36	☽ ⊥ ☉
10 03	☽ ⊻ ♄
11 29	☽ ⊻ ♃
12 03	☽ □ ♅
16 38	☽ Q ♇
19 06	☽ ☌ ♀
20 53	☽ ✶ ☿
21 45	☽ ⊥ ♃
22 39	☽ ∠ ♂
27 Friday	
01 55	☽ ∠ ☉
08 37	☿ ∠ ♆
12 31	☽ ∥ ♀
13 25	☽ ∠ ♃
13 25	☿ ⊥ ♅
17 18	☽ ⋕ ♅
19 25	☽ Q ☿
23 40	☽ □ ♇
23 46	☽ ⊻ ♆
28 Saturday	
01 53	☽ ✶ ♂
06 00	☽ ✶ ☉
09 25	♃ ✶ ♅
10 07	♀ ⊥ ♃
10 25	☽ ⊥ ♆
12 31	♀ ⋕ ♅
14 12	☽ ☌ ♄
16 14	☽ △ ♅
16 20	☽ ✶ ♃
29 Sunday	
00 28	☽ Q ♂
02 42	☽ ∠ ♆
04 45	☽ ⊻ ♀
05 20	☽ Q ☉
07 05	☽ □ ☿
14 36	☽ Q ♃
16 53	☽ ⊥ ♀
19 48	☽ ⚼ ♅
30 Monday	
06 29	☽ △ ♇
06 37	☽ ✶ ♆
11 29	☽ ∠ ♀
11 42	☽ □ ♂
12 48	☽ ± ♅
16 09	☽ ∥ ♀
16 42	♀ ∠ ♂
17 49	☽ □ ☉
18 55	☉ ⊥ ♇
22 16	☽ ⊻ ♄

OCTOBER 1957

Date	Sidereal time h m s	Sun ☉ ° ′ ″	Moon ☽ ° ′ ″	Moon ☽ 24.00 ° ′ ″	Mercury ☿ ° ′	Venus ♀ ° ′	Mars ♂ ° ′	Jupiter ♃ ° ′	Saturn ♄ ° ′	Uranus ♅ ° ′	Neptune ♆ ° ′	Pluto ♇ ° ′
01	12 39 48	08 ♎ 02 24	16 ♑ 43 34	22 ♑ 52 09	21 ♍ 56	20 ♏ 09	04 ♎ 44	11 ♎ 14	09 ♐ 40	10 ♌ 42	01 ♏ 26	01 ♍ 22
02	12 43 45	09 01 27	28 ♑ 57 26	04 ♒ 59 58	23 28	21 18	05 23	11 27	09 44	10 44	01 29	01 23
03	12 47 41	10 00 31	11 ♒ 00 14	16 ♒ 58 43	25 04	22 28	06 02	11 40	09 49	10 46	01 31	01 25
04	12 51 38	10 59 36	22 ♒ 55 53	28 ♒ 52 09	26 42	23 37	06 41	11 53	09 54	10 49	01 33	01 26
05	12 55 34	11 58 44	04 ♓ 47 53	10 ♓ 43 27	28 ♍ 22	24 46	07 20	12 06	09 58	10 51	01 35	01 28
06	12 59 31	12 57 53	16 ♓ 39 10	22 ♓ 35 20	00 ♎ 04	25 55	08 00	12 19	10 03	10 53	01 37	01 30
07	13 03 27	13 57 05	28 ♓ 32 11	04 ♈ 30 00	01 47	27 04	08 39	12 32	10 08	10 55	01 39	01 31
08	13 07 24	14 56 18	10 ♈ 28 58	16 ♈ 29 19	03 30	28 13	09 18	12 45	10 13	10 57	01 41	01 33
09	13 11 21	15 55 33	22 ♈ 31 16	28 ♈ 35 00	05 15	29 ♏ 22	09 57	12 58	10 18	10 59	01 44	01 34
10	13 15 17	16 54 51	04 ♉ 40 44	10 ♉ 48 41	07 00	00 ♐ 31	10 36	13 11	10 23	11 01	01 46	01 36
11	13 19 14	17 54 10	16 ♉ 59 04	23 ♉ 12 09	08 45	01 39	11 15	13 24	10 28	11 03	01 48	01 37
12	13 23 10	18 53 32	29 ♉ 28 11	05 ♊ 47 26	10 30	02 48	11 55	13 37	10 33	11 05	01 50	01 39
13	13 27 07	19 52 56	12 ♊ 10 10	18 ♊ 36 43	12 15	03 56	12 34	13 50	10 39	11 07	01 52	01 40
14	13 31 03	20 52 22	25 ♊ 07 21	01 ♋ 42 22	14 00	05 04	13 13	14 03	10 44	11 09	01 54	01 42
15	13 35 00	21 51 51	08 ♋ 22 01	15 ♋ 06 31	15 44	06 12	13 53	14 16	10 49	11 11	01 57	01 43
16	13 38 56	22 51 22	21 ♋ 56 05	28 ♋ 50 47	17 29	07 20	14 32	14 29	10 55	11 12	01 59	01 44
17	13 42 53	23 50 55	05 ♌ 50 41	12 ♌ 55 40	19 12	08 28	15 11	14 42	11 00	11 14	02 01	01 46
18	13 46 50	24 50 31	20 ♌ 05 32	27 ♌ 19 58	20 55	09 36	15 51	14 55	11 06	11 16	02 03	01 47
19	13 50 46	25 50 09	04 ♍ 38 29	12 ♍ 00 27	22 38	10 44	16 30	15 08	11 12	11 17	02 06	01 48
20	13 54 43	26 49 49	19 ♍ 25 09	26 ♍ 51 43	24 20	11 51	17 10	15 21	11 17	11 19	02 08	01 50
21	13 58 39	27 49 31	04 ♎ 19 12	11 ♎ 46 34	26 02	12 59	17 49	15 34	11 23	11 20	02 10	01 51
22	14 02 36	28 49 15	19 ♎ 12 49	26 ♎ 36 56	27 43	14 06	18 29	15 46	11 29	11 22	02 12	01 52
23	14 06 32	29 ♎ 49 02	03 ♏ 57 56	11 ♏ 14 59	29 ♎ 23	15 13	19 08	15 59	11 35	11 23	02 15	01 53
24	14 10 29	00 ♏ 48 50	18 ♏ 27 18	25 ♏ 34 16	01 ♏ 03	16 20	19 48	16 12	11 41	11 24	02 17	01 55
25	14 14 25	01 48 41	02 ♐ 35 26	09 ♐ 30 27	02 42	17 26	20 27	16 25	11 46	11 26	02 19	01 56
26	14 18 22	02 48 33	16 ♐ 19 10	23 ♐ 01 33	04 21	18 33	21 07	16 38	11 52	11 27	02 21	01 57
27	14 22 19	03 48 27	29 ♐ 37 42	06 ♑ 07 50	05 59	19 39	21 47	16 50	11 59	11 28	02 24	01 58
28	14 26 15	04 48 22	12 ♑ 32 17	18 ♑ 51 25	07 37	20 46	22 26	17 03	12 05	11 29	02 26	01 59
29	14 30 12	05 48 20	25 ♑ 05 43	01 ♒ 15 40	09 14	21 52	23 06	17 16	12 11	11 30	02 28	02 00
30	14 34 08	06 48 18	07 ♒ 21 51	13 ♒ 24 47	10 51	22 58	23 46	17 28	12 17	11 31	02 30	02 01
31	14 38 05	07 ♏ 48 19	19 ♒ 25 05	25 ♒ 23 19	12 ♏ 27	24 ♐ 03	24 ♎ 26	17 ♎ 41	12 ♐ 23	11 ♌ 32	02 ♏ 32	02 ♍ 02

Date	Moon True ☊ ° ′	Moon Mean ☊ ° ′	Moon ☽ Latitude ° ′
01	11 ♏ 05	12 ♏ 14	04 N 45
02	11 R 03	12 10	05 06
03	11 00	12 07	05 12
04	10 56	12 04	05 06
05	10 51	12 01	04 46
06	10 45	11 58	04 14
07	10 41	11 55	03 31
08	10 37	11 51	02 38
09	10 35	11 48	01 38
10	10 33	11 45	00 N 32
11	10 D 33	11 42	00 S 35
12	10 34	11 39	01 42
13	10 36	11 36	02 46
14	10 37	11 32	03 42
15	10 38	11 29	04 28
16	10 39	11 26	05 00
17	10 R 39	11 23	05 15
18	10 37	11 20	05 12
19	10 36	11 16	04 49
20	10 34	11 13	04 07
21	10 32	11 10	03 07
22	10 31	11 07	01 55
23	10 31	11 04	00 S 36
24	10 D 31	11 01	00 N 44
25	10 31	10 57	01 59
26	10 32	10 54	03 06
27	10 33	10 51	04 00
28	10 33	10 48	04 40
29	10 34	10 45	05 06
30	10 R 34	10 42	05 17
31	10 ♏ 34	10 ♏ 38	05 N 14

DECLINATIONS

Date	Sun ☉ ° ′	Moon ☽ ° ′	Mercury ☿ ° ′	Venus ♀ ° ′	Mars ♂ ° ′	Jupiter ♃ ° ′	Saturn ♄ ° ′	Uranus ♅ ° ′	Neptune ♆ ° ′	Pluto ♇ ° ′
01	03 S 11	17 S 41	04 N 49	19 S 19	01 S 07	03 S 26	20 S 26	18 N 09	10 S 23	21 N 19
02	03 35	15 23	04 16	19 41	01 23	03 31	20 27	18 08	10 24	21 19
03	03 58	12 27	03 41	20 03	01 39	03 36	20 28	18 07	10 24	21 19
04	04 21	09 03	03 04	20 24	01 54	03 41	20 28	18 07	10 25	21 18
05	04 44	05 19	02 25	20 44	02 10	03 46	20 29	18 06	10 26	21 18
06	05 07	01 S 22	01 45	21 05	02 26	03 51	20 30	18 06	10 27	21 18
07	05 30	02 N 38	01 03	21 24	02 42	03 56	20 31	18 05	10 27	21 17
08	05 53	06 34	00 N 21	21 43	02 57	04 01	20 32	18 05	10 28	21 17
09	06 16	10 17	00 S 23	22 02	03 13	04 06	20 33	18 04	10 29	21 17
10	06 39	13 35	01 07	22 20	03 29	04 11	20 34	18 03	10 30	21 17
11	07 01	16 21	01 51	22 38	03 45	04 16	20 35	18 03	10 30	21 16
12	07 24	18 22	02 35	22 55	04 00	04 22	20 36	18 03	10 31	21 16
13	07 47	19 31	03 20	23 11	04 16	04 27	20 36	18 02	10 32	21 16
14	08 09	19 39	04 05	23 27	04 32	04 32	20 37	18 02	10 33	21 16
15	08 31	18 44	04 50	23 42	04 47	04 37	20 38	18 01	10 33	21 15
16	08 53	16 44	05 34	23 57	05 03	04 41	20 39	18 01	10 34	21 15
17	09 15	13 43	06 18	24 11	05 18	04 46	20 40	18 00	10 35	21 15
18	09 37	09 51	07 02	24 25	05 34	04 51	20 41	18 00	10 36	21 15
19	09 59	05 19	07 46	24 38	05 49	04 56	20 42	17 59	10 37	21 15
20	10 21	00 N 24	08 29	24 50	06 05	05 01	20 43	17 59	10 37	21 14
21	10 42	04 S 35	09 12	25 02	06 20	05 06	20 44	17 59	10 38	21 14
22	11 03	09 18	09 54	25 13	06 36	05 11	20 45	17 58	10 39	21 14
23	11 25	13 25	10 36	25 23	06 51	05 16	20 46	17 58	10 40	21 14
24	11 46	16 37	11 17	25 33	07 07	05 21	20 47	17 58	10 40	21 14
25	12 06	18 44	11 57	25 42	07 22	05 26	20 48	17 57	10 41	21 14
26	12 27	19 39	12 37	25 50	07 37	05 31	20 48	17 57	10 42	21 14
27	12 47	19 26	13 16	25 58	07 52	05 35	20 49	17 57	10 43	21 14
28	13 08	18 12	13 55	26 05	08 07	05 40	20 50	17 56	10 43	21 13
29	13 28	16 06	14 32	26 12	08 23	05 45	20 51	17 56	10 44	21 13
30	13 47	13 19	15 09	26 18	08 38	05 50	20 52	17 56	10 45	21 13
31	14 S 07	10 S 02	15 S 45	26 S 23	08 S 53	05 S 55	20 S 53	17 N 56	10 S 46	21 N 13

ZODIAC SIGN ENTRIES

Date	h	m	Planets
02	14	04	☽ ♒
05	02	17	☽ ♓
06	11	09	☿ ♎
07	14	57	☽ ♈
10	01	16	♀ ♐
10	02	48	☽ ♉
12	13	01	☽ ♊
14	20	54	☽ ♋
17	01	59	☽ ♌
19	04	23	☽ ♍
21	05	03	☽ ♎
23	05	31	☽ ♏
23	16	24	☉ ♏
23	20	50	☿ ♏
25	07	33	☽ ♐
27	12	41	☽ ♑
29	21	32	☽ ♒

LATITUDES

Date	Mercury ☿ ° ′	Venus ♀ ° ′	Mars ♂ ° ′	Jupiter ♃ ° ′	Saturn ♄ ° ′	Uranus ♅ ° ′	Neptune ♆ ° ′	Pluto ♇ ° ′
01	01 N 46	01 S 35	00 N 50	01 N 06	01 N 29	00 N 37	01 N 42	11 N 06
04	01 54	01 46	00 49	01 06	01 29	00 37	01 42	11 06
07	01 55	01 57	00 48	01 06	01 28	00 37	01 42	11 07
10	01 49	02 07	00 47	01 06	01 28	00 37	01 42	11 08
13	01 38	02 17	00 46	01 06	01 27	00 37	01 42	11 09
16	01 24	02 27	00 44	01 06	01 27	00 37	01 42	11 10
19	01 07	02 36	00 43	01 06	01 26	00 37	01 42	11 11
22	00 49	02 44	00 42	01 06	01 26	00 37	01 42	11 12
25	00 29	02 52	00 41	01 07	01 25	00 38	01 42	11 13
28	00 N 09	02 59	00 39	01 07	01 25	00 38	01 42	11 14
31	00 S 11	03 S 05	00 N 38	01 N 07	01 N 25	00 N 38	01 N 42	11 N 15

DATA

Julian Date	2436113
Delta T	+32 seconds
Ayanamsa	23° 16' 12"
Synetic vernal point	05° ♓ 50' 47"
True obliquity of ecliptic	23° 26' 35"

MOON'S PHASES, APSIDES AND POSITIONS ☽

Date	h	m	Phase	Longitude ° ′	Eclipse Indicator
08	21	42	○	15 ♈ 20	
16	13	44	☾	22 ♋ 56	
23	04	43	●	29 ♎ 31	Total
30	10	48	☽	06 ♒ 45	

Day	h	m		
05	21	28	Apogee	
21	12	58	Perigee	
06	20	13	0N	
14	03	17	Max dec	19° N 44'
20	13	57	0S	
26	19	13	Max dec	19° S 42'

LONGITUDES

Date	Chiron ⚷ ° ′	Ceres ⚳ ° ′	Pallas ⚴ ° ′	Juno ⚵ ° ′	Vesta ⚶ ° ′	Black Moon Lilith ⚸ ° ′
01	12 ♒ 29	20 ♋ 16	08 ♉ 41	15 ♊ 04	15 ♈ 17	14 ♓ 17
11	12 ♒ 19	22 ♋ 59	06 ♉ 28	17 ♊ 01	12 ♈ 43	15 ♓ 24
21	12 ♒ 15	25 ♋ 21	03 ♉ 37	18 ♊ 10	10 ♈ 19	16 ♓ 31
31	12 ♒ 19	27 ♋ 17	00 ♉ 25	18 ♊ 25	08 ♈ 23	17 ♓ 38

All ephemeris data is given at 12.00 UT and the Moon's longitude is additionally given for 24.00 UT

ASPECTARIAN

h m Aspects

01 Tuesday
00 18 ☽ ⊼ ♅
01 12 ☽ □ ♃
05 37 ☽ Q ♆
05 58 ☽ # ♅
09 55 ☽ ⊥ ♄
11 17 ☽ ⚼ ♇
19 22 ☽ ⚹ ♀
23 36 ☽ △ ☿

02 Wednesday
03 37 ☽ ∠ ♄
04 56 ☽ ± ♇
07 12 ☉ ∥ ♃
16 50 ☽ ⊼ ♇
17 00 ☽ □ ♆
21 33 ☽ Q ♀

03 Thursday
01 30 ☽ △ ♂
05 11 ☉ # ☿
06 54 ☉ ⚹ ♄
09 36 ☽ ⚹ ♄
09 49 ☽ ⚼ ☿
09 50 ☽ △ ☉
11 32 ☽ ☍ ♅
13 22 ☽ △ ♃
14 53 ☿ # ♃
18 47 ☿ ⊥ ♆
22 46 ☿ ∠ ♅

04 Friday
02 45 ☽ ∥ ♆
06 46 ☽ ± ☿
07 24 ☉ ⚹ ♅
09 21 ☽ ⚼ ♂
09 53 ☽ Q ♄
13 32 ☽ □ ♀
17 42 ♀ ∥ ♄
18 45 ☽ ⚼ ☉
20 09 ☽ ⚼ ♃
20 51 ☽ ⊼ ☿

05 Saturday
04 36 ☽ ± ♂
05 15 ☽ ☍ ♇
05 28 ☽ △ ♆
06 07 ☿ Q ♄
14 36 ☽ ± ☉
14 42 ☽ ± ♃
15 15 ☽ ∥ ☉
16 00 ☉ ☌ ♃
16 52 ♂ ⊥ ♇
17 27 ☽ ⊼ ♂
18 23 ☿ # ♂
21 18 ☽ ∥ ♃
22 33 ☽ □ ♄

06 Sunday
00 17 ☽ ⊼ ♅
03 05 ☽ ⊼ ♃
03 52 ☽ ⊼ ☉
06 01 ☽ ∥ ♂
09 19 ☽ # ☿
11 56 ☽ ⚼ ♆
12 28 ☽ ± ♅

07 Monday
03 41 ♀ # ♇
03 54 ☽ ∥ ☿
06 10 ☽ ± ♆
06 43 ☽ ⚼ ♅
08 22 ☿ ⊻ ♇
08 44 ☽ △ ♀
10 14 ☿ ⊻ ♆
12 22 ☽ # ♂
18 01 ☽ ⊼ ♇
18 18 ☽ ⊼ ♆
19 38 ☽ ☍ ☿
20 01 ☽ # ♃

08 Tuesday
00 02 ♀ ∠ ♃
06 06 ☽ ± ♇
07 17 ☽ # ☉
09 29 ☽ ☍ ♂
11 28 ☽ △ ♄
12 57 ☽ △ ♅
16 38 ☽ ☍ ♃
18 03 ☽ ⚼ ♀
21 42 ☽ ☍ ☉

09 Wednesday
00 08 ☽ ⚼ ♇
13 25 ☽ # ♆
13 51 ☽ ± ♀
17 33 ☽ ⚼ ♄

10 Thursday
02 48 ♂ ⚹ ♄
02 57 ☽ ⊼ ♀
04 05 ☉ ∠ ♇
05 56 ☽ △ ♇
06 15 ☽ ☍ ♆
11 25 ☽ ± ♄
17 18 ☽ ⊼ ☿
20 19 ☿ ⊥ ♇
23 15 ☽ ⊼ ♄

11 Friday
00 15 ☽ ⊼ ♂
00 27 ☽ □ ♅
04 16 ♂ ⚹ ♅
04 56 ☽ ⊼ ♃
06 57 ☽ ± ☿
11 17 ♀ □ ♇
12 33 ☽ ± ♂
13 56 ☽ ⊼ ☉
15 06 ♀ ⊻ ♆
16 46 ☽ ± ♃

12 Saturday
02 30 ☽ ± ☉
03 11 ☽ ⚼ ☿
06 51 ☽ ⚼ ♂
07 21 ☽ ∥ ♅
10 21 ☽ ⚼ ♃
11 16 ☽ Q ♅
12 47 ☿ ⚹ ♄
16 09 ☽ □ ♇
16 31 ☽ ⊼ ♆
18 57 ☽ ☍ ♀
20 10 ☿ ⚹ ♅
21 07 ☽ ⚼ ☉

13 Sunday
03 54 ☽ ± ♆
09 07 ☽ ☍ ♄
10 02 ☽ ⚹ ♅
12 11 ☽ △ ☿
12 47 ☽ △ ♂
15 10 ☽ △ ♃
18 50 ☿ ☌ ♂
20 48 ☽ ⚼ ♆

14 Monday
01 59 ☽ Q ♇
03 32 ☽ △ ☉
11 57 ♂ ∥ ♃
12 51 ☿ ☌ ♃
13 53 ☽ ∠ ♅

15 Tuesday
00 00 ☽ ⚹ ♇
00 24 ☽ △ ♆
04 07 ☿ ∥ ♃
06 16 ☽ ⊥ ♅
07 46 ☽ ⊼ ♀
10 03 ☿ ∥ ♂
16 25 ☽ ⊼ ♄
17 02 ☽ ⊻ ♅
19 29 ☽ ± ♀
22 05 ☽ ∥ ♅
22 19 ☽ □ ♂
22 41 ☽ □ ♃

16 Wednesday
01 40 ☿ ∠ ♇
02 52 ☽ ∠ ♇
03 02 ☽ □ ☿
03 08 ☽ ± ♄
09 30 ♂ ☌ ♃
12 46 ☽ ⚼ ♀
13 44 ☽ □ ☉
18 38 ☽ ⊥ ♇
18 58 ☽ ⚼ ♄
20 45 ☉ Q ♅

17 Thursday
02 03 ♀ ⊥ ♆
05 00 ☽ ⊻ ♇
05 26 ☽ □ ♆
06 33 ☽ Q ♃
07 14 ☽ Q ♂
14 38 ☽ Q ☿
16 51 ☽ △ ♀
20 49 ☽ △ ♄
21 10 ☽ ☌ ♅
22 57 ☽ Q ☉

18 Friday
03 13 ☽ ⚹ ♃
04 34 ☽ ⚹ ♂
07 43 ☽ # ♆
11 56 ☽ Q ♆
13 11 ☽ # ☉
13 34 ☽ ⚹ ☿
20 28 ☽ ⚹ ☉

19 Saturday
01 08 ☽ # ☿
04 30 ☽ ∠ ♃
06 37 ☽ ∠ ♂
07 21 ☽ ☌ ♇
07 49 ☽ ⚹ ♆
09 36 ☽ # ♂
13 53 ☽ # ♃
17 32 ☽ ∠ ☿
19 26 ☽ ⊥ ♃
21 21 ☿ Q ♅
21 32 ☉ ∠ ♄
22 00 ☽ ⊥ ♂
22 44 ☽ □ ♀
22 45 ☽ □ ♄
22 50 ☽ ∠ ☉
22 50 ♀ ☌ ♄
22 51 ☽ ⊻ ♅
23 30 ♂ ∠ ♇

20 Sunday
00 14 ♀ △ ♅
05 19 ☽ ⊻ ♃
07 47 ☉ ∠ ♀
08 11 ☽ ⊻ ♂
08 17 ☽ ∠ ♆
08 36 ☽ ⊥ ♅
10 01 ☽ ⊥ ☿
14 26 ☽ ⊥ ☉
20 57 ☽ ⊻ ☿
20 59 ♄ △ ♅
22 51 ☽ ⊥ ♆
23 08 ☽ ∠ ♅

21 Monday
00 48 ☽ ⊻ ☉
04 00 ☽ Q ♄
06 11 ☽ Q ♀
07 20 ☉ ∥ ♆
08 01 ☽ ⊻ ♇
08 32 ☽ ⊻ ♆
14 36 ☽ ∥ ♃
17 21 ☿ ∠ ♄
17 41 ☽ ⊥ ♇
21 11 ☽ ∥ ♂
23 19 ☽ ⚹ ♅
23 26 ☽ ⚹ ♄

22 Tuesday
03 04 ☽ ⚹ ♀
06 22 ☽ ☌ ♃
08 13 ☽ ∠ ♇
10 45 ☽ ☌ ♂
15 53 ☽ ∥ ☿
18 44 ☽ Q ♅
19 25 ☽ ∥ ♆
22 37 ☽ ∥ ☉
23 51 ☽ ∠ ♄

23 Wednesday
03 33 ☽ ☌ ☿
04 43 ☽ ☌ ☉
05 21 ☽ ∠ ♀
08 36 ☽ ⚹ ♇
09 10 ☽ ☌ ♆
14 16 ☿ ∥ ♆
14 40 ☽ ⊥ ♄
21 21 ☽ ⊥ ♀

24 Thursday
00 15 ☽ □ ♅
00 38 ☽ ⊻ ♄
03 28 ☉ ☌ ☿
04 24 ☽ Q ♇
08 09 ☽ ⊻ ♀
08 10 ☽ ⊻ ♃
08 37 ♀ ⚹ ♃
14 22 ☽ ⊻ ♂
18 24 ☽ ⊥ ♃

25 Friday
00 17 ☿ ∠ ♀
00 37 ☿ ⚹ ♇
00 59 ☽ ⊥ ♂
01 37 ☽ # ♅
06 14 ☿ ☌ ♆
09 14 ♀ ∠ ♆
09 57 ☽ ∠ ♃
10 33 ☽ ⊻ ☉
10 52 ☽ □ ♇
11 32 ☽ ⊻ ♆
12 13 ☽ ⊻ ☿
14 55 ☉ ⚹ ♇
17 12 ☽ ∠ ♂
21 45 ☽ ⊥ ☉
21 57 ☽ ⊥ ♆
23 12 ☉ ∥ ☿

26 Saturday
00 03 ☽ ⊥ ☿
00 39 ☉ ☌ ♆
03 23 ☽ △ ♅
04 05 ☽ ☌ ♄
12 33 ☽ ⚹ ♃
13 51 ☽ ∠ ♆
14 52 ☽ ∠ ☉
16 20 ☽ ☌ ♀
18 09 ☽ ∠ ☿
21 01 ☽ ⚹ ♂

27 Sunday
06 14 ☽ ⚼ ♅
10 32 ☽ Q ♃
11 49 ☿ ⊥ ♄
16 18 ☽ △ ♇
17 06 ☽ ⚹ ♆
20 03 ☽ Q ♂
20 20 ☽ ⚹ ☉
22 47 ☽ ± ♅

28 Monday
01 26 ☽ ⚹ ☿
03 58 ♃ ∠ ♇
10 01 ☽ ⊼ ♅
11 07 ☽ ⊻ ♄
15 30 ☽ # ♅
15 35 ☽ Q ♆
20 27 ☽ ⚼ ♇
20 42 ☽ □ ♃
20 47 ☽ Q ☉
22 36 ☽ ⊥ ♄

29 Tuesday
03 27 ☽ Q ☿
05 10 ☽ ⊻ ♀
07 56 ☽ □ ♂
13 46 ☽ ± ♇
16 05 ☽ ∠ ♄
17 54 ☽ ⊥ ♀
22 01 ☉ ⊥ ♄
23 34 ☽ ∥ ☿

30 Wednesday
01 28 ☽ ⊼ ♇
02 24 ☽ □ ♆
02 58 ♂ Q ♅
08 38 ☽ ∥ ☉
10 48 ☽ □ ☉
13 18 ☽ ∠ ♀
19 57 ☽ □ ☿
20 15 ☽ ☍ ♅
21 50 ☽ ⚹ ♄
22 14 ☿ □ ♅

31 Thursday
06 55 ☽ ∥ ♆
08 28 ☽ △ ♃
11 02 ☿ ⊻ ♄
19 15 ☽ ∥ ♂
22 04 ☽ Q ♄
22 15 ☽ ⚹ ♀
22 39 ☽ △ ♂

NOVEMBER 1957

Date	Sidereal time h m s	Sun ☉	Moon ☽	Moon ☽ 24.00	Mercury ☿	Venus ♀	Mars ♂	Jupiter ♃	Saturn ♄	Uranus ♅	Neptune ♆	Pluto ♇
01	14 42 01	08 ♏ 48 21	01 ♓ 20 02	07 ♓ 15 48	14 ♏ 02	25 ♐ 09	25 ♎ 05	17 ♎ 54	12 ♐ 29	11 ♌ 33	02 ♏ 35	02 ♍ 03
02	14 45 58	09 48 24	13 ♓ 11 10	19 ♓ 06 37	15 38	26 14	25 45	18 06	12 36	11 34	02 37	02 04
03	14 49 54	10 48 30	25 ♓ 02 38	00 ♈ 59 40	17 13	27 19	26 25	18 19	12 42	11 35	02 39	02 05
04	14 53 51	11 48 37	06 ♈ 58 07	12 ♈ 58 21	18 47	28 24	27 05	18 31	12 49	11 35	02 41	02 06
05	14 57 48	12 48 45	19 ♈ 00 42	25 ♈ 05 27	20 21	29 ♐ 28	27 45	18 43	12 55	11 36	02 44	02 07
06	15 01 44	13 48 55	01 ♉ 12 50	07 ♉ 23 03	21 54	00 ♑ 33	28 25	18 56	13 01	11 37	02 46	02 08
07	15 05 41	14 49 07	13 ♉ 36 18	19 ♉ 52 41	23 28	01 37	29 05	19 08	13 08	11 37	02 48	02 09
08	15 09 37	15 49 21	26 ♉ 12 19	02 ♊ 35 14	25 01	02 41	29 ♎ 45	19 21	13 14	11 38	02 50	02 09
09	15 13 34	16 49 37	09 ♊ 01 31	15 ♊ 31 10	26 33	03 44	00 ♏ 25	19 33	13 21	11 38	02 52	02 10
10	15 17 30	17 49 54	22 ♊ 04 12	28 ♊ 40 34	28 05	04 47	01 05	19 45	13 28	11 39	02 55	02 11
11	15 21 27	18 50 14	05 ♋ 20 17	12 ♋ 03 18	29 ♏ 37	05 50	01 45	19 57	13 34	11 39	02 57	02 12
12	15 25 23	19 50 35	18 ♋ 49 34	25 ♋ 39 02	01 ♐ 09	06 53	02 25	20 09	13 41	11 39	02 59	02 12
13	15 29 20	20 50 58	02 ♌ 31 08	09 ♌ 27 17	02 40	07 55	03 05	20 21	13 48	11 40	03 01	02 13
14	15 33 17	21 51 23	16 ♌ 25 53	23 ♌ 27 17	04 11	08 58	03 46	20 33	13 54	11 40	03 03	02 14
15	15 37 13	22 51 50	00 ♍ 31 20	07 ♍ 37 47	05 41	09 59	04 26	20 45	14 01	11 40	03 05	02 14
16	15 41 10	23 52 19	14 ♍ 46 25	21 ♍ 56 54	07 12	11 01	05 06	20 57	14 08	11 40	03 08	02 15
17	15 45 06	24 52 50	29 ♍ 08 50	06 ♎ 21 48	08 41	12 02	05 46	21 09	14 15	11 R 40	03 10	02 15
18	15 49 03	25 53 23	13 ♎ 35 18	20 ♎ 48 47	10 11	13 02	06 27	21 21	14 22	11 40	03 12	02 16
19	15 52 59	26 53 57	28 ♎ 01 38	05 ♏ 13 15	11 40	14 03	07 07	21 33	14 28	11 40	03 14	02 16
20	15 56 56	27 54 33	12 ♏ 22 59	19 ♏ 30 11	13 09	15 03	07 48	21 44	14 35	11 40	03 16	02 17
21	16 00 52	28 55 10	26 ♏ 34 16	03 ♐ 34 39	14 37	16 02	08 28	21 56	14 42	11 40	03 18	02 17
22	16 04 49	29 ♏ 55 50	10 ♐ 30 49	17 ♐ 22 22	16 05	17 01	09 08	22 08	14 49	11 39	03 20	02 18
23	16 08 46	00 ♐ 56 30	24 ♐ 08 58	00 ♑ 50 22	17 33	18 00	09 49	22 19	14 56	11 39	03 22	02 18
24	16 12 42	01 57 12	07 ♑ 26 27	13 ♑ 57 13	19 00	18 58	10 29	22 31	15 03	11 39	03 24	02 18
25	16 16 39	02 57 55	20 ♑ 22 45	26 ♑ 43 15	20 26	19 56	11 10	22 42	15 10	11 38	03 26	02 19
26	16 20 35	03 58 39	02 ♒ 58 58	09 ♒ 10 18	21 51	20 53	11 51	22 53	15 17	11 38	03 28	02 19
27	16 24 32	04 59 24	15 ♒ 17 39	21 ♒ 21 32	23 16	21 50	12 31	23 04	15 24	11 37	03 30	02 19
28	16 28 28	06 00 10	27 ♒ 22 28	03 ♓ 21 02	24 40	22 46	13 12	23 16	15 31	11 37	03 32	02 19
29	16 32 25	07 00 57	09 ♓ 17 51	15 ♓ 13 31	26 02	23 41	13 52	23 27	15 38	11 36	03 34	02 19
30	16 36 21	08 ♐ 01 45	21 ♓ 08 40	27 ♓ 03 56	27 ♐ 24	24 ♑ 36	14 ♏ 33	23 ♎ 38	15 ♐ 45	11 ♌ 35	03 ♏ 36	02 ♍ 19

Date	Moon True ☊	Moon Mean ☊	Moon ☽ Latitude
01	10 ♏ 33	10 ♏ 35	04 N 57
02	10 R 33	10 32	04 28
03	10 D 33	10 29	03 47
04	10 33	10 26	02 56
05	10 34	10 22	01 57
06	10 34	10 19	00 N 52
07	10 R 34	10 16	00 S 17
08	10 34	10 13	01 26
09	10 33	10 10	02 32
10	10 32	10 07	03 31
11	10 31	10 03	04 19
12	10 30	10 00	04 55
13	10 29	09 57	05 14
14	10 29	09 54	05 15
15	10 D 29	09 51	04 58
16	10 29	09 47	04 22
17	10 30	09 44	03 30
18	10 32	09 41	02 24
19	10 33	09 38	01 S 09
20	10 R 33	09 35	00 N 10
21	10 32	09 32	01 27
22	10 31	09 28	02 38
23	10 28	09 25	03 38
24	10 25	09 22	04 24
25	10 21	09 19	04 56
26	10 18	09 16	05 12
27	10 15	09 13	05 13
28	10 14	09 09	05 01
29	10 D 14	09 06	04 35
30	10 ♏ 14	09 ♏ 03	03 N 58

DECLINATIONS

Date	Sun ☉	Moon ☽	Mercury ☿	Venus ♀	Mars ♂	Jupiter ♃	Saturn ♄	Uranus ♅	Neptune ♆	Pluto ♇
01	14 S 26	06 S 22	16 S 20	26 S 27	09 S 08	05 S 59	20 S 54	17 N 56	10 S 47	21 N 13
02	14 45	02 S 29	16 55	26 31	09 23	06 04	20 55	17 55	10 47	21 13
03	15 04	01 N 30	17 28	26 34	09 37	06 09	20 56	17 55	10 48	21 13
04	15 23	05 28	18 01	26 37	09 52	06 13	20 57	17 55	10 49	21 13
05	15 41	09 15	18 33	26 39	10 07	06 18	20 58	17 55	10 49	21 13
06	15 59	12 42	19 04	26 40	10 21	06 23	20 59	17 55	10 50	21 13
07	16 17	15 39	19 34	26 41	10 36	06 27	21 00	17 55	10 51	21 13
08	16 35	17 55	20 03	26 40	10 51	06 32	21 01	17 55	10 51	21 13
09	16 52	19 19	20 31	26 40	11 05	06 36	21 02	17 54	10 52	21 14
10	17 09	19 42	20 57	26 38	11 20	06 41	21 02	17 54	10 53	21 14
11	17 26	19 01	21 23	26 36	11 34	06 46	21 03	17 54	10 54	21 14
12	17 42	17 15	21 48	26 34	11 48	06 50	21 04	17 54	10 55	21 14
13	17 58	14 30	22 12	26 30	12 03	06 54	21 05	17 54	10 55	21 14
14	18 14	10 54	22 35	26 26	12 17	06 59	21 06	17 54	10 56	21 14
15	18 29	06 39	22 57	26 22	12 31	07 03	21 07	17 54	10 57	21 14
16	18 45	01 N 58	23 17	26 17	12 45	07 08	21 08	17 54	10 57	21 14
17	18 59	02 S 52	23 36	26 11	12 58	07 12	21 09	17 54	10 58	21 14
18	19 14	07 34	23 54	26 04	13 12	07 16	21 10	17 54	10 59	21 15
19	19 28	11 51	24 11	25 57	13 26	07 21	21 11	17 54	11 00	21 15
20	19 42	15 24	24 27	25 50	13 40	07 25	21 11	17 54	11 00	21 15
21	19 55	17 58	24 41	25 42	13 53	07 29	21 12	17 55	11 01	21 16
22	20 08	19 25	24 54	25 33	14 06	07 33	21 13	17 55	11 02	21 16
23	20 21	19 41	25 06	25 24	14 20	07 38	21 14	17 55	11 02	21 16
24	20 33	18 50	25 17	25 14	14 33	07 42	21 15	17 55	11 03	21 16
25	20 45	17 01	25 26	25 04	14 46	07 46	21 16	17 55	11 04	21 17
26	20 57	14 22	25 34	24 53	14 59	07 50	21 17	17 55	11 04	21 17
27	21 08	11 15	25 40	24 42	15 12	07 54	21 17	17 55	11 05	21 17
28	21 19	07 40	25 45	24 31	15 25	07 58	21 18	17 56	11 05	21 17
29	21 29	03 50	25 49	24 18	15 37	08 02	21 19	17 56	11 06	21 18
30	21 S 39	00 N 08	25 S 51	24 S 06	15 S 50	08 S 06	21 S 20	17 N 56	11 S 07	21 N 18

ZODIAC SIGN ENTRIES

Date	h	m	Planets
01	09	18	☽ ♓
03	22	00	☽ ♈
05	23	46	♀ ♑
06	09	38	☽ ♉
08	19	09	☽ ♊
08	21	04	♂ ♏
11	02	24	☽ ♋
11	18	00	☿ ♐
13	07	36	☽ ♌
15	11	07	☽ ♍
17	13	25	☽ ♎
19	15	17	☽ ♏
21	17	52	☽ ♐
22	13	39	☉ ♐
23	22	29	☽ ♑
26	06	16	☽ ♒
28	17	16	☽ ♓

LATITUDES

Date	Mercury ☿	Venus ♀	Mars ♂	Jupiter ♃	Saturn ♄	Uranus ♅	Neptune ♆	Pluto ♇
01	00 S 18	03 S 06	00 N 38	01 N 07	01 N 24	00 N 38	01 N 42	11 N 16
04	00 38	03 11	00 36	01 07	01 24	00 38	01 42	11 17
07	00 57	03 15	00 35	01 07	01 24	00 38	01 42	11 18
10	01 15	03 17	00 34	01 08	01 23	00 38	01 42	11 19
13	01 32	03 18	00 32	01 08	01 23	00 38	01 42	11 20
16	01 48	03 18	00 31	01 08	01 23	00 38	01 42	11 21
19	02 01	03 16	00 29	01 08	01 22	00 39	01 42	11 23
22	02 12	03 13	00 28	01 09	01 22	00 39	01 42	11 24
25	02 20	03 08	00 26	01 09	01 22	00 39	01 42	11 25
28	02 25	03 02	00 25	01 10	01 22	00 39	01 42	11 26
31	02 S 25	02 S 53	00 N 23	01 N 10	01 N 21	00 N 39	01 N 42	11 N 28

DATA

Julian Date	2436144
Delta T	+32 seconds
Ayanamsa	23° 16' 15"
Synetic vernal point	05° ♓ 50' 44"
True obliquity of ecliptic	23° 26' 34"

MOON'S PHASES, APSIDES AND POSITIONS ☽

Date	h	m	Phase	Longitude	Eclipse Indicator
07	14	32	○	14 ♉ 55	total
14	21	59	☾	22 ♌ 17	
21	16	19	●	29 ♏ 06	
29	06	58	☽	06 ♓ 48	

Day	h	m		
02	11	26	Apogee	
18	11	08	Perigee	
30	06	55	Apogee	
03	02	59	0N	
10	08	53	Max dec	19° N 43'
16	21	47	0S	
23	05	27	Max dec	19° S 44'
30	11	11	0N	

LONGITUDES

Date	Chiron ⚷	Ceres ⚳	Pallas ⚴	Juno ⚵	Vesta ⚶	Black Moon Lilith ⚸
01	12 ♒ 19	27 ♋ 27	00 ♉ 06	18 ♊ 23	08 ♈ 13	17 ♓ 44
11	12 ♒ 30	28 ♋ 50	26 ♈ 59	17 ♊ 37	07 ♈ 01	18 ♓ 51
21	12 ♒ 47	29 ♋ 38	24 ♈ 21	16 ♊ 02	06 ♈ 34	19 ♓ 58
31	13 ♒ 10	29 ♋ 46	22 ♈ 30	13 ♊ 54	06 ♈ 52	21 ♓ 05

All ephemeris data is given at 12.00 UT and the Moon's longitude is additionally given for 24.00 UT

ASPECTARIAN

h m Aspects

01 Friday
08 51 ♀ ✶ ♂
12 12 ☿ Q ♇
13 28 ☽ ☍ ♇
14 22 ☽ ‖ ♃
14 32 ☽ △ ♆
15 12 ☽ ⚼ ♃

02 Saturday
00 57 ☽ Q ♀
04 31 ☽ △ ☉
06 47 ☽ ⚼ ♂
08 43 ☽ ⊼ ♅
09 46 ☽ ± ♃
10 48 ☽ □ ♄
17 43 ☽ △ ☿
19 26 ♀ ⚼ ♅
20 53 ☽ ± ♅
21 00 ☽ ⚼ ♆
22 08 ☽ ⊼ ♃

03 Sunday
02 06 ☽ ± ♂
13 41 ☽ ⚼ ☉
14 56 ☽ ⊼ ♂
15 06 ☽ ⚼ ♅
15 16 ☽ ± ♆
17 03 ☽ □ ♀

04 Monday
02 13 ☽ ⊼ ♇
03 23 ☽ ⊼ ♆
04 38 ☽ ⚼ ☿
06 42 ☉ □ ♅
07 21 ☿ ⊻ ♃
07 35 ☿ # ♅
09 28 ☽ ± ☉
14 16 ☽ ± ♇
16 49 ☽ # ♃
21 15 ☽ △ ♅
22 34 ☽ ⊼ ☉
23 47 ☽ △ ♄

05 Tuesday
01 21 ☽ ± ☿
08 14 ☽ ⚼ ♇
11 25 ☽ ☍ ♃
14 46 ☉ ⊻ ♄
15 02 ☽ ⊼ ☿
18 12 ☽ # ♂
19 10 ♂ ∠ ♄
22 38 ☽ # ♆

06 Wednesday
05 42 ☽ ⚼ ♄
06 13 ☽ ☍ ♂
10 34 ☽ △ ♀
13 47 ☽ △ ♇
15 02 ☽ ☍ ♆
19 38 ☉ Q ♇
22 43 ♀ Q ♃
23 24 ☽ ± ♄

07 Thursday
08 11 ☽ □ ♅
11 05 ☽ ⊼ ♄
14 32 ☽ ☍ ☉
18 18 ☽ ⚼ ♀
18 44 ☽ # ☉
22 46 ☽ ⊼ ♃

08 Friday
00 06 ♀ △ ♇
09 26 ☽ ☍ ☿
10 21 ☽ ± ♃
11 54 ☽ ‖ ♅
12 58 ☽ ± ♀
13 31 ♂ ‖ ♆
15 45 ♀ ✶ ♆
17 58 ☿ ⊥ ♃
18 28 ☽ Q ♅
19 03 ☽ ⊼ ♂
23 12 ☽ □ ♇

09 Saturday
00 30 ☽ ⊼ ♆
01 16 ☽ ⊼ ♀
03 32 ☽ ⚼ ♃
06 53 ☽ ± ♂
11 43 ☽ ± ♆
16 51 ☽ ✶ ♅
20 04 ☽ ☍ ♄

10 Sunday
00 27 ☽ ⚼ ♂
03 36 ☽ ⊼ ☉
04 22 ☽ ⚼ ♆
07 42 ☽ △ ♃
08 33 ☽ Q ♇
15 29 ☽ ± ☉
16 38 ☿ ‖ ♄
20 20 ☽ ∠ ♅

11 Monday
00 22 ☽ ⊼ ☿
02 56 ☿ # ♇
05 13 ☽ △ ♂
06 21 ☽ ✶ ♇
07 39 ♀ ± ♅
07 42 ☽ △ ♆
09 05 ☽ ⚼ ☉
12 34 ☽ ⊥ ♅
12 34 ☽ ± ☿
12 59 ☽ ☍ ♀
23 05 ☿ ⊥ ♀
23 17 ☽ ⊻ ♅

12 Tuesday
02 49 ☽ ⊼ ♄
04 09 ♂ ✶ ♇
04 41 ☽ ‖ ♅
06 39 ☽ ⚼ ☿
07 39 ☽ # ☉
09 08 ☽ ∠ ♇
13 31 ☽ ± ♄
13 56 ☽ △ ☉
14 23 ☽ □ ♃
21 19 ☉ ⊻ ♃

13 Wednesday
00 59 ☽ ⊥ ♇
04 53 ☿ □ ♇
05 27 ☽ ⚼ ♄
06 02 ☉ # ♅
09 19 ♂ ☌ ♆
11 28 ☽ ⊻ ♇
12 16 ☽ △ ☿
12 51 ☽ □ ♆
13 02 ☽ □ ♂
17 46 ☿ ⊻ ♆
22 07 ☽ ⊼ ♀
22 15 ☽ Q ♃

14 Thursday
00 06 ☿ ⊻ ♂
03 48 ☽ ☌ ♅
03 54 ☽ # ♂
07 38 ☽ △ ♄
09 16 ☽ ± ♀
11 47 ☽ # ♆
19 10 ☽ ✶ ♃
19 56 ☽ Q ♆
21 34 ☽ Q ♂
21 59 ☽ □ ☉

15 Friday
01 52 ☽ ⚼ ♀
09 50 ☽ # ♃
13 16 ☿ ∠ ♃
14 54 ☽ ☌ ♇
16 21 ☽ ✶ ♆
18 56 ☽ ✶ ♂
20 58 ☽ ∠ ♃
21 46 ☽ □ ☿

16 Saturday
05 12 ☽ △ ♀
06 46 ☽ Q ☉
06 47 ☽ ⊻ ♅
10 55 ☽ □ ♄
12 19 ☽ ⊥ ♃
16 51 ☽ ⊥ ♅
17 38 ☽ ∠ ♆
21 21 ☽ ∠ ♂
22 29 ☽ ⊻ ♃

17 Sunday
03 28 ♀ ⊼ ♅
04 21 ☽ ✶ ☉
06 25 ♅ St R
07 26 ☽ Q ☿
07 52 ☽ ∠ ♅
08 41 ☽ ⊥ ♆
13 06 ☽ ⊥ ♂
17 10 ☽ ⊻ ♇
17 12 ☽ Q ♄
18 42 ☽ ⊻ ♆
19 45 ☿ ⊥ ♆
23 33 ☽ ⊻ ♂

18 Monday
03 09 ☽ ⊥ ♇
05 42 ☽ ✶ ☿
07 11 ☽ ∠ ☉
08 49 ☽ ✶ ♅
10 25 ☽ ‖ ♃
11 01 ☽ □ ♀
13 17 ☽ ✶ ♄
18 06 ☽ ∠ ♇
23 15 ☽ ⊥ ☉

19 Tuesday
01 04 ☽ ☌ ♃
04 45 ☽ Q ♅
06 55 ☽ ‖ ♆
09 29 ☽ ∠ ☿
09 59 ☽ ⊻ ☉
11 54 ☿ △ ♅
14 26 ☽ ∠ ♄
19 05 ☽ ✶ ♇
19 12 ☽ Q ♀
20 42 ☽ ☌ ♆
22 42 ☽ ‖ ♂
23 39 ♀ ⊻ ♄

20 Wednesday
02 13 ☽ ⊥ ☿
03 55 ☽ ☌ ♂
05 35 ☽ ⊥ ♄
07 03 ☉ ⊥ ♃
10 47 ☽ □ ♅
13 26 ☽ ⊻ ☿
15 11 ☽ Q ♇
15 44 ☽ ⊻ ♄
16 49 ☽ ✶ ♀
17 36 ♀ Q ♆

21 Thursday
04 00 ☽ ⊻ ♃
07 12 ☿ ⊥ ♂
11 14 ☽ # ♅
13 24 ☿ ☌ ♄
14 22 ☽ ⊥ ♃
16 19 ☽ ☌ ☉
20 13 ☽ ∠ ♀
21 47 ☽ □ ♇
22 15 ♂ ⊥ ♄
23 33 ☽ ⊻ ♆

22 Friday
06 03 ☽ ∠ ♃
09 29 ☽ ⊻ ♂
09 57 ☽ ⊥ ♆
12 57 ☽ ⊥ ♀
13 59 ☽ △ ♅
18 42 ♀ ⚼ ♇
19 35 ☽ ☌ ♄
20 30 ☽ ⊥ ♂
22 55 ☽ ☌ ☿

23 Saturday
00 16 ☽ ⊻ ♀
01 44 ☽ ∠ ♆
08 42 ☽ ✶ ♃
13 15 ☽ ∠ ♂
16 28 ☽ ⚼ ♅

24 Sunday
01 11 ☽ ⊻ ☉
01 57 ☿ ∠ ♆
02 39 ☽ △ ♇
04 37 ☽ ✶ ♆
06 34 ☽ Q ♃
08 43 ☽ ± ♅
09 07 ☿ ‖ ♀
10 36 ☿ ⊻ ♀
13 01 ☽ ⊥ ☉
17 54 ☽ ✶ ♂
19 43 ☽ ⊼ ♅
20 21 ☉ □ ♇

25 Monday
01 34 ☽ # ♅
02 10 ☽ ⊻ ♄
02 44 ☽ Q ♆
06 15 ☽ ⚼ ♇
07 05 ☽ ∠ ☉
11 05 ☽ ☌ ♀
12 07 ☽ ⊻ ☿
13 30 ☽ ⊥ ♄
16 26 ☽ □ ♃
17 33 ☽ Q ♂
23 13 ☽ ± ♇
23 35 ☉ ⊻ ♆

26 Tuesday
00 55 ☽ ⊥ ☿
04 30 ♂ □ ♅
06 46 ☽ ∠ ♄
07 41 ☽ ‖ ♂
10 43 ☽ ⊼ ♇
12 57 ☽ □ ♆
14 05 ☽ ✶ ☉
20 28 ☽ ∠ ☿
23 36 ♀ ⊥ ♄

27 Wednesday
01 09 ♄ # ♇
04 47 ☽ ☍ ♅
06 14 ☽ □ ♂
08 10 ☿ ✶ ♃
12 13 ☽ ✶ ♄
13 14 ☽ ‖ ♆
15 39 ☽ Q ☉

28 Thursday
02 01 ☽ ⊻ ♀
03 39 ☽ △ ♃
05 52 ☽ ✶ ☿
08 59 ☉ # ♇
10 09 ☽ ‖ ♃
10 42 ☉ ‖ ♄
12 18 ☽ Q ♄
15 01 ☽ ⊥ ♀
21 56 ☽ ☍ ♇

29 Friday
00 25 ☽ △ ♆
03 59 ♀ □ ♃
06 58 ☽ □ ☉
09 08 ☽ Q ☿
10 15 ☽ ⚼ ♃
10 40 ☽ ∠ ♀
16 39 ☽ ⊼ ♅
21 44 ☿ ⚼ ♅
21 49 ☽ △ ♂

30 Saturday
00 58 ☽ □ ♄
03 56 ♂ Q ♇
04 45 ☽ ± ♃
04 48 ☽ ± ♅
06 50 ☽ ⚼ ♆
06 59 ♃ Q ♅
17 07 ☽ ⊼ ♃
19 36 ☽ ✶ ♀
23 01 ☽ ⚼ ♅

DECEMBER 1957

Date	Sidereal time h m s	LONGITUDES Sun ☉ ° ′ ″	Moon ☽ ° ′ ″	Moon ☽ 24.00 ° ′ ″	Mercury ☿ ° ′	Venus ♀ ° ′	Mars ♂ ° ′	Jupiter ♃ ° ′	Saturn ♄ ° ′	Uranus ♅ ° ′	Neptune ♆ ° ′	Pluto ♇ ° ′
01	16 40 18	09 ♐ 02 34	02 ♈ 59 56	08 ♈ 57 15	28 ♐ 44	25 ♑ 31	15 ♏ 14	23 ♎ 49	15 ♐ 52	11 ♌ 34	03 ♏ 38	02 ♍ 20
02	16 44 15	10 03 24	14 ♈ 56 28	20 ♈ 58 08	00 ♑ 02	26 25	15 55	23 59	15 59	11 R 34	03 40	02 20
03	16 48 11	11 04 15	27 ♈ 02 44	03 ♉ 10 42	01 19	27 18	16 35	24 10	16 06	11 33	03 42	02 20
04	16 52 08	12 05 06	09 ♉ 22 26	15 ♉ 38 15	02 33	28 10	17 16	24 21	16 13	11 32	03 44	02 R 20
05	16 56 04	13 05 59	21 ♉ 58 23	28 ♉ 23 00	03 45	29 02	17 57	24 31	16 21	11 31	03 46	02 20
06	17 00 01	14 06 52	04 ♊ 52 10	11 ♊ 25 54	04 53	29 ♑ 53	18 38	24 42	16 28	11 30	03 47	02 20
07	17 03 57	15 07 47	18 ♊ 04 04	24 ♊ 46 31	05 59	00 ♒ 43	19 19	24 52	16 35	11 29	03 49	02 20
08	17 07 54	16 08 43	01 ♋ 32 57	08 ♋ 23 05	07 01	01 32	20 00	25 03	16 42	11 28	03 51	02 19
09	17 11 50	17 09 39	15 ♋ 16 30	22 ♋ 12 48	07 58	02 21	20 41	25 13	16 49	11 27	03 53	02 19
10	17 15 47	18 10 37	29 ♋ 11 31	06 ♌ 12 13	08 50	03 08	21 22	25 23	16 56	11 25	03 54	02 19
11	17 19 44	19 11 35	13 ♌ 14 27	20 ♌ 17 48	09 36	03 55	22 03	25 33	17 03	11 24	03 56	02 19
12	17 23 40	20 12 35	27 ♌ 21 53	04 ♍ 26 23	10 16	04 41	22 44	25 43	17 10	11 23	03 58	02 19
13	17 27 37	21 13 36	11 ♍ 30 59	18 ♍ 35 27	10 48	05 26	23 25	25 53	17 17	11 21	04 00	02 18
14	17 31 33	22 14 38	25 ♍ 39 33	02 ♎ 43 09	11 12	06 10	24 06	26 03	17 24	11 20	04 01	02 18
15	17 35 30	23 15 41	09 ♎ 46 03	16 ♎ 48 09	11 27	06 52	24 47	26 12	17 31	11 19	04 03	02 18
16	17 39 26	24 16 44	23 ♎ 49 16	00 ♏ 49 15	11 R 32	07 34	25 29	26 22	17 39	11 17	04 04	02 18
17	17 43 23	25 17 49	07 ♏ 47 56	14 ♏ 45 06	11 26	08 14	26 10	26 31	17 46	11 16	04 06	02 17
18	17 47 19	26 18 55	21 ♏ 40 30	28 ♏ 33 53	11 09	08 54	26 51	26 41	17 53	11 14	04 08	02 17
19	17 51 16	27 20 01	05 ♐ 24 57	12 ♐ 13 24	10 40	09 32	27 33	26 50	18 00	11 12	04 09	02 16
20	17 55 13	28 21 09	18 ♐ 58 53	25 ♐ 41 08	09 59	10 08	28 14	26 59	18 07	11 11	04 11	02 16
21	17 59 09	29 ♐ 22 16	02 ♑ 19 51	08 ♑ 54 46	09 07	10 44	28 55	27 08	18 14	11 09	04 12	02 15
22	18 03 06	00 ♑ 23 25	15 ♑ 25 43	21 ♑ 52 33	08 06	11 18	29 ♏ 37	27 17	18 21	11 07	04 14	02 15
23	18 07 02	01 24 33	28 ♑ 15 12	04 ♒ 33 42	06 55	11 50	00 ♐ 18	27 26	18 28	11 05	04 15	02 14
24	18 10 59	02 25 42	10 ♒ 48 08	16 ♒ 58 41	05 38	12 21	01 00	27 34	18 35	11 04	04 16	02 14
25	18 14 55	03 26 51	23 ♒ 05 37	29 ♒ 09 16	04 17	12 51	01 41	27 43	18 42	11 02	04 18	02 13
26	18 18 52	04 28 00	05 ♓ 10 02	11 ♓ 08 26	02 55	13 18	02 23	27 51	18 49	11 00	04 19	02 12
27	18 22 48	05 29 09	17 ♓ 04 57	23 ♓ 00 11	01 34	13 44	03 04	27 59	18 56	10 58	04 20	02 12
28	18 26 45	06 30 18	28 ♓ 54 45	04 ♈ 49 17	00 ♑ 17	14 08	03 46	28 08	19 02	10 56	04 22	02 11
29	18 30 42	07 31 27	10 ♈ 44 29	16 ♈ 41 00	29 ♐ 06	14 31	04 27	28 16	19 09	10 54	04 23	02 10
30	18 34 38	08 32 36	22 ♈ 39 33	28 ♈ 40 47	28 03	14 51	05 09	28 24	19 16	10 52	04 24	02 10
31	18 38 35	09 ♑ 33 44	04 ♉ 45 21	10 ♉ 53 53	27 ♐ 10	15 ♒ 10	05 ♐ 51	28 ♎ 31	19 ♐ 23	10 ♌ 50	04 ♏ 25	02 ♍ 09

Date	Moon True ☊ ° ′	Moon Mean ☊ ° ′	Moon ☽ Latitude ° ′
01	10 ♏ 16	09 ♏ 00	03 N 11
02	10 D 18	08 57	02 15
03	10 19	08 53	01 12
04	10 20	08 50	00 N 05
05	10 R 19	08 47	01 S 03
06	10 17	08 44	02 10
07	10 13	08 41	03 11
08	10 08	08 38	04 03
09	10 01	08 34	04 42
10	09 55	08 31	05 05
11	09 50	08 28	05 10
12	09 46	08 25	04 56
13	09 44	08 22	04 24
14	09 D 43	08 19	03 36
15	09 44	08 15	02 35
16	09 45	08 12	01 25
17	09 46	08 09	00 S 11
18	09 R 46	08 06	01 N 04
19	09 43	08 03	02 14
20	09 38	07 59	03 15
21	09 30	07 56	04 05
22	09 21	07 53	04 40
23	09 11	07 50	05 00
24	09 01	07 47	05 06
25	08 52	07 44	04 57
26	08 46	07 40	04 34
27	08 41	07 37	04 00
28	08 39	07 34	03 16
29	08 D 39	07 31	02 24
30	08 40	07 28	01 25
31	08 ♏ 40	07 ♏ 25	00 N 21

Date	DECLINATIONS Sun ☉ ° ′	Moon ☽ ° ′	Mercury ☿ ° ′	Venus ♀ ° ′	Mars ♂ ° ′	Jupiter ♃ ° ′	Saturn ♄ ° ′	Uranus ♅ ° ′	Neptune ♆ ° ′	Pluto ♇ ° ′
01	21 S 49	04 N 06	25 S 51	23 S 53	16 S 02	08 S 10	21 S 21	17 N 56	11 S 07	21 N 18
02	21 58	07 58	25 51	23 39	16 15	08 14	21 21	17 57	11 08	21 19
03	22 06	11 33	25 49	23 26	16 27	08 17	21 22	17 57	11 09	21 19
04	22 15	14 42	25 45	23 11	16 39	08 21	21 23	17 57	11 09	21 20
05	22 22	17 15	25 40	22 57	16 51	08 25	21 24	17 57	11 10	21 20
06	22 30	18 59	25 34	22 42	17 03	08 29	21 25	17 58	11 10	21 20
07	22 37	19 44	25 26	22 27	17 15	08 32	21 25	17 58	11 11	21 21
08	22 43	19 23	25 17	22 11	17 26	08 36	21 26	17 58	11 11	21 21
09	22 49	17 54	25 07	21 55	17 38	08 40	21 27	17 59	11 12	21 22
10	22 55	15 21	24 56	21 39	17 49	08 43	21 28	17 59	11 13	21 22
11	23 00	11 54	24 44	21 23	18 00	08 47	21 28	18 00	11 13	21 22
12	23 05	07 45	24 31	21 06	18 11	08 50	21 29	18 00	11 14	21 23
13	23 09	03 N 10	24 17	20 50	18 22	08 54	21 30	18 00	11 14	21 23
14	23 13	01 S 35	24 02	20 33	18 33	08 57	21 30	18 01	11 15	21 24
15	23 16	06 15	23 46	20 15	18 44	09 00	21 31	18 01	11 15	21 24
16	23 19	10 34	23 31	19 58	18 54	09 04	21 32	18 02	11 16	21 25
17	23 22	14 17	23 15	19 41	19 04	09 07	21 33	18 02	11 16	21 25
18	23 23	17 09	22 58	19 23	19 15	09 10	21 33	18 03	11 17	21 26
19	23 25	19 01	22 42	19 05	19 25	09 13	21 34	18 03	11 17	21 26
20	23 26	19 45	22 26	18 48	19 34	09 16	21 35	18 04	11 18	21 27
21	23 26	19 21	22 10	18 30	19 44	09 20	21 35	18 04	11 18	21 27
22	23 27	17 55	21 54	18 12	19 54	09 23	21 36	18 05	11 18	21 28
23	23 26	15 36	21 38	17 54	20 03	09 26	21 36	18 05	11 19	21 29
24	23 25	12 37	21 23	17 37	20 12	09 28	21 37	18 06	11 19	21 29
25	23 24	09 09	21 09	17 19	20 21	09 31	21 38	18 06	11 20	21 30
26	23 22	05 22	20 56	17 01	20 30	09 34	21 38	18 07	11 20	21 30
27	23 20	01 S 25	20 44	16 44	20 39	09 37	21 39	18 07	11 20	21 31
28	23 17	02 N 34	20 34	16 26	20 47	09 40	21 40	18 08	11 21	21 31
29	23 14	06 27	20 26	16 09	20 56	09 42	21 40	18 08	11 21	21 32
30	23 10	10 08	20 19	15 52	21 04	09 45	21 41	18 09	11 22	21 33
31	23 S 06	13 N 26	20 S 15	15 S 36	21 S 12	09 S 48	21 S 41	18 N 10	11 S 22	21 N 33

ZODIAC SIGN ENTRIES

Date	h	m	Planets
01	05	56	☽ ♈
02	11	19	☿ ♑
03	17	48	☽ ♉
06	03	00	☽ ♊
06	15	26	♀ ♒
08	09	16	☽ ♋
10	13	23	☽ ♌
12	16	28	☽ ♍
14	19	23	☽ ♎
16	22	35	☽ ♏
19	02	30	☽ ♐
21	07	47	☽ ♑
22	02	49	☉ ♑
23	01	29	♂ ♐
23	15	19	☽ ♒
26	01	41	☽ ♓
28	14	13	☽ ♈
28	17	30	☿ ♐
31	02	37	☽ ♉

LATITUDES

Date	Mercury ☿ ° ′	Venus ♀ ° ′	Mars ♂ ° ′	Jupiter ♃ ° ′	Saturn ♄ ° ′	Uranus ♅ ° ′	Neptune ♆ ° ′	Pluto ♇ ° ′
01	02 S 25	02 S 53	00 N 23	01 N 10	01 N 21	00 N 39	01 N 42	11 N 28
04	02 20	02 43	00 21	01 10	01 21	00 39	01 43	11 29
07	02 08	02 30	00 20	01 10	01 21	00 39	01 43	11 30
10	01 47	02 16	00 18	01 11	01 21	00 39	01 43	11 31
13	01 17	01 58	00 16	01 11	01 21	00 39	01 43	11 32
16	00 S 34	01 39	00 14	01 12	01 21	00 40	01 43	11 34
19	00 N 19	01 16	00 13	01 12	01 20	00 40	01 43	11 35
22	01 18	00 51	00 11	01 13	01 20	00 40	01 43	11 36
25	02 13	00 S 22	00 09	01 13	01 20	00 40	01 44	11 37
28	02 52	00 N 09	00 07	01 14	01 20	00 40	01 44	11 38
31	03 N 10	00 N 44	00 N 05	01 N 14	01 N 20	00 N 40	01 N 44	11 N 39

DATA

Julian Date	2436174
Delta T	+32 seconds
Ayanamsa	23° 16' 19"
Synetic vernal point	05° ♓ 50' 41"
True obliquity of ecliptic	23° 26' 34"

MOON'S PHASES, APSIDES AND POSITIONS ☽

Date	h	m	Phase	Longitude ° ′	Eclipse Indicator
07	06	16	○	14 ♊ 53	
14	05	45	☾	21 ♍ 59	
21	06	12	●	29 ♐ 07	
29	04	52	☽	07 ♈ 13	

Day	h	m		
14	05	14	Perigee	
28	04	19	Apogee	
07	16	35	Max dec	19° N 45'
14	04	02	0S	
20	15	23	Max dec	19° S 45'
27	20	29	0N	

LONGITUDES

Date	Chiron ⚷ ° ′	Ceres ⚳ ° ′	Pallas ⚴ ° ′	Juno ⚵ ° ′	Vesta ⚶ ° ′	Black Moon Lilith ⚸ ° ′
01	13 ♒ 10	29 ♋ 46	22 ♈ 30	13 ♊ 54	06 ♈ 52	21 ♓ 05
11	13 ♒ 38	29 ♋ 12	21 ♈ 36	11 ♊ 37	07 ♈ 51	22 ♓ 12
21	14 ♒ 11	27 ♋ 57	21 ♈ 41	09 ♊ 39	09 ♈ 27	23 ♓ 19
31	14 ♒ 49	26 ♋ 06	22 ♈ 41	08 ♊ 20	11 ♈ 33	24 ♓ 26

All ephemeris data is given at 12.00 UT and the Moon's longitude is additionally given for 24.00 UT

ASPECTARIAN

h m	Aspects
01 Sunday	
01 07	☽ ± ♆
02 17	☽ □ ☿
05 16	☉ ∠ ♃
06 04	☽ ꝑ ♂
10 39	☽ ⊼ ♇
13 17	☽ ⊼ ♆
21 51	☽ Q ♀
22 44	☽ ± ♇
02 Monday	
01 18	☽ △ ☉
01 19	☽ ± ♂
02 27	☉ ⊥ ♆
05 14	☽ △ ♅
09 46	♀ ± ♇
13 46	☽ # ♃
14 03	☽ ⊼ ♂
14 07	☽ △ ♄
15 21	♂ ⊻ ♄
16 46	☽ ꝑ ♇
03 Tuesday	
06 15	☽ ☍ ♃
09 11	☽ # ♆
09 54	☽ ꝑ ☉
12 32	☽ □ ♀
20 02	☽ ꝑ ♄
21 18	☽ △ ☿
22 21	☽ △ ♇
23 06	☉ △ ♅
23 49	☿ ∠ ♂
04 Wednesday	
01 03	☽ ☍ ♆
05 05	☽ ± ☉
07 41	☿ △ ♇
10 30	♇ St R
13 39	☽ ± ♄
16 09	☽ □ ♅
17 40	☽ ⊼ ☉
05 Thursday	
01 15	☽ ⊼ ♄
03 58	☽ ☍ ♂
05 17	☽ ꝑ ☿
07 24	☽ # ♂
12 19	☿ ⚹ ♆
16 51	☽ ⊼ ♃
20 29	☽ ∥ ♅
23 53	☽ ± ☿
06 Friday	
02 05	☽ Q ♅
02 08	☽ △ ♀
04 12	☽ ± ♃
07 19	☽ □ ♇
10 00	☽ ⊼ ♆
12 03	☽ ⊼ ☿
20 58	☽ ꝑ ♃
21 02	☽ ± ♆
07 Saturday	
00 06	☽ ⚹ ♅
01 00	☿ ± ♅
01 12	☉ ∥ ♀
06 16	☽ ☍ ☉
07 29	☽ ꝑ ♀
09 18	☽ ☍ ♄
13 21	☽ ꝑ ♆
14 22	☽ ⊼ ♂
16 03	☽ Q ♇
08 Sunday	
00 20	☽ △ ♃
00 40	☽ ± ♀
01 39	☽ ± ♂
03 01	☽ ∠ ♅
11 59	☽ ⊼ ♀
13 03	☿ Q ♃
13 22	☽ ⚹ ♇
16 03	☽ △ ♆
17 29	♀ ∠ ♄
18 23	☽ ꝑ ♂
18 52	☽ ⊥ ♅
22 20	☽ ☍ ☿
09 Monday	
02 46	☉ ☌ ♄
05 21	☽ ⊻ ♅
11 01	☽ ∥ ♅
11 16	♀ ⊼ ♇
14 42	☽ ⊼ ♄
14 49	☽ # ♂
15 32	☽ ⊼ ☉
15 33	☽ ∠ ♇
21 50	☽ △ ♂
10 Tuesday	
01 09	☽ ± ♄
02 42	☽ ± ☉
05 23	☽ □ ♃
07 04	☽ ⊥ ♇
08 12	☉ ∠ ♀
16 44	☽ ꝑ ♄
17 21	☽ ⊻ ♇
19 10	☽ ☍ ♀
19 22	☽ ꝑ ☉
20 06	☽ □ ♆
11 Wednesday	
04 35	♀ ∥ ♄
05 27	☽ ⊼ ☿
05 46	☉ ∠ ♆
08 52	☽ ☌ ♅
10 37	♂ # ♅
12 32	☽ Q ♃
12 33	♀ □ ♆
12 51	♀ # ♇
16 09	☽ # ♆
16 14	☽ ± ☿
18 32	☽ △ ♄
22 55	☽ △ ☉
12 Thursday	
02 49	☽ Q ♆
03 44	☽ □ ♂
06 06	☽ # ♃
08 16	☽ ꝑ ☿
09 10	☽ ⚹ ♃
20 23	☽ ☌ ♇
23 13	☽ ⚹ ♆
13 Friday	
01 06	☽ ⊼ ♀
07 12	♀ Q ♂
10 45	☽ △ ☿
10 55	☽ ∠ ♃
11 44	☽ ⊻ ♅
11 50	☽ Q ♂
11 51	☽ ± ♀
21 52	☽ □ ♄
21 53	☽ ⊥ ♅
14 Saturday	
00 42	☽ ∠ ♆
02 21	☽ ⊥ ♃
03 57	☽ ꝑ ♀
05 45	☽ □ ☉
09 13	☽ ⚹ ♂
12 40	☽ ⊻ ♃
13 09	☽ ∠ ♅
16 01	☽ ⊥ ♆
22 05	☿ ⊼ ♅
23 17	☽ ⊻ ♇
15 Sunday	
02 14	☽ ⊻ ♆
04 43	☽ Q ♄
06 48	☽ △ ♀
09 30	☽ ⊥ ♇
12 02	☽ ∠ ♂
14 37	☽ ⚹ ♅
14 45	☽ Q ☉
14 54	☽ □ ☿
16 Monday	
00 50	☽ ∠ ♇
01 21	☽ ⚹ ♄
03 12	☽ ∥ ♃
04 11	☽ ⊥ ♂
11 04	☿ St R
11 05	☽ Q ♅
12 51	☽ ⚹ ☉
14 59	☽ ⊻ ♂
16 10	☽ ∥ ♆
16 24	☽ ☌ ♃
21 45	☽ Q ☿
17 Tuesday	
02 31	☽ ⚹ ♇
03 00	☉ ∥ ☿
03 15	☽ ∠ ♄
05 37	☽ ☌ ♆
12 48	☽ □ ♀
16 39	☽ ∠ ☉
17 57	☽ □ ♅
18 10	☽ ⚹ ☿
18 53	☽ ⊥ ♄
19 01	☿ ∠ ♂
23 11	☽ Q ♇
18 Wednesday	
04 02	♂ ⊻ ♃
05 21	☽ ⊻ ♄
05 31	☿ ⊼ ♅
09 27	☽ ⊥ ☉
10 07	☉ ꝑ ♅
19 23	♀ ∥ ♂
19 33	☽ ∠ ☿
20 43	☽ ⊻ ☉
20 48	☽ ⊻ ♃
21 29	☽ ☌ ♂
21 32	☽ Q ♀
21 49	☽ # ♅
22 02	☉ ⚹ ♃
19 Thursday	
06 17	♀ Q ♂
06 29	☽ □ ♇
07 25	☽ ⊥ ♃
09 47	☽ ⊻ ♆
10 44	☽ ⊥ ☿
13 11	☽ ∥ ♀
19 35	☽ ⚹ ♀
20 21	☽ ⊥ ♆
20 50	☽ ⊻ ☿
22 11	☽ △ ♅
22 50	☽ ∥ ♂
23 26	☽ ∠ ♃
20 Friday	
03 12	☉ ⊻ ♂
09 17	☿ ⊻ ♀
10 26	☽ ☌ ♄
12 21	☽ ∠ ♆
23 32	☽ ∠ ♀
21 Saturday	
00 51	☽ ꝑ ♅
01 04	☽ ∥ ♂
02 30	☽ ⚹ ♃
05 30	☽ ⊻ ♂
06 12	☽ ☌ ☉
11 49	☿ Q ♃
11 52	☽ △ ♇
15 24	☽ ⚹ ♆
16 33	☽ ⊥ ♀
16 58	☽ ⊥ ♂
17 07	☽ ± ♅
23 31	☽ ☌ ☿
22 Sunday	
00 33	☽ Q ♃
04 02	☽ ⊻ ♀
04 04	☽ ⊼ ♅
04 50	♀ ☍ ♅
07 26	☽ ∥ ♀
09 53	☽ # ♅
10 24	☽ ∠ ♂
13 29	☽ Q ♆
15 22	☽ ꝑ ♇
17 28	☽ ⊻ ♄
21 47	♀ # ♅
23 Monday	
04 47	☽ ⊥ ♄
08 12	☽ ± ♇
10 25	☽ □ ♃
14 35	☿ ∥ ♄
16 07	☽ ⚹ ♂
18 31	☽ ⊻ ☉
19 33	☽ ⊼ ♇
19 37	☿ ⊥ ♂
22 00	☽ ∠ ♄
23 25	☽ □ ♆
24 Tuesday	
02 32	☿ ⊥ ♀
02 51	☿ # ♇
03 00	☽ ⊻ ☿
07 01	☽ ⊥ ☉
07 19	☉ △ ♇
12 30	☽ ☍ ♅
13 27	☽ ⊥ ☿
15 08	☽ ☌ ♀
16 30	☽ Q ♂
21 17	☽ ∥ ♆
22 36	☿ ± ♅
25 Wednesday	
02 03	☽ ∠ ☉
03 17	☽ ⚹ ♄
05 16	☽ ∠ ☿
09 32	☽ ∥ ♃
11 51	☿ ⚹ ♆
20 25	☉ ☌ ☿
21 15	☽ △ ♃
26 Thursday	
03 13	☽ Q ♄
06 05	☽ □ ♂
06 05	☽ ☍ ♇
06 09	♂ □ ♇
07 57	☽ ⚹ ☿
08 24	☉ ⚹ ♆
10 18	☽ △ ♆
10 28	☽ ⚹ ☉
18 14	☿ ⊻ ♂
23 41	☽ ⊼ ♅
27 Friday	
00 05	☉ ± ♅
00 34	☿ △ ♇
03 38	☽ ꝑ ♃
04 59	☽ ⊻ ♀
05 36	☽ Q ☿
11 46	☽ ± ♅
12 54	☽ Q ☉
15 46	☽ □ ♄
16 35	☽ ꝑ ♆
17 34	☽ ⊥ ♀
18 41	☿ ∥ ♂
22 04	☽ ± ♃
28 Saturday	
05 58	☽ ꝑ ♅
10 23	☽ ⊼ ♃
10 53	☽ ± ♆
12 29	☽ ∠ ♀
14 31	☽ □ ☿
18 38	☽ ⊼ ♇
22 28	☽ △ ♂
23 05	☽ ⊼ ♆
29 Sunday	
03 34	♂ ⊥ ♃
04 52	☽ □ ☉
05 23	☿ ∠ ♀
06 48	☽ ± ♇
09 15	♂ ⊻ ♆
12 19	☽ △ ♅
19 51	☽ ⚹ ♀
30 Monday	
00 58	☽ ꝑ ♇
04 48	☿ ⚹ ♃
05 08	☽ △ ♄
06 40	☽ ꝑ ♂
09 25	☽ # ♃
20 36	☽ Q ♀
20 36	☽ # ♆
21 59	☽ △ ☿
22 38	☉ ⊥ ♀
23 33	☽ ☍ ♃
31 Tuesday	
01 44	☽ ± ♂
06 52	☽ △ ♇
11 16	☽ ꝑ ♄
11 21	☽ ☍ ♆
14 16	☽ ⊼ ♂
21 52	♄ ∠ ♆
22 15	☽ △ ☉
23 50	☽ □ ♅

JANUARY 1958

Date	Sidereal time h m s	Sun ☉ ° ' "	Moon ☽ ° ' "	Moon ☽ 24.00 ° ' "	Mercury ☿ ° '	Venus ♀ ° '	Mars ♂ ° '	Jupiter ♃ ° '	Saturn ♄ ° '	Uranus ♅ ° '	Neptune ♆ ° '	Pluto ♇ ° '
01	18 42 31	10 ♑ 34 53	17 ♉ 06 56	23 ♉ 25 01	26 ♐ 26	15 ♒ 26	06 ♐ 33	28 ♎ 39	19 ♐ 30	10 ♌ 47	04 ♏ 26	02 ♍ 08
02	18 46 28	11 36 02	29 ♉ 48 33	06 ♊ 17 51	25 R 53	15 40	07 14	28 46	19 37	10 R 45	04 28	02 R 07
03	18 50 24	12 37 10	12 ♊ 53 07	19 ♊ 34 27	25 30	15 52	07 56	28 54	19 43	10 43	04 29	02 06
04	18 54 21	13 38 18	26 ♊ 21 45	03 ♋ 14 49	25 18	16 02	08 38	29 01	19 50	10 41	04 30	02 05
05	18 58 17	14 39 26	10 ♋ 13 17	17 ♋ 16 36	25 D 14	16 10	09 20	29 08	19 57	10 39	04 31	02 05
06	19 02 14	15 40 34	24 ♋ 24 10	01 ♌ 35 12	25 20	16 15	10 02	29 15	20 03	10 36	04 32	02 04
07	19 06 11	16 41 42	08 ♌ 48 52	16 ♌ 04 19	25 34	16 18	10 44	29 22	20 10	10 34	04 33	02 03
08	19 10 07	17 42 50	23 ♌ 20 40	00 ♍ 37 04	25 55	16 R 18	11 26	29 29	20 17	10 32	04 34	02 02
09	19 14 04	18 43 57	07 ♍ 52 45	15 ♍ 07 04	26 23	16 16	12 08	29 35	20 23	10 29	04 35	02 01
10	19 18 00	19 45 05	22 ♍ 19 24	29 ♍ 29 20	26 57	16 11	12 50	29 42	20 30	10 27	04 36	02 00
11	19 21 57	20 46 13	06 ♎ 36 33	13 ♎ 40 49	27 37	16 04	13 32	29 48	20 36	10 25	04 36	01 59
12	19 25 53	21 47 20	20 ♎ 42 02	27 ♎ 40 09	28 22	15 54	14 14	29 ♎ 54	20 42	10 22	04 37	01 58
13	19 29 50	22 48 28	04 ♏ 35 13	11 ♏ 27 18	29 ♐ 11	15 42	14 56	00 ♏ 00	20 49	10 20	04 38	01 56
14	19 33 46	23 49 35	18 ♏ 16 30	25 ♏ 02 54	00 ♑ 04	15 27	15 38	00 06	20 55	10 17	04 39	01 55
15	19 37 43	24 50 43	01 ♐ 46 36	08 ♐ 27 39	01 01	15 10	16 20	00 11	21 02	10 15	04 39	01 54
16	19 41 40	25 51 50	15 ♐ 06 07	21 ♐ 42 00	02 01	14 51	17 02	00 17	21 08	10 12	04 40	01 53
17	19 45 36	26 52 57	28 ♐ 15 15	04 ♑ 45 50	03 04	14 29	17 45	00 22	21 14	10 10	04 41	01 52
18	19 49 33	27 54 03	11 ♑ 13 40	17 ♑ 38 40	04 10	14 05	18 27	00 27	21 20	10 07	04 41	01 51
19	19 53 29	28 55 09	24 ♑ 00 45	00 ♒ 19 49	05 18	13 39	19 09	00 32	21 26	10 05	04 42	01 50
20	19 57 26	29 ♑ 56 15	06 ♒ 35 51	12 ♒ 48 49	06 28	13 11	19 52	00 37	21 32	10 02	04 43	01 48
21	20 01 22	00 ♒ 57 19	18 ♒ 58 44	25 ♒ 05 42	07 40	12 41	20 34	00 41	21 39	09 59	04 43	01 47
22	20 05 19	01 58 23	01 ♓ 09 50	07 ♓ 11 21	08 54	12 09	21 16	00 46	21 45	09 57	04 44	01 46
23	20 09 15	02 59 26	13 ♓ 10 31	19 ♓ 07 40	10 09	11 36	21 59	00 50	21 50	09 54	04 44	01 45
24	20 13 12	04 00 28	25 ♓ 03 12	00 ♈ 57 34	11 26	11 02	22 41	00 54	21 56	09 52	04 44	01 43
25	20 17 09	05 01 29	06 ♈ 51 17	12 ♈ 44 55	12 45	10 26	23 24	00 58	22 02	09 49	04 45	01 42
26	20 21 05	06 02 29	18 ♈ 39 05	24 ♈ 34 25	14 04	09 50	24 06	01 02	22 08	09 46	04 45	01 41
27	20 25 02	07 03 28	00 ♉ 31 38	06 ♉ 31 23	15 25	09 13	24 49	01 06	22 14	09 44	04 45	01 39
28	20 28 58	08 04 26	12 ♉ 34 24	18 ♉ 41 24	16 47	08 36	25 32	01 09	22 19	09 41	04 46	01 38
29	20 32 55	09 05 22	24 ♉ 53 02	01 ♊ 09 58	18 10	07 59	26 14	01 12	22 25	09 39	04 46	01 37
30	20 36 51	10 06 17	07 ♊ 32 46	14 ♊ 01 58	19 34	07 22	26 57	01 15	22 31	09 36	04 46	01 35
31	20 40 48	11 ♒ 07 12	20 ♊ 37 57	27 ♊ 21 00	20 ♑ 59	06 ♒ 46	27 ♐ 39	01 ♏ 18	22 ♐ 36	09 ♌ 33	04 ♏ 46	01 ♍ 34

Date	Moon True ☊ ° '	Moon Mean ☊ ° '	Moon ☽ Latitude ° '	DECLINATIONS Sun ☉ ° '	Moon ☽ ° '	Mercury ☿ ° '	Venus ♀ ° '	Mars ♂ ° '	Jupiter ♃ ° '	Saturn ♄ ° '	Uranus ♅ ° '	Neptune ♆ ° '	Pluto ♇ ° '
01	08 ♏ 40	07 ♏ 21	01 S 45	23 S 01	16 N 14	20 S 13	15 S 19	21 S 20	09 S 50	21 S 42	18 N 10	11 S 22	21 N 34
02	08 R 38	07 18	01 50	22 56	18 19	20 12	15 03	21 28	09 53	21 42	18 11	11 23	21 35
03	08 33	07 15	02 52	22 51	19 30	20 14	14 47	21 35	09 55	21 43	18 12	11 23	21 35
04	08 26	07 12	03 45	22 45	19 38	20 18	14 32	21 42	09 57	21 43	18 12	11 23	21 36
05	08 16	07 09	04 27	22 38	18 36	20 23	14 16	21 49	10 00	21 44	18 13	11 24	21 36
06	08 05	07 05	04 54	22 31	16 25	20 29	14 02	21 56	10 02	21 44	18 13	11 24	21 37
07	07 54	07 02	05 02	22 24	13 12	20 37	13 47	22 03	10 04	21 45	18 14	11 24	21 38
08	07 44	06 59	04 51	22 16	09 09	20 45	13 34	22 09	10 06	21 45	18 15	11 24	21 38
09	07 36	06 56	04 21	22 08	04 N 34	20 54	13 20	22 16	10 09	21 46	18 15	11 25	21 39
10	07 30	06 53	03 35	21 59	00 S 15	21 04	13 08	22 22	10 11	21 46	18 16	11 25	21 40
11	07 28	06 50	02 36	21 50	05 01	21 14	12 56	22 28	10 13	21 47	18 17	11 25	21 40
12	07 27	06 46	01 28	21 41	09 26	21 23	12 44	22 34	10 15	21 47	18 17	11 25	21 41
13	07 D 27	06 43	00 S 15	21 31	13 17	21 33	12 33	22 39	10 16	21 48	18 18	11 25	21 42
14	07 R 27	06 40	00 N 57	21 20	16 22	21 43	12 23	22 44	10 18	21 48	18 19	11 26	21 42
15	07 25	06 37	02 05	21 10	18 29	21 52	12 13	22 49	10 20	21 48	18 19	11 26	21 43
16	07 20	06 34	03 05	20 59	19 33	22 01	12 04	22 54	10 22	21 49	18 20	11 26	21 44
17	07 12	06 31	03 54	20 47	19 32	22 10	11 56	22 59	10 23	21 49	18 21	11 26	21 45
18	07 01	06 27	04 31	20 35	18 28	22 17	11 48	23 04	10 25	21 50	18 21	11 26	21 45
19	06 48	06 24	04 53	20 23	16 30	22 24	11 41	23 08	10 27	21 50	18 22	11 26	21 46
20	06 33	06 21	05 00	20 10	13 47	22 30	11 35	23 12	10 28	21 50	18 23	11 27	21 47
21	06 19	06 18	04 53	19 57	10 30	22 36	11 30	23 16	10 29	21 51	18 24	11 27	21 47
22	06 06	06 15	04 32	19 43	06 49	22 40	11 26	23 19	10 31	21 51	18 24	11 27	21 48
23	05 55	06 11	04 00	19 30	02 S 55	22 44	11 22	23 23	10 32	21 51	18 25	11 27	21 49
24	05 47	06 08	03 17	19 15	01 N 05	22 46	11 19	23 26	10 33	21 52	18 26	11 27	21 49
25	05 43	06 05	02 26	19 01	04 57	22 47	11 17	23 29	10 34	21 52	18 27	11 27	21 50
26	05 40	06 02	01 29	18 46	08 41	22 48	11 15	23 32	10 36	21 52	18 27	11 27	21 51
27	05 D 40	05 59	00 N 27	18 31	12 05	22 47	11 14	23 34	10 37	21 53	18 28	11 27	21 52
28	05 R 39	05 56	00 S 37	18 15	15 02	22 45	11 14	23 37	10 38	21 53	18 29	11 27	21 52
29	05 39	05 52	01 40	17 59	17 22	22 42	11 15	23 39	10 38	21 53	18 29	11 27	21 53
30	05 37	05 49	02 40	17 43	18 56	22 37	11 16	23 41	10 39	21 54	18 30	11 27	21 53
31	05 ♏ 32	05 ♏ 46	03 S 34	17 S 26	19 N 33	22 S 32	11 S 18	23 S 42	10 S 40	21 S 54	18 N 31	11 S 27	21 N 54

ZODIAC SIGN ENTRIES

Date	h	m	Planets
02	12	21	☽ ♊
04	18	22	☽ ♋
06	21	21	☽ ♌
08	22	59	☽ ♍
11	00	52	☽ ♎
13	04	02	☽ ♏
13	12	52	♃ ♏
14	10	03	☿ ♑
15	08	49	☽ ♐
17	15	13	☽ ♑
19	23	22	☽ ♒
20	13	28	☉ ♒
22	09	41	☽ ♓
24	22	03	☽ ♈
27	10	56	☽ ♉
29	21	47	☽ ♊

LATITUDES

Date	Mercury ☿ ° '	Venus ♀ ° '	Mars ♂ ° '	Jupiter ♃ ° '	Saturn ♄ ° '	Uranus ♅ ° '	Neptune ♆ ° '	Pluto ♇ ° '
01	03 N 11	00 N 56	00 N 04	01 N 14	01 N 20	00 N 40	01 N 44	11 N 40
04	03 04	01 35	00 N 02	01 15	01 20	00 40	01 44	11 41
07	02 46	02 16	00 00	01 16	01 20	00 40	01 44	11 42
10	02 21	03 00	00 S 02	01 16	01 20	00 40	01 44	11 43
13	01 53	03 45	00 04	01 17	01 20	00 40	01 45	11 44
16	01 24	04 31	00 06	01 17	01 20	00 40	01 45	11 44
19	00 56	05 15	00 08	01 18	01 20	00 40	01 45	11 45
22	00 29	05 57	00 10	01 18	01 20	00 40	01 45	11 46
25	00 N 03	06 35	00 12	01 19	01 20	00 40	01 45	11 47
28	00 S 22	07 07	00 15	01 20	01 20	00 41	01 46	11 47
31	00 S 44	07 N 31	00 S 17	01 N 21	01 N 20	00 N 41	01 N 46	11 N 48

DATA

Julian Date	2436205
Delta T	+32 seconds
Ayanamsa	23° 16' 24"
Synetic vernal point	05° ♓ 50' 35"
True obliquity of ecliptic	23° 26' 33"

MOON'S PHASES, APSIDES AND POSITIONS ☽

Date	h	m	Phase	Longitude ° '	Eclipse Indicator
05	20	09	○	15 ♋ 00	
12	14	01	☾	21 ♎ 52	
19	22	08	●	29 ♑ 21	
28	02	16	☽	07 ♉ 40	

Day	h	m		
08	23	54	Perigee	
25	00	32	Apogee	
04	02	50	Max dec	19° N 43'
10	10	46	0S	
16	23	34	Max dec	19° S 40'
24	05	39	0N	
31	13	58	Max dec	19° N 33'

LONGITUDES

Date	Chiron ⚷ ° '	Ceres ⚳ ° '	Pallas ⚴ ° '	Juno ⚵ ° '	Vesta ⚶ ° '	Black Moon Lilith ⚸ ° '
01	14 ♒ 52	25 ♋ 53	22 ♈ 50	08 ♊ 15	11 ♈ 48	24 ♓ 32
11	15 ♒ 33	23 ♋ 37	24 ♈ 44	07 ♊ 54	14 ♈ 23	25 ♓ 39
21	16 ♒ 16	21 ♋ 16	27 ♈ 21	08 ♊ 26	17 ♈ 20	26 ♓ 46
31	17 ♒ 01	19 ♋ 16	00 ♉ 34	09 ♊ 49	20 ♈ 35	27 ♓ 53

All ephemeris data is given at 12.00 UT and the Moon's longitude is additionally given for 24.00 UT

ASPECTARIAN

h m Aspects

01 Wednesday
01 37 ☽ ⚼ ☿
04 21 ☽ # ♀
04 58 ☽ ± ♄
08 42 ☽ □ ♀
13 50 ☉ Q ♃
16 36 ☽ ⊼ ♄
16 47 ☉ ⊼ ♅
18 03 ☽ ± ☿

02 Thursday
04 56 ☽ ⊼ ☿
05 29 ☽ ⚼ ☉
10 02 ☽ Q ♅
10 03 ☽ ⊼ ♃
10 07 ☽ ∥ ♅
16 17 ☽ □ ♇
19 50 ☿ ⚼ ♅
20 38 ☽ ⊼ ♆
21 17 ☽ ± ♃
23 38 ☽ ± ☉

03 Friday
02 30 ☽ ☍ ♂
07 38 ☽ ± ♆
08 05 ☽ ✱ ♅
11 29 ☽ ⊼ ☉
12 40 ♂ # ♇
13 51 ☽ ⚼ ♃
17 27 ☽ △ ♀
23 51 ☽ ⚼ ♆

04 Saturday
00 22 ☽ ☍ ♄
00 56 ☽ Q ♇
10 08 ☽ ☍ ☿
10 48 ☽ ∠ ♅
15 53 ♂ ∥ ♄
16 41 ☽ △ ♃
20 15 ☽ ⚼ ♀
21 59 ☽ ✱ ♇

05 Sunday
02 11 ☽ △ ♆
02 27 ☽ ⊥ ♅
08 40 ☿ St D
10 24 ☽ ⊼ ♂
11 54 ☽ ± ♀
12 43 ☽ ⊻ ♅
17 22 ☽ ∥ ♅
20 09 ☽ ☍ ☉
21 09 ☽ ± ♂
22 11 ☽ ⊼ ♀
23 39 ☽ ∠ ♇

06 Monday
04 38 ☽ ⊼ ♄
13 06 ☽ ⚼ ♂
13 34 ☽ ⊼ ☿
14 46 ☽ ⊥ ♇
14 47 ☽ ± ♄
20 11 ☽ □ ♃
23 44 ☽ ± ☿

07 Tuesday
00 46 ☽ ⊻ ♇
02 11 ☉ ⊻ ♀
04 55 ☽ □ ♆
05 40 ♂ ⊥ ♆
05 54 ☽ ⚼ ♄
06 49 ♂ △ ♅
07 45 ☽ # ♀
08 27 ☉ Q ♆
12 32 ☿ ⚼ ♅
14 24 ☉ ⊥ ♂
14 54 ☽ ☌ ♅
14 57 ☽ ⚼ ☿
15 20 ☽ △ ♂
20 06 ☉ ⚼ ♇
23 08 ☽ # ♆

08 Wednesday
00 23 ☽ ☍ ♀
02 01 ☽ ⊼ ☉
02 15 ☽ Q ♃
02 46 ♀ St R
06 42 ☽ # ♃
06 54 ☽ △ ♄
10 43 ☽ Q ♆
12 39 ☽ ± ☉
16 22 ☽ △ ☿
22 12 ☽ ✱ ♃
23 09 ☉ ∥ ♂

09 Thursday
02 19 ☽ ☌ ♇
04 38 ☽ ⚼ ☉
06 32 ☽ ✱ ♆
16 19 ☽ ⊻ ♅
19 24 ☽ □ ♂
23 12 ☽ ∠ ♃

10 Friday
01 50 ☽ ⊼ ♀
02 14 ☽ ⊥ ♅
07 23 ☽ △ ☉
07 26 ☽ ∠ ♆
08 55 ☽ □ ♄
11 46 ☽ ± ♀
14 18 ☽ ⊥ ♃
17 13 ☽ ∠ ♅
20 06 ☽ □ ☿
22 30 ☽ ⊥ ♆

11 Saturday
00 26 ☽ ⊻ ♃
02 44 ☽ ⚼ ♀
02 59 ☽ Q ♂
04 12 ☽ ⊻ ♇
07 33 ☉ ⊻ ♄
08 37 ☽ ⊻ ♆
14 19 ☽ ⊥ ♇
15 24 ☽ Q ♄
18 25 ☽ ✱ ♅
20 39 ☉ ∥ ♄

12 Sunday
00 21 ☽ ✱ ♂
03 53 ☽ △ ♀
04 09 ☽ Q ☿
05 36 ☽ ∠ ♇
11 08 ☉ # ♇
12 01 ☽ ✱ ♄
14 01 ☽ □ ☉
14 51 ☽ Q ♅
16 44 ☽ ∥ ♃
23 50 ☽ ∥ ♆

13 Monday
02 01 ☽ ✱ ☿
03 29 ☽ ∠ ♂
03 58 ☽ ☌ ♃
07 15 ☽ ∥ ♀
07 24 ☽ ✱ ♇
09 00 ☉ ∥ ☿
12 05 ☽ ☌ ♆
14 09 ☽ ∠ ♄
14 36 ♂ ∠ ♃
19 59 ☽ ⊥ ♂
22 00 ☽ □ ♅
23 44 ☽ Q ☉

14 Tuesday
04 21 ☽ Q ♇
05 57 ☽ ∠ ☿
06 03 ☽ ⊥ ♄
07 05 ☽ ⊻ ♂
07 07 ☽ □ ♀
07 29 ♀ ✱ ♂
10 39 ☿ # ♇
12 32 ☿ ✱ ♃
16 43 ☽ ⊻ ♄
19 33 ☿ ∠ ♀
22 38 ☽ ✱ ☉
23 02 ☽ ⊥ ☿

15 Wednesday
01 24 ☿ ∥ ♄
09 08 ☽ ⊻ ♃
09 46 ☽ # ♅
10 33 ☽ ⊻ ☿
12 14 ☽ □ ♇
14 26 ☽ Q ♀
17 10 ☽ ⊻ ♆
19 58 ☽ ⊥ ♃

16 Thursday
03 10 ☽ △ ♅
03 42 ☽ ∠ ☉
03 59 ☽ ⊥ ♆
08 48 ☿ △ ♇
11 32 ☽ ✱ ♀
12 19 ☽ ∠ ♃
12 30 ☉ ± ♇
15 43 ☽ ☌ ♂
20 18 ☽ ∠ ♆
21 23 ☽ ⊥ ☉
23 03 ☽ ☌ ♄

17 Friday
06 21 ☽ ⚼ ♅
09 16 ☽ ⊻ ☉
14 11 ☽ ∠ ♀
15 55 ☽ ✱ ♃
18 38 ☽ △ ♇
21 14 ☉ ⊥ ♄
21 41 ☽ ☌ ☿
22 51 ☽ ± ♅
23 51 ☽ ✱ ♆

18 Saturday
06 20 ☽ ⊥ ♀
09 57 ☽ ⊼ ♅
11 01 ☿ ± ♅
13 44 ☽ # ♅
14 18 ☽ Q ♃
17 09 ☽ ⊻ ♀
22 13 ☽ Q ♆
22 29 ☽ ⚼ ♇
23 18 ☿ ✱ ♆

19 Sunday
02 18 ☽ ⊻ ♂
07 06 ☽ ⊻ ♄
14 17 ☽ ⊥ ♂
15 26 ☽ ± ♇
18 33 ☽ ⊥ ♄
22 08 ☽ ☌ ☉

20 Monday
00 28 ☽ □ ♃
02 50 ☽ ⊼ ♇
06 48 ♂ ∠ ♆
08 22 ☽ □ ♆
08 28 ☽ ∠ ♂
11 43 ☽ ⊻ ☿
11 53 ☽ ∠ ♄
18 36 ☽ ☍ ♅
22 06 ☿ ⊥ ♀

21 Tuesday
00 13 ☽ ☌ ♀
00 32 ☽ ⊥ ☿
04 49 ☽ ∥ ♀
05 16 ☉ □ ♃
05 24 ☽ ∥ ♆
12 02 ☽ ∥ ♃
15 18 ☽ ✱ ♂
17 15 ☽ ✱ ♄
20 02 ☽ ∠ ☿

22 Wednesday
05 31 ♀ ∥ ♆
07 10 ☉ ⊼ ♇
11 12 ☽ △ ♃
13 11 ☽ ☍ ♇
13 45 ☽ ⊻ ☉
16 27 ☽ Q ♂
17 10 ☽ Q ♄
19 05 ☽ △ ♆

23 Thursday
02 49 ☽ ⊥ ☉
05 13 ☽ ✱ ☿
05 27 ☽ ⊼ ♅
06 28 ♂ ☌ ♄
07 25 ☿ ⊼ ♅
08 59 ☽ ⊻ ♀
17 23 ☽ ⚼ ♃
17 28 ☽ ± ♅
20 30 ☽ ⊥ ♀
22 37 ☽ ∠ ☉

24 Friday
01 14 ☽ ⚼ ♆
05 38 ☽ □ ♄
06 44 ☿ ⊻ ♀
06 54 ☽ □ ♂
08 19 ☽ Q ☿
11 37 ☽ ⚼ ♅
11 42 ☽ ± ♃
13 53 ☽ ∠ ♀
19 30 ☽ ± ♆
23 58 ☽ ⊼ ♃

25 Saturday
01 31 ☽ ⊼ ♇
05 25 ☉ □ ♆
07 42 ☽ ⊼ ♆
07 55 ☽ ✱ ☉
13 43 ☽ ± ♇
16 24 ☿ Q ♃
18 01 ☽ △ ♅
18 57 ☽ ✱ ♀

26 Sunday
01 30 ☽ □ ☿
08 00 ☽ ⚼ ♇
10 39 ☽ Q ☉
14 33 ♀ ☍ ♅
18 08 ☽ Q ♀
19 07 ☽ △ ♄
23 45 ☽ △ ♂

27 Monday
01 14 ☽ # ♃
01 14 ♀ ∠ ♂
05 49 ☽ # ♀
07 21 ☽ # ♆
09 18 ♂ ⚼ ♅
13 09 ☽ ☍ ♃
14 16 ☽ △ ♇
15 58 ☉ # ♅
16 26 ☉ ∠ ♄
20 29 ☽ ☍ ♆

28 Tuesday
01 31 ☽ ⚼ ♄
02 16 ☽ □ ☉
04 31 ☽ □ ♀
06 19 ☽ □ ♅
07 42 ☽ ⚼ ♂
09 27 ☿ ⚼ ♇
11 41 ☿ Q ♆
19 26 ☽ ± ♄
19 47 ☉ ☌ ♀
21 19 ☽ △ ☿

29 Wednesday
02 28 ☽ ± ♂
07 12 ☽ ⊼ ♄
14 45 ☽ ⊼ ♂
17 16 ☽ Q ♅
18 53 ☽ # ☉

30 Thursday
00 08 ☽ ⊼ ♃
00 32 ☉ ☍ ♅
00 49 ☽ □ ♇
03 44 ☽ ∥ ♅
04 28 ♄ # ♇
05 43 ☽ ⚼ ☿
06 48 ☽ ⊼ ♆
07 16 ♀ ∠ ♄
11 28 ☽ ± ♃
11 41 ☽ △ ♀
15 48 ☽ ✱ ♅
17 10 ☽ △ ☉
17 59 ☽ ± ♆

31 Friday
00 29 ☽ ± ☿
04 08 ☽ ⚼ ♃
10 04 ☽ Q ♇
10 27 ☽ ⚼ ♆
12 42 ☽ ⊼ ☿
13 57 ☽ ⚼ ♀
15 34 ☽ ☍ ♄
19 01 ☽ ∠ ♅
22 37 ☽ ⚼ ☉

FEBRUARY 1958

	Sidereal time	LONGITUDES Sun ☉	Moon ☽	Moon ☽ 24.00	Mercury ☿	Venus ♀	Mars ♂	Jupiter ♃	Saturn ♄	Uranus ♅	Neptune ♆	Pluto ♇
Date	h m s	° ' "	° ' "	° ' "	° '	° '	° '	° '	° '	° '	° '	° '
01	20 44 44	12 ♒ 08 04	04 ♋ 11 13	11 ♋ 08 33	22 ♑ 25	06 ♒ 10	28 ♐ 22	01 ♏ 21	22 ♐ 41	09 ♌ 31	04 ♏ 47	01 ♍ 32
02	20 48 41	13 08 56	18 ♋ 12 45	25 ♋ 23 21	23 52	05 R 35	29 05	01 24	22 47	09 R 28	04 47	01 R 31
03	20 52 38	14 09 46	02 ♌ 39 41	10 ♌ 00 55	25 19	05 01	29 ♐ 48	01 26	22 52	09 25	04 47	01 30
04	20 56 34	15 10 35	17 ♌ 26 02	24 ♌ 53 54	26 48	04 29	00 ♑ 30	01 28	22 57	09 23	04 47	01 28
05	21 00 31	16 11 23	02 ♍ 23 21	09 ♍ 53 08	28 17	03 59	01 13	01 30	23 03	09 20	04 R 47	01 27
06	21 04 27	17 12 10	17 ♍ 22 07	24 ♍ 49 12	29 ♑ 47	03 30	01 56	01 32	23 08	09 18	04 47	01 25
07	21 08 24	18 12 56	02 ♎ 13 28	09 ♎ 34 06	01 ♒ 18	03 03	02 39	01 34	23 13	09 15	04 47	01 24
08	21 12 20	19 13 40	16 ♎ 50 31	24 ♎ 02 15	02 50	02 38	03 22	01 35	23 18	09 12	04 47	01 22
09	21 16 17	20 14 24	01 ♏ 09 04	08 ♏ 10 49	04 23	02 16	04 05	01 36	23 23	09 10	04 46	01 21
10	21 20 13	21 15 07	15 ♏ 07 31	21 ♏ 59 17	05 56	01 55	04 48	01 37	23 27	09 07	04 46	01 19
11	21 24 10	22 15 49	28 ♏ 46 18	05 ♐ 28 48	07 31	01 37	05 31	01 38	23 32	09 05	04 46	01 18
12	21 28 07	23 16 29	12 ♐ 07 03	18 ♐ 41 20	09 06	01 22	06 14	01 39	23 37	09 02	04 46	01 16
13	21 32 03	24 17 09	25 ♐ 11 56	01 ♑ 39 05	10 42	01 09	06 57	01 39	23 41	09 00	04 46	01 15
14	21 36 00	25 17 47	08 ♑ 03 04	14 ♑ 24 03	12 19	00 58	07 40	01 40	23 46	08 57	04 45	01 13
15	21 39 56	26 18 25	20 ♑ 42 14	26 ♑ 57 45	13 57	00 50	08 23	01 40	23 51	08 55	04 45	01 12
16	21 43 53	27 19 01	03 ♒ 10 45	09 ♒ 21 20	15 35	00 45	09 06	01 R 40	23 55	08 52	04 45	01 10
17	21 47 49	28 19 35	15 ♒ 29 34	21 ♒ 35 35	17 15	00 42	09 50	01 40	23 59	08 50	04 44	01 09
18	21 51 46	29 ♒ 20 08	27 ♒ 39 27	03 ♓ 41 17	18 55	00 D 41	10 33	01 39	24 03	08 47	04 44	01 07
19	21 55 42	00 ♓ 20 39	09 ♓ 41 14	15 ♓ 39 25	20 36	00 43	11 16	01 38	24 08	08 45	04 43	01 06
20	21 59 39	01 21 09	21 ♓ 36 05	27 ♓ 31 26	22 18	00 47	11 59	01 38	24 12	08 43	04 43	01 04
21	22 03 36	02 21 37	03 ♈ 25 45	09 ♈ 19 24	24 02	00 53	12 43	01 37	24 16	08 40	04 42	01 03
22	22 07 32	03 22 03	15 ♈ 12 44	21 ♈ 06 12	25 46	01 02	13 26	01 35	24 20	08 38	04 42	01 01
23	22 11 29	04 22 28	27 ♈ 00 16	02 ♉ 55 28	27 31	01 13	14 09	01 34	24 24	08 36	04 41	01 00
24	22 15 25	05 22 51	08 ♉ 52 23	14 ♉ 51 36	29 ♒ 17	01 26	14 53	01 32	24 27	08 33	04 41	00 58
25	22 19 22	06 23 11	20 ♉ 53 45	26 ♉ 59 30	01 ♓ 04	01 41	15 36	01 31	24 31	08 31	04 40	00 57
26	22 23 18	07 23 30	03 ♊ 09 30	09 ♊ 24 24	02 52	01 58	16 19	01 29	24 35	08 29	04 39	00 55
27	22 27 15	08 23 47	15 ♊ 44 50	22 ♊ 11 22	04 41	02 17	17 03	01 27	24 38	08 27	04 38	00 54
28	22 31 11	09 ♓ 24 02	28 ♊ 44 32	05 ♋ 24 46	06 ♓ 31	02 ♒ 38	17 ♑ 46	01 ♏ 24	24 ♐ 41	08 ♌ 25	04 ♏ 38	00 ♍ 52

	Moon True ☊	Moon Mean ☊	Moon ☽ Latitude
Date	° '	° '	° '
01	05 ♏ 24	05 ♏ 43	04 S 18
02	05 R 14	05 40	04 48
03	05 03	05 36	05 00
04	04 51	05 33	04 54
05	04 40	05 30	04 27
06	04 31	05 27	03 41
07	04 25	05 24	02 41
08	04 22	05 21	01 31
09	04 21	05 17	00 S 17
10	04 D 21	05 14	00 N 57
11	04 R 21	05 11	02 06
12	04 19	05 08	03 06
13	04 15	05 05	03 56
14	04 09	05 02	04 32
15	03 59	04 58	04 54
16	03 48	04 55	05 02
17	03 35	04 52	04 56
18	03 22	04 49	04 36
19	03 11	04 46	04 04
20	03 01	04 42	03 22
21	02 54	04 39	02 30
22	02 50	04 36	01 33
23	02 48	04 33	00 N 31
24	02 D 48	04 30	00 S 33
25	02 49	04 27	01 36
26	02 50	04 23	02 36
27	02 R 49	04 20	03 30
28	02 ♏ 47	04 ♏ 17	04 S 15

DECLINATIONS

Date	Sun ☉	Moon ☽	Mercury ☿	Venus ♀	Mars ♂	Jupiter ♃	Saturn ♄	Uranus ♅	Neptune ♆	Pluto ♇
	° '	° '	° '	° '	° '	° '	° '	° '	° '	° '
01	17 S 10	19 N 05	22 S 25	11 S 21	23 S 44	10 S 41	21 S 54	18 N 32	11 S 27	21 N 55
02	16 52	17 27	22 17	11 23	23 45	10 41	21 54	18 32	11 27	21 56
03	16 35	14 41	22 07	11 27	23 46	10 42	21 55	18 33	11 27	21 57
04	16 17	10 57	21 57	11 31	23 46	10 43	21 55	18 34	11 27	21 57
05	15 59	06 28	21 45	11 35	23 47	10 43	21 55	18 34	11 27	21 58
06	15 41	01 N 35	21 31	11 40	23 47	10 44	21 55	18 35	11 27	21 59
07	15 22	03 S 21	21 17	11 45	23 47	10 44	21 56	18 36	11 27	21 59
08	15 04	08 02	21 01	11 50	23 47	10 44	21 56	18 36	11 27	22 00
09	14 44	12 08	20 44	11 55	23 47	10 44	21 56	18 37	11 27	22 01
10	14 25	15 28	20 25	12 01	23 46	10 45	21 56	18 38	11 27	22 01
11	14 06	17 50	20 05	12 07	23 45	10 45	21 56	18 38	11 27	22 02
12	13 46	19 10	19 44	12 13	23 44	10 45	21 56	18 39	11 26	22 03
13	13 26	19 26	19 21	12 19	23 43	10 45	21 57	18 40	11 26	22 03
14	13 05	18 40	18 57	12 25	23 41	10 44	21 57	18 40	11 26	22 04
15	12 45	17 00	18 32	12 31	23 39	10 44	21 57	18 41	11 26	22 05
16	12 24	14 33	18 05	12 37	23 37	10 44	21 57	18 42	11 26	22 05
17	12 03	11 29	17 37	12 43	23 35	10 44	21 57	18 42	11 26	22 06
18	11 42	07 58	17 07	12 48	23 33	10 43	21 57	18 43	11 25	22 07
19	11 21	04 10	16 37	12 54	23 30	10 43	21 58	18 44	11 25	22 07
20	11 00	00 S 15	16 04	13 00	23 27	10 42	21 58	18 44	11 25	22 08
21	10 38	03 N 40	15 31	13 05	23 24	10 42	21 58	18 45	11 25	22 08
22	10 16	07 25	14 56	13 10	23 20	10 41	21 58	18 45	11 24	22 09
23	09 54	10 53	14 19	13 15	23 17	10 41	21 58	18 46	11 24	22 10
24	09 32	13 57	13 41	13 20	23 13	10 40	21 58	18 47	11 24	22 10
25	09 10	16 27	13 02	13 24	23 09	10 39	21 58	18 47	11 24	22 11
26	08 48	18 15	12 22	13 28	23 04	10 38	21 58	18 48	11 23	22 12
27	08 25	19 12	11 40	13 32	23 00	10 37	21 58	18 48	11 23	22 12
28	08 S 03	19 N 11	10 S 57	13 S 35	22 S 55	10 S 36	21 S 58	18 N 49	11 S 23	22 N 13

ZODIAC SIGN ENTRIES

Date	h	m	Planets
01	04	41	☽ ♋
03	07	38	☽ ♌
03	18	57	♂ ♑
05	08	11	☽ ♍
06	15	21	☿ ♒
07	08	23	☽ ♎
09	10	03	☽ ♏
11	14	11	☽ ♐
13	20	55	☽ ♑
16	05	51	☽ ♒
18	16	39	☽ ♓
19	03	48	☉ ♓
21	05	02	☽ ♈
23	18	05	☽ ♉
24	21	44	☿ ♓
26	05	52	☽ ♊
28	14	17	☽ ♋

LATITUDES

Date	Mercury ☿	Venus ♀	Mars ♂	Jupiter ♃	Saturn ♄	Uranus ♅	Neptune ♆	Pluto ♇
	° '	° '	° '	° '	° '	° '	° '	° '
01	00 S 51	07 N 37	00 S 18	01 N 21	01 N 20	00 N 41	01 N 46	11 N 48
04	01 10	07 51	00 20	01 22	01 21	00 41	01 46	11 49
07	01 27	07 56	00 22	01 22	01 21	00 41	01 46	11 49
10	01 41	07 55	00 25	01 23	01 21	00 41	01 46	11 50
13	01 52	07 47	00 27	01 24	01 21	00 41	01 47	11 50
16	02 01	07 33	00 30	01 24	01 21	00 41	01 47	11 51
19	02 06	07 16	00 32	01 25	01 21	00 41	01 47	11 51
22	02 07	06 55	00 35	01 26	01 21	00 40	01 47	11 51
25	02 05	06 33	00 37	01 26	01 22	00 40	01 47	11 51
28	01 58	06 09	00 40	01 27	01 22	00 40	01 47	11 51
31	01 S 47	05 N 43	00 S 42	01 N 28	01 N 22	00 N 40	01 N 48	11 N 51

DATA	
Julian Date	2436236
Delta T	+32 seconds
Ayanamsa	23° 16' 29"
Synetic vernal point	05° ♓ 50' 30"
True obliquity of ecliptic	23° 26' 33"

MOON'S PHASES, APSIDES AND POSITIONS ☽

Date	h	m	Phase	Longitude ° '	Eclipse Indicator
04	08	05	○	15 ♌ 01	
10	23	34	☾	21 ♏ 44	
18	15	38	●	29 ♒ 29	
26	20	52	☽	07 ♊ 46	

Day	h	m		
05	22	53	Perigee	
21	15	17	Apogee	
06	19	41	0S	
13	05	57	Max dec	19° S 28'
20	13	29	0N	
27	23	42	Max dec	19° N 19'

LONGITUDES

Date	Chiron ⚷	Ceres ⚳	Pallas ⚴	Juno ⚵	Vesta ⚶	Black Moon Lilith ⚸
	° '	° '	° '	° '	° '	° '
01	17 ♒ 05	18 ♋ 56	00 ♉ 55	09 ♊ 59	20 ♈ 55	27 ♓ 59
11	17 ♒ 50	17 ♋ 19	04 ♉ 43	12 ♊ 08	24 ♈ 25	29 ♓ 06
21	18 ♒ 35	16 ♋ 21	08 ♉ 56	14 ♊ 52	28 ♈ 08	00 ♈ 13
31	19 ♒ 18	16 ♋ 07	13 ♉ 32	18 ♊ 05	01 ♉ 59	01 ♈ 20

ASPECTARIAN

h m	Aspects
01 Saturday	
01 14	☽ ☍ ♂
05 16	☽ ± ♀
07 02	☽ △ ♃
07 23	☽ ⚹ ♇
10 50	☽ ⊥ ♅
13 01	☽ △ ♆
15 18	☽ ⚻ ♀
15 39	☽ ± ☉
16 57	☿ ⚺ ♄
21 11	☽ ⚺ ♅
22 12	☽ ∥ ♅
02 Sunday	
02 46	☽ ⚻ ☉
09 09	☽ ∠ ♇
18 39	☽ # ☉
19 42	☽ ⚻ ♄
21 27	♀ ⊥ ♂
22 31	☽ ☍ ☿
03 Monday	
00 11	☽ ⊥ ♇
05 43	☽ ± ♄
07 03	☽ ⚻ ♂
09 59	☽ □ ♃
10 05	☽ ⚺ ♇
13 46	♀ ∥ ♆
14 47	☿ ± ♇
15 28	☽ □ ♆
15 44	☽ ☍ ♀
17 23	☽ ± ♂
20 34	☽ ⚼ ♄
22 50	♀ □ ♆
23 00	☽ ☌ ♅
04 Tuesday	
08 05	☽ ☍ ☉
08 44	☽ # ♀
08 44	☽ ⚼ ♂
09 02	☽ # ♆
10 50	☿ # ♇
11 32	♃ ⚹ ♇
13 20	☽ # ♃
15 17	☽ Q ♃
15 39	☿ ∥ ♄
20 36	☽ Q ♆
20 56	☽ △ ♄
05 Wednesday	
03 11	♆ St R
04 42	☽ ⚻ ☿
10 02	☽ △ ♂
10 29	☽ ☌ ♇
10 35	☽ ⚹ ♃
14 28	☽ ⚻ ♀
14 29	☉ ∠ ♂
15 22	☽ ± ☿
15 50	☽ ⚹ ♆
19 17	♂ △ ♇
21 56	♂ ⚹ ♃
23 05	☽ ⚺ ♅
23 46	☽ ± ♀
06 Thursday	
00 50	☿ ⊥ ♄
07 24	☽ ⚼ ☿
08 41	☽ ⊥ ♅
10 39	☽ ∠ ♃
11 43	☽ ⚻ ☉
13 46	☽ ⚼ ♀
15 52	☽ ∠ ♆
21 19	☽ □ ♄
22 05	☽ ± ☉
23 07	☽ ∠ ♅
07 Friday	
01 11	☽ ⊥ ♃
06 24	☽ ⊥ ♆
10 20	☽ △ ☿
10 39	☽ ⚺ ♇
10 55	☽ ⚺ ♃
12 44	☽ □ ♂
13 18	☽ △ ♀
13 23	☿ ⚻ ♇
13 44	☽ ⚼ ☉
16 04	☿ □ ♃
16 10	☽ ⚺ ♆
20 24	♀ ⚺ ♂
20 26	☽ ⊥ ♇
23 27	☽ ⚹ ♅
08 Saturday	
02 47	☽ Q ♄
07 02	♂ ± ♅
09 29	☿ ☌ ♀
11 13	☽ ∠ ♇
16 16	☽ △ ☉
19 15	☽ Q ♅
19 55	☽ Q ♂
22 49	☽ ⚹ ♄
09 Sunday	
03 18	☿ ⚺ ♂
03 21	☽ ∥ ♃
07 38	☽ ∥ ♆
10 35	☽ ∥ ♀
12 20	☽ ⚹ ♇
12 46	☽ ☌ ♃
13 50	☽ □ ♀
17 15	☽ ⚹ ♂
18 02	☿ □ ♆
18 10	☽ ☌ ♆
18 11	☽ □ ☿
10 Monday	
00 24	☽ ∠ ♄
01 39	☽ □ ♅
04 28	☽ ∥ ☉
08 53	☽ Q ♇
11 08	♂ ⚹ ♆
16 05	☽ ⊥ ♄
20 11	☽ Q ♀
20 36	☽ ∠ ♂
23 34	☽ □ ☉
11 Tuesday	
02 40	☽ ⚺ ♄
05 28	☽ Q ☿
10 46	♀ □ ♃
13 24	☽ ⊥ ♂
16 30	☽ □ ♇
16 59	☽ ⚹ ♀
17 07	☽ ⚺ ♃
22 43	☽ ⚺ ♆
12 Wednesday	
00 07	☽ # ♅
00 45	☽ ⚺ ♂
03 54	☽ ⊥ ♃
04 18	☿ ∠ ♄
05 47	☽ ⚹ ☿
06 26	☽ △ ♅
09 33	☽ ⊥ ♆
10 21	☽ Q ☉
11 06	☿ ☍ ♅
19 37	☽ ∠ ♀
20 16	☽ ∠ ♃
20 44	☉ ⚹ ♄
22 54	♀ ⚻ ♇
13 Thursday	
01 58	☽ ∠ ♆
05 12	☽ ∥ ☿
09 12	☽ ☌ ♄
09 47	☽ ⚼ ♅
10 10	☽ ⚹ ☉
11 54	☽ ⊥ ♀
13 04	☽ ∠ ☿
22 54	☽ ⚺ ♀
23 13	☽ △ ♇
14 Friday	
00 01	☽ ⚹ ♃
01 03	☽ ∥ ☿
02 28	☽ ± ♅
05 49	☽ ⚹ ♆
08 16	☽ ⊥ ☿
11 14	☽ ☌ ♂
11 59	☽ # ♅
13 42	☽ ⚻ ♅
16 36	☽ ∠ ☉
21 14	☽ ⚺ ☿
22 36	☽ Q ♃
15 Saturday	
03 26	☽ ⚼ ♇
03 42	☿ # ♅
04 28	☽ Q ♆
11 11	☽ ⊥ ☉
14 54	♃ St R
18 03	☽ ⚺ ♄
20 35	☽ ± ♇
23 34	☿ ⊥ ♂
23 41	☽ ⚺ ☉
16 Sunday	
00 54	☉ ∥ ♀
02 53	♄ ⚼ ♅
04 37	♂ ⚻ ♅
05 39	☽ ⊥ ♄
07 19	☽ ☌ ♀
08 08	☽ ⚻ ♇
09 04	☽ □ ♃
15 02	☽ □ ♆
23 01	☽ ☍ ♅
23 13	☽ ∠ ♄
17 Monday	
00 14	☽ ⚺ ♂
03 07	☽ ∥ ♀
07 14	☽ ∥ ☉
12 22	☽ ∥ ♆
12 42	☽ ⊥ ♂
15 59	☽ ☌ ☿
17 20	☽ ∥ ♃
18 Tuesday	
04 50	☽ ⚹ ♄
06 17	♀ St D
07 33	☽ ∠ ♂
15 38	☽ ☌ ☉
18 01	☽ ⚺ ♀
18 52	☽ ☍ ♇
19 56	☽ △ ♃
19 Wednesday	
02 04	☽ △ ♆
04 50	☽ Q ♄
06 01	☽ ⊥ ♀
07 27	☉ ∥ ♆
10 08	☽ ⚻ ♅
15 22	☽ ⚹ ♂
21 20	☉ ⚺ ♀
22 09	☽ ± ♅
20 Thursday	
00 11	☽ ∠ ♀
01 58	☽ ⚼ ♃
05 28	☉ ☍ ♇
08 11	☽ ⚼ ♆
13 40	☽ ⚺ ☿
16 15	☽ ⚼ ♅
17 09	☽ Q ♂
17 17	☽ □ ♄
18 26	☉ △ ♃
20 08	☽ ± ♃
21 Friday	
02 24	☽ ± ♆
03 54	☽ ⊥ ☿
06 47	☽ ⚹ ♀
07 10	☽ ⚻ ♇
07 40	☉ ∥ ♃
08 18	☽ ⚻ ♃
09 37	☽ ⚺ ☉
14 36	☽ ⚻ ♆
15 24	☿ ⚹ ♄
19 21	☽ ± ♇
22 38	☽ △ ♅
22 59	☽ ⊥ ☉
22 Saturday	
01 22	☽ ∠ ☿
07 30	☽ Q ♀
08 08	☽ □ ♂
10 05	♀ ⚻ ♇
13 39	☽ ⚼ ♇
17 07	♂ Q ♃
19 02	☽ ∠ ☉
23 Sunday	
05 38	☽ # ☉
06 40	☽ △ ♄
10 27	☽ # ♃
13 13	☽ ⚹ ☿
15 47	☽ # ♆
19 21	☉ △ ♆
20 05	☽ △ ♇
20 42	☽ □ ♀
21 14	☽ ☍ ♃
24 Monday	
03 33	☽ ☍ ♆
04 19	☽ ⚹ ☉
06 40	☽ # ♀
10 15	☽ # ☿
11 22	☽ □ ♅
13 11	☽ ⚼ ♄
17 41	☽ Q ☿
21 22	♀ □ ♃
25 Tuesday	
00 11	☿ ∥ ♀
00 48	☽ △ ♂
01 34	☿ ∠ ♂
06 34	☽ Q ☉
07 15	☽ ± ♄
10 27	☿ ☍ ♇
15 17	☉ Q ♄
17 53	☿ △ ♃
19 11	☽ ⚻ ♄
21 51	☿ ⚺ ♀
23 03	☽ Q ♅
23 05	♂ ⚼ ♇
26 Wednesday	
07 40	☽ □ ♇
08 13	☽ ⚼ ♂
08 45	☽ ⚻ ♃
09 39	☽ △ ♀
11 20	☽ □ ☿
14 53	☽ ⚻ ♆
20 17	☽ ± ♃
20 52	☽ □ ☉
22 12	☽ ⚹ ♅
22 47	♂ Q ♆
23 07	☽ ∥ ♅
27 Thursday	
02 22	☽ ± ♆
02 35	☽ ± ♂
08 12	☉ ⊥ ♀
11 31	☿ △ ♆
13 10	☉ ⚻ ♅
13 18	☽ ⚼ ♃
14 35	☽ ⚻ ♂
14 58	☽ ⚼ ♀
17 53	☽ Q ♇
19 16	☽ ⚼ ♆
21 34	☿ ∥ ♆
28 Friday	
02 17	☽ ∠ ♅
04 34	☽ ☍ ♄
08 04	☽ ± ♀
14 25	☿ Q ♄
15 51	☽ ⚹ ♇
16 48	☽ △ ♃
18 37	☽ ⊥ ♅
19 14	☽ ⚻ ♀
22 35	☽ △ ♆
23 08	☽ ∥ ♅
23 30	☿ ∥ ♃

All ephemeris data is given at 12.00 UT and the Moon's longitude is additionally given for 24.00 UT

MARCH 1958

LONGITUDES

Date	Sidereal time h m s	Sun ☉ ° ' "	Moon ☽ ° ' "	Moon ☽ 24.00 ° ' "	Mercury ☿ ° '	Venus ♀ ° '	Mars ♂ ° '	Jupiter ♃ ° '	Saturn ♄ ° '	Uranus ♅ ° '	Neptune ♆ ° '	Pluto ♇ ° '
01	22 35 08	10 ♓ 24 15	12 ♋ 12 20	19 ♋ 07 25	08 ♓ 22	03 ♒ 01	18 ♑ 30	01 ♏ 22	24 ♐ 45	08 ♌ 23	04 ♏ 37	00 ♍ 51
02	22 39 05	11 24 26	26 ♋ 09 58	03 ♌ 19 47	10 14	03 26	19 13	01 R 19	24 48	08 R 20	04 R 36	00 R 49
03	22 43 01	12 24 34	10 ♌ 36 24	17 ♌ 59 11	12 06	03 52	19 57	01 16	24 51	08 18	04 35	00 48
04	22 46 58	13 24 41	25 ♌ 27 14	02 ♍ 59 31	14 00	04 20	20 40	01 13	24 54	08 16	04 34	00 46
05	22 50 54	14 24 46	10 ♍ 34 50	18 ♍ 11 52	15 55	04 49	21 24	01 10	24 57	08 14	04 34	00 45
06	22 54 51	15 24 49	25 ♍ 49 16	03 ♎ 25 44	17 50	05 20	22 08	01 06	25 00	08 12	04 33	00 43
07	22 58 47	16 24 50	10 ♎ 59 59	18 ♎ 30 55	19 46	05 53	22 51	01 03	25 03	08 11	04 32	00 42
08	23 02 44	17 24 49	25 ♎ 57 33	03 ♏ 19 08	21 43	06 27	23 35	00 59	25 06	08 09	04 31	00 40
09	23 06 40	18 24 47	10 ♏ 35 03	17 ♏ 44 57	23 40	07 02	24 18	00 55	25 08	08 07	04 30	00 39
10	23 10 37	19 24 43	24 ♏ 48 34	01 ♐ 45 54	25 38	07 39	25 02	00 51	25 11	08 05	04 29	00 38
11	23 14 34	20 24 37	08 ♐ 36 59	15 ♐ 22 02	27 35	08 17	25 46	00 47	25 13	08 03	04 28	00 36
12	23 18 30	21 24 30	22 ♐ 01 19	28 ♐ 35 10	29 ♓ 33	08 56	26 30	00 43	25 16	08 02	04 27	00 35
13	23 22 27	22 24 21	05 ♑ 03 58	11 ♑ 28 06	01 ♈ 30	09 36	27 13	00 38	25 18	08 00	04 26	00 33
14	23 26 23	23 24 11	17 ♑ 47 58	24 ♑ 03 59	03 27	10 17	27 57	00 33	25 20	07 58	04 24	00 32
15	23 30 20	24 23 59	00 ♒ 16 32	06 ♒ 25 59	05 23	11 00	28 41	00 28	25 22	07 57	04 23	00 31
16	23 34 16	25 23 45	12 ♒ 32 41	18 ♒ 36 57	07 17	11 43	29 ♑ 25	00 23	25 24	07 55	04 22	00 29
17	23 38 13	26 23 29	24 ♒ 39 05	00 ♓ 39 21	09 10	12 28	00 ♒ 09	00 18	25 26	07 54	04 21	00 28
18	23 42 09	27 23 11	06 ♓ 38 00	12 ♓ 35 17	11 01	13 13	00 53	00 13	25 28	07 52	04 20	00 27
19	23 46 06	28 22 52	18 ♓ 31 26	24 ♓ 26 40	12 50	13 59	01 37	00 07	25 29	07 51	04 19	00 25
20	23 50 03	29 ♓ 22 30	00 ♈ 21 12	06 ♈ 15 17	14 35	14 46	02 21	00 ♏ 02	25 31	07 50	04 17	00 24
21	23 53 59	00 ♈ 22 06	12 ♈ 09 10	18 ♈ 03 06	16 17	15 34	03 04	29 ♎ 56	25 32	07 48	04 16	00 23
22	23 57 56	01 21 41	23 ♈ 57 22	29 ♈ 52 18	17 55	16 23	03 48	29 50	25 34	07 47	04 15	00 21
23	00 01 52	02 21 13	05 ♉ 48 15	11 ♉ 45 34	19 29	17 12	04 32	29 44	25 35	07 46	04 13	00 20
24	00 05 49	03 20 43	17 ♉ 44 41	23 ♉ 46 01	20 58	18 02	05 16	29 38	25 36	07 45	04 12	00 19
25	00 09 45	04 20 11	29 ♉ 50 03	05 ♊ 57 17	22 22	18 53	06 00	29 32	25 37	07 43	04 11	00 18
26	00 13 42	05 19 37	12 ♊ 08 14	18 ♊ 23 23	23 41	19 45	06 44	29 25	25 38	07 42	04 09	00 16
27	00 17 38	06 19 00	24 ♊ 43 18	01 ♋ 08 27	24 53	20 37	07 28	29 19	25 39	07 41	04 08	00 15
28	00 21 35	07 18 21	07 ♋ 39 20	14 ♋ 16 21	26 00	21 30	08 12	29 12	25 40	07 40	04 07	00 14
29	00 25 32	08 17 40	20 ♋ 59 51	27 ♋ 50 06	27 00	22 23	08 56	29 06	25 41	07 40	04 05	00 13
30	00 29 28	09 16 57	04 ♌ 47 12	11 ♌ 51 10	27 53	23 18	09 40	28 59	25 41	07 39	04 04	00 12
31	00 33 25	10 ♈ 16 11	19 ♌ 01 48	26 ♌ 18 44	28 ♈ 40	24 ♒ 12	10 ♒ 25	28 ♎ 52	25 ♐ 42	07 ♌ 38	04 ♏ 02	00 ♍ 11

Date	Moon True ☊ ° '	Moon Mean ☊ ° '	Moon ☽ Latitude ° '
01	02 ♏ 43	04 ♏ 14	04 S 48
02	02 R 37	04 11	05 05
03	02 30	04 08	05 04
04	02 22	04 04	04 43
05	02 15	04 01	04 02
06	02 09	03 58	03 03
07	02 05	03 55	01 51
08	02 03	03 52	00 S 33
09	02 D 03	03 48	00 N 46
10	02 04	03 45	02 00
11	02 06	03 42	03 05
12	02 R 06	03 39	03 58
13	02 05	03 36	04 37
14	02 03	03 33	05 01
15	01 58	03 29	05 11
16	01 53	03 26	05 05
17	01 46	03 23	04 47
18	01 39	03 20	04 15
19	01 33	03 17	03 33
20	01 29	03 14	02 42
21	01 25	03 10	01 44
22	01 24	03 07	00 N 41
23	01 D 23	03 04	00 S 24
24	01 24	03 01	01 29
25	01 26	02 58	02 30
26	01 28	02 54	03 26
27	01 29	02 51	04 13
28	01 R 30	02 48	04 49
29	01 29	02 45	05 10
30	01 27	02 42	05 15
31	01 ♏ 25	02 ♏ 39	05 S 01

DECLINATIONS

Date	Sun ☉ ° '	Moon ☽ ° '	Mercury ☿ ° '	Venus ♀ ° '	Mars ♂ ° '	Jupiter ♃ ° '	Saturn ♄ ° '	Uranus ♅ ° '	Neptune ♆ ° '	Pluto ♇ ° '
01	07 S 40	18 N 06	10 S 12	13 S 38	22 S 50	10 S 35	21 S 58	18 N 49	11 S 22	22 N 13
02	07 17	15 55	09 27	13 41	22 45	10 34	21 58	18 50	11 22	22 14
03	06 54	12 42	08 40	13 43	22 40	10 33	21 59	18 50	11 22	22 14
04	06 31	08 35	07 51	13 45	22 34	10 32	21 59	18 51	11 22	22 15
05	06 08	03 N 52	07 02	13 47	22 28	10 30	21 59	18 51	11 21	22 15
06	05 45	01 S 08	06 12	13 48	22 22	10 29	21 59	18 52	11 21	22 16
07	05 22	06 04	05 20	13 48	22 16	10 28	21 59	18 52	11 20	22 16
08	04 58	10 32	04 27	13 49	22 09	10 26	21 59	18 53	11 20	22 17
09	04 35	14 16	03 34	13 48	22 02	10 25	21 59	18 53	11 20	22 17
10	04 12	17 02	02 40	13 48	21 55	10 23	21 59	18 54	11 19	22 18
11	03 48	18 41	01 45	13 47	21 48	10 21	21 59	18 54	11 19	22 18
12	03 24	19 14	00 S 50	13 45	21 41	10 20	21 59	18 55	11 19	22 19
13	03 01	18 44	00 N 06	13 43	21 33	10 18	21 59	18 55	11 18	22 19
14	02 37	17 17	01 02	13 40	21 26	10 16	21 59	18 55	11 18	22 20
15	02 13	15 02	01 58	13 37	21 18	10 14	21 59	18 56	11 17	22 20
16	01 50	12 09	02 54	13 34	21 09	10 12	21 59	18 56	11 17	22 21
17	01 26	08 48	03 49	13 30	21 01	10 10	21 59	18 57	11 16	22 21
18	01 02	05 07	04 44	13 25	20 52	10 08	21 59	18 57	11 16	22 22
19	00 39	01 S 16	05 38	13 20	20 44	10 06	21 59	18 57	11 16	22 22
20	00 S 15	02 N 37	06 30	13 14	20 35	10 04	21 59	18 58	11 15	22 22
21	00 N 09	06 24	07 21	13 08	20 25	10 02	21 59	18 58	11 15	22 23
22	00 32	09 56	08 10	13 02	20 16	10 00	21 59	18 58	11 14	22 23
23	00 56	13 05	08 57	12 55	20 06	09 58	21 59	18 58	11 14	22 23
24	01 20	15 42	09 42	12 47	19 57	09 56	21 59	18 59	11 13	22 24
25	01 43	17 40	10 25	12 39	19 47	09 53	21 59	18 59	11 13	22 24
26	02 07	18 51	11 05	12 31	19 37	09 51	21 59	18 59	11 12	22 24
27	02 31	19 07	11 42	12 22	19 26	09 49	21 59	18 59	11 12	22 25
28	02 54	18 25	12 16	12 12	19 16	09 46	21 58	19 00	11 11	22 25
29	03 17	16 42	12 47	12 02	19 05	09 44	21 58	19 00	11 11	22 25
30	03 41	13 58	13 14	11 52	18 54	09 41	21 58	19 00	11 10	22 26
31	04 N 04	10 N 21	13 N 39	11 S 41	18 S 43	09 S 39	21 S 58	19 N 00	11 S 10	22 N 26

ZODIAC SIGN ENTRIES

Date	h m	Planets
02	18 27	☽ ♌
04	19 15	☽ ♍
06	18 35	☽ ♎
08	18 34	☽ ♏
10	20 56	☽ ♐
12	17 31	☿ ♈
13	02 36	☽ ♑
15	11 28	☽ ♒
17	07 11	♂ ♒
17	22 41	☽ ♓
20	11 17	☽ ♈
20	19 13	♃ ♎
21	03 06	☉ ♈
23	00 16	☽ ♉
25	12 20	☽ ♊
27	21 53	☽ ♋
30	03 46	☽ ♌

LATITUDES

Date	Mercury ☿ ° '	Venus ♀ ° '	Mars ♂ ° '	Jupiter ♃ ° '	Saturn ♄ ° '	Uranus ♅ ° '	Neptune ♆ ° '	Pluto ♇ ° '
01	01 S 55	06 N 00	00 S 41	01 N 27	01 N 22	00 N 40	01 N 47	11 N 51
04	01 42	05 35	00 43	01 28	01 22	00 40	01 48	11 51
07	01 24	05 09	00 46	01 28	01 22	00 40	01 48	11 51
10	01 01	04 43	00 49	01 29	01 23	00 40	01 48	11 51
13	00 S 32	04 17	00 51	01 29	01 23	00 40	01 48	11 51
16	00 00	03 52	00 54	01 30	01 23	00 40	01 48	11 51
19	00 N 36	03 27	00 57	01 30	01 23	00 40	01 48	11 50
22	01 14	03 02	00 59	01 31	01 23	00 40	01 48	11 50
25	01 50	02 38	01 02	01 31	01 24	00 40	01 48	11 50
28	02 23	02 15	01 05	01 31	01 24	00 40	01 49	11 50
31	02 N 50	01 N 53	01 S 08	01 N 32	01 N 24	00 N 40	01 N 49	11 N 49

DATA

Julian Date	2436264
Delta T	+32 seconds
Ayanamsa	23° 16' 33"
Synetic vernal point	05° ♓ 50' 27"
True obliquity of ecliptic	23° 26' 34"

MOON'S PHASES, APSIDES AND POSITIONS ☽

Date	h m	Phase	Longitude ° '	Eclipse Indicator
05	18 28	○	14 ♍ 41	
12	10 48	☾	21 ♐ 21	
20	09 50	●	29 ♓ 17	
28	11 19	☽	07 ♋ 17	

Day	h m		
06	08 45	Perigee	
20	19 44	Apogee	
06	06 34	0S	
12	12 07	Max dec	19° S 14'
19	19 50	0N	
27	06 57	Max dec	19° N 09'

LONGITUDES

Date	Chiron ⚷ ° '	Ceres ⚳ ° '	Pallas ⚴ ° '	Juno ⚵ ° '	Vesta ⚶ ° '	Black Moon Lilith ⚸ ° '
01	19 ♒ 09	16 ♋ 06	12 ♉ 35	17 ♊ 24	01 ♉ 12	01 ♈ 06
11	19 ♒ 51	16 ♋ 26	17 ♉ 25	20 ♊ 54	05 ♉ 10	02 ♈ 13
21	20 ♒ 29	17 ♋ 25	22 ♉ 32	24 ♊ 43	09 ♉ 15	03 ♈ 20
31	21 ♒ 04	19 ♋ 00	27 ♉ 52	28 ♊ 45	13 ♉ 24	04 ♈ 26

All ephemeris data is given at 12.00 UT and the Moon's longitude is additionally given for 24.00 UT

ASPECTARIAN

h m	Aspects
01 Saturday	
04 10	☽ △ ☿
05 17	☽ ⊻ ♅
08 35	☽ △ ☉
12 12	☿ ⊼ ♅
18 20	☽ ∠ ♇
22 50	☿ ⊥ ♀
23 31	☽ ☍ ♂
02 Sunday	
09 41	☽ ⊼ ♄
09 44	☽ ⊥ ♇
10 10	☽ ⚼ ☿
12 26	☽ ⚼ ☉
19 48	☽ ⊻ ♇
19 49	☽ ± ♄
20 37	☽ □ ♃
03 Monday	
00 32	☽ ☍ ♀
02 06	☽ □ ♆
03 30	☽ ± ☿
04 35	☽ ± ☉
05 10	☽ # ♀
08 14	☽ ☌ ♅
10 46	☽ ⚼ ♄
14 49	☽ ⊼ ☿
15 10	☽ ⊼ ☉
20 12	☉ ☌ ☿
20 18	☽ # ♆
04 Tuesday	
01 10	☽ # ♃
02 01	☽ Q ♃
03 56	☽ ⊼ ♂
07 24	☽ Q ♆
11 07	☽ △ ♄
14 03	☽ ± ♂
15 22	☿ ± ♅
16 45	☽ # ☿
20 28	☽ ☌ ♇
21 09	☽ ✶ ♃
23 40	♀ □ ♆
23 46	☽ # ☉
05 Wednesday	
02 30	☽ ✶ ♆
02 36	☽ ⊼ ♀
05 04	☽ ⚼ ♂
08 00	☉ ± ♅
08 19	☽ ⊻ ♅
12 24	☽ ± ♀
15 04	☿ ⚼ ♃
17 45	☽ ⊥ ♅
18 28	☽ ☍ ☉
20 46	☽ ∠ ♃
21 37	☽ ☍ ☿
06 Thursday	
02 08	☽ ∠ ♆
03 04	☽ ⚼ ♀
05 53	☽ △ ♂
07 54	☽ ∠ ♅
10 42	☽ □ ♄
10 53	☽ ⊥ ♃
16 17	☽ ⊥ ♆
19 43	☽ ⊻ ♇
20 18	☽ ⊻ ♃
07 Friday	
01 45	☽ ⊻ ♆
03 35	☽ △ ♀
03 43	☉ ⚼ ♃
05 11	☽ ⊥ ♇
07 32	☽ ✶ ♅
08 45	☽ ∥ ☉
08 54	☽ ∥ ☿
09 03	☿ ⚼ ♆
09 10	♂ # ♇
10 34	☉ ∥ ☿
15 16	☽ Q ♄
19 29	☽ ∠ ♇
21 15	☽ ⊼ ☉
08 Saturday	
02 38	☽ Q ♅
04 07	☽ ⊼ ☿
07 21	☿ ∠ ♀
07 35	☽ ± ☉
07 57	☽ □ ♂
10 36	☽ ✶ ♄
11 24	☽ ∥ ♃
15 17	☽ ± ☿
16 40	☽ ∥ ♆
19 39	☽ ✶ ♇
20 09	☽ ☌ ♃
23 17	☽ ⚼ ☉
09 Sunday	
01 57	☽ ☌ ♆
05 19	☿ ⚼ ♅
05 52	☽ □ ♀
07 55	☽ □ ♅
08 19	☽ ⚼ ☿
08 41	☽ ∥ ♀
11 15	☽ ∠ ♄
15 01	☽ Q ♂
15 26	☽ Q ♇
22 53	♂ ± ♇
10 Monday	
00 29	☿ ✶ ♂
00 31	♂ ∥ ♄
02 07	☽ △ ☉
02 23	☽ ⊥ ♄
02 50	☿ ± ♃
06 24	☿ □ ♄
12 25	☽ ✶ ♂
12 38	☽ ⊻ ♄
13 30	☽ Q ♀
13 36	☉ ⚼ ♆
13 38	☽ △ ☿
17 00	♂ ⊻ ♄
22 00	☽ □ ♇
22 22	☽ ⊻ ♃
11 Tuesday	
04 00	♀ ☍ ♅
04 43	☽ ⊻ ♆
08 47	☽ ⊥ ♃
11 01	☽ △ ♅
11 22	☽ ✶ ♀
15 16	☽ ⊥ ♆
16 01	☽ ∠ ♂
17 14	☽ # ♅
22 36	☿ ± ♆
12 Wednesday	
00 40	☽ ∠ ♃
07 20	☽ ∠ ♆
09 04	☽ ⊥ ♂
10 48	☽ □ ☉
13 49	☽ ⚼ ♅
15 39	☽ ∠ ♀
17 55	☽ ☌ ♄
20 38	☽ ⊻ ♂
13 Thursday	
00 28	☿ ⊼ ♇
01 41	☿ ⊼ ♃
03 39	☽ △ ♇
03 49	☽ ✶ ♃
04 13	☽ □ ☿
06 19	☽ ± ♅
07 05	☽ # ♅
09 07	☽ ⊥ ♀
10 49	☽ ✶ ♆
17 28	☽ ⊼ ♅
20 58	☽ ⊻ ♀
22 51	☽ Q ☉
14 Friday	
01 54	☉ ⚼ ♅
02 07	☽ Q ♃
07 42	☽ ⚼ ♇
09 21	☽ Q ♆
13 35	♀ ∠ ♄
20 15	☽ Q ☿
20 54	♃ ✶ ♇
23 39	☽ ✶ ☉
23 46	☿ ⊼ ♆
15 Saturday	
00 52	☽ ± ♇
02 29	☽ ⊻ ♄
08 43	☽ ☌ ♂
12 23	☽ □ ♃
12 27	☽ ⊼ ♇
13 37	☉ ± ♃
14 08	☽ ⊥ ♄
16 36	☉ # ☿
20 00	☽ □ ♆
23 47	☽ ✶ ☿
16 Sunday	
00 36	☽ ∥ ♀
01 59	☿ ± ♇
02 56	☽ ☍ ♅
07 24	☽ ∠ ☉
07 46	☽ ∠ ♄
10 16	☽ ☌ ♀
12 06	☉ □ ♄
18 33	☽ ∥ ♆
19 53	☿ △ ♅
17 Monday	
02 24	☽ ∥ ♃
02 46	☽ ⊥ ☉
10 52	☽ ∠ ☿
13 34	☽ ✶ ♄
15 47	☽ ⊻ ☉
16 33	♂ □ ♃
22 07	♂ ⊼ ♇
23 13	☽ △ ♃
23 36	☽ ☍ ♇
23 42	☽ ⊻ ♂
18 Tuesday	
07 23	☽ △ ♆
08 10	☽ ⊥ ☿
12 31	☽ ⊥ ♂
13 40	☽ Q ♄
13 59	☽ # ☿
14 29	☽ ⊼ ♅
22 26	☽ ⊻ ☿
19 Wednesday	
02 11	☽ ⊻ ♀
02 34	☽ ± ♅
05 10	☽ ⚼ ♃
07 51	♂ ⊥ ♄
07 52	☽ ∠ ♂
10 18	☉ ± ♆
13 35	☽ ⚼ ♆
15 10	☽ ⊥ ♀
16 18	☽ ∥ ☉
20 45	☽ ⚼ ♅
22 43	☽ # ☉
23 15	☽ ± ♃
20 Thursday	
02 09	☽ □ ♄
06 12	☿ Q ♂
07 49	☽ ± ♆
09 50	☽ ☌ ☉
10 44	☽ ∠ ♀
11 21	☽ ⊼ ♃
12 06	☽ ⊼ ♇
16 19	☽ ✶ ♂
16 47	☿ ✶ ♀
19 59	☽ ⊼ ♆
23 15	☿ ⚼ ♇
21 Friday	
00 16	☽ ± ♇
02 24	☉ ⊼ ♃
03 10	☽ △ ♅
12 12	☉ ⊼ ♇
18 20	☽ Q ♂
18 33	☽ ⚼ ♇
19 28	☽ ✶ ♀
20 12	☽ ∥ ☿
21 47	☽ ☌ ☿
22 Saturday	
12 31	☽ # ♃
14 39	☉ ∠ ♀
15 16	☽ △ ♄
21 32	☽ # ♆
21 39	☽ Q ♀
23 50	☽ ☍ ♃
23 Sunday	
00 57	☽ △ ♇
01 57	♂ □ ♆
04 23	☽ ⊻ ☉
08 49	☽ ☍ ♆
09 17	☽ □ ♂
10 42	☽ # ♀
15 57	☽ □ ♅
17 37	☽ ⊥ ☉
21 39	☽ ⚼ ♄
24 Monday	
12 38	☽ □ ♀
13 19	☽ ∠ ☉
15 43	☽ ± ♄
18 51	☿ # ♃
19 18	☽ ⊻ ☿
25 Tuesday	
03 40	☽ ⊼ ♄
03 54	☽ Q ♅
08 16	☉ ⊼ ♆
08 45	☽ ⊥ ☿
11 24	☽ ⊼ ♃
12 54	☽ □ ♇
20 31	☽ ⊼ ♆
21 37	☽ ✶ ☉
23 04	☽ ± ♃
26 Wednesday	
00 52	☽ △ ♂
03 26	☽ ✶ ♅
04 32	☽ ∠ ☿
08 10	☽ ± ♆
16 22	☽ ⚼ ♃
16 32	☿ # ♆
17 13	☽ ∥ ♅
22 49	☽ Q ☉
23 46	☽ Q ♇
27 Thursday	
01 26	☽ ⚼ ♆
03 40	☽ △ ♀
07 30	☽ ⚼ ♂
08 10	☽ ∠ ♅
10 32	☉ ± ♇
12 21	☽ ✶ ☿
13 45	☽ ☍ ♄
18 58	♂ ☍ ♅
20 21	☽ ∥ ♅
20 32	☽ △ ♃
22 20	☽ ✶ ♇
28 Friday	
01 00	☽ ⊥ ♅
01 22	☽ ± ♂
04 28	☿ △ ♄
05 30	☽ △ ♆
09 44	☽ ⚼ ♀
09 49	☿ # ♀
11 19	☽ □ ☉
12 02	☽ ⊻ ♅
12 41	☽ Q ☿
13 04	☽ ⊼ ♂
20 48	☉ △ ♅
29 Saturday	
01 43	☽ ∠ ♇
03 13	☽ ± ♀
14 38	☽ ⊼ ♀
17 40	☽ ⊥ ♇
20 14	☽ ⊼ ♄
23 10	♂ # ♅
23 18	☽ □ ☿
30 Sunday	
02 04	☽ □ ♃
04 07	☽ ⊻ ♇
06 40	☽ ± ♄
10 46	☽ □ ♆
16 44	☽ ∥ ☿
16 52	☽ ☌ ♅
20 14	☽ △ ☉
20 47	☽ ☍ ♂
22 02	☽ ⚼ ♄
31 Monday	
03 23	☽ # ♀
07 03	☽ # ♆
08 26	☽ Q ♃
16 12	☽ # ♃
16 58	☽ Q ♆
18 03	☿ ☍ ♃
21 07	☽ ☍ ♀
21 21	♂ ∠ ♄
22 59	☽ △ ♄
23 02	☽ ⚼ ☉

APRIL 1958

LONGITUDES

Date	Sidereal time h m s	Sun ☉ ° ' "	Moon ☽ ° ' "	Moon ☽ 24.00 ° ' "	Mercury ☿ ° '	Venus ♀ ° '	Mars ♂ ° '	Jupiter ♃ ° '	Saturn ♄ ° '	Uranus ♅ ° '	Neptune ♆ ° '	Pluto ♇ ° '
01	00 37 21	11 ♈ 15 22	03 ♍ 41 26	11 ♍ 09 08	29 ♈ 19	25 ♒ 07	11 ♒ 09	28 ♎ 45	25 ♐ 42	07 ♌ 37	04 ♏ 01	00 ♍ 10
02	00 41 18	12 14 32	18 ♍ 40 56	26 ♍ 15 46	29 ♈ 52	26 07	11 53	28 R 38	25 42	07 R 36	03 R 59	00 R 09
03	00 45 14	13 13 39	03 ♎ 52 27	11 ♎ 29 45	00 ♉ 17	26 59	12 37	28 31	25 42	07 36	03 58	00 08
04	00 49 11	14 12 44	19 ♎ 06 22	26 ♎ 41 06	00 35	27 56	13 21	28 24	25 42	07 35	03 56	00 07
05	00 53 07	15 11 48	04 ♏ 12 46	11 ♏ 40 21	00 46	28 53	14 05	28 16	25 R 42	07 35	03 55	00 06
06	00 57 04	16 10 49	19 ♏ 02 57	26 ♏ 19 52	00 50	29 ♒ 50	14 49	28 09	25 42	07 34	03 53	00 05
07	01 01 01	17 09 48	03 ♐ 30 34	10 ♐ 34 42	00 R 47	00 ♓ 48	15 33	28 02	25 42	07 34	03 52	00 04
08	01 04 57	18 08 46	17 ♐ 32 05	24 ♐ 22 40	00 38	01 47	16 18	27 54	25 42	07 33	03 50	00 03
09	01 08 54	19 07 42	01 ♑ 06 36	07 ♑ 44 05	00 23	02 46	17 02	27 47	25 41	07 33	03 49	00 02
10	01 12 50	20 06 37	14 ♑ 15 27	20 ♑ 41 04	00 ♉ 02	03 45	17 46	27 39	25 41	07 33	03 47	00 01
11	01 16 47	21 05 29	27 ♑ 01 24	03 ♒ 16 55	29 ♈ 36	04 45	18 30	27 32	25 40	07 32	03 45	00 ♍ 00
12	01 20 43	22 04 20	09 ♒ 28 08	15 ♒ 35 32	29 05	05 45	19 14	27 24	25 40	07 32	03 44	29 ♌ 59
13	01 24 40	23 03 09	21 ♒ 39 38	27 ♒ 40 56	28 30	06 45	19 59	27 16	25 39	07 32	03 42	29 58
14	01 28 36	24 01 56	03 ♓ 39 55	09 ♓ 37 02	27 51	07 46	20 43	27 09	25 38	07 32	03 41	29 58
15	01 32 33	25 00 42	15 ♓ 32 43	21 ♓ 27 23	27 11	08 47	21 27	27 01	25 37	07 D 32	03 39	29 57
16	01 36 30	25 59 25	27 ♓ 21 25	03 ♈ 15 10	26 28	09 48	22 11	26 53	25 36	07 32	03 37	29 56
17	01 40 26	26 58 07	09 ♈ 08 57	15 ♈ 03 07	25 45	10 50	22 55	26 46	25 35	07 32	03 36	29 55
18	01 44 23	27 56 47	20 ♈ 57 55	26 ♈ 53 39	25 01	11 51	23 40	26 38	25 33	07 32	03 34	29 55
19	01 48 19	28 55 25	02 ♉ 50 36	08 ♉ 48 59	24 19	12 54	24 24	26 30	25 32	07 32	03 33	29 54
20	01 52 16	29 ♈ 54 01	14 ♉ 49 05	20 ♉ 51 09	23 37	13 56	25 08	26 23	25 30	07 33	03 31	29 54
21	01 56 12	00 ♉ 52 35	26 ♉ 55 27	03 ♊ 02 14	22 58	14 59	25 52	26 15	25 29	07 33	03 29	29 53
22	02 00 09	01 51 08	09 ♊ 11 47	15 ♊ 24 23	22 22	16 02	26 37	26 07	25 27	07 33	03 28	29 53
23	02 04 05	02 49 38	21 ♊ 40 19	27 ♊ 59 55	21 49	17 05	27 21	26 00	25 25	07 34	03 26	29 52
24	02 08 02	03 48 06	04 ♋ 23 27	10 ♋ 51 15	21 20	18 08	28 05	25 52	25 24	07 34	03 24	29 51
25	02 11 59	04 46 32	17 ♋ 23 34	24 ♋ 00 42	20 55	19 12	28 49	25 45	25 22	07 35	03 23	29 51
26	02 15 55	05 44 56	00 ♌ 42 53	07 ♌ 30 18	20 35	20 16	29 ♒ 33	25 37	25 20	07 35	03 21	29 50
27	02 19 52	06 43 17	14 ♌ 23 05	21 ♌ 21 17	20 19	21 20	00 ♓ 17	25 30	25 18	07 36	03 19	29 50
28	02 23 48	07 41 37	28 ♌ 24 51	05 ♍ 33 39	20 08	22 24	01 02	25 22	25 15	07 37	03 18	29 50
29	02 27 45	08 39 54	12 ♍ 47 22	20 ♍ 05 37	20 01	23 29	01 46	25 15	25 13	07 37	03 16	29 49
30	02 31 41	09 ♉ 38 10	27 ♍ 27 51	04 ♎ 53 23	20 ♈ 00	24 ♓ 34	02 ♓ 30	25 ♎ 08	25 ♐ 11	07 ♌ 38	03 ♏ 15	29 ♌ 49

Date	Moon True ☊ ° '	Moon Mean ☊ ° '	Moon ☽ Latitude ° '
01	01 ♏ 22	02 ♏ 35	04 S 27
02	01 R 20	02 32	03 34
03	01 18	02 29	02 26
04	01 17	02 26	01 S 07
05	01 D 16	02 23	00 N 16
06	01 17	02 20	01 37
07	01 18	02 16	02 49
08	01 19	02 13	03 49
09	01 20	02 10	04 34
10	01 20	02 07	05 03
11	01 R 20	02 04	05 16
12	01 20	02 00	05 14
13	01 19	01 57	04 58
14	01 18	01 54	04 28
15	01 17	01 51	03 48
16	01 16	01 48	02 58
17	01 16	01 45	02 00
18	01 16	01 41	00 N 57
19	01 D 16	01 38	00 S 09
20	01 16	01 35	01 15
21	01 16	01 32	02 18
22	01 R 16	01 29	03 16
23	01 16	01 25	04 05
24	01 15	01 22	04 44
25	01 15	01 19	05 09
26	01 15	01 16	05 18
27	01 D 15	01 13	05 10
28	01 15	01 10	04 43
29	01 16	01 06	03 58
30	01 ♏ 16	01 ♏ 03	02 S 57

DECLINATIONS

Date	Sun ☉ ° '	Moon ☽ ° '	Mercury ☿ ° '	Venus ♀ ° '	Mars ♂ ° '	Jupiter ♃ ° '	Saturn ♄ ° '	Uranus ♅ ° '	Neptune ♆ ° '	Pluto ♇ ° '
01	04 N 27	06 N 00	13 N 59	11 S 29	18 S 32	09 S 36	21 S 58	19 N 00	11 S 09	22 N 26
02	04 50	01 N 11	14 16	11 17	18 21	09 34	21 58	19 01	11 09	22 26
03	05 13	03 S 46	14 30	11 05	18 09	09 31	21 58	19 01	11 08	22 26
04	05 36	08 31	14 39	10 52	17 57	09 28	21 58	19 01	11 08	22 27
05	05 59	12 40	14 45	10 39	17 46	09 26	21 58	19 01	11 07	22 27
06	06 22	15 56	14 47	10 25	17 34	09 23	21 58	19 01	11 07	22 27
07	06 45	18 05	14 46	10 11	17 21	09 21	21 58	19 01	11 06	22 27
08	07 07	19 03	14 40	09 56	17 09	09 18	21 58	19 01	11 05	22 27
09	07 29	18 52	14 31	09 41	16 57	09 15	21 58	19 01	11 05	22 28
10	07 52	17 39	14 19	09 26	16 44	09 12	21 58	19 01	11 04	22 28
11	08 14	15 35	14 03	09 10	16 31	09 10	21 58	19 01	11 04	22 28
12	08 36	12 50	13 44	08 54	16 18	09 07	21 57	19 01	11 03	22 28
13	08 58	09 35	13 22	08 37	16 05	09 04	21 57	19 01	11 03	22 28
14	09 19	06 00	12 58	08 20	15 52	09 02	21 57	19 01	11 02	22 28
15	09 41	02 S 12	12 31	08 03	15 39	08 59	21 57	19 01	11 02	22 28
16	10 02	01 N 40	12 03	07 45	15 25	08 56	21 57	19 01	11 01	22 28
17	10 24	05 28	11 34	07 27	15 11	08 53	21 57	19 01	11 00	22 28
18	10 45	09 04	11 03	07 08	14 58	08 51	21 57	19 01	11 00	22 28
19	11 06	12 19	10 33	06 50	14 44	08 48	21 57	19 01	10 59	22 28
20	11 26	15 06	10 02	06 30	14 30	08 45	21 57	19 01	10 59	22 28
21	11 47	17 14	09 32	06 11	14 16	08 42	21 57	19 01	10 58	22 28
22	12 07	18 36	09 03	05 51	14 02	08 40	21 56	19 01	10 58	22 28
23	12 27	19 06	08 35	05 31	13 47	08 37	21 56	19 01	10 57	22 28
24	12 47	18 39	08 08	05 11	13 33	08 34	21 56	19 01	10 57	22 28
25	13 07	17 13	07 44	04 50	13 18	08 32	21 56	19 00	10 56	22 28
26	13 26	14 50	07 21	04 29	13 04	08 29	21 56	19 00	10 55	22 28
27	13 46	11 35	07 01	04 08	12 49	08 27	21 56	19 00	10 55	22 28
28	14 05	07 36	06 43	03 46	12 34	08 24	21 56	19 00	10 54	22 28
29	14 23	03 05	06 28	03 25	12 19	08 21	21 56	19 00	10 54	22 28
30	14 N 42	01 S 42	06 N 16	03 S 03	12 S 04	08 S 19	21 S 55	18 N 59	10 S 53	22 N 28

ZODIAC SIGN ENTRIES

Date	h	m	Planets
01	06	01	☽ ♍
02	19	17	☿ ♉
03	05	54	☽ ♎
05	05	16	☽ ♏
06	16	00	♀ ♓
07	06	07	☽ ♐
09	10	00	☽ ♑
10	13	52	☿ ♈
11	14	58	♇ ♌
11	17	41	☽ ♒
14	04	38	☽ ♓
16	17	23	☽ ♈
19	06	16	☽ ♉
20	14	27	☉ ♉
21	18	03	☽ ♊
24	03	46	☽ ♋
26	10	44	☽ ♌
27	02	31	♂ ♓
28	14	41	☽ ♍
30	16	06	☽ ♎

LATITUDES

Date	Mercury ☿ ° '	Venus ♀ ° '	Mars ♂ ° '	Jupiter ♃ ° '	Saturn ♄ ° '	Uranus ♅ ° '	Neptune ♆ ° '	Pluto ♇ ° '
01	02 N 57	01 N 46	01 S 09	01 N 32	01 N 24	00 N 40	01 N 49	11 N 49
04	03 11	01 25	01 11	01 32	01 24	00 40	01 49	11 48
07	03 13	01 04	01 14	01 32	01 25	00 40	01 49	11 48
10	03 01	00 45	01 17	01 32	01 25	00 39	01 49	11 47
13	02 36	00 27	01 20	01 32	01 25	00 39	01 49	11 47
16	01 58	00 N 09	01 23	01 32	01 25	00 39	01 49	11 46
19	01 12	00 S 07	01 25	01 32	01 25	00 39	01 49	11 45
22	00 N 22	00 22	01 28	01 32	01 25	00 39	01 49	11 45
25	00 S 28	00 37	01 31	01 32	01 26	00 39	01 49	11 44
28	01 14	00 50	01 33	01 31	01 26	00 39	01 49	11 43
31	01 S 53	01 S 02	01 S 36	01 N 31	01 N 26	00 N 39	01 N 49	11 N 42

DATA

Julian Date	2436295
Delta T	+32 seconds
Ayanamsa	23° 16' 35"
Synetic vernal point	05° ♓ 50' 25"
True obliquity of ecliptic	23° 26' 34"

MOON'S PHASES, APSIDES AND POSITIONS ☽

Date	h	m	Phase	Longitude ° '	Eclipse Indicator
04	03	45	○	13 ♎ 52	
10	23	50	☾	20 ♑ 36	
19	03	23	●	28 ♈ 34	Annular
26	21	36	☽	06 ♌ 08	

Day	h	m		
03	20	39	Perigee	
16	22	58	Apogee	
02	17	45	0S	
08	19	54	Max dec	19° S 07'
16	01	40	0N	
23	12	40	Max dec	19° N 06'
30	03	33	0S	

LONGITUDES

Date	Chiron ⚷ ° '	Ceres ⚳ ° '	Pallas ⚴ ° '	Juno ⚵ ° '	Vesta ⚶ ° '	Black Moon Lilith ⚸ ° '
01	21 ♒ 07	19 ♋ 12	28 ♉ 25	29 ♊ 10	13 ♉ 49	04 ♈ 33
11	21 ♒ 37	21 ♋ 20	03 ♊ 57	03 ♋ 24	18 ♉ 03	05 ♈ 39
21	22 ♒ 02	23 ♋ 55	09 ♊ 38	07 ♋ 45	22 ♉ 20	06 ♈ 46
31	22 ♒ 22	26 ♋ 51	15 ♊ 25	12 ♋ 12	26 ♉ 38	07 ♈ 53

ASPECTARIAN

h m Aspects

01 Tuesday
01 16 ☉ ⚹ ♂
04 03 ☽ ⚹ ♃
04 36 ☽ △ ☿
06 17 ☽ ☌ ♇
12 31 ☽ ⚹ ♆
14 42 ☽ ± ☉
18 20 ☽ ⊻ ♅
19 20 ☽ ∥ ☉

02 Wednesday
00 36 ☽ ⊼ ♂
01 01 ☽ ⊼ ☉
03 04 ♀ ⚹ ♄
03 56 ☽ ⊥ ♅
04 02 ☽ ∠ ♃
05 43 ☽ ⚼ ☿
10 39 ☽ ± ♂
12 29 ☽ ∠ ♆
18 13 ☽ ⊥ ♃
18 13 ☽ ∠ ♅
20 28 ☽ ± ☿
23 07 ☽ □ ♄

03 Thursday
00 25 ☽ ⊼ ♀
01 38 ☽ ⚼ ♂
02 42 ☽ ⊥ ♆
02 47 ☿ △ ♇
03 37 ☽ ⊻ ♃
05 52 ♀ ∥ ♆
06 06 ☽ ⊻ ♇
06 12 ☽ ⊼ ☿
10 30 ☽ ± ♀
12 08 ☽ ⊻ ♆
15 33 ☽ ⊥ ♇
17 51 ☽ ⚹ ♅
19 45 ☽ # ☉

04 Friday
01 37 ☽ ⚼ ♀
02 28 ☽ △ ♂
03 29 ☽ Q ♄
03 45 ☽ ☍ ☉
05 42 ☽ ∠ ♇
12 45 ☽ Q ♅
17 08 ☽ ∥ ♃
19 39 ♄ St R
22 27 ☽ ⚹ ♄
22 28 ♀ △ ♃

05 Saturday
00 25 ☽ ∥ ♀
02 32 ☽ ∥ ♆
02 36 ☽ ☌ ♃
02 55 ☽ △ ♀
05 26 ☽ ⚹ ♇
06 27 ☽ ☍ ☿
09 31 ☉ ⚼ ♇
11 31 ☽ ☌ ♆
17 24 ☽ □ ♅
22 26 ☽ ∠ ♄

06 Sunday
00 40 ☽ Q ♇
02 38 ☽ # ☿
04 44 ☽ □ ♂
06 59 ☽ ⊼ ☉
13 04 ☽ ⊥ ♄
14 25 ☿ St R
17 31 ☽ ± ☉
17 48 ♀ ☍ ♇
22 58 ☽ ⊻ ♄

07 Monday
02 54 ☽ ⊻ ♃
03 11 ☽ ∥ ♂
06 13 ☽ □ ♇
07 08 ☽ □ ♀
07 28 ☽ ⊼ ☿
09 34 ☽ ⚼ ☉
11 39 ☿ ⚹ ♀
12 05 ☽ Q ♂
12 36 ☽ ⊻ ♆
12 52 ☽ ⊥ ♃
17 30 ☽ ± ☿
18 51 ☽ △ ♅
22 45 ☽ ⊥ ♆

08 Tuesday
04 04 ☽ ∠ ♃
08 45 ☽ ⚼ ☿
09 43 ☽ ⚹ ♂
10 14 ☽ # ♅
13 09 ☽ △ ☉
14 16 ☽ ∠ ♆
16 13 ☽ Q ♀
20 47 ☽ ⚼ ♅

09 Wednesday
02 20 ☽ ☌ ♄
05 36 ☽ # ♅
06 06 ☽ ⚹ ♃
10 04 ☽ △ ♇
10 43 ☽ △ ☿
12 47 ☽ ± ♅
13 45 ☽ ∠ ♂
15 13 ☽ ⚹ ♀
16 51 ☽ ⚹ ♆
23 39 ☽ ⊼ ♅

10 Thursday
03 36 ☽ Q ♃
07 07 ☽ ⊥ ♂
12 49 ♀ △ ♆
12 56 ☿ △ ♇
13 24 ☽ ⚼ ♇
14 50 ☽ Q ♆
17 34 ☿ Q ♂
18 56 ☽ ⊻ ♂
21 04 ☽ ∠ ♀
23 50 ☽ □ ☉

11 Friday
01 05 ☽ ∥ ♂
06 16 ☽ ± ♇
09 26 ☽ ⊻ ♄
12 46 ♀ ∥ ♃
12 57 ☽ □ ♃
15 34 ☽ ⊥ ♀
16 43 ☽ □ ☿
17 41 ☽ ⊼ ♇
20 53 ☽ ⊥ ♄

12 Saturday
00 53 ☽ □ ♆
03 44 ☽ # ☿
04 07 ☽ ⊻ ♀
08 14 ☽ ☍ ♅
13 17 ☽ Q ☉
14 19 ☽ ∠ ♄
23 16 ☉ # ♀

13 Sunday
01 34 ☽ ∥ ♆
02 15 ☽ Q ☿
08 27 ☽ ☌ ♂
15 01 ☽ ⚹ ☉
15 36 ☽ ∥ ♃
15 54 ☽ # ☉
18 26 ☉ # ♃
19 12 ☽ ∥ ♀
19 55 ☽ ⚹ ♄
23 04 ☽ △ ♃

14 Monday
00 57 ☽ ⚹ ☿
04 34 ☽ ☍ ♇
06 39 ♀ ⊼ ♅
09 01 ♀ Q ♄
12 01 ☽ △ ♆
19 47 ☽ ⊼ ♅
19 59 ☽ Q ♄
21 01 ☽ ☌ ♀
23 47 ☽ ∠ ☉

15 Tuesday
04 56 ☽ ⚼ ♃
05 33 ☽ ∠ ☿
07 55 ☽ ± ♅
08 27 ♅ St D
18 17 ☽ ⚼ ♆
18 33 ☿ ☍ ♃
19 40 ☽ ⊥ ☉
22 47 ☽ ⊥ ☿
23 00 ☽ ± ♃

16 Wednesday
00 47 ☽ ⊻ ♂
02 11 ☽ ⚼ ♅
02 31 ☉ △ ♄
08 25 ☽ □ ♄
08 58 ☽ ⊻ ☉
10 18 ☽ ⊻ ☿
11 04 ☽ ⊼ ♃
12 32 ☽ ± ♆
13 48 ☽ ⊥ ♂
17 15 ☽ ⊼ ♇
18 45 ☉ ☌ ☿

17 Thursday
00 44 ☽ ⊼ ♆
05 27 ☽ ± ♇
07 32 ☉ ☍ ♃
08 43 ☽ △ ♅
09 20 ☽ ∠ ♂
10 54 ☿ ∠ ♀
15 44 ☽ ⊻ ♀
17 44 ☿ △ ♄
23 44 ☽ ⚼ ♇

18 Friday
00 00 ☽ # ♀
05 05 ☽ ⊥ ♀
07 25 ♀ ⚼ ♃
10 30 ☽ # ♃
14 42 ☿ # ♆
17 49 ☽ ⚹ ♂
19 45 ☽ ☌ ☿
20 39 ☉ ∥ ☿
21 17 ☽ △ ♄
23 21 ☽ ☍ ♃

19 Saturday
00 21 ☽ ∥ ☿
01 04 ☽ ∠ ♀
01 27 ☽ ∥ ☉
01 50 ☽ # ♆
03 23 ☽ ☌ ☉
05 00 ☉ # ♆
06 05 ☽ △ ♇
10 33 ☿ ⚹ ♂
13 24 ☽ ☍ ♆
19 37 ☽ Q ♂
21 27 ☽ □ ♅

20 Sunday
03 00 ♀ ± ♅
03 24 ☽ ⚼ ♄
06 51 ☽ # ♂
10 04 ☽ ⚹ ♀
11 48 ☉ △ ♇
21 19 ☽ ± ♄
23 44 ♂ ⚹ ♄

21 Monday
04 35 ☽ ⊻ ☿
09 10 ☽ ⊼ ♄
09 17 ☽ Q ♅
09 48 ☽ □ ♂
10 41 ☽ ⊼ ♃
12 07 ☽ Q ♀
15 50 ☽ ⊥ ☿
17 49 ☽ □ ♇
20 26 ☽ ⊻ ☉
22 21 ☽ ± ♃
22 32 ♂ △ ♃

22 Tuesday
00 51 ☽ ⊼ ♆
08 37 ☽ ∠ ☿
08 49 ☽ ⚹ ♅
09 10 ☽ ⊥ ☉
12 31 ☽ ± ♆
15 42 ☽ ⚼ ♃
17 02 ☿ ⊥ ♀

23 Wednesday
02 16 ☽ ∥ ♅
02 25 ☽ □ ♀
04 02 ☽ ∠ ☉
04 44 ☽ Q ♇
05 49 ☽ ⚼ ♆
09 38 ☿ # ♃
12 16 ☽ ⚹ ☿
13 42 ☽ ∠ ♅
19 07 ☽ ☍ ♄
20 08 ☽ △ ♃
23 08 ☽ ∥ ♅
23 26 ☽ △ ♂

24 Thursday
02 31 ☉ ☍ ♆
03 30 ☽ ⚹ ♇
06 43 ☽ ⊥ ♅
10 06 ☽ Q ☿
10 10 ☽ △ ♆
10 49 ☽ ⚹ ☉
17 52 ♀ ⚼ ♆
17 56 ☽ ⊻ ♅

25 Friday
05 04 ☽ ⚼ ♂
07 21 ☽ ∠ ♇
10 47 ☽ Q ☉
15 35 ☽ △ ♀
18 14 ☽ □ ☿
20 00 ☉ # ♂
22 26 ☽ ± ♂
22 59 ♀ ± ♃
23 42 ☽ ⊥ ♇

26 Saturday
02 24 ☽ ⊼ ♄
02 59 ☽ □ ♃
09 49 ☽ ⊼ ♂
10 27 ☽ ⊻ ♇
13 05 ☽ ± ♄
16 40 ☽ □ ♆
17 28 ☿ ⊻ ♀
20 45 ☽ ⚼ ♀
21 12 ♂ ☍ ♇
21 36 ☽ □ ☉
21 58 ☽ ∥ ☉

27 Sunday
00 09 ☽ ☌ ♅
02 51 ☽ # ♂
04 54 ☽ ⚼ ♄
10 28 ☽ Q ♃
13 47 ☽ ± ♀
16 17 ☽ # ♆
22 04 ☽ △ ☿
23 55 ☽ Q ♆

28 Monday
00 57 ☽ ⊼ ♀
06 40 ☽ △ ♄
06 54 ☽ ⚹ ♃
07 25 ☽ # ♃
09 57 ☉ □ ♅
14 23 ☽ ☌ ♇
16 37 ♀ ⚼ ♅
16 39 ☽ ☍ ♂
17 10 ☽ ∥ ☿
20 12 ☽ ⚹ ♆
23 10 ☽ ⚼ ☿

29 Tuesday
03 26 ☽ ⊻ ♅
04 41 ☽ △ ☉
07 50 ☽ ∠ ♃
10 12 ☽ # ♀
13 23 ☽ ⊥ ♅
14 01 ☽ ± ☿
21 00 ☽ ∠ ♆
21 05 ♃ ⚹ ♄
22 32 ☽ ⊥ ♃
23 51 ☽ ⊼ ☿

30 Wednesday
04 09 ☽ ∠ ♅
06 55 ☽ ☍ ♀
06 57 ☿ St D
07 05 ☽ ⚼ ☉
08 14 ☽ ⊻ ♃
08 18 ☽ □ ♄
11 38 ☽ ⊥ ♆
15 48 ☽ ⊻ ♇
18 14 ☽ ∥ ♀
20 34 ☽ ⊼ ♂
21 20 ☽ ⊻ ♆
22 41 ☽ ± ☉
23 22 ♀ ⊼ ♃

All ephemeris data is given at 12.00 UT and the Moon's longitude is additionally given for 24.00 UT

MAY 1958

LONGITUDES

Date	Sidereal time h m s	Sun ☉ ° ' "	Moon ☽ ° ' "	Moon ☽ 24.00 ° ' "	Mercury ☿ ° '	Venus ♀ ° '	Mars ♂ ° '	Jupiter ♃ ° '	Saturn ♄ ° '	Uranus ♅ ° '	Neptune ♆ ° '	Pluto ♇ ° '
01	02 35 38	10 ♉ 36 23	12 ♎ 21 24	19 ♎ 51 01	20 ♈ 03	25 ♓ 39	03 ♓ 14	25 ♎ 01	25 ♐ 08	07 ♌ 39	03 ♏ 13	29 ♌ 49
02	02 39 34	11 34 34	27 ♎ 21 13	04 ♏ 50 59	20 D 12	26 44	03 58	24 R 53	25 R 06	07 40	03 R 11	29 R 48
03	02 43 31	12 32 44	12 ♏ 19 14	19 ♏ 44 57	20 24	27 49	04 42	24 46	25 03	07 41	03 10	29 48
04	02 47 28	13 30 52	27 ♏ 07 11	04 ♐ 25 03	20 42	28 ♓ 54	05 26	24 39	25 01	07 42	03 08	29 48
05	02 51 24	14 28 58	11 ♐ 37 47	18 ♐ 44 49	21 04	00 ♈ 00	06 11	24 33	24 58	07 43	03 06	29 48
06	02 55 21	15 27 03	25 ♐ 45 39	02 ♑ 40 02	21 31	01 06	06 55	24 26	24 55	07 44	03 05	29 47
07	02 59 17	16 25 07	09 ♑ 27 49	16 ♑ 09 00	22 01	02 12	07 39	24 19	24 52	07 45	03 03	29 47
08	03 03 14	17 23 09	22 ♑ 43 43	29 ♑ 12 15	22 36	03 18	08 23	24 12	24 49	07 46	03 02	29 47
09	03 07 10	18 21 09	05 ♒ 34 56	11 ♒ 52 12	23 14	04 24	09 07	24 06	24 46	07 48	03 00	29 47
10	03 11 07	19 19 08	18 ♒ 04 33	24 ♒ 12 32	23 57	05 31	09 51	24 00	24 43	07 49	02 59	29 47
11	03 15 03	20 17 06	00 ♓ 16 42	06 ♓ 17 39	24 43	06 37	10 35	23 53	24 40	07 50	02 57	29 47
12	03 19 00	21 15 03	12 ♓ 16 00	18 ♓ 12 18	25 32	07 44	11 19	23 47	24 36	07 52	02 55	29 D 47
13	03 22 57	22 12 58	24 ♓ 07 09	00 ♈ 01 08	26 25	08 51	12 03	23 41	24 33	07 53	02 54	29 47
14	03 26 53	23 10 52	05 ♈ 54 45	11 ♈ 48 31	27 21	09 58	12 47	23 35	24 30	07 55	02 52	29 47
15	03 30 50	24 08 45	17 ♈ 42 55	23 ♈ 38 23	28 20	11 05	13 31	23 29	24 26	07 56	02 51	29 47
16	03 34 46	25 06 36	29 ♈ 35 18	05 ♉ 34 03	29 ♈ 23	12 12	14 15	23 24	24 23	07 58	02 49	29 47
17	03 38 43	26 04 26	11 ♉ 34 56	17 ♉ 38 14	00 ♉ 28	13 20	14 58	23 18	24 19	07 59	02 48	29 47
18	03 42 39	27 02 15	23 ♉ 44 10	29 ♉ 52 58	01 36	14 27	15 42	23 13	24 16	08 01	02 46	29 48
19	03 46 36	28 00 02	06 ♊ 04 46	12 ♊ 19 43	02 47	15 35	16 26	23 07	24 12	08 03	02 45	29 48
20	03 50 32	28 57 48	18 ♊ 37 54	24 ♊ 59 24	04 00	16 43	17 10	23 02	24 08	08 05	02 44	29 48
21	03 54 29	29 ♉ 55 33	01 ♋ 24 16	07 ♋ 52 34	05 17	17 50	17 54	22 57	24 04	08 07	02 42	29 48
22	03 58 26	00 ♊ 53 16	14 ♋ 24 18	20 ♋ 59 31	06 35	18 58	18 37	22 53	24 01	08 08	02 41	29 49
23	04 02 22	01 50 58	27 ♋ 38 14	04 ♌ 20 27	07 57	20 06	19 21	22 48	23 57	08 10	02 39	29 49
24	04 06 19	02 48 38	11 ♌ 06 11	17 ♌ 55 27	09 21	21 14	20 04	22 43	23 53	08 12	02 38	29 49
25	04 10 15	03 46 16	24 ♌ 48 15	01 ♍ 44 31	10 47	22 23	20 48	22 39	23 49	08 14	02 37	29 50
26	04 14 12	04 43 53	08 ♍ 44 14	15 ♍ 47 18	12 16	23 31	21 31	22 35	23 45	08 16	02 35	29 50
27	04 18 08	05 41 28	22 ♍ 53 34	00 ♎ 02 51	13 47	24 39	22 15	22 31	23 41	08 19	02 34	29 51
28	04 22 05	06 39 02	07 ♎ 14 51	14 ♎ 29 14	15 21	25 48	22 58	22 27	23 36	08 21	02 33	29 51
29	04 26 01	07 36 35	21 ♎ 45 33	29 ♎ 03 17	16 57	26 56	23 42	22 23	23 32	08 23	02 31	29 52
30	04 29 58	08 34 06	06 ♏ 21 49	13 ♏ 40 28	18 36	28 05	24 25	22 20	23 28	08 25	02 30	29 52
31	04 33 55	09 ♊ 31 36	20 ♏ 58 29	28 ♏ 15 05	20 ♉ 17	29 ♈ 14	25 ♓ 08	22 ♎ 16	23 ♐ 24	08 ♌ 27	02 ♏ 29	29 ♌ 53

Date	Moon True ☊ ° '	Moon Mean ☊ ° '	Moon ☽ Latitude ° '
01	01 ♏ 17	01 ♏ 00	01 S 43
02	01 D 17	00 57	00 S 22
03	01 R 17	00 54	01 N 01
04	01 17	00 51	02 19
05	01 15	00 47	03 26
06	01 13	00 44	04 18
07	01 12	00 41	04 54
08	01 10	00 38	05 13
09	01 09	00 35	05 16
10	01 08	00 31	05 03
11	01 D 09	00 28	04 37
12	01 10	00 25	03 59
13	01 11	00 22	03 11
14	01 13	00 19	02 15
15	01 14	00 16	01 14
16	01 15	00 12	00 N 09
17	01 R 14	00 09	00 S 57
18	01 13	00 06	02 01
19	01 10	00 03	03 00
20	01 06	00 ♏ 00	03 51
21	01 01	29 ♎ 57	04 32
22	00 57	29 53	05 00
23	00 52	29 50	05 12
24	00 49	29 47	05 08
25	00 48	29 44	04 46
26	00 D 47	29 41	04 07
27	00 48	29 37	03 13
28	00 50	29 34	02 05
29	00 51	29 31	00 S 49
30	00 R 51	29 28	00 N 30
31	00 ♏ 50	29 ♎ 25	01 N 47

DECLINATIONS

Date	Sun ☉ ° '	Moon ☽ ° '	Mercury ☿ ° '	Venus ♀ ° '	Mars ♂ ° '	Jupiter ♃ ° '	Saturn ♄ ° '	Uranus ♅ ° '	Neptune ♆ ° '	Pluto ♇ ° '
01	15 N 00	06 S 28	06 N 05	02 S 41	11 S 49	08 S 16	21 S 55	18 N 59	10 S 53	22 N 28
02	15 18	10 52	05 58	02 18	11 34	08 14	21 55	18 59	10 52	22 28
03	15 36	14 34	05 53	01 56	11 18	08 11	21 55	18 59	10 52	22 27
04	15 54	17 16	05 51	01 33	11 03	08 09	21 55	18 58	10 51	22 27
05	16 11	18 47	05 51	01 10	10 47	08 07	21 55	18 58	10 51	22 27
06	16 28	19 04	05 53	00 47	10 32	08 04	21 55	18 58	10 50	22 27
07	16 45	18 13	05 58	00 S 24	10 16	08 02	21 55	18 57	10 49	22 27
08	17 01	16 23	06 05	00 00	10 01	08 00	21 54	18 57	10 49	22 26
09	17 18	13 46	06 14	00 N 23	09 45	07 57	21 54	18 57	10 48	22 26
10	17 34	10 36	06 25	00 47	09 29	07 55	21 54	18 56	10 48	22 26
11	17 49	07 03	06 38	01 11	09 13	07 53	21 54	18 56	10 47	22 26
12	18 05	03 S 17	06 53	01 35	08 57	07 51	21 54	18 56	10 47	22 26
13	18 20	00 N 35	07 10	01 59	08 41	07 49	21 54	18 55	10 46	22 25
14	18 34	04 25	07 29	02 23	08 25	07 47	21 54	18 55	10 46	22 25
15	18 49	08 06	07 49	02 47	08 09	07 45	21 53	18 54	10 45	22 25
16	19 03	11 28	08 11	03 11	07 53	07 43	21 53	18 54	10 45	22 24
17	19 16	14 25	08 34	03 35	07 37	07 41	21 53	18 53	10 44	22 24
18	19 30	16 46	08 59	03 59	07 20	07 39	21 53	18 53	10 44	22 24
19	19 43	18 22	09 25	04 24	07 04	07 38	21 53	18 52	10 44	22 23
20	19 56	19 07	09 52	04 48	06 48	07 36	21 53	18 52	10 43	22 23
21	20 08	18 54	10 20	05 12	06 31	07 34	21 52	18 51	10 43	22 23
22	20 20	17 41	10 50	05 37	06 15	07 33	21 52	18 51	10 42	22 22
23	20 32	15 32	11 20	06 01	05 59	07 31	21 52	18 50	10 42	22 22
24	20 43	12 30	11 52	06 25	05 42	07 30	21 52	18 50	10 41	22 22
25	20 54	08 46	12 24	06 50	05 26	07 28	21 52	18 49	10 41	22 21
26	21 05	04 N 29	12 57	07 14	05 09	07 27	21 52	18 49	10 40	22 21
27	21 15	00 S 08	13 30	07 38	04 53	07 26	21 51	18 48	10 40	22 20
28	21 25	04 48	14 05	08 02	04 36	07 24	21 51	18 48	10 40	22 20
29	21 35	09 15	14 39	08 26	04 20	07 23	21 51	18 47	10 39	22 20
30	21 44	13 10	15 14	08 50	04 03	07 22	21 51	18 47	10 39	22 19
31	21 N 53	16 S 16	15 N 50	09 N 13	03 S 47	07 S 21	21 S 51	18 N 46	10 S 38	22 N 19

ZODIAC SIGN ENTRIES

Date	h m	Planets
02	16 14	☽ ♏
04	16 43	☽ ♐
05	11 59	♀ ♈
06	19 21	☽ ♑
09	01 29	☽ ♒
11	11 27	☽ ♓
13	23 58	☽ ♈
16	12 50	☽ ♉
17	01 53	☿ ♉
19	00 14	☽ ♊
21	09 23	☽ ♋
21	13 51	☉ ♊
23	16 15	☽ ♌
25	21 00	☽ ♍
27	23 55	☽ ♎
30	01 33	☽ ♏

LATITUDES

Date	Mercury ☿ ° '	Venus ♀ ° '	Mars ♂ ° '	Jupiter ♃ ° '	Saturn ♄ ° '	Uranus ♅ ° '	Neptune ♆ ° '	Pluto ♇ ° '
01	01 S 53	01 S 02	01 S 36	01 N 31	01 N 26	00 N 39	01 N 49	11 N 42
04	02 25	01 13	01 39	01 31	01 26	00 39	01 49	11 42
07	02 49	01 23	01 41	01 30	01 26	00 39	01 49	11 41
10	03 05	01 32	01 44	01 30	01 26	00 39	01 49	11 40
13	03 15	01 40	01 47	01 29	01 26	00 38	01 49	11 39
16	03 17	01 47	01 49	01 29	01 26	00 38	01 49	11 38
19	03 13	01 53	01 52	01 28	01 26	00 38	01 49	11 38
22	03 03	01 58	01 54	01 27	01 26	00 38	01 49	11 37
25	02 48	02 02	01 56	01 27	01 26	00 38	01 49	11 36
28	02 28	02 05	01 59	01 26	01 26	00 38	01 48	11 35
31	02 S 04	02 S 07	02 S 01	01 N 25	01 N 26	00 N 38	01 N 48	11 N 34

DATA

Julian Date	2436325
Delta T	+32 seconds
Ayanamsa	23° 16' 38"
Synetic vernal point	05° ♓ 50' 22"
True obliquity of ecliptic	23° 26' 33"

MOON'S PHASES, APSIDES AND POSITIONS ☽

Date	h m	Phase	Longitude ° '	Eclipse Indicator
03	12 23	○	12 ♏ 34	partial
10	14 38	☾	19 ♒ 25	
18	19 00	●	27 ♉ 19	
26	04 38	☽	04 ♍ 26	

Day	h m		
02	06 04	Perigee	
14	11 27	Apogee	
30	07 35	Perigee	
06	05 41	Max dec	19° S 07'
13	08 22	0N	
20	18 45	Max dec	19° N 09'
27	11 21	0S	

LONGITUDES

Date	Chiron ⚷ ° '	Ceres ⚳ ° '	Pallas ⚴ ° '	Juno ⚵ ° '	Vesta ⚶ ° '	Black Moon Lilith ⚸ ° '
01	22 ♒ 22	26 ♋ 51	15 ♊ 25	12 ♋ 12	26 ♉ 38	07 ♈ 53
11	22 ♒ 35	00 ♌ 06	21 ♊ 17	16 ♋ 43	00 ♊ 58	08 ♈ 59
21	22 ♒ 42	03 ♌ 35	27 ♊ 12	21 ♋ 17	05 ♊ 19	10 ♈ 06
31	22 ♒ 43	07 ♌ 18	03 ♋ 09	25 ♋ 52	09 ♊ 40	11 ♈ 12

ASPECTARIAN

h m	Aspects
01 Thursday	
00 55	☉ ℙ ♄
01 16	♀ □ ♄
01 29	☽ ⊥ ♇
04 18	☉ ∠ ♀
04 26	☽ ⚹ ♅
06 44	☽ ± ♂
09 00	☽ ⚻ ☉
10 07	☽ # ☿
11 24	♂ △ ♆
13 15	☽ Q ♄
15 56	☽ ∠ ♇
21 24	☽ ∥ ♃
21 54	☽ ℙ ♂
23 42	☽ Q ♅
02 Friday	
00 25	☽ ☍ ☿
08 06	☽ ☌ ♃
08 24	☽ ⚹ ♄
10 55	☽ ⚻ ♀
11 59	☽ ∥ ♆
15 50	☽ ∥ ♂
15 55	☽ ⚹ ♇
21 16	☽ ± ♀
21 19	☽ ☌ ♆
21 55	♀ ± ♆
23 08	☽ △ ♂
03 Saturday	
04 32	☽ □ ♅
08 22	☽ ∠ ♄
11 10	☽ Q ♇
12 23	☽ ☍ ☉
12 52	☽ ℙ ♀
21 02	☽ # ☉
22 50	☽ ⊥ ♄
04 Sunday	
01 19	☽ ⚻ ☿
08 01	☽ ⚺ ♃
08 34	☽ ⚺ ♄
11 18	☽ ± ☿
15 10	☽ △ ♀
16 23	☽ □ ♇
17 45	☽ ⊥ ♃
21 52	☽ ⚺ ♆
05 Monday	
02 26	☽ □ ♂
02 29	☽ ℙ ☿
04 30	☿ ∠ ♂
05 28	☽ △ ♅
07 03	♂ ∥ ♆
07 27	♀ ⚻ ♇
07 48	☽ ⊥ ♆
08 32	☽ ∠ ♃
17 08	☽ ⚻ ☉
17 42	☽ # ♅
22 54	☽ ∠ ♆
06 Tuesday	
04 03	☽ ± ☉
04 27	☽ △ ☿
06 48	☽ ℙ ♅
09 44	☽ ⚹ ♃
10 27	☽ Q ♂
10 33	☽ ☌ ♄
12 12	♂ Q ♄
18 01	☽ # ♅
18 59	☽ △ ♇
20 45	☽ ℙ ☉
22 04	☽ □ ♀
22 23	☽ ± ♅
07 Wednesday	
00 42	☽ ⚹ ♆
06 28	☽ Q ♃
08 35	☽ ⚹ ♂
08 57	☽ ⚻ ♅
15 38	♂ ⚻ ♅
21 32	☽ ℙ ♇
22 00	☽ Q ♆
08 Thursday	
01 28	☽ △ ☉
05 36	☽ # ☉
06 14	♀ ⚻ ♆
09 08	☽ Q ♀
11 45	☽ □ ☿
13 16	☽ ∠ ♂
13 57	☽ ± ♇
14 42	☽ □ ♃
15 50	☽ ⚺ ♄
09 Friday	
01 05	☽ ⚻ ♇
02 58	☽ ⊥ ♄
07 03	☽ ⊥ ♂
07 08	☽ □ ♆
09 34	☽ ⚹ ♀
11 35	♂ ℙ ♃
16 13	☽ ☍ ♅
19 08	☽ ⚺ ♂
19 56	☽ ∠ ♄
21 47	☉ ± ♄
23 25	☽ Q ☿
10 Saturday	
10 37	☽ ∥ ♆
13 19	☿ ☍ ♃
14 38	☽ □ ☉
17 13	☽ ∠ ♀
17 51	♀ ± ♇
20 28	☽ ∥ ♂
23 29	☽ △ ♃
11 Sunday	
00 14	☽ ⚹ ☿
00 57	☽ ⚹ ♄
06 31	☽ ∥ ♃
10 36	☿ △ ♄
11 01	☽ ☍ ♇
12 45	☽ ⊥ ♀
14 33	☽ # ☿
17 18	☽ △ ♆
21 58	♇ St D
12 Monday	
00 41	☽ Q ♄
01 57	☽ ⚺ ♀
03 07	☽ ⚻ ♅
05 03	☽ ℙ ♃
05 24	☽ Q ☉
08 15	☽ ∠ ☿
09 57	☽ ☌ ♂
14 49	♀ △ ♅
15 13	☽ ± ♅
21 37	☽ # ♀
23 03	☽ ± ♃
23 24	☽ ℙ ♆
13 Tuesday	
03 52	☽ ⊥ ☿
07 47	☽ ⚹ ☉
09 29	☽ ℙ ♅
11 07	☽ ⚻ ♃
12 53	☽ □ ♄
17 04	☽ ⚺ ☿
17 38	☽ ± ♆
21 38	☽ ∥ ♀
23 31	☽ ⚻ ♇
14 Wednesday	
05 50	☽ ⚻ ♆
11 44	☽ ± ♇
16 05	☽ △ ♅
17 02	☽ ∠ ☉
21 07	☽ ☌ ♀
21 08	☉ ⚻ ♃
15 Thursday	
02 54	☽ ⚺ ♂
06 03	☽ ℙ ♇
07 33	☿ # ♃
09 42	☽ # ♃
09 57	☽ ∥ ☿
12 21	☽ # ♂
12 57	☽ ⊥ ☉
15 53	☽ ⊥ ♂
18 52	☉ ⚻ ♄
21 16	☉ ∥ ♅
23 36	☽ ☍ ♃
16 Friday	
00 37	☿ # ♂
01 33	☽ △ ♄
01 56	☿ ∠ ♂
02 10	☽ ⚺ ☉
02 30	♂ ± ♅
06 39	☽ # ♆
11 16	☽ ∠ ♂
11 32	☽ ☌ ☿
12 24	☽ △ ♇
18 29	☽ ☍ ♆
21 12	☿ △ ♇
17 Saturday	
04 19	♂ ∥ ♃
04 50	☽ □ ♅
07 31	☽ ℙ ♄
09 53	☉ Q ♅
15 49	☽ ⚺ ♀
19 10	☽ ⚹ ♂
18 Sunday	
01 17	☽ ± ♄
04 54	☽ ⊥ ♀
10 59	☽ ⚻ ♃
13 01	☽ ⚻ ♄
16 29	☽ Q ♅
19 00	☽ ☌ ☉
19 16	♀ ℙ ♇
20 15	☽ Q ♂
22 37	☽ ± ♃
23 50	☽ □ ♇
19 Monday	
00 17	☽ ∠ ♀
04 56	☽ ⚺ ☿
05 35	☽ ⚻ ♆
11 26	☿ ☍ ♆
15 48	☽ ⚹ ♅
15 55	☽ ℙ ♃
17 08	☽ ± ♆
17 45	☽ ⊥ ☿
20 Tuesday	
00 27	☽ ∥ ♅
07 59	☽ ⚹ ♀
08 20	♂ ± ♃
09 03	☽ □ ♂
10 17	☽ ℙ ♆
10 25	☽ Q ♇
12 47	☽ ∠ ☿
13 44	☉ ± ♃
20 16	☽ △ ♃
20 25	☽ ∠ ♅
22 21	☽ ☍ ♄
21 Wednesday	
05 59	♂ ℙ ♆
08 30	☉ Q ♂
08 48	☽ Q ♀
08 59	☉ □ ♇
09 01	☽ ⚹ ♇
09 01	☽ ⚺ ☉
13 16	☽ ∥ ♅
13 19	☽ ⊥ ♅
14 25	☽ △ ♆
15 08	♀ ⚺ ♂
20 00	☽ ⚹ ☿
21 04	☽ ⊥ ☉
22 Thursday	
00 28	☽ ⚺ ♅
05 49	☿ # ♆
12 44	☽ ∠ ♇
14 55	☽ ∠ ☉
20 08	☽ △ ♂
20 30	☽ Q ☿
21 07	☽ □ ♀
23 Friday	
03 19	☽ □ ♃
05 07	☽ ⊥ ♇
05 23	☽ ⚻ ♄
10 31	♀ # ♂
15 55	☽ ⚺ ♇
16 03	☿ □ ♅
16 07	☽ ± ♄
20 08	☽ ⚹ ☉
20 59	☽ □ ♆
24 Saturday	
00 42	☽ ℙ ♂
04 26	☿ ℙ ♄
06 52	☽ ☌ ♅
07 42	☉ ⚻ ♆
08 05	☽ ℙ ♄
08 31	☽ □ ☿
11 20	☽ Q ♃
15 51	☽ ∥ ☿
17 32	☽ ± ♂
19 02	☽ Q ☉
25 Sunday	
00 11	☽ # ♆
04 38	☽ ⚻ ♂
04 43	☽ Q ♆
07 24	☽ △ ♀
08 16	☽ ⚹ ♃
10 17	☽ △ ♄
17 26	♀ ☍ ♃
19 31	☽ # ♃
20 42	☽ ☌ ♇
22 10	☽ ∥ ♀
26 Monday	
01 29	☽ ⚹ ♆
04 38	☽ □ ☉
08 06	☽ # ♂
10 02	☽ ∠ ♃
11 12	☽ ⚺ ♅
11 35	☽ ℙ ♀
16 33	♀ △ ♄
18 44	☽ △ ☿
21 27	☽ ⊥ ♅
27 Tuesday	
00 29	♀ # ♃
01 17	☽ ⊥ ♃
03 02	☽ ∠ ♆
04 14	☽ ± ♀
10 52	☽ ☍ ♂
11 22	☽ ⚺ ♃
12 42	☽ ∠ ♅
13 19	☽ □ ♄
15 13	☽ ⚻ ♀
18 10	☽ ⊥ ♆
20 01	♂ ⚻ ♃
23 05	☽ ℙ ☿
23 40	☽ ⚺ ♇
28 Wednesday	
04 11	☽ ⚺ ♆
09 41	☽ ⊥ ♇
10 56	☽ △ ☉
11 04	☽ ∥ ♂
13 50	☽ ⚹ ♅
15 55	☽ ± ☿
19 12	☽ Q ♄
29 Thursday	
00 36	☽ ∠ ♇
00 59	♂ ℙ ♅
01 47	☽ ∥ ♃
03 05	☽ ⚻ ☿
06 45	☉ ℙ ♃
06 59	☽ # ♀
07 14	♂ □ ♄
09 43	☽ Q ♅
13 02	☽ ☌ ♃
13 30	☽ ℙ ☉
14 55	☽ ⚹ ♄
15 21	☽ ⚻ ♂
20 08	☽ ∥ ♆
20 19	☿ ± ♄
21 15	☽ ☍ ♀
30 Friday	
01 20	☽ ⚹ ♇
01 44	☽ ± ♂
05 20	☽ ± ☉
05 40	☽ ☌ ♆
08 05	☉ ⚹ ♅
10 24	☉ ± ♆
15 23	☽ □ ♅
15 26	☽ ∠ ♄
15 52	☽ ⚻ ☉
17 16	☽ ℙ ♂
21 02	☽ Q ♇
23 21	♄ ℙ ♅
31 Saturday	
06 09	☽ ⊥ ♄
06 11	☉ # ♄
06 58	☽ # ☿
10 42	☽ ☍ ☿
14 08	☽ ⚺ ♃
15 58	☽ ⚺ ♄
19 13	☽ △ ♂
23 59	☽ ⊥ ♃

All ephemeris data is given at 12.00 UT and the Moon's longitude is additionally given for 24.00 UT

JUNE 1958

Date	Sidereal time h m s	Sun ☉	Moon ☽	Moon ☽ 24.00	Mercury ☿	Venus ♀	Mars ♂	Jupiter ♃	Saturn ♄	Uranus ♅	Neptune ♆	Pluto ♇
		° ' "	° ' "	° ' "	° '	° '	° '	° '	° '	° '	° '	° '
01	04 37 51	10 ♊ 29 05	05 ♐ 29 29	12 ♐ 40 54	22 ♉ 00	00 ♉ 23	25 ♓ 52	22 ♎ 13	23 ♐ 20	08 ♌ 30	02 ♏ 28	29 ♌ 53
02	04 41 48	11 26 33	19 ♐ 48 36	26 ♐ 51 54	23 46	01 32	26 35	22 R 10	23 R 15	08 32	02 R 27	29 54
03	04 45 44	12 24 00	03 ♑ 50 16	10 ♑ 43 13	25 34	02 41	27 18	22 07	23 11	08 34	02 25	29 54
04	04 49 41	13 21 26	17 ♑ 30 28	24 ♑ 11 48	27 24	03 50	28 01	22 04	23 07	08 37	02 24	29 55
05	04 53 37	14 18 51	00 ♒ 47 10	07 ♒ 16 38	29 ♉ 17	04 59	28 44	22 02	23 03	08 39	02 23	29 56
06	04 57 34	15 16 15	13 ♒ 40 26	19 ♒ 58 49	01 ♊ 12	06 08	29 ♓ 27	22 00	22 58	08 42	02 22	29 57
07	05 01 30	16 13 39	26 ♒ 12 13	02 ♓ 21 05	03 10	07 17	00 ♈ 10	21 57	22 54	08 44	02 21	29 57
08	05 05 27	17 11 02	08 ♓ 25 58	14 ♓ 27 27	05 09	08 27	00 53	21 55	22 49	08 47	02 20	29 58
09	05 09 24	18 08 25	20 ♓ 26 09	26 ♓ 22 42	07 11	09 36	01 36	21 54	22 45	08 50	02 19	29 ♌ 59
10	05 13 20	19 05 47	02 ♈ 17 46	08 ♈ 11 59	09 14	10 46	02 19	21 52	22 41	08 52	02 18	00 ♍ 00
11	05 17 17	20 03 08	14 ♈ 06 02	20 ♈ 00 30	11 20	11 55	03 01	21 51	22 36	08 55	02 17	00 01
12	05 21 13	21 00 29	25 ♈ 56 01	01 ♉ 53 08	13 26	13 05	03 44	21 49	22 32	08 58	02 16	00 02
13	05 25 10	21 57 50	07 ♉ 52 23	13 ♉ 54 16	15 35	14 15	04 26	21 48	22 27	09 01	02 15	00 02
14	05 29 06	22 55 10	19 ♉ 59 11	26 ♉ 07 29	17 44	15 25	05 09	21 47	22 23	09 03	02 14	00 03
15	05 33 03	23 52 30	02 ♊ 19 29	08 ♊ 35 24	19 55	16 35	05 51	21 46	22 18	09 06	02 13	00 04
16	05 36 59	24 49 49	14 ♊ 55 23	21 ♊ 19 29	22 06	17 45	06 34	21 46	22 14	09 09	02 12	00 05
17	05 40 56	25 47 07	27 ♊ 47 42	04 ♋ 19 57	24 18	18 55	07 16	21 46	22 10	09 12	02 11	00 06
18	05 44 53	26 44 25	10 ♋ 56 06	17 ♋ 35 57	26 30	20 05	07 58	21 45	22 05	09 15	02 10	00 07
19	05 48 49	27 41 43	24 ♋ 19 16	01 ♌ 05 45	28 ♊ 41	21 15	08 40	21 D 45	22 01	09 18	02 10	00 08
20	05 52 46	28 39 00	07 ♌ 55 08	14 ♌ 47 06	00 ♋ 53	22 25	09 22	21 46	21 56	09 21	02 09	00 10
21	05 56 42	29 ♊ 36 16	21 ♌ 41 23	28 ♌ 37 42	03 04	23 35	10 04	21 46	21 52	09 24	02 08	00 11
22	06 00 39	00 ♋ 33 31	05 ♍ 35 49	12 ♍ 35 30	05 13	24 46	10 46	21 46	21 48	09 27	02 08	00 12
23	06 04 35	01 30 46	19 ♍ 36 36	26 ♍ 38 56	07 22	25 56	11 27	21 47	21 43	09 30	02 07	00 13
24	06 08 32	02 28 00	03 ♎ 42 21	10 ♎ 46 44	09 30	27 06	12 09	21 48	21 39	09 33	02 07	00 14
25	06 12 28	03 25 13	17 ♎ 51 56	24 ♎ 57 48	11 35	28 17	12 50	21 49	21 35	09 36	02 06	00 15
26	06 16 25	04 22 26	02 ♏ 04 08	09 ♏ 10 42	13 40	29 ♉ 27	13 32	21 50	21 30	09 39	02 05	00 17
27	06 20 22	05 19 39	16 ♏ 17 12	23 ♏ 23 18	15 43	00 ♊ 38	14 13	21 52	21 26	09 43	02 05	00 18
28	06 24 18	06 16 50	00 ♐ 28 36	07 ♐ 32 39	17 43	01 48	14 54	21 53	21 22	09 46	02 04	00 19
29	06 28 15	07 14 02	14 ♐ 34 58	21 ♐ 35 00	19 42	02 59	15 35	21 55	21 18	09 49	02 04	00 20
30	06 32 11	08 ♋ 11 13	28 ♐ 32 14	05 ♑ 26 10	21 ♋ 39	04 ♊ 10	16 ♈ 16	21 ♎ 57	21 ♐ 14	09 ♌ 52	02 ♏ 03	00 ♍ 22

Date	Moon True ☊	Moon Mean ☊	Moon ☽ Latitude
	° '	° '	° '
01	00 ♏ 46	29 ♎ 22	02 N 57
02	00 R 41	29 18	03 55
03	00 35	29 15	04 37
04	00 28	29 12	05 02
05	00 22	29 09	05 09
06	00 16	29 06	05 01
07	00 12	29 03	04 38
08	00 10	28 59	04 03
09	00 D 10	28 56	03 18
10	00 11	28 53	02 25
11	00 12	28 50	01 26
12	00 13	28 47	00 N 23
13	00 R 13	28 43	00 S 41
14	00 10	28 40	01 44
15	00 06	28 37	02 44
16	00 00	28 34	03 36
17	29 ♎ 51	28 31	04 19
18	29 42	28 28	04 49
19	29 32	28 24	05 04
20	29 24	28 21	05 02
21	29 17	28 18	04 42
22	29 12	28 15	04 06
23	29 09	28 12	03 15
24	29 D 09	28 09	02 12
25	29 09	28 05	01 S 00
26	29 R 09	28 02	00 N 16
27	29 08	27 59	01 30
28	29 05	27 56	02 39
29	28 59	27 53	03 37
30	28 ♎ 51	27 ♎ 49	04 N 22

DECLINATIONS

Date	Sun ☉	Moon ☽	Mercury ☿	Venus ♀	Mars ♂	Jupiter ♃	Saturn ♄	Uranus ♅	Neptune ♆	Pluto ♇
	° '	° '	° '	° '	° '	° '	° '	° '	° '	° '
01	22 N 01	18 S 19	16 N 25	09 N 37	03 S 30	07 S 20	21 S 51	18 N 45	10 S 38	22 N 18
02	22 09	19 09	17 01	10 01	03 14	07 19	21 50	18 45	10 38	22 18
03	22 17	18 47	17 36	10 24	02 57	07 19	21 50	18 44	10 37	22 17
04	22 24	17 19	18 12	10 47	02 41	07 18	21 50	18 43	10 37	22 17
05	22 31	14 57	18 46	11 10	02 24	07 17	21 50	18 43	10 37	22 16
06	22 38	11 55	19 21	11 33	02 08	07 16	21 50	18 42	10 36	22 16
07	22 44	08 25	19 54	11 56	01 51	07 16	21 50	18 41	10 36	22 15
08	22 49	04 39	20 27	12 18	01 35	07 15	21 49	18 40	10 36	22 15
09	22 55	00 S 45	20 59	12 40	01 18	07 15	21 49	18 40	10 35	22 14
10	23 00	03 N 08	21 29	13 02	01 02	07 15	21 49	18 39	10 35	22 14
11	23 04	06 53	21 58	13 24	00 46	07 14	21 49	18 38	10 35	22 13
12	23 08	10 23	22 26	13 46	00 29	07 14	21 49	18 38	10 34	22 13
13	23 12	13 29	22 51	14 07	00 S 13	07 14	21 48	18 37	10 34	22 12
14	23 15	16 04	23 14	14 28	00 N 03	07 14	21 48	18 36	10 34	22 12
15	23 18	17 57	23 36	14 49	00 19	07 14	21 48	18 35	10 34	22 11
16	23 20	19 00	23 54	15 09	00 36	07 14	21 48	18 35	10 33	22 10
17	23 22	19 06	24 10	15 29	00 52	07 14	21 48	18 34	10 33	22 10
18	23 24	18 12	24 24	15 49	01 08	07 14	21 48	18 33	10 33	22 09
19	23 25	16 16	24 35	16 09	01 24	07 14	21 47	18 32	10 33	22 08
20	23 26	13 26	24 43	16 28	01 40	07 15	21 47	18 31	10 32	22 08
21	23 26	09 49	24 48	16 47	01 56	07 15	21 47	18 30	10 32	22 07
22	23 26	05 39	24 50	17 05	02 12	07 15	21 47	18 30	10 32	22 07
23	23 26	01 N 08	24 49	17 23	02 27	07 16	21 47	18 29	10 32	22 06
24	23 25	03 S 29	24 46	17 41	02 43	07 17	21 47	18 28	10 32	22 05
25	23 24	07 56	24 40	17 58	02 59	07 17	21 46	18 27	10 32	22 05
26	23 22	11 57	24 32	18 15	03 14	07 18	21 46	18 26	10 31	22 04
27	23 20	15 16	24 21	18 31	03 30	07 19	21 46	18 25	10 31	22 03
28	23 18	17 40	24 08	18 47	03 45	07 20	21 46	18 25	10 31	22 03
29	23 15	18 57	23 52	19 03	04 01	07 20	21 46	18 24	10 31	22 02
30	23 N 11	19 S 04	23 N 35	19 N 18	04 N 16	07 S 21	21 S 46	18 N 23	10 S 31	22 N 01

ZODIAC SIGN ENTRIES

Date	h m	Planets
01	02 54	☽ ♐
01	04 07	♀ ♉
03	05 23	☽ ♑
05	10 34	☽ ♒
05	20 59	☿ ♊
07	06 21	♂ ♈
07	19 24	☽ ♓
10	07 20	☽ ♈
10	18 50	♇ ♍
12	20 12	☽ ♉
15	07 31	☽ ♊
17	16 04	☽ ♋
19	22 04	☽ ♌
20	02 20	☿ ♋
21	21 57	☉ ♋
22	02 22	☽ ♍
24	05 42	☽ ♎
26	08 30	☽ ♏
26	23 08	♀ ♊
28	11 12	☽ ♐
30	14 32	☽ ♑

LATITUDES

Date	Mercury ☿	Venus ♀	Mars ♂	Jupiter ♃	Saturn ♄	Uranus ♅	Neptune ♆	Pluto ♇
	° '	° '	° '	° '	° '	° '	° '	° '
01	01 S 55	02 S 07	02 S 01	01 N 25	01 N 26	00 N 38	01 N 48	11 N 34
04	01 26	02 08	02 04	01 24	01 26	00 38	01 48	11 33
07	00 54	02 08	02 06	01 23	01 26	00 38	01 48	11 32
10	00 S 21	02 07	02 08	01 23	01 25	00 38	01 48	11 32
13	00 N 11	02 05	02 10	01 22	01 25	00 38	01 48	11 31
16	00 42	02 03	02 11	01 21	01 25	00 38	01 48	11 30
19	01 08	02 00	02 13	01 20	01 25	00 38	01 48	11 30
22	01 30	01 56	02 15	01 19	01 25	00 38	01 47	11 29
25	01 44	01 51	02 16	01 19	01 24	00 38	01 47	11 28
28	01 53	01 46	02 18	01 18	01 24	00 38	01 47	11 28
31	01 N 54	01 S 41	02 S 19	01 N 17	01 N 24	00 N 37	01 N 47	11 N 27

DATA

Julian Date	2436356
Delta T	+32 seconds
Ayanamsa	23° 16' 42"
Synetic vernal point	05° ♓ 50' 18"
True obliquity of ecliptic	23° 26' 32"

MOON'S PHASES, APSIDES AND POSITIONS ☽

Date	h m	Phase	Longitude ° '	Eclipse Indicator
01	20 55	○	10 ♐ 50	
09	06 59	☾	17 ♓ 56	
17	07 59	●	25 ♊ 38	
24	09 45	☽	02 ♎ 23	

Day	h m		
11	04 37	Apogee	
26	08 58	Perigee	
02	16 23	Max dec	19° S 10'
09	16 37	0N	
17	02 29	Max dec	19° N 11'
23	17 51	0S	
30	02 22	Max dec	19° S 10'

LONGITUDES

Date	Chiron ⚷	Ceres ⚳	Pallas ⚴	Juno ⚵	Vesta ⚶	Black Moon Lilith ⚸
	° '	° '	° '	° '	° '	° '
01	22 ♒ 43	07 ♌ 40	03 ♋ 45	26 ♋ 19	10 ♊ 06	11 ♈ 19
11	22 ♒ 37	11 ♌ 34	09 ♋ 42	00 ♌ 55	14 ♊ 27	11 ♈ 25
21	22 ♒ 25	15 ♌ 37	15 ♋ 37	05 ♌ 30	18 ♊ 46	13 ♈ 32
31	22 ♒ 07	19 ♌ 47	21 ♋ 31	10 ♌ 05	23 ♊ 04	14 ♈ 38

ASPECTARIAN

h m	Aspects
01 Sunday	
01 40	♀ △ ♇
02 42	☽ □ ♇
02 47	☽ ⊼ ♀
06 59	☽ ⊻ ♆
13 36	☽ ± ♀
14 52	☽ ∠ ♃
14 56	☿ ⊼ ♃
16 56	☽ ⊥ ♆
17 01	☽ △ ♅
20 28	☽ ⋕ ♅
20 55	☽ ☍ ☉
02 Monday	
05 27	☿ ⊼ ♄
05 58	☽ ꝑ ♀
07 31	♂ ± ♆
08 00	☽ ∠ ♆
15 59	☽ ⚹ ♃
17 49	☽ ☌ ♄
18 20	☽ ꝑ ♅
19 41	☽ ⊼ ☿
03 Tuesday	
00 08	☽ □ ♂
05 13	☽ △ ♇
06 49	♀ ☍ ♆
07 29	☽ ± ☿
09 33	☽ ⚹ ♆
09 48	☽ △ ♀
09 48	☽ ± ♅
12 29	☽ Q ♃
12 33	☉ ∥ ♇
13 07	☽ ⋕ ♅
20 16	☽ ⊼ ♅
04 Wednesday	
01 31	☿ Q ♅
01 32	♀ ⋕ ♆
01 33	☽ ꝑ ☿
03 02	☽ ⋕ ☿
04 05	☽ ⊼ ☉
06 30	☽ Q ♆
07 24	☽ ꝑ ♇
09 12	☽ Q ♂
15 33	☽ ± ☉
20 09	☽ □ ♃
20 24	☿ ± ♃
22 00	☽ ⊻ ♄
22 35	♀ ⊥ ♂
23 31	☽ ± ♇
05 Thursday	
00 45	☿ ⚹ ♂
08 02	☽ ⚹ ♂
08 47	☽ △ ☿
08 49	☽ ⊥ ♄
09 06	☽ ꝑ ☉
09 22	☿ ∥ ♅
10 26	☽ ⊼ ♇
14 56	☽ □ ♆
20 10	☿ □ ♇
20 29	☽ □ ♀
06 Friday	
01 21	☽ ∠ ♄
02 37	☽ ☍ ♅
13 34	☽ ∠ ♂
14 20	☽ ⋕ ♀
15 16	☽ △ ☉
21 17	☽ ∥ ♆
07 Saturday	
02 11	☿ ⊼ ♆
03 49	☽ △ ♃
04 43	♂ ⊼ ♇
05 38	☽ ⚹ ♄
07 49	☽ ⊥ ♂
10 03	☽ Q ♀
19 19	☽ ☍ ♇
19 28	☽ ∥ ♃
20 12	☽ ⊻ ♂
23 51	♀ ꝑ ♄
23 59	☽ △ ♆
08 Sunday	
04 14	☽ □ ☿
04 54	☽ Q ♄
09 01	☽ ꝑ ♃
12 02	☽ ⚹ ♀
12 42	☽ ⊼ ♅
15 38	☉ ꝑ ♆
19 16	♀ □ ♅
09 Monday	
00 42	☽ ± ♅
02 54	☽ ± ♃
05 44	☽ ꝑ ♆
06 59	☽ □ ☉
07 58	☉ Q ♇
08 20	☽ ∥ ♂
08 42	☿ ꝑ ♃
14 56	☽ ⊼ ♃
16 38	☽ □ ♄
18 52	☽ ꝑ ♅
21 19	☽ ∠ ♀
23 34	☽ Q ☿
23 50	☽ ⋕ ♂
23 51	☽ ± ♆
10 Tuesday	
01 11	☿ ± ♆
07 20	☽ ⊼ ♇
07 41	☿ ⚹ ♅
11 36	♂ ⊼ ♆
12 00	☽ ⊼ ♆
12 02	☽ ☌ ♂
17 34	☽ ⊥ ♀
19 32	☽ ± ♇
22 37	☽ Q ☉
11 Wednesday	
01 25	☽ △ ♅
04 01	☿ ⋕ ♄
05 08	☽ ⚹ ☿
07 05	☽ ⊻ ♀
13 51	☽ ꝑ ♇
14 21	☽ ⋕ ♃
12 Thursday	
00 31	☿ ∥ ♇
01 09	☽ ⚹ ☉
03 07	☿ ⊻ ♀
03 41	☽ ☍ ♃
05 09	☽ △ ♄
13 24	☽ ⋕ ♆
18 10	☽ ∠ ☿
20 16	☽ △ ♇
13 Friday	
00 45	☽ ☍ ♆
04 42	☽ ⊻ ♂
08 01	☉ △ ♃
10 02	☽ ∠ ☉
11 10	☽ ꝑ ♄
14 17	☽ □ ♅
16 09	☽ ⊥ ☿
17 26	☽ ⊥ ♂
18 12	☽ ∥ ♀
23 27	☉ ☍ ♄
14 Saturday	
02 02	☽ ☌ ♀
02 15	☿ Q ♂
04 57	☽ ± ♄
05 27	☽ ⊥ ☉
06 26	☿ ꝑ ♆
06 37	☽ ⊻ ☿
12 20	☽ ∠ ♂
12 54	☉ ∥ ☿
15 31	☿ Q ♇
15 32	☽ ⊼ ♃
16 40	☽ ⊼ ♄
18 14	☽ ⊻ ☉
15 Sunday	
01 52	☽ Q ♅
03 13	☽ ± ♃
06 47	♀ ± ♄
07 39	☽ □ ♇
11 48	☽ ⊼ ♆
18 02	☉ ∠ ♅
19 11	☽ ⚹ ♂
19 35	♂ ± ♇
20 32	☽ ꝑ ♃
23 17	☽ ± ♆
23 49	☽ ∥ ♅
16 Monday	
01 02	☽ ⚹ ♅
08 19	☿ △ ♃
13 23	☿ ☍ ♄
16 17	☽ ꝑ ♆
17 50	☽ ⊻ ♀
17 58	☽ Q ♇
19 14	☽ Q ♂
17 Tuesday	
00 49	☽ △ ♃
01 37	☽ ☍ ♄
04 12	☽ ☌ ☿
05 19	☽ ∠ ♅
06 08	☽ ⊥ ♀
07 59	☽ ☌ ☉
10 54	☿ ∠ ♅
16 16	☽ ⚹ ♇
20 05	☽ △ ♆
21 58	☽ ⊥ ♅
18 Wednesday	
00 19	☽ ∠ ♀
02 14	☿ ⊥ ♀
05 15	☽ ∥ ♅
06 19	☽ □ ♂
08 56	☽ ⊻ ♅
16 43	☉ ☌ ☿
19 34	☽ ∠ ♇
19 Thursday	
01 42	♃ St D
06 00	☽ ⚹ ♀
07 26	☽ □ ♃
07 55	☽ ⊼ ♄
11 41	☽ ⊥ ♇
13 06	☽ ∥ ♀
18 26	☽ ⊻ ☉
18 31	☽ ± ♄
21 15	☽ ⊻ ☿
22 20	☽ ⊻ ♇
22 28	♀ ⊼ ♃
20 Friday	
01 52	☽ □ ♆
02 49	♀ ⊼ ♄
04 01	☿ ⚹ ♇
05 16	☽ Q ♀
05 50	☽ ⊥ ☉
09 50	☽ ⊥ ☿
10 18	☽ ꝑ ♄
11 17	♂ △ ♅
14 31	☽ ☌ ♅
14 40	☽ △ ♂
15 13	☽ Q ♃
22 46	☽ ∠ ☉
21 Saturday	
01 55	☿ △ ♆
04 31	☽ ∠ ☿
07 34	☽ ⋕ ♆
09 19	☽ Q ♆
12 08	☽ ⚹ ♃
12 18	☽ △ ♄
13 56	☉ ⊥ ♀
15 35	☽ □ ♀
15 50	☿ ⊥ ♅
18 09	☽ ꝑ ♂
22 Sunday	
02 41	☽ ⚹ ☉
02 42	☽ ☌ ♇
02 44	☉ ⚹ ♇
03 03	☽ ⋕ ♃
06 02	☽ ⚹ ♆
10 29	☽ ± ♂
11 14	☽ ⚹ ☿
14 01	☽ ∠ ♃
18 10	♃ ⚹ ♄
18 38	☽ ⊻ ♅
21 20	☽ ⊼ ♂
23 Monday	
00 49	☽ Q ☉
04 57	☽ ⊥ ♅
05 25	☽ ∥ ♂
05 28	☽ ⊥ ♃
07 45	☽ ∠ ♆
11 31	☽ Q ☿
15 35	☽ □ ♄
15 43	☽ ⊻ ♃
20 22	☽ ∠ ♅
23 05	☽ ⊥ ♆
23 45	☽ △ ♀
24 Tuesday	
03 05	☉ △ ♆
06 06	☽ ⊻ ♇
07 45	☽ ⋕ ♂
09 17	☽ ⊻ ♆
09 45	☽ □ ☉
12 42	☿ ⊻ ♅
14 14	♀ ∠ ♂
16 18	☽ ⊥ ♇
21 35	♀ Q ♅
21 57	☽ ⚹ ♅
22 02	☽ Q ♄
23 32	☽ □ ☿
25 Wednesday	
02 25	♀ ± ♃
03 03	☽ ☍ ♂
03 32	☽ ꝑ ♀
07 35	☽ ∠ ♇
08 23	☽ ∥ ♃
16 55	☉ ⊥ ♅
18 15	☽ ⚹ ♄
18 21	☽ Q ♅
18 42	☽ ☌ ♃
20 08	☽ ± ♀
26 Thursday	
03 04	☽ ∥ ♆
07 11	☽ ⊼ ♀
08 58	☽ ⚹ ♇
09 40	☿ □ ♂
12 02	☽ ☌ ♆
16 10	☽ △ ☉
19 27	☽ ∠ ♄
27 Friday	
00 51	☽ □ ♅
03 43	♀ ∥ ♅
05 07	♀ □ ♇
05 15	☽ Q ♇
07 07	☿ ∠ ♇
08 20	☽ ⊼ ♂
09 49	☿ ∠ ♀
10 34	☽ ⊥ ♄
10 52	☽ △ ☿
18 59	☽ ± ♂
19 19	☽ ꝑ ☉
20 39	☽ ⊻ ♄
21 26	☽ ⊻ ♃
28 Saturday	
07 36	☽ ⊥ ♃
10 59	☽ ꝑ ♂
11 39	☽ ± ☉
11 44	☽ □ ♇
14 28	☽ ☍ ♀
14 42	☽ ⊻ ♆
16 26	☽ ꝑ ☿
17 21	♀ ⊼ ♆
22 34	☽ ⊼ ☉
22 54	☽ ∠ ♃
22 59	☽ ⋕ ♅
29 Sunday	
00 53	☽ ⊥ ♆
02 59	♂ ꝑ ♇
03 50	☽ △ ♅
10 15	☽ ± ☿
13 49	☽ △ ♂
16 15	☽ ∠ ♆
22 12	☽ ⊼ ☿
23 27	☽ ☌ ♄
30 Monday	
00 36	☽ ⚹ ♃
05 38	☽ ꝑ ♅
06 53	☿ ⊼ ♄
15 10	☽ △ ♇
15 44	☿ □ ♃
18 06	☽ ⚹ ♆
21 19	☽ ± ♅
21 26	☽ Q ♃
22 42	☽ ⊼ ♀

All ephemeris data is given at 12.00 UT and the Moon's longitude is additionally given for 24.00 UT

JULY 1958

LONGITUDES

Date	Sidereal time h m s	Sun ☉	Moon ☽	Moon ☽ 24.00	Mercury ☿	Venus ♀	Mars ♂	Jupiter ♃	Saturn ♄	Uranus ♅	Neptune ♆	Pluto ♇
01	06 36 08	09 ♋ 08 24	12 ♑ 16 18	19 ♑ 02 14	23 ♋ 34	05 ♊ 21	16 ♈ 57	21 ♎ 59	21 ♐ 10	09 ♌ 56	02 ♏ 03	00 ♍ 23
02	06 40 04	10 05 35	25 ♑ 43 38	02 ♒ 20 13	25 27	06 32	17 38	22 01	21 R 06	09 59	02 R 03	00 25
03	06 44 01	11 02 46	08 ♒ 51 50	15 ♒ 18 28	27 18	07 42	18 19	22 04	21 01	10 02	02 02	00 26
04	06 47 57	11 59 57	21 ♒ 40 08	27 ♒ 57 02	29 ♋ 07	08 53	18 59	22 07	20 57	10 06	02 02	00 27
05	06 51 54	12 57 08	04 ♓ 09 25	10 ♓ 17 37	00 ♌ 54	10 04	19 40	22 09	20 54	10 09	02 02	00 29
06	06 55 51	13 54 20	16 ♓ 22 06	22 ♓ 23 21	02 39	11 15	20 20	22 12	20 50	10 13	02 01	00 30
07	06 59 47	14 51 31	28 ♓ 21 56	04 ♈ 18 27	04 22	12 27	21 00	22 15	20 46	10 16	02 01	00 32
08	07 03 44	15 48 43	10 ♈ 13 33	16 ♈ 07 53	06 03	13 38	21 40	22 19	20 42	10 19	02 01	00 33
09	07 07 40	16 45 56	22 ♈ 02 09	27 ♈ 57 01	07 41	14 49	22 20	22 22	20 38	10 23	02 01	00 35
10	07 11 37	17 43 08	03 ♉ 53 10	09 ♉ 51 17	09 18	16 00	23 00	22 26	20 35	10 26	02 01	00 36
11	07 15 33	18 40 22	15 ♉ 51 59	21 ♉ 55 53	10 52	17 12	23 40	22 30	20 31	10 30	02 00	00 38
12	07 19 30	19 37 35	28 ♉ 03 30	04 ♊ 15 20	12 25	18 23	24 19	22 34	20 27	10 33	02 00	00 39
13	07 23 26	20 34 49	10 ♊ 31 48	16 ♊ 53 11	13 55	19 34	24 59	22 38	20 24	10 37	02 00	00 41
14	07 27 23	21 32 04	23 ♊ 19 44	29 ♊ 51 33	15 24	20 46	25 38	22 42	20 20	10 40	02 D 00	00 43
15	07 31 20	22 29 19	06 ♋ 28 37	13 ♋ 10 49	16 50	21 58	26 17	22 46	20 17	10 44	02 00	00 44
16	07 35 16	23 26 35	19 ♋ 57 53	26 ♋ 49 29	18 14	23 09	26 56	22 51	20 14	10 48	02 00	00 46
17	07 39 13	24 23 51	03 ♌ 45 10	10 ♌ 44 24	19 35	24 21	27 35	22 56	20 10	10 51	02 01	00 48
18	07 43 09	25 21 07	17 ♌ 46 36	24 ♌ 51 09	20 55	25 32	28 13	23 01	20 07	10 55	02 01	00 49
19	07 47 06	26 18 24	01 ♍ 57 26	09 ♍ 04 51	22 12	26 44	28 52	23 06	20 04	10 58	02 01	00 51
20	07 51 02	27 15 41	16 ♍ 12 53	23 ♍ 21 01	23 27	27 56	29 ♈ 30	23 11	20 01	11 02	02 01	00 53
21	07 54 59	28 12 58	00 ♎ 28 49	07 ♎ 35 59	24 40	29 ♊ 08	00 ♉ 08	23 16	19 58	11 06	02 01	00 54
22	07 58 55	29 ♋ 10 15	14 ♎ 42 13	21 ♎ 47 19	25 50	00 ♋ 20	00 46	23 22	19 55	11 09	02 01	00 56
23	08 02 52	00 ♌ 07 32	28 ♎ 51 08	05 ♏ 53 34	26 58	01 32	01 23	23 28	19 52	11 13	02 02	00 58
24	08 06 49	01 04 50	12 ♏ 54 31	19 ♏ 53 56	28 03	02 44	02 01	23 33	19 49	11 17	02 02	01 00
25	08 10 45	02 02 09	26 ♏ 51 43	03 ♐ 47 47	29 ♌ 05	03 56	02 38	23 39	19 47	11 20	02 02	01 01
26	08 14 42	02 59 27	10 ♐ 42 00	17 ♐ 34 13	00 ♍ 05	05 08	03 15	23 45	19 44	11 24	02 03	01 03
27	08 18 38	03 56 47	24 ♐ 24 14	01 ♑ 11 51	01 01	06 20	03 52	23 52	19 41	11 28	02 03	01 05
28	08 22 35	04 54 06	07 ♑ 56 48	14 ♑ 38 51	01 55	07 32	04 29	23 58	19 39	11 31	02 04	01 07
29	08 26 31	05 51 27	21 ♑ 17 45	27 ♑ 53 14	02 45	08 44	05 06	24 05	19 37	11 35	02 04	01 09
30	08 30 28	06 48 47	04 ♒ 25 08	10 ♒ 53 14	03 32	09 56	05 42	24 11	19 34	11 39	02 05	01 10
31	08 34 24	07 ♌ 46 09	17 ♒ 17 27	23 ♒ 37 43	04 ♍ 15	11 ♋ 09	06 ♉ 18	24 ♎ 18	19 ♐ 32	11 ♌ 42	02 ♏ 05	01 ♍ 12

Date	Moon True ☊	Moon Mean ☊	Moon ☽ Latitude
01	28 ♎ 40	27 ♎ 46	04 N 50
02	28 R 29	27 43	05 02
03	28 18	27 40	04 57
04	28 08	27 37	04 38
05	28 01	27 34	04 05
06	27 55	27 30	03 22
07	27 52	27 27	02 30
08	27 51	27 24	01 32
09	27 D 51	27 21	00 N 31
10	27 R 51	27 18	00 S 32
11	27 50	27 14	01 34
12	27 47	27 11	02 33
13	27 42	27 08	03 26
14	27 33	27 05	04 09
15	27 23	27 02	04 41
16	27 11	26 59	04 58
17	26 59	26 55	04 59
18	26 48	26 52	04 41
19	26 39	26 49	04 06
20	26 33	26 46	03 16
21	26 29	26 43	02 13
22	26 28	26 40	01 S 02
23	26 D 27	26 36	00 N 13
24	26 R 27	26 33	01 26
25	26 26	26 30	02 33
26	26 22	26 27	03 31
27	26 15	26 24	04 16
28	26 06	26 20	04 46
29	25 55	26 17	05 00
30	25 43	26 14	04 58
31	25 ♎ 31	26 ♎ 11	04 N 40

DECLINATIONS

Date	Sun ☉	Moon ☽	Mercury ☿	Venus ♀	Mars ♂	Jupiter ♃	Saturn ♄	Uranus ♅	Neptune ♆	Pluto ♇
01	23 N 08	18 S 04	23 N 15	19 N 33	04 N 31	07 S 23	21 S 45	18 N 22	10 S 31	22 N 01
02	23 04	16 03	22 54	19 47	04 47	07 24	21 45	18 21	10 31	22 00
03	22 59	13 16	22 32	20 00	05 02	07 25	21 45	18 20	10 31	21 59
04	22 54	09 54	22 07	20 14	05 17	07 26	21 45	18 19	10 31	21 59
05	22 49	06 11	21 42	20 26	05 31	07 27	21 45	18 18	10 31	21 58
06	22 43	02 S 17	21 15	20 39	05 46	07 29	21 45	18 17	10 30	21 57
07	22 37	01 N 39	20 47	20 50	06 01	07 30	21 45	18 16	10 30	21 57
08	22 30	05 28	20 18	21 01	06 16	07 32	21 45	18 15	10 30	21 56
09	22 23	09 04	19 48	21 12	06 30	07 33	21 44	18 14	10 30	21 55
10	22 16	12 19	19 17	21 22	06 44	07 35	21 44	18 13	10 30	21 54
11	22 08	15 05	18 45	21 31	06 59	07 36	21 44	18 13	10 30	21 54
12	22 00	17 15	18 13	21 40	07 13	07 38	21 44	18 12	10 30	21 53
13	21 52	18 38	17 40	21 49	07 27	07 40	21 44	18 11	10 31	21 52
14	21 43	19 07	17 07	21 56	07 41	07 42	21 44	18 10	10 31	21 52
15	21 34	18 36	16 34	22 04	07 55	07 43	21 44	18 09	10 31	21 51
16	21 24	17 02	16 00	22 10	08 09	07 45	21 44	18 08	10 31	21 50
17	21 14	14 28	15 26	22 16	08 22	07 47	21 44	18 07	10 31	21 49
18	21 04	11 02	14 52	22 21	08 36	07 49	21 44	18 06	10 31	21 49
19	20 54	06 57	14 18	22 26	08 49	07 52	21 43	18 05	10 31	21 48
20	20 43	02 N 26	13 45	22 30	09 02	07 54	21 43	18 04	10 31	21 47
21	20 31	02 S 13	13 11	22 34	09 16	07 56	21 43	18 03	10 31	21 46
22	20 20	06 45	12 37	22 37	09 29	07 58	21 43	18 02	10 31	21 46
23	20 08	10 52	12 04	22 39	09 41	08 01	21 43	18 01	10 32	21 45
24	19 55	14 21	11 32	22 41	09 54	08 03	21 43	18 00	10 32	21 44
25	19 43	16 58	10 59	22 42	10 07	08 05	21 43	17 59	10 32	21 44
26	19 30	18 34	10 28	22 42	10 19	08 08	21 43	17 58	10 32	21 43
27	19 16	19 03	09 57	22 42	10 32	08 10	21 43	17 57	10 32	21 42
28	19 03	18 26	09 27	22 41	10 44	08 13	21 43	17 56	10 32	21 41
29	18 49	16 49	08 58	22 39	10 56	08 15	21 43	17 55	10 33	21 41
30	18 34	14 20	08 29	22 37	11 08	08 18	21 43	17 54	10 33	21 40
31	18 N 20	11 S 12	08 N 02	22 N 34	11 N 20	08 S 21	21 S 43	17 N 53	10 S 33	21 N 39

ZODIAC SIGN ENTRIES

Date	h	m	Planets
02	19	44	☽ ♒
04	23	46	☿ ♌
05	03	57	☽ ♓
07	15	18	☽ ♈
10	04	09	☽ ♉
12	15	47	☽ ♊
15	00	15	☽ ♋
17	05	31	☽ ♌
19	08	42	☽ ♍
21	07	03	♂ ♉
21	11	11	☽ ♎
22	05	26	♀ ♋
23	08	50	☉ ♌
23	13	57	☽ ♏
25	17	25	☽ ♐
26	10	08	☿ ♍
27	21	53	☽ ♑
30	03	52	☽ ♒

LATITUDES

Date	Mercury ☿	Venus ♀	Mars ♂	Jupiter ♃	Saturn ♄	Uranus ♅	Neptune ♆	Pluto ♇
01	01 N 54	01 S 41	02 S 19	01 N 17	01 N 24	00 N 37	01 N 47	11 N 27
04	01 50	01 35	02 20	01 16	01 23	00 37	01 47	11 26
07	01 39	01 28	02 21	01 15	01 23	00 37	01 47	11 26
10	01 24	01 21	02 22	01 15	01 22	00 37	01 46	11 25
13	01 04	01 14	02 23	01 14	01 22	00 37	01 46	11 25
16	00 40	01 06	02 24	01 13	01 21	00 37	01 46	11 25
19	00 N 12	00 58	02 24	01 12	01 21	00 37	01 46	11 24
22	00 S 18	00 50	02 25	01 12	01 21	00 37	01 46	11 24
25	00 51	00 41	02 25	01 11	01 20	00 37	01 46	11 24
28	01 27	00 33	02 25	01 10	01 20	00 37	01 45	11 23
31	02 S 03	00 S 25	02 S 25	01 N 09	01 N 19	00 N 37	01 N 45	11 N 23

DATA

Julian Date	2436386
Delta T	+32 seconds
Ayanamsa	23° 16' 47"
Synetic vernal point	05° ♓ 50' 13"
True obliquity of ecliptic	23° 26' 32"

MOON'S PHASES, APSIDES AND POSITIONS ☽

Date	h	m	Phase	Longitude	Eclipse Indicator
01	06	04	○	08 ♑ 54	
09	00	21	☾	16 ♈ 18	
16	18	33	●	23 ♋ 42	
23	14	20	☽	00 ♏ 13	
30	16	47	○	07 ♒ 00	

Day	h	m		
08	23	07	Apogee	
21	10	49	Perigee	
07	01	56	0N	
14	11	52	Max dec	19° N 07'
21	00	33	0S	
27	10	29	Max dec	19° S 03'

LONGITUDES

Date	Chiron ⚷	Ceres ⚳	Pallas ⚴	Juno ⚵	Vesta ⚶	Black Moon Lilith ☾
01	22 ♒ 07	19 ♌ 47	21 ♋ 31	10 ♌ 05	23 ♊ 04	14 ♈ 38
11	21 ♒ 45	24 ♌ 03	27 ♋ 21	14 ♌ 38	27 ♊ 20	15 ♈ 45
21	21 ♒ 18	28 ♌ 24	03 ♌ 08	19 ♌ 09	01 ♋ 33	16 ♈ 51
31	20 ♒ 49	02 ♍ 49	08 ♌ 49	23 ♌ 39	05 ♋ 43	17 ♈ 58

ASPECTARIAN

h m	Aspects
01 Tuesday	
06 04	☽ ☍ ☉
06 51	☽ # ♅
07 51	☽ ⊼ ♅
10 13	☽ ± ♀
15 08	☽ Q ♆
17 31	☽ ꝑ ♇
20 44	☽ □ ♂
22 27	☿ ⊥ ♇
23 19	☉ ∥ ☿
02 Wednesday	
03 43	☽ ⊻ ♄
03 43	☽ ꝑ ♀
05 19	☽ □ ♃
09 04	☉ ⊻ ♅
09 37	☽ ± ♇
11 26	☽ ☍ ☿
14 27	☽ ⊥ ♄
20 30	☽ ⊼ ♇
22 29	♀ ꝑ ♃
23 27	☽ □ ♆
03 Thursday	
06 47	☽ ∠ ♄
07 02	☽ Q ♂
08 27	☿ ± ♄
09 39	☽ △ ♀
14 11	☽ ☍ ♅
16 22	☽ ⊼ ☉
18 40	♀ ± ♆
04 Friday	
04 30	☽ ± ☉
06 39	☽ ⚹ ♂
07 52	☽ ∥ ♆
10 40	☽ ⚹ ♄
12 50	☽ △ ♃
20 27	☿ ∥ ♇
23 01	☽ ꝑ ☉
05 Saturday	
04 01	☽ ∥ ♃
04 39	☽ ⊼ ☿
04 52	☽ ☍ ♇
06 11	☿ ⊻ ♇
07 52	☽ △ ♆
08 59	☿ # ♄
09 33	☽ Q ♄
13 03	☽ ∠ ♂
13 40	♀ ⚹ ♅
15 51	☽ # ♂
17 52	☽ ꝑ ♃
18 15	☽ ± ☿
23 47	☽ ⊼ ♅
06 Sunday	
00 15	♃ Q ♅
00 49	☽ □ ♀
03 20	☿ □ ♆
06 42	☽ △ ☉
07 44	☽ ⊥ ♂
11 40	☽ ± ♃
11 41	☽ ± ♅
13 18	☽ ꝑ ♆
14 58	☽ ꝑ ☿
20 22	☽ ⊻ ♂
20 50	☽ □ ♄
23 41	☽ ⊼ ♃
07 Monday	
04 04	♂ △ ♄
05 44	☽ ꝑ ♅
07 17	☽ ± ♆
09 52	☿ ∥ ♀
16 22	☽ ⊼ ♇
16 39	☽ Q ♀
19 22	☽ ⊼ ♆
08 Tuesday	
02 07	☽ △ ☿
04 32	☽ ± ♇
05 20	☉ ∠ ♇
07 14	☿ ꝑ ♄
12 12	☽ △ ♅
17 31	☽ ∥ ♂
19 41	☽ ⚹ ♀
22 51	☽ ꝑ ♇
09 Wednesday	
00 21	☽ □ ☉
01 37	☽ # ♃
09 10	☽ △ ♄
12 39	☽ ☌ ♂
12 41	☽ ☍ ♃
13 15	♂ ☍ ♃
22 18	☽ # ♆
10 Thursday	
05 22	☽ △ ♇
05 32	☽ ∠ ♀
08 13	☽ ☍ ♆
15 23	☽ ꝑ ♄
16 01	☽ Q ☉
11 Friday	
00 33	☽ □ ☿
01 14	☽ □ ♅
01 39	☽ ⊥ ♀
05 56	☿ Q ♃
06 00	☿ ☌ ♅
08 15	♀ ꝑ ♆
09 19	☽ ± ♄
14 55	☽ ⊻ ♀
18 02	☽ ⚹ ☉
21 10	☽ ⊼ ♄
21 45	♃ Q ♅
12 Saturday	
01 11	☽ ⊼ ♃
04 17	☽ ⊻ ♂
12 58	☽ Q ♅
12 59	☽ ± ♃
13 07	☿ ∥ ♅
16 39	☽ ⊥ ♂
17 04	☽ □ ♇
17 13	☽ Q ☿
17 38	♀ Q ♇
19 40	☽ ⊼ ♆
22 30	♀ # ♄
22 47	☽ ∥ ☿
13 Sunday	
01 46	☽ ∠ ☉
02 26	☽ ∥ ♅
06 26	☽ ꝑ ♃
07 12	☽ ± ♆
07 38	☉ ⊼ ♄
10 54	☽ ∠ ♂
10 54	☉ ∥ ♇
12 10	☽ ⚹ ♅
16 46	☉ ∥ ♀
19 16	☽ ⚹ ☿
20 17	☽ ⊥ ☉
22 10	♀ ∥ ♇
14 Monday	
00 13	☽ ꝑ ♆
03 24	☽ Q ♇
03 47	♀ ☍ ♄
05 49	♆ St D
06 28	☽ ☍ ♄
06 46	☽ ☌ ♀
08 25	☽ ⊻ ☉
10 01	☉ # ♄
10 50	☽ △ ♃
13 11	♂ # ♃
16 21	☽ ∠ ♅
16 28	☽ ⚹ ♂
15 Tuesday	
01 35	☽ ⚹ ♇
02 34	☽ ∠ ☿
03 55	☽ △ ♆
08 50	☽ ⊥ ♅
15 25	☽ Q ♂
19 40	☽ ⊻ ♅
19 48	☉ □ ♃
20 43	☽ ⊥ ☿
20 59	☽ ∥ ♅
16 Wednesday	
04 34	☽ ∠ ♇
05 31	♀ △ ♃
08 36	☽ ⊻ ☿
12 27	☽ ⊼ ♄
17 06	☽ □ ♃
18 07	☽ ⊻ ♀
18 33	☽ ☌ ☉
20 26	☽ ⊥ ♇
22 55	☽ ± ♄
17 Thursday	
00 47	☽ □ ♂
01 48	☽ ∥ ☿
05 33	☽ ⊥ ♀
06 53	☽ ⊻ ♇
08 59	☽ □ ♆
14 26	☽ ꝑ ♄
17 08	☉ ⊻ ♀
19 30	☿ Q ♆
22 00	☿ △ ♄
22 14	☉ ⊥ ♇
22 31	☽ ∠ ♀
18 Friday	
00 15	☽ ☌ ♅
00 24	☽ Q ♃
15 17	☽ # ♆
15 48	☽ Q ♆
15 58	☽ △ ♄
17 52	☽ ☌ ☿
19 51	♀ ∠ ♅
20 56	☽ ⚹ ♃
19 Saturday	
01 46	☽ ⊻ ☉
01 57	☽ ∥ ♂
02 23	☽ ⚹ ♀
06 14	☉ ± ♄
06 32	☽ △ ♂
06 56	☽ # ♃
10 08	☽ ☌ ♇
12 06	☽ ⚹ ♆
12 38	☽ ⊥ ☉
22 24	☽ ∠ ♃
20 Sunday	
00 28	☽ Q ♀
03 15	☽ ⊻ ♅
04 52	☽ ∠ ☉
06 18	☿ ⚹ ♃
08 59	☽ ꝑ ♂
13 21	☽ ∠ ♆
13 23	☽ ⊥ ♅
13 38	☽ ⊥ ♃
18 22	☽ □ ♄
23 48	☽ ⊻ ♃
21 Monday	
00 49	☽ ± ♂
01 19	☽ ⊻ ☿
04 29	☽ ⊥ ♆
04 35	☽ ∠ ♅
07 55	☽ ⚹ ☉
09 31	☽ □ ♀
11 23	☽ ⊼ ♂
12 20	☽ ⊥ ☿
12 43	☽ ⊻ ♇
14 36	☽ ⊻ ♆
22 51	☽ ⊥ ♇
22 Tuesday	
00 34	☽ Q ♄
04 53	☽ ∠ ☿
05 36	☽ Q ☉
05 59	☽ ⚹ ♅
14 05	☽ ∠ ♇
18 53	☽ ∥ ♃
18 56	♂ △ ♇
20 48	☽ ⚹ ♄
23 Wednesday	
00 28	♀ ⚹ ♇
02 23	☽ Q ♅
02 46	☽ ☌ ♃
04 24	☽ # ♂
06 16	♀ ⚹ ♂
08 31	☽ ⚹ ☿
09 52	☽ ∥ ♆
14 20	☽ □ ☉
15 36	☽ ⚹ ♇
16 31	☽ ☍ ♂
16 59	☽ △ ♀
17 25	☽ ☌ ♆
18 43	☽ # ☿
22 06	♀ △ ♆
22 13	☽ ∠ ♄
24 Thursday	
06 42	☽ Q ☿
09 12	☽ □ ♅
09 45	☉ ⊻ ♇
12 09	☽ Q ♇
12 45	♂ ☍ ♆
13 34	☽ ⊥ ♄
21 02	☽ ꝑ ♀
23 50	☽ ⊻ ♄
25 Friday	
06 26	☽ ⊻ ♃
12 06	☉ □ ♆
14 01	☽ ± ♀
16 08	☽ □ ☿
16 52	☽ ⊥ ♃
19 13	☽ □ ♇
20 58	☽ ⊻ ♆
21 37	☽ △ ☉
22 27	☽ ⊼ ♂
26 Saturday	
00 42	♄ # ♇
01 01	☽ # ♅
01 23	☽ ⊼ ♀
07 23	☽ ⊥ ♆
08 35	☽ ∠ ♃
08 50	☿ # ♆
09 22	☽ ± ♂
13 13	☽ △ ♅
16 40	☿ ∥ ♂
17 43	♀ ⊥ ♅
23 05	☽ ∠ ♆
27 Sunday	
01 42	☽ ꝑ ☉
01 49	☽ ꝑ ♂
03 44	☽ ☌ ♄
06 44	☉ □ ♂
11 02	☽ ⚹ ♃
12 53	♂ # ♆
13 46	☿ ☌ ♇
15 38	☽ ꝑ ♅
18 43	☽ ± ☉
23 49	☽ △ ♇
28 Monday	
00 31	☽ △ ☿
01 31	☽ ⚹ ♆
05 32	☽ △ ♂
05 54	☉ ꝑ ♄
06 10	☽ ⊼ ☉
07 40	☽ ± ♅
08 27	☽ Q ♃
11 11	☽ ☍ ♀
16 14	☿ ⚹ ♆
18 06	♂ ꝑ ♄
18 25	☽ ⊼ ♅
21 21	☽ # ♅
22 57	☽ Q ♆
29 Tuesday	
02 40	☽ ꝑ ♇
05 10	☽ ꝑ ☿
08 57	☽ ⊻ ♄
17 05	☽ □ ♃
19 01	☽ ± ♇
19 49	☽ ⊥ ♄
22 34	☽ ± ☿
30 Wednesday	
06 01	☽ ⊼ ♇
07 41	☽ □ ♆
10 16	☽ ⊼ ☿
12 17	☽ ∠ ♄
14 29	☽ □ ♂
16 47	☽ ☍ ☉
20 55	☿ # ♃
23 17	☽ ⊼ ♀
31 Thursday	
01 29	☽ ☍ ♅
11 08	☽ # ♂
11 42	☽ ± ♀
16 13	☽ ⚹ ♄
16 34	☽ ∥ ♆
23 49	♀ ⊻ ♅

All ephemeris data is given at 12.00 UT and the Moon's longitude is additionally given for 24.00 UT

AUGUST 1958

Date	Sidereal time h m s	Sun ☉ ° ' "	Moon ☽ ° ' "	Moon ☽ 24.00 ° ' "	Mercury ☿ ° '	Venus ♀ ° '	Mars ♂ ° '	Jupiter ♃ ° '	Saturn ♄ ° '	Uranus ♅ ° '	Neptune ♆ ° '	Pluto ♇ ° '
01	08 38 21	08 ♌ 43 32	29 ♒ 54 04	06 ♓ 06 34	04 ♍ 55	12 ♋ 21	06 ♉ 54	24 ♎ 25	19 ♐ 30	11 ♌ 46	02 ♏ 06	01 ♍ 14
02	08 42 18	09 40 55	12 ♓ 15 25	18 ♓ 20 51	05 31	13 33	07 30	24 32	19 R 28	11 50	02 06	01 16
03	08 46 14	10 38 20	24 ♓ 23 11	00 ♈ 22 48	06 04	14 46	08 05	24 39	19 26	11 54	02 07	01 18
04	08 50 11	11 35 45	06 ♈ 20 10	12 ♈ 15 48	06 31	15 58	08 40	24 46	19 24	11 57	02 08	01 20
05	08 54 07	12 33 12	18 ♈ 10 14	24 ♈ 04 05	06 55	17 11	09 15	24 54	19 22	12 01	02 08	01 22
06	08 58 04	13 30 40	29 ♈ 58 00	05 ♉ 52 36	07 14	18 23	09 50	25 01	19 20	12 05	02 09	01 24
07	09 02 00	14 28 10	11 ♉ 48 36	17 ♉ 46 41	07 28	19 36	10 25	25 09	19 19	12 08	02 10	01 26
08	09 05 57	15 25 40	23 ♉ 47 30	29 ♉ 51 44	07 37	20 49	10 59	25 17	19 17	12 12	02 11	01 27
09	09 09 53	16 23 13	06 ♊ 00 01	12 ♊ 12 56	07 41	22 02	11 33	25 25	19 16	12 16	02 11	01 29
10	09 13 50	17 20 46	18 ♊ 31 01	24 ♊ 54 42	07 R 40	23 14	12 07	25 33	19 15	12 20	02 12	01 31
11	09 17 47	18 18 21	01 ♋ 24 20	08 ♋ 00 10	07 34	24 27	12 40	25 41	19 13	12 23	02 13	01 33
12	09 21 43	19 15 57	14 ♋ 42 18	21 ♋ 30 41	07 22	25 40	13 13	25 49	19 12	12 27	02 14	01 35
13	09 25 40	20 13 35	28 ♋ 25 07	05 ♌ 25 16	07 04	26 53	13 46	25 58	19 11	12 31	02 15	01 37
14	09 29 36	21 11 13	12 ♌ 30 36	19 ♌ 40 30	06 41	28 06	14 19	26 06	19 10	12 34	02 16	01 39
15	09 33 33	22 08 54	26 ♌ 54 13	04 ♍ 10 53	06 13	29 ♋ 19	14 51	26 15	19 09	12 38	02 17	01 41
16	09 37 29	23 06 35	11 ♍ 29 37	18 ♍ 49 32	05 40	00 ♌ 32	15 23	26 24	19 09	12 42	02 18	01 43
17	09 41 26	24 04 17	26 ♍ 09 45	03 ♎ 29 25	05 01	01 45	15 55	26 33	19 08	12 45	02 19	01 45
18	09 45 22	25 02 01	10 ♎ 47 50	18 ♎ 04 21	04 19	02 58	16 26	26 42	19 07	12 49	02 20	01 47
19	09 49 19	25 59 46	25 ♎ 18 26	02 ♏ 29 41	03 33	04 12	16 57	26 51	19 07	12 53	02 21	01 49
20	09 53 16	26 57 31	09 ♏ 37 48	16 ♏ 42 34	02 43	05 25	17 28	27 00	19 06	12 56	02 22	01 51
21	09 57 12	27 55 18	23 ♏ 43 53	00 ♐ 41 42	01 52	06 38	17 58	27 09	19 06	13 00	02 24	01 53
22	10 01 09	28 53 06	07 ♐ 36 00	14 ♐ 26 49	00 59	07 51	18 28	27 19	19 06	13 04	02 25	01 55
23	10 05 05	29 ♌ 50 56	21 ♐ 14 13	27 ♐ 58 15	00 ♍ 06	09 05	18 58	27 28	19 06	13 07	02 26	01 57
24	10 09 02	00 ♍ 48 46	04 ♑ 38 58	11 ♑ 16 26	29 ♌ 13	10 18	19 27	27 38	19 D 06	13 11	02 27	01 59
25	10 12 58	01 46 38	17 ♑ 50 41	24 ♑ 21 44	28 22	11 32	19 56	27 47	19 06	13 14	02 29	02 01
26	10 16 55	02 44 31	00 ♒ 49 36	07 ♒ 14 19	27 35	12 45	20 25	27 57	19 06	13 18	02 30	02 03
27	10 20 51	03 42 25	13 ♒ 35 52	19 ♒ 54 18	26 51	13 59	20 53	28 07	19 06	13 21	02 31	02 05
28	10 24 48	04 40 21	26 ♒ 09 37	02 ♓ 21 54	26 12	15 12	21 21	28 17	19 07	13 25	02 33	02 07
29	10 28 45	05 38 18	08 ♓ 31 13	14 ♓ 37 41	25 39	16 26	21 48	28 27	19 07	13 28	02 34	02 09
30	10 32 41	06 36 16	20 ♓ 41 28	26 ♓ 42 45	25 13	17 39	22 15	28 37	19 08	13 32	02 35	02 11
31	10 36 38	07 ♍ 34 17	02 ♈ 41 48	08 ♈ 38 55	24 ♌ 54	18 ♌ 53	22 ♉ 42	28 ♎ 47	19 ♐ 08	13 ♌ 35	02 ♏ 37	02 ♍ 13

Date	Moon True ☊ ° '	Moon Mean ☊ ° '	Moon ☽ Latitude ° '
01	25 ♎ 20	26 ♎ 08	04 N 09
02	25 R 11	26 05	03 26
03	25 05	26 01	02 35
04	25 01	25 58	01 38
05	24 59	25 55	00 N 36
06	24 D 59	25 52	00 S 27
07	25 00	25 49	01 28
08	24 R 59	25 46	02 27
09	24 58	25 42	03 20
10	24 54	25 39	04 05
11	24 47	25 36	04 39
12	24 39	25 33	04 59
13	24 29	25 30	05 03
14	24 19	25 26	04 49
15	24 10	25 23	04 16
16	24 02	25 20	03 27
17	23 57	25 17	02 23
18	23 54	25 14	01 S 10
19	23 D 54	25 11	00 N 08
20	23 54	25 07	01 23
21	23 55	25 04	02 33
22	23 R 55	25 01	03 33
23	23 53	24 58	04 19
24	23 48	24 55	04 50
25	23 42	24 52	05 05
26	23 34	24 48	05 04
27	23 25	24 45	04 48
28	23 16	24 42	04 18
29	23 09	24 39	03 37
30	23 02	24 36	02 46
31	22 ♎ 58	24 ♎ 32	01 N 47

DECLINATIONS

Date	Sun ☉ ° '	Moon ☽ ° '	Mercury ☿ ° '	Venus ♀ ° '	Mars ♂ ° '	Jupiter ♃ ° '	Saturn ♄ ° '	Uranus ♅ ° '	Neptune ♆ ° '	Pluto ♇ ° '
01	18 N 05	07 S 37	07 N 36	22 N 30	11 N 32	08 S 24	21 S 43	17 N 51	10 S 33	21 N 38
02	17 50	03 S 47	07 12	22 26	11 43	08 26	21 43	17 50	10 34	21 38
03	17 34	00 N 09	06 48	22 21	11 55	08 29	21 43	17 49	10 34	21 37
04	17 18	04 01	06 27	22 16	12 06	08 32	21 43	17 48	10 34	21 36
05	17 02	07 41	06 07	22 10	12 17	08 35	21 43	17 47	10 34	21 35
06	16 46	11 03	05 49	22 03	12 28	08 38	21 43	17 46	10 35	21 35
07	16 30	13 59	05 33	21 56	12 39	08 41	21 43	17 45	10 35	21 34
08	16 13	16 21	05 19	21 48	12 50	08 44	21 43	17 44	10 35	21 33
09	15 56	18 01	05 08	21 39	13 00	08 47	21 43	17 43	10 36	21 32
10	15 38	18 52	04 59	21 30	13 11	08 50	21 44	17 42	10 36	21 32
11	15 21	18 47	04 53	21 20	13 21	08 53	21 44	17 41	10 36	21 31
12	15 03	17 40	04 49	21 09	13 31	08 57	21 44	17 40	10 37	21 30
13	14 45	15 32	04 48	20 58	13 41	09 00	21 44	17 39	10 37	21 30
14	14 26	12 26	04 51	20 46	13 51	09 03	21 44	17 38	10 38	21 29
15	14 08	08 32	04 56	20 34	14 01	09 06	21 44	17 37	10 38	21 28
16	13 49	04 N 04	05 04	20 21	14 10	09 10	21 44	17 36	10 38	21 27
17	13 30	00 S 40	05 16	20 07	14 20	09 13	21 44	17 35	10 39	21 27
18	13 11	05 21	05 30	19 53	14 29	09 17	21 44	17 34	10 39	21 26
19	12 51	09 40	05 48	19 39	14 38	09 20	21 45	17 33	10 40	21 25
20	12 32	13 23	06 08	19 23	14 47	09 24	21 45	17 32	10 40	21 25
21	12 12	16 14	06 30	19 08	14 56	09 27	21 45	17 31	10 40	21 24
22	11 52	18 05	06 55	18 51	15 04	09 31	21 45	17 30	10 41	21 23
23	11 32	18 51	07 21	18 34	15 13	09 34	21 45	17 29	10 41	21 22
24	11 11	18 32	07 49	18 17	15 21	09 38	21 45	17 28	10 42	21 22
25	10 51	17 12	08 18	17 59	15 29	09 41	21 46	17 27	10 42	21 21
26	10 30	15 01	08 47	17 41	15 37	09 45	21 46	17 26	10 43	21 20
27	10 09	12 08	09 15	17 22	15 45	09 49	21 46	17 25	10 43	21 20
28	09 48	08 45	09 44	17 02	15 53	09 52	21 46	17 24	10 44	21 19
29	09 27	05 02	10 11	16 42	16 01	09 56	21 47	17 23	10 44	21 18
30	09 05	01 09	10 37	16 22	16 08	10 00	21 47	17 22	10 45	21 18
31	08 N 44	02 N 43	11 N 00	16 N 01	16 N 15	10 S 04	21 S 47	17 N 21	10 S 45	21 N 17

ZODIAC SIGN ENTRIES

Date	h	m	Planets
01	12	11	☽ ♓
03	23	14	☽ ♈
06	12	04	☽ ♉
09	00	16	☽ ♊
11	09	25	☽ ♋
13	14	43	☽ ♌
15	17	07	☽ ♍
16	01	28	♀ ♌
17	18	17	☽ ♎
19	19	50	☽ ♏
21	22	48	☽ ♐
23	14	31	☿ ♌
23	15	46	☉ ♍
24	03	38	☽ ♑
26	10	28	☽ ♒
28	19	25	☽ ♓
31	06	35	☽ ♈

LATITUDES

Date	Mercury ☿ ° '	Venus ♀ ° '	Mars ♂ ° '	Jupiter ♃ ° '	Saturn ♄ ° '	Uranus ♅ ° '	Neptune ♆ ° '	Pluto ♇ ° '
01	02 S 16	00 S 22	02 S 25	01 N 09	01 N 19	00 N 37	01 N 45	11 N 23
04	02 52	00 13	02 25	01 08	01 18	00 37	01 45	11 23
07	03 28	00 S 05	02 25	01 08	01 18	00 37	01 45	11 23
10	04 00	00 N 03	02 24	01 07	01 17	00 37	01 45	11 23
13	04 25	00 11	02 24	01 07	01 17	00 37	01 45	11 23
16	04 41	00 19	02 23	01 06	01 16	00 37	01 44	11 23
19	04 44	00 27	02 22	01 05	01 15	00 37	01 44	11 23
22	04 30	00 34	02 21	01 05	01 15	00 38	01 44	11 23
25	03 59	00 41	02 19	01 04	01 14	00 38	01 44	11 23
28	03 15	00 47	02 18	01 04	01 14	00 38	01 44	11 23
31	02 S 21	00 N 54	02 S 16	01 N 03	01 N 13	00 N 38	01 N 44	11 N 24

DATA

Julian Date	2436417
Delta T	+33 seconds
Ayanamsa	23° 16' 52"
Synetic vernal point	05° ♓ 50' 08"
True obliquity of ecliptic	23° 26' 32"

MOON'S PHASES, APSIDES AND POSITIONS ☽

Date	h	m	Phase	Longitude ° '	Eclipse Indicator
07	17	50	☾	14 ♉ 42	
15	03	33	●	21 ♌ 49	
21	19	45	☽	28 ♏ 14	
29	05	53	○	05 ♓ 24	

Day	h	m		
05	17	34	Apogee	
17	14	21	Perigee	
03	11	08	0N	
10	21	51	Max dec	18° N 57'
17	08	41	0S	
23	16	47	Max dec	18° S 52'
30	19	07	0N	

LONGITUDES

Date	Chiron ⚷ ° '	Ceres ⚳ ° '	Pallas ⚴ ° '	Juno ⚵ ° '	Vesta ⚶ ° '	Black Moon Lilith ⚸ ° '
01	20 ♒ 46	03 ♍ 16	09 ♌ 23	24 ♌ 06	06 ♋ 08	18 ♈ 04
11	20 ♒ 15	07 ♍ 45	15 ♌ 00	28 ♌ 33	10 ♋ 13	19 ♈ 11
21	19 ♒ 44	12 ♍ 16	20 ♌ 30	02 ♍ 56	14 ♋ 13	20 ♈ 17
31	19 ♒ 14	16 ♍ 49	25 ♌ 55	07 ♍ 17	18 ♋ 06	21 ♈ 24

All ephemeris data is given at 12.00 UT and the Moon's longitude is additionally given for 24.00 UT

ASPECTARIAN

h m Aspects

01 Friday
01 24 ☽ △ ♃
01 57 ☽ Q ♂
06 35 ☽ bQ ♀
07 05 ☽ ∥ ♃
12 08 ☽ # ☿
14 35 ☽ ☍ ♇
15 04 ☽ Q ♄
16 14 ☽ △ ♆
22 13 ☽ ☍ ☿

02 Saturday
02 13 ☽ ✶ ♂
06 32 ☽ ⊼ ☉
06 37 ☽ bQ ♃
10 44 ☉ ∥ ♅
11 10 ☽ ⊼ ♅
14 50 ☽ △ ♀
19 19 ☽ ± ☉
21 33 ☽ bQ ♆
23 02 ☽ ± ♅

03 Sunday
00 29 ☽ ± ♃
02 10 ☽ □ ♄
09 16 ☽ ∠ ♂
12 32 ☽ ⊼ ♃
14 43 ☽ bQ ☉
15 27 ☽ ± ♆
17 02 ☽ bQ ♅

04 Monday
01 53 ☽ ⊼ ♇
03 30 ☽ ⊼ ♆
04 14 ☽ ⊥ ♂
12 24 ☽ ⊼ ☿
14 01 ☽ ± ♇
16 58 ☽ ⊻ ♂
19 19 ♀ ∠ ♇
21 36 ☉ ☌ ♅
23 26 ☽ △ ♅
23 35 ☽ △ ☉

05 Tuesday
00 59 ☽ ± ☿
02 24 ☽ ∥ ☿
08 19 ☽ bQ ♇
09 45 ☽ □ ♀
14 26 ☽ △ ♄
18 15 ☽ # ♃
19 51 ☽ bQ ☿
21 56 ☉ Q ♃

06 Wednesday
01 50 ☽ ☍ ♃
08 29 ☽ # ♆
14 55 ☽ △ ♇
16 27 ☽ ☍ ♆
20 52 ☽ bQ ♄
23 53 ☽ ∥ ♂

07 Thursday
02 32 ☽ Q ♀
03 04 ☽ △ ☿
06 26 ♀ ⊼ ♄
09 02 ☽ ☌ ♂
12 40 ☽ □ ♅
15 02 ☽ ± ♄
17 50 ☽ □ ☉

08 Friday
03 03 ☽ ⊼ ♄
05 24 ☽ ✶ ♀
10 39 ☽ ∥ ☉
14 59 ☽ ⊼ ♃
23 41 ♀ # ♄

09 Saturday
00 44 ☽ Q ♅
02 57 ☽ ± ♃
03 10 ☽ □ ♇
04 34 ☽ ⊼ ♆
06 49 ☽ ∥ ♅
08 36 ☽ Q ☉
14 12 ☽ ∠ ♀
15 17 ☽ □ ☿
16 15 ☽ ± ♆
18 47 ☿ St R
20 38 ☽ bQ ♃
23 14 ☽ ⊻ ♂

10 Sunday
00 09 ☽ ✶ ♅
01 08 ☿ ∠ ♀
06 11 ♀ ∥ ♇
09 19 ☽ ⊥ ♀
09 31 ☽ bQ ♆
09 36 ☽ ✶ ☉
11 12 ☽ ⊥ ♂
13 22 ☽ ☍ ♄
13 54 ☽ Q ♇
21 48 ☽ ⊻ ♀
22 22 ♂ □ ♅

11 Monday
01 19 ☽ Q ☿
01 20 ☽ △ ♃
04 34 ☽ ∠ ♅
04 49 ☽ ∠ ♂
12 16 ☽ ✶ ♇
13 29 ☽ △ ♆
15 45 ☽ ∠ ☉
19 51 ♀ Q ♂
21 07 ☽ ⊥ ♅
23 03 ☽ ✶ ☿

12 Tuesday
02 58 ♀ ± ♄
07 58 ☽ ⊻ ♅
09 15 ☽ ⊥ ☉
09 15 ☽ ✶ ♂
10 23 ♀ ⊥ ♇
10 28 ☉ △ ♄
11 14 ♂ ± ♄
12 01 ☽ ∥ ♅
15 21 ☽ ∠ ♇
15 29 ♀ □ ♃
19 56 ☽ ⊼ ♄
20 40 ☽ ⊻ ☉

13 Wednesday
01 13 ☽ ∠ ☿
06 25 ☽ ± ♄
07 09 ☽ ⊥ ♇
07 14 ☽ Q ♂
07 43 ☽ □ ♃
09 06 ☽ ☌ ♀
12 35 ☉ Q ♆
16 27 ☽ ⊥ ☿
17 31 ☽ ⊻ ♇
18 35 ☽ □ ♆
19 34 ☽ ∥ ☉
21 53 ☽ bQ ♄

14 Thursday
02 20 ☽ ∥ ♂
02 26 ☽ ⊻ ☿
12 06 ☽ ☌ ♅
14 43 ☽ Q ♃
15 09 ☽ □ ♂
23 09 ☽ △ ♄
23 36 ☽ # ♆

15 Friday
01 00 ☽ Q ♆
03 33 ☽ ☌ ☉
08 43 ☽ # ♃
10 54 ☽ ✶ ♃
16 21 ☽ ⊻ ♀
18 01 ☉ ∥ ♂
19 55 ☽ ☌ ♇
20 53 ☽ ✶ ♆

16 Saturday
00 18 ☿ ⊥ ♀
02 48 ☽ ☌ ☿
03 08 ☽ ⊥ ♀
06 58 ☽ ∥ ☿
11 50 ☽ ∠ ♃
13 59 ☽ ⊻ ♅
18 37 ☽ △ ♂
19 13 ☽ ∠ ♀
21 31 ☽ ∠ ♆
23 50 ☽ ⊥ ♅

17 Sunday
00 30 ☽ □ ♄
02 43 ☽ ⊥ ♃
08 20 ☽ ⊻ ☉
11 59 ♀ ⊻ ♇
12 15 ☽ ⊥ ♆
12 38 ☽ ⊻ ♃
14 37 ☽ ∠ ♅
18 51 ☽ ⊥ ☉
20 04 ☽ bQ ♂
21 10 ☽ ⊻ ♇
21 59 ☽ ✶ ♀
22 05 ☽ ⊻ ♆
23 16 ♀ □ ♆

18 Monday
01 51 ☽ ⊻ ☿
05 58 ☽ Q ♄
07 02 ☽ ⊥ ♇
10 39 ☽ ∠ ☉
11 15 ☽ ⊥ ☿
11 23 ☽ ± ♂
12 55 ☽ # ☿
15 20 ☽ ✶ ♅
19 31 ☽ Q ♀
21 38 ☽ ⊼ ♂
21 54 ☽ ∠ ♇

19 Tuesday
01 21 ☽ ∠ ☿
01 44 ☽ ✶ ♄
04 16 ☿ ⊻ ♀
09 59 ☽ ∥ ♃
10 27 ♀ bQ ♄
11 17 ☽ Q ♅
13 14 ☽ ✶ ☉
14 35 ☽ ☌ ♃
17 58 ☽ ∥ ♆
22 54 ☽ ✶ ♇
23 47 ☽ ☌ ♆

20 Wednesday
01 01 ☽ ✶ ☿
02 42 ☽ ∠ ♄
04 14 ☽ □ ♀
06 32 ☽ # ☉
10 47 ☽ Q ☉
13 09 ☉ ✶ ♃
17 37 ☽ □ ♅
17 53 ☽ ⊥ ♄
19 10 ☽ Q ♇
20 08 ☽ Q ☿
21 38 ☿ ✶ ♆
23 23 ☽ # ♂

21 Thursday
01 47 ☽ ☍ ♂
04 05 ☽ ⊻ ♄
11 24 ☿ ☌ ♇
17 57 ☽ ⊻ ♃
19 45 ☽ □ ☉

22 Friday
01 11 ☽ □ ☿
02 06 ☽ □ ♇
02 51 ☽ # ♅
02 58 ☽ ⊻ ♆
04 27 ☽ ⊥ ♃
12 29 ☽ △ ♀
13 25 ☽ ⊥ ♆
20 20 ☽ ∠ ♃
21 36 ☽ △ ♅

23 Saturday
02 42 ☽ # ♀
05 15 ☽ ∠ ♆
07 49 ☽ ⊼ ♂
08 12 ☽ ☌ ♄
15 10 ☉ ☌ ☿
17 33 ☽ bQ ♀
18 28 ♂ ⊼ ♄
18 53 ☽ ± ♂
23 14 ☽ ✶ ♃

24 Sunday
00 19 ☽ bQ ♅
00 31 ♄ St D
02 50 ☽ △ ☿
04 33 ☽ △ ☉
07 11 ☽ △ ♇
08 02 ☽ ✶ ♆
11 19 ☽ ± ♀
11 38 ☽ bQ ♂
16 35 ☽ ± ♅
20 18 ☽ # ♀
21 07 ☽ Q ♃
23 17 ☽ ⊼ ♀

25 Monday
03 32 ☽ ⊼ ♅
04 19 ☽ bQ ☿
05 49 ☽ Q ♆
08 37 ☽ # ♅
09 53 ☽ bQ ☉
10 29 ☽ bQ ♇
14 18 ☽ ⊻ ♄
15 59 ☽ △ ♂
18 12 ☉ ☌ ♇
19 50 ☽ ± ☿
21 22 ☉ # ♆

26 Tuesday
01 22 ☽ ⊥ ♄
02 29 ☿ ✶ ♃
03 06 ☽ ± ♇
03 48 ☽ ± ☉
05 46 ☉ ✶ ♆
06 18 ☽ ⊼ ☿
06 28 ☽ # ♂
06 35 ☽ □ ♃
14 17 ☽ ⊼ ♇
15 07 ☽ □ ♆
15 52 ☽ ⊼ ☉
18 07 ☽ ∠ ♄
23 16 ♀ ☌ ♅

27 Wednesday
07 39 ♀ ∥ ♅
11 32 ☽ ☍ ♅
12 48 ☽ ☍ ♀
22 20 ☽ ∥ ♆
22 29 ☽ ✶ ♄

28 Thursday
02 24 ☽ □ ♂
04 02 ☽ # ☉
04 27 ☽ ∥ ♃
06 06 ☽ # ☿
07 40 ☉ # ♃
12 05 ☽ ☍ ☿
14 04 ☉ ∥ ☿
16 09 ☽ △ ♃
20 45 ☿ # ♃
21 34 ☽ Q ♄
23 33 ☽ ☍ ♇

29 Friday
00 22 ☽ △ ♆
05 53 ☽ ☍ ☉
12 28 ♀ Q ♃
14 36 ☽ Q ♂
21 46 ☽ ⊼ ♅
21 49 ☽ bQ ♃

30 Saturday
05 18 ☽ ⊼ ♀
05 50 ☽ bQ ♆
08 54 ☽ □ ♄
09 41 ☽ ± ♅
15 13 ☽ ✶ ♂
15 53 ☽ ± ♃
18 34 ☽ ± ♀
20 27 ☿ # ♆
20 46 ☽ ⊼ ☿
23 41 ♀ ∥ ♂
23 47 ☽ ± ♆

31 Sunday
03 43 ☽ bQ ♅
04 03 ☽ ⊼ ♃
08 29 ☽ ± ☿
11 02 ☽ ⊼ ♇
11 50 ☽ ⊼ ♆
14 40 ☽ bQ ♀
17 04 ♀ △ ♄
22 27 ☽ ∠ ♂
22 42 ☽ ⊼ ☉
23 10 ☽ ± ♇

SEPTEMBER 1958

LONGITUDES

Date	Sidereal time h m s	Sun ☉ ° ′ ″	Moon ☽ ° ′ ″	Moon ☽ 24.00 ° ′ ″	Mercury ☿ ° ′	Venus ♀ ° ′	Mars ♂ ° ′	Jupiter ♃ ° ′	Saturn ♄ ° ′	Uranus ♅ ° ′	Neptune ♆ ° ′	Pluto ⯓ ° ′
01	10 40 34	08 ♍ 32 19	14 ♈ 34 25	20 ♈ 28 44	24 ♌ 44	20 ♌ 07	23 ♉ 08	28 ♎ 58	19 ♐ 09	13 ♌ 39	02 ♏ 38	02 ♍ 15
02	10 44 31	09 30 23	26 ♈ 22 17	02 ♉ 15 32	24 D 41	21 21	23 34	29 08	19 10	13 42	02 40	02 17
03	10 48 27	10 28 29	08 ♉ 09 03	14 ♉ 03 21	24 47	22 34	23 59	29 19	19 11	13 46	02 41	02 19
04	10 52 24	11 26 37	19 ♉ 59 04	25 ♉ 56 47	25 01	23 48	24 24	29 29	19 12	13 49	02 43	02 21
05	10 56 20	12 24 47	01 ♊ 57 08	08 ♊ 00 46	25 25	25 02	24 48	29 40	19 13	13 53	02 44	02 23
06	11 00 17	13 22 58	14 ♊ 08 19	20 ♊ 20 24	25 56	26 16	25 12	29 ♎ 51	19 15	13 56	02 46	02 25
07	11 04 14	14 21 12	26 ♊ 37 35	03 ♋ 00 26	26 36	27 30	25 35	00 ♏ 01	19 16	13 59	02 48	02 27
08	11 08 10	15 19 28	09 ♋ 29 24	16 ♋ 04 51	27 24	28 44	25 58	00 12	19 17	14 03	02 49	02 29
09	11 12 07	16 17 46	22 ♋ 47 04	29 ♋ 36 10	28 19	29 ♌ 58	26 20	00 23	19 19	14 06	02 51	02 31
10	11 16 03	17 16 06	06 ♌ 32 09	13 ♌ 34 50	29 ♌ 23	01 ♍ 12	26 42	00 34	19 21	14 09	02 52	02 33
11	11 20 00	18 14 28	20 ♌ 43 51	27 ♌ 58 39	00 ♍ 32	02 26	27 04	00 45	19 22	14 12	02 54	02 35
12	11 23 56	19 12 52	05 ♍ 18 33	12 ♍ 42 41	01 48	03 41	27 24	00 57	19 24	14 16	02 56	02 37
13	11 27 53	20 11 18	20 ♍ 10 05	27 ♍ 39 40	03 10	04 55	27 45	01 08	19 26	14 19	02 58	02 39
14	11 31 49	21 09 46	05 ♎ 10 20	12 ♎ 40 59	04 37	06 09	28 04	01 19	19 28	14 22	02 59	02 41
15	11 35 46	22 08 15	20 ♎ 10 33	27 ♎ 38 04	06 09	07 23	28 23	01 31	19 30	14 25	03 01	02 42
16	11 39 43	23 06 46	05 ♏ 02 38	12 ♏ 23 32	07 44	08 38	28 42	01 42	19 32	14 28	03 03	02 44
17	11 43 39	24 05 19	19 ♏ 40 09	26 ♏ 52 03	09 23	09 52	28 59	01 54	19 35	14 31	03 05	02 46
18	11 47 36	25 03 54	03 ♐ 58 55	11 ♐ 00 33	11 05	11 06	29 16	02 05	19 37	14 34	03 07	02 48
19	11 51 32	26 02 30	17 ♐ 56 54	24 ♐ 48 00	12 49	12 21	29 33	02 17	19 39	14 37	03 09	02 50
20	11 55 29	27 01 08	01 ♑ 33 56	08 ♑ 14 53	14 35	13 35	29 ♉ 49	02 29	19 42	14 40	03 10	02 52
21	11 59 25	27 59 47	14 ♑ 51 04	21 ♑ 22 43	16 23	14 50	00 ♊ 04	02 41	19 45	14 43	03 12	02 54
22	12 03 22	28 58 29	27 ♑ 50 05	04 ♒ 13 28	18 11	16 04	00 19	02 52	19 47	14 46	03 14	02 56
23	12 07 18	29 ♍ 57 12	10 ♒ 33 08	16 ♒ 49 20	20 01	17 19	00 33	03 04	19 50	14 49	03 16	02 57
24	12 11 15	00 ♎ 55 56	23 ♒ 02 19	29 ♒ 12 22	21 51	18 33	00 46	03 16	19 53	14 52	03 18	02 59
25	12 15 12	01 54 43	05 ♓ 19 42	11 ♓ 24 33	23 41	19 48	00 58	03 28	19 56	14 55	03 20	03 01
26	12 19 08	02 53 31	17 ♓ 27 09	23 ♓ 27 43	25 31	21 02	01 10	03 40	19 59	14 58	03 22	03 03
27	12 23 05	03 52 21	29 ♓ 26 30	05 ♈ 23 43	27 21	22 17	01 21	03 53	20 03	15 00	03 24	03 05
28	12 27 01	04 51 13	11 ♈ 19 37	17 ♈ 14 27	29 ♍ 11	23 32	01 31	04 05	20 06	15 03	03 26	03 06
29	12 30 58	05 50 07	23 ♈ 08 31	29 ♈ 02 06	01 ♎ 00	24 46	01 41	04 17	20 09	15 06	03 28	03 08
30	12 34 54	06 ♎ 49 04	04 ♉ 55 33	10 ♉ 49 12	02 ♎ 49	26 ♍ 01	01 ♊ 50	04 ♏ 29	20 ♐ 13	15 ♌ 08	03 ♏ 30	03 ♍ 10

Moon Nodes and Latitude / DECLINATIONS

Date	Moon True ☊ ° ′	Moon Mean ☊ ° ′	Moon ☽ Latitude ° ′	Sun ☉ ° ′	Moon ☽ ° ′	Mercury ☿ ° ′	Venus ♀ ° ′	Mars ♂ ° ′	Jupiter ♃ ° ′	Saturn ♄ ° ′	Uranus ♅ ° ′	Neptune ♆ ° ′	Pluto ⯓ ° ′
01	22 ♎ 56	24 ♎ 29	00 N 45	08 N 22	06 N 26	11 N 22	15 N 40	16 N 23	10 S 08	21 S 47	17 N 20	10 S 46	21 N 17
02	22 D 56	24 26	00 S 19	08 00	09 53	11 40	15 18	16 30	10 10	21 48	17 19	10 47	21 16
03	22 57	24 23	01 22	07 38	12 56	11 56	14 56	16 36	10 15	21 48	17 18	10 47	21 15
04	22 58	24 20	02 22	07 16	15 28	12 09	14 33	16 43	10 18	21 48	17 17	10 48	21 15
05	22 59	24 17	03 17	06 54	17 21	12 18	14 10	16 50	10 23	21 48	17 16	10 48	21 14
06	23 R 00	24 13	04 03	06 32	18 28	12 24	13 47	16 56	10 27	21 49	17 15	10 49	21 13
07	22 59	24 10	04 40	06 10	18 44	12 26	13 23	17 03	10 31	21 49	17 14	10 49	21 13
08	22 57	24 07	05 03	05 47	18 03	12 24	12 59	17 09	10 35	21 49	17 13	10 50	21 12
09	22 53	24 04	05 12	05 24	16 23	12 19	12 34	17 15	10 39	21 50	17 12	10 51	21 12
10	22 48	24 01	05 03	05 02	13 45	12 09	12 09	17 21	10 43	21 50	17 11	10 51	21 11
11	22 43	23 57	04 36	04 39	10 13	11 57	11 44	17 27	10 47	21 50	17 10	10 52	21 10
12	22 38	23 54	03 50	04 16	06 00	11 41	11 18	17 32	10 51	21 51	17 10	10 52	21 10
13	22 34	23 51	02 48	03 53	01 N 19	11 21	10 52	17 38	10 55	21 51	17 09	10 53	21 09
14	22 31	23 48	01 34	03 30	03 S 29	10 58	10 26	17 43	10 59	21 51	17 08	10 54	21 09
15	22 30	23 45	00 S 13	03 07	08 05	10 32	10 00	17 48	11 03	21 52	17 07	10 54	21 08
16	22 D 31	23 42	01 N 08	02 44	12 08	10 03	09 33	17 53	11 07	21 52	17 06	10 55	21 08
17	22 32	23 38	02 24	02 21	15 21	09 32	09 06	17 58	11 11	21 52	17 05	10 56	21 07
18	22 33	23 35	03 29	01 58	17 32	08 58	08 38	18 03	11 15	21 53	17 04	10 56	21 07
19	22 34	23 32	04 19	01 34	18 35	08 21	08 11	18 08	11 20	21 53	17 03	10 57	21 06
20	22 R 35	23 29	04 54	01 11	18 32	07 43	07 43	18 13	11 24	21 53	17 03	10 58	21 06
21	22 34	23 26	05 12	00 48	17 27	07 03	07 15	18 17	11 28	21 54	17 02	10 58	21 05
22	22 31	23 23	05 14	00 24	15 28	06 22	06 47	18 21	11 32	21 54	17 01	10 59	21 05
23	22 28	23 19	04 59	00 N 01	12 47	05 39	06 18	18 26	11 36	21 55	17 00	11 00	21 04
24	22 25	23 16	04 31	00 S 22	09 34	04 56	05 50	18 30	11 40	21 55	16 59	11 00	21 04
25	22 22	23 13	03 51	00 46	05 58	04 11	05 21	18 34	11 45	21 55	16 58	11 01	21 03
26	22 19	23 10	03 01	01 09	02 S 11	03 26	04 52	18 37	11 49	21 56	16 58	11 02	21 03
27	22 17	23 07	02 03	01 32	01 N 39	02 40	04 23	18 41	11 53	21 56	16 57	11 02	21 02
28	22 15	23 03	01 N 00	01 56	05 21	01 54	03 54	18 45	11 57	21 57	16 56	11 03	21 02
29	22 D 15	23 00	00 S 05	02 19	08 55	01 07	03 24	18 48	12 01	21 57	16 55	11 04	21 01
30	22 ♎ 15	22 ♎ 57	01 S 10	02 S 42	12 N 04	00 N 21	02 N 55	18 N 52	12 S 06	21 S 57	16 N 55	11 S 05	21 N 01

ZODIAC SIGN ENTRIES

Date	h	m	Planets
02	19	24	☽ ♉
05	08	07	☽ ♊
07	08	52	♃ ♏
07	18	22	☽ ♋
09	12	35	♀ ♍
10	00	42	☽ ♌
11	01	10	☿ ♍
12	03	19	☽ ♍
14	03	44	☽ ♎
16	03	49	☽ ♏
18	05	16	☽ ♐
20	09	13	☽ ♑
21	05	26	♂ ♊
22	16	03	☽ ♒
23	13	09	☉ ♎
25	01	33	☽ ♓
27	13	07	☽ ♈
28	22	45	☿ ♎
30	01	58	☽ ♉

LATITUDES

Date	Mercury ☿ ° ′	Venus ♀ ° ′	Mars ♂ ° ′	Jupiter ♃ ° ′	Saturn ♄ ° ′	Uranus ♅ ° ′	Neptune ♆ ° ′	Pluto ⯓ ° ′
01	02 S 02	00 N 56	02 S 15	01 N 03	01 N 13	00 N 38	01 N 44	11 N 24
04	01 06	01 01	02 13	01 02	01 12	00 38	01 43	11 24
07	00 S 14	01 06	02 11	01 02	01 12	00 38	01 43	11 24
10	00 N 30	01 08	02 08	01 02	01 11	00 38	01 43	11 25
13	01 05	01 15	02 05	01 02	01 11	00 38	01 43	11 25
16	01 30	01 18	02 02	01 01	01 10	00 38	01 43	11 26
19	01 45	01 21	01 58	01 00	01 10	00 38	01 43	11 26
22	01 51	01 24	01 54	01 00	01 09	00 38	01 43	11 27
25	01 50	01 25	01 50	00 59	01 08	00 38	01 43	11 28
28	01 43	01 26	01 45	00 59	01 08	00 38	01 43	11 28
31	01 N 32	01 N 27	01 S 40	00 N 59	01 N 07	00 N 38	01 N 43	11 N 29

DATA

Julian Date	2436448
Delta T	+33 seconds
Ayanamsa	23° 16′ 55″
Synetic vernal point	05° ♓ 50′ 05″
True obliquity of ecliptic	23° 26′ 33″

MOON'S PHASES, APSIDES AND POSITIONS ☽

Date	h	m	Phase	Longitude ° ′	Eclipse Indicator
06	10	24	☾	13 ♊ 19	
13	12	02	●	20 ♍ 11	
20	03	18	☽	26 ♐ 40	
27	21	44	○	04 ♈ 16	

Day	h	m		
02	10	28	Apogee	
14	16	35	Perigee	
29	22	05	Apogee	
07	06	59	Max dec	18° N 46′
13	18	35	0S	
19	22	39	Max dec	18° S 42′
27	01	37	0N	

LONGITUDES

Date	Chiron ⚷ ° ′	Ceres ⚳ ° ′	Pallas ⚴ ° ′	Juno ⚵ ° ′	Vesta ⚶ ° ′	Black Moon Lilith ⚸ ° ′
01	19 ♒ 11	17 ♍ 16	26 ♌ 27	07 ♍ 43	18 ♋ 29	21 ♈ 30
11	18 ♒ 44	21 ♍ 51	01 ♍ 44	11 ♍ 59	22 ♋ 14	22 ♈ 37
21	18 ♒ 20	26 ♍ 25	06 ♍ 55	16 ♍ 11	25 ♋ 49	23 ♈ 43
31	18 ♒ 01	01 ♎ 00	11 ♍ 58	20 ♍ 17	29 ♋ 13	24 ♈ 50

All ephemeris data is given at 12.00 UT and the Moon's longitude is additionally given for 24.00 UT

ASPECTARIAN

h m	Aspects
01 Monday	
02 18	☽ ♇ ☿
10 07	☽ △ ♅
11 55	☽ ± ☉
17 23	☽ ⊥ ♂
17 27	☽ ♇ ⯓
21 19	☽ △ ♄
22 26	♀ Q ♆
23 53	☽ ∥ ☉
02 Tuesday	
00 34	☽ △ ♀
06 03	☽ ⊻ ♂
07 42	☿ St D
07 52	☽ ♇ ☉
08 33	☽ △ ☿
14 17	☽ ⧺ ♃
17 43	☽ ☍ ♃
18 40	☽ ⧺ ♆
03 Wednesday	
00 05	☽ △ ⯓
00 51	☽ ☍ ♆
02 57	☽ ∥ ☿
03 55	☽ ♇ ♄
17 09	☽ △ ☉
22 15	☽ ± ♄
23 28	☽ □ ♅
04 Thursday	
03 57	☽ ∥ ♀
10 25	☽ ⊼ ♄
20 35	☽ □ ♀
21 12	☽ ☌ ♂
22 26	☽ □ ☿
05 Friday	
04 00	☽ ∥ ♂
05 09	♀ □ ♂
07 22	☽ ⊼ ♃
10 49	☽ ∥ ♅
11 51	☽ Q ♅
12 52	☽ □ ⯓
13 34	☽ ⊼ ♆
19 28	☽ ± ♃
23 52	☿ ☌ ♀
06 Saturday	
01 27	☽ ± ♆
10 24	☽ □ ☉
11 34	☽ Q ☿
11 36	☽ ⚹ ♅
12 17	☽ Q ♀
13 23	☽ ♇ ♃
19 03	☽ ♇ ♆
21 54	☽ ☍ ♄
07 Sunday	
00 11	☽ Q ⯓
02 24	☉ ⊻ ♅
09 58	☽ ⊻ ♂
11 56	☽ ⚹ ☿
13 50	☽ ⚹ ♀
16 29	☽ ∠ ♅
18 30	☽ △ ♃
21 38	☽ ⊥ ♂
22 59	☽ ⚹ ⯓
23 37	☽ △ ♆
23 40	☽ Q ☉
08 Monday	
08 23	☉ ∠ ♃
09 20	☽ ⊥ ♅
14 47	☽ ∠ ♂
17 42	☽ ∠ ☿
20 21	☽ ⊻ ♅
20 33	☽ ∠ ♀
23 28	☽ ⚹ ☉
09 Tuesday	
01 55	☽ ∥ ♅
01 59	☽ ∥ ♂
02 34	☽ ∠ ⯓
03 04	♂ ∥ ♅
05 48	☽ ⊼ ♄
11 07	☽ ⊥ ☿
14 19	☽ ⊥ ♀
16 29	☽ ± ♄
18 28	☽ ⚹ ♂
18 36	☽ ⊥ ⯓
21 34	♀ ⚹ ♃
22 33	☽ ⊻ ☿
10 Wednesday	
01 33	☽ □ ♃
01 53	☽ ⊻ ♀
04 05	☽ ∠ ☉
05 06	☽ ⊻ ⯓
05 40	☽ □ ♆
08 13	☽ ♇ ♄
10 45	☿ ∥ ♀
15 49	☽ Q ♂
20 41	☽ ⊥ ☉
11 Thursday	
00 09	☽ ∥ ☿
01 01	☽ ☌ ♅
01 07	☽ ∥ ♀
03 24	☉ ∠ ♆
07 32	☽ ⊻ ☉
08 01	☽ ⧺ ♆
08 36	☽ ⧺ ♃
08 40	☽ Q ♃
09 44	☽ △ ♄
12 17	☽ Q ♆
14 45	♀ ☌ ⯓
17 03	☿ ⚹ ♃
21 12	♀ ⚹ ♆
22 45	☽ □ ♂
12 Friday	
04 47	☽ ⚹ ♃
05 44	☽ ☌ ☿
07 35	☽ ☌ ⯓
08 07	☽ ⚹ ♆
09 06	☽ ☌ ♀
16 45	☉ □ ♄
21 52	☽ ∥ ☉
23 06	♃ ∥ ♆
13 Saturday	
02 33	☽ ⊻ ♅
02 41	☿ ☌ ⯓
05 26	☽ ∠ ♃
08 19	☿ ⚹ ♆
08 27	☽ ∠ ♆
09 56	♀ ⧺ ♃
10 49	☽ □ ♄
11 19	♀ ⧺ ♆
12 02	☽ ☌ ☉
12 14	☽ ⊥ ♅
15 15	☉ ⊥ ♅
20 03	☽ ⊥ ♃
22 54	☽ ⊥ ♆
14 Sunday	
00 24	☽ △ ♂
02 42	☽ ∠ ♅
05 46	☽ ⊻ ♃
08 00	☽ ⊻ ⯓
08 30	☽ ⊻ ♆
11 01	☽ ⊻ ☿
11 26	☿ ⧺ ♃
12 04	☽ ⧺ ☉
13 42	☽ ⊻ ♀
15 40	☽ Q ♄
16 20	☿ ⧺ ♆
17 36	☽ ⊥ ⯓
21 40	☽ ⊥ ☿
15 Monday	
00 09	☽ ⊥ ♀
00 53	☽ ♇ ♂
02 45	☽ ⚹ ♅
08 02	☽ ∠ ⯓
10 55	☽ ⚹ ♄
13 44	☽ ∠ ☿
15 22	☽ ⊻ ☉
15 37	☽ ± ♂
15 52	☽ ∠ ♀
21 45	☽ ⧺ ♀
22 04	☽ Q ♅
16 Tuesday	
00 34	☽ ⧺ ☿
01 30	☽ ⊼ ♂
01 43	☽ ⊥ ☉
04 19	☽ ∥ ♆
05 28	☽ ∥ ♃
06 30	☽ ☌ ♃
08 15	☽ ⚹ ⯓
08 45	☽ ☌ ♆
11 11	☽ ∠ ♄
16 56	☽ ⚹ ☿
17 21	☽ ∠ ☉
18 23	☽ ⚹ ♀
17 Wednesday	
01 55	☽ ⊥ ♄
03 28	☽ □ ♅
03 53	☽ Q ⯓
11 51	☽ ⊻ ♄
15 13	☽ Q ☿
15 59	☽ Q ♀
19 53	☽ ⚹ ☉
18 Thursday	
03 53	☽ ☍ ♂
05 50	☽ ⧺ ♅
08 45	☽ ⊻ ♃
10 00	☽ □ ⯓
10 31	☽ ⊻ ♆
13 12	☿ ☌ ♀
17 38	☽ Q ☉
19 06	☽ ⊥ ♃
20 46	☽ ⊥ ♆
21 28	☽ ⧺ ♂
19 Friday	
01 21	☽ □ ♀
01 50	☽ □ ☿
06 13	☽ △ ♅
10 50	☽ ∠ ♃
12 20	☽ ∠ ♆
14 59	☽ ☌ ♄
19 27	☉ ⊥ ♃
20 Saturday	
03 18	☽ □ ☉
08 37	☽ ♇ ♅
08 49	☽ ⊼ ♂
12 33	☿ ∥ ♀
13 12	☿ ⊻ ♅
13 39	☽ ⚹ ♃
14 20	☽ △ ⯓
14 53	☽ ⚹ ♆
15 55	☉ ⊥ ♆
19 46	☽ ± ♂
21 02	☽ ⧺ ♂
21 Sunday	
00 49	☽ ± ♅
09 52	♀ ⊻ ♅
11 41	☽ Q ♃
11 46	☽ ⊼ ♅
11 57	☽ △ ♀
12 24	☽ ♇ ♂
12 39	☽ Q ♆
15 15	☽ △ ☿
17 36	☽ ♇ ⯓
18 03	☽ ⧺ ♅
21 01	☽ ⊻ ♄
22 Monday	
07 20	☿ ∠ ♃
08 10	☽ ⊥ ♄
10 18	☽ ± ⯓
12 38	☿ ∠ ♆
14 18	☽ △ ☉
16 43	☽ △ ♂
18 43	☽ ♇ ♀
19 20	♃ ⚹ ⯓
21 34	☽ ⊼ ⯓
21 36	☽ □ ♃
22 10	☽ □ ♆
23 44	☽ ♇ ☿
23 Tuesday	
01 07	☽ ∠ ♄
08 31	☉ ∠ ♅
09 38	☿ □ ♄
13 36	☽ ± ♀
19 45	☽ ± ☿
20 11	☽ ☍ ♅
21 01	☽ ∥ ♃
21 08	☽ ♇ ☉
22 50	☿ ⊥ ♅
24 Wednesday	
01 42	☽ ∥ ♆
02 22	☽ ⊼ ♀
05 33	♀ ∠ ♃
05 53	☽ ⚹ ♄
06 40	☉ △ ♂
07 01	♀ ∠ ♆
09 17	☽ ⊼ ☿
15 59	☽ ± ☉
16 12	♃ ☌ ♆
25 Thursday	
03 18	☽ □ ♂
04 43	☽ ⊼ ☉
05 19	☽ Q ♄
07 27	☽ ☍ ⯓
08 04	☽ △ ♆
08 18	☽ △ ♃
14 51	♀ □ ♄
16 36	☽ ⧺ ♀
26 Friday	
02 13	☽ ⧺ ☿
07 01	☽ ⊼ ♅
10 24	♀ ⊥ ♅
13 50	☽ ♇ ♆
14 29	☽ ♇ ♃
15 28	☽ Q ♂
15 54	☉ ⊻ ⯓
17 05	☽ □ ♄
17 51	☽ ∥ ☉
19 01	☽ ± ♅
19 59	☽ ☍ ♀
27 Saturday	
00 03	☉ ⊻ ♆
07 03	☽ ☍ ☿
07 53	☽ ± ♆
08 48	☽ ± ♃
11 10	☽ ⧺ ☉
12 07	☉ ⊻ ♃
12 38	☿ ⊥ ♆
13 08	☽ ♇ ♅
15 54	☽ ⚹ ♂
17 19	☽ ∥ ☿
19 20	☽ ⊼ ⯓
19 43	☿ ⊥ ♃
20 00	☽ ⊼ ♆
21 05	☽ ⊼ ♃
21 44	☽ ☍ ☉
28 Sunday	
03 20	☽ ∥ ♀
07 29	☽ ± ⯓
11 22	☉ ⧺ ☿
19 35	☽ △ ♅
22 41	☽ ∠ ♂
23 43	☿ ∠ ♅
29 Monday	
01 47	☽ ♇ ⯓
05 53	☽ △ ♄
15 42	☽ ⊼ ♀
17 14	☽ ⊥ ♂
21 45	☿ △ ♂
30 Tuesday	
04 02	☽ ⧺ ♆
05 22	☽ ± ♀
05 37	☽ ⊻ ♂
06 56	☽ ⊼ ☿
08 24	☽ △ ⯓
09 05	☽ ☍ ♆
11 05	☽ ☍ ♃
12 12	☽ ⧺ ♃
12 35	☽ ♇ ♄
16 12	☽ ⊼ ☉
16 39	☿ ⊻ ⯓
17 37	☉ ⧺ ♀
21 15	☿ ⊻ ♆
21 22	☽ ± ☿

OCTOBER 1958

Date	Sidereal time h m s	Sun ☉	Moon ☽	Moon ☽ 24.00	Mercury ☿	Venus ♀	Mars ♂	Jupiter ♃	Saturn ♄	Uranus ♅	Neptune ♆	Pluto ♇
01	12 38 51	07 ♎ 48 02	16 ♉ 43 26	22 ♉ 38 40	04 ♎ 37	27 ♍ 16	01 ♊ 58	04 ♏ 42	20 ♐ 16	15 ♌ 11	03 ♏ 32	03 ♍ 12
02	12 42 47	08 47 03	28 ♉ 35 21	04 ♊ 33 57	06 25	28 30	02 05	04 54	20 20	15 14	03 34	03 13
03	12 46 44	09 46 06	10 ♊ 34 58	16 ♊ 38 54	08 12	29 ♍ 45	02 11	05 06	20 23	15 16	03 36	03 15
04	12 50 41	10 45 11	22 ♊ 46 17	28 ♊ 57 39	09 58	01 ♎ 00	02 17	05 19	20 27	15 19	03 38	03 17
05	12 54 37	11 44 19	05 ♋ 13 33	11 ♋ 34 30	11 43	02 15	02 21	05 31	20 31	15 21	03 41	03 18
06	12 58 34	12 43 29	18 ♋ 00 58	24 ♋ 33 24	13 28	03 30	02 25	05 44	20 35	15 23	03 43	03 20
07	13 02 30	13 42 41	01 ♌ 12 10	07 ♌ 57 34	15 12	04 45	02 28	05 56	20 39	15 26	03 45	03 21
08	13 06 27	14 41 55	14 ♌ 49 47	21 ♌ 48 51	16 55	06 00	02 31	06 09	20 43	15 28	03 47	03 23
09	13 10 23	15 41 12	28 ♌ 54 40	06 ♍ 06 59	18 37	07 14	02 32	06 22	20 47	15 30	03 49	03 25
10	13 14 20	16 40 31	13 ♍ 25 20	20 ♍ 49 07	20 19	08 29	02 R 32	06 34	20 51	15 33	03 51	03 26
11	13 18 16	17 39 53	28 ♍ 17 32	05 ♎ 49 37	22 00	09 44	02 32	06 47	20 56	15 35	03 53	03 28
12	13 22 13	18 39 16	13 ♎ 24 19	21 ♎ 00 26	23 40	10 59	02 30	07 00	21 00	15 37	03 56	03 29
13	13 26 10	19 38 41	28 ♎ 36 46	06 ♏ 12 06	25 19	12 14	02 28	07 13	21 05	15 39	03 58	03 31
14	13 30 06	20 38 09	13 ♏ 45 16	21 ♏ 15 09	26 58	13 29	02 25	07 26	21 09	15 41	04 00	03 32
15	13 34 03	21 37 39	28 ♏ 40 49	06 ♐ 01 27	28 ♎ 35	14 45	02 20	07 38	21 14	15 43	04 02	03 34
16	13 37 59	22 37 10	13 ♐ 16 24	20 ♐ 25 12	00 ♏ 13	16 00	02 15	07 51	21 18	15 45	04 04	03 35
17	13 41 56	23 36 43	27 ♐ 27 34	04 ♑ 23 21	01 49	17 15	02 09	08 04	21 23	15 47	04 07	03 37
18	13 45 52	24 36 18	11 ♑ 12 33	17 ♑ 55 20	03 25	18 30	02 03	08 17	21 28	15 49	04 09	03 38
19	13 49 49	25 35 55	24 ♑ 31 55	01 ♒ 02 38	05 00	19 45	01 55	08 30	21 33	15 51	04 11	03 39
20	13 53 45	26 35 33	07 ♒ 27 52	13 ♒ 48 03	06 35	21 00	01 46	08 43	21 38	15 52	04 13	03 41
21	13 57 42	27 35 13	20 ♒ 03 38	26 ♒ 15 05	08 09	22 15	01 37	08 56	21 43	15 54	04 15	03 42
22	14 01 38	28 34 55	02 ♓ 22 54	08 ♓ 27 33	09 43	23 30	01 27	09 09	21 48	15 56	04 18	03 43
23	14 05 35	29 ♎ 34 38	14 ♓ 29 28	20 ♓ 29 06	11 16	24 45	01 15	09 22	21 53	15 57	04 20	03 45
24	14 09 32	00 ♏ 34 23	26 ♓ 26 52	02 ♈ 23 09	12 48	26 01	01 04	09 35	21 58	15 59	04 22	03 46
25	14 13 28	01 34 10	08 ♈ 18 19	14 ♈ 12 43	14 20	27 16	00 51	09 48	22 04	16 01	04 24	03 47
26	14 17 25	02 33 59	20 ♈ 06 39	26 ♈ 00 26	15 51	28 31	00 37	10 01	22 09	16 02	04 27	03 48
27	14 21 21	03 33 50	01 ♉ 54 20	07 ♉ 48 38	17 22	29 ♎ 46	00 23	10 14	22 14	16 03	04 29	03 50
28	14 25 18	04 33 43	13 ♉ 43 36	19 ♉ 39 29	18 52	01 ♏ 01	00 ♊ 08	10 28	22 20	16 05	04 31	03 51
29	14 29 14	05 33 38	25 ♉ 36 32	01 ♊ 35 02	20 22	02 17	29 ♉ 52	10 41	22 25	16 06	04 33	03 52
30	14 33 11	06 33 34	07 ♊ 35 15	13 ♊ 37 29	21 51	03 32	29 36	10 54	22 31	16 07	04 36	03 53
31	14 37 07	07 ♏ 33 33	19 ♊ 42 00	25 ♊ 49 10	23 ♏ 19	04 ♏ 47	29 ♉ 18	11 ♏ 07	22 ♐ 37	16 ♌ 09	04 ♏ 38	03 ♍ 54

Date	Moon True ☊	Moon Mean ☊	Moon ☽ Latitude
01	22 ♎ 16	22 ♎ 54	02 S 12
02	22 D 18	22 51	03 08
03	22 19	22 48	03 57
04	22 20	22 44	04 37
05	22 20	22 41	05 04
06	22 R 20	22 38	05 17
07	22 20	22 35	05 14
08	22 19	22 32	04 54
09	22 18	22 29	04 15
10	22 18	22 25	03 20
11	22 17	22 22	02 10
12	22 17	22 19	00 S 49
13	22 D 17	22 16	00 N 35
14	22 17	22 13	01 57
15	22 17	22 09	03 09
16	22 R 17	22 06	04 08
17	22 17	22 03	04 49
18	22 17	22 00	05 12
19	22 17	21 57	05 18
20	22 D 17	21 54	05 07
21	22 17	21 50	04 42
22	22 18	21 47	04 04
23	22 19	21 44	03 15
24	22 19	21 41	02 19
25	22 20	21 38	01 17
26	22 20	21 35	00 N 12
27	22 R 20	21 31	00 S 53
28	22 20	21 28	01 56
29	22 18	21 25	02 54
30	22 16	21 22	03 45
31	22 ♎ 14	21 ♎ 19	04 S 27

DECLINATIONS

Date	Sun ☉	Moon ☽	Mercury ☿	Venus ♀	Mars ♂	Jupiter ♃	Saturn ♄	Uranus ♅	Neptune ♆	Pluto ♇
01	03 S 06	14 N 44	00 S 26	02 N 25	18 N 55	12 S 10	21 S 58	16 N 54	11 S 05	21 N 01
02	03 29	16 47	01 13	01 55	18 58	12 14	21 58	16 53	11 06	21 00
03	03 52	18 07	01 59	01 26	19 01	12 18	21 59	16 52	11 07	21 00
04	04 15	18 38	02 46	00 56	19 04	12 23	21 59	16 52	11 07	20 59
05	04 39	18 17	03 32	00 N 26	19 07	12 27	22 00	16 51	11 08	20 59
06	05 02	17 00	04 18	00 S 04	19 10	12 31	22 00	16 50	11 09	20 59
07	05 25	14 47	05 04	00 34	19 12	12 35	22 00	16 50	11 10	20 58
08	05 48	11 42	05 49	01 04	19 15	12 40	22 01	16 49	11 10	20 58
09	06 10	07 52	06 33	01 34	19 17	12 44	22 01	16 48	11 11	20 58
10	06 33	03 N 26	07 18	02 04	19 19	12 48	22 02	16 48	11 12	20 57
11	06 56	01 S 18	08 01	02 34	19 21	12 52	22 02	16 47	11 13	20 57
12	07 19	06 03	08 45	03 04	19 23	12 57	22 03	16 47	11 13	20 57
13	07 41	10 26	09 27	03 34	19 25	13 01	22 03	16 46	11 14	20 57
14	08 04	14 07	10 09	04 04	19 27	13 05	22 04	16 45	11 15	20 56
15	08 26	16 47	10 51	04 33	19 28	13 09	22 04	16 45	11 15	20 56
16	08 48	18 18	11 31	05 03	19 30	13 14	22 05	16 44	11 16	20 56
17	09 10	18 36	12 11	05 33	19 31	13 18	22 05	16 44	11 17	20 56
18	09 32	17 47	12 51	06 02	19 32	13 22	22 05	16 43	11 18	20 55
19	09 54	16 00	13 29	06 31	19 33	13 26	22 06	16 43	11 19	20 55
20	10 15	13 27	14 07	07 01	19 34	13 30	22 06	16 42	11 19	20 55
21	10 37	10 20	14 45	07 30	19 35	13 35	22 07	16 42	11 20	20 55
22	10 58	06 50	15 21	07 59	19 36	13 39	22 07	16 41	11 21	20 55
23	11 19	03 S 06	15 56	08 27	19 36	13 43	22 08	16 41	11 22	20 55
24	11 40	00 N 43	16 31	08 56	19 36	13 47	22 08	16 40	11 22	20 54
25	12 01	04 29	17 05	09 24	19 37	13 51	22 09	16 40	11 23	20 54
26	12 22	08 03	17 38	09 52	19 37	13 56	22 09	16 40	11 24	20 54
27	12 42	11 19	18 10	10 20	19 37	14 00	22 10	16 39	11 25	20 54
28	13 03	14 07	18 42	10 48	19 36	14 04	22 10	16 39	11 25	20 54
29	13 23	16 21	19 12	11 15	19 36	14 08	22 11	16 38	11 26	20 54
30	13 42	17 52	19 41	11 43	19 36	14 12	22 11	16 38	11 27	20 54
31	14 S 02	18 N 36	20 S 10	12 S 09	19 N 35	14 S 16	22 S 11	16 N 38	11 S 28	20 N 54

ZODIAC SIGN ENTRIES

Date	h	m	Planets
02	14	50	☽ ♊
03	16	44	♀ ♎
05	02	00	☽ ♋
07	09	51	☽ ♌
09	13	49	☽ ♍
11	14	44	☽ ♎
13	14	11	☽ ♏
15	14	09	☽ ♐
16	08	52	☿ ♏
17	16	23	☽ ♑
19	22	04	☽ ♒
22	07	19	☽ ♓
23	22	11	☉ ♏
24	19	10	☽ ♈
27	08	07	☽ ♉
27	16	26	♀ ♏
29	00	01	♂ ♉
29	20	49	☽ ♊

LATITUDES

Date	Mercury ☿	Venus ♀	Mars ♂	Jupiter ♃	Saturn ♄	Uranus ♅	Neptune ♆	Pluto ♇
01	01 N 32	01 N 27	01 S 40	00 N 59	01 N 07	00 N 38	01 N 43	11 N 29
04	01 17	01 27	01 35	00 58	01 07	00 39	01 43	11 30
07	01 00	01 26	01 29	00 58	01 06	00 39	01 42	11 31
10	00 42	01 25	01 23	00 58	01 06	00 39	01 42	11 32
13	00 22	01 23	01 16	00 58	01 05	00 39	01 42	11 33
16	00 N 02	01 21	01 09	00 57	01 05	00 39	01 42	11 34
19	00 S 19	01 18	01 01	00 57	01 05	00 39	01 42	11 35
22	00 39	01 14	00 53	00 57	01 04	00 39	01 42	11 36
25	00 59	01 10	00 44	00 57	01 04	00 39	01 42	11 37
28	01 19	01 06	00 35	00 57	01 03	00 39	01 42	11 38
31	01 S 37	01 N 01	00 S 26	00 N 56	01 N 03	00 N 40	01 N 42	11 N 39

DATA

Julian Date	2436478
Delta T	+33 seconds
Ayanamsa	23° 16' 57"
Synetic vernal point	05° ♓ 50' 03"
True obliquity of ecliptic	23° 26' 33"

MOON'S PHASES, APSIDES AND POSITIONS ☽

Date	h	m	Phase	Longitude	Eclipse Indicator
06	01	20	☾	12 ♋ 17	
12	20	52	●	19 ♎ 01	Total
19	14	07	☽	25 ♑ 41	
27	15	41	○	03 ♉ 43	

Day	h	m		
13	02	22	Perigee	
27	00	03	Apogee	
04	14	24	Max dec	18° N 39'
11	05	29	0S	
17	06	08	Max dec	18° S 38'
24	07	30	0N	
31	20	35	Max dec	18° N 40'

LONGITUDES

Date	Chiron ⚷	Ceres ⚳	Pallas ⚴	Juno ⚵	Vesta ⚶	Black Moon Lilith ⚸
01	18 ♒ 01	01 ♎ 00	11 ♍ 58	20 ♍ 17	29 ♋ 13	24 ♈ 50
11	17 ♒ 48	05 ♎ 33	16 ♍ 53	24 ♍ 18	02 ♌ 23	25 ♈ 56
21	17 ♒ 40	10 ♎ 05	21 ♍ 40	28 ♍ 18	05 ♌ 15	27 ♈ 03
31	17 ♒ 39	14 ♎ 34	26 ♍ 17	01 ♎ 59	07 ♌ 47	28 ♈ 09

All ephemeris data is given at 12.00 UT and the Moon's longitude is additionally given for 24.00 UT

ASPECTARIAN

h m Aspects

01 Wednesday
01 52 ☽ Ᵽ ♀
05 31 ☽ ± ☉
06 59 ☽ ± ♄
08 51 ☽ □ ♅
13 04 ☿ ⊻ ♃
17 26 ♀ ⊥ ♆
18 56 ☽ Ᵽ ☿
19 14 ☽ ⊼ ♄

02 Thursday
00 09 ☉ Q ♄
01 26 ☽ Ᵽ ☉
11 49 ☽ △ ♀
13 25 ☽ ∥ ♅
19 05 ☽ ☌ ♂
21 02 ♀ ⊥ ♃
21 20 ☽ □ ♇
21 21 ☽ Q ♅
22 02 ☽ ⊼ ♆
22 57 ☉ ⊥ ♇

03 Friday
00 53 ☽ ⊼ ♃
01 24 ☿ # ♀
06 26 ☽ △ ☿
10 03 ☽ ± ♆
10 14 ☽ △ ☉
13 03 ☽ ± ♃
14 43 ☿ Q ♄
21 19 ☽ ✶ ♅
22 13 ♀ ∠ ♅

04 Saturday
02 29 ☿ ⊥ ♇
03 54 ☽ Ᵽ ♆
07 07 ☽ Ᵽ ♃
07 27 ☽ ☍ ♄
09 04 ☽ Q ♇

05 Sunday
02 39 ☽ ∠ ♅
05 41 ☽ □ ♀
06 30 ☽ ⊻ ♂
08 20 ☽ ✶ ♇
09 02 ☽ △ ♆
11 10 ♃ ∠ ♄
12 31 ☉ ☌ ☿
12 34 ☽ △ ♃
14 15 ♀ △ ♂
17 58 ☽ ⊥ ♂
19 50 ☽ ⊥ ♅

06 Monday
01 20 ☽ □ ☉
02 13 ☽ □ ☿
07 07 ☽ ⊻ ♅
08 45 ♀ ⊻ ♇
10 54 ☽ ∠ ♂
12 35 ☽ ∠ ♇
14 05 ☽ ∥ ♅
16 15 ♀ ⊻ ♆
16 45 ☽ ⊼ ♄
19 05 ☽ Q ♀

07 Tuesday
03 45 ☽ ± ♄
05 04 ☽ ⊥ ♇
12 59 ☽ Q ☉
14 17 ☽ ✶ ♂
15 18 ☿ ✶ ♅
15 51 ☽ ⊻ ♇
16 05 ☽ Q ☿
16 33 ☽ □ ♆
18 57 ☽ ✶ ♀
19 58 ☽ Ᵽ ♄
20 34 ☽ □ ♃

08 Wednesday
05 20 ☽ # ♃
10 46 ☉ ∥ ☿
11 27 ☽ Q ♂
11 45 ☽ ✶ ☉
13 06 ☽ ☌ ♅
15 34 ☽ # ♆
15 40 ♀ ⊻ ♃
16 06 ☽ ✶ ☿
20 27 ☿ Ᵽ ♂
22 11 ☽ △ ♄
23 38 ☽ ∠ ♀
23 58 ☽ Q ♆

09 Thursday
04 13 ☽ Q ♃
07 26 ☉ ✶ ♅
09 00 ☿ ∠ ♇
15 12 ☽ ∠ ☉
16 16 ☽ ⊥ ♀
18 03 ☽ □ ♂
18 16 ☽ # ☿
19 32 ☽ ☌ ♇
20 13 ☽ ✶ ♆
20 42 ☽ # ☉
20 54 ☽ ∠ ☿

10 Friday
00 35 ☽ ✶ ♃
03 10 ☽ ⊻ ♀
07 11 ☽ ⊥ ☉
09 46 ♂ St R
13 38 ☽ ⊥ ☿
15 28 ☽ ⊻ ♅
17 40 ☽ ⊻ ☉
18 23 ☽ # ♀
19 28 ♀ Q ♄
20 06 ☿ ✶ ♄
20 51 ☽ ∠ ♆

11 Saturday
00 07 ☽ □ ♄
00 37 ☽ ⊻ ☿
01 12 ☽ ⊥ ♅
01 24 ☽ ∠ ♃
06 33 ♀ ⊥ ♇
08 42 ☉ Ᵽ ♂
11 21 ☽ ⊥ ♆
15 40 ☽ ∠ ♅
16 02 ☽ ⊥ ♃
18 45 ☽ △ ♂
19 03 ☽ ∥ ♀
20 15 ☽ ⊻ ♇
20 57 ☽ ⊻ ♆

12 Sunday
01 43 ☽ ⊻ ♃
05 00 ☽ Q ♄
05 48 ☽ ⊥ ♇
07 50 ☽ ☌ ♀
07 52 ☉ ∠ ♇
15 30 ☽ ✶ ♅
18 27 ☽ Ᵽ ♂
19 14 ☽ ∥ ☉
20 02 ☽ ∠ ♇
20 52 ☽ ☌ ☉

13 Monday
00 03 ☽ ✶ ♄
05 14 ☽ ∥ ☿
06 10 ☽ ☌ ☿
08 37 ☽ ± ♂
10 29 ☽ Q ♅
16 48 ☽ ∥ ♆
18 04 ☽ ⊼ ♂
19 45 ☽ ✶ ♇
20 28 ☽ ☌ ♆
23 52 ☽ ∠ ♄

14 Tuesday
01 48 ☽ ☌ ♃
04 15 ☿ ± ♂
04 31 ☽ ∥ ♃
11 33 ☽ ⊻ ♀
14 14 ☽ ⊥ ♄
14 51 ☽ Q ♇
15 05 ☽ □ ♅
22 01 ☽ ⊥ ♀
22 53 ☿ Q ♅
23 48 ☽ ⊻ ☉
23 54 ☽ ⊻ ♄

15 Wednesday
01 33 ☉ ✶ ♄
10 10 ☽ ⊥ ☉
11 30 ☽ # ♅
11 50 ☽ ⊻ ☿
13 53 ☽ ∠ ♀
17 56 ☽ ☍ ♂
19 58 ☽ □ ♇
20 45 ☽ ⊻ ♆
22 51 ☽ ⊥ ☿

16 Thursday
01 56 ☽ ∠ ☉
02 53 ☽ ⊻ ♃
02 54 ☿ ∥ ♆
06 40 ☽ ⊥ ♆
07 14 ♀ ✶ ♅
12 59 ☽ ⊥ ♃
15 39 ☽ ∠ ☿
16 09 ☽ △ ♅
16 59 ☽ ✶ ♀
21 45 ☽ ∠ ♆

17 Friday
01 35 ☽ ☌ ♄
04 23 ☽ ∠ ♃
04 55 ☽ ✶ ☉
10 28 ♀ Ᵽ ♂
15 23 ☽ Q ♀
16 44 ☿ ⊼ ♂
17 45 ☽ Ᵽ ♅
20 03 ☽ ⊼ ♂
20 31 ☽ ✶ ☿
22 40 ☽ △ ♇
23 33 ☽ ✶ ♆

18 Saturday
03 15 ☽ Q ☉
06 27 ☽ ± ♂
06 45 ☽ ✶ ♃
09 31 ☽ ± ♅
14 41 ♀ ∠ ♇
15 16 ☿ ✶ ♇
20 14 ☽ ⊼ ♅
20 31 ☽ Q ☿
20 50 ☽ Q ♆
22 19 ☽ Ᵽ ♂
23 13 ☿ ☌ ♆

19 Sunday
01 18 ☽ Ᵽ ♇
02 23 ☽ □ ♀
03 41 ☽ # ♅
04 32 ☽ Q ♃
06 32 ☽ ⊻ ♄
09 43 ☿ ∥ ♃
14 07 ☽ □ ☉
17 34 ☽ ⊥ ♄
17 45 ☽ ± ♇
18 44 ☉ ± ♂

20 Monday
01 28 ☽ △ ♂
04 54 ☽ ⊼ ♇
05 54 ☽ □ ♆
07 19 ☽ ∥ ☿
10 07 ☽ □ ☿
10 25 ☽ ∠ ♄
11 33 ☽ ∥ ♃
12 45 ☿ ∠ ♄
14 24 ☽ □ ♃

21 Tuesday
00 59 ♀ ✶ ♄
04 00 ☽ ☍ ♅
04 44 ☽ ∥ ♆
10 11 ☽ ∥ ☉
15 13 ☽ ✶ ♄
16 42 ☽ △ ♀
19 51 ☉ Q ♅

22 Wednesday
01 58 ☿ ☌ ♃
03 53 ☽ △ ☉
05 18 ☽ ∥ ♀
10 11 ☽ □ ♂
14 39 ☽ ☍ ♇
14 49 ☽ Q ♄
15 47 ☽ △ ♆

23 Thursday
01 29 ☽ Ᵽ ♀
01 37 ☽ △ ♃
04 37 ☽ △ ☿
12 11 ☽ Ᵽ ☉
14 29 ☉ ∥ ♆
14 56 ☽ ⊼ ♅
20 19 ♀ ± ♂
21 23 ☽ Q ♂
21 32 ☽ ± ♀
21 43 ☽ Ᵽ ♆

24 Friday
02 55 ☽ □ ♄
03 00 ☽ ± ♅
07 53 ☽ ± ☉
08 11 ☽ Ᵽ ♃
11 01 ☽ ⊼ ♀
15 08 ☽ Ᵽ ☿
15 53 ☽ ± ♆
18 22 ☿ # ♅
21 06 ☽ ⊼ ☉
21 09 ☽ ✶ ♂
21 11 ☽ Ᵽ ♅
21 40 ☉ ⊼ ♂

25 Saturday
02 42 ☽ ± ♃
02 49 ☽ ⊼ ♇
04 04 ☽ ⊼ ♆
12 03 ☽ ± ☿
15 01 ☽ ± ♇
15 06 ☽ ⊼ ♃

26 Sunday
02 03 ☽ ⊼ ☿
02 37 ♀ Q ♅
03 02 ☽ ∠ ♂
03 41 ☽ △ ♅
09 20 ☽ Ᵽ ♇
11 18 ☿ Q ♇
14 58 ☿ □ ♅
16 11 ☽ △ ♄
17 03 ☿ ⊥ ♄
21 00 ☽ ⊥ ♂

27 Monday
03 15 ☽ # ♀
07 08 ☽ ☍ ♀
08 58 ☽ ⊻ ♂
12 47 ☽ # ♆
15 41 ☽ ☍ ☉
15 55 ☽ △ ♇
17 15 ☽ ☍ ♆
18 25 ☉ ✶ ♇
21 48 ♀ ⊼ ♂
22 55 ☽ Ᵽ ♄

28 Tuesday
00 57 ☽ # ☉
05 15 ☽ ☍ ♃
10 55 ☉ ☌ ♆
11 31 ☽ # ♃
16 47 ☽ □ ♅
17 19 ☽ ± ♄
23 54 ☽ ☍ ☿

29 Wednesday
05 32 ☽ ⊼ ♄
15 53 ☽ ∥ ♅
20 22 ☽ ☌ ♂
21 40 ♀ ∥ ♆

30 Thursday
02 57 ☽ ⊼ ♀
04 35 ☽ □ ♇
05 04 ☽ Q ♅
06 00 ☽ ⊼ ♆
07 23 ☿ # ♂
09 46 ☽ ⊼ ☉
16 19 ☽ ± ♀
18 00 ☽ ± ♆
18 43 ☽ ⊼ ♃
18 51 ♀ ✶ ♇
22 47 ☽ ± ☉
23 38 ☿ ⊻ ♄

31 Friday
04 59 ☽ ✶ ♅
06 49 ☽ ± ♃
08 59 ♀ ☌ ♆
11 52 ☽ Ᵽ ♆
12 11 ☽ Ᵽ ♀
13 20 ☉ ∠ ♄
16 20 ☽ Q ♇
17 46 ☽ ☍ ♄
18 07 ☽ Ᵽ ☉
20 05 ☽ ⊼ ☿

NOVEMBER 1958

LONGITUDES

Date	Sidereal time h m s	Sun ☉ ° ' "	Moon ☽ ° ' "	Moon ☽ 24.00 ° ' "	Mercury ☿ ° '	Venus ♀ ° '	Mars ♂ ° '	Jupiter ♃ ° '	Saturn ♄ ° '	Uranus ♅ ° '	Neptune ♆ ° '	Pluto ♇ ° '
01	14 41 04	08 ♏ 33 34	01 ♋ 59 17	08 ♋ 12 43	24 ♏ 47	06 ♏ 02	29 ♉ 01	11 ♏ 20	22 ♐ 42	16 ♌ 10	04 ♏ 40	03 ♍ 55
02	14 45 01	09 33 37	14 ♋ 29 50	20 ♋ 51 00	26 15	07 18	28 R 42	11 33	22 48	16 11	04 42	03 56
03	14 48 57	10 33 42	27 ♋ 16 36	03 ♌ 47 00	27 42	08 33	28 23	11 47	22 54	16 12	04 45	03 57
04	14 52 54	11 33 50	10 ♌ 22 33	17 ♌ 03 34	29 ♏ 08	09 48	28 04	12 00	23 00	16 13	04 47	03 58
05	14 56 50	12 33 59	23 ♌ 50 20	00 ♍ 43 01	00 ♐ 33	11 03	27 44	12 13	23 06	16 14	04 49	03 59
06	15 00 47	13 34 10	07 ♍ 41 44	14 ♍ 46 30	01 58	12 19	27 24	12 26	23 12	16 15	04 51	04 00
07	15 04 43	14 34 24	21 ♍ 57 09	29 ♍ 13 27	03 23	13 34	27 03	12 39	23 18	16 16	04 53	04 01
08	15 08 40	15 34 39	06 ♎ 34 55	14 ♎ 00 58	04 46	14 49	26 42	12 52	23 24	16 17	04 56	04 02
09	15 12 36	16 34 57	21 ♎ 30 48	29 ♎ 03 31	06 09	16 05	26 20	13 06	23 30	16 17	04 58	04 03
10	15 16 33	17 35 16	06 ♏ 38 02	14 ♏ 13 12	07 30	17 20	25 59	13 19	23 36	16 18	05 00	04 03
11	15 20 30	18 35 37	21 ♏ 47 47	29 ♏ 20 34	08 51	18 35	25 37	13 32	23 42	16 19	05 02	04 04
12	15 24 26	19 36 00	06 ♐ 50 22	14 ♐ 16 05	10 10	19 51	25 15	13 45	23 48	16 19	05 05	04 05
13	15 28 23	20 36 25	21 ♐ 36 45	28 ♐ 51 35	11 28	21 06	24 53	13 58	23 55	16 20	05 07	04 06
14	15 32 19	21 36 51	05 ♑ 59 58	13 ♑ 01 27	12 45	22 22	24 31	14 11	24 01	16 20	05 09	04 06
15	15 36 16	22 37 18	19 ♑ 55 49	26 ♑ 43 02	14 00	23 37	24 08	14 25	24 07	16 21	05 11	04 07
16	15 40 12	23 37 47	03 ♒ 23 11	09 ♒ 56 31	15 14	24 52	23 46	14 38	24 14	16 21	05 13	04 08
17	15 44 09	24 38 17	16 ♒ 23 25	22 ♒ 44 20	16 25	26 08	23 25	14 51	24 20	16 21	05 15	04 08
18	15 48 05	25 38 48	28 ♒ 59 48	05 ♓ 10 24	17 34	27 23	23 03	15 04	24 27	16 21	05 17	04 09
19	15 52 02	26 39 21	11 ♓ 16 44	17 ♓ 19 25	18 41	28 38	22 42	15 17	24 33	16 22	05 20	04 09
20	15 55 59	27 39 54	23 ♓ 19 05	29 ♓ 16 20	19 44	29 ♏ 54	22 20	15 30	24 40	16 22	05 22	04 10
21	15 59 55	28 40 29	05 ♈ 11 47	11 ♈ 05 58	20 44	01 ♐ 09	22 00	15 43	24 46	16 22	05 24	04 10
22	16 03 52	29 ♏ 41 06	16 ♈ 59 26	22 ♈ 52 39	21 41	02 25	21 39	15 56	24 53	16 R 22	05 26	04 11
23	16 07 48	00 ♐ 41 43	28 ♈ 46 07	04 ♉ 40 12	22 33	03 40	21 19	16 09	24 59	16 22	05 28	04 11
24	16 11 45	01 42 22	10 ♉ 35 17	16 ♉ 31 41	23 20	04 55	21 00	16 22	25 06	16 22	05 30	04 12
25	16 15 41	02 43 02	22 ♉ 30 42	28 ♉ 29 32	24 02	06 11	20 41	16 35	25 13	16 21	05 32	04 12
26	16 19 38	03 43 43	04 ♊ 31 26	10 ♊ 35 32	24 37	07 26	20 22	16 48	25 20	16 21	05 34	04 12
27	16 23 34	04 44 26	16 ♊ 41 59	22 ♊ 50 56	25 06	08 41	20 04	17 01	25 26	16 21	05 36	04 13
28	16 27 31	05 45 10	29 ♊ 02 28	05 ♋ 16 41	25 27	09 57	19 47	17 14	25 33	16 21	05 38	04 13
29	16 31 28	06 45 56	11 ♋ 33 43	17 ♋ 53 39	25 39	11 12	19 30	17 26	25 40	16 20	05 40	04 13
30	16 35 24	07 ♐ 46 42	24 ♋ 16 36	00 ♌ 42 43	25 ♐ 42	12 ♐ 28	19 ♉ 14	17 ♏ 39	25 ♐ 47	16 ♌ 20	05 ♏ 42	04 ♍ 13

MOON AND DECLINATIONS

Date	Moon True ☊ ° '	Moon Mean ☊ ° '	Moon ☽ Latitude ° '	Sun ☉ ° '	Moon ☽ ° '	Mercury ☿ ° '	Venus ♀ ° '	Mars ♂ ° '	Jupiter ♃ ° '	Saturn ♄ ° '	Uranus ♅ ° '	Neptune ♆ ° '	Pluto ♇ ° '
01	22 ♎ 11	21 ♎ 15	04 S 57	14 S 21	18 N 29	20 S 37	12 S 36	19 N 34	14 S 20	22 S 12	16 N 37	11 S 28	20 N 54
02	22 R 09	21 12	05 13	14 41	17 20	21 04	13 02	19 33	14 25	22 12	16 37	11 29	20 54
03	22 08	21 09	05 15	15 00	15 33	21 29	13 28	19 32	14 29	22 13	16 37	11 30	20 54
04	22 07	21 06	05 00	15 18	12 49	21 53	13 54	19 31	14 33	22 13	16 37	11 31	20 54
05	22 D 08	21 03	04 29	15 37	09 20	22 17	14 19	19 30	14 37	22 14	16 36	11 31	20 54
06	22 09	21 00	03 42	15 55	05 15	22 39	14 44	19 28	14 41	22 14	16 36	11 32	20 54
07	22 10	20 56	02 39	16 13	00 N 45	23 00	15 09	19 27	14 45	22 14	16 36	11 33	20 54
08	22 11	20 53	01 25	16 30	03 S 55	23 19	15 33	19 25	14 49	22 15	16 36	11 33	20 54
09	22 12	20 50	00 S 04	16 48	08 27	23 38	15 57	19 23	14 53	22 15	16 36	11 34	20 54
10	22 R 11	20 47	01 N 19	17 05	12 30	23 55	16 20	19 21	14 57	22 16	16 35	11 35	20 54
11	22 10	20 44	02 36	17 22	15 42	24 12	16 43	19 19	15 01	22 16	16 35	11 36	20 54
12	22 06	20 41	03 41	17 38	17 49	24 26	17 06	19 17	15 05	22 17	16 35	11 36	20 54
13	22 02	20 37	04 31	17 54	18 40	24 40	17 28	19 15	15 09	22 17	16 35	11 37	20 54
14	21 58	20 34	05 02	18 10	18 17	24 52	17 49	19 13	15 13	22 17	16 35	11 38	20 54
15	21 54	20 31	05 13	18 26	16 48	25 03	18 10	19 11	15 17	22 18	16 35	11 38	20 54
16	21 51	20 28	05 07	18 41	14 25	25 12	18 31	19 08	15 20	22 18	16 35	11 39	20 54
17	21 49	20 25	04 45	18 56	11 23	25 20	18 51	19 06	15 24	22 19	16 35	11 40	20 55
18	21 D 49	20 21	04 10	19 10	07 55	25 27	19 11	19 03	15 28	22 19	16 35	11 41	20 55
19	21 50	20 18	03 24	19 25	04 11	25 32	19 30	19 01	15 32	22 19	16 35	11 41	20 55
20	21 51	20 15	02 30	19 38	00 S 21	25 36	19 48	18 58	15 36	22 20	16 35	11 42	20 55
21	21 53	20 12	01 30	19 52	03 N 27	25 38	20 06	18 56	15 40	22 20	16 35	11 43	20 55
22	21 54	20 09	00 N 27	20 05	07 05	25 38	20 24	18 53	15 44	22 21	16 35	11 43	20 56
23	21 R 54	20 06	00 S 37	20 18	10 27	25 37	20 41	18 51	15 47	22 21	16 35	11 44	20 56
24	21 52	20 02	01 40	20 30	13 25	25 35	20 57	18 48	15 51	22 21	16 35	11 45	20 56
25	21 48	19 59	02 39	20 42	15 50	25 30	21 12	18 46	15 54	22 21	16 35	11 45	20 56
26	21 42	19 56	03 31	20 54	17 36	25 25	21 27	18 44	15 58	22 22	16 35	11 46	20 57
27	21 35	19 53	04 14	21 05	18 34	25 17	21 42	18 41	16 02	22 22	16 35	11 47	20 57
28	21 26	19 50	04 45	21 16	18 41	25 08	21 56	18 39	16 05	22 22	16 35	11 47	20 57
29	21 18	19 46	05 04	21 27	17 54	24 57	22 09	18 37	16 09	22 23	16 35	11 48	20 57
30	21 ♎ 10	19 ♎ 43	05 S 08	21 S 37	16 N 13	24 S 44	22 S 21	18 N 35	16 S 13	22 S 23	16 N 36	11 S 48	20 N 58

ZODIAC SIGN ENTRIES

Date	h	m	Planets
01	08	09	☽ ♋
03	17	02	☽ ♌
05	02	36	☿ ♐
05	22	45	☽ ♍
08	01	16	☽ ♎
10	01	30	☽ ♏
12	01	03	☽ ♐
14	01	54	☽ ♑
16	05	53	☽ ♒
18	13	56	☽ ♓
20	13	59	♀ ♐
21	01	28	☽ ♈
22	19	29	☉ ♐
23	14	30	☽ ♉
26	03	00	☽ ♊
28	13	51	☽ ♋
30	22	41	☽ ♌

LATITUDES

Date	Mercury ☿ ° '	Venus ♀ ° '	Mars ♂ ° '	Jupiter ♃ ° '	Saturn ♄ ° '	Uranus ♅ ° '	Neptune ♆ ° '	Pluto ♇ ° '
01	01 S 42	00 N 59	00 S 23	00 N 56	01 N 03	00 N 40	01 N 42	11 N 39
04	01 58	00 54	00 13	00 56	01 02	00 40	01 42	11 40
07	02 12	00 48	00 S 03	00 56	01 02	00 40	01 42	11 42
10	02 24	00 42	00 N 07	00 56	01 02	00 40	01 42	11 43
13	02 32	00 36	00 16	00 56	01 01	00 40	01 42	11 44
16	02 36	00 29	00 26	00 56	01 01	00 40	01 43	11 45
19	02 35	00 22	00 35	00 56	01 01	00 40	01 43	11 47
22	02 28	00 15	00 44	00 56	01 00	00 41	01 43	11 48
25	02 12	00 08	00 52	00 56	01 00	00 41	01 43	11 49
28	01 46	00 N 01	01 00	00 56	01 00	00 41	01 43	11 50
31	01 S 08	00 S 06	01 N 08	00 N 56	00 N 59	00 N 41	01 N 43	11 N 52

DATA

Julian Date	2436509
Delta T	+33 seconds
Ayanamsa	23° 17' 00"
Synetic vernal point	05° ♓ 49' 59"
True obliquity of ecliptic	23° 26' 32"

MOON'S PHASES, APSIDES AND POSITIONS ☽

Date	h	m	Phase	Longitude ° '	Eclipse Indicator
04	14	19	☾	11 ♌ 40	
11	06	34	●	18 ♏ 22	
18	04	59	☽	25 ♒ 21	
26	10	17	○	03 ♊ 39	

Day	h	m	
10	14	22	Perigee
23	05	02	Apogee

Day	h	m		
07	15	53	0S	
13	16	16	Max dec	18° S 42'
20	14	14	0N	
28	03	04	Max dec	18° N 45'

LONGITUDES

Date	Chiron ⚷ ° '	Ceres ⚳ ° '	Pallas ⚴ ° '	Juno ⚵ ° '	Vesta ⚶ ° '	Black Moon Lilith ⚸ ° '
01	17 ♒ 40	15 ♎ 00	26 ♍ 44	02 ♎ 21	08 ♌ 01	28 ♈ 16
11	17 ♒ 46	19 ♎ 25	01 ♎ 10	05 ♎ 57	10 ♌ 04	29 ♈ 22
21	17 ♒ 58	23 ♎ 45	05 ♎ 24	09 ♎ 23	11 ♌ 37	00 ♉ 08
31	18 ♒ 17	28 ♎ 00	09 ♎ 25	12 ♎ 36	12 ♌ 35	01 ♉ 35

ASPECTARIAN

01 Saturday
00 49 ☽ ⚼ ♃
06 21 ☽ ⊻ ♂
09 22 ☽ ± ☿
10 22 ☉ ∥ ♃
10 24 ☽ ∠ ♅
15 44 ☽ ⚹ ♇
17 12 ☽ △ ♆
17 42 ☽ ⊥ ♂
20 42 ☽ △ ♀

02 Sunday
01 46 ☽ △ ☉
02 48 ☿ # ♇
03 46 ☽ ⊥ ♅
05 00 ☽ ⚼ ☿
06 18 ☽ △ ♃
10 32 ☽ ∠ ♂
15 12 ☽ ⊻ ♅
20 25 ☽ ∠ ♇
22 31 ♀ ∠ ♄
23 59 ☽ ∥ ♅

03 Monday
03 47 ☽ ⊼ ♄
12 52 ☽ △ ☿
13 15 ☽ ⊥ ♇
14 01 ☽ ⚹ ♂
15 02 ☽ ± ♄
16 57 ☽ # ☉
21 28 ☿ ☍ ♂
22 03 ☽ # ♃

04 Tuesday
00 19 ☽ ⊻ ♇
01 48 ☽ □ ♆
04 26 ☽ # ♀
07 39 ☽ ⚼ ♄
10 51 ☽ □ ♀
11 27 ☽ Q ♂
14 19 ☽ □ ☉
14 58 ☽ □ ♃
21 34 ☽ # ♆
22 31 ☽ ☌ ♅

05 Wednesday
01 13 ☉ ☌ ♃
08 49 ☿ ∥ ♄
10 12 ☽ Q ♆
10 41 ☽ △ ♄
18 39 ☽ □ ♂
22 02 ☽ Q ♀
23 19 ☽ Q ♃

06 Thursday
00 39 ☽ Q ☉
01 04 ☽ □ ☿
05 40 ☽ ☌ ♇
07 07 ☽ ⚹ ♆
08 05 ♀ ∥ ♃
14 50 ♀ ☌ ♃
20 11 ☽ ⚹ ♃
20 36 ☽ ⚹ ♀
22 44 ☽ ⚹ ☉

07 Friday
02 29 ☽ ⊻ ♅
08 34 ☽ ∠ ♆
10 56 ☽ Q ☿
12 31 ☽ ⊥ ♅
14 14 ☽ □ ♄
20 13 ☽ △ ♂
21 34 ☽ ∠ ♃
23 06 ☿ □ ♇
23 29 ☽ ⊥ ♆
23 57 ☽ ∠ ♀

08 Saturday
01 30 ☽ ∠ ☉
03 21 ☽ ∠ ♅
07 51 ☽ ⊻ ♇
08 45 ☽ ⚹ ☿
09 18 ☽ ⊻ ♆
12 29 ☽ ⊥ ♃
14 52 ☿ ⊻ ♆
15 58 ☽ ⊥ ♀
17 12 ☽ ⊥ ☉
17 35 ☽ ⊥ ♇
19 15 ☉ # ♅
19 50 ☽ Q ♄
20 04 ☽ ⚼ ♂
22 19 ☽ ⊻ ♃
22 57 ☉ Q ♇

09 Sunday
02 31 ☽ ⊻ ♀
03 33 ☽ ⊻ ☉
03 38 ☽ ⚹ ♅
04 53 ☉ □ ♅
08 03 ☽ ∠ ♇
10 10 ☽ ± ♂
11 19 ♀ Q ♇
11 21 ☽ ∠ ☿
15 11 ☽ ⚹ ♄
16 02 ♀ □ ♅
19 30 ☽ ⊼ ♂
22 47 ☽ Q ♅

10 Monday
03 05 ☽ ⊥ ☿
06 12 ☽ ∥ ♆
07 55 ☽ ⚹ ♇
09 25 ☽ ☌ ♆
12 14 ☉ ⊥ ♄
13 31 ☽ ⊻ ☿
15 08 ☽ ∠ ♄
17 28 ♀ ⊥ ♄
22 43 ☽ ☌ ♃

11 Tuesday
02 55 ☽ Q ♇
03 18 ☽ □ ♅
03 49 ♀ # ♅
05 28 ☽ ⊥ ♄
05 59 ☽ ∥ ♃
06 28 ☽ ☌ ♀
06 34 ☽ ☌ ☉
12 20 ☉ ☌ ♀
15 03 ☽ ⊻ ♄
17 55 ☽ ☍ ♂
20 27 ☽ # ♅
23 54 ☽ ∥ ♀

12 Wednesday
07 34 ☽ □ ♇
08 40 ☽ ∥ ☉
09 10 ☽ ⊻ ♆
17 53 ☽ ☌ ☿
18 51 ☽ ⊥ ♆
23 20 ☽ ⊻ ♃

13 Thursday
03 21 ☽ △ ♅
05 06 ☿ ⊥ ♆
09 16 ☽ ⊥ ♃
09 32 ☽ ∠ ♆
10 14 ☽ ⊻ ☉
11 05 ☽ ⊻ ♀
15 49 ☽ ☌ ♄
17 15 ☽ ⊼ ♂
20 52 ☽ ⊥ ☉
21 56 ☽ ⊥ ♀

14 Friday
00 22 ☽ ∠ ♃
02 59 ☽ ± ♂
04 08 ☽ ⚼ ♅
08 48 ☽ △ ♇
10 33 ☽ ⚹ ♆
13 07 ☽ ∠ ☉
14 10 ☽ ∥ ☉
14 32 ☽ ∠ ♀
17 49 ☽ ⚼ ♂
19 23 ☽ ± ♅
19 28 ☽ ∥ ♀

15 Saturday
00 41 ☽ ⊻ ☿
02 14 ☽ ⚹ ♃
05 44 ☽ ⊼ ♅
07 11 ☽ Q ♆
10 34 ☽ ⚼ ♇
12 09 ☽ ⊥ ☿
13 03 ♂ ⊼ ♄
14 33 ☽ # ♅
17 07 ☽ ⚹ ☉
19 09 ☽ ⚹ ♀
19 13 ☽ △ ♂
19 27 ☽ ⊻ ♄
19 47 ♀ ☍ ♂
21 32 ☿ ⊻ ♃
22 33 ♀ ⊻ ♄
23 38 ☽ Q ♃

16 Sunday
02 31 ☽ ± ♇
03 46 ☽ ∥ ♃
05 43 ☽ ∠ ☿
06 15 ☽ ⊥ ♄
13 21 ☽ ⊼ ♇
14 32 ☉ ☍ ♂
15 21 ☽ □ ♆
16 25 ☽ Q ☉
19 02 ☽ Q ♀
22 46 ☽ ∠ ♄

17 Monday
03 54 ☉ ⊻ ♄
09 04 ☽ □ ♃
09 57 ☽ ∥ ♆
10 40 ☿ △ ♅
11 56 ☽ ☍ ♅
12 03 ☽ ⚹ ☿

18 Tuesday
00 55 ☽ □ ♂
01 51 ☉ # ♂
03 10 ☽ ⚹ ♄
03 50 ♀ # ♂
04 59 ☽ □ ☉
08 33 ☽ □ ♀
10 33 ☉ ∥ ♀
13 13 ☽ Q ☿
22 00 ☽ ☍ ♇

19 Wednesday
00 16 ☽ △ ♆
02 37 ☽ Q ♄
10 53 ☽ Q ♂
20 05 ☽ △ ♃
22 05 ☽ ⊼ ♅

20 Thursday
04 08 ☽ □ ☿
06 03 ☽ ⚼ ♆
10 05 ☽ ± ♅
10 06 ☽ ⚹ ♂
14 43 ☽ □ ♄
21 34 ☽ △ ☉

21 Friday
00 13 ☽ ± ♆
02 45 ☽ ⚼ ♃
02 50 ☽ △ ♀
03 23 ☿ ∠ ♆
04 14 ☽ ⚼ ♅
09 55 ☽ ⊼ ♇
12 25 ☽ ⊼ ♆
15 33 ☽ ∠ ♂
21 22 ☽ ± ♃
22 07 ☽ ± ♇

22 Saturday
04 48 ♅ St R
06 52 ☽ ⚼ ☉
09 21 ☽ ⊥ ♂
09 48 ☽ ⊼ ♃
10 43 ☽ △ ♅
11 33 ☿ ⊼ ♂
12 57 ☽ ⚼ ♀
16 28 ☽ ⚼ ♇
21 05 ☿ ⊥ ♃
21 15 ☽ ⊻ ♂
22 20 ☽ △ ☿

23 Sunday
02 56 ☽ ± ☉
04 14 ☽ △ ♄
09 29 ☽ ± ♀
16 17 ☽ ⊼ ☉
16 18 ♃ Q ♇
21 55 ☽ # ♆
22 02 ♀ □ ♇
23 02 ☽ △ ♇
23 09 ☽ ⊼ ♀

24 Monday
01 39 ☽ ☍ ♆
07 07 ☽ ⚼ ☿
10 54 ♀ # ♇
11 00 ☽ ⚼ ♄
11 23 ♃ □ ♅
23 23 ♀ ⊻ ♆
23 40 ☽ □ ♅
23 53 ☽ ☍ ♃

25 Tuesday
02 30 ☽ ± ☿
05 21 ☽ ± ♄
08 27 ☽ ☌ ♂
12 48 ☽ # ♃
15 15 ☽ ⊼ ☿
17 30 ☽ ⊼ ♄
21 00 ☽ ∥ ♅

26 Wednesday
10 17 ☽ ☍ ☉
11 22 ☽ □ ♇
11 40 ☽ Q ♅
14 05 ☽ ⊼ ♆
17 40 ☉ # ♇
18 26 ☽ ☍ ♀
23 23 ☉ □ ♇

27 Thursday
01 58 ☽ ± ♆
11 19 ☽ ⚹ ♅
12 37 ☽ ⊼ ♃
17 51 ☽ ∥ ♂
18 26 ☽ ⊻ ♂
19 39 ☽ ⚼ ♆
22 46 ☽ Q ♇

28 Friday
00 32 ☽ ± ♃
04 52 ☽ ☍ ☿
05 11 ☽ ☍ ♄
05 50 ☽ ⊥ ♂
09 10 ☉ ⊻ ♆
14 12 ☽ ∥ ♂
16 26 ☽ ∠ ♅
18 15 ☽ ⚼ ♃
21 58 ☽ ⚹ ♇
22 48 ☽ ∠ ♂

29 Saturday
00 43 ☽ △ ♆
02 01 ♂ ± ♄
02 03 ☽ ⊼ ☉
03 50 ☿ ± ♂
09 40 ☽ ⊥ ♅
11 15 ☽ ⊼ ♀
14 29 ☽ ± ☉
21 03 ☽ ⊻ ♅
21 09 ♀ ⊥ ♆
23 20 ☽ △ ♃
23 52 ☽ ± ♀

30 Sunday
02 30 ☽ ∠ ♇
02 43 ☽ ⚹ ♂
07 15 ☿ St R
07 30 ☽ ∥ ♅
08 57 ☽ ⚼ ☉
12 01 ☽ # ♃
14 39 ☽ ⊼ ☿
14 50 ☽ ⊼ ♄
15 35 ♀ ∥ ♄
18 36 ☽ ⚼ ♀
19 22 ☽ ⊥ ♇

All ephemeris data is given at 12.00 UT and the Moon's longitude is additionally given for 24.00 UT

DECEMBER 1958

LONGITUDES

Date	Sidereal time h m s	Sun ☉ ° ' "	Moon ☽ ° ' "	Moon ☽ 24.00 ° ' "	Mercury ☿ ° '	Venus ♀ ° '	Mars ♂ ° '	Jupiter ♃ ° '	Saturn ♄ ° '	Uranus ♅ ° '	Neptune ♆ ° '	Pluto ♇ ° '
01	16 39 21	08 ♐ 47 31	07 ♌ 12 08	13 ♌ 45 02	25 ♐ 35	13 ♐ 43	18 ♉ 59	17 ♏ 52	25 ♐ 54	16 ♌ 19	05 ♏ 44	04 ♍ 14
02	16 43 17	09 48 20	20 ♌ 21 36	27 ♌ 02 01	25 R 17	14 58	18 R 44	18 05	26 01	16 R 19	05 46	04 14
03	16 47 14	10 49 11	03 ♍ 46 29	10 ♍ 35 10	24 48	16 14	18 30	18 17	26 07	16 18	05 48	04 14
04	16 51 10	11 50 04	17 ♍ 28 15	24 ♍ 25 49	24 08	17 29	18 17	18 30	26 14	16 18	05 50	04 14
05	16 55 07	12 50 58	01 ♎ 27 57	08 ♎ 34 35	23 17	18 45	18 04	18 43	26 21	16 17	05 52	04 14
06	16 59 03	13 51 53	15 ♎ 45 35	23 ♎ 00 43	22 16	20 00	17 53	18 55	26 28	16 16	05 54	04 14
07	17 03 00	14 52 49	00 ♏ 19 33	07 ♏ 41 33	21 06	21 15	17 42	19 08	26 35	16 15	05 56	04 R 14
08	17 06 57	15 53 47	15 ♏ 06 02	22 ♏ 32 09	19 49	22 31	17 32	19 20	26 42	16 15	05 57	04 14
09	17 10 53	16 54 46	29 ♏ 58 58	07 ♐ 25 26	18 28	23 46	17 22	19 33	26 49	16 14	05 59	04 14
10	17 14 50	17 55 46	14 ♐ 50 29	22 ♐ 13 01	17 05	25 02	17 14	19 45	26 56	16 13	06 01	04 14
11	17 18 46	18 56 47	29 ♐ 32 00	06 ♑ 46 28	15 43	26 17	17 06	19 58	27 03	16 12	06 03	04 13
12	17 22 43	19 57 48	13 ♑ 55 36	20 ♑ 58 45	14 25	27 33	16 59	20 10	27 10	16 11	06 05	04 13
13	17 26 39	20 58 51	27 ♑ 55 26	04 ♒ 45 19	13 13	28 ♐ 48	16 53	20 22	27 17	16 09	06 06	04 13
14	17 30 36	21 59 53	11 ♒ 28 19	18 ♒ 04 27	12 09	00 ♑ 03	16 48	20 35	27 24	16 08	06 08	04 13
15	17 34 32	23 00 57	24 ♒ 33 57	00 ♓ 57 09	11 15	01 19	16 44	20 47	27 32	16 07	06 10	04 13
16	17 38 29	24 02 00	07 ♓ 14 30	13 ♓ 26 31	10 31	02 34	16 40	20 59	27 39	16 06	06 11	04 12
17	17 42 26	25 03 04	19 ♓ 33 50	25 ♓ 37 05	09 59	03 50	16 38	21 11	27 46	16 04	06 13	04 12
18	17 46 22	26 04 09	01 ♈ 36 58	07 ♈ 34 10	09 37	05 05	16 36	21 23	27 53	16 03	06 15	04 12
19	17 50 19	27 05 14	13 ♈ 29 23	19 ♈ 23 19	09 26	06 20	16 35	21 35	28 00	16 02	06 16	04 11
20	17 54 15	28 06 19	25 ♈ 16 37	01 ♉ 09 56	09 D 26	07 36	16 D 35	21 47	28 07	16 00	06 18	04 11
21	17 58 12	29 ♐ 07 24	07 ♉ 03 51	12 ♉ 58 57	09 35	08 51	16 35	21 58	28 14	15 59	06 19	04 11
22	18 02 08	00 ♑ 08 30	18 ♉ 55 42	24 ♉ 54 34	09 52	10 07	16 37	22 10	28 21	15 57	06 21	04 10
23	18 06 05	01 09 36	00 ♊ 55 56	07 ♊ 00 07	10 18	11 22	16 39	22 22	28 28	15 56	06 22	04 10
24	18 10 01	02 10 42	13 ♊ 07 21	19 ♊ 17 50	10 50	12 37	16 42	22 34	28 35	15 54	06 24	04 09
25	18 13 58	03 11 49	25 ♊ 31 41	01 ♋ 48 56	11 29	13 53	16 45	22 45	28 42	15 52	06 25	04 09
26	18 17 55	04 12 56	08 ♋ 09 35	14 ♋ 33 34	12 14	15 08	16 50	22 57	28 49	15 51	06 27	04 08
27	18 21 51	05 14 03	21 ♋ 00 48	27 ♋ 31 09	13 04	16 23	16 55	23 08	28 56	15 49	06 28	04 07
28	18 25 48	06 15 10	04 ♌ 04 28	10 ♌ 40 36	13 58	17 39	17 00	23 19	29 04	15 47	06 29	04 07
29	18 29 44	07 16 18	17 ♌ 19 26	24 ♌ 00 49	14 56	18 54	17 07	23 31	29 11	15 45	06 31	04 06
30	18 33 41	08 17 26	00 ♍ 44 41	07 ♍ 30 58	15 58	20 09	17 14	23 42	29 18	15 43	06 32	04 05
31	18 37 37	09 ♑ 18 35	14 ♍ 19 38	21 ♍ 10 41	17 ♐ 03	21 ♑ 25	17 ♉ 22	23 ♏ 53	29 ♐ 25	15 ♌ 41	06 ♏ 33	04 ♍ 05

Moon

Date	Moon True ☊ ° '	Moon Mean ☊ ° '	Moon ☽ Latitude ° '
01	21 ♎ 04	19 ♎ 40	04 S 56
02	21 R 00	19 37	04 29
03	20 58	19 34	03 47
04	20 D 58	19 31	02 51
05	20 58	19 27	01 43
06	20 59	19 24	00 S 28
07	20 R 59	19 21	00 N 50
08	20 57	19 18	02 06
09	20 52	19 15	03 13
10	20 45	19 12	04 08
11	20 35	19 08	04 45
12	20 25	19 05	05 03
13	20 15	19 02	05 02
14	20 07	18 59	04 44
15	20 01	18 56	04 12
16	19 57	18 52	03 28
17	19 56	18 49	02 35
18	19 D 56	18 46	01 36
19	19 56	18 43	00 N 34
20	19 R 56	18 40	00 S 28
21	19 55	18 37	01 30
22	19 51	18 33	02 28
23	19 44	18 30	03 20
24	19 34	18 27	04 03
25	19 23	18 24	04 36
26	19 09	18 21	04 55
27	18 56	18 18	05 01
28	18 43	18 14	04 50
29	18 33	18 11	04 24
30	18 25	18 08	03 43
31	18 ♎ 21	18 ♎ 05	02 S 49

DECLINATIONS

Date	Sun ☉	Moon ☽	Mercury ☿	Venus ♀	Mars ♂	Jupiter ♃	Saturn ♄	Uranus ♅	Neptune ♆	Pluto ♇
01	21 S 46	13 N 42	24 S 30	22 S 33	18 N 33	16 S 16	22 S 23	16 N 36	11 S 49	20 N 58
02	21 55	10 27	24 14	22 44	18 31	16 20	22 24	16 36	11 50	20 58
03	22 04	06 36	23 55	22 55	18 30	16 23	22 24	16 36	11 50	20 59
04	22 13	02 N 20	23 35	23 05	18 28	16 27	22 24	16 36	11 51	20 59
05	22 21	02 S 10	23 14	23 14	18 27	16 30	22 25	16 37	11 51	21 00
06	22 28	06 38	22 50	23 22	18 25	16 34	22 25	16 37	11 52	21 00
07	22 35	10 48	22 26	23 30	18 24	16 37	22 25	16 37	11 53	21 00
08	22 42	14 22	22 00	23 37	18 23	16 40	22 25	16 38	11 53	21 01
09	22 48	17 00	21 34	23 43	18 22	16 44	22 26	16 38	11 54	21 01
10	22 54	18 29	21 08	23 48	18 22	16 47	22 26	16 38	11 54	21 02
11	22 59	18 42	20 43	23 53	18 21	16 50	22 26	16 39	11 55	21 02
12	23 04	17 42	20 20	23 57	18 21	16 54	22 26	16 39	11 55	21 02
13	23 08	15 39	19 59	24 01	18 21	16 57	22 27	16 39	11 56	21 03
14	23 12	12 47	19 40	24 03	18 21	17 00	22 27	16 40	11 56	21 03
15	23 15	09 23	19 25	24 05	18 21	17 03	22 27	16 40	11 57	21 04
16	23 18	05 38	19 13	24 06	18 21	17 06	22 27	16 40	11 57	21 04
17	23 21	01 S 45	19 05	24 06	18 22	17 10	22 27	16 41	11 58	21 05
18	23 23	02 N 07	19 00	24 06	18 23	17 13	22 27	16 41	11 58	21 05
19	23 25	05 51	18 59	24 05	18 24	17 16	22 28	16 42	11 59	21 06
20	23 26	09 20	19 00	24 03	18 25	17 19	22 28	16 42	11 59	21 06
21	23 26	12 27	19 04	24 00	18 26	17 22	22 28	16 43	12 00	21 07
22	23 27	15 05	19 11	23 57	18 27	17 25	22 28	16 43	12 00	21 08
23	23 26	17 05	19 20	23 53	18 29	17 28	22 28	16 44	12 01	21 08
24	23 25	18 21	19 30	23 48	18 31	17 31	22 28	16 44	12 01	21 09
25	23 24	18 46	19 42	23 42	18 33	17 33	22 29	16 45	12 02	21 09
26	23 22	18 17	19 55	23 36	18 35	17 36	22 29	16 45	12 02	21 10
27	23 20	16 51	20 08	23 29	18 37	17 39	22 29	16 46	12 02	21 10
28	23 18	14 32	20 23	23 21	18 40	17 42	22 29	16 46	12 03	21 11
29	23 15	11 27	20 37	23 13	18 42	17 45	22 29	16 47	12 03	21 12
30	23 11	07 43	20 52	23 03	18 45	17 47	22 29	16 48	12 03	21 12
31	23 S 07	03 N 34	21 S 07	22 S 53	18 N 48	17 S 50	22 S 29	16 N 48	12 S 04	21 N 13

ZODIAC SIGN ENTRIES

Date	h	m	Planets
03	05	18	☽ ♍
05	09	31	☽ ♎
07	11	28	☽ ♏
09	12	02	☽ ♐
11	12	46	☽ ♑
13	15	38	☽ ♒
14	10	55	♀ ♑
15	22	12	☽ ♓
18	08	45	☽ ♈
20	21	38	☽ ♉
22	08	40	☉ ♑
23	10	09	☽ ♊
25	20	33	☽ ♋
28	04	33	☽ ♌
30	10	41	☽ ♍

LATITUDES

Date	Mercury ☿	Venus ♀	Mars ♂	Jupiter ♃	Saturn ♄	Uranus ♅	Neptune ♆	Pluto ♇
01	01 S 08	00 S 06	01 N 08	00 N 56	00 N 59	00 N 41	01 N 43	11 N 52
04	00 S 17	00 13	01 15	00 56	00 59	00 41	01 43	11 53
07	00 N 43	00 21	01 21	00 56	00 59	00 41	01 43	11 54
10	01 41	00 28	01 26	00 56	00 59	00 41	01 43	11 55
13	02 26	00 34	01 32	00 56	00 58	00 41	01 43	11 57
16	02 50	00 41	01 36	00 56	00 58	00 41	01 43	11 58
19	02 55	00 48	01 40	00 56	00 58	00 42	01 44	11 59
22	02 47	00 54	01 44	00 56	00 58	00 42	01 44	12 00
25	02 29	00 59	01 47	00 56	00 58	00 42	01 44	12 02
28	02 07	01 05	01 49	00 56	00 57	00 42	01 44	12 03
31	01 N 42	01 S 10	01 N 52	00 N 56	00 N 57	00 N 42	01 N 44	12 N 04

DATA

Julian Date	2436539
Delta T	+33 seconds
Ayanamsa	23° 17' 05"
Synetic vernal point	05° ♓ 49' 55"
True obliquity of ecliptic	23° 26' 32"

MOON'S PHASES, APSIDES AND POSITIONS ☽

Date	h	m	Phase	Longitude ° '	Eclipse Indicator
04	01	24	☾	11 ♍ 23	
10	17	23	●	18 ♐ 09	
17	23	52	☽	25 ♓ 33	
26	03	54	○	03 ♋ 52	

Day	h	m		
08	23	45	Perigee	
20	20	31	Apogee	
05	00	32	0S	
11	04	05	Max dec	18° S 46'
17	22	50	0N	
25	11	08	Max dec	18° N 46'

LONGITUDES

Date	Chiron ⚷ ° '	Ceres ⚳ ° '	Pallas ⚴ ° '	Juno ⚵ ° '	Vesta ⚶ ° '	Black Moon Lilith ⚸ ° '
01	18 ♒ 17	28 ♎ 00	09 ♎ 25	12 ♎ 36	12 ♌ 35	01 ♉ 35
11	18 ♒ 41	02 ♏ 06	13 ♎ 10	15 ♎ 35	12 ♌ 51	02 ♉ 42
21	19 ♒ 11	06 ♏ 04	16 ♎ 37	18 ♎ 17	12 ♌ 23	03 ♉ 48
31	19 ♒ 44	09 ♏ 51	19 ♎ 43	20 ♎ 39	11 ♌ 09	04 ♉ 55

All ephemeris data is given at 12.00 UT and the Moon's longitude is additionally given for 24.00 UT

ASPECTARIAN

h m Aspects

01 Monday
00 43 ☽ Q ♂
01 45 ☽ ± ☿
02 06 ☽ ± ♄
06 31 ☽ ⊻ ♇
09 18 ☽ □ ♆
15 10 ☽ △ ☉
18 06 ☽ bQ ☿
18 50 ☽ bQ ♄

02 Tuesday
01 12 ☽ △ ♀
02 25 ☽ # ♆
04 40 ☽ ☌ ♅
07 48 ☽ □ ♃
09 07 ☽ □ ♂
18 10 ☽ Q ♆
20 36 ☽ △ ☿
22 15 ☽ △ ♄

03 Wednesday
12 48 ☽ ☌ ♇
13 24 ♀ △ ♅
15 35 ☽ ⚹ ♆
16 31 ☽ Q ♃
23 34 ♂ ☍ ♃

04 Thursday
01 24 ☽ □ ☉
02 31 ☿ ⊥ ♃
05 10 ☿ ± ♂
09 58 ☽ ⊻ ♅
11 57 ☉ ⊥ ♆
12 02 ☽ □ ♀
13 23 ☽ △ ♂
13 49 ☽ ⚹ ♃
17 50 ☽ ∠ ♆
20 19 ☽ ⊥ ♅
22 52 ☽ □ ☿

05 Friday
00 58 ♀ ⊼ ♂
03 13 ☽ □ ♄
09 16 ☽ ⊥ ♆
10 52 ☽ Q ☉
11 17 ♀ ⊻ ♃
11 41 ☽ ∠ ♅
12 04 ☿ ∥ ♀
14 41 ☽ bQ ♂
15 52 ☽ ∠ ♃
16 41 ☽ ⊻ ♇
19 27 ☽ ⊻ ♆
21 47 ☽ Q ♀

06 Saturday
01 10 ☉ ∥ ♄
02 47 ☽ ⊥ ♇
03 28 ☽ Q ☿
05 37 ☽ ± ♂
07 12 ☽ ⊥ ♃
08 36 ☽ ⚹ ☉
09 50 ☽ Q ♄
12 51 ☽ ⚹ ♅
15 28 ☽ ⊼ ♂
15 45 ♇ St R
17 19 ☽ ⊻ ♃
17 46 ☽ ∠ ♇
19 42 ☽ ⚹ ♀
22 00 ☽ ⚹ ☿

07 Sunday
05 01 ☉ ∥ ☿
05 30 ♀ ∠ ♆
05 50 ☽ ⚹ ♄
08 37 ☽ Q ♅
10 29 ☿ ☌ ♀
11 13 ☽ ∠ ☉
12 41 ☿ ∥ ♄
14 04 ♃ # ♅
15 20 ☿ ∠ ♆
18 22 ☽ ⚹ ♇
18 44 ☽ ∥ ♆
20 41 ☽ ∠ ☿
21 09 ☽ ☌ ♆
22 34 ☽ ∠ ♀

08 Monday
02 57 ☽ ⊥ ☉
06 28 ☽ ∠ ♄
10 06 ☽ ⊥ ☿
13 23 ☽ ⊻ ☉
13 50 ☽ Q ♇
13 51 ☽ □ ♅
14 30 ☽ ⊥ ♀
15 53 ☽ ☍ ♂
18 57 ☽ ☌ ♃
19 00 ☽ ⊻ ☿
19 32 ☿ ⊻ ♃
20 02 ☉ △ ♅
21 07 ☽ ⊥ ♄

09 Tuesday
01 04 ☽ ⊻ ♀
05 10 ♀ ± ♂
06 52 ☽ ⊻ ♄
07 58 ☽ # ♅
08 56 ☽ ∥ ♃
18 51 ☽ □ ♇
21 29 ☉ ⊼ ♂
21 42 ☽ ⊻ ♆

10 Wednesday
03 36 ☉ ☌ ☿
07 25 ☽ ⊥ ♆
09 02 ☽ # ♂
09 16 ☿ ⊼ ♂
14 13 ☽ △ ♅
15 20 ☽ ☌ ☿
15 51 ☽ ⊼ ♂
17 23 ☽ ☌ ☉
18 10 ☿ # ♇
20 06 ☽ ⊻ ♃
22 04 ☽ ∠ ♆

11 Thursday
01 32 ☽ ± ♂
03 31 ☿ △ ♅
04 36 ♀ ⊥ ♃
06 03 ☽ ⊥ ♃
06 10 ☽ ☌ ♀
07 53 ☽ ☌ ♄
14 44 ☽ bQ ♅
16 12 ☽ bQ ♂
19 46 ☽ △ ♇
21 07 ☽ ∠ ♃
22 49 ☽ ⚹ ♆
23 42 ☽ # ♂

12 Friday
04 12 ♀ ☌ ♄
05 42 ☽ ± ♅
12 46 ☽ ⊻ ☿
15 48 ☽ ⊼ ♅
17 09 ☽ △ ♂
18 01 ☉ ⊻ ♃
19 03 ☽ Q ♆
21 00 ☽ bQ ♇
22 09 ☽ ⊥ ☿
22 26 ☽ ∥ ♃
22 46 ☽ ⚹ ♃
23 04 ☽ ⊻ ☉

13 Saturday
01 32 ☽ # ♅
10 14 ☽ ⊥ ☉
10 53 ☽ ⊻ ♄
12 28 ☽ ∠ ☿
12 31 ☽ ± ♇
13 41 ☽ ⊻ ♀
15 00 ☉ ∠ ♆
19 55 ☽ Q ♃
21 29 ☽ ⊥ ♄
23 03 ☽ ⊼ ♇

14 Sunday
01 19 ☽ ⊥ ♀
02 25 ☽ □ ♆
03 20 ☽ ∠ ☉
12 25 ☿ ⊥ ♆
13 08 ☽ ⚹ ☿
13 42 ☽ ∠ ♄
18 15 ☽ ∥ ♆
19 10 ☽ ∠ ♀
20 27 ☽ ☍ ♅
21 37 ☽ □ ♂

15 Monday
04 52 ☽ □ ♃
05 43 ☉ ± ♂
08 19 ♀ bQ ♅
08 52 ☽ ⚹ ☉
09 42 ☽ Q ☿
17 35 ☽ ⚹ ♄
19 36 ♀ bQ ♂

16 Tuesday
02 05 ☽ ⚹ ♀
06 11 ☽ ☍ ♇
07 06 ☽ Q ♂
09 29 ☽ Q ☉
09 58 ☽ △ ♆
16 40 ☽ Q ♄
18 02 ☽ □ ☿

17 Wednesday
03 49 ☽ Q ♀
05 09 ☽ ⊼ ♅
06 15 ☽ ⚹ ♂
15 15 ☽ △ ♃
15 16 ☽ bQ ♆
16 57 ☽ ± ♅
19 09 ♀ △ ♇
23 52 ☽ □ ☉

18 Thursday
04 26 ☽ □ ♄
09 14 ☽ ± ♆
10 52 ☽ bQ ♅
11 58 ☽ ∠ ♂
17 11 ☽ ⊼ ♇
19 48 ☽ □ ♀
21 20 ☽ ⊼ ♆
21 46 ☽ bQ ♃

19 Friday
03 52 ☽ △ ☿
05 19 ☽ ± ♇
06 06 ☽ ⊥ ♂
10 37 ♀ ⚹ ♆
16 19 ☽ ± ♃
17 09 ☽ △ ♅
17 26 ♀ ∠ ♃
18 17 ☽ ⊻ ♂
23 35 ☽ bQ ♇

20 Saturday
01 28 ☿ St D
02 25 ☉ ⊥ ♃
04 45 ☽ ⊼ ♃
06 45 ♂ St D
10 16 ☽ bQ ☿
12 17 ☉ ☌ ♄
17 51 ☽ △ ♄
18 19 ☽ △ ☉

21 Sunday
04 47 ☽ ± ☿
06 08 ☽ △ ♇
08 14 ☽ # ♆
10 29 ☽ ☍ ♆
16 04 ☽ △ ♀
17 12 ☽ ⊼ ☿

22 Monday
00 38 ☽ bQ ♄
03 39 ☽ bQ ☉
05 41 ☿ ⊻ ♀
06 01 ☽ □ ♅
07 19 ☽ ☌ ♂
09 05 ♀ ± ♅
18 37 ☽ ☍ ♃
18 57 ☽ ± ♄
23 26 ☽ ± ☉

23 Tuesday
01 51 ☽ bQ ♀
06 38 ☉ bQ ♅
06 59 ☽ ∥ ♅
07 03 ☽ ⊼ ♄
12 30 ☽ ⊼ ☉
17 52 ☽ # ♃
17 55 ☽ Q ♅
18 23 ☽ □ ♇
21 47 ☽ ± ♀
22 47 ☽ ⊼ ♆
23 58 ☉ bQ ♂

24 Wednesday
07 18 ☽ ☍ ☿
10 35 ☽ ± ♆
10 55 ☽ ⊼ ♀
17 15 ☽ ∥ ♂
17 24 ☽ ⚹ ♅
18 59 ☽ ⊻ ♂
21 24 ♃ ⊥ ♄

25 Thursday
04 05 ☽ bQ ♆
05 30 ☽ Q ♇
06 35 ☽ ⊼ ♃
06 39 ☽ ⊥ ♂
18 08 ☽ ☍ ♄
18 16 ☽ ± ♃
22 11 ☽ ∠ ♅
23 57 ☽ ∠ ♂

26 Friday
03 02 ☽ ∥ ♂
03 54 ☽ ☍ ☉
04 24 ☽ ⚹ ♇
08 45 ☽ △ ♆
10 04 ☉ △ ♇
11 35 ☽ bQ ♃
15 09 ☽ ⊥ ♅
18 25 ☿ ⊥ ♆
20 10 ☽ ⊼ ☿

27 Saturday
00 45 ☽ # ♃
01 16 ♀ ⊼ ♅
02 22 ☽ ⊻ ♅
02 29 ☽ ☍ ♀
04 20 ☽ ⚹ ♂
08 08 ☽ ± ☿
08 30 ☽ ∠ ♇
13 01 ☽ ∥ ♅
15 59 ☽ △ ♃
22 48 ♀ △ ♂

28 Sunday
01 06 ☽ ⊥ ♇
01 57 ☽ bQ ☿
02 40 ☽ Q ♂
02 45 ☽ ⊼ ♄
12 04 ☽ ⊻ ♇
13 49 ☽ ± ♄
16 18 ☽ ⊼ ☉
16 24 ☽ □ ♆
17 40 ☉ ⚹ ♆

29 Monday
03 26 ☉ ∥ ♀
04 05 ☽ ± ☉
04 26 ♀ Q ♆
06 17 ☽ bQ ♄
07 22 ☽ △ ☿
07 41 ☽ # ♆
09 11 ☽ ☌ ♅
11 37 ☽ □ ♂
15 08 ☽ ⊼ ♀
15 49 ♀ bQ ♇
21 37 ☽ bQ ☉
23 15 ☽ □ ♃

30 Tuesday
00 54 ☽ Q ♆
02 59 ☽ ± ♀
06 28 ☿ △ ♅
09 24 ☽ △ ♄
17 56 ☽ ☌ ♇
20 37 ☽ bQ ♀
22 17 ☽ ⚹ ♆
23 39 ☉ ∠ ♃

31 Wednesday
02 27 ☽ △ ☉
07 38 ☽ Q ♃
14 23 ☽ ⊻ ♅
17 12 ☽ □ ☿
17 23 ☽ △ ♂
19 44 ☿ ⊼ ♂
20 39 ☉ ± ♅
21 55 ☿ # ♇

JANUARY 1959

LONGITUDES

Date	Sidereal time h m s	Sun ☉ ° ' "	Moon ☽ ° ' "	Moon ☽ 24.00 ° ' "	Mercury ☿ ° '	Venus ♀ ° '	Mars ♂ ° '	Jupiter ♃ ° '	Saturn ♄ ° '	Uranus ♅ ° '	Neptune ♆ ° '	Pluto ♇ ° '
01	18 41 34	10 ♑ 19 44	28 ♍ 04 08	05 ♎ 00 03	18 ♐ 11	22 ♑ 40	17 ♉ 30	24 ♏ 04	29 ♐ 32	15 ♌ 39	06 ♏ 34	04 ♍ 04
02	18 45 30	11 20 53	11 ♎ 58 26	18 ♎ 59 19	19 21	23 55	17 40	24 15	29 39	15 R 37	06 36	04 R 03
03	18 49 27	12 22 03	26 ♎ 02 40	03 ♏ 08 26	20 33	25 11	17 49	24 26	29 46	15 35	06 37	04 02
04	18 53 24	13 23 13	10 ♏ 16 26	17 ♏ 26 27	21 47	26 26	18 00	24 36	29 ♐ 53	15 33	06 38	04 01
05	18 57 20	14 24 23	24 ♏ 38 07	01 ♐ 50 59	23 03	27 41	18 11	24 47	00 ♑ 00	15 31	06 39	04 01
06	19 01 17	15 25 34	09 ♐ 04 30	16 ♐ 18 00	24 21	28 ♑ 56	18 22	24 58	00 06	15 29	06 40	04 00
07	19 05 13	16 26 44	23 ♐ 30 45	00 ♑ 41 58	25 40	00 ♒ 12	18 35	25 08	00 13	15 27	06 41	03 59
08	19 09 10	17 27 55	07 ♑ 50 51	14 ♑ 56 35	27 00	01 27	18 47	25 19	00 20	15 25	06 42	03 58
09	19 13 06	18 29 05	21 ♑ 57 48	28 ♑ 55 49	28 21	02 42	19 01	25 29	00 27	15 22	06 43	03 57
10	19 17 03	19 30 15	05 ♒ 48 05	12 ♒ 34 52	29 ♐ 43	03 57	19 15	25 39	00 34	15 20	06 44	03 56
11	19 20 59	20 31 25	19 ♒ 15 52	25 ♒ 50 58	01 ♑ 07	05 13	19 29	25 49	00 41	15 18	06 45	03 55
12	19 24 56	21 32 34	02 ♓ 20 12	08 ♓ 43 41	02 31	06 28	19 44	25 59	00 48	15 16	06 46	03 54
13	19 28 53	22 33 43	15 ♓ 01 42	21 ♓ 14 38	03 56	07 43	19 59	26 09	00 54	15 13	06 47	03 53
14	19 32 49	23 34 51	27 ♓ 22 58	03 ♈ 27 15	05 22	08 58	20 15	26 19	01 01	15 11	06 48	03 52
15	19 36 46	24 35 58	09 ♈ 28 04	15 ♈ 26 06	06 49	10 14	20 32	26 29	01 08	15 08	06 48	03 51
16	19 40 42	25 37 05	21 ♈ 22 01	27 ♈ 16 31	08 16	11 29	20 49	26 38	01 15	15 06	06 49	03 50
17	19 44 39	26 38 11	03 ♉ 10 20	09 ♉ 04 08	09 44	12 44	21 06	26 48	01 21	15 04	06 50	03 49
18	19 48 35	27 39 16	14 ♉ 58 39	20 ♉ 54 30	11 12	13 59	21 24	26 57	01 28	15 01	06 51	03 47
19	19 52 32	28 40 20	26 ♉ 52 20	02 ♊ 52 45	12 42	15 14	21 43	27 07	01 34	14 59	06 51	03 46
20	19 56 28	29 ♑ 41 24	08 ♊ 56 14	15 ♊ 03 15	14 12	16 29	22 01	27 16	01 41	14 56	06 52	03 45
21	20 00 25	00 ♒ 42 26	21 ♊ 14 13	27 ♊ 29 24	15 42	17 44	22 21	27 25	01 47	14 54	06 53	03 44
22	20 04 22	01 43 28	03 ♋ 49 01	10 ♋ 13 12	17 13	18 59	22 40	27 34	01 54	14 51	06 53	03 43
23	20 08 18	02 44 29	16 ♋ 41 56	23 ♋ 15 10	18 45	20 14	23 00	27 43	02 00	14 49	06 54	03 41
24	20 12 15	03 45 29	29 ♋ 52 42	06 ♌ 34 18	20 18	21 29	23 21	27 51	02 07	14 46	06 54	03 40
25	20 16 11	04 46 28	13 ♌ 19 39	20 ♌ 08 23	21 51	22 44	23 41	28 00	02 13	14 43	06 55	03 39
26	20 20 08	05 47 27	27 ♌ 00 03	03 ♍ 54 17	23 24	23 59	24 02	28 08	02 19	14 41	06 55	03 37
27	20 24 04	06 48 25	10 ♍ 50 38	17 ♍ 48 43	24 58	25 14	24 24	28 17	02 26	14 38	06 55	03 36
28	20 28 01	07 49 22	24 ♍ 48 11	01 ♎ 48 42	26 33	26 29	24 46	28 25	02 32	14 36	06 56	03 35
29	20 31 57	08 50 18	08 ♎ 50 02	15 ♎ 51 56	28 09	27 44	25 08	28 33	02 38	14 33	06 56	03 33
30	20 35 54	09 51 13	22 ♎ 54 16	29 ♎ 56 54	29 ♑ 45	28 ♒ 59	25 30	28 41	02 44	14 30	06 56	03 32
31	20 39 51	10 ♒ 52 08	06 ♏ 59 42	14 ♏ 02 35	01 ♒ 22	00 ♓ 14	25 ♉ 53	28 ♏ 49	02 ♑ 50	14 ♌ 28	06 ♏ 57	03 ♍ 31

DECLINATIONS

Date	Moon True ☊ ° '	Moon Mean ☊ ° '	Moon ☽ Latitude ° '	Sun ☉ ° '	Moon ☽ ° '	Mercury ☿ ° '	Venus ♀ ° '	Mars ♂ ° '	Jupiter ♃ ° '	Saturn ♄ ° '	Uranus ♅ ° '	Neptune ♆ ° '	Pluto ♇ ° '
01	18 ♎ 19	18 ♎ 02	01 S 45	23 S 02	00 S 50	21 S 22	22 S 43	18 N 51	17 S 53	22 S 29	16 N 49	12 S 04	21 N 13
02	18 D 19	17 58	00 S 33	22 57	05 15	21 36	22 32	18 54	17 55	22 29	16 49	12 05	21 14
03	18 R 18	17 55	00 N 41	22 52	09 26	21 50	22 20	18 58	17 58	22 29	16 50	12 05	21 15
04	18 17	17 52	01 53	22 46	13 06	22 04	22 07	19 01	18 00	22 29	16 51	12 05	21 15
05	18 14	17 49	02 59	22 40	16 02	22 16	21 54	19 05	18 03	22 30	16 51	12 06	21 16
06	18 07	17 46	03 54	22 33	17 57	22 29	21 40	19 09	18 06	22 30	16 52	12 06	21 17
07	17 58	17 43	04 34	22 26	18 44	22 40	21 25	19 13	18 08	22 30	16 53	12 06	21 17
08	17 46	17 39	04 56	22 18	18 17	22 51	21 10	19 17	18 10	22 30	16 53	12 06	21 18
09	17 33	17 36	04 59	22 10	16 43	23 00	20 54	19 21	18 13	22 30	16 54	12 07	21 19
10	17 20	17 33	04 45	22 01	14 13	23 09	20 37	19 25	18 15	22 30	16 55	12 07	21 19
11	17 09	17 30	04 15	21 52	11 00	23 17	20 20	19 29	18 17	22 30	16 55	12 07	21 20
12	17 00	17 27	03 32	21 43	07 20	23 24	20 02	19 34	18 20	22 30	16 56	12 07	21 21
13	16 54	17 24	02 40	21 33	03 S 26	23 29	19 44	19 38	18 22	22 30	16 57	12 08	21 21
14	16 50	17 20	01 41	21 23	00 N 30	23 34	19 25	19 43	18 24	22 30	16 58	12 08	21 22
15	16 49	17 17	00 N 39	21 12	04 21	23 37	19 06	19 48	18 26	22 30	16 58	12 08	21 23
16	16 D 49	17 14	00 S 24	21 01	07 58	23 39	18 46	19 52	18 29	22 30	16 59	12 08	21 23
17	16 R 49	17 11	01 26	20 50	11 14	23 40	18 25	19 57	18 31	22 30	17 00	12 08	21 24
18	16 48	17 08	02 24	20 38	14 03	23 40	18 04	20 02	18 33	22 29	17 01	12 09	21 25
19	16 44	17 04	03 15	20 26	16 17	23 39	17 43	20 07	18 35	22 29	17 01	12 09	21 26
20	16 39	17 01	03 59	20 13	17 51	23 36	17 21	20 12	18 37	22 29	17 02	12 09	21 26
21	16 30	16 58	04 33	20 00	18 37	23 32	16 58	20 17	18 39	22 29	17 03	12 09	21 27
22	16 19	16 55	04 54	19 47	18 29	23 27	16 35	20 23	18 41	22 29	17 04	12 09	21 28
23	16 06	16 52	05 01	19 33	17 25	23 20	16 12	20 28	18 43	22 29	17 04	12 09	21 28
24	15 53	16 49	04 53	19 19	15 24	23 12	15 48	20 33	18 44	22 29	17 05	12 09	21 29
25	15 42	16 45	04 28	19 04	12 32	23 03	15 24	20 38	18 46	22 29	17 06	12 09	21 30
26	15 32	16 42	03 47	18 50	08 57	22 52	14 59	20 44	18 48	22 29	17 07	12 10	21 31
27	15 24	16 39	02 52	18 34	04 50	22 40	14 34	20 49	18 50	22 29	17 07	12 10	21 31
28	15 20	16 36	01 47	18 19	00 N 26	22 27	14 08	20 55	18 52	22 29	17 08	12 10	21 32
29	15 18	16 33	00 S 34	18 03	04 S 02	22 12	13 43	21 00	18 53	22 29	17 09	12 10	21 33
30	15 D 18	16 30	00 N 40	17 47	08 17	21 56	13 16	21 06	18 55	22 29	17 10	12 10	21 33
31	15 ♎ 19	16 ♎ 26	01 N 53	17 S 30	12 S 04	21 S 38	12 S 50	21 N 11	18 S 57	22 S 28	17 N 10	12 S 10	21 N 34

ZODIAC SIGN ENTRIES

Date	h	m	Planets
01	15	21	☽ ♎
03	18	42	☽ ♏
05	13	33	♄ ♑
05	20	56	☽ ♐
07	08	16	♀ ♒
07	22	50	☽ ♑
10	01	52	☽ ♒
10	16	47	☿ ♑
12	07	39	☽ ♓
14	17	09	☽ ♈
17	05	33	☽ ♉
19	18	16	☽ ♊
20	19	19	☉ ♒
22	04	47	☽ ♋
24	12	13	☽ ♌
26	17	13	☽ ♍
28	20	54	☽ ♎
30	15	41	☿ ♒
31	00	05	☽ ♏
31	07	28	♀ ♓

LATITUDES

Date	Mercury ☿ ° '	Venus ♀ ° '	Mars ♂ ° '	Jupiter ♃ ° '	Saturn ♄ ° '	Uranus ♅ ° '	Neptune ♆ ° '	Pluto ♇ ° '
01	01 N 34	01 S 12	01 N 52	00 N 56	00 N 57	00 N 42	01 N 44	12 N 04
04	01 08	01 16	01 54	00 57	00 57	00 42	01 44	12 05
07	00 42	01 20	01 56	00 57	00 57	00 42	01 45	12 06
10	00 N 17	01 24	01 57	00 57	00 57	00 42	01 45	12 07
13	00 S 06	01 27	01 58	00 57	00 57	00 42	01 45	12 08
16	00 28	01 29	01 59	00 58	00 57	00 42	01 45	12 09
19	00 49	01 31	02 00	00 58	00 57	00 42	01 45	12 10
22	01 07	01 32	02 00	00 58	00 56	00 43	01 46	12 11
25	01 24	01 33	02 00	00 58	00 56	00 43	01 46	12 12
28	01 38	01 33	02 01	00 58	00 56	00 43	01 46	12 13
31	01 S 49	01 S 32	02 N 01	00 N 59	00 N 56	00 N 43	01 N 46	12 N 13

DATA

Julian Date	2436570
Delta T	+33 seconds
Ayanamsa	23° 17' 10"
Synetic vernal point	05° ♓ 49' 50"
True obliquity of ecliptic	23° 26' 31"

MOON'S PHASES, APSIDES AND POSITIONS ☽

Date	h	m	Phase	Longitude ° '	Eclipse Indicator
02	10	51	☾	11 ♎ 18	
09	05	34	●	18 ♑ 13	
16	21	27	☽	26 ♈ 01	
24	19	33	○	04 ♌ 05	
31	19	06	☾	11 ♏ 10	

Day	h	m		
05	20	25	Perigee	
17	17	00	Apogee	
31	05	31	Perigee	
01	07	29	0S	
07	15	13	Max dec	18° S 44'
14	08	54	0N	
21	20	43	Max dec	18° N 40'
28	14	18	0S	

LONGITUDES

Date	Chiron ⚷ ° '	Ceres ⚳ ° '	Pallas ⚴ ° '	Juno ⚵ ° '	Vesta ⚶ ° '	Black Moon Lilith ⚸ ° '
01	19 ♒ 48	10 ♏ 13	20 ♎ 00	20 ♎ 52	10 ♌ 59	05 ♉ 01
11	20 ♒ 26	13 ♏ 46	22 ♎ 39	22 ♎ 49	09 ♌ 01	06 ♉ 08
21	21 ♒ 06	17 ♏ 03	24 ♎ 47	24 ♎ 19	06 ♌ 33	07 ♉ 14
31	21 ♒ 48	20 ♏ 01	26 ♎ 21	25 ♎ 18	03 ♌ 55	08 ♉ 21

All ephemeris data is given at 12.00 UT and the Moon's longitude is additionally given for 24.00 UT

ASPECTARIAN

h m	Aspects
01 Thursday	
00 40	☽ ∠ ♆
00 52	☽ ⊥ ♅
01 39	☽ △ ♀
04 56	☽ ⚹ ♃
14 33	☽ □ ♄
16 21	☽ ⊥ ♆
16 29	☽ ∠ ♅
19 46	☽ ꝗ ♂
22 22	☽ ⊻ ♇
02 Friday	
02 44	☽ ⊻ ♆
03 19	☽ Q ☿
07 15	☽ ∠ ♃
08 42	☽ ⊥ ♇
10 51	☽ □ ☉
11 27	☽ ± ♂
16 27	♀ ∥ ♄
18 14	☽ ⚹ ♅
19 17	♀ ⚹ ♃
21 47	☽ Q ♄
21 51	☽ ⊼ ♂
22 52	☽ ⊥ ♃
03 Saturday	
00 06	☽ ∠ ♇
01 47	☽ ⚹ ☿
09 13	☽ ⊻ ♃
10 23	☽ □ ♀
14 37	☽ Q ♅
18 21	☽ ⚹ ♄
19 53	☽ Q ☉
04 Sunday	
01 30	☽ ⚹ ♇
04 54	☽ ∥ ♆
05 35	☽ ∠ ☿
05 52	☽ ☌ ♆
08 57	☿ ∠ ♆
15 02	☿ ∥ ♀
17 37	☽ ⚹ ☉
19 38	☽ Q ♀
19 47	☽ ∠ ♄
20 49	☽ □ ♅
21 37	☽ Q ♇
22 07	☽ ⊥ ☿
05 Monday	
01 06	☽ ☍ ♂
09 07	☽ ⊻ ☿
10 55	☽ ⊥ ♄
12 15	☽ ☌ ♃
17 34	☽ ⚹ ♀
18 09	♀ ± ♇
20 33	☽ ∠ ☉
20 47	☽ ⋕ ♅
20 59	☽ ⊻ ♄
06 Tuesday	
03 34	☽ □ ♇
08 00	☽ ⊻ ♆
12 35	☿ ± ♂
12 38	☽ ⊥ ☉
13 18	☉ ⊼ ♅
13 53	☿ ∥ ♄
14 35	☽ ∥ ♃
17 35	☉ ∥ ☿
17 58	☽ ⊥ ♆
20 51	☽ ∠ ♀
22 37	☽ △ ♅
23 21	☽ ⊻ ☉
23 35	☉ ∥ ♄
07 Wednesday	
01 03	☿ ⊻ ♃
03 40	☽ ⊼ ♂
08 57	☽ ∠ ♆
12 36	♀ ⊻ ♄
13 15	☽ ⊥ ♀
13 48	☽ ± ♂
14 44	☽ ⊻ ♃
15 57	☽ ☌ ☿
23 18	☽ ☌ ♄
23 33	☽ ꝗ ♅
23 35	♀ ⋕ ♇
08 Thursday	
00 13	☽ ⊻ ♀
00 53	☽ ⊥ ♃
05 04	☽ ꝗ ♂
05 29	☽ △ ♇
10 04	☽ ⚹ ♆
14 26	☽ ∥ ♃
14 38	☽ ± ♅
16 12	☽ ∠ ♃
23 26	♄ ꝗ ♅
09 Friday	
00 46	☽ ⊼ ♅
05 34	☽ ☌ ☉
06 25	☽ Q ♆
06 50	☽ ꝗ ♇
06 51	☽ △ ♂
09 54	☽ ⋕ ♅
17 38	☉ Q ♆
18 07	☽ ⚹ ♃
22 17	☽ ± ♇
22 46	☉ ꝗ ♇
10 Saturday	
00 12	☽ ⊻ ☿
02 46	☽ ⊻ ♄
04 01	☉ △ ♂
08 26	☽ ☌ ♀
08 44	☽ ⊼ ♇
11 32	♀ ⊼ ♇
11 51	☽ ⊥ ☿
13 21	☽ ⊥ ♄
13 39	☽ □ ♆
15 18	☽ Q ♃
22 19	☿ ꝗ ♅
11 Sunday	
03 54	☿ ☌ ♄
04 08	☽ ∥ ♆
04 53	☽ ☍ ♅
05 29	☽ ∠ ♄
05 39	☽ ∠ ☿
12 24	☽ □ ♂
14 28	☽ ⊻ ☉
12 Monday	
00 06	☽ □ ♃
01 48	☿ ⊥ ♃
02 22	☽ ⊥ ☉
09 06	☽ ⚹ ♄
12 23	☽ ⚹ ☿
14 55	☽ ☍ ♇
17 49	♀ □ ♆
18 54	♀ ⊥ ♄
20 18	☽ △ ♆
20 34	☽ ∠ ☉
20 35	☽ ⊻ ♀
22 20	☽ Q ♂
13 Tuesday	
07 54	☽ Q ♄
09 13	☽ ⊥ ♀
11 08	☿ △ ♇
12 22	☽ ⊼ ♅
13 58	☽ Q ☿
18 08	♀ ⋕ ♂
21 35	♀ Q ♃
21 47	☽ ⚹ ♂
23 55	☽ ± ♅
14 Wednesday	
01 04	☽ ꝗ ♆
03 53	☽ ⚹ ☉
04 33	☽ ∠ ♀
09 48	☿ ꝗ ♂
09 53	☽ △ ♃
13 58	☉ ⋕ ♇
17 30	☽ ꝗ ♅
18 44	☽ ± ♆
19 14	☽ □ ♄
15 Thursday	
00 48	☽ ⊼ ♇
03 57	☽ ∠ ♂
05 44	☽ Q ☉
05 57	☽ □ ☿
06 40	☽ ⊼ ♆
11 59	☿ ⚹ ♆
12 45	☽ ± ♇
13 42	☽ ⚹ ♀
16 06	☽ ꝗ ♃
22 26	☽ ⊥ ♂
23 22	☽ △ ♅
16 Friday	
06 52	☽ ꝗ ♇
10 30	☽ ± ♃
10 51	☽ ⊻ ♂
16 48	☽ Q ♀
21 27	☽ □ ☉
22 51	☽ ⊼ ♃
17 Saturday	
01 20	☿ ± ♅
06 28	♀ ∥ ♃
08 16	☽ △ ♄
13 18	☽ △ ♇
16 32	☉ ⚹ ♃
19 19	☽ ⋕ ♆
19 28	☽ ☍ ♆
18 Sunday	
03 15	☽ △ ☿
09 45	☽ □ ♀
12 05	☽ □ ♅
15 02	☽ ꝗ ♄
15 07	☉ ± ♇
19 Monday	
01 21	☽ ☌ ♂
01 29	☿ ∠ ♃
07 12	♀ ☍ ♅
09 22	☽ ± ♄
12 29	☽ ☍ ♃
13 53	☽ ꝗ ☿
15 56	☽ △ ☉
21 29	☽ ⊼ ♄
21 58	☽ ∥ ♅
20 Tuesday	
00 09	☽ Q ♅
01 45	☽ □ ♇
04 48	☽ ⋕ ♀
07 54	☽ ⊼ ♆
10 20	☽ ± ☿
13 11	☉ ⋕ ♂
16 06	♀ ∠ ♄
19 44	☽ ± ♆
23 29	☿ ⊼ ♅
23 44	☽ ⚹ ♅
23 46	☽ ⊼ ☿
21 Wednesday	
00 18	☽ ꝗ ☉
04 28	☽ △ ♀
07 30	♀ ⋕ ♅
12 57	☽ Q ♇
13 14	☽ ꝗ ♆
14 11	☽ ⊻ ♂
16 00	☽ ⋕ ♃
19 16	☽ ± ☉
23 44	☽ ⋕ ♃
22 Thursday	
00 00	☽ ⊼ ♃
01 59	☽ ⊥ ♂
04 31	☽ ∠ ♅
07 42	☽ ⊼ ☉
08 21	☽ ☍ ♄
11 31	☽ ± ♃
11 48	☽ ⚹ ♇
12 22	☽ ꝗ ♀
16 37	☉ ⊻ ♄
17 46	☽ △ ♆
19 26	☽ ∠ ♂
21 25	☽ ⊥ ♅
23 Friday	
04 33	☽ ꝗ ♃
06 59	☽ ± ♀
08 32	☽ ⊻ ♅
11 00	☿ ꝗ ♇
14 14	☿ Q ♆
15 39	☽ ∠ ♇
16 17	☽ ☍ ☿
16 56	☽ ∥ ♅
19 11	☽ ⊼ ♀
23 51	☽ ⚹ ♂
24 Saturday	
07 11	☽ ⋕ ♀
08 01	☽ ⊥ ♇
08 19	☽ △ ♃
09 54	☉ ⊼ ♇
16 03	☽ ⊼ ♄
18 48	☽ ⊻ ♇
19 33	☽ ☍ ☉
22 04	☽ Q ♂
25 Sunday	
00 36	☽ □ ♆
02 52	☽ ± ♄
14 28	☽ ☌ ♅
14 48	☽ ⋕ ♆
18 55	☽ ꝗ ♄
26 Monday	
04 55	☽ ⊼ ☿
06 13	☽ ☍ ♀
06 42	☽ □ ♂
08 22	☽ Q ♆
13 19	♀ □ ♂
14 00	☽ □ ♃
14 05	☉ ∥ ♃
16 43	☽ ± ☿
21 20	☽ △ ♄
23 30	☽ ☌ ♇
27 Tuesday	
00 37	☿ △ ♂
04 28	☽ ⊼ ☉
05 14	☽ ⚹ ♆
10 18	☽ ꝗ ☿
14 48	☉ □ ♆
15 39	☽ ± ☉
18 31	☽ ⊻ ♅
21 27	☽ Q ♃
28 Wednesday	
04 48	☽ ⊥ ♅
07 04	☽ ∠ ♆
07 17	☿ ⊻ ♀
08 20	☿ ∥ ♄
08 20	☽ ꝗ ☉
11 56	☽ △ ♂
15 10	☽ ⊼ ♀
15 23	☽ △ ☿
18 15	☽ ⚹ ♃
20 11	☽ ∠ ♅
22 30	☽ ⊥ ♆
29 Thursday	
01 20	☽ □ ♄
02 27	☽ ± ♀
03 00	☽ ⊻ ♇
03 14	☿ ± ♇
06 41	☉ ⊥ ♄
08 45	☽ ⊻ ♆
12 00	☽ △ ☉
13 14	☽ ⊥ ♇
14 17	☽ ꝗ ♂
18 35	☿ ⚹ ♃
19 19	☽ ꝗ ♀
20 08	☽ ∠ ♃
21 44	☽ ⚹ ♅
30 Friday	
04 34	☽ ∠ ♇
05 28	♀ □ ♃
06 03	☽ ± ♂
08 17	☽ Q ♄
11 37	☽ ⊥ ♃
16 33	☽ ⊼ ♂
18 07	☽ Q ♅
21 56	☽ ⊻ ♃
23 22	☽ △ ♀
31 Saturday	
01 10	☽ □ ☿
04 52	☽ ⚹ ♄
06 05	☽ ⚹ ♇
10 31	☉ Q ♃
11 55	☽ ☌ ♆
12 39	☽ ∥ ♆
16 34	☿ ⋕ ♇
16 43	☽ ∥ ♀
19 06	☽ □ ☉

FEBRUARY 1959

Date	Sidereal time h m s	Sun ☉ ° ' "	Moon ☽ ° ' "	Moon ☽ 24.00 ° ' "	Mercury ☿ ° '	Venus ♀ ° '	Mars ♂ ° '	Jupiter ♃ ° '	Saturn ♄ ° '	Uranus ♅ ° '	Neptune ♆ ° '	Pluto ⯓ ° '
01	20 43 47	11 ♒ 53 02	21 ♏ 05 27	28 ♏ 08 10	03 ♒ 00	01 ♓ 29	26 ♉ 16	28 ♏ 57	02 ♑ 56	14 ♌ 25	06 ♏ 57	03 ♍ 29
02	20 47 44	12 53 56	05 ♐ 10 36	12 ♐ 12 31	04 38	02 44	26 40	29 04	03 02	14 R 23	06 57	03 R 28
03	20 51 40	13 54 49	19 ♐ 13 41	26 ♐ 13 48	06 17	03 59	27 04	29 12	03 08	14 20	06 57	03 27
04	20 55 37	14 55 41	03 ♑ 12 30	10 ♑ 09 25	07 57	05 14	27 28	29 19	03 14	14 17	06 57	03 25
05	20 59 33	15 56 32	17 ♑ 04 07	23 ♑ 56 10	09 37	06 28	27 52	29 26	03 20	14 15	06 58	03 24
06	21 03 30	16 57 22	00 ♒ 45 08	07 ♒ 30 37	11 18	07 43	28 17	29 33	03 26	14 12	06 58	03 22
07	21 07 26	17 58 10	14 ♒ 12 16	20 ♒ 49 47	13 01	08 58	28 42	29 40	03 32	14 09	06 58	03 21
08	21 11 23	18 58 58	27 ♒ 22 55	03 ♓ 51 33	14 43	10 13	29 07	29 47	03 37	14 07	06 R 58	03 19
09	21 15 20	19 59 44	10 ♓ 15 37	16 ♓ 35 10	16 27	11 27	29 32	29 ♏ 53	03 43	14 04	06 58	03 18
10	21 19 16	21 00 29	22 ♓ 50 20	29 ♓ 01 20	18 11	12 42	29 ♉ 58	00 ♐ 00	03 49	14 02	06 57	03 16
11	21 23 13	22 01 12	05 ♈ 08 29	11 ♈ 12 10	19 57	13 56	00 ♊ 24	00 06	03 54	13 59	06 57	03 15
12	21 27 09	23 01 54	17 ♈ 12 49	23 ♈ 10 58	21 43	15 11	00 50	00 12	04 00	13 56	06 57	03 13
13	21 31 06	24 02 34	29 ♈ 07 11	05 ♉ 02 02	23 30	16 26	01 17	00 18	04 05	13 54	06 57	03 12
14	21 35 02	25 03 12	10 ♉ 56 11	16 ♉ 50 17	25 17	17 40	01 43	00 24	04 10	13 51	06 57	03 10
15	21 38 59	26 03 49	22 ♉ 44 59	28 ♉ 40 58	27 06	18 55	02 10	00 29	04 16	13 49	06 57	03 09
16	21 42 55	27 04 24	04 ♊ 38 54	10 ♊ 39 26	28 ♒ 55	20 09	02 37	00 35	04 21	13 46	06 56	03 07
17	21 46 52	28 04 57	16 ♊ 43 10	22 ♊ 50 41	00 ♓ 45	21 24	03 05	00 40	04 26	13 44	06 56	03 06
18	21 50 49	29 ♒ 05 29	29 ♊ 02 31	05 ♋ 19 07	02 35	22 38	03 32	00 45	04 31	13 41	06 56	03 04
19	21 54 45	00 ♓ 05 59	11 ♋ 40 51	18 ♋ 08 00	04 27	23 52	04 00	00 50	04 36	13 39	06 55	03 03
20	21 58 42	01 06 27	24 ♋ 40 46	01 ♌ 19 11	06 18	25 07	04 28	00 55	04 41	13 36	06 55	03 01
21	22 02 38	02 06 53	08 ♌ 03 12	14 ♌ 52 37	08 11	26 21	04 56	01 00	04 46	13 34	06 54	03 00
22	22 06 35	03 07 18	21 ♌ 47 09	28 ♌ 46 21	10 03	27 35	05 24	01 04	04 51	13 31	06 54	02 58
23	22 10 31	04 07 40	05 ♍ 49 43	12 ♍ 56 37	11 56	28 ♓ 49	05 53	01 09	04 55	13 29	06 53	02 57
24	22 14 28	05 08 01	20 ♍ 06 25	27 ♍ 18 25	13 49	00 ♈ 03	06 21	01 13	05 00	13 26	06 53	02 55
25	22 18 24	06 08 21	04 ♎ 31 54	11 ♎ 46 11	15 42	01 18	06 50	01 17	05 04	13 24	06 52	02 54
26	22 22 21	07 08 39	19 ♎ 00 37	26 ♎ 14 37	17 34	02 32	07 19	01 21	05 09	13 22	06 52	02 52
27	22 26 18	08 08 55	03 ♏ 27 39	10 ♏ 39 16	19 25	03 46	07 48	01 24	05 13	13 19	06 51	02 51
28	22 30 14	09 ♓ 09 10	17 ♏ 49 05	24 ♏ 56 47	21 ♓ 16	05 ♈ 00	08 ♊ 18	01 ♐ 28	05 ♑ 18	13 ♌ 17	06 ♏ 50	02 ♍ 49

Date	Moon True ☊ ° '	Moon Mean ☊ ° '	Moon ☽ Latitude ° '
01	15 ♎ 19	16 ♎ 23	02 N 59
02	15 R 16	16 20	03 54
03	15 12	16 17	04 34
04	15 05	16 14	04 58
05	14 55	16 10	05 04
06	14 45	16 07	04 53
07	14 35	16 04	04 25
08	14 25	16 01	03 44
09	14 18	15 58	02 52
10	14 13	15 55	01 52
11	14 11	15 51	00 N 49
12	14 D 10	15 48	00 S 16
13	14 11	15 45	01 20
14	14 13	15 42	02 19
15	14 14	15 39	03 13
16	14 R 13	15 35	03 59
17	14 11	15 32	04 35
18	14 07	15 29	04 59
19	14 01	15 26	05 09
20	13 54	15 23	05 04
21	13 47	15 20	04 42
22	13 40	15 16	04 04
23	13 34	15 13	03 10
24	13 30	15 10	02 04
25	13 28	15 07	00 S 49
26	13 D 28	15 04	00 N 30
27	13 29	15 01	01 47
28	13 ♎ 31	14 ♎ 57	02 N 57

DECLINATIONS

Date	Sun ☉ ° '	Moon ☽ ° '	Mercury ☿ ° '	Venus ♀ ° '	Mars ♂ ° '	Jupiter ♃ ° '	Saturn ♄ ° '	Uranus ♅ ° '	Neptune ♆ ° '	Pluto ⯓ ° '
01	17 S 14	15 S 09	21 S 19	12 S 23	21 N 16	18 S 58	22 S 28	17 N 11	12 S 10	21 N 35
02	16 57	17 20	20 58	11 55	21 22	19 00	22 28	17 12	12 10	21 36
03	16 39	18 27	20 36	11 28	21 28	19 01	22 28	17 13	12 10	21 36
04	16 22	18 26	20 13	11 00	21 33	19 03	22 28	17 14	12 10	21 37
05	16 04	17 19	19 48	10 32	21 39	19 04	22 28	17 14	12 10	21 38
06	15 45	15 13	19 22	10 03	21 44	19 05	22 28	17 15	12 10	21 38
07	15 27	12 20	18 54	09 34	21 50	19 07	22 27	17 16	12 10	21 39
08	15 08	08 52	18 25	09 05	21 55	19 08	22 27	17 17	12 10	21 40
09	14 49	05 04	17 54	08 36	22 01	19 09	22 27	17 17	12 10	21 41
10	14 30	01 S 07	17 22	08 07	22 06	19 11	22 27	17 18	12 09	21 41
11	14 10	02 N 47	16 48	07 37	22 11	19 12	22 27	17 19	12 09	21 42
12	13 51	06 31	16 13	07 07	22 17	19 13	22 27	17 20	12 09	21 43
13	13 31	09 55	15 37	06 37	22 22	19 14	22 27	17 20	12 09	21 43
14	13 10	12 54	14 59	06 07	22 28	19 15	22 26	17 21	12 09	21 44
15	12 50	15 21	14 19	05 37	22 33	19 16	22 26	17 22	12 09	21 45
16	12 29	17 09	13 38	05 06	22 38	19 17	22 26	17 23	12 09	21 45
17	12 09	18 13	12 56	04 35	22 43	19 18	22 26	17 23	12 09	21 46
18	11 47	18 28	12 13	04 05	22 48	19 19	22 26	17 24	12 08	21 47
19	11 26	17 48	11 28	03 34	22 54	19 20	22 26	17 25	12 08	21 47
20	11 05	16 01	10 42	03 03	22 59	19 21	22 25	17 25	12 08	21 48
21	10 43	13 42	09 54	02 32	23 04	19 22	22 25	17 26	12 08	21 49
22	10 22	10 24	09 06	02 01	23 09	19 23	22 25	17 27	12 08	21 49
23	10 00	06 25	08 17	01 30	23 14	19 24	22 25	17 28	12 07	21 50
24	09 38	02 N 01	07 26	00 58	23 18	19 24	22 25	17 28	12 07	21 51
25	09 16	02 S 33	06 35	00 S 27	23 23	19 25	22 24	17 29	12 07	21 51
26	08 53	06 59	05 43	00 N 04	23 28	19 26	22 24	17 30	12 07	21 52
27	08 31	11 00	04 50	00 36	23 32	19 26	22 24	17 30	12 06	21 52
28	08 S 08	14 S 19	03 S 57	01 N 07	23 N 37	19 S 27	22 S 24	17 N 31	12 S 06	21 N 53

ZODIAC SIGN ENTRIES

Date	h	m	Planets	
02	03	11	☽	♐
04	06	29	☽	♑
06	10	40	☽	♒
08	16	50	☽	♓
10	13	46	♃	♐
10	13	57	♂	♊
11	01	55	☽	♈
13	13	47	☽	♉
16	02	39	☽	♊
17	02	15	☿	♓
18	13	51	☽	♋
19	09	38	☉	♓
20	21	38	☽	♌
23	02	06	☽	♍
24	10	53	♀	♈
25	04	29	☽	♎
27	06	14	☽	♏

LATITUDES

Date	Mercury ☿ ° '	Venus ♀ ° '	Mars ♂ ° '	Jupiter ♃ ° '	Saturn ♄ ° '	Uranus ♅ ° '	Neptune ♆ ° '	Pluto ⯓ ° '
01	01 S 52	01 S 32	02 N 01	00 N 59	00 N 56	00 N 43	01 N 46	12 N 14
04	02 00	01 30	02 01	00 59	00 56	00 43	01 46	12 14
07	02 04	01 28	02 00	00 59	00 56	00 43	01 46	12 15
10	02 05	01 26	02 00	01 00	00 56	00 43	01 47	12 15
13	02 02	01 22	02 00	01 00	00 56	00 43	01 47	12 16
16	01 54	01 18	01 59	01 00	00 56	00 43	01 47	12 16
19	01 42	01 14	01 59	01 01	00 56	00 43	01 47	12 16
22	01 24	01 09	01 58	01 01	00 56	00 43	01 47	12 17
25	01 01	01 03	01 58	01 01	00 56	00 43	01 48	12 17
28	00 S 32	00 57	01 57	01 02	00 56	00 43	01 48	12 17
31	00 N 02	00 S 50	01 N 56	01 N 02	00 N 56	00 N 43	01 N 48	12 N 17

DATA

Julian Date	2436601
Delta T	+33 seconds
Ayanamsa	23° 17' 14"
Synetic vernal point	05° ♓ 49' 46"
True obliquity of ecliptic	23° 26' 32"

MOON'S PHASES, APSIDES AND POSITIONS ☽

Date	h	m	Phase	Longitude ° '	Eclipse Indicator
07	19	22	●	18 ♒ 17	
15	19	20	☽	26 ♉ 22	
23	08	54	○	04 ♍ 00	

Day	h	m		
14	14	21	Apogee	
26	09	36	Perigee	
03	23	41	Max dec	18° S 35'
10	18	50	0N	
18	06	32	Max dec	18° N 29'
24	22	39	0S	

LONGITUDES

Date	Chiron ⚷ ° '	Ceres ⚳ ° '	Pallas ⚴ ° '	Juno ⚵ ° '	Vesta ⚶ ° '	Black Moon Lilith ⚸ ° '
01	21 ♒ 52	20 ♏ 18	26 ♎ 29	25 ♎ 22	03 ♌ 39	08 ♉ 28
11	22 ♒ 36	22 ♏ 52	27 ♎ 17	25 ♎ 44	01 ♌ 11	09 ♉ 34
21	23 ♒ 19	25 ♏ 00	27 ♎ 20	25 ♎ 30	29 ♋ 13	10 ♉ 41
31	24 ♒ 01	26 ♏ 39	26 ♎ 32	24 ♎ 38	27 ♋ 55	11 ♉ 47

All ephemeris data is given at 12.00 UT and the Moon's longitude is additionally given for 24.00 UT

ASPECTARIAN

h m	Aspects
01 Sunday	
00 41	☽ □ ♅
02 29	☽ Q ⯓
06 36	☽ ∠ ♄
11 10	☿ ⊻ ♄
11 49	☽ Q ☿
14 09	☿ # ♂
15 21	☉ # ♅
19 10	☿ ⊼ ⯓
21 05	☽ ☍ ♂
22 02	☽ ⊥ ♄
23 20	♀ ∥ ♆
02 Monday	
01 30	☽ ☌ ♃
04 08	☽ Q ☉
07 26	☽ □ ♀
07 26	☽ ∥ ☉
08 20	☽ ⊻ ♄
09 05	☽ □ ⯓
10 06	☽ # ♅
10 57	☽ ✱ ☿
15 02	☽ ⊻ ♆
18 29	♀ ✱ ♄
03 Tuesday	
01 16	☽ ⊥ ♆
01 53	♀ ☍ ⯓
02 12	☽ ✱ ☉
03 39	☽ △ ♅
15 59	☽ ∠ ☿
16 40	☽ ∠ ♆
17 10	☽ Q ♀
18 28	♂ ± ♄
21 30	☉ ☍ ♅
21 44	☿ □ ♆
04 Wednesday	
01 49	☽ ⊼ ♂
05 14	☽ ⊻ ♃
05 17	☽ ♇ ♅
05 55	☽ ∠ ☉
09 32	☽ ⊥ ☿
12 03	☽ ☌ ♄
12 22	☽ △ ⯓
12 27	☽ ± ♂
15 40	☽ ⊥ ♃
15 49	☽ ✱ ♀
18 28	☽ ✱ ♆
20 44	☽ ± ♅
21 18	☽ ⊻ ☿
22 39	☽ ⊥ ☉
05 Thursday	
04 28	☽ ♇ ♂
07 06	☽ ⊼ ♅
07 23	☽ ∠ ♃
07 42	☿ ⊥ ♄
08 09	♂ ∥ ⯓
09 53	☽ ⊻ ☉
13 11	☽ # ♅
14 18	☽ ♇ ⯓
15 18	☽ Q ♆
20 27	☽ ∠ ♀
21 23	♀ △ ♆
23 41	♄ △ ⯓
06 Friday	
05 59	☽ ∥ ☉
06 03	☽ ± ⯓
07 30	☽ △ ♂
09 51	☽ ✱ ♃
13 53	☽ ⊥ ♀
15 40	☿ Q ♃
16 38	☽ ⊼ ⯓
16 47	☽ ⊻ ♄
23 01	☽ □ ♆
07 Saturday	
01 28	☿ ∥ ♃
01 38	☽ ⊻ ♀
03 33	☽ ⊥ ♄
07 24	☽ Q ♃
09 32	☽ ☌ ☿
11 55	☽ ☍ ♅
13 17	☽ ∥ ♆
13 33	♆ St R
19 22	☽ ☌ ☉
19 52	☽ ∠ ♄
08 Sunday	
02 37	☉ ∠ ♄
03 42	☿ ☍ ♅
10 22	☽ ∥ ♀
15 18	☽ □ ♂
16 27	☽ □ ♃
22 59	☽ ☍ ⯓
23 39	☽ ✱ ♄
09 Monday	
05 48	☽ △ ♆
14 30	☽ ☌ ♀
19 11	☽ ⊼ ♅
22 25	☽ Q ♄
10 Tuesday	
01 37	☽ ⊻ ☿
02 18	☽ Q ♂
06 36	☽ ± ♅
08 10	☽ ⊻ ☉
10 18	☽ ♇ ♆
14 01	♂ ☍ ♃
14 36	☿ # ♅
15 02	☽ ⊥ ☿
20 48	☽ ⊥ ☉
20 59	☿ ∠ ♄
23 58	☽ ♇ ♅
11 Wednesday	
02 01	☽ △ ♃
02 21	☽ ✱ ♂
03 47	☽ ± ♆
08 17	☽ ⊼ ⯓
09 32	☽ □ ♄
11 33	☽ ∠ ☿
12 47	♀ ⊼ ♅
15 35	☽ ⊼ ♆
16 03	☽ ∠ ☉
20 06	☽ ± ⯓
12 Thursday	
05 29	☽ △ ♅
07 28	☽ ⊻ ♀
07 56	☽ ♇ ♃
09 08	☽ ∠ ♂
14 01	☽ ♇ ⯓
15 37	☽ # ♀
20 54	☽ ⊥ ♀
22 37	☽ ✱ ☿
13 Friday	
00 47	☽ ✱ ☉
02 10	☽ ± ♃
03 55	☽ ⊥ ♂
04 49	♀ Q ♄
14 24	☽ ⊼ ♃
16 32	☽ ⊻ ♂
17 14	☽ ∠ ♀
20 15	☽ △ ⯓
22 09	☽ △ ♄
14 Saturday	
03 15	☽ Q ☿
03 22	☽ Q ☉
03 53	☽ ☍ ♆
04 54	☉ ☌ ☿
05 35	☽ # ♆
07 06	♂ # ♄
14 09	☽ # ☉
17 55	☽ □ ♅
18 33	♂ Q ♅
15 Sunday	
03 18	☽ ✱ ♀
03 34	☽ # ☿
04 52	☽ ♇ ♄
19 20	☽ □ ☉
22 23	☽ □ ☿
23 14	☽ ± ♄
16 Monday	
03 46	☽ ☍ ♃
04 50	♀ ± ♅
06 14	☽ Q ♅
06 24	☽ Q ♀
07 46	☽ ☌ ♂
08 57	☽ □ ⯓
11 23	☽ ⊼ ♄
15 52	☽ ∥ ♅
16 35	☽ ⊼ ♆
17 Tuesday	
02 19	♀ Q ♂
04 31	☽ ± ♆
06 07	☽ ✱ ♅
10 56	☿ □ ♃
11 57	☉ ∥ ♆
13 04	♂ □ ⯓
20 34	☽ Q ⯓
22 12	☽ □ ♀
22 13	☽ ♇ ♆
22 23	♀ ♇ ♆
18 Wednesday	
11 19	☽ ∠ ♅
12 06	☽ △ ☉
14 21	☿ ∥ ♆
15 19	☽ ⊼ ♃
18 11	☿ ☍ ⯓
19 43	☽ ✱ ⯓
19 59	☽ △ ☿
20 56	☽ ⊻ ♂
22 33	☽ ☍ ♄
19 Thursday	
02 49	☽ ± ♃
03 02	☽ △ ♆
04 19	☿ □ ♂
04 26	☽ ⊥ ♅
08 43	☽ ⊥ ♂
13 33	☉ ∥ ☿
14 06	☿ ✱ ♄
15 39	☽ ⊻ ♅
18 55	☽ ♇ ☉
19 14	☽ ∥ ♅
19 48	☽ ♇ ♃
23 49	☽ ∠ ⯓
20 Friday	
02 06	☽ ∠ ♂
04 49	☽ ♇ ☿
07 07	☉ □ ♃
12 51	☽ ± ☉
12 52	☽ △ ♀
16 15	☽ ⊥ ⯓
19 45	☿ △ ♆
23 21	☽ △ ♃
23 50	☽ ± ☿
21 Saturday	
00 34	☽ ⊼ ☉
01 30	♂ ⊼ ♄
03 01	☽ ⊻ ⯓
06 07	☽ ⊼ ♄
06 16	☽ ✱ ♂
09 58	☽ □ ♆
12 15	☽ ⊼ ☿
16 49	☽ ± ♄
18 24	☽ ♇ ♀
21 40	☽ ☌ ♅
22 Sunday	
00 06	☽ # ♆
04 08	☽ Q ♂
08 30	☉ ☍ ⯓
08 37	☽ ♇ ♄
11 37	☽ ± ♀
12 14	☽ # ☉
17 21	☽ Q ♆
22 22	☽ # ☿
22 56	☽ ⊼ ♀
23 Monday	
04 00	☽ □ ♃
05 34	♀ ♇ ♅
07 07	☽ ☌ ⯓
08 54	☽ ☍ ☉
10 27	☽ △ ♄
12 05	☽ □ ♂
13 48	☽ ✱ ♆
23 52	☽ ☍ ☿
24 Tuesday	
00 52	☽ ⊻ ♅
07 18	☿ ⊼ ♅
08 30	☉ ✱ ♄
10 30	☽ Q ♃
10 53	☽ ⊥ ♅
14 57	☽ ∠ ♆
18 15	☽ # ♀
25 Wednesday	
01 51	☽ ∠ ♅
02 08	☽ ∥ ♀
03 51	♀ ± ♆
05 56	☽ ⊥ ♆
06 07	☽ ☍ ♀
06 35	☽ ✱ ♃
09 17	☽ ⊻ ⯓
11 45	♀ △ ♃
12 54	☽ □ ♄
13 34	♂ ⊼ ♆
14 52	☽ ⊼ ☉
15 53	☽ ⊻ ♆
15 57	☽ △ ♂
19 13	☽ ⊥ ⯓
26 Thursday	
01 33	☽ ± ☉
02 40	☽ ✱ ♅
05 17	☉ △ ♆
06 09	☽ ∥ ☿
06 27	☿ Q ♄
07 34	☽ ∠ ♃
09 15	☽ ⊼ ☿
10 07	☽ ∠ ⯓
17 35	☽ ♇ ☉
17 41	☽ ♇ ♂
18 30	♀ ⊼ ⯓
18 54	☽ Q ♄
20 12	☉ □ ♂
20 40	☽ ± ☿
22 05	☽ ∥ ☉
22 30	☽ Q ♅
22 33	☽ ⊥ ♃
27 Friday	
08 34	☽ ⊻ ♃
09 09	☽ ± ♂
10 42	☿ ± ♅
10 58	☽ ✱ ⯓
12 33	☽ ⊼ ♀
13 51	☽ ♇ ☿
14 57	☽ ✱ ♄
17 39	☽ ☌ ♆
19 25	☽ ∥ ♆
19 30	☽ ⊼ ♂
20 24	☽ △ ☉
23 30	☽ ± ♀
28 Saturday	
04 25	☽ □ ♅
06 59	☽ Q ⯓
16 00	☽ ♇ ♀
16 11	☽ ∠ ♄
18 14	♀ □ ♄
18 39	☽ △ ☿
19 25	☿ ♇ ♆

MARCH 1959

LONGITUDES

Date	Sidereal time h m s	Sun ☉ ° ' "	Moon ☽ ° ' "	Moon ☽ 24.00 ° ' "	Mercury ☿ ° '	Venus ♀ ° '	Mars ♂ ° '	Jupiter ♃ ° '	Saturn ♄ ° '	Uranus ♅ ° '	Neptune ♆ ° '	Pluto ♇ ° '
01	22 34 11	10 ♓ 09 24	02 ♐ 02 08	09 ♐ 04 57	23 ♓ 05	06 ♈ 14	08 ♊ 47	01 ♐ 31	05 ♑ 22	13 ♌ 15	06 ♏ 50	02 ♍ 48
02	22 38 07	11 09 36	16 ♐ 05 04	23 ♐ 02 24	24 53	07 28	09 17	01 34	05 26	13 R 13	06 R 49	02 R 46
03	22 42 04	12 09 47	29 ♐ 56 51	06 ♑ 48 20	26 38	08 41	09 47	01 37	05 30	13 10	06 48	02 45
04	22 46 00	13 09 56	13 ♑ 36 48	20 ♑ 22 11	28 ♓ 21	09 55	10 17	01 40	05 34	13 08	06 47	02 43
05	22 49 57	14 10 03	27 ♑ 04 26	03 ♒ 43 29	00 ♈ 01	11 09	10 47	01 43	05 38	13 06	06 47	02 42
06	22 53 53	15 10 09	10 ♒ 19 15	16 ♒ 51 42	01 36	12 23	11 17	01 45	05 42	13 04	06 46	02 40
07	22 57 50	16 10 13	23 ♒ 20 48	29 ♒ 46 29	03 08	13 36	11 47	01 47	05 46	13 02	06 45	02 39
08	23 01 47	17 10 16	06 ♓ 08 46	12 ♓ 27 40	04 35	14 50	12 18	01 49	05 50	13 00	06 44	02 37
09	23 05 43	18 10 16	18 ♓ 43 13	24 ♓ 55 30	05 56	16 04	12 49	01 51	05 53	12 58	06 43	02 36
10	23 09 40	19 10 15	01 ♈ 04 39	07 ♈ 10 50	07 12	17 17	13 19	01 53	05 57	12 56	06 42	02 34
11	23 13 36	20 10 12	13 ♈ 14 16	19 ♈ 15 13	08 21	18 31	13 50	01 54	06 00	12 54	06 41	02 33
12	23 17 33	21 10 06	25 ♈ 14 00	01 ♉ 10 59	09 24	19 44	14 21	01 56	06 04	12 52	06 40	02 31
13	23 21 29	22 09 59	07 ♉ 06 32	13 ♉ 01 08	10 19	20 58	14 52	01 57	06 07	12 50	06 39	02 30
14	23 25 26	23 09 49	18 ♉ 55 15	24 ♉ 49 24	11 06	22 11	15 24	01 58	06 10	12 48	06 38	02 28
15	23 29 22	24 09 37	00 ♊ 44 09	06 ♊ 40 03	11 46	23 24	15 55	01 58	06 13	12 46	06 37	02 27
16	23 33 19	25 09 24	12 ♊ 37 42	18 ♊ 37 42	12 17	24 37	16 27	01 59	06 17	12 45	06 36	02 26
17	23 37 16	26 09 07	24 ♊ 40 39	00 ♋ 47 10	12 40	25 51	16 58	01 59	06 19	12 43	06 35	02 24
18	23 41 12	27 08 49	06 ♋ 57 49	13 ♋ 13 08	12 55	27 04	17 30	01 59	06 22	12 41	06 33	02 23
19	23 45 09	28 08 28	19 ♋ 33 39	25 ♋ 59 47	13 01	28 17	18 02	01 R 59	06 25	12 40	06 32	02 22
20	23 49 05	29 ♓ 08 05	02 ♌ 31 54	09 ♌ 10 16	13 R 00	29 ♈ 30	18 34	01 59	06 28	12 38	06 31	02 20
21	23 53 02	00 ♈ 07 40	15 ♌ 55 03	22 ♌ 46 16	12 50	00 ♉ 43	19 06	01 59	06 30	12 37	06 30	02 19
22	23 56 58	01 07 13	29 ♌ 43 50	06 ♍ 47 27	12 32	01 56	19 38	01 58	06 33	12 35	06 29	02 18
23	00 00 55	02 06 43	13 ♍ 56 45	21 ♍ 11 10	12 08	03 08	20 10	01 57	06 35	12 34	06 27	02 16
24	00 04 51	03 06 11	28 ♍ 30 00	05 ♎ 52 27	11 37	04 21	20 42	01 56	06 38	12 32	06 26	02 15
25	00 08 48	04 05 37	13 ♎ 17 36	20 ♎ 44 30	11 00	05 34	21 15	01 55	06 40	12 31	06 25	02 14
26	00 12 45	05 05 01	28 ♎ 12 08	05 ♏ 39 32	10 19	06 46	21 47	01 54	06 42	12 30	06 23	02 13
27	00 16 41	06 04 24	13 ♏ 05 45	20 ♏ 29 53	09 34	07 59	22 19	01 53	06 44	12 28	06 22	02 11
28	00 20 38	07 03 44	27 ♏ 51 10	05 ♐ 08 56	08 45	09 11	22 52	01 51	06 46	12 27	06 21	02 10
29	00 24 34	08 03 03	12 ♐ 22 37	19 ♐ 31 49	07 55	10 24	23 25	01 49	06 48	12 26	06 19	02 09
30	00 28 31	09 02 20	26 ♐ 36 13	03 ♑ 35 39	07 04	11 36	23 58	01 47	06 49	12 25	06 18	02 08
31	00 32 27	10 ♈ 01 35	10 ♑ 30 02	17 ♑ 19 23	06 ♈ 13	12 ♉ 49	24 ♊ 30	01 ♐ 45	06 ♑ 51	12 ♌ 24	06 ♏ 17	02 ♍ 07

Moon / DECLINATIONS

Date	Moon True ☊ ° '	Moon Mean ☊ ° '	Moon ☽ Latitude ° '	Sun ☉ ° '	Moon ☽ ° '	Mercury ☿ ° '	Venus ♀ ° '	Mars ♂ ° '	Jupiter ♃ ° '	Saturn ♄ ° '	Uranus ♅ ° '	Neptune ♆ ° '	Pluto ♇ ° '
01	13 ♎ 32	14 ♎ 54	03 N 55	07 S 46	16 S 44	03 S 04	01 N 38	23 N 41	19 S 27	22 S 24	17 N 31	12 S 06	21 N 54
02	13 R 32	14 51	04 38	07 23	18 06	02 11	02 09	23 46	19 28	22 24	17 32	12 06	21 54
03	13 31	14 48	05 04	07 00	18 22	01 18	02 41	23 50	19 28	22 23	17 33	12 05	21 55
04	13 28	14 45	05 13	06 37	17 33	00 S 26	03 12	23 54	19 29	22 23	17 33	12 05	21 55
05	13 24	14 41	05 04	06 14	15 46	00 N 25	03 43	23 58	19 29	22 23	17 34	12 05	21 56
06	13 20	14 38	04 39	05 51	13 10	01 15	04 14	24 02	19 30	22 23	17 35	12 05	21 57
07	13 15	14 35	04 00	05 27	09 57	02 04	04 45	24 06	19 30	22 23	17 35	12 04	21 57
08	13 11	14 32	03 09	05 04	06 19	02 51	05 16	24 10	19 30	22 22	17 36	12 04	21 58
09	13 08	14 29	02 10	04 41	02 S 28	03 36	05 46	24 14	19 31	22 22	17 36	12 03	21 58
10	13 06	14 26	01 N 06	04 17	01 N 26	04 18	06 17	24 18	19 31	22 22	17 37	12 03	21 59
11	13 D 05	14 22	00 S 01	03 54	05 13	04 58	06 47	24 21	19 31	22 22	17 37	12 03	21 59
12	13 06	14 19	01 07	03 30	08 44	05 35	07 17	24 25	19 31	22 22	17 38	12 02	22 00
13	13 07	14 16	02 09	03 07	11 51	06 08	07 48	24 28	19 31	22 22	17 38	12 02	22 00
14	13 09	14 13	03 05	02 43	14 29	06 38	08 18	24 31	19 31	22 21	17 39	12 01	22 01
15	13 10	14 10	03 54	02 19	16 29	07 04	08 48	24 34	19 31	22 21	17 39	12 01	22 01
16	13 11	14 07	04 33	01 56	17 48	07 26	09 17	24 37	19 31	22 21	17 40	12 01	22 02
17	13 12	14 03	05 00	01 32	18 20	07 44	09 47	24 40	19 31	22 21	17 40	12 00	22 02
18	13 R 12	14 00	05 15	01 08	18 01	07 58	10 16	24 43	19 31	22 21	17 41	12 00	22 02
19	13 11	13 57	05 15	00 44	16 49	08 07	10 45	24 46	19 31	22 21	17 41	11 59	22 03
20	13 09	13 54	04 59	00 S 21	14 45	08 12	11 13	24 48	19 31	22 20	17 42	11 59	22 03
21	13 07	13 51	04 27	00 N 03	11 49	08 13	11 42	24 51	19 31	22 20	17 42	11 59	22 04
22	13 06	13 47	03 38	00 27	08 10	08 09	12 10	24 53	19 30	22 20	17 42	11 58	22 04
23	13 04	13 44	02 35	00 50	03 N 56	08 00	12 38	24 55	19 30	22 20	17 43	11 58	22 04
24	13 04	13 41	01 S 20	01 14	00 S 38	07 48	13 05	24 57	19 30	22 20	17 43	11 57	22 05
25	13 D 03	13 38	00 N 01	01 38	05 14	07 31	13 33	24 59	19 30	22 20	17 43	11 57	22 05
26	13 03	13 35	01 23	02 01	09 32	07 11	14 00	25 01	19 29	22 20	17 44	11 56	22 05
27	13 04	13 32	02 39	02 25	13 14	06 48	14 26	25 02	19 29	22 20	17 44	11 56	22 06
28	13 05	13 28	03 44	02 48	16 03	06 22	14 52	25 04	19 28	22 19	17 44	11 55	22 06
29	13 05	13 25	04 33	03 12	17 46	05 53	15 18	25 05	19 28	22 19	17 45	11 55	22 06
30	13 05	13 22	05 05	03 35	18 19	05 23	15 44	25 07	19 27	22 19	17 45	11 54	22 07
31	13 ♎ 06	13 ♎ 19	05 N 17	03 N 58	17 S 45	04 N 52	16 N 09	25 N 08	19 S 27	22 S 19	17 N 45	11 S 54	22 N 07

ZODIAC SIGN ENTRIES

Date	h m	Planets
01	08 33	☽ ♐
03	12 05	☽ ♑
05	11 52	☿ ♈
05	17 16	☽ ♒
08	00 25	☽ ♓
10	09 53	☽ ♈
12	21 37	☽ ♉
15	10 31	☽ ♊
17	22 28	☽ ♋
20	07 22	☽ ♌
20	21 55	♀ ♉
21	08 55	☉ ♈
22	12 28	☽ ♍
24	14 27	☽ ♎
26	14 53	☽ ♏
28	15 31	☽ ♐
30	17 49	☽ ♑

LATITUDES

Date	Mercury ☿ ° '	Venus ♀ ° '	Mars ♂ ° '	Jupiter ♃ ° '	Saturn ♄ ° '	Uranus ♅ ° '	Neptune ♆ ° '	Pluto ♇ ° '
01	00 S 21	00 S 55	01 N 57	01 N 02	00 N 56	00 N 43	01 N 48	12 N 17
04	00 N 14	00 48	01 56	01 02	00 56	00 43	01 48	12 17
07	00 54	00 41	01 55	01 03	00 56	00 42	01 48	12 17
10	01 35	00 33	01 54	01 03	00 56	00 42	01 48	12 17
13	02 14	00 25	01 54	01 03	00 57	00 42	01 48	12 17
16	02 48	00 16	01 53	01 04	00 57	00 42	01 49	12 17
19	03 14	00 S 08	01 52	01 04	00 57	00 42	01 49	12 16
22	03 28	00 N 01	01 51	01 04	00 57	00 42	01 49	12 16
25	03 26	00 11	01 50	01 05	00 57	00 42	01 49	12 16
28	03 09	00 20	01 49	01 05	00 57	00 42	01 49	12 15
31	02 N 36	00 N 29	01 N 48	01 N 05	00 N 57	00 N 42	01 N 49	12 N 15

DATA

Julian Date	2436629
Delta T	+33 seconds
Ayanamsa	23° 17' 17"
Synetic vernal point	05° ♓ 49' 42"
True obliquity of ecliptic	23° 26' 32"

MOON'S PHASES, APSIDES AND POSITIONS ☽

Date	h	m	Phase	Longitude ° '	Eclipse Indicator
02	02	54	☾	10 ♐ 47	
09	10	51	●	18 ♓ 07	
17	15	10	☽	26 ♊ 17	
24	20	02	○	03 ♎ 26	partial
31	11	06	☾	09 ♑ 59	

Day	h	m	
14	09	31	Apogee
26	09	33	Perigee

Day	h	m		
03	05	42	Max dec	18° S 25'
10	03	08	0N	
17	15	17	Max dec	18° N 20'
24	08	45	0S	
30	11	37	Max dec	18° S 19'

LONGITUDES

Date	Chiron ⚷ ° '	Ceres ⚳ ° '	Pallas ⚴ ° '	Juno ⚵ ° '	Vesta ⚶ ° '	Black Moon Lilith ⚸ ° '
01	23 ♒ 53	26 ♏ 21	26 ♎ 46	24 ♎ 51	28 ♋ 07	11 ♉ 34
11	24 ♒ 33	27 ♏ 33	25 ♎ 17	23 ♎ 31	27 ♋ 26	12 ♉ 41
21	25 ♒ 12	27 ♏ 07	23 ♎ 02	21 ♎ 41	27 ♋ 32	13 ♉ 47
31	25 ♒ 47	28 ♏ 01	20 ♎ 13	19 ♎ 29	28 ♋ 21	14 ♉ 54

ASPECTARIAN

h m Aspects

01 Sunday
07 27 ☽ ⊥ ♄
11 07 ☽ ☌ ♃
13 17 ☽ □ ♇
17 42 ☽ ⊻ ♄
19 49 ☽ △ ♀
20 09 ☽ ⊻ ♆
23 30 ☽ # ♅
23 34 ♀ ⊼ ♆
23 55 ☽ ☍ ♂

02 Monday
02 54 ☽ □ ☉
06 23 ☽ ⊥ ♆
07 05 ☽ △ ♅
12 31 ☿ # ♀
21 52 ☽ ∠ ♆

03 Tuesday
05 25 ☽ □ ☿
08 55 ☽ bQ ♅
12 24 ☽ Q ☉
13 00 ♀ ± ♇
14 56 ☽ ⊻ ♃
16 52 ☽ △ ♇
21 46 ☽ ☌ ♄
23 59 ☽ ✶ ♆

04 Wednesday
00 37 ☽ ± ♅
01 29 ☽ ⊥ ♃
04 50 ☽ □ ♀
05 53 ☽ ⊼ ♂
09 01 ☿ bQ ♅
10 33 ☿ Q ♂
11 09 ☽ ✶ ☉
11 09 ☽ ⊼ ♅
11 19 ☉ ⊼ ♅
12 01 ☽ # ♅
16 54 ☽ ± ♂
17 26 ☽ ∠ ♃
17 32 ☽ Q ☿
19 16 ☽ bQ ♇
21 10 ☽ Q ♆
23 46 ♀ ✶ ♂

05 Thursday
09 35 ☽ bQ ♂
11 19 ☽ ± ♇
16 04 ☽ ∠ ☉
16 07 ☽ Q ♀
18 01 ☽ ✶ ☿
20 23 ☽ ✶ ♃
22 07 ☽ ⊼ ♇
23 18 ☿ ± ♆

06 Friday
03 33 ☽ ⊻ ♄
05 31 ☽ □ ♆
09 43 ☽ ⊥ ☉
13 50 ☽ △ ♂
14 15 ☿ △ ♃
14 33 ☽ ⊥ ♄
16 09 ☽ ✶ ♀
17 00 ☽ ☍ ♅
18 18 ☽ Q ♃
20 41 ☽ ∥ ♆
21 37 ☽ ⊻ ☉

07 Saturday
01 01 ♀ △ ♅
01 05 ☽ ∠ ☿
04 17 ☿ ⊼ ♇
07 12 ☽ ∠ ♄
19 58 ☽ ⊥ ☿
22 51 ☽ ∠ ♀

08 Sunday
03 49 ☽ □ ♃
05 21 ☽ ☍ ♇
06 52 ☉ # ♀
08 40 ☽ ⊻ ☿
11 24 ☽ ✶ ♄
13 06 ☽ △ ♆
17 39 ☽ ⊥ ♀
17 55 ☽ # ♀
20 48 ☽ ∥ ☉

09 Monday
00 11 ☽ □ ♂
00 59 ☽ ⊼ ♅
04 51 ☉ Q ♄
06 08 ☽ # ☿
06 20 ☽ ⊻ ♀
07 48 ♂ ± ♆
10 24 ☽ Q ♄
10 51 ☽ ☌ ☉
11 05 ☿ □ ♄
12 28 ☽ ± ♅
17 46 ☽ bQ ♆
18 44 ♂ ✶ ♅

10 Tuesday
02 24 ☿ ⊼ ♆
03 50 ♀ bQ ♃
05 52 ☽ bQ ♅
06 23 ☉ ± ♅
11 16 ☽ ± ♆
11 35 ☉ # ☿
12 30 ☽ Q ♂
13 34 ☽ △ ♃
14 55 ☽ ⊼ ♇
17 25 ♀ bQ ♇
21 37 ☽ □ ♄
23 02 ☽ ⊼ ♆

11 Wednesday
01 20 ☽ ☌ ☿
02 43 ☽ ± ♇
04 17 ☽ # ☉
10 03 ☽ ∥ ☿
11 19 ☽ △ ♅
13 15 ☽ ✶ ♂
16 10 ☿ ± ♇
19 19 ☽ bQ ♃
20 34 ☽ bQ ♇
23 43 ☽ ☌ ♀

12 Thursday
00 14 ☽ ∥ ♀
03 06 ☽ ⊻ ☉
13 24 ☽ ± ♃
16 15 ☽ ⊥ ☉
20 41 ☽ ∠ ♂
23 48 ☉ bQ ♆

13 Friday
01 31 ☽ ⊼ ♃
02 41 ☽ △ ♇
09 59 ☽ △ ♄
11 04 ☽ ☍ ♆
12 08 ☽ ∠ ☉
13 27 ☽ # ♆
15 45 ☽ ⊥ ♂
19 00 ☽ ⊻ ☿
23 36 ☽ □ ♅

14 Saturday
04 30 ☽ ⊻ ♂
08 03 ☽ ⊥ ☿
16 36 ☽ bQ ♄
19 24 ☽ ⊻ ♀
21 25 ☽ ✶ ☉

15 Sunday
03 30 ☽ ∠ ☿
08 59 ☽ ⊥ ♀
10 58 ☽ ± ♄
12 05 ☽ Q ♅
14 30 ☽ ☍ ♃
15 28 ☽ □ ♇
23 09 ☽ ⊼ ♄
23 52 ☽ ⊼ ♆
23 59 ☽ Q ☉

16 Monday
05 16 ☽ ∠ ♀
08 35 ☽ ∥ ♅
11 17 ☽ ✶ ☿
11 56 ☽ ± ♆
12 14 ☽ ✶ ♅
19 59 ☽ ☌ ♂

17 Tuesday
03 33 ☽ Q ♇
05 52 ☽ bQ ♆
11 59 ☽ Q ☿
14 33 ☽ ✶ ♀
14 49 ♀ ± ♃
15 10 ☽ □ ☉
15 14 ☿ △ ♅
17 58 ☽ ∠ ♅

18 Wednesday
02 21 ☽ ⊼ ♃
03 08 ☽ ✶ ♇
10 51 ☽ ☍ ♄
11 13 ☽ △ ♆
11 28 ☽ ⊥ ♅
13 59 ☽ ± ♃
16 29 ☽ Q ♀
20 52 ☽ ∥ ♅
21 01 ☉ ⊻ ♀
22 10 ♃ St R
22 58 ☽ ⊻ ♅
23 33 ☽ □ ☿

19 Thursday
00 43 ☉ bQ ♅
07 09 ☽ bQ ♃
07 52 ☽ ∠ ♇
08 59 ☽ ⊻ ♂
18 34 ☿ St R
20 42 ☽ ⊥ ♂

20 Friday
00 39 ☽ ⊥ ♇
05 16 ☽ △ ☉
05 53 ☽ □ ♀
11 00 ☽ △ ♃
11 39 ☽ ⊻ ♇
13 57 ☽ ∠ ♂
19 09 ☽ ⊼ ♄
19 13 ☽ □ ♆

21 Saturday
05 56 ☽ ± ♄
06 09 ☽ ☌ ♅
06 37 ☽ △ ☿
09 10 ♄ ✶ ♆
10 30 ☽ bQ ☉
10 53 ☽ # ♆
12 47 ☽ ∥ ♀
17 49 ☽ ✶ ♂
20 46 ☉ ± ♆
21 50 ☽ bQ ♄

22 Sunday
02 00 ♀ # ♆
02 58 ☽ Q ♆
03 28 ☽ ± ☉
08 20 ☽ bQ ☿
08 36 ☿ △ ♅
12 08 ☽ ∥ ☿
12 50 ♀ ⊼ ♃
14 33 ☽ ⊼ ☉
15 22 ☽ Q ♂
15 49 ☽ □ ♃
16 06 ☽ △ ♀
16 22 ☽ ☌ ♇
19 06 ♀ △ ♇
23 17 ☽ ± ☿
23 27 ☽ ✶ ♆
23 37 ☽ △ ♄

23 Monday
01 59 ♄ ± ♅
08 19 ☉ △ ♃
09 04 ☽ ⊼ ☿
09 42 ☽ ⊻ ♅
15 47 ☉ ⊼ ♇
16 35 ♂ Q ♇
19 36 ☽ bQ ♀
19 39 ☽ ⊥ ♅
21 58 ☽ Q ♃
22 43 ☽ □ ♂

24 Tuesday
00 26 ☽ ∠ ♆
03 06 ☽ ∥ ☉
10 26 ☽ ∠ ♅
11 44 ☽ ± ♀
15 09 ☽ ⊥ ♆
15 26 ☽ # ☉
17 36 ☽ ✶ ♃
18 06 ☽ ⊻ ♇
20 02 ☽ ☍ ☉
22 23 ☽ ⊼ ♀

25 Wednesday
00 53 ☽ ⊻ ♆
01 15 ☽ □ ♄
03 50 ☽ ⊥ ♇
08 28 ☽ ☍ ☿
10 45 ☽ ✶ ♅
17 51 ☽ ∠ ♃
18 20 ☽ ∠ ♇
19 18 ♂ bQ ♆
23 34 ☽ # ☿

26 Thursday
01 17 ☽ △ ♂
04 32 ♀ ☍ ♆
06 03 ☽ Q ♅
06 21 ☽ Q ♄
08 18 ☽ ⊥ ♃
10 25 ♀ △ ♄
12 18 ♀ ∠ ♂
17 56 ☽ ⊻ ♃
18 26 ☽ ✶ ♇
22 00 ☿ Q ♂
23 52 ☽ ⊼ ☉

27 Friday
01 10 ☽ ☌ ♆
01 42 ☽ ✶ ♄
02 20 ☽ bQ ♂
02 56 ☽ ∥ ♆
03 01 ☽ ☍ ♀
06 34 ☽ ⊼ ☿
10 14 ☽ ± ☉
11 00 ☽ □ ♅
13 46 ☽ Q ♇
15 47 ☽ ± ☿
17 26 ☽ ± ♂
19 00 ☉ ⊼ ♆
22 54 ☽ # ♀

28 Saturday
01 52 ☽ bQ ☉
02 02 ☽ ∠ ♄
03 33 ☽ ⊼ ♂
04 30 ☉ □ ♄
05 39 ☽ bQ ☿
06 51 ☿ ⊻ ♀
16 47 ☽ ⊥ ♄
18 33 ☽ ☌ ♃
19 04 ☽ □ ♇

29 Sunday
01 57 ☽ ⊻ ♆
02 42 ☽ ⊻ ♄
04 17 ☽ △ ☉
05 00 ☽ △ ☿
05 22 ☿ ± ♇
08 24 ☽ ⊼ ♀
10 17 ☉ ☌ ☿
11 41 ☽ # ♅
11 55 ☽ ⊥ ♆
12 06 ☽ △ ♅
14 19 ☉ ± ♇
19 21 ☽ ± ♀

30 Monday
03 00 ☽ ∠ ♆
07 19 ☽ ☍ ♂
12 00 ☽ bQ ♀
13 23 ☽ bQ ♅
18 47 ☿ □ ♄
20 51 ☽ ⊻ ♃
21 28 ☽ △ ♇

31 Tuesday
03 53 ♀ □ ♅
04 39 ☽ ✶ ♆
04 52 ☽ ± ♅
04 59 ☽ □ ☿
05 00 ☿ ⊥ ♀
05 38 ☽ ☌ ♄
07 12 ☽ ⊥ ♃
10 29 ☿ ⊼ ♆
11 06 ☽ □ ☉
11 54 ☽ # ♅
15 19 ☽ ⊼ ♅
16 26 ☽ △ ♀
22 57 ☽ ∠ ♃
23 36 ☽ bQ ♇

All ephemeris data is given at 12.00 UT and the Moon's longitude is additionally given for 24.00 UT

APRIL 1959

LONGITUDES

Date	Sidereal time h m s	Sun ☉ ° ' "	Moon ☽ ° ' "	Moon ☽ 24.00 ° ' "	Mercury ☿ ° '	Venus ♀ ° '	Mars ♂ ° '	Jupiter ♃ ° '	Saturn ♄ ° '	Uranus ♅ ° '	Neptune ♆ ° '	Pluto ♇ ° '
01	00 36 24	11 ♈ 00 49	24 ♑ 03 46	00 ♒ 43 22	05 ♈ 24	14 ♉ 01	25 ♊ 03	01 ♐ 42	06 ♑ 53	12 ♌ 23	06 ♏ 15	02 ♍ 05
02	00 40 20	12 00 01	07 ♒ 18 21	13 ♒ 48 57	04 R 36	15 13	25 36	01 R 40	06 54	12 R 22	06 R 14	02 R 04
03	00 44 17	12 59 10	20 ♒ 15 25	26 ♒ 38 00	03 52	16 25	26 09	01 37	06 55	12 21	06 12	02 03
04	00 48 14	13 58 18	02 ♓ 56 58	09 ♓ 12 34	03 11	17 37	26 43	01 34	06 57	12 20	06 11	02 02
05	00 52 10	14 57 25	15 ♓ 25 03	21 ♓ 34 41	02 34	18 49	27 16	01 31	06 58	12 19	06 09	02 01
06	00 56 07	15 56 29	27 ♓ 41 41	03 ♈ 46 17	02 02	20 01	27 49	01 27	06 59	12 18	06 08	02 00
07	01 00 03	16 55 31	09 ♈ 48 45	15 ♈ 49 16	01 35	21 13	28 23	01 24	07 00	12 18	06 06	01 59
08	01 04 00	17 54 32	21 ♈ 48 06	27 ♈ 45 29	01 13	22 24	28 56	01 20	07 01	12 17	06 05	01 58
09	01 07 56	18 53 30	03 ♉ 41 40	09 ♉ 36 56	00 56	23 36	29 ♊ 30	01 16	07 01	12 16	06 03	01 57
10	01 11 53	19 52 26	15 ♉ 31 32	21 ♉ 25 49	00 45	24 48	00 ♋ 03	01 12	07 02	12 16	06 02	01 56
11	01 15 49	20 51 20	27 ♉ 20 06	03 ♊ 14 45	00 39	25 59	00 37	01 08	07 03	12 15	06 00	01 55
12	01 19 46	21 50 12	09 ♊ 10 08	15 ♊ 06 42	00 D 39	27 10	01 10	01 04	07 03	12 15	05 59	01 54
13	01 23 43	22 49 02	21 ♊ 04 53	27 ♊ 05 09	00 44	28 22	01 44	00 59	07 03	12 15	05 57	01 54
14	01 27 39	23 47 50	03 ♋ 08 00	09 ♋ 13 56	00 54	29 ♉ 33	02 18	00 55	07 04	12 14	05 55	01 53
15	01 31 36	24 46 35	15 ♋ 23 29	21 ♋ 37 10	01 09	00 ♊ 44	02 52	00 50	07 04	12 14	05 54	01 52
16	01 35 32	25 45 19	27 ♋ 55 31	04 ♌ 19 00	01 28	01 55	03 26	00 45	07 04	12 14	05 52	01 51
17	01 39 29	26 44 00	10 ♌ 48 06	17 ♌ 23 13	01 53	03 06	04 00	00 40	07 04	12 13	05 51	01 50
18	01 43 25	27 42 38	24 ♌ 04 42	00 ♍ 52 49	02 21	04 17	04 34	00 35	07 R 04	12 13	05 49	01 50
19	01 47 22	28 41 15	07 ♍ 47 40	14 ♍ 49 18	02 54	05 28	05 08	00 29	07 03	12 13	05 47	01 49
20	01 51 18	29 ♈ 39 49	21 ♍ 57 32	29 ♍ 12 05	03 31	06 38	05 42	00 24	07 03	12 D 13	05 46	01 48
21	01 55 15	00 ♉ 38 21	06 ♎ 32 27	13 ♎ 57 57	04 12	07 49	06 16	00 18	07 03	12 13	05 44	01 48
22	01 59 12	01 36 51	21 ♎ 27 46	29 ♎ 00 53	04 56	08 59	06 51	00 12	07 02	12 13	05 42	01 47
23	02 03 08	02 35 19	06 ♏ 36 11	14 ♏ 12 28	05 44	10 10	07 25	00 07	07 02	12 14	05 41	01 46
24	02 07 05	03 33 45	21 ♏ 48 28	29 ♏ 22 59	06 35	11 20	07 59	00 ♐ 01	07 01	12 14	05 39	01 46
25	02 11 01	04 32 10	06 ♐ 54 49	14 ♐ 22 53	07 30	12 30	08 34	29 ♏ 54	07 00	12 14	05 38	01 45
26	02 14 58	05 30 33	21 ♐ 46 17	29 ♐ 04 14	08 27	13 40	09 08	29 48	06 59	12 14	05 36	01 45
27	02 18 54	06 28 55	06 ♑ 16 10	13 ♑ 21 41	09 27	14 50	09 43	29 42	06 58	12 15	05 34	01 44
28	02 22 51	07 27 14	20 ♑ 20 35	27 ♑ 12 48	10 31	16 00	10 17	29 35	06 57	12 15	05 33	01 44
29	02 26 47	08 25 33	03 ♒ 58 27	10 ♒ 37 44	11 37	17 10	10 52	29 29	06 56	12 16	05 31	01 43
30	02 30 44	09 ♉ 23 49	17 ♒ 10 57	23 ♒ 38 31	12 ♈ 45	18 ♊ 19	11 ♋ 26	29 ♏ 22	06 ♑ 55	12 ♌ 16	05 ♏ 29	01 ♍ 43

Date	Moon True ☊ ° '	Moon Mean ☊ ° '	Moon ☽ Latitude ° '
01	13 ♎ 06	13 ♎ 16	05 N 12
02	13 R 05	13 13	04 50
03	13 05	13 09	04 14
04	13 D 05	13 06	03 25
05	13 06	13 03	02 28
06	13 06	13 00	01 25
07	13 06	12 57	00 N 18
08	13 R 06	12 53	00 S 48
09	13 05	12 50	01 52
10	13 05	12 47	02 51
11	13 04	12 44	03 42
12	13 03	12 41	04 24
13	13 01	12 38	04 54
14	13 00	12 34	05 12
15	13 00	12 31	05 17
16	12 D 59	12 28	05 06
17	13 00	12 25	04 41
18	13 01	12 22	03 59
19	13 02	12 19	03 03
20	13 03	12 15	01 54
21	13 04	12 12	00 S 36
22	13 R 04	12 09	00 N 46
23	13 03	12 06	02 07
24	13 01	12 03	03 18
25	12 58	11 59	04 15
26	12 55	11 56	04 54
27	12 52	11 53	05 13
28	12 50	11 50	05 12
29	12 50	11 47	04 54
30	12 ♎ 50	11 ♎ 44	04 N 20

DECLINATIONS

Date	Sun ☉ ° '	Moon ☽ ° '	Mercury ☿ ° '	Venus ♀ ° '	Mars ♂ ° '	Jupiter ♃ ° '	Saturn ♄ ° '	Uranus ♅ ° '	Neptune ♆ ° '	Pluto ♇ ° '
01	04 N 22	16 S 11	04 N 20	16 N 34	25 N 09	19 S 26	22 S 19	17 N 46	11 S 53	22 N 07
02	04 45	13 46	03 48	16 58	25 10	19 26	22 19	17 46	11 53	22 08
03	05 08	10 43	03 17	17 22	25 10	19 25	22 19	17 46	11 52	22 08
04	05 31	07 14	02 46	17 46	25 11	19 24	22 19	17 46	11 52	22 08
05	05 54	03 S 28	02 17	18 09	25 11	19 24	22 19	17 46	11 51	22 08
06	06 16	00 N 23	01 49	18 31	25 12	19 23	22 19	17 47	11 51	22 08
07	06 39	04 10	01 24	18 54	25 12	19 22	22 19	17 47	11 50	22 09
08	07 02	07 45	01 00	19 15	25 12	19 21	22 18	17 47	11 50	22 09
09	07 24	11 00	00 39	19 36	25 12	19 21	22 18	17 47	11 49	22 09
10	07 46	13 46	00 20	19 57	25 11	19 20	22 18	17 47	11 49	22 09
11	08 09	15 58	00 N 04	20 17	25 11	19 19	22 18	17 47	11 48	22 09
12	08 31	17 29	00 S 09	20 37	25 10	19 18	22 18	17 47	11 48	22 09
13	08 52	18 15	00 20	20 56	25 10	19 17	22 18	17 48	11 47	22 10
14	09 14	18 12	00 29	21 15	25 09	19 16	22 18	17 48	11 47	22 10
15	09 36	17 18	00 34	21 33	25 08	19 15	22 18	17 48	11 46	22 10
16	09 57	15 34	00 38	21 50	25 07	19 14	22 18	17 48	11 45	22 10
17	10 19	13 01	00 38	22 07	25 05	19 13	22 18	17 48	11 45	22 10
18	10 40	09 44	00 37	22 24	25 04	19 11	22 18	17 48	11 44	22 10
19	11 01	05 49	00 33	22 40	25 02	19 10	22 18	17 48	11 44	22 10
20	11 21	01 N 26	00 26	22 55	25 01	19 09	22 18	17 48	11 43	22 10
21	11 42	03 S 09	00 18	23 09	24 59	19 08	22 18	17 48	11 43	22 10
22	12 02	07 39	00 S 08	23 23	24 57	19 07	22 18	17 48	11 42	22 10
23	12 22	11 44	00 N 05	23 37	24 54	19 05	22 18	17 47	11 42	22 10
24	12 42	15 02	00 19	23 49	24 52	19 04	22 18	17 47	11 41	22 10
25	13 02	17 16	00 36	24 02	24 49	19 03	22 18	17 47	11 41	22 10
26	13 22	18 18	00 54	24 13	24 47	19 02	22 18	17 47	11 40	22 10
27	13 41	18 05	01 13	24 24	24 44	19 00	22 18	17 47	11 40	22 10
28	14 00	16 45	01 35	24 34	24 41	18 59	22 18	17 47	11 39	22 10
29	14 19	14 30	01 58	24 44	24 38	18 57	22 18	17 47	11 38	22 10
30	14 N 38	11 S 33	02 N 22	24 N 52	24 N 35	18 S 56	22 S 18	17 N 47	11 S 38	22 N 10

ZODIAC SIGN ENTRIES

Date	h	m	Planets
01	22	41	☽ ♒
04	06	23	☽ ♓
06	16	33	☽ ♈
09	04	32	☽ ♉
10	09	46	♂ ♋
11	17	25	☽ ♊
14	05	48	☽ ♋
14	21	08	♀ ♊
16	15	55	☽ ♌
18	22	27	☽ ♍
20	20	17	☉ ♉
21	01	19	☽ ♎
23	01	34	☽ ♏
24	14	10	♃ ♏
25	00	59	☽ ♐
27	01	32	☽ ♑
29	04	55	☽ ♒

LATITUDES

Date	Mercury ☿ ° '	Venus ♀ ° '	Mars ♂ ° '	Jupiter ♃ ° '	Saturn ♄ ° '	Uranus ♅ ° '	Neptune ♆ ° '	Pluto ♇ ° '
01	02 N 23	00 N 33	01 N 48	01 N 05	00 N 57	00 N 42	01 N 49	12 N 15
04	01 38	00 42	01 47	01 06	00 57	00 42	01 49	12 14
07	00 50	00 52	01 46	01 06	00 57	00 42	01 49	12 14
10	00 N 02	01 01	01 45	01 06	00 57	00 42	01 49	12 13
13	00 S 41	01 10	01 44	01 06	00 57	00 42	01 49	12 13
16	01 19	01 19	01 43	01 06	00 57	00 41	01 49	12 12
19	01 51	01 28	01 42	01 06	00 57	00 41	01 49	12 11
22	02 16	01 37	01 41	01 07	00 57	00 41	01 49	12 11
25	02 35	01 45	01 40	01 07	00 57	00 41	01 50	12 10
28	02 48	01 52	01 39	01 07	00 57	00 41	01 50	12 09
31	02 S 55	01 N 59	01 N 38	01 N 07	00 N 57	00 N 41	01 N 50	12 N 08

DATA

Julian Date	2436660
Delta T	+33 seconds
Ayanamsa	23° 17' 20"
Synetic vernal point	05° ♓ 49' 40"
True obliquity of ecliptic	23° 26' 32"

MOON'S PHASES, APSIDES AND POSITIONS ☽

Date	h	m	Phase	Longitude ° '	Eclipse Indicator
08	03	29	●	17 ♈ 34	Annular
16	07	33	☽	25 ♋ 34	
23	05	13	○	02 ♏ 19	
29	20	38	☾	08 ♒ 47	

Day	h	m		
10	23	05	Apogee	
23	18	28	Perigee	
06	09	39	0N	
13	22	37	Max dec	18° N 20'
20	19	35	0S	
26	19	42	Max dec	18° S 22'

LONGITUDES

Date	Chiron ⚷ ° '	Ceres ⚳ ° '	Pallas ⚴ ° '	Juno ⚵ ° '	Vesta ⚶ ° '	Black Moon Lilith ⚸ ° '
01	25 ♒ 51	27 ♏ 58	19 ♎ 55	19 ♎ 15	28 ♋ 28	15 ♉ 00
11	26 ♒ 22	27 ♏ 08	16 ♎ 48	16 ♎ 54	00 ♌ 00	16 ♉ 07
21	26 ♒ 48	25 ♏ 41	13 ♎ 50	14 ♎ 39	02 ♌ 06	17 ♉ 14
31	27 ♒ 10	23 ♏ 46	11 ♎ 20	12 ♎ 43	04 ♌ 40	18 ♉ 21

All ephemeris data is given at 12.00 UT and the Moon's longitude is additionally given for 24.00 UT

ASPECTARIAN

h m Aspects

01 Wednesday
01 40 ☽ Q ♆
08 04 ☽ # ♀
10 53 ☽ Q ☿
11 22 ☉ ∥ ☿
13 51 ☽ ⊼ ♂
15 38 ☽ ± ♇
21 37 ☽ Q ☉

02 Thursday
01 09 ☽ ± ♂
01 44 ☽ ⚹ ♃
02 28 ☽ ⊼ ♇
07 20 ☽ ⚹ ☿
10 02 ☽ □ ♆
11 15 ☽ ⊻ ♄
18 20 ☽ bQ ♂
20 42 ☉ △ ♅
21 18 ☽ ☍ ♅
21 21 ☽ ⚹ ☉
22 19 ☽ ⊥ ♄
23 40 ☽ Q ♃

03 Friday
03 24 ☽ ∥ ♆
04 06 ☽ □ ♀
09 32 ☽ ∠ ☿
15 08 ☽ ∠ ♄
23 36 ☽ △ ♂

04 Saturday
01 35 ☽ ⊥ ☿
03 47 ☽ ∠ ☉
09 22 ☽ □ ♃
10 16 ☽ ☍ ♇
12 25 ☽ ⊻ ☿
12 38 ♀ ∥ ♅
17 38 ☽ Q ♀
18 10 ☽ △ ♆
19 24 ☿ ∠ ♀
19 39 ☽ ⚹ ♄
22 06 ☽ # ☉
22 27 ☽ ⊥ ☉

05 Sunday
06 00 ☽ ⊼ ♅
11 02 ☽ ⊻ ☉
14 20 ♂ ∠ ♅
17 38 ☽ ± ♅
18 54 ☽ Q ♄
19 19 ☽ ⚹ ♀
20 30 ☽ # ☿
23 09 ☽ bQ ♆

06 Monday
05 12 ☉ Q ♂
11 14 ☽ bQ ♅
12 16 ☽ □ ♂
13 40 ☿ ⊼ ♇
16 47 ☽ ± ♆
19 23 ☽ △ ♃
20 06 ☽ ∥ ☿
20 14 ☽ ☌ ☿
20 29 ☽ ⊼ ♇
23 53 ☉ bQ ♃

07 Tuesday
04 03 ☽ ∠ ♀
04 39 ☽ ⊼ ♆
06 23 ☽ □ ♄
08 22 ☽ ± ♇
13 25 ☉ bQ ♇
16 57 ☽ △ ♅
23 58 ☽ ⊥ ♀

08 Wednesday
01 05 ☽ bQ ♃
01 28 ☿ △ ♃
01 45 ☽ Q ♂
02 19 ☽ bQ ♇
03 29 ☽ ☌ ☉
03 58 ♀ bQ ♄
06 23 ☽ ∥ ☉
13 21 ☽ ⊻ ♀
18 35 ♀ # ♃
19 05 ☽ ± ♃

09 Thursday
03 05 ☽ ⚹ ♂
06 32 ☽ ⊻ ☿
07 08 ☽ ⊼ ♃
07 58 ♀ ⊥ ♂
08 29 ☽ △ ♇
16 46 ☽ ☍ ♆
18 27 ☽ ⊥ ☿
18 41 ☽ # ♆
18 45 ☽ △ ♄

10 Friday
05 23 ☽ □ ♅
10 59 ☽ ∠ ♂
12 27 ☽ ∠ ☿
21 38 ☽ ⊻ ☉

11 Saturday
01 14 ☽ bQ ♄
06 12 ☽ ⊥ ♂
08 57 ☽ ☌ ♀
10 56 ☽ ⊥ ☉
13 41 ☿ □ ♂
17 56 ☽ Q ♅
18 43 ☽ ⚹ ☿
18 59 ☽ ⊻ ♂
19 32 ☽ ± ♄
19 41 ☽ ☍ ♃
21 18 ☽ □ ♇

12 Sunday
01 53 ☿ St D
05 33 ☽ ⊼ ♆
06 51 ☽ ∠ ☉
07 43 ☽ ⊼ ♄
07 54 ♂ ⊼ ♃
16 40 ♃ ⊥ ♄
17 40 ☽ ± ♆
18 13 ☽ ⚹ ♅
19 03 ☽ ∥ ♅
19 04 ☽ Q ☿

13 Monday
09 37 ☽ Q ♇
11 44 ☽ bQ ♆
15 47 ☽ ⚹ ☉
18 26 ♂ ⚹ ♇

14 Tuesday
00 18 ☽ ∠ ♅
04 07 ☽ ⊻ ♀
07 30 ☽ □ ☿
07 38 ☽ ⊼ ♃
09 31 ☽ ⚹ ♇
10 16 ☽ ☌ ♂
13 35 ☿ △ ♃
17 17 ☽ ⊥ ♀
17 29 ☽ △ ♆
17 43 ☽ Q ☉
18 07 ☽ ⊥ ♅
19 24 ☽ ± ♃
19 45 ☽ ☍ ♄

15 Wednesday
01 44 ☽ ∥ ♅
01 53 ♀ Q ♅
05 52 ☽ ⊻ ♅
12 44 ☽ ∠ ♀
12 51 ☽ bQ ♃
13 18 ☉ ± ♃
13 53 ♀ ☍ ♃
14 51 ☽ ∠ ♇
16 53 ☉ ⊥ ♀
18 40 ♀ ± ♄
23 10 ☿ ⚹ ♀

16 Thursday
07 33 ☽ □ ☉
08 05 ☽ ⊥ ♇
10 39 ♀ □ ♇
15 31 ♄ St R
17 18 ☽ △ ♃
18 53 ☽ △ ☿
19 23 ☽ ⊻ ♇
20 17 ☽ ⚹ ♀
22 50 ☽ ⊻ ♂

17 Friday
02 52 ☽ □ ♆
05 06 ☽ ⊼ ♄
10 05 ☿ ⊼ ♇
10 27 ☽ ⊥ ♂
14 37 ☽ ☌ ♅
15 36 ♀ ∥ ♇
16 09 ☽ ± ♄
20 38 ☽ Q ♀
21 56 ☽ # ♆
23 28 ☽ bQ ☿

18 Saturday
03 28 ♀ # ♄
03 34 ☽ ∠ ♂
06 17 ☽ ∥ ☉
08 24 ☽ bQ ♄
11 32 ☽ Q ♆
16 12 ☽ ± ☿
18 56 ☽ △ ☉
23 08 ♀ ⊻ ♂
23 24 ☽ □ ♃

19 Sunday
01 39 ☽ ☌ ♇
03 10 ☽ ⊼ ☿
07 13 ☽ ⚹ ♂
07 36 ☽ □ ♀
08 33 ☽ ⚹ ♆
10 44 ☽ △ ♄
18 31 ♀ ⊼ ♆
19 35 ☽ ⊻ ♅
22 49 ☽ bQ ☉

20 Monday
04 34 ☽ Q ♂
05 44 ☽ ⊥ ♅
06 04 ☽ Q ♃
06 58 ♅ St D
10 00 ☽ ∠ ♆
14 20 ♂ △ ♆
15 03 ☽ ± ☉
17 26 ☽ # ☿
20 22 ♀ ⊼ ♄
20 44 ☽ ∠ ♅
21 37 ☽ ∥ ☿

21 Tuesday
00 54 ☽ ⊥ ♆
01 40 ☽ ⊼ ☉
01 53 ☽ ⚹ ♃
04 16 ☽ ⊻ ♇
04 29 ☉ ⊼ ♃
07 59 ☽ ☍ ☿
09 52 ♂ ⊥ ♅
10 42 ☽ ⊻ ♆
11 33 ☽ □ ♂
12 49 ☽ □ ♄
13 05 ☉ # ♆
13 06 ♂ ± ♃
14 02 ☽ ⊥ ♇
14 15 ☽ △ ♀
21 12 ☽ ⚹ ♅

22 Wednesday
02 04 ☽ ∠ ♃
04 32 ☽ ∠ ♇
16 05 ☉ △ ♇
16 21 ☽ ⊥ ♃
16 22 ☽ bQ ♀
16 24 ☽ Q ♅
17 41 ☽ Q ♄
19 56 ♂ ☍ ♄

23 Thursday
01 48 ☽ ⊻ ♃
04 22 ☽ ⚹ ♇
05 13 ☽ ☍ ☉
07 49 ☽ ± ♀
10 33 ☽ ⊼ ☿
10 33 ☽ ☌ ♆
10 34 ☿ ⊼ ♆
11 47 ☽ ∥ ♆
12 40 ☽ ⚹ ♄
13 20 ☽ △ ♂
16 38 ☽ # ☉
18 05 ☽ ⊼ ♀
20 34 ☽ ± ☿
20 53 ☽ □ ♅
23 18 ☽ Q ♇

24 Friday
11 38 ☽ bQ ☿
12 20 ☽ ∠ ♄
13 56 ☽ bQ ♂
18 23 ♀ ± ♆
23 21 ☿ □ ♄

25 Saturday
00 54 ☽ ☌ ♃
02 35 ☽ ⊥ ♄
03 46 ☽ □ ♇
04 47 ☽ ± ♂
06 27 ♀ ⚹ ♅
07 56 ☽ ⊼ ☉
09 57 ☽ ⊻ ♆
12 08 ☽ ⊻ ♄
12 59 ☽ △ ☿
14 45 ☽ ⊼ ♂
18 13 ☽ ± ☉
18 36 ☿ ± ♇
19 33 ☽ ⊥ ♆
20 32 ☽ △ ♅
20 40 ☽ # ♅
21 44 ☽ ☍ ♀

26 Sunday
09 48 ☽ bQ ☉
10 05 ☽ ∠ ♆
14 09 ☉ ☍ ♆
20 59 ☽ bQ ♅

27 Monday
01 07 ☽ ⊻ ♃
04 26 ☽ △ ♇
10 50 ☽ ⚹ ♆
11 03 ☽ ⊥ ♃
11 58 ☽ ± ♅
12 23 ☽ △ ☉
13 10 ☽ ☌ ♄
17 48 ☽ □ ☿
18 03 ☽ ☍ ♂
19 25 ☽ # ♅
22 06 ☽ ⊼ ♅
23 49 ☉ △ ♄

28 Tuesday
00 55 ☿ □ ♂
02 10 ☽ ∠ ♃
03 50 ☽ ⊼ ♀
05 46 ☽ bQ ♇
07 11 ☽ Q ♆
15 09 ☽ ± ♀
21 23 ☽ ± ♇

29 Wednesday
01 03 ♀ ∥ ♂
03 32 ☽ Q ☿
04 04 ☽ ⚹ ♃
07 59 ☽ ⊼ ♇
08 28 ☽ bQ ♀
13 29 ☽ # ☉
14 46 ☽ □ ♆
17 18 ☽ ⊻ ♄
20 38 ☽ □ ☉

30 Thursday
00 59 ☽ ⊼ ♂
01 26 ☽ Q ♃
01 50 ☿ △ ♅
02 59 ☽ ☍ ♅
03 05 ☽ ⚹ ☿
04 10 ☽ ⊥ ♄
11 23 ☽ ∥ ♆
12 30 ☽ ± ♂
14 19 ☽ △ ♀
20 45 ☽ ∠ ♄

MAY 1959

LONGITUDES

Date	Sidereal time h m s	Sun ☉ ° ' "	Moon ☽ ° ' "	Moon ☽ 24.00 ° ' "	Mercury ☿ ° '	Venus ♀ ° '	Mars ♂ ° '	Jupiter ♃ ° '	Saturn ♄ ° '	Uranus ♅ ° '	Neptune ♆ ° '	Pluto ♇ ° '
01	02 34 41	10 ♉ 22 05	00 ♓ 00 52	06 ♓ 18 27	13 ♈ 56	19 ♊ 29	12 ♋ 01	29 ♏ 15	06 ♑ 53	12 ♌ 17	05 ♏ 28	01 ♍ 42
02	02 38 37	11 20 18	12 ♓ 31 46	18 ♓ 41 18	15 10	20 38	12 36	29 R 08	06 R 52	12 17	05 R 26	01 R 42
03	02 42 34	12 18 31	24 ♓ 47 33	00 ♈ 50 58	16 26	21 48	13 10	29 01	06 50	12 18	05 24	01 42
04	02 46 30	13 16 41	06 ♈ 51 59	12 ♈ 51 02	17 44	22 57	13 45	28 54	06 49	12 19	05 23	01 42
05	02 50 27	14 14 50	18 ♈ 48 29	24 ♈ 44 42	19 05	24 06	14 20	28 47	06 47	12 20	05 21	01 41
06	02 54 23	15 12 58	00 ♉ 40 01	06 ♉ 34 42	20 27	25 15	14 55	28 40	06 45	12 20	05 20	01 41
07	02 58 20	16 11 04	12 ♉ 29 04	18 ♉ 23 22	21 52	26 24	15 30	28 33	06 43	12 21	05 18	01 41
08	03 02 16	17 09 08	24 ♉ 17 50	00 ♊ 12 42	23 19	27 32	16 05	28 25	06 41	12 22	05 16	01 41
09	03 06 13	18 07 11	06 ♊ 08 14	12 ♊ 04 38	24 49	28 41	16 40	28 18	06 39	12 23	05 15	01 40
10	03 10 10	19 05 12	18 ♊ 02 11	24 ♊ 01 07	26 20	29 ♊ 49	17 15	28 11	06 37	12 24	05 13	01 40
11	03 14 06	20 03 11	00 ♋ 01 44	06 ♋ 04 20	27 53	00 ♋ 58	17 50	28 03	06 34	12 25	05 12	01 40
12	03 18 03	21 01 09	12 ♋ 09 15	18 ♋ 16 50	29 ♈ 29	02 06	18 25	27 56	06 32	12 27	05 10	01 40
13	03 21 59	21 59 05	24 ♋ 27 28	00 ♌ 41 34	01 ♉ 06	03 14	19 00	27 48	06 30	12 28	05 09	01 40
14	03 25 56	22 56 59	06 ♌ 59 31	13 ♌ 21 48	02 46	04 22	19 35	27 41	06 27	12 29	05 07	01 D 40
15	03 29 52	23 54 51	19 ♌ 48 49	26 ♌ 21 01	04 27	05 29	20 11	27 33	06 25	12 30	05 06	01 40
16	03 33 49	24 52 42	02 ♍ 58 48	09 ♍ 42 30	06 11	06 37	20 46	27 25	06 22	12 32	05 04	01 40
17	03 37 45	25 50 31	16 ♍ 32 26	23 ♍ 28 48	07 56	07 44	21 21	27 18	06 19	12 33	05 02	01 40
18	03 41 42	26 48 18	00 ♎ 31 41	07 ♎ 41 02	09 44	08 51	21 57	27 10	06 16	12 35	05 01	01 41
19	03 45 39	27 46 03	14 ♎ 56 38	22 ♎ 18 06	11 34	09 58	22 32	27 02	06 13	12 36	04 59	01 41
20	03 49 35	28 43 47	29 ♎ 44 50	07 ♏ 16 01	13 26	11 05	23 07	26 55	06 10	12 38	04 58	01 41
21	03 53 32	29 ♉ 41 29	14 ♏ 50 43	22 ♏ 27 45	15 20	12 12	23 43	26 47	06 07	12 39	04 57	01 41
22	03 57 28	00 ♊ 39 10	00 ♐ 05 53	07 ♐ 43 45	17 16	13 19	24 18	26 40	06 04	12 41	04 55	01 42
23	04 01 25	01 36 50	15 ♐ 20 00	22 ♐ 53 19	19 13	14 25	24 54	26 32	06 01	12 43	04 54	01 42
24	04 05 21	02 34 28	00 ♑ 22 30	07 ♑ 46 29	21 13	15 31	25 29	26 24	05 58	12 44	04 52	01 42
25	04 09 18	03 32 06	15 ♑ 04 24	22 ♑ 15 35	23 15	16 37	26 05	26 17	05 55	12 46	04 51	01 42
26	04 13 14	04 29 42	29 ♑ 19 36	06 ♒ 16 14	25 18	17 43	26 40	26 09	05 51	12 48	04 50	01 43
27	04 17 11	05 27 17	13 ♒ 05 26	19 ♒ 47 22	27 24	18 48	27 16	26 02	05 48	12 50	04 48	01 43
28	04 21 08	06 24 52	26 ♒ 22 18	02 ♓ 50 40	29 ♉ 30	19 54	27 51	25 54	05 44	12 52	04 47	01 43
29	04 25 04	07 22 25	09 ♓ 12 57	15 ♓ 29 42	01 ♊ 38	20 59	28 27	25 47	05 41	12 54	04 45	01 44
30	04 29 01	08 19 58	21 ♓ 41 33	27 ♓ 49 05	03 48	22 04	29 03	25 40	05 37	12 56	04 44	01 44
31	04 32 57	09 ♊ 17 30	03 ♈ 52 58	09 ♈ 53 47	05 ♊ 58	23 ♋ 08	29 ♋ 39	25 ♏ 32	05 ♑ 33	12 ♌ 58	04 ♏ 43	01 ♍ 45

Date	Moon True ☊ ° '	Moon Mean ☊ ° '	Moon ☽ Latitude ° '
01	12 ♎ 51	11 ♎ 40	03 N 34
02	12 D 52	11 37	02 39
03	12 54	11 34	01 38
04	12 55	11 31	00 N 33
05	12 R 55	11 28	00 S 32
06	12 54	11 24	01 36
07	12 50	11 21	02 35
08	12 46	11 18	03 27
09	12 40	11 15	04 10
10	12 33	11 12	04 43
11	12 27	11 09	05 03
12	12 21	11 05	05 10
13	12 16	11 02	05 03
14	12 13	10 59	04 42
15	12 12	10 56	04 06
16	12 D 12	10 53	03 17
17	12 13	10 50	02 15
18	12 14	10 46	01 S 03
19	12 R 14	10 43	00 N 15
20	12 13	10 40	01 33
21	12 10	10 37	02 47
22	12 05	10 34	03 49
23	11 58	10 30	04 35
24	11 51	10 27	05 01
25	11 43	10 24	05 07
26	11 37	10 21	04 53
27	11 33	10 18	04 22
28	11 31	10 15	03 38
29	11 D 30	10 11	02 45
30	11 31	10 08	01 45
31	11 ♎ 32	10 ♎ 05	00 N 41

DECLINATIONS

Date	Sun ☉ ° '	Moon ☽ ° '	Mercury ☿ ° '	Venus ♀ ° '	Mars ♂ ° '	Jupiter ♃ ° '	Saturn ♄ ° '	Uranus ♅ ° '	Neptune ♆ ° '	Pluto ♇ ° '
01	14 N 56	08 S 07	02 N 48	25 N 01	24 N 31	18 S 54	22 S 18	17 N 46	11 S 37	22 N 10
02	15 14	04 03	03 16	25 08	24 28	18 53	22 19	17 46	11 37	22 09
03	15 32	00 S 34	03 44	25 15	24 24	18 52	22 19	17 46	11 36	22 09
04	15 50	03 N 14	04 14	25 21	24 20	18 50	22 19	17 46	11 36	22 09
05	16 07	06 52	04 45	25 27	24 16	18 49	22 19	17 45	11 35	22 09
06	16 24	10 13	05 18	25 31	24 12	18 47	22 19	17 45	11 35	22 09
07	16 41	13 08	05 51	25 36	24 07	18 45	22 19	17 45	11 34	22 09
08	16 57	15 30	06 26	25 39	24 03	18 44	22 19	17 45	11 34	22 08
09	17 14	17 14	07 01	25 42	23 58	18 42	22 19	17 44	11 33	22 08
10	17 30	18 13	07 37	25 44	23 53	18 41	22 19	17 44	11 33	22 08
11	17 45	18 23	08 15	25 45	23 48	18 39	22 19	17 44	11 32	22 08
12	18 01	17 44	08 53	25 46	23 43	18 37	22 20	17 43	11 32	22 08
13	18 16	16 15	09 31	25 46	23 38	18 36	22 20	17 43	11 31	22 07
14	18 31	13 58	10 11	25 45	23 33	18 34	22 20	17 42	11 31	22 07
15	18 45	10 58	10 51	25 44	23 27	18 33	22 20	17 42	11 30	22 07
16	18 59	07 21	11 32	25 42	23 21	18 31	22 20	17 42	11 30	22 06
17	19 13	03 N 15	12 13	25 39	23 15	18 29	22 20	17 41	11 29	22 06
18	19 27	01 S 10	12 54	25 36	23 09	18 28	22 20	17 41	11 29	22 06
19	19 40	05 40	13 36	25 32	23 03	18 26	22 20	17 40	11 28	22 06
20	19 53	09 56	14 18	25 28	22 57	18 24	22 21	17 40	11 28	22 05
21	20 05	13 38	15 00	25 22	22 50	18 23	22 21	17 39	11 27	22 05
22	20 17	16 26	15 42	25 16	22 44	18 21	22 21	17 39	11 27	22 04
23	20 29	18 05	16 24	25 10	22 37	18 19	22 21	17 38	11 26	22 04
24	20 41	18 25	17 05	25 03	22 30	18 18	22 21	17 38	11 26	22 04
25	20 52	17 31	17 46	24 55	22 23	18 16	22 21	17 37	11 25	22 03
26	21 03	15 31	18 26	24 47	22 16	18 14	22 21	17 37	11 25	22 03
27	21 13	12 42	19 06	24 38	22 09	18 13	22 22	17 36	11 25	22 03
28	21 23	09 18	19 44	24 28	22 01	18 11	22 22	17 36	11 24	22 02
29	21 33	05 34	20 21	24 18	21 54	18 10	22 22	17 35	11 24	22 02
30	21 42	01 S 42	20 57	24 07	21 46	18 08	22 22	17 34	11 23	22 01
31	21 N 51	02 N 10	21 N 31	23 N 56	21 N 38	18 S 06	22 S 23	17 N 34	11 S 23	22 N 01

ZODIAC SIGN ENTRIES

Date	h	m	Planets
01	11	58	☽ ♓
03	22	19	☽ ♈
06	10	39	☽ ♉
08	23	34	☽ ♊
10	15	45	♀ ♋
11	11	57	☽ ♋
12	19	48	☿ ♉
13	22	40	☽ ♌
16	06	38	☽ ♍
18	11	06	☽ ♎
20	12	24	☽ ♏
21	19	42	☉ ♊
22	11	51	☽ ♐
24	11	24	☽ ♑
26	13	09	☽ ♒
28	17	35	☿ ♊
28	18	42	☽ ♓
31	04	18	☽ ♈

LATITUDES

Date	Mercury ☿ ° '	Venus ♀ ° '	Mars ♂ ° '	Jupiter ♃ ° '	Saturn ♄ ° '	Uranus ♅ ° '	Neptune ♆ ° '	Pluto ♇ ° '
01	02 S 55	01 N 59	01 N 38	01 N 07	00 N 57	00 N 41	01 N 50	12 N 08
04	02 57	02 06	01 37	01 07	00 57	00 41	01 49	12 07
07	02 53	02 12	01 35	01 06	00 57	00 41	01 49	12 06
10	02 43	02 17	01 34	01 06	00 57	00 41	01 49	12 06
13	02 29	02 22	01 33	01 06	00 57	00 41	01 49	12 05
16	02 11	02 26	01 32	01 06	00 57	00 40	01 49	12 04
19	01 47	02 29	01 31	01 05	00 57	00 40	01 49	12 03
22	01 21	02 31	01 30	01 05	00 57	00 40	01 49	12 02
25	00 51	02 32	01 29	01 05	00 57	00 40	01 49	12 01
28	00 S 19	02 32	01 29	01 05	00 57	00 40	01 49	12 00
31	00 N 13	02 N 31	01 N 26	01 N 04	00 N 57	00 N 40	01 N 49	11 N 59

DATA

Julian Date	2436690
Delta T	+33 seconds
Ayanamsa	23° 17' 23"
Synetic vernal point	05° ♓ 49' 37"
True obliquity of ecliptic	23° 26' 32"

MOON'S PHASES, APSIDES AND POSITIONS ☽

Date	h	m	Phase	Longitude ° '	Eclipse Indicator
07	20	11	●	16 ♉ 31	
15	20	09	☽	24 ♌ 14	
22	12	56	○	00 ♐ 41	
29	08	14	☾	07 ♓ 13	

Day	h	m		
08	03	55	Apogee	
22	04	57	Perigee	
03	15	34	0N	
11	05	13	Max dec	18° N 25'
18	05	43	0S	
24	06	20	Max dec	18° S 28'
30	22	28	0N	

LONGITUDES

Date	Chiron ⚷ ° '	Ceres ⚳ ° '	Pallas ⚴ ° '	Juno ⚵ ° '	Vesta ⚶ ° '	Black Moon Lilith ⚸ ° '
01	27 ♒ 10	23 ♏ 46	11 ♎ 20	12 ♎ 43	04 ♌ 40	18 ♉ 21
11	27 ♒ 26	21 ♏ 34	09 ♎ 31	11 ♎ 13	07 ♌ 37	19 ♉ 27
21	27 ♒ 36	19 ♏ 21	08 ♎ 31	10 ♎ 17	10 ♌ 55	20 ♉ 34
31	27 ♒ 39	17 ♏ 23	08 ♎ 18	09 ♎ 54	14 ♌ 31	21 ♉ 41

ASPECTARIAN

h m Aspects

01 Friday
06 04 ☽ ⚼ ♂
08 38 ☽ Q ☉
09 45 ☽ ∠ ☿
10 34 ☽ □ ♃
15 13 ☽ ☍ ♇
16 39 ♀ Q ♇
17 41 ☿ ⚼ ♃
22 21 ☽ △ ♆
23 06 ♂ ⚺ ♅

02 Saturday
01 05 ☽ ✶ ♄
04 47 ☽ ⊥ ☿
07 51 ♀ ⚼ ♆
09 30 ☽ ✶ ☉
11 32 ☽ ⚻ ♅
12 08 ☽ △ ♂
17 42 ☽ ⚺ ☿
18 24 ☽ # ☿
23 14 ☽ ± ♅

03 Sunday
00 19 ☽ Q ♄
03 23 ☽ ⚼ ♆
05 29 ☽ □ ♀
11 49 ☉ □ ♅
16 55 ☿ ⚼ ♇
16 58 ☽ ⚼ ♅
17 24 ☽ ∠ ☉
20 18 ☽ △ ♃
21 07 ☽ ± ♆

04 Monday
01 41 ☽ ⚻ ♇
09 02 ☽ ⚻ ♆
11 53 ☽ □ ♄
12 54 ☽ ⊥ ☉
13 39 ☽ ± ♇
17 14 ♂ ⚼ ♃
19 31 ☽ II ☿
21 03 ☽ Q ♀
22 56 ☽ △ ♅

05 Tuesday
01 59 ☽ ⚼ ♃
02 00 ☽ ⚺ ☉
02 31 ☽ □ ♂
07 43 ☽ ⚼ ♇
12 37 ☽ ☌ ☿
17 23 ☉ ✶ ♂
19 58 ☽ ± ♃
23 50 ☽ ✶ ♀

06 Wednesday
07 59 ☽ ⚻ ♃
14 04 ☽ △ ♇
16 48 ☽ Q ♂
21 26 ☽ ☍ ♆
22 43 ☽ # ♆

07 Thursday
00 19 ☽ △ ♄
09 33 ☽ ∠ ♀
11 44 ☽ □ ♅
18 26 ☽ ✶ ♂
20 11 ☽ ☌ ☉
22 21 ☿ ± ♃

08 Friday
05 48 ☽ ⊥ ♀
06 42 ☽ ⚼ ♄
08 25 ♀ ∠ ♅
09 45 ☽ ⚺ ☿
19 17 ☽ ⚺ ♀
20 17 ☽ ☍ ♃
23 39 ☽ ⊥ ☿

09 Saturday
00 20 ☽ Q ♅
00 55 ☽ ± ♄
02 28 ☽ ∠ ♂
02 58 ☽ □ ♇
04 47 ♀ ⚻ ♃
10 12 ☽ ⚻ ♆
12 04 ☽ II ☉
12 27 ♂ ∠ ♇
13 02 ☽ ⚻ ♄
20 30 ☽ ∠ ☿
21 37 ☽ ⊥ ♂
22 08 ☽ II ♅
22 18 ☽ ± ♆

10 Sunday
00 39 ☽ ✶ ♅
10 20 ☽ ⚺ ♂
14 18 ☽ ⚺ ☉
15 17 ☽ Q ♇
16 23 ☽ ⚼ ♆

11 Monday
03 22 ☽ ⊥ ☉
06 48 ☽ ∠ ♅
07 05 ☽ ✶ ☿
08 06 ☽ ⚻ ♃
09 16 ☉ II ♅
14 03 ☽ ☌ ♀
14 22 ☿ ⚻ ♃
15 16 ☽ ✶ ♇
19 55 ☽ ± ♃
22 14 ☽ △ ♆
22 51 ☽ ∠ ☉

12 Tuesday
00 43 ☽ ⊥ ♅
00 57 ☽ ☍ ♄
03 01 ♀ ✶ ♇
06 33 ☽ II ☉
10 28 ☽ Q ☿
12 17 ☽ II ♅
12 34 ☽ ⚺ ♅
13 30 ☽ ⚼ ♃
20 51 ☽ ∠ ♇

13 Wednesday
00 18 ☉ ⚼ ♄
00 53 ☽ ☌ ♂
06 48 ☽ ✶ ☉
14 21 ☽ ⊥ ♇
18 23 ☽ △ ♃
20 18 ☿ △ ♇
21 50 ♇ St D
22 56 ♀ ± ♃

14 Thursday
00 01 ♄ ± ♅
01 52 ☽ ⚺ ♇
02 43 ☽ □ ☿
06 31 ☽ ⚺ ♀
07 48 ☽ Q ☉
08 27 ☽ □ ♆
10 59 ☽ ⚻ ♄
17 11 ☉ # ♃
18 59 ☽ ⊥ ♀
22 16 ☽ ± ♄
22 22 ☽ ☌ ♅

15 Friday
03 44 ♀ △ ♆
08 08 ☽ # ♆
12 42 ☽ ⚺ ♂
12 45 ☽ II ☿
13 22 ☽ ∠ ♀
14 56 ☽ ⚼ ♄
18 02 ☽ Q ♆
20 09 ☽ □ ☉
20 49 ☿ ☍ ♆

16 Saturday
00 14 ☽ ⊥ ♂
02 03 ☽ □ ♃
06 53 ♀ ☍ ♄
09 39 ☽ ☌ ♇
10 08 ♀ ⊥ ♅
10 50 ☿ # ♆
14 29 ☿ △ ♄
15 44 ☽ ✶ ♆
17 13 ☽ ∠ ♂
18 02 ☽ △ ♄
18 35 ☽ △ ☿
19 06 ☽ ✶ ♀

17 Sunday
04 31 ☿ ✶ ♀
05 00 ☽ ⚺ ♅
09 51 ☽ Q ♃
15 30 ☽ ⊥ ♅
18 02 ☽ Q ♀
18 04 ☽ ∠ ♆
20 43 ☽ ✶ ♂

18 Monday
00 42 ☽ ⚼ ☿
05 13 ☽ △ ☉
06 21 ☽ ✶ ♃
06 59 ☽ ∠ ♅
09 27 ☽ ⊥ ♆
13 56 ☽ ⚺ ♇
16 02 ☿ Q ♂
18 00 ☽ Q ♂
18 11 ☽ ± ☿
19 32 ☽ ⚺ ♆
20 00 ☉ ☍ ♃
21 37 ☽ □ ♄
23 59 ☽ ⊥ ♇

19 Tuesday
03 07 ☽ □ ♀
05 38 ☽ ⚻ ☿
07 16 ☽ ∠ ♃
08 08 ☽ ✶ ♅
08 10 ☽ ⚼ ☉
14 51 ☽ ∠ ♇
21 52 ☽ ⊥ ♃
23 54 ☽ ± ☉

20 Wednesday
00 53 ☽ □ ♂
01 35 ☿ □ ♅
03 04 ☽ Q ♄
03 45 ☽ Q ♅
07 29 ☽ ⚺ ♃
10 15 ☽ ⚻ ☉
15 06 ☽ ✶ ♇
20 20 ☽ ☌ ♆
21 22 ☽ II ♆
22 14 ☽ ✶ ♄

21 Thursday
03 57 ♀ ⚼ ♃
07 29 ☽ △ ♀
08 32 ☽ □ ♅
10 10 ☽ Q ♇
12 52 ☽ ☍ ☿
21 52 ☽ ∠ ♄
22 03 ♀ ⚺ ♅
22 13 ☉ ± ♄

22 Friday
02 06 ☽ # ☿
02 31 ☽ △ ♂
06 38 ☽ ☌ ♃
08 58 ☽ ⚼ ♀
11 57 ☽ ⊥ ♄
12 44 ☉ Q ♅
12 56 ☽ ☍ ☉
14 30 ☽ □ ♇
19 34 ☽ ⚺ ♆
21 21 ☽ ⚺ ♄

23 Saturday
00 14 ☽ ± ♀
03 04 ☽ ⚼ ♂
03 31 ☽ # ♅
05 00 ☽ ⊥ ♆
07 51 ☽ △ ♅
10 26 ☽ ⚻ ♀
13 59 ☉ □ ♇
17 52 ☽ ± ♂
19 04 ☽ II ♃
19 06 ☽ ⚻ ☿
19 13 ☽ ∠ ♆

24 Sunday
03 49 ☽ ⚻ ♂
05 41 ☽ ⚺ ♃
06 08 ☽ ± ☿
07 45 ☽ ⚼ ♅
09 02 ☿ ⚼ ♄
14 08 ☽ △ ♇
15 15 ☽ ⊥ ♃
15 48 ☽ ⚻ ☉
18 42 ☽ II ♃
19 16 ☽ ✶ ♆
20 43 ♂ ⊥ ♇
21 01 ☽ ☌ ♄
22 20 ☽ ± ♅
22 58 ☽ ⚼ ☿

25 Monday
02 14 ☽ ± ☉
05 48 ☽ ∠ ♃
07 00 ☿ II ♅
08 11 ☽ ⚻ ♅
09 05 ☽ # ☿
10 07 ☽ # ♅
13 55 ♀ ∠ ♇
14 42 ☽ ⚼ ♇
14 46 ☽ ☍ ♀
14 56 ☽ Q ♆
17 56 ♂ # ♄
18 10 ☽ ⚼ ☉
18 48 ♂ △ ♃

26 Tuesday
03 59 ☽ △ ☿
05 15 ☿ # ♃
05 50 ☽ ± ♇
06 38 ☽ ✶ ♃
07 16 ☽ ☍ ♂
16 06 ☽ ⚻ ♇
20 04 ☉ ⚻ ♆
21 15 ☿ ☍ ♃
21 28 ☽ □ ♆
21 35 ☽ △ ☉
23 14 ☽ ⚺ ♄

27 Wednesday
03 09 ☽ Q ♃
09 43 ☽ ⊥ ♄
09 56 ☿ ✶ ♂
11 32 ☽ ☍ ♅
20 02 ☉ ⚻ ♄
21 29 ☽ II ♆
23 08 ☽ ⚻ ♀

28 Thursday
01 46 ☽ ∠ ♄
09 12 ♂ II ♇
11 03 ☽ ± ♀
11 09 ☽ □ ♃
14 33 ☿ ± ♄
14 52 ☽ ⚻ ♂
18 55 ☽ □ ☿
21 55 ☽ ☍ ♇

29 Friday
02 34 ☽ ± ♂
03 32 ☿ Q ♅
03 36 ☽ △ ♆
05 19 ☽ ⚼ ♀
05 21 ☽ ✶ ♄
08 14 ☽ □ ☉
13 00 ☿ □ ♇
19 02 ☽ ⚻ ♅
20 29 ☽ ⚼ ♂

30 Saturday
04 08 ☽ Q ♄
06 37 ☽ ± ♅
08 12 ☽ ⚼ ♆
12 15 ☽ Q ☿
12 47 ☽ △ ♀
17 50 ☉ II ♂
19 41 ☽ △ ♃
21 51 ☽ Q ☉
22 17 ☿ ⚻ ♆

31 Sunday
00 15 ☽ ⚼ ♅
01 47 ☽ ± ♆
03 10 ☽ △ ♂
07 33 ☿ ⚻ ♄
07 45 ☽ ⚻ ♇
13 39 ☽ ⚻ ♆
15 19 ☽ □ ♄
16 19 ☿ II ♂
17 05 ☽ ✶ ☿
19 42 ☽ ± ♇
23 44 ☽ ✶ ☉

All ephemeris data is given at 12.00 UT and the Moon's longitude is additionally given for 24.00 UT

JUNE 1959

LONGITUDES

Date	Sidereal time h m s	Sun ☉ ° ' "	Moon ☽ ° ' "	Moon ☽ 24.00 ° ' "	Mercury ☿ ° '	Venus ♀ ° '	Mars ♂ ° '	Jupiter ♃ ° '	Saturn ♄ ° '	Uranus ♅ ° '	Neptune ♆ ° '	Pluto ⯓ ° '
01	04 36 54	10 ♊ 15 01	15 ♈ 52 08	21 ♈ 48 36	08 ♊ 09	24 ♋ 13	00 ♌ 14	25 ♏ 25	05 ♑ 30	13 ♌ 00	04 ♏ 42	01 ♍ 45
02	04 40 50	11 12 31	27 ♈ 43 43	03 ♉ 37 57	10 21	25 17	00 50	25 R 18	05 R 26	13 02	04 R 40	01 46
03	04 44 47	12 10 00	09 ♉ 31 47	15 ♉ 25 35	12 33	26 21	01 26	25 11	05 22	13 04	04 39	01 46
04	04 48 43	13 07 29	21 ♉ 19 43	27 ♉ 14 31	14 45	27 25	02 02	25 04	05 18	13 07	04 38	01 47
05	04 52 40	14 04 56	03 ♊ 10 15	09 ♊ 07 09	16 57	28 29	02 38	24 57	05 14	13 09	04 37	01 48
06	04 56 37	15 02 23	15 ♊ 05 26	21 ♊ 05 16	19 09	29 ♋ 32	03 14	24 50	05 10	13 11	04 36	01 48
07	05 00 33	15 59 49	27 ♊ 06 48	03 ♋ 10 13	21 19	00 ♌ 35	03 49	24 44	05 06	13 13	04 34	01 49
08	05 04 30	16 57 14	09 ♋ 15 37	15 ♋ 23 12	23 29	01 38	04 25	24 37	05 02	13 16	04 33	01 50
09	05 08 26	17 54 39	21 ♋ 33 06	27 ♋ 45 29	25 37	02 40	05 01	24 30	04 58	13 18	04 32	01 51
10	05 12 23	18 52 02	04 ♌ 00 34	10 ♌ 18 35	27 44	03 42	05 38	24 24	04 54	13 21	04 31	01 51
11	05 16 19	19 49 24	16 ♌ 39 47	23 ♌ 04 27	29 ♊ 49	04 44	06 14	24 18	04 50	13 23	04 30	01 52
12	05 20 16	20 46 45	29 ♌ 32 53	06 ♍ 05 24	01 ♋ 52	05 46	06 50	24 11	04 46	13 26	04 29	01 53
13	05 24 12	21 44 05	12 ♍ 42 21	19 ♍ 24 02	03 53	06 47	07 26	24 05	04 41	13 28	04 28	01 54
14	05 28 09	22 41 24	26 ♍ 10 45	03 ♎ 02 46	05 53	07 48	08 02	23 59	04 37	13 31	04 27	01 55
15	05 32 06	23 38 42	10 ♎ 00 16	17 ♎ 03 20	07 50	08 49	08 38	23 54	04 33	13 34	04 26	01 56
16	05 36 02	24 36 00	24 ♎ 11 57	01 ♏ 25 56	09 45	09 49	09 14	23 48	04 28	13 36	04 25	01 57
17	05 39 59	25 33 16	08 ♏ 44 58	16 ♏ 08 30	11 38	10 49	09 51	23 42	04 24	13 39	04 24	01 58
18	05 43 55	26 30 32	23 ♏ 35 51	01 ♐ 06 09	13 28	11 48	10 27	23 37	04 20	13 42	04 24	01 59
19	05 47 52	27 27 47	08 ♐ 38 19	16 ♐ 11 13	15 16	12 47	11 03	23 32	04 15	13 45	04 23	02 00
20	05 51 48	28 25 01	23 ♐ 43 35	01 ♑ 14 10	17 02	13 46	11 39	23 26	04 11	13 48	04 22	02 01
21	05 55 45	29 ♊ 22 15	08 ♑ 41 43	16 ♑ 05 07	18 46	14 44	12 16	23 21	04 07	13 51	04 21	02 02
22	05 59 41	00 ♋ 19 29	23 ♑ 23 20	00 ♒ 35 36	20 27	15 42	12 52	23 16	04 02	13 53	04 20	02 03
23	06 03 38	01 16 42	07 ♒ 41 09	14 ♒ 39 43	22 06	16 40	13 28	23 12	03 58	13 56	04 20	02 04
24	06 07 35	02 13 55	21 ♒ 31 02	28 ♒ 15 06	23 42	17 37	14 05	23 07	03 53	13 59	04 19	02 05
25	06 11 31	03 11 08	04 ♓ 52 04	11 ♓ 22 14	25 16	18 33	14 41	23 03	03 49	14 02	04 18	02 06
26	06 15 28	04 08 21	17 ♓ 46 02	24 ♓ 04 00	26 48	19 30	15 18	22 58	03 45	14 05	04 18	02 08
27	06 19 24	05 05 33	00 ♈ 16 42	06 ♈ 24 46	28 17	20 25	15 54	22 54	03 40	14 08	04 17	02 09
28	06 23 21	06 02 46	12 ♈ 28 54	18 ♈ 29 45	29 ♋ 44	21 20	16 31	22 50	03 36	14 12	04 16	02 10
29	06 27 17	06 59 59	24 ♈ 27 59	00 ♉ 24 16	01 ♌ 08	22 15	17 07	22 47	03 31	14 15	04 16	02 11
30	06 31 14	07 ♋ 57 12	06 ♉ 19 15	12 ♉ 13 30	02 ♌ 30	23 ♌ 09	17 ♌ 44	22 ♏ 43	03 ♑ 27	14 ♌ 18	04 ♏ 15	02 ♍ 13

MOON NODES, MOON LATITUDE AND DECLINATIONS

Date	Moon True ☊ ° '	Moon Mean ☊ ° '	Moon ☽ Latitude ° '	Sun ☉ ° '	Moon ☽ ° '	Mercury ☿ ° '	Venus ♀ ° '	Mars ♂ ° '	Jupiter ♃ ° '	Saturn ♄ ° '	Uranus ♅ ° '	Neptune ♆ ° '	Pluto ⯓ ° '
01	11 ♎ 32	10 ♎ 02	00 S 23	21 N 59	05 N 53	22 N 03	23 N 44	21 N 30	18 S 05	22 S 23	17 N 33	11 S 23	22 N 00
02	11 R 30	09 59	01 26	22 07	09 20	22 33	23 32	21 22	18 03	22 23	17 33	11 22	22 00
03	11 27	09 56	02 24	22 15	12 23	23 01	23 19	21 14	18 02	22 23	17 32	11 22	21 59
04	11 20	09 52	03 16	22 23	14 56	23 26	23 05	21 05	18 00	22 23	17 31	11 21	21 59
05	11 12	09 49	03 59	22 30	16 52	23 49	22 51	20 57	17 59	22 23	17 31	11 21	21 58
06	11 01	09 46	04 33	22 36	18 05	24 10	22 37	20 48	17 57	22 24	17 30	11 21	21 58
07	10 49	09 43	04 54	22 42	18 31	24 27	22 22	20 39	17 56	22 24	17 29	11 20	21 57
08	10 37	09 40	05 02	22 48	18 05	24 42	22 06	20 30	17 54	22 24	17 29	11 20	21 57
09	10 26	09 36	04 57	22 54	16 50	24 54	21 51	20 21	17 53	22 24	17 28	11 20	21 56
10	10 17	09 33	04 37	22 59	14 45	25 04	21 34	20 12	17 52	22 25	17 27	11 19	21 56
11	10 10	09 30	04 04	23 03	11 58	25 10	21 17	20 03	17 50	22 25	17 26	11 19	21 55
12	10 06	09 27	03 18	23 07	08 33	25 14	21 00	19 53	17 49	22 25	17 26	11 19	21 55
13	10 05	09 24	02 20	23 11	04 38	25 15	20 43	19 43	17 48	22 25	17 25	11 19	21 54
14	10 D 04	09 21	01 S 13	23 14	00 N 24	25 14	20 25	19 34	17 46	22 25	17 24	11 18	21 54
15	10 05	09 17	00 00	23 17	03 S 58	25 10	20 06	19 24	17 45	22 26	17 23	11 18	21 53
16	10 R 04	09 14	01 N 14	23 20	08 14	25 04	19 48	19 14	17 44	22 26	17 22	11 18	21 52
17	10 02	09 11	02 26	23 22	12 07	24 55	19 28	19 04	17 43	22 26	17 22	11 17	21 52
18	09 57	09 08	03 29	23 24	15 18	24 45	19 09	18 54	17 42	22 26	17 21	11 17	21 51
19	09 49	09 05	04 18	23 25	17 30	24 33	18 49	18 43	17 41	22 26	17 20	11 17	21 51
20	09 40	09 02	04 50	23 26	18 28	24 19	18 29	18 33	17 39	22 27	17 19	11 17	21 50
21	09 29	08 58	05 01	23 26	18 09	24 03	18 09	18 22	17 38	22 27	17 18	11 17	21 49
22	09 18	08 55	04 52	23 26	16 37	23 45	17 48	18 12	17 37	22 27	17 17	11 16	21 49
23	09 09	08 52	04 25	23 26	14 06	23 26	17 27	18 01	17 36	22 27	17 17	11 16	21 48
24	09 02	08 49	03 43	23 25	10 49	23 06	17 06	17 50	17 36	22 27	17 16	11 16	21 47
25	08 57	08 46	02 50	23 24	07 05	22 45	16 44	17 39	17 35	22 28	17 15	11 16	21 47
26	08 55	08 42	01 50	23 23	03 S 09	22 22	16 23	17 28	17 34	22 28	17 14	11 16	21 46
27	08 D 54	08 39	00 N 46	23 21	00 N 48	21 59	16 01	17 16	17 33	22 28	17 13	11 15	21 45
28	08 R 54	08 36	00 S 19	23 18	04 38	21 34	15 38	17 05	17 32	22 28	17 12	11 15	21 45
29	08 54	08 33	01 21	23 15	08 13	21 09	15 16	16 54	17 31	22 28	17 11	11 15	21 44
30	08 ♎ 52	08 ♎ 30	02 S 20	23 N 12	11 N 26	20 N 43	14 N 54	16 N 42	17 S 31	22 S 29	17 N 10	11 S 15	21 N 43

ZODIAC SIGN ENTRIES

Date	h	m	Planets
01	02	26	♂ ♌
02	16	37	☽ ♉
05	05	35	☽ ♊
06	22	43	♀ ♌
07	17	44	☽ ♋
10	04	19	☽ ♌
11	14	11	☿ ♋
12	12	50	☽ ♍
14	18	42	☽ ♎
16	21	38	☽ ♏
18	22	14	☽ ♐
20	22	01	☽ ♑
22	03	50	☉ ♋
22	23	00	☽ ♒
25	03	09	☽ ♓
27	11	28	☽ ♈
28	16	31	☿ ♌
29	23	11	☽ ♉

LATITUDES

Date	Mercury ☿ ° '	Venus ♀ ° '	Mars ♂ ° '	Jupiter ♃ ° '	Saturn ♄ ° '	Uranus ♅ ° '	Neptune ♆ ° '	Pluto ⯓ ° '
01	00 N 23	02 N 30	01 N 26	01 N 04	00 N 57	00 N 40	01 N 49	11 N 59
04	00 53	02 27	01 25	01 04	00 57	00 40	01 49	11 58
07	01 18	02 24	01 24	01 03	00 57	00 40	01 49	11 58
10	01 38	02 19	01 22	01 03	00 57	00 40	01 49	11 57
13	01 52	02 12	01 21	01 02	00 56	00 40	01 48	11 56
16	01 59	02 05	01 20	01 01	00 56	00 40	01 48	11 55
19	01 59	01 56	01 19	01 01	00 56	00 40	01 48	11 54
22	01 53	01 45	01 18	01 00	00 56	00 40	01 48	11 54
25	01 41	01 33	01 16	00 59	00 56	00 39	01 48	11 53
28	01 24	01 19	01 15	00 59	00 55	00 39	01 48	11 52
31	01 N 01	01 N 04	01 N 14	00 N 58	00 N 55	00 N 39	01 N 48	11 N 52

DATA

Julian Date	2436721
Delta T	+33 seconds
Ayanamsa	23° 17' 27"
Synetic vernal point	05° ♓ 49' 33"
True obliquity of ecliptic	23° 26' 31"

MOON'S PHASES, APSIDES AND POSITIONS ☽

Date	h	m	Phase	Longitude ° '	Eclipse Indicator
06	11	53	●	15 ♊ 02	
14	05	23	☽	22 ♍ 26	
20	20	00	○	28 ♐ 44	
27	22	12	☾	05 ♈ 30	

Day	h	m		
04	07	50	Apogee	
19	13	12	Perigee	
07	12	05	Max dec	18° N 31'
14	14	13	0S	
20	18	01	Max dec	18° S 31'
27	07	04	0N	

LONGITUDES

Date	Chiron ⚷ ° '	Ceres ⚳ ° '	Pallas ⚴ ° '	Juno ⚵ ° '	Vesta ⚶ ° '	Black Moon Lilith ⚸ ° '
01	27 ♒ 39	17 ♏ 12	08 ♎ 19	09 ♎ 54	14 ♌ 53	21 ♉ 47
11	27 ♒ 37	15 ♏ 43	08 ♎ 54	10 ♎ 07	18 ♌ 44	22 ♉ 54
21	27 ♒ 28	14 ♏ 49	10 ♎ 09	10 ♎ 51	22 ♌ 49	24 ♉ 01
31	27 ♒ 13	14 ♏ 33	11 ♎ 56	12 ♎ 01	27 ♌ 02	25 ♉ 08

All ephemeris data is given at 12.00 UT and the Moon's longitude is additionally given for 24.00 UT

ASPECTARIAN

h m	Aspects
01 Monday	
01 09	☽ ♇ ♃
06 12	☽ △ ♅
08 15	☉ ∥ ☿
10 04	☿ ∥ ⯓
13 47	☽ ♇ ⯓
14 52	☉ ∥ ⯓
19 06	☽ ± ♃
22 52	☉ ± ♆
02 Tuesday	
03 42	☿ # ♄
06 05	☽ ∠ ☿
06 33	☽ □ ♀
07 07	☽ ⊼ ♃
08 39	☽ ∠ ☉
10 33	☿ ∠ ♀
12 18	♀ △ ♃
15 27	☿ ± ♆
18 39	☽ □ ♂
20 12	☽ △ ⯓
22 50	♀ ⊥ ⯓
03 Wednesday	
02 06	☽ ☍ ♆
03 34	☽ # ♆
03 34	☽ △ ♄
04 30	☉ ☌ ☿
04 33	☽ ⊥ ☉
04 33	☽ ⊥ ☿
17 43	☿ ⚹ ♅
17 51	☽ ⊻ ☉
19 14	☽ □ ♅
19 34	☽ ⊻ ☿
22 47	☽ Q ♀
22 48	☿ ∥ ♀
04 Thursday	
02 01	♂ ⊻ ⯓
09 13	☽ Q ♂
09 55	☽ ♇ ♄
11 35	☉ ⚹ ♅
14 29	☉ # ♄
19 31	☽ ☍ ♃
05 Friday	
01 34	☽ ⚹ ♀
04 05	☽ ± ♄
07 54	☽ Q ♅
09 13	☽ □ ⯓
10 51	☽ ⚹ ♂
14 54	☽ ⊼ ♆
16 09	☽ ⊼ ♄
22 07	☿ ∠ ♂
22 36	☽ ∥ ♅
06 Saturday	
02 59	☽ ± ♆
08 10	☽ ⚹ ♅
08 28	☽ # ♃
10 46	☽ ∠ ♀
11 53	☽ ☌ ☉
12 47	☉ ∥ ♀
16 54	☿ ♇ ♆
18 37	☽ ∠ ♂
19 20	☿ Q ⯓
21 00	☽ ♇ ♆
21 27	☽ Q ⯓
21 55	☽ ☌ ☿
07 Sunday	
06 29	☽ ⊥ ♀
07 18	☽ ⊼ ♃
08 51	♀ # ♄
13 29	☽ ⊥ ♂
14 13	☽ ∠ ♅
19 06	☽ ± ♃
19 32	☽ ⊻ ♀
21 20	☽ ⚹ ⯓
08 Monday	
01 59	☽ ⊻ ♂
02 45	☽ △ ♆
03 44	☽ ☍ ♄
08 04	☽ ⊥ ♅
12 41	☽ ♇ ♃
16 44	♀ ⊻ ⯓
16 48	☽ # ♃
17 06	♂ □ ♆
19 53	☽ ⊻ ♅
09 Tuesday	
00 07	☿ ⊼ ♃
01 43	☽ ∥ ♅
02 50	☽ ∠ ⯓
03 05	♀ ∥ ⯓
04 20	☽ ⊻ ☉
09 56	♂ ⊼ ♄
16 57	☽ ⊥ ☉
17 40	☽ △ ♃
20 19	☽ ⊥ ⯓
21 29	☽ ⊻ ☿
10 Wednesday	
07 52	☽ ⊻ ⯓
11 21	☽ ⊥ ☿
11 22	☽ ☌ ♀
11 30	☿ ⊥ ♀
11 42	☽ ∠ ☉
12 58	☽ □ ♆
13 41	☽ ⊼ ♄
15 14	☽ ☌ ♂
19 15	☿ ∠ ♅
11 Thursday	
01 03	☽ ± ♄
04 05	☉ ♇ ♆
05 48	☽ ☌ ♅
06 38	♀ □ ♆
07 50	☽ ∠ ☿
13 11	☉ Q ⯓
14 01	♀ ⊼ ♄
16 51	☽ # ♆
17 19	☿ ± ♃
17 54	☽ ♇ ♄
18 24	☽ ⚹ ☉
18 47	☿ ⊥ ♂
22 55	☽ Q ♆
12 Friday	
02 09	☽ □ ♃
12 12	☿ ⚹ ⯓
16 18	☽ ☌ ⯓
17 03	☽ ⚹ ☿
17 38	☉ ∠ ♀
18 25	☽ Q ☉
21 03	☽ ⚹ ♆
21 31	☽ △ ♄
13 Saturday	
00 22	☽ ⊻ ♀
01 59	☽ ⊻ ♂
10 54	☽ Q ♃
12 09	☽ ⊥ ♀
13 22	☽ ⊥ ♂
13 23	☽ ⊻ ♅
18 44	☽ Q ☿
18 53	☿ △ ♆
21 15	☿ ☍ ♄
14 Sunday	
00 06	☽ ∠ ♆
00 10	☽ ⊥ ♅
05 23	☽ □ ☉
05 33	☽ ∠ ♀
06 11	☽ ∠ ♂
08 10	☽ ⚹ ♃
15 59	☽ ⊥ ♆
16 07	☽ ∠ ♅
22 03	☽ ⊻ ⯓
15 Monday	
01 38	♀ ☌ ♂
02 25	☽ ⊻ ♆
02 39	☽ □ ♄
07 40	☽ □ ☿
08 26	☽ ⊥ ⯓
08 35	☿ ⊥ ♅
09 33	☽ ⚹ ♂
09 48	☽ ⚹ ♀
10 06	☽ ∠ ♃
11 19	☉ ∠ ♂
17 39	☉ ⊼ ♃
18 06	☽ ⚹ ♅
23 48	☽ ∠ ⯓
16 Tuesday	
00 34	☿ ♇ ♃
01 19	☽ ⊥ ♃
02 34	☿ ⊻ ♂
06 50	☽ Q ♂
07 43	☽ Q ♀
09 08	☽ Q ♄
11 20	☽ ⊻ ♃
12 43	☽ △ ☉
13 43	☿ ⊻ ♀
14 21	☽ Q ♅
17 Wednesday	
00 52	☽ ⚹ ⯓
02 47	♀ ± ♄
04 54	☽ ☌ ♆
04 55	☽ ⚹ ♄
06 36	☽ ∥ ♆
10 02	♄ ⚹ ♆
13 51	☽ □ ♂
15 09	☽ ♇ ☉
15 36	☽ □ ♀
17 22	☽ △ ☿
20 00	☽ □ ♅
20 29	☽ Q ⯓
18 Thursday	
05 11	☽ ∠ ♄
06 42	☽ ± ☉
07 52	♂ ± ♄
12 02	☽ ☌ ♃
15 05	☿ ⊻ ♅
16 59	☽ ⊼ ☉
19 32	☽ ⊥ ♄
20 52	☽ ♇ ☿
19 Friday	
01 25	☽ □ ⯓
05 04	☽ ⊻ ♄
05 14	☽ ⊻ ♆
09 43	☽ # ♅
13 09	☽ ± ☿
14 46	☽ ⊥ ♆
14 49	☽ ∥ ♃
16 00	☽ △ ♂
19 04	☽ △ ♀
20 09	☽ △ ♅
23 57	☽ ⊼ ☿
20 Saturday	
02 46	♀ ∥ ♂
05 04	☽ ∠ ♆
11 33	☽ ⊻ ♃
11 38	☿ ∠ ⯓
12 35	☽ # ♀
12 39	♀ ☌ ♅
16 52	☽ ♇ ♂
17 10	☽ # ♂
20 00	☽ ☍ ☉
20 07	☽ ♇ ♅
20 37	☽ ♇ ♀
21 04	☽ ⊥ ♃
22 00	☉ ∠ ♅
21 Sunday	
01 16	☽ △ ⯓
01 21	☽ # ♂
04 39	☽ ☌ ♄
05 00	☽ ⚹ ♆
07 54	☽ ± ♂
10 37	☽ ± ♅
11 27	☽ ∠ ♃
11 39	☉ ± ♃
12 05	☽ ± ♀
12 09	☽ # ♀
18 01	☽ ⊼ ♂
20 22	☽ ⊼ ♅
22 13	☽ ∥ ♃
22 30	☽ ⊼ ♀
22 Monday	
00 26	☽ Q ♆
01 34	☽ ♇ ⯓
03 21	☽ # ♅
06 32	☽ ☍ ☿
11 49	☽ ⚹ ♃
16 25	☽ ± ⯓
23 Tuesday	
00 23	☽ ⊼ ☉
00 42	♀ # ♃
02 28	☽ ⊼ ⯓
05 43	☽ ⊻ ♄
06 18	☽ □ ♆
07 47	☽ Q ♃
11 15	☽ ± ☉
12 17	☉ ∥ ☿
15 53	☽ ⊥ ♄
22 24	☽ ☍ ♂
22 47	☽ ☍ ♅
24 Wednesday	
00 21	♀ ∥ ♅
03 35	☿ △ ♃
03 55	☽ ♇ ☉
04 38	☽ ☍ ♀
07 24	☽ ∠ ♄
08 05	♂ ☌ ♅
08 16	☉ ⚹ ⯓
08 53	☽ ∥ ♆
14 49	☽ □ ♃
16 23	☽ ⊼ ☿
25 Thursday	
04 35	☽ ± ☿
06 58	☽ ☍ ⯓
08 41	☽ △ ☉
10 05	☽ ⚹ ♄
10 58	☽ △ ♆
18 09	♀ ♇ ♄
21 44	♂ # ♃
23 18	☽ ♇ ☿
26 Friday	
01 13	☿ ⊥ ⯓
02 46	☉ ☍ ♄
05 03	☽ ⊼ ♅
06 12	☿ # ♄
07 06	☽ ⊼ ♂
08 12	☽ Q ♄
14 53	☽ ♇ ♆
15 32	☽ ⊼ ♀
15 51	☉ △ ♆
16 25	☽ ± ♅
19 02	☽ ± ♂
21 51	☽ △ ♃
27 Saturday	
03 56	☽ ± ♀
07 37	☽ △ ☿
08 08	☽ ± ♆
09 47	☽ ♇ ♅
13 17	☽ ♇ ♂
15 39	☽ ⊼ ⯓
18 35	☽ □ ♄
19 49	☽ ⊼ ♆
19 49	♂ ∥ ♅
22 12	☽ □ ☉
22 52	☽ ♇ ♀
28 Sunday	
01 36	☿ ∥ ⯓
02 52	☽ ♇ ♃
03 27	☽ ± ⯓
15 25	☽ △ ♅
20 28	☽ △ ♂
20 38	☽ ± ♃
21 21	☽ ♇ ⯓
29 Monday	
07 10	☽ △ ♀
08 37	☽ ⊼ ♃
12 26	♀ Q ♆
13 10	☽ Q ☉
30 Tuesday	
01 10	♀ □ ♃
03 15	☽ □ ☿
03 39	☽ △ ⯓
06 12	☽ △ ♄
06 24	☉ ♇ ♃
06 44	☿ ⊻ ⯓
07 49	☽ ☍ ♆
10 34	☽ # ♆
15 36	☽ ⚹ ☉
21 09	☉ ⊥ ♅

JULY 1959

LONGITUDES

Date	Sidereal time h m s	Sun ☉ ° ' "	Moon ☽ ° ' "	Moon ☽ 24.00 ° ' "	Mercury ☿ ° '	Venus ♀ ° '	Mars ♂ ° '	Jupiter ♃ ° '	Saturn ♄ ° '	Uranus ♅ ° '	Neptune ♆ ° '	Pluto ♇ ° '
01	06 35 10	08 ♋ 54 25	18 ♉ 07 36	24 ♉ 02 04	03 ♌ 49	24 ♌ 02	18 ♌ 20	22 ♏ 40	03 ♑ 23	14 ♌ 21	04 ♏ 15	02 ♍ 14
02	06 39 07	09 51 38	29 ♉ 57 20	05 ♊ 53 50	05 06	24 55	18 57	22 R 36	03 R 18	14 24	04 R 14	02 15
03	06 43 04	10 48 51	11 ♊ 51 55	17 ♊ 51 52	06 20	25 48	19 34	22 33	03 14	14 27	04 14	02 17
04	06 47 00	11 46 05	23 ♊ 53 57	29 ♊ 58 22	07 32	26 39	20 10	22 30	03 09	14 31	04 14	02 18
05	06 50 57	12 43 18	06 ♋ 05 14	12 ♋ 14 40	08 40	27 31	20 47	22 27	03 05	14 34	04 13	02 19
06	06 54 53	13 40 32	18 ♋ 26 45	24 ♋ 41 30	09 46	28 21	21 24	22 25	03 01	14 37	04 13	02 21
07	06 58 50	14 37 45	00 ♌ 58 58	07 ♌ 19 10	10 49	29 ♌ 11	22 01	22 22	02 57	14 41	04 13	02 22
08	07 02 46	15 34 58	13 ♌ 42 07	20 ♌ 07 50	11 49	00 ♍ 00	22 38	22 20	02 52	14 44	04 12	02 24
09	07 06 43	16 32 12	26 ♌ 36 24	03 ♍ 07 52	12 45	00 48	23 14	22 18	02 48	14 47	04 12	02 25
10	07 10 39	17 29 25	09 ♍ 42 21	16 ♍ 19 57	13 39	01 36	23 51	22 16	02 44	14 51	04 12	02 27
11	07 14 36	18 26 38	23 ♍ 00 50	29 ♍ 45 10	14 29	02 22	24 28	22 15	02 40	14 54	04 12	02 28
12	07 18 33	19 23 51	06 ♎ 33 05	13 ♎ 24 46	15 16	03 08	25 05	22 13	02 35	14 58	04 11	02 30
13	07 22 29	20 21 05	20 ♎ 20 20	27 ♎ 19 52	15 59	03 53	25 42	22 12	02 31	15 01	04 11	02 31
14	07 26 26	21 18 18	04 ♏ 23 22	11 ♏ 30 45	16 39	04 37	26 19	22 11	02 27	15 04	04 11	02 33
15	07 30 22	22 15 31	18 ♏ 41 49	25 ♏ 56 17	17 14	05 20	26 56	22 10	02 23	15 08	04 11	02 35
16	07 34 19	23 12 44	03 ♐ 13 39	10 ♐ 33 21	17 46	06 03	27 33	22 09	02 19	15 11	04 11	02 36
17	07 38 15	24 09 57	17 ♐ 54 38	25 ♐ 16 41	18 13	06 44	28 10	22 08	02 15	15 15	04 D 11	02 38
18	07 42 12	25 07 11	02 ♑ 38 33	09 ♑ 59 16	18 36	07 24	28 47	22 08	02 11	15 19	04 11	02 40
19	07 46 08	26 04 25	17 ♑ 17 50	24 ♑ 33 16	18 54	08 03	29 ♌ 24	22 08	02 08	15 22	04 11	02 41
20	07 50 05	27 01 39	01 ♒ 44 42	08 ♒ 51 22	19 08	08 41	00 ♍ 01	22 D 07	02 04	15 26	04 11	02 43
21	07 54 02	27 58 54	15 ♒ 52 36	22 ♒ 47 55	19 17	09 18	00 39	22 08	02 00	15 29	04 12	02 45
22	07 57 58	28 56 09	29 ♒ 37 02	06 ♓ 19 46	19 22	09 53	01 16	22 08	01 56	15 33	04 12	02 46
23	08 01 55	29 ♋ 53 25	12 ♓ 56 07	19 ♓ 26 16	19 R 21	10 27	01 53	22 08	01 53	15 36	04 12	02 48
24	08 05 51	00 ♌ 50 42	25 ♓ 50 28	02 ♈ 09 08	19 16	11 00	02 30	22 09	01 49	15 40	04 12	02 50
25	08 09 48	01 47 59	08 ♈ 22 42	14 ♈ 31 45	19 05	11 32	03 08	22 10	01 46	15 44	04 12	02 52
26	08 13 44	02 45 18	20 ♈ 36 51	26 ♈ 38 39	18 49	12 02	03 45	22 11	01 42	15 47	04 13	02 53
27	08 17 41	03 42 37	02 ♉ 37 48	08 ♉ 34 58	18 29	12 31	04 22	22 12	01 39	15 51	04 13	02 55
28	08 21 37	04 39 58	14 ♉ 30 49	20 ♉ 25 58	18 04	12 58	05 00	22 13	01 35	15 55	04 13	02 57
29	08 25 34	05 37 19	26 ♉ 21 04	02 ♊ 16 43	17 35	13 23	05 37	22 15	01 32	15 58	04 14	02 59
30	08 29 31	06 34 42	08 ♊ 13 27	14 ♊ 11 47	17 01	13 47	06 14	22 17	01 29	16 02	04 14	03 01
31	08 33 27	07 ♌ 32 06	20 ♊ 12 12	26 ♊ 15 04	16 ♌ 24	14 ♍ 10	06 ♍ 52	22 ♏ 19	01 ♑ 26	16 ♌ 06	04 ♏ 15	03 ♍ 02

Date	Moon True ☊ ° '	Moon Mean ☊ ° '	Moon ☽ Latitude ° '
01	08 ♎ 48	08 ♎ 27	03 S 12
02	08 R 40	08 23	03 55
03	08 31	08 20	04 29
04	08 19	08 17	04 51
05	08 06	08 14	05 00
06	07 52	08 11	04 55
07	07 40	08 08	04 36
08	07 29	08 04	04 03
09	07 22	08 01	03 17
10	07 17	07 58	02 20
11	07 14	07 55	01 15
12	07 D 14	07 52	00 S 04
13	07 R 14	07 48	01 N 09
14	07 14	07 45	02 19
15	07 11	07 42	03 21
16	07 07	07 39	04 11
17	07 00	07 36	04 46
18	06 51	07 33	05 01
19	06 40	07 29	04 57
20	06 30	07 26	04 34
21	06 21	07 23	03 54
22	06 14	07 20	03 01
23	06 10	07 17	02 00
24	06 07	07 13	00 N 55
25	06 D 07	07 10	00 S 12
26	06 07	07 07	01 17
27	06 08	07 04	02 17
28	06 R 07	07 01	03 10
29	06 05	06 58	03 55
30	06 01	06 54	04 30
31	05 ♎ 54	06 ♎ 51	04 S 54

DECLINATIONS

Date	Sun ☉ ° '	Moon ☽ ° '	Mercury ☿ ° '	Venus ♀ ° '	Mars ♂ ° '	Jupiter ♃ ° '	Saturn ♄ ° '	Uranus ♅ ° '	Neptune ♆ ° '	Pluto ♇ ° '
01	23 N 09	14 N 10	20 N 17	14 N 31	16 N 30	17 S 30	22 S 29	17 N 09	11 S 15	21 N 43
02	23 05	16 19	19 50	14 08	16 18	17 30	22 29	17 08	11 15	21 42
03	23 00	17 46	19 23	13 45	16 07	17 29	22 29	17 07	11 15	21 41
04	22 55	18 28	18 56	13 22	15 55	17 28	22 29	17 06	11 15	21 41
05	22 50	18 19	18 28	12 59	15 42	17 28	22 30	17 05	11 15	21 40
06	22 44	17 18	18 00	12 36	15 30	17 28	22 30	17 04	11 15	21 39
07	22 38	15 27	17 33	12 12	15 18	17 27	22 30	17 03	11 14	21 38
08	22 32	12 50	17 05	11 49	15 06	17 27	22 30	17 03	11 14	21 38
09	22 25	09 33	16 38	11 25	14 53	17 27	22 30	17 02	11 14	21 37
10	22 18	05 46	16 11	11 02	14 40	17 26	22 31	17 00	11 14	21 36
11	22 10	01 N 38	15 45	10 39	14 28	17 26	22 31	16 59	11 14	21 36
12	22 02	02 S 39	15 19	10 15	14 15	17 26	22 31	16 58	11 14	21 35
13	21 54	06 53	14 53	09 52	14 02	17 26	22 31	16 57	11 14	21 34
14	21 45	10 48	14 29	09 28	13 49	17 26	22 31	16 56	11 14	21 33
15	21 36	14 10	14 05	09 05	13 36	17 26	22 32	16 55	11 14	21 33
16	21 27	16 41	13 42	08 42	13 23	17 26	22 32	16 54	11 14	21 32
17	21 17	18 09	13 21	08 19	13 10	17 26	22 32	16 53	11 15	21 31
18	21 07	18 24	13 00	07 56	12 56	17 26	22 32	16 52	11 15	21 30
19	20 56	17 25	12 41	07 33	12 43	17 26	22 32	16 51	11 15	21 30
20	20 45	15 19	12 24	07 11	12 30	17 26	22 33	16 50	11 15	21 29
21	20 34	12 21	12 08	06 48	12 16	17 27	22 33	16 49	11 15	21 28
22	20 22	08 46	11 54	06 26	12 02	17 27	22 33	16 48	11 15	21 27
23	20 11	04 51	11 42	06 04	11 49	17 27	22 33	16 47	11 15	21 27
24	19 58	00 S 49	11 31	05 42	11 35	17 28	22 33	16 46	11 15	21 26
25	19 46	03 N 08	11 23	05 21	11 21	17 28	22 33	16 45	11 15	21 25
26	19 33	06 52	11 17	05 00	11 07	17 29	22 34	16 44	11 16	21 24
27	19 20	10 15	11 13	04 39	10 53	17 29	22 34	16 43	11 16	21 24
28	19 06	13 10	11 11	04 19	10 39	17 30	22 34	16 41	11 16	21 23
29	18 52	15 32	11 12	03 59	10 25	17 31	22 34	16 40	11 16	21 22
30	18 38	17 14	11 15	03 39	10 11	17 31	22 34	16 39	11 16	21 21
31	18 N 23	18 N 12	11 N 20	03 N 20	09 N 56	17 S 32	22 S 34	16 N 38	11 S 16	21 N 21

ZODIAC SIGN ENTRIES

Date	h	m	Planets
02	12	05	☽ ♊
05	00	03	☽ ♋
07	10	08	☽ ♌
08	12	08	♀ ♍
09	18	15	☽ ♍
12	00	26	☽ ♎
14	04	33	☽ ♏
16	06	42	☽ ♐
18	07	42	☽ ♑
20	09	05	☽ ♒
20	11	03	♂ ♍
22	12	41	☽ ♓
23	14	45	☉ ♌
24	19	53	☽ ♈
27	06	43	☽ ♉
29	19	23	☽ ♊

LATITUDES

Date	Mercury ☿ ° '	Venus ♀ ° '	Mars ♂ ° '	Jupiter ♃ ° '	Saturn ♄ ° '	Uranus ♅ ° '	Neptune ♆ ° '	Pluto ♇ ° '
01	01 N 01	01 N 04	01 N 14	00 N 58	00 N 55	00 N 39	01 N 48	11 N 52
04	00 33	00 47	01 13	00 57	00 55	00 39	01 47	11 51
07	00 N 01	00 28	01 11	00 57	00 55	00 39	01 47	11 50
10	00 S 34	00 N 08	01 10	00 56	00 54	00 39	01 47	11 50
13	01 13	00 S 14	01 09	00 55	00 54	00 39	01 47	11 49
16	01 54	00 38	01 08	00 54	00 54	00 39	01 47	11 49
19	02 36	01 05	01 06	00 54	00 53	00 39	01 47	11 48
22	03 17	01 33	01 05	00 53	00 53	00 39	01 46	11 48
25	03 55	02 03	01 04	00 52	00 52	00 39	01 46	11 48
28	04 27	02 34	01 02	00 51	00 52	00 39	01 46	11 47
31	04 S 48	03 S 08	01 N 01	00 N 51	00 N 52	00 N 39	01 N 46	11 N 47

DATA

Julian Date	2436751
Delta T	+33 seconds
Ayanamsa	23° 17' 32"
Synetic vernal point	05° ♓ 49' 28"
True obliquity of ecliptic	23° 26' 31"

MOON'S PHASES, APSIDES AND POSITIONS ☽

Date	h	m	Phase	Longitude ° '	Eclipse Indicator
06	02	00	●	13 ♋ 17	
13	12	01	☽	20 ♎ 21	
20	03	33	○	26 ♑ 42	
27	14	22	☾	03 ♉ 48	

Day	h	m		
01	19	23	Apogee	
17	14	20	Perigee	
29	11	44	Apogee	
04	19	48	Max dec	18° N 30'
11	21	10	0S	
18	04	44	Max dec	18° S 27'
24	16	54	0N	

LONGITUDES

Date	Chiron ⚷ ° '	Ceres ⚳ ° '	Pallas ⚴ ° '	Juno ⚵ ° '	Vesta ⚶ ° '	Black Moon Lilith ⚸ ° '
01	27 ♒ 13	14 ♏ 33	11 ♎ 56	12 ♎ 01	27 ♌ 02	25 ♉ 08
11	26 ♒ 54	14 ♏ 54	14 ♎ 11	13 ♎ 35	01 ♍ 25	26 ♉ 15
21	26 ♒ 30	15 ♏ 49	16 ♎ 49	15 ♎ 29	05 ♍ 56	27 ♉ 22
31	26 ♒ 03	17 ♏ 15	19 ♎ 46	17 ♎ 40	10 ♍ 34	28 ♉ 29

All ephemeris data is given at 12.00 UT and the Moon's longitude is additionally given for 24.00 UT

ASPECTARIAN

h m Aspects

01 Wednesday
04 14 ☿ ⊼ ♄
04 17 ☽ □ ♅
12 28 ☽ □ ♂
12 30 ☽ ⚼ ♄
13 14 ♂ ⚼ ♄
15 02 ☽ ∥ ♀
19 51 ☿ □ ♆
20 26 ☽ Q ☿
21 10 ☽ ☍ ♃

02 Thursday
00 46 ☽ ∠ ☉
00 59 ☽ □ ♀
06 40 ☽ ± ♄
11 58 ☽ ∥ ♂
16 39 ☽ □ ♇
16 58 ☽ Q ♅
18 44 ☽ ⊼ ♄
20 35 ☽ ⊥ ☉
20 39 ☽ ⊼ ♆
23 37 ☽ ✶ ☿
23 58 ☽ ∥ ♅

03 Friday
02 53 ☽ Q ♂
06 10 ☽ # ♃
06 44 ☉ ∠ ♀
08 44 ☽ ± ♆
09 43 ☽ ⊻ ☉
16 10 ☽ Q ♀
17 13 ☽ ✶ ♅

04 Saturday
02 43 ☽ ⚼ ♆
04 12 ☽ ✶ ♂
04 50 ☽ Q ♇
08 59 ☽ ∠ ☿
09 15 ☽ ⊼ ♃
17 52 ☽ ✶ ♀
21 04 ☽ ± ♃
23 08 ☽ ∠ ♅

05 Sunday
04 37 ☽ ✶ ♇
04 37 ☽ ⊥ ☿
06 09 ☽ ☍ ♄
08 21 ☽ △ ♆
11 23 ☽ ∠ ♂
14 40 ☽ ⚼ ♃
16 52 ☽ ⊥ ♅
17 33 ☽ ⊻ ☿
20 26 ☿ ± ♄

06 Monday
01 26 ☽ ∠ ♀
02 00 ☽ ☌ ☉
04 35 ☽ ⊻ ♅
05 49 ☽ ⊥ ♂
09 11 ☽ # ♃
09 53 ☽ ∠ ♇
15 36 ☽ ∥ ♅
17 59 ☽ ⊻ ♂
19 37 ☽ △ ♃
20 03 ☽ ⊥ ♀

07 Tuesday
03 12 ☽ ⊥ ♇
08 20 ☽ ⊻ ♀
13 16 ☉ ⊻ ♅
13 44 ☽ ∥ ♂
14 39 ☽ ⊻ ♇
15 42 ☽ ⊼ ♄
16 56 ☿ # ♃
18 07 ☽ □ ♆
19 38 ♂ Q ♆

08 Wednesday
01 19 ♂ □ ♃
02 59 ☽ ± ♄
08 10 ☽ ☌ ☿
13 56 ☽ ☌ ♅
14 32 ☿ ∥ ♅
15 48 ☽ ⊻ ☉
17 48 ☉ # ♄
19 45 ☽ ⚼ ♄
21 02 ☽ ∥ ♀

09 Thursday
00 11 ☽ # ♆
03 51 ☽ Q ♆
03 52 ☽ ⊥ ☉
04 03 ☽ □ ♃
05 28 ☽ ☌ ♂
20 13 ☽ ☌ ♀
21 47 ☽ ∠ ☉
22 43 ☽ ☌ ♇
23 20 ☽ △ ♄
23 21 ♀ # ♆

10 Friday
01 57 ☽ ✶ ♆
10 52 ☉ ∠ ♇
13 02 ☽ Q ♃
19 39 ☽ ⊻ ☿
21 21 ☽ ⊻ ♅

11 Saturday
03 10 ☽ ✶ ☉
05 09 ☽ ∠ ♆
07 11 ☽ ⊥ ☿
08 12 ☽ ⊥ ♅
10 37 ☽ ✶ ♃
13 49 ☉ ⊥ ♂
14 44 ☽ ⊻ ♂
15 15 ♀ ☌ ♇
20 18 ♀ △ ♄
21 14 ☽ ⊥ ♆

12 Sunday
00 15 ☽ ∠ ☿
00 19 ☽ ∠ ♅
01 33 ☿ ☌ ♅
01 54 ☽ ⊥ ♂
02 14 ☽ Q ☉
04 51 ☽ ⊻ ♇
05 04 ☽ □ ♄
05 38 ☽ ⊻ ♀
07 51 ☽ ⊻ ♆
13 10 ☽ ∠ ♃
15 25 ☽ ⊥ ♇
16 48 ☽ ⊥ ♀
18 29 ☽ ∠ ♂

13 Monday
02 45 ☽ ✶ ♅
04 04 ☽ ✶ ☿
04 51 ☽ ⊥ ♃
07 08 ☽ ∠ ♇
09 21 ☽ ∠ ♀
11 49 ♄ △ ♇
12 01 ☽ □ ☉
12 19 ☽ Q ♄
15 12 ☽ ⊻ ♃
21 38 ☽ ✶ ♂
21 50 ♀ ✶ ♆
23 31 ☽ Q ♅

14 Tuesday
01 46 ☽ Q ☿
04 19 ☽ # ♀
08 44 ☽ ✶ ♄
08 53 ☽ ✶ ♇
11 40 ☽ ☌ ♆
12 25 ☽ ✶ ♀
14 51 ☽ ∥ ♆
18 56 ☽ Q ♂

15 Wednesday
05 07 ☽ Q ♇
06 02 ☽ □ ♅
07 52 ☽ # ♂
09 28 ☽ □ ☿
09 34 ☉ △ ♃
09 37 ☽ Q ♀
09 50 ☽ ∠ ♄
11 25 ☽ # ☿
17 45 ☽ ☌ ♃
17 58 ☿ ⚼ ♄
18 20 ☽ △ ☉
21 55 ☉ ∥ ♇

16 Thursday
00 41 ☽ ⊥ ♄
02 15 ☽ □ ♂
10 31 ☽ ⊻ ♄
10 58 ☽ □ ♇
13 34 ☽ ⊻ ♆
14 36 ☽ # ♅
16 51 ☽ □ ♀
16 51 ♆ St D
20 44 ☽ ⚼ ☉
21 53 ☽ ∥ ♃
23 24 ☽ ⊥ ♆

17 Friday
07 39 ☽ △ ♅
12 27 ☽ ± ☉
12 31 ☽ △ ☿
14 05 ☽ ∠ ♆
18 53 ☽ ⊻ ♃
22 54 ☽ ⊼ ☉

18 Saturday
04 39 ☽ ⊥ ♃
05 26 ☽ △ ♂
08 11 ☽ ⚼ ♅
11 16 ☽ ☌ ♄
12 02 ☽ △ ♇
13 36 ☽ ⚼ ☿
14 31 ☽ ✶ ♆
19 19 ☽ ∠ ♃
20 08 ☽ △ ♀
22 56 ☽ ± ♅

19 Sunday
04 39 ☽ ± ☿
04 48 ☿ ∥ ♂
07 02 ☽ ⚼ ♂
08 49 ☽ ⊼ ♅
10 10 ☽ Q ♆
11 38 ☽ ∥ ♃
12 39 ☽ ⚼ ♇
14 42 ☽ ⊼ ☿
19 47 ☽ # ♅
19 58 ☽ ✶ ♃
21 56 ☽ ⚼ ♀
22 33 ☽ ± ♂

20 Monday
03 33 ☽ ☍ ☉
03 35 ☽ ± ♇
03 53 ☉ ⊥ ♇
08 00 ♃ St D
08 59 ☽ ⊼ ♂
12 32 ☽ ⊻ ♄
13 38 ☽ ⊼ ♇
13 38 ☽ ± ♀
16 00 ☽ Q ♃
16 07 ☽ □ ♆
22 36 ☽ ⊥ ♄

21 Tuesday
00 14 ☽ ⊼ ♀
11 20 ☽ ☍ ♅
12 41 ☽ # ♂
13 43 ☽ # ☿
13 56 ☽ ∠ ♄
17 56 ☽ ☍ ☿
19 46 ☽ ∥ ♆
22 50 ☽ □ ♃

22 Wednesday
10 42 ☽ ⊼ ☉
15 04 ☽ ☍ ♂
16 07 ☽ ✶ ♄
17 38 ☽ ☍ ♇
20 10 ☽ △ ♆
21 02 ☿ St R
22 13 ☽ ± ☉
22 28 ♀ Q ♃

23 Thursday
03 57 ☽ # ♀
07 16 ☽ ☍ ♀
11 48 ♂ △ ♄
13 43 ☽ Q ♄
15 52 ☽ ⚼ ☉
16 56 ☽ ⊼ ♅
23 33 ☽ ⚼ ♆
23 47 ☽ ⊼ ☿

24 Friday
04 07 ☽ ± ♅
05 03 ☽ △ ♃
10 55 ☽ ± ☿
16 28 ☽ ± ♆
21 12 ☽ ⚼ ♅
22 17 ☽ △ ☉
23 19 ☽ □ ♄

25 Saturday
01 10 ♂ ☌ ♇
01 20 ☽ ⊼ ♇
01 20 ☽ ⊼ ♂
03 50 ☽ ⚼ ☿
03 56 ☽ ⊼ ♆
05 02 ☿ ∥ ♂
09 39 ☽ ⚼ ♃
11 03 ☉ ⊼ ♄
12 56 ☽ ± ♇
13 32 ☽ ± ♂
18 24 ☽ ⊼ ♀
21 29 ♂ # ♆

26 Sunday
00 48 ☽ ∥ ♀
02 26 ☽ △ ♅
03 14 ☽ ± ♃
06 36 ☽ ⚼ ♇
06 41 ☽ ± ♀
08 07 ☽ ⚼ ♂
08 33 ☽ △ ☿
15 07 ☽ ⊼ ♃
15 28 ☉ ⊻ ♇
17 11 ☿ # ♆

27 Monday
01 19 ☽ ⚼ ♀
06 01 ♂ ✶ ♆
10 02 ☽ △ ♄
12 35 ☽ △ ♇
14 22 ☽ □ ☉
15 12 ☽ ☍ ♆
15 42 ☽ △ ♂
16 34 ☽ ∥ ♂
19 24 ☽ ∥ ☿
19 55 ☽ # ♆

28 Tuesday
00 48 ☉ □ ♆
08 44 ☽ △ ♀
14 51 ☽ □ ♅
16 11 ☽ ⚼ ♄
18 56 ☽ □ ☿

29 Wednesday
03 40 ☽ ☍ ♃
05 59 ☽ Q ☉
10 21 ☽ ± ♄
11 39 ☉ ⊻ ♂
22 27 ☽ ⊼ ♄

30 Thursday
01 27 ☽ □ ♇
02 49 ☽ ∥ ♅
03 30 ☽ Q ♅
03 57 ☽ ⊼ ♆
05 50 ☽ Q ☿
07 47 ☽ □ ♂
08 24 ☽ ✶ ☉
16 03 ☽ ± ♆
17 32 ☽ # ♃
19 57 ☿ # ♆
23 33 ☽ □ ♀

31 Friday
03 46 ☽ ✶ ♅
04 47 ☽ ✶ ☿
09 28 ☉ ± ♄
10 05 ☽ ⚼ ♆
10 44 ☿ ⚼ ♄
13 40 ☽ Q ♇
16 12 ☽ ⊼ ♃
17 02 ☽ ∠ ☉
18 33 ☽ ∥ ☉
21 46 ☽ Q ♂
22 04 ☿ ☌ ♅

AUGUST 1959

Date	Sidereal time h m s	Sun ☉ ° ' "	Moon ☽ ° ' "	Moon ☽ 24.00 ° ' "	Mercury ☿ ° '	Venus ♀ ° '	Mars ♂ ° '	Jupiter ♃ ° '	Saturn ♄ ° '	Uranus ♅ ° '	Neptune ♆ ° '	Pluto ♇ ° '
01	08 37 24	08 ♌ 29 30	02 ♋ 20 45	08 ♋ 29 32	15 ♌ 43	14 ♍ 30	07 ♍ 29	22 ♏ 21	01 ♑ 23	16 ♌ 09	04 ♏ 15	03 ♍ 04
02	08 41 20	09 26 56	14 ♋ 41 37	20 ♋ 57 09	15 R 00	14 49	08 07	22 23	01 R 20	16 13	04 16	03 06
03	08 45 17	10 24 22	27 ♋ 16 13	03 ♌ 38 51	14 16	15 06	08 45	22 26	01 17	16 17	04 16	03 08
04	08 49 13	11 21 50	10 ♌ 05 00	16 ♌ 34 37	13 30	15 21	09 22	22 28	01 14	16 21	04 17	03 10
05	08 53 10	12 19 18	23 ♌ 07 34	29 ♌ 43 44	12 43	15 34	10 00	22 31	01 11	16 24	04 17	03 12
06	08 57 06	13 16 48	06 ♍ 22 57	13 ♍ 05 04	11 57	15 45	10 37	22 34	01 08	16 28	04 18	03 14
07	09 01 03	14 14 18	19 ♍ 49 56	26 ♍ 37 25	11 13	15 54	11 15	22 37	01 06	16 32	04 19	03 16
08	09 05 00	15 11 49	03 ♎ 27 23	10 ♎ 19 45	10 31	16 00	11 53	22 40	01 03	16 35	04 19	03 18
09	09 08 56	16 09 21	17 ♎ 14 24	24 ♎ 11 15	09 52	16 05	12 31	22 44	01 01	16 39	04 20	03 20
10	09 12 53	17 06 54	01 ♏ 10 15	08 ♏ 11 18	09 17	16 07	13 08	22 48	00 58	16 43	04 21	03 21
11	09 16 49	18 04 28	15 ♏ 14 18	22 ♏ 19 06	08 46	16 R 07	13 46	22 51	00 56	16 47	04 22	03 23
12	09 20 46	19 02 02	29 ♏ 25 31	06 ♐ 33 19	08 21	16 05	14 24	22 55	00 54	16 50	04 23	03 25
13	09 24 42	19 59 38	13 ♐ 42 11	20 ♐ 51 45	08 01	16 00	15 02	23 00	00 52	16 54	04 23	03 27
14	09 28 39	20 57 14	28 ♐ 01 33	05 ♑ 11 05	07 48	15 53	15 40	23 04	00 50	16 58	04 24	03 29
15	09 32 35	21 54 51	12 ♑ 19 47	19 ♑ 27 04	07 41	15 43	16 18	23 08	00 48	17 01	04 25	03 31
16	09 36 32	22 52 30	26 ♑ 32 18	03 ♒ 34 53	07 D 42	15 31	16 56	23 13	00 46	17 05	04 26	03 33
17	09 40 29	23 50 09	10 ♒ 34 13	17 ♒ 29 48	07 50	15 17	17 34	23 18	00 44	17 09	04 27	03 35
18	09 44 25	24 47 50	24 ♒ 21 09	01 ♓ 07 53	08 05	15 00	18 12	23 23	00 42	17 13	04 28	03 37
19	09 48 22	25 45 32	07 ♓ 49 45	14 ♓ 26 35	08 28	14 41	18 50	23 28	00 41	17 16	04 29	03 39
20	09 52 18	26 43 15	20 ♓ 58 17	27 ♓ 24 56	08 58	14 20	19 28	23 33	00 39	17 20	04 30	03 41
21	09 56 15	27 41 00	03 ♈ 46 39	10 ♈ 03 40	09 36	13 56	20 06	23 39	00 38	17 24	04 31	03 43
22	10 00 11	28 38 46	16 ♈ 16 19	22 ♈ 25 00	10 21	13 31	20 44	23 44	00 36	17 27	04 33	03 45
23	10 04 08	29 ♌ 36 34	28 ♈ 30 09	04 ♉ 32 17	11 13	13 03	21 23	23 50	00 35	17 31	04 34	03 47
24	10 08 04	00 ♍ 34 24	10 ♉ 31 57	16 ♉ 29 43	12 12	12 34	22 01	23 56	00 34	17 35	04 35	03 49
25	10 12 01	01 32 15	22 ♉ 26 12	28 ♉ 22 01	13 17	12 03	22 39	24 02	00 33	17 38	04 36	03 51
26	10 15 58	02 30 09	04 ♊ 17 46	10 ♊ 14 05	14 29	11 31	23 17	24 08	00 32	17 42	04 37	03 53
27	10 19 54	03 28 04	16 ♊ 11 34	22 ♊ 10 48	15 47	10 57	23 56	24 14	00 31	17 46	04 39	03 55
28	10 23 51	04 26 01	28 ♊ 12 19	04 ♋ 16 39	17 11	10 22	24 34	24 21	00 30	17 49	04 40	03 57
29	10 27 47	05 23 59	10 ♋ 24 15	16 ♋ 35 33	18 39	09 46	25 13	24 27	00 30	17 53	04 41	03 59
30	10 31 44	06 22 00	22 ♋ 50 52	29 ♋ 10 30	20 13	09 10	25 51	24 34	00 29	17 56	04 43	04 01
31	10 35 40	07 ♍ 20 02	05 ♌ 34 39	12 ♌ 03 25	21 ♌ 50	08 ♍ 33	26 ♍ 30	24 ♏ 41	00 ♑ 28	18 ♌ 00	04 ♏ 44	04 ♍ 03

Date	Moon True ☊ ° '	Moon Mean ☊ ° '	Moon Latitude ° '	DECLINATIONS Sun ☉ ° '	Moon ☽ ° '	Mercury ☿ ° '	Venus ♀ ° '	Mars ♂ ° '	Jupiter ♃ ° '	Saturn ♄ ° '	Uranus ♅ ° '	Neptune ♆ ° '	Pluto ♇ ° '
01	05 ♎ 45	06 ♎ 48	05 S 04	18 N 09	18 N 21	11 N 28	03 N 02	09 N 42	17 S 33	22 S 35	16 N 37	11 S 17	21 N 20
02	05 R 36	06 45	05 01	17 53	17 38	11 38	02 44	09 27	17 34	22 35	16 36	11 17	21 19
03	05 26	06 42	04 43	17 38	16 04	11 49	02 26	09 13	17 34	22 35	16 35	11 17	21 18
04	05 16	06 39	04 11	17 22	13 41	12 03	02 09	08 58	17 35	22 35	16 34	11 17	21 17
05	05 09	06 35	03 25	17 06	10 35	12 19	01 53	08 44	17 36	22 35	16 33	11 18	21 17
06	05 03	06 32	02 28	16 50	06 53	12 35	01 37	08 29	17 37	22 35	16 31	11 18	21 16
07	05 00	06 29	01 21	16 34	02 N 47	12 53	01 22	08 14	17 38	22 36	16 30	11 18	21 15
08	04 59	06 26	00 S 08	16 17	01 S 30	13 12	01 08	07 59	17 39	22 36	16 29	11 19	21 14
09	04 D 59	06 23	01 N 06	16 00	05 46	13 32	00 55	07 45	17 40	22 36	16 28	11 19	21 14
10	05 00	06 19	02 17	15 42	09 45	13 52	00 42	07 30	17 42	22 36	16 27	11 19	21 13
11	05 01	06 16	03 20	15 25	13 13	14 12	00 31	07 15	17 43	22 36	16 26	11 19	21 12
12	05 R 01	06 13	04 12	15 07	15 56	14 32	00 20	07 00	17 44	22 36	16 25	11 20	21 11
13	04 59	06 10	04 48	14 49	17 41	14 51	00 10	06 45	17 45	22 37	16 24	11 20	21 11
14	04 55	06 07	05 07	14 31	18 19	15 10	00 N 01	06 30	17 47	22 37	16 23	11 20	21 10
15	04 49	06 04	05 06	14 12	17 47	15 27	00 S 06	06 14	17 48	22 37	16 21	11 21	21 09
16	04 43	06 00	04 47	13 54	16 09	15 43	00 13	05 59	17 49	22 37	16 20	11 21	21 08
17	04 36	05 57	04 11	13 35	13 34	15 58	00 19	05 44	17 51	22 37	16 19	11 22	21 08
18	04 30	05 54	03 20	13 15	10 15	16 11	00 23	05 29	17 52	22 37	16 18	11 22	21 07
19	04 26	05 51	02 19	12 56	06 29	16 22	00 26	05 13	17 54	22 38	16 17	11 22	21 06
20	04 24	05 48	01 12	12 37	02 S 28	16 31	00 28	04 58	17 55	22 38	16 16	11 23	21 06
21	04 23	05 45	00 N 03	12 17	01 N 33	16 38	00 29	04 43	17 57	22 38	16 15	11 23	21 05
22	04 D 23	05 41	01 S 04	11 57	05 24	16 42	00 29	04 27	17 58	22 38	16 14	11 24	21 04
23	04 25	05 38	02 08	11 37	08 57	16 44	00 27	04 12	18 00	22 38	16 12	11 24	21 03
24	04 26	05 35	03 05	11 16	12 03	16 43	00 25	03 56	18 02	22 38	16 11	11 25	21 03
25	04 28	05 32	03 53	10 56	14 38	16 39	00 21	03 41	18 03	22 38	16 10	11 25	21 02
26	04 R 28	05 29	04 31	10 35	16 34	16 32	00 16	03 25	18 05	22 39	16 09	11 25	21 01
27	04 27	05 25	04 57	10 14	17 48	16 22	00 09	03 09	18 07	22 39	16 08	11 26	21 01
28	04 25	05 22	05 11	09 53	18 15	16 09	00 S 02	02 54	18 09	22 39	16 07	11 26	21 00
29	04 22	05 19	05 11	09 32	17 52	15 54	00 N 06	02 38	18 10	22 39	16 06	11 27	20 59
30	04 18	05 16	04 57	09 11	16 37	15 35	00 16	02 22	18 12	22 39	16 05	11 27	20 58
31	04 ♎ 13	05 ♎ 13	04 S 28	08 N 49	14 N 33	15 N 13	00 N 26	02 N 07	18 S 14	22 S 39	16 N 04	11 S 28	20 N 58

ZODIAC SIGN ENTRIES

Date	h	m	Planets
01	07	24	☽ ♋
03	17	09	☽ ♌
06	00	29	☽ ♍
08	05	56	☽ ♎
10	10	00	☽ ♏
12	12	58	☽ ♐
14	15	18	☽ ♑
16	17	53	☽ ♒
18	21	59	☽ ♓
21	04	51	☽ ♈
23	14	58	☽ ♉
23	21	44	☉ ♍
26	03	18	☽ ♊
28	15	33	☽ ♋
31	01	33	☽ ♌

LATITUDES

Date	Mercury ☿ ° '	Venus ♀ ° '	Mars ♂ ° '	Jupiter ♃ ° '	Saturn ♄ ° '	Uranus ♅ ° '	Neptune ♆ ° '	Pluto ♇ ° '
01	04 S 53	03 S 20	01 N 01	00 N 50	00 N 51	00 N 39	01 N 46	11 N 47
04	04 55	03 56	00 59	00 50	00 51	00 39	01 46	11 47
07	04 42	04 33	00 58	00 49	00 51	00 39	01 45	11 47
10	04 13	05 11	00 57	00 48	00 50	00 39	01 45	11 47
13	03 32	05 48	00 55	00 48	00 50	00 39	01 45	11 46
16	02 43	06 26	00 54	00 47	00 49	00 39	01 45	11 46
19	01 50	07 01	00 52	00 46	00 49	00 39	01 45	11 47
22	00 59	07 32	00 51	00 46	00 48	00 39	01 45	11 47
25	00 S 11	07 59	00 50	00 45	00 48	00 39	01 44	11 47
28	00 N 30	08 20	00 48	00 44	00 48	00 39	01 44	11 47
31	01 N 02	08 S 32	00 N 47	00 N 44	00 N 47	00 N 39	01 N 44	11 N 47

DATA

Julian Date	2436782
Delta T	+33 seconds
Ayanamsa	23° 17' 36"
Synetic vernal point	05° ♓ 49' 23"
True obliquity of ecliptic	23° 26' 31"

MOON'S PHASES, APSIDES AND POSITIONS ☽

Date	h	m	Phase	Longitude ° '	Eclipse Indicator
04	14	34	●	11 ♌ 28	
11	17	10	☽	18 ♏ 17	
18	12	51	○	24 ♒ 50	
26	08	03	☾	02 ♊ 21	

Day	h	m		
13	16	08	Perigee	
26	06	14	Apogee	
01	04	20	Max dec	18° N 24'
08	03	39	0S	
14	13	05	Max dec	18° S 19'
21	02	41	0N	
28	13	07	Max dec	18° N 15'

LONGITUDES

Date	Chiron ⚷ ° '	Ceres ⚳ ° '	Pallas ⚴ ° '	Juno ⚵ ° '	Vesta ⚶ ° '	Black Moon Lilith ⚸ ° '
01	26 ♒ 00	17 ♏ 25	20 ♎ 05	17 ♎ 54	11 ♍ 02	28 ♉ 35
11	25 ♒ 30	19 ♏ 20	23 ♎ 20	20 ♎ 22	15 ♍ 47	29 ♉ 42
21	25 ♒ 00	21 ♏ 38	26 ♎ 49	23 ♎ 01	20 ♍ 38	00 ♊ 49
31	24 ♒ 30	24 ♏ 16	00 ♏ 28	25 ♎ 51	25 ♍ 33	01 ♊ 56

All ephemeris data is given at 12.00 UT and the Moon's longitude is additionally given for 24.00 UT

ASPECTARIAN

h m	Aspects
01 Saturday	
04 07	☽ ± ♃
08 59	☽ ∠ ☿
09 39	☽ ∠ ♅
10 06	☽ ☍ ♄
12 19	☽ ⊥ ☉
12 19	☽ Q ♀
12 24	☉ ⊥ ♀
13 26	☽ ✶ ♇
15 44	☽ △ ♆
21 48	☽ Ᵽ ♃
22 35	☽ ✶ ♂
02 Sunday	
01 00	☽ ⊻ ☉
01 37	☽ ⊥ ☿
03 18	☽ ⊥ ♅
03 58	☽ ∥ ☉
12 14	☽ ✶ ♀
12 34	☽ ⊻ ☿
13 40	☽ # ♃
14 57	☽ ⊻ ♅
16 29	☿ ⊻ ♀
18 34	☽ ∠ ♇
03 Monday	
02 47	☽ △ ♃
04 58	☽ ∠ ♂
05 26	☽ ∥ ♅
11 36	♄ Ᵽ ♅
11 44	☽ ⊥ ♇
17 14	☉ # ♃
17 27	☽ ∠ ♀
19 32	☽ ⊼ ♄
22 50	☽ ⊥ ♂
23 04	☽ ⊻ ♇
04 Tuesday	
01 11	☽ □ ♆
06 43	☽ ± ♄
10 36	☽ ⊥ ♀
10 36	☽ ⊻ ♂
14 34	☽ ☌ ☉
17 58	☽ ☌ ☿
21 54	☽ ⊻ ♀
23 19	☽ Ᵽ ♄
23 37	☽ ☌ ♅
05 Wednesday	
00 15	☽ ∥ ☿
06 52	☽ # ♆
10 28	☽ Q ♆
10 53	☽ □ ♃
17 33	☉ ☌ ☿
06 Thursday	
01 17	☽ ∥ ♂
02 35	☽ △ ♄
06 19	☽ ☌ ♇
08 15	☽ ✶ ♆
09 33	♂ Q ♃
19 32	☽ Q ♃
19 59	☽ ☌ ♂
21 28	☽ ⊻ ☿
07 Friday	
01 18	☽ ⊻ ☉
04 56	☽ ☌ ♀
06 06	☽ ⊻ ♅
07 36	☽ ⊥ ☿
11 05	☽ ∠ ♆
11 22	☿ ⊻ ♂
12 46	☽ ⊥ ☉
16 48	☽ ⊥ ♅
16 57	☉ ∥ ♅
16 57	☽ ✶ ♃
20 28	☽ ∥ ♀
22 43	☽ ∠ ☿
08 Saturday	
02 59	☽ ⊥ ♆
05 51	☽ ∠ ☉
07 48	☽ □ ♄
08 43	☽ ∠ ♅
10 05	☽ # ♀
11 43	☽ ⊻ ♇
13 31	☽ ⊻ ♆
19 24	☽ ∠ ♃
22 13	☽ ⊥ ♇
23 45	☽ ✶ ☿
09 Sunday	
03 24	☽ ⊻ ♂
04 29	☿ ⊥ ♀
08 32	☉ Ᵽ ♄
09 58	☉ ⊻ ♀
09 59	☽ ⊻ ♀
09 59	☽ ✶ ☉
10 59	☽ ✶ ♅
11 07	☽ ⊥ ♃
13 53	☽ ∠ ♇
14 18	☽ ⊥ ♂
15 03	☽ Q ♄
19 39	☽ Q ☿
20 24	☽ ⊥ ♀
21 32	☽ ⊻ ♃
22 58	☽ # ♂
10 Monday	
01 17	☉ ☌ ♅
06 33	☽ ∠ ♂
07 46	☽ Q ♅
08 13	☽ Q ☉
11 40	☽ ✶ ♄
11 54	☽ ∠ ♀
15 45	☽ ✶ ♇
17 27	☽ ☌ ♆
22 21	☽ ∥ ♆
23 15	♀ St R
11 Tuesday	
01 21	☽ □ ☿
09 23	☽ ✶ ♂
12 15	☽ Q ♇
13 11	☽ ∠ ♄
13 29	☽ ✶ ♀
14 37	☽ □ ♅
17 10	☽ □ ☉
20 53	☽ # ☿
12 Wednesday	
00 58	☽ ☌ ♃
04 23	☽ ⊥ ♄
04 47	☽ # ☉
06 40	☽ Q ♂
09 44	☽ Q ♀
10 35	☿ ⊥ ♂
14 28	☽ ⊻ ♄
17 19	☽ # ♅
18 45	☽ □ ♇
20 21	☽ ⊻ ♆
13 Thursday	
02 39	☽ △ ☿
06 26	☽ ⊥ ♆
10 42	☉ ∥ ☿
13 33	☽ ∥ ♃
14 20	☽ □ ♂
15 49	☽ □ ♀
17 23	☽ △ ♅
21 33	☽ ∠ ♆
23 18	☽ △ ☉
14 Friday	
03 21	☽ Ᵽ ☿
03 39	☽ ⊻ ♃
13 45	☽ ⊥ ♃
16 41	☽ ☌ ♄
18 15	☽ ± ☿
18 29	♀ ☌ ♂
18 38	☽ Ᵽ ♅
21 11	☽ △ ♇
22 42	☽ ✶ ♆
15 Saturday	
02 15	☽ Ᵽ ☉
04 15	☽ ⊼ ☿
04 55	☽ ∠ ♃
09 48	☽ ± ♅
11 43	☽ ∥ ♃
17 38	☽ △ ♀
18 28	☽ ± ☉
18 54	☽ Q ♆
19 00	☽ △ ♂
19 56	☽ ⊼ ♅
22 05	☿ St D
22 27	☽ Ᵽ ♇
16 Sunday	
00 54	☉ Q ♆
05 20	☽ ⊼ ☉
06 20	☽ ✶ ♃
09 50	☽ # ♅
13 44	☽ ± ♇
16 07	☽ # ☿
18 31	♂ ⊻ ♅
18 40	☽ Ᵽ ♀
19 10	☽ ⊻ ♄
21 18	☉ □ ♃
21 37	☽ Ᵽ ♂
23 59	☽ ⊼ ♇
17 Monday	
01 29	☽ □ ♆
02 53	☽ Q ♃
05 25	☽ ⊥ ♄
07 13	☽ ☍ ☿
09 49	☽ ± ♀
11 53	☽ # ☉
13 48	☽ ± ♂
19 59	☽ ⊼ ♀
20 55	☽ ∠ ♄
23 27	☽ ☍ ♅
18 Tuesday	
00 42	☽ ⊼ ♂
04 25	☽ ∥ ♆
10 17	☽ □ ♃
12 51	☽ ☍ ☉
23 13	☽ ✶ ♄
19 Wednesday	
00 53	☿ ∥ ♅
04 29	☽ ☍ ♇
05 59	☽ △ ♆
13 11	☽ ⊼ ☿
18 32	☿ ⊥ ♀
20 07	☽ # ♂
20 46	☽ Q ♄
20 Thursday	
00 07	☽ ☍ ♀
00 29	☽ ± ☿
05 16	☽ ⊼ ♅
09 05	☽ ☍ ♂
09 17	☽ Ᵽ ♆
13 29	♂ ∠ ♆
16 24	☽ ± ♅
16 49	☽ △ ♃
17 49	☽ Ᵽ ☿
23 34	☽ ⊼ ☉
23 47	☽ ∥ ♀
21 Friday	
02 04	☽ ± ♆
05 35	☽ # ♀
06 03	☽ □ ♄
09 22	☽ Ᵽ ♅
11 48	☽ ± ☉
11 54	☽ ⊼ ♇
13 25	☽ ⊼ ♆
21 21	☽ Ᵽ ♃
23 23	☽ ± ♇
23 47	☽ △ ☿
22 Saturday	
06 17	☽ ∥ ♂
06 29	☽ Ᵽ ☉
06 50	☽ ⊼ ♀
14 19	☽ △ ♅
14 52	☽ ± ♃
16 51	☽ Ᵽ ♇
18 05	☽ ± ♀
21 11	☽ ⊼ ♂
23 Sunday	
02 42	☽ ⊼ ♃
09 39	☽ ± ♂
11 09	☽ Ᵽ ♀
14 23	☽ △ ☉
16 07	☽ △ ♄
22 32	☽ △ ♇
24 Monday	
00 04	☽ ☍ ♆
02 31	☉ # ♆
04 33	☽ Ᵽ ♂
06 12	☽ ∥ ☉
06 38	☽ # ♆
11 49	☉ △ ♄
15 40	☽ □ ☿
15 55	☽ △ ♀
17 44	☿ ⊻ ♀
22 07	☽ Ᵽ ♄
25 Tuesday	
02 15	☽ □ ♅
12 28	☽ △ ♂
12 46	♀ Q ♃
15 15	☽ ☍ ♃
16 16	☽ ± ♄
26 Wednesday	
04 24	☽ ⊼ ♄
06 10	☽ ∥ ♅
07 55	☽ Q ☿
08 03	☽ □ ☉
11 10	☽ □ ♇
11 36	☽ ∥ ☿
12 40	☽ ⊼ ♆
14 51	☽ Q ♅
27 Thursday	
00 48	☽ ± ♆
01 55	☽ □ ♀
04 57	♂ ⊥ ♅
07 14	☿ Ᵽ ♄
11 05	☽ ✶ ☿
15 10	☽ ✶ ♅
18 56	☽ Ᵽ ♆
23 30	☽ Q ☉
23 31	☽ Q ♇
23 42	☉ ☌ ♇
28 Friday	
00 29	☽ # ♃
01 54	♂ ✶ ♃
04 15	☽ ⊼ ♃
04 22	☽ □ ♂
12 18	☽ Q ♀
16 17	☽ ± ♃
16 24	☿ ∥ ♅
16 33	☽ ☍ ♄
17 55	☉ ✶ ♆
20 56	☽ ∠ ☿
21 10	☽ ∠ ♅
23 03	☿ ☌ ♅
23 24	☽ ✶ ♇
23 59	☽ # ♃
29 Saturday	
00 47	☽ △ ♆
01 22	☽ ✶ ☉
10 08	☽ Ᵽ ♃
10 49	☽ ✶ ♀
14 53	☽ ⊥ ♅
17 00	☽ ⊥ ☿
17 45	☽ Q ♂
19 12	♃ ⊥ ♄
30 Sunday	
02 33	☽ ⊻ ♅
02 46	☿ ⊥ ♂
04 40	☽ ∠ ♇
06 14	☽ ⊻ ☿
08 56	☽ ∠ ☉
14 23	☽ ∠ ♀
15 18	☽ △ ♃
18 01	☽ ✶ ♂
19 22	☽ ∥ ♅
21 51	☽ ⊥ ♇
31 Monday	
02 27	☽ ⊼ ♄
03 25	☽ ⊥ ☉
03 55	☽ ∥ ☿
06 36	☽ ⊥ ♀
09 09	☽ ⊻ ♇
10 25	☽ □ ♆
13 40	☽ ± ♄
15 32	☽ ⊻ ☉
17 16	☽ ⊻ ♀
23 32	☽ ∠ ♂

SEPTEMBER 1959

LONGITUDES

Date	Sidereal time h m s	Sun ☉ ° ' "	Moon ☽ ° ' "	Moon ☽ 24.00 ° ' "	Mercury ☿ ° '	Venus ♀ ° '	Mars ♂ ° '	Jupiter ♃ ° '	Saturn ♄ ° '	Uranus ♅ ° '	Neptune ♆ ° '	Pluto ♇ ° '
01	10 39 37	08 ♍ 18 06	18 ♌ 36 52	25 ♌ 14 55	23 ♌ 32	07 ♍ 56	27 ♍ 08	24 ♏ 48	00 ♑ 28	18 ♌ 04	04 ♏ 45	04 ♍ 05
02	10 43 33	09 16 12	01 ♍ 57 27	08 ♍ 44 14	25 16	07 R 19	27 47	24 55	00 R 28	18 07	04 47	04 07
03	10 47 30	10 14 19	15 ♍ 35 01	22 ♍ 29 27	27 03	06 42	28 25	25 03	00 28	18 11	04 48	04 09
04	10 51 27	11 12 28	29 ♍ 27 08	06 ♎ 27 40	28 ♌ 53	06 06	29 04	25 10	00 28	18 14	04 50	04 11
05	10 55 23	12 10 39	13 ♎ 30 35	20 ♎ 35 28	00 ♍ 44	05 30	29 ♍ 43	25 18	00 D 28	18 18	04 51	04 13
06	10 59 20	13 08 51	27 ♎ 41 50	04 ♏ 49 17	02 37	04 55	00 ♎ 21	25 25	00 28	18 21	04 53	04 15
07	11 03 16	14 07 05	11 ♏ 57 23	19 ♏ 05 45	04 31	04 22	01 00	25 33	00 28	18 25	04 54	04 17
08	11 07 13	15 05 21	26 ♏ 14 01	03 ♐ 21 52	06 26	03 50	01 39	25 41	00 28	18 28	04 56	04 19
09	11 11 09	16 03 38	10 ♐ 28 58	17 ♐ 35 02	08 21	03 19	02 18	25 49	00 28	18 31	04 58	04 21
10	11 15 06	17 01 56	24 ♐ 39 48	01 ♑ 43 01	10 16	02 50	02 57	25 57	00 29	18 35	04 59	04 23
11	11 19 02	18 00 16	08 ♑ 44 24	15 ♑ 43 44	12 11	02 23	03 35	26 06	00 30	18 38	05 01	04 25
12	11 22 59	18 58 37	22 ♑ 40 46	29 ♑ 35 17	14 06	01 58	04 14	26 14	00 30	18 42	05 02	04 27
13	11 26 56	19 57 00	06 ♒ 27 01	13 ♒ 15 47	16 00	01 34	04 53	26 23	00 31	18 45	05 04	04 29
14	11 30 52	20 55 25	20 ♒ 01 21	26 ♒ 43 34	17 54	01 14	05 32	26 31	00 32	18 48	05 06	04 31
15	11 34 49	21 53 51	03 ♓ 22 15	09 ♓ 57 17	19 47	00 55	06 11	26 40	00 33	18 51	05 08	04 33
16	11 38 45	22 52 19	16 ♓ 28 34	22 ♓ 56 03	21 40	00 39	06 51	26 49	00 34	18 55	05 09	04 35
17	11 42 42	23 50 49	29 ♓ 19 43	05 ♈ 39 37	23 31	00 25	07 30	26 58	00 35	18 58	05 11	04 37
18	11 46 38	24 49 21	11 ♈ 55 51	18 ♈ 08 33	25 22	00 13	08 09	27 07	00 36	19 01	05 13	04 39
19	11 50 35	25 47 55	24 ♈ 17 55	00 ♉ 24 14	27 11	00 ♍ 04	08 48	27 17	00 38	19 04	05 15	04 41
20	11 54 31	26 46 31	06 ♉ 27 45	12 ♉ 28 52	29 ♍ 00	29 ♌ 58	09 27	27 26	00 39	19 07	05 17	04 43
21	11 58 28	27 45 09	18 ♉ 27 57	24 ♉ 25 26	00 ♎ 48	29 54	10 06	27 36	00 41	19 11	05 18	04 45
22	12 02 25	28 43 49	00 ♊ 21 48	06 ♊ 17 34	02 35	29 52	10 46	27 45	00 42	19 14	05 20	04 47
23	12 06 21	29 ♍ 42 32	12 ♊ 13 14	18 ♊ 09 24	04 20	29 D 53	11 25	27 55	00 44	19 17	05 22	04 49
24	12 10 18	00 ♎ 41 16	24 ♊ 06 36	00 ♋ 05 26	06 05	29 ♌ 56	12 05	28 05	00 46	19 20	05 24	04 50
25	12 14 14	01 40 03	06 ♋ 06 29	12 ♋ 10 19	07 49	00 ♍ 01	12 44	28 14	00 48	19 23	05 26	04 52
26	12 18 11	02 38 53	18 ♋ 17 30	24 ♋ 28 33	09 32	00 09	13 23	28 24	00 50	19 26	05 28	04 54
27	12 22 07	03 37 44	00 ♌ 43 58	07 ♌ 04 13	11 14	00 18	14 03	28 35	00 52	19 29	05 30	04 56
28	12 26 04	04 36 38	13 ♌ 29 39	20 ♌ 00 36	12 55	00 30	14 43	28 45	00 54	19 32	05 32	04 58
29	12 30 00	05 35 34	26 ♌ 37 15	03 ♍ 19 45	14 35	00 44	15 22	28 55	00 57	19 34	05 34	05 00
30	12 33 57	06 ♎ 34 32	10 ♍ 08 05	17 ♍ 02 07	16 ♎ 15	01 ♍ 00	16 ♎ 02	29 ♏ 05	00 ♑ 59	19 ♌ 37	05 ♏ 36	05 ♍ 01

Date	Moon True ☊ ° '	Moon Mean ☊ ° '	Moon ☽ Latitude ° '
01	04 ♎ 09	05 ♎ 10	03 S 44
02	04 R 06	05 06	02 47
03	04 04	05 03	01 40
04	04 03	05 00	00 S 25
05	04 D 03	04 57	00 N 52
06	04 04	04 54	02 07
07	04 05	04 51	03 14
08	04 06	04 47	04 10
09	04 07	04 44	04 49
10	04 R 07	04 41	05 12
11	04 06	04 38	05 15
12	04 05	04 35	04 59
13	04 03	04 31	04 27
14	04 01	04 28	03 40
15	03 59	04 25	02 41
16	03 58	04 22	01 35
17	03 58	04 19	00 N 26
18	03 D 58	04 16	00 S 44
19	03 58	04 12	01 50
20	03 59	04 09	02 51
21	04 00	04 06	03 43
22	04 01	04 03	04 24
23	04 01	04 00	04 55
24	04 01	03 57	05 13
25	04 R 01	03 53	05 17
26	04 01	03 50	05 08
27	04 01	03 47	04 44
28	04 D 01	03 44	04 05
29	04 01	03 41	03 13
30	04 ♎ 01	03 ♎ 37	02 S 09

DECLINATIONS

Date	Sun ☉ ° '	Moon ☽ ° '	Mercury ☿ ° '	Venus ♀ ° '	Mars ♂ ° '	Jupiter ♃ ° '	Saturn ♄ ° '	Uranus ♅ ° '	Neptune ♆ ° '	Pluto ♇ ° '
01	08 N 27	11 N 42	14 N 48	00 N 37	01 N 51	18 S 16	22 S 40	16 N 03	11 S 28	20 N 57
02	08 06	08 10	14 20	00 49	01 35	18 18	22 40	16 02	11 29	20 56
03	07 44	04 N 09	13 50	01 02	01 19	18 20	22 40	16 00	11 29	20 56
04	07 22	00 S 10	13 18	01 15	01 03	18 22	22 40	15 59	11 30	20 55
05	07 00	04 32	12 43	01 29	00 48	18 24	22 40	15 58	11 31	20 54
06	06 37	08 41	12 06	01 44	00 32	18 26	22 40	15 57	11 31	20 54
07	06 15	12 21	11 28	01 58	00 N 16	18 28	22 40	15 56	11 32	20 53
08	05 53	15 16	10 48	02 13	00 00	18 30	22 41	15 55	11 32	20 53
09	05 30	17 15	10 06	02 28	00 S 16	18 32	22 41	15 54	11 33	20 52
10	05 07	18 09	09 23	02 44	00 32	18 34	22 41	15 53	11 33	20 51
11	04 45	17 55	08 39	02 59	00 48	18 36	22 41	15 52	11 34	20 51
12	04 22	16 37	07 54	03 14	01 04	18 39	22 41	15 51	11 35	20 50
13	03 59	14 21	07 09	03 29	01 20	18 41	22 41	15 50	11 35	20 50
14	03 36	11 20	06 22	03 44	01 35	18 43	22 42	15 49	11 36	20 49
15	03 13	07 46	05 36	03 58	01 51	18 45	22 42	15 48	11 36	20 48
16	02 50	03 S 52	04 49	04 12	02 07	18 47	22 42	15 47	11 37	20 48
17	02 27	00 N 08	04 01	04 25	02 23	18 50	22 42	15 46	11 38	20 47
18	02 03	04 03	03 14	04 39	02 39	18 52	22 42	15 45	11 38	20 47
19	01 40	07 43	02 26	04 51	02 55	18 54	22 42	15 44	11 39	20 46
20	01 17	10 59	01 39	05 03	03 11	18 57	22 42	15 43	11 39	20 46
21	00 54	13 45	00 51	05 14	03 27	18 59	22 43	15 42	11 40	20 45
22	00 30	15 55	00 N 04	05 25	03 43	19 01	22 43	15 41	11 41	20 45
23	00 N 07	17 23	00 S 43	05 35	03 59	19 04	22 43	15 40	11 41	20 44
24	00 S 16	18 06	01 30	05 45	04 14	19 06	22 43	15 39	11 42	20 44
25	00 40	18 01	02 17	05 53	04 30	19 08	22 43	15 39	11 43	20 43
26	01 03	17 07	03 03	06 01	04 46	19 11	22 43	15 38	11 43	20 43
27	01 27	15 23	03 49	06 09	05 02	19 13	22 43	15 37	11 44	20 42
28	01 50	12 51	04 34	06 15	05 18	19 15	22 43	15 36	11 45	20 42
29	02 13	09 37	05 19	06 21	05 33	19 18	22 44	15 35	11 45	20 41
30	02 S 37	05 N 47	06 S 03	06 N 26	05 S 49	19 S 20	22 S 44	15 N 34	11 S 46	20 N 41

ZODIAC SIGN ENTRIES

Date	h	m	Planets
02	08	31	☽ ♍
04	12	56	☽ ♎
05	02	28	☿ ♍
05	22	46	♂ ♎
06	15	53	☽ ♏
08	18	20	☽ ♐
10	21	04	☽ ♑
13	00	43	☽ ♒
15	05	54	☽ ♓
17	13	16	☽ ♈
19	23	12	☽ ♉
20	03	01	♀ ♌
21	01	20	☿ ♎
22	11	16	☽ ♊
23	19	08	☉ ♎
24	23	49	☽ ♋
25	08	15	♀ ♍
27	10	36	☽ ♌
29	18	04	☽ ♍

LATITUDES

Date	Mercury ☿ ° '	Venus ♀ ° '	Mars ♂ ° '	Jupiter ♃ ° '	Saturn ♄ ° '	Uranus ♅ ° '	Neptune ♆ ° '	Pluto ♇ ° '
01	01 N 11	08 S 35	00 N 46	00 N 43	00 N 47	00 N 39	01 N 44	11 N 47
04	01 32	08 37	00 45	00 43	00 46	00 39	01 44	11 48
07	01 44	08 31	00 43	00 42	00 46	00 39	01 44	11 48
10	01 48	08 17	00 42	00 42	00 46	00 40	01 44	11 48
13	01 46	07 57	00 40	00 41	00 45	00 40	01 44	11 49
16	01 38	07 31	00 39	00 41	00 45	00 40	01 44	11 49
19	01 26	07 02	00 37	00 40	00 44	00 40	01 43	11 50
22	01 11	06 30	00 36	00 39	00 44	00 40	01 43	11 50
25	00 54	05 57	00 34	00 39	00 43	00 40	01 43	11 51
28	00 35	05 23	00 33	00 38	00 43	00 40	01 43	11 52
31	00 N 15	04 S 49	00 N 31	00 N 38	00 N 42	00 N 40	01 N 43	11 N 52

DATA

Julian Date	2436813
Delta T	+33 seconds
Ayanamsa	23° 17' 40"
Synetic vernal point	05° ♓ 49' 20"
True obliquity of ecliptic	23° 26' 32"

MOON'S PHASES, APSIDES AND POSITIONS ☽

Date	h	m	Phase	Longitude ° '	Eclipse Indicator
03	01	56	●	09 ♍ 50	
09	22	07	☽	16 ♐ 28	
17	00	52	○	23 ♓ 24	
25	02	22	☾	01 ♋ 16	

Day	h	m		
07	16	28	Perigee	
23	01	25	Apogee	
04	11	05	0S	
10	19	07	Max dec	18° S 12'
17	11	15	0N	
24	21	32	Max dec	18° N 10'

LONGITUDES

Date	Chiron ⚷ ° '	Ceres ⚳ ° '	Pallas ⚴ ° '	Juno ⚵ ° '	Vesta ⚶ ° '	Black Moon Lilith ⚸ ° '
01	24 ♒ 27	24 ♏ 32	00 ♏ 51	26 ♎ 09	26 ♍ 03	02 ♊ 03
11	23 ♒ 59	27 ♏ 28	04 ♏ 41	29 ♎ 08	01 ♎ 02	03 ♊ 10
21	23 ♒ 34	00 ♐ 37	08 ♏ 38	02 ♏ 15	06 ♎ 06	04 ♊ 17
31	23 ♒ 12	03 ♐ 58	12 ♏ 41	05 ♏ 27	11 ♎ 12	05 ♊ 24

All ephemeris data is given at 12.00 UT and the Moon's longitude is additionally given for 24.00 UT

ASPECTARIAN

h m	Aspects
01 Tuesday	
01 00	☿ Q ♆
06 16	☽ ⚼ ♄
06 23	☉ ☌ ♀
10 59	☽ ☌ ♅
13 38	☽ # ♆
16 49	☽ ⊥ ♂
19 31	☽ Q ♆
22 14	☽ ☌ ☿
23 18	☽ □ ♃
02 Wednesday	
04 10	☽ ⊻ ♂
06 54	☿ □ ♃
09 20	☽ △ ♄
12 33	☽ ∥ ☉
15 51	☽ ☌ ♇
17 01	☽ ⚹ ♆
21 05	☽ ☌ ♀
03 Thursday	
01 56	☽ ☌ ☉
07 31	☽ Q ♃
16 32	☽ ⊻ ♅
19 22	☽ ∠ ♆
04 Friday	
02 08	♀ ∥ ♂
02 51	♂ ⊥ ♆
02 59	☽ ⊥ ♅
04 33	☽ ∥ ♀
04 33	☽ ⚹ ♃
04 51	☽ ∥ ♂
10 53	☽ ⊻ ☿
10 56	☽ ⊥ ♆
11 18	☽ ☌ ♂
13 44	☽ □ ♄
15 38	☿ ⊻ ♂
16 35	☽ # ♂
18 17	☽ # ♀
18 31	☽ ∠ ♅
20 08	☽ ⊻ ♇
21 14	☽ ⊻ ♆
22 43	☽ ⊥ ☿
22 54	☽ ⊻ ♀
05 Saturday	
01 02	♄ St D
06 24	☽ ⊥ ♇
06 29	☽ ∠ ♃
08 22	☿ △ ♄
08 43	☽ ⊥ ♀
09 34	☽ ⊻ ☉
16 22	☽ ∠ ☿
20 09	☽ ⚹ ♅
20 23	☽ Q ♄
20 30	☽ ⊥ ☉
21 43	☽ ∠ ♇
21 54	☽ ⊥ ♃
23 23	☽ ∠ ♀
06 Sunday	
00 50	☽ # ☉
08 07	☽ ⊻ ♃
12 49	☽ ∠ ☉
13 39	♀ ⚹ ♆
15 53	♂ □ ♄
16 30	☽ Q ♅
16 39	☽ ⚹ ♄
16 42	☽ ⊻ ♂
19 48	☉ Q ♃
21 34	☽ ⚹ ☿
23 05	☽ ⚹ ♇
23 42	☽ ⚹ ♀
07 Monday	
00 07	☽ ☌ ♆
03 16	☽ ⊥ ♂
06 15	☽ ∥ ♆
06 48	☽ # ☿
09 03	☿ ☌ ♇
09 44	☿ # ♆
10 27	☿ ☌ ♀
14 56	♀ ☌ ♇
15 54	☽ ⚹ ☉
16 55	☿ ⚹ ♆
17 54	☽ ∠ ♄
19 07	☽ ∠ ♂
19 08	☽ Q ♀
19 18	☽ Q ♇
20 51	☽ Q ☿
22 53	☽ □ ♅
08 Tuesday	
09 02	☽ ⊥ ♄
11 04	☽ ☌ ♃
13 33	☽ Q ☉
18 38	☽ # ♅
19 08	☽ ⊻ ♄
21 33	☽ ⚹ ♂
09 Wednesday	
00 20	☽ □ ♀
01 39	☽ □ ♇
02 40	☽ ⊻ ♆
07 50	☽ □ ☿
12 48	☽ ⊥ ♆
18 45	☽ Q ♂
22 07	☽ □ ☉
10 Thursday	
01 39	☽ △ ♅
04 03	☽ ∠ ♆
09 36	♀ ⊻ ♂
14 13	☽ ⊻ ♃
21 54	☽ ☌ ♄
11 Friday	
00 32	☽ ⊥ ♃
01 28	☽ △ ♀
02 46	☽ □ ♂
03 14	☽ ⚼ ♅
04 36	☽ △ ♇
05 37	☽ ⚹ ♆
13 50	♂ ∠ ♅
16 05	☽ ∠ ♃
18 43	☽ ± ♅
18 51	☽ △ ☿
12 Saturday	
02 14	☽ Q ♆
02 24	☽ ⚼ ♀
04 33	☉ ⊻ ♅
05 05	☽ ⊼ ♅
05 07	☽ △ ☉
06 25	☽ ⚼ ♇
13 52	☿ Q ♃
17 32	☽ ± ♀
18 14	☽ ⚹ ♃
20 25	♂ ⊻ ♇
21 13	☽ # ♅
22 03	☽ ± ♇
13 Sunday	
00 56	☽ ⚼ ☿
01 37	☽ ⊻ ♄
03 41	☽ ⊼ ♀
08 33	☽ ⊼ ♇
09 08	☽ △ ♂
09 10	☽ ⚼ ☉
09 34	☽ □ ♆
12 07	☽ ⊥ ♄
15 02	☉ ∠ ♆
15 25	☽ Q ♃
18 56	♂ ⊻ ♆
19 16	☽ ± ☿
14 Monday	
02 14	☽ ± ☉
04 00	☽ ∠ ♄
07 08	☉ ∥ ♀
07 37	☽ ⊼ ☿
09 49	☽ ☍ ♅
10 04	☽ ∥ ♆
12 58	☽ ⚼ ♂
13 44	☽ ⊼ ☉
23 46	☽ □ ♃
23 48	☿ ⊻ ♅
15 Tuesday	
05 57	☽ ± ♂
06 53	☽ ⚹ ♄
07 39	☽ ☍ ♀
14 09	☽ ☍ ♇
15 12	☽ △ ♆
16 25	☿ ∠ ♆
17 24	☽ ⊼ ♂
16 Wednesday	
04 47	☽ Q ♄
04 55	☽ # ☿
06 49	♀ ⊥ ♂
10 09	☽ # ♀
16 32	☽ ⊼ ♅
18 50	☽ ⚼ ♆
18 57	☽ # ☉
19 16	♀ △ ♄
21 52	☽ ∥ ♂
23 15	☽ ☍ ☿
17 Thursday	
00 52	☽ ☍ ☉
02 27	☿ ∥ ♀
03 46	☽ ± ♅
07 30	☽ △ ♃
11 44	☽ ± ♆
14 01	☽ ⊼ ♀
14 08	☉ # ♂
14 23	☽ □ ♄
20 49	☽ ⚼ ♅
21 05	☉ ☌ ☿
22 03	☽ ⊼ ♇
23 07	☽ ⊼ ♆
18 Friday	
00 49	☽ ∥ ☉
01 14	☽ ± ♀
02 43	☽ # ♂
04 21	☽ ☍ ♂
07 25	☿ ⊥ ♅
07 47	☽ ∥ ☿
09 32	☽ ± ♇
12 22	☽ ⚼ ♃
16 01	☽ ∥ ♀
17 07	☉ ⊥ ♅
18 16	☽ ⚼ ♀
19 Saturday	
01 08	☿ # ♂
01 46	☽ △ ♅
02 58	☽ ⚼ ♇
06 01	☽ ± ♃
13 17	☿ ⚹ ♃
15 12	☽ ⊼ ☉
17 55	☽ ⊼ ♃
18 40	☽ ⊼ ☿
23 14	☽ △ ♀
20 Sunday	
00 28	☽ △ ♄
04 03	☽ ± ☉
08 31	☽ △ ♇
08 35	☽ ± ☿
09 38	☽ ☍ ♆
15 44	☿ ⊥ ♆
17 24	☽ # ♆
18 18	☽ ⊼ ♂
23 32	☽ ⚼ ☉
21 Monday	
00 18	☿ ⊻ ♀
05 42	☽ ⚼ ☿
06 24	☽ ⚼ ♄
06 59	☽ ± ♂
07 18	☉ ⚹ ♃
09 36	☉ ∥ ☿
10 24	☿ □ ♄
13 26	☽ □ ♅
22 Tuesday	
00 33	☽ ± ♄
02 10	☽ ⚼ ♂
06 39	☽ ☍ ♃
08 24	☽ △ ☉
09 05	☽ ∥ ♅
11 00	☽ □ ♀
12 32	♂ ⊥ ♇
12 42	☽ ⊼ ♄
17 15	♀ St D
17 16	☽ △ ☿
20 57	☽ □ ♇
22 06	☽ ⊼ ♆
23 39	☉ # ☿
23 Wednesday	
01 57	☽ Q ♅
03 25	☉ ⊥ ♆
10 16	☽ ± ♆
10 17	☽ △ ♂
11 08	☿ ∠ ♅
16 17	☉ ⊻ ♀
18 32	☿ ⊻ ♇
23 28	☽ Q ♀
24 Thursday	
02 20	☽ ⚹ ♅
02 23	☿ ⊻ ♆
04 31	☽ ⚼ ♆
09 26	☽ Q ♇
09 42	☿ ⊥ ♀
13 59	☉ □ ♄
20 05	☽ ⊼ ♃
23 45	☽ ⚹ ♀
25 Friday	
01 23	☽ ☍ ♄
02 22	☽ □ ☉
08 14	☽ ± ♃
08 33	☽ ∠ ♅
09 32	☽ ⚹ ♇
10 39	☽ △ ♆
15 58	☽ □ ☿
26 Saturday	
01 51	☽ □ ♂
02 18	☽ ⚼ ♃
02 26	☽ ⊥ ♅
05 46	☽ ∠ ♀
12 50	♂ ∠ ♃
14 14	☽ ⊻ ♅
15 09	☽ ∠ ♇
16 59	☽ Q ☉
23 30	☽ ⊥ ♀
27 Sunday	
07 38	☿ ⊥ ♇
07 49	☽ △ ♃
08 33	☽ ⊥ ♇
08 42	☽ Q ☿
09 17	☽ ∥ ♅
11 11	☽ ⊻ ♀
12 15	☽ ⊼ ♄
14 39	☽ Q ♂
17 58	☽ ⚹ ☉
19 59	☽ ⊻ ♇
21 04	☽ □ ♆
23 39	☽ ± ♄
28 Monday	
09 50	☉ ∠ ♅
10 46	☽ ⚹ ☿
14 22	☽ ⚹ ♂
16 28	☽ ⚼ ♄
20 45	☽ # ♆
20 51	☉ ⊻ ♇
23 09	☽ ☌ ♅
29 Tuesday	
00 11	☽ ∠ ☉
01 14	☿ ∠ ♃
06 28	☽ Q ♆
11 18	☉ ⊻ ♆
16 11	☽ □ ♃
17 46	☽ ⊥ ☉
18 05	☽ ∠ ☿
19 04	☽ ∠ ♂
19 32	☽ ☌ ♀
19 46	☽ △ ♄
23 57	☿ ∥ ♂
30 Wednesday	
02 59	☽ ☌ ♇
04 00	☽ ⚹ ♆
05 15	☽ ⊻ ☉
06 51	☿ ☌ ♂
07 53	☿ ∠ ♀
08 13	☽ ∥ ♀
09 32	♀ △ ♄
10 35	♀ ∠ ♂
10 37	☽ # ☿
11 48	☽ # ♂
11 48	☽ ⊥ ♂
12 13	☽ ⊥ ☿
22 47	☽ ⊻ ♂

OCTOBER 1959

Date	Sidereal time h m s	Sun ☉	Moon ☽	Moon ☽ 24.00	Mercury ☿	Venus ♀	Mars ♂	Jupiter ♃	Saturn ♄	Uranus ♅	Neptune ♆	Pluto ⯓
01	12 37 54	07 ♎ 33 32	24 ♍ 01 37	01 ♎ 06 12	17 ♎ 53	01 ♍ 19	16 ♎ 42	29 ♏ 16	01 ♑ 01	19 ♌ 40	05 ♏ 38	05 ♍ 03
02	12 41 50	08 32 34	08 ♎ 15 21	15 ♎ 28 29	19 30	01 39	17 21	29 27	01 04	19 43	05 40	05 05
03	12 45 47	09 31 39	22 ♎ 44 52	00 ♏ 03 43	21 07	02 00	18 01	29 37	01 07	19 46	05 42	05 07
04	12 49 43	10 30 45	07 ♏ 24 12	14 ♏ 45 29	22 43	02 24	18 41	29 48	01 09	19 48	05 44	05 08
05	12 53 40	11 29 54	22 ♏ 06 41	29 ♏ 27 01	24 18	02 49	19 21	29 ♏ 59	01 12	19 51	05 46	05 10
06	12 57 36	12 29 04	06 ♐ 45 44	14 ♐ 02 08	25 52	03 16	20 01	00 ♐ 10	01 15	19 54	05 48	05 12
07	13 01 33	13 28 16	21 ♐ 15 41	28 ♐ 25 53	27 26	03 45	20 41	00 21	01 18	19 56	05 50	05 13
08	13 05 29	14 27 30	05 ♑ 32 24	12 ♑ 34 56	28 ♎ 59	04 15	21 20	00 32	01 21	19 59	05 52	05 15
09	13 09 26	15 26 46	19 ♑ 33 21	26 ♑ 27 33	00 ♏ 30	04 47	22 01	00 43	01 25	20 01	05 55	05 17
10	13 13 23	16 26 03	03 ♒ 17 32	10 ♒ 03 21	02 02	05 20	22 41	00 54	01 28	20 04	05 57	05 18
11	13 17 19	17 25 22	16 ♒ 45 05	23 ♒ 22 51	03 32	05 54	23 21	01 06	01 31	20 06	05 59	05 20
12	13 21 16	18 24 43	29 ♒ 56 49	06 ♓ 27 07	05 02	06 30	24 01	01 17	01 35	20 08	06 01	05 21
13	13 25 12	19 24 05	12 ♓ 53 57	19 ♓ 17 26	06 31	07 07	24 41	01 28	01 38	20 11	06 03	05 23
14	13 29 09	20 23 30	25 ♓ 37 46	01 ♈ 55 05	07 59	07 46	25 21	01 40	01 42	20 13	06 05	05 25
15	13 33 05	21 22 56	08 ♈ 09 32	14 ♈ 21 17	09 26	08 25	26 01	01 52	01 46	20 15	06 08	05 26
16	13 37 02	22 22 24	20 ♈ 30 29	26 ♈ 37 17	10 53	09 06	26 42	02 03	01 49	20 17	06 10	05 28
17	13 40 58	23 21 55	02 ♉ 41 51	08 ♉ 44 22	12 19	09 48	27 22	02 15	01 53	20 20	06 12	05 29
18	13 44 55	24 21 27	14 ♉ 45 03	20 ♉ 44 07	13 44	10 31	28 02	02 27	01 57	20 22	06 14	05 31
19	13 48 52	25 21 02	26 ♉ 41 48	02 ♊ 38 25	15 08	11 15	28 43	02 39	02 01	20 24	06 16	05 32
20	13 52 48	26 20 39	08 ♊ 34 16	14 ♊ 29 43	16 31	12 00	29 ♎ 23	02 51	02 05	20 26	06 19	05 33
21	13 56 45	27 20 18	20 ♊ 25 09	26 ♊ 21 00	17 53	12 46	00 ♏ 04	03 03	02 10	20 28	06 21	05 35
22	14 00 41	28 19 59	02 ♋ 17 43	08 ♋ 15 47	19 15	13 33	00 45	03 15	02 14	20 30	06 23	05 36
23	14 04 38	29 ♎ 19 42	14 ♋ 15 45	20 ♋ 18 09	20 35	14 21	01 25	03 27	02 18	20 32	06 25	05 37
24	14 08 34	00 ♏ 19 28	26 ♋ 23 33	02 ♌ 32 32	21 54	15 09	02 06	03 39	02 23	20 33	06 27	05 39
25	14 12 31	01 19 16	08 ♌ 45 39	15 ♌ 03 29	23 12	15 59	02 46	03 52	02 27	20 35	06 30	05 40
26	14 16 27	02 19 06	21 ♌ 26 33	27 ♌ 55 21	24 29	16 49	03 27	04 04	02 32	20 37	06 32	05 41
27	14 20 24	03 18 59	04 ♍ 30 19	11 ♍ 11 47	25 45	17 41	04 08	04 16	02 36	20 39	06 34	05 43
28	14 24 21	04 18 53	18 ♍ 00 01	24 ♍ 55 06	26 59	18 32	04 49	04 29	02 41	20 40	06 36	05 44
29	14 28 17	05 18 50	01 ♎ 57 00	09 ♎ 05 31	28 11	19 25	05 30	04 41	02 46	20 42	06 39	05 45
30	14 32 14	06 18 49	16 ♎ 20 14	23 ♎ 40 36	29 ♏ 22	20 18	06 11	04 54	02 51	20 43	06 41	05 46
31	14 36 10	07 ♏ 18 49	01 ♏ 05 50	08 ♏ 34 59	00 ♐ 30	21 ♍ 12	06 ♏ 52	05 ♐ 07	02 ♑ 56	20 ♌ 45	06 ♏ 43	05 ♍ 47

Date	Moon True ☊	Moon Mean ☊	Moon ☽ Latitude
01	04 ♎ 02	03 ♎ 34	00 S 55
02	04 R 02	03 31	00 N 23
03	04 01	03 28	01 42
04	04 01	03 25	02 55
05	04 00	03 22	03 56
06	03 59	03 18	04 42
07	03 58	03 15	05 09
08	03 58	03 12	05 17
09	03 D 58	03 09	05 05
10	03 58	03 06	04 36
11	03 59	03 02	03 53
12	04 00	02 59	02 58
13	04 01	02 56	01 54
14	04 02	02 53	00 N 46
15	04 R 02	02 50	00 S 23
16	04 02	02 47	01 30
17	04 00	02 43	02 32
18	03 57	02 40	03 26
19	03 54	02 37	04 11
20	03 50	02 34	04 44
21	03 46	02 31	05 06
22	03 43	02 28	05 14
23	03 41	02 24	05 09
24	03 40	02 21	04 50
25	03 D 40	02 18	04 17
26	03 41	02 15	03 31
27	03 43	02 12	02 33
28	03 44	02 08	01 25
29	03 45	02 05	00 S 10
30	03 R 44	02 02	01 N 08
31	03 ♎ 42	01 ♎ 59	02 N 24

DECLINATIONS

Date	Sun ☉	Moon ☽	Mercury ☿	Venus ♀	Mars ♂	Jupiter ♃	Saturn ♄	Uranus ♅	Neptune ♆	Pluto ⯓
01	03 S 00	01 N 32	06 S 47	06 N 30	06 S 05	19 S 23	22 S 44	15 N 33	11 S 47	20 N 40
02	03 23	02 S 55	07 31	06 34	06 20	19 25	22 44	15 32	11 48	20 40
03	03 47	07 16	08 14	06 36	06 36	19 28	22 44	15 32	11 48	20 39
04	04 10	11 14	08 56	06 39	06 52	19 30	22 44	15 31	11 49	20 39
05	04 33	14 29	09 38	06 40	07 07	19 32	22 44	15 30	11 50	20 39
06	04 56	16 48	10 19	06 40	07 23	19 35	22 44	15 29	11 50	20 38
07	05 19	18 01	10 59	06 40	07 38	19 37	22 45	15 28	11 51	20 38
08	05 42	18 03	11 39	06 39	07 54	19 40	22 45	15 28	11 52	20 38
09	06 05	16 59	12 18	06 38	08 09	19 42	22 45	15 27	11 52	20 37
10	06 28	14 56	12 56	06 36	08 25	19 45	22 45	15 26	11 53	20 37
11	06 50	12 07	13 34	06 33	08 40	19 47	22 45	15 25	11 54	20 37
12	07 13	08 43	14 11	06 29	08 55	19 50	22 45	15 25	11 55	20 36
13	07 36	04 57	14 47	06 24	09 11	19 52	22 45	15 24	11 55	20 36
14	07 58	01 S 02	15 22	06 19	09 26	19 55	22 45	15 23	11 56	20 36
15	08 20	02 N 53	15 57	06 14	09 41	19 57	22 45	15 23	11 57	20 35
16	08 43	06 38	16 31	06 07	09 56	20 00	22 45	15 22	11 58	20 35
17	09 05	10 02	17 03	06 00	10 11	20 02	22 46	15 21	11 58	20 35
18	09 27	12 59	17 35	05 53	10 26	20 05	22 46	15 21	11 59	20 35
19	09 48	15 21	18 06	05 44	10 41	20 07	22 46	15 20	12 00	20 34
20	10 10	17 03	18 36	05 36	10 56	20 09	22 46	15 19	12 00	20 34
21	10 32	18 01	19 05	05 26	11 11	20 12	22 46	15 19	12 01	20 34
22	10 53	18 12	19 34	05 16	11 25	20 14	22 46	15 18	12 02	20 34
23	11 14	17 34	20 01	05 05	11 40	20 17	22 46	15 18	12 03	20 34
24	11 35	16 08	20 27	04 54	11 54	20 19	22 46	15 17	12 03	20 33
25	11 56	13 56	20 52	04 43	12 09	20 22	22 46	15 17	12 04	20 33
26	12 17	11 01	21 15	04 30	12 23	20 24	22 46	15 16	12 05	20 33
27	12 37	07 29	21 38	04 17	12 38	20 26	22 46	15 16	12 06	20 33
28	12 58	03 N 26	21 59	04 04	12 52	20 29	22 46	15 15	12 06	20 33
29	13 18	00 S 56	22 19	03 50	13 06	20 31	22 46	15 15	12 07	20 33
30	13 38	05 22	22 38	03 36	13 20	20 34	22 46	15 14	12 08	20 33
31	13 S 57	09 S 37	22 S 56	03 N 21	13 S 34	20 S 36	22 S 46	15 N 14	12 S 09	20 N 32

ZODIAC SIGN ENTRIES

Date	h m	Planets
01	22 08	☽ ♎
03	23 54	☽ ♏
05	14 40	♃ ♐
06	00 54	☽ ♐
08	02 38	☽ ♑
09	04 02	☿ ♏
10	06 12	☽ ♒
12	12 06	☽ ♓
14	20 20	☽ ♈
17	06 40	☽ ♉
19	18 40	☽ ♊
21	09 40	♂ ♏
22	07 22	☽ ♋
24	04 11	☉ ♏
24	19 03	☽ ♌
27	03 48	☽ ♍
29	08 41	☽ ♎
31	01 16	☿ ♐
31	10 14	☽ ♏

LATITUDES

Date	Mercury ☿	Venus ♀	Mars ♂	Jupiter ♃	Saturn ♄	Uranus ♅	Neptune ♆	Pluto ⯓
01	00 N 15	04 S 49	00 N 31	00 N 38	00 N 42	00 N 40	01 N 43	11 N 52
04	00 S 06	04 16	00 30	00 37	00 42	00 40	01 43	11 53
07	00 28	03 43	00 28	00 37	00 42	00 40	01 43	11 54
10	00 49	03 11	00 27	00 37	00 41	00 40	01 43	11 55
13	01 10	02 41	00 25	00 36	00 41	00 40	01 43	11 56
16	01 30	02 12	00 23	00 36	00 40	00 41	01 43	11 57
19	01 49	01 44	00 22	00 35	00 40	00 41	01 43	11 58
22	02 06	01 18	00 20	00 35	00 40	00 41	01 43	11 59
25	02 21	00 53	00 18	00 34	00 39	00 41	01 43	12 00
28	02 34	00 30	00 17	00 34	00 39	00 41	01 43	12 01
31	02 S 44	00 S 09	00 N 15	00 N 34	00 N 38	00 N 41	01 N 43	12 N 02

DATA

Julian Date	2436843
Delta T	+33 seconds
Ayanamsa	23° 17' 42"
Synetic vernal point	05° ♓ 49' 18"
True obliquity of ecliptic	23° 26' 32"

MOON'S PHASES, APSIDES AND POSITIONS ☽

Date	h m	Phase	Longitude	Eclipse Indicator
02	12 31	●	08 ♎ 34	Total
09	04 22	☽	15 ♑ 08	
16	15 59	○	22 ♈ 32	
24	20 22	☾	00 ♌ 40	
31	22 41	●	07 ♏ 46	

Day	h m		
04	21 06	Perigee	
20	19 10	Apogee	
01	20 18	0S	
08	00 46	Max dec	18° S 11'
14	18 15	0N	
22	05 16	Max dec	18° N 13'
29	06 59	0S	

LONGITUDES

Date	Chiron ⚷	Ceres ⚳	Pallas ⚴	Juno ⚵	Vesta ⚶	Black Moon Lilith ⚸
01	23 ♒ 12	03 ♐ 58	12 ♏ 41	05 ♏ 27	11 ♎ 12	05 ♊ 24
11	22 ♒ 56	07 ♐ 29	16 ♏ 50	08 ♏ 44	16 ♎ 21	06 ♊ 31
21	22 ♒ 46	11 ♐ 07	21 ♏ 00	12 ♏ 04	21 ♎ 33	07 ♊ 38
31	22 ♒ 41	14 ♐ 53	25 ♏ 18	15 ♏ 27	26 ♎ 45	08 ♊ 45

ASPECTARIAN

h m	Aspects
01 Thursday	
00 03	☽ ⊻ ☿
00 15	☽ Q ♃
01 34	☿ # ♀
03 05	☉ ⊥ ♀
04 31	☽ ⊻ ♅
04 33	☽ # ☉
06 11	☽ ∠ ♆
14 48	☽ ⊥ ♅
21 01	☽ ✶ ♃
21 32	☽ ⊥ ♆
23 54	☽ □ ♄
02 Friday	
00 38	☽ ⊻ ♀
05 18	☿ Q ♄
06 03	☽ ∠ ♅
06 41	☽ ⊻ ⯓
07 39	☽ ⊻ ♆
10 57	☽ ⊥ ♀
12 31	☽ ☌ ☉
14 48	☽ ∥ ☉
15 09	☿ ✶ ♅
16 43	☽ ⊥ ⯓
20 39	☿ ∠ ⯓
22 25	☽ ∠ ♃
03 Saturday	
02 17	☽ ∠ ♀
03 50	☽ ☌ ♂
06 00	☽ Q ♄
07 04	☽ ✶ ♅
07 39	☽ ∠ ⯓
07 59	☽ ∥ ♂
08 13	☽ # ♀
08 59	☽ ☌ ☿
12 38	♀ # ♂
13 27	☽ ⊥ ♃
18 42	☽ ∥ ☿
23 25	☽ ⊻ ♃
04 Sunday	
01 46	☽ ✶ ♄
02 49	☽ Q ♅
03 36	☽ ✶ ♀
08 18	☽ ✶ ⯓
09 16	☽ ☌ ♆
15 56	☽ ∥ ♆
17 26	☽ ⊻ ☉
23 45	☽ Q ♀
05 Monday	
02 20	☽ ∠ ♄
03 41	☉ ⊥ ⯓
03 55	☽ Q ⯓
03 56	☽ ⊥ ☉
06 29	☿ ⊥ ♃
06 36	♂ Q ♄
07 16	☽ ⊻ ♂
08 18	☽ □ ♅
16 00	☽ ⊻ ☿
17 04	☽ ⊥ ♄
17 32	☽ ⊥ ♂
19 41	☽ ∠ ☉
21 07	☽ # ♅
06 Tuesday	
01 02	☽ ☌ ♃
02 56	☽ ⊻ ♄
03 00	☽ ⊥ ☿
06 05	☽ □ ♀
07 28	♂ ✶ ♅
08 59	☽ ∠ ♂
09 25	☽ □ ⯓
10 25	☽ ⊻ ♆
18 58	♂ ∠ ⯓
19 35	☽ ∠ ☿
20 20	☽ ⊥ ♆
22 07	☽ ✶ ☉
07 Wednesday	
09 47	☽ △ ♅
10 59	☽ ✶ ♂
11 18	☽ ∠ ♆
19 33	☽ Q ☉
23 34	☽ ✶ ☿
08 Thursday	
03 25	☽ ⊻ ♃
04 54	☽ ☌ ♄
08 06	☽ Q ♂
09 44	☽ △ ♀
11 03	☽ ♇ ♅
11 30	☽ △ ⯓
12 34	☽ ✶ ♆
13 42	☽ ⊥ ♃
19 54	☿ ∥ ♆
22 23	☽ Q ☿
09 Friday	
02 26	☽ ± ♅
04 22	☽ □ ☉
05 17	☽ ∠ ♃
09 09	☽ Q ♆
12 24	☽ ♇ ♀
12 48	☽ ⊼ ♅
13 15	☽ ♇ ⯓
15 45	☿ ⊻ ♃
16 28	☽ □ ♂
20 07	☉ ∠ ♃
10 Saturday	
02 47	☿ ✶ ♄
04 44	☽ ± ♀
04 58	☽ ± ⯓
06 58	☽ # ♅
07 44	☽ ✶ ♃
08 46	☽ ⊻ ♄
09 29	☽ □ ☿
10 57	♀ ☌ ⯓
12 32	☿ Q ♅
15 34	☽ ⊼ ⯓
15 45	☽ ⊼ ♀
16 42	☽ □ ♆
19 19	☉ # ♀
19 25	☽ ⊥ ♄
11 Sunday	
02 26	☽ ∥ ☿
05 20	☽ Q ♃
11 35	☽ ∠ ♄
13 18	☽ △ ☉
13 38	☽ ∥ ♆
15 24	♀ ✶ ♆
18 04	☽ ☍ ♅
12 Monday	
00 34	☽ △ ♂
10 44	☽ ∥ ♂
14 30	☽ □ ♃
15 01	☽ ✶ ♄
17 25	☿ ✶ ⯓
18 54	☽ ♇ ☉
20 53	☽ ∥ ☉
22 00	☽ ☍ ⯓
22 35	☽ △ ☿
23 13	☽ △ ♆
13 Tuesday	
00 41	☽ ☍ ♀
02 45	☽ # ♀
04 23	☿ ☌ ♆
05 40	☽ ♇ ♂
13 01	☽ ± ☉
13 23	☽ Q ♄
18 08	☉ Q ♄
23 27	☽ ± ♂
14 Wednesday	
01 14	☽ ⊼ ☉
01 43	☽ ⊼ ♅
03 22	☽ ♇ ♆
05 31	☿ ✶ ♀
06 19	☽ ♇ ☿
07 35	☉ ✶ ♅
11 27	☽ ⊼ ♂
12 26	☉ ∠ ⯓
12 35	☿ # ♅
13 07	☽ ± ♅
17 51	♃ ⊻ ♄
20 31	☽ ± ♆
23 38	☽ □ ♄
23 42	☽ △ ♃
15 Thursday	
01 43	☽ ± ☿
03 48	♂ ⊥ ♃
06 23	☽ ♇ ♅
06 44	☽ ⊼ ⯓
08 04	☽ ⊼ ♆
12 32	☽ ⊼ ♀
14 47	☽ ⊼ ☿
18 21	☽ ± ⯓
16 Friday	
00 50	☽ ± ♀
05 09	☽ ♇ ♃
08 46	☽ ∥ ♀
11 34	☽ △ ♅
11 54	☽ ♇ ⯓
15 59	☽ ☍ ☉
19 28	☽ ♇ ♀
23 04	☽ ± ♃
17 Saturday	
00 51	☽ ☍ ♂
04 05	☽ # ☉
10 23	☽ △ ♄
11 06	☽ ⊼ ♃
13 13	☽ # ♂
17 32	☽ △ ⯓
18 58	☽ ☍ ♆
18 Sunday	
03 00	☽ △ ♀
03 20	☽ # ♆
09 41	☽ ☍ ☿
16 26	☽ ♇ ♄
23 17	☽ □ ♅
19 Monday	
09 02	☽ ⊼ ☉
10 38	☽ ± ♄
11 46	☽ ∥ ♅
16 19	☽ ⊼ ♂
22 15	☽ ± ☉
22 49	☽ ⊼ ♄
20 Tuesday	
00 13	☽ ☍ ♃
05 10	☽ ± ♂
05 53	☽ □ ⯓
07 24	☽ ⊼ ♆
11 43	☽ Q ♅
18 08	☽ ♇ ☉
19 25	☽ □ ♀
19 36	☽ ± ♆
22 33	☿ ∠ ♄
21 Wednesday	
00 30	☽ ♇ ♂
03 14	☉ ⊥ ♃
06 12	☽ ⊼ ☿
06 29	☿ Q ⯓
12 05	☽ ✶ ♅
13 53	☽ ♇ ♆
18 25	☽ Q ⯓
19 56	☽ ± ☿
22 Thursday	
03 17	☽ △ ☉
08 41	☽ △ ♂
10 23	☽ Q ♀
11 52	☽ ☍ ♄
13 57	☽ ⊼ ♃
16 26	☽ ♇ ☿
18 27	☽ ∠ ♅
18 40	☽ ✶ ⯓
20 15	☽ △ ♆
23 Friday	
02 13	☽ ± ♃
10 55	☿ □ ♅
12 11	☽ ✶ ♀
12 32	☽ ⊥ ♅
14 08	☉ ∠ ♀
20 29	☽ ♇ ♃
24 Saturday	
00 28	☽ ⊻ ♅
00 40	☽ ∠ ⯓
02 06	☽ △ ☿
04 19	☿ ∥ ♃
18 20	☿ # ⯓
18 23	☽ ⊥ ⯓
19 53	☽ ∠ ♀
20 22	☽ □ ☉
22 16	☽ ∥ ♅
23 11	♂ ✶ ♄
23 45	☽ ⊼ ♄
23 47	☽ □ ♂
25 Sunday	
02 24	☽ △ ♃
03 34	♂ ∥ ♆
05 02	♂ Q ♅
06 02	☽ ⊻ ⯓
07 38	☽ □ ♆
11 24	☽ ± ♄
14 31	☽ ⊥ ♀
21 39	☉ ∥ ♆
26 Monday	
02 11	☽ # ♂
02 43	☽ ⊻ ♀
03 17	☽ # ☉
03 54	☽ # ♆
04 37	☽ ♇ ♄
09 43	☽ Q ☉
10 27	☽ ☌ ♅
12 01	☽ Q ♂
17 29	☉ ✶ ♄
17 46	☽ Q ♆
18 17	☽ □ ☿
19 21	☉ Q ♅
27 Tuesday	
08 32	☽ △ ♄
09 40	☽ ✶ ☉
11 18	☽ ✶ ♂
11 34	☽ □ ♃
13 45	☉ ∥ ♂
14 11	☽ ☌ ⯓
15 44	☽ ✶ ♆
19 04	♂ ⊻ ♃
28 Wednesday	
05 47	☿ ⊥ ♄
06 11	☽ Q ☿
08 11	☽ ∥ ♀
13 01	☽ ☌ ♀
14 29	☽ ∠ ☉
15 20	☽ ∠ ♂
16 40	☽ ⊻ ♅
17 02	☉ ⊻ ♃
18 18	☽ ∠ ♆
19 55	☽ Q ♃
29 Thursday	
03 02	☽ ⊥ ♅
05 00	☽ ✶ ☿
07 11	☽ ⊥ ☉
07 37	☽ ⊥ ♂
09 47	☽ ⊥ ♆
13 23	☽ □ ♄
16 42	☽ ✶ ♃
18 06	☽ ⊻ ☉
18 17	☽ ⊻ ♂
18 20	☽ ∠ ♅
18 25	☽ ⊻ ⯓
19 56	☽ ⊻ ♆
21 16	♂ ✶ ⯓
22 43	☉ ✶ ⯓
30 Friday	
01 32	♃ # ⯓
01 46	☉ ☌ ♂
02 53	☽ # ♀
04 27	☽ ⊥ ⯓
08 28	☽ ∠ ☿
17 55	☽ ∠ ♃
18 56	☽ ⊻ ♀
19 12	☽ ✶ ♅
19 17	☽ ∠ ⯓
19 26	☽ Q ♄
21 10	☉ ☌ ♆
22 25	☿ ∥ ♄
23 26	♀ ⊻ ♅
31 Saturday	
00 28	☽ ⊥ ☿
05 19	☽ ⊥ ♀
06 43	♂ ☌ ♆
08 45	☽ ⊥ ♃
10 58	☽ ⊻ ☿
14 39	☽ Q ♅
14 57	☽ ✶ ♄
18 32	☽ ⊻ ♃
19 33	☽ ✶ ⯓
20 44	☽ ∠ ♀
21 03	☽ ☌ ♆
21 41	☽ ☌ ♂
22 41	☽ ☌ ☉

All ephemeris data is given at 12.00 UT and the Moon's longitude is additionally given for 24.00 UT

NOVEMBER 1959

Date	Sidereal time h m s	Sun ☉ ° ' "	Moon ☽ ° ' "	Moon ☽ 24.00 ° ' "	Mercury ☿ ° '	Venus ♀ ° '	Mars ♂ ° '	Jupiter ♃ ° '	Saturn ♄ ° '	Uranus ♅ ° '	Neptune ♆ ° '	Pluto ♇ ° '
01	14 40 07	08 ♏ 18 52	16 ♏ 06 58	23 ♏ 40 36	01 ♐ 37	22 ♍ 07	07 ♏ 33	05 ♐ 19	03 ♑ 01	20 ♌ 46	06 ♏ 45	05 ♍ 49
02	14 44 03	09 18 57	01 ♐ 14 37	08 ♐ 47 48	02 41	23 02	08 14	05 32	03 06	20 48	06 48	05 50
03	14 48 00	10 19 03	16 ♐ 18 55	23 ♐ 46 52	03 42	23 58	08 55	05 45	03 11	20 49	06 50	05 51
04	14 51 56	11 19 12	01 ♑ 10 43	08 ♑ 29 40	04 40	24 55	09 36	05 57	03 16	20 50	06 52	05 52
05	14 55 53	12 19 21	15 ♑ 43 06	22 ♑ 50 38	05 35	25 52	10 17	06 10	03 21	20 51	06 54	05 53
06	14 59 50	13 19 33	29 ♑ 52 00	06 ♒ 47 09	06 26	26 49	10 58	06 23	03 27	20 53	06 57	05 54
07	15 03 46	14 19 45	13 ♒ 36 10	20 ♒ 19 15	07 13	27 47	11 40	06 36	03 32	20 54	06 59	05 55
08	15 07 43	15 20 00	26 ♒ 56 40	03 ♓ 28 48	07 55	28 46	12 21	06 49	03 37	20 55	07 01	05 56
09	15 11 39	16 20 15	09 ♓ 56 02	16 ♓ 18 49	08 32	29 ♍ 45	13 02	07 02	03 43	20 56	07 03	05 57
10	15 15 36	17 20 32	22 ♓ 37 33	28 ♓ 52 41	09 02	00 ♎ 44	13 44	07 15	03 48	20 57	07 05	05 57
11	15 19 32	18 20 51	05 ♈ 04 38	11 ♈ 13 46	09 26	01 44	14 25	07 28	03 54	20 58	07 08	05 58
12	15 23 29	19 21 11	17 ♈ 20 28	23 ♈ 25 02	09 43	02 45	15 07	07 41	04 00	20 59	07 10	05 59
13	15 27 25	20 21 32	29 ♈ 27 46	05 ♉ 28 55	09 52	03 45	15 48	07 54	04 05	20 59	07 12	06 00
14	15 31 22	21 21 56	11 ♉ 28 44	17 ♉ 27 25	09 R 52	04 47	16 30	08 08	04 11	21 00	07 14	06 01
15	15 35 19	22 22 20	23 ♉ 25 10	29 ♉ 22 10	09 43	05 48	17 12	08 21	04 17	21 01	07 17	06 01
16	15 39 15	23 22 47	05 ♊ 18 34	11 ♊ 14 36	09 24	06 51	17 53	08 34	04 23	21 01	07 19	06 02
17	15 43 12	24 23 15	17 ♊ 10 25	23 ♊ 06 15	08 55	07 53	18 35	08 47	04 29	21 02	07 21	06 03
18	15 47 08	25 23 45	29 ♊ 02 21	04 ♋ 58 59	08 15	08 56	19 17	09 00	04 35	21 02	07 23	06 04
19	15 51 05	26 24 17	10 ♋ 56 26	16 ♋ 55 04	07 25	09 59	19 59	09 14	04 41	21 03	07 25	06 04
20	15 55 01	27 24 50	22 ♋ 55 17	28 ♋ 57 28	06 25	11 03	20 41	09 27	04 47	21 03	07 27	06 05
21	15 58 58	28 25 25	05 ♌ 02 08	11 ♌ 09 45	05 17	12 06	21 22	09 40	04 53	21 04	07 29	06 05
22	16 02 54	29 ♏ 26 01	17 ♌ 20 53	23 ♌ 36 03	04 02	13 11	22 04	09 54	04 59	21 04	07 32	06 06
23	16 06 51	00 ♐ 26 40	29 ♌ 55 51	06 ♍ 20 50	02 43	14 15	22 46	10 07	05 05	21 04	07 34	06 06
24	16 10 48	01 27 20	12 ♍ 51 33	19 ♍ 28 29	01 21	15 20	23 28	10 21	05 12	21 04	07 36	06 07
25	16 14 44	02 28 01	26 ♍ 12 06	03 ♎ 02 44	00 ♐ 00	16 25	24 11	10 34	05 18	21 05	07 38	06 07
26	16 18 41	03 28 45	10 ♎ 00 35	17 ♎ 05 45	28 ♏ 41	17 31	24 53	10 48	05 24	21 05	07 40	06 08
27	16 22 37	04 29 30	24 ♎ 18 04	01 ♏ 37 14	27 29	18 36	25 35	11 01	05 31	21 R 05	07 42	06 08
28	16 26 34	05 30 16	09 ♏ 02 40	16 ♏ 33 34	26 25	19 42	26 17	11 14	05 37	21 05	07 44	06 08
29	16 30 30	06 31 04	24 ♏ 08 55	01 ♐ 47 28	25 30	20 49	26 59	11 28	05 44	21 04	07 46	06 09
30	16 34 27	07 ♐ 31 53	09 ♐ 27 51	17 ♐ 08 35	24 ♏ 46	21 ♎ 55	27 ♏ 42	11 ♐ 41	05 ♑ 50	21 ♌ 04	07 ♏ 48	06 ♍ 09

Date	Moon True ☊ ° '	Moon Mean ☊ ° '	Moon ☽ Latitude ° '	DECLINATIONS Sun ☉ ° '	Moon ☽	Mercury ☿	Venus ♀	Mars ♂	Jupiter ♃	Saturn ♄	Uranus ♅	Neptune ♆	Pluto ♇
01	03 ♎ 38	01 ♎ 56	03 N 31	14 S 17	13 S 18	23 S 12	03 N 06	13 S 48	20 S 38	22 S 46	15 N 13	12 S 09	20 N 32
02	03 R 33	01 53	04 23	14 36	16 07	23 27	02 50	14 02	20 41	22 46	15 13	12 10	20 32
03	03 28	01 49	04 57	14 55	17 49	23 40	02 34	14 16	20 43	22 46	15 13	12 11	20 32
04	03 23	01 46	05 10	15 14	18 16	23 51	02 18	14 29	20 45	22 46	15 12	12 12	20 32
05	03 18	01 43	05 03	15 32	17 30	24 01	02 01	14 43	20 48	22 46	15 12	12 12	20 32
06	03 16	01 40	04 38	15 50	15 39	24 10	01 44	14 56	20 50	22 46	15 12	12 13	20 32
07	03 15	01 37	03 57	16 08	12 57	24 16	01 26	15 10	20 52	22 46	15 11	12 14	20 32
08	03 D 15	01 34	03 04	16 26	09 39	24 21	01 08	15 23	20 55	22 46	15 11	12 14	20 32
09	03 17	01 30	02 03	16 44	05 56	24 23	00 50	15 36	20 57	22 46	15 11	12 15	20 32
10	03 18	01 27	00 N 58	17 01	02 S 03	24 23	00 31	15 49	20 59	22 46	15 10	12 16	20 32
11	03 R 19	01 24	00 S 10	17 18	01 N 52	24 21	00 N 12	16 02	21 01	22 46	15 10	12 16	20 32
12	03 18	01 21	01 15	17 34	05 39	24 17	00 S 07	16 15	21 04	22 46	15 10	12 17	20 32
13	03 14	01 18	02 16	17 50	09 09	24 10	00 27	16 27	21 06	22 46	15 10	12 18	20 33
14	03 09	01 14	03 11	18 06	12 15	24 00	00 47	16 40	21 08	22 46	15 09	12 19	20 33
15	03 01	01 11	03 56	18 22	14 49	23 48	01 07	16 52	21 10	22 46	15 09	12 19	20 33
16	02 51	01 08	04 31	18 37	16 44	23 32	01 28	17 04	21 12	22 46	15 09	12 20	20 33
17	02 41	01 05	04 54	18 52	17 56	23 13	01 48	17 17	21 15	22 45	15 09	12 21	20 33
18	02 30	01 02	05 05	19 07	18 21	22 51	02 09	17 29	21 17	22 45	15 09	12 21	20 33
19	02 21	00 59	05 02	19 21	17 58	22 26	02 30	17 40	21 19	22 45	15 09	12 22	20 33
20	02 13	00 55	04 46	19 35	16 48	21 57	02 52	17 52	21 21	22 45	15 09	12 23	20 33
21	02 08	00 52	04 17	19 49	14 51	21 26	03 13	18 04	21 23	22 45	15 09	12 23	20 34
22	02 05	00 49	03 36	20 02	12 13	20 52	03 35	18 15	21 25	22 45	15 09	12 24	20 34
23	02 04	00 46	02 43	20 15	08 57	20 17	03 57	18 26	21 27	22 45	15 09	12 25	20 34
24	02 D 04	00 43	01 41	20 27	05 11	19 41	04 19	18 38	21 29	22 45	15 09	12 25	20 34
25	02 05	00 40	00 S 31	20 39	01 N 02	19 05	04 41	18 49	21 31	22 45	15 09	12 26	20 35
26	02 R 05	00 36	00 N 42	20 51	03 S 19	18 30	05 03	19 00	21 33	22 44	15 09	12 27	20 35
27	02 03	00 33	01 56	21 02	07 38	17 58	05 26	19 10	21 35	22 44	15 09	12 27	20 35
28	01 59	00 30	03 03	21 13	11 37	17 29	05 48	19 21	21 37	22 44	15 09	12 28	20 35
29	01 51	00 27	04 00	21 24	14 56	17 04	06 10	19 31	21 39	22 44	15 09	12 29	20 36
30	01 ♎ 42	00 ♎ 24	04 N 40	21 S 34	17 S 15	16 S 43	06 S 33	19 S 41	21 S 41	22 S 44	15 N 09	12 S 29	20 N 36

ZODIAC SIGN ENTRIES

Date	h	m	Planets
02	10	02	☽ ♐
04	10	05	☽ ♑
06	12	14	☽ ♒
08	17	35	☽ ♓
09	18	11	♀ ♎
11	02	10	☽ ♈
13	13	04	☽ ♉
16	01	16	☽ ♊
18	13	56	☽ ♋
21	02	04	☽ ♌
23	01	27	☉ ♐
23	12	08	☽ ♍
25	11	53	☿ ♏
25	18	41	☽ ♎
27	21	22	☽ ♏
29	21	12	☽ ♐

LATITUDES

Date	Mercury ☿ ° '	Venus ♀ ° '	Mars ♂ ° '	Jupiter ♃ ° '	Saturn ♄ ° '	Uranus ♅ ° '	Neptune ♆ ° '	Pluto ♇ ° '
01	02 S 46	00 S 02	00 N 14	00 N 34	00 N 38	00 N 41	01 N 43	12 N 03
04	02 50	00 N 18	00 13	00 33	00 38	00 41	01 43	12 04
07	02 47	00 36	00 11	00 33	00 38	00 41	01 43	12 05
10	02 37	00 53	00 09	00 33	00 37	00 42	01 43	12 06
13	02 16	01 08	00 07	00 32	00 37	00 42	01 43	12 07
16	01 42	01 22	00 06	00 32	00 37	00 42	01 43	12 09
19	00 S 54	01 35	00 04	00 32	00 36	00 42	01 43	12 10
22	00 N 05	01 46	00 N 02	00 31	00 36	00 42	01 43	12 11
25	01 06	01 55	00 00	00 31	00 36	00 42	01 43	12 13
28	01 55	02 04	00 S 01	00 31	00 35	00 42	01 43	12 14
31	02 N 27	02 N 11	00 S 03	00 N 31	00 N 35	00 N 43	01 N 43	12 N 15

DATA

Julian Date	2436874
Delta T	+33 seconds
Ayanamsa	23° 17' 45"
Synetic vernal point	05° ♓ 49' 15"
True obliquity of ecliptic	23° 26' 31"

MOON'S PHASES, APSIDES AND POSITIONS ☽

Date	h	m	Phase	Longitude ° '	Eclipse Indicator
07	13	24	☽	14 ♒ 23	
15	09	42	○	22 ♉ 17	
23	13	03	☾	00 ♍ 29	
30	08	46	●	07 ♐ 24	

Day	h	m		
02	01	01	Perigee	
17	06	22	Apogee	
30	12	20	Perigee	
04	08	35	Max dec	18° S 17'
11	00	29	0N	
18	12	35	Max dec	18° N 21'
25	17	44	0S	

LONGITUDES

Date	Chiron ⚷ ° '	Ceres ⚳ ° '	Pallas ⚴ ° '	Juno ⚵ ° '	Vesta ⚶ ° '	Black Moon Lilith ⚸ ° '
01	22 ♒ 41	15 ♐ 16	25 ♏ 43	15 ♏ 47	27 ♎ 17	08 ♊ 52
11	22 ♒ 43	19 ♐ 07	00 ♐ 01	19 ♏ 11	02 ♏ 30	09 ♊ 59
21	22 ♒ 52	23 ♐ 03	04 ♐ 19	22 ♏ 36	07 ♏ 44	11 ♊ 06
31	23 ♒ 07	27 ♐ 02	08 ♐ 37	25 ♏ 59	12 ♏ 57	12 ♊ 13

All ephemeris data is given at 12.00 UT and the Moon's longitude is additionally given for 24.00 UT

ASPECTARIAN

h m	Aspects
01 Sunday	
02 05	♀ ∠ ♆
04 00	☽ ∥ ♆
14 42	☽ Q ♇
15 02	☽ ∠ ♄
16 02	☽ ∥ ♂
19 24	☽ □ ♅
20 19	☽ ∥ ☉
22 09	☽ ✶ ♀
02 Monday	
03 20	☽ # ♅
05 23	☽ ⊥ ♄
14 27	☽ ☌ ☿
14 57	☽ ⊻ ♄
18 25	☽ Q ♀
18 54	☽ ☌ ♃
19 17	☽ □ ♇
20 50	☽ ⊻ ♆
22 29	☿ ⊻ ♄
23 38	☽ ⊻ ♂
03 Tuesday	
01 44	☽ ⊻ ☉
04 29	♀ Q ♃
06 25	☽ ⊥ ♆
06 42	♀ ∠ ♂
09 39	☽ ⊥ ♂
12 00	☽ ⊥ ☉
19 14	☽ △ ♅
20 53	☽ ∠ ♆
04 Wednesday	
00 30	♃ □ ♇
00 48	☽ ∠ ♂
01 08	☽ □ ♀
03 32	☽ ∠ ☉
10 09	☉ # ♅
15 26	☽ ☌ ♄
18 06	☽ ⊻ ☿
19 38	☽ bQ ♅
19 41	☽ △ ♇
19 56	☽ ⊻ ♃
21 21	☽ ✶ ♆
05 Thursday	
02 31	☽ ✶ ♂
04 40	☽ ⊥ ☿
05 55	☽ ✶ ☉
06 00	☽ ⊥ ♃
10 34	☽ ± ♅
17 22	☽ Q ♆
20 18	☿ □ ♇
20 39	☽ ⊼ ♅
20 41	☽ bQ ♇
20 43	☽ ∠ ☿
21 18	☽ ∠ ♃
23 37	☽ Q ♂
06 Friday	
03 37	☽ Q ☉
06 23	☽ △ ♀
10 03	☿ ☌ ♃
10 13	☽ ∥ ☉
12 03	☽ ± ♇
13 30	♀ ⊥ ♅
16 36	☽ # ♅
18 13	☽ ⊻ ♄
18 27	☽ ∥ ♂
22 28	☽ ⊼ ♇
23 29	☽ ✶ ♃
07 Saturday	
00 05	☽ ✶ ☿
00 18	☽ □ ♆
04 11	☿ ⊻ ♆
04 46	☽ ⊥ ♄
08 23	☽ □ ♂
10 26	☽ bQ ♀
13 24	☽ □ ☉
14 51	♂ # ♅
17 35	☽ ∥ ♆
20 51	☽ ∠ ♄
21 03	☽ Q ♃
22 35	☽ Q ☿
08 Sunday	
01 03	☽ ☍ ♅
03 48	☽ ± ♀
15 35	☽ ⊼ ♀
09 Monday	
00 21	☽ ✶ ♄
04 33	☽ ☍ ♇
06 30	☽ □ ♃
06 37	☽ △ ♆
09 15	☽ □ ☿
14 47	♃ ⊻ ♆
18 10	☽ △ ♂
22 57	☽ Q ♄
10 Tuesday	
01 05	☽ △ ☉
08 48	☽ ⊼ ♅
10 59	☽ bQ ♆
20 18	☽ ± ♅
20 51	♀ ⊥ ♆
22 09	☽ # ♀
11 Wednesday	
00 24	☽ bQ ♂
02 29	☽ ∥ ♀
02 55	☉ Q ♇
04 19	☽ ± ♆
04 57	☽ ☍ ♀
08 21	☽ bQ ☉
09 42	☽ □ ♄
13 43	☽ bQ ♅
13 45	☽ ⊼ ♇
16 00	☽ ⊼ ♆
16 44	☽ △ ♃
18 54	☽ ± ♂
20 44	☽ △ ☿
12 Thursday	
01 28	☽ ± ♇
02 32	☉ ∠ ♄
03 28	☽ ± ☉
07 21	☽ ⊼ ♂
16 19	☽ ⊼ ☉
19 11	☽ △ ♅
19 12	☽ bQ ♇
22 45	☽ bQ ♃
13 Friday	
02 48	☽ bQ ☿
16 57	☽ ± ♃
20 36	♀ □ ♄
20 49	☽ ± ☿
21 18	☽ △ ♄
21 21	☽ ⊼ ♀
14 Saturday	
00 35	☿ St R
01 03	☽ △ ♇
03 13	☉ □ ♅
03 29	☽ ☍ ♆
05 10	☽ ⊼ ♃
08 48	☽ ⊼ ☿
10 28	☽ ± ♀
12 30	☽ # ♆
22 42	☽ ☍ ♂
15 Sunday	
03 36	☽ bQ ♄
06 14	☽ bQ ♀
07 09	☽ □ ♅
09 42	☽ ☍ ☉
15 43	☽ ∥ ♅
16 50	♀ ∠ ♅
17 07	♀ ⊻ ♇
21 53	☽ ± ♄
16 Monday	
10 06	☽ ⊼ ♄
13 28	☽ □ ♇
15 24	☽ △ ♀
16 04	☽ ⊼ ♆
17 12	♂ Q ♇
18 20	☽ # ♂
18 43	☽ ☍ ♃
19 31	☽ Q ♅
19 59	☽ ☍ ☿
23 14	♀ ⊻ ♆
17 Tuesday	
04 14	☽ ± ♆
15 02	☽ ⊼ ♂
15 46	☿ ☌ ♃
19 49	☽ ✶ ♅
22 30	☽ bQ ♆
18 Wednesday	
01 55	☽ Q ♇
02 46	☿ ✶ ♀
03 56	☽ ± ♂
03 57	☽ ⊼ ☉
14 15	♀ ✶ ♃
17 12	☽ ± ☉
17 55	☿ ∥ ♄
23 14	☽ bQ ♂
23 17	☽ ☍ ♄
19 Thursday	
00 01	♂ ∠ ♄
02 09	☽ ∠ ♅
02 11	☽ ✶ ♇
04 53	☽ △ ♆
05 24	☽ ⊼ ☿
08 30	☽ ⊼ ♃
09 53	☽ □ ♀
11 57	☿ ⊻ ♆
13 01	☽ bQ ☉
16 37	☽ ± ☿
18 37	☽ # ♂
20 15	☽ ⊥ ♅
20 47	☽ ± ♃
20 Friday	
07 14	☽ △ ♂
08 17	☽ ⊻ ♅
08 19	☽ ∠ ♇
09 15	☽ bQ ☿
15 06	☽ bQ ♃
19 33	☿ □ ♇
21 45	☽ △ ☉
21 Saturday	
01 12	♂ □ ♅
01 20	☽ Q ♀
02 14	☽ ⊥ ♇
08 53	☽ ∥ ♅
11 33	♀ ⊥ ♇
11 42	☽ ⊼ ♄
12 27	☽ △ ☿
14 02	☿ ∥ ♃
14 04	☽ ⊻ ♇
16 50	☽ □ ♆
19 25	☿ ⊻ ♄
21 16	☽ △ ♃
23 33	☽ ± ♄
22 Sunday	
00 09	☉ ⊥ ♄
03 09	☽ ✶ ♀
10 27	☽ # ♆
17 07	☽ bQ ♄
19 09	☽ ☌ ♅
21 37	☽ □ ♂
23 Monday	
00 36	☿ # ♇
03 43	☽ Q ♆
10 36	☽ ∠ ♀
13 03	☽ □ ☉
13 06	☉ ∥ ☿
16 44	☽ □ ☿
21 44	☽ △ ♄
23 33	☽ ☌ ♇
24 Tuesday	
02 17	☽ ✶ ♆
04 56	☽ ⊥ ♀
07 18	☽ □ ♃
09 19	☽ Q ♂
10 56	☉ ☌ ☿
16 44	☽ # ♀
16 55	☽ ⊻ ♀
21 57	☿ ∠ ♀
22 41	☽ Q ☿
25 Wednesday	
00 57	☽ Q ☉
02 06	☉ # ♇
02 52	☽ ⊻ ♅
05 38	☽ ∠ ♆
08 13	☽ ✶ ♂
13 33	☽ ⊥ ♅
16 15	☽ Q ♃
18 05	☽ ✶ ☿
20 29	☿ ∥ ♂
21 34	☽ ⊥ ♆
23 41	☿ ⊥ ♄
23 52	☽ ✶ ☉
26 Thursday	
04 02	☽ □ ♄
05 15	☽ ∠ ♅
05 20	☽ ⊻ ♇
07 59	☽ ⊻ ♆
11 46	☽ ∠ ♂
13 21	☽ ✶ ♃
15 37	☽ ⊥ ♇
17 45	☽ ∠ ☿
22 27	☽ ∥ ♀
27 Friday	
01 45	☽ ☌ ♀
03 24	☽ ∠ ☉
03 46	☽ ⊥ ♂
04 49	♅ St R
06 39	☽ ✶ ♅
06 45	☽ ∠ ♇
07 41	☽ ⊥ ☿
10 41	☽ Q ♄
14 13	☽ ⊻ ♂
14 52	☽ ∠ ♃
16 52	☽ ⊻ ☿
19 24	☽ ⊥ ☉
28 Saturday	
02 22	☽ Q ♅
05 47	☽ ⊥ ♃
05 53	☽ ⊻ ☉
06 26	☽ ✶ ♄
07 19	☽ ✶ ♇
09 53	☽ ☌ ♆
13 46	☿ ☌ ♂
15 00	☉ ⊻ ♄
15 35	☽ ⊻ ♃
17 39	☽ ∥ ♆
20 10	☿ ⊥ ♀
29 Sunday	
02 31	☽ Q ♇
03 06	☉ □ ♇
06 19	☽ ⊻ ♀
06 34	☽ ∠ ♄
07 09	☽ □ ♅
13 50	☽ # ♅
14 01	☽ ☌ ☿
16 31	☽ ⊥ ♀
16 41	☽ ☌ ♂
17 44	♀ ✶ ♅
19 18	♀ ∠ ♇
20 50	☽ ⊥ ♄
22 53	♀ ⊥ ♂
30 Monday	
06 14	☽ ∥ ☿
06 17	☽ ⊻ ♄
06 49	☽ □ ♇
07 43	☽ ∠ ♀
08 46	☽ ☌ ☉
09 24	☽ ⊻ ♆
15 32	☽ ☌ ♃
18 38	☉ ⊻ ♆
18 48	☽ ⊥ ♆

DECEMBER 1959

Date	Sidereal time h m s	Sun ☉ ° ' "	Moon ☽ ° ' "	Moon ☽ 24.00 ° ' "	Mercury ☿ ° '	Venus ♀ ° '	Mars ♂ ° '	Jupiter ♃ ° '	Saturn ♄ ° '	Uranus ♅ ° '	Neptune ♆ ° '	Pluto ♇ ° '
01	16 38 23	08 ♐ 32 44	24 ♐ 48 11	02 ♑ 25 10	24 ♏ 13	23 ♎ 02	28 ♏ 24	11 ♐ 55	05 ♑ 57	21 ♌ 04	07 ♏ 50	06 ♍ 09
02	16 42 20	09 33 36	09 ♑ 58 14	17 ♑ 26 13	23 R 52	24 09	29 07	12 09	06 03	21 R 04	07 52	06 09
03	16 46 17	10 34 28	24 ♑ 48 10	02 ♒ 03 22	23 43	25 16	29 ♏ 49	12 22	06 10	21 04	07 54	06 10
04	16 50 13	11 35 22	09 ♒ 11 23	16 ♒ 11 58	23 D 44	26 23	00 ♐ 32	12 36	06 16	21 03	07 56	06 10
05	16 54 10	12 36 16	23 ♒ 05 06	29 ♒ 50 57	23 55	27 31	01 14	12 49	06 23	21 03	07 58	06 10
06	16 58 06	13 37 11	06 ♓ 29 49	13 ♓ 02 08	24 15	28 39	01 57	13 03	06 30	21 02	08 00	06 10
07	17 02 03	14 38 07	19 ♓ 28 25	25 ♓ 49 12	24 44	29 ♎ 47	02 39	13 16	06 36	21 02	08 02	06 10
08	17 05 59	15 39 03	02 ♈ 05 07	08 ♈ 16 44	25 21	00 ♏ 55	03 22	13 30	06 43	21 01	08 04	06 10
09	17 09 56	16 40 00	14 ♈ 24 41	20 ♈ 29 30	26 04	02 03	04 05	13 43	06 50	21 00	08 06	06 R 10
10	17 13 52	17 40 58	26 ♈ 31 46	02 ♉ 31 57	26 53	03 12	04 48	13 57	06 57	21 00	08 07	06 10
11	17 17 49	18 41 56	08 ♉ 30 32	14 ♉ 27 56	27 48	04 21	05 30	14 10	07 04	20 59	08 09	06 10
12	17 21 46	19 42 55	20 ♉ 24 29	26 ♉ 20 33	28 47	05 30	06 13	14 24	07 11	20 58	08 11	06 10
13	17 25 42	20 43 55	02 ♊ 16 22	08 ♊ 12 11	29 ♏ 50	06 39	06 56	14 37	07 17	20 57	08 13	06 10
14	17 29 39	21 44 56	14 ♊ 08 12	20 ♊ 04 35	00 ♐ 57	07 48	07 39	14 51	07 24	20 56	08 15	06 10
15	17 33 35	22 45 57	26 ♊ 01 30	01 ♋ 59 05	02 06	08 57	08 22	15 04	07 31	20 55	08 16	06 09
16	17 37 32	23 46 59	07 ♋ 57 29	13 ♋ 56 51	03 19	10 07	09 05	15 18	07 38	20 54	08 18	06 09
17	17 41 28	24 48 02	19 ♋ 57 20	25 ♋ 59 09	04 33	11 17	09 48	15 31	07 45	20 53	08 20	06 09
18	17 45 25	25 49 05	02 ♌ 02 30	08 ♌ 07 40	05 50	12 27	10 32	15 45	07 52	20 52	08 21	06 09
19	17 49 21	26 50 10	14 ♌ 14 55	20 ♌ 24 38	07 09	13 37	11 15	15 58	07 59	20 51	08 23	06 08
20	17 53 18	27 51 15	26 ♌ 37 10	02 ♍ 52 57	08 29	14 47	11 58	16 12	08 06	20 50	08 25	06 08
21	17 57 15	28 52 20	09 ♍ 12 27	15 ♍ 36 09	09 50	15 57	12 41	16 25	08 13	20 48	08 26	06 08
22	18 01 11	29 ♐ 53 27	22 ♍ 04 31	28 ♍ 38 05	11 13	17 07	13 25	16 38	08 20	20 47	08 28	06 07
23	18 05 08	00 ♑ 54 34	05 ♎ 17 17	12 ♎ 02 33	12 36	18 18	14 08	16 52	08 27	20 46	08 29	06 07
24	18 09 04	01 55 42	18 ♎ 54 13	25 ♎ 52 31	14 01	19 29	14 51	17 05	08 34	20 44	08 31	06 06
25	18 13 01	02 56 51	02 ♏ 57 33	10 ♏ 09 16	15 26	20 40	15 35	17 18	08 41	20 43	08 32	06 06
26	18 16 57	03 58 00	17 ♏ 27 21	24 ♏ 51 20	16 53	21 50	16 18	17 32	08 48	20 41	08 34	06 05
27	18 20 54	04 59 10	02 ♐ 20 29	09 ♐ 53 52	18 20	23 02	17 02	17 45	08 55	20 40	08 35	06 05
28	18 24 50	06 00 21	17 ♐ 30 18	25 ♐ 08 30	19 47	24 13	17 45	17 58	09 03	20 38	08 37	06 04
29	18 28 47	07 01 32	02 ♑ 47 02	10 ♑ 24 25	21 15	25 24	18 29	18 11	09 10	20 37	08 38	06 04
30	18 32 44	08 02 43	17 ♑ 59 15	25 ♑ 30 11	22 44	26 35	19 13	18 25	09 17	20 35	08 39	06 03
31	18 36 40	09 ♑ 03 54	02 ♒ 56 04	10 ♒ 15 54	24 ♐ 13	27 ♏ 47	19 ♐ 57	18 ♐ 38	09 ♑ 24	20 ♌ 33	08 ♏ 41	06 ♍ 02

Date	Moon True ☊ ° '	Moon Mean ☊ ° '	Moon ☽ Latitude ° '	DECLINATIONS Sun ☉ ° '	Moon ☽ ° '	Mercury ☿ ° '	Venus ♀ ° '	Mars ♂ ° '	Jupiter ♃ ° '	Saturn ♄ ° '	Uranus ♅ ° '	Neptune ♆ ° '	Pluto ♇ ° '
01	01 ♎ 32	00 ♎ 20	05 N 00	21 S 44	18 S 20	16 S 27	06 S 56	19 S 52	21 S 43	22 S 43	15 N 09	12 S 30	20 N 36
02	01 R 21	00 17	04 59	21 53	18 06	16 16	07 18	20 01	21 45	22 43	15 09	12 30	20 37
03	01 13	00 14	04 38	22 02	16 37	16 10	07 41	20 11	21 47	22 43	15 09	12 31	20 37
04	01 06	00 11	03 59	22 11	14 07	16 08	08 04	20 21	21 48	22 43	15 09	12 32	20 37
05	01 02	00 08	03 07	22 19	10 53	16 10	08 26	20 30	21 50	22 43	15 10	12 32	20 38
06	01 01	00 05	02 06	22 26	07 10	16 15	08 49	20 39	21 52	22 42	15 10	12 33	20 38
07	01 D 01	00 ♎ 01	01 N 01	22 33	03 S 14	16 24	09 11	20 48	21 54	22 42	15 10	12 33	20 38
08	01 R 01	29 ♍ 58	00 S 06	22 40	00 N 45	16 35	09 34	20 57	21 56	22 42	15 10	12 34	20 39
09	01 00	29 55	01 10	22 46	04 36	16 49	09 56	21 06	21 57	22 42	15 10	12 34	20 39
10	00 57	29 52	02 11	22 52	08 12	17 05	10 19	21 14	21 59	22 41	15 11	12 35	20 40
11	00 52	29 49	03 04	22 58	11 26	17 22	10 41	21 23	22 01	22 41	15 11	12 36	20 40
12	00 43	29 46	03 50	23 03	14 10	17 41	11 03	21 31	22 02	22 41	15 11	12 36	20 40
13	00 32	29 42	04 24	23 07	16 18	18 00	11 25	21 39	22 04	22 41	15 12	12 37	20 41
14	00 18	29 39	04 48	23 11	17 44	18 21	11 47	21 46	22 05	22 40	15 12	12 37	20 41
15	00 04	29 36	04 59	23 15	18 24	18 41	12 09	21 54	22 07	22 40	15 12	12 38	20 42
16	29 ♍ 49	29 33	04 56	23 18	18 16	19 02	12 31	22 01	22 09	22 40	15 13	12 38	20 42
17	29 35	29 30	04 41	23 20	17 19	19 24	12 52	22 08	22 10	22 39	15 13	12 39	20 43
18	29 23	29 26	04 13	23 23	15 36	19 45	13 13	22 15	22 12	22 39	15 13	12 39	20 43
19	29 15	29 23	03 33	23 24	13 09	20 06	13 34	22 22	22 13	22 39	15 14	12 40	20 44
20	29 09	29 20	02 42	23 25	10 06	20 26	13 55	22 28	22 15	22 38	15 14	12 40	20 44
21	29 06	29 17	01 43	23 26	06 32	20 46	14 15	22 35	22 16	22 38	15 15	12 41	20 45
22	29 D 05	29 14	00 S 37	23 27	02 N 35	21 06	14 36	22 41	22 17	22 38	15 15	12 41	20 45
23	29 R 05	29 11	00 N 33	23 26	01 S 36	21 25	14 56	22 47	22 19	22 37	15 16	12 42	20 46
24	29 05	29 07	01 43	23 26	05 49	21 43	15 16	22 52	22 20	22 37	15 16	12 42	20 46
25	29 02	29 04	02 49	23 25	09 51	22 00	15 35	22 58	22 21	22 36	15 17	12 42	20 47
26	28 57	29 01	03 46	23 23	13 26	22 17	15 54	23 03	22 23	22 36	15 17	12 43	20 48
27	28 49	28 58	04 29	23 21	16 14	22 33	16 13	23 08	22 24	22 36	15 18	12 43	20 48
28	28 39	28 55	04 55	23 18	17 58	22 47	16 31	23 13	22 25	22 35	15 18	12 44	20 49
29	28 28	28 51	04 59	23 15	18 25	23 01	16 50	23 17	22 27	22 35	15 19	12 44	20 49
30	28 16	28 48	04 43	23 12	17 33	23 14	17 07	23 21	22 28	22 34	15 19	12 44	20 50
31	28 ♍ 06	28 ♍ 45	04 N 07	23 S 08	15 S 29	23 S 26	17 S 25	23 S 25	22 S 29	22 S 34	15 N 20	12 S 45	20 N 51

ZODIAC SIGN ENTRIES

Date	h	m	Planets
01	20	11	☽ ♑
03	18	09	♂ ♐
03	20	35	☽ ♒
06	00	16	☽ ♓
07	16	41	♀ ♏
08	07	59	☽ ♈
10	18	56	☽ ♉
13	07	24	☽ ♊
13	15	42	☿ ♐
15	20	00	☽ ♋
18	07	58	☽ ♌
20	18	29	☽ ♍
22	14	34	☉ ♑
23	02	29	☽ ♎
25	07	01	☽ ♏
27	08	16	☽ ♐
29	07	38	☽ ♑
31	07	15	☽ ♒

LATITUDES

Date	Mercury ☿ ° '	Venus ♀ ° '	Mars ♂ ° '	Jupiter ♃ ° '	Saturn ♄ ° '	Uranus ♅ ° '	Neptune ♆ ° '	Pluto ♇ ° '
01	02 N 27	02 N 11	00 S 03	00 N 31	00 N 35	00 N 43	01 N 43	12 N 15
04	02 40	02 17	00 05	00 30	00 35	00 43	01 43	12 17
07	02 38	02 21	00 07	00 30	00 35	00 43	01 43	12 18
10	02 27	02 25	00 09	00 30	00 34	00 43	01 44	12 19
13	02 10	02 27	00 11	00 30	00 34	00 43	01 44	12 20
16	01 49	02 28	00 12	00 29	00 34	00 43	01 44	12 22
19	01 26	02 28	00 14	00 29	00 34	00 43	01 44	12 23
22	01 02	02 27	00 16	00 29	00 33	00 43	01 44	12 24
25	00 39	02 25	00 18	00 29	00 33	00 43	01 44	12 26
28	00 N 16	02 22	00 20	00 29	00 33	00 44	01 44	12 27
31	00 S 07	02 N 19	00 S 22	00 N 28	00 N 33	00 N 44	01 N 44	12 N 28

DATA

Julian Date	2436904
Delta T	+33 seconds
Ayanamsa	23° 17' 49"
Synetic vernal point	05° ♓ 49' 11"
True obliquity of ecliptic	23° 26' 30"

MOON'S PHASES, APSIDES AND POSITIONS ☽

Date	h	m	Phase	Longitude ° '	Eclipse Indicator
07	02	12	☽	14 ♓ 13	
15	04	49	○	22 ♊ 28	
23	03	28	☾	00 ♎ 33	
29	19	09	●	07 ♑ 20	

Day	h	m		
14	06	43	Apogee	
29	01	04	Perigee	
01	19	34	Max dec	18° S 24'
08	07	29	0N	
15	20	01	Max dec	18° N 27'
23	02	54	0S	
29	08	15	Max dec	18° S 26'

LONGITUDES

Date	Chiron ⚷ ° '	Ceres ⚳ ° '	Pallas ⚴ ° '	Juno ⚵ ° '	Vesta ⚶ ° '	Black Moon Lilith ⚸ ° '
01	23 ♒ 07	27 ♐ 02	08 ♐ 37	25 ♏ 59	12 ♏ 57	12 ♊ 13
11	23 ♒ 27	01 ♑ 03	12 ♐ 54	29 ♏ 21	18 ♏ 09	13 ♊ 21
21	23 ♒ 53	05 ♑ 05	17 ♐ 09	02 ♐ 39	23 ♏ 19	14 ♊ 28
31	24 ♒ 23	09 ♑ 07	21 ♐ 21	05 ♐ 53	28 ♏ 25	15 ♊ 35

All ephemeris data is given at 12.00 UT and the Moon's longitude is additionally given for 24.00 UT

ASPECTARIAN

h m	Aspects
01 Tuesday	
06 09	☽ △ ♅
08 54	☽ ∠ ♆
09 00	☽ ⚹ ♀
09 05	☉ ∥ ♃
11 07	☽ ⊻ ☿
17 56	☽ ⊻ ♂
20 18	☽ ⊥ ☿
02 Wednesday	
03 53	☽ ⊥ ♂
05 25	☽ Q ♀
05 43	☽ ☌ ♄
05 47	☽ ⚻ ♅
05 55	☽ △ ♇
07 17	☿ ⊻ ♀
08 38	☽ ⚹ ♆
09 48	♀ Q ♄
10 17	☽ ∠ ☿
11 18	☽ ⊻ ☉
14 24	♄ ⚻ ♅
15 32	☽ ⊻ ♃
18 58	☽ ∠ ♂
20 10	☽ ± ♅
21 37	☽ ⊥ ☉
03 Thursday	
01 21	☽ ⊥ ♃
03 59	☽ Q ♆
05 53	☽ ⊼ ♅
06 03	☽ ⚻ ♇
10 13	☽ ⚹ ☿
11 22	♄ △ ♇
12 49	☽ □ ♀
13 22	☽ ∠ ☉
16 17	☽ ∠ ♃
17 13	☽ ∥ ☿
20 42	☽ ⚹ ♂
20 51	☽ ± ♇
21 26	☿ St D
04 Friday	
01 48	♂ ⊥ ♄
03 01	☽ # ♅
06 07	☽ Q ☿
06 53	☽ ⊼ ♇
07 02	☽ ⊻ ♄
09 52	☽ □ ♆
16 24	☽ ⚹ ☉
17 18	☽ ⊥ ♄
17 54	☽ ⚹ ♃
17 59	☽ Q ♂
05 Saturday	
00 18	☽ ∥ ♆
08 25	☽ ☍ ♅
09 00	☽ ∠ ♄
13 29	☽ □ ☿
14 54	☽ Q ☉
15 06	☽ Q ♃
18 30	☉ ☌ ♃
20 04	♀ ∠ ♃
20 33	☽ △ ♀
06 Sunday	
02 35	☽ ∥ ♀
03 18	☽ □ ♂
06 50	☉ ∠ ♀
08 10	♂ # ♇
11 24	☽ ☍ ♇
12 00	☽ ⚹ ♄
14 45	☽ △ ♆
21 15	☉ ⊥ ♆
07 Monday	
00 14	☽ □ ♃
02 12	☽ □ ☉
02 23	☽ ⚻ ♀
10 22	☽ Q ♄
14 55	☽ ⊼ ♅
18 43	☽ ⚻ ♆
20 55	☽ ± ♀
22 25	☽ △ ☿
08 Tuesday	
02 17	☽ ± ♅
09 31	☽ ⊼ ♀
11 57	☽ ± ♆
14 38	☽ △ ♂
18 12	☉ ∥ ♄
19 36	☽ ⚻ ♅
19 54	☽ ⊼ ♇
20 28	♇ St R
21 03	☽ □ ♄
23 36	☽ ⊼ ♆
09 Wednesday	
05 01	☽ ⚻ ☿
07 36	☽ ± ♇
10 37	☽ △ ♃
16 51	☽ △ ☉
21 47	☽ ⚻ ♂
23 57	☽ ± ☿
10 Thursday	
01 00	☽ △ ♅
01 20	☽ ⚻ ♇
07 48	♀ Q ♅
12 46	☽ ⊼ ☿
16 48	☽ ± ♂
16 55	☽ ⚻ ♃
11 Friday	
01 27	☽ ⚻ ☉
02 45	☽ ☍ ♀
05 23	☽ # ♀
05 35	☽ ⊼ ♂
07 17	☽ △ ♇
09 04	☽ △ ♄
09 50	♃ ⊥ ♆
11 17	☽ ☍ ♆
11 18	☽ ± ♃
21 14	☽ ± ☉
21 41	☽ # ♆
23 38	☽ ⊼ ♃
12 Saturday	
10 02	♂ □ ♇
10 28	☽ ⊼ ☉
13 08	☽ □ ♅
15 36	☽ ⚻ ♄
22 40	☽ ∥ ♅
13 Sunday	
01 59	♀ ⚹ ♇
06 34	☽ ☍ ☿
10 00	☽ ± ♄
17 10	☉ △ ♅
19 52	☽ □ ♇
21 48	☽ ⊼ ♀
22 03	☽ ☍ ♂
22 15	☽ ⊼ ♄
14 Monday	
00 03	☽ ⊼ ♆
01 30	☽ Q ♅
02 05	♂ ⊻ ♄
02 56	♀ ⚹ ♄
04 07	♀ ⊻ ♂
11 15	☽ ± ♀
12 13	☽ ± ♆
13 28	☽ ☍ ♃
21 28	♀ ☌ ♆
22 44	☿ ⊥ ♄
15 Tuesday	
01 43	☽ ⚹ ♅
04 49	☽ ☍ ☉
06 26	☽ ⚻ ♆
07 23	☽ ⚻ ♀
08 14	☽ Q ♇
08 33	♂ ⊻ ♆
14 59	♀ ⊥ ♃
16 Wednesday	
00 17	☉ ∠ ♆
01 36	☽ ⊼ ☿
07 53	☽ ∠ ♅
08 23	☽ ⚹ ♇
11 21	☽ ☍ ♄
12 41	☽ △ ♆
14 25	☽ ⊼ ♂
15 01	☽ ± ☿
16 47	☽ △ ♀
20 43	♀ ∥ ♆
17 Thursday	
01 54	☽ ⊥ ♅
02 59	☽ ⊼ ♃
03 11	☽ ± ♂
11 07	☽ ⚻ ☿
13 51	☽ ⊻ ♅
14 23	☽ ∠ ♇
15 11	☽ ± ♃
19 58	♂ ∥ ♃
22 16	☽ ⚻ ♂
22 32	☽ ⊼ ☉
18 Friday	
08 15	☽ ⊥ ♇
09 23	☽ ⚻ ♃
11 31	☽ ± ☉
16 08	☽ ∥ ♅
17 42	☿ □ ♇
20 05	☽ ⊻ ♇
20 23	☽ △ ☿
23 36	☽ ⊼ ♄
19 Saturday	
00 29	☽ □ ♆
05 45	☽ △ ♂
06 51	☽ ⚻ ☉
08 47	☽ # ♀
10 37	☽ □ ♀
11 29	☽ ± ♄
15 25	☽ △ ♃
16 12	☽ # ♆
20 Sunday	
00 50	☽ ☌ ♅
04 39	☿ ⊻ ♄
05 09	☽ ⚻ ♄
10 47	☿ ⊻ ♆
11 36	☽ Q ♆
14 35	☽ △ ☉
21 Monday	
01 01	☽ Q ♀
06 10	☽ ☌ ♇
10 07	☽ △ ♄
10 17	☿ # ♇
10 33	☽ ⚹ ♆
13 19	☽ □ ☿
18 56	☽ □ ♂
23 45	♀ ⊻ ♃
23 49	♂ ∥ ♄
22 Tuesday	
01 45	☽ □ ♃
01 55	☽ ⚹ ♀
09 38	☽ ⊻ ♅
14 34	☽ ∠ ♆
20 37	☽ ⊥ ♅
23 Wednesday	
02 35	☽ Q ☿
03 28	☽ □ ☉
06 00	☽ Q ♂
06 58	☽ ⊥ ♆
08 05	☽ ∠ ♀
08 11	♀ Q ♇
11 13	☽ Q ♃
12 51	☽ ∠ ♅
13 28	☽ ⊻ ♇
17 42	☽ □ ♄
17 43	☽ ⊻ ♆
21 22	♄ ⚹ ♆
24 Thursday	
00 07	☽ ⊥ ♇
00 17	♂ ⊥ ♆
01 38	☽ ⊥ ♀
02 29	☽ ⚹ ☿
04 33	☽ ⚹ ♂
08 47	☽ ⚹ ♃
12 40	♀ # ♅
13 05	☽ ⊻ ♀
13 55	☽ Q ☉
15 10	☽ ⚹ ♅
15 48	☽ ∠ ♇
20 35	☿ ⊥ ♆
25 Friday	
01 18	☽ Q ♄
07 17	☽ ∠ ☿
07 47	☽ ∠ ♂
10 53	☽ ∠ ♃
11 35	☽ Q ♅
11 59	☽ ⚹ ☉
13 07	♀ □ ♅
16 44	☿ ☌ ♂
17 15	☽ ⚹ ♇
21 20	☽ ☌ ♆
21 39	☽ ⚹ ♄
23 38	☽ ⊥ ♂
26 Saturday	
00 00	☽ ⊥ ☿
02 07	☽ ⊥ ♃
06 49	☽ ∥ ♆
10 01	☽ ⊻ ♂
10 57	☽ ⊻ ☿
12 07	☽ ⊻ ♃
13 02	☽ Q ♇
14 39	☽ ∠ ☉
17 15	☽ □ ♅
19 45	☽ ☌ ♀
21 34	☿ ∥ ♃
22 23	☽ ∠ ♄
27 Sunday	
00 45	☿ ☌ ♃
03 00	☽ # ♅
06 15	☽ ⊥ ☉
11 50	☽ ∥ ♀
12 56	☽ ⊥ ♄
16 31	☽ ⊻ ☉
16 35	☿ ∥ ♄
17 57	☽ □ ♇
21 57	☽ ⊻ ♆
22 33	☽ ⊻ ♄
28 Monday	
03 34	☉ ⚻ ♅
07 26	☽ ⊥ ♆
08 13	♀ ∠ ♄
12 25	☽ ☌ ♂
12 45	☽ ☌ ♃
13 28	☉ △ ♇
15 58	☽ ☌ ☿
16 55	☽ △ ♅
21 37	☽ ∠ ♆
22 03	♂ ☌ ♃
23 26	☽ ⊻ ♀
29 Tuesday	
01 43	☿ △ ♅
06 18	☉ ∥ ♂
09 38	☽ ⊥ ♀
16 26	☽ ⚻ ♅
17 09	☽ △ ♇
19 09	☽ ☌ ☉
21 13	☽ ⚹ ♆
22 07	☽ ☌ ♄
30 Wednesday	
01 00	☽ ∠ ♀
06 37	☽ ± ♅
08 47	☉ ∥ ☿
12 41	☽ ⊻ ♃
14 03	☽ ⊻ ♂
16 07	☽ ⊼ ♅
16 15	☽ Q ♆
16 52	☽ ⚻ ♇
17 25	☽ ∥ ♀
20 23	☽ ⊻ ☿
22 24	☽ ⊥ ♃
31 Thursday	
00 08	☽ ⊥ ♂
02 45	☉ ⚹ ♆
02 56	☽ ⚹ ♀
03 16	☿ ∠ ♆
07 06	☽ ⊥ ☿
07 18	☽ ± ♇
11 19	☿ ∥ ♂
13 09	☽ ∠ ♃
13 28	☽ # ♅
15 26	☽ ∠ ♂
17 03	☽ ⊼ ♇
20 51	☉ ☌ ♄
21 24	☽ □ ♆
22 39	☽ ⊻ ♄
22 47	☽ ⊻ ☉
23 26	☽ ∠ ☿

JANUARY 1960

LONGITUDES

Date	Sidereal time h m s	Sun ☉ ° ' "	Moon ☽ ° ' "	Moon ☽ 24.00 ° ' "	Mercury ☿ ° '	Venus ♀ ° '	Mars ♂ ° '	Jupiter ♃ ° '	Saturn ♄ ° '	Uranus ♅ ° '	Neptune ♆ ° '	Pluto ♇ ° '
01	18 40 37	10 ♑ 05 05	17 ♒ 28 57	24 ♒ 34 45	25 ♐ 42	28 ♏ 58	20 ♐ 40	18 ♐ 51	09 ♑ 31	20 ♌ 31	08 ♏ 42	06 ♍ 02
02	18 44 33	11 06 16	01 ♓ 33 00	08 ♓ 23 41	27 12	00 ♐ 10	21 24	19 04	09 38	20 R 30	08 43	06 R 01
03	18 48 30	12 07 26	15 ♓ 06 54	21 ♓ 42 59	28 ♐ 43	01 21	22 08	19 17	09 45	20 28	08 45	06 00
04	18 52 26	13 08 37	28 ♓ 12 21	04 ♈ 35 30	00 ♑ 14	02 33	22 52	19 30	09 52	20 26	08 46	05 59
05	18 56 23	14 09 47	10 ♈ 53 03	17 ♈ 05 38	01 45	03 45	23 36	19 43	09 59	20 24	08 47	05 58
06	19 00 19	15 10 56	23 ♈ 13 53	29 ♈ 18 28	03 17	04 57	24 20	19 56	10 06	20 22	08 48	05 58
07	19 04 16	16 12 05	05 ♉ 20 02	11 ♉ 19 13	04 49	06 09	25 04	20 09	10 13	20 20	08 49	05 57
08	19 08 13	17 13 14	17 ♉ 16 37	23 ♉ 12 46	06 21	07 21	25 48	20 22	10 20	20 18	08 50	05 56
09	19 12 09	18 14 23	29 ♉ 08 11	05 ♊ 03 20	07 54	08 33	26 32	20 34	10 28	20 16	08 51	05 55
10	19 16 06	19 15 31	10 ♊ 58 37	16 ♊ 54 24	09 27	09 45	27 16	20 47	10 35	20 14	08 53	05 54
11	19 20 02	20 16 39	22 ♊ 51 00	28 ♊ 48 40	11 01	10 58	28 00	21 00	10 42	20 12	08 54	05 53
12	19 23 59	21 17 46	04 ♋ 47 37	10 ♋ 48 02	12 35	12 10	28 44	21 12	10 49	20 09	08 55	05 52
13	19 27 55	22 18 53	16 ♋ 50 03	22 ♋ 53 48	14 10	13 22	29 ♐ 29	21 25	10 56	20 07	08 55	05 51
14	19 31 52	23 19 59	28 ♋ 59 22	05 ♌ 06 53	15 45	14 35	00 ♑ 13	21 38	11 03	20 05	08 56	05 50
15	19 35 48	24 21 05	11 ♌ 16 24	17 ♌ 28 03	17 20	15 47	00 57	21 50	11 10	20 03	08 57	05 49
16	19 39 45	25 22 11	23 ♌ 41 58	29 ♌ 58 17	18 56	17 00	01 42	22 03	11 17	20 00	08 58	05 48
17	19 43 42	26 23 16	06 ♍ 17 11	12 ♍ 38 52	20 33	18 13	02 26	22 15	11 23	19 58	08 59	05 47
18	19 47 38	27 24 21	19 ♍ 03 36	25 ♍ 31 38	22 10	19 25	03 11	22 27	11 30	19 56	09 00	05 46
19	19 51 35	28 25 26	02 ♎ 03 16	08 ♎ 38 48	23 48	20 38	03 55	22 40	11 37	19 53	09 00	05 45
20	19 55 31	29 ♑ 26 30	15 ♎ 18 33	22 ♎ 02 49	25 26	21 51	04 40	22 52	11 44	19 51	09 01	05 43
21	19 59 28	00 ♒ 27 34	28 ♎ 51 50	05 ♏ 45 50	27 05	23 04	05 24	23 04	11 51	19 49	09 02	05 42
22	20 03 24	01 28 37	12 ♏ 44 54	19 ♏ 49 05	28 ♑ 44	24 17	06 09	23 16	11 58	19 46	09 02	05 41
23	20 07 21	02 29 40	26 ♏ 58 15	04 ♐ 12 08	00 ♒ 24	25 30	06 53	23 28	12 05	19 44	09 03	05 40
24	20 11 17	03 30 43	11 ♐ 30 19	18 ♐ 52 13	02 04	26 43	07 38	23 40	12 11	19 41	09 04	05 39
25	20 15 14	04 31 45	26 ♐ 17 02	03 ♑ 43 54	03 46	27 56	08 23	23 52	12 18	19 39	09 04	05 37
26	20 19 11	05 32 47	11 ♑ 11 46	18 ♑ 39 31	05 27	29 ♐ 09	09 08	24 04	12 25	19 36	09 05	05 36
27	20 23 07	06 33 48	26 ♑ 06 01	03 ♒ 30 07	07 10	00 ♑ 22	09 52	24 15	12 32	19 34	09 05	05 35
28	20 27 04	07 34 48	10 ♒ 50 46	18 ♒ 07 00	08 53	01 35	10 37	24 27	12 38	19 31	09 06	05 34
29	20 31 00	08 35 47	25 ♒ 18 01	02 ♓ 23 10	10 36	02 48	11 22	24 39	12 45	19 29	09 06	05 32
30	20 34 57	09 36 45	09 ♓ 22 00	16 ♓ 14 13	12 21	04 02	12 07	24 50	12 52	19 26	09 06	05 31
31	20 38 53	10 ♒ 37 42	22 ♓ 59 43	29 ♓ 38 35	14 ♒ 05	05 ♑ 15	12 ♑ 52	25 ♐ 02	12 ♑ 58	19 ♌ 23	09 ♏ 07	05 ♍ 30

Date	Moon True ☊ ° '	Moon Mean ☊ ° '	Moon ☽ Latitude ° '
01	27 ♍ 58	28 ♍ 42	03 N 16
02	27 R 53	28 39	02 14
03	27 50	28 36	01 N 07
04	27 D 50	28 32	00 S 02
05	27 50	28 29	01 09
06	27 R 50	28 26	02 10
07	27 48	28 23	03 05
08	27 44	28 20	03 51
09	27 37	28 17	04 26
10	27 27	28 13	04 50
11	27 15	28 10	05 01
12	27 02	28 07	04 59
13	26 49	28 04	04 44
14	26 37	28 01	04 16
15	26 27	27 57	03 35
16	26 20	27 54	02 44
17	26 15	27 51	01 44
18	26 13	27 48	00 S 38
19	26 D 13	27 45	00 N 31
20	26 14	27 42	01 41
21	26 14	27 38	02 46
22	26 R 14	27 35	03 43
23	26 11	27 32	04 28
24	26 06	27 29	04 57
25	25 59	27 26	05 06
26	25 51	27 23	04 56
27	25 43	27 19	04 25
28	25 36	27 16	03 37
29	25 30	27 13	02 35
30	25 27	27 10	01 26
31	25 ♍ 26	27 ♍ 07	00 N 13

DECLINATIONS

Date	Sun ☉ ° '	Moon ☽ ° '	Mercury ☿ ° '	Venus ♀ ° '	Mars ♂ ° '	Jupiter ♃ ° '	Saturn ♄ ° '	Uranus ♅ ° '	Neptune ♆ ° '	Pluto ♇ ° '
01	23 S 03	12 S 29	23 S 36	17 S 42	23 S 29	22 S 30	22 S 34	15 N 20	12 S 45	20 N 51
02	22 59	08 50	23 46	17 58	23 33	22 31	22 33	15 21	12 45	20 52
03	22 53	04 50	23 54	18 15	23 36	22 32	22 33	15 22	12 46	20 53
04	22 48	00 S 45	24 01	18 30	23 39	22 33	22 32	15 22	12 46	20 53
05	22 41	03 N 15	24 07	18 46	23 42	22 35	22 32	15 23	12 47	20 54
06	22 35	07 00	24 12	19 01	23 45	22 36	22 31	15 24	12 47	20 54
07	22 28	10 24	24 15	19 15	23 47	22 37	22 31	15 24	12 47	20 55
08	22 20	13 18	24 17	19 29	23 49	22 38	22 30	15 25	12 47	20 56
09	22 12	15 38	24 18	19 42	23 51	22 39	22 30	15 26	12 48	20 56
10	22 04	17 19	24 17	19 55	23 53	22 39	22 29	15 26	12 48	20 57
11	21 55	18 14	24 15	20 08	23 55	22 40	22 29	15 27	12 48	20 58
12	21 45	18 22	24 12	20 20	23 56	22 41	22 28	15 28	12 49	20 59
13	21 36	17 41	24 07	20 31	23 57	22 42	22 28	15 28	12 49	20 59
14	21 26	16 11	24 01	20 42	23 57	22 43	22 27	15 29	12 49	21 00
15	21 15	13 56	23 54	20 52	23 58	22 44	22 27	15 30	12 49	21 01
16	21 04	11 02	23 45	21 02	23 58	22 45	22 26	15 31	12 49	21 01
17	20 53	07 35	23 35	21 11	23 58	22 45	22 26	15 31	12 50	21 02
18	20 41	03 N 45	23 23	21 20	23 58	22 46	22 25	15 32	12 50	21 03
19	20 29	00 S 20	23 10	21 28	23 57	22 47	22 25	15 33	12 50	21 03
20	20 16	04 29	22 55	21 35	23 56	22 48	22 24	15 34	12 50	21 04
21	20 03	08 29	22 39	21 42	23 55	22 48	22 24	15 35	12 50	21 05
22	19 50	12 07	22 21	21 48	23 54	22 49	22 23	15 35	12 50	21 06
23	19 36	15 08	22 02	21 54	23 52	22 50	22 22	15 36	12 51	21 06
24	19 22	17 16	21 41	21 59	23 50	22 50	22 22	15 37	12 51	21 07
25	19 08	18 17	21 19	22 03	23 48	22 51	22 21	15 38	12 51	21 08
26	18 53	18 04	20 55	22 07	23 46	22 51	22 21	15 39	12 51	21 09
27	18 38	16 36	20 30	22 10	23 43	22 52	22 20	15 39	12 51	21 09
28	18 23	14 02	20 03	22 12	23 41	22 52	22 20	15 40	12 51	21 10
29	18 07	10 39	19 35	22 14	23 37	22 53	22 19	15 41	12 51	21 11
30	17 51	06 44	19 05	22 15	23 34	22 54	22 18	15 42	12 51	21 12
31	17 S 34	02 S 35	18 S 33	22 S 16	23 S 31	22 S 54	22 S 18	15 N 43	12 S 51	21 N 12

ZODIAC SIGN ENTRIES

Date	h	m	Planets
02	08	43	♀ ♐
02	09	19	☽ ♓
04	08	24	☿ ♑
04	15	21	☽ ♈
07	01	22	☽ ♉
09	13	45	☽ ♊
12	02	23	☽ ♋
14	04	59	♂ ♑
14	13	59	☽ ♌
17	00	03	☽ ♍
19	08	14	☽ ♎
21	01	10	☉ ♒
21	13	59	☽ ♏
23	06	16	☿ ♒
23	17	03	☽ ♐
25	18	00	☽ ♑
27	04	46	♀ ♑
27	18	19	☽ ♒
29	19	56	☽ ♓

LATITUDES

Date	Mercury ☿ ° '	Venus ♀ ° '	Mars ♂ ° '	Jupiter ♃ ° '	Saturn ♄ ° '	Uranus ♅ ° '	Neptune ♆ ° '	Pluto ♇ ° '
01	00 S 14	02 N 17	00 S 23	00 N 28	00 N 33	00 N 44	01 N 45	12 N 28
04	00 35	02 13	00 24	00 28	00 32	00 44	01 45	12 30
07	00 54	02 07	00 26	00 28	00 32	00 44	01 45	12 31
10	01 11	02 01	00 28	00 28	00 32	00 44	01 45	12 32
13	01 27	01 54	00 30	00 28	00 32	00 44	01 45	12 33
16	01 40	01 47	00 32	00 28	00 32	00 44	01 45	12 34
19	01 51	01 39	00 34	00 27	00 31	00 44	01 46	12 35
22	01 59	01 31	00 36	00 27	00 31	00 44	01 46	12 36
25	02 04	01 23	00 38	00 27	00 31	00 44	01 46	12 36
28	02 05	01 14	00 40	00 27	00 31	00 44	01 46	12 37
31	02 S 02	01 N 05	00 S 42	00 N 27	00 N 31	00 N 44	01 N 46	12 N 38

DATA

Julian Date	2436935
Delta T	+33 seconds
Ayanamsa	23° 17' 54"
Synetic vernal point	05° ♓ 49' 05"
True obliquity of ecliptic	23° 26' 30"

MOON'S PHASES, APSIDES AND POSITIONS ☽

Date	h	m	Phase	Longitude ° '	Eclipse Indicator
05	18	53	☽	14 ♈ 27	
13	23	51	○	22 ♋ 49	
21	15	01	☾	00 ♏ 35	
28	06	15	●	07 ♒ 20	

Day	h	m		
10	13	03	Apogee	
26	09	42	Perigee	
04	16	24	ON	
12	03	55	Max dec	18° N 25'
19	10	02	OS	
25	19	45	Max dec	18° S 21'

LONGITUDES

Date	Chiron ⚷ ° '	Ceres ⚳ ° '	Pallas ⚴ ° '	Juno ⚵ ° '	Vesta ⚶ ° '	Black Moon Lilith ⚸ ° '
01	24 ♒ 27	09 ♑ 32	21 ♐ 46	06 ♐ 12	28 ♏ 56	15 ♊ 42
11	25 ♒ 01	13 ♑ 33	25 ♐ 53	09 ♐ 20	03 ♐ 58	16 ♊ 49
21	25 ♒ 39	17 ♑ 33	29 ♐ 54	12 ♐ 20	08 ♐ 54	17 ♊ 56
31	26 ♒ 19	21 ♑ 31	03 ♑ 49	15 ♐ 10	13 ♐ 43	19 ♊ 03

All ephemeris data is given at 12.00 UT and the Moon's longitude is additionally given for 24.00 UT

ASPECTARIAN

h m	Aspects
01 Friday	
00 12	☽ Q ♀
07 20	♂ △ ♅
08 41	☽ ⊥ ♄
09 29	☽ ⊥ ☉
10 04	☽ ∥ ♆
14 20	☽ ⚹ ♃
17 06	☽ ☍ ♅
17 39	☽ ⚹ ♂
02 Saturday	
00 00	☽ ∠ ♄
01 52	☽ ∠ ☉
03 36	☽ ⚹ ☿
09 22	☽ □ ♀
11 09	☽ Q ♃
15 24	☽ Q ♂
19 48	☽ ☍ ♇
03 Sunday	
00 36	☽ △ ♆
02 19	☽ ⚹ ♄
03 07	☽ Q ☿
06 11	☽ ⚹ ☉
16 05	♃ ∥ ♄
19 41	☽ □ ♃
21 41	☽ ⊼ ♅
04 Monday	
00 10	☽ Q ♄
01 31	☽ □ ♂
03 45	☽ bQ ♆
05 50	☽ Q ☉
08 42	☽ ± ♅
16 17	☽ □ ☿
20 34	☽ ± ♆
21 00	☽ △ ♀
05 Tuesday	
01 33	☽ bQ ♅
02 38	☽ ⊼ ♇
07 58	☽ ⊼ ♆
10 16	☽ □ ♄
14 06	☽ ± ♇
17 15	♀ ⊥ ♄
17 23	☉ ± ♅
18 20	♂ ∠ ♆
18 53	☽ □ ☉
06 Wednesday	
04 52	☽ bQ ♀
05 25	☽ △ ♃
06 24	☽ △ ♅
07 33	☽ bQ ♇
09 10	☉ ∥ ♃
14 18	☽ △ ♂
07 Thursday	
00 22	☉ ∥ ♄
00 32	☽ ± ♀
07 58	♀ □ ♇
10 48	☽ △ ☿
11 37	☽ bQ ♃
13 13	☽ △ ♇
13 49	☽ ⊼ ♀
19 00	☽ ☍ ♆
19 58	☿ bQ ♅
21 54	☽ △ ♄
22 06	☽ bQ ♂
08 Friday	
05 32	☿ △ ♇
06 01	☽ ± ♃
06 03	♃ △ ♅
07 26	☽ # ♆
11 53	☽ △ ☉
17 26	☽ ± ♂
18 05	☽ □ ♅
18 21	☽ ⊼ ♃
21 28	☽ bQ ☿
09 Saturday	
04 28	☽ bQ ♄
06 22	☽ ⊼ ♂
09 32	☽ ∥ ♅
18 09	♀ ⊻ ♆
18 27	☽ ± ☿
21 06	☽ bQ ☉
22 54	☽ ± ♄
10 Sunday	
01 44	☽ □ ♇
02 59	☿ ⚹ ♆
06 27	☽ Q ♅
07 44	☽ ⊼ ♆
08 27	☽ ⊼ ☿
09 15	☽ ☍ ♀
11 11	☽ ⊼ ♄
17 03	☽ ± ☉
19 54	☽ ± ♆
11 Monday	
06 04	♀ ⊻ ♄
06 19	☽ ⊼ ☉
06 38	☿ ☌ ♄
06 39	☽ ⚹ ♅
08 12	☽ ☍ ♃
08 22	☿ ⊻ ♀
10 05	☉ ⊼ ♅
14 05	☽ Q ♇
14 06	☽ bQ ♆
23 03	☽ ☍ ♂
12 Tuesday	
02 06	☉ bQ ♇
02 43	☉ Q ♆
09 23	☉ ⊻ ♃
12 43	☽ ∠ ♅
14 09	☽ ⚹ ♇
20 14	☽ △ ♆
13 Wednesday	
00 08	☽ ☍ ♄
04 22	☽ ⊼ ♀
05 54	☽ ☍ ☿
06 38	☽ ⊥ ♅
11 22	☿ ± ♅
17 36	☽ ± ♀
18 30	☽ ⊻ ♅
19 57	☽ ∠ ♇
21 14	☽ ⊼ ♃
23 51	☽ ☍ ☉
14 Thursday	
09 17	☽ ± ♃
13 18	☽ bQ ♀
13 40	☽ ⊥ ♇
14 34	☽ ⊼ ♂
19 10	♀ ⊥ ♆
20 19	☽ ∥ ♅
15 Friday	
00 56	☿ ∥ ♂
01 23	☽ ⊻ ♇
03 03	☽ ± ♂
03 13	☽ bQ ♃
07 29	☽ □ ♆
11 47	☽ ⊼ ♄
21 39	☽ bQ ♂
21 42	☽ △ ♀
21 51	☽ # ♆
23 31	☽ ± ♄
16 Saturday	
01 29	☽ ⊼ ☿
04 55	☽ ☌ ♅
08 46	☽ △ ♃
10 32	♀ # ♇
14 29	☉ ∥ ♀
14 44	☽ ± ☿
15 29	☽ ⊼ ☉
16 59	☽ bQ ♄
17 22	☉ # ♇
18 16	☽ Q ♆
17 Sunday	
03 32	☿ ⊼ ♅
03 57	☽ ± ☉
04 14	☽ △ ♂
10 24	☽ bQ ☿
11 03	☽ ☌ ♇
15 24	☿ bQ ♇
17 06	☽ ⚹ ♆
18 27	☿ Q ♆
21 44	☽ △ ♄
22 28	☽ bQ ☉
18 Monday	
12 45	☽ □ ♀
13 37	☽ ⊻ ♅
16 50	☿ ⊻ ♃
18 25	☽ □ ♃
18 37	☽ △ ☿
21 10	☽ ∠ ♆
21 41	♀ △ ♅
19 Tuesday	
00 42	☽ ⊥ ♅
04 47	☽ △ ☉
13 45	☽ ⊥ ♆
15 36	☽ □ ♂
17 10	☽ ∠ ♅
18 43	☽ ⊻ ♇
18 56	☉ ⊥ ♃
20 Wednesday	
00 40	☽ ⊻ ♆
01 11	☽ Q ♀
03 53	☽ Q ♃
05 32	☽ □ ♄
05 34	☽ ⊥ ♇
17 50	♂ bQ ♅
18 33	☉ ± ♇
20 05	☽ ⚹ ♅
21 39	☽ ∠ ♇
22 46	☿ ∥ ♃
21 Thursday	
00 48	☽ ⚹ ♀
01 39	☽ ⚹ ♃
01 50	☽ Q ♂
08 26	☽ □ ☿
12 00	♀ ☌ ♃
13 44	☽ Q ♄
15 01	☽ □ ☉
17 08	☽ Q ♅
21 31	♂ △ ♇
23 53	☽ ⚹ ♇
22 Friday	
00 01	☽ ⚹ ♂
04 12	☽ ∠ ♃
05 30	☽ ∠ ♀
05 39	☽ ☌ ♆
07 14	♀ ∠ ♆
09 23	☿ ∥ ♄
10 39	☽ ⚹ ♄
17 13	☽ ∥ ♆
19 41	☽ Q ☿
19 48	☽ ⊥ ♃
20 23	☽ Q ♇
20 45	☿ ⊥ ♃
22 17	☽ ⊥ ♀
23 53	☽ □ ♅
23 Saturday	
00 18	☽ Q ☉
01 34	☿ ± ♇
03 01	☽ ∠ ♂
06 03	☽ ⊻ ♃
09 19	☽ ⊻ ♀
12 11	☽ ∠ ♄
16 26	☽ # ♅
18 27	☽ ⚹ ☿
18 52	☽ ⊥ ♂
19 44	☿ ∥ ♀
21 52	☽ ⚹ ☉
24 Sunday	
02 24	☽ □ ♇
03 13	☽ ⊥ ♄
05 19	☽ ⊻ ♂
08 00	☽ ⊻ ♆
13 08	☽ ⊻ ♄
17 49	☽ ⊥ ♆
22 15	☽ ∠ ☿
25 Monday	
00 16	☽ ∠ ☉
01 17	☽ △ ♅
08 02	☽ ☌ ♃
08 25	☽ ∠ ♆
14 41	☽ ⊥ ☿
14 54	☽ ☌ ♀
15 53	☽ ⊥ ☉
20 38	☿ ⊥ ♀
23 00	☿ # ♇
26 Tuesday	
01 26	☽ bQ ♅
01 35	☽ ⊻ ☿
02 15	☽ ⊻ ☉
03 01	☽ △ ♇
08 30	☽ ☌ ♂
08 36	☽ ⚹ ♆
10 25	♂ ⚹ ♆
13 18	☉ ⊼ ♇
13 59	☽ ☌ ♄
14 03	☿ ⊼ ♇
14 23	♃ ∠ ♆
15 10	☉ ☌ ☿
15 51	☽ ± ♅
27 Wednesday	
01 29	☽ ⊼ ♅
03 07	☽ bQ ♇
03 54	☽ Q ♆
08 59	☽ ⊻ ♃
17 38	☽ ± ♇
18 49	☽ ⊥ ♃
19 32	☽ ⊻ ♀
21 52	☽ # ♅
28 Thursday	
03 22	☽ ⊼ ♇
06 11	☽ ⊥ ♀
06 15	☽ ☌ ☉
08 21	☽ ☌ ☿
09 08	☽ □ ♆
09 41	☽ ∠ ♃
11 08	☉ ⊥ ♀
11 37	☽ ⊻ ♂
14 58	☽ ⊻ ♄
15 00	☿ □ ♆
20 57	☿ ∠ ♃
20 59	☽ ∥ ♆
22 02	☽ ⊥ ♂
22 20	☽ ∠ ♀
29 Friday	
00 58	☽ ⊥ ♄
02 17	☽ ☍ ♅
10 53	☽ ⚹ ♃
13 54	☽ ∠ ♂
16 10	☽ ∠ ♄
30 Saturday	
00 00	☉ □ ♆
01 56	☽ ⚹ ♀
05 22	☽ ☍ ♇
06 33	☿ ⊻ ♂
07 34	☽ Q ♃
11 33	☽ △ ♆
12 28	☽ ⊻ ☉
17 04	☽ ⚹ ♂
17 56	☽ ⊻ ☿
18 08	☽ ⚹ ♄
18 28	☉ ∠ ♃
19 35	☿ ⊻ ♄
19 41	♀ bQ ♅
23 47	☽ ⊥ ☉
31 Sunday	
00 47	☽ Q ♀
05 36	☽ ⊼ ♅
06 03	☽ ⊥ ☿
14 00	☽ bQ ♆
15 34	☽ Q ♂
15 34	☽ Q ♄
15 42	☽ □ ♃
15 48	♂ ☌ ♄
16 17	☽ ± ♅
16 42	♀ △ ♇
17 07	☽ ∠ ☉

FEBRUARY 1960

LONGITUDES

Date	Sidereal time h m s	Sun ☉ ° ' "	Moon ☽ ° ' "	Moon ☽ 24.00 ° ' "	Mercury ☿ ° '	Venus ♀ ° '	Mars ♂ ° '	Jupiter ♃ ° '	Saturn ♄ ° '	Uranus ♅ ° '	Neptune ♆ ° '	Pluto ♇ ° '
01	20 42 50	11 ♒ 38 38	06 ♈ 11 01	12 ♈ 37 19	15 ♒ 51	06 ♑ 28	13 ♑ 38	25 ♐ 13	13 ♑ 05	19 ♌ 21	09 ♏ 07	05 ♍ 28
02	20 46 46	12 39 32	18 ♈ 57 55	25 ♈ 13 19	17 37	07 42	14 22	25 24	13 11	19 R 18	09 07	05 R 27
03	20 50 43	13 40 25	01 ♉ 24 05	07 ♉ 30 49	19 23	08 55	15 07	25 35	13 18	19 16	09 08	05 25
04	20 54 40	14 41 16	13 ♉ 34 07	19 ♉ 34 39	21 10	10 08	15 52	25 46	13 24	19 13	09 08	05 24
05	20 58 36	15 42 07	25 ♉ 33 02	01 ♊ 29 53	22 57	11 22	16 37	25 57	13 30	19 10	09 08	05 23
06	21 02 33	16 42 55	07 ♊ 25 48	13 ♊ 21 22	24 45	12 35	17 22	26 08	13 37	19 08	09 08	05 21
07	21 06 29	17 43 43	19 ♊ 17 06	25 ♊ 13 32	26 33	13 49	18 08	26 19	13 43	19 05	09 08	05 20
08	21 10 26	18 44 29	01 ♋ 11 04	07 ♋ 10 10	28 ♒ 20	15 02	18 53	26 30	13 49	19 02	09 08	05 18
09	21 14 22	19 45 13	13 ♋ 11 08	19 ♋ 14 18	00 ♓ 08	16 15	19 38	26 41	13 56	19 00	09 08	05 17
10	21 18 19	20 45 56	25 ♋ 19 56	01 ♌ 28 12	01 55	17 29	20 23	26 51	14 02	18 57	09 R 08	05 15
11	21 22 15	21 46 38	07 ♌ 39 18	13 ♌ 53 19	03 42	18 43	21 09	27 02	14 08	18 55	09 08	05 14
12	21 26 12	22 47 18	20 ♌ 10 19	26 ♌ 30 23	05 28	19 56	21 54	27 12	14 14	18 52	09 08	05 12
13	21 30 09	23 47 56	02 ♍ 53 30	09 ♍ 19 41	07 13	21 10	22 39	27 22	14 20	18 49	09 08	05 11
14	21 34 05	24 48 33	15 ♍ 48 55	22 ♍ 21 10	08 56	22 23	23 25	27 32	14 26	18 47	09 08	05 09
15	21 38 02	25 49 09	28 ♍ 56 27	05 ♎ 34 43	10 37	23 37	24 10	27 42	14 32	18 44	09 08	05 08
16	21 41 58	26 49 44	12 ♎ 16 00	19 ♎ 00 16	12 17	24 51	24 56	27 52	14 38	18 41	09 08	05 06
17	21 45 55	27 50 17	25 ♎ 47 33	02 ♏ 37 49	13 53	26 04	25 41	28 02	14 44	18 39	09 07	05 05
18	21 49 51	28 50 49	09 ♏ 31 06	16 ♏ 27 20	15 26	27 18	26 27	28 12	14 49	18 36	09 07	05 03
19	21 53 48	29 ♒ 51 20	23 ♏ 26 29	00 ♐ 28 26	16 55	28 32	27 12	28 21	14 55	18 34	09 07	05 02
20	21 57 44	00 ♓ 51 50	07 ♐ 33 04	14 ♐ 40 08	18 19	29 ♑ 45	27 58	28 31	15 01	18 31	09 06	05 00
21	22 01 41	01 52 18	21 ♐ 49 22	29 ♐ 00 24	19 38	00 ♒ 59	28 44	28 40	15 06	18 29	09 06	04 59
22	22 05 38	02 52 45	06 ♑ 12 46	13 ♑ 25 58	20 51	02 13	29 ♑ 29	28 50	15 12	18 26	09 06	04 57
23	22 09 34	03 53 11	20 ♑ 39 25	27 ♑ 52 27	21 57	03 27	00 ♒ 15	28 59	15 17	18 24	09 05	04 56
24	22 13 31	04 53 35	05 ♒ 04 24	12 ♒ 14 35	22 57	04 40	01 01	29 08	15 23	18 21	09 05	04 54
25	22 17 27	05 53 58	19 ♒ 22 19	26 ♒ 26 57	23 48	05 54	01 46	29 17	15 28	18 19	09 04	04 53
26	22 21 24	06 54 19	03 ♓ 27 53	10 ♓ 24 36	24 31	07 08	02 32	29 26	15 33	18 16	09 04	04 51
27	22 25 20	07 54 38	17 ♓ 16 42	24 ♓ 03 51	25 05	08 22	03 18	29 34	15 39	18 14	09 03	04 49
28	22 29 17	08 54 56	00 ♈ 45 49	07 ♈ 22 31	25 30	09 36	04 04	29 43	15 44	18 11	09 03	04 48
29	22 33 13	09 ♓ 55 11	13 ♈ 53 56	20 ♈ 20 12	25 ♓ 46	10 ♒ 49	04 ♒ 49	29 ♐ 51	15 ♑ 49	18 ♌ 09	09 ♏ 02	04 ♍ 46

Date	Moon True ☊ ° '	Moon Mean ☊ ° '	Moon ☽ Latitude ° '
01	25 ♍ 26	27 ♍ 03	00 S 58
02	25 D 27	27 00	02 04
03	25 29	26 57	03 02
04	25 30	26 54	03 51
05	25 R 29	26 51	04 29
06	25 27	26 48	04 55
07	25 23	26 44	05 09
08	25 17	26 41	05 09
09	25 11	26 38	04 55
10	25 04	26 35	04 29
11	24 58	26 32	03 49
12	24 53	26 29	02 58
13	24 49	26 25	01 57
14	24 47	26 22	00 S 49
15	24 D 47	26 19	00 N 23
16	24 48	26 16	01 34
17	24 50	26 13	02 42
18	24 51	26 09	03 41
19	24 52	26 06	04 29
20	24 R 52	26 03	05 00
21	24 51	26 00	05 14
22	24 49	25 57	05 08
23	24 46	25 54	04 43
24	24 43	25 50	04 00
25	24 41	25 47	03 02
26	24 39	25 44	01 54
27	24 38	25 41	00 N 41
28	24 D 38	25 38	00 S 34
29	24 ♍ 38	25 ♍ 35	01 S 45

DECLINATIONS

Date	Sun ☉	Moon ☽	Mercury ☿	Venus ♀	Mars ♂	Jupiter ♃	Saturn ♄	Uranus ♅	Neptune ♆	Pluto ♇
01	17 S 18	01 N 34	18 S 00	22 S 16	23 S 27	22 S 54	22 S 17	15 N 43	12 S 51	21 N 13
02	17 01	05 31	17 26	22 15	23 23	22 55	22 17	15 44	12 51	21 14
03	16 43	09 06	16 50	22 13	23 18	22 55	22 16	15 45	12 51	21 14
04	16 26	12 14	16 13	22 11	23 14	22 56	22 16	15 46	12 51	21 15
05	16 08	14 47	15 34	22 08	23 09	22 56	22 15	15 47	12 51	21 16
06	15 50	16 42	14 54	22 05	23 04	22 56	22 14	15 48	12 51	21 17
07	15 31	17 53	14 13	22 01	22 58	22 57	22 14	15 48	12 51	21 17
08	15 13	18 17	13 30	21 56	22 53	22 57	22 13	15 49	12 51	21 18
09	14 54	17 53	12 46	21 51	22 47	22 57	22 13	15 50	12 51	21 19
10	14 34	16 40	12 01	21 45	22 41	22 58	22 12	15 51	12 51	21 20
11	14 15	14 40	11 15	21 38	22 35	22 58	22 11	15 52	12 51	21 20
12	13 55	11 57	10 29	21 31	22 28	22 58	22 11	15 53	12 51	21 21
13	13 35	08 38	09 41	21 23	22 21	22 58	22 10	15 53	12 51	21 22
14	13 15	04 51	08 53	21 14	22 14	22 59	22 10	15 54	12 51	21 22
15	12 55	00 N 46	08 05	21 05	22 07	22 59	22 09	15 55	12 51	21 23
16	12 34	03 S 24	07 17	20 55	21 59	22 59	22 08	15 56	12 51	21 24
17	12 14	07 27	06 29	20 45	21 52	22 59	22 08	15 57	12 51	21 24
18	11 53	11 10	05 41	20 34	21 44	22 59	22 07	15 57	12 50	21 25
19	11 31	14 18	04 55	20 22	21 35	23 00	22 07	15 58	12 50	21 26
20	11 10	16 38	04 09	20 10	21 27	23 00	22 06	15 59	12 50	21 27
21	10 49	17 58	03 25	19 57	21 18	23 00	22 05	16 00	12 50	21 27
22	10 27	18 10	02 43	19 44	21 09	23 00	22 05	16 01	12 50	21 28
23	10 05	17 11	02 03	19 30	21 00	23 00	22 04	16 01	12 50	21 29
24	09 43	15 07	01 26	19 15	20 51	23 00	22 04	16 02	12 49	21 29
25	09 21	12 07	00 52	19 00	20 41	23 00	22 03	16 03	12 49	21 30
26	08 59	08 28	00 S 21	18 44	20 32	23 00	22 03	16 04	12 49	21 30
27	08 36	04 24	00 N 06	18 28	20 22	23 00	22 02	16 04	12 49	21 31
28	08 14	00 13	00 28	18 11	20 11	23 00	22 01	16 05	12 48	21 32
29	07 S 51	03 N 53	00 N 47	17 S 54	20 S 01	23 S 01	22 S 01	16 N 06	12 S 48	21 N 32

ZODIAC SIGN ENTRIES

Date	h	m	Planets
01	00	39	☽ ♈
03	09	16	☽ ♉
05	20	58	☽ ♊
08	09	37	☽ ♋
09	10	13	☿ ♓
10	21	08	☽ ♌
13	06	35	☽ ♍
15	13	55	☽ ♎
17	19	24	☽ ♏
19	15	26	☉ ♓
19	23	12	☽ ♐
20	16	47	♀ ♒
22	01	39	☽ ♑
23	04	11	♂ ♒
24	03	32	☽ ♒
26	06	04	☽ ♓
28	10	37	☽ ♈

LATITUDES

Date	Mercury ☿	Venus ♀	Mars ♂	Jupiter ♃	Saturn ♄	Uranus ♅	Neptune ♆	Pluto ♇
01	02 S 01	01 N 02	00 S 42	00 N 27	00 N 31	00 N 44	01 N 46	12 N 38
04	01 52	00 52	00 44	00 27	00 31	00 44	01 47	12 39
07	01 38	00 43	00 46	00 27	00 30	00 44	01 47	12 40
10	01 19	00 33	00 48	00 27	00 30	00 44	01 47	12 40
13	00 53	00 24	00 50	00 26	00 30	00 44	01 47	12 41
16	00 S 21	00 15	00 51	00 26	00 30	00 44	01 47	12 41
19	00 N 17	00 N 05	00 53	00 26	00 30	00 44	01 47	12 41
22	01 00	00 S 04	00 55	00 26	00 30	00 44	01 48	12 42
25	01 44	00 12	00 57	00 26	00 30	00 44	01 48	12 42
28	02 28	00 21	00 58	00 26	00 30	00 44	01 48	12 42
31	03 N 05	00 S 29	01 S 00	00 N 26	00 N 30	00 N 44	01 N 48	12 N 42

DATA

Julian Date	2436966
Delta T	+33 seconds
Ayanamsa	23° 17' 59"
Synetic vernal point	05° ♓ 49' 01"
True obliquity of ecliptic	23° 26' 31"

MOON'S PHASES, APSIDES AND POSITIONS ☽

Date	h	m	Phase	Longitude ° '	Eclipse Indicator
04	14	26	☽	14 ♉ 47	
12	17	24	○	23 ♌ 01	
19	23	47	☾	00 ♐ 21	
26	18	23	●	07 ♓ 10	

Day	h	m		
07	05	47	Apogee	
23	02	40	Perigee	
01	02	50	0N	
08	12	12	Max dec	18° N 17'
15	16	26	0S	
22	04	03	Max dec	18° S 14'
28	13	14	0N	

LONGITUDES

Date	Chiron ⚷ ° '	Ceres ⚳ ° '	Pallas ⚴ ° '	Juno ⚵ ° '	Vesta ⚶ ° '	Black Moon Lilith ⚸ ° '
01	26 ♒ 23	21 ♑ 54	04 ♑ 12	15 ♐ 26	14 ♐ 12	19 ♊ 10
11	27 ♒ 04	25 ♑ 48	07 ♑ 57	18 ♐ 04	18 ♐ 52	20 ♊ 17
21	27 ♒ 46	29 ♑ 36	11 ♑ 32	20 ♐ 28	23 ♐ 21	21 ♊ 25
31	28 ♒ 27	03 ♒ 19	14 ♑ 55	22 ♐ 35	27 ♐ 38	22 ♊ 32

ASPECTARIAN

h m Aspects

01 Monday
00 41 ☽ ∠ ☿
03 47 ♂ ± ♅
06 21 ☽ ± ♆
08 37 ☽ ⅊ ♅
10 41 ☽ ⊼ ♇
12 35 ☽ □ ♀
17 27 ☽ ⊼ ♆
21 49 ☽ ± ♇
23 02 ☽ ✶ ☉

02 Tuesday
00 58 ☽ □ ♄
02 44 ☽ □ ♂
09 01 ☽ ✶ ☿
12 38 ☽ △ ♅
14 49 ☽ ⅊ ♇
23 53 ☽ Q ☉

03 Wednesday
00 32 ☽ △ ♃
01 58 ☉ ⊻ ♄
06 24 ♄ ± ♅
10 18 ☿ ☍ ♅
10 39 ☿ ⊥ ♄
11 58 ☽ Q ☿
16 10 ♀ ✶ ♆
19 52 ☽ △ ♇
20 24 ☉ ∥ ☿

04 Thursday
03 11 ☽ ☍ ♆
04 25 ☽ △ ♀
06 22 ☽ ⅊ ♃
11 40 ☽ △ ♄
14 26 ☽ □ ☉
16 54 ☽ △ ♂
17 23 ☽ # ♆
23 14 ☽ □ ♅

05 Friday
00 35 ☽ ± ♃
04 15 ☿ ⊥ ♂
04 23 ☿ # ♅
05 52 ☽ □ ☿
12 50 ☽ ⊼ ♃
13 49 ☽ ⅊ ♀
18 01 ☽ ⅊ ♄
18 39 ☽ # ☿
23 29 ☽ ∥ ♅

06 Saturday
01 05 ☽ ⅊ ♂
01 41 ☽ # ☉
07 48 ☽ □ ♇
10 05 ☽ ± ♀
11 23 ☽ Q ♅
12 22 ☽ ± ♄
14 38 ☉ # ♅
15 27 ☽ ⊼ ♆
20 32 ☽ ± ♂
22 17 ♀ ± ♅
23 38 ☽ ⊼ ♀

07 Sunday
00 38 ☽ ⊼ ♄
03 36 ☽ ± ♆
08 33 ☽ △ ☉
08 39 ☿ ✶ ♃
09 30 ☽ ⊼ ♂
10 04 ♀ ☌ ♄
11 36 ☽ ✶ ♅
18 44 ♂ ∥ ♃
20 09 ☽ Q ♇
21 48 ☽ ⅊ ♆

08 Monday
02 25 ☽ ☍ ♃
05 16 ☽ △ ☿
16 46 ♂ ⊼ ♅
17 36 ☽ ⅊ ☉
17 43 ☽ ∠ ♅
18 47 ☉ ☍ ♅
18 51 ☿ ∠ ♄
20 15 ☽ ✶ ♇

09 Tuesday
01 01 ☉ ⊻ ♂
03 56 ☽ △ ♆
09 10 ☿ ∥ ♆
11 37 ☽ ⊥ ♅
13 14 ☽ ± ☉
13 29 ☽ ☍ ♄
16 33 ☽ ⅊ ☿
16 33 ☉ ⊥ ♄
18 47 ☽ ☍ ♀
23 29 ☽ ⊻ ♅

10 Wednesday
00 04 ♆ St R
01 38 ☽ ☍ ♂
02 02 ☽ ∠ ♇
02 12 ☽ ⊼ ☉
07 51 ♂ ⅊ ♇
13 21 ☽ ± ☿
15 01 ☽ ⊼ ♃
19 40 ☽ ⊥ ♇
22 53 ☽ ∥ ♅

11 Thursday
02 54 ☽ ± ♃
03 03 ☽ ⊼ ☿
07 19 ☽ ⊻ ♇
11 46 ♂ Q ♆
12 22 ☿ ∠ ♀
14 52 ☽ □ ♆
15 47 ♀ ⊼ ♅
16 43 ☽ # ☉
20 32 ☽ ⅊ ♃

12 Friday
00 34 ☽ ⊼ ♄
04 38 ☽ # ♆
08 30 ☿ ☍ ♇
09 31 ☽ ☌ ♅
11 30 ☽ ⊼ ♀
12 07 ☽ ± ♄
15 30 ☽ ⊼ ♂
17 12 ♀ ⅊ ♇
17 24 ☽ ☍ ☉

13 Saturday
00 05 ☽ ± ♀
01 11 ☽ Q ♆
01 29 ☽ △ ♃
02 33 ☽ # ☿
03 33 ☽ ± ♂
05 16 ☽ ⅊ ♄
11 30 ♀ Q ♆
15 19 ♀ # ♇
16 16 ☽ ☌ ♇
18 45 ☽ ⅊ ♀
21 19 ☽ ☍ ☿
21 27 ☽ ⅊ ♂
22 59 ☿ ∠ ♂
23 38 ☽ ✶ ♆

14 Sunday
09 26 ☽ △ ♄
14 48 ☿ △ ♆
17 26 ☽ ⊻ ♅
21 27 ☿ Q ♃

15 Monday
01 18 ☽ △ ♀
02 47 ☽ △ ♂
03 15 ☽ ∠ ♆
04 22 ☽ ⊥ ♅
04 36 ♂ ∥ ♄
05 51 ☽ ⊼ ☉
09 44 ☽ □ ♃
16 50 ☉ ∥ ♆
17 39 ☽ ± ☉
19 35 ☽ ⊥ ♆
20 39 ☽ ∠ ♅
23 10 ☽ ⊻ ♇

16 Tuesday
06 23 ☽ ⊻ ♆
09 56 ☽ ⊥ ♇
11 09 ☽ ⅊ ☉
12 01 ☽ ⊼ ☿
16 15 ☽ □ ♄
16 27 ♀ ☌ ♂
18 31 ☽ Q ♃
23 24 ☽ ✶ ♅

17 Wednesday
00 10 ☽ ± ☿
01 56 ☽ ∠ ♇
07 06 ☽ ∥ ☿
11 48 ☽ □ ♂
12 32 ☽ □ ♀
15 53 ☽ △ ☉
16 00 ☽ ✶ ♃
17 35 ☉ ✶ ♃
18 08 ☽ ⅊ ☿
20 30 ☽ Q ♅

18 Thursday
00 15 ☽ Q ♄
01 52 ☿ ✶ ♄
04 15 ☽ ✶ ♇
11 18 ☽ ☌ ♆
16 34 ☽ ∥ ☉
18 27 ☽ ∠ ♃
21 02 ☽ Q ♂
21 15 ☽ ✶ ♄
22 58 ☽ Q ♀
23 28 ☽ △ ☿

19 Friday
00 10 ☽ ∥ ♆
01 01 ☽ Q ♇
03 39 ☽ □ ♅
08 13 ♀ ⊻ ♃
10 07 ☽ ⊥ ♃
13 37 ☉ ∠ ♄
18 48 ☽ ✶ ♂
20 30 ☽ ⊻ ♃
21 31 ☽ ✶ ♀
21 38 ♀ ± ♇
23 08 ☽ ∠ ♄
23 47 ☽ □ ☉

20 Saturday
04 15 ☽ # ♅
07 42 ☽ □ ♇
13 18 ♂ # ♇
14 29 ☽ ⊥ ♄
14 38 ☽ ⊻ ♆
15 32 ☿ ⊼ ♅
21 39 ☽ ∠ ♂

21 Sunday
00 40 ☽ ⊻ ♄
00 44 ☽ ⊥ ♆
01 17 ☽ ∠ ♀
06 25 ☽ △ ♅
07 58 ☽ □ ☿
08 29 ☽ Q ☉
09 54 ♂ ⊻ ♃
13 36 ☽ ⊥ ♂
15 49 ☽ ∠ ♆
17 47 ☽ ⊥ ♀
19 43 ♂ ± ♇
23 34 ☽ ☌ ♃

22 Monday
00 11 ☽ ⊻ ♂
04 43 ☽ ⊻ ♀
06 02 ☽ ✶ ☉
07 23 ☽ ⅊ ♅
09 54 ☽ △ ♇
16 46 ☽ Q ☿
16 47 ☽ ✶ ♆
22 19 ☽ ± ♅

23 Tuesday
03 02 ☽ ☌ ♄
08 15 ☽ ⊼ ♅
08 50 ☽ ∠ ☉
10 47 ☽ ⅊ ♇
12 43 ☽ Q ♆
14 20 ☽ ✶ ☿

24 Wednesday
00 55 ☽ ⊥ ☉
01 44 ☽ ± ♇
01 59 ☽ ⊻ ♃
02 48 ☽ # ♅
04 51 ☽ ☌ ♂
11 16 ☽ ☌ ♀
11 41 ☽ ⊻ ☉
11 43 ☽ ⊼ ♇
12 06 ☽ ⊥ ♃
12 12 ☉ ☍ ♇
16 23 ♀ ⊼ ♇
17 07 ☽ ∠ ☿
18 42 ☽ □ ♆
22 12 ♀ ⊥ ♃

25 Thursday
03 20 ☽ ∠ ♃
05 23 ☽ ⊻ ♄
06 56 ☽ ∥ ♆
09 11 ☽ ⊥ ☿
10 13 ☽ ☍ ♅
11 42 ☉ ⊻ ♀
15 34 ☽ ⊥ ♄
19 56 ☽ ⊻ ☿
20 25 ☿ ⅊ ♆

26 Friday
03 34 ☿ ± ♅
05 00 ☽ ✶ ♃
06 59 ☽ ∠ ♄
08 28 ☽ ∥ ☉
10 19 ☽ ⊻ ♂
14 23 ☽ ☍ ♇
18 23 ☽ ☌ ☉
18 56 ☽ ⊻ ♀
21 15 ☽ ⊥ ♂
21 39 ☽ △ ♆

27 Saturday
01 55 ☽ Q ♃
06 23 ☽ ⊥ ♀
09 07 ☽ ✶ ♄
13 40 ☽ ⊼ ♅
13 54 ☽ ∠ ♂
23 50 ☽ ∠ ♀
23 58 ☽ ⅊ ♆

28 Sunday
00 15 ☽ ± ♅
01 22 ♀ □ ♆
02 18 ☽ ☌ ☿
06 31 ☽ Q ♄
10 05 ☽ □ ♃
10 37 ☽ # ☿
15 02 ☉ △ ♆
16 07 ☽ ± ♆
16 19 ☽ ∥ ☿
16 22 ☽ ⅊ ♅
18 20 ☽ ✶ ♂
19 17 ☽ ⊼ ♇

29 Monday
03 02 ☽ ⊼ ♆
04 03 ☽ ⊻ ☉
05 44 ☽ ✶ ♀
06 15 ☽ ± ♇
10 28 ♂ ⊼ ♇
10 43 ☿ ∠ ♀
15 35 ☽ □ ♄
16 04 ☽ ⊥ ☉
17 47 ☽ Q ♂
19 53 ☽ △ ♅
22 55 ☽ ⅊ ♇

All ephemeris data is given at 12.00 UT and the Moon's longitude is additionally given for 24.00 UT

MARCH 1960

Date	Sidereal time h m s	Sun ☉ ° ' "	Moon ☽ ° ' "	Moon ☽ 24.00 ° ' "	Mercury ☿ ° '	Venus ♀ ° '	Mars ♂ ° '	Jupiter ♃ ° '	Saturn ♄ ° '	Uranus ♅ ° '	Neptune ♆ ° '	Pluto ⯓ ° '
01	22 37 10	10 ♓ 55 25	26 ♈ 41 30	02 ♉ 58 07	25 ♓ 52	12 ♒ 03	05 ♒ 35	00 ♑ 00	15 ♑ 54	18 ♌ 06	09 ♏ 01	04 ♍ 45
02	22 41 07	11 55 36	09 ♉ 10 26	15 ♉ 18 52	25 R 49	13 17	06 21	00 08	15 59	18 R 04	09 R 01	04 R 43
03	22 45 03	12 55 46	21 ♉ 23 53	27 ♉ 26 00	25 36	14 31	07 07	00 16	16 04	18 02	09 00	04 42
04	22 49 00	13 55 54	03 ♊ 25 48	09 ♊ 23 49	25 14	15 45	07 53	00 24	16 09	17 59	08 59	04 40
05	22 52 56	14 55 59	15 ♊ 20 41	21 ♊ 16 58	24 44	16 59	08 39	00 31	16 13	17 57	08 58	04 39
06	22 56 53	15 56 03	27 ♊ 13 15	03 ♋ 10 08	24 06	18 12	09 25	00 39	16 18	17 55	08 58	04 37
07	23 00 49	16 56 04	09 ♋ 08 11	15 ♋ 07 55	23 22	19 26	10 11	00 47	16 23	17 53	08 57	04 36
08	23 04 46	17 56 03	21 ♋ 09 51	27 ♋ 14 27	22 32	20 40	10 57	00 54	16 27	17 51	08 56	04 34
09	23 08 42	18 56 00	03 ♌ 22 08	09 ♌ 33 17	21 38	21 54	11 43	01 01	16 32	17 48	08 55	04 33
10	23 12 39	19 55 55	15 ♌ 48 13	22 ♌ 07 09	20 42	23 08	12 29	01 08	16 36	17 46	08 54	04 31
11	23 16 36	20 55 48	28 ♌ 30 18	04 ♍ 57 46	19 43	24 22	13 15	01 15	16 41	17 44	08 53	04 30
12	23 20 32	21 55 38	11 ♍ 29 35	18 ♍ 05 44	18 43	25 36	14 01	01 22	16 45	17 42	08 52	04 28
13	23 24 29	22 55 27	24 ♍ 46 06	01 ♎ 30 31	17 46	26 49	14 47	01 29	16 49	17 40	08 51	04 27
14	23 28 25	23 55 14	08 ♎ 18 47	15 ♎ 10 35	16 51	28 03	15 33	01 35	16 53	17 38	08 50	04 26
15	23 32 22	24 54 58	22 ♎ 05 37	29 ♎ 03 30	15 58	29 ♒ 17	16 19	01 41	16 57	17 36	08 49	04 24
16	23 36 18	25 54 41	06 ♏ 03 52	13 ♏ 06 20	15 10	00 ♓ 31	17 05	01 48	17 01	17 34	08 48	04 23
17	23 40 15	26 54 23	20 ♏ 10 28	27 ♏ 15 52	14 27	01 45	17 51	01 54	17 05	17 32	08 47	04 21
18	23 44 11	27 54 02	04 ♐ 22 11	11 ♐ 29 01	13 49	02 59	18 37	01 59	17 09	17 31	08 46	04 20
19	23 48 08	28 53 40	18 ♐ 36 01	25 ♐ 42 52	13 16	04 13	19 24	02 05	17 12	17 29	08 45	04 18
20	23 52 05	29 ♓ 53 16	02 ♑ 49 14	09 ♑ 54 51	12 50	05 27	20 10	02 11	17 16	17 27	08 44	04 17
21	23 56 01	00 ♈ 52 51	16 ♑ 59 24	24 ♑ 02 40	12 30	06 41	20 56	02 16	17 19	17 25	08 42	04 16
22	23 59 58	01 52 23	01 ♒ 04 21	08 ♒ 04 12	12 16	07 55	21 42	02 21	17 23	17 24	08 41	04 14
23	00 03 54	02 51 54	15 ♒ 01 59	21 ♒ 57 28	12 08	09 08	22 28	02 26	17 26	17 22	08 40	04 13
24	00 07 51	03 51 23	28 ♒ 50 22	05 ♓ 40 29	12 D 06	10 22	23 15	02 31	17 30	17 21	08 39	04 12
25	00 11 47	04 50 50	12 ♓ 27 34	19 ♓ 11 25	12 10	11 36	24 01	02 36	17 33	17 19	08 37	04 11
26	00 15 44	05 50 15	25 ♓ 51 50	02 ♈ 28 41	12 19	12 50	24 47	02 41	17 36	17 18	08 36	04 09
27	00 19 40	06 49 39	09 ♈ 01 49	15 ♈ 31 09	12 34	14 04	25 33	02 45	17 39	17 16	08 35	04 08
28	00 23 37	07 49 00	21 ♈ 56 38	28 ♈ 18 19	12 54	15 18	26 20	02 49	17 42	17 15	08 33	04 07
29	00 27 34	08 48 19	04 ♉ 36 14	10 ♉ 50 31	13 18	16 32	27 06	02 53	17 45	17 13	08 32	04 06
30	00 31 30	09 47 36	17 ♉ 01 21	23 ♉ 08 59	13 47	17 46	27 52	02 57	17 47	17 12	08 31	04 04
31	00 35 27	10 ♈ 46 50	29 ♉ 13 43	05 ♊ 15 53	14 ♓ 21	19 ♓ 00	28 ♒ 38	03 ♑ 01	17 ♑ 50	17 ♌ 11	08 ♏ 29	04 ♍ 03

Date	Moon True ☊ ° '	Moon Mean ☊ ° '	Moon ☽ Latitude ° '
01	24 ♍ 40	25 ♍ 31	02 S 48
02	24 D 41	25 28	03 42
03	24 42	25 25	04 25
04	24 43	25 22	04 55
05	24 R 43	25 19	05 13
06	24 43	25 15	05 16
07	24 42	25 12	05 07
08	24 41	25 09	04 44
09	24 40	25 06	04 08
10	24 40	25 03	03 19
11	24 39	25 00	02 20
12	24 39	24 56	01 S 12
13	24 D 39	24 53	00 N 01
14	24 39	24 50	01 15
15	24 R 39	24 47	02 26
16	24 39	24 44	03 30
17	24 39	24 40	04 22
18	24 38	24 37	04 58
19	24 38	24 34	05 15
20	24 D 38	24 31	05 14
21	24 38	24 28	04 53
22	24 39	24 25	04 15
23	24 40	24 21	03 22
24	24 40	24 18	02 18
25	24 41	24 15	01 N 07
26	24 R 41	24 12	00 S 07
27	24 41	24 09	01 19
28	24 40	24 06	02 25
29	24 38	24 02	03 23
30	24 36	23 59	04 10
31	24 ♍ 33	23 ♍ 56	04 S 45

DECLINATIONS

Date	Sun ☉ ° '	Moon ☽ ° '	Mercury ☿ ° '	Venus ♀ ° '	Mars ♂ ° '	Jupiter ♃ ° '	Saturn ♄ ° '	Uranus ♅ ° '	Neptune ♆ ° '	Pluto ⯓ ° '
01	07 S 28	07 N 41	01 N 01	17 S 36	19 S 50	23 S 01	22 S 00	16 N 07	12 S 48	21 N 33
02	07 05	11 02	01 10	17 18	19 40	23 01	22 00	16 07	12 48	21 34
03	06 42	13 51	01 14	16 59	19 28	23 01	21 59	16 08	12 47	21 34
04	06 19	16 01	01 13	16 40	19 17	23 01	21 59	16 09	12 47	21 35
05	05 56	17 27	01 08	16 20	19 06	23 01	21 58	16 09	12 47	21 35
06	05 33	18 08	00 57	16 00	18 54	23 01	21 58	16 10	12 46	21 36
07	05 10	18 02	00 43	15 39	18 42	23 01	21 57	16 11	12 46	21 36
08	04 46	17 06	00 24	15 18	18 30	23 01	21 57	16 11	12 46	21 37
09	04 23	15 23	00 N 02	14 57	18 18	23 01	21 56	16 12	12 46	21 38
10	03 59	12 56	00 S 23	14 35	18 06	23 01	21 56	16 12	12 45	21 38
11	03 36	09 48	00 50	14 12	17 53	23 00	21 55	16 13	12 45	21 38
12	03 12	06 08	01 20	13 50	17 40	23 00	21 55	16 14	12 44	21 39
13	02 49	02 N 05	01 50	13 27	17 27	23 00	21 54	16 14	12 44	21 40
14	02 25	02 S 09	02 21	13 03	17 14	23 00	21 54	16 15	12 44	21 40
15	02 01	06 20	02 52	12 39	17 01	23 00	21 53	16 16	12 43	21 41
16	01 38	10 14	03 22	12 15	16 47	23 00	21 53	16 16	12 43	21 41
17	01 14	13 35	03 51	11 51	16 34	23 00	21 52	16 17	12 43	21 42
18	00 50	16 09	04 19	11 26	16 20	23 00	21 52	16 17	12 42	21 42
19	00 26	17 43	04 45	11 01	16 06	23 00	21 51	16 18	12 42	21 42
20	00 S 03	18 11	05 09	10 35	15 52	23 00	21 51	16 18	12 41	21 43
21	00 N 21	17 31	05 31	10 10	15 38	23 00	21 50	16 19	12 41	21 43
22	00 45	15 46	05 50	09 43	15 23	23 00	21 50	16 19	12 41	21 44
23	01 08	13 06	06 07	09 17	15 09	23 00	21 50	16 20	12 40	21 44
24	01 32	09 43	06 22	08 51	14 54	23 00	21 49	16 20	12 40	21 44
25	01 56	05 51	06 34	08 24	14 39	23 00	21 49	16 21	12 39	21 45
26	02 19	01 S 45	06 43	07 57	14 24	23 00	21 49	16 21	12 39	21 45
27	02 43	02 N 23	06 50	07 30	14 09	23 00	21 48	16 21	12 38	21 46
28	03 06	06 18	06 55	07 02	13 53	23 00	21 48	16 22	12 38	21 46
29	03 29	09 52	06 57	06 35	13 38	22 59	21 47	16 22	12 37	21 46
30	03 53	12 55	06 57	06 07	13 23	22 59	21 47	16 23	12 37	21 46
31	04 N 16	15 N 21	06 S 55	05 S 39	13 S 07	22 S 59	21 S 47	16 N 23	12 S 36	21 N 47

ZODIAC SIGN ENTRIES

Date	h	m	Planets
01	13	10	♃ ♑
01	18	18	☽ ♉
04	05	08	☽ ♊
06	17	37	☽ ♋
09	05	25	☽ ♌
11	14	47	☽ ♍
13	21	19	☽ ♎
16	01	37	☽ ♏
16	01	53	♀ ♓
18	04	37	☽ ♐
20	07	14	☽ ♑
20	14	43	☉ ♈
22	10	10	☽ ♒
24	14	02	☽ ♓
26	19	29	☽ ♈
29	03	13	☽ ♉
31	13	32	☽ ♊

LATITUDES

Date	Mercury ☿ ° '	Venus ♀ ° '	Mars ♂ ° '	Jupiter ♃ ° '	Saturn ♄ ° '	Uranus ♅ ° '	Neptune ♆ ° '	Pluto ⯓ ° '
01	02 N 53	00 S 26	01 S 00	00 N 26	00 N 30	00 N 44	01 N 48	12 N 42
04	03 23	00 34	01 03	00 26	00 29	00 44	01 48	12 42
07	03 39	00 42	01 03	00 26	00 29	00 44	01 48	12 42
10	03 36	00 49	01 05	00 26	00 29	00 44	01 49	12 42
13	03 15	00 56	01 06	00 26	00 29	00 44	01 49	12 42
16	02 41	01 02	01 08	00 26	00 29	00 44	01 49	12 42
19	01 58	01 07	01 09	00 25	00 29	00 44	01 49	12 42
22	01 13	01 12	01 11	00 25	00 29	00 44	01 49	12 41
25	00 N 28	01 17	01 12	00 25	00 29	00 44	01 49	12 41
28	00 S 13	01 21	01 13	00 25	00 29	00 44	01 49	12 41
31	00 S 49	01 S 24	01 S 15	00 N 25	00 N 29	00 N 44	01 N 49	12 N 40

DATA

Julian Date	2436995
Delta T	+33 seconds
Ayanamsa	23° 18' 02"
Synetic vernal point	05° ♓ 48' 58"
True obliquity of ecliptic	23° 26' 31"

MOON'S PHASES, APSIDES AND POSITIONS ☽

Date	h	m	Phase	Longitude ° '	Eclipse Indicator
05	11	06	☽	14 ♊ 54	
13	08	26	○	22 ♍ 47	total
20	06	40	☾	29 ♐ 40	
27	07	37	●	06 ♈ 39	Partial

Day	h	m		
06	02	08	Apogee	
19	07	20	Perigee	
06	20	36	Max dec	18° N 11'
13	23	53	0S	
20	09	47	Max dec	18° S 11'
26	22	06	0N	

LONGITUDES

Date	Chiron ⚷ ° '	Ceres ⚳ ° '	Pallas ⚴ ° '	Juno ⚵ ° '	Vesta ⚶ ° '	Black Moon Lilith ⚸ ° '
01	28 ♒ 23	02 ♒ 57	14 ♑ 58	22 ♐ 24	27 ♐ 13	22 ♊ 57
11	29 ♒ 03	06 ♒ 33	17 ♑ 45	24 ♐ 14	01 ♑ 15	23 ♊ 32
21	29 ♒ 42	10 ♒ 01	20 ♑ 39	25 ♐ 42	05 ♑ 00	24 ♊ 40
31	00 ♓ 17	13 ♒ 18	23 ♑ 13	26 ♐ 45	08 ♑ 24	25 ♊ 47

All ephemeris data is given at 12.00 UT and the Moon's longitude is additionally given for 24.00 UT

ASPECTARIAN

h m	Aspects
01 Tuesday	
06 28	☽ Q ♀
10 25	☽ ∠ ☉
10 26	☽ ⊻ ☿
10 45	☽ ⋕ ☉
15 10	☿ St R
18 22	☽ △ ♃
21 53	☽ ⊥ ☿
02 Wednesday	
03 24	☽ △ ⯓
03 29	♂ ⊥ ♃
06 10	☽ □ ♂
11 41	☽ ☍ ♆
15 09	☽ ∠ ☿
17 36	☉ Q ♃
17 51	☽ ⚹ ☉
20 55	☽ □ ♀
23 46	☽ ♇ ♃
03 Thursday	
01 24	☽ △ ♄
02 21	☽ ⋕ ♆
05 22	☽ □ ♅
17 45	☽ ± ♃
19 39	☽ Q ☉
20 08	☽ ⚹ ☿
04 Friday	
04 21	♀ ∠ ♃
05 51	☽ ⊼ ♃
06 59	☉ ⊥ ♂
07 23	☽ ♇ ♄
13 49	☽ ∥ ♅
14 29	☽ □ ⯓
17 08	☽ Q ♅
19 22	☽ Q ☿
19 46	☽ ⋕ ♀
20 20	♀ ⊻ ♄
21 34	☽ △ ♂
23 10	☽ ⊼ ♆
05 Saturday	
01 36	☽ ± ♄
11 06	☽ □ ☉
11 15	☽ ± ♆
13 47	☽ ⊼ ♄
15 41	☽ △ ♀
17 15	☽ ⚹ ♅
22 03	♂ □ ♆
06 Sunday	
00 26	♀ ⋕ ♅
02 44	☽ Q ⯓
05 25	☽ ♇ ♆
05 56	☽ ♇ ♂
06 03	☽ □ ☿
06 29	♀ ☍ ♅
06 53	☿ ∠ ♂
10 44	☿ ⊥ ♀
17 08	☿ ♇ ♆
18 53	☿ ± ♅
19 00	☽ ☍ ♃
21 35	☉ ⚹ ♄
23 27	☽ ∠ ♅
07 Monday	
01 21	☽ ± ♂
01 28	☽ ♇ ♀
02 54	☽ ⚹ ⯓
11 37	☽ △ ♆
14 14	☽ ⊼ ♂
17 29	☽ ⊥ ♅
21 36	☽ ± ♀
08 Tuesday	
02 35	☽ ☍ ♄
05 00	☽ △ ☉
05 25	☽ ⊻ ♅
08 51	☽ ∠ ⯓
09 53	☉ ⊼ ♅
10 54	☽ ⊼ ♀
14 33	☽ △ ☿
09 Wednesday	
02 01	☽ ∥ ♅
02 35	☽ ⊥ ⯓
07 22	☽ ⊼ ♃
09 07	☿ ⊻ ♀
13 12	☽ ♇ ☉
14 17	☽ ⊻ ⯓
17 50	☽ ⋕ ♀
17 55	☽ ♇ ☿
19 10	☽ ± ♃
22 45	☽ □ ♆
10 Thursday	
01 04	♀ ⊥ ♄
05 13	☽ ☍ ♂
08 07	☽ ± ☉
10 02	☽ ± ☿
12 39	☽ ♇ ♃
13 30	☽ ⋕ ♆
13 32	☽ ⊼ ♄
15 45	☽ ☌ ♅
20 32	☽ ⊼ ☉
20 38	☽ ⊼ ☿
21 16	☉ ☌ ☿
11 Friday	
00 59	☽ ± ♄
03 24	☽ ☍ ♀
08 58	☽ Q ♆
17 10	☽ △ ♃
17 56	☽ ♇ ♄
18 27	☿ ⊥ ♂
23 07	☽ ☌ ⯓
12 Saturday	
07 12	☽ ⚹ ♆
16 53	☽ ⊼ ♂
21 37	☽ △ ♄
23 16	☽ ⊻ ♅
13 Sunday	
00 16	☽ ☍ ☿
04 24	☽ ± ♂
07 25	☽ ⋕ ☉
08 26	☽ ☍ ☉
10 02	☽ ⊥ ♅
10 22	☽ ∠ ♆
13 18	☽ ⋕ ☿
14 43	☿ ⊼ ♅
16 03	☽ ⊼ ♀
21 28	☽ ♇ ♂
14 Monday	
00 02	☽ □ ♃
02 01	☽ ∠ ♅
02 22	☽ ⊥ ♆
03 46	☽ ± ♀
05 10	☽ ⊻ ⯓
05 22	☉ ± ♅
10 01	☉ ♇ ♆
11 03	☿ ⚹ ♄
12 55	☽ ⊻ ♆
13 19	☽ ∥ ☿
13 23	☽ ∥ ☉
13 41	☉ ∥ ☿
15 42	☽ ⊥ ⯓
21 07	☽ ♇ ♀
15 Tuesday	
01 24	☽ △ ♂
02 00	☽ ⊼ ☿
03 03	☽ □ ♄
04 14	☽ ⚹ ♅
06 54	☿ ⊻ ♂
07 21	☽ ∠ ⯓
07 48	☽ Q ♃
07 55	♀ ∥ ♆
11 48	☽ ± ☿
17 15	☽ ⊼ ☉
16 Wednesday	
00 54	☽ Q ♅
01 27	♂ ∠ ♃
01 35	☽ △ ♀
02 26	☽ ♇ ☿
04 21	☽ ± ☉
04 38	☽ ⚹ ♃
09 07	☽ ⚹ ⯓
09 39	♂ ⊻ ♄
10 12	☽ Q ♄
16 40	☽ ☌ ♆
20 54	☽ ♇ ☉
17 Thursday	
00 21	☽ ∥ ♀
02 36	♂ ☍ ♅
02 44	☽ △ ☿
05 14	☽ ∥ ♆
05 32	☽ Q ⯓
06 24	☽ ∠ ♃
06 44	☽ ⚹ ♄
07 33	☽ □ ♅
07 50	☽ □ ♂
15 01	♀ ⚹ ♃
18 48	♀ ∠ ♄
21 45	☽ ⊥ ♃
18 Friday	
00 15	☽ △ ☉
05 48	☿ Q ♃
07 57	☽ ⊻ ♃
08 14	☽ ∠ ♄
09 26	☽ □ ♀
11 56	☽ □ ⯓
13 42	☽ ⋕ ♅
14 00	☽ ∥ ♂
16 01	☽ Q ♂
16 31	♂ ⋕ ♅
19 24	☽ ⊻ ♆
23 29	☽ ⊥ ♄
19 Saturday	
03 20	☽ □ ☿
05 31	☽ ⊥ ♆
09 38	☽ ⊻ ♄
10 07	☽ △ ♅
13 25	☽ ⚹ ♂
13 48	♀ ☍ ⯓
18 40	☽ Q ♀
20 01	☉ Q ♄
20 40	☽ ∠ ♆
20 Sunday	
06 40	☽ □ ☉
08 44	☽ Q ☿
10 55	☽ ☌ ♃
11 23	☽ ♇ ♅
14 28	☽ △ ⯓
16 11	☽ ∠ ♂
16 52	☽ ⚹ ♀
21 58	☽ ⚹ ♆
21 Monday	
02 35	☽ ± ♅
04 32	☽ ⚹ ☿
08 18	☽ ⊥ ♂
12 34	☽ ☌ ♄
12 44	☽ ⊼ ♅
15 27	☽ Q ☉
15 51	☽ ♇ ⯓
18 18	☽ Q ♆
19 05	☽ ⊻ ♂
20 44	☽ ∠ ♀
22 Tuesday	
05 35	☽ ∠ ☿
05 43	☽ ⋕ ♅
07 10	☽ ± ⯓
13 28	☽ ⚹ ☉
13 34	☽ ⊥ ♀
14 13	☽ ⊻ ♃
15 49	♄ ⊼ ♅
16 24	☽ ∥ ♂
17 25	☽ ⊼ ⯓
20 27	♃ ♇ ♅
20 48	☽ ⊥ ☿
23 Wednesday	
00 18	☉ ♇ ♅
00 34	☽ ⊥ ♃
00 47	☉ □ ♃
00 52	☽ ⊻ ♀
01 02	☽ □ ♆
02 53	♀ △ ♆
07 02	☽ ⊻ ☿
07 17	☉ ± ♆
15 20	☽ ∥ ♆
16 02	☽ ☍ ♅
16 11	☽ ⊻ ♄
16 11	☽ ∠ ♃
17 17	☽ ∠ ☉
24 Thursday	
01 39	☽ ☌ ♂
02 38	☽ ⊥ ♄
08 05	☿ St D
10 09	☽ ⊥ ☉
18 22	☽ ∥ ♀
18 26	☽ ∠ ♄
18 30	☽ ⚹ ♃
20 04	☉ ⊼ ⯓
20 22	♂ ⊥ ♄
21 23	☽ ☍ ⯓
21 29	☽ ⊻ ☉
25 Friday	
05 13	☽ △ ♆
07 56	☽ ∥ ☿
10 20	☽ ☌ ♀
11 28	☽ ☌ ☿
15 50	☽ Q ♃
20 38	☽ ⊼ ♅
21 06	☽ ⚹ ♄
26 Saturday	
00 13	☿ ☌ ♀
07 22	☽ ± ♅
07 56	☽ ♇ ♆
08 58	☽ ⋕ ☉
09 56	☽ ⊻ ♂
18 47	☽ Q ♄
21 28	☽ ⊥ ♂
23 38	☽ ♇ ♅
27 Sunday	
00 12	☽ ± ♆
00 26	☽ □ ♃
03 02	☽ ⊼ ⯓
07 37	☽ ☌ ☉
11 10	☽ ⊼ ♆
14 02	☽ ± ⯓
14 12	☽ ∥ ☉
14 59	☽ ∠ ♂
18 41	☽ ⊻ ☿
22 17	☽ ⊻ ♀
28 Monday	
02 06	♀ Q ♃
03 14	☽ △ ♅
04 01	☽ □ ♄
06 07	☽ ⊥ ☿
06 42	☽ ♇ ⯓
10 40	☽ ⊥ ♀
15 59	☽ ⋕ ☿
16 12	☽ ⋕ ♀
17 40	☿ ∥ ♀
20 47	☽ ⚹ ♂
23 35	☽ ∠ ☿
29 Tuesday	
05 30	☽ ∠ ♀
05 35	☉ ⊼ ♆
08 42	☽ △ ♃
11 01	☽ △ ⯓
19 32	☽ ☍ ♆
20 46	☽ ⊻ ☉
21 12	☽ Q ♂
30 Wednesday	
01 12	♀ ⊼ ♅
05 26	☽ ⚹ ☿
09 24	☽ ⊥ ☉
09 25	☽ ⋕ ♆
12 21	☽ □ ♅
12 31	♀ ⚹ ♄
13 30	☽ △ ♄
13 36	☽ ⚹ ♀
13 49	☽ ♇ ♃
15 43	☽ ⋕ ♂
18 39	☉ ± ⯓
31 Thursday	
04 35	☽ ∠ ☉
06 01	☽ Q ☿
07 36	☽ ± ♃
10 45	☽ □ ♂
13 54	♄ ⋕ ⯓
15 54	☽ Q ♀
19 11	☽ ♇ ♄
19 33	☽ ⊼ ♃
21 34	☽ □ ⯓
23 49	☽ Q ♅

APRIL 1960

Date	Sidereal time h m s	Sun ☉ ° ′ ″	Moon ☽ ° ′ ″	Moon ☽ 24.00 ° ′ ″	Mercury ☿ ° ′	Venus ♀ ° ′	Mars ♂ ° ′	Jupiter ♃ ° ′	Saturn ♄ ° ′	Uranus ♅ ° ′	Neptune ♆ ° ′	Pluto ♇ ° ′
01	00 39 23	11 ♈ 46 03	11 ♊ 15 54	17 ♊ 14 13	14 ♓ 58	20 ♓ 14	29 ♒ 25	03 ♑ 04	17 ♑ 53	17 ♌ 10	08 ♏ 28	04 ♍ 02
02	00 43 20	12 45 13	23 ♊ 11 18	29 ♊ 07 41	15 40	21 27	00 ♓ 11	03 08	17 55	17 R 08	08 R 27	04 R 01
03	00 47 16	13 44 21	05 ♋ 03 56	11 ♋ 00 36	16 25	22 41	00 57	03 11	17 57	17 07	08 25	04 00
04	00 51 13	14 43 27	16 ♋ 58 16	22 ♋ 57 33	17 13	23 55	01 43	03 14	18 00	17 06	08 24	03 59
05	00 55 09	15 42 30	28 ♋ 59 02	05 ♌ 03 19	18 05	25 09	02 30	03 17	18 02	17 05	08 22	03 58
06	00 59 06	16 41 31	11 ♌ 10 56	17 ♌ 22 27	19 00	26 23	03 16	03 19	18 04	17 04	08 21	03 56
07	01 03 03	17 40 30	23 ♌ 38 20	29 ♌ 59 04	19 58	27 37	04 02	03 22	18 06	17 03	08 19	03 55
08	01 06 59	18 39 26	06 ♍ 24 58	12 ♍ 56 22	20 59	28 ♓ 51	04 48	03 24	18 08	17 02	08 18	03 54
09	01 10 56	19 38 20	19 ♍ 33 26	26 ♍ 16 16	22 02	00 ♈ 05	05 35	03 26	18 10	17 02	08 16	03 53
10	01 14 52	20 37 12	03 ♎ 04 50	09 ♎ 58 57	23 08	01 18	06 21	03 28	18 11	17 01	08 15	03 52
11	01 18 49	21 36 02	16 ♎ 58 20	24 ♎ 02 34	24 16	02 32	07 07	03 30	18 13	17 00	08 13	03 52
12	01 22 45	22 34 50	01 ♏ 11 05	08 ♏ 23 14	25 27	03 46	07 53	03 31	18 15	17 00	08 12	03 51
13	01 26 42	23 33 36	15 ♏ 38 17	22 ♏ 55 24	26 40	05 00	08 40	03 33	18 16	16 59	08 10	03 50
14	01 30 38	24 32 20	00 ♐ 13 45	07 ♐ 32 30	27 55	06 14	09 26	03 34	18 17	16 58	08 08	03 49
15	01 34 35	25 31 03	14 ♐ 50 50	22 ♐ 07 59	29 ♓ 13	07 27	10 12	03 35	18 19	16 58	08 07	03 48
16	01 38 32	26 29 43	29 ♐ 23 18	06 ♑ 36 11	00 ♈ 32	08 41	10 58	03 36	18 20	16 57	08 05	03 47
17	01 42 28	27 28 22	13 ♑ 46 12	20 ♑ 52 57	01 53	09 55	11 44	03 36	18 21	16 57	08 04	03 46
18	01 46 25	28 27 00	27 ♑ 56 12	04 ♒ 55 48	03 17	11 09	12 31	03 37	18 22	16 57	08 02	03 46
19	01 50 21	29 ♈ 25 36	11 ♒ 51 41	18 ♒ 43 49	04 42	12 23	13 17	03 37	18 23	16 56	08 00	03 45
20	01 54 18	00 ♉ 24 10	25 ♒ 32 16	02 ♓ 17 07	06 09	13 37	14 03	03 R 37	18 23	16 56	07 59	03 44
21	01 58 14	01 22 42	08 ♓ 58 28	15 ♓ 36 27	07 38	14 50	14 49	03 37	18 24	16 56	07 57	03 43
22	02 02 11	02 21 13	22 ♓ 11 10	28 ♓ 42 43	09 09	16 04	15 35	03 36	18 24	16 56	07 56	03 43
23	02 06 07	03 19 42	05 ♈ 11 14	11 ♈ 36 46	10 41	17 18	16 22	03 36	18 25	16 56	07 54	03 42
24	02 10 04	04 18 09	17 ♈ 59 25	24 ♈ 19 14	12 16	18 32	17 08	03 35	18 25	16 D 56	07 52	03 42
25	02 14 01	05 16 34	00 ♉ 36 18	06 ♉ 50 39	13 52	19 46	17 54	03 34	18 26	16 56	07 51	03 41
26	02 17 57	06 14 58	13 ♉ 02 22	19 ♉ 11 32	15 30	20 59	18 40	03 33	18 26	16 56	07 49	03 40
27	02 21 54	07 13 20	25 ♉ 18 16	01 ♊ 22 42	17 10	22 13	19 26	03 32	18 26	16 56	07 47	03 40
28	02 25 50	08 11 40	07 ♊ 25 00	13 ♊ 25 24	18 51	23 27	20 12	03 30	18 R 26	16 56	07 46	03 39
29	02 29 47	09 09 58	19 ♊ 24 09	25 ♊ 21 32	20 34	24 41	20 58	03 29	18 26	16 57	07 44	03 39
30	02 33 43	10 ♉ 08 14	01 ♋ 17 56	07 ♋ 13 45	22 ♈ 19	25 ♈ 54	21 ♓ 44	03 ♑ 27	18 ♑ 25	16 ♌ 57	07 ♏ 43	03 ♍ 38

Date	Moon True ☊ ° ′	Moon Mean ☊ ° ′	Moon ☽ Latitude ° ′
01	24 ♍ 31	23 ♍ 53	05 S 07
02	24 R 29	23 50	05 15
03	24 28	23 46	05 10
04	24 D 28	23 43	04 51
05	24 28	23 40	04 20
06	24 30	23 37	03 37
07	24 31	23 34	02 42
08	24 33	23 31	01 38
09	24 34	23 27	00 S 27
10	24 R 34	23 24	00 N 47
11	24 32	23 21	02 00
12	24 29	23 18	03 07
13	24 26	23 15	04 04
14	24 21	23 12	04 45
15	24 17	23 08	05 08
16	24 14	23 05	05 11
17	24 11	23 02	04 54
18	24 11	22 59	04 20
19	24 D 11	22 56	03 31
20	24 12	22 52	02 30
21	24 14	22 49	01 22
22	24 15	22 46	00 N 11
23	24 R 14	22 43	00 S 59
24	24 12	22 40	02 05
25	24 07	22 37	03 04
26	24 01	22 33	03 53
27	23 54	22 30	04 31
28	23 45	22 27	04 56
29	23 37	22 24	05 07
30	23 ♍ 30	22 ♍ 21	05 S 05

DECLINATIONS

Date	Sun ☉ ° ′	Moon ☽ ° ′	Mercury ☿ ° ′	Venus ♀ ° ′	Mars ♂ ° ′	Jupiter ♃ ° ′	Saturn ♄ ° ′	Uranus ♅ ° ′	Neptune ♆ ° ′	Pluto ♇ ° ′
01	04 N 39	17 N 04	06 S 51	05 S 11	12 S 51	22 S 59	21 S 47	16 N 23	12 S 36	21 N 47
02	05 02	18 01	06 45	04 42	12 35	22 59	21 46	16 24	12 36	21 47
03	05 25	18 11	06 36	04 14	12 19	22 59	21 46	16 24	12 35	21 48
04	05 48	17 33	06 26	03 45	12 03	22 59	21 46	16 24	12 35	21 48
05	06 11	16 07	06 14	03 17	11 47	22 59	21 45	16 25	12 34	21 48
06	06 34	13 57	06 00	02 48	11 30	22 59	21 45	16 25	12 34	21 48
07	06 56	11 05	05 44	02 19	11 14	22 59	21 45	16 25	12 33	21 49
08	07 19	07 38	05 26	01 50	10 57	22 59	21 45	16 25	12 33	21 49
09	07 41	03 N 43	05 07	01 21	10 41	22 59	21 44	16 26	12 32	21 49
10	08 03	00 S 31	04 46	00 52	10 24	22 59	21 44	16 26	12 32	21 49
11	08 25	04 49	04 24	00 S 23	10 07	22 59	21 44	16 26	12 31	21 49
12	08 47	08 57	04 00	00 N 06	09 50	22 59	21 44	16 26	12 31	21 49
13	09 09	12 38	03 35	00 35	09 33	22 59	21 44	16 26	12 30	21 49
14	09 31	15 33	03 08	01 04	09 16	22 59	21 44	16 26	12 29	21 50
15	09 52	17 29	02 40	01 33	08 59	22 59	21 43	16 26	12 29	21 50
16	10 13	18 15	02 11	02 02	08 42	22 59	21 43	16 27	12 28	21 50
17	10 35	17 51	01 41	02 32	08 24	22 59	21 43	16 27	12 28	21 50
18	10 55	16 20	01 09	03 01	08 07	22 59	21 43	16 27	12 27	21 50
19	11 16	13 52	00 36	03 30	07 49	22 59	21 43	16 27	12 27	21 50
20	11 37	10 39	00 S 01	03 59	07 32	22 59	21 43	16 27	12 26	21 50
21	11 57	06 56	00 N 34	04 28	07 14	22 59	21 43	16 27	12 26	21 50
22	12 18	02 S 56	01 10	04 56	06 57	22 59	21 43	16 27	12 25	21 50
23	12 38	01 N 09	01 48	05 25	06 39	22 59	21 43	16 27	12 25	21 50
24	12 57	05 08	02 26	05 54	06 21	22 59	21 43	16 27	12 24	21 50
25	13 17	08 48	03 06	06 22	06 03	23 00	21 43	16 27	12 24	21 50
26	13 36	12 03	03 46	06 50	05 46	23 00	21 43	16 27	12 23	21 50
27	13 55	14 42	04 28	07 19	05 28	23 00	21 43	16 27	12 23	21 50
28	14 14	16 41	05 10	07 47	05 10	23 00	21 43	16 27	12 22	21 50
29	14 33	17 55	05 53	08 15	04 52	23 00	21 43	16 26	12 22	21 50
30	14 N 52	18 N 21	06 N 37	08 N 42	04 S 34	23 S 00	21 S 43	16 N 26	12 S 21	21 N 50

ZODIAC SIGN ENTRIES

Date	h m	Planets
02	06 24	♂ ♓
03	01 46	☽ ♋
05	14 01	☽ ♌
08	00 02	☽ ♍
09	10 32	♀ ♈
10	06 36	☽ ♎
12	10 01	☽ ♏
14	11 37	☽ ♐
16	02 22	☿ ♈
16	13 01	☽ ♑
18	15 32	☽ ♒
20	02 06	☉ ♉
20	19 55	☽ ♓
23	02 23	☽ ♈
25	10 50	☽ ♉
27	21 16	☽ ♊
30	09 22	☽ ♋

LATITUDES

Date	Mercury ☿ ° ′	Venus ♀ ° ′	Mars ♂ ° ′	Jupiter ♃ ° ′	Saturn ♄ ° ′	Uranus ♅ ° ′	Neptune ♆ ° ′	Pluto ♇ ° ′
01	01 S 00	01 S 25	01 S 15	00 N 25	00 N 29	00 N 44	01 N 49	12 N 40
04	01 30	01 28	01 16	00 25	00 29	00 44	01 49	12 39
07	01 55	01 30	01 17	00 25	00 28	00 44	01 50	12 39
10	02 14	01 31	01 18	00 25	00 28	00 43	01 50	12 38
13	02 28	01 32	01 19	00 25	00 28	00 43	01 50	12 38
16	02 37	01 32	01 20	00 25	00 28	00 43	01 50	12 37
19	02 41	01 31	01 21	00 24	00 28	00 43	01 50	12 36
22	02 40	01 30	01 22	00 24	00 28	00 43	01 50	12 35
25	02 34	01 28	01 23	00 24	00 28	00 43	01 50	12 35
28	02 24	01 26	01 24	00 24	00 28	00 43	01 50	12 34
31	02 S 09	01 S 23	01 S 24	00 N 24	00 N 28	00 N 43	01 N 50	12 N 33

DATA

Julian Date	2437026
Delta T	+33 seconds
Ayanamsa	23° 18' 04"
Synetic vernal point	05° ♓ 48' 56"
True obliquity of ecliptic	23° 26' 31"

MOON'S PHASES, APSIDES AND POSITIONS ☽

Date	h m	Phase	Longitude ° ′	Eclipse Indicator
04	07 05	☽	14 ♋ 31	
11	20 27	○	21 ♎ 57	
18	12 57	☾	28 ♑ 29	
25	21 44	●	05 ♉ 40	

Day	h m		
02	22 20	Apogee	
14	18 55	Perigee	
30	16 01	Apogee	
03	04 53	Max dec	18° N 13'
10	09 09	0S	
16	15 40	Max dec	18° S 16'
23	05 11	0N	
30	12 56	Max dec	18° N 21'

LONGITUDES

Date	Chiron ⚷ ° ′	Ceres ⚳ ° ′	Pallas ⚴ ° ′	Juno ⚵ ° ′	Vesta ⚶ ° ′	Black Moon Lilith ⚸ ° ′
01	00 ♓ 20	13 ♒ 37	23 ♑ 27	26 ♐ 50	08 ♑ 43	25 ♊ 53
11	00 ♓ 52	16 ♒ 42	25 ♑ 36	27 ♐ 21	11 ♑ 37	27 ♊ 01
21	01 ♓ 19	19 ♒ 33	27 ♑ 19	27 ♐ 21	14 ♑ 01	28 ♊ 08
31	01 ♓ 42	22 ♒ 07	28 ♑ 30	26 ♐ 48	15 ♑ 49	29 ♊ 15

ASPECTARIAN

01 Friday
01 17 ☽ ∥ ♅
06 24 ☽ ⊼ ♆
13 06 ☽ ✶ ☉
13 14 ☽ ± ♄
16 00 ☿ Q ♃
18 25 ☽ ± ♆
19 53 ☽ □ ☿
23 49 ☽ ✶ ♅

02 Saturday
01 20 ☽ ⊼ ♄
02 43 ☉ ⧺ ♀
08 06 ☽ □ ♀
09 38 ☽ Q ♇
11 23 ♂ ∥ ♆
12 31 ☽ bQ ♆
15 27 ☽ Q ☉

03 Sunday
03 06 ☽ △ ♂
06 04 ☽ ∠ ♅
08 11 ☽ ☍ ♃
09 50 ☽ ✶ ♇
18 45 ☽ △ ♆
20 18 ♀ ± ♅

04 Monday
00 12 ☽ ⊥ ♅
01 56 ♀ bQ ♆
07 05 ☽ □ ☉
08 37 ☿ ⊼ ♅
11 28 ☽ bQ ♂
12 16 ☽ ⊻ ♅
12 33 ☽ △ ☿
14 04 ☽ ☍ ♄
16 01 ☽ ∠ ♇

05 Tuesday
03 30 ☽ △ ♀
06 43 ☽ ± ♂
08 03 ☽ ∥ ♅
09 58 ☽ ⊥ ♇
10 26 ☿ ✶ ♄
13 45 ☉ ⧺ ☿
19 25 ☽ ⊼ ♂
20 32 ☽ ⊼ ♃
20 46 ☽ bQ ☿
21 50 ☽ ⊻ ♇

06 Wednesday
05 34 ♂ ∠ ♄
06 28 ☽ □ ♆
08 22 ☽ ± ♃
12 26 ☽ bQ ♀
13 57 ♂ ✶ ♃
15 50 ☽ ± ☿
21 06 ☉ △ ♅
23 24 ☽ ☌ ♅
23 36 ☽ △ ☉

07 Thursday
00 23 ☽ ⧺ ♆
01 22 ☽ ⊼ ♄
01 52 ☽ bQ ♃
04 23 ☽ ⊼ ☿
07 43 ☽ ± ♀
08 38 ♂ ☍ ♇
10 50 ☽ ⧺ ♂
12 53 ☽ ± ♄
17 05 ☽ Q ♆
20 21 ☽ ⊼ ♀
22 44 ☉ □ ♄

08 Friday
05 53 ☽ bQ ♄
06 23 ☽ △ ♃
06 27 ☽ bQ ☉
07 21 ☽ ☌ ♇
08 49 ☽ ☍ ♂
13 54 ☽ ∥ ☉
15 28 ☽ ✶ ♆
18 01 ☉ bQ ♇

09 Saturday
00 25 ☽ ± ☉
02 56 ☽ ⧺ ☿
07 26 ☽ ⊻ ♅
09 29 ☽ △ ♄
12 10 ☽ ⊼ ☉
13 44 ♀ Q ♄
16 50 ☽ ☍ ☿
18 13 ☽ ⊥ ♅
18 39 ☽ ∠ ♆

10 Sunday
03 16 ☽ ⧺ ♀
08 35 ☽ ☍ ♀
09 27 ☿ ± ♅
10 08 ☽ ∠ ♅
10 32 ☽ ⊥ ♆
12 41 ☽ □ ♃
13 23 ☽ ⊻ ♇
13 48 ☽ ∥ ♀
14 17 ☿ bQ ♆
18 02 ☽ ⊼ ♂
20 59 ☽ ⊻ ♆
23 48 ☽ ⊥ ♇

11 Monday
01 42 ♀ bQ ♅
05 01 ☽ ± ♂
05 56 ♀ ± ♆
09 50 ☽ ∥ ☿
12 03 ☽ ✶ ♅
14 08 ☽ □ ♄
15 13 ☽ ∠ ♇
19 43 ☽ Q ♃
20 27 ☽ ☍ ☉
21 15 ☽ bQ ♂

12 Tuesday
01 30 ☽ ⊼ ☿
07 08 ♀ □ ♃
08 20 ☽ Q ♅
10 51 ☽ ⧺ ☉
12 29 ☽ ± ☿
13 29 ♀ ⊼ ♇
15 55 ☽ ✶ ♃
16 26 ☽ ✶ ♇
16 43 ☽ ⊼ ♀
17 02 ☽ ∥ ♂
20 27 ☽ Q ♄
21 10 ♂ △ ♆
23 39 ☽ ☌ ♆
23 48 ☽ △ ♂

13 Wednesday
03 37 ☽ ± ♀
04 50 ☽ bQ ☿
11 04 ☽ ∥ ♆
12 19 ☽ Q ♇
14 13 ☽ □ ♅
16 20 ☽ ✶ ♄
16 48 ☽ ∠ ♃
19 51 ☽ bQ ♀
23 24 ☉ ∠ ♂

14 Thursday
01 59 ☽ ⊼ ☉
02 56 ☉ ⧺ ♂
07 37 ☽ ⊥ ♃
07 51 ☽ △ ☿
12 33 ☽ ± ☉
17 02 ☽ ∠ ♄
17 29 ☽ ⊻ ♃
17 53 ☽ □ ♇
21 27 ☽ ⧺ ♅
22 45 ☽ △ ♀

15 Friday
00 58 ☽ ⊻ ♆
03 56 ☽ □ ♂
04 22 ☽ bQ ☉
07 49 ☽ ⊥ ♄
10 48 ☽ ⊥ ♆
15 29 ☽ △ ♅
17 42 ☽ ⊻ ♄

16 Saturday
00 32 ♀ ⊼ ♆
01 36 ☽ ∠ ♆
06 52 ☽ △ ☉
08 14 ☿ Q ♄
11 16 ☽ Q ♂
14 06 ☽ □ ☿
15 30 ☿ ⧺ ♀
16 16 ☽ bQ ♅
18 59 ☽ ☌ ♃
19 18 ☽ △ ♇

17 Sunday
02 27 ☽ ✶ ♆
04 56 ☽ □ ♀
07 16 ☽ ± ♅
08 24 ☽ ✶ ♂
09 12 ♀ ± ♇
13 02 ☿ bQ ♅
14 54 ☿ ± ♆
17 21 ☽ ⊼ ♅
19 43 ☽ ☌ ♄
20 25 ☽ bQ ♇
22 35 ☽ Q ♆
23 26 ☽ Q ☿

18 Monday
10 37 ☽ ⧺ ♅
11 14 ☽ ∠ ♂
11 42 ☽ ± ♇
12 57 ☽ □ ☉
14 16 ☽ Q ♀
17 40 ☿ □ ♃
20 06 ☿ ⊼ ♇
21 44 ☽ ⊻ ♃
21 59 ☽ ⊼ ♇
22 11 ☽ ✶ ☿

19 Tuesday
03 36 ☽ ⊥ ♂
05 20 ☽ □ ♆
08 06 ☽ ⊥ ♃
12 59 ☽ ✶ ♀
14 37 ☽ ⊻ ♂
20 52 ☽ ☍ ♅
22 28 ☽ Q ☉
23 12 ☽ ∥ ♆
23 23 ☽ ⊻ ♄
23 48 ☽ ∠ ♃

20 Wednesday
03 20 ☽ ∠ ☿
04 54 ♃ St R
05 53 ☽ ⧺ ☉
09 58 ☽ ⊥ ♄
18 00 ☽ ∠ ♀
21 12 ☽ ⊥ ☿
21 19 ☽ ✶ ☉

21 Thursday
01 59 ☽ ∠ ♄
02 23 ☽ ✶ ♃
02 35 ☽ ☍ ♇
09 17 ☽ ⊻ ☿
09 59 ☽ ∥ ♂
10 10 ☽ △ ♆
11 02 ♀ ⊻ ♂
11 44 ☽ ⊥ ♀
17 00 ☿ ⊼ ♆
23 14 ☽ ☌ ♂
23 42 ☽ ⊻ ♀

22 Friday
00 00 ☽ Q ♃
01 22 ☽ ⧺ ♀
02 25 ☽ ⊼ ♅
02 28 ☽ ∠ ☉
05 06 ☽ ✶ ♄
12 31 ♂ Q ♃
13 21 ☽ bQ ♆
13 22 ☽ ± ♅
20 16 ☽ ⊥ ☉
20 47 ☿ ± ♇
20 57 ☽ ⧺ ☿
21 02 ☉ ⧺ ♆

23 Saturday
03 09 ☽ Q ♄
04 48 ♀ △ ♅
05 54 ☽ ± ♆
05 57 ☽ bQ ♅
08 16 ☽ ⊻ ☉
09 03 ☽ □ ♃
09 15 ☽ ⊼ ♇
16 31 ☽ ∥ ☿
17 03 ☽ ⊼ ♆
18 35 ☉ △ ♃
20 25 ☽ ± ♇
21 07 ☉ △ ♇
23 42 ☽ ☌ ☿

24 Sunday
05 47 ♂ ⊼ ♅
07 48 ♅ St D
09 53 ♀ □ ♄
10 00 ☽ △ ♅
10 16 ☽ ⊻ ♂
12 49 ☽ □ ♄
13 08 ☽ ☌ ♀
13 20 ☽ bQ ♇
15 09 ♀ bQ ♇
17 31 ☽ ∥ ♀
19 10 ☽ ⧺ ♂
22 22 ☽ ⊥ ♂

25 Monday
02 16 ♀ ⧺ ♂
16 41 ☽ ∠ ♂
17 41 ☽ △ ♃
17 54 ☽ △ ♇
21 44 ☽ ☌ ☉

26 Tuesday
01 54 ☽ ☍ ♆
04 33 ♂ ✶ ♄
14 47 ☽ ⧺ ♆
17 32 ☽ ⊻ ☿
19 35 ☽ □ ♅
22 30 ☽ △ ♄
22 44 ☽ bQ ♃
23 42 ☽ ✶ ♂

27 Wednesday
03 14 ☽ ∥ ☉
05 15 ☽ ⊻ ♀
07 07 ☽ ⊥ ☿
08 45 ☿ △ ♅
14 06 ♄ St R
16 23 ☽ ± ♃
18 24 ☽ ⊥ ♀

28 Thursday
00 56 ☽ Q ♂
01 38 ☉ ☍ ♆
03 45 ☽ ∠ ☿
04 04 ☽ bQ ♄
04 14 ☽ ⊼ ♃
04 31 ☽ □ ♇
06 02 ☿ □ ♄
07 04 ☽ Q ♅
08 27 ☽ ∥ ♅
09 15 ☿ bQ ♇
11 55 ☿ ⧺ ♂
12 41 ☽ ⊼ ♆
13 41 ☽ ⊻ ☉
14 17 ☽ ∠ ♀
22 00 ☽ ± ♄

29 Friday
00 39 ☽ ± ♆
02 44 ☽ ⊥ ☉
07 03 ☽ ✶ ♅
10 02 ☽ ⊼ ♄
14 45 ☽ ✶ ☿
15 22 ☽ □ ♂
16 31 ☽ Q ♇
18 42 ☽ bQ ♆
21 46 ☿ ⊻ ♂
22 27 ☽ ∠ ☉
23 51 ☽ ✶ ♀

30 Saturday
13 19 ☽ ∠ ♅
16 20 ☽ ☍ ♃
16 44 ☽ ✶ ♇
19 12 ☽ Q ☿

All ephemeris data is given at 12.00 UT and the Moon's longitude is additionally given for 24.00 UT

MAY 1960

LONGITUDES

Date	Sidereal time h m s	Sun ☉ ° ' "	Moon ☽ ° ' "	Moon ☽ 24.00 ° ' "	Mercury ☿ ° '	Venus ♀ ° '	Mars ♂ ° '	Jupiter ♃ ° '	Saturn ♄ ° '	Uranus ♅ ° '	Neptune ♆ ° '	Pluto ♇ ° '
01	02 37 40	11 ♉ 06 28	13 ♋ 09 24	19 ♋ 05 24	24 ♈ 06	27 ♈ 08	22 ♓ 30	03 ♑ 25	18 ♑ 25	16 ♌ 57	07 ♏ 41	03 ♍ 38
02	02 41 36	12 04 41	25 ♋ 02 16	01 ♌ 00 33	25 55	28 22	23 16	03 R 23	18 R 25	16 58	07 R 39	03 R 38
03	02 45 33	13 02 51	07 ♌ 00 52	13 ♌ 03 49	27 45	29 ♈ 36	24 02	03 20	18 24	16 58	07 38	03 37
04	02 49 30	14 00 59	19 ♌ 10 01	25 ♌ 20 07	29 ♈ 38	00 ♉ 49	24 48	03 18	18 23	16 59	07 36	03 37
05	02 53 26	14 59 05	01 ♍ 34 42	07 ♍ 54 23	01 ♉ 32	02 03	25 34	03 15	18 23	16 59	07 34	03 37
06	02 57 23	15 57 10	14 ♍ 19 41	20 ♍ 51 06	03 27	03 17	26 20	03 12	18 22	17 00	07 33	03 36
07	03 01 19	16 55 12	27 ♍ 29 01	04 ♎ 13 41	05 25	04 31	27 06	03 09	18 21	17 01	07 31	03 36
08	03 05 16	17 53 12	11 ♎ 05 17	18 ♎ 03 47	07 24	05 44	27 52	03 06	18 20	17 01	07 30	03 36
09	03 09 12	18 51 11	25 ♎ 09 00	02 ♏ 20 33	09 25	06 58	28 37	03 02	18 19	17 02	07 28	03 36
10	03 13 09	19 49 08	09 ♏ 37 50	17 ♏ 00 07	11 28	08 12	29 ♓ 23	02 59	18 18	17 03	07 26	03 36
11	03 17 05	20 47 03	24 ♏ 26 25	01 ♐ 55 40	13 32	09 25	00 ♈ 09	02 55	18 16	17 04	07 25	03 35
12	03 21 02	21 44 57	09 ♐ 26 41	16 ♐ 58 14	15 38	10 39	00 55	02 51	18 15	17 05	07 23	03 35
13	03 24 59	22 42 50	24 ♐ 29 06	01 ♑ 58 07	17 45	11 53	01 40	02 47	18 14	17 06	07 22	03 35
14	03 28 55	23 40 41	09 ♑ 24 15	16 ♑ 46 35	19 53	13 06	02 26	02 42	18 12	17 07	07 20	03 35
15	03 32 52	24 38 31	24 ♑ 04 24	01 ♒ 17 09	22 03	14 20	03 12	02 38	18 10	17 08	07 19	03 D 35
16	03 36 48	25 36 20	08 ♒ 24 29	15 ♒ 26 12	24 13	15 34	03 57	02 33	18 09	17 09	07 17	03 35
17	03 40 45	26 34 07	22 ♒ 22 17	29 ♒ 12 48	26 24	16 47	04 43	02 29	18 07	17 10	07 15	03 35
18	03 44 41	27 31 54	05 ♓ 57 59	12 ♓ 38 04	28 ♉ 35	18 01	05 28	02 24	18 05	17 11	07 14	03 35
19	03 48 38	28 29 39	19 ♓ 13 23	25 ♓ 44 18	00 ♊ 47	19 15	06 14	02 19	18 03	17 13	07 12	03 36
20	03 52 34	29 ♉ 27 23	02 ♈ 11 09	08 ♈ 34 19	02 58	20 28	06 59	02 13	18 01	17 14	07 11	03 36
21	03 56 31	00 ♊ 25 06	14 ♈ 54 08	21 ♈ 10 56	05 09	21 42	07 45	02 08	17 59	17 16	07 09	03 36
22	04 00 28	01 22 48	27 ♈ 24 58	03 ♉ 36 31	07 19	22 56	08 30	02 03	17 56	17 17	07 08	03 36
23	04 04 24	02 20 29	09 ♉ 45 48	15 ♉ 53 00	09 29	24 09	09 16	01 57	17 54	17 19	07 07	03 36
24	04 08 21	03 18 09	21 ♉ 58 18	28 ♉ 01 50	11 37	25 23	10 01	01 51	17 52	17 20	07 05	03 37
25	04 12 17	04 15 48	04 ♊ 03 45	10 ♊ 04 12	13 43	26 37	10 46	01 45	17 49	17 22	07 04	03 37
26	04 16 14	05 13 25	16 ♊ 03 18	22 ♊ 01 13	15 48	27 51	11 31	01 39	17 46	17 23	07 02	03 37
27	04 20 10	06 11 01	27 ♊ 58 08	03 ♋ 54 16	17 51	29 ♉ 04	12 17	01 33	17 44	17 25	07 01	03 38
28	04 24 07	07 08 36	09 ♋ 49 50	15 ♋ 45 09	19 52	00 ♊ 18	13 02	01 27	17 41	17 27	07 00	03 38
29	04 28 03	08 06 10	21 ♋ 40 31	27 ♋ 36 19	21 51	01 32	13 47	01 21	17 38	17 29	06 58	03 38
30	04 32 00	09 03 43	03 ♌ 32 58	09 ♌ 30 57	23 48	02 45	14 32	01 14	17 35	17 30	06 57	03 38
31	04 35 56	10 ♊ 01 14	15 ♌ 30 46	21 ♌ 32 58	25 ♊ 42	03 ♊ 59	15 ♈ 17	01 ♑ 07	17 ♑ 32	17 ♌ 32	06 ♏ 56	03 ♍ 39

Date	Moon True ☊ ° '	Moon Mean ☊ ° '	Moon Latitude ° '
01	23 ♍ 24	22 ♍ 18	04 S 50
02	23 R 21	22 14	04 22
03	23 19	22 11	03 43
04	23 D 19	22 08	02 53
05	23 20	22 05	01 54
06	23 21	22 02	00 S 48
07	23 R 21	21 58	00 N 22
08	23 20	21 55	01 34
09	23 16	21 52	02 42
10	23 10	21 49	03 42
11	23 02	21 46	04 28
12	22 54	21 43	04 56
13	22 45	21 39	05 04
14	22 38	21 36	04 52
15	22 32	21 33	04 20
16	22 29	21 30	03 32
17	22 27	21 27	02 33
18	22 D 27	21 24	01 27
19	22 28	21 20	00 N 17
20	22 R 28	21 17	00 S 52
21	22 25	21 14	01 56
22	22 21	21 11	02 54
23	22 13	21 08	03 43
24	22 03	21 04	04 21
25	21 51	21 01	04 47
26	21 38	20 58	04 59
27	21 25	20 55	04 59
28	21 13	20 52	04 45
29	21 03	20 49	04 20
30	20 56	20 45	03 42
31	20 ♍ 51	20 ♍ 42	02 S 55

DECLINATIONS

Date	Sun ☉ ° '	Moon ☽ ° '	Mercury ☿ ° '	Venus ♀ ° '	Mars ♂ ° '	Jupiter ♃ ° '	Saturn ♄ ° '	Uranus ♅ ° '	Neptune ♆ ° '	Pluto ♇ ° '
01	15 N 10	17 N 59	07 N 21	09 N 10	04 S 16	23 S 00	21 S 43	16 N 26	12 S 21	21 N 50
02	15 28	16 50	08 06	09 37	03 58	23 00	21 43	16 26	12 20	21 50
03	15 45	14 55	08 52	10 04	03 40	23 00	21 43	16 26	12 19	21 49
04	16 03	12 20	09 38	10 31	03 22	23 00	21 43	16 26	12 19	21 49
05	16 20	09 08	10 25	10 58	03 04	23 01	21 44	16 25	12 18	21 49
06	16 37	05 26	11 12	11 24	02 46	23 01	21 44	16 25	12 18	21 49
07	16 54	01 N 21	11 59	11 50	02 27	23 01	21 44	16 25	12 17	21 49
08	17 10	02 S 57	12 46	12 16	02 09	23 01	21 44	16 25	12 17	21 49
09	17 26	07 13	13 34	12 42	01 51	23 01	21 44	16 24	12 16	21 48
10	17 42	11 12	14 21	13 07	01 33	23 01	21 44	16 24	12 16	21 48
11	17 57	14 33	15 08	13 32	01 15	23 01	21 45	16 24	12 15	21 48
12	18 12	16 59	15 55	13 56	00 57	23 02	21 45	16 23	12 15	21 48
13	18 27	18 16	16 41	14 21	00 39	23 02	21 45	16 23	12 14	21 47
14	18 42	18 16	17 26	14 45	00 21	23 02	21 45	16 23	12 14	21 47
15	18 56	17 02	18 10	15 08	00 S 03	23 02	21 46	16 22	12 13	21 47
16	19 10	14 45	18 53	15 31	00 N 15	23 02	21 46	16 22	12 13	21 47
17	19 23	11 38	19 35	15 54	00 33	23 02	21 46	16 22	12 13	21 46
18	19 37	07 59	20 15	16 17	00 51	23 03	21 47	16 21	12 12	21 46
19	19 50	04 S 00	20 53	16 39	01 09	23 03	21 47	16 21	12 12	21 46
20	20 02	00 N 05	21 29	17 00	01 27	23 03	21 47	16 20	12 11	21 45
21	20 14	04 05	22 03	17 21	01 45	23 03	21 47	16 20	12 11	21 45
22	20 26	07 51	22 35	17 42	02 03	23 03	21 48	16 19	12 10	21 45
23	20 38	11 13	23 04	18 03	02 21	23 03	21 48	16 19	12 10	21 44
24	20 49	14 03	23 31	18 22	02 39	23 04	21 48	16 18	12 09	21 44
25	21 00	16 16	23 55	18 42	02 57	23 04	21 49	16 18	12 09	21 43
26	21 10	17 45	24 16	19 00	03 14	23 04	21 49	16 17	12 08	21 43
27	21 21	18 27	24 35	19 19	03 32	23 04	21 50	16 17	12 08	21 43
28	21 30	18 20	24 51	19 37	03 50	23 04	21 50	16 16	12 08	21 42
29	21 40	17 25	25 04	19 54	04 07	23 04	21 50	16 16	12 07	21 42
30	21 49	15 45	25 15	20 11	04 25	23 05	21 51	16 15	12 07	21 41
31	21 N 57	13 N 24	25 N 23	20 N 27	04 N 42	23 S 05	21 S 51	16 N 14	12 S 06	21 N 41

ZODIAC SIGN ENTRIES

Date	h	m	Planets
02	21	59	☽ ♌
03	19	56	♀ ♉
04	16	45	☿ ♉
05	08	59	☽ ♍
07	16	30	☽ ♎
09	20	07	☽ ♏
11	07	19	♂ ♈
11	20	55	☽ ♐
13	20	50	☽ ♑
15	21	51	☽ ♒
18	01	23	☽ ♓
19	03	27	☿ ♊
20	07	55	☽ ♈
21	01	34	☉ ♊
22	17	00	☽ ♉
25	03	55	☽ ♊
27	16	06	☽ ♋
28	06	11	♀ ♊
30	04	50	☽ ♌

LATITUDES

Date	Mercury ☿ ° '	Venus ♀ ° '	Mars ♂ ° '	Jupiter ♃ ° '	Saturn ♄ ° '	Uranus ♅ ° '	Neptune ♆ ° '	Pluto ♇ ° '
01	02 S 09	01 S 23	01 S 24	00 N 24	00 N 28	00 N 43	01 N 50	12 N 33
04	01 49	01 20	01 25	00 24	00 28	00 43	01 50	12 32
07	01 26	01 16	01 25	00 23	00 27	00 43	01 50	12 31
10	00 58	01 11	01 25	00 23	00 27	00 43	01 50	12 30
13	00 S 28	01 07	01 26	00 23	00 27	00 42	01 50	12 29
16	00 N 03	01 01	01 26	00 23	00 27	00 42	01 50	12 29
19	00 35	00 56	01 26	00 23	00 27	00 42	01 50	12 28
22	01 04	00 50	01 26	00 22	00 27	00 42	01 50	12 27
25	01 29	00 44	01 26	00 22	00 27	00 42	01 49	12 26
28	01 48	00 37	01 26	00 22	00 26	00 42	01 49	12 25
31	02 N 01	00 S 30	01 S 26	00 N 21	00 N 26	00 N 42	01 N 49	12 N 24

DATA

Julian Date	2437056
Delta T	+33 seconds
Ayanamsa	23° 18' 07"
Synetic vernal point	05° ♓ 48' 53"
True obliquity of ecliptic	23° 26' 31"

MOON'S PHASES, APSIDES AND POSITIONS ☽

Date	h	m	Phase	Longitude ° '	Eclipse Indicator
04	01	00	☽	13 ♌ 34	
11	05	42	○	20 ♏ 32	
17	19	54	☾	26 ♒ 53	
25	12	26	●	04 ♊ 17	

Day	h	m		
12	18	16	Perigee	
28	04	39	Apogee	
07	19	34	0S	
13	23	58	Max dec	18° S 25'
20	11	31	0N	
27	20	41	Max dec	18° N 30'

LONGITUDES

Date	Chiron ⚷ ° '	Ceres ⚳ ° '	Pallas ⚴ ° '	Juno ⚵ ° '	Vesta ⚶ ° '	Black Moon Lilith ⚸ ° '
01	01 ♓ 42	22 ♒ 07	28 ♑ 30	26 ♐ 48	15 ♑ 49	29 ♊ 15
11	01 ♓ 59	24 ♒ 22	29 ♑ 07	25 ♐ 42	16 ♑ 55	00 ♋ 22
21	02 ♓ 11	26 ♒ 15	29 ♑ 05	24 ♐ 21	17 ♑ 14	01 ♋ 30
31	02 ♓ 17	27 ♒ 43	28 ♑ 22	22 ♐ 06	16 ♑ 44	02 ♋ 37

All ephemeris data is given at 12.00 UT and the Moon's longitude is additionally given for 24.00 UT

ASPECTARIAN

01 Sunday
00 56 ☽ △ ♆
02 55 ☽ Q ♀
07 29 ☽ ✱ ☉
07 32 ☽ ⊥ ♅
17 22 ♂ ⚼ ♆
19 41 ☽ ⊻ ♅
22 38 ☽ ☍ ♄
23 04 ☽ ∠ ♇

02 Monday
02 12 ♂ ± ♅
08 12 ☽ △ ♂
09 54 ☽ Q ☉
14 05 ☽ □ ☿
17 13 ☽ ⊥ ♇
17 47 ☽ ∥ ♅
19 28 ☽ □ ♀

03 Tuesday
03 46 ☽ ∥ ☉
04 41 ☽ ⊼ ♃
05 14 ☽ ⊻ ♇
13 13 ☽ □ ♆
16 18 ☽ ⚼ ♂
16 36 ☽ ± ♃

04 Wednesday
01 00 ☽ □ ☉
07 42 ☽ ☌ ♅
10 18 ☽ ⚼ ♃
10 29 ☽ ⊼ ♄
10 54 ♀ ⊥ ♂
11 14 ☽ ± ♂
12 08 ☽ ⋕ ♆
22 10 ☽ ± ♄
23 41 ☽ ⊼ ♂

05 Thursday
00 24 ☽ ∥ ♀
00 29 ☽ Q ♆
04 37 ☽ ∥ ☿
11 53 ☽ △ ☿
12 51 ☿ ⊥ ♂
13 00 ☽ △ ♀
15 10 ☽ △ ♃
15 26 ☽ ⚼ ♄
15 52 ☽ ☌ ♇
19 32 ☉ ∥ ♅
23 21 ☽ ✱ ♆

06 Friday
06 03 ☿ ☌ ♀
08 55 ☿ △ ♃
10 31 ♀ △ ♃
13 51 ☿ △ ♇
15 15 ☽ △ ☉
16 56 ☽ ⊻ ♅
18 22 ♀ △ ♇
19 26 ☽ △ ♄
20 03 ☽ ⚼ ♀
20 57 ☽ ⚼ ☿

07 Saturday
01 59 ☿ ∥ ♀
03 03 ☽ ∠ ♆
03 55 ☽ ⊥ ♅
05 06 ☽ ⋕ ♂
11 16 ☽ ☍ ♂
14 01 ☽ ± ♀
14 13 ☉ □ ♅
16 03 ☽ ± ☿
19 11 ☽ ⊥ ♆
20 05 ☽ ∠ ♅
20 32 ☽ ⚼ ☉
21 11 ☿ ⋕ ♆
22 03 ☽ □ ♃
22 54 ☽ ⊻ ♇

08 Sunday
01 44 ☽ ⊼ ♀
04 29 ☽ ⊼ ☿
05 45 ☽ ⊻ ♆
07 52 ☽ ∥ ♂
09 25 ☽ ⊥ ♇
12 45 ♀ ⋕ ♆
13 02 ☿ ☍ ♆
13 29 ☽ ± ☉
16 53 ☉ ⚼ ♃
22 14 ☽ ✱ ♅
22 52 ☉ △ ♄

09 Monday
00 27 ☽ □ ♄
00 34 ☽ ⊼ ☉
00 55 ☽ ∠ ♇
05 05 ☽ Q ♃
18 08 ☽ ⊼ ♂
18 30 ☽ Q ♅
21 34 ♀ ☍ ♆

10 Tuesday
01 06 ☽ ✱ ♃
02 04 ☽ ✱ ♇
04 39 ☽ ± ♂
06 32 ☽ Q ♄
08 25 ☽ ☌ ♆
09 26 ☽ ☍ ♀
15 29 ☽ ☍ ☿
19 04 ☽ ∥ ♆
20 11 ☽ ⚼ ♂
21 43 ☽ Q ♇

11 Wednesday
00 05 ☽ □ ♅
01 31 ☽ ∠ ♃
02 04 ☽ ✱ ♄
02 54 ☽ ⋕ ♀
05 42 ☽ ☍ ☉
15 48 ♂ Q ♄
15 57 ☽ ⊥ ♃
18 52 ☽ ⋕ ☿
21 39 ☽ △ ♂

12 Thursday
01 31 ☽ ⊻ ♃
02 08 ☽ ∠ ♄
02 39 ☽ □ ♇
05 02 ☽ ⋕ ♅
08 43 ☽ ⊻ ♆
14 06 ☽ ⊼ ♀
16 28 ☽ ⊥ ♄
16 56 ☿ ∠ ♂
18 16 ☽ ⊥ ♆
23 29 ☽ ⊼ ☿

13 Friday
00 11 ☽ △ ♅
00 31 ☽ ± ♀
02 01 ☽ ⊻ ♄
02 29 ♂ ± ♆
02 56 ☿ ∥ ♅
04 31 ☿ □ ♅
08 37 ☽ ∠ ♆
08 58 ☽ ⊼ ☉
10 38 ☽ ± ☿
12 17 ☿ ⚼ ♃
16 10 ☽ ⚼ ♀
17 16 ☿ △ ♄
19 14 ☽ ± ☉

14 Saturday
00 09 ☽ □ ♂
00 13 ☽ ⚼ ♅
01 14 ☽ ☌ ♃
01 38 ♂ ⚼ ♅
02 36 ☽ △ ♇
03 29 ☽ ⚼ ☿
08 39 ☽ ✱ ♆
10 45 ☽ ⚼ ☉
14 46 ☽ ± ♅
18 33 ☽ △ ♀
19 53 ♂ □ ♃
23 55 ☽ ⋕ ☿

15 Sunday
00 34 ☽ ⊼ ♅
00 49 ♇ St D
02 18 ☽ ☌ ♄
02 58 ☽ ⚼ ♇
04 10 ☽ Q ♆
06 59 ☽ Q ♂
08 04 ☽ △ ☿
13 00 ☽ △ ☉
17 50 ☽ ± ♇
20 03 ☽ ⋕ ♅

16 Monday
00 27 ♂ ⊼ ♇
02 11 ☽ ⊻ ♃
03 52 ☽ ⊼ ♇
04 03 ☽ ✱ ♂
05 48 ☽ ⋕ ♀
10 06 ☽ □ ♆
12 15 ☽ ⊥ ♃

17 Tuesday
01 24 ☽ □ ♀
02 22 ☉ ∥ ☿
02 58 ☽ ☍ ♅
03 34 ☽ ∠ ♃
04 38 ☽ ⊻ ♄
07 07 ☽ ∠ ♂
07 58 ☽ ∥ ♆
09 55 ☉ ± ♃
12 49 ☿ ± ♃
13 20 ♄ ⋕ ♇
15 02 ☽ ⊥ ♄
15 17 ☉ ☌ ☿
19 34 ♀ □ ♅
19 54 ☽ □ ☉
20 24 ☽ □ ☿
23 47 ☽ ⊥ ♂

18 Wednesday
00 37 ♀ ⚼ ♃
05 41 ☽ ✱ ♃
06 52 ☽ ∠ ♄
07 46 ☽ ☍ ♇
11 04 ☽ ⊻ ♂
12 06 ☽ Q ♀
13 12 ♀ △ ♄
14 16 ☽ △ ♆
16 49 ♀ ∥ ♅

19 Thursday
03 06 ☽ Q ♃
06 37 ☽ Q ☉
08 19 ☽ ⊼ ♅
09 51 ☽ ✱ ♄
11 02 ☽ Q ☿
12 03 ☽ ✱ ♀
17 28 ☽ ⚼ ♆
19 21 ☽ ± ♅

20 Friday
03 34 ☽ ⋕ ♂
04 08 ☿ ⊼ ♃
06 30 ☽ ✱ ☉
07 57 ☽ Q ♄
09 39 ♃ ⚼ ♅
10 08 ☽ ± ♆
12 04 ☽ □ ♃
12 06 ☽ ⚼ ♅
12 28 ☿ ⚼ ♄
13 46 ☽ ✱ ☿
14 38 ☽ ⊼ ♇
17 55 ♂ ⊼ ♆
18 49 ☽ ∠ ♀
18 53 ☿ □ ♇
20 48 ☽ ∥ ♂
21 22 ☽ ⊼ ♆
21 35 ☽ ☌ ♂
23 11 ☿ ∥ ♇

21 Saturday
00 44 ☿ ⋕ ♄
01 56 ☽ ± ♇
13 04 ☽ ∠ ☉
13 12 ☿ Q ♅
13 41 ☽ ⊥ ♀
16 30 ☽ △ ♅
17 51 ☽ □ ♄
19 03 ☽ ⚼ ♇

22 Sunday
00 08 ☽ ∠ ☿
02 25 ☽ ⊻ ♀
07 45 ☽ ⊥ ☉
09 55 ☿ ⊼ ♆
20 19 ☽ ⊻ ☉
20 54 ☽ △ ♃
21 10 ☽ ⊥ ☿
23 59 ☽ △ ♇

23 Monday
03 07 ☉ ⊼ ♃
06 50 ☽ ☍ ♆
08 14 ☿ ✱ ♂
10 57 ☽ ⊻ ♂
11 19 ☽ ⊻ ☿
11 27 ☿ ⋕ ♃
17 10 ♀ ∠ ♂
19 29 ☽ ⋕ ♆
23 04 ♂ ± ♇
23 29 ☽ ⊥ ♂

24 Tuesday
01 22 ☉ ⚼ ♄
01 59 ☽ ⚼ ♃
02 50 ☽ □ ♅
03 55 ☽ △ ♄
14 44 ☿ ± ♄
18 25 ☽ ∠ ♂
19 31 ☽ ☌ ♀
19 37 ☽ ± ♃
19 44 ☉ □ ♇
20 29 ♀ ± ♃

25 Wednesday
04 32 ☿ ± ♆
07 27 ☽ ⊼ ♃
09 32 ☽ ⚼ ♄
11 07 ☽ □ ♇
12 26 ☽ ∥ ♅
12 26 ☽ ☌ ☉
14 36 ☽ Q ♅
17 58 ☽ ⊼ ♆

26 Thursday
02 18 ☽ ✱ ♂
03 27 ☽ ± ♄
05 57 ☽ ± ♆
11 24 ☽ ☌ ☿
14 41 ☽ ✱ ♅
15 26 ☽ ⊼ ♄
16 15 ☉ Q ♅
23 12 ☽ Q ♇

27 Friday
00 01 ☽ ⚼ ♆
04 03 ☽ Q ♂
06 46 ☿ ✱ ♅
10 33 ☿ ⊼ ♄
14 29 ☽ ⊻ ♀
19 11 ☽ ☍ ♃
21 01 ☽ ∠ ♅
23 27 ☽ ✱ ♇

28 Saturday
04 01 ☽ ⊥ ♀
06 05 ☽ ⊻ ☉
06 16 ☽ △ ♆
08 18 ☉ ⊼ ♆
15 17 ☽ ⊥ ♅
18 55 ☽ □ ♂
19 18 ☽ ⊥ ☉

29 Sunday
00 22 ☽ ∠ ♀
03 28 ☽ ⊻ ♅
03 51 ☽ ☍ ♄
05 51 ☽ ∠ ♇
08 43 ♀ ⊼ ♃
09 24 ☿ Q ♇
12 26 ☽ ⊻ ☿
13 26 ☿ ⚼ ♆
15 09 ☽ ∠ ☉
17 13 ☉ ∥ ♇

30 Monday
00 05 ☽ ⊥ ♇
02 57 ☽ ⊥ ☿
05 49 ☽ ∥ ♅
07 22 ☽ ⊼ ♃
08 54 ♀ ⚼ ♄
10 13 ☽ ✱ ♀
12 12 ☽ ⊻ ♇
18 44 ☉ ⋕ ♄
18 50 ☽ □ ♆
19 21 ☽ ± ♃

31 Tuesday
00 03 ☽ ✱ ☉
00 33 ☽ ∠ ☿
05 36 ♀ □ ♇
11 30 ☽ △ ♂
12 13 ♄ ⊼ ♅
13 02 ☽ Q ♀
13 13 ☽ ⚼ ♃
16 01 ☽ ⊼ ♄
16 03 ☽ ☌ ♅
23 01 ☽ ⋕ ♆

JUNE 1960

LONGITUDES

Date	Sidereal time h m s	Sun ☉ ° ' "	Moon ☽ ° ' "	Moon ☽ 24.00 ° ' "	Mercury ☿ ° '	Venus ♀ ° '	Mars ♂ ° '	Jupiter ♃ ° '	Saturn ♄ ° '	Uranus ♅ ° '	Neptune ♆ ° '	Pluto ♇ ° '
01	04 39 53	10 ♊ 58 43	27 ♌ 38 09	03 ♍ 46 55	27 ♊ 33	05 ♊ 13	16 ♈ 02	01 ♑ 01	17 ♑ 29	17 ♌ 34	06 ♏ 54	03 ♍ 40
02	04 43 50	11 56 12	09 ♍ 59 54	16 ♍ 17 43	29 ♊ 22	06 26	16 47	00 R 54	17 R 26	17 36	06 R 53	03 40
03	04 47 46	12 53 39	22 ♍ 40 59	29 ♍ 10 16	01 ♋ 08	07 40	17 31	00 47	17 23	17 38	06 52	03 41
04	04 51 43	13 51 05	05 ♎ 46 07	12 ♎ 28 56	02 52	08 53	18 16	00 40	17 20	17 40	06 50	03 42
05	04 55 39	14 48 30	19 ♎ 19 04	26 ♎ 16 39	04 33	10 07	19 01	00 33	17 16	17 42	06 49	03 42
06	04 59 36	15 45 53	03 ♏ 21 42	10 ♏ 34 01	06 11	11 21	19 46	00 26	17 13	17 45	06 48	03 43
07	05 03 32	16 43 16	17 ♏ 53 08	25 ♏ 18 24	07 46	12 34	20 30	00 19	17 10	17 47	06 47	03 43
08	05 07 29	17 40 38	02 ♐ 48 56	10 ♐ 23 35	09 19	13 48	21 15	00 12	17 06	17 49	06 46	03 44
09	05 11 25	18 37 58	17 ♐ 01 06	25 ♐ 40 05	10 49	15 02	21 59	00 ♑ 04	17 03	17 51	06 45	03 45
10	05 15 22	19 35 18	03 ♑ 19 04	10 ♑ 56 39	12 16	16 15	22 44	29 ♐ 57	16 59	17 54	06 44	03 46
11	05 19 19	20 32 38	18 ♑ 31 27	26 ♑ 02 17	13 40	17 29	23 28	29 49	16 55	17 56	06 42	03 46
12	05 23 15	21 29 56	03 ♒ 28 09	10 ♒ 48 13	15 01	18 43	24 13	29 42	16 52	17 58	06 41	03 47
13	05 27 12	22 27 15	18 ♒ 01 57	25 ♒ 09 00	16 19	19 56	24 57	29 34	16 48	18 01	06 40	03 48
14	05 31 08	23 24 32	02 ♓ 09 12	09 ♓ 02 35	17 35	21 10	25 41	29 27	16 44	18 03	06 39	03 49
15	05 35 05	24 21 50	15 ♓ 49 22	22 ♓ 29 49	18 47	22 24	26 25	29 19	16 40	18 06	06 38	03 50
16	05 39 01	25 19 07	29 ♓ 04 21	05 ♈ 33 25	19 56	23 37	27 09	29 12	16 36	18 08	06 37	03 51
17	05 42 58	26 16 24	11 ♈ 57 29	18 ♈ 17 03	21 02	24 51	27 54	29 04	16 32	18 11	06 36	03 52
18	05 46 54	27 13 40	24 ♈ 32 38	00 ♉ 44 42	22 05	26 05	28 38	28 56	16 28	18 14	06 36	03 53
19	05 50 51	28 10 57	06 ♉ 53 43	13 ♉ 00 06	23 05	27 19	29 ♈ 21	28 49	16 24	18 16	06 35	03 54
20	05 54 48	29 ♊ 08 13	19 ♉ 04 13	25 ♉ 06 26	24 01	28 32	00 ♉ 05	28 41	16 20	18 19	06 34	03 55
21	05 58 44	00 ♋ 05 29	01 ♊ 07 03	07 ♊ 06 21	24 53	29 ♊ 46	00 49	28 33	16 16	18 22	06 33	03 56
22	06 02 41	01 02 44	13 ♊ 04 32	19 ♊ 01 51	25 42	01 ♋ 00	01 33	28 26	16 12	18 24	06 32	03 57
23	06 06 37	01 59 59	24 ♊ 58 28	00 ♋ 54 35	26 28	02 13	02 17	28 18	16 08	18 27	06 31	03 58
24	06 10 34	02 57 14	06 ♋ 50 22	12 ♋ 46 00	27 09	03 27	03 00	28 10	16 03	18 30	06 31	03 59
25	06 14 30	03 54 29	18 ♋ 41 40	24 ♋ 37 35	27 47	04 41	03 44	28 03	15 59	18 33	06 30	04 00
26	06 18 27	04 51 43	00 ♌ 33 59	06 ♌ 31 09	28 21	05 55	04 27	27 55	15 55	18 36	06 29	04 02
27	06 22 23	05 48 57	12 ♌ 29 22	18 ♌ 29 01	28 50	07 08	05 10	27 48	15 50	18 39	06 29	04 03
28	06 26 20	06 46 11	24 ♌ 30 27	00 ♍ 34 07	29 16	08 22	05 54	27 40	15 46	18 42	06 28	04 04
29	06 30 17	07 43 24	06 ♍ 40 30	12 ♍ 50 05	29 36	09 36	06 37	27 33	15 42	18 45	06 27	04 05
30	06 34 13	08 ♋ 40 36	19 ♍ 03 25	25 ♍ 21 03	29 ♋ 53	10 ♋ 50	07 ♉ 20	27 ♐ 25	15 ♑ 37	18 ♌ 48	06 ♏ 27	04 ♍ 07

Date	Moon True ☊ ° '	Moon Mean ☊ ° '	Moon ☽ Latitude ° '
01	20 ♍ 49	20 ♍ 39	01 S 59
02	20 D 49	20 36	00 S 57
03	20 R 49	20 33	00 N 10
04	20 48	20 29	01 18
05	20 46	20 26	02 25
06	20 42	20 23	03 25
07	20 34	20 20	04 13
08	20 25	20 17	04 47
09	20 15	20 14	05 00
10	20 04	20 10	04 52
11	19 54	20 07	04 24
12	19 47	20 04	03 38
13	19 42	20 01	02 39
14	19 40	19 58	01 31
15	19 D 39	19 55	00 N 20
16	19 R 39	19 51	00 S 50
17	19 39	19 48	01 55
18	19 36	19 45	02 53
19	19 32	19 42	03 42
20	19 24	19 39	04 20
21	19 14	19 35	04 46
22	19 01	19 32	04 59
23	18 48	19 29	04 59
24	18 35	19 26	04 46
25	18 23	19 23	04 20
26	18 12	19 20	03 43
27	18 05	19 16	02 56
28	18 00	19 13	02 01
29	17 58	19 10	01 00
30	17 ♍ 57	19 ♍ 07	00 N 06

DECLINATIONS

Date	Sun ☉ ° '	Moon ☽ ° '	Mercury ☿ ° '	Venus ♀ ° '	Mars ♂ ° '	Jupiter ♃ ° '	Saturn ♄ ° '	Uranus ♅ ° '	Neptune ♆ ° '	Pluto ♇ ° '
01	22 N 06	10 N 26	25 N 29	20 N 43	04 N 59	23 S 05	21 S 52	16 N 14	12 S 06	21 N 40
02	22 13	06 56	25 32	20 58	05 17	23 05	21 52	16 13	12 06	21 40
03	22 21	03 N 03	25 34	21 12	05 34	23 05	21 53	16 13	12 05	21 39
04	22 28	01 S 06	25 32	21 26	05 51	23 05	21 53	16 12	12 05	21 39
05	22 35	05 20	25 29	21 39	06 08	23 06	21 54	16 11	12 04	21 38
06	22 41	09 26	25 24	21 52	06 25	23 06	21 54	16 11	12 04	21 38
07	22 47	13 06	25 17	22 04	06 42	23 06	21 55	16 10	12 04	21 37
08	22 52	16 02	25 09	22 16	06 59	23 06	21 55	16 09	12 03	21 37
09	22 57	17 55	24 58	22 27	07 16	23 06	21 56	16 08	12 03	21 36
10	23 02	18 32	24 47	22 37	07 32	23 06	21 56	16 08	12 03	21 36
11	23 06	17 48	24 34	22 47	07 49	23 06	21 57	16 07	12 02	21 35
12	23 10	15 51	24 20	22 55	08 05	23 06	21 57	16 06	12 02	21 35
13	23 14	12 55	24 04	23 04	08 22	23 07	21 58	16 05	12 02	21 34
14	23 17	09 17	23 48	23 11	08 38	23 07	21 58	16 05	12 02	21 33
15	23 19	05 17	23 30	23 18	08 54	23 07	21 59	16 04	12 01	21 33
16	23 22	01 S 08	23 12	23 25	09 10	23 07	21 59	16 03	12 01	21 32
17	23 23	02 N 58	22 53	23 30	09 26	23 07	22 00	16 02	12 01	21 32
18	23 25	06 49	22 34	23 35	09 42	23 07	22 01	16 01	12 01	21 31
19	23 26	10 19	22 14	23 39	09 58	23 07	22 01	16 00	12 00	21 31
20	23 26	13 19	21 54	23 43	10 14	23 07	22 02	16 00	12 00	21 30
21	23 26	15 43	21 33	23 45	10 29	23 07	22 02	15 59	12 00	21 29
22	23 26	17 26	21 13	23 48	10 45	23 07	22 03	15 58	12 00	21 28
23	23 26	18 22	20 52	23 49	11 00	23 07	22 03	15 57	11 59	21 28
24	23 25	18 31	20 31	23 50	11 15	23 07	22 04	15 56	11 59	21 27
25	23 23	17 51	20 10	23 50	11 30	23 07	22 05	15 55	11 59	21 26
26	23 21	16 24	19 50	23 49	11 45	23 07	22 05	15 54	11 59	21 26
27	23 19	14 14	19 30	23 48	12 00	23 07	22 06	15 53	11 59	21 25
28	23 16	11 27	19 10	23 45	12 15	23 07	22 06	15 52	11 59	21 24
29	23 13	08 08	18 51	23 43	12 29	23 07	22 07	15 51	11 58	21 24
30	23 N 09	04 N 25	18 N 33	23 N 39	12 N 44	23 S 07	22 S 08	15 N 50	11 S 58	21 N 23

ZODIAC SIGN ENTRIES

Date	h	m	Planets
01	16	38	☽ ♍
02	20	31	☿ ♋
04	01	31	☽ ♎
06	06	20	☽ ♏
08	07	31	☽ ♐
10	01	52	♃ ♐
10	06	48	☽ ♑
12	06	23	☽ ♒
14	08	17	☽ ♓
16	13	42	☽ ♈
18	22	33	☽ ♉
20	09	05	♂ ♉
21	09	42	☉ ♋
21	09	46	☽ ♊
21	16	34	♀ ♋
23	22	10	☽ ♋
26	10	51	☽ ♌
28	22	53	☽ ♍

LATITUDES

Date	Mercury ☿ ° '	Venus ♀ ° '	Mars ♂ ° '	Jupiter ♃ ° '	Saturn ♄ ° '	Uranus ♅ ° '	Neptune ♆ ° '	Pluto ♇ ° '
01	02 N 04	00 S 28	01 S 26	00 N 21	00 N 26	00 N 42	01 N 49	12 N 24
04	02 08	00 21	01 25	00 21	00 26	00 42	01 49	12 23
07	02 05	00 14	01 25	00 21	00 26	00 42	01 49	12 22
10	01 55	00 S 07	01 24	00 20	00 26	00 41	01 49	12 21
13	01 38	00 00	01 24	00 20	00 26	00 41	01 49	12 20
16	01 15	00 N 07	01 23	00 20	00 25	00 41	01 49	12 19
19	00 47	00 14	01 22	00 19	00 25	00 41	01 49	12 19
22	00 N 13	00 21	01 21	00 19	00 25	00 41	01 48	12 18
25	00 S 26	00 28	01 20	00 18	00 25	00 41	01 48	12 17
28	01 09	00 35	01 19	00 18	00 24	00 41	01 48	12 16
31	01 S 55	00 N 41	01 S 18	00 N 17	00 N 24	00 N 41	01 N 48	12 N 16

DATA

Julian Date	2437087
Delta T	+33 seconds
Ayanamsa	23° 18' 11"
Synetic vernal point	05° ♓ 48' 48"
True obliquity of ecliptic	23° 26' 30"

MOON'S PHASES, APSIDES AND POSITIONS ☽

Date	h	m	Phase	Longitude ° '	Eclipse Indicator
02	16	01	☽	12 ♍ 06	
09	13	02	○	18 ♐ 40	
16	04	35	☾	25 ♓ 01	
24	03	27	●	02 ♋ 37	

Day	h	m		
10	01	52	Perigee	
24	09	47	Apogee	
04	05	46	0S	
10	10	51	Max dec	18° S 32'
16	18	32	0N	
24	04	07	Max dec	18° N 33'

LONGITUDES

Date	Chiron ⚷ ° '	Ceres ⚳ ° '	Pallas ⚴ ° '	Juno ⚵ ° '	Vesta ⚶ ° '	Black Moon Lilith ⚸ ° '
01	02 ♓ 17	27 ♒ 50	28 ♑ 15	21 ♐ 53	16 ♑ 39	02 ♋ 44
11	02 ♓ 17	28 ♒ 46	26 ♑ 47	19 ♐ 39	15 ♑ 18	03 ♋ 51
21	02 ♓ 10	29 ♒ 30	24 ♑ 43	17 ♐ 24	13 ♑ 18	04 ♋ 58
31	01 ♓ 58	28 ♒ 59	22 ♑ 11	15 ♐ 21	10 ♑ 57	06 ♋ 05

All ephemeris data is given at 12.00 UT and the Moon's longitude is additionally given for 24.00 UT

ASPECTARIAN

h m	Aspects
01 Wednesday	
02 02	☽ Q ☉
03 52	☽ ± ♄
06 38	☽ Q ♆
11 48	☽ ⚹ ☿
18 33	☽ △ ♃
19 04	☽ ⚼ ♂
19 16	♀ Q ♅
21 27	☽ ⚼ ♄
22 38	☿ Q ♂
23 47	☽ ☌ ♇
02 Thursday	
00 07	☉ ± ♄
04 23	☽ □ ♀
06 01	☽ ⚹ ♆
13 35	☽ ± ♂
15 03	☽ Q ☿
16 01	☽ □ ☉
20 35	♀ ⊼ ♆
21 48	☽ ∥ ♂
03 Friday	
01 43	☽ ⊼ ♂
02 06	☽ △ ♄
02 31	☽ ⊻ ♅
07 29	☿ ☍ ♃
07 45	♂ □ ♄
10 28	☽ ∠ ♆
11 13	☉ ± ♆
13 47	☽ ⊥ ♅
15 47	♂ △ ♅
04 Saturday	
02 49	☽ □ ♃
03 04	☽ ⊥ ♆
05 58	☽ □ ☿
06 23	☽ ∠ ♅
08 15	☽ ⊻ ♇
09 14	☿ ∠ ♅
13 56	☽ ⊻ ♆
18 10	☽ △ ♀
19 03	☽ ⊥ ♇
23 48	☿ ⚹ ♇
05 Sunday	
01 44	♂ ⚼ ♇
03 31	☽ △ ☉
08 27	☽ □ ♄
09 11	☽ ⚹ ♅
10 04	♀ ∥ ♇
10 41	☽ Q ♃
10 56	☽ ∠ ♇
11 27	☽ ☍ ♂
16 57	☽ # ♂
22 59	☽ ⚼ ♀
06 Monday	
05 53	☽ Q ♅
07 06	☽ ⚹ ♃
07 19	☽ ⚼ ☉
09 35	♀ ± ♄
12 35	☽ ⚹ ♇
15 06	☽ Q ♄
15 38	☽ ± ♀
16 04	♀ # ♄
17 19	☽ △ ☿
17 45	☽ ☌ ♆
21 09	☿ △ ♆
23 26	☽ ± ☉
07 Tuesday	
02 31	☽ ⊼ ♀
04 48	☽ ∥ ♆
07 50	☽ ∠ ♃
08 28	☽ Q ♇
09 58	☽ ⊼ ☉
10 49	☽ ⚹ ♄
11 50	☽ □ ♅
16 00	♀ ± ♆
16 29	☽ ⊼ ♂
20 51	☽ ⚼ ☿
22 19	☽ ⊥ ♃
22 24	☉ ⊼ ♄
08 Wednesday	
02 39	☽ ± ♂
07 51	☽ ⊻ ♃
10 52	☽ ∠ ♄
12 53	☽ ± ☿
13 08	☽ # ♅
13 28	☽ □ ♇
15 38	☉ ⚹ ♅
17 44	☽ ⚼ ♂
18 15	☽ ⊻ ♆
23 26	☽ ⊼ ☿
09 Thursday	
01 04	☽ ⊥ ♄
03 43	☽ ⊥ ♆
06 54	☽ ☍ ♀
10 28	☽ ⊻ ♄
11 45	☽ △ ♅
13 02	☽ ☍ ☉
17 50	☽ ∠ ♆
18 33	☽ △ ♂
10 Friday	
05 39	☿ ⊥ ♅
06 45	☽ ☌ ♃
11 20	☽ ⚼ ♅
12 42	☽ △ ♇
17 21	☽ ⚹ ♆
11 Saturday	
01 31	♀ ⊼ ♄
01 32	☽ ± ♅
03 31	☽ ☍ ☿
09 28	☽ ☌ ♄
10 12	☽ ⊼ ♀
11 03	☽ ⊼ ♅
12 09	☉ # ♃
12 17	☽ Q ♆
12 24	☽ ⚼ ♇
15 26	☽ ⊼ ☉
20 18	☽ □ ♂
20 37	☽ ± ♀
21 03	♀ ⚹ ♅
12 Sunday	
01 41	☽ ± ☉
02 48	☽ ± ♇
05 57	☽ ⊻ ♃
09 26	☽ # ♅
12 26	☽ ⚼ ♀
12 31	☽ ⊼ ♇
15 36	☽ ⊥ ♃
16 43	☉ ⚼ ♆
17 15	☽ □ ♆
17 17	☽ ⚼ ☉
19 23	☉ Q ♇
13 Monday	
03 05	☽ Q ♂
06 17	☽ ∠ ♃
08 52	☽ ⊼ ☿
09 57	☽ ⊻ ♄
11 58	☽ ☍ ♅
15 30	☽ △ ♀
18 11	☽ ∥ ♆
19 55	☽ ± ☿
19 58	☽ △ ☉
19 59	☽ ⊥ ♄
20 32	☿ ☍ ♄
20 38	♀ # ♃
14 Tuesday	
00 18	☽ ⚹ ♂
07 23	☽ ⚹ ♃
11 17	☽ ∠ ♄
12 48	☽ ⚼ ☿
14 53	☽ ☍ ♇
15 46	☽ # ♂
19 49	☽ △ ♆
21 24	♀ ⚼ ♆
21 44	☿ ⊻ ♅
15 Wednesday	
00 50	♀ Q ♇
03 45	☽ ∠ ♂
04 05	☽ Q ♃
13 02	☿ ∠ ♇
13 30	☽ ⚹ ♄
16 05	☽ ⊼ ♅
17 48	☽ △ ☿
17 57	☉ ∥ ♀
20 44	☽ ⊥ ♂
22 26	☽ ⚼ ♆
23 58	☿ ∥ ♀
16 Thursday	
01 02	☽ □ ♀
01 10	☉ ∥ ☿
02 57	☽ ± ♅
04 35	☽ □ ☉
08 17	☽ ⊻ ♂
11 08	☽ Q ♄
12 13	☽ □ ♃
14 51	☽ ± ♆
19 05	☿ # ♃
19 32	☽ ⚼ ♅
20 50	☽ ⊼ ♇
17 Friday	
01 58	☽ ⊼ ♆
08 03	☽ ± ♇
13 52	☽ Q ♀
16 44	☽ Q ☉
20 38	☽ □ ♄
23 51	☽ △ ♅
18 Saturday	
01 07	☽ ⚼ ♇
06 51	☽ □ ☿
15 17	☽ ⚹ ♀
17 37	☽ ⚹ ☉
20 23	☽ ☌ ♂
20 25	☽ △ ♃
20 45	♂ △ ♃
19 Sunday	
06 08	☽ △ ♇
09 14	☽ ∥ ♂
11 23	☽ ☍ ♆
20 54	☽ Q ☿
23 50	☽ ∠ ♀
20 Monday	
00 55	☽ # ♆
01 24	☽ ∠ ☉
01 27	☽ ⚼ ♃
01 57	☉ ☍ ♃
03 03	☿ # ♄
06 37	☽ △ ♄
10 30	☽ □ ♅
14 35	♀ ☍ ♃
19 06	☽ ± ♃
19 40	☽ ⊥ ♀
20 46	☽ ⊥ ☉
22 36	☽ ⚹ ☿
21 Tuesday	
06 56	☽ ⊼ ♃
08 59	☽ ⊻ ♀
09 46	☽ ⊻ ☉
11 22	☽ ⊻ ♂
12 18	☽ ⚼ ♄
15 02	☽ ∥ ♅
17 05	☿ ∥ ♇
17 39	☽ □ ♇
22 33	☽ Q ♅
22 52	☽ ⊼ ♆
22 Wednesday	
00 10	☽ ⊥ ♂
06 14	☽ ± ♄
06 54	☽ ∠ ☿
10 55	☽ ± ♆
16 24	☉ ☌ ♀
18 15	☽ ⊼ ♄
19 27	☽ ∠ ♂
22 47	☽ ⚹ ♅
23 Thursday	
02 17	☽ ⊥ ☿
05 02	☽ ⚼ ♆
05 55	☽ Q ♇
14 27	♀ ⚹ ♂
15 12	☽ ⊻ ☿
18 39	☽ ☍ ♃
24 Friday	
03 27	☽ ☌ ☉
03 44	☽ ⚹ ♂
04 21	☽ ☌ ♀
05 13	☽ ∠ ♅
06 13	☽ ⚹ ♇
11 20	☽ △ ♆
13 00	♀ ∠ ♅
17 01	☉ ⚹ ♂
22 35	♀ ⚹ ♇
23 31	☽ ⊥ ♅
25 Saturday	
02 31	☉ ∠ ♅
05 36	☽ Q ♂
06 33	☽ ☍ ♄
11 42	☽ ⊻ ♅
12 38	☽ ∠ ♇
14 31	☉ ⚹ ♇
20 49	☿ ⊼ ♃
21 25	☿ ⊥ ♇
21 30	♂ △ ♇
26 Sunday	
06 43	☽ ⊼ ♃
06 52	☽ ⊥ ♇
07 18	☽ ☌ ☿
18 15	☽ ∥ ♅
18 42	☽ ± ♃
18 59	☽ ⊻ ♇
20 20	☽ □ ♂
21 25	☽ ⊻ ☉
23 11	♀ △ ♆
23 56	☽ □ ♆
27 Monday	
00 01	☽ ⊻ ♀
09 42	♂ # ♆
10 32	☽ ⊥ ☉
12 36	☽ ⚼ ♃
13 27	☽ ⊥ ♀
18 41	☽ ⊼ ♄
28 Tuesday	
00 23	☽ ☌ ♅
04 28	☉ △ ♆
06 05	☽ ∠ ☉
06 06	☽ ∥ ♂
06 35	☽ ± ♄
07 49	☽ # ♆
09 29	☽ ∠ ♀
11 55	☽ Q ♆
18 12	☽ △ ♃
21 43	☽ ⊻ ☿
29 Wednesday	
00 19	☽ ⚼ ♄
06 49	♂ ☍ ♆
06 55	☽ ☌ ♇
09 51	☽ ⊥ ☿
11 34	☽ ⚹ ♆
11 52	☽ △ ♂
14 13	☽ ⚹ ☉
18 20	☽ ⚹ ♀
30 Thursday	
03 48	☽ ∠ ☿
05 26	☽ △ ♄
11 30	☽ ⊻ ♅
15 21	☽ Q ☉
16 34	☽ ∠ ♆
18 39	☽ ⚼ ♂
19 59	☽ Q ♀
23 00	☽ ⊥ ♅

JULY 1960

LONGITUDES

Date	Sidereal time h m s	Sun ☉ ° ' "	Moon ☽ ° ' "	Moon ☽ 24.00 ° ' "	Mercury ☿ ° '	Venus ♀ ° '	Mars ♂ ° '	Jupiter ♃ ° '	Saturn ♄ ° '	Uranus ♅ ° '	Neptune ♆ ° '	Pluto ♇ ° '
01	06 38 10	09 ♋ 37 49	01 ♎ 43 32	08 ♎ 11 25	00 ♌ 05	12 ♋ 03	08 ♉ 03	27 ♐ 18	15 ♑ 33	18 ♌ 51	06 ♏ 26	04 ♍ 08
02	06 42 06	10 35 01	14 ♎ 45 13	21 ♎ 25 23	00 12	13 17	08 46	27 R 11	15 R 29	18 54	06 R 26	04 09
03	06 46 03	11 32 12	28 ♎ 12 20	05 ♏ 06 18	00 15	14 31	09 29	27 03	15 24	18 57	06 25	04 11
04	06 49 59	12 29 23	12 ♏ 07 26	19 ♏ 15 41	00 R 12	15 45	10 12	26 56	15 20	19 00	06 25	04 12
05	06 53 56	13 26 35	26 ♏ 30 48	03 ♐ 52 20	00 ♌ 06	16 58	10 54	26 49	15 15	19 04	06 24	04 13
06	06 57 52	14 23 45	11 ♐ 19 37	18 ♐ 51 42	29 ♋ 54	18 12	11 37	26 42	15 11	19 07	06 24	04 15
07	07 01 49	15 20 56	26 ♐ 27 31	04 ♑ 05 46	29 39	19 26	12 20	26 35	15 07	19 10	06 24	04 16
08	07 05 46	16 18 07	11 ♑ 45 04	19 ♑ 24 01	29 19	20 40	13 02	26 28	15 02	19 13	06 23	04 18
09	07 09 42	17 15 18	27 ♑ 01 11	04 ♒ 35 15	28 54	21 54	13 44	26 21	14 58	19 16	06 23	04 19
10	07 13 39	18 12 29	12 ♒ 05 02	19 ♒ 29 32	28 27	23 07	14 27	26 15	14 53	19 20	06 23	04 21
11	07 17 35	19 09 41	26 ♒ 47 58	03 ♓ 59 45	27 55	24 21	15 09	26 08	14 49	19 23	06 23	04 22
12	07 21 32	20 06 52	11 ♓ 04 34	18 ♓ 02 14	27 21	25 35	15 51	26 02	14 45	19 26	06 22	04 24
13	07 25 28	21 04 05	24 ♓ 52 47	01 ♈ 36 26	26 45	26 49	16 33	25 55	14 40	19 30	06 22	04 25
14	07 29 25	22 01 18	08 ♈ 13 28	14 ♈ 44 18	26 06	28 03	17 15	25 49	14 36	19 33	06 22	04 27
15	07 33 21	22 58 31	21 ♈ 09 24	27 ♈ 29 17	25 27	29 ♋ 16	17 57	25 43	14 31	19 37	06 22	04 28
16	07 37 18	23 55 45	03 ♉ 44 30	09 ♉ 55 37	24 47	00 ♌ 30	18 39	25 37	14 27	19 40	06 22	04 30
17	07 41 15	24 53 00	16 ♉ 03 09	22 ♉ 07 40	24 07	01 44	19 20	25 31	14 23	19 44	06 22	04 32
18	07 45 11	25 50 15	28 ♉ 09 39	04 ♊ 09 35	23 28	02 58	20 02	25 26	14 18	19 47	06 D 22	04 33
19	07 49 08	26 47 32	10 ♊ 07 55	16 ♊ 05 02	22 50	04 12	20 43	25 20	14 14	19 51	06 22	04 35
20	07 53 04	27 44 49	22 ♊ 01 20	27 ♊ 57 07	22 15	05 26	21 25	25 14	14 10	19 54	06 22	04 37
21	07 57 01	28 42 06	03 ♋ 52 43	09 ♋ 48 22	21 42	06 39	22 06	25 09	14 06	19 58	06 22	04 38
22	08 00 57	29 ♋ 39 24	15 ♋ 44 19	21 ♋ 40 49	21 13	07 53	22 47	25 04	14 02	20 01	06 22	04 40
23	08 04 54	00 ♌ 36 43	27 ♋ 38 02	03 ♌ 36 12	20 48	09 07	23 28	24 59	13 57	20 05	06 22	04 42
24	08 08 50	01 34 03	09 ♌ 35 30	15 ♌ 36 08	20 28	10 21	24 09	24 54	13 53	20 08	06 23	04 43
25	08 12 47	02 31 23	21 ♌ 38 21	27 ♌ 42 21	20 12	11 35	24 50	24 49	13 49	20 12	06 23	04 45
26	08 16 44	03 28 43	03 ♍ 48 25	09 ♍ 56 49	20 02	12 49	25 31	24 45	13 45	20 15	06 23	04 47
27	08 20 40	04 26 04	16 ♍ 07 52	22 ♍ 21 55	19 58	14 03	26 11	24 40	13 41	20 19	06 23	04 49
28	08 24 37	05 23 26	28 ♍ 39 19	05 ♎ 00 26	19 D 59	15 17	26 52	24 36	13 37	20 23	06 24	04 51
29	08 28 33	06 20 48	11 ♎ 25 40	17 ♎ 55 25	20 07	16 31	27 32	24 32	13 33	20 26	06 24	04 52
30	08 32 30	07 18 11	24 ♎ 30 03	01 ♏ 09 56	20 20	17 44	28 12	24 28	13 29	20 30	06 24	04 54
31	08 36 26	08 ♌ 15 34	07 ♏ 55 20	14 ♏ 46 29	20 ♋ 40	18 ♌ 58	28 ♉ 52	24 ♐ 24	13 ♑ 26	20 ♌ 34	06 ♏ 25	04 ♍ 56

Date	Moon True ☊ ° '	Moon Mean ☊ ° '	Moon ☽ Latitude ° '
01	17 ♍ 57	19 ♍ 04	01 N 13
02	17 R 58	19 01	02 18
03	17 56	18 57	03 17
04	17 53	18 54	04 07
05	17 48	18 51	04 43
06	17 40	18 48	05 02
07	17 32	18 45	05 00
08	17 23	18 41	04 37
09	17 15	18 38	03 54
10	17 09	18 35	02 56
11	17 05	18 32	01 46
12	17 03	18 29	00 N 32
13	17 D 03	18 26	00 S 42
14	17 04	18 22	01 51
15	17 05	18 19	02 52
16	17 R 04	18 16	03 44
17	17 02	18 13	04 23
18	16 58	18 10	04 51
19	16 51	18 07	05 05
20	16 43	18 03	05 05
21	16 34	18 00	04 53
22	16 25	17 57	04 28
23	16 17	17 54	03 51
24	16 10	17 51	03 04
25	16 06	17 47	02 08
26	16 03	17 44	01 S 06
27	16 D 02	17 41	00 N 01
28	16 03	17 38	01 08
29	16 04	17 35	02 14
30	16 06	17 32	03 14
31	16 ♍ 06	17 ♍ 28	04 N 05

DECLINATIONS

Date	Sun ☉ ° '	Moon ☽ ° '	Mercury ☿ ° '	Venus ♀ ° '	Mars ♂ ° '	Jupiter ♃ ° '	Saturn ♄ ° '	Uranus ♅ ° '	Neptune ♆ ° '	Pluto ♇ ° '
01	23 N 06	00 N 25	18 N 15	23 N 35	12 N 58	23 S 07	22 S 08	15 N 49	11 S 58	21 N 22
02	23 01	03 S 42	17 59	23 30	13 12	23 07	22 09	15 49	11 58	21 22
03	22 56	07 46	17 43	23 24	13 26	23 07	22 09	15 48	11 58	21 21
04	22 51	11 33	17 28	23 18	13 40	23 07	22 10	15 47	11 58	21 20
05	22 46	14 47	17 15	23 11	13 54	23 07	22 10	15 46	11 58	21 20
06	22 40	17 09	17 03	23 03	14 07	23 07	22 11	15 44	11 58	21 19
07	22 34	18 24	16 52	22 55	14 21	23 07	22 12	15 43	11 58	21 18
08	22 27	18 19	16 42	22 46	14 34	23 07	22 12	15 42	11 58	21 17
09	22 20	16 55	16 34	22 36	14 47	23 07	22 13	15 41	11 57	21 17
10	22 12	14 22	16 28	22 25	15 00	23 07	22 13	15 40	11 57	21 16
11	22 04	10 55	16 23	22 14	15 13	23 07	22 14	15 39	11 57	21 15
12	21 56	06 55	16 19	22 03	15 26	23 07	22 14	15 38	11 57	21 14
13	21 47	02 S 41	16 18	21 50	15 38	23 07	22 15	15 37	11 57	21 14
14	21 38	01 N 34	16 17	21 37	15 51	23 07	22 16	15 36	11 57	21 13
15	21 29	05 35	16 19	21 24	16 03	23 07	22 16	15 35	11 57	21 12
16	21 19	09 15	16 22	21 10	16 15	23 07	22 17	15 34	11 57	21 11
17	21 09	12 26	16 26	20 55	16 27	23 07	22 17	15 33	11 58	21 11
18	20 59	15 02	16 31	20 39	16 38	23 07	22 18	15 32	11 58	21 10
19	20 48	16 57	16 38	20 24	16 50	23 07	22 18	15 31	11 58	21 09
20	20 37	18 07	16 46	20 07	17 01	23 07	22 19	15 29	11 58	21 08
21	20 25	18 30	16 56	19 49	17 13	23 07	22 19	15 28	11 58	21 07
22	20 13	18 05	17 06	19 32	17 24	23 07	22 20	15 27	11 58	21 07
23	20 01	16 51	17 17	19 13	17 35	23 07	22 21	15 26	11 58	21 06
24	19 49	14 54	17 28	18 55	17 45	23 07	22 21	15 25	11 58	21 05
25	19 36	12 16	17 40	18 35	17 56	23 07	22 22	15 24	11 58	21 04
26	19 23	09 05	17 53	18 15	18 06	23 07	22 22	15 23	11 58	21 04
27	19 09	05 29	18 05	17 55	18 17	23 07	22 23	15 22	11 59	21 03
28	18 55	01 N 35	18 17	17 34	18 27	23 07	22 23	15 20	11 59	21 02
29	18 41	02 S 28	18 31	17 13	18 36	23 07	22 24	15 19	11 59	21 01
30	18 27	06 29	18 43	16 51	18 46	23 07	22 24	15 18	11 59	21 00
31	18 N 12	10 S 17	18 N 55	16 N 29	18 N 56	23 S 07	22 S 25	15 N 17	11 S 59	21 N 00

ZODIAC SIGN ENTRIES

Date	h	m	Planets
01	01	13	☿ ♌
01	08	46	☽ ♎
03	15	08	☽ ♏
05	17	42	☽ ♐
06	01	23	☿ ♋
07	17	34	☽ ♑
09	16	43	☽ ♒
11	17	19	☽ ♓
13	21	07	☽ ♈
16	02	11	♀ ♌
16	04	48	☽ ♉
18	15	40	☽ ♊
21	04	09	☽ ♋
22	20	37	☉ ♌
23	16	46	☽ ♌
26	04	31	☽ ♍
28	14	33	☽ ♎
30	21	55	☽ ♏

LATITUDES

Date	Mercury ☿ ° '	Venus ♀ ° '	Mars ♂ ° '	Jupiter ♃ ° '	Saturn ♄ ° '	Uranus ♅ ° '	Neptune ♆ ° '	Pluto ♇ ° '
01	01 S 55	00 N 41	01 S 18	00 N 17	00 N 24	00 N 41	01 N 48	12 N 16
04	02 42	00 47	01 16	00 17	00 24	00 41	01 48	12 15
07	03 26	00 53	01 15	00 17	00 24	00 41	01 48	12 14
10	04 06	00 59	01 14	00 16	00 23	00 41	01 47	12 14
13	04 36	01 04	01 12	00 16	00 23	00 41	01 47	12 13
16	04 53	01 08	01 10	00 15	00 23	00 41	01 47	12 13
19	04 56	01 13	01 08	00 15	00 23	00 41	01 47	12 12
22	04 44	01 16	01 07	00 14	00 22	00 41	01 47	12 12
25	04 18	01 20	01 05	00 14	00 22	00 41	01 47	12 11
28	03 41	01 22	01 03	00 13	00 22	00 41	01 46	12 11
31	02 S 58	01 N 25	01 S 00	00 N 13	00 N 21	00 N 41	01 N 46	12 N 11

DATA

Julian Date	2437117
Delta T	+33 seconds
Ayanamsa	23° 18' 16"
Synetic vernal point	05° ♓ 48' 44"
True obliquity of ecliptic	23° 26' 30"

MOON'S PHASES, APSIDES AND POSITIONS ☽

Date	h	m	Phase	Longitude ° '	Eclipse Indicator
02	03	48	☽	10 ♎ 15	
08	19	37	○	16 ♑ 36	
15	15	43	☾	23 ♈ 07	
23	18	31	●	00 ♌ 52	
31	12	38	☽	08 ♏ 17	

Day	h	m		
08	11	28	Perigee	
21	13	50	Apogee	
01	14	29	0S	
07	22	44	Max dec	18° S 32'
14	03	04	0N	
21	11	21	Max dec	18° N 30'
28	21	24	0S	

LONGITUDES

Date	Chiron ⚷ ° '	Ceres ⚳ ° '	Pallas ⚴ ° '	Juno ⚵ ° '	Vesta ⚶ ° '	Black Moon Lilith ⚸ ° '
01	01 ♓ 58	28 ♒ 59	22 ♑ 11	15 ♐ 21	10 ♑ 57	06 ♋ 05
11	01 ♓ 41	28 ♒ 13	19 ♑ 27	13 ♐ 41	08 ♑ 35	07 ♋ 12
21	01 ♓ 19	26 ♒ 53	16 ♑ 44	12 ♐ 29	06 ♑ 32	08 ♋ 19
31	00 ♓ 54	25 ♒ 05	14 ♑ 28	11 ♐ 52	05 ♑ 05	09 ♋ 27

All ephemeris data is given at 12.00 UT and the Moon's longitude is additionally given for 24.00 UT

ASPECTARIAN

h m	Aspects
01 Friday	
01 00	☉ # ♃
03 46	☽ □ ♃
08 53	☽ ✱ ☿
09 36	☽ ⊥ ♆
12 39	☽ ± ♂
15 59	☽ ∠ ♅
16 30	☽ ⊻ ♇
20 46	☽ ⊻ ♆
02 Saturday	
00 26	☽ ⊼ ♂
03 36	☽ ⊥ ♇
03 48	☽ □ ☉
04 09	♀ ⊥ ♅
07 19	☽ Q ☿
09 04	☽ □ ♀
12 45	☽ Q ♃
13 18	☽ □ ♄
19 31	☽ ✱ ♅
19 57	☽ ∠ ♇
03 Sunday	
10 00	☽ ✱ ♃
13 15	☿ St R
15 34	☽ □ ☿
16 49	☽ Q ♅
21 01	☽ Q ♄
22 25	☽ ✱ ♇
04 Monday	
02 16	☽ ☌ ♆
04 23	♀ ☍ ♄
08 33	☽ ☍ ♂
11 41	☽ ∠ ♃
12 40	☽ △ ☉
14 48	☽ ∥ ♆
17 23	☽ ✱ ♄
18 41	☽ △ ♀
18 53	☽ Q ♇
23 37	☽ □ ♅
05 Tuesday	
01 45	☉ ⊥ ♅
02 40	☽ ⊥ ♃
04 18	☽ # ♂
12 30	☽ ⊻ ♃
15 23	☽ Ꝑ ☉
17 48	☽ △ ☿
18 06	☽ ∠ ♄
20 40	☽ # ♅
21 44	☽ Ꝑ ♀
22 59	♀ # ♃
06 Wednesday	
00 35	☽ □ ♇
04 05	☽ ⊻ ♆
06 59	☽ ± ☉
08 35	☽ ⊥ ♄
10 43	☽ # ☿
12 29	☽ ⊼ ♂
13 32	☽ ± ♀
13 43	☽ ⊥ ♆
14 24	♂ Ꝑ ♃
17 14	☽ ⊼ ☉
17 38	☽ Ꝑ ☿
18 08	☽ ⊻ ♄
22 31	☽ ± ♂
23 55	☽ ⊼ ♀
07 Thursday	
00 26	☽ △ ♅
04 01	☽ ∠ ♆
06 25	☉ ☍ ♄
06 33	♀ ⊻ ♅
07 39	☽ ± ☿
08 44	♀ ∠ ♇
12 12	☽ ☌ ♃
13 26	☽ Ꝑ ♂
16 55	☽ ⊼ ☿
08 Friday	
00 09	☽ Ꝑ ♅
00 17	☽ △ ♇
03 36	☽ ✱ ♆
14 06	☽ △ ♂
14 19	☽ ± ♅
17 08	☽ ☌ ♄
19 37	☽ ☍ ☉
22 25	☽ Q ♆
23 46	☽ ⊼ ♅
23 51	☽ Ꝑ ♇
09 Saturday	
03 13	☽ ☍ ♀
10 58	☽ ⊻ ♃
14 03	☽ ± ♇
14 54	☽ ☍ ☿
16 15	☽ # ☿
20 23	☽ ⊥ ♃
23 35	☽ ⊼ ♇
10 Sunday	
00 51	☽ # ♅
02 52	☽ □ ♆
07 09	☽ # ♂
08 41	☉ # ♄
10 40	☽ ∠ ♃
16 00	☽ □ ♂
16 30	☽ ⊻ ♄
16 37	☿ ⊥ ♇
22 36	☽ ⊼ ☉
23 47	☽ ☍ ♅
11 Monday	
01 43	♂ △ ♄
02 13	☽ ⊥ ♄
05 14	☽ ∥ ♆
07 35	☽ ⊼ ♀
09 06	☽ ± ☉
10 55	☽ ✱ ♃
13 00	♀ # ♄
13 48	☽ ⊼ ☿
16 59	☽ ∠ ♄
17 21	☉ ∠ ♇
18 00	☉ ⊻ ♅
18 27	☽ ± ♀
23 07	☽ Q ♂
23 26	☽ ± ☿
12 Tuesday	
00 39	☽ ☍ ♇
01 09	☽ Ꝑ ☉
02 44	☿ Q ♂
04 01	☽ △ ♆
06 51	☽ Q ♃
11 05	☽ Ꝑ ♀
14 06	☽ Ꝑ ☿
18 16	☽ ✱ ♄
20 02	♀ ⊼ ♃
20 39	☽ ✱ ♂
13 Wednesday	
02 30	☽ ⊼ ♅
04 47	☽ △ ☉
05 49	☽ Ꝑ ♆
10 02	♂ ∥ ♅
11 09	☿ ☌ ♀
13 06	☽ ± ♅
13 50	☽ □ ♃
15 09	☽ Q ♄
15 10	☽ △ ☿
15 46	☽ △ ♀
21 47	☽ ± ♆
14 Thursday	
00 33	☽ ∠ ♂
05 07	☽ ⊼ ♇
05 17	☽ Ꝑ ♅
05 39	☉ ∥ ♀
08 37	☽ ⊼ ♆
16 05	☽ ± ♇
17 52	☽ ⊥ ♂
20 02	♀ ⊥ ♇
23 40	☽ □ ♄
15 Friday	
00 25	☿ ⊼ ♃
05 38	☽ ⊻ ♂
08 50	☽ Ꝑ ♇
09 05	☽ △ ♅
15 43	☽ □ ☉
19 42	☽ □ ☿
20 34	☽ △ ♃
16 Saturday	
05 05	☽ □ ♀
08 58	♀ ∥ ♇
13 28	☽ △ ♇
17 05	☽ ☍ ♆
18 17	♀ Q ♂
17 Sunday	
00 34	☉ ☌ ☿
01 15	☽ Ꝑ ♃
04 40	☽ Q ☿
05 15	☽ Q ☉
08 07	☽ # ♆
08 09	♀ ± ♃
08 38	☿ ∥ ♂
08 44	☽ △ ♄
08 49	☉ ∥ ♇
17 33	♂ ± ♃
18 47	☽ ± ♃
18 52	☽ ☌ ♂
19 17	☽ □ ♅
20 05	☽ Q ♀
18 Monday	
02 34	☉ ⊼ ♃
02 40	♂ □ ♅
03 07	☽ ✱ ☿
06 36	☽ ⊼ ♃
06 58	♆ St D
06 58	☽ ✱ ☉
14 17	☽ Ꝑ ♄
17 25	☽ ∥ ♅
22 42	☽ ✱ ♀
19 Tuesday	
00 49	☽ □ ♇
04 26	☽ ⊼ ♆
07 07	☽ ∥ ☿
07 22	☽ Q ♅
07 36	☽ ∠ ☿
08 13	☽ ± ♄
10 04	☽ ∥ ♂
15 38	☽ ∠ ☉
16 30	☽ ± ♆
19 42	♀ ⊻ ♇
20 13	☽ ⊼ ♄
20 Wednesday	
00 52	☽ ⊥ ☿
07 41	☽ ✱ ♅
08 24	☽ ∠ ♀
10 40	☽ Ꝑ ♆
10 41	☽ ⊻ ♂
11 24	☽ ⊥ ☉
12 26	☽ ⊻ ☿
13 11	☽ Q ♇
18 28	☽ ☍ ♃
23 35	☽ ⊥ ♂
21 Thursday	
00 36	☽ ⊻ ☉
04 09	☿ ✱ ♂
04 43	☽ ⊥ ♀
06 20	♀ □ ♆
10 21	☉ ⊥ ♇
13 33	☽ ✱ ♇
14 12	☽ ∠ ♅
17 02	☽ △ ♆
18 17	☽ ⊻ ♀
18 55	☽ ∠ ♂
22 Friday	
08 30	☽ ⊥ ♅
08 33	☽ ☍ ♄
13 21	♄ ± ♅
19 57	☽ ∠ ♇
20 41	☽ ⊻ ♅
22 41	☽ ☌ ☿
23 Saturday	
01 16	☽ ∥ ♂
03 06	☽ ✱ ♂
06 03	☽ ∥ ☿
06 42	☽ ⊼ ♃
14 09	☽ ⊥ ♇
18 31	☽ ☌ ☉
18 41	☽ ± ♃
20 33	☉ ± ♃
24 Sunday	
02 14	☽ ⊻ ♇
03 45	♀ Ꝑ ♃
04 42	☽ Q ♂
05 33	☽ □ ♆
06 20	☽ ∥ ♅
12 37	☽ Ꝑ ♃
13 42	☽ ☌ ♀
20 32	☽ ⊼ ♄
25 Monday	
08 25	☽ ± ♄
09 08	☽ ☌ ♅
09 13	☽ ⊻ ☿
11 40	♂ ⊼ ♃
12 51	☿ ⊻ ♅
14 26	☽ # ♆
17 26	☽ Q ♆
18 16	☽ △ ♃
18 42	☽ □ ♂
20 54	☽ ⊥ ☿
26 Tuesday	
02 07	☽ Ꝑ ♄
11 18	☽ ⊻ ☉
13 55	☽ ☌ ♇
14 23	☽ ∠ ☿
17 03	☽ ✱ ♆
19 09	♀ ∥ ♂
27 Wednesday	
00 01	☽ ⊥ ☉
04 34	☿ ∥ ♀
05 21	♀ ⊼ ♄
07 18	☽ △ ♄
07 31	☽ ⊻ ♀
18 24	☿ St D
18 54	☽ ∠ ☉
19 23	☽ ✱ ☿
20 07	☽ ⊻ ♅
20 22	☽ ⊥ ♀
21 50	☉ ⊻ ♇
22 08	☽ ∠ ♆
28 Thursday	
04 19	☽ □ ♃
07 39	☽ ⊥ ♅
08 24	☽ △ ♂
15 18	☽ ⊥ ♆
15 24	☽ ∠ ♀
18 21	☽ Q ☿
23 43	☽ ⊻ ♇
29 Friday	
00 46	☽ ∠ ♅
01 45	☽ ✱ ☉
02 36	☽ ⊻ ♆
10 58	☽ ⊥ ♇
13 19	☉ □ ♆
14 02	☽ Q ♃
14 10	☽ Ꝑ ♂
15 56	☽ □ ♄
16 50	☉ ∥ ♂
21 30	☉ ∥ ☿
22 23	☽ ✱ ♀
30 Saturday	
01 47	☽ Q ☉
03 37	☽ ∠ ♇
04 16	☽ □ ☿
04 41	☽ ✱ ♅
07 36	☽ ± ♂
11 56	☽ ✱ ♃
19 03	☽ ⊼ ♂
21 20	♂ Ꝑ ♄
22 24	☽ Q ♀
31 Sunday	
00 31	☽ Q ♄
02 27	☽ Q ♅
03 20	☿ ⊻ ♅
06 42	☽ ✱ ♇
09 20	☽ ☌ ♆
12 38	☽ □ ☉
14 36	☽ ∠ ♃
19 46	☿ ∥ ♂
20 26	♀ ± ♄
21 37	☽ ✱ ♄
23 45	☽ ∥ ♆

AUGUST 1960

LONGITUDES

Date	Sidereal time h m s	Sun ☉ ° ' "	Moon ☽ ° ' "	Moon ☽ 24.00 ° ' "	Mercury ☿ ° '	Venus ♀ ° '	Mars ♂ ° '	Jupiter ♃ ° '	Saturn ♄ ° '	Uranus ♅ ° '	Neptune ♆ ° '	Pluto ⯓ ° '
01	08 40 23	09 ♌ 12 58	21 ♏ 43 33	28 ♏ 46 31	21 ♋ 07	20 ♌ 12	29 ♉ 33	24 ♐ 21	13 ♑ 22	20 ♌ 37	06 ♏ 25	04 ♍ 58
02	08 44 19	10 10 23	05 ♐ 55 17	13 ♐ 09 33	21 39	21 26	00 ♊ 12	24 R 17	13 R 18	20 41	06 26	05 00
03	08 48 16	11 07 48	20 ♐ 28 53	27 ♐ 52 38	22 18	22 40	00 52	24 14	13 15	20 45	06 26	05 02
04	08 52 13	12 05 14	05 ♑ 20 02	12 ♑ 50 08	23 03	23 54	01 32	24 11	13 11	20 48	06 27	05 04
05	08 56 09	13 02 41	20 ♑ 21 51	27 ♑ 54 03	23 55	25 08	02 11	24 08	13 07	20 52	06 27	05 06
06	09 00 06	14 00 08	05 ♒ 25 34	12 ♒ 55 11	24 52	26 22	02 51	24 05	13 04	20 56	06 28	05 07
07	09 04 02	14 57 37	20 ♒ 21 50	27 ♒ 44 30	25 55	27 36	03 30	24 03	13 01	21 00	06 28	05 09
08	09 07 59	15 55 06	05 ♓ 02 18	12 ♓ 14 33	27 05	28 ♌ 50	04 09	24 00	12 57	21 03	06 29	05 11
09	09 11 55	16 52 37	19 ♓ 20 43	26 ♓ 20 26	28 20	00 ♍ 03	04 48	23 58	12 54	21 07	06 30	05 13
10	09 15 52	17 50 09	03 ♈ 13 30	09 ♈ 59 56	29 ♋ 40	01 17	05 27	23 56	12 51	21 11	06 31	05 15
11	09 19 48	18 47 42	16 ♈ 39 49	23 ♈ 13 24	01 ♌ 05	02 31	06 06	23 55	12 47	21 14	06 31	05 17
12	09 23 45	19 45 17	29 ♈ 41 01	06 ♉ 03 04	02 35	03 45	06 45	23 53	12 44	21 18	06 32	05 19
13	09 27 42	20 42 53	12 ♉ 20 03	18 ♉ 32 29	04 10	04 59	07 23	23 51	12 41	21 22	06 33	05 21
14	09 31 38	21 40 31	24 ♉ 40 53	00 ♊ 45 50	05 49	06 13	08 01	23 50	12 38	21 26	06 34	05 23
15	09 35 35	22 38 10	06 ♊ 47 54	12 ♊ 47 39	07 32	07 27	08 40	23 49	12 36	21 29	06 35	05 25
16	09 39 31	23 35 51	18 ♊ 45 35	24 ♊ 42 16	09 18	08 41	09 18	23 48	12 33	21 33	06 36	05 27
17	09 43 28	24 33 33	00 ♋ 38 10	06 ♋ 33 46	11 08	09 55	09 56	23 48	12 30	21 37	06 37	05 29
18	09 47 24	25 31 17	12 ♋ 29 29	18 ♋ 25 43	13 00	11 09	10 34	23 47	12 27	21 41	06 37	05 31
19	09 51 21	26 29 03	24 ♋ 22 51	00 ♌ 21 11	14 54	12 22	11 11	23 47	12 25	21 44	06 38	05 33
20	09 55 17	27 26 49	06 ♌ 21 02	12 ♌ 22 39	16 50	13 36	11 49	23 47	12 22	21 48	06 40	05 35
21	09 59 14	28 24 38	18 ♌ 26 16	24 ♌ 32 06	18 48	14 50	12 26	23 D 47	12 20	21 52	06 41	05 37
22	10 03 11	29 ♌ 22 28	00 ♍ 40 20	06 ♍ 51 07	20 46	16 04	13 03	23 47	12 18	21 56	06 42	05 39
23	10 07 07	00 ♍ 20 19	13 ♍ 04 39	19 ♍ 21 02	22 45	17 18	13 40	23 47	12 15	21 59	06 43	05 41
24	10 11 04	01 18 12	25 ♍ 40 25	02 ♎ 02 58	24 45	18 32	14 17	23 48	12 13	22 03	06 44	05 43
25	10 15 00	02 16 06	08 ♎ 28 47	14 ♎ 58 01	26 45	19 46	14 54	23 49	12 11	22 07	06 45	05 45
26	10 18 57	03 14 01	21 ♎ 30 48	28 ♎ 07 15	28 ♌ 44	21 00	15 30	23 50	12 09	22 10	06 46	05 47
27	10 22 53	04 11 58	04 ♏ 47 31	11 ♏ 31 40	00 ♍ 44	22 14	16 06	23 51	12 07	22 14	06 48	05 49
28	10 26 50	05 09 56	18 ♏ 19 49	25 ♏ 11 58	02 42	23 28	16 42	23 52	12 06	22 18	06 49	05 51
29	10 30 46	06 07 55	02 ♐ 08 10	09 ♐ 08 19	04 41	24 42	17 18	23 54	12 04	22 21	06 50	05 53
30	10 34 43	07 05 56	16 ♐ 12 18	23 ♐ 19 55	06 38	25 55	17 54	23 56	12 02	22 25	06 51	05 55
31	10 38 40	08 ♍ 03 58	00 ♑ 30 51	07 ♑ 44 44	08 ♍ 34	27 ♍ 09	18 ♊ 30	23 ♐ 58	12 ♑ 01	22 ♌ 29	06 ♏ 53	05 ♍ 58

Date	Moon True ☊ ° '	Moon Mean ☊ ° '	Moon ☽ Latitude ° '
01	16 ♍ 05	17 ♍ 25	04 N 44
02	16 R 03	17 22	05 07
03	16 00	17 19	05 11
04	15 55	17 16	04 54
05	15 51	17 13	04 17
06	15 46	17 09	03 22
07	15 43	17 06	02 14
08	15 42	17 03	00 N 58
09	15 D 41	17 00	00 S 20
10	15 42	16 57	01 35
11	15 44	16 53	02 42
12	15 45	16 50	03 38
13	15 46	16 47	04 23
14	15 R 46	16 44	04 54
15	15 46	16 41	05 11
16	15 44	16 38	05 14
17	15 41	16 34	05 04
18	15 38	16 31	04 41
19	15 35	16 28	04 06
20	15 32	16 25	03 19
21	15 30	16 22	02 24
22	15 28	16 18	01 21
23	15 28	16 15	00 S 13
24	15 D 28	16 12	00 N 56
25	15 29	16 09	02 04
26	15 30	16 06	03 07
27	15 31	16 03	04 00
28	15 32	15 59	04 42
29	15 32	15 56	05 08
30	15 R 32	15 53	05 17
31	15 ♍ 31	15 ♍ 50	05 N 06

DECLINATIONS

Date	Sun ☉ ° '	Moon ☽ ° '	Mercury ☿ ° '	Venus ♀ ° '	Mars ♂ ° '	Jupiter ♃ ° '	Saturn ♄ ° '	Uranus ♅ ° '	Neptune ♆ ° '	Pluto ⯓ ° '
01	17 N 57	13 S 37	19 N 06	16 N 06	19 N 05	23 S 07	22 S 25	15 N 16	11 S 59	20 N 59
02	17 42	16 16	19 17	15 43	19 14	23 07	22 26	15 14	12 00	20 58
03	17 26	17 56	19 26	15 19	19 23	23 07	22 26	15 13	12 00	20 57
04	17 10	18 26	19 35	14 55	19 32	23 07	22 26	15 12	12 00	20 56
05	16 54	17 40	19 42	14 31	19 41	23 07	22 27	15 11	12 00	20 56
06	16 38	15 38	19 47	14 06	19 49	23 07	22 27	15 10	12 01	20 55
07	16 21	12 35	19 51	13 41	19 57	23 07	22 28	15 09	12 01	20 54
08	16 04	08 46	19 53	13 15	20 06	23 07	22 28	15 07	12 01	20 53
09	15 47	04 32	19 53	12 49	20 13	23 07	22 29	15 06	12 01	20 53
10	15 29	00 S 10	19 51	12 23	20 21	23 07	22 29	15 05	12 02	20 52
11	15 12	04 N 03	19 47	11 56	20 29	23 07	22 29	15 04	12 02	20 51
12	14 54	07 57	19 39	11 30	20 36	23 07	22 30	15 03	12 02	20 50
13	14 35	11 22	19 30	11 02	20 43	23 07	22 30	15 01	12 03	20 49
14	14 17	14 12	19 17	10 35	20 51	23 07	22 31	15 00	12 03	20 49
15	13 58	16 20	19 02	10 07	20 57	23 08	22 31	14 59	12 03	20 48
16	13 39	17 45	18 45	09 39	21 04	23 08	22 31	14 58	12 04	20 47
17	13 20	18 22	18 24	09 11	21 11	23 08	22 32	14 57	12 04	20 46
18	13 01	18 12	18 01	08 42	21 17	23 08	22 32	14 55	12 04	20 46
19	12 41	17 13	17 35	08 13	21 23	23 08	22 32	14 54	12 05	20 45
20	12 22	15 28	17 07	07 44	21 29	23 08	22 33	14 53	12 05	20 44
21	12 02	13 01	16 36	07 15	21 35	23 08	22 33	14 52	12 06	20 43
22	11 42	09 59	16 03	06 46	21 41	23 08	22 33	14 51	12 06	20 43
23	11 21	06 27	15 29	06 16	21 46	23 09	22 34	14 49	12 06	20 42
24	11 01	02 N 35	14 52	05 47	21 52	23 09	22 34	14 48	12 07	20 41
25	10 40	01 S 28	14 13	05 17	21 57	23 09	22 34	14 47	12 07	20 40
26	10 19	05 30	13 33	04 47	22 02	23 09	22 35	14 46	12 08	20 40
27	09 58	09 21	12 52	04 16	22 07	23 09	22 35	14 45	12 08	20 39
28	09 37	12 46	12 10	03 46	22 12	23 10	22 35	14 43	12 09	20 38
29	09 16	15 33	11 26	03 16	22 16	23 10	22 35	14 42	12 09	20 38
30	08 54	17 28	10 42	02 45	22 21	23 10	22 36	14 41	12 09	20 37
31	08 N 33	18 S 20	09 N 57	02 N 14	22 N 25	23 S 10	22 S 36	14 N 40	12 S 10	20 N 36

ZODIAC SIGN ENTRIES

Date	h m	Planets
02	02 04	☽ ♐
02	04 32	♂ ♊
04	03 25	☽ ♑
06	03 21	☽ ♒
08	03 42	☽ ♓
09	10 54	♀ ♍
10	06 21	☽ ♈
10	17 49	☿ ♌
12	12 36	☽ ♉
14	22 29	☽ ♊
17	10 43	☽ ♋
19	23 18	☽ ♌
22	10 41	☽ ♍
23	03 34	☉ ♍
24	20 09	☽ ♎
27	03 11	☿ ♍
27	03 24	☽ ♏
29	08 19	☽ ♐
31	11 09	☽ ♑

LATITUDES

Date	Mercury ☿ ° '	Venus ♀ ° '	Mars ♂ ° '	Jupiter ♃ ° '	Saturn ♄ ° '	Uranus ♅ ° '	Neptune ♆ ° '	Pluto ⯓ ° '
01	02 S 43	01 N 25	01 S 00	00 N 13	00 N 21	00 N 41	01 N 46	12 N 11
04	01 55	01 27	00 57	00 12	00 21	00 41	01 46	12 10
07	01 08	01 28	00 55	00 12	00 21	00 41	01 46	12 10
10	00 S 23	01 28	00 52	00 11	00 20	00 41	01 46	12 10
13	00 N 17	01 27	00 50	00 11	00 20	00 41	01 45	12 10
16	00 51	01 26	00 47	00 10	00 20	00 41	01 45	12 10
19	01 16	01 25	00 44	00 10	00 19	00 41	01 45	12 10
22	01 34	01 23	00 41	00 09	00 19	00 41	01 45	12 10
25	01 44	01 20	00 38	00 09	00 19	00 41	01 45	12 10
28	01 46	01 17	00 35	00 08	00 18	00 41	01 45	12 10
31	01 N 43	01 N 13	00 S 32	00 N 08	00 N 18	00 N 41	01 N 45	12 N 10

DATA

Julian Date	2437148
Delta T	+33 seconds
Ayanamsa	23° 18' 20"
Synetic vernal point	05° ♓ 48' 39"
True obliquity of ecliptic	23° 26' 31"

MOON'S PHASES, APSIDES AND POSITIONS ☽

Date	h m	Phase	Longitude ° '	Eclipse Indicator
07	02 41	○	14 ♒ 35	
14	05 37	☾	21 ♉ 25	
22	09 15	●	29 ♌ 16	
29	19 22	☽	06 ♐ 26	

Day	h m		
05	19 41	Perigee	
18	01 03	Apogee	
04	09 31	Max dec	18° S 27'
10	12 56	0N	
17	18 38	Max dec	18° N 24'
25	03 22	0S	
31	17 46	Max dec	18° S 22'

LONGITUDES

Date	Chiron ⚷ ° '	Ceres ⚳ ° '	Pallas ⚴ ° '	Juno ⚵ ° '	Vesta ⚶ ° '	Black Moon Lilith ⚸ ° '
01	00 ♓ 51	24 ♒ 53	14 ♑ 07	11 ♐ 50	04 ♑ 59	09 ♋ 33
11	00 ♓ 23	22 ♒ 45	12 ♑ 16	11 ♐ 50	04 ♑ 22	10 ♋ 40
21	29 ♒ 53	20 ♒ 33	11 ♑ 03	12 ♐ 22	04 ♑ 32	11 ♋ 48
31	29 ♒ 24	18 ♒ 31	10 ♑ 28	13 ♐ 25	05 ♑ 28	12 ♋ 55

ASPECTARIAN

h m	Aspects
01 Monday	
03 47	☽ Q ⯓
06 13	☽ ⊥ ♃
09 08	☽ □ ♀
10 06	☽ □ ♅
10 54	☽ △ ☿
15 02	☉ ♇ ♃
16 28	☽ ⊻ ♃
20 35	♀ ☌ ♅
23 15	☽ ∠ ♄
02 Tuesday	
01 52	☽ # ♅
01 57	☽ ☍ ♂
07 06	☽ # ♀
10 27	☽ □ ⯓
12 51	☽ ⊻ ♆
13 16	☽ ♇ ☿
14 18	☽ ⊥ ♄
19 34	☽ △ ☉
20 26	☿ ⊻ ♀
22 48	☽ ⊥ ♆
03 Wednesday	
00 11	☽ ⊻ ♄
04 20	☽ # ☉
04 50	☽ ± ☿
12 26	☽ △ ♅
13 33	☽ ∠ ♆
15 07	☽ ⊼ ☿
15 53	☽ △ ♀
18 05	☽ ☌ ♃
18 14	♀ ∥ ♅
21 48	☽ ♇ ☉
04 Thursday	
05 36	☽ ⊼ ♂
11 34	☽ △ ⯓
12 46	☽ ♇ ♅
13 17	☽ ± ☉
13 47	☽ ⚹ ♆
15 41	☽ ± ♂
17 19	♀ △ ♃
18 13	☽ ♇ ♀
22 43	♀ Q ♆
23 32	☽ ⊼ ☉
05 Friday	
00 30	☽ ☌ ♄
03 12	☽ ± ♅
06 43	☽ ♇ ♂
08 57	☽ Q ♆
09 52	☽ ± ♀
11 34	☽ ♇ ⯓
12 48	☽ ⊼ ♅
13 51	☉ ⊼ ♄
17 35	☿ ⊼ ♃
17 59	☽ ⊻ ♃
18 01	☽ ☍ ☿
20 16	☽ ⊼ ♀
06 Saturday	
00 00	☿ ∥ ♂
00 25	☽ # ☉
01 56	☽ ± ⯓
03 31	☽ ⊥ ♃
07 42	☽ △ ♂
11 31	☽ ⊼ ⯓
13 40	☽ □ ♆
16 20	☽ # ♅
17 50	☽ ∠ ♃
07 Sunday	
00 11	☽ ⊻ ♄
02 41	☽ ☍ ☉
02 57	☽ # ♀
09 49	☽ ⊥ ♄
13 01	☽ ☍ ♅
15 49	☽ ∥ ♆
17 58	☽ ⚹ ♃
19 44	♀ ♇ ♄
21 47	☽ ⊼ ☿
08 Monday	
00 23	☽ ∠ ♄
00 50	☽ ☍ ♀
08 28	☽ ± ☿
10 28	☽ □ ♂
12 15	☽ ☍ ⯓
13 36	☽ Q ♃
14 24	☽ △ ♆
09 Tuesday	
00 49	☽ ♇ ☿
01 08	☽ ⚹ ♄
06 23	☉ Q ♂
07 30	☽ ⊼ ☉
15 02	☽ ⊼ ♅
15 41	☽ ♇ ♆
18 12	☽ Q ♂
18 29	☽ ± ☉
19 54	☽ □ ♃
21 28	☽ Q ♄
10 Wednesday	
01 24	☽ ± ♅
04 13	♂ □ ⯓
04 39	☿ ⊥ ⯓
05 05	☽ △ ☿
07 14	☽ ± ♆
08 16	☽ ⊼ ♀
11 16	☽ ♇ ☉
15 35	☽ ⊼ ⯓
16 07	☽ ⚹ ♂
16 40	☿ ± ♃
17 14	☽ ♇ ♅
17 48	☽ ⊼ ♆
19 54	☽ ± ♀
11 Thursday	
02 17	☽ ± ⯓
05 02	☽ □ ♄
07 08	♀ # ♆
11 55	☉ ± ♄
13 43	☽ ♇ ♀
16 11	☽ △ ☉
18 37	☽ ♇ ⯓
20 24	☽ △ ♅
20 31	☽ ∠ ♂
23 11	☉ ∥ ♅
12 Friday	
01 14	☽ △ ♃
04 04	♂ ⊼ ♆
11 53	♂ ± ♄
14 05	☽ ⊥ ♂
18 13	☽ □ ☿
20 28	☽ △ ♀
22 38	☽ △ ⯓
13 Saturday	
00 56	☽ ☍ ♆
02 02	☽ ⊻ ♂
05 21	☽ ♇ ♃
09 50	☽ ∥ ♀
12 41	☽ △ ♄
17 18	☽ # ♆
19 23	♀ ☌ ⯓
22 39	☽ ± ♃
14 Sunday	
05 24	☉ ☌ ♅
05 36	☽ □ ♅
05 37	☽ □ ☉
05 39	☿ ⊻ ⯓
06 12	♂ ∥ ⯓
10 02	☽ Q ☿
10 21	☽ ⊼ ♃
12 46	☽ ∥ ☉
17 48	☽ ♇ ♄
18 51	♀ ⚹ ♆
20 06	☽ ∥ ♅
22 36	☿ □ ♆
15 Monday	
07 52	☿ ⊻ ♀
09 14	☽ □ ⯓
11 34	☽ ⊼ ♆
11 36	☽ ± ♄
13 27	☽ □ ♀
13 43	☽ ⚹ ☿
15 56	☽ ☌ ♂
17 24	☽ Q ♅
20 21	☽ Q ☉
23 33	☽ ⊼ ♄
23 35	☽ ± ♆
16 Tuesday	
05 19	☿ ♇ ♃
11 47	☿ ⚹ ♂
17 05	☉ △ ♃
17 40	☽ ⚹ ♅
17 43	☽ ♇ ♆
21 30	☽ Q ⯓
22 10	☽ ☍ ♃
22 37	☽ ⚹ ☉
22 46	♂ Q ♅
17 Wednesday	
01 13	☽ ∠ ☿
05 51	☽ Q ♀
12 42	♀ □ ♂
13 06	☽ ∥ ☿
13 15	☉ Q ♆
21 51	☽ ⚹ ⯓
22 47	☽ ⊥ ☿
18 Thursday	
00 07	☽ △ ♆
00 10	☽ ∠ ♅
05 16	☿ ⊼ ♄
07 40	☽ ∠ ☉
07 52	☽ ⊻ ♂
08 57	☽ ⚹ ♀
11 56	☽ ☍ ♄
13 13	☽ ⊻ ☿
18 29	☽ ⊥ ♅
20 41	☽ ⊥ ♂
23 52	☽ ∥ ☿
19 Friday	
03 27	☽ ⊥ ☉
04 16	☽ ∠ ⯓
06 39	☽ ⊻ ♅
10 47	☽ ⊼ ♃
12 45	♀ △ ♄
15 50	☽ ∠ ♂
16 36	☽ ⊻ ☉
18 43	☽ ∠ ♀
22 25	☽ ⊥ ⯓
22 51	☽ ± ♃
20 Saturday	
10 14	☉ ♇ ♄
10 28	☽ ⊻ ⯓
12 37	☽ □ ♆
14 48	☽ ⊥ ♀
16 42	♃ St D
16 50	☽ ♇ ♃
18 25	☽ ∥ ♅
23 28	☽ ⚹ ♂
23 57	☽ ⊼ ♄
21 Sunday	
04 05	☽ ⊻ ♀
06 29	☿ ± ♄
07 30	☉ # ♆
08 24	♂ ⊼ ♄
11 48	☽ ± ♄
12 50	☽ ☌ ☿
18 47	☽ ☌ ♅
19 47	☽ # ♆
21 23	☽ ∥ ☉
21 43	♂ ± ♆
22 31	☽ △ ♃
22 Monday	
00 18	☽ Q ♆
00 26	☽ Q ♂
05 26	☽ ♇ ♄
09 15	☽ ☌ ☉
21 42	☽ ☌ ⯓
23 43	☽ ⚹ ♆
23 Tuesday	
02 24	☿ ☌ ♅
10 26	☽ △ ♄
13 12	☽ □ ♂
13 18	☽ ∥ ♀
20 58	☽ ☌ ♀
24 Wednesday	
00 28	☿ △ ♃
04 31	☽ ∠ ♆
05 06	☽ ⊻ ♅
08 27	☽ □ ♃
09 56	☽ ⊻ ☿
11 45	☿ Q ♆
14 23	☿ ∥ ♅
16 30	☽ ⊥ ♅
21 33	☽ ⊥ ♆
23 20	☽ ⊥ ☿
23 28	☽ ⊻ ☉
25 Thursday	
06 55	☽ ⊻ ⯓
08 47	☽ ⊻ ♆
09 26	☽ ∠ ♅
11 35	☽ ⊥ ☉
14 31	☿ Q ♂
17 12	☿ ♇ ♄
18 05	☽ ⊥ ⯓
18 11	☽ Q ♃
18 51	☽ □ ♄
19 10	☽ ∠ ☿
26 Friday	
00 27	☽ △ ♂
05 31	☽ ∠ ☉
08 07	☽ # ♀
10 40	☽ ∠ ⯓
10 58	☽ ⊻ ♀
13 12	☽ ⚹ ♅
16 13	☽ ⚹ ♃
22 59	☽ ⊥ ♀
27 Saturday	
03 20	♀ ∠ ♆
03 26	☽ ⚹ ☿
03 38	☽ Q ♄
05 04	☽ ♇ ♂
10 51	☽ ⚹ ☉
11 00	☽ Q ♅
12 05	♀ ⊻ ♅
13 51	☽ ⚹ ⯓
15 35	☽ ☌ ♆
15 47	☽ # ☉
16 48	☽ ∠ ♀
19 15	☽ ∠ ♃
21 55	☽ ± ♂
28 Sunday	
01 02	☽ ⚹ ♄
04 32	☽ Q ☿
07 19	☽ ∥ ♆
08 17	☽ # ☿
09 01	☽ ⊼ ♂
09 48	☽ Q ☉
11 10	☽ Q ⯓
11 12	☽ ⊥ ♃
12 39	☿ # ♆
18 58	☽ □ ♅
20 11	♀ □ ♃
21 42	☽ ⊻ ♃
21 51	☽ ⚹ ♀
29 Monday	
03 15	☽ ∠ ♄
04 01	☽ # ♅
05 47	☉ ☌ ⯓
17 04	☽ □ ☿
18 28	☽ □ ⯓
18 44	☽ ⊥ ♄
19 22	☽ □ ☉
20 05	☽ ⊻ ♆
20 34	☽ Q ♀
30 Tuesday	
03 10	☿ ☌ ⯓
04 57	☽ ⊻ ♄
05 50	☉ ⚹ ♆
06 19	☽ ⊥ ♆
14 50	☿ ⚹ ♆
15 00	☽ ☍ ♂
21 32	☽ ∠ ♆
22 31	☽ △ ♅
23 33	☉ ☌ ☿
31 Wednesday	
01 02	☽ ☌ ♃
05 52	☽ □ ♀
21 04	☽ △ ⯓
22 35	☽ ⚹ ♆
23 36	☽ ♇ ♅

All ephemeris data is given at 12.00 UT and the Moon's longitude is additionally given for 24.00 UT

SEPTEMBER 1960

Date	Sidereal time h m s	Sun ☉ ° ' "	Moon ☽ ° ' "	Moon ☽ 24.00 ° ' "	Mercury ☿ ° '	Venus ♀ ° '	Mars ♂ ° '	Jupiter ♃ ° '	Saturn ♄ ° '	Uranus ♅ ° '	Neptune ♆ ° '	Pluto ♇ ° '
01	10 42 36	09 ♍ 02 01	15 ♑ 01 03	22 ♑ 19 14	10 ♍ 29	28 ♍ 23	19 ♊ 05	24 ♐ 00	11 ♑ 59	22 ♌ 32	06 ♏ 54	06 ♍ 00
02	10 46 33	10 00 06	29 ♑ 38 38	06 ♒ 58 30	12 24	29 ♍ 37	19 40	24 02	11 R 58	22 36	06 55	06 02
03	10 50 29	10 58 12	14 ♒ 18 03	21 ♒ 36 31	14 17	00 ♎ 51	20 16	24 04	11 57	22 39	06 57	06 04
04	10 54 26	11 56 20	28 ♒ 53 04	06 ♓ 06 56	16 09	02 05	20 50	24 07	11 56	22 43	06 58	06 06
05	10 58 22	12 54 29	13 ♓ 17 23	20 ♓ 23 48	18 00	03 18	21 25	24 10	11 55	22 47	07 00	06 08
06	11 02 19	13 52 40	27 ♓ 25 35	04 ♈ 22 19	19 49	04 32	21 59	24 13	11 54	22 50	07 01	06 10
07	11 06 15	14 50 53	11 ♈ 13 40	17 ♈ 59 25	21 38	05 46	22 34	24 16	11 53	22 54	07 03	06 12
08	11 10 12	15 49 08	24 ♈ 39 29	01 ♉ 13 52	23 25	07 00	23 08	24 20	11 52	22 57	07 04	06 14
09	11 14 09	16 47 25	07 ♉ 42 44	14 ♉ 06 16	25 12	08 14	23 42	24 23	11 51	23 01	07 06	06 16
10	11 18 05	17 45 43	20 ♉ 24 49	26 ♉ 38 45	26 57	09 27	24 15	24 27	11 51	23 04	07 07	06 18
11	11 22 02	18 44 04	02 ♊ 48 31	08 ♊ 54 36	28 ♍ 41	10 41	24 49	24 31	11 50	23 08	07 09	06 20
12	11 25 58	19 42 27	14 ♊ 57 32	20 ♊ 57 54	00 ♎ 24	11 55	25 22	24 35	11 50	23 11	07 11	06 22
13	11 29 55	20 40 53	26 ♊ 56 15	02 ♋ 53 10	02 05	13 09	25 55	24 39	11 50	23 15	07 12	06 24
14	11 33 51	21 39 20	08 ♋ 49 14	14 ♋ 45 01	03 46	14 23	26 28	24 43	11 49	23 18	07 14	06 26
15	11 37 48	22 37 49	20 ♋ 41 06	26 ♋ 37 59	05 25	15 36	27 00	24 48	11 49	23 21	07 16	06 28
16	11 41 44	23 36 21	02 ♌ 36 12	08 ♌ 36 13	07 04	16 50	27 33	24 53	11 D 49	23 25	07 17	06 30
17	11 45 41	24 34 55	14 ♌ 38 29	20 ♌ 43 23	08 41	18 04	28 05	24 58	11 49	23 28	07 19	06 32
18	11 49 38	25 33 30	26 ♌ 51 16	03 ♍ 02 25	10 18	19 18	28 36	25 03	11 50	23 31	07 21	06 34
19	11 53 34	26 32 08	09 ♍ 17 06	15 ♍ 35 29	11 53	20 31	29 08	25 08	11 50	23 35	07 23	06 36
20	11 57 31	27 30 48	21 ♍ 57 42	28 ♍ 23 48	13 28	21 45	29 ♊ 39	25 13	11 50	23 38	07 25	06 38
21	12 01 27	28 29 30	04 ♎ 53 50	11 ♎ 27 44	15 01	22 59	00 ♋ 10	25 19	11 51	23 41	07 26	06 39
22	12 05 24	29 ♍ 28 13	18 ♎ 05 25	24 ♎ 46 44	16 34	24 12	00 41	25 25	11 51	23 44	07 28	06 41
23	12 09 20	00 ♎ 26 59	01 ♏ 31 32	08 ♏ 19 35	18 05	25 26	01 11	25 31	11 52	23 48	07 30	06 43
24	12 13 17	01 25 46	15 ♏ 10 41	22 ♏ 04 34	19 36	26 40	01 42	25 37	11 53	23 51	07 32	06 45
25	12 17 13	02 24 36	29 ♏ 00 59	05 ♐ 59 39	21 05	27 54	02 11	25 43	11 54	23 54	07 34	06 47
26	12 21 10	03 23 27	13 ♐ 00 20	20 ♐ 02 44	22 34	29 ♎ 07	02 41	25 49	11 55	23 57	07 36	06 49
27	12 25 07	04 22 19	27 ♐ 06 37	04 ♑ 11 43	24 01	00 ♏ 21	03 10	25 56	11 56	24 00	07 38	06 51
28	12 29 03	05 21 14	11 ♑ 17 46	18 ♑ 24 30	25 28	01 34	03 39	26 02	11 57	24 03	07 40	06 53
29	12 33 00	06 20 10	25 ♑ 31 38	02 ♒ 38 52	26 54	02 48	04 08	26 09	11 58	24 06	07 42	06 54
30	12 36 56	07 ♎ 19 08	09 ♒ 45 55	16 ♒ 52 25	28 ♎ 18	04 ♏ 02	04 ♋ 37	26 ♐ 16	12 ♑ 00	24 ♌ 09	07 ♏ 44	06 ♍ 56

Date	Moon True ☊ ° '	Moon Mean ☊ ° '	Moon Latitude ° '
01	15 ♍ 31	15 ♍ 47	04 N 36
02	15 R 30	15 44	03 48
03	15 30	15 40	02 45
04	15 29	15 37	01 31
05	15 29	15 34	00 N 12
06	15 D 29	15 31	01 S 06
07	15 29	15 28	02 18
08	15 R 29	15 24	03 21
09	15 29	15 21	04 11
10	15 29	15 18	04 47
11	15 29	15 15	05 10
12	15 29	15 12	05 17
13	15 D 29	15 09	05 11
14	15 29	15 05	04 51
15	15 30	15 02	04 19
16	15 30	14 59	03 36
17	15 31	14 56	02 43
18	15 32	14 53	01 42
19	15 33	14 50	00 S 35
20	15 R 33	14 46	00 N 35
21	15 32	14 43	01 45
22	15 31	14 40	02 50
23	15 29	14 37	03 48
24	15 27	14 34	04 33
25	15 25	14 30	05 03
26	15 23	14 27	05 15
27	15 22	14 24	05 09
28	15 D 22	14 21	04 44
29	15 23	14 18	04 01
30	15 ♍ 24	14 ♍ 15	03 N 04

DECLINATIONS

Date	Sun ☉ ° '	Moon ☽ ° '	Mercury ☿ ° '	Venus ♀ ° '	Mars ♂ ° '	Jupiter ♃ ° '	Saturn ♄ ° '	Uranus ♅ ° '	Neptune ♆ ° '	Pluto ♇ ° '
01	08 N 11	18 S 01	09 N 11	01 N 44	22 N 29	23 S 11	22 S 36	14 N 39	12 S 10	20 N 35
02	07 49	16 31	08 25	01 13	22 33	23 11	22 36	14 38	12 11	20 35
03	07 27	13 55	07 38	00 42	22 36	23 11	22 37	14 36	12 11	20 34
04	07 05	10 27	06 51	00 N 11	22 40	23 11	22 37	14 35	12 12	20 33
05	06 43	06 23	06 04	00 S 19	22 44	23 12	22 37	14 34	12 12	20 33
06	06 21	02 S 02	05 17	00 50	22 47	23 12	22 37	14 33	12 13	20 32
07	05 58	02 N 19	04 30	01 21	22 50	23 12	22 37	14 32	12 14	20 31
08	05 36	06 26	03 42	01 52	22 53	23 12	22 38	14 31	12 14	20 31
09	05 13	10 07	02 55	02 23	22 56	23 13	22 38	14 30	12 15	20 30
10	04 50	13 14	02 08	02 54	22 59	23 13	22 38	14 28	12 15	20 29
11	04 27	15 40	01 21	03 25	23 01	23 13	22 38	14 27	12 16	20 29
12	04 05	17 20	00 N 35	03 55	23 04	23 14	22 38	14 26	12 16	20 28
13	03 42	18 14	00 S 12	04 26	23 06	23 14	22 38	14 25	12 17	20 28
14	03 19	18 18	00 58	04 57	23 08	23 14	22 38	14 24	12 17	20 27
15	02 55	17 34	01 43	05 27	23 11	23 15	22 39	14 23	12 18	20 26
16	02 32	16 04	02 29	05 57	23 13	23 15	22 39	14 22	12 19	20 26
17	02 09	13 51	03 14	06 28	23 14	23 15	22 39	14 21	12 19	20 25
18	01 46	10 58	03 58	06 58	23 16	23 15	22 39	14 20	12 20	20 25
19	01 23	07 33	04 42	07 28	23 18	23 16	22 39	14 19	12 20	20 24
20	00 59	03 N 44	05 26	07 58	23 19	23 16	22 39	14 17	12 21	20 23
21	00 36	00 S 26	06 09	08 27	23 21	23 16	22 39	14 16	12 22	20 23
22	00 N 13	04 28	06 51	08 57	23 22	23 17	22 39	14 15	12 22	20 22
23	00 S 11	08 27	07 33	09 26	23 23	23 17	22 39	14 14	12 23	20 22
24	00 34	12 02	08 14	09 55	23 24	23 17	22 39	14 13	12 24	20 21
25	00 57	15 01	08 55	10 24	23 25	23 18	22 39	14 12	12 24	20 21
26	01 21	17 09	09 35	10 53	23 26	23 18	22 39	14 11	12 25	20 20
27	01 44	18 16	10 15	11 21	23 27	23 18	22 39	14 10	12 26	20 20
28	02 08	18 15	10 53	11 50	23 28	23 19	22 39	14 09	12 26	20 19
29	02 31	17 05	11 31	12 18	23 29	23 19	22 39	14 08	12 27	20 19
30	02 S 54	14 S 51	12 S 09	12 S 45	23 N 29	23 S 19	22 S 39	14 N 07	12 S 28	20 N 18

ZODIAC SIGN ENTRIES

Date	h	m	Planets
02	12	35	☽ ♒
02	19	29	♀ ♎
04	13	51	☽ ♓
06	16	26	☽ ♈
08	21	44	☽ ♉
11	06	31	☽ ♊
12	06	29	☿ ♎
13	18	10	☽ ♋
16	06	46	☽ ♌
18	18	07	☽ ♍
21	02	58	☽ ♎
21	04	06	♂ ♋
23	00	59	☉ ♎
23	09	18	☽ ♏
25	13	42	☽ ♐
27	05	13	♀ ♏
27	16	54	☽ ♑
29	19	32	☽ ♒

LATITUDES

Date	Mercury ☿ ° '	Venus ♀ ° '	Mars ♂ ° '	Jupiter ♃ ° '	Saturn ♄ ° '	Uranus ♅ ° '	Neptune ♆ ° '	Pluto ♇ ° '
01	01 N 40	01 N 11	00 S 31	00 N 08	00 N 18	00 N 41	01 N 44	12 N 10
04	01 30	01 06	00 28	00 07	00 17	00 41	01 44	12 11
07	01 17	01 01	00 24	00 07	00 17	00 41	01 44	12 11
10	01 00	00 56	00 20	00 07	00 17	00 41	01 44	12 11
13	00 42	00 49	00 17	00 06	00 17	00 41	01 44	12 12
16	00 N 21	00 43	00 13	00 06	00 16	00 41	01 44	12 12
19	00 00	00 36	00 09	00 05	00 16	00 41	01 44	12 13
22	00 S 22	00 28	00 S 04	00 05	00 16	00 41	01 44	12 13
25	00 45	00 21	00 00	00 05	00 15	00 41	01 44	12 14
28	01 07	00 13	00 N 05	00 04	00 15	00 41	01 43	12 15
31	01 S 29	00 N 04	00 N 09	00 N 04	00 N 15	00 N 41	01 N 43	12 N 16

DATA

Julian Date	2437179
Delta T	+34 seconds
Ayanamsa	23° 18' 24"
Synetic vernal point	05° ♓ 48' 36"
True obliquity of ecliptic	23° 26' 31"

MOON'S PHASES, APSIDES AND POSITIONS ☽

Date	h	m	Phase	Longitude ° '	Eclipse Indicator
05	11	19	○	12 ♓ 53	total
12	22	19	☾	20 ♊ 08	
20	23	12	●	27 ♍ 58	Partial
28	01	13	☽	04 ♑ 55	

Day	h	m		
02	21	26	Perigee	
14	17	43	Apogee	
29	22	28	Perigee	
06	23	06	ON	
14	02	16	Max dec	18° N 22'
21	10	02	0S	
27	23	40	Max dec	18° S 24'

LONGITUDES

Date	Chiron ⚷ ° '	Ceres ⚳ ° '	Pallas ⚴ ° '	Juno ⚵ ° '	Vesta ⚶ ° '	Black Moon Lilith ⚸ ° '
01	29 ♒ 21	18 ♒ 19	10 ♑ 27	13 ♐ 32	05 ♑ 36	13 ♋ 01
11	28 ♒ 52	16 ♒ 41	10 ♑ 34	15 ♐ 05	07 ♑ 16	14 ♋ 08
21	28 ♒ 26	15 ♒ 34	11 ♑ 14	17 ♐ 01	09 ♑ 30	15 ♋ 15
31	28 ♒ 04	15 ♒ 02	12 ♑ 25	19 ♐ 18	12 ♑ 05	16 ♋ 22

ASPECTARIAN

h m Aspects

01 Thursday
01 26 ☽ △ ☉
03 24 ☽ △ ☿
07 01 ☽ ☌ ♄
14 31 ☽ ± ♅
15 08 ♀ ⊥ ♅
18 24 ☽ Q ♆
18 58 ☽ ⊼ ♂
21 51 ☽ ⚼ ♇

02 Friday
00 24 ☽ ⊼ ♅
02 47 ☽ ⊻ ♃
03 51 ☽ ⚼ ☉
05 14 ☽ ± ♂
06 39 ☿ △ ♄
07 46 ☽ ⚼ ☿
11 57 ☽ △ ♀
12 38 ☽ ± ♇
12 38 ☽ ⊥ ♃
19 38 ☽ ± ☉
20 35 ☽ ⚼ ♂
22 28 ☽ ⊼ ♇
23 56 ☽ □ ♆

03 Saturday
00 41 ☽ ± ☿
03 25 ☽ ∠ ♃
06 09 ☽ ⊼ ☉
06 23 ☽ # ♅
08 09 ☽ ⊻ ♄
11 58 ☽ ⊼ ☿
13 08 ♂ # ♄
14 01 ♀ ⊥ ♆
14 46 ☽ ⚼ ♀
17 58 ☽ ⊥ ♄
22 11 ☽ △ ♂
22 39 ☉ ∥ ☿

04 Sunday
00 32 ☽ ∥ ♆
01 47 ☽ ☍ ♅
04 06 ☽ ⚹ ♃
06 56 ☽ ± ♀
08 46 ☽ ∠ ♄
11 42 ☉ △ ♄
17 47 ☽ ⊼ ♀

05 Monday
00 00 ☽ ☍ ♇
00 03 ☽ Q ♃
01 27 ☽ △ ♆
09 41 ☽ ⚹ ♄
09 56 ☽ # ☉
11 19 ☽ ☍ ☉
14 08 ☽ # ☿
21 07 ☽ ☍ ☿

06 Tuesday
02 19 ☽ □ ♂
02 45 ☽ ⚼ ♆
04 07 ☽ ⊼ ♅
05 57 ☽ Q ♄
06 29 ☽ □ ♃
13 18 ♂ ⚼ ♆
14 26 ☽ ± ♅
17 49 ☽ ∥ ♀
18 12 ☽ ± ♆

07 Wednesday
01 30 ☽ ☍ ♀
03 09 ☽ ⊼ ♇
04 39 ☽ ⊼ ♆
05 51 ☽ # ♀
06 07 ☽ ⚼ ♅
10 47 ☽ Q ♂
13 09 ☽ □ ♄
13 43 ☽ ± ♇
17 36 ☿ ∠ ♆
18 54 ☽ ⊼ ☉
20 35 ♀ ⊻ ♇
22 25 ☽ ∥ ☿

08 Thursday
03 41 ♂ ⚹ ♅
05 28 ☿ ⊻ ♅
05 48 ☽ ⚼ ♇
06 13 ☿ □ ♂
06 28 ☽ ± ☉
07 19 ☽ ∥ ☉
08 54 ☽ △ ♅
09 07 ☽ ⚹ ♂
09 25 ☽ ⊼ ☿
11 24 ☽ △ ♃
13 29 ♀ ⊻ ♆
22 03 ☽ ± ☿

09 Friday
00 09 ☽ ⚼ ☉
00 37 ☿ □ ♃
07 36 ♀ ∠ ♅
09 18 ☽ △ ♇
10 51 ☽ ☍ ♆
13 04 ☽ ⊼ ♀
13 55 ☽ ∠ ♂
15 08 ☽ ⚼ ♃
17 23 ☽ ⚼ ☿
19 45 ☽ △ ♄
21 56 ☿ # ♀

10 Saturday
01 33 ☽ ± ♀
03 52 ☽ # ♆
06 31 ☽ △ ☉
07 41 ☽ ⊥ ♂
08 13 ☽ ± ♃
13 52 ♂ Q ♇
17 07 ☽ □ ♅
19 44 ☽ ⊻ ♂
19 47 ☽ ⊼ ♃
20 37 ☽ ⚼ ♀
21 18 ♂ ☍ ♃
23 18 ☽ ∥ ♅

11 Sunday
00 23 ☽ ⚼ ♄
02 38 ☽ △ ☿
17 56 ☽ ± ♄
18 30 ☿ ⊥ ♅
18 56 ☽ □ ♇
20 33 ☽ ⊼ ♆

12 Monday
04 28 ☽ Q ♅
05 16 ☽ △ ♀
05 47 ☽ ⊼ ♄
08 27 ☽ ± ♆
10 19 ♀ □ ♄
16 07 ☉ # ♀
20 57 ♀ ⊥ ♇
22 19 ☽ □ ☉
23 17 ☿ ⊥ ♆

13 Tuesday
01 43 ♀ Q ♃
02 28 ☽ ⚼ ♆
04 32 ☽ ⚹ ♅
06 52 ☽ Q ♇
07 22 ☽ ☍ ♃
09 51 ☽ ☌ ♂

14 Wednesday
00 06 ☽ □ ☿
07 09 ☽ ⚹ ♇
08 47 ☽ △ ♆
10 57 ☽ ∠ ♅
13 50 ☽ Q ☉
18 04 ☽ ☍ ♄

15 Thursday
00 33 ☽ □ ♀
02 41 ☉ ∠ ♆
05 14 ☽ ⊥ ♅
13 34 ☽ ∠ ♇
16 17 ☽ ⚹ ☉
17 25 ☽ ⊻ ♅
18 25 ☽ Q ☿
20 22 ☽ ⊼ ♃
22 48 ♄ St D

16 Friday
01 21 ☽ ⊻ ♂
03 27 ☿ ⊻ ♇
06 58 ☉ ⊻ ♅
07 45 ☽ ⊥ ♇
08 31 ☽ ± ♃
13 16 ☉ # ☿
13 58 ☽ ⊥ ♂
15 22 ☿ ⊻ ♆
16 59 ☽ Q ♀
19 49 ☽ ⊻ ♇
21 24 ☽ □ ♆
22 20 ☽ ⚹ ☿

17 Saturday
01 04 ☽ ∠ ☉
02 38 ☽ ⚼ ♃
06 24 ☽ ⊼ ♄
07 08 ☽ ∥ ♅
08 35 ☿ ∠ ♅
08 45 ☽ ∠ ♂
18 17 ☽ ± ♄
19 31 ☽ ⚹ ♀
20 28 ☽ ⊥ ☉
22 13 ☉ □ ♃

18 Sunday
00 31 ♂ # ♃
01 16 ☽ # ♆
05 28 ☽ ☌ ♅
08 27 ☽ △ ♃
08 31 ☽ ∠ ☿
09 04 ☽ Q ♆
09 15 ☽ ⊻ ☉
11 57 ☽ ⚼ ♄
15 34 ☽ ⚹ ♂

19 Monday
04 00 ☽ ∠ ♀
04 32 ☽ ⊥ ☿
06 50 ☽ ☌ ♇
08 21 ☽ ⚹ ♆
11 07 ☿ □ ♄
12 31 ☽ # ♀
15 41 ☽ Q ♂
16 52 ☽ △ ♄
17 41 ☽ ⊻ ☿
22 55 ☿ ⊥ ♇
23 03 ☽ ⊥ ♀

20 Tuesday
03 14 ☽ # ☿
08 07 ☿ Q ♃
09 29 ♀ ∠ ♇
11 34 ☽ ⊻ ♀
12 51 ☽ ∠ ♆
15 09 ☽ ⊻ ♅
18 09 ☽ □ ♃
23 12 ☽ ☌ ☉

21 Wednesday
02 21 ☽ ⊥ ♅
02 55 ☽ □ ♂
05 37 ☽ ⊥ ♆
05 58 ☽ ∥ ☉
13 23 ☽ # ☉
15 14 ☽ ⊻ ♇
16 40 ☽ ⊻ ♆
18 58 ☽ ∠ ♅

22 Thursday
00 42 ☽ □ ♄
02 12 ☽ ⊥ ♇
02 28 ♀ ⚹ ♅
03 29 ☽ Q ♃
08 53 ☽ ☌ ☿
18 29 ☽ ∠ ♇
19 01 ☉ ⊥ ♅
22 11 ☽ ⚹ ♅

23 Friday
00 05 ☽ ☌ ♀
01 14 ☽ ⚹ ♃
05 16 ☽ ∥ ☿
09 03 ☽ Q ♄
09 57 ☽ ⊻ ☉
11 23 ☽ △ ♂
13 36 ♀ ⚹ ♃
19 16 ☽ ∥ ♀
19 34 ☽ Q ♅
21 12 ☽ ⚹ ♇
21 22 ☽ ⊥ ☉
22 34 ☽ ☌ ♆

24 Saturday
03 57 ☽ ∠ ♃
06 14 ☽ ⚹ ♄
14 21 ☽ ∠ ☉
14 36 ☽ ∥ ♆
14 38 ☉ ⊥ ♆
14 44 ☽ ⚼ ♂
18 15 ☽ Q ♇
19 47 ☽ ⊥ ♃
20 38 ☽ ⊻ ☿

25 Sunday
01 09 ☉ □ ♂
03 08 ☽ □ ♅
04 53 ☽ # ♅
06 15 ☽ ⊻ ♃
06 57 ☽ ± ♂
08 17 ☽ ⊥ ☿
08 20 ☽ ∠ ♄
09 52 ☽ ⊻ ♀
17 40 ☽ ⊼ ♂
18 17 ☽ ⚹ ☉
21 12 ☽ ⊥ ♀
23 31 ☿ ∠ ♇
23 51 ☽ ⊥ ♄

26 Monday
01 23 ☽ □ ♇
01 36 ☽ ∠ ☿
02 44 ☽ ⊻ ♆
10 08 ☽ ⊻ ♄
13 01 ☽ ⊥ ♆
14 05 ☽ ∠ ♀
16 23 ☽ Q ☉

27 Tuesday
03 43 ♀ Q ♄
04 23 ☽ ∠ ♆
06 10 ☽ ⚹ ☿
06 42 ☽ △ ♅
09 59 ☽ ☌ ♃
11 36 ☿ ⚹ ♅
18 00 ☽ ⚹ ♀
22 38 ☽ ☍ ♂

28 Wednesday
01 13 ☽ □ ☉
04 31 ☽ △ ♇
04 48 ☽ Q ☿
05 51 ☽ ⚹ ♆
08 12 ☽ ⚼ ♅
13 06 ☽ ☌ ♄
16 13 ☽ Q ♀
22 25 ☿ ⚹ ♃
23 26 ☽ ± ♅

29 Thursday
02 09 ☽ Q ♆
05 53 ☽ ⚼ ♇
09 35 ☽ ⊼ ♅
13 04 ☽ ⊻ ♃
14 34 ☽ □ ☿
20 17 ♀ ∥ ♆
21 05 ☽ ± ♇
23 16 ☽ ⊥ ♃

30 Friday
01 25 ☽ □ ♀
02 24 ☉ ⊻ ♇
03 00 ☽ ⊼ ♂
07 13 ☽ ⊼ ♇
07 34 ☽ △ ☉
08 33 ☽ □ ♆
13 28 ☽ ± ♂
14 34 ☽ ∠ ♃
15 46 ☽ ⊻ ♄
18 11 ☽ # ♅
22 20 ☉ ⊻ ♆

All ephemeris data is given at 12.00 UT and the Moon's longitude is additionally given for 24.00 UT

OCTOBER 1960

LONGITUDES

Date	Sidereal time h m s	Sun ☉ ° ' "	Moon ☽ ° ' "	Moon ☽ 24.00 ° ' "	Mercury ☿ ° '	Venus ♀ ° '	Mars ♂ ° '	Jupiter ♃ ° '	Saturn ♄ ° '	Uranus ♅ ° '	Neptune ♆ ° '	Pluto ♇ ° '
01	12 40 53	08 ♎ 18 08	23 ♒ 58 02	01 ♓ 02 23	29 ♎ 42	05 ♏ 15	05 ♋ 05	26 ♐ 23	12 ♑ 01	24 ♌ 12	07 ♏ 46	06 ♍ 58
02	12 44 49	09 17 09	08 ♓ 05 03	15 ♓ 05 40	01 ♏ 06	06 29	05 32	26 31	12 03	24 15	07 48	07 00
03	12 48 46	10 16 12	22 ♓ 03 47	28 ♓ 59 01	02 26	07 42	06 00	26 38	12 04	24 18	07 50	07 02
04	12 52 42	11 15 17	05 ♈ 50 58	12 ♈ 39 16	03 46	08 56	06 27	26 46	12 06	24 21	07 52	07 03
05	12 56 39	12 14 25	19 ♈ 23 39	26 ♈ 03 49	05 05	10 09	06 54	26 53	12 08	24 24	07 54	07 05
06	13 00 36	13 13 34	02 ♉ 39 35	09 ♉ 10 52	06 22	11 23	07 20	27 01	12 10	24 26	07 56	07 07
07	13 04 32	14 12 45	15 ♉ 37 35	21 ♉ 59 49	07 39	12 36	07 46	27 09	12 12	24 29	07 58	07 09
08	13 08 29	15 11 59	28 ♉ 17 40	04 ♊ 31 21	08 54	13 50	08 12	27 17	12 14	24 32	08 00	07 10
09	13 12 25	16 11 15	10 ♊ 41 09	16 ♊ 47 26	10 07	15 03	08 37	27 25	12 16	24 34	08 02	07 12
10	13 16 22	17 10 33	22 ♊ 50 36	28 ♊ 51 07	11 19	16 17	09 02	27 34	12 19	24 37	08 04	07 14
11	13 20 18	18 09 54	04 ♋ 49 31	10 ♋ 46 21	12 29	17 30	09 27	27 42	12 21	24 40	08 06	07 15
12	13 24 15	19 09 17	16 ♋ 42 12	22 ♋ 37 43	13 37	18 44	09 51	27 51	12 24	24 42	08 09	07 17
13	13 28 11	20 08 42	28 ♋ 33 29	04 ♌ 30 08	14 43	19 57	10 15	27 59	12 26	24 45	08 11	07 18
14	13 32 08	21 08 09	10 ♌ 28 20	16 ♌ 28 40	15 47	21 11	10 38	28 08	12 29	24 47	08 13	07 20
15	13 36 05	22 07 39	22 ♌ 31 44	28 ♌ 38 05	16 49	22 24	11 01	28 17	12 32	24 50	08 15	07 22
16	13 40 01	23 07 11	04 ♍ 48 15	11 ♍ 02 41	17 48	23 37	11 24	28 26	12 35	24 52	08 17	07 23
17	13 43 58	24 06 45	17 ♍ 21 46	23 ♍ 45 50	18 44	24 51	11 46	28 36	12 38	24 54	08 19	07 25
18	13 47 54	25 06 21	00 ♎ 15 04	06 ♎ 49 36	19 37	26 04	12 08	28 45	12 41	24 57	08 22	07 26
19	13 51 51	26 06 00	13 ♎ 29 27	20 ♎ 14 29	20 26	27 17	12 29	28 54	12 44	24 59	08 24	07 28
20	13 55 47	27 05 41	27 ♎ 04 28	03 ♏ 59 05	21 11	28 31	12 50	29 04	12 47	25 01	08 26	07 29
21	13 59 44	28 05 23	10 ♏ 57 52	18 ♏ 00 16	21 53	29 ♏ 44	13 10	29 14	12 50	25 03	08 28	07 31
22	14 03 40	29 ♎ 05 08	25 ♏ 05 42	02 ♐ 13 30	22 29	00 ♐ 57	13 30	29 23	12 54	25 05	08 30	07 32
23	14 07 37	00 ♏ 04 54	09 ♐ 22 58	16 ♐ 33 28	23 00	02 11	13 49	29 33	12 57	25 07	08 33	07 33
24	14 11 34	01 04 43	23 ♐ 44 20	00 ♑ 54 59	23 26	03 24	14 07	29 43	13 01	25 09	08 35	07 35
25	14 15 30	02 04 33	08 ♑ 04 54	15 ♑ 13 39	23 45	04 37	14 26	29 ♐ 54	13 05	25 11	08 37	07 36
26	14 19 27	03 04 24	22 ♑ 20 51	29 ♑ 26 15	23 57	05 50	14 43	00 ♑ 04	13 08	25 13	08 39	07 37
27	14 23 23	04 04 18	06 ♒ 29 37	13 ♒ 30 51	24 01	07 04	15 00	00 14	13 12	25 15	08 42	07 39
28	14 27 20	05 04 13	20 ♒ 29 51	27 ♒ 26 34	23 R 58	08 17	15 17	00 25	13 16	25 17	08 44	07 40
29	14 31 16	06 04 09	04 ♓ 20 57	11 ♓ 13 00	23 46	09 30	15 33	00 35	13 20	25 19	08 46	07 41
30	14 35 13	07 04 07	18 ♓ 02 41	24 ♓ 49 57	23 25	10 43	15 48	00 46	13 24	25 21	08 48	07 42
31	14 39 09	08 ♏ 04 07	01 ♈ 34 44	08 ♈ 16 58	22 ♏ 54	11 ♐ 56	16 ♋ 03	00 ♑ 57	13 ♑ 28	25 ♌ 22	08 ♏ 51	07 ♍ 44

Date	Moon True ☊ ° '	Moon Mean ☊ ° '	Moon ☽ Latitude ° '
01	15 ♍ 25	14 ♍ 11	01 N 56
02	15 D 26	14 08	00 N 40
03	15 R 26	14 05	00 S 36
04	15 25	14 02	01 50
05	15 23	13 59	02 56
06	15 19	13 56	03 50
07	15 14	13 52	04 32
08	15 09	13 49	04 59
09	15 04	13 46	05 11
10	15 01	13 43	05 09
11	14 58	13 40	04 54
12	14 57	13 36	04 26
13	14 D 58	13 33	03 47
14	14 59	13 30	02 57
15	15 01	13 27	02 00
16	15 02	13 24	00 S 55
17	15 R 02	13 21	00 N 13
18	15 01	13 17	01 22
19	14 58	13 14	02 28
20	14 53	13 11	03 28
21	14 47	13 08	04 17
22	14 39	13 05	04 50
23	14 32	13 02	05 07
24	14 26	12 58	05 04
25	14 22	12 55	04 42
26	14 19	12 52	04 03
27	14 D 19	12 49	03 09
28	14 19	12 46	02 05
29	14 20	12 42	00 N 54
30	14 R 21	12 39	00 S 20
31	14 ♍ 19	12 ♍ 36	01 S 31

DECLINATIONS

Date	Sun ☉ ° '	Moon ☽ ° '	Mercury ☿ ° '	Venus ♀ ° '	Mars ♂ ° '	Jupiter ♃ ° '	Saturn ♄ ° '	Uranus ♅ ° '	Neptune ♆ ° '	Pluto ♇ ° '
01	03 S 18	11 S 43	12 S 45	13 S 13	23 N 30	23 S 20	22 S 39	14 N 06	12 S 28	20 N 18
02	03 41	07 55	13 21	13 40	23 30	23 20	22 39	14 06	12 29	20 17
03	04 04	03 S 42	13 56	14 06	23 31	23 20	22 39	14 05	12 30	20 17
04	04 27	00 N 39	14 30	14 33	23 31	23 21	22 39	14 04	12 30	20 17
05	04 50	04 53	15 03	14 59	23 32	23 21	22 39	14 03	12 31	20 16
06	05 13	08 47	15 35	15 24	23 32	23 21	22 39	14 02	12 32	20 16
07	05 36	12 11	16 07	15 50	23 32	23 22	22 39	14 01	12 32	20 15
08	05 59	14 55	16 37	16 15	23 32	23 22	22 39	14 00	12 33	20 15
09	06 22	16 55	17 06	16 39	23 32	23 22	22 39	13 59	12 34	20 15
10	06 45	18 06	17 34	17 03	23 32	23 22	22 39	13 58	12 34	20 14
11	07 07	18 27	18 01	17 27	23 32	23 23	22 38	13 58	12 35	20 14
12	07 30	18 00	18 27	17 50	23 33	23 23	22 38	13 57	12 36	20 14
13	07 52	16 45	18 52	18 13	23 33	23 23	22 38	13 56	12 36	20 13
14	08 15	14 46	19 15	18 35	23 33	23 23	22 38	13 55	12 37	20 13
15	08 37	12 07	19 37	18 57	23 33	23 24	22 38	13 54	12 38	20 13
16	08 59	08 53	19 57	19 19	23 33	23 24	22 38	13 54	12 39	20 12
17	09 21	05 11	20 16	19 40	23 33	23 24	22 37	13 53	12 39	20 12
18	09 43	01 N 09	20 34	20 00	23 33	23 24	22 37	13 52	12 40	20 12
19	10 05	03 S 03	20 49	20 20	23 33	23 25	22 37	13 51	12 41	20 12
20	10 26	07 12	21 03	20 39	23 33	23 25	22 37	13 51	12 41	20 11
21	10 48	11 03	21 15	20 58	23 33	23 25	22 37	13 50	12 42	20 11
22	11 09	14 20	21 24	21 16	23 33	23 25	22 36	13 49	12 43	20 11
23	11 30	16 48	21 31	21 34	23 33	23 25	22 36	13 49	12 44	20 11
24	11 51	18 14	21 36	21 51	23 33	23 25	22 36	13 48	12 44	20 11
25	12 12	18 30	21 38	22 08	23 33	23 25	22 36	13 47	12 45	20 10
26	12 32	17 35	21 37	22 24	23 33	23 26	22 35	13 47	12 46	20 10
27	12 53	15 36	21 33	22 39	23 34	23 26	22 35	13 46	12 46	20 10
28	13 13	12 41	21 26	22 54	23 34	23 26	22 35	13 46	12 47	20 10
29	13 33	09 05	21 15	23 08	23 34	23 26	22 35	13 45	12 48	20 10
30	13 52	05 02	21 00	23 21	23 35	23 26	22 34	13 45	12 49	20 10
31	14 S 12	00 S 46	20 S 41	23 S 34	23 N 35	23 S 26	22 S 34	13 N 44	12 S 49	20 N 10

ZODIAC SIGN ENTRIES

Date	h m	Planets
01	17 17	☿ ♏
01	22 14	☽ ♓
04	01 46	☽ ♈
06	07 09	☽ ♉
08	15 16	☽ ♊
11	02 18	☽ ♋
13	14 55	☽ ♌
16	02 40	☽ ♍
18	11 32	☽ ♎
20	17 06	☽ ♏
21	17 12	♀ ♐
22	20 16	☽ ♐
23	10 02	☉ ♏
24	22 28	☽ ♑
26	03 01	♃ ♑
27	00 57	☽ ♒
29	04 26	☽ ♓
31	09 11	☽ ♈

LATITUDES

Date	Mercury ☿ ° '	Venus ♀ ° '	Mars ♂ ° '	Jupiter ♃ ° '	Saturn ♄ ° '	Uranus ♅ ° '	Neptune ♆ ° '	Pluto ♇ ° '
01	01 S 29	00 N 04	00 N 09	00 N 04	00 N 15	00 N 41	01 N 43	12 N 16
04	01 50	00 S 04	00 14	00 03	00 14	00 42	01 43	12 16
07	02 10	00 13	00 19	00 03	00 14	00 42	01 43	12 17
10	02 28	00 22	00 24	00 03	00 14	00 42	01 43	12 18
13	02 44	00 30	00 30	00 02	00 14	00 42	01 43	12 19
16	02 56	00 39	00 36	00 02	00 13	00 42	01 43	12 20
19	03 04	00 48	00 41	00 02	00 13	00 42	01 43	12 21
22	03 07	00 56	00 48	00 01	00 13	00 42	01 43	12 22
25	03 01	01 05	00 54	00 01	00 12	00 42	01 43	12 23
28	02 45	01 13	01 01	00 01	00 12	00 42	01 43	12 24
31	02 S 16	01 S 21	01 N 07	00 00	00 N 12	00 N 43	01 N 43	12 N 25

DATA

Julian Date	2437209
Delta T	+34 seconds
Ayanamsa	23° 18' 27"
Synetic vernal point	05° ♓ 48' 33"
True obliquity of ecliptic	23° 26' 31"

MOON'S PHASES, APSIDES AND POSITIONS ☽

Date	h m	Phase	Longitude ° '	Eclipse Indicator
04	22 16	○	11 ♈ 41	
12	17 25	☾	19 ♋ 23	
20	12 02	●	27 ♎ 06	
27	07 34	☽	03 ♒ 53	

Day	h m		
12	13 09	Apogee	
24	19 25	Perigee	
04	08 27	0N	
11	10 28	Max dec	18° N 28'
18	18 37	0S	
25	05 23	Max dec	18° S 33'
31	16 17	0N	

LONGITUDES

Date	Chiron ⚷ ° '	Ceres ⚳ ° '	Pallas ⚴ ° '	Juno ⚵ ° '	Vesta ⚶ ° '	Black Moon Lilith ⚸ ° '
01	28 ♒ 04	15 ♒ 02	12 ♑ 25	19 ♐ 18	12 ♑ 15	16 ♋ 22
11	27 ♒ 46	15 ♒ 06	14 ♑ 02	21 ♐ 53	15 ♑ 24	17 ♋ 29
21	27 ♒ 33	15 ♒ 45	16 ♑ 00	24 ♐ 43	18 ♑ 53	18 ♋ 36
31	27 ♒ 26	16 ♒ 57	18 ♑ 18	27 ♐ 48	22 ♑ 40	19 ♋ 43

All ephemeris data is given at 12.00 UT and the Moon's longitude is additionally given for 24.00 UT

ASPECTARIAN

h m	Aspects
01 Saturday	
00 34	☿ ∥ ♆
01 55	☽ ⊥ ♄
02 38	☽ ∥ ♀
05 12	☽ bQ ♂
05 49	☽ ∥ ☿
06 25	♀ △ ♂
06 45	☽ ∥ ♆
10 47	☽ bQ ☉
12 24	☽ ☍ ♅
16 08	☽ ⚹ ♃
17 11	☽ ∠ ♄
17 43	☿ Q ♄
22 46	☽ △ ☿
02 Sunday	
03 12	☽ ± ☉
07 18	♀ Q ♅
07 31	☽ △ ♂
09 00	☽ △ ♀
10 08	☽ ☍ ♇
11 05	☉ ∠ ♅
11 30	☽ △ ♆
12 44	☽ Q ♃
14 12	☽ ⊼ ☉
18 47	☽ ⚹ ♄
22 22	♀ ⚹ ♇
03 Monday	
03 09	☽ bQ ☿
10 09	☽ ∥ ☉
10 33	♀ # ♅
13 13	☽ bQ ♀
13 20	☽ bQ ♆
14 27	♀ ☌ ♆
15 29	☽ Q ♄
15 53	☽ ⊼ ♅
17 59	☿ # ♅
19 59	☽ □ ♃
20 22	☽ ± ☿
04 Tuesday	
02 20	☽ ± ♅
05 00	☽ ± ♆
06 23	☽ ± ♀
07 57	☽ ⊼ ☿
13 05	☽ □ ♂
14 08	☽ ⊼ ♇
15 33	☽ ⊼ ♆
17 58	☽ ⊼ ♀
18 10	☽ bQ ♅
21 11	☿ ∥ ♀
22 16	☽ ☍ ☉
23 03	☽ □ ♄
05 Wednesday	
00 44	☽ ± ♇
09 20	☉ □ ♄
11 44	☽ # ☉
16 50	☽ bQ ♇
21 01	☽ △ ♅
22 14	☽ Q ♂
22 58	♂ ⚹ ♇
06 Thursday	
01 37	☽ △ ♃
09 12	☉ ⊥ ♇
13 17	☿ Q ♅
19 33	☽ ☍ ☿
20 12	☽ △ ♇
20 54	☽ ⚹ ♂
21 43	☽ ☍ ♆
07 Friday	
01 55	♀ ∠ ♃
02 16	☿ ⚹ ♇
03 47	♀ ⚹ ♄
05 26	☽ bQ ♃
05 35	☽ △ ♄
05 46	☽ ☍ ♀
09 08	☽ ⊼ ☉
14 51	☽ # ♆
15 41	☿ △ ♂
18 18	☿ ☌ ♆
21 21	☽ ± ☉
22 30	☽ ± ♃
23 46	♂ △ ♆
08 Saturday	
01 57	☽ ∠ ♂
03 18	☽ ∥ ♅
04 47	☽ □ ♅
09 58	☽ bQ ♄
10 03	☽ ⊼ ♃
14 22	☉ Q ♃
15 58	☽ bQ ☉
19 47	☽ ⊥ ♂
09 Sunday	
03 22	☽ ± ♄
05 11	☽ □ ♇
06 49	☽ ⊼ ♆
07 07	☽ # ♀
07 50	☽ ⊻ ♂
10 46	☽ ⊼ ☿
15 07	☽ ⊼ ♄
15 43	☽ Q ♅
16 22	☽ # ☿
18 35	☽ ± ♆
21 32	☽ ⊼ ♀
23 46	☽ △ ☉
23 51	☽ ± ☿
10 Monday	
10 45	☽ ± ♀
12 27	☽ bQ ♆
15 33	☽ ⚹ ♅
16 46	☽ Q ♇
19 41	☽ bQ ☿
21 31	☽ ☍ ♃
11 Tuesday	
06 48	☽ bQ ♀
09 13	☿ ⚹ ♄
16 54	☽ ⚹ ♇
17 13	☿ ∠ ♃
18 38	☽ △ ♆
21 40	☽ ☌ ♂
21 47	☽ ∠ ♅
12 Wednesday	
01 53	♂ ∠ ♅
02 19	☽ # ☿
03 15	☽ ☍ ♄
05 06	☽ △ ☿
15 00	☽ # ♀
16 04	☽ ⊥ ♅
16 35	☽ △ ♀
17 25	☽ □ ☉
23 03	♀ Q ♇
23 19	☽ ∠ ♇
13 Thursday	
04 15	☽ ⊻ ♅
10 50	☽ ⊼ ♃
17 34	☽ ⊥ ♇
23 06	☽ ± ♃
14 Friday	
05 41	☽ ⊻ ♇
07 27	☽ □ ♆
07 45	☉ ⊻ ♀
09 05	☽ Q ☉
12 21	☽ ⊻ ♂
16 03	☽ ⊼ ♄
17 24	☽ bQ ♃
20 24	☽ ∥ ♅
23 38	☽ □ ☿
15 Saturday	
00 44	☽ ⊥ ♂
04 03	☽ ± ♄
07 48	☽ # ♆
09 29	♀ ⊥ ♃
11 08	☽ ⚹ ☉
11 43	☽ □ ♀
16 33	☽ ☌ ♅
17 47	☉ ∠ ♇
19 06	☽ ∠ ♂
19 21	☽ Q ♆
21 53	☽ bQ ♄
23 28	☽ △ ♃
16 Sunday	
11 24	☽ # ☉
14 05	☽ Q ☿
17 00	☽ ☌ ♇
18 44	☽ ⚹ ♆
18 57	☽ ∠ ☉
17 Monday	
01 04	☽ ⚹ ♂
02 31	☽ Q ♀
02 59	☽ △ ♄
06 26	☿ # ♇
13 11	♀ □ ♅
13 32	☽ ⊥ ☉
14 46	☽ ⚹ ☿
23 13	☽ ∠ ♆
18 Tuesday	
00 21	☽ Q ♂
01 42	☽ ⊻ ☉
02 10	☽ ⊻ ♅
03 29	☽ ⚹ ♀
06 59	☿ Q ♇
07 54	☉ ⚹ ♅
09 12	☽ □ ♃
13 16	☽ ⊥ ♅
15 53	☽ ⊥ ♆
20 31	☽ ∠ ☿
19 Wednesday	
01 08	☽ ⊻ ♇
01 56	♀ # ♇
02 49	☽ ⊻ ♆
05 41	☽ ∠ ♅
09 38	☽ ∠ ♀
10 09	☽ □ ♂
10 38	☽ □ ♄
11 57	☽ ⊥ ♇
13 47	☽ ⊥ ☿
17 13	♀ bQ ♂
18 10	☽ Q ♃
21 03	♀ ∠ ♄
20 Thursday	
01 05	☽ ⊻ ☿
03 13	☽ ⊥ ♀
03 57	☽ ∠ ♇
08 24	☽ ⚹ ♅
08 39	♂ ☍ ♄
12 02	☽ ☌ ☉
14 45	☽ ⊻ ♀
15 31	☽ ⚹ ♃
18 29	☽ Q ♄
21 Friday	
00 31	♀ ⊻ ♃
05 16	☽ Q ♅
06 04	☽ ⚹ ♇
07 43	☽ ☌ ♆
10 08	☽ ∥ ☉
15 13	☽ ⚹ ♄
15 51	☽ △ ♂
17 38	☽ ∠ ♃
23 28	☽ ∥ ♆
22 Saturday	
02 35	☽ Q ♇
07 24	☽ ☌ ☿
07 52	☽ # ♅
09 05	☽ ⊥ ♃
11 59	☽ □ ♅
16 45	☽ ∠ ♄
17 51	☽ bQ ♂
19 14	☽ ⊻ ☉
19 19	☽ ⊻ ♃
20 50	☉ ⚹ ♃
22 48	☽ ☌ ♀
23 Sunday	
06 03	☽ ⊥ ☉
06 14	☿ ∥ ♀
07 55	☽ ⊥ ♄
08 56	☽ □ ♇
09 19	☽ ± ♂
10 36	☽ ⊻ ♆
18 00	☽ ⊻ ♄
19 35	☽ ⊼ ♂
20 39	☽ ⊥ ♆
22 15	☽ ∠ ☉
24 Monday	
10 25	☉ Q ♄
11 28	☽ ⊻ ☿
11 44	☽ ∠ ♆
14 23	☽ △ ♅
21 45	☽ ⊥ ☿
22 07	☽ ☌ ♃
25 Tuesday	
01 11	☽ ⚹ ☉
05 40	☽ ⊻ ♀
11 12	☽ △ ♇
12 54	☽ ⚹ ♆
13 08	☽ ∠ ☿
15 33	☽ bQ ♅
16 39	☽ ⊥ ♀
20 25	☽ ☌ ♄
22 49	☽ Q ☉
22 53	☽ ☍ ♂
26 Wednesday	
06 43	☽ ± ♅
09 08	☽ Q ♆
09 13	☽ ∠ ♀
12 28	☽ bQ ♇
14 44	☽ ⚹ ☿
16 52	☽ ⊼ ♅
27 Thursday	
01 13	☽ ⊻ ♃
03 44	☽ ± ♇
04 26	☉ ∥ ♆
06 09	♀ ∥ ♄
07 34	☽ □ ☉
11 12	☽ Q ☿
11 33	☽ ⊥ ♃
13 03	☽ ⚹ ♀
13 58	☽ ⊼ ♇
14 01	☿ St R
15 00	♀ ⊥ ♄
15 46	☽ □ ♆
23 31	☽ ⊻ ♄
23 43	♀ □ ♇
28 Friday	
02 51	☽ ⊼ ♂
03 09	☽ ∠ ♃
03 50	☽ # ♅
08 27	☽ ∥ ☉
09 52	☽ ⊥ ♄
11 16	☽ ∥ ♆
11 35	☽ Q ♀
13 23	☽ ± ♂
17 55	☽ □ ☿
20 17	☽ ☍ ♅
21 10	♀ ⊻ ♆
29 Saturday	
01 30	☽ ∠ ♄
05 16	☽ bQ ♂
05 22	☽ ⚹ ♃
13 18	♀ ± ♂
15 14	☽ △ ☉
17 50	☽ ☍ ♇
19 44	☽ △ ♆
21 52	☽ □ ♀
30 Sunday	
02 33	☉ # ♅
02 35	☽ Q ♃
03 48	☽ ⚹ ♄
07 59	☽ △ ♂
18 47	☉ Q ♅
19 40	☽ bQ ☉
21 06	♀ ∥ ♃
21 10	☽ △ ☿
22 12	☽ bQ ♆
31 Monday	
00 56	☽ ⊼ ♅
01 05	☽ Q ♄
03 39	☉ ⚹ ♇
10 51	☽ □ ♃
11 38	☽ ± ♅
12 57	☽ ± ☉
14 16	☽ ± ♆
15 36	♀ # ♂
22 49	☽ bQ ☿
23 01	☽ ⊼ ♇

NOVEMBER 1960

LONGITUDES

Date	Sidereal time h m s	Sun ☉ ° ' "	Moon ☽ ° ' "	Moon ☽ 24.00 ° ' "	Mercury ☿ ° '	Venus ♀ ° '	Mars ♂ ° '	Jupiter ♃ ° '	Saturn ♄ ° '	Uranus ♅ ° '	Neptune ♆ ° '	Pluto ♇ ° '
01	14 43 06	09 ♏ 04 08	14 ♈ 56 32	21 ♈ 33 18	22 ♏ 14	13 ♐ 09	16 ♋ 18	01 ♑ 07	13 ♑ 33	25 ♌ 24	08 ♏ 53	07 ♍ 45
02	14 47 03	10 04 12	28 ♈ 07 08	04 ♉ 37 55	21 R 25	14 22	16 31	01 18	13 37	25 25	08 55	07 46
03	14 50 59	11 04 17	11 ♉ 05 29	17 ♉ 29 46	20 26	15 35	16 44	01 29	13 41	25 27	08 57	07 47
04	14 54 56	12 04 24	23 ♉ 50 40	00 ♊ 08 10	19 20	16 48	16 57	01 41	13 46	25 29	09 00	07 49
05	14 58 52	13 04 32	06 ♊ 22 16	12 ♊ 33 03	18 08	18 01	17 09	01 52	13 50	25 30	09 02	07 49
06	15 02 49	14 04 43	18 ♊ 40 38	24 ♊ 45 15	16 51	19 14	17 20	02 03	13 55	25 31	09 04	07 50
07	15 06 45	15 04 56	00 ♋ 47 10	06 ♋ 46 42	15 32	20 27	17 30	02 14	14 00	25 33	09 06	07 51
08	15 10 42	16 05 11	12 ♋ 44 15	18 ♋ 40 17	14 13	21 40	17 40	02 26	14 04	25 34	09 09	07 52
09	15 14 38	17 05 27	24 ♋ 35 19	00 ♌ 29 54	12 57	22 53	17 49	02 38	14 09	25 35	09 11	07 53
10	15 18 35	18 05 46	06 ♌ 24 38	12 ♌ 20 10	11 46	24 06	17 57	02 49	14 14	25 36	09 13	07 54
11	15 22 32	19 06 06	18 ♌ 17 09	24 ♌ 16 17	10 42	25 18	18 05	03 01	14 19	25 37	09 15	07 55
12	15 26 28	20 06 29	00 ♍ 18 13	06 ♍ 23 40	09 48	26 31	18 12	03 13	14 24	25 39	09 17	07 56
13	15 30 25	21 06 53	12 ♍ 33 16	18 ♍ 47 39	09 04	27 44	18 18	03 25	14 29	25 40	09 20	07 57
14	15 34 21	22 07 20	25 ♍ 07 23	01 ♎ 32 59	08 31	28 ♐ 57	18 23	03 37	14 34	25 41	09 22	07 58
15	15 38 18	23 07 48	08 ♎ 04 49	14 ♎ 43 12	08 10	00 ♑ 09	18 28	03 49	14 40	25 41	09 24	07 58
16	15 42 14	24 08 18	21 ♎ 28 17	28 ♎ 20 02	08 01	01 22	18 32	04 01	14 45	25 42	09 26	07 59
17	15 46 11	25 08 49	05 ♏ 18 17	12 ♏ 22 40	08 D 03	02 34	18 35	04 13	14 50	25 43	09 28	08 00
18	15 50 07	26 09 23	19 ♏ 32 38	26 ♏ 47 28	08 16	03 47	18 37	04 25	14 56	25 44	09 31	08 01
19	15 54 04	27 09 58	04 ♐ 06 18	11 ♐ 28 07	08 39	05 00	18 39	04 37	15 01	25 45	09 33	08 01
20	15 58 01	28 10 35	18 ♐ 51 53	26 ♐ 16 30	09 10	06 12	18 39	04 50	15 07	25 45	09 35	08 02
21	16 01 57	29 ♏ 11 13	03 ♑ 40 54	11 ♑ 04 05	09 50	07 24	18 R 39	05 02	15 12	25 46	09 37	08 02
22	16 05 54	00 ♐ 11 52	18 ♑ 25 08	25 ♑ 43 20	10 37	08 37	18 38	05 15	15 18	25 46	09 39	08 03
23	16 09 50	01 12 32	02 ♒ 58 03	10 ♒ 08 51	11 30	09 49	18 36	05 27	15 24	25 47	09 41	08 04
24	16 13 47	02 13 14	17 ♒ 15 27	24 ♒ 17 41	12 28	11 02	18 33	05 40	15 29	25 47	09 43	08 04
25	16 17 43	03 13 56	01 ♓ 15 32	08 ♓ 09 04	13 32	12 14	18 30	05 53	15 35	25 47	09 46	08 05
26	16 21 40	04 14 40	14 ♓ 58 25	21 ♓ 43 47	14 40	13 26	18 25	06 06	15 41	25 48	09 48	08 05
27	16 25 36	05 15 24	28 ♓ 25 22	05 ♈ 03 26	15 51	14 38	18 20	06 18	15 47	25 48	09 50	08 05
28	16 29 33	06 16 09	11 ♈ 38 10	18 ♈ 09 47	17 06	15 50	18 14	06 31	15 53	25 48	09 52	08 06
29	16 33 30	07 16 56	24 ♈ 38 28	01 ♉ 04 21	18 23	17 02	18 07	06 44	15 59	25 48	09 54	08 06
30	16 37 26	08 ♐ 17 43	07 ♉ 27 34	13 ♉ 48 10	19 ♏ 42	18 ♑ 14	17 ♋ 59	06 ♑ 57	16 ♑ 05	25 ♌ 48	09 ♏ 56	08 ♍ 06

DECLINATIONS

Date	Moon True ☊ ° '	Moon Mean ☊ ° '	Moon ☽ Latitude ° '	Sun ☉ ° '	Moon ☽ ° '	Mercury ☿ ° '	Venus ♀ ° '	Mars ♂ ° '	Jupiter ♃ ° '	Saturn ♄ ° '	Uranus ♅ ° '	Neptune ♆ ° '	Pluto ♇ ° '
01	14 ♍ 16	12 ♍ 33	02 S 36	14 S 31	03 N 29	20 S 18	23 S 46	23 N 36	23 S 26	22 S 34	13 N 43	12 S 50	20 N 09
02	14 R 09	12 30	03 32	14 50	07 30	19 51	23 57	23 37	23 26	22 33	13 43	12 51	20 09
03	14 00	12 27	04 16	15 09	11 06	19 24	24 08	23 37	23 26	22 33	13 42	12 51	20 09
04	13 50	12 23	04 46	15 28	14 07	18 44	24 18	23 38	23 26	22 32	13 42	12 52	20 09
05	13 39	12 20	05 01	15 46	16 25	18 06	24 27	23 39	23 26	22 32	13 42	12 53	20 09
06	13 28	12 17	05 03	16 04	17 56	17 25	24 36	23 40	23 26	22 32	13 41	12 54	20 09
07	13 18	12 14	04 50	16 22	18 36	16 42	24 44	23 41	23 26	22 31	13 41	12 54	20 09
08	13 10	12 11	04 25	16 39	18 26	15 59	24 51	23 43	23 26	22 31	13 40	12 55	20 09
09	13 05	12 07	03 48	16 56	17 28	15 17	24 57	23 44	23 25	22 30	13 40	12 56	20 09
10	13 02	12 04	03 02	17 13	15 44	14 37	25 03	23 45	23 25	22 30	13 40	12 56	20 09
11	13 02	12 01	02 08	17 30	13 19	14 00	25 08	23 47	23 25	22 30	13 39	12 57	20 09
12	13 D 02	11 58	01 07	17 46	10 19	13 27	25 13	23 49	23 25	22 29	13 39	12 58	20 09
13	13 02	11 55	00 S 03	18 02	06 49	13 00	25 16	23 50	23 25	22 29	13 39	12 58	20 09
14	13 R 02	11 52	01 N 04	18 18	02 N 55	12 37	25 19	23 52	23 25	22 28	13 38	12 59	20 09
15	12 59	11 48	02 09	18 33	01 S 14	12 20	25 21	23 54	23 24	22 28	13 38	13 00	20 10
16	12 54	11 45	03 09	18 49	05 27	12 09	25 22	23 57	23 24	22 27	13 38	13 01	20 10
17	12 46	11 42	04 00	19 03	09 31	12 03	25 23	23 59	23 23	22 27	13 38	13 01	20 10
18	12 36	11 39	04 37	19 18	13 10	12 02	25 23	24 01	23 23	22 26	13 37	13 02	20 10
19	12 24	11 36	04 58	19 32	16 05	12 06	25 22	24 04	23 23	22 26	13 37	13 03	20 10
20	12 12	11 33	04 59	19 45	18 00	12 14	25 20	24 07	23 23	22 25	13 37	13 03	20 10
21	12 02	11 29	04 40	19 59	18 43	12 26	25 18	24 10	23 22	22 25	13 37	13 04	20 10
22	11 54	11 26	04 03	20 12	18 10	12 41	25 15	24 13	23 22	22 24	13 37	13 04	20 11
23	11 49	11 23	03 10	20 24	16 25	12 59	25 11	24 16	23 22	22 24	13 37	13 05	20 11
24	11 47	11 20	02 06	20 36	13 40	13 19	25 06	24 19	23 21	22 23	13 37	13 06	20 11
25	11 D 46	11 17	00 N 55	20 48	10 10	13 42	25 01	24 22	23 21	22 23	13 37	13 06	20 11
26	11 R 46	11 13	00 S 17	21 00	06 11	14 05	24 55	24 26	23 20	22 22	13 36	13 07	20 12
27	11 46	11 10	01 27	21 11	01 S 57	14 30	24 48	24 30	23 20	22 21	13 36	13 08	20 12
28	11 43	11 07	02 31	21 21	02 N 17	14 56	24 41	24 33	23 19	22 21	13 36	13 08	20 12
29	11 37	11 04	03 26	21 32	06 21	15 23	24 33	24 37	23 19	22 20	13 36	13 09	20 12
30	11 ♍ 29	11 ♍ 01	04 S 09	21 S 42	10 N 04	15 S 50	24 S 24	24 N 41	23 S 18	22 S 19	13 N 36	13 S 10	20 N 13

ZODIAC SIGN ENTRIES

Date	h	m	Planets
02	15	27	☽ ♉
04	23	44	☽ ♊
07	10	26	☽ ♋
09	22	59	☽ ♌
12	11	24	☽ ♍
14	21	07	☽ ♎
15	08	57	♀ ♑
17	02	53	☽ ♏
19	05	17	☽ ♐
21	06	02	☽ ♑
22	07	18	☉ ♐
23	07	04	☽ ♒
25	09	49	☽ ♓
27	14	51	☽ ♈
29	22	00	☽ ♉

LATITUDES

Date	Mercury ☿ ° '	Venus ♀ ° '	Mars ♂ ° '	Jupiter ♃ ° '	Saturn ♄ ° '	Uranus ♅ ° '	Neptune ♆ ° '	Pluto ♇ ° '
01	02 S 02	01 S 24	01 N 10	00 00	00 N 12	00 N 43	01 N 43	12 N 26
04	01 13	01 31	01 17	00 00	00 11	00 43	01 43	12 27
07	00 S 13	01 38	01 24	00 00	00 11	00 43	01 43	12 28
10	00 N 47	01 45	01 32	00 S 01	00 11	00 43	01 43	12 29
13	01 37	01 51	01 40	00 01	00 11	00 43	01 43	12 31
16	02 09	01 56	01 48	00 01	00 10	00 43	01 43	12 32
19	02 24	02 01	01 56	00 01	00 10	00 43	01 43	12 33
22	02 26	02 05	02 05	00 02	00 10	00 43	01 43	12 35
25	02 19	02 09	02 14	00 02	00 10	00 44	01 43	12 36
28	02 05	02 12	02 23	00 02	00 09	00 44	01 43	12 37
31	01 N 47	02 S 14	02 N 31	00 S 03	00 N 09	00 N 44	01 N 43	12 N 39

DATA

Julian Date	2437240
Delta T	+34 seconds
Ayanamsa	23° 18' 29"
Synetic vernal point	05° ♓ 48' 30"
True obliquity of ecliptic	23° 26' 31"

LONGITUDES

Date	Chiron ⚷ ° '	Ceres ⚳ ° '	Pallas ⚴ ° '	Juno ⚵ ° '	Vesta ⚶ ° '	Black Moon Lilith ⚸ ° '
01	27 ♒ 26	17 ♒ 05	18 ♑ 32	28 ♐ 07	23 ♑ 04	19 ♋ 50
11	27 ♒ 25	18 ♒ 48	21 ♑ 07	01 ♑ 24	27 ♑ 06	20 ♋ 57
21	27 ♒ 31	20 ♒ 55	23 ♑ 53	04 ♑ 51	01 ♒ 28	22 ♋ 04
31	27 ♒ 43	23 ♒ 23	26 ♑ 50	08 ♑ 27	05 ♒ 44	23 ♋ 11

MOON'S PHASES, APSIDES AND POSITIONS ☽

Date	h	m	Phase	Longitude ° '	Eclipse Indicator
03	11	58	○	11 ♉ 04	
11	13	47	☾	19 ♌ 11	
18	23	46	●	26 ♏ 39	
25	15	42	☽	03 ♓ 23	

Day	h	m		
09	09	22	Apogee	
21	04	07	Perigee	
07	19	07	Max dec	18° N 39'
15	04	58	0S	
21	13	28	Max dec	18° S 44'
27	23	00	0N	

ASPECTARIAN

01 Tuesday
00 33 ☽ ⊼ ☉
01 03 ☽ ⊼ ♆
03 47 ☽ bQ ♅
07 18 ☉ ☌ ♆
08 27 ☽ △ ♀
09 28 ☽ □ ♄
09 50 ☽ ± ♇
14 13 ☽ ± ☿
14 29 ☽ □ ♂
19 54 ☿ # ♇
20 10 ♀ ⊻ ♄

02 Wednesday
00 29 ☽ ⊼ ☿
02 12 ☽ bQ ♇
07 03 ☽ △ ♅
14 32 ☽ bQ ♀
17 57 ☽ △ ♃
23 08 ♀ ⊥ ♆

03 Thursday
00 00 ☽ Q ♂
05 50 ☽ △ ♇
08 01 ☽ ☍ ♆
08 54 ☽ ± ♀
11 58 ☽ ☍ ☉
16 53 ☽ △ ♄
21 18 ☽ ⊼ ♀
22 15 ☽ bQ ♃
22 45 ☽ ⚹ ♂

04 Friday
01 24 ☽ # ♆
02 21 ☿ Q ♇
04 10 ☽ ☍ ☿
08 21 ☽ ∥ ♅
15 06 ☽ □ ♅
15 22 ♀ ⊼ ♂
15 32 ☽ ± ♃
21 26 ☽ bQ ♄

05 Saturday
03 03 ☽ # ☉
03 11 ☽ ⊼ ♃
03 43 ☽ ∠ ♂
13 07 ☿ ⊻ ♀
14 49 ☽ □ ♇
14 51 ☽ ± ♄
17 10 ☽ ⊼ ♆
21 24 ☽ ⊥ ♂

06 Sunday
01 53 ☽ Q ♅
02 11 ☽ ⊼ ☉
02 36 ☽ ⊼ ♄
04 19 ☿ △ ♂
04 54 ☽ ± ♆
05 23 ☽ # ☿
07 46 ☉ ⚹ ♄
08 46 ☽ ⊼ ☿
08 49 ☿ ∠ ♃
09 19 ☽ ⊻ ♂
13 13 ☽ ☍ ♀
15 00 ☽ ± ☉
19 26 ☽ ± ☿
22 40 ☽ bQ ♆

07 Monday
01 33 ☽ ⚹ ♅
02 10 ☽ Q ♇
10 28 ☽ bQ ☉
11 33 ☽ bQ ☿
14 57 ☽ ☍ ♃
16 40 ☉ ☌ ☿
20 02 ☉ ∥ ☿
22 15 ☿ ⊥ ♀

08 Tuesday
02 11 ☽ ⚹ ♇
04 44 ☽ △ ♆
07 37 ☽ ∠ ♅
14 32 ☿ ⚹ ♄
14 42 ☽ △ ☿
14 43 ☽ ☍ ♄
19 24 ☽ △ ☉
22 06 ☽ ☌ ♂

09 Wednesday
01 50 ☽ ⊥ ♅
08 08 ☽ ⊼ ♀
08 33 ☽ ∠ ♇
14 02 ☽ ⊻ ♅
19 05 ☽ # ☉
21 43 ☽ ± ♀

10 Thursday
02 50 ☽ ⊥ ♇
03 48 ☉ ∠ ♃
04 35 ☽ ⊼ ♃
08 08 ☉ △ ♂
12 21 ☉ ⊥ ♀
14 31 ♀ ∠ ♆
15 02 ☽ ⊻ ♇
16 58 ☽ ± ♃
17 42 ☽ □ ♆
18 03 ☽ bQ ♀
21 55 ☽ □ ☿

11 Friday
03 57 ☽ ⊼ ♄
03 57 ☽ # ☿
09 00 ☽ ∥ ♅
11 27 ☽ bQ ♃
11 35 ☽ ⊻ ♂
13 47 ☽ □ ☉
15 12 ☽ # ♆
16 07 ☽ ± ♄
16 21 ☿ ∠ ♀
18 24 ♀ △ ♅
23 44 ☽ ⊥ ♂

12 Saturday
02 43 ☽ ☌ ♅
02 59 ☿ # ♅
03 39 ☽ △ ♀
06 00 ☽ Q ♆
07 21 ☽ Q ☿
07 45 ☉ Q ♇
10 12 ☽ bQ ♄
17 46 ☽ ∠ ♂
17 50 ☽ △ ♃

13 Sunday
03 02 ☽ ☌ ♇
03 06 ☿ ☌ ♆
04 43 ☽ Q ☉
05 34 ☽ ⚹ ☿
05 43 ☽ ⚹ ♆
13 01 ☿ ∥ ♆
15 45 ☽ △ ♄
23 08 ☽ ⚹ ♂

14 Monday
05 51 ☽ ⚹ ☉
09 05 ☽ ∠ ☿
10 34 ☽ ∠ ♆
13 02 ☽ ⊻ ♅
19 54 ☽ □ ♀
21 55 ☽ Q ♂

15 Tuesday
00 15 ☽ ⊥ ♅
01 24 ☽ ⊥ ☿
03 24 ☽ ⊥ ♆
04 03 ☽ □ ♃
11 48 ☽ ⊻ ♇
12 06 ☽ ∠ ☉
12 10 ☽ ⊻ ☿
14 25 ☽ ⊻ ♆
16 45 ☽ ∠ ♅
22 40 ☽ ⊥ ♇
23 58 ☽ □ ♄

16 Wednesday
05 37 ☽ ⊥ ☉
06 47 ☽ □ ♂
07 55 ☽ Q ♀
12 58 ☽ Q ♃
14 40 ☽ ∠ ♇
17 04 ☽ ⊻ ☉
19 26 ☽ ⚹ ♅
19 27 ☿ St D

17 Thursday
06 53 ☽ ⚹ ♀
07 45 ☽ Q ♄
10 06 ☽ ⚹ ♃
16 07 ☽ Q ♅
16 36 ☽ ⚹ ♇
16 44 ☽ ☌ ☿
19 07 ☽ ☌ ♆

18 Friday
01 45 ☉ □ ♅
04 08 ☽ ∥ ☿
04 14 ☽ ⚹ ♄
10 27 ☽ △ ♂
10 37 ☽ ∠ ♀
11 02 ☽ ∥ ♆
11 47 ☽ ∠ ♃
12 46 ☽ Q ♇
15 21 ☽ # ♅
22 16 ☽ □ ♅
23 46 ☽ ☌ ☉

19 Saturday
02 52 ☽ ⊥ ♀
02 53 ☽ ⊥ ♃
03 11 ♀ ☌ ♃
05 16 ☽ ∠ ♄
11 15 ☽ bQ ♂
12 52 ☽ ⊻ ♃
13 35 ☽ ⊻ ♀
18 24 ☽ □ ♇
19 39 ☽ ⊻ ☿
20 04 ☽ ⊥ ♄
20 54 ☽ ⊻ ♆

20 Sunday
01 55 ☽ ± ♂
05 46 ☽ ⊥ ☿
05 52 ☽ ⊻ ♄
06 40 ☽ ⊥ ♆
11 39 ☽ ⊼ ♂
17 03 ♂ St R
20 58 ☽ ∠ ☿
21 17 ☽ ∠ ♆
23 10 ☽ △ ♅

21 Monday
04 11 ☽ ⊻ ☉
04 25 ☿ ☌ ♆
07 34 ☉ ⊥ ♃
14 14 ☽ ☌ ♃
14 37 ☽ ⊥ ☉
18 35 ☽ ☌ ♀
19 05 ☽ △ ♇
21 40 ☽ ⚹ ♆
22 31 ☽ ⚹ ☿
23 31 ☽ bQ ♅

22 Tuesday
00 41 ♀ △ ♇
06 21 ☽ ∠ ☉
06 52 ☽ ☌ ♄
10 05 ☉ # ♇
12 21 ☽ ☍ ♂
14 13 ☽ ± ♅
14 36 ☉ ∠ ♄
17 19 ☽ Q ♆
19 18 ☽ Q ☿
19 36 ☽ bQ ♇

23 Wednesday
00 05 ☽ ⊼ ♅
08 51 ☽ ⚹ ☉
09 18 ♀ ⚹ ♆
10 29 ☽ ± ♇
16 13 ☽ ⊻ ♃
19 36 ☿ ∥ ♆
20 30 ☽ ⊼ ♇
23 16 ☽ □ ♆

24 Thursday
00 30 ☽ ⊻ ♀
02 25 ☽ ⊥ ♃
03 18 ☽ □ ☿
06 28 ☽ Q ☉
07 10 ♀ bQ ♅
08 59 ☽ ⊻ ♄
11 34 ☽ ⊥ ♀
12 25 ☽ # ♅
14 12 ☽ ⊼ ♂
14 18 ☽ ∥ ☿
16 12 ☽ ∥ ♆
17 53 ☽ ∠ ♃
19 15 ☽ ⊥ ♄

25 Friday
00 24 ☽ ± ♂
02 34 ☽ ☍ ♅
04 23 ☽ ∠ ♀
06 41 ☿ # ♅
10 50 ☽ ∠ ♄
15 42 ☽ □ ☉
15 51 ☽ bQ ♂
17 47 ☉ bQ ♂
20 09 ☽ ⚹ ♃
23 53 ☽ ☍ ♇

26 Saturday
02 51 ☽ △ ♆
09 01 ☽ ⚹ ♀
11 24 ☽ △ ☿
13 16 ☽ ⚹ ♄
17 37 ☽ Q ♃
18 04 ☽ △ ♂

27 Sunday
05 32 ☽ bQ ♆
07 17 ☽ ⊼ ♅
08 28 ☽ Q ♀
10 28 ☿ ⚹ ♄
10 50 ☽ Q ♄
16 50 ☽ bQ ☿
18 06 ☽ ± ♅
21 48 ☽ ± ♆

28 Monday
01 23 ☽ △ ☉
02 30 ☽ □ ♃
05 32 ☽ ⊼ ♇
08 45 ☽ ⊼ ♆
10 28 ☽ bQ ♅
10 54 ☽ ± ☿
12 57 ♀ ☌ ♄
16 31 ☽ ± ♇
19 34 ☉ ⊻ ♃
19 51 ☽ □ ♄
20 30 ☽ □ ♀
23 07 ☽ ⊼ ☿

29 Tuesday
00 01 ☽ □ ♂
03 05 ♀ # ♂
07 15 ☽ bQ ☉
07 25 ☿ △ ♂
09 08 ☽ bQ ♇
14 10 ☽ △ ♅

30 Wednesday
01 27 ☽ ± ☉
07 20 ♀ ☍ ♂
07 33 ☉ □ ♇
09 14 ☽ Q ♂
11 02 ☽ △ ♃
13 13 ☽ △ ♇
13 43 ☽ ⊼ ☉
16 41 ☽ ☍ ♆
19 12 ☿ Q ♇

All ephemeris data is given at 12.00 UT and the Moon's longitude is additionally given for 24.00 UT

DECEMBER 1960

LONGITUDES

Date	Sidereal time h m s	Sun ☉ ° ' "	Moon ☽ ° ' "	Moon ☽ 24.00 ° ' "	Mercury ☿ ° '	Venus ♀ ° '	Mars ♂ ° '	Jupiter ♃ ° '	Saturn ♄ ° '	Uranus ♅ ° '	Neptune ♆ ° '	Pluto ♇ ° '
01	16 41 23	09 ♐ 18 32	20 ♉ 06 14	26 ♉ 21 47	21 ♏ 04	19 ♑ 26	17 ♋ 50	07 ♑ 10	16 ♑ 11	25 ♌ 48	09 ♏ 58	08 ♍ 07
02	16 45 19	10 19 22	02 ♊ 34 52	08 ♊ 45 29	22 27	20 38	17 R 40	07 23	16 17	25 R 48	10 00	08 07
03	16 49 16	11 20 13	14 ♊ 53 41	20 ♊ 59 30	23 51	21 50	17 30	07 36	16 23	25 48	10 02	08 07
04	16 53 12	12 21 05	27 ♊ 03 04	03 ♋ 04 27	25 17	23 01	17 19	07 50	16 30	25 48	10 04	08 07
05	16 57 09	13 21 58	09 ♋ 03 51	15 ♋ 01 29	26 44	24 13	17 07	08 03	16 36	25 48	10 06	08 08
06	17 01 05	14 22 52	20 ♋ 57 36	26 ♋ 52 33	28 11	25 25	16 54	08 16	16 42	25 48	10 08	08 08
07	17 05 02	15 23 47	02 ♌ 46 42	08 ♌ 40 30	29 ♏ 40	26 36	16 40	08 29	16 49	25 47	10 10	08 08
08	17 08 59	16 24 44	14 ♌ 34 27	20 ♌ 29 04	01 ♐ 09	27 48	16 25	08 43	16 55	25 47	10 12	08 08
09	17 12 55	17 25 41	26 ♌ 24 59	02 ♍ 22 47	02 38	28 ♑ 59	16 08	08 56	17 02	25 46	10 14	08 08
10	17 16 52	18 26 40	08 ♍ 23 10	14 ♍ 26 48	04 08	00 ♒ 10	15 54	09 10	17 08	25 46	10 15	08 R 08
11	17 20 48	19 27 40	20 ♍ 34 22	26 ♍ 46 35	05 38	01 21	15 37	09 23	17 15	25 45	10 17	08 08
12	17 24 45	20 28 41	03 ♎ 04 05	09 ♎ 27 31	07 09	02 32	15 20	09 37	17 21	25 45	10 19	08 08
13	17 28 41	21 29 43	15 ♎ 57 27	22 ♎ 34 21	08 40	03 43	15 01	09 50	17 28	25 44	10 21	08 08
14	17 32 38	22 30 46	29 ♎ 18 33	06 ♏ 10 17	10 12	04 54	14 43	10 04	17 35	25 43	10 23	08 08
15	17 36 34	23 31 50	13 ♏ 09 33	20 ♏ 16 12	11 43	06 05	14 23	10 17	17 41	25 43	10 24	08 08
16	17 40 31	24 32 56	27 ♏ 29 48	04 ♐ 49 46	13 15	07 16	14 03	10 31	17 48	25 42	10 26	08 07
17	17 44 28	25 34 02	12 ♐ 15 13	19 ♐ 45 08	14 47	08 27	13 42	10 45	17 55	25 41	10 28	08 07
18	17 48 24	26 35 08	27 ♐ 18 19	04 ♑ 53 25	16 19	09 37	13 21	10 58	18 01	25 40	10 30	08 07
19	17 52 21	27 36 15	12 ♑ 29 06	20 ♑ 04 01	17 52	10 48	12 59	11 12	18 08	25 39	10 31	08 07
20	17 56 17	28 37 23	27 ♑ 36 53	05 ♒ 06 35	19 24	11 58	12 37	11 26	18 15	25 38	10 33	08 06
21	18 00 14	29 ♐ 38 31	12 ♒ 32 09	19 ♒ 52 49	20 57	13 08	12 15	11 40	18 22	25 37	10 35	08 06
22	18 04 10	00 ♑ 39 39	27 ♒ 08 01	04 ♓ 17 24	22 30	14 18	11 52	11 53	18 29	25 36	10 36	08 06
23	18 08 07	01 40 48	11 ♓ 20 46	18 ♓ 18 06	24 03	15 28	11 29	12 07	18 36	25 35	10 38	08 05
24	18 12 03	02 41 56	25 ♓ 09 32	01 ♈ 55 17	25 37	16 38	11 06	12 21	18 43	25 33	10 39	08 05
25	18 16 00	03 43 04	08 ♈ 35 38	15 ♈ 10 57	27 10	17 48	10 42	12 35	18 49	25 32	10 41	08 04
26	18 19 57	04 44 13	21 ♈ 41 36	28 ♈ 07 59	28 ♐ 44	18 57	10 18	12 49	18 56	25 31	10 42	08 04
27	18 23 53	05 45 21	04 ♉ 30 29	10 ♉ 49 28	00 ♑ 18	20 07	09 54	13 03	19 03	25 29	10 44	08 03
28	18 27 50	06 46 30	17 ♉ 05 17	23 ♉ 18 13	01 53	21 16	09 31	13 17	19 10	25 28	10 45	08 03
29	18 31 46	07 47 38	29 ♉ 28 35	05 ♊ 36 36	03 27	22 25	09 07	13 31	19 17	25 27	10 47	08 02
30	18 35 43	08 48 47	11 ♊ 42 29	17 ♊ 46 27	05 02	23 34	08 43	13 45	19 24	25 25	10 48	08 02
31	18 39 39	09 ♑ 49 55	23 ♊ 48 39	29 ♊ 49 14	06 ♑ 38	24 ♒ 43	08 ♋ 19	13 ♑ 58	19 ♑ 31	25 ♌ 24	10 ♏ 50	08 ♍ 01

Date	Moon True ☊ ° '	Moon Mean ☊ ° '	Moon ☽ Latitude ° '
01	11 ♍ 17	10 ♍ 58	04 S 40
02	11 R 04	10 54	04 56
03	10 49	10 51	04 59
04	10 34	10 48	04 47
05	10 21	10 45	04 23
06	10 11	10 42	03 48
07	10 03	10 39	03 03
08	09 58	10 35	02 10
09	09 56	10 32	01 11
10	09 D 55	10 29	00 S 08
11	09 R 55	10 26	00 N 56
12	09 55	10 23	01 59
13	09 52	10 19	02 59
14	09 48	10 16	03 50
15	09 40	10 13	04 30
16	09 31	10 10	04 55
17	09 20	10 07	05 01
18	09 08	10 04	04 47
19	08 58	10 00	04 12
20	08 50	09 57	03 20
21	08 45	09 54	02 14
22	08 42	09 51	01 N 01
23	08 D 42	09 48	00 S 14
24	08 42	09 45	01 27
25	08 R 42	09 41	02 32
26	08 41	09 38	03 28
27	08 37	09 35	04 12
28	08 31	09 32	04 43
29	08 22	09 29	05 00
30	08 11	09 25	05 03
31	07 ♍ 59	09 ♍ 22	04 S 53

DECLINATIONS

Date	Sun ☉ ° '	Moon ☽ ° '	Mercury ☿ ° '	Venus ♀ ° '	Mars ♂ ° '	Jupiter ♃ ° '	Saturn ♄ ° '	Uranus ♅ ° '	Neptune ♆ ° '	Pluto ♇ ° '
01	21 S 51	13 N 17	16 S 18	24 S 14	24 N 45	23 S 17	22 S 19	13 N 36	13 S 10	20 N 13
02	22 00	15 50	16 46	24 04	24 49	23 17	22 18	13 37	13 11	20 13
03	22 09	17 38	17 13	23 53	24 54	23 16	22 17	13 37	13 11	20 14
04	22 17	18 37	17 41	23 41	24 58	23 15	22 17	13 37	13 12	20 14
05	22 24	18 45	18 08	23 29	25 03	23 15	22 16	13 37	13 13	20 14
06	22 32	18 03	18 35	23 16	25 07	23 14	22 15	13 37	13 13	20 15
07	22 39	16 34	19 01	23 02	25 12	23 13	22 15	13 37	13 14	20 15
08	22 45	14 23	19 27	22 48	25 16	23 13	22 14	13 37	13 14	20 15
09	22 51	11 36	19 52	22 33	25 21	23 12	22 13	13 37	13 15	20 16
10	22 56	08 18	20 16	22 18	25 26	23 11	22 12	13 38	13 15	20 16
11	23 01	04 36	20 40	22 02	25 31	23 10	22 12	13 38	13 16	20 17
12	23 06	00 N 36	21 02	21 45	25 35	23 09	22 11	13 38	13 16	20 17
13	23 10	03 S 32	21 24	21 28	25 40	23 08	22 10	13 38	13 17	20 18
14	23 14	07 38	21 45	21 10	25 45	23 07	22 09	13 39	13 17	20 18
15	23 17	11 29	22 05	20 52	25 50	23 06	22 08	13 39	13 18	20 18
16	23 20	14 49	22 24	20 33	25 54	23 05	22 08	13 39	13 18	20 19
17	23 22	17 17	22 42	20 13	25 59	23 04	22 07	13 40	13 19	20 19
18	23 24	18 38	22 59	19 53	26 04	23 03	22 06	13 40	13 19	20 20
19	23 25	18 40	23 14	19 33	26 08	23 02	22 05	13 40	13 20	20 20
20	23 26	17 22	23 29	19 12	26 12	23 01	22 04	13 41	13 20	20 21
21	23 26	14 54	23 42	18 50	26 17	23 00	22 03	13 41	13 21	20 21
22	23 26	11 31	23 54	18 28	26 21	22 59	22 03	13 42	13 21	20 22
23	23 26	07 32	24 05	18 06	26 25	22 58	22 02	13 42	13 22	20 22
24	23 25	03 S 15	24 15	17 43	26 29	22 57	22 01	13 42	13 22	20 23
25	23 23	01 N 05	24 24	17 20	26 33	22 56	22 00	13 43	13 23	20 24
26	23 21	05 14	24 31	16 56	26 36	22 54	21 59	13 43	13 23	20 24
27	23 19	09 04	24 37	16 32	26 40	22 53	21 58	13 44	13 23	20 25
28	23 16	12 25	24 41	16 08	26 43	22 52	21 57	13 44	13 24	20 25
29	23 13	15 09	24 45	15 43	26 46	22 51	21 56	13 45	13 24	20 26
30	23 09	17 11	24 46	15 18	26 49	22 49	21 55	13 45	13 25	20 27
31	23 S 05	18 N 26	24 S 47	14 S 52	26 N 52	22 S 48	21 S 55	13 N 46	13 S 25	20 N 27

ZODIAC SIGN ENTRIES

Date	h	m	Planets
02	07	01	☽ ♊
04	17	52	☽ ♋
07	06	21	☽ ♌
07	17	30	☿ ♐
09	19	13	☽ ♍
10	08	34	♀ ♒
12	06	10	☽ ♎
14	13	13	☽ ♏
16	16	07	☽ ♐
18	16	16	☽ ♑
20	15	49	☽ ♒
21	20	26	☉ ♑
22	16	47	☽ ♓
24	20	34	☽ ♈
27	03	30	☽ ♉
27	07	21	☿ ♑
29	13	01	☽ ♊

LATITUDES

Date	Mercury ☿ ° '	Venus ♀ ° '	Mars ♂ ° '	Jupiter ♃ ° '	Saturn ♄ ° '	Uranus ♅ ° '	Neptune ♆ ° '	Pluto ♇ ° '
01	01 N 47	02 S 14	02 N 31	00 S 03	00 N 09	00 N 44	01 N 43	12 N 39
04	01 27	02 15	02 40	00 03	00 09	00 44	01 43	12 40
07	01 05	02 15	02 49	00 03	00 09	00 44	01 44	12 41
10	00 43	02 14	02 57	00 03	00 08	00 44	01 44	12 43
13	00 N 21	02 12	03 06	00 04	00 08	00 44	01 44	12 44
16	00 00	02 10	03 13	00 04	00 08	00 44	01 44	12 46
19	00 S 21	02 06	03 21	00 04	00 08	00 45	01 44	12 47
22	00 41	02 01	03 27	00 05	00 07	00 45	01 44	12 48
25	00 59	01 56	03 33	00 05	00 07	00 45	01 44	12 49
28	01 16	01 49	03 38	00 05	00 07	00 45	01 44	12 51
31	01 S 31	01 S 41	03 N 42	00 S 05	00 N 07	00 N 45	01 N 45	12 N 52

DATA

Julian Date	2437270
Delta T	+34 seconds
Ayanamsa	23° 18' 33"
Synetic vernal point	05° ♓ 48' 26"
True obliquity of ecliptic	23° 26' 31"

MOON'S PHASES, APSIDES AND POSITIONS ☽

Date	h	m	Phase	Longitude ° '	Eclipse Indicator
03	04	24	○	11 ♊ 01	
11	09	38	☾	19 ♍ 22	
18	10	47	●	26 ♐ 32	
25	02	30	☽	03 ♈ 19	

Day	h	m		
07	03	16	Apogee	
19	10	25	Perigee	
05	03	44	Max dec	18° N 48'
12	15	33	0S	
19	00	38	Max dec	18° S 49'
25	05	59	0N	

LONGITUDES

Date	Chiron ⚷ ° '	Ceres ⚳ ° '	Pallas ⚴ ° '	Juno ⚵ ° '	Vesta ⚶ ° '	Black Moon Lilith ⚸ ° '
01	27 ♒ 43	23 ♒ 23	26 ♑ 50	08 ♑ 27	05 ♒ 44	23 ♋ 11
11	28 ♒ 00	26 ♒ 09	29 ♑ 56	12 ♑ 10	10 ♒ 15	24 ♋ 18
21	28 ♒ 23	29 ♒ 10	03 ♒ 08	15 ♑ 59	14 ♒ 52	25 ♋ 25
31	28 ♒ 51	02 ♓ 24	06 ♒ 25	19 ♑ 53	19 ♒ 34	26 ♋ 31

All ephemeris data is given at 12.00 UT and the Moon's longitude is additionally given for 24.00 UT

ASPECTARIAN

h m	Aspects
01 Thursday	
04 20	♅ St R
04 28	☽ △ ♄
07 43	☽ ⚹ ♂
10 35	☽ △ ♀
11 07	☽ # ♆
14 04	☽ ☍ ☿
14 47	☽ ∥ ♅
16 01	☽ ⚼ ♃
19 24	♀ ± ♅
22 56	☽ □ ♅
02 Friday	
04 05	☉ ⊻ ♆
09 29	☽ ⚼ ♄
09 39	☽ ± ♃
10 47	☿ ∠ ♃
11 05	☉ ⊥ ♄
12 10	☽ ∠ ♂
18 33	☽ ⚼ ♀
21 30	☽ ⊼ ♃
22 45	☽ □ ♇
03 Saturday	
02 28	☽ ⊼ ♆
02 54	☽ # ☿
03 07	☽ ± ♄
04 24	☽ ☍ ☉
05 27	☽ ⊥ ♂
09 52	☽ Q ♅
14 02	☽ ± ♀
14 14	☽ ± ♆
14 58	☽ ⊼ ♄
15 14	☉ ± ♂
16 11	♀ Q ♆
17 02	☽ ⊻ ♂
04 Sunday	
03 09	☽ ⊼ ♀
08 01	☽ ⊼ ☿
08 03	☽ ⚼ ♆
09 31	☽ ⚹ ♅
10 10	☽ Q ♇
11 51	☉ ∥ ♄
14 00	♀ ⚼ ♇
20 35	☿ □ ♅
21 34	☽ ± ☿
05 Monday	
09 55	☽ ☍ ♃
10 07	☽ ⚹ ♇
14 05	☽ △ ♆
15 29	☽ ∠ ♅
18 06	☽ ⚼ ☿
20 47	♃ △ ♇
21 28	☽ ⊼ ☉
06 Tuesday	
02 55	☽ # ☿
03 19	☽ ☍ ♄
03 55	☽ ☌ ♂
09 38	☽ ⊥ ♅
10 43	☽ ± ☉
15 49	♀ ∥ ♃
16 24	☽ ∠ ♇
19 39	♀ ⊼ ♅
21 48	☽ ⊻ ♅
22 02	☽ ☍ ♀
07 Wednesday	
01 31	♂ ☍ ♄
04 45	☽ △ ☿
06 42	☽ ⚼ ☉
10 41	☽ ⊥ ♇
22 54	☽ ⊻ ♇
23 51	☽ ⊼ ♃
08 Thursday	
03 04	☽ □ ♆
06 42	☉ ⊥ ♆
12 12	☉ ⊼ ♂
12 17	☽ ± ♃
15 41	☽ ⊻ ♂
15 51	☿ ⚼ ♂
15 53	☉ ∥ ♀
16 05	☽ △ ☉
16 49	☽ ⊼ ♄
19 07	☽ ∥ ♅
22 29	☽ # ♆
09 Friday	
01 25	☉ ⊻ ♄
01 28	☿ ∠ ♄
03 36	☽ ⊥ ♂
05 05	☽ ± ♄
06 54	☽ ⚼ ♃
10 42	☽ ☌ ♅
15 40	☽ Q ♆
17 40	☿ ⊥ ♃
17 45	☽ ⊼ ♀
21 22	☽ ∠ ♂
23 24	☽ ⚼ ♄
10 Saturday	
02 18	☽ □ ☿
04 39	♇ St R
07 06	☽ ± ♀
11 30	☽ ☌ ♇
12 12	☿ # ♇
13 34	☽ △ ♃
15 44	☽ ⚹ ♆
20 56	♀ ∥ ♄
11 Sunday	
02 32	☽ ⚹ ♂
02 52	☽ ⚼ ♀
05 26	☽ △ ♄
09 38	☽ □ ☉
18 47	☽ Q ☿
21 10	☽ ∠ ♆
22 02	☽ ⊻ ♅
12 Monday	
01 19	☽ Q ♂
03 43	♀ ± ♇
09 30	☽ ⊥ ♅
10 54	☽ △ ♀
14 22	☽ ⊥ ♆
20 44	☽ ⚹ ☿
21 31	☽ ⊻ ♇
23 03	☽ Q ☉
13 Tuesday	
00 30	☽ □ ♃
01 38	☽ ⊻ ♆
02 23	☽ ∠ ♅
03 09	☉ ∥ ♃
03 28	☿ □ ♇
08 39	☽ ⊥ ♇
10 20	☽ □ ♂
14 23	☿ ∥ ♀
14 47	☽ □ ♄
16 38	☿ ± ♂
22 54	☽ ⚹ ☉
14 Wednesday	
01 00	☽ ∠ ♇
03 46	☽ ∠ ☿
05 39	☽ ⚹ ♅
09 35	☿ ⊻ ♃
09 46	☽ Q ♃
14 59	☿ ⊻ ♆
21 38	☽ ⊥ ☿
22 43	☽ □ ♀
23 03	☽ Q ♄
15 Thursday	
02 40	☽ Q ♅
03 23	☽ ⚹ ♇
03 27	☽ ∠ ☉
07 01	☽ ⚹ ♃
07 17	☽ ☌ ♆
09 15	☽ ⊻ ☿
11 27	☿ ⊥ ♄
14 02	☽ △ ♂
16 12	☿ ∥ ♄
19 43	☽ ⚹ ♄
19 58	☽ ⊥ ☉
23 45	☽ Q ♇
16 Friday	
00 29	☽ ∥ ♆
02 23	♃ ⚹ ♆
03 02	☽ # ♅
06 46	☽ ⊻ ☉
07 59	☽ Q ♀
08 41	☽ ∠ ♃
09 02	☽ □ ♅
14 30	☽ ⚼ ♂
20 46	☽ ∠ ♄
22 15	☿ ⊼ ♂
23 40	☽ ⊥ ♃
17 Saturday	
04 46	♀ # ♇
04 50	☽ ± ♂
05 20	☽ ⚹ ♀
05 20	☽ □ ♇
05 25	♀ ⊼ ♇
05 52	♃ ⚼ ♅
09 07	☽ ⊻ ♆
09 32	☽ ⊻ ♃
09 33	☉ ∠ ♆
11 27	☽ ⊥ ♄
14 17	☽ ⊼ ♂
14 41	☉ △ ♅
16 31	☽ ☌ ☿
18 46	☽ ⊥ ♆
21 08	☽ ⊻ ♄
18 Sunday	
07 23	☽ ∠ ♀
09 07	☽ ∠ ♆
09 24	☽ △ ♅
10 47	☽ ☌ ☉
14 47	☿ ⊥ ♆
18 53	☿ ∥ ♃
22 50	☽ ⊥ ♀
19 Monday	
05 05	☽ △ ♇
06 21	♀ □ ♆
08 54	☽ ⚹ ♆
09 06	☽ ⊻ ♀
09 06	☽ ⚼ ♅
09 57	☽ ☌ ♃
12 47	☽ ☍ ♂
16 39	☿ ⊻ ♄
21 01	☽ ☌ ♄
21 28	☽ ⊻ ☿
22 26	♀ ⊻ ♃
23 20	☽ ± ♅
20 Tuesday	
03 56	☽ Q ♆
04 50	☽ ⚼ ♇
07 06	☉ ∥ ☿
08 04	☽ ⊥ ☿
08 51	☽ ⊼ ♅
13 44	☽ ⊻ ☉
19 11	☽ ± ♇
22 15	♀ ⊼ ♂
21 Wednesday	
00 02	☽ ⊥ ☉
00 07	☽ ∠ ☿
04 49	☽ ⊼ ♇
08 49	☽ □ ♆
10 33	☽ ⊻ ♃
11 33	☽ ⊼ ♂
13 03	☽ ☌ ♀
15 41	☽ ∠ ☉
20 30	☽ ⊥ ♃
21 05	☽ ± ♂
21 13	☽ # ♅
21 35	☽ ⊻ ♄
23 37	☽ ∥ ♆
22 Thursday	
03 24	☽ ⚹ ☿
07 33	☽ ⊥ ♄
09 27	☽ ☍ ♅
11 05	♂ ☍ ♃
11 34	☽ ⚼ ♂
11 35	☽ ∠ ♃
18 21	☽ ⚹ ☉
22 43	☽ ∠ ♄
23 Friday	
01 52	☽ Q ☿
06 26	☽ ☍ ♇
10 46	☽ △ ♆
12 14	☽ △ ♂
13 21	☽ ⚹ ♃
16 19	☽ Q ☉
19 44	☽ ⊻ ♀
24 Saturday	
00 37	☽ ⚹ ♄
07 09	☽ ⊥ ♀
10 33	☽ Q ♃
11 12	☿ △ ♅
12 42	☽ ⊼ ♅
12 43	☿ ∠ ♆
12 53	☽ ⚼ ♆
12 54	☽ □ ☿
19 07	♀ ± ♂
21 55	☽ Q ♄
23 20	☽ ± ♅
25 Sunday	
00 34	☽ ∠ ♀
02 30	☽ □ ☉
04 56	☽ ± ♆
11 03	☽ ⊼ ♇
13 03	♂ △ ♆
15 31	☽ ⚼ ♅
15 43	☽ □ ♂
15 48	☽ ⊼ ♆
19 23	☽ □ ♃
21 58	☽ ± ♇
22 34	♂ ∠ ♅
26 Monday	
06 26	☽ ⚹ ♀
06 52	☽ □ ♄
08 25	♀ ⊥ ♃
11 42	♀ ⊻ ♄
14 32	☽ ⚼ ♇
19 05	☽ △ ♅
23 57	☽ Q ♂
27 Tuesday	
02 58	☽ △ ☿
07 01	☽ Q ♀
14 34	☽ △ ☉
18 43	☽ △ ♇
21 56	☽ ⚹ ♂
23 51	☽ ☍ ♆
28 Wednesday	
04 33	☽ △ ♃
11 32	☽ ⚼ ☿
16 03	☽ △ ♄
20 03	☽ # ♆
20 53	☽ □ ♀
21 51	☽ ⚼ ☉
22 58	☽ ∥ ♅
29 Thursday	
01 54	☽ ∠ ♂
04 10	☽ □ ♅
07 29	☽ ± ☿
10 05	☽ ⚼ ♃
16 56	☽ # ♀
16 56	☽ ± ☉
17 39	☉ △ ♇
18 53	☽ ⊥ ♂
20 56	☽ ⊼ ☿
21 30	☽ ⚼ ♄
30 Friday	
04 02	☽ ± ♃
04 45	☽ □ ♇
05 46	☽ ⊼ ☉
06 17	☽ ⊻ ♂
10 13	☽ ⊼ ♆
10 21	☉ ☍ ♂
13 50	♄ ± ♅
14 20	♀ ⚼ ♂
15 22	☽ Q ♅
15 23	☽ ± ♄
16 06	☽ ⊼ ♃
22 06	☽ ± ♆
31 Saturday	
03 23	☽ ⊼ ♄
13 59	☽ △ ♀
15 09	☽ ⚹ ♅
16 02	☽ ⚼ ♆
16 24	☽ Q ♇

JANUARY 1961

LONGITUDES

Date	Sidereal time h m s	Sun ☉ ° ' "	Moon ☽ ° ' "	Moon ☽ 24.00 ° ' "	Mercury ☿ ° '	Venus ♀ ° '	Mars ♂ ° '	Jupiter ♃ ° '	Saturn ♄ ° '	Uranus ♅ ° '	Neptune ♆ ° '	Pluto ♇ ° '
01	18 43 36	10 ♑ 51 04	05 ♋ 48 21	11 ♋ 46 11	08 ♑ 13	25 ♒ 51	07 ♋ 56	14 ♑ 12	19 ♑ 38	25 ♌ 22	10 ♏ 51	08 ♍ 00
02	18 47 32	11 52 13	17 ♋ 42 52	23 ♋ 38 35	09 49	27 00	07 R 32	14 26	19 46	25 R 20	10 52	08 R 00
03	18 51 29	12 53 21	29 ♋ 33 32	05 ♌ 27 58	11 25	28 08	07 09	14 40	19 53	25 19	10 54	07 59
04	18 55 26	13 54 30	11 ♌ 22 10	17 ♌ 16 25	13 02	29 ♒ 16	06 46	14 54	20 00	25 17	10 55	07 58
05	18 59 22	14 55 39	23 ♌ 11 05	29 ♌ 06 35	14 39	00 ♓ 24	06 24	15 08	20 07	25 15	10 56	07 57
06	19 03 19	15 56 47	05 ♍ 03 20	11 ♍ 01 51	16 16	01 32	06 01	15 22	20 14	25 13	10 57	07 57
07	19 07 15	16 57 56	17 ♍ 02 38	23 ♍ 06 16	17 54	02 39	05 40	15 36	20 21	25 11	10 58	07 56
08	19 11 12	17 59 05	29 ♍ 13 20	05 ♎ 24 26	19 32	03 46	05 18	15 50	20 28	25 09	11 00	07 55
09	19 15 08	19 00 14	11 ♎ 40 12	18 ♎ 01 13	21 11	04 54	04 57	16 04	20 35	25 08	11 01	07 54
10	19 19 05	20 01 22	24 ♎ 28 04	01 ♏ 01 17	22 50	06 00	04 37	16 18	20 42	25 06	11 02	07 53
11	19 23 01	21 02 31	07 ♏ 41 18	14 ♏ 28 28	24 30	07 07	04 17	16 32	20 49	25 04	11 03	07 52
12	19 26 58	22 03 40	21 ♏ 23 01	28 ♏ 25 00	26 10	08 13	03 58	16 46	20 56	25 02	11 04	07 51
13	19 30 55	23 04 49	05 ♐ 34 17	12 ♐ 50 30	27 50	09 20	03 39	17 00	21 04	24 59	11 05	07 50
14	19 34 51	24 05 58	20 ♐ 13 08	27 ♐ 41 21	29 ♑ 31	10 26	03 21	17 14	21 11	24 57	11 06	07 49
15	19 38 48	25 07 06	05 ♑ 14 12	12 ♑ 50 29	01 ♒ 12	11 31	03 04	17 28	21 18	24 55	11 07	07 48
16	19 42 44	26 08 15	20 ♑ 28 56	28 ♑ 08 09	02 53	12 37	02 47	17 42	21 25	24 53	11 08	07 47
17	19 46 41	27 09 22	05 ♒ 46 44	13 ♒ 23 22	04 35	13 42	02 31	17 56	21 32	24 51	11 08	07 46
18	19 50 37	28 10 29	20 ♒ 56 47	28 ♒ 25 56	06 17	14 47	02 16	18 10	21 39	24 49	11 09	07 45
19	19 54 34	29 ♑ 11 35	05 ♓ 49 52	13 ♓ 07 55	08 00	15 51	02 02	18 23	21 46	24 46	11 10	07 44
20	19 58 30	00 ♒ 12 41	20 ♓ 19 34	27 ♓ 24 31	09 42	16 55	01 48	18 37	21 53	24 44	11 11	07 43
21	20 02 27	01 13 45	04 ♈ 22 39	11 ♈ 14 01	11 25	17 59	01 36	18 51	22 00	24 42	11 12	07 42
22	20 06 24	02 14 48	17 ♈ 58 48	24 ♈ 37 15	13 07	19 03	01 23	19 05	22 07	24 39	11 12	07 40
23	20 10 20	03 15 51	01 ♉ 09 45	07 ♉ 36 44	14 50	20 06	01 12	19 19	22 14	24 37	11 13	07 39
24	20 14 17	04 16 52	13 ♉ 58 39	20 ♉ 15 58	16 32	21 09	01 02	19 33	22 21	24 35	11 13	07 38
25	20 18 13	05 17 52	26 ♉ 29 12	02 ♊ 38 48	18 13	22 12	00 52	19 46	22 28	24 32	11 14	07 37
26	20 22 10	06 18 52	08 ♊ 45 15	14 ♊ 49 00	19 54	23 14	00 43	20 00	22 35	24 30	11 15	07 35
27	20 26 06	07 19 50	20 ♊ 50 26	26 ♊ 49 58	21 34	24 16	00 35	20 14	22 42	24 27	11 15	07 34
28	20 30 03	08 20 47	02 ♋ 47 57	08 ♋ 44 43	23 12	25 17	00 28	20 27	22 49	24 25	11 16	07 33
29	20 33 59	09 21 43	14 ♋ 40 35	20 ♋ 35 47	24 49	26 18	00 22	20 41	22 56	24 22	11 16	07 32
30	20 37 56	10 22 38	26 ♋ 30 38	02 ♌ 25 20	26 24	27 19	00 16	20 55	23 03	24 20	11 17	07 30
31	20 41 53	11 ♒ 23 31	08 ♌ 20 09	14 ♌ 15 19	27 ♒ 56	28 ♓ 19	00 ♋ 12	21 ♑ 08	23 ♑ 10	24 ♌ 17	11 ♏ 17	07 ♍ 29

Date	Moon True ☊ ° '	Moon Mean ☊ ° '	Moon ☽ Latitude ° '
01	07 ♍ 47	09 ♍ 19	04 S 29
02	07 R 37	09 16	03 54
03	07 28	09 13	03 08
04	07 22	09 10	02 15
05	07 18	09 06	01 15
06	07 17	09 03	00 S 12
07	07 D 17	09 00	00 N 52
08	07 19	08 57	01 56
09	07 20	08 54	02 55
10	07 R 20	08 51	03 47
11	07 19	08 47	04 29
12	07 15	08 44	04 58
13	07 10	08 41	05 09
14	07 03	08 38	05 01
15	06 57	08 35	04 33
16	06 50	08 31	03 45
17	06 45	08 28	02 40
18	06 43	08 25	01 25
19	06 42	08 22	00 N 05
20	06 D 42	08 19	01 S 14
21	06 44	08 16	02 25
22	06 45	08 12	03 26
23	06 46	08 09	04 15
24	06 R 45	08 06	04 49
25	06 43	08 03	05 08
26	06 40	08 00	05 13
27	06 35	07 56	05 03
28	06 30	07 53	04 41
29	06 25	07 50	04 06
30	06 21	07 47	03 21
31	06 ♍ 17	07 ♍ 44	02 S 28

DECLINATIONS

Date	Sun ☉ ° '	Moon ☽ ° '	Mercury ☿ ° '	Venus ♀ ° '	Mars ♂ ° '	Jupiter ♃ ° '	Saturn ♄ ° '	Uranus ♅ ° '	Neptune ♆ ° '	Pluto ♇ ° '
01	23 S 00	18 N 50	24 S 46	14 S 26	26 N 55	22 S 47	21 S 54	13 N 47	13 S 25	20 N 28
02	22 55	18 24	24 44	14 00	26 57	22 45	21 53	13 47	13 26	20 29
03	22 49	17 11	24 40	13 33	27 00	22 44	21 52	13 48	13 26	20 29
04	22 43	15 13	24 35	13 07	27 02	22 42	21 51	13 48	13 26	20 30
05	22 36	12 36	24 28	12 40	27 04	22 41	21 50	13 49	13 27	20 31
06	22 29	09 28	24 20	12 12	27 05	22 39	21 49	13 50	13 27	20 31
07	22 22	05 55	24 10	11 45	27 07	22 38	21 48	13 50	13 27	20 32
08	22 14	02 N 05	23 59	11 17	27 08	22 36	21 47	13 51	13 28	20 33
09	22 06	01 S 56	23 46	10 49	27 09	22 35	21 46	13 52	13 28	20 33
10	21 57	05 58	23 32	10 20	27 11	22 33	21 45	13 52	13 28	20 34
11	21 48	09 50	23 16	09 52	27 11	22 32	21 44	13 53	13 29	20 35
12	21 38	13 19	22 58	09 23	27 12	22 30	21 43	13 54	13 29	20 35
13	21 28	16 10	22 39	08 54	27 13	22 28	21 42	13 54	13 29	20 36
14	21 18	18 04	22 19	08 25	27 13	22 27	21 41	13 55	13 29	20 37
15	21 07	18 48	21 57	07 56	27 13	22 25	21 40	13 56	13 30	20 37
16	20 55	18 11	21 33	07 27	27 13	22 23	21 39	13 57	13 30	20 38
17	20 44	16 14	21 08	06 57	27 13	22 21	21 38	13 57	13 30	20 39
18	20 32	13 11	20 41	06 28	27 13	22 20	21 37	13 58	13 30	20 40
19	20 19	09 18	20 13	05 58	27 13	22 18	21 36	13 59	13 30	20 40
20	20 06	04 58	19 43	05 28	27 13	22 16	21 35	14 00	13 31	20 41
21	19 53	00 S 29	19 12	04 58	27 12	22 14	21 33	14 01	13 31	20 42
22	19 40	03 N 52	18 39	04 28	27 12	22 12	21 32	14 01	13 31	20 43
23	19 26	07 54	18 05	03 59	27 11	22 11	21 31	14 02	13 31	20 43
24	19 11	11 26	17 29	03 29	27 11	22 09	21 30	14 03	13 31	20 44
25	18 57	14 23	16 53	02 59	27 10	22 07	21 29	14 04	13 31	20 45
26	18 42	16 37	16 15	02 29	27 09	22 05	21 28	14 05	13 31	20 46
27	18 26	18 05	15 36	01 59	27 08	22 03	21 27	14 06	13 31	20 46
28	18 11	18 44	14 56	01 29	27 07	22 01	21 26	14 06	13 32	20 47
29	17 55	18 33	14 16	00 59	27 06	21 59	21 25	14 07	13 32	20 48
30	17 38	17 34	13 35	00 29	27 05	21 57	21 24	14 08	13 32	20 48
31	17 S 22	15 N 48	12 S 53	00 N 01	27 N 04	21 S 55	21 S 23	14 N 09	13 S 32	20 N 49

ZODIAC SIGN ENTRIES

Date	h m	Planets
01	00 22	☽ ♋
03	12 54	☽ ♌
05	03 31	♀ ♓
06	01 48	☽ ♍
08	13 31	☽ ♎
10	22 09	☽ ♏
13	02 40	☽ ♐
14	18 58	☿ ♒
15	03 41	☽ ♑
17	02 55	☽ ♒
19	02 32	☽ ♓
20	07 01	☉ ♒
21	04 26	☽ ♈
23	09 51	☽ ♉
25	18 50	☽ ♊
28	06 22	☽ ♋
30	19 05	☽ ♌

LATITUDES

Date	Mercury ☿ ° '	Venus ♀ ° '	Mars ♂ ° '	Jupiter ♃ ° '	Saturn ♄ ° '	Uranus ♅ ° '	Neptune ♆ ° '	Pluto ♇ ° '
01	01 S 35	01 S 38	03 N 43	00 S 05	00 N 07	00 N 45	01 N 45	12 N 52
04	01 47	01 28	03 46	00 06	00 06	00 45	01 45	12 54
07	01 56	01 18	03 48	00 06	00 06	00 45	01 45	12 55
10	02 03	01 06	03 49	00 06	00 06	00 45	01 45	12 56
13	02 06	00 54	03 49	00 07	00 06	00 46	01 45	12 57
16	02 06	00 40	03 49	00 07	00 06	00 46	01 45	12 58
19	02 01	00 25	03 48	00 07	00 05	00 46	01 46	12 59
22	01 51	00 S 09	03 46	00 08	00 05	00 46	01 46	13 00
25	01 35	00 N 08	03 43	00 08	00 05	00 46	01 46	13 01
28	01 13	00 26	03 41	00 08	00 05	00 46	01 46	13 02
31	00 S 44	00 N 45	03 N 38	00 S 08	00 N 05	00 N 46	01 N 46	13 N 02

DATA

Julian Date	2437301
Delta T	+34 seconds
Ayanamsa	23° 18' 39"
Synetic vernal point	05° ♓ 48' 21"
True obliquity of ecliptic	23° 26' 30"

MOON'S PHASES, APSIDES AND POSITIONS ☽

Date	h m	Phase	Longitude ° '	Eclipse Indicator
01	23 06	○	11 ♋ 19	
10	03 02	☾	19 ♎ 39	
16	21 30	●	26 ♑ 32	
23	16 13	☽	03 ♉ 27	
31	18 47	○	11 ♌ 41	

Day	h m		
03	12 39	Apogee	
16	22 51	Perigee	
30	12 17	Apogee	
01	11 40	Max dec	18° N 50'
09	00 31	0S	
15	13 08	Max dec	18° S 48'
21	14 37	0N	
28	18 46	Max dec	18° N 46'

LONGITUDES

Date	Chiron ⚷ ° '	Ceres ⚳ ° '	Pallas ⚴ ° '	Juno ⚵ ° '	Vesta ⚶ ° '	Black Moon Lilith ⚸ ° '
01	28 ♒ 54	02 ♓ 44	06 ♒ 44	20 ♑ 17	20 ♒ 03	26 ♋ 38
11	29 ♒ 26	06 ♓ 10	10 ♒ 05	24 ♑ 16	24 ♒ 49	27 ♋ 45
21	00 ♓ 02	09 ♓ 44	13 ♒ 28	28 ♑ 17	29 ♒ 38	28 ♋ 52
31	00 ♓ 39	13 ♓ 25	16 ♒ 52	02 ♒ 21	04 ♓ 29	29 ♋ 58

All ephemeris data is given at 12.00 UT and the Moon's longitude is additionally given for 24.00 UT

ASPECTARIAN

h m	Aspects
01 Sunday	
00 51	☉ ᵽ ♅
01 56	♀ ☍ ♅
06 59	♀ ⊥ ♄
07 06	♂ ✱ ♇
08 29	☿ ☍ ♂
08 49	☿ △ ♇
11 11	☉ ∠ ♀
11 56	☉ ✱ ♆
16 08	☽ ☌ ♂
16 25	☽ ✱ ♇
17 36	☽ ☍ ☿
21 09	☽ ∠ ♅
22 10	☽ △ ♆
23 06	☽ ☍ ☉
23 14	☽ ᵽ ♀
02 Monday	
05 15	☽ ☍ ♃
15 16	☽ ⊥ ♅
16 11	☽ ☍ ♄
19 21	☽ ± ♀
19 40	☿ ᵽ ♅
22 40	☽ ∠ ♇
23 23	♀ # ♅
03 Tuesday	
03 24	☽ ⊻ ♅
03 59	☿ ✱ ♆
08 48	☽ ⊼ ♀
16 55	☽ ⊥ ♇
18 30	♀ ∥ ♆
04 Wednesday	
02 56	☽ ⊻ ♂
05 06	☽ ⊻ ♇
11 04	☽ □ ♆
14 45	☽ ⊥ ♂
15 17	☉ ∥ ♃
15 55	☽ ⊼ ☿
17 39	☽ ⊼ ☉
19 20	☽ ⊼ ♃
05 Thursday	
01 33	☽ ∥ ♅
04 50	☽ # ♆
04 59	♀ ∠ ♃
05 42	☽ ⊼ ♄
06 02	☽ ± ☿
06 59	☽ ± ☉
07 46	☽ ± ♃
08 29	☽ ∠ ♂
11 29	☽ # ♀
16 11	☽ ☌ ♅
18 00	☽ ± ♄
18 25	☉ ☌ ♃
20 27	☿ ☌ ♃
23 06	☉ ☌ ☿
23 40	☽ Q ♆
06 Friday	
02 22	☽ ᵽ ♃
02 56	☽ ᵽ ☉
03 10	☽ ᵽ ☿
04 08	☽ ☍ ♀
12 21	☽ ᵽ ♄
13 53	☽ ✱ ♂
17 48	☽ ☌ ♇
23 52	☽ ✱ ♆
07 Saturday	
00 13	☿ ∠ ♀
09 05	☽ △ ♃
11 50	☽ △ ☉
13 11	☽ Q ♂
13 58	☽ △ ☿
18 37	☽ △ ♄
08 Sunday	
04 04	☽ ⊻ ♅
05 40	☽ ∠ ♆
06 32	☿ ± ♅
15 46	☽ ⊥ ♅
21 43	☽ ⊼ ♀
23 13	☽ ⊥ ♆
23 29	☽ □ ♂
09 Monday	
02 37	☿ ☌ ♄
04 48	☽ ⊻ ♇
09 04	☽ ∠ ♅
10 22	☽ ± ♀
10 45	☽ ⊻ ♆
13 03	♀ △ ♂
14 47	☉ ± ♅
16 14	☽ ⊥ ♇
20 29	☽ □ ♃
10 Tuesday	
03 02	☽ □ ☉
04 43	♀ ∠ ♄
04 57	☽ □ ♄
04 58	☽ ᵽ ♀
08 32	☽ □ ☿
09 04	☽ ∠ ♇
12 43	☿ ᵽ ♇
13 09	☽ ✱ ♅
14 51	☿ Q ♆
11 Wednesday	
06 03	☽ △ ♂
06 10	☉ ☌ ♄
06 15	☽ Q ♃
10 53	☽ Q ♅
10 53	☽ △ ♀
12 12	☽ ∥ ♀
12 19	☽ ✱ ♇
14 02	☽ Q ♄
14 36	☽ Q ☉
17 58	☽ ☌ ♆
20 01	☿ ⊼ ♅
21 42	☽ Q ☿
23 00	☉ ∥ ♄
12 Thursday	
03 52	☽ ✱ ♃
04 05	♀ ☍ ♇
07 55	☽ ᵽ ♂
09 22	☽ Q ♇
11 14	☽ ✱ ♄
13 14	☽ ∥ ♆
13 15	☽ ✱ ☉
16 26	☽ # ♅
18 13	☽ □ ♅
21 16	☽ ✱ ☿
23 00	☽ ± ♂
13 Friday	
05 56	☽ ∠ ♃
06 22	☉ ᵽ ♇
08 52	☽ ⊼ ♂
11 59	☉ Q ♆
12 49	☽ ∠ ♄
15 45	☽ □ ♇
16 29	☽ ∠ ☉
18 44	☽ □ ♀
21 07	☽ ⊻ ♆
21 07	☽ ⊥ ♃
14 Saturday	
01 32	☽ ∠ ☿
02 23	☿ ∥ ♃
03 45	☽ ⊥ ♄
06 56	☽ ⊥ ♆
07 05	☽ ⊻ ♃
08 19	☽ ⊥ ☉
13 34	☽ ⊻ ♄
17 59	☽ ⊥ ☿
18 42	☽ ⊻ ☉
19 36	☽ △ ♅
21 28	☽ ∠ ♆
15 Sunday	
02 13	☽ Q ♀
02 54	♀ △ ♆
04 47	☽ ⊻ ☿
07 29	☉ ⊼ ♅
08 37	☽ ☍ ♂
16 03	☽ △ ♇
19 23	☽ ᵽ ♅
20 31	☿ ± ♇
21 17	☽ ✱ ♆
22 41	☽ ✱ ♀
16 Monday	
06 10	☿ ∥ ♄
07 34	☽ ☌ ♃
09 30	☽ ± ♅
10 47	☿ ⊼ ♂
13 28	☽ ☌ ♄
15 36	☽ ᵽ ♇
16 09	☽ Q ♆
18 53	☽ ⊼ ♅
21 30	☽ ☌ ☉
17 Tuesday	
00 02	☽ ∠ ♀
05 42	☽ ± ♇
06 58	☽ ⊼ ♂
09 53	☽ ☌ ☿
15 08	☽ ⊼ ♇
15 15	☽ ⊥ ♀
16 15	☽ ± ♂
20 27	☽ □ ♆
21 20	☉ # ♇
18 Wednesday	
01 27	☽ ⊻ ♀
06 15	☽ ᵽ ♂
06 28	☽ # ♅
07 30	☽ ⊻ ♃
09 45	☽ ∥ ♆
13 08	☽ ⊻ ♄
13 14	☿ # ♇
17 13	☽ ⊥ ♃
18 10	☽ ☍ ♅
22 50	☽ ⊥ ♄
19 Thursday	
00 26	☽ ⊻ ☉
02 19	☉ ∥ ☿
05 55	☽ △ ♂
07 58	☽ ∠ ♃
08 19	☿ ⊼ ♇
10 53	☽ ⊥ ☉
12 27	☿ ± ♂
13 33	☽ ∠ ♄
15 06	☽ ☍ ♇
16 00	☽ ⊻ ☿
20 46	☽ △ ♆
20 Friday	
02 48	☽ ∠ ☉
03 14	☽ ⊥ ☿
05 51	☽ ☌ ♀
08 56	☽ ∥ ♀
09 06	☽ ✱ ♃
14 39	☽ ✱ ♄
19 25	☽ ⊼ ♅
20 25	☽ ∠ ☿
21 53	♃ ± ♅
21 55	☽ ᵽ ♆
21 Saturday	
05 39	☽ ± ♅
05 48	☽ Q ♃
06 08	☽ ✱ ☉
07 15	☽ □ ♂
08 51	☿ □ ♆
11 21	☽ Q ♄
13 25	☽ ± ♆
17 46	☽ ⊼ ♇
19 07	☉ ⊼ ♂
21 16	☽ ᵽ ♅
22 43	☉ ± ♇
23 56	☽ ⊼ ♆
22 Sunday	
02 06	☽ ✱ ☿
04 20	☽ ± ♇
04 48	☽ Q ☉
09 17	☿ ⊥ ♀
12 57	♀ ✱ ♃
14 01	☽ □ ♃
14 05	☽ ⊻ ♀
14 30	☽ Q ♂
15 05	☽ # ♀
19 32	☽ □ ♄
20 27	☽ ᵽ ♇
23 Monday	
00 02	☽ △ ♅
01 54	☽ ⊥ ♀
02 51	☽ Q ☿
12 05	☽ ✱ ♂
16 13	☽ □ ☉
19 58	☽ ∠ ♀
24 Tuesday	
00 03	☽ △ ♇
05 34	☿ ᵽ ♂
06 47	☽ ☍ ♆
15 51	☽ ∠ ♂
17 37	☽ □ ☿
22 49	☽ △ ♃
25 Wednesday	
02 57	☽ ✱ ♀
04 10	☽ △ ♄
04 25	☽ # ♆
08 14	☽ □ ♅
08 54	☽ ⊥ ♂
09 09	☽ ∥ ♅
19 11	♀ ✱ ♄
20 25	☽ ⊻ ♂
26 Thursday	
04 26	☽ Q ♀
04 28	☽ ᵽ ♃
06 45	☽ △ ☉
08 34	☽ # ☿
09 41	☽ ᵽ ♄
09 43	☽ □ ♇
12 16	♄ ᵽ ♇
13 37	☿ ⊻ ♃
16 55	☽ ⊼ ♆
19 22	☽ Q ♅
20 31	☉ ± ♂
22 35	☽ ± ♃
27 Friday	
03 41	☽ ± ♄
04 50	☽ ± ♆
10 45	☽ ⊼ ♃
13 41	☽ △ ☿
15 15	☽ ᵽ ☉
15 46	☽ ⊼ ♄
16 15	♀ ⊼ ♅
17 32	☉ ⊼ ♇
19 10	☽ # ☉
19 12	☽ ✱ ♅
19 29	☽ □ ♀
21 27	☽ Q ♇
22 51	☽ ᵽ ♆
28 Saturday	
05 53	☿ ⊻ ♄
07 21	☽ ☌ ♂
11 00	☽ ± ☉
21 34	☽ ✱ ♇
29 Sunday	
00 14	☽ ⊼ ☉
00 38	☽ ᵽ ☿
01 18	☽ ∠ ♅
05 06	☽ △ ♆
05 25	☿ ☍ ♅
11 09	♀ ᵽ ♆
16 51	☿ # ♅
19 28	☽ ⊥ ♅
21 42	☽ ± ☿
30 Monday	
00 25	☽ ☍ ♃
03 53	☽ ∠ ♇
04 55	☽ ☍ ♄
07 35	☽ ⊻ ♅
10 10	☽ # ☉
11 44	☽ ⊼ ☿
13 39	☿ ∥ ♆
13 47	☽ △ ♀
19 35	☽ ⊻ ♂
21 17	☿ ⊥ ♃
22 07	☽ ⊥ ♇
31 Tuesday	
07 41	☽ ⊥ ♂
09 23	☉ □ ♆
10 16	☽ ⊻ ♇
17 59	☽ □ ♆
18 47	☽ ☍ ☉
23 01	☽ ᵽ ♀

FEBRUARY 1961

LONGITUDES

Date	Sidereal time h m s	Sun ☉ ° ' "	Moon ☽ ° ' "	Moon ☽ 24.00 ° ' "	Mercury ☿ ° '	Venus ♀ ° '	Mars ♂ ° '	Jupiter ♃ ° '	Saturn ♄ ° '	Uranus ♅ ° '	Neptune ♆ ° '	Pluto ♇ ° '
01	20 45 49	12 ♒ 24 24	20 ♌ 11 03	26 ♌ 07 36	29 ♒ 25	29 ♓ 19	00 ♋ 08	21 ♑ 22	23 ♑ 17	24 ♌ 15	11 ♏ 17	07 ♍ 28
02	20 49 46	13 25 16	02 ♍ 05 13	08 ♍ 04 11	00 ♓ 51	00 ♈ 18	00 R 05	21 35	23 24	24 R 12	11 18	07 R 26
03	20 53 42	14 26 06	14 ♍ 04 48	20 ♍ 07 21	02 12	01 16	00 02	21 49	23 31	24 09	11 18	07 25
04	20 57 39	15 26 56	26 ♍ 12 12	02 ♎ 19 43	03 28	02 15	00 01	22 02	23 37	24 07	11 18	07 23
05	21 01 35	16 27 44	08 ♎ 30 17	14 ♎ 44 18	04 38	03 12	00 00	22 16	23 44	24 04	11 18	07 22
06	21 05 32	17 28 32	21 ♎ 02 12	27 ♎ 24 24	05 42	04 09	00 D 00	22 29	23 51	24 02	11 19	07 21
07	21 09 28	18 29 18	03 ♏ 51 19	10 ♏ 23 23	06 39	05 06	00 00	22 42	23 57	23 59	11 19	07 19
08	21 13 25	19 30 04	17 ♏ 00 56	23 ♏ 44 18	07 28	06 02	00 02	22 56	24 04	23 56	11 19	07 18
09	21 17 22	20 30 49	00 ♐ 33 43	07 ♐ 29 19	08 08	06 57	00 04	23 09	24 11	23 54	11 19	07 16
10	21 21 18	21 31 32	14 ♐ 31 10	21 ♐ 39 07	08 38	07 52	00 07	23 22	24 17	23 51	11 19	07 15
11	21 25 15	22 32 15	28 ♐ 52 56	06 ♑ 12 09	08 59	08 46	00 10	23 35	24 24	23 49	11 R 19	07 13
12	21 29 11	23 32 56	13 ♑ 36 11	21 ♑ 04 16	09 09	09 40	00 14	23 48	24 31	23 46	11 19	07 12
13	21 33 08	24 33 37	28 ♑ 35 28	06 ♒ 08 43	09 R 09	10 33	00 19	24 01	24 37	23 43	11 19	07 10
14	21 37 04	25 34 16	13 ♒ 42 55	21 ♒ 16 51	08 58	11 25	00 25	24 14	24 43	23 41	11 19	07 09
15	21 41 01	26 34 53	28 ♒ 49 20	06 ♓ 19 14	08 37	12 16	00 31	24 27	24 50	23 38	11 19	07 07
16	21 44 57	27 35 30	13 ♓ 45 29	21 ♓ 07 09	08 06	13 07	00 38	24 40	24 56	23 35	11 19	07 06
17	21 48 54	28 36 04	28 ♓ 23 28	05 ♈ 33 49	07 26	13 56	00 46	24 53	25 03	23 33	11 18	07 04
18	21 52 51	29 ♒ 36 37	12 ♈ 37 45	19 ♈ 35 00	06 38	14 45	00 54	25 06	25 09	23 30	11 18	07 03
19	21 56 47	00 ♓ 37 08	26 ♈ 25 28	03 ♉ 09 12	05 43	15 33	01 03	25 18	25 15	23 28	11 18	07 01
20	22 00 44	01 37 37	09 ♉ 46 22	16 ♉ 17 15	04 43	16 20	01 12	25 31	25 21	23 25	11 18	07 00
21	22 04 40	02 38 04	22 ♉ 42 14	29 ♉ 01 45	03 39	17 07	01 22	25 44	25 28	23 22	11 17	06 58
22	22 08 37	03 38 29	05 ♊ 16 18	11 ♊ 26 26	02 33	17 52	01 33	25 56	25 34	23 20	11 17	06 57
23	22 12 33	04 38 53	17 ♊ 32 40	23 ♊ 35 36	01 27	18 36	01 44	26 09	25 40	23 17	11 17	06 55
24	22 16 30	05 39 14	29 ♊ 35 45	05 ♋ 33 42	00 ♓ 22	19 19	01 56	26 21	25 46	23 15	11 16	06 53
25	22 20 26	06 39 34	11 ♋ 29 57	17 ♋ 25 02	29 ♒ 20	20 01	02 08	26 33	25 52	23 12	11 16	06 52
26	22 24 23	07 39 52	23 ♋ 19 26	29 ♋ 13 34	28 22	20 42	02 21	26 46	25 58	23 10	11 15	06 50
27	22 28 20	08 40 07	05 ♌ 07 53	11 ♌ 02 46	27 29	21 22	02 34	26 58	26 03	23 07	11 15	06 49
28	22 32 16	09 ♓ 40 21	16 ♌ 58 33	22 ♌ 55 35	26 ♒ 42	22 ♈ 00	02 ♋ 48	27 ♑ 10	26 ♑ 09	23 ♌ 05	11 ♏ 14	06 ♍ 47

Date	Moon True ☊ ° '	Moon Mean ☊ ° '	Moon ☽ Latitude ° '
01	06 ♍ 15	07 ♍ 41	01 S 27
02	06 R 13	07 37	00 S 23
03	06 D 14	07 34	00 N 43
04	06 15	07 31	01 48
05	06 16	07 28	02 49
06	06 18	07 25	03 43
07	06 19	07 22	04 27
08	06 20	07 18	04 59
09	06 R 19	07 15	05 15
10	06 18	07 12	05 14
11	06 17	07 09	04 53
12	06 15	07 06	04 12
13	06 14	07 02	03 14
14	06 13	06 59	02 02
15	06 12	06 56	00 N 41
16	06 D 12	06 53	00 S 42
17	06 13	06 50	02 00
18	06 13	06 47	03 09
19	06 14	06 43	04 04
20	06 14	06 40	04 45
21	06 14	06 37	05 09
22	06 R 15	06 34	05 18
23	06 15	06 31	05 12
24	06 14	06 28	04 52
25	06 D 14	06 24	04 20
26	06 15	06 21	03 37
27	06 15	06 18	02 45
28	06 ♍ 15	06 ♍ 15	01 S 45

DECLINATIONS

Date	Sun ☉ ° '	Moon ☽ ° '	Mercury ☿ ° '	Venus ♀ ° '	Mars ♂ ° '	Jupiter ♃ ° '	Saturn ♄ ° '	Uranus ♅ ° '	Neptune ♆ ° '	Pluto ♇ ° '
01	17 S 05	13 N 23	12 S 12	00 N 31	27 N 03	21 S 53	21 S 22	14 N 10	13 S 32	20 N 50
02	16 48	10 23	11 30	01 00	27 02	21 51	21 21	14 11	13 32	20 51
03	16 30	06 56	10 49	01 30	27 01	21 49	21 19	14 12	13 32	20 51
04	16 12	03 N 10	10 09	01 59	27 00	21 47	21 18	14 12	13 32	20 52
05	15 54	00 S 47	09 30	02 29	26 58	21 45	21 17	14 13	13 32	20 53
06	15 36	04 46	08 53	02 58	26 57	21 43	21 16	14 14	13 32	20 54
07	15 17	08 36	08 17	03 27	26 56	21 41	21 15	14 15	13 32	20 54
08	14 58	12 08	07 45	03 56	26 55	21 39	21 14	14 16	13 32	20 55
09	14 39	15 08	07 15	04 24	26 53	21 37	21 13	14 17	13 32	20 56
10	14 20	17 21	06 48	04 53	26 52	21 35	21 12	14 18	13 32	20 57
11	14 00	18 34	06 25	05 21	26 51	21 32	21 11	14 19	13 32	20 57
12	13 40	18 34	06 06	05 49	26 50	21 30	21 10	14 19	13 32	20 58
13	13 20	17 17	05 52	06 17	26 48	21 28	21 09	14 20	13 32	20 59
14	13 00	14 46	05 42	06 45	26 47	21 26	21 07	14 21	13 32	21 00
15	12 39	11 15	05 37	07 12	26 45	21 24	21 06	14 22	13 31	21 00
16	12 19	07 02	05 37	07 39	26 44	21 22	21 05	14 23	13 31	21 01
17	11 58	02 S 29	05 42	08 06	26 43	21 19	21 04	14 24	13 31	21 02
18	11 37	02 N 05	05 51	08 33	26 41	21 17	21 03	14 25	13 31	21 02
19	11 15	06 24	06 04	08 59	26 40	21 15	21 02	14 26	13 31	21 03
20	10 54	10 14	06 22	09 25	26 38	21 13	21 01	14 26	13 31	21 04
21	10 32	13 28	06 42	09 50	26 37	21 11	21 00	14 27	13 31	21 05
22	10 10	15 58	07 05	10 15	26 36	21 08	20 59	14 28	13 31	21 05
23	09 48	17 41	07 30	10 40	26 34	21 06	20 58	14 29	13 30	21 06
24	09 26	18 35	07 57	11 04	26 33	21 04	20 57	14 30	13 30	21 07
25	09 04	18 38	08 24	11 28	26 31	21 02	20 56	14 31	13 30	21 07
26	08 42	17 52	08 51	11 52	26 29	21 00	20 55	14 31	13 30	21 08
27	08 19	16 19	09 18	12 15	26 28	20 57	20 54	14 32	13 30	21 09
28	07 S 57	14 N 05	09 S 44	12 N 37	26 N 26	20 S 55	20 S 53	14 N 33	13 S 29	21 N 09

ZODIAC SIGN ENTRIES

Date	h	m	Planets
01	21	39	☿ ♓
02	04	46	♀ ♈
02	07	48	☽ ♍
04	19	27	☽ ♎
05	00	22	♂ ♊
07	04	51	☽ ♏
07	05	26	♂ ♋
09	11	01	☽ ♐
11	13	51	☽ ♑
13	14	14	☽ ♒
15	13	53	☽ ♓
17	14	41	☽ ♈
18	21	16	☉ ♓
19	18	21	☽ ♉
22	01	51	☽ ♊
24	12	49	☽ ♋
24	20	22	☿ ♒
27	01	34	☽ ♌

LATITUDES

Date	Mercury ☿ ° '	Venus ♀ ° '	Mars ♂ ° '	Jupiter ♃ ° '	Saturn ♄ ° '	Uranus ♅ ° '	Neptune ♆ ° '	Pluto ♇ ° '
01	00 S 33	00 N 51	03 N 37	00 S 09	00 N 04	00 N 46	01 N 46	13 N 03
04	00 N 06	01 12	03 33	00 09	00 04	00 46	01 47	13 03
07	00 50	01 33	03 29	00 09	00 04	00 46	01 47	13 04
10	01 39	01 55	03 26	00 10	00 04	00 46	01 47	13 05
13	02 27	02 18	03 22	00 10	00 04	00 46	01 47	13 05
16	03 08	02 41	03 18	00 10	00 03	00 46	01 47	13 06
19	03 36	03 06	03 14	00 11	00 03	00 46	01 48	13 06
22	03 44	03 31	03 10	00 11	00 03	00 46	01 48	13 06
25	03 32	03 56	03 06	00 11	00 03	00 46	01 48	13 07
28	03 04	04 22	03 02	00 12	00 03	00 46	01 48	13 07
31	02 N 26	04 N 48	02 N 58	00 S 12	00 N 02	00 N 46	01 N 48	13 N 07

DATA

Julian Date	2437332
Delta T	+34 seconds
Ayanamsa	23° 18' 44"
Synetic vernal point	05° ♓ 48' 16"
True obliquity of ecliptic	23° 26' 31"

MOON'S PHASES, APSIDES AND POSITIONS ☽

Date	h	m	Phase	Longitude ° '	Eclipse Indicator
08	16	49	☾	19 ♏ 42	
15	08	10	●	26 ♒ 25	Total
22	08	34	☽	03 ♊ 30	

Day	h	m		
14	11	20	Perigee	
26	21	04	Apogee	
05	07	17	0S	
12	00	08	Max dec	18° S 43'
18	00	56	0N	
25	01	33	Max dec	18° N 42'

LONGITUDES

Date	Chiron ⚷ ° '	Ceres ⚳ ° '	Pallas ⚴ ° '	Juno ⚵ ° '	Vesta ⚶ ° '	Black Moon Lilith ⚸ ° '
01	00 ♓ 43	13 ♓ 47	17 ♒ 12	02 ♒ 46	04 ♓ 58	00 ♌ 05
11	01 ♓ 23	17 ♓ 35	20 ♒ 36	06 ♒ 51	09 ♓ 50	01 ♌ 12
21	02 ♓ 03	21 ♓ 26	23 ♒ 58	10 ♒ 57	14 ♓ 42	02 ♌ 18
31	02 ♓ 44	25 ♓ 20	27 ♒ 18	15 ♒ 02	19 ♓ 33	03 ♌ 25

ASPECTARIAN

h m	Aspects
01 Wednesday	
01 49	☽ ∠ ♂
04 54	☽ ∥ ♅
06 26	☿ ⊻ ♀
09 32	☿ ⊥ ♄
10 38	☽ # ♆
13 35	♄ Q ♆
14 26	☽ ⊼ ♃
18 19	☽ ⊼ ♄
18 53	☽ ± ♀
20 10	☽ ☌ ♅
23 22	☿ △ ♂
02 Thursday	
00 59	☽ # ☿
02 46	☽ ± ♃
06 23	☽ Q ♆
06 32	☽ ± ♄
06 51	♀ □ ♂
07 58	☽ ✶ ♂
08 05	☽ ⊼ ♀
09 10	☽ ☍ ☿
09 45	♀ ± ♅
21 12	☽ bQ ♃
22 43	☽ ☌ ♇
03 Friday	
00 46	☽ bQ ♄
06 27	☽ ✶ ♆
07 56	☽ Q ♂
12 46	☽ ⊼ ☉
04 Saturday	
01 46	☽ ± ☉
01 49	☉ bQ ♂
03 38	☽ △ ♃
06 52	☽ △ ♄
07 54	☽ ⊻ ♅
12 12	☽ ∠ ♆
18 26	☽ ∥ ♀
19 27	☽ □ ♂
19 39	☽ ⊥ ♅
21 05	☽ bQ ☉
05 Sunday	
00 50	☽ ☍ ♀
03 44	☽ ⊼ ☿
05 48	☽ ⊥ ♆
09 48	☽ ⊻ ♇
13 06	☽ ∠ ♅
16 31	☽ ± ☿
17 25	☽ ⊻ ♆
21 21	☽ ⊥ ♇
22 21	♃ bQ ♇
23 37	☽ # ♀
06 Monday	
02 51	♂ St D
04 38	☽ △ ☉
11 19	☽ bQ ☿
14 28	☽ ∠ ♇
14 47	☽ □ ♃
17 21	☽ □ ♄
17 38	☽ ✶ ♅
22 47	♀ Q ♃
07 Tuesday	
04 51	☽ △ ♂
10 14	☽ ∥ ☿
14 29	☽ ⊼ ♀
15 55	☽ Q ♅
16 04	♄ ⊼ ♅
17 25	♀ ± ♆
17 31	☽ △ ☿
18 22	☽ ✶ ♇
08 Wednesday	
00 47	☽ Q ♃
01 41	☽ ☌ ♆
02 19	☽ ± ♀
02 59	☽ Q ♄
06 52	☿ ☍ ♇
08 25	☽ bQ ♂
13 02	♀ Q ♄
16 05	☽ Q ♇
16 49	☽ □ ☉
19 44	☽ bQ ♀
22 33	☽ ∥ ♆
22 44	☽ ✶ ♃
09 Thursday	
00 19	☽ □ ♅
00 33	☽ ± ♂
00 41	☽ ✶ ♄
04 35	☽ # ♅
08 13	☽ ∥ ☉
11 08	☽ ⊼ ♂
13 12	☿ ∠ ♃
19 59	♀ ⊼ ♇
23 36	☽ □ ♇
23 52	☽ △ ♀
10 Friday	
01 20	☽ ∠ ♃
01 38	☽ □ ☿
02 50	☽ Q ☉
03 01	☽ ∠ ♄
06 29	♃ Q ♆
06 33	☽ ⊻ ♆
14 28	☉ # ♅
16 44	☽ ⊥ ♆
16 53	☽ ⊥ ♃
18 25	☽ ⊥ ♄
11 Saturday	
00 41	☽ ✶ ☉
03 05	☽ ⊻ ♃
03 37	☽ △ ♅
04 31	☽ ⊻ ♄
07 46	☽ ∠ ♆
08 48	☽ Q ☿
11 33	♆ St R
12 57	♀ bQ ♅
14 08	☽ ☍ ♂
19 39	☿ ⊻ ♀
12 Sunday	
01 38	☽ △ ♇
03 13	☽ ∠ ☉
04 12	☽ bQ ♅
04 45	☽ ✶ ☿
05 13	☽ □ ♀
08 18	☽ ✶ ♆
08 29	♃ ⊼ ♅
16 55	☉ ☍ ♅
18 41	☽ ± ♅
18 49	☽ ⊥ ☉
19 42	☉ ⊻ ♃
21 19	☿ # ♀
22 19	☉ ∥ ♆
23 32	☿ St R
13 Monday	
01 47	☽ bQ ♇
03 35	☽ Q ♆
04 16	☽ ⊼ ♅
04 37	☽ ☌ ♃
04 57	☽ ∠ ☿
05 07	☽ ⊻ ☉
05 37	☽ ☌ ♄
11 55	☽ Q ♀
13 30	☉ ⊻ ♄
14 46	☽ ⊼ ♂
16 06	☽ ± ♇
19 12	☽ ⊥ ☿
21 13	☿ ∠ ♃
14 Tuesday	
00 21	☽ ± ♂
01 36	☽ ⊼ ♇
04 36	☽ ⊻ ☿
08 08	☽ ✶ ♀
08 12	☽ □ ♆
09 17	♀ ⊼ ♆
14 43	☽ bQ ♂
15 10	☽ # ♅
21 10	☽ ∥ ♆
15 Wednesday	
02 08	☽ ∥ ☉
03 46	☽ ☍ ♅
04 56	☽ ⊻ ♃
05 36	☽ ⊻ ♄
08 10	☽ ☌ ☉
09 22	☽ ∠ ♀
14 38	☽ ⊥ ♃
14 44	☽ △ ♂
15 14	☽ ⊥ ♄
16 Thursday	
00 38	☽ ⊥ ♀
01 16	☽ ☍ ♇
03 12	☽ ☌ ☿
05 17	☽ ∠ ♃
05 47	☽ ∠ ♄
08 02	☽ △ ♆
08 56	☽ # ♀
10 53	☽ ⊻ ♀
11 37	♀ ± ♇
19 29	☽ ∥ ☿
17 Friday	
04 01	☽ ⊼ ♅
04 09	☿ ⊥ ♀
06 06	☽ ✶ ♃
06 25	☽ ✶ ♄
08 33	☽ bQ ♆
12 22	☽ ⊻ ☉
13 55	☽ ± ♅
15 59	☽ □ ♂
23 10	☽ ⊥ ☉
23 34	☽ ± ♆
23 44	☿ ☍ ♇
18 Saturday	
02 22	☽ ⊻ ☿
02 26	☽ Q ♃
02 31	☽ ⊼ ♇
02 36	☽ Q ♄
04 59	☽ bQ ♅
09 44	☽ ⊼ ♆
12 00	☽ ⊥ ☿
12 43	☽ ± ♇
15 40	☽ ∠ ☉
15 52	☽ ☌ ♀
22 04	♄ # ♇
22 56	☽ Q ♂
19 Sunday	
00 02	♃ ☌ ♄
02 37	☽ ∠ ☿
04 16	☽ bQ ♇
04 49	☉ ∠ ♀
06 47	☽ △ ♅
09 55	☽ □ ♄
10 00	☽ □ ♃
10 00	☽ # ☿
20 04	☽ ✶ ☉
20 19	☽ ✶ ♂
20 Monday	
00 04	☉ △ ♂
03 28	☽ ✶ ☿
04 49	☉ ⊥ ♄
05 48	☽ ∥ ♀
06 57	☽ △ ♇
08 42	☉ ⊥ ♃
14 47	☽ ☍ ♆
16 06	☽ # ☉
19 40	☽ Q ☉
23 50	☽ Q ☿
21 Tuesday	
00 00	☽ ∠ ♂
00 52	☽ ⊻ ♀
12 24	☽ # ♆
12 49	☽ ⊥ ♀
13 15	☽ □ ♅
17 07	☽ ⊥ ♂
17 15	☽ △ ♄
17 49	☽ △ ♃
20 43	☽ ∥ ♅
23 35	☉ ☌ ☿
22 Wednesday	
04 43	☽ ⊻ ♂
07 04	☽ ∠ ♀
07 11	☽ □ ☿
07 55	☿ ∠ ♀
08 34	☽ □ ☉
09 30	☉ # ♀
15 14	☽ □ ♇
22 22	☽ bQ ♄
23 12	☽ bQ ♃
23 14	☿ ⊥ ♃
23 41	☽ ⊼ ♆
23 44	☽ Q ♅
23 Thursday	
06 37	☿ △ ♂
07 42	☿ ⊥ ♄
11 28	☽ ± ♆
14 13	☽ ✶ ♀
14 55	♃ # ♇
16 13	☽ ± ♄
17 14	☽ ± ♃
23 21	☽ ✶ ♅
24 Friday	
02 36	☽ Q ♇
04 16	☽ ⊼ ♄
05 20	☽ bQ ♆
05 23	☽ ⊼ ♃
13 25	☽ △ ☿
15 41	☽ Q ♀
16 46	☽ ☌ ♂
25 Saturday	
01 19	☽ △ ☉
02 39	☽ ✶ ♇
05 21	☽ ∠ ♅
11 31	☽ △ ♆
16 47	☉ ☍ ♇
17 18	☽ bQ ☿
17 48	♀ Q ♂
23 31	☽ ⊥ ♅
26 Sunday	
06 21	☽ □ ♀
07 25	☉ ∥ ☿
08 59	☽ ∠ ♇
10 12	☽ ± ☿
10 32	☽ bQ ☉
11 40	☽ ⊻ ♅
17 24	☽ ☍ ♄
19 06	☽ ☍ ♃
21 31	☽ ⊼ ☿
27 Monday	
03 15	☽ ⊥ ♇
06 32	☽ ± ☉
06 42	☽ ⊻ ♂
15 24	☽ ⊻ ♇
19 07	☽ ⊥ ♂
19 51	☽ ⊼ ☉
28 Tuesday	
00 21	☿ ⊻ ♃
00 23	☽ □ ♆
04 05	♀ bQ ♇
07 30	☽ ∥ ♅
13 42	☽ ∠ ♂
17 22	☽ # ♆
22 42	☽ △ ♀
23 20	☽ ∥ ♀

All ephemeris data is given at 12.00 UT and the Moon's longitude is additionally given for 24.00 UT

MARCH 1961

LONGITUDES

Date	Sidereal time h m s	Sun ☉ ° ' "	Moon ☽ ° ' "	Moon ☽ 24.00 ° ' "	Mercury ☿ ° '	Venus ♀ ° '	Mars ♂ ° '	Jupiter ♃ ° '	Saturn ♄ ° '	Uranus ♅ ° '	Neptune ♆ ° '	Pluto ♇ ° '
01	22 36 13	10 ♓ 40 33	28 ♌ 54 09	04 ♍ 54 30	26 ♒ 01	22 ♈ 37	03 ♋ 02	27 ♑ 22	26 ♑ 15	23 ♌ 02	11 ♏ 14	06 ♍ 46
02	22 40 09	11 40 43	10 ♍ 56 52	17 ♍ 01 30	25 R 27	23 13	03 17	27 34	26 21	23 R 00	11 R 13	06 R 44
03	22 44 06	12 40 51	23 ♍ 08 34	29 ♍ 18 16	25 00	23 48	03 33	27 46	26 26	22 57	11 12	06 43
04	22 48 02	13 40 58	05 ♎ 30 45	11 ♎ 46 11	24 40	24 21	03 48	27 57	26 32	22 55	11 12	06 41
05	22 51 59	14 41 02	18 ♎ 04 44	24 ♎ 26 33	24 27	24 52	04 04	28 09	26 38	22 52	11 11	06 40
06	22 55 55	15 41 05	00 ♏ 51 48	07 ♏ 20 36	24 21	25 22	04 21	28 21	26 43	22 50	11 10	06 38
07	22 59 52	16 41 06	13 ♏ 53 08	20 ♏ 29 32	24 D 21	25 50	04 38	28 32	26 48	22 48	11 09	06 36
08	23 03 49	17 41 06	27 ♏ 09 55	03 ♐ 54 25	24 28	26 17	04 55	28 44	26 54	22 45	11 09	06 35
09	23 07 45	18 41 04	10 ♐ 43 07	17 ♐ 36 05	24 40	26 42	05 13	28 55	26 59	22 43	11 08	06 33
10	23 11 42	19 41 01	24 ♐ 33 19	01 ♑ 34 46	24 58	27 05	05 32	29 07	27 04	22 41	11 07	06 32
11	23 15 38	20 40 56	08 ♑ 40 18	15 ♑ 49 43	25 21	27 27	05 50	29 18	27 09	22 38	11 06	06 30
12	23 19 35	21 40 49	23 ♑ 02 44	00 ♒ 18 56	25 49	27 46	06 09	29 29	27 15	22 36	11 05	06 29
13	23 23 31	22 40 41	07 ♒ 37 49	14 ♒ 58 46	26 21	28 03	06 29	29 40	27 20	22 34	11 04	06 27
14	23 27 28	23 40 30	22 ♒ 21 05	29 ♒ 44 00	26 58	28 19	06 48	29 ♑ 51	27 25	22 32	11 03	06 26
15	23 31 24	24 40 18	07 ♓ 06 39	14 ♓ 28 11	27 39	28 32	07 09	00 ♒ 02	27 29	22 30	11 02	06 25
16	23 35 21	25 40 04	21 ♓ 47 42	29 ♓ 04 21	28 24	28 43	07 29	00 13	27 34	22 28	11 01	06 23
17	23 39 18	26 39 49	06 ♈ 17 18	13 ♈ 25 52	29 ♒ 12	28 52	07 50	00 23	27 39	22 26	11 00	06 22
18	23 43 14	27 39 31	20 ♈ 29 24	27 ♈ 27 25	00 ♓ 04	28 59	08 11	00 34	27 44	22 24	10 59	06 20
19	23 47 11	28 39 10	04 ♉ 19 34	11 ♉ 05 37	00 59	29 04	08 33	00 44	27 48	22 22	10 58	06 19
20	23 51 07	29 ♓ 38 48	17 ♉ 45 30	24 ♉ 19 15	01 56	29 06	08 54	00 55	27 53	22 20	10 57	06 17
21	23 55 04	00 ♈ 38 24	00 ♊ 47 04	07 ♊ 09 13	02 56	29 R 05	09 17	01 05	27 57	22 18	10 56	06 16
22	23 59 00	01 37 57	13 ♊ 26 05	19 ♊ 38 06	03 59	29 02	09 39	01 15	28 02	22 16	10 55	06 15
23	00 02 57	02 37 28	25 ♊ 45 48	01 ♋ 49 44	05 05	28 57	10 02	01 25	28 06	22 14	10 54	06 13
24	00 06 53	03 36 57	07 ♋ 50 29	13 ♋ 48 42	06 13	28 49	10 25	01 35	28 10	22 12	10 52	06 12
25	00 10 50	04 36 23	19 ♋ 44 58	25 ♋ 39 56	07 23	28 39	10 48	01 45	28 15	22 11	10 51	06 11
26	00 14 47	05 35 47	01 ♌ 34 13	07 ♌ 28 24	08 35	28 26	11 12	01 54	28 19	22 09	10 50	06 09
27	00 18 43	06 35 09	13 ♌ 23 04	19 ♌ 18 46	09 49	28 11	11 35	02 04	28 23	22 07	10 49	06 08
28	00 22 40	07 34 28	25 ♌ 15 59	01 ♍ 15 14	11 05	27 53	12 00	02 13	28 27	22 06	10 47	06 07
29	00 26 36	08 33 46	07 ♍ 16 53	13 ♍ 21 20	12 23	27 33	12 24	02 23	28 30	22 04	10 46	06 05
30	00 30 33	09 33 01	19 ♍ 28 53	25 ♍ 39 48	13 43	27 10	12 49	02 32	28 34	22 02	10 45	06 04
31	00 34 29	10 ♈ 32 13	01 ♎ 54 16	08 ♎ 12 25	15 ♓ 05	26 ♈ 46	13 ♋ 13	02 ♒ 41	28 ♑ 38	22 ♌ 01	10 ♏ 43	06 ♍ 03

Date	Moon True ☊ ° '	Moon Mean ☊ ° '	Moon ☽ Latitude ° '
01	06 ♍ 15	06 ♍ 12	00 S 41
02	06 R 15	06 08	00 N 26
03	06 15	06 05	01 32
04	06 14	06 02	02 35
05	06 13	05 59	03 32
06	06 12	05 56	04 19
07	06 11	05 53	04 53
08	06 10	05 49	05 13
09	06 09	05 46	05 16
10	06 D 09	05 43	05 01
11	06 10	05 40	04 28
12	06 11	05 37	03 37
13	06 12	05 34	02 32
14	06 13	05 30	01 N 16
15	06 R 13	05 27	00 S 05
16	06 13	05 24	01 25
17	06 11	05 21	02 39
18	06 08	05 18	03 41
19	06 05	05 14	04 28
20	06 02	05 11	04 59
21	05 58	05 08	05 13
22	05 56	05 05	05 12
23	05 55	05 02	04 56
24	05 D 55	04 59	04 27
25	05 55	04 55	03 47
26	05 57	04 52	02 58
27	05 59	04 49	02 01
28	06 00	04 46	00 S 59
29	06 R 01	04 43	00 N 07
30	06 00	04 39	01 13
31	05 ♍ 57	04 ♍ 36	02 N 17

DECLINATIONS

Date	Sun ☉ ° '	Moon ☽ ° '	Mercury ☿ ° '	Venus ♀ ° '	Mars ♂ ° '	Jupiter ♃ ° '	Saturn ♄ ° '	Uranus ♅ ° '	Neptune ♆ ° '	Pluto ♇ ° '
01	07 S 34	11 N 13	10 S 09	12 N 59	26 N 25	20 S 53	20 S 52	14 N 34	13 S 29	21 N 10
02	07 11	07 52	10 32	13 21	26 23	20 51	20 51	14 35	13 29	21 10
03	06 48	04 08	10 53	13 42	26 21	20 48	20 50	14 35	13 29	21 11
04	06 25	00 N 11	11 13	14 02	26 20	20 46	20 49	14 36	13 28	21 12
05	06 02	03 S 49	11 30	14 22	26 18	20 44	20 48	14 37	13 28	21 12
06	05 39	07 44	11 46	14 41	26 16	20 42	20 47	14 38	13 28	21 13
07	05 15	11 20	11 59	15 00	26 14	20 40	20 46	14 39	13 28	21 13
08	04 52	14 27	12 10	15 18	26 12	20 37	20 45	14 39	13 27	21 14
09	04 29	16 50	12 19	15 35	26 10	20 35	20 44	14 40	13 27	21 15
10	04 05	18 19	12 25	15 51	26 08	20 33	20 43	14 41	13 27	21 15
11	03 42	18 43	12 30	16 07	26 06	20 31	20 42	14 42	13 26	21 16
12	03 18	17 54	12 33	16 22	26 04	20 29	20 41	14 42	13 26	21 16
13	02 54	15 55	12 34	16 36	26 02	20 26	20 40	14 43	13 26	21 17
14	02 31	12 52	12 32	16 49	26 00	20 24	20 39	14 44	13 25	21 17
15	02 07	08 59	12 29	17 01	25 57	20 22	20 38	14 44	13 25	21 18
16	01 43	04 S 33	12 24	17 12	25 55	20 20	20 38	14 45	13 25	21 18
17	01 20	00 N 04	12 17	17 22	25 52	20 18	20 37	14 46	13 24	21 19
18	00 56	04 36	12 09	17 31	25 50	20 16	20 36	14 46	13 24	21 19
19	00 32	08 45	11 59	17 39	25 47	20 13	20 35	14 47	13 24	21 20
20	00 S 08	12 20	11 47	17 46	25 45	20 11	20 34	14 47	13 23	21 20
21	00 N 15	15 12	11 33	17 52	25 42	20 09	20 33	14 48	13 23	21 21
22	00 39	17 15	11 18	17 57	25 39	20 07	20 33	14 49	13 22	21 21
23	01 03	18 26	11 02	18 00	25 36	20 05	20 32	14 49	13 22	21 22
24	01 26	18 46	10 44	18 02	25 33	20 03	20 31	14 50	13 22	21 22
25	01 50	18 14	10 24	18 02	25 30	20 01	20 30	14 50	13 21	21 22
26	02 13	16 55	10 03	18 01	25 27	19 59	20 29	14 51	13 21	21 23
27	02 37	14 52	09 40	17 59	25 24	19 57	20 29	14 51	13 20	21 23
28	03 00	12 11	09 17	17 55	25 21	19 55	20 28	14 52	13 20	21 23
29	03 24	08 57	08 51	17 50	25 17	19 53	20 27	14 52	13 19	21 24
30	03 47	05 17	08 25	17 43	25 14	19 51	20 27	14 53	13 19	21 24
31	04 N 10	01 N 20	07 S 57	17 N 35	25 N 10	19 S 49	20 S 26	14 N 53	13 S 19	21 N 24

ZODIAC SIGN ENTRIES

Date	h	m	Planets
01	14	12	☽ ♍
04	01	21	☽ ♎
06	10	24	☽ ♏
08	17	04	☽ ♐
10	21	19	☽ ♑
12	23	29	☽ ♒
15	00	26	☽ ♓
15	08	01	♃ ♒
17	01	32	☽ ♈
18	10	16	☿ ♓
19	04	25	☽ ♉
20	20	32	☉ ♈
21	10	32	☽ ♊
23	20	22	☽ ♋
26	08	48	☽ ♌
28	21	30	☽ ♍
31	08	21	☽ ♎

LATITUDES

Date	Mercury ☿ ° '	Venus ♀ ° '	Mars ♂ ° '	Jupiter ♃ ° '	Saturn ♄ ° '	Uranus ♅ ° '	Neptune ♆ ° '	Pluto ♇ ° '
01	02 N 52	04 N 31	03 N 00	00 S 12	00 N 03	00 N 46	01 N 48	13 N 07
04	02 13	04 57	02 56	00 12	00 02	00 46	01 48	13 07
07	01 31	05 23	02 52	00 13	00 02	00 46	01 48	13 07
10	00 50	05 49	02 49	00 13	00 02	00 46	01 49	13 07
13	00 N 11	06 14	02 45	00 13	00 02	00 46	01 49	13 07
16	00 S 24	06 37	02 41	00 14	00 01	00 46	01 49	13 07
19	00 55	06 59	02 38	00 14	00 01	00 46	01 49	13 06
22	01 21	07 18	02 34	00 14	00 01	00 46	01 49	13 06
25	01 43	07 33	02 31	00 15	00 01	00 46	01 49	13 05
28	02 01	07 44	02 27	00 15	00 01	00 46	01 49	13 05
31	02 S 15	07 N 48	02 N 24	00 S 16	00 N 00	00 N 45	01 N 49	13 N 05

DATA

Julian Date	2437360
Delta T	+34 seconds
Ayanamsa	23° 18' 47"
Synetic vernal point	05° ♓ 48' 13"
True obliquity of ecliptic	23° 26' 32"

MOON'S PHASES, APSIDES AND POSITIONS ☽

Date	h	m	Phase	Longitude ° '	Eclipse Indicator
02	13	35	○	11 ♍ 45	partial
10	02	57	☾	19 ♐ 18	
16	18	51	●	25 ♓ 57	
24	02	48	☽	03 ♋ 14	

Day	h	m		
14	18	12	Perigee	
26	14	25	Apogee	
04	13	08	0S	
11	08	01	Max dec	18° S 44'
17	11	38	0N	
24	08	59	Max dec	18° N 46'
31	19	53	0S	

LONGITUDES

Date	Chiron ⚷ ° '	Ceres ⚳ ° '	Pallas ⚴ ° '	Juno ⚵ ° '	Vesta ⚶ ° '	Black Moon Lilith ⚸ ° '
01	02 ♓ 36	24 ♓ 33	26 ♒ 38	14 ♒ 13	18 ♓ 35	03 ♌ 12
11	03 ♓ 15	28 ♓ 29	29 ♒ 55	18 ♒ 17	23 ♓ 25	04 ♌ 18
21	03 ♓ 53	02 ♈ 26	03 ♓ 07	22 ♒ 20	28 ♓ 51	05 ♌ 25
31	04 ♓ 29	06 ♈ 24	06 ♓ 14	26 ♒ 19	03 ♈ 00	06 ♌ 32

ASPECTARIAN

h m Aspects

01 Wednesday
00 15 ☽ ☌ ♅
04 27 ☿ ⊻ ♄
06 31 ☽ ☍ ☿
06 38 ☽ ⊼ ♄
08 52 ☽ ⊼ ♃
12 39 ☽ Q ♆
18 45 ☽ ± ♄
19 10 ☽ # ☿
20 27 ☽ ⚹ ♂
21 04 ☽ ± ♃

02 Thursday
01 01 ☉ △ ♆
03 12 ☉ ∠ ♄
03 18 ♀ △ ♅
03 39 ☽ ☌ ♇
06 19 ☽ bQ ♀
10 46 ♃ ∥ ♄
12 32 ☽ ⚹ ♆
12 48 ☽ bQ ♄
13 35 ☽ ☍ ☉
15 15 ☽ bQ ♃
17 04 ☽ # ☉
20 46 ☽ Q ♂
20 52 ♀ # ♆

03 Friday
01 00 ☽ ± ♀
11 38 ☽ ⊻ ♅
13 20 ☽ ⊼ ♀
14 24 ☉ ∠ ♃
15 32 ☽ ⊼ ☿
17 58 ☽ ∠ ♆
18 29 ☽ △ ♄
21 09 ☽ △ ♃
23 17 ☽ ⊥ ♅

04 Saturday
02 52 ☽ ± ☿
08 38 ☽ □ ♂
11 23 ☽ ⊥ ♆
14 15 ☽ ⊻ ♇
16 36 ☽ ∠ ♅
19 50 ☽ bQ ☿
22 03 ☿ ⚹ ♀
22 53 ☽ ⊻ ♆

05 Sunday
01 43 ☽ ⊥ ♇
05 00 ☽ ⊼ ☉
17 21 ☽ ± ☉
18 45 ☽ ∠ ♇
21 01 ☽ ⚹ ♅
23 54 ☽ △ ☿

06 Monday
00 13 ☽ ∥ ☉
01 20 ☽ ☍ ♀
04 12 ☽ □ ♄
07 14 ☽ □ ♃
07 25 ♀ ∥ ♅
11 38 ☽ bQ ☉
18 37 ☽ △ ♂
19 20 ☽ Q ♅
22 40 ☽ ⚹ ♇
23 16 ☿ St D

07 Tuesday
07 01 ☽ ☌ ♆
13 42 ☽ Q ♄
16 54 ☽ Q ♃
16 55 ☽ ∥ ☿
17 31 ☽ △ ☉
20 34 ☽ Q ♇
22 41 ☽ bQ ♂

08 Wednesday
03 50 ☽ ∥ ♆
04 06 ☽ □ ♅
07 06 ☽ □ ☿
10 22 ☽ ⊼ ♀
11 31 ☽ ⚹ ♄
13 51 ☽ # ♅
14 50 ☽ ⚹ ♃
15 13 ☽ ± ♂
20 37 ☽ # ♀
21 25 ☽ ± ♀

09 Thursday
02 07 ☽ ⊼ ♂
04 42 ☽ □ ♇
12 43 ☽ ⊻ ♆
13 46 ☽ bQ ♀
14 14 ☽ ∠ ♄
15 28 ☽ Q ☿
17 41 ☽ ∠ ♃
23 10 ☽ ⊥ ♆

10 Friday
02 57 ☽ □ ☉
05 58 ☽ ⊥ ♄
08 47 ☽ △ ♅
09 29 ☽ ⊥ ♃
10 41 ♀ □ ♄
12 43 ☽ ⚹ ☿
14 40 ☽ ∠ ♆
16 20 ☽ ⊻ ♄
16 27 ☽ △ ♀
19 54 ☽ ⊻ ♃

11 Saturday
07 07 ☽ ☍ ♂
08 21 ☽ △ ♇
10 16 ☽ bQ ♅
12 01 ☽ Q ☉
14 54 ☽ ∠ ☿
16 05 ☽ ⚹ ♆

12 Sunday
01 19 ☽ ± ♅
06 27 ☽ ⊥ ☿
09 25 ☽ bQ ♇
09 34 ☽ ⚹ ☉
11 16 ☽ ⊼ ♅
12 04 ☽ Q ♆
15 00 ☉ ⊥ ♀
16 45 ☽ ⊻ ☿
18 59 ☽ ☌ ♄
19 58 ☽ □ ♀
22 46 ☽ ☌ ♃

13 Monday
00 15 ☽ ± ♇
05 52 ☽ # ♀
09 26 ☉ ⊼ ♅
10 04 ☽ ⊼ ♂
10 05 ☽ ⊼ ♇
10 38 ♂ ⚹ ♇
12 05 ☽ ∠ ☉
17 37 ☽ □ ♆
20 06 ☽ ± ♂
22 21 ☽ # ♅

14 Tuesday
02 01 ☽ Q ♀
03 50 ☽ ⊥ ☉
08 07 ☽ ∥ ♆
11 06 ☽ bQ ♂
12 18 ☽ ☍ ♅
14 14 ☽ ∥ ☿
14 19 ☽ ⊻ ☉
19 52 ☽ ☌ ☿
20 16 ☽ ⊻ ♄
21 51 ☽ ⚹ ♀

15 Wednesday
00 20 ☽ ⊻ ♃
05 44 ☿ ⊻ ♄
06 05 ☽ ⊥ ♄
10 13 ☽ ⊥ ♃
10 52 ☽ ☍ ♇
12 03 ☽ △ ♂
18 24 ☽ △ ♆
20 49 ☽ ∠ ♄
22 37 ☽ ∠ ♀

16 Thursday
01 05 ☽ ∠ ♃
10 33 ♂ ∠ ♅
13 06 ☽ ⊼ ♅
13 33 ☽ ⊥ ♀
18 51 ☽ ☌ ☉
18 57 ☽ bQ ♆
20 23 ☉ bQ ♆
21 34 ☽ ⚹ ♄
22 58 ☽ ± ♅
23 31 ☽ ⊻ ☿
23 33 ☽ ⊻ ♀

17 Friday
00 19 ☿ ⚹ ♀
02 03 ☽ ⚹ ♃
04 06 ☽ ∥ ☉
08 56 ♃ ± ♇
09 52 ☽ ± ♆
10 05 ☽ ⊥ ☿
12 07 ☽ ⊼ ♇
13 54 ☽ bQ ♅
14 39 ☽ □ ♂
17 40 ☽ Q ♄
18 02 ☽ # ☉
19 54 ☽ ⊼ ♆
22 11 ☽ ± ♇
22 22 ☽ Q ♃

18 Saturday
02 09 ☽ ∠ ☿
13 27 ☽ bQ ♇
13 51 ☉ ⚹ ♄
15 15 ☽ △ ♅
22 03 ☽ Q ♂

19 Sunday
00 33 ☽ □ ♄
01 18 ☽ ⊻ ☉
02 45 ☽ ☌ ♀
04 24 ☿ ⊻ ♃
05 11 ☉ ± ♅
05 38 ☽ □ ♃
05 42 ☽ ⚹ ☿
12 37 ☽ ⊥ ☉
15 30 ☽ △ ♇
19 40 ☽ ⚹ ♂
22 17 ☉ ⊻ ♀
23 46 ☽ ☍ ♆

20 Monday
04 33 ☽ Q ☿
05 55 ☽ ∠ ☉
08 11 ☽ # ☿
20 02 ☽ # ♆
20 12 ♀ St R
20 19 ☽ □ ♅
23 34 ☽ ∠ ♂

21 Tuesday
06 42 ☽ △ ♄
08 10 ☽ ∥ ♅
08 50 ☽ ⊻ ♀
11 42 ☽ ⚹ ☉
12 34 ☽ △ ♃
16 24 ☽ □ ☿
16 49 ☽ ⊥ ♂
20 04 ☽ ⊥ ♀
22 18 ☽ □ ♇

22 Wednesday
00 49 ☉ ⚹ ♃
04 32 ☽ ⊻ ♂
05 56 ☽ Q ♅
07 10 ☽ ⊼ ♆
11 13 ☽ bQ ♄
12 25 ☽ Q ☉
12 56 ☿ ⊥ ♄
13 09 ☽ ∠ ♀
17 30 ☽ bQ ♃
18 42 ☽ ± ♆

23 Thursday
00 21 ☽ ∥ ♀
04 46 ☽ ± ♄
05 06 ☽ ⚹ ♅
08 58 ☽ Q ♇
11 18 ☽ ± ♃
12 15 ☽ bQ ♆
16 38 ☽ ⊼ ♄
18 14 ☽ ⚹ ♀
23 20 ☽ ⊼ ♃

24 Friday
02 48 ☽ □ ☉
08 24 ☽ △ ☿
08 43 ☽ ⚹ ♇
10 44 ☽ ∠ ♅
11 45 ☿ ☍ ♇
17 20 ☽ ☌ ♂
17 54 ☽ Q ♀
18 04 ☽ △ ♆

25 Saturday
04 47 ☽ ⊥ ♅
05 41 ♀ Q ♂
14 53 ☽ ∠ ♇
15 02 ♂ △ ♆
16 49 ☽ ∥ ♀
16 54 ☽ ⊻ ♅
17 50 ☉ ± ♆
17 55 ☽ bQ ☿
20 33 ☿ ⊥ ♃

26 Sunday
05 20 ☽ ☍ ♄
05 45 ☽ □ ♀
09 08 ☽ ⊥ ♇
12 41 ☽ ☍ ♃
14 17 ☽ ± ☿
20 56 ☽ △ ☉
21 18 ☽ ⊻ ♇
21 37 ♀ □ ♄

27 Monday
01 15 ☉ ⊼ ♇
03 54 ☽ ⊼ ☿
06 47 ☽ □ ♆
08 14 ☽ ⊻ ♂
12 09 ☽ ∥ ♅
20 49 ☽ ⊥ ♂

28 Tuesday
00 37 ☉ bQ ♅
02 21 ☽ # ♆
05 38 ☽ ☌ ♅
06 05 ☽ bQ ☉
06 33 ☿ △ ♆
15 35 ☽ ∠ ♂
17 07 ☽ △ ♀
18 25 ☽ ⊼ ♄
19 03 ☽ Q ♆

29 Wednesday
01 46 ☽ ± ☉
02 07 ☽ ⊼ ♃
06 27 ☽ ± ♄
09 38 ☽ ☌ ♇
12 22 ☿ △ ♂
12 42 ☽ # ☿
14 12 ☽ ± ♃
14 20 ☿ ∠ ♀
14 46 ☽ ⊼ ☉
18 53 ☽ ⚹ ♆
22 06 ☽ bQ ♀
22 28 ☽ ⚹ ♂
23 19 ☽ ☍ ☿

30 Thursday
00 22 ☽ bQ ♄
08 09 ☽ bQ ♃
09 16 ☿ ∠ ♄
15 12 ☽ ± ♀
16 58 ☽ ⊻ ♅
20 26 ☽ ∥ ☉
22 42 ☽ Q ♂

31 Friday
00 08 ☽ ∠ ♆
02 27 ☽ ⊼ ♀
04 33 ☽ ⊥ ♅
05 41 ☽ △ ♄
13 30 ☽ △ ♃
14 26 ☉ Q ♄
16 25 ☉ ⊼ ♆
17 22 ☽ ⊥ ♆
19 54 ☽ ⊻ ♇
21 43 ☽ ∠ ♅

All ephemeris data is given at 12.00 UT and the Moon's longitude is additionally given for 24.00 UT

APRIL 1961

LONGITUDES

Date	Sidereal time h m s	Sun ☉ ° ' "	Moon ☽ ° ' "	Moon ☽ 24.00 ° ' "	Mercury ☿ ° '	Venus ♀ ° '	Mars ♂ ° '	Jupiter ♃ ° '	Saturn ♄ ° '	Uranus ♅ ° '	Neptune ♆ ° '	Pluto ♇ ° '
01	00 38 26	11 ♈ 31 24	14 ♎ 34 20	21 ♎ 00 02	16 ♓ 28	26 ♈ 19	13 ♋ 38	02 ♒ 50	28 ♑ 41	22 ♌ 00	10 ♏ 42	06 ♍ 02
02	00 42 22	12 30 33	27 ♎ 29 29	04 ♏ 02 35	17 53	25 R 50	14 04	02 59	28 45	21 R 58	10 R 41	06 R 01
03	00 46 19	13 29 40	10 ♏ 39 14	17 ♏ 19 16	19 20	25 19	14 29	03 08	28 48	21 57	10 39	05 59
04	00 50 16	14 28 45	24 ♏ 02 30	00 ♐ 48 46	20 48	24 47	14 55	03 16	28 52	21 55	10 38	05 58
05	00 54 12	15 27 48	07 ♐ 37 51	14 ♐ 29 35	22 18	24 13	15 21	03 25	28 55	21 54	10 36	05 57
06	00 58 09	16 26 49	21 ♐ 23 46	28 ♐ 20 14	23 49	23 38	15 47	03 33	28 58	21 53	10 35	05 56
07	01 02 05	17 25 49	05 ♑ 18 50	12 ♑ 19 24	25 22	23 02	16 14	03 41	29 01	21 52	10 33	05 55
08	01 06 02	18 24 46	19 ♑ 21 47	26 ♑ 25 51	26 57	22 26	16 40	03 49	29 04	21 51	10 32	05 54
09	01 09 58	19 23 43	03 ♒ 31 25	10 ♒ 38 19	28 ♓ 33	21 48	17 07	03 57	29 07	21 50	10 31	05 53
10	01 13 55	20 22 37	17 ♒ 46 19	24 ♒ 55 08	00 ♈ 11	21 10	17 34	04 05	29 10	21 49	10 29	05 52
11	01 17 51	21 21 30	02 ♓ 04 27	09 ♓ 13 55	01 50	20 33	18 01	04 13	29 13	21 48	10 28	05 51
12	01 21 48	22 20 21	16 ♓ 23 03	23 ♓ 31 24	03 31	19 55	18 29	04 20	29 15	21 47	10 26	05 50
13	01 25 45	23 19 10	00 ♈ 38 24	07 ♈ 43 30	05 13	19 18	18 56	04 28	29 18	21 46	10 24	05 49
14	01 29 41	24 17 57	14 ♈ 46 07	21 ♈ 45 41	06 57	18 41	19 24	04 35	29 21	21 45	10 23	05 48
15	01 33 38	25 16 42	28 ♈ 41 38	05 ♉ 33 29	08 43	18 05	19 52	04 42	29 23	21 44	10 21	05 47
16	01 37 34	26 15 25	12 ♉ 20 50	19 ♉ 03 20	10 30	17 31	20 20	04 49	29 25	21 44	10 20	05 46
17	01 41 31	27 14 06	25 ♉ 40 45	02 ♊ 12 59	12 18	16 58	20 48	04 56	29 27	21 43	10 18	05 45
18	01 45 27	28 12 46	08 ♊ 40 00	15 ♊ 01 54	14 08	16 26	21 17	05 02	29 29	21 42	10 17	05 44
19	01 49 24	29 ♈ 11 23	21 ♊ 18 55	27 ♊ 31 20	16 00	15 56	21 45	05 09	29 31	21 42	10 15	05 44
20	01 53 20	00 ♉ 09 58	03 ♋ 39 32	09 ♋ 43 58	17 54	15 28	22 14	05 15	29 33	21 41	10 13	05 43
21	01 57 17	01 08 30	15 ♋ 45 12	21 ♋ 43 46	19 49	15 01	22 43	05 21	29 35	21 41	10 12	05 42
22	02 01 14	02 07 01	27 ♋ 40 19	03 ♌ 35 29	21 45	14 37	23 12	05 27	29 37	21 41	10 10	05 41
23	02 05 10	03 05 30	09 ♌ 29 55	15 ♌ 24 19	23 43	14 15	23 42	05 33	29 39	21 40	10 09	05 41
24	02 09 07	04 03 56	21 ♌ 19 20	27 ♌ 15 38	25 43	13 56	24 11	05 39	29 40	21 40	10 07	05 40
25	02 13 03	05 02 20	03 ♍ 13 51	09 ♍ 14 35	27 44	13 38	24 40	05 44	29 41	21 40	10 05	05 39
26	02 17 00	06 00 42	15 ♍ 18 24	21 ♍ 25 50	29 ♈ 47	13 23	25 10	05 50	29 43	21 39	10 04	05 39
27	02 20 56	06 59 02	27 ♍ 37 18	03 ♎ 53 11	01 ♉ 51	13 11	25 40	05 55	29 44	21 39	10 02	05 38
28	02 24 53	07 57 20	10 ♎ 13 45	16 ♎ 39 14	03 56	13 01	26 10	06 00	29 45	21 39	10 00	05 38
29	02 28 49	08 55 36	23 ♎ 09 42	29 ♎ 45 07	06 03	12 53	26 40	06 05	29 46	21 D 39	09 59	05 37
30	02 32 46	09 ♉ 53 50	06 ♏ 25 24	13 ♏ 10 17	08 ♉ 10	12 ♈ 48	27 ♋ 10	06 ♒ 10	29 ♑ 47	21 ♌ 39	09 ♏ 57	05 ♍ 36

Date	Moon True ☊ ° '	Moon Mean ☊ ° '	Moon ☽ Latitude ° '
01	05 ♍ 53	04 ♍ 33	03 N 15
02	05 R 48	04 30	04 04
03	05 41	04 27	04 41
04	05 35	04 24	05 04
05	05 29	04 20	05 10
06	05 25	04 17	04 58
07	05 23	04 14	04 29
08	05 D 22	04 11	03 44
09	05 23	04 08	02 44
10	05 24	04 05	01 34
11	05 25	04 01	00 N 18
12	05 R 24	03 58	00 S 59
13	05 22	03 55	02 12
14	05 17	03 52	03 16
15	05 10	03 49	04 07
16	05 02	03 45	04 43
17	04 53	03 42	05 03
18	04 44	03 39	05 06
19	04 37	03 36	04 54
20	04 31	03 33	04 28
21	04 27	03 30	03 51
22	04 26	03 26	03 04
23	04 D 26	03 23	02 10
24	04 27	03 20	01 10
25	04 28	03 17	00 S 07
26	04 R 27	03 14	00 N 58
27	04 24	03 11	02 01
28	04 19	03 07	02 59
29	04 12	03 04	03 49
30	04 ♍ 03	03 ♍ 01	04 N 29

DECLINATIONS

Date	Sun ☉ ° '	Moon ☽ ° '	Mercury ☿ ° '	Venus ♀ ° '	Mars ♂ ° '	Jupiter ♃ ° '	Saturn ♄ ° '	Uranus ♅ ° '	Neptune ♆ ° '	Pluto ♇ ° '
01	04 N 33	02 S 45	07 S 28	17 N 25	25 N 07	19 S 47	20 S 25	14 N 54	13 S 18	21 N 25
02	04 57	06 47	06 57	17 14	25 03	19 45	20 25	14 54	13 18	21 25
03	05 20	10 34	06 26	17 01	24 59	19 43	20 24	14 55	13 17	21 25
04	05 43	13 52	05 53	16 47	24 55	19 42	20 23	14 55	13 17	21 26
05	06 05	16 29	05 19	16 31	24 51	19 40	20 23	14 55	13 16	21 26
06	06 28	18 12	04 44	16 14	24 47	19 38	20 22	14 56	13 16	21 26
07	06 51	18 51	04 07	15 56	24 43	19 36	20 22	14 56	13 15	21 26
08	07 13	18 21	03 30	15 37	24 39	19 34	20 21	14 56	13 15	21 27
09	07 35	16 42	02 51	15 17	24 34	19 33	20 21	14 57	13 14	21 27
10	07 58	14 01	02 11	14 55	24 30	19 31	20 20	14 57	13 14	21 27
11	08 20	10 27	01 31	14 33	24 25	19 29	20 20	14 57	13 13	21 27
12	08 42	06 17	00 49	14 10	24 20	19 28	20 19	14 58	13 13	21 27
13	09 04	01 S 46	00 S 06	13 47	24 15	19 26	20 19	14 58	13 12	21 28
14	09 25	02 N 48	00 N 38	13 23	24 10	19 25	20 18	14 58	13 12	21 28
15	09 47	07 09	01 22	12 59	24 05	19 23	20 18	14 58	13 11	21 28
16	10 08	11 03	02 08	12 35	24 00	19 22	20 17	14 58	13 11	21 28
17	10 29	14 17	02 55	12 11	23 55	19 20	20 17	14 59	13 10	21 28
18	10 50	16 43	03 42	11 47	23 50	19 19	20 17	14 59	13 10	21 28
19	11 11	18 16	04 30	11 23	23 44	19 17	20 16	14 59	13 09	21 28
20	11 32	18 55	05 19	11 00	23 38	19 16	20 16	14 59	13 09	21 28
21	11 52	18 41	06 08	10 37	23 33	19 15	20 16	14 59	13 08	21 28
22	12 13	17 37	06 58	10 15	23 27	19 13	20 15	14 59	13 08	21 28
23	12 33	15 47	07 49	09 54	23 21	19 12	20 15	14 59	13 07	21 28
24	12 53	13 18	08 40	09 33	23 15	19 11	20 15	14 59	13 07	21 28
25	13 12	10 13	09 31	09 13	23 09	19 10	20 15	14 59	13 06	21 28
26	13 32	06 41	10 23	08 54	23 02	19 08	20 14	14 59	13 06	21 28
27	13 51	02 N 47	11 14	08 36	22 56	19 07	20 14	14 59	13 05	21 28
28	14 10	01 S 19	12 06	08 19	22 49	19 06	20 14	14 59	13 05	21 28
29	14 29	05 27	12 58	08 03	22 43	19 05	20 14	14 59	13 04	21 28
30	14 N 47	09 S 26	13 N 49	07 N 48	22 N 36	19 S 04	20 S 14	14 N 59	13 S 04	21 N 28

ZODIAC SIGN ENTRIES

Date	h	m	Planets
02	16	36	☽ ♏
04	22	34	☽ ♐
07	02	52	☽ ♑
09	06	03	☽ ♒
10	09	22	☿ ♈
11	08	31	☽ ♓
13	10	55	☽ ♈
15	14	16	☽ ♉
17	19	55	☽ ♊
20	04	50	☽ ♋
20	07	55	☉ ♉
22	16	43	☽ ♌
25	05	31	☽ ♍
26	14	34	☿ ♉
27	16	34	☽ ♎
30	00	27	☽ ♏

LATITUDES

Date	Mercury ☿ ° '	Venus ♀ ° '	Mars ♂ ° '	Jupiter ♃ ° '	Saturn ♄ ° '	Uranus ♅ ° '	Neptune ♆ ° '	Pluto ♇ ° '
01	02 S 18	07 N 48	02 N 23	00 S 16	00 00	00 N 45	01 N 50	13 N 05
04	02 26	07 44	02 20	00 16	00 00	00 45	01 50	13 04
07	02 29	07 32	02 17	00 17	00 00	00 45	01 50	13 04
10	02 28	07 12	02 13	00 17	00 00	00 45	01 50	13 03
13	02 22	06 45	02 10	00 18	00 S 01	00 45	01 50	13 03
16	02 12	06 11	02 07	00 18	00 01	00 45	01 50	13 02
19	01 57	05 33	02 05	00 19	00 01	00 45	01 50	13 01
22	01 37	04 52	02 02	00 19	00 01	00 45	01 50	13 00
25	01 14	04 10	01 59	00 20	00 02	00 45	01 50	13 00
28	00 47	03 27	01 56	00 20	00 02	00 44	01 50	12 59
31	00 S 16	02 N 46	01 N 54	00 S 21	00 S 02	00 N 44	01 N 50	12 N 58

DATA

Julian Date	2437391
Delta T	+34 seconds
Ayanamsa	23° 18' 49"
Synetic vernal point	05° ♓ 48' 11"
True obliquity of ecliptic	23° 26' 32"

MOON'S PHASES, APSIDES AND POSITIONS ☽

Date	h	m	Phase	Longitude ° '	Eclipse Indicator
01	05	47	○	11 ♎ 16	
08	10	16	☾	18 ♑ 21	
15	05	37	●	25 ♈ 01	
22	21	49	☽	02 ♌ 31	
30	18	40	○	10 ♏ 10	

Day	h	m		
11	07	24	Perigee	
23	10	02	Apogee	
07	13	45	Max dec	18° S 51'
13	21	13	0N	
20	17	29	Max dec	18° N 57'
28	04	25	0S	

LONGITUDES

Date	Chiron ⚷ ° '	Ceres ⚳ ° '	Pallas ⚴ ° '	Juno ⚵ ° '	Vesta ⚶ ° '	Black Moon Lilith ⚸ ° '
01	04 ♓ 33	06 ♈ 48	06 ♓ 33	26 ♒ 43	03 ♈ 28	06 ♌ 38
11	05 ♓ 05	10 ♈ 44	09 ♓ 32	00 ♓ 38	08 ♈ 11	07 ♌ 45
21	05 ♓ 33	14 ♈ 40	12 ♓ 22	04 ♓ 28	12 ♈ 51	08 ♌ 52
31	05 ♓ 58	18 ♈ 33	15 ♓ 03	08 ♓ 12	17 ♈ 26	09 ♌ 58

All ephemeris data is given at 12.00 UT and the Moon's longitude is additionally given for 24.00 UT

ASPECTARIAN

h m	Aspects
01 Saturday	
04 44	☽ ⊻ ♆
05 47	☽ ☍ ☉
07 14	☽ ⊥ ♇
10 12	☽ □ ♂
15 59	☽ ⚻ ☿
23 48	☽ # ☉
02 Sunday	
00 02	☽ ∠ ♇
00 04	☉ ± ♇
01 49	☽ ✶ ♅
04 31	☽ ± ☿
09 04	☽ ☍ ♀
12 57	☽ ∥ ☿
13 50	☿ ∠ ♃
14 19	☽ □ ♄
22 10	☽ □ ♃
23 06	☽ bQ ☿
23 51	☽ Q ♅
03 Monday	
03 33	☽ ✶ ♇
11 58	☿ ⊥ ♀
12 00	☽ ☌ ♆
17 32	☽ ⚻ ☉
19 09	☽ △ ♂
23 08	☽ Q ♄
04 Tuesday	
01 11	☽ Q ♇
05 09	☽ ± ☉
05 30	☽ △ ☿
07 00	☽ Q ♃
07 22	☽ ∥ ♆
08 14	☽ □ ♅
13 16	☽ ⚻ ♀
16 24	☉ # ☿
20 35	☽ ✶ ♄
20 46	☽ # ♅
22 24	☽ bQ ☉
22 46	☽ bQ ♂
23 29	☽ ± ♀
05 Wednesday	
04 30	☽ ✶ ♃
05 50	☿ ⚻ ♅
07 06	☉ □ ♂
09 03	☽ □ ♇
10 29	☉ Q ♃
12 21	☽ # ♀
14 41	☽ bQ ♀
15 07	☽ ± ♂
17 12	☽ ⊻ ♆
23 02	☽ ∠ ♄
06 Thursday	
01 56	☽ ⚻ ♂
02 45	☽ △ ☉
03 39	☽ ⊥ ♆
07 01	☽ ∠ ♃
09 57	☿ ⊻ ♀
12 51	☽ △ ♅
14 44	☽ ⊥ ♄
15 44	☽ △ ♀
16 43	☽ □ ☿
17 02	♂ ⊥ ♅
19 14	☽ ∠ ♆
22 45	☽ ⊥ ♃
07 Friday	
01 08	☽ ⊻ ♄
09 11	☽ ⊻ ♃
13 02	☽ △ ♇
14 39	☽ bQ ♅
14 49	☿ bQ ♆
20 58	☽ ✶ ♆
08 Saturday	
03 32	☽ Q ☿
06 01	☽ ± ♅
07 16	☽ ☍ ♂
10 16	☽ □ ☉
14 36	☽ bQ ♇
16 13	☽ ⚻ ♅
16 59	☽ □ ♀
17 23	☽ Q ♆
09 Sunday	
01 20	☿ ± ♅
02 31	☽ ✶ ☿
04 32	☽ ☌ ♄
05 51	☽ ± ♇
10 57	♀ △ ♅
12 44	☽ ☌ ♃
15 58	☽ ⚻ ♇
19 01	☽ Q ☉
20 42	☿ ✶ ♄
22 09	☽ Q ♀
23 46	☽ □ ♆
10 Monday	
03 47	☽ # ♀
04 35	☽ # ♅
07 05	☽ ∠ ☿
10 07	♀ ∥ ♅
11 39	☽ ⚻ ♂
16 42	☽ ✶ ☉
17 28	☽ ✶ ♀
17 47	☽ ∥ ♆
18 47	☽ ☍ ♅
22 03	☽ ± ♂
23 40	☉ bQ ♇
23 51	☉ ☌ ♀
11 Tuesday	
00 07	♀ bQ ♇
00 10	☽ ⊥ ☿
07 11	☽ ⊻ ♄
11 33	☽ ⊻ ☿
13 38	☽ bQ ♂
15 37	☽ ⊻ ♃
17 17	☽ ⊥ ♄
17 34	☽ ∠ ♀
18 19	☽ ☍ ♇
19 43	☽ ∠ ☉
22 31	☉ △ ♅
23 33	☽ # ☉
12 Wednesday	
01 46	☽ ⊥ ♃
02 02	☽ △ ♆
08 02	☽ ⊥ ♀
08 25	☽ ∠ ♄
11 55	☽ ⊥ ☉
15 38	☽ △ ♂
17 00	☽ ∠ ♃
17 41	☽ ⊻ ♀
21 03	☽ ⚻ ♅
22 45	☽ ⊻ ☉
13 Thursday	
00 32	☿ ✶ ♃
00 48	☿ ± ♆
03 11	☽ bQ ♆
07 09	☽ ± ♅
09 44	☽ ✶ ♄
18 22	☽ ± ♆
18 31	☽ ✶ ♃
19 58	♀ □ ♂
20 12	☿ ⚻ ♇
20 24	☽ # ☿
20 45	☽ ⚻ ♇
20 49	☽ ☌ ☿
22 19	☽ ∥ ☿
22 22	☽ bQ ♅
14 Friday	
04 32	☽ ⚻ ♆
06 08	☽ Q ♄
06 56	☽ ± ♇
09 17	☿ bQ ♅
15 08	☽ Q ♃
18 26	☽ ☌ ♀
20 13	☽ □ ♂
22 20	☽ bQ ♇
23 49	♀ # ♆
23 58	☽ △ ♅
15 Saturday	
05 37	☽ ☌ ☉
13 12	☽ □ ♄
22 35	☽ □ ♃
16 Sunday	
00 23	☽ △ ♇
04 38	☽ Q ♂
05 26	☽ ∥ ☉
08 12	☽ ⊻ ☿
08 26	☽ ☍ ♆
09 52	☿ ⚻ ♆
20 33	☽ ⊥ ☿
20 52	☽ ⊻ ♀
21 36	☽ ∥ ♀
17 Monday	
00 36	☿ Q ♄
02 50	☽ ✶ ♂
03 10	☽ # ♆
04 48	☽ □ ♅
04 49	☿ ± ♇
07 15	☽ ⊥ ♀
09 24	♂ ∠ ♇
13 17	♀ Q ♃
15 04	☽ ⊻ ☉
15 27	☽ ∠ ☿
18 06	☽ ∥ ♅
18 56	☽ △ ♄
23 04	☽ ∠ ♀
18 Tuesday	
03 01	☽ ⊥ ☉
05 10	☽ △ ♃
06 32	☽ □ ♇
07 22	☽ ∠ ♂
13 57	☽ Q ♅
15 01	☽ ⚻ ♆
21 16	☽ ∠ ☉
23 00	☽ bQ ♄
19 Wednesday	
00 05	☽ ✶ ☿
00 58	☽ ⊥ ♂
02 06	☽ ✶ ♀
02 20	☽ ± ♆
09 01	♂ ⊻ ♅
09 44	☽ bQ ♃
11 17	☿ ☌ ♀
12 44	☽ ✶ ♅
12 53	☽ ⊻ ♂
16 16	☽ ± ♄
16 38	☽ Q ♇
18 37	☉ ∥ ♀
19 35	☽ bQ ♆
20 30	☉ □ ♄
20 Thursday	
00 19	☽ Q ♀
03 16	☽ Q ☿
03 18	☽ ± ♃
03 25	☿ Q ♃
03 57	☽ ⚻ ♄
04 34	☽ ✶ ☉
15 10	☽ ⚻ ♃
16 03	☽ ✶ ♇
17 58	☽ ∠ ♅
21 Friday	
00 57	☽ △ ♆
06 19	☽ Q ☉
10 35	☽ □ ♀
11 51	☽ ⊥ ♅
21 43	☽ □ ☿
21 55	☽ ∠ ♇
22 59	☿ bQ ♇
23 54	☽ ⊻ ♅
22 Saturday	
02 35	☽ ☌ ♂
11 04	☿ △ ♅
15 57	☽ ☍ ♄
16 05	☽ ⊥ ♇
21 49	☽ □ ☉
23 Sunday	
03 55	☽ ☍ ♃
04 15	☽ ⊻ ♇
11 32	☿ □ ♂
13 18	☽ □ ♆
20 24	☽ ∥ ♅
21 24	☽ △ ♀
24 Monday	
10 54	♄ ± ♇
12 42	☽ ☌ ♅
13 32	☽ # ♆
15 09	☽ ∥ ☉
16 25	♃ ⚻ ♇
18 02	☽ ⊻ ♂
22 42	☽ △ ☿
25 Tuesday	
01 42	☽ Q ♆
02 59	☽ bQ ♀
04 51	☉ # ♆
04 53	☽ ⚻ ♄
05 49	☿ ∥ ♀
06 39	☽ ⊥ ♂
15 56	☽ △ ☉
15 59	☽ ∥ ☿
16 51	☽ ☌ ♇
16 56	☽ ± ♄
17 03	☽ ⚻ ♃
19 50	☽ ∥ ♀
20 37	☽ ± ♀
26 Wednesday	
01 24	☽ ∠ ♂
01 39	☽ ✶ ♆
03 02	☉ △ ♇
05 05	☽ ± ♃
07 01	☉ □ ♃
08 17	☽ ⚻ ♀
10 45	☽ bQ ☿
10 50	☽ bQ ♄
11 13	☿ □ ♄
22 54	☽ bQ ♃
27 Thursday	
00 09	☽ bQ ☉
00 26	☽ ⊻ ♅
07 01	☽ ∠ ♆
07 53	☽ ± ☿
08 04	☽ ✶ ♂
12 04	☽ ⊥ ♅
16 04	☽ △ ♄
19 00	☽ ± ☉
21 43	☽ ⚻ ☿
28 Friday	
00 15	☽ ⊥ ♆
03 19	☽ ⊻ ♇
03 58	☽ △ ♃
05 15	☽ ∠ ♅
07 22	☽ ⚻ ☉
07 57	☽ Q ♂
11 35	☽ ⊻ ♆
14 37	☽ ⊥ ♇
17 09	☽ ☍ ♀
29 Saturday	
07 09	☿ △ ♇
07 20	☽ ∠ ♇
07 49	♅ St D
09 14	☽ ✶ ♅
12 25	☿ □ ♃
15 03	☿ # ♆
18 39	☽ □ ♂
30 Sunday	
00 03	☽ □ ♄
02 35	☽ # ♀
07 02	☽ Q ♅
10 33	☽ ✶ ♇
11 31	☽ □ ♃
13 21	☉ ☍ ♆
15 43	☽ ☍ ☿
18 17	☽ ☌ ♆
18 40	☽ ☍ ☉
23 17	☽ ⚻ ♀

MAY 1961

LONGITUDES

Date	Sidereal time h m s	Sun ☉ ° ' "	Moon ☽ ° ' "	Moon ☽ 24.00 ° ' "	Mercury ☿ ° '	Venus ♀ ° '	Mars ♂ ° '	Jupiter ♃ ° '	Saturn ♄ ° '	Uranus ♅ ° '	Neptune ♆ ° '	Pluto ♇ ° '
01	02 36 43	10 ♉ 52 03	19 ♏ 59 28	26 ♏ 52 32	10 ♉ 19	12 ♈ 45	27 ♋ 40	06 ♒ 14	29 ♑ 48	21 ♌ 39	09 ♏ 56	05 ♍ 36
02	02 40 39	11 50 14	03 ♐ 49 01	10 ♐ 48 23	12 28	12 D 44	28 11	06 18	29 49	21 40	09 R 54	05 R 36
03	02 44 36	12 48 23	17 ♐ 50 05	24 ♐ 53 36	14 38	12 46	28 41	06 23	29 49	21 40	09 52	05 35
04	02 48 32	13 46 31	01 ♑ 58 22	09 ♑ 03 56	16 47	12 51	29 12	06 27	29 50	21 40	09 51	05 35
05	02 52 29	14 44 37	16 ♑ 09 52	23 ♑ 15 46	18 57	12 57	29 ♋ 43	06 31	29 50	21 40	09 49	05 34
06	02 56 25	15 42 42	00 ♒ 21 21	07 ♒ 26 23	21 07	13 06	00 ♌ 14	06 34	29 51	21 41	09 47	05 34
07	03 00 22	16 40 45	14 ♒ 30 40	21 ♒ 34 05	23 16	13 17	00 45	06 38	29 51	21 41	09 46	05 34
08	03 04 18	17 38 48	28 ♒ 36 30	05 ♓ 37 50	25 24	13 30	01 16	06 41	29 51	21 42	09 44	05 33
09	03 08 15	18 36 48	12 ♓ 37 59	19 ♓ 36 52	27 31	13 45	01 47	06 44	29 51	21 42	09 43	05 33
10	03 12 12	19 34 48	26 ♓ 34 20	03 ♈ 30 14	29 ♉ 36	14 02	02 19	06 47	29 R 51	21 43	09 41	05 33
11	03 16 08	20 32 46	10 ♈ 24 22	17 ♈ 16 29	01 ♊ 40	14 21	02 50	06 50	29 51	21 43	09 39	05 33
12	03 20 05	21 30 43	24 ♈ 06 20	00 ♉ 53 36	03 42	14 41	03 22	06 53	29 51	21 44	09 38	05 33
13	03 24 01	22 28 38	07 ♉ 37 59	14 ♉ 19 12	05 41	15 04	03 54	06 55	29 51	21 45	09 36	05 33
14	03 27 58	23 26 32	20 ♉ 56 56	27 ♉ 30 57	07 39	15 28	04 25	06 57	29 50	21 45	09 35	05 32
15	03 31 54	24 24 25	04 ♊ 01 03	10 ♊ 27 05	09 33	15 54	04 57	06 59	29 50	21 46	09 33	05 32
16	03 35 51	25 22 16	16 ♊ 48 59	23 ♊ 06 46	11 25	16 21	05 29	07 01	29 49	21 47	09 31	05 32
17	03 39 47	26 20 05	29 ♊ 20 31	05 ♋ 30 24	13 14	16 50	06 01	07 03	29 48	21 48	09 30	05 D 32
18	03 43 44	27 17 53	11 ♋ 36 43	17 ♋ 39 46	15 00	17 21	06 34	07 04	29 48	21 49	09 28	05 32
19	03 47 41	28 15 40	23 ♋ 39 58	29 ♋ 37 47	16 43	17 53	07 06	07 05	29 47	21 50	09 27	05 32
20	03 51 37	29 ♉ 13 25	05 ♌ 33 47	11 ♌ 28 31	18 23	18 26	07 38	07 06	29 46	21 51	09 25	05 32
21	03 55 34	00 ♊ 11 08	17 ♌ 22 37	23 ♌ 16 44	20 00	19 00	08 11	07 07	29 45	21 52	09 24	05 33
22	03 59 30	01 08 49	29 ♌ 11 33	05 ♍ 07 45	21 33	19 36	08 43	07 08	29 43	21 54	09 22	05 33
23	04 03 27	02 06 29	11 ♍ 06 01	17 ♍ 07 03	23 04	20 13	09 16	07 09	29 42	21 55	09 21	05 33
24	04 07 23	03 04 08	23 ♍ 11 30	29 ♍ 20 00	24 31	20 51	09 49	07 09	29 41	21 56	09 19	05 33
25	04 11 20	04 01 45	05 ♎ 33 06	11 ♎ 51 20	25 54	21 30	10 22	07 09	29 39	21 58	09 18	05 33
26	04 15 16	04 59 20	18 ♎ 15 07	24 ♎ 44 46	27 14	22 11	10 55	07 R 09	29 38	21 59	09 16	05 34
27	04 19 13	05 56 55	01 ♏ 20 29	08 ♏ 02 21	28 31	22 52	11 28	07 09	29 36	22 00	09 15	05 34
28	04 23 10	06 54 27	14 ♏ 50 16	21 ♏ 44 02	29 ♊ 44	23 34	12 01	07 09	29 34	22 02	09 14	05 34
29	04 27 06	07 51 59	28 ♏ 43 16	05 ♐ 47 26	00 ♋ 54	24 18	12 34	07 08	29 32	22 03	09 12	05 35
30	04 31 03	08 49 30	12 ♐ 55 53	20 ♐ 07 52	02 00	25 02	13 07	07 07	29 31	22 05	09 11	05 35
31	04 34 59	09 ♊ 46 59	27 ♐ 22 34	04 ♑ 39 05	03 ♋ 02	25 ♈ 47	13 ♌ 40	07 ♒ 06	29 ♑ 29	22 ♌ 07	09 ♏ 09	05 ♍ 35

Date	Moon True ☊ ° '	Moon Mean ☊ ° '	Moon ☽ Latitude ° '
01	03 ♍ 52	02 ♍ 58	04 N 54
02	03 R 41	02 55	05 03
03	03 31	02 51	04 53
04	03 23	02 48	04 26
05	03 18	02 45	03 43
06	03 15	02 42	02 45
07	03 15	02 39	01 38
08	03 D 15	02 36	00 N 25
09	03 R 14	02 32	00 S 50
10	03 13	02 29	02 00
11	03 09	02 26	03 03
12	03 02	02 23	03 55
13	02 52	02 20	04 33
14	02 40	02 17	04 55
15	02 27	02 13	05 01
16	02 15	02 10	04 51
17	02 04	02 07	04 28
18	01 55	02 04	03 52
19	01 48	02 01	03 07
20	01 45	01 57	02 14
21	01 43	01 54	01 15
22	01 D 43	01 51	00 S 13
23	01 R 43	01 48	00 N 50
24	01 42	01 45	01 51
25	01 39	01 42	02 49
26	01 34	01 38	03 40
27	01 26	01 35	04 21
28	01 16	01 32	04 49
29	01 05	01 29	05 01
30	00 53	01 26	04 54
31	00 ♍ 43	01 ♍ 23	04 N 29

DECLINATIONS

Date	Sun ☉ ° '	Moon ☽ ° '	Mercury ☿ ° '	Venus ♀ ° '	Mars ♂ ° '	Jupiter ♃ ° '	Saturn ♄ ° '	Uranus ♅ ° '	Neptune ♆ ° '	Pluto ♇ ° '
01	15 N 05	13 S 01	14 N 39	07 N 35	22 N 29	19 S 03	20 S 14	14 N 59	13 S 03	21 N 28
02	15 23	15 58	15 29	07 22	22 22	19 02	20 14	14 59	13 03	21 28
03	15 41	18 01	16 18	07 11	22 15	19 01	20 14	14 59	13 02	21 28
04	15 59	18 59	17 06	07 00	22 08	19 01	20 14	14 59	13 01	21 28
05	16 16	18 47	17 52	06 51	22 00	19 00	20 14	14 59	13 01	21 28
06	16 33	17 23	18 37	06 43	21 53	18 59	20 14	14 59	13 01	21 27
07	16 49	14 55	19 21	06 36	21 45	18 58	20 14	14 59	13 00	21 27
08	17 06	11 34	20 02	06 30	21 38	18 58	20 14	14 58	13 00	21 27
09	17 22	07 35	20 41	06 25	21 30	18 57	20 14	14 58	12 59	21 27
10	17 38	03 S 12	21 18	06 22	21 22	18 57	20 14	14 58	12 59	21 27
11	17 53	01 N 19	21 53	06 19	21 14	18 56	20 14	14 58	12 58	21 26
12	18 09	05 42	22 25	06 17	21 06	18 56	20 14	14 57	12 58	21 26
13	18 24	09 46	22 54	06 16	20 57	18 55	20 14	14 57	12 57	21 26
14	18 38	13 15	23 22	06 16	20 49	18 55	20 14	14 57	12 57	21 26
15	18 52	16 02	23 46	06 17	20 40	18 54	20 15	14 57	12 56	21 25
16	19 06	17 57	24 08	06 19	20 32	18 54	20 15	14 56	12 56	21 25
17	19 20	18 59	24 27	06 22	20 23	18 54	20 15	14 56	12 55	21 25
18	19 33	19 05	24 44	06 25	20 14	18 54	20 15	14 56	12 55	21 25
19	19 46	18 18	24 59	06 29	20 05	18 54	20 16	14 55	12 54	21 24
20	19 59	16 43	25 11	06 34	19 56	18 54	20 16	14 55	12 54	21 24
21	20 12	14 26	25 20	06 40	19 47	18 53	20 16	14 54	12 53	21 24
22	20 23	11 33	25 28	06 47	19 38	18 54	20 17	14 54	12 53	21 23
23	20 35	08 10	25 33	06 54	19 28	18 54	20 17	14 54	12 52	21 23
24	20 46	04 24	25 37	07 01	19 19	18 54	20 17	14 53	12 52	21 23
25	20 57	00 N 23	25 38	07 10	19 09	18 54	20 18	14 53	12 52	21 22
26	21 08	03 S 46	25 38	07 19	18 59	18 54	20 18	14 52	12 51	21 22
27	21 18	07 52	25 36	07 28	18 49	18 54	20 19	14 52	12 51	21 21
28	21 28	11 41	25 32	07 38	18 39	18 55	20 19	14 51	12 50	21 21
29	21 37	14 59	25 27	07 49	18 29	18 55	20 19	14 50	12 50	21 20
30	21 46	17 29	25 20	08 00	18 19	18 55	20 20	14 50	12 49	21 20
31	21 N 55	18 S 56	25 N 13	08 N 11	18 N 09	18 S 56	20 S 20	14 N 49	12 S 49	21 N 20

ZODIAC SIGN ENTRIES

Date	h	m	Planets
02	05	25	☽ ♐
04	08	40	☽ ♑
06	01	13	♂ ♌
06	11	24	☽ ♒
08	14	23	☽ ♓
10	16	34	☿ ♊
10	17	56	☽ ♈
12	22	25	☽ ♉
15	04	34	☽ ♊
17	13	17	☽ ♋
20	00	45	☽ ♌
21	07	22	☉ ♊
22	13	38	☽ ♍
25	01	18	☽ ♎
27	09	34	☽ ♏
28	17	23	☿ ♋
29	14	11	☽ ♐
31	16	20	☽ ♑

LATITUDES

Date	Mercury ☿ ° '	Venus ♀ ° '	Mars ♂ ° '	Jupiter ♃ ° '	Saturn ♄ ° '	Uranus ♅ ° '	Neptune ♆ ° '	Pluto ♇ ° '
01	00 S 16	02 N 46	01 N 54	00 S 21	00 S 02	00 N 44	01 N 50	12 N 58
04	00 N 15	02 06	01 51	00 21	00 03	00 44	01 50	12 57
07	00 47	01 28	01 48	00 22	00 03	00 44	01 50	12 56
10	01 16	00 54	01 46	00 22	00 03	00 44	01 50	12 55
13	01 41	00 N 22	01 43	00 23	00 03	00 44	01 50	12 54
16	02 00	00 S 08	01 41	00 24	00 03	00 44	01 50	12 53
19	02 12	00 34	01 38	00 24	00 04	00 44	01 50	12 52
22	02 18	00 58	01 36	00 25	00 04	00 44	01 50	12 51
25	02 16	01 19	01 34	00 26	00 04	00 43	01 50	12 50
28	02 06	01 38	01 31	00 26	00 05	00 43	01 50	12 49
31	01 N 48	01 S 54	01 N 29	00 S 27	00 S 05	00 N 43	01 N 50	12 N 48

DATA

Julian Date	2437421
Delta T	+34 seconds
Ayanamsa	23° 18' 52"
Synetic vernal point	05° ♓ 48' 08"
True obliquity of ecliptic	23° 26' 31"

MOON'S PHASES, APSIDES AND POSITIONS ☽

Date	h	m	Phase	Longitude ° '	Eclipse Indicator
07	15	57	☾	16 ♒ 50	
14	16	54	●	23 ♉ 38	
22	16	18	☽	01 ♍ 19	
30	04	37	○	08 ♐ 32	

Day	h	m		
06	12	15	Perigee	
21	05	12	Apogee	
04	19	48	Max dec	19° S 03'
11	05	02	0N	
18	02	36	Max dec	19° N 09'
25	14	13	0S	

LONGITUDES

Date	Chiron ⚷ ° '	Ceres ⚳ ° '	Pallas ⚴ ° '	Juno ⚵ ° '	Vesta ⚶ ° '	Black Moon Lilith ⚸ ° '
01	05 ♓ 58	18 ♈ 33	15 ♓ 03	08 ♓ 12	17 ♈ 26	09 ♌ 58
11	06 ♓ 17	22 ♈ 23	17 ♓ 31	11 ♓ 47	21 ♈ 57	11 ♌ 05
21	06 ♓ 31	26 ♈ 10	19 ♓ 46	15 ♓ 13	26 ♈ 22	12 ♌ 11
31	06 ♓ 39	29 ♈ 51	21 ♓ 45	18 ♓ 27	00 ♉ 41	13 ♌ 18

All ephemeris data is given at 12.00 UT and the Moon's longitude is additionally given for 24.00 UT

ASPECTARIAN

h m	Aspects
01 Monday	
04 06	☉ ∥ ♅
07 44	☿ ☍ ♆
07 49	☽ Q ♇
08 09	☽ Q ♄
09 49	☽ ± ♀
12 15	☽ ∥ ♆
14 55	☽ □ ♅
19 27	☽ Q ♃
21 34	☿ ∥ ♅
23 17	☉ ☌ ☿
02 Tuesday	
01 30	☽ ⚼ ♀
01 54	☽ △ ♂
03 20	☽ # ♅
04 16	♀ St D
05 05	☽ ⚹ ♄
05 46	☽ # ☿
06 09	☽ # ☉
07 34	☉ ∥ ☿
15 03	☽ □ ♇
15 05	☿ ⊻ ♀
16 18	☽ ⚹ ♃
22 26	☽ ⊻ ♆
03 Wednesday	
02 47	☽ ⊼ ☉
03 20	☽ △ ♀
04 40	☽ ⚼ ♂
05 32	☽ ⊼ ☿
06 52	☽ ∠ ♄
08 40	☽ ⊥ ♆
11 06	☉ ⊻ ♀
13 47	☽ ± ☉
17 37	☽ ± ☿
18 03	☽ ∠ ♃
18 31	☽ △ ♅
20 34	☽ ± ♂
22 11	☽ ⊥ ♄
23 56	☽ ∠ ♆
04 Thursday	
06 11	☽ ⚼ ☉
07 08	☽ ⊼ ♂
08 22	☽ ⊻ ♄
09 24	☽ ⊥ ♃
11 38	☽ ⚼ ☿
13 17	☽ ∥ ♃
18 00	☿ Q ♂
18 06	☽ △ ♇
19 36	☽ ⊻ ♃
19 57	☽ ⚼ ♅
05 Friday	
01 17	☽ ⚹ ♆
02 51	☽ ∥ ♃
05 22	♂ ⊥ ♇
06 31	☽ □ ♀
09 25	☽ △ ☉
11 10	☽ ± ♅
11 58	☿ ⊥ ♀
17 34	☽ △ ☿
17 52	♂ ☍ ♄
19 27	☽ ⚼ ♇
21 19	☽ ⊼ ♅
21 32	☽ Q ♆
23 43	☽ # ☿
06 Saturday	
10 40	☽ ± ♇
11 08	☽ ☌ ♄
11 47	☽ ☍ ♂
13 16	☽ Q ♀
18 18	☿ □ ♅
20 20	☽ # ☉
20 49	☽ ⊼ ♇
22 34	☽ ☌ ♃
23 45	☿ # ♃
07 Sunday	
03 57	☽ □ ♆
09 53	☽ ⚹ ♀
11 33	☽ # ♅
15 57	☽ □ ☉
08 Monday	
00 12	☽ ☍ ♅
02 28	☽ ∥ ♆
05 33	☽ □ ☿
11 48	☽ ∠ ♀
14 08	☽ ⊻ ♄
16 43	☽ ⊼ ♂
19 08	☿ # ♄
23 52	☽ ☍ ♇
09 Tuesday	
00 23	☽ ⊥ ♄
00 55	☽ Q ☉
01 51	☽ ⊻ ♃
03 22	☽ ± ♂
03 28	☽ ⊥ ♀
07 00	☽ △ ♆
12 11	☽ ⊥ ♃
13 57	☽ ⊻ ♀
15 49	☽ ∠ ♄
16 20	♄ St R
17 49	☽ Q ☿
18 37	☽ # ♀
19 25	☽ ⚼ ♂
21 01	♂ ∥ ♇
23 03	☽ ⚹ ☉
10 Wednesday	
03 36	☽ ⊼ ♅
03 43	☽ ∠ ♃
04 16	☿ ∠ ♀
08 45	☽ ⚼ ♆
13 58	☽ ± ♅
14 05	☿ ∥ ♂
14 53	☿ △ ♄
17 41	☽ ⚹ ♄
17 48	☿ ∥ ♇
18 10	☽ ⚹ ☿
22 19	☽ △ ♂
11 Thursday	
00 17	☽ ± ♆
02 55	☽ ∠ ☉
03 33	☽ ⊼ ♇
04 27	☉ ⊥ ♀
05 35	☽ ⚼ ♅
05 46	☽ ⚹ ♃
10 42	☽ ⊼ ♆
13 59	☽ ± ♇
14 31	☽ Q ♄
19 03	☽ ☌ ♀
19 46	☽ ⊥ ☉
12 Friday	
00 51	☽ ∠ ☿
02 47	☽ Q ♃
03 53	☉ Q ♂
05 44	☽ ⚼ ♇
06 37	☿ ⚹ ♂
07 05	☽ ⊻ ☉
07 49	☽ △ ♅
15 14	☽ ∥ ♀
17 33	☉ □ ♅
19 26	☽ ⊥ ☿
22 09	☽ □ ♄
13 Saturday	
05 03	☽ □ ♂
07 56	☽ ⊻ ☿
08 16	☽ △ ♇
10 11	☿ □ ♇
10 43	☽ □ ♃
15 31	☽ ☍ ♆
14 Sunday	
01 45	☽ ⊻ ♀
03 17	☿ △ ♃
09 40	☽ # ♆
12 59	☽ ⊥ ♀
13 28	☽ □ ♅
14 48	☽ Q ♂
16 54	☽ ☌ ☉
15 Monday	
01 46	☽ ∥ ♅
04 15	☽ △ ♄
06 02	☽ ∠ ♀
11 57	☿ ⊼ ♆
13 49	☽ ⚹ ♂
14 47	☿ Q ♅
14 50	☽ □ ♇
15 19	☉ # ♃
17 32	☽ △ ♃
22 17	☽ ⊼ ♆
22 44	☽ Q ♅
16 Tuesday	
00 05	☽ ☌ ☿
08 13	☽ ⚼ ♄
09 33	☽ ± ♆
11 06	☽ ⚹ ♀
14 17	♂ ⊻ ♇
19 17	☽ ∠ ♂
21 28	☽ ⚹ ♅
21 56	☽ ⚼ ♃
17 Wednesday	
00 49	☽ Q ♇
01 20	☽ ± ♄
02 41	☽ ⚼ ♆
04 24	♇ St D
05 42	☽ ⊻ ☉
09 00	☽ # ♃
10 59	☽ Q ♀
12 54	☽ ⊼ ♄
13 23	☽ ⊥ ♂
15 19	☽ ± ♃
18 18	☽ ⊥ ☉
18 Thursday	
00 04	☽ ⚹ ♇
01 36	☽ ⊻ ♂
02 33	☽ ∠ ♅
03 03	☽ ⊼ ♃
07 47	☽ △ ♆
08 49	♂ # ♄
09 06	☿ ⚼ ♄
13 28	☽ ∠ ☉
18 22	☿ ± ♆
19 50	☽ ⊻ ☿
20 21	☽ ⊥ ♅
20 34	☽ # ♃
23 53	☽ □ ♀
19 Friday	
05 44	☽ ∠ ♇
08 20	☽ ⊻ ♅
09 48	☽ ⊥ ☿
11 42	♂ ☍ ♃
22 03	☽ ⚹ ☉
23 49	☽ ⊥ ♇
20 Saturday	
00 17	☽ ☍ ♄
06 53	☽ ∠ ☿
08 33	☉ ∥ ♂
11 57	☽ ⊻ ♇
12 58	☿ ⚹ ♀
15 08	☽ ☍ ♃
16 25	☽ ☌ ♂
19 49	☽ □ ♆
21 Sunday	
00 30	☽ Q ☉
01 09	☉ △ ♄
07 32	☽ ∥ ♅
15 29	☽ △ ♀
17 00	♀ Q ♃
18 09	☽ ⚹ ☿
21 10	☽ ☌ ♅
21 40	☉ # ♄
22 Monday	
01 26	☽ # ♆
08 19	☽ Q ♆
13 04	☽ ⊼ ♄
16 18	☽ □ ☉
17 22	☿ ⚹ ♅
21 12	☿ ⚼ ♃
22 07	☽ Q ☿
23 32	☽ ⚼ ♀
23 Tuesday	
00 51	☽ ☌ ♇
01 10	☽ ± ♄
04 03	☽ ⊼ ♃
08 09	☽ ⊻ ♂
08 29	☽ ⚹ ♆
15 22	♂ □ ♆
16 06	☽ ± ♃
18 34	☽ ± ♀
19 11	☽ ⚼ ♄
19 59	☿ Q ♇
20 05	☽ ∥ ♀
20 43	☽ ⊥ ♂
22 19	☿ ± ♄
24 Wednesday	
00 42	♀ ⚼ ♇
07 08	☽ ⊼ ♀
08 53	☿ ⚼ ♆
09 32	☽ ⊻ ♅
09 57	☽ ⚼ ♃
14 13	☽ ∠ ♆
14 56	☽ □ ☿
15 20	☽ ∠ ♂
20 25	☿ ∠ ♂
21 18	☽ ⊥ ♅
25 Thursday	
00 39	☽ △ ♄
07 41	☽ ⊥ ♆
08 50	☽ △ ☉
12 00	☽ ⊻ ♇
14 42	☽ ∠ ♅
15 04	☽ △ ♃
18 36	♃ St R
19 08	☽ ⊻ ♆
21 35	☽ ⚹ ♂
23 26	☽ ⊥ ♇
26 Friday	
04 53	♀ △ ♅
15 29	☽ ⚼ ☉
16 17	☽ ∠ ♇
18 56	☽ ⚹ ♅
19 41	☽ ☍ ♀
21 00	☽ Q ♂
27 Saturday	
00 02	♂ # ♃
02 21	☉ □ ♇
06 21	☽ △ ☿
08 51	☽ □ ♄
09 17	☽ ± ☉
09 35	☽ # ♀
16 48	☽ Q ♅
19 28	☉ ∥ ♇
19 35	☽ ⚹ ♇
20 55	☽ ⊼ ☉
22 25	☽ □ ♃
28 Sunday	
02 08	☽ ☌ ♆
06 49	☽ □ ♂
08 48	☿ ⊼ ♄
11 48	☽ ⚼ ☿
12 50	♄ ± ♇
16 46	☽ Q ♄
16 47	☽ Q ♇
17 49	☉ △ ♃
19 51	☽ ∥ ♆
29 Monday	
00 32	☽ □ ♅
04 00	☽ ⊼ ♀
04 52	☽ ± ☿
05 52	☽ Q ♃
10 51	☽ # ♅
13 24	☽ ⚹ ♄
14 50	☽ ± ♀
16 01	☽ ⊼ ☿
17 01	☿ ± ♃
23 39	☽ □ ♇
30 Tuesday	
02 15	☽ ⚹ ♃
04 37	☽ ☍ ☉
05 43	☽ ⊻ ♆
06 52	☽ ⚼ ♀
12 19	☽ △ ♂
14 38	☽ ∠ ♄
15 45	☽ ⊥ ♆
20 42	☉ ⊼ ♆
22 16	☽ # ♂
31 Wednesday	
03 16	☽ △ ♅
03 17	☽ ∠ ♃
05 34	☽ ⊥ ♄
06 41	☽ ∠ ♆
09 13	☽ △ ♀
12 06	☽ ∥ ♃
14 14	☽ ⚼ ♂
15 28	☽ ⊻ ♄
18 09	☽ ⊥ ♃
20 29	☉ Q ♅
22 01	☽ ☍ ☿

JUNE 1961

LONGITUDES

Date	Sidereal time h m s	Sun ☉	Moon ☽	Moon ☽ 24.00	Mercury ☿	Venus ♀	Mars ♂	Jupiter ♃	Saturn ♄	Uranus ♅	Neptune ♆	Pluto ♇
01	04 38 56	10 ♊ 44 28	11 ♑ 56 35	19 ♑ 14 14	04 ♋ 01	26 ♈ 33	14 ♌ 14	07 ♒ 05	29 ♑ 26	22 ♌ 08	09 ♏ 08	05 ♍ 36
02	04 42 52	11 41 56	26 ♑ 31 16	03 ♒ 47 01	04 55	27 20	14 47	07 R 04	29 R 24	22 10	09 R 07	05 36
03	04 46 49	12 39 23	11 ♒ 00 54	18 ♒ 12 29	05 46	28 07	15 21	07 02	29 22	22 12	09 06	05 37
04	04 50 45	13 36 49	25 ♒ 21 25	02 ♓ 27 29	06 33	28 56	15 54	07 00	29 20	22 14	09 04	05 37
05	04 54 42	14 34 14	09 ♓ 30 30	16 ♓ 30 26	07 16	29 ♈ 45	16 28	06 58	29 17	22 16	09 03	05 38
06	04 58 39	15 31 39	23 ♓ 27 14	00 ♈ 20 58	07 55	00 ♉ 34	17 02	06 56	29 15	22 18	09 02	05 38
07	05 02 35	16 29 04	07 ♈ 11 38	13 ♈ 59 19	08 29	01 25	17 36	06 54	29 12	22 20	09 00	05 39
08	05 06 32	17 26 27	20 ♈ 44 02	27 ♈ 25 50	09 00	02 16	18 10	06 51	29 10	22 22	08 59	05 40
09	05 10 28	18 23 51	04 ♉ 04 43	10 ♉ 40 40	09 26	03 08	18 44	06 49	29 07	22 24	08 58	05 40
10	05 14 25	19 21 13	17 ♉ 13 40	23 ♉ 43 39	09 47	04 00	19 18	06 46	29 04	22 26	08 57	05 41
11	05 18 21	20 18 35	00 ♊ 10 35	06 ♊ 34 23	10 04	04 53	19 52	06 43	29 01	22 28	08 56	05 42
12	05 22 18	21 15 57	12 ♊ 55 01	19 ♊ 12 29	10 16	05 47	20 26	06 40	28 58	22 30	08 55	05 43
13	05 26 14	22 13 18	25 ♊ 26 45	01 ♋ 37 52	10 24	06 41	21 00	06 36	28 55	22 32	08 54	05 43
14	05 30 11	23 10 38	07 ♋ 45 57	13 ♋ 51 06	10 27	07 35	21 35	06 33	28 52	22 35	08 53	05 44
15	05 34 08	24 07 58	19 ♋ 53 33	25 ♋ 53 30	10 R 26	08 30	22 09	06 29	28 49	22 37	08 51	05 45
16	05 38 04	25 05 16	01 ♌ 51 19	07 ♌ 47 19	10 20	09 26	22 44	06 25	28 46	22 39	08 50	05 46
17	05 42 01	26 02 34	13 ♌ 41 58	19 ♌ 35 42	10 10	10 22	23 18	06 21	28 42	22 42	08 49	05 47
18	05 45 57	26 59 52	25 ♌ 29 05	01 ♍ 22 38	09 56	11 18	23 53	06 17	28 39	22 44	08 49	05 48
19	05 49 54	27 57 08	07 ♍ 16 59	13 ♍ 12 46	09 38	12 15	24 28	06 12	28 35	22 47	08 48	05 49
20	05 53 50	28 54 24	19 ♍ 10 37	25 ♍ 11 14	09 16	13 12	25 02	06 08	28 32	22 49	08 47	05 50
21	05 57 47	29 ♊ 51 39	01 ♎ 15 15	07 ♎ 23 21	08 51	14 10	25 37	06 03	28 28	22 52	08 46	05 51
22	06 01 43	00 ♋ 48 53	13 ♎ 36 09	19 ♎ 54 15	08 22	15 08	26 12	05 58	28 25	22 54	08 45	05 52
23	06 05 40	01 46 07	26 ♎ 18 11	02 ♏ 48 23	07 52	16 07	26 47	05 53	28 21	22 57	08 44	05 53
24	06 09 37	02 43 20	09 ♏ 25 12	16 ♏ 08 52	07 19	17 05	27 22	05 48	28 17	23 00	08 43	05 54
25	06 13 33	03 40 33	22 ♏ 59 26	29 ♏ 56 49	06 45	18 05	27 57	05 43	28 14	23 02	08 42	05 55
26	06 17 30	04 37 45	07 ♐ 00 46	14 ♐ 10 48	06 09	19 04	28 32	05 37	28 10	23 05	08 42	05 56
27	06 21 26	05 34 57	21 ♐ 26 18	28 ♐ 46 29	05 34	20 04	29 07	05 31	28 06	23 08	08 41	05 57
28	06 25 23	06 32 08	06 ♑ 10 25	13 ♑ 37 04	04 59	21 04	29 ♌ 43	05 26	28 02	23 11	08 40	05 59
29	06 29 19	07 29 19	21 ♑ 05 22	28 ♑ 34 12	04 24	22 05	00 ♍ 18	05 20	27 58	23 14	08 40	06 00
30	06 33 16	08 ♋ 26 31	06 ♒ 02 30	13 ♒ 29 17	03 ♋ 51	23 ♉ 06	00 ♍ 53	05 ♒ 14	27 ♑ 54	23 ♌ 16	08 ♏ 39	06 ♍ 01

Date	Moon True ☊	Moon Mean ☊	Moon ☽ Latitude
01	00 ♍ 34	01 ♍ 19	03 N 47
02	00 R 28	01 16	02 49
03	00 25	01 13	01 41
04	00 24	01 10	00 N 27
05	00 D 24	01 07	00 S 48
06	00 R 24	01 03	01 59
07	00 23	01 00	03 02
08	00 19	00 57	03 54
09	00 13	00 54	04 32
10	00 04	00 51	04 55
11	29 ♌ 53	00 48	05 02
12	29 41	00 44	04 54
13	29 29	00 41	04 32
14	29 19	00 38	03 58
15	29 11	00 35	03 13
16	29 05	00 32	02 20
17	29 01	00 28	01 21
18	29 00	00 25	00 S 19
19	29 D 00	00 22	00 N 44
20	29 01	00 19	01 46
21	29 R 01	00 16	02 44
22	29 00	00 13	03 36
23	28 58	00 09	04 18
24	28 53	00 06	04 49
25	28 46	00 03	05 05
26	28 38	00 ♍ 00	05 03
27	28 30	29 ♌ 57	04 43
28	28 22	29 54	04 03
29	28 16	29 50	03 06
30	28 ♌ 12	29 ♌ 47	01 N 57

DECLINATIONS

Date	Sun ☉	Moon ☽	Mercury ☿	Venus ♀	Mars ♂	Jupiter ♃	Saturn ♄	Uranus ♅	Neptune ♆	Pluto ♇
01	22 N 04	19 S 08	25 N 04	08 N 23	17 N 58	18 S 56	20 S 21	14 N 49	12 S 49	21 N 19
02	22 11	18 05	24 53	08 35	17 48	18 57	20 22	14 48	12 48	21 19
03	22 19	15 51	24 42	08 48	17 37	18 57	20 22	14 48	12 48	21 18
04	22 26	12 39	24 30	09 01	17 26	18 58	20 23	14 47	12 48	21 18
05	22 33	08 45	24 17	09 14	17 15	18 59	20 23	14 46	12 47	21 17
06	22 39	04 S 25	24 03	09 28	17 04	18 59	20 24	14 46	12 47	21 17
07	22 45	00 N 04	23 49	09 42	16 53	19 00	20 24	14 45	12 46	21 16
08	22 51	04 29	23 34	09 56	16 42	19 01	20 25	14 44	12 46	21 16
09	22 56	08 37	23 19	10 11	16 31	19 02	20 26	14 44	12 46	21 15
10	23 01	12 16	23 03	10 26	16 20	19 03	20 26	14 43	12 45	21 15
11	23 05	15 16	22 47	10 41	16 08	19 04	20 27	14 42	12 45	21 14
12	23 09	17 29	22 31	10 56	15 57	19 05	20 28	14 41	12 45	21 13
13	23 13	18 50	22 15	11 11	15 45	19 06	20 29	14 41	12 44	21 13
14	23 16	19 15	21 58	11 26	15 33	19 07	20 29	14 40	12 44	21 12
15	23 19	18 47	21 42	11 42	15 21	19 08	20 30	14 39	12 44	21 12
16	23 21	17 29	21 26	11 58	15 09	19 09	20 31	14 38	12 44	21 11
17	23 23	15 25	21 10	12 13	14 57	19 11	20 32	14 37	12 43	21 10
18	23 24	12 44	20 54	12 29	14 45	19 12	20 32	14 37	12 43	21 10
19	23 26	09 31	20 39	12 45	14 33	19 13	20 33	14 36	12 43	21 09
20	23 26	05 55	20 24	13 01	14 21	19 14	20 34	14 35	12 43	21 09
21	23 27	02 N 01	20 10	13 17	14 08	19 16	20 35	14 34	12 42	21 08
22	23 26	02 S 03	19 57	13 33	13 56	19 17	20 36	14 33	12 42	21 07
23	23 26	06 08	19 44	13 49	13 43	19 19	20 36	14 32	12 42	21 07
24	23 25	10 03	19 32	14 05	13 30	19 20	20 37	14 31	12 42	21 06
25	23 23	13 36	19 22	14 21	13 18	19 22	20 38	14 31	12 41	21 05
26	23 22	16 30	19 12	14 37	13 05	19 23	20 39	14 30	12 41	21 05
27	23 19	18 28	19 04	14 52	12 52	19 25	20 40	14 29	12 41	21 04
28	23 17	19 15	18 56	15 08	12 39	19 26	20 41	14 28	12 41	21 03
29	23 14	18 43	18 50	15 24	12 26	19 28	20 42	14 27	12 41	21 02
30	23 N 10	16 S 53	18 N 46	15 N 39	12 N 12	19 S 30	20 S 42	14 N 26	12 S 41	21 N 02

ZODIAC SIGN ENTRIES

Date	h	m	Planets
02	17	45	☽ ♒
04	19	50	☽ ♓
05	19	25	♀ ♉
06	23	23	☽ ♈
09	04	38	☽ ♉
11	11	40	☽ ♊
13	20	50	☽ ♋
16	08	16	☽ ♌
18	21	12	☽ ♍
21	09	32	☽ ♎
21	15	30	☉ ♋
23	18	51	☽ ♏
26	00	05	☽ ♐
28	02	00	☽ ♑
28	23	47	♂ ♍
30	02	18	☽ ♒

LATITUDES

Date	Mercury ☿	Venus ♀	Mars ♂	Jupiter ♃	Saturn ♄	Uranus ♅	Neptune ♆	Pluto ♇
01	01 N 41	01 S 59	01 N 28	00 S 27	00 S 05	00 N 43	01 N 50	12 N 48
04	01 13	02 13	01 26	00 28	00 05	00 43	01 49	12 47
07	00 N 39	02 25	01 24	00 28	00 05	00 43	01 49	12 46
10	00 S 02	02 35	01 21	00 29	00 06	00 43	01 49	12 45
13	00 48	02 43	01 19	00 29	00 06	00 43	01 49	12 44
16	01 37	02 49	01 17	00 30	00 06	00 43	01 49	12 43
19	02 27	02 54	01 15	00 31	00 07	00 43	01 49	12 43
22	03 14	02 57	01 13	00 31	00 07	00 43	01 49	12 42
25	03 55	02 59	01 11	00 32	00 07	00 43	01 49	12 41
28	04 25	03 00	01 09	00 32	00 07	00 42	01 48	12 40
31	04 S 42	02 S 59	01 N 07	00 S 33	00 S 08	00 N 42	01 N 48	12 N 39

DATA

Julian Date	2437452
Delta T	+34 seconds
Ayanamsa	23° 18' 56"
Synetic vernal point	05° ♓ 48' 04"
True obliquity of ecliptic	23° 26' 31"

LONGITUDES

Date	Chiron ⚷	Ceres ⚳	Pallas ⚴	Juno ⚵	Vesta ⚶	Black Moon Lilith ⚸
01	06 ♓ 40	00 ♉ 13	21 ♓ 56	18 ♓ 46	01 ♉ 06	13 ♌ 24
11	06 ♓ 42	03 ♉ 47	23 ♓ 33	21 ♓ 43	05 ♉ 17	14 ♌ 31
21	06 ♓ 38	07 ♉ 14	24 ♓ 48	24 ♓ 23	09 ♉ 20	15 ♌ 37
31	06 ♓ 28	10 ♉ 32	25 ♓ 37	26 ♓ 41	13 ♉ 13	16 ♌ 44

All ephemeris data is given at 12.00 UT and the Moon's longitude is additionally given for 24.00 UT

MOON'S PHASES, APSIDES AND POSITIONS ☽

Date	h	m	Phase	Longitude	Eclipse Indicator
05	21	19	☾	14 ♓ 57	
13	05	16	●	21 ♊ 57	
21	09	01	☽	29 ♍ 45	
28	12	38	○	06 ♑ 34	

Day	h	m		
02	02	49	Perigee	
17	22	24	Apogee	
30	01	25	Perigee	
01	04	02	Max dec	19° S 13'
07	11	38	0N	
14	11	19	Max dec	19° N 15'
21	23	57	0S	
28	14	29	Max dec	19° S 16'

ASPECTARIAN

h m Aspects

01 Thursday
01 33 ☽ △ ♇
04 01 ☽ ⚺ ♃
04 05 ☽ ⚼ ♅
05 39 ☽ ± ♂
07 23 ☽ ✶ ♆
09 53 ☽ ⚻ ☉
15 55 ☽ ⚻ ♂
18 55 ☽ ± ♅
19 37 ☽ ∥ ♃
20 27 ☽ ± ☉

02 Friday
02 15 ☽ ⚼ ♇
03 06 ☽ Q ♆
04 49 ☽ ⚻ ♅
12 19 ☽ ⚼ ☉
13 24 ☽ □ ♀
16 20 ☽ ⋕ ♂
16 45 ☽ ☌ ♄
17 06 ☽ ± ♇

03 Saturday
02 46 ☽ ⚻ ☿
03 02 ☽ ⚻ ♇
05 24 ☽ ☌ ♃
07 19 ☿ ✶ ♇
08 48 ☽ □ ♆
13 20 ☽ ± ☿
14 56 ☽ △ ☉
19 31 ☽ ☍ ♂
20 42 ☽ ⋕ ♅
21 01 ☽ Q ♀

04 Sunday
05 15 ☽ ⚼ ☿
06 44 ☽ ☍ ♅
11 03 ☽ ∥ ♆
18 23 ☽ ✶ ♀
18 41 ☽ ⚺ ♄
23 18 ♀ □ ♄

05 Monday
02 14 ☿ ⚻ ♃
04 50 ☽ ⊥ ♄
05 13 ☉ ⚼ ♄
05 23 ☽ ☍ ♇
07 41 ☽ ⚺ ♃
07 59 ☽ △ ☿
09 17 ☽ ⋕ ♀
11 13 ☽ △ ♆
11 43 ☿ ∠ ♅
17 55 ☽ ⊥ ♃
20 10 ☽ ∠ ♄
21 19 ☽ □ ☉
21 32 ☽ ∠ ♀
23 45 ☉ ± ♆

06 Tuesday
00 26 ☽ ⚻ ♂
09 23 ☽ ∠ ♃
09 59 ☽ ⚻ ♅
11 14 ☽ ± ♂
13 00 ☽ ⚼ ♆
14 04 ☽ ⊥ ♀
20 26 ☽ ± ♅
21 32 ☉ ∠ ♀
22 03 ☽ ✶ ♄

07 Wednesday
01 12 ☽ ⚺ ♀
03 35 ☽ ⚼ ♂
04 40 ☽ ± ♆
06 53 ☽ Q ☉
09 17 ☽ ⚻ ♇
11 29 ☽ ✶ ♃
12 14 ☽ ⚼ ♅
14 23 ☽ □ ☿
15 11 ☽ ⚻ ♆
19 03 ☽ Q ♄
19 52 ☽ ± ♇

08 Thursday
05 41 ☽ ✶ ☉
07 13 ☽ △ ♂
08 40 ☽ Q ♃
11 42 ☿ △ ♆
11 52 ☽ ⚼ ♇
14 55 ☽ △ ♅
23 36 ☽ Q ☿

09 Friday
03 04 ☽ □ ♄
10 10 ☽ ☌ ♀
10 40 ☽ ∠ ☉
14 54 ☽ △ ♇
16 57 ☽ □ ♃
20 52 ☽ ☍ ♆
22 00 ☽ ✶ ☿
22 35 ☽ ∥ ♀

10 Saturday
04 20 ☽ ⊥ ☉
08 23 ☉ ✶ ♂
14 41 ☉ ∥ ☿
15 36 ☽ ⋕ ♆
15 59 ☽ □ ♂
16 13 ☽ ⚺ ☉
21 37 ☽ □ ♅

11 Sunday
02 17 ☽ ∠ ☿
07 04 ☽ ∥ ♅
09 51 ☽ △ ♄
19 25 ☽ ⊥ ☿
19 47 ☽ ∥ ♂
21 29 ☽ ⚺ ♀
22 22 ☽ □ ♇

12 Monday
00 13 ☽ △ ♃
03 07 ☽ Q ♂
04 25 ☽ ⚻ ♆
06 55 ☽ ⚺ ☿
07 24 ☽ Q ♅
09 40 ☽ ⊥ ♀
10 11 ♀ △ ♇
14 00 ☽ ⚼ ♄
15 47 ☽ ± ♆
21 23 ☉ ⚼ ♃

13 Tuesday
03 02 ☽ ✶ ♂
04 10 ☽ ∠ ♀
04 38 ☽ ⚼ ♃
05 16 ☽ ☌ ☉
06 23 ☽ ✶ ♅
07 09 ☽ ± ♄
08 40 ☽ Q ♇
09 00 ☽ ⚼ ♆
10 11 ♀ □ ♃
18 42 ☽ ⚻ ♄
20 18 ☉ ✶ ♅
21 25 ☽ ⋕ ♃
21 57 ☽ ± ♃

14 Wednesday
04 37 ☉ ± ♄
08 01 ☽ ✶ ♇
09 33 ☽ ∠ ♂
09 37 ☽ ⚻ ♃
11 37 ☽ ✶ ♀
11 38 ☽ ∠ ♅
14 11 ☽ △ ♆
17 06 ☿ St R
17 17 ☽ ☌ ☿

15 Thursday
00 19 ☽ ⋕ ♃
02 17 ☉ Q ♇
04 11 ☽ ⊥ ♂
05 14 ☉ ⚼ ♆
05 28 ☽ ⊥ ♅
13 19 ☽ Q ♀
13 43 ☽ ∠ ♇
16 44 ☽ ⚺ ♂
17 27 ☽ ⚺ ♅
21 02 ♀ ☍ ♆
21 12 ☽ ⚺ ☉

16 Friday
05 48 ☽ ☍ ♄
07 47 ☽ ⊥ ♇
08 45 ♂ ☌ ♅
10 19 ☽ ⊥ ☉
19 55 ☽ ⚺ ♇
21 10 ☽ ☍ ♃

17 Saturday
02 07 ☽ □ ♆
04 38 ☽ □ ♀
04 56 ☽ ⚺ ☿
06 08 ☽ ∠ ☉
07 52 ☿ ✶ ♀
10 54 ☿ ∥ ♇
16 56 ☽ ⊥ ☿
16 58 ☽ ∥ ♂
19 43 ☽ ∥ ♅

18 Sunday
06 23 ☽ ☌ ♅
08 34 ☽ ☌ ♂
10 54 ☽ ∠ ☿
12 07 ☽ ⋕ ♆
13 47 ☽ ∥ ♀
14 42 ☽ Q ♆
15 21 ☽ ✶ ☉
18 25 ☽ ⚻ ♄

19 Monday
05 55 ♂ ∥ ♅
06 34 ☽ ± ♄
08 24 ♀ ⋕ ♆
09 01 ☽ ☌ ♇
09 50 ☽ ⚻ ♃
15 03 ☽ ✶ ♆
16 23 ☿ ∠ ♂
16 37 ☽ ✶ ☿
17 53 ☽ Q ☉
20 52 ☿ ⋕ ♄
21 54 ☽ ± ♃
22 56 ☽ △ ♀

20 Tuesday
00 42 ☽ ⚼ ♄
03 09 ☉ ⚻ ♄
15 53 ☽ ⚼ ♃
16 03 ☽ Q ☿
19 19 ☽ ⚺ ♅
21 11 ☽ ∠ ♆

21 Wednesday
00 18 ☽ ⚺ ♂
06 33 ☽ △ ♄
07 16 ☽ ⊥ ♅
07 32 ☽ ⚼ ♀
09 01 ☽ □ ☉
12 45 ☽ ⊥ ♂
14 58 ☽ ⊥ ♆
16 23 ☿ △ ♆
16 23 ☉ ± ♃
21 00 ☽ ⚺ ♇
21 20 ☽ △ ♃

22 Thursday
00 58 ☽ ∠ ♅
02 17 ☽ □ ☿
02 39 ☽ ⚺ ♆
02 39 ☽ ± ♀
07 09 ☽ ∠ ♂
08 39 ☽ ⊥ ♇
15 11 ☽ ⚻ ♀

23 Friday
01 50 ☽ ∠ ♇
05 43 ☽ ✶ ♅
06 56 ♀ ∥ ♂
08 15 ☿ ∠ ♅
10 01 ♂ Q ♆
12 41 ♃ ⚻ ♇
12 56 ☽ ✶ ♂
15 47 ☽ □ ♄
22 54 ☽ △ ☉

24 Saturday
03 58 ☽ Q ♅
05 30 ☽ □ ♃
05 38 ☽ ✶ ♇
08 21 ☽ △ ☿
10 45 ☽ ☌ ♆
11 54 ☽ Q ♂

25 Sunday
00 12 ☽ Q ♄
02 44 ☽ ☍ ♀
03 07 ☽ Q ♇
03 53 ☽ ⚼ ☉
05 29 ☽ ∥ ♆
09 54 ☽ ⋕ ♂
09 55 ☽ ⚼ ☿
12 05 ☽ □ ♅
12 17 ☿ ⋕ ♃
13 14 ☽ Q ♃
18 09 ☽ ⋕ ♀
18 50 ☽ ⋕ ♅
20 42 ☽ ± ☉
20 57 ☽ □ ♂
21 01 ☽ ✶ ♄
22 12 ♂ ⚻ ♄

26 Monday
00 50 ☽ ± ☿
01 46 ♀ ∥ ♅
07 41 ☽ ⚻ ☉
09 40 ☽ ✶ ♃
10 11 ☽ □ ♇
10 37 ☽ ⚻ ☿
14 50 ☽ ⚺ ♆
20 35 ☿ ✶ ♇
22 16 ☽ ∠ ♄

27 Tuesday
00 51 ☽ ⊥ ♆
09 35 ☽ ⚻ ♀
10 30 ☽ ∠ ♃
10 39 ☉ ⚻ ♃
11 42 ☉ ☌ ☿
13 05 ☽ ⊥ ♄
13 55 ☿ ⚻ ♃
14 47 ☽ △ ♅
15 41 ☽ ∠ ♆
19 29 ☿ ∠ ♀
20 08 ☽ ± ♀
21 37 ☉ ✶ ♇
22 51 ☽ ⚺ ♄
23 23 ☽ ⋕ ☿

28 Wednesday
01 05 ☽ △ ♂
01 08 ☽ ⊥ ♃
07 58 ♂ ⋕ ♆
10 08 ☽ ☍ ☿
10 48 ☽ ⚺ ♃
11 41 ☽ △ ♇
11 49 ☽ ⚼ ♀
12 38 ☽ ☍ ☉
15 15 ☽ ⚼ ♅
16 02 ☽ ✶ ♆

29 Thursday
02 19 ☽ ⚼ ♂
05 47 ☽ ± ♅
09 19 ☽ ⋕ ☿
11 19 ☽ Q ♆
11 51 ☽ ⚼ ♇
13 42 ☽ △ ♀
15 26 ☽ ⚻ ♅
17 22 ☽ ± ♂
22 59 ☽ ☌ ♄

30 Friday
02 19 ☽ ± ♇
03 23 ☽ ⚻ ♂
07 32 ☉ ∠ ♅
08 36 ☽ ⚻ ☿
10 42 ☽ ☌ ♃
11 58 ☽ ⚻ ♇
16 08 ☽ ⚻ ☉
16 12 ☽ □ ♆
16 28 ♀ □ ♅
17 10 ☉ △ ♆
17 56 ☽ ± ☿
22 04 ☽ ⋕ ♀

JULY 1961

LONGITUDES

Date	Sidereal time h m s	Sun ☉ ° ' "	Moon ☽ ° ' "	Moon ☽ 24.00 ° ' "	Mercury ☿ ° '	Venus ♀ ° '	Mars ♂ ° '	Jupiter ♃ ° '	Saturn ♄ ° '	Uranus ♅ ° '	Neptune ♆ ° '	Pluto ♇ ° '
01	06 37 12	09 ♋ 23 42	20 ♒ 53 40	28 ♒ 14 52	03 ♋ 20	24 ♉ 07	01 ♍ 29	05 ♒ 08	27 ♑ 50	23 ♌ 19	08 ♏ 38	06 ♍ 02
02	06 41 09	10 20 53	05 ♓ 32 19	12 ♓ 45 30	02 R 52	25 08	02 04	05 R 01	27 R 46	23 22	08 R 38	06 04
03	06 45 06	11 18 05	19 ♓ 54 08	26 ♓ 57 59	02 27	26 10	02 40	04 55	27 42	23 25	08 37	06 05
04	06 49 02	12 15 16	03 ♈ 56 59	10 ♈ 51 08	02 05	27 12	03 15	04 48	27 38	23 28	08 37	06 06
05	06 52 59	13 12 28	17 ♈ 40 32	24 ♈ 25 17	01 48	28 14	03 51	04 42	27 33	23 31	08 36	06 08
06	06 56 55	14 09 41	01 ♉ 05 36	07 ♉ 41 39	01 34	29 ♉ 17	04 27	04 35	27 29	23 34	08 36	06 09
07	07 00 52	15 06 53	14 ♉ 13 40	20 ♉ 41 52	01 25	00 ♊ 20	05 03	04 28	27 25	23 37	08 35	06 10
08	07 04 48	16 04 06	27 ♉ 06 26	03 ♊ 27 35	01 21	01 23	05 38	04 21	27 21	23 41	08 35	06 12
09	07 08 45	17 01 20	09 ♊ 45 29	16 ♊ 00 20	01 D 22	02 26	06 14	04 14	27 16	23 44	08 35	06 13
10	07 12 41	17 58 34	22 ♊ 12 18	28 ♊ 21 32	01 28	03 29	06 50	04 07	27 12	23 47	08 34	06 15
11	07 16 38	18 55 48	04 ♋ 28 11	10 ♋ 32 26	01 39	04 33	07 26	04 00	27 08	23 50	08 34	06 16
12	07 20 35	19 53 02	16 ♋ 34 27	22 ♋ 34 25	01 56	05 37	08 02	03 53	27 03	23 53	08 34	06 18
13	07 24 31	20 50 17	28 ♋ 32 33	04 ♌ 29 05	02 18	06 41	08 39	03 45	26 59	23 57	08 33	06 19
14	07 28 28	21 47 32	10 ♌ 24 17	16 ♌ 18 26	02 45	07 46	09 15	03 38	26 54	24 00	08 33	06 21
15	07 32 24	22 44 47	22 ♌ 11 53	28 ♌ 04 58	03 18	08 50	09 51	03 31	26 50	24 03	08 33	06 22
16	07 36 21	23 42 02	03 ♍ 58 08	09 ♍ 51 48	03 56	09 55	10 27	03 23	26 46	24 07	08 33	06 24
17	07 40 17	24 39 18	15 ♍ 46 27	21 ♍ 42 36	04 39	11 00	11 04	03 15	26 41	24 10	08 33	06 26
18	07 44 14	25 36 34	27 ♍ 40 47	03 ♎ 41 34	05 27	12 05	11 40	03 08	26 37	24 13	08 33	06 27
19	07 48 10	26 33 49	09 ♎ 45 33	15 ♎ 53 18	06 21	13 10	12 17	03 00	26 32	24 17	08 33	06 29
20	07 52 07	27 31 06	22 ♎ 05 26	28 ♎ 22 30	07 20	14 16	12 53	02 52	26 28	24 20	08 33	06 30
21	07 56 04	28 28 22	04 ♏ 45 04	11 ♏ 13 37	08 24	15 22	13 30	02 45	26 23	24 24	08 D 33	06 32
22	08 00 00	29 ♋ 25 39	17 ♏ 48 34	24 ♏ 30 16	09 33	16 27	14 06	02 37	26 19	24 27	08 33	06 34
23	08 03 57	00 ♌ 22 56	01 ♐ 18 56	08 ♐ 14 40	10 46	17 33	14 43	02 29	26 15	24 31	08 33	06 36
24	08 07 53	01 20 13	15 ♐ 17 22	22 ♐ 26 49	12 05	18 40	15 20	02 22	26 10	24 34	08 33	06 37
25	08 11 50	02 17 31	29 ♐ 42 35	07 ♑ 04 03	13 28	19 46	15 57	02 14	26 06	24 38	08 33	06 39
26	08 15 46	03 14 50	14 ♑ 30 25	22 ♑ 00 44	14 55	20 53	16 33	02 06	26 01	24 41	08 33	06 41
27	08 19 43	04 12 08	29 ♑ 33 57	07 ♒ 08 54	16 27	21 59	17 10	01 58	25 57	24 45	08 33	06 43
28	08 23 39	05 09 28	14 ♒ 44 22	22 ♒ 19 09	18 04	23 06	17 47	01 50	25 53	24 48	08 33	06 44
29	08 27 36	06 06 49	29 ♒ 52 06	07 ♓ 22 11	19 44	24 13	18 24	01 43	25 48	24 52	08 34	06 46
30	08 31 33	07 04 10	14 ♓ 48 26	22 ♓ 10 06	21 28	25 20	19 01	01 35	25 44	24 55	08 34	06 48
31	08 35 29	08 ♌ 01 32	29 ♓ 26 33	06 ♈ 37 19	23 ♋ 15	26 ♊ 27	19 ♍ 38	01 ♒ 27	25 ♑ 40	24 ♌ 59	08 ♏ 34	06 ♍ 50

Date	Moon True ☊ ° '	Moon Mean ☊ ° '	Moon ☽ Latitude ° '
01	28 ♌ 11	29 ♌ 44	00 N 39
02	28 D 11	29 41	00 S 40
03	28 12	29 38	01 55
04	28 13	29 34	03 01
05	28 R 13	29 31	03 56
06	28 11	29 28	04 36
07	28 08	29 25	05 01
08	28 03	29 22	05 10
09	27 57	29 19	05 03
10	27 50	29 15	04 42
11	27 43	29 12	04 09
12	27 37	29 09	03 24
13	27 32	29 06	02 31
14	27 29	29 03	01 32
15	27 27	29 00	00 S 29
16	27 D 27	28 56	00 N 36
17	27 29	28 53	01 39
18	27 30	28 50	02 38
19	27 32	28 47	03 32
20	27 33	28 44	04 16
21	27 R 33	28 40	04 50
22	27 32	28 37	05 10
23	27 29	28 34	05 14
24	27 26	28 31	04 59
25	27 23	28 28	04 26
26	27 19	28 25	03 34
27	27 17	28 21	02 27
28	27 15	28 18	01 N 09
29	27 D 15	28 15	00 S 14
30	27 16	28 12	01 36
31	27 ♌ 17	28 ♌ 09	02 S 49

DECLINATIONS

Date	Sun ☉ ° '	Moon ☽ ° '	Mercury ☿ ° '	Venus ♀ ° '	Mars ♂ ° '	Jupiter ♃ ° '	Saturn ♄ ° '	Uranus ♅ ° '	Neptune ♆ ° '	Pluto ♇ ° '
01	23 N 07	13 S 55	18 N 42	15 N 54	11 N 59	19 S 31	20 S 43	14 N 25	12 S 40	21 N 01
02	23 02	10 06	18 40	16 10	11 46	19 33	20 44	14 24	12 40	21 00
03	22 58	05 46	18 40	16 25	11 32	19 35	20 45	14 23	12 40	21 00
04	22 53	01 S 12	18 40	16 39	11 19	19 36	20 46	14 22	12 40	20 59
05	22 47	03 N 18	18 43	16 54	11 05	19 38	20 47	14 21	12 40	20 58
06	22 41	07 32	18 46	17 08	10 52	19 40	20 48	14 20	12 40	20 57
07	22 35	11 19	18 51	17 23	10 38	19 42	20 49	14 19	12 40	20 57
08	22 28	14 29	18 57	17 36	10 24	19 44	20 50	14 18	12 40	20 56
09	22 21	16 55	19 04	17 50	10 10	19 45	20 51	14 17	12 40	20 55
10	22 14	18 31	19 12	18 04	09 56	19 47	20 52	14 15	12 40	20 54
11	22 06	19 13	19 21	18 17	09 42	19 49	20 53	14 14	12 40	20 54
12	21 58	19 02	19 31	18 29	09 28	19 51	20 54	14 13	12 40	20 53
13	21 50	17 59	19 41	18 42	09 14	19 53	20 54	14 12	12 39	20 52
14	21 41	16 10	19 52	18 54	09 00	19 55	20 55	14 11	12 39	20 51
15	21 31	13 40	20 04	19 06	08 45	19 57	20 56	14 10	12 39	20 51
16	21 22	10 36	20 16	19 18	08 31	19 59	20 57	14 09	12 39	20 50
17	21 12	07 08	20 28	19 29	08 17	20 01	20 58	14 08	12 39	20 49
18	21 01	03 N 21	20 39	19 40	08 02	20 03	20 59	14 07	12 40	20 48
19	20 51	00 S 37	20 51	19 50	07 48	20 04	21 00	14 05	12 40	20 47
20	20 40	04 38	21 02	20 00	07 33	20 06	21 01	14 04	12 40	20 47
21	20 28	08 33	21 13	20 10	07 18	20 08	21 02	14 03	12 40	20 46
22	20 16	12 11	21 22	20 19	07 04	20 10	21 03	14 02	12 40	20 45
23	20 04	15 18	21 31	20 28	06 49	20 12	21 04	14 01	12 40	20 44
24	19 52	17 40	21 39	20 36	06 34	20 14	21 05	14 00	12 40	20 43
25	19 39	19 00	21 45	20 44	06 19	20 16	21 06	13 58	12 40	20 43
26	19 26	19 06	21 50	20 51	06 04	20 18	21 06	13 57	12 40	20 42
27	19 13	17 51	21 53	20 58	05 49	20 20	21 07	13 56	12 40	20 41
28	18 59	15 19	21 54	21 05	05 34	20 22	21 08	13 55	12 40	20 40
29	18 45	11 45	21 53	21 11	05 19	20 23	21 09	13 54	12 41	20 39
30	18 30	07 27	21 49	21 16	05 04	20 25	21 10	13 52	12 41	20 39
31	18 N 16	02 S 48	21 N 44	21 N 21	04 N 49	20 S 27	21 S 11	13 N 51	12 S 41	20 N 38

ZODIAC SIGN ENTRIES

Date	h m	Planets
02	02 52	☽ ♓
04	05 12	☽ ♈
06	10 01	☽ ♉
07	04 32	♀ ♊
08	17 27	☽ ♊
11	03 13	☽ ♋
13	14 56	☽ ♌
16	03 55	☽ ♍
18	16 39	☽ ♎
21	03 05	☽ ♏
23	02 24	☉ ♌
23	09 42	☽ ♐
25	12 29	☽ ♑
27	12 41	☽ ♒
29	12 13	☽ ♓
31	12 56	☽ ♈

LATITUDES

Date	Mercury ☿ ° '	Venus ♀ ° '	Mars ♂ ° '	Jupiter ♃ ° '	Saturn ♄ ° '	Uranus ♅ ° '	Neptune ♆ ° '	Pluto ♇ ° '
01	04 S 42	02 S 59	01 N 07	00 S 33	00 S 08	00 N 42	01 N 48	12 N 39
04	04 45	02 57	01 04	00 34	00 08	00 42	01 48	12 39
07	04 35	02 55	01 02	00 34	00 08	00 42	01 48	12 38
10	04 14	02 51	01 00	00 35	00 08	00 42	01 48	12 37
13	03 44	02 46	00 58	00 35	00 09	00 42	01 48	12 37
16	03 07	02 41	00 56	00 36	00 09	00 42	01 47	12 36
19	02 27	02 34	00 54	00 36	00 09	00 42	01 47	12 36
22	01 44	02 27	00 52	00 36	00 10	00 42	01 47	12 35
25	01 01	02 19	00 50	00 37	00 10	00 42	01 47	12 35
28	00 S 20	02 11	00 48	00 37	00 10	00 42	01 47	12 34
31	00 N 18	02 S 02	00 N 46	00 S 38	00 S 10	00 N 42	01 N 46	12 N 34

DATA

Julian Date	2437482
Delta T	+34 seconds
Ayanamsa	23° 19' 01"
Synetic vernal point	05° ♓ 47' 59"
True obliquity of ecliptic	23° 26' 31"

MOON'S PHASES, APSIDES AND POSITIONS ☽

Date	h m	Phase	Longitude ° '	Eclipse Indicator
05	03 32	☾	12 ♈ 52	
12	19 11	●	20 ♋ 10	
20	23 13	☽	27 ♎ 58	
27	19 50	○	04 ♒ 31	

Day	h m		
15	11 27	Apogee	
28	08 43	Perigee	
04	18 20	0N	
11	18 53	Max dec	19° N 16'
19	08 16	0S	
26	01 44	Max dec	19° S 13'

LONGITUDES

Date	Chiron ⚷ ° '	Ceres ⚳ ° '	Pallas ⚴ ° '	Juno ⚵ ° '	Vesta ⚶ ° '	Black Moon Lilith ⚸ ° '
01	06 ♓ 28	10 ♉ 32	25 ♓ 37	26 ♓ 41	13 ♉ 13	16 ♌ 44
11	06 ♓ 14	13 ♉ 40	25 ♓ 57	28 ♓ 32	16 ♉ 55	17 ♌ 50
21	05 ♓ 54	16 ♉ 34	25 ♓ 45	29 ♓ 51	20 ♉ 23	18 ♌ 57
31	05 ♓ 31	19 ♉ 14	24 ♓ 58	00 ♈ 34	23 ♉ 37	20 ♌ 03

ASPECTARIAN

h m	Aspects
01 Saturday	
02 28	☽ ± ☉
07 59	☽ ⅊ ☿
08 26	☽ # ♅
15 58	☽ ☍ ♅
17 38	☽ □ ♀
18 06	☽ ⅊ ☉
20 18	☽ ∥ ♆
23 16	☽ ⊻ ♄
02 Sunday	
01 26	☽ # ♂
06 02	☽ ☍ ♂
07 43	☽ △ ☿
09 05	☽ ⊥ ♄
11 09	☽ ⊻ ♃
12 52	☽ ☍ ♇
17 07	☽ △ ♆
20 33	☽ △ ☉
21 02	☽ ⊥ ♃
23 57	☽ ∠ ♄
03 Monday	
01 36	☽ Q ♀
06 47	☿ ∗ ♂
12 01	☽ ∠ ♃
16 47	☿ ⊥ ♀
17 59	☽ ⊼ ♅
18 18	☽ ⅊ ♆
23 28	☽ ∗ ♀
04 Tuesday	
01 11	☽ ∗ ♄
04 16	☽ ± ♅
08 52	☽ □ ☿
09 41	☽ ± ♆
10 45	☽ ⊼ ♂
13 28	☽ ∗ ♃
15 44	☽ ⊼ ♇
19 52	☽ ⅊ ♅
20 05	☽ ⊼ ♆
21 17	♀ △ ♄
21 38	☽ ± ♂
21 49	☽ Q ♄
05 Wednesday	
01 22	♂ ± ♄
02 13	☽ ± ♇
03 32	☽ ∠ ♀
03 32	☽ □ ☉
03 55	☉ ∠ ♀
10 17	☽ Q ♃
14 11	☽ ⅊ ♂
15 41	☽ Q ☿
18 08	☽ ⅊ ♇
20 47	☽ ⊥ ♀
22 26	☽ △ ♅
06 Thursday	
05 32	☽ □ ♄
08 27	☽ ⊻ ♀
12 51	☽ ∗ ☿
14 05	☽ Q ☉
16 40	♂ ⊼ ♃
18 17	☽ □ ♃
18 22	☽ △ ♂
21 12	☽ △ ♇
07 Friday	
01 38	☽ ☍ ♆
07 38	☽ ∥ ♂
13 46	☽ ∗ ☉
16 01	☽ ∠ ☿
21 33	☽ # ♆
08 Saturday	
05 32	☽ □ ♅
08 43	☽ ⊥ ☿
10 22	☽ ∥ ♅
11 29	☿ ⊻ ♀
12 27	☽ △ ♄
19 36	☿ St D
20 00	☽ ⊻ ☿
20 05	☽ ∠ ☉
20 47	☽ ☌ ♀
09 Sunday	
01 34	☽ △ ♃
04 57	☽ □ ♂
05 14	☽ □ ♇
09 44	☽ ⊼ ♆
11 14	♂ ☌ ♇
14 37	☽ ⊥ ☉
15 47	☽ Q ♅
16 47	☽ ⅊ ♄
21 14	☽ ± ♆
10 Monday	
02 13	☽ ∥ ♀
03 07	☽ ⊻ ☉
06 04	☽ ⅊ ♃
06 50	☉ ⊥ ♅
10 04	☽ ± ♄
14 39	☽ ⅊ ♆
15 05	☽ ∗ ♅
15 59	☽ Q ♇
17 23	☽ Q ♂
21 41	☽ ⊼ ♄
23 25	☽ ± ♃
11 Tuesday	
00 48	♀ △ ♃
06 22	☽ ☌ ☿
11 05	☽ ⊼ ♃
12 11	☽ ⊻ ♀
15 33	☽ ∗ ♇
18 10	☽ ∗ ♂
20 05	☽ △ ♆
20 40	☽ ∠ ♅
12 Wednesday	
01 11	☽ ⊥ ♀
03 30	♄ # ♇
14 38	☽ ⊥ ♅
19 11	☽ ☌ ☉
20 52	☽ ∠ ♀
21 27	☽ ∠ ♇
13 Thursday	
00 19	☽ ∥ ♀
01 37	☽ ∠ ♂
02 42	☽ ⊻ ♅
03 34	♀ □ ♇
08 35	♂ ∗ ♆
08 52	☽ ☍ ♄
15 36	☽ ⊥ ♇
19 52	☽ ⊻ ☿
20 43	☽ ⊥ ♂
22 25	☽ ☍ ♃
14 Friday	
00 28	☉ ∠ ♇
03 45	☽ ⊻ ♇
06 06	☽ ∗ ♀
08 15	☽ □ ♆
08 30	☽ ⊥ ☿
09 31	☽ ⊻ ♂
18 03	☿ # ♃
15 Saturday	
00 46	♂ ± ♃
03 39	☽ ∠ ☿
05 37	♀ ⊼ ♆
07 32	☽ ∥ ♅
08 57	☽ Q ♀
13 13	☽ ⊻ ☉
15 48	☽ ☌ ♅
18 59	☿ ⊼ ♃
20 19	☽ # ♆
20 52	☽ Q ♆
21 24	☽ ⊼ ♄
16 Sunday	
02 32	☽ ⊥ ☉
09 33	☽ ± ♄
10 49	☽ ⊼ ♃
11 55	☽ ∗ ☿
16 58	☽ ☌ ♇
21 19	☽ ∗ ♆
22 29	☽ ∠ ☉
22 54	☽ ± ♃
22 56	☉ ⊻ ♅
17 Monday	
01 20	☽ □ ♀
01 55	☽ ☌ ♂
03 45	☽ ⅊ ♄
03 47	☽ ∥ ♂
13 54	☽ Q ☿
15 07	♀ □ ♂
16 58	☽ ⅊ ♃
18 Tuesday	
02 14	♀ ⅊ ♄
03 42	☽ ∠ ♆
05 02	☽ ⊻ ♅
07 29	☽ ∗ ☉
09 52	☽ △ ♄
10 02	♂ ⅊ ♄
15 12	♀ Q ♅
16 41	☉ # ♄
17 07	☽ ⊥ ♅
21 43	☽ ⊥ ♆
22 46	☽ △ ♃
19 Wednesday	
04 43	☽ □ ☿
05 13	☿ ∥ ♇
05 31	☽ ⊻ ♇
09 27	☽ Q ☉
09 36	☽ ⊻ ♆
11 03	☽ ∠ ♅
11 24	☉ ☍ ♄
11 44	☉ ∥ ☿
15 21	☿ ∗ ♇
17 12	☽ ⊻ ♂
17 21	☽ ⊥ ♇
19 21	☽ △ ♀
19 34	☉ ∥ ♇
20 Thursday	
05 30	☽ ⊥ ♂
09 27	☿ # ♄
10 53	☽ ∠ ♇
16 20	☽ ∗ ♅
18 04	♀ ± ♆
18 48	♆ St D
20 19	☽ □ ♄
23 13	☽ □ ☉
23 38	☽ ∠ ♂
21 Friday	
02 59	☽ ⅊ ♀
04 42	☽ # ♂
07 41	♀ # ♃
08 17	☽ □ ♃
13 32	☉ ∠ ♂
15 04	☽ Q ♅
15 10	☿ △ ♆
15 20	☽ ∗ ♇
19 03	☽ ☌ ♆
19 25	☽ △ ☿
21 21	☽ ± ♀
22 Saturday	
04 57	☽ ∗ ♂
05 42	☽ Q ♄
09 12	☉ ∥ ♀
09 20	☽ ⊼ ♀
10 03	☿ ∠ ♅
13 22	☽ Q ♇
15 26	☽ ∥ ♆
17 01	☽ Q ♃
22 42	☉ # ♃
23 57	☽ □ ♅
23 Sunday	
01 15	☽ ⅊ ☿
01 33	☽ # ♅
03 08	☽ ∗ ♄
03 32	☽ Q ♂
10 15	☽ △ ☉
10 38	♀ ⅊ ♃
14 02	☽ ∗ ♃
17 27	☉ ⊥ ♇
18 37	☽ ± ☿
21 11	☽ □ ♇
24 Monday	
00 31	☽ ⊻ ♆
05 03	☽ ∠ ♄
05 59	☽ ⊼ ☿
10 45	☽ ⊥ ♆
12 04	☽ □ ♂
13 54	☽ ⅊ ☉
15 27	☽ ∠ ♃
18 09	☽ ☍ ♀
20 10	☽ ⊥ ♄
25 Tuesday	
01 50	☽ ∠ ♆
03 35	☽ △ ♅
05 59	☽ ± ☉
06 05	☽ ⊻ ♄
06 19	☽ ⊥ ♃
08 40	♀ ∥ ♇
10 36	☉ ☍ ♃
16 05	☽ ⊻ ♃
16 31	☽ ⊼ ☉
18 41	♀ ± ♄
23 21	☽ △ ♇
26 Wednesday	
02 24	☽ ∗ ♆
04 12	☽ ⅊ ♅
12 45	☽ ☍ ☿
15 26	☽ △ ♂
18 43	☽ ± ♅
21 40	☽ Q ♆
23 00	☽ ⊼ ♀
23 30	☽ ⅊ ♇
27 Thursday	
04 19	☽ ⊼ ♅
05 30	♂ ⅊ ♃
06 17	☽ ☌ ♄
09 18	☽ ± ♀
13 49	☽ ± ♇
15 47	☽ ☌ ♃
16 18	☽ ⅊ ♂
19 50	☽ ☍ ☉
23 20	☽ ⊼ ♇
28 Friday	
00 40	☽ ⅊ ♀
02 14	☽ □ ♆
05 32	☿ ∗ ♂
07 08	☽ ± ♂
17 02	☽ ⊼ ♂
17 54	☽ ⊼ ☿
21 53	♀ ⅊ ♆
22 14	☽ # ♅
23 12	☿ ⊥ ♅
29 Saturday	
02 18	☽ △ ♀
04 00	☽ ☍ ♅
04 14	♀ # ♄
04 35	☽ ± ☿
05 34	☽ ⊻ ♄
06 17	☽ ∥ ♆
14 55	☽ ⊻ ♃
15 05	☽ ⊥ ♄
20 46	☽ ⅊ ☿
22 40	☽ ⊼ ☉
23 04	☽ ☍ ♇
30 Sunday	
00 11	♀ Q ♇
00 27	☽ ⊥ ♃
01 55	☽ △ ♆
02 39	♀ ∗ ♅
05 01	☉ ⊻ ♇
05 27	☽ ∠ ♄
09 00	☽ ± ☉
14 52	☽ ∠ ♃
16 40	☿ ∠ ♇
16 46	♀ ± ♃
19 09	☽ ☍ ♂
20 00	♀ ⊼ ♄
31 Monday	
00 20	☽ △ ☿
00 40	☽ ⅊ ☉
01 08	☽ # ♂
02 18	☽ ⅊ ♆
04 36	☽ ⊼ ♅
05 47	☽ ∗ ♄
06 39	☽ □ ♀
14 34	☽ ± ♅
15 19	☽ ∗ ♃
17 13	☽ ± ♆

All ephemeris data is given at 12.00 UT and the Moon's longitude is additionally given for 24.00 UT

AUGUST 1961

Date	Sidereal time h m s	Sun ☉ ° ' "	Moon ☽ ° ' "	Moon ☽ 24.00 ° ' "	Mercury ☿ ° '	Venus ♀ ° '	Mars ♂ ° '	Jupiter ♃ ° '	Saturn ♄ ° '	Uranus ♅ ° '	Neptune ♆ ° '	Pluto ♇ ° '
01	08 39 26	08 ♌ 58 55	13 ♈ 42 06	20 ♈ 40 46	25 ♋ 06	27 ♊ 35	20 ♍ 16	01 ♒ 20	25 ♑ 35	25 ♌ 03	08 ♏ 35	06 ♍ 52
02	08 43 22	09 56 20	27 ♈ 33 17	04 ♉ 19 45	26 59	28 42	20 53	01 R 12	25 R 31	25 06	08 35	06 54
03	08 47 19	10 53 46	11 ♉ 00 21	17 ♉ 35 21	28 ♋ 55	29 ♊ 50	21 30	01 04	25 27	25 10	08 36	06 55
04	08 51 15	11 51 13	24 ♉ 05 03	00 ♊ 29 49	00 ♌ 53	00 ♋ 58	22 07	00 57	25 23	25 14	08 36	06 57
05	08 55 12	12 48 41	06 ♊ 50 01	13 ♊ 06 03	02 53	02 06	22 45	00 49	25 19	25 17	08 36	06 59
06	08 59 08	13 46 10	19 ♊ 18 19	25 ♊ 27 11	04 54	03 14	23 22	00 42	25 15	25 21	08 37	07 01
07	09 03 05	14 43 41	01 ♋ 33 02	07 ♋ 36 15	06 57	04 22	24 00	00 35	25 11	25 25	08 38	07 03
08	09 07 02	15 41 13	13 ♋ 37 10	19 ♋ 36 06	09 00	05 31	24 37	00 27	25 07	25 28	08 38	07 05
09	09 10 58	16 38 46	25 ♋ 33 24	01 ♌ 29 20	11 04	06 39	25 15	00 20	25 03	25 32	08 39	07 07
10	09 14 55	17 36 21	07 ♌ 24 13	13 ♌ 18 19	13 07	07 48	25 53	00 13	24 59	25 36	08 39	07 09
11	09 18 51	18 33 56	19 ♌ 11 56	25 ♌ 05 20	15 11	08 57	26 30	00 ♒ 06	24 55	25 39	08 40	07 11
12	09 22 48	19 31 33	00 ♍ 58 49	06 ♍ 52 40	17 14	10 06	27 08	29 ♑ 59	24 51	25 43	08 41	07 13
13	09 26 44	20 29 11	12 ♍ 47 11	18 ♍ 42 43	19 17	11 15	27 46	29 52	24 47	25 47	08 42	07 15
14	09 30 41	21 26 50	24 ♍ 39 35	00 ♎ 38 10	21 19	12 24	28 24	29 46	24 43	25 51	08 42	07 17
15	09 34 37	22 24 30	06 ♎ 38 50	12 ♎ 41 59	23 20	13 33	29 02	29 39	24 40	25 54	08 43	07 19
16	09 38 34	23 22 11	18 ♎ 48 03	24 ♎ 57 28	25 20	14 42	29 ♍ 40	29 32	24 36	25 58	08 44	07 21
17	09 42 31	24 19 53	01 ♏ 10 42	07 ♏ 28 10	27 19	15 52	00 ♎ 18	29 26	24 33	26 02	08 45	07 23
18	09 46 27	25 17 36	13 ♏ 50 20	20 ♏ 17 38	29 ♌ 17	17 01	00 56	29 20	24 29	26 06	08 46	07 25
19	09 50 24	26 15 20	26 ♏ 50 27	03 ♐ 29 08	01 ♍ 13	18 11	01 34	29 13	24 26	26 09	08 47	07 27
20	09 54 20	27 13 06	10 ♐ 13 59	17 ♐ 05 11	03 09	19 21	02 12	29 07	24 22	26 13	08 48	07 29
21	09 58 17	28 10 52	24 ♐ 02 51	01 ♑ 06 55	05 02	20 31	02 51	29 02	24 19	26 17	08 49	07 31
22	10 02 13	29 ♌ 08 40	08 ♑ 17 15	15 ♑ 33 30	06 55	21 41	03 29	28 56	24 16	26 21	08 50	07 33
23	10 06 10	00 ♍ 06 29	22 ♑ 55 10	00 ♒ 21 36	08 46	22 51	04 07	28 50	24 13	26 24	08 51	07 35
24	10 10 06	01 04 19	07 ♒ 51 59	15 ♒ 25 19	10 36	24 01	04 46	28 45	24 09	26 28	08 52	07 37
25	10 14 03	02 02 11	23 ♒ 00 33	00 ♓ 36 31	12 25	25 11	05 24	28 39	24 06	26 32	08 53	07 39
26	10 18 00	03 00 03	08 ♓ 12 02	15 ♓ 45 52	14 12	26 21	06 03	28 34	24 03	26 36	08 54	07 41
27	10 21 56	03 57 58	23 ♓ 16 56	00 ♈ 44 08	15 58	27 32	06 41	28 29	24 01	26 39	08 55	07 43
28	10 25 53	04 55 54	08 ♈ 06 35	15 ♈ 23 31	17 42	28 43	07 20	28 24	23 58	26 43	08 56	07 45
29	10 29 49	05 53 52	22 ♈ 34 20	29 ♈ 38 37	19 25	29 ♋ 53	07 59	28 19	23 55	26 47	08 58	07 47
30	10 33 46	06 51 51	06 ♉ 36 07	13 ♉ 26 46	21 07	01 ♌ 04	08 37	28 15	23 52	26 50	08 59	07 49
31	10 37 42	07 ♍ 49 53	20 ♉ 10 37	26 ♉ 47 51	22 ♍ 48	02 ♌ 15	09 ♎ 16	28 ♑ 10	23 ♑ 50	26 ♌ 54	09 ♏ 00	07 ♍ 51

Date	Moon True ☊ ° '	Moon Mean ☊ ° '	Moon ☽ Latitude ° '
01	27 ♌ 18	28 ♌ 06	03 S 50
02	27 D 19	28 02	04 35
03	27 19	27 59	05 04
04	27 R 19	27 56	05 16
05	27 18	27 53	05 12
06	27 16	27 50	04 54
07	27 15	27 46	04 22
08	27 13	27 43	03 39
09	27 11	27 40	02 47
10	27 10	27 37	01 48
11	27 10	27 34	00 S 44
12	27 D 10	27 31	00 N 21
13	27 10	27 27	01 26
14	27 11	27 24	02 27
15	27 11	27 21	03 22
16	27 12	27 18	04 10
17	27 12	27 15	04 46
18	27 12	27 12	05 09
19	27 12	27 08	05 18
20	27 12	27 05	05 10
21	27 12	27 02	04 44
22	27 12	26 59	04 00
23	27 13	26 56	03 00
24	27 13	26 52	01 46
25	27 13	26 49	00 N 23
26	27 R 13	26 46	01 S 01
27	27 13	26 43	02 20
28	27 12	26 40	03 28
29	27 11	26 37	04 22
30	27 10	26 33	04 57
31	27 ♌ 09	26 ♌ 30	05 S 15

DECLINATIONS

Date	Sun ☉ ° '	Moon ☽ ° '	Mercury ☿ ° '	Venus ♀ ° '	Mars ♂ ° '	Jupiter ♃ ° '	Saturn ♄ ° '	Uranus ♅ ° '	Neptune ♆ ° '	Pluto ♇ ° '
01	18 N 01	01 N 53	21 N 36	21 N 26	04 N 34	20 S 29	21 S 12	13 N 50	12 S 41	20 N 37
02	17 46	06 19	21 25	21 30	04 18	20 31	21 13	13 49	12 41	20 36
03	17 30	10 19	21 11	21 33	04 03	20 32	21 14	13 48	12 41	20 35
04	17 14	13 41	20 55	21 36	03 48	20 34	21 14	13 46	12 42	20 35
05	16 58	16 19	20 36	21 39	03 32	20 36	21 15	13 45	12 42	20 34
06	16 42	18 08	20 15	21 41	03 17	20 38	21 16	13 44	12 42	20 33
07	16 25	19 04	19 51	21 42	03 02	20 39	21 17	13 43	12 42	20 32
08	16 08	19 07	19 25	21 43	02 46	20 41	21 18	13 41	12 42	20 31
09	15 51	18 18	18 56	21 43	02 31	20 43	21 18	13 40	12 43	20 31
10	15 34	16 41	18 25	21 43	02 15	20 44	21 19	13 39	12 43	20 30
11	15 16	14 22	17 53	21 42	01 59	20 46	21 20	13 38	12 43	20 29
12	14 58	11 27	17 18	21 40	01 44	20 47	21 21	13 36	12 44	20 28
13	14 40	08 05	16 42	21 38	01 28	20 49	21 22	13 35	12 44	20 27
14	14 21	04 22	16 04	21 35	01 13	20 50	21 22	13 34	12 44	20 27
15	14 03	00 N 28	15 24	21 32	00 57	20 52	21 23	13 32	12 44	20 26
16	13 44	03 S 31	14 44	21 28	00 41	20 53	21 24	13 31	12 45	20 25
17	13 25	07 25	14 02	21 24	00 25	20 55	21 24	13 30	12 45	20 24
18	13 06	11 04	13 20	21 19	00 N 10	20 56	21 25	13 29	12 45	20 23
19	12 46	14 18	12 37	21 14	00 S 06	20 57	21 26	13 27	12 46	20 23
20	12 26	16 52	11 53	21 08	00 22	20 59	21 26	13 26	12 46	20 22
21	12 06	18 35	11 08	21 01	00 38	21 00	21 27	13 25	12 46	20 21
22	11 46	19 11	10 23	20 54	00 53	21 01	21 28	13 24	12 47	20 20
23	11 26	18 33	09 38	20 46	01 09	21 03	21 28	13 22	12 47	20 20
24	11 06	16 36	08 52	20 37	01 25	21 04	21 29	13 21	12 48	20 19
25	10 45	13 29	08 06	20 28	01 41	21 05	21 30	13 20	12 48	20 18
26	10 24	09 26	07 20	20 19	01 57	21 06	21 30	13 19	12 48	20 17
27	10 03	04 S 49	06 34	20 09	02 13	21 07	21 31	13 17	12 49	20 17
28	09 42	00 N 02	05 48	19 58	02 28	21 08	21 31	13 16	12 49	20 16
29	09 21	04 44	05 02	19 47	02 44	21 09	21 32	13 15	12 50	20 15
30	09 00	09 02	04 16	19 35	03 00	21 10	21 32	13 13	12 50	20 14
31	08 N 38	12 N 44	03 N 30	19 N 22	03 S 16	21 S 11	21 S 33	13 N 12	12 S 51	20 N 14

ZODIAC SIGN ENTRIES

Date	h	m	Planets
02	16	19	☽ ♉
03	15	28	♀ ♋
04	01	15	☿ ♌
04	23	04	☽ ♊
07	08	56	☽ ♋
09	20	59	☽ ♌
12	08	54	♃ ♑
12	10	00	☽ ♍
14	22	44	☽ ♎
17	00	41	♂ ♎
17	09	44	☽ ♏
18	20	52	☿ ♍
19	17	44	☽ ♐
21	22	07	☽ ♑
23	09	19	☉ ♍
23	23	25	☽ ♒
25	23	02	☽ ♓
27	22	49	☽ ♈
29	14	18	♀ ♌
30	00	37	☽ ♉

LATITUDES

Date	Mercury ☿ ° '	Venus ♀ ° '	Mars ♂ ° '	Jupiter ♃ ° '	Saturn ♄ ° '	Uranus ♅ ° '	Neptune ♆ ° '	Pluto ♇ ° '
01	00 N 29	01 S 59	00 N 46	00 S 38	00 S 10	00 N 42	01 N 46	12 N 34
04	00 59	01 50	00 44	00 38	00 11	00 42	01 46	12 33
07	01 21	01 40	00 42	00 38	00 11	00 42	01 46	12 33
10	01 37	01 30	00 40	00 39	00 11	00 42	01 46	12 33
13	01 44	01 20	00 38	00 39	00 11	00 42	01 46	12 33
16	01 45	01 10	00 36	00 39	00 12	00 42	01 46	12 33
19	01 41	00 59	00 34	00 39	00 12	00 42	01 45	12 33
22	01 31	00 49	00 32	00 40	00 12	00 42	01 45	12 33
25	01 18	00 38	00 30	00 40	00 12	00 42	01 45	12 33
28	01 01	00 28	00 28	00 40	00 12	00 42	01 45	12 33
31	00 N 42	00 S 18	00 N 27	00 S 40	00 S 13	00 N 42	01 N 45	12 N 33

DATA

Julian Date	2437513
Delta T	+34 seconds
Ayanamsa	23° 19' 06"
Synetic vernal point	05° ♓ 47' 54"
True obliquity of ecliptic	23° 26' 32"

LONGITUDES

Date	Chiron ⚷ ° '	Ceres ⚳ ° '	Pallas ⚴ ° '	Juno ⚵ ° '	Vesta ⚶ ° '	Black Moon Lilith ⚸ ° '
01	05 ♓ 29	19 ♉ 29	24 ♓ 51	00 ♈ 36	23 ♉ 55	20 ♌ 10
11	05 ♓ 02	21 ♉ 49	23 ♓ 26	00 ♈ 32	26 ♉ 49	21 ♌ 16
21	04 ♓ 34	23 ♉ 47	21 ♓ 30	29 ♓ 44	29 ♉ 21	22 ♌ 23
31	04 ♓ 04	25 ♉ 20	19 ♓ 09	28 ♓ 14	01 ♊ 28	23 ♌ 29

All ephemeris data is given at 12.00 UT and the Moon's longitude is additionally given for 24.00 UT

MOON'S PHASES, APSIDES AND POSITIONS ☽

Date	h	m	Phase	Longitude ° '	Eclipse Indicator
03	11	47	☾	10 ♉ 53	
11	10	36	●	18 ♌ 31	Annular
19	10	51	☽	26 ♏ 13	
26	03	13	○	02 ♓ 39	total

Day	h	m		
11	17	08	Apogee	
25	18	49	Perigee	
01	02	17	0N	
08	01	17	Max dec	19° N 12'
15	14	47	0S	
22	11	59	Max dec	19° S 11'
28	11	52	0N	

ASPECTARIAN

01 Tuesday
00 23 ☽ ⊼ ♇
01 41 ☽ Q ♄
01 49 ☉ □ ♆
03 18 ☽ ⊼ ♆
03 24 ☽ △ ☉
05 45 ☽ bQ ♅
10 34 ☽ ± ♇
11 20 ☿ ⊻ ♅
11 22 ☽ Q ♃
15 30 ☽ Q ♀
18 08 ☿ ☍ ♄
23 48 ☽ ⊼ ♂

02 Wednesday
01 28 ☽ ∥ ♂
02 05 ☽ bQ ♇
04 06 ☿ ∥ ♀
07 41 ☽ △ ♅
08 27 ☽ □ ♄
10 46 ☽ ± ♂
10 50 ☽ □ ☿
14 13 ☽ ⚹ ♀
18 23 ☽ □ ♃

03 Thursday
03 30 ☽ bQ ♂
04 37 ☽ △ ♇
07 38 ☽ ☍ ♆
08 32 ☿ # ♄
11 47 ☽ □ ☉
19 37 ☽ ∠ ♀

04 Friday
00 40 ☽ Q ☿
04 24 ☽ # ♆
08 11 ☽ △ ♂
11 05 ♃ ± ♇
11 38 ♀ ⊼ ♃
12 41 ☽ ∥ ♅
12 43 ☿ ☍ ♃
12 51 ☿ ⊥ ♇
13 48 ☽ ⊥ ♀
14 08 ☽ □ ♅
14 19 ☿ ⊻ ♀
14 24 ☽ △ ♄
16 17 ♃ # ♇
23 40 ☽ Q ☉

05 Saturday
00 44 ☽ △ ♃
02 09 ☽ ⊻ ♀
03 06 ☽ ⚹ ☿
12 18 ☽ □ ♇
12 33 ☿ # ♃
15 12 ☿ ∥ ♇
15 23 ☽ ⊼ ♆
16 29 ♄ ⊼ ♅
18 29 ☽ ∥ ☉
18 36 ☽ bQ ♄

06 Sunday
00 24 ☽ ⚹ ☉
00 25 ☽ Q ♅
02 55 ☽ ± ♆
05 05 ☽ bQ ♃
11 53 ☽ ± ♄
13 24 ☽ ∠ ☿
20 21 ☽ □ ♂
20 25 ☽ bQ ♆
21 32 ♂ ∠ ♆
22 25 ☽ ± ♃
23 11 ☽ Q ♇
23 32 ☽ ⊼ ♄
23 51 ☽ ⚹ ♅

07 Monday
08 06 ☽ ∠ ☉
10 06 ☽ ⊼ ♃
10 34 ☽ ⊥ ☿
13 14 ☿ ⊻ ♇
18 10 ☽ ☌ ♀
22 56 ☽ ⚹ ♇

08 Tuesday
00 53 ☽ ⊻ ☿
02 03 ☽ △ ♆
03 28 ☽ ⊥ ☉
05 41 ☽ ∠ ♅
07 43 ☿ □ ♆
09 54 ☽ Q ♂
16 30 ☽ ⊻ ☉
22 26 ☿ ∠ ♂
23 48 ☽ ⊥ ♅

09 Wednesday
04 48 ♂ △ ♄
05 02 ☽ ∠ ♇
10 58 ☽ ☍ ♄
11 21 ☽ ⚹ ♂
11 57 ☽ ⊻ ♅
21 34 ☽ ☍ ♃
21 54 ♀ ⚹ ♇
23 16 ☽ ⊥ ♇

10 Thursday
00 02 ♂ ⊻ ♅
11 29 ☽ ⊻ ♇
12 54 ☽ ⊻ ♀
14 33 ☽ □ ♆
19 28 ☽ ∠ ♂

11 Friday
02 06 ☽ ☌ ☿
02 18 ☽ ∥ ☉
02 24 ☽ ⊥ ♀
05 45 ☿ ⊥ ♀
06 07 ♀ △ ♆
10 36 ☽ ☌ ☉
14 49 ☽ ⊥ ♂
18 37 ☽ ∥ ♅
22 43 ☽ ∠ ♀
23 35 ☽ ⊼ ♄

12 Saturday
01 14 ☽ ☌ ♅
02 05 ☽ # ♆
03 14 ☽ Q ♆
03 44 ☽ ⊻ ♂
10 00 ☽ ⊼ ♃
11 44 ☽ ± ♄
22 05 ☽ ± ♃

13 Sunday
00 43 ☽ ☌ ♇
01 48 ♀ ∠ ♅
03 41 ☽ ⚹ ♆
05 56 ☽ bQ ♄
08 32 ☽ ⚹ ♀
16 11 ☽ bQ ♃

14 Monday
03 53 ☽ ⊻ ☿
04 57 ☽ ⊻ ☉
10 05 ☽ ∠ ♆
11 25 ☽ Q ♀
12 08 ☽ △ ♄
14 24 ☽ ⊻ ♅
14 56 ☉ ☌ ☿
18 06 ☽ ⊥ ☉
18 26 ☽ ⊥ ☿
19 56 ☽ ☌ ♂
22 09 ☽ △ ♃

15 Tuesday
02 30 ☽ ⊥ ♅
04 09 ☽ ⊥ ♆
06 43 ☿ ⊥ ♂
08 50 ☽ ∥ ♂
13 20 ☽ ⊻ ♇
13 39 ☽ ∠ ☉
16 01 ☽ ∠ ☿
16 07 ☽ ⊻ ♆
19 58 ☽ # ♂
20 30 ☽ ∠ ♅

16 Wednesday
01 15 ☽ ⊥ ♇
03 07 ☽ □ ♀
03 26 ☿ ⊼ ♄
07 56 ♂ △ ♃
18 57 ☽ ∠ ♇
19 54 ☿ ☌ ♅
21 40 ☽ ⚹ ☉
23 15 ☽ □ ♄

17 Thursday
02 02 ☽ ⚹ ♅
03 10 ☽ ⚹ ☿
05 01 ☿ Q ♆
05 04 ☉ ∥ ♅
08 40 ☽ □ ♃
09 38 ☉ ⊥ ♂
10 13 ☽ ⊻ ♂
10 26 ♀ # ♄
16 58 ☉ ⊼ ♄
22 18 ☽ ⊥ ♂
22 39 ☽ Q ☉
23 52 ☽ ⚹ ♇

18 Friday
01 08 ☽ Q ♅
02 26 ☽ ☌ ♆
06 20 ☽ Q ☿
06 56 ☿ ∥ ♅
09 28 ☽ Q ♄
12 33 ☿ ⊼ ♃
16 07 ☽ ∠ ♂
18 27 ☽ Q ♃
18 31 ☽ △ ♀
22 24 ☽ Q ♇

19 Saturday
00 06 ☽ ∥ ♆
01 15 ☽ # ☉
01 23 ☽ # ☿
02 25 ☿ ± ♄
02 37 ☉ ∥ ☿
05 26 ☽ # ♅
07 02 ☿ # ♆
07 37 ☽ ⚹ ♄
09 21 ☉ ☌ ♅
10 45 ☽ □ ♅
10 51 ☽ □ ☉
12 19 ☉ # ♆
16 18 ☽ ⚹ ♃
18 27 ☿ ⊻ ♂
20 59 ☽ ⚹ ♂
21 16 ☽ □ ☿

20 Sunday
00 33 ☽ bQ ♀
01 14 ☉ Q ♆
07 07 ☽ □ ♇
09 27 ☽ ⊻ ♆
10 29 ☽ ∠ ♄
12 30 ♂ ⊥ ♅
17 59 ☽ ± ♀
18 47 ☽ ∠ ♃
19 19 ☽ Q ♂
20 01 ☽ ⊥ ♆

21 Monday
02 10 ☽ ⊥ ♄
05 22 ☽ ⊼ ♀
07 01 ♀ ⊥ ♅
10 16 ☽ ⊥ ♃
10 39 ♂ ⊥ ♆
11 36 ☽ ∠ ♆
11 48 ☿ ± ♃
12 27 ☽ ⊻ ♄
14 10 ♀ # ♃
15 50 ☽ △ ♅
19 33 ☽ △ ☉
20 25 ☽ ⊻ ♃

22 Tuesday
03 13 ♀ Q ♂
03 36 ☽ □ ♂
03 46 ☿ ∠ ♀
07 07 ☉ ⊼ ♃
09 23 ☽ △ ☿
10 46 ☽ △ ♇
12 54 ☽ ⚹ ♆
17 05 ☽ bQ ♅
20 17 ☿ ☌ ♇
22 22 ☽ bQ ☉

23 Wednesday
06 28 ♀ ∠ ♇
07 54 ☽ ± ♅
08 38 ☽ Q ♆
11 27 ☽ bQ ♇
11 52 ☽ ☍ ♀
12 58 ☿ ⚹ ♆
13 34 ☽ bQ ☿
14 04 ☽ ± ☉
14 05 ☽ ☌ ♄
14 23 ☉ ± ♄
17 34 ☿ bQ ♄
17 40 ☽ ⊼ ♅
21 29 ☽ ☌ ♃

24 Thursday
00 24 ☽ ⊼ ☉
02 00 ☽ ± ♇
06 04 ☽ ± ☿
06 50 ☽ △ ♂
11 36 ☽ ⊼ ♇
13 35 ☽ □ ♆
14 49 ♀ ☍ ♄
16 57 ☽ ⊼ ☿

25 Friday
07 42 ☽ bQ ♂
13 00 ☽ # ♅
13 44 ☽ ⊻ ♄
15 43 ☽ ⊼ ♀
16 23 ☽ ∥ ♆
17 35 ☽ ☍ ♅
20 52 ☽ ⊻ ♃
22 33 ☽ ± ♂
23 10 ☽ ⊥ ♄

26 Saturday
01 59 ☽ ± ♀
03 13 ☽ ☍ ☉
03 54 ☿ bQ ♃
06 09 ☽ # ☉
06 17 ☽ ⊥ ♃
08 26 ☽ ⊼ ♂
11 11 ☽ ☍ ♇
13 06 ☽ △ ♆
13 21 ☽ ∠ ♄
15 44 ♀ ∥ ♇
17 04 ♀ ⊻ ♅
17 25 ☽ bQ ♀
20 28 ☽ ∠ ♃
22 46 ☽ ☍ ☿

27 Sunday
01 20 ☽ # ☿
13 01 ☽ bQ ♆
13 10 ☽ ⚹ ♄
17 26 ☽ ⊼ ♅
19 25 ☽ △ ♀
20 19 ☽ ⚹ ♃
23 51 ☉ ± ♃

28 Monday
00 15 ☽ ∥ ♂
03 10 ☽ ± ♅
03 34 ☽ ± ♆
06 09 ♀ ☍ ♃
06 27 ☽ ⊼ ☉
08 30 ☽ Q ♄
10 40 ☽ ☍ ♂
11 25 ☽ ⊼ ♇
13 22 ☽ ⊼ ♆
15 44 ☽ Q ♃
16 58 ☽ ± ☉
17 57 ☽ bQ ♅
21 18 ☽ ± ♇

29 Tuesday
01 02 ☽ # ♂
04 31 ♂ ⊻ ♇
06 00 ☽ ⊼ ☿
08 59 ☽ bQ ☉
12 22 ☽ bQ ♇
13 12 ☉ ⊥ ♀
13 22 ☽ ∥ ☿
14 16 ☽ □ ♄
17 29 ☽ ± ☿
19 09 ☽ △ ♅
21 41 ☽ □ ♃

30 Wednesday
01 33 ☽ □ ♀
11 03 ☽ bQ ☿
11 46 ☽ ∥ ☉
12 29 ☽ △ ☉
14 08 ☽ △ ♇
15 42 ☽ ⊼ ♂
16 09 ☽ ☍ ♆

31 Thursday
01 40 ♂ ⊻ ♆
02 47 ☽ ± ♂
03 48 ♀ ⊥ ♇
12 08 ☽ Q ♀
12 38 ☉ ☌ ♇
12 51 ☽ # ♆
15 29 ☽ ∥ ♅
17 25 ☽ △ ☿
17 34 ☿ # ♂
18 35 ☽ △ ♄
19 47 ☽ bQ ♂

SEPTEMBER 1961

LONGITUDES

Date	Sidereal time h m s	Sun ☉ ° ' "	Moon ☽ ° ' "	Moon ☽ 24.00 ° ' "	Mercury ☿ ° '	Venus ♀ ° '	Mars ♂ ° '	Jupiter ♃ ° '	Saturn ♄ ° '	Uranus ♅ ° '	Neptune ♆ ° '	Pluto ♇ ° '
01	10 41 39	08 ♍ 47 56	03 ♊ 18 48	09 ♊ 43 49	24 ♍ 27	03 ♌ 26	09 ♎ 55	28 ♑ 06	23 ♑ 47	26 ♌ 58	09 ♏ 01	07 ♍ 53
02	10 45 35	09 46 02	16 ♊ 03 21	22 ♊ 17 56	26 06	04 37	10 34	28 R 02	23 R 45	27 02	09 03	07 55
03	10 49 32	10 44 09	28 ♊ 28 04	04 ♋ 34 17	27 43	05 48	11 13	27 58	23 43	27 05	09 04	07 58
04	10 53 29	11 42 18	10 ♋ 37 09	16 ♋ 37 11	29 ♍ 18	06 59	11 52	27 54	23 40	27 09	09 05	08 00
05	10 57 25	12 40 30	22 ♋ 34 56	28 ♋ 30 54	00 ♎ 53	08 11	12 31	27 51	23 38	27 13	09 07	08 02
06	11 01 22	13 38 43	04 ♌ 25 32	10 ♌ 19 20	02 26	09 22	13 10	27 48	23 36	27 16	09 08	08 04
07	11 05 18	14 36 58	16 ♌ 12 40	22 ♌ 05 58	03 58	10 34	13 49	27 44	23 34	27 20	09 10	08 06
08	11 09 15	15 35 15	27 ♌ 59 35	03 ♍ 53 50	05 29	11 45	14 29	27 41	23 32	27 23	09 11	08 08
09	11 13 11	16 33 34	09 ♍ 49 03	15 ♍ 45 28	06 59	12 57	15 08	27 39	23 30	27 27	09 13	08 10
10	11 17 08	17 31 54	21 ♍ 43 23	27 ♍ 43 01	08 28	14 09	15 47	27 36	23 29	27 31	09 14	08 12
11	11 21 04	18 30 16	03 ♎ 44 37	09 ♎ 48 22	09 55	15 21	16 27	27 33	23 27	27 34	09 16	08 14
12	11 25 01	19 28 41	15 ♎ 54 31	22 ♎ 03 16	11 21	16 33	17 06	27 31	23 26	27 38	09 17	08 16
13	11 28 58	20 27 06	28 ♎ 14 51	04 ♏ 29 27	12 46	17 45	17 46	27 29	23 24	27 41	09 19	08 18
14	11 32 54	21 25 34	10 ♏ 47 21	17 ♏ 08 45	14 10	18 57	18 25	27 27	23 23	27 45	09 21	08 20
15	11 36 51	22 24 03	23 ♏ 33 55	00 ♐ 03 06	15 32	20 09	19 05	27 26	23 22	27 48	09 22	08 22
16	11 40 47	23 22 34	06 ♐ 36 33	13 ♐ 14 30	16 53	21 21	19 45	27 24	23 20	27 52	09 24	08 24
17	11 44 44	24 21 07	19 ♐ 57 10	26 ♐ 44 45	18 13	22 33	20 24	27 23	23 19	27 55	09 26	08 26
18	11 48 40	25 19 41	03 ♑ 37 23	10 ♑ 35 09	19 31	23 46	21 04	27 22	23 18	27 59	09 27	08 28
19	11 52 37	26 18 17	17 ♑ 38 04	24 ♑ 46 02	20 48	24 58	21 44	27 21	23 18	28 02	09 29	08 30
20	11 56 33	27 16 54	01 ♒ 58 50	09 ♒ 16 09	22 03	26 11	22 24	27 20	23 17	28 05	09 31	08 32
21	12 00 30	28 15 34	16 ♒ 37 29	24 ♒ 02 15	23 17	27 23	23 04	27 20	23 16	28 09	09 33	08 34
22	12 04 27	29 ♍ 14 14	01 ♓ 29 41	08 ♓ 58 55	24 28	28 36	23 44	27 19	23 16	28 12	09 34	08 36
23	12 08 23	00 ♎ 12 57	16 ♓ 28 56	23 ♓ 58 43	25 39	29 ♌ 49	24 24	27 19	23 15	28 16	09 36	08 38
24	12 12 20	01 11 41	01 ♈ 27 10	08 ♈ 53 11	26 47	01 ♍ 02	25 04	27 D 19	23 15	28 19	09 38	08 40
25	12 16 16	02 10 28	16 ♈ 15 44	23 ♈ 33 53	27 53	02 14	25 44	27 19	23 14	28 22	09 40	08 42
26	12 20 13	03 09 16	00 ♉ 46 47	07 ♉ 53 48	28 57	03 27	26 24	27 20	23 14	28 25	09 42	08 44
27	12 24 09	04 08 07	14 ♉ 54 24	21 ♉ 48 17	29 ♎ 59	04 40	27 05	27 21	23 14	28 29	09 44	08 46
28	12 28 06	05 07 00	28 ♉ 35 17	05 ♊ 15 26	00 ♏ 59	05 54	27 45	27 21	23 D 14	28 32	09 46	08 47
29	12 32 02	06 05 55	11 ♊ 48 52	18 ♊ 15 53	01 55	07 07	28 25	27 22	23 14	28 35	09 47	08 49
30	12 35 59	07 ♎ 04 53	24 ♊ 36 53	00 ♋ 52 22	02 ♏ 49	08 ♍ 20	29 ♎ 06	27 ♑ 24	23 ♑ 15	28 ♌ 38	09 ♏ 49	08 ♍ 51

Date	Moon True ☊ ° '	Moon Mean ☊ ° '	Moon ☽ Latitude ° '
01	27 ♌ 08	26 ♌ 27	05 S 16
02	27 D 09	26 24	05 00
03	27 09	26 21	04 31
04	27 10	26 17	03 51
05	27 12	26 14	03 01
06	27 13	26 11	02 03
07	27 14	26 08	01 S 01
08	27 R 14	26 05	00 N 04
09	27 14	26 02	01 09
10	27 12	25 58	02 11
11	27 10	25 55	03 08
12	27 06	25 52	03 57
13	27 03	25 49	04 36
14	26 59	25 46	05 02
15	26 56	25 43	05 14
16	26 53	25 39	05 10
17	26 52	25 36	04 49
18	26 D 53	25 33	04 12
19	26 54	25 30	03 19
20	26 55	25 27	02 13
21	26 56	25 23	00 N 56
22	26 R 57	25 20	00 S 25
23	26 55	25 17	01 45
24	26 52	25 14	02 58
25	26 48	25 11	03 57
26	26 43	25 08	04 40
27	26 37	25 04	05 05
28	26 32	25 01	05 11
29	26 27	24 58	05 00
30	26 ♌ 25	24 ♌ 55	04 S 35

DECLINATIONS

Date	Sun ☉ ° '	Moon ☽ ° '	Mercury ☿ ° '	Venus ♀ ° '	Mars ♂ ° '	Jupiter ♃ ° '	Saturn ♄ ° '	Uranus ♅ ° '	Neptune ♆ ° '	Pluto ♇ ° '
01	08 N 16	15 N 39	02 N 45	19 N 09	03 S 32	21 S 12	21 S 33	13 N 11	12 S 51	20 N 13
02	07 54	17 44	01 59	18 56	03 48	21 13	21 34	13 10	12 52	20 12
03	07 33	18 55	01 14	18 42	04 04	21 13	21 34	13 08	12 52	20 11
04	07 10	19 11	00 N 29	18 27	04 19	21 14	21 35	13 07	12 53	20 11
05	06 48	18 35	00 S 15	18 12	04 35	21 15	21 35	13 06	12 53	20 10
06	06 26	17 10	00 59	17 57	04 51	21 16	21 36	13 05	12 53	20 09
07	06 03	15 01	01 43	17 41	05 07	21 16	21 36	13 04	12 54	20 09
08	05 41	12 14	02 26	17 24	05 23	21 17	21 36	13 02	12 55	20 08
09	05 18	08 57	03 09	17 07	05 38	21 18	21 37	13 01	12 55	20 07
10	04 56	05 18	03 51	16 49	05 54	21 18	21 37	13 00	12 56	20 07
11	04 33	01 N 23	04 33	16 31	06 10	21 19	21 38	12 59	12 56	20 06
12	04 10	02 S 37	05 14	16 13	06 26	21 19	21 38	12 57	12 57	20 05
13	03 47	06 33	05 55	15 54	06 41	21 19	21 38	12 56	12 57	20 05
14	03 24	10 17	06 35	15 34	06 57	21 20	21 38	12 55	12 58	20 04
15	03 01	13 36	07 14	15 14	07 13	21 20	21 39	12 54	12 58	20 03
16	02 38	16 19	07 53	14 54	07 28	21 20	21 39	12 53	12 59	20 03
17	02 15	18 15	08 30	14 33	07 44	21 21	21 39	12 51	12 59	20 02
18	01 51	19 12	09 08	14 12	07 59	21 21	21 39	12 50	13 00	20 01
19	01 28	18 59	09 44	13 50	08 15	21 21	21 40	12 49	13 01	20 01
20	01 05	17 34	10 19	13 28	08 30	21 21	21 40	12 48	13 01	20 00
21	00 42	14 58	10 54	13 06	08 46	21 21	21 40	12 47	13 02	20 00
22	00 N 18	11 20	11 28	12 43	09 01	21 21	21 40	12 46	13 02	19 59
23	00 S 05	06 57	12 00	12 20	09 16	21 21	21 40	12 45	13 03	19 59
24	00 29	02 S 08	12 32	11 57	09 32	21 21	21 40	12 43	13 04	19 58
25	00 52	02 N 45	13 03	11 33	09 47	21 21	21 41	12 42	13 04	19 57
26	01 15	07 22	13 32	11 08	10 02	21 21	21 41	12 41	13 05	19 57
27	01 39	11 27	14 00	10 44	10 17	21 21	21 41	12 40	13 05	19 56
28	02 02	14 47	14 27	10 19	10 32	21 21	21 41	12 39	13 06	19 56
29	02 25	17 15	14 53	09 54	10 47	21 21	21 41	12 38	13 07	19 55
30	02 S 49	18 N 46	15 S 17	09 N 29	11 S 02	21 S 20	21 S 41	12 N 37	13 S 07	19 N 55

ZODIAC SIGN ENTRIES

Date	h m	Planets
01	05 52	☽ ♊
03	15 00	☽ ♋
04	22 32	☿ ♎
06	03 01	☽ ♌
08	16 05	☽ ♍
11	04 33	☽ ♎
13	15 23	☽ ♏
15	23 54	☽ ♐
18	05 42	☽ ♑
20	08 43	☽ ♒
22	09 36	☽ ♓
23	06 42	☉ ♎
23	15 43	♀ ♍
24	09 40	☽ ♈
26	10 42	☽ ♉
27	12 16	☿ ♏
28	14 31	☽ ♊
30	22 19	☽ ♋

LATITUDES

Date	Mercury ☿ ° '	Venus ♀ ° '	Mars ♂ ° '	Jupiter ♃ ° '	Saturn ♄ ° '	Uranus ♅ ° '	Neptune ♆ ° '	Pluto ♇ ° '
01	00 N 35	00 S 14	00 N 26	00 S 40	00 S 13	00 N 42	01 N 45	12 N 33
04	00 N 14	00 S 04	00 24	00 40	00 13	00 42	01 45	12 33
07	00 S 09	00 N 05	00 22	00 40	00 13	00 42	01 44	12 34
10	00 32	00 15	00 20	00 40	00 13	00 42	01 44	12 34
13	00 57	00 24	00 18	00 40	00 14	00 42	01 44	12 34
16	01 21	00 32	00 17	00 40	00 14	00 42	01 44	12 35
19	01 44	00 41	00 15	00 40	00 14	00 42	01 44	12 35
22	02 08	00 48	00 13	00 40	00 14	00 42	01 44	12 36
25	02 29	00 56	00 11	00 40	00 14	00 42	01 44	12 36
28	02 49	01 02	00 09	00 40	00 14	00 42	01 44	12 37
31	03 S 06	01 N 09	00 N 07	00 S 40	00 S 15	00 N 43	01 N 44	12 N 38

DATA

Julian Date	2437544
Delta T	+34 seconds
Ayanamsa	23° 19' 09"
Synetic vernal point	05° ♓ 47' 51"
True obliquity of ecliptic	23° 26' 32"

MOON'S PHASES, APSIDES AND POSITIONS ☽

Date	h m	Phase	Longitude ° '	Eclipse Indicator
01	23 05	☾	09 ♊ 15	
10	02 50	●	17 ♍ 10	
17	20 23	☽	24 ♐ 42	
24	11 33	○	01 ♈ 11	

Day	h m		
07	20 39	Apogee	
23	04 09	Perigee	
04	07 21	Max dec	19° N 12'
11	20 21	0S	
18	19 56	Max dec	19° S 15'
24	22 27	0N	

LONGITUDES

Date	Chiron ⚷ ° '	Ceres ⚳ ° '	Pallas ⚴ ° '	Juno ⚵ ° '	Vesta ⚶ ° '	Black Moon Lilith ⚸ ° '
01	04 ♓ 01	25 ♉ 27	18 ♓ 54	28 ♓ 03	01 ♊ 39	23 ♌ 36
11	03 ♓ 33	26 ♉ 27	16 ♓ 18	25 ♓ 55	03 ♊ 12	24 ♌ 42
21	03 ♓ 06	26 ♉ 53	13 ♓ 44	23 ♓ 29	04 ♊ 10	25 ♌ 49
31	02 ♓ 42	26 ♉ 42	11 ♓ 24	21 ♓ 07	04 ♊ 29	26 ♌ 55

All ephemeris data is given at 12.00 UT and the Moon's longitude is additionally given for 24.00 UT

ASPECTARIAN

h m	Aspects
01 Friday	
00 15	☽ □ ♅
02 26	☽ △ ♃
02 31	☿ △ ♄
05 35	☿ ∠ ♆
11 46	☉ bQ ♄
12 14	☽ ✶ ♀
17 40	☉ ✶ ♆
20 34	☽ □ ♇
22 12	☽ bQ ♄
22 41	☽ ⊼ ♆
23 05	☽ □ ☉
02 Saturday	
01 01	☽ △ ♂
06 17	☽ bQ ♃
10 01	☽ Q ♅
10 04	☽ ± ♆
15 14	☽ ± ♄
19 32	☽ ∠ ♀
23 26	☽ ± ♃
23 41	♅ Q ♆
03 Sunday	
02 21	☿ ⊻ ♅
02 46	☽ ⊼ ♄
03 25	☽ bQ ♆
07 05	☽ Q ♇
07 10	☽ ∥ ♀
09 17	☽ ✶ ♅
10 18	☽ □ ☿
11 02	☽ ⊼ ♃
12 34	☽ Q ☉
14 54	☽ ⊥ ♀
15 45	☿ △ ♃
04 Monday	
04 00	☽ ⊻ ♀
06 46	☽ ✶ ♇
08 57	☽ △ ♆
14 21	☽ ✶ ☉
14 38	☽ □ ♂
15 04	☽ ∠ ♅
23 28	♂ ∠ ♅
05 Tuesday	
00 06	☉ ⊻ ♂
03 25	☽ Q ☿
08 52	♀ ⊻ ♇
09 13	☽ ⊥ ♅
12 54	☽ ∠ ♇
14 07	☽ ☍ ♄
16 03	☉ bQ ♃
21 24	☽ ⊻ ♅
21 44	☽ ∥ ♀
22 36	☽ ☍ ♃
23 13	☽ ∠ ☉
06 Wednesday	
05 00	☽ Q ♂
07 11	☽ ⊥ ♇
07 14	♀ □ ♆
07 21	☽ ✶ ☿
19 08	☽ ⊥ ☉
19 25	☽ ⊻ ♇
21 36	☽ □ ♆
23 05	☿ ⊥ ♆
23 11	☽ ☌ ♀
07 Thursday	
01 30	☿ ⊥ ♅
06 51	☽ ✶ ♂
08 27	☽ ⊻ ☉
18 28	☽ ∠ ☿
22 35	♂ ⊥ ♇
08 Friday	
02 57	☽ ⊼ ♄
05 31	☽ ∥ ♅
06 38	☽ # ♆
10 21	☽ Q ♆
10 46	☽ ☌ ♅
11 23	☽ ⊼ ♃
15 08	☽ ± ♄
15 12	☽ ∠ ♂
15 29	☽ ⊥ ☿
23 29	☉ # ♂
23 32	☽ ± ♃
09 Saturday	
05 26	☽ ⊻ ☿
08 39	☽ ☌ ♇
09 21	☽ bQ ♄
10 32	☽ ⊥ ♂
10 46	☽ ✶ ♆
17 41	☽ bQ ♃
19 02	☽ ⊻ ♀
23 22	☽ ⊻ ♂
10 Sunday	
02 50	☽ ☌ ☉
07 37	☿ ⊻ ♇
08 23	☽ # ♂
08 29	☽ ⊥ ♀
14 33	☽ ∥ ☉
15 31	☽ △ ♄
17 03	☽ ∠ ♆
19 37	☽ # ☿
23 39	☽ ⊻ ♅
23 43	☽ △ ♃
11 Monday	
01 00	☿ ⊻ ♆
04 29	☽ ∠ ♀
08 59	♃ ⊼ ♅
11 03	☽ ⊥ ♆
11 39	☽ ⊥ ♅
11 59	☉ # ☿
20 55	☽ ⊻ ♇
22 57	☽ ⊻ ♆
12 Tuesday	
01 51	☽ ☌ ☿
05 32	☽ ∠ ♅
08 46	☽ ⊥ ♇
13 23	☽ ✶ ♀
14 28	☽ ☌ ♂
19 35	☽ ⊻ ☉
20 33	☽ # ☉
22 58	♅ # ♆
13 Wednesday	
02 24	☽ ∠ ♇
02 38	☽ □ ♄
07 11	☽ ∥ ☿
08 14	☽ ⊥ ☉
10 32	☽ □ ♃
10 36	☿ ∠ ♅
10 55	☽ ✶ ♅
12 50	♀ ✶ ♂
12 53	☽ ∥ ♂
15 11	☽ Q ♀
14 Thursday	
03 00	☽ ∠ ☉
07 19	☽ ✶ ♇
09 15	☽ ☌ ♆
10 01	☽ Q ♅
13 07	☽ Q ♄
15 05	☿ ⊥ ♇
19 09	☽ ⊻ ☿
20 48	☽ Q ♃
15 Friday	
03 11	☽ ⊻ ♂
04 58	☽ □ ♀
06 01	☽ Q ♇
06 39	☽ # ♅
07 09	☽ ∥ ♆
07 46	☽ ⊥ ☿
08 48	♄ bQ ♇
09 39	☽ ✶ ☉
10 32	☿ ∥ ♂
11 37	☽ ✶ ♄
14 58	☽ ⊥ ♂
19 09	☽ ✶ ♃
19 54	☽ □ ♅
16 Saturday	
00 10	☽ # ♀
02 22	☽ ∠ ☿
08 25	☽ ∠ ♂
09 34	☽ Q ☉
11 08	☉ △ ♄
15 08	☽ ∠ ♄
15 16	☽ □ ♇
17 04	☽ ⊻ ♆
22 28	☽ ∠ ♃
17 Sunday	
03 54	☽ ⊥ ♆
07 19	☽ ⊥ ♄
08 34	☽ ✶ ☿
12 51	☽ ✶ ♂
13 53	☉ ∠ ♆
14 32	☽ ⊥ ♃
17 04	☽ △ ♀
17 58	☽ ⊻ ♄
19 56	☽ ∠ ♆
20 23	☽ □ ☉
18 Monday	
01 06	☽ ⊻ ♃
02 07	☽ △ ♅
03 04	♀ ⊼ ♄
07 57	☽ Q ☿
10 59	☽ Q ♂
20 23	☽ △ ♇
21 43	☽ bQ ♀
22 05	☽ ✶ ♆
19 Tuesday	
04 09	☽ bQ ♅
14 28	☽ ± ♀
17 51	☽ □ ☿
18 30	☽ Q ♆
19 15	☽ □ ♂
19 27	☽ ± ♅
21 31	☽ ☌ ♄
21 54	☽ bQ ♇
20 Wednesday	
01 28	☽ ⊼ ♀
03 38	☽ △ ☉
04 17	☽ ☌ ♃
05 31	☽ ⊼ ♅
12 55	☽ ± ♇
13 16	☉ △ ♃
22 49	☽ ⊼ ♇
21 Thursday	
00 25	☽ □ ♆
02 49	☿ ☌ ♂
06 08	☽ bQ ☉
09 05	☉ ⊻ ♅
10 45	♀ ⊼ ♃
11 52	☿ □ ♄
15 08	♀ Q ♆
16 20	♀ # ♆
17 55	☿ ∠ ♇
19 16	♂ □ ♄
21 46	☽ ± ☉
22 45	☽ ⊻ ♄
22 55	☽ △ ♂
23 43	☽ △ ☿
22 Friday	
01 28	☽ ∥ ♆
02 32	☽ # ♀
03 11	☽ # ♅
03 46	♀ ☌ ♅
05 17	☽ ⊻ ♃
06 41	☽ ☍ ♅
06 56	☽ ☍ ♀
07 00	♂ ∠ ♆
08 07	☽ ⊼ ☉
08 25	☽ ⊥ ♄
09 15	♀ ∥ ♅
11 19	☽ ∥ ☿
14 56	☽ ⊥ ♃
22 50	☽ ∠ ♄
23 25	☽ ☍ ♇
23 Saturday	
00 08	☽ bQ ♂
00 22	☽ ∥ ♂
00 58	♀ ± ♄
00 58	☽ △ ♆
01 53	☽ bQ ☿
05 20	☽ ∠ ♃
15 12	☽ ± ♂
15 29	♃ St D
17 29	☽ ± ☿
20 28	☿ # ♀
22 50	☽ ✶ ♄
24 Sunday	
01 02	☽ bQ ♆
01 16	☽ ⊼ ♂
03 53	☽ ⊼ ☿
05 21	☽ ✶ ♃
06 56	☽ ⊼ ♅
11 15	☽ ⊼ ♀
11 33	☽ ☍ ☉
15 31	☽ ± ♆
16 38	☽ ± ♅
18 07	☽ Q ♄
19 31	☽ ∥ ☉
20 26	☿ # ♅
21 47	☽ ± ♀
23 35	☿ □ ♃
23 40	☽ ⊼ ♇
25 Monday	
00 42	☽ Q ♃
01 14	☽ ⊼ ♆
01 53	☽ # ☉
05 15	☉ ⊻ ♀
07 16	☽ bQ ♅
09 26	☽ ± ♇
13 12	☿ ∥ ♆
13 45	☽ bQ ♀
23 15	☿ ✶ ♅
23 28	☽ □ ♄
26 Tuesday	
00 15	☽ bQ ♇
04 21	☽ ☍ ♂
06 14	☽ □ ♃
08 03	☽ △ ♅
08 43	☽ ☍ ☿
09 30	♀ ± ♃
16 17	☽ ⊼ ☉
16 55	☽ △ ♀
27 Wednesday	
01 27	☽ △ ♇
01 41	☉ ⊥ ♆
03 06	☽ ☍ ♆
03 12	☽ ± ☉
04 14	☽ # ♂
07 55	☽ ∥ ♀
19 32	♄ St D
19 54	☽ bQ ☉
20 04	☽ ∥ ♅
20 50	☉ ⊥ ♅
21 36	♂ □ ♃
23 06	☽ # ♆
28 Thursday	
02 31	☽ △ ♄
04 13	♀ # ♂
08 50	☽ # ☿
09 48	☽ △ ♃
10 26	☽ ⊼ ♂
11 54	☽ □ ♅
16 37	☽ ⊼ ☿
21 46	☽ ± ♂
29 Friday	
00 41	☽ △ ☉
02 30	☽ □ ♀
04 19	☽ ± ☿
05 26	☽ bQ ♄
06 29	☽ □ ♇
08 16	☽ ⊼ ♆
13 02	☽ bQ ♃
15 08	☽ bQ ♂
18 08	♂ ✶ ♅
19 24	☽ ± ♆
20 53	☽ Q ♅
22 05	☽ ± ♄
22 13	☽ bQ ☿
30 Saturday	
05 53	☽ ± ♃
09 23	☽ ⊼ ♄
10 14	♀ bQ ♄
12 24	☽ bQ ♆
15 38	☽ Q ♀
16 17	☽ Q ♇
17 19	☽ ⊼ ♃
19 43	☽ ✶ ♅
21 04	☽ △ ♂
22 30	♀ ☌ ♇

OCTOBER 1961

LONGITUDES

Date	Sidereal time h m s	Sun ☉ ° ' "	Moon ☽ ° ' "	Moon ☽ 24.00 ° ' "	Mercury ☿ ° '	Venus ♀ ° '	Mars ♂ ° '	Jupiter ♃ ° '	Saturn ♄ ° '	Uranus ♅ ° '	Neptune ♆ ° '	Pluto ♇ ° '
01	12 39 56	08 ♎ 03 52	07 ♋ 02 52	13 ♋ 09 00	03 ♏ 40	09 ♍ 33	29 ♎ 46	27 ♑ 25	23 ♑ 15	28 ♌ 41	09 ♏ 51	08 ♍ 53
02	12 43 52	09 02 55	19 ♋ 11 23	25 ♋ 10 40	04 27	10 47	00 ♏ 27	27 27	23 15	28 44	09 53	08 55
03	12 47 49	10 01 59	01 ♌ 07 30	07 ♌ 02 32	05 11	12 00	01 08	27 29	23 16	28 47	09 55	08 57
04	12 51 45	11 01 06	12 ♌ 56 23	18 ♌ 49 39	05 51	13 13	01 48	27 31	23 16	28 50	09 57	08 59
05	12 55 42	12 00 14	24 ♌ 42 54	00 ♍ 36 40	06 26	14 27	02 29	27 33	23 17	28 53	09 59	09 00
06	12 59 38	12 59 26	06 ♍ 31 25	12 ♍ 27 37	06 56	15 41	03 10	27 35	23 18	28 56	10 01	09 02
07	13 03 35	13 58 39	18 ♍ 25 39	24 ♍ 25 50	07 21	16 54	03 51	27 38	23 19	28 59	10 03	09 04
08	13 07 31	14 57 54	00 ♎ 28 28	06 ♎ 33 46	07 41	18 08	04 32	27 41	23 20	29 02	10 06	09 06
09	13 11 28	15 57 12	12 ♎ 41 56	18 ♎ 53 05	07 54	19 22	05 13	27 43	23 21	29 05	10 08	09 07
10	13 15 25	16 56 31	25 ♎ 07 19	01 ♏ 24 39	08 01	20 36	05 54	27 47	23 22	29 08	10 10	09 09
11	13 19 21	17 55 53	07 ♏ 45 09	14 ♏ 08 46	08 R 00	21 49	06 35	27 50	23 23	29 11	10 12	09 11
12	13 23 18	18 55 16	20 ♏ 35 31	27 ♏ 05 22	07 52	23 03	07 16	27 54	23 25	29 13	10 14	09 12
13	13 27 14	19 54 42	03 ♐ 38 16	10 ♐ 14 15	07 36	24 17	07 57	27 57	23 26	29 16	10 16	09 14
14	13 31 11	20 54 09	16 ♐ 53 16	23 ♐ 35 22	07 12	25 31	08 38	28 01	23 28	29 19	10 18	09 16
15	13 35 07	21 53 39	00 ♑ 20 34	07 ♑ 08 56	06 39	26 45	09 20	28 05	23 30	29 21	10 20	09 17
16	13 39 04	22 53 10	14 ♑ 00 29	20 ♑ 55 17	05 58	28 00	10 01	28 09	23 32	29 24	10 22	09 19
17	13 43 00	23 52 42	27 ♑ 53 22	04 ♒ 54 44	05 09	29 ♍ 14	10 43	28 14	23 33	29 27	10 25	09 21
18	13 46 57	24 52 17	11 ♒ 59 21	19 ♒ 07 05	04 12	00 ♎ 28	11 24	28 19	23 35	29 29	10 27	09 22
19	13 50 54	25 51 53	26 ♒ 17 45	03 ♓ 31 04	03 08	01 42	12 06	28 23	23 38	29 32	10 29	09 24
20	13 54 50	26 51 30	10 ♓ 46 38	18 ♓ 03 56	01 59	02 56	12 47	28 28	23 40	29 34	10 31	09 25
21	13 58 47	27 51 10	25 ♓ 21 13	02 ♈ 41 11	00 ♏ 45	04 11	13 29	28 33	23 42	29 36	10 33	09 27
22	14 02 43	28 50 51	09 ♈ 59 35	17 ♈ 16 43	29 ♎ 30	05 25	14 10	28 39	23 44	29 39	10 36	09 28
23	14 06 40	29 ♎ 50 34	24 ♈ 31 42	01 ♉ 43 39	28 15	06 40	14 52	28 44	23 47	29 41	10 38	09 30
24	14 10 36	00 ♏ 50 19	08 ♉ 51 44	15 ♉ 55 14	27 03	07 54	15 34	28 50	23 49	29 43	10 40	09 31
25	14 14 33	01 50 06	22 ♉ 53 32	29 ♉ 46 07	25 55	09 08	16 16	28 56	23 52	29 45	10 42	09 32
26	14 18 29	02 49 56	06 ♊ 32 40	13 ♊ 12 59	24 54	10 23	16 58	29 02	23 55	29 48	10 44	09 34
27	14 22 26	03 49 47	19 ♊ 47 04	26 ♊ 15 00	24 02	11 38	17 40	29 08	23 58	29 50	10 47	09 35
28	14 26 23	04 49 41	02 ♋ 37 03	08 ♋ 53 35	23 19	12 52	18 22	29 14	24 01	29 52	10 49	09 37
29	14 30 19	05 49 36	15 ♋ 05 03	21 ♋ 11 59	22 48	14 07	19 04	29 21	24 04	29 54	10 51	09 38
30	14 34 16	06 49 34	27 ♋ 14 59	03 ♌ 14 43	22 28	15 22	19 46	29 27	24 07	29 56	10 53	09 39
31	14 38 12	07 ♏ 49 34	09 ♌ 11 51	15 ♌ 07 05	22 ♎ 19	16 ♎ 36	20 ♏ 28	29 ♑ 34	24 ♑ 10	29 ♌ 58	10 ♏ 56	09 ♍ 40

Date	Moon True ☊ ° '	Moon Mean ☊ ° '	Moon ☽ Latitude ° '
01	26 ♌ 24	24 ♌ 52	03 S 57
02	26 D 24	24 49	03 13
03	26 26	24 45	02 13
04	26 27	24 42	01 13
05	26 28	24 39	00 S 10
06	26 R 28	24 36	00 N 54
07	26 25	24 33	01 56
08	26 20	24 29	02 53
09	26 13	24 26	03 43
10	26 05	24 23	04 23
11	25 55	24 20	04 51
12	25 45	24 17	05 05
13	25 37	24 14	05 03
14	25 30	24 10	04 45
15	25 26	24 07	04 11
16	25 23	24 04	03 23
17	25 D 23	24 01	02 22
18	25 24	23 58	01 N 11
19	25 R 24	23 55	00 S 05
20	25 23	23 51	01 21
21	25 20	23 48	02 33
22	25 14	23 45	03 34
23	25 06	23 42	04 22
24	24 55	23 39	04 51
25	24 44	23 35	05 03
26	24 34	23 32	04 57
27	24 25	23 29	04 34
28	24 18	23 26	03 59
29	24 14	23 23	03 12
30	24 12	23 20	02 18
31	24 ♌ 11	23 ♌ 16	01 S 19

DECLINATIONS

Date	Sun ☉ ° '	Moon ☽ ° '	Mercury ☿ ° '	Venus ♀ ° '	Mars ♂ ° '	Jupiter ♃ ° '	Saturn ♄ ° '	Uranus ♅ ° '	Neptune ♆ ° '	Pluto ♇ ° '
01	03 S 12	19 N 19	15 S 39	09 N 03	11 S 17	21 S 20	21 S 41	12 N 36	13 S 08	19 N 54
02	03 35	18 57	16 00	08 37	11 32	21 20	21 41	12 35	13 09	19 54
03	03 58	17 45	16 19	08 10	11 47	21 19	21 41	12 34	13 09	19 53
04	04 22	15 46	16 36	07 44	12 01	21 19	21 41	12 33	13 10	19 53
05	04 45	13 08	16 51	07 17	12 16	21 19	21 41	12 32	13 11	19 53
06	05 08	09 57	17 03	06 50	12 30	21 18	21 41	12 31	13 11	19 52
07	05 31	06 21	17 13	06 23	12 45	21 18	21 40	12 30	13 12	19 52
08	05 54	02 N 27	17 20	05 56	12 59	21 17	21 40	12 29	13 13	19 51
09	06 17	01 S 36	17 24	05 28	13 14	21 16	21 40	12 28	13 13	19 51
10	06 39	05 38	17 25	05 00	13 28	21 16	21 40	12 27	13 14	19 50
11	07 02	09 30	17 23	04 32	13 42	21 15	21 40	12 26	13 15	19 50
12	07 25	13 00	17 17	04 04	13 56	21 14	21 40	12 25	13 15	19 50
13	07 47	15 55	17 07	03 36	14 10	21 14	21 40	12 24	13 16	19 49
14	08 10	18 04	16 52	03 07	14 24	21 13	21 39	12 23	13 17	19 49
15	08 32	19 15	16 33	02 39	14 38	21 12	21 39	12 22	13 17	19 49
16	08 54	19 20	16 10	02 10	14 51	21 11	21 39	12 21	13 18	19 48
17	09 16	18 16	15 42	01 42	15 05	21 10	21 39	12 20	13 19	19 48
18	09 38	16 04	15 10	01 13	15 18	21 09	21 38	12 20	13 19	19 48
19	10 00	12 50	14 33	00 44	15 32	21 08	21 38	12 19	13 20	19 47
20	10 21	08 47	13 53	00 N 15	15 45	21 07	21 38	12 18	13 21	19 47
21	10 43	04 S 11	13 10	00 S 14	15 58	21 06	21 37	12 17	13 21	19 47
22	11 04	00 N 40	12 25	00 43	16 11	21 05	21 37	12 16	13 22	19 47
23	11 25	05 27	11 39	01 12	16 24	21 04	21 37	12 16	13 23	19 46
24	11 46	09 51	10 54	01 41	16 37	21 03	21 36	12 15	13 24	19 46
25	12 07	13 37	10 10	02 10	16 50	21 01	21 36	12 14	13 24	19 46
26	12 27	16 32	09 29	02 39	17 02	21 00	21 35	12 13	13 25	19 46
27	12 48	18 30	08 52	03 09	17 15	20 59	21 35	12 13	13 26	19 46
28	13 08	19 26	08 20	03 38	17 27	20 58	21 34	12 12	13 26	19 45
29	13 28	19 24	07 53	04 06	17 39	20 56	21 34	12 11	13 27	19 45
30	13 48	18 27	07 32	04 35	17 51	20 55	21 33	12 11	13 28	19 45
31	14 S 07	16 N 41	07 S 17	05 S 04	18 S 03	20 S 53	21 S 33	12 N 10	13 S 28	19 N 45

ZODIAC SIGN ENTRIES

Date	h	m	Planets
01	20	02	♂ ♏
03	09	43	☽ ♌
05	22	45	☽ ♍
08	11	04	☽ ♎
10	21	19	☽ ♏
13	05	21	☽ ♐
15	11	24	☽ ♑
17	15	37	☽ ♒
18	02	58	♀ ♎
19	18	10	☽ ♓
21	19	36	☽ ♈
22	02	29	☿ ♎
23	15	47	☉ ♏
23	21	07	☽ ♉
26	00	24	☽ ♊
28	07	03	☽ ♋
30	17	30	☽ ♌

LATITUDES

Date	Mercury ☿ ° '	Venus ♀ ° '	Mars ♂ ° '	Jupiter ♃ ° '	Saturn ♄ ° '	Uranus ♅ ° '	Neptune ♆ ° '	Pluto ♇ ° '
01	03 S 06	01 N 09	00 N 07	00 S 40	00 S 15	00 N 43	01 N 44	12 N 38
04	03 19	01 14	00 05	00 40	00 15	00 43	01 43	12 39
07	03 26	01 19	00 03	00 40	00 15	00 43	01 43	12 39
10	03 26	01 23	00 N 02	00 40	00 15	00 43	01 43	12 40
13	03 14	01 27	00 00	00 40	00 15	00 43	01 43	12 41
16	02 49	01 30	00 S 02	00 40	00 15	00 43	01 43	12 42
19	02 07	01 32	00 04	00 40	00 16	00 43	01 43	12 43
22	01 12	01 34	00 06	00 40	00 16	00 43	01 43	12 44
25	00 S 10	01 35	00 08	00 40	00 16	00 43	01 43	12 45
28	00 N 47	01 35	00 09	00 40	00 16	00 43	01 43	12 46
31	01 N 31	01 N 35	00 S 11	00 S 40	00 S 16	00 N 44	01 N 43	12 N 48

DATA

Julian Date	2437574
Delta T	+34 seconds
Ayanamsa	23° 19' 11"
Synetic vernal point	05° ♓ 47' 48"
True obliquity of ecliptic	23° 26' 32"

MOON'S PHASES, APSIDES AND POSITIONS ☽

Date	h	m	Phase	Longitude ° '	Eclipse Indicator
01	14	10	☾	08 ♋ 09	
09	18	52	●	16 ♎ 14	
17	04	34	☽	23 ♑ 34	
23	21	30	○	00 ♉ 14	
31	08	58	☾	07 ♌ 42	

Day	h	m		
05	08	02	Apogee	
21	07	01	Perigee	
01	14	19	Max dec	19° N 19'
09	02	37	0S	
16	01	55	Max dec	19° S 27'
22	08	41	0N	
28	23	01	Max dec	19° N 33'

LONGITUDES

Date	Chiron ⚷ ° '	Ceres ⚳ ° '	Pallas ⚴ ° '	Juno ⚵ ° '	Vesta ⚶ ° '	Black Moon Lilith ⚸ ° '
01	02 ♓ 42	26 ♉ 42	11 ♓ 24	21 ♓ 07	04 ♊ 29	26 ♌ 55
11	02 ♓ 22	25 ♉ 53	09 ♓ 30	19 ♓ 30	04 ♊ 03	28 ♌ 02
21	02 ♓ 07	24 ♉ 27	08 ♓ 09	17 ♓ 51	02 ♊ 54	29 ♌ 08
31	01 ♓ 57	22 ♉ 31	07 ♓ 26	17 ♓ 24	01 ♊ 04	00 ♍ 15

All ephemeris data is given at 12.00 UT and the Moon's longitude is additionally given for 24.00 UT

ASPECTARIAN

h m Aspects

01 Sunday
04 56 ☽ △ ☿
14 10 ☽ □ ☉
15 36 ☽ ⚹ ♇
17 27 ☽ ⚹ ♀
17 31 ☽ △ ♆
18 07 ♀ ⚹ ♆

02 Monday
01 07 ☽ ∠ ♅
08 38 ☉ ⊻ ♇
19 08 ☽ ⊥ ♅
20 08 ☽ ☍ ♄
21 29 ☽ ∠ ♇

03 Tuesday
02 43 ☽ ∠ ♀
04 36 ☽ ☍ ♃
05 11 ☽ Q ☉
07 16 ☽ ⊻ ♅
09 13 ☉ ⊻ ♆
12 00 ☽ □ ♂
15 42 ☽ ⊥ ♇
20 44 ☽ □ ☿
21 35 ♀ bQ ♃
23 02 ☽ ⊥ ♀

04 Wednesday
03 55 ☽ ⊻ ♇
03 59 ☽ # ☿
05 54 ☽ □ ♆
07 44 ☽ ⚹ ☉
12 39 ☽ ⊻ ♀

05 Thursday
02 51 ☽ Q ♂
09 05 ☽ ⊼ ♄
11 23 ☽ Q ☿
11 40 ☽ # ♆
16 55 ☽ ∥ ♅
17 05 ☽ ∠ ☉
17 47 ☽ ⊼ ♃
18 28 ☽ # ♂
18 41 ☽ Q ♆
20 32 ☽ ☌ ♅
21 19 ☽ ± ♄

06 Friday
04 46 ☽ ⚹ ♂
06 01 ☽ ± ♃
12 30 ♂ # ♅
12 52 ☽ ⚹ ☿
13 02 ☽ ⊥ ☉
15 36 ☽ bQ ♄
17 06 ☽ ☌ ♇
19 06 ☽ ⚹ ♆

07 Saturday
00 18 ☽ bQ ♃
02 15 ☽ ⊻ ☉
08 36 ☽ ☌ ♀
11 48 ☽ ∥ ♀
12 16 ☉ ∠ ♅
12 53 ☽ ∠ ♂
16 50 ☽ # ☉
20 06 ☽ ∠ ☿
21 47 ☽ △ ♄

08 Sunday
01 17 ☽ ∠ ♆
06 26 ☽ △ ♃
07 55 ☽ ⊥ ♂
09 08 ☽ ⊻ ♅
12 52 ☉ # ♀
14 26 ☽ ⊥ ☿
15 13 ☉ ⊥ ♇
19 10 ☽ ⊥ ♆
20 28 ☽ ⊻ ♂
21 02 ☽ ⊥ ♅

09 Monday
02 29 ☽ ⊻ ☿
05 00 ☽ ⊻ ♇
06 58 ☽ ⊻ ♆
11 32 ♂ ∥ ♆
14 42 ☽ ∠ ♅
16 43 ☽ ⊥ ♇
18 52 ☽ ☌ ☉

10 Tuesday
02 21 ☽ ⊻ ♀
08 35 ☽ # ♀
08 38 ☽ □ ♄
10 08 ☽ ∠ ♇
15 07 ☽ ⊥ ♀
17 06 ☽ □ ♃
18 50 ☽ ∥ ☉
19 41 ☽ ⚹ ♅
22 41 ☿ St R

11 Wednesday
01 14 ♀ ∠ ♂
09 40 ☽ ☌ ♂
10 04 ☽ ∠ ♀
12 29 ☽ ☌ ☿
14 42 ☽ ⚹ ♇
16 37 ☽ ☌ ♆
18 28 ☽ Q ♅
18 51 ☽ Q ♄

12 Thursday
03 13 ☽ Q ♃
07 45 ☽ # ♅
08 38 ☽ ⊻ ☉
08 55 ☿ ∠ ♀
13 09 ☽ Q ♇
13 55 ☽ ∥ ♆
17 03 ☽ ⚹ ♀
17 14 ☽ ⚹ ♄
19 07 ♀ △ ♄
19 41 ☽ ∥ ♂
20 40 ☽ ⊥ ☉

13 Friday
01 32 ☽ ⚹ ♃
03 43 ☿ ☌ ♂
03 58 ☽ □ ♅
14 31 ☽ ∠ ☉
17 20 ☽ Q ♀
19 02 ☽ ⊻ ☿
20 17 ☽ ⊻ ♂
20 45 ☽ ∠ ♄
22 12 ☽ □ ♇
22 58 ☽ ∥ ☿

14 Saturday
00 05 ☽ ⊻ ♆
05 00 ☽ ∠ ♃
05 35 ☽ ⊥ ☿
07 35 ♀ ∠ ♆
07 44 ☽ ⊥ ♂
10 57 ☽ ⊥ ♆
13 03 ☽ ⊥ ♄
19 46 ☽ ⚹ ☉
21 11 ☽ ∠ ☿
21 15 ☽ ⊥ ♃
23 48 ☽ ⊻ ♄

15 Sunday
00 44 ☽ ∠ ♂
03 06 ☽ ∠ ♆
05 00 ☽ □ ♀
07 59 ☽ ⊻ ♃
10 15 ☽ △ ♅
10 36 ♂ ⚹ ♇
18 46 ☽ Q ☉
22 38 ☽ ⚹ ☿

16 Monday
03 47 ☽ △ ♇
04 40 ☽ ⚹ ♂
05 38 ☽ ⚹ ♆
12 41 ☽ bQ ♅
15 24 ♀ △ ♃
18 31 ☽ Q ☿

17 Tuesday
01 05 ♂ ☌ ♆
02 33 ☽ Q ♆
02 37 ☽ Q ♂
03 58 ☉ □ ♄
04 20 ☽ ± ♅
04 32 ☽ ☌ ♄
04 34 ☽ □ ☉
05 54 ☽ bQ ♇
11 04 ☿ ⊥ ♀
12 35 ☽ ☌ ♃
14 31 ☽ △ ♀
14 40 ☽ ⊼ ♅
16 17 ♀ ⊻ ♅
21 21 ☽ ± ♇
23 32 ☉ ∠ ♇
23 38 ☽ □ ☿

18 Wednesday
07 34 ☽ ⊼ ♇
07 36 ☿ ∥ ♂
09 23 ☽ □ ♆
10 57 ☽ □ ♂
15 06 ♂ Q ♅
17 51 ☽ ∥ ♂
18 25 ☽ bQ ♀
18 57 ♂ Q ♄
21 10 ☽ ∥ ☿

19 Thursday
07 32 ☽ ⊻ ♄
08 39 ☽ ∥ ♆
10 55 ☽ ± ♀
11 14 ☽ △ ☉
15 19 ☽ # ♅
15 30 ☽ ⊻ ♃
17 23 ☽ ☍ ♅
17 33 ☽ ⊥ ♄
21 50 ☽ ⊼ ♀
22 32 ☽ △ ☿

20 Friday
01 31 ☽ ⊥ ♃
02 25 ☿ ⊻ ♀
03 50 ☽ ∥ ☉
08 30 ☽ ∠ ♄
09 45 ☽ ☍ ♇
11 34 ☽ △ ♆
13 55 ☽ bQ ☉
15 29 ☽ △ ♂
16 28 ☽ ∠ ♃
21 26 ☽ bQ ☿

21 Saturday
05 48 ☽ ± ☉
05 48 ☿ ∥ ♆
09 15 ☽ ⚹ ♄
11 04 ☽ ± ☿
12 18 ☽ bQ ♆
16 22 ☽ ⊼ ☉
17 15 ☽ ⚹ ♃
17 21 ☽ bQ ♂
18 58 ☽ ⊼ ♅
19 32 ♀ ⊥ ♆
20 08 ☽ ⊼ ☿

22 Sunday
03 06 ☽ ± ♆
03 47 ☽ ☍ ♀
04 50 ☽ ± ♅
04 59 ☽ Q ♄
05 46 ☽ ∥ ♀
06 40 ☉ □ ♃
08 52 ☽ ± ♂
09 22 ☿ ⚹ ♅
11 08 ☽ ⊼ ♇
12 16 ☽ # ♀
12 59 ☽ ⊼ ♆
13 05 ☽ Q ♃
16 30 ♀ ⊥ ♅
16 40 ☿ # ♅
18 59 ☉ ☌ ☿
19 14 ☽ ⊼ ♂
19 41 ☽ bQ ♅
21 02 ☽ ± ♇

23 Monday
03 17 ☿ □ ♃
07 59 ☉ ⚹ ♅
10 45 ☽ □ ♄
11 57 ☽ bQ ♇
17 07 ☉ ∥ ☿
17 43 ☽ ☍ ☿
19 03 ☽ □ ♃
20 36 ☽ △ ♅
21 30 ☽ ☍ ☉

24 Tuesday
10 13 ☽ ⊼ ♀
13 07 ☽ △ ♇
15 04 ☽ ☍ ♆
17 16 ☽ # ☿
21 23 ☽ ± ♀

25 Wednesday
00 00 ☽ ☍ ♂
00 49 ☽ # ☉
02 41 ☽ ∥ ♅
10 32 ☽ # ♆
13 42 ☽ △ ♄
14 23 ☽ bQ ♀
16 53 ☽ ⊼ ☿
19 53 ♀ ⊻ ♇
20 11 ☉ # ♅
22 36 ☽ △ ♃

26 Thursday
00 01 ☽ □ ♅
02 40 ☽ ± ☿
04 53 ☽ ⊼ ☉
14 36 ♂ Q ♃
16 15 ☽ bQ ♄
16 25 ☽ ± ☉
17 25 ☽ □ ♇
17 37 ☽ bQ ☿
17 39 ☽ # ♂
19 08 ♀ ⊻ ♆
19 33 ☽ ⊼ ♆
19 35 ☽ △ ♀
20 33 ☿ ∠ ♇

27 Friday
01 35 ☽ bQ ♃
06 28 ☽ ± ♆
07 53 ☽ ⊼ ♂
08 24 ☽ Q ♅
08 38 ☽ ± ♄
10 06 ☽ bQ ☉
13 39 ♀ ⊥ ♂
13 59 ☿ □ ♄
18 14 ☽ ± ♃
19 25 ☽ △ ☿
19 35 ☽ ± ♂
19 45 ☽ ⊼ ♄
23 09 ☽ bQ ♆

28 Saturday
02 32 ☽ Q ♇
05 33 ☽ ⊼ ♃
06 46 ☽ ⚹ ♅
13 30 ☽ bQ ♂
16 34 ☽ △ ☉

29 Sunday
01 24 ☽ ⚹ ♇
03 45 ☽ △ ♆
09 54 ☽ □ ♀
10 54 ☉ ∥ ♆
11 38 ☽ ∠ ♅
20 16 ☽ △ ♂

30 Monday
02 42 ☽ □ ☿
03 32 ♀ ∠ ♅
05 23 ☽ ⊥ ♅
05 44 ☽ ☍ ♄
06 50 ☽ ∠ ♇
16 27 ☽ ☍ ♃
17 22 ☽ ⊻ ♅
17 47 ♀ ⊥ ♇
20 25 ☽ # ♂

31 Tuesday
00 51 ☽ ⊥ ♇
01 39 ☽ Q ♀
08 58 ☽ □ ☉
12 58 ☽ ⊻ ♇
14 15 ☽ Q ☿
15 31 ☽ □ ♆
18 01 ☿ St D

NOVEMBER 1961

Date	Sidereal time h m s	LONGITUDES Sun ☉ ° ' "	Moon ☽ ° ' "	Moon ☽ 24.00 ° ' "	Mercury ☿ ° '	Venus ♀ ° '	Mars ♂ ° '	Jupiter ♃ ° '	Saturn ♄ ° '	Uranus ♅ ° '	Neptune ♆ ° '	Pluto ♇ ° '
01	14 42 09	08 ♏ 49 36	21 ♌ 01 07	26 ♌ 54 38	22 ♎ 22	17 ♎ 51	21 ♏ 11	29 ♑ 41	24 ♑ 13	00 ♍ 00	10 ♏ 58	09 ♍ 42
02	14 46 05	09 49 40	02 ♍ 48 19	08 ♍ 42 50	22 D 35	19 06	21 53	29 48	24 17	00 02	11 00	09 43
03	14 50 02	10 49 47	14 ♍ 38 48	20 ♍ 36 47	22 59	20 21	22 35	29 ♑ 55	24 20	00 03	11 02	09 44
04	14 53 58	11 49 55	26 ♍ 37 18	02 ♎ 40 48	23 33	21 35	23 18	00 ♒ 03	24 24	00 05	11 05	09 45
05	14 57 55	12 50 05	08 ♎ 47 42	14 ♎ 58 18	24 15	22 50	24 00	00 10	24 27	00 07	11 07	09 46
06	15 01 52	13 50 18	21 ♎ 12 50	27 ♎ 31 27	25 04	24 05	24 43	00 18	24 31	00 08	11 09	09 48
07	15 05 48	14 50 32	03 ♏ 54 13	10 ♏ 21 06	26 01	25 20	25 25	00 26	24 35	00 10	11 11	09 49
08	15 09 45	15 50 48	16 ♏ 52 00	23 ♏ 26 46	27 03	26 35	26 08	00 34	24 39	00 12	11 14	09 50
09	15 13 41	16 51 06	00 ♐ 05 08	06 ♐ 46 52	28 11	27 50	26 51	00 42	24 43	00 13	11 16	09 51
10	15 17 38	17 51 26	13 ♐ 31 37	20 ♐ 19 05	29 ♎ 23	29 ♎ 05	27 34	00 50	24 47	00 14	11 18	09 52
11	15 21 34	18 51 47	27 ♐ 08 57	04 ♑ 00 56	00 ♏ 39	00 ♏ 20	28 17	00 59	24 51	00 16	11 20	09 53
12	15 25 31	19 52 10	10 ♑ 54 45	17 ♑ 50 11	01 58	01 35	28 59	01 07	24 55	00 17	11 23	09 54
13	15 29 27	20 52 34	24 ♑ 47 04	01 ♒ 45 14	03 20	02 50	29 ♏ 42	01 16	24 59	00 19	11 25	09 55
14	15 33 24	21 52 59	08 ♒ 44 36	15 ♒ 45 06	04 44	04 05	00 ♐ 25	01 25	25 04	00 20	11 27	09 55
15	15 37 21	22 53 26	22 ♒ 46 40	29 ♒ 49 14	06 11	05 21	01 08	01 34	25 08	00 21	11 29	09 56
16	15 41 17	23 53 54	06 ♓ 52 45	13 ♓ 57 06	07 39	06 36	01 52	01 43	25 12	00 22	11 31	09 57
17	15 45 14	24 54 23	21 ♓ 02 08	28 ♓ 07 36	09 08	07 51	02 35	01 52	25 17	00 23	11 34	09 58
18	15 49 10	25 54 54	05 ♈ 13 14	12 ♈ 18 40	10 38	09 06	03 18	02 02	25 22	00 24	11 36	09 59
19	15 53 07	26 55 25	19 ♈ 23 28	26 ♈ 27 07	12 09	10 21	04 01	02 11	25 26	00 25	11 38	09 59
20	15 57 03	27 55 59	03 ♉ 29 05	10 ♉ 28 47	13 41	11 36	04 45	02 21	25 31	00 26	11 40	10 00
21	16 01 00	28 56 33	17 ♉ 25 38	24 ♉ 19 05	15 14	12 52	05 28	02 31	25 36	00 27	11 42	10 01
22	16 04 56	29 ♏ 57 09	01 ♊ 08 38	07 ♊ 53 48	16 47	14 07	06 11	02 40	25 41	00 28	11 45	10 01
23	16 08 53	00 ♐ 57 46	14 ♊ 34 16	21 ♊ 09 46	18 20	15 22	06 55	02 50	25 46	00 28	11 47	10 02
24	16 12 50	01 58 25	27 ♊ 40 10	04 ♋ 05 26	19 53	16 37	07 38	03 00	25 51	00 29	11 49	10 03
25	16 16 46	02 59 06	10 ♋ 25 40	16 ♋ 41 04	21 27	17 53	08 22	03 11	25 56	00 30	11 51	10 03
26	16 20 43	03 59 48	22 ♋ 51 56	28 ♋ 58 39	23 01	19 08	09 06	03 21	26 02	00 30	11 53	10 04
27	16 24 39	05 00 31	05 ♌ 01 42	11 ♌ 01 38	24 35	20 23	09 49	03 31	26 07	00 31	11 55	10 04
28	16 28 36	06 01 16	16 ♌ 59 01	22 ♌ 54 30	26 09	21 39	10 33	03 42	26 12	00 31	11 57	10 05
29	16 32 32	07 02 02	28 ♌ 48 46	04 ♍ 42 29	27 43	22 54	11 17	03 53	26 18	00 32	11 59	10 05
30	16 36 29	08 ♐ 02 50	10 ♍ 36 22	16 ♍ 31 07	29 ♏ 17	24 ♏ 09	12 ♐ 01	04 ♒ 03	26 ♑ 23	00 ♍ 32	12 ♏ 02	10 ♍ 06

Date	Moon True ☊ ° '	Moon Mean ☊ ° '	Moon ☽ Latitude ° '
01	24 ♌ 12	23 ♌ 13	00 S 17
02	24 R 11	23 10	00 N 46
03	24 09	23 07	01 46
04	24 05	23 04	02 43
05	23 58	23 01	03 33
06	23 49	22 57	04 14
07	23 37	22 54	04 43
08	23 23	22 51	04 58
09	23 10	22 48	04 58
10	22 58	22 45	04 42
11	22 48	22 41	04 09
12	22 41	22 38	03 21
13	22 37	22 35	02 21
14	22 35	22 32	01 N 13
15	22 D 35	22 29	00 S 01
16	22 R 35	22 26	01 15
17	22 33	22 22	02 24
18	22 29	22 19	03 25
19	22 22	22 16	04 13
20	22 12	22 13	04 45
21	22 00	22 10	05 00
22	21 47	22 06	04 57
23	21 35	22 03	04 37
24	21 24	22 00	04 03
25	21 15	21 57	03 18
26	21 09	21 54	02 24
27	21 06	21 51	01 24
28	21 05	21 47	00 S 22
29	21 D 05	21 44	00 N 41
30	21 ♌ 05	21 ♌ 41	01 N 42

DECLINATIONS

Date	Sun ☉ ° '	Moon ☽ ° '	Mercury ☿ ° '	Venus ♀ ° '	Mars ♂ ° '	Jupiter ♃ ° '	Saturn ♄ ° '	Uranus ♅ ° '	Neptune ♆ ° '	Pluto ♇ ° '
01	14 S 27	14 N 14	07 S 07	05 S 33	18 S 15	20 S 52	21 S 32	12 N 09	13 S 29	19 N 45
02	14 46	11 11	07 04	06 02	18 26	20 50	21 32	12 09	13 30	19 45
03	15 05	07 41	07 06	06 30	18 38	20 49	21 31	12 08	13 31	19 45
04	15 23	03 N 50	07 12	06 59	18 49	20 47	21 31	12 08	13 31	19 45
05	15 42	00 S 13	07 23	07 27	19 00	20 46	21 30	12 07	13 32	19 45
06	16 00	04 21	07 38	07 55	19 11	20 44	21 30	12 06	13 33	19 45
07	16 17	08 23	07 57	08 23	19 22	20 42	21 29	12 06	13 33	19 44
08	16 35	12 06	08 19	08 51	19 33	20 40	21 28	12 05	13 34	19 44
09	16 52	15 19	08 43	09 19	19 43	20 39	21 28	12 05	13 35	19 44
10	17 09	17 46	09 10	09 46	19 54	20 37	21 27	12 04	13 35	19 44
11	17 26	19 16	09 39	10 13	20 04	20 35	21 26	12 04	13 36	19 44
12	17 42	19 39	10 09	10 40	20 14	20 33	21 26	12 04	13 37	19 44
13	17 59	18 51	10 40	11 07	20 24	20 31	21 25	12 03	13 37	19 45
14	18 14	16 55	11 12	11 33	20 34	20 29	21 24	12 03	13 38	19 45
15	18 30	13 56	11 45	12 00	20 43	20 27	21 23	12 02	13 39	19 45
16	18 45	10 09	12 18	12 25	20 52	20 25	21 23	12 02	13 39	19 45
17	19 00	05 46	12 52	12 51	21 02	20 23	21 22	12 02	13 40	19 45
18	19 14	01 S 04	13 25	13 16	21 10	20 21	21 21	12 01	13 41	19 45
19	19 28	03 N 41	13 59	13 41	21 19	20 19	21 20	12 01	13 41	19 45
20	19 42	08 13	14 33	14 06	21 28	20 17	21 19	12 01	13 42	19 45
21	19 56	12 14	15 06	14 30	21 36	20 14	21 18	12 01	13 43	19 45
22	20 09	15 33	15 39	14 54	21 44	20 12	21 18	12 00	13 43	19 46
23	20 21	17 58	16 11	15 18	21 52	20 10	21 17	12 00	13 44	19 46
24	20 34	19 22	16 43	15 41	22 00	20 08	21 16	12 00	13 45	19 46
25	20 45	19 45	17 15	16 04	22 08	20 05	21 15	12 00	13 45	19 46
26	20 57	19 08	17 46	16 26	22 15	20 03	21 14	12 00	13 46	19 46
27	21 08	17 39	18 16	16 48	22 22	20 00	21 13	11 59	13 46	19 47
28	21 19	15 24	18 45	17 10	22 29	19 58	21 12	11 59	13 47	19 47
29	21 29	12 32	19 13	17 31	22 36	19 55	21 11	11 59	13 48	19 47
30	21 S 39	09 N 10	19 S 41	17 S 51	22 S 42	19 S 53	21 S 10	11 N 59	13 S 48	19 N 47

ZODIAC SIGN ENTRIES

Date	h	m	Planets
01	16	01	♅ ♍
02	06	17	☽ ♍
04	02	49	♃ ♒
04	18	42	☽ ♎
07	04	40	☽ ♏
09	11	51	☽ ♐
10	23	53	☿ ♏
11	05	33	♀ ♏
11	16	59	☽ ♑
13	20	59	☽ ♒
13	21	50	♂ ♐
16	00	18	☽ ♓
18	03	10	☽ ♈
20	06	03	☽ ♉
22	09	59	☽ ♊
22	13	08	☉ ♐
24	16	20	☽ ♋
27	02	01	☽ ♌
29	14	25	☽ ♍
30	22	54	☿ ♐

LATITUDES

Date	Mercury ☿ ° '	Venus ♀ ° '	Mars ♂ ° '	Jupiter ♃ ° '	Saturn ♄ ° '	Uranus ♅ ° '	Neptune ♆ ° '	Pluto ♇ ° '
01	01 N 42	01 N 34	00 S 12	00 S 40	00 S 16	00 N 44	01 N 43	12 N 48
04	02 05	01 33	00 14	00 40	00 16	00 44	01 43	12 49
07	02 15	01 31	00 15	00 39	00 17	00 44	01 43	12 50
10	02 14	01 29	00 17	00 39	00 17	00 44	01 43	12 52
13	02 05	01 26	00 19	00 39	00 17	00 44	01 43	12 53
16	01 52	01 22	00 21	00 39	00 17	00 44	01 43	12 54
19	01 34	01 18	00 22	00 39	00 17	00 44	01 43	12 56
22	01 15	01 14	00 24	00 39	00 17	00 45	01 43	12 57
25	00 55	01 09	00 26	00 39	00 18	00 45	01 43	12 58
28	00 34	01 03	00 28	00 39	00 18	00 45	01 43	13 00
31	00 N 13	00 N 57	00 S 29	00 S 39	00 S 18	00 N 45	01 N 43	13 N 01

DATA

Julian Date	2437605
Delta T	+34 seconds
Ayanamsa	23° 19' 15"
Synetic vernal point	05° ♓ 47' 45"
True obliquity of ecliptic	23° 26' 32"

MOON'S PHASES, APSIDES AND POSITIONS ☽

Date	h	m	Phase	Longitude ° '	Eclipse Indicator
08	09	58	●	15 ♏ 46	
15	12	12	☽	22 ♒ 54	
22	09	44	○	29 ♉ 51	
30	06	18	☾	07 ♍ 48	

Day	h	m		
02	01	45	Apogee	
17	05	30	Perigee	
29	22	28	Apogee	
05	10	42	0S	
12	07	54	Max dec	19° S 40'
18	17	20	0N	
25	08	56	Max dec	19° N 45'

LONGITUDES

Date	Chiron ⚷ ° '	Ceres ⚳ ° '	Pallas ⚴ ° '	Juno ⚵ ° '	Vesta ⚶ ° '	Black Moon Lilith ⚸ ° '
01	01 ♓ 57	22 ♉ 18	07 ♓ 24	17 ♓ 24	00 ♊ 51	00 ♍ 21
11	01 ♓ 53	20 ♉ 02	07 ♓ 21	17 ♓ 54	28 ♉ 28	01 ♍ 28
21	01 ♓ 56	17 ♉ 43	07 ♓ 54	19 ♓ 15	25 ♉ 51	02 ♍ 34
31	02 ♓ 04	15 ♉ 39	08 ♓ 59	21 ♓ 21	23 ♉ 21	03 ♍ 41

All ephemeris data is given at 12.00 UT and the Moon's longitude is additionally given for 24.00 UT

ASPECTARIAN

h m	Aspects
01 Wednesday	
04 47	☽ ⚹ ♀
10 18	☽ ⋕ ☉
12 21	☽ □ ♂
14 46	☽ ⚹ ☿
18 13	☽ ⋕ ♆
18 33	☽ ⊼ ♄
02 Thursday	
00 56	☽ Q ☉
04 14	☽ Q ♆
04 49	☽ ∥ ♅
05 50	☽ ⊼ ♃
06 10	♂ Q ♇
06 20	☽ ☌ ♅
06 50	☽ ± ♄
09 15	☉ ⚹ ♇
14 56	☽ ∠ ♀
18 09	☽ ± ♃
22 01	☽ ∠ ☿
03 Friday	
01 12	☽ ⚼ ♄
02 03	☽ ☌ ♇
03 17	☽ Q ♂
03 34	☽ ⚹ ☉
04 41	☽ ⚹ ♆
11 19	☽ ⊥ ♀
12 34	☽ ⚼ ♃
15 44	☽ ⋕ ☿
16 56	☽ ⊥ ☿
17 14	☉ ☌ ♆
18 42	☽ ⋕ ♀
04 Saturday	
00 48	☽ ⊻ ♀
04 57	☽ ⚹ ♂
05 32	☽ ⊻ ☿
07 32	☽ △ ♄
10 55	☽ ∠ ♆
12 27	☽ ∠ ☉
18 13	☉ Q ♅
18 52	☽ △ ♃
18 53	☽ ⊻ ♅
21 11	♃ ⊼ ♅
05 Sunday	
02 15	☉ Q ♄
04 46	☽ ⊥ ♆
06 22	☿ ∥ ♀
06 44	☽ ⊥ ♅
07 49	☽ ⊥ ☉
12 26	☽ ∠ ♂
13 55	☽ ⊻ ♇
16 32	☽ ⊻ ♆
18 55	☿ □ ♄
20 33	☽ ⊻ ☉
06 Monday	
00 18	☽ ∠ ♅
01 34	☽ ⊥ ♇
04 07	☿ ∠ ♇
04 35	♂ ⚹ ♄
06 55	☽ ⊥ ♂
18 05	☽ ☌ ♀
18 19	☽ □ ♄
18 50	☽ ∠ ♇
19 04	☽ ⊻ ♂
19 55	☽ ☌ ☿
20 38	♀ □ ♄
07 Tuesday	
01 45	♀ ∠ ♇
04 58	☽ ⚹ ♅
05 25	☽ □ ♃
09 10	☽ ∥ ☿
12 03	☽ ∥ ♀
15 56	♀ ⊻ ♂
23 01	☽ ⚹ ♇
08 Wednesday	
01 35	☽ ☌ ♆
03 23	☽ Q ♅
04 12	☽ Q ♄
09 58	☽ ☌ ☉
11 54	☽ ⋕ ♅
15 09	☽ Q ♃
21 04	☽ Q ♇
22 24	☽ ∥ ♆
09 Thursday	
02 14	☽ ⚹ ♄
05 50	☽ ☌ ♂
07 31	☽ ⊻ ♀
08 14	☽ ⊻ ☿
12 14	☽ □ ♅
13 07	☽ ⚹ ♃
14 26	♂ ⋕ ♇
19 26	☽ ⊥ ♀
20 03	☽ ⊥ ☿
10 Friday	
04 12	☽ ∥ ☉
05 18	☽ ∠ ♄
05 29	☽ □ ♇
08 02	☽ ⊻ ♆
13 05	☽ ∠ ♀
13 40	☽ ∠ ☿
16 08	☽ ∠ ♃
18 42	☽ ⊥ ♆
20 16	☽ ⊻ ☉
21 20	☽ ⊥ ♄
11 Saturday	
04 45	☿ ⚹ ♅
07 40	☽ ⊥ ☉
07 56	☽ ⊻ ♄
08 10	☽ ⊥ ♃
10 35	☽ ∠ ♆
10 36	♀ ⚹ ♅
14 05	☽ ⊻ ♂
15 18	☉ Q ♃
17 28	☽ △ ♅
18 08	☽ ⚹ ♀
18 46	☽ ⚹ ☿
18 46	☽ ⊻ ♃
18 56	☿ □ ♃
12 Sunday	
00 40	☽ ∠ ☉
01 08	☽ ⊥ ♂
01 58	♀ □ ♃
02 58	♄ ⚼ ♇
10 14	☽ △ ♇
12 48	☽ ⚹ ♆
17 06	☽ Q ♀
17 38	☽ ∠ ♂
17 52	☽ Q ☿
19 36	☽ ⚼ ♅
13 Monday	
04 44	☽ ⚹ ☉
09 38	☽ Q ♆
11 11	☽ ± ♅
12 13	☽ ⚼ ♇
12 21	☽ ☌ ♄
20 56	☽ ⚹ ♂
21 32	☽ ⊼ ♅
23 02	☽ ∥ ☉
23 17	☽ ☌ ♃
14 Tuesday	
02 57	♂ ∥ ♃
03 01	☽ Q ☉
03 14	☽ □ ♀
03 43	☽ ± ♇
04 21	☽ □ ☿
08 46	♂ □ ♅
13 00	☉ Q ♇
14 02	☽ ⊼ ♇
16 39	☽ □ ♆
18 39	☽ Q ♂
15 Wednesday	
12 12	☽ □ ☉
14 03	☽ ∥ ♆
14 37	♀ ⋕ ♅
16 02	☽ ⊻ ♄
23 34	☽ ∥ ♀
16 Thursday	
00 33	☿ ⋕ ♅
00 36	☽ ∥ ☿
00 36	☽ ⋕ ♅
00 55	☽ ☍ ♅
02 19	☽ ⊥ ♄
03 01	☽ □ ♂
03 08	☽ ⊻ ♃
06 02	♂ ⚹ ♃
11 28	☽ △ ♀
13 26	☽ ⊥ ♃
13 27	☽ △ ☿
17 13	☽ ☍ ♇
17 41	☽ ∠ ♄
19 54	☽ △ ♆
17 Friday	
04 52	☽ ∠ ♃
09 58	☿ ∥ ♀
15 22	☽ ⚼ ♀
17 51	☽ ⚼ ☿
19 03	☽ △ ☉
19 14	☽ ⚹ ♄
21 23	☽ ⚼ ♆
21 45	☉ ⚹ ♄
18 Saturday	
01 28	☿ ⚹ ♇
03 50	☽ ⊼ ♅
06 32	☽ ⚹ ♃
08 04	☽ ± ♀
08 35	☽ △ ♂
10 53	☽ ± ☿
12 38	☽ ± ♆
14 00	☽ ± ♅
15 39	☽ Q ♄
19 12	☽ ⊼ ♀
20 03	☽ ⊼ ♇
22 16	☽ ⊼ ☿
22 22	☽ ⚼ ☉
22 49	☽ ⊼ ♆
23 03	☿ ∥ ♆
19 Sunday	
03 05	☽ Q ♃
03 36	☿ ☌ ♆
05 00	♀ ⚹ ♇
05 15	☽ ⚼ ♅
06 14	☽ ± ♇
11 20	☽ ⚼ ♂
12 01	♀ ∥ ♆
14 18	♂ ∥ ♄
14 48	☽ ± ☉
16 10	☿ Q ♅
21 31	☽ ⚼ ♇
22 20	☽ □ ♄
20 Monday	
01 47	☽ ⊼ ☉
03 28	☽ ± ♂
06 47	☽ △ ♅
09 16	☿ Q ♄
10 02	☽ □ ♃
13 17	♀ ☌ ♆
14 16	☽ ⊼ ♂
17 33	☉ ⋕ ♇
23 11	☽ △ ♇
21 Tuesday	
02 05	☽ ☍ ♆
03 19	☽ ☍ ♀
04 01	♀ Q ♅
07 43	☽ ☍ ☿
10 31	☽ ∥ ♅
21 59	☽ ⋕ ♆
22 Wednesday	
02 20	☽ △ ♄
03 13	♀ Q ♄
06 02	☽ ⋕ ♀
09 44	☽ ☍ ☉
10 48	☽ □ ♅
13 02	☽ ⋕ ☿
14 44	☽ △ ♃
17 39	☉ ∥ ♃
21 28	☽ ☍ ♂
23 Thursday	
00 13	☉ □ ♅
03 49	☽ □ ♇
05 06	☽ ⚼ ♄
06 57	☽ ⊼ ♆
13 36	☽ ⊼ ♀
17 50	☽ ± ♆
18 00	☽ ⚼ ♃
19 05	☽ Q ♅
19 45	☽ ⊼ ☿
21 31	☽ ± ♄
24 Friday	
01 41	☽ ± ♀
08 15	☽ ± ☿
08 37	☽ ⊼ ♄
10 25	☽ ⚼ ♆
10 45	☽ ± ♃
12 42	☽ Q ♇
17 15	☽ ⚹ ♅
20 10	☽ ⚼ ♀
20 43	☽ ⊼ ☉
22 06	☽ ⊼ ♃
25 Saturday	
03 24	☽ ⚼ ☿
07 15	☿ Q ♃
07 50	☽ ⊼ ♂
09 01	☽ ± ☉
11 17	☽ ⚹ ♇
14 43	☽ △ ♆
17 31	☉ ⚹ ♃
20 00	☽ ± ♂
21 16	☿ Q ♇
21 43	☽ ∠ ♅
26 Sunday	
03 48	☽ ⚼ ☉
03 55	☽ △ ♀
12 21	☽ △ ☿
14 33	☽ ⚼ ♂
15 12	☽ ⊥ ♅
16 18	☽ ∠ ♇
18 14	☽ ☍ ♄
27 Monday	
03 02	☽ ⊻ ♅
05 35	☽ ⋕ ☿
08 58	☽ ☍ ♃
10 06	☽ ⊥ ♇
11 57	☽ △ ☉
20 16	♂ □ ♇
20 32	☽ ⋕ ♀
22 03	☉ ∥ ♄
22 05	☽ ⊻ ♇
22 12	☽ △ ♂
28 Tuesday	
01 50	☽ □ ♆
12 50	☿ ⚹ ♄
13 17	♀ Q ♃
20 24	♀ Q ♇
22 33	☽ □ ♀
29 Wednesday	
02 02	☽ ⋕ ♆
06 50	☽ ⊼ ♄
09 26	☽ □ ☿
12 25	♂ ∠ ♄
14 24	☽ Q ♆
15 30	☽ ☌ ♅
16 06	☽ ∥ ♅
19 08	☽ ± ♄
22 28	☽ ⊼ ♃
30 Thursday	
06 18	☽ □ ☉
10 52	☽ ± ♃
10 57	☽ ☌ ♇
12 25	♂ ⊻ ♆
13 36	☽ ⚼ ♄
14 54	☽ ⚹ ♆
15 03	☽ □ ♂
15 31	☽ Q ♀
16 58	♃ ± ♇
17 50	☿ ⋕ ♇
21 42	☿ ∥ ♃

DECEMBER 1961

LONGITUDES

Date	Sidereal time h m s	Sun ☉ ° ' "	Moon ☽ ° ' "	Moon ☽ 24.00 ° ' "	Mercury ☿ ° '	Venus ♀ ° '	Mars ♂ ° '	Jupiter ♃ ° '	Saturn ♄ ° '	Uranus ♅ ° '	Neptune ♆ ° '	Pluto ♇ ° '
01	16 40 25	09 ♐ 03 40	22 ♍ 27 26	28 ♍ 25 58	00 ♐ 51	25 ♏ 25	12 ♐ 45	04 ♒ 14	26 ♑ 29	00 ♍ 32	12 ♏ 04	10 ♍ 06
02	16 44 22	10 04 30	04 ♎ 27 22	10 ♎ 32 13	02 25	26 40	13 29	04 25	26 34	00 33	12 06	10 06
03	16 48 19	11 05 22	16 ♎ 41 04	22 ♎ 54 21	03 59	27 55	14 13	04 36	26 40	00 33	12 08	10 07
04	16 52 15	12 06 16	29 ♎ 12 29	05 ♏ 35 43	05 34	29 ♏ 11	14 57	04 47	26 46	00 33	12 10	10 07
05	16 56 12	13 07 11	12 ♏ 04 14	18 ♏ 38 05	07 08	00 ♐ 26	15 41	04 59	26 51	00 33	12 12	10 07
06	17 00 08	14 08 07	25 ♏ 17 14	02 ♐ 01 29	08 42	01 42	16 25	05 10	26 57	00 R 33	12 14	10 07
07	17 04 05	15 09 04	08 ♐ 50 32	15 ♐ 44 00	10 16	02 57	17 09	05 22	27 03	00 33	12 16	10 07
08	17 08 01	16 10 02	22 ♐ 41 22	29 ♐ 42 05	11 50	04 12	17 54	05 33	27 09	00 33	12 18	10 08
09	17 11 58	17 11 01	06 ♑ 45 32	13 ♑ 51 07	13 24	05 28	18 38	05 45	27 15	00 33	12 20	10 08
10	17 15 54	18 12 01	20 ♑ 58 12	28 ♑ 06 12	14 58	06 43	19 22	05 57	27 21	00 32	12 21	10 08
11	17 19 51	19 13 02	05 ♒ 14 37	12 ♒ 22 56	16 32	07 59	20 07	06 08	27 27	00 32	12 23	10 08
12	17 23 48	20 14 03	19 ♒ 30 48	26 ♒ 37 51	18 06	09 14	20 51	06 20	27 33	00 32	12 25	10 R 08
13	17 27 44	21 15 04	03 ♓ 43 52	10 ♓ 48 39	19 40	10 30	21 36	06 32	27 39	00 31	12 27	10 08
14	17 31 41	22 16 07	17 ♓ 52 03	24 ♓ 53 57	21 15	11 45	22 21	06 44	27 46	00 31	12 29	10 08
15	17 35 37	23 17 09	01 ♈ 54 16	08 ♈ 52 56	22 49	13 01	23 05	06 57	27 52	00 30	12 31	10 08
16	17 39 34	24 18 12	15 ♈ 49 50	22 ♈ 44 52	24 24	14 16	23 50	07 09	27 58	00 30	12 33	10 08
17	17 43 30	25 19 15	29 ♈ 37 54	06 ♉ 28 47	25 58	15 32	24 35	07 21	28 04	00 29	12 34	10 07
18	17 47 27	26 20 19	13 ♉ 17 20	20 ♉ 03 20	27 33	16 47	25 19	07 33	28 11	00 29	12 36	10 07
19	17 51 23	27 21 23	26 ♉ 46 34	03 ♊ 26 47	29 ♐ 08	18 03	26 04	07 46	28 17	00 28	12 38	10 07
20	17 55 20	28 22 27	10 ♊ 03 47	16 ♊ 37 21	00 ♑ 43	19 18	26 49	07 59	28 24	00 27	12 40	10 07
21	17 59 17	29 ♐ 23 32	23 ♊ 07 19	29 ♊ 33 33	02 19	20 34	27 34	08 11	28 30	00 26	12 41	10 07
22	18 03 13	00 ♑ 24 38	05 ♋ 55 57	12 ♋ 14 32	03 54	21 49	28 19	08 24	28 37	00 26	12 43	10 06
23	18 07 10	01 25 44	18 ♋ 29 18	24 ♋ 40 24	05 30	23 04	29 04	08 37	28 43	00 25	12 45	10 06
24	18 11 06	02 26 50	00 ♌ 48 00	06 ♌ 52 22	07 06	24 20	29 ♐ 49	08 49	28 50	00 24	12 46	10 06
25	18 15 03	03 27 57	12 ♌ 53 48	18 ♌ 52 42	08 42	25 35	00 ♑ 34	09 02	28 57	00 23	12 48	10 05
26	18 18 59	04 29 04	24 ♌ 49 29	00 ♍ 44 41	10 19	26 51	01 19	09 15	29 03	00 22	12 49	10 05
27	18 22 56	05 30 12	06 ♍ 38 49	12 ♍ 32 28	11 55	28 06	02 05	09 28	29 10	00 20	12 51	10 04
28	18 26 52	06 31 20	18 ♍ 26 14	24 ♍ 20 47	13 32	29 ♐ 22	02 50	09 41	29 17	00 19	12 52	10 04
29	18 30 49	07 32 29	00 ♎ 16 45	06 ♎ 14 48	15 09	00 ♑ 37	03 35	09 54	29 24	00 18	12 54	10 03
30	18 34 45	08 33 38	12 ♎ 15 35	18 ♎ 19 46	16 47	01 53	04 20	10 08	29 30	00 17	12 55	10 03
31	18 38 42	09 ♑ 34 47	24 ♎ 27 57	00 ♏ 40 44	18 ♑ 24	03 ♑ 08	05 ♑ 06	10 ♒ 21	29 ♑ 37	00 ♍ 15	12 ♏ 57	10 ♍ 02

Date	Moon True ☊ ° '	Moon Mean ☊ ° '	Moon ☽ Latitude ° '
01	21 ♌ 04	21 ♌ 38	02 N 39
02	21 R 01	21 35	03 29
03	20 56	21 32	04 11
04	20 48	21 28	04 42
05	20 38	21 25	05 00
06	20 26	21 22	05 02
07	20 14	21 19	04 47
08	20 04	21 16	04 16
09	19 55	21 12	03 28
10	19 49	21 09	02 27
11	19 46	21 06	01 17
12	19 45	21 03	00 N 01
13	19 D 45	21 00	01 S 14
14	19 46	20 57	02 25
15	19 R 46	20 53	03 26
16	19 44	20 50	04 15
17	19 39	20 47	04 48
18	19 32	20 44	05 04
19	19 24	20 41	05 04
20	19 14	20 38	04 47
21	19 05	20 34	04 15
22	18 57	20 31	03 30
23	18 50	20 28	02 36
24	18 46	20 25	01 36
25	18 44	20 22	00 S 32
26	18 D 44	20 18	00 N 33
27	18 46	20 15	01 36
28	18 47	20 12	02 34
29	18 49	20 09	03 27
30	18 R 49	20 06	04 11
31	18 ♌ 48	20 ♌ 03	04 N 44

DECLINATIONS

Date	Sun ☉ ° '	Moon ☽ ° '	Mercury ☿ ° '	Venus ♀ ° '	Mars ♂ ° '	Jupiter ♃ ° '	Saturn ♄ ° '	Uranus ♅ ° '	Neptune ♆ ° '	Pluto ♇ ° '
01	21 S 49	05 N 25	20 S 08	18 S 11	22 S 49	19 S 50	21 S 09	11 N 59	13 S 49	19 N 48
02	21 58	01 N 26	20 33	18 31	22 55	19 48	21 08	11 59	13 50	19 48
03	22 06	02 S 41	20 58	18 50	23 01	19 45	21 07	11 59	13 50	19 48
04	22 15	06 48	21 22	19 09	23 06	19 42	21 06	11 59	13 51	19 49
05	22 23	10 42	21 45	19 27	23 12	19 40	21 05	11 59	13 51	19 49
06	22 30	14 12	22 06	19 44	23 17	19 37	21 04	11 59	13 52	19 49
07	22 37	17 03	22 27	20 01	23 22	19 34	21 03	11 59	13 52	19 50
08	22 43	18 59	22 47	20 18	23 26	19 31	21 02	11 59	13 53	19 50
09	22 49	19 48	23 05	20 33	23 31	19 28	21 01	11 59	13 54	19 50
10	22 55	19 23	23 22	20 49	23 35	19 25	21 00	11 59	13 54	19 51
11	23 00	17 43	23 38	21 03	23 39	19 22	20 58	11 59	13 55	19 51
12	23 05	14 57	23 53	21 17	23 43	19 19	20 57	12 00	13 55	19 52
13	23 09	11 18	24 07	21 31	23 46	19 16	20 56	12 00	13 56	19 52
14	23 13	07 01	24 19	21 43	23 50	19 13	20 55	12 00	13 56	19 52
15	23 16	02 S 24	24 30	21 56	23 53	19 10	20 54	12 00	13 57	19 53
16	23 19	02 N 19	24 40	22 07	23 55	19 07	20 53	12 00	13 57	19 53
17	23 22	06 51	24 49	22 18	23 58	19 04	20 51	12 01	13 58	19 54
18	23 23	10 59	24 56	22 28	24 00	19 01	20 50	12 01	13 58	19 54
19	23 25	14 31	25 02	22 38	24 02	18 58	20 49	12 01	13 59	19 55
20	23 26	17 14	25 06	22 46	24 04	18 55	20 48	12 02	13 59	19 55
21	23 26	19 02	25 09	22 54	24 05	18 51	20 46	12 02	14 00	19 56
22	23 26	19 49	25 11	23 02	24 06	18 48	20 45	12 02	14 00	19 56
23	23 26	19 35	25 11	23 09	24 07	18 45	20 44	12 03	14 01	19 57
24	23 25	18 26	25 10	23 15	24 08	18 41	20 42	12 03	14 01	19 58
25	23 24	16 26	25 07	23 20	24 09	18 38	20 41	12 03	14 01	19 58
26	23 22	13 46	25 03	23 25	24 09	18 35	20 40	12 04	14 02	19 59
27	23 20	10 33	24 57	23 28	24 09	18 31	20 39	12 04	14 02	19 59
28	23 17	06 56	24 49	23 32	24 08	18 28	20 37	12 05	14 03	20 00
29	23 14	03 N 03	24 41	23 34	24 08	18 24	20 36	12 05	14 03	20 00
30	23 10	01 S 00	24 30	23 36	24 07	18 21	20 34	12 06	14 03	20 01
31	23 S 06	05 S 05	24 S 18	23 S 37	24 S 06	18 S 17	20 S 33	12 N 06	14 S 04	20 N 02

ZODIAC SIGN ENTRIES

Date	h	m	Planets
02	03	08	☽ ♎
04	13	30	☽ ♏
05	03	40	♀ ♐
06	20	25	☽ ♐
09	00	31	☽ ♑
11	03	11	☽ ♒
13	05	41	☽ ♓
15	08	44	☽ ♈
17	12	39	☽ ♉
19	17	47	☽ ♊
20	01	04	☿ ♑
22	00	50	☽ ♋
22	02	19	☉ ♑
24	10	26	☽ ♌
24	17	50	♂ ♑
26	22	29	☽ ♍
29	00	07	♀ ♑
29	11	26	☽ ♎
31	22	42	☽ ♏

LATITUDES

Date	Mercury ☿ ° '	Venus ♀ ° '	Mars ♂ ° '	Jupiter ♃ ° '	Saturn ♄ ° '	Uranus ♅ ° '	Neptune ♆ ° '	Pluto ♇ ° '
01	00 N 13	00 N 57	00 S 29	00 S 39	00 S 18	00 N 45	01 N 43	13 N 01
04	00 S 08	00 51	00 31	00 39	00 18	00 45	01 43	13 03
07	00 28	00 45	00 33	00 39	00 18	00 45	01 44	13 04
10	00 47	00 38	00 34	00 39	00 18	00 45	01 44	13 05
13	01 05	00 31	00 36	00 39	00 19	00 45	01 44	13 07
16	01 21	00 24	00 37	00 39	00 19	00 46	01 44	13 08
19	01 35	00 17	00 39	00 39	00 19	00 46	01 44	13 10
22	01 48	00 09	00 41	00 39	00 19	00 46	01 44	13 11
25	01 58	00 N 02	00 42	00 39	00 19	00 46	01 44	13 12
28	02 05	00 S 05	00 44	00 40	00 19	00 46	01 44	13 14
31	02 S 09	00 S 12	00 S 45	00 S 40	00 S 20	00 N 46	01 N 45	13 N 15

DATA

Julian Date	2437635
Delta T	+34 seconds
Ayanamsa	23° 19' 19"
Synetic vernal point	05° ♓ 47' 41"
True obliquity of ecliptic	23° 26' 32"

MOON'S PHASES, APSIDES AND POSITIONS ☽

Date	h	m	Phase	Longitude ° '	Eclipse Indicator
07	23	52	●	15 ♐ 39	
14	20	05	☽	22 ♓ 37	
22	00	42	○	29 ♊ 56	
30	03	57	☾	08 ♎ 13	

Day	h	m		
12	00	19	Perigee	
27	19	23	Apogee	
02	20	22	0S	
09	15	58	Max dec	19° S 49'
16	00	11	0N	
22	18	34	Max dec	19° N 51'
30	06	07	0S	

LONGITUDES

Date	Chiron ⚷ ° '	Ceres ⚳ ° '	Pallas ⚴ ° '	Juno ⚵ ° '	Vesta ⚶ ° '	Black Moon Lilith ⚸ ° '
01	02 ♓ 04	15 ♉ 39	08 ♓ 59	21 ♓ 21	23 ♉ 21	03 ♍ 41
11	02 ♓ 19	14 ♉ 02	10 ♓ 32	24 ♓ 06	21 ♉ 14	04 ♍ 47
21	02 ♓ 39	13 ♉ 01	12 ♓ 30	27 ♓ 26	19 ♉ 44	05 ♍ 54
31	03 ♓ 03	12 ♉ 40	14 ♓ 50	01 ♈ 14	18 ♉ 57	07 ♍ 00

All ephemeris data is given at 12.00 UT and the Moon's longitude is additionally given for 24.00 UT

ASPECTARIAN

h m	Aspects
01 Friday	
03 37	☽ Q ☿
05 24	☽ bQ ♃
07 08	☿ □ ♅
18 38	☽ ⚹ ♀
20 09	☽ △ ♄
21 17	☽ ∠ ♆
22 07	☽ Q ☉
02 Saturday	
04 13	☽ ⊻ ♅
05 42	☽ Q ♂
07 22	☽ ⚹ ☿
09 08	♃ # ♇
10 00	♀ ⚹ ♄
11 56	☽ △ ♃
12 42	☉ □ ♇
15 15	☽ ⊥ ♆
16 08	☽ ⊥ ♅
23 09	☽ ⊻ ♇
03 Sunday	
00 06	☽ ⚹ ☉
03 06	☽ ⊻ ♆
03 50	☽ ∠ ♀
06 53	☽ ⚹ ♂
09 47	☽ ∠ ♅
10 53	☽ ⊥ ♇
17 07	☽ ∠ ☿
20 37	☿ ∥ ♄
22 40	☿ ⚹ ♃
23 15	☽ ⊥ ♀
04 Monday	
02 59	☉ ∠ ♄
04 13	☽ ∠ ♇
07 19	☽ □ ♄
07 40	☽ ∠ ☉
11 56	☽ ⊻ ♀
12 45	☽ ⊥ ☿
13 23	☉ ⊻ ♆
13 29	☽ ∠ ♂
14 32	☽ ⚹ ♅
22 39	☽ □ ♃
05 Tuesday	
01 35	☽ ⊻ ☿
02 03	☽ ⊥ ☉
07 20	☽ ⊥ ♂
08 24	☽ ⚹ ♇
12 14	☽ ☌ ♆
12 53	☽ Q ♅
14 05	☽ ⊻ ☉
14 08	♀ □ ♅
17 09	☽ Q ♄
19 01	☽ ⊻ ♂
20 23	☽ # ♅
06 Wednesday	
03 03	♀ ∥ ♃
04 27	♅ St R
06 18	☽ Q ♇
08 08	☽ Q ♃
09 30	☽ ∥ ♆
15 00	☽ ⚹ ♄
19 07	♀ # ♇
21 23	☽ □ ♅
07 Thursday	
00 35	☽ ☌ ♀
05 48	☽ ⚹ ♃
09 55	☿ □ ♇
14 15	☽ □ ♇
14 48	☽ ☌ ☿
17 38	☽ ∠ ♄
17 59	☽ ⊻ ♆
23 52	☽ ☌ ☉
08 Friday	
03 17	☽ ☌ ♂
04 25	☽ ⊥ ♆
05 55	☉ ∥ ☿
08 16	☽ ∠ ♃
09 20	☽ ⊥ ♄
17 15	☿ ∠ ♄
19 17	☿ ⊻ ♆
19 42	☽ ⊻ ♄
19 54	☽ ∠ ♆
22 46	☽ ∥ ♃
23 55	☽ ⊥ ♃
09 Saturday	
01 26	☽ △ ♅
01 31	♂ ⊥ ♆
09 36	☽ ⊻ ♀
10 16	☽ ⊻ ♃
17 42	☽ △ ♇
18 21	♀ ⚹ ♃
20 45	☽ ⊥ ♀
21 27	☽ ⚹ ♆
10 Sunday	
00 37	☽ ⊻ ☿
02 51	☽ bQ ♅
06 59	☽ ⊻ ☉
09 10	☽ ⊻ ♂
10 54	☽ ∥ ♃
11 59	☽ ⊥ ☿
13 23	☽ ∠ ♀
15 50	☉ ⊥ ♆
17 43	☽ Q ♆
17 51	☽ ⊥ ☉
18 00	☽ ± ♅
19 00	☽ bQ ♇
19 49	☽ ⊥ ♂
22 48	☽ ☌ ♄
11 Monday	
04 05	☽ ⊼ ♅
04 29	♀ ∥ ♄
05 00	☽ ∠ ☿
10 08	☽ ± ♇
10 09	☽ ∠ ☉
11 46	☽ ∠ ♂
13 22	☿ ∥ ♂
13 32	☽ ☌ ♃
17 03	☽ ⚹ ♀
20 13	☽ ⊼ ♇
12 Tuesday	
00 02	☽ □ ♆
09 20	☽ ⚹ ☿
11 55	♇ St R
13 18	☽ ⚹ ☉
14 23	☽ ⚹ ♂
15 11	☽ Q ♀
16 58	☿ ⊥ ♆
19 17	☽ ∥ ♆
13 Wednesday	
01 39	☽ ⊻ ♄
05 02	♀ □ ♇
06 35	☽ ☍ ♅
07 44	☽ # ♅
08 05	☽ Q ☿
09 17	♂ ∠ ♃
11 07	☽ Q ☉
11 46	☽ Q ♂
11 52	☽ ⊥ ♄
14 06	♂ ⊥ ♄
16 49	☽ ⊻ ♃
20 26	☉ ∠ ♃
22 36	☉ ⊥ ♄
22 51	☽ ☍ ♇
14 Thursday	
00 35	☽ □ ♀
02 49	☽ △ ♆
03 09	☽ ⊥ ♃
03 15	☽ ∠ ♄
18 29	☉ ☌ ♂
18 29	☽ □ ☿
18 42	☽ ∠ ♃
20 03	☽ □ ♂
20 05	☽ □ ☉
20 24	☿ ⊥ ♄
20 39	☿ ∠ ♃
15 Friday	
02 14	♀ ⊻ ♆
04 27	☽ bQ ♆
05 01	☽ ⚹ ♄
08 55	♀ ∠ ♄
09 23	♅ Q ♆
09 36	☽ ⊼ ♅
19 43	☿ ☌ ♂
19 54	☽ ± ♅
19 56	☽ ± ♆
20 47	☽ ⚹ ♃
16 Saturday	
01 48	☽ Q ♄
02 09	☽ ⊼ ♇
06 18	☽ ⊼ ♆
08 04	☉ ☌ ☿
09 02	☽ △ ♀
11 26	☽ bQ ♅
12 31	☽ ± ♇
17 50	☽ Q ♃
17 Sunday	
02 41	☽ △ ♂
03 53	☽ △ ☉
04 08	☽ bQ ♇
04 47	☽ △ ☿
09 16	☽ □ ♄
13 30	☽ △ ♅
13 43	☽ bQ ♀
18 Monday	
01 44	☽ □ ♃
06 25	☽ △ ♇
06 28	☽ bQ ♂
07 08	☽ ± ♀
08 17	☽ bQ ☉
10 32	☽ bQ ☿
10 47	☽ ☍ ♆
12 44	☿ ∠ ♆
18 34	☽ ∥ ♅
18 49	☽ ⊼ ♀
22 12	☿ ⊻ ♄
23 19	☽ ± ♂
19 Tuesday	
01 31	☽ ± ☉
04 37	☽ ± ☿
08 02	☽ # ♆
10 40	☽ ⊼ ♂
13 08	☽ ⊼ ☉
14 44	☽ △ ♄
16 48	☽ ⊼ ☿
18 37	☽ □ ♅
18 38	☉ ∠ ♆
23 28	♀ ⊥ ♆
20 Wednesday	
07 57	☿ △ ♅
08 08	☽ △ ♃
12 06	☽ □ ♇
12 35	☉ ⊻ ♄
16 45	☽ ⊼ ♆
18 08	☽ bQ ♄
21 Thursday	
03 21	☽ Q ♅
03 47	☽ ± ♆
06 45	☽ ☍ ♀
09 06	☽ # ♃
09 48	☿ ⊥ ♃
10 51	☽ ± ♄
12 07	☽ bQ ♃
16 00	♂ ∠ ♆
20 31	☽ bQ ♆
20 47	☽ ☍ ♂
21 17	☽ Q ♇
22 07	☽ ⊼ ♄
22 Friday	
00 42	☽ ☍ ☉
01 38	☽ ⚹ ♅
05 13	☽ ± ♃
07 37	☽ ☍ ☿
12 20	☉ △ ♅
16 45	☽ ⊼ ♃
19 55	☽ ⚹ ♇
23 10	♂ ⊻ ♄
23 Saturday	
00 56	☽ △ ♆
04 40	♀ ⊥ ♄
06 05	☽ ∠ ♅
21 54	☽ ⊼ ♀
23 28	☽ ⊥ ♅
24 Sunday	
00 18	♀ ∠ ♃
00 49	☽ ∠ ♇
07 40	☽ # ♃
08 06	☽ ☍ ♄
09 56	☽ ⊼ ♂
10 58	☽ ± ♀
11 12	☽ ⊻ ♅
15 32	☽ ⊼ ☉
18 29	☽ ⊥ ♇
22 34	☽ ± ♂
23 14	☉ ⊥ ♃
25 Monday	
02 21	☽ ⊼ ☿
04 10	☽ ☍ ♃
04 31	☽ ± ☉
05 59	♂ △ ♅
06 24	☽ ⊻ ♇
06 51	☽ bQ ♀
11 48	☽ □ ♆
16 11	☽ ± ☿
17 43	☽ bQ ♂
17 46	☿ ⊻ ♃
26 Tuesday	
00 13	☽ bQ ☉
02 23	☉ ∥ ♀
08 34	☿ △ ♇
09 51	☽ # ♆
13 08	☽ bQ ☿
16 35	☽ △ ♀
20 39	☽ ⊼ ♄
23 12	☽ ☌ ♅
27 Wednesday	
00 11	☽ Q ♆
01 08	☽ ∥ ♅
02 04	☽ △ ♂
06 57	♀ ∠ ♆
08 58	☽ ± ♄
09 27	☽ △ ☉
17 51	☽ ⊼ ♃
18 58	☽ ☌ ♇
28 Thursday	
00 27	☽ △ ☿
00 39	☽ ⚹ ♆
01 59	☿ ⚹ ♆
03 28	☽ bQ ♄
06 18	☽ ± ♃
10 14	♀ ⊻ ♄
29 Friday	
00 56	☽ bQ ♃
05 55	♀ △ ♅
07 11	☽ ∠ ♆
10 12	☽ △ ♄
12 02	☽ ⊻ ♅
12 47	☽ □ ♀
14 07	☿ bQ ♅
19 06	☽ □ ♂
30 Saturday	
00 05	☽ ⊥ ♅
01 20	☽ ⊥ ♆
02 31	♂ ⊥ ♃
03 27	♃ ⊼ ♇
03 57	☽ □ ☉
07 36	☽ ⊻ ♇
07 41	☽ △ ♃
13 19	☽ ⊻ ♆
17 58	☽ ∠ ♅
19 30	☽ ⊥ ♇
22 19	☽ □ ☿
31 Sunday	
04 46	☽ Q ♀
09 10	☽ Q ♂
13 06	☽ ∠ ♇
18 34	☽ Q ☉
22 04	☽ □ ♄
22 39	☉ △ ♇
23 10	☽ ⚹ ♅

JANUARY 1962

Date	Sidereal time h m s	Sun ☉ ° ' "	Moon ☽ ° ' "	Moon ☽ 24.00 ° ' "	Mercury ☿ ° '	Venus ♀ ° '	Mars ♂ ° '	Jupiter ♃ ° '	Saturn ♄ ° '	Uranus ♅ ° '	Neptune ♆ ° '	Pluto ♇ ° '
01	18 42 39	10 ♑ 35 57	06 ♏ 58 37	13 ♏ 22 04	20 ♑ 02	04 ♑ 24	05 ♑ 51	10 ♒ 34	29 ♑ 44	00 ♍ 14	12 ♏ 58	10 ♍ 02
02	18 46 35	11 37 08	19 ♏ 51 26	26 ♏ 26 59	21 39	05 39	06 37	10 48	29 51	00 R 13	13 00	10 R 01
03	18 50 32	12 38 18	03 ♐ 08 50	09 ♐ 57 00	23 17	06 55	07 22	11 01	29 ♑ 58	00 11	13 01	10 00
04	18 54 28	13 39 29	16 ♐ 51 19	23 ♐ 51 30	24 55	08 10	08 08	11 15	00 ♒ 05	00 10	13 02	10 00
05	18 58 25	14 40 40	00 ♑ 57 04	08 ♑ 07 28	26 33	09 26	08 53	11 28	00 12	00 08	13 04	09 59
06	19 02 21	15 41 51	15 ♑ 21 59	22 ♑ 39 48	28 10	10 41	09 39	11 42	00 19	00 06	13 05	09 58
07	19 06 18	16 43 02	00 ♒ 00 03	07 ♒ 21 50	29 ♑ 47	11 57	10 25	11 55	00 26	00 05	13 06	09 57
08	19 10 14	17 44 12	14 ♒ 44 14	22 ♒ 06 24	01 ♒ 24	13 12	11 10	12 09	00 33	00 03	13 07	09 57
09	19 14 11	18 45 23	29 ♒ 27 31	06 ♓ 46 52	03 00	14 28	11 56	12 23	00 40	00 01	13 08	09 56
10	19 18 08	19 46 32	14 ♓ 03 48	21 ♓ 17 50	04 36	15 43	12 42	12 36	00 47	00 ♍ 00	13 10	09 55
11	19 22 04	20 47 41	28 ♓ 28 31	05 ♈ 35 34	06 10	16 59	13 28	12 50	00 54	29 ♌ 58	13 11	09 54
12	19 26 01	21 48 50	12 ♈ 38 47	19 ♈ 38 00	07 44	18 14	14 14	13 04	01 01	29 56	13 12	09 53
13	19 29 57	22 49 57	26 ♈ 33 12	03 ♉ 24 22	09 15	19 30	15 00	13 18	01 08	29 54	13 13	09 52
14	19 33 54	23 51 05	10 ♉ 11 33	16 ♉ 54 48	10 45	20 45	15 46	13 32	01 15	29 52	13 14	09 51
15	19 37 50	24 52 11	23 ♉ 34 15	00 ♊ 09 58	12 13	22 01	16 32	13 46	01 22	29 50	13 15	09 50
16	19 41 47	25 53 17	06 ♊ 42 05	13 ♊ 10 42	13 37	23 16	17 18	14 00	01 29	29 48	13 16	09 49
17	19 45 43	26 54 22	19 ♊ 35 57	25 ♊ 57 55	14 59	24 32	18 04	14 14	01 37	29 46	13 17	09 48
18	19 49 40	27 55 26	02 ♋ 16 43	08 ♋ 32 27	16 16	25 47	18 50	14 28	01 44	29 44	13 18	09 47
19	19 53 37	28 56 30	14 ♋ 45 15	20 ♋ 55 14	17 29	27 02	19 36	14 42	01 51	29 42	13 19	09 46
20	19 57 33	29 ♑ 57 33	27 ♋ 02 32	03 ♌ 07 20	18 36	28 18	20 22	14 56	01 58	29 40	13 19	09 45
21	20 01 30	00 ♒ 58 35	09 ♌ 09 47	15 ♌ 10 06	19 37	29 ♑ 33	21 08	15 10	02 05	29 37	13 20	09 44
22	20 05 26	01 59 36	21 ♌ 08 32	27 ♌ 05 21	20 31	00 ♒ 49	21 54	15 24	02 12	29 35	13 21	09 43
23	20 09 23	03 00 37	03 ♍ 00 51	08 ♍ 55 23	21 18	02 04	22 41	15 38	02 19	29 33	13 22	09 42
24	20 13 19	04 01 38	14 ♍ 49 20	20 ♍ 43 06	21 56	03 19	23 27	15 52	02 27	29 31	13 22	09 40
25	20 17 16	05 02 37	26 ♍ 37 09	02 ♎ 31 58	22 24	04 35	24 13	16 07	02 34	29 28	13 23	09 39
26	20 21 12	06 03 36	08 ♎ 28 03	14 ♎ 25 58	22 42	05 50	25 00	16 21	02 41	29 26	13 24	09 38
27	20 25 09	07 04 35	20 ♎ 26 16	26 ♎ 29 32	22 49	07 06	25 46	16 35	02 48	29 24	13 24	09 37
28	20 29 06	08 05 32	02 ♏ 36 20	08 ♏ 47 16	22 R 45	08 21	26 32	16 49	02 55	29 21	13 25	09 35
29	20 33 02	09 06 29	15 ♏ 02 52	21 ♏ 23 42	22 30	09 36	27 19	17 04	03 02	29 19	13 25	09 34
30	20 36 59	10 07 26	27 ♏ 50 14	04 ♐ 22 54	22 03	10 52	28 05	17 18	03 09	29 17	13 26	09 33
31	20 40 55	11 ♒ 08 22	11 ♐ 02 02	17 ♐ 47 51	21 ♒ 26	12 ♒ 07	28 ♑ 52	17 ♒ 32	03 ♒ 16	29 ♌ 14	13 ♏ 26	09 ♍ 32

Date	Moon True ☊ ° '	Moon Mean ☊ ° '	Moon ☽ Latitude ° '
01	18 ♌ 44	19 ♌ 59	05 N 05
02	18 R 40	19 56	05 11
03	18 34	19 53	05 02
04	18 28	19 50	04 34
05	18 23	19 47	03 50
06	18 18	19 44	02 50
07	18 15	19 40	01 38
08	18 14	19 37	00 N 19
09	18 D 14	19 34	01 S 01
10	18 16	19 31	02 17
11	18 17	19 28	03 23
12	18 18	19 24	04 16
13	18 R 19	19 21	04 53
14	18 18	19 18	05 12
15	18 16	19 15	05 13
16	18 13	19 12	04 59
17	18 09	19 09	04 29
18	18 06	19 05	03 46
19	18 03	19 02	02 54
20	18 01	18 59	01 53
21	17 59	18 56	00 S 49
22	17 D 59	18 53	00 N 17
23	18 00	18 50	01 22
24	18 01	18 46	02 23
25	18 02	18 43	03 18
26	18 04	18 40	04 05
27	18 05	18 37	04 42
28	18 05	18 34	05 06
29	18 R 05	18 30	05 17
30	18 05	18 27	05 13
31	18 ♌ 04	18 ♌ 24	04 N 53

DECLINATIONS

Date	Sun ☉ ° '	Moon ☽ ° '	Mercury ☿ ° '	Venus ♀ ° '	Mars ♂ ° '	Jupiter ♃ ° '	Saturn ♄ ° '	Uranus ♅ ° '	Neptune ♆ ° '	Pluto ♇ ° '
01	23 S 01	09 S 02	24 S 05	23 S 37	24 S 04	18 S 14	20 S 32	12 N 07	14 S 04	20 N 02
02	22 56	12 42	23 50	23 36	24 03	18 10	20 30	12 07	14 05	20 03
03	22 50	15 51	23 33	23 35	24 01	18 06	20 29	12 08	14 05	20 04
04	22 44	18 14	23 15	23 33	23 58	18 03	20 27	12 08	14 05	20 04
05	22 38	19 36	22 55	23 30	23 56	17 59	20 26	12 09	14 06	20 05
06	22 31	19 44	22 34	23 27	23 53	17 55	20 25	12 10	14 06	20 05
07	22 24	18 33	22 11	23 23	23 50	17 51	20 23	12 10	14 06	20 06
08	22 16	16 07	21 47	23 18	23 47	17 48	20 22	12 11	14 07	20 07
09	22 08	12 37	21 23	23 12	23 43	17 44	20 20	12 11	14 07	20 08
10	21 59	08 23	20 54	23 06	23 40	17 40	20 19	12 12	14 07	20 08
11	21 50	03 S 43	20 26	22 59	23 36	17 36	20 17	12 13	14 08	20 09
12	21 40	01 N 04	19 56	22 51	23 31	17 32	20 16	12 13	14 08	20 10
13	21 31	05 42	19 26	22 43	23 27	17 28	20 14	12 14	14 08	20 10
14	21 20	09 56	18 54	22 34	23 22	17 24	20 13	12 15	14 08	20 11
15	21 09	13 36	18 22	22 24	23 17	17 20	20 11	12 16	14 09	20 12
16	20 58	16 31	17 49	22 13	23 11	17 16	20 10	12 16	14 09	20 12
17	20 47	18 34	17 15	22 02	23 06	17 12	20 08	12 17	14 09	20 13
18	20 35	19 39	16 42	21 50	23 00	17 08	20 07	12 18	14 09	20 14
19	20 22	19 45	16 08	21 38	22 54	17 04	20 05	12 19	14 09	20 15
20	20 10	18 54	15 35	21 25	22 47	17 00	20 04	12 19	14 10	20 15
21	19 57	17 11	15 03	21 11	22 41	16 56	20 02	12 20	14 10	20 16
22	19 43	14 44	14 31	20 56	22 34	16 52	20 01	12 21	14 10	20 17
23	19 29	11 41	14 02	20 41	22 27	16 48	19 59	12 22	14 10	20 18
24	19 15	08 11	13 34	20 26	22 19	16 44	19 58	12 23	14 10	20 18
25	19 00	04 23	13 09	20 09	22 11	16 39	19 56	12 23	14 10	20 19
26	18 46	00 N 24	12 47	19 52	22 04	16 35	19 55	12 24	14 11	20 20
27	18 30	03 S 38	12 28	19 35	21 55	16 31	19 53	12 25	14 11	20 21
28	18 15	07 35	12 12	19 17	21 47	16 27	19 51	12 26	14 11	20 21
29	17 59	11 17	12 01	18 58	21 38	16 22	19 50	12 27	14 11	20 22
30	17 43	14 35	11 53	18 39	21 29	16 18	19 48	12 28	14 11	20 23
31	17 S 26	17 S 16	11 S 50	18 S 20	21 S 20	16 S 14	19 S 47	12 N 29	14 S 11	20 N 24

ZODIAC SIGN ENTRIES

Date	h	m	Planets
03	06	23	☽ ♐
03	19	01	♄ ♒
05	10	24	☽ ♑
07	12	00	☽ ♒
07	15	08	☿ ♒
09	12	53	☽ ♓
10	05	53	♅ ♌
11	14	34	☽ ♈
13	18	01	☽ ♉
15	23	42	☽ ♊
18	07	39	☽ ♋
20	12	58	☉ ♒
20	17	50	☽ ♌
21	20	31	♀ ♒
23	05	53	☽ ♍
25	18	52	☽ ♎
28	06	54	☽ ♏
30	15	59	☽ ♐

LATITUDES

Date	Mercury ☿ ° '	Venus ♀ ° '	Mars ♂ ° '	Jupiter ♃ ° '	Saturn ♄ ° '	Uranus ♅ ° '	Neptune ♆ ° '	Pluto ♇ ° '
01	02 S 09	00 S 15	00 S 46	00 S 40	00 S 20	00 N 46	01 N 45	13 N 15
04	02 08	00 22	00 47	00 40	00 20	00 46	01 45	13 17
07	02 02	00 29	00 48	00 40	00 20	00 46	01 45	13 18
10	01 50	00 35	00 50	00 40	00 20	00 47	01 45	13 19
13	01 33	00 42	00 51	00 40	00 20	00 47	01 45	13 20
16	01 07	00 48	00 52	00 40	00 21	00 47	01 45	13 21
19	00 S 34	00 54	00 54	00 40	00 21	00 47	01 46	13 22
22	00 N 08	00 59	00 55	00 41	00 21	00 47	01 46	13 23
25	00 57	01 04	00 56	00 41	00 21	00 47	01 46	13 24
28	01 50	01 09	00 57	00 41	00 21	00 47	01 46	13 25
31	02 N 40	01 S 13	00 S 58	00 S 41	00 S 22	00 N 47	01 N 46	13 N 26

LONGITUDES

Date	Chiron ⚷ ° '	Ceres ⚳ ° '	Pallas ⚴ ° '	Juno ⚵ ° '	Vesta ⚶ ° '	Black Moon Lilith ⚸ ° '
01	03 ♓ 06	12 ♉ 40	15 ♓ 05	01 ♈ 38	18 ♉ 55	07 ♍ 07
11	03 ♓ 35	13 ♉ 04	17 ♓ 45	05 ♈ 53	18 ♉ 57	08 ♍ 13
21	04 ♓ 06	14 ♉ 00	20 ♓ 41	10 ♈ 08	19 ♉ 41	09 ♍ 20
31	04 ♓ 44	15 ♉ 38	23 ♓ 51	15 ♈ 20	21 ♉ 02	10 ♍ 26

DATA

Julian Date	2437666
Delta T	+34 seconds
Ayanamsa	23° 19' 24"
Synetic vernal point	05° ♓ 47' 36"
True obliquity of ecliptic	23° 26' 32"

MOON'S PHASES, APSIDES AND POSITIONS ☽

Date	h	m	Phase	Longitude ° '	Eclipse Indicator
06	12	35	●	15 ♑ 43	
13	05	01	☽	22 ♈ 30	
20	18	16	○	00 ♌ 13	
28	23	36	☾	08 ♏ 35	

Day	h	m	
08	13	43	Perigee
24	12	46	Apogee
06	02	30	Max dec 19° S 50'
12	06	38	0N
19	02	25	Max dec 19° N 50'
26	14	22	0S

All ephemeris data is given at 12.00 UT and the Moon's longitude is additionally given for 24.00 UT

ASPECTARIAN

h m	Aspects
01 Monday	
06 34	☽ ⚹ ♀
09 44	☽ ⚹ ♂
11 09	☉ ⊻ ♃
12 48	☿ ∥ ♂
14 17	☽ Q ☿
16 00	♀ ⊥ ♃
17 45	☽ ⚹ ♇
18 53	☽ □ ♃
19 25	☽ ⚹ ☉
21 52	☽ Q ♅
23 17	☽ ☌ ♆
02 Tuesday	
08 00	☽ # ♅
08 17	☽ Q ♄
13 37	☽ ∠ ♀
15 24	☽ ∠ ♂
15 45	☽ ⚹ ☿
15 57	☽ Q ♇
21 55	☽ ∥ ♆
03 Wednesday	
01 20	☽ ∠ ☉
04 30	☽ Q ♃
06 16	☽ ⚹ ♄
06 44	☽ □ ♅
07 37	☽ ⊥ ♀
08 39	☽ ⊥ ♂
08 47	☿ ∥ ♀
18 41	☽ ⊥ ☉
19 21	☽ ⊻ ♀
19 55	☽ ⊻ ♂
21 04	☉ ⚹ ♆
22 19	☽ ∠ ☿
04 Thursday	
00 05	☽ □ ♇
01 03	☿ ± ♅
02 06	☽ ⚹ ♃
05 23	☽ ⊻ ♆
06 01	☽ ⊻ ☉
08 55	☽ ∠ ♄
09 40	☽ ∥ ♃
09 53	♀ ☌ ♂
13 09	☿ ⚼ ♇
13 50	☿ Q ♆
15 46	☽ ⊥ ♆
16 01	☽ ⊥ ☿
05 Friday	
00 29	☽ ⊥ ♄
01 09	♄ ⊼ ♅
03 36	☽ ⊻ ☿
04 19	☽ ∠ ♃
07 07	☽ ∠ ♆
10 38	☽ △ ♅
10 43	☽ ⊻ ♄
19 41	☽ ⊥ ♃
22 23	♀ △ ♇
22 27	☉ ⚼ ♅
06 Saturday	
02 01	☽ ☌ ♂
03 04	☽ △ ♇
03 32	☽ ☌ ♀
05 50	☽ ⊻ ♃
08 13	☽ ⚹ ♆
11 34	☽ ⚼ ♅
12 35	☽ ☌ ☉
16 11	☉ ∥ ☿
21 54	♂ △ ♇
07 Sunday	
02 20	☽ ± ♅
03 46	☽ ⚼ ♇
03 59	☽ Q ♆
11 21	♀ ⊻ ♃
11 37	☽ ☌ ☿
12 08	☽ ⊼ ♅
12 42	☽ ☌ ♄
16 14	☿ ⊼ ♅
18 27	☽ ± ♇
20 22	☽ ∥ ♃
22 16	☿ ☌ ♄
08 Monday	
04 12	☽ ⊼ ♇
05 53	☽ ⊻ ♂
07 43	☽ ☌ ♃
09 17	☽ ⊻ ♀
09 22	☽ □ ♆
10 20	♀ ⚹ ♆
16 11	☽ ⊥ ♂
17 15	☽ ⊻ ☉
19 57	☽ ⊥ ♀
09 Tuesday	
02 28	☽ ∥ ♆
03 45	☽ ⊥ ☉
07 39	☽ ∠ ♂
12 01	☽ ∠ ♀
12 55	☽ ☍ ♅
13 59	☽ ⊻ ♄
14 36	☽ # ♅
18 31	☽ ⊻ ☿
19 34	☽ ∠ ☉
22 23	♀ ⚼ ♅
23 54	☽ ⊥ ♄
10 Wednesday	
01 45	☿ ± ♇
05 10	☽ ☍ ♇
05 35	☽ ⊥ ☿
07 47	♂ ⊻ ♃
09 33	☽ ⊻ ♃
09 37	☽ ⚹ ♂
10 30	☽ △ ♆
14 52	☽ ∠ ♄
15 00	☽ ⚹ ♀
19 39	☽ ⊥ ♃
22 11	☽ ⚹ ☉
22 18	☽ ∠ ☿
11 Thursday	
02 49	♂ ⚹ ♆
06 40	☽ Q ♂
10 55	☽ ∠ ♃
11 30	☽ ⚼ ♆
12 56	☽ Q ♀
14 30	☽ ⊼ ♅
16 07	☽ ⚹ ♄
19 19	☿ ∥ ♄
19 50	☽ Q ☉
12 Friday	
00 36	☽ ± ♅
01 31	☿ # ♇
02 36	☽ ⚹ ☿
02 42	☽ ± ♆
07 18	☽ ⊼ ♇
12 38	☽ Q ♄
12 44	☽ ⚹ ♃
12 57	☽ ⊼ ♆
14 52	☽ □ ♂
15 54	☽ ⚼ ♅
17 33	☽ ± ♇
22 33	☽ □ ♀
13 Saturday	
01 40	☽ Q ☿
02 40	♃ □ ♆
05 01	☽ □ ☉
09 04	☽ ⚼ ♇
09 09	♂ ⚼ ♅
09 47	☽ Q ♃
17 50	☽ △ ♅
20 05	☽ □ ♄
21 40	☿ ⊼ ♇
14 Sunday	
11 24	☽ △ ♇
12 21	☉ ± ♅
13 07	☽ □ ☿
17 25	☽ ☍ ♆
18 03	☽ □ ♃
22 32	☽ △ ♂
15 Monday	
02 38	☽ ∥ ♅
08 20	♄ # ♇
08 53	☽ △ ♀
11 15	☉ ⚼ ♇
14 33	☽ △ ☉
15 58	☽ # ♆
21 04	☉ Q ♆
23 22	☽ □ ♅
16 Tuesday	
02 20	☽ △ ♄
03 23	☽ ⚼ ♂
05 47	☿ □ ♆
15 12	☽ ⚼ ♀
17 46	☽ □ ♇
19 25	☽ # ♃
19 51	☿ ☌ ♃
20 24	☽ ⚼ ☉
21 02	☽ ± ♂
21 53	♀ ± ♅
22 51	☽ # ☿
17 Wednesday	
00 10	☽ ⊼ ♆
01 46	☽ △ ♃
02 21	☽ △ ☿
06 21	☽ ⚼ ♄
08 34	☽ Q ♅
08 56	☽ ⊼ ♂
09 46	☽ ± ♀
11 24	☽ ± ♆
14 22	☿ ∥ ♃
14 40	☽ ± ☉
17 13	♀ ⚼ ♇
22 18	☽ ⊼ ♀
23 26	☽ ± ♄
18 Thursday	
02 34	♀ Q ♆
02 59	☽ ⊼ ☉
03 28	☽ Q ♇
04 24	☽ ⚼ ♆
06 32	☽ ⚼ ♃
07 10	☽ ⚹ ♅
09 51	☽ ⚼ ☿
10 57	☽ ⊼ ♄
19 Friday	
00 04	☽ ± ♃
02 23	☽ ⚹ ♇
05 00	☽ ± ☿
09 12	☽ △ ♆
11 53	☽ ⊼ ♃
11 53	☽ ∠ ♅
17 50	☽ ⊼ ☿
22 03	☽ ☍ ♂
20 Saturday	
01 57	☉ # ♇
05 12	☉ ⊼ ♅
05 23	☽ ⊥ ♅
07 30	☽ ∠ ♇
14 45	☽ ☍ ♀
17 09	☽ ⊻ ♅
18 16	☽ ☍ ☉
21 49	☽ ☍ ♄
21 Sunday	
00 18	☉ ∥ ♄
01 13	☽ ⊥ ♇
13 08	☽ ⊻ ♇
13 18	♀ ⊼ ♅
14 52	☽ # ♃
20 21	☽ □ ♆
22 Monday	
00 14	☽ ☍ ♃
10 40	☽ ☍ ☿
13 39	☽ ⊼ ♂
14 08	☽ # ☿
16 46	☽ # ♆
17 39	☉ ☌ ♄
23 Tuesday	
02 36	☽ ± ♂
05 00	☽ ☌ ♅
05 11	☿ ∥ ♆
06 59	☽ ∥ ♅
08 39	☽ Q ♆
09 51	☽ ⊼ ♀
10 35	☽ ⊼ ♄
11 59	☽ ⊼ ☉
17 24	♀ ☌ ♄
22 08	☽ ⚼ ♂
22 54	☽ ± ♄
23 29	☽ ± ♀
24 Wednesday	
01 20	☽ ± ☉
01 33	☽ ☌ ♇
03 49	☉ ± ♇
09 03	☽ ⚹ ♆
13 51	♂ ± ♅
14 11	☽ ⊼ ♃
17 23	☽ ⚼ ♄
18 34	♀ ± ♇
19 59	☽ ⚼ ♀
21 22	☽ ⚼ ☉
22 17	♀ # ♇
25 Thursday	
02 39	☽ ± ♃
03 06	☽ ⊼ ☿
06 47	☽ △ ♂
15 35	☽ ∠ ♆
15 44	☽ ± ☿
17 47	☽ ⊻ ♅
21 18	☽ ⚼ ♃
26 Friday	
00 11	☽ △ ♄
01 04	♂ ⚼ ♇
05 54	☽ ⊥ ♅
06 04	☽ △ ♀
06 41	☽ △ ☉
08 47	♀ ∥ ♄
09 50	☽ ⊥ ♆
10 25	☽ ⚼ ☿
14 21	☽ ⊻ ♇
21 56	☽ ⊻ ♆
23 58	☽ ∠ ♅
27 Saturday	
00 39	♂ Q ♆
02 23	☽ ⊥ ♇
04 09	☽ △ ♃
10 19	☉ ☌ ♀
15 18	☿ # ♅
15 30	☿ St R
16 44	☽ △ ☿
20 16	☽ ∠ ♇
23 17	☽ □ ♂
28 Sunday	
05 39	☽ ⚹ ♅
12 37	☽ □ ♄
23 36	☽ □ ☉
29 Monday	
00 24	☽ □ ♀
01 32	☽ ⚹ ♇
04 53	☽ Q ♅
08 54	☽ ☌ ♆
11 20	♀ ⊼ ♇
12 33	☽ Q ♂
15 54	☽ □ ♃
16 44	☽ ∥ ☿
20 03	☽ # ♅
22 41	☉ ⊼ ♇
23 26	☽ Q ♄
30 Tuesday	
00 18	☽ Q ♇
01 38	☽ □ ☿
08 49	☽ ∥ ♆
12 30	☽ ⚹ ♂
12 34	☽ Q ☉
14 06	☽ Q ♀
14 39	☽ □ ♅
21 51	☽ ⚹ ♄
31 Wednesday	
01 55	☽ Q ♃
02 09	☽ ∥ ♃
09 16	☽ Q ☿
09 18	☽ □ ♇
12 12	☽ ⚹ ☉
13 34	☽ ∥ ☉
14 08	☽ ⚹ ♀
16 18	☽ ⊻ ♆
17 21	☽ ∠ ♂
22 29	☽ ∥ ♀
22 52	♂ ⊼ ♅
23 45	☽ ⚹ ♃

FEBRUARY 1962

LONGITUDES

Date	Sidereal time h m s	Sun ☉ ° ′ ″	Moon ☽ ° ′ ″	Moon ☽ 24.00 ° ′ ″	Mercury ☿ ° ′	Venus ♀ ° ′	Mars ♂ ° ′	Jupiter ♃ ° ′	Saturn ♄ ° ′	Uranus ♅ ° ′	Neptune ♆ ° ′	Pluto ♇ ° ′
01	20 44 52	12 ♒ 09 17	24 ♐ 40 30	01 ♑ 39 55	20 ♒ 39	13 ♒ 22	29 ♑ 38	17 ♒ 47	03 ♒ 24	29 ♌ 12	13 ♏ 27	09 ♍ 30
02	20 48 48	13 10 11	08 ♑ 45 56	15 ♑ 58 11	19 R 43	14 38	00 ♒ 25	18 01	03 31	29 R 09	13 27	09 R 29
03	20 52 45	14 11 04	23 ♑ 16 09	00 ♒ 39 08	18 40	15 53	01 12	18 15	03 38	29 07	13 28	09 28
04	20 56 41	15 11 57	08 ♒ 06 16	15 ♒ 36 35	17 31	17 08	01 58	18 30	03 45	29 04	13 28	09 26
05	21 00 38	16 12 48	23 ♒ 09 01	00 ♓ 42 23	16 20	18 24	02 45	18 44	03 52	29 02	13 28	09 25
06	21 04 35	17 13 38	08 ♓ 15 34	15 ♓ 47 25	15 07	19 39	03 32	18 58	03 59	28 59	13 29	09 23
07	21 08 31	18 14 26	23 ♓ 16 53	00 ♈ 42 59	13 56	20 54	04 18	19 13	04 06	28 56	13 29	09 22
08	21 12 28	19 15 13	08 ♈ 04 55	15 ♈ 21 59	12 47	22 09	05 05	19 27	04 13	28 54	13 29	09 21
09	21 16 24	20 15 59	22 ♈ 33 41	29 ♈ 39 39	11 42	23 25	05 52	19 41	04 20	28 51	13 29	09 19
10	21 20 21	21 16 43	06 ♉ 39 41	13 ♉ 33 43	10 43	24 40	06 38	19 56	04 27	28 49	13 29	09 18
11	21 24 17	22 17 25	20 ♉ 21 47	27 ♉ 04 03	09 50	25 55	07 25	20 10	04 34	28 46	13 29	09 16
12	21 28 14	23 18 06	03 ♊ 40 45	10 ♊ 12 11	09 05	27 10	08 12	20 25	04 41	28 43	13 29	09 15
13	21 32 10	24 18 45	16 ♊ 38 42	23 ♊ 00 39	08 27	28 26	08 59	20 39	04 48	28 41	13 29	09 13
14	21 36 07	25 19 22	29 ♊ 18 26	05 ♋ 32 27	07 58	29 ♒ 41	09 46	20 53	04 54	28 38	13 R 29	09 12
15	21 40 04	26 19 58	11 ♋ 43 04	17 ♋ 50 39	07 36	00 ♓ 56	10 33	21 08	05 01	28 36	13 29	09 10
16	21 44 00	27 20 32	23 ♋ 55 35	29 ♋ 58 11	07 22	02 11	11 19	21 22	05 08	28 33	13 29	09 09
17	21 47 57	28 21 04	05 ♌ 58 47	11 ♌ 57 41	07 16	03 26	12 06	21 36	05 15	28 30	13 29	09 07
18	21 51 53	29 ♒ 21 35	17 ♌ 55 10	23 ♌ 51 29	07 D 16	04 41	12 53	21 51	05 22	28 28	13 29	09 06
19	21 55 50	00 ♓ 22 04	29 ♌ 46 56	05 ♍ 41 44	07 24	05 56	13 40	22 05	05 28	28 25	13 29	09 04
20	21 59 46	01 22 31	11 ♍ 36 10	17 ♍ 30 28	07 37	07 11	14 27	22 19	05 35	28 22	13 29	09 03
21	22 03 43	02 22 57	23 ♍ 24 55	29 ♍ 19 47	07 57	08 26	15 14	22 33	05 42	28 20	13 29	09 01
22	22 07 39	03 23 22	05 ♎ 15 22	11 ♎ 11 58	08 22	09 42	16 01	22 48	05 48	28 17	13 29	08 59
23	22 11 36	04 23 44	17 ♎ 09 56	23 ♎ 09 38	08 52	10 57	16 48	23 02	05 55	28 15	13 28	08 58
24	22 15 33	05 24 06	29 ♎ 11 26	05 ♏ 15 46	09 27	12 12	17 35	23 16	06 01	28 12	13 28	08 56
25	22 19 29	06 24 26	11 ♏ 23 04	17 ♏ 33 46	10 07	13 27	18 22	23 30	06 08	28 09	13 27	08 55
26	22 23 26	07 24 44	23 ♏ 48 21	00 ♐ 07 18	10 50	14 42	19 09	23 44	06 14	28 07	13 27	08 53
27	22 27 22	08 25 01	06 ♐ 31 05	13 ♐ 00 08	11 37	15 56	19 56	23 59	06 21	28 04	13 26	08 52
28	22 31 19	09 ♓ 25 17	19 ♐ 34 53	26 ♐ 15 42	12 ♒ 28	17 ♓ 11	20 ♒ 43	24 ♒ 13	06 ♒ 27	28 ♌ 02	13 ♏ 26	08 ♍ 50

Date	Moon True ☊ ° ′	Moon Mean ☊ ° ′	Moon ☽ Latitude ° ′
01	18 ♌ 03	18 ♌ 21	04 N 15
02	18 R 03	18 18	03 22
03	18 02	18 15	02 13
04	18 02	18 11	00 N 55
05	18 D 02	18 08	00 S 28
06	18 02	18 05	01 50
07	18 R 02	18 02	03 04
08	18 02	17 59	04 04
09	18 02	17 55	04 47
10	18 01	17 52	05 12
11	18 01	17 49	05 18
12	18 D 01	17 46	05 06
13	18 02	17 43	04 39
14	18 02	17 40	03 59
15	18 03	17 36	03 09
16	18 04	17 33	02 10
17	18 05	17 30	01 07
18	18 05	17 27	00 S 01
19	18 R 05	17 24	01 N 04
20	18 04	17 21	02 07
21	18 02	17 17	03 04
22	18 00	17 14	03 53
23	17 57	17 11	04 32
24	17 55	17 08	04 59
25	17 52	17 05	05 14
26	17 51	17 01	05 14
27	17 50	16 58	04 59
28	17 ♌ 50	16 ♌ 55	04 N 29

DECLINATIONS

Date	Sun ☉	Moon ☽	Mercury ☿	Venus ♀	Mars ♂	Jupiter ♃	Saturn ♄	Uranus ♅	Neptune ♆	Pluto ♇
01	17 S 09	19 S 05	11 S 51	17 S 59	21 S 11	16 S 10	19 S 45	12 N 29	14 S 11	20 N 24
02	16 52	19 48	11 56	17 39	21 01	16 05	19 44	12 30	14 11	20 25
03	16 35	19 15	12 05	17 18	20 52	16 01	19 42	12 31	14 11	20 26
04	16 17	17 21	12 17	16 56	20 42	15 56	19 40	12 32	14 11	20 27
05	15 59	14 15	12 31	16 34	20 31	15 52	19 39	12 33	14 11	20 27
06	15 40	10 11	12 49	16 11	20 21	15 48	19 37	12 34	14 11	20 28
07	15 22	05 29	13 07	15 48	20 10	15 43	19 36	12 35	14 11	20 29
08	15 03	00 S 32	13 27	15 24	19 59	15 39	19 34	12 36	14 11	20 30
09	14 44	04 N 20	13 48	15 01	19 48	15 34	19 32	12 37	14 11	20 31
10	14 25	08 50	14 08	14 36	19 36	15 30	19 31	12 38	14 11	20 31
11	14 05	12 44	14 29	14 11	19 25	15 25	19 29	12 39	14 11	20 32
12	13 45	15 52	14 48	13 46	19 13	15 21	19 28	12 39	14 11	20 33
13	13 25	18 08	15 07	13 21	19 01	15 17	19 26	12 40	14 11	20 33
14	13 05	19 27	15 25	12 55	18 49	15 12	19 25	12 41	14 11	20 34
15	12 44	19 48	15 41	12 29	18 36	15 07	19 23	12 42	14 11	20 35
16	12 24	19 11	15 56	12 02	18 24	15 03	19 21	12 43	14 11	20 36
17	12 03	17 42	16 10	11 35	18 11	14 58	19 20	12 44	14 11	20 36
18	11 42	15 27	16 22	11 08	17 58	14 54	19 18	12 45	14 11	20 37
19	11 21	12 33	16 32	10 41	17 45	14 49	19 17	12 46	14 11	20 38
20	10 59	09 10	16 41	10 13	17 31	14 45	19 15	12 47	14 11	20 39
21	10 38	05 25	16 48	09 45	17 18	14 40	19 14	12 48	14 10	20 39
22	10 16	01 N 28	16 54	09 17	17 04	14 36	19 12	12 49	14 10	20 40
23	09 54	02 S 33	16 58	08 48	16 50	14 31	19 11	12 50	14 10	20 41
24	09 32	06 31	17 01	08 19	16 36	14 26	19 09	12 50	14 10	20 41
25	09 10	10 16	17 02	07 50	16 22	14 22	19 08	12 51	14 10	20 42
26	08 47	13 39	17 01	07 21	16 07	14 17	19 06	12 52	14 10	20 43
27	08 25	16 29	16 59	06 52	15 53	14 13	19 05	12 53	14 10	20 43
28	08 S 02	18 S 34	16 S 56	06 S 22	15 S 38	14 S 08	19 S 03	12 N 54	14 S 09	20 N 44

ZODIAC SIGN ENTRIES

Date	h	m	Planets
01	21	10	☽ ♑
01	23	06	♂ ♒
03	22	57	☽ ♒
05	22	53	☽ ♓
07	22	50	☽ ♈
10	00	35	☽ ♉
12	05	18	☽ ♊
14	13	20	☽ ♋
14	18	09	♀ ♓
17	00	04	☽ ♌
19	03	15	☉ ♓
19	12	27	☽ ♍
22	01	22	☽ ♎
24	13	36	☽ ♏
26	23	46	☽ ♐

LATITUDES

Date	Mercury ☿	Venus ♀	Mars ♂	Jupiter ♃	Saturn ♄	Uranus ♅	Neptune ♆	Pluto ♇
01	02 N 55	01 S 14	00 S 59	00 S 41	00 S 22	00 N 47	01 N 46	13 N 26
04	03 28	01 17	01 00	00 41	00 22	00 47	01 47	13 27
07	03 41	01 20	01 01	00 42	00 22	00 47	01 47	13 28
10	03 33	01 23	01 02	00 42	00 22	00 47	01 47	13 28
13	03 08	01 25	01 02	00 42	00 23	00 47	01 47	13 29
16	02 35	01 26	01 03	00 42	00 23	00 47	01 47	13 29
19	01 57	01 27	01 04	00 43	00 23	00 47	01 47	13 30
22	01 19	01 27	01 05	00 43	00 23	00 47	01 48	13 30
25	00 42	01 27	01 05	00 43	00 24	00 47	01 48	13 31
28	00 N 08	01 26	01 06	00 43	00 24	00 47	01 48	13 31
31	00 S 22	01 S 24	01 S 07	00 S 44	00 S 24	00 N 47	01 N 48	13 N 31

DATA

Julian Date	2437697
Delta T	+34 seconds
Ayanamsa	23° 19′ 29″
Synetic vernal point	05° ♓ 47′ 31″
True obliquity of ecliptic	23° 26′ 32″

MOON'S PHASES, APSIDES AND POSITIONS ☽

Date	h	m	Phase	Longitude ° ′	Eclipse Indicator
05	00	10	●	15 ♒ 43	Total
11	15	43	☽	22 ♉ 27	
19	13	18	○	00 ♍ 25	
27	15	50	☾	08 ♐ 35	

Day	h	m		
05	21	33	Perigee	
20	20	34	Apogee	
02	13	51	Max dec	19° S 48′
08	14	35	0N	
15	08	27	Max dec	19° N 48′
22	20	46	0S	

LONGITUDES

Date	Chiron ⚷ ° ′	Ceres ⚳ ° ′	Pallas ⚴ ° ′	Juno ⚵ ° ′	Vesta ⚶ ° ′	Black Moon Lilith ⚸ ° ′
01	04 ♓ 48	15 ♉ 49	24 ♓ 10	15 ♈ 50	21 ♉ 12	10 ♍ 33
11	05 ♓ 26	17 ♉ 54	27 ♓ 33	20 ♈ 57	23 ♉ 09	11 ♍ 40
21	06 ♓ 05	20 ♉ 23	01 ♈ 06	26 ♈ 15	25 ♉ 32	12 ♍ 46
31	06 ♓ 44	23 ♉ 13	04 ♈ 47	01 ♉ 42	28 ♉ 18	13 ♍ 53

ASPECTARIAN

h m	Aspects
01 Thursday	
00 57	☽ ∠ ♄
02 54	☽ ⊥ ♆
05 24	☽ ∗ ☿
10 06	☽ ⊥ ♂
13 27	♀ □ ♆
16 37	☽ ∠ ☉
16 44	☽ ⊥ ♄
18 30	☽ ∠ ♆
19 00	☽ ∠ ♀
19 45	☽ △ ♅
21 03	☽ ⊻ ♂
02 Friday	
02 08	☽ ∠ ♃
03 04	☽ ⊻ ♄
05 37	☽ ∠ ☿
05 51	☽ ‖ ♄
07 04	♄ ± ♇
09 07	☽ ⊥ ☉
11 45	☽ ⊥ ♀
13 12	☽ △ ♇
17 32	☽ ⊥ ♃
18 48	☉ □ ♆
19 43	☽ ⊥ ☿
19 50	☽ ∗ ♆
19 55	☽ ⊻ ☉
20 58	☽ ⊓ ♅
22 42	☽ ⊻ ♀
22 47	☽ ‖ ♄
03 Saturday	
03 38	☽ ⊻ ♃
04 58	☽ ⊻ ☿
11 44	☽ ± ♅
13 56	☽ ⊓ ♇
15 35	☽ Q ♆
19 14	☿ ☌ ♃
21 29	☽ ⊼ ♅
04 Sunday	
01 35	☽ ☌ ♂
04 30	☽ ± ♇
04 56	☽ ☌ ♄
14 08	☽ ⊼ ♇
15 49	☿ ☌ ♀
16 30	☽ ‖ ♀
20 35	☽ □ ♆
22 24	☽ ‖ ☉
05 Monday	
00 10	☽ ☌ ☉
00 19	☽ ‖ ♃
01 57	☽ ☌ ☿
03 45	☽ ☌ ♀
04 52	☽ ☌ ♃
12 24	☽ ‖ ♆
13 18	☉ ☌ ☿
14 30	☿ ǂ ♅
20 01	♀ ☌ ♃
20 04	♂ ǂ ♇
21 18	☽ ☍ ♅
22 05	☽ ‖ ☿
22 34	☽ ǂ ♅
23 32	☉ ‖ ♃
06 Tuesday	
04 04	☽ ⊻ ♂
05 09	☽ ⊻ ♄
07 54	♂ ± ♇
13 48	☽ ☍ ♇
14 07	☽ ⊥ ♂
14 46	☽ ⊥ ♄
20 19	☽ △ ♆
22 07	☽ ⊻ ☿
07 Wednesday	
03 20	☽ ⊻ ☉
04 30	♂ ☌ ♄
05 14	☽ ∠ ♄
05 16	☽ ∠ ♂
05 22	☽ ⊻ ♃
07 01	☽ ⊥ ☿
07 50	☽ ⊻ ♀
13 39	☽ ⊥ ☉
15 09	☽ ⊥ ♃
17 57	♀ ‖ ♃
18 22	☽ ⊥ ♀
20 23	☽ ⊓ ♆
20 26	☽ ∠ ☿
21 06	☽ ⊼ ♅
21 09	☿ □ ♆
08 Thursday	
05 17	☽ ∠ ☉
05 38	☽ ∗ ♄
05 58	☽ ∠ ♃
06 49	☽ ± ♅
06 50	☽ ∗ ♂
10 21	☽ ∠ ♀
11 01	☽ ± ♆
14 04	☽ ⊼ ♇
18 07	☉ ☌ ♃
19 10	☽ ∗ ☿
20 53	☽ ⊼ ♆
21 32	☽ ⊓ ♅
23 56	☽ ± ♇
09 Friday	
01 31	☽ Q ♄
03 42	☽ Q ♂
07 07	☽ ∗ ♃
07 52	☽ ∗ ☉
13 34	☽ ∗ ♀
13 47	☽ Q ☿
14 57	☽ ⊓ ♇
22 36	☽ △ ♅
10 Saturday	
03 44	☽ Q ♃
05 44	☽ Q ☉
08 09	☽ □ ♄
11 58	☽ □ ♂
12 01	☽ Q ♀
15 25	☿ ‖ ♆
16 33	☽ △ ♇
18 35	☽ □ ☿
21 42	☉ ‖ ☿
23 52	☽ ☍ ♆
11 Sunday	
01 25	♂ ‖ ♄
02 47	☿ ‖ ♀
04 16	☉ ‖ ♆
11 24	☽ ‖ ♅
11 39	☽ □ ♃
12 07	♀ ‖ ♆
15 43	☽ □ ☉
20 41	☽ ǂ ☉
21 11	☽ ǂ ♀
22 21	☽ ǂ ♆
22 58	☽ □ ♀
12 Monday	
02 03	☽ ǂ ☿
03 01	☽ □ ♅
06 09	☿ ⊼ ♇
07 37	☽ ǂ ♃
13 50	☽ △ ♄
16 56	☉ ‖ ♀
20 50	☽ △ ♂
21 26	☽ △ ☿
22 12	☽ □ ♇
13 Tuesday	
02 43	☿ ☌ ♂
06 06	☽ ⊼ ♆
12 04	☽ Q ♅
16 42	♀ ☍ ♅
17 21	☽ ± ♆
17 58	☽ ⊓ ♄
19 07	♂ ⊼ ♇
19 40	☽ △ ♃
20 04	♆ St R
21 55	☿ ‖ ♃
14 Wednesday	
00 00	☽ ǂ ♂
00 20	☽ ⊓ ☿
02 45	☽ ⊓ ♂
03 44	☽ △ ☉
07 58	☽ Q ♇
10 26	☽ ⊓ ♆
10 43	☽ ∗ ♅
10 51	☽ ǂ ♄
11 13	☽ ± ♄
12 48	☽ △ ♀
16 56	☽ ± ☿
21 08	☽ ± ♂
22 52	☽ ⊼ ♄
15 Thursday	
00 09	♀ ǂ ♅
00 55	☽ ⊓ ♃
04 11	☽ ⊼ ☿
07 03	☽ ∗ ♇
09 33	☽ ⊼ ♂
11 11	☽ ⊓ ☉
14 32	☉ ǂ ♅
15 28	☽ △ ♆
15 39	☽ ∠ ♅
18 48	☽ ± ♃
21 11	☽ ⊓ ♀
16 Friday	
06 26	☽ ± ☉
06 50	☽ ⊼ ♃
07 39	☽ ǂ ♄
09 17	☽ ⊥ ♅
12 26	☽ ∠ ♇
16 59	☽ ± ♀
19 23	☽ ⊼ ☉
21 08	☽ ⊻ ♅
17 Saturday	
04 33	☽ ǂ ♂
06 18	☽ ⊥ ♇
06 19	☽ ⊼ ♀
10 31	☽ ☍ ♄
14 33	☽ ☍ ☿
15 31	☉ ☍ ♅
18 17	☽ ⊻ ♇
21 36	☿ St D
18 Sunday	
01 09	☽ ☍ ♂
03 04	☽ □ ♆
03 48	☽ ǂ ☿
17 06	☽ ǂ ♃
20 05	☽ ☍ ♃
23 07	☽ ǂ ♆
19 Monday	
02 08	♀ ⊻ ♄
06 20	♂ □ ♆
09 15	☽ ☌ ♅
10 25	☽ ‖ ♅
13 18	☽ ☍ ☉
15 27	☽ Q ♆
22 00	☽ ǂ ☉
23 39	☽ ⊼ ♄
20 Tuesday	
01 58	☽ ☍ ♀
03 44	☽ ⊼ ☿
03 46	☽ ǂ ♀
06 49	☽ ☌ ♇
11 57	☽ ± ♄
15 49	☽ ∗ ♆
16 13	☽ ± ☿
18 12	☽ ⊼ ♂
22 56	☿ ⊻ ♀
21 Wednesday	
06 25	☽ ⊓ ♄
07 15	☽ ± ♂
10 13	☽ ⊼ ♃
11 02	☽ ⊓ ☿
21 56	☽ ⊻ ♅
22 16	☽ ∠ ♆
22 39	☽ ± ♃
22 49	♀ ☍ ♇
22 Thursday	
02 49	☽ ⊓ ♂
07 52	☽ ⊼ ☉
10 03	☽ ⊥ ♅
13 07	☽ △ ♄
16 28	☽ ⊥ ♆
17 14	☽ ⊓ ♃
18 33	☽ △ ☿
19 32	☽ ⊻ ♇
21 07	☽ ± ☉
22 01	☽ ⊼ ♀
23 Friday	
01 04	☿ ‖ ♂
04 09	☽ ∠ ♅
04 34	☽ ⊻ ♆
07 35	☽ ⊥ ♇
11 13	☽ △ ♂
11 30	☽ ± ♀
15 45	☿ ⊼ ♇
16 53	☽ ⊓ ☉
23 59	☽ △ ♃
24 Saturday	
01 35	☽ ∠ ♇
07 34	☽ ⊓ ♀
08 25	♀ ⊥ ♄
10 02	☽ ∗ ♅
22 03	☽ ‖ ♀
25 Sunday	
01 23	☽ △ ☉
01 37	☽ □ ♄
04 37	☉ ⊻ ♄
05 23	☽ ‖ ☉
07 11	☽ ∗ ♇
09 22	☽ □ ☿
09 36	☽ Q ♅
12 12	♀ △ ♆
16 02	☽ ☌ ♆
16 28	☽ △ ♀
26 Monday	
02 28	☽ □ ♂
06 08	☽ ǂ ♅
06 25	☽ Q ♇
11 52	☽ □ ♃
12 50	☽ Q ♄
15 58	☽ ‖ ♆
16 51	☽ ‖ ♃
20 10	☽ □ ♅
22 11	☽ Q ☿
27 Tuesday	
06 51	☽ ‖ ♂
11 41	☽ ∗ ♄
14 48	☽ Q ♂
15 50	☽ □ ☉
16 21	☽ □ ♇
16 56	☽ ‖ ☿
22 06	☽ ∗ ☿
22 18	☽ Q ♃
22 21	☉ ☍ ♇
28 Wednesday	
00 48	☽ ⊻ ♆
05 25	♃ ‖ ♆
07 12	☽ □ ♀
11 44	☽ ⊥ ♆
14 11	☽ ∗ ♂
15 24	☽ ∠ ♄
19 42	☽ ‖ ♄
20 29	☽ ∗ ♃

All ephemeris data is given at 12.00 UT and the Moon's longitude is additionally given for 24.00 UT

MARCH 1962

LONGITUDES

Date	Sidereal time h m s	Sun ☉ ° ' "	Moon ☽ ° ' "	Moon ☽ 24.00 ° ' "	Mercury ☿ ° '	Venus ♀ ° '	Mars ♂ ° '	Jupiter ♃ ° '	Saturn ♄ ° '	Uranus ♅ ° '	Neptune ♆ ° '	Pluto ♇ ° '
01	22 35 15	10 ♓ 25 31	03 ♑ 02 54	09 ♑ 56 40	13 ♒ 22	18 ♓ 26	21 ♒ 30	24 ♒ 27	06 ♒ 34	27 ♌ 59	13 ♏ 25	08 ♍ 49
02	22 39 12	11 25 44	16 ♑ 57 07	24 ♑ 04 12	14 19	19 41	22 17	24 41	06 40	27 R 56	13 R 25	08 R 47
03	22 43 08	12 25 55	01 ♒ 17 41	08 ♒ 37 13	15 19	20 56	23 04	24 55	06 46	27 54	13 24	08 45
04	22 47 05	13 26 05	16 ♒ 02 13	23 ♒ 31 57	16 21	22 11	23 51	25 09	06 52	27 51	13 24	08 44
05	22 51 02	14 26 13	01 ♓ 05 27	08 ♓ 41 38	17 26	23 26	24 38	25 23	06 59	27 49	13 23	08 42
06	22 54 58	15 26 19	16 ♓ 19 18	23 ♓ 57 09	18 33	24 41	25 25	25 37	07 05	27 46	13 22	08 41
07	22 58 55	16 26 23	01 ♈ 33 52	09 ♈ 08 10	19 42	25 56	26 12	25 51	07 11	27 44	13 22	08 39
08	23 02 51	17 26 25	16 ♈ 38 51	24 ♈ 04 51	20 53	27 10	27 00	26 05	07 17	27 42	13 21	08 38
09	23 06 48	18 26 25	01 ♉ 25 17	08 ♉ 39 25	22 06	28 25	27 47	26 19	07 23	27 39	13 20	08 36
10	23 10 44	19 26 23	15 ♉ 46 47	22 ♉ 47 03	23 21	29 ♓ 40	28 34	26 33	07 29	27 37	13 19	08 35
11	23 14 41	20 26 19	29 ♉ 40 08	06 ♊ 26 05	24 38	00 ♈ 55	29 ♒ 21	26 46	07 35	27 34	13 19	08 33
12	23 18 37	21 26 12	13 ♊ 05 08	19 ♊ 37 35	25 56	02 09	00 ♓ 08	27 00	07 40	27 32	13 18	08 32
13	23 22 34	22 26 04	26 ♊ 03 52	02 ♋ 24 29	27 16	03 24	00 55	27 14	07 46	27 30	13 17	08 30
14	23 26 31	23 25 53	08 ♋ 39 59	14 ♋ 50 55	28 ♒ 38	04 39	01 42	27 27	07 52	27 27	13 16	08 29
15	23 30 27	24 25 40	20 ♋ 57 53	27 ♋ 01 26	00 ♓ 01	05 53	02 29	27 41	07 58	27 25	13 15	08 27
16	23 34 24	25 25 25	03 ♌ 02 09	09 ♌ 00 34	01 26	07 08	03 16	27 55	08 03	27 23	13 14	08 26
17	23 38 20	26 25 07	14 ♌ 57 11	20 ♌ 52 29	02 52	08 22	04 03	28 08	08 09	27 21	13 13	08 24
18	23 42 17	27 24 47	26 ♌ 46 53	02 ♍ 40 49	04 19	09 37	04 50	28 22	08 14	27 18	13 12	08 23
19	23 46 13	28 24 26	08 ♍ 34 38	14 ♍ 28 38	05 48	10 51	05 37	28 35	08 20	27 16	13 11	08 21
20	23 50 10	29 ♓ 24 02	20 ♍ 23 08	26 ♍ 18 23	07 18	12 06	06 24	28 48	08 25	27 14	13 10	08 20
21	23 54 06	00 ♈ 23 36	02 ♎ 14 37	08 ♎ 12 03	08 50	13 20	07 12	29 02	08 30	27 12	13 09	08 18
22	23 58 03	01 23 07	14 ♎ 10 52	20 ♎ 11 16	10 23	14 35	07 59	29 15	08 36	27 10	13 08	08 17
23	00 02 00	02 22 37	26 ♎ 13 25	02 ♏ 17 32	11 57	15 49	08 46	29 28	08 41	27 08	13 07	08 16
24	00 05 56	03 22 05	08 ♏ 23 48	14 ♏ 32 26	13 33	17 04	09 33	29 41	08 46	27 06	13 06	08 14
25	00 09 53	04 21 32	20 ♏ 43 41	26 ♏ 57 47	15 10	18 18	10 20	29 ♒ 54	08 51	27 04	13 05	08 13
26	00 13 49	05 20 56	03 ♐ 15 03	09 ♐ 35 46	16 48	19 32	11 07	00 ♓ 08	08 56	27 02	13 03	08 11
27	00 17 46	06 20 19	16 ♐ 00 15	22 ♐ 28 53	18 28	20 46	11 54	00 21	09 01	27 00	13 02	08 10
28	00 21 42	07 19 40	29 ♐ 01 59	05 ♑ 39 54	20 09	22 01	12 41	00 33	09 06	26 59	13 01	08 09
29	00 25 39	08 18 59	12 ♑ 22 58	19 ♑ 11 27	21 51	23 15	13 28	00 46	09 10	26 57	13 00	08 07
30	00 29 35	09 18 16	26 ♑ 05 36	03 ♒ 05 33	23 35	24 29	14 15	00 59	09 15	26 55	12 58	08 06
31	00 33 32	10 ♈ 17 32	10 ♒ 11 20	17 ♒ 22 51	25 ♓ 20	25 ♈ 43	15 ♓ 02	01 ♓ 12	09 ♒ 20	26 ♌ 53	12 ♏ 57	08 ♍ 05

DECLINATIONS

Date	Moon True ☊ ° '	Moon Mean ☊ ° '	Moon ☽ Latitude ° '	Sun ☉ ° '	Moon ☽ ° '	Mercury ☿ ° '	Venus ♀ ° '	Mars ♂ ° '	Jupiter ♃ ° '	Saturn ♄ ° '	Uranus ♅ ° '	Neptune ♆ ° '	Pluto ♇ ° '
01	17 ♌ 52	16 ♌ 52	03 N 43	07 S 40	19 S 42	16 S 51	05 S 53	15 S 23	14 S 04	19 S 02	12 N 55	14 S 09	20 N 45
02	17 D 53	16 49	02 43	07 17	19 41	16 44	05 23	15 08	13 59	19 00	12 56	14 09	20 45
03	17 54	16 46	01 30	06 54	18 24	16 36	04 53	14 53	13 54	18 59	12 57	14 09	20 46
04	17 55	16 42	00 N 10	06 31	15 52	16 27	04 23	14 37	13 50	18 57	12 57	14 08	20 47
05	17 R 55	16 39	01 S 12	06 08	12 13	16 16	03 52	14 22	13 45	18 56	12 58	14 08	20 47
06	17 53	16 36	02 30	05 44	07 42	16 04	03 22	14 06	13 40	18 54	12 59	14 08	20 48
07	17 49	16 33	03 37	05 21	02 S 42	15 50	02 51	13 50	13 36	18 53	13 00	14 08	20 49
08	17 45	16 30	04 28	04 58	02 N 25	15 35	02 21	13 34	13 31	18 51	13 01	14 07	20 49
09	17 41	16 27	05 01	04 34	07 16	15 19	01 50	13 18	13 27	18 50	13 02	14 07	20 50
10	17 37	16 23	05 13	04 11	11 34	15 01	01 20	13 02	13 22	18 48	13 03	14 07	20 50
11	17 34	16 20	05 06	03 47	15 06	14 42	00 49	12 46	13 17	18 47	13 04	14 07	20 51
12	17 32	16 17	04 43	03 24	17 42	14 21	00 S 18	12 29	13 13	18 46	13 04	14 06	20 52
13	17 D 32	16 14	04 05	03 00	19 18	14 00	00 N 13	12 13	13 08	18 44	13 05	14 06	20 52
14	17 33	16 11	03 17	02 37	19 53	13 37	00 43	11 56	13 04	18 43	13 06	14 06	20 53
15	17 34	16 07	02 20	02 13	19 30	13 12	01 14	11 39	12 59	18 41	13 07	14 06	20 53
16	17 36	16 04	01 19	01 49	18 12	12 47	01 45	11 22	12 55	18 40	13 07	14 05	20 54
17	17 37	16 01	00 S 15	01 25	16 07	12 20	02 16	11 05	12 50	18 39	13 08	14 05	20 54
18	17 R 36	15 58	00 N 50	01 02	13 22	11 51	02 46	10 48	12 45	18 37	13 09	14 04	20 55
19	17 34	15 55	01 52	00 38	10 05	11 22	03 17	10 31	12 41	18 36	13 09	14 04	20 55
20	17 30	15 52	02 49	00 S 14	06 23	10 51	03 48	10 13	12 36	18 35	13 10	14 04	20 56
21	17 23	15 48	03 38	00 N 09	02 N 27	10 19	04 18	09 56	12 32	18 34	13 11	14 03	20 56
22	17 16	15 45	04 19	00 33	01 S 37	09 46	04 48	09 38	12 27	18 32	13 11	14 03	20 57
23	17 07	15 42	04 48	00 57	05 39	09 12	05 19	09 21	12 23	18 31	13 12	14 03	20 57
24	16 58	15 39	05 04	01 20	09 30	08 36	05 49	09 03	12 18	18 30	13 13	14 02	20 57
25	16 50	15 36	05 07	01 44	13 00	08 00	06 19	08 45	12 14	18 29	13 13	14 02	20 58
26	16 44	15 33	04 55	02 08	15 59	07 22	06 49	08 27	12 09	18 27	13 14	14 01	20 58
27	16 40	15 29	04 29	02 31	18 15	06 43	07 19	08 09	12 05	18 26	13 15	14 01	20 59
28	16 38	15 26	03 48	02 55	19 38	06 03	07 48	07 52	12 01	18 25	13 15	14 01	20 59
29	16 D 37	15 23	02 54	03 18	19 59	05 21	08 18	07 33	11 56	18 24	13 16	14 00	21 00
30	16 38	15 20	01 48	03 41	18 10	04 39	08 47	07 15	11 52	18 23	13 16	14 00	21 00
31	16 ♌ 39	15 ♌ 17	00 N 35	04 N 05	17 S 08	03 S 55	09 N 16	06 S 57	11 S 47	18 S 22	13 N 17	13 S 59	21 N 00

ZODIAC SIGN ENTRIES

Date	h	m	Planets
01	06	38	☽ ♑
03	09	52	☽ ♒
05	10	16	☽ ♓
07	09	32	☽ ♈
09	09	40	☽ ♉
10	18	28	♀ ♈
11	12	35	☽ ♊
12	07	58	♂ ♓
13	19	25	☽ ♋
15	11	43	☿ ♓
16	05	56	☽ ♌
18	18	33	☽ ♍
21	02	30	☉ ♈
21	07	28	☽ ♎
23	19	29	☽ ♏
25	22	07	♃ ♓
26	05	49	☽ ♐
28	13	46	☽ ♑
30	18	43	☽ ♒

LATITUDES

Date	Mercury ☿ ° '	Venus ♀ ° '	Mars ♂ ° '	Jupiter ♃ ° '	Saturn ♄ ° '	Uranus ♅ ° '	Neptune ♆ ° '	Pluto ♇ ° '
01	00 S 02	01 S 25	01 S 06	00 S 44	00 S 24	00 N 47	01 N 48	13 N 31
04	00 32	01 23	01 07	00 44	00 24	00 47	01 48	13 31
07	00 58	01 21	01 07	00 44	00 25	00 47	01 48	13 31
10	01 21	01 18	01 08	00 45	00 25	00 47	01 49	13 31
13	01 41	01 15	01 08	00 45	00 25	00 47	01 49	13 31
16	01 56	01 11	01 08	00 45	00 26	00 47	01 49	13 31
19	02 08	01 06	01 08	00 46	00 26	00 47	01 49	13 31
22	02 16	01 01	01 09	00 46	00 26	00 47	01 49	13 31
25	02 20	00 56	01 09	00 47	00 27	00 47	01 49	13 30
28	02 19	00 50	01 09	00 47	00 27	00 47	01 49	13 30
31	02 S 15	00 S 44	01 S 09	00 S 47	00 S 27	00 N 47	01 N 49	13 N 29

DATA

Julian Date	2437725
Delta T	+34 seconds
Ayanamsa	23° 19' 32"
Synetic vernal point	05° ♓ 47' 28"
True obliquity of ecliptic	23° 26' 33"

MOON'S PHASES, APSIDES AND POSITIONS ☽

Date	h	m	Phase	Longitude ° '	Eclipse Indicator
06	10	31	●	15 ♓ 23	
13	04	39	☽	22 ♊ 08	
21	07	55	○	00 ♎ 13	
29	04	11	☾	08 ♑ 00	

Day	h	m		
06	09	46	Perigee	
19	21	05	Apogee	
01	23	47	Max dec	19° S 50'
08	00	38	ON	
14	14	11	Max dec	19° N 53'
22	02	28	OS	
29	07	16	Max dec	20° S 00'

LONGITUDES

Date	Chiron ⚷ ° '	Ceres ⚳ ° '	Pallas ⚴ ° '	Juno ⚵ ° '	Vesta ⚶ ° '	Black Moon Lilith ⚸ ° '
01	06 ♓ 37	22 ♉ 37	04 ♈ 02	00 ♉ 36	27 ♉ 43	13 ♍ 40
11	07 ♓ 16	25 ♉ 41	07 ♈ 50	06 ♉ 10	00 ♊ 44	14 ♍ 46
21	07 ♓ 53	29 ♉ 00	11 ♈ 45	11 ♉ 49	04 ♊ 01	15 ♍ 53
31	08 ♓ 29	02 ♊ 32	15 ♈ 46	17 ♉ 32	07 ♊ 32	16 ♍ 59

All ephemeris data is given at 12.00 UT and the Moon's longitude is additionally given for 24.00 UT

ASPECTARIAN

h m Aspects

01 Thursday
03 05 ☽ △ ♅
03 09 ☽ ∠ ☿
03 12 ☽ Q ☉
03 51 ☽ ∠ ♆
07 35 ☽ ⊥ ♄
13 28 ☿ □ ♆
18 11 ☽ ⊻ ♄
18 24 ☽ ∠ ♂
18 31 ☽ Q ♀
20 05 ☽ ⊥ ☿
22 01 ☽ △ ♇
23 20 ☽ ∠ ♃

02 Friday
01 50 ☽ ⚹ ☉
05 11 ☽ ⚼ ♅
05 58 ☽ ⚹ ♆
07 10 ☽ ⊻ ☿
10 48 ☽ ⊥ ♂
14 59 ☽ ⊥ ♃
17 05 ☽ ⚹ ♀
20 24 ☽ ± ♅
21 32 ☽ ⊻ ♂
23 30 ☽ ⚼ ♇

03 Saturday
01 14 ☽ ⊻ ♃
02 14 ☽ Q ♆
03 48 ☽ ∥ ♄
05 08 ☽ ∠ ☉
06 24 ☽ ⊼ ♅
14 24 ☽ ± ♇
20 20 ☽ ∠ ♀
21 00 ☉ ⊥ ♄
21 03 ☽ ☌ ♄
21 03 ☽ ⊥ ☉

04 Sunday
00 12 ☽ ⊼ ♇
05 28 ♀ ∠ ♄
07 06 ☽ ∥ ☿
07 30 ☽ ⊻ ☉
07 45 ☽ □ ♆
11 04 ☉ △ ♆
12 15 ☽ ⊥ ♀
12 32 ☽ ☌ ☿
21 39 ☽ ∥ ♂
22 44 ☽ ⊻ ♀

05 Monday
00 10 ☽ ∥ ♆
01 12 ☽ ☌ ♂
02 28 ☽ ∥ ♃
02 48 ☽ ☌ ♃
06 49 ☽ ☍ ♅
07 30 ☽ ⋕ ♅
10 58 ☿ ⊥ ♀
21 21 ☽ ⊻ ♄

06 Tuesday
00 00 ☽ ☍ ♇
06 52 ☽ ⊥ ♄
07 22 ☽ △ ♆
08 58 ♂ ∥ ♆
10 31 ☽ ☌ ☉
15 46 ☽ ⊻ ☿
20 26 ♂ ☌ ♃
21 07 ☽ ∠ ♄
22 23 ☽ ∥ ☉

07 Wednesday
01 59 ☽ ⊥ ☿
02 19 ☽ ☌ ♀
02 51 ☽ ⊻ ♃
03 05 ☽ ⊻ ♂
05 58 ☽ ⊼ ♅
06 57 ☽ ⚼ ♆
10 11 ♀ ⊻ ♃
11 10 ☽ ∥ ♀
12 27 ☽ ⊥ ♃
13 04 ☽ ⊥ ♂
15 25 ☽ ± ♅
17 22 ☽ ∠ ☿
20 57 ☽ ⚹ ♄
21 10 ☽ ± ♆
23 13 ☽ ⊼ ♇

08 Thursday
02 38 ♀ ⊻ ♂
02 57 ☽ ∠ ♃
04 08 ☽ ∠ ♂
05 41 ☽ ⚼ ♅
06 43 ☽ ⊼ ♆
08 46 ☽ ± ♇
11 44 ☽ ⋕ ♀
13 22 ☽ ⊻ ☉
16 16 ☽ Q ♄
18 12 ♂ ∥ ♃
19 25 ☽ ⚹ ☿
21 42 ♀ ⊼ ♅
23 15 ☽ ⚼ ♇
23 27 ☽ ⋕ ☉
23 45 ☽ ⊥ ☉

09 Friday
03 30 ☽ ⚹ ♃
05 41 ☽ ⚹ ♂
05 50 ☽ △ ♅
06 37 ☽ ⊻ ♀
08 21 ♂ ☍ ♅
10 28 ♀ ⚼ ♆
15 35 ☽ ∠ ☉
16 50 ☽ Q ☿
17 25 ☽ ⊥ ♀
21 56 ☽ □ ♄
23 37 ☽ Q ♃
23 53 ☽ △ ♇

10 Saturday
02 41 ☽ Q ♂
07 51 ☽ ☍ ♆
09 56 ☽ ∠ ♀
11 11 ♂ ⋕ ♅
18 43 ☽ ⚹ ☉
20 38 ☽ ⋕ ♂
21 22 ☽ ∥ ♅
23 17 ☽ ⋕ ♃

11 Sunday
02 18 ☽ □ ☿
04 41 ☽ ⋕ ♆
06 50 ☽ □ ♃
08 20 ☽ □ ♅
09 15 ☽ ⋕ ☿
11 24 ☽ □ ♂
14 24 ☽ ⚹ ♀
17 17 ☽ Q ☉

12 Monday
02 09 ☽ △ ♄
03 46 ☽ □ ♇
12 23 ☽ ⊼ ♆
14 09 ☽ Q ♀
16 27 ☽ Q ♅
23 23 ☽ ± ♆

13 Tuesday
01 52 ☽ ⋕ ♄
04 39 ☽ □ ☉
05 03 ☿ ∥ ♆
05 47 ☽ ⚼ ♄
08 00 ♀ ⊥ ♃
11 06 ☿ ☌ ♃
12 49 ☽ Q ♇
13 47 ♀ ± ♅
14 14 ☽ △ ♃
14 32 ☽ △ ☿
14 41 ☽ ⚹ ♅
15 52 ☿ ☍ ♅
16 10 ☽ ⚼ ♆
20 56 ☉ ∠ ♄
21 46 ☽ △ ♂
22 52 ☽ ± ♄

14 Wednesday
03 21 ♃ ⋕ ♅
03 25 ☽ □ ♀
10 27 ☽ ⊼ ♄
11 38 ☽ ⚹ ♇
11 56 ♃ ☍ ♅
14 09 ☿ ⊥ ♀
19 19 ☽ ∠ ♅
19 29 ☽ ⚼ ♃
20 55 ☽ △ ♆
22 50 ☽ ⚼ ☿

15 Thursday
04 41 ☽ ⚼ ♂
12 54 ☽ ⊥ ♅
13 27 ☽ ± ♃
16 54 ☽ ∠ ♇
17 22 ☿ ⋕ ♅
18 49 ☽ ± ☿
19 28 ☽ △ ☉
23 41 ☽ ± ♂

16 Friday
00 45 ☽ ⊻ ♅
01 34 ☽ ⊼ ♃
03 02 ☿ ∥ ♃
04 53 ☽ ⋕ ♄
08 21 ☽ ⊼ ☿
10 47 ☽ ⊥ ♇
12 30 ☽ ⊼ ♂
13 51 ☉ ⋕ ♀
14 03 ♀ ± ♆
21 10 ☽ △ ♀
22 09 ☽ ☍ ♄
22 48 ☽ ⊻ ♇

17 Saturday
04 12 ☽ ⚼ ☉
07 17 ♀ ⚹ ♄
08 30 ☽ □ ♆
12 34 ♀ ⊼ ♇

18 Sunday
00 06 ☽ ± ☉
06 18 ☽ ⋕ ♆
07 05 ☽ ⚼ ♀
09 33 ☉ ⊼ ♅
13 04 ☽ ☌ ♅
13 24 ☽ ⊼ ☉
13 45 ☽ ∥ ♅
15 16 ☽ ☍ ♃
16 51 ☽ ⋕ ♃
20 59 ☽ Q ♆

19 Monday
01 25 ☽ ⋕ ☿
03 32 ☽ ± ♀
05 32 ☽ ☍ ☿
05 34 ☽ ☍ ♂
05 59 ☿ ☌ ♂
06 47 ☉ ⚼ ♆
08 46 ☽ ⋕ ♂
11 29 ☽ ⊼ ♄
11 33 ☽ ☌ ♇
17 11 ☽ ⊼ ♀
17 27 ♄ ⊼ ♇
17 30 ☉ ⊻ ♃
21 22 ☽ ⚹ ♆
23 47 ☽ ± ♄

20 Tuesday
14 37 ♀ ⚼ ♅
18 12 ☽ ⚼ ♄

21 Wednesday
01 51 ☽ ⊻ ♅
02 07 ☽ ∥ ♀
03 45 ☽ ∠ ♆
03 54 ☿ ☍ ♇
04 20 ♀ ⊥ ♂
05 23 ☽ ⊼ ♃
06 36 ☿ ⊻ ♄
07 55 ☽ ☍ ☉
08 27 ♀ ⊼ ♆
13 56 ☽ ⊥ ♅
17 43 ☽ ± ♃
21 53 ☽ ⊥ ♆
22 40 ☽ ⊼ ♂

22 Thursday
00 11 ☽ ⊻ ♇
00 21 ☽ ∥ ☉
00 42 ☽ △ ♄
03 14 ☽ ⊼ ☿
04 15 ♀ ∠ ♃
05 03 ☽ ⋕ ☉
06 22 ♀ ± ♇
07 59 ☽ ∠ ♅
09 54 ☽ ⊻ ♆
11 34 ☽ ± ♂
12 08 ☽ ⚼ ♃
12 12 ☽ ⊥ ♇
12 53 ☽ ☍ ♀
17 03 ☽ ± ☿
21 05 ♂ ☍ ♇
23 34 ☿ ∥ ♂

23 Friday
06 07 ☽ ∠ ♇
06 46 ☽ ⚼ ♂
09 13 ♂ ⊻ ♄
09 40 ☽ ⋕ ♀
13 40 ☽ ⚼ ☿
13 48 ☽ ⚹ ♅
18 33 ☽ △ ♃

24 Saturday
01 15 ☽ ⊼ ☉
05 21 ☿ △ ♆
05 45 ☉ ± ♅
07 04 ☽ ∥ ☿
09 20 ☽ ∥ ♂
11 41 ☽ ⚹ ♇
12 44 ☽ □ ♄
13 23 ☽ Q ♅
14 04 ☽ ± ☉
14 24 ☽ △ ♂
21 10 ☽ ☌ ♆
23 34 ☽ △ ☿

25 Sunday
06 36 ☽ ∥ ♃
06 46 ☽ ⊼ ♀
07 08 ☿ ⊥ ♄
09 07 ☽ ⚼ ☉
11 00 ☽ Q ♇
13 41 ☽ ⋕ ♅
19 38 ☽ ± ♀
19 46 ☽ ∥ ♆
23 52 ☽ Q ♄

26 Monday
00 10 ☽ □ ♅
05 57 ☽ □ ♃
14 42 ☽ ⚼ ♀
16 19 ☽ △ ☉
21 20 ☽ □ ♇
22 49 ☽ ⚹ ♄
23 32 ☿ ⋕ ♀

27 Tuesday
03 49 ☽ □ ♂
06 28 ☽ ⊻ ♆
12 07 ☉ ⊥ ♃
14 24 ☽ ∥ ♄
16 25 ☽ Q ♃
16 58 ♀ Q ♄
17 15 ☽ □ ☿
17 38 ☽ ⊥ ♆
21 47 ☽ △ ♀

28 Wednesday
02 55 ☽ ∠ ♄
04 36 ☉ ± ♆
08 15 ☽ △ ♅
10 09 ☽ ∠ ♆
13 41 ♀ ⋕ ♂
14 49 ☽ ⚹ ♃
15 11 ☽ Q ♂
19 25 ☽ ⊥ ♄
22 04 ♂ △ ♆

29 Thursday
04 11 ☽ □ ☉
04 25 ☽ △ ♇
06 15 ☽ ⊻ ♄
06 50 ☽ Q ☿
07 25 ☉ ⊼ ♇
09 36 ♀ ⚼ ♇
11 14 ☽ ⚼ ♅
13 05 ☽ ⚹ ♆
14 02 ☽ ⚹ ♂
18 05 ☽ ∠ ♃

30 Friday
03 02 ☽ ± ♅
06 50 ☽ ⚼ ♇
07 01 ☽ ⚹ ☿
08 57 ☽ □ ♀
10 03 ☽ ⊥ ♃
10 04 ☽ Q ♆
10 38 ☉ ⚹ ♄
13 25 ☽ ⊼ ♅
14 15 ☽ Q ☉
17 45 ☽ ∠ ♂
20 32 ☽ ⊻ ♃
21 44 ☿ ∠ ♄
22 18 ☽ ± ♇
23 07 ☽ ∥ ♄

31 Saturday
08 27 ☽ ⊼ ♇
08 40 ☉ ⋕ ☿
09 56 ☽ ⊥ ♂
10 33 ☽ ☌ ♄
12 11 ☽ ⚹ ☉
12 16 ☽ ∠ ☿
16 37 ☽ □ ♆
18 28 ☽ Q ♀
20 33 ☽ ⊻ ♂
22 16 ♂ ⊥ ♄

APRIL 1962

LONGITUDES

Date	Sidereal time h m s	Sun ☉ ° ' "	Moon ☽ ° ' "	Moon ☽ 24.00 ° ' "	Mercury ☿ ° '	Venus ♀ ° '	Mars ♂ ° '	Jupiter ♃ ° '	Saturn ♄ ° '	Uranus ♅ ° '	Neptune ♆ ° '	Pluto ⯓ ° '
01	00 37 29	11 ♈ 16 46	24 ♒ 39 50	02 ♓ 01 53	27 ♓ 06	26 ♈ 58	15 ♓ 49	01 ♓ 25	09 ♒ 24	26 ♌ 52	12 ♏ 56	08 ♍ 04
02	00 41 25	12 15 58	09 ♓ 28 23	16 ♓ 58 30	28 ♓ 54	28 12	16 36	01 37	09 29	26 R 50	12 R 55	08 R 02
03	00 45 22	13 15 08	24 ♓ 31 18	02 ♈ 05 39	00 ♈ 43	29 ♈ 26	17 22	01 50	09 33	26 48	12 53	08 01
04	00 49 18	14 14 16	09 ♈ 40 18	17 ♈ 14 00	02 34	00 ♉ 40	18 09	02 02	09 38	26 47	12 52	08 00
05	00 53 15	15 13 22	24 ♈ 45 26	02 ♉ 13 24	04 26	01 54	18 56	02 15	09 42	26 45	12 50	07 59
06	00 57 11	16 12 26	09 ♉ 36 44	16 ♉ 54 31	06 20	03 08	19 43	02 27	09 46	26 44	12 49	07 58
07	01 01 08	17 11 28	24 ♉ 05 58	01 ♊ 10 31	08 15	04 22	20 30	02 39	09 50	26 43	12 48	07 56
08	01 05 04	18 10 28	08 ♊ 07 49	14 ♊ 57 45	10 11	05 36	21 17	02 51	09 54	26 41	12 46	07 55
09	01 09 01	19 09 25	21 ♊ 40 19	28 ♊ 15 46	12 09	06 50	22 04	03 03	09 58	26 40	12 45	07 54
10	01 12 58	20 08 21	04 ♋ 44 26	11 ♋ 06 48	14 08	08 04	22 50	03 15	10 02	26 39	12 43	07 53
11	01 16 54	21 07 13	17 ♋ 23 23	23 ♋ 34 50	16 08	09 18	23 37	03 27	10 06	26 37	12 42	07 52
12	01 20 51	22 06 04	29 ♋ 41 46	05 ♌ 44 51	18 10	10 31	24 24	03 39	10 10	26 36	12 40	07 51
13	01 24 47	23 04 53	11 ♌ 44 47	17 ♌ 42 13	20 13	11 45	25 11	03 51	10 13	26 35	12 39	07 50
14	01 28 44	24 03 39	23 ♌ 37 47	29 ♌ 32 07	22 17	12 59	25 57	04 03	10 17	26 34	12 37	07 49
15	01 32 40	25 02 22	05 ♍ 25 47	11 ♍ 19 18	24 22	14 13	26 44	04 14	10 20	26 33	12 36	07 48
16	01 36 37	26 01 04	17 ♍ 13 12	23 ♍ 07 52	26 28	15 26	27 31	04 26	10 24	26 32	12 34	07 47
17	01 40 33	26 59 44	29 ♍ 03 44	05 ♎ 01 06	28 ♈ 34	16 40	28 17	04 37	10 27	26 31	12 33	07 46
18	01 44 30	27 58 21	11 ♎ 00 16	17 ♎ 01 28	00 ♉ 42	17 54	29 04	04 48	10 30	26 30	12 31	07 45
19	01 48 27	28 56 56	23 ♎ 04 52	29 ♎ 10 37	02 49	19 07	29 ♓ 50	05 00	10 33	26 30	12 29	07 44
20	01 52 23	29 ♈ 55 30	05 ♏ 18 49	11 ♏ 29 33	04 56	20 21	00 ♈ 37	05 11	10 36	26 29	12 28	07 44
21	01 56 20	00 ♉ 54 02	17 ♏ 42 53	23 ♏ 58 51	07 04	21 34	01 23	05 22	10 39	26 28	12 26	07 43
22	02 00 16	01 52 31	00 ♐ 17 30	06 ♐ 38 53	09 10	22 48	02 10	05 33	10 42	26 27	12 25	07 42
23	02 04 13	02 51 00	13 ♐ 03 04	19 ♐ 30 09	11 16	24 01	02 56	05 44	10 45	26 27	12 23	07 41
24	02 08 09	03 49 26	26 ♐ 00 14	02 ♑ 33 28	13 21	25 15	03 43	05 54	10 47	26 26	12 21	07 40
25	02 12 06	04 47 51	09 ♑ 10 00	15 ♑ 50 02	15 24	26 28	04 29	06 05	10 50	26 26	12 20	07 40
26	02 16 02	05 46 14	22 ♑ 33 46	29 ♑ 21 23	17 26	27 41	05 15	06 15	10 52	26 25	12 18	07 39
27	02 19 59	06 44 36	06 ♒ 13 04	13 ♒ 08 59	19 26	28 ♉ 55	06 02	06 26	10 55	26 25	12 17	07 38
28	02 23 56	07 42 56	20 ♒ 09 13	27 ♒ 13 47	21 23	00 ♊ 08	06 48	06 36	10 57	26 25	12 15	07 38
29	02 27 52	08 41 14	04 ♓ 22 35	11 ♓ 35 27	23 17	01 21	07 34	06 46	10 59	26 24	12 13	07 37
30	02 31 49	09 ♉ 39 32	18 ♓ 52 00	26 ♓ 11 46	25 ♉ 09	02 ♊ 34	08 ♈ 21	06 ♓ 56	11 ♒ 02	26 ♌ 24	12 ♏ 12	07 ♍ 37

DECLINATIONS

Date	Moon True ☊ ° '	Moon Mean ☊ ° '	Moon ☽ Latitude ° '	Sun ☉ ° '	Moon ☽ ° '	Mercury ☿ ° '	Venus ♀ ° '	Mars ♂ ° '	Jupiter ♃ ° '	Saturn ♄ ° '	Uranus ♅ ° '	Neptune ♆ ° '	Pluto ⯓ ° '
01	16 ♌ 39	15 ♌ 13	00 S 43	04 N 28	13 S 59	03 S 11	09 N 45	06 S 39	11 S 43	18 S 20	13 N 18	13 S 59	21 N 01
02	16 R 37	15 10	01 59	04 51	09 52	02 25	10 13	06 21	11 39	18 19	13 18	13 58	21 01
03	16 33	15 07	03 09	05 14	05 S 04	01 38	10 42	06 02	11 34	18 18	13 19	13 58	21 01
04	16 26	15 04	04 05	05 37	00 N 05	00 51	11 10	05 44	11 30	18 17	13 19	13 58	21 02
05	16 17	15 01	04 44	06 00	05 11	00 S 02	11 38	05 25	11 26	18 16	13 20	13 57	21 02
06	16 08	14 58	05 03	06 23	09 54	00 N 48	12 05	05 07	11 21	18 15	13 20	13 57	21 02
07	15 59	14 54	05 02	06 45	13 55	01 38	12 32	04 48	11 17	18 14	13 20	13 56	21 02
08	15 51	14 51	04 42	07 08	17 01	02 29	12 59	04 30	11 13	18 13	13 21	13 56	21 03
09	15 46	14 48	04 07	07 30	19 04	03 21	13 26	04 11	11 09	18 12	13 21	13 55	21 03
10	15 42	14 45	03 21	07 52	20 01	04 14	13 52	03 53	11 05	18 11	13 22	13 55	21 03
11	15 41	14 42	02 25	08 14	19 55	05 07	14 18	03 34	11 00	18 10	13 22	13 54	21 03
12	15 D 41	14 39	01 24	08 37	18 51	06 01	14 44	03 16	10 56	18 10	13 22	13 54	21 03
13	15 42	14 35	00 S 21	08 58	16 56	06 55	15 09	02 57	10 52	18 09	13 23	13 53	21 04
14	15 R 42	14 32	00 N 42	09 20	14 19	07 50	15 34	02 38	10 48	18 08	13 23	13 53	21 04
15	15 40	14 29	01 43	09 42	11 07	08 45	15 59	02 20	10 44	18 07	13 23	13 52	21 04
16	15 35	14 26	02 39	10 03	07 30	09 40	16 23	02 01	10 40	18 06	13 24	13 52	21 04
17	15 28	14 23	03 29	10 24	03 N 34	10 34	16 47	01 42	10 36	18 05	13 24	13 52	21 04
18	15 19	14 19	04 10	10 45	00 S 32	11 29	17 10	01 23	10 32	18 05	13 24	13 51	21 04
19	15 07	14 16	04 39	11 06	04 39	12 23	17 33	01 05	10 28	18 04	13 25	13 51	21 04
20	14 54	14 13	04 57	11 27	08 38	13 16	17 55	00 46	10 24	18 03	13 25	13 50	21 05
21	14 41	14 10	05 00	11 47	12 18	14 08	18 17	00 27	10 20	18 03	13 25	13 50	21 05
22	14 29	14 07	04 50	12 08	15 29	14 59	18 39	00 S 09	10 16	18 02	13 25	13 49	21 05
23	14 19	14 04	04 24	12 28	18 00	15 49	18 59	00 N 10	10 13	18 01	13 25	13 49	21 05
24	14 12	14 00	03 45	12 48	19 38	16 38	19 20	00 28	10 09	18 01	13 26	13 48	21 05
25	14 07	13 57	02 53	13 07	20 14	17 25	19 40	00 47	10 05	18 00	13 26	13 48	21 05
26	14 05	13 54	01 51	13 27	19 43	18 09	19 59	01 06	10 01	18 00	13 26	13 47	21 05
27	14 D 05	13 51	00 N 42	13 46	18 03	18 52	20 18	01 24	09 58	17 59	13 26	13 47	21 05
28	14 R 05	13 48	00 S 32	14 05	15 17	19 33	20 37	01 43	09 54	17 59	13 26	13 46	21 05
29	14 04	13 44	01 45	14 24	11 32	20 11	20 54	02 01	09 51	17 58	13 26	13 46	21 05
30	14 ♌ 01	13 ♌ 41	02 S 53	14 N 42	07 S 03	20 N 47	21 N 12	02 N 20	09 S 47	17 S 58	13 N 26	13 S 45	21 N 05

ZODIAC SIGN ENTRIES

Date	h	m	Planets
01	20	42	☽ ♓
03	02	32	☽ ♈
03	20	41	♀ ♉
03	23	05	☿ ♈
05	20	25	☽ ♉
07	22	00	☽ ♊
10	03	12	☽ ♋
12	12	36	☽ ♌
15	00	57	☽ ♍
17	13	54	☽ ♎
18	04	10	☿ ♉
19	16	58	♂ ♈
20	01	37	☽ ♏
20	13	51	☉ ♉
22	11	27	☽ ♐
24	19	20	☽ ♑
27	01	08	☽ ♒
28	09	23	♀ ♊
29	04	40	☽ ♓

LATITUDES

Date	Mercury ☿ ° '	Venus ♀ ° '	Mars ♂ ° '	Jupiter ♃ ° '	Saturn ♄ ° '	Uranus ♅ ° '	Neptune ♆ ° '	Pluto ⯓ ° '
01	02 S 12	00 S 42	01 S 08	00 S 48	00 S 27	00 N 47	01 N 49	13 N 29
04	02 02	00 35	01 08	00 48	00 28	00 47	01 50	13 29
07	01 47	00 28	01 08	00 49	00 28	00 47	01 50	13 28
10	01 27	00 21	01 08	00 49	00 28	00 46	01 50	13 28
13	01 03	00 13	01 07	00 50	00 29	00 46	01 50	13 27
16	00 36	00 S 05	01 07	00 50	00 29	00 46	01 50	13 26
19	00 S 05	00 N 02	01 06	00 51	00 29	00 46	01 50	13 25
22	00 N 28	00 10	01 06	00 51	00 30	00 46	01 50	13 25
25	01 00	00 18	01 05	00 52	00 30	00 46	01 50	13 24
28	01 29	00 26	01 05	00 53	00 31	00 46	01 50	13 23
31	01 N 55	00 N 34	01 S 04	00 S 53	00 S 31	00 N 46	01 N 50	13 N 22

DATA

Julian Date	2437756
Delta T	+34 seconds
Ayanamsa	23° 19' 35"
Synetic vernal point	05° ♓ 47' 25"
True obliquity of ecliptic	23° 26' 33"

MOON'S PHASES, APSIDES AND POSITIONS ☽

Date	h	m	Phase	Longitude ° '	Eclipse Indicator
04	19	45	●	14 ♈ 33	
11	19	50	☽	21 ♋ 26	
20	00	33	○	29 ♎ 28	
27	12	59	☾	06 ♒ 47	

Day	h	m		
03	20	47	Perigee	
16	06	40	Apogee	
04	11	38	0N	
10	21	28	Max dec	20° N 06'
18	08	56	0S	
25	13	14	Max dec	20° S 15'

LONGITUDES

Date	Chiron ⚷ ° '	Ceres ⚳ ° '	Pallas ⚴ ° '	Juno ⚵ ° '	Vesta ⚶ ° '	Black Moon Lilith ⚸ ° '
01	08 ♓ 33	02 ♊ 53	16 ♈ 10	18 ♉ 07	07 ♊ 25	17 ♍ 06
11	09 ♓ 06	06 ♊ 36	20 ♈ 17	23 ♉ 54	11 ♊ 36	18 ♍ 13
21	09 ♓ 35	10 ♊ 27	24 ♈ 29	29 ♉ 42	15 ♊ 27	19 ♍ 19
31	10 ♓ 01	14 ♊ 26	28 ♈ 46	05 ♊ 31	19 ♊ 26	20 ♍ 26

All ephemeris data is given at 12.00 UT and the Moon's longitude is additionally given for 24.00 UT

ASPECTARIAN

h m Aspects

01 Sunday
05 20 ☽ ⊥ ☿
05 43 ☿ ⊻ ♀
08 48 ☿ ⊼ ♅
10 08 ♀ △ ♅
11 58 ☽ ∥ ♆
14 50 ☽ ∠ ☉
15 35 ☽ ☍ ♅
16 06 ☽ ✶ ♀
16 21 ☽ # ♅
16 32 ☽ ⊻ ☿
22 57 ☿ ♇ ♆
23 09 ☽ ☌ ♃

02 Monday
01 46 ☉ ♇ ♅
02 01 ☽ ∥ ♃
06 29 ☽ ⊥ ☉
09 42 ☽ ☍ ⯓
10 15 ☽ # ♀
12 01 ☽ ⊻ ♄
16 48 ☽ ⊻ ☉
17 30 ☽ △ ♆
18 30 ☽ ∠ ♀
21 40 ☽ ⊥ ♄

03 Tuesday
00 01 ☽ ☌ ♂
03 17 ☉ ⊼ ♆
07 00 ☽ ∥ ♂
10 07 ☽ ⊥ ♀
11 14 ☽ # ☉
12 03 ☽ ∠ ♄
15 37 ☽ ⊼ ♅
17 20 ☽ ♇ ♆
20 28 ☽ ⊻ ♀
23 11 ☽ ☌ ☿
23 44 ☽ ⊻ ♃

04 Wednesday
01 06 ☽ ± ♅
04 16 ☿ ⊻ ♃
06 17 ☉ ± ⯓
06 57 ☽ ∥ ☿
07 33 ☽ ± ♆
09 21 ☽ ⊼ ⯓
09 22 ☽ ⊥ ♃
11 56 ☽ ✶ ♄
14 45 ☿ ± ♅
15 04 ☽ # ☿
15 20 ☽ ♇ ♅
16 04 ☉ # ♂
17 03 ☽ ⊼ ♆
18 51 ☽ ± ⯓
19 45 ☽ ☌ ☉
23 51 ☽ ∠ ♃

05 Thursday
02 12 ☽ ⊻ ♂
03 05 ♀ # ♃
07 05 ☽ Q ♄
09 10 ☽ ♇ ⯓
12 18 ☽ ⊥ ♂
13 05 ☽ # ♂
15 12 ☽ △ ♅
16 16 ☽ ∥ ☉
20 01 ♀ ✶ ♃

06 Friday
00 12 ☽ ✶ ♃
00 31 ☽ ☌ ♀
03 36 ☽ ∠ ♂
05 51 ☽ ⊻ ☿
09 18 ☽ △ ⯓
12 15 ☽ □ ♄
17 07 ☽ ⊥ ☿
17 14 ☽ ☍ ♆
18 05 ☿ ± ♆
20 00 ☽ # ♃
20 03 ☽ Q ♃
23 38 ☽ ⊻ ☉

07 Saturday
02 01 ☽ ∥ ♀
05 37 ☽ ✶ ♂
08 11 ☽ ∥ ♅
08 16 ☿ ⊼ ⯓
10 20 ☽ ∠ ☿
10 22 ☽ ⊥ ☉
12 06 ☽ # ♆
16 24 ☽ □ ♅
17 42 ☿ ⊥ ♃

08 Sunday
02 11 ☉ ∠ ♃
02 45 ☽ □ ♃
02 47 ☽ ∠ ☉
03 07 ☽ Q ♂
07 11 ☽ ⊻ ♀
08 27 ☿ ✶ ♄
11 38 ☽ □ ⯓
15 07 ☽ △ ♄
16 11 ☽ ✶ ☿
18 40 ☽ ⊥ ♀
20 07 ☽ ⊼ ♆
23 30 ☽ Q ♅

09 Monday
00 23 ☽ # ♄
06 14 ☿ ♇ ♅
06 45 ☽ ± ♆
07 07 ☽ ✶ ☉
07 38 ♀ ∥ ♅
12 19 ☽ ∠ ♀
12 45 ☽ □ ♂
17 16 ☽ Q ☿
18 00 ☽ ♇ ♄
19 12 ☿ ⊼ ♆
19 40 ☽ Q ⯓
21 03 ☽ ✶ ♅
23 02 ☽ ♇ ♆

10 Tuesday
00 13 ♀ ∠ ♂
04 55 ☿ # ♂
06 46 ☽ Q ☉
08 35 ♀ △ ⯓
09 05 ☿ ± ⯓
09 11 ☽ △ ♃
10 40 ☽ ± ♄
14 13 ♀ # ♆
17 53 ☽ ✶ ⯓
18 54 ☽ ✶ ♀
22 00 ☽ ⊼ ♄

11 Wednesday
00 59 ☽ ∠ ♅
03 02 ☽ △ ♆
09 08 ☽ □ ☿
14 05 ☽ ♇ ♃
18 14 ☽ ⊥ ♅
19 50 ☽ □ ☉
20 23 ☽ Q ♀
22 36 ☽ ∠ ⯓

12 Thursday
00 54 ☽ △ ♂
04 30 ♀ □ ♄
05 56 ☽ ⊻ ♅
07 55 ☽ ± ♃
13 30 ☉ Q ♄
16 15 ☽ ⊥ ⯓
18 20 ☿ ∠ ♃
19 58 ☽ ⊼ ♃
21 49 ☽ # ♄

13 Friday
04 10 ☽ ⊻ ⯓
06 02 ☉ ♇ ⯓
08 38 ☽ ♇ ♂
08 55 ☽ ☍ ♄
12 01 ☽ □ ♀
13 24 ♂ ∠ ♄
13 48 ☽ □ ♆

14 Saturday
02 37 ☽ ∥ ♀
04 02 ☽ ± ♂
05 03 ♀ ☍ ♆
08 41 ☽ △ ☿
11 58 ☿ Q ♄
12 57 ☽ △ ☉
15 27 ☽ # ♆
17 03 ☽ ⊼ ♂
17 58 ☽ ☌ ♅
18 08 ☿ ♇ ⯓
19 21 ☽ ∥ ♅

15 Sunday
02 11 ☽ Q ♆
06 30 ♂ ⊼ ♅
09 32 ☽ ☍ ♃
14 44 ☽ # ♃
16 49 ☽ ☌ ⯓
20 51 ☽ ∥ ☉
21 45 ☽ ♇ ☿
22 02 ☽ ⊼ ♄
22 14 ☽ ♇ ☉

16 Monday
00 47 ☽ ∥ ☿
02 29 ☉ ☌ ☿
02 34 ☽ ✶ ♆
07 58 ☽ △ ♀
10 19 ☽ ± ♄
12 49 ☿ △ ♅
13 45 ♂ ♇ ♆
18 12 ☽ ± ☉
20 01 ☽ ± ☿

17 Tuesday
00 30 ☉ △ ♅
04 39 ☽ ♇ ♄
04 47 ☉ ∥ ☿
06 51 ☿ ⊻ ♂
06 52 ☽ ⊻ ♅
07 27 ☽ ⊼ ☉
08 56 ☽ ∠ ♆
10 20 ☽ ☍ ♂
10 48 ☽ ⊼ ☿
10 50 ♀ Q ♃
12 42 ☿ # ♃
17 52 ☽ ♇ ♀
18 58 ☽ ⊥ ♅
23 16 ☉ # ♃
23 22 ☽ ⊼ ♃
23 54 ☽ # ♂

18 Wednesday
03 02 ☽ ⊥ ♆
05 30 ☽ ⊻ ⯓
10 59 ☽ △ ♄
11 36 ☽ ± ♃
13 00 ☽ ∠ ♅
13 59 ☽ ± ♀
15 01 ☽ ⊻ ♆
16 40 ☽ ∥ ♂
17 29 ☽ ⊥ ⯓

19 Thursday
03 17 ☽ ⊼ ♀
05 48 ☽ ♇ ♃
11 20 ☽ ∠ ⯓
18 43 ☽ ✶ ♅

20 Friday
00 33 ☽ ☍ ☉
02 12 ☽ ⊼ ♂
11 07 ☽ ☍ ☿
11 44 ☽ △ ♃
14 42 ☽ ± ♂
14 57 ☿ ✶ ♃
16 04 ☿ ∥ ♅
16 41 ☽ ✶ ⯓
18 09 ☽ Q ♅
20 37 ♀ # ♄
22 19 ☽ □ ♄
23 05 ☽ ∥ ♃

21 Saturday
01 51 ☽ ☌ ♆
03 28 ☿ # ♆
08 05 ☽ # ☉
09 17 ☽ ♇ ♂
15 50 ☽ Q ⯓
17 53 ☿ ⊥ ♂
19 21 ☿ △ ⯓
19 54 ☽ # ♅
20 12 ☽ ☍ ♀
22 53 ☽ ∥ ♆

22 Sunday
04 43 ☽ □ ♅
06 19 ☽ # ☿
08 58 ☽ Q ♄
15 15 ☽ ⊼ ☉
15 46 ☽ △ ♂
20 57 ♂ ± ♅
22 04 ☽ □ ♃

23 Monday
01 58 ☽ □ ⯓
03 29 ☽ ± ☉
05 49 ☿ □ ♄
07 40 ☽ ✶ ♄
08 02 ☽ ⊼ ☿
10 45 ☽ ⊻ ♆
12 19 ☽ ∥ ♄
21 22 ☽ ± ☿
21 40 ☽ ♇ ☉
21 54 ☽ ⊥ ♆
22 35 ☉ ⊻ ♂

24 Tuesday
00 39 ☿ ☍ ♆
04 36 ☽ # ♀
08 05 ☽ Q ♃
10 28 ☽ ⊼ ♀
11 36 ☽ ∠ ♄
12 48 ☽ △ ♅
14 29 ☽ ∠ ♆
17 07 ☽ ♇ ☿
22 35 ☽ ± ♀

25 Wednesday
02 59 ☽ □ ♂
03 27 ☽ △ ☉
04 07 ☽ ⊥ ♄
06 20 ☽ ✶ ♃
09 17 ☽ △ ⯓
11 15 ♀ □ ♅
15 01 ☽ ⊻ ♄
16 05 ☽ ♇ ♅
16 34 ☽ ♇ ♀
17 42 ☽ ✶ ♆

26 Thursday
01 15 ☽ △ ☿
06 42 ☿ # ♄
07 06 ☽ # ♀
08 12 ☽ ± ♅
09 39 ☽ ∠ ♃
10 40 ☉ ∥ ♅
12 09 ☽ ♇ ⯓
13 18 ☽ Q ♂
15 05 ☽ Q ♆
18 50 ☽ ⊼ ♅
21 58 ☽ △ ♀
22 49 ☿ Q ♃

27 Friday
01 45 ☽ ⊥ ♃
02 37 ☉ ✶ ♃
04 01 ☽ ± ⯓
04 55 ☽ # ☿
11 39 ☽ ✶ ♂
12 23 ☽ ⊻ ♃
12 36 ☉ # ♆
12 43 ☽ ∥ ♄
12 59 ☽ □ ☉
14 28 ☽ ⊼ ⯓
19 26 ♂ ± ♆
20 10 ☽ ☌ ♄
22 29 ☽ □ ♆

28 Saturday
04 04 ♂ ⊻ ♃
09 54 ☉ △ ⯓
14 25 ☽ □ ☿
14 58 ☽ ∠ ♂
19 37 ☽ # ☉
20 49 ☿ ∠ ♂
22 08 ☽ Q ☉
22 24 ☽ ∥ ♆
22 37 ☽ ☍ ♅

29 Sunday
00 32 ☽ # ♅
06 28 ☽ □ ♀
07 02 ☽ ⊥ ♂
13 28 ♂ ⊼ ⯓
16 03 ☽ ☌ ♃
17 24 ☽ ☍ ⯓
17 38 ☽ ⊻ ♂
19 42 ☽ ✶ ☉
21 37 ☽ ∥ ♃
23 02 ☽ ⊻ ♄

30 Monday
01 01 ☽ △ ♆
01 12 ☽ Q ☿
02 06 ♀ ∥ ⯓
08 58 ☽ ⊥ ♄
15 03 ☽ Q ♀
22 10 ☽ ∠ ☉
23 45 ☽ ✶ ☿
23 45 ☽ ∠ ♄

MAY 1962

LONGITUDES

Date	Sidereal time h m s	Sun ☉ ° ' "	Moon ☽ ° ' "	Moon ☽ 24.00 ° ' "	Mercury ☿ ° '	Venus ♀ ° '	Mars ♂ ° '	Jupiter ♃ ° '	Saturn ♄ ° '	Uranus ♅ ° '	Neptune ♆ ° '	Pluto ♇ ° '
01	02 35 45	10 ♉ 37 47	03 ♈ 34 05	10 ♈ 58 10	26 ♉ 58	03 ♊ 48	09 ♈ 07	07 ♓ 06	11 ♒ 04	26 ♌ 24	12 ♏ 10	07 ♍ 36
02	02 39 42	11 36 01	18 ♈ 23 06	25 ♈ 47 52	28 ♉ 43	05 01	09 53	07 16	11 05	26 R 24	12 R 08	07 R 36
03	02 43 38	12 34 13	03 ♉ 11 25	10 ♉ 32 39	00 ♊ 25	06 14	10 39	07 26	11 07	26 24	12 07	07 35
04	02 47 35	13 32 24	17 ♉ 50 32	25 ♉ 04 08	02 03	07 27	11 25	07 36	11 09	26 D 24	12 05	07 35
05	02 51 31	14 30 33	02 ♊ 12 37	09 ♊ 15 18	03 38	08 40	12 11	07 45	11 10	26 24	12 04	07 34
06	02 55 28	15 28 40	16 ♊ 11 43	23 ♊ 01 32	05 09	09 53	12 57	07 55	11 12	26 24	12 02	07 34
07	02 59 25	16 26 46	29 ♊ 44 37	06 ♋ 21 02	06 36	11 06	13 43	08 04	11 14	26 24	12 00	07 33
08	03 03 21	17 24 50	12 ♋ 50 56	19 ♋ 14 40	07 59	12 19	14 29	08 13	11 15	26 24	11 59	07 33
09	03 07 18	18 22 51	25 ♋ 32 40	01 ♌ 45 27	09 18	13 32	15 15	08 22	11 16	26 24	11 57	07 33
10	03 11 14	19 20 51	07 ♌ 53 36	13 ♌ 57 46	10 33	14 45	16 01	08 31	11 17	26 24	11 55	07 32
11	03 15 11	20 18 49	19 ♌ 58 37	25 ♌ 56 49	11 43	15 57	16 47	08 40	11 19	26 25	11 54	07 32
12	03 19 07	21 16 45	01 ♍ 53 04	07 ♍ 48 03	12 50	17 10	17 32	08 48	11 20	26 25	11 52	07 32
13	03 23 04	22 14 40	13 ♍ 42 25	19 ♍ 36 48	13 52	18 23	18 18	08 57	11 20	26 26	11 51	07 32
14	03 27 00	23 12 32	25 ♍ 31 46	01 ♎ 27 54	14 50	19 36	19 04	09 05	11 21	26 26	11 49	07 32
15	03 30 57	24 10 23	07 ♎ 25 40	13 ♎ 25 32	15 43	20 48	19 49	09 14	11 22	26 27	11 47	07 31
16	03 34 54	25 08 12	19 ♎ 27 51	25 ♎ 32 58	16 32	22 01	20 35	09 22	11 22	26 28	11 46	07 31
17	03 38 50	26 06 00	01 ♏ 41 06	07 ♏ 52 28	17 17	23 13	21 20	09 30	11 23	26 28	11 44	07 31
18	03 42 47	27 03 46	14 ♏ 07 09	20 ♏ 25 15	17 56	24 26	22 06	09 38	11 23	26 29	11 43	07 31
19	03 46 43	28 01 31	26 ♏ 46 43	03 ♐ 11 33	18 32	25 38	22 51	09 45	11 24	26 30	11 41	07 D 31
20	03 50 40	28 59 14	09 ♐ 39 39	16 ♐ 10 54	19 02	26 51	23 37	09 53	11 24	26 31	11 40	07 31
21	03 54 36	29 ♉ 56 56	22 ♐ 45 11	29 ♐ 22 41	19 28	28 03	24 22	10 00	11 24	26 32	11 38	07 31
22	03 58 33	00 ♊ 54 37	06 ♑ 02 17	12 ♑ 44 54	19 48	29 ♊ 16	25 07	10 07	11 R 24	26 32	11 37	07 31
23	04 02 29	01 52 17	19 ♑ 30 04	26 ♑ 17 45	20 04	00 ♋ 28	25 52	10 15	11 24	26 33	11 35	07 32
24	04 06 26	02 49 55	03 ♒ 07 54	10 ♒ 00 31	20 16	01 40	26 38	10 22	11 24	26 34	11 34	07 32
25	04 10 23	03 47 33	16 ♒ 55 34	23 ♒ 53 04	20 22	02 52	27 23	10 28	11 23	26 36	11 32	07 32
26	04 14 19	04 45 10	00 ♓ 52 59	07 ♓ 55 18	20 R 24	04 04	28 08	10 35	11 23	26 37	11 31	07 32
27	04 18 16	05 42 45	14 ♓ 59 55	22 ♓ 06 41	20 21	05 16	28 53	10 42	11 22	26 38	11 29	07 32
28	04 22 12	06 40 20	29 ♓ 15 23	06 ♈ 25 42	20 14	06 29	29 ♈ 38	10 48	11 22	26 39	11 28	07 33
29	04 26 09	07 37 54	13 ♈ 37 16	20 ♈ 49 33	20 02	07 41	00 ♉ 23	10 54	11 21	26 40	11 26	07 33
30	04 30 05	08 35 27	28 ♈ 02 00	05 ♉ 13 57	19 47	08 53	01 08	11 00	11 20	26 42	11 25	07 33
31	04 34 02	09 ♊ 33 00	12 ♉ 24 42	19 ♉ 33 32	19 ♊ 27	10 ♋ 04	01 ♉ 53	11 ♓ 06	11 ♒ 19	26 ♌ 43	11 ♏ 23	07 ♍ 33

MOON NODES AND LATITUDE

Date	Moon True ☊ ° '	Moon Mean ☊ ° '	Moon ☽ Latitude ° '
01	13 ♌ 55	13 ♌ 38	03 S 50
02	13 R 47	13 35	04 32
03	13 36	13 32	04 56
04	13 25	13 29	05 00
05	13 13	13 25	04 44
06	13 03	13 22	04 12
07	12 56	13 19	03 27
08	12 51	13 16	02 31
09	12 48	13 13	01 30
10	12 47	13 10	00 S 26
11	12 D 47	13 06	00 N 38
12	12 R 47	13 03	01 39
13	12 45	13 00	02 36
14	12 41	12 57	03 26
15	12 35	12 54	04 07
16	12 26	12 50	04 38
17	12 15	12 47	04 56
18	12 03	12 44	05 01
19	11 50	12 41	04 51
20	11 38	12 38	04 26
21	11 29	12 35	03 47
22	11 22	12 31	02 55
23	11 18	12 28	01 53
24	11 16	12 25	00 N 43
25	11 D 16	12 22	00 S 30
26	11 16	12 19	01 43
27	11 R 16	12 16	02 49
28	11 14	12 12	03 47
29	11 10	12 09	04 30
30	11 03	12 06	04 56
31	10 ♌ 55	12 ♌ 03	05 S 04

DECLINATIONS

Date	Sun ☉ ° '	Moon ☽ ° '	Mercury ☿ ° '	Venus ♀ ° '	Mars ♂ ° '	Jupiter ♃ ° '	Saturn ♄ ° '	Uranus ♅ ° '	Neptune ♆ ° '	Pluto ♇ ° '
01	15 N 01	02 S 06	21 N 20	21 N 28	02 N 38	09 S 43	17 S 57	13 N 26	13 S 45	21 N 05
02	15 19	03 N 01	21 51	21 45	02 57	09 40	17 57	13 26	13 44	21 05
03	15 37	07 57	22 20	22 00	03 15	09 37	17 56	13 26	13 44	21 04
04	15 54	12 21	22 46	22 15	03 33	09 33	17 56	13 26	13 43	21 04
05	16 12	15 57	23 09	22 29	03 51	09 30	17 56	13 26	13 43	21 04
06	16 29	18 33	23 30	22 43	04 10	09 27	17 55	13 26	13 42	21 04
07	16 45	20 00	23 49	22 56	04 28	09 23	17 55	13 26	13 42	21 04
08	17 02	20 19	24 05	23 08	04 46	09 20	17 55	13 26	13 41	21 04
09	17 18	19 34	24 19	23 20	05 04	09 17	17 55	13 26	13 41	21 04
10	17 34	17 53	24 31	23 31	05 22	09 14	17 54	13 25	13 40	21 03
11	17 50	15 26	24 41	23 41	05 40	09 11	17 54	13 25	13 40	21 03
12	18 05	12 21	24 48	23 51	05 57	09 08	17 54	13 25	13 39	21 03
13	18 20	08 48	24 54	24 00	06 15	09 05	17 54	13 25	13 39	21 03
14	18 35	04 56	24 57	24 09	06 33	09 02	17 54	13 25	13 38	21 03
15	18 49	00 N 50	24 59	24 16	06 50	08 59	17 54	13 24	13 38	21 02
16	19 03	03 S 20	24 59	24 23	07 08	08 56	17 54	13 24	13 37	21 02
17	19 17	07 26	24 57	24 30	07 25	08 53	17 54	13 24	13 37	21 02
18	19 30	11 17	24 53	24 35	07 43	08 50	17 54	13 23	13 36	21 01
19	19 43	14 43	24 48	24 40	08 00	08 48	17 54	13 23	13 36	21 01
20	19 56	17 31	24 41	24 44	08 17	08 45	17 54	13 23	13 35	21 01
21	20 09	19 28	24 33	24 48	08 34	08 43	17 54	13 23	13 35	21 00
22	20 21	20 23	24 24	24 51	08 51	08 40	17 54	13 22	13 35	21 00
23	20 32	20 10	24 13	24 53	09 08	08 38	17 55	13 22	13 34	21 00
24	20 44	18 46	24 01	24 54	09 25	08 35	17 55	13 22	13 34	20 59
25	20 55	16 15	23 47	24 55	09 41	08 33	17 55	13 21	13 33	20 59
26	21 05	12 46	23 33	24 54	09 58	08 31	17 55	13 21	13 33	20 59
27	21 16	08 31	23 18	24 54	10 14	08 28	17 55	13 20	13 33	20 58
28	21 26	03 S 46	23 01	24 52	10 30	08 26	17 56	13 20	13 32	20 58
29	21 35	01 N 14	22 44	24 50	10 47	08 24	17 56	13 19	13 32	20 57
30	21 44	06 10	22 26	24 47	11 03	08 22	17 56	13 19	13 31	20 57
31	21 N 53	10 N 44	22 N 07	24 N 43	11 N 19	08 S 20	17 S 57	13 N 18	13 S 31	20 N 57

ZODIAC SIGN ENTRIES

Date	h	m	Planets
01	06	12	☽ ♈
03	06	05	☿ ♊
03	06	49	☽ ♉
05	08	16	☽ ♊
07	12	28	☽ ♋
09	20	35	☽ ♌
12	08	11	☽ ♍
14	21	03	☽ ♎
17	08	43	☽ ♏
19	18	02	☽ ♐
21	13	17	☉ ♊
22	01	08	☽ ♑
23	02	46	♀ ♋
24	06	31	☽ ♒
26	10	29	☽ ♓
28	13	15	☽ ♈
28	23	47	♂ ♉
30	15	17	☽ ♉

LATITUDES

Date	Mercury ☿ ° '	Venus ♀ ° '	Mars ♂ ° '	Jupiter ♃ ° '	Saturn ♄ ° '	Uranus ♅ ° '	Neptune ♆ ° '	Pluto ♇ ° '
01	01 N 55	00 N 34	01 S 04	00 S 53	00 S 31	00 N 46	01 N 50	13 N 22
04	02 14	00 42	01 03	00 54	00 31	00 46	01 50	13 21
07	02 26	00 50	01 02	00 54	00 32	00 45	01 50	13 20
10	02 31	00 57	01 01	00 55	00 32	00 45	01 50	13 19
13	02 27	01 04	01 00	00 56	00 33	00 45	01 50	13 18
16	02 14	01 11	00 59	00 57	00 33	00 45	01 50	13 17
19	01 52	01 18	00 58	00 57	00 33	00 45	01 50	13 16
22	01 21	01 24	00 56	00 58	00 34	00 45	01 50	13 15
25	00 N 42	01 30	00 55	00 59	00 34	00 45	01 50	13 14
28	00 S 04	01 35	00 54	01 00	00 35	00 45	01 50	13 13
31	00 S 55	01 N 40	00 S 52	01 S 00	00 S 35	00 N 45	01 N 50	13 N 12

DATA

Julian Date	2437786
Delta T	+34 seconds
Ayanamsa	23° 19' 37"
Synetic vernal point	05° ♓ 47' 22"
True obliquity of ecliptic	23° 26' 33"

MOON'S PHASES, APSIDES AND POSITIONS ☽

Date	h	m	Phase	Longitude ° '	Eclipse Indicator
04	04	25	●	13 ♉ 14	
11	12	44	☽	20 ♌ 21	
19	14	32	○	28 ♏ 08	
26	19	05	☾	05 ♓ 02	

Day	h	m		
02	01	37	Perigee	
13	23	08	Apogee	
29	12	54	Perigee	
01	21	51	ON	
08	06	45	Max dec	20° N 20'
15	16	49	0S	
22	19	29	Max dec	20° S 26'
29	06	07	0N	

LONGITUDES

Date	Chiron ⚷ ° '	Ceres ⚳ ° '	Pallas ⚴ ° '	Juno ⚵ ° '	Vesta ⚶ ° '	Black Moon Lilith ⚸ ° '
01	10 ♓ 01	14 ♊ 26	28 ♈ 46	05 ♊ 31	19 ♊ 26	20 ♍ 26
11	10 ♓ 22	18 ♊ 30	03 ♉ 06	11 ♊ 20	23 ♊ 31	21 ♍ 33
21	10 ♓ 37	22 ♊ 39	07 ♉ 31	17 ♊ 08	27 ♊ 41	22 ♍ 40
31	10 ♓ 48	26 ♊ 52	11 ♉ 59	22 ♊ 54	01 ♋ 56	23 ♍ 46

ASPECTARIAN

01 Tuesday
00 20 ☽ ⚻ ♅
00 24 ☿ ∥ ♇
01 36 ☽ ⚼ ♆
04 29 ☿ □ ♅
09 36 ☽ # ♂
10 06 ☽ ± ♅
12 24 ☽ ✶ ♀
13 51 ☽ ⊥ ☉
16 13 ☽ ± ♆
17 49 ☽ ⚺ ♃
18 32 ☽ ⚻ ♇
21 29 ☽ ☌ ♂
22 59 ☉ □ ♄

02 Wednesday
00 10 ☽ ✶ ♄
00 15 ☽ ⚺ ☉
00 22 ☿ ∥ ♀
00 41 ☽ ⚼ ♅
01 55 ☽ ⚻ ♆
03 26 ☽ ∠ ☿
03 38 ☽ ⊥ ♃
04 15 ☽ ± ♇
11 38 ☽ ∥ ♂
14 52 ☽ ∠ ♀
18 22 ☽ ∠ ♃
18 48 ☽ ⚼ ♇
19 38 ☽ Q ♄
19 56 ☽ ⊥ ☿

03 Thursday
00 58 ☽ △ ♅
01 01 ☉ ☍ ♆
06 46 ☽ ⊥ ♀
06 55 ☽ ⚺ ☿
17 24 ☽ ⚺ ♀
19 00 ☽ ✶ ♃
19 09 ☽ △ ♇
20 33 ☽ # ♃

04 Friday
00 51 ☽ ⚺ ♂
00 58 ☽ □ ♄
02 33 ☽ ☍ ♆
03 19 ♂ ✶ ♄
04 25 ☽ ☌ ☉
08 56 ♅ St D
09 13 ♃ ☍ ♇
11 13 ♂ ⚼ ♅
11 16 ☽ ⊥ ♂
14 29 ♀ □ ♇
14 56 ☽ Q ♃
15 19 ♀ □ ♃
18 35 ☽ ∥ ♅
20 22 ☽ # ♆

05 Saturday
02 13 ☽ □ ♅
03 04 ☽ ∠ ♂
08 11 ♂ ⚻ ♆
14 02 ☽ ∥ ☉
14 42 ☽ ☌ ☿
21 07 ☽ □ ♇
21 32 ☽ □ ♃

06 Sunday
00 02 ☽ ☌ ♀
03 20 ☽ △ ♄
04 48 ☽ ⚻ ♆
05 13 ☽ # ♄
06 02 ☽ ✶ ♂
08 52 ☽ Q ♅
10 39 ☽ ⚺ ☉
15 12 ☽ ± ♆
21 58 ☽ ⊥ ☉

07 Monday
04 21 ☽ Q ♂
04 30 ☽ Q ♇
05 41 ☽ ⚼ ♄
06 00 ☽ ✶ ♅
06 58 ♂ ± ♇
07 06 ☽ ⚼ ♆
14 35 ♀ △ ♄
15 19 ☽ ∠ ☉
21 58 ☽ ± ♄

08 Tuesday
01 37 ♂ ⊥ ♃
01 56 ☽ ⚺ ☿
02 12 ☽ ✶ ♇
03 20 ☽ △ ♃
04 27 ☿ □ ♇
05 28 ♀ ⚻ ♆
09 02 ☽ ⚻ ♄
09 19 ☽ ∠ ♅
10 23 ☽ △ ♆
10 54 ☽ ⚺ ♀
14 21 ☽ ⊥ ☿
15 15 ☽ □ ♂
16 48 ☿ □ ♃
21 15 ☽ ✶ ☉
23 20 ☽ ⊥ ♀

09 Wednesday
02 12 ☽ ⊥ ♅
06 16 ☽ ∠ ♇
07 47 ☽ ⚼ ♃
09 20 ☽ ∠ ☿
13 39 ☽ ⚺ ♅
18 22 ☽ ∠ ♀
22 07 ☽ Q ☉
23 35 ☽ ⊥ ♇

10 Thursday
01 21 ☽ ± ♃
05 22 ♀ Q ♅
11 18 ☽ ⚺ ♇
11 41 ☽ # ♄
13 15 ☽ ⚻ ♃
15 07 ☽ ∥ ☉
17 48 ☽ ✶ ☿
18 43 ☽ ☍ ♄
19 56 ☽ □ ♆

11 Friday
03 04 ☽ ✶ ♀
03 16 ☿ △ ♄
05 10 ☽ △ ♂
12 44 ☽ □ ☉
15 34 ☿ ⚻ ♆
19 10 ☉ # ♄
20 19 ☽ Q ☿
22 15 ☉ Q ♃

12 Saturday
00 57 ☽ ☌ ♅
02 19 ☽ # ♆
04 07 ☽ ∥ ♅
05 53 ☽ Q ♀
07 56 ☽ Q ♆
13 25 ☽ ⚼ ♂
23 27 ☽ ☌ ♇

13 Sunday
01 33 ♀ ± ♆
02 13 ☽ ☍ ♃
07 11 ☽ ⚻ ♄
07 33 ♀ ✶ ♂
08 13 ☽ ✶ ♆
08 57 ☽ ± ♂
10 14 ☽ # ♃
12 22 ☽ □ ☿
19 23 ☽ ± ♄
21 58 ☽ ⚻ ♂
22 35 ☽ □ ♀

14 Monday
01 54 ☿ Q ♅
02 52 ☽ ∥ ♂
06 53 ☽ △ ☉
13 40 ☽ ⚼ ♄
13 51 ☽ ⚺ ♅
14 36 ☽ ∠ ♆

15 Tuesday
01 59 ☽ ⊥ ♅
08 43 ☽ ⊥ ♆
12 12 ☽ ⚺ ♇
15 39 ☽ ⚻ ♃
15 48 ☽ ⚼ ☉
19 53 ☽ △ ♄
20 04 ☽ ∠ ♅
20 43 ☽ ⚺ ♆

16 Wednesday
00 12 ☽ ⊥ ♇
03 46 ☽ ± ♃
05 48 ☽ △ ☿
11 18 ☽ ± ☉
14 21 ☽ ☍ ♂
17 36 ☽ △ ♀
18 03 ☽ ∠ ♇
21 46 ☽ ⚼ ♃

17 Thursday
00 09 ☽ ⚻ ☉
01 48 ☽ ✶ ♅
11 58 ☽ # ♂
13 13 ☽ ⚼ ☿
20 45 ☽ ∥ ♃
21 22 ☉ □ ♅
23 19 ☽ ✶ ♇

18 Friday
01 10 ☽ Q ♅
02 02 ☽ ⚼ ♀
03 17 ☽ △ ♃
03 42 ☿ ± ♆
06 46 ☽ □ ♄
07 24 ☽ ☌ ♆
07 36 ☽ ± ☿
19 40 ☽ ⚻ ☿
21 06 ☽ ± ♀
22 18 ☽ Q ♇

19 Saturday
01 28 ♂ ⚼ ♇
02 17 ☽ # ♅
03 46 ☽ ∥ ♆
04 08 ☽ ⚻ ♂
09 38 ♇ St D
09 38 ♀ Q ♇
09 38 ☽ ⚻ ♀
11 28 ☽ □ ♅
14 32 ☽ ☍ ☉
16 08 ☽ ± ♂
16 54 ☽ Q ♄

20 Sunday
03 01 ♀ ⚼ ♄
05 12 ♂ Q ♄
05 14 ♀ ✶ ♅
05 47 ☿ ∥ ♀
08 02 ☽ □ ♇
08 21 ♀ ⚼ ♆
09 56 ☽ ⚼ ♂
12 25 ☽ □ ♃
15 12 ☽ ✶ ♄
15 41 ☽ ⚺ ♆
16 01 ☽ ∥ ♄

21 Monday
02 41 ☽ ⊥ ♆
05 48 ☽ ☍ ☿
15 06 ☽ △ ♂
18 37 ☽ ∠ ♄
18 51 ☽ △ ♅
19 02 ☽ ∠ ♆
21 37 ☽ Q ♃
22 35 ☽ ☍ ♀
22 37 ♂ # ♃
23 22 ♄ St R

22 Tuesday
02 03 ☽ ⚻ ☉
08 14 ☽ # ☉
10 51 ☽ ⊥ ♄
12 12 ♂ ∠ ♃
13 41 ☽ ± ☉
14 40 ☽ △ ♇
19 23 ☽ ✶ ♃
21 35 ☽ ⚺ ♄
21 52 ☽ ⚼ ♅
21 57 ☽ ✶ ♆
22 38 ☽ # ☉

23 Wednesday
06 59 ☽ ⚼ ☉
12 18 ☉ ⊥ ♂
13 02 ☽ ⚻ ☿
13 52 ☽ ± ♅
17 21 ☽ ⚼ ♇
19 12 ☽ Q ♆
22 14 ☽ ∠ ♃
23 47 ☽ ± ☿
23 55 ☽ □ ♂

24 Thursday
00 29 ☽ ⚻ ♅
09 11 ☽ ⚻ ♀
09 11 ☽ ± ♇
10 18 ♂ △ ♅
11 26 ☽ △ ☉
14 10 ☽ ⊥ ♃
15 46 ☽ ⚼ ☿
19 41 ☽ ⚻ ♇
20 41 ☽ ± ♀
21 18 ☽ ∥ ♄

25 Friday
00 43 ☽ ⚺ ♃
02 24 ☽ ☌ ♄
02 40 ☽ □ ♆
09 10 ☽ Q ♂
13 47 ☽ ⚼ ♀
17 58 ☽ △ ☿
21 24 ☉ ∥ ♇

26 Saturday
04 41 ☽ ☍ ♅
07 01 ☽ ✶ ♂
07 03 ☽ ∥ ♆
08 20 ☽ # ♅
09 09 ☿ St R
17 57 ☽ △ ♀
19 05 ☽ □ ☉
23 21 ☽ ☍ ♇

27 Sunday
03 15 ☽ # ♂
04 39 ☽ ☌ ♃
05 52 ☽ ⚺ ♄
06 04 ☽ △ ♆
10 00 ☽ ∠ ♂
12 13 ☽ ∥ ♃
13 57 ♀ ± ♄
16 01 ☽ ⊥ ♄
20 58 ☽ □ ☿

28 Monday
02 02 ☽ ⊥ ♂
03 45 ☽ Q ☉
07 09 ☽ ∠ ♄
07 19 ☽ ⚼ ♆
07 38 ☽ ⚻ ♅
12 40 ☽ ⚺ ♂
17 42 ☽ ± ♅
22 22 ☽ ± ♆

29 Tuesday
01 11 ☽ □ ♀
01 18 ☽ ✶ ☉
01 52 ☽ ⚻ ♇
02 49 ☽ Q ☿
07 26 ☽ ⚺ ♃
07 34 ☉ ⚺ ♀
08 13 ☽ ✶ ♄
08 22 ☽ ⚻ ♆
08 45 ☽ ⚼ ♅
09 24 ♀ ✶ ♇
09 52 ☉ □ ♇
11 53 ☽ ± ♇
17 30 ☽ ⊥ ♃
22 31 ☽ ✶ ☿

30 Wednesday
02 52 ☽ ⚼ ♇
04 05 ☽ ∠ ☉
04 11 ☽ Q ♄
08 36 ☽ ∠ ♃
09 46 ☽ △ ♅
09 54 ☽ Q ♀
17 26 ☽ ☌ ♂
20 08 ☽ ⊥ ☉
23 00 ☽ ∠ ☿
23 10 ☽ # ♃

31 Thursday
03 53 ☽ △ ♇
06 52 ☽ ⚺ ☉
07 44 ☽ ✶ ♀
09 48 ☽ ✶ ♃
10 11 ☽ □ ♄
10 17 ☽ ☍ ♆
13 42 ☽ ⊥ ☿
15 28 ☽ ∥ ♂
23 32 ☽ ⚺ ☿

All ephemeris data is given at 12.00 UT and the Moon's longitude is additionally given for 24.00 UT

JUNE 1962

LONGITUDES

Date	Sidereal time h m s	Sun ☉ ° ' "	Moon ☽ ° ' "	Moon ☽ 24.00 ° ' "	Mercury ☿ ° '	Venus ♀ ° '	Mars ♂ ° '	Jupiter ♃ ° '	Saturn ♄ ° '	Uranus ♅ ° '	Neptune ♆ ° '	Pluto ♇ ° '
01	04 37 58	10 ♊ 30 31	26 ♉ 39 42	03 ♊ 42 31	19 ♊ 05	11 ♋ 16	02 ♉ 37	11 ♓ 12	11 ♒ 18	26 ♌ 45	11 ♏ 22	07 ♍ 34
02	04 41 55	11 28 02	10 ♊ 41 21	17 ♊ 35 38	18 R 39	12 28	03 22	11 17	11 R 17	26 46	11 R 21	07 34
03	04 45 52	12 25 31	24 ♊ 24 58	01 ♋ 08 59	18 11	13 40	04 07	11 23	11 16	26 48	11 19	07 35
04	04 49 48	13 23 00	07 ♋ 47 32	14 ♋ 20 31	17 40	14 52	04 51	11 28	11 15	26 49	11 18	07 35
05	04 53 45	14 20 27	20 ♋ 48 00	27 ♋ 10 11	17 08	16 03	05 36	11 33	11 14	26 51	11 17	07 36
06	04 57 41	15 17 54	03 ♌ 27 20	09 ♌ 39 49	16 35	17 15	06 21	11 38	11 12	26 53	11 15	07 36
07	05 01 38	16 15 19	15 ♌ 48 06	21 ♌ 52 41	16 02	18 27	07 05	11 43	11 11	26 54	11 14	07 37
08	05 05 34	17 12 43	27 ♌ 54 08	03 ♍ 53 05	15 28	19 38	07 49	11 47	11 09	26 56	11 13	07 37
09	05 09 31	18 10 06	09 ♍ 50 08	15 ♍ 45 57	14 55	20 50	08 34	11 52	11 07	26 58	11 12	07 38
10	05 13 27	19 07 28	21 ♍ 41 11	27 ♍ 36 28	14 24	22 01	09 18	11 56	11 06	27 00	11 10	07 39
11	05 17 24	20 04 49	03 ♎ 32 27	09 ♎ 29 43	13 54	23 12	10 02	12 00	11 04	27 02	11 09	07 39
12	05 21 21	21 02 09	15 ♎ 28 53	21 ♎ 30 28	13 27	24 24	10 46	12 04	11 02	27 04	11 08	07 40
13	05 25 17	21 59 28	27 ♎ 34 57	03 ♏ 42 46	13 02	25 35	11 30	12 07	11 00	27 06	11 07	07 41
14	05 29 14	22 56 46	09 ♏ 54 18	16 ♏ 09 50	12 40	26 46	12 14	12 11	10 57	27 08	11 06	07 42
15	05 33 10	23 54 03	22 ♏ 29 35	28 ♏ 53 42	12 22	27 57	12 58	12 14	10 55	27 10	11 05	07 42
16	05 37 07	24 51 20	05 ♐ 22 13	11 ♐ 55 07	12 08	29 ♋ 08	13 42	12 17	10 53	27 12	11 04	07 43
17	05 41 03	25 48 36	18 ♐ 32 19	25 ♐ 13 36	11 58	00 ♌ 19	14 26	12 20	10 50	27 14	11 02	07 44
18	05 45 00	26 45 51	01 ♑ 58 44	08 ♑ 47 27	11 52	01 30	15 10	12 23	10 48	27 17	11 01	07 45
19	05 48 56	27 43 06	15 ♑ 39 25	22 ♑ 34 15	11 D 50	02 41	15 54	12 25	10 45	27 19	11 00	07 46
20	05 52 53	28 40 21	29 ♑ 31 37	06 ♒ 31 09	11 54	03 52	16 37	12 28	10 43	27 21	10 59	07 47
21	05 56 50	29 ♊ 37 35	13 ♒ 32 30	20 ♒ 35 19	12 01	05 02	17 21	12 30	10 40	27 24	10 59	07 48
22	06 00 46	00 ♋ 34 49	27 ♒ 39 19	04 ♓ 44 12	12 14	06 13	18 04	12 32	10 37	27 26	10 58	07 49
23	06 04 43	01 32 02	11 ♓ 49 42	18 ♓ 55 35	12 31	07 24	18 48	12 34	10 34	27 28	10 57	07 50
24	06 08 39	02 29 16	26 ♓ 01 37	03 ♈ 07 34	12 53	08 34	19 31	12 35	10 31	27 31	10 56	07 51
25	06 12 36	03 26 30	10 ♈ 13 11	17 ♈ 18 14	13 19	09 45	20 15	12 37	10 28	27 33	10 55	07 52
26	06 16 32	04 23 43	24 ♈ 22 27	01 ♉ 25 33	13 50	10 55	20 58	12 38	10 25	27 36	10 54	07 53
27	06 20 29	05 20 57	08 ♉ 27 12	15 ♉ 27 05	14 26	12 05	21 41	12 39	10 22	27 39	10 53	07 54
28	06 24 25	06 18 11	22 ♉ 24 52	29 ♉ 20 11	15 07	13 15	22 24	12 40	10 19	27 41	10 53	07 55
29	06 28 22	07 15 24	06 ♊ 12 42	13 ♊ 02 04	15 51	14 26	23 07	12 41	10 16	27 44	10 52	07 56
30	06 32 19	08 ♋ 12 38	19 ♊ 47 58	26 ♊ 30 08	16 ♊ 41	15 ♌ 36	23 ♉ 50	12 ♓ 41	10 ♒ 12	27 ♌ 47	10 ♏ 51	07 ♍ 58

Moon

Date	Moon True ☊ ° '	Moon Mean ☊ ° '	Moon ☽ Latitude ° '
01	10 ♌ 45	12 ♌ 00	04 S 52
02	10 R 36	11 56	04 24
03	10 27	11 53	03 40
04	10 21	11 50	02 45
05	10 17	11 47	01 42
06	10 15	11 44	00 S 36
07	10 D 15	11 41	00 N 30
08	10 16	11 37	01 34
09	10 17	11 34	02 32
10	10 R 17	11 31	03 24
11	10 16	11 28	04 07
12	10 13	11 25	04 40
13	10 08	11 22	05 01
14	10 02	11 18	05 08
15	09 54	11 15	05 01
16	09 46	11 12	04 38
17	09 39	11 09	04 00
18	09 33	11 06	03 09
19	09 29	11 02	02 05
20	09 27	10 59	00 N 54
21	09 D 27	10 56	00 S 22
22	09 28	10 53	01 37
23	09 29	10 50	02 47
24	09 30	10 47	03 46
25	09 R 30	10 43	04 32
26	09 29	10 40	05 01
27	09 26	10 37	05 12
28	09 22	10 34	05 04
29	09 17	10 31	04 38
30	09 ♌ 12	10 ♌ 28	03 S 57

DECLINATIONS

Date	Sun ☉ ° '	Moon ☽ ° '	Mercury ☿ ° '	Venus ♀ ° '	Mars ♂ ° '	Jupiter ♃ ° '	Saturn ♄ ° '	Uranus ♅ ° '	Neptune ♆ ° '	Pluto ♇ ° '
01	22 N 02	14 N 40	21 N 48	24 N 39	11 N 35	08 S 18	17 S 57	13 N 18	13 S 30	20 N 56
02	22 10	17 42	21 28	24 34	11 50	08 16	17 57	13 17	13 30	20 56
03	22 17	19 40	21 09	24 28	12 06	08 14	17 58	13 17	13 30	20 55
04	22 25	20 28	20 49	24 21	12 21	08 13	17 58	13 16	13 29	20 55
05	22 31	20 09	20 29	24 14	12 37	08 11	17 59	13 16	13 29	20 54
06	22 38	18 48	20 10	24 06	12 52	08 09	17 59	13 15	13 28	20 54
07	22 44	16 35	19 52	23 58	13 07	08 08	18 00	13 14	13 28	20 53
08	22 50	13 40	19 34	23 49	13 22	08 06	18 01	13 14	13 28	20 53
09	22 55	10 14	19 17	23 39	13 37	08 05	18 01	13 13	13 27	20 52
10	23 00	06 25	19 01	23 28	13 51	08 04	18 02	13 12	13 27	20 52
11	23 04	02 N 22	18 47	23 17	14 06	08 02	18 02	13 12	13 27	20 51
12	23 08	01 S 47	18 34	23 05	14 20	08 01	18 03	13 11	13 26	20 50
13	23 12	05 56	18 23	22 53	14 34	08 00	18 04	13 10	13 26	20 50
14	23 15	09 55	18 13	22 40	14 48	07 59	18 05	13 09	13 26	20 49
15	23 18	13 33	18 05	22 26	15 02	07 58	18 05	13 09	13 25	20 49
16	23 21	16 38	17 59	22 12	15 16	07 57	18 06	13 08	13 25	20 48
17	23 23	18 57	17 54	21 57	15 29	07 56	18 07	13 07	13 25	20 47
18	23 24	20 17	17 52	21 42	15 43	07 55	18 08	13 06	13 25	20 47
19	23 25	20 27	17 51	21 26	15 56	07 55	18 08	13 06	13 24	20 46
20	23 26	19 23	17 52	21 09	16 09	07 54	18 09	13 05	13 24	20 46
21	23 27	17 07	17 55	20 52	16 22	07 54	18 10	13 04	13 24	20 45
22	23 26	13 49	18 00	20 34	16 35	07 53	18 11	13 03	13 23	20 44
23	23 26	09 42	18 06	20 16	16 47	07 53	18 12	13 02	13 23	20 44
24	23 25	05 03	18 14	19 57	17 00	07 52	18 13	13 01	13 23	20 43
25	23 24	00 S 07	18 23	19 38	17 12	07 52	18 14	13 00	13 23	20 42
26	23 22	04 N 47	18 33	19 18	17 24	07 52	18 15	12 59	13 23	20 42
27	23 20	09 24	18 45	18 58	17 36	07 52	18 16	12 59	13 22	20 41
28	23 18	13 29	18 58	18 37	17 48	07 52	18 17	12 58	13 22	20 40
29	23 15	16 47	19 12	18 16	17 59	07 52	18 18	12 57	13 22	20 39
30	23 N 11	19 N 06	19 N 26	17 N 55	18 N 11	07 S 52	18 S 19	12 N 56	13 S 22	20 N 39

ZODIAC SIGN ENTRIES

Date	h	m	Planets
01	17	40	☽ ♊
03	21	56	☽ ♋
06	05	23	☽ ♌
08	16	12	☽ ♍
11	04	51	☽ ♎
13	16	45	☽ ♏
16	02	03	☽ ♐
17	05	31	♀ ♌
18	08	30	☽ ♑
20	12	49	☽ ♒
21	21	24	☉ ♋
22	15	59	☽ ♓
24	18	43	☽ ♈
26	21	34	☽ ♉
29	01	09	☽ ♊

LATITUDES

Date	Mercury ☿ ° '	Venus ♀ ° '	Mars ♂ ° '	Jupiter ♃ ° '	Saturn ♄ ° '	Uranus ♅ ° '	Neptune ♆ ° '	Pluto ♇ ° '
01	01 S 12	01 N 41	00 S 52	01 S 01	00 S 35	00 N 44	01 N 50	13 N 12
04	02 04	01 45	00 50	01 01	00 36	00 44	01 50	13 11
07	02 52	01 49	00 49	01 02	00 36	00 44	01 49	13 10
10	03 32	01 51	00 47	01 03	00 36	00 44	01 49	13 09
13	04 01	01 53	00 45	01 04	00 37	00 44	01 49	13 08
16	04 19	01 54	00 43	01 05	00 37	00 44	01 49	13 07
19	04 24	01 55	00 42	01 06	00 38	00 44	01 49	13 06
22	04 18	01 55	00 40	01 07	00 38	00 44	01 49	13 05
25	04 03	01 53	00 38	01 08	00 38	00 44	01 49	13 04
28	03 40	01 51	00 36	01 08	00 39	00 44	01 48	13 04
31	03 S 11	01 N 49	00 S 34	01 S 09	00 S 39	00 N 44	01 N 48	13 N 03

DATA

Julian Date	2437817
Delta T	+34 seconds
Ayanamsa	23° 19' 42"
Synetic vernal point	05° ♓ 47' 18"
True obliquity of ecliptic	23° 26' 32"

MOON'S PHASES, APSIDES AND POSITIONS ☽

Date	h	m	Phase	Longitude ° '	Eclipse Indicator
02	13	27	●	11 ♊ 31	
10	06	21	☽	18 ♍ 54	
18	02	02	○	26 ♐ 22	
24	23	42	☾	02 ♈ 57	

Day	h	m		
10	17	39	Apogee	
23	19	46	Perigee	
04	16	52	Max dec	20° N 30'
12	01	43	0S	
19	03	19	Max dec	20° S 32'
25	12	36	0N	

LONGITUDES

Date	Chiron ⚷ ° '	Ceres ⚳ ° '	Pallas ⚴ ° '	Juno ⚵ ° '	Vesta ⚶ ° '	Black Moon Lilith ⚸ ° '
01	10 ♓ 49	27 ♊ 18	12 ♉ 26	23 ♊ 28	02 ♋ 21	23 ♍ 53
11	10 ♓ 53	01 ♋ 34	16 ♉ 58	29 ♊ 11	06 ♋ 39	25 ♍ 00
21	10 ♓ 52	05 ♋ 53	21 ♉ 33	04 ♋ 51	11 ♋ 00	26 ♍ 07
31	10 ♓ 45	10 ♋ 14	26 ♉ 10	10 ♋ 26	15 ♋ 23	27 ♍ 14

All ephemeris data is given at 12.00 UT and the Moon's longitude is additionally given for 24.00 UT

ASPECTARIAN

h m	Aspects
01 Friday	
00 03	☉ ∥ ☿
03 06	☽ ∥ ♅
04 24	☽ # ♆
06 06	☽ Q ♃
10 21	♀ △ ♃
11 17	☽ ∠ ♀
12 08	☽ □ ♅
12 42	♀ ⊼ ♄
13 50	♀ △ ♆
21 40	♀ ∠ ♅
22 43	☽ ⊻ ♂
02 Saturday	
04 03	☽ ⊥ ♀
06 37	☽ □ ♇
07 05	☉ □ ♃
07 39	☉ △ ♄
08 59	☉ ⊼ ♆
09 36	☽ ⊥ ♂
12 10	♃ ⊻ ♄
13 02	☽ △ ♄
13 03	☽ □ ♃
13 08	☽ ⊼ ♆
13 27	☽ ☌ ☉
14 30	☽ # ♄
15 22	☽ ⊻ ♀
17 34	☿ ∠ ♂
19 05	☽ Q ♅
23 30	♃ △ ♆
23 33	☽ ± ♆
03 Sunday	
01 23	☽ ☌ ☿
02 07	☽ ∠ ♂
14 04	☽ Q ♇
15 17	☽ ⚼ ♄
15 22	☽ ⚼ ♆
16 14	☽ ✶ ♅
04 Monday	
04 48	☿ ∥ ♇
06 22	☽ ✶ ♂
07 24	☽ ± ♄
11 38	☽ ✶ ♇
18 18	☽ ⊼ ♄
18 24	☽ △ ♆
18 45	☽ △ ♃
19 23	☽ ∠ ♅
23 03	☽ ⊻ ☉
05 Tuesday	
02 16	☽ ☌ ♀
05 25	☿ ± ♆
05 27	☽ ⊻ ☿
05 40	☽ Q ♂
11 04	☽ ⊥ ☉
12 06	☽ ⊥ ♅
15 22	☽ ∠ ♇
16 12	☽ ⊥ ☿
22 54	☽ ⚼ ♃
23 25	☽ ⊻ ♅
06 Wednesday	
01 09	☉ Q ♅
02 51	☿ ⊻ ♀
05 27	☽ ∠ ☉
08 27	☽ ⊥ ♇
08 34	☽ ∠ ☿
16 13	☽ ± ♃
17 55	☽ □ ♂
20 00	☽ ⊻ ♇
21 41	☽ # ♄
07 Thursday	
02 58	☽ ☍ ♄
03 04	☽ □ ♆
03 56	☽ ⊼ ♃
08 22	☉ ☌ ☿
12 25	☽ ✶ ☿
12 58	☽ ✶ ☉
17 46	☽ ⊻ ♀
23 28	♂ ∥ ♅
08 Friday	
05 26	♂ △ ♇
06 59	☽ ⊥ ♀
10 04	☽ ☌ ♅
11 10	☽ Q ☿
12 02	☉ ± ♆
13 32	☽ # ♆
14 06	☽ ∥ ♂
14 37	☽ Q ♆
14 51	☽ Q ☉
15 16	☽ ∥ ♅
21 26	♂ # ♆
21 52	♀ Q ♂
09 Saturday	
03 01	☽ ∠ ♀
07 33	☽ ☌ ♇
09 15	☽ △ ♂
10 06	☿ Q ♅
13 20	☿ ⊥ ♀
14 36	☽ ⊼ ♄
14 44	☽ ✶ ♆
14 54	♀ ⊥ ♅
16 07	☽ ☍ ♃
18 49	☿ ⊥ ♂
21 51	☽ □ ☿
10 Sunday	
01 51	☽ # ♃
02 43	☽ ± ♄
06 21	☽ □ ☉
12 45	☽ ✶ ♀
17 39	☽ ⚼ ♂
20 54	☽ ⚼ ♄
21 05	☽ ∠ ♆
22 48	☽ ⊻ ♅
11 Monday	
00 46	♀ ∠ ♇
10 58	☽ ⊥ ♅
13 04	☽ ± ♂
15 15	☽ ⊥ ♆
15 44	☽ Q ♀
20 18	☽ ⊻ ♇
12 Tuesday	
01 57	☽ ⊼ ♂
03 06	☽ △ ♄
03 18	☽ ⊻ ♆
05 07	☽ ⊼ ♃
05 08	☽ ∠ ♅
07 28	☉ ∥ ♀
08 04	☽ △ ☿
08 22	☽ ⊥ ♇
17 10	☽ ± ♃
20 00	♂ □ ♄
23 31	♂ ☍ ♆
13 Wednesday	
00 01	☽ △ ☉
02 19	☽ ∠ ♇
07 38	☽ □ ♀
11 03	☽ ✶ ♅
11 05	☽ ⚼ ♃
12 51	☽ ⚼ ☿
14 Thursday	
00 14	☽ ∥ ♃
05 55	☽ ± ☿
07 43	☽ ✶ ♇
07 54	☽ ⚼ ☉
09 49	♂ ✶ ♃
10 30	☽ Q ♅
14 01	☽ □ ♄
14 17	☽ ☌ ♆
16 24	☽ △ ♃
16 46	☽ ☍ ♂
17 11	☽ ⊼ ☿
19 36	♀ ⊻ ♅
20 43	♀ ⚼ ♃
21 53	☿ ⊻ ♂
15 Friday	
02 30	☿ ∠ ♀
02 36	☽ ± ☉
06 44	☽ Q ♇
09 11	☽ # ♅
10 38	☿ # ♄
11 08	☽ ∥ ♆
14 52	☽ ⊼ ☉
20 48	☽ □ ♅
22 46	☿ □ ♃
23 17	☽ △ ♀
23 47	☽ # ♂
16 Saturday	
00 01	☽ Q ♄
16 19	☽ □ ♇
22 05	☽ ✶ ♄
22 25	☽ ⊻ ♆
17 Sunday	
00 13	☽ ☍ ☿
00 18	☽ # ☿
00 43	☽ □ ♃
02 07	☽ ∥ ♄
04 09	☽ ⊼ ♂
05 37	☽ ⚼ ♀
09 18	☽ ⊥ ♆
10 03	☉ Q ♇
12 45	☉ ⚼ ♄
15 37	☽ ± ♂
17 43	☉ ⚼ ♆
23 23	☽ ± ♀
18 Monday	
01 03	☽ ∠ ♄
01 26	☽ ∠ ♆
02 02	☽ ☍ ☉
03 38	☽ △ ♅
08 36	☽ ⚼ ♂
09 10	☽ Q ♃
11 04	☽ ⊼ ♀
16 58	☽ ⊥ ♄
17 05	♀ ⊥ ♇
22 11	☽ △ ♇
19 Tuesday	
01 26	☉ ✶ ♅
03 28	☽ ⊻ ♄
03 54	☽ ✶ ♆
05 20	☽ ⊼ ☿
06 09	☽ ⚼ ♅
06 20	☽ ✶ ♃
07 45	☿ St D
12 26	☽ △ ♂
15 48	☽ ± ☿
21 51	☽ ± ♅
20 Wednesday	
00 21	☽ ⚼ ♇
00 44	☽ Q ♆
07 26	☽ ⚼ ☿
08 15	☽ ⊼ ♅
08 26	☽ ∠ ♃
10 25	☽ ⊼ ☉
15 52	☽ ± ♇
20 08	☽ ☍ ♀
21 29	☽ ± ☉
23 56	☽ ⊥ ♃
21 Thursday	
02 10	☽ ⊼ ♇
02 20	☽ ∥ ♄
04 55	☽ # ☿
07 07	☽ ☌ ♄
07 37	☽ □ ♆
09 23	☽ △ ☿
10 13	☽ ⊻ ♃
13 59	☽ ⚼ ☉
17 42	☽ # ♂
18 50	☽ □ ♂
21 54	♀ ∥ ♇
22 Friday	
11 37	☽ ☍ ♅
14 40	☽ ∥ ♆
16 48	☽ # ♅
17 19	☽ △ ☉
18 36	♀ ± ♃
23 Saturday	
03 02	☽ Q ♂
03 49	☽ ⊼ ♀
05 13	☽ ☍ ♇
09 53	☽ ⊻ ♄
10 30	☽ △ ♆
13 11	☽ □ ☿
13 15	☽ ☌ ♃
14 53	☽ ± ♀
15 44	☿ □ ♃
20 00	☽ ⊥ ♄
21 01	♀ ⊻ ♇
21 40	☽ ∥ ♃
24 Sunday	
00 25	☽ ✶ ♂
03 16	☉ ⊥ ♀
07 28	☽ ⚼ ♀
09 34	☿ # ♄
11 09	☽ ∠ ♄
11 50	☽ ⚼ ♆
14 31	☽ ⊼ ♅
20 27	☽ Q ☿
23 42	☽ □ ☉
25 Monday	
00 42	☽ ± ♅
03 02	☽ ± ♆
03 08	☽ ∠ ♂
08 00	☽ ⊼ ♇
11 07	☽ △ ♀
12 26	☽ ✶ ♄
13 11	☽ ⊼ ♆
15 58	☽ ⚼ ♅
16 04	☽ ⊻ ♃
17 26	☽ ✶ ☿
18 11	☽ ± ♇
19 11	☽ ⊥ ♂
26 Tuesday	
02 14	☽ ⊥ ♃
02 21	♀ ☍ ♄
05 54	☽ ⊻ ♂
08 24	☽ Q ☉
08 42	☽ Q ♄
09 28	☽ ⚼ ♇
11 45	♀ □ ♆
12 38	☉ ± ♄
17 30	☽ △ ♅
17 33	☽ ∠ ♃
19 55	☽ ∠ ☿
27 Wednesday	
03 45	☽ # ♃
06 18	☽ ✶ ☉
11 03	☽ △ ♇
11 58	☽ ⊥ ☿
15 16	☽ □ ♄
16 10	☽ ☍ ♆
18 47	☽ □ ♀
19 12	☽ ✶ ♃
21 25	☿ ∥ ♀
22 45	☽ ⊻ ☿
23 45	♀ ⊼ ♃
28 Thursday	
08 43	☽ ∥ ♅
09 56	☽ ∠ ☉
11 19	☽ # ♆
11 59	☽ ☌ ♂
15 54	☽ Q ♃
21 10	☽ □ ♅
29 Friday	
02 42	☽ ⊥ ☉
04 46	☽ Q ♀
07 56	☿ Q ♅
10 16	♀ # ♄
13 58	☽ ⊻ ☉
15 02	☽ □ ♇
19 05	☽ △ ♄
20 10	☽ ⊼ ♆
23 23	☽ □ ♃
30 Saturday	
00 05	☽ ∥ ♂
00 08	☽ ∥ ♀
00 20	♀ ∥ ♂
02 26	☽ # ♄
03 49	☽ ✶ ♀
04 50	☽ Q ♅
05 31	☉ ✶ ♇
06 04	☽ ☌ ☿
06 46	☽ ± ♆
16 49	☿ ± ♆
17 36	☽ ∥ ☿
19 38	☽ ⊻ ♂
21 38	☽ ⚼ ♄
22 49	☽ ⚼ ♆
23 02	☽ Q ♇

JULY 1962

LONGITUDES

Date	Sidereal time h m s	Sun ☉ ° ' "	Moon ☽ ° ' "	Moon ☽ 24.00 ° ' "	Mercury ☿ ° '	Venus ♀ ° '	Mars ♂ ° '	Jupiter ♃ ° '	Saturn ♄ ° '	Uranus ♅ ° '	Neptune ♆ ° '	Pluto ♇ ° '
01	06 36 15	09 ♋ 09 52	03 ♋ 08 21	09 ♋ 42 26	17 ♊ 34	16 ♌ 46	24 ♉ 33	12 ♓ 41	10 ♒ 09	27 ♌ 49	10 ♏ 50	07 ♍ 59
02	06 40 12	10 07 06	16 ♋ 12 18	22 ♋ 37 53	18 32	17 55	25 16	12 R 41	10 R 05	27 52	10 R 50	08 00
03	06 44 08	11 04 19	28 ♋ 59 15	05 ♌ 16 30	19 34	19 05	25 59	12 41	10 02	27 55	10 49	08 01
04	06 48 05	12 01 33	11 ♌ 29 49	17 ♌ 39 27	20 41	20 15	26 42	12 41	09 58	27 58	10 48	08 03
05	06 52 01	12 58 46	23 ♌ 45 43	29 ♌ 48 58	21 51	21 25	27 24	12 40	09 54	28 01	10 48	08 04
06	06 55 58	13 55 59	05 ♍ 49 39	11 ♍ 48 13	23 06	22 34	28 07	12 40	09 51	28 04	10 47	08 05
07	06 59 54	14 53 12	17 ♍ 45 11	23 ♍ 41 06	24 24	23 44	28 50	12 39	09 47	28 07	10 47	08 07
08	07 03 51	15 50 25	29 ♍ 36 31	05 ♎ 32 02	25 47	24 53	29 ♉ 32	12 38	09 43	28 09	10 46	08 08
09	07 07 48	16 47 37	11 ♎ 28 15	17 ♎ 25 46	27 13	26 02	00 ♊ 14	12 36	09 39	28 13	10 46	08 09
10	07 11 44	17 44 50	23 ♎ 25 11	29 ♎ 27 04	28 ♊ 43	27 12	00 57	12 35	09 35	28 16	10 46	08 11
11	07 15 41	18 42 02	05 ♏ 31 59	11 ♏ 40 28	00 ♋ 17	28 21	01 39	12 33	09 31	28 19	10 45	08 12
12	07 19 37	19 39 15	17 ♏ 53 01	24 ♏ 10 02	01 55	29 ♌ 30	02 21	12 31	09 27	28 22	10 45	08 14
13	07 23 34	20 36 27	00 ♐ 31 54	06 ♐ 58 55	03 37	00 ♍ 39	03 03	12 29	09 23	28 25	10 45	08 15
14	07 27 30	21 33 40	13 ♐ 31 15	20 ♐ 09 03	05 21	01 47	03 45	12 27	09 19	28 28	10 44	08 17
15	07 31 27	22 30 53	26 ♐ 52 18	03 ♑ 40 53	07 09	02 56	04 27	12 25	09 15	28 31	10 44	08 18
16	07 35 23	23 28 06	10 ♑ 34 36	17 ♑ 33 07	09 00	04 05	05 09	12 22	09 11	28 34	10 44	08 20
17	07 39 20	24 25 19	24 ♑ 36 02	01 ♒ 42 49	10 54	05 13	05 51	12 19	09 07	28 38	10 44	08 21
18	07 43 17	25 22 33	08 ♒ 52 54	16 ♒ 05 37	12 51	06 21	06 32	12 16	09 02	28 41	10 43	08 23
19	07 47 13	26 19 47	23 ♒ 20 20	00 ♓ 36 19	14 50	07 30	07 14	12 13	08 58	28 44	10 43	08 25
20	07 51 10	27 17 01	07 ♓ 52 54	15 ♓ 09 24	16 51	08 38	07 56	12 10	08 54	28 48	10 43	08 26
21	07 55 06	28 14 17	22 ♓ 25 13	29 ♓ 39 47	18 54	09 46	08 37	12 06	08 49	28 51	10 43	08 28
22	07 59 03	29 ♋ 11 33	06 ♈ 52 33	14 ♈ 03 07	20 58	10 54	09 19	12 02	08 45	28 54	10 43	08 30
23	08 02 59	00 ♌ 08 50	21 ♈ 11 05	28 ♈ 16 09	23 04	12 01	10 00	11 59	08 41	28 58	10 D 43	08 31
24	08 06 56	01 06 08	05 ♉ 18 05	12 ♉ 16 42	25 10	13 09	10 41	11 54	08 36	29 01	10 43	08 33
25	08 10 52	02 03 27	19 ♉ 11 52	26 ♉ 03 29	27 17	14 16	11 22	11 50	08 32	29 04	10 43	08 35
26	08 14 49	03 00 47	02 ♊ 51 29	09 ♊ 35 52	29 ♋ 24	15 24	12 03	11 46	08 28	29 08	10 43	08 36
27	08 18 46	03 58 08	16 ♊ 17 37	22 ♊ 53 44	01 ♌ 31	16 31	12 44	11 41	08 23	29 11	10 43	08 38
28	08 22 42	04 55 30	29 ♊ 27 16	05 ♋ 57 15	03 38	17 38	13 25	11 36	08 19	29 15	10 43	08 40
29	08 26 39	05 52 52	12 ♋ 23 45	18 ♋ 46 49	05 44	18 45	14 06	11 32	08 14	29 18	10 44	08 42
30	08 30 35	06 50 16	25 ♋ 06 33	01 ♌ 23 02	07 49	19 52	14 47	11 26	08 10	29 22	10 44	08 44
31	08 34 32	07 ♌ 47 41	07 ♌ 36 24	13 ♌ 46 48	09 ♌ 54	20 ♍ 58	15 ♊ 28	11 ♓ 21	08 ♒ 05	29 ♌ 25	10 ♏ 44	08 ♍ 45

Date	Moon True ☊ ° '	Moon Mean ☊ ° '	Moon ☽ Latitude ° '
01	09 ♌ 08	10 ♌ 24	03 S 04
02	09 R 05	10 21	02 02
03	09 03	10 18	00 S 55
04	09 D 03	10 15	00 N 13
05	09 03	10 12	01 20
06	09 05	10 08	02 22
07	09 06	10 05	03 17
08	09 08	10 02	04 03
09	09 09	09 59	04 39
10	09 R 09	09 56	05 04
11	09 08	09 53	05 15
12	09 06	09 49	05 12
13	09 04	09 46	04 54
14	09 01	09 43	04 20
15	08 59	09 40	03 32
16	08 57	09 37	02 32
17	08 56	09 33	01 N 18
18	08 D 56	09 30	00 00
19	08 56	09 27	01 S 19
20	08 57	09 24	02 34
21	08 58	09 21	03 38
22	08 58	09 18	04 29
23	08 59	09 14	05 02
24	08 R 59	09 11	05 16
25	08 59	09 08	05 12
26	08 58	09 05	04 50
27	08 57	09 02	04 12
28	08 57	08 59	03 22
29	08 57	08 55	02 22
30	08 56	08 52	01 16
31	08 ♌ 56	08 ♌ 49	00 S 07

DECLINATIONS

Date	Sun ☉ ° '	Moon ☽ ° '	Mercury ☿ ° '	Venus ♀ ° '	Mars ♂ ° '	Jupiter ♃ ° '	Saturn ♄ ° '	Uranus ♅ ° '	Neptune ♆ ° '	Pluto ♇ ° '
01	23 N 08	20 N 20	19 N 42	17 N 32	18 N 22	07 S 52	18 S 20	12 N 55	13 S 22	20 N 38
02	23 03	20 26	19 57	17 10	18 33	07 52	18 21	12 54	13 22	20 37
03	22 59	19 28	20 14	16 47	18 44	07 52	18 22	12 53	13 21	20 37
04	22 54	17 33	20 30	16 24	18 54	07 53	18 23	12 52	13 21	20 36
05	22 49	14 52	20 47	16 00	19 05	07 53	18 24	12 51	13 21	20 35
06	22 43	11 35	21 03	15 36	19 15	07 54	18 25	12 50	13 21	20 34
07	22 37	07 52	21 20	15 12	19 25	07 55	18 26	12 49	13 21	20 34
08	22 30	03 N 53	21 35	14 47	19 35	07 55	18 28	12 48	13 21	20 33
09	22 23	00 S 15	21 51	14 22	19 45	07 56	18 29	12 47	13 21	20 32
10	22 16	04 22	22 05	13 56	19 54	07 57	18 30	12 45	13 21	20 31
11	22 08	08 25	22 18	13 30	20 03	07 58	18 31	12 44	13 21	20 31
12	22 00	12 10	22 31	13 04	20 12	07 59	18 32	12 43	13 21	20 30
13	21 52	15 29	22 41	12 38	20 21	08 00	18 33	12 42	13 21	20 29
14	21 43	18 07	22 51	12 11	20 30	08 01	18 35	12 41	13 21	20 28
15	21 34	19 52	22 58	11 44	20 39	08 02	18 36	12 40	13 21	20 27
16	21 24	20 31	23 03	11 17	20 47	08 03	18 37	12 39	13 21	20 27
17	21 14	19 55	23 07	10 49	20 55	08 05	18 38	12 38	13 21	20 26
18	21 04	18 02	23 08	10 22	21 03	08 06	18 39	12 36	13 21	20 25
19	20 53	14 59	23 06	09 54	21 11	08 08	18 41	12 35	13 21	20 24
20	20 42	10 59	23 02	09 26	21 18	08 09	18 42	12 34	13 21	20 23
21	20 31	06 21	22 55	08 57	21 26	08 11	18 43	12 33	13 21	20 22
22	20 19	01 S 23	22 46	08 29	21 33	08 12	18 44	12 32	13 21	20 22
23	20 07	03 N 36	22 34	08 00	21 40	08 14	18 46	12 31	13 21	20 21
24	19 55	08 19	22 19	07 31	21 46	08 16	18 47	12 29	13 21	20 20
25	19 42	12 31	22 01	07 02	21 53	08 18	18 48	12 28	13 21	20 19
26	19 29	15 49	21 41	06 33	21 59	08 20	18 49	12 27	13 21	20 18
27	19 16	18 33	21 19	06 03	22 05	08 22	18 51	12 26	13 21	20 18
28	19 02	20 05	20 54	05 34	22 11	08 24	18 52	12 24	13 21	20 17
29	18 48	20 30	20 27	05 04	22 17	08 26	18 53	12 23	13 21	20 16
30	18 34	19 52	19 58	04 34	22 23	08 28	18 54	12 22	13 21	20 15
31	18 N 19	18 N 15	19 N 27	04 N 05	22 N 28	08 S 30	18 S 56	12 N 21	13 S 21	20 N 14

ZODIAC SIGN ENTRIES

Date	h	m	Planets
01	06	19	☽ ♋
03	13	55	☽ ♌
06	00	22	☽ ♍
08	12	48	☽ ♎
09	03	50	♂ ♊
11	01	05	☽ ♏
11	07	36	☿ ♋
12	22	32	♀ ♍
13	11	00	☽ ♐
15	17	32	☽ ♑
17	21	07	☽ ♒
19	23	00	☽ ♓
22	00	34	☽ ♈
23	08	18	☉ ♌
24	02	57	☽ ♉
26	06	57	☽ ♊
26	18	50	☿ ♌
28	13	00	☽ ♋
30	21	21	☽ ♌

LATITUDES

Date	Mercury ☿ ° '	Venus ♀ ° '	Mars ♂ ° '	Jupiter ♃ ° '	Saturn ♄ ° '	Uranus ♅ ° '	Neptune ♆ ° '	Pluto ♇ ° '
01	03 S 11	01 N 49	00 S 34	01 S 09	00 S 39	00 N 44	01 N 48	13 N 03
04	02 37	01 45	00 32	01 10	00 39	00 43	01 48	13 02
07	02 00	01 40	00 30	01 11	00 40	00 43	01 48	13 01
10	01 21	01 35	00 28	01 12	00 40	00 43	01 48	13 00
13	00 42	01 29	00 25	01 13	00 40	00 43	01 48	13 00
16	00 S 05	01 24	00 23	01 14	00 41	00 43	01 48	12 59
19	00 N 29	01 13	00 21	01 15	00 41	00 43	01 47	12 58
22	00 58	01 05	00 18	01 15	00 41	00 43	01 47	12 58
25	01 21	00 55	00 16	01 16	00 42	00 43	01 47	12 57
28	01 36	00 44	00 14	01 17	00 42	00 43	01 47	12 57
31	01 N 45	00 N 33	00 S 11	01 S 18	00 S 42	00 N 43	01 N 47	12 N 56

DATA

Julian Date	2437847
Delta T	+34 seconds
Ayanamsa	23° 19' 47"
Synetic vernal point	05° ♓ 47' 13"
True obliquity of ecliptic	23° 26' 32"

MOON'S PHASES, APSIDES AND POSITIONS ☽

Date	h	m	Phase	Longitude ° '	Eclipse Indicator
01	23	52	●	09 ♋ 38	
09	23	39	☽	17 ♎ 15	
17	11	41	○	24 ♑ 25	
24	04	18	☾	00 ♉ 37	
31	12	24	●	07 ♌ 49	Annular

Day	h	m		
08	12	18	Apogee	
20	10	12	Perigee	
02	02	08	Max dec	20° N 32'
09	10	33	0S	
16	12	43	Max dec	20° S 31'
22	18	37	0N	
29	09	30	Max dec	20° N 31'

LONGITUDES

Date	Chiron ⚷ ° '	Ceres ⚳ ° '	Pallas ⚴ ° '	Juno ⚵ ° '	Vesta ⚶ ° '	Black Moon Lilith ⚸ ° '
01	10 ♓ 45	10 ♋ 14	26 ♉ 10	10 ♋ 26	15 ♋ 23	27 ♍ 14
11	10 ♓ 33	14 ♋ 36	00 ♊ 49	15 ♋ 57	19 ♋ 48	28 ♍ 21
21	10 ♓ 16	18 ♋ 59	05 ♊ 30	21 ♋ 24	24 ♋ 13	29 ♍ 27
31	09 ♓ 55	23 ♋ 22	10 ♊ 12	26 ♋ 44	28 ♋ 40	00 ♎ 34

All ephemeris data is given at 12.00 UT and the Moon's longitude is additionally given for 24.00 UT

ASPECTARIAN

h m Aspects

01 Sunday
02 21 ☽ ⚹ ♅
07 03 ☽ ⊥ ♂
07 24 ♂ # ♄
09 15 ☽ ∠ ♀
13 49 ☽ ± ♄
16 28 ♂ Q ♃
20 51 ☽ ⚹ ♇
23 52 ☽ ☌ ☉

02 Monday
00 24 ☽ ∠ ♂
00 45 ☽ ⊼ ♄
02 04 ☽ △ ♆
03 18 ☽ ⊥ ♀
05 30 ☽ △ ♃
05 48 ☽ ∠ ♅
08 59 ♃ St R
11 18 ☉ ⊼ ♄
15 31 ☽ ⊻ ♀
16 42 ☽ ⊻ ☿
22 36 ☽ ⊥ ♅
23 49 ☽ ∥ ☿

03 Tuesday
00 43 ☽ ∠ ♇
03 15 ☉ ∠ ♂
04 57 ☽ ⊥ ☿
05 42 ☉ △ ♆
05 59 ☽ ⚹ ♂
09 32 ☽ ⚼ ♃
09 58 ☽ ⊻ ♅
17 47 ☽ ⊥ ♇
21 42 ☽ ∥ ♂
23 40 ☽ ∠ ☿

04 Wednesday
02 43 ☽ ± ♃
02 55 ☽ # ♄
05 19 ☽ ⊻ ♇
06 15 ☽ Q ♂
09 03 ☽ ☍ ♄
10 40 ☽ □ ♆
13 07 ☽ ⊻ ☉
14 18 ☽ ⊼ ♃
19 51 ☿ ∥ ♇

05 Thursday
00 57 ☽ ∥ ♀
01 48 ☽ ⊥ ☉
04 23 ☉ △ ♃
06 53 ☽ ☌ ♀
07 49 ☽ ⚹ ☿
12 50 ☉ ∠ ♅
19 40 ☽ □ ♂
20 27 ☽ ☌ ♅
21 04 ☽ ∠ ☉
21 58 ☽ Q ♆
23 29 ☽ # ♆

06 Friday
03 14 ☽ ∥ ♅
09 53 ♂ □ ♅
10 21 ☽ Q ☿
16 32 ☽ ☌ ♇
20 01 ☽ ⊼ ♄
21 57 ☽ ⚹ ♆

07 Saturday
01 43 ☽ ☍ ♃
05 43 ☽ ⚹ ☉
08 03 ☽ ± ♄
11 43 ☽ # ♃
18 27 ☿ ⚼ ♄

08 Sunday
01 24 ☽ ⊻ ♀
02 09 ☽ ⚼ ♄
03 12 ☽ □ ☿
04 14 ☽ ∠ ♆
08 06 ☽ Q ☉
09 03 ☽ ⊻ ♅
11 50 ☽ △ ♂
11 56 ☿ ⚼ ♆
14 52 ☽ ⊥ ♀
18 06 ☿ Q ♇
21 15 ☽ ⊥ ♅
22 27 ☽ ⊥ ♆

09 Monday
05 17 ☽ ⊻ ♇
08 21 ☽ △ ♄
10 35 ☽ ⊻ ♆
11 02 ☽ ∠ ♀
14 17 ☽ ⊼ ♃
15 31 ☽ ∠ ♅
17 25 ☽ ⊥ ♇
20 04 ☽ ⚼ ♂
23 39 ☽ □ ☉

10 Tuesday
02 20 ☽ ± ♃
04 27 ☿ ⚹ ♅
11 31 ☽ ∠ ♇
15 14 ☽ ± ♂
20 16 ☽ ⚼ ♃
20 19 ☽ ⚹ ♀
21 41 ☽ ⚹ ♅

11 Wednesday
00 07 ☽ △ ☿
00 24 ☉ ∥ ☿
03 53 ☽ ⊼ ♂
09 13 ☽ ∥ ♃
11 13 ♀ ☌ ♅
17 15 ☽ ⚹ ♇
19 46 ☽ □ ♄
20 27 ♀ Q ♆
20 51 ♀ # ♆
21 23 ☽ Q ♅
22 12 ☽ ☌ ♆
22 23 ☽ Q ♀

12 Thursday
01 41 ☽ △ ♃
09 52 ☽ ⚼ ☿
15 41 ☽ △ ☉
15 43 ☽ # ♅
16 30 ☽ Q ♇
17 27 ☽ # ♀
20 04 ☽ ∥ ♆
22 40 ☿ ⊻ ♂

13 Friday
05 39 ☽ ± ☿
06 07 ☽ Q ♄
07 50 ♀ ∥ ♅
08 01 ☽ □ ♅
09 01 ☿ ± ♄
12 14 ☽ □ ♀
16 59 ☽ ☍ ♂
18 38 ☽ ⊼ ☿
22 12 ☽ ⚼ ☉

14 Saturday
02 22 ☽ □ ♇
04 21 ☽ ⚹ ♄
06 55 ☽ ⊻ ♆
07 11 ♂ ∥ ♇
10 03 ☽ □ ♃
16 00 ☽ ± ☉
17 10 ☽ ∥ ♄
17 50 ☽ ⊥ ♆

15 Sunday
03 39 ☽ ⊼ ☉
07 22 ☽ ∠ ♄
09 59 ☽ ∠ ♆
12 08 ☉ ⊥ ♅
14 56 ☽ △ ♅
18 14 ☽ Q ♃
23 11 ☽ ⊥ ♄
23 40 ☽ △ ♀

16 Monday
02 04 ☽ ⊼ ♂
03 13 ☿ ⚹ ♇
04 27 ☽ # ♇
08 06 ☽ △ ♇
08 26 ☉ ∠ ♇
08 51 ☽ ☍ ☿
09 36 ☽ ⊻ ♄
12 16 ☽ ⚹ ♆
13 03 ☽ ± ♂
14 10 ☿ ⊼ ♄
15 05 ☽ ⚹ ♃
17 12 ☽ ⚼ ♅
21 27 ☽ # ♇

17 Tuesday
03 54 ☽ ⚼ ♀
05 17 ☽ ⚼ ♂
08 38 ☽ ± ♅
08 49 ☽ Q ♆
09 47 ☿ △ ♆
09 53 ☽ ⚼ ♇
11 41 ☽ ☍ ☉
16 35 ☽ ∠ ♃
18 50 ☽ ⊼ ♅
20 29 ☽ ± ♀

18 Wednesday
01 06 ☽ ± ♇
05 07 ☿ △ ♃
05 41 ☽ ∥ ♄
06 12 ☿ ⊥ ♂
07 25 ☽ ⊼ ♀
07 39 ☽ ⊥ ♃
07 54 ☽ △ ♂
11 10 ☽ ⊼ ♇
12 16 ☽ ☌ ♄
13 18 ☉ ∥ ♂
15 04 ☽ □ ♆
17 38 ☽ ⊻ ♃
19 39 ☽ ⊼ ☿
21 57 ♀ □ ♂
22 27 ☿ ∠ ♅

19 Thursday
05 08 ♅ Q ♆
07 11 ☽ ± ☿
17 17 ☽ ⊼ ☉
20 57 ☽ ☍ ♅
22 28 ☽ ∥ ♆

20 Friday
00 26 ☽ ⚼ ☿
03 01 ☽ # ♅
03 53 ☽ ± ☉
07 51 ♀ ☌ ♇
09 06 ☉ ⚼ ♃
12 05 ☽ □ ♂
12 55 ☽ ☍ ♇
13 20 ☽ ☍ ♀
13 40 ☽ ⊻ ♄
16 41 ☽ △ ♆
17 19 ♀ ⊼ ♄
19 02 ☽ ☌ ♃
19 46 ☽ ⚼ ☉
21 20 ☽ # ♀
23 31 ☽ ⊥ ♄

21 Saturday
02 53 ☽ ∥ ♃
05 13 ☽ △ ☿
06 25 ♂ □ ♇
14 19 ☽ ∠ ♄
17 27 ☽ ⚼ ♆
18 27 ♂ △ ♄
19 18 ☽ Q ♂
22 19 ☽ △ ☉
22 41 ☽ ⊼ ♅

22 Sunday
04 17 ☉ ⊻ ♅
06 40 ☉ ∥ ♇
08 15 ♀ ⚹ ♆
08 24 ☽ ± ♆
08 42 ☽ ± ♅
14 42 ☽ ⊼ ♇
15 07 ☽ ⚹ ♄
16 16 ☽ ⚹ ♂
18 25 ☽ ⊼ ♆
19 17 ☽ ⊼ ♀
20 36 ☽ ⊻ ♃
23 48 ☽ ⚼ ♅

23 Monday
00 46 ☽ ± ♇
00 48 ♀ # ♃
06 13 ☽ ± ♀
06 37 ☽ ⊥ ♃
08 08 ♆ St D
10 49 ☿ ⊥ ♅
11 05 ♀ ☍ ♃
11 09 ☽ Q ♄
15 43 ☽ □ ☿
15 57 ☽ ⚼ ♇
17 21 ☿ ∠ ♇
18 47 ☽ ∠ ♂
21 45 ☽ ∠ ♃
22 44 ☽ ⚼ ♀

24 Tuesday
01 14 ☽ △ ♅
04 18 ☽ □ ☉
08 11 ☽ ∥ ♀
10 54 ☽ ⊥ ♂
11 44 ☽ # ♃
13 04 ♂ ⊼ ♆
17 35 ☽ △ ♇
17 38 ☽ □ ♄
20 46 ☿ ∠ ♂
21 18 ☽ ☍ ♆
21 44 ☽ ⊻ ♂
23 18 ☽ ⚹ ♃

25 Wednesday
01 13 ♄ ⊼ ♇
02 42 ☽ △ ♀
03 58 ☽ Q ☿
07 09 ☿ ⚼ ♃
11 42 ☽ ∥ ♅
13 37 ☽ Q ☉
17 12 ♀ ± ♄
17 15 ☽ # ♆
20 00 ☿ ∥ ♂
20 04 ☽ Q ♃

26 Thursday
01 30 ☉ ⊥ ♇
02 42 ♂ □ ♃
04 45 ☽ ⚹ ☿
05 23 ☽ □ ♅
08 54 ☿ ⊻ ♅
12 18 ☽ ⚹ ☉
21 55 ☽ △ ♄
22 15 ☽ □ ♇

27 Friday
02 01 ☽ ⊼ ♆
03 47 ☽ □ ♃
05 17 ☽ ☌ ♂
12 02 ☿ ∠ ♀
12 28 ☽ □ ♀
12 31 ☽ ∠ ☿
12 48 ☽ ± ♆
13 39 ☽ Q ♅
15 36 ☽ # ♄
17 15 ☽ ∠ ☉
20 08 ☽ ∥ ☉

28 Saturday
00 49 ☽ ⚼ ♄
00 56 ☿ ⊥ ♇
05 10 ☽ ⚼ ♆
06 52 ☽ Q ♇
08 00 ☽ ⊥ ☿
10 57 ☽ ⊥ ☉
11 37 ☽ ⚹ ♅
17 14 ☽ ± ♄
17 43 ☽ ∥ ♇
21 11 ☽ ⊻ ☿
22 54 ☽ ⊻ ☉

29 Sunday
00 29 ☽ Q ♀
03 48 ☉ ± ♃
04 17 ☽ ⊼ ♄
04 22 ☉ # ♄
05 05 ☽ ⚹ ♇
08 51 ☽ ∥ ☿
08 53 ☽ △ ♆
09 47 ☿ ± ♃
10 23 ☽ △ ♃
15 14 ☉ ☌ ☿
15 23 ☽ ⊻ ♂
15 36 ☽ ∠ ♅
21 36 ☿ ∥ ♇

30 Monday
01 05 ☽ ⚹ ♀
02 02 ☽ ∥ ♇
03 19 ☽ ⊥ ♂
07 49 ☽ ∥ ☿
08 40 ☽ ⊥ ♅
09 22 ☽ ∠ ♇
14 31 ☽ ⚼ ♃
15 50 ☿ ☍ ♄
20 10 ☽ ⊻ ♅
21 27 ☽ ∠ ♂
22 39 ☿ ⊻ ♇

31 Tuesday
02 37 ☽ ⊥ ♇
03 37 ☽ # ♄
07 41 ☽ ± ♃
08 32 ☽ ∠ ♀
11 03 ☽ ∥ ☉
12 24 ☽ ☌ ☉
12 56 ☽ ☍ ♄
14 14 ☽ ⊻ ♇
17 19 ☽ ☌ ☿
18 04 ☽ □ ♆
18 49 ☉ ☍ ♄
19 13 ☽ ⊼ ♃
21 50 ☿ □ ♆

AUGUST 1962

Date	Sidereal time h m s	LONGITUDES Sun ☉ ° ′ ″	Moon ☽ ° ′ ″	Moon ☽ 24.00 ° ′ ″	Mercury ☿ ° ′	Venus ♀ ° ′	Mars ♂ ° ′	Jupiter ♃ ° ′	Saturn ♄ ° ′	Uranus ♅ ° ′	Neptune ♆ ° ′	Pluto ⯓ ° ′
01	08 38 28	08 ♌ 45 06	19 ♌ 54 23	25 ♌ 59 22	11 ♌ 57	22 ♍ 05	16 ♊ 08	11 ♓ 16	08 ♒ 01	29 ♌ 29	10 ♏ 44	08 ♍ 47
02	08 42 25	09 42 32	02 ♍ 01 59	08 ♍ 02 28	13 59	23 11	16 49	11 R 10	07 R 56	29 33	10 45	08 49
03	08 46 21	10 39 59	14 ♍ 01 09	19 ♍ 58 21	16 00	24 17	17 29	11 05	07 52	29 36	10 45	08 51
04	08 50 18	11 37 26	25 ♍ 54 26	01 ♎ 49 47	18 00	25 23	18 10	10 59	07 47	29 40	10 45	08 53
05	08 54 15	12 34 55	07 ♎ 44 52	13 ♎ 40 09	19 58	26 29	18 50	10 53	07 43	29 43	10 46	08 55
06	08 58 11	13 32 24	19 ♎ 36 06	25 ♎ 33 15	21 54	27 35	19 30	10 47	07 39	29 47	10 46	08 57
07	09 02 08	14 29 54	01 ♏ 32 08	07 ♏ 33 19	23 49	28 40	20 10	10 40	07 34	29 51	10 47	08 59
08	09 06 04	15 27 25	13 ♏ 37 21	19 ♏ 44 47	25 43	29 ♍ 46	20 50	10 34	07 30	29 54	10 47	09 00
09	09 10 01	16 24 56	25 ♏ 56 12	02 ♐ 12 05	27 35	00 ♎ 51	21 30	10 28	07 25	29 ♌ 58	10 48	09 02
10	09 13 57	17 22 29	08 ♐ 32 58	14 ♐ 59 16	29 ♌ 26	01 56	22 10	10 21	07 21	00 ♍ 02	10 48	09 04
11	09 17 54	18 20 02	21 ♐ 31 23	28 ♐ 09 37	01 ♍ 15	03 01	22 50	10 14	07 17	00 05	10 49	09 06
12	09 21 50	19 17 36	04 ♑ 54 09	11 ♑ 45 06	03 02	04 05	23 29	10 07	07 12	00 09	10 50	09 08
13	09 25 47	20 15 12	18 ♑ 42 24	25 ♑ 45 53	04 48	05 09	24 09	10 01	07 08	00 13	10 50	09 10
14	09 29 44	21 12 48	02 ♒ 55 12	10 ♒ 09 51	06 33	06 13	24 48	09 54	07 04	00 16	10 51	09 12
15	09 33 40	22 10 25	17 ♒ 29 13	24 ♒ 52 29	08 16	07 17	25 28	09 46	07 00	00 20	10 52	09 14
16	09 37 37	23 08 04	02 ♓ 18 46	09 ♓ 47 04	09 58	08 21	26 07	09 39	06 55	00 24	10 52	09 16
17	09 41 33	24 05 44	17 ♓ 16 21	24 ♓ 45 32	11 38	09 24	26 46	09 32	06 51	00 28	10 53	09 18
18	09 45 30	25 03 25	02 ♈ 13 35	09 ♈ 39 31	13 16	10 27	27 25	09 24	06 47	00 31	10 54	09 20
19	09 49 26	26 01 08	17 ♈ 02 27	24 ♈ 21 37	14 54	11 30	28 04	09 17	06 43	00 35	10 55	09 22
20	09 53 23	26 58 52	01 ♉ 36 24	08 ♉ 46 19	16 29	12 33	28 43	09 10	06 39	00 39	10 56	09 24
21	09 57 19	27 56 38	15 ♉ 51 01	22 ♉ 50 20	18 04	13 35	29 ♊ 22	09 02	06 35	00 43	10 57	09 26
22	10 01 16	28 54 26	29 ♉ 44 10	06 ♊ 32 36	19 37	14 38	00 ♋ 01	08 54	06 31	00 46	10 58	09 28
23	10 05 13	29 ♌ 52 16	13 ♊ 15 44	19 ♊ 53 47	21 08	15 39	00 39	08 47	06 27	00 50	10 59	09 30
24	10 09 09	00 ♍ 50 08	26 ♊ 27 01	02 ♋ 55 43	22 38	16 41	01 18	08 39	06 23	00 54	11 00	09 33
25	10 13 06	01 48 01	09 ♋ 20 12	15 ♋ 40 48	24 07	17 42	01 56	08 31	06 19	00 58	11 01	09 35
26	10 17 02	02 45 56	21 ♋ 57 51	28 ♋ 11 39	25 34	18 43	02 34	08 23	06 16	01 01	11 02	09 37
27	10 20 59	03 43 52	04 ♌ 22 31	10 ♌ 30 43	27 00	19 44	03 13	08 15	06 12	01 05	11 03	09 39
28	10 24 55	04 41 50	16 ♌ 36 33	22 ♌ 40 15	28 24	20 44	03 51	08 07	06 08	01 09	11 04	09 41
29	10 28 52	05 39 50	28 ♌ 42 04	04 ♍ 42 13	29 ♍ 47	21 44	04 29	07 59	06 05	01 13	11 05	09 43
30	10 32 48	06 37 52	10 ♍ 40 56	16 ♍ 38 27	01 ♎ 08	22 44	05 07	07 52	06 01	01 16	11 06	09 45
31	10 36 45	07 ♍ 35 55	22 ♍ 34 58	28 ♍ 30 44	02 ♎ 28	23 ♎ 43	05 ♋ 44	07 ♓ 44	05 ♒ 58	01 ♍ 20	11 ♏ 07	09 ♍ 47

Date	Moon True ☊ ° ′	Moon Mean ☊ ° ′	Moon ☽ Latitude ° ′
01	08 ♌ 56	08 ♌ 46	01 N 01
02	08 R 56	08 43	02 05
03	08 56	08 39	03 03
04	08 56	08 36	03 52
05	08 55	08 33	04 32
06	08 55	08 30	04 59
07	08 55	08 27	05 14
08	08 54	08 24	05 16
09	08 D 55	08 20	05 03
10	08 55	08 17	04 36
11	08 56	08 14	03 54
12	08 57	08 11	02 58
13	08 57	08 08	01 50
14	08 58	08 05	00 N 33
15	08 R 58	08 01	00 S 47
16	08 57	07 58	02 06
17	08 56	07 55	03 16
18	08 54	07 52	04 13
19	08 52	07 49	04 53
20	08 50	07 45	05 13
21	08 49	07 42	05 13
22	08 D 49	07 39	04 55
23	08 49	07 36	04 20
24	08 50	07 33	03 33
25	08 52	07 30	02 36
26	08 53	07 26	01 32
27	08 54	07 23	00 S 25
28	08 R 54	07 20	00 N 42
29	08 52	07 17	01 47
30	08 50	07 14	02 46
31	08 ♌ 45	07 ♌ 11	03 N 37

DECLINATIONS

Date	Sun ☉ ° ′	Moon ☽ ° ′	Mercury ☿ ° ′	Venus ♀ ° ′	Mars ♂ ° ′	Jupiter ♃ ° ′	Saturn ♄ ° ′	Uranus ♅ ° ′	Neptune ♆ ° ′	Pluto ⯓ ° ′
01	18 N 04	15 N 48	18 N 54	03 N 35	22 N 33	08 S 33	18 S 57	12 N 19	13 S 22	20 N 14
02	17 49	12 42	18 20	03 05	22 38	08 35	18 58	12 18	13 22	20 13
03	17 34	09 06	17 44	02 35	22 43	08 37	18 59	12 17	13 22	20 12
04	17 18	05 11	17 07	02 05	22 47	08 40	19 01	12 16	13 22	20 11
05	17 02	01 N 05	16 29	01 35	22 51	08 42	19 02	12 14	13 22	20 10
06	16 46	03 S 03	15 49	01 04	22 56	08 45	19 03	12 13	13 22	20 09
07	16 29	07 05	15 09	00 34	23 00	08 47	19 04	12 12	13 23	20 09
08	16 12	10 54	14 28	00 N 04	23 03	08 50	19 05	12 11	13 23	20 08
09	15 55	14 20	13 46	00 S 26	23 07	08 53	19 07	12 09	13 23	20 07
10	15 38	17 11	13 04	00 56	23 10	08 55	19 08	12 08	13 23	20 06
11	15 20	19 17	12 21	01 27	23 13	08 58	19 09	12 07	13 24	20 05
12	15 02	20 23	11 38	01 57	23 16	09 01	19 10	12 05	13 24	20 04
13	14 44	20 19	10 55	02 27	23 19	09 04	19 11	12 04	13 24	20 04
14	14 26	18 58	10 11	02 57	23 21	09 06	19 12	12 03	13 24	20 03
15	14 07	16 20	09 27	03 27	23 24	09 09	19 14	12 01	13 25	20 02
16	13 48	12 36	08 43	03 57	23 26	09 12	19 15	12 00	13 25	20 01
17	13 29	08 02	07 59	04 27	23 28	09 15	19 16	11 59	13 25	20 00
18	13 10	02 S 59	07 15	04 56	23 30	09 18	19 17	11 57	13 26	20 00
19	12 51	02 N 11	06 31	05 26	23 31	09 21	19 18	11 56	13 26	19 59
20	12 31	07 08	05 47	05 56	23 33	09 24	19 19	11 55	13 26	19 58
21	12 11	11 35	05 03	06 25	23 34	09 27	19 20	11 53	13 27	19 57
22	11 51	15 18	04 20	06 54	23 35	09 30	19 21	11 52	13 27	19 56
23	11 31	18 05	03 36	07 23	23 36	09 33	19 22	11 51	13 27	19 56
24	11 11	19 51	02 53	07 52	23 36	09 36	19 23	11 49	13 28	19 55
25	10 50	20 32	02 11	08 21	23 37	09 39	19 24	11 48	13 28	19 54
26	10 29	20 08	01 28	08 50	23 37	09 42	19 25	11 47	13 28	19 53
27	10 08	18 46	00 47	09 18	23 37	09 45	19 26	11 45	13 29	19 52
28	09 47	16 32	00 N 05	09 47	23 37	09 48	19 27	11 44	13 29	19 52
29	09 26	13 36	00 S 36	10 15	23 37	09 51	19 28	11 43	13 30	19 51
30	09 05	10 07	01 16	10 43	23 37	09 54	19 29	11 41	13 30	19 50
31	08 N 43	06 N 16	01 S 56	11 S 10	23 N 36	09 S 57	19 S 30	11 N 40	13 S 30	19 N 49

ZODIAC SIGN ENTRIES

Date	h	m	Planets
02	07	57	☽ ♍
04	20	17	☽ ♎
07	08	56	☽ ♏
08	17	13	♀ ♎
09	19	48	☽ ♐
10	01	19	♅ ♍
10	19	29	☿ ♍
12	03	18	☽ ♑
14	07	07	☽ ♒
16	08	17	☽ ♓
18	08	25	☽ ♈
20	09	20	☽ ♉
22	11	37	♂ ♋
22	12	28	☽ ♊
23	15	12	☉ ♍
24	18	34	☽ ♋
27	03	30	☽ ♌
29	14	36	☽ ♍
29	15	48	☿ ♎

LATITUDES

Date	Mercury ☿ ° ′	Venus ♀ ° ′	Mars ♂ ° ′	Jupiter ♃ ° ′	Saturn ♄ ° ′	Uranus ♅ ° ′	Neptune ♆ ° ′	Pluto ⯓ ° ′
01	01 N 46	00 N 29	00 S 10	01 S 18	00 S 42	00 N 43	01 N 47	12 N 56
04	01 46	00 16	00 08	01 19	00 43	00 43	01 46	12 56
07	01 40	00 N 03	00 05	01 19	00 43	00 43	01 46	12 56
10	01 29	00 S 11	00 S 03	01 20	00 43	00 43	01 46	12 55
13	01 15	00 26	00 00	01 21	00 43	00 43	01 46	12 55
16	00 57	00 41	00 N 03	01 21	00 43	00 43	01 46	12 55
19	00 36	00 58	00 06	01 22	00 44	00 43	01 46	12 55
22	00 N 14	01 14	00 08	01 22	00 44	00 43	01 45	12 55
25	00 S 10	01 31	00 11	01 22	00 44	00 43	01 45	12 55
28	00 36	01 49	00 14	01 23	00 44	00 43	01 45	12 55
31	01 S 02	02 S 07	00 N 17	01 S 23	00 S 44	00 N 43	01 N 45	12 N 55

DATA

Julian Date	2437878
Delta T	+34 seconds
Ayanamsa	23° 19′ 52″
Synetic vernal point	05° ♓ 47′ 08″
True obliquity of ecliptic	23° 26′ 33″

MOON'S PHASES, APSIDES AND POSITIONS ☽

Date	h	m	Phase	Longitude ° ′	Eclipse Indicator
08	15	55	☽	15 ♏ 37	
15	20	09	○	22 ♒ 30	
22	10	26	☾	28 ♉ 51	
30	03	09	●	06 ♍ 16	

Day	h	m		
05	05	48	Apogee	
17	08	30	Perigee	
05	18	17	0S	
12	22	44	Max dec	20° S 31′
19	01	51	0N	
25	15	07	Max dec	20° N 32′

LONGITUDES

Date	Chiron ⚷ ° ′	Ceres ⚳ ° ′	Pallas ⚴ ° ′	Juno ⚵ ° ′	Vesta ⚶ ° ′	Black Moon Lilith ⚸ ° ′
01	09 ♓ 52	23 ♋ 48	10 ♊ 41	27 ♋ 16	29 ♋ 06	00 ♎ 41
11	09 ♓ 27	28 ♋ 10	15 ♊ 22	02 ♌ 30	03 ♌ 33	01 ♎ 48
21	09 ♓ 00	02 ♌ 30	20 ♊ 03	07 ♌ 38	07 ♌ 59	02 ♎ 55
31	08 ♓ 32	06 ♌ 49	24 ♊ 40	12 ♌ 39	12 ♌ 24	04 ♎ 02

ASPECTARIAN

h m	Aspects
01 Wednesday	
03 45	☽ ⊥ ♀
04 11	☽ ⚹ ♂
04 20	☿ ⊼ ♃
10 11	☿ # ♄
12 56	☉ ⊻ ⯓
16 43	☽ ⊻ ♀
02 Thursday	
05 13	☽ Q ♂
05 27	☽ Q ♆
06 56	♀ ♇ ♄
07 01	☽ ☌ ♅
07 11	☽ # ♆
09 28	♂ ± ♆
14 45	☽ ∥ ♅
23 43	☽ ⊼ ♄
03 Friday	
01 36	☽ ☌ ⯓
04 41	☽ ⊻ ☉
05 26	☽ ⚹ ♆
06 08	☽ ☍ ♃
11 41	☽ ± ♄
14 07	☉ □ ♆
14 58	☽ # ♃
16 26	♂ Q ♅
16 48	☽ ⊻ ☿
17 48	☽ ⊥ ☉
19 24	☽ □ ♂
21 21	☉ ⊼ ♃
23 38	☉ ∥ ☿
04 Saturday	
05 44	☽ ♇ ♄
07 21	☽ ⊥ ☿
10 51	☽ ☌ ♀
11 42	☽ ∠ ♆
13 35	☽ ∠ ☉
15 06	☿ ⚹ ♂
19 39	☽ ⊻ ♅
20 02	♀ ∠ ♆
05 Sunday	
05 14	☽ ∠ ☿
05 57	☽ ⊥ ♆
07 52	☽ ⊥ ♅
08 45	☽ ∥ ♀
11 56	☽ △ ♄
14 22	☽ ⊻ ⯓
18 07	☽ ⊻ ♆
18 18	☽ ⊼ ♃
22 40	☽ ⚹ ☉
06 Monday	
01 45	☽ # ♀
02 12	☽ ∠ ♅
02 33	☽ ⊥ ⯓
02 50	☿ ⊥ ♀
06 21	☽ ± ♃
11 47	☽ △ ♂
13 41	♃ △ ♆
17 33	☽ ⚹ ☿
20 47	☽ ∠ ⯓
07 Tuesday	
00 21	☽ ♇ ♃
01 01	☽ Q ☉
05 42	☽ ⊻ ♀
08 36	☽ ⚹ ♅
18 54	☽ ⊥ ♀
19 41	☽ ♇ ♂
22 10	☽ Q ☿
22 36	☽ ∥ ♃
23 57	☽ □ ♄
08 Wednesday	
02 51	☽ ⚹ ⯓
06 02	☽ △ ♃
06 24	☽ ☌ ♆
08 36	☽ Q ♅
14 28	☽ ∠ ♀
14 32	☽ ± ♂
15 19	♀ ⊻ ♅
15 55	☽ □ ☉
20 31	☽ # ♅
09 Thursday	
02 30	☽ Q ⯓
02 56	☽ ⊼ ♂
05 03	☽ ∥ ♆
08 36	☽ # ☿
11 01	☽ Q ♄
15 44	☽ □ ☿
19 46	☽ □ ♅
22 19	☽ ⚹ ♀
23 29	☽ # ☉
10 Friday	
01 08	☿ # ♆
03 47	☿ Q ♆
09 45	☽ ⚹ ♄
12 59	☽ □ ⯓
15 21	☽ □ ♃
16 14	☽ ⊻ ♆
18 01	♂ ♇ ♄
20 07	☿ ☌ ♅
22 57	☽ Q ♀
11 Saturday	
03 22	☽ ⊥ ♆
05 42	☽ △ ☉
10 02	☽ ∥ ♄
13 22	☽ ∠ ♄
14 30	☽ ☍ ♂
19 47	☽ ∠ ♆
20 23	☿ ∥ ♅
12 Sunday	
00 02	☽ Q ♃
02 23	☽ # ⯓
03 31	☽ △ ♅
05 29	☽ ⊥ ♄
08 12	☽ △ ☿
10 26	☽ □ ♀
10 51	☽ ♇ ☉
16 02	☽ ⊻ ♄
19 28	☽ △ ⯓
21 05	☽ ⚹ ♃
22 24	☽ ⚹ ♆
13 Monday	
03 46	☽ ± ☉
04 46	♀ ⊥ ♆
05 58	☽ ♇ ♅
14 09	☽ ♇ ☿
14 50	☽ ⊼ ☉
19 03	☽ Q ♆
19 07	☽ # ⯓
21 20	☽ ♇ ⯓
21 25	☽ ± ♅
21 43	☽ ⊼ ♂
22 38	☽ ∠ ♃
14 Tuesday	
00 20	☿ ⊻ ♀
07 30	☽ ± ☿
07 34	☽ ⊼ ♅
08 18	☽ ± ♂
09 02	☽ ∥ ♄
09 26	☉ ∠ ♀
12 28	☽ ± ⯓
13 11	♀ ⊥ ♅
13 36	☽ ⊥ ♃
17 43	☿ Q ♂
17 56	☽ △ ♀
18 51	☽ ⊼ ☿
18 51	☽ ☌ ♄
18 51	☿ ⊼ ♄
22 26	☽ ⊼ ⯓
23 27	☽ ⊻ ♃
23 57	☽ ♇ ♂
15 Wednesday	
01 08	☽ □ ♆
05 43	♀ △ ♄
20 09	☽ ☍ ☉
20 25	☽ ♇ ♀
21 01	☿ # ♃
16 Thursday	
01 33	☽ △ ♂
01 59	☿ ☌ ⯓
02 57	♂ ♇ ♆
04 18	☽ # ☉
07 17	☽ ∥ ♆
07 55	☿ ☍ ♃
08 54	☽ ☍ ♅
12 04	☽ ± ♀
15 26	☽ # ♅
19 22	☽ ⊻ ♄
22 26	☽ ⊼ ♀
23 12	☽ ☍ ⯓
23 42	☽ ☌ ♃
17 Friday	
01 12	☿ ⚹ ♆
01 46	☽ △ ♆
01 50	☽ ☍ ☿
04 57	☽ ⊥ ♄
06 01	☽ ∥ ♃
09 37	♀ ⊻ ⯓
12 21	☽ # ☿
14 34	♀ ⊼ ♃
17 15	☉ # ♆
19 18	☽ ∠ ♄
23 41	☽ ⊼ ☉
18 Saturday	
01 49	☽ ♇ ♆
03 42	☽ ∥ ♀
03 55	☽ □ ♂
05 07	☿ ± ♄
08 49	♂ Q ⯓
09 15	☽ ⊼ ♅
09 59	☽ ± ☉
16 19	☽ ± ♆
18 57	☽ ± ♅
19 19	☽ ⚹ ♄
22 15	♀ ⊻ ♆
22 41	♃ ☍ ⯓
23 30	☽ ⊻ ♃
23 30	☽ ⊼ ⯓
19 Sunday	
01 31	☽ ♇ ☉
02 02	☽ ⊼ ♆
02 19	☽ ☍ ♀
08 04	☽ ⊼ ☿
09 10	☽ ⊥ ♃
09 16	☽ ± ⯓
09 37	☽ ♇ ♅
10 21	☽ Q ♂
14 43	☽ Q ♄
19 05	☽ ± ☿
23 46	☽ ∠ ♃
20 Monday	
00 03	☽ ♇ ⯓
03 47	☽ △ ☉
05 18	☽ # ♀
06 08	☽ ∥ ☿
06 59	☽ ⚹ ♂
09 08	☿ # ♀
10 24	☽ △ ♅
11 47	☽ ♇ ☿
20 23	☽ □ ♄
23 55	☽ # ♃
21 Tuesday	
00 32	☽ ⚹ ♃
01 06	☽ △ ⯓
03 39	☽ ☍ ♆
07 51	☽ ⊼ ♀
09 21	☽ ∠ ♂
13 48	☽ ∥ ♅
15 19	☽ ∥ ☉
16 15	☽ △ ☿
18 55	☽ ± ♀
20 48	☽ Q ♃
23 22	☽ # ♆
22 Wednesday	
01 32	☽ ⊥ ♂
10 26	☽ □ ☉
11 02	☉ ∥ ♅
11 47	☽ ♇ ♀
12 30	☽ ⊻ ♂
13 20	☉ Q ♆
13 50	☽ □ ♅
17 46	♀ ± ♃
23 54	☽ △ ♄
23 Thursday	
04 03	☽ □ ♃
05 06	♂ ± ♄
05 16	☽ □ ⯓
07 54	☽ ⊼ ♆
08 25	♀ ⊥ ⯓
16 27	♀ ∠ ♅
16 40	☽ △ ♀
16 47	☿ ♇ ♄
18 42	☽ ± ♆
19 31	♂ ⚹ ♅
20 58	☽ Q ☉
22 07	☽ Q ♅
24 Friday	
02 46	☽ ♇ ♄
04 03	☽ # ♄
04 06	☽ □ ☿
11 09	☽ ♇ ♆
13 19	☽ ∥ ⯓
13 40	☉ ☌ ♅
14 01	☽ Q ⯓
19 14	☽ ± ♄
20 16	☽ ⚹ ♅
20 46	☽ ⚹ ☉
21 26	☽ ☌ ♂
25 Saturday	
06 22	☽ ⊼ ♄
10 28	☽ △ ♃
12 27	☽ ⚹ ⯓
15 10	☽ △ ♆
17 56	☽ Q ☿
22 01	☉ ⚹ ♂
26 Sunday	
00 36	☽ ∠ ♅
03 18	☽ ∠ ☉
05 15	☽ □ ♀
14 42	☽ ♇ ♃
17 06	☽ ∠ ⯓
17 55	☽ ⊥ ♅
18 01	☽ ∥ ⯓
19 44	☿ ∠ ♆
19 50	☽ ⚹ ☿
22 01	☽ ⊥ ☉
27 Monday	
02 20	☽ # ♄
05 34	☽ ⊻ ♅
07 55	☽ ± ♃
09 37	☽ ⊻ ♂
10 34	☽ ⊥ ⯓
10 38	☽ ⊻ ☉
15 32	☽ ☍ ♄
19 08	☽ Q ♀
19 30	☽ ⊼ ♃
21 58	☽ ⊥ ♂
22 20	☽ ⊻ ⯓
28 Tuesday	
01 04	☽ □ ♆
04 52	☽ ∠ ☿
11 14	☉ # ♃
12 19	☉ # ♀
13 21	♀ ∥ ♃
16 40	☽ ∠ ♂
20 54	☽ ⚹ ♀
29 Wednesday	
00 57	☽ ⊥ ☿
12 45	☽ # ♆
12 46	☽ Q ♆
14 26	☽ ⊻ ☿
17 02	☽ ☌ ♅
21 40	☉ ⊼ ♄
30 Thursday	
00 12	☽ ⚹ ♂
01 26	☽ ∥ ♅
02 41	☽ ⊼ ♄
03 09	☽ ☌ ☉
05 32	☽ ∠ ♀
06 23	☽ ☍ ♃
08 31	☽ # ♀
10 07	☽ ☌ ⯓
12 51	☽ ⚹ ♆
13 20	☽ # ♃
14 32	☿ ⊻ ♅
14 40	☽ ± ♄
14 42	♀ ♇ ♃
19 20	☽ ∥ ☉
31 Friday	
01 18	☽ ⊥ ♀
01 40	☽ Q ♂
08 44	☽ ♇ ♄
14 31	☽ ⊻ ♀
14 48	☉ ☍ ♃
19 11	☽ ∠ ♆
19 41	♂ ⊼ ♄

All ephemeris data is given at 12.00 UT and the Moon's longitude is additionally given for 24.00 UT

SEPTEMBER 1962

Date	Sidereal time h m s	LONGITUDES Sun ☉ ° ' "	Moon ☽ ° ' "	Moon ☽ 24.00 ° ' "	Mercury ☿ ° '	Venus ♀ ° '	Mars ♂ ° '	Jupiter ♃ ° '	Saturn ♄ ° '	Uranus ♅ ° '	Neptune ♆ ° '	Pluto ♇ ° '
01	10 40 42	08 ♍ 33 59	04 ♎ 26 01	10 ♎ 21 03	03 ♎ 46	24 ♎ 42	06 ♋ 22	07 ♓ 36	05 ♒ 54	01 ♍ 24	11 ♏ 09	09 ♍ 49
02	10 44 38	09 32 05	16 ♎ 16 09	22 ♎ 11 38	05 02	25 41	07 00	07 R 28	05 R 51	01 28	11 10	09 51
03	10 48 35	10 30 13	28 ♎ 07 52	04 ♏ 05 12	06 17	26 39	07 37	07 20	05 48	01 31	11 11	09 53
04	10 52 31	11 28 22	10 ♏ 04 06	16 ♏ 04 58	07 30	27 37	08 15	07 12	05 44	01 35	11 12	09 55
05	10 56 28	12 26 33	22 ♏ 08 18	28 ♏ 14 37	08 41	28 34	08 52	07 04	05 41	01 39	11 14	09 57
06	11 00 24	13 24 45	04 ♐ 24 25	10 ♐ 38 14	09 50	29 ♎ 31	09 29	06 56	05 38	01 42	11 15	09 59
07	11 04 21	14 22 58	16 ♐ 56 38	23 ♐ 20 07	10 57	00 ♏ 28	10 06	06 48	05 35	01 46	11 17	10 02
08	11 08 17	15 21 14	29 ♐ 49 13	06 ♑ 24 21	12 02	01 24	10 43	06 41	05 32	01 50	11 18	10 04
09	11 12 14	16 19 30	13 ♑ 05 56	19 ♑ 54 16	13 04	02 19	11 19	06 33	05 30	01 54	11 19	10 06
10	11 16 11	17 17 48	26 ♑ 49 34	03 ♒ 51 50	14 04	03 14	11 56	06 25	05 27	01 57	11 21	10 08
11	11 20 07	18 16 08	11 ♒ 01 01	18 ♒ 16 46	15 02	04 09	12 33	06 18	05 24	02 01	11 22	10 10
12	11 24 04	19 14 29	25 ♒ 38 38	03 ♓ 05 53	15 56	05 03	13 09	06 10	05 22	02 04	11 24	10 12
13	11 28 00	20 12 52	10 ♓ 37 39	18 ♓ 12 49	16 48	05 56	13 45	06 03	05 19	02 08	11 25	10 14
14	11 31 57	21 11 17	25 ♓ 50 11	03 ♈ 28 26	17 37	06 49	14 21	05 55	05 17	02 12	11 27	10 16
15	11 35 53	22 09 44	11 ♈ 06 10	18 ♈ 42 04	18 22	07 41	14 57	05 48	05 14	02 15	11 29	10 18
16	11 39 50	23 08 12	26 ♈ 14 51	03 ♉ 43 23	19 03	08 33	15 33	05 41	05 12	02 19	11 30	10 20
17	11 43 46	24 06 43	11 ♉ 06 43	18 ♉ 24 04	19 41	09 24	16 09	05 34	05 10	02 22	11 32	10 22
18	11 47 43	25 05 16	25 ♉ 34 55	02 ♊ 38 54	20 14	10 14	16 45	05 27	05 08	02 26	11 34	10 24
19	11 51 40	26 03 51	09 ♊ 35 53	16 ♊ 25 55	20 43	11 04	17 20	05 20	05 06	02 30	11 35	10 26
20	11 55 36	27 02 28	23 ♊ 09 11	29 ♊ 46 00	21 07	11 53	17 56	05 13	05 04	02 33	11 37	10 28
21	11 59 33	28 01 08	06 ♋ 16 45	12 ♋ 41 55	21 26	12 41	18 31	05 06	05 02	02 37	11 39	10 30
22	12 03 29	28 59 50	19 ♋ 02 00	25 ♋ 17 33	21 38	13 28	19 06	04 59	05 00	02 40	11 40	10 32
23	12 07 26	29 ♍ 58 34	01 ♌ 29 05	07 ♌ 37 08	21 45	14 15	19 41	04 53	04 59	02 43	11 42	10 34
24	12 11 22	00 ♎ 57 20	13 ♌ 42 13	19 ♌ 44 47	21 R 46	15 00	20 16	04 47	04 57	02 47	11 44	10 36
25	12 15 19	01 56 08	25 ♌ 45 17	01 ♍ 44 08	21 40	15 45	20 51	04 40	04 56	02 50	11 46	10 38
26	12 19 15	02 54 59	07 ♍ 41 42	13 ♍ 38 17	21 26	16 29	21 25	04 34	04 54	02 54	11 48	10 40
27	12 23 12	03 53 51	19 ♍ 34 10	25 ♍ 29 39	21 06	17 12	22 00	04 28	04 53	02 57	11 49	10 42
28	12 27 09	04 52 46	01 ♎ 24 55	07 ♎ 20 12	20 37	17 54	22 34	04 22	04 52	03 00	11 51	10 44
29	12 31 05	05 51 43	13 ♎ 15 42	19 ♎ 11 34	20 02	18 35	23 08	04 17	04 51	03 04	11 53	10 46
30	12 35 02	06 ♎ 50 41	25 ♎ 08 01	01 ♏ 05 15	19 ♎ 19	19 ♏ 15	23 ♋ 42	04 ♓ 11	04 ♒ 50	03 ♍ 07	11 ♏ 55	10 ♍ 48

Date	Moon True ☊ ° '	Moon Mean ☊ ° '	Moon ☽ Latitude ° '	DECLINATIONS Sun ☉ ° '	Moon ☽ ° '	Mercury ☿ ° '	Venus ♀ ° '	Mars ♂ ° '	Jupiter ♃ ° '	Saturn ♄ ° '	Uranus ♅ ° '	Neptune ♆ ° '	Pluto ♇ ° '
01	08 ♌ 40	07 ♌ 07	04 N 18	08 N 22	02 N 11	02 S 35	11 S 38	23 N 35	10 00	19 S 31	11 N 39	13 S 31	19 N 49
02	08 R 35	07 04	04 49	08 00	01 S 57	03 14	12 05	23 35	10 03	19 32	11 37	13 31	19 48
03	08 30	07 01	05 07	07 38	06 02	03 51	12 32	23 34	10 06	19 32	11 36	13 32	19 47
04	08 25	06 58	05 12	07 16	09 54	04 28	12 59	23 32	10 09	19 33	11 35	13 32	19 46
05	08 22	06 55	05 03	06 54	13 25	05 04	13 25	23 31	10 12	19 34	11 34	13 33	19 46
06	08 20	06 51	04 40	06 31	16 26	05 40	13 52	23 29	10 15	19 35	11 32	13 33	19 45
07	08 D 19	06 48	04 04	06 09	18 45	06 14	14 17	23 28	10 18	19 36	11 31	13 34	19 44
08	08 20	06 45	03 14	05 46	20 12	06 47	14 43	23 26	10 21	19 36	11 30	13 34	19 43
09	08 22	06 42	02 13	05 24	20 35	07 20	15 08	23 24	10 24	19 37	11 28	13 34	19 43
10	08 23	06 39	01 N 02	05 01	19 46	07 51	15 33	23 22	10 27	19 38	11 27	13 35	19 42
11	08 R 23	06 36	00 S 14	04 38	17 42	08 21	15 58	23 19	10 30	19 39	11 26	13 35	19 41
12	08 22	06 32	01 32	04 16	14 25	08 49	16 22	23 17	10 33	19 39	11 24	13 36	19 41
13	08 19	06 29	02 46	03 53	10 08	09 17	16 46	23 14	10 35	19 40	11 23	13 37	19 40
14	08 14	06 26	03 48	03 30	05 S 09	09 42	17 10	23 12	10 38	19 40	11 22	13 37	19 39
15	08 08	06 23	04 35	03 07	00 N 10	10 06	17 33	23 09	10 41	19 41	11 21	13 38	19 39
16	08 01	06 20	05 02	02 43	05 26	10 28	17 56	23 06	10 43	19 42	11 19	13 38	19 38
17	07 55	06 17	05 08	02 20	10 17	10 49	18 19	23 03	10 46	19 42	11 18	13 39	19 37
18	07 50	06 13	04 54	01 57	14 24	11 07	18 41	22 59	10 49	19 43	11 17	13 39	19 37
19	07 47	06 10	04 22	01 34	17 34	11 23	19 02	22 56	10 51	19 43	11 15	13 40	19 36
20	07 46	06 07	03 37	01 11	19 39	11 36	19 24	22 52	10 54	19 44	11 14	13 40	19 35
21	07 D 46	06 04	02 41	00 47	20 36	11 47	19 45	22 49	10 56	19 44	11 13	13 41	19 35
22	07 47	06 01	01 39	00 24	20 27	11 55	20 05	22 45	10 58	19 45	11 12	13 41	19 34
23	07 48	05 57	00 S 34	00 N 01	19 17	12 00	20 25	22 41	11 01	19 45	11 11	13 42	19 34
24	07 R 48	05 54	00 N 32	00 S 23	17 13	12 02	20 45	22 37	11 03	19 45	11 09	13 43	19 33
25	07 46	05 51	01 35	00 46	14 26	12 00	21 04	22 33	11 05	19 46	11 08	13 43	19 32
26	07 42	05 48	02 33	01 10	11 03	11 54	21 22	22 29	11 07	19 46	11 07	13 44	19 32
27	07 35	05 45	03 24	01 33	07 16	11 44	21 40	22 24	11 09	19 46	11 06	13 44	19 31
28	07 26	05 42	04 06	01 56	03 N 12	11 29	21 58	22 20	11 11	19 47	11 05	13 45	19 31
29	07 15	05 38	04 38	02 20	00 S 58	11 11	22 15	22 15	11 13	19 47	11 03	13 46	19 30
30	07 ♌ 04	05 ♌ 35	04 N 57	02 S 43	05 S 07	10 S 48	22 S 32	22 N 11	11 S 15	19 S 47	11 N 02	13 S 46	19 N 30

ZODIAC SIGN ENTRIES

Date	h	m	Planets
01	03	01	☽ ♎
03	15	46	☽ ♏
06	03	26	☽ ♐
07	00	11	♀ ♏
08	12	20	☽ ♑
10	17	26	☽ ♒
12	19	02	☽ ♓
14	18	33	☽ ♈
16	18	00	☽ ♉
18	19	29	☽ ♊
21	00	26	☽ ♋
23	09	07	☽ ♌
23	12	35	☉ ♎
25	20	31	☽ ♍
28	09	08	☽ ♎
30	21	49	☽ ♏

LATITUDES

Date	Mercury ☿ ° '	Venus ♀ ° '	Mars ♂ ° '	Jupiter ♃ ° '	Saturn ♄ ° '	Uranus ♅ ° '	Neptune ♆ ° '	Pluto ♇ ° '
01	01 S 11	02 S 13	00 N 18	01 S 23	00 S 44	00 N 43	01 N 45	12 N 55
04	01 38	02 31	00 21	01 23	00 44	00 43	01 45	12 55
07	02 04	02 50	00 24	01 23	00 44	00 43	01 45	12 56
10	02 30	03 09	00 27	01 23	00 45	00 43	01 44	12 56
13	02 54	03 28	00 31	01 23	00 45	00 43	01 44	12 56
16	03 15	03 47	00 34	01 23	00 45	00 43	01 44	12 57
19	03 33	04 06	00 37	01 23	00 45	00 43	01 44	12 57
22	03 46	04 24	00 40	01 23	00 45	00 43	01 44	12 58
25	03 50	04 43	00 44	01 23	00 45	00 43	01 44	12 58
28	03 43	05 00	00 47	01 23	00 45	00 43	01 44	12 59
31	03 S 20	05 S 17	00 N 51	01 S 22	00 S 45	00 N 43	01 N 44	13 N 00

DATA

Julian Date	2437909
Delta T	+34 seconds
Ayanamsa	23° 19' 55"
Synetic vernal point	05° ♓ 47' 05"
True obliquity of ecliptic	23° 26' 34"

MOON'S PHASES, APSIDES AND POSITIONS ☽

Date	h	m	Phase	Longitude ° '	Eclipse Indicator
07	06	44	☽	14 ♐ 10	
14	04	11	○	20 ♓ 52	
20	19	36	☾	27 ♊ 21	
28	19	39	●	05 ♎ 12	

Day	h	m		
01	19	36	Apogee	
14	16	20	Perigee	
29	01	03	Apogee	
02	00	41	0S	
09	08	02	Max dec	20° S 36'
15	11	13	0N	
21	20	31	Max dec	20° N 40'
29	06	26	0S	

LONGITUDES

Date	Chiron ⚷ ° '	Ceres ⚳ ° '	Pallas ⚴ ° '	Juno ⚵ ° '	Vesta ⚶ ° '	Black Moon Lilith ⚸ ° '
01	08 ♓ 29	07 ♌ 14	25 ♊ 07	13 ♌ 08	12 ♌ 50	04 ♎ 09
11	08 ♓ 00	11 ♌ 30	29 ♊ 38	18 ♌ 00	17 ♌ 13	05 ♎ 16
21	07 ♓ 33	15 ♌ 33	03 ♋ 59	22 ♌ 43	21 ♌ 43	06 ♎ 23
31	07 ♓ 08	19 ♌ 46	08 ♋ 26	27 ♌ 17	25 ♌ 51	07 ♎ 30

All ephemeris data is given at 12.00 UT and the Moon's longitude is additionally given for 24.00 UT

ASPECTARIAN

h m	Aspects
01 Saturday	
05 49	☽ ⊻ ♅
10 01	☽ # ☿
10 29	☽ ☌ ☿
12 44	♀ # ♅
13 27	☽ ⊥ ♆
14 53	♀ ∠ ♇
14 58	☽ △ ♄
16 09	☽ □ ♂
18 03	☽ ⊥ ♅
18 20	☽ ⊼ ♃
21 08	☽ ⊻ ☉
22 57	☽ ⊻ ♇
02 Sunday	
01 38	☽ ⊻ ♆
06 22	☽ ± ♃
10 23	☽ ⊥ ☉
11 09	☽ ⊥ ♇
12 23	☽ ∠ ♅
14 25	☿ ⊥ ♆
20 10	☉ ☌ ♇
20 45	☽ ∥ ☿
03 Monday	
00 24	☽ bQ ♃
02 49	♂ △ ♃
02 52	☿ △ ♄
05 26	☽ ∠ ♇
06 13	☽ ∠ ☉
08 45	☽ ☌ ♀
18 52	☽ ✱ ♅
20 52	☽ # ☉
04 Tuesday	
03 22	☽ □ ♄
05 18	☉ ✱ ♆
06 17	☽ ⊻ ☿
06 19	☽ △ ♃
06 36	☿ ⊼ ♃
08 09	☽ △ ♂
11 42	☽ ✱ ♇
13 39	☽ ∥ ♃
13 48	☿ ⊥ ♅
14 17	☽ ☌ ♆
15 03	☽ ✱ ☉
18 17	☉ ± ♄
19 04	☽ Q ♅
19 37	☽ ⊥ ☿
23 00	☽ # ♅
05 Wednesday	
11 38	☽ Q ♇
11 59	☽ ∥ ♀
12 52	☽ ∥ ♆
15 03	☽ Q ♄
15 22	☽ ∠ ☿
15 35	☽ bQ ♂
16 56	☽ Q ☉
18 43	♀ ∥ ♆
19 58	☿ □ ♂
06 Thursday	
01 42	☽ ⊻ ♀
06 44	☽ □ ♅
10 07	☽ ± ♂
14 20	☽ ⊥ ♀
14 22	☽ ✱ ♄
15 29	☿ ⊻ ♇
16 50	☽ □ ♃
22 17	☽ ⊼ ♂
22 48	☽ □ ♇
23 29	☽ ✱ ☿
07 Friday	
01 12	☽ ⊻ ♆
06 44	☽ □ ☉
08 58	☽ ∠ ♀
09 04	♂ ✱ ♇
09 47	☉ # ☿
12 38	☽ ⊥ ♆
18 50	☽ ∠ ♄
19 21	☿ ⊻ ♆
08 Saturday	
00 00	☽ ∥ ♄
00 19	☽ Q ☿
02 11	☽ # ♇
02 36	☽ Q ♃
05 30	☽ ∠ ♆
11 29	☽ ⊥ ♄
15 07	☽ ✱ ♀
15 42	☽ △ ♅
22 24	☽ ⊻ ♄
09 Sunday	
00 01	♀ ✱ ♅
00 22	☽ ✱ ♃
01 12	☿ ± ♃
06 38	☽ △ ♇
08 41	☽ ☍ ♂
08 50	☽ ✱ ♆
11 57	☽ □ ☿
12 00	♂ △ ♆
14 20	☽ Q ♀
18 09	☽ △ ☉
18 45	☽ bQ ♅
10 Monday	
02 44	☽ ∠ ♃
05 59	☽ Q ♆
09 04	☽ bQ ♇
10 29	☽ ± ♅
13 11	☽ # ♇
14 14	☽ ∥ ♄
18 06	☽ ⊥ ♃
20 48	☽ ⊼ ♅
22 02	☽ bQ ☉
23 42	☽ □ ♀
11 Tuesday	
00 29	☽ ± ♇
02 38	☽ ☌ ♄
04 10	☽ ⊻ ♃
10 35	☽ ⊼ ♇
12 36	☽ □ ♆
14 14	☽ ± ☉
14 39	☽ ⊼ ♂
19 06	☽ △ ☿
12 Wednesday	
00 12	☽ ∥ ♀
00 50	☽ ⊼ ☉
00 58	☽ ± ♂
16 13	☽ bQ ♂
17 01	☽ ∥ ♆
19 19	☿ ⊥ ♇
19 58	♀ □ ♄
21 04	☽ bQ ☿
22 24	☽ ☍ ♅
13 Thursday	
03 34	☽ ⊻ ♄
04 04	☽ △ ♀
04 46	☽ ☌ ♃
05 27	☽ # ♅
09 42	☽ ∥ ♃
11 22	☽ ☍ ♇
12 18	☽ ± ☿
13 05	☽ ⊥ ♄
13 16	☽ △ ♆
14 22	♄ # ♇
14 26	☉ bQ ♄
14 31	♀ △ ♃
15 59	☽ ∥ ☿
17 10	☽ △ ♂
22 20	☽ ⊼ ☿
22 26	☿ ∠ ♅
14 Friday	
03 16	☽ ∠ ♄
04 11	☽ ☍ ☉
05 18	☽ bQ ♀
12 58	☽ bQ ♆
20 08	☽ # ☉
20 18	☽ ± ♀
22 02	☽ ⊼ ♅
15 Saturday	
02 48	☽ ✱ ♄
03 08	☽ ± ♆
03 43	☽ ⊻ ♃
06 18	☽ ⊼ ♀
07 30	☽ ± ♅
10 44	☽ ⊼ ♇
12 35	☽ ⊼ ♆
13 05	☽ ⊥ ♃
18 20	☽ □ ♂
20 13	☽ ± ♇
21 40	☽ Q ♄
21 45	☽ bQ ♅
16 Sunday	
00 02	☽ ☍ ☿
00 20	☽ ∥ ☉
03 12	☽ ∠ ♃
06 42	☽ ⊼ ☉
10 32	☽ bQ ♇
16 57	☽ ± ☉
21 46	☽ △ ♅
17 Monday	
00 13	☽ Q ♂
02 21	☽ □ ♄
03 02	☽ ✱ ♃
08 01	☿ ∥ ♃
08 30	☽ bQ ☉
09 02	☽ ☍ ♀
10 47	☽ △ ♇
12 41	☽ ☍ ♆
14 37	☽ # ♃
15 04	☽ # ☿
17 29	☽ ∥ ♅
20 38	☽ ✱ ♂
22 31	☽ Q ♃
18 Tuesday	
02 43	☽ ⊼ ☿
07 14	☽ # ♆
11 06	☽ △ ☉
13 09	☽ ± ☿
17 00	♀ ✱ ♇
19 52	☿ bQ ♃
22 55	☽ ∠ ♂
23 41	☽ □ ♅
19 Wednesday	
01 13	☿ # ♅
04 13	☽ △ ♄
04 40	☽ □ ♃
05 03	☽ bQ ☿
11 52	☉ ∠ ♀
13 28	☽ □ ♇
14 43	☽ ⊼ ♀
15 10	☽ ⊥ ♂
15 29	☽ ⊼ ♆
18 54	♂ ∠ ♅
20 Thursday	
01 14	☉ ∠ ♆
01 58	☽ ± ♀
02 05	☽ ± ♆
02 14	☽ ⊻ ♂
03 57	♀ ☌ ♆
06 28	☽ bQ ♄
07 10	☽ # ♀
07 19	☽ Q ♅
08 14	☽ △ ☿
11 02	☽ ∥ ♇
13 13	☽ # ♄
18 16	☽ bQ ♆
19 10	☽ bQ ♀
19 36	☽ □ ☉
21 39	☽ Q ♇
22 41	☽ ± ♄
21 Friday	
01 03	♀ # ♇
05 11	☽ ✱ ♅
09 42	☽ ⊼ ♄
09 50	☽ △ ♃
11 31	♀ ∥ ♄
19 54	☽ ✱ ♇
22 02	☽ △ ♆
22 Saturday	
00 45	☽ △ ♀
07 39	♃ ⊻ ♄
07 48	☽ Q ☉
09 23	☽ ∠ ♅
12 08	☽ ☌ ♂
13 49	☽ bQ ♃
17 03	☽ □ ☿
19 31	☽ # ♀
23 Sunday	
00 30	☽ ∠ ♇
02 43	☽ ⊥ ♅
04 31	☽ # ♄
06 59	☽ ± ♃
07 40	☽ ∥ ♇
08 49	☽ ✱ ☉
14 26	☽ ⊻ ♅
18 02	☽ ⊥ ♇
18 35	☽ ⊼ ♃
18 48	☽ ☍ ♄
18 52	♂ bQ ♃
24 Monday	
01 52	☿ St R
04 14	☽ Q ☿
04 16	♀ Q ♅
05 51	☽ ⊻ ♇
08 06	☽ □ ♆
14 45	☽ □ ♀
16 51	☽ ∠ ☉
25 Tuesday	
01 42	☽ ⊻ ♂
03 55	☽ ✱ ☿
12 24	☽ ⊥ ☉
14 18	☽ ⊥ ♂
17 19	☽ # ♆
20 03	☽ Q ♆
26 Wednesday	
01 31	☽ ⊻ ☉
02 17	☽ ☌ ♅
05 07	☽ Q ♀
05 45	☽ ☍ ♃
06 07	☽ # ☿
06 23	☽ ⊼ ♄
09 06	♃ # ♅
09 18	☽ ∠ ♂
09 31	☽ ∠ ☿
11 27	☉ ⊻ ♅
11 33	☽ # ♃
11 35	☽ ∥ ♅
12 31	☿ □ ♂
18 01	☽ ☌ ♇
18 28	☽ ± ♄
20 18	☽ ✱ ♆
27 Thursday	
03 14	☽ ⊥ ☿
06 54	☽ ✱ ♀
12 38	☽ bQ ♄
14 59	☽ ⊻ ☿
17 09	☽ ✱ ♂
17 36	☉ Q ♂
28 Friday	
00 45	☉ ⊼ ♃
02 44	☽ ∠ ♆
11 40	☉ △ ♄
15 12	☽ ∠ ♀
15 14	☽ ⊻ ♅
17 57	☽ ⊼ ♃
18 41	☽ # ☉
18 42	☽ Q ♂
18 59	☽ △ ♄
19 39	☽ ☌ ☉
21 01	☽ ⊥ ♆
29 Saturday	
03 27	☽ ⊥ ♅
06 00	☽ ± ♃
06 56	☽ ⊻ ♇
09 03	☿ ∥ ♃
09 13	☽ ⊻ ♆
10 33	☽ ⊥ ♀
12 17	♀ # ♂
12 38	☉ ⊥ ♆
19 07	☽ ⊥ ♇
20 34	☿ # ♅
20 36	☽ ∥ ☉
21 45	☽ ∠ ♅
23 25	☽ ⊻ ♀
30 Sunday	
00 05	☽ bQ ♃
00 56	☽ ☌ ☿
08 57	☽ □ ♂
13 01	☿ ⊻ ♀
13 21	☽ ∠ ♇
16 22	☿ bQ ♃

OCTOBER 1962

Date	Sidereal time h m s	LONGITUDES Sun ☉ ° ′ ″	Moon ☽ ° ′ ″	Moon ☽ 24.00 ° ′ ″	Mercury ☿ ° ′	Venus ♀ ° ′	Mars ♂ ° ′	Jupiter ♃ ° ′	Saturn ♄ ° ′	Uranus ♅ ° ′	Neptune ♆ ° ′	Pluto ♇ ° ′
01	12 38 58	07 ♎ 49 42	07 ♏ 03 27	13 ♏ 02 53	18 ♎ 29	19 ♏ 54	24 ♋ 16	04 ♓ 06	04 ♒ 49	03 ♍ 10	11 ♏ 57	10 ♍ 50
02	12 42 55	08 48 44	19 ♏ 03 47	25 ♏ 06 28	17 R 33	20 32	24 49	04 R 01	04 R 48	03 14	11 59	10 52
03	12 46 51	09 47 49	01 ♐ 11 16	07 ♐ 18 32	16 31	21 08	25 23	03 56	04 48	03 17	12 01	10 54
04	12 50 48	10 46 55	13 ♐ 28 42	19 ♐ 42 11	15 25	21 43	25 56	03 51	04 47	03 20	12 03	10 55
05	12 54 44	11 46 03	25 ♐ 59 27	02 ♑ 21 01	14 16	22 17	26 29	03 46	04 47	03 23	12 05	10 57
06	12 58 41	12 45 13	08 ♑ 47 22	15 ♑ 18 59	13 06	22 50	27 02	03 41	04 46	03 26	12 07	10 59
07	13 02 38	13 44 24	21 ♑ 56 22	28 ♑ 39 54	11 56	23 21	27 35	03 37	04 46	03 29	12 09	11 01
08	13 06 34	14 43 38	05 ♒ 29 58	12 ♒ 26 49	10 50	23 51	28 07	03 33	04 46	03 32	12 11	11 03
09	13 10 31	15 42 53	19 ♒ 30 34	26 ♒ 41 10	09 47	24 19	28 40	03 29	04 46	03 35	12 13	11 05
10	13 14 27	16 42 10	03 ♓ 58 23	11 ♓ 21 47	08 52	24 45	29 12	03 25	04 D 46	03 38	12 15	11 06
11	13 18 24	17 41 28	18 ♓ 50 41	26 ♓ 24 12	08 03	25 10	29 ♋ 44	03 21	04 46	03 41	12 17	11 08
12	13 22 20	18 40 48	04 ♈ 01 13	11 ♈ 40 27	07 25	25 33	00 ♌ 16	03 18	04 46	03 44	12 19	11 10
13	13 26 17	19 40 11	19 ♈ 20 30	26 ♈ 59 53	06 56	25 55	00 48	03 15	04 46	03 47	12 21	11 11
14	13 30 13	20 39 35	04 ♉ 37 09	12 ♉ 10 55	06 38	26 14	01 19	03 12	04 47	03 50	12 23	11 13
15	13 34 10	21 39 02	19 ♉ 39 57	27 ♉ 03 12	06 31	26 32	01 51	03 09	04 47	03 53	12 26	11 15
16	13 38 07	22 38 30	04 ♊ 19 50	11 ♊ 29 17	06 D 35	26 47	02 22	03 06	04 48	03 56	12 28	11 17
17	13 42 03	23 38 02	18 ♊ 31 12	25 ♊ 25 28	06 50	27 01	02 53	03 04	04 49	03 58	12 30	11 18
18	13 46 00	24 37 35	02 ♋ 12 10	08 ♋ 51 32	07 15	27 13	03 23	03 01	04 50	04 01	12 32	11 20
19	13 49 56	25 37 11	15 ♋ 23 59	21 ♋ 49 59	07 50	27 22	03 54	02 59	04 51	04 04	12 34	11 21
20	13 53 53	26 36 48	28 ♋ 10 06	04 ♌ 24 58	08 33	27 29	04 24	02 57	04 52	04 06	12 36	11 23
21	13 57 49	27 36 29	10 ♌ 35 13	16 ♌ 41 31	09 25	27 34	04 54	02 55	04 53	04 09	12 38	11 25
22	14 01 46	28 36 11	22 ♌ 44 29	28 ♌ 44 45	10 24	27 37	05 24	02 54	04 54	04 12	12 41	11 26
23	14 05 42	29 ♎ 35 56	04 ♍ 42 54	10 ♍ 39 29	11 29	27 R 37	05 54	02 53	04 55	04 14	12 43	11 28
24	14 09 39	00 ♏ 35 42	16 ♍ 35 00	22 ♍ 29 56	12 40	27 36	06 23	02 52	04 57	04 17	12 45	11 29
25	14 13 36	01 35 31	28 ♍ 24 39	04 ♎ 19 32	13 56	27 31	06 52	02 51	04 58	04 19	12 47	11 31
26	14 17 32	02 35 22	10 ♎ 14 52	16 ♎ 10 56	15 16	27 24	07 21	02 50	05 00	04 21	12 49	11 32
27	14 21 29	03 35 15	22 ♎ 07 56	28 ♎ 06 04	16 40	27 15	07 50	02 49	05 02	04 24	12 52	11 34
28	14 25 25	04 35 10	04 ♏ 05 28	10 ♏ 06 17	18 07	27 04	08 18	02 49	05 03	04 26	12 54	11 35
29	14 29 22	05 35 07	16 ♏ 08 37	22 ♏ 12 36	19 36	26 50	08 47	02 D 49	05 05	04 28	12 56	11 36
30	14 33 18	06 35 06	28 ♏ 18 21	04 ♐ 26 01	21 07	26 33	09 14	02 49	05 07	04 31	12 58	11 38
31	14 37 15	07 ♏ 35 07	10 ♐ 35 45	16 ♐ 47 44	22 ♎ 40	26 ♏ 15	09 ♌ 42	02 ♓ 49	05 ♒ 10	04 ♍ 33	13 ♏ 01	11 ♍ 39

Date	Moon True ☊ ° ′	Moon Mean ☊ ° ′	Moon ☽ Latitude ° ′
01	06 ♌ 52	05 ♌ 32	05 N 03
02	06 R 42	05 29	04 56
03	06 33	05 26	04 36
04	06 27	05 22	04 03
05	06 24	05 19	03 17
06	06 23	05 16	02 21
07	06 D 23	05 13	01 16
08	06 23	05 10	00 N 05
09	06 R 22	05 07	01 S 09
10	06 20	05 03	02 21
11	06 15	05 00	03 25
12	06 07	04 57	04 16
13	05 57	04 54	04 49
14	05 46	04 51	05 01
15	05 36	04 48	04 52
16	05 27	04 44	04 24
17	05 20	04 41	03 41
18	05 17	04 38	02 45
19	05 15	04 35	01 43
20	05 D 15	04 32	00 S 37
21	05 R 15	04 28	00 N 28
22	05 14	04 25	01 31
23	05 11	04 22	02 29
24	05 05	04 19	03 20
25	04 56	04 16	04 02
26	04 45	04 13	04 33
27	04 31	04 09	04 52
28	04 16	04 06	04 59
29	04 02	04 03	04 53
30	03 48	04 00	04 33
31	03 ♌ 37	03 ♌ 57	04 N 00

DECLINATIONS

Date	Sun ☉ ° ′	Moon ☽ ° ′	Mercury ☿ ° ′	Venus ♀ ° ′	Mars ♂ ° ′	Jupiter ♃ ° ′	Saturn ♄ ° ′	Uranus ♅ ° ′	Neptune ♆ ° ′	Pluto ♇ ° ′
01	03 S 06	09 S 05	10 S 20	22 S 48	22 N 06	11 S 17	19 S 47	11 N 01	13 S 47	19 N 29
02	03 30	12 44	09 48	23 04	22 01	11 19	19 48	11 00	13 47	19 29
03	03 53	15 54	09 12	23 19	21 56	11 21	19 48	10 59	13 48	19 28
04	04 16	18 24	08 33	23 33	21 51	11 22	19 48	10 58	13 49	19 28
05	04 39	20 06	07 51	23 47	21 46	11 24	19 48	10 57	13 49	19 27
06	05 02	20 48	07 08	24 00	21 41	11 26	19 48	10 55	13 50	19 27
07	05 25	20 24	06 23	24 13	21 36	11 27	19 48	10 54	13 51	19 26
08	05 48	18 49	05 38	24 25	21 31	11 28	19 48	10 53	13 51	19 26
09	06 11	16 04	04 55	24 36	21 25	11 30	19 48	10 52	13 52	19 25
10	06 34	12 14	04 14	24 47	21 20	11 31	19 48	10 51	13 53	19 25
11	06 57	07 33	03 37	24 57	21 15	11 32	19 48	10 50	13 53	19 25
12	07 19	02 S 19	03 03	25 06	21 09	11 33	19 48	10 49	13 54	19 24
13	07 42	03 N 07	02 34	25 15	21 03	11 34	19 48	10 48	13 54	19 24
14	08 04	08 20	02 11	25 23	20 58	11 35	19 48	10 47	13 55	19 23
15	08 26	12 57	01 53	25 30	20 52	11 36	19 48	10 46	13 56	19 23
16	08 49	16 41	01 42	25 36	20 46	11 37	19 48	10 45	13 56	19 23
17	09 11	19 17	01 35	25 42	20 41	11 38	19 48	10 44	13 57	19 22
18	09 33	20 40	01 34	25 46	20 35	11 39	19 47	10 43	13 58	19 22
19	09 54	20 51	01 39	25 50	20 29	11 39	19 47	10 42	13 58	19 22
20	10 16	19 55	01 48	25 53	20 23	11 40	19 47	10 41	13 59	19 21
21	10 37	18 02	02 02	25 54	20 17	11 40	19 47	10 40	14 00	19 21
22	10 59	15 22	02 20	25 55	20 11	11 41	19 46	10 39	14 00	19 21
23	11 20	12 06	02 42	25 55	20 05	11 41	19 46	10 38	14 01	19 20
24	11 41	08 22	03 07	25 54	19 59	11 41	19 46	10 38	14 02	19 20
25	12 02	04 20	03 34	25 52	19 54	11 41	19 45	10 37	14 03	19 20
26	12 22	00 N 08	04 05	25 48	19 48	11 41	19 45	10 36	14 03	19 20
27	12 43	04 S 06	04 37	25 44	19 42	11 41	19 44	10 35	14 04	19 20
28	13 03	08 11	05 11	25 38	19 36	11 41	19 44	10 34	14 05	19 19
29	13 23	12 00	05 46	25 31	19 30	11 41	19 44	10 33	14 05	19 19
30	13 43	15 21	06 22	25 23	19 24	11 41	19 43	10 33	14 06	19 19
31	14 S 03	18 S 05	07 S 00	25 S 13	19 N 18	11 S 41	19 S 43	10 N 32	14 S 07	19 N 19

ZODIAC SIGN ENTRIES

Date	h	m	Planets
03	09	40	☽ ♐
05	19	35	☽ ♑
08	02	22	☽ ♒
10	05	29	☽ ♓
11	23	54	♂ ♌
12	05	41	☽ ♈
14	04	43	☽ ♉
16	04	50	☽ ♊
18	08	05	☽ ♋
20	15	30	☽ ♌
23	02	31	☽ ♍
23	21	40	☉ ♏
25	15	14	☽ ♎
28	03	49	☽ ♏
30	15	19	☽ ♐

LATITUDES

Date	Mercury ☿ ° ′	Venus ♀ ° ′	Mars ♂ ° ′	Jupiter ♃ ° ′	Saturn ♄ ° ′	Uranus ♅ ° ′	Neptune ♆ ° ′	Pluto ♇ ° ′
01	03 S 20	05 S 17	00 N 51	01 S 22	00 S 45	00 N 43	01 N 44	13 N 00
04	02 42	05 33	00 55	01 22	00 45	00 43	01 43	13 00
07	01 48	05 48	00 58	01 22	00 45	00 43	01 43	13 01
10	00 S 47	06 01	01 02	01 21	00 45	00 44	01 43	13 02
13	00 N 12	06 12	01 06	01 21	00 45	00 44	01 43	13 03
16	01 00	06 20	01 10	01 20	00 45	00 44	01 43	13 04
19	01 35	06 26	01 14	01 20	00 45	00 44	01 43	13 05
22	01 56	06 28	01 18	01 19	00 45	00 44	01 43	13 06
25	02 05	06 26	01 23	01 19	00 45	00 44	01 43	13 07
28	02 05	06 18	01 27	01 18	00 45	00 44	01 43	13 08
31	01 N 58	06 S 05	01 N 32	01 S 18	00 S 45	00 N 44	01 N 43	13 N 09

DATA

Julian Date	2437939
Delta T	+34 seconds
Ayanamsa	23° 19′ 57″
Synetic vernal point	05° ♓ 47′ 02″
True obliquity of ecliptic	23° 26′ 34″

MOON'S PHASES, APSIDES AND POSITIONS ☽

Date	h	m	Phase	Longitude ° ′	Eclipse Indicator
06	19	54	☽	13 ♑ 05	
13	12	33	○	19 ♈ 42	
20	08	47	☾	26 ♋ 29	
28	13	05	●	04 ♏ 38	

Day	h	m		
13	03	29	Perigee	
26	03	41	Apogee	
06	15	42	Max dec	20° S 49′
12	22	14	0N	
19	03	38	Max dec	20° N 55′
26	12	43	0S	

LONGITUDES

Date	Chiron ⚷ ° ′	Ceres ⚳ ° ′	Pallas ⚴ ° ′	Juno ⚵ ° ′	Vesta ⚶ ° ′	Black Moon Lilith ⚸ ° ′
01	07 ♓ 08	19 ♌ 46	08 ♋ 08	27 ♌ 17	25 ♌ 51	07 ♎ 30
11	06 ♓ 46	23 ♌ 45	11 ♋ 59	01 ♍ 39	00 ♍ 04	08 ♎ 37
21	06 ♓ 29	27 ♌ 35	15 ♋ 26	05 ♍ 49	04 ♍ 11	09 ♎ 44
31	06 ♓ 17	01 ♍ 15	18 ♋ 20	09 ♍ 46	08 ♍ 11	10 ♎ 51

All ephemeris data is given at 12.00 UT and the Moon's longitude is additionally given for 24.00 UT

ASPECTARIAN

h m	Aspects
01 Monday	
04 09	☽ ⚹ ♅
05 24	♄ ± ♇
06 06	☽ △ ♃
07 30	☽ □ ♄
13 41	☽ ⊻ ☉
18 58	☽ ∥ ☿
19 35	☽ ⚹ ♇
19 45	☿ ∠ ♅
21 50	☽ ☌ ♆
02 Tuesday	
00 18	☽ # ♅
02 16	☽ ∥ ♃
02 46	☽ ⊥ ☉
04 19	☽ Q ♅
09 12	☽ ⊻ ☿
15 04	☽ ☌ ♀
19 26	☽ Q ♄
19 34	☽ Q ♇
19 36	☽ ∥ ♆
20 13	☽ ⊥ ☿
22 16	☽ ∠ ☉
22 39	☉ ⊥ ♅
23 59	☽ △ ♂
03 Wednesday	
03 37	☿ ⊥ ♇
12 35	☽ ∠ ☿
14 55	☉ ± ♃
16 08	☽ □ ♅
17 21	☽ □ ♃
19 04	☽ ⚹ ♄
04 Thursday	
06 19	☽ ⚹ ☉
06 50	☽ ꝑ ♂
07 02	☽ □ ♇
07 36	☿ ⊥ ♀
09 13	☽ ⊻ ♆
11 33	♂ ∠ ♇
15 26	☽ ⚹ ☿
15 33	☉ ⊻ ♇
20 51	☽ ⊥ ♆
05 Friday	
00 09	☽ ∠ ♄
01 01	☽ ± ♂
01 23	☽ # ♇
04 00	☽ Q ♃
04 37	☽ ⊻ ♀
06 45	☽ ∥ ♄
07 25	☽ Q ☉
12 28	☽ Q ☿
12 59	☽ ⊼ ♂
14 04	☽ ∠ ♆
16 33	☽ ⊥ ♀
17 16	☽ ⊥ ♄
19 55	☉ ⊻ ♆
06 Saturday	
02 00	☽ △ ♅
02 34	☽ ⚹ ♃
04 32	☽ ⊻ ♄
09 15	♀ Q ♄
10 09	☽ ∠ ♀
15 49	☉ ☌ ☿
16 04	☽ △ ♇
18 09	☽ ⚹ ♆
19 17	☽ □ ☿
19 26	♀ Q ♇
19 54	☽ □ ☉
07 Sunday	
05 45	☽ ꝑ ♅
06 02	☽ ∠ ♃
07 29	♂ ⊥ ♅
07 45	☿ ⊻ ♆
13 28	♂ ± ♃
14 38	☽ ⚹ ♀
15 58	☽ Q ♆
19 19	☽ ꝑ ♇
21 57	☽ ± ♅
22 06	☽ ⊥ ♃
22 30	☽ ☍ ♂
23 22	☽ ∥ ♄
08 Monday	
04 40	☽ # ♇
07 19	☿ ⊻ ♇
08 27	☉ ∥ ☿
08 34	☽ ⊼ ♅
08 37	☽ ⊻ ♃
10 43	☽ ☌ ♄
11 12	☽ ± ♇
12 37	☽ Q ♀
13 55	♃ ☍ ♅
20 34	☽ △ ☿
21 37	☽ ⊼ ♇
22 32	☿ Q ♂
23 34	☽ □ ♆
09 Tuesday	
05 06	☽ △ ☉
16 24	♄ St D
16 42	☿ ⊥ ♅
20 04	☿ ∠ ♀
20 11	☿ ± ♃
20 17	☽ ꝑ ☿
20 19	☽ □ ♀
10 Wednesday	
02 30	☽ ∥ ♆
03 52	☽ ⊼ ♂
08 01	☽ ꝑ ☉
10 17	☽ ± ☿
11 06	☽ ☌ ♃
11 27	☽ ☍ ♅
13 17	☽ ⊻ ♄
14 05	☽ ± ♂
15 57	☽ ∥ ♃
19 31	☽ ⊼ ☿
19 33	☽ # ♅
22 04	☉ ⊥ ♇
23 02	☽ ⊥ ♄
23 36	☽ ☍ ♇
23 43	☽ ± ☉
11 Thursday	
01 28	☽ △ ♆
05 11	☽ ꝑ ♂
10 02	☽ ⊼ ☉
13 28	☽ ∠ ♄
14 42	☽ ∥ ☉
22 19	☽ △ ♀
12 Friday	
01 25	☽ ꝑ ♆
03 16	☉ ꝑ ♃
05 53	☽ △ ♂
08 22	☽ ∥ ☿
10 52	☽ ⊻ ♃
11 33	☽ ⊼ ♅
13 10	☽ ⚹ ♄
13 28	☉ ∠ ♅
15 37	☽ ± ♆
17 08	☽ ☍ ☿
20 15	☽ ⊥ ♃
21 00	☽ ± ♅
22 30	☽ ꝑ ♀
23 13	☽ ⊼ ♇
13 Saturday	
01 02	☽ ⊼ ♆
07 59	☽ Q ♄
08 38	☽ ± ♇
09 47	☽ # ☿
10 17	☽ ∠ ♃
11 08	☽ ꝑ ♅
12 33	☽ ☍ ☉
12 55	☽ ± ♀
20 50	☉ ⊥ ♀
22 31	☽ ⊼ ♀
22 45	☽ ꝑ ♇
14 Sunday	
06 37	☽ □ ♂
09 45	☽ ⚹ ♃
10 39	☽ # ☉
10 45	☽ △ ♅
12 15	☽ □ ♄
15 09	☽ ⊼ ☿
22 29	☽ △ ♇
15 Monday	
00 11	☽ ∥ ♅
00 22	☽ ☍ ♆
00 35	☽ ± ☿
04 32	☽ # ♃
04 45	☽ Q ♃
12 18	☽ Q ♂
14 59	☽ ꝑ ☿
15 05	☿ St D
15 26	☽ ⊼ ☉
17 42	☽ # ♆
23 21	☽ ☍ ♀
16 Tuesday	
01 55	☽ ± ☉
08 37	☽ ⚹ ♂
09 58	☽ □ ♃
11 20	☽ □ ♅
12 47	☽ △ ♄
15 48	☽ △ ☿
17 56	☽ ꝑ ☉
23 40	☽ □ ♇
17 Wednesday	
01 41	☽ ⊼ ♆
10 51	☽ ∠ ♂
11 58	☽ ± ♆
13 03	☽ ∥ ♇
14 14	☽ ꝑ ♄
18 00	☽ Q ♅
18 40	☽ # ♄
19 53	♂ ⊼ ♃
21 33	☽ △ ☉
18 Thursday	
03 01	☽ ⊼ ♀
03 07	☽ ⊥ ♂
03 41	☽ ꝑ ♆
05 59	☽ ± ♄
06 53	☽ Q ♇
09 44	☽ ∥ ♂
13 27	☽ △ ♃
13 50	☽ ± ♀
14 12	☽ ⊻ ♂
15 16	☽ ⚹ ♅
16 43	☽ ⊼ ♄
21 27	☽ □ ☿
19 Friday	
04 33	☽ ⚹ ♇
06 21	☽ ꝑ ♀
06 46	☽ △ ♆
16 47	☽ ꝑ ♃
18 50	☽ ∠ ♅
20 37	♂ ⊻ ♅
20 Saturday	
02 08	☽ ∥ ♂
06 18	☉ ∠ ♇
08 36	☽ ∠ ♇
08 43	☽ Q ☿
08 47	☽ □ ☉
09 41	☽ ± ♃
10 42	☽ △ ♀
11 53	☽ ⊥ ♅
14 16	☽ # ♄
20 31	☽ ∥ ♇
21 09	☽ ⊼ ♃
23 08	☿ ± ♃
23 27	☽ ⊻ ♅
21 Sunday	
00 29	☽ ☌ ♂
00 53	☽ ☍ ♄
01 54	☽ ⊥ ♇
09 32	☽ ⚹ ☿
10 42	♂ ☍ ♄
11 05	☉ ⊻ ♀
13 37	☽ ⊻ ♇
15 01	☉ # ♅
16 02	☽ □ ♆
22 44	☽ Q ☉
22 Monday	
06 59	☿ ⊥ ♅
13 44	♂ ⊥ ♇
17 49	☽ ∠ ☿
21 45	☽ □ ♀
22 26	☽ # ♆
23 Tuesday	
00 47	☽ ⚹ ☉
03 56	☽ Q ♆
04 14	♀ St R
08 18	☽ ☍ ♃
11 02	☽ ☌ ♅
11 24	☿ ⊻ ♇
12 25	☽ ⊼ ♄
13 44	☽ ⊥ ☿
14 29	☽ ⊻ ♂
14 47	☽ # ♃
16 40	☽ # ☉
21 41	☽ ∥ ♅
24 Wednesday	
00 33	☽ ± ♄
01 39	☽ ☌ ♇
03 08	☽ ⊥ ♂
03 11	☽ ⊻ ☿
04 13	☽ ⚹ ♆
09 49	☽ ∠ ☉
10 00	☽ Q ♀
10 29	☿ ∠ ♀
12 14	☉ ∥ ♃
13 34	☿ ⊻ ♆
18 50	☽ ꝑ ♄
22 10	☽ ∠ ♂
25 Thursday	
05 45	☽ ⊥ ☉
10 12	☽ ⚹ ♀
10 44	☽ ∠ ♆
15 53	☽ # ☿
19 03	☽ ⊻ ☉
20 59	☽ ⊼ ♃
26 Friday	
00 01	☽ ⊻ ♅
01 20	☽ △ ♄
05 03	☽ ⊥ ♆
05 53	☽ ⚹ ♂
09 08	☽ ± ♃
12 13	☽ ⊥ ♅
14 37	☽ ⊻ ♇
16 19	☽ ∠ ♀
17 14	☽ ⊻ ♆
17 46	☉ △ ♃
23 29	♂ # ♄
23 30	☽ ☌ ☿
27 Saturday	
02 46	☽ ⊥ ♇
03 19	☽ ꝑ ♃
06 28	☽ ∠ ♅
07 10	☽ Q ♂
10 15	☽ ⊥ ♀
15 26	☽ ∥ ☿
20 55	☽ ∠ ♇
22 09	☽ ⊻ ♀
28 Sunday	
03 09	☿ ⊥ ♇
07 12	☿ ꝑ ♃
08 14	☉ ⚹ ♅
09 27	☽ △ ♃
12 41	☽ ⚹ ♅
13 05	☽ ☌ ☉
13 56	☽ □ ♄
20 46	☽ □ ♂
23 41	☉ □ ♄
29 Monday	
02 42	☽ # ♅
02 58	☽ ⚹ ♇
05 37	☽ ☌ ♆
09 58	☿ ∠ ♅
09 58	☽ ∥ ♃
10 35	♃ St D
12 39	☽ Q ♅
19 49	☽ ⊻ ☿
22 32	☽ ∥ ☉
30 Tuesday	
01 46	☽ Q ♄
02 34	☽ ∥ ♆
02 47	☽ Q ♇
04 36	☿ ⊥ ♀
08 39	☽ ☌ ♀
09 20	☽ ⊥ ☿
14 46	☿ Q ♂
20 51	☽ □ ♃
31 Wednesday	
00 11	☽ □ ♅
01 23	☽ ⚹ ♄
05 29	☽ ∠ ☿
05 38	☽ ⊻ ☉
07 38	♂ ∥ ♇
10 12	☽ △ ♂
14 03	☽ □ ♇
16 42	☽ ⊻ ♆
17 04	☉ ∥ ♆
18 18	☽ ⊥ ☉

NOVEMBER 1962

LONGITUDES

Date	Sidereal time h m s	Sun ☉ ° ' "	Moon ☽ ° ' "	Moon ☽ 24.00 ° ' "	Mercury ☿ ° '	Venus ♀ ° '	Mars ♂ ° '	Jupiter ♃ ° '	Saturn ♄ ° '	Uranus ♅ ° '	Neptune ♆ ° '	Pluto ♇ ° '
01	14 41 11	08 ♏ 35 09	23 ♐ 02 12	29 ♐ 19 26	24 ♎ 14	25 ♏ 54	10 ♌ 10	02 ♓ 50	05 ♒ 12	04 ♍ 35	13 ♏ 03	11 ♍ 40
02	14 45 08	09 35 13	05 ♑ 39 42	12 ♑ 03 20	25 49	25 R 31	10 37	02 51	05 14	04 37	13 05	11 42
03	14 49 05	10 35 19	18 ♑ 30 43	25 ♑ 02 14	27 25	25 05	11 03	02 52	05 16	04 39	13 07	11 43
04	14 53 01	11 35 27	01 ♒ 38 15	08 ♒ 19 10	29 ♎ 02	24 38	11 30	02 53	05 19	04 41	13 10	11 44
05	14 56 58	12 35 35	15 ♒ 05 19	21 ♒ 57 00	00 ♏ 39	24 09	11 56	02 54	05 22	04 43	13 12	11 45
06	15 00 54	13 35 46	28 ♒ 54 25	05 ♓ 57 41	02 17	23 39	12 22	02 56	05 24	04 45	13 14	11 47
07	15 04 51	14 35 58	13 ♓ 06 45	20 ♓ 21 25	03 54	23 06	12 48	02 57	05 27	04 47	13 16	11 48
08	15 08 47	15 36 11	27 ♓ 41 18	05 ♈ 05 49	05 32	22 33	13 13	02 59	05 30	04 49	13 19	11 49
09	15 12 44	16 36 25	12 ♈ 34 11	20 ♈ 05 25	07 09	21 58	13 38	03 02	05 33	04 50	13 21	11 50
10	15 16 40	17 36 42	27 ♈ 38 23	05 ♉ 11 51	08 47	21 23	14 03	03 04	05 36	04 52	13 23	11 51
11	15 20 37	18 37 00	12 ♉ 44 30	20 ♉ 15 02	10 24	20 47	14 27	03 06	05 39	04 54	13 25	11 52
12	15 24 34	19 37 19	27 ♉ 42 11	05 ♊ 04 51	12 02	20 10	14 51	03 09	05 42	04 55	13 28	11 53
13	15 28 30	20 37 41	12 ♊ 22 02	19 ♊ 33 00	13 39	19 33	15 15	03 12	05 46	04 57	13 30	11 54
14	15 32 27	21 38 04	26 ♊ 37 10	03 ♋ 34 11	15 16	18 57	15 39	03 15	05 49	04 58	13 32	11 55
15	15 36 23	22 38 29	10 ♋ 23 56	17 ♋ 06 28	16 52	18 21	16 02	03 19	05 53	05 00	13 34	11 56
16	15 40 20	23 38 56	23 ♋ 41 58	00 ♌ 10 50	18 29	17 46	16 24	03 22	05 56	05 01	13 37	11 57
17	15 44 16	24 39 24	06 ♌ 33 30	12 ♌ 50 30	20 05	17 11	16 47	03 26	06 00	05 02	13 39	11 58
18	15 48 13	25 39 55	19 ♌ 02 29	25 ♌ 10 04	21 41	16 38	17 09	03 30	06 04	05 04	13 41	11 59
19	15 52 09	26 40 27	01 ♍ 13 55	07 ♍ 14 43	23 17	16 06	17 30	03 34	06 08	05 05	13 43	11 59
20	15 56 06	27 41 01	13 ♍ 13 07	19 ♍ 09 46	24 53	15 35	17 51	03 38	06 12	05 06	13 45	12 00
21	16 00 03	28 41 37	25 ♍ 05 16	01 ♎ 00 13	26 28	15 06	18 12	03 43	06 16	05 07	13 48	12 01
22	16 03 59	29 ♏ 42 14	06 ♎ 55 07	12 ♎ 50 29	28 03	14 39	18 33	03 47	06 20	05 08	13 50	12 02
23	16 07 56	00 ♐ 42 53	18 ♎ 46 44	24 ♎ 44 15	29 ♏ 38	14 14	18 52	03 52	06 24	05 09	13 52	12 02
24	16 11 52	01 43 34	00 ♏ 43 21	06 ♏ 44 20	01 ♐ 13	13 52	19 12	03 57	06 28	05 10	13 54	12 03
25	16 15 49	02 44 16	12 ♏ 47 23	18 ♏ 52 41	02 48	13 31	19 31	04 02	06 32	05 11	13 56	12 04
26	16 19 45	03 45 00	25 ♏ 00 22	01 ♐ 10 31	04 22	13 13	19 50	04 08	06 37	05 12	13 58	12 04
27	16 23 42	04 45 45	07 ♐ 23 12	13 ♐ 38 26	05 57	12 57	20 08	04 13	06 41	05 13	14 01	12 05
28	16 27 38	05 46 31	19 ♐ 56 16	26 ♐ 16 42	07 31	12 43	20 26	04 19	06 46	05 14	14 03	12 05
29	16 31 35	06 47 19	02 ♑ 39 48	09 ♑ 05 35	09 05	12 32	20 43	04 25	06 51	05 14	14 05	12 06
30	16 35 32	07 ♐ 48 07	15 ♑ 34 08	22 ♑ 05 31	10 ♐ 39	12 ♏ 24	20 ♌ 59	04 ♓ 31	06 ♒ 55	05 ♍ 15	14 ♏ 07	12 ♍ 06

Date	Moon True ☊ ° '	Moon Mean ☊ ° '	Moon ☽ Latitude ° '
01	03 ♌ 29	03 ♌ 54	03 N 16
02	03 R 24	03 50	02 21
03	03 22	03 47	01 18
04	03 D 21	03 44	00 N 09
05	03 R 21	03 41	01 S 02
06	03 20	03 38	02 11
07	03 18	03 34	03 14
08	03 13	03 31	04 06
09	03 05	03 28	04 42
10	02 55	03 25	05 00
11	02 44	03 22	04 57
12	02 33	03 19	04 33
13	02 23	03 15	03 52
14	02 16	03 12	02 57
15	02 12	03 09	01 53
16	02 10	03 06	00 S 45
17	02 D 10	03 03	00 N 23
18	02 10	03 00	01 29
19	02 R 10	02 56	02 28
20	02 09	02 53	03 20
21	02 05	02 50	04 03
22	01 59	02 47	04 35
23	01 51	02 44	04 56
24	01 40	02 40	05 04
25	01 28	02 37	04 58
26	01 17	02 34	04 38
27	01 06	02 31	04 06
28	00 57	02 28	03 21
29	00 51	02 25	02 25
30	00 ♌ 47	02 ♌ 21	01 N 21

DECLINATIONS

Date	Sun ☉ ° '	Moon ☽ ° '	Mercury ☿ ° '	Venus ♀ ° '	Mars ♂ ° '	Jupiter ♃ ° '	Saturn ♄ ° '	Uranus ♅ ° '	Neptune ♆ ° '	Pluto ♇ ° '
01	14 S 22	20 00	07 S 38	25 S 03	19 N 12	11 S 40	19 S 42	10 N 31	14 S 07	19 N 19
02	14 41	20 59	08 16	24 51	19 06	11 40	19 41	10 30	14 08	19 19
03	15 00	20 53	08 55	24 37	19 00	11 39	19 41	10 30	14 09	19 18
04	15 19	19 39	09 34	24 23	18 54	11 39	19 40	10 29	14 09	19 18
05	15 37	17 18	10 13	24 07	18 48	11 38	19 40	10 28	14 10	19 18
06	15 55	13 54	10 52	23 50	18 43	11 37	19 39	10 28	14 11	19 18
07	16 13	09 37	11 30	23 32	18 37	11 37	19 38	10 27	14 11	19 18
08	16 31	04 S 41	12 09	23 13	18 31	11 36	19 38	10 26	14 12	19 18
09	16 48	00 N 38	12 46	22 53	18 26	11 35	19 37	10 26	14 13	19 18
10	17 05	05 58	13 24	22 32	18 20	11 34	19 36	10 25	14 13	19 18
11	17 22	10 57	14 01	22 10	18 15	11 33	19 35	10 25	14 14	19 18
12	17 39	15 13	14 37	21 47	18 09	11 31	19 35	10 24	14 15	19 18
13	17 55	18 27	15 13	21 24	18 04	11 30	19 34	10 24	14 15	19 18
14	18 11	20 27	15 48	21 00	17 58	11 29	19 33	10 23	14 16	19 18
15	18 26	21 10	16 22	20 36	17 53	11 27	19 32	10 23	14 17	19 18
16	18 41	20 38	16 56	20 11	17 48	11 26	19 31	10 22	14 17	19 18
17	18 56	19 01	17 28	19 47	17 43	11 24	19 30	10 22	14 18	19 18
18	19 10	16 31	18 00	19 23	17 38	11 23	19 29	10 21	14 19	19 18
19	19 25	13 21	18 31	18 59	17 33	11 21	19 28	10 21	14 19	19 18
20	19 39	09 41	19 01	18 35	17 28	11 20	19 27	10 20	14 20	19 18
21	19 52	05 40	19 31	18 11	17 24	11 18	19 26	10 20	14 20	19 19
22	20 05	01 N 28	19 59	17 49	17 19	11 16	19 25	10 20	14 21	19 19
23	20 18	02 S 48	20 26	17 27	17 15	11 14	19 24	10 19	14 22	19 19
24	20 31	06 59	20 52	17 05	17 11	11 12	19 23	10 19	14 22	19 19
25	20 43	10 57	21 18	16 45	17 06	11 10	19 22	10 19	14 23	19 19
26	20 54	14 31	21 42	16 25	17 02	11 08	19 21	10 19	14 24	19 19
27	21 05	17 30	22 05	16 07	16 58	11 06	19 20	10 18	14 24	19 20
28	21 16	19 44	22 27	15 49	16 55	11 03	19 19	10 18	14 25	19 20
29	21 27	21 00	22 48	15 33	16 51	11 01	19 18	10 18	14 25	19 20
30	21 S 37	21 S 12	23 S 08	15 S 17	16 N 48	10 S 59	19 S 16	10 N 18	14 S 26	19 N 20

ZODIAC SIGN ENTRIES

Date	h	m	Planets
02	01	17	☽ ♑
04	09	02	☽ ♒
05	02	20	☿ ♏
06	13	52	☽ ♓
08	15	45	☽ ♈
10	15	45	☽ ♉
12	15	43	☽ ♊
14	17	49	☽ ♋
16	23	40	☽ ♌
19	09	33	☽ ♍
21	21	58	☽ ♎
22	19	02	☉ ♐
23	17	31	☿ ♐
24	10	33	☽ ♏
26	21	43	☽ ♐
29	07	00	☽ ♑

LATITUDES

Date	Mercury ☿ ° '	Venus ♀ ° '	Mars ♂ ° '	Jupiter ♃ ° '	Saturn ♄ ° '	Uranus ♅ ° '	Neptune ♆ ° '	Pluto ♇ ° '
01	01 N 54	05 S 59	01 N 33	01 S 18	00 S 45	00 N 44	01 N 43	13 N 10
04	01 41	05 37	01 38	01 17	00 45	00 44	01 43	13 11
07	01 24	05 09	01 43	01 16	00 45	00 45	01 43	13 12
10	01 05	04 35	01 48	01 16	00 45	00 45	01 43	13 13
13	00 46	03 55	01 53	01 15	00 45	00 45	01 43	13 15
16	00 25	03 11	01 58	01 15	00 45	00 45	01 43	13 16
19	00 N 05	02 25	02 04	01 14	00 45	00 45	01 43	13 17
22	00 S 15	01 39	02 09	01 14	00 45	00 45	01 43	13 19
25	00 35	00 53	02 15	01 13	00 45	00 45	01 43	13 20
28	00 54	00 S 10	02 21	01 12	00 45	00 45	01 43	13 22
31	01 S 11	00 N 30	02 N 27	01 S 12	00 S 45	00 N 46	01 N 43	13 N 23

DATA

Julian Date	2437970
Delta T	+34 seconds
Ayanamsa	23° 20' 00"
Synetic vernal point	05° ♓ 46' 59"
True obliquity of ecliptic	23° 26' 34"

MOON'S PHASES, APSIDES AND POSITIONS ☽

Date	h	m	Phase	Longitude ° '	Eclipse Indicator
05	07	15	☽	12 ♒ 24	
11	22	03	○	19 ♉ 02	
19	02	09	☾	26 ♌ 16	
27	06	29	●	04 ♐ 32	

Day	h	m		
10	13	53	Perigee	
22	15	48	Apogee	
02	21	56	Max dec	21° S 04'
09	09	11	0N	
15	13	18	Max dec	21° N 10'
22	20	17	0S	
30	04	10	Max dec	21° S 15'

LONGITUDES

Date	Chiron ⚷ ° '	Ceres ⚳ ° '	Pallas ⚴ ° '	Juno ⚵ ° '	Vesta ⚶ ° '	Black Moon Lilith ° '
01	06 ♓ 16	01 ♍ 37	18 ♋ 35	10 ♍ 08	08 ♍ 34	10 ♎ 58
11	06 ♓ 10	05 ♍ 02	20 ♋ 41	13 ♍ 47	12 ♍ 35	12 ♎ 02
21	06 ♓ 09	08 ♍ 11	21 ♋ 52	17 ♍ 07	16 ♍ 01	13 ♎ 02
31	06 ♓ 15	11 ♍ 01	21 ♋ 58	20 ♍ 04	19 ♍ 24	14 ♎ 19

All ephemeris data is given at 12.00 UT and the Moon's longitude is additionally given for 24.00 UT

ASPECTARIAN

h m	Aspects
01 Thursday	
01 09	☽ # ♂
02 03	☽ # ♇
04 19	☽ ⊥ ♆
06 32	☽ ∠ ♄
07 22	☽ ∥ ♄
07 46	☽ Q ♃
13 09	☽ ∠ ☉
14 37	☽ ⚹ ☿
16 13	☽ ⚼ ♂
17 19	☽ ⊻ ♀
21 36	☽ ∠ ♆
23 48	☽ ⊥ ♄
02 Friday	
04 24	☽ ⊥ ♀
06 40	☽ ⚹ ♃
08 16	☿ ⊻ ♀
09 57	☽ ± ♂
10 01	☽ △ ♅
11 11	☽ ⊻ ♄
16 39	☽ Q ☿
20 00	☽ ⚹ ☉
20 50	☽ ∠ ♀
21 38	☽ ⊼ ♂
23 21	☽ △ ♇
03 Saturday	
01 16	☿ ∠ ♇
01 58	☽ ⚹ ♆
10 48	☽ ∠ ♃
14 07	☽ ⚼ ♅
20 08	☽ Q ☉
23 42	☽ ⚹ ♀
04 Sunday	
00 12	☽ Q ♆
03 05	☽ ⚼ ♇
03 21	☽ ⊥ ♃
06 38	☽ □ ☿
06 38	☽ ± ♅
08 07	☉ □ ♂
11 44	☽ ∥ ♄
14 15	☽ ⊻ ♃
15 33	☉ ⚹ ♇
16 22	☽ # ♇
17 30	☽ ⊼ ♅
18 39	☽ ☌ ♄
19 23	☽ ± ♇
20 41	☽ Q ♀
21 19	☽ # ♂
05 Monday	
01 32	♂ ⊻ ♇
06 06	☽ ⊼ ♇
06 15	☽ ☍ ♂
07 15	☽ □ ☉
08 39	☽ □ ♆
21 26	☿ # ♅
23 40	☽ ∥ ☉
06 Tuesday	
03 02	☉ ☌ ♆
03 16	☽ □ ♀
06 06	♀ Q ♇
10 16	☽ ∥ ♆
18 30	☽ △ ☿
18 52	☽ ☌ ♃
21 48	☿ △ ♃
21 54	♀ Q ♄
21 58	☽ ☍ ♅
23 06	☽ ⊻ ♄
07 Wednesday	
01 20	☽ ∥ ♃
03 14	☽ ∥ ☿
07 38	☽ # ♅
09 13	☽ ⊥ ♄
09 48	☽ ☍ ♇
11 28	☽ ⊼ ♂
12 16	☽ △ ♆
14 40	☽ △ ☉
15 51	☿ ∥ ♃
21 43	☽ ± ♂
22 49	☽ ⚼ ☿
08 Thursday	
00 12	☽ ∠ ♄
01 11	☿ ⚹ ♅
03 55	☽ △ ♀
11 33	☿ □ ♄
12 53	☽ ⚼ ♂
13 01	☽ ⚼ ♆
15 22	☽ ± ☿
17 05	☽ ⚼ ☉
17 39	♂ □ ♆
20 37	☽ ⊻ ♃
23 34	☽ ⊼ ♅
09 Friday	
00 41	☽ ⚹ ♄
02 15	☽ ⊼ ☿
03 21	☽ ⚼ ♀
03 36	☽ ± ♆
06 18	☽ ⊥ ♃
08 38	☽ ± ☉
09 13	☽ ± ♅
10 49	☽ ⊼ ♇
13 15	☽ ⊼ ♆
13 45	☽ △ ♂
17 14	☽ ± ♀
17 41	☉ Q ♅
18 55	☽ ⊼ ☉
19 59	☽ Q ♄
20 25	☽ ± ♇
20 44	☽ ∠ ♃
23 37	☽ ⚼ ♅
10 Saturday	
02 26	☽ ⊼ ♀
10 45	☽ ⚼ ♇
20 38	☽ ⚹ ♃
23 30	☽ △ ♅
11 Sunday	
00 41	☽ □ ♄
07 50	☽ ☍ ☿
09 16	☽ ∥ ♅
10 36	☽ △ ♇
13 05	☽ ☍ ♆
14 49	☽ □ ♂
15 04	☽ # ♃
15 47	☽ Q ♃
20 49	☿ ∥ ♆
22 03	☽ ☍ ☉
12 Monday	
00 21	☽ ☍ ♀
06 02	☽ # ♆
07 38	☽ # ☿
09 51	☿ ⚹ ♇
20 06	☉ ☌ ♀
20 36	☽ Q ♂
20 53	☽ □ ♃
23 46	☽ □ ♅
13 Tuesday	
01 04	☽ △ ♄
06 50	☽ # ☉
08 41	☽ ∥ ♂
09 43	☿ ☌ ♆
11 14	☽ □ ♇
13 53	☽ ⊼ ♆
14 24	☽ ⊼ ☿
16 56	☽ ⚹ ♂
20 24	☽ ∥ ♇
22 15	☉ # ♂
23 17	☽ # ♄
23 31	☽ ⊼ ♀
23 57	☽ ± ♆
14 Wednesday	
01 43	☽ ± ☿
02 06	☽ ⚼ ♄
02 52	☽ ⊼ ☉
05 46	☽ Q ♅
09 16	☽ ± ♀
13 52	☽ ± ☉
15 18	☽ ⚼ ♆
17 31	☽ ± ♄
17 40	☽ Q ♇
19 05	☽ ⚼ ☿
19 07	☽ ∠ ♂
19 28	☿ □ ♂
20 34	☽ # ♀
23 30	☽ △ ♃
15 Thursday	
00 08	☽ ⚼ ♀
02 28	☽ ⚹ ♅
04 00	☽ ⊼ ♄
06 44	☽ ⚼ ☉
11 19	☽ ⊥ ♂
13 50	☿ Q ♅
14 44	☽ ⚹ ♇
17 40	☽ △ ♆
22 21	☽ ⊻ ♂
16 Friday	
01 10	☽ △ ☿
01 38	☽ △ ♀
02 14	☽ ⚼ ♃
04 05	☿ ☌ ♀
05 16	☽ ∠ ♅
11 54	☽ △ ☉
17 24	♄ ± ♇
18 00	☽ ∠ ♇
18 48	☽ ± ♃
19 14	☉ Q ♇
19 20	☉ Q ♄
21 51	☽ ⊥ ♅
23 07	☽ # ♀
17 Saturday	
06 01	☽ # ♄
06 04	☽ ⊼ ♃
08 35	☽ ∥ ♇
09 07	☽ ⊻ ♅
10 52	☽ ⊥ ♇
10 56	☽ ☍ ♄
12 48	☽ # ☉
17 50	♀ Q ♅
21 31	☿ # ♂
22 19	☽ ⊻ ♇
22 27	♀ □ ♂
18 Sunday	
00 59	☽ # ☿
01 35	☽ □ ♆
01 51	☽ ∥ ♂
05 18	♀ ∥ ♄
07 31	☽ □ ♀
08 12	☽ ☌ ♂
16 27	♀ # ♇
17 57	☽ □ ☿
19 29	☉ ∥ ♀
19 Monday	
00 43	☉ # ♇
02 09	☽ □ ☉
05 07	☽ # ♆
12 58	☽ Q ♆
16 40	☽ ☍ ♃
17 26	☉ ∥ ♄
17 28	☽ Q ♀
19 41	☽ ☌ ♅
21 49	☽ ⊼ ♄
22 43	☿ Q ♇
20 Tuesday	
00 08	☿ ∥ ♀
01 13	☿ Q ♄
01 28	☽ # ♃
07 50	☽ ∥ ♅
09 33	☽ ☌ ♇
09 55	☽ ± ♄
11 12	☽ Q ☿
13 05	☽ ⚹ ♆
16 35	☽ ⚹ ♀
17 26	☽ Q ☉
21 39	☽ ⊻ ♂
21 Wednesday	
02 02	☿ # ♇
04 12	☽ ⚼ ♄
08 37	☿ ∥ ♄
10 09	☽ ⊥ ♂
15 14	☽ ⚹ ☿
19 32	☽ ∠ ♆
20 00	☽ ⚹ ☉
21 48	☽ ∠ ♀
22 Thursday	
04 57	☽ ∠ ♂
05 36	☽ ⊼ ♃
08 23	☽ ⊻ ♅
10 48	☽ △ ♄
13 51	☽ ⊥ ♆
15 24	☽ ⊥ ♀
17 51	☽ ± ♃
20 34	☽ ⊥ ♅
22 22	☽ ⊻ ♇
22 51	☉ ∥ ☿
23 Friday	
02 03	☽ ⊻ ♆
02 21	☽ ∠ ☿
03 08	☽ ⊻ ♀
05 14	☽ ∠ ☉
10 30	☽ ⊥ ♇
12 11	☽ ⚼ ♃
12 12	☽ ⚹ ♂
14 47	☽ ∠ ♅
23 17	☽ ⊥ ☿
24 Saturday	
01 04	☽ ⊥ ☉
04 30	♀ # ♂
04 38	☽ ∠ ♇
09 25	♀ ☌ ♆
12 59	☽ Q ♂
13 08	☽ ⊻ ☿
14 11	☽ ⊻ ☉
18 30	☽ △ ♃
20 54	☽ ⚹ ♅
23 32	☽ □ ♄
25 Sunday	
08 03	☽ # ♅
09 36	☉ ☌ ☿
10 34	☽ ⚹ ♇
13 23	☽ ∥ ♃
13 24	☽ ☌ ♀
14 16	☽ ☌ ♆
20 41	☽ Q ♅
26 Monday	
01 36	☽ □ ♂
08 04	☿ □ ♃
10 10	☽ Q ♇
11 07	☽ ∥ ♆
11 14	☽ Q ♄
21 49	☉ □ ♃
27 Tuesday	
00 49	☿ □ ♅
01 15	☽ ∥ ♀
05 51	☽ □ ♃
06 29	☽ ☌ ☉
07 24	☽ # ♂
07 49	☽ □ ♅
08 49	☽ ☌ ☿
10 39	☽ ⚹ ♄
17 12	♄ # ♇
21 01	☽ □ ♇
22 29	☽ ⊻ ♀
22 53	☉ □ ♅
23 59	☿ ⚹ ♄
28 Wednesday	
00 44	☽ ⊻ ♆
06 45	☽ ∥ ♄
06 55	☽ # ♇
09 44	☽ ⊥ ♀
12 12	☽ ⊥ ♆
12 57	☽ △ ♂
15 29	☽ ∠ ♄
16 33	☽ Q ♃
29 Thursday	
02 30	☽ ∠ ♀
05 15	☽ ∠ ♆
08 34	☽ ⊥ ♄
13 23	☉ ⚹ ♄
15 18	☽ ⚹ ♃
16 49	☽ △ ♅
17 50	☽ ⚼ ♂
19 51	☽ ⊻ ♄
20 22	☽ ⊻ ☉
30 Friday	
01 38	☽ ⊻ ☿
05 36	☽ △ ♇
06 11	☽ ⚹ ♀
08 27	☽ ⊥ ☉
09 18	☽ ⚹ ♆
10 55	☽ ± ♂
14 16	☽ ⊥ ☿
19 20	☽ ∠ ♃
20 38	☽ ⚼ ♅
22 12	☽ ⊼ ♂

DECEMBER 1962

Date	Sidereal time h m s	Sun ☉ ° ' "	Moon ☽ ° ' "	Moon ☽ 24.00 ° ' "	Mercury ☿ ° '	Venus ♀ ° '	Mars ♂ ° '	Jupiter ♃ ° '	Saturn ♄ ° '	Uranus ♅ ° '	Neptune ♆ ° '	Pluto ♇ ° '
01	16 39 28	08 ♐ 48 57	28 ♑ 39 53	05 ♒ 17 20	12 ♐ 13	12 ♏ 18	21 ♌ 16	04 ♓ 37	07 ♒ 00	05 ♍ 16	14 ♏ 09	12 ♍ 07
02	16 43 25	09 49 48	11 ♒ 58 03	18 ♒ 42 11	13 47	12 R 14	21 31	04 44	07 05	05 16	14 11	12 07
03	16 47 21	10 50 39	25 ♒ 29 53	02 ♓ 21 20	15 21	12 D 13	21 47	04 50	07 10	05 17	14 13	12 08
04	16 51 18	11 51 32	09 ♓ 16 36	16 ♓ 15 45	16 55	12 14	22 01	04 57	07 15	05 17	14 15	12 08
05	16 55 14	12 52 25	23 ♓ 18 46	00 ♈ 25 33	18 28	12 18	22 15	05 04	07 20	05 17	14 17	12 08
06	16 59 11	13 53 19	07 ♈ 35 52	14 ♈ 49 22	20 02	12 24	22 29	05 11	07 25	05 18	14 19	12 09
07	17 03 07	14 54 13	22 ♈ 05 34	29 ♈ 23 54	21 36	12 32	22 42	05 18	07 30	05 18	14 21	12 09
08	17 07 04	15 55 08	06 ♉ 43 37	14 ♉ 03 55	23 10	12 43	22 54	05 26	07 36	05 18	14 23	12 09
09	17 11 01	16 56 04	21 ♉ 23 54	28 ♉ 42 37	24 44	12 55	23 06	05 33	07 41	05 18	14 25	12 09
10	17 14 57	17 57 01	05 ♊ 59 10	13 ♊ 12 39	26 18	13 10	23 17	05 41	07 46	05 18	14 27	12 09
11	17 18 54	18 57 59	20 ♊ 22 14	27 ♊ 27 12	27 52	13 27	23 28	05 49	07 52	05 R 18	14 29	12 10
12	17 22 50	19 58 58	04 ♋ 26 58	11 ♋ 21 05	29 ♐ 25	13 46	23 38	05 57	07 57	05 18	14 31	12 10
13	17 26 47	20 59 57	18 ♋ 09 16	24 ♋ 51 23	00 ♑ 59	14 07	23 48	06 05	08 03	05 18	14 33	12 10
14	17 30 43	22 00 58	01 ♌ 27 24	07 ♌ 57 29	02 33	14 30	23 56	06 13	08 09	05 18	14 35	12 10
15	17 34 40	23 01 59	14 ♌ 21 52	20 ♌ 40 53	04 07	14 55	24 05	06 22	08 14	05 18	14 37	12 R 10
16	17 38 36	24 03 01	26 ♌ 55 00	03 ♍ 04 43	05 41	15 22	24 12	06 30	08 20	05 17	14 39	12 10
17	17 42 33	25 04 04	09 ♍ 10 34	15 ♍ 13 09	07 14	15 50	24 19	06 39	08 26	05 17	14 40	12 10
18	17 46 30	26 05 08	21 ♍ 13 05	27 ♍ 11 01	08 48	16 20	24 25	06 48	08 32	05 17	14 42	12 10
19	17 50 26	27 06 13	03 ♎ 07 33	09 ♎ 03 20	10 21	16 51	24 31	06 57	08 38	05 16	14 44	12 09
20	17 54 23	28 07 19	14 ♎ 58 57	20 ♎ 55 01	11 54	17 24	24 35	07 06	08 44	05 16	14 46	12 09
21	17 58 19	29 ♐ 08 25	26 ♎ 52 04	02 ♏ 50 37	13 27	17 59	24 39	07 15	08 50	05 15	14 47	12 09
22	18 02 16	00 ♑ 09 32	08 ♏ 51 09	14 ♏ 54 05	15 00	18 34	24 42	07 25	08 56	05 15	14 49	12 09
23	18 06 12	01 10 40	20 ♏ 59 47	27 ♏ 08 33	16 31	19 12	24 45	07 34	09 02	05 14	14 51	12 09
24	18 10 09	02 11 49	03 ♐ 20 39	09 ♐ 36 16	18 02	19 50	24 47	07 44	09 08	05 13	14 53	12 08
25	18 14 05	03 12 58	15 ♐ 55 31	22 ♐ 18 28	19 33	20 30	24 48	07 54	09 14	05 13	14 54	12 08
26	18 18 02	04 14 07	28 ♐ 45 08	05 ♑ 15 28	21 02	21 11	24 R 48	08 04	09 21	05 12	14 56	12 08
27	18 21 59	05 15 17	11 ♑ 49 23	18 ♑ 26 45	22 30	21 53	24 47	08 14	09 27	05 11	14 57	12 07
28	18 25 55	06 16 28	25 ♑ 07 24	01 ♒ 51 11	23 56	22 36	24 46	08 24	09 33	05 10	14 59	12 07
29	18 29 52	07 17 38	08 ♒ 37 53	15 ♒ 27 20	25 21	23 21	24 44	08 34	09 40	05 09	15 01	12 06
30	18 33 48	08 18 48	22 ♒ 19 20	29 ♒ 13 41	26 43	24 06	24 41	08 45	09 46	05 08	15 02	12 06
31	18 37 45	09 ♑ 19 58	06 ♓ 10 13	13 ♓ 08 44	28 ♑ 03	24 ♏ 52	24 ♌ 37	08 ♓ 55	09 ♒ 53	05 ♍ 07	15 ♏ 04	12 ♍ 05

Date	Moon True ☊ ° '	Moon Mean ☊ ° '	Moon ☽ Latitude ° '
01	00 ♌ 46	02 ♌ 18	00 N 11
02	00 D 47	02 15	01 S 00
03	00 48	02 12	02 09
04	00 48	02 09	03 12
05	00 R 48	02 06	04 05
06	00 46	02 02	04 44
07	00 41	01 59	05 05
08	00 35	01 56	05 06
09	00 28	01 53	04 48
10	00 21	01 50	04 11
11	00 15	01 46	03 19
12	00 10	01 43	02 15
13	00 08	01 40	01 S 05
14	00 D 07	01 37	00 N 07
15	00 08	01 34	01 17
16	00 09	01 31	02 21
17	00 11	01 27	03 17
18	00 12	01 24	04 03
19	00 R 12	01 21	04 38
20	00 11	01 18	05 02
21	00 08	01 15	05 12
22	00 03	01 11	05 09
23	29 ♋ 58	01 08	04 52
24	29 53	01 05	04 21
25	29 49	01 02	03 37
26	29 45	00 59	02 42
27	29 42	00 56	01 37
28	29 41	00 52	00 N 25
29	29 D 41	00 49	00 S 49
30	29 43	00 46	02 02
31	29 ♋ 44	00 ♌ 43	03 S 08

DECLINATIONS

Date	Sun ☉ ° '	Moon ☽ ° '	Mercury ☿ ° '	Venus ♀ ° '	Mars ♂ ° '	Jupiter ♃ ° '	Saturn ♄ ° '	Uranus ♅ ° '	Neptune ♆ ° '	Pluto ♇ ° '
01	21 S 46	20 S 15	23 S 26	15 S 03	16 N 44	10 S 56	19 S 15	10 N 17	14 S 27	19 N 21
02	21 56	18 10	23 44	14 50	16 41	10 54	19 14	10 17	14 27	19 21
03	22 04	15 03	24 00	14 38	16 38	10 51	19 13	10 17	14 28	19 21
04	22 13	11 04	24 15	14 28	16 36	10 48	19 11	10 17	14 28	19 21
05	22 21	06 24	24 28	14 18	16 33	10 46	19 10	10 17	14 29	19 22
06	22 28	01 S 20	24 41	14 09	16 31	10 43	19 09	10 17	14 29	19 22
07	22 35	03 N 53	24 52	14 02	16 28	10 40	19 08	10 17	14 30	19 22
08	22 42	08 56	25 02	13 56	16 26	10 37	19 06	10 17	14 31	19 23
09	22 48	13 28	25 10	13 51	16 25	10 34	19 05	10 17	14 31	19 23
10	22 54	17 11	25 17	13 46	16 23	10 31	19 03	10 17	14 32	19 24
11	22 59	19 48	25 23	13 43	16 21	10 28	19 02	10 17	14 32	19 24
12	23 04	21 07	25 27	13 41	16 20	10 25	19 01	10 17	14 33	19 24
13	23 08	21 09	25 30	13 40	16 19	10 22	18 59	10 17	14 33	19 25
14	23 12	19 57	25 31	13 39	16 19	10 19	18 58	10 17	14 34	19 25
15	23 16	17 45	25 31	13 40	16 18	10 15	18 56	10 17	14 34	19 26
16	23 19	14 45	25 29	13 41	16 18	10 12	18 55	10 17	14 35	19 26
17	23 21	11 10	25 26	13 43	16 18	10 09	18 53	10 18	14 35	19 27
18	23 23	07 12	25 22	13 46	16 18	10 05	18 52	10 18	14 36	19 27
19	23 25	03 N 01	25 16	13 50	16 18	10 02	18 50	10 18	14 36	19 27
20	23 26	01 S 16	25 08	13 54	16 19	09 58	18 49	10 18	14 37	19 28
21	23 26	05 30	24 59	13 59	16 20	09 55	18 47	10 18	14 37	19 28
22	23 27	09 34	24 48	14 05	16 21	09 51	18 46	10 19	14 38	19 29
23	23 26	13 19	24 36	14 11	16 22	09 47	18 44	10 19	14 38	19 30
24	23 25	16 33	24 22	14 17	16 24	09 44	18 42	10 19	14 39	19 30
25	23 24	19 06	24 07	14 24	16 26	09 40	18 41	10 20	14 39	19 31
26	23 22	20 44	23 51	14 32	16 28	09 36	18 39	10 20	14 39	19 31
27	23 20	21 19	23 33	14 40	16 31	09 32	18 37	10 20	14 40	19 32
28	23 18	20 42	23 14	14 48	16 34	09 28	18 36	10 21	14 40	19 32
29	23 15	18 54	22 53	14 57	16 37	09 24	18 34	10 21	14 41	19 33
30	23 11	15 59	22 32	15 06	16 40	09 20	18 32	10 21	14 41	19 34
31	23 S 07	12 S 09	22 S 09	15 S 15	16 N 44	09 S 16	18 S 31	10 N 22	14 S 42	19 N 34

ZODIAC SIGN ENTRIES

Date	h	m	Planets
01	14	26	☽ ♒
03	19	53	☽ ♓
05	23	17	☽ ♈
08	00	59	☽ ♉
10	02	07	☽ ♊
12	04	21	☽ ♋
12	20	51	☿ ♑
14	09	20	☽ ♌
16	17	59	☽ ♍
19	05	41	☽ ♎
21	18	18	☽ ♏
22	08	15	☉ ♑
24	05	33	☽ ♐
26	14	19	☽ ♑
28	20	42	☽ ♒
31	01	20	☽ ♓

LATITUDES

Date	Mercury ☿ ° '	Venus ♀ ° '	Mars ♂ ° '	Jupiter ♃ ° '	Saturn ♄ ° '	Uranus ♅ ° '	Neptune ♆ ° '	Pluto ♇ ° '
01	01 S 11	00 N 30	02 N 27	01 S 12	00 S 45	00 N 46	01 N 43	13 N 23
04	01 27	01 06	02 34	01 11	00 45	00 46	01 43	13 24
07	01 41	01 39	02 40	01 11	00 45	00 46	01 43	13 26
10	01 54	02 07	02 47	01 10	00 45	00 46	01 44	13 27
13	02 03	02 32	02 54	01 10	00 45	00 46	01 44	13 29
16	02 10	02 53	03 01	01 09	00 45	00 46	01 44	13 30
19	02 14	03 10	03 08	01 09	00 45	00 46	01 44	13 32
22	02 13	03 25	03 15	01 08	00 46	00 47	01 44	13 33
25	02 08	03 36	03 22	01 08	00 46	00 47	01 44	13 34
28	01 56	03 45	03 30	01 08	00 46	00 47	01 44	13 36
31	01 S 38	03 N 51	03 N 37	01 S 07	00 S 46	00 N 47	01 N 44	13 N 37

DATA

Julian Date	2438000
Delta T	+34 seconds
Ayanamsa	23° 20' 05"
Synetic vernal point	05° ♓ 46' 55"
True obliquity of ecliptic	23° 26' 33"

MOON'S PHASES, APSIDES AND POSITIONS ☽

Date	h	m	Phase	Longitude ° '	Eclipse Indicator
04	16	48	☽	12 ♓ 04	
11	09	27	○	18 ♊ 52	
18	22	42	☾	26 ♍ 32	
26	22	59	●	04 ♑ 42	

Day	h	m		
08	16	45	Perigee	
20	10	45	Apogee	
06	18	07	0N	
13	00	21	Max dec	21° N 18'
20	04	55	0S	
27	11	48	Max dec	21° S 19'

LONGITUDES

Date	Chiron ⚷ ° '	Ceres ⚳ ° '	Pallas ⚴ ° '	Juno ⚵ ° '	Vesta ⚶ ° '	Black Moon Lilith ⚸ ° '
01	06 ♓ 15	11 ♍ 01	21 ♋ 58	20 ♍ 04	19 ♍ 24	14 ♎ 19
11	06 ♓ 26	13 ♍ 26	20 ♋ 51	22 ♍ 37	22 ♍ 29	15 ♎ 26
21	06 ♓ 43	15 ♍ 24	18 ♋ 36	24 ♍ 41	25 ♍ 13	16 ♎ 34
31	07 ♓ 05	16 ♍ 48	15 ♋ 30	26 ♍ 10	27 ♍ 29	17 ♎ 41

All ephemeris data is given at 12.00 UT and the Moon's longitude is additionally given for 24.00 UT

ASPECTARIAN

h m	Aspects
01 Saturday	
02 25	☽ ∠ ☉
04 05	☽ Q ♀
07 24	☽ Q ♆
09 00	☽ ∠ ☿
09 11	☽ ⚼ ♇
10 26	☿ □ ♇
11 55	☽ ⊥ ♃
13 05	☽ ± ♅
13 09	☿ ⊻ ♀
22 53	☽ ⊻ ♃
23 57	☽ ⊼ ♅
23 58	☽ ⋕ ♇
02 Sunday	
01 08	☽ ∥ ♄
01 30	☽ ± ♇
03 11	☽ ☌ ♄
07 51	☽ ✱ ☉
12 17	☽ ⊼ ♇
12 29	☽ □ ♀
15 40	☽ ✱ ☿
15 58	☽ □ ♆
18 19	☿ ⊻ ♆
03 Monday	
00 30	☽ ⋕ ♂
05 19	☽ ☍ ♂
06 57	☽ Q ☉
11 26	♀ St D
14 52	☽ ∥ ♀
15 40	☽ Q ☿
15 51	☽ ∥ ♆
04 Tuesday	
04 27	☽ ☌ ♃
05 05	☽ ☍ ♅
08 28	☽ ⊻ ♄
10 09	♀ ∥ ♆
13 24	☽ ∥ ♃
16 13	☽ ⋕ ♅
16 48	☽ □ ☉
16 55	☽ ☍ ♇
17 07	☽ △ ♀
18 32	☉ □ ♇
18 53	☽ ⊥ ♄
20 35	☽ △ ♆
21 23	☉ ⊻ ♀
05 Wednesday	
02 45	☽ □ ☿
09 08	☿ ⊥ ♀
10 11	☽ ⊼ ♂
10 20	☽ ∠ ♄
18 47	☽ ⚼ ♀
20 29	☽ ± ♂
22 07	☽ ⚼ ♆
06 Thursday	
07 56	☽ ⊻ ♃
08 09	☽ ⊼ ♅
09 59	☽ ± ♀
11 42	☽ ✱ ♄
11 48	☽ ⚼ ♂
13 12	☽ ± ♆
16 26	☿ ⊥ ♆
18 01	☽ ⊥ ♃
18 09	☽ ± ♅
19 34	☽ ⊼ ♇
20 03	☽ ⊼ ♀
22 34	☉ ⊻ ♆
23 12	☽ ⊼ ♆
23 14	☽ △ ☉
07 Friday	
05 30	☽ ± ♇
07 43	☽ Q ♄
09 02	☽ ∠ ♃
09 03	☽ ⚼ ♅
11 01	♃ ☍ ♅
11 06	☽ △ ☿
13 01	☽ △ ♂
20 19	☽ ⚼ ♇
08 Saturday	
01 47	☽ ⚼ ☉
02 41	☿ ∠ ♄
07 24	☿ △ ♂
09 40	☽ △ ♅
09 51	☽ ✱ ♃
13 26	☽ □ ♄
14 38	☽ ⚼ ☿
16 20	☿ Q ♃
17 36	☽ ± ☉
18 47	☽ ∥ ♅
20 26	☽ ⋕ ♃
20 52	☽ △ ♇
21 55	☽ ☍ ♀
09 Sunday	
00 33	☽ ☍ ♆
04 09	☽ ⊼ ☉
05 39	☽ Q ♃
07 07	☽ ± ☿
14 05	☽ ⋕ ♀
14 50	☽ □ ♂
18 07	☽ ⊼ ☿
18 11	☽ ⋕ ♆
10 Monday	
06 15	☽ ∥ ♂
10 52	☽ □ ♅
11 29	☽ □ ♃
14 59	☽ △ ♄
20 55	☽ Q ♂
22 15	☽ □ ♇
11 Tuesday	
00 10	☽ ⊼ ♀
02 06	☽ ⊼ ♆
03 50	☽ ⋕ ♄
05 13	♅ St R
07 30	☽ ∥ ♇
09 27	☽ ☍ ☉
10 26	☽ ± ♀
12 12	☽ ± ♆
16 14	☽ ⚼ ♄
16 57	☽ Q ♅
17 18	☽ ✱ ♂
23 25	☿ ∠ ♀
12 Wednesday	
02 01	☽ ⚼ ♀
02 17	☽ ☍ ☿
03 30	☽ ⚼ ♆
04 38	☽ Q ♇
04 42	☉ ⊥ ♀
07 41	☽ ± ♄
13 28	☿ ∠ ♆
13 29	☽ ✱ ♅
14 37	☽ △ ♃
18 07	☽ ⊼ ♄
19 21	☽ ∠ ♂
13 Thursday	
01 02	☉ ⊥ ♆
01 25	☽ ✱ ♇
04 40	☽ △ ♀
05 36	☽ △ ♆
11 21	☽ ⊥ ♂
15 49	☽ ∠ ♅
17 16	☽ ⚼ ♃
17 29	☽ ⊼ ☉
22 12	☽ ⊻ ♂
14 Friday	
04 10	☽ ∠ ♇
05 12	☽ ± ☉
05 19	☿ ⊥ ♄
08 03	☽ ⊥ ♅
09 43	☽ ± ♃
14 17	☽ ⊼ ☿
16 56	♀ ☌ ♆
18 51	☽ ∥ ♇
19 04	☽ ⊻ ♅
20 40	☽ ⊥ ♇
20 52	☽ ⊼ ♃
21 46	♇ St R
22 47	♃ ⋕ ♅
23 08	☽ ⚼ ☉
15 Saturday	
00 10	☽ ⋕ ♄
00 26	☽ ☍ ♄
02 55	☽ ± ☿
07 51	☽ ⊻ ♇
12 28	☽ □ ♆
13 05	☽ □ ♀
17 21	☉ ∠ ♄
22 17	☽ ⚼ ☿
16 Sunday	
00 17	☽ ∥ ♂
05 58	☽ △ ☉
06 03	☿ △ ♅
06 42	☽ ☌ ♂
13 12	☽ ⋕ ♆
16 04	☉ △ ♂
19 23	☽ ⋕ ♀
23 10	☽ Q ♆
17 Monday	
00 27	☉ Q ♃
01 03	☽ Q ♀
01 57	☿ ✱ ♃
04 20	☽ ☌ ♅
06 57	☽ ☍ ♃
07 37	☽ △ ☿
10 31	☽ ⊼ ♄
17 28	☽ ∥ ♅
17 55	☽ ☌ ♇
18 29	☽ ⋕ ♃
22 31	☽ ± ♄
22 56	☽ ✱ ♆
18 Tuesday	
01 47	☽ ✱ ♀
07 33	☿ ⊻ ♄
16 41	☽ ⚼ ♄
18 29	☽ ⊻ ♂
22 08	☿ ⚼ ♂
22 42	☽ □ ☉
19 Wednesday	
05 08	☽ ∠ ♆
06 40	☽ ⊥ ♂
09 18	☽ ∠ ♀
16 20	☽ ⊻ ♅
19 50	☽ ⊼ ♃
23 14	☽ △ ♄
23 23	☽ ⊥ ♆
20 Thursday	
01 00	☽ ∠ ♂
04 24	☽ ⊥ ♀
04 29	☽ ⊥ ♅
04 50	☽ □ ☿
06 07	♀ Q ♅
06 17	☽ ⊻ ♇
08 08	☽ ± ♃
11 33	☽ ⊻ ♆
14 31	☽ Q ☉
15 48	☿ △ ♇
17 08	☽ ⊻ ♀
18 25	☽ ⊥ ♇
22 40	☽ ∠ ♅
21 Friday	
02 35	☽ ⚼ ♃
07 31	☽ ✱ ♂
12 34	☽ ∠ ♇
17 00	☽ ✱ ☉
22 35	☽ Q ☿
22 Saturday	
03 47	☉ ∠ ♆
04 49	☽ ✱ ♅
07 42	☽ Q ♂
09 05	☽ △ ♃
09 15	☿ ✱ ♆
12 09	☽ □ ♄
13 40	☽ ∥ ♃
16 33	☽ ⋕ ♅
18 33	☽ ✱ ♇
23 52	☽ ☌ ♆
23 Sunday	
01 39	☽ ∠ ☉
01 56	☽ ✱ ☿
04 37	☽ Q ♅
08 16	☽ ☌ ♀
18 09	☽ Q ♇
18 12	☽ ∥ ♀
19 22	☽ □ ♂
20 55	☽ ⊥ ☉
21 18	☽ ∥ ♆
23 53	☽ Q ♄
24 Monday	
09 35	☽ ⊻ ☉
10 46	☽ ⋕ ♂
11 20	☽ ∠ ☿
15 36	☽ □ ♅
20 32	☽ □ ♃
23 12	☽ ✱ ♄
23 53	☿ ± ♂
25 Tuesday	
04 49	☽ □ ♇
06 54	☽ ⊥ ☿
07 32	☽ ∥ ♄
10 04	☽ ⊻ ♆
12 38	☉ ⊥ ♄
16 55	☽ ⋕ ♇
19 43	☽ ⊻ ☿
21 06	☽ ⊻ ♀
21 23	☽ ⊥ ♆
22 34	☿ ⚼ ♅
26 Wednesday	
03 44	☽ ∠ ♄
04 39	☽ △ ♂
06 11	♂ St R
06 56	☽ Q ♃
08 55	☽ ⊥ ♀
14 11	☽ ∠ ♆
16 36	☿ ✱ ♀
20 33	☽ ⊥ ♄
22 59	☽ ☌ ☉
23 52	☽ △ ♅
27 Thursday	
02 28	☽ ∠ ♀
05 21	☽ ✱ ♃
07 38	☽ ⊻ ♄
08 18	☽ ⚼ ♂
10 16	☉ △ ♅
12 13	♀ ∥ ♆
12 32	☽ △ ♇
17 42	☽ ✱ ♆
28 Friday	
00 36	☽ ± ♂
01 42	☿ ∠ ♃
03 07	☽ ⚼ ♅
06 17	☉ ∥ ☿
07 14	☽ ✱ ♀
08 52	☽ ∠ ♃
09 37	☽ ☌ ☿
11 22	☽ ⊼ ♂
15 20	☽ Q ♆
15 33	☽ ⚼ ♇
19 13	☽ ± ♅
29 Saturday	
01 08	☽ ⊥ ♃
01 41	☿ ⊼ ♂
04 57	☽ ⋕ ♇
05 51	☽ Q ♀
05 51	☽ ⊼ ♅
07 32	☽ ± ♇
09 27	☽ ⊻ ☉
11 53	☽ ⊻ ♃
13 50	☽ ☌ ♄
15 14	☽ ∥ ♄
18 07	☽ ⊼ ♇
20 52	☽ ⊥ ☉
23 14	☽ □ ♆
30 Sunday	
07 10	☽ ⋕ ♂
12 02	♀ Q ♇
13 52	☽ ∠ ☉
15 16	☽ □ ♀
16 05	☽ ☍ ♂
17 44	☿ Q ♆
17 49	☽ ∥ ♀
18 42	☿ ⚼ ♇
20 28	☽ ⊻ ☿
20 44	☽ ∥ ♆
31 Monday	
00 12	☉ ✱ ♃
04 52	♀ □ ♂
07 58	☽ ⊥ ☿
10 11	☽ ☍ ♅
16 48	☽ ☌ ♃
17 53	☽ ✱ ☉
18 14	☉ ⚼ ♂
18 26	☽ ⊻ ♄
21 51	☽ ⋕ ♅
22 11	☽ ☍ ♇

JANUARY 1963

LONGITUDES

Date	Sidereal time h m s	Sun ☉ ° ' "	Moon ☽ ° ' "	Moon ☽ 24.00 ° ' "	Mercury ☿ ° '	Venus ♀ ° '	Mars ♂ ° '	Jupiter ♃ ° '	Saturn ♄ ° '	Uranus ♅ ° '	Neptune ♆ ° '	Pluto ♇ ° '
01	18 41 41	10 ♑ 21 08	20 ♓ 09 04	27 ♓ 11 03	29 ♑ 19	25 ♏ 39	24 ♌ 32	09 ♓ 06	09 ♒ 59	05 ♍ 06	15 ♏ 05	12 ♍ 05
02	18 45 38	11 22 18	04 ♈ 14 28	11 ♈ 19 07	00 ♒ 32	26 27	24 R 27	09 16	10 06	05 R 04	15 07	12 R 04
03	18 49 34	12 23 27	18 ♈ 24 47	25 ♈ 31 11	01 41	27 16	24 21	09 27	10 13	05 03	15 08	12 04
04	18 53 31	13 24 37	02 ♉ 38 01	09 ♉ 44 58	02 45	28 06	24 14	09 38	10 19	05 02	15 09	12 03
05	18 57 28	14 25 45	16 ♉ 51 40	23 ♉ 57 40	03 43	28 57	24 06	09 49	10 26	05 01	15 11	12 02
06	19 01 24	15 26 54	01 ♊ 02 34	08 ♊ 05 54	04 34	29 ♏ 48	23 57	10 01	10 33	04 59	15 12	12 02
07	19 05 21	16 28 02	15 ♊ 07 11	22 ♊ 05 58	05 18	00 ♐ 40	23 48	10 12	10 39	04 58	15 13	12 01
08	19 09 17	17 29 10	29 ♊ 01 48	05 ♋ 54 16	05 54	01 33	23 37	10 23	10 46	04 56	15 15	12 00
09	19 13 14	18 30 18	12 ♋ 43 00	19 ♋ 27 41	06 20	02 26	23 26	10 35	10 53	04 55	15 16	11 59
10	19 17 10	19 31 25	26 ♋ 08 05	02 ♌ 44 01	06 37	03 21	23 15	10 46	11 00	04 53	15 17	11 59
11	19 21 07	20 32 33	09 ♌ 15 24	15 ♌ 42 13	06 R 42	04 15	23 02	10 58	11 07	04 52	15 18	11 58
12	19 25 03	21 33 39	22 ♌ 04 33	28 ♌ 22 33	06 36	05 11	22 48	11 10	11 14	04 50	15 20	11 57
13	19 29 00	22 34 46	04 ♍ 36 27	10 ♍ 46 32	06 19	06 07	22 34	11 22	11 20	04 48	15 21	11 56
14	19 32 57	23 35 53	16 ♍ 53 10	22 ♍ 56 46	05 49	07 03	22 19	11 34	11 27	04 47	15 22	11 55
15	19 36 53	24 36 59	28 ♍ 57 47	04 ♎ 56 44	05 08	08 01	22 04	11 46	11 34	04 45	15 23	11 54
16	19 40 50	25 38 05	10 ♎ 54 10	16 ♎ 50 37	04 17	08 58	21 47	11 58	11 41	04 43	15 24	11 53
17	19 44 46	26 39 11	22 ♎ 46 42	28 ♎ 42 59	03 16	09 56	21 30	12 10	11 48	04 41	15 25	11 52
18	19 48 43	27 40 16	04 ♏ 40 05	10 ♏ 38 35	02 08	10 55	21 12	12 22	11 55	04 39	15 26	11 51
19	19 52 39	28 41 22	16 ♏ 39 04	22 ♏ 42 05	00 ♒ 54	11 54	20 54	12 35	12 02	04 37	15 27	11 50
20	19 56 36	29 ♑ 42 27	28 ♏ 48 10	04 ♐ 57 49	29 ♑ 37	12 54	20 35	12 47	12 10	04 35	15 28	11 49
21	20 00 32	00 ♒ 43 31	11 ♐ 11 27	17 ♐ 29 28	28 20	13 54	20 15	13 00	12 17	04 33	15 29	11 48
22	20 04 29	01 44 36	23 ♐ 52 10	00 ♑ 19 46	27 04	14 54	19 55	13 12	12 24	04 31	15 29	11 47
23	20 08 26	02 45 39	06 ♑ 52 27	13 ♑ 30 14	25 51	15 55	19 34	13 25	12 31	04 29	15 30	11 46
24	20 12 22	03 46 42	20 ♑ 13 06	27 ♑ 00 53	24 44	16 56	19 13	13 38	12 38	04 27	15 31	11 45
25	20 16 19	04 47 45	03 ♒ 53 20	10 ♒ 50 08	23 44	17 58	18 51	13 50	12 45	04 25	15 32	11 44
26	20 20 15	05 48 46	17 ♒ 50 50	24 ♒ 54 56	22 52	19 00	18 29	14 03	12 52	04 23	15 33	11 42
27	20 24 12	06 49 47	02 ♓ 01 53	09 ♓ 11 05	22 08	20 02	18 06	14 16	12 59	04 21	15 33	11 41
28	20 28 08	07 50 46	16 ♓ 21 53	23 ♓ 33 41	21 34	21 05	17 43	14 29	13 07	04 18	15 34	11 40
29	20 32 05	08 51 45	00 ♈ 45 52	07 ♈ 57 51	21 08	22 08	17 20	14 43	13 14	04 16	15 35	11 39
30	20 36 01	09 52 42	15 ♈ 09 06	22 ♈ 19 09	20 51	23 11	16 56	14 56	13 21	04 14	15 35	11 38
31	20 39 58	10 ♒ 53 38	29 ♈ 27 35	06 ♉ 34 04	20 ♑ 42	24 ♐ 15	16 ♌ 33	15 ♓ 09	13 ♒ 28	04 ♍ 11	15 ♏ 36	11 ♍ 36

Date	Moon True ☊ ° '	Moon Mean ☊ ° '	Moon ☽ Latitude ° '
01	29 ♋ 45	00 ♌ 40	04 S 03
02	29 D 46	00 37	04 45
03	29 R 46	00 33	05 10
04	29 45	00 30	05 15
05	29 44	00 27	05 02
06	29 42	00 24	04 30
07	29 40	00 21	03 42
08	29 39	00 17	02 41
09	29 38	00 14	01 32
10	29 37	00 11	00 S 19
11	29 D 37	00 08	00 N 53
12	29 38	00 05	02 01
13	29 38	00 ♌ 02	03 02
14	29 39	29 ♋ 58	03 53
15	29 40	29 55	04 33
16	29 40	29 52	05 00
17	29 40	29 49	05 15
18	29 R 40	29 46	05 16
19	29 40	29 43	05 03
20	29 D 40	29 39	04 37
21	29 40	29 36	03 58
22	29 41	29 33	03 06
23	29 41	29 30	02 03
24	29 41	29 27	00 N 52
25	29 R 41	29 23	00 S 23
26	29 41	29 20	01 39
27	29 40	29 17	02 50
28	29 39	29 14	03 51
29	29 38	29 11	04 37
30	29 37	29 08	05 07
31	29 ♋ 36	29 ♋ 04	05 S 17

DECLINATIONS

Date	Sun ☉ ° '	Moon ☽ ° '	Mercury ☿ ° '	Venus ♀ ° '	Mars ♂ ° '	Jupiter ♃ ° '	Saturn ♄ ° '	Uranus ♅ ° '	Neptune ♆ ° '	Pluto ♇ ° '
01	23 S 02	07 S 38	21 S 45	15 S 25	16 N 47	09 S 12	18 S 29	10 N 22	14 S 42	19 N 35
02	22 57	02 S 40	21 21	15 34	16 51	09 08	18 27	10 23	14 42	19 35
03	22 52	02 N 27	20 56	15 44	16 56	09 03	18 25	10 23	14 43	19 36
04	22 46	07 27	20 31	15 55	17 00	08 59	18 24	10 24	14 43	19 37
05	22 40	12 03	20 06	16 05	17 05	08 55	18 22	10 24	14 43	19 37
06	22 33	15 58	19 41	16 15	17 10	08 51	18 20	10 25	14 44	19 38
07	22 26	18 56	19 17	16 26	17 16	08 46	18 18	10 25	14 44	19 39
08	22 18	20 45	18 53	16 36	17 21	08 42	18 16	10 26	14 44	19 39
09	22 10	21 18	18 31	16 47	17 27	08 37	18 15	10 27	14 45	19 40
10	22 01	20 37	18 10	16 57	17 33	08 33	18 13	10 27	14 45	19 41
11	21 52	18 48	17 52	17 08	17 40	08 28	18 11	10 28	14 45	19 41
12	21 43	16 04	17 35	17 18	17 46	08 24	18 09	10 28	14 46	19 42
13	21 33	12 39	17 22	17 29	17 53	08 19	18 07	10 29	14 46	19 43
14	21 23	08 45	17 11	17 39	18 00	08 14	18 05	10 30	14 46	19 43
15	21 12	04 35	17 02	17 50	18 07	08 10	18 03	10 30	14 46	19 44
16	21 01	00 N 18	16 57	18 00	18 14	08 05	18 02	10 31	14 47	19 45
17	20 50	03 S 59	16 55	18 10	18 21	08 00	18 00	10 32	14 47	19 46
18	20 38	08 07	16 56	18 20	18 29	07 56	17 58	10 33	14 47	19 46
19	20 26	11 58	16 59	18 29	18 36	07 51	17 56	10 33	14 47	19 47
20	20 13	15 23	17 05	18 39	18 44	07 46	17 54	10 34	14 48	19 48
21	20 00	18 12	17 12	18 48	18 52	07 41	17 52	10 35	14 48	19 49
22	19 46	20 12	17 21	18 57	19 00	07 36	17 50	10 36	14 48	19 49
23	19 33	21 13	17 31	19 06	19 08	07 31	17 48	10 36	14 48	19 50
24	19 19	21 04	17 42	19 15	19 16	07 26	17 46	10 37	14 48	19 51
25	19 04	19 40	17 53	19 23	19 24	07 21	17 44	10 38	14 48	19 52
26	18 49	17 03	18 05	19 31	19 32	07 16	17 42	10 39	14 49	19 52
27	18 34	13 24	18 17	19 38	19 40	07 11	17 40	10 40	14 49	19 53
28	18 19	08 55	18 29	19 45	19 48	07 06	17 38	10 41	14 49	19 54
29	18 03	03 S 56	18 41	19 52	19 56	07 01	17 36	10 41	14 49	19 55
30	17 47	01 N 15	18 52	19 59	20 04	06 56	17 34	10 42	14 49	19 55
31	17 S 30	06 N 21	19 S 03	20 S 05	20 N 12	06 S 50	17 S 32	10 N 43	14 S 49	19 N 56

ZODIAC SIGN ENTRIES

Date	h	m	Planets
02	01	10	☿ ♒
02	04	48	☽ ♈
04	07	34	☽ ♉
06	10	14	☽ ♊
06	17	35	♀ ♐
08	13	41	☽ ♋
10	19	01	☽ ♌
13	03	07	☽ ♍
15	14	05	☽ ♎
18	02	35	☽ ♏
20	04	59	☿ ♑
20	14	20	☽ ♐
20	18	54	☉ ♒
22	23	23	☽ ♑
25	05	14	☽ ♒
27	08	35	☽ ♓
29	10	44	☽ ♈
31	12	55	☽ ♉

LATITUDES

Date	Mercury ☿ ° '	Venus ♀ ° '	Mars ♂ ° '	Jupiter ♃ ° '	Saturn ♄ ° '	Uranus ♅ ° '	Neptune ♆ ° '	Pluto ♇ ° '
01	01 S 30	03 N 52	03 N 39	01 S 07	00 S 46	00 N 47	01 N 44	13 N 38
04	01 00	03 56	03 47	01 07	00 46	00 47	01 45	13 39
07	00 S 21	03 57	03 54	01 06	00 46	00 47	01 45	13 40
10	00 N 28	03 56	04 00	01 06	00 46	00 47	01 45	13 41
13	01 23	03 54	04 07	01 06	00 46	00 47	01 45	13 43
16	02 18	03 51	04 13	01 05	00 46	00 47	01 45	13 44
19	03 03	03 46	04 18	01 05	00 47	00 48	01 45	13 45
22	03 28	03 39	04 23	01 05	00 47	00 48	01 46	13 46
25	03 31	03 32	04 27	01 05	00 47	00 48	01 46	13 47
28	03 16	03 24	04 30	01 04	00 47	00 48	01 46	13 48
31	02 N 50	03 N 14	04 N 32	01 S 04	00 S 47	00 N 48	01 N 46	13 N 49

DATA

Julian Date	2438031
Delta T	+35 seconds
Ayanamsa	23° 20' 10"
Synetic vernal point	05° ♓ 46' 49"
True obliquity of ecliptic	23° 26' 34"

MOON'S PHASES, APSIDES AND POSITIONS ☽

Date	h	m	Phase	Longitude ° '	Eclipse Indicator
03	01	02	☽	11 ♈ 55	
09	23	08	○	18 ♋ 59	
17	20	34	☾	27 ♎ 01	
25	13	42	●	04 ♒ 52	Annular

Day	h	m		
04	08	29	Perigee	
17	08	02	Apogee	
29	07	17	Perigee	
03	00	33	0N	
09	10	26	Max dec	21° N 18'
16	13	38	0S	
23	21	00	Max dec	21° S 18'
30	06	13	0N	

LONGITUDES

Date	Chiron ⚷ ° '	Ceres ⚳ ° '	Pallas ⚴ ° '	Juno ⚵ ° '	Vesta ⚶ ° '	Black Moon Lilith ⚸ ° '
01	07 ♓ 08	16 ♍ 54	15 ♋ 10	26 ♍ 17	27 ♍ 41	17 ♎ 47
11	07 ♓ 35	17 ♍ 36	11 ♋ 51	27 ♍ 06	29 ♍ 22	18 ♎ 55
21	08 ♓ 06	17 ♍ 35	08 ♋ 58	27 ♍ 12	00 ♎ 25	20 ♎ 02
31	08 ♓ 39	16 ♍ 51	07 ♋ 02	26 ♍ 34	00 ♎ 44	21 ♎ 09

All ephemeris data is given at 12.00 UT and the Moon's longitude is additionally given for 24.00 UT

ASPECTARIAN

h m	Aspects
01 Tuesday	
01 02	☽ ∠ ☿
02 25	☉ ⊻ ♄
03 19	☽ △ ♆
03 54	☽ ∥ ♃
04 49	☽ ⊥ ♄
07 39	☿ ± ♅
16 03	☽ Q ☉
19 27	☽ ⊼ ♂
20 19	☽ ∠ ♄
21 57	☽ △ ♀
02 Wednesday	
04 58	☽ bQ ♆
05 08	☽ ⚹ ☿
05 36	☽ ± ♂
13 25	☽ ⊼ ♅
20 16	☽ ± ♆
20 39	☽ ⊻ ♃
20 46	☽ bQ ♂
21 39	☉ ∠ ♀
22 01	☽ ⚹ ♄
23 34	☽ ± ♅
03 Thursday	
00 58	☽ bQ ♀
01 02	☽ □ ☉
01 16	☽ ⊼ ♇
03 19	☽ Q ☿
04 18	☉ △ ♇
06 27	☽ ⊼ ♆
06 56	☽ ⊥ ♃
11 24	☽ ± ♇
14 46	☽ bQ ♅
17 07	☽ ± ♀
18 28	☽ Q ♄
21 56	☽ △ ♂
22 20	☽ ∠ ♃
04 Friday	
02 35	☽ bQ ♇
03 53	☽ ⊼ ♀
12 12	☽ □ ☿
16 02	☽ △ ♅
19 14	♀ Q ♄
19 37	☽ # ♃
23 58	☽ ⚹ ♃
05 Saturday	
01 04	☽ □ ♄
03 05	☽ ∥ ♅
03 52	☽ △ ♇
07 35	☽ △ ☉
09 09	☽ ☍ ♆
15 43	☿ ⊥ ♃
20 30	☽ Q ♃
06 Sunday	
00 07	☽ □ ♂
03 50	☽ # ♆
06 04	☉ ⚹ ♆
09 45	☽ ☍ ♀
10 55	☽ bQ ☉
14 07	☽ # ♀
15 09	☿ # ♇
18 20	☽ △ ☿
18 42	☽ □ ♅
21 02	☽ ∥ ♂
07 Monday	
00 47	☿ ⊼ ♅
03 25	☽ ± ☉
03 28	☽ □ ♃
04 18	☽ △ ♄
06 10	☽ # ♄
06 23	☽ Q ♂
06 42	☽ □ ♇
12 11	☽ ⊼ ♆
14 30	☽ ⊼ ☉
15 02	☽ # ☿
19 39	☽ ∥ ♇
21 20	☽ bQ ☿
22 30	☽ ± ♆
08 Tuesday	
01 28	☽ Q ♅
02 45	☽ ⚹ ♂
06 18	☽ bQ ♄
13 34	☽ ± ☿
13 42	☽ Q ♇
14 07	☽ bQ ♆
14 46	☉ ± ♂
16 41	☽ ⊼ ♀
17 01	☿ ± ♇
22 06	☽ ± ♄
22 17	☽ ⚹ ♅
09 Wednesday	
00 25	☽ ⊼ ☿
03 55	☽ ± ♀
04 33	☽ ∠ ♂
08 10	☽ △ ♃
08 44	☽ ⊼ ♄
10 43	☽ ⚹ ♇
16 32	☽ △ ♆
20 16	☽ ⊥ ♂
20 59	☽ bQ ♀
23 08	☽ ☍ ☉
10 Thursday	
00 47	☽ ∠ ♅
06 52	☽ ⊻ ♂
08 52	☿ ∥ ♄
11 20	☽ bQ ♃
13 31	☽ ∠ ♇
16 59	☽ ⊥ ♅
20 22	☉ bQ ♅
11 Friday	
01 54	☽ ∥ ♇
02 06	☽ △ ♀
03 55	☽ ⊻ ♅
03 58	☽ ± ♃
05 56	☽ ⊥ ♇
07 17	☽ ☍ ☿
11 46	☿ St R
15 13	☽ ⊼ ♃
15 28	☽ ☍ ♄
17 01	☽ ⊻ ♇
18 09	☽ # ♄
22 08	☽ # ☿
22 27	☽ ∥ ♂
23 16	☽ □ ♆
12 Saturday	
00 30	☿ # ♂
02 31	☽ # ♀
03 19	♀ □ ♅
10 56	☽ ⊼ ☉
13 22	☽ ☌ ♂
21 39	☽ # ♆
23 22	☽ ± ☉
13 Sunday	
04 23	☿ ∥ ♀
06 25	♃ ⊻ ♄
09 33	☽ Q ♆
11 50	☉ ⊼ ♂
12 23	☽ ☌ ♅
15 09	☽ □ ♀
15 12	☽ ⊼ ☿
15 29	☿ ⚹ ♀
18 17	☽ bQ ☉
14 Monday	
01 14	☽ ⊼ ♄
01 22	☽ ☍ ♃
01 34	☽ ∥ ♅
02 15	☽ ☌ ♇
02 28	☽ ± ☿
07 35	☿ ± ♇
09 00	☽ ⚹ ♆
13 08	☽ ± ♄
15 05	☽ # ♃
19 24	☽ bQ ☿
19 37	☿ ⊥ ♃
22 32	☽ ⊻ ♂
15 Tuesday	
02 31	☽ △ ☉
03 33	♂ # ♄
05 35	☽ Q ♀
07 11	☽ bQ ♄
10 14	☽ ⊥ ♂
14 51	☽ ∠ ♆
23 34	☽ ⊻ ♅
23 36	☽ △ ☿
23 58	☿ ⊼ ♅
16 Wednesday	
03 54	☽ ∠ ♂
04 00	♃ ☍ ♇
07 46	☽ ⚹ ♀
08 58	☽ ⊥ ♆
11 38	☽ ⊥ ♅
13 36	☽ △ ♄
13 59	☽ ⊻ ♇
14 10	☽ ⊼ ♃
15 27	♀ ∥ ♄
21 05	☽ ⊻ ♆
17 Thursday	
02 06	☽ ⊥ ♇
02 31	☽ ± ♃
05 46	☽ ∠ ♅
09 29	☽ ⚹ ♂
16 46	☽ ∠ ♀
17 05	☉ bQ ♇
20 16	☽ ∠ ♇
20 34	☽ □ ☉
21 01	☽ bQ ♃
23 56	♄ ⊼ ♇
18 Friday	
03 06	☉ ∠ ♃
06 15	☉ Q ♆
07 21	☽ □ ☿
09 08	☽ Q ♂
10 54	☽ ∥ ♃
11 58	☽ ⚹ ♅
12 33	☽ ⊥ ♀
19 Saturday	
01 40	☽ ⊻ ♀
02 24	☽ ⚹ ♇
02 42	☽ □ ♄
02 56	☽ # ♅
03 44	☽ △ ♃
09 36	☽ ☌ ♆
10 27	♀ □ ♇
10 27	☉ ± ♅
11 56	☽ Q ♅
12 05	☽ Q ☉
15 46	♀ ⚹ ♄
16 03	☽ Q ☿
20 13	☽ □ ♂
20 Sunday	
02 13	☽ Q ♇
07 34	☽ ∥ ♆
08 33	♀ □ ♃
11 07	☉ ☌ ☿
13 27	☽ ⚹ ☿
13 56	☽ ⚹ ☉
14 41	☽ Q ♄
23 15	☽ □ ♅
21 Monday	
02 19	☽ ∥ ☿
06 02	☿ ∠ ♀
07 40	☿ ± ♅
08 52	☽ ∥ ♄
13 10	☽ □ ♇
14 06	☽ ⚹ ♄
15 30	☽ □ ♃
15 42	☽ ∠ ☿
17 23	☿ ∠ ♃
17 37	☽ ☌ ♀
18 42	☽ ∥ ♀
19 22	☽ # ♂
20 11	☽ ⊻ ♆
21 25	☽ ∠ ☉
22 Tuesday	
03 49	☿ Q ♆
04 46	☽ △ ♂
06 24	☽ # ♇
06 30	☽ ∥ ☉
07 12	☉ # ♇
07 12	☽ ⊥ ☿
07 32	☽ ⊥ ♆
15 48	☽ ⊥ ☉
17 25	☿ bQ ♇
17 26	☽ ⊻ ☿
18 38	☽ ∠ ♄
23 Wednesday	
00 19	☽ ∠ ♆
01 50	☽ Q ♃
02 04	♀ ⊻ ♆
03 50	☽ ⊻ ☉
07 39	☽ △ ♅
07 54	☽ bQ ♂
11 20	☽ ⊥ ♄
20 51	☽ △ ♇
22 19	☽ ⊻ ♄
23 48	☽ ± ♂
24 Thursday	
00 02	☽ ⚹ ♃
03 36	☽ ⚹ ♆
05 41	☽ ⊻ ♀
10 15	☽ ⊼ ♂
10 38	☽ bQ ♅
14 49	☉ # ♂
16 20	☉ ∥ ♀
17 13	☽ ⊥ ♀
19 25	☽ ☌ ☿
19 33	☿ ∥ ♄
23 31	☽ bQ ♇
25 Friday	
00 54	☽ Q ♆
02 29	☽ ± ♅
03 04	☽ ∠ ♃
03 20	☉ ⊼ ♅
09 11	☿ ⊥ ♀
09 35	☽ # ♇
10 16	☽ ∠ ♀
12 55	☽ ⊼ ♅
13 42	☽ ☌ ☉
14 49	☽ # ♂
15 03	☽ ∥ ♀
15 11	☽ ± ♇
18 57	☽ ⊥ ♃
19 17	☽ ∥ ☉
26 Saturday	
01 31	☽ ⊼ ♇
03 05	♀ △ ♂
03 25	☽ ☌ ♄
04 14	☽ ∥ ☿
05 25	☽ ⊻ ♃
06 51	☽ ∥ ♄
08 04	☽ □ ♆
09 35	☉ ± ♇
13 03	☽ ☍ ♂
14 07	☽ ⚹ ♀
20 05	☽ ⊻ ☿
27 Sunday	
03 25	☽ ∥ ♆
05 45	☽ ⊥ ☿
12 01	☽ Q ♀
15 52	☽ ☍ ♅
20 13	☽ ∠ ☿
20 40	☽ ⊻ ☉
28 Monday	
02 40	☉ ∥ ☿
03 02	☽ # ♅
04 10	☽ ☍ ♇
06 31	☽ ⊻ ♄
07 28	☽ ⊥ ☉
08 49	☽ ☌ ♃
10 40	☽ △ ♆
14 12	☽ ⊼ ♂
16 37	☽ ⊥ ♄
19 33	☿ ⊻ ♀
20 24	☽ ⚹ ☿
20 29	☽ □ ♀
21 10	☽ ∥ ♃
23 09	♀ ⊥ ♆
23 38	☽ ∠ ☉
23 57	☽ ± ♂
29 Tuesday	
06 52	♂ ∥ ♇
07 22	☉ ⊥ ♃
07 44	☽ ∠ ♄
11 41	☽ bQ ♆
14 33	☽ bQ ♂
15 51	☽ Q ☿
17 49	☽ ⊼ ♅
22 08	♀ # ♇
30 Wednesday	
02 31	☽ ⚹ ☉
02 42	☽ ± ♆
03 48	☽ ± ♅
06 07	☽ ⊼ ♇
08 58	☽ ⚹ ♄
11 37	☽ ⊻ ♃
12 44	☽ ⊼ ♆
14 55	☽ △ ♂
16 08	☽ ± ♇
18 48	☽ bQ ♅
21 25	☽ □ ☿
21 49	☽ ⊥ ♃
31 Thursday	
00 07	☽ Q ☉
02 32	☽ △ ♀
05 14	☽ Q ♄
07 12	☽ bQ ♇
08 30	☉ ∥ ♄
13 11	☽ ∠ ♃
14 23	☽ # ♃
19 57	☽ △ ♅

FEBRUARY 1963

LONGITUDES

Date	Sidereal time h m s	Sun ☉ ° ' "	Moon ☽ ° ' "	Moon ☽ 24.00 ° ' "	Mercury ☿ ° '	Venus ♀ ° '	Mars ♂ ° '	Jupiter ♃ ° '	Saturn ♄ ° '	Uranus ♅ ° '	Neptune ♆ ° '	Pluto ♇ ° '
01	20 43 55	11 ♒ 54 32	13 ♉ 38 18	20 ♉ 40 05	20 ♑ 42	25 ♐ 19	16 ♌ 09	15 ♓ 22	13 ♒ 35	04 ♍ 09	15 ♏ 36	11 ♍ 35
02	20 47 51	12 55 26	27 ♉ 39 14	04 ♊ 35 37	20 D 49	26 23	15 R 45	15 36	13 42	04 R 07	15 37	11 R 34
03	20 51 48	13 56 18	11 ♊ 29 08	18 ♊ 19 43	21 02	27 27	15 21	15 49	13 50	04 04	15 37	11 32
04	20 55 44	14 57 08	25 ♊ 07 20	01 ♋ 51 55	21 23	28 32	14 57	16 02	13 57	04 02	15 38	11 31
05	20 59 41	15 57 57	08 ♋ 33 27	15 ♋ 11 53	21 49	29 ♐ 37	14 33	16 16	14 04	03 59	15 38	11 30
06	21 03 37	16 58 45	21 ♋ 47 12	28 ♋ 19 22	22 20	00 ♑ 42	14 09	16 30	14 11	03 57	15 38	11 28
07	21 07 34	17 59 32	04 ♌ 48 21	11 ♌ 14 10	22 57	01 47	13 45	16 43	14 18	03 54	15 39	11 27
08	21 11 30	19 00 17	17 ♌ 36 47	23 ♌ 56 14	23 38	02 53	13 22	16 57	14 26	03 52	15 39	11 25
09	21 15 27	20 01 01	00 ♍ 12 33	06 ♍ 25 50	24 23	03 59	12 58	17 11	14 33	03 49	15 39	11 24
10	21 19 24	21 01 43	12 ♍ 36 10	18 ♍ 43 43	25 12	05 05	12 35	17 24	14 40	03 47	15 39	11 23
11	21 23 20	22 02 25	24 ♍ 48 40	00 ♎ 51 15	26 05	06 11	12 12	17 38	14 47	03 44	15 40	11 21
12	21 27 17	23 03 05	06 ♎ 51 46	12 ♎ 50 32	27 01	07 18	11 50	17 52	14 54	03 42	15 40	11 20
13	21 31 13	24 03 44	18 ♎ 47 56	24 ♎ 44 24	28 00	08 24	11 27	18 06	15 01	03 39	15 40	11 18
14	21 35 10	25 04 21	00 ♏ 40 23	06 ♏ 36 23	29 ♑ 01	09 31	11 06	18 20	15 09	03 37	15 40	11 17
15	21 39 06	26 04 58	12 ♏ 32 58	18 ♏ 30 40	00 ♒ 05	10 38	10 44	18 34	15 16	03 34	15 40	11 15
16	21 43 03	27 05 33	24 ♏ 30 04	00 ♐ 31 48	01 11	11 45	10 23	18 48	15 23	03 31	15 R 40	11 14
17	21 46 59	28 06 07	06 ♐ 36 26	12 ♐ 44 37	02 20	12 53	10 03	19 02	15 30	03 29	15 40	11 12
18	21 50 56	29 ♒ 06 40	18 ♐ 56 53	25 ♐ 13 50	03 30	14 00	09 43	19 16	15 37	03 26	15 40	11 11
19	21 54 53	00 ♓ 07 12	01 ♑ 35 58	08 ♑ 03 43	04 43	15 08	09 24	19 30	15 44	03 23	15 40	11 09
20	21 58 49	01 07 43	14 ♑ 37 29	21 ♑ 17 31	05 57	16 16	09 05	19 44	15 51	03 21	15 40	11 08
21	22 02 46	02 08 12	28 ♑ 04 00	04 ♒ 56 56	07 13	17 24	08 47	19 58	15 58	03 18	15 40	11 06
22	22 06 42	03 08 39	11 ♒ 56 11	19 ♒ 01 28	08 30	18 32	08 30	20 13	16 05	03 16	15 39	11 04
23	22 10 39	04 09 05	26 ♒ 12 18	03 ♓ 28 04	09 49	19 41	08 13	20 27	16 12	03 13	15 39	11 03
24	22 14 35	05 09 30	10 ♓ 47 59	18 ♓ 11 09	11 09	20 49	07 57	20 41	16 19	03 10	15 39	11 01
25	22 18 32	06 09 52	25 ♓ 36 39	03 ♈ 03 04	12 31	21 58	07 42	20 55	16 26	03 08	15 38	11 00
26	22 22 28	07 10 13	10 ♈ 29 41	17 ♈ 55 20	13 54	23 06	07 28	21 10	16 33	03 05	15 38	10 58
27	22 26 25	08 10 32	25 ♈ 19 02	02 ♉ 39 55	15 18	24 15	07 14	21 24	16 40	03 02	15 38	10 57
28	22 30 22	09 ♓ 10 49	09 ♉ 57 15	17 ♉ 10 27	16 ♒ 44	25 ♑ 24	07 ♌ 01	21 ♓ 39	16 ♒ 46	03 ♍ 00	15 ♏ 37	10 ♍ 55

Moon

Date	Moon True ☊ ° '	Moon Mean ☊ ° '	Moon ☽ Latitude ° '
01	29 ♋ 36	29 ♋ 01	05 S 07
02	29 D 37	28 58	04 40
03	29 38	28 55	03 56
04	29 39	28 52	03 00
05	29 40	28 49	01 54
06	29 41	28 45	00 S 43
07	29 R 41	28 42	00 N 28
08	29 40	28 39	01 37
09	29 38	28 36	02 40
10	29 35	28 33	03 35
11	29 31	28 29	04 18
12	29 27	28 26	04 50
13	29 23	28 23	05 08
14	29 20	28 20	05 14
15	29 18	28 17	05 05
16	29 17	28 14	04 44
17	29 D 17	28 10	04 10
18	29 19	28 07	03 23
19	29 20	28 04	02 26
20	29 22	28 01	01 20
21	29 23	27 58	00 N 07
22	29 R 22	27 55	01 S 08
23	29 20	27 51	02 21
24	29 16	27 48	03 26
25	29 10	27 45	04 19
26	29 04	27 42	04 54
27	28 59	27 39	05 09
28	28 ♋ 54	27 ♋ 35	05 S 04

DECLINATIONS

Date	Sun ☉ ° '	Moon ☽ ° '	Mercury ☿ ° '	Venus ♀ ° '	Mars ♂ ° '	Jupiter ♃ ° '	Saturn ♄ ° '	Uranus ♅ ° '	Neptune ♆ ° '	Pluto ♇ ° '
01	17 S 13	11 N 03	19 S 14	20 S 11	20 N 20	06 S 45	17 S 30	10 N 44	14 S 49	19 N 57
02	16 56	15 06	19 23	20 16	20 27	06 40	17 28	10 45	14 49	19 58
03	16 39	18 16	19 32	20 21	20 35	06 35	17 26	10 46	14 49	19 59
04	16 21	20 22	19 40	20 25	20 42	06 29	17 24	10 47	14 49	19 59
05	16 03	21 16	19 47	20 29	20 50	06 24	17 22	10 48	14 50	20 00
06	15 45	20 58	19 54	20 33	20 57	06 19	17 20	10 48	14 50	20 01
07	15 26	19 31	19 59	20 36	21 03	06 13	17 18	10 49	14 50	20 02
08	15 08	17 06	20 03	20 38	21 10	06 08	17 16	10 50	14 50	20 02
09	14 49	13 54	20 06	20 41	21 17	06 03	17 14	10 51	14 50	20 03
10	14 29	10 08	20 09	20 42	21 23	05 57	17 12	10 52	14 50	20 04
11	14 10	06 01	20 10	20 43	21 29	05 52	17 10	10 53	14 50	20 05
12	13 50	01 N 43	20 09	20 44	21 35	05 46	17 08	10 54	14 50	20 05
13	13 30	02 S 36	20 08	20 44	21 40	05 41	17 06	10 55	14 50	20 06
14	13 10	06 48	20 06	20 44	21 45	05 35	17 04	10 56	14 50	20 07
15	12 50	10 45	20 02	20 43	21 50	05 30	17 02	10 57	14 50	20 08
16	12 29	14 18	19 57	20 41	21 55	05 24	17 00	10 58	14 49	20 08
17	12 08	17 19	19 51	20 39	22 00	05 19	16 58	10 59	14 49	20 09
18	11 47	19 36	19 43	20 37	22 04	05 13	16 56	11 00	14 49	20 10
19	11 26	21 00	19 35	20 34	22 08	05 08	16 54	11 01	14 49	20 11
20	11 04	21 19	19 25	20 30	22 12	05 02	16 52	11 02	14 49	20 11
21	10 43	20 26	19 13	20 26	22 15	04 56	16 50	11 03	14 49	20 12
22	10 21	18 18	19 01	20 21	22 18	04 51	16 48	11 04	14 49	20 13
23	09 59	15 00	18 47	20 16	22 21	04 45	16 46	11 05	14 49	20 14
24	09 37	10 42	18 32	20 10	22 24	04 40	16 44	11 06	14 49	20 14
25	09 15	05 42	18 15	20 04	22 26	04 34	16 42	11 06	14 49	20 15
26	08 53	00 S 21	17 58	19 57	22 28	04 28	16 40	11 07	14 48	20 16
27	08 30	04 N 59	17 39	19 50	22 30	04 23	16 38	11 08	14 48	20 16
28	08 S 08	09 N 59	17 S 19	19 S 42	22 N 31	04 S 17	16 S 36	11 N 09	14 S 48	20 N 17

ZODIAC SIGN ENTRIES

Date	h m	Planets
02	16 03	☽ ♊
04	20 40	☽ ♋
05	20 36	♀ ♑
07	03 06	☽ ♌
09	11 36	☽ ♍
11	22 18	☽ ♎
14	10 38	☽ ♏
15	10 08	☿ ♒
16	22 57	☽ ♐
19	09 00	☽ ♑
19	09 09	☉ ♓
21	15 23	☽ ♒
23	18 17	☽ ♓
25	19 05	☽ ♈
27	19 38	☽ ♉

LATITUDES

Date	Mercury ☿ ° '	Venus ♀ ° '	Mars ♂ ° '	Jupiter ♃ ° '	Saturn ♄ ° '	Uranus ♅ ° '	Neptune ♆ ° '	Pluto ♇ ° '
01	02 N 39	03 N 11	04 N 33	01 S 04	00 S 47	00 N 48	01 N 46	13 N 49
04	02 06	03 01	04 33	01 04	00 48	00 48	01 46	13 50
07	01 32	02 50	04 33	01 04	00 48	00 48	01 46	13 51
10	00 58	02 39	04 32	01 04	00 48	00 48	01 47	13 51
13	00 N 26	02 27	04 30	01 03	00 48	00 48	01 47	13 52
16	00 S 03	02 14	04 27	01 03	00 48	00 48	01 47	13 53
19	00 30	02 02	04 23	01 03	00 49	00 48	01 47	13 53
22	00 54	01 49	04 19	01 03	00 49	00 48	01 47	13 54
25	01 15	01 36	04 14	01 03	00 49	00 48	01 48	13 54
28	01 34	01 23	04 08	01 03	00 50	00 48	01 48	13 54
31	01 S 48	01 N 10	04 N 02	01 S 03	00 S 50	00 N 48	01 N 48	13 N 54

DATA

Julian Date	2438062
Delta T	+35 seconds
Ayanamsa	23° 20' 15"
Synetic vernal point	05° ♓ 46' 45"
True obliquity of ecliptic	23° 26' 34"

MOON'S PHASES, APSIDES AND POSITIONS ☽

Date	h m	Phase	Longitude ° '	Eclipse Indicator
01	08 50	☽	11 ♉ 46	
08	14 52	○	19 ♌ 08	
16	17 38	☾	27 ♏ 20	
24	02 06	●	04 ♓ 45	

Day	h m		
14	04 17	Apogee	
26	00 06	Perigee	
05	17 52	Max dec	21° N 18'
12	21 29	0S	
20	06 40	Max dec	21° S 21'
26	13 34	0N	

LONGITUDES

Date	Chiron ⚷ ° '	Ceres ⚳ ° '	Pallas ⚴ ° '	Juno ⚵ ° '	Vesta ⚶ ° '	Black Moon Lilith ⚸ ° '
01	08 ♓ 43	16 ♍ 44	06 ♋ 54	26 ♍ 28	00 ♎ 43	21 ♎ 16
11	09 ♓ 19	15 ♍ 14	06 ♋ 14	25 ♍ 05	00 ♎ 10	22 ♎ 23
21	09 ♓ 57	13 ♍ 14	06 ♋ 40	23 ♍ 06	28 ♍ 49	23 ♎ 30
31	10 ♓ 35	10 ♍ 56	08 ♋ 04	20 ♍ 43	26 ♍ 48	24 ♎ 38

ASPECTARIAN

h m Aspects

01 Friday
01 59 ☿ St D
04 27 ☉ ⚻ ♇
05 53 ☽ ⚼ ♀
08 31 ☽ △ ♇
08 50 ☽ □ ☉
10 18 ☽ ∥ ♅
11 55 ☽ □ ♄
15 00 ☽ ⚹ ♃
15 21 ☽ ☍ ♆
16 10 ☽ □ ♂
22 28 ☽ ± ♀

02 Saturday
00 07 ☽ △ ☿
09 37 ☽ ⚻ ♀
10 12 ☽ ⋕ ♆
11 54 ☽ Q ♃
14 04 ♃ △ ♆
18 04 ♂ ⚻ ♃
20 10 ♂ □ ♆
22 14 ☽ Q ♂
23 08 ☽ □ ♅
23 55 ☽ ⋕ ☉

03 Sunday
02 20 ☽ ⚼ ☿
05 01 ☽ ⋕ ♄
09 03 ☉ ☌ ♄
12 06 ☽ □ ♇
16 08 ☽ △ ♄
16 38 ☽ △ ☉
18 22 ☽ ± ☿
18 34 ☽ ⚹ ♂
19 14 ☽ ⚻ ♆
19 43 ☽ □ ♃

04 Monday
01 50 ☽ ⋕ ☿
05 11 ☽ ⚻ ☿
05 49 ☽ ± ♆
06 32 ☽ ∥ ♇
06 33 ☽ Q ♅
11 57 ☉ ☍ ♂
12 59 ☽ ⋕ ♀
18 35 ☽ ☍ ♀
18 45 ☽ ∥ ♂
18 51 ☽ ⚼ ♄
19 48 ☽ Q ♇
20 20 ☽ ∠ ♂
21 17 ☽ ⚼ ☉
21 48 ☽ ⚼ ♆
22 26 ♀ ∠ ♄

05 Tuesday
03 49 ☽ ⚹ ♅
04 03 ☉ □ ♆
11 00 ♀ ⚼ ♂
11 07 ☽ ± ♄
11 59 ☽ ⊥ ♂
14 45 ☽ ± ☉
17 17 ☽ ⚹ ♇
21 08 ☉ ⚺ ♃
22 02 ☽ ⚻ ♄
22 30 ☽ ⚺ ♂

06 Wednesday
00 48 ☽ △ ♆
02 11 ☽ △ ♃
02 30 ☽ ⚻ ☉
06 50 ☽ ∠ ♅
10 19 ♂ ☍ ♄
10 41 ♀ ∠ ♆
12 30 ☽ ∥ ♂
13 03 ☽ ☍ ☿
20 34 ☽ ∠ ♇
20 44 ☽ ⋕ ♀
23 16 ☽ ⊥ ♅

07 Thursday
05 25 ☽ ∥ ♇
05 53 ☽ ⚻ ♀
06 10 ☽ ⚼ ♃
06 16 ☽ ⋕ ☿
10 20 ☽ ⚺ ♅
13 11 ☽ ⊥ ♇
18 04 ☽ ± ♀
23 14 ☽ ± ♃

08 Friday
00 22 ☽ ⚺ ♇
04 14 ☽ ☌ ♂
05 17 ☿ ⋕ ♇
05 56 ☽ ☍ ♄
08 17 ☽ □ ♆
10 33 ☽ ⋕ ♄
10 43 ☽ ⚻ ♃
12 33 ☽ ⚼ ♀
14 52 ☽ ☍ ☉

09 Saturday
00 08 ☽ ⚻ ☿
04 59 ☽ ⋕ ☉
05 32 ☽ ⋕ ♆
08 43 ♀ △ ♅
10 45 ☉ ∥ ♆
12 22 ☽ ± ☿
18 38 ☽ Q ♆
18 56 ☽ ☌ ♅
19 58 ☽ △ ♀

10 Sunday
06 59 ☽ ⚼ ☿
07 33 ☽ ∥ ♅
09 37 ☽ ☌ ♇
11 58 ☽ ⚺ ♂
16 04 ☽ ⚻ ♄
17 58 ☽ ⚹ ♆
20 55 ♀ Q ♃
21 35 ☽ ☍ ♃
23 22 ☽ ⊥ ♂

11 Monday
03 58 ☽ ± ♄
06 02 ☽ ⚻ ☉
12 17 ♀ ± ♂
12 52 ☽ ⋕ ♃
14 44 ☽ △ ☿
16 36 ☽ ∠ ♂
18 51 ☿ ⚼ ♇
18 59 ☽ ± ☉
21 58 ☽ ⚼ ♄
23 37 ☽ ∠ ♆

12 Tuesday
05 41 ☽ ⚺ ♅
10 25 ♂ ± ♃
12 57 ☽ □ ♀
14 36 ☽ ⚼ ☉
17 37 ☽ ⊥ ♆
17 39 ☽ ⊥ ♅
20 56 ☽ ⚺ ♇
21 39 ☽ ⚹ ♂

13 Wednesday
04 01 ☿ Q ♆
04 07 ☿ ± ♅
04 19 ☽ △ ♄
05 41 ☽ ⚺ ♆
06 23 ♅ Q ♆
08 59 ☽ ⊥ ♇
10 33 ☽ ⚻ ♃
11 42 ☽ ∠ ♅
21 07 ☽ Q ♂
22 50 ♂ ⚺ ♇
22 55 ☽ ± ♃
23 37 ☽ △ ☉

14 Thursday
02 54 ☿ ⋕ ♇
02 55 ♀ ⊥ ♄
03 08 ☽ ∠ ♇
04 57 ☽ Q ♀
05 04 ☽ ∥ ♃
08 20 ☽ □ ☿
17 29 ☽ ⚼ ♃
17 55 ☽ ⚹ ♅

15 Friday
07 44 ☽ ⚹ ♀
08 27 ☽ □ ♂
09 23 ☽ ⚹ ♇
13 15 ☽ ⋕ ♅
13 39 ♀ ⚻ ♂
17 31 ☽ □ ♄
18 03 ☽ Q ♅
18 17 ☽ ☌ ♆

16 Saturday
00 16 ☽ Q ☿
00 21 ☽ △ ♃
00 25 ☽ ∥ ☉
00 55 ♀ △ ♇
06 05 ♆ St R
09 28 ☽ Q ♇
15 49 ☽ ∥ ♆
16 58 ☽ ∠ ♀
17 38 ☽ □ ☉

17 Sunday
02 41 ☽ ⚹ ☿
05 49 ☽ Q ♄
05 51 ☽ □ ♅
08 59 ☽ ∥ ♄
12 36 ☽ ⊥ ♀
18 34 ☽ △ ♂
20 59 ☽ □ ♇

18 Monday
01 30 ☽ ⚺ ♀
05 31 ☽ ⚹ ♄
05 40 ☽ ⚺ ♆
08 09 ☽ Q ☉
10 39 ☿ ⚻ ♅
11 03 ☽ ∠ ☿
12 37 ☽ □ ♃
13 24 ☽ ∥ ☿
17 12 ☽ ⊥ ♆
19 53 ☽ ⋕ ♇
21 48 ♄ □ ♆
22 45 ☽ ⚼ ♂

19 Tuesday
02 52 ☽ ∥ ♀
06 00 ☽ ⊥ ☿
06 53 ☿ ∠ ♃
08 48 ☉ ∠ ♀
08 59 ☽ ⚹ ☉
10 15 ☽ ∠ ♆
10 22 ☽ ∠ ♄
15 17 ☽ ± ♂
15 20 ☽ △ ♅
18 24 ☽ ⚺ ☿
20 28 ☿ ± ♇
23 08 ♀ ⚹ ♆
23 10 ☽ Q ♃

20 Wednesday
02 07 ♀ ⚺ ♄
02 07 ☽ ⚻ ♂
03 13 ☽ ⊥ ♄
05 38 ☽ △ ♇
13 53 ☽ ⚹ ♆
14 14 ☽ ⚺ ♄
14 57 ☽ ∠ ☉
14 59 ☉ ⋕ ♅
15 15 ☽ ☌ ♀
18 42 ☽ ⚼ ♅
21 23 ☽ ⚹ ♃

21 Thursday
08 20 ☽ ⊥ ☉
08 33 ☽ ⚼ ♇
10 40 ☽ ± ♅
11 17 ☽ Q ♆
11 59 ☽ ∥ ♀
15 27 ☽ ⋕ ♇
19 41 ☽ ⚺ ☉
21 07 ☽ ⚻ ♅

22 Friday
00 14 ☽ ± ♇
00 15 ☽ ∠ ♃
04 44 ☽ ∥ ☿
05 31 ☽ ☌ ☿
06 14 ☽ ☍ ♂
06 20 ♀ ⚼ ♅
10 32 ☽ ⚻ ♇
12 02 ☿ ☍ ♂
14 38 ☉ ☍ ♅
15 56 ☽ ⊥ ♃
18 19 ☽ □ ♆
19 06 ☽ ☌ ♄

23 Saturday
00 03 ☽ ∥ ♄
00 09 ☽ ⚺ ♀
02 14 ☽ ⚺ ♃
11 03 ☽ ⊥ ♀
13 07 ☽ ∥ ♆
21 47 ♀ ⋕ ♇
23 33 ☽ ☍ ♅

24 Sunday
02 06 ☽ ☌ ☉
03 10 ☽ ∠ ♀
07 26 ☽ ⚻ ♂
08 30 ♀ ⚹ ♃
09 45 ☿ ⚻ ♇
09 59 ☽ ⋕ ♅
12 22 ☽ ☍ ♇
12 38 ☽ ⚺ ☿
17 03 ☽ ± ♂
17 52 ☽ ∥ ☉
19 53 ☽ △ ♆
21 02 ☽ ⚺ ♄
23 22 ☽ ⊥ ☿

25 Monday
04 19 ☽ ☌ ♃
05 37 ☽ ⚹ ♀
06 50 ☽ ⊥ ♄
07 23 ☽ ⚼ ♂
15 23 ☽ ∠ ☿
17 17 ☽ ∥ ♃
20 07 ☽ ⚼ ♆
21 28 ☽ ∠ ♄

26 Tuesday
00 05 ☽ ⚻ ♅
02 35 ☽ Q ♀
06 15 ☽ ⚺ ☉
07 11 ☽ △ ♂
09 44 ☽ ± ♅
10 37 ☽ ± ♆
12 46 ☽ ⚻ ♇
16 38 ☽ ⊥ ☉
17 39 ☉ ⚻ ♂
18 04 ☽ ⚹ ☿
20 18 ☽ ⚻ ♆
21 51 ☽ ⚹ ♄
22 27 ☽ ± ♇

27 Wednesday
00 14 ☽ ⚼ ♅
05 32 ☽ ⚺ ♃
08 16 ☽ ∠ ☉
09 14 ☽ ⋕ ♃
10 07 ☽ □ ♀
13 01 ☽ ⚼ ♇
14 02 ☿ ⊥ ♃
15 27 ☽ ⊥ ♃
15 35 ☽ Q ☿
17 30 ☽ Q ♄
17 30 ☿ □ ♆

28 Thursday
00 35 ☽ △ ♅
03 27 ☽ ⋕ ☉
06 27 ☽ ∠ ♃
07 13 ☽ □ ♂
10 38 ☽ ⚹ ☉
12 50 ☿ ☌ ♄
13 36 ☽ △ ♇
18 06 ☽ ∥ ♅
21 24 ☽ ☍ ♆
22 33 ♀ ⚼ ♇
23 25 ☽ □ ♄

All ephemeris data is given at 12.00 UT and the Moon's longitude is additionally given for 24.00 UT

MARCH 1963

LONGITUDES

Date	Sidereal time h m s	Sun ☉ ° ' "	Moon ☽ ° ' "	Moon ☽ 24.00 ° ' "	Mercury ☿ ° '	Venus ♀ ° '	Mars ♂ ° '	Jupiter ♃ ° '	Saturn ♄ ° '	Uranus ♅ ° '	Neptune ♆ ° '	Pluto ♇ ° '
01	22 34 18	10 ♓ 11 04	24 ♉ 19 05	01 ♊ 22 54	18 ♒ 10	26 ♑ 33	06 ♌ 49	21 ♓ 53	16 ♒ 53	02 ♍ 57	15 ♏ 37	10 ♍ 54
02	22 38 15	11 11 17	08 ♊ 21 44	15 ♊ 15 35	19 38	27 42	06 R 37	22 07	17 00	02 R 55	15 R 36	10 R 52
03	22 42 11	12 11 28	22 ♊ 04 33	28 ♊ 48 47	21 08	28 ♑ 52	06 27	22 22	17 07	02 52	15 36	10 50
04	22 46 08	13 11 37	05 ♋ 28 31	12 ♋ 04 00	22 38	00 ♒ 01	06 17	22 36	17 14	02 49	15 35	10 49
05	22 50 04	14 11 43	18 ♋ 35 33	25 ♋ 03 25	24 09	01 10	06 08	22 51	17 20	02 47	15 35	10 47
06	22 54 01	15 11 48	01 ♌ 27 54	07 ♌ 49 15	25 42	02 20	06 00	23 05	17 27	02 44	15 34	10 46
07	22 57 57	16 11 50	14 ♌ 07 43	20 ♌ 23 32	27 16	03 30	05 52	23 20	17 34	02 42	15 34	10 44
08	23 01 54	17 11 51	26 ♌ 36 51	02 ♍ 47 53	28 ♒ 50	04 39	05 46	23 34	17 40	02 39	15 33	10 43
09	23 05 51	18 11 49	08 ♍ 56 45	15 ♍ 03 35	00 ♓ 26	05 49	05 40	23 49	17 47	02 37	15 32	10 41
10	23 09 47	19 11 45	21 ♍ 08 32	27 ♍ 11 41	02 03	06 59	05 35	24 03	17 53	02 34	15 32	10 39
11	23 13 44	20 11 40	03 ♎ 13 13	09 ♎ 13 14	03 42	08 09	05 30	24 18	18 00	02 32	15 31	10 38
12	23 17 40	21 11 32	15 ♎ 11 56	21 ♎ 09 31	05 21	09 19	05 27	24 32	18 06	02 29	15 30	10 36
13	23 21 37	22 11 23	27 ♎ 06 13	03 ♏ 02 18	07 02	10 29	05 24	24 47	18 13	02 27	15 29	10 35
14	23 25 33	23 11 12	08 ♏ 58 05	14 ♏ 53 56	08 43	11 40	05 22	25 01	18 19	02 24	15 29	10 33
15	23 29 30	24 10 59	20 ♏ 50 16	26 ♏ 47 32	10 26	12 50	05 21	25 16	18 25	02 22	15 28	10 32
16	23 33 26	25 10 45	02 ♐ 46 14	08 ♐ 46 54	12 10	14 00	05 20	25 30	18 31	02 20	15 27	10 30
17	23 37 23	26 10 29	14 ♐ 50 07	20 ♐ 56 28	13 56	15 11	05 D 20	25 45	18 38	02 17	15 26	10 29
18	23 41 20	27 10 11	27 ♐ 06 35	03 ♑ 21 05	15 42	16 21	05 21	25 59	18 44	02 15	15 25	10 27
19	23 45 16	28 09 51	09 ♑ 40 35	16 ♑ 05 41	17 30	17 32	05 23	26 14	18 50	02 13	15 24	10 26
20	23 49 13	29 ♓ 09 30	22 ♑ 36 54	29 ♑ 14 44	19 19	18 43	05 25	26 28	18 56	02 10	15 23	10 24
21	23 53 09	00 ♈ 09 07	05 ♒ 59 33	12 ♒ 51 36	21 09	19 53	05 28	26 43	19 02	02 08	15 22	10 23
22	23 57 06	01 08 42	19 ♒ 50 59	26 ♒ 57 37	23 01	21 04	05 32	26 58	19 08	02 06	15 21	10 21
23	00 01 02	02 08 16	04 ♓ 11 12	11 ♓ 31 14	24 54	22 15	05 36	27 12	19 14	02 04	15 20	10 20
24	00 04 59	03 07 47	18 ♓ 56 59	26 ♓ 27 28	26 48	23 26	05 42	27 27	19 20	02 02	15 19	10 19
25	00 08 55	04 07 16	04 ♈ 01 34	11 ♈ 38 00	28 ♓ 43	24 37	05 47	27 41	19 26	02 00	15 18	10 17
26	00 12 52	05 06 44	19 ♈ 15 21	26 ♈ 52 14	00 ♈ 40	25 48	05 54	27 56	19 31	01 57	15 17	10 16
27	00 16 49	06 06 09	04 ♉ 27 17	11 ♉ 59 13	02 37	26 59	06 01	28 10	19 37	01 55	15 16	10 14
28	00 20 45	07 05 32	19 ♉ 26 56	26 ♉ 49 31	04 36	28 10	06 08	28 24	19 43	01 53	15 14	10 13
29	00 24 42	08 04 53	04 ♊ 06 17	11 ♊ 16 44	06 36	29 ♒ 21	06 17	28 39	19 48	01 51	15 13	10 12
30	00 28 38	09 04 12	18 ♊ 20 36	25 ♊ 17 50	08 37	00 ♓ 33	06 26	28 53	19 54	01 49	15 12	10 10
31	00 32 35	10 ♈ 03 28	02 ♋ 08 30	08 ♋ 52 51	10 ♈ 39	01 ♓ 44	06 ♌ 35	29 ♓ 08	19 ♒ 59	01 ♍ 48	15 ♏ 11	10 ♍ 09

Date	Moon True ☊ ° '	Moon Mean ☊ ° '	Moon ☽ Latitude ° '
01	28 ♋ 51	27 ♋ 32	04 S 40
02	28 R 50	27 29	04 00
03	28 D 50	27 26	03 06
04	28 52	27 23	02 03
05	28 53	27 20	00 S 55
06	28 R 53	27 16	00 N 14
07	28 52	27 13	01 21
08	28 49	27 10	02 24
09	28 43	27 07	03 18
10	28 35	27 04	04 03
11	28 25	27 00	04 37
12	28 15	26 57	04 57
13	28 04	26 54	05 05
14	27 55	26 51	05 00
15	27 48	26 48	04 41
16	27 43	26 45	04 10
17	27 40	26 41	03 28
18	27 39	26 38	02 36
19	27 D 39	26 35	01 35
20	27 40	26 32	00 N 27
21	27 R 40	26 29	00 S 44
22	27 38	26 26	01 55
23	27 33	26 22	03 01
24	27 26	26 19	03 57
25	27 17	26 16	04 38
26	27 07	26 13	04 59
27	26 57	26 10	05 00
28	26 48	26 06	04 39
29	26 42	26 03	04 01
30	26 38	26 00	03 08
31	26 ♋ 37	25 ♋ 57	02 S 06

DECLINATIONS

Date	Sun ☉ ° '	Moon ☽ ° '	Mercury ☿ ° '	Venus ♀ ° '	Mars ♂ ° '	Jupiter ♃ ° '	Saturn ♄ ° '	Uranus ♅ ° '	Neptune ♆ ° '	Pluto ♇ ° '
01	07 S 45	14 N 19	16 S 57	19 S 33	22 N 32	04 S 11	16 S 34	11 N 10	14 S 48	20 N 18
02	07 22	17 46	16 34	19 24	22 33	04 06	16 32	11 11	14 48	20 19
03	06 59	20 07	16 10	19 14	22 34	04 00	16 30	11 12	14 48	20 19
04	06 36	21 17	15 45	19 04	22 35	03 54	16 28	11 13	14 47	20 20
05	06 13	21 14	15 18	18 54	22 35	03 48	16 26	11 14	14 47	20 20
06	05 50	20 04	14 50	18 42	22 35	03 43	16 24	11 15	14 47	20 21
07	05 27	17 53	14 21	18 31	22 35	03 37	16 22	11 16	14 47	20 22
08	05 03	14 54	13 51	18 18	22 35	03 31	16 20	11 17	14 46	20 22
09	04 40	11 17	13 19	18 06	22 34	03 25	16 18	11 18	14 46	20 23
10	04 17	07 14	12 46	17 52	22 33	03 20	16 17	11 18	14 46	20 24
11	03 53	02 N 57	12 12	17 38	22 32	03 14	16 15	11 19	14 46	20 24
12	03 30	01 S 25	11 36	17 24	22 31	03 08	16 13	11 20	14 45	20 25
13	03 06	05 42	11 00	17 09	22 29	03 02	16 11	11 21	14 45	20 25
14	02 42	09 45	10 22	16 54	22 28	02 57	16 09	11 22	14 45	20 26
15	02 19	13 27	09 43	16 38	22 26	02 51	16 07	11 23	14 45	20 27
16	01 55	16 37	09 02	16 22	22 24	02 45	16 05	11 24	14 44	20 27
17	01 31	19 08	08 21	16 05	22 22	02 39	16 04	11 24	14 44	20 28
18	01 08	20 49	07 38	15 48	22 19	02 34	16 02	11 25	14 44	20 28
19	00 44	21 31	06 54	15 30	22 17	02 28	16 00	11 26	14 43	20 29
20	00 S 20	21 06	06 09	15 12	22 14	02 22	15 58	11 27	14 43	20 29
21	00 N 04	19 30	05 22	14 54	22 11	02 16	15 56	11 28	14 43	20 30
22	00 27	16 41	04 35	14 35	22 08	02 11	15 55	11 28	14 42	20 30
23	00 51	12 47	03 46	14 15	22 05	02 05	15 53	11 29	14 42	20 31
24	01 15	08 01	02 57	13 55	22 01	01 59	15 51	11 30	14 42	20 31
25	01 38	02 S 39	02 06	13 35	21 58	01 54	15 50	11 31	14 41	20 32
26	02 02	02 N 55	01 15	13 15	21 54	01 48	15 48	11 31	14 41	20 32
27	02 25	08 18	00 S 22	12 54	21 50	01 42	15 46	11 32	14 40	20 32
28	02 49	13 07	00 N 31	12 32	21 46	01 36	15 44	11 33	14 40	20 33
29	03 12	17 02	01 25	12 11	21 42	01 31	15 43	11 34	14 40	20 33
30	03 36	19 48	02 20	11 49	21 38	01 25	15 41	11 34	14 39	20 34
31	03 N 59	21 N 20	03 N 16	11 S 26	21 N 33	01 S 19	15 S 40	11 N 35	14 S 39	20 N 34

ZODIAC SIGN ENTRIES

Date	h	m	Planets
01	21	39	☽ ♊
04	02	08	☽ ♋
04	11	41	♀ ♒
06	09	15	☽ ♌
08	18	34	☽ ♍
09	05	26	☿ ♓
11	05	35	☽ ♎
13	17	51	☽ ♏
16	06	27	☽ ♐
18	17	35	☽ ♑
21	01	21	☽ ♒
21	08	20	☉ ♈
23	05	04	☽ ♓
25	05	38	☽ ♈
26	03	52	☿ ♈
27	04	57	☽ ♉
29	05	13	☽ ♊
30	01	00	♀ ♓
31	08	13	☽ ♋

LATITUDES

Date	Mercury ☿ ° '	Venus ♀ ° '	Mars ♂ ° '	Jupiter ♃ ° '	Saturn ♄ ° '	Uranus ♅ ° '	Neptune ♆ ° '	Pluto ♇ ° '
01	01 S 39	01 N 19	04 N 06	01 S 03	00 S 50	00 N 48	01 N 48	13 N 54
04	01 53	01 06	04 00	01 03	00 50	00 48	01 48	13 54
07	02 03	00 53	03 54	01 03	00 50	00 48	01 48	13 55
10	02 10	00 41	03 47	01 03	00 51	00 48	01 48	13 55
13	02 13	00 29	03 41	01 03	00 51	00 48	01 48	13 55
16	02 13	00 17	03 34	01 03	00 51	00 48	01 49	13 55
19	02 08	00 N 05	03 27	01 03	00 52	00 48	01 49	13 54
22	01 58	00 S 06	03 21	01 03	00 52	00 48	01 49	13 54
25	01 44	00 17	03 14	01 03	00 52	00 48	01 49	13 54
28	01 26	00 27	03 08	01 04	00 53	00 48	01 49	13 54
31	01 S 02	00 S 37	03 N 01	01 S 04	00 S 53	00 N 48	01 N 49	13 N 53

DATA

Julian Date	2438090
Delta T	+35 seconds
Ayanamsa	23° 20' 18"
Synetic vernal point	05° ♓ 46' 41"
True obliquity of ecliptic	23° 26' 35"

MOON'S PHASES, APSIDES AND POSITIONS ☽

Date	h	m	Phase	Longitude ° '	Eclipse Indicator
02	17	17	☽	11 ♊ 25	
10	07	49	○	19 ♍ 01	
18	12	08	☾	27 ♐ 10	
25	12	10	●	04 ♈ 08	

Day	h	m		
13	19	51	Apogee	
26	07	42	Perigee	
04	23	10	Max dec	21° N 24'
12	04	13	0S	
19	15	25	Max dec	21° S 32'
25	23	26	0N	

LONGITUDES

Date	Chiron ⚷ ° '	Ceres ⚳ ° '	Pallas ⚴ ° '	Juno ⚵ ° '	Vesta ⚶ ° '	Black Moon Lilith ⚸ ° '
01	10 ♓ 28	11 ♍ 24	07 ♋ 43	21 ♍ 13	27 ♍ 15	24 ♎ 24
11	11 ♓ 06	09 ♍ 06	09 ♋ 45	18 ♍ 43	24 ♍ 51	25 ♎ 31
21	11 ♓ 44	07 ♍ 04	12 ♋ 26	16 ♍ 19	22 ♍ 15	26 ♎ 39
31	12 ♓ 20	05 ♍ 33	15 ♋ 38	14 ♍ 16	19 ♍ 50	27 ♎ 46

ASPECTARIAN

h m	Aspects
01 Friday	
00 31	☽ □ ☿
07 49	☽ ✶ ♃
08 08	☽ Q ☉
08 15	♂ ⚼ ♃
12 50	☽ Q ♂
14 57	☽ ⋕ ♆
16 07	☽ △ ♀
20 03	♀ ± ♅
02 Saturday	
02 39	☽ □ ♅
02 40	☽ ⋕ ♄
03 50	☽ ⋕ ☿
04 30	☉ ☍ ♇
04 34	☽ Q ♃
09 02	☽ ✶ ♂
10 00	♀ Q ♆
14 38	☿ ∥ ♄
16 20	☽ □ ♇
17 17	☽ □ ☉
20 14	☽ ⚼ ♀
03 Sunday	
00 36	☽ ⊼ ♆
02 22	☽ ⋕ ♀
03 11	☽ △ ♄
09 52	☽ Q ♅
10 07	☽ △ ☿
10 54	☽ ∠ ♂
11 09	☽ ± ♆
12 31	☽ □ ♃
13 31	☽ ± ♀
14 57	☽ ∥ ♇
17 14	☉ ± ♂
04 Monday	
00 02	☽ Q ♇
01 13	☽ ⊼ ♀
02 45	☽ ⊥ ♂
03 12	☽ ⚼ ♆
06 05	☽ ⚼ ♄
07 13	☽ ✶ ♅
11 29	☿ ⊻ ♃
13 27	☽ ⊻ ♂
16 25	☽ ⚼ ☿
21 41	☽ ✶ ♇
22 33	☽ ± ♄
05 Tuesday	
03 14	☽ △ ☉
06 27	☽ △ ♆
09 40	☽ ⊼ ♄
10 30	☽ ∠ ♅
11 05	☽ ± ☿
20 02	☽ △ ♃
23 43	☽ ⊼ ☿
06 Wednesday	
01 20	☽ ∠ ♇
03 10	☽ ⊥ ♅
07 43	☽ ∥ ♇
09 25	☽ ⚼ ☉
13 48	☽ ☍ ♀
14 23	☽ ⊻ ♅
14 57	☿ ∥ ♆
18 12	☽ ⊥ ♇
20 07	♀ ⊼ ♅
20 27	☽ ☌ ♂
20 55	☉ △ ♆
07 Thursday	
00 45	☽ ⚼ ♃
03 52	☽ ± ☉
05 32	☽ ⋕ ♀
05 33	☽ ⊻ ♇
14 44	☽ □ ♆
16 18	☽ ⊼ ☉
18 14	☽ ± ♃
18 37	☽ ☍ ♄
08 Friday	
01 02	☽ ⋕ ♄
06 00	☽ ⊼ ♃
12 53	☽ ⋕ ♆
13 07	♀ ± ♇
16 57	☽ ☍ ☿
20 37	☽ ⋕ ☿
23 41	☽ ☌ ♅
09 Saturday	
00 44	☉ ⊻ ♄
01 27	☽ Q ♆
05 15	☽ ⊼ ♀
05 38	☽ ⊻ ♂
06 17	♃ ⊥ ♄
09 01	♀ ☍ ♂
11 55	☽ ∥ ♅
15 24	☽ ☌ ♇
17 17	☽ ⊥ ♂
18 13	☽ ± ♀
10 Sunday	
00 56	☽ ✶ ♆
05 31	☽ ⊼ ♄
07 49	☽ ☍ ☉
10 53	☽ ∠ ♂
13 51	☽ ⚼ ♀
17 29	☽ ± ♄
17 53	☽ ☍ ♃
19 21	☿ ☍ ♅
11 Monday	
06 19	☽ ⋕ ☉
06 37	☽ ∠ ♆
10 24	☽ ⋕ ♃
10 38	☽ ⊻ ♅
11 33	☽ ⚼ ♄
13 06	☽ ⊼ ☿
16 32	☽ ✶ ♂
19 02	☉ ⚼ ♂
22 35	☽ ⊥ ♅
22 55	☽ △ ♀
12 Tuesday	
00 35	☽ ⊥ ♆
02 48	☽ ⊻ ♇
03 02	☽ ± ☿
12 37	☽ ⊻ ♆
13 19	☿ ⊼ ♂
14 49	☽ ⊥ ♇
16 30	☽ Q ♂
16 35	☽ ∠ ♅
17 38	♀ ∠ ♃
17 54	☽ △ ♄
21 21	☽ ∥ ♃
22 25	☿ ⋕ ♅
22 35	☽ ∥ ☉
13 Wednesday	
00 04	☽ ⚼ ☿
01 10	☽ ⊼ ☉
07 12	☽ ⊼ ♃
08 56	☽ ∠ ♇
13 51	♀ ⊼ ♇
14 24	☽ ± ☉
16 40	☉ ∥ ♃
19 35	☽ ± ♃
22 46	☽ ✶ ♅
14 Thursday	
04 44	☽ □ ♂
10 16	☽ ⚼ ☉
11 25	☽ △ ☿
14 10	☽ ⚼ ♃
15 12	☽ ✶ ♇
15 14	☽ ∥ ☿
18 03	☽ □ ♀
22 11	☽ ⋕ ♅
22 58	☽ Q ♅
15 Friday	
01 09	☽ ☌ ♆
07 04	☽ □ ♄
13 15	☿ ☍ ♇
15 24	☽ Q ♇
18 22	☉ ⊥ ♄
19 22	☽ △ ☉
21 06	☽ △ ♃
21 16	☽ ∥ ♆
16 Saturday	
00 29	☿ ± ♂
07 41	☽ ∥ ♄
10 02	☽ ∥ ♀
10 18	☽ Q ♀
11 07	☽ □ ♅
17 08	☽ △ ♂
17 21	♂ St D
19 34	☽ Q ♄
22 23	☉ ☌ ♃
17 Sunday	
03 24	☽ □ ♇
09 54	☽ □ ☿
12 45	☽ ✶ ♀
13 11	☽ ⊻ ♆
14 20	♀ ∥ ♄
17 06	♀ □ ♆
19 32	☽ ✶ ♄
22 50	☽ ⚼ ♂
18 Monday	
00 57	☽ ⊥ ♆
05 46	☽ ⋕ ♇
08 10	☿ △ ♆
09 47	☽ □ ♃
12 08	☽ □ ☉
16 20	☽ ± ♂
18 22	☽ ∠ ♆
21 02	☽ ∠ ♀
21 52	☽ △ ♅
19 Tuesday	
00 50	☽ ∠ ♄
02 47	☽ Q ☿
03 51	☽ ⊼ ♂
13 12	☿ ⊻ ♀
13 25	☽ △ ♇
15 51	☽ ⊥ ♀
17 58	☽ ⊥ ♄
20 42	☽ Q ♃
22 42	☽ ✶ ♆
20 Wednesday	
01 08	☽ Q ☉
02 02	☽ ⚼ ♅
04 08	☽ ⊻ ♀
04 58	☽ ✶ ☿
05 12	☽ ⊻ ♄
06 39	☿ ⊻ ♄
16 58	♀ ☌ ♄
17 04	☽ ⚼ ♇
18 27	☽ ± ♅
19 08	☽ ✶ ♃
20 39	☽ Q ♆
23 22	☽ ⋕ ♇
21 Thursday	
00 48	☽ ✶ ☉
02 49	☿ ⚼ ♂
05 11	☽ ⊼ ♅
09 10	☽ ± ♇
11 05	☽ ☍ ♂
12 20	☽ ∠ ☿
17 08	☉ ⚼ ♆
19 41	☽ ⊼ ♇
22 12	☽ ∠ ♃
22 Friday	
02 07	♀ ∥ ♆
04 19	☽ □ ♆
05 11	☽ ∠ ☉
06 25	☽ ⊥ ☿
09 08	♀ ⊥ ♃
10 46	☽ ☌ ♄
13 55	☽ ⊥ ♃
14 16	☽ ☌ ♀
17 22	☽ ∥ ♄
18 11	☽ ⊻ ☿
21 37	☽ ⊥ ☉
23 Saturday	
00 12	☽ ⊻ ♃
01 00	☽ ∥ ♆
02 58	☽ ∥ ♀
08 22	☽ ⊻ ☉
08 30	☽ ☍ ♅
10 15	☉ ⊼ ♅
14 21	☽ ⊼ ♂
16 31	☿ ⊥ ♄
18 55	☽ ⋕ ♅
22 03	☽ ☍ ♇
24 Sunday	
00 13	☽ ± ♂
06 09	☽ △ ♆
12 37	☽ ⊻ ♄
14 49	☽ ⚼ ♂
19 48	☽ ⊻ ♀
21 16	☿ ☌ ♃
22 16	☽ ⊥ ♄
25 Monday	
01 47	☽ ☌ ♃
02 22	☽ ☌ ☿
06 06	☽ ⚼ ♆
06 09	☽ ⊥ ♀
08 47	☽ ⊼ ♅
12 10	☽ ☌ ☉
12 38	☽ ∠ ♄
14 48	☽ △ ♂
14 48	☽ ∥ ☿
15 21	☽ ∥ ♃
16 06	☽ ⋕ ☉
18 15	☽ ± ♅
18 42	☿ ∥ ♃
20 14	☉ ∠ ♄
20 18	☽ ± ♆
20 57	☉ ⋕ ☿
21 34	☽ ∠ ♀
21 52	☽ ⊼ ♇
26 Tuesday	
00 27	☉ ⋕ ♃
05 45	☽ ⊼ ♆
05 45	☽ ⋕ ☿
07 15	☽ ⋕ ♃
07 18	☽ ± ♇
07 20	☿ ⚼ ♆
07 53	☽ ∥ ☉
08 23	☽ ⚼ ♅
12 26	☽ ✶ ♄
21 27	☽ ⚼ ♇
23 11	☽ ✶ ♀
27 Wednesday	
01 53	☽ ⊻ ♃
03 37	☿ ⊼ ♅
07 29	☽ Q ♄
08 00	☽ △ ♅
08 40	☽ ⊻ ☿
09 31	☉ △ ♂
11 32	☽ ⊥ ♃
14 30	☽ □ ♂
14 48	☽ ⊻ ☉
19 38	☽ ⊥ ☿
19 49	☽ Q ♀
21 12	☽ △ ♇
23 02	☿ ⊥ ♀
28 Thursday	
01 03	☽ ⊥ ☉
02 07	☽ ∠ ♃
03 45	☽ ∥ ♅
05 13	☽ ☍ ♆
09 09	☽ ⋕ ♀
12 17	☽ ∠ ☿
12 26	☽ □ ♄
13 24	☿ ∠ ♄
16 35	☽ ∠ ☉
18 00	♀ ⊻ ♃
19 41	☽ Q ♂
20 48	☽ ⋕ ♆
29 Friday	
02 50	☽ ✶ ♃
03 16	☽ ⋕ ♄
03 28	☽ □ ♀
06 43	☉ ± ♅
07 50	☿ △ ♂
08 17	☽ □ ♅
14 06	☿ ⋕ ♃
15 39	☽ ✶ ♂
16 50	☽ ✶ ☿
19 07	☽ ✶ ☉
22 09	☽ □ ♇
23 08	☽ Q ♃
30 Saturday	
02 43	☿ ± ♅
06 39	☽ ⊼ ♆
14 32	☽ Q ♅
14 41	☽ △ ♄
15 03	☉ ± ♆
16 34	☽ Q ☿
16 54	☽ ± ♆
17 02	☽ Q ☉
17 21	☽ ∠ ♂
18 48	☿ ± ♆
21 31	☽ ∥ ♇
22 25	☉ ☌ ☿
31 Sunday	
03 05	♀ ⋕ ♅
04 59	☽ Q ♇
06 10	☿ ⊼ ♇
06 36	☽ □ ♃
08 33	☽ ⚼ ♆
09 13	☽ ⊥ ♂
11 12	☽ △ ♀
11 23	☽ ✶ ♅
13 11	♀ ☍ ♅
14 10	☉ ⊼ ♇
17 05	☽ ⚼ ♄
19 09	☽ ∥ ♂
19 59	☽ ⊻ ♂

All ephemeris data is given at 12.00 UT and the Moon's longitude is additionally given for 24.00 UT

APRIL 1963

LONGITUDES

Date	Sidereal time h m s	Sun ☉ ° ' "	Moon ☽ ° ' "	Moon ☽ 24.00 ° ' "	Mercury ☿ ° '	Venus ♀ ° '	Mars ♂ ° '	Jupiter ♃ ° '	Saturn ♄ ° '	Uranus ♅ ° '	Neptune ♆ ° '	Pluto ⯓ ° '
01	00 36 31	11 ♈ 02 42	15 ♋ 31 12	22 ♋ 03 58	12 ♈ 42	02 ♓ 55	06 ♌ 45	29 ♓ 22	20 ♒ 05	01 ♍ 46	15 ♏ 09	10 ♍ 08
02	00 40 28	12 01 53	28 ♋ 31 37	04 ♌ 54 36	14 45	04 07	06 56	29 37	20 10	01 R 44	15 R 08	10 R 06
03	00 44 24	13 01 03	11 ♌ 13 26	17 ♌ 28 34	16 49	05 18	07 07	29 ♓ 51	20 15	01 42	15 07	10 05
04	00 48 21	14 00 09	23 ♌ 40 29	29 ♌ 49 36	18 53	06 29	07 19	00 ♈ 05	20 21	01 40	15 05	10 04
05	00 52 18	14 59 14	05 ♍ 56 18	12 ♍ 00 56	20 57	07 41	07 31	00 19	20 26	01 39	15 04	10 03
06	00 56 14	15 58 16	18 ♍ 03 47	24 ♍ 05 09	23 01	08 52	07 44	00 34	20 31	01 37	15 03	10 01
07	01 00 11	16 57 16	00 ♎ 05 16	06 ♎ 04 18	25 05	10 04	07 57	00 48	20 36	01 35	15 01	10 00
08	01 04 07	17 56 15	12 ♎ 02 28	17 ♎ 59 55	27 08	11 15	08 11	01 02	20 41	01 34	15 00	09 59
09	01 08 04	18 55 11	23 ♎ 56 48	29 ♎ 53 17	29 ♈ 09	12 27	08 25	01 16	20 46	01 32	14 59	09 58
10	01 12 00	19 54 05	05 ♏ 49 31	11 ♏ 45 42	01 ♉ 10	13 39	08 40	01 31	20 51	01 31	14 57	09 57
11	01 15 57	20 52 57	17 ♏ 42 02	23 ♏ 38 47	03 08	14 50	08 55	01 45	20 55	01 29	14 56	09 56
12	01 19 53	21 51 47	29 ♏ 36 11	05 ♐ 34 36	05 04	16 02	09 11	01 59	21 00	01 28	14 54	09 54
13	01 23 50	22 50 35	11 ♐ 34 23	17 ♐ 35 56	06 58	17 14	09 27	02 13	21 05	01 27	14 53	09 53
14	01 27 47	23 49 22	23 ♐ 39 43	29 ♐ 46 14	08 49	18 26	09 43	02 27	21 09	01 25	14 51	09 52
15	01 31 43	24 48 07	05 ♑ 56 02	12 ♑ 09 39	10 37	19 37	10 00	02 41	21 14	01 24	14 50	09 51
16	01 35 40	25 46 50	18 ♑ 27 40	24 ♑ 50 41	12 21	20 49	10 17	02 55	21 18	01 23	14 48	09 50
17	01 39 36	26 45 32	01 ♒ 19 16	07 ♒ 53 57	14 01	22 01	10 35	03 09	21 22	01 22	14 47	09 49
18	01 43 33	27 44 12	14 ♒ 35 13	21 ♒ 23 26	15 38	23 13	10 53	03 23	21 27	01 20	14 45	09 48
19	01 47 29	28 42 50	28 ♒ 18 53	05 ♓ 21 40	17 10	24 25	11 12	03 37	21 31	01 19	14 44	09 47
20	01 51 26	29 ♈ 41 26	12 ♓ 31 43	19 ♓ 48 45	18 38	25 37	11 31	03 50	21 35	01 18	14 42	09 47
21	01 55 22	00 ♉ 40 01	27 ♓ 12 16	04 ♈ 41 29	20 01	26 49	11 50	04 04	21 39	01 17	14 41	09 46
22	01 59 19	01 38 34	12 ♈ 15 25	19 ♈ 52 53	21 20	28 01	12 10	04 18	21 43	01 16	14 39	09 45
23	02 03 16	02 37 05	27 ♈ 32 31	05 ♉ 12 52	22 34	29 ♓ 13	12 30	04 31	21 47	01 16	14 37	09 44
24	02 07 12	03 35 35	12 ♉ 52 26	20 ♉ 29 46	23 43	00 ♈ 25	12 50	04 45	21 50	01 15	14 36	09 43
25	02 11 09	04 34 02	28 ♉ 03 31	05 ♊ 31 32	24 46	01 37	13 11	04 59	21 54	01 14	14 34	09 42
26	02 15 05	05 32 28	12 ♊ 55 48	20 ♊ 12 37	25 45	02 49	13 32	05 12	21 58	01 13	14 33	09 42
27	02 19 02	06 30 51	27 ♊ 22 29	04 ♋ 25 06	26 38	04 01	13 54	05 26	22 01	01 13	14 31	09 41
28	02 22 58	07 29 13	11 ♋ 20 26	18 ♋ 08 34	27 27	05 13	14 15	05 39	22 05	01 12	14 29	09 40
29	02 26 55	08 27 32	24 ♋ 49 48	01 ♌ 24 30	28 09	06 26	14 38	05 52	22 08	01 12	14 28	09 40
30	02 30 51	09 ♉ 25 50	07 ♌ 53 07	14 ♌ 16 13	28 ♉ 47	07 ♈ 38	15 ♌ 00	06 ♈ 06	22 ♒ 11	01 ♍ 11	14 ♏ 26	09 ♍ 39

Date	Moon True ☊ ° '	Moon Mean ☊ ° '	Moon ☽ Latitude ° '
01	26 ♋ 36	25 ♋ 54	00 S 58
02	26 R 37	25 51	00 N 10
03	26 36	25 47	01 17
04	26 33	25 44	02 18
05	26 28	25 41	03 12
06	26 19	25 38	03 56
07	26 08	25 35	04 30
08	25 55	25 32	04 51
09	25 41	25 28	04 59
10	25 27	25 25	04 55
11	25 14	25 22	04 37
12	25 03	25 19	04 08
13	24 55	25 16	03 27
14	24 50	25 12	02 36
15	24 47	25 09	01 38
16	24 D 47	25 06	00 N 33
17	24 R 47	25 03	00 S 35
18	24 46	25 00	01 43
19	24 44	24 57	02 47
20	24 39	24 53	03 44
21	24 32	24 50	04 27
22	24 22	24 47	04 54
23	24 11	24 44	05 00
24	24 00	24 41	04 45
25	23 51	24 38	04 09
26	23 44	24 34	03 17
27	23 39	24 31	02 14
28	23 37	24 28	01 S 05
29	23 D 36	24 25	00 N 06
30	23 ♋ 37	24 ♋ 22	01 N 15

DECLINATIONS

Date	Sun ☉ ° '	Moon ☽ ° '	Mercury ☿ ° '	Venus ♀ ° '	Mars ♂ ° '	Jupiter ♃ ° '	Saturn ♄ ° '	Uranus ♅ ° '	Neptune ♆ ° '	Pluto ⯓ ° '
01	04 N 22	21 N 34	04 N 12	11 S 04	21 N 29	01 S 14	15 S 38	11 N 36	14 S 39	20 N 34
02	04 45	20 38	05 08	10 41	21 24	01 08	15 36	11 36	14 38	20 35
03	05 08	18 38	06 05	10 17	21 19	01 02	15 35	11 37	14 38	20 35
04	05 31	15 48	07 01	09 54	21 14	00 57	15 33	11 37	14 37	20 35
05	05 54	12 18	07 58	09 30	21 09	00 51	15 32	11 38	14 37	20 36
06	06 17	08 21	08 54	09 06	21 04	00 45	15 30	11 39	14 36	20 36
07	06 40	04 N 05	09 50	08 41	20 58	00 40	15 29	11 39	14 36	20 36
08	07 02	00 S 18	10 45	08 16	20 53	00 34	15 27	11 40	14 36	20 37
09	07 25	04 39	11 39	07 51	20 47	00 29	15 26	11 40	14 35	20 37
10	07 47	08 50	12 31	07 26	20 42	00 23	15 25	11 41	14 35	20 37
11	08 09	12 40	13 23	07 01	20 36	00 17	15 23	11 41	14 34	20 37
12	08 31	16 02	14 13	06 35	20 30	00 12	15 22	11 42	14 34	20 38
13	08 53	18 46	15 01	06 09	20 23	00 06	15 21	11 42	14 33	20 38
14	09 15	20 41	15 47	05 43	20 17	00 S 01	15 19	11 42	14 33	20 38
15	09 36	21 41	16 31	05 17	20 11	00 N 05	15 18	11 43	14 32	20 38
16	09 58	21 37	17 13	04 51	20 04	00 10	15 17	11 43	14 32	20 38
17	10 19	20 26	17 53	04 24	19 58	00 15	15 15	11 44	14 32	20 38
18	10 40	18 06	18 30	03 57	19 51	00 21	15 14	11 44	14 31	20 39
19	11 01	14 40	19 04	03 30	19 44	00 26	15 13	11 44	14 31	20 39
20	11 22	10 18	19 36	03 04	19 37	00 32	15 12	11 45	14 30	20 39
21	11 42	05 S 12	20 05	02 36	19 30	00 37	15 11	11 45	14 30	20 39
22	12 03	00 N 20	20 32	02 09	19 23	00 42	15 10	11 45	14 29	20 39
23	12 23	05 56	20 56	01 42	19 16	00 48	15 09	11 46	14 29	20 39
24	12 43	11 11	21 17	01 15	19 08	00 53	15 07	11 46	14 28	20 39
25	13 03	15 41	21 36	00 47	19 01	00 58	15 06	11 46	14 28	20 39
26	13 22	19 05	21 52	00 S 20	18 53	01 04	15 05	11 46	14 27	20 39
27	13 42	21 11	22 06	00 N 08	18 45	01 09	15 04	11 47	14 27	20 39
28	14 01	21 53	22 17	00 36	18 37	01 14	15 03	11 47	14 26	20 39
29	14 20	21 16	22 26	01 03	18 29	01 19	15 03	11 47	14 26	20 39
30	14 N 38	19 N 30	22 N 32	01 N 31	18 N 21	01 N 24	15 S 02	11 N 47	14 S 25	20 N 39

ZODIAC SIGN ENTRIES

Date	h	m	Planets
02	14	45	☽ ♌
04	03	19	♃ ♈
05	00	20	☽ ♍
07	11	49	☽ ♎
09	22	03	☿ ♉
10	00	14	☽ ♏
12	12	48	☽ ♐
15	00	27	☽ ♑
17	09	34	☽ ♒
19	14	53	☽ ♓
20	19	36	☉ ♉
21	16	30	☽ ♈
23	15	51	☽ ♉
24	03	39	♀ ♈
25	15	06	☽ ♊
27	16	27	☽ ♋
29	21	25	☽ ♌

LATITUDES

Date	Mercury ☿ ° '	Venus ♀ ° '	Mars ♂ ° '	Jupiter ♃ ° '	Saturn ♄ ° '	Uranus ♅ ° '	Neptune ♆ ° '	Pluto ⯓ ° '
01	00 S 54	00 S 40	02 N 59	01 S 04	00 S 53	00 N 48	01 N 49	13 N 53
04	00 S 25	00 49	02 53	01 04	00 54	00 48	01 49	13 52
07	00 N 08	00 58	02 47	01 04	00 54	00 48	01 49	13 52
10	00 41	01 05	02 41	01 04	00 55	00 47	01 50	13 51
13	01 15	01 12	02 36	01 05	00 55	00 47	01 50	13 51
16	01 46	01 19	02 30	01 05	00 55	00 47	01 50	13 50
19	02 12	01 24	02 25	01 05	00 56	00 47	01 50	13 49
22	02 31	01 29	02 19	01 05	00 56	00 47	01 50	13 49
25	02 43	01 33	02 14	01 06	00 57	00 47	01 50	13 48
28	02 46	01 37	02 10	01 06	00 57	00 47	01 50	13 47
31	02 N 39	01 S 40	02 N 05	01 S 06	00 S 58	00 N 47	01 N 50	13 N 46

DATA

Julian Date	2438121
Delta T	+35 seconds
Ayanamsa	23° 20' 21"
Synetic vernal point	05° ♓ 46' 38"
True obliquity of ecliptic	23° 26' 35"

MOON'S PHASES, APSIDES AND POSITIONS ☽

Date	h	m	Phase	Longitude ° '	Eclipse Indicator
01	03	15	☽	10 ♋ 41	
09	00	57	○	18 ♎ 28	
17	02	52	☾	26 ♑ 23	
23	20	29	●	02 ♉ 58	
30	15	08	☽	09 ♌ 33	

Day	h	m		
10	02	16	Apogee	
23	18	28	Perigee	
01	04	40	Max dec	21° N 38'
08	10	23	0S	
15	22	43	Max dec	21° S 47'
22	10	35	0N	
28	12	26	Max dec	21° N 53'

LONGITUDES

Date	Chiron ⚷ ° '	Ceres ⚳ ° '	Pallas ⚴ ° '	Juno ⚵ ° '	Vesta ⚶ ° '	Black Moon Lilith ⚸ ° '
01	12 ♓ 23	05 ♍ 26	15 ♋ 58	14 ♍ 05	19 ♍ 37	27 ♎ 52
11	12 ♓ 56	04 ♍ 38	19 ♋ 34	12 ♍ 37	17 ♍ 47	29 ♎ 00
21	13 ♓ 27	04 ♍ 33	23 ♋ 27	11 ♍ 47	16 ♍ 40	00 ♏ 07
31	13 ♓ 53	05 ♍ 09	27 ♋ 34	11 ♍ 34	16 ♍ 22	01 ♏ 14

All ephemeris data is given at 12.00 UT and the Moon's longitude is additionally given for 24.00 UT

ASPECTARIAN

h m	Aspects
01 Monday	
02 15	☽ ⚹ ⯓
03 15	☽ □ ☉
05 56	☽ □ ☿
09 22	☽ ± ♄
11 20	☽ △ ♆
14 15	☽ ∠ ♅
16 49	☽ ♇ ♀
17 18	☽ ∥ ♂
19 42	☉ ∥ ☿
20 24	☽ ⊼ ♄
02 Tuesday	
05 38	☽ ∠ ⯓
06 48	☽ ⊥ ♅
11 08	☽ ± ♀
12 43	☽ ∥ ⯓
14 04	☽ △ ♃
16 25	☿ ⊼ ♆
18 00	☽ ⊻ ♅
22 28	☽ ⊥ ⯓
23 34	☽ ⊼ ♀
03 Wednesday	
03 36	☿ ± ⯓
04 04	☽ ☌ ♂
09 50	☽ ⊻ ⯓
10 41	☿ ♇ ♅
15 43	☽ △ ☉
19 05	☽ ♇ ♃
19 26	☽ □ ♆
04 Thursday	
00 52	☽ △ ☿
05 30	☽ ☍ ♄
12 28	♃ ♇ ♆
12 49	☽ ± ♃
13 50	☽ # ♄
20 34	☽ # ♆
23 18	☽ ♇ ☉
05 Friday	
00 45	☽ ⊼ ♃
03 35	☽ ☌ ♅
05 39	☿ ⚹ ♄
06 22	☽ Q ♆
07 59	♀ ⊼ ♂
12 02	☽ ♇ ☿
13 57	☉ ⊼ ♆
15 10	☽ ⊻ ♂
15 48	☽ ☍ ♀
16 15	☽ ∥ ♅
18 33	☽ ± ☉
20 05	☽ ☌ ⯓
06 Saturday	
03 14	☽ ⊥ ♂
06 01	☽ ⚹ ♆
07 09	☽ # ♀
07 28	☽ ⊼ ☉
09 22	☽ ∥ ☿
09 30	☽ ± ☿
13 14	☉ ± ⯓
15 29	☿ # ♀
16 55	☽ ⊼ ♄
21 28	☽ ∠ ♂
22 48	☽ ∥ ☉
23 55	☽ ⊼ ☿
07 Sunday	
03 18	☉ ♇ ♅
04 58	☽ ± ♄
10 47	♀ ☍ ⯓
11 04	☿ ♇ ⯓
11 29	☿ ∠ ♀
11 52	☽ ∠ ♆
13 27	☽ ☍ ♃
15 00	☽ ⊻ ♅
23 08	☽ ♇ ♄
08 Monday	
03 01	☽ ⊥ ♅
04 05	☽ ⚹ ♂
05 54	☽ ⊥ ♆
07 11	☽ # ♃
07 52	☽ ⊻ ⯓
10 15	☽ ⊼ ♀
13 28	☽ ∥ ♃
17 57	☽ ⊻ ♆
19 55	☽ ⊥ ⯓
21 05	☽ ∠ ♅
23 40	☽ ± ♀
09 Tuesday	
00 57	☽ ☍ ☉
04 44	☽ Q ♂
05 32	☽ △ ♄
12 42	☿ ∥ ♅
14 03	☽ ∠ ⯓
19 52	☽ ♇ ♀
10 Wednesday	
00 40	☽ ☍ ☿
01 29	☉ # ♀
03 06	☽ ⊼ ♃
03 18	☽ ⚹ ♅
04 35	☽ ∥ ♀
05 15	☽ # ☉
12 14	♃ ⊼ ♅
15 28	☽ ± ♃
16 11	☿ △ ♅
16 46	☿ ⊻ ♃
17 51	☽ □ ♂
20 19	☽ ⚹ ⯓
11 Thursday	
03 30	☽ Q ♅
04 54	♂ ∥ ⯓
05 34	☽ # ♅
05 34	☽ △ ♀
06 25	☽ ☌ ♆
09 17	☿ Q ♄
10 02	☽ ♇ ♃
13 04	☉ ⚹ ♄
13 46	♀ △ ♆
13 56	♀ ± ♂
16 52	☉ ⊥ ♀
18 15	☽ # ☿
18 33	☽ □ ♄
19 00	☽ ⊼ ☉
20 31	☽ Q ⯓
12 Friday	
01 01	☽ ∥ ♆
06 54	☽ ∥ ♄
08 11	☽ ± ☉
15 44	☽ □ ♅
16 52	☽ △ ♃
22 13	☿ # ♆
13 Saturday	
01 04	☽ ⊼ ☿
03 53	☽ ♇ ☉
06 59	☽ Q ♄
07 39	☽ △ ♂
08 39	☽ □ ⯓
15 18	☽ ± ☿
18 35	☽ ⊻ ♆
21 44	☿ # ♄
14 Sunday	
00 30	☽ □ ♀
06 20	☽ # ♂
06 28	☽ ⊥ ♆
06 28	☿ ⊥ ♃
07 01	☽ ⚹ ♄
11 08	☽ # ⯓
12 21	☽ △ ☉
12 21	☽ ♇ ☿
14 08	☽ ♇ ♂
15 Monday	
00 08	☽ ∠ ♆
00 25	♂ ⊻ ⯓
01 55	☿ △ ⯓
02 13	☿ □ ♂
03 12	☽ △ ♅
05 34	☽ □ ♃
08 10	☽ ± ♂
12 34	☽ ∠ ♄
13 18	☉ ♇ ⯓
15 37	☽ Q ♀
19 34	☽ △ ⯓
20 02	☽ ⊼ ♂
22 30	☽ △ ☿
16 Tuesday	
05 04	☽ ⚹ ♆
05 58	☽ ⊥ ♄
08 03	☽ ♇ ♅
16 43	☽ Q ♃
16 55	☽ ⚹ ♀
17 23	☽ ⊻ ♄
22 13	♀ ⊻ ♄
23 58	☽ ♇ ⯓
17 Wednesday	
00 59	☽ ± ♅
02 52	☽ □ ☉
03 37	☽ Q ♆
08 59	☽ # ⯓
12 04	☽ ⊼ ♅
15 25	☽ ⚹ ♃
16 35	☽ ± ⯓
18 14	☽ # ♂
23 00	☿ ☍ ♆
23 26	☽ ∠ ♀
18 Thursday	
03 27	☽ ⊼ ⯓
05 14	☽ ☍ ♂
09 13	☽ # ☿
12 18	☽ □ ♆
14 06	☽ □ ☿
14 12	☽ Q ☉
17 07	☽ ⊥ ♀
18 50	☽ ∠ ♃
19 Friday	
00 09	☽ ☌ ♄
04 38	☽ ⊻ ♀
08 36	☽ ∥ ♄
10 46	☽ ⊥ ♃
12 44	☽ ⚹ ☉
13 00	☽ ∥ ♆
17 09	☽ ☍ ♅
21 11	☽ ⊻ ♃
20 Saturday	
01 02	☽ Q ☿
04 35	☽ # ♅
06 57	☽ # ☉
07 25	☽ ☍ ⯓
10 17	☽ ⊼ ♂
12 49	☿ ∥ ♂
15 36	☽ △ ♆
15 50	☽ ∠ ☉
16 08	☿ ∠ ♃
20 25	☽ ± ♂
23 09	☽ ⚹ ☿
21 Sunday	
02 57	☽ ⊻ ♄
07 37	☽ ⊥ ☉
11 19	☽ ☌ ♀
11 24	☽ ♇ ♂
12 34	♀ ♇ ♂
12 43	☽ ⊥ ♄
15 04	☉ ∥ ♅
15 58	☽ ♇ ♆
17 57	☽ ⊻ ☉
18 34	☽ ⊼ ♅
23 11	☽ ☌ ♃
22 Monday	
00 24	☽ ∥ ♀
01 44	☽ ∠ ☿
03 05	☉ △ ♅
03 11	☽ ∠ ♄
04 07	☽ ± ♅
05 35	♀ ⊥ ♄
06 18	☽ ± ♆
07 38	☽ # ♃
08 02	☽ ⊼ ⯓
11 51	☽ △ ♂
13 37	☽ ∥ ♃
15 46	☽ ⊼ ♆
17 17	☽ ⊥ ☿
17 30	☽ ± ⯓
18 20	☽ ♇ ♅
19 02	☿ ∥ ⯓
19 10	☽ # ♀
19 37	☿ □ ♄
23 Tuesday	
02 56	☽ ⚹ ♄
03 32	☽ ⊻ ☿
07 37	☽ ♇ ⯓
14 51	☽ ⊻ ♀
17 49	☽ △ ♅
19 57	♀ ♇ ♆
20 29	☽ ☌ ☉
21 47	☽ Q ♄
23 05	☽ ⊻ ♃
24 Wednesday	
01 01	☽ ⊥ ♀
07 03	☽ △ ⯓
08 37	☽ ⊥ ♃
11 57	☽ □ ♂
14 42	☽ ☍ ♆
14 52	☽ ∥ ♅
16 20	☽ ∠ ♀
18 27	☉ Q ♄
20 14	☽ ∥ ☉
22 59	☽ ∠ ♃
25 Thursday	
02 11	☽ □ ♄
03 43	♀ # ♃
04 23	♀ ⊼ ♅
04 59	☽ # ♆
06 23	☽ ☌ ☿
08 38	☽ # ♄
17 04	☽ □ ♅
17 07	☽ Q ♂
18 11	☽ ⚹ ♀
23 10	☽ ⊻ ☉
23 16	☽ ⚹ ♃
26 Friday	
01 10	☉ ⊻ ♃
06 44	☽ □ ⯓
09 34	☽ ⊥ ☉
10 19	☽ ∥ ♂
13 01	☽ ⚹ ♂
14 38	☽ ⊼ ♆
15 23	☽ Q ♀
19 08	☽ Q ♃
22 21	☽ Q ♅
27 Saturday	
00 32	☽ ± ♆
01 27	☽ ∠ ☉
02 58	☽ △ ♄
04 13	☽ ∥ ⯓
10 41	☽ ⊻ ☿
12 31	☽ Q ⯓
14 38	☽ ∠ ♂
15 37	☽ ♇ ♆
18 30	☽ ⚹ ♅
21 31	☽ ⊥ ☿
28 Sunday	
00 23	☽ □ ♀
01 58	☽ □ ♃
04 33	☽ ♇ ♄
04 47	☽ ⚹ ☉
06 29	☽ ⊥ ♂
09 05	☽ ⚹ ⯓
14 03	☽ ∠ ☿
17 16	☽ ⊻ ♂
17 31	☽ △ ♆
20 22	☽ ± ♄
20 33	☽ ∠ ♅
22 28	♀ ☌ ♃
29 Monday	
02 09	♂ □ ♆
03 31	☽ Q ☉
07 07	☽ ⊼ ♄
11 42	☽ ∠ ⯓
12 39	☽ ⊥ ♅
18 22	☽ ⚹ ☿
19 49	☉ # ♆
22 08	☽ ∥ ⯓
23 36	☽ ⊻ ♅
30 Tuesday	
02 42	♀ ∠ ♄
03 09	♀ ± ♅
04 08	☽ ⊥ ⯓
05 07	♀ ∥ ♃
08 37	☽ △ ♃
11 28	☽ △ ♀
15 08	☽ □ ☉
15 18	☽ ⊻ ⯓
17 21	☉ △ ⯓
17 40	☽ Q ☿
23 48	☽ ∥ ♂

MAY 1963

LONGITUDES

Date	Sidereal time h m s	Sun ☉ ° ' "	Moon ☽ ° ' "	Moon ☽ 24.00 ° ' "	Mercury ☿ ° '	Venus ♀ ° '	Mars ♂ ° '	Jupiter ♃ ° '	Saturn ♄ ° '	Uranus ♅ ° '	Neptune ♆ ° '	Pluto ♇ ° '
01	02 34 48	10 ♉ 24 05	20 ♌ 34 19	26 ♌ 48 00	29 ♉ 19	08 ♈ 50	15 ♌ 23	06 ♈ 19	22 ♒ 14	01 ♍ 11	14 ♏ 24	09 ♍ 38
02	02 38 45	11 22 18	02 ♍ 57 52	09 ♍ 04 26	29 ♉ 45	10 02	15 46	06 32	22 17	01 R 10	14 R 23	09 R 38
03	02 42 41	12 20 29	15 ♍ 08 15	21 ♍ 09 48	00 ♊ 06	11 14	16 09	06 45	22 20	01 10	14 21	09 37
04	02 46 38	13 18 38	27 ♍ 09 33	03 ♎ 07 54	00 22	12 27	16 32	06 58	22 23	01 10	14 20	09 37
05	02 50 34	14 16 45	09 ♎ 05 14	15 ♎ 01 52	00 32	13 39	16 56	07 11	22 26	01 09	14 18	09 36
06	02 54 31	15 14 51	20 ♎ 58 06	26 ♎ 54 10	00 37	14 51	17 20	07 24	22 29	01 09	14 16	09 36
07	02 58 27	16 12 54	02 ♏ 50 19	08 ♏ 46 43	00 R 36	16 03	17 45	07 37	22 31	01 09	14 15	09 35
08	03 02 24	17 10 56	14 ♏ 43 35	20 ♏ 41 05	00 31	17 16	18 09	07 50	22 34	01 09	14 13	09 35
09	03 06 20	18 08 56	26 ♏ 39 22	02 ♐ 38 38	00 21	18 28	18 34	08 03	22 36	01 D 09	14 11	09 34
10	03 10 17	19 06 55	08 ♐ 39 04	14 ♐ 40 54	00 ♊ 06	19 40	18 59	08 16	22 39	01 09	14 10	09 34
11	03 14 14	20 04 52	20 ♐ 44 21	26 ♐ 49 42	29 ♉ 47	20 53	19 25	08 28	22 41	01 09	14 08	09 34
12	03 18 10	21 02 48	02 ♑ 57 16	09 ♑ 07 24	29 25	22 05	19 50	08 41	22 43	01 09	14 07	09 34
13	03 22 07	22 00 43	15 ♑ 20 27	21 ♑ 36 51	28 59	23 17	20 16	08 53	22 45	01 09	14 05	09 33
14	03 26 03	22 58 36	27 ♑ 57 01	04 ♒ 21 25	28 30	24 30	20 42	09 06	22 47	01 10	14 03	09 33
15	03 30 00	23 56 28	10 ♒ 50 30	17 ♒ 24 42	27 58	25 42	21 09	09 18	22 49	01 10	14 02	09 33
16	03 33 56	24 54 18	24 ♒ 04 26	00 ♓ 50 03	27 25	26 55	21 35	09 30	22 51	01 10	14 00	09 33
17	03 37 53	25 52 08	07 ♓ 41 49	14 ♓ 39 56	26 51	28 07	22 02	09 42	22 53	01 11	13 59	09 33
18	03 41 49	26 49 56	21 ♓ 44 25	28 ♓ 55 08	26 16	29 ♈ 20	22 29	09 54	22 54	01 11	13 57	09 32
19	03 45 46	27 47 43	06 ♈ 11 47	13 ♈ 33 51	25 41	00 ♉ 32	22 56	10 06	22 56	01 12	13 55	09 32
20	03 49 43	28 45 29	21 ♈ 00 38	28 ♈ 31 14	25 06	01 45	23 23	10 18	22 57	01 12	13 54	09 32
21	03 53 39	29 ♉ 43 14	06 ♉ 04 35	13 ♉ 39 28	24 33	02 57	23 51	10 30	22 58	01 13	13 52	09 32
22	03 57 36	00 ♊ 40 58	21 ♉ 14 36	28 ♉ 49 28	24 02	04 10	24 19	10 42	23 00	01 13	13 51	09 D 32
23	04 01 32	01 38 40	06 ♊ 20 24	13 ♊ 48 36	23 33	05 22	24 47	10 54	23 01	01 14	13 49	09 32
24	04 05 29	02 36 22	21 ♊ 12 13	28 ♊ 30 22	23 06	06 35	25 15	11 05	23 02	01 15	13 48	09 32
25	04 09 25	03 34 02	05 ♋ 42 23	12 ♋ 47 46	22 43	07 47	25 43	11 17	23 03	01 16	13 46	09 32
26	04 13 22	04 31 40	19 ♋ 46 15	26 ♋ 37 44	22 23	09 00	26 12	11 28	23 04	01 17	13 45	09 33
27	04 17 18	05 29 17	03 ♌ 22 18	10 ♌ 00 10	22 07	10 13	26 41	11 40	23 04	01 18	13 43	09 33
28	04 21 15	06 26 53	16 ♌ 31 39	22 ♌ 57 12	21 55	11 25	27 10	11 51	23 05	01 19	13 42	09 33
29	04 25 12	07 24 27	29 ♌ 17 17	05 ♍ 32 28	21 48	12 38	27 39	12 02	23 05	01 20	13 40	09 33
30	04 29 08	08 21 59	11 ♍ 43 35	17 ♍ 50 22	21 44	13 50	28 08	12 13	23 06	01 21	13 39	09 33
31	04 33 05	09 ♊ 19 31	23 ♍ 54 15	29 ♍ 55 32	21 ♉ 45	15 ♉ 03	28 ♌ 38	12 ♈ 24	23 ♒ 06	01 ♍ 22	13 ♏ 37	09 ♍ 34

Date	Moon True ☊ ° '	Moon Mean ☊ ° '	Moon ☽ Latitude ° '
01	23 ♋ 36	24 ♋ 18	02 N 18
02	23 R 34	24 15	03 13
03	23 30	24 12	03 57
04	23 23	24 09	04 31
05	23 13	24 06	04 53
06	23 02	24 03	05 02
07	22 50	23 59	04 57
08	22 37	23 56	04 40
09	22 26	23 53	04 10
10	22 17	23 50	03 30
11	22 10	23 47	02 39
12	22 06	23 44	01 40
13	22 04	23 40	00 N 36
14	22 D 04	23 37	00 S 31
15	22 04	23 34	01 39
16	22 05	23 31	02 43
17	22 R 04	23 28	03 39
18	22 02	23 24	04 25
19	21 57	23 21	04 55
20	21 51	23 18	05 06
21	21 43	23 15	04 57
22	21 36	23 12	04 26
23	21 29	23 09	03 38
24	21 24	23 05	02 35
25	21 21	23 02	01 23
26	21 20	22 59	00 S 08
27	21 D 20	22 56	01 N 05
28	21 21	22 53	02 12
29	21 22	22 49	03 11
30	21 R 23	22 46	03 59
31	21 ♋ 21	22 ♋ 43	04 N 35

DECLINATIONS

Date	Sun ☉ ° '	Moon ☽ ° '	Mercury ☿ ° '	Venus ♀ ° '	Mars ♂ ° '	Jupiter ♃ ° '	Saturn ♄ ° '	Uranus ♅ ° '	Neptune ♆ ° '	Pluto ♇ ° '
01	14 N 57	16 N 49	22 N 35	01 N 59	18 N 13	01 N 30	15 S 01	11 N 47	14 S 25	20 N 39
02	15 15	13 25	22 36	02 26	18 04	01 35	15 00	11 47	14 24	20 39
03	15 33	09 30	22 35	02 54	17 56	01 40	14 59	11 47	14 24	20 39
04	15 50	05 17	22 31	03 21	17 47	01 45	14 58	11 47	14 23	20 39
05	16 08	00 N 53	22 26	03 49	17 39	01 50	14 58	11 47	14 23	20 39
06	16 25	03 S 31	22 17	04 17	17 30	01 55	14 57	11 47	14 22	20 39
07	16 41	07 48	22 07	04 44	17 21	02 00	14 56	11 47	14 22	20 39
08	16 58	11 48	21 54	05 11	17 12	02 05	14 55	11 47	14 21	20 39
09	17 14	15 21	21 40	05 39	17 03	02 10	14 55	11 47	14 21	20 38
10	17 30	18 18	21 23	06 06	16 54	02 15	14 54	11 47	14 21	20 38
11	17 46	20 29	21 05	06 33	16 44	02 19	14 54	11 47	14 20	20 38
12	18 01	21 44	20 45	07 00	16 35	02 24	14 53	11 47	14 20	20 38
13	18 16	21 58	20 23	07 27	16 25	02 29	14 53	11 47	14 19	20 38
14	18 31	21 05	20 00	07 54	16 16	02 34	14 52	11 47	14 19	20 37
15	18 46	19 06	19 37	08 20	16 06	02 38	14 52	11 47	14 18	20 37
16	19 00	16 03	19 12	08 47	15 56	02 43	14 51	11 47	14 18	20 37
17	19 14	12 04	18 47	09 13	15 46	02 48	14 51	11 46	14 17	20 37
18	19 27	07 20	18 22	09 39	15 36	02 52	14 50	11 46	14 17	20 36
19	19 40	02 S 03	17 57	10 05	15 26	02 57	14 50	11 46	14 16	20 36
20	19 53	03 N 28	17 32	10 31	15 15	03 01	14 50	11 46	14 16	20 36
21	20 06	08 53	17 08	10 56	15 05	03 06	14 50	11 46	14 15	20 36
22	20 18	13 47	16 46	11 21	14 55	03 10	14 49	11 45	14 15	20 35
23	20 30	17 48	16 24	11 46	14 44	03 15	14 49	11 45	14 15	20 35
24	20 41	20 34	16 04	12 11	14 33	03 19	14 49	11 45	14 14	20 34
25	20 52	21 56	15 45	12 35	14 23	03 24	14 49	11 44	14 14	20 34
26	21 03	21 51	15 29	13 00	14 12	03 28	14 49	11 44	14 13	20 34
27	21 13	20 27	15 14	13 23	14 01	03 32	14 49	11 44	14 13	20 33
28	21 23	17 59	15 02	13 47	13 50	03 36	14 49	11 43	14 12	20 33
29	21 33	14 42	14 51	14 10	13 38	03 41	14 49	11 43	14 12	20 33
30	21 42	10 51	14 43	14 33	13 27	03 45	14 49	11 42	14 12	20 33
31	21 N 51	06 N 38	14 N 38	14 N 56	13 N 16	03 N 49	14 S 49	11 N 42	14 S 11	20 N 32

ZODIAC SIGN ENTRIES

Date	h	m	Planets
02	06	13	☽ ♍
03	04	17	☿ ♊
04	17	42	☽ ♎
07	06	16	☽ ♏
09	18	42	☽ ♐
10	20	39	☿ ♉
12	06	13	☽ ♑
14	15	51	☽ ♒
16	22	32	☽ ♓
19	01	21	♀ ♉
19	01	48	☽ ♈
21	02	21	☽ ♉
21	18	58	☉ ♊
23	01	53	☽ ♊
25	02	29	☽ ♋
27	05	58	☽ ♌
29	13	22	☽ ♍

LATITUDES

Date	Mercury ☿ ° '	Venus ♀ ° '	Mars ♂ ° '	Jupiter ♃ ° '	Saturn ♄ ° '	Uranus ♅ ° '	Neptune ♆ ° '	Pluto ♇ ° '
01	02 N 39	01 S 40	02 N 05	01 S 06	00 S 58	00 N 47	01 N 50	13 N 46
04	02 21	01 42	02 00	01 07	00 58	00 47	01 50	13 45
07	01 52	01 43	01 56	01 07	00 59	00 46	01 50	13 44
10	01 14	01 44	01 51	01 07	00 59	00 46	01 50	13 43
13	00 N 28	01 43	01 47	01 08	00 59	00 46	01 50	13 42
16	00 S 24	01 43	01 43	01 08	01 00	00 46	01 50	13 41
19	01 16	01 41	01 39	01 09	01 00	00 46	01 50	13 40
22	02 05	01 39	01 35	01 09	01 01	00 46	01 50	13 39
25	02 47	01 36	01 31	01 09	01 02	00 46	01 50	13 38
28	03 20	01 33	01 27	01 10	01 02	00 46	01 50	13 37
31	03 S 42	01 S 29	01 N 24	01 S 11	01 S 03	00 N 45	01 N 50	13 N 36

DATA

Julian Date	2438151
Delta T	+35 seconds
Ayanamsa	23° 20' 25"
Synetic vernal point	05° ♓ 46' 35"
True obliquity of ecliptic	23° 26' 35"

MOON'S PHASES, APSIDES AND POSITIONS ☽

Date	h	m	Phase	Longitude ° '	Eclipse Indicator
08	17	23	○	17 ♏ 24	
16	13	36	☾	24 ♒ 58	
23	04	00	●	01 ♊ 19	
30	04	55	☽	08 ♍ 05	

Day	h	m		
07	04	04	Apogee	
22	03	52	Perigee	
05	16	48	0S	
13	05	02	Max dec	22° S 01'
19	20	57	0N	
25	22	28	Max dec	22° N 04'

LONGITUDES

Date	Chiron ⚷ ° '	Ceres ⚳ ° '	Pallas ⚴ ° '	Juno ⚵ ° '	Vesta ⚶ ° '	Black Moon Lilith ⚸ ° '
01	13 ♓ 53	05 ♍ 09	27 ♋ 33	11 ♍ 34	16 ♍ 22	01 ♏ 14
11	14 ♓ 16	06 ♍ 22	01 ♌ 47	11 ♍ 58	16 ♍ 53	02 ♏ 21
21	14 ♓ 33	08 ♍ 09	06 ♌ 09	12 ♍ 53	18 ♍ 08	03 ♏ 29
31	14 ♓ 46	10 ♍ 23	10 ♌ 36	14 ♍ 17	20 ♍ 02	04 ♏ 36

ASPECTARIAN

01 Wednesday
- 00 17 ☽ □ ♆
- 01 47 ☽ ☌ ♂
- 03 43 ♀ ± ♆
- 13 27 ☽ ⚼ ♃
- 15 13 ☽ ☍ ♄
- 17 15 ☉ # ♄
- 18 56 ☽ ⚼ ♀

02 Thursday
- 00 35 ☽ ∥ ☉
- 01 16 ☽ # ♄
- 03 57 ♀ ⊼ ♇
- 05 22 ☽ # ♆
- 05 31 ☽ □ ☿
- 07 10 ☽ ± ♃
- 08 30 ☽ ☌ ♅
- 10 52 ☽ Q ♆
- 14 20 ☽ ± ♀
- 19 08 ☽ ⊼ ♃
- 22 17 ☽ ∥ ♅

03 Friday
- 01 05 ☽ ☌ ♇
- 03 26 ☽ ⊼ ♀
- 05 58 ☽ △ ☉
- 10 27 ☽ ⚹ ♆
- 14 04 ☽ ⊻ ♂

04 Saturday
- 01 14 ☉ ⊥ ♃
- 02 24 ☽ ⊼ ♄
- 02 26 ☽ ⊥ ♂
- 14 28 ☽ ± ♄
- 14 31 ☽ ⚼ ☉
- 16 20 ☽ ∠ ♆
- 18 32 ☽ △ ☿
- 20 02 ☽ ⊻ ♅
- 21 06 ☽ ∠ ♂
- 21 33 ☽ ∥ ♀

05 Sunday
- 06 57 ☽ ∥ ♃
- 08 06 ♃ ± ♅
- 08 06 ☽ ⊥ ♅
- 08 06 ☽ ☍ ♃
- 08 39 ☽ ⚼ ♄
- 10 14 ☽ ± ☉
- 10 25 ☽ ⊥ ♆
- 12 29 ☉ ☍ ♆
- 13 02 ☽ ⊻ ♇
- 22 15 ☽ ☍ ♀
- 22 30 ☽ ⊻ ♆
- 23 25 ☽ ⊼ ☉

06 Monday
- 00 42 ♀ ⊼ ♆
- 01 07 ☽ ⚼ ☿
- 01 09 ☽ ⊥ ♇
- 02 16 ☽ ∠ ♅
- 03 02 ☽ # ♃
- 04 24 ☽ ⚹ ♂
- 15 04 ☽ △ ♄
- 16 38 ☽ # ♀
- 19 20 ☽ ∠ ♇
- 19 23 ☽ ± ☿
- 22 15 ♃ ∠ ♄
- 22 30 ☿ St R

07 Tuesday
- 02 43 ♀ ± ♇
- 03 17 ☿ ∠ ♀
- 05 32 ☽ Q ♂
- 07 30 ☽ ⊼ ☿
- 08 35 ☽ ⚹ ♅
- 13 51 ♀ ⚼ ♅
- 21 50 ☽ ⊼ ♃

08 Wednesday
- 01 38 ☽ ⚹ ♇
- 04 00 ☉ ⊻ ♀
- 08 49 ☽ Q ♅
- 10 10 ☽ ± ♃
- 10 59 ☽ ☌ ♆
- 11 59 ☽ # ♅
- 17 23 ☽ ☍ ☉
- 17 41 ☽ ⊼ ♀
- 19 10 ☽ □ ♂

09 Thursday
- 01 10 ☉ ∥ ♂
- 01 48 ☽ Q ♇
- 03 50 ☽ □ ♄
- 04 37 ☽ ⚼ ♃
- 04 54 ☽ ∥ ♆
- 07 07 ☽ ± ♀
- 08 52 ☽ ∥ ♄
- 10 16 ♅ St D
- 15 11 ♀ △ ♂
- 19 16 ☽ ☍ ☿
- 21 00 ☽ □ ♅

10 Friday
- 00 29 ☽ # ♂
- 02 24 ♃ ± ♆
- 03 10 ☽ ⚼ ♀
- 04 13 ☽ # ☉
- 06 31 ☉ □ ♂
- 11 12 ☽ △ ♃
- 13 50 ☽ □ ♇
- 15 59 ☽ Q ♄
- 22 57 ☽ ⊻ ♆

11 Saturday
- 09 17 ☽ △ ♂
- 10 35 ☽ ⊼ ☉
- 10 49 ☽ ⊥ ♆
- 12 18 ☽ △ ♀
- 14 12 ☽ # ♇
- 15 51 ☽ ⚹ ♄
- 19 28 ☽ # ☿
- 23 26 ☽ ± ☉

12 Sunday
- 04 30 ☽ ∠ ♆
- 05 17 ☽ ⊼ ☿
- 08 29 ☽ △ ♅
- 15 49 ☽ ⚼ ♂
- 16 38 ☽ ± ☿
- 18 32 ☽ ⚼ ☉
- 19 42 ☿ ∥ ♇
- 21 18 ☽ ∠ ♄
- 23 20 ☽ □ ♃

13 Monday
- 00 50 ☽ △ ♇
- 01 00 ♀ ⚹ ♄
- 07 29 ☿ ⊥ ♀
- 09 28 ☽ ⚼ ☿
- 09 35 ☽ ⚹ ♆
- 09 52 ☽ ± ♂
- 13 34 ☽ ⚼ ♅
- 14 43 ☽ ⊥ ♄
- 21 47 ☽ ⊼ ♂

14 Tuesday
- 01 49 ☽ △ ☉
- 02 12 ☽ ⊻ ♄
- 04 47 ☽ □ ♀
- 05 35 ☽ ⚼ ♇
- 06 44 ☽ ± ♅
- 07 07 ☉ □ ♄
- 08 26 ☽ Q ♆
- 10 21 ☽ Q ♃
- 12 59 ☽ △ ☿
- 13 02 ♀ ⚼ ♇
- 18 02 ☽ ⊼ ♅
- 19 00 ☽ # ♇
- 22 30 ☽ ± ♇

15 Wednesday
- 05 52 ☽ # ☿
- 09 07 ☽ ⚹ ♃
- 09 37 ☽ ⊼ ♇
- 14 52 ☽ # ☉
- 17 47 ☽ Q ♀
- 17 50 ☽ □ ♆
- 23 18 ☉ ∠ ♃

16 Thursday
- 07 23 ☽ ☍ ♂
- 09 48 ☽ ☌ ♄
- 12 47 ☽ ∠ ♃
- 12 52 ☽ # ♂
- 13 36 ☽ □ ☉
- 16 49 ♃ ⊼ ♇
- 17 33 ☽ ⚹ ♀
- 17 43 ☽ □ ☿
- 18 51 ☿ ⊻ ♀
- 19 34 ☉ ∥ ☿
- 19 50 ☽ ∥ ♄
- 23 17 ☽ ∥ ♆

17 Friday
- 00 36 ☽ ☍ ♅
- 04 56 ☽ ⊥ ♃
- 13 37 ☽ # ♅
- 15 12 ☽ ☍ ♇
- 15 32 ☽ ⊻ ♃
- 22 14 ☽ ∠ ♀
- 22 48 ☽ △ ♆
- 23 25 ☽ Q ☉
- 23 49 ☽ Q ☿

18 Saturday
- 01 37 ☽ # ♀
- 03 10 ☉ ☌ ☿
- 13 17 ☽ ⊼ ♂
- 13 58 ☽ ⊻ ♄
- 14 55 ☽ ⊥ ♀
- 19 17 ☽ ⚹ ☿
- 21 08 ☽ ⚹ ☉
- 23 38 ☽ ± ♂

19 Sunday
- 00 00 ☽ ⊥ ♄
- 00 02 ☽ ⚼ ♆
- 01 50 ☽ ⊻ ♀
- 03 46 ☽ ⊼ ♅
- 08 05 ☽ # ♃
- 11 41 ♂ ☍ ♄
- 13 38 ☽ ± ♅
- 14 49 ☽ ± ♆
- 14 50 ☽ ∠ ♄
- 14 56 ☽ ⚼ ♂
- 17 28 ☽ ⊼ ♇
- 18 29 ☽ ☌ ♃
- 19 03 ☽ ∠ ☿
- 23 30 ☽ ∠ ☉

20 Monday
- 00 34 ☽ ⊼ ♆
- 01 09 ♀ △ ♅
- 03 12 ☽ ± ♇
- 04 16 ☽ ⚼ ♅
- 05 45 ☿ ∠ ♃
- 09 03 ☽ ⊥ ☿
- 10 02 ☽ ∥ ♃
- 15 00 ☽ ⊥ ☉
- 15 07 ☽ ⚹ ♄
- 15 56 ☽ △ ♂
- 17 39 ☽ ⚼ ♇
- 18 19 ☽ ⊻ ☿

21 Tuesday
- 01 13 ☽ ⊻ ☉
- 04 17 ☽ △ ♅
- 06 37 ☽ ☌ ♀
- 10 15 ☽ Q ♄
- 15 39 ♇ St D
- 17 29 ☽ △ ♇
- 19 06 ☽ ⊻ ♃
- 22 32 ☽ ∥ ♀

22 Wednesday
- 00 19 ☽ ☍ ♆
- 01 39 ☽ ∥ ♅
- 04 43 ☽ ⊥ ♃
- 05 03 ☿ □ ♂
- 14 30 ☽ # ♆
- 14 47 ☽ □ ♄
- 16 16 ☽ ☌ ☿
- 17 01 ☽ □ ♂
- 17 38 ☽ # ♄
- 17 54 ☽ ∥ ♂
- 19 09 ☽ ∠ ♃

23 Thursday
- 00 15 ♂ # ♄
- 01 41 ☉ □ ♅
- 03 39 ☽ ∥ ☿
- 03 51 ☽ □ ♅
- 04 00 ☽ ☌ ☉
- 04 47 ♀ Q ♄
- 10 19 ☽ ⊻ ♀
- 10 49 ♀ ∥ ♅
- 17 07 ☽ □ ♇
- 19 24 ☽ ⚹ ♃
- 20 47 ☽ ⊥ ♀
- 22 34 ☉ ∥ ♇
- 22 41 ☽ Q ♂

24 Friday
- 00 00 ☽ ⊼ ♆
- 08 49 ☽ Q ♅
- 09 42 ☽ ± ♆
- 12 00 ☽ ∥ ♇
- 12 40 ☽ ∠ ♀
- 13 25 ☽ ∥ ☉
- 14 59 ☽ △ ♄
- 15 02 ☽ ⊻ ☿
- 15 07 ☽ Q ♃
- 16 15 ☿ □ ♄
- 18 51 ☽ ⚹ ♂
- 22 24 ☽ Q ♇

25 Saturday
- 00 27 ☽ ⚼ ♆
- 00 39 ☽ ⊥ ☿
- 04 34 ☽ ⚹ ♅
- 08 10 ☽ ⊻ ☉
- 15 18 ☽ ∠ ☿
- 15 50 ☽ ⚹ ♀
- 15 57 ☽ ⚼ ♄
- 18 28 ☽ ⚹ ♇
- 18 59 ☽ ⊥ ☉
- 20 46 ☽ ∠ ♂
- 21 33 ☽ □ ♃

26 Sunday
- 01 38 ☽ △ ♆
- 05 58 ☽ ∠ ♅
- 07 18 ☽ ± ♄
- 08 15 ♂ # ♆
- 11 33 ☽ ∠ ☉
- 12 46 ☽ ⊥ ♂
- 14 21 ☽ Q ♀
- 16 28 ☽ ⚹ ☿
- 17 44 ☽ ⊼ ♄
- 20 20 ☽ ∠ ♇
- 21 38 ☽ ⊥ ♅
- 22 48 ♀ △ ♇
- 23 39 ☽ ⊻ ♂

27 Monday
- 02 25 ☽ ∥ ☉
- 08 17 ☽ ⊻ ♅
- 10 20 ♂ ⚼ ♃
- 10 43 ☽ ∥ ♇
- 12 19 ☽ ⊥ ♇
- 13 19 ☽ Q ☿
- 16 06 ☽ ⚹ ☉
- 23 10 ☽ ⊻ ♇

28 Tuesday
- 01 38 ☽ □ ♀
- 03 15 ☽ △ ♃
- 06 46 ☽ □ ♆
- 13 44 ♀ ∥ ♂
- 15 51 ☽ Q ☉
- 21 57 ☽ □ ☿
- 21 58 ♀ ⊻ ♃

29 Wednesday
- 00 15 ☽ ☍ ♄
- 07 39 ☽ ⚼ ♃
- 08 45 ☽ ☌ ♂
- 10 53 ☽ ∥ ☿
- 11 15 ☽ # ♄
- 13 44 ♀ # ♆
- 15 08 ☽ ∥ ♀
- 15 17 ☽ # ♆
- 15 54 ☽ ☌ ♅
- 16 33 ☽ Q ♆
- 19 15 ☽ ∥ ♂
- 19 44 ☿ # ♄

30 Thursday
- 01 09 ☽ ± ♃
- 04 55 ☽ □ ☉
- 06 51 ☽ ∥ ♅
- 07 47 ☽ ☌ ♇
- 08 11 ♀ ☍ ♆
- 12 59 ☽ ⊼ ♃
- 15 45 ☽ ⚹ ♆
- 16 36 ☽ △ ♀
- 18 51 ☿ St D
- 20 17 ☿ ∥ ♀

31 Friday
- 04 18 ♀ # ♄
- 07 43 ☽ △ ☿
- 10 25 ☽ ⊼ ♄
- 17 58 ☉ □ ♇
- 21 22 ☽ ∠ ♆
- 21 48 ☽ ⊻ ♂
- 22 22 ☽ ± ♄

All ephemeris data is given at 12.00 UT and the Moon's longitude is additionally given for 24.00 UT

JUNE 1963

LONGITUDES

Date	Sidereal time h m s	Sun ☉ ° ' "	Moon ☽ ° ' "	Moon ☽ 24.00 ° ' "	Mercury ☿ ° '	Venus ♀ ° '	Mars ♂ ° '	Jupiter ♃ ° '	Saturn ♄ ° '	Uranus ♅ ° '	Neptune ♆ ° '	Pluto ♇ ° '
01	04 37 01	10 ♊ 17 01	05 ♎ 54 45	11 ♎ 52 26	21 ♉ 50	16 ♉ 16	29 ♌ 07	12 ♈ 35	23 ♒ 06	01 ♍ 23	13 ♏ 36	09 ♍ 34
02	04 40 58	11 14 29	17 ♎ 49 04	23 ♎ 45 06	22 D 00	17 28	29 ♌ 37	12 45	23 07	01 24	13 R 34	09 34
03	04 44 54	12 11 57	29 ♎ 40 58	05 ♏ 37 02	22 15	18 41	00 ♍ 07	12 56	23 R 07	01 26	13 33	09 35
04	04 48 51	13 09 23	11 ♏ 33 39	17 ♏ 31 07	22 34	19 54	00 37	13 07	23 07	01 27	13 32	09 35
05	04 52 47	14 06 48	23 ♏ 29 43	29 ♏ 29 40	22 57	21 06	01 07	13 17	23 06	01 28	13 30	09 36
06	04 56 44	15 04 12	05 ♐ 31 11	11 ♐ 34 28	23 24	22 19	01 38	13 27	23 06	01 30	13 29	09 36
07	05 00 41	16 01 36	17 ♐ 39 41	23 ♐ 47 01	23 56	23 32	02 08	13 37	23 06	01 31	13 28	09 37
08	05 04 37	16 58 58	29 ♐ 56 36	06 ♑ 08 37	24 32	24 45	02 39	13 48	23 05	01 33	13 26	09 37
09	05 08 34	17 56 20	12 ♑ 23 14	18 ♑ 40 38	25 12	25 57	03 10	13 58	23 05	01 34	13 25	09 38
10	05 12 30	18 53 41	25 ♑ 01 00	01 ♒ 24 32	25 56	27 10	03 41	14 07	23 04	01 36	13 24	09 38
11	05 16 27	19 51 01	07 ♒ 51 27	14 ♒ 21 59	26 44	28 23	04 12	14 17	23 03	01 38	13 23	09 39
12	05 20 23	20 48 21	20 ♒ 56 21	27 ♒ 34 47	27 36	29 ♉ 36	04 43	14 27	23 03	01 40	13 21	09 40
13	05 24 20	21 45 41	04 ♓ 17 30	11 ♓ 04 38	28 32	00 ♊ 49	05 15	14 36	23 02	01 41	13 20	09 40
14	05 28 16	22 42 59	17 ♓ 56 21	24 ♓ 52 43	29 ♉ 31	02 02	05 46	14 46	23 01	01 43	13 19	09 41
15	05 32 13	23 40 18	01 ♈ 53 41	08 ♈ 59 11	00 ♊ 34	03 14	06 18	14 55	23 00	01 45	13 18	09 42
16	05 36 10	24 37 36	16 ♈ 08 59	23 ♈ 22 44	01 40	04 27	06 50	15 04	22 58	01 47	13 17	09 42
17	05 40 06	25 34 54	00 ♉ 39 58	08 ♉ 00 05	02 50	05 40	07 21	15 13	22 57	01 49	13 16	09 43
18	05 44 03	26 32 11	15 ♉ 22 22	22 ♉ 46 00	04 03	06 53	07 54	15 22	22 56	01 51	13 14	09 44
19	05 47 59	27 29 29	00 ♊ 10 05	07 ♊ 33 41	05 20	08 06	08 26	15 31	22 54	01 53	13 13	09 45
20	05 51 56	28 26 46	14 ♊ 55 50	22 ♊ 15 35	06 40	09 19	08 58	15 39	22 53	01 55	13 12	09 46
21	05 55 52	29 ♊ 24 03	29 ♊ 32 06	06 ♋ 44 33	08 03	10 32	09 31	15 48	22 51	01 57	13 11	09 47
22	05 59 49	00 ♋ 21 19	13 ♋ 52 17	20 ♋ 54 46	09 30	11 45	10 03	15 56	22 49	01 59	13 10	09 48
23	06 03 45	01 18 35	27 ♋ 51 34	04 ♌ 42 27	10 59	12 58	10 36	16 05	22 47	02 02	13 09	09 49
24	06 07 42	02 15 50	11 ♌ 27 17	18 ♌ 06 05	12 32	14 11	11 09	16 13	22 45	02 04	13 08	09 50
25	06 11 39	03 13 04	24 ♌ 38 59	01 ♍ 06 13	14 09	15 24	11 42	16 21	22 43	02 06	13 07	09 51
26	06 15 35	04 10 19	07 ♍ 28 07	13 ♍ 45 05	15 48	16 37	12 15	16 28	22 41	02 09	13 07	09 52
27	06 19 32	05 07 32	19 ♍ 57 34	26 ♍ 06 06	17 30	17 50	12 48	16 36	22 39	02 11	13 06	09 53
28	06 23 28	06 04 45	02 ♎ 11 11	08 ♎ 13 24	19 16	19 03	13 21	16 44	22 37	02 13	13 05	09 54
29	06 27 25	07 01 57	14 ♎ 13 19	20 ♎ 11 29	21 04	20 17	13 55	16 51	22 34	02 16	13 04	09 55
30	06 31 21	07 ♋ 59 09	26 ♎ 08 30	02 ♏ 04 52	22 ♊ 55	21 ♊ 30	14 ♍ 28	16 ♈ 58	22 ♒ 32	02 ♍ 18	13 ♏ 03	09 ♍ 56

Date	Moon True ☊ ° '	Moon Mean ☊ ° '	Moon ☽ Latitude ° '
01	21 ♋ 18	22 ♋ 40	04 N 58
02	21 R 14	22 37	05 09
03	21 08	22 34	05 06
04	21 01	22 30	04 50
05	20 54	22 27	04 21
06	20 48	22 24	03 41
07	20 43	22 21	02 50
08	20 40	22 18	01 50
09	20 38	22 15	00 N 45
10	20 D 38	22 11	00 S 24
11	20 39	22 08	01 33
12	20 40	22 05	02 38
13	20 42	22 02	03 37
14	20 43	21 59	04 24
15	20 R 42	21 55	04 57
16	20 41	21 52	05 13
17	20 39	21 49	05 09
18	20 36	21 46	04 45
19	20 33	21 43	04 02
20	20 30	21 40	03 03
21	20 28	21 36	01 53
22	20 27	21 33	00 S 36
23	20 D 27	21 30	00 N 41
24	20 28	21 27	01 53
25	20 29	21 24	02 58
26	20 30	21 21	03 51
27	20 31	21 17	04 32
28	20 32	21 14	05 00
29	20 R 32	21 11	05 14
30	20 ♋ 31	21 ♋ 08	05 N 14

DECLINATIONS

Date	Sun ☉ ° '	Moon ☽ ° '	Mercury ☿ ° '	Venus ♀ ° '	Mars ♂ ° '	Jupiter ♃ ° '	Saturn ♄ ° '	Uranus ♅ ° '	Neptune ♆ ° '	Pluto ♇ ° '
01	22 N 00	02 N 13	14 N 34	15 N 18	13 N 04	03 N 53	14 S 49	11 N 41	14 S 11	20 N 31
02	22 08	02 S 14	14 33	15 40	12 53	03 57	14 49	11 41	14 10	20 31
03	22 16	06 35	14 33	16 02	12 41	04 01	14 49	11 40	14 10	20 30
04	22 23	10 42	14 36	16 23	12 29	04 05	14 49	11 40	14 10	20 30
05	22 30	14 26	14 41	16 43	12 17	04 09	14 49	11 39	14 09	20 29
06	22 36	17 36	14 48	17 04	12 05	04 12	14 50	11 39	14 09	20 29
07	22 43	20 03	14 57	17 24	11 53	04 16	14 50	11 38	14 09	20 28
08	22 48	21 36	15 08	17 43	11 41	04 20	14 50	11 38	14 08	20 28
09	22 54	22 07	15 20	18 02	11 29	04 24	14 51	11 37	14 08	20 27
10	22 59	21 31	15 34	18 21	11 17	04 27	14 51	11 37	14 08	20 27
11	23 03	19 48	15 49	18 39	11 04	04 31	14 51	11 36	14 07	20 26
12	23 07	17 01	16 06	18 57	10 52	04 34	14 52	11 35	14 07	20 26
13	23 11	13 18	16 24	19 14	10 40	04 38	14 52	11 35	14 07	20 25
14	23 15	08 49	16 43	19 31	10 27	04 41	14 53	11 34	14 06	20 24
15	23 17	03 S 47	17 03	19 47	10 14	04 45	14 53	11 33	14 06	20 24
16	23 20	01 N 32	17 24	20 02	10 01	04 48	14 54	11 32	14 06	20 23
17	23 22	06 53	17 46	20 17	09 49	04 51	14 55	11 32	14 05	20 23
18	23 24	11 54	18 08	20 32	09 36	04 54	14 55	11 31	14 05	20 22
19	23 25	16 15	18 31	20 46	09 23	04 58	14 56	11 30	14 05	20 21
20	23 26	19 33	18 55	21 00	09 09	05 01	14 57	11 29	14 04	20 21
21	23 26	21 34	19 18	21 13	08 56	05 04	14 57	11 29	14 04	20 20
22	23 27	22 07	19 42	21 25	08 43	05 07	14 58	11 28	14 04	20 19
23	23 26	21 16	20 06	21 37	08 30	05 10	14 59	11 27	14 04	20 19
24	23 25	19 10	20 29	21 48	08 16	05 13	15 00	11 26	14 03	20 18
25	23 24	16 06	20 53	21 58	08 03	05 15	15 00	11 25	14 03	20 17
26	23 23	12 21	21 15	22 08	07 49	05 18	15 01	11 24	14 03	20 17
27	23 21	08 09	21 37	22 18	07 36	05 21	15 02	11 23	14 03	20 16
28	23 18	03 N 43	21 58	22 26	07 22	05 24	15 03	11 23	14 03	20 15
29	23 15	00 S 47	22 18	22 34	07 08	05 26	15 04	11 22	14 02	20 15
30	23 N 12	05 S 12	22 N 37	22 N 42	06 N 54	05 N 29	15 S 05	11 N 21	14 S 02	20 N 14

ZODIAC SIGN ENTRIES

Date	h	m	Planets
01	00	09	☽ ♎
03	06	30	♂ ♍
03	12	39	☽ ♏
06	01	01	☽ ♐
08	12	07	☽ ♑
10	21	22	☽ ♒
12	19	57	♀ ♊
13	04	21	☽ ♓
14	23	21	☿ ♊
15	08	46	☽ ♈
17	10	54	☽ ♉
19	11	44	☽ ♊
21	12	46	☽ ♋
22	03	04	☉ ♋
23	15	44	☽ ♌
25	21	56	☽ ♍
28	07	41	☽ ♎
30	19	48	☽ ♏

LATITUDES

Date	Mercury ☿ ° '	Venus ♀ ° '	Mars ♂ ° '	Jupiter ♃ ° '	Saturn ♄ ° '	Uranus ♅ ° '	Neptune ♆ ° '	Pluto ♇ ° '
01	03 S 47	01 S 28	01 N 23	01 S 11	01 S 03	00 N 45	01 N 50	13 N 35
04	03 56	01 23	01 19	01 12	01 04	00 45	01 49	13 34
07	03 56	01 18	01 16	01 12	01 04	00 45	01 49	13 33
10	03 47	01 13	01 12	01 13	01 05	00 45	01 49	13 32
13	03 31	01 07	01 09	01 13	01 05	00 45	01 49	13 31
16	03 10	01 01	01 06	01 14	01 06	00 45	01 49	13 30
19	02 43	00 54	01 02	01 15	01 06	00 45	01 49	13 29
22	02 12	00 47	00 59	01 15	01 07	00 45	01 49	13 28
25	01 38	00 40	00 56	01 16	01 08	00 45	01 49	13 27
28	01 03	00 33	00 53	01 17	01 08	00 44	01 48	13 26
31	00 S 27	00 S 26	00 N 50	01 S 17	01 S 09	00 N 44	01 N 48	13 N 25

DATA

Julian Date	2438182
Delta T	+35 seconds
Ayanamsa	23° 20' 29"
Synetic vernal point	05° ♓ 46' 31"
True obliquity of ecliptic	23° 26' 35"

MOON'S PHASES, APSIDES AND POSITIONS ☽

Date	h	m	Phase	Longitude ° '	Eclipse Indicator
07	08	31	○	15 ♐ 53	
14	20	53	☾	23 ♓ 04	
21	11	46	●	29 ♊ 23	
28	20	24	☽	06 ♎ 25	

Day	h	m		
03	13	47	Apogee	
19	07	30	Perigee	
01	23	57	0S	
09	11	21	Max dec	22° S 07'
16	05	08	0N	
22	09	08	Max dec	22° N 08'
29	07	50	0S	

LONGITUDES

Date	Chiron ⚷ ° '	Ceres ⚳ ° '	Pallas ⚴ ° '	Juno ⚵ ° '	Vesta ⚶ ° '	Black Moon Lilith ⚸ ° '
01	14 ♓ 47	10 ♍ 38	11 ♌ 03	14 ♍ 27	20 ♍ 15	04 ♏ 43
11	14 ♓ 53	13 ♍ 18	15 ♌ 34	16 ♍ 17	22 ♍ 46	05 ♏ 50
21	14 ♓ 55	16 ♍ 18	20 ♌ 07	18 ♍ 27	25 ♍ 45	06 ♏ 57
31	14 ♓ 50	19 ♍ 34	24 ♌ 43	20 ♍ 53	29 ♍ 08	08 ♏ 04

ASPECTARIAN

h m	Aspects
01 Saturday	
01 38	☽ ⚼ ♀
02 54	☽ ⊻ ♅
03 10	☽ ∥ ♃
10 20	☽ ⊥ ♂
13 53	☽ ⚼ ☿
14 58	☽ ⊥ ♅
15 23	☽ ⊥ ♆
16 25	☽ ⚼ ♄
19 21	☽ ⊻ ♇
21 34	☽ △ ☉
21 45	☽ ± ♀
02 Sunday	
01 38	☽ ☍ ♃
03 27	☽ ⊻ ♆
05 15	☽ ∠ ♂
07 28	☽ ⊥ ♇
08 17	☽ ± ☿
09 09	☽ ∠ ♅
11 14	☽ ⊼ ♀
20 38	☽ ⊼ ☿
21 30	☽ # ♃
22 42	☽ △ ♄
03 Monday	
01 40	☽ ∠ ♇
06 32	☽ ⚼ ☉
09 39	♄ St R
12 55	☽ ⚹ ♂
15 32	☽ ⚹ ♅
17 46	♀ ⊥ ♃
04 Tuesday	
02 20	☽ ± ☉
08 01	☽ ⚹ ♇
10 33	☉ ⚹ ♃
14 13	☽ Q ♂
15 10	☽ ⊼ ♃
15 30	☽ ⊼ ☉
15 49	☽ Q ♅
15 57	☽ ☌ ♆
17 56	☽ # ♅
21 05	☉ ⊼ ♆
22 35	☽ # ♂
05 Wednesday	
03 25	☽ ± ♃
06 40	☽ ☍ ♀
08 11	☽ Q ♇
10 08	☽ ∥ ♆
10 52	☽ ☍ ☿
11 13	☽ □ ♄
13 51	☽ # ☿
14 46	☽ ∥ ♄
20 44	☿ □ ♄
21 43	☽ ⚼ ♃
06 Thursday	
03 55	☽ □ ♂
03 59	☽ □ ♅
04 22	♅ Q ♆
05 26	♂ Q ♆
05 33	♂ ☌ ♅
06 59	☽ # ♀
15 26	♃ ⊼ ♆
16 39	☿ # ♄
20 06	☽ □ ♇
23 04	☽ Q ♄
07 Friday	
03 26	♀ □ ♄
03 44	☽ ⊻ ♆
03 56	☽ △ ♃
08 31	☽ ☍ ☉
15 32	☽ ⊥ ♆
17 15	☽ # ♇
22 39	☽ ⚹ ♄
08 Saturday	
00 46	☽ ⊼ ♀
00 55	☽ ⊼ ☿
03 31	☿ ☌ ♀
09 05	☽ ∠ ♆
13 13	☽ ± ☿
13 44	☽ ± ♀
15 07	☽ △ ♅
17 28	☽ △ ♂
19 29	♂ ∥ ♅
09 Sunday	
03 44	☽ ∠ ♄
06 42	☽ △ ♇
07 34	☽ ⚼ ☿
08 58	☽ ⚼ ♀
13 58	☽ ⚹ ♆
15 03	☽ □ ♃
20 01	☽ ⚼ ♅
20 57	☽ ⊥ ♄
23 28	☽ ⊼ ☉
23 29	☽ ⚼ ♂
10 Monday	
08 20	☽ ⊻ ♄
11 17	☽ ⚼ ♇
11 45	☽ ± ☉
12 43	☽ Q ♆
13 06	☽ ± ♅
13 51	☽ △ ☿
16 29	☽ △ ♀
17 13	☽ ± ♂
11 Tuesday	
00 20	☉ ± ♆
00 23	☽ ⊼ ♅
01 30	☽ Q ♃
04 10	☽ ± ♇
04 45	☽ # ♇
04 55	☽ ⊼ ♂
05 58	☽ ⚼ ☉
06 18	☉ Q ♅
15 19	☽ ⊼ ♇
21 56	☽ # ♀
22 10	☽ □ ♆
12 Wednesday	
00 00	☽ ⚹ ♃
08 33	♀ ∠ ♃
11 44	☽ △ ☉
15 49	☽ ☌ ♄
17 59	☽ # ☿
13 Thursday	
00 55	☽ □ ☿
02 30	☽ ∥ ♄
03 32	☽ ∠ ♃
05 11	☽ □ ♀
07 11	☽ ∥ ♆
07 21	☽ ☍ ♅
13 45	☽ ☍ ♂
19 44	☽ ⊥ ♃
21 32	☽ ☍ ♇
21 40	☽ # ♅
14 Friday	
03 15	☽ # ♂
03 56	☽ △ ♆
05 46	♀ □ ♅
06 24	☽ ⊻ ♃
11 12	☽ Q ☿
15 58	☽ Q ♀
18 53	☿ ∠ ♃
19 16	☉ △ ♄
20 46	☽ ⊻ ♄
20 53	☽ □ ☉
15 Saturday	
05 52	☽ ⚼ ♆
07 04	☽ ⊥ ♄
07 38	☽ # ♃
09 32	☽ ⚹ ☿
11 45	☽ ⊼ ♅
14 30	☽ ⚹ ♀
19 45	☽ ⊼ ♂
21 08	☽ ± ♆
21 56	☽ ± ♅
22 19	☽ ∠ ♄
16 Sunday	
01 12	☽ ⊼ ♇
05 42	☽ Q ☉
06 14	☽ ± ♂
07 12	☽ ⊼ ♆
10 11	☽ ☌ ♃
11 16	☽ ± ♇
12 56	☽ ∠ ☿
13 03	☽ ⚼ ♅
14 33	☿ □ ♅
18 00	☽ ∠ ♀
21 47	☽ ⚼ ♂
23 17	☉ Q ♂
23 19	☽ ⚹ ♄
17 Monday	
02 12	☽ ⚼ ♇
02 43	☽ ∥ ♃
03 03	☽ ⚹ ☉
05 08	☽ ⊥ ☿
10 13	☽ ⊥ ♀
13 53	☽ △ ♅
15 52	☽ ⊻ ☿
19 00	☽ Q ♄
20 04	♀ ∥ ♇
20 56	☽ ⊻ ♀
23 22	☽ △ ♂
18 Tuesday	
01 09	☽ ∥ ♂
02 49	☽ △ ♇
05 20	☽ ∠ ☉
08 32	☽ ☍ ♆
10 05	☽ ∥ ♅
12 00	☽ ⊻ ♃
20 58	☽ ⊥ ☉
21 50	☽ ⊥ ♃
23 28	☽ # ♆
19 Wednesday	
00 14	☽ □ ♄
04 12	☽ # ♄
07 22	☽ ⊻ ☉
12 34	☽ ∠ ♃
12 41	☉ Q ♃
14 47	☽ □ ♅
18 34	☉ Q ♇
21 11	☽ ☌ ☿
23 34	♀ □ ♂
20 Thursday	
01 55	☽ □ ♂
02 02	☽ ☌ ♀
03 34	☽ □ ♇
05 44	☽ ∥ ☿
06 03	☉ ⚼ ♆
09 11	☽ ⊼ ♆
13 12	☽ ⚹ ♃
18 59	☽ ± ♆
19 34	☽ ∥ ♇
20 11	☽ Q ♅
20 52	♀ □ ♇
21 Friday	
00 59	☽ △ ♄
05 20	☽ ∥ ♀
07 49	♃ ± ♇
08 31	☽ Q ♂
09 05	☽ Q ♇
09 06	☽ Q ♃
09 46	☽ ⚼ ♆
11 46	☽ ☌ ☉
16 01	☽ ⚹ ♅
22 Saturday	
00 14	♂ ☌ ♇
01 49	☽ ⚼ ♄
03 46	☽ ⊻ ☿
05 07	☽ ⚹ ♇
05 18	☽ ⚹ ♂
05 18	♂ ± ♃
08 05	☽ ⊻ ♀
10 49	☽ △ ♆
15 04	☽ ⊥ ☿
15 33	☽ □ ♃
16 58	☿ □ ♇
17 00	☽ ± ♄
17 19	☽ ∠ ♅
19 13	☽ ⊥ ♀
23 Sunday	
02 18	☿ □ ♂
03 15	☽ ⊼ ♄
06 37	☽ ∥ ♀
06 42	☽ ∠ ♇
07 55	☽ ∠ ♂
08 21	☽ ∠ ☿
08 49	☽ ⊥ ♅
12 13	☽ ∠ ♀
15 38	♀ ⊼ ♆
18 29	☽ ⊻ ☉
19 18	☽ ⊻ ♅
22 26	☽ ⊥ ♇
24 Monday	
00 18	☽ ⊥ ♂
00 28	☽ ∥ ♇
00 31	☽ ∥ ☿
00 48	☿ ∥ ♇
05 52	☽ ⊥ ☉
06 49	☉ ⚹ ♅
09 05	☽ ⊻ ♇
11 25	☽ ⊻ ♂
14 12	☽ ⚹ ☿
15 01	☽ □ ♆
17 24	☽ ⚹ ♀
20 39	☽ △ ♃
21 00	☿ ⊼ ♆
23 18	☽ ∠ ☉
25 Tuesday	
08 27	☽ ☍ ♄
15 09	☽ Q ☿
17 38	☽ Q ♀
19 20	☽ # ♄
26 Wednesday	
00 01	☽ Q ♆
00 35	☽ ⚼ ♃
01 30	☽ # ♆
01 55	☽ ☌ ♅
05 16	☽ ⚹ ☉
08 46	♀ ⚹ ♃
16 33	☽ ☌ ♇
17 33	☽ ∥ ♅
17 47	☽ ± ♃
21 32	☽ ☌ ♂
22 24	☿ ⚹ ♃
22 45	☽ ⚹ ♆
27 Thursday	
05 26	☽ ⊼ ♃
06 03	☽ Q ☉
06 28	☽ □ ☿
07 26	☽ □ ♀
15 13	☽ ∥ ♂
17 13	☽ ⊼ ♄
28 Friday	
00 30	♂ ⚹ ♆
03 06	☽ ∥ ♃
03 13	☿ ☌ ♀
03 54	☽ ∠ ♆
04 57	☽ ± ♄
09 36	☿ ± ♆
12 05	☽ ⊻ ♅
12 27	♀ ± ♆
20 24	☽ □ ☉
21 43	☽ ⊥ ♆
22 45	☽ ⚼ ♄
29 Saturday	
00 03	☽ ⊥ ♅
01 11	☿ Q ♅
03 22	☽ ⊻ ♇
09 41	☽ ⊻ ♆
11 21	☽ ⊻ ♂
11 47	♀ Q ♅
15 24	☽ ⊥ ♇
17 20	☽ ☍ ♃
18 08	☽ ∠ ♅
30 Sunday	
00 00	☽ ⊥ ♂
01 00	☉ ⚼ ♄
01 33	☽ △ ♀
04 17	☽ △ ☿
04 44	☽ △ ♄
07 07	☿ △ ♄
09 34	☽ ∠ ♇
13 32	☽ # ♃
19 03	☽ ∠ ♂
21 00	☽ # ♂
23 34	☿ ∥ ♀

All ephemeris data is given at 12.00 UT and the Moon's longitude is additionally given for 24.00 UT

LONGITUDES

Date	Sidereal time h m s	Sun ☉ ° ' "	Moon ☽ ° ' "	Moon ☽ 24.00 ° ' "	Mercury ☿ ° '	Venus ♀ ° '	Mars ♂ ° '	Jupiter ♃ ° '	Saturn ♄ ° '	Uranus ♅ ° '	Neptune ♆ ° '	Pluto ♇ ° '
01	06 35 18	08 ♋ 56 21	08 ♏ 01 09	13 ♏ 57 49	24 ♊ 49	22 ♊ 43	15 ♍ 02	17 ♈ 05	22 ♒ 29	02 ♍ 21	13 ♏ 03	09 ♍ 57
02	06 39 14	09 53 32	19 ♏ 55 23	25 ♏ 54 15	26 46	23 56	15 36	17 12	22 R 27	02 24	13 R 02	09 58
03	06 43 11	10 50 44	01 ♐ 54 49	07 ♐ 57 29	28 ♊ 45	25 09	16 09	17 19	22 24	02 26	13 01	10 00
04	06 47 08	11 47 54	14 ♐ 02 32	20 ♐ 10 15	00 ♋ 46	26 22	16 43	17 26	22 21	02 29	13 00	10 01
05	06 51 04	12 45 05	26 ♐ 20 53	02 ♑ 34 37	02 49	27 36	17 17	17 32	22 18	02 32	13 00	10 02
06	06 55 01	13 42 16	08 ♑ 51 37	15 ♑ 11 59	04 54	28 ♊ 49	17 52	17 39	22 15	02 34	12 59	10 04
07	06 58 57	14 39 27	21 ♑ 35 47	28 ♑ 03 06	07 00	00 ♋ 02	18 26	17 45	22 12	02 37	12 59	10 05
08	07 02 54	15 36 38	04 ♒ 33 54	11 ♒ 08 13	09 08	01 15	19 00	17 51	22 09	02 40	12 58	10 06
09	07 06 50	16 33 49	17 ♒ 45 59	24 ♒ 27 09	11 17	02 29	19 35	17 57	22 06	02 43	12 58	10 08
10	07 10 47	17 31 00	01 ♓ 11 39	07 ♓ 59 24	13 26	03 42	20 09	18 02	22 03	02 46	12 57	10 09
11	07 14 43	18 28 11	14 ♓ 50 17	21 ♓ 44 10	15 35	04 55	20 44	18 08	22 00	02 49	12 57	10 10
12	07 18 40	19 25 23	28 ♓ 40 56	05 ♈ 40 25	17 45	06 09	21 19	18 13	21 56	02 52	12 56	10 12
13	07 22 37	20 22 36	12 ♈ 42 24	19 ♈ 46 41	19 54	07 22	21 54	18 18	21 53	02 55	12 56	10 13
14	07 26 33	21 19 49	26 ♈ 53 01	04 ♉ 01 05	22 03	08 36	22 28	18 23	21 49	02 58	12 55	10 15
15	07 30 30	22 17 03	11 ♉ 10 34	18 ♉ 21 06	24 11	09 49	23 04	18 28	21 46	03 01	12 55	10 16
16	07 34 26	23 14 17	25 ♉ 32 14	02 ♊ 43 31	26 18	11 03	23 39	18 33	21 42	03 04	12 55	10 18
17	07 38 23	24 11 32	09 ♊ 54 28	17 ♊ 04 33	28 ♋ 25	12 16	24 14	18 37	21 39	03 07	12 54	10 19
18	07 42 19	25 08 48	24 ♊ 13 14	01 ♋ 19 58	00 ♌ 29	13 30	24 49	18 41	21 35	03 10	12 54	10 21
19	07 46 16	26 06 05	08 ♋ 24 13	15 ♋ 25 28	02 33	14 43	25 25	18 46	21 31	03 13	12 54	10 22
20	07 50 12	27 03 22	22 ♋ 23 16	29 ♋ 17 10	04 35	15 57	26 00	18 50	21 27	03 16	12 54	10 24
21	07 54 09	28 00 39	06 ♌ 06 50	12 ♌ 51 59	06 36	17 10	26 36	18 53	21 23	03 19	12 54	10 26
22	07 58 06	28 57 57	19 ♌ 32 25	26 ♌ 08 00	08 35	18 24	27 12	18 57	21 19	03 23	12 54	10 27
23	08 02 02	29 ♋ 55 15	02 ♍ 38 44	09 ♍ 04 40	10 32	19 38	27 48	19 00	21 15	03 26	12 53	10 29
24	08 05 59	00 ♌ 52 34	15 ♍ 25 56	21 ♍ 42 47	12 28	20 51	28 23	19 04	21 11	03 29	12 53	10 31
25	08 09 55	01 49 53	27 ♍ 55 30	04 ♎ 04 27	14 22	22 05	28 59	19 07	21 07	03 33	12 53	10 32
26	08 13 52	02 47 13	10 ♎ 10 02	16 ♎ 12 45	16 14	23 19	29 ♍ 36	19 09	21 03	03 36	12 D 53	10 34
27	08 17 48	03 44 32	22 ♎ 13 04	28 ♎ 11 34	18 05	24 32	00 ♎ 12	19 12	20 59	03 39	12 53	10 36
28	08 21 45	04 41 53	04 ♏ 08 46	10 ♏ 05 17	19 53	25 46	00 48	19 14	20 55	03 43	12 53	10 38
29	08 25 41	05 39 14	16 ♏ 01 40	21 ♏ 58 32	21 41	27 00	01 24	19 17	20 51	03 46	12 54	10 39
30	08 29 38	06 36 35	27 ♏ 56 26	03 ♐ 55 57	23 26	28 13	02 01	19 19	20 46	03 49	12 54	10 41
31	08 33 35	07 ♌ 33 57	09 ♐ 57 36	16 ♐ 01 55	25 ♌ 10	29 ♋ 27	02 ♎ 37	19 ♈ 21	20 ♒ 42	03 ♍ 53	12 ♏ 54	10 ♍ 43

Date	Moon True ☊ ° '	Moon Mean ☊ ° '	Moon ☽ Latitude ° '
01	20 ♋ 30	21 ♋ 05	05 N 01
02	20 R 29	21 01	04 35
03	20 27	20 58	03 57
04	20 26	20 55	03 08
05	20 25	20 52	02 09
06	20 25	20 49	01 N 04
07	20 D 25	20 46	00 S 07
08	20 25	20 42	01 18
09	20 25	20 39	02 26
10	20 25	20 36	03 27
11	20 26	20 33	04 18
12	20 26	20 30	04 55
13	20 26	20 27	05 14
14	20 26	20 23	05 15
15	20 26	20 20	04 57
16	20 26	20 17	04 20
17	20 26	20 14	03 26
18	20 27	20 11	02 20
19	20 27	20 07	01 S 06
20	20 R 27	20 04	00 N 11
21	20 27	20 01	01 26
22	20 26	19 58	02 34
23	20 25	19 55	03 33
24	20 24	19 52	04 19
25	20 22	19 48	04 52
26	20 21	19 45	05 11
27	20 20	19 42	05 16
28	20 20	19 39	05 07
29	20 D 20	19 36	04 45
30	20 21	19 33	04 11
31	20 ♋ 22	19 ♋ 29	03 N 25

DECLINATIONS

Date	Sun ☉ ° '	Moon ☽ ° '	Mercury ☿ ° '	Venus ♀ ° '	Mars ♂ ° '	Jupiter ♃ ° '	Saturn ♄ ° '	Uranus ♅ ° '	Neptune ♆ ° '	Pluto ♇ ° '
01	23 N 09	09 S 25	22 N 54	22 N 49	06 N 40	05 N 31	15 S 06	11 N 20	14 S 02	20 N 13
02	23 04	13 18	23 09	22 55	06 26	05 34	15 07	11 19	14 02	20 12
03	23 00	16 40	23 23	23 00	06 12	05 36	15 08	11 18	14 02	20 12
04	22 55	19 23	23 34	23 05	05 58	05 38	15 09	11 17	14 02	20 11
05	22 50	21 14	23 44	23 09	05 44	05 41	15 10	11 16	14 01	20 10
06	22 44	22 05	23 51	23 13	05 30	05 43	15 11	11 15	14 01	20 09
07	22 38	21 49	23 55	23 16	05 16	05 45	15 12	11 14	14 01	20 09
08	22 32	20 23	23 57	23 18	05 01	05 47	15 14	11 13	14 01	20 08
09	22 25	17 49	23 56	23 19	04 47	05 49	15 15	11 12	14 01	20 07
10	22 18	14 17	23 52	23 20	04 33	05 51	15 16	11 11	14 01	20 06
11	22 10	09 56	23 45	23 20	04 18	05 53	15 17	11 10	14 01	20 06
12	22 02	05 S 02	23 36	23 19	04 04	05 54	15 18	11 08	14 01	20 05
13	21 54	00 N 12	23 24	23 18	03 49	05 56	15 20	11 07	14 01	20 04
14	21 45	05 28	23 09	23 16	03 34	05 58	15 21	11 06	14 01	20 03
15	21 36	10 29	22 52	23 13	03 20	05 59	15 22	11 05	14 01	20 02
16	21 27	14 57	22 32	23 10	03 05	06 01	15 23	11 04	14 00	20 02
17	21 17	18 32	22 10	23 06	02 50	06 02	15 25	11 03	14 00	20 01
18	21 06	20 59	21 46	23 01	02 35	06 04	15 26	11 02	14 00	20 00
19	20 56	22 04	21 20	22 56	02 20	06 05	15 27	11 01	14 00	19 59
20	20 45	21 46	20 52	22 50	02 06	06 06	15 29	10 59	14 00	19 58
21	20 34	20 08	20 23	22 43	01 51	06 08	15 30	10 58	14 00	19 58
22	20 22	17 24	19 51	22 36	01 36	06 09	15 32	10 57	14 00	19 57
23	20 10	13 50	19 19	22 28	01 21	06 10	15 33	10 56	14 01	19 56
24	19 58	09 44	18 45	22 19	01 06	06 11	15 34	10 55	14 01	19 55
25	19 45	05 18	18 10	22 10	00 50	06 11	15 36	10 53	14 01	19 54
26	19 32	00 N 45	17 34	22 00	00 35	06 12	15 37	10 52	14 01	19 53
27	19 19	03 S 46	16 57	21 49	00 20	06 13	15 39	10 51	14 01	19 53
28	19 06	08 05	16 19	21 38	00 N 05	06 14	15 40	10 50	14 01	19 52
29	18 52	12 05	15 40	21 26	00 S 10	06 14	15 42	10 48	14 01	19 51
30	18 37	15 38	15 01	21 13	00 26	06 15	15 43	10 47	14 01	19 50
31	18 N 23	18 S 34	14 N 21	21 N 00	00 S 41	06 N 15	15 S 45	10 N 46	14 S 01	19 N 49

ZODIAC SIGN ENTRIES

Date	h	m	Planets
03	08	11	☽ ♐
04	03	00	☿ ♋
05	19	03	☽ ♑
07	11	18	♀ ♋
08	03	36	☽ ♒
10	09	53	☽ ♓
12	14	16	☽ ♈
14	17	15	☽ ♉
16	19	27	☽ ♊
18	06	19	☿ ♌
18	21	45	☽ ♋
21	01	15	☽ ♌
23	07	06	☽ ♍
23	13	59	☉ ♌
25	16	02	☽ ♎
27	04	14	♂ ♎
28	03	38	☽ ♏
30	16	08	☽ ♐
31	22	38	♀ ♌

LATITUDES

Date	Mercury ☿ ° '	Venus ♀ ° '	Mars ♂ ° '	Jupiter ♃ ° '	Saturn ♄ ° '	Uranus ♅ ° '	Neptune ♆ ° '	Pluto ♇ ° '
01	00 S 27	00 S 26	00 N 50	01 S 17	01 S 09	00 N 44	01 N 48	13 N 25
04	00 N 08	00 19	00 48	01 18	01 09	00 44	01 48	13 25
07	00 39	00 11	00 45	01 19	01 10	00 44	01 48	13 24
10	01 06	00 S 04	00 42	01 20	01 10	00 44	01 48	13 23
13	01 27	00 N 04	00 39	01 21	01 10	00 44	01 48	13 22
16	01 40	00 11	00 37	01 21	01 11	00 44	01 47	13 21
19	01 47	00 18	00 34	01 22	01 11	00 44	01 47	13 21
22	01 48	00 25	00 31	01 23	01 12	00 44	01 47	13 20
25	01 43	00 32	00 29	01 24	01 12	00 44	01 47	13 20
28	01 32	00 39	00 26	01 25	01 12	00 44	01 47	13 19
31	01 N 18	00 N 45	00 N 24	01 S 25	01 S 13	00 N 44	01 N 47	13 N 19

DATA

Julian Date	2438212
Delta T	+35 seconds
Ayanamsa	23° 20' 34"
Synetic vernal point	05° ♓ 46' 26"
True obliquity of ecliptic	23° 26' 35"

MOON'S PHASES, APSIDES AND POSITIONS ☽

Date	h	m	Phase	Longitude ° '	Eclipse Indicator
06	21	55	○	14 ♑ 06	partial
14	01	57	☾	20 ♈ 56	
20	20	43	●	27 ♋ 24	Total
28	13	13	☽	04 ♏ 45	

Day	h	m		
01	05	30	Apogee	
16	18	13	Perigee	
28	23	36	Apogee	
06	18	24	Max dec	22° S 08'
13	11	06	ON	
19	18	34	Max dec	22° N 08'
26	15	55	OS	

LONGITUDES

Date	Chiron ⚷ ° '	Ceres ⚳ ° '	Pallas ⚴ ° '	Juno ⚵ ° '	Vesta ⚶ ° '	Black Moon Lilith ⚸ ° '
01	14 ♓ 50	19 ♍ 34	24 ♌ 43	20 ♍ 53	29 ♍ 08	08 ♏ 04
11	14 ♓ 40	23 ♍ 04	29 ♌ 19	23 ♍ 33	02 ♎ 50	09 ♏ 11
21	14 ♓ 26	26 ♍ 45	03 ♍ 57	26 ♍ 25	06 ♎ 50	10 ♏ 18
31	14 ♓ 06	00 ♎ 36	08 ♍ 35	29 ♍ 26	11 ♎ 04	11 ♏ 26

All ephemeris data is given at 12.00 UT and the Moon's longitude is additionally given for 24.00 UT

ASPECTARIAN

h m Aspects

01 Monday
00 30 ☽ ⚹ ♅
07 41 ♀ △ ♄
11 19 ☽ Ᵽ ♀
14 01 ☽ △ ☉
15 55 ☽ ⚹ ♇
16 20 ☽ Ᵽ ☿
22 08 ☽ ☌ ♆
23 26 ☽ # ♅

02 Tuesday
00 50 ☽ Q ♅
02 51 ☽ ⚹ ♂
05 47 ☉ ∥ ☿
06 29 ☽ ⊼ ♃
07 33 ☽ ± ♀
14 01 ☽ ± ☿
14 06 ☉ ⚹ ♇
16 08 ☽ Q ♇
16 53 ☽ ∥ ♆
17 03 ☽ □ ♄
18 39 ☽ ± ♃
20 58 ☽ ⊼ ♀
22 50 ☽ Ᵽ ☉

03 Wednesday
00 29 ☽ ∥ ♄
02 54 ☿ Q ♇
03 20 ☿ Ᵽ ♆
04 08 ☽ Q ♂
04 25 ☽ ⊼ ☿
11 22 ☉ ∥ ♀
12 49 ☽ Ᵽ ♃
13 03 ☽ □ ♅
18 20 ☽ ± ☉
19 17 ☿ Q ♃

04 Thursday
04 03 ☽ □ ♇
04 46 ☽ Q ♄
07 13 ☽ ⊼ ☉
09 58 ☽ ⊻ ♆
17 31 ☽ □ ♂
18 42 ☽ △ ♃
20 57 ☽ # ♇
21 43 ☽ ⊥ ♆

05 Friday
04 11 ☽ ⚹ ♄
08 19 ♃ Ᵽ ♅
08 35 ☿ ⚹ ♅
14 40 ☽ ☍ ♀
15 11 ☽ ∠ ♆
17 18 ♂ ∥ ♃
18 06 ☉ △ ♆
19 51 ♀ Ᵽ ♆
20 52 ♀ Q ♇
23 57 ☽ △ ♅

06 Saturday
00 45 ♂ ⊼ ♃
02 56 ☽ ☍ ☿
08 57 ☽ ∠ ♄
14 17 ☽ △ ♇
19 49 ☽ ⚹ ♆
21 55 ☽ ☍ ☉

07 Sunday
01 56 ☽ ⊥ ♄
03 04 ☿ Q ♂
04 31 ☽ Ᵽ ♅
04 44 ☽ □ ♃
05 45 ♀ Q ♃
05 48 ☽ △ ♂
13 08 ☽ ⊻ ♄
14 14 ☿ Ᵽ ♄
18 18 ☽ Q ♆
18 30 ☽ Ᵽ ♇
21 23 ☽ ± ♅

08 Monday
05 17 ☽ ⊼ ♀
08 30 ☽ ⊼ ♅
10 55 ☽ Ᵽ ♂
11 09 ☽ ± ♇
14 22 ☽ Q ♃
14 53 ☽ # ♇
17 26 ☽ ± ♀
21 58 ☽ ⊼ ☿
22 08 ☽ ⊼ ♇
23 00 ☿ ⚹ ♇

09 Tuesday
00 59 ☉ ± ♄
03 19 ☽ □ ♆
04 05 ☽ ± ♂
09 40 ☽ ⊼ ☉
10 57 ☽ ± ☿
11 26 ☽ Ᵽ ♀
12 19 ☽ ⚹ ♃
15 25 ☽ ⊼ ♂
16 48 ♀ ⚹ ♅
19 46 ☽ ☌ ♄
21 17 ☽ ± ☉

10 Wednesday
05 55 ☽ ∥ ♄
06 09 ☽ Ᵽ ☿
06 42 ☿ △ ♆
13 35 ☽ ∥ ♆
14 31 ☽ Ᵽ ☉
14 47 ☽ ☍ ♅
15 17 ☽ ∠ ♃
16 53 ☽ △ ♀
18 30 ☉ ∠ ♅

11 Thursday
02 29 ☉ □ ♃
03 49 ☽ ☍ ♇
05 34 ☽ # ♅
07 14 ☽ ⊥ ♃
08 41 ☽ △ ♆
13 33 ☽ △ ☿
16 24 ☿ ± ♄
17 46 ☽ ⊻ ♃
18 48 ☽ △ ☉
22 42 ☽ ☍ ♂

12 Friday
00 24 ☽ ⊻ ♄
07 52 ☽ # ♃
10 43 ☽ Ᵽ ♆
10 43 ☽ ⊥ ♄
13 17 ☿ ∠ ♅
16 44 ☽ # ♂
17 27 ☿ □ ♃
19 12 ☽ ⊼ ♅

13 Saturday
02 02 ☽ □ ♀
02 06 ☽ ∠ ♄
02 09 ☽ ± ♆
02 50 ♀ Ᵽ ♄
05 30 ☽ ± ♅
07 46 ☽ ⊼ ♇
11 34 ♂ ⊼ ♄
12 23 ☽ ⊼ ♆
17 59 ☽ ± ♇
20 52 ☽ Ᵽ ♅
21 31 ☉ ☌ ☿
21 34 ☽ ☌ ♃
23 06 ☿ ∥ ♀

14 Sunday
01 57 ☽ □ ☉
02 23 ☽ □ ☿
03 30 ☽ ⚹ ♄
03 43 ☽ ∥ ♂
04 15 ☽ ⊼ ♂
09 14 ☽ Ᵽ ♇
09 31 ☿ ⊼ ♄
11 28 ☽ Q ♀
14 20 ☽ ∥ ♃
14 48 ☽ ± ♂
18 33 ☿ ⚹ ♂
22 15 ☽ △ ♅
23 37 ☽ Q ♄
23 40 ☉ ⊼ ♄

15 Monday
06 33 ☽ Ᵽ ♂
09 31 ☽ ⚹ ♀
10 24 ☽ Q ☉
10 29 ☽ △ ♇
13 59 ☽ Q ☿
14 55 ☽ ☍ ♆
15 02 ☽ ∥ ♅
21 04 ♀ ⚹ ♇

16 Tuesday
00 16 ☽ ⊻ ♃
00 24 ☿ ∠ ♇
05 38 ☽ □ ♄
06 37 ☽ # ♆
07 53 ☽ ⚹ ☉
08 42 ☽ △ ♂
10 20 ☽ ⊥ ♃
12 55 ☽ ∠ ♀
13 30 ☽ ⚹ ☿
14 41 ☽ # ♄
20 48 ☿ ⊥ ♅

17 Wednesday
00 36 ☽ □ ♅
01 26 ☽ ∠ ♃
05 21 ☽ ⊥ ♀
10 41 ♀ Q ♂
10 43 ☽ ∠ ☉
12 42 ☽ □ ♇
14 38 ☉ ⚹ ♂
16 19 ☽ ⊻ ♀
17 01 ☽ ⊼ ♆
18 51 ☽ ∠ ☿

18 Thursday
00 29 ♀ △ ♆
00 52 ☽ ∥ ♇
02 40 ☽ ⚹ ♃
02 52 ☽ ⊥ ☉
03 04 ☽ ± ♆
06 51 ☽ Q ♅
07 35 ☽ △ ♄
12 32 ☽ ⊥ ☿
13 03 ☽ □ ♂
13 36 ☽ ∥ ☉
13 40 ☽ ⊻ ☉
17 12 ☉ ∠ ♇
18 12 ☽ Ᵽ ♆
18 58 ☽ Q ♇
21 48 ☽ ∥ ☿
22 58 ☽ Q ♃

19 Friday
00 23 ☽ ⊻ ☿
03 09 ☽ ⚹ ♅
08 48 ☽ Ᵽ ♄
15 22 ☽ ⚹ ♇
19 40 ☽ △ ♆
20 02 ☿ ⊻ ♅
20 56 ☽ Q ♂
23 49 ☽ ☌ ♀

20 Saturday
00 06 ☽ ± ♄
02 51 ♀ ± ♄
04 52 ☽ ∠ ♅
05 49 ☽ □ ♃
09 47 ☿ ⊥ ♇
10 23 ☽ ⊼ ♄
17 14 ☽ ∠ ♇
17 42 ☉ ⊥ ♅
18 34 ☽ ⚹ ♂
20 31 ☽ ⊥ ♅
20 43 ☽ ☌ ☉
21 33 ☉ ∥ ☿

21 Sunday
06 33 ☽ ∥ ☉
07 04 ☽ ⊻ ♅
08 24 ☽ ∥ ☿
09 01 ☽ ⊥ ♇
13 00 ☽ ☌ ☿
13 50 ☽ ∥ ♇
19 40 ☽ ⊻ ♇
22 11 ☽ ∠ ♂

22 Monday
00 03 ☽ □ ♆
07 50 ☿ ∥ ♇
09 44 ☽ ⊻ ♀
10 56 ☽ △ ♃
11 36 ♀ ∠ ♅
15 08 ☽ ⊥ ♂
15 13 ☽ ☍ ♄
16 37 ♂ ± ♄
21 44 ☽ ⊥ ♀
23 18 ♀ □ ♃

23 Tuesday
01 07 ☽ # ♄
02 37 ☽ ⊻ ♂
06 34 ☽ ⊻ ☉
08 45 ☽ Q ♆
10 58 ☽ # ♆
11 22 ☿ ⊻ ♇
13 28 ☽ ☌ ♅
14 32 ☽ Ᵽ ♃
15 57 ♂ ∠ ♆
16 04 ☽ ∠ ♀
18 35 ☽ ⊥ ☉

24 Wednesday
02 41 ☽ ☌ ♇
05 18 ☽ ∥ ♅
05 22 ☽ ⊻ ☿
07 11 ☽ ⚹ ♆
07 29 ☽ ± ♃
12 55 ☽ ∠ ☉
17 22 ☿ □ ♆
17 58 ☉ ∥ ♇
18 13 ♀ ⊼ ♄
18 48 ☽ ⊥ ☿
18 56 ☽ ⊼ ♃
22 56 ☽ ⊼ ♄
23 28 ☽ ⚹ ♀

25 Thursday
05 05 ☿ ∠ ♂
07 14 ☽ ∥ ♃
10 27 ☽ ± ♄
11 56 ☽ ∠ ♆
14 11 ☽ ☌ ♂
15 18 ☽ ∠ ☿
19 10 ♆ St D
20 15 ☽ ⚹ ☉
23 00 ☽ ⊻ ♅

26 Friday
01 21 ☽ Q ♀
03 56 ☽ Ᵽ ♄
05 32 ☽ ⊥ ♆
10 52 ☽ ⊥ ♅
12 48 ☽ ⊻ ♇
12 52 ☽ ∥ ♂
17 24 ☽ ⊻ ♆
18 39 ☽ # ♂
21 57 ☽ Q ☉

27 Saturday
00 44 ☽ ⊥ ♇
02 14 ☽ ⚹ ☿
04 50 ☽ ∠ ♅
05 56 ☽ ☍ ♃
09 32 ☽ △ ♄
09 39 ☉ ⊻ ♅
17 11 ☽ □ ♀
18 48 ☽ ∠ ♇

28 Sunday
01 30 ☽ # ♃
03 11 ☿ △ ♃
04 53 ☽ ⊻ ♂
06 38 ☽ Q ☿
09 13 ♀ ∠ ♇
10 10 ☉ ⊥ ♇
11 07 ☽ ⚹ ♅
13 13 ☽ □ ☉
17 39 ☽ ⊥ ♂

29 Monday
01 07 ☽ ⚹ ♇
01 11 ☿ ☍ ♄
04 07 ☽ # ♅
05 40 ☽ ☌ ♆
11 06 ☿ # ♄
11 28 ☽ Q ♅
12 48 ☽ ∠ ♂
18 35 ☽ ⊼ ♃
21 39 ☽ □ ♄

30 Tuesday
00 37 ☽ ∥ ♆
01 23 ☽ □ ☿
01 24 ☽ Q ♇
03 49 ♀ ⊥ ♅
06 42 ☽ ± ♃
08 17 ☽ # ☿
12 38 ☽ △ ♀
12 39 ☽ ∥ ♄
20 36 ☽ ⚹ ♂
23 51 ☽ □ ♅

31 Wednesday
00 48 ☽ Ᵽ ♃
06 50 ☽ △ ☉
09 31 ☽ Q ♄
10 27 ☽ # ☉
13 30 ☽ □ ♇
17 49 ☽ ⊻ ♆
21 43 ☽ Q ♂
21 54 ☽ Ᵽ ♀

AUGUST 1963

Date	Sidereal time h m s	Sun ☉ ° ' "	Moon ☽ ° ' "	Moon ☽ 24.00 ° ' "	Mercury ☿ ° '	Venus ♀ ° '	Mars ♂ ° '	Jupiter ♃ ° '	Saturn ♄ ° '	Uranus ♅ ° '	Neptune ♆ ° '	Pluto ♇ ° '
01	08 37 31	08 ♌ 31 20	22 ♐ 09 21	28 ♐ 20 19	26 ♌ 52	00 ♌ 41	03 ♎ 14	19 ♈ 22	20 ♒ 38	03 ♍ 56	12 ♏ 54	10 ♍ 45
02	08 41 28	09 28 43	04 ♑ 35 13	10 ♑ 54 20	28 ♌ 32	01 55	03 51	19 24	20 R 33	04 00	12 54	10 47
03	08 45 24	10 26 07	17 ♑ 17 54	23 ♑ 46 05	00 ♍ 11	03 09	04 27	19 25	20 29	04 03	12 55	10 48
04	08 49 21	11 23 32	00 ♒ 18 58	06 ♒ 56 33	01 48	04 23	05 04	19 26	20 25	04 07	12 55	10 50
05	08 53 17	12 20 58	13 ♒ 38 42	20 ♒ 25 17	03 23	05 37	05 41	19 27	20 20	04 10	12 55	10 52
06	08 57 14	13 18 25	27 ♒ 16 00	04 ♓ 10 32	04 57	06 50	06 18	19 28	20 16	04 14	12 56	10 54
07	09 01 10	14 15 53	11 ♓ 08 28	18 ♓ 09 22	06 29	08 04	06 55	19 28	20 11	04 18	12 56	10 56
08	09 05 07	15 13 21	25 ♓ 12 45	02 ♈ 18 05	07 59	09 18	07 33	19 29	20 07	04 21	12 56	10 58
09	09 09 03	16 10 52	09 ♈ 24 52	16 ♈ 32 37	09 28	10 32	08 10	19 29	20 02	04 25	12 57	11 00
10	09 13 00	17 08 23	23 ♈ 40 51	00 ♉ 49 07	10 55	11 46	08 47	19 R 29	19 58	04 28	12 57	11 02
11	09 16 57	18 05 56	07 ♉ 57 02	15 ♉ 04 16	12 21	13 00	09 25	19 29	19 53	04 32	12 58	11 04
12	09 20 53	19 03 30	22 ♉ 10 29	29 ♉ 15 27	13 44	14 14	10 02	19 28	19 49	04 36	12 58	11 06
13	09 24 50	20 01 06	06 ♊ 18 55	13 ♊ 20 41	15 06	15 29	10 40	19 27	19 44	04 39	12 59	11 08
14	09 28 46	20 58 44	20 ♊ 20 36	27 ♊ 18 28	16 26	16 43	11 17	19 27	19 40	04 43	13 00	11 10
15	09 32 43	21 56 23	04 ♋ 14 09	11 ♋ 07 29	17 44	17 57	11 55	19 26	19 35	04 47	13 00	11 12
16	09 36 39	22 54 03	17 ♋ 58 18	24 ♋ 46 26	19 00	19 11	12 33	19 24	19 31	04 50	13 01	11 14
17	09 40 36	23 51 45	01 ♌ 31 42	08 ♌ 13 55	20 15	20 25	13 11	19 23	19 26	04 54	13 02	11 16
18	09 44 32	24 49 28	14 ♌ 52 57	21 ♌ 28 36	21 27	21 39	13 49	19 21	19 22	04 58	13 02	11 18
19	09 48 29	25 47 13	28 ♌ 00 46	04 ♍ 29 19	22 37	22 54	14 27	19 19	19 17	05 02	13 03	11 20
20	09 52 26	26 44 59	10 ♍ 54 13	17 ♍ 15 24	23 45	24 08	15 05	19 17	19 13	05 05	13 04	11 22
21	09 56 22	27 42 46	23 ♍ 32 56	29 ♍ 46 54	24 51	25 22	15 44	19 15	19 09	05 09	13 05	11 24
22	10 00 19	28 40 35	05 ♎ 57 26	12 ♎ 04 46	25 55	26 36	16 22	19 13	19 04	05 13	13 06	11 26
23	10 04 15	29 ♌ 38 24	18 ♎ 09 11	24 ♎ 10 59	26 55	27 51	17 00	19 10	19 00	05 17	13 07	11 28
24	10 08 12	00 ♍ 36 15	00 ♏ 10 34	06 ♏ 08 24	27 54	29 ♌ 05	17 39	19 07	18 55	05 20	13 07	11 30
25	10 12 08	01 34 08	12 ♏ 04 58	18 ♏ 00 48	28 49	00 ♍ 19	18 17	19 04	18 51	05 24	13 08	11 32
26	10 16 05	02 32 01	23 ♏ 56 28	29 ♏ 52 34	29 ♍ 42	01 34	18 56	19 01	18 47	05 28	13 09	11 34
27	10 20 01	03 29 56	05 ♐ 49 43	11 ♐ 48 32	00 ♎ 32	02 48	19 35	18 57	18 42	05 32	13 10	11 36
28	10 23 58	04 27 52	17 ♐ 49 41	23 ♐ 53 47	01 18	04 02	20 14	18 54	18 38	05 35	13 12	11 38
29	10 27 55	05 25 50	00 ♑ 01 26	06 ♑ 13 12	02 01	05 17	20 53	18 50	18 34	05 39	13 13	11 40
30	10 31 51	06 23 49	12 ♑ 29 39	18 ♑ 51 14	02 40	06 31	21 31	18 46	18 30	05 43	13 14	11 42
31	10 35 48	07 ♍ 21 49	25 ♑ 18 21	01 ♒ 51 19	03 ♎ 15	07 ♍ 45	22 ♎ 10	18 ♈ 42	18 ♒ 26	05 ♍ 47	13 ♏ 15	11 ♍ 45

DECLINATIONS

Date	Moon True ☊ ° '	Moon Mean ☊ ° '	Moon ☽ Latitude ° '	Sun ☉ ° '	Moon ☽ ° '	Mercury ☿ ° '	Venus ♀ ° '	Mars ♂ ° '	Jupiter ♃ ° '	Saturn ♄ ° '	Uranus ♅ ° '	Neptune ♆ ° '	Pluto ♇ ° '
01	20 ♋ 24	19 ♋ 26	02 N 30	18 N 08	20 S 43	13 N 41	20 N 46	00 S 56	06 N 16	15 S 46	10 N 45	14 S 01	19 N 48
02	20 D 25	19 23	01 26	17 53	21 56	13 01	20 32	01 12	06 16	15 48	10 43	14 01	19 47
03	20 26	19 20	00 N 17	17 38	22 02	12 20	20 17	01 27	06 16	15 49	10 42	14 01	19 47
04	20 R 25	19 17	00 S 54	17 22	20 58	11 39	20 01	01 42	06 16	15 51	10 41	14 02	19 46
05	20 24	19 13	02 04	17 06	18 43	10 58	19 45	01 58	06 16	15 52	10 39	14 02	19 45
06	20 22	19 10	03 09	16 50	15 23	10 17	19 28	02 13	06 16	15 54	10 38	14 02	19 44
07	20 19	19 07	04 03	16 33	11 08	09 36	19 11	02 28	06 16	15 55	10 37	14 02	19 43
08	20 15	19 04	04 44	16 16	06 15	08 55	18 53	02 44	06 16	15 57	10 35	14 02	19 42
09	20 12	19 01	05 08	15 59	00 S 59	08 14	18 35	02 59	06 16	15 58	10 34	14 03	19 42
10	20 09	18 58	05 13	15 42	04 N 21	07 33	18 16	03 15	06 16	16 00	10 33	14 03	19 41
11	20 08	18 54	04 58	15 24	09 27	06 53	17 57	03 30	06 15	16 01	10 31	14 03	19 40
12	20 D 08	18 51	04 26	15 07	14 02	06 12	17 37	03 46	06 15	16 03	10 30	14 03	19 39
13	20 08	18 48	03 37	14 49	17 48	05 32	17 16	04 01	06 15	16 04	10 29	14 03	19 38
14	20 10	18 45	02 36	14 30	20 30	04 53	16 56	04 17	06 14	16 06	10 27	14 04	19 37
15	20 11	18 42	01 26	14 12	21 56	04 14	16 34	04 32	06 13	16 07	10 26	14 04	19 36
16	20 12	18 38	00 S 12	13 53	22 02	03 35	16 12	04 48	06 13	16 09	10 25	14 04	19 36
17	20 R 11	18 35	01 N 01	13 34	20 49	02 57	15 50	05 03	06 12	16 10	10 23	14 04	19 35
18	20 09	18 32	02 10	13 15	18 27	02 20	15 27	05 19	06 11	16 12	10 22	14 05	19 34
19	20 05	18 29	03 11	12 56	15 09	01 43	15 04	05 34	06 10	16 13	10 21	14 05	19 33
20	20 00	18 26	04 01	12 36	11 12	01 07	14 41	05 49	06 09	16 14	10 19	14 05	19 32
21	19 54	18 23	04 38	12 16	06 49	00 N 32	14 17	06 05	06 08	16 16	10 18	14 06	19 31
22	19 47	18 19	05 01	11 56	02 N 14	00 S 03	13 52	06 20	06 07	16 17	10 17	14 06	19 31
23	19 40	18 16	05 09	11 36	02 S 21	00 36	13 28	06 36	06 05	16 19	10 15	14 06	19 30
24	19 35	18 13	05 04	11 16	06 47	01 08	13 03	06 51	06 04	16 20	10 14	14 07	19 29
25	19 31	18 10	04 46	10 55	10 55	01 39	12 37	07 07	06 03	16 22	10 12	14 07	19 28
26	19 29	18 07	04 16	10 34	14 38	02 09	12 11	07 22	06 01	16 23	10 11	14 07	19 27
27	19 D 29	18 04	03 34	10 14	17 46	02 38	11 45	07 37	06 00	16 24	10 10	14 08	19 27
28	19 30	18 00	02 43	09 52	20 11	03 05	11 19	07 53	05 58	16 26	10 08	14 08	19 26
29	19 31	17 57	01 43	09 31	21 43	03 31	10 52	08 08	05 57	16 27	10 07	14 08	19 25
30	19 32	17 54	00 38	09 10	22 14	03 55	10 25	08 23	05 55	16 28	10 05	14 09	19 24
31	19 ♋ 32	17 ♋ 51	00 S 31	08 N 48	21 S 35	04 S 17	09 N 58	08 S 38	05 N 53	16 S 30	10 N 04	14 S 09	19 N 23

ZODIAC SIGN ENTRIES

Date	h	m	Planets
02	03	12	☽ ♑
03	09	20	☿ ♍
04	11	25	☽ ♒
06	16	46	☽ ♓
08	20	07	☽ ♈
10	22	37	☽ ♉
13	01	16	☽ ♊
15	04	39	☽ ♋
17	09	17	☽ ♌
19	15	40	☽ ♍
22	00	25	☽ ♎
23	20	58	☉ ♍
24	11	39	☽ ♏
25	05	49	♀ ♍
26	20	33	☿ ♎
27	00	15	☽ ♐
29	11	57	☽ ♑
31	20	37	☽ ♒

LATITUDES

Date	Mercury ☿ ° '	Venus ♀ ° '	Mars ♂ ° '	Jupiter ♃ ° '	Saturn ♄ ° '	Uranus ♅ ° '	Neptune ♆ ° '	Pluto ♇ ° '
01	01 N 12	00 N 47	00 N 23	01 S 26	01 S 13	00 N 44	01 N 47	13 N 18
04	00 53	00 53	00 20	01 27	01 13	00 44	01 46	13 18
07	00 30	00 58	00 18	01 27	01 13	00 44	01 46	13 18
10	00 N 05	01 03	00 16	01 28	01 14	00 44	01 46	13 17
13	00 S 22	01 08	00 13	01 29	01 14	00 44	01 46	13 17
16	00 50	01 12	00 11	01 30	01 14	00 44	01 46	13 17
19	01 19	01 15	00 09	01 31	01 14	00 43	01 45	13 17
22	01 49	01 18	00 06	01 31	01 15	00 43	01 45	13 17
25	02 19	01 21	00 04	01 32	01 15	00 43	01 45	13 16
28	02 48	01 23	00 02	01 33	01 15	00 43	01 45	13 16
31	03 S 16	01 N 24	00 00	01 S 34	01 S 15	00 N 43	01 N 45	13 N 17

DATA

Julian Date	2438243
Delta T	+35 seconds
Ayanamsa	23° 20' 39"
Synetic vernal point	05° ♓ 46' 21"
True obliquity of ecliptic	23° 26' 35"

MOON'S PHASES, APSIDES AND POSITIONS ☽

Date	h	m	Phase	Longitude ° '	Eclipse Indicator
05	09	31	○	12 ♒ 15	
12	06	21	☾	18 ♉ 50	
19	07	34	●	25 ♌ 37	
27	06	54	☽	03 ♐ 18	

Day	h	m		
11	00	30	Perigee	
25	18	30	Apogee	
03	02	23	Max dec	22° S 08'
09	16	25	0N	
16	01	40	Max dec	22° N 10'
22	23	39	0S	
30	10	56	Max dec	22° S 14'

LONGITUDES

Date	Chiron ⚷ ° '	Ceres ⚳ ° '	Pallas ⚴ ° '	Juno ⚵ ° '	Vesta ⚶ ° '	Black Moon Lilith ⚸ ° '
01	14 ♓ 04	00 ♎ 59	09 ♍ 03	29 ♍ 45	11 ♎ 30	11 ♏ 32
11	13 ♓ 41	04 ♎ 58	13 ♍ 41	02 ♎ 55	15 ♎ 57	12 ♏ 39
21	13 ♓ 15	09 ♎ 04	18 ♍ 19	06 ♎ 11	20 ♎ 34	13 ♏ 47
31	12 ♓ 47	13 ♎ 15	22 ♍ 57	09 ♎ 32	25 ♎ 21	14 ♏ 54

ASPECTARIAN

h m	Aspects
01 Thursday	
00 09	☿ # ♆
00 41	☽ # ♇
05 38	☽ ⊥ ♆
06 33	☽ △ ♃
09 02	☽ ⚹ ♄
12 38	☽ # ♀
14 53	☽ ꝗ ☉
17 28	☽ ± ♀
20 17	☿ ⊥ ♂
22 35	☽ △ ☿
23 09	☽ ∠ ♆
02 Friday	
06 20	☽ ⊼ ♀
09 42	☽ ± ☉
10 30	☽ □ ♂
10 52	☽ △ ♅
13 50	☽ ∠ ♄
18 42	♂ ⊻ ♅
22 04	☽ ⊼ ☉
23 47	☽ △ ♇
03 Saturday	
03 46	☽ ⚹ ♆
06 46	☽ ⊥ ♄
07 28	☽ ꝗ ☿
15 17	☽ ꝗ ♅
15 57	☽ □ ♃
17 53	☽ ⊻ ♄
21 40	☉ ⊻ ♇
22 47	☿ Q ♆
04 Sunday	
02 06	☽ Q ♆
02 35	☽ ± ☿
03 47	☽ ꝗ ♇
06 38	♀ ⊻ ♅
07 58	☽ ± ♅
15 04	☽ ⊼ ☿
18 56	☽ ⊼ ♅
20 08	☽ ☍ ♀
20 14	☽ ± ♇
21 03	☽ △ ♂
21 14	♀ ⊥ ♇
23 46	♂ ꝗ ♄
05 Monday	
00 55	☽ Q ♃
01 10	☽ # ♀
02 24	☽ # ♇
07 02	☽ ⊼ ♇
09 31	☽ ☍ ☉
10 42	☽ □ ♆
12 22	♀ ∥ ♇
15 02	♀ ⚹ ♂
22 18	☽ ⚹ ♃
23 20	☿ ∥ ♅
23 47	☽ ☌ ♄
06 Tuesday	
00 30	☿ ☌ ♅
01 04	☽ ꝗ ♂
01 36	☽ # ☉
02 23	☉ □ ♆
04 28	☿ ꝗ ♃
08 43	☽ ∥ ♄
17 32	☽ ± ♂
20 07	☽ ∥ ♆
07 Wednesday	
00 09	☽ ☍ ♅
00 31	☽ ∠ ♃
03 00	☽ ☍ ☿
04 24	☽ ⊼ ♂
06 13	☽ ⊼ ♀
11 39	☽ ☍ ♇
12 23	♂ ⊥ ♆
14 43	☽ # ♅
15 04	☽ △ ♆
16 00	☽ ⊥ ♃
17 31	☽ ± ♀
17 45	☽ ⊼ ☉
21 08	☽ # ☿
23 46	☿ ⊻ ♂
08 Thursday	
02 15	☽ ⊻ ♃
03 23	☽ ⊻ ♄
04 44	☽ ± ☉
10 19	☽ ꝗ ♀
11 53	☽ # ♃
13 31	☽ ⊥ ♄
16 37	☽ ꝗ ♆
21 06	☽ ꝗ ☉
09 Friday	
03 24	☽ ∥ ♂
03 32	☽ ⊼ ♅
04 40	☽ ∠ ♄
07 50	☽ ± ♆
09 48	☽ ☍ ♂
12 06	☽ ⊼ ☿
13 32	☉ # ♄
13 41	☽ ± ♅
14 04	☽ △ ♀
14 40	☽ ⊼ ♇
15 30	♃ St R
17 57	☽ ⊼ ♆
21 10	♀ ⊻ ♇
23 21	☽ ± ☿
10 Saturday	
00 13	☽ △ ☉
00 48	☽ ± ♇
04 54	☽ ꝗ ♅
04 56	☽ ☌ ♃
05 47	☽ ⚹ ♄
06 45	☽ # ♂
13 53	☿ ☌ ♇
14 41	♃ ꝗ ♅
15 58	☽ ꝗ ♇
16 11	☽ ꝗ ☿
20 48	☽ ∥ ♃
11 Sunday	
01 07	☽ ∥ ☿
01 51	☽ Q ♄
06 14	☽ △ ♅
11 10	♀ □ ♆
14 34	☽ ⊼ ♂
17 15	☽ △ ♇
17 18	☽ ∥ ♅
20 12	☽ △ ☿
20 27	☽ ☍ ♆
21 20	☽ □ ♀
22 44	☿ ⚹ ♆
12 Monday	
01 09	☽ ± ♂
06 21	☽ □ ☉
07 25	☿ ± ♃
07 26	☽ ⊻ ♃
08 02	☽ □ ♄
10 25	☿ ∥ ♃
12 07	☽ # ♆
17 04	☽ ꝗ ♂
17 34	☽ ⊥ ♃
17 52	☽ ∥ ☉
22 10	☉ △ ♃
13 Tuesday	
00 09	☽ # ♄
05 32	☉ ☍ ♄
06 42	☽ Q ♀
08 35	☽ ∥ ♀
08 50	☽ ∠ ♃
09 10	☽ □ ♅
11 45	♂ ⊥ ♅
15 07	☽ Q ☉
19 46	☽ △ ♂
20 14	☽ □ ♇
23 23	☽ ⊼ ♆
14 Wednesday	
03 01	☽ ∥ ♇
04 35	☽ □ ☿
05 10	☽ ⚹ ♀
06 47	♂ ⊻ ♇
09 41	☽ ± ♆
10 27	☽ ⚹ ♃
10 50	☽ △ ♄
13 10	☽ ⚹ ☉
16 06	☽ Q ♅
15 Thursday	
01 12	☽ ꝗ ♆
03 14	☽ Q ♇
03 56	☿ # ♂
07 08	☽ Q ♃
09 33	☽ ∠ ♀
12 37	☽ ꝗ ♄
12 57	☽ ⚹ ♅
14 52	☽ Q ☿
17 03	☽ ∠ ☉
22 07	☉ # ♆
16 Friday	
00 09	☽ ⚹ ♇
02 02	☽ □ ♂
02 46	☽ ⊥ ♀
03 18	☽ △ ♆
04 13	☽ ± ♄
09 59	☽ ⊥ ☉
14 00	☽ ⚹ ☿
14 21	☽ ⊻ ♀
14 31	☽ □ ♃
14 42	☽ ⊼ ♄
15 18	☽ ∠ ♅
15 50	♀ # ♄
16 14	♀ △ ♃
18 05	♀ ☍ ♄
19 31	☿ ⊼ ♃
21 13	☿ ⊼ ♄
21 21	☽ ⊻ ☉
17 Saturday	
02 37	☽ ∠ ♇
05 58	♂ ⊻ ♆
07 18	☽ ⊥ ♅
11 21	☽ Q ♂
18 03	☽ ⊻ ♅
18 41	☽ ⊥ ♇
19 18	☽ ∠ ☿
18 Sunday	
01 57	☽ ∥ ♇
05 30	☽ ⊻ ♇
08 40	☽ □ ♆
09 58	☽ ⚹ ♂
13 08	☽ ⊥ ☿
18 55	♃ ⚹ ♄
20 06	☽ ☍ ♄
20 06	☽ △ ♃
19 Monday	
01 08	☽ ⊻ ☿
01 37	☽ ☌ ♀
04 55	☽ # ♄
07 34	☽ ☌ ☉
12 37	☽ ∥ ♀
14 48	☽ ∠ ♂
17 37	☽ Q ♆
18 50	☽ # ♆
23 39	☽ ꝗ ♃
20 Tuesday	
01 04	☽ ☌ ♅
03 07	☽ ∥ ☉
08 25	☽ ⊥ ♂
12 52	☽ ☌ ♇
16 05	☽ ⚹ ♆
16 29	☽ ± ♃
16 56	☽ ∥ ♅
20 19	☽ ⊻ ♂
21 Wednesday	
03 38	☽ ⊼ ♄
03 49	☽ ⊼ ♃
14 44	☽ ☌ ☿
15 02	☽ ± ♄
15 37	☽ ∥ ♃
15 40	☽ # ♂
15 52	☽ ⊻ ♀
16 17	♂ # ♃
18 06	☿ ± ♄
20 41	☽ ⊻ ☉
20 43	☽ ∠ ♆
22 56	♀ # ♆
22 Thursday	
04 45	☽ ⊥ ♀
08 21	☽ ꝗ ♄
09 17	☽ ⊥ ☉
10 33	☽ ⊻ ♅
14 13	☽ ⊥ ♆
22 11	☽ # ☿
22 21	☽ ⊥ ♅
22 45	☽ ⊻ ♇
23 Friday	
00 19	☽ ∠ ♀
01 32	☽ ∥ ☿
02 01	☽ ⊻ ♆
04 27	☽ ∠ ☉
09 36	☽ ☌ ♂
10 38	☽ ⊥ ♇
13 40	☽ △ ♄
14 00	☽ ☍ ♃
16 14	☽ ∠ ♅
24 Saturday	
04 37	☽ ∠ ♇
06 11	♂ ⊥ ♇
07 02	☽ ⊻ ☿
08 04	☽ # ♃
09 33	☽ ⚹ ♀
12 25	☽ ∥ ♂
12 56	☽ ⚹ ☉
17 55	☿ ∠ ♆
20 07	☽ ⊥ ☿
22 26	☽ ⚹ ♅
25 Sunday	
01 10	☉ Q ♆
07 43	☽ # ♅
10 53	☽ ⚹ ♇
11 59	☽ # ☉
12 32	☽ Q ♀
14 09	☽ ☌ ♆
15 16	☽ Q ☉
15 48	☽ ∠ ☿
21 30	☽ # ♀
22 49	☽ Q ♅
26 Monday	
01 17	☽ ⊻ ♂
01 37	☽ □ ♄
02 04	☽ ⊼ ♃
04 08	♀ Q ♆
06 48	♂ △ ♄
08 29	☽ ∥ ♆
11 15	☽ Q ♇
14 08	☽ ⊥ ♂
14 10	☽ ± ♃
14 41	♂ ☍ ♃
27 Tuesday	
00 32	☽ ⚹ ☿
00 54	☽ ∥ ♄
05 11	☽ □ ♀
06 54	☽ □ ☉
08 15	☽ ꝗ ♃
09 21	☽ ∠ ♂
11 23	☽ □ ♅
13 46	☽ Q ♄
16 47	☉ ∥ ♅
22 43	☉ ꝗ ♃
23 37	☽ □ ♇
28 Wednesday	
02 22	☽ Q ☿
02 45	☽ ⊻ ♆
03 39	☽ # ♇
09 24	♀ ꝗ ♃
13 36	☽ ⚹ ♄
14 07	☽ △ ♃
14 43	☽ ⊥ ♆
17 02	☽ ⚹ ♂
29 Thursday	
02 51	♂ ∠ ♅
08 27	☽ ∠ ♆
16 05	☽ □ ☿
17 51	☽ Q ♂
17 53	☉ ☌ ♅
18 50	☽ ∠ ♄
19 39	♀ ☌ ♅
22 58	☽ △ ♅
23 18	☽ △ ♀
23 22	☽ △ ☉
30 Friday	
01 30	☉ ☌ ♀
10 30	☽ △ ♇
12 00	☽ ⊥ ♄
12 20	♀ ∠ ♂
13 24	☽ ⚹ ♆
21 40	☉ ∠ ♂
23 16	☽ ⊻ ♄
23 47	☽ □ ♃
31 Saturday	
03 33	☽ ꝗ ♅
05 53	☽ □ ♂
05 58	♀ ∥ ♅
06 07	☽ ꝗ ☉
06 47	☽ ꝗ ♀
11 53	☽ Q ♆
14 39	☽ ꝗ ♇
18 31	☉ # ♂
19 09	☿ ꝗ ♄
20 15	☽ ± ♅
23 59	☽ ± ☉

All ephemeris data is given at 12.00 UT and the Moon's longitude is additionally given for 24.00 UT

SEPTEMBER 1963

Date	Sidereal time h m s	Sun ☉	Moon ☽	Moon ☽ 24.00	Mercury ☿	Venus ♀	Mars ♂	Jupiter ♃	Saturn ♄	Uranus ♅	Neptune ♆	Pluto ♇
01	10 39 44	08 ♍ 19 51	08 ♒ 30 18	15 ♒ 15 23	03 ♎ 46	09 ♍ 00	22 ♎ 50	18 ♈ 38	18 ♒ 22	05 ♍ 50	13 ♏ 16	11 ♍ 47
02	10 43 41	09 17 54	22 ♒ 06 29	29 ♒ 03 23	04 12	10 14	23 29	18 R 33	18 R 18	05 54	13 17	11 49
03	10 47 37	10 15 58	06 ♓ 05 43	13 ♓ 12 56	04 34	11 29	24 08	18 28	18 14	05 58	13 18	11 51
04	10 51 34	11 14 05	20 ♓ 24 23	27 ♓ 39 18	04 51	12 43	24 47	18 24	18 10	06 02	13 20	11 53
05	10 55 30	12 12 13	04 ♈ 56 49	12 ♈ 16 01	05 02	13 58	25 27	18 19	18 06	06 05	13 21	11 55
06	10 59 27	13 10 22	19 ♈ 35 59	26 ♈ 55 48	05 08	15 12	26 06	18 13	18 02	06 09	13 22	11 57
07	11 03 24	14 08 34	04 ♉ 14 37	11 ♉ 31 43	05 R 07	16 27	26 46	18 08	17 58	06 13	13 24	11 59
08	11 07 20	15 06 48	18 ♉ 46 26	25 ♉ 58 15	05 01	17 41	27 25	18 03	17 54	06 17	13 25	12 01
09	11 11 17	16 05 04	03 ♊ 06 47	10 ♊ 11 46	04 48	18 56	28 05	17 57	17 51	06 20	13 26	12 03
10	11 15 13	17 03 22	17 ♊ 13 03	24 ♊ 10 34	04 29	20 10	28 45	17 51	17 47	06 24	13 28	12 06
11	11 19 10	18 01 42	01 ♋ 04 21	07 ♋ 54 29	04 03	21 25	29 ♎ 25	17 45	17 43	06 28	13 29	12 08
12	11 23 06	19 00 05	14 ♋ 41 03	21 ♋ 24 13	03 31	22 39	00 ♏ 05	17 39	17 40	06 32	13 31	12 10
13	11 27 03	19 58 29	28 ♋ 04 08	04 ♌ 40 54	02 52	23 54	00 45	17 33	17 36	06 35	13 32	12 12
14	11 30 59	20 56 56	11 ♌ 14 40	17 ♌ 45 31	02 07	25 09	01 25	17 27	17 33	06 39	13 34	12 14
15	11 34 56	21 55 24	24 ♌ 13 31	00 ♍ 38 44	01 16	26 23	02 05	17 20	17 30	06 43	13 35	12 16
16	11 38 53	22 53 55	07 ♍ 01 12	13 ♍ 20 55	00 ♎ 21	27 38	02 45	17 14	17 27	06 46	13 37	12 18
17	11 42 49	23 52 27	19 ♍ 37 54	25 ♍ 52 10	29 ♍ 21	28 ♍ 52	03 25	17 07	17 23	06 50	13 38	12 20
18	11 46 46	24 51 02	02 ♎ 03 46	08 ♎ 12 43	28 19	00 ♎ 07	04 06	17 00	17 20	06 54	13 40	12 22
19	11 50 42	25 49 38	14 ♎ 19 07	20 ♎ 23 05	27 15	01 22	04 46	16 53	17 17	06 57	13 41	12 24
20	11 54 39	26 48 16	26 ♎ 24 47	02 ♏ 24 26	26 12	02 36	05 27	16 46	17 14	07 01	13 43	12 26
21	11 58 35	27 46 56	08 ♏ 22 17	14 ♏ 18 41	25 09	03 51	06 07	16 39	17 11	07 04	13 45	12 28
22	12 02 32	28 45 38	20 ♏ 14 00	26 ♏ 08 40	24 10	05 06	06 48	16 31	17 09	07 08	13 46	12 30
23	12 06 28	29 ♍ 44 21	02 ♐ 03 11	07 ♐ 58 04	23 16	06 21	07 29	16 24	17 06	07 11	13 48	12 32
24	12 10 25	00 ♎ 43 07	13 ♐ 53 55	19 ♐ 51 19	22 27	07 35	08 10	16 17	17 03	07 15	13 50	12 34
25	12 14 22	01 41 54	25 ♐ 50 57	01 ♑ 53 27	21 46	08 50	08 51	16 09	17 01	07 19	13 52	12 36
26	12 18 18	02 40 43	07 ♑ 59 31	14 ♑ 09 49	21 13	10 05	09 32	16 01	16 58	07 22	13 53	12 38
27	12 22 15	03 39 33	20 ♑ 24 59	26 ♑ 45 40	20 50	11 19	10 13	15 54	16 56	07 26	13 55	12 40
28	12 26 11	04 38 25	03 ♒ 12 24	09 ♒ 45 41	20 36	12 34	10 54	15 46	16 54	07 29	13 57	12 42
29	12 30 08	05 37 19	16 ♒ 25 53	23 ♒ 13 14	20 D 33	13 49	11 35	15 38	16 51	07 32	13 59	12 44
30	12 34 04	06 ♎ 36 15	00 ♓ 07 50	07 ♓ 09 34	20 ♍ 40	15 ♎ 03	12 ♏ 16	15 ♈ 30	16 ♒ 49	07 ♍ 36	14 ♏ 01	12 ♍ 46

Date	Moon True ☊	Moon Mean ☊	Moon Latitude
01	19 ♋ 30	17 ♋ 48	01 S 40
02	19 R 27	17 44	02 46
03	19 21	17 41	03 43
04	19 13	17 38	04 28
05	19 05	17 35	04 56
06	18 57	17 32	05 06
07	18 50	17 29	04 55
08	18 45	17 25	04 25
09	18 43	17 22	03 39
10	18 D 42	17 19	02 40
11	18 42	17 16	01 33
12	18 43	17 13	00 S 22
13	18 R 43	17 10	00 N 50
14	18 41	17 06	01 57
15	18 36	17 03	02 57
16	18 28	17 00	03 47
17	18 18	16 57	04 25
18	18 07	16 54	04 50
19	17 54	16 50	05 01
20	17 42	16 47	04 58
21	17 31	16 44	04 42
22	17 23	16 41	04 14
23	17 17	16 38	03 35
24	17 13	16 35	02 47
25	17 12	16 31	01 51
26	17 D 12	16 28	00 N 49
27	17 R 12	16 25	00 S 17
28	17 11	16 22	01 24
29	17 08	16 19	02 28
30	17 ♋ 03	16 ♋ 16	03 S 26

DECLINATIONS

Date	Sun ☉	Moon ☽	Mercury ☿	Venus ♀	Mars ♂	Jupiter ♃	Saturn ♄	Uranus ♅	Neptune ♆	Pluto ♇
01	08 N 27	19 S 45	04 S 37	09 N 30	08 S 54	05 N 51	16 S 31	10 N 03	14 S 09	19 N 23
02	08 05	16 46	04 55	09 02	09 09	05 49	16 32	10 01	14 10	19 22
03	07 43	12 44	05 11	08 34	09 24	05 47	16 34	10 00	14 10	19 21
04	07 21	07 55	05 24	08 06	09 39	05 45	16 35	09 59	14 11	19 20
05	06 59	02 S 34	05 35	07 37	09 54	05 43	16 36	09 57	14 11	19 19
06	06 37	02 N 57	05 42	07 08	10 09	05 41	16 37	09 56	14 12	19 19
07	06 14	08 18	05 46	06 39	10 24	05 39	16 38	09 54	14 12	19 18
08	05 52	13 10	05 48	06 10	10 39	05 37	16 40	09 53	14 12	19 17
09	05 29	17 12	05 45	05 41	10 54	05 34	16 41	09 52	14 13	19 16
10	05 07	20 10	05 39	05 11	11 09	05 32	16 42	09 50	14 13	19 16
11	04 44	21 53	05 29	04 41	11 24	05 30	16 43	09 49	14 14	19 15
12	04 21	22 17	05 15	04 12	11 38	05 27	16 44	09 48	14 14	19 14
13	03 58	21 22	04 57	03 42	11 53	05 25	16 45	09 46	14 15	19 14
14	03 35	19 17	04 35	03 12	12 08	05 22	16 46	09 45	14 15	19 13
15	03 12	16 14	04 08	02 41	12 22	05 19	16 47	09 44	14 16	19 12
16	02 49	12 27	03 39	02 11	12 37	05 17	16 48	09 42	14 16	19 12
17	02 26	08 10	03 06	01 41	12 51	05 14	16 49	09 41	14 17	19 11
18	02 03	03 N 37	02 30	01 10	13 05	05 11	16 50	09 40	14 17	19 10
19	01 40	01 S 01	01 51	00 40	13 20	05 09	16 51	09 38	14 18	19 09
20	01 16	05 33	01 12	00 N 09	13 34	05 06	16 52	09 37	14 18	19 09
21	00 53	09 50	00 S 31	00 S 21	13 48	05 03	16 53	09 36	14 19	19 08
22	00 30	13 43	00 N 09	00 52	14 02	05 00	16 54	09 34	14 20	19 08
23	00 N 06	17 04	00 49	01 22	14 16	04 57	16 54	09 33	14 20	19 07
24	00 S 17	19 43	01 26	01 53	14 30	04 54	16 55	09 32	14 21	19 06
25	00 41	21 32	02 01	02 23	14 43	04 51	16 56	09 30	14 21	19 06
26	01 04	22 24	02 32	02 54	14 57	04 48	16 57	09 29	14 22	19 05
27	01 27	22 10	03 00	03 24	15 11	04 45	16 57	09 28	14 22	19 04
28	01 51	20 48	03 23	03 55	15 24	04 42	16 58	09 26	14 23	19 04
29	02 14	18 16	03 41	04 25	15 38	04 39	16 59	09 25	14 23	19 03
30	02 S 37	14 S 39	03 N 54	04 S 55	15 S 51	04 N 36	16 S 59	09 N 24	14 S 24	19 N 03

ZODIAC SIGN ENTRIES

Date	h	m	Planets
03	01	37	☽ ♓
05	03	52	☽ ♈
07	05	02	☽ ♉
09	06	45	☽ ♊
11	10	08	☽ ♋
12	09	11	♂ ♏
13	15	30	☽ ♌
15	22	47	☽ ♍
16	20	29	☿ ♍
18	08	00	☽ ♎
18	09	43	♀ ♎
20	19	10	☽ ♏
23	07	50	☽ ♐
23	18	24	☉ ♎
25	20	15	☽ ♑
28	06	03	☽ ♒
30	11	47	☽ ♓

LATITUDES

Date	Mercury ☿	Venus ♀	Mars ♂	Jupiter ♃	Saturn ♄	Uranus ♅	Neptune ♆	Pluto ♇
01	03 S 24	01 N 24	00 S 01	01 S 34	01 S 15	00 N 43	01 N 45	13 N 17
04	03 47	01 25	00 03	01 34	01 15	00 44	01 45	13 17
07	04 04	01 25	00 05	01 35	01 15	00 44	01 44	13 17
10	04 13	01 24	00 07	01 36	01 15	00 44	01 44	13 17
13	04 09	01 23	00 09	01 36	01 15	00 44	01 44	13 18
16	03 49	01 21	00 11	01 36	01 15	00 44	01 44	13 18
19	03 13	01 19	00 13	01 37	01 15	00 44	01 44	13 18
22	02 21	01 16	00 15	01 37	01 15	00 44	01 44	13 19
25	01 22	01 13	00 17	01 37	01 15	00 44	01 44	13 19
28	00 S 23	01 09	00 19	01 38	01 15	00 44	01 44	13 20
31	00 N 29	01 N 04	00 S 21	01 S 38	01 S 15	00 N 44	01 N 43	13 N 21

DATA

Julian Date	2438274
Delta T	+35 seconds
Ayanamsa	23° 20' 43"
Synetic vernal point	05° ♓ 46' 17"
True obliquity of ecliptic	23° 26' 36"

MOON'S PHASES, APSIDES AND POSITIONS ☽

Date	h	m	Phase	Longitude	Eclipse Indicator
03	19	34	○	10 ♓ 34	
10	11	42	☾	17 ♊ 03	
17	20	51	●	24 ♍ 14	
26	00	38	☽	02 ♑ 13	

Day	h	m		
06	15	42	Perigee	
22	12	40	Apogee	
05	23	11	ON	
12	06	57	Max dec	22° N 18'
19	06	43	OS	
26	19	19	Max dec	22° S 27'

LONGITUDES

Date	Chiron ⚷	Ceres ⚳	Pallas ⚴	Juno ⚵	Vesta ⚶	Black Moon Lilith ⚸
01	12 ♓ 44	13 ♎ 40	23 ♍ 25	09 ♎ 52	25 ♎ 50	15 ♏ 00
11	12 ♓ 16	17 ♎ 55	28 ♍ 02	13 ♎ 17	00 ♏ 44	16 ♏ 07
21	11 ♓ 49	22 ♎ 13	02 ♎ 39	16 ♎ 45	05 ♏ 45	17 ♏ 14
31	11 ♓ 23	26 ♎ 34	07 ♎ 14	20 ♎ 16	10 ♏ 52	18 ♏ 21

All ephemeris data is given at 12.00 UT and the Moon's longitude is additionally given for 24.00 UT

ASPECTARIAN

h m Aspects

01 Sunday
01 03 ☽ ± ♀
03 08 ☽ △ ☿
07 05 ☽ ± ♇
07 11 ☽ ⊼ ♅
08 39 ☽ Q ♃
11 40 ☽ ⊼ ☉
12 58 ☽ ⊼ ♀
15 38 ☽ # ♇
17 51 ☽ ⊼ ♇
20 30 ☽ □ ♆

02 Monday
05 22 ☽ ☌ ♄
05 50 ☽ ✶ ♃
06 47 ☽ bQ ☿
08 12 ♀ # ♂
13 28 ☽ ∥ ♄
14 30 ☽ △ ♂
22 50 ☽ ± ☿

03 Tuesday
04 03 ☽ ∥ ♆
07 34 ☽ ∠ ♃
09 21 ☽ ⊼ ☿
11 47 ☽ ☍ ♅
17 23 ☽ bQ ♂
19 22 ♀ ☌ ♇
19 34 ☽ ☍ ☉
21 44 ☽ ☍ ♇
21 57 ☽ ☍ ♀
22 42 ☽ ⊥ ♃

04 Wednesday
00 10 ☽ △ ♆
02 04 ☽ # ♅
04 06 ☽ ∥ ♂
06 06 ♀ ± ♃
08 17 ☽ ⊻ ♄
08 40 ☽ ⊻ ♃
09 11 ☽ ± ♂
11 03 ☽ # ♀
14 47 ☽ # ☉
18 12 ☽ ⊥ ♄
19 37 ☽ ⊼ ♂
21 56 ☽ # ♃
23 05 ☽ ∥ ☿
23 58 ♀ ✶ ♆

05 Thursday
01 08 ☽ bQ ♆
04 40 ☉ ☌ ♇
08 58 ☽ ∠ ♄
12 08 ☽ ☍ ☿
13 53 ☽ ⊼ ♅
14 27 ☉ ± ♃
15 57 ☽ ± ♆
16 24 ♂ # ♅
23 27 ☽ ⊼ ♇
23 46 ☽ ± ♅

06 Friday
00 44 ☽ ⊼ ☉
01 47 ☽ ⊼ ♆
04 08 ☽ ⊼ ♀
09 18 ☽ ± ♇
09 27 ☽ ✶ ♄
09 29 ☿ # ♃
09 46 ☽ ☌ ♃
11 15 ☽ ± ☉
14 33 ☽ bQ ♅
14 52 ☽ ± ♀
17 00 ☉ ✶ ♆
23 09 ☿ St R
23 09 ☽ ☍ ♂
23 59 ☽ ∥ ♃

07 Saturday
00 04 ☽ bQ ♇
00 20 ☽ # ☿
03 02 ☽ bQ ☉
03 10 ☽ ∥ ☉
05 01 ☽ Q ♄
05 03 ☽ ∥ ♀
06 58 ☽ bQ ♀
13 26 ☽ ⊼ ☿
15 15 ☽ △ ♅
19 32 ☽ ∥ ♅
20 38 ♂ ∠ ♇
22 30 ☽ # ♂
23 16 ☽ ± ☿

08 Sunday
00 47 ☽ △ ♇
03 06 ☽ ☍ ♆
05 30 ☽ △ ☉
10 01 ☽ △ ♀
10 34 ☽ □ ♄
10 48 ☽ ⊻ ♃
14 03 ☽ bQ ☿
16 02 ♀ ⊼ ♄
16 08 ♃ ± ♇
16 55 ☉ # ☿
17 44 ☽ # ♆
18 28 ♀ ⊼ ♃
20 43 ☽ ⊥ ♃

09 Monday
03 08 ☽ ⊼ ♂
06 04 ☉ ∥ ♃
07 45 ☿ # ♀
08 31 ☽ # ♄
11 44 ☽ ∠ ♃
13 43 ☽ ± ♂
14 48 ☽ △ ☿
17 29 ☽ □ ♅
17 29 ♀ ∥ ♃

10 Tuesday
03 13 ☽ □ ♇
03 36 ☽ ∥ ♇
05 33 ☽ ⊼ ♆
05 46 ☽ bQ ♂
11 42 ☽ □ ☉
12 58 ☽ △ ♄
13 05 ☽ ✶ ♃
15 52 ☽ ± ♆
17 35 ☽ □ ♀

11 Wednesday
00 27 ☽ Q ♅
03 00 ☉ ∥ ♀
04 54 ☉ ⊼ ♄
05 54 ☉ ⊼ ♃
07 29 ☽ bQ ♆
08 57 ☽ △ ♂
09 43 ☽ Q ♃
10 03 ☿ # ♃
10 21 ☽ Q ♇
14 53 ☽ bQ ♄
17 02 ☽ □ ☿
21 21 ☽ Q ☉
21 30 ☽ ✶ ♅

12 Thursday
04 08 ☽ Q ♀
06 14 ♃ ✶ ♄
06 40 ☽ ± ♄
07 31 ☽ ✶ ♇
09 55 ☽ △ ♆
17 15 ☽ □ ♃
17 17 ☽ ⊼ ♄
20 18 ☽ ✶ ☉
23 39 ☽ Q ☿

13 Friday
00 16 ☽ ∠ ♅
03 43 ☽ ✶ ♀
06 37 ♀ ± ♄
10 25 ☽ ∠ ♇
16 35 ☽ ⊥ ♅
17 06 ☽ □ ♂
20 14 ☽ ✶ ☿
21 09 ☿ bQ ♄

14 Saturday
01 32 ☽ ∠ ☉
02 48 ☽ ⊥ ♇
03 33 ☽ ⊻ ♅
09 46 ☽ ∠ ♀
12 37 ☽ ∥ ♇
13 49 ☽ ⊻ ♇
16 16 ☽ □ ♆
19 22 ☽ ⊥ ☉
22 10 ☽ ∠ ☿
23 14 ☿ ⊻ ♂
23 17 ♀ ⊥ ♂
23 20 ☽ △ ♃
23 34 ☽ ☍ ♄

15 Sunday
03 53 ☽ Q ♂
04 06 ☽ ⊥ ♀
07 22 ☽ ⊻ ☉
08 06 ☽ # ♄
13 49 ☽ ⊥ ☿
16 28 ☽ ⊻ ♀

16 Monday
00 18 ☽ ⊻ ☿
00 57 ☽ # ♆
01 47 ☽ Q ♆
03 03 ☽ bQ ♃
03 31 ☽ ✶ ♂
11 07 ☽ # ♂
11 32 ☽ ☌ ♅
19 54 ☽ ± ♃
22 02 ☽ ☌ ♇

17 Tuesday
00 32 ☽ ✶ ♆
00 40 ☉ ± ♄
03 44 ☽ ∥ ♅
07 13 ☽ ⊼ ♃
07 19 ♀ ∠ ♆
07 43 ☽ ⊼ ♄
09 33 ☽ ∠ ♂
17 05 ☿ ☌ ♀
19 11 ☽ ± ♄
20 51 ☽ ☌ ☉

18 Wednesday
03 44 ☽ ∥ ♃
03 52 ☽ ⊥ ♂
04 15 ☿ ∠ ♆
05 18 ☽ ☌ ☿
05 23 ☽ ∠ ♆
07 48 ☽ ☌ ♀
12 32 ☽ bQ ♄
15 02 ☿ ⊥ ♂
16 12 ☽ ⊻ ♂
18 44 ☽ # ☿
20 52 ☽ ∥ ☉
21 28 ☽ ⊻ ♅
22 57 ☽ ⊥ ♆

19 Thursday
02 12 ☽ ∥ ♀
08 13 ☽ ⊻ ♇
09 18 ☽ ⊥ ♅
10 21 ☽ # ♀
10 46 ☽ ⊻ ♆
15 05 ☽ # ☉
15 50 ☽ ∥ ☿
17 01 ☽ ☍ ♃
17 50 ☽ △ ♄
20 06 ☽ ⊥ ♇

20 Friday
03 11 ☽ ∠ ♅
04 47 ☉ ☌ ☿
05 08 ♀ bQ ♄
05 17 ☉ # ☿
09 33 ☽ # ♃
11 36 ☽ ⊻ ☿
12 51 ☽ ⊻ ☉
14 03 ☽ ∠ ♇
22 38 ☽ ⊥ ☿

21 Saturday
01 51 ☽ ⊻ ♀
01 56 ☽ ⊥ ☉
07 12 ☽ ☌ ♂
09 22 ☽ ✶ ♅
10 34 ☽ # ♅
15 19 ☽ ∠ ☿
15 20 ☽ ⊥ ♀
15 22 ☿ ∥ ♀
20 18 ☽ ✶ ♇
21 42 ☽ ∠ ☉
22 53 ☽ ☌ ♆

22 Sunday
02 11 ☉ # ♀
04 33 ☽ ⊼ ♃
05 46 ☽ □ ♄
09 45 ☽ Q ♅
11 41 ☽ ∠ ♀
12 22 ☉ ∠ ♆
14 13 ☽ ∥ ♂
16 03 ☽ ∥ ♆
16 36 ☽ ± ♃
19 24 ☽ ✶ ☿
19 43 ☉ ∥ ☿
20 42 ☽ Q ♇

23 Monday
00 45 ♂ ✶ ♅
06 53 ☽ ✶ ☉
10 41 ☽ bQ ♃
10 48 ☽ ∥ ♄
16 48 ☿ ± ♄
18 04 ☽ Q ☿
18 09 ☽ Q ♄
19 34 ♂ ∥ ♆
21 44 ☽ ✶ ♀
22 29 ☽ □ ♅
23 41 ☽ ⊻ ♂

24 Tuesday
00 19 ☿ ∠ ♂
05 11 ♀ ⊻ ♅
05 53 ☽ # ♇
09 19 ☽ □ ♇
09 24 ☽ Q ☉
11 52 ☽ ⊻ ♆
12 34 ☽ ⊥ ♂
16 45 ☽ △ ♃
16 51 ♀ ⊥ ♆
18 21 ☽ ✶ ♄
23 59 ☽ ⊥ ♆

25 Wednesday
00 48 ☽ Q ♀
04 15 ☽ □ ☿
07 45 ☽ ∠ ♂
12 31 ♀ ⊻ ♂
18 01 ☽ ∠ ♆
19 22 ☉ bQ ♄

26 Thursday
00 12 ☽ ∠ ♄
00 38 ☽ □ ☉
10 46 ☽ △ ♅
15 10 ☽ ✶ ♂
16 32 ☽ □ ♀
17 48 ☽ ⊥ ♄
21 05 ☽ △ ♇
23 30 ☽ ✶ ♆

27 Friday
03 26 ☽ □ ♃
05 21 ☽ ⊻ ♄
12 46 ☽ △ ☿
15 36 ☽ Q ♂
15 50 ☽ bQ ♅
22 27 ☽ Q ♆

28 Saturday
01 45 ☽ bQ ♇
08 48 ☽ ± ♅
13 01 ☽ Q ♃
14 46 ♀ ⊻ ♇
14 51 ☽ △ ☉
16 22 ☽ bQ ☿
18 27 ☽ ± ♇
19 53 ☽ ⊼ ♅

29 Sunday
02 49 ☽ □ ♂
05 22 ☽ ⊼ ♇
05 34 ☽ # ♇
06 30 ♀ ⊥ ♅
06 50 ☽ △ ♀
07 37 ☽ □ ♆
08 04 ☿ St D
08 37 ☽ ± ☿
08 46 ☉ ⊥ ♂
10 36 ☽ ✶ ♃
12 45 ☽ ☌ ♄
15 21 ♀ ⊻ ♆
19 20 ☽ ⊼ ☿
20 01 ☽ bQ ☉
21 19 ☽ ∥ ♄
22 12 ♀ # ♃

30 Monday
05 08 ☽ ∥ ♂
11 52 ☽ bQ ♀
12 38 ☽ ∠ ♃
12 53 ☽ ± ☉
13 24 ☽ ∥ ♆
19 46 ♀ ☍ ♃
23 53 ☽ ⊼ ☉

OCTOBER 1963

LONGITUDES

Date	Sidereal time	Sun ☉	Moon ☽	Moon ☽ 24.00	Mercury ☿	Venus ♀	Mars ♂	Jupiter ♃	Saturn ♄	Uranus ♅	Neptune ♆	Pluto ♇
	h m s	° ′ ″	° ′ ″	° ′ ″	° ′	° ′	° ′	° ′	° ′	° ′	° ′	° ′
01	12 38 01	07 ♎ 35 13	14 ♓ 18 08	21 ♓ 33 03	20 ♍ 57	16 ♎ 18	12 ♏ 57	15 ♈ 22	16 ♒ 47	07 ♍ 39	14 ♏ 03	12 ♍ 48
02	12 41 57	08 34 12	28 ♓ 53 33	06 ♈ 18 45	21 23	17 33	13 39	15 R 14	16 R 45	07 43	14 05	12 50
03	12 45 54	09 33 13	13 ♈ 47 34	21 ♈ 18 47	22 00	18 48	14 20	15 06	16 43	07 46	14 06	12 52
04	12 49 51	10 32 17	28 ♈ 51 08	06 ♉ 23 22	22 45	20 02	15 02	14 58	16 42	07 49	14 08	12 54
05	12 53 47	11 31 23	13 ♉ 54 13	21 ♉ 22 35	23 38	21 17	15 43	14 50	16 40	07 53	14 10	12 56
06	12 57 44	12 30 30	28 ♉ 47 28	06 ♊ 08 06	24 39	22 32	16 25	14 42	16 39	07 56	14 12	12 58
07	13 01 40	13 29 41	13 ♊ 23 51	20 ♊ 34 19	25 47	23 47	17 07	14 34	16 37	07 59	14 14	13 00
08	13 05 37	14 28 53	27 ♊ 39 15	04 ♋ 38 35	27 00	25 01	17 49	14 26	16 36	08 02	14 16	13 02
09	13 09 33	15 28 08	11 ♋ 32 22	18 ♋ 20 46	28 19	26 16	18 31	14 18	16 34	08 06	14 18	13 03
10	13 13 30	16 27 26	25 ♋ 04 03	01 ♌ 42 32	29 ♍ 43	27 31	19 12	14 10	16 33	08 09	14 20	13 05
11	13 17 26	17 26 45	08 ♌ 16 32	14 ♌ 46 25	01 ♎ 11	28 ♎ 46	19 54	14 02	16 32	08 12	14 22	13 07
12	13 21 23	18 26 07	21 ♌ 12 31	27 ♌ 35 11	02 41	00 ♏ 01	20 37	13 54	16 31	08 15	14 24	13 09
13	13 25 20	19 25 31	03 ♍ 54 43	10 ♍ 11 22	04 15	01 15	21 19	13 46	16 30	08 18	14 27	13 11
14	13 29 16	20 24 58	16 ♍ 25 24	22 ♍ 37 00	05 51	02 30	22 01	13 38	16 30	08 21	14 29	13 12
15	13 33 13	21 24 26	28 ♍ 46 21	04 ♎ 53 35	07 29	03 45	22 43	13 30	16 29	08 24	14 31	13 14
16	13 37 09	22 23 57	10 ♎ 58 51	17 ♎ 02 14	09 08	05 00	23 26	13 22	16 28	08 27	14 33	13 16
17	13 41 06	23 23 29	23 ♎ 03 53	29 ♎ 03 53	10 48	06 15	24 08	13 14	16 28	08 30	14 35	13 18
18	13 45 02	24 23 04	05 ♏ 02 24	10 ♏ 59 34	12 29	07 29	24 51	13 06	16 27	08 33	14 37	13 19
19	13 48 59	25 22 41	16 ♏ 55 34	22 ♏ 50 39	14 11	08 44	25 33	12 59	16 27	08 36	14 39	13 21
20	13 52 55	26 22 20	28 ♏ 45 05	04 ♐ 39 10	15 53	09 59	26 16	12 51	16 27	08 39	14 41	13 23
21	13 56 52	27 22 00	10 ♐ 33 17	16 ♐ 27 51	17 35	11 14	26 59	12 43	16 27	08 41	14 44	13 24
22	14 00 49	28 21 43	22 ♐ 23 19	28 ♐ 20 13	19 17	12 29	27 41	12 36	16 D 27	08 44	14 46	13 26
23	14 04 45	29 ♎ 21 27	04 ♑ 19 06	10 ♑ 20 34	21 00	13 43	28 24	12 28	16 27	08 47	14 48	13 28
24	14 08 42	00 ♏ 21 13	16 ♑ 25 15	22 ♑ 33 47	22 42	14 58	29 07	12 21	16 27	08 50	14 50	13 29
25	14 12 38	01 21 01	28 ♑ 46 51	05 ♒ 05 06	24 23	16 13	29 ♏ 50	12 14	16 28	08 52	14 52	13 31
26	14 16 35	02 20 50	11 ♒ 29 08	17 ♒ 59 34	26 05	17 28	00 ♐ 33	12 07	16 28	08 55	14 54	13 32
27	14 20 31	03 20 41	24 ♒ 36 53	01 ♓ 21 30	27 46	18 43	01 16	12 00	16 29	08 57	14 57	13 34
28	14 24 28	04 20 34	08 ♓ 13 41	15 ♓ 13 32	29 ♎ 27	19 57	01 59	11 53	16 29	09 00	14 59	13 35
29	14 28 24	05 20 28	22 ♓ 20 58	29 ♓ 35 41	01 ♏ 07	21 12	02 43	11 46	16 30	09 02	15 01	13 37
30	14 32 21	06 20 24	06 ♈ 57 09	14 ♈ 24 34	02 47	22 27	03 26	11 39	16 31	09 05	15 03	13 38
31	14 36 18	07 ♏ 20 22	21 ♈ 56 57	29 ♈ 33 06	04 ♏ 27	23 ♏ 42	04 ♐ 09	11 ♈ 33	16 ♒ 32	09 ♍ 07	15 ♏ 06	13 ♍ 39

Date	Moon True ☊	Moon Mean ☊	Moon ☽ Latitude
	° ′	° ′	° ′
01	16 ♋ 55	16 ♋ 12	04 S 14
02	16 R 45	16 09	04 46
03	16 34	16 06	05 00
04	16 22	16 03	04 53
05	16 13	16 00	04 26
06	16 05	15 56	03 42
07	16 01	15 53	02 43
08	15 59	15 50	01 35
09	15 D 58	15 47	00 S 23
10	15 R 58	15 44	00 N 48
11	15 57	15 41	01 55
12	15 54	15 37	02 55
13	15 48	15 34	03 44
14	15 39	15 31	04 22
15	15 28	15 28	04 47
16	15 14	15 25	04 59
17	15 00	15 22	04 57
18	14 46	15 18	04 42
19	14 33	15 15	04 14
20	14 23	15 12	03 36
21	14 16	15 09	02 48
22	14 11	15 06	01 53
23	14 09	15 02	00 N 52
24	14 D 09	14 59	00 S 12
25	14 09	14 56	01 17
26	14 R 09	14 53	02 20
27	14 07	14 50	03 18
28	14 02	14 47	04 06
29	13 56	14 43	04 42
30	13 47	14 40	05 01
31	13 ♋ 37	14 ♋ 37	05 S 00

DECLINATIONS

Date	Sun ☉	Moon ☽	Mercury ☿	Venus ♀	Mars ♂	Jupiter ♃	Saturn ♄	Uranus ♅	Neptune ♆	Pluto ♇
	° ′	° ′	° ′	° ′	° ′	° ′	° ′	° ′	° ′	° ′
01	03 S 01	10 S 05	04 N 02	05 S 25	16 S 04	04 N 33	17 S 00	09 N 23	14 S 25	19 N 02
02	03 24	04 S 49	04 05	05 55	16 17	04 30	17 00	09 21	14 25	19 02
03	03 47	00 N 50	04 03	06 25	16 30	04 27	17 01	09 20	14 26	19 01
04	04 10	06 30	03 56	06 55	16 43	04 24	17 02	09 19	14 26	19 01
05	04 34	11 46	03 45	07 25	16 56	04 21	17 02	09 18	14 27	19 00
06	04 57	16 17	03 29	07 54	17 08	04 17	17 02	09 17	14 28	19 00
07	05 20	19 43	03 10	08 23	17 21	04 14	17 03	09 15	14 28	18 59
08	05 43	21 50	02 46	08 53	17 33	04 11	17 03	09 14	14 29	18 59
09	06 05	22 33	02 20	09 22	17 45	04 08	17 04	09 13	14 29	18 58
10	06 28	21 54	01 50	09 50	17 58	04 05	17 04	09 12	14 30	18 58
11	06 51	20 03	01 18	10 19	18 09	04 02	17 04	09 11	14 31	18 57
12	07 14	17 11	00 44	10 47	18 21	03 59	17 04	09 10	14 31	18 57
13	07 36	13 33	00 N 08	11 15	18 33	03 56	17 05	09 08	14 32	18 56
14	07 59	09 23	00 S 30	11 43	18 45	03 53	17 05	09 07	14 33	18 56
15	08 21	04 53	01 09	12 10	18 56	03 50	17 05	09 06	14 33	18 56
16	08 43	00 N 14	01 50	12 37	19 07	03 47	17 05	09 05	14 34	18 55
17	09 05	04 S 22	02 31	13 04	19 18	03 44	17 05	09 04	14 35	18 55
18	09 27	08 47	03 13	13 31	19 29	03 41	17 05	09 03	14 35	18 54
19	09 49	12 50	03 55	13 57	19 40	03 38	17 05	09 02	14 36	18 54
20	10 11	16 23	04 38	14 23	19 51	03 35	17 05	09 01	14 36	18 54
21	10 32	19 16	05 21	14 49	20 01	03 32	17 05	09 00	14 37	18 53
22	10 54	21 21	06 04	15 14	20 12	03 29	17 05	08 59	14 38	18 53
23	11 15	22 30	06 47	15 39	20 22	03 27	17 05	08 58	14 38	18 53
24	11 36	22 38	07 30	16 03	20 32	03 24	17 05	08 57	14 39	18 53
25	11 57	21 40	08 12	16 27	20 42	03 21	17 05	08 56	14 40	18 52
26	12 17	19 35	08 54	16 51	20 51	03 19	17 05	08 55	14 40	18 52
27	12 38	16 25	09 36	17 14	21 01	03 16	17 05	08 54	14 41	18 52
28	12 58	12 17	10 18	17 37	21 10	03 13	17 04	08 53	14 42	18 51
29	13 18	07 21	10 59	17 59	21 19	03 11	17 04	08 52	14 42	18 51
30	13 38	01 51	11 39	18 21	21 28	03 09	17 04	08 51	14 43	18 51
31	13 S 58	03 N 55	12 S 19	18 S 42	21 S 37	03 N 06	17 S 03	08 N 51	14 S 44	18 N 51

ZODIAC SIGN ENTRIES

Date	h m	Planets
02	13 48	☽ ♈
04	13 50	☽ ♉
06	13 58	☽ ♊
08	16 01	☽ ♋
10	16 44	☿ ♎
10	20 54	☽ ♌
12	11 50	♀ ♏
13	04 34	☽ ♍
15	14 24	☽ ♎
18	01 53	☽ ♏
20	14 32	☽ ♐
23	03 21	☽ ♑
24	03 29	☉ ♏
25	14 20	☽ ♒
25	17 31	♂ ♐
27	21 36	☽ ♓
28	19 54	☿ ♏
30	00 40	☽ ♈

LATITUDES

Date	Mercury ☿	Venus ♀	Mars ♂	Jupiter ♃	Saturn ♄	Uranus ♅	Neptune ♆	Pluto ♇
	° ′	° ′	° ′	° ′	° ′	° ′	° ′	° ′
01	00 N 29	01 N 04	00 S 21	01 S 38	01 S 15	00 N 44	01 N 43	13 N 21
04	01 09	01 00	00 23	01 38	01 15	00 44	01 43	13 21
07	01 37	00 54	00 25	01 38	01 15	00 44	01 43	13 22
10	01 53	00 48	00 27	01 38	01 15	00 44	01 43	13 23
13	01 59	00 42	00 29	01 38	01 15	00 44	01 43	13 24
16	01 57	00 36	00 30	01 37	01 14	00 44	01 43	13 25
19	01 49	00 29	00 32	01 37	01 14	00 44	01 43	13 26
22	01 36	00 22	00 34	01 37	01 14	00 44	01 43	13 27
25	01 21	00 15	00 35	01 36	01 14	00 45	01 43	13 28
28	01 03	00 08	00 37	01 36	01 14	00 45	01 43	13 29
31	00 N 44	00 00	00 S 38	01 S 35	01 S 14	00 N 45	01 N 43	13 N 30

DATA

Julian Date	2438304
Delta T	+35 seconds
Ayanamsa	23° 20' 45"
Synetic vernal point	05° ♓ 46' 14"
True obliquity of ecliptic	23° 26' 37"

MOON'S PHASES, APSIDES AND POSITIONS ☽

Date	h m	Phase	Longitude ° ′	Eclipse Indicator
03	04 44	○	09 ♈ 15	
09	19 27	☾	15 ♋ 47	
17	12 43	●	23 ♎ 25	
25	17 20	☽	01 ♒ 34	

Day	h m		
04	15 11	Perigee	
20	02 41	Apogee	
03	08 30	0N	
09	12 21	Max dec	22° N 33'
16	13 14	0S	
24	02 52	Max dec	22° S 43'
30	19 44	0N	

LONGITUDES

Date	Chiron ⚷	Ceres ⚳	Pallas ⚴	Juno ⚵	Vesta ⚶	Black Moon Lilith ⚸
	° ′	° ′	° ′	° ′	° ′	° ′
01	11 ♓ 23	26 ♎ 34	07 ♎ 14	20 ♎ 16	10 ♏ 52	18 ♏ 21
11	11 ♓ 00	00 ♏ 57	11 ♎ 47	23 ♎ 46	16 ♏ 03	19 ♏ 28
21	10 ♓ 41	05 ♏ 20	16 ♎ 19	27 ♎ 17	21 ♏ 18	20 ♏ 35
31	10 ♓ 27	09 ♏ 43	20 ♎ 48	00 ♏ 48	26 ♏ 37	21 ♏ 42

ASPECTARIAN

h m Aspects

01 Tuesday
00 48 ☽ ☍ ♅
03 49 ☽ ⊥ ♃
04 40 ☽ ± ♀
06 26 ♂ ⚹ ♇
09 30 ☽ ☍ ♇
09 39 ☽ △ ♂
11 34 ☽ △ ♆
13 44 ☉ ⚺ ♅
13 46 ☽ ⚺ ♃
15 22 ☽ ⋕ ♅
15 39 ☽ ⚻ ♀
16 07 ☽ ⚺ ♄
21 05 ♀ △ ♄
23 19 ☽ ☍ ☿
23 33 ☉ ⊥ ♆

02 Wednesday
02 00 ☽ ⊥ ♄
07 33 ☽ ‖ ♀
11 35 ☽ ⚼ ♂
12 18 ☽ ⚼ ♆
13 23 ☽ ⋕ ♃
15 10 ☽ ⋕ ☿
16 38 ☽ ∠ ♄
17 43 ☽ ‖ ☉

03 Thursday
02 18 ☽ ⚻ ♅
02 50 ☽ ± ♂
02 52 ☽ ± ♆
03 36 ♂ ☌ ♆
04 44 ☽ ☍ ☉
10 31 ☽ ⚻ ♇
11 57 ☽ ± ♅
12 30 ☽ ⚻ ♆
12 55 ☽ ⚻ ♂
13 30 ♀ ⊥ ♇
14 05 ☽ ☌ ♃
16 41 ☽ ⚹ ♄
20 07 ☽ ± ♇
20 42 ☽ ☍ ♀

04 Friday
01 08 ☉ ⋕ ☿
01 19 ☽ ‖ ☿
01 20 ☽ ⋕ ☉
01 45 ☽ ⚻ ☿
02 22 ☽ ⚼ ♅
03 05 ☽ ‖ ♃
10 18 ♂ ⚻ ♃
10 29 ☽ ⚼ ♇
10 31 ☿ ± ♄
11 45 ☽ Q ♄
11 49 ☽ ± ☿
14 01 ☽ ⋕ ♀

05 Saturday
00 08 ☉ ⋕ ♃
00 28 ☽ ‖ ♅
02 20 ☽ △ ♅
03 02 ☽ ⚼ ☿
07 56 ☽ ⚻ ☉
10 27 ☽ △ ♇
12 26 ☽ ☍ ♆
13 29 ☽ ⚺ ♃
15 03 ☽ ☍ ♂
16 25 ☽ □ ♄
18 13 ☽ ± ☉
23 02 ☽ ⊥ ♃

06 Sunday
00 24 ♂ ‖ ♄
00 56 ☽ ⚻ ♀
01 39 ☽ ⋕ ♆
04 46 ☽ △ ☿
09 46 ☽ ⚼ ☉
11 32 ☽ ± ♀
13 28 ☽ ∠ ♃
16 37 ☽ ⋕ ♄
17 33 ☽ ⋕ ♂
19 27 ♂ □ ♄
20 02 ♀ ∠ ♅
23 29 ☉ ⚺ ♇

07 Monday
03 01 ☽ □ ♅
03 38 ☽ ⚼ ♀
06 05 ☽ ‖ ♇
11 20 ☽ □ ♇
12 10 ☽ △ ☉
13 24 ☽ ⚻ ♆
13 56 ☽ ⚹ ♃
17 22 ☽ △ ♄
18 31 ☽ ⚻ ♂
23 28 ☽ ± ♆

08 Tuesday
00 37 ☉ ⊥ ♅
05 08 ☽ ± ♂
06 44 ☉ ⚺ ♆
07 06 ☽ △ ♀
09 14 ☽ Q ♅
09 56 ☽ Q ♃
10 47 ☽ □ ☿
10 59 ☉ ☍ ♃
14 46 ☽ ⚼ ♆
17 47 ☽ Q ♇
18 44 ☽ ⚼ ♄
21 18 ☽ ⚼ ♂

09 Wednesday
05 15 ♀ ⋕ ♅
05 57 ☽ ⚹ ♅
10 19 ☽ ± ♄
11 09 ♃ ⚻ ♆
14 40 ☽ ⚹ ♇
16 48 ☽ □ ♃
16 52 ☽ △ ♆
19 27 ☽ □ ☉
20 51 ☽ ⚻ ♄
21 22 ☽ Q ☿

10 Thursday
00 58 ☽ △ ♂
05 27 ☿ ∠ ♆
08 32 ☽ ∠ ♅
14 19 ☉ △ ♄
14 36 ♃ ± ♅
16 52 ☽ □ ♀
17 27 ☽ ∠ ♇
21 25 ☽ ⚹ ☿
23 18 ♀ ∠ ♇

11 Friday
00 51 ☽ ⊥ ♅
06 23 ☽ Q ☉
09 52 ☽ ⊥ ♇
11 51 ☽ ⚺ ♅
15 31 ♂ ± ♃
17 43 ☿ ⚼ ♄
20 57 ☽ ⚺ ♇
22 07 ☽ ‖ ♇
22 31 ☽ △ ♃
22 40 ♂ Q ♅
23 17 ☽ □ ♆

12 Saturday
03 15 ☽ ☍ ♄
03 36 ☽ ⋕ ♂
04 32 ☽ ∠ ☿
05 23 ☽ Q ♀
06 23 ☽ ⚹ ☉
10 49 ☽ □ ♂
12 49 ☽ ⋕ ♄
23 44 ☽ ⊥ ☿

13 Sunday
02 20 ☽ ⚼ ♃
05 50 ☉ ⊥ ♇
05 57 ☽ ⋕ ♆
06 24 ☽ ⚹ ♀
09 12 ☽ Q ♆
12 44 ☽ ⚺ ☿
13 04 ☽ ∠ ☉
19 16 ☽ ± ♃
20 25 ☽ ☌ ♅
22 56 ☽ Q ♂

14 Monday
00 13 ☽ ⋕ ♀
05 47 ☽ ☌ ♇
06 40 ☽ ⚻ ♃
07 48 ☽ ⊥ ☉
08 14 ☽ ⚹ ♆
12 08 ☽ ⚻ ♄
13 26 ☽ ‖ ♅
14 19 ☽ ∠ ♀
19 03 ☽ ⋕ ☉
20 24 ☽ ⚺ ☉
23 29 ☽ ⚹ ♂
23 45 ☽ ± ♄

15 Tuesday
09 46 ☽ ⊥ ♀
11 05 ♂ ⋕ ♇
13 27 ☽ ∠ ♆
17 18 ☽ ⚼ ♄
17 32 ☽ ‖ ♃
18 15 ☿ ∠ ♂
22 52 ☽ ⚺ ♀

16 Wednesday
01 52 ☿ ⚺ ♅
03 23 ☿ ⊥ ♆
04 51 ☽ ⋕ ☿
06 39 ☽ ∠ ♂
06 59 ☽ ⚺ ♅
07 11 ☽ ⊥ ♆
07 46 ☽ ☌ ☿
16 32 ☽ ⚺ ♇
16 40 ☽ ☍ ♃
18 54 ☽ ⊥ ♅
19 05 ☽ ⚺ ♆
22 52 ☽ △ ♄

17 Thursday
00 33 ☽ ‖ ☿
01 34 ☽ ⊥ ♂
02 54 ♃ ⚻ ♇
04 28 ☽ ⊥ ♇
08 40 ☽ ⋕ ♃
10 49 ☉ ⋕ ♅
12 43 ☽ ☌ ☉
12 52 ☽ ∠ ♅
14 16 ☽ ⚺ ♂
14 45 ☉ ∠ ♅
22 29 ☽ ∠ ♇

18 Friday
13 32 ☽ ⋕ ♅
16 12 ☽ ‖ ☉
17 30 ☽ ☌ ♀
19 06 ☽ ⚹ ♅
20 09 ☿ ☍ ♃

19 Saturday
00 04 ☿ ⚺ ♇
02 51 ☿ ⋕ ♃
04 05 ☽ ⚻ ♃
04 45 ☽ ⚹ ♇
04 52 ♂ Q ♇
05 31 ☽ ⚺ ☿
07 23 ☽ ☌ ♆
09 12 ♀ ⚹ ♅
11 02 ☽ □ ♄
16 06 ☽ ± ♃
18 02 ☿ ⊥ ♅
18 49 ☿ ⚺ ♆
19 28 ☽ Q ♅
19 42 ☽ ⊥ ☿
20 10 ☽ ‖ ♀
23 30 ☽ ‖ ♆

20 Sunday
02 50 ☉ ⚺ ♂
05 08 ☽ Q ♇
06 37 ☽ ☌ ♂
06 43 ☽ ⚺ ☉
10 11 ☽ ⚼ ♃
17 04 ☽ ∠ ☿
17 24 ☽ ‖ ♄
20 00 ☿ △ ♄
20 02 ☽ ⊥ ☉
23 35 ☽ Q ♄

21 Monday
00 44 ♀ ‖ ♆
08 12 ☽ □ ♅
08 31 ☽ ⋕ ♇
13 32 ☽ ⚺ ♀
16 01 ☽ ∠ ☉
16 21 ☽ △ ♃
16 21 ♄ St D
17 48 ☽ □ ♇
20 18 ☽ ‖ ♂
20 30 ☽ ⚺ ♆
23 58 ☽ ⚹ ♄

22 Tuesday
03 09 ☽ ⊥ ♀
04 40 ☽ ⚹ ☿
08 42 ☽ ⊥ ♆
09 20 ♂ ⚼ ♃
13 44 ☉ ∠ ♇
14 02 ☿ ⊥ ♇
14 06 ♀ ⚻ ♃
23 23 ☽ ⚺ ♂
23 28 ☽ ∠ ♀

23 Wednesday
01 09 ☽ ⚹ ☉
02 55 ☽ ∠ ♆
06 15 ☽ ∠ ♄
06 49 ♀ ⚹ ♇
08 55 ☽ Q ☿
12 11 ☽ ⊥ ♂
20 56 ☽ △ ♅

24 Thursday
00 13 ☽ ⊥ ♄
03 16 ☽ Q ☉
04 03 ☽ □ ♃
06 12 ☽ △ ♇
07 11 ☽ ∠ ♂
08 49 ☽ ⚹ ♀
08 52 ☽ ⚹ ♆
09 19 ♀ ☌ ♆
12 04 ☽ ⚺ ♄
22 22 ☿ ⊥ ♂

25 Friday
02 12 ☽ □ ☿
02 30 ☽ ⚼ ♅
04 25 ☿ ∠ ♅
08 19 ☽ Q ♆
10 48 ☽ Q ♀
11 29 ☽ ⚼ ♇
14 08 ☽ ⚹ ♂
14 45 ☽ Q ♃
16 44 ♀ □ ♄
17 20 ☽ □ ☉
19 50 ☽ ± ♅
23 50 ☽ ‖ ♂

26 Saturday
04 36 ☽ ± ♇
07 11 ☽ ⚻ ♅
12 20 ☿ ⋕ ♅
13 09 ☽ ⚹ ♃
14 06 ☽ Q ♂
15 49 ☽ ⚻ ♇
18 11 ☽ ⋕ ♇
18 21 ☽ □ ♆
21 13 ☽ ☌ ♄
23 25 ♀ ± ♃

27 Sunday
00 11 ☽ □ ♀
02 15 ♀ ‖ ♄
07 04 ☽ ‖ ♀
07 36 ☽ ‖ ♄
16 13 ☽ ∠ ♃
18 27 ☽ △ ☿
22 43 ☽ ‖ ♆
23 28 ☿ ∠ ♇

28 Monday
00 30 ☽ □ ♂
04 43 ☽ △ ☉
07 57 ☽ ⊥ ♃
08 38 ☽ ‖ ☉
13 20 ☽ ☍ ♅
18 14 ☽ ⚺ ♃
20 52 ☽ ‖ ☿
21 13 ☽ ☍ ♇
23 37 ☽ △ ♆

29 Tuesday
00 07 ☽ ⚼ ☿
02 09 ☽ ⚺ ♄
04 56 ☽ ⋕ ♅
08 23 ☽ ⚼ ☉
08 47 ♀ Q ♅
09 54 ☽ △ ♀
12 15 ☽ ⊥ ♄
17 13 ☽ ± ☿

30 Wednesday
00 26 ☽ ± ☉
00 44 ☽ ⚼ ♆
03 08 ☽ ∠ ♄
04 22 ☽ ⚻ ☿
05 59 ☽ △ ♂
06 27 ☽ ⋕ ♃
10 56 ☽ ⚻ ☉
12 53 ☽ ⚼ ♀
15 24 ☽ ± ♆
15 27 ☽ ⚻ ♅
19 32 ☽ ☌ ♃
22 47 ☽ ⚻ ♇

31 Thursday
01 04 ☽ ⚻ ♆
01 07 ☽ ± ♅
03 23 ☽ ⚹ ♄
04 33 ☿ ⚺ ♂
04 38 ☽ ± ♀
07 21 ☽ ⚼ ♂
08 22 ☽ ± ♇
08 39 ☽ ‖ ♃
15 01 ☽ ⚻ ♀
15 27 ☽ ⚼ ♅
22 06 ♀ ⋕ ♇
22 17 ☽ ± ♂
22 24 ☽ Q ♄
22 37 ☽ ⚼ ♇

All ephemeris data is given at 12.00 UT and the Moon's longitude is additionally given for 24.00 UT

NOVEMBER 1963

Date	Sidereal time h m s	Sun ☉ ° ' "	Moon ☽ ° ' "	Moon ☽ 24.00 ° ' "	Mercury ☿ ° '	Venus ♀ ° '	Mars ♂ ° '	Jupiter ♃ ° '	Saturn ♄ ° '	Uranus ♅ ° '	Neptune ♆ ° '	Pluto ♇ ° '
01	14 40 14	08 ♏ 20 21	07 ♉ 11 41	14 ♉ 51 15	06 ♏ 06	24 ♏ 56	04 ♐ 53	11 ♈ 26	16 ♒ 33	09 ♍ 10	15 ♏ 08	13 ♍ 41
02	14 44 11	09 20 23	22 ♉ 30 22	00 ♊ 07 37	07 44	26 11	05 36	11 R 20	16 34	09 12	15 10	13 42
03	14 48 07	10 20 26	07 ♊ 41 45	15 ♊ 11 37	09 22	27 26	06 20	11 14	16 35	09 14	15 12	13 44
04	14 52 04	11 20 32	22 ♊ 36 17	29 ♊ 55 04	11 00	28 41	07 03	11 08	16 37	09 16	15 14	13 45
05	14 56 00	12 20 39	07 ♋ 07 28	14 ♋ 13 12	12 37	29 ♏ 56	07 47	11 02	16 38	09 19	15 17	13 46
06	14 59 57	13 20 49	21 ♋ 12 12	28 ♋ 04 30	14 14	01 ♐ 10	08 31	10 57	16 40	09 21	15 19	13 47
07	15 03 53	14 21 00	04 ♌ 50 21	11 ♌ 30 02	15 51	02 25	09 15	10 51	16 41	09 23	15 21	13 49
08	15 07 50	15 21 14	18 ♌ 03 58	24 ♌ 32 33	17 27	03 40	09 59	10 46	16 43	09 25	15 23	13 50
09	15 11 47	16 21 30	00 ♍ 56 16	07 ♍ 15 35	19 03	04 55	10 42	10 41	16 45	09 27	15 26	13 51
10	15 15 43	17 21 48	13 ♍ 30 59	19 ♍ 42 55	20 38	06 09	11 26	10 35	16 47	09 29	15 28	13 52
11	15 19 40	18 22 07	25 ♍ 51 47	01 ♎ 58 00	22 13	07 24	12 11	10 31	16 49	09 31	15 30	13 53
12	15 23 36	19 22 29	08 ♎ 01 56	14 ♎ 03 54	23 48	08 39	12 55	10 26	16 51	09 33	15 32	13 54
13	15 27 33	20 22 52	20 ♎ 04 12	26 ♎ 03 05	25 22	09 54	13 39	10 21	16 53	09 34	15 35	13 55
14	15 31 29	21 23 17	02 ♏ 00 47	07 ♏ 57 31	26 56	11 09	14 23	10 17	16 56	09 36	15 37	13 57
15	15 35 26	22 23 44	13 ♏ 53 29	19 ♏ 48 53	28 ♏ 30	12 23	15 07	10 13	16 58	09 38	15 39	13 58
16	15 39 22	23 24 13	25 ♏ 43 53	01 ♐ 38 42	00 ♐ 03	13 38	15 52	10 09	17 01	09 40	15 41	13 59
17	15 43 19	24 24 43	07 ♐ 33 33	13 ♐ 28 39	01 37	14 53	16 36	10 05	17 03	09 41	15 44	13 59
18	15 47 16	25 25 15	19 ♐ 24 15	25 ♐ 20 41	03 10	16 08	17 21	10 02	17 06	09 43	15 46	14 00
19	15 51 12	26 25 48	01 ♑ 18 14	07 ♑ 17 18	04 43	17 22	18 05	09 58	17 09	09 44	15 48	14 01
20	15 55 09	27 26 23	13 ♑ 18 15	19 ♑ 21 34	06 15	18 37	18 50	09 55	17 12	09 46	15 50	14 02
21	15 59 05	28 26 59	25 ♑ 27 41	01 ♒ 37 09	07 48	19 52	19 35	09 52	17 15	09 47	15 53	14 03
22	16 03 02	29 ♏ 27 36	07 ♒ 50 27	14 ♒ 08 10	09 20	21 07	20 19	09 50	17 18	09 48	15 55	14 04
23	16 06 58	00 ♐ 28 15	20 ♒ 30 49	26 ♒ 58 56	10 52	22 21	21 04	09 47	17 21	09 50	15 57	14 05
24	16 10 55	01 28 54	03 ♓ 33 00	10 ♓ 13 27	12 24	23 36	21 49	09 45	17 24	09 51	15 59	14 05
25	16 14 51	02 29 34	17 ♓ 00 38	23 ♓ 54 46	13 55	24 51	22 34	09 42	17 28	09 52	16 01	14 06
26	16 18 48	03 30 16	00 ♈ 55 58	08 ♈ 04 08	15 27	26 06	23 19	09 41	17 31	09 53	16 04	14 07
27	16 22 45	04 30 58	15 ♈ 19 01	22 ♈ 40 09	16 58	27 20	24 04	09 39	17 35	09 54	16 06	14 07
28	16 26 41	05 31 42	00 ♉ 06 50	07 ♉ 38 12	18 29	28 35	24 49	09 37	17 39	09 55	16 08	14 08
29	16 30 38	06 32 27	15 ♉ 13 11	22 ♉ 50 33	20 00	29 ♐ 50	25 34	09 36	17 42	09 56	16 10	14 09
30	16 34 34	07 ♐ 33 13	00 ♊ 28 59	08 ♊ 07 09	21 ♐ 30	01 ♑ 04	26 ♐ 19	09 ♈ 35	17 ♒ 46	09 ♍ 57	16 ♏ 12	14 ♍ 09

Date	Moon True ☊ ° '	Moon Mean ☊ ° '	Moon ☽ Latitude ° '
01	13 ♋ 27	14 ♋ 34	04 S 38
02	13 R 18	14 31	03 55
03	13 11	14 27	02 57
04	13 07	14 24	01 47
05	13 06	14 21	00 S 32
06	13 D 06	14 18	00 N 43
07	13 06	14 15	01 54
08	13 R 07	14 12	02 56
09	13 06	14 08	03 47
10	13 02	14 05	04 26
11	12 56	14 02	04 52
12	12 48	13 59	05 05
13	12 39	13 56	05 03
14	12 28	13 53	04 48
15	12 18	13 49	04 21
16	12 09	13 46	03 43
17	12 01	13 43	02 54
18	11 56	13 40	01 59
19	11 54	13 37	00 N 57
20	11 D 53	13 33	00 S 08
21	11 54	13 30	01 13
22	11 55	13 27	02 16
23	11 57	13 24	03 14
24	11 R 57	13 21	04 04
25	11 56	13 18	04 42
26	11 53	13 14	05 06
27	11 49	13 11	05 11
28	11 44	13 08	04 55
29	11 38	13 05	04 20
30	11 ♋ 33	13 ♋ 02	03 S 25

DECLINATIONS

Date	Sun ☉ ° '	Moon ☽ ° '	Mercury ☿ ° '	Venus ♀ ° '	Mars ♂ ° '	Jupiter ♃ ° '	Saturn ♄ ° '	Uranus ♅ ° '	Neptune ♆ ° '	Pluto ♇ ° '
01	14 S 17	09 N 32	12 S 58	19 S 03	21 S 45	03 N 04	17 S 03	08 N 50	14 S 44	18 N 51
02	14 36	14 36	13 36	19 23	21 53	03 02	17 03	08 49	14 45	18 51
03	14 55	18 41	14 14	19 43	22 02	02 59	17 02	08 48	14 46	18 50
04	15 14	21 27	14 51	20 02	22 09	02 57	17 02	08 47	14 46	18 50
05	15 33	22 43	15 28	20 21	22 17	02 55	17 01	08 46	14 47	18 50
06	15 51	22 29	16 03	20 39	22 25	02 53	17 01	08 46	14 48	18 50
07	16 09	20 54	16 38	20 56	22 32	02 51	17 00	08 45	14 48	18 50
08	16 27	18 12	17 12	21 13	22 39	02 49	17 00	08 44	14 49	18 50
09	16 44	14 41	17 45	21 30	22 46	02 47	16 59	08 43	14 49	18 50
10	17 01	10 35	18 17	21 45	22 53	02 46	16 58	08 43	14 50	18 50
11	17 18	06 07	18 48	22 01	22 59	02 44	16 58	08 42	14 51	18 50
12	17 34	01 N 29	19 18	22 15	23 05	02 42	16 57	08 41	14 51	18 50
13	17 51	03 S 10	19 48	22 29	23 11	02 41	16 56	08 41	14 52	18 50
14	18 07	07 40	20 16	22 42	23 17	02 39	16 55	08 40	14 53	18 50
15	18 22	11 51	20 43	22 55	23 22	02 38	16 55	08 40	14 53	18 50
16	18 38	15 35	21 10	23 07	23 28	02 37	16 54	08 39	14 54	18 50
17	18 53	18 42	21 35	23 18	23 33	02 36	16 53	08 38	14 55	18 50
18	19 07	21 03	21 59	23 28	23 37	02 34	16 52	08 38	14 55	18 50
19	19 22	22 29	22 22	23 38	23 42	02 33	16 51	08 37	14 56	18 50
20	19 35	22 54	22 44	23 47	23 46	02 32	16 50	08 37	14 57	18 50
21	19 49	22 05	23 05	23 56	23 50	02 31	16 49	08 36	14 57	18 50
22	20 02	20 03	23 25	24 03	23 54	02 31	16 48	08 36	14 58	18 50
23	20 15	17 43	23 43	24 10	23 58	02 30	16 47	08 35	14 58	18 50
24	20 28	14 00	24 01	24 17	24 01	02 29	16 46	08 35	14 59	18 50
25	20 40	09 28	24 17	24 22	24 04	02 29	16 45	08 35	15 00	18 51
26	20 51	04 S 18	24 32	24 27	24 07	02 28	16 44	08 34	15 00	18 51
27	21 03	01 N 15	24 45	24 31	24 10	02 28	16 43	08 34	15 01	18 51
28	21 14	06 54	24 58	24 34	24 12	02 27	16 42	08 33	15 01	18 51
29	21 24	12 16	25 09	24 37	24 14	02 27	16 40	08 33	15 02	18 51
30	21 S 34	16 N 55	25 S 18	24 S 38	24 S 16	02 N 27	16 S 39	08 N 33	15 S 03	18 N 51

ZODIAC SIGN ENTRIES

Date	h	m	Planets
01	00	42	☽ ♉
02	23	48	☽ ♊
05	00	08	☽ ♋
05	13	25	♀ ♐
07	03	24	☽ ♌
09	10	14	☽ ♍
11	20	07	☽ ♎
14	07	57	☽ ♏
16	11	07	☿ ♐
16	20	40	☽ ♐
19	09	23	☽ ♑
21	20	51	☽ ♒
23	00	49	☉ ♐
24	05	32	☽ ♓
26	10	25	☽ ♈
28	11	49	☽ ♉
29	15	21	♀ ♑
30	11	14	☽ ♊

LATITUDES

Date	Mercury ☿ ° '	Venus ♀ ° '	Mars ♂ ° '	Jupiter ♃ ° '	Saturn ♄ ° '	Uranus ♅ ° '	Neptune ♆ ° '	Pluto ♇ ° '
01	00 N 37	00 S 03	00 S 39	01 S 35	01 S 14	00 N 45	01 N 43	13 N 31
04	00 N 17	00 10	00 41	01 35	01 13	00 45	01 43	13 32
07	00 S 03	00 18	00 42	01 34	01 13	00 45	01 43	13 33
10	00 23	00 25	00 44	01 33	01 13	00 45	01 43	13 34
13	00 42	00 33	00 45	01 33	01 13	00 45	01 43	13 36
16	01 01	00 40	00 46	01 32	01 13	00 45	01 43	13 37
19	01 19	00 48	00 48	01 31	01 13	00 45	01 43	13 38
22	01 35	00 55	00 49	01 30	01 13	00 46	01 43	13 40
25	01 49	01 02	00 50	01 29	01 12	00 46	01 43	13 41
28	02 01	01 08	00 51	01 28	01 12	00 46	01 43	13 42
31	02 S 11	01 S 14	00 S 53	01 S 28	01 S 12	00 N 46	01 N 43	13 N 44

DATA

Julian Date	2438335
Delta T	+35 seconds
Ayanamsa	23° 20' 48"
Synetic vernal point	05° ♓ 46' 11"
True obliquity of ecliptic	23° 26' 36"

MOON'S PHASES, APSIDES AND POSITIONS ☽

Date	h	m	Phase	Longitude ° '	Eclipse Indicator
01	13	55	○	08 ♉ 25	
08	06	37	☾	15 ♌ 08	
16	06	50	●	23 ♏ 11	
24	07	56	☽	01 ♓ 19	
30	23	54	○	08 ♊ 03	

Day	h	m		
02	00	34	Perigee	
16	06	34	Apogee	
30	13	13	Perigee	
05	20	02	Max dec	22° N 48'
12	19	36	0S	
20	09	28	Max dec	22° S 55'
27	06	40	0N	

LONGITUDES

Date	Chiron ⚷ ° '	Ceres ⚳ ° '	Pallas ⚴ ° '	Juno ⚵ ° '	Vesta ⚶ ° '	Black Moon Lilith ⚸ ° '
01	10 ♓ 25	10 ♏ 10	21 ♎ 15	01 ♏ 09	27 ♏ 09	21 ♏ 49
11	10 ♓ 17	14 ♏ 32	25 ♎ 41	04 ♏ 37	02 ♐ 30	22 ♏ 56
21	10 ♓ 14	18 ♏ 53	00 ♏ 03	08 ♏ 02	07 ♐ 53	24 ♏ 03
31	10 ♓ 17	23 ♏ 12	04 ♏ 19	11 ♏ 22	13 ♐ 18	25 ♏ 10

ASPECTARIAN

h m	Aspects
01 Friday	
00 44	♂ Q ♄
08 11	☽ ⚻ ♂
08 53	☽ ∥ ♅
10 04	☽ ☍ ☿
13 55	☽ ☍ ☉
15 05	☽ △ ♅
18 36	☽ ⚺ ♃
22 11	☽ △ ♇
02 Saturday	
00 28	☽ ☍ ♆
02 30	♀ Q ♇
02 40	☽ □ ♄
03 57	☽ ⊥ ♃
06 12	☽ # ☿
08 30	☉ ⚹ ♅
12 02	☽ # ☉
12 46	☽ # ♆
14 38	♀ bQ ♃
17 59	☽ ∠ ♃
18 18	☽ ☍ ♀
23 02	☉ ∥ ♆
03 Sunday	
01 28	☽ # ♄
09 43	☽ ☍ ♂
09 56	☿ ⚹ ♅
13 04	☽ ∥ ♇
14 28	☽ □ ♅
15 00	☽ ⚻ ☿
16 31	☽ ⚻ ☉
17 37	☽ ⚹ ♃
20 29	☽ # ♀
21 39	☽ □ ♇
04 Monday	
00 03	☽ ⚻ ♆
01 48	☽ ± ☿
02 16	☽ △ ♄
02 51	☽ ± ☉
07 28	☉ ⚻ ♃
08 37	☿ ∥ ♆
09 47	☽ ± ♆
12 51	☽ Q ♃
13 50	☿ ⚻ ♃
18 15	☽ bQ ☿
18 34	☽ bQ ☉
19 40	☽ Q ♅
22 50	☽ # ♂
22 53	☽ ⚻ ♀
05 Tuesday	
00 34	☽ bQ ♆
01 08	☉ ☌ ☿
02 50	☽ bQ ♄
03 03	☽ Q ♇
09 48	☽ ± ♀
13 10	☽ ⚻ ♂
15 41	☽ ⚹ ♅
17 55	☽ ± ♄
18 33	☽ □ ♃
18 54	☉ ∥ ☿
21 29	☽ △ ☉
22 29	☽ △ ☿
23 15	☽ ⚹ ♇
23 52	☽ ± ♂
06 Wednesday	
01 51	☽ △ ♆
02 29	☽ bQ ♀
04 09	☽ ⚻ ♄
05 14	☿ ⚹ ♇
13 34	☽ # ♂
16 14	☽ bQ ♂
17 29	☽ ∠ ♅
22 49	☉ ⚹ ♇
07 Thursday	
01 17	☽ ∠ ♇
04 28	☿ ☌ ♆
07 15	☽ △ ♀
09 23	☽ ⊥ ♅
11 35	☽ # ♀
16 44	♂ □ ♅
17 20	☽ ⊥ ♇
20 11	☽ ⚺ ♅
20 23	☽ △ ♂
22 45	☽ △ ♃
08 Friday	
00 51	☿ □ ♄
02 15	☿ ± ♃
03 27	☿ ∥ ♄
04 14	☽ ⚺ ♇
06 37	☽ □ ☉
07 05	☽ □ ♆
07 06	☽ ∥ ♇
09 31	☽ ☍ ♄
10 42	☽ □ ☿
12 56	☉ ☌ ♆
18 21	☽ # ☿
20 51	☽ # ♄
23 38	☽ # ☉
09 Saturday	
02 11	☽ bQ ♃
08 47	♀ Q ♄
11 03	♂ △ ♃
11 06	☽ # ♆
16 44	☽ Q ♆
18 59	☉ ± ♃
19 02	☽ ± ♃
19 02	☽ Q ☉
20 21	☽ □ ♀
21 39	☉ □ ♄
10 Sunday	
01 16	☽ Q ☿
04 14	☽ ☌ ♅
06 25	☽ ⚻ ♃
07 45	☽ □ ♂
08 09	☉ ∥ ♄
12 41	☽ ☌ ♇
15 46	☽ ⚹ ♆
18 20	☽ ⚻ ♄
20 05	☽ ⚹ ☉
22 14	☽ ∥ ♅
11 Monday	
01 08	☿ Q ♅
03 49	☽ ⚹ ☿
06 01	☽ ± ♄
11 00	☽ Q ♀
13 17	☿ # ♇
21 01	☽ Q ♂
21 09	☽ ∠ ♆
23 44	☽ bQ ♄
12 Tuesday	
04 06	☽ ∠ ☉
05 38	☽ ∥ ♃
13 22	☽ ⚹ ♀
13 44	☽ ∠ ☿
15 00	☽ ⊥ ♆
15 01	☽ ⚺ ♅
16 44	☽ ☍ ♃
22 20	☽ ⚹ ♂
23 36	☽ ⊥ ☉
23 42	☽ ⚺ ♇
13 Wednesday	
02 59	☽ ⊥ ♅
02 59	☽ ⚺ ♆
05 37	☽ △ ♄
05 39	♀ □ ♅
09 30	☽ # ♃
10 23	☽ ⊥ ☿
11 42	☽ ⊥ ♇
11 53	☿ bQ ♃
12 41	☽ ⚺ ☉
20 24	♀ △ ♃
20 38	☿ Q ♇
21 03	☽ ∠ ♅
21 15	♂ □ ♇
22 48	☽ ∠ ♀
14 Thursday	
00 14	☽ ⚺ ☿
05 48	☽ ∠ ♇
06 21	☽ ∠ ♂
17 17	☉ Q ♅
17 34	☽ # ♅
19 03	☽ ⊥ ♀
15 Friday	
01 43	☽ ⊥ ♂
03 22	☽ ⚹ ♅
04 36	☽ ⚻ ♃
08 36	☽ ⚺ ♀
12 08	☽ ⚹ ♇
14 40	☽ ⚺ ♂
15 35	☽ ☌ ♆
16 41	☽ ± ♃
18 15	☽ □ ♄
16 Saturday	
03 43	☽ Q ♅
06 07	♂ ⚺ ♆
06 50	☽ ☌ ☉
07 17	☽ ∥ ♆
10 50	☽ bQ ♃
12 30	☽ Q ♇
18 38	♀ □ ♇
21 23	☽ ∥ ♄
22 06	☽ ☌ ☿
17 Sunday	
06 54	☽ Q ♄
07 15	☉ # ♇
13 04	☽ # ♇
13 39	☽ ∥ ☉
16 19	☽ □ ♅
17 06	☽ △ ♃
18 Monday	
01 03	☽ □ ♇
03 13	☉ bQ ♃
03 36	♂ ⚹ ♄
04 35	☽ ☌ ♀
04 37	☽ ⚺ ♆
04 49	♀ ⚺ ♆
07 19	☽ ⚹ ♄
07 33	☽ ☌ ♂
16 47	☽ ⊥ ♆
19 Tuesday	
01 17	☽ ⚺ ☉
02 08	☉ Q ♇
07 30	♀ ⚹ ♄
07 43	☽ ∥ ☿
10 59	☽ ∠ ♆
13 42	☽ ∠ ♄
14 28	☽ ⊥ ☉
19 01	☿ Q ♄
19 51	☽ ⚺ ☿
20 Wednesday	
04 56	☽ △ ♅
05 17	☽ □ ♃
07 17	♀ ∥ ♂
07 47	☽ ⊥ ♄
09 36	☽ ⊥ ☿
10 07	☽ ∠ ☉
13 27	☽ △ ♇
17 03	☽ ⚹ ♆
18 35	☽ ∥ ☿
19 45	☽ ⚺ ♄
22 09	♀ ☌ ♂
23 40	☽ ⚺ ♂
23 45	☽ ⚺ ♀
21 Thursday	
06 01	☽ ∠ ☿
10 40	☽ bQ ♅
10 52	♃ ± ♆
12 14	☽ ⊥ ♂
12 53	☽ ⊥ ♀
16 42	☽ Q ♃
16 44	☽ Q ♆
18 22	☽ ⚹ ☉
19 01	☽ bQ ♇
22 Friday	
04 13	☽ ± ♅
06 51	☽ ∠ ♂
08 18	☽ ∠ ♀
12 26	☽ ± ♇
15 15	☽ ⚹ ☿
15 46	☽ ⚻ ♅
15 47	☽ ⚹ ♃
16 20	☽ ∥ ☉
18 44	♃ ⚻ ♅
19 31	☿ △ ♃
19 31	☽ Q ☉
19 33	☿ □ ♅
23 52	☽ ⚻ ♇
23 Saturday	
03 23	☽ # ♇
03 24	☽ □ ♆
03 55	♀ ⊥ ♆
06 03	☽ ☌ ♄
13 06	☽ ⚹ ♂
15 48	☽ ⚹ ♀
16 58	☽ Q ☿
18 39	☽ ∥ ♄
19 55	☽ ∠ ♃
24 Sunday	
06 11	☽ ∥ ♆
07 56	☽ □ ☉
12 21	☽ ⊥ ♃
12 31	☽ Q ♂
12 43	☿ ∥ ♂
16 06	☽ Q ♀
17 46	♂ ⊥ ♆
23 07	☽ ⚺ ♃
23 21	☽ ☍ ♅
25 Monday	
05 53	☽ □ ☿
06 52	☽ ☍ ♇
10 16	☽ △ ♆
12 48	☽ ⚺ ♄
14 50	☿ □ ♇
16 19	☽ # ♅
22 13	☽ □ ♂
23 16	☽ ⊥ ♄
26 Tuesday	
00 17	☿ ∥ ♀
02 56	☽ □ ♀
12 13	☽ bQ ♆
14 42	☽ ∠ ♄
16 41	☽ △ ☉
20 03	☽ # ♃
21 54	☿ ⚺ ♆
27 Wednesday	
02 39	☽ ☌ ♃
03 03	☽ ⚻ ♅
03 21	☽ ± ♆
10 02	☽ ⚻ ♇
12 58	☽ ± ♅
13 17	☽ ⚻ ♆
15 01	☽ △ ☿
15 44	☽ ⚹ ♄
17 06	☽ ∥ ♃
19 23	☽ bQ ☉
19 52	☽ ± ♇
22 09	☿ ⚹ ♄
28 Thursday	
03 01	☽ △ ♂
03 38	☽ bQ ♅
09 19	☽ △ ♀
10 26	☽ bQ ♇
11 00	☽ ± ☉
11 14	☽ Q ♄
14 53	☉ Q ♄
17 59	☽ bQ ☿
19 12	☽ ∥ ♅
21 16	☽ ⚻ ☉
29 Friday	
03 07	☽ ⚺ ♃
03 39	☽ △ ♅
04 16	☽ bQ ♂
09 51	☽ ± ☿
10 18	☽ △ ♇
11 19	☽ bQ ♀
12 36	☽ ⊥ ♃
13 30	☽ ☍ ♆
15 56	☽ □ ♄
19 12	☽ ± ♂
20 21	☽ ⚻ ☿
30 Saturday	
01 45	☽ # ♆
02 44	☽ ∠ ♃
02 45	☽ ± ♀
05 07	☽ ⚻ ♂
10 32	☽ # ♄
13 00	☽ ⚻ ♀
14 36	♀ ∠ ♆
23 27	☿ ⊥ ♆
23 54	☽ ☍ ☉

All ephemeris data is given at 12.00 UT and the Moon's longitude is additionally given for 24.00 UT

DECEMBER 1963

LONGITUDES

Date	Sidereal time h m s	Sun ☉	Moon ☽	Moon ☽ 24.00	Mercury ☿	Venus ♀	Mars ♂	Jupiter ♃	Saturn ♄	Uranus ♅	Neptune ♆	Pluto ♇
01	16 38 31	08 ♐ 34 00	15 ♊ 43 41	23 ♊ 17 20	23 ♐ 00	02 ♑ 19	27 ♐ 04	09 ♈ 34	17 ♒ 50	09 ♍ 58	16 ♏ 14	14 ♍ 10
02	16 42 27	09 34 48	00 ♋ 46 57	08 ♋ 11 34	24 30	03 34	27 49	09 R 33	17 54	09 59	16 16	14 10
03	16 46 24	10 35 38	15 ♋ 30 22	22 ♋ 42 48	25 59	04 48	28 35	09 33	17 58	10 00	16 18	14 11
04	16 50 20	11 36 29	29 ♋ 48 26	06 ♌ 47 06	27 28	06 03	29 ♐ 20	09 32	18 02	10 00	16 21	14 11
05	16 54 17	12 37 21	13 ♌ 38 46	20 ♌ 23 34	28 ♐ 57	07 17	00 ♑ 06	09 D 32	18 06	10 01	16 23	14 11
06	16 58 14	13 38 15	27 ♌ 01 43	03 ♍ 33 36	00 ♑ 25	08 32	00 51	09 32	18 11	10 02	16 25	14 12
07	17 02 10	14 39 09	09 ♍ 59 38	16 ♍ 20 17	01 52	09 47	01 37	09 33	18 15	10 02	16 27	14 12
08	17 06 07	15 40 05	22 ♍ 36 04	28 ♍ 47 32	03 18	11 01	02 22	09 33	18 20	10 03	16 29	14 12
09	17 10 03	16 41 03	04 ♎ 55 12	10 ♎ 59 38	04 43	12 16	03 08	09 34	18 24	10 03	16 31	14 13
10	17 14 00	17 42 01	17 ♎ 01 19	23 ♎ 00 46	06 07	13 30	03 53	09 35	18 29	10 03	16 33	14 13
11	17 17 56	18 43 01	28 ♎ 58 27	04 ♏ 54 48	07 30	14 45	04 39	09 36	18 34	10 04	16 35	14 13
12	17 21 53	19 44 01	10 ♏ 50 14	16 ♏ 45 07	08 51	16 00	05 25	09 37	18 38	10 04	16 37	14 13
13	17 25 49	20 45 03	22 ♏ 39 48	28 ♏ 34 34	10 10	17 14	06 11	09 39	18 43	10 04	16 39	14 13
14	17 29 46	21 46 06	04 ♐ 29 45	10 ♐ 25 35	11 27	18 29	06 57	09 41	18 48	10 04	16 41	14 14
15	17 33 43	22 47 09	16 ♐ 22 18	22 ♐ 20 09	12 42	19 43	07 43	09 43	18 53	10 04	16 42	14 14
16	17 37 39	23 48 13	28 ♐ 19 22	04 ♑ 20 08	13 53	20 58	08 29	09 45	18 58	10 R 04	16 44	14 14
17	17 41 36	24 49 18	10 ♑ 21 42	16 ♑ 27 15	15 01	22 12	09 15	09 47	19 03	10 04	16 46	14 R 14
18	17 45 32	25 50 24	22 ♑ 34 03	28 ♑ 43 21	16 05	23 27	10 01	09 50	19 09	10 04	16 48	14 14
19	17 49 29	26 51 30	04 ♒ 55 25	11 ♒ 10 30	17 05	24 41	10 47	09 53	19 14	10 04	16 50	14 14
20	17 53 25	27 52 36	17 ♒ 28 56	23 ♒ 51 00	17 59	25 55	11 33	09 56	19 19	10 04	16 52	14 14
21	17 57 22	28 53 43	00 ♓ 17 01	06 ♓ 47 17	18 47	27 10	12 19	09 59	19 25	10 03	16 54	14 13
22	18 01 18	29 ♐ 54 50	13 ♓ 22 08	20 ♓ 01 48	19 28	28 24	13 05	10 02	19 30	10 03	16 55	14 13
23	18 05 15	00 ♑ 55 57	26 ♓ 46 33	03 ♈ 36 33	20 01	29 ♑ 39	13 52	10 06	19 36	10 03	16 57	14 13
24	18 09 12	01 57 04	10 ♈ 31 56	17 ♈ 32 41	20 25	00 ♒ 53	14 38	10 10	19 41	10 02	16 59	14 13
25	18 13 08	02 58 11	24 ♈ 38 43	01 ♉ 49 49	20 40	02 07	15 24	10 14	19 47	10 02	17 01	14 13
26	18 17 05	03 59 18	09 ♉ 05 38	16 ♉ 25 41	20 R 44	03 22	16 11	10 18	19 53	10 01	17 02	14 12
27	18 21 01	05 00 25	23 ♉ 49 17	01 ♊ 15 42	20 38	04 36	16 57	10 22	19 58	10 01	17 04	14 12
28	18 24 58	06 01 33	08 ♊ 44 01	16 ♊ 13 16	20 19	05 50	17 43	10 27	20 04	10 00	17 06	14 12
29	18 28 54	07 02 40	23 ♊ 42 24	01 ♋ 10 21	19 49	07 04	18 30	10 32	20 10	09 59	17 07	14 11
30	18 32 51	08 03 48	08 ♋ 36 05	15 ♋ 58 37	19 07	08 19	19 16	10 36	20 16	09 59	17 09	14 11
31	18 36 47	09 ♑ 04 56	23 ♋ 17 03	00 ♌ 30 36	18 ♑ 14	09 ♒ 33	20 ♑ 03	10 ♈ 42	20 ♒ 22	09 ♍ 58	17 ♏ 10	14 ♍ 10

MOON NODES / LATITUDE AND DECLINATIONS

Date	Moon True ☊	Moon Mean ☊	Moon ☽ Latitude	Sun ☉	Moon ☽	Mercury ☿	Venus ♀	Mars ♂	Jupiter ♃	Saturn ♄	Uranus ♅	Neptune ♆	Pluto ♇
01	11 ♋ 30	12 ♋ 59	02 S 16	21 S 44	20 N 25	25 S 26	24 S 39	24 S 17	02 N 27	16 S 38	08 N 33	15 S 03	18 N 52
02	11 R 28	12 55	00 S 58	21 53	22 28	25 33	24 40	24 18	02 27	16 37	08 32	15 04	18 52
03	11 D 27	12 52	00 N 22	22 02	22 55	25 39	24 39	24 19	02 27	16 35	08 32	15 04	18 52
04	11 28	12 49	01 39	22 11	21 48	25 43	24 38	24 20	02 27	16 34	08 32	15 05	18 53
05	11 30	12 46	02 47	22 19	19 24	25 46	24 36	24 21	02 27	16 33	08 32	15 05	18 53
06	11 31	12 43	03 44	22 26	16 01	25 47	24 33	24 21	02 28	16 31	08 31	15 06	18 53
07	11 32	12 39	04 28	22 34	11 57	25 46	24 30	24 21	02 28	16 30	08 31	15 07	18 53
08	11 R 32	12 36	04 57	22 40	07 29	25 45	24 26	24 20	02 29	16 29	08 31	15 07	18 54
09	11 31	12 33	05 12	22 47	02 N 49	25 42	24 21	24 20	02 29	16 27	08 31	15 08	18 54
10	11 28	12 30	05 13	22 52	01 S 52	25 37	24 15	24 19	02 30	16 26	08 31	15 08	18 54
11	11 25	12 27	05 00	22 58	06 26	25 31	24 08	24 18	02 31	16 24	08 31	15 09	18 55
12	11 21	12 24	04 34	23 03	10 44	25 24	24 01	24 16	02 31	16 23	08 31	15 09	18 55
13	11 17	12 20	03 57	23 07	14 38	25 15	23 53	24 15	02 32	16 21	08 31	15 10	18 56
14	11 14	12 17	03 09	23 11	17 57	25 05	23 45	24 13	02 33	16 20	08 31	15 10	18 56
15	11 12	12 14	02 13	23 15	20 33	24 53	23 35	24 10	02 34	16 18	08 31	15 11	18 56
16	11 10	12 11	01 10	23 18	22 16	24 41	23 25	24 08	02 35	16 16	08 31	15 11	18 57
17	11 09	12 08	00 N 04	23 21	22 58	24 27	23 14	24 05	02 37	16 15	08 31	15 12	18 57
18	11 D 10	12 05	01 S 03	23 23	22 35	24 12	23 03	24 02	02 38	16 13	08 31	15 12	18 58
19	11 11	12 01	02 08	23 24	21 06	23 56	22 51	23 59	02 39	16 11	08 31	15 13	18 58
20	11 12	11 58	03 08	23 26	18 35	23 40	22 38	23 55	02 41	16 10	08 31	15 13	18 59
21	11 13	11 55	04 00	23 26	15 07	23 22	22 24	23 51	02 42	16 08	08 31	15 14	18 59
22	11 14	11 52	04 41	23 27	10 51	23 05	22 10	23 47	02 44	16 06	08 32	15 14	19 00
23	11 15	11 49	05 07	23 26	05 59	22 47	21 55	23 43	02 46	16 04	08 32	15 15	19 00
24	11 R 14	11 45	05 17	23 26	00 S 42	22 29	21 40	23 38	02 47	16 03	08 32	15 15	19 01
25	11 14	11 42	05 09	23 25	04 N 45	22 12	21 24	23 33	02 49	16 01	08 32	15 16	19 01
26	11 13	11 39	04 41	23 23	10 05	21 54	21 07	23 28	02 51	15 59	08 32	15 16	19 02
27	11 12	11 36	03 54	23 21	14 57	21 38	20 50	23 22	02 53	15 57	08 33	15 16	19 02
28	11 12	11 33	02 51	23 18	18 57	21 22	20 32	23 17	02 55	15 55	08 33	15 17	19 03
29	11 11	11 30	01 36	23 15	21 42	21 08	20 14	23 10	02 57	15 53	08 33	15 17	19 04
30	11 11	11 26	00 14	23 12	22 56	20 54	19 55	23 04	02 59	15 52	08 34	15 18	19 04
31	11 ♋ 11	11 ♋ 23	01 N 07	23 S 08	22 N 32	20 S 42	19 S 35	22 S 58	03 N 02	15 S 50	08 N 34	15 S 18	19 N 05

ZODIAC SIGN ENTRIES

Date	h	m	Planets
02	10	44	☽ ♋
04	12	20	☽ ♌
05	09	03	♂ ♑
06	05	17	☿ ♑
06	17	26	☽ ♍
09	02	21	☽ ♎
11	14	04	☽ ♏
14	02	53	☽ ♐
16	15	21	☽ ♑
19	02	29	☽ ♒
21	11	28	☽ ♓
22	14	02	☉ ♑
23	17	41	☽ ♈
23	18	53	♀ ♒
25	20	57	☽ ♉
27	21	58	☽ ♊
29	22	07	☽ ♋
31	23	09	☽ ♌

LATITUDES

Date	Mercury ☿	Venus ♀	Mars ♂	Jupiter ♃	Saturn ♄	Uranus ♅	Neptune ♆	Pluto ♇
01	02 S 11	01 S 14	00 S 53	01 S 28	01 S 12	00 N 46	01 N 43	13 N 44
04	02 18	01 20	00 54	01 27	01 12	00 46	01 43	13 45
07	02 21	01 25	00 55	01 26	01 12	00 46	01 43	13 47
10	02 19	01 30	00 56	01 25	01 12	00 46	01 43	13 48
13	02 12	01 34	00 57	01 24	01 12	00 47	01 43	13 50
16	01 58	01 38	00 58	01 23	01 12	00 47	01 43	13 51
19	01 36	01 41	00 59	01 22	01 11	00 47	01 44	13 53
22	01 04	01 43	00 59	01 21	01 11	00 47	01 44	13 54
25	00 S 21	01 45	01 00	01 20	01 11	00 47	01 44	13 56
28	00 N 33	01 46	01 01	01 20	01 11	00 47	01 44	13 57
31	01 N 31	01 S 47	01 S 02	01 S 19	01 S 11	00 N 47	01 N 44	13 N 59

DATA

Julian Date	2438365
Delta T	+35 seconds
Ayanamsa	23° 20' 53"
Synetic vernal point	05° ♓ 46' 07"
True obliquity of ecliptic	23° 26' 36"

MOON'S PHASES, APSIDES AND POSITIONS ☽

Date	h	m	Phase	Longitude	Eclipse Indicator
07	21	34	☾	15 ♍ 03	
16	02	06	●	23 ♐ 23	
23	19	54	☽	01 ♈ 16	
30	11	04	○	08 ♋ 01	total

Day	h	m	
13	09	31	Apogee
29	00	15	Perigee

Day	h	m		
03	06	30	Max dec	22° N 57'
10	02	24	0S	
17	15	45	Max dec	22° S 59'
24	15	05	0N	
30	18	04	Max dec	22° N 59'

LONGITUDES

Date	Chiron ⚷	Ceres ⚳	Pallas ⚴	Juno ⚵	Vesta ⚶	Black Moon Lilith ⚸
01	10 ♓ 17	23 ♏ 12	04 ♏ 19	11 ♏ 22	13 ♐ 18	25 ♏ 10
11	10 ♓ 25	27 ♏ 27	08 ♏ 30	14 ♏ 38	18 ♐ 42	26 ♏ 17
21	10 ♓ 39	01 ♐ 39	12 ♏ 33	17 ♏ 46	24 ♐ 07	27 ♏ 24
31	10 ♓ 59	05 ♐ 44	16 ♏ 28	20 ♏ 46	29 ♐ 31	28 ♏ 30

ASPECTARIAN

h m	Aspects
01 Sunday	
00 11	☽ ∥ ♇
02 17	☽ ⚹ ♃
02 54	☽ □ ♅
09 31	☽ □ ♇
12 48	☽ ⚻ ♆
15 21	☽ △ ♄
21 15	☽ Q ♃
22 21	☽ ± ♆
22 34	♀ ∠ ♄
02 Monday	
00 49	☽ ☍ ☿
02 13	☽ # ☉
07 00	☽ ☍ ♂
07 30	☽ Q ♅
11 21	☉ △ ♃
12 47	☽ bQ ♆
14 14	☽ Q ♇
15 26	☽ bQ ♄
16 53	☽ ☍ ♀
21 38	☉ □ ♅
03 Tuesday	
02 13	☽ □ ♃
02 56	☽ ⚹ ♅
03 19	☽ ⚻ ☉
06 09	☽ ± ♄
09 48	☽ ⚹ ♇
13 20	☽ △ ♆
13 56	☽ ± ☉
16 06	☽ ⚻ ♄
04 Wednesday	
03 51	☽ ∠ ♅
06 09	☽ bQ ☉
07 02	☽ # ☉
07 34	☽ ⚻ ☿
10 56	☽ ∠ ♇
11 09	☽ ⚻ ♂
19 01	☽ ± ☿
19 12	☽ ⊥ ♅
22 02	☽ ± ♂
23 47	☽ ⚻ ♀
05 Thursday	
02 26	☽ ⊥ ♇
04 47	☽ △ ♃
05 37	☽ ⚺ ♅
10 03	☽ △ ☉
10 13	♃ St D
11 19	☽ ± ♀
12 35	☽ bQ ☿
12 58	☽ ⚺ ♇
14 42	☽ bQ ♂
16 08	☽ ∥ ♇
16 51	☽ □ ♆
19 57	☽ ☍ ♄
06 Friday	
05 00	☽ bQ ♀
07 28	☽ bQ ♃
08 43	☽ # ♄
17 42	☽ # ♆
18 58	☽ △ ☿
19 26	☽ △ ♂
23 58	☽ ± ♃
07 Saturday	
01 19	☉ □ ♇
01 37	☽ Q ♆
03 17	☿ ☌ ♂
04 57	☿ ∠ ♆
06 33	♂ ∠ ♆
07 29	♀ □ ♃
11 09	☽ ⚻ ♃
11 33	☽ △ ♀
12 05	☽ ☌ ♅
17 00	♀ △ ♅
19 57	☽ ☌ ♇
21 34	☽ □ ☉
08 Sunday	
00 14	☽ ⚹ ♆
03 45	☽ ⚻ ♄
06 35	☽ ∥ ♅
12 33	☿ ∠ ♄
15 21	☽ ± ♄
09 Monday	
05 18	☽ ∠ ♆
07 48	☉ ⚺ ♆
08 15	☽ □ ♂
09 00	☽ bQ ♄
11 30	☽ Q ☉
11 33	☽ □ ☿
13 43	☽ ∥ ♃
14 52	♀ ⊥ ♄
15 58	♀ ∥ ♂
21 11	☽ ☍ ♃
21 36	♂ ∠ ♄
22 08	☽ ⚺ ♅
23 05	☽ ⊥ ♆
10 Tuesday	
04 11	☽ □ ♀
06 24	☽ ⚺ ♇
10 04	☽ ⊥ ♅
11 03	☽ ⚺ ♆
13 29	☽ ⚹ ☉
14 56	☽ △ ♄
15 15	☽ # ♃
18 23	☽ ⊥ ♇
22 25	☽ Q ♂
11 Wednesday	
01 44	♀ △ ♇
04 05	☽ Q ☿
04 07	☽ ∠ ♅
07 57	☉ ⚹ ♄
12 30	☽ ∠ ♇
20 31	☽ Q ♀
22 29	☽ ∠ ☉
23 22	☽ # ♅
12 Thursday	
00 16	☽ ⚹ ♂
07 27	☽ ⚹ ☿
09 32	☽ ⚻ ♃
10 26	☽ ⚹ ♅
18 26	☽ ⊥ ☉
18 52	☽ ⚹ ♇
21 44	☽ ± ♃
23 41	☽ ⚹ ♀
23 45	☽ ☌ ♆
13 Friday	
00 17	♀ ⚹ ♆
02 18	☿ □ ♃
03 56	☽ □ ♄
07 45	☽ ⚺ ☉
08 47	☽ ∠ ♂
10 07	☿ △ ♅
10 47	☽ Q ♅
15 36	☽ ∥ ♆
16 03	☽ bQ ♃
17 43	☽ ∠ ☿
19 14	☽ Q ♇
23 47	☽ ∥ ♄
14 Saturday	
04 18	☽ ⊥ ♂
09 42	☽ ∠ ♀
14 10	☽ ⊥ ☿
16 42	☽ Q ♄
17 18	☽ ⚺ ♂
18 43	♀ ⚺ ♄
20 18	☽ # ♇
22 31	☽ △ ♃
23 17	☽ □ ♅
15 Sunday	
03 44	☽ ⚺ ☿
06 02	☽ ⊥ ♀
07 41	☽ □ ♇
10 07	☉ ⊥ ♆
12 41	☽ ⚺ ♆
16 02	☿ ⊥ ♄
17 06	☽ ⚹ ♄
19 31	☽ ⚺ ♀
16 Monday	
00 47	☽ ⊥ ♆
02 06	☽ ☌ ☉
05 11	♅ St R
18 51	☽ ∠ ♆
19 06	☿ △ ♇
23 21	☽ ∠ ♄
17 Tuesday	
01 06	☉ ∥ ♀
07 08	♇ St R
09 36	☽ ☌ ♂
10 50	☽ □ ♃
11 23	☽ △ ♅
17 20	☽ ⊥ ♄
19 37	☽ △ ♇
22 04	☽ ☌ ☿
18 Wednesday	
00 39	☽ ⚹ ♆
05 14	☽ ⚺ ♄
06 08	♂ □ ♃
13 49	♂ △ ♅
13 54	☽ ☌ ♀
16 53	☽ bQ ♅
18 58	☽ ⚺ ☉
22 19	☽ Q ♃
19 Thursday	
00 11	☽ Q ♆
00 59	☽ bQ ♇
05 42	☿ ⚹ ♆
07 02	☿ ∥ ♂
07 39	☽ ⊥ ☉
10 21	☽ ± ♅
18 21	☽ ± ♇
19 22	♀ bQ ♅
21 34	☽ ⚹ ♃
21 53	☽ ⚻ ♅
23 58	☽ ⚺ ♂
20 Friday	
02 29	☽ ∠ ☉
05 49	☽ ⚻ ♇
08 46	☽ # ♇
10 49	☽ □ ♆
12 08	☽ ⊥ ♂
13 00	☽ ⚺ ☿
15 30	☽ ☌ ♄
21 Saturday	
01 05	☽ ⊥ ☿
02 05	☽ ∠ ♃
05 29	☽ ∥ ♄
05 35	☽ ⚺ ♀
06 08	☽ ∠ ♂
06 46	☉ ∥ ☿
09 12	☽ ⚹ ☉
11 17	☽ ∥ ♆
17 54	☽ ⊥ ♀
18 50	☽ ∠ ☿
18 52	☽ ⊥ ♃
22 Sunday	
04 35	♀ Q ♃
05 55	☽ ⚺ ♃
05 58	☽ ☍ ♅
09 08	☽ Q ☉
11 28	☽ ⚹ ♂
12 04	☽ ∠ ♀
13 33	☽ ☍ ♇
13 55	☿ ⚺ ♄
16 37	♃ ⚻ ♅
18 26	☽ △ ♆
22 16	♀ Q ♆
23 08	☽ ⚺ ♄
23 29	☽ ⚹ ☿
23 47	☽ # ♅
23 Monday	
02 36	♂ ⊥ ♄
03 46	♀ bQ ♇
09 54	☽ ⊥ ♄
10 17	☽ Q ♂
17 34	☽ ⚹ ♀
19 54	☽ □ ☉
21 07	☽ bQ ♆
21 31	☽ Q ☿
23 08	♂ △ ♇
24 Tuesday	
01 48	☽ ∠ ♄
02 42	☽ # ♃
11 09	☽ ⚻ ♅
11 22	☽ ☌ ♃
12 43	☉ ∠ ♆
12 46	☽ ± ♆
16 26	☽ Q ♀
18 19	☽ ⚻ ♇
19 26	☽ □ ♂
21 26	☽ ± ♅
23 04	☽ ⚻ ♆
25 Wednesday	
03 26	☽ ∥ ♃
03 44	☽ ⚹ ♄
04 31	☽ ± ♇
05 12	☽ □ ☿
12 39	☽ bQ ♅
19 38	☽ bQ ♇
26 Thursday	
00 00	☽ Q ♄
01 39	☽ □ ♀
02 56	☽ △ ☉
04 52	☽ ∥ ♅
09 39	☿ St R
13 31	☽ △ ♅
13 59	☽ ⚺ ♃
20 22	☽ △ ♇
23 51	☽ ⊥ ♃
27 Friday	
00 14	☽ △ ♂
00 42	♀ ± ♅
01 01	☽ ☍ ♆
05 22	☽ bQ ☉
05 43	☽ □ ♄
06 53	☽ △ ☿
11 05	☉ ∠ ♄
13 44	☽ # ♆
14 31	☽ ∠ ♃
15 43	♂ ⚹ ♆
17 26	☽ # ♄
20 59	☽ ± ☉
21 57	☉ ∥ ♂
28 Saturday	
01 49	☽ bQ ♂
06 40	☽ bQ ☿
06 56	☽ △ ♀
07 20	☽ ⚻ ☉
12 42	☽ ∥ ♇
14 02	☽ □ ♅
14 46	☽ ⚹ ♃
17 03	☽ ± ♂
20 42	☽ ± ☿
20 45	☽ □ ♇
22 57	☿ ⚺ ♄
23 04	☽ # ♀
29 Sunday	
01 25	☽ ⚻ ♆
03 12	☽ ⚻ ♂
05 59	☽ ⚻ ☿
06 17	☽ △ ♄
06 20	☽ # ☿
08 57	☉ ⚺ ♀
09 09	☽ bQ ♀
10 06	☽ Q ♃
11 03	☽ ± ♆
18 52	☽ Q ♅
30 Monday	
00 55	☽ ± ♀
01 33	☽ bQ ♆
01 38	☽ Q ♇
06 34	☽ bQ ♄
09 33	♀ ± ♇
09 34	☿ ☌ ♂
11 04	☽ ☍ ☉
11 29	☽ ⚻ ♀
14 14	☽ ⚹ ♅
15 16	☽ □ ♃
21 04	☽ ⚹ ♇
21 16	☽ ± ♄
31 Tuesday	
01 56	☽ △ ♆
04 12	☽ ☍ ☿
06 22	☽ ☍ ♂
07 10	☽ ⚻ ♄
14 46	☽ ∠ ♅
20 01	♀ ⚻ ♅
21 46	☽ ∠ ♇
23 16	♂ ⚺ ♄

All ephemeris data is given at 12.00 UT and the Moon's longitude is additionally given for 24.00 UT

JANUARY 1964

LONGITUDES

Date	Sidereal time h m s	Sun ☉ ° ' "	Moon ☽ ° ' "	Moon ☽ 24.00 ° ' "	Mercury ☿ ° '	Venus ♀ ° '	Mars ♂ ° '	Jupiter ♃ ° '	Saturn ♄ ° '	Uranus ♅ ° '	Neptune ♆ ° '	Pluto ♇ ° '
01	18 40 44	10 ♑ 06 04	07 ♌ 38 39	14 ♌ 40 41	17 ♑ 11	10 ♒ 47	20 ♑ 50	10 ♈ 47	20 ♒ 28	09 ♍ 57	17 ♏ 12	14 ♍ 10
02	18 44 41	11 07 12	21 ♌ 36 23	28 ♌ 25 36	16 R 00	12 01	21 36	10 52	20 34	09 R 56	17 13	14 R 09
03	18 48 37	12 08 21	05 ♍ 08 17	11 ♍ 44 34	14 43	13 15	22 23	10 58	20 40	09 55	17 15	14 09
04	18 52 34	13 09 30	18 ♍ 14 39	24 ♍ 38 52	13 23	14 29	23 10	11 04	20 47	09 54	17 16	14 08
05	18 56 30	14 10 38	00 ♎ 57 38	07 ♎ 11 26	12 02	15 43	23 56	11 10	20 53	09 53	17 18	14 08
06	19 00 27	15 11 47	13 ♎ 20 47	19 ♎ 26 14	10 43	16 57	24 43	11 16	20 59	09 52	17 19	14 07
07	19 04 23	16 12 57	25 ♎ 28 23	01 ♏ 27 49	09 27	18 11	25 30	11 22	21 05	09 51	17 21	14 07
08	19 08 20	17 14 06	07 ♏ 25 07	13 ♏ 20 53	08 18	19 25	26 17	11 29	21 12	09 49	17 22	14 06
09	19 12 16	18 15 15	19 ♏ 15 40	25 ♏ 10 01	07 17	20 39	27 04	11 35	21 18	09 48	17 23	14 05
10	19 16 13	19 16 25	01 ♐ 04 28	06 ♐ 59 28	06 25	21 52	27 50	11 42	21 25	09 47	17 24	14 04
11	19 20 10	20 17 34	12 ♐ 55 29	18 ♐ 52 55	05 43	23 06	28 37	11 49	21 31	09 45	17 26	14 04
12	19 24 06	21 18 44	24 ♐ 52 09	00 ♑ 53 31	05 10	24 20	29 ♑ 24	11 56	21 38	09 44	17 27	14 03
13	19 28 03	22 19 53	06 ♑ 57 16	13 ♑ 03 40	04 47	25 34	00 ♒ 11	12 04	21 44	09 42	17 28	14 02
14	19 31 59	23 21 02	19 ♑ 12 55	25 ♑ 25 10	04 34	26 47	00 58	12 11	21 51	09 41	17 29	14 01
15	19 35 56	24 22 10	01 ♒ 40 34	07 ♒ 59 11	04 D 30	28 01	01 45	12 19	21 58	09 39	17 31	14 00
16	19 39 52	25 23 18	14 ♒ 21 06	20 ♒ 46 21	04 34	29 ♒ 14	02 32	12 27	22 05	09 38	17 32	13 59
17	19 43 49	26 24 26	27 ♒ 14 58	03 ♓ 46 55	04 46	00 ♓ 28	03 20	12 35	22 11	09 36	17 33	13 59
18	19 47 45	27 25 32	10 ♓ 22 15	17 ♓ 00 54	05 06	01 41	04 07	12 43	22 18	09 34	17 34	13 58
19	19 51 42	28 26 38	23 ♓ 42 53	00 ♈ 28 10	05 32	02 55	04 54	12 51	22 25	09 33	17 35	13 57
20	19 55 39	29 ♑ 27 43	07 ♈ 16 43	14 ♈ 08 29	06 03	04 08	05 41	12 59	22 32	09 31	17 36	13 56
21	19 59 35	00 ♒ 28 47	21 ♈ 03 26	28 ♈ 01 28	06 41	05 21	06 28	13 08	22 39	09 29	17 37	13 55
22	20 03 32	01 29 51	05 ♉ 02 30	12 ♉ 06 24	07 23	06 35	07 15	13 17	22 45	09 27	17 38	13 53
23	20 07 28	02 30 53	19 ♉ 12 58	26 ♉ 21 57	08 10	07 48	08 02	13 25	22 52	09 25	17 39	13 52
24	20 11 25	03 31 54	03 ♊ 33 05	10 ♊ 45 58	09 00	09 01	08 50	13 34	22 59	09 23	17 39	13 51
25	20 15 21	04 32 54	18 ♊ 00 10	25 ♊ 15 10	09 55	10 14	09 37	13 43	23 06	09 21	17 40	13 50
26	20 19 18	05 33 53	02 ♋ 30 23	09 ♋ 45 12	10 52	11 27	10 24	13 53	23 13	09 19	17 41	13 49
27	20 23 14	06 34 51	16 ♋ 58 56	24 ♋ 10 53	11 53	12 40	11 11	14 02	23 20	09 17	17 42	13 48
28	20 27 11	07 35 49	01 ♌ 20 21	08 ♌ 26 41	12 56	13 53	11 59	14 11	23 27	09 15	17 43	13 47
29	20 31 08	08 36 45	15 ♌ 29 14	22 ♌ 27 26	14 02	15 05	12 46	14 21	23 35	09 13	17 43	13 46
30	20 35 04	09 37 40	29 ♌ 20 50	06 ♍ 09 03	15 10	16 18	13 33	14 31	23 42	09 11	17 44	13 44
31	20 39 01	10 ♒ 38 34	12 ♍ 51 50	19 ♍ 29 03	16 ♑ 20	17 ♓ 31	14 ♒ 21	14 ♈ 41	23 ♒ 49	09 ♍ 09	17 ♏ 45	13 ♍ 43

DECLINATIONS

Date	Moon True ☊ ° '	Moon Mean ☊ ° '	Moon ☽ Latitude ° '	Sun ☉ ° '	Moon ☽ ° '	Mercury ☿ ° '	Venus ♀ ° '	Mars ♂ ° '	Jupiter ♃ ° '	Saturn ♄ ° '	Uranus ♅ ° '	Neptune ♆ ° '	Pluto ♇ ° '
01	11 ♋ 11	11 ♋ 20	02 N 22	23 S 04	20 N 39	20 S 31	19 S 15	22 S 51	03 N 04	15 S 48	08 N 34	15 S 18	19 N 05
02	11 D 12	11 17	03 26	22 59	17 33	20 22	18 54	22 44	03 06	15 46	08 35	15 19	19 06
03	11 R 11	11 14	04 17	22 53	13 37	20 14	18 33	22 36	03 09	15 44	08 35	15 19	19 07
04	11 11	11 10	04 53	22 48	09 08	20 07	18 11	22 29	03 11	15 42	08 36	15 20	19 07
05	11 11	11 07	05 13	22 41	04 N 24	20 02	17 49	22 21	03 14	15 40	08 36	15 20	19 08
06	11 11	11 04	05 17	22 35	00 S 24	19 58	17 26	22 13	03 17	15 38	08 36	15 20	19 09
07	11 D 11	11 01	05 08	22 28	05 04	19 56	17 03	22 04	03 19	15 36	08 37	15 21	19 09
08	11 11	10 58	04 45	22 20	09 29	19 55	16 39	21 56	03 22	15 34	08 37	15 21	19 10
09	11 12	10 55	04 10	22 12	13 32	19 56	16 15	21 47	03 25	15 32	08 38	15 21	19 11
10	11 13	10 51	03 25	22 03	17 02	19 58	15 51	21 38	03 28	15 29	08 38	15 22	19 12
11	11 14	10 48	02 31	21 55	19 51	20 01	15 26	21 29	03 31	15 27	08 39	15 22	19 12
12	11 15	10 45	01 30	21 45	21 51	20 06	15 00	21 19	03 34	15 25	08 40	15 22	19 13
13	11 15	10 42	00 N 24	21 36	22 52	20 11	14 35	21 09	03 37	15 23	08 40	15 23	19 14
14	11 R 15	10 39	00 S 44	21 25	22 48	20 18	14 09	20 59	03 40	15 21	08 41	15 23	19 14
15	11 14	10 36	01 51	21 15	21 35	20 25	13 42	20 49	03 43	15 19	08 41	15 23	19 15
16	11 13	10 32	02 53	21 04	19 17	20 33	13 15	20 38	03 47	15 17	08 42	15 23	19 15
17	11 11	10 29	03 48	20 53	16 00	20 41	12 48	20 27	03 50	15 14	08 43	15 24	19 16
18	11 08	10 26	04 32	20 41	11 52	20 49	12 21	20 16	03 53	15 12	08 43	15 24	19 17
19	11 06	10 23	05 01	20 29	07 06	20 58	11 53	20 05	03 57	15 10	08 44	15 24	19 18
20	11 04	10 20	05 15	20 16	01 S 56	21 06	11 25	19 54	04 00	15 08	08 45	15 24	19 18
21	11 03	10 16	05 11	20 03	03 N 25	21 15	10 56	19 42	04 04	15 06	08 45	15 24	19 19
22	11 D 03	10 13	04 49	19 50	08 40	21 22	10 27	19 30	04 07	15 03	08 46	15 25	19 20
23	11 03	10 10	04 08	19 36	13 33	21 30	09 58	19 18	04 11	15 01	08 47	15 25	19 21
24	11 05	10 07	03 13	19 22	17 43	21 37	09 29	19 06	04 15	14 59	08 48	15 25	19 21
25	11 06	10 04	02 04	19 08	20 51	21 43	09 00	18 53	04 19	14 57	08 48	15 25	19 22
26	11 07	10 01	00 S 47	18 53	22 38	21 49	08 30	18 41	04 22	14 54	08 49	15 25	19 23
27	11 R 07	09 57	00 N 32	18 38	22 54	21 54	08 00	18 28	04 26	14 52	08 50	15 25	19 24
28	11 06	09 54	01 49	18 22	21 38	21 58	07 30	18 15	04 30	14 50	08 51	15 26	19 24
29	11 03	09 51	02 58	18 07	19 01	22 01	07 00	18 01	04 34	14 47	08 52	15 26	19 25
30	10 59	09 48	03 54	17 51	15 21	22 03	06 30	17 48	04 38	14 45	08 53	15 26	19 26
31	10 ♋ 55	09 ♋ 45	04 N 36	17 S 34	10 N 59	22 S 04	05 S 59	17 S 34	04 N 42	14 S 43	08 N 53	15 S 26	19 N 27

ZODIAC SIGN ENTRIES

Date	h	m	Planets
03	02	48	☽ ♍
05	10	10	☽ ♎
07	21	04	☽ ♏
10	09	49	☽ ♐
12	22	14	☽ ♑
13	06	13	♂ ♒
15	08	48	☽ ♒
17	02	54	♀ ♓
17	17	04	☽ ♓
19	23	10	☽ ♈
21	00	41	☉ ♒
22	03	23	☽ ♉
24	06	05	☽ ♊
26	07	51	☽ ♋
28	09	45	☽ ♌
30	13	09	☽ ♍

LATITUDES

Date	Mercury ☿ ° '	Venus ♀ ° '	Mars ♂ ° '	Jupiter ♃ ° '	Saturn ♄ ° '	Uranus ♅ ° '	Neptune ♆ ° '	Pluto ♇ ° '
01	01 N 50	01 S 47	01 S 02	01 S 18	01 S 11	00 N 47	01 N 44	13 N 59
04	02 40	01 46	01 02	01 18	01 11	00 47	01 44	14 00
07	03 11	01 45	01 03	01 17	01 11	00 48	01 44	14 02
10	03 20	01 43	01 03	01 16	01 11	00 48	01 45	14 03
13	03 10	01 40	01 04	01 15	01 11	00 48	01 45	14 04
16	02 49	01 37	01 04	01 14	01 11	00 48	01 45	14 06
19	02 22	01 33	01 04	01 14	01 11	00 48	01 45	14 07
22	01 52	01 28	01 05	01 13	01 11	00 48	01 45	14 08
25	01 21	01 22	01 05	01 12	01 11	00 48	01 45	14 09
28	00 51	01 16	01 05	01 11	01 12	00 48	01 46	14 10
31	00 N 23	01 S 08	01 S 05	01 S 11	01 S 12	00 N 48	01 N 46	14 N 11

DATA

Julian Date	2438396
Delta T	+35 seconds
Ayanamsa	23° 20' 59"
Synetic vernal point	05° ♓ 46' 01"
True obliquity of ecliptic	23° 26' 36"

MOON'S PHASES, APSIDES AND POSITIONS ☽

Date	h	m	Phase	Longitude ° '	Eclipse Indicator
06	15	58	☾	15 ♎ 22	
14	20	43	●	23 ♑ 43	Partial
22	05	29	☽	01 ♉ 13	
28	23	23	○	08 ♌ 05	

Day	h	m		
09	23	48	Apogee	
26	01	31	Perigee	
06	10	02	0S	
13	22	30	Max dec	22° S 58'
20	20	43	0N	
27	04	04	Max dec	22° N 59'

LONGITUDES

Date	Chiron ⚷ ° '	Ceres ⚳ ° '	Pallas ⚴ ° '	Juno ⚵ ° '	Vesta ⚶ ° '	Black Moon Lilith ⚸ ° '
01	11 ♓ 01	06 ♐ 09	16 ♏ 51	21 ♏ 03	00 ♑ 03	28 ♏ 37
11	11 ♓ 26	10 ♐ 07	20 ♏ 33	23 ♏ 52	05 ♑ 24	29 ♏ 44
21	11 ♓ 54	13 ♐ 57	24 ♏ 13	26 ♏ 27	10 ♑ 46	00 ♐ 51
31	12 ♓ 26	17 ♐ 38	27 ♏ 15	28 ♏ 48	16 ♑ 03	01 ♐ 57

All ephemeris data is given at 12.00 UT and the Moon's longitude is additionally given for 24.00 UT

ASPECTARIAN

h m	Aspects
01 Wednesday	
05 46	☽ ⊥ ♅
08 26	☉ △ ♅
11 47	☿ ⚹ ♆
12 02	♀ ⚹ ♃
12 53	☽ ⊥ ♇
13 16	☽ # ☿
15 54	☽ ⊻ ♅
16 30	☽ ⊼ ☉
17 22	☽ △ ♃
17 50	☽ ☍ ♀
22 42	♀ # ♇
23 07	☽ ⊻ ♇
02 Thursday	
01 03	☽ ∥ ♇
01 21	☽ # ♀
03 04	☽ ⊼ ☿
03 36	☽ ± ☉
04 22	☽ □ ♆
05 35	☉ □ ♃
10 11	☽ ☍ ♄
12 00	☽ ⊼ ♂
12 38	☽ ± ☿
19 32	☽ ⚼ ♃
20 34	☽ ⚼ ☉
23 11	☽ ± ♂
23 33	☽ # ♄
03 Friday	
02 08	☽ # ♆
03 10	☽ ⚼ ☿
11 41	☽ ± ♃
12 12	☽ Q ♆
12 52	☿ ⊥ ♄
16 18	☽ ⚼ ♂
20 39	☽ ☌ ♅
22 26	☿ △ ♇
22 39	☽ ⊼ ♃
04 Saturday	
01 48	☿ ⊻ ♀
01 48	☽ △ ☉
03 51	☽ △ ☿
04 18	☽ ⊼ ♀
04 25	☽ ☌ ♇
05 24	♀ ⊼ ♇
10 11	☽ ⚹ ♆
14 17	☉ ☌ ☿
14 48	☽ ∥ ♅
16 37	☽ ± ♀
16 46	☽ ⊼ ♄
21 48	☽ △ ♂
05 Sunday	
04 10	☽ ± ♄
10 54	☉ △ ♇
11 29	☽ ⚼ ♀
14 34	☽ ∠ ♆
17 47	☽ ∥ ♃
21 33	☽ ⚼ ♄
06 Monday	
02 32	☿ □ ♃
05 12	☽ ⊻ ♅
06 26	☉ ⊥ ♄
07 20	☽ □ ☿
07 54	☽ ☍ ♃
08 02	☽ ⊥ ♆
09 44	☿ ⊥ ♀
13 31	☽ ⊻ ♇
15 58	☽ □ ☉
16 20	♂ ⚼ ♅
16 56	☽ ⊥ ♅
19 22	♀ □ ♆
19 50	☽ ⊻ ♆
19 53	☽ △ ♀
07 Tuesday	
01 20	☽ ⊥ ♇
02 50	☽ # ♃
03 12	☽ △ ♄
03 36	♃ ± ♆
04 18	☿ △ ♅
10 45	☽ ∠ ♅
12 03	☽ □ ♂
15 36	☽ Q ☿
19 16	☽ ∠ ♇
08 Wednesday	
07 08	☽ # ♅
07 11	☽ Q ☉
13 39	☽ ⚹ ☿
15 07	☉ ⚹ ♆
16 51	☽ ⚹ ♅
20 17	☽ ⊼ ♃
09 Thursday	
01 30	☽ ⚹ ♇
02 52	☽ Q ♂
08 11	☽ ☌ ♆
08 34	☽ ± ♃
09 46	☽ ⚹ ☉
15 08	☽ □ ♀
16 11	☽ □ ♄
17 09	☽ Q ♅
17 42	☽ ∠ ☿
10 Friday	
00 01	☽ ∥ ♆
01 03	☽ ∥ ♄
01 51	☽ Q ♇
02 10	♀ ☌ ♄
03 02	☽ ⚼ ♃
04 20	☽ ∥ ♀
04 58	☽ ⚹ ♂
06 42	☿ ∠ ♀
10 46	☽ ⊥ ☿
12 16	☿ ∠ ♄
19 06	☽ ∠ ☉
22 12	☽ ⊻ ☿
11 Saturday	
05 03	☽ Q ♄
05 36	☽ □ ♅
05 44	☽ # ♇
07 54	☽ Q ♀
09 45	☽ △ ♃
10 23	♀ ∥ ♄
13 30	☽ ∠ ♂
13 40	☽ ∥ ☿
14 17	☽ □ ♇
15 01	☽ ⊥ ☉
15 45	♀ ∥ ♆
21 06	☽ ⊻ ♆
12 Sunday	
01 13	♂ ⚼ ♇
04 13	☽ ⊻ ☉
04 59	☽ ∥ ♂
05 27	☽ ⚹ ♄
08 52	☽ ⊥ ♂
09 09	☽ ⊥ ♆
10 40	☽ ∥ ☉
10 48	☽ ⚹ ♀
13 24	♂ Q ♆
20 24	☉ ⊻ ♄
21 40	☽ ⊻ ♂
13 Monday	
03 07	☽ ∠ ♆
07 25	♂ Q ♃
07 50	☽ ☌ ☿
11 34	☽ ∠ ♄
17 24	☽ △ ♅
18 54	♄ ∥ ♆
19 53	☽ ∠ ♀
22 09	☽ □ ♃
14 Tuesday	
01 53	☽ △ ♇
05 24	☽ ⊥ ♄
08 38	☽ ⚹ ♆
15 23	☽ ⊥ ♀
17 09	☽ ⊻ ♄
20 43	☽ ☌ ☉
20 44	♀ ∠ ♃
22 33	☽ ⚼ ♅
15 Wednesday	
04 13	☽ ⊻ ♀
06 54	☽ ⚼ ♇
07 51	☽ Q ♆
09 22	☽ Q ♃
11 43	☿ St D
12 10	☽ ☌ ♂
15 46	☽ ± ♅
16 51	☽ ∥ ☉
17 23	☽ ⊻ ☿
18 33	☉ ⚼ ♅
22 14	☽ ∥ ♂
16 Thursday	
00 01	☽ ± ♇
00 54	☽ ∥ ☿
03 07	☽ ⊼ ♅
04 49	☽ ⊥ ☿
08 23	☽ ⚹ ♃
11 19	☽ ⊼ ♇
12 16	☽ # ♇
17 57	☽ □ ♆
19 00	☿ ∥ ♂
21 53	☽ ∠ ☿
17 Friday	
02 33	☽ ☌ ♄
10 19	☽ ⊻ ☉
12 37	☽ ∠ ♃
15 47	☽ ∥ ♆
16 46	☽ ∥ ♄
18 32	☽ ☌ ♀
20 08	♂ ± ♅
22 17	☽ ⊥ ☉
23 52	☽ ⊻ ♂
18 Saturday	
01 52	☉ ∥ ☿
02 08	☽ ⚹ ☿
05 17	☽ ⊥ ♃
09 08	☽ ∥ ♀
10 33	☽ ☍ ♅
11 30	☽ ⊥ ♂
16 02	☽ ∠ ☉
16 17	☽ ⊻ ♃
18 29	☽ ☍ ♇
19 Sunday	
00 31	☽ Q ☿
01 00	☽ △ ♆
04 06	☽ # ♅
04 45	☽ ∠ ♂
09 40	☽ ⊻ ♄
20 26	☽ ⊥ ♄
21 06	☽ ⚹ ☉
23 34	☉ ⚼ ♇
20 Monday	
02 39	☽ # ♃
03 45	☽ ⚼ ♆
05 56	☽ ⊻ ♀
09 01	☽ ⚹ ♂
09 46	☽ □ ☿
12 27	☽ ∠ ♄
15 13	☉ Q ♆
15 55	☽ ⊼ ♅
17 30	☽ ⊥ ♀
19 34	☽ ± ♆
19 55	☽ Q ☉
22 06	☽ ☌ ♃
23 37	☽ ⊼ ♇
21 Tuesday	
02 21	☽ ± ♅
06 02	☽ ⊼ ♆
07 15	☽ Q ♂
10 01	☽ ± ♇
10 40	☽ ∠ ♀
14 46	☽ ⚹ ♄
14 58	☽ ∥ ♃
17 54	☽ ⚼ ♅
22 Wednesday	
01 30	☽ ⚼ ♇
05 29	☽ □ ☉
05 54	☉ Q ♃
11 31	☽ Q ♄
12 29	☽ ∥ ♅
14 52	☽ ⚹ ♀
15 59	☽ □ ♂
16 13	☽ △ ☿
19 29	☽ △ ♅
19 46	☽ # ♀
23 Thursday	
01 51	☿ ∠ ♄
02 07	☽ ⊻ ♃
03 00	☽ △ ♇
03 37	♀ ⊥ ♃
07 01	♂ ± ♇
07 18	♂ # ♇
09 21	☽ ☍ ♆
12 21	☽ ⊥ ♃
13 04	☽ Q ♀
18 12	☽ □ ♄
19 02	☽ ⚼ ☿
19 52	☽ # ♄
22 09	☽ # ♆
24 Friday	
01 32	♀ ⊻ ♂
03 36	☽ ∠ ♃
08 43	☉ ± ♅
11 02	☽ ± ☿
11 27	☿ ⚹ ♀
11 58	☽ △ ☉
13 06	☉ # ♇
19 09	♀ ☍ ♅
20 56	☽ # ♂
21 17	☽ △ ♂
21 40	☽ ⊼ ☿
21 41	☽ □ ♅
21 55	☿ △ ♅
21 56	☽ □ ♀
22 43	☽ # ☉
23 29	☽ ∥ ♇
25 Saturday	
04 25	♂ ⊼ ♅
04 50	☽ ⚹ ♃
05 06	☽ □ ♇
11 27	☽ ⊼ ♆
14 45	☽ ⚼ ☉
20 31	☽ △ ♄
20 59	♀ # ♅
21 24	☽ ± ♆
21 57	☽ # ☿
23 34	☽ ⚼ ♂
26 Sunday	
00 55	☽ Q ♃
03 27	☽ Q ♅
03 45	♃ ⊼ ♇
06 46	☽ ± ☉
10 52	☽ Q ♇
12 18	☽ ⚼ ♆
15 19	☽ ± ♂
17 27	☽ ⊼ ☉
21 32	☽ ⚼ ♄
23 16	☽ ⚹ ♅
27 Monday	
01 50	☽ ⊼ ♂
01 54	☉ ⊥ ♀
02 53	☽ ☍ ☿
04 10	☽ △ ♀
06 43	☽ ⚹ ♇
07 03	☽ □ ♃
12 36	☽ ± ♄
13 11	☽ △ ♆
22 41	☽ ⊼ ♄
28 Tuesday	
00 09	☽ ∠ ♅
07 29	☽ ⚼ ♀
07 42	☽ ∠ ♇
07 55	☽ # ☿
10 03	♀ ☍ ♇
15 13	☽ ⊥ ♅
16 13	☉ ± ♇
19 07	♀ ⊻ ♃
22 51	☽ ⊥ ♇
23 23	☽ ☍ ☉
29 Wednesday	
00 05	☽ ± ♀
01 20	☽ ⊻ ♅
06 10	☿ △ ♇
07 04	☽ ☍ ♂
08 57	☽ ∥ ♇
09 03	☽ ⊻ ♇
09 18	☽ ⊼ ☿
10 02	☽ △ ♃
11 16	☽ ⊼ ♀
15 50	☽ □ ♆
19 06	☽ # ☉
19 41	☽ # ♂
19 57	☿ □ ♃
20 30	☽ ± ☿
30 Thursday	
01 49	☉ ⊼ ♅
02 04	☽ ☍ ♄
11 34	☽ # ♆
12 18	☽ ⚼ ♃
13 34	☽ ⚼ ☿
15 34	☽ # ♄
17 26	♂ ⊼ ♇
23 16	☽ Q ♆
31 Friday	
04 24	☽ ± ♃
05 21	☽ ☌ ♅
07 41	☽ ⊼ ☉
13 32	☽ ☌ ♇
14 09	☉ ∥ ♂
14 50	☽ ⊼ ♂
15 19	☽ ⊼ ♃
16 36	♀ △ ♆
18 53	☽ △ ☿
19 24	☽ ± ☉
20 50	☽ ⚹ ♆
21 15	☽ ☍ ♀
22 39	☽ ∥ ♅

FEBRUARY 1964

Date	Sidereal time h m s	Sun ☉ ° ' "	Moon ☽ ° ' "	Moon ☽ 24.00 ° ' "	Mercury ☿ ° '	Venus ♀ ° '	Mars ♂ ° '	Jupiter ♃ ° '	Saturn ♄ ° '	Uranus ♅ ° '	Neptune ♆ ° '	Pluto ⯓ ° '
01	20 42 57	11 ♒ 39 28	26 ♍ 00 40	02 ♎ 26 48	17 ♑ 32	18 ♓ 43	15 ♒ 08	14 ♈ 51	23 ♒ 56	09 ♍ 07	17 ♏ 45	13 ♍ 42
02	20 46 54	12 40 20	08 ♎ 47 39	15 ♎ 03 30	18 46	19 56	15 55	15 01	24 03	09 R 04	17 46	13 R 40
03	20 50 50	13 41 12	21 ♎ 14 45	27 ♎ 21 51	20 01	21 08	16 43	15 11	24 10	09 02	17 46	13 39
04	20 54 47	14 42 03	03 ♏ 25 21	09 ♏ 25 46	21 18	22 20	17 30	15 21	24 17	09 00	17 47	13 38
05	20 58 43	15 42 53	15 ♏ 23 45	21 ♏ 19 55	22 36	23 33	18 17	15 32	24 25	08 57	17 47	13 37
06	21 02 40	16 43 42	27 ♏ 14 54	03 ♐ 09 21	23 56	24 45	19 05	15 42	24 32	08 55	17 48	13 35
07	21 06 37	17 44 30	09 ♐ 03 55	14 ♐ 59 13	25 17	25 57	19 52	15 53	24 39	08 53	17 48	13 34
08	21 10 33	18 45 17	20 ♐ 55 53	26 ♐ 54 28	26 39	27 09	20 40	16 04	24 46	08 50	17 48	13 32
09	21 14 30	19 46 04	02 ♑ 55 31	08 ♑ 59 32	28 02	28 21	21 27	16 15	24 53	08 48	17 49	13 31
10	21 18 26	20 46 49	15 ♑ 06 56	21 ♑ 18 05	29 ♑ 26	29 ♓ 33	22 14	16 26	25 01	08 45	17 49	13 30
11	21 22 23	21 47 33	27 ♑ 33 19	03 ♒ 52 49	00 ♒ 52	00 ♈ 44	23 02	16 37	25 08	08 43	17 49	13 28
12	21 26 19	22 48 16	10 ♒ 16 44	16 ♒ 45 08	02 18	01 56	23 49	16 48	25 15	08 40	17 50	13 27
13	21 30 16	23 48 57	23 ♒ 17 58	29 ♒ 55 07	03 46	03 07	24 37	16 59	25 22	08 38	17 50	13 25
14	21 34 12	24 49 37	06 ♓ 36 24	13 ♓ 21 31	05 14	04 19	25 24	17 11	25 30	08 35	17 50	13 24
15	21 38 09	25 50 16	20 ♓ 10 11	27 ♓ 01 59	06 43	05 30	26 12	17 22	25 37	08 33	17 50	13 22
16	21 42 06	26 50 53	03 ♈ 56 33	10 ♈ 53 28	08 14	06 41	26 59	17 34	25 44	08 30	17 50	13 21
17	21 46 02	27 51 28	17 ♈ 52 20	24 ♈ 52 44	09 45	07 53	27 47	17 45	25 51	08 28	17 50	13 19
18	21 49 59	28 52 01	01 ♉ 54 21	08 ♉ 56 51	11 17	09 04	28 34	17 57	25 59	08 25	17 50	13 18
19	21 53 55	29 ♒ 52 33	15 ♉ 59 57	23 ♉ 03 27	12 50	10 14	29 ♒ 21	18 09	26 06	08 23	17 R 50	13 16
20	21 57 52	00 ♓ 53 03	00 ♊ 07 08	07 ♊ 10 50	14 24	11 25	00 ♓ 09	18 21	26 13	08 20	17 50	13 15
21	22 01 48	01 53 31	14 ♊ 14 25	21 ♊ 17 45	15 59	12 36	00 56	18 33	26 20	08 17	17 50	13 14
22	22 05 45	02 53 57	28 ♊ 20 40	05 ♋ 23 02	17 35	13 46	01 44	18 45	26 28	08 15	17 50	13 12
23	22 09 41	03 54 21	12 ♋ 24 38	19 ♋ 25 15	19 12	14 57	02 31	18 57	26 35	08 12	17 50	13 10
24	22 13 38	04 54 44	26 ♋ 24 37	03 ♌ 22 26	20 50	16 07	03 18	19 10	26 42	08 10	17 50	13 09
25	22 17 35	05 55 04	10 ♌ 18 20	17 ♌ 11 58	22 28	17 17	04 06	19 22	26 49	08 07	17 49	13 07
26	22 21 31	06 55 23	24 ♌ 02 57	00 ♍ 50 52	24 08	18 27	04 53	19 34	26 56	08 04	17 49	13 05
27	22 25 28	07 55 39	07 ♍ 35 23	14 ♍ 16 08	25 49	19 37	05 41	19 47	27 04	08 02	17 49	13 04
28	22 29 24	08 55 52	20 ♍ 52 51	27 ♍ 25 18	27 31	20 46	06 28	20 00	27 11	07 59	17 49	13 02
29	22 33 21	09 ♓ 56 08	03 ♎ 53 21	10 ♎ 16 56	29 ♒ 13	21 ♈ 56	07 ♓ 15	20 ♈ 12	27 ♒ 18	07 ♍ 56	17 ♏ 48	13 ♍ 01

Date	Moon True ☊ ° '	Moon Mean ☊ ° '	Moon ☽ Latitude ° '
01	10 ♋ 49	09 ♋ 42	05 N 02
02	10 R 45	09 38	05 12
03	10 41	09 35	05 07
04	10 39	09 32	04 48
05	10 38	09 29	04 17
06	10 D 39	09 26	03 35
07	10 40	09 22	02 44
08	10 42	09 19	01 46
09	10 43	09 16	00 N 42
10	10 R 44	09 13	00 S 24
11	10 42	09 10	01 30
12	10 38	09 07	02 33
13	10 33	09 03	03 30
14	10 26	09 00	04 16
15	10 18	08 57	04 49
16	10 10	08 54	05 06
17	10 03	08 51	05 05
18	09 58	08 48	04 46
19	09 55	08 44	04 09
20	09 54	08 41	03 17
21	09 D 54	08 38	02 14
22	09 55	08 35	01 S 02
23	09 R 56	08 32	00 N 13
24	09 54	08 28	01 27
25	09 51	08 25	02 35
26	09 45	08 22	03 33
27	09 36	08 19	04 18
28	09 26	08 16	04 48
29	09 ♋ 15	08 ♋ 13	05 N 03

DECLINATIONS

Date	Sun ☉ ° '	Moon ☽ ° '	Mercury ☿ ° '	Venus ♀ ° '	Mars ♂ ° '	Jupiter ♃ ° '	Saturn ♄ ° '	Uranus ♅ ° '	Neptune ♆ ° '	Pluto ⯓ ° '
01	17 S 18	06 N 12	22 S 04	05 S 28	17 S 20	04 N 46	14 S 40	08 N 54	15 S 26	19 N 28
02	17 00	01 N 18	22 03	04 57	17 06	04 50	14 38	08 55	15 26	19 28
03	16 43	03 S 32	22 01	04 26	16 52	04 54	14 36	08 56	15 26	19 29
04	16 26	08 08	21 58	03 55	16 37	04 58	14 33	08 57	15 26	19 30
05	16 08	12 21	21 53	03 24	16 23	05 03	14 31	08 58	15 26	19 31
06	15 50	16 03	21 48	02 53	16 08	05 07	14 29	08 59	15 27	19 31
07	15 31	19 07	21 41	02 22	15 53	05 11	14 26	09 00	15 27	19 32
08	15 12	21 22	21 33	01 50	15 38	05 15	14 24	09 00	15 27	19 33
09	14 53	22 42	21 24	01 19	15 23	05 20	14 22	09 01	15 27	19 34
10	14 34	22 59	21 13	00 47	15 07	05 24	14 19	09 02	15 27	19 35
11	14 15	22 08	21 01	00 S 16	14 52	05 29	14 17	09 03	15 27	19 35
12	13 55	20 08	20 48	00 N 16	14 36	05 33	14 14	09 04	15 27	19 36
13	13 35	17 03	20 34	00 47	14 20	05 38	14 12	09 05	15 27	19 37
14	13 15	13 03	20 18	01 19	14 04	05 42	14 10	09 06	15 27	19 38
15	12 55	08 19	20 01	01 50	13 48	05 47	14 07	09 07	15 27	19 39
16	12 34	03 S 07	19 43	02 22	13 31	05 51	14 05	09 08	15 27	19 39
17	12 13	02 N 19	19 23	02 53	13 15	05 56	14 02	09 09	15 27	19 40
18	11 52	07 40	19 03	03 25	12 58	06 01	14 00	09 10	15 27	19 41
19	11 31	12 39	18 40	03 56	12 42	06 05	13 58	09 11	15 26	19 42
20	11 10	16 58	18 17	04 27	12 25	06 10	13 55	09 12	15 26	19 42
21	10 48	20 18	17 52	04 58	12 08	06 15	13 53	09 13	15 26	19 43
22	10 27	22 24	17 26	05 29	11 51	06 19	13 50	09 14	15 26	19 44
23	10 05	23 05	16 58	06 00	11 34	06 24	13 48	09 15	15 26	19 45
24	09 43	22 18	16 30	06 31	11 16	06 29	13 46	09 16	15 26	19 45
25	09 21	20 09	15 59	07 01	10 59	06 34	13 43	09 17	15 26	19 46
26	08 58	16 52	15 28	07 32	10 42	06 39	13 41	09 18	15 26	19 47
27	08 36	12 43	14 55	08 02	10 24	06 44	13 38	09 19	15 26	19 47
28	08 13	08 02	14 21	08 32	10 06	06 48	13 36	09 20	15 26	19 48
29	07 S 51	03 N 05	13 S 46	09 N 02	09 S 49	06 N 53	13 S 34	09 N 21	15 S 25	19 N 49

ZODIAC SIGN ENTRIES

Date	h	m	Planets	
01	19	25	☽	♎
04	05	12	☽	♏
06	17	35	☽	♐
09	06	11	☽	♑
10	21	09	♀	♈
10	21	30	☿	♒
11	16	39	☽	♒
14	00	09	☽	♓
16	05	10	☽	♈
18	08	45	☽	♉
19	14	57	☉	♓
20	07	33	♂	♓
20	11	48	☽	♊
22	14	49	☽	♋
24	18	11	☽	♌
26	22	30	☽	♍
29	04	46	☽	♎
29	22	50	☿	♓

LATITUDES

Date	Mercury ☿ ° '	Venus ♀ ° '	Mars ♂ ° '	Jupiter ♃ ° '	Saturn ♄ ° '	Uranus ♅ ° '	Neptune ♆ ° '	Pluto ⯓ ° '
01	00 N 14	01 S 06	01 S 05	01 S 11	01 S 12	00 N 48	01 N 46	14 N 11
04	00 S 12	00 58	01 05	01 10	01 12	00 48	01 46	14 12
07	00 36	00 49	01 05	01 09	01 12	00 49	01 46	14 13
10	00 58	00 40	01 05	01 09	01 12	00 49	01 46	14 14
13	01 17	00 30	01 04	01 08	01 12	00 49	01 46	14 14
16	01 33	00 19	01 04	01 07	01 12	00 49	01 47	14 15
19	01 47	00 S 08	01 04	01 07	01 13	00 49	01 47	14 16
22	01 57	00 N 03	01 03	01 06	01 13	00 49	01 47	14 16
25	02 05	00 15	01 03	01 06	01 13	00 49	01 47	14 17
28	02 08	00 28	01 02	01 05	01 13	00 49	01 47	14 17
31	02 S 08	00 N 40	01 S 02	01 S 05	01 S 14	00 N 49	01 N 48	14 N 17

DATA

Julian Date	2438427
Delta T	+35 seconds
Ayanamsa	23° 21' 04"
Synetic vernal point	05° ♓ 45' 56"
True obliquity of ecliptic	23° 26' 37"

MOON'S PHASES, APSIDES AND POSITIONS ☽

Date	h	m	Phase	Longitude ° '	Eclipse Indicator
05	12	42	☾	15 ♏ 45	
13	13	01	●	23 ♒ 52	
20	13	24	☽	00 ♊ 57	
27	12	39	○	07 ♍ 57	

Day	h	m		
06	19	49	Apogee	
21	08	00	Perigee	
02	18	23	0S	
10	06	01	Max dec	23° S 01'
17	01	48	0N	
23	11	08	Max dec	23° N 05'

LONGITUDES

Date	Chiron ⚷ ° '	Ceres ⚳ ° '	Pallas ⚴ ° '	Juno ⚵ ° '	Vesta ⚶ ° '	Black Moon Lilith ⚸ ° '
01	12 ♓ 29	18 ♐ 00	27 ♏ 33	29 ♏ 01	16 ♑ 34	02 ♐ 04
11	12 ♓ 04	21 ♐ 28	00 ♐ 25	01 ♐ 20	21 ♑ 47	03 ♐ 11
21	13 ♓ 41	24 ♐ 42	02 ♐ 53	01 ♐ 42	26 ♑ 55	04 ♐ 18
31	14 ♓ 18	27 ♐ 41	04 ♐ 53	03 ♐ 58	01 ♒ 57	05 ♐ 24

ASPECTARIAN

h m Aspects

01 Saturday
00 53 ♂ ⚹ ♃
02 26 ☽ ± ♂
08 07 ☽ ⊼ ♄
13 18 ☽ ♇ ☉
16 02 ☽ # ♀
16 21 ☿ ⚹ ♆
18 58 ☽ ∥ ♃
19 22 ☽ ± ♄
20 10 ☽ ♇ ♂
20 41 ☿ ⊥ ♄

02 Sunday
00 35 ☽ ∠ ♆
12 30 ☽ ♇ ♄
12 32 ☽ ⊻ ♅
17 04 ♀ # ♃
17 40 ☽ ⊥ ♆
18 49 ♃ ± ♅
20 04 ☽ △ ☉
21 19 ☽ ⊻ ⯓
23 59 ☽ ⊥ ♅

03 Monday
00 05 ☽ ☍ ♃
02 36 ☽ △ ♂
05 14 ☽ ⊻ ♆
08 54 ☽ ⊥ ⯓
09 20 ☽ □ ☿
11 13 ☉ ⊼ ⯓
11 46 ☽ ⊼ ♀
16 09 ☽ ∥ ♀
17 26 ☽ ∠ ♅
17 47 ☽ △ ♄
19 05 ☽ # ♃

04 Tuesday
00 49 ☽ ± ♀
02 31 ☽ ∠ ⯓
16 28 ☽ # ♅
20 35 ♂ □ ♆
20 41 ☽ ♇ ♀
23 06 ☽ ⚹ ♅

05 Wednesday
01 10 ☽ Q ☿
06 41 ☉ ⚹ ♃
08 24 ☽ ⚹ ⯓
12 16 ☽ ⊼ ♃
12 42 ☽ □ ☉
16 50 ☽ ☌ ♆
18 16 ☽ □ ♂
23 12 ☽ Q ♅

06 Thursday
00 35 ☽ ± ♃
01 25 ☽ ∥ ♄
04 25 ☽ ⚹ ☿
06 21 ☽ △ ♀
06 26 ☽ □ ♄
07 09 ♀ ⊻ ♄
07 45 ☽ ∥ ♆
08 38 ☽ Q ⯓
10 31 ☽ ∥ ☉
11 46 ☿ ♇ ♅
12 30 ☽ ∥ ♂
19 08 ☽ ♇ ♃
23 46 ☿ ⊻ ♄

07 Friday
04 37 ☽ Q ☉
07 22 ♀ ⊥ ♂
09 24 ☽ Q ♂
11 37 ☽ □ ♅
13 25 ☉ □ ♆
14 47 ☽ ∠ ☿
15 57 ☽ # ⯓
17 46 ☉ ∥ ♆
19 21 ☽ Q ♄
21 06 ☽ □ ⯓

08 Saturday
02 01 ☽ △ ♃
05 42 ☽ ⊻ ♆
07 12 ☽ ⚹ ☉
11 21 ☽ ⊥ ☿
11 25 ☽ ⚹ ♂
14 09 ☽ ∥ ☿
17 47 ☽ ⊥ ♆
19 48 ☽ ⚹ ♄

09 Sunday
00 58 ☽ ⊻ ☿
01 52 ☽ □ ♀
05 39 ♂ ∥ ♆
11 47 ☽ ∠ ♆
15 59 ☽ ∠ ☉
19 28 ☽ ∠ ♂
20 10 ☿ ♇ ⯓
23 35 ☽ △ ♅

10 Monday
01 54 ☽ ∠ ♄
08 50 ☽ △ ⯓
11 17 ☽ ⊥ ☉
14 21 ☽ ⊥ ♂
14 36 ☽ □ ♃
17 14 ☽ Q ♀
17 16 ☽ ⚹ ♆
18 27 ☿ Q ♆
19 39 ☽ ⊥ ♄
23 20 ☿ ⚹ ♀
23 58 ☽ ⊻ ☉

11 Tuesday
02 45 ☽ ⊻ ♂
04 40 ☽ ♇ ♅
07 19 ☽ ⊻ ♄
09 10 ☉ ∥ ♄
13 44 ☽ ♇ ⯓
16 19 ☽ Q ♆
18 41 ☽ ⚹ ♀
19 06 ☽ ☌ ☿
20 47 ♀ ⊥ ♄
21 46 ☽ ± ♅

12 Wednesday
01 35 ☽ Q ♃
04 31 ☽ ∥ ☿
06 43 ☽ ± ⯓
09 01 ☽ ⊼ ♅
16 43 ☽ # ⯓
17 53 ☽ ⊼ ⯓
17 59 ☿ ± ♅

13 Thursday
00 16 ☽ ⚹ ♃
01 34 ☽ ∠ ♀
01 59 ☽ □ ♆
06 02 ♀ ♇ ♆
13 01 ☽ ☌ ☉
14 33 ☽ ☌ ♂
15 49 ☽ ☌ ♄
19 38 ☽ ⊥ ♀
22 17 ☽ ∥ ♆

14 Friday
02 05 ♂ ∥ ♄
03 57 ☽ ∠ ♃
05 47 ☽ ∥ ♄
06 00 ☽ ∥ ♂
07 30 ☽ ⊻ ♀
09 14 ☽ ⊻ ☿
10 51 ☽ ∥ ☉
11 00 ☿ Q ♃
15 16 ♂ ☌ ♄
15 32 ☽ ☍ ♅
20 15 ☽ ⊥ ♃
21 15 ☽ ⊥ ☿

15 Saturday
00 03 ☽ ☍ ⯓
06 00 ☉ ☌ ♄
07 00 ☽ ⊻ ♃
07 54 ☽ △ ♆
08 11 ☽ # ♅
15 03 ☽ ∠ ☿
21 37 ☽ ⊻ ♄
22 15 ☿ ± ⯓
22 42 ☽ ⊻ ☉
23 11 ☽ ⊻ ♂
23 45 ☽ # ♃

16 Sunday
08 09 ☽ ⊥ ♄
09 57 ☽ ⊥ ☉
10 05 ☽ ♇ ♆
10 14 ☽ ⊥ ♂
15 03 ☽ # ♀
16 19 ☿ ⊼ ♅
16 27 ☿ # ⯓
17 12 ☽ ☌ ♀
19 52 ☽ ⊼ ♅
20 19 ☽ ⚹ ☿
23 50 ☽ ∠ ♄

17 Monday
01 38 ☽ ± ♆
02 43 ☽ ∠ ☉
02 43 ☽ ∠ ♂
02 57 ☉ ☌ ♂
04 12 ☽ ⊼ ⯓
06 10 ☽ ± ♅
11 48 ☽ ☌ ♃
11 56 ☽ ⊼ ♆
14 29 ☽ ± ⯓
14 48 ☽ ∥ ♀
19 27 ☽ Q ☿
21 33 ☽ ♇ ♅
21 42 ♃ ⊼ ♆
23 29 ♀ ⊼ ♅

18 Tuesday
01 47 ☽ ⚹ ♄
04 21 ☽ ∥ ♃
05 51 ☽ ♇ ⯓
05 58 ☽ ⚹ ♂
06 25 ☽ ⚹ ☉
14 27 ♆ St R
19 00 ☽ ∥ ♅
22 26 ☽ Q ♄
23 04 ☽ △ ♅

19 Wednesday
01 18 ☽ ⊻ ♀
03 38 ☽ Q ♂
04 27 ☽ Q ☉
05 57 ☽ □ ☿
06 42 ☽ # ☉
07 22 ☽ △ ⯓
12 13 ☽ # ♂
12 27 ☽ ⊥ ♀
15 07 ☽ ☍ ♆
15 43 ☽ ⊻ ♃
18 37 ☿ ⊼ ⯓
18 46 ☽ # ♄

20 Thursday
02 03 ☽ ⊥ ♃
02 59 ☽ # ♆
05 08 ☽ ∠ ♀
05 19 ☽ □ ♄
07 28 ♀ ∠ ♄
12 03 ☽ □ ♂
13 24 ☽ □ ☉
17 34 ☽ ∠ ♃
19 41 ☽ # ☿
20 29 ♀ ± ♆

21 Friday
01 55 ☽ □ ♅
07 08 ☽ ∥ ⯓
08 57 ☽ ⚹ ♀
10 16 ☽ □ ⯓
15 20 ☽ △ ☿
18 07 ☽ ⊼ ♆
19 26 ☽ ⚹ ♃

22 Saturday
00 29 ♀ ⊼ ⯓
04 19 ☽ ± ♆
07 13 ☽ Q ♀
08 26 ☽ Q ♅
08 46 ☽ △ ♄
15 44 ☿ □ ♆
16 10 ☽ Q ♃
16 51 ☽ Q ⯓
18 06 ☽ △ ♂
19 39 ☽ ♇ ♆
20 09 ☽ ♇ ☿
20 22 ☽ △ ☉
21 22 ♀ ± ♅

23 Sunday
04 50 ☽ ⚹ ♅
07 57 ☿ ⚹ ♃
10 34 ☽ ♇ ♄
13 18 ☽ ⚹ ⯓
13 31 ☽ ± ☿
13 31 ☉ ∠ ♃
16 44 ☽ □ ♀
21 16 ☽ ♇ ♂
21 16 ☽ △ ♆
23 22 ☽ □ ♃
23 59 ☽ ♇ ☉

24 Monday
01 08 ☽ ⊼ ☿
02 07 ☽ ± ♄
06 26 ☽ ∠ ♅
10 08 ♃ ± ⯓
10 22 ♀ ∥ ♃
12 30 ☽ ⊼ ♄
13 38 ☽ ± ♂
14 59 ☽ ∠ ⯓
16 38 ☽ ± ☉
21 52 ☽ ⊥ ♅

25 Tuesday
00 36 ☽ ⊼ ♂
03 48 ☽ ⊼ ☉
06 29 ☽ ⊥ ⯓
08 13 ☽ ⊻ ♅
15 15 ☽ ∥ ⯓
15 47 ☉ # ♅
16 53 ☽ ⊻ ⯓
23 05 ♀ ⊼ ♆
23 08 ♂ ∠ ♃

26 Wednesday
01 05 ☽ □ ♆
01 16 ☽ △ ♀
04 02 ☽ △ ♃
12 10 ☽ ☍ ☿
13 35 ☿ ∥ ♆
17 08 ☽ ☍ ♄
20 46 ☽ # ♆
21 52 ☽ # ☿

27 Thursday
00 57 ♀ ± ⯓
06 11 ☽ ♇ ♀
06 55 ☽ ♇ ♃
06 56 ☽ # ♄
08 22 ☽ ☍ ♂
08 50 ☽ Q ♆
12 39 ☽ ☍ ☉
12 47 ☽ ☌ ♅
14 17 ☉ ☍ ♅
16 18 ♀ ☌ ♃
21 49 ☽ ☌ ⯓
23 18 ☽ ± ♃
23 51 ☽ ± ♀

28 Friday
00 56 ☽ # ♂
03 23 ☉ # ♀
05 33 ☽ ∥ ♅
06 25 ☽ ⚹ ♆
07 00 ☿ ☌ ♄
09 44 ☽ ∥ ♀
10 21 ☽ ⊼ ♃
10 58 ☽ # ☉
11 47 ☽ ⊼ ♀
17 54 ☽ ∥ ♃
23 40 ☽ ⊼ ♄

29 Saturday
02 00 ☽ ⊼ ☿
09 59 ☽ ∠ ♆
10 53 ☽ ± ♄
14 53 ☽ ± ☿
18 43 ☽ ⊼ ♂
19 34 ☽ ⊻ ♅
20 22 ☿ ∥ ♄

All ephemeris data is given at 12.00 UT and the Moon's longitude is additionally given for 24.00 UT

MARCH 1964

LONGITUDES

Date	Sidereal time h m s	Sun ☉ ° ' "	Moon ☽ ° ' "	Moon ☽ 24.00 ° ' "	Mercury ☿ ° '	Venus ♀ ° '	Mars ♂ ° '	Jupiter ♃ ° '	Saturn ♄ ° '	Uranus ♅ ° '	Neptune ♆ ° '	Pluto ♇ ° '
01	22 37 17	10 ♓ 56 19	16 ♎ 36 06	22 ♎ 50 58	00 ♓ 57	23 ♈ 05	08 ♓ 03	20 ♈ 25	27 ♒ 25	07 ♍ 54	17 ♏ 48	12 ♍ 59
02	22 41 14	11 56 29	29 ♎ 01 44	05 ♏ 08 43	02 42	24 15	08 50	20 38	27 32	07 R 51	17 R 47	12 R 58
03	22 45 10	12 56 38	11 ♏ 12 17	17 ♏ 12 53	04 28	25 24	09 37	20 51	27 39	07 49	17 47	12 56
04	22 49 07	13 56 44	23 ♏ 11 03	29 ♏ 07 19	06 15	26 32	10 25	21 04	27 46	07 46	17 47	12 54
05	22 53 04	14 56 50	05 ♐ 02 20	10 ♐ 56 43	08 03	27 41	11 12	21 16	27 53	07 43	17 46	12 53
06	22 57 00	15 56 53	16 ♐ 51 08	22 ♐ 46 18	09 52	28 50	11 59	21 30	28 00	07 41	17 45	12 51
07	23 00 57	16 56 56	28 ♐ 42 52	04 ♑ 41 33	11 42	29 ♈ 58	12 46	21 43	28 07	07 38	17 45	12 50
08	23 04 53	17 56 56	10 ♑ 43 00	16 ♑ 47 51	13 34	01 ♉ 06	13 34	21 56	28 14	07 36	17 44	12 48
09	23 08 50	18 56 55	22 ♑ 56 43	29 ♑ 10 06	15 26	02 14	14 21	22 09	28 21	07 33	17 44	12 47
10	23 12 46	19 56 52	05 ♒ 28 30	11 ♒ 52 16	17 20	03 22	15 08	22 22	28 28	07 30	17 43	12 45
11	23 16 43	20 56 48	18 ♒ 21 41	24 ♒ 56 54	19 14	04 30	15 55	22 36	28 35	07 28	17 42	12 43
12	23 20 39	21 56 41	01 ♓ 37 56	08 ♓ 24 39	21 10	05 38	16 43	22 49	28 42	07 25	17 42	12 42
13	23 24 36	22 56 33	15 ♓ 16 49	22 ♓ 14 00	23 06	06 45	17 30	23 03	28 49	07 23	17 41	12 40
14	23 28 33	23 56 23	29 ♓ 15 42	06 ♈ 21 15	25 04	07 52	18 17	23 16	28 56	07 20	17 40	12 39
15	23 32 29	24 56 11	13 ♈ 29 57	20 ♈ 41 01	27 02	08 59	19 04	23 30	29 03	07 18	17 39	12 37
16	23 36 26	25 55 56	27 ♈ 53 39	05 ♉ 07 05	29 ♓ 01	10 06	19 51	23 43	29 09	07 15	17 38	12 36
17	23 40 22	26 55 40	12 ♉ 20 34	19 ♉ 33 26	01 ♈ 00	11 12	20 38	23 57	29 16	07 13	17 38	12 34
18	23 44 19	27 55 21	26 ♉ 45 09	03 ♊ 55 12	03 00	12 18	21 25	24 11	29 23	07 11	17 37	12 33
19	23 48 15	28 55 01	11 ♊ 03 16	18 ♊ 09 05	05 00	13 24	22 12	24 24	29 29	07 08	17 36	12 31
20	23 52 12	29 ♓ 54 37	25 ♊ 12 28	02 ♋ 13 20	07 01	14 30	22 59	24 38	29 36	07 06	17 35	12 30
21	23 56 08	00 ♈ 54 12	09 ♋ 11 40	16 ♋ 07 28	09 01	15 36	23 46	24 52	29 43	07 03	17 34	12 28
22	00 00 05	01 53 44	23 ♋ 00 46	29 ♋ 51 37	11 00	16 41	24 33	25 06	29 49	07 01	17 33	12 27
23	00 04 02	02 53 14	06 ♌ 40 01	13 ♌ 26 00	12 59	17 46	25 20	25 20	29 ♒ 56	06 59	17 32	12 25
24	00 07 58	03 52 42	20 ♌ 09 31	26 ♌ 50 33	14 57	18 51	26 07	25 34	00 ♓ 02	06 57	17 31	12 24
25	00 11 55	04 52 07	03 ♍ 28 59	10 ♍ 04 44	16 53	19 55	26 54	25 47	00 08	06 54	17 30	12 22
26	00 15 51	05 51 30	16 ♍ 37 40	23 ♍ 07 38	18 47	20 59	27 41	26 01	00 15	06 52	17 29	12 21
27	00 19 48	06 50 51	29 ♍ 34 32	05 ♎ 58 13	20 39	22 03	28 28	26 15	00 21	06 50	17 27	12 20
28	00 23 44	07 50 10	12 ♎ 18 36	18 ♎ 35 39	22 29	23 07	29 ♓ 14	26 30	00 27	06 48	17 26	12 18
29	00 27 41	08 49 27	24 ♎ 49 22	00 ♏ 59 47	24 15	24 10	00 ♈ 01	26 44	00 34	06 46	17 25	12 17
30	00 31 37	09 48 42	07 ♏ 07 01	13 ♏ 11 17	25 58	25 13	00 48	26 58	00 40	06 44	17 24	12 15
31	00 35 34	10 ♈ 47 55	19 ♏ 12 48	25 ♏ 11 54	27 ♈ 36	26 ♉ 16	01 ♈ 35	27 ♈ 12	00 ♓ 46	06 ♍ 42	17 ♏ 23	12 ♍ 14

Date	Moon True ☊ ° '	Moon Mean ☊ ° '	Moon ☽ Latitude ° '
01	09 ♋ 04	08 ♋ 09	05 N 01
02	08 R 55	08 06	04 46
03	08 47	08 03	04 17
04	08 43	08 00	03 38
05	08 40	07 57	02 49
06	08 D 40	07 54	01 54
07	08 40	07 50	00 N 53
08	08 R 41	07 47	00 S 11
09	08 40	07 44	01 15
10	08 37	07 41	02 17
11	08 31	07 38	03 14
12	08 23	07 34	04 02
13	08 12	07 31	04 38
14	08 00	07 28	04 58
15	07 48	07 25	05 00
16	07 38	07 22	04 43
17	07 29	07 19	04 08
18	07 24	07 15	03 17
19	07 21	07 12	02 14
20	07 D 20	07 09	01 S 04
21	07 R 20	07 06	00 N 10
22	07 20	07 03	01 22
23	07 17	06 59	02 28
24	07 12	06 56	03 26
25	07 04	06 53	04 11
26	06 54	06 50	04 42
27	06 41	06 47	04 58
28	06 27	06 44	04 58
29	06 14	06 40	04 45
30	06 02	06 37	04 18
31	05 ♋ 52	06 ♋ 34	03 N 39

DECLINATIONS

Date	Sun ☉ ° '	Moon ☽ ° '	Mercury ☿ ° '	Venus ♀ ° '	Mars ♂ ° '	Jupiter ♃ ° '	Saturn ♄ ° '	Uranus ♅ ° '	Neptune ♆ ° '	Pluto ♇ ° '
01	07 S 28	01 S 53	13 S 09	09 N 32	09 S 31	06 N 58	13 S 31	09 N 22	15 S 25	19 N 50
02	07 05	06 41	12 31	10 01	09 13	07 03	13 29	09 23	15 25	19 50
03	06 42	11 07	11 51	10 31	08 55	07 08	13 27	09 24	15 25	19 51
04	06 19	15 03	11 10	11 00	08 37	07 13	13 24	09 25	15 25	19 52
05	05 56	18 22	10 28	11 29	08 19	07 18	13 22	09 26	15 25	19 52
06	05 33	20 55	09 45	11 57	08 00	07 23	13 20	09 27	15 24	19 53
07	05 09	22 33	09 01	12 26	07 42	07 28	13 17	09 28	15 24	19 54
08	04 46	23 12	08 15	12 54	07 24	07 33	13 15	09 29	15 24	19 54
09	04 22	22 44	07 28	13 21	07 05	07 38	13 13	09 30	15 24	19 55
10	03 59	21 07	06 40	13 49	06 47	07 43	13 10	09 31	15 23	19 56
11	03 35	18 24	05 50	14 16	06 28	07 48	13 08	09 31	15 23	19 56
12	03 12	14 40	05 00	14 43	06 09	07 53	13 06	09 32	15 23	19 57
13	02 48	10 04	04 08	15 09	05 51	07 59	13 03	09 33	15 23	19 57
14	02 24	04 S 51	03 16	15 36	05 32	08 04	13 01	09 34	15 22	19 58
15	02 01	00 N 44	02 22	16 01	05 13	08 09	12 59	09 35	15 22	19 59
16	01 37	06 19	01 28	16 27	04 55	08 14	12 56	09 36	15 22	19 59
17	01 13	11 37	00 S 33	16 52	04 36	08 19	12 54	09 37	15 22	20 00
18	00 50	16 14	00 N 23	17 17	04 17	08 24	12 52	09 38	15 21	20 00
19	00 26	19 53	01 19	17 41	03 58	08 29	12 50	09 39	15 21	20 01
20	00 S 02	22 18	02 16	18 05	03 39	08 34	12 48	09 40	15 21	20 01
21	00 N 22	23 17	03 12	18 29	03 20	08 40	12 45	09 40	15 20	20 02
22	00 45	22 50	04 09	18 52	03 01	08 45	12 43	09 41	15 20	20 02
23	01 09	21 00	05 06	19 14	02 43	08 50	12 41	09 42	15 20	20 03
24	01 33	18 01	06 02	19 37	02 24	08 55	12 39	09 43	15 19	20 03
25	01 56	14 07	06 57	19 59	02 05	09 00	12 37	09 44	15 19	20 04
26	02 20	09 36	07 52	20 20	01 46	09 05	12 35	09 45	15 19	20 04
27	02 43	04 N 43	08 45	20 41	01 27	09 10	12 32	09 45	15 18	20 05
28	03 07	00 S 17	09 38	21 01	01 08	09 16	12 30	09 46	15 18	20 05
29	03 30	05 12	10 28	21 21	00 49	09 21	12 28	09 47	15 18	20 06
30	03 53	09 50	11 17	21 41	00 30	09 26	12 26	09 48	15 17	20 06
31	04 N 16	14 S 01	12 N 03	22 N 00	00 S 11	09 N 31	12 S 24	09 N 48	15 S 17	20 N 06

ZODIAC SIGN ENTRIES

Date	h	m	Planets
02	13	54	☽ ♏
05	01	47	☽ ♐
07	12	38	♀ ♉
07	14	35	☽ ♑
10	01	35	☽ ♒
12	09	05	☽ ♓
14	13	15	☽ ♈
16	15	30	☽ ♉
16	23	54	☿ ♈
18	17	26	☽ ♊
20	14	10	☉ ♈
20	20	11	☽ ♋
23	00	15	☽ ♌
24	04	18	♄ ♓
25	05	42	☽ ♍
27	12	48	☽ ♎
29	11	24	♂ ♈
29	22	03	☽ ♏

LATITUDES

Date	Mercury ☿ ° '	Venus ♀ ° '	Mars ♂ ° '	Jupiter ♃ ° '	Saturn ♄ ° '	Uranus ♅ ° '	Neptune ♆ ° '	Pluto ♇ ° '
01	02 S 09	00 N 36	01 S 02	01 S 05	01 S 13	00 N 49	01 N 47	14 N 17
04	02 06	00 49	01 01	01 05	01 14	00 49	01 48	14 17
07	01 59	01 02	01 01	01 04	01 14	00 49	01 48	14 17
10	01 48	01 15	01 00	01 04	01 14	00 49	01 48	14 17
13	01 31	01 28	00 59	01 04	01 15	00 49	01 48	14 17
16	01 10	01 41	00 58	01 03	01 15	00 49	01 48	14 17
19	00 44	01 55	00 57	01 03	01 15	00 49	01 48	14 17
22	00 S 13	02 08	00 56	01 03	01 16	00 49	01 49	14 17
25	00 N 21	02 20	00 55	01 02	01 16	00 48	01 49	14 17
28	00 57	02 33	00 54	01 02	01 17	00 48	01 49	14 16
31	01 N 32	02 N 45	00 S 53	01 S 02	01 S 17	00 N 48	01 N 49	14 N 16

DATA

Julian Date	2438456
Delta T	+35 seconds
Ayanamsa	23° 21' 07"
Synetic vernal point	05° ♓ 45' 52"
True obliquity of ecliptic	23° 26' 38"

MOON'S PHASES, APSIDES AND POSITIONS ☽

Date	h	m	Phase	Longitude ° '	Eclipse Indicator
06	10	00	☾	15 ♐ 52	
14	02	14	●	23 ♓ 32	
20	20	39	☽	00 ♋ 16	
28	02	48	○	07 ♎ 27	

Day	h	m		
05	16	47	Apogee	
17	15	25	Perigee	
01	02	50	0S	
08	14	05	Max dec	23° S 12'
15	08	54	0N	
21	16	17	Max dec	23° N 19'
28	10	38	0S	

LONGITUDES

Date	Chiron ⚷ ° '	Ceres ⚳ ° '	Pallas ⚴ ° '	Juno ⚵ ° '	Vesta ⚶ ° '	Black Moon Lilith ⚸ ° '
01	14 ♓ 14	27 ♐ 23	04 ♐ 42	03 ♐ 51	01 ♒ 27	05 ♐ 18
11	14 ♓ 52	00 ♑ 05	06 ♐ 12	04 ♐ 43	06 ♒ 24	06 ♐ 24
21	15 ♓ 29	02 ♑ 26	07 ♐ 05	05 ♐ 05	11 ♒ 13	07 ♐ 31
31	16 ♓ 05	04 ♑ 22	07 ♐ 14	04 ♐ 54	15 ♒ 53	08 ♐ 37

All ephemeris data is given at 12.00 UT and the Moon's longitude is additionally given for 24.00 UT

ASPECTARIAN

h m Aspects

01 Sunday
00 19 ☽ ⊼ ☉
02 52 ☽ ⊥ ♆
03 34 ♀ ∥ ♅
03 58 ☽ ⚼ ♄
05 08 ☽ ⊻ ♇
06 48 ☽ ± ♂
06 52 ☽ ⊥ ♅
07 44 ♂ ☍ ♅
08 09 ♀ ⚼ ♅
09 06 ♀ ∠ ♂
10 34 ☽ ⚼ ☿
11 25 ♀ # ♂
12 42 ☽ ± ☉
14 17 ☽ ⊻ ♆
16 33 ☽ ⊥ ♇
19 26 ☽ ☍ ♃
23 21 ♂ # ♅

02 Monday
00 03 ☽ ∠ ♅
01 13 ☽ ⚼ ♂
01 44 ☽ ☍ ♀
07 34 ☽ ⚼ ☉
09 04 ☽ △ ♄
09 55 ☽ ∠ ♇
13 35 ☉ # ♃
13 58 ☽ ∥ ☉
14 00 ☽ # ♃
20 24 ☽ △ ☿

03 Tuesday
00 35 ☽ ∥ ♂
02 24 ☽ # ♅
05 17 ☽ ⚹ ♅
08 10 ☽ # ♀
08 38 ☽ △ ♂
11 46 ☉ ☍ ♇
15 26 ☽ ⚹ ♇
15 40 ☽ ∥ ☿
15 47 ☽ △ ☉

04 Wednesday
01 08 ☽ ☌ ♆
01 37 ☽ ∥ ♄
05 09 ☽ Q ♅
07 38 ☽ ⊼ ♃
09 07 ☿ ∠ ♃
14 21 ☽ ∥ ♆
15 28 ☽ Q ♇
15 39 ☿ # ♀
19 30 ☽ ⊼ ♀
19 58 ☽ ± ♃
21 22 ☽ □ ♄

05 Thursday
07 46 ☿ ☍ ♅
08 58 ☽ ± ♀
14 33 ☽ ⚼ ♃
15 59 ♀ ⚼ ♇
16 44 ♀ ⚹ ♄
17 26 ☽ □ ♅
19 13 ☽ □ ☿
22 02 ☉ ⊥ ♃

06 Friday
01 17 ☽ # ♇
01 25 ☽ □ ♂
03 54 ☽ □ ♇
05 12 ☽ ⚼ ♀
10 00 ☽ □ ☉
10 16 ☽ Q ♄
13 50 ☽ ⊻ ♆
21 35 ☽ △ ♃
21 50 ☿ # ♅

07 Saturday
01 59 ☽ ⊥ ♆
10 48 ☽ ⚹ ♄
13 36 ♂ ☍ ♇
14 22 ☽ Q ☿
14 48 ☽ △ ♀
16 26 ☽ Q ♂
20 06 ☽ ∠ ♆

08 Sunday
01 39 ☽ Q ☉
02 11 ♂ # ♃
02 21 ☿ ☍ ♇
05 49 ☽ △ ♅
07 00 ☉ △ ♆
12 01 ☿ ☌ ♂
16 07 ☽ △ ♇
17 03 ☽ ∠ ♄
18 01 ☽ ⚹ ♂
18 39 ☽ ⚹ ☿

09 Monday
01 50 ☽ ⚹ ♆
03 31 ☽ ⚹ ☉
04 55 ♀ # ♄
07 15 ☿ # ♃
10 26 ☽ □ ♃
10 51 ☽ ⊥ ♄
11 14 ☽ ⚼ ♅
21 19 ☽ ⚼ ♇
22 19 ☿ ⊥ ♃
22 32 ☽ ⊻ ♄

10 Tuesday
01 04 ☽ Q ♆
01 10 ☽ ∠ ♂
04 30 ☽ ± ♅
04 58 ☽ ∠ ☿
06 22 ☿ ∥ ♂
07 38 ☽ □ ♀
10 55 ☽ ∠ ☉
14 24 ☽ ± ♇
15 49 ☽ ⊼ ♅
16 54 ☿ △ ♆
19 21 ☽ ⊥ ♂
21 22 ☽ Q ♃
23 43 ☽ # ♇

11 Wednesday
00 09 ♃ ⚼ ♅
00 53 ☽ ⊥ ☿
01 36 ☽ ⊼ ♇
05 11 ☽ ⊥ ☉
07 13 ☽ ⊻ ♂
10 48 ☽ □ ♆
13 53 ☽ ⊻ ☿
17 07 ☽ ⊻ ☉
19 52 ☽ ⚹ ♃
20 01 ☿ ∠ ♀
20 16 ☽ Q ♀

12 Thursday
06 43 ☽ ☌ ♄
07 49 ☽ ∥ ♆
11 44 ☽ # ♀
16 37 ♂ ⊥ ♃
19 44 ☽ ⚹ ♀
20 44 ☽ ∥ ♄
22 14 ☽ ☍ ♅
23 08 ☽ ∠ ♃

13 Friday
07 28 ☽ ☍ ♇
07 57 ☉ ☌ ☿
11 09 ☿ ⊻ ♃
14 27 ☽ # ♅
15 06 ☽ ⊥ ♃
15 06 ☉ ⊻ ♃
16 04 ☽ ☌ ♂
16 09 ☽ △ ♆
17 33 ♂ △ ♆
21 44 ☽ # ♃

14 Saturday
00 02 ♀ # ♆
00 08 ☽ ∠ ♀
01 04 ♀ △ ♅
01 36 ☽ ⊻ ♃
02 14 ☽ ☌ ☉
03 41 ☽ ☌ ☿
08 45 ☽ ∥ ♂
11 26 ☽ ⊻ ♄
16 48 ☽ ⊥ ♀
17 46 ☽ ⚼ ♆
20 14 ☽ ∥ ☿
21 40 ☽ ⊥ ♄
23 25 ☽ ∥ ☉

15 Sunday
01 37 ☽ ⊼ ♅
03 47 ☽ ⊻ ♀
08 55 ☽ ± ♆
10 32 ☽ ⊼ ♇
11 40 ☽ ± ♅
12 55 ☽ ∠ ♄
17 07 ☽ # ☉
18 02 ☽ # ☿
18 56 ☽ ⊼ ♆
20 33 ☽ ± ♇
21 51 ☽ ⊻ ♂

16 Monday
02 39 ☽ ⚼ ♅
04 50 ☉ ∥ ☿
04 57 ☽ ☌ ♃
06 12 ☽ # ♂
08 25 ☽ ⊥ ♂
08 30 ☽ ⊻ ☉
11 30 ☽ ⚼ ♇
13 49 ☿ ⊻ ♄
14 07 ☽ ⚹ ♄
14 09 ☽ ⊻ ☿
19 12 ☽ ⊥ ☉
20 33 ☽ ∥ ♃

17 Tuesday
00 13 ☽ ∠ ♂
01 43 ☽ ⊥ ☿
02 39 ☽ ∥ ♅
03 30 ☽ △ ♅
09 57 ☽ ☌ ♀
10 12 ☽ Q ♄
11 16 ☽ ∠ ☉
12 23 ☽ △ ♇
13 37 ♀ Q ♄
18 16 ☽ # ♄
19 04 ☽ ∠ ☿
20 47 ☽ ☍ ♆

18 Wednesday
02 36 ☽ ⚹ ♂
07 04 ☽ # ♆
07 19 ☿ ⚼ ♆
07 38 ☽ ⊻ ♃
14 06 ☽ ⚹ ☉
16 26 ☽ □ ♄
17 06 ♀ △ ♇
17 49 ☽ ⊥ ♃
18 52 ☽ ∥ ♀
20 00 ☉ # ☿
23 49 ☽ Q ♂

19 Thursday
00 10 ☽ ⚹ ☿
05 25 ☽ □ ♅
09 10 ☽ ∠ ♃
11 45 ☽ Q ☉
12 59 ☽ ∥ ♇
14 28 ☽ □ ♇
16 18 ☽ ⊻ ♀
18 06 ☿ ⊥ ♄
23 03 ☽ ⊼ ♆
23 43 ☽ Q ☿

20 Friday
03 19 ☽ ⊥ ♀
03 34 ☉ ⊻ ♄
08 00 ☽ □ ♂
09 14 ☽ ± ♆
11 00 ☽ ⚹ ♃
11 49 ☽ Q ♅
12 59 ☿ ⊼ ♅
19 34 ☽ △ ♄
19 57 ☽ ∠ ♀
20 39 ☽ □ ☉
21 01 ☽ Q ♇

21 Saturday
00 36 ☽ ⚼ ♆
07 55 ☽ Q ♃
08 19 ☽ ⚹ ♅
11 38 ☽ □ ☿
14 30 ☿ # ♂
17 39 ☽ ⚹ ♇
21 37 ☽ ⚼ ♄

22 Sunday
00 02 ☽ ⚹ ♀
02 29 ☽ △ ♆
03 21 ☿ ⊥ ♀
10 16 ☽ ∠ ♅
13 25 ☽ ± ♄
14 52 ☽ △ ♂
15 42 ☽ □ ♃
18 29 ☿ ± ♆
19 45 ☽ ∠ ♇
22 47 ☽ Q ♀

23 Monday
00 01 ☽ ⊼ ♄
02 00 ☽ ⊥ ♅
03 30 ☉ ⚼ ♆
04 48 ☽ △ ☉
05 12 ☿ ⊼ ♇
06 53 ♀ ☍ ♆
11 32 ♂ ⊻ ♃
11 34 ☽ ⊥ ♇
11 55 ☿ ± ♅
12 33 ☽ ⊻ ♅
18 54 ☽ ⚼ ♂
20 38 ☽ ∥ ♇
22 11 ☽ ⊻ ♇

24 Tuesday
01 07 ☽ △ ☿
01 29 ☽ ∥ ♀
07 17 ☽ □ ♆
09 27 ☽ □ ♀
09 32 ☽ ⚼ ☉
11 55 ☽ ± ♂
13 07 ☿ ∠ ♄
21 38 ☉ ∠ ♀
21 52 ☽ △ ♃
23 22 ☽ ⊼ ♂

25 Wednesday
02 59 ☽ ± ☉
05 06 ☽ # ♆
05 54 ☽ ☍ ♄
08 37 ☽ ⚼ ☿
14 43 ☽ ⊼ ☉
15 39 ☽ Q ♆
16 48 ☉ # ♂
17 53 ♀ ∥ ♇
18 12 ☽ ☌ ♅
19 34 ☿ ⊼ ♆
20 23 ☽ # ♄

26 Thursday
01 32 ☽ ⚼ ♃
03 46 ☽ ± ☿
04 10 ☽ ☌ ♇
06 28 ☿ ± ♇
11 17 ☽ ∥ ♅
13 34 ☽ ⚹ ♆
14 32 ☽ ∥ ♃
16 39 ☽ ⊼ ☿
18 22 ☽ ± ♃
19 19 ☽ ∥ ☿
20 46 ☽ △ ♀
22 34 ☉ ⊥ ♄

27 Friday
05 42 ☽ ⊼ ♃
09 47 ☽ ☍ ♂
11 38 ☉ ⊼ ♅
13 28 ☽ ⊼ ♄
17 23 ☽ ∠ ♆
20 54 ☽ ∥ ☉

28 Saturday
00 42 ☿ ∥ ♃
00 50 ☽ ± ♄
01 35 ☽ ⊻ ♅
02 48 ☽ ☍ ☉
03 06 ☿ ⚼ ♅
03 19 ☽ ⚼ ♀
04 45 ☽ # ♂
10 21 ☽ ⊥ ♆
11 59 ☽ ⊻ ♇
12 55 ☽ ⊥ ♅
15 49 ☽ ∥ ♂
16 04 ☿ ∥ ♅
18 03 ☽ ⚼ ♄
21 46 ☽ ⊻ ♆
22 00 ☽ ± ♀
23 25 ☽ ⊥ ♇

29 Sunday
02 52 ☽ # ☉
06 06 ☽ ∠ ♅
09 11 ☿ ⊻ ♀
10 37 ☽ ⊼ ♀
10 43 ☽ ☍ ☿
15 46 ☽ ☍ ♃
16 45 ☽ ∠ ♇
22 47 ☽ ⊼ ♂
23 15 ☽ △ ♄

30 Monday
07 16 ♂ ⊻ ♄
09 49 ☽ # ♃
11 14 ☽ ⚹ ♅
11 20 ☽ ± ♂
11 48 ☽ # ♅
17 47 ☽ ⊼ ☉
21 48 ☽ # ☿
22 08 ☽ ⚹ ♇

31 Tuesday
02 29 ☽ ∥ ♄
04 55 ☿ ☌ ♃
06 22 ☽ ⚼ ♂
06 33 ☿ ⚼ ♇
06 45 ☽ ± ☉
08 21 ☽ ☌ ♆
10 58 ☽ Q ♅
15 23 ♃ ⚼ ♇
20 00 ☽ ∥ ♆
22 02 ☽ Q ♇
22 35 ☿ # ♄

APRIL 1964

LONGITUDES

Date	Sidereal time h m s	Sun ☉ ° ' "	Moon ☽ ° ' "	Moon ☽ 24.00 ° ' "	Mercury ☿ ° '	Venus ♀ ° '	Mars ♂ ° '	Jupiter ♃ ° '	Saturn ♄ ° '	Uranus ♅ ° '	Neptune ♆ ° '	Pluto ♇ ° '
01	00 39 31	11 ♈ 47 06	01 ♐ 08 59	07 ♐ 04 29	29 ♈ 11	27 ♉ 18	02 ♈ 21	27 ♈ 26	00 ♓ 52	06 ♍ 39	17 ♏ 22	12 ♍ 13
02	00 43 27	12 46 15	12 ♐ 58 54	18 ♐ 52 48	00 ♉ 41	28 20	03 08	27 40	00 58	06 R 38	17 R 20	12 R 11
03	00 47 24	13 45 23	24 ♐ 46 46	00 ♑ 41 27	02 06	29 ♉ 22	03 54	27 54	01 04	06 36	17 19	12 10
04	00 51 20	14 44 29	06 ♑ 37 31	12 ♑ 35 38	03 25	00 ♊ 23	04 41	28 09	01 10	06 34	17 18	12 09
05	00 55 17	15 43 33	18 ♑ 36 31	24 ♑ 40 50	04 40	01 24	05 28	28 23	01 16	06 32	17 16	12 07
06	00 59 13	16 42 35	00 ♒ 49 17	07 ♒ 02 30	05 48	02 24	06 14	28 37	01 22	06 30	17 15	12 06
07	01 03 10	17 41 35	13 ♒ 21 04	19 ♒ 45 31	06 51	03 24	07 00	28 51	01 28	06 28	17 14	12 05
08	01 07 06	18 40 34	26 ♒ 16 17	02 ♓ 53 40	07 47	04 24	07 47	29 06	01 33	06 26	17 12	12 04
09	01 11 03	19 39 31	09 ♓ 37 52	16 ♓ 28 53	08 38	05 23	08 33	29 20	01 39	06 25	17 11	12 03
10	01 15 00	20 38 26	23 ♓ 26 36	00 ♈ 30 38	09 22	06 22	09 20	29 34	01 45	06 23	17 10	12 01
11	01 18 56	21 37 19	07 ♈ 40 30	14 ♈ 55 30	09 59	07 21	10 06	29 ♈ 49	01 50	06 21	17 08	12 00
12	01 22 53	22 36 10	22 ♈ 14 47	29 ♈ 37 22	10 30	08 19	10 52	00 ♉ 03	01 56	06 20	17 07	11 59
13	01 26 49	23 34 59	07 ♉ 02 12	14 ♉ 28 11	10 55	09 16	11 38	00 17	02 01	06 18	17 05	11 58
14	01 30 46	24 33 46	21 ♉ 54 15	29 ♉ 19 23	11 13	10 13	12 25	00 32	02 06	06 17	17 04	11 57
15	01 34 42	25 32 31	06 ♊ 42 40	14 ♊ 03 18	11 25	11 10	13 11	00 46	02 12	06 15	17 02	11 56
16	01 38 39	26 31 14	21 ♊ 20 37	28 ♊ 34 08	11 31	12 06	13 57	01 01	02 17	06 14	17 01	11 55
17	01 42 35	27 29 55	05 ♋ 43 31	12 ♋ 48 31	11 R 30	13 01	14 43	01 15	02 22	06 13	16 59	11 54
18	01 46 32	28 28 34	19 ♋ 49 04	26 ♋ 45 10	11 24	13 56	15 29	01 29	02 27	06 11	16 58	11 53
19	01 50 29	29 ♈ 27 10	03 ♌ 36 54	10 ♌ 24 24	11 12	14 50	16 15	01 44	02 32	06 10	16 56	11 52
20	01 54 25	00 ♉ 25 44	17 ♌ 07 50	23 ♌ 47 25	10 54	15 44	17 01	01 58	02 37	06 09	16 55	11 51
21	01 58 22	01 24 16	00 ♍ 23 20	06 ♍ 55 45	10 32	16 37	17 47	02 13	02 42	06 08	16 53	11 50
22	02 02 18	02 22 45	13 ♍ 24 53	19 ♍ 50 50	10 05	17 29	18 33	02 27	02 47	06 07	16 52	11 49
23	02 06 15	03 21 13	26 ♍ 13 47	02 ♎ 33 49	09 35	18 21	19 19	02 41	02 52	06 06	16 50	11 48
24	02 10 11	04 19 38	08 ♎ 51 02	15 ♎ 05 30	09 01	19 12	20 04	02 56	02 56	06 05	16 48	11 47
25	02 14 08	05 18 01	21 ♎ 17 20	27 ♎ 26 35	08 25	20 02	20 50	03 10	03 01	06 04	16 47	11 47
26	02 18 04	06 16 23	03 ♏ 33 21	09 ♏ 37 45	07 47	20 52	21 36	03 25	03 05	06 03	16 45	11 46
27	02 22 01	07 14 43	15 ♏ 39 55	21 ♏ 40 01	07 07	21 41	22 21	03 39	03 10	06 02	16 44	11 45
28	02 25 58	08 13 00	27 ♏ 38 15	03 ♐ 34 53	06 27	22 29	23 07	03 53	03 14	06 01	16 42	11 44
29	02 29 54	09 11 17	09 ♐ 30 11	15 ♐ 24 30	05 48	23 16	23 52	04 08	03 19	06 00	16 40	11 44
30	02 33 51	10 ♉ 09 31	21 ♐ 18 14	27 ♐ 11 48	05 ♉ 09	24 ♊ 02	24 ♈ 38	04 ♉ 22	03 ♓ 23	05 ♍ 59	16 ♏ 39	11 ♍ 43

Moon

Date	Moon True ☊ ° '	Moon Mean ☊ ° '	Moon ☽ Latitude ° '
01	05 ♋ 45	06 ♋ 31	02 N 52
02	05 R 41	06 28	01 57
03	05 40	06 25	00 N 57
04	05 D 39	06 21	00 S 05
05	05 R 39	06 18	01 08
06	05 39	06 15	02 09
07	05 36	06 12	03 06
08	05 31	06 09	03 55
09	05 24	06 05	04 32
10	05 14	06 02	04 55
11	05 03	05 59	05 01
12	04 52	05 56	04 48
13	04 42	05 53	04 16
14	04 34	05 50	03 26
15	04 29	05 46	02 22
16	04 27	05 43	01 S 09
17	04 D 26	05 40	00 N 07
18	04 26	05 37	01 21
19	04 R 26	05 34	02 29
20	04 25	05 31	03 27
21	04 22	05 27	04 13
22	04 16	05 24	04 44
23	04 08	05 21	05 01
24	03 57	05 18	05 03
25	03 47	05 15	04 50
26	03 36	05 11	04 24
27	03 27	05 08	03 46
28	03 19	05 05	02 59
29	03 14	05 02	02 04
30	03 ♋ 11	04 ♋ 59	01 03

DECLINATIONS

Date	Sun ☉ ° '	Moon ☽ ° '	Mercury ☿ ° '	Venus ♀ ° '	Mars ♂ ° '	Jupiter ♃ ° '	Saturn ♄ ° '	Uranus ♅ ° '	Neptune ♆ ° '	Pluto ♇ ° '
01	04 N 40	17 S 35	12 N 48	22 N 18	00 N 08	09 N 36	12 S 22	09 N 49	15 S 16	20 N 07
02	05 03	20 25	13 30	22 36	00 27	09 41	12 20	09 50	15 16	20 07
03	05 26	22 23	14 09	22 54	00 46	09 47	12 18	09 50	15 16	20 07
04	05 49	23 22	14 46	23 11	01 05	09 52	12 16	09 51	15 15	20 08
05	06 11	23 17	15 20	23 27	01 23	09 57	12 14	09 52	15 15	20 08
06	06 34	22 05	15 51	23 43	01 42	10 02	12 12	09 52	15 15	20 08
07	06 57	19 47	16 19	23 59	02 01	10 07	12 10	09 53	15 14	20 09
08	07 19	16 26	16 44	24 14	02 20	10 12	12 08	09 54	15 14	20 09
09	07 41	12 09	17 05	24 28	02 38	10 17	12 06	09 54	15 13	20 09
10	08 04	07 07	17 24	24 42	02 57	10 22	12 05	09 55	15 13	20 10
11	08 26	01 S 34	17 39	24 55	03 16	10 28	12 03	09 55	15 12	20 10
12	08 48	04 N 12	17 52	25 08	03 34	10 33	12 01	09 56	15 12	20 10
13	09 10	09 50	18 00	25 20	03 53	10 38	11 59	09 57	15 12	20 10
14	09 31	14 56	18 06	25 32	04 11	10 43	11 57	09 57	15 11	20 10
15	09 53	19 06	18 08	25 43	04 30	10 48	11 56	09 58	15 11	20 11
16	10 14	22 01	18 07	25 53	04 48	10 53	11 54	09 58	15 10	20 11
17	10 35	23 26	18 03	26 03	05 06	10 58	11 52	09 59	15 10	20 11
18	10 56	23 19	17 55	26 13	05 24	11 03	11 51	09 59	15 09	20 11
19	11 17	21 46	17 45	26 22	05 42	11 08	11 49	09 59	15 09	20 11
20	11 37	18 59	17 31	26 30	06 01	11 13	11 47	10 00	15 08	20 11
21	11 58	15 16	17 15	26 38	06 19	11 18	11 46	10 00	15 08	20 11
22	12 18	10 54	16 56	26 45	06 36	11 23	11 44	10 01	15 08	20 12
23	12 38	06 06	16 34	26 52	06 54	11 28	11 43	10 01	15 07	20 12
24	12 58	01 N 08	16 11	26 58	07 12	11 33	11 41	10 01	15 07	20 12
25	13 17	03 S 49	15 45	27 04	07 30	11 38	11 39	10 02	15 06	20 12
26	13 37	08 34	15 18	27 09	07 47	11 43	11 38	10 02	15 06	20 12
27	13 56	12 55	14 50	27 14	08 05	11 48	11 37	10 02	15 05	20 12
28	14 15	16 44	14 21	27 18	08 22	11 53	11 35	10 03	15 05	20 12
29	14 34	19 50	13 52	27 22	08 39	11 57	11 34	10 03	15 04	20 12
30	14 N 52	22 S 06	13 N 23	27 N 25	08 N 57	12 N 02	11 S 32	10 N 03	15 S 04	20 N 12

ZODIAC SIGN ENTRIES

Date	h	m	Planets	
01	09	41	☽	♐
02	00	57	☿	♉
03	22	36	☽	♑
04	03	03	♀	♊
06	10	24	☽	♒
08	18	47	☽	♓
10	23	08	☽	♈
12	06	52	♃	♉
13	00	37	☽	♉
15	01	06	☽	♊
17	02	23	☽	♋
19	05	40	☽	♌
20	01	27	☉	♉
21	11	17	☽	♍
23	19	08	☽	♎
26	05	01	☽	♏
28	16	46	☽	♐

LATITUDES

Date	Mercury ☿ ° '	Venus ♀ ° '	Mars ♂ ° '	Jupiter ♃ ° '	Saturn ♄ ° '	Uranus ♅ ° '	Neptune ♆ ° '	Pluto ♇ ° '
01	01 N 43	02 N 49	00 S 53	01 S 02	01 S 17	00 N 48	01 N 49	14 N 16
04	02 14	03 01	00 51	01 01	01 17	00 48	01 49	14 15
07	02 40	03 12	00 50	01 01	01 18	00 48	01 49	14 15
10	02 56	03 22	00 49	01 01	01 18	00 48	01 49	14 14
13	03 03	03 32	00 47	01 01	01 19	00 48	01 49	14 14
16	02 59	03 41	00 46	01 01	01 19	00 48	01 49	14 13
19	02 42	03 48	00 44	01 00	01 20	00 48	01 49	14 12
22	02 12	03 55	00 43	01 00	01 20	00 48	01 50	14 11
25	01 31	04 01	00 41	01 00	01 21	00 48	01 50	14 10
28	00 N 43	04 05	00 40	01 00	01 21	00 47	01 50	14 09
31	00 S 08	04 N 07	00 S 38	01 S 00	01 S 22	00 N 47	01 N 50	14 N 09

DATA

Julian Date	2438487
Delta T	+35 seconds
Ayanamsa	23° 21' 10"
Synetic vernal point	05° ♓ 45' 50"
True obliquity of ecliptic	23° 26' 38"

MOON'S PHASES, APSIDES AND POSITIONS ☽

Date	h	m	Phase	Longitude ° '	Eclipse Indicator
05	05	45	☾	15 ♑ 28	
12	12	37	●	22 ♈ 38	
19	04	09	☽	29 ♋ 08	
26	17	50	○	06 ♏ 31	

Day	h	m		
02	11	49	Apogee	
14	09	30	Perigee	
30	02	13	Apogee	
04	22	07	Max dec	23° S 28'
11	18	34	ON	
17	22	05	Max dec	23° N 34'
24	17	27	OS	

LONGITUDES

Date	Chiron ⚷ ° '	Ceres ⚳ ° '	Pallas ⚴ ° '	Juno ⚵ ° '	Vesta ⚶ ° '	Black Moon Lilith ⚸ ° '
01	16 ♓ 08	04 ♑ 32	07 ♐ 12	04 ♐ 51	16 ♒ 21	08 ♐ 44
11	16 ♓ 42	05 ♑ 57	06 ♐ 29	04 ♐ 04	20 ♒ 50	09 ♐ 51
21	17 ♓ 13	06 ♑ 49	04 ♐ 58	02 ♐ 44	25 ♒ 07	10 ♐ 57
31	17 ♓ 40	07 ♑ 06	02 ♐ 44	00 ♐ 58	29 ♒ 39	12 ♐ 04

ASPECTARIAN

h m	Aspects
01 Wednesday	
01 50	☉ ± ♆
02 24	☽ ⚼ ☉
03 29	☽ ☍ ♀
04 21	☽ ⊼ ♃
07 26	☽ ⊼ ☿
11 26	☽ □ ♄
12 09	♂ ⚼ ♆
14 36	☽ △ ♂
16 01	♀ ⊻ ♃
16 43	☽ ± ♃
21 22	☽ ± ☿
22 09	☉ ⊼ ♇
23 07	☽ □ ♅
02 Thursday	
08 34	☉ ± ♅
09 01	☽ # ♇
10 23	☽ □ ♇
11 21	☽ ⚼ ♃
11 32	☽ △ ☉
17 10	☿ ✶ ♄
18 15	☽ ⚼ ☿
20 51	☽ ⊻ ♆
03 Friday	
00 17	☽ Q ♄
09 02	☽ ⊥ ♆
18 29	☽ △ ♃
22 11	☽ ⊼ ♀
04 Saturday	
00 52	☽ ✶ ♄
01 21	☽ # ♀
03 16	☽ ∠ ♆
04 44	☽ △ ☿
07 48	☽ □ ♂
09 13	♃ ∥ ♅
11 28	☽ ± ♀
11 52	☽ △ ♅
23 05	☽ △ ♇
05 Sunday	
05 45	☽ □ ☉
06 30	☽ # ♀
07 12	☽ ⚼ ♀
07 18	☽ ∠ ♄
08 27	☿ # ♆
08 37	♀ □ ♄
09 21	☽ ✶ ♆
17 46	☽ ⚼ ♅
22 15	☽ Q ♂
06 Monday	
01 15	☽ ⊥ ♄
02 37	☉ ∠ ♄
04 46	☽ ⚼ ♇
07 38	☽ □ ♃
08 57	☽ Q ♆
11 23	☽ ± ♅
13 04	☽ ⊻ ♄
15 20	☽ △ ♀
19 56	♂ ⊼ ♅
20 10	☽ Q ☉
22 11	☽ ± ♇
22 31	☽ □ ☿
22 56	☽ ⊼ ♅
23 08	☽ ✶ ♂
07 Tuesday	
00 55	☉ ⊼ ♆
03 22	☿ △ ♅
08 51	☽ # ♇
09 36	☽ ⊼ ♇
18 43	☽ Q ♃
19 16	☽ □ ♆
20 50	☽ ✶ ☉
21 20	☉ ± ♇
08 Wednesday	
03 59	♂ ⊥ ♄
05 11	☽ ∠ ♂
10 20	☽ # ☿
10 57	☿ ⊻ ♂
11 03	☽ Q ☿
17 14	☽ ✶ ♃
19 18	☽ ∥ ♆
21 39	☽ ☌ ♄
22 37	☽ ⊥ ♂
09 Thursday	
02 28	☽ ∠ ☉
03 52	☽ □ ♀
06 18	☽ ☍ ♅
09 59	☽ ⊻ ♂
10 07	☽ ✶ ☿
10 12	♀ ⊥ ♃
12 16	☽ ∥ ♄
16 15	☽ ☍ ♇
19 37	☽ ⊥ ☉
20 24	☽ ∠ ♃
21 08	☽ # ♃
23 06	☽ # ♅
10 Friday	
01 12	☽ △ ♆
06 50	☽ ⊻ ☉
08 02	☽ # ☉
12 14	☽ ⊥ ♃
12 20	♀ □ ♅
13 39	☽ ∠ ☿
13 42	☽ Q ♀
19 07	☿ ⊻ ♂
22 36	☽ ⊻ ♃
11 Saturday	
02 10	☽ ⊻ ♄
02 45	☽ ⚼ ♆
05 14	☽ # ♂
05 35	☽ ⊥ ☿
05 42	☉ ⚼ ♅
09 49	☽ ⊼ ♅
11 25	☽ ✶ ♀
12 16	☽ ⊥ ♄
16 00	☽ ⊻ ☿
16 15	☽ ☌ ♂
17 44	☽ ± ♆
19 10	☽ ⊼ ♇
19 45	☽ ± ♅
12 Sunday	
03 14	☽ ∠ ♄
03 36	☽ ⊼ ♆
05 02	☽ ± ♇
09 14	☽ ∥ ♂
10 30	☽ ⚼ ♅
12 37	☽ ☌ ☉
13 52	☽ ∠ ♀
19 16	♂ ± ♆
19 43	☽ ⚼ ♇
13 Monday	
00 54	☽ ☌ ♃
03 50	☽ ✶ ♄
05 29	☽ ⊥ ♀
08 49	☽ ∥ ☉
10 49	☽ △ ♅
12 29	☽ ∥ ♅
15 35	☽ ∥ ♃
15 51	☽ ⊻ ♀
18 25	☽ ☌ ☿
19 50	☽ ⊻ ♂
19 57	☽ △ ♇
21 40	☽ # ♄
21 56	♂ ⊼ ♇
23 20	☽ Q ♄
14 Tuesday	
04 12	☽ ☍ ♆
06 03	☽ ⊥ ♂
08 03	♂ ± ♅
13 17	☽ # ♆
16 36	☽ ⊻ ☉
21 23	☽ ∠ ♂
15 Wednesday	
02 11	☽ ⊻ ♃
03 00	☽ ⊥ ☉
04 37	☽ □ ♄
05 46	☽ ∥ ☿
11 16	☽ □ ♅
12 06	☽ ⊥ ♃
17 42	☉ ∥ ♅
18 42	☽ ∠ ☉
19 38	☽ ∥ ♇
19 40	☿ ⊻ ♀
19 45	☽ ⊻ ☿
19 45	☽ ☌ ♀
20 31	☽ □ ♇
23 09	☽ ✶ ♂
16 Thursday	
03 04	☽ ∠ ♃
04 52	☽ ⊼ ♆
05 39	☽ ⊥ ☿
07 25	♀ □ ♇
14 45	☽ ± ♆
16 47	☽ Q ♅
20 04	☽ Q ♂
20 35	☽ ∠ ☿
21 13	☽ ✶ ☉
21 30	☉ ⚼ ♇
21 51	☿ St R
17 Friday	
02 14	☽ Q ♇
04 21	☽ ✶ ♃
05 44	☽ ⚼ ♆
06 19	☽ △ ♄
12 49	☽ ✶ ♅
18 51	☽ Q ☉
21 44	☽ ✶ ☿
22 26	☽ ✶ ♇
18 Saturday	
00 58	☽ Q ♃
01 13	☽ ⊻ ♀
04 08	☽ □ ♂
07 06	☽ △ ♆
07 55	☽ ⚼ ♄
12 13	☽ ⊥ ♀
14 22	☽ ∠ ♅
18 06	☽ Q ☿
22 26	☉ ∥ ♃
23 33	☽ ± ♄
19 Sunday	
00 13	☽ ∠ ♇
04 09	☽ □ ☉
04 55	☽ ∠ ♀
05 58	☽ ⊥ ♅
08 38	☽ □ ♃
10 06	☽ ⊼ ♄
15 57	☽ ⊥ ♇
16 29	☽ ⊻ ♅
20 Monday	
01 08	☽ □ ☿
02 35	☽ ⊻ ♇
02 42	☽ ∥ ♇
08 50	♂ ⊼ ♆
09 19	☽ ✶ ♀
11 36	☽ □ ♆
11 47	☽ △ ♂
22 38	☉ # ♄
22 53	☽ ∥ ☿
21 Tuesday	
08 32	☽ Q ♀
09 10	♂ ∠ ♄
10 27	☿ ⊥ ♀
12 50	☽ # ♆
13 38	♂ ± ♇
14 00	☽ △ ☉
15 24	☽ △ ♃
16 15	☽ ☍ ♄
16 39	☽ ⚼ ♂
19 07	♀ ⊼ ♆
20 13	☽ Q ♆
22 31	☽ ☌ ♅
22 Wednesday	
05 06	☽ ∥ ☉
06 03	☽ △ ☿
07 35	☽ # ♄
09 03	☽ ☌ ♇
09 30	☽ ∥ ♃
10 17	☽ ± ♂
10 28	♀ ∠ ♃
14 20	☉ ☌ ♃
16 32	☽ ∥ ♅
18 24	☽ ✶ ♆
19 40	☽ ⚼ ♃
19 59	☽ ⚼ ☉
20 08	☽ □ ♀
22 10	☽ ⊼ ♂
22 46	☉ ✶ ♄
23 Thursday	
08 18	☽ ∥ ♂
09 01	☽ ⚼ ☿
11 48	☉ ∠ ♀
12 53	☽ ± ♃
14 18	☽ ± ☉
22 35	☽ ∠ ♆
24 Friday	
00 29	☽ ⊼ ♃
00 39	☽ ⊼ ♄
01 21	☽ ± ☿
02 38	☽ ⊼ ☉
06 42	☽ ⊻ ♅
12 10	☽ ± ♄
12 19	☽ ⊼ ☿
13 05	♃ ✶ ♄
15 45	☽ ⊥ ♆
17 38	☽ ⊻ ♇
18 11	☽ ⊥ ♅
25 Saturday	
03 17	☽ ⊻ ♆
05 12	☽ ⊥ ♇
05 37	☽ ⚼ ♄
09 24	☽ △ ♀
11 03	☽ ☍ ♂
11 33	☽ ∠ ♅
18 20	♃ # ♄
18 59	♂ ⚼ ♅
22 41	☽ ∠ ♇
26 Sunday	
06 25	☉ △ ♅
07 42	☽ # ♂
11 05	☽ △ ♄
11 42	☽ ☍ ♃
16 53	☽ ⚼ ♀
16 54	☽ ✶ ♅
17 50	☽ ☍ ☉
19 50	☽ # ♅
19 54	☽ ☍ ☿
22 57	☿ # ♆
27 Monday	
04 13	☽ ✶ ♇
04 33	☽ ∥ ♄
05 25	☽ # ♃
10 10	☉ ☌ ☿
12 02	☽ ± ♀
14 07	☽ ☌ ♆
16 43	☽ Q ♅
18 32	☽ # ☉
19 14	☿ ∠ ♀
22 17	☽ # ☿
28 Tuesday	
00 53	☽ ⊼ ♀
01 07	☽ ∥ ♆
02 17	☽ ⊼ ♂
04 10	☽ Q ♇
15 10	☉ ∥ ☿
15 11	☽ ± ♂
18 34	♀ ± ♆
23 22	☽ □ ♄
29 Wednesday	
00 53	☽ ⊼ ♃
04 22	☿ △ ♅
04 53	☽ ⊼ ☿
04 55	☽ □ ♅
10 38	☽ ⚼ ♂
11 18	☽ ⊼ ☉
13 18	☽ ± ♃
15 15	☽ # ♇
16 25	☽ ± ☿
16 31	☽ □ ♇
30 Thursday	
00 35	☽ ± ☉
02 32	☽ ⊻ ♆
07 59	☽ ⚼ ♃
09 47	☽ ⚼ ☿
10 42	♀ Q ♅
12 09	☽ Q ♄
14 44	☽ ⊥ ♆
17 56	☽ ☍ ♀
19 14	☽ △ ♂
20 33	☽ ⚼ ☉

All ephemeris data is given at 12.00 UT and the Moon's longitude is additionally given for 24.00 UT

MAY 1964

LONGITUDES

Date	Sidereal time h m s	Sun ☉ ° ' "	Moon ☽ ° ' "	Moon ☽ 24.00 ° ' "	Mercury ☿ ° '	Venus ♀ ° '	Mars ♂ ° '	Jupiter ♃ ° '	Saturn ♄ ° '	Uranus ♅ ° '	Neptune ♆ ° '	Pluto ♇ ° '
01	02 37 47	11 ♉ 07 44	03 ♑ 05 40	09 ♑ 00 23	04 ♉ 32	24 ♊ 47	25 ♈ 23	04 ♉ 36	03 ♓ 27	05 ♍ 59	16 ♏ 37	11 ♍ 42
02	02 41 44	12 05 56	14 ♑ 56 30	20 ♑ 54 36	03 R 58	25 32	26 09	04 51	03 31	05 R 58	16 R 36	11 R 42
03	02 45 40	13 04 06	26 ♑ 55 18	02 ♒ 59 14	03 26	26 15	26 54	05 05	03 35	05 58	16 34	11 41
04	02 49 37	14 02 15	09 ♒ 07 03	15 ♒ 19 23	02 57	26 58	27 39	05 19	03 39	05 57	16 32	11 41
05	02 53 33	15 00 22	21 ♒ 36 51	28 ♒ 00 01	02 32	27 39	28 25	05 34	03 43	05 57	16 31	11 40
06	02 57 30	15 58 27	04 ♓ 29 24	11 ♓ 05 25	02 11	28 19	29 10	05 48	03 47	05 56	16 29	11 39
07	03 01 27	16 56 31	17 ♓ 48 25	24 ♓ 38 35	01 55	28 58	29 ♈ 55	06 02	03 50	05 56	16 27	11 39
08	03 05 23	17 54 34	01 ♈ 35 56	08 ♈ 40 21	01 42	29 ♊ 36	00 ♉ 40	06 16	03 54	05 56	16 26	11 39
09	03 09 20	18 52 36	15 ♈ 51 29	23 ♈ 07 36	01 34	00 ♋ 13	01 25	06 31	03 57	05 55	16 24	11 38
10	03 13 16	19 50 36	00 ♉ 31 33	07 ♉ 58 52	01 31	00 49	02 10	06 45	04 01	05 55	16 23	11 38
11	03 17 13	20 48 34	15 ♉ 29 40	23 ♉ 02 47	01 D 33	01 23	02 55	06 59	04 04	05 55	16 21	11 37
12	03 21 09	21 46 32	00 ♊ 37 01	08 ♊ 11 07	01 39	01 56	03 40	07 13	04 07	05 55	16 19	11 37
13	03 25 06	22 44 27	15 ♊ 43 55	23 ♊ 14 17	01 50	02 28	04 25	07 27	04 10	05 D 55	16 18	11 37
14	03 29 02	23 42 21	00 ♋ 41 17	08 ♋ 04 05	02 05	02 58	05 10	07 41	04 14	05 55	16 16	11 36
15	03 32 59	24 40 14	15 ♋ 22 03	22 ♋ 34 41	02 25	03 26	05 54	07 56	04 16	05 55	16 14	11 36
16	03 36 56	25 38 04	29 ♋ 41 41	06 ♌ 42 54	02 49	03 53	06 39	08 10	04 19	05 55	16 13	11 36
17	03 40 52	26 35 53	13 ♌ 38 17	20 ♌ 27 55	03 17	04 18	07 24	08 24	04 22	05 55	16 11	11 36
18	03 44 49	27 33 40	27 ♌ 11 59	03 ♍ 50 42	03 50	04 42	08 08	08 38	04 25	05 56	16 10	11 36
19	03 48 45	28 31 25	10 ♍ 24 23	16 ♍ 53 20	04 26	05 04	08 53	08 52	04 28	05 56	16 08	11 36
20	03 52 42	29 ♉ 29 09	23 ♍ 17 53	29 ♍ 38 22	05 06	05 24	09 37	09 05	04 30	05 56	16 06	11 35
21	03 56 38	00 ♊ 26 51	05 ♎ 55 09	12 ♎ 08 33	05 51	05 42	10 22	09 19	04 32	05 57	16 05	11 35
22	04 00 35	01 24 32	18 ♎ 18 52	24 ♎ 26 24	06 38	05 58	11 06	09 33	04 35	05 57	16 03	11 35
23	04 04 31	02 22 11	00 ♏ 31 27	06 ♏ 34 16	07 29	06 12	11 51	09 47	04 37	05 58	16 02	11 D 35
24	04 08 28	03 19 48	12 ♏ 35 06	18 ♏ 34 12	08 24	06 25	12 35	10 01	04 39	05 58	16 00	11 35
25	04 12 25	04 17 25	24 ♏ 31 49	00 ♐ 28 11	09 22	06 35	13 19	10 14	04 41	05 59	15 59	11 35
26	04 16 21	05 15 00	06 ♐ 23 31	12 ♐ 18 06	10 23	06 42	14 03	10 28	04 43	05 59	15 57	11 36
27	04 20 18	06 12 34	18 ♐ 12 11	24 ♐ 06 02	11 27	06 48	14 47	10 42	04 45	06 00	15 56	11 36
28	04 24 14	07 10 07	00 ♑ 00 00	05 ♑ 54 23	12 35	06 51	15 31	10 55	04 47	06 01	15 54	11 36
29	04 28 11	08 07 39	11 ♑ 49 32	17 ♑ 45 52	13 45	06 R 52	16 15	11 09	04 49	06 02	15 53	11 36
30	04 32 07	09 05 09	23 ♑ 43 47	29 ♑ 43 44	14 58	06 51	16 59	11 22	04 50	06 03	15 51	11 36
31	04 36 04	10 ♊ 02 39	05 ♒ 46 11	11 ♒ 51 37	16 ♉ 15	06 ♋ 47	17 ♉ 43	11 ♉ 36	04 ♓ 52	06 ♍ 04	15 ♏ 50	11 ♍ 37

Moon True ☊, Moon Mean ☊, Moon Latitude, DECLINATIONS

Date	Moon True ☊ ° '	Moon Mean ☊ ° '	Moon Latitude ° '	Sun ☉ ° '	Moon ☽ ° '	Mercury ☿ ° '	Venus ♀ ° '	Mars ♂ ° '	Jupiter ♃ ° '	Saturn ♄ ° '	Uranus ♅ ° '	Neptune ♆ ° '	Pluto ♇ ° '
01	03 ♋ 10	04 ♋ 56	00 00	15 N 10	23 S 24	12 N 54	27 N 27	09 N 14	12 N 07	11 S 31	10 N 03	15 S 03	20 N 12
02	03 D 11	04 52	01 S 03	15 28	23 39	12 26	27 29	09 31	12 12	11 30	10 03	15 03	20 12
03	03 12	04 49	02 05	15 46	22 49	12 00	27 31	09 48	12 17	11 28	10 04	15 02	20 12
04	03 13	04 46	03 02	16 03	20 54	11 35	27 32	10 04	12 21	11 27	10 04	15 02	20 12
05	03 R 13	04 43	03 52	16 20	17 57	11 11	27 33	10 21	12 26	11 26	10 04	15 02	20 12
06	03 11	04 40	04 31	16 37	14 04	10 50	27 33	10 38	12 31	11 25	10 04	15 01	20 12
07	03 07	04 37	04 58	16 54	09 24	10 30	27 33	10 54	12 36	11 24	10 04	15 01	20 11
08	03 02	04 33	05 09	17 10	04 S 05	10 13	27 32	11 10	12 40	11 23	10 04	15 00	20 11
09	02 56	04 30	05 01	17 26	01 N 36	09 59	27 31	11 27	12 45	11 21	10 04	15 00	20 11
10	02 49	04 27	04 34	17 42	07 23	09 47	27 30	11 43	12 50	11 20	10 04	14 59	20 11
11	02 43	04 24	03 48	17 58	12 51	09 37	27 28	11 59	12 54	11 19	10 04	14 59	20 11
12	02 38	04 21	02 45	18 13	17 36	09 30	27 25	12 14	12 59	11 18	10 04	14 58	20 11
13	02 35	04 17	01 30	18 28	21 11	09 25	27 23	12 30	13 04	11 17	10 04	14 58	20 10
14	02 34	04 14	00 S 10	18 42	23 16	09 23	27 19	12 46	13 08	11 16	10 04	14 57	20 10
15	02 D 35	04 11	01 N 09	18 56	23 42	09 23	27 16	13 01	13 13	11 16	10 04	14 57	20 10
16	02 36	04 08	02 22	19 10	22 32	09 26	27 12	13 16	13 17	11 15	10 04	14 56	20 10
17	02 37	04 05	03 25	19 24	20 01	09 30	27 08	13 32	13 22	11 14	10 04	14 56	20 10
18	02 38	04 02	04 15	19 37	16 26	09 37	27 03	13 47	13 26	11 13	10 04	14 56	20 09
19	02 R 37	03 58	04 49	19 50	12 08	09 46	26 58	14 01	13 31	11 12	10 04	14 55	20 09
20	02 35	03 55	05 08	20 03	07 23	09 57	26 53	14 16	13 35	11 12	10 03	14 55	20 09
21	02 32	03 52	05 12	20 15	02 N 25	10 10	26 47	14 31	13 39	11 11	10 03	14 54	20 08
22	02 27	03 49	05 00	20 27	02 S 33	10 25	26 42	14 45	13 44	11 10	10 03	14 54	20 08
23	02 22	03 46	04 36	20 38	07 21	10 42	26 35	14 59	13 48	11 10	10 03	14 53	20 08
24	02 17	03 43	03 59	20 50	11 49	11 00	26 29	15 13	13 52	11 09	10 03	14 53	20 07
25	02 12	03 39	03 12	21 00	15 48	11 20	26 22	15 27	13 57	11 08	10 02	14 53	20 07
26	02 09	03 36	02 17	21 11	19 08	11 41	26 15	15 41	14 01	11 08	10 02	14 52	20 07
27	02 07	03 33	01 16	21 21	21 40	12 03	26 07	15 55	14 05	11 07	10 02	14 52	20 06
28	02 06	03 30	00 N 12	21 31	23 15	12 27	25 59	16 08	14 09	11 07	10 01	14 51	20 06
29	02 D 06	03 27	00 S 53	21 40	23 48	12 52	25 51	16 22	14 14	11 07	10 01	14 51	20 06
30	02 08	03 23	01 56	21 49	23 16	13 18	25 42	16 35	14 18	11 06	10 01	14 51	20 05
31	02 ♋ 09	03 ♋ 20	02 S 55	21 N 58	21 S 40	13 N 45	25 N 33	16 N 48	14 N 22	11 S 06	10 N 00	14 S 50	20 N 04

ZODIAC SIGN ENTRIES

Date	h	m	Planets
01	05	42	☽ ♑
03	18	06	☽ ♒
06	03	43	☽ ♓
07	14	41	♂ ♉
08	09	16	☽ ♈
09	03	16	♀ ♋
10	11	09	☽ ♉
12	11	01	☽ ♊
14	10	53	☽ ♋
16	12	31	☽ ♌
18	17	02	☽ ♍
21	00	41	☽ ♎
21	00	50	☉ ♊
23	10	58	☽ ♏
25	23	03	☽ ♐
28	12	00	☽ ♑
31	00	32	☽ ♒

LATITUDES

Date	Mercury ☿ ° '	Venus ♀ ° '	Mars ♂ ° '	Jupiter ♃ ° '	Saturn ♄ ° '	Uranus ♅ ° '	Neptune ♆ ° '	Pluto ♇ ° '
01	00 S 08	04 N 07	00 S 38	01 S 00	01 S 22	00 N 47	01 N 50	14 N 09
04	00 59	04 08	00 37	01 00	01 23	00 47	01 50	14 08
07	01 44	04 06	00 35	01 00	01 23	00 47	01 50	14 07
10	02 22	04 05	00 33	01 00	01 24	00 47	01 50	14 06
13	02 52	03 57	00 31	01 00	01 24	00 47	01 50	14 05
16	03 13	03 49	00 30	01 00	01 25	00 47	01 50	14 03
19	03 26	03 38	00 28	01 00	01 25	00 47	01 50	14 02
22	03 30	03 23	00 26	01 00	01 26	00 46	01 50	14 01
25	03 28	03 05	00 24	01 00	01 27	00 46	01 49	14 00
28	03 19	02 43	00 22	01 00	01 27	00 46	01 49	13 59
31	03 S 04	02 N 17	00 S 20	01 S 00	01 S 28	00 N 46	01 N 49	13 N 58

DATA

Julian Date	2438517
Delta T	+35 seconds
Ayanamsa	23° 21' 13"
Synetic vernal point	05° ♓ 45' 46"
True obliquity of ecliptic	23° 26' 37"

MOON'S PHASES, APSIDES AND POSITIONS ☽

Date	h	m	Phase	Longitude ° '	Eclipse Indicator
04	22	20	☾	14 ♒ 27	
11	21	02	●	21 ♉ 10	
18	12	42	☽	27 ♌ 35	
26	09	29	○	05 ♐ 09	

Day	h	m		
12	16	10	Perigee	
27	08	44	Apogee	
02	05	36	Max dec	23° S 41'
09	05	20	0N	
15	06	14	Max dec	23° N 45'
21	23	39	0S	
29	12	18	Max dec	23° S 48'

LONGITUDES

Date	Chiron ⚷ ° '	Ceres ⚳ ° '	Pallas ⚴ ° '	Juno ⚵ ° '	Vesta ⚶ ° '	Black Moon Lilith ⚸ ° '
01	17 ♓ 40	07 ♑ 06	02 ♐ 44	00 ♐ 58	29 ♒ 10	12 ♐ 06
11	18 ♓ 03	06 ♑ 47	29 ♏ 57	28 ♏ 52	02 ♓ 58	13 ♐ 10
21	18 ♓ 22	05 ♑ 26	26 ♏ 56	26 ♏ 38	06 ♓ 26	14 ♐ 17
31	18 ♓ 36	04 ♑ 20	24 ♏ 27	24 ♏ 29	09 ♓ 32	15 ♐ 24

All ephemeris data is given at 12.00 UT and the Moon's longitude is additionally given for 24.00 UT

ASPECTARIAN

h m Aspects

01 Friday
03 10 ☉ # ♆
09 00 ☽ ∠ ♆
10 06 ☿ ☌ ♃
12 43 ☽ ⚹ ♄
14 48 ☽ △ ☿
15 08 ☽ △ ♃
17 51 ☽ △ ♅

02 Saturday
02 06 ☉ △ ♇
05 27 ☽ △ ♇
05 45 ☽ △ ☉
15 19 ☽ ⚹ ♆
19 14 ☽ ∠ ♄
22 59 ☿ ∥ ♃

03 Sunday
00 07 ☽ Ᵽ ♅
05 14 ♂ Ᵽ ♇
05 41 ☿ ⚹ ♄
10 35 ☽ ⊼ ♀
11 32 ☽ Ᵽ ♇
11 57 ☽ □ ♂
13 19 ☽ ⊥ ♄
15 15 ☽ Q ♆
18 01 ☽ ± ♅
23 12 ☽ ± ♀

04 Monday
00 23 ☽ □ ☿
01 15 ☽ ⊻ ♄
04 26 ☽ □ ♃
05 17 ☽ ± ♇
05 49 ☽ ⊼ ♅
11 04 ♂ ∥ ♅
16 58 ☽ ⊼ ♇
17 50 ☽ Ᵽ ♀
18 33 ☽ # ♇
19 45 ☿ # ♄
22 20 ☽ □ ☉

05 Tuesday
01 27 ☽ Q ♂
02 18 ☽ □ ♆
10 01 ☽ Q ☿
15 45 ☽ Q ♃
21 55 ☽ # ☉
23 58 ☽ △ ♀

06 Wednesday
01 33 ☽ ⚹ ♂
06 37 ☽ ∥ ♆
06 44 ☉ Q ♄
07 52 ☽ ⚹ ☿
10 41 ☽ ☌ ♄
10 59 ☽ Q ☉
14 26 ☽ ⚹ ♃
14 39 ☽ ☍ ♅
19 59 ☿ ∥ ♂
20 16 ☽ # ♃

07 Thursday
00 18 ☉ ☍ ♆
01 01 ☽ ☍ ♇
01 34 ♃ △ ♅
02 08 ☽ ∥ ♄
05 03 ☽ # ♂
06 14 ☽ # ☿
06 33 ☽ ∠ ♂
08 45 ☽ # ♅
09 37 ☽ △ ♆
10 21 ☽ ⚹ ☉
10 26 ☽ ∠ ☿
17 48 ☽ ∠ ♃
23 21 ☽ ⊥ ♂

08 Friday
01 59 ☽ ⊥ ☿
08 25 ☽ □ ♀
09 41 ☽ ⊥ ♃
10 19 ☽ ⊻ ♂
11 43 ☽ Ᵽ ♆
12 11 ☽ ⊻ ☿
13 20 ♀ Q ♇
14 24 ☽ ∠ ☉
15 56 ☽ ⊻ ♄
19 22 ☽ ⊼ ♅
20 05 ☽ ⊻ ♃

09 Saturday
02 07 ☽ ⊥ ♄
02 44 ☿ ∥ ♅
02 55 ☽ ± ♆
04 55 ♂ # ♄
04 58 ☽ ⊼ ♇
05 27 ☽ ± ♅
06 41 ☽ ⊥ ☉
12 54 ☽ ⊼ ♆
14 56 ☽ ± ♇
16 05 ☽ Q ♀
16 27 ☿ ☌ ♂
17 08 ☽ ∠ ♄
17 21 ☽ ⊻ ☉
20 21 ☽ Ᵽ ♅

10 Sunday
05 41 ☽ Ᵽ ♇
12 29 ☽ ⚹ ♀
13 36 ☽ ☌ ☿
14 48 ☽ ☌ ♂
16 08 ☿ St D
17 39 ☽ ⚹ ♄
20 41 ☽ △ ♅
21 57 ☽ ∥ ☿
22 11 ☽ ☌ ♃
23 32 ☽ ∥ ♅

11 Monday
05 05 ☽ # ♄
05 50 ☽ △ ♇
07 47 ☽ ∥ ♂
10 29 ♀ Ᵽ ♆
12 15 ☽ ∥ ♃
12 55 ☽ Q ♄
13 21 ☽ ☍ ♆
13 28 ☽ ∠ ♀
20 06 ☿ ⚹ ♀
21 02 ☽ ☌ ☉
22 10 ☽ # ♆

12 Tuesday
04 18 ☽ ⊥ ♀
13 39 ☽ ⊻ ☿
14 10 ☽ ⊻ ♀
15 48 ☽ ∥ ☉
17 05 ☽ ⊻ ♂
17 34 ☽ □ ♄
20 24 ☽ □ ♅
22 38 ☽ ⊻ ♃
23 16 ☽ ⊥ ☿

13 Wednesday
03 06 ☽ ⊥ ♂
03 41 ♂ ⚹ ♄
04 15 ☽ ∥ ♇
05 27 ☽ □ ♇
08 19 ☽ ⊥ ♃
11 27 ♅ St D
12 54 ☽ ⊼ ♆
13 47 ☽ ∠ ☿
18 11 ☽ ∠ ♂
22 28 ☽ ± ♆
22 55 ☽ ∠ ♃
23 58 ☽ ⊻ ☉

14 Thursday
01 05 ☽ Q ♅
10 15 ☽ Q ♇
10 18 ☽ ⊥ ☉
12 56 ☽ Ᵽ ♆
14 18 ☽ ⚹ ☿
15 48 ☽ ☌ ♀
17 45 ☽ △ ♄
19 39 ☽ ⚹ ♂
20 29 ☽ ⚹ ♅
23 34 ☽ ⚹ ♃

15 Friday
01 58 ☽ ∠ ☉
05 48 ☽ ⚹ ♇
10 23 ☽ Q ☿
12 18 ♂ △ ♅
13 26 ☽ △ ♆
16 26 ☽ Q ♂
18 30 ☽ Ᵽ ♄
19 42 ☽ Q ♃
21 13 ☽ ∠ ♅

16 Saturday
04 38 ☽ ⚹ ☉
06 46 ☽ ∠ ♇
09 40 ☽ ± ♄
12 23 ☽ ⊥ ♅
13 48 ♂ ∥ ♃
17 29 ☽ □ ☿
19 22 ☽ ⊻ ♀
19 55 ☽ ⊼ ♄
22 05 ☽ ⊥ ♇
22 38 ☽ ⊻ ♅

17 Sunday
00 34 ☽ □ ♂
02 36 ☽ Q ☉
02 44 ☽ □ ♃
06 02 ☽ ⊥ ♀
08 27 ☽ ⊻ ♇
10 51 ☽ ∥ ♇
16 10 ♀ △ ♄
16 17 ☽ ∥ ☉
16 27 ☽ □ ♆
22 15 ☽ ∠ ♀

18 Monday
12 42 ☽ □ ☉
20 47 ☽ # ♆

19 Tuesday
00 32 ☽ △ ☿
00 33 ☽ Q ♆
01 05 ☽ ☍ ♄
01 57 ☽ ⚹ ♀
02 21 ☽ ∥ ♂
03 48 ☽ ☌ ♅
04 44 ☽ ∥ ♃
09 02 ☽ △ ♂
09 06 ☽ △ ♃
10 54 ♂ ☌ ♃
12 54 ☿ ⚹ ♄
14 11 ☽ ☌ ♇
16 48 ☽ # ♄
22 34 ☽ ⚹ ♆
22 39 ☽ ∥ ♅
23 40 ☽ ∥ ☿

20 Wednesday
00 40 ☽ Q ♀
05 40 ☽ Ᵽ ☿
08 45 ☉ ⊥ ♀
13 31 ☽ Ᵽ ♃
14 39 ☽ Ᵽ ♂
23 14 ☉ ∥ ♇
23 33 ☿ ∥ ♅
23 39 ☽ ± ☿

21 Thursday
00 40 ☽ △ ☉
02 46 ☽ ∠ ♆
04 44 ☿ ⚹ ♀
06 56 ☽ ± ♃
08 50 ☽ ± ♂
09 21 ☽ ⊼ ♄
11 34 ☽ □ ♀
11 51 ☽ ⊼ ☿
12 03 ☽ ⊻ ♅
15 12 ☿ △ ♅
18 40 ☽ ⊼ ♃
20 00 ☽ ⊥ ♆
20 56 ☽ ± ♄
21 06 ☽ ⊼ ♂
22 56 ☽ ⊻ ♇
23 37 ☽ ⊥ ♅

22 Friday
07 36 ☽ ⊻ ♆
07 58 ☽ Ᵽ ☉
10 02 ♀ ⚹ ♅
10 35 ☽ ⊥ ♇
14 29 ☽ Ᵽ ♄
17 10 ☽ ∠ ♅
21 27 ♇ St D

23 Saturday
02 16 ♂ # ♆
03 06 ☽ ± ☉
03 49 ♂ △ ♇
04 14 ☽ ∠ ♇
15 58 ☽ ⊼ ☉
20 08 ☽ △ ♄
22 48 ☽ ⚹ ♅
23 29 ☽ △ ♀

24 Sunday
02 11 ☽ # ♅
02 57 ☽ ☍ ☿
06 46 ☽ ☍ ♃
07 02 ☽ # ☿
08 15 ☽ ∥ ♄
10 01 ☽ ⚹ ♇
11 59 ☽ ☍ ♂
18 50 ☽ ☌ ♆
22 48 ☽ Q ♅
23 08 ☿ # ♄

25 Monday
00 09 ☽ # ♃
05 58 ☽ Ᵽ ♀
06 05 ☽ ∥ ♆
09 36 ☽ # ♂
10 06 ☽ Q ♇
22 19 ☉ □ ♄

26 Tuesday
00 22 ☽ ± ♀
08 36 ☽ □ ♄
09 29 ☽ ☍ ☉
11 11 ☽ □ ♅
12 39 ☽ ⊼ ♀
14 28 ☿ ☌ ♃
20 18 ☽ # ♇
20 26 ☽ ⊼ ♃
20 54 ☽ ⊼ ☿
22 34 ☽ □ ♇

27 Wednesday
04 35 ☽ ⊼ ♂
06 46 ☉ □ ♅
07 23 ☽ ⊻ ♆
08 13 ☽ # ☉
08 53 ☽ ± ♃
10 20 ☽ ± ☿
15 03 ☿ △ ♇
17 36 ☽ ± ♂
19 34 ☽ ⊥ ♆
21 17 ☽ Q ♄

28 Thursday
03 33 ☽ Ᵽ ♃
03 50 ☉ ⊻ ♀
06 33 ☽ Ᵽ ☿
13 08 ☽ Ᵽ ♂
13 50 ☽ ∠ ♆
21 44 ☽ ⚹ ♄

29 Friday
00 03 ♂ ☍ ♆
00 14 ☽ △ ♅
01 57 ☽ ☍ ♀
03 51 ☽ ⊼ ☉
10 29 ♀ St R
10 36 ☽ △ ♃
11 33 ☽ △ ♇
16 20 ☽ △ ☿
17 04 ☽ ± ☉
20 10 ☽ ⚹ ♆
21 32 ☽ △ ♂

30 Saturday
04 10 ☽ ∠ ♄
06 36 ☽ Ᵽ ♅
06 55 ♂ Q ♄
12 47 ☽ Ᵽ ☉
17 46 ☽ Ᵽ ♇
20 14 ☽ Q ♆
22 15 ☽ ⊥ ♄

31 Sunday
00 39 ☽ ± ♅
04 25 ☿ ☍ ♆
08 45 ☽ # ☉
10 12 ☽ ⊻ ♄
11 41 ☽ ± ♇
12 34 ☽ ⊼ ♅
13 04 ♃ △ ♇
14 00 ☽ ⊼ ♀
21 09 ☽ △ ☉
23 31 ☽ ⊼ ♇
23 37 ☿ Q ♄
23 42 ☽ □ ♃

JUNE 1964

Date	Sidereal time h m s	Sun ☉ ° ' "	Moon ☽ ° ' "	Moon ☽ 24.00 ° ' "	Mercury ☿ ° '	Venus ♀ ° '	Mars ♂ ° '	Jupiter ♃ ° '	Saturn ♄ ° '	Uranus ♅ ° '	Neptune ♆ ° '	Pluto ♇ ° '
01	04 40 00	11 ♊ 00 09	18 ♒ 00 34	24 ♒ 13 31	17 ♉ 33	06 ♋ 41	18 ♉ 27	11 ♉ 49	04 ♓ 53	06 ♍ 05	15 ♏ 48	11 ♍ 37
02	04 43 57	11 57 37	00 ♓ 31 01	06 ♓ 53 33	18 55	06 R 33	19 11	12 03	04 54	06 06	15 R 47	11 37
03	04 47 54	12 55 04	13 ♓ 21 36	19 ♓ 55 34	20 20	06 22	19 54	12 16	04 56	06 07	15 45	11 37
04	04 51 50	13 52 31	26 ♓ 35 51	03 ♈ 22 41	21 47	06 09	20 38	12 29	04 57	06 08	15 44	11 38
05	04 55 47	14 49 58	10 ♈ 16 16	17 ♈ 16 37	23 17	05 53	21 22	12 43	04 58	06 09	15 43	11 38
06	04 59 43	15 47 23	24 ♈ 23 36	01 ♉ 36 57	24 50	05 35	22 05	12 56	04 59	06 10	15 41	11 39
07	05 03 40	16 44 48	08 ♉ 56 11	16 ♉ 20 40	26 25	05 15	22 49	13 09	04 59	06 12	15 40	11 39
08	05 07 36	17 42 13	23 ♉ 49 34	01 ♊ 21 57	28 03	04 52	23 32	13 22	05 00	06 13	15 39	11 40
09	05 11 33	18 39 36	08 ♊ 56 41	16 ♊ 32 37	29 ♉ 44	04 27	24 16	13 35	05 01	06 14	15 37	11 40
10	05 15 29	19 36 59	24 ♊ 08 31	01 ♋ 43 12	01 ♊ 27	04 01	24 59	13 48	05 01	06 16	15 36	11 41
11	05 19 26	20 34 22	09 ♋ 15 29	16 ♋ 44 21	03 13	03 32	25 42	14 01	05 02	06 17	15 35	11 41
12	05 23 23	21 31 43	24 ♋ 08 50	01 ♌ 28 11	05 02	03 02	26 25	14 14	05 02	06 19	15 33	11 42
13	05 27 19	22 29 04	08 ♌ 41 46	15 ♌ 49 11	06 53	02 30	27 09	14 26	05 02	06 20	15 32	11 43
14	05 31 16	23 26 24	22 ♌ 50 08	29 ♌ 44 32	08 46	01 57	27 52	14 39	05 02	06 22	15 31	11 43
15	05 35 12	24 23 42	06 ♍ 32 22	13 ♍ 13 49	10 42	01 22	28 35	14 52	05 R 02	06 24	15 30	11 44
16	05 39 09	25 21 00	19 ♍ 49 08	26 ♍ 18 37	12 41	00 46	29 ♉ 18	15 04	05 02	06 25	15 29	11 45
17	05 43 05	26 18 17	02 ♎ 42 41	09 ♎ 01 44	14 41	00 ♋ 10	00 ♊ 01	15 17	05 02	06 27	15 27	11 45
18	05 47 02	27 15 33	15 ♎ 16 16	21 ♎ 26 44	16 43	29 ♊ 33	00 43	15 29	05 02	06 29	15 26	11 46
19	05 50 58	28 12 48	27 ♎ 33 37	03 ♏ 37 24	18 48	28 55	01 26	15 42	05 01	06 31	15 25	11 47
20	05 54 55	29 ♊ 10 03	09 ♏ 38 33	15 ♏ 37 29	20 54	28 17	02 09	15 54	05 01	06 33	15 24	11 48
21	05 58 52	00 ♋ 07 17	21 ♏ 34 40	27 ♏ 30 29	23 02	27 40	02 52	16 06	05 00	06 35	15 23	11 49
22	06 02 48	01 04 30	03 ♐ 25 19	09 ♐ 19 32	25 11	27 03	03 34	16 18	05 00	06 37	15 22	11 50
23	06 06 45	02 01 43	15 ♐ 13 28	21 ♐ 07 26	27 20	26 26	04 17	16 30	04 59	06 39	15 21	11 51
24	06 10 41	02 58 56	27 ♐ 01 44	02 ♑ 56 41	29 ♊ 31	25 50	04 59	16 42	04 58	06 41	15 20	11 52
25	06 14 38	03 56 08	08 ♑ 52 32	14 ♑ 49 34	01 ♋ 42	25 16	05 42	16 54	04 57	06 43	15 19	11 53
26	06 18 34	04 53 20	20 ♑ 48 04	26 ♑ 48 18	03 53	24 42	06 24	17 06	04 56	06 45	15 18	11 54
27	06 22 31	05 50 32	02 ♒ 50 33	08 ♒ 55 06	06 05	24 10	07 06	17 18	04 55	06 47	15 17	11 55
28	06 26 27	06 47 43	15 ♒ 02 14	21 ♒ 12 17	08 15	23 40	07 49	17 29	04 54	06 50	15 16	11 56
29	06 30 24	07 44 55	27 ♒ 25 31	03 ♓ 42 18	10 26	23 11	08 31	17 41	04 52	06 52	15 16	11 57
30	06 34 21	08 ♋ 42 07	10 ♓ 02 56	16 ♓ 27 45	12 ♋ 35	22 ♊ 45	09 ♊ 13	17 ♉ 52	04 ♓ 51	06 ♍ 54	15 ♏ 15	11 ♍ 58

Date	Moon True ☊ ° '	Moon Mean ☊ ° '	Moon ☽ Latitude ° '
01	02 ♋ 11	03 ♋ 17	03 S 47
02	02 D 12	03 14	04 29
03	02 R 12	03 11	04 59
04	02 12	03 08	05 15
05	02 11	03 04	05 13
06	02 09	03 01	04 53
07	02 07	02 58	04 13
08	02 05	02 55	03 16
09	02 04	02 52	02 05
10	02 03	02 48	00 S 44
11	02 D 03	02 45	00 N 40
12	02 04	02 42	02 00
13	02 04	02 39	03 10
14	02 05	02 36	04 06
15	02 06	02 33	04 47
16	02 06	02 29	05 10
17	02 R 06	02 26	05 18
18	02 06	02 23	05 09
19	02 05	02 20	04 47
20	02 05	02 17	04 12
21	02 05	02 14	03 27
22	02 05	02 10	02 33
23	02 D 04	02 07	01 32
24	02 05	02 04	00 N 28
25	02 R 04	02 01	00 S 38
26	02 04	01 58	01 42
27	02 04	01 54	02 43
28	02 04	01 51	03 37
29	02 03	01 48	04 21
30	02 ♋ 02	01 ♋ 45	04 S 54

DECLINATIONS

Date	Sun ☉ ° '	Moon ☽ ° '	Mercury ☿ ° '	Venus ♀ ° '	Mars ♂ ° '	Jupiter ♃ ° '	Saturn ♄ ° '	Uranus ♅ ° '	Neptune ♆ ° '	Pluto ♇ ° '
01	22 N 06	19 S 02	14 N 13	25 N 24	17 N 00	14 N 26	11 S 06	10 00	14 S 50	20 N 04
02	22 14	15 29	14 42	25 15	17 13	14 30	11 05	09 59	14 49	20 04
03	22 21	11 09	15 11	25 05	17 26	14 34	11 05	09 59	14 49	20 03
04	22 28	06 10	15 41	24 55	17 38	14 38	11 05	09 59	14 49	20 03
05	22 35	00 S 44	16 12	24 44	17 50	14 42	11 05	09 58	14 48	20 02
06	22 41	04 N 55	16 43	24 33	18 02	14 46	11 05	09 58	14 48	20 02
07	22 47	10 28	17 14	24 22	18 14	14 50	11 04	09 57	14 48	20 01
08	22 52	15 34	17 46	24 10	18 25	14 54	11 04	09 57	14 47	20 01
09	22 58	19 45	18 17	23 58	18 37	14 58	11 04	09 56	14 47	20 00
10	23 02	22 35	18 48	23 46	18 48	15 01	11 04	09 55	14 46	19 59
11	23 07	23 47	19 19	23 33	18 59	15 05	11 04	09 55	14 46	19 59
12	23 10	23 15	19 50	23 20	19 09	15 09	11 05	09 54	14 46	19 58
13	23 14	21 08	20 20	23 07	19 20	15 13	11 05	09 54	14 46	19 58
14	23 17	17 47	20 49	22 54	19 31	15 16	11 05	09 53	14 45	19 57
15	23 20	13 33	21 17	22 40	19 41	15 20	11 05	09 52	14 45	19 57
16	23 22	08 47	21 44	22 26	19 51	15 24	11 05	09 52	14 45	19 56
17	23 24	03 N 47	22 10	22 12	20 01	15 27	11 06	09 51	14 44	19 55
18	23 25	01 S 15	22 34	21 58	20 10	15 31	11 06	09 50	14 44	19 55
19	23 26	06 08	22 56	21 44	20 20	15 34	11 06	09 49	14 44	19 54
20	23 26	10 43	23 17	21 29	20 29	15 38	11 07	09 49	14 43	19 53
21	23 27	14 50	23 35	21 15	20 38	15 41	11 07	09 48	14 43	19 53
22	23 26	18 21	23 51	21 01	20 47	15 45	11 07	09 47	14 43	19 52
23	23 26	21 06	24 05	20 47	20 56	15 48	11 08	09 46	14 43	19 51
24	23 25	22 57	24 16	20 34	21 04	15 51	11 08	09 46	14 42	19 51
25	23 23	23 46	24 24	20 20	21 12	15 55	11 09	09 45	14 42	19 50
26	23 21	23 31	24 30	20 07	21 20	15 58	11 10	09 44	14 42	19 49
27	23 19	22 10	24 32	19 55	21 28	16 01	11 10	09 43	14 42	19 49
28	23 16	19 47	24 32	19 43	21 36	16 04	11 11	09 42	14 42	19 48
29	23 13	16 27	24 29	19 31	21 43	16 07	11 12	09 41	14 41	19 47
30	23 N 09	12 S 20	24 N 24	19 N 20	21 N 50	16 N 11	11 S 12	09 N 40	14 S 41	19 N 47

ZODIAC SIGN ENTRIES

Date	h m	Planets
02	11 01	☽ ♓
04	18 03	☽ ♈
06	21 20	☽ ♉
08	21 50	☽ ♊
09	15 45	☿ ♊
10	21 16	☽ ♋
12	21 35	☽ ♌
15	00 27	☽ ♍
17	06 54	☽ ♎
17	11 43	♂ ♊
17	18 17	♀ ♊
19	16 49	☽ ♏
21	08 57	☉ ♋
22	05 03	☽ ♐
24	17 17	☿ ♋
24	18 02	☽ ♑
27	06 22	☽ ♒
29	16 56	☽ ♓

LATITUDES

Date	Mercury ☿ ° '	Venus ♀ ° '	Mars ♂ ° '	Jupiter ♃ ° '	Saturn ♄ ° '	Uranus ♅ ° '	Neptune ♆ ° '	Pluto ♇ ° '
01	02 S 58	02 N 08	00 S 20	01 S 00	01 S 28	00 N 46	01 N 49	13 N 58
04	02 37	01 37	00 18	01 00	01 29	00 46	01 49	13 57
07	02 11	01 01	00 16	01 00	01 30	00 46	01 49	13 55
10	01 41	00 N 23	00 14	01 00	01 30	00 46	01 49	13 54
13	01 09	00 S 18	00 12	01 00	01 31	00 46	01 49	13 53
16	00 35	01 00	00 10	01 01	01 32	00 45	01 49	13 52
19	00 S 02	01 43	00 08	01 01	01 32	00 45	01 49	13 51
22	00 N 30	02 24	00 05	01 01	01 33	00 45	01 49	13 50
25	00 58	03 01	00 03	01 01	01 34	00 45	01 48	13 49
28	01 21	03 35	00 S 01	01 01	01 34	00 45	01 48	13 48
31	01 N 38	04 S 04	00 N 01	01 S 02	01 S 35	00 N 45	01 N 48	13 N 47

DATA

Julian Date	2438548
Delta T	+35 seconds
Ayanamsa	23° 21' 18"
Synetic vernal point	05° ♓ 45' 41"
True obliquity of ecliptic	23° 26' 37"

MOON'S PHASES, APSIDES AND POSITIONS ☽

Date	h m	Phase	Longitude ° '	Eclipse Indicator
03	11 07	☾	12 ♓ 53	
10	04 22	●	19 ♊ 19	Partial
16	23 02	☽	25 ♍ 47	
25	01 08	○	03 ♑ 30	total

Day	h m		
10	01 54	Perigee	
23	11 20	Apogee	
05	15 08	0N	
11	16 28	Max dec	23° N 49'
18	05 58	0S	
25	18 25	Max dec	23° S 49'

LONGITUDES

Date	Chiron ⚷ ° '	Ceres ⚳ ° '	Pallas ⚴ ° '	Juno ⚵ ° '	Vesta ⚶ ° '	Black Moon Lilith ⚸ ° '
01	18 ♓ 37	04 ♑ 09	23 ♏ 47	24 ♏ 14	09 ♓ 49	15 ♐ 30
11	18 ♓ 45	02 ♑ 11	21 ♏ 22	22 ♏ 20	12 ♓ 26	16 ♐ 37
21	18 ♓ 48	00 ♑ 00	19 ♏ 38	20 ♏ 51	14 ♓ 32	17 ♐ 43
31	18 ♓ 45	27 ♐ 51	18 ♏ 39	19 ♏ 52	16 ♓ 03	18 ♐ 50

ASPECTARIAN

h m	Aspects
01 Monday	
01 43	☽ ± ♀
03 33	☽ # ♇
07 43	☽ □ ♆
11 01	☽ □ ☿
12 54	☽ □ ♂
19 03	☽ bQ ♀
02 Tuesday	
00 34	☿ ∥ ♃
01 36	☽ # ♂
03 24	☉ □ ♇
11 05	☽ Q ♃
14 48	☉ ⊻ ♃
15 57	☽ ∥ ♆
16 11	☽ # ☿
17 44	☽ # ♃
18 04	☿ # ♆
20 18	☽ ☌ ♄
21 14	☿ ☌ ♂
22 31	☽ ☍ ♅
23 12	☽ △ ♀
03 Wednesday	
01 17	☽ Q ♂
01 32	☽ Q ☿
08 48	☽ ☍ ♇
09 57	☽ ✶ ♃
11 07	☽ □ ☉
12 19	☽ ∥ ♄
16 24	☽ △ ♆
17 51	☽ # ♅
04 Thursday	
00 40	☽ ✶ ♂
02 17	☽ ✶ ☿
02 56	☿ ∠ ♀
13 07	♀ ✶ ♅
13 37	☽ ∠ ♃
19 20	☽ bQ ♆
22 04	☽ Q ☉
05 Friday	
00 28	♀ ∠ ♂
02 46	☽ ⊻ ♄
04 32	☽ □ ♀
04 50	☽ ⊼ ♅
04 50	☽ ∠ ♂
05 43	☽ ⊥ ♃
08 08	☽ ∠ ☿
11 02	☽ ± ♆
13 12	☽ ⊥ ♄
14 21	☽ ⊼ ♇
15 15	☽ ± ♅
16 16	☽ ⊻ ♃
20 24	☽ ✶ ☉
21 13	☽ ⊥ ♂
21 19	☽ ⊼ ♆
06 Saturday	
00 37	☽ ± ♇
01 29	☽ ⊥ ☿
04 34	☽ ∠ ♄
06 35	☽ bQ ♅
07 55	☽ ⊻ ♂
09 29	☉ ⊼ ♆
10 40	☽ Q ♀
12 49	☽ ⊻ ☿
15 46	☽ bQ ♇
22 51	♃ # ♆
23 23	☽ ∠ ☉
07 Sunday	
05 33	☽ ✶ ♄
06 06	☽ ✶ ♀
07 31	☽ △ ♅
09 42	☽ ∥ ♅
14 41	☽ # ♄
15 09	☽ ⊥ ☉
16 25	☽ △ ♇
18 56	☽ ☌ ♃
22 53	☽ ☍ ♆
08 Monday	
01 03	☽ Q ♄
01 31	☽ ⊻ ☉
03 55	♀ △ ♄
05 50	☽ ∠ ♀
08 07	☽ # ♆
08 36	☽ ∥ ♃
11 31	☽ ☌ ♂
19 35	☽ ☌ ☿
19 50	☽ ⊥ ♀
21 27	☿ ⊥ ♀
09 Tuesday	
01 30	☽ ∥ ☿
04 26	☽ ∥ ♂
05 06	☽ ⊻ ♀
05 47	☽ □ ♄
07 43	☽ □ ♅
13 45	☽ ∥ ♇
16 18	☽ □ ♇
19 26	☽ ⊻ ♃
22 32	☽ ⊼ ♆
10 Wednesday	
04 22	☽ ☌ ☉
05 02	☽ ⊥ ♃
07 59	☽ ± ♆
11 28	☿ ∥ ♂
12 11	☽ Q ♅
13 24	☽ ⊻ ♂
17 54	☉ ⊥ ♃
18 08	☽ ∥ ☉
19 28	☽ ∠ ♃
20 46	☽ Q ♇
22 12	☽ bQ ♆
23 22	☽ ⊥ ♂
11 Thursday	
01 06	☽ ⊻ ☿
03 10	☽ ☌ ♀
05 02	☽ ∥ ♀
05 15	☽ △ ♄
07 15	☽ ✶ ♅
11 56	☽ ⊥ ☿
14 26	☽ ∠ ♂
15 20	☿ ⊻ ♀
15 53	☽ ✶ ♇
19 44	☽ ✶ ♃
22 07	☽ △ ♆
12 Friday	
04 23	☽ ∠ ☿
05 19	☽ bQ ♄
07 23	☽ ∠ ♅
07 27	☽ ⊻ ☉
10 01	☽ ∥ ♀
12 03	☿ □ ♄
12 42	☉ ± ♆
13 15	☽ ∥ ☉
15 27	☽ Q ♃
15 54	☽ ✶ ♂
16 10	☽ ∠ ♇
17 55	☽ ⊥ ☉
18 47	☿ ∥ ♇
20 00	☽ ± ♄
22 07	☽ ⊥ ♅
23 45	♀ ⊥ ♂
13 Saturday	
02 04	☽ ⊻ ♀
02 17	☉ ∥ ♀
04 56	☿ □ ♅
05 54	☽ ⊼ ♄
07 01	☽ ⊥ ♇
08 04	☽ ⊻ ♅
08 31	☽ ✶ ☿
09 50	☽ ∠ ☉
11 41	☽ ⊥ ♀
12 47	☽ Q ♂
17 04	☽ ⊻ ♇
17 41	☽ ∥ ☿
21 21	☽ ∥ ♇
21 48	☽ □ ♃
23 30	☽ □ ♆
14 Sunday	
01 08	☽ ∥ ♂
02 18	☽ ∠ ♀
07 54	☽ Q ☿
13 07	☽ ✶ ☉
21 11	☽ □ ♂
15 Monday	
02 28	☽ ∥ ♃
03 13	☽ ✶ ♀
03 26	♄ St R
05 33	☽ # ♆
06 37	☽ Q ♆
09 20	☽ ☍ ♄
11 43	☽ Q ☉
11 44	☽ ☌ ♅
11 59	☉ Q ♅
20 44	☽ □ ☿
21 18	☽ ☌ ♇
23 43	☽ Q ♀
16 Tuesday	
00 38	☽ # ♄
00 39	☿ □ ♇
03 12	☽ △ ♃
04 05	☽ ✶ ♆
06 43	☽ ∥ ♅
23 02	☽ □ ☉
23 51	♂ ∥ ♇
17 Wednesday	
00 00	♀ bQ ♆
06 37	☽ △ ♂
07 21	☽ bQ ♃
07 25	☽ □ ♀
07 46	☽ ∠ ♆
08 34	♀ ∠ ♃
13 18	☿ ∥ ♀
14 45	♀ ⊻ ♂
16 24	☽ ⊼ ♄
19 06	☽ ⊻ ♅
19 51	☿ ⊻ ♃
21 04	☿ ⊼ ♆
18 Thursday	
00 41	☽ ± ♃
00 48	☽ ⊥ ♆
03 21	♀ Q ♇
03 50	☽ ± ♄
05 15	☽ ⊻ ♇
06 37	☽ ⊥ ♅
06 58	♃ ☍ ♆
12 19	☽ ⊻ ♆
12 25	☽ ⊼ ♃
12 56	☽ bQ ♂
15 23	☽ △ ☿
16 51	☽ ⊥ ♇
21 14	☽ bQ ♄
19 Friday	
00 06	☽ ∠ ♅
07 34	☽ ± ♂
10 28	☽ ∠ ♇
13 24	☽ △ ☉
14 33	☽ △ ♀
20 08	☽ ⊼ ♂
22 40	☉ ☌ ♀
20 Saturday	
02 46	☽ △ ♄
02 57	☽ bQ ☿
05 48	☽ ✶ ♅
07 06	☽ # ♅
14 10	☽ ∥ ♄
16 19	☽ ✶ ♇
17 38	☿ ± ♆
18 57	☽ bQ ♀
21 51	☽ bQ ☉
23 32	☽ ☌ ♆
21 Sunday	
00 13	☉ ∥ ☿
00 28	☿ ⊥ ♃
00 46	☽ ☍ ♃
00 49	☽ ± ☿
04 08	☉ Q ♇
05 56	☽ Q ♅
11 17	☽ ∥ ♆
12 10	☽ ± ♀
15 35	☽ ⊼ ☿
16 32	☽ Q ♇
17 30	☽ # ♃
17 35	☽ ± ☉
18 31	☉ bQ ♆
23 42	☽ ⊼ ♀
22 Monday	
05 38	☿ Q ♅
06 49	☽ ⊼ ☉
12 19	☽ ☍ ♂
15 12	☽ □ ♄
18 30	☽ □ ♅
19 15	☉ ∠ ♃
23 Tuesday	
00 20	☽ # ♇
03 11	♀ ∥ ♂
04 09	☿ ☌ ♀
05 07	☽ □ ♇
09 10	☽ # ♀
10 09	☽ # ♂
12 15	☽ ⊻ ♆
14 39	☽ ⊼ ♃
24 Wednesday	
00 27	☽ ⊥ ♆
03 04	☽ ± ♃
03 46	☽ Q ♄
09 42	☽ ☍ ♀
11 22	♂ □ ♄
15 47	☿ Q ♇
18 12	☽ ☍ ☿
18 42	☽ ∠ ♆
20 54	☿ bQ ♆
21 33	☽ # ☉
21 39	☽ bQ ♃
25 Thursday	
01 08	☽ ☍ ☉
04 05	☽ ✶ ♄
05 10	☽ ⊼ ♂
07 38	☽ △ ♅
14 23	☿ ∠ ♃
17 41	♃ Q ♄
18 03	☽ ± ♂
18 04	☽ △ ♇
26 Friday	
00 58	☽ ✶ ♆
04 27	☽ △ ♃
09 54	♀ Q ♅
10 16	☽ ∠ ♄
13 07	☉ △ ♄
13 17	☽ bQ ♂
13 55	☽ bQ ♅
16 30	☽ # ☉
19 28	☽ ⊼ ♀
23 21	☿ △ ♄
27 Saturday	
00 12	☽ bQ ♇
00 42	♂ □ ♅
00 59	☽ Q ♆
04 13	☽ ⊥ ♄
06 55	☽ ± ♀
07 25	☉ ☌ ☿
07 55	☽ ± ♅
16 06	☽ ⊻ ♄
18 05	☽ ± ♇
18 26	☽ ⊼ ☉
19 43	☽ # ♂
19 48	☽ ⊼ ☿
19 50	☽ ⊼ ♅
20 00	☿ ✶ ♅
20 57	☽ △ ♂
28 Sunday	
00 00	☽ bQ ♀
01 01	♀ ∥ ♇
04 43	☿ ⊻ ♂
05 54	☽ ⊼ ♇
07 15	☽ ± ☉
10 09	☽ ± ☿
11 49	☽ # ♇
12 28	☽ □ ♆
12 34	☽ # ♀
12 53	☉ ✶ ♅
16 51	☽ □ ♃
18 16	♀ ⊥ ♃
29 Monday	
02 14	☽ bQ ☉
04 08	☽ △ ♀
07 21	☽ bQ ☿
14 05	☽ # ♃
22 49	☽ ∥ ♆
30 Tuesday	
02 11	☽ ☌ ♄
04 00	☽ Q ♃
05 05	☿ ✶ ♇
06 03	☽ ☍ ♅
09 15	☽ △ ☉
10 20	☽ □ ♂
15 36	☽ ☍ ♇
17 43	☽ △ ☿
18 00	☽ ∥ ♄
21 43	☽ △ ♆

All ephemeris data is given at 12.00 UT and the Moon's longitude is additionally given for 24.00 UT

JULY 1964

LONGITUDES

Date	Sidereal time h m s	Sun ☉ ° ' "	Moon ☽ ° ' "	Moon ☽ 24.00 ° ' "	Mercury ☿ ° '	Venus ♀ ° '	Mars ♂ ° '	Jupiter ♃ ° '	Saturn ♄ ° '	Uranus ♅ ° '	Neptune ♆ ° '	Pluto ♇ ° '
01	06 38 17	09 ♋ 39 18	22 ♓ 57 05	29 ♓ 31 12	14 ♋ 43	22 ♊ 20	09 ♊ 55	18 ♉ 04	04 ♓ 49	06 ♍ 57	15 ♏ 14	11 ♍ 59
02	06 42 14	10 36 30	06 ♈ 10 24	12 ♈ 54 53	16 50	21 R 58	10 37	18 15	04 R 48	06 59	15 R 13	12 00
03	06 46 10	11 33 42	19 ♈ 44 52	26 ♈ 40 25	18 56	21 38	11 19	18 26	04 46	07 02	15 12	12 02
04	06 50 07	12 30 55	03 ♉ 41 33	10 ♉ 48 10	21 00	21 20	12 01	18 37	04 44	07 04	15 12	12 03
05	06 54 03	13 28 07	18 ♉ 00 03	25 ♉ 16 50	23 02	21 04	12 43	18 48	04 42	07 07	15 11	12 04
06	06 58 00	14 25 20	02 ♊ 38 03	10 ♊ 03 03	25 03	20 51	13 25	18 59	04 40	07 09	15 10	12 05
07	07 01 56	15 22 34	17 ♊ 31 04	25 ♊ 01 11	27 02	20 41	14 06	19 10	04 38	07 12	15 10	12 07
08	07 05 53	16 19 48	02 ♋ 32 27	10 ♋ 03 47	28 ♋ 59	20 32	14 48	19 21	04 36	07 15	15 09	12 08
09	07 09 50	17 17 02	17 ♋ 34 07	25 ♋ 02 20	00 ♌ 54	20 26	15 30	19 32	04 34	07 17	15 09	12 09
10	07 13 46	18 14 16	02 ♌ 27 27	09 ♌ 48 30	02 48	20 23	16 11	19 42	04 32	07 20	15 08	12 11
11	07 17 43	19 11 30	17 ♌ 04 39	24 ♌ 15 15	04 39	20 21	16 53	19 52	04 29	07 23	15 08	12 12
12	07 21 39	20 08 44	01 ♍ 19 44	08 ♍ 17 45	06 29	20 D 22	17 34	20 03	04 27	07 26	15 07	12 13
13	07 25 36	21 05 58	15 ♍ 09 07	21 ♍ 53 46	08 16	20 26	18 16	20 13	04 24	07 29	15 07	12 15
14	07 29 32	22 03 12	28 ♍ 31 50	05 ♎ 03 30	10 02	20 31	18 57	20 23	04 21	07 31	15 06	12 16
15	07 33 29	23 00 27	11 ♎ 29 08	17 ♎ 49 06	11 46	20 39	19 38	20 33	04 19	07 34	15 06	12 18
16	07 37 25	23 57 41	24 ♎ 03 56	00 ♏ 14 07	13 28	20 49	20 19	20 43	04 16	07 37	15 05	12 19
17	07 41 22	24 54 55	06 ♏ 20 15	12 ♏ 22 54	15 08	21 01	21 00	20 53	04 13	07 40	15 05	12 21
18	07 45 19	25 52 10	18 ♏ 22 39	24 ♏ 20 05	16 46	21 15	21 41	21 02	04 10	07 43	15 05	12 22
19	07 49 15	26 49 25	00 ♐ 15 48	06 ♐ 10 20	18 22	21 30	22 22	21 12	04 07	07 46	15 05	12 24
20	07 53 12	27 46 40	12 ♐ 04 12	17 ♐ 57 56	19 56	21 48	23 03	21 21	04 04	07 49	15 04	12 26
21	07 57 08	28 43 56	23 ♐ 52 00	29 ♐ 46 48	21 29	22 08	23 44	21 31	04 00	07 53	15 04	12 27
22	08 01 05	29 ♋ 41 12	05 ♑ 42 45	11 ♑ 40 12	22 59	22 29	24 25	21 40	03 57	07 56	15 04	12 29
23	08 05 01	00 ♌ 38 28	17 ♑ 39 28	23 ♑ 40 51	24 27	22 52	25 06	21 49	03 54	07 59	15 04	12 31
24	08 08 58	01 35 45	29 ♑ 44 33	05 ♒ 50 49	25 54	23 17	25 46	21 58	03 50	08 02	15 04	12 32
25	08 12 54	02 33 03	11 ♒ 59 48	18 ♒ 11 39	27 18	23 43	26 27	22 07	03 47	08 05	15 04	12 34
26	08 16 51	03 30 21	24 ♒ 26 31	00 ♓ 44 29	28 ♌ 41	24 11	27 08	22 15	03 43	08 08	15 04	12 36
27	08 20 48	04 27 40	07 ♓ 05 39	13 ♓ 30 06	00 ♍ 01	24 41	27 48	22 24	03 40	08 12	15 D 04	12 37
28	08 24 44	05 25 00	19 ♓ 57 56	26 ♓ 29 11	01 20	25 12	28 29	22 32	03 36	08 15	15 04	12 39
29	08 28 41	06 22 21	03 ♈ 03 58	09 ♈ 42 21	02 36	25 44	29 09	22 41	03 32	08 18	15 04	12 41
30	08 32 37	07 19 43	16 ♈ 24 24	23 ♈ 10 11	03 50	26 18	29 ♊ 49	22 49	03 29	08 22	15 04	12 43
31	08 36 34	08 ♌ 17 06	29 ♈ 59 47	06 ♉ 53 14	05 ♍ 02	26 ♊ 53	00 ♋ 30	22 ♉ 57	03 ♓ 25	08 ♍ 25	15 ♏ 04	12 ♍ 44

Date	Moon True ☊ ° '	Moon Mean ☊ ° '	Moon ☽ Latitude ° '
01	02 ♋ 02	01 ♋ 42	05 S 13
02	02 R 01	01 39	05 16
03	02 D 01	01 35	05 02
04	02 02	01 32	04 30
05	02 03	01 29	03 41
06	02 04	01 26	02 36
07	02 04	01 23	01 S 20
08	02 R 05	01 20	00 N 03
09	02 04	01 16	01 25
10	02 03	01 13	02 40
11	02 01	01 10	03 44
12	01 59	01 07	04 32
13	01 57	01 04	05 02
14	01 54	01 00	05 15
15	01 53	00 57	05 11
16	01 D 53	00 54	04 52
17	01 53	00 51	04 21
18	01 54	00 48	03 38
19	01 56	00 45	02 46
20	01 58	00 41	01 47
21	01 59	00 38	00 N 45
22	01 R 59	00 35	00 S 20
23	01 58	00 32	01 25
24	01 55	00 29	02 26
25	01 52	00 26	03 22
26	01 47	00 22	04 08
27	01 42	00 19	04 44
28	01 36	00 16	05 05
29	01 32	00 13	05 12
30	01 29	00 10	05 01
31	01 ♋ 27	00 ♋ 06	04 S 34

DECLINATIONS

Date	Sun ☉ ° '	Moon ☽ ° '	Mercury ☿ ° '	Venus ♀ ° '	Mars ♂ ° '	Jupiter ♃ ° '	Saturn ♄ ° '	Uranus ♅ ° '	Neptune ♆ ° '	Pluto ♇ ° '
01	23 N 06	07 S 36	24 N 15	19 N 10	21 N 57	16 N 14	11 S 13	09 N 39	14 S 41	19 N 46
02	23 01	02 S 24	24 04	19 00	22 04	16 17	11 14	09 39	14 41	19 45
03	22 56	03 N 03	23 51	18 51	22 11	16 20	11 15	09 38	14 41	19 44
04	22 51	08 31	23 35	18 43	22 17	16 23	11 15	09 37	14 41	19 44
05	22 46	13 40	23 17	18 35	22 23	16 26	11 16	09 36	14 40	19 43
06	22 40	18 08	22 57	18 28	22 29	16 28	11 17	09 35	14 40	19 42
07	22 33	21 32	22 35	18 22	22 35	16 31	11 18	09 34	14 40	19 41
08	22 27	23 28	22 11	18 16	22 40	16 34	11 19	09 33	14 40	19 40
09	22 20	23 41	21 45	18 11	22 46	16 37	11 20	09 32	14 40	19 40
10	22 12	22 13	21 18	18 07	22 51	16 40	11 21	09 31	14 40	19 39
11	22 04	19 16	20 50	18 03	22 56	16 42	11 22	09 29	14 40	19 38
12	21 56	15 14	20 20	18 00	23 00	16 45	11 23	09 28	14 40	19 37
13	21 47	10 30	19 49	17 58	23 05	16 48	11 25	09 27	14 40	19 36
14	21 38	05 24	19 17	17 56	23 09	16 50	11 26	09 26	14 39	19 36
15	21 29	00 N 14	18 44	17 54	23 13	16 53	11 27	09 25	14 39	19 35
16	21 19	04 S 48	18 11	17 54	23 17	16 55	11 28	09 24	14 39	19 34
17	21 09	09 32	17 36	17 53	23 21	16 58	11 29	09 23	14 39	19 33
18	20 59	13 49	17 01	17 54	23 24	17 00	11 31	09 22	14 39	19 32
19	20 48	17 30	16 26	17 54	23 27	17 03	11 32	09 21	14 39	19 32
20	20 37	20 28	15 50	17 55	23 30	17 05	11 33	09 19	14 39	19 31
21	20 25	22 34	15 13	17 57	23 33	17 07	11 35	09 18	14 39	19 30
22	20 13	23 40	14 37	17 59	23 35	17 10	11 36	09 17	14 39	19 29
23	20 01	23 41	14 00	18 01	23 38	17 12	11 37	09 16	14 39	19 28
24	19 48	22 36	13 23	18 03	23 40	17 14	11 39	09 15	14 39	19 27
25	19 36	20 26	12 46	18 06	23 42	17 16	11 40	09 13	14 39	19 26
26	19 22	17 17	12 09	18 09	23 44	17 18	11 42	09 12	14 39	19 26
27	19 09	13 18	11 32	18 12	23 45	17 20	11 43	09 11	14 39	19 25
28	18 55	08 39	10 56	18 15	23 46	17 22	11 44	09 10	14 39	19 24
29	18 41	03 S 33	10 19	18 18	23 47	17 24	11 46	09 08	14 40	19 23
30	18 27	01 N 49	09 43	18 22	23 48	17 26	11 47	09 07	14 40	19 22
31	18 N 12	07 N 12	09 N 07	18 N 25	23 N 49	17 N 28	11 S 49	09 N 06	14 S 40	19 N 21

ZODIAC SIGN ENTRIES

Date	h	m	Planets
02	00	52	☽ ♈
04	05	42	☽ ♉
06	07	43	☽ ♊
08	07	57	☽ ♋
09	00	38	☿ ♌
10	08	01	☽ ♌
12	09	44	☽ ♍
14	14	41	☽ ♎
16	23	32	☽ ♏
19	11	28	☽ ♐
22	00	27	☽ ♑
22	19	53	☉ ♌
24	12	30	☽ ♒
26	22	36	☽ ♓
27	11	35	☿ ♍
29	06	25	☽ ♈
30	18	23	♂ ♋
31	12	00	☽ ♉

LATITUDES

Date	Mercury ☿ ° '	Venus ♀ ° '	Mars ♂ ° '	Jupiter ♃ ° '	Saturn ♄ ° '	Uranus ♅ ° '	Neptune ♆ ° '	Pluto ♇ ° '
01	01 N 38	04 S 04	00 N 01	01 S 02	01 S 35	00 N 45	01 N 48	13 N 47
04	01 48	04 27	00 03	01 02	01 36	00 45	01 48	13 46
07	01 52	04 46	00 05	01 02	01 36	00 45	01 48	13 45
10	01 49	05 00	00 07	01 02	01 37	00 45	01 48	13 45
13	01 41	05 09	00 09	01 03	01 37	00 45	01 47	13 44
16	01 27	05 15	00 11	01 03	01 38	00 44	01 47	13 43
19	01 10	05 17	00 14	01 03	01 38	00 44	01 47	13 42
22	00 48	05 16	00 16	01 04	01 39	00 44	01 47	13 42
25	00 N 23	05 12	00 18	01 04	01 39	00 44	01 47	13 41
28	00 S 05	05 07	00 20	01 04	01 40	00 44	01 47	13 40
31	00 S 35	04 S 59	00 N 22	01 S 05	01 S 40	00 N 44	01 N 46	13 N 40

DATA

Julian Date	2438578
Delta T	+35 seconds
Ayanamsa	23° 21' 23"
Synetic vernal point	05° ♓ 45' 36"
True obliquity of ecliptic	23° 26' 37"

MOON'S PHASES, APSIDES AND POSITIONS ☽

Date	h	m	Phase	Longitude ° '	Eclipse Indicator
02	20	31	☾	10 ♈ 57	
09	11	31	●	17 ♋ 16	Partial
16	11	47	☽	23 ♎ 57	
24	15	58	○	01 ♒ 45	

Day	h	m		
08	10	49	Perigee	
20	20	42	Apogee	
02	22	37	0N	
09	03	08	Max dec	23° N 49'
15	13	05	0S	
23	00	29	Max dec	23° S 49'
30	03	56	0N	

LONGITUDES

Date	Chiron ⚷ ° '	Ceres ⚳ ° '	Pallas ⚴ ° '	Juno ⚵ ° '	Vesta ⚶ ° '	Black Moon Lilith ⚸ ° '
01	18 ♓ 45	27 ♐ 51	18 ♏ 39	19 ♏ 52	16 ♓ 03	18 ♐ 50
11	18 ♓ 37	25 ♐ 55	18 ♏ 26	19 ♏ 25	16 ♓ 52	19 ♐ 56
21	18 ♓ 24	24 ♐ 26	18 ♏ 56	19 ♏ 30	16 ♓ 56	21 ♐ 03
31	18 ♓ 06	23 ♐ 29	20 ♏ 04	20 ♏ 06	16 ♓ 13	22 ♐ 09

All ephemeris data is given at 12.00 UT and the Moon's longitude is additionally given for 24.00 UT

ASPECTARIAN

h m	Aspects
01 Wednesday	
01 52	☽ # ♅
02 50	☽ ⚹ ♃
10 54	☽ □ ♀
17 47	☿ △ ♆
21 36	☽ Q ♂
02 Thursday	
01 17	☽ ⚼ ♆
06 40	☽ ∠ ♃
08 18	☿ ⊥ ♂
09 32	☽ ⊻ ♄
12 55	☉ ⊻ ♂
13 28	☽ ⚻ ♅
17 26	☽ ± ♆
18 36	☽ Q ♀
20 14	☽ ⊥ ♄
20 22	☽ ⚹ ♂
20 31	☽ □ ☉
22 24	☽ ⚻ ♇
22 59	☽ ⊥ ♃
03 Friday	
00 10	☽ ± ♅
04 03	☽ ⚻ ♆
05 47	☿ ⚹ ♃
08 59	☽ ± ♇
09 41	☽ ⊻ ♃
10 19	☽ □ ☿
12 02	☽ ∠ ♄
15 12	☽ ⚹ ♀
15 59	☽ ⚼ ♅
21 34	☿ ⚼ ♄
23 55	☉ ⚹ ♇
23 59	☽ ∠ ♂
04 Saturday	
00 37	☽ ⚼ ♇
06 12	☽ Q ☉
13 02	♂ □ ♇
13 46	☽ ⚹ ♄
15 27	☿ ⊻ ♀
16 09	☽ ⊥ ♂
16 23	☽ ∠ ♀
16 57	☽ ǁ ♅
17 44	☽ △ ♅
22 28	☽ Q ☿
05 Sunday	
00 34	☽ # ♄
00 41	♀ ± ♆
00 52	☿ ∠ ♅
02 06	☽ △ ♇
02 45	☽ ⊻ ♂
03 56	☽ ⚹ ☉
07 13	☽ ⊥ ♀
07 19	☽ ☍ ♆
09 51	☽ Q ♄
13 21	☽ ☌ ♃
17 00	☽ ⊻ ♀
17 03	☽ # ♆
21 39	☽ ⚹ ☿
06 Monday	
02 25	☽ ǁ ♃
06 25	☽ ∠ ☉
13 56	☽ ǁ ♀
15 18	☽ □ ♄
19 21	☽ □ ♅
21 55	☽ ǁ ♇
22 01	☽ ⊥ ☉
07 Tuesday	
01 51	☽ ∠ ☿
03 18	☽ □ ♇
06 16	☽ ☌ ♂
06 41	☉ △ ♆
07 59	☿ ⊥ ♀
08 14	☽ ⚻ ♆
08 20	☽ ⊻ ☉
09 14	☉ ǁ ♂
12 04	☿ ǁ ♂
12 57	☿ ∠ ♇
14 10	☉ ǁ ☿
14 41	☽ ⊻ ♃
17 00	☽ ☌ ♀
17 50	☽ ± ♆
18 28	☽ ⊥ ☿
20 55	☽ ǁ ☿
21 46	☽ ǁ ☉
23 04	☽ ǁ ♂
08 Wednesday	
00 19	☽ Q ♅
00 23	☽ ⊥ ♃
05 29	☽ ⊻ ☿
07 21	☿ ± ♄
08 09	☽ Q ♇
08 12	☽ ⚼ ♆
14 55	☽ ∠ ♃
15 17	☽ △ ♄
19 31	☽ ⚹ ♅
09 Thursday	
00 00	♂ ⚻ ♆
03 20	☽ ⚹ ♇
03 54	☿ ∠ ♂
08 07	☽ △ ♆
08 31	☽ ⊻ ♂
11 31	☽ ☌ ☉
15 10	☽ ⚹ ♃
15 11	☽ ⚼ ♄
16 35	☽ ⊻ ♀
16 58	☿ ⊥ ♅
18 36	☽ ⊥ ♂
19 36	☽ ∠ ♅
20 38	☿ Q ♃
10 Friday	
02 11	☽ ⊥ ♀
03 26	☽ ∠ ♇
04 53	☽ ǁ ♂
05 39	☽ ± ♄
09 50	☽ ∠ ♂
10 10	☽ ⊥ ♅
10 45	☽ Q ♃
12 11	☽ ǁ ☉
12 38	☽ ☌ ☿
15 21	☽ ⚻ ♄
16 44	☽ ∠ ♀
18 04	☽ ⊥ ♇
19 58	☽ ⊻ ♅
22 23	☽ ǁ ☿
11 Saturday	
03 55	☽ ⊻ ♇
08 46	☽ □ ♆
09 31	☽ ǁ ♇
09 51	☿ ⚻ ♄
11 39	☽ ⚹ ♂
13 01	♀ St D
15 46	☽ ⊻ ☉
16 43	☽ □ ♃
17 28	☽ ⚹ ♀
19 04	☉ ⚼ ♄
19 59	☽ ǁ ♀
21 13	☿ ∠ ♀
12 Sunday	
02 33	☽ ⊥ ☉
03 40	☽ ǁ ♃
08 35	☿ ⊥ ♇
08 51	☽ Q ♂
08 58	☉ ⚹ ♃
13 48	☽ Q ♀
15 04	☽ Q ♆
15 04	☽ # ♆
17 19	☽ ☍ ♄
18 00	☉ ⊻ ♀
19 02	☽ ∠ ☉
22 11	☽ ⊻ ☿
22 32	☽ ☌ ♅
13 Monday	
00 59	☿ ⊻ ♅
06 53	☽ ☌ ♇
07 34	☽ # ♄
10 13	☽ ⊥ ☿
11 56	☽ ⚹ ♆
17 00	☽ ǁ ♅
17 48	☽ □ ♂
21 06	☽ △ ♃
21 26	☽ □ ♀
21 55	☿ ǁ ♇
23 23	☽ ⚹ ☉
14 Tuesday	
04 41	☽ ∠ ☿
14 52	☽ ∠ ♆
22 40	☽ ⚻ ♄
22 57	☽ Q ☉
15 Wednesday	
00 29	☉ ∠ ♅
00 46	☽ ⚼ ♃
04 39	☽ ⊻ ♅
07 31	☽ ⊥ ♆
09 48	☽ ± ♄
12 37	☽ ⚹ ☿
13 32	☽ ⊻ ♇
15 57	☽ ⊥ ♅
17 52	☽ ± ♃
18 49	☽ ⊻ ♆
19 35	☿ ⊻ ♇
16 Thursday	
00 56	☽ ⊥ ♇
02 48	☽ ⚼ ♄
04 22	☽ △ ♂
05 28	☽ ⚻ ♃
05 39	☽ △ ♀
09 12	☽ ∠ ♅
11 47	☽ □ ☉
15 08	☽ Q ☿
18 20	☽ ∠ ♇
17 Friday	
00 04	☿ ǁ ♀
06 10	♂ ⊻ ♃
07 50	☽ △ ♄
11 12	☽ # ♅
11 18	☽ ⚼ ♂
11 21	☽ ⚼ ♀
11 21	☿ □ ♆
12 15	♀ ☌ ♂
14 39	☽ ⚹ ♅
14 46	♂ ± ♆
19 55	♀ ± ♆
22 40	☽ ǁ ♄
23 58	☽ ⚹ ♇
18 Saturday	
05 24	☽ ☌ ♆
05 35	☽ ± ♀
06 17	☽ ± ♂
08 16	☽ □ ☿
12 40	☿ ǁ ♃
14 43	☽ Q ♅
17 07	☽ ǁ ♆
17 26	☽ ☍ ♃
17 53	☽ ⚻ ♀
19 04	☽ ⚻ ♂
19 Sunday	
00 06	☽ Q ♇
04 25	☽ △ ☉
05 36	☽ # ☿
08 42	☽ # ♃
14 55	☽ # ♀
19 47	☽ □ ♄
20 Monday	
02 56	☉ ∠ ♇
03 19	☽ □ ♅
03 33	☽ # ♇
12 44	☽ □ ♇
12 57	☉ ⊥ ♀
13 16	☽ # ☉
13 34	☽ ⚼ ☉
18 07	☽ ⊻ ♆
18 44	☉ ± ♄
21 Tuesday	
06 19	☽ ⊥ ♆
06 25	☽ △ ☿
07 09	☽ ⚻ ♃
08 14	☽ Q ♄
08 22	☽ ☍ ♀
09 30	☽ ± ☉
11 43	☽ ☍ ♂
12 35	☿ □ ♃
19 30	☽ ± ♃
22 45	☽ ⚻ ☉
22 Wednesday	
00 35	☽ ∠ ♆
01 27	☿ ⚹ ♀
08 28	☽ ⚹ ♄
09 06	☽ # ♂
10 21	☿ # ♆
13 57	☽ ⚼ ♃
16 29	☽ △ ♅
17 14	☽ ⚼ ☿
23 Thursday	
01 40	☽ △ ♇
06 49	☽ ⚹ ♆
13 49	☽ ± ☿
14 07	☽ # ♂
14 28	☽ ∠ ♄
20 24	☽ △ ♃
22 39	☽ ⚼ ♅
22 45	☽ ⚻ ♀
24 Friday	
03 23	☽ ⚻ ☿
03 41	☽ ⚻ ♂
06 42	☽ Q ♆
07 38	☽ ⚼ ♇
08 00	☿ ⚹ ♂
08 16	☽ ⊥ ♄
11 04	☽ ± ♀
15 58	☽ ☍ ☉
16 14	☽ ± ♂
16 32	☽ ± ♅
20 02	☽ ⊻ ♄
21 57	♂ Q ♅
23 38	☉ ⊥ ♅
25 Saturday	
01 23	☽ ± ♇
03 26	☉ ⊥ ♂
04 21	☽ ⚻ ♅
05 23	☽ ⚼ ♀
10 53	☽ ⚼ ♂
13 06	☽ ⚻ ♇
17 57	☽ □ ♆
19 38	☽ # ☉
20 20	☽ # ♇
26 Sunday	
05 56	☉ ǁ ♇
06 07	☽ # ♀
07 46	☽ □ ♃
11 30	☽ △ ♀
11 50	☽ # ♃
17 09	☉ ⚻ ♄
17 25	☽ △ ♂
21 03	☽ ☍ ☿
27 Monday	
04 16	☽ ǁ ♆
05 16	☿ # ♄
05 34	☽ ☌ ♄
06 38	☽ ⚻ ☉
07 00	♆ St D
10 07	☉ Q ♃
14 05	☽ ☍ ♅
18 16	☽ Q ♃
18 50	☽ ± ☉
20 27	☽ ǁ ♄
22 23	☽ ☍ ♇
22 52	☽ # ☿
28 Tuesday	
02 54	☽ △ ♆
09 29	☽ # ♅
12 54	☽ ⚼ ☉
14 45	♂ ⊥ ♃
16 48	☽ ⚹ ♃
22 03	☽ □ ♀
29 Wednesday	
04 29	☽ □ ♂
06 32	☽ ⚼ ♆
11 04	☽ ⚻ ☿
12 51	☽ ⊻ ♄
18 27	☽ △ ☉
20 00	☉ ⊥ ♇
20 26	☽ ∠ ♃
20 58	☿ Q ♆
21 31	☽ ⚻ ♅
22 51	☽ ± ♆
23 02	☽ ± ☿
23 39	☽ ⊥ ♄
30 Thursday	
05 23	☽ ⚻ ♇
05 24	☿ ☍ ♄
08 04	☽ Q ♀
08 20	☽ ± ♅
09 36	☽ ⚻ ♆
12 44	☽ ⊥ ♃
14 39	☽ Q ♂
14 59	♀ Q ♅
15 40	☽ ∠ ♄
16 07	☽ ± ♇
16 44	☽ ⚼ ☿
18 21	☉ ǁ ♀
20 40	♂ ⚼ ♆
23 29	☽ ⊻ ♃
31 Friday	
00 23	☽ ⚼ ♅
06 17	☽ ⚹ ♀
08 02	☽ ⚼ ♇
12 55	☽ ⚹ ♂
13 04	☿ ǁ ♅
15 31	☉ ⊻ ♅
17 56	☽ ⚹ ♄
19 55	☽ ǁ ☿
20 40	☽ ǁ ♅
21 18	♂ Q ♇
21 35	☽ △ ☿

AUGUST 1964

Date	Sidereal time h m s	Sun ☉ ° ' "	Moon ☽ ° ' "	Moon ☽ 24.00 ° ' "	Mercury ☿ ° '	Venus ♀ ° '	Mars ♂ ° '	Jupiter ♃ ° '	Saturn ♄ ° '	Uranus ♅ ° '	Neptune ♆ ° '	Pluto ♇ ° '
01	08 40 30	09 ♌ 14 30	13 ♉ 50 32	20 ♉ 51 40	06 ♍ 11	27 ♊ 29	01 ♋ 10	23 ♉ 05	03 ♓ 21	08 ♍ 28	15 ♏ 04	12 ♍ 46
02	08 44 27	10 11 56	27 ♉ 56 33	05 ♊ 05 02	07 18	28 06	01 50	23 12	03 R 17	08 32	15 04	12 48
03	08 48 23	11 09 22	12 ♊ 16 52	19 ♊ 31 45	08 22	28 45	02 30	23 20	03 13	08 35	15 04	12 50
04	08 52 20	12 06 50	26 ♊ 49 13	04 ♋ 08 46	09 24	29 ♊ 25	03 10	23 27	03 09	08 39	15 05	12 52
05	08 56 17	13 04 19	11 ♋ 29 43	18 ♋ 51 22	10 24	00 ♋ 05	03 50	23 35	03 05	08 42	15 05	12 54
06	09 00 13	14 01 50	26 ♋ 12 53	03 ♌ 33 25	11 20	00 47	04 30	23 42	03 01	08 46	15 05	12 56
07	09 04 10	14 59 21	10 ♌ 52 03	18 ♌ 07 55	12 13	01 30	05 10	23 49	02 57	08 49	15 06	12 57
08	09 08 06	15 56 54	25 ♌ 20 11	02 ♍ 28 05	13 04	02 14	05 49	23 56	02 52	08 53	15 06	12 59
09	09 12 03	16 54 27	09 ♍ 30 57	16 ♍ 28 15	13 51	02 58	06 29	24 02	02 48	08 56	15 06	13 01
10	09 15 59	17 52 01	23 ♍ 19 37	00 ♎ 04 48	14 34	03 44	07 09	24 09	02 44	09 00	15 07	13 03
11	09 19 56	18 49 37	06 ♎ 43 43	13 ♎ 16 24	15 14	04 30	07 48	24 15	02 40	09 03	15 07	13 05
12	09 23 52	19 47 13	19 ♎ 43 03	26 ♎ 03 59	15 50	05 17	08 28	24 21	02 35	09 07	15 08	13 07
13	09 27 49	20 44 50	02 ♏ 19 34	08 ♏ 30 19	16 22	06 05	09 07	24 28	02 31	09 11	15 08	13 09
14	09 31 46	21 42 29	14 ♏ 36 45	20 ♏ 39 30	16 50	06 54	09 47	24 33	02 26	09 14	15 09	13 11
15	09 35 42	22 40 08	26 ♏ 39 10	02 ♐ 36 25	17 13	07 43	10 26	24 39	02 22	09 18	15 10	13 13
16	09 39 39	23 37 48	08 ♐ 31 55	14 ♐ 26 19	17 32	08 33	11 05	24 45	02 18	09 22	15 10	13 15
17	09 43 35	24 35 29	20 ♐ 20 15	26 ♐ 14 14	17 46	09 24	11 44	24 50	02 13	09 25	15 11	13 17
18	09 47 32	25 33 12	02 ♑ 09 15	08 ♑ 05 27	17 54	10 16	12 23	24 55	02 09	09 29	15 12	13 19
19	09 51 28	26 30 56	14 ♑ 03 31	20 ♑ 03 53	17 58	11 08	13 02	25 00	02 04	09 33	15 12	13 21
20	09 55 25	27 28 40	26 ♑ 06 58	02 ♒ 13 08	17 R 55	12 00	13 41	25 05	02 00	09 36	15 13	13 23
21	09 59 21	28 26 26	08 ♒ 22 39	14 ♒ 35 44	17 47	12 54	14 20	25 10	01 55	09 40	15 14	13 25
22	10 03 18	29 ♌ 24 14	20 ♒ 52 33	27 ♒ 13 10	17 34	13 48	14 59	25 15	01 51	09 44	15 15	13 27
23	10 07 15	00 ♍ 22 02	03 ♓ 37 35	10 ♓ 05 45	17 14	14 42	15 38	25 19	01 46	09 47	15 16	13 30
24	10 11 11	01 19 53	16 ♓ 37 35	23 ♓ 12 54	16 48	15 37	16 17	25 23	01 41	09 51	15 16	13 32
25	10 15 08	02 17 44	29 ♓ 51 33	06 ♈ 33 18	16 17	16 33	16 55	25 27	01 37	09 55	15 17	13 34
26	10 19 04	03 15 38	13 ♈ 17 57	20 ♈ 05 17	15 41	17 29	17 34	25 31	01 32	09 59	15 18	13 36
27	10 23 01	04 13 33	26 ♈ 55 06	03 ♉ 47 13	14 59	18 25	18 12	25 35	01 28	10 02	15 19	13 38
28	10 26 57	05 11 30	10 ♉ 41 28	17 ♉ 37 44	14 12	19 22	18 51	25 38	01 23	10 06	15 20	13 40
29	10 30 54	06 09 28	24 ♉ 35 54	01 ♊ 35 54	13 22	20 20	19 29	25 41	01 19	10 10	15 21	13 42
30	10 34 50	07 07 29	08 ♊ 37 37	15 ♊ 40 59	12 28	21 18	20 07	25 45	01 14	10 14	15 22	13 44
31	10 38 47	08 ♍ 05 32	22 ♊ 45 54	29 ♊ 52 13	11 ♍ 32	22 ♋ 16	20 ♋ 46	25 ♉ 47	01 ♓ 10	10 ♍ 17	15 ♏ 23	13 ♍ 46

Date	Moon True ☊ ° '	Moon Mean ☊ ° '	Moon ☽ Latitude ° '
01	01 ♋ 27	00 ♋ 03	03 S 50
02	01 D 27	00 ♋ 00	02 52
03	01 29	29 ♊ 57	01 43
04	01 30	29 54	00 S 25
05	01 R 29	29 51	00 N 54
06	01 27	29 47	02 10
07	01 23	29 44	03 17
08	01 17	29 41	04 10
09	01 10	29 38	04 47
10	01 03	29 35	05 05
11	00 56	29 32	05 07
12	00 50	29 28	04 52
13	00 47	29 25	04 23
14	00 45	29 22	03 43
15	00 D 45	29 19	02 53
16	00 46	29 16	01 57
17	00 47	29 12	00 N 56
18	00 R 47	29 09	00 S 07
19	00 46	29 06	01 11
20	00 43	29 03	02 12
21	00 38	29 00	03 07
22	00 30	28 57	03 55
23	00 20	28 53	04 32
24	00 10	28 50	04 55
25	00 00	28 47	05 04
26	29 ♊ 50	28 44	04 55
27	29 43	28 41	04 31
28	29 38	28 37	03 50
29	29 36	28 34	02 55
30	29 D 36	28 31	01 49
31	29 ♊ 36	28 ♊ 28	00 S 36

DECLINATIONS

Date	Sun ☉	Moon ☽	Mercury ☿	Venus ♀	Mars ♂	Jupiter ♃	Saturn ♄	Uranus ♅	Neptune ♆	Pluto ♇
01	17 N 57	12 N 20	08 N 32	18 N 29	23 N 49	17 N 30	11 S 51	09 N 05	14 S 40	19 N 20
02	17 41	16 54	07 57	18 33	23 50	17 32	11 52	09 03	14 40	19 20
03	17 26	20 34	07 23	18 36	23 50	17 34	11 54	09 02	14 40	19 19
04	17 10	22 59	06 50	18 40	23 50	17 35	11 55	09 01	14 40	19 18
05	16 54	23 51	06 17	18 43	23 49	17 37	11 57	08 59	14 40	19 17
06	16 37	23 02	05 46	18 47	23 49	17 39	11 59	08 58	14 40	19 16
07	16 21	20 40	05 15	18 50	23 48	17 40	12 00	08 57	14 41	19 15
08	16 04	17 00	04 45	18 53	23 47	17 42	12 02	08 55	14 41	19 14
09	15 46	12 26	04 17	18 57	23 46	17 43	12 03	08 54	14 41	19 13
10	15 29	07 19	03 49	19 00	23 45	17 45	12 05	08 53	14 41	19 13
11	15 11	02 N 01	03 24	19 02	23 43	17 46	12 07	08 51	14 41	19 12
12	14 53	03 S 12	02 59	19 05	23 42	17 48	12 08	08 50	14 42	19 11
13	14 35	08 10	02 37	19 07	23 40	17 49	12 10	08 49	14 42	19 10
14	14 16	12 40	02 16	19 09	23 38	17 50	12 12	08 47	14 42	19 09
15	13 58	16 36	01 57	19 11	23 36	17 51	12 14	08 46	14 42	19 08
16	13 39	19 48	01 40	19 13	23 33	17 53	12 15	08 44	14 42	19 07
17	13 20	22 10	01 26	19 14	23 30	17 54	12 17	08 43	14 43	19 07
18	13 00	23 33	01 14	19 15	23 28	17 55	12 18	08 42	14 43	19 06
19	12 41	23 52	01 04	19 16	23 25	17 56	12 20	08 40	14 43	19 05
20	12 21	23 05	00 58	19 16	23 22	17 57	12 22	08 39	14 44	19 04
21	12 01	21 11	00 54	19 16	23 18	17 58	12 24	08 37	14 44	19 03
22	11 41	18 15	00 54	19 16	23 15	17 59	12 25	08 36	14 44	19 02
23	11 21	14 24	00 57	19 15	23 11	18 00	12 27	08 35	14 44	19 01
24	11 00	09 49	01 03	19 14	23 07	18 01	12 29	08 33	14 45	19 01
25	10 40	04 S 42	01 12	19 13	23 03	18 01	12 31	08 32	14 45	19 00
26	10 19	00 N 43	01 26	19 11	22 59	18 02	12 32	08 30	14 45	18 59
27	09 58	06 10	01 42	19 09	22 55	18 03	12 34	08 29	14 46	18 58
28	09 37	11 24	02 02	19 06	22 50	18 04	12 36	08 28	14 46	18 57
29	09 15	16 06	02 25	19 03	22 45	18 04	12 37	08 26	14 46	18 56
30	08 54	19 57	02 51	19 00	22 41	18 05	12 39	08 25	14 47	18 56
31	08 N 32	22 N 39	03 N 20	18 N 56	22 N 36	18 N 05	12 S 41	08 N 23	14 S 47	18 N 55

ZODIAC SIGN ENTRIES

Date	h	m	Planets
02	15	28	☽ ♊
04	17	13	☽ ♋
05	08	53	♀ ♋
06	18	11	☽ ♌
08	19	50	☽ ♍
10	23	51	☽ ♎
13	07	31	☽ ♏
15	18	44	☽ ♐
18	07	38	☽ ♑
20	19	39	☽ ♒
23	02	51	☉ ♍
23	05	13	☽ ♓
25	12	15	☽ ♈
27	17	24	☽ ♉
29	21	16	☽ ♊

LATITUDES

Date	Mercury ☿	Venus ♀	Mars ♂	Jupiter ♃	Saturn ♄	Uranus ♅	Neptune ♆	Pluto ♇
01	00 S 46	04 S 56	00 N 23	01 S 05	01 S 41	00 N 44	01 N 46	13 N 40
04	01 18	04 47	00 25	01 05	01 41	00 44	01 46	13 39
07	01 52	04 36	00 28	01 06	01 41	00 44	01 46	13 39
10	02 27	04 24	00 30	01 06	01 42	00 44	01 46	13 38
13	03 00	04 11	00 32	01 06	01 42	00 44	01 46	13 38
16	03 32	03 58	00 34	01 07	01 42	00 44	01 45	13 38
19	04 00	03 43	00 37	01 07	01 43	00 44	01 45	13 38
22	04 22	03 29	00 39	01 08	01 43	00 44	01 45	13 37
25	04 34	03 14	00 41	01 08	01 43	00 44	01 45	13 37
28	04 32	02 58	00 43	01 09	01 43	00 44	01 45	13 37
31	04 S 14	02 S 43	00 N 46	01 S 09	01 S 44	00 N 44	01 N 45	13 N 37

DATA

Julian Date	2438609
Delta T	+36 seconds
Ayanamsa	23° 21' 28"
Synetic vernal point	05° ♓ 45' 31"
True obliquity of ecliptic	23° 26' 38"

MOON'S PHASES, APSIDES AND POSITIONS ☽

Date	h	m	Phase	Longitude ° '	Eclipse Indicator
01	03	29	☾	08 ♉ 54	
07	19	17	●	15 ♌ 17	
15	03	19	☽	22 ♏ 19	
23	05	25	○	00 ♓ 06	
30	09	15	☾	07 ♊ 01	

Day	h	m		
05	14	44	Perigee	
17	12	10	Apogee	
05	12	27	Max dec	23° N 51'
11	21	11	0S	
19	07	05	Max dec	23° S 54'
26	08	53	0N	

LONGITUDES

Date	Chiron ⚷ ° '	Ceres ⚳ ° '	Pallas ⚴ ° '	Juno ⚵ ° '	Vesta ⚶ ° '	Black Moon Lilith ⚸ ° '
01	18 ♓ 04	23 ♐ 25	20 ♏ 13	20 ♏ 11	16 ♓ 06	22 ♐ 16
11	17 ♓ 42	23 ♐ 07	21 ♏ 55	21 ♏ 18	14 ♓ 34	23 ♐ 22
21	17 ♓ 18	23 ♐ 24	24 ♏ 05	22 ♏ 58	12 ♓ 26	24 ♐ 29
31	16 ♓ 51	24 ♐ 17	26 ♏ 37	24 ♏ 40	09 ♓ 58	25 ♐ 35

All ephemeris data is given at 12.00 UT and the Moon's longitude is additionally given for 24.00 UT

ASPECTARIAN

h m Aspects

01 Saturday
02 43 ☽ △ ♅
03 29 ☽ □ ☉
09 33 ☽ ∠ ♀
09 38 ☽ # ♄
10 09 ☽ △ ♇
14 06 ☽ ☍ ♆
14 34 ☽ Q ♄
16 11 ☽ ∠ ♂
23 47 ☽ # ♆

02 Sunday
01 39 ☽ ⊥ ♀
03 55 ☽ ☌ ♃
08 15 ☽ ⊥ ♂
12 17 ☽ ⊻ ♀
12 28 ☽ Q ☉
15 40 ☽ ∥ ♃
16 20 ☽ ∥ ☉
18 52 ☽ ⊻ ♂
20 57 ☽ □ ♄
22 01 ☽ ∥ ♀

03 Monday
01 27 ☉ ∥ ♃
02 56 ☽ ∥ ♇
04 59 ☽ □ ☿
05 50 ☽ □ ♅
10 00 ☽ ⚹ ☉
12 55 ☽ □ ♇
16 38 ☽ ⊼ ♆
17 10 ☿ ☌ ♅

04 Tuesday
02 33 ☽ ± ♆
06 25 ☽ ⊻ ♃
11 26 ♂ △ ♄
11 43 ☽ Q ♅
12 31 ☽ ∠ ☉
13 02 ☽ Q ☿
14 00 ♀ ⊥ ♃
16 22 ☽ ⊥ ♃
16 27 ☽ ☌ ♀
17 21 ☽ ⅌ ♆
18 39 ☽ Q ♇
22 19 ☽ △ ♄
22 54 ☽ ☌ ♂

05 Wednesday
04 17 ☽ ⊥ ☉
07 12 ☽ ∠ ♃
07 25 ☉ ⊻ ♇
07 26 ☽ ⚹ ♅
08 43 ☽ ∥ ♂
10 04 ☽ ⚹ ☿
11 48 ♀ ⅌ ♆
14 17 ☽ ⚹ ♇
14 45 ☽ ⊻ ☉
16 23 ☽ ∥ ♂
17 51 ☽ △ ♆
22 41 ☽ ⅌ ♄

06 Thursday
07 52 ☽ ⚹ ♃
07 59 ☽ ∠ ♅
12 12 ☽ ∠ ☿
13 18 ☽ ± ♄
14 48 ☽ ∠ ♇
17 01 ♀ Q ♇
19 51 ☽ ⊻ ♀
22 44 ☽ ⊥ ♅
23 03 ☽ ⊼ ♄

07 Friday
02 11 ☽ ⊻ ♂
03 38 ☽ Q ♃
03 53 ☽ ⊥ ☿
05 34 ☽ ⊥ ♇
06 10 ☽ ⊥ ♀
08 37 ☽ ⊻ ♅
12 30 ☽ ⊥ ♂
14 22 ☽ ⊻ ☿
14 39 ☉ □ ♆
15 27 ☽ ⊻ ♇
18 58 ☽ □ ♆
19 17 ☽ ☌ ☉
21 47 ☽ ∠ ♀
22 11 ☽ ∥ ♇

08 Saturday
00 41 ☽ ∥ ♀
04 07 ☽ ∠ ♂
07 58 ☽ ∥ ♃
09 38 ☽ □ ♃
09 54 ☿ ☌ ♇
17 39 ☽ ∥ ☉

09 Sunday
00 14 ☽ ⚹ ♀
00 37 ☽ ☍ ♄
00 38 ☽ # ♆
01 05 ☽ Q ♆
06 34 ☽ ⚹ ♂
07 06 ♀ △ ♄
11 00 ☽ ☌ ♅
13 47 ☽ # ♄
18 03 ☽ ☌ ♇
19 52 ☽ ☌ ☿
21 38 ☽ ⚹ ♆
21 56 ☽ Q ♀

10 Monday
01 43 ☽ ⊻ ☉
04 17 ☽ Q ♂
04 49 ☽ ∥ ♅
13 02 ☽ ⊥ ☉
13 28 ☽ △ ♃

11 Tuesday
00 04 ☽ ∠ ♆
04 40 ☽ ⊼ ♄
05 17 ☽ ∥ ☿
06 20 ☽ ∠ ☉
07 42 ☽ □ ♀
07 48 ☿ ⚹ ♆
14 04 ☽ □ ♂
15 30 ☽ ± ♄
16 16 ☽ ⊻ ♅
16 22 ☽ ⊥ ♆
16 39 ☽ ⅌ ♃
23 41 ☽ ⊻ ♇

12 Wednesday
03 22 ☽ ⊥ ♅
03 26 ☽ ⊻ ♆
04 24 ☽ ⊻ ☿
08 02 ☽ ⅌ ♄
09 26 ☽ ± ♃
10 53 ☽ ⊥ ♇
11 03 ☽ # ☿
12 08 ☽ ⚹ ☉
16 10 ☽ ⊥ ☿
20 20 ☽ ∠ ♅
20 50 ☽ ⊼ ♃

13 Thursday
02 58 ☉ # ♆
03 58 ☽ ∠ ♇
10 05 ☽ ∠ ☿
12 22 ☽ △ ♄
12 53 ☽ Q ☉
14 17 ♂ ⚹ ♅
15 16 ☽ # ♅
19 47 ☽ △ ♀

14 Friday
01 23 ☽ ⚹ ♅
01 57 ☽ △ ♂
02 34 ♂ ∠ ♃
09 11 ☽ ⚹ ♇
09 18 ☽ ∥ ♄
09 47 ♀ ∥ ♇
13 04 ☽ ☌ ♆
16 33 ☽ ⚹ ☿
20 38 ☽ # ☉
23 54 ☽ ∥ ♆

15 Saturday
01 13 ☽ Q ♅
03 19 ☽ □ ☉
03 32 ☽ ⅌ ♀
07 57 ☽ ☍ ♃
09 07 ☽ Q ♇
09 24 ☽ ⅌ ♂
17 20 ☽ Q ☿
20 47 ☽ # ♃
21 13 ☉ ∠ ♀
22 58 ☽ ± ♀
23 26 ☽ □ ♄

16 Sunday
04 36 ☽ ± ♂
06 23 ☽ # ♇
07 03 ☽ # ♀
12 03 ☽ ⊼ ♀
13 41 ☽ □ ♅
17 29 ☽ ⊼ ♂
21 37 ☽ □ ♇

17 Monday
01 30 ☽ ⊻ ♆
06 41 ☽ □ ☿
11 45 ☽ Q ♄
12 36 ♀ ⚹ ♅
13 43 ☽ ⊥ ♆
18 43 ☉ □ ♃
21 13 ☽ ⊼ ♃
21 25 ☽ △ ☉

18 Tuesday
01 36 ♀ ∠ ♃
08 01 ☽ ∠ ♆
09 29 ☽ ± ♃
09 49 ☽ # ♂
11 59 ☽ ⚹ ♄

19 Wednesday
02 53 ☽ △ ♅
03 48 ☽ ⅌ ♃
05 39 ☽ ☍ ♀
06 27 ☽ ⅌ ☉
09 50 ☽ ☍ ♂
10 35 ☽ △ ♇
14 14 ☿ St R
14 18 ☽ ⚹ ♆
17 59 ☽ ∠ ♄
19 48 ☽ △ ☿

20 Thursday
00 16 ♂ ⚹ ♇
02 01 ☽ ± ☉
06 21 ☽ # ♂
09 00 ☽ ⅌ ♅
09 57 ☽ △ ♃
10 52 ☉ # ♄
11 45 ☽ ⊥ ♄
14 11 ☽ Q ♆
14 55 ☽ ⊼ ☉
16 30 ☽ ⅌ ♇
23 29 ☽ ⊻ ♄

21 Friday
01 16 ☽ ⅌ ☿
02 47 ☽ ± ♅
10 09 ☽ ± ♇
14 31 ☽ ⊼ ♅
18 30 ☽ ± ☿
21 24 ☽ ⊼ ♀
21 46 ☽ ⊼ ♇

22 Saturday
00 08 ☽ ⊼ ♂
01 14 ☽ □ ♆
02 42 ♀ ⚹ ♇
04 30 ☽ # ♀
05 49 ☽ ⊼ ☿
06 14 ☽ # ♇
09 47 ☽ ± ♀
12 13 ☽ ± ♂
13 50 ☽ # ♃
20 19 ☽ □ ♃
21 52 ♂ △ ♆

23 Sunday
04 06 ☽ ⅌ ♀
05 25 ☽ ☍ ☉
06 06 ☽ ⅌ ♂
08 33 ☽ ☌ ♄
10 03 ☽ ∥ ♆
22 32 ☽ ∥ ♄
23 30 ☽ ☍ ♅

24 Monday
02 51 ♀ △ ♆
05 36 ☽ # ☉
06 02 ☽ Q ♃
06 18 ☽ ☍ ♇
07 50 ☉ ∠ ♂
09 31 ☽ △ ♆
10 01 ☽ △ ♀
11 20 ☽ △ ♂
12 19 ☽ ☍ ☿
18 08 ☽ # ♅
20 18 ☉ ☍ ♄
23 12 ☿ ⚹ ♂

25 Tuesday
01 50 ♂ ⅌ ♄
04 01 ☽ ⚹ ♃
07 51 ☿ ⚹ ♀
12 46 ☽ ⅌ ♆
13 40 ♀ ⅌ ♄
15 08 ☽ ⊻ ♄
16 43 ☽ ⊼ ☉

26 Wednesday
01 49 ☽ ⊥ ♄
03 00 ☽ # ☿
04 16 ☽ ± ☉
04 54 ☽ ± ♆
06 04 ☽ ⊼ ♅
07 02 ☽ ∠ ♃
12 32 ☽ ⊼ ♇
13 07 ☉ Q ♆
15 17 ☽ ∥ ☿
15 33 ☽ ⊼ ♆
16 01 ☽ ⊼ ☿
16 46 ☽ ± ♅
17 42 ☽ ∠ ♄
18 38 ♀ ☌ ♂
19 55 ☽ □ ♂
19 57 ☽ □ ♀
21 27 ☽ ⅌ ☉
23 03 ☽ ⊥ ♃
23 10 ☽ ± ♇

27 Thursday
00 48 ☿ ⚹ ♆
02 05 ☽ ± ☿
08 42 ☽ ⅌ ♅
09 38 ☽ ⊻ ♃
15 00 ☽ ⅌ ♇
17 05 ☽ ⅌ ☿
19 54 ☽ ⚹ ♄
22 23 ☽ ∥ ♅

28 Friday
01 44 ☽ △ ☉
04 09 ☽ ∥ ☉
05 00 ☽ Q ♂
05 49 ☽ Q ♀
10 59 ☽ △ ♅
16 39 ☽ Q ♄
17 10 ☽ △ ♇
17 45 ☽ △ ☿
17 51 ☽ # ♄
20 03 ☽ ☍ ♆

29 Saturday
02 47 ☽ ⚹ ♂
02 54 ☿ ☌ ♇
04 07 ☽ ⚹ ♀
04 53 ☽ # ♆
13 53 ☽ ☌ ♃
23 27 ☽ □ ♄
23 34 ☽ ∥ ♃

30 Sunday
05 01 ☽ ∥ ♇
05 32 ☽ ∥ ♀
05 44 ☽ ∠ ♂
07 44 ☽ ∠ ♀
09 15 ☽ □ ☉
14 44 ☽ □ ♅
18 08 ☽ □ ☿
20 43 ☽ □ ♇
21 47 ☽ ⊥ ♂
23 29 ☽ ⊼ ♆

31 Monday
00 11 ☽ ⊥ ♀
08 27 ☽ ⊻ ♂
09 40 ☽ ± ♆
11 06 ☽ ⊻ ♀
11 27 ☽ ∥ ♂
17 08 ☽ ⊻ ♃
17 27 ♀ ∥ ♇
18 02 ☽ Q ☉
21 23 ☽ Q ♅
22 37 ☉ ∥ ♅
22 43 ☽ Q ☿

SEPTEMBER 1964

LONGITUDES

Date	Sidereal time h m s	Sun ☉ ° ' "	Moon ☽ ° ' "	Moon ☽ 24.00 ° ' "	Mercury ☿ ° '	Venus ♀ ° '	Mars ♂ ° '	Jupiter ♃ ° '	Saturn ♄ ° '	Uranus ♅ ° '	Neptune ♆ ° '	Pluto ♇ ° '
01	10 42 44	09 ♍ 03 37	06 ♋ 59 45	14 ♋ 08 15	10 ♍ 35	23 ♋ 15	21 ♋ 24	25 ♉ 50	01 ♓ 05	10 ♍ 21	15 ♏ 25	13 ♍ 48
02	10 46 40	10 01 43	21 ♋ 17 23	28 ♋ 26 46	09 R 38	24 15	22 02	25 53	01 R 01	10 25	15 26	13 50
03	10 50 37	10 59 52	05 ♌ 35 55	12 ♌ 44 18	08 42	25 14	22 40	25 55	00 56	10 29	15 27	13 53
04	10 54 33	11 58 02	19 ♌ 51 20	26 ♌ 56 22	07 49	26 14	23 18	25 57	00 52	10 33	15 28	13 55
05	10 58 30	12 56 15	03 ♍ 58 47	10 ♍ 57 56	06 59	27 15	23 56	25 59	00 48	10 36	15 29	13 57
06	11 02 26	13 54 29	17 ♍ 53 14	24 ♍ 44 12	06 14	28 16	24 33	26 01	00 43	10 40	15 31	13 59
07	11 06 23	14 52 44	01 ♎ 30 22	08 ♎ 11 26	05 36	29 ♋ 17	25 11	26 02	00 39	10 44	15 32	14 01
08	11 10 19	15 51 02	14 ♎ 47 11	21 ♎ 17 32	05 04	00 ♌ 18	25 49	26 04	00 35	10 48	15 33	14 03
09	11 14 16	16 49 20	27 ♎ 42 32	04 ♏ 02 20	04 40	01 20	26 26	26 05	00 30	10 51	15 35	14 05
10	11 18 13	17 47 41	10 ♏ 17 11	16 ♏ 27 29	04 24	02 22	27 04	26 06	00 26	10 55	15 36	14 07
11	11 22 09	18 46 03	22 ♏ 33 39	28 ♏ 36 13	04 18	03 25	27 41	26 07	00 22	10 59	15 37	14 10
12	11 26 06	19 44 27	04 ♐ 35 45	10 ♐ 32 54	04 D 20	04 27	28 19	26 07	00 18	11 03	15 39	14 12
13	11 30 02	20 42 53	16 ♐ 28 19	22 ♐ 22 40	04 32	05 31	28 56	26 08	00 14	11 06	15 40	14 14
14	11 33 59	21 41 20	28 ♐ 16 40	04 ♑ 11 00	04 53	06 34	29 ♋ 33	26 08	00 09	11 10	15 42	14 16
15	11 37 55	22 39 49	10 ♑ 06 20	16 ♑ 03 22	05 23	07 38	00 ♌ 10	26 R 08	00 05	11 14	15 43	14 18
16	11 41 52	23 38 19	22 ♑ 02 42	28 ♑ 04 56	06 01	08 42	00 47	26 07	00 ♓ 01	11 18	15 45	14 20
17	11 45 48	24 36 51	04 ♒ 10 36	10 ♒ 20 11	06 49	09 46	01 24	26 07	29 ♒ 58	11 21	15 46	14 22
18	11 49 45	25 35 25	16 ♒ 34 05	22 ♒ 52 35	07 44	10 50	02 01	26 06	29 54	11 25	15 48	14 24
19	11 53 42	26 34 01	29 ♒ 15 56	05 ♓ 44 14	08 47	11 55	02 38	26 06	29 50	11 29	15 49	14 26
20	11 57 38	27 32 38	12 ♓ 17 28	18 ♓ 55 34	09 56	13 00	03 15	26 04	29 46	11 32	15 51	14 28
21	12 01 35	28 31 17	25 ♓ 38 17	02 ♈ 25 21	11 12	14 05	03 51	26 03	29 42	11 36	15 53	14 31
22	12 05 31	29 ♍ 29 58	09 ♈ 16 20	16 ♈ 10 48	12 33	15 11	04 28	26 02	29 39	11 40	15 54	14 33
23	12 09 28	00 ♎ 28 41	23 ♈ 08 15	00 ♉ 08 09	13 59	16 16	05 04	26 00	29 35	11 43	15 56	14 35
24	12 13 24	01 27 26	07 ♉ 09 57	14 ♉ 13 11	15 30	17 22	05 41	25 58	29 32	11 47	15 58	14 37
25	12 17 21	02 26 14	21 ♉ 17 22	28 ♉ 22 05	17 04	18 29	06 17	25 56	29 28	11 50	15 59	14 39
26	12 21 17	03 25 03	05 ♊ 26 59	12 ♊ 31 48	18 42	19 35	06 53	25 54	29 25	11 54	16 01	14 41
27	12 25 14	04 23 55	19 ♊ 36 18	26 ♊ 40 19	20 22	20 42	07 29	25 52	29 22	11 58	16 03	14 43
28	12 29 11	05 22 50	03 ♋ 43 43	10 ♋ 46 24	22 04	21 49	08 05	25 49	29 18	12 01	16 05	14 45
29	12 33 07	06 21 46	17 ♋ 48 17	24 ♋ 49 17	23 48	22 56	08 41	25 46	29 15	12 05	16 07	14 47
30	12 37 04	07 ♎ 20 46	01 ♌ 49 16	08 ♌ 48 07	25 ♍ 33	24 ♌ 03	09 ♌ 17	25 ♉ 43	29 ♒ 12	12 ♍ 08	16 ♏ 08	14 ♍ 49

Moon nodes and latitude / DECLINATIONS

Date	Moon True ☊ ° '	Moon Mean ☊ ° '	Moon ☽ Latitude ° '	Sun ☉ ° '	Moon ☽ ° '	Mercury ☿ ° '	Venus ♀ ° '	Mars ♂ ° '	Jupiter ♃ ° '	Saturn ♄ ° '	Uranus ♅ ° '	Neptune ♆ ° '	Pluto ♇ ° '
01	29 ♊ 36	28 ♊ 25	00 N 39	08 N 10	23 N 55	03 N 50	18 N 51	22 N 30	18 N 06	12 S 42	08 N 22	14 S 48	18 N 54
02	29 R 34	28 22	01 53	07 49	23 37	04 23	18 46	22 25	18 06	12 44	08 20	14 48	18 53
03	29 30	28 18	02 59	07 27	21 46	04 56	18 41	22 20	18 07	12 45	08 19	14 48	18 52
04	29 23	28 15	03 53	07 04	18 33	05 30	18 35	22 14	18 07	12 47	08 18	14 49	18 52
05	29 14	28 12	04 33	06 42	14 17	06 03	18 29	22 08	18 07	12 49	08 16	14 49	18 51
06	29 03	28 09	04 56	06 20	09 19	06 36	18 22	22 02	18 08	12 50	08 15	14 50	18 50
07	28 51	28 06	05 01	05 57	04 N 00	07 07	18 15	21 56	18 08	12 52	08 13	14 50	18 49
08	28 39	28 03	04 50	05 35	01 S 22	07 36	18 07	21 50	18 08	12 53	08 12	14 50	18 49
09	28 30	27 59	04 24	05 12	06 33	08 03	17 59	21 44	18 08	12 55	08 10	14 51	18 48
10	28 22	27 56	03 45	04 50	11 20	08 27	17 51	21 37	18 08	12 56	08 09	14 51	18 47
11	28 17	27 53	02 57	04 27	15 34	08 47	17 42	21 31	18 08	12 58	08 08	14 52	18 46
12	28 15	27 50	02 02	04 04	19 04	09 04	17 32	21 24	18 08	12 59	08 06	14 52	18 46
13	28 14	27 47	01 N 02	03 41	21 44	09 17	17 22	21 17	18 08	13 01	08 05	14 53	18 45
14	28 D 14	27 43	00 00	03 18	23 26	09 26	17 11	21 10	18 08	13 02	08 03	14 53	18 44
15	28 R 14	27 40	01 S 03	02 55	24 06	09 31	17 00	21 03	18 08	13 04	08 02	14 54	18 43
16	28 12	27 37	02 03	02 32	23 39	09 31	16 49	20 56	18 08	13 05	08 00	14 54	18 43
17	28 08	27 34	02 58	02 08	22 06	09 27	16 37	20 49	18 08	13 06	07 59	14 55	18 42
18	28 01	27 31	03 46	01 45	19 28	09 19	16 24	20 41	18 07	13 08	07 58	14 55	18 41
19	27 52	27 28	04 24	01 22	15 51	09 07	16 11	20 34	18 07	13 09	07 56	14 56	18 41
20	27 41	27 24	04 50	00 59	11 25	08 51	15 58	20 26	18 07	13 10	07 55	14 56	18 40
21	27 28	27 21	05 00	00 35	06 19	08 32	15 44	20 18	18 06	13 12	07 54	14 57	18 39
22	27 16	27 18	04 54	00 N 12	00 S 50	08 09	15 30	20 10	18 06	13 13	07 52	14 57	18 38
23	27 04	27 15	04 30	00 S 11	04 N 48	07 42	15 15	20 02	18 05	13 14	07 51	14 58	18 38
24	26 56	27 12	03 50	00 35	10 17	07 13	15 00	19 54	18 05	13 15	07 49	14 58	18 37
25	26 49	27 09	02 56	00 58	15 16	06 41	14 44	19 46	18 04	13 17	07 48	14 59	18 37
26	26 46	27 05	01 50	01 22	19 25	06 07	14 28	19 37	18 03	13 18	07 47	14 59	18 36
27	26 45	27 02	00 S 38	01 45	22 25	05 30	14 11	19 29	18 03	13 19	07 45	15 00	18 35
28	26 D 45	26 59	00 N 37	02 08	24 00	04 52	13 54	19 21	18 02	13 20	07 44	15 00	18 35
29	26 R 45	26 56	01 49	02 32	24 04	04 12	13 37	19 12	18 01	13 21	07 43	15 01	18 34
30	26 ♊ 43	26 ♊ 53	02 N 54	02 S 55	22 N 35	03 N 30	13 N 19	19 N 03	18 N 00	13 S 22	07 N 41	15 S 01	18 N 34

ZODIAC SIGN ENTRIES

Date	h	m	Planets
01	00	13	☽ ♋
03	02	36	☽ ♌
05	05	12	☽ ♍
07	09	19	☽ ♎
08	04	53	♀ ♌
09	16	19	☽ ♏
12	02	47	☽ ♐
14	15	30	☽ ♑
15	05	22	♂ ♌
16	21	04	♄ ♒
17	03	47	☽ ♒
19	13	22	☽ ♓
21	19	44	☽ ♈
23	00	17	☉ ♎
23	23	46	☽ ♉
26	02	46	☽ ♊
28	05	39	☽ ♋
30	08	52	☽ ♌

LATITUDES

Date	Mercury ☿ ° '	Venus ♀ ° '	Mars ♂ ° '	Jupiter ♃ ° '	Saturn ♄ ° '	Uranus ♅ ° '	Neptune ♆ ° '	Pluto ♇ ° '
01	04 S 04	02 S 37	00 N 47	01 S 09	01 S 44	00 N 44	01 N 45	13 N 37
04	03 23	02 22	00 49	01 10	01 44	00 44	01 44	13 37
07	02 31	02 06	00 51	01 10	01 44	00 44	01 44	13 38
10	01 33	01 50	00 54	01 11	01 44	00 44	01 44	13 38
13	00 S 37	01 35	00 56	01 11	01 44	00 44	01 44	13 38
16	00 N 14	01 19	00 58	01 11	01 44	00 44	01 44	13 39
19	00 54	01 04	01 01	01 12	01 44	00 44	01 44	13 39
22	01 24	00 50	01 03	01 12	01 44	00 44	01 44	13 39
25	01 43	00 35	01 06	01 13	01 44	00 44	01 43	13 40
28	01 52	00 21	01 08	01 13	01 44	00 44	01 43	13 41
31	01 N 53	00 S 08	01 N 10	01 S 13	01 S 43	00 N 44	01 N 43	13 N 41

DATA

Julian Date	2438640
Delta T	+36 seconds
Ayanamsa	23° 21' 32"
Synetic vernal point	05° ♓ 45' 27"
True obliquity of ecliptic	23° 26' 39"

MOON'S PHASES, APSIDES AND POSITIONS ☽

Date	h	m	Phase	Longitude ° '	Eclipse Indicator
06	04	34	●	13 ♍ 36	
13	21	24	☽	21 ♐ 06	
21	17	31	○	28 ♓ 45	
28	15	01	☾	05 ♋ 30	

Day	h	m		
02	02	12	Perigee	
14	06	46	Apogee	
27	04	48	Perigee	
01	19	29	Max dec	23° N 59'
08	05	50	0S	
15	14	30	Max dec	24° S 06'
22	15	32	0N	
29	00	50	Max dec	24° N 14'

LONGITUDES

Date	Chiron ⚷ ° '	Ceres ⚳ ° '	Pallas ⚴ ° '	Juno ⚵ ° '	Vesta ⚶ ° '	Black Moon Lilith ⚸ ° '
01	16 ♓ 48	24 ♐ 23	26 ♏ 53	24 ♏ 53	09 ♓ 43	25 ♐ 42
11	16 ♓ 20	25 ♐ 48	29 ♏ 45	27 ♏ 05	07 ♓ 15	26 ♐ 48
21	15 ♓ 52	27 ♐ 38	02 ♐ 53	29 ♏ 34	05 ♓ 09	27 ♐ 55
31	15 ♓ 26	29 ♐ 51	06 ♐ 12	02 ♐ 17	03 ♓ 38	29 ♐ 01

ASPECTARIAN

h m Aspects

01 Tuesday
00 54 ☽ bQ ♆
02 06 ☽ △ ♄
03 15 ☽ Q ♇
03 17 ☽ ⊥ ♃
15 43 ☽ ✶ ☉
17 25 ☿ ☌ ♅
17 39 ☽ ✶ ☿
17 40 ☽ ✶ ♅
18 29 ☽ ∠ ♃
23 28 ☽ ✶ ♇

02 Wednesday
02 09 ☽ △ ♆
03 12 ☽ bQ ♄
07 02 ☉ ☌ ☿
13 18 ☽ ☌ ♂
16 50 ☿ ∠ ♀
17 16 ☽ ∠ ☿
17 19 ☽ ☌ ♀
18 13 ☽ ± ♄
18 44 ☽ ∠ ☉
18 57 ☽ ∠ ♅
19 43 ☽ ✶ ♃
22 17 ☉ ☌ ♅

03 Thursday
00 42 ☽ ∠ ♇
04 13 ☽ ⊼ ♄
05 19 ♀ ± ♄
06 07 ☽ ∥ ♂
07 26 ☽ ⊥ ☿
10 07 ☽ ⊥ ♅
10 55 ☽ ⊥ ☉
15 50 ☽ ⊥ ♇
15 54 ☽ Q ♃
16 54 ☽ ⊻ ☿
18 12 ♀ ∠ ♅
20 14 ☽ ⊻ ♅
21 44 ☽ ⊻ ☉

04 Friday
01 57 ☽ ⊻ ♇
04 17 ☿ ∠ ♂
04 35 ☽ □ ♆
04 54 ♀ ✶ ♃
10 01 ☽ ∥ ♇
11 46 ☽ ∥ ♀
14 41 ☽ ∥ ♃
18 06 ☽ ⊻ ♂
22 21 ☽ □ ♃
23 38 ☽ ⊻ ♀

05 Saturday
04 46 ☽ ⊥ ♂
06 35 ☽ ☍ ♄
09 15 ☽ # ♆
10 39 ☽ ⊥ ♀
11 10 ☽ Q ♆
16 53 ☽ ☌ ☿
19 23 ☽ # ♄
20 53 ☽ ∠ ♂
23 26 ☽ ☌ ♅

06 Sunday
03 20 ☽ ∠ ♀
04 34 ☽ ☌ ☉
05 00 ☉ ∥ ☿
05 12 ☽ ☌ ♇
07 52 ☽ ✶ ♆
13 56 ☉ ☌ ♇
16 59 ☽ ∥ ♅
17 32 ♂ ± ♄
23 17 ☽ ∥ ☿

07 Monday
00 15 ☽ ✶ ♂
02 17 ☽ △ ♃
02 38 ☽ ∥ ☉
05 39 ♀ ∠ ♇
07 43 ☽ ✶ ♀
10 16 ☽ ∠ ♆
10 29 ☽ ⊼ ♄
16 45 ☿ ⊥ ♀
19 01 ☽ ⊻ ☿
21 10 ☽ ± ♄
22 42 ☽ Q ♂

08 Tuesday
02 27 ☽ ⊥ ♆
04 30 ☉ ✶ ♆
04 41 ☽ ⊻ ♅
05 12 ☽ bQ ♃
05 27 ☽ ⊥ ☿
07 05 ☽ Q ♀
09 57 ♀ ∥ ♃
10 39 ☽ ⊻ ♇
11 08 ♂ ∠ ♅
13 25 ☽ ⊻ ♆
13 26 ☽ bQ ♄
14 06 ☽ ⊻ ☉
15 42 ☽ ⊥ ♅
17 55 ♀ ⊼ ♄
21 25 ☽ ∠ ☿
21 44 ☽ ⊥ ♇
21 44 ☽ ± ♃
21 51 ♂ ✶ ♃

09 Wednesday
02 06 ☽ ⊥ ☉
06 02 ☽ # ☉
08 30 ☽ ∠ ♅
08 56 ☽ ⊼ ♃
09 30 ☽ □ ♂
14 37 ☽ ∠ ♇
17 15 ☽ △ ♄
18 35 ☿ ∥ ♅
19 28 ☽ □ ♀
19 49 ☽ # ♅
19 56 ☽ # ☿
20 26 ☽ ∠ ☉

10 Thursday
00 54 ☽ ✶ ☿
13 14 ☽ ✶ ♅
19 28 ☽ ✶ ♇
20 44 ☽ ∥ ♄
22 20 ☽ ☌ ♆
23 45 ☽ Q ☿

11 Friday
03 53 ☽ ✶ ☉
07 45 ☽ ∥ ♆
12 50 ☽ Q ♅
17 48 ☿ St D
19 03 ☽ ☍ ♃
19 09 ☽ Q ♇
22 44 ☽ △ ♂

12 Saturday
01 21 ☽ # ♀
03 26 ☽ □ ♄
05 07 ☽ # ♃
05 46 ☽ Q ☉
08 52 ☿ ⊻ ♀
09 40 ☽ # ♇
11 28 ☽ □ ☿
11 42 ☽ △ ♀
13 09 ☿ ⊥ ♂

13 Sunday
01 04 ☽ □ ♅
02 14 ♀ ⊥ ♅
06 34 ☽ bQ ♂
07 26 ☽ □ ♇
07 37 ☽ # ♂
10 22 ☽ ⊻ ♆
15 32 ☽ Q ♄
21 00 ☽ bQ ♀
21 24 ☽ □ ☉
22 35 ☽ ⊥ ♆

14 Monday
00 13 ♂ ∠ ♇
01 51 ☽ ± ♂
07 38 ☽ ⊼ ♃
14 44 ☽ ⊼ ♂
15 48 ☽ ✶ ♄
16 55 ☽ ∠ ♆
17 07 ☽ ± ♀
19 03 ♃ St R
19 50 ☽ ± ♃

15 Tuesday
01 58 ☽ △ ☿
06 29 ☽ ⊼ ♀
09 12 ♂ ⊼ ♄
14 04 ☽ bQ ♃
14 17 ☽ △ ♅
20 29 ☽ △ ♇
21 56 ☉ ∠ ♀
22 00 ☽ ∠ ♄
23 16 ♀ Q ♃
23 21 ☽ ✶ ♆

16 Wednesday
03 42 ♀ ⊥ ♇
09 50 ☽ bQ ☿
15 27 ☽ △ ☉
15 56 ☽ ⊥ ♄
20 07 ☽ △ ♃
20 30 ☽ bQ ♅
23 21 ☽ Q ♆

17 Thursday
02 31 ☽ bQ ♇
03 45 ☽ ⊻ ♄
04 54 ☽ ± ☿
06 16 ☽ ☍ ♂
14 19 ☽ ± ♅
17 32 ☽ ⊼ ☿
20 12 ☽ ± ♇
23 30 ☽ bQ ☉
23 55 ☽ ☍ ♀

18 Friday
01 32 ☽ # ♂
02 02 ☽ ⊼ ♅
07 50 ☽ ⊼ ♇
10 31 ☽ □ ♆
17 46 ☽ # ♇
18 15 ☽ ± ☉
21 41 ☽ # ♃

19 Saturday
00 30 ☉ △ ♃
01 40 ♀ ⊻ ♅
04 41 ☿ ⊥ ♂
06 04 ☽ □ ♃
06 32 ☽ ⊼ ☉
09 53 ☽ # ♀
13 03 ☽ ☌ ♄
17 22 ☽ ∥ ♆
18 34 ☽ ⊼ ♂

20 Sunday
02 59 ☽ ∥ ♄
06 10 ☽ ± ♂
07 16 ☽ ☍ ☿
10 37 ☽ ☍ ♅
13 24 ☽ ⊼ ♀
15 14 ☽ Q ♃
15 59 ☽ ☍ ♇
18 28 ☽ △ ♆
23 17 ☽ bQ ♂

21 Monday
01 11 ☽ # ☿
01 12 ☽ ± ♀
04 49 ☽ # ♅
12 44 ☽ ✶ ♃
17 31 ☽ ☍ ☉
18 39 ☽ bQ ♀
19 11 ☽ ⊻ ♄
19 40 ☿ ☌ ♅
21 18 ☽ bQ ♆
21 36 ♀ ⊻ ♇

22 Tuesday
03 12 ☽ △ ♂
05 41 ☽ ⊥ ♄
13 06 ☽ ± ♆
14 53 ☽ # ☉
15 03 ☽ ∠ ♃
15 24 ☉ ⊼ ♄
16 06 ☽ ∥ ☉
16 11 ☽ ⊼ ♅
18 21 ☽ ⊼ ☿
21 11 ☽ ⊼ ♇
21 18 ☽ ∠ ♄
23 09 ☽ △ ♀
23 33 ☽ ⊼ ♆

23 Wednesday
02 37 ☽ ± ♅
04 05 ☿ ∥ ♅
04 21 ♀ □ ♆
05 56 ☽ ± ☿
06 37 ☽ ⊥ ♃
07 35 ☽ ± ♇
16 55 ☽ ⊻ ♃
18 11 ☽ bQ ♅
21 44 ☿ ☌ ♇
23 01 ☽ ✶ ♄
23 04 ☽ bQ ♇
23 14 ☽ bQ ☿
23 29 ☉ ∠ ♆
23 32 ☽ ∥ ☿

24 Thursday
01 06 ☽ ∥ ♅
01 32 ☽ ⊼ ☉
09 21 ☽ □ ♂
12 32 ☽ ± ☉
14 25 ♀ # ♆
16 32 ♂ ⊥ ♅
19 19 ☿ ✶ ♆
19 24 ☽ Q ♄
19 53 ☽ △ ♅

25 Friday
00 42 ☽ △ ♇
02 01 ☽ # ♄
02 59 ☽ ☍ ♆
03 56 ☽ △ ☿
04 59 ☽ bQ ☉
06 50 ☽ □ ♀
09 27 ☽ ∥ ♀
10 33 ☽ # ♆
17 18 ☽ Q ♂
19 52 ☽ ☌ ♃

26 Saturday
01 49 ☽ □ ♄
03 32 ☽ ∥ ♃
06 50 ☽ ∥ ♇
08 18 ☽ △ ☉
13 23 ☽ ∥ ♂
14 33 ☽ ✶ ♂
15 56 ☽ Q ♀
22 59 ☽ □ ♅

27 Sunday
03 41 ☽ □ ♇
05 57 ☽ ⊼ ♆
13 27 ☽ □ ☿
14 01 ☽ ✶ ♀
16 09 ☽ ± ♆
17 07 ☽ ∠ ♂
22 35 ☽ ⊻ ♃

28 Monday
01 47 ♂ Q ♃
01 50 ☿ ⊻ ♀
04 30 ☽ △ ♄
05 40 ☽ Q ♅
07 29 ☽ bQ ♆
08 45 ☽ ⊥ ♃
09 05 ☽ ⊥ ♂
10 17 ☉ ± ♄
10 20 ☽ Q ♇
15 01 ☽ □ ☉
17 42 ☽ ∠ ♀
19 45 ☽ ⊻ ♂

29 Tuesday
00 02 ☽ ∠ ♃
00 18 ☽ Q ☿
02 11 ☽ ✶ ♅
05 58 ☽ bQ ♄
06 49 ☽ ✶ ♇
09 06 ☽ △ ♆
09 49 ☿ ∠ ♂
10 22 ☽ ⊥ ♀
15 54 ♂ ⊥ ♇
21 17 ☽ ± ♄
21 31 ☽ ⊻ ♀
23 42 ☽ ✶ ☿

30 Wednesday
00 04 ☽ Q ☉
01 34 ☽ ✶ ♃
03 56 ☽ ∠ ♅
07 31 ☽ ⊼ ♄
07 49 ♀ # ♄
08 33 ☽ ∠ ♇
14 16 ☿ △ ♃
19 27 ☽ ⊥ ♅
22 06 ☽ Q ♃
22 13 ☽ ✶ ☉

All ephemeris data is given at 12.00 UT and the Moon's longitude is additionally given for 24.00 UT

OCTOBER 1964

Date	Sidereal time h m s	Sun ☉ ° ' "	Moon ☽ ° ' "	Moon ☽ 24.00 ° ' "	Mercury ☿ ° '	Venus ♀ ° '	Mars ♂ ° '	Jupiter ♃ ° '	Saturn ♄ ° '	Uranus ♅ ° '	Neptune ♆ ° '	Pluto ♇ ° '
01	12 41 00	08 ♎ 19 47	15 ♌ 45 38	22 ♌ 41 37	27 ♍ 19	25 ♌ 11	09 ♌ 53	25 ♉ 40	29 ♒ 09	12 ♍ 12	16 ♏ 10	14 ♍ 51
02	12 44 57	09 18 50	29 35 48	06 ♍ 27 51	29 ♍ 06	26 18	10 29	25 R 36	29 R 06	12 15	16 12	14 53
03	12 48 53	10 17 56	13 ♍ 17 29	20 ♍ 04 20	00 ♎ 53	27 26	11 04	25 33	29 03	12 19	16 14	14 55
04	12 52 50	11 17 04	26 ♍ 48 03	03 ♎ 28 20	02 40	28 34	11 40	25 29	29 01	12 22	16 16	14 57
05	12 56 46	12 16 14	10 ♎ 04 52	16 ♎ 37 26	04 27	29 ♌ 42	12 15	25 25	28 58	12 25	16 18	14 59
06	13 00 43	13 15 26	23 ♎ 05 50	29 ♎ 29 59	06 14	00 ♍ 51	12 51	25 21	28 55	12 29	16 20	15 01
07	13 04 40	14 14 40	05 ♏ 49 51	12 ♏ 05 31	08 01	01 59	13 26	25 17	28 53	12 32	16 22	15 03
08	13 08 36	15 13 56	18 ♏ 17 07	24 ♏ 24 54	09 48	03 08	14 01	25 12	28 50	12 36	16 24	15 04
09	13 12 33	16 13 14	00 ♐ 29 10	06 ♐ 30 20	11 34	04 17	14 36	25 07	28 48	12 39	16 26	15 06
10	13 16 29	17 12 34	12 ♐ 28 50	18 ♐ 25 13	13 19	05 26	15 11	25 02	28 46	12 42	16 28	15 08
11	13 20 26	18 11 56	24 ♐ 20 02	00 ♑ 13 53	15 04	06 35	15 46	24 57	28 44	12 45	16 30	15 10
12	13 24 22	19 11 19	06 ♑ 07 27	12 ♑ 01 23	16 49	07 45	16 21	24 52	28 42	12 49	16 32	15 12
13	13 28 19	20 10 44	17 ♑ 56 22	23 ♑ 53 06	18 32	08 54	16 55	24 47	28 40	12 52	16 34	15 14
14	13 32 15	21 10 11	29 ♑ 52 15	05 ♒ 54 30	20 15	10 04	17 30	24 41	28 38	12 55	16 36	15 16
15	13 36 12	22 09 40	12 ♒ 00 30	18 ♒ 10 48	21 58	11 14	18 04	24 36	28 36	12 58	16 38	15 17
16	13 40 09	23 09 11	24 ♒ 25 59	00 ♓ 46 28	23 39	12 24	18 38	24 30	28 34	13 01	16 40	15 19
17	13 44 05	24 08 43	07 ♓ 12 38	13 ♓ 44 44	25 21	13 34	19 13	24 24	28 33	13 04	16 42	15 21
18	13 48 02	25 08 17	20 ♓ 22 54	27 ♓ 07 07	27 01	14 44	19 47	24 17	28 31	13 07	16 44	15 23
19	13 51 58	26 07 53	03 ♈ 57 16	10 ♈ 53 01	28 ♎ 41	15 55	20 21	24 11	28 30	13 10	16 46	15 24
20	13 55 55	27 07 31	17 ♈ 53 58	24 ♈ 59 33	00 ♏ 20	17 05	20 55	24 05	28 28	13 13	16 49	15 26
21	13 59 51	28 07 11	02 ♉ 09 04	09 ♉ 21 47	01 58	18 16	21 28	23 58	28 27	13 16	16 51	15 28
22	14 03 48	29 ♎ 06 53	16 ♉ 36 53	23 ♉ 53 32	03 36	19 27	22 02	23 52	28 26	13 19	16 53	15 29
23	14 07 44	00 ♏ 06 37	01 ♊ 10 55	08 ♊ 28 17	05 14	20 38	22 35	23 45	28 25	13 22	16 55	15 31
24	14 11 41	01 06 23	15 ♊ 44 55	23 ♊ 00 11	06 50	21 49	23 09	23 38	28 24	13 25	16 57	15 33
25	14 15 38	02 06 11	00 ♋ 13 36	07 ♋ 24 45	08 27	23 00	23 42	23 31	28 23	13 28	16 59	15 34
26	14 19 34	03 06 02	14 ♋ 33 17	21 ♋ 39 00	10 02	24 11	24 15	23 24	28 23	13 31	17 02	15 36
27	14 23 31	04 05 55	28 ♋ 41 44	05 ♌ 41 24	11 37	25 23	24 48	23 16	28 22	13 33	17 04	15 37
28	14 27 27	05 05 50	12 ♌ 37 58	19 ♌ 31 24	13 12	26 34	25 21	23 09	28 22	13 36	17 06	15 39
29	14 31 24	06 05 47	26 ♌ 21 43	03 ♍ 08 56	14 46	27 46	25 54	23 01	28 21	13 39	17 08	15 40
30	14 35 20	07 05 47	09 ♍ 53 03	16 ♍ 34 05	16 20	28 ♍ 57	26 27	22 54	28 21	13 41	17 10	15 42
31	14 39 17	08 ♏ 05 48	23 ♍ 12 01	29 ♍ 46 49	17 ♏ 53	00 ♎ 09	26 ♌ 59	22 ♉ 46	28 ♒ 21	13 ♍ 44	17 ♏ 13	15 ♍ 43

Date	Moon True ☊ ° '	Moon Mean ☊ ° '	Moon ☽ Latitude ° '
01	26 ♊ 39	26 ♊ 49	03 N 49
02	26 R 31	26 46	04 29
03	26 22	26 43	04 54
04	26 10	26 40	05 01
05	25 58	26 37	04 52
06	25 46	26 34	04 28
07	25 35	26 30	03 51
08	25 27	26 27	03 03
09	25 22	26 24	02 08
10	25 19	26 21	01 08
11	25 18	26 18	00 N 05
12	25 D 18	26 15	00 S 58
13	25 19	26 11	01 58
14	25 R 18	26 08	02 54
15	25 16	26 05	03 43
16	25 12	26 02	04 22
17	25 05	25 59	04 50
18	24 57	25 55	05 04
19	24 47	25 52	05 01
20	24 37	25 49	04 40
21	24 28	25 46	04 02
22	24 21	25 43	03 08
23	24 16	25 40	02 01
24	24 14	25 36	00 S 46
25	24 D 14	25 33	00 N 32
26	24 15	25 30	01 47
27	24 16	25 27	02 55
28	24 R 16	25 24	03 51
29	24 14	25 21	04 33
30	24 10	25 17	04 59
31	24 ♊ 04	25 ♊ 14	05 N 08

DECLINATIONS

Date	Sun ☉ ° '	Moon ☽ ° '	Mercury ☿ ° '	Venus ♀ ° '	Mars ♂ ° '	Jupiter ♃ ° '	Saturn ♄ ° '	Uranus ♅ ° '	Neptune ♆ ° '	Pluto ♇ ° '
01	03 S 18	19 N 45	02 N 48	13 N 00	18 N 54	18 N 00	13 S 23	07 N 40	15 S 02	18 N 33
02	03 42	15 49	02 04	12 42	18 46	17 59	13 24	07 39	15 02	18 32
03	04 05	11 05	01 20	12 23	18 37	17 58	13 25	07 37	15 03	18 32
04	04 28	05 53	00 N 35	12 03	18 28	17 57	13 26	07 36	15 04	18 31
05	04 51	00 N 29	00 S 10	11 43	18 18	17 56	13 27	07 35	15 04	18 31
06	05 14	04 S 49	00 55	11 23	18 09	17 55	13 28	07 34	15 05	18 30
07	05 37	09 50	01 41	11 02	18 00	17 53	13 28	07 32	15 05	18 30
08	06 00	14 20	02 27	10 41	17 51	17 52	13 29	07 31	15 06	18 29
09	06 23	18 10	03 13	10 20	17 41	17 51	13 30	07 30	15 07	18 29
10	06 46	21 11	03 58	09 58	17 32	17 50	13 31	07 29	15 07	18 28
11	07 08	23 14	04 44	09 36	17 22	17 49	13 31	07 27	15 08	18 28
12	07 31	24 16	05 29	09 14	17 13	17 47	13 32	07 26	15 08	18 27
13	07 53	24 11	06 14	08 51	17 03	17 46	13 33	07 25	15 09	18 27
14	08 16	23 01	06 58	08 28	16 53	17 45	13 33	07 24	15 10	18 26
15	08 38	20 45	07 42	08 04	16 43	17 43	13 34	07 23	15 10	18 26
16	09 00	17 30	08 25	07 41	16 34	17 42	13 34	07 21	15 11	18 26
17	09 22	13 21	09 08	07 17	16 24	17 40	13 35	07 20	15 11	18 25
18	09 44	08 28	09 51	06 53	16 14	17 39	13 35	07 19	15 12	18 25
19	10 06	03 S 02	10 32	06 28	16 04	17 37	13 36	07 18	15 13	18 24
20	10 27	02 N 42	11 14	06 04	15 54	17 35	13 36	07 17	15 13	18 24
21	10 48	08 26	11 54	05 39	15 44	17 34	13 36	07 16	15 14	18 24
22	11 10	13 48	12 34	05 14	15 34	17 32	13 37	07 15	15 15	18 23
23	11 31	18 26	13 13	04 48	15 24	17 30	13 37	07 14	15 15	18 23
24	11 52	21 56	13 52	04 23	15 14	17 29	13 37	07 12	15 16	18 23
25	12 12	23 59	14 29	03 57	15 04	17 27	13 37	07 11	15 16	18 22
26	12 33	24 26	15 06	03 31	14 53	17 25	13 37	07 10	15 17	18 22
27	12 53	23 17	15 42	03 05	14 43	17 23	13 38	07 09	15 18	18 22
28	13 13	20 43	16 18	02 39	14 33	17 22	13 38	07 08	15 18	18 22
29	13 33	17 00	16 52	02 12	14 23	17 20	13 38	07 07	15 19	18 21
30	13 53	12 28	17 26	01 46	14 12	17 18	13 38	07 06	15 20	18 21
31	14 S 13	07 N 25	17 S 58	01 N 19	14 N 02	17 N 16	13 S 38	07 N 05	15 S 20	18 N 21

ZODIAC SIGN ENTRIES

Date	h	m	Planets
02	12	42	☽ ♍
03	00	12	☿ ♎
04	17	44	☽ ♎
05	18	10	♀ ♍
07	00	57	☽ ♏
09	11	02	☽ ♐
11	23	32	☽ ♑
14	12	15	☽ ♒
16	22	33	☽ ♓
19	05	05	☽ ♈
20	07	11	☿ ♏
21	08	24	☽ ♉
23	09	21	☉ ♏
23	10	03	☽ ♊
25	11	37	☽ ♋
27	14	14	☽ ♌
29	18	25	☽ ♍
31	08	54	♀ ♎

LATITUDES

Date	Mercury ☿ ° '	Venus ♀ ° '	Mars ♂ ° '	Jupiter ♃ ° '	Saturn ♄ ° '	Uranus ♅ ° '	Neptune ♆ ° '	Pluto ♇ ° '
01	01 N 53	00 S 08	01 N 10	01 S 13	01 S 43	00 N 44	01 N 43	13 N 41
04	01 48	00 N 05	01 13	01 14	01 43	00 44	01 43	13 42
07	01 38	00 17	01 16	01 14	01 43	00 44	01 43	13 43
10	01 24	00 29	01 18	01 14	01 43	00 44	01 43	13 43
13	01 08	00 40	01 21	01 14	01 43	00 44	01 43	13 44
16	00 49	00 50	01 23	01 14	01 42	00 44	01 43	13 45
19	00 30	01 00	01 26	01 15	01 42	00 45	01 43	13 46
22	00 N 10	01 08	01 29	01 15	01 42	00 45	01 43	13 47
25	00 S 10	01 16	01 31	01 15	01 41	00 45	01 43	13 48
28	00 31	01 24	01 34	01 15	01 41	00 45	01 43	13 49
31	00 S 50	01 N 30	01 N 37	01 S 15	01 S 41	00 N 45	01 N 42	13 N 51

DATA

Julian Date	2438670
Delta T	+36 seconds
Ayanamsa	23° 21' 35"
Synetic vernal point	05° ♓ 45' 24"
True obliquity of ecliptic	23° 26' 39"

MOON'S PHASES, APSIDES AND POSITIONS ☽

Date	h	m	Phase	Longitude ° '	Eclipse Indicator
05	16	20	●	12 ♎ 27	
13	16	56	☽	20 ♑ 23	
21	04	45	○	27 ♈ 49	
27	21	59	☾	04 ♌ 31	

Day	h	m		
12	02	45	Apogee	
23	22	01	Perigee	
05	14	11	0S	
12	22	29	Max dec	24° S 22'
20	00	46	0N	
26	06	32	Max dec	24° N 28'

LONGITUDES

Date	Chiron ⚷ ° '	Ceres ⚳ ° '	Pallas ⚴ ° '	Juno ⚵ ° '	Vesta ⚶ ° '	Black Moon Lilith ⚸ ° '
01	15 ♓ 26	29 ♐ 51	06 ♐ 12	02 ♐ 17	03 ♓ 38	29 ♐ 01
11	15 ♓ 03	03 ♑ 23	09 ♐ 42	05 ♐ 11	02 ♓ 51	00 ♑ 08
21	14 ♓ 42	05 ♑ 12	13 ♐ 20	08 ♐ 15	02 ♓ 51	01 ♑ 14
31	14 ♓ 27	08 ♑ 16	17 ♐ 04	11 ♐ 28	03 ♓ 34	02 ♑ 20

ASPECTARIAN

h m Aspects

01 Thursday
00 03 ☽ ⊥ ♇
00 58 ☉ # ☿
01 25 ☽ ☌ ♂
05 11 ☽ ∠ ☿
05 49 ☽ ⚺ ♅
10 25 ☽ ⚺ ♇
12 43 ☽ □ ♆
17 55 ☽ ∥ ♂
20 00 ☽ ∥ ♇
21 54 ♀ □ ♃
23 01 ☽ ⊥ ☿
23 29 ☽ ∥ ♃

02 Friday
02 06 ☽ ∠ ☉
05 05 ☽ □ ♃
05 45 ☽ ☌ ♀
11 00 ☽ ⚺ ☿
11 09 ☽ ☍ ♄
12 08 ☿ ⚻ ♄
16 08 ☽ # ♆
18 59 ☽ ⊥ ☉
20 03 ☽ Q ♆

03 Saturday
00 35 ☽ # ♄
05 17 ☽ ∥ ♀
06 19 ☽ ⚺ ☉
07 55 ☽ ⚺ ♂
10 16 ☽ ☌ ♅
10 21 ☉ ⊥ ♆
14 52 ☽ ☌ ♇
16 51 ☿ ∠ ♆
17 12 ☽ ⚹ ♆
17 41 ☉ bQ ♃
18 59 ☽ ⊥ ♂

04 Sunday
01 28 ♂ ∥ ♇
04 10 ☽ ∥ ♅
09 39 ☽ △ ♃
11 45 ☽ ∠ ♂
15 28 ☽ ⚺ ♀
15 57 ☽ ⚻ ♄
17 54 ☽ # ☉
20 02 ☽ ∠ ♆
20 56 ♀ ☍ ♄

05 Monday
00 11 ☽ ☌ ☿
02 44 ☽ ± ♄
03 18 ☽ ⊥ ♀
11 06 ☉ ⚹ ♂
12 24 ☽ ⊥ ♆
12 37 ☽ bQ ♃
13 16 ☽ # ☿
15 24 ☽ ∥ ☿
15 58 ☉ ⚺ ♅
16 10 ☽ ⚹ ♂
16 18 ☽ ⚺ ♅
16 20 ☽ ☌ ☉
18 41 ☿ ± ♄
19 05 ☽ bQ ♄
19 35 ♂ ⚺ ♅
21 00 ☽ ⚺ ♇
21 17 ☽ ∠ ♀
23 26 ☽ ⚺ ♆

06 Tuesday
03 23 ☽ ⊥ ♅
05 04 ☽ ± ♃
08 06 ☽ ⊥ ♇
14 03 ☽ ∥ ☉
15 25 ☽ Q ♂
16 11 ☽ ⚻ ♃
20 14 ☽ ∠ ♅
22 53 ☽ △ ♄

07 Wednesday
00 49 ☽ # ♅
01 00 ☽ ∠ ♇
03 30 ☉ bQ ♄
03 59 ☽ ⚹ ♀
10 49 ☿ ⊥ ♀
16 53 ☽ ⚺ ☿
17 42 ☽ # ♀

08 Thursday
00 55 ☽ ⚹ ♅
03 18 ☽ □ ♂
05 16 ☽ Q ♀
05 33 ☽ ⚺ ☉
05 45 ☽ ⚹ ♇
06 21 ☽ ⊥ ☿
06 58 ♂ ∥ ♃
07 12 ☽ ∥ ♄
08 02 ☉ ⚺ ♇
08 19 ☽ ☌ ♆
16 26 ☽ ∥ ♆
17 14 ☿ bQ ♃
18 16 ☽ ⊥ ☉
20 16 ☿ ⊥ ♆

09 Friday
00 24 ☽ Q ♅
01 28 ☽ ☍ ♃
02 55 ☽ ∠ ☿
05 17 ☽ Q ♇
08 40 ☽ □ ♄
08 50 ☽ # ♂
09 50 ☽ # ♃
13 35 ☽ ∠ ☉
14 11 ☽ # ♇
15 04 ♀ Q ♆
17 13 ☉ ⚺ ♆
20 22 ☽ □ ♀

10 Saturday
03 14 ☿ ⚺ ♅
09 58 ♂ ⚺ ♇
12 27 ☽ □ ♅
13 59 ☽ ⚹ ☿
17 22 ☽ □ ♇
17 44 ☽ △ ♂
17 54 ☿ bQ ♄
20 04 ☽ ⚺ ♆
20 37 ☽ Q ♄
22 25 ☽ ⚹ ☉

11 Sunday
08 15 ☽ ⊥ ♆
13 15 ☽ ⚻ ♃
13 20 ☿ ⚺ ♇
18 32 ☽ Q ☿
20 55 ☽ ⚹ ♄

12 Monday
01 02 ☽ Q ☉
01 23 ☽ ± ♃
01 46 ☽ bQ ♂
02 18 ☉ ⊥ ♅
02 18 ☿ ⚹ ♂
02 37 ☽ ∠ ♆
04 53 ☉ ± ♃
07 15 ☉ # ♅
08 02 ☿ ⚺ ♆
15 40 ☽ △ ♀
19 34 ☽ bQ ♃
20 16 ♂ □ ♆
21 02 ☽ ± ♂

13 Tuesday
01 40 ☽ △ ♅
03 21 ☽ ∠ ♄
06 30 ☽ △ ♇
09 12 ☽ ⚹ ♆
09 50 ☽ ⚻ ♂
13 25 ☽ □ ☿
15 11 ☿ ± ♃
16 40 ☿ ⊥ ♅
16 56 ☽ □ ☉
21 30 ☽ ⊥ ♄

14 Wednesday
01 21 ☽ bQ ♀
01 42 ☽ △ ♃
08 04 ☽ bQ ♅
09 27 ☽ Q ♆
09 31 ☽ ⚺ ♄
12 47 ☽ bQ ♇
14 13 ☉ ⊥ ♇
18 24 ☉ # ♀
21 15 ☽ ± ♀

15 Thursday
01 42 ☿ # ♅
02 03 ☽ ± ♅
02 20 ☿ ⊥ ♇
06 39 ☽ ± ♇
10 19 ☽ ⚻ ♀
13 53 ☽ ⚻ ♅
18 25 ☽ ⚻ ♇
18 44 ☉ ☌ ☿
20 05 ☿ # ♀
21 02 ☽ □ ♆

16 Friday
00 21 ☽ ☍ ♂
05 50 ☽ # ♇
09 21 ☽ △ ☉
10 18 ☽ △ ☿
10 45 ☽ # ♃
12 07 ☽ □ ♃
18 08 ☽ # ♂
19 50 ☽ ☌ ♄
23 14 ☿ ⚻ ♃

17 Saturday
01 22 ♀ ☌ ♅
01 58 ☽ ∥ ♆
08 34 ♀ ∥ ♅
10 48 ☽ ∥ ♄
15 52 ☽ bQ ☉
17 27 ☉ ⚻ ♃
18 37 ☽ bQ ☿
21 28 ☽ Q ♃
22 49 ☽ ☍ ♅

18 Sunday
00 49 ☽ ☍ ♀
02 57 ☽ ☍ ♇
03 58 ☉ ∥ ☿
05 25 ☽ △ ♆
06 20 ☽ ∥ ☿
06 28 ☽ ∥ ☉
09 35 ☽ ± ☉
10 52 ☽ ⚻ ♂
13 18 ☽ ± ☿
17 15 ☽ # ♅
18 56 ☽ ⚹ ♃
19 48 ☽ # ♀
21 10 ☽ ⚻ ☉
22 03 ☽ ± ♂

19 Monday
01 22 ♀ ☌ ♇
01 28 ☽ ⚻ ☿
02 27 ☽ ⚺ ♄
04 27 ☿ ∠ ♅
08 11 ☽ bQ ♆
09 23 ☿ △ ♄
12 56 ☽ ⊥ ♄
14 31 ☽ bQ ♂
21 00 ☽ ∠ ♃
23 50 ☽ ± ♆

20 Tuesday
03 59 ☽ ⚻ ♅
04 27 ☽ ∠ ♄
06 09 ♀ ⚹ ♆
07 47 ☽ ⚻ ♇
10 08 ☽ ⚻ ♆
10 30 ☽ ⚻ ♀
12 18 ☽ ⊥ ♃
13 31 ☿ ∠ ♇
14 15 ☽ ± ♅
17 19 ☽ △ ♂
18 00 ☽ ± ♇
21 35 ☽ ± ♀
22 23 ☽ ⚺ ♃

21 Wednesday
01 02 ☽ ∥ ♀
04 45 ☽ ☍ ☉
05 29 ☽ bQ ♅
05 49 ☽ ⚹ ♄
07 03 ☽ ∥ ♅
09 10 ☽ bQ ♇
11 40 ☽ ☍ ☿
14 01 ☽ bQ ♀
15 51 ☉ ∠ ♅
19 54 ☉ △ ♄
23 02 ☽ # ☉

22 Thursday
01 47 ☽ Q ♄
05 24 ☽ # ☿
06 32 ☽ △ ♅
10 08 ☽ △ ♇
11 05 ☽ # ♄
12 26 ☽ ☍ ♆
17 05 ☽ △ ♀
18 58 ☽ # ♆
20 17 ☽ ∥ ♂
21 17 ☽ □ ♂
21 37 ☿ Q ♂
23 51 ☽ ☌ ♃

23 Friday
06 48 ☽ ∥ ♃
07 27 ☽ □ ♄
10 06 ☽ ⚻ ☉
11 43 ☽ ∥ ♇
19 29 ☽ ⚻ ☿
20 42 ☽ ± ☉
22 02 ☉ ∠ ♇

24 Saturday
02 55 ☿ ∥ ♄
04 06 ☽ Q ♂
06 36 ☽ ± ☿
07 24 ♂ # ♆
08 08 ☽ □ ♅
10 25 ☿ ∠ ♀
11 40 ☽ □ ♇
12 38 ☽ bQ ☉
14 00 ☽ ⚻ ♆
22 55 ☽ □ ♀
23 20 ☽ bQ ☿
23 57 ☽ ± ♆

25 Sunday
00 44 ☽ ⚹ ♂
00 56 ☽ ⚺ ♃
05 14 ♂ □ ♃
08 57 ☽ △ ♄
10 49 ☽ ⊥ ♃
14 04 ☽ Q ♅
14 57 ☽ bQ ♆
15 22 ☽ △ ☉
17 35 ☽ Q ♇
21 28 ♀ △ ♃

26 Monday
01 44 ☽ ∠ ♃
02 44 ☽ ∠ ♂
03 27 ☽ △ ☿
05 29 ☿ # ♂
07 39 ☽ Q ♀
10 01 ☽ bQ ♄
10 14 ☽ ⚹ ♅
13 45 ☽ ⚹ ♇
14 37 ♀ ⚺ ♂
16 11 ☽ △ ♆
18 30 ☽ ⊥ ♂
19 21 ☿ ∥ ♆

27 Tuesday
01 14 ☽ ± ♄
02 50 ☽ ⚹ ♃
05 06 ☽ ⚺ ♂
05 49 ☽ ⚹ ♀
11 26 ☽ ⚻ ♄
11 45 ☽ ∠ ♅
15 18 ☽ ∠ ♇
21 59 ☽ □ ☉
23 11 ☽ Q ♃

28 Wednesday
03 16 ☽ ⊥ ♅
06 49 ☽ ⊥ ♇
09 59 ☽ ∠ ♀
13 07 ☽ □ ☿
13 41 ☽ ⚺ ♅
17 15 ☽ ⚺ ♇
18 17 ☿ ⚹ ♅
19 47 ☽ □ ♆

29 Thursday
03 09 ☽ ⊥ ♀
03 56 ☽ ∥ ♇
06 11 ☽ □ ♃
07 42 ☽ Q ☉
10 07 ☽ ∥ ♃
11 09 ☽ ☌ ♂
12 42 ☽ # ☿
14 42 ☽ ⚺ ♀
15 31 ☽ ☍ ♄
17 19 ☉ ∥ ♄
21 21 ☽ # ♆
23 49 ♀ ⚻ ♄

30 Friday
00 48 ☽ Q ☿
02 05 ☿ ⚹ ♇
02 53 ☽ ∥ ♂
03 34 ☽ Q ♆
05 21 ☽ # ☉
06 11 ☽ # ♄
06 37 ☽ ⚹ ☉
06 40 ☿ # ♃
18 50 ☽ ☌ ♅
22 27 ☽ ☌ ♇

31 Saturday
01 06 ☽ ⚹ ☿
01 08 ☽ ⚹ ♆
01 19 ☿ ☌ ♆
03 38 ☉ # ♂
11 14 ☽ △ ♃
11 48 ☽ ∠ ☉
13 30 ☽ ∥ ♅
19 11 ☽ ⚺ ♂
21 22 ☽ ⚻ ♄

All ephemeris data is given at 12.00 UT and the Moon's longitude is additionally given for 24.00 UT

NOVEMBER 1964

Date	Sidereal time h m s	Sun ☉ ° ' "	Moon ☽ ° ' "	Moon ☽ 24.00 ° ' "	Mercury ☿ ° '	Venus ♀ ° '	Mars ♂ ° '	Jupiter ♃ ° '	Saturn ♄ ° '	Uranus ♅ ° '	Neptune ♆ ° '	Pluto ♇ ° '
01	14 43 13	09 ♏ 05 52	06 ♎ 18 28	12 ♎ 46 56	19 ♏ 26	01 ♎ 21	27 ♌ 32	22 ♉ 39	28 ♒ 21	13 ♍ 46	17 ♏ 15	15 ♍ 45
02	14 47 10	10 05 57	19 ♎ 12 10	25 ♎ 34 10	20 58	02 33	28 04	22 R 31	28 D 21	13 49	17 17	15 46
03	14 51 07	11 06 05	01 ♏ 52 54	08 ♏ 08 24	22 30	03 45	28 36	22 23	28 21	13 51	17 19	15 47
04	14 55 03	12 06 15	14 ♏ 20 43	20 ♏ 29 55	24 02	04 58	29 08	22 15	28 21	13 54	17 22	15 49
05	14 59 00	13 06 26	26 ♏ 36 09	02 ♐ 39 35	25 33	06 10	29 ♌ 40	22 07	28 21	13 56	17 24	15 50
06	15 02 56	14 06 39	08 ♐ 40 27	14 ♐ 39 02	27 03	07 22	00 ♍ 11	21 59	28 22	13 58	17 26	15 51
07	15 06 53	15 06 54	20 ♐ 35 39	26 ♐ 30 41	28 ♏ 34	08 35	00 43	21 51	28 22	14 01	17 28	15 53
08	15 10 49	16 07 10	02 ♑ 24 35	08 ♑ 17 48	00 ♐ 04	09 48	01 14	21 43	28 23	14 03	17 31	15 54
09	15 14 46	17 07 28	14 ♑ 10 52	20 ♑ 04 20	01 33	11 00	01 46	21 35	28 24	14 05	17 33	15 55
10	15 18 42	18 07 48	25 ♑ 58 48	01 ♒ 54 53	03 02	12 13	02 17	21 27	28 25	14 07	17 35	15 56
11	15 22 39	19 08 09	07 ♒ 53 11	13 ♒ 54 23	04 31	13 26	02 47	21 18	28 25	14 09	17 37	15 58
12	15 26 36	20 08 31	19 ♒ 59 06	26 ♒ 07 58	05 59	14 39	03 18	21 10	28 27	14 11	17 40	15 59
13	15 30 32	21 08 55	02 ♓ 21 36	08 ♓ 40 32	07 26	15 52	03 49	21 02	28 28	14 13	17 42	16 00
14	15 34 29	22 09 20	15 ♓ 05 18	21 ♓ 36 17	08 53	17 05	04 19	20 54	28 29	14 15	17 44	16 01
15	15 38 25	23 09 46	28 ♓ 13 51	04 ♈ 58 10	10 20	18 18	04 49	20 46	28 30	14 17	17 46	16 02
16	15 42 22	24 10 14	11 ♈ 49 19	18 ♈ 47 12	11 46	19 31	05 19	20 37	28 32	14 19	17 49	16 03
17	15 46 18	25 10 43	25 ♈ 51 35	03 ♉ 02 01	13 11	20 44	05 49	20 29	28 33	14 21	17 51	16 04
18	15 50 15	26 11 14	10 ♉ 17 56	17 ♉ 38 33	14 36	21 57	06 19	20 21	28 35	14 23	17 53	16 05
19	15 54 11	27 11 46	25 ♉ 03 01	02 ♊ 30 19	15 59	23 11	06 48	20 13	28 37	14 24	17 55	16 06
20	15 58 08	28 12 19	09 ♊ 59 25	17 ♊ 29 15	17 22	24 24	07 18	20 05	28 39	14 26	17 57	16 07
21	16 02 05	29 ♏ 12 54	24 ♊ 58 43	02 ♋ 26 51	18 44	25 38	07 47	19 57	28 41	14 28	18 00	16 08
22	16 06 01	00 ♐ 13 31	09 ♋ 52 41	17 ♋ 15 27	20 04	26 51	08 16	19 49	28 43	14 29	18 02	16 09
23	16 09 58	01 14 09	24 ♋ 34 26	01 ♌ 49 07	21 23	28 05	08 45	19 41	28 45	14 31	18 04	16 10
24	16 13 54	02 14 49	08 ♌ 59 05	16 ♌ 04 04	22 41	29 ♎ 19	09 13	19 34	28 47	14 32	18 06	16 10
25	16 17 51	03 15 31	23 ♌ 03 55	29 ♌ 58 35	23 57	00 ♏ 33	09 41	19 26	28 50	14 34	18 09	16 11
26	16 21 47	04 16 14	06 ♍ 48 07	13 ♍ 32 38	25 11	01 46	10 10	19 18	28 52	14 35	18 11	16 12
27	16 25 44	05 16 59	20 ♍ 12 19	26 ♍ 47 21	26 23	03 00	10 38	19 11	28 55	14 36	18 13	16 13
28	16 29 40	06 17 45	03 ♎ 18 00	09 ♎ 44 31	27 32	04 14	11 05	19 03	28 57	14 38	18 15	16 13
29	16 33 37	07 18 33	16 ♎ 07 09	22 ♎ 26 10	28 38	05 28	11 33	18 56	29 00	14 39	18 17	16 14
30	16 37 34	08 ♐ 19 22	28 ♎ 41 50	04 ♏ 54 22	29 ♐ 41	06 ♏ 42	12 ♍ 00	18 ♉ 48	29 ♒ 03	14 ♍ 40	18 ♏ 19	16 ♍ 14

Date	Moon True ☊ ° '	Moon Mean ☊ ° '	Moon ☽ Latitude ° '
01	23 ♊ 56	25 ♊ 11	05 N 01
02	23 R 48	25 08	04 39
03	23 40	25 05	04 03
04	23 33	25 01	03 16
05	23 28	24 58	02 21
06	23 25	24 55	01 19
07	23 24	24 52	00 N 15
08	23 D 24	24 49	00 S 49
09	23 25	24 46	01 51
10	23 27	24 42	02 49
11	23 29	24 39	03 40
12	23 29	24 36	04 22
13	23 R 29	24 33	04 53
14	23 27	24 30	05 10
15	23 23	24 26	05 12
16	23 19	24 23	04 57
17	23 14	24 20	04 24
18	23 10	24 17	03 34
19	23 07	24 14	02 28
20	23 05	24 11	01 S 12
21	23 D 05	24 07	00 N 10
22	23 06	24 04	01 31
23	23 07	24 01	02 45
24	23 08	23 58	03 47
25	23 09	23 55	04 34
26	23 R 10	23 52	05 03
27	23 09	23 48	05 16
28	23 08	23 45	05 11
29	23 06	23 42	04 51
30	23 ♊ 03	23 ♊ 39	04 N 17

DECLINATIONS

Date	Sun ☉ ° '	Moon ☽ ° '	Mercury ☿ ° '	Venus ♀ ° '	Mars ♂ ° '	Jupiter ♃ ° '	Saturn ♄ ° '	Uranus ♅ ° '	Neptune ♆ ° '	Pluto ♇ ° '
01	14 S 32	02 N 06	18 S 30	00 N 52	13 N 52	17 N 14	13 S 38	07 N 04	15 S 21	18 N 21
02	14 51	03 S 13	19 01	00 N 25	13 42	17 12	13 38	07 04	15 21	18 21
03	15 10	08 20	19 31	00 S 02	13 31	17 10	13 37	07 03	15 22	18 20
04	15 28	13 01	20 00	00 29	13 21	17 08	13 37	07 02	15 23	18 20
05	15 47	17 07	20 28	00 56	13 11	17 06	13 37	07 01	15 23	18 20
06	16 05	20 27	20 55	01 23	13 01	17 04	13 37	07 00	15 24	18 20
07	16 22	22 51	21 21	01 51	12 50	17 02	13 36	06 59	15 25	18 20
08	16 40	24 14	21 46	02 18	12 40	17 00	13 36	06 58	15 25	18 20
09	16 57	24 32	22 10	02 45	12 30	16 58	13 36	06 58	15 26	18 20
10	17 14	23 43	22 33	03 13	12 20	16 56	13 35	06 57	15 27	18 20
11	17 31	21 50	22 55	03 40	12 09	16 54	13 35	06 56	15 27	18 19
12	17 47	18 57	23 15	04 07	11 59	16 52	13 34	06 55	15 28	18 19
13	18 03	15 11	23 35	04 35	11 49	16 50	13 34	06 54	15 28	18 19
14	18 19	10 38	23 53	05 02	11 39	16 48	13 33	06 54	15 29	18 19
15	18 34	05 S 29	24 10	05 29	11 29	16 46	13 33	06 53	15 30	18 19
16	18 49	00 N 07	24 26	05 56	11 19	16 44	13 32	06 52	15 30	18 19
17	19 04	05 53	24 40	06 23	11 09	16 42	13 32	06 52	15 31	18 19
18	19 18	11 31	24 53	06 50	10 59	16 40	13 31	06 51	15 31	18 19
19	19 32	16 38	25 05	07 17	10 49	16 38	13 30	06 50	15 32	18 20
20	19 46	20 46	25 16	07 44	10 39	16 36	13 29	06 50	15 32	18 20
21	19 59	23 31	25 25	08 11	10 29	16 34	13 29	06 49	15 33	18 20
22	20 12	24 36	25 33	08 37	10 19	16 32	13 28	06 49	15 34	18 20
23	20 25	23 55	25 39	09 03	10 09	16 30	13 27	06 48	15 34	18 20
24	20 37	21 40	25 44	09 29	09 59	16 28	13 26	06 48	15 35	18 20
25	20 49	18 08	25 48	09 55	09 49	16 26	13 25	06 47	15 36	18 20
26	21 00	13 43	25 50	10 21	09 40	16 24	13 24	06 47	15 36	18 20
27	21 11	08 43	25 51	10 47	09 31	16 22	13 23	06 46	15 37	18 20
28	21 22	03 N 27	25 50	11 12	09 21	16 20	13 22	06 46	15 37	18 21
29	21 32	01 S 52	25 48	11 37	09 12	16 18	13 21	06 45	15 38	18 21
30	21 S 42	07 S 00	25 S 44	12 S 02	09 N 02	16 N 17	13 S 20	06 N 45	15 S 39	18 N 21

ZODIAC SIGN ENTRIES

Date	h	m	Planets
01	00	24	☽ ♎
03	08	25	☽ ♏
05	18	43	☽ ♐
06	03	20	♂ ♍
08	07	06	☽ ♑
08	11	02	☿ ♐
10	20	08	☽ ♒
13	07	28	☽ ♓
15	15	10	☽ ♈
17	18	57	☽ ♉
19	19	58	☽ ♊
21	20	04	☽ ♋
22	06	39	☉ ♐
23	20	59	☽ ♌
25	01	25	♀ ♏
26	00	02	☽ ♍
28	05	54	☽ ♎
30	14	31	☽ ♏
30	19	30	☿ ♑

LATITUDES

Date	Mercury ☿ ° '	Venus ♀ ° '	Mars ♂ ° '	Jupiter ♃ ° '	Saturn ♄ ° '	Uranus ♅ ° '	Neptune ♆ ° '	Pluto ♇ ° '
01	00 S 57	01 N 32	01 N 38	01 S 15	01 S 41	00 N 45	01 N 42	13 N 51
04	01 16	01 37	01 41	01 14	01 40	00 45	01 42	13 52
07	01 33	01 42	01 44	01 14	01 40	00 45	01 42	13 54
10	01 49	01 45	01 47	01 14	01 40	00 45	01 42	13 55
13	02 03	01 48	01 50	01 14	01 39	00 45	01 42	13 56
16	02 15	01 50	01 53	01 13	01 39	00 46	01 42	13 58
19	02 24	01 51	01 56	01 13	01 39	00 46	01 42	13 59
22	02 29	01 52	01 59	01 13	01 38	00 46	01 42	14 00
25	02 30	01 51	02 03	01 12	01 38	00 46	01 43	14 01
28	02 25	01 50	02 06	01 12	01 38	00 46	01 43	14 03
31	02 S 13	01 N 48	02 N 09	01 S 11	01 S 38	00 N 46	01 N 43	14 N 05

DATA

Julian Date	2438701
Delta T	+36 seconds
Ayanamsa	23° 21' 39"
Synetic vernal point	05° ♓ 45' 21"
True obliquity of ecliptic	23° 26' 39"

MOON'S PHASES, APSIDES AND POSITIONS ☽

Date	h	m	Phase	Longitude ° '	Eclipse Indicator
04	07	16	●	11 ♏ 54	
12	12	20	☽	20 ♒ 09	
19	15	43	○	27 ♉ 21	
26	07	10	☾	04 ♍ 04	

Day	h	m		
08	22	04	Apogee	
20	23	59	Perigee	
01	21	27	0S	
09	06	19	Max dec	24° S 34'
16	11	32	0N	
22	14	36	Max dec	24° N 36'
29	03	33	0S	

LONGITUDES

Date	Chiron ⚷ ° '	Ceres ⚳ ° '	Pallas ⚴ ° '	Juno ⚵ ° '	Vesta ⚶ ° '	Black Moon Lilith ⚸ ° '
01	14 ♓ 25	08 ♑ 35	17 ♐ 26	11 ♐ 48	03 ♓ 41	02 ♑ 27
11	14 ♓ 15	11 ♑ 51	21 ♐ 16	15 ♐ 09	05 ♓ 08	03 ♑ 34
21	14 ♓ 10	15 ♑ 17	25 ♐ 10	18 ♐ 35	07 ♓ 09	04 ♑ 40
31	14 ♓ 11	18 ♑ 52	29 ♐ 05	22 ♐ 07	09 ♓ 39	05 ♑ 46

ASPECTARIAN

01 Sunday
01 58 ☽ ☌ ♀
04 30 ☽ ∠ ♆
04 51 ☿ # ♇
05 36 ☽ ⊥ ☉
06 39 ☽ ⊥ ♂
08 05 ☽ ∠ ☿
08 23 ☽ ± ♄
14 26 ☽ Ᵽ ♃
17 35 ☽ ⊻ ☉
18 03 ☽ ∥ ♀
20 43 ♄ St D
21 10 ☽ ⊥ ♆

02 Monday
00 02 ☽ ∠ ♂
00 19 ☽ # ♀
01 03 ☽ Ᵽ ♄
01 53 ☽ ⊻ ♅
03 00 ☽ ⊥ ☿
05 33 ☽ ⊻ ♇
06 26 ♀ ∠ ♆
07 01 ☽ ± ♃
08 24 ☽ ⊻ ♆
10 12 ☉ Q ♂
13 09 ☽ ⊥ ♅
15 47 ☽ ⊻ ☿
16 50 ☽ ⊥ ♇
18 10 ☽ ⊼ ♃
21 49 ♂ # ♄

03 Tuesday
00 32 ♂ ☍ ♄
05 16 ☽ △ ♄
05 28 ☽ ⚹ ♂
05 51 ☽ # ♅
06 13 ☽ ∠ ♅
09 55 ☽ ∠ ♇
10 15 ☿ ☍ ♃
15 58 ☽ ⊻ ♀
23 46 ♀ ± ♄

04 Wednesday
04 34 ☉ ∥ ♆
04 44 ☽ ⊥ ♀
05 30 ☽ Q ♂
07 16 ☽ ☌ ☉
11 07 ☽ ⚹ ♅
13 43 ☽ # ♂
14 52 ☽ ⚹ ♇
15 16 ☽ ∥ ♄
17 53 ☽ ☌ ♆
18 06 ♀ ⊥ ♂

05 Thursday
00 08 ☽ ∠ ♀
01 20 ☽ ∥ ♆
02 59 ☽ ∥ ☉
03 16 ☽ ☍ ♃
09 37 ☽ ☌ ☿
10 41 ☽ Q ♅
11 53 ☽ # ♃
14 26 ☽ Q ♇
15 28 ☽ □ ♄
18 20 ☿ Q ♅
18 20 ☽ □ ♂
20 05 ☽ # ♇

06 Friday
05 00 ♀ Ᵽ ♃
08 36 ☉ ⚹ ♅
09 06 ☽ ⚹ ♀
16 50 ☽ ∥ ☿
22 40 ☽ □ ♅
23 55 ☽ ⊻ ☉

07 Saturday
00 56 ☿ Q ♇
02 28 ☽ □ ♇
03 28 ☽ Q ♄
05 40 ☽ ⊻ ♆
08 55 ☿ □ ♄
11 58 ☽ Q ♀
13 09 ☽ ⊥ ☉
14 31 ☽ ⊼ ♃
17 51 ☽ ⊥ ♆

08 Sunday
02 33 ☽ ± ♃
03 48 ☽ ⚹ ♄
06 31 ☽ ⊻ ☿
06 39 ☉ ⚹ ♇
09 08 ☽ ∠ ☉
09 30 ☽ △ ♂
12 12 ☽ ∠ ♆
20 31 ☽ ⊥ ☿
20 40 ☽ Ᵽ ♃

09 Monday
04 47 ☽ □ ♀
10 24 ☽ ∠ ♄
11 48 ☽ △ ♅
13 15 ☉ # ♃
15 33 ☽ △ ♇
17 08 ☿ □ ♂
17 30 ☽ Ᵽ ♂
17 32 ☽ ∠ ☿
18 33 ☽ ⚹ ☉
18 53 ☽ ⚹ ♆
22 28 ☉ ☌ ♆
23 07 ♀ ⊥ ♆

10 Tuesday
02 07 ☉ ⊥ ♀
02 53 ☽ △ ♃
04 44 ☽ ⊥ ♄
12 38 ☽ ± ♂
16 55 ☽ ⊻ ♄
18 23 ☽ Ᵽ ♅
19 19 ☽ Q ♆
21 10 ☽ Q ☉
22 03 ☽ Ᵽ ♇

11 Wednesday
01 18 ☽ ⊼ ♂
01 49 ☽ ∥ ☿
04 16 ☽ ⚹ ☿
11 55 ♀ Ᵽ ♄
12 33 ☽ ± ♅
16 09 ☽ ± ♇

12 Thursday
00 17 ☽ △ ♀
00 32 ☽ ⊼ ♅
02 48 ♀ ⊻ ♅
04 06 ☽ ⊼ ♇
07 25 ☽ □ ♆
07 31 ☽ Q ☿
12 20 ☽ □ ☉
14 18 ☽ □ ♃
16 25 ☽ # ♇
19 30 ☽ ∥ ☉
21 20 ♀ ± ♃

13 Friday
02 03 ☽ # ♃
04 30 ☽ ☌ ♄
08 49 ☽ Ᵽ ♀
09 35 ☉ ☍ ♃
10 19 ☽ ∥ ♆
14 46 ♀ ⊻ ♇
14 53 ☽ ☍ ♂
21 00 ☽ ∥ ♄
22 55 ☽ □ ☿

14 Saturday
00 32 ☽ Q ♃
03 44 ☽ ± ♀
06 47 ☽ # ♂
10 27 ☽ ☍ ♅
13 06 ☉ # ♇
13 43 ☽ ☍ ♇
16 04 ☽ ⊼ ♀
16 55 ☽ △ ♆
22 36 ☽ ⚹ ♃

15 Sunday
01 21 ♀ ⊻ ♆
02 05 ☽ △ ☉
05 41 ☽ # ♅
11 58 ☽ ∥ ♀
12 30 ☽ ⊻ ♄
20 08 ☽ Ᵽ ♆
23 12 ☽ ⊥ ♄

16 Monday
00 11 ☽ ⊼ ♂
01 16 ☽ ∠ ♃
07 01 ☽ Ᵽ ☉
11 06 ☽ ± ♂
11 53 ☽ △ ☿
11 59 ☽ ± ♆
14 58 ☽ ∠ ♄
16 20 ☽ ⊼ ♅
16 48 ☽ ⊥ ♃
19 19 ☽ ⊼ ♇
22 21 ☽ ⊼ ♆
23 48 ☽ ± ☉

17 Tuesday
02 30 ☽ ☍ ♀
02 39 ☽ ± ♅
03 00 ☽ ⊻ ♃
03 09 ☽ Ᵽ ♂
04 13 ♀ ⊥ ♅
05 35 ☽ ± ♇
07 38 ♀ ⊼ ♃
10 46 ☽ ⊼ ☉
13 21 ♂ Q ♆
14 17 ☽ # ♀
14 47 ♀ ∠ ♂
16 04 ☽ ∥ ♅
16 20 ☽ Ᵽ ☿
16 32 ☽ ⚹ ♄
17 52 ☽ Ᵽ ♅
20 44 ☽ Ᵽ ♇

18 Wednesday
05 12 ☽ △ ♂
08 15 ☿ □ ♅
08 54 ☽ ± ☿
09 41 ☽ ∥ ♂
12 28 ☽ Q ♄
12 42 ♀ # ♅
14 30 ♀ ⊥ ♇
16 42 ☉ Q ♅
18 42 ☽ △ ♅
19 46 ☽ ⊼ ☿
20 56 ☽ # ♄
21 28 ☽ △ ♇

19 Thursday
00 25 ☽ ☍ ♆
04 15 ☽ ☌ ♃
06 30 ☽ # ♆
08 43 ☽ ⊼ ♀
11 59 ☽ ∥ ♃
13 57 ☿ □ ♇
15 43 ☽ ☍ ☉
17 45 ☽ □ ♄
19 15 ☽ ± ♀
21 01 ☽ ∥ ♇
23 06 ☿ Q ♄

20 Friday
05 03 ☽ # ☉
07 32 ☽ □ ♂
09 50 ☉ Q ♇
10 59 ☽ Ᵽ ♀
19 08 ☽ □ ♅
21 49 ☽ □ ♇
22 39 ☿ ⊻ ♆
22 48 ☉ □ ♄

21 Saturday
00 47 ☽ ⊼ ♆
01 00 ☽ ☍ ☿
04 01 ☽ ⊻ ♃
10 25 ☽ ± ♆
13 08 ☽ △ ♀
13 20 ☽ Q ♂
13 33 ☽ ⊥ ♃
17 57 ☽ △ ♄
19 18 ☽ ⊼ ☉

22 Sunday
00 03 ☽ Q ♅
00 55 ☽ Ᵽ ♆
02 44 ☽ Q ♇
03 54 ☽ ∠ ♃
05 40 ☽ ± ☉
07 54 ☿ ⊼ ♃
09 18 ☽ ⚹ ♂
18 14 ☽ Ᵽ ♄
19 30 ☽ ⚹ ♅
21 20 ☽ Ᵽ ☉
22 12 ☽ ⚹ ♇

23 Monday
01 18 ☽ △ ♆
04 03 ☽ ⚹ ♃
06 15 ☽ ⊼ ☿
08 59 ☽ ± ♄
10 35 ☽ ∠ ♂
17 07 ☽ ± ☿
18 20 ☽ □ ♀
18 55 ☽ ⊼ ♄
20 11 ☽ ∠ ♅
22 55 ☽ ∠ ♇
23 41 ☽ Q ♃
23 52 ☽ △ ☉

24 Tuesday
01 24 ♀ △ ♄
02 00 ☽ ⊥ ♂
09 36 ☽ Ᵽ ☿
11 15 ☽ ⊥ ♅
12 24 ☽ ⊻ ♂
14 00 ☽ ⊥ ♇
16 29 ♀ ∠ ♅
19 31 ☽ # ☉
21 25 ☽ ⊻ ♅

25 Wednesday
00 11 ☽ ⊻ ♇
03 29 ☽ Q ♀
03 31 ☽ □ ♆
05 48 ☽ □ ♃
08 09 ♀ # ♂
10 49 ☽ ∥ ♇
13 41 ☽ △ ☿
15 45 ☿ ⊥ ♆
21 47 ☽ ∥ ♃
22 01 ☽ ☍ ♄

26 Thursday
00 40 ♀ ∠ ♇
02 13 ☽ # ♆
02 16 ☽ ⚹ ♀
07 10 ☽ □ ☉
10 54 ☽ Q ♆
13 33 ☽ # ♄
14 07 ☿ ± ♃
18 10 ☽ ☌ ♂

27 Friday
01 53 ☽ ☌ ♅
03 04 ☽ # ♀
04 47 ☽ ☌ ♇
07 37 ☽ ∠ ♀
08 11 ☽ ∥ ♂
08 23 ☽ ⚹ ♆
10 09 ☽ △ ♃
18 04 ☽ Q ☉
20 58 ☽ ∥ ♅

28 Saturday
00 21 ☽ □ ☿
01 41 ☽ ⊥ ♀
03 57 ☽ ⊼ ♄
11 55 ☽ ∠ ♆
13 23 ☽ Ᵽ ♃
13 55 ☽ ⊻ ♀
15 05 ☽ ± ♄
18 02 ☽ ⚹ ☉

29 Sunday
03 04 ☽ ⊻ ♂
04 45 ☽ ⊥ ♆
06 02 ☽ ± ♃
07 59 ☽ Ᵽ ♄
09 13 ☽ ⊻ ♅
12 13 ☽ ⊻ ♇
13 04 ☽ Q ☿
14 48 ☽ ⊥ ♂
16 07 ☽ ⊻ ♆
17 16 ☽ ⊼ ♃
20 33 ☿ ⚹ ♄
20 36 ☽ ⊥ ♅
23 37 ☽ ⊥ ♇

30 Monday
00 47 ☽ ∠ ☉
08 37 ☽ ∠ ♂
10 47 ☽ # ♅
12 41 ☽ △ ♄
13 52 ☽ ∠ ♅
14 04 ☽ ⚹ ☿
16 54 ☽ ∠ ♇
19 37 ☽ ⊥ ☉
21 38 ☽ # ♂

All ephemeris data is given at 12.00 UT and the Moon's longitude is additionally given for 24.00 UT

DECEMBER 1964

LONGITUDES

Date	Sidereal time h m s	Sun ☉ ° ' "	Moon ☽ ° ' "	Moon ☽ 24.00 ° ' "	Mercury ☿ ° '	Venus ♀ ° '	Mars ♂ ° '	Jupiter ♃ ° '	Saturn ♄ ° '	Uranus ♅ ° '	Neptune ♆ ° '	Pluto ♇ ° '
01	16 41 30	09 ♐ 20 13	11 ♏ 04 01	17 ♏ 11 01	00 ♑ 40	07 ♏ 56	12 ♍ 27	18 ♉ 41	29 ♒ 06	14 ♍ 41	18 ♏ 21	16 ♍ 15
02	16 45 27	10 21 05	23 ♏ 15 36	29 ♏ 17 57	01 35	09 10	12 54	18 R 34	29 09	14 42	18 24	16 16
03	16 49 23	11 21 58	05 ♐ 18 18	11 ♐ 16 53	02 24	10 24	13 20	18 27	29 12	14 43	18 26	16 16
04	16 53 20	12 22 52	17 ♐ 13 55	23 ♐ 09 38	03 08	11 39	13 46	18 20	29 15	14 44	18 28	16 17
05	16 57 16	13 23 48	29 ♐ 04 18	04 ♑ 58 13	03 45	12 53	14 13	18 14	29 19	14 45	18 30	16 17
06	17 01 13	14 24 44	10 ♑ 51 39	16 ♑ 44 58	04 15	14 07	14 38	18 07	29 22	14 46	18 32	16 17
07	17 05 09	15 25 41	22 ♑ 38 30	28 ♑ 32 40	04 37	15 21	15 04	18 01	29 26	14 46	18 34	16 18
08	17 09 06	16 26 40	04 ♒ 27 52	10 ♒ 24 34	04 50	16 36	15 29	17 54	29 29	14 47	18 36	16 18
09	17 13 03	17 27 38	16 ♒ 23 15	22 ♒ 24 25	04 R 53	17 50	15 54	17 48	29 33	14 48	18 38	16 18
10	17 16 59	18 28 38	28 ♒ 28 35	04 ♓ 36 19	04 45	19 05	16 19	17 42	29 37	14 48	18 40	16 19
11	17 20 56	19 29 38	10 ♓ 48 08	17 ♓ 04 36	04 27	20 19	16 43	17 36	29 41	14 49	18 42	16 19
12	17 24 52	20 30 38	23 ♓ 26 13	29 ♓ 53 28	03 56	21 33	17 07	17 31	29 44	14 49	18 44	16 19
13	17 28 49	21 31 39	06 ♈ 26 50	13 ♈ 06 39	03 15	22 48	17 31	17 25	29 49	14 50	18 46	16 19
14	17 32 45	22 32 40	19 ♈ 53 12	26 ♈ 46 39	02 22	24 02	17 54	17 20	29 53	14 50	18 48	16 20
15	17 36 42	23 33 42	03 ♉ 47 03	10 ♉ 54 16	01 19	25 17	18 18	17 15	29 ♒ 57	14 50	18 50	16 20
16	17 40 38	24 34 45	18 ♉ 08 01	25 ♉ 27 48	00 ♑ 08	26 32	18 40	17 10	00 ♓ 01	14 50	18 52	16 20
17	17 44 35	25 35 48	02 ♊ 53 00	10 ♊ 22 47	28 ♐ 50	27 46	19 03	17 05	00 06	14 51	18 54	16 20
18	17 48 32	26 36 51	17 ♊ 56 08	25 ♊ 31 57	27 28	29 ♏ 01	19 25	17 00	00 10	14 51	18 56	16 20
19	17 52 28	27 37 55	03 ♋ 09 02	10 ♋ 46 08	26 06	00 ♐ 15	19 47	16 56	00 14	14 51	18 58	16 R 20
20	17 56 25	28 39 00	18 ♋ 22 00	25 ♋ 55 26	24 44	01 30	20 09	16 51	00 19	14 R 51	18 59	16 20
21	18 00 21	29 ♐ 40 05	03 ♌ 25 21	10 ♌ 50 45	23 27	02 45	20 30	16 47	00 24	14 51	19 01	16 20
22	18 04 18	00 ♑ 41 11	18 ♌ 10 52	25 ♌ 25 04	22 16	04 00	20 51	16 43	00 29	14 51	19 03	16 20
23	18 08 14	01 42 17	02 ♍ 32 54	09 ♍ 34 05	21 14	05 14	21 11	16 39	00 33	14 51	19 05	16 19
24	18 12 11	02 43 24	16 ♍ 28 33	23 ♍ 16 21	20 21	06 29	21 31	16 36	00 38	14 50	19 07	16 19
25	18 16 07	03 44 31	29 ♍ 57 37	06 ♎ 32 40	19 39	07 44	21 51	16 32	00 43	14 50	19 08	16 19
26	18 20 04	04 45 40	13 ♎ 01 51	19 ♎ 25 35	19 08	08 59	22 10	16 29	00 48	14 50	19 10	16 19
27	18 24 01	05 46 48	25 ♎ 44 20	01 ♏ 58 34	18 47	10 13	22 29	16 26	00 53	14 49	19 12	16 19
28	18 27 57	06 47 58	08 ♏ 08 47	14 ♏ 15 29	18 36	11 28	22 48	16 24	00 59	14 49	19 13	16 18
29	18 31 54	07 49 07	20 ♏ 19 08	26 ♏ 20 12	18 D 35	12 43	23 06	16 21	01 04	14 48	19 15	16 18
30	18 35 50	08 50 18	02 ♐ 19 07	08 ♐ 16 18	18 43	13 58	23 23	16 19	01 09	14 48	19 17	16 18
31	18 39 47	09 ♑ 51 28	14 ♐ 12 06	20 ♐ 06 54	19 ♐ 00	15 ♐ 13	23 ♍ 41	16 ♉ 16	01 ♓ 15	14 ♍ 47	19 ♏ 18	16 ♍ 17

Date	Moon True ☊ ° '	Moon Mean ☊ ° '	Moon ☽ Latitude ° '
01	23 ♊ 01	23 ♊ 36	03 N 32
02	22 R 59	23 32	02 37
03	22 58	23 29	01 36
04	22 57	23 26	00 N 32
05	22 D 57	23 23	00 S 34
06	22 58	23 20	01 38
07	22 58	23 17	02 38
08	22 59	23 13	03 31
09	23 00	23 10	04 16
10	23 00	23 07	04 50
11	23 01	23 04	05 11
12	23 R 01	23 01	05 18
13	23 01	22 58	05 09
14	23 01	22 54	04 44
15	23 D 01	22 51	04 01
16	23 01	22 48	03 02
17	23 01	22 45	01 50
18	23 01	22 42	00 S 28
19	23 R 01	22 38	00 N 56
20	23 01	22 35	02 16
21	23 00	22 32	03 26
22	22 59	22 29	04 21
23	22 58	22 26	04 58
24	22 58	22 23	05 16
25	22 57	22 19	05 15
26	22 D 57	22 16	04 58
27	22 58	22 13	04 27
28	22 59	22 10	03 44
29	23 00	22 07	02 52
30	23 02	22 04	01 52
31	23 ♊ 03	22 ♊ 00	00 N 49

DECLINATIONS

Date	Sun ☉ ° '	Moon ☽ ° '	Mercury ☿ ° '	Venus ♀ ° '	Mars ♂ ° '	Jupiter ♃ ° '	Saturn ♄ ° '	Uranus ♅ ° '	Neptune ♆ ° '	Pluto ♇ ° '
01	21 S 51	11 S 47	25 S 39	12 S 27	08 N 53	16 N 15	13 S 19	06 N 45	15 S 39	18 N 21
02	22 00	16 03	25 33	12 51	08 44	16 13	13 18	06 44	15 40	18 22
03	22 09	19 37	25 26	13 15	08 35	16 12	13 16	06 44	15 40	18 22
04	22 17	22 18	25 17	13 39	08 26	16 10	13 15	06 44	15 41	18 22
05	22 25	24 00	25 06	14 03	08 17	16 08	13 14	06 43	15 41	18 22
06	22 32	24 38	24 55	14 26	08 08	16 07	13 13	06 43	15 42	18 23
07	22 39	24 08	24 43	14 48	07 59	16 05	13 11	06 43	15 42	18 23
08	22 45	22 34	24 29	15 11	07 50	16 04	13 10	06 42	15 43	18 23
09	22 51	19 59	24 14	15 33	07 42	16 02	13 09	06 42	15 44	18 24
10	22 57	16 32	23 59	15 55	07 33	16 01	13 07	06 42	15 44	18 24
11	23 02	12 18	23 42	16 16	07 25	15 59	13 06	06 42	15 45	18 24
12	23 06	07 28	23 25	16 37	07 17	15 58	13 04	06 42	15 45	18 25
13	23 10	02 S 11	23 07	16 57	07 09	15 57	13 03	06 42	15 46	18 25
14	23 14	03 N 24	22 48	17 17	07 01	15 55	13 01	06 42	15 46	18 25
15	23 17	09 00	22 29	17 37	06 53	15 54	13 00	06 42	15 47	18 26
16	23 20	14 19	22 09	17 56	06 45	15 53	12 58	06 41	15 47	18 26
17	23 22	18 57	21 50	18 15	06 37	15 52	12 56	06 41	15 48	18 27
18	23 24	22 26	21 30	18 33	06 30	15 51	12 55	06 41	15 48	18 27
19	23 25	24 20	21 12	18 51	06 23	15 50	12 53	06 41	15 49	18 28
20	23 26	24 26	20 54	19 08	06 15	15 49	12 51	06 41	15 49	18 28
21	23 27	22 44	20 39	19 24	06 08	15 48	12 50	06 42	15 49	18 29
22	23 27	19 31	20 25	19 40	06 01	15 47	12 48	06 42	15 50	18 29
23	23 26	15 12	20 14	19 56	05 54	15 46	12 46	06 42	15 50	18 30
24	23 25	10 11	20 05	20 11	05 48	15 45	12 44	06 42	15 51	18 30
25	23 23	04 N 50	19 59	20 26	05 41	15 45	12 42	06 42	15 51	18 31
26	23 21	00 S 34	19 56	20 39	05 35	15 44	12 41	06 42	15 52	18 31
27	23 19	05 48	19 55	20 53	05 29	15 44	12 39	06 42	15 52	18 32
28	23 16	10 41	19 57	21 06	05 23	15 43	12 37	06 43	15 52	18 32
29	23 13	15 04	20 01	21 18	05 17	15 43	12 35	06 43	15 53	18 33
30	23 09	18 48	20 06	21 29	05 11	15 42	12 33	06 43	15 53	18 33
31	23 S 05	21 S 42	20 S 14	21 S 40	05 N 06	15 N 42	12 S 31	06 N 43	15 S 54	18 N 34

ZODIAC SIGN ENTRIES

Date	h	m	Planets
03	01	24	☽ ♐
05	13	53	☽ ♑
08	02	57	☽ ♒
10	15	00	☽ ♓
13	00	12	☽ ♈
15	05	33	☽ ♉
16	05	39	♄ ♓
16	14	31	☿ ♐
17	07	21	☽ ♊
19	07	02	♀ ♐
19	07	02	☽ ♋
21	06	31	☽ ♌
21	19	50	☉ ♑
23	07	41	☽ ♍
25	12	04	☽ ♎
27	20	11	☽ ♏
30	07	20	☽ ♐

LATITUDES

Date	Mercury ☿ ° '	Venus ♀ ° '	Mars ♂ ° '	Jupiter ♃ ° '	Saturn ♄ ° '	Uranus ♅ ° '	Neptune ♆ ° '	Pluto ♇ ° '
01	02 S 13	01 N 48	02 N 09	01 S 11	01 S 38	00 N 46	01 N 43	14 N 05
04	01 52	01 46	02 13	01 10	01 37	00 46	01 43	14 06
07	01 21	01 43	02 17	01 10	01 37	00 46	01 43	14 08
10	00 S 37	01 39	02 20	01 09	01 37	00 47	01 43	14 09
13	00 N 17	01 34	02 24	01 08	01 36	00 47	01 43	14 11
16	01 17	01 29	02 28	01 07	01 36	00 47	01 43	14 12
19	02 11	01 24	02 32	01 07	01 36	00 47	01 43	14 14
22	02 48	01 18	02 36	01 06	01 36	00 47	01 43	14 15
25	03 04	01 12	02 40	01 05	01 35	00 47	01 43	14 17
28	03 01	01 05	02 45	01 04	01 35	00 47	01 43	14 18
31	02 N 46	00 N 58	02 N 49	01 S 03	01 S 35	00 N 47	01 N 44	14 N 20

DATA

Julian Date	2438731
Delta T	+36 seconds
Ayanamsa	23° 21' 43"
Synetic vernal point	05° ♓ 45' 17"
True obliquity of ecliptic	23° 26' 39"

MOON'S PHASES, APSIDES AND POSITIONS ☽

Date	h	m	Phase	Longitude ° '	Eclipse Indicator
04	01	18	●	11 ♐ 56	Partial
12	06	01	☽	20 ♓ 15	
19	02	41	○	27 ♊ 14	total
25	19	27	☾	04 ♎ 03	

Day	h	m		
06	12	04	Apogee	
19	11	15	Perigee	
06	13	22	Max dec	24° S 38'
13	21	28	0N	
20	01	13	Max dec	24° N 37'
26	09	29	0S	

LONGITUDES

Date	Chiron ⚷ ° '	Ceres ⚳ ° '	Pallas ⚴ ° '	Juno ⚵ ° '	Vesta ⚶ ° '	Black Moon Lilith ⚸ ° '
01	14 ♓ 11	18 ♑ 52	29 ♐ 05	22 ♐ 07	09 ♓ 39	05 ♑ 46
11	14 ♓ 18	22 ♑ 33	03 ♑ 02	25 ♐ 42	12 ♓ 33	06 ♑ 53
21	14 ♓ 30	26 ♑ 19	06 ♑ 59	29 ♐ 19	15 ♓ 47	07 ♑ 59
31	14 ♓ 47	00 ♒ 09	10 ♑ 55	02 ♑ 58	19 ♓ 17	09 ♑ 06

All ephemeris data is given at 12.00 UT and the Moon's longitude is additionally given for 24.00 UT

ASPECTARIAN

h m Aspects

01 Tuesday
05 13 ☽ ☌ ♀
08 19 ☽ ⊻ ☉
14 48 ☽ ⚹ ♂
15 50 ☽ ∥ ♀
19 06 ☽ ⚹ ♅
20 09 ☽ ∥ ♄
21 46 ☽ ∠ ☿
22 10 ☽ ⚹ ♇

02 Wednesday
02 21 ☽ ☌ ♆
02 49 ☽ ☍ ♃
09 39 ☽ ∥ ♆
13 02 ☽ # ♃
15 22 ☽ Q ♂
16 57 ☽ ⊥ ☿
18 50 ☽ Q ♅
21 56 ☽ Q ♇
23 45 ☽ □ ♄

03 Thursday
02 57 ☽ # ♇
05 47 ☽ ⊻ ☿
13 04 ♀ ∥ ♄
16 01 ♃ ☍ ♆
23 26 ☽ ⊻ ♀

04 Friday
01 18 ☽ ☌ ☉
04 45 ☽ □ ♂
06 57 ☽ □ ♅
10 04 ☽ □ ♇
11 45 ☽ ∥ ☉
12 03 ☽ Q ♄
12 56 ☽ ⊥ ♀
14 13 ☽ ⊼ ♃
14 30 ☽ ⊻ ♆
18 28 ☿ ᑫ ♃

05 Saturday
01 02 ☿ ∠ ♆
02 15 ☽ ± ♃
02 41 ☽ ⊥ ♆
09 18 ☽ ∠ ♀
12 29 ☽ ⚹ ♄
20 23 ☽ ᑫ ♃
21 02 ☽ ∠ ♆
21 59 ☽ ☌ ☿

06 Sunday
19 02 ♂ ☌ ♅
19 11 ☽ ∠ ♄
19 25 ☽ ⚹ ♀
19 55 ☽ ⊻ ☉
19 57 ☽ △ ♅
19 59 ☽ △ ♂
20 18 ☉ □ ♅
21 11 ☉ □ ♂
23 04 ☽ △ ♇

07 Monday
00 32 ♀ ⚹ ♅
02 39 ☽ △ ♃
03 19 ♀ ⚹ ♂
03 41 ☽ ⚹ ♆
09 18 ☽ ⊥ ☉
13 36 ☽ ⊥ ♄
19 35 ☉ ⊻ ♀
22 43 ☽ Q ♀

08 Tuesday
01 52 ☽ ⊻ ♄
02 30 ☽ ᑫ ♅
03 38 ☽ ᑫ ♂
04 09 ☽ Q ♆
05 19 ☽ ∠ ☉
05 36 ☽ ᑫ ♇
06 17 ♀ ⚹ ♇
08 38 ☉ □ ♇
09 51 ☽ ∥ ☉
12 45 ☽ ⊻ ☿
20 44 ☽ ± ♅
22 30 ☽ ± ♂
23 47 ☽ ± ♇

09 Wednesday
00 57 ☽ ⊥ ☿
07 04 ☿ St R
08 48 ☽ ⊼ ♅
10 59 ☽ ⊼ ♂
11 25 ♀ ☍ ♃
11 50 ☽ ⊼ ♇
14 12 ☉ Q ♄
14 21 ☽ ⚹ ☉
14 48 ☽ □ ♃
15 14 ☽ □ ♀
16 30 ☽ □ ♆
18 56 ☽ ∠ ☿
19 22 ☉ ⊼ ♃
23 45 ☽ # ♇
23 58 ♀ ∥ ♆

10 Thursday
03 55 ♀ ☌ ♆
12 09 ♂ ☌ ♇
14 15 ☽ ☌ ♄
15 11 ☽ # ♃
15 28 ☽ ∥ ♀
16 17 ☽ Q ☉
16 43 ☉ ⊻ ♆
16 49 ☽ ∥ ♆
18 19 ♀ # ♃
22 51 ☿ ∠ ♀

11 Friday
00 02 ☽ ⚹ ☿
02 02 ☽ Q ♃
07 47 ☽ ∥ ♄
19 42 ☽ ☍ ♅
22 25 ☽ Q ☿
22 34 ☽ ☍ ♇
23 41 ☽ ☍ ♂

12 Saturday
00 54 ☽ ⚹ ♃
03 08 ☽ △ ♆
06 01 ☽ □ ☉
08 06 ☽ △ ♀
12 56 ☽ # ♂
15 38 ☽ # ♅
19 21 ☿ ∠ ♆
23 47 ☽ ⊻ ♄

13 Sunday
04 42 ☽ ∠ ♃
06 29 ☽ □ ☿
07 07 ☽ ᑫ ♆
07 17 ♂ △ ♃
08 15 ☉ ∥ ☿
10 50 ☽ ⊥ ♄
14 42 ☽ ᑫ ♀
20 55 ☽ ⊥ ♃
23 25 ☽ ± ♆

14 Monday
03 04 ☽ ⊼ ♅
03 06 ☽ ∠ ♄
05 43 ☽ ⊼ ♇
07 31 ☽ ⊻ ♃
08 25 ☽ ⊼ ♂
08 26 ☽ ± ♀
10 05 ☽ ⊼ ♆
13 00 ☿ ᑫ ♃
13 40 ☽ ± ♅
16 16 ☽ ± ♇
17 01 ☽ △ ☉
19 14 ☽ ± ♂
19 59 ☽ ⊼ ♀

15 Tuesday
02 04 ☽ ∥ ♅
03 05 ☽ ∥ ♂
05 03 ☉ ± ♃
05 16 ☽ ᑫ ♅
05 25 ☽ ⚹ ♄
07 49 ☽ ᑫ ♇
08 07 ☽ △ ☿
11 08 ☽ ᑫ ♂
12 22 ☿ ⊥ ♀
20 41 ☽ ᑫ ☉

16 Wednesday
01 49 ☽ Q ♄
05 44 ☽ # ♄
06 33 ☽ △ ♅
07 26 ☽ ᑫ ☿
09 01 ☽ △ ♇
10 24 ☽ ☌ ♃
12 47 ☽ ± ☉
12 55 ☽ △ ♂
13 13 ☽ ☍ ♆
14 02 ☿ ⚹ ♄
18 06 ♀ Q ♅
19 00 ☉ ⊥ ♆
19 09 ☽ # ♆
19 36 ☽ ∥ ♃
21 03 ☽ ± ☿
23 21 ☽ ⊼ ☉
23 22 ♂ ∥ ♅

17 Thursday
01 28 ♂ ⚹ ♆
02 59 ☽ ☍ ♀
06 00 ☽ ⊼ ☿
07 29 ☽ □ ♄
07 43 ☽ # ♀
09 08 ☽ ∥ ♇
21 52 ☿ ⊻ ♀
22 50 ♀ Q ♇

18 Friday
04 14 ♀ # ♇
05 17 ☽ # ☿
07 06 ☽ □ ♅
09 27 ☽ □ ♇
10 32 ☽ ⊻ ♃
13 35 ☽ ⊼ ♆
14 24 ☽ □ ♂
18 38 ♇ St R
19 58 ☽ ⊥ ♃
20 36 ☉ ☌ ☿
21 43 ☽ # ☉
23 04 ☽ ± ♆

19 Saturday
01 48 ☽ ☍ ☿
02 41 ☽ ☍ ☉
07 02 ☽ ⊼ ♀
07 24 ☽ △ ♄
10 05 ☽ ∠ ♃
11 31 ☽ Q ♅
11 40 ♀ □ ♄
13 17 ☽ ᑫ ♆
13 51 ☽ Q ♇
17 20 ☽ ± ♀
19 28 ☽ Q ♂

20 Sunday
06 26 ☽ ⚹ ♅
06 46 ♅ St R
07 09 ☽ ᑫ ♄
07 30 ☿ ⊥ ♆
08 20 ♃ # ♆
08 47 ☽ ⚹ ♇
08 47 ☽ ᑫ ♀
09 37 ☽ ⚹ ♃
12 59 ☽ △ ♆
14 53 ☽ ⚹ ♂
21 18 ☽ ⊼ ☿
21 30 ☽ ± ♄

21 Monday
04 34 ☽ # ☉
04 36 ☽ Q ♃
05 16 ♀ Q ♂
05 32 ☽ ⊼ ☉
06 07 ☽ ± ☿
06 16 ☽ ∠ ♅
07 07 ☽ ⊼ ♄
08 38 ☽ ∠ ♇
10 49 ☽ △ ♀
15 25 ☽ ∠ ♂
15 53 ☽ ± ☉
19 30 ☽ ᑫ ☿
20 45 ☽ ⊥ ♅
23 10 ☽ ⊥ ♇

22 Tuesday
02 04 ☿ ± ♃
05 48 ☽ # ☿
06 23 ☽ ⊥ ♂
06 31 ☽ ⊻ ♅
06 37 ☉ ⚹ ♄
07 36 ☽ ᑫ ☉
08 57 ☽ ⊻ ♇
09 36 ☽ □ ♃
11 05 ☽ # ♀
13 26 ☽ □ ♆
16 30 ☽ ⊻ ♂
18 13 ☽ ∥ ♇
18 17 ☽ △ ☿

23 Wednesday
08 37 ☽ ☍ ♄
08 42 ☽ # ♆
09 03 ☽ ∥ ♃
10 28 ☽ △ ☉
10 56 ☉ ᑫ ♃
12 57 ☿ □ ♂
17 01 ☽ □ ♀
19 45 ☽ Q ♆
23 59 ☽ # ♄

24 Thursday
05 43 ☿ ∥ ♀
09 08 ☽ ☌ ♅
11 44 ☽ ☌ ♇
12 13 ☽ △ ♃
16 38 ☽ ⚹ ♆
18 28 ☽ □ ☿
21 07 ☽ ☌ ♂

25 Friday
03 37 ☽ Q ♀
03 43 ☽ ∥ ♅
08 08 ☽ ∥ ♂
13 23 ☽ ⊼ ♄
14 51 ☽ ᑫ ♃
19 27 ☽ □ ☉
19 37 ☽ ∠ ♆
21 39 ☉ ∠ ♆

26 Saturday
00 24 ☽ ± ♄
01 27 ☽ Q ☿
03 41 ☽ ⚹ ♀
07 18 ☽ ± ♃
09 57 ☿ ⊻ ♆
12 15 ☽ ⊥ ♆
15 21 ☽ ⊻ ♅
17 13 ☽ ᑫ ♄
18 08 ☽ ⊻ ♇
18 26 ☽ ⊼ ♃
23 06 ☽ ⚹ ☿
23 32 ☽ ⊻ ♆

27 Sunday
02 39 ☽ ⊥ ♅
04 52 ☿ Q ♄
05 28 ☽ ⊥ ♇
05 38 ☽ ⊻ ♂
07 56 ☽ Q ☉
10 32 ☽ # ♂
10 55 ☽ ∠ ♀
16 19 ☽ # ♅
17 24 ☽ ⊥ ♂
19 50 ☽ ∠ ♅
21 58 ☽ △ ♄
22 42 ☽ ∠ ♇

28 Monday
03 14 ☽ ∠ ☿
06 12 ☽ ⊥ ♀
09 08 ☽ ⚹ ☉
11 18 ☽ ∠ ♂
19 15 ☽ ⊻ ♀
20 42 ☽ ⊥ ☿
22 05 ☽ ∥ ♄

29 Tuesday
01 06 ☽ ⚹ ♅
02 16 ☿ St D
04 02 ☽ ⚹ ♇
04 09 ☽ ☍ ♃
08 33 ☽ ⊻ ☿
09 53 ☽ ☌ ♆
15 49 ☽ # ♃
16 52 ☽ ∥ ♆
17 26 ☽ ∠ ☉
17 40 ☽ ⚹ ♂

30 Wednesday
00 56 ☽ Q ♅
03 55 ☽ Q ♇
09 38 ☽ □ ♄
10 19 ☽ # ♇
13 09 ☽ ⊥ ☉
18 20 ☽ Q ♂
22 21 ☽ ∥ ☿

31 Thursday
01 16 ♃ △ ♇
02 23 ☽ ⊻ ☉
03 51 ♀ □ ♅
11 39 ☽ ∥ ♀
13 11 ☽ □ ♅
14 18 ☽ ☌ ♀
16 11 ☽ ⊼ ♃
16 14 ☽ □ ♇
22 03 ☽ ☌ ☿
22 19 ☽ Q ♄
22 23 ☽ ⊻ ♆

JANUARY 1965

Date	Sidereal time h m s	Sun ☉ ° ' "	Moon ☽ ° ' "	Moon ☽ 24.00 ° ' "	Mercury ☿ ° '	Venus ♀ ° '	Mars ♂ ° '	Jupiter ♃ ° '	Saturn ♄ ° '	Uranus ♅ ° '	Neptune ♆ ° '	Pluto ⯓ ° '
01	18 43 43	10 ♑ 52 39	26 ♐ 01 01	01 ♑ 54 45	19 ♐ 24	16 ♐ 28	23 ♍ 57	16 ♉ 14	01 ♓ 20	14 ♍ 47	19 ♏ 20	16 ♍ 17
02	18 47 40	11 53 50	07 ♑ 48 23	13 ♑ 42 10	19 55	17 43	24 14	16 R 12	01 26	14 R 46	19 21	16 R 16
03	18 51 36	12 55 01	19 ♑ 36 22	25 ♑ 31 14	20 32	18 58	24 29	16 11	01 31	14 45	19 23	16 16
04	18 55 33	13 56 11	01 ♒ 26 59	07 ♒ 23 53	21 15	20 13	24 45	16 10	01 37	14 44	19 25	16 15
05	18 59 30	14 57 22	13 ♒ 22 11	19 ♒ 22 08	22 02	21 28	25 00	16 09	01 43	14 43	19 26	16 15
06	19 03 26	15 58 32	25 ♒ 24 02	01 ♓ 28 12	22 54	22 43	25 14	16 08	01 48	14 43	19 27	16 14
07	19 07 23	16 59 42	07 ♓ 34 56	13 ♓ 44 36	23 50	23 58	25 28	16 07	01 54	14 42	19 29	16 14
08	19 11 19	18 00 52	19 ♓ 57 34	26 ♓ 14 13	24 49	25 13	25 41	16 07	02 00	14 41	19 30	16 13
09	19 15 16	19 02 01	02 ♈ 34 57	09 ♈ 00 11	25 52	26 28	25 54	16 06	02 06	14 39	19 32	16 12
10	19 19 12	20 03 10	15 ♈ 30 18	22 ♈ 05 42	26 57	27 43	26 06	16 D 06	02 12	14 38	19 33	16 12
11	19 23 09	21 04 18	28 ♈ 46 44	05 ♉ 33 41	28 05	28 ♐ 57	26 18	16 06	02 18	14 37	19 34	16 11
12	19 27 05	22 05 25	12 ♉ 26 47	19 ♉ 26 11	29 ♐ 15	00 ♑ 12	26 29	16 07	02 24	14 36	19 36	16 10
13	19 31 02	23 06 32	26 ♉ 31 53	03 ♊ 43 45	00 ♑ 27	01 27	26 40	16 07	02 30	14 35	19 37	16 09
14	19 34 59	24 07 38	11 ♊ 01 31	18 ♊ 24 41	01 40	02 42	26 50	16 08	02 37	14 33	19 38	16 08
15	19 38 55	25 08 43	25 ♊ 52 39	03 ♋ 24 33	02 56	03 57	26 59	16 09	02 43	14 32	19 39	16 08
16	19 42 52	26 09 48	10 ♋ 59 23	18 ♋ 36 02	04 13	05 13	27 08	16 10	02 49	14 31	19 40	16 07
17	19 46 48	27 10 53	26 ♋ 13 15	03 ♌ 49 43	05 31	06 28	27 16	16 11	02 56	14 29	19 41	16 06
18	19 50 45	28 11 56	11 ♌ 24 08	18 ♌ 55 16	06 51	07 43	27 24	16 13	03 02	14 28	19 43	16 05
19	19 54 41	29 ♑ 13 00	26 ♌ 21 58	03 ♍ 43 14	08 12	08 58	27 31	16 15	03 09	14 26	19 44	16 04
20	19 58 38	00 ♒ 14 02	10 ♍ 58 18	18 ♍ 06 33	09 34	10 13	27 37	16 17	03 15	14 24	19 45	16 03
21	20 02 34	01 15 04	25 ♍ 07 35	02 ♎ 01 12	10 57	11 28	27 43	16 19	03 22	14 23	19 46	16 02
22	20 06 31	02 16 06	08 ♎ 47 27	15 ♎ 26 27	12 21	12 43	27 48	16 21	03 28	14 21	19 47	16 01
23	20 10 28	03 17 07	21 ♎ 58 31	28 ♎ 24 06	13 46	13 58	27 52	16 24	03 35	14 19	19 48	16 00
24	20 14 24	04 18 08	04 ♏ 43 41	10 ♏ 57 50	15 11	15 13	27 56	16 27	03 41	14 18	19 48	15 59
25	20 18 21	05 19 08	17 ♏ 07 12	23 ♏ 12 22	16 38	16 28	27 58	16 30	03 48	14 16	19 49	15 58
26	20 22 17	06 20 08	29 ♏ 14 01	05 ♐ 12 47	18 05	17 43	28 01	16 33	03 55	14 14	19 50	15 57
27	20 26 14	07 21 07	11 ♐ 09 16	17 ♐ 04 03	19 33	18 58	28 02	16 36	04 02	14 12	19 51	15 56
28	20 30 10	08 22 05	22 ♐ 57 44	28 ♐ 50 47	21 02	20 13	28 03	16 40	04 09	14 10	19 52	15 54
29	20 34 07	09 23 03	04 ♑ 43 43	10 ♑ 36 56	22 32	21 28	28 R 03	16 43	04 15	14 08	19 53	15 53
30	20 38 03	10 24 00	16 ♑ 30 51	22 ♑ 25 46	24 03	22 43	28 02	16 47	04 22	14 06	19 53	15 52
31	20 42 00	11 ♒ 24 56	28 ♑ 22 00	04 ♒ 19 49	25 ♑ 34	23 ♑ 58	28 ♍ 00	16 ♉ 51	04 ♓ 29	14 ♍ 04	19 ♏ 54	15 ♍ 51

DECLINATIONS

Date	Moon True ☊ ° '	Moon Mean ☊ ° '	Moon ☽ Latitude ° '	Sun ☉ ° '	Moon ☽ ° '	Mercury ☿ ° '	Venus ♀ ° '	Mars ♂ ° '	Jupiter ♃ ° '	Saturn ♄ ° '	Uranus ♅ ° '	Neptune ♆ ° '	Pluto ⯓ ° '
01	23 ♊ 03	21 ♊ 57	00 S 16	23 S 00	23 S 39	20 S 22	21 S 50	05 N 01	15 N 42	12 S 29	06 N 44	15 S 54	18 N 35
02	23 R 02	21 54	01 21	22 55	24 33	20 32	22 00	04 55	15 41	12 27	06 44	15 54	18 35
03	23 00	21 51	02 22	22 49	24 21	20 42	22 09	04 51	15 41	12 25	06 44	15 55	18 36
04	22 57	21 48	03 16	22 43	23 02	20 54	22 17	04 46	15 41	12 23	06 45	15 55	18 37
05	22 54	21 44	04 03	22 36	20 41	21 05	22 25	04 41	15 41	12 21	06 45	15 56	18 37
06	22 50	21 41	04 39	22 29	17 26	21 17	22 32	04 37	15 41	12 18	06 45	15 56	18 38
07	22 46	21 38	05 03	22 22	13 25	21 29	22 38	04 33	15 41	12 16	06 46	15 56	18 38
08	22 42	21 35	05 14	22 14	08 47	21 40	22 44	04 29	15 41	12 14	06 46	15 57	18 39
09	22 40	21 32	05 10	22 06	03 S 42	21 51	22 49	04 25	15 41	12 12	06 47	15 57	18 40
10	22 39	21 29	04 50	21 57	01 N 39	22 02	22 53	04 22	15 42	12 10	06 47	15 57	18 41
11	22 D 39	21 25	04 14	21 48	07 05	22 13	22 56	04 19	15 42	12 07	06 48	15 57	18 41
12	22 40	21 22	03 23	21 38	12 21	22 23	22 59	04 16	15 42	12 05	06 48	15 58	18 42
13	22 42	21 19	02 19	21 28	17 08	22 32	23 01	04 13	15 43	12 03	06 49	15 58	18 43
14	22 43	21 16	01 S 04	21 17	21 03	22 40	23 03	04 10	15 43	12 01	06 49	15 58	18 43
15	22 R 43	21 13	00 N 17	21 07	23 40	22 47	23 04	04 08	15 44	11 58	06 50	15 59	18 44
16	22 42	21 10	01 38	20 55	24 37	22 54	23 04	04 06	15 44	11 56	06 50	15 59	18 45
17	22 39	21 06	02 53	20 44	23 44	23 00	23 03	04 04	15 45	11 54	06 51	15 59	18 45
18	22 35	21 03	03 55	20 32	21 07	23 04	23 02	04 03	15 46	11 51	06 52	15 59	18 46
19	22 29	21 00	04 40	20 19	17 07	23 08	23 00	04 02	15 47	11 49	06 52	16 00	18 47
20	22 23	20 57	05 05	20 06	12 09	23 11	22 57	04 00	15 48	11 47	06 53	16 00	18 48
21	22 18	20 54	05 11	19 53	06 41	23 12	22 53	04 00	15 48	11 44	06 54	16 00	18 48
22	22 14	20 50	04 58	19 39	01 N 05	23 12	22 49	03 59	15 49	11 42	06 54	16 00	18 49
23	22 11	20 47	04 30	19 26	04 S 23	23 11	22 44	03 59	15 50	11 39	06 55	16 00	18 50
24	22 D 11	20 44	03 50	19 11	09 30	23 09	22 39	03 59	15 51	11 37	06 56	16 01	18 51
25	22 11	20 41	02 59	18 57	14 05	23 06	22 32	03 59	15 53	11 34	06 56	16 01	18 52
26	22 13	20 38	02 02	18 42	18 01	23 02	22 25	04 00	15 54	11 32	06 57	16 01	18 52
27	22 14	20 35	01 N 00	18 26	21 08	22 56	22 18	04 01	15 55	11 30	06 58	16 01	18 53
28	22 R 15	20 31	00 S 04	18 11	23 19	22 49	22 10	04 02	15 56	11 27	06 59	16 01	18 54
29	22 14	20 28	01 07	17 55	24 29	22 40	22 01	04 03	15 58	11 25	06 59	16 01	18 55
30	22 12	20 25	02 07	17 38	24 32	22 31	21 51	04 05	15 59	11 22	07 00	16 02	18 55
31	22 ♊ 05	20 ♊ 22	03 S 02	17 S 22	23 S 28	22 S 20	21 S 41	04 N 07	16 N 00	11 S 19	07 N 01	16 S 02	18 N 56

ZODIAC SIGN ENTRIES

Date	h m	Planets
01	20 06	☽ ♑
04	09 04	☽ ♒
06	21 06	☽ ♓
09	07 08	☽ ♈
11	14 10	☽ ♉
12	08 00	♀ ♑
13	03 12	☿ ♑
13	17 48	☽ ♊
15	18 35	☽ ♋
17	17 57	☽ ♌
19	17 55	☽ ♍
20	06 29	☉ ♒
21	20 28	☽ ♎
24	03 01	☽ ♏
26	13 32	☽ ♐
29	02 21	☽ ♑
31	15 18	☽ ♒

LATITUDES

Date	Mercury ☿ ° '	Venus ♀ ° '	Mars ♂ ° '	Jupiter ♃ ° '	Saturn ♄ ° '	Uranus ♅ ° '	Neptune ♆ ° '	Pluto ⯓ ° '
01	02 N 39	00 N 55	02 N 50	01 S 03	01 S 35	00 N 47	01 N 44	14 N 20
04	02 16	00 48	02 55	01 02	01 35	00 48	01 44	14 22
07	01 50	00 40	03 00	01 01	01 35	00 48	01 44	14 23
10	01 22	00 33	03 04	01 01	01 35	00 48	01 44	14 24
13	00 55	00 25	03 09	01 00	01 34	00 48	01 44	14 26
16	00 29	00 17	03 14	00 59	01 34	00 48	01 44	14 27
19	00 N 03	00 09	03 19	00 58	01 34	00 48	01 45	14 28
22	00 S 20	00 N 01	03 23	00 57	01 34	00 48	01 45	14 29
25	00 42	00 S 07	03 28	00 56	01 34	00 48	01 45	14 30
28	01 00	00 14	03 33	00 55	01 34	00 48	01 45	14 31
31	01 S 19	00 S 22	03 N 38	00 S 54	01 S 34	00 N 49	01 N 45	14 N 32

DATA

Julian Date	2438762
Delta T	+36 seconds
Ayanamsa	23° 21' 49"
Synetic vernal point	05° ♓ 45' 11"
True obliquity of ecliptic	23° 26' 39"

MOON'S PHASES, APSIDES AND POSITIONS ☽

Date	h	m	Phase	Longitude ° '	Eclipse Indicator
02	21	07	●	12 ♑ 17	
10	21	00	☽	20 ♈ 26	
17	13	37	○	27 ♋ 15	
24	11	07	☾	04 ♏ 16	

Day	h	m		
02	14	24	Apogee	
17	00	38	Perigee	
29	18	37	Apogee	
02	19	27	Max dec	24° S 37'
10	04	42	ON	
16	12	33	Max dec	24° N 37'
22	16	41	OS	
30	01	08	Max dec	24° S 39'

LONGITUDES

Date	Chiron ⚷ ° '	Ceres ⚳ ° '	Pallas ⚴ ° '	Juno ⚵ ° '	Vesta ⚶ ° '	Black Moon Lilith ⚸ ° '
01	14 ♓ 49	00 ♒ 33	11 ♑ 19	03 ♑ 20	19 ♓ 39	09 ♑ 13
11	15 ♓ 12	04 ♒ 27	15 ♑ 13	06 ♑ 59	23 ♓ 25	10 ♑ 10
21	15 ♓ 39	08 ♒ 22	19 ♑ 03	10 ♑ 37	27 ♓ 21	11 ♑ 26
31	16 ♓ 09	12 ♒ 18	22 ♑ 50	14 ♑ 14	01 ♈ 26	12 ♑ 32

All ephemeris data is given at 12.00 UT and the Moon's longitude is additionally given for 24.00 UT

ASPECTARIAN

h m Aspects

01 Friday
02 50 ☽ ∥ ☉
04 20 ☽ ± ♃
07 42 ☽ □ ♂
07 43 ☿ Q ♄
07 49 ♀ ⊼ ♃
08 14 ☿ ⊻ ♆
08 28 ♀ □ ⯓
10 36 ☽ ⊥ ♆
22 36 ☽ ♇ ♃
22 55 ☽ ✶ ♄

02 Saturday
04 58 ☽ ∠ ♆
21 07 ☽ ☌ ☉

03 Sunday
02 09 ☽ △ ♅
05 04 ☽ △ ♃
05 13 ☽ △ ⯓
05 41 ☽ ∠ ♄
10 32 ☽ ⊻ ♀
11 33 ☽ ✶ ♆
14 00 ☽ ⊻ ☿
20 15 ♀ ⊻ ♆
22 08 ☽ △ ♂
23 36 ♀ Q ♄

04 Monday
00 06 ☽ ⊥ ♄
00 09 ☽ ⊥ ♀
02 55 ☽ ⊥ ☿
08 33 ☽ ♇ ♅
11 36 ☽ ♇ ⯓
11 55 ☽ Q ♆
12 20 ☽ ⊻ ♄
16 10 ☽ ∥ ☉
20 19 ☽ ∥ ♀
20 29 ☽ ∠ ♀
22 20 ☽ ∠ ☿

05 Tuesday
02 41 ☽ ± ♅
05 05 ☽ ♇ ♂
05 44 ☽ ± ⯓
06 38 ☉ △ ♅
08 47 ☽ ∥ ☿
14 43 ☽ ⊼ ♅
15 08 ☿ ± ♃
15 28 ☽ ⊻ ☉
17 33 ☽ □ ♃
17 45 ☽ ⊼ ⯓
23 29 ☽ ± ♂

06 Wednesday
00 09 ☽ □ ♆
00 59 ♀ ± ♃
03 54 ☽ # ⯓
04 34 ☽ ⊥ ☉
06 02 ☽ ✶ ♀
06 38 ☽ ✶ ☿
07 40 ☉ ∥ ♀
11 40 ☽ ⊼ ♂
15 35 ☉ △ ♃
18 05 ☉ △ ⯓
21 28 ☽ ∥ ♆
22 58 ☽ # ♃

07 Thursday
00 02 ☽ ∠ ☉
00 46 ☽ ☌ ♄
01 44 ☿ ☌ ♀
05 13 ☽ Q ♃
08 17 ☽ Q ☿
08 28 ☽ Q ♀
09 38 ☉ ∠ ♄
18 15 ☽ ∥ ♄

08 Friday
01 49 ☽ ☍ ♅
04 35 ☽ ✶ ♃
04 48 ☽ ☍ ⯓
07 55 ☽ ✶ ☉
11 07 ☽ △ ♆
17 46 ♀ ⊥ ♆
21 43 ☽ # ♅
22 08 ☽ □ ☿
23 05 ♀ □ ♂
23 09 ☽ ☍ ♂
23 09 ☽ □ ♀

09 Saturday
04 15 ☿ ⊥ ♆
08 41 ☽ # ♂
08 50 ☽ Q ☉
09 13 ☽ ∠ ♃
11 05 ☽ ⊻ ♄
13 05 ☿ □ ♂
15 39 ☽ ♇ ♆
22 24 ☽ ⊥ ♄
23 52 ☉ ✶ ♆

10 Sunday
02 03 ☽ ⊥ ♃
05 13 ☉ ∥ ☿
08 24 ☽ ± ♆
09 33 ♃ St D
10 25 ☽ ⊼ ♅
13 06 ☽ ⊻ ♃
13 15 ☽ ⊼ ⯓
15 08 ☽ ∠ ♄
19 24 ☽ ⊼ ♆
21 00 ☽ □ ☉
21 21 ☽ ± ♅
23 52 ☽ ∥ ♂

11 Monday
00 10 ☽ ± ⯓
07 30 ☽ ⊼ ♂
10 38 ☽ △ ☿
10 43 ☽ ∥ ♅
12 21 ☽ △ ♀
13 30 ☽ ♇ ♅
16 16 ☽ ♇ ⯓
18 18 ☽ ✶ ♄
18 20 ☽ ± ♂

12 Tuesday
10 19 ☽ ♇ ♂
10 47 ☽ # ♄
15 23 ☽ ♇ ☿
15 24 ☽ Q ♄
15 43 ☽ △ ♅
17 14 ☽ ♇ ♀
18 19 ☽ ☌ ♃
18 24 ☽ △ ⯓

13 Wednesday
00 17 ☽ ☍ ♆
04 30 ☽ ∥ ♃
05 28 ♀ ♇ ♃
05 47 ☽ △ ☉
05 49 ☽ # ♆
08 10 ☽ ± ☿
10 01 ☽ ± ♀
12 13 ☽ △ ♂
19 09 ☽ ⊼ ☿
20 56 ☽ ∥ ⯓
21 01 ☽ ⊼ ♀
22 03 ☽ □ ♄

14 Thursday
01 24 ☿ ♇ ♃
08 40 ☽ ♇ ☉
09 59 ♀ ✶ ♄
13 40 ☽ # ☉
17 45 ☽ □ ♅
18 43 ♃ △ ⯓
20 19 ☽ □ ⯓
20 19 ☽ ⊻ ♃

15 Friday
00 23 ☽ ± ☉
01 48 ☽ # ☿
01 59 ☽ ⊼ ♆
04 58 ☽ # ♀
06 01 ☽ ⊥ ♃
07 32 ☿ ✶ ♄
10 45 ☽ ⊼ ☉
11 38 ☽ ± ♆
13 47 ☽ □ ♂
20 25 ☽ ∠ ♃
22 35 ☽ Q ♅
22 59 ☽ △ ♄

16 Saturday
00 17 ☽ ☍ ☿
01 08 ☽ Q ⯓
01 32 ♀ ∠ ♆
01 59 ☽ ♇ ♆
02 02 ☽ ☍ ♀
17 33 ☽ ✶ ♅
18 36 ☽ Q ♂
20 04 ☽ ✶ ⯓
20 11 ☽ ✶ ♃
20 33 ☿ ∠ ♆
22 51 ☽ ♇ ♄

17 Sunday
01 42 ☽ △ ♆
05 20 ☉ ⊥ ♄
13 07 ☽ ± ♄
13 37 ☽ ☍ ☉
13 40 ☽ ✶ ♂
14 23 ☉ △ ♂
15 07 ☽ Q ♃
17 08 ☽ ∠ ♅
19 41 ☽ ∠ ⯓
20 01 ☽ # ♀
20 14 ☽ # ☿
22 39 ☽ ⊼ ♄

18 Monday
01 04 ☿ ∥ ♀
04 05 ☽ ⊼ ☿
05 37 ☽ ⊼ ♀
07 20 ☽ ⊥ ♅
09 54 ☽ ⊥ ⯓
13 36 ☽ ∠ ♂
14 32 ☽ ± ☿
16 00 ☽ ± ♀
16 19 ☽ # ☉
16 51 ☽ ⊻ ♅
19 27 ☽ ⊻ ⯓
19 41 ☽ □ ♃

19 Tuesday
01 17 ☽ □ ♆
02 50 ☽ ∥ ⯓
04 06 ☽ ⊥ ♂
06 22 ☽ ♇ ☿
07 45 ☽ ♇ ♀
13 52 ☽ ⊻ ♂
16 59 ☽ ⊼ ☉
16 59 ☉ ♇ ♅
17 44 ☽ # ♆
18 47 ☽ ∥ ♃
23 08 ☽ ☍ ♄

20 Wednesday
03 33 ☽ ± ☉
06 38 ☽ Q ♆
09 25 ☽ △ ☿
10 37 ☽ △ ♀
13 42 ☽ # ♄
17 44 ☽ ☌ ♅
19 41 ☽ ♇ ☉
20 31 ☽ ☌ ⯓
20 56 ☽ △ ♃

21 Thursday
02 48 ☽ ✶ ♆
06 57 ☉ ♇ ⯓
11 07 ☽ ∥ ♅
16 30 ☽ ☌ ♂
22 48 ☽ ♇ ♃
23 30 ☽ △ ☉
23 33 ☽ ∥ ♂

22 Friday
00 14 ☉ Q ♆
02 28 ☽ ⊼ ♄
04 51 ☽ ∠ ♆
13 14 ☽ ± ♄
14 49 ☽ ± ♃
19 09 ☽ □ ☿
19 47 ☽ □ ♀
20 59 ☽ ⊥ ♆
22 00 ☽ ⊻ ♅

23 Saturday
01 02 ☽ ⊻ ⯓
01 43 ☽ ⊼ ♃
05 41 ☽ ♇ ♄
07 58 ☽ ⊻ ♆
08 57 ☽ ⊥ ♅
10 13 ☽ # ♂
12 03 ☽ ⊥ ⯓
18 44 ♀ △ ♅
19 49 ☉ ⊻ ♄
21 16 ☿ △ ♅
23 03 ☽ ⊻ ♂
23 40 ☽ # ♅

24 Sunday
01 42 ☽ ∠ ♅
04 53 ☽ ∠ ⯓
08 41 ☽ Q ☿
08 47 ☽ Q ♀
10 00 ☽ △ ♄
10 28 ☽ ⊥ ♂
11 07 ☽ □ ☉
15 09 ☿ ☌ ♀
22 37 ☽ ∥ ♄

25 Monday
01 03 ☿ △ ⯓
02 32 ♀ △ ⯓
03 52 ☽ ∠ ♂
06 25 ☽ ✶ ♅
09 38 ☿ △ ♃
09 44 ☽ ✶ ⯓
10 34 ☽ ✶ ♀
10 46 ☽ ☍ ♃
10 55 ☽ ✶ ☿
12 35 ♀ △ ♃
17 19 ☽ ☌ ♆
19 44 ☉ # ⯓
22 27 ☽ # ♃
23 14 ☽ ∥ ♆

26 Tuesday
01 21 ☽ Q ☉
06 01 ☽ Q ♅
09 26 ☽ Q ⯓
09 33 ☽ ✶ ♂
16 24 ☽ ∥ ☉
18 02 ☽ # ⯓
19 48 ☽ ∠ ♀
20 48 ☽ ∠ ☿
21 29 ☽ □ ♄

27 Wednesday
02 40 ☿ ∠ ♄
03 36 ☽ ✶ ☉
09 44 ☽ Q ♂
13 20 ♀ ∠ ♄
16 06 ☽ ⊥ ♀
16 48 ☿ ✶ ♆
17 34 ☽ ⊥ ☿
18 09 ☽ □ ♅
21 40 ☽ □ ⯓
22 46 ☽ ∥ ♀
23 06 ☽ ⊼ ♃

28 Thursday
05 09 ♀ ✶ ♆
05 41 ☽ ⊻ ♆
05 44 ☽ ∥ ☿
05 45 ☽ ⊻ ♀
07 23 ☉ ± ♅
07 31 ☽ ⊻ ☿
10 19 ☽ Q ♄
11 23 ☽ ± ♃
12 54 ☽ ∠ ☉
17 55 ☽ ⊥ ♆
22 22 ☽ □ ♂
22 38 ♂ St R

29 Friday
05 50 ☽ ♇ ♃
09 00 ☽ ⊥ ☉
11 02 ☽ ✶ ♄
12 18 ☽ ∠ ♆
22 23 ☽ ⊻ ☉
23 38 ☉ ± ⯓

30 Saturday
07 06 ☽ △ ♅
10 41 ☽ △ ⯓
12 34 ☽ △ ♃
17 52 ☽ ∠ ♄
18 51 ☽ ✶ ♆

31 Sunday
02 04 ☽ ☌ ♀
05 30 ☽ ☌ ☿
11 16 ☽ △ ♂
12 15 ☽ ⊥ ♄
13 24 ☽ ♇ ♅
16 59 ☽ ♇ ⯓
19 07 ☽ Q ♆

FEBRUARY 1965

LONGITUDES

Date	Sidereal time h m s	Sun ☉ ° ′ ″	Moon ☽ ° ′ ″	Moon ☽ 24.00 ° ′ ″	Mercury ☿ ° ′	Venus ♀ ° ′	Mars ♂ ° ′	Jupiter ♃ ° ′	Saturn ♄ ° ′	Uranus ♅ ° ′	Neptune ♆ ° ′	Pluto ♇ ° ′
01	20 45 57	12 ♒ 25 51	10 ♒ 19 23	16 ♒ 20 55	27 ♑ 06	25 ♑ 13	27 ♍ 58	16 ♉ 56	04 ♓ 36	14 ♍ 02	19 ♏ 55	15 ♍ 50
02	20 49 53	13 26 44	22 ♒ 24 33	28 ♒ 30 25	28 ♑ 38	26 28	27 R 55	17 00	04 43	14 R 00	19 55	15 R 48
03	20 53 50	14 27 37	04 ♓ 38 37	10 ♓ 49 17	00 ♒ 12	27 43	27 51	17 05	04 50	13 58	19 56	15 47
04	20 57 46	15 28 29	17 ♓ 02 31	23 ♓ 18 25	01 46	28 ♑ 58	27 46	17 10	04 57	13 55	19 56	15 46
05	21 01 43	16 29 19	29 ♓ 37 09	05 ♈ 58 51	03 21	00 ♒ 13	27 41	17 15	05 04	13 53	19 57	15 44
06	21 05 39	17 30 07	12 ♈ 23 41	18 ♈ 51 53	04 56	01 29	27 35	17 20	05 11	13 51	19 57	15 43
07	21 09 36	18 30 54	25 ♈ 23 38	01 ♉ 59 13	06 32	02 44	27 28	17 25	05 18	13 49	19 58	15 42
08	21 13 32	19 31 40	08 ♉ 38 51	15 ♉ 22 47	08 10	03 59	27 20	17 31	05 26	13 46	19 58	15 40
09	21 17 29	20 32 24	22 ♉ 11 16	29 ♉ 04 30	09 48	05 14	27 11	17 37	05 33	13 44	19 59	15 39
10	21 21 26	21 33 07	06 ♊ 02 37	13 ♊ 05 41	11 26	06 29	27 02	17 43	05 40	13 42	19 59	15 37
11	21 25 22	22 33 48	20 ♊ 13 40	27 ♊ 26 25	13 06	07 44	26 52	17 49	05 47	13 39	19 59	15 36
12	21 29 19	23 34 27	04 ♋ 43 38	12 ♋ 04 52	14 46	08 59	26 41	17 55	05 54	13 37	20 00	15 35
13	21 33 15	24 35 05	19 ♋ 29 29	26 ♋ 56 41	16 27	10 14	26 29	18 01	06 01	13 35	20 00	15 33
14	21 37 12	25 35 41	04 ♌ 25 33	11 ♌ 55 01	18 09	11 29	26 16	18 08	06 09	13 32	20 00	15 32
15	21 41 08	26 36 16	19 ♌ 23 58	26 ♌ 51 11	19 52	12 44	26 03	18 15	06 16	13 30	20 00	15 30
16	21 45 05	27 36 49	04 ♍ 15 33	11 ♍ 35 57	21 36	13 59	25 49	18 22	06 23	13 27	20 00	15 29
17	21 49 01	28 37 20	18 ♍ 51 25	26 ♍ 01 09	23 21	15 14	25 35	18 29	06 30	13 25	20 00	15 27
18	21 52 58	29 ♒ 37 50	03 ♎ 04 29	10 ♎ 01 00	25 06	16 28	25 19	18 36	06 38	13 22	20 00	15 26
19	21 56 55	00 ♓ 38 18	16 ♎ 50 26	23 ♎ 32 44	26 53	17 43	25 03	18 43	06 45	13 20	20 00	15 24
20	22 00 51	01 38 45	00 ♏ 08 02	06 ♏ 36 36	28 ♒ 40	18 58	24 46	18 51	06 52	13 17	20 R 00	15 23
21	22 04 48	02 39 11	12 ♏ 58 49	19 ♏ 15 11	00 ♓ 29	20 13	24 29	18 58	07 00	13 15	20 00	15 21
22	22 08 44	03 39 36	25 ♏ 26 18	01 ♐ 32 48	02 18	21 28	24 11	19 06	07 07	13 12	20 00	15 20
23	22 12 41	04 39 59	07 ♐ 35 21	13 ♐ 34 39	04 08	22 43	23 52	19 14	07 14	13 09	20 00	15 18
24	22 16 37	05 40 21	19 ♐ 31 24	25 ♐ 26 19	05 59	23 58	23 33	19 22	07 22	13 07	20 00	15 17
25	22 20 34	06 40 41	01 ♑ 20 03	07 ♑ 13 16	07 51	25 13	23 13	19 30	07 29	13 04	20 00	15 15
26	22 24 30	07 41 00	13 ♑ 06 34	19 ♑ 00 32	09 44	26 28	22 53	19 39	07 36	13 02	20 00	15 13
27	22 28 27	08 41 18	24 ♑ 55 41	00 ♒ 52 28	11 37	27 43	22 32	19 47	07 44	12 59	20 00	15 12
28	22 32 24	09 ♓ 41 34	06 ♒ 51 18	12 ♒ 52 32	13 ♓ 31	28 ♒ 58	22 ♍ 10	19 ♉ 56	07 ♓ 51	12 ♍ 56	19 ♏ 59	15 ♍ 10

DECLINATIONS

Date	Moon True ☊ ° ′	Moon Mean ☊ ° ′	Moon ☽ Latitude ° ′	Sun ☉ ° ′	Moon ☽ ° ′	Mercury ☿ ° ′	Venus ♀ ° ′	Mars ♂ ° ′	Jupiter ♃ ° ′	Saturn ♄ ° ′	Uranus ♅ ° ′	Neptune ♆ ° ′	Pluto ♇ ° ′
01	21 ♊ 57	20 ♊ 19	03 S 49	17 S 05	21 S 20	22 S 08	21 S 30	04 N 10	16 N 02	11 S 17	07 N 02	16 S 02	18 N 57
02	21 R 48	20 15	04 27	16 47	18 15	21 54	21 18	04 12	16 03	11 14	07 03	16 02	18 58
03	21 37	20 12	04 52	16 30	14 20	21 39	21 06	04 15	16 05	11 12	07 04	16 02	18 59
04	21 27	20 09	05 05	16 12	09 47	21 23	20 53	04 18	16 07	11 09	07 04	16 02	18 59
05	21 17	20 06	05 02	15 54	04 S 46	21 05	20 39	04 22	16 08	11 07	07 05	16 02	19 00
06	21 10	20 03	04 45	15 36	00 N 31	20 46	20 25	04 25	16 10	11 04	07 06	16 02	19 01
07	21 05	20 00	04 13	15 17	05 54	20 25	20 10	04 30	16 12	11 01	07 07	16 02	19 02
08	21 02	19 56	03 27	14 58	11 07	20 03	19 55	04 34	16 13	10 59	07 08	16 02	19 03
09	21 D 02	19 53	02 28	14 39	15 56	19 40	19 39	04 38	16 15	10 56	07 09	16 02	19 03
10	21 02	19 50	01 19	14 19	20 01	19 15	19 23	04 43	16 17	10 54	07 10	16 02	19 04
11	21 03	19 47	00 S 04	14 00	23 01	18 49	19 06	04 49	16 19	10 51	07 11	16 02	19 05
12	21 R 02	19 44	01 N 12	13 40	24 34	18 22	18 48	04 54	16 21	10 48	07 12	16 02	19 06
13	20 59	19 41	02 26	13 20	24 26	17 53	18 30	05 00	16 23	10 46	07 13	16 02	19 07
14	20 53	19 37	03 30	12 59	22 33	17 23	18 11	05 06	16 25	10 43	07 14	16 02	19 07
15	20 45	19 34	04 19	12 39	19 06	16 51	17 52	05 12	16 27	10 40	07 15	16 02	19 08
16	20 35	19 31	04 51	12 18	14 27	16 18	17 32	05 18	16 29	10 38	07 16	16 02	19 09
17	20 24	19 28	05 03	11 57	09 03	15 43	17 12	05 25	16 32	10 35	07 16	16 02	19 10
18	20 14	19 25	04 55	11 36	03 N 18	15 07	16 51	05 32	16 34	10 32	07 17	16 02	19 10
19	20 05	19 21	04 31	11 15	02 S 27	14 30	16 30	05 39	16 36	10 30	07 18	16 02	19 11
20	19 58	19 18	03 52	10 53	07 54	13 51	16 09	05 46	16 38	10 27	07 19	16 02	19 12
21	19 54	19 15	03 03	10 32	12 50	13 11	15 47	05 54	16 41	10 24	07 20	16 02	19 13
22	19 53	19 12	02 06	10 10	17 05	12 30	15 24	06 01	16 43	10 22	07 21	16 02	19 13
23	19 D 52	19 09	01 05	09 48	20 31	11 47	15 01	06 09	16 45	10 19	07 22	16 02	19 14
24	19 52	19 06	00 N 02	09 26	23 00	11 02	14 38	06 17	16 48	10 16	07 23	16 02	19 15
25	19 R 52	19 02	01 S 00	09 04	24 27	10 17	14 14	06 25	16 50	10 14	07 24	16 02	19 16
26	19 50	18 59	02 00	08 41	24 47	09 30	13 50	06 34	16 53	10 11	07 25	16 02	19 16
27	19 45	18 56	02 54	08 19	24 00	08 42	13 25	06 42	16 55	10 08	07 26	16 02	19 17
28	19 ♊ 37	18 ♊ 53	03 S 41	07 S 56	22 S 08	07 S 53	13 S 00	06 N 50	16 N 58	10 S 05	07 N 27	16 S 02	19 N 18

ZODIAC SIGN ENTRIES

Date	h	m	Planets
03	02	56	☽ ♓
03	09	02	☿ ♒
05	07	41	♀ ♒
05	12	43	☽ ♈
07	20	24	☽ ♉
10	01	36	☽ ♊
12	04	14	☽ ♋
14	04	54	☽ ♌
16	05	05	☽ ♍
18	06	45	☽ ♎
18	20	00	☉ ♓
20	11	45	☽ ♏
21	05	40	☿ ♓
22	20	57	☽ ♐
25	09	17	☽ ♑
27	22	14	☽ ♒

LATITUDES

Date	Mercury ☿ ° ′	Venus ♀ ° ′	Mars ♂ ° ′	Jupiter ♃ ° ′	Saturn ♄ ° ′	Uranus ♅ ° ′	Neptune ♆ ° ′	Pluto ♇ ° ′
01	01 S 24	00 S 24	03 N 39	00 S 54	01 S 34	00 N 49	01 N 45	14 N 33
04	01 39	00 31	03 43	00 53	01 34	00 49	01 45	14 34
07	01 50	00 38	03 48	00 53	01 34	00 49	01 45	14 35
10	01 59	00 45	03 52	00 52	01 34	00 49	01 46	14 35
13	02 04	00 51	03 55	00 51	01 34	00 49	01 46	14 36
16	02 06	00 57	03 58	00 50	01 34	00 49	01 46	14 37
19	02 04	01 02	04 01	00 50	01 34	00 49	01 46	14 37
22	01 58	01 07	04 03	00 49	01 34	00 49	01 47	14 38
25	01 47	01 11	04 04	00 48	01 34	00 49	01 47	14 38
28	01 31	01 15	04 04	00 47	01 35	00 49	01 47	14 39
31	01 S 10	01 S 18	04 N 04	00 S 47	01 S 35	00 N 49	01 N 47	14 N 39

DATA

Julian Date	2438793
Delta T	+36 seconds
Ayanamsa	23° 21′ 55″
Synetic vernal point	05° ♓ 45′ 05″
True obliquity of ecliptic	23° 26′ 39″

MOON'S PHASES, APSIDES AND POSITIONS ☽

Date	h	m	Phase	Longitude ° ′	Eclipse Indicator
01	16	36	●	12 ♒ 37	
09	08	53	☽	20 ♉ 25	
16	00	27	○	27 ♌ 08	
23	05	39	☾	04 ♐ 24	

Day	h	m		
14	10	48	Perigee	
26	09	59	Apogee	
06	09	40	ON	
12	22	10	Max dec	24° N 43′
19	01	40	0S	
26	07	19	Max dec	24° S 49′

LONGITUDES

Date	Chiron ⚷ ° ′	Ceres ⚳ ° ′	Pallas ⚴ ° ′	Juno ⚵ ° ′	Vesta ⚶ ° ′	Black Moon Lilith ⚸ ° ′
01	16 ♓ 12	12 ♒ 42	23 ♑ 12	14 ♑ 36	01 ♈ 51	12 ♑ 39
11	16 ♓ 46	16 ♒ 38	26 ♑ 53	18 ♑ 09	06 ♈ 04	13 ♑ 45
21	17 ♓ 21	20 ♒ 33	00 ♒ 27	21 ♑ 38	10 ♈ 21	14 ♑ 52
31	17 ♓ 57	24 ♒ 26	03 ♒ 53	25 ♑ 02	14 ♈ 44	15 ♑ 59

All ephemeris data is given at 12.00 UT and the Moon's longitude is additionally given for 24.00 UT

ASPECTARIAN

h m Aspects

01 Monday
00 26 ☽ ⊻ ♄
03 35 ☽ ∥ ☿
07 26 ☽ ± ♅
10 27 ☽ ∥ ♀
11 00 ☽ ± ♇
12 16 ♃ ⋕ ♆
16 36 ☽ ☌ ☉
17 15 ☽ Ᵽ ♂
19 22 ☽ ⊼ ♅
22 56 ☽ ⊼ ♇

02 Tuesday
00 05 ☉ Ᵽ ♂
01 11 ☿ △ ♂
01 14 ☽ □ ♃
07 01 ☽ ⋕ ♇
07 05 ☽ □ ♆
11 02 ☽ ± ♂
13 22 ☿ ⊥ ♄
17 26 ☿ Ᵽ ♅
20 55 ☽ ⊻ ♀
22 17 ☽ ∥ ☉
22 47 ☽ ⊼ ♂

03 Wednesday
00 35 ☉ ⊼ ♅
01 55 ☽ ⋕ ♃
02 02 ☽ ⊻ ☿
02 09 ☽ ∥ ♆
10 00 ☽ ⊥ ♀
12 23 ☽ ☌ ♄
12 52 ☽ Q ♃
14 18 ♀ △ ♂
15 27 ☽ ⊥ ☿
20 55 ☿ Ᵽ ♇

04 Thursday
05 03 ☽ ∥ ♄
05 26 ☽ ∠ ♀
06 01 ☽ ☍ ♅
08 43 ☽ ⊻ ☉
09 32 ☽ ☍ ♇
11 04 ♀ Ᵽ ♅
11 23 ☽ ∠ ☿
11 33 ♀ ⊥ ♄
12 14 ☽ ✶ ♃
14 43 ☿ Q ♆
17 34 ☽ △ ♆
18 38 ☉ ⊼ ♇
18 41 ☉ ⋕ ♃
21 15 ☽ ⊥ ☉

05 Friday
01 06 ☉ ∥ ♆
01 10 ☽ ⋕ ♅
08 21 ☽ ☍ ♂
13 16 ☽ ✶ ♀
13 52 ☽ ⋕ ♂
15 51 ☽ ∠ ☉
17 00 ☽ ∠ ♃
20 02 ☽ ✶ ☿
21 42 ♀ Ᵽ ♇
22 04 ☽ Ᵽ ♆
22 23 ☽ ⊻ ♄

06 Saturday
07 38 ☉ □ ♃
09 44 ☽ ⊥ ♄
10 00 ☽ ⊥ ♃
14 14 ☽ Q ♀
14 42 ☽ ⊼ ♅
14 54 ☽ ± ♆
16 06 ☿ ⊻ ♄
18 10 ☽ ⊼ ♇
21 14 ☽ ⊻ ♃
21 17 ♀ Q ♆
21 37 ☽ Q ☿
22 17 ☽ ✶ ☉

07 Sunday
01 47 ☽ ± ♅
02 01 ☽ ⊼ ♆
02 34 ☽ ∠ ♄
05 14 ☽ ± ♇
05 37 ☽ ∥ ♂
15 44 ☽ ⊼ ♂
17 31 ☽ ∥ ♅
18 13 ☽ Ᵽ ♅
21 38 ☽ Ᵽ ♇
22 06 ☽ Q ☉

08 Monday
02 31 ☽ ± ♂
02 43 ☽ □ ♀
06 10 ☽ ✶ ♄
06 25 ☿ ± ♅
11 00 ☽ □ ☿
11 20 ☽ ⋕ ♄
18 31 ☽ Ᵽ ♂
21 07 ☽ △ ♅
22 33 ☉ □ ♆

09 Tuesday
00 30 ☽ △ ♇
03 45 ☽ Q ♄
03 53 ☽ ☌ ♃
05 43 ☽ ⋕ ☉
08 07 ☽ ☍ ♆
08 53 ☽ □ ☉
09 55 ☿ ± ♇
12 34 ☽ ⋕ ♆
13 44 ☽ ∥ ♃
14 46 ☿ ∥ ♀
18 45 ♀ ⊻ ♄
20 38 ☽ △ ♂

10 Wednesday
01 16 ☉ ± ♂
05 53 ☽ ∥ ♇
07 34 ☽ ⋕ ☿
08 08 ☽ ⋕ ♀
11 21 ☽ □ ♄
12 49 ☽ △ ♀
19 48 ☿ Ᵽ ♂
22 13 ☿ ⋕ ♇
22 25 ☽ △ ☿

11 Thursday
00 59 ☽ □ ♅
04 15 ☽ □ ♇
07 55 ☽ ⊻ ♃
10 41 ♀ ± ♅
11 36 ☽ ⊼ ♆
12 57 ♀ ⋕ ♇
16 12 ☽ △ ☉
16 34 ☽ Ᵽ ♀
18 01 ☽ ⊥ ♃
19 52 ☿ ⊼ ♅
21 36 ☽ ± ♆
22 54 ☽ □ ♂

12 Friday
02 47 ☽ Ᵽ ☿
06 54 ☽ Q ♅
08 52 ☽ ± ♀
09 00 ☽ ∠ ♃
10 07 ☽ Q ♇
12 26 ☽ Ᵽ ♆
13 57 ☽ △ ♄
18 45 ☽ Ᵽ ☉
19 28 ☽ ± ☿
19 35 ☽ ⊼ ♀
23 18 ♀ ± ♇
23 22 ☿ ⊼ ♇

13 Saturday
02 27 ☽ ✶ ♅
04 00 ☽ Q ♂
05 39 ☽ ✶ ♇
06 28 ☽ ⊼ ☿
08 49 ♃ Q ♄
09 37 ☽ ✶ ♃
10 26 ☽ ± ☉
12 49 ☽ △ ♆
14 30 ☽ Ᵽ ♄
20 48 ☽ ⊼ ☉
23 06 ☽ ✶ ♂

14 Sunday
02 35 ☽ ∠ ♅
05 04 ☽ Q ♃
05 05 ☽ ± ♄
05 46 ☽ ∠ ♇
08 40 ♀ Ᵽ ♂
11 38 ☿ □ ♃
14 47 ☽ ⊼ ♄
16 58 ☽ ⊥ ♅
20 10 ☽ ⊥ ♇
22 49 ☽ ∠ ♂

15 Monday
00 19 ☽ ☍ ♀
01 17 ☉ ⊼ ♂
02 33 ☽ ⊻ ♅
05 46 ☽ ⊻ ♇
10 08 ☽ □ ♃
11 49 ☽ ∥ ♇
12 51 ☽ ☍ ☿
12 58 ☽ □ ♆
13 02 ☽ ⊥ ♂
13 50 ☿ □ ♆
14 15 ☿ ± ♂
19 28 ☽ ⋕ ♀
22 33 ☽ ⊻ ♂

16 Tuesday
00 27 ☽ ☍ ☉
01 54 ☽ ⋕ ☿
02 10 ☽ ∥ ♃
02 16 ♀ ⊼ ♅
04 09 ☿ ⋕ ♃
04 21 ☽ ⋕ ♆
15 30 ☽ ☍ ♄
18 07 ☽ Q ♆
22 32 ☽ ⋕ ☉
22 47 ☿ ∥ ♆

17 Wednesday
03 00 ☽ ☌ ♅
05 21 ☽ ⋕ ♄
05 25 ☽ ⊼ ♀
06 22 ☽ ☌ ♇
11 22 ☽ △ ♃
13 55 ☽ ✶ ♆
16 19 ♀ ⊼ ♇
16 20 ☽ ± ♀
19 27 ☽ ∥ ♅
20 33 ☽ ⊼ ☿
23 03 ☽ ☌ ♂

18 Thursday
02 55 ☽ ∥ ♂
05 40 ☽ ⊼ ☉
08 09 ☽ ± ☿
09 00 ☽ Ᵽ ♀
12 54 ☽ Ᵽ ♃
14 31 ☿ ⊼ ♂
15 19 ☽ ∠ ♆
16 44 ☽ ± ☉
18 10 ☽ ⊼ ♄

19 Friday
01 57 ☽ Ᵽ ☿
04 40 ☽ ± ♃
04 43 ☽ ± ♄
05 49 ☽ ⊻ ♅
06 04 ♀ ⋕ ♃
07 00 ☽ ⊥ ♆
09 28 ☽ ⊻ ♇
09 42 ☽ Ᵽ ☉
13 44 ☽ △ ♀
15 22 ☽ ⊼ ♃
16 25 ☽ ⊥ ♅
17 39 ☽ ⊻ ♆
20 08 ☽ ⊥ ♇
20 51 ☽ Ᵽ ♄

20 Saturday
01 23 ♆ St R
02 11 ☽ ⋕ ♂
02 25 ☽ ⊻ ♂
08 37 ☽ ∠ ♅
08 51 ♀ ± ♂
08 54 ☽ △ ☿
09 14 ♀ □ ♃
09 23 ☽ ⋕ ♅
12 27 ☽ ∠ ♇
13 09 ☽ ⊥ ♂
15 01 ☽ △ ☉
19 00 ♀ ∥ ♆
23 56 ☽ ∥ ♄

21 Sunday
00 37 ☽ △ ♄
01 14 ☽ ∥ ☉
05 32 ☽ ∠ ♂
07 52 ♀ □ ♆
12 30 ☽ ✶ ♅
13 36 ☽ ∥ ☿
16 31 ☽ ✶ ♇
21 32 ☉ ∥ ♄
23 35 ☽ ☍ ♃

22 Monday
01 27 ☽ ☌ ♆
02 48 ☽ ∥ ♀
03 25 ☽ □ ♀
05 40 ☽ ∥ ♆
09 36 ☽ ✶ ♂
09 43 ☽ ⋕ ♃
11 32 ☽ Q ♅
15 41 ☽ Q ♇

23 Tuesday
02 16 ☽ ⋕ ♇
03 54 ☽ □ ☿
05 39 ☽ □ ☉
08 40 ☽ Q ♂
11 17 ☽ □ ♄
19 00 ☽ Q ♀
20 21 ♃ Q ♄
23 07 ☽ □ ♅

24 Wednesday
03 07 ☉ ☌ ☿
03 26 ☽ □ ♇
05 34 ♀ ⊼ ♂
11 41 ☽ ⊼ ♃
12 58 ☽ ⊻ ♆
19 56 ☽ □ ♂
21 12 ☽ Q ☉
22 05 ☽ ✶ ♀
22 45 ☽ Q ☿
23 58 ☽ Q ♄

25 Thursday
00 00 ☽ ± ♃
01 09 ☽ ⊥ ♆
06 54 ☿ ☌ ♄
07 11 ☿ Q ♃
13 53 ☿ ∥ ♄
18 32 ☽ Ᵽ ♃
19 28 ☽ ∠ ♆
23 55 ☽ ✶ ☉

26 Friday
00 40 ☽ ✶ ♄
03 48 ☽ ✶ ☿
08 15 ☽ ∠ ♀
09 49 ☉ ☌ ♄
10 55 ☉ Q ♃
11 50 ☽ △ ♅
16 18 ☽ △ ♇

27 Saturday
01 27 ☽ △ ♃
02 00 ☽ ✶ ♆
04 44 ☽ ⊥ ♀
07 17 ☽ △ ♂
07 30 ☽ ∠ ♄
09 16 ☽ ∠ ☉
16 05 ☽ ∠ ☿
18 09 ☽ Ᵽ ♅
18 18 ☽ ⊻ ♀
22 37 ☽ Ᵽ ♇

28 Sunday
01 51 ☽ ⊥ ♄
02 15 ☽ Q ♆
04 48 ☿ ☍ ♅
05 05 ☽ ⊥ ☉
09 12 ☉ ∥ ☿
12 10 ☽ ± ♅
12 37 ☽ Ᵽ ♂
13 35 ☽ ⊥ ☿
14 00 ☽ ⊻ ♄
16 37 ☽ ± ♇
18 11 ☽ ⊻ ☉
21 19 ♃ ☍ ♆

MARCH 1965

LONGITUDES

Date	Sidereal time h m s	Sun ☉ ° ′ ″	Moon ☽ ° ′ ″	Moon ☽ 24.00 ° ′ ″	Mercury ☿ ° ′	Venus ♀ ° ′	Mars ♂ ° ′	Jupiter ♃ ° ′	Saturn ♄ ° ′	Uranus ♅ ° ′	Neptune ♆ ° ′	Pluto ♇ ° ′
01	22 36 20	10 ♓ 41 48	18 ♒ 56 26	25 ♒ 03 14	15 ♓ 26	00 ♓ 13	21 ♍ 49	20 ♉ 05	07 ♓ 58	12 ♍ 54	19 ♏ 59	15 ♍ 09
02	22 40 17	11 42 00	01 ♓ 13 03	07 ♓ 26 00	17 21	01 28	21 R 27	20 13	08 05	12 R 51	19 R 59	15 R 07
03	22 44 13	12 42 11	13 ♓ 42 07	20 ♓ 01 24	19 17	02 43	21 04	20 23	08 13	12 48	19 58	15 06
04	22 48 10	13 42 20	26 ♓ 23 47	02 ♈ 49 12	21 13	03 57	20 41	20 32	08 20	12 46	19 58	15 04
05	22 52 06	14 42 27	09 ♈ 17 34	15 ♈ 48 47	23 08	05 12	20 18	20 41	08 27	12 43	19 57	15 02
06	22 56 03	15 42 32	22 ♈ 22 47	28 ♈ 59 29	25 04	06 27	19 55	20 50	08 35	12 41	19 57	15 01
07	22 59 59	16 42 35	05 ♉ 38 52	12 ♉ 20 54	26 58	07 42	19 31	21 00	08 42	12 38	19 57	14 59
08	23 03 56	17 42 35	19 ♉ 05 37	25 ♉ 53 04	28 ♓ 53	08 57	19 08	21 10	08 49	12 35	19 56	14 58
09	23 07 53	18 42 34	02 ♊ 43 18	09 ♊ 36 25	00 ♈ 45	10 12	18 44	21 19	08 56	12 33	19 55	14 56
10	23 11 49	19 42 31	16 ♊ 32 29	23 ♊ 31 33	02 37	11 26	18 21	21 29	09 04	12 30	19 55	14 54
11	23 15 46	20 42 25	00 ♋ 33 39	07 ♋ 38 43	04 26	12 41	17 57	21 39	09 11	12 27	19 54	14 53
12	23 19 42	21 42 18	14 ♋ 46 39	21 ♋ 57 12	06 13	13 56	17 33	21 49	09 18	12 25	19 54	14 51
13	23 23 39	22 42 07	29 ♋ 10 04	06 ♌ 24 46	07 57	15 11	17 10	22 00	09 25	12 22	19 53	14 50
14	23 27 35	23 41 55	13 ♌ 40 46	20 ♌ 57 23	09 38	16 25	16 46	22 10	09 33	12 20	19 52	14 48
15	23 31 32	24 41 41	28 ♌ 13 50	05 ♍ 29 16	11 14	17 40	16 23	22 20	09 40	12 17	19 51	14 47
16	23 35 28	25 41 24	12 ♍ 42 51	19 ♍ 53 42	12 47	18 55	16 00	22 31	09 47	12 15	19 51	14 45
17	23 39 25	26 41 05	27 ♍ 00 58	04 ♎ 03 56	14 14	20 09	15 38	22 41	09 54	12 12	19 50	14 43
18	23 43 21	27 40 44	11 ♎ 01 57	17 ♎ 54 31	15 36	21 24	15 15	22 52	10 01	12 10	19 49	14 42
19	23 47 18	28 40 21	24 ♎ 41 15	01 ♏ 21 56	16 53	22 39	14 53	23 03	10 08	12 07	19 48	14 40
20	23 51 15	29 ♓ 39 57	07 ♏ 56 33	14 ♏ 25 09	18 03	23 53	14 32	23 14	10 15	12 05	19 47	14 39
21	23 55 11	00 ♈ 39 30	20 ♏ 47 58	27 ♏ 05 20	19 07	25 08	14 10	23 25	10 22	12 02	19 46	14 37
22	23 59 08	01 39 02	03 ♐ 17 41	09 ♐ 25 31	20 04	26 23	13 50	23 36	10 29	12 00	19 46	14 36
23	00 03 04	02 38 32	15 ♐ 29 26	21 ♐ 30 04	20 54	27 37	13 29	23 47	10 36	11 57	19 45	14 34
24	00 07 01	03 38 01	27 ♐ 28 04	03 ♑ 24 06	21 36	28 ♓ 52	13 10	23 59	10 43	11 55	19 44	14 33
25	00 10 57	04 37 27	09 ♑ 18 54	15 ♑ 13 07	22 11	00 ♈ 07	12 51	24 10	10 50	11 53	19 43	14 31
26	00 14 54	05 36 52	21 ♑ 07 28	27 ♑ 02 33	22 39	01 21	12 32	24 22	10 57	11 50	19 42	14 30
27	00 18 50	06 36 15	02 ♒ 59 02	08 ♒ 57 28	22 59	02 36	12 14	24 33	11 04	11 48	19 40	14 28
28	00 22 47	07 35 36	14 ♒ 58 24	21 ♒ 02 17	23 11	03 50	11 57	24 45	11 11	11 46	19 39	14 27
29	00 26 44	08 34 56	27 ♒ 09 32	03 ♓ 20 28	23 15	05 05	11 40	24 57	11 18	11 43	19 38	14 26
30	00 30 40	09 34 13	09 ♓ 35 21	15 ♓ 54 20	23 R 13	06 19	11 24	25 08	11 24	11 41	19 37	14 24
31	00 34 37	10 ♈ 33 29	22 ♓ 17 31	28 ♓ 44 54	23 ♈ 03	07 ♈ 34	11 ♍ 09	25 ♉ 20	11 ♓ 31	11 ♍ 39	19 ♏ 36	14 ♍ 23

Date	Moon True ☊ ° ′	Moon Mean ☊ ° ′	Moon ☽ Latitude ° ′
01	19 ♊ 27	18 ♊ 50	04 S 19
02	19 R 14	18 47	04 46
03	19 00	18 43	04 59
04	18 46	18 40	04 57
05	18 33	18 37	04 41
06	18 23	18 34	04 09
07	18 15	18 31	03 24
08	18 11	18 27	02 27
09	18 09	18 24	01 21
10	18 D 09	18 21	00 S 08
11	18 R 09	18 18	01 N 05
12	18 07	18 15	02 16
13	18 04	18 12	03 19
14	17 58	18 08	04 09
15	17 49	18 05	04 43
16	17 38	18 02	04 59
17	17 25	17 59	04 56
18	17 14	17 56	04 35
19	17 04	17 53	03 59
20	16 56	17 49	03 10
21	16 51	17 46	02 13
22	16 48	17 43	01 11
23	16 47	17 40	00 N 07
24	16 D 47	17 37	00 S 56
25	16 R 47	17 33	01 57
26	16 46	17 30	02 52
27	16 43	17 27	03 40
28	16 37	17 24	04 18
29	16 29	17 21	04 46
30	16 19	17 18	05 01
31	16 ♊ 07	17 ♊ 14	05 S 01

DECLINATIONS

Date	Sun ☉ ° ′	Moon ☽ ° ′	Mercury ☿ ° ′	Venus ♀ ° ′	Mars ♂ ° ′	Jupiter ♃ ° ′	Saturn ♄ ° ′	Uranus ♅ ° ′	Neptune ♆ ° ′	Pluto ♇ ° ′
01	07 S 33	19 S 15	07 S 03	12 S 35	06 N 59	17 N 00	10 S 03	07 N 28	16 S 01	19 N 19
02	07 11	15 29	06 11	12 10	07 08	17 03	10 00	07 29	16 01	19 19
03	06 48	11 00	05 19	11 44	07 16	17 06	09 57	07 30	16 01	19 20
04	06 25	05 59	04 26	11 17	07 25	17 08	09 55	07 31	16 01	19 21
05	06 01	00 S 37	03 32	10 51	07 34	17 11	09 52	07 33	16 01	19 21
06	05 38	04 N 51	02 38	10 24	07 42	17 14	09 49	07 34	16 01	19 22
07	05 15	10 12	01 43	09 57	07 51	17 16	09 47	07 35	16 00	19 23
08	04 51	15 08	00 S 48	09 30	08 00	17 19	09 44	07 36	16 00	19 24
09	04 28	19 23	00 N 07	09 02	08 08	17 22	09 41	07 37	16 00	19 24
10	04 05	22 37	01 02	08 34	08 17	17 25	09 39	07 37	16 00	19 25
11	03 41	24 32	01 56	08 06	08 25	17 27	09 36	07 38	16 00	19 25
12	03 17	24 52	02 50	07 38	08 33	17 30	09 33	07 39	15 59	19 26
13	02 54	23 34	03 42	07 09	08 41	17 33	09 31	07 40	15 59	19 27
14	02 30	20 42	04 34	06 41	08 49	17 36	09 28	07 41	15 59	19 27
15	02 06	16 31	05 24	06 12	08 57	17 39	09 26	07 42	15 59	19 28
16	01 43	11 24	06 12	05 43	09 05	17 42	09 23	07 43	15 58	19 29
17	01 19	05 N 43	06 58	05 13	09 12	17 44	09 20	07 44	15 58	19 29
18	00 55	00 S 09	07 41	04 44	09 19	17 47	09 18	07 45	15 58	19 30
19	00 32	05 52	08 22	04 15	09 26	17 50	09 15	07 46	15 58	19 30
20	00 S 08	11 09	09 00	03 45	09 33	17 53	09 13	07 47	15 57	19 31
21	00 N 16	15 49	09 36	03 15	09 39	17 56	09 10	07 48	15 57	19 31
22	00 39	19 39	10 07	02 46	09 46	17 59	09 07	07 49	15 57	19 32
23	01 03	22 32	10 36	02 16	09 52	18 02	09 05	07 50	15 56	19 32
24	01 27	24 22	11 00	01 46	09 57	18 05	09 02	07 51	15 56	19 33
25	01 50	25 03	11 21	01 16	10 03	18 08	09 00	07 52	15 56	19 33
26	02 14	24 37	11 39	00 46	10 08	18 11	08 57	07 53	15 56	19 34
27	02 37	23 03	11 52	00 S 16	10 13	18 14	08 55	07 53	15 55	19 34
28	03 01	20 28	12 01	00 N 14	10 17	18 17	08 52	07 54	15 55	19 35
29	03 24	16 56	12 06	00 44	10 22	18 20	08 50	07 55	15 55	19 35
30	03 48	12 37	12 07	01 15	10 26	18 23	08 47	07 56	15 54	19 36
31	04 N 11	07 S 40	12 N 04	01 N 45	10 N 29	18 N 26	08 S 45	07 N 57	15 S 54	19 N 36

ZODIAC SIGN ENTRIES

Date	h	m	Planets
01	07	55	♀ ♓
02	09	38	☽ ♓
04	18	45	☽ ♈
07	01	49	☽ ♉
09	02	19	☿ ♈
09	07	14	☽ ♊
11	11	03	☽ ♋
13	13	23	☽ ♌
15	14	55	☽ ♍
17	17	04	☽ ♎
19	21	32	☽ ♏
20	20	05	☉ ♈
22	05	37	☽ ♐
24	17	07	☽ ♑
25	09	54	♀ ♈
27	05	59	☽ ♒
29	17	32	☽ ♓

LATITUDES

Date	Mercury ☿ ° ′	Venus ♀ ° ′	Mars ♂ ° ′	Jupiter ♃ ° ′	Saturn ♄ ° ′	Uranus ♅ ° ′	Neptune ♆ ° ′	Pluto ♇ ° ′
01	01 S 25	01 S 16	04 N 04	00 S 47	01 S 35	00 N 49	01 N 47	14 N 39
04	01 02	01 19	04 03	00 47	01 35	00 49	01 47	14 39
07	00 34	01 22	04 02	00 46	01 35	00 49	01 47	14 39
10	00 S 01	01 24	03 59	00 45	01 35	00 49	01 47	14 39
13	00 N 36	01 25	03 56	00 45	01 36	00 49	01 48	14 40
16	01 15	01 26	03 52	00 44	01 36	00 49	01 48	14 39
19	01 53	01 27	03 47	00 44	01 36	00 49	01 48	14 39
22	02 28	01 26	03 41	00 43	01 36	00 49	01 48	14 39
25	02 56	01 25	03 35	00 42	01 37	00 49	01 48	14 39
28	03 14	01 24	03 28	00 42	01 37	00 49	01 48	14 39
31	03 N 21	01 S 22	03 N 21	00 S 41	01 S 37	00 N 49	01 N 48	14 N 38

DATA

Julian Date	2438821
Delta T	+36 seconds
Ayanamsa	23° 21′ 58″
Synetic vernal point	05° ♓ 45′ 02″
True obliquity of ecliptic	23° 26′ 40″

MOON'S PHASES, APSIDES AND POSITIONS ☽

Date	h	m	Phase	Longitude ° ′	Eclipse Indicator
03	09	56	●	12 ♓ 37	
10	17	52	☽	19 ♊ 57	
17	11	24	○	26 ♍ 40	
25	01	37	☾	04 ♑ 12	

Day	h	m		
14	09	14	Perigee	
26	05	35	Apogee	
05	14	44	0N	
12	05	03	Max dec	24° N 57′
18	11	25	0S	
25	14	37	Max dec	25° S 04′

LONGITUDES

Date	Chiron ⚷ ° ′	Ceres ⚳ ° ′	Pallas ⚴ ° ′	Juno ⚵ ° ′	Vesta ⚶ ° ′	Black Moon Lilith ⚸ ° ′
01	17 ♓ 50	23 ♒ 40	03 ♒ 12	24 ♑ 22	13 ♈ 51	15 ♑ 45
11	18 ♓ 27	27 ♒ 30	06 ♒ 31	27 ♑ 40	18 ♈ 15	16 ♑ 52
21	19 ♓ 04	01 ♓ 17	09 ♒ 38	00 ♒ 49	22 ♈ 42	17 ♑ 59
31	19 ♓ 40	04 ♓ 58	12 ♒ 34	03 ♒ 49	27 ♈ 10	19 ♑ 05

All ephemeris data is given at 12.00 UT and the Moon's longitude is additionally given for 24.00 UT

ASPECTARIAN

h m	Aspects
01 Monday	
00 03	☿ # ♅
00 05	☽ ⊼ ♅
03 47	☽ ⊻ ☿
04 31	☽ ⊼ ♇
06 00	☽ ± ♂
08 24	☿ ☍ ♇
11 34	☽ # ♇
13 30	☿ # ♂
14 03	☽ □ ♆
14 16	☽ □ ♃
17 01	☉ # ♅
17 29	☽ ⊼ ♂
02 Tuesday	
02 45	☽ # ♃
08 52	☽ ∥ ♆
12 31	☽ ☌ ♀
14 12	☉ # ♂
03 Wednesday	
01 24	☽ ☌ ♄
01 41	☽ Q ♃
07 57	☽ ∥ ♀
09 56	☽ ☌ ☉
10 18	☽ ☍ ♅
14 23	☉ ☍ ♅
14 39	☽ ☍ ♇
17 13	☽ ∥ ♄
20 34	☿ △ ♆
23 54	☽ △ ♆
04 Thursday	
00 30	☽ ☌ ☿
00 49	☽ ✱ ♃
01 34	☽ ☍ ♂
02 47	☿ ✱ ♃
04 51	☽ # ♅
05 31	☽ # ♂
06 35	☿ ☍ ♂
09 51	☽ ∥ ☉
19 08	♂ △ ♃
20 26	☽ ∥ ☿
05 Friday	
03 37	☽ ⊻ ♀
03 59	☽ ⚼ ♆
05 14	☽ ∠ ♃
08 01	♂ ∥ ♅
10 26	☽ ⊻ ♄
15 54	☽ ⊥ ♀
18 18	☽ ⊼ ♅
19 45	☉ ☍ ♇
20 35	☽ ± ♆
21 36	☽ ⊥ ♄
22 03	☽ ⊥ ♃
22 34	☽ ⊼ ♇
22 48	☽ ⊻ ☉
06 Saturday	
03 39	☽ # ☿
05 16	☽ ± ♅
07 34	☽ ⊼ ♆
07 38	☽ ⊼ ♂
09 10	☽ ⊻ ♃
09 31	☽ ± ♇
09 47	♂ ✱ ♆
10 08	☽ ∠ ♀
10 41	☽ ⊥ ☉
14 12	☽ ∠ ♄
15 14	☽ # ☉
17 42	☽ ⊻ ☿
18 15	☽ ± ♂
21 35	☽ ⚼ ♅
07 Sunday	
00 03	☽ ∥ ♅
01 02	☽ ∥ ♂
01 49	☽ ⚼ ♇
04 20	☽ ∠ ☉
06 23	☽ ⊥ ☿
10 02	☽ ⚼ ♂
10 06	☽ # ♄
10 57	☽ # ♀
16 04	☽ ✱ ♀
17 32	☽ ✱ ♄
21 59	♀ ∥ ♄
08 Monday	
00 28	☽ △ ♅
01 12	☽ ∠ ☿
04 40	☽ △ ♇
09 21	♀ ☌ ♄
09 21	☽ ✱ ☉
12 04	☽ △ ♂
13 29	☽ ☍ ♆
15 05	☽ Q ♄
15 37	☽ Q ♀
15 42	☽ ☌ ♃
16 31	☽ # ♆
16 44	♀ Q ♃
23 49	☽ ∥ ♃
09 Tuesday	
08 01	☽ ✱ ☿
08 12	☽ Q ☉
12 04	☽ ∥ ♇
12 29	☉ ☍ ♂
22 56	☽ □ ♄
10 Wednesday	
02 18	☽ □ ♀
05 02	☽ □ ♅
08 10	☽ Q ☿
09 11	☽ □ ♇
15 01	☽ □ ♂
16 53	☉ △ ♆
17 48	☽ ⊼ ♆
17 52	☽ □ ☉
20 36	☽ ⊻ ♃
23 27	♀ # ♂
11 Thursday	
04 04	☽ ± ♆
06 59	☽ ⊥ ♃
07 47	♀ ☍ ♅
11 50	☽ Q ♅
15 56	☽ Q ♇
18 15	☿ ⚼ ♆
19 22	☽ ⚼ ♆
19 31	☽ □ ☿
20 53	☽ Q ♂
22 27	☽ ∠ ♃
12 Friday	
02 43	☽ △ ♄
08 03	☽ ✱ ♅
10 27	☽ △ ♀
10 35	♀ # ♅
12 08	☽ ✱ ♇
15 24	☉ ✱ ♃
16 32	☽ ✱ ♂
20 33	☽ △ ♆
20 43	☉ # ☿
21 13	☿ ∠ ♃
23 55	☽ ✱ ♃
13 Saturday	
00 27	☽ △ ☉
04 03	☽ ⚼ ♄
05 25	♀ ☍ ♇
09 02	☽ ∠ ♅
13 06	☽ ∠ ♇
13 50	☽ ⚼ ♀
16 50	☽ ∠ ♂
19 07	☽ ± ♄
20 06	☽ Q ♃
23 54	☽ ⊥ ♅
14 Sunday	
03 10	☽ ⚼ ☉
03 58	☽ ⊥ ♇
04 27	☽ △ ☿
05 07	☽ ⊼ ♄
06 07	☽ ± ♀
07 20	☽ ⊥ ♂
09 47	☽ ⊻ ♅
10 42	☿ ⊻ ♄
13 51	☽ ⊻ ♇
16 57	☽ ⊼ ♀
16 58	☽ ⊻ ♂
17 10	♀ ☍ ♂
19 07	☽ ± ☉
19 50	☽ ∥ ♇
22 12	☽ □ ♆
15 Monday	
02 10	☽ □ ♃
05 44	☽ ⊼ ☉
06 10	☽ ∥ ♃
08 19	☽ ⚼ ☿
14 43	☽ # ♆
16 Tuesday	
00 59	☽ ± ☿
02 53	☿ # ♀
03 46	☿ ⊼ ♅
03 55	☽ Q ♆
07 05	☽ ☍ ♄
11 13	☽ ☌ ♅
12 07	☽ ⊼ ☿
15 23	☽ ☌ ♇
17 21	☽ ☌ ♂
20 46	☽ # ♄
21 46	☽ ∥ ♂
23 20	☽ ☍ ♀
23 54	☽ ✱ ♆
17 Wednesday	
03 39	☽ ∥ ♅
04 36	☽ △ ♃
05 15	☿ ± ♆
05 47	♀ △ ♆
07 27	☽ ∥ ☿
11 24	☽ ☍ ☉
14 13	☽ # ♀
20 11	☿ ⊼ ♇
18 Thursday	
01 18	☽ ∠ ♆
06 28	☽ ⚼ ♃
07 01	☿ ⊼ ♂
07 19	☽ # ☉
08 01	♂ # ♄
10 14	☽ ⊼ ♄
13 57	☽ ⊻ ♅
14 18	☿ ∥ ♅
15 01	☽ ∥ ☉
16 50	☽ ⊥ ♆
18 22	☽ ⊻ ♇
19 09	☽ ⊻ ♂
20 17	☿ ⊥ ♄
20 46	☽ ± ♄
20 48	☽ ☍ ☿
22 19	☽ ± ♃
19 Friday	
00 24	☽ ⊥ ♅
03 21	☽ ⊻ ♆
04 53	☽ ⊥ ♇
05 26	☽ ⊥ ♂
05 38	☽ ∥ ♀
08 00	☽ ⊼ ♀
09 03	☽ ⊼ ♃
12 49	☽ ⚼ ♄
15 54	☿ ⊥ ♃
16 20	☽ ∠ ♅
19 43	☽ ⊼ ☉
19 49	☽ ± ♀
20 25	☽ # ♅
20 55	☽ ∠ ♇
21 05	☽ ∠ ♂
21 08	♀ ✱ ♃
20 Saturday	
00 37	☽ # ☿
02 58	☽ ∥ ♄
03 23	♂ ☌ ♇
04 17	☽ # ♂
07 29	☽ ± ☉
12 29	☿ ± ♅
13 56	☽ ⚼ ♀
16 18	☽ △ ♄
19 29	☿ # ♄
19 37	☽ ✱ ♅
23 52	☽ ✱ ♂
21 Sunday	
00 24	☽ ✱ ♇
01 30	☽ ⚼ ☉
08 33	☽ ⊼ ☿
10 04	☽ ☌ ♆
12 45	☽ ∥ ♆
15 33	☿ ∥ ♂
17 03	☽ ☍ ♃
18 08	☽ Q ♅
20 55	☽ ± ☿
21 10	☽ △ ♀
21 58	☽ Q ♂
23 05	☽ Q ♇
22 Monday	
00 42	☽ # ♃
04 00	☿ ⊼ ♆
07 23	☿ ± ♂
08 32	☽ △ ☉
11 07	☽ # ♇
15 43	☽ ⚼ ☿
23 Tuesday	
02 14	☽ □ ♄
02 26	☿ ± ♇
05 01	☽ □ ♅
08 08	☽ □ ♂
10 11	☽ □ ♇
20 28	☽ ⊻ ♆
23 30	☽ △ ☿
24 Wednesday	
04 52	☽ ⊼ ♃
08 30	☽ ⊥ ♆
14 33	☽ Q ♄
15 09	☽ □ ♀
17 09	☽ ± ♃
20 34	☉ # ♀
25 Thursday	
01 37	☽ □ ☉
02 40	☽ ∠ ♆
11 42	☽ ⚼ ♃
14 02	☉ ⚼ ♆
15 07	☽ ✱ ♄
17 11	☽ △ ♅
18 59	☽ △ ♂
22 34	☽ △ ♇
26 Friday	
07 59	☽ Q ♀
09 06	☽ ✱ ♆
15 12	☽ □ ☿
17 31	☽ Q ☉
18 40	☽ △ ♃
21 53	☽ ∠ ♄
23 33	☽ ⚼ ♅
27 Saturday	
00 40	☽ ⚼ ♂
04 56	☽ ⚼ ♇
09 22	☽ Q ♆
11 07	☽ ✱ ♀
16 14	☽ ⊥ ♄
17 39	☽ ± ♅
18 23	☽ ± ♂
19 56	☽ ✱ ☉
23 01	☽ ± ♇
28 Sunday	
04 20	☽ Q ☿
04 22	☽ ⊻ ♄
05 37	☽ ⊼ ♅
06 06	☽ ⊼ ♂
10 58	☽ ⊼ ♇
18 33	☽ # ♇
20 32	☽ ∠ ♀
21 16	☽ □ ♆
29 Monday	
03 16	☽ # ♃
03 35	♀ ⚼ ♆
04 20	☽ ✱ ☿
04 23	☽ ∠ ☉
06 03	♂ ☌ ♅
07 36	☽ □ ♃
14 53	☿ St R
16 10	☽ ⊥ ♀
18 04	☽ ∥ ♆
23 27	☽ ⊥ ☉
30 Tuesday	
05 03	☽ ⊻ ♀
09 23	☽ ∠ ☿
11 25	♂ ☍ ♄
11 58	☽ ⊻ ☉
14 33	☽ # ☿
15 23	☽ ☍ ♂
15 30	☽ ☌ ♄
15 59	☽ ☍ ♅
18 52	☽ Q ♃
21 08	☽ ☍ ♇
22 47	☽ # ♂
31 Wednesday	
02 18	☽ ⊥ ☿
05 17	☉ ∠ ♃
06 55	☽ ∥ ♄
06 58	☽ △ ♆
10 42	☽ # ♅
13 23	☽ ⊻ ☿
17 46	☽ ✱ ♃
23 22	☉ ⊼ ♂

APRIL 1965

LONGITUDES

Date	Sidereal time h m s	Sun ☉ ° ' "	Moon ☽ ° ' "	Moon ☽ 24.00 ° ' "	Mercury ☿ ° '	Venus ♀ ° '	Mars ♂ ° '	Jupiter ♃ ° '	Saturn ♄ ° '	Uranus ♅ ° '	Neptune ♆ ° '	Pluto ♇ ° '
01	00 38 33	11 ♈ 32 42	05 ♈ 16 23	11 ♈ 51 49	22 ♈ 46	08 ♈ 48	10 ♍ 54	25 ♉ 32	11 ♓ 38	11 ♍ 37	19 ♏ 35	14 ♍ 21
02	00 42 30	12 31 54	18 ♈ 31 00	25 ♈ 13 40	22 R 23	10 03	10 R 40	25 44	11 44	11 R 35	19 R 34	14 R 20
03	00 46 26	13 31 03	01 ♉ 59 30	08 ♉ 48 13	21 55	11 17	10 27	25 56	11 51	11 33	19 32	14 19
04	00 50 23	14 30 11	15 ♉ 39 29	22 ♉ 32 59	21 21	12 32	10 15	26 09	11 58	11 30	19 31	14 17
05	00 54 19	15 29 16	29 ♉ 28 26	06 ♊ 25 35	20 44	13 46	10 03	26 21	12 04	11 28	19 30	14 16
06	00 58 16	16 28 19	13 ♊ 24 14	20 ♊ 24 10	20 02	15 01	09 52	26 33	12 11	11 26	19 29	14 15
07	01 02 13	17 27 20	27 ♊ 25 14	04 ♋ 27 19	19 18	16 15	09 42	26 46	12 17	11 24	19 27	14 13
08	01 06 09	18 26 18	11 ♋ 30 17	18 ♋ 34 01	18 33	17 29	09 33	26 58	12 24	11 22	19 26	14 12
09	01 10 06	19 25 14	25 ♋ 38 22	02 ♌ 43 11	17 46	18 44	09 25	27 11	12 30	11 21	19 25	14 11
10	01 14 02	20 24 08	09 ♌ 48 15	16 ♌ 53 20	17 00	19 58	09 17	27 23	12 36	11 19	19 23	14 10
11	01 17 59	21 23 00	23 ♌ 58 06	01 ♍ 02 13	16 14	21 12	09 10	27 36	12 43	11 17	19 22	14 08
12	01 21 55	22 21 49	08 ♍ 05 16	15 ♍ 06 47	15 30	22 27	09 04	27 49	12 49	11 15	19 20	14 07
13	01 25 52	23 20 36	22 ♍ 06 17	29 ♍ 03 18	14 49	23 41	08 59	28 01	12 55	11 13	19 19	14 06
14	01 29 48	24 19 20	05 ♎ 57 20	12 ♎ 47 56	14 11	24 55	08 54	28 14	13 01	11 12	19 18	14 05
15	01 33 45	25 18 03	19 ♎ 34 41	26 ♎ 17 15	13 36	26 10	08 51	28 27	13 07	11 10	19 16	14 04
16	01 37 42	26 16 44	02 ♏ 55 21	09 ♏ 28 48	13 06	27 24	08 48	28 40	13 13	11 08	19 15	14 03
17	01 41 38	27 15 22	15 ♏ 57 31	22 ♏ 21 31	12 40	28 38	08 45	28 53	13 19	11 07	19 13	14 02
18	01 45 35	28 13 59	28 ♏ 40 54	04 ♐ 55 51	12 19	29 ♈ 52	08 44	29 06	13 25	11 05	19 12	14 01
19	01 49 31	29 ♈ 12 35	11 ♐ 06 40	17 ♐ 13 41	12 03	01 ♉ 06	08 43	29 19	13 31	11 04	19 10	14 00
20	01 53 28	00 ♉ 11 08	23 ♐ 17 21	29 ♐ 18 08	11 52	02 21	08 D 43	29 32	13 37	11 02	19 09	13 59
21	01 57 24	01 09 40	05 ♑ 16 35	11 ♑ 13 15	11 46	03 35	08 44	29 45	13 43	11 01	19 07	13 58
22	02 01 21	02 08 10	17 ♑ 08 46	23 ♑ 03 44	11 D 45	04 49	08 46	29 ♉ 59	13 48	11 00	19 06	13 57
23	02 05 17	03 06 39	28 ♑ 58 48	04 ♒ 54 37	11 49	06 03	08 48	00 ♊ 12	13 54	10 58	19 04	13 56
24	02 09 14	04 05 06	10 ♒ 51 49	16 ♒ 51 01	11 59	07 17	08 51	00 25	13 59	10 57	19 03	13 55
25	02 13 11	05 03 31	22 ♒ 52 49	28 ♒ 57 47	12 13	08 31	08 54	00 39	14 05	10 56	19 01	13 54
26	02 17 07	06 01 55	05 ♓ 06 26	11 ♓ 19 14	12 31	09 45	08 59	00 52	14 10	10 55	18 59	13 53
27	02 21 04	07 00 17	17 ♓ 36 34	23 ♓ 58 45	12 54	10 59	09 04	01 05	14 16	10 53	18 58	13 52
28	02 25 00	07 58 37	00 ♈ 26 00	06 ♈ 58 26	13 22	12 14	09 10	01 19	14 21	10 52	18 56	13 51
29	02 28 57	08 56 56	13 ♈ 36 07	20 ♈ 18 55	13 53	13 28	09 16	01 32	14 26	10 51	18 55	13 51
30	02 32 53	09 ♉ 55 13	27 ♈ 06 41	03 ♉ 59 07	14 ♈ 29	14 ♉ 42	09 ♍ 23	01 ♊ 46	14 ♓ 32	10 ♍ 50	18 ♏ 53	13 ♍ 50

Date	Moon True ☊ ° '	Moon Mean ☊ ° '	Moon ☽ Latitude ° '
01	15 ♊ 55	17 ♊ 11	04 S 46
02	15 R 45	17 08	04 15
03	15 36	17 05	03 30
04	15 30	17 02	02 32
05	15 27	16 59	01 24
06	15 25	16 55	00 S 11
07	15 D 26	16 52	01 N 04
08	15 27	16 49	02 15
09	15 R 27	16 46	03 18
10	15 25	16 43	04 09
11	15 21	16 39	04 45
12	15 15	16 36	05 04
13	15 07	16 33	05 04
14	14 59	16 30	04 46
15	14 50	16 27	04 12
16	14 43	16 24	03 25
17	14 38	16 20	02 28
18	14 34	16 17	01 25
19	14 33	16 14	00 N 19
20	14 D 33	16 11	00 S 47
21	14 35	16 08	01 50
22	14 36	16 04	02 47
23	14 37	16 01	03 38
24	14 R 37	15 58	04 19
25	14 35	15 55	04 49
26	14 32	15 52	05 07
27	14 27	15 49	05 10
28	14 21	15 45	04 59
29	14 15	15 42	04 31
30	14 ♊ 10	15 ♊ 39	03 S 48

DECLINATIONS

Date	Sun ☉ ° '	Moon ☽ ° '	Mercury ☿ ° '	Venus ♀ ° '	Mars ♂ ° '	Jupiter ♃ ° '	Saturn ♄ ° '	Uranus ♅ ° '	Neptune ♆ ° '	Pluto ♇ ° '
01	04 N 34	02 S 17	11 N 57	02 N 15	10 N 32	18 N 29	08 S 42	07 N 58	15 S 53	19 N 37
02	04 57	03 N 19	11 46	02 45	10 35	18 32	08 40	07 58	15 53	19 37
03	05 20	08 53	11 31	03 15	10 38	18 35	08 38	07 59	15 53	19 37
04	05 43	14 06	11 13	03 45	10 41	18 38	08 35	08 00	15 52	19 38
05	06 06	18 40	10 51	04 14	10 43	18 41	08 33	08 01	15 52	19 38
06	06 29	22 14	10 27	04 44	10 44	18 44	08 30	08 01	15 52	19 38
07	06 51	24 29	10 01	05 14	10 46	18 47	08 28	08 02	15 51	19 39
08	07 14	25 11	09 32	05 43	10 47	18 50	08 26	08 03	15 51	19 39
09	07 36	24 16	09 02	06 13	10 48	18 53	08 23	08 04	15 50	19 39
10	07 58	21 48	08 31	06 42	10 48	18 56	08 21	08 04	15 50	19 40
11	08 20	18 01	08 00	07 11	10 48	18 59	08 19	08 05	15 50	19 40
12	08 42	13 14	07 28	07 40	10 48	19 02	08 17	08 06	15 49	19 40
13	09 04	07 47	06 57	08 09	10 48	19 05	08 14	08 06	15 49	19 41
14	09 26	02 N 01	06 27	08 38	10 47	19 08	08 12	08 07	15 48	19 41
15	09 47	03 S 46	05 59	09 06	10 46	19 11	08 10	08 07	15 48	19 41
16	10 09	09 16	05 31	09 34	10 45	19 14	08 08	08 08	15 48	19 41
17	10 30	14 15	05 06	10 02	10 43	19 17	08 06	08 09	15 47	19 41
18	10 51	18 30	04 43	10 30	10 41	19 20	08 04	08 09	15 47	19 42
19	11 12	21 48	04 22	10 57	10 39	19 23	08 01	08 10	15 46	19 42
20	11 32	24 04	04 04	11 25	10 37	19 26	07 59	08 10	15 46	19 42
21	11 53	25 10	03 48	11 52	10 34	19 29	07 57	08 11	15 45	19 42
22	12 13	25 07	03 35	12 18	10 31	19 32	07 55	08 11	15 45	19 42
23	12 33	23 55	03 25	12 45	10 28	19 35	07 53	08 12	15 45	19 42
24	12 53	21 39	03 17	13 11	10 24	19 38	07 51	08 12	15 44	19 42
25	13 13	18 26	03 12	13 37	10 21	19 40	07 49	08 13	15 44	19 42
26	13 32	14 23	03 09	14 02	10 17	19 43	07 47	08 13	15 43	19 42
27	13 51	09 39	03 09	14 27	10 13	19 46	07 45	08 13	15 43	19 43
28	14 10	04 S 24	03 11	14 52	10 08	19 49	07 44	08 14	15 42	19 43
29	14 29	01 N 12	03 15	15 16	10 04	19 52	07 42	08 14	15 42	19 43
30	14 N 48	06 N 54	03 N 22	15 N 40	09 N 59	19 N 55	07 S 40	08 N 14	15 S 42	19 N 43

ZODIAC SIGN ENTRIES

Date	h	m	Planets
01	02	19	☽ ♈
03	08	29	☽ ♉
05	12	55	☽ ♊
07	16	24	☽ ♋
09	19	24	☽ ♌
11	22	14	☽ ♍
14	01	38	☽ ♎
16	06	42	☽ ♏
18	14	31	♀ ♉
18	14	31	☽ ♐
20	07	26	☉ ♉
21	01	24	☽ ♑
22	14	32	♃ ♊
23	14	04	☽ ♒
26	02	02	☽ ♓
28	11	12	☽ ♈
30	17	04	☽ ♉

LATITUDES

Date	Mercury ☿ ° '	Venus ♀ ° '	Mars ♂ ° '	Jupiter ♃ ° '	Saturn ♄ ° '	Uranus ♅ ° '	Neptune ♆ ° '	Pluto ♇ ° '
01	03 N 19	01 S 21	03 N 19	00 S 41	01 S 38	00 N 49	01 N 48	14 N 38
04	03 07	01 19	03 11	00 41	01 38	00 49	01 49	14 38
07	02 39	01 15	03 03	00 40	01 38	00 49	01 49	14 37
10	02 00	01 12	02 56	00 40	01 39	00 48	01 49	14 37
13	01 13	01 08	02 48	00 39	01 39	00 48	01 49	14 36
16	00 N 23	01 03	02 40	00 39	01 40	00 48	01 49	14 35
19	00 S 26	00 58	02 32	00 39	01 40	00 48	01 49	14 34
22	01 09	00 52	02 24	00 38	01 41	00 48	01 49	14 34
25	01 46	00 47	02 17	00 38	01 41	00 48	01 49	14 33
28	02 17	00 40	02 10	00 37	01 42	00 48	01 49	14 32
31	02 S 40	00 S 34	02 N 03	00 S 37	01 S 42	00 N 48	01 N 49	14 N 31

DATA

Julian Date	2438852
Delta T	+36 seconds
Ayanamsa	23° 22' 01"
Synetic vernal point	05° ♓ 44' 59"
True obliquity of ecliptic	23° 26' 40"

MOON'S PHASES, APSIDES AND POSITIONS ☽

Date	h	m	Phase	Longitude ° '	Eclipse Indicator
02	00	21	●	12 ♈ 03	
09	00	40	☽	18 ♋ 57	
15	23	02	○	25 ♎ 45	
23	21	07	☾	03 ♒ 29	

Day	h	m		
09	10	46	Perigee	
23	01	25	Apogee	
01	21	49	ON	
08	10	24	Max dec	25° N 11'
14	20	18	OS	
21	22	47	Max dec	25° S 17'
29	06	55	ON	

LONGITUDES

Date	Chiron ⚷ ° '	Ceres ⚳ ° '	Pallas ⚴ ° '	Juno ⚵ ° '	Vesta ⚶ ° '	Black Moon Lilith ⚸ ° '
01	19 ♓ 43	05 ♓ 20	12 ♒ 50	04 ♒ 06	27 ♈ 37	19 ♑ 12
11	20 ♓ 17	08 ♓ 55	15 ♒ 29	06 ♒ 52	02 ♉ 06	20 ♑ 19
21	20 ♓ 48	12 ♓ 22	17 ♒ 51	09 ♒ 23	06 ♉ 35	21 ♑ 25
31	21 ♓ 17	15 ♓ 41	19 ♒ 52	11 ♒ 36	11 ♉ 03	22 ♑ 32

ASPECTARIAN

h m	Aspects
01 Thursday	
02 39	☽ # ☉
09 07	♄ ☍ ♅
10 44	☽ ⚼ ♆
12 08	☽ # ♀
13 35	☉ ⊼ ♅
14 21	☉ ⊻ ♄
19 07	☽ ☌ ♀
21 44	☽ ∠ ♃
22 04	☽ ⊼ ♂
23 31	☽ ⊼ ♅
23 41	☽ ⊻ ♄
02 Friday	
00 21	☽ ☌ ☉
03 05	☽ ± ♆
04 29	☽ ⊼ ♇
08 44	☽ ± ♂
09 18	☽ ∥ ♀
10 19	☽ ± ♅
10 36	☽ ⊥ ♄
13 52	☽ ⊼ ♆
14 13	☽ ⊥ ♃
15 15	☽ ± ♇
18 43	☽ ☌ ☿
19 29	☽ ∥ ☉
22 10	♀ ⊼ ♂
03 Saturday	
00 35	☽ ⚼ ♂
01 06	☽ ⊻ ♃
02 22	☽ ⚼ ♅
02 49	☽ ∠ ♄
03 57	♀ ∠ ♃
07 16	☽ ⚼ ♇
08 04	☽ ∥ ♅
10 53	☽ # ♄
12 31	☉ ± ♆
16 46	♀ ⊼ ♅
19 53	☽ ∥ ♂
23 11	☽ ∥ ☿
23 58	♀ ⊻ ♄
04 Sunday	
02 40	☽ △ ♂
04 46	☽ △ ♅
05 29	☽ ✶ ♄
05 59	☽ ⊻ ♀
06 53	☉ ⊼ ♇
09 37	☽ △ ♇
09 50	☽ ⊻ ☉
17 30	☽ ⊥ ♀
18 43	☽ ☍ ♆
20 47	☽ # ♆
21 05	☽ ⊥ ☉
21 30	☽ ⊻ ☿
05 Monday	
02 34	☽ Q ♄
06 30	☽ ☌ ♃
06 49	♀ ± ♆
07 28	☽ ⊥ ☿
10 40	☽ ∠ ♀
12 04	☽ ∥ ♃
13 53	☽ ∠ ☉
17 48	☽ ∥ ♇
20 23	☿ ∥ ♂
21 27	♀ ⊼ ♇
22 18	☽ ∠ ☿
22 20	☿ ⊥ ♃
23 33	☉ ± ♂
06 Tuesday	
04 43	☿ ± ♇
06 00	☽ □ ♂
08 38	☽ □ ♅
09 53	☽ □ ♄
13 27	☽ □ ♇
15 02	☽ ✶ ♀
17 40	☽ ✶ ☉
22 24	☽ ⊼ ♆
22 49	☽ ✶ ☿
07 Wednesday	
02 38	♀ ± ♂
07 02	☿ ⊼ ♆
08 39	☽ ± ♆
10 50	☉ ± ♅
10 51	☽ ⊻ ♃
12 29	☽ Q ♂
13 33	☽ Q ♀
15 23	☽ Q ♅
15 44	☽ Q ☉
18 17	☽ Q ☿
20 11	☽ Q ♇
21 15	☽ ⊥ ♃
23 59	☽ ⚼ ♆
08 Thursday	
08 42	☽ ✶ ♂
09 49	♀ ± ♅
10 46	☉ ⊥ ♄
11 47	☽ ✶ ♅
12 48	☽ ∠ ♃
13 26	☉ ☌ ☿
13 31	☽ △ ♄
16 05	☿ ⊥ ♄
16 35	☽ ✶ ♇
23 09	☽ □ ♀
23 20	☽ □ ☿
09 Friday	
00 33	☿ ☌ ♀
00 40	☽ □ ☉
01 27	☽ △ ♆
07 08	♀ ⊥ ♄
09 56	☽ ∠ ♂
11 45	☉ ⊼ ♆
13 11	☽ ∠ ♅
14 39	☽ ✶ ♃
15 11	☽ ⚼ ♄
18 00	☽ ∠ ♇
10 Saturday	
00 57	♀ ⊼ ♆
01 03	☽ ⊥ ♂
01 42	☿ ± ♅
04 24	☽ ⊥ ♅
06 14	☉ ± ♇
06 32	☽ ± ♄
09 13	☽ ⊥ ♇
11 07	☽ ⊻ ♂
11 17	☽ Q ♃
14 33	☽ ⊻ ♅
15 41	♀ ± ♇
16 47	☽ ⊼ ♄
18 35	☉ ∥ ♅
19 22	☽ ⊻ ♇
20 16	☿ # ♄
23 33	☽ △ ☿
11 Sunday	
02 25	☽ ∥ ♇
02 42	☉ ∥ ☿
04 12	☽ □ ♆
06 34	☽ ∥ ♃
06 52	☽ △ ♀
07 17	☽ △ ☉
08 07	☿ ∥ ♅
10 27	☉ # ♄
18 15	☽ □ ♃
18 41	☉ ⊥ ♃
21 07	♀ ⊥ ♃
23 31	☽ # ♆
23 43	☽ ⚼ ☿
12 Monday	
04 21	☉ ☌ ♀
07 15	☿ ∥ ♀
10 40	☽ ⚼ ☉
10 44	☽ Q ♆
10 48	☽ ⚼ ♀
13 39	☽ ☌ ♂
14 18	☽ ± ☿
17 23	☽ ☌ ♅
20 08	☽ ☍ ♄
22 17	☽ ☌ ♇
22 59	☽ ∥ ♂
13 Tuesday	
00 04	☽ ⊼ ☿
03 12	☽ ± ☉
03 40	☽ ± ♀
05 25	☿ ± ♂
06 51	☽ ∥ ☉
07 13	☽ ✶ ♆
09 37	♀ ∥ ♅
10 03	☽ # ♄
10 34	☽ ∥ ♀
10 39	☽ ∥ ♅
14 18	☽ ⊼ ☉
14 59	☽ ⊼ ♀
15 50	☽ ∥ ☿
16 12	♀ # ♄
17 22	♀ ⚼ ♂
22 22	☽ △ ♃
14 Wednesday	
02 26	☉ ⚼ ♂
09 06	☽ ∠ ♆
16 02	☿ ⊼ ♇
17 08	☽ ⊻ ♂
21 09	☽ ⊻ ♅
15 Thursday	
00 29	☽ ⊼ ♄
00 51	☽ ⊥ ♆
00 58	☽ ⚼ ♃
01 50	☽ ☍ ☿
02 15	☽ ⊻ ♇
03 38	☽ ⊥ ♂
07 43	☽ ⊥ ♅
11 11	☽ ± ♄
11 27	☽ ⊻ ♆
12 07	♀ ⚼ ♅
12 52	☽ ⊥ ♇
17 01	☿ ∠ ♃
17 12	☽ ± ♃
19 35	☽ ∠ ♂
20 42	☽ # ☿
23 02	☽ ☍ ☉
23 45	☽ ∠ ♅
16 Friday	
00 58	☽ ☍ ♀
03 25	☽ ⚼ ♄
04 09	☽ ⊼ ♃
04 26	☿ ± ♆
04 59	☽ ∠ ♇
06 52	☽ # ♅
06 54	☽ ∥ ♄
07 03	☿ ⊻ ♄
08 39	☉ ⚼ ♅
10 22	♄ # ♅
13 29	☽ # ♀
16 17	☽ # ☉
18 45	☽ # ♂
22 42	☽ ✶ ♂
17 Saturday	
03 02	☽ ✶ ♅
05 22	♀ ∠ ♄
06 05	☽ ⊼ ☿
07 03	☽ △ ♄
08 25	☽ ✶ ♇
16 55	☽ ± ☿
17 51	♀ ⊻ ♃
18 05	☽ ☌ ♆
19 33	♀ ⚼ ♇
20 08	☽ ∥ ♆
20 58	☽ Q ♂
18 Sunday	
01 24	☽ Q ♅
01 46	☉ ∥ ♂
06 55	☽ Q ♇
09 28	☽ ⚼ ☿
11 04	☽ ⊼ ☉
12 49	☽ ☍ ♃
14 31	☽ ⊼ ♀
17 04	☉ ∠ ♄
17 32	☽ # ♃
19 54	☽ # ♇
21 04	♀ ∥ ♂
23 34	☽ ± ☉
19 Monday	
03 21	☽ ± ♀
06 47	☉ ⚼ ♇
07 21	☽ □ ♂
11 54	☽ □ ♅
13 48	☽ △ ☿
15 25	☉ ⊻ ♃
16 45	☽ □ ♄
17 38	☽ □ ♇
18 35	☽ ⚼ ☉
21 56	♂ St D
22 54	☽ ⚼ ♀
20 Tuesday	
03 48	☽ ⊻ ♆
15 41	☽ ⊥ ♆
21 Wednesday	
00 42	☽ ⊼ ♃
03 00	☽ △ ☉
04 46	☽ Q ♄
08 11	☽ △ ♀
09 41	☽ ∠ ♆
12 59	☽ ± ♃
16 08	☉ ∥ ♀
18 59	☽ △ ♂
23 34	☽ △ ♅
22 Thursday	
01 04	☽ □ ☿
04 00	☿ St D
05 10	☽ ✶ ♄
05 31	☽ △ ♇
07 31	☽ ⚼ ♃
15 57	☽ ✶ ♆
23 Friday	
01 27	☽ ⚼ ♂
05 55	☽ ⚼ ♅
11 50	☽ ∠ ♄
11 54	☽ ⚼ ♇
13 44	☽ Q ☿
14 31	☽ △ ♃
16 13	☽ Q ♆
18 58	♄ ☍ ♇
19 46	☽ ± ♂
21 07	☽ □ ☉
24 Saturday	
00 06	☽ ± ♅
03 58	☽ □ ♀
06 04	☽ ± ♇
06 10	☽ ⊥ ♄
07 55	☽ ⊼ ♂
12 10	☽ ⊼ ♅
14 16	☽ ✶ ☿
18 07	☽ ⊼ ♇
18 20	☽ ⊻ ♄
25 Sunday	
03 18	☽ # ♇
03 39	☽ # ♃
04 20	☽ □ ♆
12 23	☽ Q ☉
19 57	♀ △ ♂
20 01	☽ Q ♀
20 45	☽ ∠ ☿
26 Monday	
03 35	☽ □ ♃
04 34	☽ ∥ ♆
04 38	♃ ∥ ♇
13 44	☽ # ♀
13 57	☽ ✶ ☉
14 49	☽ ⊥ ☿
16 16	☽ # ☉
19 33	☽ ☍ ♂
21 59	☽ ✶ ♀
23 12	☽ ☍ ♅
27 Tuesday	
02 44	☽ ⊻ ☿
04 54	☽ ☍ ♇
05 35	☽ ☌ ♄
09 17	☽ # ♂
10 04	♀ △ ♅
14 34	☽ △ ♆
14 51	☽ Q ♃
15 10	☿ ± ♆
18 44	☽ # ♅
20 57	☽ ∥ ♄
20 59	☽ ∠ ☉
28 Wednesday	
05 26	☽ ∠ ♀
13 39	☽ ✶ ♃
15 05	☽ ⊥ ☉
17 15	☽ # ☿
18 26	☽ ⚼ ♆
23 44	☽ ⊥ ♀
29 Thursday	
02 55	☽ ⊻ ☉
04 06	☽ ⊼ ♂
07 03	☽ ⊼ ♅
10 04	☿ ⊼ ♇
10 46	☽ ± ♆
11 43	☽ ⊻ ♀
12 26	☽ ⊼ ♇
12 33	☽ ☌ ☿
13 31	☽ ⊻ ♄
15 01	☽ ± ♂
17 22	☽ ∠ ♃
17 50	☽ ± ♅
19 25	♀ △ ♇
20 48	☽ ∥ ☿
20 50	☉ △ ♂
21 29	☽ ⊼ ♆
23 09	☽ ± ♇
30 Friday	
00 18	☽ ⊥ ♄
03 56	☿ ⊻ ♀
07 09	☽ ⚼ ♂
08 30	♀ ✶ ♄
09 36	☽ ⊥ ♃
09 46	☽ ⚼ ♅
13 06	♀ # ♆
13 46	☿ ⊻ ♄
15 01	☽ ⚼ ♇
15 15	☽ # ♄
16 16	☽ ∠ ♄
17 44	☽ ∥ ♅
20 16	☽ ⊻ ♃

All ephemeris data is given at 12.00 UT and the Moon's longitude is additionally given for 24.00 UT

MAY 1965

Date	Sidereal time h m s	Sun ☉ ° ' "	Moon ☽ ° ' "	Moon ☽ 24.00 ° ' "	Mercury ☿ ° '	Venus ♀ ° '	Mars ♂ ° '	Jupiter ♃ ° '	Saturn ♄ ° '	Uranus ♅ ° '	Neptune ♆ ° '	Pluto ♇ ° '
01	02 36 50	10 ♉ 53 28	10 ♉ 55 49	17 ♉ 56 21	15 ♈ 09	15 ♉ 56	09 ♍ 31	01 ♊ 59	14 ♓ 37	10 ♍ 49	18 ♏ 51	13 ♍ 49
02	02 40 46	11 51 42	25 ♉ 00 10	02 ♊ 06 45	15 52	17 10	09 39	02 13	14 42	10 R 49	18 R 50	13 R 48
03	02 44 43	12 49 54	09 ♊ 15 28	16 ♊ 25 45	16 39	18 24	09 48	02 27	14 47	10 48	18 48	13 48
04	02 48 40	13 48 05	23 ♊ 37 01	00 ♋ 48 43	17 29	19 38	09 57	02 40	14 52	10 47	18 47	13 47
05	02 52 36	14 46 13	08 ♋ 00 22	15 ♋ 11 29	18 23	20 52	10 07	02 54	14 57	10 46	18 45	13 47
06	02 56 33	15 44 19	22 ♋ 21 40	29 ♋ 30 33	19 20	22 06	10 18	03 08	15 01	10 46	18 43	13 46
07	03 00 29	16 42 24	06 ♌ 37 51	13 ♌ 43 18	20 19	23 19	10 29	03 22	15 06	10 45	18 42	13 46
08	03 04 26	17 40 26	20 ♌ 46 39	27 ♌ 47 43	21 22	24 33	10 41	03 35	15 11	10 44	18 40	13 45
09	03 08 22	18 38 27	04 ♍ 46 20	11 ♍ 42 20	22 27	25 47	10 54	03 49	15 15	10 44	18 38	13 45
10	03 12 19	19 36 25	18 ♍ 35 33	25 ♍ 25 52	23 36	27 01	11 07	04 03	15 20	10 43	18 37	13 44
11	03 16 15	20 34 22	02 ♎ 13 09	08 ♎ 57 16	24 47	28 15	11 20	04 17	15 24	10 43	18 35	13 44
12	03 20 12	21 32 17	15 ♎ 38 06	22 ♎ 15 32	26 00	29 ♉ 29	11 34	04 31	15 29	10 43	18 34	13 43
13	03 24 09	22 30 10	28 ♎ 49 31	05 ♏ 19 57	27 16	00 ♊ 43	11 48	04 45	15 33	10 42	18 32	13 43
14	03 28 05	23 28 02	11 ♏ 46 48	18 ♏ 10 04	28 35	01 56	12 03	04 59	15 37	10 42	18 30	13 43
15	03 32 02	24 25 52	24 ♏ 29 46	00 ♐ 45 59	29 ♈ 55	03 10	12 19	05 12	15 41	10 42	18 29	13 42
16	03 35 58	25 23 40	06 ♐ 58 50	13 ♐ 08 27	01 ♉ 19	04 24	12 35	05 26	15 45	10 42	18 27	13 42
17	03 39 55	26 21 28	19 ♐ 15 04	25 ♐ 18 55	02 44	05 38	12 51	05 40	15 49	10 42	18 25	13 42
18	03 43 51	27 19 14	01 ♑ 20 19	07 ♑ 19 36	04 12	06 52	13 08	05 54	15 53	10 42	18 24	13 41
19	03 47 48	28 16 59	13 ♑ 17 09	19 ♑ 13 24	05 43	08 05	13 26	06 08	15 57	10 D 42	18 22	13 41
20	03 51 44	29 ♉ 14 42	25 ♑ 08 49	01 ♒ 03 53	07 15	09 19	13 43	06 22	16 00	10 42	18 21	13 41
21	03 55 41	00 ♊ 12 25	06 ♒ 59 09	12 ♒ 55 09	08 50	10 33	14 01	06 36	16 04	10 42	18 19	13 41
22	03 59 38	01 10 06	18 ♒ 52 28	24 ♒ 51 40	10 27	11 46	14 20	06 50	16 08	10 42	18 17	13 41
23	04 03 34	02 07 46	00 ♓ 53 19	06 ♓ 58 02	12 06	13 00	14 39	07 04	16 11	10 42	18 16	13 41
24	04 07 31	03 05 26	13 ♓ 06 21	19 ♓ 18 49	13 48	14 14	14 59	07 18	16 14	10 43	18 14	13 41
25	04 11 27	04 03 04	25 ♓ 35 55	01 ♈ 58 08	15 32	15 27	15 18	07 32	16 18	10 43	18 13	13 D 41
26	04 15 24	05 00 41	08 ♈ 25 49	14 ♈ 59 17	17 18	16 41	15 39	07 46	16 21	10 43	18 11	13 41
27	04 19 20	05 58 18	21 ♈ 38 44	28 ♈ 24 17	19 06	17 55	15 59	08 00	16 24	10 44	18 10	13 41
28	04 23 17	06 55 53	05 ♉ 15 55	12 ♉ 13 27	20 57	19 08	16 20	08 14	16 27	10 44	18 08	13 41
29	04 27 13	07 53 27	19 ♉ 16 38	26 ♉ 25 01	22 49	20 22	16 42	08 28	16 30	10 45	18 07	13 41
30	04 31 10	08 51 01	03 ♊ 38 03	10 ♊ 55 04	24 44	21 35	17 03	08 42	16 33	10 45	18 05	13 41
31	04 35 07	09 ♊ 48 34	18 ♊ 15 18	25 ♊ 37 55	26 ♉ 41	22 ♊ 49	17 ♍ 25	08 ♊ 56	16 ♓ 35	10 ♍ 46	18 ♏ 04	13 ♍ 41

Date	Moon True ☊ ° '	Moon Mean ☊ ° '	Moon Latitude ° '
01	14 ♊ 05	15 ♊ 36	02 S 51
02	14 R 03	15 33	01 42
03	14 01	15 30	00 S 26
04	14 D 02	15 26	00 N 52
05	14 03	15 23	02 08
06	14 04	15 20	03 15
07	14 05	15 17	04 10
08	14 R 05	15 14	04 49
09	14 05	15 10	05 10
10	14 03	15 07	05 13
11	14 00	15 04	04 58
12	13 57	15 01	04 28
13	13 53	14 58	03 43
14	13 51	14 55	02 47
15	13 49	14 51	01 44
16	13 48	14 48	00 N 38
17	13 D 48	14 45	00 S 30
18	13 48	14 42	01 35
19	13 49	14 39	02 36
20	13 51	14 36	03 29
21	13 52	14 32	04 14
22	13 53	14 29	04 47
23	13 53	14 26	05 09
24	13 R 53	14 23	05 17
25	13 52	14 20	05 10
26	13 51	14 16	04 48
27	13 50	14 13	04 11
28	13 50	14 10	03 18
29	13 49	14 07	02 12
30	13 49	14 04	00 56
31	13 ♊ 49	14 ♊ 01	00 N 25

DECLINATIONS

Date	Sun ☉ ° '	Moon ☽ ° '	Mercury ☿ ° '	Venus ♀ ° '	Mars ♂ ° '	Jupiter ♃ ° '	Saturn ♄ ° '	Uranus ♅ ° '	Neptune ♆ ° '	Pluto ♇ ° '
01	15 N 06	12 N 24	03 N 31	16 N 04	09 N 54	19 N 58	07 S 38	08 N 15	15 S 41	19 N 43
02	15 24	17 22	03 42	16 27	09 49	20 00	07 36	08 15	15 41	19 43
03	15 42	21 25	03 55	16 50	09 43	20 03	07 35	08 15	15 40	19 43
04	15 59	24 10	04 10	17 12	09 37	20 06	07 33	08 16	15 40	19 42
05	16 16	25 20	04 27	17 34	09 31	20 09	07 31	08 16	15 39	19 42
06	16 33	24 48	04 45	17 56	09 25	20 12	07 29	08 16	15 39	19 42
07	16 50	22 39	05 05	18 17	09 19	20 14	07 28	08 16	15 38	19 42
08	17 06	19 07	05 27	18 37	09 12	20 17	07 26	08 16	15 38	19 42
09	17 23	14 34	05 50	18 57	09 06	20 20	07 25	08 17	15 38	19 42
10	17 38	09 19	06 15	19 17	08 59	20 22	07 23	08 17	15 37	19 42
11	17 54	03 N 41	06 41	19 36	08 52	20 25	07 22	08 17	15 37	19 42
12	18 09	02 S 02	07 09	19 54	08 44	20 28	07 20	08 17	15 36	19 42
13	18 24	07 35	07 37	20 12	08 37	20 30	07 19	08 17	15 36	19 41
14	18 39	12 43	08 07	20 30	08 29	20 33	07 17	08 17	15 35	19 41
15	18 53	17 13	08 38	20 46	08 22	20 36	07 16	08 17	15 35	19 41
16	19 07	20 52	09 10	21 03	08 14	20 38	07 14	08 17	15 34	19 41
17	19 21	23 31	09 43	21 18	08 05	20 41	07 13	08 17	15 34	19 41
18	19 34	25 02	10 17	21 33	07 57	20 43	07 12	08 17	15 34	19 40
19	19 47	25 24	10 52	21 48	07 49	20 46	07 10	08 17	15 33	19 40
20	20 00	24 32	11 28	22 01	07 40	20 48	07 09	08 17	15 33	19 40
21	20 12	22 37	12 04	22 15	07 31	20 51	07 08	08 17	15 32	19 39
22	20 24	19 43	12 40	22 28	07 22	20 53	07 07	08 17	15 32	19 39
23	20 36	15 58	13 18	22 40	07 13	20 56	07 06	08 17	15 31	19 39
24	20 47	11 31	13 55	22 51	07 04	20 58	07 05	08 16	15 31	19 38
25	20 58	06 30	14 33	23 02	06 55	21 00	07 04	08 16	15 31	19 38
26	21 08	01 S 04	15 12	23 12	06 45	21 03	07 03	08 16	15 30	19 38
27	21 18	04 N 34	15 50	23 21	06 36	21 05	07 02	08 16	15 30	19 37
28	21 28	10 10	16 28	23 30	06 26	21 07	07 01	08 16	15 29	19 37
29	21 38	15 26	17 07	23 38	06 16	21 10	07 00	08 15	15 29	19 37
30	21 47	19 58	17 45	23 45	06 06	21 12	06 59	08 15	15 29	19 36
31	21 N 56	23 N 20	18 N 22	23 N 52	05 N 56	21 N 14	06 S 58	08 N 15	15 S 28	19 N 36

ZODIAC SIGN ENTRIES

Date	h	m	Planets
02	20	26	☽ ♊
04	22	39	☽ ♋
07	00	50	☽ ♌
09	03	47	☽ ♍
11	08	04	☽ ♎
12	22	08	♀ ♊
13	14	10	☽ ♏
15	13	19	☿ ♉
15	22	32	☽ ♐
18	09	20	☽ ♑
20	21	50	☽ ♒
21	06	50	☉ ♊
23	10	14	☽ ♓
25	20	19	☽ ♈
28	02	48	☽ ♉
30	05	58	☽ ♊

LATITUDES

Date	Mercury ☿ ° '	Venus ♀ ° '	Mars ♂ ° '	Jupiter ♃ ° '	Saturn ♄ ° '	Uranus ♅ ° '	Neptune ♆ ° '	Pluto ♇ ° '
01	02 S 40	00 S 34	02 N 03	00 S 37	01 S 42	00 N 48	01 N 49	14 N 31
04	02 55	00 27	01 56	00 37	01 43	00 48	01 49	14 30
07	03 05	00 20	01 49	00 36	01 44	00 47	01 49	14 29
10	03 08	00 13	01 42	00 36	01 44	00 47	01 49	14 28
13	03 05	00 S 06	01 36	00 36	01 45	00 47	01 49	14 27
16	02 57	00 N 01	01 30	00 35	01 45	00 47	01 49	14 26
19	02 43	00 08	01 24	00 35	01 46	00 47	01 49	14 25
22	02 24	00 16	01 18	00 35	01 47	00 47	01 49	14 23
25	02 01	00 23	01 13	00 34	01 47	00 47	01 49	14 22
28	01 35	00 30	01 08	00 34	01 48	00 47	01 49	14 21
31	01 S 05	00 N 37	01 N 03	00 S 34	01 S 49	00 N 46	01 N 49	14 N 20

DATA

Julian Date	2438882
Delta T	+36 seconds
Ayanamsa	23° 22' 04"
Synetic vernal point	05° ♓ 44' 56"
True obliquity of ecliptic	23° 26' 40"

MOON'S PHASES, APSIDES AND POSITIONS ☽

Date	h	m	Phase	Longitude ° '	Eclipse Indicator
01	11	56	●	10 ♉ 53	
08	06	20	☽	17 ♌ 27	
15	11	52	○	24 ♏ 26	
23	14	40	☾	02 ♓ 14	
30	21	13	●	09 ♊ 13	Total

Day	h	m	
05	00	56	Perigee
20	19	27	Apogee

Day	h	m		
05	16	26	Max dec	25° N 21'
12	03	25	0S	
19	06	51	Max dec	25° S 24'
26	16	37	0N	

LONGITUDES

Date	Chiron ⚷ ° '	Ceres ⚳ ° '	Pallas ⚴ ° '	Juno ⚵ ° '	Vesta ⚶ ° '	Black Moon Lilith ⚸ ° '
01	21 ♓ 17	15 ♓ 41	19 ♒ 52	11 ♒ 36	11 ♉ 03	22 ♑ 32
11	21 ♓ 41	18 ♓ 49	21 ♒ 31	13 ♒ 28	15 ♉ 30	23 ♑ 39
21	22 ♓ 01	21 ♓ 45	22 ♒ 43	14 ♒ 55	19 ♉ 55	24 ♑ 46
31	22 ♓ 17	24 ♓ 26	23 ♒ 26	15 ♒ 53	24 ♉ 18	25 ♑ 52

All ephemeris data is given at 12.00 UT and the Moon's longitude is additionally given for 24.00 UT

ASPECTARIAN

h m	Aspects
01 Saturday	
01 04	☽ ∥ ♂
09 32	☽ △ ♂
10 22	☉ △ ♅
11 49	☽ △ ♅
11 56	☽ ☌ ☉
16 57	☽ △ ♇
18 21	☽ ⚹ ♄
19 37	☽ ⊻ ☿
21 23	☽ ☌ ♀
02 Sunday	
01 24	☽ ∥ ☉
01 32	☽ ☍ ♆
03 07	☿ ± ♂
03 26	☽ # ♆
06 23	☽ ⊥ ☿
06 53	☽ ∥ ♀
14 53	☽ Q ♄
22 28	☽ ∠ ☿
03 Monday	
00 23	☽ ☌ ♃
01 03	☽ ∥ ♇
03 02	☽ ∥ ♃
10 09	☉ # ♆
12 55	☽ □ ♂
14 34	☽ □ ♅
16 14	☿ ± ♅
18 25	☽ ⊻ ☉
19 36	☽ □ ♇
19 48	♀ ☍ ♆
21 18	☽ □ ♄
04 Tuesday	
01 07	☽ ⚹ ☿
03 56	☽ ⊼ ♆
04 43	☽ ⊻ ♀
05 10	☽ ⊥ ☉
11 40	☉ △ ♇
13 56	☽ ± ♆
15 40	☽ ⊥ ♀
18 56	☿ ∠ ♃
19 19	☽ Q ♂
20 37	☽ Q ♅
21 16	☽ ∠ ☉
22 26	☽ Q ☿
05 Wednesday	
01 37	☽ Q ♇
03 21	☽ ⊻ ♃
04 55	☽ Ꝑ ♆
08 05	☽ ∠ ♀
13 31	☽ ⊥ ♃
15 35	☽ ⚹ ♂
16 37	☽ ⚹ ♅
16 40	☉ ⚹ ♄
21 16	☿ ⊼ ♆
21 38	☽ ⚹ ♇
23 39	☽ △ ♄
06 Thursday	
00 07	☽ ⚹ ☉
04 48	☽ ∠ ♃
05 55	☽ △ ♆
06 33	☽ □ ☿
11 30	☽ ⚹ ♀
17 00	☽ ∠ ♂
17 42	☽ ∠ ♅
21 41	☽ Q ☉
22 43	☿ ± ♇
22 45	☽ ∠ ♇
07 Friday	
00 56	☽ Ꝑ ♄
06 24	☽ ⚹ ♃
08 20	☽ ⊥ ♂
08 50	☽ ⊥ ♅
09 35	☽ Q ♀
13 54	☽ ⊥ ♇
16 12	☽ ± ♄
18 37	☽ ⊻ ♂
18 58	☽ ⊻ ♅
08 Saturday	
00 03	☽ ⊻ ♇
02 25	☽ ⊼ ♄
03 02	☽ Q ♃
05 01	☽ ∥ ♃
06 20	☽ □ ☉
07 27	☿ ⊥ ♄
08 25	☽ □ ♆
08 33	☽ ∥ ♇
13 05	☽ △ ☿
14 41	☽ ∥ ♀
18 04	♂ ☌ ♅
19 04	☽ □ ♀
22 33	☽ ∥ ☉
09 Sunday	
06 47	☽ # ♆
10 20	☽ □ ♃
12 01	☉ ☍ ♆
15 13	☽ Q ♆
17 03	☽ Ꝑ ☿
22 18	☽ ☌ ♅
22 45	☽ ☌ ♂
10 Monday	
03 32	☽ ☌ ♇
06 16	☽ ☍ ♄
10 06	☽ ± ☿
12 02	☽ ⚹ ♆
13 28	☽ ∥ ♂
13 55	☽ △ ☉
16 29	☽ ∥ ♅
18 26	♀ Q ♄
20 22	☽ # ♄
21 35	☽ ⊼ ☿
11 Tuesday	
00 13	☽ ∥ ☿
04 16	☽ △ ♀
14 25	☽ ∠ ♆
15 44	☽ △ ♃
18 25	☽ Ꝑ ☉
19 31	♀ ∥ ♇
12 Wednesday	
03 09	☽ ⊻ ♅
04 33	☽ ⊻ ♂
06 25	☿ Ꝑ ♅
06 28	☽ ⊥ ♆
08 33	☽ ⊻ ♇
09 42	☽ Ꝑ ♀
11 43	☽ ⊼ ♄
11 49	☽ ± ☉
13 57	☽ ⊥ ♅
15 33	☽ ⊥ ♂
17 16	☽ ⊻ ♆
19 08	☽ Ꝑ ♃
19 23	☽ ⊥ ♇
21 04	☿ # ♄
22 38	☽ ± ♄
23 32	☽ ⊼ ☉
13 Thursday	
01 20	☿ Ꝑ ♂
03 41	☽ ± ♀
06 17	☽ ∠ ♅
08 14	☽ ∠ ♂
08 50	☽ ☍ ☿
10 47	☽ ∥ ♄
11 48	☽ ∠ ♇
11 51	☽ ± ♃
12 12	☽ # ☿
15 08	☽ # ♅
15 11	☽ Ꝑ ♄
15 50	☽ ⊼ ♀
16 32	☽ # ♂
23 06	☽ ⊼ ♃
14 Friday	
09 59	☽ ⚹ ♅
12 32	☽ ⚹ ♂
14 23	☿ Ꝑ ♇
15 37	☽ ⚹ ♇
17 32	♀ ∥ ♃
19 14	☽ △ ♄
19 34	☿ ∥ ♅
20 40	☿ ⊥ ♃
15 Saturday	
00 37	☽ ☌ ♆
01 42	☿ ∥ ♂
02 51	☽ ∥ ♆
08 35	☽ Q ♅
11 39	☽ Q ♂
11 52	☽ ☍ ☉
14 18	☽ Q ♇
22 56	☽ # ☉
23 40	☽ ⊼ ☿
16 Sunday	
01 38	♂ ∥ ♅
01 54	☿ ∠ ♄
03 31	☽ # ♇
06 27	☽ ☍ ♀
08 57	☽ ☍ ♃
10 16	☽ # ♃
12 44	☽ ± ☿
13 31	☽ # ♀
19 14	☽ □ ♅
23 09	☽ □ ♂
17 Monday	
01 05	☽ □ ♇
05 13	☽ □ ♄
08 37	☽ Ꝑ ☿
10 22	☽ ⊻ ♆
13 01	♀ ☌ ♃
22 13	☽ ⊥ ♆
18 Tuesday	
03 17	☽ ⊼ ☉
14 33	♅ St D
16 07	☽ ∠ ♆
16 19	☽ ± ☉
17 07	☽ Q ♄
18 33	☽ △ ☿
21 19	☽ ⊼ ♃
23 22	☉ ∥ ♇
19 Wednesday	
00 20	☽ ⊼ ♀
02 59	☉ Q ♄
06 47	☽ △ ♅
09 38	☽ ± ♃
12 00	☽ Ꝑ ☉
12 17	☽ △ ♂
12 49	☽ △ ♇
13 48	☽ ± ♀
17 24	☽ ⚹ ♄
19 53	☿ ⊻ ♃
22 15	☽ ⚹ ♆
20 Thursday	
04 12	☽ Ꝑ ♃
09 07	♂ ☌ ♇
10 07	☽ Ꝑ ♀
13 07	☽ Ꝑ ♅
19 10	☽ Ꝑ ♇
19 26	☽ Ꝑ ♂
21 03	☽ △ ☉
22 31	☽ Q ♆
23 57	☽ ∠ ♄
21 Friday	
07 22	☽ ± ♅
11 13	☽ △ ♃
13 25	☽ ± ♇
14 10	☽ ± ♂
15 01	♀ □ ♅
15 13	☽ # ♀
16 19	☽ □ ☿
18 16	☽ ⊥ ♄
19 31	☽ ⊼ ♅
20 02	☽ △ ♀
22 Saturday	
01 32	☽ ⊼ ♇
02 37	☽ ⊼ ♂
03 18	☽ # ♃
06 26	☽ ⊻ ♄
07 15	☽ # ☉
10 50	☽ □ ♆
12 27	☽ # ♇
15 40	☿ △ ♅
23 Sunday	
10 12	☽ Q ☿
14 33	☽ ∥ ♆
14 40	☽ □ ☉
24 Monday	
00 26	☽ □ ♃
00 59	☽ # ☿
01 16	♀ □ ♇
07 20	☽ ☍ ♅
10 19	☿ △ ♇
10 37	♂ # ♄
13 07	☽ ☍ ♇
13 34	☽ ⚹ ☿
14 25	☽ □ ♀
15 44	☽ ☍ ♂
18 06	☽ ☌ ♄
21 55	☽ △ ♆
25 Tuesday	
03 45	☽ # ♅
04 41	☽ Q ☉
05 21	♇ St D
07 58	♀ □ ♂
08 11	☿ △ ♂
08 33	☿ ⊻ ♀
09 23	☽ ∥ ♄
10 01	☽ # ♂
11 53	☽ Q ♃
19 31	☉ ∥ ♃
22 45	☿ ⚹ ♄
22 47	☽ ∠ ☿
26 Wednesday	
02 18	☽ Ꝑ ♆
04 20	☽ Q ♀
05 07	♀ □ ♄
05 10	☽ ⚹ ☉
10 46	☽ ⚹ ♃
16 13	☽ ⊼ ♅
18 06	☽ ⊥ ☿
18 53	☽ ± ♆
21 37	☽ ⊼ ♇
23 26	☿ # ♆
23 42	☿ ☍ ♆
27 Thursday	
01 32	☽ ⊼ ♂
02 31	☽ ⊻ ♄
03 09	☽ ± ♅
04 36	☽ ⚹ ♀
05 45	☽ ⊼ ♆
06 43	☽ ⊻ ☿
08 29	☽ ± ♇
10 42	☽ ∠ ☉
12 38	☽ ± ♂
13 21	☽ ⊥ ♄
14 28	☽ ∠ ♃
16 48	♀ ⊼ ♆
19 17	☽ Ꝑ ♅
20 23	☽ ∥ ♂
22 26	☽ # ♄
28 Friday	
00 29	☽ Ꝑ ♇
03 45	☽ ∥ ♅
03 52	☽ ⊥ ☉
04 58	☽ Ꝑ ♂
05 19	☽ ∠ ♄
06 38	☽ ⊥ ♃
09 51	☽ ∠ ♀
15 06	☽ ⊻ ☉
17 14	☽ ⊻ ♃
20 50	♂ ☍ ♄
21 27	☽ △ ♅
29 Saturday	
02 30	☽ △ ♇
02 52	☽ ⊥ ♀
07 16	☽ ⚹ ♄
07 30	☽ △ ♂
10 02	☽ ☍ ♆
12 15	☽ # ♆
14 01	☽ ⊻ ♀
18 54	☽ ☌ ☿
21 39	☽ ∥ ☿
30 Sunday	
03 31	☽ Q ♄
07 12	☉ ☌ ♃
09 53	☽ ∥ ♇
19 49	☽ ∥ ♃
20 30	☽ ☌ ♃
21 13	☽ ☌ ☉
23 45	☽ □ ♅
31 Monday	
00 15	☽ ∥ ☉
04 32	☽ □ ♇
09 17	☽ □ ♄
10 36	☽ □ ♂
11 41	☽ ⊼ ♆
17 23	☽ ∥ ♀
20 06	☽ ☌ ♀
21 26	☽ ± ♆

JUNE 1965

LONGITUDES

Date	Sidereal time h m s	Sun ☉ ° ' "	Moon ☽ ° ' "	Moon ☽ 24.00 ° ' "	Mercury ☿ ° '	Venus ♀ ° '	Mars ♂ ° '	Jupiter ♃ ° '	Saturn ♄ ° '	Uranus ♅ ° '	Neptune ♆ ° '	Pluto ♇ ° '
01	04 39 03	10 ♊ 46 05	03 ♋ 01 59	10 ♋ 26 37	28 ♉ 41	24 ♊ 03	17 ♍ 48	09 ♊ 10	16 ♓ 38	10 ♍ 47	18 ♏ 02	13 ♍ 42
02	04 43 00	11 43 36	17 ♋ 50 54	25 ♋ 13 58	00 ♊ 42	25 16	18 11	09 24	16 41	10 48	18 R 01	13 42
03	04 46 56	12 41 05	02 ♌ 35 03	09 ♌ 53 25	02 45	26 30	18 34	09 38	16 43	10 48	17 59	13 42
04	04 50 53	13 38 33	17 ♌ 08 29	24 ♌ 19 45	04 50	27 43	18 57	09 52	16 45	10 49	17 58	13 42
05	04 54 49	14 35 59	01 ♍ 26 51	08 ♍ 29 30	06 56	28 ♊ 57	19 21	10 06	16 48	10 50	17 56	13 43
06	04 58 46	15 33 25	15 ♍ 27 32	22 ♍ 20 54	09 04	00 ♋ 10	19 45	10 20	16 50	10 51	17 55	13 43
07	05 02 42	16 30 49	29 ♍ 09 34	05 ♎ 53 37	11 13	01 24	20 09	10 34	16 52	10 52	17 54	13 44
08	05 06 39	17 28 12	12 ♎ 33 11	19 ♎ 08 25	13 24	02 37	20 34	10 48	16 54	10 53	17 52	13 44
09	05 10 36	18 25 34	25 ♎ 39 30	02 ♏ 06 39	15 35	03 51	20 59	11 02	16 56	10 54	17 51	13 44
10	05 14 32	19 22 54	08 ♏ 30 04	14 ♏ 50 00	17 46	05 04	21 24	11 16	16 58	10 56	17 49	13 45
11	05 18 29	20 20 14	21 ♏ 06 38	27 ♏ 20 13	19 58	06 17	21 50	11 30	16 59	10 57	17 48	13 45
12	05 22 25	21 17 34	03 ♐ 30 57	09 ♐ 39 03	22 10	07 31	22 16	11 44	17 01	10 58	17 47	13 46
13	05 26 22	22 14 52	15 ♐ 44 44	21 ♐ 48 14	24 22	08 44	22 42	11 58	17 02	10 59	17 46	13 47
14	05 30 18	23 12 10	27 ♐ 49 45	03 ♑ 49 32	26 33	09 58	23 08	12 12	17 04	11 01	17 44	13 47
15	05 34 15	24 09 27	09 ♑ 47 48	15 ♑ 44 51	28 ♊ 44	11 11	23 35	12 26	17 05	11 02	17 43	13 48
16	05 38 11	25 06 44	21 ♑ 40 55	27 ♑ 36 21	00 ♋ 53	12 24	24 02	12 39	17 06	11 04	17 42	13 49
17	05 42 08	26 04 00	03 ♒ 31 27	09 ♒ 26 34	03 02	13 38	24 29	12 53	17 08	11 05	17 41	13 49
18	05 46 05	27 01 16	15 ♒ 22 07	21 ♒ 18 28	05 09	14 51	24 56	13 07	17 09	11 07	17 39	13 50
19	05 50 01	27 58 31	27 ♒ 16 06	03 ♓ 15 27	07 14	16 04	25 24	13 21	17 09	11 09	17 38	13 51
20	05 53 58	28 55 46	09 ♓ 17 02	15 ♓ 21 20	09 18	17 18	25 52	13 34	17 10	11 10	17 37	13 52
21	05 57 54	29 ♊ 53 01	21 ♓ 28 53	27 ♓ 40 13	11 19	18 31	26 20	13 48	17 11	11 12	17 36	13 52
22	06 01 51	00 ♋ 50 16	03 ♈ 55 50	10 ♈ 16 16	13 19	19 44	26 49	14 02	17 12	11 14	17 35	13 53
23	06 05 47	01 47 30	16 ♈ 42 00	23 ♈ 13 27	15 17	20 57	27 17	14 15	17 12	11 16	17 34	13 54
24	06 09 44	02 44 45	29 ♈ 51 00	06 ♉ 34 57	17 13	22 10	27 46	14 29	17 13	11 17	17 33	13 55
25	06 13 40	03 41 59	13 ♉ 25 30	20 ♉ 22 42	19 07	23 24	28 15	14 43	17 13	11 19	17 32	13 56
26	06 17 37	04 39 14	27 ♉ 26 30	04 ♊ 36 41	20 58	24 37	28 44	14 56	17 13	11 21	17 31	13 57
27	06 21 34	05 36 28	11 ♊ 52 49	19 ♊ 14 21	22 48	25 50	29 14	15 10	17 13	11 23	17 30	13 58
28	06 25 30	06 33 43	26 ♊ 40 32	04 ♋ 10 26	24 35	27 03	29 ♍ 44	15 23	17 R 13	11 25	17 29	13 59
29	06 29 27	07 30 57	11 ♋ 43 02	19 ♋ 17 11	26 20	28 16	00 ♎ 14	15 37	17 13	11 27	17 28	14 00
30	06 33 23	08 ♋ 28 11	26 ♋ 51 40	04 ♌ 25 16	28 ♋ 03	29 ♋ 30	00 ♎ 44	15 ♊ 50	17 ♓ 13	11 ♍ 30	17 ♏ 27	14 ♍ 01

Date	Moon True ☊ ° '	Moon Mean ☊ ° '	Moon ☽ Latitude ° '
01	13 ♊ 49	13 ♊ 57	01 N 45
02	13 D 49	13 54	02 58
03	13 49	13 51	03 59
04	13 49	13 48	04 44
05	13 R 49	13 45	05 10
06	13 49	13 42	05 17
07	13 D 49	13 38	05 06
08	13 49	13 35	04 38
09	13 50	13 32	03 57
10	13 50	13 29	03 04
11	13 51	13 26	02 03
12	13 51	13 22	00 N 57
13	13 R 51	13 19	00 S 10
14	13 51	13 16	01 17
15	13 50	13 13	02 19
16	13 49	13 10	03 14
17	13 47	13 07	04 01
18	13 45	13 03	04 38
19	13 44	13 00	05 03
20	13 42	12 57	05 15
21	13 41	12 54	05 13
22	13 D 41	12 51	04 56
23	13 42	12 48	04 25
24	13 43	12 44	03 39
25	13 44	12 41	02 40
26	13 45	12 38	01 29
27	13 46	12 35	00 S 10
28	13 R 45	12 32	01 N 11
29	13 44	12 28	02 28
30	13 ♊ 41	12 ♊ 25	03 N 35

DECLINATIONS

Date	Sun ☉ ° '	Moon ☽ ° '	Mercury ☿ ° '	Venus ♀ ° '	Mars ♂ ° '	Jupiter ♃ ° '	Saturn ♄ ° '	Uranus ♅ ° '	Neptune ♆ ° '	Pluto ♇ ° '
01	22 N 04	25 N 09	18 N 59	23 N 58	05 N 46	21 N 16	06 S 57	08 N 14	15 S 28	19 N 35
02	22 12	25 12	19 35	24 03	05 35	21 19	06 56	08 14	15 27	19 35
03	22 19	23 28	20 10	24 08	05 25	21 21	06 56	08 14	15 27	19 34
04	22 27	20 12	20 45	24 12	05 14	21 23	06 55	08 13	15 27	19 34
05	22 33	15 47	21 17	24 15	05 03	21 25	06 54	08 13	15 26	19 33
06	22 40	10 36	21 48	24 17	04 52	21 27	06 54	08 13	15 26	19 33
07	22 46	05 N 01	22 18	24 19	04 41	21 29	06 53	08 12	15 26	19 32
08	22 51	00 S 41	22 45	24 20	04 30	21 31	06 52	08 12	15 25	19 32
09	22 56	06 15	23 11	24 20	04 19	21 33	06 52	08 11	15 25	19 31
10	23 01	11 26	23 34	24 20	04 07	21 35	06 52	08 11	15 25	19 31
11	23 06	16 04	23 54	24 18	03 56	21 37	06 51	08 10	15 24	19 30
12	23 10	19 56	24 12	24 16	03 44	21 39	06 51	08 10	15 24	19 30
13	23 13	22 51	24 27	24 14	03 33	21 41	06 50	08 09	15 24	19 29
14	23 16	24 42	24 40	24 11	03 21	21 43	06 50	08 09	15 23	19 29
15	23 19	25 23	24 49	24 06	03 09	21 45	06 50	08 08	15 23	19 28
16	23 21	24 54	24 56	24 02	02 57	21 47	06 49	08 08	15 23	19 27
17	23 23	23 17	25 00	23 56	02 45	21 49	06 49	08 07	15 22	19 27
18	23 25	20 39	25 01	23 50	02 33	21 50	06 49	08 07	15 22	19 26
19	23 26	17 10	25 00	23 43	02 21	21 52	06 49	08 06	15 22	19 26
20	23 26	12 57	24 56	23 36	02 08	21 54	06 49	08 05	15 22	19 25
21	23 27	08 10	24 49	23 27	01 56	21 56	06 49	08 04	15 21	19 24
22	23 26	02 S 58	24 40	23 18	01 43	21 57	06 49	08 03	15 21	19 24
23	23 26	02 N 29	24 29	23 09	01 31	21 59	06 49	08 03	15 21	19 23
24	23 25	08 20	24 16	22 59	01 18	22 01	06 49	08 02	15 21	19 22
25	23 24	13 20	24 00	22 48	01 06	22 02	06 49	08 01	15 20	19 22
26	23 22	18 09	23 43	22 36	00 53	22 04	06 49	08 00	15 20	19 21
27	23 20	22 03	23 24	22 24	00 40	22 05	06 49	08 00	15 20	19 20
28	23 17	24 35	23 03	22 11	00 27	22 07	06 50	07 59	15 20	19 19
29	23 14	25 23	22 41	21 57	00 14	22 08	06 50	07 58	15 19	19 19
30	23 N 10	24 N 18	22 N 18	21 N 43	00 N 01	22 N 10	06 S 50	07 N 57	15 S 19	19 N 18

ZODIAC SIGN ENTRIES

Date	h	m	Planets
01	07	05	☽ ♋
02	03	47	☿ ♊
03	07	46	☽ ♌
05	09	33	☽ ♍
06	08	39	♀ ♋
07	13	29	☽ ♎
09	20	04	☽ ♏
12	05	10	☽ ♐
14	16	20	☽ ♑
16	02	04	☿ ♋
17	04	51	☽ ♒
19	17	29	☽ ♓
21	14	56	☉ ♋
22	04	29	☽ ♈
24	12	16	☽ ♉
26	16	18	☽ ♊
28	17	20	☽ ♋
29	01	12	♂ ♎
30	16	59	☽ ♌
30	21	59	♀ ♌

LATITUDES

Date	Mercury ☿ ° '	Venus ♀ ° '	Mars ♂ ° '	Jupiter ♃ ° '	Saturn ♄ ° '	Uranus ♅ ° '	Neptune ♆ ° '	Pluto ♇ ° '
01	00 S 54	00 N 39	01 N 01	00 S 34	01 S 49	00 N 46	01 N 49	14 N 20
04	00 S 22	00 46	00 56	00 34	01 50	00 46	01 49	14 18
07	00 N 10	00 53	00 51	00 33	01 51	00 46	01 49	14 17
10	00 41	00 59	00 47	00 33	01 51	00 46	01 49	14 16
13	01 08	01 05	00 42	00 33	01 52	00 46	01 49	14 15
16	01 30	01 10	00 38	00 33	01 53	00 46	01 48	14 14
19	01 45	01 15	00 34	00 33	01 54	00 46	01 48	14 13
22	01 54	01 20	00 30	00 32	01 54	00 46	01 48	14 12
25	01 56	01 24	00 26	00 32	01 55	00 45	01 48	14 11
28	01 52	01 27	00 22	00 32	01 56	00 45	01 48	14 10
31	01 N 43	01 N 31	00 N 18	00 S 32	01 S 57	00 N 45	01 N 48	14 N 09

DATA

Julian Date	2438913
Delta T	+36 seconds
Ayanamsa	23° 22' 09"
Synetic vernal point	05° ♓ 44' 51"
True obliquity of ecliptic	23° 26' 40"

MOON'S PHASES, APSIDES AND POSITIONS ☽

Date	h	m	Phase	Longitude ° '	Eclipse Indicator
06	12	11	☽	15 ♍ 34	
14	01	59	○	22 ♐ 48	partial
22	05	36	☾	00 ♈ 35	
29	04	52	●	07 ♋ 14	

Day	h	m		
01	18	19	Perigee	
17	09	41	Apogee	
29	23	57	Perigee	
02	00	30	Max dec	25° N 24'
08	09	06	0S	
15	13	53	Max dec	25° S 24'
23	01	10	0N	
29	10	18	Max dec	25° N 23'

LONGITUDES

Date	Chiron ⚷ ° '	Ceres ⚳ ° '	Pallas ⚴ ° '	Juno ⚵ ° '	Vesta ⚶ ° '	Black Moon Lilith ⚸ ° '
01	22 ♓ 18	24 ♓ 41	23 ♒ 28	15 ♒ 57	24 ♉ 44	25 ♑ 59
11	22 ♓ 28	27 ♓ 04	23 ♒ 34	16 ♒ 19	29 ♉ 03	27 ♑ 06
21	22 ♓ 33	29 ♓ 06	23 ♒ 04	16 ♒ 04	03 ♊ 19	28 ♑ 13
31	22 ♓ 33	00 ♈ 45	21 ♒ 58	15 ♒ 12	07 ♊ 31	29 ♑ 20

All ephemeris data is given at 12.00 UT and the Moon's longitude is additionally given for 24.00 UT

ASPECTARIAN

h m	Aspects
01 Tuesday	
03 50	☽ ⊻ ☿
05 06	☽ Q ♅
09 50	☽ Q ♇
11 29	☿ Q ♄
11 51	♀ ± ♆
12 00	☽ bQ ♆
12 19	☉ □ ♅
15 05	☽ ⊥ ☿
16 36	☽ Q ♂
22 06	☽ ⊻ ♃
02 Wednesday	
00 33	☽ ⚹ ♅
01 24	☽ ⊻ ☉
02 14	♂ ⚹ ♆
05 16	☽ ⚹ ♇
07 57	☽ ∠ ☿
07 59	☽ ⊥ ♃
10 06	☽ △ ♄
11 43	☿ ∥ ♇
11 47	☽ ⊥ ☉
12 16	☽ △ ♆
12 33	☽ ⚹ ♂
22 50	☽ ∠ ♃
03 Thursday	
00 55	☽ ∠ ♅
01 09	☽ ⊻ ♀
03 26	☽ ∠ ☉
05 21	☽ ∥ ♀
05 39	☽ ∠ ♇
10 35	☽ bQ ♄
11 51	☽ ⊥ ♀
12 18	☽ ⚹ ☿
13 39	☽ ∠ ♂
15 39	☽ ⊥ ♅
20 24	☽ ⊥ ♇
21 14	☽ ∥ ☉
23 47	☽ ⚹ ♃
04 Friday	
01 24	☽ ± ♄
01 32	☽ ⊻ ♅
04 00	☽ ∠ ♀
04 27	☽ ∥ ♃
04 52	☽ ⊥ ♂
05 47	☽ ⚹ ☉
06 18	☽ ⊻ ♇
09 03	☽ ∥ ☿
11 22	☽ ⊼ ♄
11 23	☽ Q ☿
13 22	☽ □ ♆
13 38	☉ □ ♇
15 06	☽ ⊻ ♂
15 50	☽ ∥ ♇
20 01	☽ Q ♃
05 Saturday	
03 14	☽ Q ☉
07 22	☽ ⚹ ♀
09 50	♀ Q ♅
13 42	☽ # ♆
18 16	☿ ∥ ♃
19 37	☽ Q ♆
23 00	☽ □ ☿
06 Sunday	
03 01	☽ □ ♃
04 03	☽ ☌ ♅
05 46	☽ Q ♀
08 59	☽ ☌ ♇
12 11	☽ □ ☉
14 23	☽ ☍ ♄
16 15	☽ ⚹ ♆
19 41	☽ ☌ ♂
22 25	☽ ∥ ♅
07 Monday	
03 56	☿ ☌ ♃
04 04	☽ # ♄
08 05	☿ □ ♅
13 26	☽ ∥ ♂
16 22	☽ □ ♀
18 30	♀ Q ♇
18 37	☽ ∠ ♆
21 10	☉ □ ♄
08 Tuesday	
08 47	☽ △ ♃
08 59	☽ ⊻ ♅
10 46	☽ ⊥ ♆
13 49	☽ △ ☿
14 08	☽ ⊻ ♇
15 46	☿ □ ♇
16 49	♀ bQ ♆
18 55	☉ ∥ ☿
19 54	☽ ⊥ ♅
19 55	☽ ⊼ ♄
21 29	♃ □ ♅
21 39	☽ △ ☉
21 40	☽ ⊻ ♆
21 48	☉ ⊼ ♆
09 Wednesday	
01 06	☽ ⊥ ♇
03 06	☽ ⊻ ♂
03 49	☽ # ♂
06 57	☽ ± ♄
12 28	☽ ∠ ♅
12 43	☽ bQ ♃
14 32	☽ ⊥ ♂
14 46	☽ ∥ ♄
17 43	☽ ∠ ♇
20 44	☽ # ♅
23 01	☽ bQ ☿
23 41	☽ bQ ♄
10 Thursday	
03 01	☿ □ ♄
03 38	☽ bQ ☉
04 51	☽ △ ♀
05 48	☽ ± ♃
07 55	☽ ∠ ♂
12 34	☿ ⊼ ♆
16 36	☽ ⚹ ♅
17 20	☽ ⊼ ♃
19 29	☽ ± ☿
21 57	☽ ⚹ ♇
22 00	☽ ± ☉
11 Friday	
04 06	☽ △ ♄
05 41	☽ ☌ ♆
08 21	☽ ∥ ♆
09 21	☽ ⊼ ☿
10 24	☽ ⊼ ☉
12 23	☽ bQ ♀
13 26	☽ ⚹ ♂
15 32	☽ Q ♅
19 02	☉ ☌ ☿
20 57	☽ Q ♇
12 Saturday	
07 41	☽ ± ♀
09 02	☽ # ♇
13 14	☿ □ ♂
13 31	☽ Q ♂
17 47	☿ ∥ ♀
20 41	☽ ⊼ ♀
13 Sunday	
01 15	☽ # ♃
02 37	☽ □ ♅
04 24	☽ ☍ ♃
05 24	☿ ± ♆
08 07	☽ □ ♇
14 34	☽ □ ♄
15 50	☽ # ☉
15 59	☽ ⊻ ♆
14 Monday	
01 59	☽ ☍ ☉
02 18	☽ □ ♂
03 44	☽ # ♀
03 52	☽ ⊥ ♆
08 54	☽ ☍ ☿
09 02	☉ □ ♂
11 03	☽ # ☿
21 48	☽ ∠ ♆
15 Tuesday	
01 11	☉ ± ♆
02 31	☽ Q ♄
09 07	♀ ⚹ ♅
14 30	☽ △ ♅
15 07	☽ ☍ ♀
15 26	☿ Q ♅
17 24	☽ ⊼ ♃
20 04	☽ △ ♇
16 Wednesday	
00 25	♀ Q ♂
02 44	☽ ⚹ ♄
03 57	☽ ⚹ ♆
05 45	☽ ± ♃
11 13	☽ # ☿
16 57	☽ △ ♂
18 04	♀ ⊻ ♃
19 33	☽ ⊼ ☉
20 54	☽ bQ ♅
22 20	☿ Q ♇
17 Thursday	
00 20	☽ bQ ♃
02 27	☽ bQ ♇
03 34	☽ # ♀
04 13	☽ Q ♆
08 05	☿ bQ ♆
08 47	☽ ± ☉
09 10	☽ ∠ ♄
10 47	☽ ⊼ ☿
10 50	☽ # ☉
15 11	☽ ± ♅
15 50	♀ ⚹ ♇
20 43	☽ ± ♇
18 Friday	
00 34	☽ bQ ♂
01 36	☽ ± ☿
02 20	☽ # ♃
03 22	☽ ⊼ ♅
03 26	☽ ⊥ ♄
04 38	☽ bQ ☉
07 21	☽ △ ♃
08 53	☽ ⊼ ♇
10 50	☽ ⊼ ♀
15 36	☽ ⊻ ♄
16 37	☽ □ ♆
19 31	☽ ± ♂
21 03	☽ # ♇
23 43	☽ bQ ☿
19 Saturday	
00 20	☽ ± ♀
08 06	☽ ⊼ ♂
13 33	☽ △ ☉
20 29	☽ bQ ♀
22 42	☽ ∥ ♆
20 Sunday	
09 37	♀ △ ♄
12 01	☽ △ ☿
15 45	☽ ☍ ♅
18 15	☉ Q ♅
18 21	♀ △ ♆
20 39	☽ □ ♃
21 04	☽ ☍ ♇
21 Monday	
03 35	☽ ☌ ♄
04 25	☽ △ ♆
05 33	☽ △ ♀
10 32	☿ ⚹ ♅
12 29	☽ # ♅
14 06	☉ ∥ ♀
18 25	☽ ∥ ♄
19 49	♃ □ ♇
21 48	☽ ☍ ♂
22 Tuesday	
05 36	☽ □ ☉
08 18	☽ Q ♃
09 26	☽ bQ ♆
17 47	☽ # ♂
18 57	☿ ⚹ ♇
19 10	♀ ⊥ ♃
21 47	☿ ⊻ ♃
23 Wednesday	
01 50	☽ ⊼ ♅
02 27	☽ ± ♆
06 47	☽ ⊼ ♇
07 23	☽ ⚹ ♃
07 57	☽ ∥ ♂
08 54	☽ □ ☿
12 03	☿ Q ♂
12 56	☽ ⊻ ♄
13 02	☽ ± ♅
13 36	☽ ⊼ ♆
14 47	☉ Q ♇
17 55	☽ ± ♇
18 10	☽ Q ☉
20 39	☽ □ ♀
23 58	☽ ⊥ ♄
24 Thursday	
05 34	☽ bQ ♅
06 49	☽ # ♄
07 07	☉ bQ ♆
08 06	☽ ⊼ ♂
10 19	☽ bQ ♇
11 20	☽ ∠ ♃
11 57	☿ △ ♄
12 09	☽ ∥ ♅
16 10	☿ △ ♆
16 14	☽ ∠ ♄
17 35	☽ ⚹ ☉
19 16	☽ ± ♂
23 08	☽ Q ☿
25 Friday	
03 37	☽ ⊥ ♃
08 07	☽ Q ♀
08 19	☽ △ ♅
11 41	☽ bQ ♂
12 53	☽ △ ♇
14 16	☽ ⊻ ♃
18 34	☽ ⚹ ♄
19 06	☽ ☍ ♆
21 34	☽ # ♆
21 47	☽ ∠ ☉
23 20	☽ ⚹ ☿
26 Saturday	
06 46	☽ ⚹ ♀
11 32	☿ ⊥ ♃
14 11	☽ ⊥ ☉
14 16	☽ △ ♂
14 59	☽ Q ♄
18 40	☽ ∥ ♇
27 Sunday	
00 55	☽ ⊻ ☉
04 19	☽ ∠ ☿
10 08	☽ ∠ ♀
11 11	☽ □ ♅
12 19	☽ ∥ ♃
14 28	☽ ∥ ♀
15 25	☽ □ ♇
17 27	☽ ☌ ♃
18 02	☉ ∥ ☿
20 43	☽ □ ♄
21 08	☽ ⊥ ☿
21 10	☽ ⊼ ♆
21 45	☽ ∥ ☿
22 09	☽ ∥ ☉
23 12	♀ ∠ ♅
28 Monday	
02 08	☽ ⊥ ♀
05 32	♄ St R
06 52	☽ ± ♆
08 11	☽ ⊻ ☿
12 40	☽ ⊻ ♀
16 25	☽ Q ♅
17 04	☽ □ ♂
18 38	♀ ∥ ♃
20 31	☽ Q ♇
21 17	☽ bQ ♆
29 Tuesday	
04 52	☽ ☌ ☉
11 35	☽ ⚹ ♅
13 45	☿ ∠ ♅
15 37	☽ ⚹ ♇
18 16	☽ ⊻ ♃
20 43	☽ △ ♄
21 07	☽ △ ♆
22 40	☽ Q ♂
30 Wednesday	
02 29	♀ ∠ ♇
03 55	☽ ⊥ ♃
11 25	☽ ∠ ♅
14 07	☽ ☌ ☿
15 25	☽ ∠ ♇
16 32	☽ ☌ ♀
18 21	☽ ⚹ ♂
18 24	☽ ∠ ♃
19 20	☿ ∥ ♃
20 30	☽ bQ ♄
23 35	☽ ∥ ☉

JULY 1965

LONGITUDES

Date	Sidereal time h m s	Sun ☉ ° ' "	Moon ☽ ° ' "	Moon ☽ 24.00 ° ' "	Mercury ☿ ° '	Venus ♀ ° '	Mars ♂ ° '	Jupiter ♃ ° '	Saturn ♄ ° '	Uranus ♅ ° '	Neptune ♆ ° '	Pluto ♇ ° '
01	06 37 20	09 ♋ 25 25	11 ♌ 56 50	19 ♌ 25 16	29 ♋ 44	00 ♌ 43	01 ♎ 14	16 ♊ 04	17 ♓ 13	11 ♍ 32	17 ♏ 26	14 ♍ 02
02	06 41 16	10 22 38	26 ♌ 49 35	04 ♍ 09 01	01 ♌ 22	01 56	01 45	16 17	17 R 12	11 34	17 R 25	14 03
03	06 45 13	11 19 52	11 ♍ 22 54	18 ♍ 30 47	02 59	03 09	02 15	16 30	17 12	11 36	17 25	14 04
04	06 49 09	12 17 04	25 ♍ 32 25	02 ♎ 27 39	04 33	04 22	02 46	16 43	17 11	11 39	17 24	14 06
05	06 53 06	13 14 17	09 ♎ 16 33	15 ♎ 59 16	06 05	05 35	03 17	16 57	17 11	11 41	17 23	14 07
06	06 57 03	14 11 29	22 ♎ 36 03	29 ♎ 07 15	07 35	06 48	03 49	17 10	17 10	11 43	17 22	14 08
07	07 00 59	15 08 41	05 ♏ 33 15	11 ♏ 54 29	09 02	08 01	04 20	17 23	17 09	11 46	17 22	14 09
08	07 04 56	16 05 53	18 ♏ 11 23	24 ♏ 24 27	10 27	09 14	04 52	17 36	17 08	11 48	17 21	14 11
09	07 08 52	17 03 05	00 ♐ 34 05	06 ♐ 40 44	11 50	10 27	05 24	17 49	17 07	11 51	17 20	14 12
10	07 12 49	18 00 16	12 ♐ 44 49	18 ♐ 46 43	13 11	11 40	05 56	18 02	17 06	11 53	17 20	14 13
11	07 16 45	18 57 28	24 ♐ 46 47	00 ♑ 45 22	14 29	12 53	06 28	18 15	17 04	11 56	17 19	14 15
12	07 20 42	19 54 40	06 ♑ 42 45	12 ♑ 39 13	15 45	14 06	07 01	18 28	17 03	11 58	17 19	14 16
13	07 24 38	20 51 53	18 ♑ 35 03	24 ♑ 30 28	16 58	15 19	07 33	18 41	17 02	12 01	17 18	14 17
14	07 28 35	21 49 05	00 ♒ 25 42	06 ♒ 21 01	18 09	16 31	08 06	18 54	17 00	12 04	17 18	14 19
15	07 32 32	22 46 18	12 ♒ 16 36	18 ♒ 12 44	19 17	17 44	08 39	19 06	16 59	12 07	17 17	14 20
16	07 36 28	23 43 31	24 ♒ 09 38	00 ♓ 07 36	20 23	18 57	09 12	19 19	16 57	12 09	17 17	14 22
17	07 40 25	24 40 45	06 ♓ 06 54	12 ♓ 07 52	21 25	20 10	09 45	19 32	16 55	12 12	17 16	14 23
18	07 44 21	25 37 59	18 ♓ 10 50	24 ♓ 16 11	22 25	21 23	10 18	19 44	16 53	12 15	17 16	14 25
19	07 48 18	26 35 14	00 ♈ 24 20	06 ♈ 35 43	23 22	22 35	10 52	19 57	16 51	12 18	17 16	14 26
20	07 52 14	27 32 29	12 ♈ 50 46	19 ♈ 09 58	24 16	23 48	11 26	20 09	16 49	12 21	17 15	14 28
21	07 56 11	28 29 46	25 ♈ 32 11	02 ♉ 02 42	25 06	25 01	11 59	20 21	16 47	12 24	17 15	14 29
22	08 00 07	29 ♋ 27 03	08 ♉ 37 09	15 ♉ 17 33	25 53	26 13	12 33	20 34	16 44	12 27	17 15	14 31
23	08 04 04	00 ♌ 24 21	22 ♉ 04 14	28 ♉ 57 28	26 37	27 26	13 08	20 46	16 42	12 30	17 15	14 33
24	08 08 01	01 21 40	05 ♊ 57 24	13 ♊ 04 01	27 17	28 39	13 42	20 58	16 40	12 33	17 14	14 34
25	08 11 57	02 19 00	20 ♊ 17 11	27 ♊ 36 32	27 53	29 ♌ 51	14 16	21 10	16 37	12 36	17 14	14 36
26	08 15 54	03 16 21	05 ♋ 01 33	12 ♋ 31 27	28 25	01 ♍ 04	14 51	21 22	16 35	12 39	17 14	14 38
27	08 19 50	04 13 43	20 ♋ 05 17	27 ♋ 41 56	28 53	02 17	15 26	21 34	16 32	12 42	17 14	14 39
28	08 23 47	05 11 05	05 ♌ 20 08	12 ♌ 58 30	29 17	03 29	16 01	21 46	16 29	12 45	17 14	14 41
29	08 27 43	06 08 28	20 ♌ 35 41	28 ♌ 10 18	29 36	04 42	16 36	21 58	16 26	12 48	17 14	14 43
30	08 31 40	07 05 52	05 ♍ 41 06	13 ♍ 07 00	29 ♌ 51	05 54	17 11	22 10	16 23	12 52	17 D 14	14 45
31	08 35 36	08 ♌ 03 17	20 ♍ 27 04	27 ♍ 40 37	00 ♍ 00	07 ♍ 07	17 ♎ 46	22 ♊ 21	16 ♓ 20	12 ♍ 55	17 ♏ 14	14 ♍ 46

Date	Moon True ☊ ° '	Moon Mean ☊ ° '	Moon ☽ Latitude ° '
01	13 ♊ 38	12 ♊ 22	04 N 27
02	13 R 35	12 19	05 00
03	13 32	12 16	05 13
04	13 29	12 13	05 07
05	13 28	12 09	04 42
06	13 D 29	12 06	04 04
07	13 30	12 03	03 13
08	13 31	12 00	02 14
09	13 33	11 57	01 10
10	13 33	11 53	00 N 04
11	13 R 33	11 50	01 S 01
12	13 31	11 47	02 03
13	13 26	11 44	02 59
14	13 21	11 41	03 47
15	13 14	11 38	04 25
16	13 07	11 34	04 52
17	13 00	11 31	05 07
18	12 53	11 28	05 08
19	12 48	11 25	04 55
20	12 45	11 22	04 28
21	12 44	11 19	03 47
22	12 D 45	11 15	02 54
23	12 46	11 12	01 49
24	12 47	11 09	00 S 37
25	12 R 47	11 06	00 N 40
26	12 45	11 03	01 57
27	12 41	10 59	03 07
28	12 35	10 56	04 04
29	12 27	10 53	04 44
30	12 19	10 50	05 04
31	12 ♊ 11	10 ♊ 47	05 N 03

DECLINATIONS

Date	Sun ☉ ° '	Moon ☽ ° '	Mercury ☿ ° '	Venus ♀ ° '	Mars ♂ ° '	Jupiter ♃ ° '	Saturn ♄ ° '	Uranus ♅ ° '	Neptune ♆ ° '	Pluto ♇ ° '
01	23 N 07	21 N 29	21 N 53	21 N 29	00 S 13	22 N 11	06 S 50	07 N 56	15 S 19	19 N 17
02	23 02	17 16	21 27	21 13	00 26	22 13	06 51	07 55	15 19	19 16
03	22 58	12 07	21 00	20 57	00 39	22 14	06 51	07 55	15 19	19 16
04	22 53	06 28	20 32	20 41	00 52	22 15	06 52	07 54	15 18	19 15
05	22 47	00 N 39	20 04	20 24	01 06	22 17	06 52	07 53	15 18	19 14
06	22 41	05 S 01	19 35	20 06	01 19	22 18	06 53	07 52	15 18	19 13
07	22 35	10 20	19 05	19 48	01 33	22 19	06 53	07 51	15 18	19 13
08	22 28	15 06	18 34	19 29	01 47	22 21	06 54	07 50	15 18	19 12
09	22 21	19 08	18 04	19 10	02 00	22 22	06 55	07 49	15 18	19 11
10	22 14	22 15	17 33	18 50	02 14	22 23	06 55	07 48	15 18	19 10
11	22 06	24 21	17 01	18 29	02 28	22 24	06 56	07 47	15 18	19 09
12	21 58	25 19	16 30	18 09	02 42	22 26	06 57	07 46	15 17	19 09
13	21 49	25 07	15 59	17 47	02 55	22 27	06 57	07 45	15 17	19 08
14	21 41	23 46	15 27	17 26	03 09	22 28	06 58	07 44	15 17	19 07
15	21 31	21 22	14 56	17 03	03 23	22 29	06 59	07 43	15 17	19 06
16	21 22	18 04	14 25	16 41	03 37	22 30	07 00	07 42	15 17	19 05
17	21 12	14 01	13 54	16 18	03 51	22 31	07 01	07 40	15 17	19 05
18	21 01	09 23	13 24	15 54	04 05	22 32	07 02	07 39	15 17	19 04
19	20 50	04 S 21	12 54	15 30	04 19	22 33	07 03	07 38	15 17	19 03
20	20 39	00 N 58	12 25	15 06	04 34	22 34	07 04	07 37	15 17	19 02
21	20 28	06 22	11 56	14 41	04 48	22 35	07 05	07 36	15 17	19 01
22	20 16	11 38	11 29	14 16	05 02	22 36	07 06	07 35	15 17	19 00
23	20 04	16 32	11 02	13 50	05 16	22 37	07 07	07 34	15 17	18 59
24	19 52	20 42	10 36	13 24	05 30	22 38	07 08	07 32	15 17	18 59
25	19 39	23 46	10 12	12 58	05 45	22 38	07 10	07 31	15 17	18 58
26	19 26	25 18	09 49	12 32	05 59	22 39	07 11	07 30	15 17	18 57
27	19 12	25 01	09 27	12 05	06 13	22 40	07 12	07 29	15 17	18 56
28	18 59	22 53	09 06	11 38	06 28	22 41	07 13	07 27	15 17	18 55
29	18 44	19 07	08 48	11 10	06 42	22 42	07 15	07 26	15 17	18 54
30	18 30	14 08	08 31	10 42	06 56	22 42	07 16	07 25	15 17	18 53
31	18 N 15	08 N 25	08 N 16	10 N 14	07 S 11	22 N 43	07 S 17	07 N 24	15 S 17	18 N 52

ZODIAC SIGN ENTRIES

Date	h m	Planets
01	15 55	☿ ♌
02	17 11	☽ ♍
04	19 43	☽ ♎
07	01 38	☽ ♏
09	10 53	☽ ♐
11	22 29	☽ ♑
14	11 08	☽ ♒
16	23 45	☽ ♓
19	11 13	☽ ♈
21	20 14	☽ ♉
23	01 48	☉ ♌
24	01 48	☽ ♊
25	14 51	♀ ♍
26	03 53	☽ ♋
28	03 37	☽ ♌
30	02 55	☽ ♍
31	11 24	☿ ♍

LATITUDES

Date	Mercury ☿ ° '	Venus ♀ ° '	Mars ♂ ° '	Jupiter ♃ ° '	Saturn ♄ ° '	Uranus ♅ ° '	Neptune ♆ ° '	Pluto ♇ ° '
01	01 N 43	01 N 31	00 N 18	00 S 32	01 S 57	00 N 45	01 N 48	14 N 09
04	01 27	01 33	00 15	00 32	01 57	00 45	01 48	14 08
07	01 07	01 35	00 11	00 31	01 58	00 45	01 47	14 07
10	00 43	01 36	00 08	00 31	01 59	00 45	01 47	14 06
13	00 N 14	01 37	00 05	00 31	01 59	00 45	01 47	14 05
16	00 S 18	01 37	00 N 02	00 31	02 00	00 45	01 47	14 04
19	00 53	01 36	00 S 02	00 31	02 01	00 45	01 47	14 03
22	01 30	01 35	00 05	00 31	02 02	00 45	01 47	14 03
25	02 09	01 33	00 07	00 31	02 02	00 44	01 46	14 02
28	02 48	01 30	00 10	00 31	02 03	00 44	01 46	14 01
31	03 S 26	01 N 26	00 S 13	00 S 30	02 S 03	00 N 44	01 N 46	14 N 01

DATA

Julian Date	2438943
Delta T	+36 seconds
Ayanamsa	23° 22' 15"
Synetic vernal point	05° ♓ 44' 45"
True obliquity of ecliptic	23° 26' 40"

MOON'S PHASES, APSIDES AND POSITIONS ☽

Date	h m	Phase	Longitude ° '	Eclipse Indicator
05	19 36	☽	13 ♎ 32	
13	17 01	○	21 ♑ 04	
21	17 53	☾	28 ♈ 44	
28	11 45	●	05 ♌ 10	

Day	h m		
14	16 57	Apogee	
28	09 14	Perigee	
05	14 43	0S	
12	19 42	Max dec	25° S 23'
20	07 41	0N	
26	20 30	Max dec	25° N 25'

LONGITUDES

Date	Chiron ⚷ ° '	Ceres ⚳ ° '	Pallas ⚴ ° '	Juno ⚵ ° '	Vesta ⚶ ° '	Black Moon Lilith ⚸ ° '
01	22 ♓ 33	00 ♈ 45	21 ♒ 58	15 ♒ 12	07 ♊ 31	29 ♑ 20
11	22 ♓ 27	01 ♈ 57	20 ♒ 16	13 ♒ 43	11 ♊ 38	00 ♒ 26
21	22 ♓ 16	02 ♈ 39	18 ♒ 06	11 ♒ 44	15 ♊ 38	01 ♒ 33
31	22 ♓ 00	02 ♈ 47	15 ♒ 36	09 ♒ 24	19 ♊ 32	02 ♒ 40

All ephemeris data is given at 12.00 UT and the Moon's longitude is additionally given for 24.00 UT

ASPECTARIAN

h m	Aspects
01 Thursday	
01 44	☽ ⊥ ♅
01 54	☿ ∠ ♇
05 45	☽ ⊥ ♇
07 08	☽ ∥ ♃
07 42	☽ ⊻ ☉
08 53	☽ ∥ ☿
10 49	☽ ± ♄
11 20	☽ ⊻ ♅
12 04	☽ ∥ ♀
15 21	☽ ⊻ ♇
17 57	☽ ⊥ ☉
18 41	☽ ✶ ♃
19 07	☽ ∠ ♂
20 23	♀ ∠ ♃
20 26	☽ ⊼ ♄
20 48	☽ □ ♆
02 Friday	
01 23	☽ ∥ ♇
05 39	♀ ✶ ♂
09 29	☽ ∠ ☉
10 10	☽ ⊥ ♂
10 27	☿ ∠ ♃
14 25	☽ Q ♃
17 24	♀ ⚼ ♄
20 01	☿ ✶ ♂
20 20	☽ ⊻ ♂
20 21	☽ ⊻ ☿
21 06	☽ ⊻ ♀
21 32	☽ # ♆
03 Saturday	
00 19	☿ ⚼ ♄
02 05	☽ Q ♆
07 30	☽ ⊥ ☿
07 56	☽ ⊥ ♀
11 55	☽ ✶ ☉
12 22	☽ ☌ ♅
16 31	☽ ☌ ♇
18 33	☿ ∥ ♀
19 04	♂ ∠ ♆
19 09	☉ ✶ ♅
20 44	☽ □ ♃
21 46	☽ ☍ ♄
22 07	☽ ✶ ♆
23 11	☿ ☌ ♀
04 Sunday	
00 27	☽ ∠ ♀
00 29	☽ ∠ ☿
06 02	☽ ∥ ♅
09 41	☽ Q ☉
10 21	☽ # ♄
23 53	☽ ∠ ♆
05 Monday	
01 02	☽ ☌ ♂
04 50	☽ ✶ ♀
05 30	☿ ⊥ ♅
05 39	☽ ✶ ☿
10 13	☽ # ♂
14 00	♀ ⊥ ♅
15 45	☽ ⊥ ♆
16 18	☽ ⊻ ♅
19 36	☽ ∥ ♂
19 36	☽ □ ☉
20 39	☽ ⊻ ♇
06 Tuesday	
01 57	☽ △ ♃
02 08	☽ ⊼ ♄
02 30	☽ ⊻ ♆
03 06	☽ ⊥ ♅
04 23	☽ Q ♀
05 48	☽ Q ☿
07 30	☽ ⊥ ♇
10 30	☉ ✶ ♇
11 55	♃ □ ♄
13 02	☽ ± ♄
19 35	☽ ∠ ♅
20 09	☽ ∥ ♄
21 11	☿ ⊥ ♇
07 Wednesday	
00 03	☽ ∠ ♇
00 31	☽ # ♅
05 35	☿ ∥ ♇
05 38	☽ ⚼ ♄
05 58	☽ ⚼ ♃
09 37	☽ ⊻ ♂
09 42	♃ ⊼ ♆
14 45	♀ ⊥ ♇
17 08	☽ □ ♀
19 24	☽ □ ☿
21 25	☽ ⊥ ♂
23 12	☽ ± ♃
23 46	☽ ✶ ♅
08 Thursday	
04 18	☽ ✶ ♇
07 40	☽ △ ☉
09 58	☽ △ ♄
10 23	☽ ☌ ♆
10 51	☽ ⊼ ♃
13 05	☽ ∥ ♆
15 22	☽ ∠ ♂
22 52	☽ Q ♅
23 29	☿ ± ♄
09 Friday	
03 28	☽ Q ♇
05 58	☽ # ☿
10 09	♀ ∥ ♇
10 24	☉ ∥ ♃
12 05	☿ ⊻ ♅
12 13	☽ # ♀
12 23	☽ # ♇
13 35	☉ △ ♄
15 09	☽ ⚼ ☉
19 11	☉ △ ♆
21 55	☽ ✶ ♂
10 Saturday	
00 56	♀ ± ♄
09 37	☽ △ ♀
10 17	☽ □ ♅
10 24	☽ ± ☉
11 47	☽ # ☉
12 58	☽ △ ☿
13 00	☉ ⊻ ♃
13 11	☽ # ♃
14 56	☽ □ ♇
16 33	♀ ⊻ ♅
20 38	☽ □ ♄
21 06	☽ ⊻ ♆
22 43	☽ ☍ ♃
22 48	☽ Q ♂
23 21	☽ ⊼ ☉
11 Sunday	
07 21	☿ ⊻ ♇
09 05	☽ ⊥ ♆
18 56	☽ ⚼ ♀
22 35	☽ ⚼ ☿
12 Monday	
03 08	☽ ∠ ♆
08 40	☽ Q ♄
12 38	☽ □ ♂
15 07	☽ ± ♀
15 25	♀ ⊻ ♇
18 51	☽ ± ☿
22 40	☽ △ ♅
13 Tuesday	
03 17	☽ △ ♇
04 37	☽ ⊼ ♀
08 22	☽ ⊼ ☿
08 51	☽ ✶ ♄
09 24	☽ ✶ ♆
12 12	☽ ⊼ ♃
13 05	☿ ⊼ ♄
17 01	☽ ☍ ☉
18 34	☿ □ ♆
14 Wednesday	
00 34	☽ ± ♃
05 09	☽ ⚼ ♅
09 42	☽ Q ♆
09 44	☽ ⚼ ♇
15 11	☽ ∠ ♄
19 09	☽ ⚼ ♃
19 37	☿ # ♆
21 15	♀ ⊼ ♄
23 28	☽ ± ♅
15 Thursday	
02 04	☽ # ♃
03 07	♀ □ ♆
04 00	☽ ± ♇
04 17	☽ △ ♂
07 09	☿ ✶ ♃
09 22	☽ ⊥ ♄
10 35	☽ # ☉
11 40	☽ ⊼ ♅
16 11	☽ ⊼ ♇
21 29	☽ ⊻ ♄
22 07	☽ □ ♆
16 Friday	
00 18	☽ ☍ ♀
02 03	☽ △ ♃
03 37	☽ ☍ ☿
05 07	☽ # ♇
11 03	☽ ⊼ ☉
12 05	☽ ⚼ ♂
20 43	♀ ✶ ♃
21 36	☽ # ♀
17 Saturday	
00 10	☽ ± ☉
04 53	☽ ∥ ♆
07 02	☽ ± ♂
12 42	☽ # ☿
19 37	☽ ⊼ ♂
19 44	☽ ⚼ ☉
18 Sunday	
00 11	☽ ☍ ♅
04 31	☽ ☍ ♇
09 26	☽ ☌ ♄
10 11	☽ △ ♆
15 08	☽ □ ♃
15 17	☉ ⊥ ♃
19 00	☽ ⊼ ♀
20 28	☽ # ♅
21 05	☽ ⊼ ☿
23 23	☽ ∥ ♄
19 Monday	
03 55	☽ △ ☉
08 04	☽ ± ♀
09 49	☽ ± ☿
12 05	☽ ∥ ♂
15 36	☽ ⚼ ♆
20 Tuesday	
00 54	♀ # ♆
02 51	☽ Q ♃
03 25	☽ ⚼ ♀
04 37	☽ ⚼ ☿
04 45	♂ ⊥ ♆
06 49	☉ ∠ ♅
08 58	☽ ± ♆
09 09	☽ ☍ ♂
11 02	☽ ⊼ ♅
15 05	☽ ⊼ ♇
19 32	☽ ⊻ ♄
20 23	☽ ⊼ ♆
22 29	☽ ± ♅
21 Wednesday	
02 05	☽ ✶ ♃
02 28	☽ ± ♇
04 43	☽ # ♂
06 49	☽ ⊥ ♄
10 52	☽ △ ♀
11 05	☽ △ ☿
15 15	☽ # ♄
15 25	☽ ⚼ ♅
17 19	☿ ☌ ♀
17 32	☽ ∥ ♅
17 53	☽ □ ☉
19 18	☽ ⚼ ♇
23 29	☽ ∠ ♄
22 Thursday	
06 21	☽ ∠ ♃
06 47	♂ ⊻ ♅
11 21	☽ ∥ ☿
13 41	☉ ∠ ♇
18 56	☽ △ ♅
19 25	☽ ⊼ ♂
22 38	☽ △ ♇
22 52	☽ ⊥ ♃
23 34	☽ ∥ ♀
23 Friday	
02 32	☽ ✶ ♄
03 28	☽ ☍ ♆
05 03	☽ Q ☉
05 37	☽ # ♆
06 35	☽ ± ♂
09 40	☽ ⊻ ♃
20 21	☽ □ ☿
22 16	☽ □ ♀
23 01	☽ ⚼ ♂
23 31	☽ Q ♄
24 Saturday	
01 28	☽ ∥ ♇
03 34	☽ ✶ ☉
06 58	☽ ∥ ☉
13 58	♀ ∠ ♂
19 14	☉ ⚼ ♄
23 10	☽ □ ♅
25 Sunday	
01 36	☽ △ ♂
01 53	☽ ∥ ♃
02 33	☽ □ ♇
04 24	☽ Q ☿
05 57	☽ □ ♄
06 44	☽ ∠ ☉
06 57	☽ ⊼ ♆
07 37	☽ Q ♀
09 12	☉ Q ♂
13 29	☽ ☌ ♃
16 51	☽ ± ♆
22 35	☽ ⊥ ☉
26 Monday	
00 56	☽ ✶ ☿
02 16	♂ ⊻ ♇
04 55	☽ Q ♅
05 03	☽ ✶ ♀
07 30	☽ ⚼ ♆
08 08	☽ Q ♇
08 59	☽ ⊻ ☉
27 Tuesday	
00 14	☽ ✶ ♅
01 52	☽ ∠ ☿
03 22	☽ ✶ ♇
04 19	☽ □ ♂
06 23	☽ △ ♄
07 10	☽ ∠ ♀
07 29	☽ △ ♆
14 22	☽ ⊻ ♃
16 33	☽ ⊥ ☿
22 36	☽ ⊥ ♀
23 57	☽ ⊥ ♃
28 Wednesday	
00 03	☽ ∠ ♅
02 15	☽ ⊻ ☿
03 06	☽ ∠ ♇
05 58	☽ ⚼ ♄
08 51	☽ ⊻ ♀
09 50	☽ Q ♂
11 45	☽ ☌ ☉
13 34	☽ ∥ ♃
14 14	☽ ⊥ ♅
14 17	☽ ∠ ♃
17 16	☽ ⊥ ♇
18 25	☉ ∥ ♇
18 45	♀ Q ♃
20 04	☽ ± ♄
23 42	☽ ⊻ ♅
29 Thursday	
02 43	☽ ⊻ ♇
05 27	☽ ✶ ♂
05 28	☽ ⊼ ♄
06 02	♂ ⊼ ♄
06 42	☽ □ ♆
13 07	☽ ∥ ♇
14 05	☽ ∥ ☉
14 12	☽ ✶ ♃
17 43	♆ St D
22 43	♀ Q ♆
30 Friday	
02 31	☽ ☌ ☿
05 41	☉ ⊥ ♅
06 10	☽ ∠ ♂
06 49	☽ # ♆
09 31	☽ Q ♃
11 16	☽ Q ♆
12 23	☽ ☌ ♀
14 00	☉ ∠ ♃
14 07	♂ ⊻ ♆
14 26	☽ ⊻ ☉
21 14	☽ ⊥ ♂
23 38	☽ ☌ ♅
31 Saturday	
00 48	☽ ⊥ ☉
02 40	☽ ☌ ♇
03 55	☽ ∥ ♀
05 16	☽ ☍ ♄
06 43	☽ ✶ ♆
07 24	☽ ⊻ ♂
12 39	☽ ∥ ☿
15 11	☽ □ ♃
16 08	☽ ∥ ♅
16 32	☽ # ♄
16 36	☽ ∠ ☉
16 49	☽ # ♂

AUGUST 1965

LONGITUDES

Date	Sidereal time h m s	Sun ☉ ° ' "	Moon ☽ ° ' "	Moon ☽ 24.00 ° ' "	Mercury ☿ ° '	Venus ♀ ° '	Mars ♂ ° '	Jupiter ♃ ° '	Saturn ♄ ° '	Uranus ♅ ° '	Neptune ♆ ° '	Pluto ♇ ° '
01	08 39 33	09 ♌ 00 42	04 ♎ 47 08	11 ♎ 46 23	00 ♍ 05	08 ♍ 19	18 ♎ 22	22 ♊ 33	16 ♓ 17	12 ♍ 58	17 ♏ 14	14 ♍ 48
02	08 43 30	09 58 08	18 ♎ 38 18	25 ♎ 22 59	00 ♍ 04	09 31	18 57	22 44	16 R 14	13 01	17 14	14 50
03	08 47 26	10 55 34	02 ♏ 00 42	08 ♏ 31 52	29 ♌ 59	10 44	19 33	22 56	16 11	13 05	17 14	14 52
04	08 51 23	11 53 01	14 ♏ 56 56	21 ♏ 16 26	29 R 48	11 56	20 09	23 07	16 07	13 08	17 14	14 54
05	08 55 19	12 50 29	27 ♏ 30 58	03 ♐ 41 07	29 32	13 08	20 45	23 18	16 04	13 11	17 15	14 55
06	08 59 16	13 47 57	09 ♐ 47 30	15 ♐ 50 42	29 10	14 21	21 21	23 30	16 01	13 15	17 15	14 57
07	09 03 12	14 45 26	21 ♐ 51 18	27 ♐ 49 50	28 44	15 33	21 57	23 41	15 57	13 18	17 15	14 59
08	09 07 09	15 42 57	03 ♑ 46 48	09 ♑ 42 39	28 13	16 45	22 34	23 52	15 53	13 22	17 15	15 01
09	09 11 05	16 40 28	15 ♑ 37 50	21 ♑ 32 42	27 37	17 57	23 10	24 03	15 50	13 25	17 16	15 03
10	09 15 02	17 38 00	27 ♑ 27 36	03 ♒ 22 48	26 57	19 09	23 47	24 13	15 46	13 29	17 16	15 05
11	09 18 59	18 35 33	09 ♒ 18 35	15 ♒ 15 09	26 13	20 21	24 23	24 24	15 42	13 32	17 17	15 07
12	09 22 55	19 33 07	21 ♒ 12 42	27 ♒ 11 25	25 27	21 33	25 00	24 35	15 39	13 36	17 17	15 09
13	09 26 52	20 30 42	03 ♓ 11 27	09 ♓ 12 58	24 38	22 45	25 37	24 45	15 35	13 39	17 17	15 11
14	09 30 48	21 28 19	15 ♓ 16 08	21 ♓ 21 07	23 49	23 57	26 14	24 56	15 31	13 43	17 18	15 13
15	09 34 45	22 25 56	27 ♓ 28 06	03 ♈ 37 18	22 58	25 09	26 52	25 06	15 27	13 46	17 18	15 15
16	09 38 41	23 23 36	09 ♈ 48 58	16 ♈ 03 23	22 08	26 21	27 29	25 16	15 23	13 50	17 19	15 17
17	09 42 38	24 21 16	22 ♈ 20 50	28 ♈ 41 40	21 20	27 33	28 06	25 26	15 19	13 53	17 20	15 19
18	09 46 34	25 18 59	05 ♉ 06 16	11 ♉ 34 59	20 34	28 45	28 44	25 36	15 14	13 57	17 20	15 21
19	09 50 31	26 16 42	18 ♉ 08 15	24 ♉ 46 25	19 52	29 ♍ 57	29 ♎ 22	25 46	15 10	14 01	17 21	15 23
20	09 54 28	27 14 28	01 ♊ 29 52	08 ♊ 18 55	19 14	01 ♎ 08	00 ♏ 00	25 56	15 06	14 04	17 22	15 25
21	09 58 24	28 12 16	15 ♊ 13 49	22 ♊ 14 42	18 42	02 20	00 38	26 06	15 02	14 08	17 22	15 27
22	10 02 21	29 ♌ 10 05	29 ♊ 21 38	06 ♋ 34 27	18 15	03 32	01 16	26 15	14 58	14 12	17 23	15 29
23	10 06 17	00 ♍ 07 56	13 ♋ 52 51	21 ♋ 16 22	17 55	04 43	01 54	26 25	14 53	14 15	17 24	15 31
24	10 10 14	01 05 48	28 ♋ 44 16	06 ♌ 15 40	17 42	05 55	02 32	26 34	14 49	14 19	17 25	15 33
25	10 14 10	02 03 42	13 ♌ 49 29	21 ♌ 24 30	17 37	07 07	03 10	26 43	14 44	14 23	17 25	15 35
26	10 18 07	03 01 38	28 ♌ 59 24	06 ♍ 32 51	17 D 39	08 18	03 49	26 52	14 40	14 26	17 26	15 37
27	10 22 03	03 59 35	14 ♍ 03 33	21 ♍ 30 14	17 50	09 30	04 28	27 01	14 36	14 30	17 27	15 39
28	10 26 00	04 57 34	28 ♍ 51 52	06 ♎ 07 32	18 09	10 41	05 06	27 10	14 31	14 34	17 28	15 42
29	10 29 57	05 55 34	13 ♎ 16 33	20 ♎ 18 29	18 35	11 52	05 45	27 19	14 27	14 38	17 29	15 44
30	10 33 53	06 53 35	27 ♎ 13 03	04 ♏ 00 15	19 10	13 04	06 24	27 28	14 22	14 41	17 30	15 46
31	10 37 50	07 ♍ 51 38	10 ♏ 40 13	17 ♏ 13 14	19 ♌ 53	14 ♎ 15	07 ♏ 03	27 ♊ 36	14 ♓ 17	14 ♍ 45	17 ♏ 31	15 ♍ 48

MOON NODES AND LATITUDE

Date	Moon True ☊ ° '	Moon Mean ☊ ° '	Moon ☽ Latitude ° '
01	12 ♊ 05	10 ♊ 44	04 N 42
02	12 R 01	10 40	04 06
03	11 59	10 37	03 17
04	11 D 59	10 34	02 19
05	11 59	10 31	01 17
06	12 00	10 28	00 N 12
07	12 R 00	10 25	00 S 52
08	11 57	10 21	01 53
09	11 52	10 18	02 49
10	11 45	10 15	03 37
11	11 35	10 12	04 16
12	11 24	10 09	04 43
13	11 11	10 05	04 59
14	10 59	10 02	05 01
15	10 48	09 59	04 49
16	10 39	09 56	04 24
17	10 32	09 53	03 46
18	10 29	09 50	02 56
19	10 27	09 46	01 55
20	10 D 27	09 43	00 S 47
21	10 R 27	09 40	00 N 25
22	10 26	09 37	01 38
23	10 23	09 34	02 47
24	10 17	09 31	03 46
25	10 09	09 27	04 30
26	09 59	09 24	04 55
27	09 47	09 21	05 00
28	09 37	09 18	04 44
29	09 28	09 15	04 10
30	09 21	09 11	03 23
31	09 ♊ 17	09 ♊ 08	02 N 25

DECLINATIONS

Date	Sun ☉ ° '	Moon ☽ ° '	Mercury ☿ ° '	Venus ♀ ° '	Mars ♂ ° '	Jupiter ♃ ° '	Saturn ♄ ° '	Uranus ♅ ° '	Neptune ♆ ° '	Pluto ♇ ° '
01	18 N 00	02 N 25	08 N 03	09 N 46	07 S 25	22 N 44	07 S 19	07 N 22	15 S 17	18 N 52
02	17 45	03 S 31	07 52	09 18	07 39	22 44	07 20	07 21	15 17	18 51
03	17 30	09 05	07 44	08 49	07 54	22 45	07 22	07 20	15 17	18 50
04	17 14	14 06	07 38	08 20	08 08	22 46	07 23	07 19	15 18	18 49
05	16 58	18 22	07 35	07 51	08 22	22 46	07 25	07 17	15 18	18 48
06	16 41	21 44	07 34	07 21	08 37	22 47	07 26	07 16	15 18	18 47
07	16 25	24 04	07 37	06 52	08 51	22 47	07 28	07 15	15 18	18 46
08	16 08	25 17	07 42	06 22	09 06	22 48	07 29	07 13	15 18	18 45
09	15 51	25 20	07 49	05 52	09 20	22 49	07 31	07 12	15 18	18 44
10	15 33	24 13	08 00	05 22	09 34	22 49	07 32	07 11	15 18	18 44
11	15 15	22 02	08 13	04 52	09 48	22 50	07 34	07 09	15 19	18 43
12	14 57	18 54	08 29	04 21	10 03	22 50	07 36	07 08	15 19	18 42
13	14 39	14 58	08 47	03 51	10 17	22 50	07 37	07 07	15 19	18 41
14	14 21	10 25	09 07	03 20	10 31	22 51	07 39	07 05	15 19	18 40
15	14 02	05 26	09 28	02 50	10 46	22 51	07 41	07 04	15 19	18 39
16	13 43	00 S 09	09 52	02 19	11 00	22 52	07 42	07 02	15 20	18 38
17	13 24	05 N 13	10 16	01 48	11 14	22 52	07 44	07 01	15 20	18 37
18	13 05	10 28	10 41	01 17	11 28	22 52	07 46	07 00	15 20	18 36
19	12 46	15 23	11 06	00 46	11 42	22 53	07 47	06 58	15 20	18 36
20	12 26	19 42	11 31	00 N 15	11 56	22 53	07 49	06 57	15 20	18 35
21	12 06	23 03	11 56	00 S 16	12 10	22 53	07 51	06 55	15 21	18 34
22	11 46	25 05	12 19	00 47	12 24	22 54	07 53	06 54	15 21	18 33
23	11 26	25 29	12 42	01 18	12 38	22 54	07 55	06 52	15 21	18 32
24	11 05	24 06	13 03	01 49	12 52	22 54	07 56	06 51	15 22	18 31
25	10 45	20 59	13 21	02 20	13 06	22 55	07 58	06 50	15 22	18 30
26	10 24	16 26	13 38	02 51	13 20	22 55	08 00	06 48	15 22	18 30
27	10 03	10 53	13 52	03 22	13 34	22 55	08 02	06 47	15 22	18 29
28	09 42	04 N 48	14 04	03 52	13 47	22 55	08 04	06 45	15 23	18 28
29	09 20	01 S 24	14 13	04 23	14 01	22 56	08 05	06 44	15 23	18 27
30	08 59	07 20	14 19	04 54	14 14	22 56	08 07	06 42	15 23	18 26
31	08 N 37	12 S 44	14 N 21	05 S 25	14 S 28	22 N 56	08 S 09	06 N 41	15 S 24	18 N 25

ZODIAC SIGN ENTRIES

Date	h m	Planets
01	03 54	☽ ♎
03	08 09	☿ ♌
03	08 20	☽ ♏
05	16 49	☽ ♐
08	04 22	☽ ♑
10	17 09	☽ ♒
13	05 37	☽ ♓
15	16 57	☽ ♈
18	02 27	☽ ♉
19	13 06	♀ ♎
20	09 20	☽ ♊
20	12 16	♂ ♏
22	13 04	☽ ♋
23	08 43	☉ ♍
24	14 01	☽ ♌
26	13 36	☽ ♍
28	13 52	☽ ♎
30	16 54	☽ ♏

LATITUDES

Date	Mercury ☿ ° '	Venus ♀ ° '	Mars ♂ ° '	Jupiter ♃ ° '	Saturn ♄ ° '	Uranus ♅ ° '	Neptune ♆ ° '	Pluto ♇ ° '
01	03 S 38	01 N 25	00 S 14	00 S 30	02 S 04	00 N 44	01 N 46	14 N 00
04	04 10	01 21	00 17	00 30	02 04	00 44	01 46	14 00
07	04 35	01 16	00 19	00 30	02 05	00 44	01 46	13 59
10	04 49	01 10	00 22	00 30	02 05	00 44	01 45	13 59
13	04 48	01 04	00 24	00 30	02 06	00 44	01 45	13 58
16	04 31	00 57	00 27	00 30	02 06	00 44	01 45	13 58
19	03 57	00 49	00 29	00 30	02 07	00 44	01 45	13 58
22	03 12	00 41	00 31	00 30	02 07	00 44	01 45	13 58
25	02 18	00 32	00 34	00 30	02 07	00 44	01 45	13 58
28	01 24	00 23	00 36	00 30	02 08	00 44	01 44	13 57
31	00 S 32	00 N 14	00 S 38	00 S 30	02 S 08	00 N 44	01 N 44	13 N 57

DATA

Julian Date	2438974
Delta T	+36 seconds
Ayanamsa	23° 22' 20"
Synetic vernal point	05° ♓ 44' 40"
True obliquity of ecliptic	23° 26' 41"

MOON'S PHASES, APSIDES AND POSITIONS ☽

Date	h m	Phase	Longitude ° '	Eclipse Indicator
04	05 47	☽	11 ♏ 38	
12	08 22	○	19 ♒ 24	
20	03 50	☾	26 ♉ 55	
26	18 50	●	03 ♍ 18	

Day	h m		
10	19 22	Apogee	
25	18 25	Perigee	
01	21 41	0S	
09	00 56	Max dec	25° S 27'
16	12 42	0N	
23	05 34	Max dec	25° N 33'
29	06 32	0S	

LONGITUDES

Date	Chiron ⚷ ° '	Ceres ⚳ ° '	Palias ⚴ ° '	Juno ⚵ ° '	Vesta ⚶ ° '	Black Moon Lilith ⚸ ° '
01	21 ♓ 58	02 ♈ 46	15 ♒ 20	09 ♒ 09	19 ♊ 55	02 ♒ 47
11	21 ♓ 38	02 ♈ 14	12 ♒ 44	06 ♒ 44	23 ♊ 39	03 ♒ 54
21	21 ♓ 15	01 ♈ 08	10 ♒ 16	04 ♒ 31	27 ♊ 14	05 ♒ 01
31	20 ♓ 49	29 ♓ 29	08 ♒ 09	02 ♒ 45	00 ♋ 36	06 ♒ 08

ASPECTARIAN

h m	Aspects
01 Sunday	
00 29	♂ ∥ ♄
04 00	☽ ⊻ ☿
06 34	☉ ⊥ ♇
07 40	☽ ∠ ♆
08 15	♂ # ♅
14 13	☽ ⊥ ☿
18 36	☽ ⊻ ♀
19 46	☽ ⚹ ☉
21 55	☿ St R
23 04	☽ ⊥ ♆
02 Monday	
02 08	☽ ⊻ ♅
05 18	☽ ⊻ ♇
05 46	☽ ∠ ☿
06 00	☽ ⊥ ♀
07 47	☽ ⚻ ♄
09 32	☽ ⊻ ♆
12 35	☽ ☌ ♂
12 41	☽ ⊥ ♅
15 05	♂ ⊥ ♅
15 53	☽ ⊥ ♇
18 16	☉ ± ♄
18 21	☽ ± ♄
18 21	☽ Q ☉
19 23	☽ △ ♃
21 30	♄ # ♅
23 30	☽ ∠ ♀
03 Tuesday	
01 30	☿ # ♂
04 14	☽ # ♅
04 17	☽ ∥ ♄
04 50	☽ ∠ ♅
06 06	☽ # ☿
06 26	☽ ∥ ♂
08 05	☽ ∠ ♇
08 20	☽ ⚹ ☿
10 29	☽ ⚼ ♄
10 51	☽ # ♀
23 03	☽ ⚼ ♃
04 Wednesday	
05 45	☽ ⚹ ♀
05 47	☽ □ ☉
06 11	☽ Q ☿
07 08	☉ ⊻ ♀
08 34	☽ ⚹ ♅
11 54	☽ ⚹ ♇
14 12	☽ △ ♄
16 10	☽ ± ♃
16 20	☽ ☌ ♆
18 16	☽ ∥ ♆
18 29	♀ # ♂
22 21	☽ ⊻ ♂
05 Thursday	
03 46	☽ ⚻ ♃
03 48	♃ ± ♆
04 05	☽ # ☉
06 55	☽ Q ♀
07 29	☽ Q ♅
10 26	☽ ⊥ ♂
10 51	☽ Q ♇
13 05	♀ ☌ ♅
14 45	☽ # ♇
15 48	☽ □ ☿
19 26	♂ ⊥ ♇
21 18	☉ ⊻ ♅
06 Friday	
01 15	☿ ∥ ♀
04 52	☽ ∠ ♂
08 16	♀ # ♄
16 27	♀ ∥ ♅
18 52	☽ □ ♅
20 37	☽ △ ☉
21 32	☽ # ♃
22 01	☽ □ ♀
22 15	☽ □ ♇
07 Saturday	
00 16	☽ □ ♄
00 32	♀ ☌ ♇
02 48	☽ ⊻ ♆
11 54	♂ ± ♄
12 12	☽ ⚹ ♂
14 48	☽ ⊥ ♆
15 43	☽ ☍ ♃
17 56	☉ ⊻ ♇
19 41	♀ ☍ ♄
08 Sunday	
01 15	☽ △ ☿
04 20	♀ ⊥ ♂
05 16	☽ ⚼ ☉
08 55	☽ ∠ ♆
12 13	☽ Q ♄
13 40	☽ Q ♂
16 08	☉ ⚻ ♄
22 11	♀ ⚹ ♆
09 Monday	
01 04	☽ ± ☉
06 11	☽ ⚼ ☿
07 30	☽ △ ♅
10 49	☽ △ ♇
12 24	☽ ⚹ ♄
14 18	☽ ⚻ ☉
15 19	☽ ⚹ ♆
17 15	☽ △ ♀
23 30	☽ ± ☿
10 Tuesday	
02 50	☉ □ ♆
04 08	☽ □ ♂
05 20	☽ ⚻ ♃
11 01	☽ ⚻ ☿
14 04	☽ ⚼ ♅
15 40	☽ Q ♆
17 20	☽ ⚼ ♇
17 41	☽ ± ♃
18 40	☽ ∠ ♄
11 Wednesday	
03 06	☽ ⚼ ♀
04 32	☽ # ♃
07 50	☉ # ♆
08 24	☽ ± ♅
11 36	☽ ± ♇
12 11	☽ ⚼ ♃
12 33	♂ △ ♃
12 48	☽ ⊥ ♄
20 35	☽ ⚻ ♅
23 20	☽ ± ♀
23 45	☽ ⚻ ♇
12 Thursday	
00 51	☽ ⊻ ♄
04 05	☽ □ ♆
08 22	☽ ☍ ☉
12 46	☽ ⚻ ♀
13 22	☽ # ♇
18 52	☽ △ ♃
19 32	☿ ⚹ ♂
19 59	☽ ☍ ☿
20 02	☽ △ ♂
13 Friday	
09 16	☿ ⚹ ♃
10 04	☽ ∥ ♆
13 54	☽ # ☉
14 Saturday	
03 35	☽ ⚼ ♂
08 54	☽ ☍ ♅
10 15	☿ ⊻ ♀
11 31	☽ ∥ ♂
11 53	☽ ☍ ♇
12 29	☽ ☌ ♄
16 01	☽ △ ♆
18 03	☽ # ☿
22 20	☽ ± ♂
15 Sunday	
01 17	☽ ⚻ ☉
01 28	☽ ∥ ♄
03 45	☽ ⚻ ☿
04 17	☽ # ♅
06 59	☽ ☍ ♀
07 18	☽ □ ♃
10 41	♀ □ ♃
10 45	☽ ⚻ ♂
14 03	☽ ± ☉
14 45	☽ ± ☿
19 11	☉ ☌ ☿
21 27	☽ ⚼ ♆
16 Monday	
01 13	☽ # ♀
07 09	☽ ⚼ ☿
09 01	☽ ⚼ ☉
14 54	☽ ± ♆
18 45	☽ Q ♃
19 46	☽ ⚻ ♅
22 06	☽ ∥ ♀
22 32	☽ ⚻ ♇
22 39	☽ ⊻ ♄
17 Tuesday	
02 25	☽ ⚻ ♆
07 18	☽ ± ♅
09 23	☿ ⊥ ♀
10 02	☽ ± ♇
10 02	☽ ⊥ ♄
10 11	☽ △ ☿
11 11	♄ ☍ ♇
16 07	☽ △ ☉
17 56	☽ ⚹ ♃
20 07	☽ ∥ ♅
22 52	☽ ⚻ ♀
23 28	☽ ☍ ♂
23 28	☽ # ♄
18 Wednesday	
00 26	☽ ⚼ ♅
02 57	☽ ∠ ♄
03 05	☽ ⚼ ♇
11 16	☽ ± ♀
11 24	♀ ⊻ ♂
13 05	☽ ∥ ☿
16 56	☽ # ♂
20 42	☉ ⚹ ♃
21 09	♂ ∠ ♅
22 20	☽ ∠ ♃
23 43	☽ ∥ ☉
19 Thursday	
04 26	☽ △ ♅
05 35	☽ ⚼ ♀
06 37	☽ ⚹ ♄
06 58	☽ △ ♇
10 34	☽ ☍ ♆
11 44	☽ # ♆
15 00	☽ □ ☿
15 00	☽ ⊥ ♃
20 Friday	
01 58	☽ ⊻ ♃
03 50	☽ □ ☉
04 13	☽ Q ♄
05 21	☽ ∥ ♇
09 12	☽ ⚻ ♂
11 18	☽ △ ♀
15 42	♂ ⚼ ♄
20 19	☽ ± ♂
21 43	☽ Q ☿
21 Saturday	
04 58	♂ ∠ ♇
09 01	☉ # ♂
10 06	☽ □ ♅
10 40	☽ ∥ ♃
11 39	☽ □ ♄
12 23	☽ □ ♇
12 43	☽ ⚼ ♂
12 43	♀ ∠ ♆
13 30	☿ Q ♂
13 48	☽ Q ☉
15 41	☽ ⚻ ♆
17 38	☉ ∥ ☿
17 45	☽ ⚹ ☿
22 Sunday	
01 55	☽ ± ♆
06 43	☽ ☌ ♃
07 48	☿ ∠ ♀
11 39	☽ ⚹ ☉
15 19	☽ △ ♂
16 45	☽ Q ♅
17 03	☽ ⚼ ♆
18 20	☽ ∠ ☿
18 54	☽ Q ♇
19 35	☽ □ ♀
23 Monday	
01 21	☿ # ♂
08 51	☽ ⊥ ☿
12 37	☽ ⚹ ♅
13 38	☽ △ ♄
14 11	☽ ∠ ☉
14 41	☽ ⚹ ♇
17 44	☽ △ ♆
18 27	☽ ⊻ ☿
24 Tuesday	
03 36	☽ Q ♀
05 45	☽ ⊥ ☉
08 29	☽ ⊻ ♃
12 56	☽ ∠ ♅
13 43	☽ ⚼ ♄
14 55	☽ ∠ ♇
16 02	☽ ⊻ ☉
18 11	☽ ⊥ ♃
18 20	☽ □ ♂
22 43	☽ ∥ ♃
25 Wednesday	
00 26	☽ ⚹ ♀
03 20	☽ ⊥ ♅
03 59	☽ ± ♄
05 16	☽ ⊥ ♇
08 38	☽ ∠ ♃
12 53	☽ ⊻ ♅
13 27	☽ ⚻ ♄
14 48	☽ ⊻ ♇
16 24	☿ St D
17 42	☽ □ ♆
18 00	☽ ☌ ☿
26 Thursday	
00 08	☽ Q ♂
01 53	☽ ∥ ♇
02 14	☽ ∠ ♀
08 37	☽ ⚹ ♃
16 52	☽ # ♆
17 42	☽ ⊥ ♀
18 50	☽ ☌ ☉
20 00	☽ ⚹ ♂
22 15	☽ Q ♆
23 57	☽ ∥ ☿
27 Friday	
01 16	☽ # ♂
03 52	☽ Q ♃
04 04	☽ ⊻ ♀
12 43	☽ ☌ ♅
12 51	☽ ☍ ♄
14 34	☽ ☌ ♇
15 34	☽ ∥ ☉
17 28	☽ ⚹ ♆
18 11	☽ ⊻ ☿
21 05	☽ ∠ ♂
23 19	☽ # ♄
28 Saturday	
03 36	♄ ☍ ♅
04 06	☽ ⊥ ☿
04 21	☽ ∥ ♅
09 12	☽ □ ♃
12 25	☽ ⊥ ♂
15 17	☽ # ♀
17 57	☽ ∠ ♆
19 15	☽ ∠ ☿
22 47	☽ ⊻ ☉
22 47	☽ ⊻ ♂
22 52	☉ ⚹ ♂
29 Sunday	
00 52	☉ Q ♆
04 02	♀ ⊥ ♆
08 58	☽ ⊥ ♆
09 25	☽ ☌ ♀
09 33	☽ ⊥ ☉
13 58	☽ ⚻ ♄
14 18	☽ ⊻ ♅
16 10	☽ ⊻ ♇
17 35	☉ ⊥ ♀
19 10	☽ ⊻ ♆
21 25	☽ ⚹ ☿
30 Monday	
00 10	☽ ± ♄
00 36	☽ ⊥ ♅
01 00	☽ ∥ ♀
02 02	☽ ∠ ☉
02 29	☽ ⊥ ♇
09 23	☽ # ♅
12 26	☽ △ ♃
15 22	☽ ∥ ♄
15 45	☽ ⚼ ♄
16 22	☽ ∠ ♅
18 16	☽ ∠ ♇
18 37	☽ # ☉
19 20	☽ Q ☿
22 08	☿ # ♂
31 Tuesday	
05 07	☽ ☌ ♂
06 31	☽ ⚹ ☉
12 45	♀ ⚻ ♄
15 33	☽ ⚼ ♃
18 34	☽ △ ♄
19 11	☽ ⊻ ♀
19 29	☽ ⚹ ♅
19 55	☽ # ☿
20 53	☽ ∥ ♂
21 24	☽ ⚹ ♇
22 41	♀ ⊻ ♅

All ephemeris data is given at 12.00 UT and the Moon's longitude is additionally given for 24.00 UT

SEPTEMBER 1965

LONGITUDES

Date	Sidereal time h m s	Sun ☉ ° ' "	Moon ☽ ° ' "	Moon ☽ 24.00 ° ' "	Mercury ☿ ° '	Venus ♀ ° '	Mars ♂ ° '	Jupiter ♃ ° '	Saturn ♄ ° '	Uranus ♅ ° '	Neptune ♆ ° '	Pluto ♇ ° '
01	10 41 46	08 ♍ 49 43	23 ♏ 39 43	00 ♐ 00 12	20 ♌ 44	15 ♎ 26	07 ♏ 42	27 ♊ 44	14 ♓ 13	14 ♍ 49	17 ♏ 32	15 ♍ 50
02	10 45 43	09 47 49	06 ♐ 15 16	12 ♐ 25 32	21 41	16 38	08 22	27 53	14 R 08	14 53	17 33	15 52
03	10 49 39	10 45 56	18 ♐ 31 41	24 ♐ 34 23	22 46	17 49	09 01	28 01	14 04	14 56	17 34	15 54
04	10 53 36	11 44 04	00 ♑ 34 16	06 ♑ 31 59	23 57	19 00	09 40	28 09	13 59	15 00	17 35	15 56
05	10 57 32	12 42 14	12 ♑ 28 11	18 ♑ 23 24	25 15	20 11	10 20	28 17	13 55	15 04	17 37	15 59
06	11 01 29	13 40 26	24 ♑ 18 12	00 ♒ 13 03	26 38	21 22	11 00	28 24	13 50	15 08	17 38	16 01
07	11 05 26	14 38 39	06 ♒ 08 24	12 ♒ 04 39	28 06	22 33	11 39	28 32	13 45	15 12	17 39	16 03
08	11 09 22	15 36 54	18 ♒ 02 06	24 ♒ 01 04	29 ♌ 39	23 44	12 19	28 39	13 41	15 15	17 40	16 05
09	11 13 19	16 35 10	00 ♓ 01 45	06 ♓ 04 22	01 ♍ 16	24 54	12 59	28 46	13 36	15 19	17 42	16 07
10	11 17 15	17 33 28	12 ♓ 09 01	18 ♓ 15 51	02 56	26 05	13 39	28 54	13 32	15 23	17 43	16 09
11	11 21 12	18 31 48	24 ♓ 24 57	00 ♈ 36 21	04 40	27 16	14 19	29 00	13 27	15 27	17 44	16 11
12	11 25 08	19 30 09	06 ♈ 50 08	13 ♈ 06 22	06 26	28 26	15 00	29 07	13 23	15 30	17 46	16 14
13	11 29 05	20 28 33	19 ♈ 25 06	25 ♈ 46 26	08 14	29 ♎ 37	15 40	29 14	13 18	15 34	17 47	16 16
14	11 33 01	21 26 58	02 ♉ 10 29	08 ♉ 37 22	10 03	00 ♏ 47	16 20	29 20	13 13	15 38	17 48	16 18
15	11 36 58	22 25 26	15 ♉ 07 16	21 ♉ 40 23	11 54	01 58	17 01	29 27	13 09	15 42	17 50	16 20
16	11 40 55	23 23 56	28 ♉ 16 56	04 ♊ 57 08	13 45	03 08	17 42	29 33	13 04	15 45	17 51	16 22
17	11 44 51	24 22 28	11 ♊ 41 15	18 ♊ 29 29	15 37	04 19	18 22	29 39	13 00	15 49	17 53	16 24
18	11 48 48	25 21 02	25 ♊ 22 03	02 ♋ 19 04	17 30	05 29	19 03	29 45	12 56	15 53	17 54	16 26
19	11 52 44	26 19 39	09 ♋ 20 37	16 ♋ 26 40	19 22	06 39	19 44	29 51	12 51	15 57	17 56	16 28
20	11 56 41	27 18 17	23 ♋ 37 04	00 ♌ 51 32	21 14	07 49	20 25	29 ♊ 56	12 47	16 00	17 57	16 31
21	12 00 37	28 16 58	08 ♌ 09 36	15 ♌ 30 40	23 06	08 59	21 06	00 ♋ 02	12 42	16 04	17 59	16 33
22	12 04 34	29 ♍ 15 42	22 ♌ 54 00	00 ♍ 18 43	24 58	10 09	21 48	00 07	12 38	16 08	18 01	16 35
23	12 08 30	00 ♎ 14 27	07 ♍ 43 48	15 ♍ 08 12	26 48	11 19	22 29	00 12	12 34	16 12	18 02	16 37
24	12 12 27	01 13 14	22 ♍ 30 49	29 ♍ 50 38	28 ♍ 39	12 29	23 10	00 17	12 30	16 15	18 04	16 39
25	12 16 24	02 12 03	07 ♎ 06 37	14 ♎ 17 55	00 ♎ 28	13 39	23 52	00 21	12 25	16 19	18 06	16 41
26	12 20 20	03 10 55	21 ♎ 23 48	28 ♎ 23 42	02 17	14 48	24 33	00 26	12 21	16 23	18 07	16 43
27	12 24 17	04 09 48	05 ♏ 17 14	12 ♏ 04 11	04 05	15 58	25 15	00 30	12 17	16 26	18 09	16 45
28	12 28 13	05 08 43	18 ♏ 44 33	25 ♏ 18 26	05 52	17 08	25 57	00 35	12 13	16 30	18 11	16 47
29	12 32 10	06 07 40	01 ♐ 46 05	08 ♐ 07 53	07 38	18 17	26 39	00 39	12 09	16 34	18 12	16 49
30	12 36 06	07 ♎ 06 39	14 ♐ 24 18	20 ♐ 35 52	09 ♎ 24	19 ♏ 26	27 ♏ 21	00 ♋ 42	12 ♓ 05	16 ♍ 37	18 ♏ 14	16 ♍ 51

Date	Moon True ☊ ° '	Moon Mean ☊ ° '	Moon ☽ Latitude ° '
01	09 ♊ 15	09 ♊ 05	01 N 21
02	09 D 15	09 02	00 N 16
03	09 R 15	08 59	00 S 49
04	09 14	08 56	01 50
05	09 11	08 52	02 46
06	09 06	08 49	03 34
07	08 58	08 46	04 13
08	08 47	08 43	04 41
09	08 35	08 40	04 57
10	08 21	08 37	04 59
11	08 08	08 33	04 48
12	07 55	08 30	04 23
13	07 45	08 27	03 45
14	07 38	08 24	02 55
15	07 34	08 21	01 56
16	07 32	08 17	00 S 49
17	07 D 32	08 14	00 N 22
18	07 32	08 11	01 33
19	07 R 32	08 08	02 40
20	07 29	08 05	03 39
21	07 24	08 02	04 24
22	07 17	07 58	04 53
23	07 08	07 55	05 02
24	06 57	07 52	04 52
25	06 47	07 49	04 22
26	06 39	07 46	03 36
27	06 33	07 42	02 38
28	06 29	07 39	01 33
29	06 27	07 36	00 25
30	06 ♊ 28	07 ♊ 33	00 S 42

DECLINATIONS

Date	Sun ☉ ° '	Moon ☽ ° '	Mercury ☿ ° '	Venus ♀ ° '	Mars ♂ ° '	Jupiter ♃ ° '	Saturn ♄ ° '	Uranus ♅ ° '	Neptune ♆ ° '	Pluto ♇ ° '
01	08 N 16	17 S 23	14 N 21	05 S 55	14 S 41	22 N 56	08 S 11	06 N 39	15 S 24	18 N 24
02	07 54	21 06	14 17	06 26	14 55	22 56	08 13	06 38	15 24	18 24
03	07 32	23 46	14 09	06 56	15 08	22 56	08 15	06 37	15 25	18 23
04	07 10	25 17	13 59	07 26	15 21	22 56	08 17	06 35	15 25	18 22
05	06 48	25 37	13 44	07 56	15 34	22 57	08 18	06 34	15 26	18 21
06	06 25	24 46	13 27	08 26	15 47	22 57	08 20	06 32	15 26	18 20
07	06 03	22 49	13 06	08 56	16 00	22 57	08 22	06 31	15 26	18 20
08	05 40	19 53	12 43	09 25	16 13	22 57	08 24	06 29	15 27	18 19
09	05 18	16 05	12 16	09 55	16 26	22 57	08 26	06 28	15 27	18 18
10	04 55	11 37	11 47	10 24	16 39	22 57	08 27	06 26	15 28	18 17
11	04 32	06 38	11 15	10 53	16 51	22 57	08 29	06 25	15 28	18 16
12	04 09	01 S 19	10 40	11 22	17 04	22 57	08 31	06 23	15 28	18 16
13	03 46	04 N 07	10 04	11 50	17 16	22 57	08 33	06 22	15 29	18 15
14	03 23	09 29	09 26	12 18	17 29	22 57	08 35	06 20	15 29	18 14
15	03 00	14 32	08 46	12 46	17 41	22 57	08 36	06 19	15 30	18 13
16	02 37	18 59	08 04	13 14	17 53	22 57	08 38	06 18	15 30	18 13
17	02 14	22 33	07 21	13 42	18 05	22 58	08 40	06 16	15 31	18 12
18	01 51	24 55	06 37	14 09	18 17	22 58	08 41	06 15	15 31	18 11
19	01 28	25 47	05 52	14 36	18 28	22 58	08 43	06 13	15 32	18 10
20	01 04	24 58	05 07	15 02	18 40	22 58	08 45	06 12	15 32	18 10
21	00 41	22 29	04 21	15 28	18 51	22 58	08 47	06 10	15 32	18 09
22	00 N 18	18 30	03 34	15 54	19 03	22 58	08 48	06 09	15 33	18 08
23	00 S 06	13 21	02 47	16 20	19 14	22 58	08 50	06 07	15 33	18 08
24	00 29	07 26	02 00	16 45	19 25	22 58	08 51	06 06	15 34	18 07
25	00 53	01 N 11	01 13	17 10	19 36	22 58	08 53	06 05	15 34	18 06
26	01 16	05 S 00	00 N 26	17 34	19 47	22 58	08 55	06 03	15 35	18 06
27	01 39	10 48	00 S 21	17 58	19 57	22 58	08 56	06 02	15 35	18 05
28	02 03	15 55	01 09	18 22	20 08	22 58	08 58	06 00	15 36	18 04
29	02 26	20 07	01 55	18 45	20 18	22 58	08 59	05 59	15 37	18 04
30	02 S 49	23 S 14	02 S 42	19 S 08	20 S 28	22 N 58	09 S 01	05 N 58	15 S 37	18 N 03

ZODIAC SIGN ENTRIES

Date	h	m	Planets
02	00	00	☽ ♐
04	10	51	☽ ♑
06	23	34	☽ ♒
08	17	14	☿ ♍
09	11	57	☽ ♓
11	22	50	☽ ♈
13	19	50	♀ ♏
14	07	56	☽ ♉
16	15	06	☽ ♊
18	20	01	☽ ♋
20	22	35	☽ ♌
21	04	39	♃ ♋
22	23	30	☽ ♍
23	06	06	☉ ♎
25	00	15	☽ ♎
25	05	49	☿ ♎
27	02	47	☽ ♏
29	08	42	☽ ♐

LATITUDES

Date	Mercury ☿ ° '	Venus ♀ ° '	Mars ♂ ° '	Jupiter ♃ ° '	Saturn ♄ ° '	Uranus ♅ ° '	Neptune ♆ ° '	Pluto ♇ ° '
01	00 S 15	00 N 10	00 S 38	00 S 29	02 S 08	00 N 44	01 N 44	13 N 57
04	00 N 28	00 00	00 40	00 29	02 08	00 44	01 44	13 57
07	01 02	00 S 10	00 42	00 29	02 09	00 44	01 44	13 58
10	01 27	00 21	00 44	00 29	02 09	00 44	01 44	13 58
13	01 42	00 32	00 46	00 29	02 09	00 44	01 44	13 58
16	01 49	00 43	00 48	00 29	02 09	00 44	01 43	13 58
19	01 49	00 54	00 50	00 29	02 09	00 44	01 43	13 59
22	01 42	01 06	00 51	00 29	02 09	00 44	01 43	13 59
25	01 32	01 17	00 53	00 29	02 09	00 44	01 43	14 00
28	01 18	01 29	00 54	00 29	02 09	00 44	01 43	14 00
31	01 N 01	01 S 40	00 S 56	00 S 29	02 S 09	00 N 44	01 N 43	14 N 01

DATA

Julian Date	2439005
Delta T	+36 seconds
Ayanamsa	23° 22' 24"
Synetic vernal point	05° ♓ 44' 36"
True obliquity of ecliptic	23° 26' 41"

MOON'S PHASES, APSIDES AND POSITIONS ☽

Date	h	m	Phase	Longitude ° '	Eclipse Indicator
02	19	27	☽	10 ♐ 06	
10	23	32	○	18 ♓ 01	
18	11	58	☾	25 ♊ 21	
25	03	18	●	01 ♎ 51	

Day	h	m		
07	03	56	Apogee	
22	23	13	Perigee	
05	06	42	Max dec	25° S 39'
12	17	49	0N	
19	12	37	Max dec	25° N 47'
25	16	32	0S	

LONGITUDES

Date	Chiron ⚷ ° '	Ceres ⚳ ° '	Pallas ⚴ ° '	Juno ⚵ ° '	Vesta ⚶ ° '	Black Moon Lilith ⚸ ° '
01	20 ♓ 46	29 ♓ 18	07 ♒ 58	02 ♒ 36	00 ♋ 56	06 ♒ 15
11	20 ♓ 19	27 ♓ 14	06 ♒ 24	01 ♒ 30	04 ♋ 02	07 ♒ 22
21	19 ♓ 51	25 ♓ 00	05 ♒ 25	01 ♒ 06	06 ♋ 51	08 ♒ 29
31	19 ♓ 24	22 ♓ 50	05 ♒ 02	01 ♒ 25	09 ♋ 19	09 ♒ 36

ASPECTARIAN

h m	Aspects
01 Wednesday	
00 34	☽ ☌ ♆
01 14	☽ ∥ ♆
06 06	☽ □ ☿
06 16	☽ Q ☉
07 25	☽ ⊥ ♀
08 22	☽ ± ♃
16 47	☉ # ♄
17 58	☽ Q ♅
18 01	☽ # ♇
19 47	☽ ⊼ ♃
19 54	☽ Q ♇
20 14	♀ ⊻ ♇
02 Thursday	
02 10	☽ ∠ ♀
14 21	☉ Q ♃
16 18	☽ ⊻ ♂
19 27	☽ □ ☉
21 18	♀ # ♅
03 Friday	
03 16	☽ □ ♄
03 26	☽ # ♃
04 41	☽ ⊥ ♂
04 54	☽ □ ♅
06 49	☽ □ ♇
07 05	♀ ⊻ ♆
10 07	☽ ⊻ ♆
10 26	☽ ⚹ ♀
21 18	☽ △ ☿
22 01	☽ ⊥ ♆
23 31	☽ ∠ ♂
04 Saturday	
04 29	☉ # ♀
07 05	☽ ☍ ♃
12 57	☽ Q ♀
14 49	☽ Q ♄
16 04	☽ ∠ ♆
19 20	♂ ∥ ♆
05 Sunday	
06 51	♀ ± ♄
06 56	☽ bQ ☿
07 25	☽ ⚹ ♂
12 31	☽ △ ☉
14 54	☽ ⚹ ♄
17 17	☽ △ ♅
19 08	☽ △ ♇
22 26	☽ ⚹ ♆
06 Monday	
03 33	☽ ± ☿
04 08	☉ ∥ ♅
05 22	☽ □ ♀
06 53	♀ ∥ ♄
07 00	♀ ⊥ ♅
09 11	☽ Q ♂
15 39	☉ ☍ ♄
17 23	☽ ⊼ ☿
20 25	☽ ⊼ ♃
21 08	☽ ∠ ♄
21 40	☽ bQ ☉
22 50	☽ Q ♆
23 53	☽ bQ ♅
07 Tuesday	
01 35	♀ ⊥ ♇
01 39	☽ bQ ♇
08 42	☽ ± ♃
10 46	☽ # ♃
15 15	☽ ⊥ ♄
17 31	☽ ± ☉
18 12	☽ ± ♅
19 16	☿ ⚹ ♃
19 56	☽ ± ♇
23 49	☽ □ ♂
08 Wednesday	
02 30	☉ ☌ ♅
03 05	☽ bQ ♃
03 17	☽ ⊻ ♄
06 23	☽ ⊼ ♅
06 42	☽ ⊼ ☉
08 04	☽ ⊼ ♇
11 16	☽ □ ♆
22 33	☽ # ♇
09 Thursday	
00 02	☉ ☌ ♇
00 40	☽ △ ♀
05 02	☿ Q ♂
09 28	☽ △ ♃
10 05	☽ ∥ ♂
14 51	☽ ☍ ☿
15 36	☽ ∥ ♆
10 Friday	
07 54	♂ △ ♄
09 41	☽ bQ ♀
11 03	☽ # ☿
14 42	☽ ☌ ♄
15 08	☽ △ ♂
15 58	☉ ⚹ ♆
17 29	☽ ∥ ♀
18 23	☽ ☍ ♅
19 53	☽ ☍ ♇
22 20	♂ bQ ♃
22 57	☽ △ ♆
23 32	☽ ☍ ☉
11 Saturday	
03 19	☽ ∥ ♄
05 13	☽ ± ♀
12 59	☽ # ♅
18 07	☽ ⊼ ♀
20 22	☿ # ♀
21 00	☽ □ ♃
22 04	☽ bQ ♂
22 18	☽ # ☉
12 Sunday	
02 51	☿ Q ♆
04 09	☽ bQ ♆
10 45	♀ bQ ♄
11 05	☽ ⊼ ☿
16 23	☽ ± ♂
21 27	☽ ± ♆
13 Monday	
00 26	☽ ⊻ ♄
00 29	☽ ± ☿
03 22	♀ △ ♃
04 29	☽ ⊼ ♂
04 39	☽ ⊼ ♅
06 00	☽ ⊼ ♇
07 49	☽ Q ♃
08 10	♂ ⚹ ♅
08 54	☽ ⊼ ♆
10 33	☽ ∥ ☉
11 47	☽ ⊥ ♄
14 10	☽ ⊼ ☉
16 05	☽ ± ♅
17 24	☽ ± ♇
20 24	☽ bQ ☿
21 53	☽ ∥ ♅
14 Tuesday	
02 25	☽ ± ☉
04 39	☽ ∠ ♄
06 39	☽ ⚹ ♃
07 50	☽ # ♄
08 34	♀ ∠ ♅
09 06	☽ bQ ♅
09 09	☽ ☍ ♀
10 21	♂ ⚹ ♇
10 21	☽ bQ ♇
11 46	☽ ∥ ☿
20 37	☽ bQ ☉
22 40	♀ ∠ ♇
15 Wednesday	
02 31	☽ # ♀
05 04	☽ △ ☿
05 50	☿ Q ♃
08 23	☽ ⚹ ♄
10 45	☽ ∠ ♃
13 04	☽ △ ♅
14 14	☽ △ ♇
15 40	☽ ☍ ♂
16 54	☽ # ♆
16 59	☽ ☍ ♆
17 12	☿ # ♄
16 Thursday	
02 26	☽ △ ☉
03 21	☽ ⊥ ♃
03 35	☿ ☍ ♄
05 19	☽ # ♂
06 13	☽ Q ♄
07 31	☽ ∥ ♇
14 19	☽ ⊻ ♃
17 53	♂ ☌ ♆
21 35	☽ ⊼ ♀
17 Friday	
09 19	☽ ± ♀
14 19	☽ □ ♄
14 39	☿ ☌ ♅
15 18	☽ ∥ ♃
19 20	☽ □ ♅
20 04	☽ □ ☿
20 21	☽ □ ♇
22 14	☿ ☌ ♇
22 57	☽ ⊼ ♆
18 Saturday	
00 25	☽ ⊼ ♂
01 42	♂ # ♇
02 41	☽ bQ ♀
09 27	☽ ± ♆
11 26	☽ ± ♂
11 58	☽ □ ☉
17 21	☿ ⚹ ♆
19 38	☽ ☌ ♃
19 Sunday	
00 36	☿ ∥ ♅
01 02	☽ bQ ♆
02 45	☽ Q ♅
03 41	☽ Q ♇
03 45	☽ bQ ♂
07 00	☽ △ ♀
08 07	☽ Q ☿
17 55	☽ △ ♄
19 29	☿ ⚹ ♂
21 03	☽ Q ☉
23 13	☽ ⚹ ♅
20 Monday	
00 05	☽ ⚹ ♇
02 31	☽ △ ♆
06 24	☽ △ ♂
07 26	☽ ⚹ ☿
18 34	☽ ⚹ ☉
18 53	☽ bQ ♄
22 33	☽ ⊻ ♃
21 Tuesday	
00 18	☽ ∠ ♅
01 06	☽ ∠ ♇
08 23	☽ ∥ ♃
08 29	☽ ⊥ ♃
09 38	☽ ± ♄
11 53	☽ ∠ ☿
13 28	☽ □ ♀
15 08	☽ ⊥ ♅
15 45	♀ ∥ ♆
15 55	☽ ⊥ ♇
19 24	☽ ⊼ ♄
20 58	☽ ∠ ☉
23 17	☽ ∠ ♃
22 Wednesday	
00 58	☽ ⊻ ♅
01 43	☽ ⊻ ♇
04 03	☽ □ ♆
04 41	☽ ⊥ ☿
09 15	☽ # ♂
10 07	☽ □ ♂
12 38	☽ ⊥ ☉
13 51	☽ ∥ ♇
15 49	☽ ⊻ ☿
18 49	☿ ∠ ♀
21 14	☽ Q ♀
23 02	☽ ⊻ ☉
23 40	☽ # ♀
23 45	☽ ⚹ ♃
23 Thursday	
02 15	☽ # ♆
09 15	☽ Q ♆
10 52	☉ □ ♃
16 40	☽ Q ♂
18 18	☽ ⚹ ♀
19 17	☽ Q ♃
19 47	☽ ☍ ♄
24 Friday	
01 46	☽ ☌ ♅
02 26	☽ ☌ ♇
04 44	☽ ⚹ ♆
06 26	☽ # ♄
12 12	♀ △ ♄
13 08	☽ ⚹ ♂
17 12	☽ ∥ ♅
20 50	☽ ∠ ♀
23 28	☽ ☌ ☿
25 Saturday	
00 47	☽ □ ♃
03 18	☽ ☌ ☉
05 20	☽ ∠ ♆
10 28	☿ □ ♃
11 52	☽ ∥ ☿
12 58	☽ ⊥ ♀
13 07	☽ # ☉
15 04	☽ ∠ ♂
18 57	☉ # ☿
20 10	☽ # ☿
20 19	☽ ⊥ ♆
20 25	☽ ∥ ☉
20 49	☽ ⊼ ♄
23 52	☽ ⊻ ♀
26 Sunday	
03 28	☽ ⊻ ♅
04 03	☽ ⊻ ♇
06 26	☽ ⊻ ♆
06 52	☽ ± ♄
06 56	☽ ⊥ ♂
07 17	♂ ± ♃
10 27	☉ ∠ ♆
13 41	☽ ⊥ ♅
14 16	☽ ⊥ ♇
16 10	☽ # ♅
17 41	☽ ⊻ ♂
22 09	☽ bQ ♄
23 21	☿ ∠ ♆
27 Monday	
01 49	♀ bQ ♃
03 37	☽ △ ♃
03 58	☽ ∥ ♄
05 15	☽ ∠ ♅
05 49	☽ ∠ ♇
09 34	☽ ⊻ ☿
09 53	☽ ⊻ ☉
14 27	☉ ☌ ☿
18 26	♀ # ♇
21 17	☽ ⊥ ☉
21 45	☽ ⊥ ☿
22 16	♀ ⚹ ♅
28 Tuesday	
00 19	☽ △ ♄
04 47	♀ ⚹ ♇
06 15	☽ bQ ♃
07 55	☽ ⚹ ♅
08 27	☽ ⚹ ♇
08 48	☽ ☌ ♀
10 25	☽ ∥ ♆
10 59	☽ ☌ ♆
14 45	☽ ∠ ☉
16 28	☽ ∠ ☿
22 43	☽ ± ♃
23 36	☽ # ♇
29 Wednesday	
01 56	☽ ☌ ♂
02 43	☽ ∥ ♀
05 59	☽ Q ♅
06 30	☽ Q ♇
09 53	☽ ⊼ ♃
10 22	♀ ☌ ♆
13 18	☽ ∥ ♂
20 53	☽ ⚹ ☉
30 Thursday	
00 51	☽ ⚹ ☿
07 34	☽ □ ♄
09 29	☽ # ♃
16 18	☽ □ ♅
16 45	☽ □ ♇
19 26	☽ ⊻ ♆
19 40	☉ ∥ ☿
21 54	☽ Q ☉
22 45	☽ ⊻ ♀

All ephemeris data is given at 12.00 UT and the Moon's longitude is additionally given for 24.00 UT

OCTOBER 1965

LONGITUDES

Date	Sidereal time h m s	Sun ☉ ° ′ ″	Moon ☽ ° ′ ″	Moon ☽ 24.00 ° ′ ″	Mercury ☿ ° ′	Venus ♀ ° ′	Mars ♂ ° ′	Jupiter ♃ ° ′	Saturn ♄ ° ′	Uranus ♅ ° ′	Neptune ♆ ° ′	Pluto ♇ ° ′
01	12 40 03	08 ♎ 05 39	26 ♐ 43 11	02 ♑ 46 53	11 ♎ 08	20 ♏ 36	28 ♏ 03	00 ♋ 46	12 ♓ 01	16 ♍ 41	18 ♏ 16	16 ♍ 53
02	12 43 59	09 04 41	08 ♑ 47 37	14 ♑ 46 03	12 52	21 45	28 45	00 49	11 R 57	16 44	18 18	16 55
03	12 47 56	10 03 45	20 ♑ 42 49	26 ♑ 38 35	14 35	22 54	29 ♏ 27	00 53	11 54	16 48	18 20	16 57
04	12 51 53	11 02 51	02 ♒ 33 57	08 ♒ 29 30	16 16	24 03	00 ♐ 09	00 56	11 50	16 51	18 22	16 59
05	12 55 49	12 01 59	14 ♒ 25 48	20 ♒ 23 20	17 58	25 12	00 52	00 59	11 46	16 55	18 23	17 01
06	12 59 46	13 01 08	26 ♒ 22 33	02 ♓ 23 51	19 38	26 21	01 34	01 01	11 43	16 58	18 25	17 03
07	13 03 42	14 00 19	08 ♓ 27 35	14 ♓ 34 00	21 17	27 29	02 17	01 04	11 39	17 02	18 27	17 05
08	13 07 39	14 59 32	20 ♓ 43 22	26 ♓ 55 47	22 56	28 38	02 59	01 06	11 36	17 05	18 29	17 07
09	13 11 35	15 58 47	03 ♈ 11 24	09 ♈ 30 15	24 34	29 ♏ 46	03 42	01 08	11 32	17 09	18 31	17 09
10	13 15 32	16 58 04	15 ♈ 52 20	22 ♈ 17 37	26 11	00 ♐ 55	04 25	01 10	11 29	17 12	18 33	17 11
11	13 19 28	17 57 23	28 ♈ 46 03	05 ♉ 17 33	27 48	02 03	05 07	01 12	11 26	17 16	18 35	17 13
12	13 23 25	18 56 44	11 ♉ 52 01	18 ♉ 29 21	29 ♎ 23	03 11	05 50	01 14	11 23	17 19	18 37	17 15
13	13 27 22	19 56 08	25 ♉ 09 29	01 ♊ 52 21	00 ♏ 58	04 19	06 33	01 15	11 20	17 22	18 39	17 17
14	13 31 18	20 55 33	08 ♊ 37 52	15 ♊ 26 00	02 33	05 27	07 17	01 16	11 17	17 26	18 41	17 19
15	13 35 15	21 55 01	22 ♊ 16 44	29 ♊ 10 03	04 06	06 35	08 00	01 17	11 14	17 29	18 43	17 21
16	13 39 11	22 54 32	06 ♋ 05 54	13 ♋ 04 17	05 39	07 42	08 43	01 18	11 11	17 32	18 45	17 22
17	13 43 08	23 54 04	20 ♋ 05 07	27 ♋ 08 18	07 12	08 50	09 26	01 18	11 08	17 35	18 47	17 24
18	13 47 04	24 53 39	04 ♌ 13 42	11 ♌ 21 06	08 43	09 57	10 10	01 19	11 05	17 39	18 49	17 26
19	13 51 01	25 53 16	18 ♌ 30 12	25 ♌ 40 38	10 15	11 04	10 53	01 19	11 03	17 42	18 52	17 28
20	13 54 57	26 52 56	02 ♍ 51 58	10 ♍ 03 39	11 45	12 11	11 37	01 R 19	11 00	17 45	18 54	17 30
21	13 58 54	27 52 37	17 ♍ 15 06	24 ♍ 25 40	13 15	13 18	12 20	01 19	10 58	17 48	18 56	17 31
22	14 02 51	28 52 21	01 ♎ 34 40	08 ♎ 41 27	14 44	14 25	13 04	01 19	10 56	17 51	18 58	17 33
23	14 06 47	29 ♎ 52 07	15 ♎ 45 19	22 ♎ 45 41	16 13	15 32	13 48	01 18	10 53	17 54	19 00	17 35
24	14 10 44	00 ♏ 51 55	29 ♎ 41 59	06 ♏ 33 46	17 41	16 38	14 32	01 17	10 51	17 57	19 02	17 36
25	14 14 40	01 51 45	13 ♏ 20 39	20 ♏ 02 25	19 08	17 44	15 16	01 16	10 49	18 00	19 04	17 38
26	14 18 37	02 51 37	26 ♏ 38 54	03 ♐ 10 06	20 35	18 50	16 00	01 14	10 47	18 03	19 07	17 40
27	14 22 33	03 51 31	09 ♐ 36 06	15 ♐ 57 05	22 01	19 56	16 44	01 13	10 45	18 06	19 09	17 41
28	14 26 30	04 51 28	22 ♐ 13 20	28 ♐ 25 14	23 26	21 02	17 28	01 11	10 44	18 09	19 11	17 43
29	14 30 26	05 51 24	04 ♑ 33 11	10 ♑ 37 41	24 50	22 07	18 12	01 09	10 42	18 12	19 13	17 45
30	14 34 23	06 51 23	16 ♑ 39 17	22 ♑ 38 33	26 14	23 13	18 57	01 07	10 40	18 15	19 15	17 46
31	14 38 20	07 ♏ 51 23	28 ♑ 36 04	04 ♒ 32 27	27 ♏ 37	24 ♐ 18	19 ♐ 41	01 ♋ 05	10 ♓ 39	18 ♍ 17	19 ♏ 18	17 ♍ 48

Date	Moon True ☊ ° ′	Moon Mean ☊ ° ′	Moon ☽ Latitude ° ′
01	06 ♊ 28	07 ♊ 30	01 S 46
02	06 D 29	07 27	02 44
03	06 R 28	07 23	03 34
04	06 26	07 20	04 15
05	06 21	07 17	04 44
06	06 14	07 14	05 01
07	06 05	07 11	05 06
08	05 56	07 08	04 56
09	05 46	07 04	04 32
10	05 38	07 01	03 54
11	05 31	06 58	03 04
12	05 26	06 55	02 03
13	05 24	06 52	00 S 55
14	05 D 23	06 48	00 N 18
15	05 24	06 45	01 30
16	05 25	06 42	02 39
17	05 26	06 39	03 38
18	05 R 26	06 36	04 26
19	05 24	06 33	04 57
20	05 20	06 29	05 10
21	05 15	06 26	05 03
22	05 09	06 23	04 38
23	05 04	06 20	03 56
24	04 59	06 17	03 00
25	04 55	06 14	01 55
26	04 54	06 10	00 N 45
27	04 D 53	06 07	00 S 26
28	04 54	06 04	01 34
29	04 56	06 01	02 35
30	04 58	05 58	03 29
31	04 ♊ 59	05 ♊ 54	04 S 13

DECLINATIONS

Date	Sun ☉ ° ′	Moon ☽ ° ′	Mercury ☿ ° ′	Venus ♀ ° ′	Mars ♂ ° ′	Jupiter ♃ ° ′	Saturn ♄ ° ′	Uranus ♅ ° ′	Neptune ♆ ° ′	Pluto ♇ ° ′
01	03 S 13	25 S 11	03 S 28	19 S 30	20 S 38	22 N 58	09 S 02	05 N 56	15 S 38	18 N 02
02	03 36	25 53	04 14	19 52	20 48	22 58	09 03	05 55	15 38	18 02
03	03 59	25 22	05 00	20 14	20 58	22 58	09 05	05 53	15 39	18 01
04	04 22	23 43	05 45	20 35	21 07	22 58	09 06	05 52	15 39	18 01
05	04 45	21 02	06 30	20 55	21 17	22 58	09 08	05 51	15 40	18 00
06	05 09	17 27	07 14	21 15	21 26	22 58	09 09	05 49	15 40	18 00
07	05 32	13 07	07 57	21 35	21 35	22 58	09 10	05 48	15 41	17 59
08	05 54	08 12	08 40	21 54	21 44	22 58	09 11	05 47	15 41	17 58
09	06 17	02 S 53	09 23	22 12	21 52	22 58	09 13	05 45	15 42	17 58
10	06 40	02 N 38	10 05	22 30	22 01	22 58	09 14	05 44	15 43	17 57
11	07 03	08 10	10 46	22 47	22 09	22 58	09 15	05 43	15 43	17 57
12	07 25	13 26	11 26	23 04	22 17	22 58	09 16	05 41	15 44	17 56
13	07 48	18 10	12 06	23 20	22 25	22 58	09 17	05 40	15 44	17 56
14	08 10	22 02	12 45	23 36	22 32	22 58	09 18	05 39	15 45	17 55
15	08 32	24 43	13 24	23 51	22 40	22 58	09 19	05 38	15 45	17 55
16	08 55	25 57	14 02	24 05	22 47	22 58	09 20	05 36	15 46	17 55
17	09 17	25 32	14 39	24 19	22 54	22 58	09 21	05 35	15 46	17 55
18	09 38	23 30	15 15	24 33	23 01	22 58	09 22	05 34	15 47	17 54
19	10 00	19 59	15 50	24 45	23 08	22 58	09 23	05 33	15 48	17 53
20	10 22	15 16	16 25	24 57	23 14	22 58	09 24	05 32	15 48	17 53
21	10 43	09 41	16 59	25 09	23 20	22 58	09 25	05 30	15 49	17 52
22	11 05	03 N 37	17 31	25 20	23 26	22 58	09 25	05 29	15 50	17 52
23	11 26	02 S 34	18 03	25 30	23 32	22 58	09 26	05 28	15 50	17 52
24	11 47	08 34	18 35	25 39	23 37	22 58	09 27	05 27	15 51	17 51
25	12 07	14 01	19 05	25 48	23 43	22 58	09 27	05 26	15 51	17 51
26	12 28	18 41	19 34	25 57	23 48	22 58	09 28	05 25	15 52	17 51
27	12 48	22 19	20 02	26 04	23 53	22 58	09 29	05 23	15 53	17 50
28	13 09	24 46	20 30	26 11	23 57	22 58	09 29	05 22	15 53	17 50
29	13 29	25 57	20 56	26 17	24 02	22 59	09 30	05 21	15 54	17 50
30	13 48	25 52	21 21	26 23	24 06	22 59	09 30	05 20	15 54	17 50
31	14 S 08	24 S 34	21 S 45	26 S 28	24 S 10	22 N 59	09 S 31	05 N 19	15 S 55	17 N 49

ZODIAC SIGN ENTRIES

Date	h	m	Planets	
01	18	29	☽	♑
04	06	46	♂	♐
04	06	48	☽	♒
06	19	14	☽	♓
09	05	54	☽	♈
09	16	46	♀	♐
11	14	16	☽	♉
12	21	15	☿	♏
13	20	40	☽	♊
16	01	27	☽	♋
18	04	51	☽	♌
20	07	13	☽	♍
22	09	21	☽	♎
23	15	10	☉	♏
24	12	31	☽	♏
26	18	09	☽	♐
29	03	05	☽	♑
31	14	49	☽	♒

LATITUDES

Date	Mercury ☿ ° ′	Venus ♀ ° ′	Mars ♂ ° ′	Jupiter ♃ ° ′	Saturn ♄ ° ′	Uranus ♅ ° ′	Neptune ♆ ° ′	Pluto ♇ ° ′
01	01 N 01	01 S 40	00 S 56	00 S 29	02 S 09	00 N 44	01 N 43	14 N 01
04	00 42	01 51	00 57	00 29	02 08	00 44	01 43	14 01
07	00 23	02 02	00 59	00 29	02 08	00 44	01 43	14 02
10	00 N 02	02 12	01 00	00 29	02 08	00 44	01 42	14 03
13	00 S 19	02 22	01 01	00 28	02 08	00 44	01 42	14 04
16	00 39	02 31	01 02	00 28	02 07	00 44	01 42	14 05
19	01 00	02 40	01 03	00 28	02 07	00 44	01 42	14 06
22	01 19	02 49	01 04	00 28	02 07	00 45	01 42	14 07
25	01 38	02 56	01 05	00 28	02 06	00 45	01 42	14 08
28	01 55	03 03	01 06	00 28	02 06	00 45	01 42	14 09
31	02 S 11	03 S 09	01 S 07	00 S 28	02 S 05	00 N 45	01 N 42	14 N 10

DATA

Julian Date	2439035
Delta T	+36 seconds
Ayanamsa	23° 22' 27"
Synetic vernal point	05° ♓ 44' 33"
True obliquity of ecliptic	23° 26' 42"

MOON'S PHASES, APSIDES AND POSITIONS ☽

Date	h	m	Phase	Longitude ° ′	Eclipse Indicator
02	12	37	☽	09 ♑ 06	
10	14	14	○	17 ♈ 04	
17	19	00	☾	24 ♋ 11	
24	14	11	●	00 ♏ 57	

Day	h	m		
04	19	34	Apogee	
20	10	46	Perigee	
02	13	49	Max dec	25° S 53'
10	00	36	ON	
16	18	11	Max dec	26° N 00'
23	02	01	OS	
29	22	14	Max dec	26° S 04'

LONGITUDES

Date	Chiron ⚷ ° ′	Ceres ⚳ ° ′	Pallas ⚴ ° ′	Juno ⚵ ° ′	Vesta ⚶ ° ′	Black Moon Lilith ⚸ ° ′
01	19 ♓ 24	22 ♓ 50	05 ♒ 02	01 ♒ 25	09 ♋ 19	09 ♒ 36
11	19 ♓ 00	20 ♓ 57	05 ♒ 14	02 ♒ 24	11 ♋ 22	10 ♒ 43
21	18 ♓ 38	19 ♓ 33	05 ♒ 56	04 ♒ 00	12 ♋ 56	11 ♒ 50
31	18 ♓ 21	18 ♓ 44	07 ♒ 07	06 ♒ 08	13 ♋ 55	12 ♒ 57

All ephemeris data is given at 12.00 UT and the Moon's longitude is additionally given for 24.00 UT

ASPECTARIAN

h m	Aspects
01 Friday	
03 48	☽ Q ☿
07 10	☽ ⊥ ♆
11 44	☽ ⊥ ♀
14 46	☽ ⊻ ♂
18 29	☽ Q ♄
20 02	☽ ☍ ♃
23 51	☿ ⊼ ♄
02 Saturday	
01 00	☽ ∠ ♆
03 25	☽ ⊥ ♂
04 00	☿ ⊥ ♆
07 28	☽ ∠ ♀
11 43	♂ Q ♅
12 37	☽ □ ☉
18 19	☽ ✱ ♄
18 25	♂ Q ♇
21 32	☽ □ ☿
22 34	☽ ∠ ♂
03 Sunday	
04 03	☽ △ ♅
04 24	☽ △ ♇
07 10	☽ ✱ ♆
08 59	☿ ∠ ♂
16 54	☽ ✱ ♀
04 Monday	
00 27	☽ ∠ ♄
06 48	☽ ✱ ♂
07 31	☽ Q ♆
08 40	☽ ⊼ ♃
10 33	☽ ⚼ ♅
10 50	☽ ⚼ ♇
15 44	☿ # ♅
18 35	☽ ⊥ ♄
19 42	☽ # ♃
19 49	☽ Q ♀
20 34	☿ ⊻ ♅
20 53	☽ ± ♃
22 24	☿ ⊻ ♇
05 Tuesday	
04 52	☽ ± ♅
05 06	☽ ± ♇
06 00	☉ ⊼ ♄
06 40	☽ ⊻ ♄
06 43	☽ △ ☉
07 12	♀ ± ♃
08 38	☽ Q ♂
09 24	☿ ± ♄
10 12	☽ ∥ ♂
12 46	☽ ∥ ♀
15 08	☽ ⚼ ♃
16 18	♂ ⊼ ♃
17 02	☽ ⊼ ♅
17 15	☽ ⊼ ♇
18 18	☿ ⊻ ♆
20 00	☽ □ ♆
20 17	☽ △ ☿
21 01	☉ ⊥ ♆
06 Wednesday	
08 39	☽ # ♇
11 56	☽ □ ♀
15 35	☽ ⚼ ☉
21 18	☽ △ ♃
22 16	☽ ∥ ♆
23 00	☽ □ ♂
07 Thursday	
07 02	☽ ⚼ ☿
11 01	☽ ± ☉
12 22	♀ ∥ ♂
18 15	☽ ☌ ♄
21 32	☿ ⊥ ♀
23 51	☽ ⊼ ☉
08 Friday	
03 29	☽ ± ☿
04 20	☉ # ♅
04 54	☽ ☍ ♅
04 59	☽ ☍ ♇
07 24	☽ ∥ ♄
07 39	☽ △ ♆
10 06	☽ ∥ ☿
14 21	☿ ⊥ ♅
14 49	☿ ⊥ ♇
16 56	☽ ⊼ ☿
21 51	☽ ∥ ☉
22 06	♀ Q ♅
22 36	♀ Q ♇
23 11	☽ # ♅
09 Saturday	
04 49	☽ △ ♀
06 03	☿ ∥ ♄
08 04	☽ □ ♃
12 38	☽ ⚼ ♆
13 02	☽ △ ♂
20 17	♅ ☌ ♇
10 Sunday	
03 47	☽ ⊻ ♄
05 44	☽ ± ♆
12 05	☽ ⚼ ♀
14 14	☽ ☍ ☉
14 28	☽ ⊼ ♇
14 30	☽ ⊼ ♅
15 01	☽ ⊥ ♄
16 18	☿ ⚼ ♄
17 02	☽ ⊼ ♆
17 30	☉ ⊻ ♇
17 37	♀ ⊼ ♃
18 04	☉ ⊻ ♅
18 12	☽ Q ♃
19 01	☽ ⚼ ♂
23 53	☉ ± ♄
11 Monday	
01 19	☽ ∥ ♅
01 42	☽ ± ♇
01 45	☽ ± ♅
06 29	☽ ± ♀
06 43	☽ # ☉
07 42	☽ ∠ ♄
09 57	☽ ☍ ☿
12 42	☽ ± ♂
16 30	☽ ✱ ♃
16 48	☽ # ♄
18 22	☽ ⚼ ♇
18 28	☽ ⚼ ♅
18 37	☽ ⊼ ♀
12 Tuesday	
00 22	☽ ⊼ ♂
01 18	☽ # ☿
03 07	♀ # ♃
03 48	☉ ⊻ ♆
11 07	☽ ✱ ♄
19 55	☽ ∠ ♃
21 47	☽ △ ♇
21 55	☽ △ ♅
23 12	☽ # ♆
13 Wednesday	
00 16	☽ ☍ ♆
00 29	☿ ⊥ ♂
01 51	☽ ⊼ ☉
08 43	☽ Q ♄
10 42	☽ ∥ ♇
12 10	☽ ⊥ ♃
13 30	☽ ± ☉
16 17	☿ △ ♃
22 54	☽ ⊻ ♃
23 46	☽ ⊼ ☿
14 Thursday	
05 51	☽ ☍ ♀
06 50	☽ ⚼ ☉
08 23	☿ ∠ ♇
09 28	☽ ☍ ♂
10 08	☿ ∠ ♅
11 50	☽ ± ☿
15 54	☽ # ♂
16 40	☽ □ ♄
19 04	☽ ∥ ♃
15 Friday	
01 47	☽ # ♀
03 20	☽ □ ♇
03 34	☽ □ ♅
05 44	☽ ⚼ ☿
05 45	☽ ⊼ ♆
11 19	☽ △ ☉
16 16	☽ ± ♆
16 Saturday	
03 41	☽ ☌ ♃
07 56	☽ ⚼ ♆
10 45	☽ Q ♇
11 02	☽ Q ♅
11 08	☽ △ ☿
15 01	☽ ⊼ ♀
16 46	☽ ⊼ ♂
20 43	☽ △ ♄
23 37	☉ ⊥ ♇
17 Sunday	
02 14	☽ ± ♀
03 37	☽ ± ♂
04 03	☉ ⊥ ♅
07 25	☽ ✱ ♇
07 43	☽ ✱ ♅
09 47	☽ △ ♆
17 12	☉ ∥ ♄
18 56	☽ ⚼ ♀
19 00	☽ □ ☉
19 49	☽ ⚼ ♂
22 16	☽ ⚼ ♄
18 Monday	
01 11	♂ # ♃
01 18	☉ ∠ ♀
02 54	☽ # ♀
07 04	☽ ⊻ ♃
08 58	☽ ∠ ♇
09 19	☽ ∠ ♅
13 27	☽ ± ♄
15 50	☽ # ♂
16 22	☽ ∥ ♃
17 12	☽ ⊥ ♃
20 29	☽ □ ☿
22 28	☽ △ ♀
22 32	☽ △ ♂
23 31	☽ ⊼ ♄
19 Tuesday	
00 10	☽ ⊥ ♇
00 32	☽ ⊥ ♅
00 44	♀ ☌ ♂
03 41	☽ Q ☉
08 20	☽ ∠ ♃
10 15	☽ ⊻ ♇
10 19	☿ ∥ ♆
10 39	☽ ⊻ ♅
11 29	♀ □ ♄
11 51	☉ ∠ ♂
12 36	☽ □ ♆
15 41	☉ ⚼ ♄
17 01	♂ □ ♄
19 34	♃ St R
23 20	☽ ∥ ♇
20 Wednesday	
00 26	☿ △ ♄
01 16	☽ ✱ ☉
06 12	☽ Q ☿
07 10	☽ # ☿
07 44	☿ ⊻ ♂
09 25	☽ ✱ ♃
09 30	☽ # ♆
18 44	☽ Q ♆
21 Thursday	
01 32	☽ ☍ ♄
03 22	☽ □ ♂
04 09	☽ ∠ ☉
04 33	☽ ✱ ☿
04 51	☽ □ ♀
05 26	☽ Q ♃
08 01	☽ # ☉
12 27	☽ ☌ ♇
12 55	☽ ☌ ♅
13 09	☽ # ♄
14 49	☽ ✱ ♆
15 34	☿ ⊻ ♀
20 18	☽ ⊥ ☉
22 Friday	
04 44	☽ ∥ ♅
07 07	☽ ⊻ ☉
08 33	☽ ∠ ☿
11 06	☽ Q ♂
11 32	☽ □ ♃
13 32	☽ Q ♀
16 02	☽ ∠ ♆
23 Saturday	
01 29	☽ ⊥ ☿
03 15	☿ # ♇
03 45	☽ ⊼ ♄
07 18	☽ ⊥ ♆
08 29	☽ ✱ ♂
11 35	☽ ✱ ♀
12 52	☽ ⊻ ☿
13 19	☿ ⚼ ♃
13 56	☽ ± ♄
15 07	☽ ⊻ ♇
15 41	☽ ⊻ ♅
17 34	☽ ⊻ ♆
23 23	☽ # ♅
24 Sunday	
01 26	☽ ⊥ ♇
02 01	☽ ⊥ ♅
05 21	☽ ⚼ ♄
10 49	☿ ✱ ♇
11 41	☽ ∠ ♂
14 11	☽ ☌ ☉
14 45	☽ △ ♃
15 40	☽ ∠ ♀
15 44	☽ ∥ ♄
16 43	☿ ✱ ♅
17 05	☽ ∠ ♇
17 42	☽ ∠ ♅
21 48	☉ △ ♃
25 Monday	
02 44	☽ ∥ ☉
04 21	☽ ⊥ ♂
07 32	☽ △ ♄
08 53	☽ ⊥ ♀
09 43	♀ □ ♇
10 59	☿ ☌ ♆
15 37	☽ ⊻ ♂
17 12	☽ ⚼ ♃
18 05	♀ □ ♅
19 41	☽ ✱ ♇
20 22	☽ ✱ ♅
20 34	☽ ⊻ ♀
20 55	☽ ∥ ♆
22 17	☽ ☌ ♆
23 38	☽ ☌ ☿
26 Tuesday	
07 07	☉ ∠ ♇
07 19	☽ # ♇
09 26	☽ ± ♃
16 53	☉ ∠ ♅
17 32	☽ Q ♇
17 56	☽ ∥ ☿
18 06	♀ ⊻ ♆
18 16	☽ Q ♅
20 25	☽ ⊼ ♃
27 Wednesday	
00 23	☽ ⊻ ☉
12 31	☽ ⊥ ☉
14 10	☽ □ ♄
17 21	☽ # ♃
28 Thursday	
02 13	☽ ∥ ♂
02 19	☽ ☌ ♂
03 21	☽ □ ♇
04 10	☽ □ ♅
06 09	☽ ⊻ ♆
07 04	☽ ∠ ☉
09 30	☽ ☌ ♀
14 38	☽ ⊻ ☿
17 44	☽ ⊥ ♆
20 26	♂ □ ♇
29 Friday	
00 34	☽ Q ♄
03 47	☽ ⊥ ☿
05 22	☽ ☍ ♃
11 20	☽ ∠ ♆
11 48	♂ □ ♅
14 48	☽ ✱ ☉
17 20	☿ ± ♃
23 48	☽ ∠ ☿
30 Saturday	
00 07	☽ ✱ ♄
14 14	☽ △ ♇
15 12	☽ △ ♅
16 48	☽ Q ☉
16 53	☽ ⊻ ♂
17 13	☽ ✱ ♆
22 42	♂ ⊻ ♆
31 Sunday	
02 28	☽ ⊻ ♀
05 44	☽ ⊥ ♂
06 04	☽ ∠ ♄
09 45	☽ ✱ ☿
15 46	☽ ⊥ ♀
16 55	☽ ∥ ♂
17 00	☽ ⊼ ♃
17 27	☽ Q ♆
20 29	☽ ⚼ ♇
21 31	☽ ⚼ ♅

NOVEMBER 1965

LONGITUDES

Date	Sidereal time h m s	Sun ☉ ° ' "	Moon ☽ ° ' "	Moon ☽ 24.00 ° ' "	Mercury ☿ ° '	Venus ♀ ° '	Mars ♂ ° '	Jupiter ♃ ° '	Saturn ♄ ° '	Uranus ♅ ° '	Neptune ♆ ° '	Pluto ♇ ° '
01	14 42 16	08 ♏ 51 25	10 ♒ 28 19	16 ♒ 24 18	28 ♏ 59	25 ♐ 23	20 ♐ 25	01 ♋ 03	10 ♓ 38	18 ♍ 20	19 ♏ 20	17 ♍ 49
02	14 46 13	09 51 29	22 ♒ 20 59	28 ♒ 18 58	00 ♐ 20	26 27	21 10	01 R 00	10 R 36	18 23	19 22	17 51
03	14 50 09	10 51 35	04 ♓ 16 48	10 ♓ 21 00	01 40	27 32	21 54	00 57	10 35	18 26	19 24	17 52
04	14 54 06	11 51 42	16 ♓ 26 04	22 ♓ 34 26	02 59	28 36	22 39	00 54	10 34	18 28	19 26	17 53
05	14 58 02	12 51 50	28 ♓ 46 27	05 ♈ 02 26	04 16	29 ♐ 40	23 24	00 51	10 33	18 31	19 29	17 55
06	15 01 59	13 52 00	11 ♈ 22 38	17 ♈ 47 13	05 32	00 ♑ 43	24 09	00 48	10 32	18 33	19 31	17 56
07	15 05 55	14 52 12	24 ♈ 16 14	00 ♉ 49 44	06 47	01 47	24 53	00 44	10 32	18 36	19 33	17 58
08	15 09 52	15 52 25	07 ♉ 27 37	14 ♉ 09 44	08 00	02 50	25 38	00 40	10 31	18 38	19 35	17 59
09	15 13 49	16 52 41	20 ♉ 55 54	27 ♉ 45 48	09 11	03 53	26 23	00 36	10 30	18 41	19 38	18 00
10	15 17 45	17 52 58	04 ♊ 39 09	11 ♊ 35 34	10 20	04 55	27 08	00 32	10 30	18 43	19 40	18 02
11	15 21 42	18 53 16	18 ♊ 34 41	25 ♊ 36 04	11 27	05 57	27 53	00 28	10 30	18 45	19 42	18 03
12	15 25 38	19 53 37	02 ♋ 39 19	09 ♋ 44 02	12 30	06 59	28 38	00 23	10 29	18 48	19 44	18 04
13	15 29 35	20 54 00	16 ♋ 49 50	23 ♋ 56 20	13 31	08 01	29 ♐ 24	00 18	10 29	18 50	19 47	18 05
14	15 33 31	21 54 24	01 ♌ 03 11	08 ♌ 10 04	14 29	09 02	00 ♑ 09	00 14	10 D 29	18 52	19 49	18 06
15	15 37 28	22 54 51	15 ♌ 16 39	22 ♌ 22 41	15 22	10 03	00 54	00 09	10 29	18 54	19 51	18 08
16	15 41 24	23 55 19	29 ♌ 27 52	06 ♍ 31 56	16 12	11 03	01 40	00 ♋ 03	10 30	18 56	19 53	18 09
17	15 45 21	24 55 49	13 ♍ 34 38	20 ♍ 35 43	16 56	12 03	02 25	29 ♊ 58	10 30	18 58	19 56	18 10
18	15 49 18	25 56 21	27 ♍ 34 57	04 ♎ 32 03	17 35	13 03	03 10	29 52	10 30	19 00	19 58	18 11
19	15 53 14	26 56 55	11 ♎ 26 47	18 ♎ 18 54	18 08	14 02	03 56	29 47	10 31	19 02	20 00	18 12
20	15 57 11	27 57 30	25 ♎ 08 10	01 ♏ 54 22	18 34	15 01	04 42	29 41	10 31	19 04	20 02	18 13
21	16 01 07	28 58 07	08 ♏ 37 15	15 ♏ 16 40	18 52	15 59	05 27	29 35	10 32	19 06	20 05	18 14
22	16 05 04	29 ♏ 58 46	21 ♏ 52 27	28 ♏ 24 29	19 02	16 57	06 13	29 29	10 33	19 08	20 07	18 15
23	16 09 00	00 ♐ 59 26	04 ♐ 52 41	11 ♐ 17 03	19 R 03	17 54	06 59	29 22	10 34	19 10	20 09	18 16
24	16 12 57	02 00 08	17 ♐ 37 34	23 ♐ 54 22	18 54	18 51	07 45	29 16	10 35	19 12	20 11	18 17
25	16 16 53	03 00 51	00 ♑ 07 33	06 ♑ 17 20	18 35	19 48	08 30	29 09	10 36	19 13	20 14	18 17
26	16 20 50	04 01 35	12 ♑ 23 58	18 ♑ 27 45	18 05	20 44	09 16	29 03	10 37	19 15	20 16	18 18
27	16 24 47	05 02 21	24 ♑ 29 02	00 ♒ 28 15	17 24	21 39	10 02	28 56	10 39	19 16	20 18	18 19
28	16 28 43	06 03 07	06 ♒ 25 50	12 ♒ 22 16	16 32	22 34	10 48	28 49	10 40	19 18	20 20	18 20
29	16 32 40	07 03 55	18 ♒ 18 05	24 ♒ 13 50	15 30	23 28	11 35	28 42	10 42	19 19	20 22	18 21
30	16 36 36	08 ♐ 04 43	00 ♓ 10 04	06 ♓ 07 23	14 ♐ 20	24 ♑ 21	12 ♑ 21	28 ♊ 35	10 ♓ 43	19 ♍ 21	20 ♏ 24	18 ♍ 21

ASPECTARIAN

h m	Aspects
01 Monday	
00 12	☽ ⊥ ♄
01 06	☽ ∠ ♂
05 05	☽ ± ♃
05 10	☽ # ♃
08 26	☽ □ ☉
10 51	♀ ⊥ ♆
11 48	☽ ∠ ♀
12 19	☽ ⊻ ♄
12 29	☽ ∥ ☿
13 10	☽ Q ☿
14 44	☽ ± ♇
15 47	☽ ± ♅
23 14	☽ ♇ ♃
02 Tuesday	
02 53	☽ ⊼ ♇
03 07	☿ Q ♇
03 58	☽ ⊼ ♅
05 58	☽ □ ♆
09 27	☽ ✱ ♂
12 55	☿ Q ♅
18 44	☽ # ♇
21 05	☽ ✱ ♀
23 36	☿ ⊼ ♃
03 Wednesday	
05 19	☽ △ ♃
05 34	☉ △ ♄
05 41	☽ ∥ ♆
06 03	☽ □ ☿
10 32	☽ ∥ ☉
11 08	☽ Q ♂
22 08	☿ # ♃
23 23	☽ Q ♀
04 Thursday	
00 27	☽ ☌ ♄
02 11	☽ △ ☉
11 19	♀ Q ♄
14 37	☽ ∥ ♄
14 52	☽ ☍ ♇
16 00	☽ ☍ ♅
17 54	☽ △ ♆
05 Friday	
00 56	☽ □ ♂
10 06	☽ ♇ ☉
10 25	☽ # ♅
13 52	☽ □ ♀
15 58	☽ □ ♃
22 58	☽ ♇ ♆
23 43	☽ △ ☿
06 Saturday	
04 48	☽ ± ☉
09 47	☉ ∥ ♆
10 25	☽ ⊻ ♄
13 28	♀ ☍ ♃
16 02	☽ ± ♆
17 04	☽ ⊼ ☉
21 40	☽ ⊥ ♄
07 Sunday	
00 18	☽ ⊼ ♇
01 28	☽ ⊼ ♅
01 48	☽ Q ♃
03 15	☽ ⊼ ♆
06 56	☽ ♇ ☿
07 34	☽ ∥ ♅
11 26	☽ ± ♇
12 36	☽ ± ♅
13 12	☽ △ ♂
14 18	☽ ∠ ♄
23 46	☽ ✱ ♃
08 Monday	
01 08	☽ ± ☿
02 18	☽ # ♄
02 55	☽ △ ♀
03 54	☽ ♇ ♇
05 05	☽ ♇ ♅
07 24	☉ ♇ ♃
10 25	♂ ⊥ ♆
13 04	☽ ⊼ ☿
17 29	☽ ✱ ♄
18 02	☽ ♇ ♂
09 Tuesday	
02 37	☽ ∠ ♃
04 15	☽ ☍ ☉
06 49	☽ △ ♇
08 00	☽ △ ♅
08 04	☽ ♇ ♀
08 06	☽ # ♆
09 41	☽ ☍ ♆
10 59	☽ ± ♂
12 31	☽ # ☉
14 46	☽ Q ♄
17 13	☽ ∥ ♇
18 26	☽ ⊥ ♃
22 09	☽ ⊼ ♂
22 59	☿ ∥ ♂
10 Wednesday	
01 12	☽ ± ♀
04 52	☽ ⊻ ♃
05 55	♀ ∠ ♆
12 30	☽ ⊼ ♀
15 30	☿ □ ♄
15 30	☉ ✱ ♇
22 07	☽ □ ♄
22 41	☽ ☍ ☿
11 Thursday	
01 23	☽ ∥ ♃
08 45	☉ ✱ ♅
11 05	☽ □ ♇
12 18	☽ □ ♅
12 34	☽ ⊼ ☉
13 56	☽ ⊼ ♆
15 49	☽ # ♂
19 01	☽ # ☿
23 37	☽ ± ☉
12 Friday	
00 12	☽ ± ♆
04 47	☽ ☍ ♂
07 14	♂ Q ♄
08 10	☽ ☌ ♃
08 12	☉ ☌ ♆
15 33	☽ ♇ ♆
16 05	☽ ♇ ☉
17 48	☽ Q ♇
18 39	☉ # ♇
19 02	☽ Q ♅
19 55	☽ ☍ ♀
13 Saturday	
01 17	☽ △ ♄
05 59	☽ ⊼ ☿
14 08	☽ ✱ ♇
15 23	☽ ✱ ♅
16 53	☽ ± ☿
16 59	☽ △ ♆
19 24	☽ △ ☉
14 Sunday	
01 36	☽ # ☿
02 37	☽ ♇ ♄
03 16	♄ St D
07 55	☽ # ♂
09 10	☽ ♇ ☿
10 23	☽ ⊼ ♂
10 37	☽ ⊻ ♃
14 17	♂ ☍ ♃
15 28	☽ ∠ ♇
16 46	☽ ∠ ♅
17 48	☽ ± ♄
20 41	☽ ⊥ ♃
21 04	☽ ± ♂
22 23	☽ ∥ ♃
15 Monday	
02 29	☽ ⊼ ♀
03 55	☽ ⊼ ♄
06 40	☽ ⊥ ♇
07 59	☽ ⊥ ♅
11 46	☽ ∠ ♃
12 10	☽ △ ☿
13 07	☽ ♇ ♂
13 23	☽ ± ♀
16 49	☽ ⊻ ♇
18 09	☽ ⊻ ♅
19 45	☽ □ ♆
22 39	♀ ✱ ♄
16 Tuesday	
01 29	☽ # ☉
01 54	☽ □ ☉
05 46	☽ ♇ ♀
05 57	☽ ∥ ♇
13 00	☽ ✱ ♃
14 24	☽ # ♆
14 56	☉ ± ♃
15 56	☽ △ ♂
17 Wednesday	
02 21	☽ Q ♆
06 45	☽ ☍ ♄
09 12	☽ △ ♀
09 16	☽ Q ♃
10 49	☽ Q ☉
18 02	☽ □ ☿
19 19	☽ # ♄
19 51	☽ ☌ ♇
21 15	☽ ☌ ♅
22 53	☽ ✱ ♆
18 Thursday	
08 57	☽ ✱ ☉
13 30	☽ ∥ ♅
15 55	☽ □ ♃
22 12	☽ □ ♂
19 Friday	
00 47	☽ ∠ ♆
02 25	☽ Q ☿
10 22	☽ ⊼ ♄
12 57	☽ ∠ ☉
15 16	☿ □ ♇
16 28	☽ ⊥ ♆
16 51	☽ □ ♀
20 51	☽ ± ♄
23 49	☽ ⊻ ♇
20 Saturday	
00 06	☽ ✱ ☿
01 18	☽ ⊻ ♅
03 00	☽ ⊻ ♆
05 35	☽ # ♅
05 57	☽ ⊥ ☉
07 26	☽ Q ♂
10 22	☽ ⊥ ♇
11 53	☽ ⊥ ♅
12 41	☽ ♇ ♄
17 24	☽ ⊻ ☉
19 59	☽ △ ♃
23 29	♂ ∠ ♆
21 Sunday	
00 14	☽ ∥ ♄
02 21	☽ ∠ ♇
03 04	☽ Q ♀
03 20	☽ ∠ ☿
03 54	☽ ∠ ♅
05 59	☽ ✱ ♂
15 27	☽ △ ♄
19 47	☽ ⊥ ☿
22 39	☽ ♇ ♃
22 Monday	
01 13	☉ ⊼ ♃
02 20	☽ ✱ ♀
05 23	☽ ✱ ♇
06 47	☽ ⊻ ☿
06 57	☽ ∥ ♆
06 59	☽ ✱ ♅
08 47	☽ ☌ ♆
10 43	☽ ∠ ♂
14 55	☽ ± ♃
15 42	☽ # ♇
18 26	☉ Q ♇
23 Tuesday	
01 52	☽ ⊼ ♃
02 14	☿ St R
03 25	☽ Q ♇
04 10	☽ ☌ ☉
04 18	☽ ⊥ ♂
05 05	☽ Q ♅
06 35	☽ ∥ ☉
08 02	☽ ∠ ♀
10 55	☉ ⊥ ♂
16 10	☽ ⊻ ♂
16 14	☉ Q ♅
21 05	♀ △ ♇
22 40	☽ □ ♄
24 Wednesday	
02 13	☽ # ♃
02 14	☽ ⊥ ♀
13 01	☿ ⊻ ♀
13 11	☿ ∥ ♂
13 14	☽ □ ♇
14 23	☽ ☌ ☿
14 32	☽ ⊻ ♀
14 59	☽ □ ♅
15 33	☽ ∥ ☿
15 47	☽ ∥ ♂
16 54	☽ ⊻ ♆
20 50	♀ △ ♅
25 Thursday	
02 08	☽ ∥ ♀
04 26	☽ ⊥ ♆
09 03	☽ Q ♄
10 08	☽ ☍ ♃
18 07	☽ ⊻ ☉
21 57	☽ ∠ ♆
23 30	♀ ✱ ♆
26 Friday	
02 34	☿ □ ♇
05 26	☽ ☌ ♂
06 54	☽ ⊥ ☉
08 30	☽ ✱ ♄
22 41	☽ ⊻ ☿
23 42	☽ △ ♇
27 Saturday	
01 35	☽ △ ♅
02 19	☽ ∠ ☉
03 38	☽ ✱ ♆
05 52	☽ ☌ ♀
09 57	☽ ⊥ ☿
14 20	☽ ∠ ♄
19 18	☽ ∥ ♀
20 50	☽ ⊼ ♃
28 Sunday	
01 10	☽ ∥ ♂
02 50	☽ ∠ ☿
03 43	☽ Q ♆
05 45	☽ ♇ ♇
07 32	♂ ✱ ♄
07 42	☽ ♇ ♅
08 26	☽ ⊥ ♄
08 47	☽ ± ♃
11 10	☽ ✱ ☉
11 39	☿ ⊥ ♀
13 03	☽ ∥ ☿
13 33	☽ # ♃
16 04	☿ # ♃
20 35	☽ ⊻ ♄
21 27	☽ ⊻ ♂
23 56	☽ ± ♇
29 Monday	
01 54	☽ ± ♅
02 46	☽ ∥ ☉
02 47	☽ ♇ ♃
06 49	☽ ✱ ☿
10 26	☽ ⊥ ♂
12 05	☽ ⊼ ♇
13 41	☽ Q ☉
14 04	☽ ⊼ ♅
16 12	☽ □ ♆
23 18	☽ ⊻ ♀
30 Tuesday	
03 19	☽ # ♇
04 59	☽ Q ☿
05 54	☽ ∠ ♂
08 50	☽ △ ♃
12 24	☽ ⊥ ♀
12 43	☽ ∥ ♆

Moon Nodes and Latitude

Date	Moon True ☊ ° '	Moon Mean ☊ ° '	Moon ☽ Latitude ° '
01	05 ♊ 00	05 ♊ 51	04 S 46
02	04 R 59	05 48	05 07
03	04 57	05 45	05 14
04	04 54	05 42	05 08
05	04 51	05 39	04 47
06	04 47	05 35	04 13
07	04 44	05 32	03 25
08	04 42	05 29	02 24
09	04 40	05 26	01 S 15
10	04 D 40	05 23	00 00
11	04 40	05 20	01 N 16
12	04 41	05 16	02 29
13	04 42	05 13	03 33
14	04 43	05 10	04 24
15	04 44	05 07	04 59
16	04 R 44	05 04	05 15
17	04 43	05 00	05 13
18	04 42	04 57	04 52
19	04 41	04 54	04 14
20	04 41	04 51	03 22
21	04 40	04 48	02 19
22	04 39	04 45	01 N 10
23	04 D 39	04 41	00 S 01
24	04 39	04 38	01 11
25	04 40	04 35	02 17
26	04 40	04 32	03 14
27	04 R 40	04 29	04 03
28	04 40	04 26	04 40
29	04 39	04 22	05 04
30	04 ♊ 39	04 ♊ 19	05 S 16

DECLINATIONS

Date	Sun ☉ ° '	Moon ☽ ° '	Mercury ☿ ° '	Venus ♀ ° '	Mars ♂ ° '	Jupiter ♃ ° '	Saturn ♄ ° '	Uranus ♅ ° '	Neptune ♆ ° '	Pluto ♇ ° '
01	14 S 27	22 S 12	22 S 08	26 S 32	24 S 13	22 N 59	09 S 31	05 N 18	15 S 56	17 N 49
02	14 46	18 53	22 30	26 36	24 17	22 59	09 31	05 17	15 56	17 49
03	15 05	14 48	22 51	26 39	24 20	22 59	09 32	05 16	15 57	17 49
04	15 24	10 05	23 10	26 41	24 22	22 59	09 32	05 15	15 57	17 49
05	15 42	04 S 53	23 28	26 43	24 25	22 59	09 32	05 14	15 58	17 48
06	16 00	00 N 37	23 45	26 44	24 27	22 59	09 32	05 13	15 59	17 48
07	16 18	06 14	24 01	26 45	24 30	22 59	09 32	05 12	15 59	17 48
08	16 36	11 44	24 15	26 44	24 31	22 59	09 33	05 11	16 00	17 48
09	16 53	16 47	24 28	26 43	24 33	22 59	09 33	05 10	16 01	17 48
10	17 10	21 04	24 40	26 42	24 34	22 59	09 33	05 09	16 01	17 48
11	17 27	24 13	24 50	26 40	24 35	23 00	09 33	05 08	16 02	17 48
12	17 43	25 54	24 58	26 37	24 36	23 00	09 33	05 08	16 02	17 47
13	17 59	25 54	25 05	26 33	24 37	23 00	09 32	05 07	16 03	17 47
14	18 15	24 13	25 11	26 29	24 37	23 00	09 32	05 06	16 04	17 47
15	18 30	21 00	25 15	26 25	24 37	23 00	09 32	05 05	16 04	17 47
16	18 45	16 35	25 17	26 20	24 37	23 00	09 32	05 04	16 05	17 47
17	19 00	11 16	25 17	26 14	24 36	23 00	09 32	05 04	16 05	17 47
18	19 15	05 N 25	25 16	26 07	24 35	23 00	09 31	05 03	16 06	17 47
19	19 29	00 S 38	25 12	26 00	24 34	23 00	09 31	05 02	16 06	17 47
20	19 43	06 36	25 07	25 53	24 33	23 00	09 31	05 01	16 07	17 47
21	19 56	12 11	25 00	25 45	24 31	23 00	09 30	05 01	16 08	17 47
22	20 09	17 06	24 51	25 36	24 29	23 00	09 30	05 00	16 08	17 47
23	20 22	21 08	24 39	25 27	24 27	23 00	09 29	04 59	16 09	17 48
24	20 34	24 03	24 25	25 17	24 24	23 00	09 29	04 59	16 09	17 48
25	20 46	25 43	24 09	25 07	24 22	23 01	09 28	04 58	16 10	17 48
26	20 57	26 06	23 50	24 57	24 19	23 01	09 27	04 57	16 11	17 48
27	21 09	25 12	23 29	24 46	24 15	23 01	09 27	04 57	16 11	17 48
28	21 19	23 11	23 05	24 34	24 12	23 01	09 26	04 56	16 12	17 48
29	21 30	20 10	22 39	24 22	24 08	23 01	09 25	04 56	16 12	17 48
30	21 S 40	16 S 20	22 S 11	24 S 10	24 S 04	23 N 01	09 S 25	04 N 55	16 S 13	17 N 49

ZODIAC SIGN ENTRIES

Date	h m	Planets
02	06 04	☿ ♐
03	03 23	☽ ♓
05	14 21	☽ ♈
05	19 36	♀ ♑
07	22 29	☽ ♉
10	03 54	☽ ♊
12	07 29	☽ ♋
14	07 19	♂ ♑
14	10 13	☽ ♌
16	12 54	☽ ♍
17	03 08	♃ ♊
18	16 10	☽ ♎
20	20 37	☽ ♏
22	12 29	☉ ♐
23	02 56	☽ ♐
25	11 45	☽ ♑
27	23 03	☽ ♒
30	11 40	☽ ♓

LATITUDES

Date	Mercury ☿ ° '	Venus ♀ ° '	Mars ♂ ° '	Jupiter ♃ ° '	Saturn ♄ ° '	Uranus ♅ ° '	Neptune ♆ ° '	Pluto ♇ ° '
01	02 S 15	03 S 11	01 S 07	00 S 28	02 S 05	00 N 45	01 N 42	14 N 10
04	02 27	03 15	01 08	00 28	02 05	00 45	01 42	14 12
07	02 36	03 19	01 09	00 27	02 04	00 45	01 42	14 13
10	02 41	03 21	01 09	00 27	02 04	00 45	01 42	14 14
13	02 41	03 22	01 10	00 27	02 04	00 45	01 42	14 15
16	02 34	03 21	01 10	00 27	02 03	00 45	01 42	14 17
19	02 18	03 19	01 11	00 27	02 03	00 45	01 42	14 18
22	01 51	03 16	01 11	00 26	02 02	00 46	01 42	14 20
25	01 12	03 10	01 12	00 26	02 02	00 46	01 42	14 21
28	00 S 19	03 03	01 12	00 26	02 01	00 46	01 42	14 22
31	00 N 41	02 S 54	01 S 12	00 S 25	02 S 01	00 N 46	01 N 42	14 N 24

DATA

Julian Date	2439066
Delta T	+36 seconds
Ayanamsa	23° 22' 31"
Synetic vernal point	05° ♓ 44' 29"
True obliquity of ecliptic	23° 26' 41"

MOON'S PHASES, APSIDES AND POSITIONS ☽

Date	h m	Phase	Longitude ° '	Eclipse Indicator
01	08 26	☽	08 ♒ 42	
09	04 15	○	16 ♉ 33	
16	01 54	☾	23 ♌ 30	
23	04 10	●	00 ♐ 40	Annular

Day	h m	
01	15 07	Apogee
14	07 44	Perigee
29	12 05	Apogee

Day	h m		
06	09 19	ON	
13	00 04	Max dec	26° N 07'
19	09 30	OS	
26	06 52	Max dec	26° S 07'

LONGITUDES

Date	Chiron ⚷ ° '	Ceres ⚳ ° '	Pallas ⚴ ° '	Juno ⚵ ° '	Vesta ⚶ ° '	Black Moon Lilith ⚸ ° '
01	18 ♓ 19	18 ♓ 41	07 ♒ 15	06 ♒ 23	13 ♋ 58	13 ♒ 04
11	18 ♓ 07	18 ♓ 34	08 ♒ 52	09 ♒ 03	14 ♋ 13	14 ♒ 11
21	18 ♓ 00	19 ♓ 03	10 ♒ 50	12 ♒ 08	13 ♋ 45	15 ♒ 18
31	17 ♓ 58	20 ♓ 06	13 ♒ 05	15 ♒ 35	12 ♋ 32	16 ♒ 25

All ephemeris data is given at 12.00 UT and the Moon's longitude is additionally given for 24.00 UT

DECEMBER 1965

LONGITUDES

Date	Sidereal time h m s	Sun ☉ ° ' "	Moon ☽ ° ' "	Moon ☽ 24.00 ° ' "	Mercury ☿ ° '	Venus ♀ ° '	Mars ♂ ° '	Jupiter ♃ ° '	Saturn ♄ ° '	Uranus ♅ ° '	Neptune ♆ ° '	Pluto ♇ ° '
01	16 40 33	09 ♐ 05 32	12 ♓ 06 23	18 ♓ 07 38	13 ♐ 04	25 ♑ 14	13 ♑ 07	28 ♊ 28	10 ♓ 45	19 ♍ 22	20 ♏ 27	18 ♍ 22
02	16 44 29	10 06 22	24 ♓ 11 45	00 ♈ 19 17	11 R 43	26 06	13 53	28 R 20	10 47	19 23	20 29	18 22
03	16 48 26	11 07 13	06 ♈ 30 46	12 ♈ 46 41	10 20	26 57	14 39	28 13	10 49	19 25	20 31	18 23
04	16 52 22	12 08 05	19 ♈ 07 29	25 ♈ 33 32	08 58	27 48	15 26	28 05	10 51	19 26	20 33	18 24
05	16 56 19	13 08 58	02 ♉ 05 08	08 ♉ 42 29	07 40	28 38	16 12	27 58	10 53	19 27	20 35	18 24
06	17 00 16	14 09 51	15 ♉ 25 40	22 ♉ 14 39	06 29	29 ♑ 27	16 58	27 50	10 56	19 28	20 37	18 25
07	17 04 12	15 10 46	29 ♉ 09 17	06 ♊ 09 18	05 25	00 ♒ 15	17 45	27 42	10 58	19 29	20 39	18 25
08	17 08 09	16 11 41	13 ♊ 14 17	20 ♊ 23 40	04 32	01 02	18 31	27 34	11 00	19 30	20 41	18 26
09	17 12 05	17 12 38	27 ♊ 36 51	04 ♋ 53 04	03 49	01 48	19 18	27 26	11 03	19 31	20 44	18 26
10	17 16 02	18 13 35	12 ♋ 11 31	19 ♋ 31 21	03 18	02 34	20 04	27 18	11 06	19 32	20 46	18 26
11	17 19 58	19 14 33	26 ♋ 51 43	04 ♌ 11 46	02 58	03 18	20 51	27 10	11 09	19 33	20 48	18 27
12	17 23 55	20 15 33	11 ♌ 30 43	18 ♌ 47 51	02 49	04 01	21 37	27 02	11 11	19 33	20 50	18 27
13	17 27 51	21 16 33	26 ♌ 02 31	03 ♍ 14 13	02 D 50	04 43	22 24	26 54	11 14	19 34	20 52	18 27
14	17 31 48	22 17 34	10 ♍ 22 31	17 ♍ 27 08	03 01	05 24	23 11	26 46	11 17	19 35	20 54	18 28
15	17 35 45	23 18 37	24 ♍ 27 51	01 ♎ 24 32	03 21	06 04	23 57	26 38	11 21	19 35	20 56	18 28
16	17 39 41	24 19 40	08 ♎ 17 11	15 ♎ 05 48	03 49	06 43	24 44	26 30	11 24	19 36	20 58	18 28
17	17 43 38	25 20 44	21 ♎ 50 29	28 ♎ 31 21	04 25	07 20	25 31	26 22	11 27	19 36	21 00	18 28
18	17 47 34	26 21 50	05 ♏ 08 31	11 ♏ 42 08	05 06	07 56	26 18	26 14	11 31	19 37	21 02	18 28
19	17 51 31	27 22 56	18 ♏ 12 22	24 ♏ 39 21	05 54	08 31	27 05	26 05	11 34	19 37	21 03	18 28
20	17 55 27	28 24 03	01 ♐ 03 14	07 ♐ 24 08	06 46	09 04	27 52	25 57	11 38	19 37	21 05	18 28
21	17 59 24	29 ♐ 25 10	13 ♐ 42 10	19 ♐ 57 27	07 43	09 36	28 39	25 49	11 41	19 38	21 07	18 R 28
22	18 03 20	00 ♑ 26 18	26 ♐ 10 06	02 ♑ 20 13	08 44	10 06	29 ♑ 26	25 41	11 45	19 38	21 09	18 28
23	18 07 17	01 27 27	08 ♑ 27 56	14 ♑ 33 22	09 48	10 35	00 ♒ 13	25 33	11 49	19 38	21 11	18 28
24	18 11 14	02 28 36	20 ♑ 36 41	26 ♑ 38 03	10 55	11 02	01 00	25 25	11 53	19 38	21 13	18 28
25	18 15 10	03 29 45	02 ♒ 37 40	08 ♒ 35 47	12 05	11 27	01 47	25 17	11 57	19 R 38	21 14	18 28
26	18 19 07	04 30 54	14 ♒ 32 42	20 ♒ 28 42	13 17	11 50	02 34	25 09	12 01	19 38	21 16	18 28
27	18 23 03	05 32 04	26 ♒ 24 11	02 ♓ 19 32	14 32	12 12	03 21	25 01	12 06	19 38	21 18	18 28
28	18 27 00	06 33 13	08 ♓ 15 12	14 ♓ 11 40	15 48	12 31	04 08	24 53	12 10	19 38	21 20	18 27
29	18 30 56	07 34 23	20 ♓ 09 28	26 ♓ 09 09	17 06	12 49	04 55	24 45	12 14	19 38	21 21	18 27
30	18 34 53	08 35 32	02 ♈ 11 17	08 ♈ 16 29	18 25	13 04	05 42	24 37	12 19	19 37	21 23	18 27
31	18 38 49	09 ♑ 36 41	14 ♈ 25 21	20 ♈ 38 29	19 ♐ 46	13 ♒ 18	06 ♒ 29	24 ♊ 30	12 ♓ 23	19 ♍ 37	21 ♏ 25	18 ♍ 26

Date	Moon True ☊ ° '	Moon Mean ☊ ° '	Moon ☽ Latitude ° '
01	04 ♊ 39	04 ♊ 16	05 S 14
02	04 D 39	04 13	04 59
03	04 40	04 10	04 30
04	04 41	04 06	03 47
05	04 41	04 03	02 51
06	04 42	04 00	01 45
07	04 43	03 57	00 S 31
08	04 R 43	03 54	00 N 47
09	04 42	03 51	02 03
10	04 41	03 47	03 13
11	04 39	03 44	04 10
12	04 37	03 41	04 50
13	04 35	03 38	05 12
14	04 34	03 35	05 14
15	04 D 33	03 31	04 57
16	04 34	03 28	04 23
17	04 35	03 25	03 34
18	04 37	03 22	02 35
19	04 38	03 19	01 29
20	04 39	03 16	00 N 20
21	04 R 39	03 12	00 S 50
22	04 37	03 09	01 55
23	04 34	03 06	02 55
24	04 30	03 03	03 46
25	04 24	03 00	04 26
26	04 19	02 57	04 54
27	04 13	02 53	05 09
28	04 09	02 50	05 11
29	04 05	02 47	05 00
30	04 04	02 44	04 35
31	04 ♊ 04	02 ♊ 41	03 S 58

DECLINATIONS

Date	Sun ☉ ° '	Moon ☽ ° '	Mercury ☿ ° '	Venus ♀ ° '	Mars ♂ ° '	Jupiter ♃ ° '	Saturn ♄ ° '	Uranus ♅ ° '	Neptune ♆ ° '	Pluto ♇ ° '
01	21 S 49	11 S 52	21 S 41	23 S 57	24 S 00	23 N 01	09 S 24	04 N 55	16 S 13	17 N 49
02	21 58	06 53	21 11	23 44	23 55	23 01	09 23	04 54	16 14	17 49
03	22 07	01 S 32	20 40	23 30	23 50	23 01	09 22	04 54	16 14	17 49
04	22 15	04 N 00	20 10	23 16	23 45	23 01	09 21	04 53	16 15	17 49
05	22 23	09 31	19 42	23 02	23 39	23 01	09 20	04 53	16 16	17 50
06	22 30	14 48	19 16	22 47	23 34	23 01	09 19	04 53	16 16	17 50
07	22 37	19 28	18 54	22 32	23 27	23 01	09 18	04 52	16 17	17 50
08	22 44	23 10	18 34	22 17	23 21	23 01	09 17	04 52	16 17	17 51
09	22 50	25 29	18 19	22 01	23 15	23 01	09 16	04 52	16 18	17 51
10	22 55	26 05	18 08	21 46	23 08	23 01	09 15	04 51	16 18	17 51
11	23 01	24 52	18 01	21 30	23 01	23 01	09 13	04 51	16 19	17 51
12	23 05	21 59	17 58	21 14	22 53	23 01	09 12	04 51	16 19	17 52
13	23 09	17 44	17 58	20 57	22 46	23 01	09 11	04 51	16 20	17 52
14	23 13	12 31	18 02	20 41	22 38	23 01	09 10	04 50	16 20	17 53
15	23 17	06 44	18 08	20 24	22 30	23 00	09 08	04 50	16 21	17 53
16	23 19	00 N 44	18 17	20 07	22 22	23 00	09 07	04 50	16 21	17 53
17	23 22	05 S 12	18 28	19 50	22 13	23 00	09 05	04 50	16 22	17 54
18	23 24	10 48	18 41	19 33	22 04	23 00	09 04	04 50	16 22	17 54
19	23 25	15 49	18 56	19 16	21 55	23 00	09 02	04 50	16 23	17 55
20	23 26	20 03	19 11	18 59	21 46	23 00	09 01	04 50	16 23	17 55
21	23 27	23 16	19 27	18 42	21 36	23 00	08 59	04 50	16 24	17 56
22	23 27	25 19	19 44	18 25	21 26	23 00	08 58	04 49	16 24	17 56
23	23 26	26 05	20 01	18 08	21 16	23 00	08 56	04 49	16 24	17 57
24	23 25	25 35	20 19	17 51	21 06	23 00	08 55	04 49	16 25	17 57
25	23 24	23 53	20 36	17 34	20 55	22 59	08 53	04 50	16 25	17 58
26	23 22	21 09	20 53	17 17	20 44	22 59	08 51	04 50	16 26	17 58
27	23 20	17 33	21 10	17 00	20 33	22 59	08 49	04 50	16 26	17 59
28	23 17	13 17	21 27	16 44	20 22	22 59	08 48	04 50	16 27	17 59
29	23 14	08 29	21 43	16 28	20 11	22 59	08 46	04 50	16 27	18 00
30	23 10	03 20	21 58	16 12	19 59	22 59	08 44	04 50	16 27	18 00
31	23 S 06	02 N 02	22 S 13	15 S 56	19 S 47	22 N 59	08 S 42	04 N 50	16 S 28	18 N 01

ZODIAC SIGN ENTRIES

Date	h	m	Planets
02	23	22	☽ ♈
05	08	11	☽ ♉
07	04	37	♀ ♒
07	13	27	☽ ♊
09	15	57	☽ ♋
11	17	08	☽ ♌
13	18	35	☽ ♍
15	21	33	☽ ♎
18	02	40	☽ ♏
20	10	01	☽ ♐
22	01	40	☉ ♑
22	19	27	☽ ♑
23	05	36	♂ ♒
25	06	44	☽ ♒
27	19	17	☽ ♓
30	07	40	☽ ♈

LATITUDES

Date	Mercury ☿ ° '	Venus ♀ ° '	Mars ♂ ° '	Jupiter ♃ ° '	Saturn ♄ ° '	Uranus ♅ ° '	Neptune ♆ ° '	Pluto ♇ ° '
01	00 N 41	02 S 54	01 S 12	00 S 25	02 S 01	00 N 46	01 N 42	14 N 24
04	01 39	02 43	01 12	00 25	02 00	00 46	01 42	14 26
07	02 21	02 29	01 12	00 25	02 00	00 46	01 42	14 27
10	02 44	02 14	01 12	00 24	01 59	00 46	01 42	14 29
13	02 49	01 55	01 12	00 24	01 59	00 46	01 42	14 30
16	02 41	01 34	01 12	00 24	01 58	00 47	01 42	14 32
19	02 24	01 10	01 11	00 23	01 58	00 47	01 43	14 33
22	02 03	00 43	01 11	00 23	01 57	00 47	01 43	14 35
25	01 40	00 S 13	01 11	00 22	01 57	00 47	01 43	14 36
28	01 15	00 N 20	01 10	00 22	01 57	00 47	01 43	14 38
31	00 N 50	00 N 56	01 S 10	00 S 21	01 S 56	00 N 47	01 N 43	14 N 39

DATA

Julian Date	2439096
Delta T	+37 seconds
Ayanamsa	23° 22' 35"
Synetic vernal point	05° ♓ 44' 24"
True obliquity of ecliptic	23° 26' 41"

LONGITUDES

Date	Chiron ⚷ ° '	Ceres ⚳ ° '	Pallas ⚴ ° '	Juno ⚵ ° '	Vesta ⚶ ° '	Black Moon Lilith ⚸ ° '
01	17 ♓ 58	20 ♓ 06	13 ♒ 05	15 ♒ 35	12 ♋ 32	16 ♒ 25
11	18 ♓ 02	21 ♓ 40	15 ♒ 35	19 ♒ 21	10 ♋ 38	17 ♒ 32
21	18 ♓ 12	23 ♓ 41	18 ♒ 17	23 ♒ 23	08 ♋ 15	18 ♒ 39
31	18 ♓ 27	26 ♓ 05	21 ♒ 09	27 ♒ 39	05 ♋ 37	19 ♒ 47

MOON'S PHASES, APSIDES AND POSITIONS ☽

Date	h	m	Phase	Longitude ° '	Eclipse Indicator
01	05	24	☽	08 ♓ 49	
08	17	21	○	16 ♊ 25	
15	09	52	☾	23 ♍ 13	
22	21	03	●	00 ♑ 49	
31	01	46	☽	09 ♈ 11	

Day	h	m		
11	06	07	Perigee	
27	07	24	Apogee	
03	18	44	0N	
10	08	00	Max dec	26° N 07'
16	14	57	0S	
23	14	22	Max dec	26° S 05'
31	03	00	0N	

ASPECTARIAN

h m	Aspects
01 Wednesday	
04 23	♀ ‖ ♂
05 24	☽ □ ☉
07 17	☉ ‖ ☿
07 57	☽ ∠ ♀
09 17	☽ ☌ ♄
11 25	☿ ⊻ ♂
13 43	☽ □ ☿
14 09	☽ ✶ ♂
02 Thursday	
00 10	☽ ‖ ♄
00 29	☽ ☍ ♇
02 29	☽ ☍ ♅
02 55	♀ ∠ ♄
04 39	☽ △ ♆
15 32	☽ Q ♂
16 01	☽ ✶ ♀
18 33	☿ ∠ ♀
20 03	☽ □ ♃
21 02	☽ # ♅
03 Friday	
03 44	☿ □ ♄
04 05	☉ ☌ ☿
04 35	☉ □ ♄
10 04	☽ ⚼ ♆
17 02	☽ Q ♀
18 36	☽ △ ☿
20 17	☽ ⊻ ♄
21 37	☽ △ ☉
04 Saturday	
03 21	☽ ± ♆
04 34	☽ □ ♂
06 20	☽ Q ♃
06 48	☿ ⊥ ♂
07 43	☽ ⊥ ♄
10 38	☽ ⊼ ♇
12 35	☽ ⊼ ♅
14 41	☽ ⊼ ♆
15 52	☽ ‖ ♅
19 12	♀ ⊼ ♃
20 12	☽ ⚼ ☿
21 51	☽ ± ♇
23 47	☽ ± ♅
05 Sunday	
00 35	☽ ∠ ♄
04 10	☽ ⚼ ☉
04 30	☽ ✶ ♃
05 14	☽ □ ♀
11 10	☽ # ♄
11 18	☽ ± ☿
13 29	♀ # ♃
14 24	☽ ⚼ ♇
16 19	☽ ⚼ ♅
21 16	☽ ⊼ ☿
21 57	☽ ± ☉
06 Monday	
03 57	☽ ✶ ♄
07 26	☽ ∠ ♃
09 34	☽ ⊼ ☉
14 54	☽ △ ♂
17 17	☽ △ ♇
19 09	☽ △ ♅
19 11	☽ # ♆
21 11	☽ ☍ ♆
23 10	☽ ⊥ ♃
07 Tuesday	
01 14	☽ Q ♄
03 07	☽ ‖ ♇
06 23	☉ ‖ ♀
09 01	☽ # ☿
09 31	☽ ⊻ ♃
14 00	☽ △ ♀
18 32	☽ ⚼ ♂
19 06	☉ ∠ ♀
22 04	☽ ☍ ☿
08 Wednesday	
05 54	☽ # ♀
08 13	☽ □ ♄
08 32	☽ # ☉
09 05	♂ △ ♇
10 43	☽ ± ♂
10 47	☽ ‖ ♃
13 24	☽ # ♂
16 58	☽ ⚼ ♀
17 21	☽ ☍ ☉
20 43	☽ □ ♇
21 22	☽ ⊼ ♂
22 31	☽ □ ♅
09 Thursday	
00 31	☽ ⊼ ♆
04 01	☿ ∠ ♂
08 50	☽ ± ♀
10 31	☽ ± ♆
11 43	☽ ☌ ♃
19 00	♂ △ ♅
19 18	☽ ⊼ ♀
21 52	☽ ⊼ ☿
10 Friday	
01 25	☽ ⚼ ♆
02 33	☽ Q ♇
04 21	☽ Q ♅
07 24	☽ ± ☿
10 12	☽ △ ♄
17 04	☉ □ ♇
18 48	♀ Q ♆
21 45	☽ ⚼ ☿
22 14	☽ ✶ ♇
22 37	☽ ⊼ ☉
11 Saturday	
00 01	☽ ✶ ♅
01 37	☽ ☍ ♂
02 03	☽ △ ♆
04 06	☿ ✶ ♀
08 33	♀ ± ♃
09 09	☽ ± ☉
10 18	♂ ✶ ♆
10 49	☽ ⚼ ♄
12 23	♂ # ♃
12 30	☽ ⊻ ♃
12 34	☉ ‖ ♂
12 49	☉ # ♃
16 55	♀ ⚼ ♇
19 13	☉ □ ♅
21 51	☽ △ ☿
22 14	☽ ⊥ ♃
22 46	☽ ∠ ♇
23 05	☽ ☍ ♀
12 Sunday	
00 35	☽ ∠ ♅
00 59	☽ ⚼ ☉
01 36	☽ ± ♄
04 23	☽ # ☉
04 47	☽ ‖ ♃
05 25	☽ # ♂
11 28	☽ ⊼ ♄
12 52	☽ ∠ ♃
13 33	☽ ⊥ ♇
15 22	☽ ⊥ ♅
17 10	☽ # ♀
20 41	☿ St D
23 26	☽ ⊻ ♇
13 Monday	
01 16	☽ ⊻ ♅
01 53	☉ ⊻ ♆
03 23	☽ □ ♆
03 30	☽ △ ☉
05 37	☽ ⊼ ♂
06 39	♀ ⚼ ♅
10 47	☽ # ☿
11 17	☽ ‖ ♇
13 25	☽ ✶ ♃
16 09	☽ ± ♂
18 07	♃ ± ♆
18 48	☽ # ♆
23 27	☽ □ ☿
14 Tuesday	
03 13	☽ ⊼ ♀
07 35	♀ ⊥ ♄
08 05	☽ ⚼ ♂
09 19	☽ Q ♃
09 30	☽ Q ♆
13 33	☽ ☍ ♄
13 50	☽ ± ♀
15 Wednesday	
01 43	☽ ☌ ♇
02 11	☽ # ♄
03 38	☽ ☌ ♅
05 53	☽ ⚼ ♀
05 55	☽ ✶ ♆
06 31	☽ Q ☿
09 52	☽ □ ☉
11 05	☽ △ ♂
15 42	☽ □ ♃
19 39	☽ ‖ ♅
16 Thursday	
03 54	☽ ✶ ☿
07 55	☽ ∠ ♆
09 07	☽ △ ♀
17 29	☽ ⊼ ♄
19 41	☽ Q ☉
23 47	☽ ⊥ ♆
17 Friday	
04 09	☽ ± ♄
05 59	☽ ⊻ ♇
07 27	☽ ∠ ☿
08 00	☽ ⊻ ♅
10 29	☽ ⊻ ♆
10 30	☽ # ♅
16 42	☽ ⊥ ♇
18 45	☽ ⊥ ♅
18 48	☽ ✶ ☉
19 00	☽ □ ♂
20 02	☽ △ ♃
20 18	☽ ⚼ ♄
18 Saturday	
00 25	☽ ⊥ ☿
04 24	☽ ‖ ♄
05 20	☉ ⊻ ♂
08 57	☽ ∠ ♇
09 08	☉ ☍ ♃
10 07	♂ ⊼ ♃
11 02	☽ ∠ ♅
11 55	☽ ⊻ ☿
17 20	☽ □ ♀
19 03	♂ ∠ ♄
23 01	☽ ⚼ ♃
23 42	☽ △ ♄
19 Sunday	
00 20	☽ ∠ ☉
04 05	☉ ⊥ ♆
05 51	☽ Q ♂
12 29	☽ ✶ ♇
14 37	☽ ✶ ♅
14 52	☽ ‖ ♆
15 27	☽ ± ♃
17 18	☽ ☌ ♆
18 24	☽ ⊥ ☉
23 13	☽ # ♇
20 Monday	
02 32	☽ ⊼ ♃
03 17	☿ ‖ ♀
04 11	☽ Q ♀
05 37	☽ ✶ ♂
05 51	☽ ‖ ♀
06 13	☽ ‖ ☿
06 35	☽ ⊻ ☉
10 54	☽ Q ♇
13 04	☽ Q ♅
23 10	☽ ‖ ♂
23 39	☽ ☌ ☿
21 Tuesday	
03 50	☽ ✶ ♀
05 07	♇ St R
08 08	☽ □ ♄
09 36	☽ # ♃
11 53	☽ ∠ ♂
13 36	☽ ‖ ☉
18 50	☉ Q ♄
21 08	☽ □ ♇
23 22	☽ □ ♅
22 Wednesday	
02 16	☽ ⊻ ♆
06 20	☽ ⊥ ♂
09 51	☽ ∠ ♀
11 04	☽ ☍ ♃
13 55	☽ ⊥ ♆
18 45	☽ ⊻ ♂
19 00	☽ Q ♄
21 03	☽ ☌ ☉
23 Thursday	
04 05	☽ ⊥ ♀
07 30	☽ ∠ ♆
10 42	♀ ⚼ ♃
14 53	☽ ⊻ ☿
16 19	☽ ⊻ ♀
18 38	☽ ✶ ♄
24 Friday	
03 22	♀ # ♇
03 55	☽ ⊥ ☿
07 45	☽ △ ♇
10 03	☽ △ ♅
13 12	☽ ✶ ♆
15 40	☿ ✶ ♀
21 27	☽ ⊼ ♃
23 00	♂ ± ♃
23 42	☽ ∠ ☿
25 Saturday	
00 34	☽ ∠ ♄
06 06	♅ St R
09 11	☿ □ ♄
09 20	☽ ± ♃
10 10	☽ ☌ ♂
13 14	☽ Q ♆
13 41	☽ ⚼ ♇
13 54	☽ ⊻ ☉
16 02	☽ ⚼ ♅
17 00	☽ ‖ ☉
18 43	☽ ⊥ ♄
20 49	☽ # ♃
26 Sunday	
03 06	☽ ⊥ ☉
03 13	☽ ⚼ ♃
04 22	☿ ‖ ♂
06 21	☽ ☌ ♀
06 53	☽ ⊻ ♄
07 48	☽ ± ♇
09 11	☽ ✶ ☿
10 10	☽ ± ♅
13 47	☽ ‖ ☿
15 12	☽ ‖ ♂
19 55	☽ ⊼ ♇
22 17	☽ ⊼ ♅
23 00	☽ ∠ ☉
27 Monday	
01 38	☽ □ ♆
03 07	♀ ⊻ ♄
09 13	☽ △ ♃
09 25	☽ # ♇
10 33	♂ Q ♆
12 17	☽ Q ☿
15 28	♂ ⚼ ♇
15 30	☽ ‖ ♀
18 36	☽ ‖ ♆
28 Tuesday	
03 04	☽ ⊻ ♂
06 33	☉ ∠ ♆
06 57	♀ ± ♇
08 14	☽ ✶ ☉
10 57	☉ ⊥ ♀
16 04	☽ ⊥ ♂
19 57	☽ ☌ ♄
20 51	☽ ⊻ ♀
29 Wednesday	
03 09	♂ ⚼ ♅
05 05	☽ □ ☿
08 34	☽ ☍ ♇
09 14	☽ ⊥ ♀
10 40	☽ ‖ ♄
10 43	☽ Q ☉
10 56	☽ ☍ ♅
11 29	☽ ∠ ♂
13 08	♀ ‖ ♆
14 25	☽ △ ♆
21 07	☽ □ ♃
30 Thursday	
03 39	☽ ∠ ♀
05 10	☽ # ♅
12 32	☿ □ ♇
19 25	☽ ✶ ♂
20 18	☽ ⚼ ♆
31 Friday	
01 46	☽ □ ☉
08 01	☽ ⊻ ♄
08 18	☽ Q ♃
08 40	♂ ⊥ ♄
09 27	☿ □ ♅
09 46	☽ ✶ ♀
13 56	☽ ± ♆
19 43	☽ ⊥ ♄
19 46	☽ ⊼ ♇
20 24	☽ Q ♂
22 02	☽ ⊼ ♅
23 34	☽ △ ☿

All ephemeris data is given at 12.00 UT and the Moon's longitude is additionally given for 24.00 UT

JANUARY 1966

LONGITUDES

Date	Sidereal time h m s	Sun ☉ ° ' "	Moon ☽ ° ' "	Moon ☽ 24.00 ° ' "	Mercury ☿ ° '	Venus ♀ ° '	Mars ♂ ° '	Jupiter ♃ ° '	Saturn ♄ ° '	Uranus ♅ ° '	Neptune ♆ ° '	Pluto ♇ ° '
01	18 42 46	10 ♑ 37 50	26 ♈ 56 28	03 ♉ 19 51	21 ♐ 07	13 ♒ 29	07 ♒ 17	24 ♊ 22	12 ♓ 28	19 ♍ 37	21 ♏ 26	18 ♍ 26
02	18 46 43	11 38 59	09 ♉ 49 10	16 ♉ 24 50	22 30	13 38	08 04	24 R 15	12 33	19 R 36	21 28	18 R 26
03	18 50 39	12 40 08	23 ♉ 07 13	29 ♉ 56 31	23 54	13 44	08 51	24 07	12 38	19 36	21 30	18 25
04	18 54 36	13 41 16	06 ♊ 52 50	13 ♊ 56 05	25 18	13 48	09 38	24 00	12 42	19 35	21 31	18 25
05	18 58 32	14 42 24	21 ♊ 06 01	28 ♊ 22 10	26 44	13 50	10 25	23 53	12 47	19 34	21 33	18 24
06	19 02 29	15 43 32	05 ♋ 43 52	13 ♋ 10 18	28 10	13 R 49	11 13	23 46	12 52	19 34	21 34	18 24
07	19 06 25	16 44 40	20 ♋ 40 26	28 ♋ 13 06	29 ♐ 37	13 46	12 00	23 39	12 58	19 33	21 36	18 23
08	19 10 22	17 45 48	05 ♌ 47 04	13 ♌ 21 02	01 ♑ 04	13 40	12 47	23 32	13 03	19 32	21 37	18 23
09	19 14 18	18 46 56	20 ♌ 53 44	28 ♌ 24 00	02 32	13 32	13 35	23 25	13 08	19 32	21 39	18 22
10	19 18 15	19 48 03	05 ♍ 50 44	13 ♍ 13 03	04 01	13 21	14 22	23 19	13 13	19 31	21 40	18 22
11	19 22 12	20 49 10	20 ♍ 30 14	27 ♍ 41 44	05 30	13 07	15 09	23 12	13 19	19 30	21 41	18 21
12	19 26 08	21 50 17	04 ♎ 47 15	11 ♎ 46 36	07 00	12 52	15 57	23 06	13 24	19 29	21 43	18 20
13	19 30 05	22 51 25	18 ♎ 39 50	25 ♎ 27 03	08 30	12 33	16 44	23 00	13 30	19 28	21 44	18 19
14	19 34 01	23 52 32	02 ♏ 08 32	08 ♏ 44 35	10 01	12 13	17 31	22 54	13 35	19 27	21 45	18 19
15	19 37 58	24 53 39	15 ♏ 15 37	21 ♏ 42 02	11 32	11 50	18 19	22 48	13 41	19 26	21 47	18 18
16	19 41 54	25 54 46	28 ♏ 04 16	04 ♐ 22 44	13 04	11 25	19 06	22 42	13 47	19 24	21 48	18 17
17	19 45 51	26 55 52	10 ♐ 37 50	16 ♐ 49 59	14 37	10 58	19 54	22 37	13 52	19 23	21 49	18 16
18	19 49 47	27 56 59	22 ♐ 59 31	29 ♐ 06 45	16 10	10 29	20 41	22 31	13 58	19 22	21 50	18 15
19	19 53 44	28 58 04	05 ♑ 11 58	11 ♑ 15 25	17 43	09 58	21 28	22 26	14 04	19 21	21 51	18 15
20	19 57 41	29 ♑ 59 10	17 ♑ 17 19	23 ♑ 17 51	19 17	09 26	22 16	22 21	14 10	19 19	21 52	18 14
21	20 01 37	01 ♒ 00 15	29 ♑ 17 11	05 ♒ 15 29	20 52	08 52	23 03	22 16	14 16	19 18	21 54	18 13
22	20 05 34	02 01 19	11 ♒ 12 55	17 ♒ 09 36	22 27	08 18	23 51	22 11	14 22	19 16	21 55	18 12
23	20 09 30	03 02 22	23 ♒ 05 44	29 ♒ 01 31	24 03	07 42	24 38	22 07	14 28	19 15	21 56	18 11
24	20 13 27	04 03 24	04 ♓ 57 08	10 ♓ 52 52	25 39	07 05	25 25	22 02	14 34	19 13	21 57	18 10
25	20 17 23	05 04 26	16 ♓ 49 00	22 ♓ 45 51	27 16	06 29	26 13	21 58	14 41	19 12	21 58	18 09
26	20 21 20	06 05 26	28 ♓ 43 50	04 ♈ 43 20	28 ♑ 54	05 52	27 00	21 54	14 47	19 10	21 58	18 08
27	20 25 16	07 06 25	10 ♈ 44 51	16 ♈ 48 53	00 ♒ 32	05 15	27 48	21 50	14 53	19 08	21 59	18 07
28	20 29 13	08 07 24	22 ♈ 55 59	29 ♈ 06 44	02 11	04 38	28 35	21 47	15 00	19 06	22 00	18 05
29	20 33 10	09 08 21	05 ♉ 21 44	11 ♉ 41 34	03 51	04 02	29 ♒ 22	21 43	15 06	19 05	22 01	18 04
30	20 37 06	10 09 16	18 ♉ 06 51	24 ♉ 38 10	05 31	03 27	00 ♓ 10	21 40	15 13	19 03	22 02	18 03
31	20 41 03	11 ♒ 10 11	01 ♊ 16 00	08 ♊ 00 48	07 ♒ 12	02 ♒ 53	00 ♓ 57	21 ♊ 37	15 ♓ 19	19 ♍ 01	22 ♏ 03	18 ♍ 02

Date	Moon True ☊ ° '	Moon Mean ☊ ° '	Moon ☽ Latitude ° '
01	04 ♊ 05	02 ♊ 37	03 S 09
02	04 D 06	02 34	02 08
03	04 08	02 31	01 S 00
04	04 R 08	02 28	00 N 15
05	04 07	02 25	01 31
06	04 04	02 22	02 43
07	03 59	02 18	03 45
08	03 52	02 15	04 32
09	03 45	02 12	05 00
10	03 38	02 09	05 08
11	03 33	02 06	04 55
12	03 29	02 03	04 24
13	03 28	01 59	03 38
14	03 D 28	01 56	02 41
15	03 29	01 53	01 37
16	03 30	01 50	00 N 29
17	03 R 30	01 47	00 S 38
18	03 28	01 43	01 43
19	03 23	01 40	02 41
20	03 16	01 37	03 32
21	03 06	01 34	04 13
22	02 54	01 31	04 42
23	02 42	01 28	04 59
24	02 30	01 24	05 03
25	02 19	01 21	04 53
26	02 10	01 18	04 31
27	02 04	01 15	03 57
28	02 00	01 12	03 12
29	01 59	01 09	02 17
30	01 D 59	01 05	01 13
31	02 ♊ 00	01 ♊ 02	00 S 04

DECLINATIONS

Date	Sun ☉ ° '	Moon ☽ ° '	Mercury ☿ ° '	Venus ♀ ° '	Mars ♂ ° '	Jupiter ♃ ° '	Saturn ♄ ° '	Uranus ♅ ° '	Neptune ♆ ° '	Pluto ♇ ° '
01	23 S 01	07 N 27	22 S 27	15 S 40	19 S 35	22 N 58	08 S 40	04 N 50	16 S 28	18 N 01
02	22 56	12 44	22 40	15 25	19 22	22 58	08 38	04 51	16 28	18 02
03	22 50	17 36	22 53	15 11	19 10	22 58	08 36	04 51	16 29	18 03
04	22 44	21 43	23 04	14 56	18 57	22 58	08 34	04 51	16 29	18 03
05	22 38	24 39	23 14	14 42	18 44	22 58	08 32	04 51	16 30	18 04
06	22 31	26 02	23 24	14 29	18 31	22 58	08 30	04 52	16 30	18 05
07	22 24	25 33	23 32	14 15	18 18	22 57	08 28	04 52	16 30	18 05
08	22 16	23 13	23 39	14 03	18 04	22 57	08 26	04 52	16 31	18 06
09	22 08	19 16	23 45	13 51	17 50	22 57	08 24	04 53	16 31	18 07
10	21 59	14 08	23 50	13 39	17 36	22 57	08 22	04 53	16 31	18 07
11	21 50	08 17	23 54	13 28	17 22	22 57	08 20	04 53	16 32	18 08
12	21 40	02 N 08	23 56	13 18	17 08	22 57	08 18	04 54	16 32	18 09
13	21 30	03 S 58	23 57	13 08	16 53	22 57	08 15	04 54	16 32	18 09
14	21 20	09 42	23 57	12 59	16 39	22 56	08 13	04 55	16 32	18 10
15	21 09	14 53	23 56	12 51	16 24	22 56	08 11	04 55	16 33	18 11
16	20 58	19 16	23 53	12 43	16 09	22 56	08 09	04 56	16 33	18 11
17	20 47	22 41	23 49	12 36	15 53	22 56	08 06	04 56	16 33	18 12
18	20 35	24 58	23 44	12 30	15 38	22 56	08 04	04 57	16 33	18 13
19	20 22	26 02	23 37	12 24	15 23	22 56	08 02	04 57	16 34	18 14
20	20 09	25 50	23 29	12 19	15 07	22 56	07 59	04 58	16 34	18 14
21	19 56	24 25	23 20	12 15	14 51	22 56	07 57	04 58	16 34	18 15
22	19 43	21 56	23 09	12 12	14 35	22 55	07 54	04 59	16 34	18 16
23	19 29	18 31	22 57	12 09	14 19	22 55	07 52	05 00	16 35	18 17
24	19 15	14 23	22 43	12 07	14 03	22 55	07 49	05 00	16 35	18 17
25	19 00	09 42	22 28	12 05	13 46	22 55	07 47	05 01	16 35	18 18
26	18 45	04 S 39	22 12	12 04	13 30	22 55	07 44	05 02	16 35	18 19
27	18 30	00 N 37	21 54	12 04	13 13	22 55	07 42	05 02	16 35	18 20
28	18 14	05 57	21 34	12 05	12 56	22 55	07 39	05 03	16 35	18 20
29	17 58	11 10	21 13	12 06	12 39	22 55	07 37	05 04	16 36	18 21
30	17 42	16 04	20 51	12 07	12 22	22 55	07 34	05 05	16 36	18 22
31	17 S 26	20 N 21	20 S 27	12 S 09	12 S 05	22 N 55	07 S 32	05 N 05	16 S 36	18 N 23

ZODIAC SIGN ENTRIES

Date	h	m	Planets
01	17	46	☽ ♉
04	00	06	☽ ♊
06	02	40	☽ ♋
07	18	26	☿ ♑
08	02	50	☽ ♌
10	02	34	☽ ♍
12	03	53	☽ ♎
14	08	08	☽ ♏
16	15	39	☽ ♐
19	01	45	☽ ♑
20	12	20	☉ ♒
21	13	26	☽ ♒
24	01	58	☽ ♓
26	14	33	☽ ♈
27	04	10	☿ ♒
29	01	43	☽ ♉
30	07	01	♂ ♓
31	09	43	☽ ♊

LATITUDES

Date	Mercury ☿ ° '	Venus ♀ ° '	Mars ♂ ° '	Jupiter ♃ ° '	Saturn ♄ ° '	Uranus ♅ ° '	Neptune ♆ ° '	Pluto ♇ ° '
01	00 N 42	01 N 09	01 S 10	00 S 21	01 S 56	00 N 47	01 N 43	14 N 40
04	00 N 18	01 50	01 09	00 21	01 56	00 47	01 43	14 41
07	00 S 05	02 33	01 08	00 20	01 55	00 48	01 43	14 43
10	00 27	03 18	01 08	00 20	01 55	00 48	01 43	14 44
13	00 47	04 04	01 07	00 19	01 55	00 48	01 44	14 46
16	01 06	04 49	01 06	00 19	01 54	00 48	01 44	14 47
19	01 22	05 33	01 05	00 18	01 54	00 48	01 44	14 48
22	01 36	06 12	01 04	00 17	01 54	00 48	01 44	14 49
25	01 48	06 47	01 03	00 17	01 54	00 48	01 44	14 51
28	01 57	07 14	01 02	00 16	01 53	00 48	01 44	14 52
31	02 S 02	07 N 34	01 S 01	00 S 16	01 S 53	00 N 48	01 N 45	14 N 53

DATA

Julian Date	2439127
Delta T	+37 seconds
Ayanamsa	23° 22' 41"
Synetic vernal point	05° ♓ 44' 19"
True obliquity of ecliptic	23° 26' 41"

MOON'S PHASES, APSIDES AND POSITIONS ☽

Date	h	m	Phase	Longitude ° '	Eclipse Indicator
07	05	16	○	16 ♋ 28	
13	20	00	☾	23 ♎ 12	
21	15	46	●	01 ♒ 10	
29	19	48	☽	09 ♉ 28	

Day	h	m		
08	10	27	Perigee	
23	19	39	Apogee	
06	18	05	Max dec	26° N 05'
12	20	18	0S	
19	20	09	Max dec	26° S 06'
27	09	13	0N	

LONGITUDES

Date	Chiron ⚷ ° '	Ceres ⚳ ° '	Pallas ⚴ ° '	Juno ⚵ ° '	Vesta ⚶ ° '	Black Moon Lilith ⚸ ° '
01	18 ♓ 29	26 ♓ 20	21 ♒ 27	28 ♒ 06	05 ♋ 21	19 ♒ 53
11	18 ♓ 49	29 ♓ 05	24 ♒ 28	02 ♓ 36	02 ♋ 51	21 ♒ 01
21	19 ♓ 14	02 ♈ 06	27 ♒ 35	07 ♓ 17	00 ♋ 46	22 ♒ 08
31	19 ♓ 43	05 ♈ 20	00 ♓ 48	12 ♓ 08	29 ♊ 19	23 ♒ 15

All ephemeris data is given at 12.00 UT and the Moon's longitude is additionally given for 24.00 UT

ASPECTARIAN

h m Aspects

01 Saturday
00 24 ☽ ∥ ♅
01 30 ☽ ⊼ ♆
07 10 ☽ ⚹ ♃
07 15 ☽ ± ♇
09 12 ☽ Q ♀
09 29 ☽ ± ♅
13 00 ☽ ∠ ♄
17 25 ☽ # ♄
17 40 ☿ ⊻ ♆

02 Sunday
00 11 ☽ bQ ♇
01 27 ☉ # ♃
02 22 ☽ bQ ♅
07 14 ☽ bQ ☿
07 40 ♀ ± ♅
08 34 ☽ □ ♂
10 58 ☽ ∠ ♃
15 38 ☽ △ ☉
17 01 ☽ ⚹ ♄
19 01 ☽ □ ♀

03 Monday
00 18 ☽ # ♀
01 35 ☽ ± ☿
03 10 ☽ ⊥ ♃
03 37 ☽ △ ♇
05 43 ☽ △ ♅
06 13 ☽ # ♆
09 06 ☽ ☍ ♆
09 10 ☉ ∥ ☿
10 11 ☿ ∠ ♂
10 56 ☉ ⚹ ♄
13 32 ☽ ⊼ ☿
13 46 ☽ ⊻ ♃
14 23 ☽ ∥ ♇
14 41 ☽ Q ♄
15 35 ☿ ☍ ♃
19 14 ♂ bQ ♃
20 07 ☽ # ♂
20 40 ☽ bQ ☉
23 20 ☿ # ♃

04 Tuesday
13 30 ☽ ± ☉
14 51 ☉ ⊻ ♀
17 00 ☽ △ ♂
19 01 ☽ # ☉
20 56 ☽ ∥ ♃
21 59 ☽ □ ♄
22 18 ☽ # ☿
23 49 ☽ △ ♀

05 Wednesday
00 29 ☽ ⊼ ☉
07 31 ☽ □ ♇
09 28 ☽ □ ♅
12 44 ☽ ⊼ ♆
16 20 ♀ St R
16 35 ☽ ☌ ♃
19 34 ☽ bQ ♂
22 19 ☽ ☍ ☿
22 40 ☽ ± ♆

06 Thursday
00 45 ☽ bQ ♀
01 55 ☿ ⊥ ♆
11 07 ☽ ± ♂
13 05 ☽ Q ♇
13 22 ☽ bQ ♆
14 58 ☽ Q ♅
15 22 ☽ ± ♀
21 21 ☽ ⊼ ♂
22 34 ☿ ∠ ♀
23 35 ☽ △ ♄

07 Friday
01 00 ☽ ⊼ ♀
05 16 ☽ ☍ ☉
08 21 ☽ ⚹ ♇
10 13 ☽ ⚹ ♅
13 28 ☽ △ ♆
16 42 ☽ ⊻ ♃
23 39 ☽ bQ ♄
23 42 ♂ ± ♇

08 Saturday
02 10 ☽ ⊥ ♃
03 43 ☽ ⊼ ☿
08 11 ☽ ∠ ♇
08 47 ☽ # ☿
08 48 ♂ # ♇
10 02 ☽ ∠ ♅
11 37 ☿ Q ♄
13 57 ☽ ∥ ♃
14 01 ☽ ± ♄
14 15 ☽ ± ☿
16 20 ☽ ∠ ♃
18 57 ☽ # ☉
20 47 ♂ ⊻ ♄
22 27 ☽ ⊥ ♇
23 35 ☽ ⊼ ♄
23 43 ☽ ☍ ♂

09 Sunday
00 17 ☽ ⊥ ♅
00 24 ☽ ☍ ♀
02 23 ☉ △ ♇
06 04 ☽ bQ ☿
07 59 ☽ ⊻ ♇
08 23 ☽ ⊼ ☉
09 49 ☽ ⊻ ♅
10 27 ♂ ± ♅
10 44 ♀ ☌ ♂
12 19 ♀ ± ♅
13 12 ☽ □ ♆
16 00 ☽ ⚹ ♃
17 51 ☽ ∥ ♇
18 39 ☽ ± ☉
19 32 ☽ # ♂

10 Monday
01 24 ☽ # ♆
05 16 ☉ △ ♅
08 43 ☽ △ ☿
10 11 ☽ bQ ☉
11 09 ☽ Q ♃
14 07 ☽ # ♀
18 13 ☽ Q ♆
22 00 ♀ ⊻ ♄

11 Tuesday
00 02 ☽ ⊼ ♀
00 05 ☽ ☍ ♄
02 40 ☽ ⊼ ♂
08 26 ☽ ☌ ♇
09 45 ☽ ± ♀
10 20 ☽ ☌ ♅
11 48 ☽ # ♄
12 34 ☽ △ ☉
13 09 ☽ ± ♂
13 58 ☽ ⚹ ♆
16 27 ☽ □ ♃

12 Wednesday
00 30 ☽ bQ ♀
01 15 ☽ ∥ ♅
05 06 ☽ bQ ♂
07 22 ☿ ∠ ♆
08 57 ☉ ⚹ ♆
10 09 ☿ ⊥ ♀
15 17 ☽ ∠ ♆
16 14 ☽ □ ☿

13 Thursday
01 35 ☽ △ ♀
02 55 ☽ ⊼ ♄
06 52 ☽ ⊥ ♆
08 25 ☽ △ ♂
11 24 ☽ ⊻ ♇
13 24 ☽ ⊻ ♅
13 28 ☽ ± ♄
15 01 ☉ ⊼ ♃
15 50 ☽ # ♅
17 25 ☽ ⊻ ♆
19 35 ☽ △ ♃
20 00 ☽ □ ☉
21 59 ☽ ⊥ ♇

14 Friday
00 00 ☽ ⊥ ♅
03 38 ☽ Q ☿
05 04 ♀ ± ♇
05 34 ☽ bQ ♄
05 38 ☽ ∥ ♄
14 07 ☽ ∠ ♇
16 10 ☽ ∠ ♅
21 51 ♂ ∥ ♆
22 23 ☽ bQ ♃

15 Saturday
02 28 ☽ ∥ ♀
04 13 ☽ ⚹ ☿
05 51 ☽ □ ♀
07 15 ☽ Q ☉
09 03 ☽ △ ♄
11 36 ♂ ⊼ ♇
14 50 ☽ ± ♃
15 38 ☿ ⊻ ♀
17 38 ☽ ⚹ ♇
18 02 ☽ □ ♂
19 23 ☽ ∥ ♂
19 44 ☽ ⚹ ♅
20 35 ☽ ∥ ♆

16 Sunday
00 10 ☽ ☌ ♆
01 57 ☽ ⊼ ♃
05 39 ☽ # ♇
07 34 ☽ ⚹ ☉
12 00 ☽ ∠ ☿
12 58 ☿ ⊥ ♂
14 28 ☽ Q ♀
16 12 ☽ Q ♇
18 19 ☽ Q ♅
21 00 ♂ ⊼ ♅
22 30 ☽ ∥ ☉
23 43 ☿ ⚹ ♄

17 Monday
06 23 ☽ Q ♂
07 34 ☽ ⊥ ☿
12 37 ☽ ⚹ ♀
14 11 ☽ # ♃
14 44 ☽ ∠ ☉
18 19 ☽ □ ♄
20 48 ☽ ⊻ ☿
22 09 ☽ ∥ ☿

18 Tuesday
02 47 ☽ □ ♇
04 56 ☽ □ ♅
07 11 ☽ ⚹ ♂
09 44 ☽ ⊻ ♆
09 47 ☽ ⊥ ☉
11 05 ☽ ☍ ♃
16 41 ☽ ∠ ♀
21 30 ☽ ⊥ ♆
22 36 ☽ ⊻ ☉

19 Wednesday
00 23 ☉ ± ♃
05 46 ☽ Q ♄
09 40 ☽ ⊥ ♀
14 37 ☉ ∠ ♄
14 42 ☽ ∠ ♂
15 17 ☽ ∠ ♆
19 55 ☿ △ ♇
21 03 ☽ ⊻ ♀
23 54 ♂ □ ♆

20 Thursday
05 44 ☽ ⚹ ♄
09 49 ☽ ⊥ ♂
12 26 ☿ △ ♅
13 52 ☽ △ ♇
14 21 ♂ △ ♃
16 03 ☽ △ ♅
16 36 ☽ ☌ ☿
21 10 ☽ ⚹ ♆
22 02 ☽ ⊼ ♃
22 38 ☽ ⊻ ♂

21 Friday
09 58 ☽ ± ♃
11 58 ☽ ∠ ♄
15 46 ☽ ☌ ☉
19 52 ☽ bQ ♇
21 16 ☽ Q ♆
22 03 ☽ bQ ♅

22 Saturday
00 33 ☽ ∥ ☿
03 30 ☽ # ♃
03 41 ☿ ⚹ ♆
03 56 ☽ bQ ♃
06 13 ☽ ⊥ ♄
06 23 ☽ ☌ ♀
08 10 ☿ ⊼ ♃
09 22 ♀ ⊥ ♄
13 59 ☽ ± ♇
16 08 ☽ ± ♅
18 25 ☽ ⊻ ♄

23 Sunday
02 04 ☽ ⊼ ♇
04 14 ☽ ⊼ ♅
05 19 ☽ ∥ ☉
09 38 ☽ □ ♆
10 01 ☽ △ ♃
13 31 ☽ # ♇
14 14 ☽ ⊻ ☿
15 00 ☿ # ♃
15 15 ☉ bQ ♇
15 20 ☽ ☌ ♂
23 44 ☽ ∥ ♆

24 Monday
04 17 ☽ ⊥ ☿
05 11 ☿ ⊻ ♂
09 16 ☉ Q ♆
10 01 ☽ ⊻ ☉
13 56 ☽ ∥ ♂
14 21 ♀ bQ ♃
15 45 ☉ bQ ♅
16 07 ☽ ⊻ ♀
23 18 ☽ ⊥ ☉

25 Tuesday
00 01 ☽ ∥ ♀
01 22 ☽ ∠ ☿
03 40 ☽ ⊥ ♀
07 38 ☽ ☌ ♄
14 41 ☽ ☍ ♇
14 57 ♃ ⊼ ♆
16 47 ☽ ☍ ♅
19 11 ☽ ∠ ☉
20 57 ☽ ∠ ♀
21 23 ☽ ∥ ♄
21 53 ☿ ± ♃
22 20 ☽ □ ♃
22 23 ☽ △ ♆

26 Wednesday
08 17 ☽ ⊻ ♂
08 37 ☉ ☌ ♀
10 16 ☽ # ♅
12 24 ☽ ⚹ ☿
21 10 ☽ ⊥ ♂

27 Thursday
01 35 ☽ ⚹ ♀
01 52 ☿ ∠ ♄
04 05 ☽ ⚹ ☉
04 31 ☽ bQ ♆
06 01 ☉ bQ ♃
10 12 ☽ Q ♃
16 06 ☽ Q ☿
16 20 ☽ ∠ ♂
20 16 ☽ ⊻ ♄
22 23 ☽ ± ♆

28 Friday
00 14 ☽ Q ♀
02 32 ☽ ⊼ ♇
03 08 ☉ # ♇
04 32 ☽ ⊼ ♅
06 00 ☽ Q ☉
07 57 ☽ ∥ ♅
08 11 ☽ ⊥ ♄
09 45 ☽ ⚹ ♃
10 11 ☽ ⊼ ♆
14 15 ☽ ± ♇
16 14 ☽ ± ♅
19 42 ☽ # ♄
23 44 ☽ ⚹ ♂

29 Saturday
00 59 ☿ bQ ♇
01 49 ☽ ∠ ♄
07 38 ☽ bQ ♇
08 39 ☽ □ ☿
09 33 ☽ bQ ♅
09 35 ☽ □ ♀
10 04 ♀ bQ ♅
11 00 ☉ ⊥ ♄
12 38 ♀ Q ♆
14 01 ☿ ☌ ♀
14 32 ☿ Q ♆
14 35 ☽ ∠ ♃
15 20 ☿ bQ ♅
16 25 ☽ # ♀
18 43 ☽ # ♂
19 48 ☽ □ ☉

30 Sunday
00 09 ☽ Q ♂
06 33 ☽ ⚹ ♄
07 28 ☽ ⊥ ♃
11 53 ☽ △ ♇
13 44 ☽ △ ♅
14 48 ☽ # ♆
18 32 ☽ ⊻ ♃
19 14 ☽ ☍ ♆
20 12 ☽ # ☉

31 Monday
00 24 ☽ ∥ ♇
03 59 ☿ bQ ♃
04 49 ☽ Q ♄
05 05 ♀ bQ ♇
07 15 ♀ ∥ ♂
11 24 ☽ □ ♂
12 32 ☽ # ☿
14 46 ☽ △ ♀

FEBRUARY 1966

LONGITUDES

Date	Sidereal time h m s	Sun ☉ ° ' "	Moon ☽ ° ' "	Moon ☽ 24.00 ° ' "	Mercury ☿ ° '	Venus ♀ ° '	Mars ♂ ° '	Jupiter ♃ ° '	Saturn ♄ ° '	Uranus ♅ ° '	Neptune ♆ ° '	Pluto ♇ ° '
01	20 44 59	12 ♒ 11 04	14 ♊ 52 54	21 ♊ 52 28	08 ♒ 53	02 ♒ 20	01 ♓ 45	21 ♊ 34	15 ♓ 26	18 ♍ 59	22 ♏ 03	18 ♍ 01
02	20 48 56	13 11 56	28 ♊ 59 32	06 ♋ 13 55	10 36	01 R 48	02 32	21 R 31	15 32	18 R 57	22 04	18 R 00
03	20 52 52	14 12 47	13 ♋ 35 09	21 ♋ 02 37	12 19	01 18	03 19	21 29	15 39	18 55	22 05	17 58
04	20 56 49	15 13 36	28 ♋ 35 23	06 ♌ 12 20	14 02	00 50	04 07	21 27	15 46	18 53	22 05	17 57
05	21 00 45	16 14 24	13 ♌ 52 09	21 ♌ 33 22	15 47	00 25	04 54	21 25	15 52	18 51	22 06	17 56
06	21 04 42	17 15 11	29 ♌ 14 29	06 ♍ 53 58	17 32	00 ♒ 01	05 41	21 23	15 59	18 49	22 06	17 54
07	21 08 39	18 15 57	14 ♍ 30 23	22 ♍ 02 29	19 18	29 ♑ 39	06 29	21 21	16 06	18 47	22 07	17 53
08	21 12 35	19 16 41	29 ♍ 29 09	06 ♎ 49 32	21 05	29 20	07 16	21 20	16 13	18 45	22 07	17 52
09	21 16 32	20 17 25	14 ♎ 03 02	21 ♎ 09 16	22 52	29 03	08 03	21 18	16 20	18 43	22 08	17 50
10	21 20 28	21 18 07	28 ♎ 08 06	04 ♏ 59 35	24 40	28 49	08 50	21 17	16 27	18 40	22 08	17 49
11	21 24 25	22 18 48	11 ♏ 43 57	18 ♏ 21 32	26 29	28 37	09 38	21 16	16 33	18 38	22 09	17 48
12	21 28 21	23 19 29	24 ♏ 52 49	01 ♐ 18 17	28 ♒ 18	28 28	10 25	21 16	16 40	18 36	22 09	17 46
13	21 32 18	24 20 08	07 ♐ 38 32	13 ♐ 54 08	00 ♓ 08	28 21	11 12	21 15	16 47	18 34	22 09	17 45
14	21 36 14	25 20 46	20 ♐ 05 40	26 ♐ 13 42	01 58	28 16	11 59	21 15	16 55	18 31	22 10	17 43
15	21 40 11	26 21 23	02 ♑ 18 48	08 ♑ 21 26	03 48	28 14	12 46	21 D 15	17 02	18 29	22 10	17 42
16	21 44 08	27 21 59	14 ♑ 22 04	20 ♑ 21 09	05 39	28 D 15	13 34	21 15	17 09	18 27	22 10	17 40
17	21 48 04	28 22 33	26 ♑ 19 01	02 ♒ 16 00	07 30	28 18	14 21	21 15	17 16	18 24	22 10	17 39
18	21 52 01	29 ♒ 23 06	08 ♒ 12 24	14 ♒ 08 25	09 20	28 23	15 08	21 16	17 23	18 22	22 11	17 38
19	21 55 57	00 ♓ 23 38	20 ♒ 04 18	26 ♒ 00 11	11 11	28 31	15 55	21 17	17 30	18 19	22 11	17 36
20	21 59 54	01 24 07	01 ♓ 56 16	07 ♓ 52 42	13 00	28 40	16 42	21 18	17 37	18 17	22 11	17 34
21	22 03 50	02 24 36	13 ♓ 49 36	19 ♓ 47 09	14 49	28 52	17 29	21 19	17 45	18 14	22 11	17 33
22	22 07 47	03 25 02	25 ♓ 45 32	01 ♈ 44 56	16 36	29 07	18 16	21 20	17 52	18 12	22 R 11	17 31
23	22 11 43	04 25 27	07 ♈ 45 35	13 ♈ 47 46	18 22	29 23	19 03	21 22	17 59	18 09	22 11	17 30
24	22 15 40	05 25 50	19 ♈ 51 47	25 ♈ 57 59	20 05	29 ♑ 41	19 50	21 23	18 06	18 07	22 11	17 28
25	22 19 37	06 26 11	02 ♉ 06 47	08 ♉ 18 37	21 46	00 ♒ 01	20 37	21 25	18 14	18 04	22 11	17 27
26	22 23 33	07 26 31	14 ♉ 33 57	20 ♉ 53 17	23 24	00 23	21 24	21 28	18 21	18 02	22 11	17 25
27	22 27 30	08 26 48	27 ♉ 17 10	03 ♊ 46 06	24 58	00 47	22 11	21 30	18 28	17 59	22 10	17 24
28	22 31 26	09 ♓ 27 03	10 ♊ 20 35	17 ♊ 01 07	26 ♓ 28	01 ♒ 12	22 ♓ 58	21 ♊ 32	18 ♓ 36	17 ♍ 57	22 ♏ 10	17 ♍ 22

Date	Moon True ☊ ° '	Moon Mean ☊ ° '	Moon ☽ Latitude ° '
01	01 ♊ 59	00 ♊ 59	01 N 08
02	01 R 56	00 56	02 18
03	01 51	00 53	03 21
04	01 43	00 49	04 13
05	01 33	00 46	04 47
06	01 22	00 43	05 01
07	01 11	00 40	04 53
08	01 02	00 37	04 25
09	00 55	00 34	03 40
10	00 51	00 30	02 44
11	00 49	00 27	01 39
12	00 D 49	00 24	00 N 31
13	00 R 49	00 21	00 S 36
14	00 48	00 18	01 40
15	00 45	00 15	02 38
16	00 39	00 11	03 28
17	00 29	00 08	04 08
18	00 17	00 05	04 37
19	00 03	00 ♊ 02	04 54
20	29 ♉ 48	29 ♉ 59	04 59
21	29 33	29 55	04 50
22	29 19	29 52	04 28
23	29 08	29 49	03 55
24	29 00	29 46	03 10
25	28 55	29 43	02 16
26	28 52	29 40	01 15
27	28 D 52	29 36	00 08
28	28 ♉ 52	29 ♉ 33	01 N 00

DECLINATIONS

Date	Sun ☉ ° '	Moon ☽ ° '	Mercury ☿ ° '	Venus ♀ ° '	Mars ♂ ° '	Jupiter ♃ ° '	Saturn ♄ ° '	Uranus ♅ ° '	Neptune ♆ ° '	Pluto ♇ ° '
01	17 S 09	23 N 43	20 S 02	12 S 12	11 S 48	22 N 55	07 S 29	05 N 06	16 S 36	18 N 24
02	16 52	25 44	19 35	12 15	11 31	22 55	07 26	05 07	16 36	18 24
03	16 34	26 05	19 07	12 18	11 13	22 55	07 24	05 08	16 36	18 25
04	16 16	24 34	18 37	12 22	10 56	22 55	07 21	05 08	16 36	18 26
05	15 58	21 14	18 05	12 26	10 38	22 55	07 18	05 09	16 36	18 27
06	15 40	16 26	17 33	12 30	10 20	22 55	07 16	05 10	16 37	18 28
07	15 21	10 36	16 58	12 35	10 02	22 55	07 13	05 11	16 37	18 28
08	15 03	04 N 15	16 23	12 40	09 44	22 55	07 10	05 12	16 37	18 29
09	14 44	02 S 09	15 45	12 45	09 26	22 55	07 08	05 13	16 37	18 30
10	14 24	08 16	15 07	12 50	09 08	22 55	07 05	05 14	16 37	18 31
11	14 05	13 47	14 27	12 55	08 50	22 56	07 02	05 15	16 37	18 32
12	13 45	18 29	13 45	13 00	08 32	22 56	07 00	05 15	16 37	18 32
13	13 25	22 11	13 02	13 06	08 13	22 56	06 57	05 16	16 37	18 33
14	13 05	24 44	12 18	13 11	07 55	22 56	06 54	05 17	16 37	18 34
15	12 44	26 03	11 33	13 16	07 36	22 56	06 51	05 18	16 37	18 35
16	12 23	26 07	10 46	13 21	07 18	22 56	06 48	05 19	16 37	18 36
17	12 02	24 57	09 59	13 26	06 59	22 56	06 46	05 20	16 37	18 36
18	11 41	22 41	09 10	13 31	06 41	22 57	06 43	05 21	16 37	18 37
19	11 20	19 27	08 21	13 36	06 22	22 57	06 40	05 22	16 37	18 38
20	10 59	15 26	07 31	13 41	06 03	22 57	06 37	05 23	16 37	18 39
21	10 37	10 49	06 40	13 45	05 44	22 57	06 34	05 24	16 37	18 40
22	10 15	05 47	05 49	13 50	05 26	22 58	06 32	05 25	16 37	18 40
23	09 53	00 S 31	04 57	13 54	05 07	22 58	06 29	05 26	16 37	18 41
24	09 31	04 N 50	04 06	13 58	04 48	22 58	06 26	05 27	16 37	18 42
25	09 09	10 05	03 15	14 01	04 29	22 58	06 23	05 28	16 37	18 43
26	08 47	15 01	02 25	14 04	04 10	22 59	06 20	05 29	16 37	18 43
27	08 24	19 25	01 36	14 07	03 51	22 59	06 17	05 30	16 36	18 44
28	08 S 02	23 N 00	00 S 48	14 S 10	03 S 32	22 N 59	06 S 14	05 N 31	16 S 36	18 N 45

ZODIAC SIGN ENTRIES

Date	h	m	Planets
02	13	41	☽ ♋
04	14	14	☽ ♌
06	12	46	♀ ♑
06	13	11	☽ ♍
08	12	50	☽ ♎
10	15	15	☽ ♏
12	21	33	☽ ♐
13	10	17	☿ ♓
15	07	26	☽ ♑
17	19	26	☽ ♒
19	02	38	☉ ♓
20	08	05	☽ ♓
22	20	30	☽ ♈
25	07	53	☽ ♉
25	10	55	♀ ♒
27	17	03	☽ ♊

LATITUDES

Date	Mercury ☿ ° '	Venus ♀ ° '	Mars ♂ ° '	Jupiter ♃ ° '	Saturn ♄ ° '	Uranus ♅ ° '	Neptune ♆ ° '	Pluto ♇ ° '
01	02 S 04	07 N 38	01 S 01	00 S 16	01 S 53	00 N 48	01 N 45	14 N 53
04	02 05	07 47	00 59	00 15	01 53	00 48	01 45	14 54
07	02 02	07 49	00 58	00 15	01 53	00 49	01 45	14 55
10	01 55	07 44	00 57	00 14	01 53	00 49	01 45	14 56
13	01 43	07 33	00 56	00 14	01 53	00 49	01 45	14 57
16	01 26	07 18	00 54	00 13	01 53	00 49	01 45	14 58
19	01 03	07 00	00 53	00 13	01 53	00 49	01 46	14 58
22	00 S 34	06 39	00 51	00 12	01 53	00 49	01 46	14 59
25	00 N 01	06 16	00 50	00 12	01 53	00 49	01 46	14 59
28	00 40	05 52	00 48	00 11	01 53	00 49	01 46	15 00
31	01 N 22	05 N 28	00 S 47	00 S 11	01 S 53	00 N 49	01 N 46	15 N 00

DATA

Julian Date	2439158
Delta T	+37 seconds
Ayanamsa	23° 22' 46"
Synetic vernal point	05° ♓ 44' 13"
True obliquity of ecliptic	23° 26' 42"

MOON'S PHASES, APSIDES AND POSITIONS ☽

Date	h	m	Phase	Longitude ° '	Eclipse Indicator
05	15	58	○	16 ♌ 24	
12	08	53	☾	23 ♏ 12	
20	10	49	●	01 ♓ 21	
28	10	15	☽	09 ♊ 23	

Day	h	m		
05	22	14	Perigee	
19	20	53	Apogee	
03	04	38	Max dec	26° N 11'
09	03	52	0S	
16	01	06	Max dec	26° S 15'
23	14	19	0N	

LONGITUDES

Date	Chiron ⚷ ° '	Ceres ⚳ ° '	Pallas ⚴ ° '	Juno ⚵ ° '	Vesta ⚶ ° '	Black Moon Lilith ⚸ ° '
01	19 ♓ 46	05 ♈ 41	01 ♓ 07	12 ♓ 38	29 ♊ 12	23 ♒ 22
11	20 ♓ 18	09 ♈ 07	04 ♓ 24	17 ♓ 39	28 ♊ 34	24 ♒ 29
21	20 ♓ 52	12 ♈ 43	07 ♓ 43	22 ♓ 47	28 ♊ 41	25 ♒ 36
31	21 ♓ 27	16 ♈ 27	11 ♓ 03	28 ♓ 01	29 ♊ 29	26 ♒ 43

ASPECTARIAN

h m Aspects

01 Tuesday
00 03 ☽ △ ☿
05 29 ☽ ∥ ♃
06 56 ☽ △ ☉
08 00 ☉ ± ♇
12 57 ☽ □ ♄
16 04 ☽ ⚼ ♀
17 23 ☽ □ ♇
19 03 ☽ □ ♅
20 08 ☿ ⊥ ♄
22 37 ♀ ⊻ ♂
23 26 ☽ ☌ ♃

02 Wednesday
00 19 ☽ ⊼ ♆
05 31 ☽ ⚼ ☿
06 22 ☉ ± ♅
06 50 ☽ ± ♀
10 27 ☽ ± ♆
10 34 ☽ ⚼ ☉
16 31 ☽ ⊼ ♀
18 14 ☽ △ ♂
22 33 ☽ ± ☿
23 35 ☽ Q ♇

03 Thursday
01 09 ☽ Q ♅
01 23 ☽ ⚼ ♆
02 36 ☽ ± ☉
07 20 ☿ ± ♇
09 05 ☉ ∥ ♆
09 40 ☽ ⊼ ☿
13 05 ☽ ⊼ ☉
15 22 ☽ △ ♄
19 04 ☽ ⚹ ♇
20 04 ☽ ⚼ ♂
20 20 ☿ ± ♅
20 35 ☽ ⚹ ♅

04 Friday
00 40 ☽ ⊻ ♃
01 40 ☽ △ ♆
10 11 ☽ ⊥ ♃
11 12 ☽ ± ♂
15 27 ☽ ⚼ ♄
15 27 ☽ ☍ ♀
15 29 ♀ ∠ ♄
18 52 ☽ ∠ ♇
20 05 ☿ # ♇
20 20 ☽ ∠ ♅
21 11 ☽ ⊼ ♂

05 Saturday
00 21 ☽ ∠ ♃
01 24 ☽ ∥ ♃
02 12 ☉ ⊻ ♄
05 42 ☽ ± ♄
08 58 ☽ ⊥ ♇
10 25 ☽ ⊥ ♅
13 20 ☿ ⊻ ♄
15 09 ☽ ⊼ ♄
15 22 ☽ ☍ ☿
15 58 ☽ ☍ ☉
18 20 ☽ ⊻ ♇
19 46 ☽ ⊻ ♅
23 45 ☽ ⚹ ♃

06 Sunday
00 51 ☽ □ ♆
02 38 ☽ ∥ ♇
02 54 ☉ ☌ ☿
06 20 ☽ # ☿
11 11 ☽ # ♆
13 11 ☽ ⊼ ♀
15 32 ☽ # ☉
17 00 ☿ ⊼ ♇
18 28 ☽ Q ♃
18 43 ♀ ⊥ ♂
22 21 ☽ ± ♀
22 39 ☽ ☍ ♂

07 Monday
03 10 ☉ ⊼ ♇
04 17 ☽ # ♀
05 04 ☽ Q ♆
05 07 ☿ ⊼ ♅
12 14 ☽ ⚼ ♀
14 17 ☽ # ♂
14 33 ☽ ☍ ♄
17 21 ☽ ☌ ♇
18 24 ☽ ⊼ ☉
18 47 ☽ ☌ ♅
20 38 ☽ ⊼ ☿
22 53 ☽ □ ♃
23 51 ☉ ⊼ ♅

08 Tuesday
00 08 ☽ ⚹ ♆
01 00 ☽ # ♄
02 36 ☿ ∥ ♆
04 42 ☽ ± ☉
07 34 ☽ ± ☿
08 30 ☽ ∥ ♅
11 45 ☽ △ ♀
15 17 ☿ △ ♃
20 24 ☽ ⚼ ☉

09 Wednesday
00 17 ☽ ⚼ ☿
00 30 ☽ ∠ ♆
01 27 ☽ ⊼ ♂
02 06 ☿ □ ♆
12 00 ☽ ± ♂
15 30 ☽ ⊥ ♆
15 51 ☽ ⊼ ♄
18 22 ☽ ⊻ ♇
19 50 ☽ ⊻ ♅
23 21 ☽ △ ☉
23 48 ☽ # ♅

10 Thursday
00 14 ☽ △ ♃
01 40 ☽ ⊻ ♆
02 07 ☽ ± ♄
04 09 ☽ ⚼ ♂
04 34 ☽ ⊥ ♇
05 08 ☽ △ ☿
06 02 ☽ ⊥ ♅
07 14 ☽ ∥ ♄
11 40 ☉ △ ♃
13 09 ☽ □ ♀
15 25 ☽ ∥ ♂
17 49 ☽ ⚼ ♄
20 09 ☽ ∠ ♇
21 39 ☽ ∠ ♅

11 Friday
02 16 ☽ ⚼ ♃
07 55 ☽ ∥ ♀
08 00 ☽ △ ♂
08 00 ☉ □ ♆
13 17 ☽ ∥ ☉
14 44 ☽ ∥ ☿
18 23 ☽ ± ♃
20 43 ☽ Q ♀
20 48 ☽ △ ♄
22 57 ☽ ⚹ ♇

12 Saturday
00 28 ☽ ⚹ ♅
01 53 ☽ ∥ ♆
05 19 ☽ ⊼ ♃
06 57 ☽ ☌ ♆
08 53 ☽ □ ☉
12 19 ☽ # ♇
12 20 ☉ ∥ ☿
13 55 ☿ ⊻ ♀
18 36 ☽ ⚹ ♀
19 26 ☽ □ ☿
21 06 ☽ Q ♇
22 39 ☽ Q ♅

13 Sunday
10 26 ☿ ∥ ♀
17 59 ☽ # ♃
19 16 ☽ □ ♂
21 47 ☽ Q ☉
22 51 ☽ ∠ ♀

14 Monday
05 45 ☽ □ ♄
06 06 ☉ ∥ ♀
07 24 ☽ □ ♇
08 57 ☽ □ ♅
11 42 ☽ Q ☿
14 15 ☽ ☍ ♃
16 02 ☽ ⊻ ♆
16 14 ☽ ⊥ ♀
23 12 ☽ ⚹ ☉

15 Tuesday
03 49 ☽ ⊥ ♆
03 58 ☽ ⊻ ♀
07 01 ♃ St D
08 45 ☽ Q ♂
15 30 ☽ ⚹ ☿
17 26 ☽ Q ♄
17 36 ☿ ⊥ ♀
18 41 ♀ St D
21 38 ☽ ∠ ♆

16 Wednesday
02 14 ♀ ∠ ♂
07 38 ☽ ∠ ☉
10 16 ☽ ⚹ ♂
17 37 ☽ ⚹ ♄
18 37 ☽ △ ♇
20 08 ☽ △ ♅

17 Thursday
01 49 ☽ ⊼ ♃
02 54 ☽ ∠ ☿
03 20 ☽ ⊥ ☉
03 39 ☽ ⚹ ♆
10 00 ☉ ⊻ ♀
13 54 ☽ ± ♃
16 01 ☽ ☌ ♀
16 32 ☽ ⊻ ☉
18 32 ☽ ∠ ♂

18 Friday
00 07 ☽ ∠ ♄
00 22 ☽ ⊥ ☿
00 45 ☽ ⚼ ♇
02 15 ☽ ⚼ ♅
03 51 ☽ Q ♆
08 04 ☽ ⚼ ♃
08 49 ♂ ∥ ♄
09 41 ☽ # ♃
14 00 ☽ ⊥ ♂
14 43 ☽ ⊻ ☿
18 29 ☽ ⊥ ♄
18 54 ☽ ± ♇
20 23 ☽ ± ♅

19 Saturday
03 00 ☽ ⊻ ♂
06 45 ☽ ⊻ ♄
07 01 ☽ ⊼ ♇
08 28 ☽ ⊼ ♅
14 27 ☽ △ ♃
16 16 ☽ □ ♆
17 12 ☽ # ♇

20 Sunday
04 11 ♄ ☍ ♇
05 18 ☽ ⊻ ♀
05 20 ☽ ∥ ♆
10 49 ☽ ☌ ☉
17 37 ☽ ⊥ ♀
21 15 ☽ ∥ ♀
21 56 ☿ ∠ ♀

21 Monday
12 06 ☽ ∠ ♀
13 04 ☽ ∥ ☉
13 55 ♂ ☍ ♇
14 20 ☽ ☌ ☿
14 45 ☿ ∥ ♄
19 29 ☽ ☍ ♇
19 53 ☽ ☌ ♂
19 58 ☽ ☌ ♄
20 52 ☽ ☍ ♅
21 19 ♂ ☌ ♄

22 Tuesday
03 06 ☽ □ ♃
04 49 ☽ △ ♆
08 33 ☽ ∥ ♄
09 59 ♂ ☍ ♅
10 44 ♆ St R
11 52 ☽ ∥ ☿
12 47 ♂ # ♅
13 43 ☽ # ♅
13 47 ☽ ∥ ♂
18 52 ☽ ⚹ ♀
22 55 ☿ # ♅

23 Wednesday
00 21 ☿ ☍ ♇
04 44 ☽ ⊻ ☉
05 07 ☿ ∥ ♂
06 26 ☿ ☌ ♄
09 16 ☿ ☍ ♅
10 51 ☽ ⚼ ♆
15 12 ☽ Q ♃
17 47 ☽ ⊥ ☉
19 23 ☽ Q ♀

24 Thursday
04 43 ☽ ± ♆
05 30 ☿ ☌ ♂
07 17 ☽ ⊼ ♇
08 30 ☽ ⊻ ♄
08 34 ☽ ⊼ ♅
09 10 ☽ # ☿
11 51 ☽ # ♂
11 57 ☽ ⊻ ♂
12 31 ☽ ⊻ ☿
13 13 ☽ ∠ ☉
13 30 ♄ ☍ ♅
14 47 ☽ ∥ ♅
15 01 ☽ ⚹ ♃
16 34 ☽ ⊼ ♆
19 05 ☽ ± ♇
19 09 ☽ # ♄
20 20 ☽ ± ♅
20 26 ☽ ⊥ ♄
20 50 ☉ ⊥ ♀

25 Friday
00 33 ☽ ⊥ ♂
02 12 ☽ ⊥ ☿
06 54 ☿ □ ♃
07 48 ☽ □ ♀
07 59 ☽ # ☉
12 39 ☽ ⚼ ♇
13 52 ☽ ⚼ ♅
14 11 ☽ ∠ ♄
17 55 ☿ △ ♆
19 15 ☽ ∠ ♂
20 23 ☽ ∠ ♃
21 07 ☽ ⚹ ☉
22 24 ☽ ∠ ☿

26 Saturday
07 10 ☽ # ♀
13 42 ☽ ⊥ ♃
13 52 ♂ □ ♃
17 25 ☽ △ ♇
18 34 ☽ △ ♅
19 16 ☽ ⚹ ♄
20 15 ☽ # ♆
22 04 ☽ Q ☉

27 Sunday
01 07 ☽ ⊻ ♃
01 49 ☽ ⚹ ♂
02 25 ☽ ☍ ♆
07 05 ☽ ⚹ ☿
07 59 ☽ ∥ ♇
11 44 ♂ △ ♆
17 58 ☽ Q ♄
18 42 ☽ △ ♀

28 Monday
01 34 ☽ Q ♂
08 11 ☽ Q ☿
10 15 ☽ □ ☉
11 55 ☽ ∥ ♃
22 54 ☽ ⚼ ♀

All ephemeris data is given at 12.00 UT and the Moon's longitude is additionally given for 24.00 UT

MARCH 1966

LONGITUDES

Date	Sidereal time h m s	Sun ☉ ° ' "	Moon ☽ ° ' "	Moon ☽ 24.00 ° ' "	Mercury ☿ ° '	Venus ♀ ° '	Mars ♂ ° '	Jupiter ♃ ° '	Saturn ♄ ° '	Uranus ♅ ° '	Neptune ♆ ° '	Pluto ♇ ° '
01	22 35 23	10 ♓ 27 17	23 ♊ 48 04	00 ♋ 41 46	27 ♓ 53	01 ♒ 39	23 ♓ 45	21 ♊ 35	18 ♓ 43	17 ♍ 54	22 ♏ 10	17 ♍ 21
02	22 39 19	11 27 28	07 ♋ 42 24	14 ♋ 49 59	29 ♓ 13	02 08	24 31	21 38	18 50	17 R 51	22 R 10	17 R 19
03	22 43 16	12 27 37	22 ♋ 04 21	29 ♋ 25 08	00 ♈ 27	02 38	25 18	21 41	18 58	17 49	22 09	17 17
04	22 47 12	13 27 45	06 ♌ 51 42	14 ♌ 23 14	01 34	03 10	26 05	21 44	19 05	17 46	22 09	17 16
05	22 51 09	14 27 50	21 ♌ 58 40	29 ♌ 36 45	02 34	03 43	26 52	21 48	19 12	17 44	22 09	17 14
06	22 55 06	15 27 53	07 ♍ 16 06	14 ♍ 55 16	03 27	04 18	27 38	21 51	19 20	17 41	22 08	17 13
07	22 59 02	16 27 54	22 ♍ 32 47	00 ♎ 07 15	04 11	04 54	28 25	21 55	19 27	17 38	22 08	17 11
08	23 02 59	17 27 53	07 ♎ 37 24	15 ♎ 02 10	04 47	05 31	29 12	21 59	19 35	17 36	22 08	17 09
09	23 06 55	18 27 50	22 ♎ 20 39	29 ♎ 32 14	05 15	06 09	29 ♓ 58	22 03	19 42	17 33	22 07	17 08
10	23 10 52	19 27 46	06 ♏ 36 30	13 ♏ 33 17	05 33	06 49	00 ♈ 45	22 08	19 49	17 31	22 07	17 06
11	23 14 48	20 27 40	20 ♏ 22 36	27 ♏ 04 39	05 43	07 30	01 31	22 12	19 57	17 28	22 06	17 05
12	23 18 45	21 27 33	03 ♐ 39 45	10 ♐ 08 22	05 R 44	08 12	02 18	22 17	20 04	17 25	22 05	17 03
13	23 22 41	22 27 23	16 ♐ 31 00	22 ♐ 48 15	05 36	08 54	03 04	22 22	20 12	17 23	22 05	17 01
14	23 26 38	23 27 13	29 ♐ 00 44	05 ♑ 09 04	05 20	09 38	03 51	22 27	20 19	17 20	22 04	17 00
15	23 30 35	24 27 00	11 ♑ 13 52	17 ♑ 15 46	04 56	10 23	04 37	22 32	20 26	17 18	22 04	16 58
16	23 34 31	25 26 46	23 ♑ 15 19	29 ♑ 13 06	04 25	11 09	05 23	22 38	20 34	17 15	22 03	16 57
17	23 38 28	26 26 30	05 ♒ 09 36	11 ♒ 05 18	03 47	11 56	06 10	22 43	20 41	17 12	22 02	16 55
18	23 42 24	27 26 12	17 ♒ 00 36	22 ♒ 55 54	03 04	12 43	06 56	22 49	20 48	17 10	22 01	16 54
19	23 46 21	28 25 53	28 ♒ 51 31	04 ♓ 47 43	02 17	13 32	07 42	22 55	20 56	17 07	22 01	16 52
20	23 50 17	29 ♓ 25 31	10 ♓ 44 47	16 ♓ 42 54	01 26	14 21	08 28	23 01	21 03	17 05	22 00	16 50
21	23 54 14	00 ♈ 25 08	22 ♓ 42 15	28 ♓ 43 00	00 ♈ 33	15 11	09 15	23 07	21 11	17 02	21 59	16 49
22	23 58 10	01 24 42	04 ♈ 45 17	10 ♈ 49 16	29 ♓ 39	16 01	10 01	23 13	21 18	17 00	21 58	16 47
23	00 02 07	02 24 14	16 ♈ 55 04	23 ♈ 02 51	28 45	16 53	10 47	23 20	21 25	16 57	21 57	16 46
24	00 06 04	03 23 45	29 ♈ 12 46	05 ♉ 25 01	27 52	17 45	11 33	23 26	21 32	16 55	21 56	16 44
25	00 10 00	04 23 13	11 ♉ 39 49	17 ♉ 57 25	27 01	18 37	12 19	23 33	21 40	16 52	21 55	16 43
26	00 13 57	05 22 39	24 ♉ 18 05	00 ♊ 42 07	26 13	19 31	13 05	23 40	21 47	16 50	21 54	16 41
27	00 17 53	06 22 03	07 ♊ 09 50	13 ♊ 41 34	25 29	20 24	13 51	23 47	21 54	16 47	21 53	16 40
28	00 21 50	07 21 25	20 ♊ 17 40	26 ♊ 58 27	24 50	21 19	14 36	23 54	22 02	16 45	21 52	16 38
29	00 25 46	08 20 44	03 ♋ 44 11	10 ♋ 35 07	24 15	22 14	15 22	24 02	22 09	16 42	21 51	16 37
30	00 29 43	09 20 01	17 ♋ 31 24	24 ♋ 33 06	23 46	23 09	16 08	24 09	22 16	16 40	21 50	16 35
31	00 33 39	10 ♈ 19 16	01 ♌ 40 09	08 ♌ 52 21	23 ♓ 22	24 ♒ 05	16 ♈ 54	24 ♊ 17	22 ♓ 23	16 ♍ 38	21 ♏ 49	16 ♍ 34

Date	Moon True ☊ ° '	Moon Mean ☊ ° '	Moon ☽ Latitude ° '
01	28 ♉ 51	29 ♉ 30	02 N 08
02	28 R 49	29 27	03 10
03	28 44	29 24	04 02
04	28 37	29 20	04 40
05	28 27	29 17	04 59
06	28 17	29 14	04 57
07	28 06	29 11	04 35
08	27 57	29 08	03 53
09	27 50	29 05	02 56
10	27 46	29 01	01 50
11	27 44	28 58	00 N 39
12	27 D 44	28 55	00 S 32
13	27 44	28 52	01 38
14	27 R 44	28 49	02 38
15	27 43	28 46	03 30
16	27 39	28 42	04 11
17	27 33	28 39	04 41
18	27 24	28 36	04 59
19	27 13	28 33	05 03
20	27 02	28 30	04 55
21	26 51	28 26	04 34
22	26 40	28 23	04 00
23	26 32	28 20	03 15
24	26 26	28 17	02 20
25	26 22	28 14	01 18
26	26 21	28 11	00 S 11
27	26 D 22	28 07	00 N 58
28	26 23	28 04	02 05
29	26 24	28 01	03 07
30	26 R 24	27 58	04 00
31	26 ♉ 22	27 ♉ 55	04 N 40

DECLINATIONS

Date	Sun ☉ ° '	Moon ☽ ° '	Mercury ☿ ° '	Venus ♀ ° '	Mars ♂ ° '	Jupiter ♃ ° '	Saturn ♄ ° '	Uranus ♅ ° '	Neptune ♆ ° '	Pluto ♇ ° '
01	07 S 39	25 N 26	00 S 01	14 S 12	03 S 13	23 N 00	06 S 12	05 N 32	16 S 36	18 N 46
02	07 16	26 23	00 N 44	14 14	02 54	23 00	06 09	05 33	16 36	18 46
03	06 53	25 37	01 26	14 15	02 35	23 00	06 06	05 34	16 36	18 47
04	06 30	23 04	02 06	14 16	02 16	23 01	06 03	05 35	16 36	18 48
05	06 07	18 54	02 43	14 17	01 57	23 01	06 00	05 36	16 36	18 49
06	05 44	13 26	03 16	14 17	01 38	23 01	05 57	05 37	16 36	18 49
07	05 21	07 10	03 46	14 17	01 19	23 02	05 54	05 38	16 35	18 50
08	04 57	00 N 32	04 12	14 16	01 00	23 02	05 51	05 39	16 35	18 51
09	04 34	05 S 59	04 34	14 15	00 41	23 03	05 49	05 40	16 35	18 51
10	04 10	12 00	04 52	14 14	00 21	23 03	05 46	05 41	16 35	18 52
11	03 47	17 13	05 05	14 12	00 S 02	23 03	05 43	05 42	16 35	18 53
12	03 23	21 24	05 14	14 09	00 N 17	23 04	05 40	05 43	16 34	18 53
13	03 00	24 23	05 17	14 06	00 35	23 04	05 37	05 44	16 34	18 54
14	02 36	26 05	05 17	14 03	00 54	23 05	05 34	05 45	16 34	18 55
15	02 12	26 27	05 11	13 59	01 13	23 05	05 31	05 46	16 34	18 55
16	01 49	25 34	05 01	13 55	01 32	23 06	05 28	05 47	16 34	18 56
17	01 25	23 31	04 47	13 50	01 51	23 06	05 26	05 48	16 33	18 57
18	01 01	20 29	04 29	13 45	02 10	23 07	05 23	05 49	16 33	18 57
19	00 37	16 36	04 07	13 39	02 29	23 07	05 20	05 50	16 33	18 58
20	00 S 14	12 05	03 42	13 33	02 48	23 07	05 17	05 51	16 33	18 58
21	00 N 10	07 05	03 15	13 26	03 06	23 08	05 14	05 52	16 32	18 59
22	00 34	01 S 47	02 46	13 19	03 25	23 08	05 11	05 53	16 32	18 59
23	00 57	03 N 39	02 15	13 11	03 44	23 09	05 09	05 54	16 32	19 00
24	01 21	09 01	01 43	13 02	04 02	23 09	05 06	05 55	16 32	19 01
25	01 45	14 06	01 12	12 54	04 21	23 10	05 03	05 56	16 31	19 01
26	02 08	18 40	00 40	12 44	04 39	23 10	05 00	05 57	16 31	19 02
27	02 32	22 28	00 N 09	12 35	04 57	23 11	04 57	05 58	16 31	19 02
28	02 55	25 10	00 S 20	12 25	05 16	23 11	04 55	05 59	16 30	19 03
29	03 19	26 31	00 48	12 14	05 34	23 12	04 52	06 00	16 30	19 03
30	03 42	26 16	01 14	12 03	05 52	23 12	04 49	06 01	16 30	19 04
31	04 N 05	24 N 21	01 S 38	11 S 51	06 N 10	23 N 13	04 S 46	06 N 02	16 S 29	19 N 04

ZODIAC SIGN ENTRIES

Date	h	m	Planets
01	22	48	☽ ♋
03	02	57	☿ ♈
04	00	57	☽ ♌
06	00	36	☽ ♍
07	23	48	☽ ♎
09	12	55	♂ ♈
10	00	47	☽ ♏
12	05	18	☽ ♐
14	13	55	☽ ♑
17	01	35	☽ ♒
19	14	19	☽ ♓
21	01	53	☉ ♈
22	02	33	☽ ♈
22	02	34	☿ ♓
24	13	32	☽ ♉
26	22	41	☽ ♊
29	05	23	☽ ♋
31	09	12	☽ ♌

LATITUDES

Date	Mercury ☿ ° '	Venus ♀ ° '	Mars ♂ ° '	Jupiter ♃ ° '	Saturn ♄ ° '	Uranus ♅ ° '	Neptune ♆ ° '	Pluto ♇ ° '
01	00 N 54	05 N 44	00 S 48	00 S 11	01 S 53	00 N 49	01 N 46	15 N 00
04	01 36	05 19	00 46	00 11	01 53	00 49	01 46	15 00
07	02 18	04 54	00 44	00 10	01 53	00 49	01 47	15 00
10	02 54	04 29	00 43	00 10	01 53	00 49	01 47	15 01
13	03 20	04 04	00 41	00 09	01 53	00 49	01 47	15 01
16	03 34	03 39	00 39	00 09	01 53	00 49	01 47	15 01
19	03 30	03 15	00 38	00 08	01 53	00 49	01 47	15 01
22	03 10	02 51	00 36	00 08	01 53	00 49	01 47	15 01
25	02 36	02 28	00 34	00 08	01 54	00 49	01 47	15 00
28	01 52	02 06	00 32	00 07	01 54	00 49	01 48	15 00
31	01 N 05	01 N 45	00 S 30	00 S 07	01 S 54	00 N 49	01 N 48	15 N 00

DATA

Julian Date	2439186
Delta T	+37 seconds
Ayanamsa	23° 22' 50"
Synetic vernal point	05° ♓ 44' 10"
True obliquity of ecliptic	23° 26' 42"

MOON'S PHASES, APSIDES AND POSITIONS ☽

Date	h	m	Phase	Longitude ° '	Eclipse Indicator
07	01	45	○	16 ♍ 02	
14	00	19	☾	22 ♐ 58	
22	04	46	●	01 ♈ 07	
29	20	43	☽	08 ♋ 42	

Day	h	m		
06	10	40	Perigee	
19	02	58	Apogee	
02	13	41	Max dec	26° N 23'
08	13	57	0S	
15	06	54	Max dec	26° S 29'
22	19	54	0N	
29	20	29	Max dec	26° N 37'

LONGITUDES

Date	Chiron ⚷ ° '	Ceres ⚳ ° '	Pallas ⚴ ° '	Juno ⚵ ° '	Vesta ⚶ ° '	Black Moon Lilith ⚸ ° '
01	21 ♓ 20	15 ♈ 41	10 ♓ 23	26 ♓ 58	29 ♊ 16	26 ♒ 30
11	21 ♓ 56	19 ♈ 30	13 ♓ 43	02 ♈ 18	00 ♋ 36	27 ♒ 37
21	22 ♓ 33	23 ♈ 23	17 ♓ 03	07 ♈ 43	02 ♋ 28	28 ♒ 44
31	23 ♓ 09	27 ♈ 20	20 ♓ 22	13 ♈ 13	04 ♋ 49	29 ♒ 52

All ephemeris data is given at 12.00 UT and the Moon's longitude is additionally given for 24.00 UT

ASPECTARIAN

01 Tuesday
00 36 ☽ □ ♇
01 36 ☽ □ ♅
02 56 ☽ □ ♄
08 05 ☽ ☌ ♃
09 08 ☽ ⚻ ♆
11 54 ☽ □ ♂
15 21 ☽ ± ♀
19 37 ☽ ± ♆
19 55 ☽ □ ☿

02 Wednesday
02 08 ☽ ⚻ ♀
07 56 ☽ Q ♇
08 52 ☽ Q ♅
11 05 ☽ ⚼ ♆
18 49 ☽ △ ☉
20 28 ♀ ⚼ ♇

03 Thursday
04 06 ☽ ⚹ ♇
04 59 ☽ ⚹ ♅
06 49 ☽ △ ♄
11 22 ☽ ⚺ ♃
12 08 ☽ △ ♆
17 36 ☽ △ ♂
19 37 ♀ ⚼ ♅
21 13 ☽ ⊥ ♃
21 28 ☽ ⚼ ☉

04 Friday
02 49 ☽ △ ☿
04 37 ☽ ∠ ♇
05 27 ☽ ∠ ♅
05 50 ☽ ☍ ♀
07 30 ☽ ⚼ ♄
11 48 ☽ ∠ ♃
12 26 ☽ ∥ ♃
13 02 ☽ ± ☉
16 11 ☿ ⋕ ♂
19 01 ☽ ⊥ ♇
19 07 ☽ ⚼ ♂
19 49 ☽ ⊥ ♅
22 01 ☽ ± ♄
23 17 ☽ ⚻ ☉

05 Saturday
04 32 ☽ ⚺ ♇
04 35 ☽ ⚼ ☿
05 19 ☽ ⚺ ♅
07 36 ☽ ⚻ ♄
10 09 ☽ ± ♂
11 43 ☽ ⚹ ♃
12 16 ☽ □ ♆
12 26 ☽ ∥ ♇
19 41 ☽ ± ☿
20 06 ☽ ⚻ ♂
20 19 ☉ ∥ ♄
22 43 ☽ ⋕ ♆

06 Sunday
05 40 ☽ ⚻ ☿
05 43 ♀ Q ♆
06 38 ☽ Q ♃
07 09 ☽ ⚻ ♀
08 34 ☽ ⋕ ♀
13 46 ♀ ∠ ♄
16 30 ☽ Q ♆
16 56 ☽ ± ♀
18 39 ☉ ⋕ ♅

07 Monday
01 45 ☽ ☍ ☉
03 34 ☽ ☌ ♇
04 18 ☽ ☌ ♅
07 05 ☽ ☍ ♄
07 39 ☽ ⚼ ♀
11 01 ☽ □ ♃
11 21 ☽ ⚹ ♆
16 37 ☽ ⋕ ♄
17 33 ☽ ∥ ♅
19 03 ☽ ⋕ ☉
21 48 ☽ ☍ ♂
23 33 ☽ ∥ ☿

08 Tuesday
04 47 ☉ ☍ ♇
07 17 ☽ ☍ ☿
08 28 ☽ △ ♀
10 16 ☽ ⋕ ♂
11 12 ☽ ∠ ♆
15 02 ☉ ☍ ♅
17 17 ☽ ∥ ♂

09 Wednesday
01 46 ☽ ⊥ ♆
03 26 ☽ ⚺ ♇
04 08 ☽ ⚺ ♅
05 08 ☽ ⚻ ☉
06 25 ☽ ⋕ ☿
06 59 ☽ ∥ ☉
07 36 ☽ ⚻ ♄
10 50 ☽ ⋕ ♅
11 22 ☽ ∥ ♄
11 31 ☽ △ ♃
11 38 ☽ ⚺ ♆
11 47 ☉ ⋕ ☿
13 18 ☽ ⊥ ♇
14 00 ☽ ⊥ ♅
15 47 ☽ ± ☉
17 37 ☽ ± ♄

10 Thursday
01 28 ☽ ⚻ ♂
04 21 ☽ ∠ ♇
05 02 ☽ ∠ ♅
06 05 ♃ ⚻ ♆
08 03 ☽ ⚼ ☉
08 55 ☽ ⚼ ♄
10 10 ☽ ⚻ ☿
12 15 ☽ ± ♂
12 22 ☽ □ ♀
12 54 ☽ ⚼ ♃
20 39 ☽ ± ☿
21 41 ☽ ∥ ♀
21 51 ☉ ☌ ♄

11 Friday
00 34 ♀ ⚼ ♃
04 36 ☽ ± ♃
04 47 ☽ ⚼ ♂
06 11 ☽ ⚹ ♇
06 52 ☽ ⚹ ♅
08 48 ☽ ∥ ♆
11 14 ☽ △ ♄
12 10 ☽ △ ☉
12 36 ☽ ⚼ ☿
15 04 ☽ ☌ ♆
15 15 ♄ ⋕ ♅
15 16 ☽ ⚻ ♃
20 47 ☽ ⋕ ♇
21 39 ☽ Q ♀

12 Saturday
02 19 ☿ St R
03 35 ☽ Q ♇
04 16 ☽ Q ♅
09 20 ☽ △ ♂
15 47 ☽ △ ☿
20 52 ☽ ⚹ ♀

13 Sunday
00 06 ☽ ⋕ ♃
03 03 ☉ △ ♆
09 36 ☉ □ ♃
12 58 ☽ □ ♇
13 38 ☽ □ ♅
19 04 ☽ □ ♄
22 36 ☽ ⚺ ♆
23 14 ☽ ☍ ♃

14 Monday
00 19 ☽ □ ☉
03 00 ☽ ∠ ♀
10 10 ☽ ⊥ ♆
21 37 ☽ ⊥ ♀
22 04 ☽ □ ♂

15 Tuesday
00 00 ☽ □ ☿
03 46 ☽ ∠ ♆
06 25 ☽ Q ♄
10 13 ☽ ⚺ ♀
14 38 ☽ Q ☉
18 06 ☿ ☌ ♂
23 24 ☽ △ ♇

16 Wednesday
00 01 ☽ △ ♅
05 44 ♀ ± ♇
06 32 ☽ ⚹ ♄
09 35 ☽ ⚹ ♆
10 23 ☽ Q ☿
10 44 ☽ ⚻ ♃
12 17 ☽ Q ♂
14 51 ♀ ± ♅
16 48 ☽ ⚹ ☉
21 13 ☉ ⋕ ♂
22 53 ☽ ± ♃

17 Thursday
05 28 ☽ ⚼ ♇
06 03 ☽ ⚼ ♅
09 23 ☽ ⚹ ☿
09 44 ☽ Q ♆
13 04 ☽ ∠ ♄
14 10 ☽ ⚹ ♂
15 48 ☽ ⋕ ♃
17 13 ☽ ⚼ ♃
23 38 ☽ ± ♇

18 Friday
00 12 ☽ ± ♅
01 53 ☽ ∠ ☉
02 41 ☽ ☌ ♀
07 29 ☽ ⊥ ♄
11 46 ☽ ⚻ ♇
12 18 ☽ ⚻ ♅
14 01 ☽ ∠ ☿
14 48 ♂ ⚼ ♆
19 47 ☽ ⚺ ♄
21 48 ☽ ⊥ ☉
22 00 ☽ ⋕ ♇
22 09 ☽ □ ♆
22 40 ☽ ∠ ♂
23 51 ☽ △ ♃

19 Saturday
07 07 ☽ ⊥ ☿
11 03 ☽ ⚺ ☉
12 20 ☽ ∥ ♆
18 09 ☽ ⊥ ♂
18 28 ☽ ⚺ ☿

20 Sunday
01 12 ☉ ∠ ♀
04 24 ☽ ∥ ♀
07 06 ☽ ⚺ ♂
19 47 ☽ ⚺ ♀

21 Monday
00 13 ☽ ☍ ♇
00 41 ☽ ☍ ♅
08 43 ☽ ⊥ ♀
08 55 ☽ ☌ ♄
10 34 ☽ △ ♆
11 50 ♀ ⊥ ♄
12 50 ☽ □ ♃
13 38 ☉ ☌ ☿
16 28 ☿ ∥ ♂
17 04 ☿ ∠ ♀
17 34 ☽ ⋕ ♅
20 33 ☽ ∥ ♄

22 Tuesday
02 33 ☽ ☌ ☿
04 02 ☽ ∠ ♀
04 46 ☽ ☌ ☉
05 06 ☽ ⋕ ♂
07 11 ☽ ⋕ ☿
16 23 ☽ ⚼ ♆
17 03 ☽ ⋕ ☉
23 06 ☽ ☌ ♂
23 12 ☽ ∥ ☉

23 Wednesday
00 54 ☽ Q ♃
06 22 ☽ ∥ ☿
08 52 ♀ ⚻ ♇
10 06 ☽ ± ♆
11 42 ☽ ⚻ ♇
11 55 ☽ ⚹ ♀
12 04 ☽ ⚻ ♅
12 22 ☽ ∥ ♂
13 56 ♀ ⚻ ♅
18 34 ☽ ⋕ ♄
20 54 ☽ ⚺ ♄
21 51 ☽ ⚻ ♆
22 03 ☽ ∥ ♅
23 25 ☽ ± ♇
23 46 ☽ ± ♅

24 Thursday
00 40 ☽ ⚹ ♃
08 08 ♂ Q ♃
08 43 ☽ ⊥ ♄
09 33 ☽ ⚺ ☿
13 07 ☽ Q ♀
16 53 ☽ ⚼ ♇
17 12 ☽ ⚼ ♅
20 25 ☽ ⊥ ☿
20 48 ☽ ⚺ ☉
21 40 ☉ ∥ ☿

25 Friday
02 18 ☽ ∠ ♄
05 59 ☽ ∠ ♃
06 19 ☽ ⋕ ♀
09 21 ☽ ⊥ ☉
12 38 ☽ ∠ ☿
13 19 ☽ ⚺ ♂
21 37 ☽ △ ♇
21 54 ☽ △ ♅
23 20 ☽ ⊥ ♃

26 Saturday
00 17 ☽ ⋕ ♆
01 29 ☽ ⊥ ♂
02 16 ☽ □ ♀
03 58 ☽ ∠ ☉
07 12 ☽ ⚹ ♄
07 29 ☽ ☍ ♆
10 48 ☽ ⚺ ♃
14 02 ☽ ∥ ♇
15 24 ☽ ⚹ ☿
19 33 ☽ ∠ ♂
22 22 ☿ ⊥ ♀

27 Sunday
05 54 ☽ Q ♄
09 17 ♄ △ ♆
10 24 ☽ ⚹ ☉
11 54 ♂ ⋕ ♄
12 34 ☽ Q ☿
17 26 ☽ ∥ ♃

28 Monday
00 26 ☉ ⚼ ♆
01 02 ☽ ⚹ ♂
05 23 ☽ □ ♇
05 35 ☽ □ ♅
10 10 ☽ Q ☉
13 58 ☽ △ ♀
14 51 ☽ ⚻ ♆
15 09 ☽ □ ♄
18 34 ☽ ☌ ♃
19 49 ☽ □ ☿

29 Tuesday
00 02 ☽ Q ♂
01 35 ☽ ± ♆
02 27 ♀ □ ♆
09 34 ♀ ⚺ ♄
13 33 ☽ Q ♇
13 42 ☽ Q ♅
17 29 ☽ ⚼ ♆
18 35 ☽ ⚼ ♀
20 23 ☿ □ ♃
20 43 ☽ □ ☉

30 Wednesday
02 52 ♂ ± ♆
09 28 ☽ □ ♂
10 24 ☽ ⚹ ♇
10 32 ☽ ⚹ ♅
11 19 ☽ ± ♀
19 22 ☽ △ ♆
20 11 ☽ △ ♄
22 18 ☽ ⚻ ♀
22 22 ☽ △ ☿
22 59 ☿ ⚺ ♀
23 26 ☽ ⚺ ♃
23 57 ♂ ∥ ♅

31 Thursday
01 55 ♂ ⚻ ♇
03 59 ♂ ⚻ ♅
09 39 ☽ ⊥ ♃
11 50 ☽ ∠ ♇
11 56 ☽ ∠ ♅
17 59 ♀ △ ♃
20 55 ☽ ∥ ♃
21 37 ☽ ⚼ ♄
22 56 ☽ ⚼ ☿

APRIL 1966

LONGITUDES

Date	Sidereal time h m s	Sun ☉ ° ' "	Moon ☽ ° ' "	Moon ☽ 24.00 ° ' "	Mercury ☿ ° '	Venus ♀ ° '	Mars ♂ ° '	Jupiter ♃ ° '	Saturn ♄ ° '	Uranus ♅ ° '	Neptune ♆ ° '	Pluto ♇ ° '
01	00 37 36	11 ♈ 18 28	16 ♌ 09 18	23 ♌ 30 29	23 ♓ 04	25 ♒ 01	17 ♈ 39	24 ♊ 25	22 ♓ 30	16 ♍ 35	21 ♏ 48	16 ♍ 33
02	00 41 33	12 17 38	00 ♍ 55 11	08 ♍ 22 33	22 R 52	25 58	18 25	24 33	22 37	16 R 33	21 R 47	16 R 31
03	00 45 29	13 16 46	15 ♍ 51 34	23 ♍ 21 08	22 46	26 56	19 11	24 41	22 45	16 31	21 46	16 30
04	00 49 26	14 15 51	00 ♎ 50 08	08 ♎ 17 24	22 D 44	27 54	19 56	24 49	22 52	16 29	21 44	16 28
05	00 53 22	15 14 54	15 ♎ 41 49	23 ♎ 02 22	22 49	28 52	20 42	24 57	22 59	16 26	21 43	16 27
06	00 57 19	16 13 55	00 ♏ 18 11	07 ♏ 28 30	22 59	29 ♒ 50	21 27	25 06	23 06	16 24	21 42	16 26
07	01 01 15	17 12 55	14 ♏ 32 47	21 ♏ 30 39	23 14	00 ♓ 50	22 13	25 14	23 13	16 22	21 41	16 24
08	01 05 12	18 11 52	28 ♏ 21 54	05 ♐ 06 28	23 33	01 49	22 58	25 23	23 20	16 20	21 39	16 23
09	01 09 08	19 10 48	11 ♐ 44 30	18 ♐ 16 13	23 58	02 49	23 43	25 32	23 27	16 18	21 38	16 22
10	01 13 05	20 09 42	24 ♐ 41 57	01 ♑ 02 10	24 27	03 49	24 28	25 41	23 33	16 16	21 37	16 20
11	01 17 02	21 08 34	07 ♑ 17 20	13 ♑ 28 01	25 00	04 50	25 14	25 50	23 40	16 14	21 36	16 19
12	01 20 58	22 07 25	19 ♑ 34 48	25 ♑ 38 15	25 37	05 50	25 59	25 59	23 47	16 12	21 34	16 18
13	01 24 55	23 06 14	01 ♒ 39 01	07 ♒ 37 39	26 18	06 52	26 44	26 09	23 54	16 10	21 33	16 17
14	01 28 51	24 05 01	13 ♒ 34 46	19 ♒ 30 55	27 03	07 53	27 29	26 18	24 01	16 08	21 31	16 15
15	01 32 48	25 03 46	25 ♒ 26 38	01 ♓ 22 25	27 51	08 55	28 14	26 27	24 07	16 06	21 30	16 14
16	01 36 44	26 02 30	07 ♓ 18 43	13 ♓ 15 59	28 43	09 57	28 59	26 37	24 14	16 04	21 29	16 13
17	01 40 41	27 01 11	19 ♓ 14 34	25 ♓ 14 50	29 ♓ 37	10 59	29 ♈ 44	26 47	24 21	16 03	21 27	16 12
18	01 44 37	27 59 51	01 ♈ 17 04	07 ♈ 21 31	00 ♈ 35	12 02	00 ♉ 29	26 57	24 27	16 01	21 26	16 11
19	01 48 34	28 58 29	13 ♈ 28 23	19 ♈ 37 52	01 36	13 05	01 14	27 07	24 34	15 59	21 24	16 10
20	01 52 31	29 ♈ 57 05	25 ♈ 50 05	02 ♉ 05 10	02 39	14 08	01 58	27 17	24 41	15 57	21 23	16 09
21	01 56 27	00 ♉ 55 40	08 ♉ 23 11	14 ♉ 44 14	03 45	15 12	02 43	27 27	24 47	15 56	21 21	16 08
22	02 00 24	01 54 12	21 ♉ 08 22	27 ♉ 35 38	04 54	16 15	03 28	27 37	24 53	15 54	21 20	16 07
23	02 04 20	02 52 43	04 ♊ 06 05	10 ♊ 39 47	06 05	17 19	04 13	27 47	25 00	15 53	21 18	16 06
24	02 08 17	03 51 11	17 ♊ 16 45	23 ♊ 57 03	07 18	18 23	04 57	27 58	25 06	15 51	21 17	16 05
25	02 12 13	04 49 38	00 ♋ 40 44	07 ♋ 27 50	08 33	19 27	05 42	28 08	25 12	15 50	21 15	16 04
26	02 16 10	05 48 02	14 ♋ 18 22	21 ♋ 12 20	09 51	20 32	06 26	28 19	25 19	15 48	21 14	16 03
27	02 20 06	06 46 25	28 ♋ 09 42	05 ♌ 10 23	11 11	21 37	07 11	28 30	25 25	15 47	21 12	16 02
28	02 24 03	07 44 45	12 ♌ 14 15	19 ♌ 21 04	12 33	22 41	07 55	28 41	25 31	15 46	21 11	16 01
29	02 28 00	08 43 03	26 ♌ 30 35	03 ♍ 42 25	13 58	23 47	08 39	28 52	25 37	15 44	21 09	16 00
30	02 31 56	09 ♉ 41 19	10 ♍ 56 08	18 ♍ 11 11	15 ♈ 24	24 ♓ 52	09 ♉ 24	29 ♊ 03	25 ♓ 43	15 ♍ 43	21 ♏ 07	15 ♍ 59

Date	Moon True ☊ ° '	Moon Mean ☊ ° '	Moon ☽ Latitude ° '
01	26 ♉ 18	27 ♉ 52	05 N 04
02	26 R 12	27 48	05 07
03	26 06	27 45	04 50
04	25 59	27 42	04 14
05	25 54	27 39	03 20
06	25 50	27 36	02 14
07	25 47	27 32	01 N 01
08	25 D 47	27 29	00 S 14
09	25 48	27 26	01 26
10	25 49	27 23	02 31
11	25 51	27 20	03 27
12	25 52	27 17	04 12
13	25 R 51	27 13	04 45
14	25 49	27 10	05 05
15	25 46	27 07	05 12
16	25 42	27 04	05 06
17	25 37	27 01	04 46
18	25 32	26 58	04 14
19	25 28	26 54	03 30
20	25 24	26 51	02 35
21	25 22	26 48	01 32
22	25 21	26 45	00 S 23
23	25 D 21	26 42	00 N 48
24	25 22	26 38	01 58
25	25 24	26 35	03 02
26	25 25	26 32	03 58
27	25 26	26 29	04 40
28	25 R 26	26 26	05 07
29	25 25	26 23	05 15
30	25 ♉ 24	26 ♉ 19	05 N 04

DECLINATIONS

Date	Sun ☉ ° '	Moon ☽ ° '	Mercury ☿ ° '	Venus ♀ ° '	Mars ♂ ° '	Jupiter ♃ ° '	Saturn ♄ ° '	Uranus ♅ ° '	Neptune ♆ ° '	Pluto ♇ ° '
01	04 N 28	20 N 49	02 S 00	11 S 39	06 N 28	23 N 13	04 S 43	06 N 02	16 S 29	19 N 04
02	04 52	15 56	02 19	11 27	06 46	23 14	04 41	06 03	16 29	19 05
03	05 15	10 02	02 35	11 14	07 04	23 14	04 38	06 04	16 28	19 05
04	05 38	03 N 33	02 49	11 00	07 22	23 14	04 35	06 05	16 28	19 06
05	06 00	03 S 06	03 01	10 46	07 40	23 15	04 33	06 06	16 28	19 06
06	06 23	09 30	03 10	10 32	07 57	23 15	04 30	06 07	16 27	19 06
07	06 46	15 14	03 16	10 17	08 15	23 16	04 27	06 07	16 27	19 07
08	07 08	20 02	03 20	10 02	08 32	23 16	04 25	06 08	16 27	19 07
09	07 31	23 37	03 21	09 47	08 49	23 17	04 22	06 09	16 26	19 07
10	07 53	25 51	03 20	09 31	09 07	23 17	04 19	06 10	16 26	19 08
11	08 15	26 41	03 17	09 14	09 24	23 17	04 17	06 11	16 26	19 08
12	08 37	26 10	03 11	08 57	09 41	23 18	04 14	06 11	16 25	19 08
13	08 59	24 25	03 04	08 40	09 58	23 18	04 12	06 12	16 25	19 09
14	09 21	21 37	02 54	08 23	10 15	23 19	04 09	06 13	16 24	19 09
15	09 42	17 56	02 42	08 05	10 31	23 19	04 06	06 14	16 24	19 09
16	10 04	13 33	02 29	07 46	10 48	23 19	04 04	06 14	16 24	19 09
17	10 25	08 39	02 13	07 28	11 04	23 20	04 01	06 15	16 23	19 10
18	10 46	03 S 22	01 56	07 09	11 21	23 20	03 59	06 16	16 23	19 10
19	11 07	02 N 06	01 37	06 50	11 37	23 20	03 56	06 16	16 22	19 10
20	11 27	07 35	01 16	06 30	11 53	23 21	03 54	06 17	16 22	19 10
21	11 48	12 51	00 54	06 10	12 09	23 21	03 52	06 17	16 22	19 10
22	12 08	17 41	00 30	05 50	12 25	23 21	03 49	06 18	16 21	19 11
23	12 28	21 45	00 S 05	05 29	12 40	23 22	03 47	06 19	16 21	19 11
24	12 48	24 47	00 N 22	05 08	12 56	23 22	03 44	06 19	16 20	19 11
25	13 08	26 29	00 50	04 47	13 12	23 22	03 42	06 20	16 20	19 11
26	13 28	26 37	01 19	04 26	13 27	23 22	03 40	06 20	16 20	19 11
27	13 47	25 07	01 50	04 04	13 42	23 23	03 37	06 21	16 19	19 11
28	14 06	22 02	02 21	03 43	13 57	23 23	03 35	06 21	16 19	19 11
29	14 25	17 37	02 54	03 21	14 12	23 23	03 33	06 22	16 18	19 11
30	14 N 43	12 N 09	03 N 29	02 S 58	14 N 27	23 N 23	03 S 31	06 N 22	16 S 18	19 N 11

ZODIAC SIGN ENTRIES

Date	h m	Planets
02	10 31	☽ ♍
04	10 39	☽ ♎
06	11 30	☽ ♏
06	15 53	♀ ♓
08	14 54	☽ ♐
10	22 02	☽ ♑
13	08 42	☽ ♒
15	21 13	☽ ♓
17	20 35	♂ ♉
17	21 31	☿ ♈
18	09 27	☽ ♈
20	13 12	☉ ♉
20	20 00	☽ ♉
23	04 27	☽ ♊
25	10 48	☽ ♋
27	15 09	☽ ♌
29	17 50	☽ ♍

LATITUDES

Date	Mercury ☿ ° '	Venus ♀ ° '	Mars ♂ ° '	Jupiter ♃ ° '	Saturn ♄ ° '	Uranus ♅ ° '	Neptune ♆ ° '	Pluto ♇ ° '
01	00 N 49	01 N 38	00 S 30	00 S 07	01 S 54	00 N 49	01 N 48	15 N 00
04	00 N 04	01 17	00 28	00 06	01 55	00 49	01 48	14 59
07	00 S 38	00 57	00 26	00 06	01 55	00 49	01 48	14 59
10	01 14	00 39	00 24	00 05	01 55	00 48	01 48	14 58
13	01 44	00 21	00 22	00 05	01 56	00 48	01 48	14 57
16	02 09	00 N 04	00 20	00 05	01 56	00 48	01 48	14 57
19	02 27	00 S 12	00 19	00 04	01 56	00 48	01 48	14 56
22	02 40	00 26	00 17	00 04	01 57	00 48	01 48	14 55
25	02 48	00 40	00 15	00 04	01 57	00 48	01 48	14 54
28	02 50	00 53	00 13	00 03	01 58	00 48	01 48	14 53
31	02 S 47	01 S 05	00 S 11	00 S 03	01 S 58	00 N 48	01 N 49	14 N 52

DATA

Julian Date	2439217
Delta T	+37 seconds
Ayanamsa	23° 22' 54"
Synetic vernal point	05° ♓ 44' 06"
True obliquity of ecliptic	23° 26' 43"

MOON'S PHASES, APSIDES AND POSITIONS ☽

Date	h m	Phase	Longitude ° '	Eclipse Indicator
05	11 13	○	15 ♎ 13	
12	17 28	☾	22 ♑ 21	
20	20 35	●	00 ♉ 18	
28	03 49	☽	07 ♌ 25	

Day	h m		
03	18 50	Perigee	
15	18 25	Apogee	
05	00 46	0S	
11	14 33	Max dec	26° S 42'
19	02 50	0N	
26	02 01	Max dec	26° N 46'

LONGITUDES

Date	Chiron ⚷ ° '	Ceres ⚳ ° '	Pallas ⚴ ° '	Juno ⚵ ° '	Vesta ⚶ ° '	Black Moon Lilith ⚸ ° '
01	23 ♓ 12	27 ♈ 44	20 ♓ 41	13 ♈ 46	05 ♋ 04	29 ♒ 58
11	23 ♓ 46	01 ♉ 44	23 ♓ 57	19 ♈ 21	07 ♋ 51	01 ♓ 06
21	24 ♓ 18	05 ♉ 46	27 ♓ 08	24 ♈ 59	10 ♋ 38	02 ♓ 13
31	24 ♓ 47	09 ♉ 49	00 ♈ 15	00 ♉ 40	14 ♋ 21	03 ♓ 20

ASPECTARIAN

h m Aspects

01 Friday
00 48 ☽ ∠ ♃
02 47 ☽ ⊥ ♇
02 52 ☽ ⊥ ♅
03 27 ☽ △ ☉
12 35 ☽ ± ♄
12 38 ☽ ⊻ ♇
12 43 ☽ ⊻ ♅
13 29 ☽ ± ☿
14 36 ☽ △ ♂
21 13 ☽ □ ♆
21 16 ☽ ∥ ♇
22 27 ☽ ⊼ ♄
23 07 ☽ ⊼ ☿

02 Saturday
01 36 ☽ ⚹ ♃
01 56 ☉ # ♄
03 27 ☽ ☍ ♀
05 43 ☽ ⊡ ☉
09 34 ☽ # ♆
16 15 ☽ ⊡ ♂
19 09 ☉ Q ♃
21 09 ☽ Q ♃
21 16 ☽ ± ☉

03 Sunday
02 14 ☽ Q ♆
07 13 ☽ # ♀
07 28 ☽ ± ♂
07 35 ☽ ⊼ ☉
13 01 ☽ ☌ ♇
13 03 ☽ ☌ ♅
14 16 ☿ ☌ ♄
17 36 ☽ ⊼ ♂
21 26 ☽ ⚹ ♆
22 40 ☽ ∥ ♂
23 01 ☽ ☍ ☿
23 07 ☽ ☍ ♄

04 Monday
02 16 ☽ □ ♃
02 47 ☽ ∥ ♅
04 25 ☿ St D
04 52 ☽ ∥ ☉
06 57 ☽ ⊼ ♀
08 12 ☽ # ♄
14 32 ☽ # ☿
17 15 ☽ ± ♀
20 31 ♅ ☌ ♇
21 29 ☽ ∠ ♆

05 Tuesday
08 49 ☽ ⊡ ♀
11 13 ☽ ☍ ☉
11 40 ☽ ∥ ☿
12 02 ☽ ⊥ ♆
13 12 ☽ ⊻ ♅
13 13 ☽ ⊻ ♇
17 14 ☽ ∥ ♄
18 02 ☉ ∥ ♅
20 36 ☽ ☍ ♂
21 49 ☽ ⊻ ♆
22 59 ☽ ⊥ ♅
23 01 ☽ ⊥ ♇
23 04 ☽ # ♅
23 18 ☉ ± ♆
23 22 ☽ # ☉
23 45 ☽ ⊼ ☿

06 Wednesday
00 00 ☽ ⊼ ♄
03 18 ☽ △ ♃
05 45 ☽ # ♂
09 46 ☽ ± ☿
09 59 ☽ ± ♄
11 11 ☽ △ ♀
13 50 ☽ ∠ ♅
13 52 ☽ ∠ ♇
15 57 ☽ ∥ ♀
16 03 ☉ ⊼ ♅
16 39 ☉ ⊼ ♇
19 41 ♂ ⊼ ♆

07 Thursday
01 03 ☽ ⊡ ☿
01 09 ☽ ⊡ ♄
04 36 ☽ ⊡ ♃
09 55 ☿ ☌ ♄
15 07 ☽ ⚹ ♅
15 11 ☽ ⚹ ♇
16 51 ♂ ± ♅
16 55 ☽ ⊼ ☉
17 35 ☽ ∥ ♆
18 03 ♂ ± ♇
20 09 ☽ ± ♃

08 Friday
00 16 ☽ ☌ ♆
01 59 ☽ ⊼ ♂
03 05 ☽ △ ♄
03 20 ☽ △ ☿
04 07 ☽ ± ☉
06 42 ☽ ⊼ ♃
06 58 ☽ # ♇
11 57 ☽ Q ♅
12 02 ☽ Q ♇
13 07 ☽ ± ♂
18 36 ☽ □ ♀
21 15 ☽ ⊡ ☉

09 Saturday
01 37 ♂ ⊻ ♄
06 11 ☽ ⊡ ♂
09 17 ☽ # ♃
20 20 ☽ □ ♅
20 27 ☽ □ ♇

10 Sunday
02 49 ☽ △ ☉
06 08 ☽ Q ♀
06 14 ☽ ⊻ ♆
09 04 ☿ ⊻ ♂
09 50 ☽ □ ♄
11 30 ☽ □ ☿
11 33 ☽ △ ♂
13 52 ☽ ☍ ♃
17 29 ☽ ⊥ ♆

11 Monday
05 07 ♀ # ♂
06 50 ☽ ⚹ ♀
10 40 ☽ ∠ ♆
20 35 ☽ Q ♄
22 46 ☉ ⊼ ♆
23 39 ☽ Q ☿

12 Tuesday
05 22 ☽ △ ♅
05 33 ☽ △ ♇
12 17 ♂ ⚹ ♃
13 46 ☉ ± ♅
14 43 ☽ ∠ ♀
15 55 ☽ ⚹ ♆
16 11 ☉ ± ♇
17 28 ☽ □ ☉
20 24 ☽ ⚹ ♄

13 Wednesday
00 29 ☉ # ♀
00 40 ☽ ⚹ ☿
00 52 ☽ ⊼ ♃
01 32 ☽ □ ♂
04 59 ☿ □ ♃
10 16 ☽ ⊥ ♀
11 02 ☽ ⊡ ♅
11 15 ☽ ⊡ ♇
13 00 ☽ ± ♃
15 48 ☽ Q ♆
22 33 ☽ # ♃
23 26 ☽ ⊻ ♀

14 Thursday
02 42 ☽ ∠ ♄
05 04 ☽ ± ♅
05 19 ☽ ± ♇
07 20 ☽ ⊡ ♃
08 42 ☽ ∠ ☿
08 43 ☽ Q ☉
10 02 ☉ ⊻ ♄
16 06 ☽ Q ♂
17 09 ☽ ⊼ ♅
17 24 ☽ ⊼ ♇
21 03 ☽ ⊥ ♄

15 Friday
04 02 ☽ □ ♆
04 11 ☽ ⊥ ☿
04 35 ☽ # ♇
09 18 ☽ ⊻ ♄
11 10 ☽ ⚹ ☉
14 05 ☽ △ ♃
17 15 ☽ ⊻ ☿
18 02 ☽ ⚹ ♂
20 48 ☽ ∥ ♆

16 Saturday
17 50 ☽ ☌ ♀
20 12 ☽ ∠ ☉

17 Sunday
01 01 ☽ # ♂
02 21 ☽ ∠ ♂
04 08 ☽ # ☉
04 57 ☉ ⚹ ♃
05 36 ☽ ☍ ♅
05 54 ☽ ☍ ♇
15 52 ☽ ⊥ ☉
16 25 ☽ △ ♆
17 50 ☽ ∥ ♀
21 34 ☽ ⊥ ♂
22 18 ☽ ☌ ♄
23 02 ☽ # ♅

18 Monday
00 51 ☿ ⊻ ♂
03 16 ☽ □ ♃
04 54 ☽ ⊻ ☉
09 15 ☽ ∥ ♄
10 18 ☽ ⊻ ♂
10 30 ☽ ☌ ☿
11 06 ♂ ⊥ ♄
18 44 ☽ ∥ ☿
22 09 ☽ ⊡ ♆

19 Tuesday
04 28 ♂ ⊡ ♅
09 59 ♂ ⊡ ♇
10 01 ☽ # ☿
11 10 ☽ ⊻ ♀
15 15 ☽ Q ♃
15 46 ☽ ± ♆
16 54 ☽ ⊼ ♅
17 15 ☽ ⊼ ♇
19 58 ☽ # ♄
23 57 ☽ ⊥ ♀

20 Wednesday
03 25 ☽ ⊼ ♆
04 32 ☽ ± ♅
04 53 ☽ ± ♇
06 15 ☽ ∥ ♅
07 30 ☽ # ♀
09 45 ☽ ⊻ ♄
14 49 ☽ ⚹ ♃
18 56 ☽ ∠ ♀
20 35 ☽ ☌ ☉
21 23 ☽ ⊥ ♄
21 49 ☽ ⊡ ♅
22 11 ☽ ⊡ ♇

21 Thursday
00 32 ☽ ☌ ♂
00 36 ♃ ± ♆
02 20 ☽ ⊻ ☿
03 22 ♀ # ♅
06 44 ☽ ∥ ☉
08 00 ☉ ⊥ ♄
08 31 ☽ ∥ ♂
12 03 ☉ ⊡ ♅
14 40 ☽ ∠ ♄
14 50 ☽ ⊥ ☿
16 51 ☉ ⊡ ♇
19 47 ☽ ∠ ♃

22 Friday
02 01 ☽ ⚹ ♀
02 13 ☽ △ ♅
02 36 ☽ △ ♇
04 17 ♀ ☍ ♅
05 07 ☽ # ♆
08 51 ♀ ☍ ♇
09 26 ☽ ∠ ☿
12 21 ☽ ☍ ♆
12 54 ☽ ⊥ ♃
19 03 ☽ ⚹ ♄
20 15 ☽ ∥ ♇

23 Saturday
00 12 ☽ ⊻ ♃
02 24 ☽ Q ♀
09 34 ☽ ⊻ ☉
12 13 ☽ ⊻ ♂
15 59 ☽ ⚹ ☿
16 27 ☿ ⊡ ♆
17 21 ☽ Q ♄
21 27 ☽ ⊥ ☉
23 31 ☽ ∥ ♃
23 50 ☽ ⊥ ♂

24 Sunday
09 26 ☽ □ ♅
09 50 ☽ □ ♇
14 10 ☽ □ ♀
15 04 ☽ ∠ ☉
16 01 ☽ Q ☿
17 06 ☽ ∠ ♂
19 12 ☽ ⊼ ♆

25 Monday
02 10 ☽ □ ♄
05 55 ☽ ± ♆
07 25 ☽ ☌ ♃
17 34 ☽ Q ♅
17 59 ☽ Q ♇
19 55 ☽ ⚹ ☉
21 24 ☽ ⚹ ♂
21 51 ☽ ⊡ ♆

26 Tuesday
03 23 ☽ □ ☿
07 55 ☉ ∥ ♂
14 37 ☽ ⚹ ♅
15 02 ☽ ⚹ ♇
18 33 ☽ Q ☉
19 36 ☽ Q ♂
23 45 ☽ △ ♀

27 Wednesday
00 01 ☽ △ ♆
03 08 ♀ △ ♆
07 14 ☽ △ ♄
12 35 ☽ ⊻ ♃
16 29 ☽ ∠ ♅
16 55 ☽ ∠ ♇
22 59 ☽ ⊥ ♃

28 Thursday
02 52 ☽ ∥ ♃
03 39 ☽ ⊡ ♀
03 49 ☽ □ ☉
04 16 ☽ □ ♂
07 49 ☽ ⊥ ♅
08 15 ☽ ⊥ ♇
09 04 ☽ ⊡ ♄
12 36 ☽ △ ☿
14 28 ☽ ∠ ♃
17 57 ☽ ⊻ ♅
18 23 ☽ ⊻ ♇
20 08 ☽ ± ♀
21 09 ♀ ∥ ♄
23 49 ☉ ∠ ♀

29 Friday
00 22 ☽ ± ♄
03 02 ☽ □ ♆
03 38 ♀ ∠ ♂
04 11 ☽ ∥ ♇
05 29 ☉ ☌ ♂
07 03 ☽ ⊼ ♀
10 30 ☽ ⊼ ♄
15 58 ☽ ⚹ ♃
16 32 ☽ ⊡ ☿
18 09 ☽ # ♆
23 12 ☿ # ♀

30 Saturday
01 45 ☽ ∥ ☉
02 47 ☽ ∥ ♂
07 34 ☿ ± ♆
09 00 ☽ Q ♆
09 10 ☽ ± ☿
09 18 ☽ △ ♂
09 47 ☽ △ ☉
12 11 ☽ Q ♃
13 21 ☿ # ♄
17 16 ☿ ⊼ ♅
19 55 ☽ ☌ ♅
20 13 ☽ ⊼ ☿
20 22 ☽ ☌ ♇
21 42 ☿ ⊼ ♇

All ephemeris data is given at 12.00 UT and the Moon's longitude is additionally given for 24.00 UT

MAY 1966

LONGITUDES

Date	Sidereal time h m s	Sun ☉ ° ′ ″	Moon ☽ ° ′ ″	Moon ☽ 24.00 ° ′ ″	Mercury ☿ ° ′	Venus ♀ ° ′	Mars ♂ ° ′	Jupiter ♃ ° ′	Saturn ♄ ° ′	Uranus ♅ ° ′	Neptune ♆ ° ′	Pluto ♇ ° ′
01	02 35 53	10 ♉ 39 33	25 ♍ 27 00	02 ♎ 42 53	16 ♈ 52	25 ♓ 57	10 ♉ 08	29 ♊ 14	25 ♓ 49	15 ♍ 42	21 ♏ 06	15 ♍ 59
02	02 39 49	11 37 44	09 ♎ 58 10	17 ♎ 12 06	18 22	27 03	10 52	29 25	25 55	15 R 41	21 R 04	15 R 58
03	02 43 46	12 35 54	24 ♎ 23 59	01 ♏ 33 07	19 54	28 09	11 36	29 36	26 01	15 40	21 03	15 57
04	02 47 42	13 34 02	08 ♏ 38 52	15 ♏ 40 37	21 28	29 ♓ 14	12 20	29 47	26 07	15 39	21 01	15 56
05	02 51 39	14 32 09	22 ♏ 37 55	29 ♏ 30 22	23 04	00 ♈ 21	13 04	29 ♊ 59	26 13	15 38	20 59	15 56
06	02 55 35	15 30 14	06 ♐ 17 40	12 ♐ 59 40	24 42	01 27	13 48	00 ♋ 10	26 18	15 37	20 58	15 55
07	02 59 32	16 28 17	19 ♐ 36 17	26 ♐ 07 34	26 22	02 33	14 32	00 22	26 24	15 36	20 56	15 55
08	03 03 29	17 26 19	02 ♑ 33 39	08 ♑ 54 46	28 04	03 40	15 16	00 33	26 29	15 35	20 54	15 54
09	03 07 25	18 24 19	15 ♑ 11 14	21 ♑ 23 26	29 ♈ 48	04 46	16 00	00 45	26 35	15 34	20 53	15 53
10	03 11 22	19 22 19	27 ♑ 31 47	03 ♒ 36 47	01 ♉ 33	05 53	16 43	00 57	26 40	15 34	20 51	15 53
11	03 15 18	20 20 16	09 ♒ 38 58	15 ♒ 38 51	03 21	07 00	17 27	01 08	26 46	15 33	20 50	15 52
12	03 19 15	21 18 13	21 ♒ 37 00	27 ♒ 34 02	05 11	08 07	18 11	01 20	26 51	15 32	20 48	15 52
13	03 23 11	22 16 08	03 ♓ 30 29	09 ♓ 26 57	07 02	09 15	18 54	01 32	26 56	15 32	20 46	15 51
14	03 27 08	23 14 02	15 ♓ 23 58	21 ♓ 22 06	08 56	10 22	19 38	01 44	27 02	15 31	20 45	15 51
15	03 31 04	24 11 54	27 ♓ 21 51	03 ♈ 23 41	10 51	11 29	20 21	01 56	27 07	15 31	20 43	15 51
16	03 35 01	25 09 45	09 ♈ 28 05	15 ♈ 35 24	12 48	12 37	21 05	02 09	27 12	15 30	20 41	15 50
17	03 38 58	26 07 35	21 ♈ 46 01	28 ♈ 00 13	14 47	13 45	21 48	02 21	27 17	15 30	20 40	15 50
18	03 42 54	27 05 24	04 ♉ 18 15	10 ♉ 40 17	16 48	14 53	22 31	02 33	27 22	15 30	20 38	15 50
19	03 46 51	28 03 12	17 ♉ 06 27	23 ♉ 36 46	18 51	16 00	23 15	02 45	27 27	15 29	20 37	15 49
20	03 50 47	29 00 58	00 ♊ 11 14	06 ♊ 49 47	20 56	17 08	23 58	02 58	27 31	15 29	20 35	15 49
21	03 54 44	29 ♉ 58 43	13 ♊ 32 17	20 ♊ 18 32	23 02	18 17	24 41	03 10	27 36	15 29	20 33	15 49
22	03 58 40	00 ♊ 56 27	27 ♊ 08 18	04 ♋ 01 19	25 09	19 25	25 24	03 23	27 41	15 29	20 32	15 49
23	04 02 37	01 54 09	10 ♋ 57 17	17 ♋ 55 51	27 18	20 33	26 07	03 35	27 45	15 29	20 30	15 49
24	04 06 33	02 51 49	24 ♋ 56 42	01 ♌ 59 27	29 ♉ 27	21 42	26 50	03 48	27 50	15 D 29	20 29	15 49
25	04 10 30	03 49 29	09 ♌ 03 46	16 ♌ 09 17	01 ♊ 38	22 50	27 33	04 00	27 54	15 29	20 27	15 48
26	04 14 27	04 47 06	23 ♌ 15 40	00 ♍ 22 35	03 50	23 59	28 16	04 13	27 58	15 29	20 25	15 48
27	04 18 23	05 44 42	07 ♍ 29 43	14 ♍ 36 44	06 01	25 07	28 59	04 26	28 03	15 30	20 24	15 D 48
28	04 22 20	06 42 17	21 ♍ 43 19	28 ♍ 49 11	08 13	26 16	29 ♉ 42	04 38	28 07	15 30	20 22	15 48
29	04 26 16	07 39 50	05 ♎ 54 02	12 ♎ 57 32	10 25	27 25	00 ♊ 25	04 51	28 11	15 30	20 21	15 48
30	04 30 13	08 37 22	19 ♎ 59 25	26 ♎ 59 20	12 37	28 34	01 07	05 04	28 15	15 30	20 19	15 49
31	04 34 09	09 ♊ 34 52	03 ♏ 57 02	10 ♏ 52 10	14 ♊ 47	29 ♈ 43	01 ♊ 50	05 ♋ 17	28 ♓ 19	15 ♍ 31	20 ♏ 18	15 ♍ 49

Date	Moon True ☊ ° ′	Moon Mean ☊ ° ′	Moon ☽ Latitude ° ′
01	25 ♉ 22	26 ♉ 16	04 N 33
02	25 R 20	26 13	03 45
03	25 19	26 10	02 43
04	25 18	26 07	01 31
05	25 17	26 04	00 N 15
06	25 D 18	26 00	01 S 01
07	25 18	25 57	02 11
08	25 19	25 54	03 12
09	25 20	25 51	04 02
10	25 20	25 48	04 40
11	25 21	25 44	05 05
12	25 R 21	25 41	05 16
13	25 21	25 38	05 14
14	25 20	25 35	04 58
15	25 20	25 32	04 29
16	25 20	25 29	03 48
17	25 D 20	25 25	02 55
18	25 20	25 22	01 54
19	25 20	25 19	00 S 45
20	25 R 20	25 16	00 N 27
21	25 20	25 13	01 39
22	25 20	25 09	02 47
23	25 19	25 06	03 46
24	25 18	25 03	04 33
25	25 18	25 00	05 04
26	25 17	24 57	05 16
27	25 D 17	24 54	05 09
28	25 17	24 50	04 43
29	25 18	24 47	04 00
30	25 19	24 44	03 03
31	25 ♉ 20	24 ♉ 41	01 N 55

DECLINATIONS

Date	Sun ☉ ° ′	Moon ☽ ° ′	Mercury ☿ ° ′	Venus ♀ ° ′	Mars ♂ ° ′	Jupiter ♃ ° ′	Saturn ♄ ° ′	Uranus ♅ ° ′	Neptune ♆ ° ′	Pluto ♇ ° ′
01	15 N 01	05 N 59	04 N 04	02 S 36	14 N 41	23 N 24	03 S 28	06 N 22	16 S 17	19 N 11
02	15 20	00 S 30	04 40	02 13	14 56	23 24	03 26	06 23	16 17	19 11
03	15 37	06 56	05 17	01 50	15 10	23 24	03 24	06 23	16 17	19 11
04	15 55	12 57	05 55	01 27	15 24	23 24	03 22	06 23	16 16	19 11
05	16 12	18 12	06 34	01 04	15 38	23 24	03 20	06 24	16 16	19 11
06	16 29	22 21	07 14	00 41	15 52	23 24	03 18	06 24	16 15	19 11
07	16 46	25 12	07 55	00 S 17	16 05	23 24	03 16	06 25	16 15	19 11
08	17 02	26 37	08 36	00 N 07	16 19	23 24	03 14	06 25	16 14	19 11
09	17 19	26 36	09 18	00 30	16 32	23 24	03 12	06 25	16 14	19 11
10	17 35	25 14	10 01	00 54	16 45	23 24	03 10	06 25	16 14	19 11
11	17 50	22 44	10 44	01 18	16 58	23 24	03 08	06 26	16 13	19 11
12	18 05	19 17	11 28	01 43	17 11	23 24	03 06	06 26	16 13	19 10
13	18 20	15 05	12 12	02 07	17 24	23 24	03 04	06 26	16 12	19 10
14	18 35	10 20	12 56	02 31	17 36	23 24	03 02	06 26	16 12	19 10
15	18 50	05 S 09	13 40	02 55	17 49	23 24	03 00	06 26	16 11	19 10
16	19 04	00 N 16	14 25	03 20	18 01	23 24	02 58	06 26	16 11	19 10
17	19 17	05 46	15 09	03 44	18 13	23 24	02 56	06 26	16 11	19 10
18	19 31	11 10	15 53	04 09	18 24	23 24	02 55	06 27	16 10	19 09
19	19 44	16 13	16 37	04 33	18 36	23 24	02 53	06 27	16 10	19 09
20	19 57	20 38	17 20	04 58	18 47	23 24	02 51	06 27	16 09	19 09
21	20 09	24 04	18 02	05 22	18 59	23 23	02 49	06 27	16 09	19 09
22	20 21	26 12	18 44	05 47	19 10	23 23	02 48	06 27	16 09	19 08
23	20 33	26 45	19 24	06 11	19 21	23 23	02 46	06 27	16 08	19 08
24	20 44	25 37	20 03	06 36	19 31	23 23	02 45	06 27	16 08	19 08
25	20 55	22 52	20 41	07 00	19 42	23 23	02 43	06 27	16 07	19 07
26	21 06	18 44	21 17	07 24	19 52	23 23	02 42	06 26	16 07	19 07
27	21 16	13 32	21 51	07 48	20 02	23 22	02 40	06 26	16 07	19 07
28	21 26	07 37	22 22	08 13	20 12	23 22	02 39	06 26	16 06	19 06
29	21 36	01 N 20	22 52	08 37	20 22	23 21	02 37	06 26	16 06	19 06
30	21 45	04 S 59	23 19	09 01	20 31	23 21	02 36	06 26	16 05	19 05
31	21 N 54	11 S 02	23 N 44	09 N 24	20 N 40	23 N 20	02 S 35	06 N 26	16 S 05	19 N 05

ZODIAC SIGN ENTRIES

Date	h	m	Planets
01	19	31	☽ ♎
03	21	23	☽ ♏
05	04	33	♀ ♈
05	14	52	♃ ♋
06	00	52	☽ ♐
08	07	12	☽ ♑
09	14	48	☿ ♉
10	16	52	☽ ♒
13	04	55	☽ ♓
15	17	15	☽ ♈
18	03	49	☽ ♉
20	11	40	☽ ♊
21	12	32	☉ ♊
22	17	00	☽ ♋
24	17	59	☿ ♊
24	20	37	☽ ♌
26	23	22	☽ ♍
28	22	07	♂ ♊
29	02	00	☽ ♎
31	05	11	☽ ♏
31	18	00	♀ ♉

LATITUDES

Date	Mercury ☿ ° ′	Venus ♀ ° ′	Mars ♂ ° ′	Jupiter ♃ ° ′	Saturn ♄ ° ′	Uranus ♅ ° ′	Neptune ♆ ° ′	Pluto ♇ ° ′
01	02 S 47	01 S 05	00 S 11	00 S 03	01 S 58	00 N 48	01 N 49	14 N 52
04	02 39	01 15	00 09	00 03	01 59	00 48	01 49	14 51
07	02 26	01 25	00 07	00 02	01 59	00 48	01 49	14 50
10	02 08	01 34	00 05	00 02	02 00	00 47	01 49	14 49
13	01 46	01 41	00 03	00 02	02 01	00 47	01 49	14 48
16	01 20	01 48	00 S 01	00 01	02 01	00 47	01 49	14 47
19	00 51	01 53	00 N 01	00 01	02 02	00 47	01 49	14 46
22	00 S 20	01 58	00 03	00 01	02 03	00 47	01 49	14 45
25	00 N 11	02 02	00 05	00 S 01	02 03	00 47	01 48	14 44
28	00 42	02 04	00 07	00 00	02 04	00 47	01 48	14 42
31	01 N 10	02 S 06	00 N 08	00 00	02 S 05	00 N 47	01 N 48	14 N 41

DATA

Julian Date	2439247
Delta T	+37 seconds
Ayanamsa	23° 22′ 57″
Synetic vernal point	05° ♓ 44′ 03″
True obliquity of ecliptic	23° 26′ 43″

MOON'S PHASES, APSIDES AND POSITIONS ☽

Date	h	m	Phase	Longitude ° ′	Eclipse Indicator
04	21	00	○	13 ♏ 56	
12	11	19	☾	21 ♒ 17	
20	09	42	●	28 ♉ 55	Annular
27	08	50	☽	05 ♍ 37	

Day	h	m		
01	14	27	Perigee	
13	13	01	Apogee	
27	14	06	Perigee	
02	10	09	0S	
08	23	30	Max dec	26° S 47′
16	10	51	0N	
23	07	59	Max dec	26° N 47′
29	17	02	0S	

LONGITUDES

Date	Chiron ⚷ ° ′	Ceres ⚳ ° ′	Pallas ⚴ ° ′	Juno ⚵ ° ′	Vesta ⚶ ° ′	Black Moon Lilith ⚸ ° ′
01	24 ♓ 47	09 ♉ 49	00 ♈ 15	00 ♉ 40	14 ♋ 21	03 ♓ 20
11	25 ♓ 13	13 ♉ 52	03 ♈ 16	06 ♉ 24	17 ♋ 59	04 ♓ 27
21	25 ♓ 35	17 ♉ 55	06 ♈ 10	12 ♉ 09	21 ♋ 48	05 ♓ 35
31	25 ♓ 52	21 ♉ 57	08 ♈ 54	17 ♉ 56	25 ♋ 47	06 ♓ 42

All ephemeris data is given at 12.00 UT and the Moon's longitude is additionally given for 24.00 UT

ASPECTARIAN

h m	Aspects
01 Sunday	
04 49	☽ ⚹ ♆
08 46	♀ ☌ ♄
10 32	☽ ∥ ♅
11 26	☽ ⚼ ♂
12 22	☽ ⚼ ☉
12 37	☽ ☍ ♄
12 54	☽ ☍ ♀
16 25	☉ ∠ ♄
18 19	☽ □ ♃
18 34	☽ ∥ ☿
18 38	☿ Q ♃
21 25	☽ # ♄
02 Monday	
01 22	☽ # ♀
03 06	☽ ± ♂
04 18	☽ ± ☉
05 33	☽ ∠ ♆
13 34	☽ ⊼ ♂
14 00	♂ ∠ ♄
14 57	☽ ⊼ ☉
17 58	☽ ∥ ♀
20 26	☽ ⊥ ♆
21 28	☽ ⊻ ♅
21 56	☽ ⊻ ♇
22 46	☽ ∥ ♄
03 Tuesday	
03 36	☽ ☍ ☿
05 04	☽ # ☿
06 24	☽ ⊻ ♆
07 26	☽ ⊥ ♅
07 55	☽ ⊥ ♇
09 54	☽ # ♅
14 43	☽ ⊼ ♄
18 47	☽ ⊼ ♀
20 50	☽ △ ♃
22 29	☽ ∠ ♅
22 59	☽ ∠ ♇
04 Wednesday	
00 52	☽ ± ♄
05 12	☿ ⊼ ♆
05 44	☽ ± ♀
14 37	☿ ± ♅
16 13	☽ ⚼ ♄
18 37	☽ ☍ ♂
19 03	☿ ± ♇
21 00	☽ ☍ ☉
22 21	☽ ⚼ ♀
22 37	☽ ⚼ ♃
23 10	☽ # ♂
23 56	☽ ⚹ ♅
05 Thursday	
00 27	☽ ⚹ ♇
01 47	☽ # ☉
02 24	♀ □ ♃
02 38	☽ ∥ ♆
05 34	☿ ∥ ♅
09 10	☽ ☌ ♆
12 52	☽ ⊼ ☿
14 22	☽ ± ♃
16 50	☉ # ♆
17 08	☽ # ♇
18 16	☽ △ ♄
20 42	☽ Q ♅
21 14	☽ Q ♇
06 Friday	
00 45	☽ ± ☿
01 01	☽ ⊼ ♃
01 38	☉ ∠ ♃
02 39	☽ △ ♀
14 42	☉ △ ♅
18 57	☽ ⚼ ☿
19 33	☽ # ♃
22 11	☉ △ ♇
07 Saturday	
02 14	☽ ⊼ ♂
04 43	☽ □ ♅
05 17	☽ □ ♇
05 50	☽ ⊼ ☉
12 26	☿ ⊻ ♄
13 48	☽ ± ♂
14 26	☽ ⊻ ♆
17 41	☽ ± ☉
08 Sunday	
00 36	☽ □ ♄
01 28	☽ ⊥ ♆
02 19	☽ △ ☿
04 13	♂ # ♆
07 26	☽ ⚼ ♂
08 11	☽ ☍ ♃
11 45	☽ ⚼ ☉
14 16	☽ □ ♀
18 17	☽ ∠ ♆
22 27	♂ △ ♅
09 Monday	
01 03	♂ ∠ ♃
08 39	♂ △ ♇
10 50	☽ Q ♄
12 45	☽ △ ♅
13 21	☽ △ ♇
13 39	☽ △ ♂
18 44	☽ △ ☉
22 35	☿ ⚼ ♅
22 59	☽ ⚹ ♆
10 Tuesday	
02 40	☿ ⚹ ♃
02 53	☿ ⚼ ♇
04 09	☽ Q ♀
10 18	☽ ⚹ ♄
11 18	♀ ⚼ ♆
17 58	☽ ⚼ ♅
18 36	☽ ⚼ ♇
18 50	☽ ⊼ ♃
21 18	☽ □ ☿
22 28	☽ Q ♆
11 Wednesday	
03 47	☿ ⊥ ♄
06 12	☽ ⚹ ♀
06 25	☽ # ♃
06 55	☽ ± ♃
11 48	☽ ± ♅
12 27	☽ ± ♇
16 15	☽ ∠ ♄
23 48	☽ ⊼ ♅
23 49	☉ ☍ ♆
12 Thursday	
00 27	☽ ⊼ ♇
01 13	☽ ⚼ ♃
04 38	☽ □ ♂
10 21	☽ □ ♆
10 27	☽ ⊥ ♄
11 19	☽ □ ☉
12 40	☽ # ♇
15 21	☽ ∠ ♀
15 43	☽ Q ☿
18 48	☽ # ☉
22 38	☽ ⊻ ♄
23 52	☽ # ♂
13 Friday	
05 56	☽ ∥ ♆
07 57	☽ △ ♃
11 24	☽ ⊥ ♀
19 18	☽ Q ♂
20 28	☽ ⚹ ☿
14 Saturday	
00 47	☽ ⊻ ♀
00 54	☽ # ☿
02 52	☽ Q ☉
12 15	☽ ☍ ♅
12 54	☽ ☍ ♇
21 04	☽ ⚹ ♂
22 44	☽ △ ♆
15 Sunday	
05 07	☽ ⚹ ☉
06 12	☽ # ♅
08 24	☽ ∠ ☿
11 30	☽ ☌ ♄
16 08	♀ # ♄
21 16	☽ □ ♃
21 17	☽ # ♀
21 42	☽ ∥ ♄
23 36	♂ ☍ ♆
16 Monday	
04 14	☿ ∠ ♄
04 34	☽ ⚼ ♆
04 53	☽ ∠ ♂
05 44	☽ ⊥ ☿
06 36	☿ ⊻ ♀
13 29	☽ ∠ ☉
18 49	☽ ☌ ♀
19 49	☽ ⊻ ☿
22 13	☽ ± ♆
22 29	☉ ∥ ♇
23 41	☽ ⊥ ♂
23 44	☽ # ♄
23 50	☽ ⊼ ♅
17 Tuesday	
00 29	☽ ⊼ ♇
02 26	☽ ∥ ♀
08 33	☽ ⊥ ☉
09 12	☽ Q ♃
09 52	☽ ⊼ ♆
11 29	☽ ± ♅
12 04	☽ ⊻ ♂
12 08	☽ ± ♇
14 56	☽ ∥ ♅
18 55	☉ ⊥ ♃
20 28	☿ △ ♅
21 06	☽ ⊻ ☉
22 41	☽ ⊻ ♄
18 Wednesday	
00 24	☿ △ ♇
04 46	☽ ⚼ ♅
05 24	☽ ⚼ ♇
07 03	♀ ± ♆
08 37	☽ ⚹ ♃
10 12	☽ ⊥ ♄
19 24	☉ ⚹ ♄
21 24	☿ # ♆
21 43	☿ ∠ ♃
19 Thursday	
01 06	♀ ⊼ ♅
03 16	☽ ∠ ♄
08 07	♀ ⊼ ♇
09 00	☽ △ ♅
09 37	☽ △ ♇
09 46	☽ ⊻ ♀
11 43	☽ # ♆
13 13	☽ ∠ ♃
14 19	☽ ∥ ☿
15 51	☽ ☌ ☿
18 28	☽ ☍ ♆
21 55	☽ ⊥ ♀
23 59	☽ ☌ ♂
20 Friday	
00 58	☽ ∥ ♂
03 25	☽ ∥ ♇
06 02	☽ ⊥ ♃
07 07	☽ ⚹ ♄
07 43	☽ ∥ ☉
08 06	☿ ☍ ♆
09 42	☽ ☌ ☉
15 52	☽ ∠ ♀
17 06	☽ ⊻ ♃
21 Saturday	
04 56	☽ Q ♄
06 33	☽ ∥ ♃
15 28	☽ □ ♅
16 03	☽ □ ♇
21 11	☽ ⚹ ♀
22 Sunday	
00 25	☽ ⊼ ♆
07 52	☽ ⊻ ☿
08 48	☽ ⊻ ♂
08 52	♂ ∥ ♇
10 56	☽ ± ♆
11 31	♀ ⊥ ♂
12 57	☽ □ ♄
16 19	☿ ☌ ♂
18 20	☿ ⊥ ♀
19 08	☽ ⊻ ☉
19 51	☽ ⊥ ♂
20 08	☽ Q ♀
20 17	☽ ⊥ ☿
23 03	☽ ☌ ♃
23 04	☽ Q ♅
23 38	☽ Q ♇
23 Monday	
00 22	♃ Q ♅
02 23	☿ ∥ ♇
02 35	☽ ⚼ ♆
06 20	☽ ⊥ ☉
09 04	☿ ∥ ♂
10 59	♀ ⊼ ♆
12 18	☽ ∠ ♂
14 44	☽ ∠ ☿
15 35	☿ ⊥ ♃
16 36	♅ St D
17 18	☿ ⚹ ♄
19 48	☽ ⚹ ♅
20 22	☽ ⚹ ♇
22 59	☽ ∠ ☉
24 Tuesday	
03 09	♀ ∥ ♅
04 23	☽ △ ♆
05 57	☽ □ ♀
07 37	♀ ± ♅
13 37	♃ Q ♇
14 27	♀ ± ♇
14 39	♀ Q ♃
15 24	☽ ⚹ ♂
16 57	☽ △ ♄
21 06	☽ ⚹ ☿
21 26	☽ ∠ ♅
21 59	☽ ∠ ♇
25 Wednesday	
02 28	☽ ⚹ ☉
03 18	☽ ⊻ ♃
08 26	☽ ∥ ♃
12 43	☽ ⊥ ♅
12 53	☽ Q ♂
13 16	☽ ⊥ ♇
13 37	☽ ⊥ ♃
17 47	☉ ⊻ ♃
18 32	☽ ⚼ ♄
21 09	☽ Q ☿
22 52	☽ ⊻ ♅
23 25	☽ ⊻ ♇
23 43	☽ ∥ ☉
23 53	☽ ∥ ☿
26 Thursday	
00 16	☽ Q ☉
00 53	♂ ⚹ ♄
01 32	☉ ∥ ☿
05 04	☽ ∠ ♃
06 19	☽ ∥ ♂
07 13	☽ □ ♆
09 22	♂ ⊥ ♃
09 49	☽ ± ♄
10 03	☽ ∥ ♇
13 19	☽ △ ♀
16 44	☿ ⊻ ♃
19 59	☽ ⊼ ♄
20 54	☽ □ ♂
27 Friday	
00 37	☽ # ♆
06 38	☉ ☌ ☿
06 45	☽ ⚹ ♃
08 50	☽ □ ☉
09 04	☽ □ ☿
11 13	♇ St D
13 31	☽ Q ♆
16 49	☽ ⚼ ♀
28 Saturday	
01 29	☽ ☌ ♅
02 01	☽ ☌ ♇
03 18	☽ Q ♃
09 20	☽ ± ♀
09 43	☽ ⚹ ♆
09 50	☽ ∥ ♀
16 36	☽ ∥ ♅
20 21	☽ ⊼ ♀
22 51	☽ ☍ ♄
29 Sunday	
02 12	☽ △ ♂
07 07	☽ # ♄
09 20	☿ Q ♄
10 12	☽ □ ♃
11 04	☽ ∠ ♆
15 13	☽ △ ☉
21 05	☽ △ ☿
30 Monday	
02 20	☽ ⊥ ♆
02 54	☽ ∥ ♄
04 20	☽ ⊻ ♅
04 51	☽ ⊻ ♇
05 03	☽ ⚼ ♂
05 03	♀ ⊻ ♄
12 34	☽ ⊻ ♆
13 28	☿ ∥ ♃
14 36	☽ ⊥ ♅
15 07	☽ ⊥ ♇
17 35	☽ # ♅
18 41	☽ ⚼ ☉
21 16	☽ ± ♂
31 Tuesday	
02 14	☽ ⊼ ♄
03 29	☽ ⚼ ☿
04 02	☽ ☍ ♀
04 54	☽ # ♀
06 04	☽ ∠ ♅
06 35	☽ ∠ ♇
08 09	☽ ⊼ ♂
10 11	☿ ∠ ♀
11 19	☽ ± ☉
12 38	☽ ± ♄
13 11	♃ ⚼ ♆
14 21	☽ △ ♃
20 00	☿ □ ♅
21 57	☽ ± ☿
22 29	☽ ⊼ ☉
23 19	☿ □ ♇

JUNE 1966

LONGITUDES

Date	Sidereal time h m s	Sun ☉ ° ' "	Moon ☽ ° ' "	Moon ☽ 24.00 ° ' "	Mercury ☿ ° '	Venus ♀ ° '	Mars ♂ ° '	Jupiter ♃ ° '	Saturn ♄ ° '	Uranus ♅ ° '	Neptune ♆ ° '	Pluto ♇ ° '
01	04 38 06	10 ♊ 32 21	17 ♏ 44 29	24 ♏ 33 41	16 ♊ 57	00 ♉ 52	02 ♊ 33	05 ♋ 30	28 ♓ 23	15 ♍ 31	20 ♏ 16	15 ♍ 49
02	04 42 02	11 29 50	01 ♐ 19 32	08 ♐ 01 47	19 06	02 01	03 15	05 43	28 26	15 31	20 R 15	15 49
03	04 45 59	12 27 17	14 ♐ 40 15	21 ♐ 14 48	21 13	03 10	03 58	05 56	28 30	15 32	20 13	15 49
04	04 49 56	13 24 43	27 ♐ 45 20	04 ♑ 11 49	23 19	04 19	04 40	06 09	28 34	15 33	20 12	15 49
05	04 53 52	14 22 08	10 ♑ 34 16	16 ♑ 52 46	25 23	05 29	05 22	06 22	28 37	15 33	20 10	15 50
06	04 57 49	15 19 33	23 ♑ 07 29	29 ♑ 18 36	27 25	06 38	06 05	06 35	28 41	15 34	20 09	15 50
07	05 01 45	16 16 57	05 ♒ 26 23	11 ♒ 31 12	29 ♊ 25	07 48	06 47	06 48	28 44	15 35	20 07	15 50
08	05 05 42	17 14 20	17 ♒ 33 25	23 ♒ 33 28	01 ♋ 22	08 57	07 29	07 01	28 47	15 36	20 06	15 51
09	05 09 38	18 11 42	29 ♒ 31 49	05 ♓ 28 59	03 18	10 07	08 12	07 15	28 50	15 37	20 05	15 51
10	05 13 35	19 09 04	11 ♓ 25 31	17 ♓ 21 59	05 11	11 17	08 54	07 28	28 53	15 37	20 03	15 51
11	05 17 31	20 06 25	23 ♓ 18 58	29 ♓ 17 04	07 02	12 27	09 36	07 41	28 56	15 38	20 02	15 52
12	05 21 28	21 03 46	05 ♈ 18 52	11 ♈ 18 58	08 50	13 37	10 18	07 54	28 59	15 39	20 01	15 52
13	05 25 25	22 01 07	17 ♈ 23 56	23 ♈ 32 19	10 36	14 46	11 00	08 08	29 02	15 40	19 59	15 53
14	05 29 21	22 58 27	29 ♈ 44 38	06 ♉ 01 20	12 19	15 56	11 42	08 21	29 05	15 42	19 58	15 53
15	05 33 18	23 55 46	12 ♉ 22 48	18 ♉ 49 24	14 00	17 07	12 24	08 34	29 07	15 43	19 57	15 54
16	05 37 14	24 53 05	25 ♉ 21 19	01 ♊ 58 44	15 38	18 17	13 05	08 48	29 10	15 44	19 55	15 55
17	05 41 11	25 50 24	08 ♊ 41 40	15 ♊ 30 01	17 14	19 27	13 47	09 01	29 12	15 45	19 54	15 55
18	05 45 07	26 47 43	22 ♊ 23 36	29 ♊ 22 03	18 47	20 37	14 29	09 15	29 15	15 47	19 53	15 56
19	05 49 04	27 45 01	06 ♋ 24 55	13 ♋ 31 41	20 18	21 47	15 11	09 28	29 17	15 48	19 52	15 57
20	05 53 00	28 42 18	20 ♋ 41 40	27 ♋ 54 10	21 46	22 58	15 52	09 41	29 19	15 49	19 50	15 57
21	05 56 57	29 ♊ 39 35	05 ♌ 08 27	12 ♌ 23 43	23 12	24 08	16 34	09 55	29 21	15 51	19 49	15 58
22	06 00 54	00 ♋ 36 51	19 ♌ 39 15	26 ♌ 54 20	24 35	25 19	17 15	10 08	29 23	15 52	19 48	15 59
23	06 04 50	01 34 06	04 ♍ 08 18	11 ♍ 20 38	25 55	26 29	17 57	10 22	29 25	15 54	19 47	16 00
24	06 08 47	02 31 21	18 ♍ 30 49	25 ♍ 38 30	27 13	27 40	18 38	10 35	29 27	15 56	19 46	16 01
25	06 12 43	03 28 35	02 ♎ 43 24	09 ♎ 45 18	28 27	28 ♉ 50	19 20	10 49	29 28	15 57	19 45	16 02
26	06 16 40	04 25 49	16 ♎ 44 05	23 ♎ 39 41	29 ♋ 39	00 ♊ 01	20 01	11 02	29 30	15 59	19 44	16 02
27	06 20 36	05 23 02	00 ♏ 32 06	07 ♏ 21 19	00 ♌ 48	01 12	20 42	11 16	29 31	16 01	19 43	16 03
28	06 24 33	06 20 14	14 ♏ 07 23	20 ♏ 50 21	01 54	02 22	21 23	11 29	29 33	16 03	19 42	16 04
29	06 28 29	07 17 26	27 ♏ 30 15	04 ♐ 07 07	02 57	03 33	22 05	11 43	29 34	16 04	19 41	16 05
30	06 32 26	08 ♋ 14 38	10 ♐ 40 59	17 ♐ 11 52	03 ♌ 57	04 ♊ 44	22 ♊ 46	11 ♋ 57	29 ♓ 35	16 ♍ 06	19 ♏ 40	16 ♍ 06

Date	Moon True ☊ ° '	Moon Mean ☊ ° '	Moon ☽ Latitude ° '
01	25 ♉ 21	24 ♉ 38	00 N 42
02	25 R 21	24 35	00 S 33
03	25 20	24 31	01 45
04	25 18	24 28	02 49
05	25 16	24 25	03 44
06	25 13	24 22	04 26
07	25 10	24 19	04 56
08	25 07	24 15	05 11
09	25 05	24 12	05 13
10	25 03	24 09	05 01
11	25 D 03	24 06	04 37
12	25 04	24 03	04 00
13	25 05	24 00	03 12
14	25 07	23 56	02 15
15	25 08	23 53	01 S 09
16	25 R 09	23 50	00 N 01
17	25 08	23 47	01 13
18	25 06	23 44	02 23
19	25 02	23 41	03 26
20	24 58	23 37	04 17
21	24 53	23 34	04 53
22	24 48	23 31	05 09
23	24 44	23 28	05 07
24	24 42	23 25	04 44
25	24 41	23 21	04 05
26	24 D 42	23 18	03 11
27	24 43	23 15	02 07
28	24 44	23 12	00 N 57
29	24 R 44	23 09	00 S 15
30	24 ♉ 43	23 ♉ 06	01 S 25

DECLINATIONS

Date	Sun ☉ ° '	Moon ☽ ° '	Mercury ☿ ° '	Venus ♀ ° '	Mars ♂ ° '	Jupiter ♃ ° '	Saturn ♄ ° '	Uranus ♅ ° '	Neptune ♆ ° '	Pluto ♇ ° '
01	22 N 02	16 S 27	24 N 06	09 N 48	20 N 49	23 N 20	02 S 33	06 N 25	16 S 05	19 N 05
02	22 10	20 58	24 25	10 12	20 58	23 20	02 32	06 25	16 04	19 04
03	22 18	24 18	24 42	10 35	21 07	23 19	02 31	06 25	16 04	19 04
04	22 25	26 15	24 55	10 58	21 15	23 19	02 30	06 25	16 04	19 03
05	22 32	26 45	25 06	11 21	21 24	23 18	02 28	06 24	16 03	19 03
06	22 38	25 50	25 15	11 44	21 32	23 18	02 27	06 24	16 03	19 02
07	22 44	23 41	25 20	12 07	21 40	23 17	02 26	06 24	16 02	19 02
08	22 50	20 31	25 23	12 29	21 47	23 16	02 25	06 23	16 02	19 01
09	22 55	16 31	25 24	12 51	21 55	23 16	02 24	06 23	16 02	19 01
10	23 00	11 55	25 22	13 13	22 02	23 15	02 23	06 23	16 01	19 00
11	23 05	06 53	25 18	13 35	22 09	23 14	02 22	06 22	16 01	19 00
12	23 09	01 S 35	25 12	13 57	22 16	23 14	02 21	06 22	16 01	18 59
13	23 12	03 N 52	25 04	14 18	22 22	23 13	02 20	06 21	16 00	18 59
14	23 16	09 17	24 54	14 39	22 29	23 12	02 19	06 21	16 00	18 58
15	23 18	14 27	24 42	15 00	22 35	23 12	02 19	06 20	16 00	18 57
16	23 21	19 07	24 28	15 20	22 41	23 11	02 18	06 20	16 00	18 57
17	23 23	22 58	24 13	15 40	22 47	23 10	02 17	06 19	15 59	18 56
18	23 24	25 37	23 57	16 00	22 52	23 09	02 16	06 19	15 59	18 56
19	23 26	26 43	23 39	16 19	22 57	23 08	02 16	06 18	15 59	18 55
20	23 26	26 05	23 20	16 38	23 03	23 07	02 15	06 18	15 58	18 54
21	23 27	23 43	23 00	16 57	23 07	23 07	02 15	06 17	15 58	18 54
22	23 27	19 49	22 39	17 15	23 12	23 06	02 14	06 16	15 58	18 53
23	23 26	14 45	22 17	17 33	23 17	23 05	02 14	06 16	15 58	18 52
24	23 25	08 54	21 55	17 51	23 21	23 04	02 13	06 15	15 57	18 52
25	23 24	02 N 40	21 32	18 08	23 25	23 03	02 13	06 14	15 57	18 51
26	23 22	03 S 38	21 08	18 24	23 29	23 02	02 12	06 14	15 57	18 50
27	23 20	09 40	20 44	18 41	23 32	23 01	02 12	06 13	15 57	18 50
28	23 18	15 10	20 19	18 56	23 36	23 00	02 12	06 12	15 56	18 49
29	23 15	19 51	19 55	19 12	23 39	22 59	02 12	06 11	15 56	18 48
30	23 N 11	23 S 28	19 N 30	19 N 27	23 N 42	22 N 57	02 S 11	06 N 11	15 S 56	18 N 47

ZODIAC SIGN ENTRIES

Date	h	m	Planets
02	09	38	☽ ♐
04	16	10	☽ ♑
07	01	21	☽ ♒
07	19	11	☿ ♋
09	12	57	☽ ♓
12	01	26	☽ ♈
14	12	30	☽ ♉
16	20	26	☽ ♊
19	01	05	☽ ♋
21	03	29	☽ ♌
21	20	33	☉ ♋
23	05	08	☽ ♍
25	07	23	☽ ♎
26	11	40	♀ ♊
26	19	05	☿ ♌
27	11	04	☽ ♏
29	16	31	☽ ♐

LATITUDES

Date	Mercury ☿ ° '	Venus ♀ ° '	Mars ♂ ° '	Jupiter ♃ ° '	Saturn ♄ ° '	Uranus ♅ ° '	Neptune ♆ ° '	Pluto ♇ ° '
01	01 N 18	02 S 06	00 N 09	00 00	02 S 05	00 N 46	01 N 48	14 N 41
04	01 39	02 07	00 11	00 00	02 06	00 46	01 48	14 40
07	01 54	02 07	00 13	00 N 01	02 06	00 46	01 48	14 38
10	02 02	02 06	00 15	00 01	02 07	00 46	01 48	14 37
13	02 03	02 04	00 17	00 01	02 08	00 46	01 48	14 36
16	01 57	02 01	00 18	00 02	02 09	00 46	01 48	14 35
19	01 46	01 58	00 20	00 02	02 09	00 46	01 48	14 34
22	01 28	01 54	00 22	00 02	02 10	00 46	01 48	14 33
25	01 04	01 49	00 24	00 02	02 11	00 45	01 48	14 31
28	00 36	01 44	00 26	00 03	02 12	00 45	01 47	14 30
31	00 N 03	01 S 38	00 N 28	00 N 03	02 S 13	00 N 45	01 N 47	14 N 29

DATA

Julian Date	2439278
Delta T	+37 seconds
Ayanamsa	23° 23' 02"
Synetic vernal point	05° ♓ 43' 58"
True obliquity of ecliptic	23° 26' 42"

MOON'S PHASES, APSIDES AND POSITIONS ☽

Date	h	m	Phase	Longitude ° '	Eclipse Indicator
03	07	40	○	12 ♐ 17	
11	04	58	☾	19 ♓ 50	
18	20	09	●	27 ♊ 07	
25	13	22	☽	03 ♎ 32	

Day	h	m	
10	07	48	Apogee
22	08	16	Perigee

Day	h	m		
05	08	15	Max dec	26° S 46'
12	18	58	0N	
19	15	28	Max dec	26° N 44'
25	22	07	0S	

LONGITUDES

Date	Chiron ⚷ ° '	Ceres ⚳ ° '	Pallas ⚴ ° '	Juno ⚵ ° '	Vesta ⚶ ° '	Black Moon Lilith ⚸ ° '
01	25 ♓ 54	22 ♉ 21	09 ♈ 10	18 ♉ 31	26 ♋ 12	06 ♓ 49
11	26 ♓ 05	26 ♉ 21	11 ♈ 42	24 ♉ 19	00 ♌ 20	07 ♓ 56
21	26 ♓ 11	00 ♊ 18	14 ♈ 02	00 ♊ 06	04 ♌ 36	09 ♓ 03
31	26 ♓ 14	04 ♊ 12	16 ♈ 05	05 ♊ 53	08 ♌ 58	10 ♓ 10

All ephemeris data is given at 12.00 UT and the Moon's longitude is additionally given for 24.00 UT

ASPECTARIAN

h m Aspects

01 Wednesday
04 20 ☽ ⚼ ♄
04 45 ♀ ⚼ ♅
07 41 ☉ Q ♄
08 06 ☽ ⚹ ♅
08 37 ☽ ⚹ ♇
10 12 ☽ ∥ ♆
10 22 ☽ ⊼ ☿
10 58 ♀ ⚼ ♇
16 26 ☽ ☌ ♆
16 55 ☽ ⚼ ♃

02 Thursday
01 13 ☽ ⋕ ♇
05 15 ☽ Q ♅
05 46 ☽ Q ♇
06 51 ☽ △ ♄
09 05 ☽ ± ♃
12 01 ☽ ⋕ ♂
13 21 ☽ ⊼ ♀
15 38 ☽ ☍ ♂
19 55 ☽ ⋕ ☉
19 58 ☽ ⊼ ♃

03 Friday
00 48 ☿ ⊼ ♆
01 06 ☽ ± ♀
03 55 ☽ ⋕ ♃
07 40 ☽ ☍ ☉
13 34 ☽ □ ♅
14 05 ☽ □ ♇
16 09 ☽ ⋕ ☿
18 59 ☽ ⚼ ♀
22 06 ☽ ⊻ ♆

04 Saturday
02 14 ☽ ☍ ☿
09 07 ☽ ⊥ ♆
13 30 ☽ □ ♄
17 12 ♀ ⊥ ♄

05 Sunday
01 27 ☽ △ ♀
01 38 ☽ ⊼ ♂
01 51 ☽ ∠ ♆
03 56 ☽ ☍ ♃
06 22 ♀ ⊻ ♂
13 37 ☽ ± ♂
19 48 ☽ ⊼ ☉
21 13 ☿ ± ♆
21 29 ☽ △ ♅
22 00 ☽ △ ♇
23 33 ☽ Q ♄

06 Monday
06 17 ☽ ⚹ ♆
07 49 ☽ ⚼ ♂
08 15 ☽ ± ☉
10 39 ♀ ⚹ ♃
18 10 ☉ □ ♅
19 41 ☽ ⋕ ☿
21 55 ☽ ⊼ ☿
22 49 ☽ ⚹ ♄

07 Tuesday
00 48 ☉ □ ♇
02 28 ☽ ⚼ ♅
02 59 ☽ ⚼ ♇
03 10 ☽ ⚼ ☉
03 36 ☿ □ ♄
05 31 ☽ Q ♆
11 56 ☽ ± ☿
12 57 ♂ ⊻ ♃
14 44 ☽ ⊼ ♃
14 49 ☽ △ ♂
15 33 ☽ ⋕ ♃
17 08 ☽ □ ♀
19 44 ☽ ⋕ ☉
20 10 ☽ ± ♅
20 41 ☽ ± ♇

08 Wednesday
02 49 ☽ ± ♃
03 30 ☽ ⋕ ♂
04 28 ☽ ∠ ♄
08 05 ☽ ⊼ ♅
08 35 ☽ ⊼ ♇
09 11 ☽ ⚼ ☿
11 19 ☽ △ ☉
17 04 ☽ □ ♆
21 06 ☽ ⚼ ♃
21 29 ☽ ⋕ ♇
22 30 ☽ ⊥ ♄

09 Thursday
08 51 ☽ Q ♀
10 36 ☽ ⊻ ♄
14 43 ☽ ∥ ♆
16 00 ☿ Q ♅
19 03 ☿ Q ♇
21 01 ☽ △ ☿

10 Friday
03 51 ☽ △ ♃
05 56 ☽ ⋕ ♀
06 34 ☽ □ ♂
10 24 ☿ ⚼ ♆
11 41 ☽ ⚹ ♀
20 30 ☽ ☍ ♅
20 58 ☽ ☍ ♇

11 Saturday
04 58 ☽ □ ☉
05 23 ☽ △ ♆
10 08 ☉ ⊼ ♆
14 24 ☽ ⋕ ♅
21 09 ☽ Q ♂
21 12 ☽ ∠ ♀
21 55 ☿ ☌ ♃
23 21 ☽ ☌ ♄

12 Sunday
08 32 ☽ ∥ ♄
11 27 ☽ ⚼ ♆
17 08 ☽ ⊥ ♀
17 20 ☽ □ ♃
20 09 ♀ ∠ ♄
20 11 ☽ Q ☉
20 17 ☽ □ ☿
22 35 ☽ ⚹ ♂

13 Monday
05 17 ☽ ⋕ ♄
05 18 ☽ ± ♆
06 17 ☽ ⊻ ♀
08 36 ☽ ⊼ ♅
09 01 ☽ ⊼ ♇
13 27 ♂ Q ♄
16 34 ☉ ∥ ♃
17 04 ☽ ⊼ ♆
20 23 ☽ ± ♅
20 47 ☽ ± ♇
21 17 ☿ ⊻ ♂
21 48 ☽ ⚹ ☉
22 56 ☽ ∥ ♅

14 Tuesday
05 20 ☽ Q ♃
05 46 ☽ ∠ ♂
06 50 ♀ △ ♅
10 43 ☽ ⊻ ♄
10 58 ♀ △ ♇
13 17 ☽ Q ☿
13 50 ☽ ⚼ ♅
14 12 ☽ ⚼ ♇
22 15 ☽ ⊥ ♄

15 Wednesday
00 03 ☽ ⊥ ♂
04 42 ☽ ⚹ ♃
04 59 ☽ ∠ ☉
12 01 ☽ ⊻ ♂
14 49 ☽ ∥ ♀
15 16 ☽ ∠ ♄
15 29 ☽ ⚹ ☿
18 14 ☽ △ ♅
18 35 ☽ △ ♇
19 34 ☽ ⋕ ♆
21 42 ☽ ☌ ♀
23 10 ☽ ⊥ ☉

16 Thursday
02 03 ☽ ☍ ♆
09 06 ☽ ∠ ♃
11 01 ☽ ∥ ♇
11 04 ☽ ⊻ ☉
13 25 ☿ ⚹ ♅
16 04 ☿ ⚹ ♇
18 57 ☽ ⚹ ♄
22 54 ☽ ∠ ☿

17 Friday
01 42 ☽ ⊥ ♃
10 36 ☽ ∥ ♂
12 35 ☽ ⊻ ♃
13 28 ☽ ∥ ♃
13 30 ☉ ± ♆
15 06 ☽ ∥ ☉
16 28 ☽ Q ♄
17 05 ☽ ⊥ ☿
21 01 ☽ ∥ ☿
21 09 ♀ ☍ ♆
21 28 ☽ ☌ ♂

18 Saturday
00 28 ☽ □ ♅
00 45 ☽ □ ♇
04 58 ☽ ⊻ ☿
07 39 ☽ ⊼ ♆
08 38 ☽ ⊻ ♀
10 57 ♀ ⋕ ♆
18 01 ☽ ± ♆
19 57 ☽ ⊥ ♀
20 09 ☽ ☌ ☉
23 49 ☽ □ ♄

19 Sunday
04 59 ☿ △ ♆
07 33 ☽ Q ♅
07 37 ☉ ⊥ ♀
07 48 ☽ Q ♇
09 22 ☽ ⚼ ♆
12 41 ☽ ∠ ♀
17 15 ☽ ☌ ♃

20 Monday
03 31 ☽ ⊻ ♂
03 50 ☽ ⚹ ♅
04 04 ☽ ⚹ ♇
04 25 ☉ ∥ ☿
10 20 ♂ □ ♅
10 35 ☽ △ ♆
14 00 ☽ ☌ ☿
14 04 ☽ ⊥ ♂
15 03 ♂ □ ♇
15 06 ☿ ⊥ ♂
16 07 ☽ ⚹ ♀

21 Tuesday
02 16 ☽ ⊻ ☉
02 23 ☽ △ ♄
03 58 ☿ ∥ ♃
03 59 ☉ □ ♄
04 53 ☽ ∠ ♅
04 59 ☿ ∥ ♂
05 05 ☽ ∠ ♇
05 47 ☽ ∠ ♂
08 24 ♂ ∥ ♃
12 55 ☽ ⊥ ☉
13 48 ☽ Q ♀
13 59 ☽ ∥ ☉
16 10 ☽ ∥ ♂
16 23 ☽ ∥ ♃
17 40 ☽ ∥ ☿
19 48 ☽ ⊥ ♅
20 00 ☽ ⊥ ♇
20 01 ☽ ⊻ ♃

22 Wednesday
03 16 ☽ ⚼ ♄
04 51 ☽ ∠ ☉
05 44 ☽ ⊻ ♅
05 55 ☽ ⊻ ♇
06 06 ☽ ⊥ ♃
07 41 ♀ ∠ ♃
07 50 ☽ ⚹ ♂
12 15 ☽ □ ♆
16 48 ☽ ∥ ♇
18 11 ☽ ± ♄
20 59 ☽ ⊻ ☿
21 13 ☽ ∠ ♃
22 11 ☽ □ ♀

23 Thursday
00 00 ☽ ∥ ♀
04 09 ☽ ⊼ ♄
04 42 ☽ Q ♂
06 38 ☽ ⋕ ♆
07 26 ☽ ⚹ ☉
07 57 ☽ ⊥ ☿
18 03 ☽ Q ♆
22 32 ☽ ⚹ ♃

24 Friday
00 25 ☽ ∠ ☿
04 50 ☽ Q ☉
07 39 ☽ ☌ ♅
07 48 ☽ ☌ ♇
12 13 ☽ □ ♂
14 06 ☽ ⚹ ♆
18 58 ☽ Q ♃
22 20 ☽ ∥ ♅

25 Saturday
04 04 ☽ ⚹ ☿
04 49 ☽ △ ♀
06 28 ☽ ☍ ♄
08 14 ☉ ∥ ♂
13 22 ☽ □ ☉
13 43 ☽ ⋕ ♄
15 26 ☽ ∠ ♆

26 Sunday
00 25 ☉ Q ♅
01 12 ♀ ⚹ ♄
02 01 ☉ Q ♇
02 03 ☽ □ ♃
02 17 ♂ ⊼ ♆
02 27 ☽ Q ☿
06 32 ☽ ∥ ♄
06 50 ☽ ⊥ ♆
08 44 ☿ △ ♄
08 46 ☽ ⚼ ♀
10 42 ☽ ⊻ ♅
10 48 ☽ ⊻ ♇
17 10 ☽ ⊻ ♆
17 58 ☽ △ ♂
19 24 ☉ ⚼ ♆
21 06 ☽ ⊥ ♅
21 12 ☽ ⊥ ♇
22 06 ☽ ⋕ ♅

27 Monday
01 48 ☽ ± ♀
10 13 ☽ ⊼ ♄
12 31 ☽ □ ☿
12 50 ☽ ∠ ♅
12 55 ☽ ∠ ♇
13 16 ☽ ⊼ ♀
16 34 ☿ ∠ ♅
17 26 ☿ ∠ ♇
20 47 ☽ ± ♄
21 10 ☽ △ ☉
21 34 ☽ ⚼ ♂

28 Tuesday
00 59 ♀ ∥ ♇
07 15 ☽ △ ♃
12 45 ☽ ⚼ ♄
14 23 ☽ ± ♂
15 26 ☽ ⚹ ♅
15 29 ☽ ⚹ ♇
15 38 ☽ ∥ ♆
21 56 ☽ ☌ ♆

29 Wednesday
01 41 ☽ ⊼ ♂
01 53 ☽ ⚼ ☉
06 11 ☽ ⋕ ♇
08 06 ☽ ⋕ ♀
10 33 ☽ ⚼ ♃
12 20 ☽ ⋕ ☿
13 02 ☽ Q ♅
13 03 ☽ Q ♇
15 44 ☽ △ ♄
19 23 ☽ ± ☉
22 43 ☽ △ ☿

30 Thursday
00 03 ☽ ☍ ♀
03 10 ☽ ± ♃
07 11 ☽ ⊼ ☉
08 08 ☽ ⋕ ♃
09 55 ☽ ⋕ ☉
09 58 ♅ ☌ ♇
13 58 ☽ ⋕ ♂
14 03 ☿ ∥ ♀
14 21 ☽ ⊼ ♃
22 00 ☽ □ ♇
22 01 ☽ □ ♅

JULY 1966

LONGITUDES

Date	Sidereal time h m s	Sun ☉ ° ' "	Moon ☽ ° ' "	Moon ☽ 24.00 ° ' "	Mercury ☿ ° '	Venus ♀ ° '	Mars ♂ ° '	Jupiter ♃ ° '	Saturn ♄ ° '	Uranus ♅ ° '	Neptune ♆ ° '	Pluto ♇ ° '
01	06 36 23	09 ♋ 11 49	23 ♐ 39 46	00 ♑ 04 41	04 ♌ 54	05 ♊ 55	23 ♊ 27	12 ♋ 10	29 ♓ 36	16 ♍ 08	19 ♏ 39	16 ♍ 07
02	06 40 19	10 09 01	06 ♑ 26 37	12 ♑ 45 35	05 47	07 06	24 08	12 24	29 37	16 10	19 R 38	16 08
03	06 44 16	11 06 12	19 ♑ 01 36	25 ♑ 14 44	06 37	08 17	24 49	12 37	29 38	16 12	19 37	16 10
04	06 48 12	12 03 23	01 ♒ 25 02	07 ♒ 32 38	07 24	09 28	25 30	12 51	29 39	16 14	19 36	16 11
05	06 52 09	13 00 34	13 ♒ 37 41	19 ♒ 40 24	08 07	10 39	26 11	13 04	29 39	16 17	19 35	16 12
06	06 56 05	13 57 46	25 ♒ 41 01	01 ♓ 39 51	08 46	11 50	26 51	13 18	29 40	16 19	19 35	16 13
07	07 00 02	14 54 57	07 ♓ 37 15	13 ♓ 33 38	09 21	13 02	27 32	13 31	29 40	16 21	19 34	16 14
08	07 03 58	15 52 09	19 ♓ 29 27	25 ♓ 25 11	09 51	14 13	28 13	13 45	29 41	16 23	19 33	16 15
09	07 07 55	16 49 21	01 ♈ 21 24	07 ♈ 18 40	10 18	15 24	28 54	13 59	29 41	16 26	19 32	16 17
10	07 11 52	17 46 34	13 ♈ 17 35	19 ♈ 18 46	10 41	16 36	29 ♊ 34	14 12	29 41	16 28	19 32	16 18
11	07 15 48	18 43 47	25 ♈ 22 52	01 ♉ 30 31	10 59	17 47	00 ♋ 15	14 26	29 41	16 30	19 31	16 19
12	07 19 45	19 41 00	07 ♉ 42 20	13 ♉ 58 55	11 12	18 59	00 55	14 39	29 R 41	16 33	19 30	16 21
13	07 23 41	20 38 14	20 ♉ 20 49	26 ♉ 48 33	11 21	20 10	01 36	14 53	29 41	16 35	19 30	16 22
14	07 27 38	21 35 29	03 ♊ 22 29	10 ♊ 02 58	11 25	21 22	02 16	15 06	29 41	16 38	19 29	16 23
15	07 31 34	22 32 44	16 ♊ 50 08	23 ♊ 44 02	11 R 24	22 34	02 57	15 20	29 40	16 40	19 29	16 25
16	07 35 31	23 30 00	00 ♋ 44 31	07 ♋ 51 16	11 19	23 45	03 37	15 33	29 40	16 43	19 28	16 26
17	07 39 27	24 27 16	15 ♋ 03 47	22 ♋ 21 20	11 08	24 57	04 17	15 47	29 39	16 45	19 28	16 28
18	07 43 24	25 24 33	29 ♋ 43 06	07 ♌ 08 05	10 53	26 09	04 58	16 00	29 39	16 48	19 27	16 29
19	07 47 21	26 21 50	14 ♌ 35 10	22 ♌ 03 14	10 33	27 21	05 38	16 14	29 38	16 51	19 27	16 31
20	07 51 17	27 19 07	29 ♌ 31 08	06 ♍ 57 45	10 09	28 33	06 18	16 27	29 37	16 54	19 27	16 32
21	07 55 14	28 16 25	14 ♍ 22 07	21 ♍ 43 22	09 41	29 ♊ 44	06 58	16 40	29 36	16 56	19 26	16 34
22	07 59 10	29 ♋ 13 43	29 ♍ 00 46	06 ♎ 13 47	09 09	00 ♋ 56	07 38	16 54	29 35	16 59	19 26	16 35
23	08 03 07	00 ♌ 11 01	13 ♎ 22 02	20 ♎ 25 19	08 33	02 08	08 18	17 07	29 34	17 02	19 26	16 37
24	08 07 03	01 08 19	27 ♎ 23 32	04 ♏ 16 45	07 55	03 20	08 58	17 20	29 33	17 05	19 25	16 38
25	08 11 00	02 05 38	11 ♏ 05 05	17 ♏ 48 45	07 14	04 33	09 38	17 34	29 31	17 08	19 25	16 40
26	08 14 56	03 02 57	24 ♏ 28 01	01 ♐ 03 09	06 32	05 45	10 18	17 47	29 30	17 11	19 25	16 42
27	08 18 53	04 00 17	07 ♐ 34 27	14 ♐ 02 12	05 48	06 57	10 58	18 00	29 29	17 14	19 25	16 43
28	08 22 50	04 57 37	20 ♐ 26 40	26 ♐ 48 07	05 05	08 09	11 38	18 14	29 27	17 17	19 25	16 45
29	08 26 46	05 54 58	03 ♑ 06 45	09 ♑ 22 46	04 22	09 21	12 17	18 27	29 25	17 20	19 24	16 47
30	08 30 43	06 52 19	15 ♑ 36 20	21 ♑ 47 34	03 40	10 34	12 57	18 40	29 23	17 23	19 24	16 49
31	08 34 39	07 ♌ 49 41	27 ♑ 56 37	04 ♒ 03 33	03 ♌ 01	11 ♋ 46	13 ♋ 37	18 ♋ 53	29 ♓ 21	17 ♍ 26	19 ♏ 24	16 ♍ 50

Date	Moon True ☊ ° '	Moon Mean ☊ ° '	Moon ☽ Latitude ° '
01	24 ♉ 40	23 ♉ 02	02 S 30
02	24 R 35	22 59	03 26
03	24 28	22 56	04 10
04	24 20	22 53	04 43
05	24 11	22 50	05 01
06	24 02	22 47	05 06
07	23 55	22 43	04 58
08	23 50	22 40	04 37
09	23 46	22 37	04 03
10	23 45	22 34	03 19
11	23 D 45	22 31	02 26
12	23 46	22 27	01 25
13	23 47	22 24	00 S 18
14	23 R 46	22 21	00 N 51
15	23 44	22 18	02 00
16	23 39	22 15	03 04
17	23 33	22 12	03 58
18	23 24	22 08	04 38
19	23 15	22 05	05 00
20	23 06	22 02	05 01
21	22 58	21 59	04 43
22	22 53	21 56	04 05
23	22 50	21 52	03 13
24	22 48	21 49	02 11
25	22 D 49	21 46	01 N 02
26	22 R 49	21 43	00 S 09
27	22 48	21 40	01 17
28	22 45	21 37	02 21
29	22 40	21 33	03 16
30	22 31	21 30	04 01
31	22 ♉ 21	21 ♉ 27	04 S 34

DECLINATIONS

Date	Sun ☉ ° '	Moon ☽ ° '	Mercury ☿ ° '	Venus ♀ ° '	Mars ♂ ° '	Jupiter ♃ ° '	Saturn ♄ ° '	Uranus ♅ ° '	Neptune ♆ ° '	Pluto ♇ ° '
01	23 N 08	25 S 47	19 N 05	19 N 41	23 N 44	22 N 56	02 S 11	06 N 10	15 S 56	18 N 47
02	23 03	26 43	18 41	19 55	23 47	22 55	02 11	06 09	15 56	18 46
03	22 59	26 14	18 16	20 09	23 49	22 54	02 11	06 08	15 55	18 45
04	22 54	24 26	17 52	20 22	23 51	22 53	02 11	06 07	15 55	18 44
05	22 49	21 33	17 28	20 34	23 53	22 52	02 11	06 07	15 55	18 44
06	22 43	17 46	17 05	20 46	23 55	22 50	02 11	06 06	15 55	18 43
07	22 37	13 19	16 43	20 57	23 56	22 49	02 11	06 05	15 55	18 42
08	22 30	08 24	16 21	21 08	23 58	22 48	02 11	06 04	15 55	18 41
09	22 23	03 S 11	16 00	21 19	23 59	22 46	02 11	06 03	15 54	18 41
10	22 16	02 N 11	15 40	21 28	24 00	22 45	02 11	06 02	15 54	18 40
11	22 08	07 33	15 21	21 38	24 00	22 44	02 12	06 01	15 54	18 39
12	22 00	12 45	15 03	21 46	24 01	22 42	02 12	06 00	15 54	18 38
13	21 52	17 33	14 46	21 54	24 01	22 41	02 12	05 59	15 54	18 37
14	21 43	21 40	14 31	22 02	24 01	22 40	02 13	05 58	15 54	18 36
15	21 34	24 47	14 18	22 09	24 01	22 38	02 13	05 57	15 54	18 36
16	21 25	26 30	14 06	22 15	24 00	22 37	02 13	05 56	15 54	18 35
17	21 14	26 32	13 56	22 20	23 59	22 35	02 14	05 55	15 54	18 34
18	21 04	24 44	13 47	22 26	23 59	22 34	02 14	05 54	15 54	18 33
19	20 53	21 14	13 41	22 30	23 58	22 32	02 15	05 53	15 53	18 32
20	20 42	16 20	13 37	22 34	23 56	22 31	02 15	05 52	15 53	18 31
21	20 31	10 30	13 34	22 37	23 55	22 29	02 16	05 51	15 53	18 31
22	20 19	04 N 09	13 34	22 39	23 53	22 28	02 17	05 49	15 53	18 30
23	20 07	02 S 18	13 35	22 41	23 51	22 26	02 17	05 48	15 53	18 29
24	19 55	08 31	13 39	22 43	23 49	22 24	02 18	05 47	15 53	18 28
25	19 42	14 11	13 44	22 43	23 47	22 23	02 19	05 46	15 53	18 27
26	19 29	19 03	13 51	22 43	23 45	22 21	02 20	05 45	15 53	18 26
27	19 16	22 51	14 00	22 43	23 42	22 20	02 21	05 44	15 53	18 25
28	19 02	25 27	14 11	22 41	23 39	22 18	02 21	05 42	15 53	18 24
29	18 48	26 41	14 23	22 39	23 36	22 16	02 22	05 41	15 53	18 24
30	18 34	26 31	14 36	22 37	23 33	22 14	02 23	05 40	15 53	18 23
31	18 N 19	25 S 03	14 N 51	22 N 34	23 N 30	22 N 13	02 S 24	05 N 39	15 S 53	18 N 22

ZODIAC SIGN ENTRIES

Date	h	m	Planets
01	23	51	☽ ♑
04	09	14	☽ ♒
06	20	39	☽ ♓
09	09	16	☽ ♈
11	03	15	♂ ♋
11	21	03	☽ ♉
14	05	51	☽ ♊
16	10	44	☽ ♋
18	12	27	☽ ♌
20	12	46	☽ ♍
21	17	11	♀ ♋
22	13	38	☽ ♎
23	07	23	☉ ♌
24	16	32	☽ ♏
26	22	04	☽ ♐
29	06	04	☽ ♑
31	16	02	☽ ♒

LATITUDES

Date	Mercury ☿ ° '	Venus ♀ ° '	Mars ♂ ° '	Jupiter ♃ ° '	Saturn ♄ ° '	Uranus ♅ ° '	Neptune ♆ ° '	Pluto ♇ ° '
01	00 N 03	01 S 38	00 N 28	00 N 03	02 S 13	00 N 45	01 N 47	14 N 29
04	00 S 35	01 32	00 29	00 03	02 13	00 45	01 47	14 28
07	01 15	01 25	00 31	00 04	02 14	00 45	01 47	14 27
10	01 59	01 18	00 33	00 04	02 15	00 45	01 47	14 26
13	02 43	01 11	00 35	00 04	02 16	00 45	01 47	14 25
16	03 25	01 03	00 36	00 05	02 17	00 45	01 46	14 24
19	04 04	00 55	00 38	00 05	02 17	00 45	01 46	14 24
22	04 34	00 47	00 40	00 05	02 18	00 45	01 46	14 23
25	04 53	00 39	00 41	00 05	02 19	00 44	01 46	14 22
28	04 58	00 30	00 43	00 06	02 20	00 44	01 46	14 21
31	04 S 46	00 S 22	00 N 45	00 N 06	02 S 21	00 N 44	01 N 46	14 N 21

DATA

Julian Date	2439308
Delta T	+37 seconds
Ayanamsa	23° 23' 08"
Synetic vernal point	05° ♓ 43' 52"
True obliquity of ecliptic	23° 26' 42"

MOON'S PHASES, APSIDES AND POSITIONS ☽

Date	h	m	Phase	Longitude ° '	Eclipse Indicator
02	19	36	○	10 ♑ 27	
10	21	43	☾	18 ♈ 10	
18	04	30	●	25 ♋ 07	
24	19	00	☽	01 ♏ 25	

Day	h	m		
08	01	21	Apogee	
20	01	01	Perigee	
02	15	35	Max dec	26° S 44'
10	02	16	ON	
17	00	26	Max dec	26° N 45'
23	03	23	OS	
29	21	15	Max dec	26° S 47'

LONGITUDES

Date	Chiron ⚷ ° '	Ceres ⚳ ° '	Pallas ⚴ ° '	Juno ⚵ ° '	Vesta ⚶ ° '	Black Moon Lilith ⚸ ° '
01	26 ♓ 14	04 ♊ 12	16 ♈ 05	05 ♊ 53	08 ♌ 58	10 ♓ 10
11	26 ♓ 10	08 ♊ 01	18 ♈ 49	11 ♊ 37	13 ♌ 26	11 ♓ 17
21	26 ♓ 01	11 ♊ 45	19 ♈ 11	17 ♊ 19	17 ♌ 59	12 ♓ 24
31	25 ♓ 47	15 ♊ 22	20 ♈ 06	22 ♊ 57	22 ♌ 35	13 ♓ 32

ASPECTARIAN

h m	Aspects
01 Friday	
04 28	☽ ⚼ ☿
04 33	☽ ⊻ ♆
11 34	☽ ☍ ♂
15 42	☽ ⊥ ♆
18 18	♀ ⊥ ♃
22 32	☽ ± ☿
23 07	☽ □ ♄
02 Saturday	
06 42	☿ ∥ ♇
08 35	☽ ∠ ♆
10 40	☽ ⊼ ☿
13 22	☽ ⊼ ♀
19 36	☽ ☍ ☉
23 31	☽ ☍ ♃
03 Sunday	
01 58	☽ ± ♀
06 29	☽ △ ♇
06 34	☽ △ ♅
09 19	☽ Q ♄
13 08	☽ ✶ ♆
21 04	☽ ⚼ ♀
23 49	☽ ⊼ ♂
04 Monday	
08 33	☽ ✶ ♄
11 32	☽ ⚼ ♇
11 39	☽ ⚼ ♅
12 10	☽ ± ♂
12 22	☽ Q ♆
15 47	♂ ± ♆
17 31	☽ # ♂
18 39	☉ ∥ ♃
05 Tuesday	
00 27	☽ ☍ ☿
02 01	☽ # ♃
02 12	☽ # ☉
05 13	☽ ± ♇
05 22	☽ ± ♅
05 30	☽ △ ♀
06 52	☽ ⚼ ♂
10 40	☽ ⊼ ☉
10 53	☽ ⊼ ♃
14 02	☽ ∠ ♄
14 05	☉ ☌ ♃
17 06	☽ ⊼ ♇
17 16	☽ ⊼ ♅
18 20	☽ # ♀
23 01	☽ ± ♃
23 36	☽ ± ☉
23 49	☽ □ ♆
06 Wednesday	
06 21	☽ # ♇
07 58	☽ ⊥ ♄
08 26	♀ Q ♄
14 29	☽ △ ♂
16 14	☽ # ☿
17 20	☽ ⚼ ♃
19 09	☽ ⚼ ☉
19 59	☽ ⊻ ♄
22 21	☽ ∥ ♆
07 Thursday	
15 39	☽ ⊼ ☿
08 Friday	
00 08	☽ □ ♀
00 10	☽ △ ♃
00 22	♀ ⊻ ♃
04 02	☽ △ ☉
04 20	☽ ± ☿
05 27	☽ ☍ ♇
05 42	☽ ☍ ♅
12 07	☽ △ ♆
21 59	☉ ✶ ♇
22 54	☽ # ♅
23 18	☽ ⚼ ☿
09 Saturday	
01 35	☉ ✶ ♅
06 43	☽ □ ♂
08 37	☽ ☌ ♄
10 23	☿ ⊥ ♇
16 29	☽ ∥ ♄
16 36	☽ Q ♀
18 12	☿ # ♆
18 25	☽ ⚼ ♆
20 05	☿ ⊥ ♅
10 Sunday	
05 55	♀ □ ♇
06 36	☽ △ ☿
09 15	♀ □ ♅
12 01	☽ # ♄
12 28	☽ ± ♆
13 51	☽ □ ♃
16 09	♂ □ ♄
18 01	☽ ⊼ ♇
18 21	☽ ⊼ ♅
19 19	☽ ✶ ♀
21 03	☽ Q ♂
21 43	☽ □ ☉
11 Monday	
00 25	☽ ⊼ ♆
05 07	☽ ∥ ♅
05 57	☽ ± ♇
06 18	☽ ± ♅
13 03	♄ St R
20 27	☽ ⊻ ♄
22 06	☽ ✶ ♂
23 39	☽ ⚼ ♇
12 Tuesday	
00 02	☽ ⚼ ♅
02 03	☽ Q ♃
04 02	☽ ∠ ♀
07 37	☉ △ ♆
08 06	☽ ⊥ ♄
11 57	☽ Q ☉
18 48	☽ □ ☿
22 33	♀ ⊼ ♆
22 35	☽ ∥ ☿
23 09	☽ ⊥ ♀
13 Wednesday	
01 20	☽ ∠ ♄
01 31	☽ ✶ ♃
03 29	☽ # ♆
04 30	☽ △ ♇
04 33	☽ ∠ ♂
04 54	☽ △ ♅
08 10	☉ ∥ ♀
10 25	☽ ☍ ♆
11 38	☽ ⊻ ♀
12 35	☽ ✶ ☉
17 49	☽ ∥ ♇
22 18	☽ ⊥ ♂
14 Thursday	
04 46	☽ Q ☿
05 16	☽ ✶ ♄
05 57	☽ ∠ ♃
09 53	☽ ⊻ ♂
12 15	☽ ∥ ☉
14 26	☽ ∥ ♀
18 15	☽ ∠ ☉
18 40	☽ ∥ ♃
20 14	☿ St R
22 29	☽ ⊥ ♃
15 Friday	
02 26	☽ ✶ ☿
02 54	☽ Q ♄
05 03	☽ ∥ ♂
09 18	☽ ⊻ ♃
10 40	☉ ⊻ ♀
11 15	☽ □ ♇
11 27	☽ ⊥ ☉
11 43	☽ □ ♅
16 37	☽ ⊼ ♆
22 41	☽ ⊻ ☉
22 55	☽ ☌ ♀
16 Saturday	
03 00	☽ ± ♆
04 30	☽ ∠ ☿
10 10	☽ □ ♄
17 07	☽ ☌ ♂
18 16	☽ Q ♇
18 19	☽ ⚼ ♆
18 44	☽ Q ♅
19 39	☽ ⊥ ☿
17 Sunday	
05 35	☽ ⊻ ☿
13 12	☽ ☌ ♃
14 19	☽ ✶ ♇
14 48	☽ ✶ ♅
18 06	♂ ⚼ ♆
18 21	♂ Q ♇
19 15	☽ △ ♆
22 14	♀ ± ♆
18 Monday	
04 30	☽ ☌ ☉
05 41	☽ ⊻ ♀
05 52	♂ Q ♅
07 48	☿ ∠ ♀
10 08	☿ ⊥ ♂
11 53	☽ △ ♄
14 52	☽ ∠ ♇
15 23	☽ ∠ ♅
16 17	☽ ⊥ ♀
17 57	☿ ⊥ ♅
18 19	☽ ∥ ♂
20 54	☽ ⊻ ♂
19 Tuesday	
04 12	☽ ∥ ♃
04 37	☽ ∥ ♀
05 26	☽ ⊥ ♇
05 40	☽ ☌ ☿
05 58	☽ ⊥ ♅
07 01	☽ ⊥ ♂
08 05	☽ ∠ ♀
12 05	☽ ⚼ ♄
14 01	☽ ∥ ☉
14 41	☽ ⊻ ♃
14 52	☿ ⊥ ♇
15 06	☽ ⊻ ♇
15 39	☽ ⊻ ♅
19 49	☽ □ ♆
22 01	♀ ∥ ♃
22 10	☽ ∠ ♂
20 Wednesday	
00 28	☽ ⊥ ♃
01 58	☽ ∥ ♇
02 32	☽ ± ♄
08 13	☽ ⊻ ☉
10 17	☽ ✶ ♀
12 10	☽ ⊼ ♄
13 58	☽ # ♆
15 09	☽ ∠ ♃
18 32	☽ ⊥ ☉
22 27	♃ ✶ ♇
23 27	☽ ✶ ♂
23 41	☽ ∥ ☿
21 Thursday	
00 46	☽ Q ♆
04 39	☽ ⊻ ☿
07 21	☽ Q ♀
09 19	♀ □ ♄
10 06	☽ ∠ ☉
14 04	☽ ⊥ ☿
15 35	☽ ☌ ♇
15 49	☽ ✶ ♃
16 12	☽ ☌ ♅
19 51	☽ Q ♂
20 15	☽ ✶ ♆
22 Friday	
04 16	☽ ∠ ☿
05 44	☽ ∥ ♅
11 48	☽ Q ♃
12 23	☽ ✶ ☉
12 57	☽ ☍ ♄
15 29	☽ □ ♀
18 54	☽ # ♄
20 50	☉ △ ♄
21 00	☽ ∠ ♆
23 Saturday	
00 02	♃ ✶ ♅
03 04	☽ □ ♂
04 14	☽ ✶ ☿
09 51	☽ Q ☉
11 56	☽ ∥ ♄
12 06	☽ ⊥ ♆
16 41	☿ ⊻ ♂
17 29	☿ ⊥ ♀
17 31	☽ ⊻ ♇
18 14	☽ ⊻ ♅
18 28	☽ □ ♃
22 18	☽ ⊻ ♆
23 42	☽ Q ☿
24 Sunday	
01 18	☽ # ♅
03 47	☽ ⊥ ♇
04 32	☽ ⊥ ♅
15 44	☽ ⊼ ♄
19 00	☽ □ ☉
19 24	☽ ∠ ♇
20 11	☽ ∠ ♅
23 21	☽ △ ♀
25 Monday	
01 00	☉ ∠ ♇
02 12	☽ ± ♄
05 32	☽ □ ☿
09 18	☽ △ ♂
09 30	♀ ⚼ ♆
09 59	☽ # ☿
12 55	☉ ∠ ♅
14 34	♀ Q ♇
18 06	☽ ⚼ ♄
19 57	☽ ∥ ♆
21 58	☽ ✶ ♇
22 49	☽ ✶ ♅
23 45	☽ △ ♃
26 Tuesday	
00 11	♀ Q ♅
02 53	☽ ☌ ♆
04 36	☽ ⚼ ♀
08 47	☽ # ♇
13 36	☽ ⚼ ♂
14 22	☽ # ☉
19 43	☽ Q ♇
20 36	☽ Q ♅
21 09	☽ △ ♄
21 47	☿ ⊻ ♀
22 35	☽ ± ♀
27 Wednesday	
03 26	☽ ⚼ ♃
04 54	☽ △ ☉
06 56	☽ ± ♂
08 14	☽ # ♃
08 55	☽ △ ☿
10 44	☽ ⊼ ♀
10 57	☽ # ♀
18 32	☽ # ♂
18 37	☽ ⊼ ♂
20 22	☽ ± ♃
28 Thursday	
05 04	☽ □ ♇
06 02	☽ □ ♅
07 46	☽ ⊼ ♃
10 03	☽ ⊻ ♆
11 01	☽ ⚼ ☉
11 21	☽ ⚼ ☿
13 43	☉ ☌ ☿
21 22	☽ ⊥ ♆
29 Friday	
03 27	☽ ± ☿
04 59	☽ □ ♄
05 25	☽ ± ☉
14 16	☽ ⊼ ☿
14 28	☽ ∠ ♆
17 48	☽ ⊼ ☉
30 Saturday	
01 14	☽ ☍ ♀
06 35	☽ ☍ ♂
14 20	☽ △ ♇
15 27	☽ Q ♄
15 27	☽ △ ♅
18 02	☽ ☍ ♃
19 22	☽ ✶ ♆
31 Sunday	
07 12	☉ ∥ ♇
14 46	☽ ✶ ♄
18 47	☽ Q ♆
19 39	☽ ⚼ ♇
20 50	☽ ⚼ ♅
21 27	☽ ☍ ☿

All ephemeris data is given at 12.00 UT and the Moon's longitude is additionally given for 24.00 UT

AUGUST 1966

LONGITUDES

Date	Sidereal time h m s	Sun ☉ ° ' "	Moon ☽ ° ' "	Moon ☽ 24.00 ° ' "	Mercury ☿ ° '	Venus ♀ ° '	Mars ♂ ° '	Jupiter ♃ ° '	Saturn ♄ ° '	Uranus ♅ ° '	Neptune ♆ ° '	Pluto ♇ ° '
01	08 38 36	08 ♌ 47 04	10 ♒ 18 30	16 ♒ 11 35	02 ♌ 24	12 ♋ 59	14 ♋ 16	19 ♋ 06	29 ♓ 19	17 ♍ 29	19 ♏ 24	16 ♍ 52
02	08 42 32	09 44 28	22 ♒ 12 54	28 ♒ 12 36	01 R 51	14 11	14 56	19 19	29 R 17	17 32	19 D 24	16 54
03	08 46 29	10 41 52	04 ♓ 10 52	10 ♓ 07 55	01 22	15 24	15 36	19 32	29 15	17 35	19 24	16 56
04	08 50 25	11 39 18	16 ♓ 03 59	21 ♓ 59 23	00 58	16 36	16 15	19 45	29 13	17 38	19 25	16 58
05	08 54 22	12 36 45	27 ♓ 54 27	03 ♈ 49 35	00 40	17 49	16 54	19 58	29 11	17 42	19 25	16 59
06	08 58 19	13 34 12	09 ♈ 45 14	15 ♈ 41 53	00 27	19 01	17 34	20 11	29 08	17 45	19 25	17 01
07	09 02 15	14 31 41	21 ♈ 40 04	27 ♈ 40 23	00 21	20 14	18 13	20 24	29 06	17 48	19 25	17 03
08	09 06 12	15 29 12	03 ♉ 43 24	09 ♉ 49 48	00 D 21	21 27	18 52	20 37	29 03	17 52	19 25	17 05
09	09 10 08	16 26 43	16 ♉ 00 12	22 ♉ 15 15	00 28	22 40	19 32	20 50	29 00	17 55	19 25	17 07
10	09 14 05	17 24 16	28 ♉ 35 36	05 ♊ 01 50	00 42	23 53	20 11	21 03	28 57	17 58	19 26	17 09
11	09 18 01	18 21 51	11 ♊ 34 29	18 ♊ 14 01	01 02	25 06	20 50	21 16	28 55	18 02	19 26	17 11
12	09 21 58	19 19 27	25 ♊ 00 46	01 ♋ 54 56	01 30	26 18	21 29	21 28	28 52	18 05	19 26	17 13
13	09 25 54	20 17 04	08 ♋ 56 33	16 ♋ 05 26	02 04	27 31	22 08	21 41	28 49	18 09	19 27	17 15
14	09 29 51	21 14 43	23 ♋ 21 12	00 ♌ 43 13	02 46	28 45	22 47	21 53	28 45	18 12	19 27	17 17
15	09 33 48	22 12 24	08 ♌ 10 39	15 ♌ 42 26	03 34	29 ♋ 58	23 26	22 06	28 42	18 15	19 28	17 19
16	09 37 44	23 10 05	23 ♌ 17 22	00 ♍ 54 06	04 29	01 ♌ 11	24 05	22 19	28 39	18 19	19 28	17 21
17	09 41 41	24 07 48	08 ♍ 31 13	16 ♍ 07 22	05 31	02 24	24 44	22 31	28 36	18 22	19 29	17 23
18	09 45 37	25 05 32	23 ♍ 41 18	01 ♎ 11 35	06 38	03 37	25 23	22 43	28 32	18 26	19 29	17 25
19	09 49 34	26 03 17	08 ♎ 37 28	15 ♎ 58 05	07 52	04 50	26 02	22 56	28 29	18 30	19 30	17 27
20	09 53 30	27 01 03	23 ♎ 12 50	00 ♏ 21 21	09 12	06 04	26 41	23 08	28 25	18 33	19 30	17 29
21	09 57 27	27 58 50	07 ♏ 23 27	14 ♏ 19 09	10 37	07 17	27 19	23 20	28 22	18 37	19 31	17 31
22	10 01 23	28 56 38	21 ♏ 08 34	27 ♏ 51 57	12 06	08 30	27 58	23 32	28 18	18 40	19 32	17 33
23	10 05 20	29 ♌ 54 28	04 ♐ 29 41	11 ♐ 02 07	13 41	09 44	28 37	23 45	28 14	18 44	19 32	17 35
24	10 09 17	00 ♍ 52 19	17 ♐ 29 43	23 ♐ 52 54	15 20	10 57	29 15	23 57	28 10	18 48	19 33	17 37
25	10 13 13	01 50 11	00 ♑ 12 08	06 ♑ 27 50	17 02	12 11	29 ♋ 54	24 09	28 06	18 51	19 34	17 39
26	10 17 10	02 48 04	12 ♑ 40 24	18 ♑ 50 13	18 48	13 24	00 ♌ 32	24 21	28 03	18 55	19 35	17 41
27	10 21 06	03 45 59	24 ♑ 57 37	01 ♒ 02 52	20 36	14 38	01 11	24 32	27 59	18 59	19 35	17 43
28	10 25 03	04 43 55	07 ♒ 06 16	13 ♒ 08 02	22 27	15 52	01 49	24 44	27 55	19 02	19 36	17 45
29	10 28 59	05 41 52	19 ♒ 08 23	25 ♒ 07 29	24 20	17 05	02 28	24 56	27 50	19 06	19 37	17 47
30	10 32 56	06 39 51	01 ♓ 05 31	07 ♓ 02 39	26 15	18 19	03 06	25 08	27 46	19 10	19 38	17 50
31	10 36 52	07 ♍ 37 51	12 ♓ 59 02	18 ♓ 54 52	28 ♌ 10	19 ♌ 33	03 ♌ 44	25 ♋ 19	27 ♓ 42	19 ♍ 13	19 ♏ 39	17 ♍ 52

MOON

Date	Moon True ☊ ° '	Moon Mean ☊ ° '	Moon ☽ Latitude ° '
01	22 ♉ 08	21 ♉ 24	04 S 54
02	21 R 55	21 21	05 00
03	21 42	21 18	04 53
04	21 31	21 14	04 34
05	21 21	21 11	04 02
06	21 15	21 08	03 20
07	21 11	21 05	02 29
08	21 09	21 02	01 31
09	21 D 09	20 58	00 S 27
10	21 R 09	20 55	00 N 39
11	21 08	20 52	01 46
12	21 05	20 49	02 49
13	21 00	20 46	03 44
14	20 52	20 43	04 27
15	20 42	20 39	04 53
16	20 31	20 36	05 00
17	20 20	20 33	04 45
18	20 11	20 30	04 11
19	20 04	20 27	03 21
20	20 00	20 24	02 16
21	19 58	20 20	01 N 06
22	19 D 58	20 17	00 S 06
23	19 R 58	20 14	01 16
24	19 57	20 11	02 20
25	19 54	20 08	03 16
26	19 49	20 04	04 01
27	19 40	20 01	04 34
28	19 30	19 58	04 54
29	19 17	19 55	05 01
30	19 04	19 52	04 54
31	18 ♉ 51	19 ♉ 49	04 S 35

DECLINATIONS

Date	Sun ☉ ° '	Moon ☽ ° '	Mercury ☿ ° '	Venus ♀ ° '	Mars ♂ ° '	Jupiter ♃ ° '	Saturn ♄ ° '	Uranus ♅ ° '	Neptune ♆ ° '	Pluto ♇ ° '
01	18 N 04	22 S 25	15 N 06	22 N 30	23 N 26	22 N 11	02 S 25	05 N 38	15 S 53	18 N 21
02	17 49	18 50	15 22	22 25	23 22	22 09	02 26	05 36	15 53	18 20
03	17 33	14 32	15 38	22 20	23 18	22 07	02 27	05 35	15 54	18 19
04	17 18	09 42	15 55	22 14	23 14	22 06	02 29	05 34	15 54	18 18
05	17 02	04 S 32	16 11	22 08	23 10	22 04	02 30	05 32	15 54	18 17
06	16 45	00 N 48	16 27	22 00	23 05	22 02	02 31	05 31	15 54	18 16
07	16 29	06 09	16 43	21 53	23 01	22 00	02 32	05 30	15 54	18 15
08	16 12	11 20	16 58	21 44	22 56	21 58	02 33	05 29	15 54	18 14
09	15 55	16 12	17 12	21 35	22 51	21 57	02 35	05 27	15 54	18 14
10	15 37	20 29	17 25	21 26	22 45	21 55	02 36	05 26	15 54	18 13
11	15 20	23 55	17 37	21 15	22 40	21 53	02 37	05 25	15 55	18 12
12	15 02	26 10	17 48	21 04	22 35	21 51	02 39	05 23	15 55	18 11
13	14 44	26 52	17 57	20 53	22 29	21 49	02 40	05 22	15 55	18 10
14	14 25	25 48	18 04	20 41	22 23	21 47	02 42	05 20	15 55	18 09
15	14 07	22 56	18 09	20 28	22 17	21 45	02 43	05 19	15 55	18 08
16	13 48	18 28	18 12	20 15	22 11	21 43	02 45	05 18	15 55	18 07
17	13 29	12 47	18 12	20 01	22 04	21 41	02 46	05 16	15 56	18 06
18	13 10	06 N 21	18 11	19 47	21 58	21 40	02 48	05 15	15 56	18 05
19	12 50	00 S 22	18 06	19 32	21 51	21 38	02 49	05 14	15 56	18 05
20	12 31	06 55	17 59	19 16	21 44	21 36	02 51	05 12	15 56	18 04
21	12 11	12 57	17 49	19 00	21 37	21 34	02 52	05 11	15 56	18 03
22	11 51	18 09	17 36	18 43	21 30	21 32	02 54	05 09	15 57	18 02
23	11 30	22 17	17 21	18 26	21 23	21 30	02 56	05 08	15 57	18 01
24	11 10	25 11	17 02	18 08	21 15	21 28	02 57	05 06	15 57	18 00
25	10 49	26 42	16 41	17 50	21 08	21 26	02 59	05 05	15 57	17 59
26	10 29	26 50	16 17	17 31	21 00	21 24	03 01	05 04	15 57	17 58
27	10 08	25 38	15 50	17 12	20 52	21 23	03 02	05 02	15 58	17 57
28	09 47	23 14	15 21	16 52	20 44	21 20	03 04	05 01	15 58	17 57
29	09 25	19 51	14 49	16 32	20 36	21 18	03 06	04 59	15 59	17 56
30	09 04	15 40	14 16	16 11	20 27	21 16	03 08	04 58	15 59	17 55
31	08 N 43	10 S 55	13 N 39	15 N 50	20 N 19	21 N 14	03 S 09	04 N 56	15 S 59	17 N 54

ZODIAC SIGN ENTRIES

Date	h	m	Planets
03	03	36	☽ ♓
05	16	15	☽ ♈
08	04	38	☽ ♉
10	14	38	☽ ♊
12	20	41	☽ ♋
14	22	50	☽ ♌
15	12	47	♀ ♌
16	22	35	☽ ♍
18	22	05	☽ ♎
20	23	24	☽ ♏
23	03	51	☽ ♐
23	14	18	☉ ♍
25	11	37	☽ ♑
25	15	52	♂ ♌
27	21	56	☽ ♒
30	09	48	☽ ♓

LATITUDES

Date	Mercury ☿ ° '	Venus ♀ ° '	Mars ♂ ° '	Jupiter ♃ ° '	Saturn ♄ ° '	Uranus ♅ ° '	Neptune ♆ ° '	Pluto ♇ ° '
01	04 S 39	00 S 19	00 N 45	00 N 06	02 S 21	00 N 44	01 N 46	14 N 20
04	04 08	00 11	00 47	00 06	02 22	00 44	01 45	14 20
07	03 27	00 S 03	00 49	00 07	02 22	00 44	01 45	14 19
10	02 39	00 N 06	00 50	00 07	02 23	00 44	01 45	14 19
13	01 48	00 14	00 52	00 07	02 24	00 44	01 45	14 18
16	00 58	00 21	00 54	00 08	02 24	00 44	01 45	14 18
19	00 S 13	00 29	00 55	00 08	02 25	00 44	01 44	14 17
22	00 N 27	00 36	00 57	00 08	02 25	00 44	01 44	14 17
25	01 00	00 43	00 58	00 08	02 26	00 44	01 44	14 17
28	01 24	00 49	00 59	00 09	02 26	00 44	01 44	14 17
31	01 N 39	00 N 55	01 N 01	00 N 09	02 S 27	00 N 44	01 N 44	14 N 17

DATA

Julian Date	2439339
Delta T	+37 seconds
Ayanamsa	23° 23' 13"
Synetic vernal point	05° ♓ 43' 46"
True obliquity of ecliptic	23° 26' 43"

MOON'S PHASES, APSIDES AND POSITIONS ☽

Date	h	m	Phase	Longitude ° '	Eclipse Indicator
01	09	05	○	08 ♒ 40	
09	12	56	☾	16 ♉ 29	
16	11	48	●	23 ♌ 10	
23	03	02	☽	29 ♏ 33	
31	00	14	○	07 ♓ 09	

Day	h	m		
04	15	43	Apogee	
17	06	28	Perigee	
31	23	03	Apogee	
06	08	26	0N	
13	09	55	Max dec	26° N 52'
19	10	42	0S	
26	02	11	Max dec	26° S 57'

LONGITUDES

Date	Chiron ⚷ ° '	Ceres ⚳ ° '	Pallas ⚴ ° '	Juno ⚵ ° '	Vesta ⚶ ° '	Black Moon Lilith ⚸ ° '
01	25 ♓ 46	15 ♊ 43	20 ♈ 10	23 ♊ 30	23 ♌ 03	13 ♓ 38
11	25 ♓ 27	19 ♊ 11	20 ♈ 30	29 ♊ 02	27 ♌ 44	14 ♓ 45
21	25 ♓ 05	22 ♊ 18	20 ♈ 15	04 ♋ 27	02 ♍ 27	15 ♓ 53
31	24 ♓ 40	25 ♊ 33	19 ♈ 20	09 ♋ 43	07 ♍ 14	17 ♓ 00

All ephemeris data is given at 12.00 UT and the Moon's longitude is additionally given for 24.00 UT

ASPECTARIAN

01 Monday
03 38 ☽ ⋕ ♂
04 26 ♆ St D
08 53 ☿ ∠ ♅
09 05 ☽ ☍ ☉
11 24 ☽ ⋕ ♀
13 26 ☽ ± ♇
13 47 ☽ ⋕ ♃
14 40 ☽ ± ♅
18 14 ☽ ⊼ ♀
20 16 ☽ ∠ ♄
20 40 ☽ ⊼ ♂

02 Tuesday
01 23 ☽ ⊼ ♇
02 38 ☽ ⊼ ♅
06 07 ☽ ⊼ ♃
06 24 ☽ □ ♆
07 30 ☽ ± ♀
09 18 ☽ ± ♂
09 52 ☿ ∠ ♇
14 08 ☽ ⊥ ♄
15 00 ☽ ⋕ ♇
18 20 ☽ ± ♃
18 25 ☽ ⋕ ☉
21 13 ♃ △ ♆

03 Wednesday
02 07 ☽ ⊻ ♄
03 31 ☽ ⚼ ♀
04 22 ☽ ⚼ ♂
04 44 ☽ ∥ ♆
06 28 ☽ ⋕ ☿
06 33 ☽ ⊼ ☿
12 44 ☽ ⚼ ♃
17 58 ☉ ⊥ ♇
18 12 ☽ ± ☿
20 39 ♀ ☌ ♂

04 Thursday
02 18 ☽ ⊼ ☉
10 51 ☿ ⋕ ♆
11 39 ☉ ⊥ ♅
11 49 ☽ ⚼ ☿
12 24 ☽ △ ♂
13 13 ☽ △ ♀
13 49 ☽ ☍ ♇
15 12 ☽ ☍ ♅
15 30 ☽ ± ☉
18 46 ☽ △ ♆
19 15 ♀ ⚹ ♇
19 37 ☽ △ ♃

05 Friday
07 23 ☽ ⋕ ♅
09 34 ♀ ⚹ ♅
11 21 ☽ ⚼ ☉
14 34 ☽ ☌ ♄
15 10 ♂ ⚹ ♇
17 29 ☽ △ ☿
21 11 ☽ ∥ ♄

06 Saturday
01 11 ☽ ⚼ ♆
04 44 ♀ ∥ ♃
19 24 ☽ ± ♆
19 27 ♂ ⚹ ♅
19 43 ☽ ⋕ ♄
19 44 ♀ △ ♆
20 23 ☽ △ ☉

07 Sunday
01 25 ☉ ∥ ☿
01 32 ☉ ⚼ ♄
02 42 ☽ ⊼ ♇
04 12 ☽ ⊼ ♅
04 40 ☽ □ ♂
07 29 ☽ ⊼ ♆
08 48 ☽ □ ♀
09 06 ☽ ∥ ♅
09 25 ☽ □ ♃
14 47 ☽ ± ♇
16 04 ♀ ☌ ♃
16 18 ☽ ± ♅
23 41 ☿ St D

08 Monday
02 46 ☽ ⊻ ♄
05 18 ☽ □ ☿
08 45 ☽ ⚼ ♇
10 17 ☽ ⚼ ♅
14 36 ☽ ⊥ ♄
18 33 ☽ Q ♂
21 48 ☽ Q ♃

09 Tuesday
00 29 ☽ Q ♀
08 08 ☽ ∠ ♄
08 09 ♂ △ ♆
10 29 ☽ ⋕ ♆
10 36 ☽ ∥ ☉
12 36 ☉ ⋕ ♆
12 56 ☽ □ ☉
14 09 ☽ △ ♇
15 42 ☽ △ ♅
16 49 ☽ Q ☿
17 35 ☽ ∥ ☿
18 35 ☽ ☍ ♆
19 10 ☽ ⚹ ♂
21 27 ☽ ⚹ ♃
22 50 ☽ ∥ ♇

10 Wednesday
02 08 ☽ ⚹ ♀
05 19 ☉ ⊻ ♇
12 41 ☽ ⚹ ♄
16 02 ☽ ⚹ ☿
17 39 ☽ ∥ ♀
21 02 ☽ ∥ ♃

11 Thursday
00 56 ☽ ∠ ♂
01 42 ☽ Q ☉
02 06 ☽ ∠ ♃
02 37 ☽ ∥ ♂
03 04 ☉ ⊻ ♅
09 01 ☽ ∠ ♀
10 48 ☽ Q ♄
18 12 ☽ ⊥ ♂
18 46 ☽ ⊥ ♃
20 20 ☽ ∠ ☿
22 08 ☽ □ ♇
23 41 ☽ □ ♅

12 Friday
01 10 ☽ ⚹ ☉
02 09 ☽ ⊼ ♆
02 52 ☽ ⊥ ♀
05 28 ☽ ⊻ ♂
05 39 ☽ ⊻ ♃
11 06 ♂ ☌ ♃
12 45 ☽ ± ♆
12 53 ☽ ⊥ ☿
14 29 ☽ ⊻ ♀
14 55 ☉ □ ♆
18 41 ☽ □ ♄
23 44 ☽ ⊻ ☿

13 Saturday
04 21 ☽ ⚼ ♆
05 19 ☽ ∠ ☉
05 42 ☽ Q ♇
07 13 ☽ Q ♅
18 39 ☿ ∠ ♇
21 38 ☽ ⊥ ☉

14 Sunday
01 57 ☽ ⚹ ♇
03 28 ☽ ⚹ ♅
05 35 ☽ △ ♆
08 17 ☽ ⊻ ☉
09 34 ☽ ☌ ♃
11 02 ☽ ☌ ♂
12 17 ♀ △ ♄
20 47 ☽ △ ♄
21 35 ☽ ☌ ♀

15 Monday
02 25 ☿ ∠ ♅
02 33 ☽ ∠ ♇
04 04 ☽ ∠ ♅
04 09 ☽ ☌ ☿
08 24 ☿ ∥ ♇
08 38 ☉ ⊻ ♃
16 13 ☽ ∥ ♂
17 01 ☽ ⊥ ♇
18 32 ☽ ⊥ ♅
19 15 ☽ ∥ ♃
20 47 ☽ ⚼ ♄
23 45 ☉ ± ♄

16 Tuesday
02 34 ☽ ⊻ ♇
02 54 ☽ ∥ ♀
04 06 ☽ ⊻ ♅
05 57 ☽ □ ♆
10 26 ☽ ⊻ ♃
11 00 ☽ ± ♄
11 48 ☽ ☌ ☉
13 16 ☽ ∥ ☿
13 19 ☽ ⊻ ♂
13 38 ☽ ∥ ♇
20 02 ☽ ⊥ ♃
20 25 ☽ ⊼ ♄
23 12 ☽ ⊥ ♂
23 16 ☽ ⋕ ♆

17 Wednesday
01 31 ☽ ⊻ ♀
06 54 ☽ ⊻ ☿
09 06 ☽ ∥ ☉
10 21 ☽ Q ♆
10 24 ☽ ∠ ♃
11 33 ♀ ∠ ♇
11 47 ☽ ⊥ ♀
14 00 ☽ ∠ ♂
17 05 ☽ ⊥ ☿

18 Thursday
02 01 ☽ ☌ ♇
03 15 ☽ ∠ ♀
03 38 ☽ ☌ ♅
05 19 ☽ ⚹ ♆
08 10 ♀ ∠ ♅
08 28 ☽ ∠ ☿
10 27 ☽ ⚹ ♃
14 23 ☽ ⊻ ☉
14 50 ☽ ⚹ ♂
15 58 ☽ ∥ ♅
19 43 ☽ ☍ ♄

19 Friday
00 39 ☽ ⊥ ☉
00 41 ☽ ⋕ ♄
05 19 ☽ ∠ ♆
05 19 ☽ ⚹ ♀
05 56 ☽ Q ♃
10 14 ☉ ⊻ ♂
10 40 ☽ ⚹ ☿
11 00 ☽ Q ♂
16 14 ☽ ∠ ☉
19 38 ☿ ∥ ♇
19 57 ☽ ⊥ ♆
20 54 ☽ ∥ ♄

20 Saturday
02 28 ☽ ⊻ ♇
02 40 ☽ Q ♀
04 14 ☽ ⊻ ♅
05 37 ☽ ⋕ ♅
05 50 ☽ ⊻ ♆
08 17 ☽ Q ☿
11 52 ☽ □ ♃
12 26 ☽ ⊥ ♇
14 15 ☽ ⊥ ♅
18 04 ☽ □ ♂
18 50 ☽ ⚹ ☉
20 42 ☽ ⊼ ♄

21 Sunday
03 38 ☽ ∠ ♇
05 30 ☽ ∠ ♅
06 50 ☽ ± ♄
08 58 ☽ ⋕ ☉
11 48 ☽ □ ♀
16 48 ☽ Q ☉
18 13 ☽ □ ☿
20 52 ☉ ⊼ ♄
22 17 ☽ ⚼ ♄

22 Monday
01 17 ☽ ∥ ♆
02 58 ☿ ⊥ ♇
04 12 ♂ ∥ ♃
05 38 ☽ ⚹ ♇
07 37 ☽ ⚹ ♅
09 08 ☽ ☌ ♆
09 24 ☽ ⋕ ☿
11 24 ☽ ⋕ ♇
14 46 ☽ ⋕ ♀
16 20 ☽ △ ♃
21 06 ☿ ⊥ ♅
23 14 ♂ △ ♄

23 Tuesday
00 43 ☽ △ ♄
00 48 ☽ △ ♂
03 02 ☽ □ ☉
03 04 ☽ Q ♇
05 08 ☽ Q ♅
05 34 ☿ ⚼ ♄
06 15 ☽ ⋕ ♂
06 49 ☉ ⊥ ♃
06 51 ☽ ⋕ ♃
19 54 ☽ ⚼ ♃
22 35 ☽ △ ♀

24 Wednesday
05 39 ☽ ⚼ ♂
07 21 ☽ △ ☿
12 14 ☽ □ ♇
12 51 ☽ ± ♃
14 26 ☽ □ ♅
15 51 ☽ ⊻ ♆
23 19 ♀ ∥ ♇
23 23 ☽ ± ♂

25 Thursday
00 19 ☽ ⊼ ♃
01 20 ♀ ⊥ ♇
03 10 ☽ ⊥ ♆
05 38 ☽ ⚼ ♀
08 02 ☽ □ ♄
11 23 ☽ ⊼ ♂
15 23 ☽ △ ☉
16 04 ☽ ⚼ ☿
20 21 ☽ ∠ ♆
20 39 ☿ ⊻ ♇

26 Friday
00 42 ☽ ± ♀
01 54 ♀ ⊥ ♅
05 15 ♀ ⚼ ♄
12 17 ☽ ± ☿
13 35 ☽ ⊼ ♀
13 39 ☿ ⊻ ♅
18 31 ☽ Q ♄
21 47 ☽ △ ♇
22 31 ☿ □ ♆
22 50 ☽ ⚼ ☉

27 Saturday
00 13 ☽ △ ♅
01 28 ☽ ⚹ ♆
01 58 ☽ ⊼ ☿
05 20 ☿ ⋕ ♆
11 10 ☽ ☍ ♃
17 54 ☽ ⚹ ♄
18 00 ☽ ± ☉

28 Sunday
00 57 ☽ ☍ ♂
01 05 ☽ Q ♆
03 21 ☽ ⚼ ♇
05 12 ☿ ± ♄
05 53 ☽ ⚼ ♅
06 53 ☽ ⊼ ☉
21 17 ☽ ± ♇
23 29 ☽ ∠ ♄
23 52 ☽ ± ♅

29 Monday
02 22 ☽ ⋕ ♃
06 58 ☽ ⋕ ♂
07 25 ☽ ☍ ♀
09 18 ☽ ⊼ ♇
11 55 ☽ ⊼ ♅
12 58 ☽ □ ♆
17 23 ☽ ⊥ ♄
20 22 ☿ ⊻ ♃
23 32 ☽ ⋕ ♇
23 48 ☽ ⊼ ♃

30 Tuesday
00 24 ☽ ☍ ☿
01 14 ♂ ∠ ♇
02 11 ♀ ⊻ ♇
05 21 ☽ ⊻ ♄
08 58 ☽ ⋕ ♀
10 19 ☽ ∥ ♆
12 04 ☽ ± ♃
16 16 ☽ ⊼ ♂
20 27 ☽ ⋕ ☿

31 Wednesday
00 14 ☽ ☍ ☉
01 58 ♀ ⋕ ♆
05 04 ☽ ± ♂
05 24 ♀ ⊻ ♅
06 21 ☿ ⊼ ♄
06 32 ☽ ⚼ ♃
12 31 ☉ Q ♆
14 08 ♀ □ ♆
21 54 ☽ ☍ ♇
23 16 ☽ ⋕ ☉

SEPTEMBER 1966

Date	Sidereal time h m s	Sun ☉ ° ' "	Moon ☽ ° ' "	Moon ☽ 24.00 ° ' "	Mercury ☿ ° '	Venus ♀ ° '	Mars ♂ ° '	Jupiter ♃ ° '	Saturn ♄ ° '	Uranus ♅ ° '	Neptune ♆ ° '	Pluto ♇ ° '
01	10 40 49	08 ♍ 35 53	24 ♓ 50 20	00 ♈ 45 40	00 ♍ 07	20 ♌ 46	04 ♌ 22	25 ♋ 31	27 ♓ 38	19 ♍ 17	19 ♏ 40	17 ♍ 54
02	10 44 46	09 33 57	06 ♈ 41 07	12 ♈ 36 58	02 04	22 00	05 00	25 42	27 R 34	19 21	19 41	17 56
03	10 48 42	10 32 02	18 ♈ 33 32	24 ♈ 31 14	04 01	23 14	05 39	25 53	27 29	19 25	19 42	17 58
04	10 52 39	11 30 10	00 ♉ 30 26	06 ♉ 31 38	05 58	24 28	06 17	26 05	27 25	19 28	19 43	18 00
05	10 56 35	12 28 19	12 ♉ 35 19	18 ♉ 42 01	07 55	25 42	06 55	26 16	27 21	19 32	19 44	18 02
06	11 00 32	13 26 30	24 ♉ 52 19	01 ♊ 06 46	09 51	26 56	07 33	26 27	27 16	19 36	19 45	18 05
07	11 04 28	14 24 43	07 ♊ 26 00	13 ♊ 50 33	11 47	28 10	08 11	26 38	27 12	19 39	19 47	18 07
08	11 08 25	15 22 59	20 ♊ 21 00	26 ♊ 57 50	13 42	29 ♌ 24	08 49	26 49	27 07	19 43	19 48	18 09
09	11 12 21	16 21 16	03 ♋ 41 29	10 ♋ 32 16	15 37	00 ♍ 38	09 27	27 00	27 03	19 47	19 49	18 11
10	11 16 18	17 19 36	17 ♋ 30 20	24 ♋ 35 43	17 30	01 52	10 04	27 11	26 58	19 51	19 50	18 13
11	11 20 15	18 17 57	01 ♌ 48 12	09 ♌ 07 25	19 23	03 06	10 42	27 22	26 54	19 55	19 51	18 15
12	11 24 11	19 16 21	16 ♌ 32 41	24 ♌ 03 11	21 14	04 21	11 20	27 32	26 49	19 59	19 53	18 18
13	11 28 08	20 14 47	01 ♍ 37 49	09 ♍ 15 21	23 05	05 35	11 58	27 43	26 44	20 02	19 54	18 20
14	11 32 04	21 13 14	16 ♍ 54 26	24 ♍ 33 39	24 54	06 49	12 35	27 53	26 40	20 06	19 55	18 22
15	11 36 01	22 11 44	02 ♎ 11 34	09 ♎ 46 50	26 43	08 03	13 13	28 03	26 35	20 10	19 57	18 24
16	11 39 57	23 10 15	17 ♎ 18 15	24 ♎ 44 47	28 ♍ 30	09 18	13 51	28 14	26 31	20 14	19 58	18 26
17	11 43 54	24 08 48	02 ♏ 05 34	09 ♏ 20 01	00 ♎ 16	10 32	14 28	28 24	26 26	20 17	20 00	18 28
18	11 47 50	25 07 22	16 ♏ 27 42	23 ♏ 28 25	02 01	11 47	15 05	28 34	26 21	20 21	20 01	18 31
19	11 51 47	26 05 59	00 ♐ 22 10	07 ♐ 09 04	03 46	13 01	15 43	28 44	26 17	20 25	20 02	18 33
20	11 55 44	27 04 37	13 ♐ 49 24	20 ♐ 23 31	05 29	14 15	16 20	28 54	26 12	20 29	20 04	18 35
21	11 59 40	28 03 17	26 ♐ 51 51	03 ♑ 14 54	07 11	15 30	16 58	29 03	26 07	20 32	20 06	18 37
22	12 03 37	29 ♍ 01 58	09 ♑ 33 09	15 ♑ 47 09	08 52	16 44	17 35	29 13	26 03	20 36	20 07	18 39
23	12 07 33	00 ♎ 00 41	21 ♑ 57 25	28 ♑ 04 26	10 32	17 59	18 12	29 23	25 58	20 40	20 09	18 41
24	12 11 30	00 59 26	04 ♒ 08 42	10 ♒ 10 40	12 11	19 14	18 49	29 32	25 53	20 44	20 10	18 43
25	12 15 26	01 58 13	16 ♒ 10 45	22 ♒ 09 20	13 50	20 28	19 26	29 41	25 49	20 47	20 12	18 45
26	12 19 23	02 57 01	28 ♒ 06 46	04 ♓ 03 22	15 27	21 43	20 03	29 ♋ 50	25 44	20 51	20 14	18 48
27	12 23 19	03 55 51	09 ♓ 59 26	15 ♓ 55 12	17 03	22 57	20 40	00 ♌ 00	25 40	20 55	20 15	18 50
28	12 27 16	04 54 43	21 ♓ 50 55	27 ♓ 46 48	18 39	24 12	21 17	00 08	25 35	20 59	20 17	18 52
29	12 31 13	05 53 37	03 ♈ 43 04	09 ♈ 39 55	20 13	25 27	21 54	00 17	25 31	21 02	20 19	18 54
30	12 35 09	06 ♎ 52 33	15 ♈ 37 35	21 ♈ 36 15	21 ♎ 47	26 ♍ 41	22 ♌ 31	00 ♌ 26	25 ♓ 26	21 ♍ 06	20 ♏ 20	18 ♍ 56

Date	Moon True ☊ ° '	Moon Mean ☊ ° '	Moon ☽ Latitude ° '
01	18 ♉ 40	19 ♉ 45	04 S 04
02	18 R 31	19 42	03 22
03	18 24	19 39	02 31
04	18 20	19 36	01 33
05	18 19	19 33	00 S 30
06	18 D 19	19 30	00 N 35
07	18 19	19 26	01 40
08	18 R 19	19 23	02 42
09	18 18	19 20	03 38
10	18 15	19 17	04 22
11	18 09	19 14	04 53
12	18 01	19 10	05 05
13	17 53	19 07	04 56
14	17 44	19 04	04 27
15	17 37	19 01	03 38
16	17 31	18 58	02 34
17	17 28	18 55	01 22
18	17 27	18 51	00 N 05
19	17 D 27	18 48	01 S 09
20	17 29	18 45	02 17
21	17 29	18 42	03 16
22	17 R 29	18 39	04 03
23	17 26	18 36	04 38
24	17 22	18 32	05 00
25	17 15	18 29	05 08
26	17 07	18 26	05 02
27	16 59	18 23	04 44
28	16 51	18 20	04 13
29	16 44	18 16	03 31
30	16 ♉ 38	18 ♉ 13	02 S 40

DECLINATIONS

Date	Sun ☉ ° '	Moon ☽ ° '	Mercury ☿ ° '	Venus ♀ ° '	Mars ♂ ° '	Jupiter ♃ ° '	Saturn ♄ ° '	Uranus ♅ ° '	Neptune ♆ ° '	Pluto ♇ ° '
01	08 N 21	05 S 47	13 N 02	15 N 29	20 N 10	21 N 12	03 S 11	04 N 55	15 S 59	17 N 53
02	07 59	00 S 26	12 22	15 07	20 02	21 10	03 13	04 53	16 00	17 52
03	07 37	04 N 57	11 41	14 44	19 53	21 08	03 15	04 52	16 00	17 51
04	07 15	10 12	10 59	14 21	19 44	21 06	03 17	04 50	16 00	17 51
05	06 53	15 08	10 15	13 58	19 35	21 04	03 19	04 49	16 01	17 50
06	06 31	19 33	09 31	13 34	19 26	21 02	03 20	04 47	16 01	17 49
07	06 08	23 12	08 46	13 10	19 16	21 00	03 22	04 46	16 02	17 48
08	05 46	25 47	08 00	12 46	19 07	20 58	03 24	04 44	16 02	17 47
09	05 23	27 01	07 14	12 21	18 57	20 56	03 26	04 43	16 03	17 46
10	05 00	26 38	06 27	11 56	18 48	20 54	03 28	04 41	16 03	17 46
11	04 38	24 31	05 40	11 30	18 38	20 52	03 30	04 40	16 03	17 45
12	04 15	20 43	04 52	11 05	18 28	20 50	03 32	04 38	16 03	17 44
13	03 52	15 30	04 05	10 39	18 18	20 48	03 34	04 37	16 04	17 43
14	03 29	09 16	03 17	10 12	18 08	20 46	03 36	04 35	16 04	17 43
15	03 06	02 N 28	02 29	09 45	17 58	20 44	03 37	04 34	16 05	17 42
16	02 43	04 S 25	01 42	09 18	17 47	20 42	03 39	04 32	16 05	17 41
17	02 20	10 56	00 55	08 51	17 37	20 40	03 41	04 31	16 06	17 40
18	01 56	16 41	00 N 08	08 24	17 26	20 38	03 43	04 30	16 06	17 39
19	01 33	21 21	00 S 39	07 56	17 16	20 37	03 45	04 28	16 06	17 39
20	01 10	24 44	01 26	07 28	17 05	20 35	03 47	04 27	16 07	17 38
21	00 46	26 40	02 12	07 00	16 54	20 33	03 49	04 25	16 07	17 37
22	00 N 23	27 09	02 58	06 32	16 43	20 31	03 51	04 24	16 08	17 36
23	00 00	26 14	03 43	06 03	16 32	20 29	03 53	04 22	16 08	17 36
24	00 S 24	24 05	04 29	05 34	16 21	20 27	03 54	04 21	16 09	17 35
25	00 47	20 53	05 13	05 05	16 10	20 25	03 56	04 19	16 09	17 34
26	01 10	16 51	05 57	04 36	15 59	20 24	03 58	04 18	16 10	17 33
27	01 34	12 12	06 41	04 07	15 47	20 22	04 00	04 16	16 10	17 33
28	01 57	07 06	07 24	03 38	15 36	20 20	04 02	04 15	16 11	17 32
29	02 20	01 45	08 06	03 08	15 25	20 18	04 03	04 13	16 11	17 31
30	02 S 44	03 N 42	08 S 48	02 N 39	15 N 13	20 N 17	04 S 05	04 N 12	16 S 12	17 N 31

ZODIAC SIGN ENTRIES

Date	h	m	Planets
01	10	35	☿ ♍
01	22	27	☽ ♈
04	10	59	☽ ♉
06	21	52	☽ ♊
08	23	40	♀ ♍
09	05	26	☽ ♋
11	09	01	☽ ♌
13	09	26	☽ ♍
15	08	33	☽ ♎
17	08	19	☿ ♎
17	08	34	☽ ♏
19	11	21	☽ ♐
21	17	52	☽ ♑
23	11	43	☉ ♎
24	03	48	☽ ♒
26	15	48	☽ ♓
27	13	19	♃ ♌
29	04	29	☽ ♈

LATITUDES

Date	Mercury ☿ ° '	Venus ♀ ° '	Mars ♂ ° '	Jupiter ♃ ° '	Saturn ♄ ° '	Uranus ♅ ° '	Neptune ♆ ° '	Pluto ♇ ° '
01	01 N 42	00 N 57	01 N 02	00 N 09	02 S 27	00 N 44	01 N 44	14 N 17
04	01 47	01 03	01 03	00 10	02 27	00 44	01 44	14 17
07	01 45	01 08	01 05	00 10	02 28	00 44	01 43	14 17
10	01 38	01 12	01 07	00 10	02 28	00 44	01 43	14 17
13	01 27	01 16	01 08	00 11	02 28	00 44	01 43	14 17
16	01 12	01 19	01 10	00 11	02 28	00 44	01 43	14 17
19	00 55	01 22	01 11	00 12	02 28	00 44	01 43	14 18
22	00 36	01 24	01 13	00 12	02 29	00 44	01 43	14 18
25	00 N 16	01 26	01 14	00 12	02 29	00 44	01 42	14 18
28	00 S 06	01 27	01 16	00 13	02 29	00 44	01 42	14 19
31	00 S 27	01 N 27	01 N 17	00 N 13	02 S 29	00 N 44	01 N 42	14 N 19

DATA

Julian Date	2439370
Delta T	+37 seconds
Ayanamsa	23° 23' 17"
Synetic vernal point	05° ♓ 43' 43"
True obliquity of ecliptic	23° 26' 44"

MOON'S PHASES, APSIDES AND POSITIONS ☽

Date	h	m	Phase	Longitude ° '	Eclipse Indicator
08	02	07	☾	14 ♊ 59	
14	19	13	●	21 ♍ 31	
21	14	25	☽	28 ♐ 09	
29	16	47	○	06 ♈ 05	

Day	h	m		
14	16	22	Perigee	
28	00	44	Apogee	
02	13	57	0N	
09	18	36	Max dec	27° N 05'
15	20	32	0S	
22	07	55	Max dec	27° S 10'
29	19	44	0N	

LONGITUDES

Date	Chiron ⚷ ° '	Ceres ⚳ ° '	Pallas ⚴ ° '	Juno ⚵ ° '	Vesta ⚶ ° '	Black Moon Lilith ⚸ ° '
01	24 ♓ 38	25 ♊ 51	19 ♈ 12	10 ♋ 14	07 ♍ 43	17 ♓ 06
11	24 ♓ 11	28 ♊ 40	17 ♈ 34	15 ♋ 19	12 ♍ 31	18 ♓ 13
21	23 ♓ 43	01 ♋ 11	15 ♈ 19	20 ♋ 11	17 ♍ 22	19 ♓ 20
31	23 ♓ 16	03 ♋ 19	12 ♈ 37	24 ♋ 47	22 ♍ 13	20 ♓ 27

All ephemeris data is given at 12.00 UT and the Moon's longitude is additionally given for 24.00 UT

ASPECTARIAN

01 Thursday
00 18 ☽ bQ ♂
00 41 ☽ ☍ ♅
01 31 ☽ △ ♆
02 49 ☽ ⊼ ♀
08 22 ♂ ∠ ♅
13 23 ☽ △ ♃
15 57 ☽ ⋕ ♅
16 22 ☽ ± ♀
17 38 ☽ ☌ ♄
23 39 ☽ ∥ ♄

02 Friday
00 48 ☽ ⊼ ☿
03 50 ♀ ± ♄
07 04 ☿ ⊥ ♃
07 57 ☽ bQ ♆
08 25 ☽ △ ♂
12 43 ☽ bQ ♀
15 20 ☽ ± ☿
18 21 ☽ ⊼ ☉

03 Saturday
02 11 ☽ ± ♆
04 22 ☽ ⋕ ♄
07 33 ☽ ± ☉
10 48 ☽ ⊼ ♇
11 39 ☽ ∥ ♅
13 06 ☽ bQ ☿
13 43 ☽ ⊼ ♅
14 18 ☽ ⊼ ♆
22 30 ☽ △ ♀
22 55 ☽ ± ♇
23 00 ☉ ∠ ♃
23 19 ☽ ∥ ☉

04 Sunday
01 51 ☽ ± ♅
02 59 ☽ □ ♃
03 16 ☽ bQ ☉
05 51 ☽ ⊻ ♄
15 14 ☽ ∥ ☿
17 00 ☽ bQ ♇
17 40 ☿ ⊻ ♂
17 46 ☽ ⊥ ♄
19 57 ☽ bQ ♅

05 Monday
00 09 ☽ □ ♂
00 59 ☽ △ ☿
06 34 ☽ ∥ ♀
09 46 ☿ Q ♆
11 31 ☽ ∠ ♄
11 45 ☽ △ ☉
15 21 ☽ Q ♃
16 31 ☽ ⋕ ♆
22 44 ☽ △ ♇

06 Tuesday
01 01 ♀ ⊻ ♃
01 42 ☽ △ ♅
02 03 ☽ ☍ ♆
02 09 ☽ ∥ ♇
11 17 ☽ ∥ ♂
13 22 ☽ Q ♂
15 06 ☽ ⚹ ♃
16 25 ☽ □ ♀
16 36 ☽ ⚹ ♄
18 12 ♀ ⊼ ♄
19 29 ☉ ⊥ ♂
20 56 ☽ ∥ ♃

07 Wednesday
09 55 ☿ ∠ ♃
13 29 ☽ ⚹ ♂
15 18 ☽ Q ♄
20 00 ☽ ∠ ♃
21 36 ☽ □ ☿

08 Thursday
02 07 ☽ □ ☉
06 01 ☽ Q ♀
07 57 ☽ □ ♇
10 51 ☽ □ ♅
10 59 ☽ ⊼ ♆
12 52 ☽ ⊥ ♃
18 37 ☽ ∠ ♂
21 55 ☽ ± ♆
23 54 ☽ ⊻ ♃

09 Friday
00 13 ☽ □ ♄
06 01 ☽ ⚹ ♀
08 48 ☿ ⊥ ♂
11 32 ☽ ⊥ ♂
11 50 ☽ Q ☿
13 16 ☽ Q ☉
13 59 ☽ bQ ♆
16 13 ♃ △ ♄
16 24 ☽ Q ♇
19 14 ☽ Q ♅
22 35 ☽ ⊻ ♂

10 Saturday
04 28 ♅ ⚹ ♆
07 22 ☉ ☌ ☿
10 49 ☽ ∠ ♀
11 40 ☽ ⚹ ☉
12 00 ☽ ⚹ ☿
13 13 ☽ ⚹ ♇
15 58 ☽ △ ♆
16 00 ☽ ⚹ ♅
21 20 ☿ ☌ ♇

11 Sunday
03 28 ☽ ⊥ ♀
03 53 ☽ △ ♄
04 32 ☽ ☌ ♃
09 28 ☉ ∥ ♅
10 54 ☉ ☌ ♇
14 21 ☽ ⊻ ♀
14 24 ☽ ∠ ♇
14 39 ☽ ∠ ☉
16 52 ☽ ∠ ☿
17 08 ☽ ∠ ♅
17 44 ♀ ⊥ ♃
18 14 ☿ ⚹ ♆
19 07 ☿ ☌ ♅

12 Monday
03 13 ☽ ☌ ♂
04 25 ☽ bQ ♄
05 08 ☽ ⊥ ♇
06 21 ☽ ⊥ ☉
07 51 ☽ ⊥ ♅
09 36 ☽ ⊥ ☿
11 23 ☽ ∥ ♃
14 49 ☽ ⊻ ♇
16 41 ☽ ⊻ ☉
17 21 ☽ □ ♆
17 31 ☽ ⊻ ♅
18 49 ☽ ± ♄
19 09 ☿ ∥ ♅
20 34 ☽ ⊻ ☿
23 27 ☽ ∥ ♂

13 Tuesday
02 27 ☽ ∥ ♇
03 17 ☉ ⚹ ♆
04 18 ☽ ⊼ ♄
04 30 ♂ bQ ♄
05 44 ☽ ⊻ ♃
06 31 ☉ ☌ ♅
09 39 ☽ ⋕ ♆
15 19 ☽ ⊥ ♃
18 46 ☽ ☌ ♀
21 53 ☽ Q ♆

14 Wednesday
00 25 ☉ ∥ ☿
02 53 ♂ ⊥ ♇
03 02 ☿ ⋕ ♄
04 56 ☽ ⊻ ♂
05 36 ☉ ⋕ ♄
05 37 ☽ ∠ ♃
08 18 ☽ ∥ ♀
14 17 ☽ ☌ ♇
14 45 ☽ ⊥ ♂
16 44 ☽ ⚹ ♆
17 02 ☽ ☌ ♅
19 13 ☽ ☌ ☉

15 Thursday
02 13 ☽ ☌ ☿
03 14 ☽ ☍ ♄
04 39 ☽ ∥ ♅
05 25 ☽ ⚹ ♃
05 28 ☽ ∠ ♂
07 58 ☽ ⋕ ♄
09 39 ☽ ∥ ☉
09 48 ♀ Q ♆
10 26 ☿ ☍ ♄
11 53 ☽ ∥ ☿
16 21 ☽ ∠ ♆
22 06 ☽ ⊻ ♀

16 Friday
00 35 ☽ Q ♃
03 26 ☽ ⋕ ☿
06 13 ☽ ⚹ ♂
06 18 ☽ ⋕ ☉
06 39 ☽ ⊥ ♆
07 58 ☿ ⚹ ♃
08 30 ☽ ⊥ ♀
09 17 ☽ ∥ ♄
12 26 ☽ ⋕ ♅
13 49 ☽ ⊻ ♇
16 17 ☽ ⊻ ♆
16 43 ☽ ⊻ ♅
19 09 ☿ ∠ ♂
22 07 ☽ ⊻ ☉
23 32 ☽ ⊥ ♇

17 Saturday
00 18 ☽ ∠ ♀
02 23 ☽ Q ♂
02 28 ☽ ⊥ ♅
02 47 ☽ ⊼ ♄
03 58 ♂ ∥ ♇
04 26 ♂ ⊥ ♅
04 38 ☽ ⋕ ♀
05 52 ☽ □ ♃
08 35 ☽ ⊥ ☉
08 36 ☽ ⊻ ☿
12 33 ☽ ± ♄
14 17 ☽ ∠ ♇
17 18 ☽ ∠ ♅
19 51 ☽ ⊥ ☿

18 Sunday
00 32 ☽ ∠ ☉
03 20 ☽ ⚹ ♀
03 26 ☽ bQ ♄
09 23 ☽ ∥ ♆
09 34 ☽ □ ♂
13 05 ☽ ∠ ☿
15 24 ☽ ⋕ ♂
15 29 ☽ ⚹ ♇
16 32 ☽ ⋕ ♇
18 04 ☽ ☌ ♆
18 40 ☽ ⚹ ♅

19 Monday
01 45 ☽ Q ♀
03 59 ☽ ⚹ ☉
04 54 ☽ △ ♄
07 44 ☽ ⋕ ♃
09 06 ☽ △ ♃
12 18 ☽ Q ♇
15 37 ☽ Q ♅
16 02 ☉ ☍ ♄
18 51 ☽ ⚹ ☿

20 Tuesday
02 46 ☽ Q ☉
03 56 ♀ ∠ ♃
06 07 ☿ ∠ ♆
06 27 ☉ ⋕ ☿
12 08 ☽ bQ ♃
12 52 ☽ □ ♀
16 48 ☽ △ ♂
19 39 ☽ Q ☿
20 42 ☽ □ ♇
23 25 ☽ ⊻ ♆

21 Wednesday
00 13 ☽ □ ♅
04 50 ☽ ± ♃
10 33 ☽ ⊥ ♆
10 37 ☽ □ ♄
14 25 ☽ □ ☉
16 09 ☽ ⊼ ♃
22 03 ☽ bQ ♂

22 Thursday
03 32 ☽ ∠ ♆
10 29 ☽ □ ☿
16 06 ☽ ± ♂
17 24 ☉ ⚹ ♃
20 35 ☽ Q ♄

23 Friday
03 24 ☽ △ ♀
04 18 ☽ ⊼ ♂
05 37 ☽ △ ♇
08 27 ☽ ⚹ ♆
09 28 ☽ △ ♅
17 01 ☿ ∥ ♄
19 48 ☽ ⚹ ♄
20 26 ♀ ⊻ ♂

24 Saturday
01 26 ☿ Q ♃
02 00 ♀ ☌ ♇
02 45 ☽ ☍ ♃
05 12 ☽ △ ☉
07 57 ♂ ⊻ ♇
07 57 ☿ ⋕ ♅
08 05 ☽ Q ♆
11 10 ☽ bQ ♇
12 11 ☽ bQ ♀
15 10 ☽ bQ ♅

25 Sunday
01 20 ☽ ∠ ♄
05 08 ☽ ± ♇
06 33 ☽ △ ☿
06 39 ♀ ⚹ ♆
08 11 ☽ ± ♀
09 12 ☽ ± ♅
09 27 ☿ ⋕ ♀
13 41 ♂ ⋕ ♆
13 44 ☽ bQ ☉
15 00 ☽ ⋕ ♃
17 11 ☽ ⊼ ♇
17 34 ☿ ⊥ ♆
18 33 ♀ ☌ ♅
18 54 ☽ ☍ ♂
19 15 ☽ ⊥ ♄
20 05 ☽ □ ♆
21 18 ☽ ⊼ ♅
21 37 ☽ ⊼ ♀

26 Monday
00 54 ♂ ± ♄
07 14 ☽ ⊻ ♄
08 05 ☽ ⋕ ♇
09 27 ☽ ± ☉
15 32 ☽ ⊼ ♃
15 46 ☽ ∥ ♆
16 57 ☽ ⋕ ♂
17 27 ☽ bQ ☿
18 49 ♂ □ ♆
22 39 ☽ ⊼ ☉

27 Tuesday
03 49 ☽ ± ♃
03 59 ♀ ∥ ♅
14 29 ☽ ± ☿
17 30 ♀ ⋕ ♄
22 15 ☽ bQ ♃
22 24 ♂ ⊻ ♅

28 Wednesday
04 31 ☽ ⊼ ☿
05 56 ☽ ☍ ♇
08 49 ☽ △ ♆
10 14 ☽ ☍ ♅
10 49 ☽ ⊼ ♂
10 50 ☽ ∥ ☿
15 22 ☿ ⊻ ♇
17 19 ☽ ☍ ♀
19 31 ☽ ☌ ♄
21 18 ☉ ∠ ♆
23 37 ☽ ± ♂

29 Thursday
00 58 ☽ ⋕ ♅
01 49 ☽ ∥ ♄
04 59 ☽ △ ♃
05 15 ☽ ⋕ ♀
09 34 ☽ ∥ ☉
13 09 ♀ ☍ ♄
13 22 ☿ ⊻ ♆
15 13 ☽ bQ ♆
16 47 ☽ ☍ ☉
18 47 ☽ bQ ♂

30 Friday
01 01 ☿ ⊻ ♅
07 26 ☽ ⋕ ☉
07 46 ☽ ∥ ♀
09 24 ☽ ± ♆
09 39 ☉ ⋕ ♀
13 46 ☽ ⋕ ♄
14 14 ☽ ∥ ♅
18 40 ☽ ⊼ ♇
21 29 ☽ ⊼ ♆
23 03 ☽ ⊼ ♅

OCTOBER 1966

LONGITUDES

Date	Sidereal time h m s	Sun ☉ ° ' "	Moon ☽ ° ' "	Moon ☽ 24.00 ° ' "	Mercury ☿ ° '	Venus ♀ ° '	Mars ♂ ° '	Jupiter ♃ ° '	Saturn ♄ ° '	Uranus ♅ ° '	Neptune ♆ ° '	Pluto ♇ ° '
01	12 39 06	07 ♎ 51 31	27 ♈ 36 11	03 ♉ 37 37	23 ♎ 20	27 ♍ 56	23 ♌ 08	00 ♌ 35	25 ♓ 22	21 ♍ 10	20 ♏ 22	18 ♍ 58
02	12 43 02	08 50 31	09 ♉ 40 51	15 ♉ 46 10	24 52	29 ♍ 11	23 45	00 43	25 R 17	21 13	20 24	19 00
03	12 46 59	09 49 33	21 ♉ 53 55	28 ♉ 04 28	26 23	00 ♎ 26	24 22	00 51	25 13	21 17	20 26	19 02
04	12 50 55	10 48 38	04 ♊ 18 11	10 ♊ 35 30	27 53	01 41	24 58	01 00	25 08	21 21	20 27	19 04
05	12 54 52	11 47 44	16 ♊ 56 50	23 ♊ 22 36	29 ♎ 23	02 55	25 35	01 08	25 04	21 24	20 29	19 06
06	12 58 48	12 46 54	29 ♊ 53 14	06 ♋ 29 08	00 ♏ 51	04 10	26 12	01 16	25 00	21 28	20 31	19 08
07	13 02 45	13 46 05	13 ♋ 10 38	19 ♋ 58 02	02 19	05 25	26 48	01 23	24 55	21 31	20 33	19 10
08	13 06 42	14 45 19	26 ♋ 51 31	03 ♌ 51 11	03 46	06 40	27 25	01 31	24 51	21 35	20 35	19 12
09	13 10 38	15 44 35	10 ♌ 57 00	18 ♌ 08 44	05 12	07 55	28 01	01 39	24 47	21 39	20 37	19 14
10	13 14 35	16 43 54	25 ♌ 26 02	02 ♍ 48 20	06 37	09 10	28 37	01 46	24 43	21 42	20 39	19 16
11	13 18 31	17 43 14	10 ♍ 14 55	17 ♍ 44 52	08 02	10 25	29 14	01 53	24 39	21 46	20 41	19 18
12	13 22 28	18 42 37	25 ♍ 17 09	02 ♎ 50 37	09 25	11 40	29 ♌ 50	02 01	24 35	21 49	20 43	19 20
13	13 26 24	19 42 02	10 ♎ 24 05	17 ♎ 56 19	10 47	12 55	00 ♍ 26	02 08	24 31	21 53	20 45	19 22
14	13 30 21	20 41 30	25 ♎ 26 09	02 ♏ 52 29	12 08	14 10	01 02	02 14	24 27	21 56	20 47	19 24
15	13 34 17	21 40 59	10 ♏ 14 22	17 ♏ 30 59	13 29	15 25	01 38	02 21	24 23	21 59	20 49	19 26
16	13 38 14	22 40 30	24 ♏ 41 42	01 ♐ 46 04	14 48	16 40	02 15	02 28	24 19	22 03	20 51	19 28
17	13 42 11	23 40 03	08 ♐ 43 48	15 ♐ 34 48	16 06	17 55	02 51	02 34	24 15	22 06	20 53	19 30
18	13 46 07	24 39 38	22 ♐ 19 05	28 ♐ 56 52	17 22	19 10	03 26	02 40	24 11	22 10	20 55	19 32
19	13 50 04	25 39 15	05 ♑ 28 24	11 ♑ 54 06	18 37	20 25	04 02	02 46	24 08	22 13	20 57	19 33
20	13 54 00	26 38 53	18 ♑ 14 23	24 ♑ 29 45	19 51	21 41	04 38	02 52	24 04	22 16	20 59	19 35
21	13 57 57	27 38 33	00 ♒ 40 44	06 ♒ 47 55	21 03	22 56	05 14	02 58	24 01	22 19	21 01	19 37
22	14 01 53	28 38 15	12 ♒ 51 49	18 ♒ 53 01	22 14	24 11	05 50	03 04	23 57	22 23	21 03	19 39
23	14 05 50	29 ♎ 37 59	24 ♒ 52 02	00 ♓ 49 26	23 22	25 26	06 25	03 09	23 54	22 26	21 05	19 41
24	14 09 46	00 ♏ 37 44	06 ♓ 45 40	12 ♓ 41 15	24 29	26 41	07 01	03 14	23 51	22 29	21 07	19 42
25	14 13 43	01 37 31	18 ♓ 36 36	24 ♓ 32 07	25 33	27 56	07 36	03 19	23 48	22 32	21 10	19 44
26	14 17 39	02 37 20	00 ♈ 28 12	06 ♈ 25 09	26 34	29 ♎ 12	08 12	03 24	23 45	22 35	21 12	19 46
27	14 21 36	03 37 10	12 ♈ 23 19	18 ♈ 22 56	27 33	00 ♏ 27	08 47	03 29	23 42	22 38	21 14	19 47
28	14 25 33	04 37 02	24 ♈ 24 17	00 ♉ 27 34	28 28	01 42	09 22	03 34	23 39	22 41	21 16	19 49
29	14 29 29	05 36 57	06 ♉ 33 00	12 ♉ 40 45	29 ♏ 20	02 57	09 58	03 38	23 36	22 45	21 18	19 51
30	14 33 26	06 36 53	18 ♉ 51 00	25 ♉ 03 55	00 ♐ 08	04 12	10 33	03 42	23 33	22 48	21 20	19 52
31	14 37 22	07 ♏ 36 51	01 ♊ 19 39	07 ♊ 38 21	00 ♐ 52	05 ♏ 28	11 ♍ 08	03 ♌ 46	23 ♓ 30	22 ♍ 50	21 ♏ 23	19 ♍ 54

Date	Moon True ☊ ° '	Moon Mean ☊ ° '	Moon ☽ Latitude ° '
01	16 ♉ 34	18 ♉ 10	01 S 41
02	16 R 33	18 07	00 S 37
03	16 D 32	18 04	00 N 29
04	16 34	18 01	01 35
05	16 35	17 57	02 38
06	16 37	17 54	03 35
07	16 37	17 51	04 21
08	16 R 37	17 48	04 55
09	16 35	17 45	05 11
10	16 31	17 41	05 09
11	16 28	17 38	04 46
12	16 23	17 35	04 04
13	16 20	17 32	03 04
14	16 18	17 29	01 52
15	16 16	17 26	00 N 33
16	16 D 16	17 22	00 S 46
17	16 17	17 19	02 01
18	16 19	17 16	03 06
19	16 20	17 13	03 59
20	16 21	17 10	04 38
21	16 R 22	17 07	05 04
22	16 21	17 03	05 15
23	16 19	17 00	05 12
24	16 17	16 57	04 56
25	16 15	16 54	04 27
26	16 13	16 51	03 47
27	16 11	16 47	02 56
28	16 09	16 44	01 58
29	16 09	16 41	00 S 53
30	16 D 09	16 38	00 N 15
31	16 ♉ 09	16 ♉ 35	01 N 23

DECLINATIONS

Date	Sun ☉ ° '	Moon ☽ ° '	Mercury ☿ ° '	Venus ♀ ° '	Mars ♂ ° '	Jupiter ♃ ° '	Saturn ♄ ° '	Uranus ♅ ° '	Neptune ♆ ° '	Pluto ♇ ° '
01	03 S 07	09 N 03	09 S 29	02 N 09	15 N 01	20 N 15	04 S 07	04 N 11	16 S 12	17 N 30
02	03 30	14 08	10 10	01 39	14 50	20 13	04 09	04 09	16 13	17 30
03	03 54	18 43	10 50	01 09	14 38	20 11	04 11	04 08	16 13	17 29
04	04 17	22 34	11 29	00 39	14 26	20 10	04 12	04 06	16 14	17 28
05	04 40	25 26	12 08	00 N 10	14 14	20 08	04 14	04 05	16 14	17 28
06	05 03	27 01	12 46	00 S 20	14 02	20 07	04 16	04 03	16 15	17 27
07	05 26	27 08	13 23	00 50	13 50	20 05	04 17	04 02	16 15	17 26
08	05 49	25 36	13 59	01 20	13 38	20 03	04 19	04 01	16 16	17 26
09	06 12	22 28	14 35	01 50	13 26	20 02	04 21	03 59	16 16	17 25
10	06 35	17 53	15 10	02 20	13 14	20 00	04 22	03 58	16 17	17 25
11	06 57	12 09	15 44	02 50	13 01	19 59	04 24	03 57	16 17	17 24
12	07 20	05 N 36	16 17	03 20	12 49	19 57	04 25	03 55	16 18	17 24
13	07 42	01 S 18	16 49	03 50	12 37	19 56	04 27	03 54	16 18	17 23
14	08 05	08 06	17 20	04 20	12 24	19 55	04 28	03 52	16 19	17 23
15	08 27	14 22	17 51	04 50	12 12	19 53	04 30	03 51	16 20	17 22
16	08 49	19 42	18 20	05 19	11 59	19 52	04 31	03 50	16 20	17 22
17	09 11	23 45	18 49	05 49	11 47	19 51	04 32	03 48	16 21	17 21
18	09 33	26 19	19 16	06 18	11 34	19 49	04 34	03 47	16 21	17 21
19	09 55	27 19	19 42	06 48	11 22	19 48	04 35	03 46	16 22	17 20
20	10 17	26 48	20 07	07 17	11 09	19 47	04 36	03 45	16 22	17 20
21	10 38	24 57	20 31	07 46	10 57	19 46	04 38	03 43	16 23	17 20
22	11 00	21 59	20 54	08 15	10 44	19 45	04 39	03 42	16 23	17 19
23	11 21	18 08	21 15	08 43	10 31	19 44	04 40	03 41	16 24	17 19
24	11 42	13 36	21 35	09 12	10 18	19 43	04 41	03 40	16 25	17 18
25	12 03	08 36	21 54	09 40	10 06	19 41	04 43	03 38	16 25	17 18
26	12 23	03 S 17	22 11	10 08	09 53	19 41	04 44	03 37	16 26	17 18
27	12 44	02 N 12	22 27	10 36	09 40	19 40	04 45	03 36	16 26	17 17
28	13 04	07 38	22 41	11 03	09 27	19 39	04 46	03 35	16 27	17 17
29	13 24	12 52	22 53	11 31	09 14	19 38	04 47	03 34	16 28	17 17
30	13 44	17 40	23 04	11 58	09 01	19 37	04 48	03 32	16 28	17 16
31	14 S 03	21 N 47	23 S 12	12 S 25	08 N 48	19 N 36	04 S 49	03 N 31	16 S 29	17 N 16

ZODIAC SIGN ENTRIES

Date	h m	Planets
01	16 47	☽ ♉
03	03 44	♀ ♎
04	03 43	☽ ♊
05	22 03	☿ ♏
06	12 12	☽ ♋
08	17 25	☽ ♌
10	19 27	☽ ♍
12	18 37	♂ ♍
12	19 29	☽ ♎
14	19 21	☽ ♏
16	20 59	☽ ♐
19	01 55	☽ ♑
21	10 41	☽ ♒
23	20 51	☉ ♏
23	22 20	☽ ♓
26	11 03	☽ ♈
27	03 28	♀ ♏
28	23 05	☽ ♉
30	07 38	☿ ♐
31	09 28	☽ ♊

LATITUDES

Date	Mercury ☿ ° '	Venus ♀ ° '	Mars ♂ ° '	Jupiter ♃ ° '	Saturn ♄ ° '	Uranus ♅ ° '	Neptune ♆ ° '	Pluto ♇ ° '
01	00 S 27	01 N 27	01 N 17	00 N 13	02 S 29	00 N 44	01 N 42	14 N 19
04	00 49	01 27	01 19	00 14	02 28	00 44	01 42	14 20
07	01 10	01 26	01 20	00 14	02 28	00 44	01 42	14 21
10	01 31	01 24	01 21	00 15	02 28	00 44	01 42	14 21
13	01 51	01 22	01 23	00 15	02 28	00 44	01 42	14 22
16	02 09	01 20	01 24	00 15	02 28	00 44	01 42	14 23
19	02 25	01 17	01 26	00 16	02 27	00 44	01 42	14 24
22	02 39	01 13	01 27	00 16	02 27	00 44	01 42	14 25
25	02 50	01 09	01 29	00 17	02 27	00 44	01 42	14 26
28	02 56	01 05	01 30	00 17	02 26	00 44	01 41	14 27
31	02 S 56	01 N 00	01 N 32	00 N 18	02 S 26	00 N 44	01 N 41	14 N 28

DATA

Julian Date	2439400
Delta T	+37 seconds
Ayanamsa	23° 23' 20"
Synetic vernal point	05° ♓ 43' 40"
True obliquity of ecliptic	23° 26' 44"

LONGITUDES

Date	Chiron ⚷ ° '	Ceres ⚳ ° '	Pallas ⚴ ° '	Juno ⚵ ° '	Vesta ⚶ ° '	Black Moon Lilith ⚸ ° '
01	23 ♓ 16	03 ♋ 19	12 ♈ 37	24 ♋ 47	22 ♍ 13	20 ♓ 27
11	22 ♓ 51	05 ♋ 02	09 ♈ 43	29 ♋ 04	27 ♍ 05	21 ♓ 34
21	22 ♓ 28	06 ♋ 13	06 ♈ 53	02 ♌ 59	01 ♎ 58	22 ♓ 41
31	22 ♓ 09	06 ♋ 50	04 ♈ 26	06 ♌ 27	06 ♎ 49	23 ♓ 48

MOON'S PHASES, APSIDES AND POSITIONS ☽

Date	h m	Phase	Longitude ° '	Eclipse Indicator
07	13 08	☾	13 ♋ 49	
14	03 52	●	20 ♎ 21	
21	05 34	☽	27 ♑ 23	
29	10 00	○	05 ♉ 32	

Day	h m		
13	02 46	Perigee	
25	09 49	Apogee	
07	01 36	Max dec	27° N 17'
13	07 31	0S	
19	15 35	Max dec	27° S 20'
27	02 25	0N	

ASPECTARIAN

h m Aspects

01 Saturday
02 12 ☽ ☍ ☿
02 35 ☽ △ ♂
06 43 ☽ ± ♇
06 55 ☿ ⚹ ♂
07 33 ☽ ⊻ ♄
11 07 ☽ ± ♅
12 45 ☽ ⊼ ♀
14 19 ☽ # ☿
18 00 ☽ □ ♃
19 26 ☽ ⊥ ♄

02 Sunday
00 43 ☽ ꝑ ♇
02 04 ☽ ± ♀
05 07 ☽ ꝑ ♅
05 56 ☉ ∠ ♂
10 12 ☽ ⊼ ☉
13 11 ☽ ∠ ♄
14 12 ☿ ⊥ ♇
14 19 ♄ # ♅
15 20 ☽ ∥ ♂
18 17 ☿ ⊼ ♄
21 53 ☽ ꝑ ♀
22 31 ☽ # ♆
23 04 ☽ ± ☉

03 Monday
05 14 ☽ ∥ ♇
05 59 ☽ Q ♃
06 23 ☽ △ ♇
09 07 ☽ ☍ ♆
09 24 ♀ ⊥ ♂
10 48 ☽ △ ♅
17 03 ☽ □ ♂
18 12 ☽ ꝑ ☉
18 25 ☽ ⚹ ♄
20 31 ☽ ∥ ♃
21 14 ♀ ⚹ ♃
21 56 ☽ ⊼ ☿

04 Tuesday
01 43 ☉ # ♅
02 54 ☿ ⊥ ♅
05 34 ☽ ⚹ ♃
06 23 ☽ △ ♀
06 55 ☉ ∥ ♄
11 06 ☽ ± ☿
17 23 ☽ Q ♄
17 47 ♂ ⊼ ♄

05 Wednesday
01 28 ☽ △ ☉
05 21 ☽ Q ♂
06 32 ☽ ꝑ ☿
10 27 ☽ ∠ ♃
16 03 ☽ □ ♇
18 39 ☽ ⊼ ♆
20 22 ☽ □ ♅

06 Thursday
03 02 ☽ □ ♄
03 24 ☽ ⊥ ♃
04 53 ☽ ⚹ ♂
05 48 ☽ ± ♆
14 00 ☽ △ ☿
14 06 ☿ ± ♄
14 32 ☽ ⊻ ♃
19 15 ☿ □ ♃
20 37 ☽ □ ♀
22 17 ☽ ꝑ ♆

07 Friday
01 13 ☽ Q ♇
01 27 ☉ Q ♃
05 26 ☽ Q ♅
09 26 ☽ ∠ ♂
13 08 ☽ □ ☉
14 35 ♀ ∠ ♆
22 38 ☽ ⚹ ♇

08 Saturday
01 03 ☽ △ ♆
01 25 ☿ # ♂
02 05 ☽ ⊥ ♂
02 47 ☽ ⚹ ♅
07 39 ☉ ⊥ ♆
07 50 ☽ Q ♀
08 32 ☽ △ ♄
13 00 ☽ ⊻ ♂
19 27 ☿ ∠ ♇
20 05 ☽ ☌ ♃
22 54 ☽ Q ☉

09 Sunday
00 38 ☽ ∠ ♇
01 12 ☽ □ ☿
04 42 ☽ ∠ ♅
06 24 ☽ ⚹ ♀
10 03 ☽ ꝑ ♄
15 51 ☽ ⊥ ♇
19 53 ☽ ⊥ ♅
20 36 ☽ ⚹ ☉

10 Monday
00 59 ☽ ± ♄
01 42 ☽ ∥ ♃
01 50 ☽ ⊻ ♇
04 07 ☽ □ ♆
05 51 ☽ ⊻ ♅
09 44 ☽ ∠ ♀
10 32 ☽ Q ☿
10 49 ☽ ⊼ ♄
13 24 ☿ ∠ ♅
14 11 ☽ ∥ ♇
17 26 ☽ ☌ ♂
19 10 ☽ # ♆
22 24 ☽ ⊻ ♃
22 50 ☽ # ☿
23 00 ☽ ∠ ☉

11 Tuesday
01 45 ☽ ⊥ ♀
08 03 ☽ ⚹ ☿
08 11 ☽ ⊥ ♃
08 27 ☽ ∥ ♂
09 28 ☽ Q ♆
12 18 ☽ ⊻ ♀
14 32 ☽ ⊥ ☉
22 43 ☽ ∠ ♃

12 Wednesday
00 48 ☽ ⊻ ☉
02 31 ☽ ☌ ♇
04 42 ☽ ⚹ ♆
06 11 ☽ # ☉
06 28 ☽ ☌ ♅
10 28 ☽ ∠ ☿
10 53 ☽ ☍ ♄
12 45 ☿ ∥ ♆
14 41 ☿ ꝑ ♄
16 09 ☽ # ♄
17 57 ☽ ∥ ♅
19 25 ☽ # ♀
19 31 ☽ ⊻ ♂
22 45 ☽ ⚹ ♃

13 Thursday
02 12 ☽ ⊥ ☿
03 41 ☉ ⊻ ♇
04 35 ☽ ∠ ♆
05 27 ☽ ⊥ ♂
12 40 ☽ ⊻ ☿
14 44 ♀ # ♅
16 22 ☽ ☌ ♀
17 58 ☽ Q ♃
18 55 ☽ ⊥ ♆
20 21 ☽ ∠ ♂
21 01 ☽ # ♅
21 32 ☽ ∥ ♀
23 01 ☽ ∥ ♄

14 Friday
02 19 ☽ ⊻ ♇
03 52 ☽ ☌ ☉
04 31 ☽ ⊻ ♆
06 22 ☽ ⊻ ♅
10 25 ☽ ⊼ ♄
11 56 ☽ ∥ ☉
11 57 ☽ ⊥ ♇
13 30 ♀ Q ♃
13 46 ☿ # ♇
14 09 ☉ ⊻ ♆
16 02 ☽ ⊥ ♅
18 54 ♀ ∥ ♄
20 02 ☽ ± ♄
21 25 ☽ ⚹ ♂
23 03 ☽ □ ♃

15 Saturday
00 02 ♀ ⊥ ♆
02 30 ☽ ∠ ♇
03 38 ☽ # ♂
06 40 ☽ ∠ ♅
10 36 ☽ ꝑ ♄
17 26 ☿ Q ♂
17 50 ☽ Q ♂
17 51 ☽ ☌ ☿
19 52 ☉ ⊻ ♅
20 16 ☽ ∥ ♆
21 20 ☽ ⊻ ♀

16 Sunday
00 52 ☽ # ♇
03 13 ☽ ⚹ ♇
04 40 ☽ ∥ ☿
05 32 ☽ ☌ ♆
07 32 ☽ ⚹ ♅
08 16 ☽ ⊥ ♀
08 21 ☽ ⊻ ☉
11 22 ☽ △ ♄
12 52 ☽ # ♃
19 14 ☽ ⊥ ☉
22 34 ♂ ⊻ ♃
23 30 ☽ Q ♇

17 Monday
00 59 ☽ ∠ ♀
01 17 ☽ △ ♃
01 23 ☽ □ ♂
03 58 ☽ Q ♅
09 06 ♀ ∠ ♂
11 53 ☽ ∠ ☉

18 Tuesday
01 19 ☉ ⊼ ♄
02 15 ☽ ⊻ ☿
03 38 ☽ ꝑ ♃
05 48 ☽ ⚹ ♀
07 00 ☽ □ ♇
09 29 ☽ ⊻ ♆
11 43 ☽ □ ♅
14 05 ☽ ⊥ ☿
15 21 ☽ □ ♄
16 34 ☽ ⚹ ☉
18 57 ♀ ⊻ ♇
19 55 ☽ ± ♃
20 19 ☽ ⊥ ♆

19 Wednesday
05 46 ☽ Q ♀
06 58 ☽ ⊼ ♃
08 13 ☽ ∠ ☿
09 13 ☽ △ ♂
09 34 ☉ ⊥ ♇
12 53 ☽ ∠ ♆
16 23 ☽ Q ☉
17 30 ☿ # ♃
22 19 ♀ ⊻ ♆

20 Thursday
00 22 ☽ Q ♄
06 37 ☿ ⚹ ♇
14 35 ☽ △ ♇
14 48 ☽ ꝑ ♂
15 25 ☽ ⚹ ☿
17 15 ☽ ⚹ ♆
19 18 ☽ □ ♀
19 45 ☽ △ ♅
23 08 ☽ ⚹ ♄
23 52 ♀ ⊻ ♅

21 Friday
05 34 ☽ □ ☉
09 02 ☽ ± ♂
11 12 ☿ ☌ ♆
16 30 ☽ ☍ ♃
16 35 ☽ Q ♆
17 09 ☽ Q ☿
19 44 ☽ ꝑ ♇
21 22 ☽ ⊼ ♂

22 Saturday
00 51 ☉ # ♂
01 06 ☽ ꝑ ♅
04 18 ☽ ∠ ♄
05 22 ☉ ⊥ ♅
07 54 ♀ ⊼ ♄
13 34 ☽ ± ♇
15 14 ☿ ⚹ ♅
18 39 ☽ ∥ ☿
19 02 ☽ ± ♅
22 06 ☽ ⊥ ♄

23 Sunday
01 34 ☽ ⊼ ♇
02 35 ☽ # ♃
04 23 ☽ □ ♆
07 05 ☽ ⊼ ♅
08 41 ☽ □ ☿
10 04 ☽ ⊻ ♄
13 16 ☽ △ ♀
16 36 ☽ # ♇
16 45 ♀ ⊥ ♇
18 09 ☉ ± ♄
21 31 ☽ ∥ ♆
22 28 ☽ △ ☉
22 54 ☿ △ ♄

24 Monday
04 49 ☽ ⊼ ♃
12 32 ☽ ☍ ♂
17 03 ☽ ± ♃
20 46 ☽ ∥ ☉
23 09 ☽ ꝑ ♀

25 Tuesday
04 43 ☽ # ♂
07 26 ☽ ∥ ♀
07 37 ☽ ꝑ ☉
11 25 ☽ ꝑ ♃
14 17 ☽ ☍ ♇
17 11 ☽ △ ♆
19 33 ☽ ± ♀
19 59 ☽ ☍ ♅
22 27 ☽ ☌ ♄
23 56 ♀ ⊥ ♅

26 Wednesday
02 56 ♀ # ♂
03 23 ☽ △ ☿
03 31 ☽ ± ☉
05 35 ☽ ∥ ♄
09 07 ☽ ⊼ ♀
10 29 ☽ # ♅
16 45 ☽ ⊼ ☉
17 58 ☽ △ ♃
22 08 ♀ ± ♄
23 35 ☽ ꝑ ♆

27 Thursday
04 23 ☽ ⊼ ♂
08 27 ☉ □ ♃
12 21 ☽ ꝑ ☿
17 03 ☽ ± ♂
17 43 ☽ ± ♆
18 08 ☽ ∥ ♅
23 13 ☽ # ♄

28 Friday
02 51 ☽ ⊼ ♇
05 44 ☽ ⊼ ♆
07 24 ♂ Q ♆
07 51 ☽ ± ☿
08 35 ☽ ⊼ ♅
10 30 ☽ ⊻ ♄
11 56 ☽ ꝑ ♂
14 49 ☽ ± ♇
16 58 ☉ ∠ ♇
19 50 ☽ ∥ ♂
20 32 ☽ ± ♅
20 42 ☽ ⊼ ☿
20 46 ♂ ⊥ ♃
22 21 ☽ ⊥ ♄

29 Saturday
04 07 ☽ ☍ ♀
05 00 ☽ # ♀
06 14 ☽ □ ♃
08 39 ☽ ꝑ ♇
10 00 ☽ ☍ ☉
14 21 ☽ ꝑ ♅
14 41 ☽ # ☉
16 00 ☽ ∠ ♄
19 01 ☽ △ ♂

30 Sunday
01 50 ♀ □ ♃
05 41 ☽ # ♆
09 53 ☽ ∥ ♇
13 59 ☽ △ ♇
16 50 ☽ ☍ ♆
17 33 ☽ Q ♃
19 39 ☽ △ ♅
21 03 ☽ ⚹ ♄
22 42 ☽ ∥ ♃

31 Monday
01 02 ♀ ∠ ♇
11 05 ☽ ☍ ☿
16 41 ☽ ⚹ ♃
17 44 ☉ ∠ ♅
19 56 ☽ Q ♄
20 44 ☽ ⊼ ♀
22 10 ☽ # ☿

All ephemeris data is given at 12.00 UT and the Moon's longitude is additionally given for 24.00 UT

NOVEMBER 1966

Date	Sidereal time h m s	Sun ☉ ° ' "	Moon ☽ ° ' "	Moon ☽ 24.00 ° ' "	Mercury ☿ ° '	Venus ♀ ° '	Mars ♂ ° '	Jupiter ♃ ° '	Saturn ♄ ° '	Uranus ♅ ° '	Neptune ♆ ° '	Pluto ♇ ° '
01	14 41 19	08 ♏ 36 51	14 ♊ 00 10	20 ♊ 25 18	01 ♐ 31	06 ♏ 43	11 ♍ 43	03 ♌ 50	23 ♓ 28	22 ♍ 53	21 ♏ 25	19 ♍ 56
02	14 45 15	09 36 54	26 ♊ 53 51	03 ♋ 26 02	02 05	07 58	12 18	03 54	23 R 25	22 56	21 27	19 57
03	14 49 12	10 36 58	10 ♋ 01 58	16 ♋ 41 48	02 33	09 13	12 53	03 58	23 23	22 59	21 29	19 59
04	14 53 08	11 37 05	23 ♋ 25 40	00 ♌ 13 41	02 54	10 29	13 28	04 01	23 21	23 02	21 31	20 00
05	14 57 05	12 37 13	07 ♌ 05 53	14 ♌ 02 18	03 08	11 44	14 03	04 04	23 18	23 05	21 34	20 02
06	15 01 02	13 37 24	21 ♌ 02 53	28 ♌ 07 30	03 15	12 59	14 37	04 07	23 16	23 08	21 36	20 03
07	15 04 58	14 37 37	05 ♍ 15 57	12 ♍ 27 55	03 R 12	14 15	15 12	04 10	23 14	23 10	21 38	20 05
08	15 08 55	15 37 51	19 ♍ 43 00	27 ♍ 00 41	03 01	15 30	15 46	04 12	23 12	23 13	21 40	20 06
09	15 12 51	16 38 08	04 ♎ 20 22	11 ♎ 41 20	02 40	16 45	16 21	04 15	23 10	23 16	21 43	20 07
10	15 16 48	17 38 27	19 ♎ 02 49	26 ♎ 24 01	02 09	18 01	16 55	04 17	23 09	23 18	21 45	20 09
11	15 20 44	18 38 47	03 ♏ 44 02	11 ♏ 02 04	01 28	19 16	17 30	04 19	23 07	23 21	21 47	20 10
12	15 24 41	19 39 10	18 ♏ 17 16	25 ♏ 28 53	00 ♐ 37	20 31	18 04	04 21	23 05	23 23	21 49	20 11
13	15 28 37	20 39 34	02 ♐ 36 13	09 ♐ 38 43	29 ♏ 38	21 47	18 38	04 23	23 04	23 26	21 52	20 13
14	15 32 34	21 40 00	16 ♐ 35 54	23 ♐ 27 26	28 30	23 02	19 12	04 24	23 03	23 28	21 54	20 14
15	15 36 31	22 40 27	00 ♑ 13 07	06 ♑ 52 52	27 15	24 17	19 46	04 25	23 01	23 30	21 56	20 15
16	15 40 27	23 40 56	13 ♑ 26 45	19 ♑ 54 54	25 56	25 33	20 20	04 27	23 00	23 33	21 58	20 16
17	15 44 24	24 41 26	26 ♑ 17 36	02 ♒ 35 10	24 35	26 48	20 54	04 27	22 59	23 35	22 01	20 17
18	15 48 20	25 41 57	08 ♒ 48 04	14 ♒ 56 45	23 15	28 03	21 28	04 28	22 58	23 37	22 03	20 19
19	15 52 17	26 42 30	21 ♒ 01 45	27 ♒ 03 38	21 58	29 ♏ 19	22 01	04 29	22 58	23 40	22 05	20 20
20	15 56 13	27 43 04	03 ♓ 02 59	09 ♓ 00 24	20 47	00 ♐ 34	22 35	04 29	22 57	23 42	22 07	20 21
21	16 00 10	28 43 39	14 ♓ 56 29	20 ♓ 51 49	19 43	01 50	23 08	04 R 29	22 56	23 44	22 10	20 22
22	16 04 06	29 ♏ 44 15	26 ♓ 47 00	02 ♈ 42 36	18 50	03 05	23 42	04 29	22 56	23 46	22 12	20 23
23	16 08 03	00 ♐ 44 53	08 ♈ 39 08	14 ♈ 37 08	18 07	04 20	24 15	04 29	22 55	23 48	22 14	20 24
24	16 12 00	01 45 31	20 ♈ 37 01	26 ♈ 39 15	17 35	05 36	24 48	04 28	22 55	23 50	22 16	20 25
25	16 15 56	02 46 11	02 ♉ 44 12	08 ♉ 52 11	17 16	06 51	25 21	04 27	22 55	23 52	22 19	20 26
26	16 19 53	03 46 52	15 ♉ 03 29	21 ♉ 18 17	17 07	08 06	25 54	04 26	22 55	23 54	22 21	20 27
27	16 23 49	04 47 34	27 ♉ 36 47	03 ♊ 59 02	17 D 10	09 22	26 27	04 25	22 D 55	23 55	22 23	20 28
28	16 27 46	05 48 18	10 ♊ 25 07	16 ♊ 54 59	17 23	10 37	27 00	04 24	22 55	23 57	22 25	20 28
29	16 31 42	06 49 03	23 ♊ 28 35	00 ♋ 05 48	17 46	11 53	27 33	04 23	22 55	23 59	22 27	20 29
30	16 35 39	07 ♐ 49 49	06 ♋ 46 29	13 ♋ 30 28	18 ♏ 17	13 ♐ 08	28 ♍ 05	04 ♌ 21	22 ♓ 56	24 ♍ 01	22 ♏ 30	20 ♍ 30

Date	Moon True ☊ ° '	Moon Mean ☊ ° '	Moon ☽ Latitude ° '
01	16 ♉ 10	16 ♉ 32	02 N 28
02	16 D 10	16 28	03 27
03	16 11	16 25	04 16
04	16 11	16 22	04 53
05	16 11	16 19	05 13
06	16 R 11	16 16	05 16
07	16 11	16 13	05 00
08	16 11	16 09	04 25
09	16 D 11	16 06	03 32
10	16 11	16 03	02 25
11	16 11	16 00	01 N 09
12	16 R 12	15 57	00 S 12
13	16 11	15 53	01 30
14	16 11	15 50	02 41
15	16 10	15 47	03 41
16	16 09	15 44	04 27
17	16 08	15 41	04 58
18	16 08	15 38	05 14
19	16 07	15 34	05 16
20	16 D 07	15 31	05 03
21	16 08	15 28	04 38
22	16 09	15 25	04 01
23	16 10	15 22	03 13
24	16 11	15 19	02 17
25	16 12	15 15	01 14
26	16 13	15 12	00 S 06
27	16 R 13	15 09	01 N 03
28	16 11	15 06	02 10
29	16 09	15 03	03 11
30	16 ♉ 06	14 ♉ 59	04 N 03

DECLINATIONS

Date	Sun ☉ ° '	Moon ☽ ° '	Mercury ☿ ° '	Venus ♀ ° '	Mars ♂ ° '	Jupiter ♃ ° '	Saturn ♄ ° '	Uranus ♅ ° '	Neptune ♆ ° '	Pluto ♇ ° '
01	14 S 23	24 N 57	23 S 19	12 S 51	08 N 35	19 N 35	04 S 49	03 N 30	16 S 29	17 N 16
02	14 42	26 52	23 24	13 17	08 23	19 35	04 50	03 29	16 30	17 16
03	15 01	27 20	23 26	13 43	08 10	19 34	04 51	03 28	16 30	17 15
04	15 19	26 13	23 25	14 08	07 57	19 33	04 52	03 27	16 31	17 15
05	15 38	23 33	23 22	14 34	07 44	19 33	04 53	03 26	16 32	17 15
06	15 56	19 28	23 16	14 58	07 31	19 32	04 53	03 25	16 32	17 15
07	16 14	14 14	23 08	15 23	07 18	19 32	04 54	03 24	16 33	17 14
08	16 31	08 08	22 55	15 47	07 05	19 31	04 55	03 23	16 33	17 14
09	16 49	01 N 31	22 40	16 10	06 52	19 31	04 55	03 22	16 34	17 14
10	17 06	05 S 13	22 20	16 33	06 39	19 31	04 56	03 21	16 35	17 14
11	17 23	11 41	21 57	16 56	06 26	19 30	04 56	03 20	16 35	17 14
12	17 39	17 28	21 30	17 18	06 13	19 30	04 57	03 19	16 36	17 14
13	17 55	22 09	21 00	17 40	06 00	19 30	04 57	03 18	16 36	17 14
14	18 11	25 26	20 26	18 02	05 47	19 30	04 57	03 17	16 37	17 14
15	18 27	27 07	19 50	18 22	05 34	19 29	04 58	03 16	16 37	17 13
16	18 42	27 11	19 12	18 43	05 21	19 29	04 58	03 15	16 38	17 13
17	18 57	25 46	18 32	19 03	05 09	19 29	04 58	03 14	16 39	17 13
18	19 11	23 07	17 53	19 22	04 56	19 29	04 58	03 13	16 39	17 13
19	19 26	19 28	17 15	19 41	04 43	19 29	04 58	03 12	16 40	17 13
20	19 39	15 06	16 39	19 59	04 30	19 30	04 59	03 12	16 40	17 13
21	19 53	10 12	16 07	20 17	04 17	19 30	04 59	03 11	16 41	17 13
22	20 06	04 S 58	15 39	20 34	04 04	19 30	04 59	03 10	16 41	17 13
23	20 19	00 N 28	15 16	20 50	03 52	19 30	04 59	03 09	16 42	17 13
24	20 31	05 56	14 57	21 06	03 39	19 30	04 59	03 08	16 43	17 14
25	20 43	11 16	14 44	21 22	03 26	19 31	04 58	03 08	16 43	17 14
26	20 55	16 15	14 36	21 36	03 14	19 31	04 58	03 07	16 44	17 14
27	21 06	20 39	14 33	21 50	03 01	19 32	04 58	03 06	16 44	17 14
28	21 17	24 09	14 34	22 04	02 49	19 32	04 58	03 06	16 45	17 14
29	21 27	26 28	14 40	22 17	02 36	19 33	04 57	03 05	16 45	17 14
30	21 S 37	27 N 19	14 S 48	22 S 29	02 N 23	19 N 33	04 S 57	03 N 04	16 S 46	17 N 14

ZODIAC SIGN ENTRIES

Date	h	m	Planets
02	17	43	☽ ♋
04	23	36	☽ ♌
07	03	10	☽ ♍
09	04	54	☽ ♎
11	05	53	☽ ♏
13	03	26	☿ ♏
13	07	36	☽ ♐
15	11	37	☽ ♑
17	19	03	☽ ♒
20	01	06	♀ ♐
20	05	53	☽ ♓
22	18	14	☉ ♐
22	18	31	☽ ♈
25	06	37	☽ ♉
27	16	31	☽ ♊
29	23	50	☽ ♋

LATITUDES

Date	Mercury ☿ ° '	Venus ♀ ° '	Mars ♂ ° '	Jupiter ♃ ° '	Saturn ♄ ° '	Uranus ♅ ° '	Neptune ♆ ° '	Pluto ♇ ° '
01	02 S 55	00 N 58	01 N 32	00 N 18	02 S 26	00 N 44	01 N 41	14 N 29
04	02 44	00 52	01 34	00 19	02 25	00 44	01 41	14 30
07	02 22	00 46	01 35	00 19	02 25	00 45	01 41	14 31
10	01 47	00 40	01 37	00 20	02 24	00 45	01 41	14 32
13	00 S 57	00 34	01 38	00 20	02 24	00 45	01 41	14 34
16	00 N 03	00 27	01 40	00 21	02 23	00 45	01 41	14 35
19	01 03	00 20	01 41	00 21	02 22	00 45	01 41	14 37
22	01 51	00 13	01 43	00 22	02 22	00 45	01 41	14 38
25	02 21	00 N 06	01 44	00 22	02 21	00 45	01 41	14 39
28	02 34	00 S 01	01 46	00 23	02 20	00 45	01 41	14 41
31	02 N 33	00 S 08	01 N 47	00 N 23	02 S 20	00 N 45	01 N 41	14 N 42

DATA

Julian Date	2439431
Delta T	+37 seconds
Ayanamsa	23° 23' 24"
Synetic vernal point	05° ♓ 43' 36"
True obliquity of ecliptic	23° 26' 43"

MOON'S PHASES, APSIDES AND POSITIONS ☽

Date	h	m	Phase	Longitude ° '	Eclipse Indicator
05	22	18	☾	13 ♌ 03	
12	14	26	●	19 ♏ 45	Total
20	00	20	☽	27 ♒ 14	
28	02	40	○	05 ♊ 25	

Day	h	m		
10	08	31	Perigee	
22	02	57	Apogee	
03	07	13	Max dec	27° N 22'
09	17	24	OS	
16	00	57	Max dec	27° S 21'
23	09	56	ON	
30	12	59	Max dec	27° N 19'

LONGITUDES

Date	Chiron ⚷ ° '	Ceres ⚳ ° '	Pallas ⚴ ° '	Juno ⚵ ° '	Vesta ⚶ ° '	Black Moon Lilith ⚸ ° '
01	22 ♓ 08	06 ♋ 52	04 ♈ 14	06 ♌ 46	07 ♎ 18	23 ♓ 55
11	21 ♓ 53	06 ♋ 45	02 ♈ 27	09 ♌ 38	12 ♎ 08	25 ♓ 02
21	21 ♓ 44	05 ♋ 58	01 ♈ 25	11 ♌ 52	16 ♎ 55	26 ♓ 09
31	21 ♓ 40	04 ♋ 31	01 ♈ 10	13 ♌ 21	21 ♎ 38	27 ♓ 16

All ephemeris data is given at 12.00 UT and the Moon's longitude is additionally given for 24.00 UT

ASPECTARIAN

01 Tuesday
00 59 ☽ ⊼ ☉
07 30 ☽ □ ♂
08 31 ☉ ⚼ ♄
09 19 ☽ ± ♀
13 15 ☽ ± ☉
21 06 ☽ ∠ ♃
23 06 ☽ □ ♇

02 Wednesday
01 53 ☽ ⊼ ♆
03 57 ☽ ⚼ ♀
04 39 ☽ □ ♅
05 36 ☽ □ ♄
05 40 ☿ Q ♇
07 26 ☽ ⚼ ☉
11 23 ♀ ∠ ♅
13 01 ☽ ± ♆
13 51 ☽ ⊥ ♃
18 33 ☽ Q ♂
20 23 ♀ ⚼ ♄
21 54 ☽ ⊼ ☿

03 Thursday
00 55 ☽ ⊻ ♃
05 33 ☽ ⚼ ♆
08 16 ☽ Q ♇
09 13 ☽ ± ☿
10 23 ☽ △ ♀
13 08 ☽ △ ☉
13 44 ☽ Q ♅
17 23 ☽ ⚹ ♂

04 Friday
01 55 ☽ ⚼ ☿
05 54 ☽ ⚹ ♇
08 37 ☽ △ ♆
11 18 ☽ ⚹ ♅
11 51 ☽ △ ♄
21 17 ☽ ∠ ♂

05 Saturday
05 00 ☽ △ ☿
06 42 ☽ ☌ ♃
08 23 ☽ ∠ ♇
13 16 ☽ # ☿
13 43 ☽ ∠ ♅
13 43 ☽ ⊥ ♂
14 05 ☽ ⚼ ♄
20 49 ☽ □ ♀
22 18 ☽ □ ☉

06 Sunday
00 00 ☽ ⊥ ♇
00 32 ☽ ⊻ ♂
05 17 ☽ ⊥ ♅
05 34 ☽ ± ♄
10 18 ☽ ⊻ ♇
11 40 ☽ ∥ ♃
12 56 ☽ □ ♆
15 33 ☽ ⊻ ♅
15 46 ☽ ⊼ ♄
17 54 ☿ St R
22 47 ☽ ∥ ♇

07 Monday
01 58 ☽ # ♆
03 49 ☽ # ☉
06 27 ☽ Q ♀
07 15 ☽ Q ☉
07 26 ☽ # ♀
08 34 ☽ □ ☿
10 09 ☽ ⊻ ♃
19 19 ☽ Q ♆
20 12 ☽ ⊥ ♃

08 Tuesday
04 22 ☽ ⚹ ♀
04 45 ☽ ⚹ ☉
05 13 ☽ ☌ ♂
08 21 ♄ ☍ ♅
11 10 ☽ ∠ ♃
12 38 ☽ ☌ ♇
14 06 ☽ Q ☿
14 43 ☉ ∥ ♆
15 14 ☽ ⚹ ♆
16 02 ☽ ∥ ♂
17 44 ☽ ☍ ♄
17 47 ☽ ☌ ♅
19 59 ☉ ⚹ ♂
21 43 ♀ ⚹ ♂
23 49 ☽ # ♄

09 Wednesday
00 40 ☉ ☌ ♀
05 24 ☽ ∥ ♅
07 15 ☽ ∠ ☉
07 23 ☽ ∠ ♀
09 20 ☽ ⚹ ☿
11 51 ☽ ⚹ ♃
15 53 ☽ ∠ ♆
23 02 ☽ ⊥ ☉
23 27 ☽ ⊥ ♀

10 Thursday
05 17 ☽ # ♅
06 36 ☽ ⊥ ♆
07 29 ☽ Q ♃
08 24 ☽ ⊻ ♂
09 02 ☽ ∠ ☿
09 32 ☽ ⊻ ☉
10 09 ☽ ⊻ ♀
10 56 ☽ ∥ ♄
12 17 ☿ Q ♇
13 12 ♀ ∥ ♆
13 48 ☽ ⊻ ♇
16 25 ☽ ⊻ ♆
17 00 ☽ # ♂
18 35 ☽ ⊥ ♂
18 40 ☽ ⊼ ♄
18 58 ☽ ⊻ ♅

23 07 ☽ ⊥ ☿
23 29 ☉ # ♇
23 36 ☽ ⊥ ♇

11 Friday
04 27 ☽ ± ♄
04 48 ☽ ⊥ ♅
08 28 ☽ ⊻ ☿
09 53 ☽ ∠ ♂
12 58 ☽ □ ♃
14 21 ☽ ∠ ♇
19 11 ☽ ⚼ ♄
19 36 ☽ ∠ ♅

12 Saturday
05 32 ♀ ⚹ ♇
06 57 ♀ # ♇
08 07 ☽ ∥ ♆
10 56 ☽ # ♇
11 14 ☽ ∥ ♀
11 37 ☽ ⚹ ♂
12 54 ☽ ∥ ☉
14 26 ☽ ☌ ☉
15 10 ☽ ⚹ ♇
16 04 ☽ ☌ ♀
17 54 ☽ ☌ ♆
19 59 ☽ △ ♄
20 31 ☽ ⚹ ♅
20 49 ☿ Q ♂
21 38 ☽ # ♃

13 Sunday
01 03 ☉ ⚹ ♇
06 09 ☽ ∥ ☿
07 19 ☽ ☌ ☿
08 32 ☽ Q ♂
11 20 ☽ Q ♇
13 38 ♀ ☌ ♆
15 01 ☽ △ ♃
16 49 ☽ Q ♅

14 Monday
12 13 ♀ △ ♄
16 44 ☽ □ ♂
16 54 ☽ ⚼ ♃
17 43 ☉ ☌ ♆
18 21 ☽ □ ♇
20 35 ♀ ⚹ ♅
20 48 ♂ ∠ ♃
21 17 ☽ ⊻ ♆
21 33 ☽ ⊻ ☉
23 15 ☽ □ ♄

15 Tuesday
00 03 ☽ □ ♅
00 24 ☽ ⊻ ♀
07 10 ☽ ⊻ ☿
07 55 ☽ ⊥ ♆
08 48 ☽ ± ♃
09 01 ☽ ⊥ ☉
12 08 ☽ ⊥ ♀
16 58 ☽ ⊥ ☿
19 34 ☽ ⊼ ♃
20 10 ☉ △ ♄

16 Wednesday
00 08 ☽ ∠ ♆
01 06 ☿ # ♃
02 33 ☽ ∠ ☉
06 07 ☽ ∠ ♀
07 31 ☽ Q ♄
07 37 ☉ ∥ ♀
07 49 ☽ ∠ ☿
08 39 ☉ ⚹ ♅
09 11 ♂ ☌ ♇
15 38 ☿ ☌ ♀
23 44 ☿ ∥ ♀

17 Thursday
00 41 ☽ △ ♇
01 14 ☉ ∥ ☿
01 22 ☽ △ ♂
03 54 ☽ ⚹ ♆
05 46 ☽ ⚹ ♄
06 52 ☽ △ ♅
08 42 ☽ ⚹ ☉
09 05 ☽ ⚹ ☿
10 58 ☉ ☌ ☿
13 04 ☽ ⚹ ♀

18 Friday
02 47 ☽ Q ♆
03 37 ☽ ☍ ♃
05 14 ☽ ⚼ ♇
05 27 ☿ ⚹ ♅
05 47 ☽ Q ☿
07 08 ♂ # ♄
07 15 ☽ ⚼ ♂
09 40 ☽ Q ☉
10 24 ☽ ∠ ♄
11 39 ☽ ⚼ ♅
14 43 ☽ Q ♀
17 08 ☿ △ ♄
21 17 ♀ # ♃
22 46 ☽ ± ♇

19 Saturday
01 38 ☽ ± ♂
03 58 ☽ ⊥ ♄
05 19 ☽ ± ♅
09 45 ☿ ☌ ♆
10 37 ☽ ⊼ ♇
10 50 ☽ ∥ ♀
11 14 ☿ ⚹ ♂
11 53 ☽ # ♃
12 15 ☽ ∥ ☉
13 06 ☿ # ♇
13 41 ☽ □ ☿
14 04 ☽ ⊼ ♂
14 06 ☽ □ ♆
14 53 ♂ ⚹ ♆

15 50 ☽ ⊻ ♄
17 14 ☽ ⊼ ♅
18 42 ☉ # ♃

20 Sunday
00 20 ☽ □ ☉
00 45 ☽ # ♇
02 35 ☽ ∥ ☿
03 45 ☽ ∥ ♆
06 26 ☽ □ ♀
11 21 ☿ ∥ ♆
14 53 ☽ ⊼ ♃
21 15 ☿ ⚹ ♇

21 Monday
02 59 ☽ ± ♃
03 24 ♂ ☍ ♄
10 23 ♃ St R
20 58 ☽ △ ☿
21 12 ☽ ⚼ ♃
22 25 ♀ Q ♇
23 00 ☽ ☍ ♇

22 Tuesday
02 41 ☽ △ ♆
04 12 ☽ ☌ ♄
05 26 ☽ ☍ ♂
05 52 ☽ ☍ ♅
11 55 ☽ ∥ ♄
15 07 ♂ ☌ ♅
16 07 ☽ # ♂
18 33 ☽ △ ☉
20 00 ☽ # ♅

23 Wednesday
01 24 ☽ ⚼ ☿
02 16 ☽ △ ♀
03 34 ☽ △ ♃
09 08 ☽ ⚼ ♆
14 36 ♀ △ ♃
18 38 ☽ ± ☿
23 43 ☽ ∥ ♅

24 Thursday
02 18 ☽ ∥ ♂
03 17 ☽ ± ♆
03 35 ☽ ⚼ ☉
06 10 ☽ ⊼ ☿
07 45 ☽ # ♄
11 36 ☽ ⊼ ♇
11 57 ☽ ⚼ ♀
15 18 ☽ ⊼ ♆
16 35 ☽ ⊻ ♄
16 37 ♀ Q ♅
18 25 ☽ ⊼ ♅
20 44 ☽ ⊼ ♂
23 09 ☽ ± ☉
23 32 ☽ ± ♇

25 Friday
03 47 ☉ Q ♇
04 28 ☽ ⊥ ♄
06 20 ☽ ± ♅
07 52 ☽ ± ♀
09 09 ☽ ± ♂
12 04 ☽ ⊼ ☉
15 22 ☽ □ ♃
17 17 ☽ ⚼ ♇
20 59 ☽ ⊼ ♀
22 08 ☽ ∠ ♄

26 Saturday
00 01 ☽ ⚼ ♅
03 35 ☽ ⚼ ♂
03 58 ☽ # ☿
05 10 ♀ Q ♂
14 25 ☽ # ♆
15 33 ♄ St D
15 58 ☽ ☍ ☿
17 00 ☽ ∥ ♇
17 50 ☿ St D
22 22 ☽ △ ♇

27 Sunday
01 24 ♂ ∥ ♅
02 02 ☽ ☍ ♆
02 09 ☽ Q ♃
03 04 ☽ ⚹ ♄
03 22 ☉ △ ♃
04 59 ☽ △ ♅
05 28 ☽ ∥ ♃
09 42 ☽ △ ♂
14 53 ☽ # ☉
19 55 ☽ # ♀

28 Monday
00 48 ☽ ⚹ ♃
01 45 ☽ Q ♄
02 40 ☽ ☍ ☉
12 25 ☽ ☍ ♀
15 38 ☉ Q ♅

29 Tuesday
01 13 ☽ ⊼ ☿
04 32 ☽ ∠ ♃
06 32 ☽ □ ♇
10 08 ☽ ⊼ ♆
10 59 ☽ □ ♄
12 32 ☽ ± ☿
12 55 ☽ □ ♅
19 42 ☽ □ ♂
20 52 ☽ ⊥ ♃
21 04 ☽ ± ♆

30 Wednesday
05 27 ☽ ⚼ ☿
07 40 ☽ ⊻ ♃
13 17 ☽ ⚼ ♆
14 03 ☽ ⊼ ☉
15 05 ☽ Q ♇
21 21 ☽ Q ♅

DECEMBER 1966

LONGITUDES

Date	Sidereal time h m s	Sun ☉ ° ' "	Moon ☽ ° ' "	Moon ☽ 24.00 ° ' "	Mercury ☿ ° '	Venus ♀ ° '	Mars ♂ ° '	Jupiter ♃ ° '	Saturn ♄ ° '	Uranus ♅ ° '	Neptune ♆ ° '	Pluto ♇ ° '
01	16 39 35	08 ♐ 50 37	20 ♋ 17 32	27 ♋ 07 29	18 ♏ 55	14 ♐ 23	28 ♍ 38	04 ♌ 19	22 ♓ 56	24 ♍ 02	22 ♏ 32	20 ♍ 31
02	16 43 32	09 51 26	04 ♌ 00 04	10 ♌ 55 04	19 41	15 39	29 10	04 R 17	22 57	24 04	22 34	20 31
03	16 47 29	10 52 17	17 ♌ 52 17	24 ♌ 51 28	20 33	16 54	29 ♍ 43	04 15	22 57	24 05	22 36	20 32
04	16 51 25	11 53 08	01 ♍ 52 26	08 ♍ 54 58	21 30	18 10	00 ♎ 15	04 12	22 58	24 07	22 38	20 33
05	16 55 22	12 54 01	15 ♍ 58 53	23 ♍ 03 58	22 31	19 25	00 47	04 09	22 59	24 08	22 40	20 33
06	16 59 18	13 54 56	00 ♎ 09 59	07 ♎ 16 44	23 37	20 40	01 19	04 07	23 00	24 10	22 43	20 34
07	17 03 15	14 55 52	14 ♎ 23 54	21 ♎ 31 14	24 46	21 56	01 51	04 04	23 01	24 11	22 45	20 35
08	17 07 11	15 56 49	28 ♎ 38 23	05 ♏ 44 59	25 58	23 11	02 22	04 00	23 02	24 12	22 47	20 35
09	17 11 08	16 57 47	12 ♏ 50 36	19 ♏ 54 49	27 13	24 27	02 54	03 57	23 04	24 13	22 49	20 36
10	17 15 04	17 58 46	26 ♏ 57 09	03 ♐ 57 06	28 30	25 42	03 26	03 53	23 05	24 14	22 51	20 36
11	17 19 01	18 59 47	10 ♐ 54 13	17 ♐ 48 02	29 ♏ 49	26 57	03 57	03 49	23 07	24 16	22 53	20 36
12	17 22 58	20 00 48	24 ♐ 38 06	01 ♑ 24 04	01 ♐ 09	28 13	04 28	03 45	23 08	24 17	22 55	20 37
13	17 26 54	21 01 51	08 ♑ 05 37	14 ♑ 42 31	02 31	29 ♐ 28	04 59	03 41	23 10	24 18	22 57	20 37
14	17 30 51	22 02 54	21 ♑ 14 38	27 ♑ 41 55	03 55	00 ♑ 44	05 30	03 37	23 12	24 18	22 59	20 37
15	17 34 47	23 03 57	04 ♒ 04 25	10 ♒ 22 16	05 19	01 59	06 01	03 32	23 14	24 19	23 01	20 38
16	17 38 44	24 05 01	16 ♒ 35 43	22 ♒ 45 04	06 44	03 14	06 32	03 28	23 16	24 20	23 03	20 38
17	17 42 40	25 06 05	28 ♒ 50 44	04 ♓ 53 09	08 11	04 30	07 02	03 23	23 18	24 21	23 05	20 38
18	17 46 37	26 07 10	10 ♓ 52 51	16 ♓ 50 25	09 38	05 45	07 33	03 18	23 20	24 22	23 07	20 38
19	17 50 33	27 08 15	22 ♓ 46 25	28 ♓ 41 30	11 05	07 00	08 03	03 13	23 23	24 22	23 09	20 39
20	17 54 30	28 09 21	04 ♈ 36 18	10 ♈ 31 30	12 33	08 16	08 33	03 07	23 25	24 23	23 11	20 39
21	17 58 27	29 ♐ 10 26	16 ♈ 27 44	22 ♈ 25 40	14 02	09 31	09 03	03 02	23 28	24 23	23 13	20 39
22	18 02 23	00 ♑ 11 32	28 ♈ 25 54	04 ♉ 29 03	15 31	10 47	09 32	02 56	23 30	24 24	23 15	20 39
23	18 06 20	01 12 38	10 ♉ 35 39	16 ♉ 46 14	17 01	12 02	10 02	02 50	23 33	24 24	23 17	20 39
24	18 10 16	02 13 44	23 ♉ 01 12	29 ♉ 20 56	18 31	13 17	10 32	02 44	23 36	24 25	23 19	20 R 39
25	18 14 13	03 14 51	05 ♊ 45 42	12 ♊ 15 39	20 01	14 33	11 01	02 38	23 39	24 25	23 21	20 39
26	18 18 09	04 15 58	18 ♊ 50 52	25 ♊ 31 17	21 32	15 48	11 30	02 32	23 42	24 25	23 22	20 39
27	18 22 06	05 17 05	02 ♋ 16 45	09 ♋ 06 56	23 03	17 03	11 59	02 26	23 45	24 25	23 24	20 39
28	18 26 02	06 18 12	16 ♋ 01 30	22 ♋ 59 54	24 34	18 19	12 28	02 19	23 49	24 25	23 26	20 38
29	18 29 59	07 19 20	00 ♌ 01 37	07 ♌ 06 01	26 06	19 34	12 56	02 12	23 52	24 26	23 28	20 38
30	18 33 56	08 20 28	14 ♌ 12 27	21 ♌ 20 17	27 38	20 49	13 25	02 06	23 55	24 R 26	23 29	20 38
31	18 37 52	09 ♑ 21 36	28 ♌ 28 53	05 ♍ 37 41	29 ♐ 10	22 ♑ 05	13 ♎ 53	01 ♌ 59	23 ♓ 59	24 ♍ 25	23 ♏ 31	20 ♍ 38

Date	Moon True ☊ ° '	Moon Mean ☊ ° '	Moon ☽ Latitude ° '
01	16 ♉ 02	14 ♉ 56	04 N 43
02	15 R 59	14 53	05 07
03	15 56	14 50	05 14
04	15 55	14 47	05 02
05	15 D 54	14 44	04 32
06	15 55	14 40	03 45
07	15 57	14 37	02 44
08	15 58	14 34	01 34
09	15 59	14 31	00 N 17
10	15 R 58	14 28	01 S 00
11	15 56	14 25	02 12
12	15 52	14 21	03 15
13	15 47	14 18	04 06
14	15 41	14 15	04 43
15	15 34	14 12	05 04
16	15 28	14 09	05 10
17	15 23	14 05	05 02
18	15 20	14 02	04 40
19	15 18	13 59	04 07
20	15 D 18	13 56	03 23
21	15 19	13 53	02 30
22	15 21	13 50	01 30
23	15 22	13 46	00 S 26
24	15 R 22	13 43	00 N 41
25	15 20	13 40	01 48
26	15 15	13 37	02 50
27	15 09	13 34	03 45
28	15 00	13 30	04 28
29	14 51	13 27	04 56
30	14 42	13 24	05 06
31	14 ♉ 35	13 ♉ 21	04 N 57

DECLINATIONS

Date	Sun ☉ ° '	Moon ☽ ° '	Mercury ☿ ° '	Venus ♀ ° '	Mars ♂ ° '	Jupiter ♃ ° '	Saturn ♄ ° '	Uranus ♅ ° '	Neptune ♆ ° '	Pluto ♇ ° '
01	21 S 47	26 N 35	15 S 00	22 S 40	02 N 11	19 N 34	04 S 57	03 N 04	16 S 46	17 N 14
02	21 56	24 14	15 15	22 51	01 59	19 35	04 56	03 03	16 47	17 15
03	22 05	20 27	15 32	23 01	01 46	19 35	04 56	03 03	16 47	17 15
04	22 13	15 30	15 51	23 11	01 34	19 36	04 55	03 02	16 48	17 15
05	22 21	09 42	16 11	23 19	01 22	19 37	04 55	03 02	16 49	17 15
06	22 29	03 N 23	16 33	23 27	01 09	19 38	04 54	03 01	16 49	17 15
07	22 36	03 S 09	16 56	23 35	00 57	19 39	04 54	03 01	16 50	17 16
08	22 42	09 32	17 19	23 41	00 45	19 40	04 53	03 00	16 50	17 16
09	22 48	15 25	17 42	23 47	00 33	19 41	04 52	03 00	16 51	17 16
10	22 54	20 27	18 06	23 52	00 21	19 42	04 51	02 59	16 51	17 17
11	22 59	24 16	18 30	23 57	00 N 09	19 43	04 51	02 59	16 52	17 17
12	23 04	26 35	18 54	24 00	00 S 03	19 44	04 50	02 59	16 52	17 17
13	23 09	27 18	19 18	24 03	00 15	19 45	04 49	02 58	16 53	17 17
14	23 12	26 25	19 41	24 05	00 27	19 46	04 48	02 58	16 53	17 18
15	23 16	24 10	20 04	24 07	00 38	19 47	04 47	02 58	16 54	17 18
16	23 19	20 47	20 26	24 07	00 50	19 49	04 46	02 57	16 54	17 19
17	23 21	16 35	20 47	24 07	01 01	19 50	04 45	02 57	16 54	17 19
18	23 23	11 48	21 08	24 07	01 13	19 51	04 44	02 57	16 55	17 19
19	23 25	06 39	21 28	24 05	01 24	19 53	04 43	02 57	16 55	17 20
20	23 26	01 S 16	21 48	24 03	01 36	19 54	04 42	02 56	16 56	17 20
21	23 27	04 N 10	22 06	24 00	01 47	19 56	04 40	02 56	16 56	17 21
22	23 27	09 31	22 24	23 56	01 58	19 57	04 39	02 56	16 57	17 21
23	23 26	14 36	22 40	23 51	02 09	19 59	04 38	02 56	16 57	17 22
24	23 26	19 12	22 56	23 46	02 20	20 00	04 37	02 56	16 58	17 22
25	23 24	23 02	23 10	23 40	02 31	20 02	04 35	02 56	16 58	17 23
26	23 23	25 48	23 23	23 33	02 42	20 03	04 34	02 56	16 58	17 23
27	23 20	27 10	23 36	23 26	02 53	20 05	04 32	02 56	16 59	17 24
28	23 18	26 55	23 47	23 18	03 04	20 07	04 31	02 56	16 59	17 24
29	23 15	24 58	23 57	23 09	03 14	20 08	04 29	02 56	17 00	17 25
30	23 11	21 26	24 05	22 59	03 25	20 10	04 28	02 56	17 00	17 25
31	23 S 07	16 N 38	24 S 13	22 S 49	03 S 35	20 N 12	04 S 26	02 N 56	17 S 00	17 N 26

ZODIAC SIGN ENTRIES

Date	h	m	Planets
02	05	02	☽ ♌
04	00	55	♂ ♎
04	08	48	☽ ♍
06	11	43	☽ ♎
08	14	18	☽ ♏
10	17	13	☽ ♐
11	15	27	☿ ♐
12	21	30	☽ ♑
13	22	09	♀ ♑
15	04	19	☽ ♒
17	14	17	☽ ♓
20	02	39	☽ ♈
22	07	28	☉ ♑
22	15	07	☽ ♉
25	01	14	☽ ♊
27	07	58	☽ ♋
29	11	57	☽ ♌
31	14	33	☽ ♍

LATITUDES

Date	Mercury ☿ ° '	Venus ♀ ° '	Mars ♂ ° '	Jupiter ♃ ° '	Saturn ♄ ° '	Uranus ♅ ° '	Neptune ♆ ° '	Pluto ♇ ° '
01	02 N 33	00 S 08	01 N 47	00 N 23	02 S 20	00 N 45	01 N 41	14 N 42
04	02 22	00 16	01 49	00 24	02 19	00 46	01 41	14 44
07	02 06	00 23	01 50	00 25	02 19	00 46	01 42	14 46
10	01 46	00 30	01 52	00 25	02 18	00 46	01 42	14 47
13	01 24	00 37	01 53	00 26	02 17	00 46	01 42	14 49
16	01 02	00 43	01 55	00 26	02 17	00 46	01 42	14 50
19	00 39	00 50	01 57	00 27	02 16	00 46	01 42	14 52
22	00 N 16	00 56	01 58	00 27	02 15	00 46	01 42	14 53
25	00 S 06	01 01	02 00	00 28	02 15	00 46	01 42	14 55
28	00 27	01 07	02 01	00 28	02 14	00 47	01 42	14 57
31	00 S 46	01 S 12	02 N 03	00 N 29	02 S 14	00 N 47	01 N 42	14 N 58

DATA

Julian Date	2439461
Delta T	+37 seconds
Ayanamsa	23° 23' 29"
Synetic vernal point	05° ♓ 43' 31"
True obliquity of ecliptic	23° 26' 43"

MOON'S PHASES, APSIDES AND POSITIONS ☽

Date	h	m	Phase	Longitude ° '	Eclipse Indicator
05	06	22	☾	12 ♍ 40	
12	03	13	●	19 ♐ 38	
19	21	41	☽	27 ♓ 33	
27	17	43	○	05 ♋ 32	

Day	h	m		
07	17	41	Perigee	
19	23	47	Apogee	
07	00	26	0S	
13	10	27	Max dec	27° S 18'
20	17	37	0N	
27	20	21	Max dec	27° N 17'

LONGITUDES

Date	Chiron ⚷ ° '	Ceres ⚳ ° '	Pallas ⚴ ° '	Juno ⚵ ° '	Vesta ⚶ ° '	Black Moon Lilith ⚸ ° '
01	21 ♓ 40	04 ♋ 31	01 ♈ 10	13 ♌ 21	21 ♎ 38	27 ♓ 16
11	21 ♓ 42	02 ♋ 33	01 ♈ 40	13 ♌ 59	26 ♎ 17	28 ♓ 23
21	21 ♓ 49	00 ♋ 15	02 ♈ 52	13 ♌ 43	00 ♏ 49	29 ♓ 30
31	22 ♓ 02	27 ♊ 54	04 ♈ 41	12 ♌ 34	05 ♏ 14	00 ♈ 36

All ephemeris data is given at 12.00 UT and the Moon's longitude is additionally given for 24.00 UT

ASPECTARIAN

h m	Aspects
01 Thursday	
00 30	☽ ⊼ ♀
01 35	☽ ± ☉
05 16	☽ Q ♂
09 27	☽ △ ☿
12 11	☽ ± ♀
12 23	☽ ✶ ♇
15 57	☽ △ ♆
16 39	☽ △ ♄
18 36	☽ ✶ ♅
18 45	☽ ⚼ ☉
02 Friday	
03 14	☽ ✶ ♂
05 34	☽ ⚼ ♀
12 29	☽ ☌ ♃
14 39	☽ ∠ ♇
18 51	☽ ⚼ ♄
20 48	☽ ∠ ♅
21 20	☽ ⋕ ♀
22 58	☽ △ ☉
03 Saturday	
02 58	☽ ⋕ ☉
06 15	☽ ⊥ ♇
06 20	☽ ∠ ♂
10 10	☽ △ ♀
10 25	☽ ± ♄
11 44	☿ ✶ ♇
12 23	☽ ⊥ ♅
16 34	☽ ∥ ♃
16 35	☽ ⊻ ♇
16 55	☽ □ ☿
20 09	☽ □ ♆
20 45	☽ ⊼ ♄
22 26	☽ ⊥ ♂
22 42	☽ ⊻ ♅
04 Sunday	
04 04	☽ ∥ ♇
06 10	☽ ⋕ ♆
09 07	☽ ⊻ ♂
10 33	☽ ⋕ ☿
15 58	☽ ⊻ ♃
05 Monday	
01 59	☽ Q ☿
02 09	☽ ⊥ ♃
02 58	☽ Q ♆
06 06	☉ Q ♂
06 22	☽ □ ☉
07 14	♀ ⚼ ♃
15 37	☿ ☌ ♆
17 22	☽ ∠ ♃
18 23	☽ □ ♀
19 45	☽ ☌ ♇
22 31	☿ △ ♄
23 22	☽ ✶ ♆
23 52	☽ ☍ ♄
23 59	☽ ✶ ☿
06 Tuesday	
01 50	☽ ☌ ♅
06 19	☽ ⋕ ♄
09 58	♀ □ ♇
13 20	☽ ∥ ♅
14 01	☽ ☌ ♂
15 11	☽ Q ☉
18 38	☽ ✶ ♃
20 27	☽ ∥ ♂
23 46	☿ ✶ ♅
07 Wednesday	
00 46	☽ ∠ ♆
03 29	☽ ∠ ☿
03 44	☽ Q ♀
04 10	☽ ⋕ ♂
05 32	☿ ∥ ♆
11 29	☽ ⋕ ♅
12 58	☽ ✶ ☉
14 47	☽ Q ♃
15 58	☽ ⊥ ♆
18 26	☽ ∥ ♄
20 01	☽ ⊥ ☿
22 25	☽ ⊻ ♇
08 Thursday	
01 55	☽ ✶ ♀
02 06	☽ ⊻ ♆
02 33	☽ ⊼ ♄
04 03	♀ ⊻ ♆
04 30	☽ ⊻ ♅
07 04	☽ ⊻ ☿
08 32	☽ ⊥ ♇
09 06	☿ ⋕ ♇
09 10	♀ □ ♄
12 40	☽ ± ♄
14 38	☽ ⊥ ♅
16 11	☽ ∠ ☉
18 33	☽ ⊻ ♂
21 01	☽ □ ♃
23 44	☽ ∠ ♇
09 Friday	
03 54	☽ ⚼ ♄
05 04	☽ ⊥ ♂
05 41	☽ ∠ ♀
05 52	☽ ∠ ♅
07 44	♀ □ ♅
08 34	☽ ⊥ ☉
18 19	☽ ∥ ♆
19 32	☽ ⊻ ☉
20 16	☽ ⋕ ♇
20 55	☽ ∠ ♂
22 26	☽ ⊥ ♀
23 10	☽ ∥ ☿
10 Saturday	
01 10	☽ ✶ ♇
04 59	☽ ☌ ♆
05 23	☽ △ ♄
07 22	☽ ✶ ♅
08 01	☽ ⋕ ♃
09 39	☽ ⊻ ♀
14 54	☽ ☌ ☿
21 41	☽ Q ♇
23 32	☽ ✶ ♂
23 50	☽ △ ♃
11 Sunday	
02 45	☽ ∥ ☉
03 58	☽ Q ♅
06 53	♂ ✶ ♃
08 09	☉ ⚼ ♃
09 33	☽ ∥ ♀
21 07	☽ Q ♂
12 Monday	
01 43	☽ ⚼ ♃
03 13	☽ ☌ ☉
03 45	♀ ± ♃
04 55	☽ □ ♇
08 58	☽ ⊻ ♆
09 21	☽ □ ♄
11 22	☽ □ ♅
17 30	☽ ± ♃
18 59	☽ ☌ ♀
19 36	☽ ⊥ ♆
13 Tuesday	
00 51	☽ ⊻ ☿
01 53	♀ ⊥ ♆
02 14	☉ □ ♇
04 08	☽ ⊼ ♃
06 11	☽ □ ♂
11 45	☽ ∠ ♆
12 52	☽ ⊥ ☿
13 44	☿ Q ♇
17 34	☽ Q ♄
14 Wednesday	
07 11	☽ ∠ ☿
07 12	☿ △ ♃
10 51	☽ △ ♇
13 37	☽ ⊻ ☉
15 14	☽ ✶ ♆
15 38	☽ ✶ ♄
17 41	☽ △ ♅
17 59	☿ ⋕ ♃
15 Thursday	
01 45	☽ ⊥ ☉
07 37	☽ ⊻ ♀
10 55	☉ ⊻ ♆
11 00	☽ ☍ ♃
12 25	☽ ∥ ♀
13 48	☽ Q ♆
14 39	☽ ✶ ☿
14 57	☽ ⚼ ♇
15 50	☽ △ ♂
16 01	☉ □ ♄
19 01	☽ ∥ ☉
19 56	☽ ∠ ♄
20 15	☽ ⊥ ♀
20 16	☽ ∠ ☉
22 00	☽ ⚼ ♅
16 Friday	
05 09	☿ Q ♅
06 26	☿ ✶ ♂
08 12	☽ ± ♇
13 18	☽ ⊥ ♄
14 03	☽ ∥ ☿
15 23	☽ ± ♅
15 33	☽ ∠ ♀
16 00	♀ ⊼ ♃
16 43	☽ Q ☿
17 57	☽ ⋕ ♃
18 01	☉ □ ♅
19 52	☽ ⊼ ♇
22 01	☽ ⚼ ♂
17 Saturday	
00 38	☽ □ ♆
01 03	☽ ⊻ ♄
03 07	☽ ⊼ ♅
03 57	☽ ✶ ☉
08 06	☽ ⋕ ♇
10 18	☽ ∥ ♆
16 32	☽ ± ♂
20 56	☽ ⊼ ♃
18 Sunday	
00 32	☽ ✶ ♀
05 01	☽ ⊼ ♂
05 57	☽ Q ☉
08 51	☽ ± ♃
09 08	☽ □ ☿
19 Monday	
02 50	☽ ⚼ ♃
03 29	☽ Q ♀
07 41	☽ ☍ ♇
12 46	☽ △ ♆
13 14	☽ ☌ ♄
13 34	☉ ± ♃
15 14	☽ ☍ ♅
17 31	♂ ∠ ♆
20 44	☽ ∥ ♄
21 41	☽ □ ☉
20 Tuesday	
04 36	☽ ⋕ ♅
09 01	☽ △ ♃
10 28	♀ ∠ ♆
10 37	☽ ∥ ♂
19 17	☽ ⚼ ♆
20 18	☽ □ ♀
20 21	☽ ☍ ♂
20 58	♀ □ ♂
21 Wednesday	
01 06	☽ ⋕ ♂
06 24	☽ △ ☿
06 35	☽ ∥ ♅
13 04	☉ ⊥ ♆
13 32	☽ ± ♆
14 15	☽ ⋕ ♄
20 26	☽ ⊼ ♇
22 Thursday	
01 37	☽ ⊼ ♆
02 08	☽ ⊻ ♄
03 56	☽ ⊼ ♅
08 27	☽ ± ♇
14 09	☽ ⊥ ♄
15 49	☽ △ ☉
15 55	☽ ± ♅
16 44	☽ ⚼ ☿
20 52	☽ □ ♃
23 Friday	
02 18	☽ ⚼ ♇
02 30	♀ Q ♄
07 59	☽ ∠ ♄
09 40	☽ ⚼ ♅
10 52	☽ ⊼ ♂
12 56	☽ ± ☿
15 07	☽ △ ♀
15 35	♇ St R
23 01	☽ ± ♂
23 54	☽ ⚼ ☉
23 55	☽ ⋕ ♆
24 Saturday	
00 22	☿ ⚼ ♃
02 03	☽ ∥ ♇
02 11	☽ ⊼ ☿
07 28	☽ △ ♇
07 40	☽ Q ♃
12 33	☽ ☍ ♆
13 07	☽ ✶ ♄
14 39	☽ △ ♅
16 38	☽ ∥ ♃
16 58	☽ ⚼ ♂
18 38	☽ ± ☉
22 54	☉ ⊼ ♃
23 06	☽ ⚼ ♀
25 Sunday	
06 13	☽ ✶ ♃
06 55	☽ ⊼ ☉
11 48	☽ Q ♄
12 59	☽ ⋕ ☿
14 38	☽ ⋕ ☉
16 28	☽ ⋕ ♀
17 42	☽ ± ♀
21 58	☿ □ ♇
22 05	☽ △ ♂
26 Monday	
05 53	☽ ⊼ ♀
09 38	☽ ∠ ♃
10 36	☉ ∥ ☿
15 15	☽ □ ♇
17 28	☽ ☍ ☿
20 10	☽ ⊼ ♆
20 47	☽ □ ♄
22 02	☽ □ ♅
23 56	☿ ∥ ♀
27 Tuesday	
01 42	☽ ⊥ ♃
06 54	☽ ± ♆
12 15	☽ ⊻ ♃
17 42	☿ ⊻ ♆
17 43	☽ ☍ ☉
18 04	♂ ⋕ ♅
22 47	☽ ⚼ ♆
23 10	☽ Q ♇
23 32	☿ □ ♄
28 Wednesday	
05 36	☽ □ ♂
05 46	☽ Q ♅
09 28	☿ Q ♂
09 40	☿ □ ♅
11 45	☉ ∥ ♀
16 20	☽ ☍ ♀
19 57	☽ ✶ ♇
29 Thursday	
00 46	☽ △ ♆
01 26	☽ △ ♄
02 27	☽ ✶ ♅
04 29	☽ ⊼ ☿
13 33	☿ ± ♃
13 36	☽ Q ♂
15 41	☽ ☌ ♃
15 57	☽ ± ☿
19 40	☽ ⋕ ☿
21 31	☽ ∠ ♇
30 Friday	
01 08	☽ ⋕ ☉
01 20	☽ ⊼ ☉
02 11	☽ ⋕ ♀
03 03	☽ ⚼ ♄
03 56	☽ ∠ ♅
07 30	♅ St R
08 28	♀ △ ♇
09 02	☽ ⚼ ☿
10 37	☽ ✶ ♂
12 15	☽ ± ☉
12 43	☽ ⊥ ♇
15 38	☉ ∠ ♆
18 17	☽ ± ♄
18 54	☽ ∥ ♃
19 06	☽ ⊥ ♅
22 49	☽ ⊻ ♇
31 Saturday	
00 12	☽ ⊼ ♀
03 39	☽ □ ♆
04 25	☽ ⊼ ♄
04 33	☽ ⚼ ☉
05 11	☽ ⊻ ♅
08 23	☽ ∥ ♇
10 19	☽ ⋕ ♆
11 15	☽ ± ♀
12 42	☽ ∠ ♂
13 18	☽ △ ☿
15 58	♂ Q ♃
17 30	☿ ⊥ ♆
17 50	☽ ⊻ ♃

JANUARY 1967

LONGITUDES

Date	Sidereal time h m s	Sun ☉ ° ' "	Moon ☽ ° ' "	Moon ☽ 24.00 ° ' "	Mercury ☿ ° '	Venus ♀ ° '	Mars ♂ ° '	Jupiter ♃ ° '	Saturn ♄ ° '	Uranus ♅ ° '	Neptune ♆ ° '	Pluto ♇ ° '
01	18 41 49	10 ♑ 22 45	12 ♍ 46 09	19 ♍ 53 53	00 ♑ 43	23 ♑ 20	14 ♎ 21	01 ♌ 52	24 ♓ 02	24 ♍ 25	23 ♏ 33	20 ♍ 38
02	18 45 45	11 23 53	27 ♍ 00 31	04 ♎ 05 46	02 16	24 35	14 49	01 R 45	24 06	24 R 25	23 35	20 R 37
03	18 49 42	12 25 02	11 ♎ 09 27	18 ♎ 11 25	03 49	25 50	15 17	01 38	24 10	24 25	23 36	20 37
04	18 53 38	13 26 12	25 ♎ 11 36	02 ♏ 09 56	05 23	27 06	15 44	01 30	24 14	24 25	23 38	20 37
05	18 57 35	14 27 22	09 ♏ 06 24	16 ♏ 00 58	06 57	28 21	16 11	01 23	24 18	24 24	23 39	20 36
06	19 01 31	15 28 32	22 ♏ 53 34	29 ♏ 44 10	08 32	29 ♑ 36	16 38	01 15	24 22	24 24	23 41	20 36
07	19 05 28	16 29 42	06 ♐ 32 39	13 ♐ 18 55	10 06	00 ♒ 51	17 05	01 08	24 26	24 24	23 42	20 35
08	19 09 25	17 30 52	20 ♐ 02 48	26 ♐ 44 07	11 41	02 07	17 32	01 00	24 30	24 23	23 44	20 35
09	19 13 21	18 32 03	03 ♑ 22 40	09 ♑ 58 14	13 17	03 22	17 58	00 53	24 35	24 23	23 46	20 34
10	19 17 18	19 33 13	16 ♑ 30 38	22 ♑ 59 40	14 53	04 37	18 24	00 45	24 39	24 22	23 47	20 34
11	19 21 14	20 34 23	29 ♑ 25 10	05 ♒ 47 04	16 30	05 52	18 50	00 37	24 44	24 21	23 48	20 33
12	19 25 11	21 35 32	12 ♒ 05 17	18 ♒ 19 51	18 06	07 08	19 16	00 29	24 48	24 21	23 50	20 32
13	19 29 07	22 36 41	24 ♒ 30 49	00 ♓ 38 23	19 44	08 23	19 41	00 21	24 53	24 20	23 51	20 32
14	19 33 04	23 37 50	06 ♓ 42 45	12 ♓ 44 15	21 22	09 38	20 07	00 13	24 58	24 19	23 53	20 31
15	19 37 00	24 38 57	18 ♓ 43 15	24 ♓ 40 12	23 00	10 53	20 32	00 ♌ 05	25 03	24 18	23 54	20 30
16	19 40 57	25 40 04	00 ♈ 35 35	06 ♈ 30 00	24 39	12 08	20 56	29 ♋ 57	25 07	24 17	23 55	20 30
17	19 44 54	26 41 11	12 ♈ 24 00	18 ♈ 18 17	26 18	13 23	21 20	29 49	25 12	24 16	23 56	20 29
18	19 48 50	27 42 16	24 ♈ 13 29	00 ♉ 10 18	27 58	14 39	21 45	29 41	25 17	24 15	23 58	20 28
19	19 52 47	28 43 21	06 ♉ 09 26	12 ♉ 11 36	29 ♑ 39	15 54	22 09	29 33	25 23	24 14	23 59	20 27
20	19 56 43	29 ♑ 44 25	18 ♉ 17 28	24 ♉ 27 42	01 ♒ 20	17 09	22 32	29 25	25 28	24 13	24 00	20 26
21	20 00 40	00 ♒ 45 28	00 ♊ 42 53	07 ♊ 03 36	03 01	18 24	22 55	29 17	25 33	24 12	24 01	20 26
22	20 04 36	01 46 30	13 ♊ 30 17	20 ♊ 03 17	04 43	19 39	23 18	29 09	25 38	24 11	24 02	20 25
23	20 08 33	02 47 32	26 ♊ 42 50	03 ♋ 29 01	06 26	20 54	23 41	29 01	25 44	24 09	24 03	20 24
24	20 12 29	03 48 32	10 ♋ 21 46	17 ♋ 20 47	08 09	22 09	24 03	28 53	25 49	24 08	24 05	20 23
25	20 16 26	04 49 31	24 ♋ 25 40	01 ♌ 35 47	09 52	23 24	24 25	28 45	25 55	24 07	24 06	20 22
26	20 20 23	05 50 30	08 ♌ 50 22	16 ♌ 08 33	11 36	24 39	24 47	28 37	26 00	24 05	24 07	20 21
27	20 24 19	06 51 28	23 ♌ 29 19	00 ♍ 51 39	13 21	25 54	25 09	28 29	26 06	24 04	24 08	20 20
28	20 28 16	07 52 25	08 ♍ 14 29	15 ♍ 36 51	15 06	27 09	25 30	28 21	26 12	24 02	24 08	20 19
29	20 32 12	08 53 21	22 ♍ 57 50	00 ♎ 16 38	16 51	28 24	25 50	28 13	26 18	24 01	24 09	20 18
30	20 36 09	09 54 16	07 ♎ 32 36	14 ♎ 45 14	18 36	29 ♒ 39	26 11	28 06	26 24	23 59	24 10	20 16
31	20 40 05	10 ♒ 55 10	21 ♎ 54 09	28 ♎ 59 09	20 ♒ 22	00 ♓ 53	26 ♎ 31	27 ♋ 58	26 ♓ 29	23 ♍ 57	24 ♏ 11	20 ♍ 15

Date	Moon True ☊ ° '	Moon Mean ☊ ° '	Moon ☽ Latitude ° '
01	14 ♉ 29	13 ♉ 18	04 N 29
02	14 R 26	13 15	03 46
03	14 25	13 11	02 48
04	14 D 26	13 08	01 41
05	14 26	13 05	00 N 28
06	14 R 26	13 02	00 S 45
07	14 24	12 59	01 55
08	14 19	12 56	02 58
09	14 11	12 52	03 50
10	14 01	12 49	04 28
11	13 49	12 46	04 53
12	13 36	12 43	05 02
13	13 23	12 40	04 56
14	13 13	12 36	04 37
15	13 04	12 33	04 06
16	12 59	12 30	03 25
17	12 56	12 27	02 34
18	12 55	12 24	01 38
19	12 D 55	12 21	00 S 36
20	12 R 55	12 17	00 N 28
21	12 54	12 14	01 33
22	12 50	12 11	02 34
23	12 44	12 08	03 30
24	12 35	12 05	04 15
25	12 24	12 02	04 46
26	12 12	11 58	05 00
27	12 00	11 55	04 54
28	11 49	11 52	04 29
29	11 41	11 49	03 46
30	11 36	11 46	02 46
31	11 ♉ 33	11 ♉ 42	01 N 42

DECLINATIONS

Date	Sun ☉ ° '	Moon ☽ ° '	Mercury ☿ ° '	Venus ♀ ° '	Mars ♂ ° '	Jupiter ♃ ° '	Saturn ♄ ° '	Uranus ♅ ° '	Neptune ♆ ° '	Pluto ♇ ° '
01	23 S 02	10 N 55	24 S 19	22 S 38	03 S 46	20 N 13	04 S 25	02 N 56	17 S 01	17 N 26
02	22 57	04 N 38	24 24	22 26	03 56	20 15	04 23	02 56	17 01	17 27
03	22 52	01 S 50	24 28	22 14	04 06	20 17	04 21	02 56	17 02	17 28
04	22 46	08 11	24 30	22 01	04 16	20 19	04 20	02 56	17 02	17 28
05	22 40	14 05	24 31	21 47	04 26	20 20	04 18	02 56	17 02	17 29
06	22 33	19 14	24 31	21 33	04 36	20 22	04 16	02 57	17 03	17 29
07	22 26	23 18	24 29	21 18	04 46	20 24	04 14	02 57	17 03	17 30
08	22 18	26 01	24 26	21 02	04 55	20 26	04 12	02 57	17 03	17 31
09	22 10	27 14	24 21	20 46	05 05	20 28	04 10	02 57	17 04	17 31
10	22 01	26 52	24 15	20 29	05 14	20 30	04 09	02 58	17 04	17 32
11	21 52	25 03	24 08	20 12	05 23	20 31	04 07	02 58	17 04	17 33
12	21 43	22 00	23 59	19 54	05 33	20 33	04 05	02 58	17 05	17 33
13	21 33	18 00	23 48	19 35	05 42	20 35	04 03	02 58	17 05	17 34
14	21 23	13 20	23 36	19 16	05 51	20 37	04 01	02 59	17 05	17 35
15	21 12	08 14	23 23	18 56	05 59	20 39	03 59	02 59	17 05	17 35
16	21 01	02 S 54	23 08	18 36	06 08	20 41	03 56	03 00	17 06	17 36
17	20 49	02 N 32	22 52	18 15	06 17	20 42	03 54	03 00	17 06	17 37
18	20 38	07 53	22 34	17 54	06 25	20 44	03 52	03 00	17 06	17 38
19	20 25	13 01	22 15	17 32	06 33	20 46	03 50	03 01	17 06	17 38
20	20 13	17 44	21 54	17 10	06 42	20 48	03 48	03 01	17 07	17 39
21	20 00	21 49	21 31	16 47	06 50	20 50	03 46	03 02	17 07	17 40
22	19 46	24 59	21 07	16 24	06 58	20 52	03 43	03 02	17 07	17 41
23	19 32	26 54	20 41	16 00	07 05	20 53	03 41	03 03	17 07	17 41
24	19 18	27 16	20 15	15 36	07 13	20 55	03 39	03 04	17 08	17 42
25	19 04	25 55	19 46	15 11	07 21	20 57	03 36	03 04	17 08	17 43
26	18 49	22 52	19 15	14 46	07 28	20 59	03 34	03 05	17 08	17 44
27	18 34	18 19	18 44	14 21	07 35	21 00	03 32	03 05	17 08	17 44
28	18 18	12 38	18 10	13 55	07 42	21 02	03 29	03 06	17 08	17 45
29	18 02	06 N 17	17 36	13 29	07 49	21 04	03 27	03 07	17 09	17 46
30	17 46	00 S 24	17 00	13 03	07 56	21 05	03 24	03 07	17 09	17 47
31	17 S 30	06 S 58	16 S 22	12 S 36	08 S 03	21 N 07	03 S 22	03 N 08	17 S 09	17 N 48

ZODIAC SIGN ENTRIES

Date	h	m	Planets
01	00	52	☽ ♍
02	17	04	☽ ♎
04	20	16	☽ ♏
06	19	36	♀ ♒
07	00	28	☽ ♐
09	05	53	☽ ♑
11	13	05	☽ ♒
13	22	45	☽ ♓
16	03	50	♃ ♋
16	10	48	☽ ♈
18	23	39	☽ ♉
19	17	05	☿ ♒
20	18	08	☉ ♒
21	10	38	☽ ♊
23	17	51	☽ ♋
25	21	20	☽ ♌
27	22	36	☽ ♍
29	23	33	☽ ♎
30	18	53	♀ ♓

LATITUDES

Date	Mercury ☿ ° '	Venus ♀ ° '	Mars ♂ ° '	Jupiter ♃ ° '	Saturn ♄ ° '	Uranus ♅ ° '	Neptune ♆ ° '	Pluto ♇ ° '
01	00 S 52	01 S 13	02 N 04	00 N 29	02 S 14	00 N 47	01 N 42	14 N 59
04	01 10	01 18	02 05	00 30	02 13	00 47	01 42	15 00
07	01 25	01 21	02 07	00 30	02 12	00 47	01 43	15 02
10	01 39	01 25	02 09	00 31	02 12	00 47	01 43	15 03
13	01 50	01 28	02 10	00 31	02 11	00 47	01 43	15 04
16	01 58	01 30	02 12	00 31	02 11	00 47	01 43	15 05
19	02 04	01 32	02 14	00 32	02 10	00 47	01 43	15 07
22	02 05	01 33	02 15	00 32	02 10	00 48	01 43	15 08
25	02 03	01 33	02 17	00 32	02 10	00 48	01 43	15 10
28	01 57	01 33	02 19	00 33	02 09	00 48	01 44	15 11
31	01 S 45	01 S 32	02 N 20	00 N 33	02 S 09	00 N 48	01 N 44	15 N 12

DATA

Julian Date	2439492
Delta T	+38 seconds
Ayanamsa	23° 23' 35"
Synetic vernal point	05° ♓ 43' 25"
True obliquity of ecliptic	23° 26' 43"

MOON'S PHASES, APSIDES AND POSITIONS ☽

Date	h	m	Phase	Longitude ° '	Eclipse Indicator
03	14	19	☾	12 ♎ 31	
10	18	06	●	19 ♑ 49	
18	19	41	☽	28 ♈ 02	
26	06	40	○	05 ♌ 37	

Day	h	m		
01	09	55	Perigee	
16	21	04	Apogee	
28	15	10	Perigee	
03	05	12	0S	
09	18	18	Max dec	27° S 17'
17	00	48	0N	
24	05	26	Max dec	27° N 20'
30	10	33	0S	

LONGITUDES

Date	Chiron ⚷ ° '	Ceres ⚳ ° '	Pallas ⚴ ° '	Juno ⚵ ° '	Vesta ⚶ ° '	Black Moon Lilith ⚸ ° '
01	22 ♓ 04	27 ♊ 41	04 ♈ 54	12 ♌ 25	05 ♏ 40	00 ♈ 43
11	22 ♓ 22	25 ♊ 37	07 ♈ 19	10 ♌ 26	09 ♏ 53	01 ♈ 50
21	22 ♓ 45	24 ♊ 04	10 ♈ 12	07 ♌ 58	13 ♏ 53	02 ♈ 57
31	23 ♓ 12	23 ♊ 11	13 ♈ 29	05 ♌ 24	17 ♏ 38	04 ♈ 03

All ephemeris data is given at 12.00 UT and the Moon's longitude is additionally given for 24.00 UT

ASPECTARIAN

h m	Aspects
01 Sunday	
03 49	☽ ⚼ ♀
03 49	☽ ⊥ ♃
04 19	☽ ⊥ ♂
07 40	☽ △ ☉
09 57	☽ Q ♆
14 45	☽ ⚺ ♂
16 15	♀ ⚹ ♆
18 50	☽ ∠ ♃
02 Monday	
01 13	☽ ☌ ♇
02 19	♀ ⚹ ♄
04 30	☿ ⚻ ♃
06 11	☽ ⚹ ♆
07 04	☽ ☍ ♄
07 31	☽ △ ♀
07 38	☽ ☌ ♅
08 52	♀ △ ♅
12 58	☽ # ♄
14 35	☽ # ♂
18 21	☽ ∥ ♅
19 57	☽ ⚹ ♃
22 00	☽ □ ☿
03 Tuesday	
05 43	☉ Q ♄
07 39	☽ ∠ ♆
14 19	☽ □ ☉
16 05	☽ # ♅
16 10	☽ Q ♃
19 16	☽ ☌ ♂
20 40	☽ ∥ ♂
21 22	☽ ∥ ♄
23 01	☽ ⊥ ♆
04 Wednesday	
04 09	☽ ⚺ ♇
08 30	☽ Q ☿
09 19	☽ ⚺ ♆
10 21	☽ ⚻ ♄
10 40	☽ ⚺ ♅
14 26	☽ ⊥ ♇
15 35	☽ □ ♀
19 21	♂ ∥ ♄
20 42	☽ ± ♄
20 58	☽ ⊥ ♅
22 46	☽ □ ♃
23 36	☽ Q ☉
05 Thursday	
05 56	☽ ∠ ♇
07 48	☽ ⚹ ☿
12 20	☽ ⚼ ♄
12 31	☽ ∠ ♅
22 02	☽ ⚹ ☉
06 Friday	
00 43	☽ ⚺ ♂
01 16	☽ ∥ ♆
01 50	☽ Q ♀
03 23	☽ # ♇
07 59	☽ ⚹ ♇
11 33	☽ ⊥ ♂
13 15	☽ ∠ ☿
13 23	☽ ☌ ♆
14 26	☿ ∠ ♆
14 36	☽ △ ♄
14 38	☽ ⚹ ♅
18 08	☽ # ♃
22 53	♄ ☍ ♅
07 Saturday	
00 03	☽ ∥ ♀
00 57	☽ ⚹ ♀
02 22	☽ ∠ ☉
02 32	☽ △ ♃
03 52	☽ ∠ ♂
05 01	☽ Q ♇
06 23	☽ ∥ ☉
07 08	☽ ⊥ ☿
11 44	☽ Q ♅
16 45	♀ ☍ ♃
19 08	☽ ⚺ ☿
19 34	☽ ⊥ ☉
20 47	☽ ∥ ☿
08 Sunday	
04 51	☽ ⚼ ♃
06 13	☽ ∠ ♀
07 06	☽ ⚺ ☉
07 21	☽ ⚹ ♂
12 40	☉ □ ♂
12 57	☽ □ ♇
18 37	☽ ⚺ ♆
19 46	☽ □ ♅
20 02	☽ □ ♄
20 48	☽ ± ♃
23 43	♂ ⊥ ♆
09 Monday	
00 00	☽ ⊥ ♀
00 52	☿ Q ♄
05 26	☽ ⊥ ♆
05 37	☽ Q ♂
07 31	☽ ⚻ ♃
11 59	☽ ⚺ ♀
21 48	☽ ∠ ♆
10 Tuesday	
04 52	☽ Q ♄
08 35	☽ ☌ ☿
11 22	♀ # ♃
15 37	☽ □ ♂
18 06	☽ ☌ ☉
19 29	☽ △ ♇
11 Wednesday	
01 30	☽ ⚹ ♆
02 32	☽ △ ♅
03 11	☽ ⚹ ♄
05 53	♀ ⚼ ♇
10 42	♀ Q ♆
11 30	☉ △ ♇
14 14	☽ ☍ ♃
20 37	☽ ∥ ☿
23 33	☽ ⚼ ♇
12 Thursday	
00 04	☽ Q ♆
01 30	☽ ☌ ♀
06 46	☽ ⚼ ♅
07 37	☽ ∠ ♄
13 58	☽ ∥ ☉
16 42	☽ ± ♇
21 11	☽ # ♃
13 Friday	
00 01	☽ ± ♅
01 00	☽ ⊥ ♄
01 18	☽ ⚺ ☿
02 18	☽ △ ♂
02 19	☽ ∥ ♀
04 16	☽ ⚻ ♇
07 58	☽ ⚺ ☉
10 43	☽ □ ♆
11 11	☿ □ ♂
11 39	☽ ⚻ ♅
12 43	☽ ⚺ ♄
14 22	☽ # ♇
14 44	☽ ⊥ ☿
16 59	☽ ∥ ♆
20 45	☽ ⊥ ☉
23 19	☽ ⚻ ♃
23 42	☿ △ ♇
14 Saturday	
06 01	♀ ⚼ ♅
08 43	☽ ⚼ ♂
11 02	☽ ± ♃
11 12	☽ ∠ ☿
16 10	☽ ∠ ☉
17 55	☉ ⚹ ♆
18 29	☽ ⚺ ♀
18 43	♀ ∠ ♄
15 Sunday	
03 17	☽ ± ♂
03 58	☉ △ ♅
04 47	☽ ⚼ ♃
07 53	☽ ⊥ ♀
10 59	♂ ⚺ ♇
15 36	☽ ☍ ♇
15 46	☽ ⚻ ♂
21 54	☽ ∥ ♂
22 01	☽ ⚹ ☿
22 03	☉ ⚹ ♄
22 28	☽ △ ♆
23 15	☽ ☍ ♅
16 Monday	
00 50	☽ ☌ ♄
01 05	☽ ⚹ ☉
01 16	☿ ⚹ ♆
04 10	☽ ∠ ♀
06 48	☿ △ ♅
07 18	☽ ∥ ♄
10 43	☽ △ ♃
11 33	☽ # ♅
19 15	☿ ⚹ ♄
17 Tuesday	
02 19	☽ Q ☿
03 44	☽ Q ☉
04 57	☽ ⚼ ♆
14 05	☽ ∥ ♅
14 15	☽ ⚹ ♀
18 04	☽ # ♄
23 17	☽ ± ♆
18 Wednesday	
00 17	☉ # ♃
02 11	☉ ☌ ☿
04 24	☽ ⚻ ♇
05 11	☽ # ♂
06 48	☽ ☍ ♂
08 41	♀ ± ♇
11 28	☽ ⚻ ♆
12 04	☽ ⚻ ♅
14 10	☽ ⚺ ♄
16 32	☽ ± ♇
17 28	☽ Q ♀
19 41	☽ □ ☉
20 48	☽ □ ☿
22 54	☽ □ ♃
19 Thursday	
00 09	☽ ± ♅
02 21	☽ ⊥ ♄
05 15	♀ # ♇
10 36	☽ ⚼ ♇
10 47	☿ ☍ ♃
18 08	☽ ⚼ ♅
20 28	☽ ∠ ♄
20 Friday	
05 17	☉ ☍ ♃
08 40	☽ # ♆
09 10	☽ # ♀
09 30	☽ □ ♀
10 19	☽ Q ♃
11 33	☽ ∥ ♇
15 06	♀ ∥ ♆
16 12	☽ △ ♇
20 32	☽ ⚻ ♂
23 08	☽ ☍ ♆
23 31	☽ △ ♅
21 Saturday	
01 17	☽ # ☉
02 02	☽ ⚹ ♄
05 41	☽ ∥ ♃
08 14	♀ ± ♅
08 28	☽ ± ♂
09 18	☽ ⚹ ♃
10 14	☽ # ☿
12 05	☽ △ ☉
17 03	☽ △ ☿
22 Sunday	
01 01	☽ Q ♄
02 02	☽ ⚼ ♂
11 50	♀ ⊥ ♄
13 11	☽ ∠ ♃
18 31	☽ ⚼ ☉
21 38	☿ ⚼ ♇
23 Monday	
00 26	☽ △ ♀
00 38	☽ □ ♇
01 05	☽ ⚼ ☿
01 44	☿ # ♃
02 27	♀ ⚻ ♇
05 26	☽ ⊥ ♃
06 24	☽ △ ♂
06 45	☿ Q ♆
07 14	☽ ⚻ ♆
07 25	☽ □ ♅
10 14	☽ □ ♄
12 09	☽ ± ☉
16 04	☽ ⚺ ♃
17 58	☽ ± ♆
19 34	☽ ± ☿
23 39	☽ ⚻ ☉
24 Tuesday	
05 52	☽ ⚼ ♀
07 36	☽ ⚻ ☿
08 34	☽ Q ♇
09 46	☽ ⚼ ♆
13 21	♂ ⚺ ♆
15 03	☽ Q ♅
16 47	♂ ⚺ ♅
22 56	☽ ± ♀
25 Wednesday	
01 33	☿ ⚼ ♅
05 08	☽ ⚹ ♇
10 06	☽ ⚻ ♀
11 26	☽ △ ♆
11 28	☽ ⚹ ♅
12 00	☽ □ ♂
14 31	☽ △ ♄
19 11	☽ ☌ ♃
22 52	♅ ⚹ ♆
26 Thursday	
00 29	☉ ⚼ ♇
01 26	♀ ⚻ ♅
01 32	♀ □ ♆
03 14	☿ ∠ ♄
06 14	☽ ∠ ♇
06 40	☽ ☍ ☉
12 25	☽ ∠ ♅
15 36	☽ ⚼ ♄
15 44	♀ △ ♂
17 11	☽ ☍ ☿
18 26	☉ Q ♆
18 40	☽ Q ♂
21 03	☽ ⊥ ♇
22 47	☽ ∥ ♃
27 Friday	
03 09	☽ ⊥ ♅
06 26	☽ ± ♄
06 51	☽ ⚺ ♇
09 52	☽ # ☿
10 48	☽ # ☉
12 56	☽ ⚺ ♅
13 02	☽ □ ♆
14 38	☽ ∥ ♇
14 46	☽ ⚹ ♂
16 16	♀ ⚺ ♄
16 17	☽ ⚻ ♄
16 17	☽ ☍ ♀
17 21	☽ # ♆
20 04	☽ ⚺ ♃
28 Saturday	
01 20	☿ ± ♇
01 27	☉ ∥ ☿
05 44	☽ ⊥ ♃
06 30	☽ # ♀
11 21	☽ ⚻ ☉
15 45	☽ ∠ ♂
18 21	☽ Q ♆
20 15	☽ ∠ ♃
21 51	☽ ± ☉
29 Sunday	
00 40	☽ ⚻ ☿
05 08	☿ # ♇
06 22	☽ # ♂
06 46	☽ ⊥ ♂
07 38	☽ ☌ ♇
09 03	♀ ⚻ ♃
11 47	☽ ± ☿
13 38	☽ ⚼ ☉
13 43	☽ ☌ ♅
13 57	☽ ⚹ ♆
14 50	☉ ⚼ ♅
16 49	☽ ⚺ ♂
17 30	☽ ☍ ♄
20 33	☽ ⚹ ♃
21 44	☽ ⚻ ♀
22 14	☽ # ♄
23 22	☽ ∥ ♅
30 Monday	
03 39	☿ ± ♅
04 35	☽ ⚼ ☿
06 03	☿ ∥ ♆
08 33	☽ ± ♀
11 20	☉ # ♇
14 42	☽ ∠ ♆
16 12	☽ Q ♃
16 13	☽ △ ☉
18 23	♂ ⊥ ♇
21 50	☽ # ♅
22 48	☽ ∥ ♄
31 Tuesday	
00 56	☽ ⚼ ♀
05 44	☽ ⊥ ♆
09 03	☽ △ ☿
09 14	☽ ⚺ ♇
09 41	♂ ⚻ ♄
10 32	☿ ⚻ ♇
13 50	☿ ⊥ ♄
15 28	☽ ⚺ ♅
15 52	☽ ⚺ ♆
16 12	☽ ∥ ♂
19 21	☽ ⊥ ♇
19 49	☽ ⚻ ♄
19 59	☽ ☌ ♂
22 10	☽ □ ♃

FEBRUARY 1967

LONGITUDES

Date	Sidereal time h m s	Sun ☉ ° ' "	Moon ☽ ° ' "	Moon ☽ 24.00 ° ' "	Mercury ☿ ° '	Venus ♀ ° '	Mars ♂ ° '	Jupiter ♃ ° '	Saturn ♄ ° '	Uranus ♅ ° '	Neptune ♆ ° '	Pluto ♇ ° '
01	20 44 02	11 ♒ 56 04	06 ♏ 00 08	12 ♏ 57 05	22 ♒ 07	02 ♓ 08	26 ♎ 50	27 ♋ 50	26 ♓ 35	23 ♍ 56	24 ♏ 12	20 ♍ 14
02	20 47 58	12 56 58	19 ♏ 50 06	26 ♏ 39 19	23 53	03 23	27 10	27 R 43	26 42	23 R 54	24 13	20 R 13
03	20 51 55	13 57 50	03 ♐ 24 54	10 ♐ 07 02	25 38	04 38	27 29	27 35	26 48	23 52	24 13	20 12
04	20 55 52	14 58 42	16 ♐ 45 55	23 ♐ 21 42	27 22	05 53	27 47	27 28	26 54	23 50	24 14	20 11
05	20 59 48	15 59 33	29 ♐ 54 31	06 ♑ 24 30	29 ♒ 06	07 07	28 05	27 21	27 00	23 49	24 15	20 09
06	21 03 45	17 00 22	12 ♑ 51 44	19 ♑ 16 15	00 ♓ 48	08 22	28 23	27 13	27 06	23 47	24 15	20 08
07	21 07 41	18 01 11	25 ♑ 38 05	01 ♒ 57 15	02 30	09 37	28 40	27 06	27 13	23 45	24 16	20 07
08	21 11 38	19 01 59	08 ♒ 13 44	14 ♒ 27 33	04 09	10 51	28 57	26 59	27 19	23 43	24 17	20 05
09	21 15 34	20 02 45	20 ♒ 38 41	26 ♒ 47 11	05 46	12 06	29 13	26 52	27 25	23 41	24 17	20 04
10	21 19 31	21 03 30	02 ♓ 53 07	08 ♓ 56 34	07 20	13 21	29 29	26 46	27 32	23 39	24 18	20 03
11	21 23 27	22 04 14	14 ♓ 57 41	20 ♓ 56 40	08 51	14 35	29 ♎ 45	26 39	27 38	23 37	24 18	20 01
12	21 27 24	23 04 56	26 ♓ 53 47	02 ♈ 49 21	10 18	15 50	00 ♏ 00	26 32	27 45	23 35	24 19	20 00
13	21 31 21	24 05 37	08 ♈ 43 43	14 ♈ 37 19	11 41	17 04	00 14	26 26	27 52	23 32	24 19	19 59
14	21 35 17	25 06 16	20 ♈ 30 39	26 ♈ 24 14	12 58	18 19	00 28	26 20	27 58	23 30	24 19	19 57
15	21 39 14	26 06 53	02 ♉ 18 40	08 ♉ 14 34	14 09	19 33	00 42	26 14	28 05	23 28	24 20	19 56
16	21 43 10	27 07 29	14 ♉ 12 34	20 ♉ 13 22	15 14	20 48	00 55	26 08	28 12	23 26	24 20	19 54
17	21 47 07	28 08 03	26 ♉ 17 40	02 ♊ 26 08	16 11	22 02	01 07	26 02	28 19	23 24	24 20	19 53
18	21 51 03	29 ♒ 08 35	08 ♊ 39 28	14 ♊ 58 18	17 00	23 16	01 19	25 56	28 25	23 21	24 21	19 51
19	21 55 00	00 ♓ 09 05	21 ♊ 23 13	27 ♊ 54 46	17 41	24 31	01 31	25 50	28 32	23 19	24 21	19 50
20	21 58 56	01 09 34	04 ♋ 33 19	11 ♋ 19 10	18 12	25 45	01 42	25 45	28 39	23 17	24 21	19 49
21	22 02 53	02 10 01	18 ♋ 12 27	25 ♋ 13 05	18 34	26 59	01 52	25 40	28 46	23 14	24 21	19 47
22	22 06 50	03 10 26	02 ♌ 20 50	09 ♌ 35 12	18 46	28 13	02 02	25 35	28 53	23 12	24 21	19 46
23	22 10 46	04 10 49	16 ♌ 55 31	24 ♌ 20 53	18 R 48	29 ♓ 28	02 11	25 30	29 00	23 09	24 21	19 44
24	22 14 43	05 11 10	01 ♍ 50 14	09 ♍ 22 24	18 40	00 ♈ 42	02 19	25 25	29 07	23 07	24 21	19 42
25	22 18 39	06 11 29	16 ♍ 56 06	24 ♍ 30 02	18 22	01 56	02 27	25 20	29 14	23 05	24 R 21	19 41
26	22 22 36	07 11 47	02 ♎ 02 58	09 ♎ 33 44	17 55	03 10	02 35	25 16	29 21	23 02	24 21	19 39
27	22 26 32	08 12 04	17 ♎ 01 17	24 ♎ 24 48	17 19	04 24	02 41	25 12	29 28	23 00	24 21	19 38
28	22 30 29	09 ♓ 12 18	01 ♏ 43 35	08 ♏ 57 08	16 ♓ 36	05 ♈ 38	02 ♏ 47	25 ♋ 08	29 ♓ 35	22 ♍ 57	24 ♏ 21	19 ♍ 36

Date	Moon True ☊ ° '	Moon Mean ☊ ° '	Moon ☽ Latitude ° '
01	11 ♉ 33	11 ♉ 39	00 N 29
02	11 R 33	11 36	00 S 44
03	11 32	11 33	01 53
04	11 29	11 30	02 54
05	11 23	11 27	03 46
06	11 14	11 23	04 24
07	11 03	11 20	04 49
08	10 49	11 17	04 59
09	10 35	11 14	04 55
10	10 21	11 11	04 37
11	10 08	11 08	04 07
12	09 59	11 04	03 26
13	09 52	11 01	02 37
14	09 47	10 58	01 40
15	09 46	10 55	00 S 39
16	09 D 46	10 52	00 N 24
17	09 46	10 48	01 27
18	09 R 45	10 45	02 27
19	09 43	10 42	03 22
20	09 39	10 39	04 09
21	09 32	10 36	04 43
22	09 23	10 33	05 01
23	09 13	10 29	05 00
24	09 03	10 26	04 40
25	08 54	10 23	03 59
26	08 47	10 20	03 02
27	08 42	10 17	01 53
28	08 ♉ 40	10 ♉ 14	00 N 37

DECLINATIONS

Date	Sun ☉ ° '	Moon ☽ ° '	Mercury ☿ ° '	Venus ♀ ° '	Mars ♂ ° '	Jupiter ♃ ° '	Saturn ♄ ° '	Uranus ♅ ° '	Neptune ♆ ° '	Pluto ♇ ° '
01	17 S 13	13 S 04	15 S 43	12 S 09	08 S 10	21 N 09	03 S 19	03 N 09	17 S 09	17 N 48
02	16 56	18 24	15 03	11 41	08 16	21 10	03 17	03 09	17 09	17 49
03	16 38	22 41	14 22	11 13	08 22	21 12	03 14	03 10	17 09	17 50
04	16 21	25 41	13 39	10 45	08 28	21 13	03 12	03 11	17 09	17 51
05	16 03	27 12	12 56	10 17	08 34	21 15	03 09	03 12	17 09	17 52
06	15 45	27 13	12 12	09 48	08 40	21 16	03 07	03 12	17 10	17 52
07	15 26	25 45	11 27	09 19	08 46	21 18	03 04	03 13	17 10	17 53
08	15 07	23 02	10 42	08 50	08 51	21 19	03 01	03 14	17 10	17 54
09	14 48	19 16	09 56	08 21	08 56	21 21	02 59	03 15	17 10	17 55
10	14 29	14 45	09 10	07 51	09 01	21 22	02 56	03 16	17 10	17 56
11	14 09	09 43	08 25	07 22	09 06	21 23	02 53	03 16	17 10	17 56
12	13 50	04 S 23	07 40	06 52	09 11	21 25	02 51	03 17	17 10	17 57
13	13 30	01 N 04	06 56	06 21	09 16	21 26	02 48	03 18	17 10	17 58
14	13 09	06 28	06 14	05 51	09 20	21 27	02 45	03 19	17 10	17 59
15	12 49	11 40	05 33	05 21	09 24	21 28	02 43	03 20	17 10	18 00
16	12 28	16 29	04 54	04 50	09 29	21 30	02 40	03 21	17 10	18 00
17	12 08	20 44	04 18	04 19	09 32	21 31	02 37	03 22	17 10	18 01
18	11 47	24 11	03 44	03 48	09 36	21 32	02 34	03 23	17 10	18 02
19	11 25	26 32	03 14	03 18	09 40	21 33	02 31	03 24	17 10	18 03
20	11 04	27 30	02 48	02 46	09 43	21 34	02 29	03 25	17 10	18 04
21	10 42	26 52	02 25	02 15	09 46	21 35	02 26	03 25	17 10	18 04
22	10 21	24 31	02 07	01 44	09 49	21 36	02 23	03 26	17 10	18 05
23	09 59	20 32	01 54	01 13	09 52	21 37	02 20	03 27	17 10	18 06
24	09 37	15 10	01 45	00 42	09 54	21 38	02 17	03 28	17 10	18 07
25	09 15	08 50	01 41	00 S 10	09 57	21 39	02 14	03 29	17 10	18 08
26	08 52	01 N 58	01 43	00 N 21	09 59	21 40	02 12	03 30	17 10	18 08
27	08 30	04 S 57	01 49	00 52	10 01	21 41	02 09	03 31	17 10	18 09
28	08 S 07	11 S 30	02 S 00	01 N 23	10 S 03	21 N 41	02 S 06	03 N 32	17 S 10	18 N 10

ZODIAC SIGN ENTRIES

Date	h	m	Planets	
01	01	44	☽	♏
03	05	55	☽	♐
05	12	10	☽	♑
06	00	38	☿	♓
07	20	17	☽	♒
10	06	19	☽	♓
12	12	20	♂	♏
12	18	17	☽	♈
15	07	19	☽	♉
17	19	16	☽	♊
19	08	24	☉	♓
20	03	48	☽	♋
22	08	04	☽	♌
23	22	30	♀	♈
24	09	04	☽	♍
26	08	44	☽	♎
28	09	09	☽	♏

LATITUDES

Date	Mercury ☿ ° '	Venus ♀ ° '	Mars ♂ ° '	Jupiter ♃ ° '	Saturn ♄ ° '	Uranus ♅ ° '	Neptune ♆ ° '	Pluto ♇ ° '
01	01 S 40	01 S 32	02 N 21	00 N 33	02 S 09	00 N 48	01 N 44	15 N 13
04	01 21	01 30	02 22	00 34	02 08	00 48	01 44	15 14
07	00 55	01 28	02 24	00 34	02 08	00 48	01 44	15 15
10	00 S 23	01 25	02 26	00 34	02 08	00 48	01 44	15 16
13	00 N 16	01 21	02 27	00 34	02 07	00 48	01 44	15 16
16	01 00	01 17	02 28	00 34	02 07	00 48	01 45	15 17
19	01 46	01 13	02 30	00 35	02 07	00 48	01 45	15 18
22	02 32	01 07	02 31	00 35	02 07	00 48	01 45	15 19
25	03 10	01 02	02 32	00 35	02 07	00 48	01 45	15 19
28	03 35	00 55	02 33	00 35	02 06	00 48	01 45	15 20
31	03 N 42	00 S 48	02 N 34	00 N 35	02 S 06	00 N 48	01 N 45	15 N 20

DATA

Julian Date	2439523
Delta T	+38 seconds
Ayanamsa	23° 23' 40"
Synetic vernal point	05° ♓ 43' 19"
True obliquity of ecliptic	23° 26' 44"

MOON'S PHASES, APSIDES AND POSITIONS ☽

Date	h	m	Phase	Longitude ° '	Eclipse Indicator
01	23	03	☾	12 ♏ 24	
09	10	44	●	20 ♒ 00	
17	15	56	☽	28 ♉ 18	
24	17	43	○	05 ♍ 26	

Day	h	m		
13	15	19	Apogee	
25	20	47	Perigee	
06	00	00	Max dec	27° S 24'
13	07	20	ON	
20	14	56	Max dec	27° N 31'
26	18	48	OS	

LONGITUDES

Date	Chiron ⚷ ° '	Ceres ⚳ ° '	Pallas ⚴ ° '	Juno ⚵ ° '	Vesta ⚶ ° '	Black Moon Lilith ⚸ ° '
01	23 ♓ 15	23 ♊ 08	13 ♈ 50	05 ♌ 09	17 ♏ 59	04 ♈ 10
11	23 ♓ 45	23 ♊ 01	17 ♈ 31	02 ♌ 53	21 ♏ 22	05 ♈ 17
21	24 ♓ 18	23 ♊ 35	21 ♈ 30	01 ♌ 13	24 ♏ 22	06 ♈ 24
31	24 ♓ 53	24 ♊ 48	25 ♈ 46	00 ♌ 17	26 ♏ 53	07 ♈ 30

All ephemeris data is given at 12.00 UT and the Moon's longitude is additionally given for 24.00 UT

ASPECTARIAN

h m	Aspects
01 Wednesday	
01 38	☽ ⊥ ♅
02 59	☉ ∠ ♄
04 44	☽ △ ♀
06 07	☽ ± ♄
08 29	☽ ∥ ♀
10 41	☽ ∠ ♇
17 02	☽ ∠ ♅
17 39	☉ ∥ ♆
21 43	☽ Ᵽ ♄
22 10	☽ ∥ ☿
23 03	☽ □ ☉
02 Thursday	
05 19	☽ ∥ ☉
05 59	☽ ∥ ♆
09 09	☽ # ♇
12 18	☿ ⊼ ♅
12 40	☽ ∗ ♇
16 36	☿ □ ♆
17 46	♀ ± ♃
19 08	☽ ∗ ♅
19 42	☽ ☌ ♆
20 09	☽ □ ☿
03 Friday	
00 09	☽ △ ♄
01 12	☽ ⊻ ♂
01 45	☽ △ ♃
02 48	☽ # ♃
09 12	☽ Q ☉
09 50	☽ Q ♇
12 07	☽ ⊥ ♂
14 23	☽ □ ♀
16 23	☽ Q ♅
17 24	☉ ± ♇
18 06	♂ □ ♃
04 Saturday	
04 18	☽ Ᵽ ♃
04 38	☽ ∠ ♂
05 03	☿ ⊻ ♄
08 30	☽ ∗ ☉
09 06	☽ Q ☿
13 14	☿ ⊼ ♃
18 11	☽ □ ♇
19 00	☿ △ ♂
19 21	♄ # ♅
20 28	☽ ± ♃
05 Sunday	
00 51	☽ □ ♅
01 37	☽ ⊻ ♆
02 18	☽ Q ♀
06 37	☽ □ ♄
07 20	☽ ⊼ ♃
08 35	☽ ∗ ♂
10 17	☽ ∗ ☿
12 37	☽ ⊥ ♆
14 10	☽ ∠ ☉
06 Monday	
02 45	☽ ∗ ♀
05 17	☽ ∠ ♆
07 16	☽ Q ♂
08 15	☽ ⊥ ☉
16 13	☽ Q ♄
18 21	☽ ∠ ☿
20 25	☽ ⊻ ☉
07 Tuesday	
00 41	♃ △ ♄
01 36	☽ △ ♇
05 44	☉ ± ♅
08 26	☽ △ ♅
09 25	☽ ∗ ♆
09 51	☽ ∠ ♀
13 52	☽ ⊥ ☿
14 45	☽ ☍ ♃
15 01	☽ ∗ ♄
17 53	☽ □ ♂
20 14	☿ ± ♃
08 Wednesday	
03 01	☽ ⊻ ☿
04 50	☽ ⊥ ♀
06 00	☽ Ᵽ ♇
08 15	☽ Q ♆
11 26	♀ ∥ ♂
12 56	☽ Ᵽ ♅
17 37	☽ ⊻ ♀
19 56	☽ ∠ ♄
23 16	☽ ± ♇
23 33	☽ # ♃
09 Thursday	
06 15	☽ ± ♅
07 58	♀ Ᵽ ♃
10 44	☽ ☌ ☉
10 53	☽ ⊼ ♇
12 33	☉ ⊼ ♇
13 32	☽ ⊥ ♄
17 54	☽ ⊼ ♅
19 07	☽ □ ♆
19 35	☽ # ♇
23 38	☽ ∥ ♆
10 Friday	
00 03	☽ ⊼ ♃
01 22	☽ ⊻ ♄
05 10	☽ △ ♂
11 45	☽ ± ♃
13 27	☽ ∥ ☉
16 13	☿ ∥ ♂
22 05	☽ ☌ ☿
11 Saturday	
00 34	☉ ⊥ ♄
00 51	♂ ⊥ ♅
05 27	☽ Ᵽ ♃
11 10	☽ ☌ ♀
11 34	☽ Ᵽ ♂
14 47	☽ ∥ ♂
15 51	♀ Ᵽ ♂
18 57	☽ ∥ ☿
22 08	☽ ☍ ♇
23 51	☽ ∥ ♀
12 Sunday	
03 35	☽ ⊻ ☉
05 19	☽ ☍ ♅
06 01	☽ ± ♂
06 47	☽ △ ♆
11 17	☽ △ ♃
13 44	☽ ☌ ♄
16 50	☽ ⊥ ☉
16 51	☽ # ♅
18 24	☽ ⊼ ♂
18 52	☽ ∥ ♄
23 20	☉ ⊼ ♅
13 Monday	
07 55	☿ Ᵽ ♃
12 49	☽ ∠ ☉
13 12	☽ Ᵽ ♆
17 21	☉ □ ♆
18 46	☽ ⊻ ☿
19 36	☽ # ♄
21 55	☽ ∥ ♅
14 Tuesday	
07 00	☽ ⊻ ♀
07 32	☽ ± ♆
08 29	☽ ⊥ ☿
09 29	☽ # ♀
10 52	☽ ⊼ ♇
11 04	☽ # ☿
18 05	☽ ⊼ ♅
19 46	☽ ⊼ ♆
20 40	☽ ⊥ ♀
22 14	☽ ∗ ☉
23 04	☽ ± ♇
23 45	☽ □ ♃
15 Wednesday	
01 17	☽ # ♂
03 20	☽ ⊻ ♄
04 54	☽ ∠ ☿
06 15	☽ ± ♅
08 40	☽ ☍ ♂
14 23	☉ ⊼ ♃
15 37	☽ ⊥ ♄
17 05	☽ ∠ ♀
17 12	☽ # ☉
17 18	☽ Ᵽ ♇
19 08	♀ ☍ ♇
16 Thursday	
00 25	☽ Ᵽ ♅
00 50	☽ Q ☉
09 57	☽ ∠ ♄
11 50	☽ Q ♃
14 14	☽ ∗ ☿
15 38	☽ # ♆
20 12	☽ ∥ ♇
23 21	☽ △ ♇
17 Friday	
02 38	☽ ∗ ♀
03 46	☿ ∥ ♀
06 18	☽ △ ♅
08 09	☽ ☍ ♆
09 49	☿ Ᵽ ♂
11 29	☽ ∗ ♃
15 56	☽ □ ☉
15 59	☽ ∗ ♄
16 00	☽ Q ☿
16 40	☉ ⊻ ♄
16 56	☽ ∥ ♃
21 36	☽ ⊼ ♂
18 Saturday	
04 47	☽ Q ♀
09 24	☽ ± ♂
13 31	♀ ☍ ♅
15 24	☽ Q ♄
16 19	☽ ∠ ♃
19 Sunday	
02 46	☽ Ᵽ ♂
04 23	☿ # ♅
04 44	☽ □ ☿
07 26	♀ # ♅
08 45	♀ △ ♆
09 07	☽ □ ♇
09 09	☽ ⊥ ♃
15 34	☽ □ ♅
17 28	☽ ⊼ ♆
18 23	☽ □ ♀
20 09	☽ ⊻ ♃
20 Monday	
01 15	☽ □ ♄
04 25	☽ ± ♆
05 23	☽ △ ☉
06 47	☽ △ ♂
07 33	☿ ∥ ♀
10 41	♀ ± ♂
11 59	♀ △ ♃
17 47	☽ Q ♇
20 32	☽ Ᵽ ♆
23 53	☽ Q ♅
21 Tuesday	
00 56	☉ ± ♃
03 11	♀ ∥ ♄
03 21	☉ △ ♂
10 04	☽ Ᵽ ☉
11 15	☿ ∥ ♄
12 38	☽ △ ☿
14 43	☽ ∗ ♇
20 36	☽ ∗ ♅
22 32	☽ △ ♆
22 Wednesday	
00 41	☽ ☌ ♃
02 39	☽ ± ☉
04 26	☽ △ ♀
06 09	☽ △ ♄
11 28	☽ □ ♂
13 29	☽ ⊼ ☉
14 23	☽ Ᵽ ☿
16 01	☽ ∠ ♇
21 41	☽ ∠ ♅
23 Thursday	
02 08	♀ ☌ ♄
04 26	☿ St R
05 17	☽ ± ☿
06 19	☽ ∥ ♃
06 48	☽ ⊥ ♇
07 12	☽ Ᵽ ♄
07 37	☽ Ᵽ ♀
12 23	☽ ⊥ ♅
15 02	☽ ⊼ ☿
16 33	☽ ⊻ ♇
17 20	☽ Q ♂
18 41	☉ ∥ ♂
21 54	☽ ± ♄
22 03	☽ ⊻ ♅
23 32	☽ ± ♀
23 32	☽ ∥ ♇
24 Friday	
00 01	☽ □ ♆
01 46	☽ ⊻ ♃
03 40	☽ # ♆
07 37	☽ ⊼ ♄
10 01	☽ ⊼ ♀
11 20	☽ ⊥ ♃
12 47	☽ ∗ ♂
17 43	☽ ☍ ☉
22 04	♆ St R
25 Saturday	
01 35	☽ ∠ ♃
04 44	☽ Q ♆
08 00	☽ # ♂
10 27	☽ # ☉
12 50	☽ ∠ ♂
14 13	☽ ☍ ☿
16 21	☽ ☌ ♇
21 43	☽ ☌ ♅
23 23	♀ ⊼ ♂
23 46	☽ ∗ ♆
26 Sunday	
01 16	☽ ∗ ♃
03 14	☽ ⊥ ♂
06 44	☽ ∥ ♅
07 41	☽ ☍ ♄
11 14	☽ # ♄
12 51	☽ ⊻ ♂
12 53	☽ # ☿
13 56	☽ ☍ ♀
17 12	☽ ∥ ♀
20 17	☽ Q ♃
20 39	☽ # ♀
20 48	☽ ⊼ ☉
23 40	☽ ∠ ♆
23 55	☿ Ᵽ ♂
27 Monday	
00 52	☽ ∥ ☿
02 16	☽ ∥ ♄
06 59	☽ # ♅
07 07	☽ ± ☉
12 28	☽ ⊼ ☿
14 09	☽ ⊥ ♆
16 13	☽ ⊻ ♇
21 40	☽ ⊻ ♅
21 46	☽ ± ☿
22 45	☽ Ᵽ ☉
23 54	☽ ⊻ ♆
28 Tuesday	
00 03	☽ ∥ ☉
01 13	☽ □ ♃
01 58	☽ ⊥ ♇
06 28	☽ ∥ ♂
07 27	☽ ⊥ ♅
08 27	☽ ⊼ ♄
11 48	☽ Ᵽ ☿
13 46	☽ ☌ ♂
16 45	☽ ∠ ♇
18 27	☽ ± ♄
19 04	☽ ⊼ ♀
20 56	☿ ∥ ♄
22 18	☽ ∠ ♅

MARCH 1967

Date	Sidereal time h m s	Sun ☉ ° ' "	Moon ☽ ° ' "	Moon ☽ 24.00 ° ' "	Mercury ☿ ° '	Venus ♀ ° '	Mars ♂ ° '	Jupiter ♃ ° '	Saturn ♄ ° '	Uranus ♅ ° '	Neptune ♆ ° '	Pluto ♇ ° '
01	22 34 25	10 ♓ 12 32	16 ♏ 05 10	23 ♏ 07 31	15 ♓ 46	06 ♈ 52	02 ♏ 53	25 ♋ 04	29 ♓ 43	22 ♍ 55	24 ♏ 21	19 ♍ 35
02	22 38 22	11 12 44	00 ♐ 04 11	06 ♐ 55 17	14 R 51	08 05	02 58	25 R 00	29 50	22 R 52	24 R 21	19 R 33
03	22 42 19	12 12 54	13 ♐ 41 00	20 ♐ 21 37	13 52	09 19	03 02	24 57	29 ♓ 57	22 50	24 21	19 32
04	22 46 15	13 13 03	26 ♐ 57 24	03 ♑ 28 43	12 51	10 33	03 05	24 53	00 ♈ 04	22 47	24 20	19 30
05	22 50 12	14 13 11	09 ♑ 55 53	16 ♑ 19 15	11 49	11 47	03 08	24 50	00 12	22 44	24 20	19 28
06	22 54 08	15 13 16	22 ♑ 39 06	28 ♑ 55 44	10 47	13 00	03 10	24 47	00 19	22 42	24 20	19 27
07	22 58 05	16 13 21	05 ♒ 09 26	11 ♒ 20 26	09 48	14 14	03 11	24 44	00 26	22 39	24 19	19 25
08	23 02 01	17 13 23	17 ♒ 28 57	23 ♒ 35 11	08 51	15 28	03 12	24 42	00 34	22 37	24 19	19 24
09	23 05 58	18 13 24	29 ♒ 39 17	05 ♓ 41 27	07 59	16 41	03 R 11	24 39	00 41	22 34	24 19	19 22
10	23 09 54	19 13 22	11 ♓ 41 49	17 ♓ 40 35	07 12	17 55	03 11	24 37	00 48	22 31	24 18	19 20
11	23 13 51	20 13 19	23 ♓ 37 53	29 ♓ 33 56	06 30	19 08	03 09	24 35	00 56	22 29	24 18	19 19
12	23 17 48	21 13 14	05 ♈ 28 57	11 ♈ 23 10	05 55	20 21	03 06	24 33	01 03	22 26	24 17	19 17
13	23 21 44	22 13 07	17 ♈ 16 53	23 ♈ 10 23	05 25	21 35	03 03	24 32	01 11	22 24	24 17	19 15
14	23 25 41	23 12 58	29 ♈ 04 02	04 ♉ 58 15	05 03	22 48	02 59	24 30	01 18	22 21	24 16	19 14
15	23 29 37	24 12 47	10 ♉ 53 27	16 ♉ 50 07	04 46	24 01	02 54	24 29	01 25	22 18	24 16	19 12
16	23 33 34	25 12 33	22 ♉ 48 47	28 ♉ 49 59	04 36	25 14	02 49	24 28	01 33	22 16	24 15	19 11
17	23 37 30	26 12 18	04 ♊ 54 18	11 ♊ 02 21	04 33	26 27	02 43	24 27	01 40	22 13	24 14	19 09
18	23 41 27	27 12 00	17 ♊ 14 42	23 ♊ 31 58	04 D 35	27 40	02 36	24 26	01 48	22 10	24 14	19 08
19	23 45 23	28 11 40	29 ♊ 54 44	06 ♋ 23 33	04 43	28 ♈ 53	02 28	24 26	01 55	22 08	24 13	19 06
20	23 49 20	29 ♓ 11 18	12 ♋ 58 51	19 ♋ 41 04	04 57	00 ♉ 06	02 19	24 26	02 03	22 05	24 12	19 04
21	23 53 17	00 ♈ 10 53	26 ♋ 30 27	03 ♌ 27 09	05 16	01 19	02 10	24 D 26	02 10	22 03	24 11	19 03
22	23 57 13	01 10 26	10 ♌ 31 08	17 ♌ 42 10	05 40	02 32	02 00	24 26	02 18	22 00	24 11	19 01
23	00 01 10	02 09 57	24 ♌ 59 50	02 ♍ 23 30	06 08	03 45	01 49	24 26	02 25	21 58	24 10	19 00
24	00 05 06	03 09 26	09 ♍ 52 21	17 ♍ 25 22	06 42	04 57	01 37	24 26	02 33	21 55	24 09	18 58
25	00 09 03	04 08 52	25 ♍ 01 22	02 ♎ 39 06	07 19	06 10	01 25	24 27	02 40	21 52	24 08	18 57
26	00 12 59	05 08 16	10 ♎ 17 14	17 ♎ 54 27	08 00	07 22	01 12	24 28	02 48	21 50	24 07	18 55
27	00 16 56	06 07 38	25 ♎ 29 31	03 ♏ 01 16	08 45	08 35	00 58	24 29	02 55	21 47	24 06	18 54
28	00 20 52	07 06 59	10 ♏ 28 43	17 ♏ 51 03	09 33	09 47	00 44	24 30	03 03	21 45	24 05	18 52
29	00 24 49	08 06 17	25 ♏ 07 36	02 ♐ 17 57	10 25	10 59	00 28	24 32	03 10	21 42	24 04	18 50
30	00 28 46	09 05 34	09 ♐ 21 48	16 ♐ 19 05	11 20	12 12	00 ♏ 12	24 33	03 17	21 40	24 03	18 49
31	00 32 42	10 ♈ 04 49	23 ♐ 09 48	29 ♐ 54 08	12 ♓ 18	13 ♉ 24	29 ♎ 56	24 ♋ 35	03 ♈ 25	21 ♍ 37	24 ♏ 02	18 ♍ 48

Date	Moon True ☊ ° '	Moon Mean ☊ ° '	Moon ☽ Latitude ° '
01	08 ♉ 40	10 ♉ 10	00 S 40
02	08 D 41	10 07	01 52
03	08 R 42	10 04	02 56
04	08 41	10 01	03 49
05	08 37	09 58	04 29
06	08 32	09 54	04 54
07	08 24	09 51	05 05
08	08 15	09 48	05 02
09	08 05	09 45	04 45
10	07 55	09 42	04 15
11	07 47	09 39	03 34
12	07 40	09 35	02 44
13	07 36	09 32	01 47
14	07 33	09 29	00 S 46
15	07 D 33	09 26	00 N 18
16	07 34	09 23	01 22
17	07 35	09 19	02 23
18	07 37	09 16	03 19
19	07 R 37	09 13	04 07
20	07 36	09 10	04 43
21	07 34	09 07	05 06
22	07 30	09 04	05 11
23	07 25	09 00	04 57
24	07 19	08 57	04 23
25	07 15	08 54	03 30
26	07 11	08 51	02 22
27	07 09	08 48	01 N 03
28	07 D 09	08 45	00 S 18
29	07 10	08 41	01 37
30	07 11	08 38	02 48
31	07 ♉ 12	08 ♉ 35	03 S 46

DECLINATIONS

Date	Sun ☉ ° '	Moon ☽ ° '	Mercury ☿ ° '	Venus ♀ ° '	Mars ♂ ° '	Jupiter ♃ ° '	Saturn ♄ ° '	Uranus ♅ ° '	Neptune ♆ ° '	Pluto ♇ ° '
01	07 S 45	17 S 17	02 S 15	01 N 55	10 S 04	21 N 42	02 S 03	03 N 33	17 S 10	18 N 11
02	07 22	22 00	02 34	02 26	10 06	21 43	02 00	03 34	17 10	18 11
03	06 59	25 22	02 56	02 57	10 07	21 43	01 57	03 35	17 09	18 12
04	06 36	27 14	03 21	03 28	10 08	21 44	01 54	03 36	17 09	18 13
05	06 13	27 32	03 48	03 59	10 09	21 45	01 51	03 37	17 09	18 14
06	05 50	26 23	04 17	04 30	10 09	21 45	01 48	03 38	17 09	18 14
07	05 26	23 55	04 46	05 01	10 10	21 46	01 45	03 39	17 09	18 15
08	05 03	20 23	05 16	05 32	10 10	21 46	01 42	03 41	17 09	18 16
09	04 39	16 02	05 45	06 03	10 09	21 47	01 39	03 42	17 09	18 17
10	04 16	11 06	06 14	06 33	10 09	21 47	01 36	03 43	17 08	18 17
11	03 52	05 49	06 41	07 04	10 09	21 47	01 34	03 44	17 08	18 18
12	03 29	00 S 20	07 07	07 34	10 08	21 48	01 31	03 45	17 08	18 19
13	03 05	05 N 08	07 31	08 04	10 07	21 48	01 28	03 46	17 08	18 19
14	02 42	10 26	07 53	08 34	10 05	21 48	01 25	03 47	17 08	18 20
15	02 18	15 23	08 12	09 04	10 04	21 49	01 22	03 48	17 08	18 21
16	01 54	19 48	08 30	09 33	10 02	21 49	01 19	03 49	17 07	18 21
17	01 31	23 28	08 45	10 02	10 00	21 49	01 16	03 50	17 07	18 22
18	01 07	26 08	08 57	10 32	09 58	21 49	01 13	03 51	17 07	18 23
19	00 43	27 34	09 08	11 00	09 56	21 49	01 10	03 52	17 07	18 23
20	00 S 19	27 31	09 16	11 29	09 53	21 49	01 07	03 53	17 06	18 24
21	00 N 04	25 51	09 21	11 57	09 50	21 49	01 04	03 54	17 06	18 24
22	00 28	22 35	09 25	12 25	09 47	21 49	01 01	03 55	17 06	18 25
23	00 52	17 51	09 26	12 53	09 44	21 49	00 58	03 56	17 06	18 26
24	01 15	11 55	09 25	13 20	09 41	21 49	00 55	03 57	17 06	18 26
25	01 39	05 N 11	09 22	13 47	09 37	21 49	00 52	03 58	17 05	18 27
26	02 03	01 S 54	09 17	14 14	09 33	21 49	00 49	03 59	17 05	18 27
27	02 26	08 53	09 10	14 41	09 29	21 49	00 46	04 00	17 05	18 28
28	02 50	15 15	09 01	15 07	09 25	21 48	00 43	04 01	17 04	18 28
29	03 13	20 37	08 50	15 33	09 21	21 48	00 40	04 02	17 04	18 29
30	03 36	24 37	08 37	15 58	09 16	21 48	00 38	04 03	17 04	18 29
31	04 N 00	27 S 02	08 S 23	16 N 23	09 S 11	21 N 48	00 S 35	04 N 04	17 S 04	18 N 30

ZODIAC SIGN ENTRIES

Date	h	m	Planets
02	11	53	☽ ♐
03	21	32	♄ ♈
04	17	35	☽ ♑
07	02	03	☽ ♒
09	12	41	☽ ♓
12	00	53	☽ ♈
14	13	54	☽ ♉
17	02	19	☽ ♊
19	12	10	☽ ♋
20	09	56	♀ ♉
21	07	37	☉ ♈
21	18	04	☽ ♌
23	20	08	☽ ♍
25	19	50	☽ ♎
27	19	10	☽ ♏
29	20	08	☽ ♐
31	06	10	♂ ♎

LATITUDES

Date	Mercury ☿ ° '	Venus ♀ ° '	Mars ♂ ° '	Jupiter ♃ ° '	Saturn ♄ ° '	Uranus ♅ ° '	Neptune ♆ ° '	Pluto ♇ ° '
01	03 N 39	00 S 53	02 N 33	00 N 35	02 S 06	00 N 48	01 N 45	15 N 20
04	03 40	00 46	02 34	00 35	02 06	00 48	01 46	15 20
07	03 22	00 39	02 34	00 35	02 06	00 48	01 46	15 21
10	02 50	00 31	02 35	00 35	02 06	00 48	01 46	15 21
13	02 10	00 22	02 35	00 35	02 06	00 48	01 46	15 21
16	01 26	00 14	02 34	00 35	02 06	00 48	01 46	15 21
19	00 42	00 S 05	02 33	00 35	02 06	00 48	01 46	15 21
22	00 N 02	00 N 04	02 32	00 35	02 06	00 48	01 46	15 21
25	00 S 35	00 13	02 30	00 35	02 06	00 48	01 47	15 21
28	01 06	00 23	02 28	00 35	02 06	00 48	01 47	15 20
31	01 S 33	00 N 32	02 N 25	00 N 35	02 S 06	00 N 48	01 N 47	15 N 20

DATA

Julian Date	2439551
Delta T	+38 seconds
Ayanamsa	23° 23' 44"
Synetic vernal point	05° ♓ 43' 16"
True obliquity of ecliptic	23° 26' 44"

LONGITUDES

Date	Chiron ⚷ ° '	Ceres ⚳ ° '	Pallas ⚴ ° '	Juno ⚵ ° '	Vesta ⚶ ° '	Black Moon Lilith ⚸ ° '
01	24 ♓ 46	24 ♊ 30	24 ♈ 54	00 ♌ 25	26 ♏ 25	07 ♈ 17
11	25 ♓ 21	26 ♊ 10	29 ♈ 22	00 ♌ 06	28 ♏ 29	08 ♈ 24
21	25 ♓ 57	28 ♊ 20	04 ♉ 03	00 ♌ 32	29 ♏ 54	09 ♈ 30
31	26 ♓ 33	00 ♋ 55	08 ♉ 57	01 ♌ 38	00 ♐ 35	10 ♈ 37

MOON'S PHASES, APSIDES AND POSITIONS ☽

Date	h	m	Phase	Longitude ° '	Eclipse Indicator
03	09	10	☾	12 ♐ 06	
11	04	30	●	19 ♓ 55	
19	08	31	☽	28 ♊ 03	
26	03	21	○	04 ♎ 47	

Day	h	m		
13	01	40	Apogee	
26	08	02	Perigee	
05	04	55	Max dec	27° S 36'
12	13	28	0N	
19	23	21	Max dec	27° N 44'
26	05	36	0S	

ASPECTARIAN

h m	Aspects
01 Wednesday	
01 21	☽ △ ☉
06 02	☽ ± ♀
08 44	☉ ⚼ ♃
09 39	☽ ⚼ ♄
11 25	☽ ∥ ♆
11 29	☽ △ ☿
16 07	☽ # ♇
17 42	♀ # ♄
17 55	☽ ⚹ ♇
22 47	☽ ⚼ ♀
23 36	☽ ⚹ ♅
02 Thursday	
02 06	☽ ☌ ♆
03 16	☽ △ ♃
10 21	☽ # ♃
11 35	☽ △ ♄
14 35	☽ Q ♇
17 04	☽ ⊻ ♂
20 22	☽ Q ♅
03 Friday	
03 28	☽ △ ♀
03 41	☽ ⊥ ♂
05 22	☽ ⚼ ♃
07 37	☿ # ♀
09 10	☽ □ ☉
12 18	☽ □ ☿
12 26	♀ ⚼ ♆
19 50	☽ ∠ ♂
21 24	☽ ± ♃
22 28	☽ □ ♇
04 Saturday	
04 25	☽ □ ♅
07 13	☽ ⊻ ♆
07 36	☉ ☌ ☿
08 14	☽ ⊼ ♃
17 46	☽ □ ♄
18 12	☽ ⊥ ♆
18 27	♀ ∥ ♅
18 37	☽ Q ☿
20 29	☽ Q ☉
23 19	☽ ⚹ ♂
05 Sunday	
02 28	☿ # ♅
10 53	☽ ∠ ♆
12 21	☿ ⊻ ♀
15 15	☽ ⚹ ☿
15 49	☽ □ ♀
20 44	☽ ⚹ ☉
21 47	☽ Q ♂
06 Monday	
03 41	☽ Q ♄
05 55	☽ △ ♇
12 05	☽ △ ♅
15 12	☽ ⚹ ♆
16 03	☽ ☍ ♃
17 32	☽ ∠ ☿
07 Tuesday	
02 49	☽ ⚹ ♄
03 45	☽ ∠ ☉
05 44	☽ Q ♀
08 11	☽ □ ♂
09 34	☽ ⊥ ☿
10 35	☽ ⚼ ♇
13 29	☿ ⚼ ♃
14 15	☽ Q ♆
16 49	☽ ⚼ ♅
20 21	☽ ⊻ ☿
22 41	☽ ⊥ ☉
23 03	☉ # ♀
08 Wednesday	
03 21	☽ # ♃
04 01	☽ ± ♇
05 15	☿ ⊥ ♀
06 05	☉ ∥ ☿
07 36	☽ ⚹ ♀
08 12	☽ ∠ ♄
10 18	☽ ± ♅
11 27	☽ ⊻ ☉
15 44	☽ ⊼ ♇
17 44	♂ St R
22 02	☽ ⊼ ♅
09 Thursday	
00 10	☽ # ♇
01 26	☽ □ ♆
02 04	☽ ⊥ ♄
02 08	☽ ⊼ ♃
06 14	☽ ∥ ♆
11 15	☉ ⚼ ♂
13 59	☽ ± ♃
14 04	☽ ⊻ ♄
16 29	☽ ∠ ♀
19 01	☽ △ ♂
10 Friday	
03 32	☽ ☌ ☿
07 51	☽ ⚼ ♃
12 28	☽ ⊥ ♀
14 42	☉ ☍ ♇
16 26	☽ ∥ ♂
19 41	♀ ± ♆
11 Saturday	
00 59	☽ ⚼ ♂
01 54	☽ ⊻ ♀
03 19	☽ ☍ ♇
04 30	☽ ☌ ☉
06 54	☽ # ♀
08 23	☽ ∥ ☿
09 41	☽ ☍ ♅
13 20	☽ △ ♆
13 55	☽ △ ♃
15 27	♀ ⊼ ♇
19 05	☽ ± ♂
20 38	☉ # ♅
21 09	☽ # ♅
21 12	☽ ∥ ☉
12 Sunday	
02 55	☽ ☌ ♄
06 50	☽ ∥ ♄
07 12	☽ ⊼ ♂
12 50	☽ ⊻ ☿
19 38	☿ ∠ ♀
19 44	☽ ⚼ ♆
20 00	☽ # ♄
13 Monday	
00 31	☽ ⊥ ☿
03 35	☽ # ☉
05 56	☽ ∥ ♅
14 02	☽ ± ♆
16 00	☉ ☍ ♅
16 01	☽ ⊼ ♇
18 11	☽ ∠ ☿
21 46	☽ ☌ ♀
22 22	☽ ⊼ ♅
22 59	☽ ⊻ ☉
23 28	☽ # ☿
14 Tuesday	
02 14	☽ ⊼ ♆
02 31	☽ ∥ ♀
02 44	☽ □ ♃
03 28	♀ ⊼ ♅
04 13	☽ ± ♇
10 26	☽ # ♂
10 33	☽ ± ♅
12 20	☽ ⊥ ☉
16 35	☽ ⊻ ♄
19 55	☽ ☍ ♂
22 29	☽ ⚼ ♇
23 51	☽ ⚹ ☿
15 Wednesday	
04 46	☽ ⚼ ♅
04 54	☽ ⊥ ♄
08 17	☽ ∠ ☉
13 07	☉ △ ♆
15 13	☽ Q ♃
16 42	♀ ⊼ ♆
18 21	☉ △ ♃
20 58	♀ □ ♃
21 03	☽ # ♆
23 17	☽ ∠ ♄
23 41	☽ Q ☿
16 Thursday	
03 43	☽ ∥ ♇
04 44	☽ △ ♇
08 55	☉ ⊻ ♀
10 51	♀ ± ♇
10 54	☽ △ ♅
14 52	☽ ☍ ♆
15 18	☽ ⚹ ♃
17 13	☽ ⚹ ☉
17 23	☽ ⊻ ♀
17 Friday	
00 28	☽ ∥ ♃
05 34	☽ ⚹ ♄
06 38	☽ ⊥ ♀
07 43	☽ ⊼ ♂
10 24	♀ # ♂
11 17	☽ □ ☿
14 28	☿ St D
19 03	☽ Q ☉
19 24	☽ ± ♂
20 54	☽ ∠ ♃
22 57	☉ ± ♂
18 Saturday	
02 13	☽ ∠ ♀
05 03	☉ ∥ ♄
05 17	☽ Q ♄
12 40	☽ ⚼ ♂
14 17	☽ ⊥ ♃
15 36	☽ □ ♇
21 23	☽ □ ♅
21 33	♀ ± ♅
19 Sunday	
01 18	☽ ⊼ ♆
01 42	☽ ⊻ ♃
08 31	☽ □ ☉
09 53	☽ ⚹ ♀
12 34	☽ ± ♆
15 47	☽ □ ♄
16 42	☽ △ ♂
21 04	☽ △ ☿
20 Monday	
01 16	☽ Q ♇
05 09	☽ ⚼ ♆
06 46	☽ Q ♅
10 15	☽ Q ♀
22 54	☽ ⚹ ♇
21 Tuesday	
00 45	☽ ⚼ ☿
04 12	☽ ⚹ ♅
07 57	☽ △ ♆
08 22	☽ ☌ ♃
09 10	♃ St D
11 28	♂ ⊼ ♄
16 55	☽ ± ☿
18 52	☽ △ ☉
21 08	☽ □ ♀
21 41	☽ □ ♂
21 53	☽ △ ♄
22 Wednesday	
01 00	☽ ∠ ♇
02 42	♀ ☍ ♂
03 31	☽ ⊼ ☿
06 04	☽ ∠ ♅
06 47	♀ ⊻ ♄
16 12	☽ ⊥ ♇
16 25	☽ ∥ ♃
21 09	☽ ⊥ ♅
22 10	☽ ⚼ ☉
23 25	☽ ⚼ ♄
23 Thursday	
02 09	☽ ⊻ ♇
03 36	☽ Q ♂
04 50	☉ ⊼ ♂
07 02	☽ ⊻ ♅
09 25	☽ ∥ ♇
10 38	☽ □ ♆
11 05	☽ ⊻ ♃
14 03	☽ ± ☉
14 20	☽ ± ♄
15 19	☽ # ♆
16 52	♀ ⚼ ♇
17 44	☉ # ♄
19 01	☉ ☌ ♄
20 50	☽ ⊥ ♃
22 56	☽ ⚹ ♂
24 Friday	
00 09	☽ ⊼ ♄
00 28	☽ ⊼ ☉
03 26	☽ △ ♀
06 42	☽ ☍ ☿
06 59	☽ ∥ ♀
11 19	☽ ∠ ♃
15 37	☽ Q ♆
20 19	☽ # ♂
21 16	☽ # ☿
22 35	☽ ∠ ♂
25 Saturday	
02 25	☽ ☌ ♇
05 23	☽ ⚼ ♀
07 03	☽ ☌ ♅
10 36	☽ ⚹ ♆
11 06	☽ ⚹ ♃
12 37	☽ ⊥ ♂
16 11	☽ ∥ ♅
20 47	☽ ± ♀
21 55	☽ ⊻ ♂
23 24	☽ ∥ ☉
26 Sunday	
00 08	☽ ☍ ♄
01 37	♀ ⚼ ♅
02 46	☽ # ♄
03 21	☽ ☍ ☉
05 59	☽ Q ♃
07 01	☽ ⊼ ♀
08 14	☽ ⊼ ☿
08 23	☽ ∥ ♄
10 10	☽ ∠ ♆
12 30	☽ # ☉
18 08	☽ ± ☿
19 04	☽ # ♅
27 Monday	
00 19	☽ ⊥ ♆
01 34	☽ ⊻ ♇
06 09	☽ ⊻ ♅
09 05	☽ ⚼ ☿
09 48	☽ ⊻ ♆
10 24	☽ □ ♃
11 03	☽ ⊥ ♇
13 00	☽ ∥ ☿
14 10	☽ ∥ ♂
15 38	☽ ⊥ ♅
19 31	♀ ⊥ ♄
20 35	☽ ☌ ♂
21 42	☿ ⚹ ♀
23 56	☽ ⊼ ♄
28 Tuesday	
01 22	☽ ∠ ♇
06 00	☽ ∠ ♅
06 11	☽ ⊼ ☉
09 40	☽ ± ♄
10 25	☽ △ ☿
10 34	☿ ⚼ ♃
10 47	☽ ☍ ♀
11 22	☽ # ♀
16 35	☽ ± ☉
19 33	☽ ∥ ♆
29 Wednesday	
00 25	☽ ⚼ ♄
01 39	☽ ⚹ ♇
01 44	☽ # ♇
06 21	☽ ⚹ ♅
08 24	☽ ⚼ ☉
10 15	☽ ☌ ♆
11 00	☽ △ ♃
18 13	☽ # ♃
20 46	☽ ⊻ ♂
21 32	☽ Q ♇
30 Thursday	
01 35	☽ △ ♄
02 20	☽ Q ♅
06 43	☽ ⊥ ♂
11 05	☉ ⚼ ♆
11 30	☽ △ ☉
12 20	☽ ⚼ ♃
15 37	☽ □ ☿
17 19	☽ ⊼ ♀
19 25	♀ Q ♃
21 53	☽ ∠ ♂
31 Friday	
03 56	☽ ± ♃
04 20	☽ □ ♇
04 44	☽ ± ♀
09 18	☽ □ ♅
13 33	☽ ⊻ ♆
14 32	☽ ⊼ ♃
16 29	☉ ∥ ♅
22 13	☽ ⚼ ♀
23 48	☽ ⚹ ♂

All ephemeris data is given at 12.00 UT and the Moon's longitude is additionally given for 24.00 UT

APRIL 1967

LONGITUDES

Date	Sidereal time h m s	Sun ☉ ° ' "	Moon ☽ ° ' "	Moon ☽ 24.00 ° ' "	Mercury ☿ ° '	Venus ♀ ° '	Mars ♂ ° '	Jupiter ♃ ° '	Saturn ♄ ° '	Uranus ♅ ° '	Neptune ♆ ° '	Pluto ♇ ° '
01	00 36 39	11 ♈ 04 02	06 ♑ 32 21	13 ♑ 04 45	13 ♓ 18	14 ♉ 36	29 ♎ 39	24 ♋ 37	03 ♈ 32	21 ♍ 35	24 ♏ 01	18 ♍ 46
02	00 40 35	12 03 14	19 ♑ 31 46	25 ♑ 53 47	14 21	15 48	29 R 21	24 39	03 40	21 R 33	24 R 00	18 R 45
03	00 44 32	13 02 24	02 ♒ 11 16	08 ♒ 24 40	15 27	17 00	29 03	24 42	03 47	21 30	23 59	18 43
04	00 48 28	14 01 32	14 ♒ 34 25	20 ♒ 40 58	16 35	18 12	28 44	24 44	03 55	21 28	23 58	18 42
05	00 52 25	15 00 38	26 ♒ 44 43	02 ♓ 46 05	17 46	19 24	28 24	24 47	04 02	21 26	23 57	18 40
06	00 56 21	15 59 42	08 ♓ 45 26	14 ♓ 43 07	18 58	20 35	28 04	24 50	04 09	21 23	23 55	18 39
07	01 00 18	16 58 44	20 ♓ 39 27	26 ♓ 34 45	20 13	21 47	27 44	24 53	04 17	21 21	23 54	18 38
08	01 04 15	17 57 44	02 ♈ 29 18	08 ♈ 23 24	21 30	22 58	27 23	24 56	04 24	21 19	23 53	18 36
09	01 08 11	18 56 43	14 ♈ 17 17	20 ♈ 11 13	22 49	24 10	27 02	25 00	04 32	21 16	23 52	18 35
10	01 12 08	19 55 39	26 ♈ 05 29	02 ♉ 00 20	24 09	25 21	26 40	25 04	04 39	21 14	23 51	18 34
11	01 16 04	20 54 34	07 ♉ 56 03	13 ♉ 52 55	25 32	26 33	26 18	25 07	04 46	21 12	23 49	18 32
12	01 20 01	21 53 26	19 ♉ 51 15	25 ♉ 51 22	26 56	27 44	25 56	25 11	04 53	21 10	23 48	18 31
13	01 23 57	22 52 16	01 ♊ 53 38	07 ♊ 58 24	28 22	28 ♉ 55	25 33	25 15	05 01	21 08	23 47	18 30
14	01 27 54	23 51 05	14 ♊ 06 05	20 ♊ 17 04	29 ♓ 50	00 ♊ 06	25 11	25 20	05 08	21 06	23 45	18 28
15	01 31 50	24 49 51	26 ♊ 31 48	02 ♋ 50 42	01 ♈ 20	01 17	24 48	25 24	05 15	21 04	23 44	18 27
16	01 35 47	25 48 34	09 ♋ 15 11	15 ♋ 42 40	02 51	02 28	24 25	25 29	05 22	21 02	23 43	18 26
17	01 39 44	26 47 16	22 ♋ 16 33	28 ♋ 56 09	04 24	03 39	24 03	25 34	05 29	21 00	23 41	18 25
18	01 43 40	27 45 55	05 ♌ 41 45	12 ♌ 33 33	05 59	04 50	23 40	25 39	05 37	20 58	23 40	18 24
19	01 47 37	28 44 32	19 ♌ 31 38	26 ♌ 35 58	07 35	06 00	23 17	25 44	05 44	20 56	23 38	18 22
20	01 51 33	29 ♈ 43 07	03 ♍ 46 22	11 ♍ 02 30	09 14	07 11	22 54	25 49	05 51	20 54	23 37	18 21
21	01 55 30	00 ♉ 41 40	18 ♍ 23 52	25 ♍ 49 48	10 53	08 21	22 32	25 55	05 58	20 52	23 35	18 20
22	01 59 26	01 40 10	03 ♎ 19 29	10 ♎ 51 58	12 35	09 31	22 09	26 00	06 05	20 50	23 34	18 19
23	02 03 23	02 38 38	18 ♎ 26 09	26 ♎ 00 55	14 18	10 41	21 47	26 06	06 12	20 49	23 32	18 18
24	02 07 19	03 37 05	03 ♏ 35 04	11 ♏ 07 28	16 03	11 52	21 25	26 12	06 19	20 47	23 31	18 17
25	02 11 16	04 35 29	18 ♏ 36 59	26 ♏ 02 37	17 49	13 02	21 04	26 18	06 26	20 45	23 29	18 16
26	02 15 13	05 33 52	03 ♐ 23 29	10 ♐ 38 50	19 37	14 11	20 42	26 24	06 33	20 44	23 28	18 15
27	02 19 09	06 32 13	17 ♐ 48 06	24 ♐ 50 54	21 27	15 21	20 22	26 30	06 39	20 42	23 26	18 14
28	02 23 06	07 30 33	01 ♑ 47 00	08 ♑ 36 19	23 19	16 31	20 01	26 37	06 46	20 41	23 25	18 13
29	02 27 02	08 28 51	15 ♑ 18 56	21 ♑ 55 02	25 12	17 40	19 41	26 44	06 53	20 39	23 23	18 12
30	02 30 59	09 ♉ 27 08	28 ♑ 24 54	04 ♒ 48 57	27 ♈ 07	18 ♊ 50	19 ♎ 22	26 ♋ 50	07 ♈ 00	20 ♍ 38	23 ♏ 22	18 ♍ 11

Date	Moon True ☊ ° '	Moon Mean ☊ ° '	Moon ☽ Latitude ° '
01	07 ♉ 13	08 ♉ 32	04 S 31
02	07 R 13	08 29	05 00
03	07 12	08 25	05 13
04	07 10	08 22	05 12
05	07 06	08 19	04 56
06	07 03	08 16	04 28
07	06 59	08 13	03 48
08	06 56	08 10	02 59
09	06 54	08 06	02 02
10	06 53	08 03	00 S 59
11	06 D 53	08 00	00 N 06
12	06 53	07 57	01 11
13	06 54	07 54	02 14
14	06 55	07 51	03 12
15	06 56	07 47	04 02
16	06 57	07 44	04 41
17	06 58	07 41	05 07
18	06 R 58	07 38	05 17
19	06 57	07 35	05 10
20	06 56	07 31	04 43
21	06 56	07 28	03 58
22	06 55	07 25	02 56
23	06 55	07 22	01 41
24	06 55	07 19	00 N 18
25	06 D 55	07 16	01 S 05
26	06 55	07 12	02 22
27	06 55	07 09	03 28
28	06 R 55	07 06	04 20
29	06 55	07 03	04 55
30	06 ♉ 55	07 ♉ 00	05 S 14

DECLINATIONS

Date	Sun ☉ ° '	Moon ☽ ° '	Mercury ☿ ° '	Venus ♀ ° '	Mars ♂ ° '	Jupiter ♃ ° '	Saturn ♄ ° '	Uranus ♅ ° '	Neptune ♆ ° '	Pluto ♇ ° '
01	04 N 23	27 S 47	08 S 07	16 N 47	09 S 06	21 N 47	00 S 32	04 N 05	17 S 03	18 N 30
02	04 46	26 58	07 50	17 12	09 01	21 47	00 29	04 06	17 03	18 31
03	05 09	24 46	07 30	17 35	08 56	21 46	00 26	04 07	17 03	18 31
04	05 32	21 26	07 10	17 59	08 50	21 46	00 23	04 07	17 02	18 32
05	05 55	17 14	06 47	18 22	08 44	21 45	00 20	04 08	17 02	18 32
06	06 18	12 25	06 24	18 44	08 39	21 45	00 17	04 09	17 02	18 32
07	06 40	07 12	05 59	19 06	08 33	21 44	00 14	04 10	17 01	18 33
08	07 03	01 S 45	05 32	19 27	08 27	21 44	00 12	04 11	17 01	18 33
09	07 25	03 N 46	05 04	19 48	08 21	21 43	00 09	04 12	17 01	18 33
10	07 48	09 09	04 35	20 09	08 14	21 42	00 06	04 13	17 00	18 34
11	08 10	14 15	04 05	20 29	08 08	21 42	00 S 03	04 14	17 00	18 34
12	08 32	18 51	03 33	20 48	08 02	21 41	00 00	04 14	17 00	18 34
13	08 54	22 44	03 00	21 07	07 55	21 40	00 N 03	04 15	16 59	18 35
14	09 16	25 41	02 26	21 26	07 49	21 39	00 05	04 16	16 59	18 35
15	09 37	27 26	01 51	21 43	07 43	21 39	00 08	04 17	16 59	18 35
16	09 59	27 48	01 15	22 01	07 36	21 38	00 11	04 18	16 58	18 36
17	10 20	26 39	00 S 37	22 17	07 30	21 37	00 14	04 18	16 58	18 36
18	10 41	23 58	00 N 01	22 33	07 23	21 36	00 16	04 19	16 57	18 36
19	11 02	19 52	00 41	22 49	07 17	21 35	00 19	04 20	16 57	18 36
20	11 23	14 31	01 21	23 04	07 11	21 34	00 22	04 20	16 57	18 37
21	11 43	08 14	02 03	23 18	07 05	21 33	00 24	04 21	16 56	18 37
22	12 03	01 N 22	02 45	23 32	06 59	21 32	00 27	04 22	16 56	18 37
23	12 24	05 S 40	03 29	23 45	06 53	21 31	00 30	04 22	16 56	18 37
24	12 44	12 25	04 13	23 57	06 47	21 30	00 32	04 23	16 55	18 37
25	13 03	18 24	04 58	24 09	06 41	21 28	00 35	04 24	16 55	18 37
26	13 23	23 10	05 43	24 20	06 35	21 27	00 38	04 24	16 54	18 38
27	13 42	26 21	06 30	24 31	06 30	21 26	00 40	04 25	16 54	18 38
28	14 01	27 46	07 17	24 41	06 24	21 25	00 43	04 26	16 54	18 38
29	14 20	27 27	08 05	24 50	06 19	21 24	00 45	04 26	16 53	18 38
30	14 N 39	25 S 37	08 N 53	24 N 58	06 S 14	21 N 22	00 N 48	04 N 27	16 S 53	18 N 38

ZODIAC SIGN ENTRIES

Date	h	m	Planets
01	00	11	☽ ♑
03	07	49	☽ ♒
05	18	29	☽ ♓
08	06	57	☽ ♈
10	19	56	☽ ♉
13	08	15	☽ ♊
14	09	54	♀ ♊
14	14	38	☿ ♈
15	18	37	☽ ♋
18	01	54	☽ ♌
20	05	43	☽ ♍
20	18	55	☉ ♉
22	06	41	☽ ♎
24	06	19	☽ ♏
26	06	27	☽ ♐
28	08	54	☽ ♑
30	14	57	☽ ♒

LATITUDES

Date	Mercury ☿ ° '	Venus ♀ ° '	Mars ♂ ° '	Jupiter ♃ ° '	Saturn ♄ ° '	Uranus ♅ ° '	Neptune ♆ ° '	Pluto ♇ ° '
01	01 S 41	00 N 36	02 N 24	00 N 35	02 S 07	00 N 48	01 N 47	15 N 20
04	02 01	00 45	02 21	00 35	02 07	00 48	01 47	15 20
07	02 17	00 55	02 16	00 35	02 07	00 48	01 47	15 19
10	02 28	01 04	02 12	00 35	02 07	00 48	01 47	15 19
13	02 34	01 14	02 06	00 35	02 07	00 48	01 47	15 18
16	02 36	01 23	02 00	00 35	02 08	00 48	01 47	15 17
19	02 32	01 31	01 54	00 35	02 08	00 48	01 48	15 17
22	02 24	01 40	01 47	00 35	02 08	00 48	01 48	15 16
25	02 12	01 48	01 40	00 35	02 09	00 48	01 48	15 15
28	01 55	01 56	01 32	00 35	02 09	00 48	01 48	15 14
31	01 S 33	02 N 03	01 N 24	00 N 35	02 S 09	00 N 47	01 N 48	15 N 13

DATA

Julian Date	2439582
Delta T	+38 seconds
Ayanamsa	23° 23' 48"
Synetic vernal point	05° ♓ 43' 12"
True obliquity of ecliptic	23° 26' 44"

MOON'S PHASES, APSIDES AND POSITIONS ☽

Date	h	m	Phase	Longitude ° '	Eclipse Indicator
01	20	58	☾	11 ♑ 26	
09	22	20	●	19 ♈ 22	
17	20	48	☽	27 ♋ 09	
24	12	03	○	03 ♏ 37	total

Day	h	m		
09	03	24	Apogee	
23	19	08	Perigee	
01	11	07	Max dec	27° S 47'
08	19	35	0N	
16	06	00	Max dec	27° N 51'
22	16	40	0S	
28	19	29	Max dec	27° S 51'

LONGITUDES

Date	Chiron ⚷ ° '	Ceres ⚳ ° '	Pallas ⚴ ° '	Juno ⚵ ° '	Vesta ⚶ ° '	Black Moon Lilith ⚸ ° '
01	26 ♓ 36	01 ♋ 11	09 ♉ 27	01 ♌ 46	00 ♐ 37	10 ♈ 43
11	27 ♓ 11	04 ♋ 10	14 ♉ 33	03 ♌ 29	00 ♐ 24	11 ♈ 50
21	27 ♓ 48	07 ♋ 26	19 ♉ 48	05 ♌ 40	29 ♏ 23	12 ♈ 57
31	28 ♓ 13	10 ♋ 56	25 ♉ 13	08 ♌ 13	27 ♏ 38	14 ♈ 03

ASPECTARIAN

h m Aspects

01 Saturday
00 14 ☽ ⊥ ♆
01 44 ☽ Q ☿
06 30 ☽ □ ♄
16 31 ☽ ∠ ♆
20 58 ☽ □ ☉
21 09 ☽ Q ♂

02 Sunday
01 30 ☽ ✱ ☿
03 32 ♀ # ♆
04 20 ☽ △ ♀
10 32 ☽ △ ♇
11 54 ☿ bQ ♂
15 46 ☽ △ ♅
16 03 ☽ Q ♄
20 24 ☽ ✱ ♆
21 41 ☽ ☍ ♃

03 Monday
06 08 ☽ □ ♂
08 21 ☽ ∠ ☿
09 37 ☽ Q ☉
14 56 ☽ bQ ♇
15 06 ☽ ✱ ♄
19 18 ☽ Q ♆
20 17 ☽ bQ ♅

04 Tuesday
03 25 ☽ ⊥ ☿
08 20 ☽ ± ♇
09 52 ☽ # ♃
10 50 ☽ ✱ ☉
13 44 ☽ ± ♅
16 21 ☽ ⊻ ☿
19 53 ☽ □ ♀
20 04 ☽ ⊼ ♇
20 36 ☽ ∠ ♄
21 50 ♀ △ ♇

05 Wednesday
01 30 ☽ ⊼ ♅
03 59 ♀ ∠ ♄
05 02 ☽ # ♇
06 27 ☽ # ♀
06 27 ☽ □ ♆
08 06 ☽ ⊼ ♃
13 05 ☽ ∥ ♆
14 35 ☽ ⊥ ♄
15 12 ☽ △ ♂
19 05 ☽ ∠ ☉
20 04 ☽ ± ♃
23 18 ♀ ∥ ♇

06 Thursday
02 41 ☽ ⊻ ♄
05 49 ☿ ☍ ♇
11 37 ☽ Q ♀
14 10 ☽ bQ ♃
14 43 ☽ ⊥ ☉
15 05 ☉ # ☿
20 26 ☽ bQ ♂

07 Friday
03 34 ♀ △ ♅
03 53 ☽ ⊻ ☉
05 49 ☽ ∥ ♂
07 54 ☽ ☍ ♇
11 00 ☽ ☌ ☿
13 24 ☽ ☍ ♅
14 07 ☽ ± ♂
14 11 ☽ # ☉
14 32 ☽ ✱ ♀
17 54 ☽ ∥ ☿
18 34 ☽ △ ♆
20 36 ☽ △ ♃

08 Saturday
01 22 ☽ # ♅
01 56 ☽ ⊼ ♂
08 41 ☿ ☍ ♅
10 08 ☉ ± ♆
10 20 ☿ ± ♂
15 56 ☽ ☌ ♄
17 21 ♂ ⊥ ♅
18 48 ☽ ∥ ♄
20 21 ☽ # ♄

09 Sunday
00 25 ☽ ∠ ♀
00 59 ☽ bQ ♆
03 18 ☉ ⊼ ♇
06 00 ♀ ☍ ♆
13 54 ☽ ∥ ♅
17 17 ☽ # ☿
19 16 ☽ ± ♆
20 43 ☽ ⊼ ♇
20 47 ☽ ⊥ ♀
22 20 ☽ ☌ ☉

10 Monday
02 10 ☽ ⊼ ♅
05 23 ☽ ∥ ☉
05 40 ♀ ✱ ♃
06 35 ☿ △ ♆
07 26 ☽ ⊼ ♆
07 33 ☽ ⊻ ☿
07 56 ☽ # ♂
08 54 ☽ ± ♇
09 53 ☽ □ ♃
10 20 ☽ ⊻ ♀
13 08 ☽ ☍ ♂
14 19 ☽ ± ♅
21 19 ☽ ⊥ ☿

11 Tuesday
03 07 ☽ bQ ♇
04 36 ☿ △ ♃
05 21 ☿ # ♅
05 32 ☽ ⊻ ♄
08 14 ♀ ⊼ ♂
08 30 ☽ bQ ♅
10 37 ☉ # ♂
17 47 ☽ ⊥ ♄
17 56 ☽ ∠ ☿
18 53 ☉ ⊼ ♅
22 29 ☿ ⊼ ♂
22 32 ☽ Q ♃

12 Wednesday
01 57 ☽ # ♆
09 19 ☽ △ ♇
10 28 ☽ ∥ ♇
12 04 ☽ ∠ ♄
14 37 ☽ △ ♅
16 27 ☽ ⊻ ☉
19 53 ☽ ☍ ♆
22 44 ☽ ✱ ♃
23 47 ☽ ⊼ ♂

13 Thursday
00 26 ☽ ∥ ♀
04 04 ☽ ✱ ☿
04 53 ☽ ∥ ♃
05 27 ☽ ☌ ♀
05 28 ☽ ⊥ ☉
06 27 ☉ ⊥ ♀
11 21 ☽ ± ♂
18 14 ☽ ✱ ♄

14 Friday
00 50 ☽ ∠ ☉
04 06 ♂ □ ♃
04 34 ☽ bQ ♂
04 35 ☽ ∠ ♃
06 58 ☽ Q ☿
09 41 ☉ ⊼ ♆
17 57 ☽ Q ♄
20 29 ☽ □ ♇
22 13 ☽ ⊥ ♃

15 Saturday
01 32 ☽ □ ♅
02 55 ☉ ± ♇
04 34 ♀ ± ♂
05 48 ♀ ∥ ♃
06 39 ☽ ⊼ ♆
08 28 ☽ ✱ ☉
08 37 ☿ ✱ ♀
08 47 ☽ △ ♂
09 50 ☽ ⊻ ♃
11 30 ☉ ☍ ♂
18 05 ☽ ± ♆
21 59 ☽ ⊻ ♀
22 23 ☽ □ ☿

16 Sunday
03 16 ☉ □ ♃
04 42 ☽ □ ♄
06 46 ☽ Q ♇
09 07 ☽ Q ☉
10 25 ☽ ⊥ ♀
11 01 ☽ bQ ♆
11 21 ♂ ⊥ ♇
11 37 ☽ Q ♅

17 Monday
04 44 ☽ ∠ ♀
04 58 ☽ ✱ ♇
09 41 ☽ ✱ ♅
14 33 ☽ △ ♆
15 07 ☽ □ ♂
16 55 ☉ ± ♅
17 59 ☽ ☌ ♃
20 48 ☽ □ ☉

18 Tuesday
01 48 ☿ # ♄
05 54 ☿ ☌ ♄
07 56 ☽ ∠ ♇
10 19 ☽ ✱ ♀
11 51 ☽ △ ♄
11 55 ♂ ⊻ ♆
12 28 ☽ ∠ ♅
12 35 ☽ △ ☿
20 38 ☽ ∥ ♀
21 57 ☿ ∥ ♄
22 10 ☽ Q ♂
23 42 ☽ ⊥ ♇

19 Wednesday
02 51 ☽ ∥ ♃
04 07 ☽ ⊥ ♅
05 47 ♀ ✱ ♄
09 09 ☽ Q ♀
10 02 ☽ ⊻ ♇
14 04 ☽ bQ ♄
14 23 ☽ ⊻ ♅
17 54 ☽ bQ ☿
18 07 ☽ ∥ ♇
18 13 ☽ ✱ ♂
18 59 ☽ □ ♆
22 36 ☽ ⊻ ♃

20 Thursday
01 42 ☽ # ♆
03 12 ☿ bQ ♆
04 45 ☽ △ ☉
05 24 ☽ ± ♄
08 44 ☽ ⊥ ♃
10 58 ☽ ± ☿
15 28 ☽ ⊼ ♄
18 08 ☽ □ ♀
18 40 ☽ ∠ ♂
22 10 ☽ ⊼ ☿
23 14 ♀ bQ ♂
23 42 ☽ ∠ ♃
23 45 ☽ ∥ ☉

21 Friday
00 55 ☽ Q ♆
07 18 ☽ bQ ☉
09 02 ☽ ⊥ ♂
11 54 ☽ ☌ ♇
16 00 ☽ ☌ ♅
16 14 ☽ # ♂
18 31 ☽ ⊻ ♂
20 23 ☽ ✱ ♆
22 53 ☽ ± ☉

22 Saturday
00 12 ☽ ✱ ♃
01 43 ☽ ∥ ♅
07 42 ☽ ∥ ☿
09 10 ☽ ⊼ ☉
15 06 ☽ ∥ ♄
16 26 ☽ ☍ ♄
18 15 ☽ # ♄
19 30 ☽ Q ♃
20 20 ☽ ∠ ♆
22 42 ☽ △ ♀

23 Sunday
03 37 ☽ # ☿
04 36 ☽ ☍ ☿
07 33 ☽ # ♅
10 35 ☽ ⊥ ♆
11 47 ☽ ⊻ ♇
15 45 ☽ ⊻ ♅
16 06 ☽ ∥ ♂
17 11 ☽ ☌ ♂
20 04 ☽ ⊻ ♆
21 09 ♀ ∠ ♃
21 17 ☽ ⊥ ♇

24 Monday
00 13 ☽ □ ♃
00 27 ☽ bQ ♀
01 14 ☽ ⊥ ♅
03 55 ☉ bQ ♇
11 31 ☽ ∠ ♇
12 03 ☽ ☍ ☉
13 11 ☽ # ☉
15 29 ☽ ∠ ♅
15 55 ☽ ± ♀
16 22 ☽ ⊼ ♄
17 43 ☿ ∥ ♅

25 Tuesday
02 00 ☽ ± ♄
02 17 ☽ ⊼ ♀
05 37 ☽ ∥ ♆
07 40 ☿ ± ♆
10 32 ☽ ⊼ ☿
11 26 ☽ ✱ ♇
12 58 ☽ # ♇
15 26 ☽ ✱ ♅
15 51 ☽ ⊻ ♂
16 34 ☽ bQ ♄
17 58 ☿ ⊼ ♇
19 51 ☽ ☌ ♆
21 33 ☽ ± ☿

26 Wednesday
00 30 ☽ △ ♃
01 20 ☽ ⊥ ♂
02 34 ☽ # ♃
06 52 ☽ Q ♇
10 28 ♂ ⊻ ♅
10 55 ☽ Q ♅
14 19 ☽ bQ ☿
15 43 ☽ ∠ ♂
15 50 ☽ ⊼ ☉
15 55 ☉ bQ ♅
17 14 ☽ △ ♄
19 55 ☽ # ♀

27 Thursday
00 00 ☿ ☍ ♂
01 21 ☽ bQ ♃
02 21 ☿ ⊼ ♅
02 31 ☽ ± ☉
07 30 ☽ ☍ ♀
12 00 ☿ # ♂
12 44 ☽ □ ♇
15 22 ☉ ⊻ ♄
16 14 ☽ ✱ ♂
16 37 ☽ ± ♃
16 54 ☽ □ ♅
18 48 ☽ bQ ☉
19 08 ☽ △ ☿
21 34 ☽ ⊻ ♆

28 Friday
02 58 ☽ ⊼ ♃
07 53 ☽ ⊥ ♆
12 24 ☽ Q ♂
13 18 ☿ ⊼ ♆
20 50 ☽ □ ♄
22 50 ☽ △ ☉
23 29 ☿ ± ♇
23 38 ☽ ∠ ♆

29 Saturday
16 40 ☽ ⊼ ♀
17 13 ☽ △ ♇
19 44 ☽ □ ♂
21 40 ☽ △ ♅
22 56 ♀ □ ♇

30 Sunday
02 40 ☽ ✱ ♆
04 42 ☽ ± ♀
05 36 ☽ Q ♄
05 57 ☿ ± ♅
08 20 ☿ □ ♃
09 03 ☽ ☍ ♃
09 10 ☽ □ ☿
15 53 ♀ Q ♄
17 26 ☽ # ♀
20 40 ♀ △ ♂
20 55 ☽ bQ ♇
23 09 ☽ bQ ♀

All ephemeris data is given at 12.00 UT and the Moon's longitude is additionally given for 24.00 UT

MAY 1967

LONGITUDES

Date	Sidereal time h m s	Sun ☉ ° ' "	Moon ☽ ° ' "	Moon ☽ 24.00 ° ' "	Mercury ☿ ° '	Venus ♀ ° '	Mars ♂ ° '	Jupiter ♃ ° '	Saturn ♄ ° '	Uranus ♅ ° '	Neptune ♆ ° '	Pluto ♇ ° '
01	02 34 55	10 ♉ 25 23	11 ♒ 07 36	17 ♒ 21 21	29 ♈ 04	19 ♊ 59	19 ♎ 03	26 ♋ 57	07 ♈ 06	20 ♍ 36	23 ♏ 20	18 ♍ 11
02	02 38 52	11 23 36	23 ♒ 30 44	29 ♒ 36 18	01 ♉ 02	21 08	18 R 44	27 04	07 13	20 R 35	23 R 19	18 R 10
03	02 42 48	12 21 48	05 ♓ 38 35	11 ♓ 38 09	03 02	22 17	18 26	27 12	07 20	20 33	23 17	18 09
04	02 46 45	13 19 58	17 ♓ 35 31	23 ♓ 31 13	05 04	23 26	18 09	27 19	07 26	20 32	23 15	18 08
05	02 50 42	14 18 07	29 ♓ 25 45	05 ♈ 19 34	07 07	24 35	17 52	27 26	07 33	20 31	23 14	18 07
06	02 54 38	15 16 15	11 ♈ 13 07	17 ♈ 06 48	09 12	25 43	17 37	27 34	07 39	20 30	23 12	18 07
07	02 58 35	16 14 21	23 ♈ 01 01	28 ♈ 56 07	11 18	26 52	17 21	27 42	07 46	20 29	23 11	18 06
08	03 02 31	17 12 25	04 ♉ 52 24	10 ♉ 50 10	13 25	28 00	17 07	27 50	07 52	20 28	23 09	18 05
09	03 06 28	18 10 28	16 ♉ 49 42	22 ♉ 51 14	15 34	29 ♊ 09	16 53	27 58	07 58	20 27	23 07	18 05
10	03 10 24	19 08 29	28 ♉ 55 00	05 ♊ 01 12	17 43	00 ♋ 17	16 40	28 06	08 05	20 26	23 06	18 04
11	03 14 21	20 06 29	11 ♊ 10 04	17 ♊ 21 46	19 53	01 25	16 28	28 14	08 11	20 25	23 04	18 03
12	03 18 17	21 04 27	23 ♊ 36 30	29 ♊ 54 26	22 04	02 33	16 16	28 22	08 17	20 24	23 02	18 03
13	03 22 14	22 02 23	06 ♋ 15 47	12 ♋ 40 44	24 15	03 40	16 05	28 31	08 23	20 23	23 01	18 02
14	03 26 11	23 00 18	19 ♋ 09 26	25 ♋ 42 06	26 26	04 48	15 55	28 39	08 29	20 22	22 59	18 02
15	03 30 07	23 58 11	02 ♌ 18 52	08 ♌ 59 56	28 ♉ 36	05 55	15 46	28 48	08 35	20 21	22 58	18 01
16	03 34 04	24 56 02	15 ♌ 45 24	22 ♌ 35 23	00 ♊ 46	07 02	15 38	28 57	08 41	20 21	22 56	18 01
17	03 38 00	25 53 51	29 ♌ 29 56	06 ♍ 29 04	02 55	08 09	15 30	29 06	08 47	20 20	22 54	18 01
18	03 41 57	26 51 39	13 ♍ 32 41	20 ♍ 40 40	05 04	09 16	15 24	29 15	08 53	20 20	22 53	18 00
19	03 45 53	27 49 25	27 ♍ 52 45	05 ♎ 08 35	07 10	10 23	15 18	29 24	08 59	20 19	22 51	18 00
20	03 49 50	28 47 09	12 ♎ 27 42	19 ♎ 49 31	09 15	11 29	15 13	29 33	09 05	20 19	22 49	18 00
21	03 53 46	29 ♉ 44 52	27 ♎ 13 22	04 ♏ 38 28	11 19	12 36	15 08	29 42	09 10	20 18	22 48	17 59
22	03 57 43	00 ♊ 42 33	12 ♏ 03 56	19 ♏ 28 53	13 20	13 42	15 05	29 ♋ 52	09 16	20 18	22 46	17 59
23	04 01 40	01 40 12	26 ♏ 52 21	04 ♐ 13 24	15 19	14 48	15 02	00 ♌ 01	09 21	20 17	22 45	17 59
24	04 05 36	02 37 51	11 ♐ 31 09	18 ♐ 44 46	17 15	15 54	15 00	00 11	09 27	20 17	22 43	17 59
25	04 09 33	03 35 28	25 ♐ 53 33	02 ♑ 56 54	19 09	16 59	14 59	00 21	09 32	20 17	22 41	17 59
26	04 13 29	04 33 05	09 ♑ 54 21	16 ♑ 45 36	21 01	18 04	14 D 59	00 31	09 38	20 17	22 40	17 58
27	04 17 26	05 30 40	23 ♑ 30 29	00 ♒ 08 59	22 49	19 09	15 00	00 41	09 43	20 17	22 38	17 58
28	04 21 22	06 28 14	06 ♒ 41 13	13 ♒ 07 26	24 35	20 14	15 01	00 51	09 48	20 17	22 37	17 58
29	04 25 19	07 25 47	19 ♒ 27 58	25 ♒ 43 14	26 18	21 19	15 03	01 01	09 53	20 D 17	22 35	17 58
30	04 29 15	08 23 20	01 ♓ 53 44	08 ♓ 00 02	27 59	22 23	15 06	01 11	09 58	20 17	22 34	17 D 58
31	04 33 12	09 ♊ 20 51	14 ♓ 02 41	20 ♓ 02 20	29 ♊ 36	23 ♋ 28	15 ♎ 09	01 ♌ 21	10 ♈ 03	20 ♍ 17	22 ♏ 32	17 ♍ 58

Date	Moon True ☊ ° '	Moon Mean ☊ ° '	Moon ☽ Latitude ° '
01	06 ♉ 54	06 ♉ 57	05 S 17
02	06 D 55	06 53	05 04
03	06 55	06 50	04 39
04	06 56	06 47	04 01
05	06 56	06 44	03 14
06	06 57	06 41	02 18
07	06 58	06 37	01 17
08	06 58	06 34	00 S 12
09	06 R 58	06 31	00 N 54
10	06 57	06 28	01 59
11	06 56	06 25	02 58
12	06 54	06 22	03 50
13	06 52	06 18	04 32
14	06 50	06 15	05 01
15	06 48	06 12	05 15
16	06 47	06 09	05 12
17	06 D 47	06 06	04 51
18	06 47	06 03	04 13
19	06 49	05 59	03 19
20	06 50	05 56	02 10
21	06 51	05 53	00 N 53
22	06 R 51	05 50	00 S 29
23	06 50	05 47	01 48
24	06 48	05 43	02 59
25	06 44	05 40	03 58
26	06 40	05 37	04 40
27	06 36	05 34	05 05
28	06 33	05 31	05 13
29	06 30	05 28	05 05
30	06 29	05 24	04 43
31	06 ♉ 29	05 ♉ 21	04 S 09

DECLINATIONS

Date	Sun ☉ ° '	Moon ☽ ° '	Mercury ☿ ° '	Venus ♀ ° '	Mars ♂ ° '	Jupiter ♃ ° '	Saturn ♄ ° '	Uranus ♅ ° '	Neptune ♆ ° '	Pluto ♇ ° '
01	14 N 57	22 S 31	09 N 41	25 N 06	06 S 10	21 N 21	00 N 50	04 N 27	16 S 52	18 N 38
02	15 15	18 28	10 30	25 14	06 05	21 20	00 53	04 28	16 52	18 38
03	15 33	13 46	11 19	25 20	06 01	21 18	00 55	04 28	16 52	18 38
04	15 51	08 36	12 09	25 26	05 57	21 17	00 58	04 29	16 51	18 38
05	16 08	03 S 11	12 58	25 31	05 53	21 15	01 00	04 29	16 51	18 38
06	16 25	02 N 19	13 47	25 36	05 49	21 14	01 03	04 29	16 50	18 38
07	16 42	07 46	14 36	25 40	05 46	21 12	01 05	04 30	16 50	18 38
08	16 59	12 58	15 24	25 43	05 43	21 11	01 08	04 30	16 50	18 38
09	17 15	17 44	16 12	25 45	05 40	21 09	01 10	04 31	16 49	18 38
10	17 31	21 51	16 59	25 47	05 37	21 07	01 12	04 31	16 49	18 38
11	17 46	25 04	17 45	25 48	05 35	21 06	01 14	04 31	16 48	18 38
12	18 02	27 08	18 30	25 48	05 33	21 04	01 17	04 32	16 48	18 37
13	18 17	27 50	19 13	25 48	05 31	21 02	01 19	04 32	16 48	18 37
14	18 32	27 03	19 54	25 47	05 30	21 01	01 21	04 32	16 47	18 37
15	18 46	24 45	20 34	25 46	05 29	20 59	01 23	04 32	16 47	18 37
16	19 00	21 05	21 12	25 44	05 28	20 57	01 26	04 33	16 46	18 37
17	19 14	16 12	21 47	25 41	05 27	20 55	01 28	04 33	16 46	18 37
18	19 28	10 22	22 20	25 37	05 27	20 53	01 30	04 33	16 45	18 36
19	19 41	03 N 53	22 51	25 33	05 27	20 51	01 32	04 33	16 45	18 36
20	19 54	02 S 55	23 19	25 28	05 28	20 49	01 34	04 33	16 45	18 36
21	20 06	09 40	23 44	25 22	05 28	20 47	01 36	04 34	16 44	18 36
22	20 18	15 55	24 07	25 16	05 29	20 45	01 38	04 34	16 44	18 35
23	20 30	21 13	24 27	25 10	05 30	20 43	01 40	04 34	16 43	18 35
24	20 42	25 08	24 45	25 02	05 32	20 41	01 42	04 34	16 43	18 35
25	20 53	27 20	25 00	24 54	05 33	20 39	01 44	04 34	16 43	18 35
26	21 03	27 44	25 12	24 46	05 35	20 37	01 46	04 34	16 42	18 34
27	21 14	26 25	25 22	24 37	05 38	20 35	01 48	04 34	16 42	18 34
28	21 24	23 39	25 29	24 27	05 40	20 33	01 50	04 34	16 42	18 33
29	21 33	19 49	25 35	24 16	05 43	20 31	01 52	04 34	16 41	18 33
30	21 43	15 12	25 38	24 06	05 46	20 28	01 53	04 34	16 41	18 33
31	21 N 51	10 S 06	25 N 38	23 N 54	05 S 50	20 N 26	01 N 55	04 N 34	16 S 40	18 N 32

ZODIAC SIGN ENTRIES

Date	h m	Planets
01	23 26	☿ ♉
03	00 47	☽ ♓
05	13 10	☽ ♈
08	02 09	☽ ♉
10	06 05	♀ ♋
10	14 08	☽ ♊
13	00 11	☽ ♋
15	07 49	☽ ♌
16	03 27	☿ ♊
17	12 52	☽ ♍
19	15 31	☽ ♎
21	16 30	☽ ♏
21	18 18	☉ ♊
23	08 21	♃ ♌
23	17 06	☽ ♐
25	18 58	☽ ♑
27	23 44	☽ ♒
30	08 18	☽ ♓
31	18 02	☿ ♋

LATITUDES

Date	Mercury ☿ ° '	Venus ♀ ° '	Mars ♂ ° '	Jupiter ♃ ° '	Saturn ♄ ° '	Uranus ♅ ° '	Neptune ♆ ° '	Pluto ♇ ° '
01	01 S 33	02 N 03	01 N 24	00 N 35	02 S 09	00 N 47	01 N 48	15 N 13
04	01 08	02 09	01 16	00 35	02 10	00 47	01 48	15 12
07	00 39	02 15	01 08	00 35	02 10	00 47	01 48	15 11
10	00 S 08	02 20	01 00	00 35	02 11	00 47	01 48	15 10
13	00 N 23	02 25	00 53	00 35	02 11	00 47	01 48	15 09
16	00 54	02 28	00 45	00 35	02 12	00 47	01 48	15 08
19	01 21	02 31	00 37	00 35	02 12	00 47	01 48	15 07
22	01 43	02 33	00 30	00 35	02 13	00 47	01 48	15 05
25	02 00	02 34	00 23	00 35	02 13	00 47	01 48	15 04
28	02 09	02 33	00 16	00 35	02 14	00 46	01 48	15 03
31	02 N 12	02 N 32	00 N 09	00 N 35	02 S 15	00 N 46	01 N 48	15 N 02

DATA

Julian Date	2439612
Delta T	+38 seconds
Ayanamsa	23° 23' 52"
Synetic vernal point	05° ♓ 43' 08"
True obliquity of ecliptic	23° 26' 44"

MOON'S PHASES, APSIDES AND POSITIONS ☽

Date	h m	Phase	Longitude ° '	Eclipse Indicator
01	10 32	☾	10 ♒ 22	
09	14 55	●	18 ♉ 18	Partial
17	05 18	☽	25 ♌ 38	
23	20 22	○	02 ♐ 00	
31	01 52	☾	08 ♓ 57	

Day	h m		
06	10 51	Apogee	
22	01 49	Perigee	
06	01 54	ON	
13	11 30	Max dec	27° N 50'
20	01 45	OS	
26	05 10	Max dec	27° S 48'

LONGITUDES

Date	Chiron ⚷ ° '	Ceres ⚳ ° '	Pallas ⚴ ° '	Juno ⚵ ° '	Vesta ⚶ ° '	Black Moon Lilith ⚸ ° '
01	28 ♓ 13	10 ♋ 56	25 ♉ 13	08 ♌ 13	27 ♏ 38	14 ♈ 03
11	28 ♓ 40	14 ♋ 39	00 ♊ 46	11 ♌ 04	25 ♏ 23	15 ♈ 10
21	29 ♓ 03	18 ♋ 33	06 ♊ 26	14 ♌ 10	22 ♏ 57	16 ♈ 16
31	29 ♓ 22	22 ♋ 35	12 ♊ 12	17 ♌ 28	20 ♏ 43	17 ♈ 23

All ephemeris data is given at 12.00 UT and the Moon's longitude is additionally given for 24.00 UT

ASPECTARIAN

h m	Aspects
01 Monday	
01 00	☽ Q ♆
01 30	☽ ⚼ ♅
04 16	☽ ✶ ♄
10 32	☽ □ ☉
14 00	☽ ± ♇
18 40	☽ ± ♅
19 28	☽ ǂ ♃
02 Tuesday	
00 40	♀ □ ♅
01 34	☽ ⊼ ♇
01 36	☽ Q ☿
02 54	☽ △ ♂
06 17	☽ ⊼ ♅
06 52	☽ △ ♀
09 27	☽ ∠ ♄
10 35	♀ ⊥ ♃
11 07	☽ ǂ ♇
11 36	☽ □ ♆
19 04	☽ ⊼ ♃
20 31	☽ ∥ ♆
03 Wednesday	
00 35	☽ Q ☉
03 20	☽ ⊥ ♄
03 42	☽ ǂ ☉
05 46	☽ ✶ ☿
07 04	☽ ± ♃
07 43	☽ ⚼ ♂
13 19	☿ ⚼ ♇
15 24	☽ ⊻ ♄
21 57	☽ ǂ ☿
04 Thursday	
01 15	☽ ⚼ ♃
01 17	☽ ± ♂
02 39	☽ ✶ ☉
08 23	♀ ⊼ ♆
13 06	☽ ☍ ♇
13 06	☽ ⊼ ♂
13 25	♂ ⊻ ♇
14 56	☉ ⊥ ♄
17 29	☿ ⚼ ♅
17 57	☽ ☍ ♅
18 03	☽ ∠ ☿
23 26	☽ △ ♆
05 Friday	
00 02	☽ ∥ ♂
01 06	☽ □ ♀
06 20	☽ ǂ ♅
07 55	☽ △ ♃
11 43	☽ ∠ ☉
16 10	☽ ⊥ ☿
17 14	☿ ⊻ ♄
21 27	☽ ǂ ♄
06 Saturday	
04 41	☽ ☌ ♄
05 52	☽ ⚼ ♆
06 24	☽ ∥ ♄
07 00	☽ ⊻ ☿
07 41	☽ ⊥ ☉
17 39	☽ Q ♀
20 26	☉ Q ♃
20 59	☽ ⊻ ☉
21 31	☽ ∥ ♅
07 Sunday	
00 09	☽ ± ♆
00 44	☽ ☍ ♂
02 01	☽ ⊼ ♇
03 11	☽ ǂ ♂
06 51	☽ ⊼ ♅
12 19	☽ ⊼ ♆
14 12	☽ ± ♇
19 01	☽ ± ♅
20 39	☽ ✶ ♀
21 36	☽ □ ♃
23 13	☉ ǂ ♆
08 Monday	
01 57	☿ ∠ ♀
07 41	♀ ⊻ ♃
07 43	♂ ⊥ ♆
08 24	☽ ⚼ ♇
10 07	☉ ⊼ ♂
13 11	☽ ⚼ ♅
17 18	☿ ⊥ ♄
18 05	☽ ⊻ ♄
09 Tuesday	
02 21	☽ ∥ ☿
06 04	☽ ∠ ♀
06 14	☽ ⊥ ♄
07 09	☽ ǂ ♆
08 55	☽ ☌ ☿
09 14	☽ ∥ ☉
09 37	☉ △ ♇
10 15	☽ Q ♃
11 33	♀ ± ♆
12 06	☽ ⊼ ♂
14 29	☽ △ ♇
14 55	☽ ☌ ☉
16 44	☿ Q ♃
16 50	☽ ∥ ♇
19 12	☽ △ ♅
23 50	☽ ± ♂
10 Wednesday	
00 21	☽ ∠ ♄
00 30	☽ ☍ ♆
01 21	☿ ⊼ ♂
01 53	☽ ⊥ ♀
06 41	☿ ǂ ♆
07 25	☽ ∥ ♃
10 21	☽ ✶ ♃
14 58	☽ ⊻ ♀
15 06	♄ ⚼ ♆
15 52	☿ △ ♇
17 19	☽ ⚼ ♂
11 Thursday	
06 08	☽ ✶ ♄
13 13	☉ ∥ ☿
16 03	☽ ∠ ♃
16 25	☉ ☌ ☿
17 45	☿ △ ♅
19 06	☽ ∥ ♀
19 25	☉ △ ♅
22 06	☽ △ ♂
12 Friday	
01 20	☽ □ ♇
05 35	☽ Q ♄
05 51	☽ □ ♅
06 44	☽ ⊻ ☉
08 25	☽ ⊻ ☿
09 36	☽ ⊥ ♃
10 55	☽ ⊼ ♆
14 04	☿ ± ♂
16 16	☿ ∥ ♇
19 10	☽ ⊥ ☉
21 11	☽ ⊻ ♃
22 16	☽ ⊥ ☿
22 20	☽ ± ♆
22 38	☿ ☍ ♆
13 Saturday	
02 07	☿ ∠ ♄
06 39	☽ ☌ ♀
11 35	☽ Q ♇
13 01	☉ ± ♂
13 35	☽ ∠ ☉
15 17	☽ ⚼ ♆
15 58	☽ Q ♅
16 01	☽ □ ♄
18 44	☽ ∠ ☿
14 Sunday	
06 06	☽ □ ♂
09 56	☽ ✶ ♇
11 34	☉ ☍ ♆
14 14	☽ ✶ ♅
19 01	☽ △ ♆
19 38	☽ ✶ ☉
20 46	☉ ∥ ♇
15 Monday	
01 26	☉ ∠ ♄
03 06	☽ ∥ ♀
03 57	☽ ✶ ☿
05 34	☽ ☌ ♃
13 17	☽ ∠ ♇
14 14	♀ Q ♇
14 19	☿ ✶ ♃
14 36	☽ Q ♂
17 28	☽ ∠ ♅
19 05	☽ ⊻ ♀
19 05	☽ Q ☉
23 21	☽ △ ♄
16 Tuesday	
03 01	☿ ∥ ♃
05 23	☽ ⊥ ♇
05 42	☽ Q ☿
06 45	☽ ⊥ ♀
09 30	☽ ⊥ ♅
10 33	☿ ⚼ ♂
11 26	☽ ∥ ☿
11 47	☽ ✶ ♂
12 42	☽ ∥ ♃
15 59	☽ ⊻ ♇
18 12	☿ ⊥ ♀
20 04	☽ ⊻ ♅
22 21	☽ ∥ ☉
17 Wednesday	
00 01	☽ ∠ ♀
00 34	☽ □ ♆
00 47	☽ ∥ ♇
02 01	☽ ⚼ ♄
05 18	☽ □ ☉
06 44	♀ ⚼ ♆
09 27	☽ ǂ ♆
11 18	☽ ⊻ ♃
13 43	☽ ∠ ♂
15 49	♀ Q ♅
17 42	☽ ± ♄
18 58	☽ □ ☿
21 44	☽ ⊥ ♃
18 Thursday	
02 52	♀ □ ♄
04 02	☽ ⊼ ♄
04 08	☽ ✶ ♀
05 01	☽ ⊥ ♂
07 30	☽ Q ♆
13 12	☽ ∠ ♃
15 06	☽ ⊻ ♂
19 31	☽ ☌ ♇
23 24	☽ ☌ ♅
19 Friday	
02 05	☽ Q ♀
03 39	☽ ✶ ♆
06 21	☽ ǂ ♂
09 35	☽ ∥ ♅
11 54	☽ △ ☉
14 33	☽ ✶ ♃
20 18	☽ ∥ ♄
20 Saturday	
04 25	☽ ∠ ♆
05 53	☽ △ ☿
06 25	☽ ☍ ♄
07 14	☽ ǂ ♄
09 49	☿ ✶ ♄
10 17	☽ □ ♀
10 30	☽ Q ♃
14 19	☽ ⚼ ☉
16 28	☽ ☌ ♂
17 45	☽ ǂ ♅
19 06	☽ ⊥ ♆
20 56	☽ ∥ ♂
21 01	☽ ⊻ ♇
21 Sunday	
00 47	☽ ⊻ ♅
04 50	☽ ⊻ ♆
05 59	☽ ± ☉
06 46	☽ ⊥ ♇
10 17	☽ ⚼ ☿
10 31	☽ ⊥ ♅
10 48	☉ ✶ ♃
16 04	☽ □ ♃
16 22	☽ ⊼ ☉
21 14	☉ ⚼ ♂
21 20	☽ ∠ ♇
22 Monday	
01 04	☽ ∠ ♅
03 09	☽ ± ☿
07 27	☽ ⊼ ♄
14 22	☽ ⊼ ☿
14 51	☽ △ ♀
15 23	☽ ∥ ♆
16 52	☽ ⊻ ♂
17 12	☽ ± ♄
21 34	☽ ✶ ♇
21 52	☿ ⊻ ♀
23 26	☽ ǂ ♇
23 Tuesday	
01 19	☽ ✶ ♅
02 33	☽ ⊥ ♂
05 18	☽ ☌ ♆
07 53	☽ ⚼ ♄
08 10	☿ ∠ ♃
08 16	☽ ǂ ☉
08 43	☿ △ ♂
09 32	☽ ǂ ♃
17 04	☽ Q ♇
17 08	♀ □ ♂
17 09	☽ ∠ ♂
17 09	☽ ⚼ ♀
17 12	☽ △ ♃
20 22	☽ ☍ ☉
20 50	☽ Q ♅
24 Wednesday	
08 34	☽ △ ♄
08 51	☽ ǂ ☿
09 06	☽ ± ♀
11 19	☽ ǂ ♀
11 51	☉ ∥ ♃
17 46	☽ ✶ ♂
18 08	☽ ⚼ ♃
19 50	☽ ⊼ ♀
21 04	☿ □ ♇
22 43	☽ □ ♇
22 58	☽ ☍ ☿
25 Thursday	
02 34	☽ □ ♅
06 17	☿ ∥ ♀
06 37	☽ ⊻ ♆
09 22	☽ ± ♃
13 51	☽ Q ♂
16 44	☽ ⊥ ♆
19 39	☽ ⊼ ♃
26 Friday	
02 04	☽ ⊼ ☉
02 29	☿ □ ♅
08 07	☽ ∠ ♆
09 29	♂ St D
09 50	♀ ✶ ♇
11 31	☽ □ ♄
13 12	☽ ± ☉
20 30	☿ Q ♄
20 53	☽ □ ♂
27 Saturday	
02 09	☽ △ ♇
03 34	☽ ☍ ♀
06 14	☽ △ ♅
06 14	☽ ⚼ ☉
09 32	☿ ⊼ ♆
10 27	☽ ✶ ♆
10 35	☽ ⊼ ☿
19 38	☽ Q ♄
21 58	☽ ǂ ☿
23 05	☽ ± ☿
28 Sunday	
01 08	☽ ☍ ♃
05 09	☽ ⚼ ♇
05 46	☽ ǂ ♀
08 11	☽ Q ♆
09 24	☽ ⚼ ♅
11 34	☽ △ ☉
12 54	♀ ✶ ♅
15 54	☿ ⊥ ♃
17 50	☽ ✶ ♄
18 14	☽ ⚼ ☿
21 37	♅ St D
21 50	☽ ± ♇
29 Monday	
02 10	☽ ± ♅
02 12	☽ ǂ ☉
03 36	☽ △ ♂
07 59	☽ ǂ ♃
09 09	☽ ⊼ ♇
13 33	☽ ⊼ ♅
15 52	☽ ⊼ ♀
17 57	☽ □ ♆
18 55	☽ ǂ ♇
20 32	♇ St D
22 28	☽ ∠ ♄
30 Tuesday	
03 12	☽ △ ☿
04 31	☽ ± ♀
04 39	☽ ∥ ♆
08 28	☽ ⚼ ♂
10 35	☽ ⊼ ♃
15 43	♀ △ ♆
16 06	☽ ⊥ ♄
20 23	☿ ± ♆
22 32	☽ ± ♃
23 50	☽ ⚼ ♀
31 Wednesday	
01 52	☽ □ ☉
02 13	☽ ± ♂
04 01	☽ ⊻ ♄
14 13	☽ ⊼ ♂
16 41	☽ ⚼ ♃
19 51	☽ ☍ ♇

JUNE 1967

LONGITUDES

Date	Sidereal time h m s	Sun ☉	Moon ☽	Moon ☽ 24.00	Mercury ☿	Venus ♀	Mars ♂	Jupiter ♃	Saturn ♄	Uranus ♅	Neptune ♆	Pluto ♇
01	04 37 09	10 ♊ 18 22	25 ♓ 59 35	01 ♈ 55 04	01 ♋ 10	24 ♋ 32	15 ♎ 13	01 ♌ 32	10 ♈ 08	20 ♍ 17	22 ♏ 30	17 ♍ 58
02	04 41 05	11 15 52	07 ♈ 49 25	13 ♈ 43 12	02 42	25 35	15 18	01 42	10 13	20 17	22 R 29	17 58
03	04 45 02	12 13 21	19 ♈ 37 02	25 ♈ 31 28	04 10	26 39	15 24	01 53	10 18	20 17	22 27	17 59
04	04 48 58	13 10 49	01 ♉ 27 00	07 ♉ 24 06	05 35	27 42	15 30	02 03	10 23	20 18	22 26	17 59
05	04 52 55	14 08 17	13 ♉ 23 13	19 ♉ 24 44	06 58	28 45	15 37	02 14	10 27	20 18	22 24	17 59
06	04 56 51	15 05 44	25 ♉ 28 58	01 ♊ 36 13	08 17	29 ♋ 48	15 45	02 25	10 32	20 19	22 23	17 59
07	05 00 48	16 03 10	07 ♊ 46 41	14 ♊ 00 32	09 33	00 ♌ 50	15 53	02 36	10 37	20 19	22 22	17 59
08	05 04 44	17 00 35	20 ♊ 17 53	26 ♊ 38 49	10 46	01 52	16 03	02 47	10 41	20 20	22 20	18 00
09	05 08 41	17 58 00	03 ♋ 03 19	09 ♋ 31 23	11 55	02 54	16 12	02 58	10 45	20 20	22 19	18 00
10	05 12 38	18 55 23	16 ♋ 02 57	22 ♋ 37 55	13 01	03 55	16 23	03 09	10 50	20 21	22 17	18 00
11	05 16 34	19 52 46	29 ♋ 16 13	05 ♌ 57 42	14 04	04 56	16 34	03 20	10 54	20 22	22 16	18 01
12	05 20 31	20 50 08	12 ♌ 42 17	19 ♌ 29 49	15 04	05 57	16 46	03 31	10 58	20 22	22 14	18 01
13	05 24 27	21 47 29	26 ♌ 20 13	03 ♍ 13 23	15 59	06 58	16 58	03 43	11 02	20 23	22 13	18 02
14	05 28 24	22 44 48	10 ♍ 09 13	17 ♍ 07 37	16 52	07 58	17 11	03 54	11 06	20 24	22 12	18 02
15	05 32 20	23 42 07	24 ♍ 08 30	01 ♎ 11 44	17 40	08 58	17 24	04 06	11 10	20 25	22 10	18 03
16	05 36 17	24 39 25	08 ♎ 17 12	15 ♎ 24 43	18 25	09 57	17 39	04 17	11 13	20 26	22 09	18 03
17	05 40 13	25 36 42	22 ♎ 34 04	29 ♎ 44 58	19 05	10 56	17 53	04 29	11 17	20 27	22 08	18 04
18	05 44 10	26 33 58	06 ♏ 57 05	14 ♏ 09 58	19 42	11 55	18 09	04 40	11 21	20 28	22 06	18 04
19	05 48 07	27 31 13	21 ♏ 23 08	28 ♏ 36 01	20 14	12 53	18 24	04 52	11 24	20 29	22 05	18 05
20	05 52 03	28 28 28	05 ♐ 48 00	12 ♐ 58 26	20 43	13 51	18 41	05 04	11 28	20 30	22 04	18 06
21	05 56 00	29 ♊ 25 42	20 ♐ 06 36	27 ♐ 11 51	21 07	14 48	18 57	05 16	11 31	20 32	22 03	18 06
22	05 59 56	00 ♋ 22 56	04 ♑ 13 32	11 ♑ 11 03	21 26	15 45	19 15	05 28	11 34	20 33	22 02	18 07
23	06 03 53	01 20 10	18 ♑ 03 54	24 ♑ 51 41	21 41	16 42	19 33	05 40	11 37	20 34	22 00	18 08
24	06 07 49	02 17 23	01 ♒ 34 06	08 ♒ 10 59	21 52	17 38	19 51	05 52	11 40	20 35	21 59	18 09
25	06 11 46	03 14 35	14 ♒ 42 17	21 ♒ 08 04	21 58	18 33	20 10	06 04	11 43	20 37	21 58	18 09
26	06 15 42	04 11 48	27 ♒ 28 32	03 ♓ 43 58	21 R 59	19 28	20 29	06 16	11 46	20 38	21 57	18 10
27	06 19 39	05 09 01	09 ♓ 54 46	16 ♓ 01 24	21 56	20 22	20 49	06 28	11 49	20 40	21 56	18 11
28	06 23 36	06 06 13	22 ♓ 04 23	28 ♓ 04 19	21 48	21 16	21 09	06 40	11 52	20 41	21 55	18 12
29	06 27 32	07 03 26	04 ♈ 01 48	09 ♈ 57 31	21 36	22 09	21 30	06 52	11 54	20 43	21 54	18 13
30	06 31 29	08 ♋ 00 38	15 ♈ 52 07	21 ♈ 46 16	21 ♋ 19	23 ♌ 02	21 ♎ 51	07 ♌ 05	11 ♈ 57	20 ♍ 45	21 ♏ 53	18 ♍ 14

DECLINATIONS

Date	Moon True ☊	Moon Mean ☊	Moon ☽ Latitude	Sun ☉	Moon ☽	Mercury ☿	Venus ♀	Mars ♂	Jupiter ♃	Saturn ♄	Uranus ♅	Neptune ♆	Pluto ♇
01	06 ♉ 30	05 ♉ 18	03 S 24	22 N 00	04 S 43	25 N 37	23 N 42	05 S 53	20 N 24	01 N 57	04 N 34	16 S 40	18 N 32
02	06 D 31	05 15	02 31	22 08	00 N 48	25 34	23 30	05 57	20 22	01 59	04 33	16 40	18 32
03	06 33	05 12	01 31	22 16	06 16	25 30	23 17	06 01	20 19	02 00	04 33	16 39	18 31
04	06 34	05 08	00 S 28	22 23	11 33	25 25	23 03	06 05	20 17	02 02	04 33	16 39	18 31
05	06 R 34	05 05	00 N 37	22 30	16 27	25 15	22 49	06 10	20 14	02 04	04 33	16 39	18 30
06	06 32	05 02	01 41	22 37	20 46	25 06	22 34	06 15	20 12	02 05	04 33	16 38	18 30
07	06 28	04 59	02 42	22 43	24 16	24 55	22 19	06 20	20 10	02 07	04 32	16 38	18 29
08	06 23	04 56	03 35	22 49	26 40	24 43	22 04	06 25	20 07	02 08	04 32	16 38	18 29
09	06 17	04 53	04 19	22 54	27 43	24 31	21 48	06 31	20 05	02 10	04 32	16 37	18 28
10	06 09	04 49	04 50	22 59	27 17	24 17	21 31	06 36	20 02	02 11	04 32	16 37	18 28
11	06 02	04 46	05 07	23 04	25 18	24 02	21 15	06 42	19 59	02 13	04 31	16 37	18 27
12	05 56	04 43	05 07	23 08	21 54	23 46	20 57	06 48	19 57	02 14	04 31	16 36	18 27
13	05 52	04 40	04 49	23 11	17 16	23 30	20 40	06 55	19 54	02 15	04 31	16 36	18 26
14	05 49	04 37	04 15	23 15	11 42	23 13	20 22	07 01	19 52	02 17	04 30	16 36	18 26
15	05 D 49	04 34	03 26	23 18	05 N 28	22 55	20 03	07 08	19 49	02 18	04 30	16 35	18 25
16	05 49	04 30	02 23	23 20	01 S 06	22 38	19 45	07 15	19 46	02 19	04 29	16 35	18 24
17	05 50	04 27	01 N 11	23 22	07 41	22 19	19 26	07 22	19 43	02 20	04 29	16 35	18 24
18	05 R 51	04 24	00 S 06	23 24	13 56	22 01	19 06	07 30	19 41	02 21	04 29	16 34	18 23
19	05 50	04 21	01 23	23 25	19 27	21 43	18 46	07 37	19 38	02 23	04 28	16 34	18 23
20	05 47	04 18	02 34	23 26	23 48	21 25	18 26	07 45	19 35	02 24	04 28	16 34	18 22
21	05 41	04 14	03 35	23 27	26 39	21 07	18 06	07 53	19 32	02 25	04 27	16 34	18 21
22	05 34	04 11	04 21	23 27	27 44	20 49	17 45	08 01	19 29	02 26	04 26	16 33	18 21
23	05 25	04 08	04 52	23 26	27 02	20 31	17 25	08 09	19 26	02 27	04 26	16 33	18 20
24	05 16	04 05	05 05	23 26	24 46	20 14	17 04	08 17	19 23	02 28	04 25	16 33	18 19
25	05 07	04 02	05 01	23 24	21 13	19 57	16 42	08 26	19 19	02 29	04 25	16 33	18 19
26	05 00	03 59	04 42	23 23	16 46	19 41	16 21	08 34	19 17	02 30	04 24	16 32	18 18
27	04 55	03 55	04 11	23 21	11 43	19 26	15 59	08 43	19 14	02 30	04 23	16 32	18 17
28	04 52	03 52	03 28	23 18	06 20	19 11	15 37	08 52	19 11	02 31	04 23	16 32	18 17
29	04 51	03 49	02 37	23 15	00 48	18 57	15 15	09 01	19 08	02 32	04 22	16 32	18 16
30	04 ♉ 51	03 ♉ 46	01 S 40	23 N 12	04 N 43	18 N 45	14 N 53	09 S 10	19 N 05	02 N 33	04 N 21	16 S 31	18 N 15

ZODIAC SIGN ENTRIES

Date	h	m	Planets
01	20	07	☽ ♈
04	09	04	☽ ♉
06	16	48	♀ ♌
06	20	52	☽ ♊
09	06	18	☽ ♋
11	13	19	☽ ♌
13	18	24	☽ ♍
15	21	58	☽ ♎
18	00	25	☽ ♏
20	02	20	☽ ♐
22	02	23	☉ ♋
22	04	46	☽ ♑
24	09	11	☽ ♒
26	16	49	☽ ♓
29	03	53	☽ ♈

LATITUDES

Date	Mercury ☿	Venus ♀	Mars ♂	Jupiter ♃	Saturn ♄	Uranus ♅	Neptune ♆	Pluto ♇
01	02 N 11	02 N 31	00 N 07	00 N 35	02 S 15	00 N 46	01 N 48	15 N 01
04	02 04	02 28	00 N 01	00 35	02 16	00 46	01 48	15 00
07	01 50	02 24	00 S 05	00 35	02 16	00 46	01 47	14 59
10	01 29	02 19	00 11	00 35	02 17	00 46	01 47	14 58
13	01 02	02 12	00 16	00 36	02 18	00 46	01 47	14 56
16	00 N 27	02 04	00 21	00 36	02 18	00 46	01 47	14 55
19	00 S 12	01 54	00 26	00 36	02 19	00 45	01 47	14 54
22	00 56	01 43	00 31	00 36	02 20	00 45	01 47	14 53
25	01 43	01 30	00 35	00 36	02 21	00 45	01 47	14 51
28	02 32	01 16	00 39	00 36	02 22	00 45	01 47	14 50
31	03 S 18	01 N 00	00 S 43	00 N 36	02 S 22	00 N 45	01 N 47	14 N 49

DATA

Julian Date	2439643
Delta T	+38 seconds
Ayanamsa	23° 23' 56"
Synetic vernal point	05° ♓ 43' 03"
True obliquity of ecliptic	23° 26' 44"

MOON'S PHASES, APSIDES AND POSITIONS ☽

Date	h	m	Phase	Longitude	Eclipse Indicator
08	05	13	●	16 ♊ 44	
15	11	12	☽	23 ♍ 40	
22	04	57	○	00 ♑ 06	
29	18	39	☾	07 ♈ 19	

Day	h	m		
03	01	53	Apogee	
18	20	17	Perigee	
30	19	47	Apogee	
02	08	32	ON	
09	17	06	Max dec	27° N 45'
16	08	02	0S	
22	14	28	Max dec	27° S 44'
29	15	28	ON	

LONGITUDES

Date	Chiron ⚷	Ceres ⚳	Pallas ⚴	Juno ⚵	Vesta ⚶	Black Moon Lilith ⚸
01	29 ♓ 24	22 ♋ 59	12 ♊ 47	17 ♌ 48	20 ♏ 31	17 ♈ 30
11	29 ♓ 37	27 ♋ 09	18 ♊ 39	21 ♌ 17	18 ♏ 51	18 ♈ 36
21	29 ♓ 46	01 ♌ 25	24 ♊ 35	24 ♌ 52	17 ♏ 55	19 ♈ 43
31	29 ♓ 50	05 ♌ 46	00 ♋ 34	28 ♌ 33	17 ♏ 48	20 ♈ 49

ASPECTARIAN

h m	Aspects
01 Thursday	
00 29	☽ ☍ ♅
04 59	☽ △ ♆
06 54	☽ ∥ ♂
07 27	☉ ✶ ♄
08 45	☽ △ ♀
12 40	☽ # ♅
17 05	☽ Q ☉
18 15	☿ ⊻ ♃
23 23	☽ △ ♃
02 Friday	
00 00	☽ # ♄
00 03	☽ □ ☿
11 19	☽ ⚼ ♆
16 55	☽ ☌ ♄
17 10	☽ ∥ ♄
19 37	☽ ✶ ☉
03 Saturday	
03 20	☽ ☍ ♂
04 26	☽ ∥ ♅
05 35	☽ ± ♆
08 40	☽ ⊼ ♇
10 52	☽ # ♂
13 22	☽ ⊼ ♅
17 46	☽ ⊼ ♆
17 54	☽ Q ☿
20 52	☽ ± ♇
04 Sunday	
01 34	☽ ± ♅
03 40	☽ □ ♀
04 48	☽ ∠ ☉
13 15	☽ □ ♃
15 05	☽ ⚼ ♇
18 44	☿ Q ♇
19 46	☽ ⚼ ♅
21 27	☽ ✶ ☿
05 Monday	
00 34	☽ ⊥ ☉
06 06	☽ ⊻ ♄
13 00	☽ # ♆
13 38	☽ ⊻ ☉
16 30	☽ ⊼ ♂
18 10	☽ ⊥ ♄
19 20	☽ Q ♀
19 52	☿ ⚼ ♆
21 10	☽ △ ♇
22 54	☽ ∥ ♇
06 Tuesday	
01 47	☽ △ ♅
01 50	☽ Q ♃
04 33	☽ ± ♂
05 54	☽ ☍ ♆
07 08	☽ ∠ ☿
08 36	☽ ∥ ♃
09 18	☉ ∥ ♀
12 06	☽ ∠ ♄
12 34	☿ Q ♅
21 15	☽ ✶ ♀
22 27	☽ ⚼ ♂
22 46	☽ ∥ ♀
07 Wednesday	
00 10	☽ ∥ ☉
01 47	☽ ✶ ♃
02 52	☽ ⊥ ☿
07 14	☉ △ ♂
15 48	☽ ⊻ ☿
17 08	☽ ∥ ☿
17 30	☽ ✶ ♄
08 Thursday	
03 48	☽ △ ♂
04 53	☽ ∠ ♀
05 13	☽ ☌ ☉
07 08	☽ ∠ ♃
07 37	☽ □ ♇
10 19	☿ □ ♄
12 03	☽ □ ♅
15 51	☽ ⊼ ♆
16 33	☽ Q ♄
23 28	☽ ⊥ ♀
09 Friday	
00 26	☽ ⊥ ♃
03 09	☽ ± ♆
11 41	☽ ⊻ ♀
11 50	☽ ⊻ ♃
11 56	☉ ∠ ♃
12 52	☉ □ ♇
13 56	♀ ☌ ♃
14 27	♀ ∠ ♇
16 55	♃ ∠ ♇
17 29	☽ Q ♇
19 54	☽ ⚼ ♆
21 49	☽ Q ♅
10 Saturday	
00 42	♂ ⊥ ♆
02 21	☽ □ ♄
05 57	☽ ☌ ☿
12 30	☉ ∠ ♀
12 37	☽ □ ♂
15 35	☽ ✶ ♇
17 40	☽ ⊻ ☉
19 51	☽ ✶ ♅
23 21	☽ △ ♆
11 Sunday	
01 09	♀ Q ♂
05 24	☽ ⊥ ☉
18 44	☽ ∠ ♇
19 24	☽ ☌ ♃
21 39	☽ Q ♂
22 02	♀ ∠ ♅
22 50	☽ ∠ ☉
22 56	☽ ∠ ♅
22 59	☽ ∥ ☿
23 01	☽ ☌ ♀
12 Monday	
00 14	☉ □ ♅
04 28	☽ ∥ ☉
08 54	☽ △ ♄
10 47	☽ ⊥ ♇
14 58	☽ ⊥ ♅
16 29	☽ ⊻ ☿
17 45	☽ ∥ ♀
19 17	☽ ✶ ♂
21 24	☽ ⊻ ♇
22 55	☽ ∥ ♃
13 Tuesday	
01 33	☽ ⊻ ♅
03 26	☽ ✶ ☉
03 50	☽ ⊥ ☿
04 48	☽ □ ♆
06 26	☽ ∥ ♇
11 28	☽ ⚼ ♄
15 06	☽ # ♆
20 40	☽ ∠ ☿
21 58	☽ ∠ ♂
22 27	☉ ⊼ ♆
14 Wednesday	
01 02	☽ ⊻ ♃
01 57	☽ Q ☉
03 13	☽ ± ♄
07 56	☽ ⊻ ♀
09 39	☉ ∥ ☿
11 34	☽ ⊥ ♃
12 04	☽ Q ♆
13 38	☽ ⊼ ♄
13 48	☽ ⊥ ♂
19 05	☽ ⊥ ♀
21 25	☉ Q ♄
15 Thursday	
00 16	☽ ✶ ☿
00 17	☽ ⊻ ♂
00 59	☿ □ ♂
01 34	☽ ☌ ♇
03 15	☽ ∠ ♃
05 38	☽ ☌ ♅
05 53	☽ # ♂
08 39	☽ ✶ ♆
11 12	☽ □ ☉
11 40	☽ ∠ ♀
15 38	☽ ∥ ♅
21 56	☽ Q ☿
23 39	☽ ∥ ♄
16 Friday	
00 06	☿ ✶ ♇
05 09	☽ ✶ ♃
09 54	♀ ∥ ♃
10 05	☽ ∠ ♆
15 01	☽ ✶ ♀
16 26	☽ # ♄
16 58	☽ ☍ ♄
17 Saturday	
00 17	☽ # ♅
01 13	☽ ⊥ ♆
01 39	☽ Q ♃
04 01	☽ ☌ ♂
04 27	☽ ⊻ ♇
05 53	☽ □ ☿
08 27	☽ ⊻ ♅
10 49	☽ ∥ ♂
11 16	☽ ⊻ ♆
12 40	☽ Q ♀
14 30	☽ ⊥ ♇
17 27	☽ △ ☉
18 30	☽ ⊥ ♅
21 03	♀ △ ♄
18 Sunday	
05 18	♂ ⊻ ♇
05 32	☽ ∠ ♇
08 09	☽ □ ♃
09 32	☽ ∠ ♅
15 55	♀ ⊥ ♇
19 20	☽ ⊼ ♄
20 13	☽ ⚼ ☉
20 51	☽ □ ♀
22 59	☽ ∥ ♆
19 Monday	
05 21	☽ ± ♄
06 30	☽ ✶ ♇
06 57	☽ ⊻ ♂
07 01	☽ # ♇
09 02	☽ # ♀
10 02	☽ △ ☿
10 30	☽ ✶ ♅
12 14	☽ ± ☉
12 54	☽ # ♃
13 10	☽ ☌ ♆
17 07	☽ ⊥ ♂
20 23	☽ ⚼ ♄
22 51	☽ # ☿
22 56	☽ ⊼ ☉
20 Tuesday	
00 34	☿ ✶ ♅
01 56	☉ ± ♆
02 29	☽ Q ♇
06 30	☽ Q ♅
08 23	☽ ∠ ♂
09 38	☽ # ☉
10 45	☽ △ ♃
11 51	☽ ⚼ ☿
17 27	♀ ∥ ♇
21 30	☽ △ ♄
21 Wednesday	
02 26	☽ △ ♀
03 21	☽ ± ☿
04 48	♀ ⊥ ♅
06 44	☉ ⊥ ♃
08 37	☽ □ ♇
10 01	☽ ✶ ♂
12 16	☽ ⚼ ♃
12 42	☽ □ ♅
13 44	☽ ⊼ ☿
15 16	☽ ⊻ ♆
22 Thursday	
01 25	☽ ⊥ ♆
03 44	☽ ± ♃
04 57	☽ ☍ ☉
05 38	☽ ⚼ ♀
06 48	☽ Q ♂
14 09	☽ ⊼ ♃
16 48	☽ ∠ ♆
22 13	☽ ± ♀
23 40	♃ ∠ ♅
23 Friday	
00 43	☽ □ ♄
09 25	☽ ⊼ ♀
12 07	☽ △ ♇
14 40	☽ □ ♂
16 25	☽ △ ♅
18 28	☽ ☍ ☿
18 56	☽ ✶ ♆
24 Saturday	
08 35	☽ Q ♄
13 24	☽ ⊼ ☉
14 51	☽ ⚼ ♇
16 22	☽ Q ♆
19 17	☽ ⚼ ♅
19 53	☽ ☍ ♃
22 01	☽ # ☉
25 Sunday	
01 09	☽ ± ☉
01 37	♀ ⊻ ♇
06 28	☽ ✶ ♄
07 17	☽ ± ♇
11 50	☽ ± ♅
14 48	☿ △ ♆
18 26	☽ ⊼ ♇
19 07	☽ ⚼ ☉
19 43	☽ ☍ ♀
19 45	☽ # ☿
22 26	☽ △ ♂
22 45	☽ # ♃
22 50	♀ # ♆
23 03	☽ ⊼ ♅
26 Monday	
01 33	☽ □ ♆
01 36	☽ ⊼ ☿
04 09	☽ # ♇
06 51	☿ St R
10 39	☽ ∠ ♄
12 58	☽ ± ☿
13 08	☽ ∥ ♆
14 17	☽ # ♀
27 Tuesday	
00 06	♂ ⊻ ♅
01 58	☽ △ ☉
03 49	☽ ⚼ ♂
04 00	☽ ⊥ ♄
05 10	☽ ⊼ ♃
06 13	☽ ⚼ ☿
10 29	☿ △ ♆
15 45	☽ ⊻ ♄
17 05	☽ ± ♃
20 01	♀ ⊻ ♅
21 54	☽ ± ♂
28 Wednesday	
01 09	☽ ∥ ♂
04 18	☽ ☍ ♇
06 57	♀ ✶ ♂
09 15	☽ ☍ ♅
10 07	☽ ⊼ ♂
10 16	☽ ⊼ ♀
11 11	☽ ⚼ ♃
11 12	☿ ∥ ♃
11 27	☽ △ ☿
11 41	☽ △ ♆
14 29	☉ Q ♇
20 31	☽ # ♅
23 14	☽ ± ♀
23 46	☿ ⊻ ♀
29 Thursday	
04 31	☽ # ♄
05 01	♀ □ ♆
06 04	☉ ⊻ ♃
08 01	☉ ⚼ ♆
15 53	☿ □ ♂
17 47	☽ ⚼ ♆
17 51	☽ △ ♃
18 39	☽ □ ☉
18 50	☽ ⚼ ♀
30 Friday	
02 30	☽ ∥ ♄
04 01	☽ ☌ ♄
10 27	☽ ∥ ♅
12 01	☽ ± ♆
13 51	♂ ⊻ ♆
16 49	☽ ⊼ ♇
19 39	☉ ∠ ♀
21 56	☽ ⊼ ♅
22 47	☽ □ ☿

All ephemeris data is given at 12.00 UT and the Moon's longitude is additionally given for 24.00 UT

JULY 1967

LONGITUDES

Date	Sidereal time h m s	Sun ☉ ° ' "	Moon ☽ ° ' "	Moon ☽ 24.00 ° ' "	Mercury ☿ ° '	Venus ♀ ° '	Mars ♂ ° '	Jupiter ♃ ° '	Saturn ♄ ° '	Uranus ♅ ° '	Neptune ♆ ° '	Pluto ♇ ° '
01	06 35 25	08 ♋ 57 51	27 ♈ 40 40	03 ♉ 35 56	20 ♋ 59	23 ♌ 54	22 ♎ 13	07 ♌ 17	11 ♈ 59	20 ♍ 46	21 ♏ 52	18 ♍ 15
02	06 39 22	09 55 04	09 ♉ 32 43	15 ♉ 31 37	20 R 35	24 46	22 35	07 29	12 02	20 48	21 R 51	18 16
03	06 43 18	10 52 17	21 ♉ 33 11	27 ♉ 37 55	20 07	25 37	22 57	07 42	12 04	20 50	21 50	18 17
04	06 47 15	11 49 30	03 ♊ 46 17	09 ♊ 58 37	19 36	26 27	23 20	07 54	12 06	20 52	21 49	18 18
05	06 51 11	12 46 44	16 ♊ 15 15	22 ♊ 36 21	19 03	27 16	23 43	08 07	12 08	20 54	21 48	18 19
06	06 55 08	13 43 57	29 ♊ 02 04	05 ♋ 32 23	18 28	28 05	24 07	08 19	12 10	20 56	21 47	18 20
07	06 59 05	14 41 11	12 ♋ 07 15	18 ♋ 46 29	17 51	28 53	24 31	08 32	12 12	20 58	21 46	18 21
08	07 03 01	15 38 25	25 ♋ 29 51	02 ♌ 17 00	17 14	29 ♌ 40	24 55	08 45	12 14	21 00	21 46	18 23
09	07 06 58	16 35 39	09 ♌ 07 36	16 ♌ 01 12	16 36	00 ♍ 27	25 20	08 57	12 15	21 02	21 45	18 24
10	07 10 54	17 32 53	22 ♌ 57 22	29 ♌ 55 42	15 58	01 12	25 45	09 10	12 17	21 04	21 44	18 25
11	07 14 51	18 30 07	06 ♍ 55 45	13 ♍ 57 08	15 22	01 57	26 11	09 23	12 18	21 06	21 43	18 26
12	07 18 47	19 27 20	20 ♍ 59 31	28 ♍ 02 37	14 47	02 41	26 36	09 35	12 20	21 08	21 43	18 28
13	07 22 44	20 24 34	05 ♎ 06 11	12 ♎ 10 01	14 15	03 24	27 03	09 48	12 21	21 11	21 42	18 29
14	07 26 40	21 21 48	19 ♎ 13 57	26 ♎ 17 51	13 45	04 05	27 29	10 01	12 22	21 13	21 41	18 30
15	07 30 37	22 19 01	03 ♏ 21 37	10 ♏ 25 07	13 19	04 46	27 56	10 14	12 23	21 15	21 41	18 32
16	07 34 34	23 16 15	17 ♏ 28 12	24 ♏ 30 43	12 57	05 26	28 23	10 27	12 24	21 18	21 40	18 33
17	07 38 30	24 13 29	01 ♐ 32 28	08 ♐ 33 11	12 39	06 05	28 51	10 40	12 25	21 20	21 40	18 34
18	07 42 27	25 10 43	15 ♐ 32 35	22 ♐ 30 18	12 26	06 42	29 19	10 53	12 25	21 22	21 39	18 36
19	07 46 23	26 07 58	29 ♐ 25 59	06 ♑ 19 12	12 18	07 18	29 ♎ 47	11 06	12 26	21 25	21 39	18 37
20	07 50 20	27 05 12	13 ♑ 09 34	19 ♑ 56 40	12 15	07 53	00 ♏ 16	11 19	12 27	21 27	21 38	18 39
21	07 54 16	28 02 27	26 ♑ 40 08	03 ♒ 19 37	12 D 18	08 27	00 44	11 32	12 27	21 30	21 38	18 40
22	07 58 13	28 59 43	09 ♒ 54 51	16 ♒ 25 40	12 27	09 00	01 13	11 45	12 27	21 33	21 37	18 42
23	08 02 09	29 ♋ 56 59	22 ♒ 51 57	29 ♒ 13 40	12 41	09 31	01 43	11 58	12 28	21 35	21 37	18 43
24	08 06 06	00 ♌ 54 16	05 ♓ 30 56	11 ♓ 43 53	13 01	10 00	02 12	12 11	12 28	21 38	21 37	18 45
25	08 10 03	01 51 33	17 ♓ 52 49	23 ♓ 58 05	13 27	10 28	02 42	12 24	12 R 28	21 41	21 36	18 47
26	08 13 59	02 48 51	00 ♈ 00 04	05 ♈ 59 18	13 59	10 55	03 12	12 37	12 28	21 43	21 36	18 48
27	08 17 56	03 46 11	11 ♈ 56 18	17 ♈ 51 41	14 36	11 20	03 43	12 50	12 27	21 46	21 36	18 50
28	08 21 52	04 43 31	23 ♈ 46 04	29 ♈ 40 08	15 20	11 43	04 14	13 03	12 27	21 49	21 35	18 51
29	08 25 49	05 40 52	05 ♉ 34 33	11 ♉ 30 00	16 09	12 05	04 45	13 16	12 27	21 52	21 35	18 53
30	08 29 45	06 38 14	17 ♉ 27 10	23 ♉ 26 45	17 05	12 25	05 16	13 29	12 26	21 55	21 35	18 55
31	08 33 42	07 ♌ 35 38	29 ♉ 29 23	05 ♊ 35 41	18 ♋ 05	12 ♍ 43	05 ♏ 47	13 ♌ 42	12 ♈ 26	21 ♍ 58	21 ♏ 35	18 ♍ 57

Date	Moon True ☊ ° '	Moon Mean ☊ ° '	Moon ☽ Latitude ° '
01	04 ♉ 51	03 ♉ 43	00 S 38
02	04 R 52	03 40	00 N 25
03	04 50	03 36	01 28
04	04 47	03 33	02 28
05	04 41	03 30	03 22
06	04 32	03 27	04 07
07	04 22	03 24	04 40
08	04 11	03 20	04 58
09	03 59	03 17	05 01
10	03 50	03 14	04 45
11	03 42	03 11	04 13
12	03 37	03 08	03 25
13	03 34	03 05	02 25
14	03 34	03 01	01 15
15	03 D 34	02 58	00 N 01
16	03 R 33	02 55	01 S 13
17	03 31	02 52	02 22
18	03 26	02 49	03 23
19	03 19	02 46	04 10
20	03 09	02 42	04 43
21	02 57	02 39	04 59
22	02 45	02 36	04 58
23	02 33	02 33	04 42
24	02 23	02 30	04 12
25	02 15	02 26	03 31
26	02 09	02 23	02 41
27	02 07	02 20	01 45
28	02 06	02 17	00 S 44
29	02 D 06	02 14	00 N 18
30	02 R 05	02 11	01 21
31	02 ♉ 04	02 ♉ 07	02 N 20

DECLINATIONS

Date	Sun ☉ ° '	Moon ☽ ° '	Mercury ☿ ° '	Venus ♀ ° '	Mars ♂ ° '	Jupiter ♃ ° '	Saturn ♄ ° '	Uranus ♅ ° '	Neptune ♆ ° '	Pluto ♇ ° '
01	23 N 09	10 N 03	18 N 33	14 N 30	09 S 19	19 N 02	02 N 33	04 N 21	16 S 31	18 N 14
02	23 04	15 04	18 22	14 08	09 29	18 59	02 34	04 20	16 31	18 14
03	23 00	19 34	18 13	13 45	09 38	18 56	02 35	04 19	16 31	18 13
04	22 55	23 20	18 04	13 22	09 48	18 53	02 35	04 19	16 31	18 12
05	22 50	26 05	17 57	12 59	09 57	18 50	02 36	04 18	16 30	18 11
06	22 44	27 33	17 52	12 36	10 07	18 46	02 36	04 17	16 30	18 11
07	22 38	27 32	17 47	12 13	10 17	18 43	02 37	04 16	16 30	18 10
08	22 32	25 56	17 44	11 50	10 27	18 40	02 37	04 15	16 30	18 09
09	22 25	22 48	17 43	11 27	10 37	18 36	02 38	04 14	16 30	18 08
10	22 18	18 21	17 42	11 04	10 48	18 33	02 38	04 14	16 30	18 07
11	22 10	12 53	17 43	10 41	10 58	18 30	02 38	04 13	16 30	18 07
12	22 02	06 43	17 46	10 18	11 08	18 26	02 38	04 12	16 29	18 06
13	21 54	00 N 11	17 49	09 56	11 19	18 23	02 39	04 11	16 29	18 05
14	21 45	06 S 22	17 54	09 33	11 30	18 20	02 39	04 10	16 29	18 04
15	21 36	12 37	18 00	09 10	11 40	18 16	02 39	04 09	16 29	18 03
16	21 26	18 13	18 07	08 47	11 51	18 13	02 39	04 08	16 29	18 02
17	21 16	22 48	18 15	08 25	12 02	18 09	02 39	04 07	16 29	18 02
18	21 06	26 01	18 24	08 03	12 12	18 06	02 39	04 06	16 29	18 01
19	20 56	27 37	18 33	07 41	12 23	18 02	02 39	04 05	16 29	18 00
20	20 45	27 29	18 44	07 19	12 34	17 59	02 39	04 04	16 29	17 59
21	20 34	25 43	18 54	06 57	12 45	17 55	02 39	04 03	16 29	17 58
22	20 22	22 33	19 05	06 36	12 56	17 52	02 39	04 02	16 29	17 57
23	20 10	18 20	19 17	06 15	13 07	17 48	02 39	04 01	16 29	17 56
24	19 58	13 24	19 28	05 54	13 18	17 45	02 39	04 00	16 28	17 55
25	19 45	08 02	19 40	05 33	13 30	17 41	02 38	03 59	16 28	17 55
26	19 32	02 S 28	19 51	05 13	13 41	17 37	02 38	03 57	16 28	17 54
27	19 19	03 N 07	20 02	04 54	13 52	17 34	02 38	03 56	16 28	17 53
28	19 05	08 33	20 12	04 34	14 03	17 30	02 37	03 55	16 28	17 52
29	18 51	13 40	20 21	04 16	14 14	17 26	02 37	03 54	16 28	17 51
30	18 37	18 20	20 30	03 57	14 26	17 23	02 36	03 53	16 28	17 50
31	18 N 23	22 N 19	20 N 37	03 N 40	14 S 37	17 N 19	02 N 36	03 N 52	16 S 28	17 N 49

ZODIAC SIGN ENTRIES

Date	h	m	Planets
01	16	43	☽ ♉
04	04	39	☽ ♊
06	13	47	☽ ♋
08	19	58	☽ ♌
08	22	11	♀ ♍
11	00	07	☽ ♍
13	03	20	☽ ♎
15	06	17	☽ ♏
17	09	22	☽ ♐
19	12	59	☽ ♑
19	22	56	♂ ♏
21	17	59	☽ ♒
23	13	16	☉ ♌
24	01	28	☽ ♓
26	12	00	☽ ♈
29	00	40	☽ ♉
31	13	00	☽ ♊

LATITUDES

Date	Mercury ☿ ° '	Venus ♀ ° '	Mars ♂ ° '	Jupiter ♃ ° '	Saturn ♄ ° '	Uranus ♅ ° '	Neptune ♆ ° '	Pluto ♇ ° '
01	03 S 18	01 N 00	00 S 43	00 N 36	02 S 22	00 N 45	01 N 47	14 N 49
04	03 59	00 42	00 47	00 36	02 23	00 45	01 46	14 48
07	04 30	00 23	00 51	00 36	02 24	00 45	01 46	14 47
10	04 49	00 N 01	00 54	00 36	02 25	00 45	01 46	14 46
13	04 53	00 S 22	00 57	00 37	02 26	00 44	01 46	14 45
16	04 43	00 47	01 01	00 37	02 27	00 44	01 46	14 44
19	04 20	01 14	01 04	00 37	02 27	00 44	01 46	14 43
22	03 47	01 44	01 06	00 37	02 28	00 44	01 45	14 42
25	03 07	02 15	01 09	00 37	02 29	00 44	01 45	14 41
28	02 23	02 48	01 11	00 37	02 30	00 44	01 45	14 40
31	01 S 37	03 S 23	01 S 14	00 N 38	02 S 31	00 N 44	01 N 45	14 N 40

DATA

Julian Date	2439673
Delta T	+38 seconds
Ayanamsa	23° 24' 02"
Synetic vernal point	05° ♓ 42' 58"
True obliquity of ecliptic	23° 26' 44"

MOON'S PHASES, APSIDES AND POSITIONS ☽

Date	h	m	Phase	Longitude ° '	Eclipse Indicator
07	17	00	●	14 ♋ 53	
14	15	53	☽	21 ♎ 31	
21	14	39	○	28 ♑ 09	
29	12	14	☾	05 ♉ 41	

Day	h	m		
14	19	57	Perigee	
28	14	24	Apogee	
06	23	48	Max dec	27° N 44'
13	12	41	0S	
19	22	07	Max dec	27° S 46'
26	22	34	0N	

LONGITUDES

Date	Chiron ⚷ ° '	Ceres ⚳ ° '	Pallas ⚴ ° '	Juno ⚵ ° '	Vesta ⚶ ° '	Black Moon Lilith ⚸ ° '
01	29 ♓ 50	05 ♌ 46	00 ♋ 34	28 ♌ 33	17 ♏ 48	20 ♈ 49
11	29 ♓ 48	10 ♌ 11	06 ♋ 35	02 ♍ 18	18 ♏ 29	21 ♈ 56
21	29 ♓ 41	14 ♌ 40	12 ♋ 36	06 ♍ 07	19 ♏ 54	23 ♈ 02
31	29 ♓ 29	19 ♌ 11	18 ♋ 36	09 ♍ 58	21 ♏ 57	24 ♈ 08

ASPECTARIAN

01 Saturday
00 12 ☽ ⚻ ♆
00 32 ☽ ☍ ♂
03 44 ☽ △ ♀
05 02 ☽ ± ♇
07 03 ☉ Q ♅
08 32 ☽ # ♂
10 10 ☽ ± ♅
10 25 ☽ Q ☉
23 18 ☽ ᖘ ♇
23 55 ☿ ⚹ ♅

02 Sunday
04 26 ☽ ᖘ ♅
07 39 ☽ ∥ ♀
07 47 ☽ □ ♃
10 07 ☽ Q ☿
12 49 ☽ ⚹ ☉
17 00 ☽ ⚺ ♄
19 22 ☽ # ♆

03 Monday
04 26 ☽ ∥ ♇
04 38 ☽ ∥ ☿
05 03 ☽ ⊥ ♄
05 29 ☽ △ ♇
08 25 ☽ ∥ ♃
09 16 ☽ ⚹ ☿
10 34 ☽ △ ♅
11 07 ☿ ∥ ♇
12 33 ☽ ☍ ♆
14 51 ☽ ⚻ ♂
20 20 ☽ Q ♃
20 37 ☽ □ ♀
21 08 ☿ ⊥ ♀
21 16 ☽ ∠ ☉
22 55 ☽ ∠ ♄

04 Tuesday
03 03 ☽ ± ♂
09 10 ☽ ∥ ☉
13 33 ☽ ∠ ☿
16 19 ☽ ⊥ ☉
19 12 ☉ □ ♄
20 09 ☽ ⚹ ♃
21 07 ☽ ᖘ ♂

05 Wednesday
04 07 ☽ ⚹ ♄
04 50 ☽ ⚺ ☉
06 10 ☽ ⊥ ☿
07 51 ♀ ᖘ ♄
10 00 ☽ Q ♀
15 55 ☽ □ ♇
17 05 ☽ ⚺ ☿
20 48 ☽ □ ♅
22 28 ☽ ⚻ ♆

06 Thursday
01 10 ☽ ∠ ♃
02 32 ☽ △ ♂
02 54 ☽ Q ♄
09 41 ☽ ± ♆
10 07 ☽ ⚹ ♀
16 59 ☿ ⚹ ♇
18 11 ☽ ⊥ ♃

07 Friday
01 29 ☽ Q ♇
02 16 ☽ ᖘ ♆
02 18 ♂ ⊥ ♇
05 22 ☽ ⚺ ♃
06 14 ☽ Q ♅
12 08 ☽ □ ♄
15 24 ☽ ∠ ♀
17 00 ☽ ☌ ☉
21 53 ☽ ☌ ☿
23 16 ☽ ⚹ ♇

08 Saturday
03 57 ☽ ⚹ ♅
05 21 ☽ △ ♆
08 33 ☽ ⊥ ♀
10 56 ☽ □ ♂
19 50 ☽ ⚺ ♀

09 Sunday
01 57 ☽ ∠ ♇
06 34 ☽ ∠ ♅
11 42 ☽ ☌ ♃
12 03 ☉ ☌ ☿
14 28 ☽ ∥ ☉
17 28 ☽ △ ♄
17 43 ☽ ⊥ ♇
19 34 ☽ Q ♂
22 18 ☽ ⊥ ♅

10 Monday
00 26 ☽ ⚺ ☿
01 57 ☽ ⚺ ☉
04 09 ☽ ⚺ ♇
07 58 ☿ ∠ ♀
08 44 ☽ ⚺ ♅
09 53 ☽ □ ♆
10 22 ☽ ⊥ ☿
11 03 ☽ ∥ ♃
13 06 ☽ ⊥ ☉
13 07 ☽ ∥ ♇
15 04 ☽ ∥ ☿
16 58 ☽ ⚹ ♂
19 28 ☽ ᖘ ♄
20 37 ☽ # ♆
23 58 ♀ # ♂

11 Tuesday
01 13 ☽ ∠ ☿
02 59 ☽ ☌ ♀
05 42 ☽ ∠ ☉
10 22 ☉ ⚹ ♇
10 56 ☽ ± ♄
16 15 ☽ ⚺ ♃
16 46 ☽ Q ♆
19 28 ☽ # ♂
19 29 ☽ ∠ ♂
21 12 ☽ ⚻ ♄
21 22 ☽ ∥ ♀

12 Wednesday
01 50 ☽ ⚹ ☿
02 39 ☽ ⊥ ♃
07 41 ☽ ☌ ♇
09 12 ☽ ⚹ ☉
11 20 ☽ ⊥ ♂
12 15 ☽ ☌ ♅
13 13 ☽ ⚹ ♆
18 13 ☽ ∠ ♃
21 22 ☽ ∥ ♅
21 29 ☽ Q ☿
21 52 ☽ ⚺ ♂

13 Thursday
03 03 ☽ ∥ ♄
07 05 ☽ Q ☉
08 56 ☽ ⚺ ♀
14 43 ☽ ∠ ♆
19 40 ☽ ⊥ ♀
19 45 ♂ ⊥ ♅
20 07 ☽ ⚹ ♃
22 18 ☽ # ♄

14 Friday
00 19 ☽ ☍ ♄
03 00 ☽ □ ☿
03 53 ☽ # ♅
05 59 ☽ ⊥ ♆
08 05 ☉ ⚹ ♅
10 46 ☽ ⚺ ♇
11 45 ☽ ∠ ♀
15 22 ☽ ⚺ ♅
15 53 ☽ □ ☉
16 10 ☽ ⚺ ♆
16 48 ☽ Q ♃
20 05 ☉ △ ♆
20 58 ☽ ⊥ ♇
23 17 ☽ # ♀

15 Saturday
01 36 ☽ ⊥ ♅
02 29 ☽ ☌ ♂
08 06 ☽ ∥ ♂
12 17 ☽ ∠ ♇
14 31 ☽ ⚹ ♀
16 56 ☽ ∠ ♅
22 14 ☿ ∥ ♇
23 52 ☽ □ ♃

16 Sunday
03 22 ☽ ⚻ ♄
04 10 ☽ ∥ ♆
04 29 ☽ △ ☿
11 11 ☽ # ♇
11 32 ☽ # ☿
11 56 ☽ Q ♀
11 59 ☽ # ♃
13 35 ☽ ± ♄
13 51 ☽ ⚹ ♇
18 32 ☽ ⚹ ♅
19 09 ☽ ☌ ♆
22 36 ☽ △ ☉

17 Monday
00 22 ☿ ∥ ♃
03 36 ☽ # ☉
04 56 ☽ ᖘ ♄
05 29 ☽ ᖘ ☿
07 15 ☽ ⚺ ♂
10 21 ☽ Q ♇
15 04 ☽ Q ♅
17 51 ☽ ⊥ ♂
20 07 ☽ □ ♀
20 36 ☽ ± ☿

18 Tuesday
01 03 ♀ ± ♄
02 07 ☽ ᖘ ☉
03 52 ☽ △ ♃
06 38 ☽ △ ♄
06 44 ☽ ⚻ ☿
09 49 ☽ ∠ ♂
13 23 ☿ □ ♄
17 16 ☽ □ ♇
18 43 ☽ ± ☉
22 05 ☽ □ ♅
22 31 ☽ ⚺ ♆

19 Wednesday
05 51 ☽ ⚻ ☉
06 07 ☽ ᖘ ♃
08 54 ☽ ⊥ ♆
12 38 ☽ ⚹ ♂
22 01 ☽ ± ♃

20 Thursday
00 33 ☽ ∠ ♆
02 20 ☽ △ ♀
08 42 ☽ ⚻ ♃
10 21 ☽ Q ♂
10 25 ☽ ☍ ☿
10 36 ♃ ∥ ♇
10 44 ☽ □ ♄
12 00 ☿ St D
21 43 ☽ △ ♇

21 Friday
02 44 ☽ △ ♅
03 00 ☽ ⚹ ♆
06 00 ☽ ᖘ ♀
14 39 ☽ ☍ ☉
18 48 ☽ Q ♄
19 36 ☽ □ ♂
22 52 ☽ ± ♀

22 Saturday
00 32 ☽ Q ♆
00 39 ☽ ᖘ ♇
05 49 ☽ ᖘ ♅
10 15 ☽ ⚻ ♀
13 43 ☿ □ ♄
15 25 ☽ ☍ ♃
16 40 ☽ ⚹ ♄
16 43 ☽ ⚻ ☿
17 07 ☽ ± ♇
22 24 ☽ ± ♅

23 Sunday
01 41 ☽ # ☉
04 01 ☽ ± ☿
04 15 ☽ ⚻ ♇
07 13 ☽ # ☿
09 36 ☽ ⚻ ♅
09 39 ☽ □ ♆
14 03 ☽ # ♇
14 46 ☽ # ♃
16 50 ♀ Q ♆
20 39 ☽ ∠ ♄
21 17 ☽ ᖘ ☿
21 23 ☽ ∥ ♆

24 Monday
00 24 ♅ ⚹ ♆
02 28 ☽ ⚻ ☉
05 24 ☽ △ ♂
12 24 ☽ ∥ ♂
13 49 ☽ ⊥ ♄
14 53 ☽ ± ☉
21 00 ☽ ☍ ♀

25 Tuesday
01 06 ☽ ⚻ ♃
01 25 ☽ ⚺ ♄
03 00 ☽ △ ☿
04 07 ♄ St R
09 50 ☽ ᖘ ☉
11 38 ☽ ᖘ ♂
13 02 ☽ ± ♃
13 46 ☽ ☍ ♇
17 25 ☉ ∥ ☿
19 16 ♃ △ ♄
19 19 ☽ △ ♆
19 30 ☽ ☍ ♅
23 24 ☽ # ♀

26 Wednesday
05 35 ☽ # ♅
06 11 ☽ ± ♂
07 09 ☽ ᖘ ♃
11 16 ☽ # ♄
18 07 ☽ △ ☉
18 42 ☽ ⚻ ♂

27 Thursday
01 13 ☽ ᖘ ♆
09 03 ☉ □ ♂
09 53 ☽ ∥ ♄
10 44 ☽ ⚻ ♀
11 40 ♃ ⊥ ♇
13 03 ☽ ☌ ♄
13 34 ☉ ∠ ♇
13 51 ☽ △ ♃
15 34 ☽ ∥ ♅
17 43 ♂ ∠ ♇
17 44 ☽ □ ☿
19 20 ☽ ∥ ♀
19 24 ☽ ± ♆
23 18 ☽ ± ♀

28 Friday
02 00 ☽ ⚻ ♇
07 35 ☽ ⚻ ♆
08 01 ☽ ⚻ ♅
14 13 ☽ ± ♇
18 12 ☽ ᖘ ♀
20 16 ☽ ± ♅

29 Saturday
08 34 ☽ ᖘ ♇
08 54 ☽ Q ☿
10 14 ☽ ☍ ♂
12 14 ☽ □ ☉
14 38 ☽ ᖘ ♅
14 53 ☽ # ♂

30 Sunday
01 34 ☽ △ ♀
01 54 ☽ ⚺ ♄
02 04 ☽ # ♆
03 30 ☉ ⊥ ♀
03 52 ☽ □ ♃
06 54 ☽ ∥ ♃
09 18 ☽ ∥ ♇
11 11 ☽ ⚹ ☿
13 30 ☽ ∥ ☉
13 55 ♀ ⚻ ♄
13 59 ☽ ⊥ ♄
14 56 ☽ △ ♇
18 24 ♀ ∥ ♅
19 20 ☉ ∠ ♅
20 17 ☽ ☍ ♆
20 59 ☽ △ ♅

31 Monday
00 50 ☽ ∥ ☿
03 37 ☽ Q ☉
07 55 ☽ ∠ ♄
16 27 ☽ Q ♃
19 46 ☽ ∠ ☿

All ephemeris data is given at 12.00 UT and the Moon's longitude is additionally given for 24.00 UT

AUGUST 1967

Date	Sidereal time h m s	Sun ☉ ° ' "	Moon ☽ ° ' "	Moon ☽ 24.00 ° ' "	Mercury ☿ ° '	Venus ♀ ° '	Mars ♂ ° '	Jupiter ♃ ° '	Saturn ♄ ° '	Uranus ♅ ° '	Neptune ♆ ° '	Pluto ♇ ° '
01	08 37 38	08 ♌ 33 02	11 ♊ 46 12	18 ♊ 01 27	19 ♋ 12	12 ♍ 59	06 ♏ 19	13 ♌ 56	12 ♈ 25	22 ♍ 01	21 ♏ 35	18 ♍ 58
02	08 41 35	09 30 28	24 ♊ 21 49	00 ♋ 47 39	20 23	13 13	06 51	14 09	12 R 24	22 04	21 R 35	19 00
03	08 45 32	10 27 54	07 ♋ 19 09	13 ♋ 56 23	21 40	13 25	07 23	14 22	12 23	22 07	21 35	19 02
04	08 49 28	11 25 22	20 ♋ 39 19	27 ♋ 27 47	23 02	13 35	07 56	14 35	12 22	22 10	21 D 35	19 04
05	08 53 25	12 22 51	04 ♌ 21 27	11 ♌ 19 52	24 29	13 43	08 28	14 48	12 21	22 13	21 35	19 05
06	08 57 21	13 20 21	18 ♌ 22 29	25 ♌ 28 39	26 01	13 49	09 01	15 02	12 20	22 16	21 35	19 07
07	09 01 18	14 17 52	02 ♍ 37 41	09 ♍ 48 48	27 37	13 53	09 35	15 15	12 19	22 19	21 35	19 09
08	09 05 14	15 15 23	17 ♍ 01 17	24 ♍ 14 26	29 ♋ 17	13 54	10 08	15 28	12 17	22 22	21 35	19 11
09	09 09 11	16 12 56	01 ♎ 27 33	08 ♎ 40 05	01 ♌ 00	13 R 53	10 42	15 41	12 16	22 26	21 35	19 13
10	09 13 07	17 10 29	15 ♎ 51 30	23 ♎ 01 26	02 48	13 50	11 15	15 54	12 14	22 29	21 36	19 15
11	09 17 04	18 08 03	00 ♏ 09 32	07 ♏ 15 36	04 38	13 45	11 50	16 08	12 13	22 32	21 36	19 17
12	09 21 01	19 05 38	14 ♏ 19 27	21 ♏ 20 59	06 31	13 37	12 24	16 21	12 11	22 35	21 36	19 19
13	09 24 57	20 03 14	28 ♏ 20 08	05 ♐ 16 53	08 26	13 26	12 58	16 34	12 09	22 39	21 36	19 21
14	09 28 54	21 00 51	12 ♐ 11 11	19 ♐ 03 00	10 23	13 13	13 33	16 47	12 07	22 42	21 37	19 23
15	09 32 50	21 58 29	25 ♐ 52 19	02 ♑ 39 03	12 22	12 58	14 08	17 00	12 05	22 45	21 37	19 24
16	09 36 47	22 56 08	09 ♑ 23 36	16 ♑ 04 23	14 22	12 41	14 43	17 13	12 03	22 49	21 38	19 26
17	09 40 43	23 53 48	22 ♑ 42 46	29 ♑ 18 06	16 23	12 21	15 18	17 27	12 00	22 52	21 38	19 28
18	09 44 40	24 51 29	05 ♒ 50 16	12 ♒ 19 08	18 24	11 59	15 54	17 40	11 58	22 56	21 38	19 30
19	09 48 36	25 49 11	18 ♒ 44 35	25 ♒ 06 33	20 26	11 34	16 29	17 53	11 56	22 59	21 39	19 32
20	09 52 33	26 46 55	01 ♓ 24 59	07 ♓ 39 55	22 27	11 08	17 05	18 06	11 53	23 03	21 39	19 35
21	09 56 30	27 44 40	13 ♓ 51 23	19 ♓ 59 33	24 28	10 40	17 41	18 19	11 50	23 06	21 40	19 37
22	10 00 26	28 42 26	26 ♓ 04 35	02 ♈ 06 44	26 29	10 10	18 17	18 32	11 48	23 10	21 41	19 39
23	10 04 23	29 ♌ 40 14	08 ♈ 06 19	14 ♈ 03 44	28 ♌ 29	09 38	18 54	18 45	11 45	23 13	21 41	19 41
24	10 08 19	00 ♍ 38 03	19 ♈ 59 23	25 ♈ 53 47	00 ♍ 28	09 05	19 30	18 58	11 42	23 17	21 42	19 43
25	10 12 16	01 35 54	01 ♉ 47 27	07 ♉ 40 57	02 27	08 31	20 07	19 11	11 39	23 20	21 43	19 45
26	10 16 12	02 33 47	13 ♉ 34 55	19 ♉ 29 59	04 24	07 56	20 44	19 24	11 36	23 24	21 43	19 47
27	10 20 09	03 31 42	25 ♉ 26 48	01 ♊ 26 02	06 20	07 20	21 21	19 37	11 33	23 27	21 44	19 49
28	10 24 05	04 29 39	07 ♊ 28 21	13 ♊ 34 25	08 15	06 43	21 58	19 50	11 30	23 31	21 45	19 51
29	10 28 02	05 27 37	19 ♊ 44 52	26 ♊ 00 16	10 09	06 06	22 36	20 03	11 27	23 35	21 46	19 53
30	10 31 59	06 25 37	02 ♋ 21 09	08 ♋ 47 58	12 01	05 29	23 13	20 16	11 23	23 38	21 46	19 55
31	10 35 55	07 ♍ 23 40	15 ♋ 21 05	22 ♋ 00 43	13 ♍ 53	04 ♍ 52	23 ♏ 51	20 ♌ 29	11 ♈ 20	23 ♍ 42	21 ♏ 47	19 ♍ 58

Date	Moon True ☊ ° '	Moon Mean ☊ ° '	Moon ☽ Latitude ° '
01	02 ♉ 00	02 ♉ 04	03 N 14
02	01 R 54	02 01	04 00
03	01 46	01 58	04 35
04	01 35	01 55	04 56
05	01 24	01 52	05 01
06	01 12	01 48	04 48
07	01 02	01 45	04 17
08	00 54	01 42	03 30
09	00 49	01 39	02 29
10	00 46	01 36	01 18
11	00 45	01 32	00 N 03
12	00 D 46	01 29	01 S 11
13	00 R 45	01 26	02 21
14	00 44	01 23	03 21
15	00 40	01 20	04 09
16	00 33	01 17	04 43
17	00 24	01 13	05 00
18	00 14	01 10	05 01
19	00 03	01 07	04 47
20	29 ♈ 52	01 04	04 19
21	29 43	01 01	03 39
22	29 35	00 57	02 49
23	29 31	00 54	01 52
24	29 28	00 51	00 S 51
25	29 D 28	00 48	00 N 12
26	29 29	00 45	01 15
27	29 30	00 42	02 15
28	29 R 30	00 38	03 10
29	29 29	00 35	03 57
30	29 25	00 32	04 34
31	29 ♈ 20	00 ♉ 29	04 N 58

DECLINATIONS

Date	Sun ☉ ° '	Moon ☽ ° '	Mercury ☿ ° '	Venus ♀ ° '	Mars ♂ ° '	Jupiter ♃ ° '	Saturn ♄ ° '	Uranus ♅ ° '	Neptune ♆ ° '	Pluto ♇ ° '
01	18 N 08	25 N 24	20 N 43	03 N 22	14 S 48	17 N 15	02 N 35	03 N 50	16 S 29	17 N 48
02	17 53	27 19	20 48	03 06	14 59	17 11	02 35	03 49	16 29	17 47
03	17 37	27 49	20 50	02 50	15 11	17 08	02 34	03 48	16 29	17 46
04	17 22	26 44	20 52	02 34	15 22	17 04	02 34	03 47	16 29	17 46
05	17 06	24 02	20 50	02 19	15 33	17 00	02 33	03 46	16 29	17 45
06	16 49	19 53	20 47	02 05	15 45	16 56	02 32	03 44	16 29	17 44
07	16 33	14 32	20 42	01 52	15 56	16 52	02 31	03 43	16 29	17 43
08	16 16	08 21	20 34	01 40	16 07	16 49	02 31	03 42	16 29	17 42
09	15 59	01 N 42	20 23	01 28	16 18	16 45	02 30	03 40	16 29	17 41
10	15 42	05 S 02	20 10	01 18	16 29	16 41	02 29	03 39	16 29	17 40
11	15 24	11 29	19 54	01 08	16 41	16 37	02 28	03 38	16 29	17 39
12	15 06	17 19	19 35	00 59	16 52	16 33	02 27	03 37	16 29	17 38
13	14 48	22 05	19 14	00 52	17 03	16 29	02 26	03 35	16 30	17 37
14	14 30	25 35	18 50	00 45	17 14	16 25	02 25	03 34	16 30	17 36
15	14 11	27 32	18 24	00 40	17 25	16 21	02 24	03 32	16 30	17 35
16	13 53	27 49	17 55	00 35	17 36	16 18	02 23	03 31	16 30	17 34
17	13 34	26 28	17 24	00 32	17 47	16 14	02 22	03 30	16 30	17 34
18	13 14	23 41	16 51	00 30	17 58	16 10	02 21	03 28	16 30	17 33
19	12 55	19 45	16 16	00 29	18 09	16 06	02 20	03 27	16 31	17 32
20	12 35	15 00	15 39	00 29	18 19	16 02	02 19	03 26	16 31	17 31
21	12 16	09 43	15 01	00 31	18 30	15 58	02 17	03 24	16 31	17 30
22	11 56	04 S 08	14 21	00 33	18 41	15 54	02 16	03 23	16 31	17 29
23	11 35	01 N 30	13 40	00 37	18 51	15 50	02 15	03 21	16 31	17 28
24	11 15	07 02	12 57	00 42	19 02	15 46	02 13	03 20	16 32	17 27
25	10 55	12 18	12 14	00 48	19 12	15 42	02 12	03 19	16 32	17 26
26	10 34	17 07	11 30	00 55	19 23	15 38	02 11	03 17	16 32	17 25
27	10 13	21 19	10 45	01 04	19 33	15 34	02 09	03 16	16 32	17 24
28	09 52	24 41	10 00	01 13	19 43	15 30	02 08	03 14	16 33	17 24
29	09 31	26 59	09 14	01 23	19 53	15 26	02 06	03 13	16 33	17 23
30	09 09	27 59	08 28	01 33	20 03	15 22	02 05	03 11	16 33	17 22
31	08 N 48	27 N 30	07 N 41	01 N 45	20 S 13	15 N 18	02 N 03	03 N 10	16 S 33	17 N 21

ZODIAC SIGN ENTRIES

Date	h	m	Planets
02	22	32	☽ ♋
05	04	26	☽ ♌
07	07	36	☽ ♍
08	22	09	☿ ♌
09	09	34	☽ ♎
11	11	44	☽ ♏
13	14	52	☽ ♐
15	19	18	☽ ♑
18	01	17	☽ ♒
20	09	18	☽ ♓
22	19	47	☽ ♈
23	20	12	☉ ♍
24	06	17	☿ ♍
25	08	21	☽ ♉
27	21	08	☽ ♊
30	07	34	☽ ♋

LATITUDES

Date	Mercury ☿ ° '	Venus ♀ ° '	Mars ♂ ° '	Jupiter ♃ ° '	Saturn ♄ ° '	Uranus ♅ ° '	Neptune ♆ ° '	Pluto ♇ ° '
01	01 S 22	03 S 35	01 S 15	00 N 38	02 S 31	00 N 44	01 N 45	14 N 39
04	00 S 38	04 12	01 17	00 38	02 32	00 44	01 45	14 39
07	00 N 03	04 50	01 19	00 38	02 33	00 44	01 45	14 38
10	00 38	05 29	01 21	00 38	02 33	00 44	01 44	14 38
13	01 07	06 07	01 23	00 39	02 34	00 44	01 44	14 37
16	01 27	06 43	01 24	00 39	02 35	00 44	01 44	14 37
19	01 40	07 17	01 26	00 39	02 36	00 43	01 44	14 36
22	01 46	07 46	01 27	00 39	02 36	00 43	01 44	14 36
25	01 45	08 09	01 29	00 40	02 37	00 43	01 43	14 35
28	01 38	08 26	01 30	00 40	02 38	00 43	01 43	14 35
31	01 N 27	08 S 34	01 S 31	00 N 40	02 S 38	00 N 43	01 N 43	14 N 35

DATA

Julian Date	2439704
Delta T	+38 seconds
Ayanamsa	23° 24' 07"
Synetic vernal point	05° ♓ 42' 52"
True obliquity of ecliptic	23° 26' 44"

MOON'S PHASES, APSIDES AND POSITIONS ☽

Date	h	m	Phase	Longitude ° '	Eclipse Indicator
06	02	48	●	12 ♌ 58	
12	20	44	☽	19 ♏ 27	
20	02	27	○	26 ♒ 24	
28	05	35	☾	04 ♊ 14	

Day	h	m		
09	14	23	Perigee	
25	08	33	Apogee	
03	07	47	Max dec	27° N 50'
09	18	01	0S	
16	03	58	Max dec	27° S 54'
23	05	36	0N	
30	16	27	Max dec	28° N 01'

LONGITUDES

Date	Chiron ⚷ ° '	Ceres ⚳ ° '	Pallas ⚴ ° '	Juno ⚵ ° '	Vesta ⚶ ° '	Black Moon Lilith ⚸ ° '
01	29 ♓ 28	19 ♌ 38	19 ♋ 12	10 ♍ 22	22 ♏ 11	24 ♈ 15
11	29 ♓ 11	24 ♌ 11	25 ♋ 10	14 ♍ 14	24 ♏ 50	25 ♈ 22
21	28 ♓ 50	28 ♌ 45	01 ♌ 04	18 ♍ 08	27 ♏ 57	26 ♈ 28
31	28 ♓ 27	03 ♍ 21	06 ♌ 54	22 ♍ 02	01 ♐ 26	27 ♈ 34

All ephemeris data is given at 12.00 UT and the Moon's longitude is additionally given for 24.00 UT

ASPECTARIAN

h m Aspects

01 Tuesday
00 56 ☽ ⊼ ♂
05 14 ☽ ✶ ☉
07 11 ☿ ✶ ♇
13 06 ☽ ± ♂
13 15 ☽ ✶ ♄
14 23 ☽ □ ♀
15 02 ☽ ⊥ ☿
16 14 ☽ ✶ ♃

02 Wednesday
01 50 ☽ □ ♇
03 42 ☽ ⊻ ☿
06 45 ☽ ⊼ ♆
07 03 ☽ ⚼ ♂
07 39 ☽ □ ♅
12 04 ☽ Q ♄
12 18 ☽ ∠ ☉
18 02 ☽ ± ♆
20 46 ☉ ∥ ♇
21 06 ☽ ∠ ♃
22 30 ♂ ∠ ♅

03 Thursday
01 00 ☽ Q ♀
06 22 ☽ ⊥ ☉
10 22 ☿ △ ♆
10 39 ☽ ⚼ ♆
11 28 ☽ Q ♇
12 08 ☽ △ ♂
13 57 ☽ ⊥ ♃
15 18 ♆ St D
17 06 ☽ Q ♅
18 10 ☽ ⊻ ☉
20 14 ☿ ✶ ♅
21 12 ☽ □ ♄
23 13 ☽ ✶ ♀

04 Friday
00 59 ☽ ⊻ ♃
09 10 ☽ ✶ ♇
12 50 ♀ ∥ ♄
13 39 ☽ △ ♆
14 41 ☽ ✶ ♅
16 42 ☽ ☌ ☿

05 Saturday
02 07 ☽ ∠ ♀
11 20 ☉ △ ♄
11 32 ☽ ∠ ♇
16 57 ☽ ∠ ♅
17 51 ☽ ⊥ ♀
19 23 ☽ □ ♂
22 35 ☉ ∥ ♃

06 Sunday
01 44 ☽ △ ♄
02 48 ☽ ☌ ☉
03 02 ☽ ⊥ ♇
04 13 ☽ ⊻ ♀
06 13 ☽ ☌ ♃
06 22 ☉ ⊥ ♇
07 18 ☽ ∥ ☿
08 25 ☽ ⊥ ♅
13 16 ☽ ⊻ ♇
17 26 ☽ □ ♆
18 37 ☽ ⊻ ♅
22 16 ☽ ∥ ♇

07 Monday
01 03 ☉ ⊻ ♀
01 56 ☽ ∥ ♃
02 30 ☽ ⊻ ☿
03 03 ☽ ∥ ☉
03 06 ☽ ⚼ ♄
03 11 ☽ Q ♂
03 46 ☽ # ♆
06 20 ☽ # ♂
13 51 ☽ ⊥ ☿
17 26 ☉ # ♆
18 09 ☽ ± ♄
23 37 ☽ Q ♆

08 Tuesday
00 04 ☽ ✶ ♂
04 08 ☽ ⊼ ♄
06 44 ☿ ∠ ♀
06 49 ☽ ☌ ♀
06 49 ☽ ∠ ☿
08 51 ☽ ⊻ ☉
09 22 ☽ ⊻ ♃
14 29 ♀ St R
15 36 ☽ ☌ ♇
18 49 ☉ ☌ ♃
19 30 ☽ ⊥ ♃
19 32 ☉ # ♂
19 32 ☽ ⊥ ☉
19 36 ☽ ✶ ♆
20 56 ☽ ☌ ♅

09 Wednesday
02 02 ☽ ∠ ♂
04 55 ☽ ∥ ♅
09 08 ☽ ∥ ♄
10 42 ☽ ∠ ♃
11 08 ☽ ✶ ☿
11 34 ☽ ∠ ☉
12 48 ☽ ∥ ♀
17 36 ☽ ⊥ ♂
17 37 ☉ ⊥ ♅
20 32 ☽ ∠ ♆
22 57 ☽ # ♀

10 Thursday
02 52 ☽ # ♄
04 00 ☽ ⊻ ♂
05 58 ☽ ☍ ♄
07 02 ☽ # ♅
08 38 ☽ ⊻ ♀
09 58 ☽ Q ☿
11 27 ♂ ∥ ♆
11 33 ☽ ⊥ ♆
12 05 ☽ ✶ ♃
14 22 ☽ ✶ ☉
17 41 ☽ ⊻ ♇
18 37 ☽ ⊥ ♀
21 36 ☽ ⊻ ♆
23 08 ☽ ⊻ ♅

11 Friday
03 46 ☽ ⊥ ♇
06 15 ♂ # ♃
07 22 ☿ ∠ ♇
08 31 ☽ Q ♃
09 15 ☽ ⊥ ♅
09 38 ☽ ∠ ♀
11 57 ☽ Q ☉
18 58 ☽ ∠ ♇
20 42 ☽ □ ☿

12 Saturday
00 31 ☽ ∠ ♅
03 03 ☽ # ☉
03 19 ♂ ⊼ ♄
08 21 ☽ ⊼ ♄
08 31 ☽ ∥ ♆
08 35 ☽ ☌ ♂
08 50 ☽ # ♃
10 05 ☽ ∥ ♂
10 48 ☽ ✶ ♀
13 37 ☽ # ♇
15 30 ☽ □ ♃
17 36 ☉ ⊻ ♇
18 34 ☽ ± ♄
20 32 ☽ ✶ ♇
20 44 ☽ □ ☉
22 07 ☽ # ☿

13 Sunday
00 26 ☽ ☌ ♆
00 38 ☿ ⊥ ♀
01 53 ☿ ∠ ♅
02 11 ☽ ✶ ♅
07 05 ☽ Q ♀
09 58 ☽ ⚼ ♄
10 06 ♃ # ♆
17 12 ☽ Q ♇
22 57 ☽ Q ♅
23 48 ♃ ⊥ ♅

14 Monday
02 13 ♀ ✶ ♂
08 21 ☽ △ ☿
11 52 ☽ △ ♄
13 46 ☽ □ ♀
14 29 ☽ ⊻ ♂
20 10 ☽ △ ♃

15 Tuesday
00 36 ☽ □ ♇
01 27 ☽ ⊥ ♂
03 03 ☉ □ ♆
04 30 ☽ ⊻ ♆
04 37 ☽ △ ☉
06 29 ☽ □ ♅
08 32 ☿ △ ♄
15 05 ☽ ⊥ ♆
15 06 ☽ ⚼ ☿
18 01 ☽ ∠ ♂
18 18 ☿ ⊻ ♀
23 02 ☽ ⚼ ♃

16 Wednesday
00 40 ☿ ⊥ ♇
07 04 ☽ ∠ ♆
08 46 ☉ ⊻ ♅
08 54 ♂ # ♇
09 13 ☽ ⚼ ☉
09 52 ☽ ± ☿
15 21 ☽ ± ♃
16 45 ☽ □ ♄
17 45 ☽ △ ♀
17 48 ☿ □ ♂
22 00 ☽ ✶ ♂
22 31 ☽ ⊼ ☿
23 00 ☿ # ♂

17 Thursday
02 18 ☽ ⊼ ♃
02 37 ☽ ± ☉
04 33 ☿ ∥ ♇
06 07 ☽ △ ♇
10 02 ☽ ✶ ♆
12 17 ☽ △ ♅
14 19 ☽ ⊼ ☉
17 57 ☿ ⊥ ♅
20 12 ☽ ⚼ ♀
20 45 ☽ Q ♂

18 Friday
01 15 ☽ Q ♄
02 05 ☿ ☌ ♃
07 57 ☽ Q ♆
09 33 ☽ ⚼ ♇
12 15 ☽ ± ♀
12 42 ♀ ⊼ ♄
15 53 ☽ ⚼ ♅
23 02 ☽ ⊼ ♀
23 19 ☽ ✶ ♄

19 Saturday
01 16 ☿ ⊻ ♇
02 05 ☿ # ♆
02 15 ☽ ± ♇
07 34 ☽ □ ♂
08 41 ☽ ± ♅
10 21 ☽ ☍ ♃
13 30 ☽ ⊼ ♇
15 46 ☽ ☍ ☿
17 28 ☽ □ ♆
19 28 ☿ ∥ ♃
20 01 ☽ ⊼ ♅
20 13 ☽ ∥ ♂
23 42 ☽ # ♇

20 Sunday
02 27 ☽ ☍ ☉
02 29 ☿ □ ♆
03 24 ☽ ∠ ♄
04 43 ☽ ∥ ♆
07 00 ☽ # ♃
08 28 ☽ # ☿
14 27 ☉ ⚼ ♄
19 11 ☿ ⊻ ♅
20 32 ☽ ⊥ ♄
23 55 ☽ # ☉

21 Monday
06 02 ☽ ☍ ♀
08 06 ☽ ⊻ ♄
17 45 ♂ ± ♄
19 52 ☽ △ ♂
20 52 ☽ ⊼ ♃
23 17 ☽ ☍ ♇

22 Tuesday
03 18 ☽ △ ♆
06 12 ☽ ☍ ♅
08 54 ☽ ± ♃
12 58 ☽ ⊼ ☿
15 15 ☽ # ♅
15 38 ☿ ⚼ ♄
17 40 ☽ ⊼ ☉
19 59 ☽ # ♄

23 Wednesday
03 04 ☽ # ♀
03 07 ☽ ⚼ ♂
03 07 ☽ ⚼ ♃
03 11 ♂ □ ♃
03 18 ☽ ± ☿
06 41 ☽ ± ☉
08 11 ☽ ∥ ♀
09 09 ☽ ⚼ ♆
09 59 ♀ Q ♆
14 57 ☽ ⊼ ♀
15 10 ☽ ∥ ♄
19 18 ☽ ☌ ♄
19 55 ☽ ∥ ♅
22 10 ☽ ± ♂

24 Thursday
01 01 ☽ ⚼ ☿
02 24 ☽ ⚼ ☉
02 31 ☽ ± ♀
03 18 ☽ ± ♆
09 54 ☽ △ ♃
10 58 ☽ ⊼ ♂
11 26 ☽ ⊼ ♇
15 28 ☽ ⊼ ♆
15 51 ☉ ☌ ☿
18 43 ☽ ⊼ ♅
19 57 ☽ ⚼ ♀
20 41 ♂ ✶ ♇
23 40 ☽ ± ♇

25 Friday
05 55 ☽ ∥ ☉
06 59 ☽ ± ♅
11 34 ☽ △ ☉
11 45 ☽ ∥ ☿
13 36 ☽ △ ☿
18 02 ☽ ⚼ ♇
20 05 ♀ Q ♂

26 Saturday
01 03 ☽ △ ♀
01 24 ☽ ⚼ ♅
04 26 ☽ ∥ ♃
08 00 ☽ ⊻ ♄
08 58 ☽ # ♆
13 38 ☽ ∥ ♇
20 07 ☽ ⊥ ♄

27 Sunday
00 02 ☽ □ ♃
00 37 ☽ △ ♇
00 56 ☽ # ♂
02 32 ☿ ± ♄
03 17 ☽ ☍ ♂
04 30 ☽ ☍ ♆
07 58 ☽ △ ♅
14 12 ☽ ∠ ♄
21 27 ☿ ☌ ♀

28 Monday
03 13 ♂ ☌ ♆
05 35 ☽ □ ☉
10 35 ☽ □ ♀
12 44 ☽ Q ♃
13 49 ☽ □ ☿
14 10 ♃ ⊻ ♇
19 39 ☉ ∥ ☿
19 54 ☽ ✶ ♄

29 Tuesday
07 04 ☿ Q ♆
11 34 ☉ ± ♄
12 16 ☽ □ ♇
12 36 ☽ ✶ ♃
15 53 ☽ ⊼ ♆
17 46 ☽ ⊼ ♂
19 04 ☽ Q ♄
19 24 ☽ □ ♅
19 44 ☽ Q ☉
19 58 ☽ Q ♀
20 32 ☿ Q ♂
21 40 ☉ ☌ ♀

30 Wednesday
03 21 ☽ ± ♆
04 05 ☿ ⊼ ♄
05 48 ☽ ± ♂
06 52 ☽ Q ☿
15 59 ♀ ± ♄
17 33 ☽ ∠ ♃
17 35 ☽ ✶ ♀
20 13 ☽ ✶ ☉
20 16 ☽ ⚼ ♆
22 25 ☽ Q ♇
23 29 ☽ ⚼ ♂

31 Thursday
04 41 ☽ □ ♄
05 18 ☽ Q ♅
05 47 ♂ ✶ ♅
08 53 ☽ ✶ ☿
10 24 ☽ ⊥ ♃
19 47 ☽ ∠ ♀
20 21 ☽ ✶ ♇
21 25 ☽ ⊻ ♃
23 37 ☽ △ ♆

SEPTEMBER 1967

LONGITUDES

Date	Sidereal time h m s	Sun ☉ ° ' "	Moon ☽ ° ' "	Moon ☽ 24.00 ° ' "	Mercury ☿ ° '	Venus ♀ ° '	Mars ♂ ° '	Jupiter ♃ ° '	Saturn ♄ ° '	Uranus ♅ ° '	Neptune ♆ ° '	Pluto ♇ ° '
01	10 39 52	08 ♍ 21 44	28 ♋ 46 59	05 ♌ 39 49	15 ♍ 43	04 ♍ 15	24 ♏ 29	20 ♌ 42	11 ♈ 16	23 ♍ 46	21 ♏ 48	20 ♍ 00
02	10 43 48	09 19 49	12 ♌ 39 01	19 ♌ 44 10	17 32	03 R 39	25 07	20 55	11 R 13	23 49	21 49	20 02
03	10 47 45	10 17 57	26 ♌ 54 46	04 ♍ 10 04	19 19	03 03	25 45	21 07	11 09	23 53	21 50	20 04
04	10 51 41	11 16 06	11 ♍ 29 17	18 ♍ 51 29	21 06	02 29	26 23	21 20	11 05	23 57	21 51	20 06
05	10 55 38	12 14 17	26 ♍ 15 40	03 ♎ 40 51	22 51	01 56	27 02	21 33	11 02	24 01	21 52	20 08
06	10 59 34	13 12 30	11 ♎ 05 02	18 ♎ 30 18	24 35	01 25	27 41	21 46	10 58	24 04	21 53	20 10
07	11 03 31	14 10 44	25 ♎ 52 49	03 ♏ 12 51	26 18	00 55	28 19	21 58	10 54	24 08	21 54	20 13
08	11 07 28	15 09 00	10 ♏ 29 47	17 ♏ 43 10	27 59	00 26	28 58	22 11	10 50	24 12	21 55	20 15
09	11 11 24	16 07 17	24 ♏ 51 37	01 ♐ 57 55	29 ♍ 40	00 ♍ 00	29 ♏ 38	22 24	10 46	24 15	21 57	20 17
10	11 15 21	17 05 36	08 ♐ 58 56	15 ♐ 55 36	01 ♎ 19	29 ♌ 36	00 ♐ 17	22 36	10 42	24 19	21 58	20 19
11	11 19 17	18 03 56	22 ♐ 47 56	29 ♐ 36 02	02 57	29 13	00 56	22 49	10 38	24 23	21 59	20 21
12	11 23 14	19 02 18	06 ♑ 19 59	12 ♑ 59 55	04 34	28 53	01 36	23 01	10 34	24 27	22 00	20 24
13	11 27 10	20 00 42	19 ♑ 36 00	26 ♑ 08 21	06 10	28 36	02 15	23 14	10 29	24 31	22 01	20 26
14	11 31 07	20 59 07	02 ♒ 37 07	09 ♒ 02 27	07 45	28 20	02 55	23 26	10 25	24 34	22 03	20 28
15	11 35 03	21 57 33	15 ♒ 24 29	21 ♒ 43 20	09 19	28 07	03 35	23 38	10 21	24 38	22 04	20 30
16	11 39 00	22 56 02	27 ♒ 59 07	04 ♓ 11 59	10 51	27 56	04 15	23 51	10 16	24 42	22 05	20 32
17	11 42 57	23 54 32	10 ♓ 22 01	16 ♓ 29 23	12 23	27 48	04 56	24 03	10 12	24 46	22 07	20 34
18	11 46 53	24 53 04	22 ♓ 34 13	28 ♓ 36 41	13 53	27 42	05 36	24 15	10 08	24 49	22 08	20 37
19	11 50 50	25 51 37	04 ♈ 37 00	10 ♈ 35 22	15 23	27 39	06 16	24 27	10 03	24 53	22 09	20 39
20	11 54 46	26 50 13	16 ♈ 32 03	22 ♈ 27 20	16 51	27 D 38	06 57	24 39	09 59	24 57	22 11	20 41
21	11 58 43	27 48 51	28 ♈ 21 33	04 ♉ 15 04	18 19	27 40	07 38	24 51	09 54	25 01	22 12	20 43
22	12 02 39	28 47 31	10 ♉ 08 18	16 ♉ 01 40	19 45	27 43	08 18	25 03	09 50	25 05	22 14	20 45
23	12 06 36	29 ♍ 46 13	21 ♉ 55 41	27 ♉ 50 50	21 10	27 49	08 59	25 15	09 45	25 08	22 15	20 47
24	12 10 32	00 ♎ 44 57	03 ♊ 47 41	09 ♊ 46 47	22 34	27 58	09 40	25 27	09 40	25 12	22 17	20 50
25	12 14 29	01 43 44	15 ♊ 48 43	21 ♊ 54 06	23 57	28 08	10 22	25 39	09 36	25 16	22 18	20 52
26	12 18 26	02 42 32	28 ♊ 03 32	04 ♋ 17 35	25 18	28 21	11 03	25 51	09 31	25 20	22 20	20 54
27	12 22 22	03 41 24	10 ♋ 36 49	17 ♋ 01 45	26 39	28 36	11 44	26 02	09 26	25 23	22 22	20 56
28	12 26 19	04 40 17	23 ♋ 32 51	00 ♌ 10 29	27 58	28 52	12 26	26 14	09 22	25 27	22 23	20 58
29	12 30 15	05 39 13	06 ♌ 54 55	13 ♌ 46 18	29 ♎ 16	29 11	13 08	26 26	09 17	25 31	22 25	21 00
30	12 34 12	06 ♎ 38 10	20 ♌ 44 38	27 ♌ 49 46	00 ♏ 32	29 ♌ 32	13 ♐ 49	26 ♌ 37	09 ♈ 12	25 ♍ 35	22 ♏ 27	21 ♍ 03

Date	Moon True ☊ ° '	Moon Mean ☊ ° '	Moon ☽ Latitude ° '
01	29 ♈ 14	00 ♉ 26	05 N 07
02	29 R 06	00 23	04 59
03	28 58	00 19	04 32
04	28 51	00 16	03 47
05	28 45	00 13	02 46
06	28 42	00 10	01 34
07	28 41	00 07	00 N 15
08	28 D 41	00 03	01 S 04
09	28 42	00 ♉ 00	02 17
10	28 44	29 ♈ 57	03 21
11	28 R 44	29 54	04 12
12	28 42	29 51	04 48
13	28 39	29 48	05 07
14	28 34	29 44	05 10
15	28 29	29 41	04 57
16	28 22	29 38	04 31
17	28 16	29 35	03 51
18	28 11	29 32	03 02
19	28 07	29 29	02 05
20	28 05	29 25	01 S 03
21	28 D 04	29 22	00 N 02
22	28 05	29 19	01 06
23	28 07	29 16	02 07
24	28 08	29 13	03 04
25	28 10	29 09	03 53
26	28 11	29 06	04 33
27	28 R 11	29 03	05 01
28	28 10	29 00	05 14
29	28 07	28 57	05 12
30	28 ♈ 05	28 ♈ 54	04 N 51

DECLINATIONS

Date	Sun ☉ ° '	Moon ☽ ° '	Mercury ☿ ° '	Venus ♀ ° '	Mars ♂ ° '	Jupiter ♃ ° '	Saturn ♄ ° '	Uranus ♅ ° '	Neptune ♆ ° '	Pluto ♇ ° '
01	08 N 26	25 N 25	06 N 54	01 N 57	20 S 23	15 N 14	02 N 02	03 N 08	16 S 34	17 N 20
02	08 04	21 47	06 08	02 10	20 32	15 10	02 00	03 07	16 34	17 19
03	07 43	16 48	05 21	02 23	20 42	15 06	01 59	03 05	16 34	17 18
04	07 20	10 45	04 34	02 37	20 51	15 02	01 57	03 04	16 35	17 17
05	06 58	04 N 02	03 47	02 51	21 01	14 58	01 55	03 02	16 35	17 16
06	06 36	02 S 57	03 00	03 05	21 10	14 54	01 54	03 01	16 35	17 16
07	06 14	09 46	02 14	03 20	21 19	14 50	01 52	03 00	16 36	17 15
08	05 51	15 59	01 27	03 34	21 28	14 46	01 50	02 58	16 36	17 14
09	05 29	21 13	00 N 41	03 49	21 37	14 42	01 49	02 57	16 36	17 13
10	05 06	25 07	00 S 04	04 03	21 45	14 38	01 47	02 55	16 37	17 12
11	04 43	27 27	00 50	04 17	21 54	14 34	01 45	02 54	16 37	17 11
12	04 20	28 05	01 35	04 31	22 02	14 30	01 44	02 52	16 37	17 11
13	03 57	27 04	02 20	04 45	22 11	14 26	01 42	02 51	16 38	17 10
14	03 34	24 36	03 04	04 58	22 19	14 22	01 40	02 49	16 38	17 09
15	03 11	20 57	03 48	05 11	22 27	14 18	01 38	02 48	16 38	17 08
16	02 48	16 24	04 31	05 24	22 34	14 14	01 36	02 46	16 39	17 07
17	02 25	11 15	05 14	05 36	22 42	14 10	01 35	02 45	16 39	17 06
18	02 02	05 44	05 56	05 47	22 50	14 06	01 33	02 43	16 40	17 06
19	01 39	00 S 05	06 38	05 58	22 57	14 02	01 31	02 42	16 40	17 05
20	01 15	05 N 32	07 19	06 08	23 04	13 59	01 29	02 40	16 40	17 04
21	00 52	10 55	08 00	06 18	23 11	13 55	01 27	02 39	16 41	17 03
22	00 29	15 54	08 39	06 27	23 18	13 51	01 25	02 37	16 41	17 03
23	00 N 05	20 18	09 19	06 35	23 24	13 47	01 23	02 36	16 42	17 02
24	00 S 18	23 56	09 57	06 43	23 31	13 43	01 22	02 34	16 42	17 01
25	00 41	26 33	10 35	06 50	23 37	13 39	01 20	02 33	16 43	17 00
26	01 05	27 59	11 12	06 56	23 43	13 35	01 18	02 31	16 43	17 00
27	01 28	28 01	11 48	07 01	23 49	13 31	01 16	02 30	16 44	16 59
28	01 51	26 33	12 24	07 06	23 54	13 28	01 14	02 28	16 44	16 58
29	02 15	23 34	12 58	07 10	24 00	13 24	01 12	02 27	16 44	16 57
30	02 S 38	19 N 11	13 S 32	07 N 13	24 S 05	13 N 20	01 N 10	02 N 25	16 S 45	16 N 57

ZODIAC SIGN ENTRIES

Date	h	m	Planets
01	14	08	☽ ♌
03	17	07	☽ ♍
05	18	03	☽ ♎
07	18	44	☽ ♏
09	11	58	♀ ♌
09	16	53	☿ ♎
09	20	40	☽ ♐
10	01	44	♂ ♐
12	00	43	☽ ♑
14	07	08	☽ ♒
16	15	53	☽ ♓
19	02	46	☽ ♈
21	15	20	☽ ♉
23	17	38	☉ ♎
24	04	21	☽ ♊
26	15	45	☽ ♋
28	23	41	☽ ♌
30	01	46	☿ ♏

LATITUDES

Date	Mercury ☿ ° '	Venus ♀ ° '	Mars ♂ ° '	Jupiter ♃ ° '	Saturn ♄ ° '	Uranus ♅ ° '	Neptune ♆ ° '	Pluto ♇ ° '
01	01 N 23	08 S 35	01 S 32	00 N 40	02 S 38	00 N 43	01 N 43	14 N 35
04	01 07	08 32	01 33	00 41	02 39	00 43	01 43	14 35
07	00 49	08 21	01 34	00 41	02 39	00 43	01 43	14 35
10	00 29	08 04	01 35	00 41	02 40	00 43	01 43	14 35
13	00 N 08	07 41	01 35	00 42	02 40	00 43	01 42	14 35
16	00 S 14	07 14	01 36	00 42	02 41	00 43	01 42	14 35
19	00 37	06 44	01 37	00 42	02 41	00 43	01 42	14 36
22	01 00	06 12	01 37	00 43	02 41	00 43	01 42	14 36
25	01 23	05 38	01 37	00 43	02 41	00 43	01 42	14 36
28	01 46	05 05	01 38	00 44	02 42	00 43	01 42	14 37
31	02 S 07	04 S 32	01 S 38	00 N 44	02 S 42	00 N 43	01 N 42	14 N 37

DATA

Julian Date	2439735
Delta T	+38 seconds
Ayanamsa	23° 24' 12"
Synetic vernal point	05° ♓ 42' 48"
True obliquity of ecliptic	23° 26' 45"

LONGITUDES

Date	Chiron ⚷ ° '	Ceres ⚳ ° '	Pallas ⚴ ° '	Juno ⚵ ° '	Vesta ⚶ ° '	Black Moon Lilith ⚸ ° '
01	28 ♓ 24	03 ♍ 48	07 ♌ 29	22 ♍ 25	01 ♐ 48	27 ♈ 41
11	27 ♓ 58	08 ♍ 23	13 ♌ 11	26 ♍ 18	05 ♐ 38	28 ♈ 48
21	27 ♓ 31	12 ♍ 58	18 ♌ 46	00 ♎ 10	09 ♐ 44	29 ♈ 54
31	27 ♓ 04	17 ♍ 31	24 ♌ 11	04 ♎ 00	14 ♐ 03	01 ♉ 00

All ephemeris data is given at 12.00 UT and the Moon's longitude is additionally given for 24.00 UT

MOON'S PHASES, APSIDES AND POSITIONS ☽

Date	h	m	Phase	Longitude ° '	Eclipse Indicator
04	11	37	●	11 ♍ 15	
11	03	06	☽	17 ♐ 42	
18	16	59	○	25 ♓ 05	
26	21	44	☾	03 ♋ 06	

Day	h	m		
06	07	31	Perigee	
21	23	56	Apogee	
06	01	51	0S	
12	09	03	Max dec	28° S 06'
19	12	19	0N	
27	00	39	Max dec	28° N 11'

ASPECTARIAN

h m Aspects

01 Friday
01 40 ☽ ∠ ☉
03 05 ☽ ✶ ♅
04 01 ☽ △ ♂
11 06 ☽ ⊥ ♀
15 54 ☽ ∠ ☿
18 44 ☽ ⊥ ☉
19 45 ♀ ∥ ♄
21 09 ☽ ⊻ ♀
22 52 ☽ ∠ ♇

02 Saturday
05 25 ☽ ∠ ♅
05 54 ☽ ⊻ ☉
09 33 ☽ △ ♄
09 48 ☽ ⊥ ☿
14 21 ☽ ⊥ ♇
18 24 ☽ # ♂
20 49 ☽ ⊥ ♅
21 29 ☽ ⊻ ☿

03 Sunday
00 19 ☉ Q ♆
00 32 ☽ ⊻ ♇
02 11 ☽ ☌ ♃
03 31 ☽ □ ♆
06 56 ☽ ⊻ ♅
09 48 ☽ ∥ ♇
09 59 ☽ □ ♂
10 44 ☽ ⚼ ♄
12 58 ☽ # ♆
19 12 ☽ ∥ ♃
21 47 ☽ ☌ ♀
22 15 ☿ ☌ ♇

04 Monday
01 34 ☽ ± ♄
01 44 ♂ ⚼ ♄
07 50 ☉ ⊼ ♄
09 19 ☽ Q ♆
11 21 ☽ ⊼ ♄
11 37 ☽ ☌ ☉
15 45 ☿ ⊻ ♃
16 57 ☽ Q ♂
22 25 ☿ ✶ ♆

05 Tuesday
01 06 ☽ ∥ ☉
02 03 ☽ ☌ ♇
04 15 ☽ ⊻ ♃
04 53 ☽ ✶ ♆
05 44 ☽ ☌ ☿
08 20 ☽ ☌ ♅
12 57 ☽ ∥ ☿
13 18 ☽ ✶ ♂
14 07 ☽ ⊥ ♃
15 25 ☽ ∥ ♅
15 56 ☽ ∥ ♀
19 17 ☽ ∥ ♄
20 52 ☽ ⊻ ♀

06 Wednesday
04 38 ☿ ☌ ♅
04 53 ☽ ∠ ♃
05 10 ☽ ∠ ♆
05 17 ♀ ∥ ♅
06 14 ☽ ⊥ ♀
08 23 ☽ # ♄
09 57 ☿ ∥ ♀
11 35 ☿ ∥ ♅
11 47 ☽ ☍ ♄
12 09 ☽ # ☿
12 13 ☽ # ♅
12 29 ☽ # ♀
14 40 ☽ ∠ ♂
15 39 ☽ ⊻ ☉
19 46 ☽ ⊥ ♆
20 19 ☽ ∠ ♀

07 Thursday
00 01 ☽ # ☉
02 04 ☽ ⊥ ☉
02 45 ☽ ⊻ ♇
03 32 ♃ □ ♆
05 31 ☽ ⊻ ♆
05 33 ☽ ✶ ♃
05 56 ☽ ⊥ ♂
09 08 ☽ ⊻ ♅
12 32 ☽ ⊥ ♇
12 46 ☽ ⊻ ☿
16 10 ☽ ⊻ ♂
17 46 ☽ ∠ ☉
18 59 ☽ ⊥ ♅
19 58 ☽ ✶ ♀
23 33 ☿ ∥ ♄
23 52 ☽ ⊥ ☿

08 Friday
01 26 ☽ Q ♃
03 19 ☽ ∠ ♇
07 06 ☽ # ♃
09 50 ☽ ∠ ♅
12 33 ☽ ⊼ ♄
14 35 ☽ ∥ ♆
15 07 ☽ Q ♀
15 11 ☿ ⊥ ♃
16 40 ☽ ∠ ☿
17 16 ☽ # ♇
20 17 ☽ ✶ ☉
22 28 ☽ ± ♄

09 Saturday
04 16 ☽ ✶ ♇
07 04 ☽ ☌ ♆
07 46 ☽ □ ♃
10 57 ☽ ✶ ♅
11 09 ☿ ✶ ♂
13 29 ☽ ⚼ ♄
14 10 ☽ ∥ ♂
15 53 ☿ ⊻ ♀
17 53 ☽ Q ☉
20 22 ♀ □ ♂
20 25 ☽ □ ♀
20 25 ☽ ☌ ♂
21 10 ☽ ✶ ☿

10 Sunday
00 34 ☽ Q ♇
07 25 ☽ Q ♅
14 56 ☽ △ ♄
20 29 ☽ Q ☿

11 Monday
03 06 ☽ □ ☉
07 43 ☽ □ ♇
10 34 ☽ ⊻ ♆
12 01 ☽ △ ♃
14 48 ☽ □ ♅
21 09 ☽ ⊥ ♆
23 03 ☽ △ ♀

12 Tuesday
03 07 ☽ ⊻ ♂
04 47 ☉ ∥ ♀
08 25 ☽ □ ☿
13 12 ☽ ∠ ♆
14 23 ☽ ⊥ ♂
15 04 ☽ ⚼ ♃
15 59 ☿ ⊥ ♀
16 24 ☿ # ♄
19 33 ☽ □ ♄

13 Wednesday
01 18 ☽ ⚼ ♀
07 30 ☽ ∠ ♂
07 36 ☽ ± ♃
12 49 ☽ △ ☉
13 31 ☽ △ ♇
16 26 ☽ ✶ ♆
17 22 ☽ ± ♀
18 33 ♂ Q ♇
18 45 ☽ ⊼ ♃
21 03 ☽ △ ♅
22 41 ☉ ☌ ♇

14 Thursday
01 07 ☿ ∠ ♆
04 09 ☿ # ♅
04 12 ☽ ⊼ ♀
04 15 ☽ Q ♄
07 03 ☉ Q ♂
12 36 ☽ ✶ ♂
14 40 ☽ Q ♆
17 19 ☽ ⚼ ♇
18 48 ☽ ⚼ ☉
22 55 ☽ △ ☿
22 56 ☉ # ☿

15 Friday
00 04 ☿ ∠ ♃
01 04 ☽ ⚼ ♅
02 30 ☽ ✶ ♄
03 17 ☽ ∥ ♂
10 17 ☽ ± ♇
12 21 ☽ Q ♂
13 08 ☽ ± ☉
14 41 ☉ ✶ ♆
18 09 ☽ ± ♅
21 42 ☽ ⊼ ♇

16 Saturday
00 41 ☽ □ ♆
01 30 ☽ ⊼ ☉
03 20 ☿ ☍ ♄
03 55 ☽ ☍ ♃
05 39 ☽ ⊼ ♅
06 49 ☽ ∠ ♄
07 20 ☽ ⚼ ☿
08 27 ☽ # ♇
10 48 ☽ ∥ ♆
11 55 ☽ ☍ ♀
14 33 ☉ ∥ ♅
22 30 ☽ # ♃

17 Sunday
00 04 ☽ ⊥ ♄
00 48 ☽ □ ♂
03 09 ☽ ± ☿
11 41 ☽ ⊻ ♄
16 18 ☉ ⊻ ♃
16 30 ☽ ⊼ ☿
18 13 ☿ ∠ ♀

18 Monday
05 00 ☿ # ♀
08 07 ☽ ☍ ♇
10 25 ☉ ☌ ♅
11 08 ☽ △ ♆
11 14 ☽ ∥ ☿
11 47 ☽ # ♀
15 23 ☽ ⊼ ♃
16 30 ☽ ☍ ♅
16 59 ☽ ☍ ☉
22 09 ☽ ⊼ ♀

19 Tuesday
00 53 ☽ # ♅
03 32 ☽ ± ♃
04 52 ☽ # ☉
05 53 ☽ # ♄
10 04 ☽ ± ♀
15 31 ☽ △ ♂
17 06 ☽ ⚼ ♆
18 43 ☽ ∥ ♄
18 50 ☽ ∥ ☉
20 48 ☉ ∥ ♄
21 53 ☽ ⚼ ♃
22 51 ☽ ☌ ♄
23 43 ☽ ∥ ♅

20 Wednesday
00 49 ☿ ⊥ ♆
04 08 ☽ ⚼ ♀
09 34 ♀ St D
11 17 ☽ ± ♆
12 06 ♂ Q ♅
12 44 ☽ ☍ ☿
14 43 ☽ ∥ ♀
20 26 ☽ ⊼ ♇
20 55 ☽ # ☿
23 28 ☽ ⊼ ♆
23 38 ☽ ⚼ ♂

21 Thursday
04 45 ☽ △ ♃
05 10 ☽ ⊼ ♅
08 03 ☉ ⊻ ♀
08 39 ☽ ± ♇
10 34 ☽ △ ♀
10 47 ☽ ⊼ ☉
16 06 ♃ ⚼ ♄
17 26 ☽ ± ♅
19 04 ☽ ± ♂

22 Friday
00 07 ☽ ± ☉
01 56 ☽ ∥ ♃
03 02 ☽ ⚼ ♇
08 02 ☽ ⊼ ♂
11 22 ☽ ⊻ ♄
11 52 ☽ ⚼ ♅
15 52 ♃ ⊻ ♅
16 03 ☽ # ♆
17 52 ☽ ∥ ♇
20 07 ☽ ⚼ ☉
23 31 ☽ ⊥ ♄

23 Saturday
05 28 ☿ ⊻ ♇
09 41 ☽ △ ♇
10 14 ☽ ⊼ ☿
12 40 ☽ ☍ ♆
17 41 ☽ ∠ ♄
18 33 ☽ △ ♅
18 52 ☽ □ ♃

24 Sunday
00 03 ☽ ± ☿
00 05 ☽ □ ♀
05 19 ☽ △ ☉
07 01 ☿ ⊻ ♆
08 49 ☽ # ♂
11 58 ♂ △ ♄
20 34 ☽ ⚼ ☿
23 43 ☽ ✶ ♄

25 Monday
00 30 ☽ ☍ ♂
07 38 ☽ Q ♃
09 32 ☉ ⊥ ♃
12 39 ☽ Q ♀
22 00 ☽ □ ♇
23 20 ☽ Q ♄

26 Tuesday
00 49 ☽ ⊼ ♆
06 00 ☽ △ ☿
06 40 ☽ □ ♅
07 38 ☽ ✶ ♃
12 23 ☿ ⊻ ♅
12 32 ☽ ± ♆
12 34 ☽ ✶ ♀
21 44 ☽ □ ☉
23 11 ☿ ✶ ♃

27 Wednesday
00 32 ☉ # ♄
05 50 ☽ ⚼ ♆
08 49 ☽ Q ♇
09 48 ☽ □ ♄
12 49 ☽ ∠ ♃
14 14 ☽ ⊼ ♂
15 24 ☿ ∠ ♂
17 15 ☽ Q ♅
17 20 ☿ ⊥ ♇
17 43 ☽ ∠ ♀

28 Thursday
02 04 ☽ ± ♂
05 50 ☽ ⊥ ♃
07 16 ☽ ✶ ♇
09 53 ☽ △ ♆
10 16 ☽ Q ☉
10 44 ☽ ⊥ ♀
15 29 ☽ ✶ ♅
16 58 ☽ ⊻ ♃
19 03 ☉ ⊥ ♀
19 27 ☽ ⚼ ♂
20 54 ☽ □ ☿
21 53 ☽ ⊻ ♀

29 Friday
09 14 ☽ # ♂
09 36 ☽ ✶ ☉
09 56 ☿ ✶ ♀
10 23 ☽ ∠ ♇
16 09 ☽ △ ♄
18 21 ☽ ∠ ♅
23 24 ☉ # ♅
23 27 ☽ △ ♂

30 Saturday
02 11 ☽ ⊥ ♇
04 11 ☿ # ♃
07 51 ☽ Q ☿
10 00 ☽ ⊥ ♅
12 31 ☽ ⊻ ♇
13 38 ☽ ∠ ☉
14 54 ☽ □ ♆
17 51 ☽ ⚼ ♄
20 15 ☽ ⊻ ♅
22 06 ☽ ☌ ♃
22 09 ☽ ∥ ♇
22 58 ☽ # ♆

OCTOBER 1967

LONGITUDES

Date	Sidereal time h m s	Sun ☉ ° ' "	Moon ☽ ° ' "	Moon ☽ 24.00 ° ' "	Mercury ☿ ° '	Venus ♀ ° '	Mars ♂ ° '	Jupiter ♃ ° '	Saturn ♄ ° '	Uranus ♅ ° '	Neptune ♆ ° '	Pluto ♇ ° '
01	12 38 08	07 ♎ 37 11	05 ♍ 01 21	12 ♍ 18 52	01 ♏ 47	29 ♌ 54	14 ♐ 31	26 ♌ 48	09 ♈ 08	25 ♍ 38	22 ♏ 28	21 ♍ 05
02	12 42 05	08 36 13	19 ♍ 41 37	27 ♍ 08 46	03 01	00 ♍ 18	15 13	27 00	09 R 03	25 42	22 30	21 07
03	12 46 01	09 35 17	04 ♎ 39 18	12 ♎ 12 09	04 13	00 44	15 55	27 11	08 58	25 46	22 32	21 09
04	12 49 58	10 34 23	19 ♎ 46 07	27 ♎ 20 04	05 23	01 12	16 37	27 22	08 53	25 50	22 34	21 11
05	12 53 55	11 33 32	04 ♏ 52 50	12 ♏ 23 21	06 31	01 41	17 20	27 33	08 49	25 53	22 35	21 13
06	12 57 51	12 32 42	19 ♏ 50 38	27 ♏ 13 52	07 38	02 11	18 02	27 44	08 44	25 57	22 37	21 15
07	13 01 48	13 31 55	04 ♐ 32 20	11 ♐ 45 32	08 42	02 44	18 45	27 55	08 39	26 01	22 39	21 17
08	13 05 44	14 31 09	18 ♐ 53 04	25 ♐ 54 44	09 44	03 17	19 27	28 06	08 35	26 04	22 41	21 19
09	13 09 41	15 30 25	02 ♑ 50 26	09 ♑ 40 12	10 43	03 52	20 10	28 17	08 30	26 08	22 43	21 21
10	13 13 37	16 29 42	16 ♑ 24 11	23 ♑ 02 34	11 39	04 28	20 53	28 28	08 25	26 12	22 45	21 23
11	13 17 34	17 29 02	29 ♑ 35 38	06 ♒ 03 43	12 33	05 06	21 36	28 39	08 21	26 15	22 47	21 25
12	13 21 30	18 28 23	12 ♒ 27 09	18 ♒ 46 18	13 23	05 45	22 19	28 49	08 16	26 19	22 48	21 27
13	13 25 27	19 27 46	25 ♒ 01 33	01 ♓ 13 15	14 09	06 25	23 02	29 00	08 12	26 22	22 50	21 29
14	13 29 24	20 27 10	07 ♓ 21 46	13 ♓ 27 28	14 52	07 06	23 45	29 10	08 07	26 26	22 52	21 31
15	13 33 20	21 26 37	19 ♓ 30 40	25 ♓ 31 42	15 30	07 48	24 28	29 20	08 02	26 29	22 54	21 33
16	13 37 17	22 26 05	01 ♈ 30 50	07 ♈ 28 24	16 03	08 31	25 12	29 30	07 58	26 33	22 56	21 35
17	13 41 13	23 25 35	13 ♈ 24 40	19 ♈ 19 53	16 31	09 15	25 55	29 41	07 54	26 36	22 58	21 37
18	13 45 10	24 25 07	25 ♈ 14 20	01 ♉ 08 18	16 53	10 01	26 39	29 ♌ 51	07 49	26 40	23 00	21 39
19	13 49 06	25 24 41	07 ♉ 02 02	12 ♉ 55 49	17 09	10 47	27 23	00 ♍ 00	07 45	26 43	23 02	21 41
20	13 53 03	26 24 18	18 ♉ 49 57	24 ♉ 44 45	17 18	11 34	28 06	00 10	07 40	26 47	23 04	21 43
21	13 56 59	27 23 56	00 ♊ 40 33	06 ♊ 37 41	17 R 20	12 22	28 50	00 20	07 36	26 50	23 06	21 45
22	14 00 56	28 23 37	12 ♊ 36 33	18 ♊ 37 31	17 14	13 11	29 ♐ 34	00 30	07 32	26 53	23 09	21 46
23	14 04 53	29 ♎ 23 20	24 ♊ 41 02	00 ♋ 47 32	16 59	14 01	00 ♑ 18	00 39	07 28	26 57	23 11	21 48
24	14 08 49	00 ♏ 23 05	06 ♋ 57 28	13 ♋ 11 18	16 36	14 51	01 02	00 49	07 23	27 00	23 13	21 50
25	14 12 46	01 22 52	19 ♋ 29 30	25 ♋ 52 33	16 03	15 43	01 46	00 58	07 19	27 03	23 15	21 52
26	14 16 42	02 22 42	02 ♌ 20 52	08 ♌ 54 53	15 22	16 35	02 30	01 07	07 15	27 07	23 17	21 54
27	14 20 39	03 22 34	15 ♌ 34 56	22 ♌ 21 18	14 32	17 28	03 15	01 16	07 11	27 10	23 19	21 55
28	14 24 35	04 22 28	29 ♌ 14 12	06 ♍ 13 41	13 33	18 21	03 59	01 25	07 07	27 13	23 21	21 57
29	14 28 32	05 22 24	13 ♍ 19 42	20 ♍ 32 03	12 27	19 15	04 43	01 34	07 04	27 16	23 23	21 59
30	14 32 28	06 22 22	27 ♍ 50 20	05 ♎ 14 02	11 16	20 10	05 28	01 43	07 00	27 20	23 26	22 01
31	14 36 25	07 ♏ 22 22	12 ♎ 42 24	20 ♎ 14 33	10 ♏ 00	21 ♍ 06	06 ♑ 12	01 ♍ 51	06 ♈ 56	27 ♍ 23	23 ♏ 28	22 ♍ 02

Date	Moon True ☊ ° '	Moon Mean ☊ ° '	Moon ☽ Latitude ° '
01	28 ♈ 02	28 ♈ 50	04 N 12
02	27 R 59	28 47	03 16
03	27 57	28 44	02 05
04	27 56	28 41	00 N 45
05	27 D 56	28 38	00 S 38
06	27 57	28 35	01 58
07	27 58	28 31	03 09
08	27 59	28 28	04 06
09	28 00	28 25	04 47
10	28 00	28 22	05 11
11	28 R 00	28 19	05 17
12	27 59	28 15	05 07
13	27 58	28 12	04 43
14	27 57	28 09	04 05
15	27 56	28 06	03 18
16	27 55	28 03	02 22
17	27 55	28 00	01 20
18	27 55	27 56	00 S 15
19	27 D 55	27 53	00 N 51
20	27 55	27 50	01 54
21	27 55	27 47	02 52
22	27 R 55	27 44	03 44
23	27 55	27 41	04 26
24	27 55	27 37	04 57
25	27 54	27 34	05 14
26	27 54	27 31	05 17
27	27 D 54	27 28	05 03
28	27 55	27 25	04 32
29	27 55	27 21	03 43
30	27 56	27 18	02 40
31	27 ♈ 57	27 ♈ 15	01 N 24

DECLINATIONS

Date	Sun ☉ ° '	Moon ☽ ° '	Mercury ☿ ° '	Venus ♀ ° '	Mars ♂ ° '	Jupiter ♃ ° '	Saturn ♄ ° '	Uranus ♅ ° '	Neptune ♆ ° '	Pluto ♇ ° '
01	03 S 01	13 N 35	14 S 05	07 N 16	24 S 10	13 N 16	01 N 08	02 N 24	16 S 45	16 N 56
02	03 25	07 05	14 37	07 18	24 15	13 13	01 07	02 22	16 46	16 55
03	03 48	00 N 04	15 08	07 19	24 20	13 09	01 05	02 21	16 46	16 55
04	04 11	07 S 02	15 37	07 19	24 24	13 05	01 03	02 19	16 47	16 54
05	04 34	13 45	16 06	07 19	24 28	13 01	01 01	02 18	16 47	16 54
06	04 57	19 36	16 34	07 18	24 32	12 58	00 59	02 16	16 48	16 53
07	05 21	24 09	17 00	07 16	24 36	12 54	00 57	02 15	16 48	16 52
08	05 43	27 04	17 25	07 14	24 39	12 51	00 55	02 13	16 49	16 52
09	06 06	28 12	17 48	07 11	24 43	12 47	00 54	02 12	16 49	16 51
10	06 29	27 35	18 10	07 07	24 46	12 43	00 52	02 11	16 50	16 50
11	06 52	25 24	18 31	07 02	24 49	12 40	00 50	02 09	16 50	16 50
12	07 15	21 59	18 50	06 57	24 51	12 36	00 48	02 08	16 51	16 49
13	07 37	17 37	19 07	06 52	24 53	12 33	00 46	02 06	16 51	16 49
14	08 00	12 36	19 22	06 45	24 56	12 29	00 45	02 05	16 52	16 48
15	08 22	07 11	19 35	06 38	24 57	12 26	00 43	02 04	16 52	16 48
16	08 44	01 S 34	19 45	06 31	24 59	12 22	00 41	02 02	16 53	16 47
17	09 06	04 N 04	19 54	06 23	25 00	12 19	00 40	02 01	16 53	16 47
18	09 28	09 32	19 59	06 14	25 02	12 16	00 38	01 59	16 54	16 46
19	09 50	14 40	20 02	06 05	25 02	12 12	00 36	01 58	16 55	16 46
20	10 11	19 15	20 02	05 55	25 03	12 09	00 35	01 57	16 55	16 45
21	10 33	23 06	19 58	05 44	25 03	12 06	00 33	01 55	16 56	16 45
22	10 54	26 01	19 51	05 33	25 04	12 03	00 31	01 54	16 56	16 44
23	11 16	27 46	19 40	05 21	25 03	11 59	00 30	01 53	16 57	16 44
24	11 37	28 09	19 25	05 09	25 03	11 56	00 28	01 51	16 57	16 44
25	11 57	27 13	19 06	04 56	25 02	11 53	00 27	01 50	16 58	16 43
26	12 18	24 47	18 42	04 43	25 02	11 50	00 25	01 49	16 58	16 43
27	12 39	20 59	18 14	04 29	25 00	11 47	00 24	01 48	16 59	16 42
28	12 59	15 59	17 41	04 15	24 59	11 44	00 22	01 46	16 59	16 42
29	13 19	09 59	17 05	04 01	24 57	11 41	00 21	01 45	17 00	16 42
30	13 39	03 18	16 25	03 45	24 55	11 38	00 20	01 44	17 01	16 41
31	13 S 59	03 S 44	15 S 42	03 N 30	24 S 53	11 N 35	00 N 18	01 N 43	17 S 01	16 N 41

ZODIAC SIGN ENTRIES

Date	h	m	Planets
01	03	38	☽ ♍
01	18	07	♀ ♍
03	04	34	☽ ♎
05	04	14	☽ ♏
07	04	32	☽ ♐
09	07	04	☽ ♑
11	12	45	☽ ♒
13	21	38	☽ ♓
16	08	58	☽ ♈
18	21	41	☽ ♉
19	10	52	♃ ♍
21	10	38	☽ ♊
23	02	14	♂ ♑
23	22	27	☽ ♋
24	02	44	☉ ♏
26	07	40	☽ ♌
28	13	19	☽ ♍
30	15	31	☽ ♎

LATITUDES

Date	Mercury ☿ ° '	Venus ♀ ° '	Mars ♂ ° '	Jupiter ♃ ° '	Saturn ♄ ° '	Uranus ♅ ° '	Neptune ♆ ° '	Pluto ♇ ° '
01	02 S 07	04 S 32	01 S 38	00 N 44	02 S 42	00 N 43	01 N 42	14 N 37
04	02 27	04 00	01 38	00 45	02 42	00 43	01 41	14 38
07	02 44	03 28	01 38	00 45	02 42	00 43	01 41	14 38
10	02 59	02 58	01 38	00 46	02 42	00 43	01 41	14 39
13	03 10	02 28	01 38	00 46	02 42	00 43	01 41	14 40
16	03 15	02 01	01 38	00 47	02 41	00 43	01 41	14 41
19	03 12	01 34	01 37	00 47	02 41	00 43	01 41	14 42
22	03 00	01 09	01 37	00 48	02 41	00 43	01 41	14 43
25	02 33	00 45	01 36	00 48	02 41	00 44	01 41	14 44
28	01 52	00 23	01 36	00 49	02 40	00 44	01 41	14 45
31	00 S 56	00 S 02	01 S 35	00 N 49	02 S 40	00 N 44	01 N 41	14 N 46

DATA

Julian Date	2439765
Delta T	+38 seconds
Ayanamsa	23° 24' 15"
Synetic vernal point	05° ♓ 42' 45"
True obliquity of ecliptic	23° 26' 45"

MOON'S PHASES, APSIDES AND POSITIONS ☽

Date	h	m	Phase	Longitude ° '	Eclipse Indicator
03	20	24	●	09 ♎ 56	
10	12	11	☽	16 ♑ 30	
18	10	11	○	24 ♈ 21	total
26	12	04	☾	02 ♌ 23	

Day	h	m		
04	14	10	Perigee	
19	07	13	Apogee	
03	12	13	0S	
09	15	10	Max dec	28° S 13'
16	18	38	0N	
24	07	29	Max dec	28° N 14'
30	23	19	0S	

LONGITUDES

Date	Chiron ⚷ ° '	Ceres ⚳ ° '	Pallas ⚴ ° '	Juno ⚵ ° '	Vesta ⚶ ° '	Black Moon Lilith ⚸ ° '
01	27 ♓ 04	17 ♍ 31	24 ♌ 11	04 ♎ 00	14 ♐ 03	01 ♉ 00
11	26 ♓ 38	22 ♍ 01	29 ♌ 26	07 ♎ 47	18 ♐ 33	02 ♉ 07
21	26 ♓ 14	26 ♍ 29	04 ♍ 28	11 ♎ 10	23 ♐ 12	03 ♉ 13
31	25 ♓ 54	00 ♎ 52	09 ♍ 16	15 ♎ 10	27 ♐ 59	04 ♉ 20

ASPECTARIAN

h m Aspects

01 Sunday
03 14 ☽ ☌ ♀
05 56 ☽ ⊥ ☉
06 07 ☽ ⚹ ☿
08 18 ☉ ∠ ♆
08 52 ☽ ± ♄
08 56 ☿ ⊥ ♅
10 10 ☽ # ☿
13 13 ☽ ∥ ♃
16 36 ☽ ⊻ ☉
18 44 ☽ ⊼ ♄
21 00 ☽ Q ♆

02 Monday
04 23 ☽ □ ♂
09 02 ☽ ∠ ☿
11 16 ☽ ∥ ♀
14 18 ☽ ☌ ♇
16 33 ☽ ⚹ ♆
21 43 ☽ ☌ ♅
22 03 ☉ ☍ ♄
23 55 ☽ ⊻ ♃

03 Tuesday
00 02 ☽ # ☉
00 49 ☽ ⊥ ☿
04 17 ☽ ∥ ♅
05 33 ☽ ⊻ ♀
08 35 ☽ ∥ ♄
09 38 ☽ ⊥ ♃
10 46 ☽ Q ♂
11 14 ☽ ⊻ ☿
15 25 ☽ ⊥ ♀
15 50 ☽ # ♄
16 35 ☽ ∠ ♆
18 50 ☽ ☍ ♄
20 04 ☽ # ♅
20 24 ☽ ☌ ☉

04 Wednesday
00 07 ☽ ∠ ♃
01 45 ☽ ∥ ☉
06 09 ☽ ∠ ♀
06 46 ☽ ⚹ ♂
06 54 ☽ ⊥ ♆
12 57 ☽ # ♀
14 15 ☽ ⊻ ♇
16 26 ☽ ⊻ ♆
21 39 ☽ ⊻ ♅
23 47 ☽ ⊥ ♇

05 Thursday
00 13 ☽ ⚹ ♃
05 18 ☿ ∠ ♇
06 44 ☽ ⚹ ♀
07 13 ☽ ⊥ ♅
07 44 ☽ ∠ ♂
09 17 ☽ # ♃
14 08 ☽ ∠ ♇
14 50 ☽ ☌ ☿
18 15 ☽ ⊼ ♄
19 33 ☽ Q ♃
21 38 ☽ ∠ ♅
21 52 ☽ ∥ ☿
22 49 ☽ ⊥ ♂
23 25 ☽ ⊻ ☉
23 54 ☽ ∥ ♆

06 Friday
00 17 ☽ # ♇
02 34 ☽ Q ♀
03 48 ☽ ± ♄
08 56 ☽ ⊻ ♂
09 45 ☽ ⊥ ☉
14 17 ☽ ⚹ ♇
16 30 ☽ ☌ ♆
17 52 ☉ ∠ ♃
18 16 ☽ ꝑ ♄
21 57 ☽ ⚹ ♅

07 Saturday
01 00 ☽ □ ♃
01 08 ☿ ∥ ♆
01 25 ☽ ∠ ☉
05 08 ☿ # ♇
08 54 ☽ □ ♀
09 19 ♀ ± ♄
09 56 ☽ Q ♇
11 10 ☿ ⊼ ♄
14 57 ☽ ∥ ♂
17 46 ☽ Q ♅
18 47 ☽ △ ♄
19 26 ☽ ⊻ ☿

08 Sunday
04 05 ☽ ⚹ ☉
06 15 ☽ ⊥ ☿
13 01 ☽ ☌ ♂
16 09 ☽ □ ♇
18 29 ☽ ⊻ ♆
22 44 ☽ ∠ ☿
23 09 ☿ Q ♃

09 Monday
00 20 ☽ □ ♅
02 02 ☽ Q ☉
03 59 ☽ △ ♃
04 49 ☽ ⊥ ♆
13 52 ☽ △ ♀
20 34 ☽ ∠ ♆
21 53 ☽ □ ♄
23 14 ☿ ∠ ♅

10 Tuesday
02 54 ☽ ⚹ ☿
06 40 ☽ ꝑ ♃
12 11 ☽ □ ☉
17 48 ☽ ꝑ ♀
18 15 ☉ ⊥ ♆
20 32 ☽ ⊻ ♂
21 01 ☽ △ ♇
23 06 ☽ ± ♃
23 29 ☽ ⚹ ♆

11 Wednesday
01 56 ♆ # ♇
02 05 ☽ Q ☿
05 48 ♂ □ ♇
05 50 ☽ △ ♅
06 04 ☽ Q ♄
08 07 ☽ ⊥ ♂
10 13 ☽ ⊼ ♃
11 02 ☽ ± ♀
16 44 ☽ ∥ ♂
21 12 ☉ # ♀
21 38 ☽ Q ♆
22 44 ☽ ⊼ ♀

12 Thursday
00 42 ☽ ꝑ ♇
01 46 ☽ ∠ ♂
04 11 ☽ ⚹ ♄
09 50 ☽ ꝑ ♅
13 52 ☽ □ ☿
17 42 ☽ ± ♇

13 Friday
00 24 ☽ △ ☉
03 01 ☽ ± ♅
04 41 ☽ ∥ ☿
05 11 ☽ ⊼ ♇
05 15 ♂ ⊻ ♆
07 47 ☽ □ ♆
07 56 ☽ ⚹ ♂
08 30 ☽ ∠ ♄
13 06 ☿ ± ♄
14 37 ☽ ⊼ ♅
15 48 ☽ ∥ ♆
16 02 ☽ # ♇
19 47 ☽ ☍ ♃

14 Saturday
01 48 ☽ ⊥ ♄
07 56 ☽ ꝑ ☉
08 39 ☽ Q ♂
11 26 ☽ ☍ ♀
12 32 ☽ # ♃
13 28 ☽ ⊻ ♄

15 Sunday
03 13 ☽ ± ☉
03 37 ☽ △ ☿
07 12 ☽ ∥ ☉
14 23 ☽ # ♀
14 45 ☉ ⊻ ♇
16 05 ☽ ☍ ♇
16 11 ☽ ⊼ ☉
18 47 ☽ △ ♆
19 25 ♀ ⊼ ♄
22 32 ☽ □ ♂

16 Monday
01 59 ☽ ☍ ♅
07 55 ☽ ⊼ ♃
09 59 ☽ # ♅
11 02 ☽ ꝑ ☿
15 44 ☽ # ♄
18 07 ♄ ꝑ ♆
20 09 ☽ ± ♃
21 30 ☽ ∥ ♄

17 Tuesday
00 38 ☉ ⊻ ♆
00 55 ☽ ☌ ♄
00 59 ☽ ꝑ ♆
03 02 ☽ ⊼ ♀
03 14 ☽ ∥ ♅
05 56 ☽ ± ☿
14 36 ☽ ꝑ ♃
15 59 ☽ ± ♀
18 31 ☽ ⊼ ☿
19 14 ☽ ± ♆
21 45 ☽ ∥ ♀

18 Wednesday
04 41 ☽ ⊼ ♇
07 27 ☽ ⊼ ♆
10 11 ☽ ☍ ☉
11 30 ☽ ꝑ ♀
11 40 ☽ # ☉
12 34 ♂ □ ♅
14 55 ☽ ⊼ ♅
15 03 ☽ △ ♂
16 55 ☽ ± ♇
21 30 ☽ △ ♃

19 Thursday
00 22 ☽ ∥ ♃
03 11 ☽ ± ♅
11 17 ☽ ꝑ ♇
13 26 ☽ ⊻ ♄
20 10 ☽ △ ♀
20 16 ♀ Q ♆
21 35 ☽ ꝑ ♅
22 34 ☽ ∥ ♇
23 22 ☽ # ♆
23 35 ☽ ꝑ ♂

20 Friday
01 34 ☽ ⊥ ♄
08 53 ☽ ☍ ☿
16 26 ☽ # ☿
17 52 ☽ △ ♇
19 05 ☽ ± ♂
19 45 ☽ ∠ ♄
20 38 ☽ ☍ ♆
21 34 ☉ ⊻ ♅

21 Saturday
04 12 ☽ △ ♅
04 46 ☽ ⊼ ☉
05 15 ☿ St R
08 02 ☽ ⊼ ♂
08 14 ☉ ∠ ♀
11 18 ☽ □ ♃
18 00 ☽ ± ☉
20 35 ☉ ⊥ ♇
21 27 ♂ ⊥ ♆

22 Sunday
01 52 ☽ ⚹ ♄
03 06 ☽ # ♂
13 14 ☽ □ ♀
13 43 ☽ ꝑ ☉
21 05 ☽ ⊼ ☿
23 54 ☽ Q ♃

23 Monday
01 43 ☽ Q ♄
06 18 ☽ □ ♇
08 44 ☽ ± ☿
09 01 ☽ ⊼ ♆
16 29 ☽ □ ♅
20 52 ☽ ± ♆
22 04 ☽ △ ☉
23 44 ☽ ☍ ♂
23 53 ☽ ⚹ ♃

24 Tuesday
01 56 ☽ ꝑ ☿
02 43 ♂ △ ♃
03 27 ☽ Q ♀
12 50 ☽ □ ♄
14 26 ☽ ꝑ ♆
17 34 ☽ Q ♇
22 58 ☿ ∠ ♂

25 Wednesday
00 06 ☉ ⚹ ♃
03 32 ☽ Q ♅
04 18 ☽ ⚹ ♀
05 13 ☽ ∠ ♃
05 46 ☽ △ ☿
07 40 ☉ # ♃
16 29 ☽ ⚹ ♇
17 29 ☿ ⚹ ♀
19 06 ☽ △ ♆
22 25 ☽ ⊥ ♃

26 Thursday
02 16 ☽ ⚹ ♅
09 42 ☽ ⊻ ♃
10 05 ☽ # ♂
10 29 ☽ ∠ ♀
12 04 ☽ □ ☉
12 18 ☽ ⊼ ♂
20 21 ☽ ∠ ♇
20 56 ☽ △ ♄
23 40 ☉ ⚹ ♂
23 55 ☽ ± ♂

27 Friday
03 17 ♃ ± ♄
04 05 ☽ ⊥ ♀
05 51 ☽ ∠ ♅
06 41 ☉ ⊥ ♅
10 14 ☽ □ ☿
12 37 ☽ ⊥ ♇
15 35 ☽ ⊻ ♀
17 01 ☽ ꝑ ♂
21 57 ☽ ⊥ ♅
23 05 ☽ Q ☉
23 16 ☽ ⊻ ♇
23 39 ☽ ꝑ ♄

28 Saturday
01 44 ☽ □ ♆
03 28 ☽ # ☿
07 33 ☽ # ♆
08 29 ☽ ⊻ ♅
08 50 ☽ ∥ ♇
14 40 ☿ Q ♃
15 15 ☽ ± ♄
15 43 ☽ Q ☿
15 48 ☽ ☌ ♃
20 37 ☽ △ ♂
21 31 ☽ ⚹ ☉
22 14 ☿ ± ♄
23 47 ☽ # ☉

29 Sunday
01 28 ☽ ⊼ ♄
05 30 ☽ ∥ ♃
08 44 ☽ Q ♆
10 38 ☽ ⚹ ☿
14 51 ☿ ∥ ♆
15 34 ☿ ∠ ♅
22 33 ☽ ☌ ♀

30 Monday
00 36 ☽ ∠ ☉
02 12 ☿ # ♇
02 25 ☽ ☌ ♇
04 45 ☽ ⚹ ♆
09 37 ☽ ∠ ☿
10 21 ☽ ∥ ♀
11 10 ☽ ☌ ♅
16 26 ☽ ⊥ ☉
17 26 ☽ ∥ ♅
18 22 ☽ ⊻ ♃
22 15 ☽ ∥ ♄
23 07 ☽ ⊥ ☿

31 Tuesday
00 24 ☽ # ♄
01 01 ☽ □ ♂
02 04 ☉ ⊼ ♄
02 47 ☽ ☍ ♄
02 50 ☽ ⊻ ☉
03 44 ☉ ∠ ♇
04 09 ☽ ⊥ ♃
05 09 ☽ # ♅
05 11 ☽ ∠ ♆
08 00 ☽ ⊻ ☿
11 13 ☽ # ♀
18 41 ☽ ∠ ♃
19 36 ☽ ⊥ ♆

All ephemeris data is given at 12.00 UT and the Moon's longitude is additionally given for 24.00 UT

NOVEMBER 1967

LONGITUDES

Date	Sidereal time h m s	Sun ☉ ° ' "	Moon ☽ ° ' "	Moon ☽ 24.00 ° ' "	Mercury ☿ ° '	Venus ♀ ° '	Mars ♂ ° '	Jupiter ♃ ° '	Saturn ♄ ° '	Uranus ♅ ° '	Neptune ♆ ° '	Pluto ♇ ° '
01	14 40 22	08 ♏ 22 25	27 ♎ 49 27	05 ♏ 25 56	08 ♏ 42	22 ♍ 02	06 ♑ 57	02 ♍ 00	06 ♈ 52	27 ♍ 26	23 ♏ 30	22 ♍ 04
02	14 44 18	09 22 29	13 ♏ 02 48	20 ♏ 38 47	07 R 25	22 58	07 42	02 08	06 R 49	27 29	23 32	22 06
03	14 48 15	10 22 36	28 ♏ 12 40	05 ♐ 43 18	06 11	23 55	08 27	02 16	06 45	27 32	23 34	22 07
04	14 52 11	11 22 44	13 ♐ 09 38	20 ♐ 30 46	05 02	24 53	09 12	02 24	06 42	27 35	23 37	22 09
05	14 56 08	12 22 54	27 ♐ 46 01	04 ♑ 54 48	04 00	25 51	09 57	02 32	06 38	27 38	23 39	22 10
06	15 00 04	13 23 05	11 ♑ 56 48	18 ♑ 51 50	03 07	26 50	10 42	02 40	06 35	27 41	23 41	22 12
07	15 04 01	14 23 18	25 ♑ 39 55	02 ♒ 21 09	02 25	27 49	11 27	02 48	06 32	27 44	23 43	22 13
08	15 07 57	15 23 32	08 ♒ 55 49	15 ♒ 24 17	01 54	28 49	12 12	02 56	06 29	27 47	23 45	22 15
09	15 11 54	16 23 48	21 ♒ 46 59	28 ♒ 04 25	01 35	29 ♍ 49	12 57	03 03	06 26	27 49	23 48	22 16
10	15 15 51	17 24 05	04 ♓ 17 05	10 ♓ 25 34	01 27	00 ♎ 49	13 42	03 10	06 23	27 52	23 50	22 18
11	15 19 47	18 24 24	16 ♓ 30 25	22 ♓ 32 10	01 D 30	01 50	14 28	03 17	06 20	27 55	23 52	22 19
12	15 23 44	19 24 44	28 ♓ 31 22	04 ♈ 28 31	01 45	02 51	15 13	03 24	06 17	27 58	23 54	22 21
13	15 27 40	20 25 05	10 ♈ 24 07	16 ♈ 18 37	02 09	03 53	15 58	03 31	06 14	28 00	23 57	22 22
14	15 31 37	21 25 28	22 ♈ 12 27	28 ♈ 06 00	02 42	04 55	16 44	03 38	06 12	28 03	23 59	22 23
15	15 35 33	22 25 53	03 ♉ 59 37	09 ♉ 53 37	03 24	05 58	17 29	03 45	06 09	28 06	24 01	22 24
16	15 39 30	23 26 19	15 ♉ 48 20	21 ♉ 44 00	04 13	07 01	18 15	03 51	06 07	28 08	24 03	22 26
17	15 43 26	24 26 47	27 ♉ 40 52	03 ♊ 39 10	05 08	08 04	19 01	03 57	06 04	28 11	24 06	22 27
18	15 47 23	25 27 16	09 ♊ 39 07	15 ♊ 40 55	06 09	09 08	19 46	04 04	06 02	28 13	24 08	22 28
19	15 51 20	26 27 47	21 ♊ 44 47	27 ♊ 50 53	07 15	10 11	20 32	04 10	06 00	28 16	24 10	22 29
20	15 55 16	27 28 20	03 ♋ 59 28	10 ♋ 10 44	08 25	11 16	21 18	04 15	05 58	28 18	24 12	22 31
21	15 59 13	28 28 54	16 ♋ 24 57	22 ♋ 42 20	09 39	12 20	22 04	04 21	05 56	28 20	24 15	22 32
22	16 03 09	29 ♏ 29 30	29 ♋ 03 10	05 ♌ 27 43	10 55	13 25	22 49	04 27	05 54	28 23	24 17	22 33
23	16 07 06	00 ♐ 30 08	11 ♌ 56 19	18 ♌ 29 13	12 15	14 30	23 35	04 32	05 52	28 25	24 19	22 34
24	16 11 02	01 30 47	25 ♌ 06 44	01 ♍ 49 08	13 37	15 36	24 21	04 37	05 51	28 27	24 21	22 35
25	16 14 59	02 31 28	08 ♍ 36 40	15 ♍ 29 30	15 00	16 42	25 07	04 42	05 49	28 29	24 24	22 36
26	16 18 55	03 32 10	22 ♍ 27 46	29 ♍ 31 30	16 26	17 48	25 53	04 47	05 48	28 31	24 26	22 37
27	16 22 52	04 32 54	06 ♎ 40 37	13 ♎ 54 55	17 52	18 54	26 39	04 52	05 47	28 33	24 28	22 38
28	16 26 49	05 33 40	21 ♎ 14 03	28 ♎ 37 28	19 20	20 00	27 26	04 56	05 45	28 35	24 30	22 39
29	16 30 45	06 34 28	06 ♏ 04 31	13 ♏ 34 22	20 49	21 07	28 12	05 01	05 44	28 37	24 33	22 40
30	16 34 42	07 ♐ 35 16	21 ♏ 06 00	28 ♏ 38 22	22 ♏ 18	22 ♎ 14	28 ♑ 58	05 ♍ 05	05 ♈ 43	28 ♍ 39	24 ♏ 35	22 ♍ 41

Date	Moon True ☊ ° '	Moon Mean ☊ ° '	Moon ☽ Latitude ° '
01	27 ♈ 57	27 ♈ 12	00 N 01
02	27 R 57	27 09	01 S 23
03	27 56	27 06	02 40
04	27 54	27 02	03 45
05	27 52	26 59	04 34
06	27 51	26 56	05 04
07	27 49	26 53	05 16
08	27 48	26 50	05 11
09	27 D 48	26 46	04 49
10	27 49	26 43	04 15
11	27 50	26 40	03 29
12	27 52	26 37	02 35
13	27 54	26 34	01 35
14	27 55	26 31	00 S 31
15	27 R 55	26 27	00 N 33
16	27 53	26 24	01 37
17	27 51	26 21	02 36
18	27 47	26 18	03 29
19	27 42	26 15	04 13
20	27 36	26 12	04 47
21	27 31	26 08	05 07
22	27 27	26 05	05 12
23	27 24	26 02	05 02
24	27 22	25 59	04 37
25	27 D 22	25 56	03 55
26	27 23	25 52	02 59
27	27 25	25 49	01 51
28	27 26	25 46	00 N 34
29	27 R 26	25 43	00 S 47
30	27 ♈ 24	25 ♈ 40	02 S 05

DECLINATIONS

Date	Sun ☉ ° '	Moon ☽ ° '	Mercury ☿ ° '	Venus ♀ ° '	Mars ♂ ° '	Jupiter ♃ ° '	Saturn ♄ ° '	Uranus ♅ ° '	Neptune ♆ ° '	Pluto ♇ ° '
01	14 S 18	10 S 42	14 S 58	03 N 14	24 S 51	11 N 32	00 N 17	01 N 41	17 S 02	16 N 41
02	14 37	17 04	14 14	02 57	24 48	11 30	00 16	01 40	17 02	16 40
03	14 56	22 22	13 30	02 41	24 45	11 27	00 14	01 39	17 03	16 40
04	15 15	26 07	12 48	02 23	24 42	11 24	00 13	01 38	17 03	16 40
05	15 33	28 00	12 09	02 06	24 38	11 21	00 12	01 37	17 04	16 40
06	15 52	27 58	11 35	01 48	24 34	11 19	00 11	01 36	17 04	16 39
07	16 10	26 12	11 06	01 30	24 30	11 16	00 10	01 34	17 05	16 39
08	16 27	23 01	10 42	01 11	24 26	11 14	00 09	01 33	17 06	16 39
09	16 45	18 48	10 23	00 52	24 21	11 11	00 08	01 32	17 06	16 39
10	17 02	13 54	10 11	00 33	24 16	11 09	00 07	01 31	17 07	16 39
11	17 19	08 32	10 04	00 N 13	24 11	11 06	00 06	01 30	17 07	16 38
12	17 35	02 S 58	10 03	00 S 07	24 06	11 04	00 05	01 29	17 08	16 38
13	17 51	02 N 40	10 06	00 27	24 00	11 02	00 04	01 28	17 08	16 38
14	18 07	08 10	10 14	00 48	23 54	10 59	00 03	01 27	17 09	16 38
15	18 23	13 23	10 26	01 08	23 48	10 57	00 02	01 26	17 10	16 38
16	18 38	18 07	10 42	01 29	23 42	10 55	00 01	01 25	17 10	16 38
17	18 53	22 11	11 01	01 50	23 35	10 53	00 N 01	01 24	17 11	16 38
18	19 08	25 21	11 22	02 12	23 28	10 51	00 00	01 23	17 11	16 38
19	19 22	27 24	11 46	02 33	23 21	10 49	00 S 01	01 22	17 12	16 38
20	19 36	28 09	12 11	02 55	23 13	10 47	00 01	01 21	17 12	16 38
21	19 50	27 30	12 38	03 17	23 06	10 45	00 02	01 20	17 13	16 38
22	20 03	25 27	13 06	03 39	22 58	10 44	00 02	01 19	17 13	16 38
23	20 16	22 03	13 35	04 01	22 49	10 42	00 03	01 19	17 14	16 38
24	20 28	17 30	14 04	04 24	22 41	10 40	00 03	01 18	17 14	16 38
25	20 40	11 59	14 34	04 46	22 32	10 39	00 04	01 17	17 15	16 38
26	20 52	05 N 44	15 05	05 09	22 23	10 37	00 04	01 16	17 16	16 38
27	21 03	00 S 57	15 35	05 32	22 14	10 35	00 04	01 15	17 16	16 38
28	21 14	07 46	16 05	05 54	22 05	10 34	00 05	01 14	17 17	16 38
29	21 25	14 17	16 35	06 17	21 55	10 33	00 05	01 14	17 17	16 38
30	21 S 35	20 S 03	17 S 05	06 S 40	21 S 45	10 N 31	00 S 05	01 N 13	17 S 18	16 N 38

ZODIAC SIGN ENTRIES

Date	h	m	Planets
01	15	26	☽ ♏
03	14	51	☽ ♐
05	15	44	☽ ♑
07	19	45	☽ ♒
09	16	32	♀ ♎
10	03	42	☽ ♓
12	14	58	☽ ♈
15	03	52	☽ ♉
17	16	40	☽ ♊
20	04	13	☽ ♋
22	13	47	☽ ♌
23	00	04	☉ ♐
24	20	46	☽ ♍
27	00	48	☽ ♎
29	02	13	☽ ♏

LATITUDES

Date	Mercury ☿ ° '	Venus ♀ ° '	Mars ♂ ° '	Jupiter ♃ ° '	Saturn ♄ ° '	Uranus ♅ ° '	Neptune ♆ ° '	Pluto ♇ ° '
01	00 S 36	00 N 04	01 S 35	00 N 50	02 S 40	00 N 44	01 N 41	14 N 46
04	00 N 25	00 23	01 34	00 50	02 39	00 44	01 41	14 47
07	01 18	00 41	01 33	00 51	02 39	00 44	01 41	14 49
10	01 55	00 57	01 32	00 51	02 38	00 44	01 41	14 50
13	02 15	01 11	01 31	00 52	02 38	00 44	01 41	14 51
16	02 22	01 25	01 30	00 53	02 37	00 44	01 41	14 53
19	02 18	01 37	01 29	00 54	02 37	00 44	01 41	14 54
22	02 07	01 47	01 28	00 54	02 36	00 44	01 41	14 55
25	01 51	01 56	01 27	00 55	02 35	00 44	01 41	14 57
28	01 32	02 04	01 26	00 56	02 35	00 44	01 41	14 58
31	01 N 12	02 N 11	01 S 24	00 N 56	02 S 34	00 N 45	01 N 41	15 N 00

DATA

Julian Date	2439796
Delta T	+38 seconds
Ayanamsa	23° 24' 19"
Synetic vernal point	05° ♓ 42' 41"
True obliquity of ecliptic	23° 26' 45"

MOON'S PHASES, APSIDES AND POSITIONS ☽

Date	h	m	Phase	Longitude ° '	Eclipse Indicator
02	05	48	●	09 ♏ 07	Total
09	01	00	☽	15 ♒ 56	
17	04	53	○	24 ♉ 09	
25	00	23	☾	02 ♍ 02	

Day	h	m		
02	01	42	Perigee	
15	07	56	Apogee	
30	13	24	Perigee	
05	23	31	Max dec	28° S 13'
13	00	37	ON	
20	13	02	Max dec	28° N 10'
27	08	38	0S	

LONGITUDES

Date	Chiron ⚷ ° '	Ceres ⚳ ° '	Pallas ⚴ ° '	Juno ⚵ ° '	Vesta ⚶ ° '	Black Moon Lilith ⚸ ° '
01	25 ♓ 52	01 ♎ 18	09 ♍ 44	15 ♎ 32	28 ♐ 28	04 ♉ 26
11	25 ♓ 36	05 ♎ 35	14 ♍ 14	19 ♎ 05	03 ♑ 22	05 ♉ 33
21	25 ♓ 24	09 ♎ 45	18 ♍ 25	22 ♎ 31	08 ♑ 21	06 ♉ 39
31	25 ♓ 18	13 ♎ 46	22 ♍ 28	25 ♎ 49	13 ♑ 24	07 ♉ 46

All ephemeris data is given at 12.00 UT and the Moon's longitude is additionally given for 24.00 UT

ASPECTARIAN

h m	Aspects
01 Wednesday	
02 13	☽ ⊻ ♀
02 52	☽ ⊻ ♇
05 09	☽ ⊻ ♆
07 14	☽ Q ♂
09 37	♂ □ ♄
11 23	☽ ⊻ ♅
12 20	☽ ⊥ ♀
12 23	☽ ⊥ ♇
13 03	♀ ☌ ♇
15 00	☽ # ♃
15 28	☉ ☌ ☿
18 39	☽ ✶ ♃
20 52	☽ ⊥ ♅
02 Thursday	
01 53	☽ ∥ ☉
02 04	☽ ∥ ☿
02 13	☽ ⊼ ♄
02 36	☽ ∠ ♇
03 08	☉ ∥ ☿
03 08	☽ ✶ ♂
03 28	☽ ∠ ♀
03 49	☽ ☌ ☿
05 48	☽ ☌ ☉
06 04	☿ ∠ ♀
08 41	☿ ✶ ♂
10 23	☽ # ♇
11 06	☽ ∠ ♅
11 38	☽ ± ♄
11 51	☽ ∥ ♆
13 44	☽ Q ♃
18 04	☿ ∠ ♇
03 Friday	
00 09	☿ ⊼ ♄
01 47	☽ ⚼ ♄
02 19	☽ ✶ ♇
02 55	♀ ✶ ♆
04 02	☽ ∠ ♂
04 37	☽ ☌ ♆
04 44	☽ ✶ ♀
10 55	☽ ✶ ♅
16 20	♂ ∠ ♆
18 32	☽ □ ♃
19 06	☽ ⊥ ♂
21 27	☽ Q ♇
23 48	☽ ⊻ ☿
04 Saturday	
01 10	☽ Q ♀
01 36	☽ △ ♄
01 42	☽ ∥ ♂
05 15	☽ ⊻ ♂
06 12	☽ Q ♅
08 47	☽ ⊥ ☿
08 54	☽ ⊻ ☉
19 22	☽ ⊥ ☉
22 26	☽ ∠ ☿
05 Sunday	
02 43	☽ □ ♇
05 09	☽ ⊻ ♆
08 35	☽ □ ♀
11 19	☽ ∠ ☉
11 46	☽ □ ♅
15 09	☽ ⊥ ♆
17 52	☉ ± ♄
18 18	☉ ∠ ♅
20 04	☽ △ ♃
20 59	☿ ⊥ ♅
21 49	☽ ✶ ☿
06 Monday	
02 52	☽ □ ♄
06 24	☽ ∠ ♆
09 43	☽ ☌ ♂
14 40	☽ ✶ ☉
15 58	☿ ⊥ ♀
17 11	☽ Q ☿
22 01	☽ ⚼ ♃
07 Tuesday	
00 18	☿ ✶ ♃
02 00	☿ # ♃
05 13	♀ ∥ ♅
05 53	☽ △ ♇
08 32	☽ ✶ ♆
09 48	♀ ☌ ♅
10 00	☽ Q ♄
13 24	☽ Q ☉
14 02	☽ ± ♃
15 42	☽ △ ♅
16 09	☽ △ ♀
23 14	☉ Q ♃
23 37	☽ □ ☿
08 Wednesday	
00 56	☽ ⊼ ♃
02 12	☽ ∥ ♂
06 10	☽ Q ♆
07 31	☽ ✶ ♄
08 54	☽ ⚼ ♇
18 24	☽ ⊻ ♂
19 08	☽ ⚼ ♅
21 47	☽ ⚼ ♀
09 Thursday	
01 00	☽ □ ☉
01 36	☽ ± ♇
03 48	☉ # ♇
06 19	☽ ⊥ ♂
11 20	☽ ∠ ♄
12 05	☽ ± ♅
12 56	☽ ⊼ ♇
15 50	☽ □ ♆
16 11	☽ ± ♀
20 38	☽ ∥ ♆
21 50	☽ ∥ ☉
22 54	☽ # ♇
23 34	☽ ⊼ ♅
10 Friday	
00 31	☽ ∠ ♂
04 28	☽ ⊥ ♄
04 29	☿ Q ♂
04 41	☽ ⊼ ♀
06 32	☽ △ ☿
09 49	☽ ☍ ♃
16 04	☽ ⊻ ♄
16 16	☿ St D
19 09	☉ ∥ ♆
11 Saturday	
00 35	☽ # ♃
03 17	☿ ⊻ ♀
05 10	☽ ∥ ☿
07 41	☽ ✶ ♂
12 00	☽ ⚼ ☿
16 07	☽ △ ☉
21 16	♀ ∥ ♄
23 35	☽ ☍ ♇
12 Sunday	
02 43	☽ △ ♆
06 17	☽ ± ☿
09 12	☽ Q ♂
09 21	♀ # ♄
10 52	☽ ☍ ♅
18 19	☽ # ♅
18 41	☽ ⊼ ☿
21 33	☽ ☍ ♀
21 56	☽ ⊼ ♃
23 27	☽ ∥ ♀
13 Monday	
00 19	☽ # ♄
00 55	☽ ∥ ♄
00 58	☽ ⚼ ☉
01 57	☽ # ♀
02 29	♀ ⊻ ♃
03 36	☽ ☌ ♄
06 54	☽ ∥ ♅
09 02	☽ ⚼ ♆
10 12	☽ ± ♃
20 55	☽ ± ☉
14 Tuesday	
00 05	☽ □ ♂
03 22	☽ ± ♆
04 39	☽ ⚼ ♃
06 46	☉ ⚼ ♄
10 15	☽ ⊼ ☉
12 22	☽ ⊼ ♇
15 38	☽ ⊼ ♆
21 41	☽ # ☿
23 57	☽ ⊼ ♅
15 Wednesday	
00 36	☽ ± ♇
00 43	☽ ∥ ♃
10 42	☽ ☍ ☿
11 26	☉ ✶ ♇
11 29	☽ △ ♃
12 12	☽ ± ♅
16 12	♀ ☍ ♄
16 23	☽ ⊻ ♄
16 24	☽ ⊼ ♀
18 58	☽ ⚼ ♇
16 Thursday	
00 20	☿ ✶ ♃
04 08	☽ ∥ ♇
04 32	☽ ⊥ ♄
05 47	☽ ± ♀
06 34	☽ ⚼ ♅
06 55	☽ # ♆
07 16	♀ # ♅
09 50	☿ ⊥ ♅
15 00	☽ # ☉
17 17	☽ △ ♂
22 43	☽ ∠ ♄
17 Friday	
01 26	☽ △ ♇
01 47	☽ ⚼ ♀
03 19	☉ ☌ ♆
03 29	☿ # ♃
04 45	☽ ☍ ♆
04 53	☽ ☍ ☉
10 05	♂ ⚼ ♃
13 00	☽ △ ♅
21 21	☽ # ♂
18 Saturday	
00 43	☽ □ ♃
01 35	☽ ⚼ ♂
04 20	☽ ⊼ ☿
04 48	☽ ✶ ♄
09 35	☿ ⊼ ♄
10 51	☽ △ ♀
12 10	♀ ∠ ♆
17 27	☽ ± ☿
20 45	☽ ± ♂
19 Sunday	
04 37	☽ Q ♄
09 27	☽ ⊼ ♂
11 12	♀ ⊥ ♃
12 49	☽ Q ♃
13 05	☽ ⚼ ☿
13 28	☽ □ ♇
16 47	☽ ⊼ ♆
17 15	☿ ∠ ♇
22 07	☽ ⊼ ☉
20 Monday	
00 51	☽ □ ♅
04 36	☽ ± ♆
10 54	☽ ± ☉
12 31	☽ ✶ ♃
15 50	☽ □ ♄
21 31	☽ △ ☿
22 09	☽ ⚼ ♆
21 Tuesday	
00 40	☽ Q ♇
03 25	☽ □ ♀
05 52	☽ ⚼ ☉
08 27	☉ ✶ ♅
11 51	☽ Q ♅
17 39	☽ ∠ ♃
23 28	☽ ☍ ♂
23 41	☽ ✶ ♇
22 Wednesday	
02 58	☽ △ ♆
03 08	♂ △ ♇
07 30	☿ Q ♂
10 43	☽ ✶ ♅
10 51	☽ ⊥ ♃
12 54	☽ △ ☉
16 51	☽ Q ♀
22 10	☽ ⊻ ♃
23 Thursday	
00 48	☽ △ ♄
03 54	☽ ∠ ♇
05 26	☿ ± ♄
07 03	☽ # ♂
12 38	☽ □ ☿
14 43	☽ ∠ ♅
17 09	☽ ✶ ♀
20 30	☽ ⊥ ♇
20 39	♂ Q ♄
21 40	☽ # ☉
24 Friday	
04 18	☽ ⚼ ♄
07 11	☽ ⊥ ♅
07 26	☽ ⊻ ♇
09 06	☿ ∠ ♅
10 33	☽ ⊼ ♂
10 38	☽ □ ♆
12 06	♂ ✶ ♆
13 11	☽ # ♆
16 02	☽ ∥ ♇
18 00	☽ ⊻ ♅
20 28	☽ ± ♄
21 58	☽ ± ♂
22 42	☽ ∠ ♀
25 Saturday	
00 23	☽ □ ☉
00 58	☽ Q ☿
02 03	☽ # ☿
05 04	☽ ☌ ♃
07 06	☽ ⊼ ♄
14 48	☽ ⚼ ♂
15 58	☽ ⊥ ♀
17 21	☽ ∥ ♃
18 38	☽ Q ♆
26 Sunday	
00 26	☽ ✶ ☿
03 17	☽ ⊻ ♀
10 17	☽ Q ☉
12 16	☽ ☌ ♇
14 02	☽ # ♀
15 22	☽ ✶ ♆
18 11	☽ △ ♂
18 16	☿ Q ♃
22 20	☽ ☌ ♅
27 Monday	
02 23	♀ ⊥ ♆
04 10	☽ ∥ ♅
04 55	☽ ∠ ☿
08 10	☽ ✶ ☉
08 22	☽ # ♄
08 53	☽ ∥ ♄
08 57	☽ ⊻ ♃
10 30	☽ ☍ ♄
13 03	☽ # ♅
14 01	☉ Q ♇
16 39	☽ ∠ ♆
18 59	☽ ⊥ ♃
20 02	☉ □ ♃
21 35	☽ ⊥ ☿
28 Tuesday	
05 02	☽ ∥ ♀
07 32	☽ ⊥ ♆
08 33	☽ ⊻ ☿
09 50	☽ ☌ ♀
09 52	☽ ∠ ♃
10 24	♀ ∠ ♃
10 49	☽ ∠ ☉
14 18	☽ ⊻ ♇
16 29	☉ △ ♄
17 20	☽ ⊻ ♆
22 03	☽ # ♃
22 37	☽ □ ♂
23 58	☽ ⊻ ♅
29 Wednesday	
00 03	☽ ⊥ ♇
02 30	☽ ⊥ ☉
09 40	☽ ⊥ ♅
10 17	☽ ✶ ♃
10 45	☿ ⚼ ♄
11 27	☽ ⊼ ♄
12 52	☽ ⊻ ☉
14 18	☿ # ♇
14 33	☽ ∠ ♇
21 03	☽ ± ♄
21 17	☽ # ♇
21 55	☽ ∥ ☿
23 58	☽ ∥ ♆
30 Thursday	
00 06	☽ ∠ ♅
01 54	♂ △ ♅
05 03	☽ Q ♂
05 34	☽ Q ♃
07 35	☿ ⊻ ♀
11 24	☽ ⚼ ♄
13 57	☽ ⊻ ♀
14 08	☽ ☌ ☿
14 31	☽ ✶ ♇
15 53	♂ ± ♃
17 33	☽ ☌ ♆
18 01	☿ ✶ ♇
19 36	☽ ∥ ☉
19 55	☽ ∥ ♂
21 35	♀ ⊻ ♇
22 34	☿ ∥ ♆

DECEMBER 1967

LONGITUDES

Date	Sidereal time h m s	Sun ☉ ° ' "	Moon ☽ ° ' "	Moon ☽ 24.00 ° ' "	Mercury ☿ ° '	Venus ♀ ° '	Mars ♂ ° '	Jupiter ♃ ° '	Saturn ♄ ° '	Uranus ♅ ° '	Neptune ♆ ° '	Pluto ♇ ° '
01	16 38 38	08 ♐ 36 06	06 ♐ 10 17	13 ♐ 40 33	23 ♏ 48	23 ♎ 21	29 ♑ 44	05 ♍ 09	05 ♈ 42	28 ♍ 41	24 ♏ 37	22 ♍ 41
02	16 42 35	09 36 58	21 ♐ 07 59	28 ♐ 31 31	25 19	24 29	00 ♒ 30	05 13	05 R 41	28 43	24 39	22 42
03	16 46 31	10 37 50	05 ♑ 50 08	13 ♑ 03 00	26 50	25 37	01 17	05 16	05 41	28 45	24 41	22 43
04	16 50 28	11 38 44	20 ♑ 09 30	27 ♑ 09 09	28 22	26 44	02 03	05 20	05 40	28 46	24 44	22 44
05	16 54 24	12 39 38	04 ♒ 01 41	10 ♒ 47 03	29 ♏ 54	27 52	02 50	05 23	05 40	28 48	24 46	22 44
06	16 58 21	13 40 33	17 ♒ 25 19	23 ♒ 56 44	01 ♐ 26	29 ♎ 01	03 36	05 26	05 39	28 50	24 48	22 45
07	17 02 18	14 41 29	00 ♓ 21 42	06 ♓ 40 40	02 58	00 ♏ 09	04 22	05 29	05 39	28 51	24 50	22 46
08	17 06 14	15 42 26	12 ♓ 54 11	19 ♓ 02 52	04 30	01 18	05 09	05 32	05 39	28 53	24 52	22 46
09	17 10 11	16 43 23	25 ♓ 07 22	01 ♈ 08 20	06 03	02 27	05 55	05 34	05 D 39	28 54	24 54	22 47
10	17 14 07	17 44 21	07 ♈ 06 28	13 ♈ 02 23	07 35	03 36	06 42	05 37	05 39	28 56	24 57	22 48
11	17 18 04	18 45 19	18 ♈ 56 46	24 ♈ 50 13	09 08	04 45	07 28	05 39	05 39	28 57	24 59	22 48
12	17 22 00	19 46 18	00 ♉ 43 20	06 ♉ 36 38	10 41	05 54	08 15	05 41	05 39	28 58	25 01	22 49
13	17 25 57	20 47 18	12 ♉ 30 38	18 ♉ 25 47	12 14	07 03	09 02	05 43	05 40	29 00	25 03	22 49
14	17 29 53	21 48 18	24 ♉ 22 28	00 ♊ 21 02	13 47	08 13	09 48	05 44	05 40	29 01	25 05	22 49
15	17 33 50	22 49 19	06 ♊ 21 47	12 ♊ 24 55	15 21	09 23	10 35	05 46	05 41	29 02	25 07	22 50
16	17 37 47	23 50 21	18 ♊ 30 39	24 ♊ 39 06	16 54	10 33	11 22	05 47	05 42	29 03	25 09	22 50
17	17 41 43	24 51 23	00 ♋ 50 21	07 ♋ 04 29	18 28	11 43	12 08	05 48	05 42	29 04	25 11	22 50
18	17 45 40	25 52 26	13 ♋ 21 30	19 ♋ 41 25	20 01	12 53	12 55	05 49	05 43	29 05	25 13	22 51
19	17 49 36	26 53 30	26 ♋ 04 13	02 ♌ 29 54	21 35	14 03	13 42	05 50	05 44	29 06	25 15	22 51
20	17 53 33	27 54 35	08 ♌ 58 29	15 ♌ 29 58	23 09	15 14	14 28	05 50	05 45	29 07	25 17	22 51
21	17 57 29	28 55 40	22 ♌ 04 24	28 ♌ 41 50	24 43	16 24	15 15	05 50	05 47	29 08	25 19	22 51
22	18 01 26	29 ♐ 56 46	05 ♍ 22 21	12 ♍ 06 04	26 17	17 35	16 02	05 R 50	05 48	29 08	25 21	22 51
23	18 05 22	00 ♑ 57 52	18 ♍ 53 05	25 ♍ 43 33	27 52	18 46	16 48	05 50	05 50	29 09	25 23	22 52
24	18 09 19	01 58 59	02 ♎ 37 34	09 ♎ 35 14	29 ♐ 26	19 57	17 35	05 50	05 51	29 10	25 25	22 52
25	18 13 16	03 00 07	16 ♎ 36 35	23 ♎ 41 37	01 ♑ 01	21 08	18 22	05 50	05 53	29 10	25 27	22 52
26	18 17 12	04 01 16	00 ♏ 50 14	08 ♏ 02 13	02 36	22 19	19 09	05 49	05 55	29 11	25 29	22 R 52
27	18 21 09	05 02 25	15 ♏ 17 14	22 ♏ 34 51	04 12	23 30	19 56	05 48	05 57	29 11	25 30	22 52
28	18 25 05	06 03 35	29 ♏ 54 28	07 ♐ 15 21	05 47	24 42	20 42	05 47	05 59	29 12	25 32	22 52
29	18 29 02	07 04 46	14 ♐ 36 40	21 ♐ 57 31	07 23	25 53	21 29	05 46	06 01	29 12	25 34	22 52
30	18 32 58	08 05 56	29 ♐ 16 55	06 ♑ 33 53	08 59	27 05	22 16	05 44	06 03	29 12	25 36	22 51
31	18 36 55	09 ♑ 07 07	13 ♑ 47 28	20 ♑ 56 48	10 ♑ 36	28 ♏ 16	23 ♒ 03	05 ♍ 42	06 ♈ 05	29 ♍ 13	25 ♏ 38	22 ♍ 51

Date	Moon True ☊ ° '	Moon Mean ☊ ° '	Moon ☽ Latitude ° '
01	27 ♈ 20	25 ♈ 37	03 S 15
02	27 R 14	25 33	04 10
03	27 07	25 30	04 48
04	26 59	25 27	05 07
05	26 53	25 24	05 07
06	26 48	25 21	04 49
07	26 45	25 18	04 18
08	26 43	25 14	03 34
09	26 D 44	25 11	02 42
10	26 45	25 08	01 44
11	26 46	25 05	00 S 42
12	26 R 46	25 02	00 N 21
13	26 45	24 58	01 24
14	26 41	24 55	02 23
15	26 34	24 52	03 16
16	26 25	24 49	04 01
17	26 15	24 46	04 35
18	26 03	24 43	04 56
19	25 52	24 39	05 04
20	25 42	24 36	04 56
21	25 34	24 33	04 32
22	25 29	24 30	03 54
23	25 26	24 27	03 02
24	25 D 25	24 24	01 58
25	25 26	24 20	00 N 47
26	25 R 26	24 17	00 S 29
27	25 24	24 14	01 43
28	25 20	24 11	02 52
29	25 13	24 08	03 50
30	25 04	24 04	04 32
31	24 ♈ 52	24 ♈ 01	04 S 56

DECLINATIONS

Date	Sun ☉ ° '	Moon ☽ ° '	Mercury ☿ ° '	Venus ♀ ° '	Mars ♂ ° '	Jupiter ♃ ° '	Saturn ♄ ° '	Uranus ♅ ° '	Neptune ♆ ° '	Pluto ♇ ° '
01	21 S 45	24 S 32	17 S 34	07 S 03	21 S 35	10 N 30	00 S 05	01 N 12	17 S 18	16 N 38
02	21 54	27 19	18 03	07 26	21 25	10 29	00 05	01 12	17 19	16 38
03	22 03	28 07	18 31	07 49	21 14	10 28	00 05	01 11	17 19	16 39
04	22 11	26 59	18 59	08 12	21 03	10 27	00 05	01 10	17 20	16 39
05	22 19	24 13	19 26	08 34	20 52	10 26	00 05	01 10	17 20	16 39
06	22 27	20 12	19 52	08 57	20 41	10 25	00 05	01 09	17 21	16 39
07	22 34	15 22	20 17	09 20	20 29	10 24	00 05	01 08	17 21	16 39
08	22 41	10 01	20 42	09 43	20 17	10 23	00 05	01 08	17 22	16 40
09	22 47	04 S 25	21 05	10 05	20 05	10 23	00 05	01 07	17 22	16 40
10	22 53	01 N 14	21 28	10 28	19 53	10 22	00 05	01 07	17 23	16 40
11	22 58	06 46	21 49	10 50	19 41	10 21	00 04	01 06	17 23	16 40
12	23 03	12 04	22 10	11 13	19 28	10 21	00 04	01 06	17 24	16 41
13	23 08	16 55	22 29	11 35	19 15	10 21	00 04	01 05	17 24	16 41
14	23 12	21 10	22 48	11 57	19 02	10 20	00 03	01 05	17 25	16 41
15	23 15	24 35	23 05	12 19	18 49	10 20	00 03	01 04	17 25	16 42
16	23 18	26 57	23 21	12 40	18 36	10 20	00 02	01 04	17 26	16 42
17	23 21	28 01	23 36	13 02	18 22	10 20	00 02	01 04	17 26	16 42
18	23 23	27 41	23 50	13 23	18 09	10 19	00 S 01	01 03	17 26	16 43
19	23 25	25 54	24 03	13 44	17 55	10 19	00 00	01 03	17 27	16 43
20	23 26	22 46	24 14	14 05	17 40	10 20	00 00	01 03	17 27	16 44
21	23 26	18 26	24 24	14 25	17 26	10 20	00 N 01	01 02	17 28	16 44
22	23 27	13 10	24 33	14 46	17 12	10 20	00 02	01 02	17 28	16 44
23	23 27	07 11	24 40	15 06	16 57	10 20	00 03	01 02	17 29	16 45
24	23 26	00 N 46	24 47	15 25	16 42	10 21	00 03	01 02	17 29	16 45
25	23 25	05 S 49	24 51	15 45	16 27	10 21	00 04	01 02	17 29	16 46
26	23 23	12 13	24 55	16 04	16 12	10 21	00 05	01 01	17 30	16 46
27	23 21	18 04	24 57	16 23	15 57	10 22	00 06	01 01	17 30	16 47
28	23 18	22 56	24 58	16 41	15 41	10 23	00 07	01 01	17 31	16 47
29	23 15	26 22	24 57	16 59	15 26	10 23	00 08	01 01	17 31	16 48
30	23 12	27 58	24 55	17 17	15 10	10 24	00 09	01 01	17 31	16 48
31	23 S 08	27 S 38	24 S 51	17 S 34	14 S 54	10 N 25	00 N 11	01 N 01	17 S 32	16 N 49

ZODIAC SIGN ENTRIES

Date	h	m	Planets
01	02	10	☽ ♐
01	20	12	♂ ♒
03	02	25	☽ ♑
05	04	57	☽ ♒
05	13	41	☿ ♐
07	08	48	♀ ♏
07	11	19	☽ ♓
09	21	43	☽ ♈
12	10	32	☽ ♉
14	23	18	☽ ♊
17	10	23	☽ ♋
19	19	21	☽ ♌
22	02	21	☽ ♍
22	13	16	☉ ♑
24	07	27	☽ ♎
24	20	33	☿ ♑
26	10	36	☽ ♏
28	12	09	☽ ♐
30	13	11	☽ ♑

LATITUDES

Date	Mercury ☿ ° '	Venus ♀ ° '	Mars ♂ ° '	Jupiter ♃ ° '	Saturn ♄ ° '	Uranus ♅ ° '	Neptune ♆ ° '	Pluto ♇ ° '
01	01 N 12	02 N 11	01 S 24	00 N 56	02 S 34	00 N 45	01 N 41	15 N 00
04	00 50	02 16	01 23	00 57	02 33	00 45	01 41	15 01
07	00 29	02 21	01 21	00 58	02 32	00 45	01 41	15 03
10	00 N 07	02 24	01 20	00 59	02 32	00 45	01 41	15 05
13	00 S 13	02 26	01 18	01 00	02 31	00 45	01 41	15 06
16	00 33	02 26	01 16	01 00	02 30	00 45	01 41	15 08
19	00 52	02 26	01 15	01 01	02 29	00 45	01 41	15 09
22	01 09	02 25	01 13	01 02	02 29	00 45	01 41	15 11
25	01 25	02 23	01 11	01 03	02 28	00 46	01 41	15 13
28	01 39	02 20	01 09	01 04	02 27	00 46	01 41	15 14
31	01 S 50	02 N 16	01 S 07	01 N 04	02 S 26	00 N 46	01 N 41	15 N 16

DATA

Julian Date	2439826
Delta T	+38 seconds
Ayanamsa	23° 24' 24"
Synetic vernal point	05° ♓ 42' 36"
True obliquity of ecliptic	23° 26' 44"

MOON'S PHASES, APSIDES AND POSITIONS ☽

Date	h	m	Phase	Longitude ° '	Eclipse Indicator
01	16	10	●	08 ♐ 47	
08	17	57	☽	15 ♓ 58	
16	23	21	○	24 ♊ 19	
24	10	48	☾	01 ♎ 56	
31	03	38	●	08 ♑ 46	

Day	h	m		
12	18	14	Apogee	
28	19	06	Perigee	
03	09	40	Max dec	28° S 07'
10	06	46	ON	
17	18	29	Max dec	28° N 05'
24	14	48	OS	
30	19	46	Max dec	28° S 05'

LONGITUDES

Date	Chiron ⚷ ° '	Ceres ⚳ ° '	Pallas ⚴ ° '	Juno ⚵ ° '	Vesta ⚶ ° '	Black Moon Lilith ⚸ ° '
01	25 ♓ 18	13 ♎ 46	22 ♍ 13	25 ♎ 49	13 ♑ 24	07 ♉ 46
11	25 ♓ 18	17 ♎ 37	25 ♍ 36	28 ♎ 57	18 ♑ 30	08 ♉ 52
21	25 ♓ 23	21 ♎ 15	28 ♍ 28	01 ♏ 53	23 ♑ 38	09 ♉ 59
31	25 ♓ 34	24 ♎ 37	00 ♎ 45	04 ♏ 35	28 ♑ 48	11 ♉ 05

All ephemeris data is given at 12.00 UT and the Moon's longitude is additionally given for 24.00 UT

ASPECTARIAN

h m Aspects

01 Friday
00 03 ☽ ⚹ ♅
00 15 ☉ ∥ ♂
00 16 ☽ ⊥ ♀
01 12 ☽ ⚹ ♂
09 38 ☽ Q ♇
10 21 ☽ □ ♃
11 15 ☽ △ ♄
15 46 ☽ ∠ ♀
16 10 ☽ ☌ ☉
19 14 ☽ Q ♅

02 Saturday
01 12 ☿ ☌ ♆
02 27 ☽ ∠ ♂
14 33 ☽ □ ♇
15 49 ♀ ⊻ ♆
17 43 ☽ ⊻ ♆
17 46 ☽ ⊥ ♂
17 52 ☽ ⚹ ♀
19 33 ☽ ⊻ ☿

03 Sunday
00 20 ☽ □ ♅
03 31 ☽ ⊥ ♆
04 05 ☽ ⊻ ♂
06 29 ☽ ⊥ ☿
11 04 ☽ △ ♃
11 44 ☽ □ ♄
14 49 ☉ Q ♅
15 11 ☽ Q ♀
16 34 ☉ ∠ ♀
18 24 ☽ ∠ ♆
20 34 ☽ ⊻ ☉
23 09 ☽ ∠ ☿

04 Monday
07 24 ☽ ⊥ ☉
12 18 ☽ ⚼ ♃
16 24 ☽ △ ♇
18 00 ☽ Q ♄
18 36 ☿ ⚹ ♅
19 50 ☽ ⚹ ♆

05 Tuesday
00 00 ☽ ∠ ☉
00 17 ☽ □ ♀
02 51 ☽ △ ♅
03 50 ☽ ± ♃
03 51 ☽ ⚹ ☿
09 46 ☽ ☌ ♂
14 24 ☽ ⊼ ♃
14 53 ☽ ⚹ ♄
16 51 ☽ Q ♆
18 35 ☽ ⚼ ♇
23 43 ☽ ∥ ☉

06 Wednesday
03 48 ☽ Q ☿
04 38 ☽ ⚹ ☉
05 28 ☽ ⚼ ♅
06 29 ♀ ⊥ ♇
08 04 ♀ ⊻ ♅
09 20 ☽ ∥ ♂
10 47 ☽ ± ♇
13 39 ☽ ∥ ☿
17 55 ☽ ∠ ♄
21 48 ☽ ⊼ ♇
21 57 ☽ ± ♅

07 Thursday
01 37 ☽ □ ♆
02 31 ☽ ∥ ♆
04 31 ☽ Q ☉
05 53 ☽ ⋕ ♇
09 10 ☽ ⊼ ♅
10 40 ☽ ⊥ ♄
11 34 ☽ △ ♀
17 36 ☽ □ ☿
19 53 ☿ ∥ ♂
20 06 ☽ ⊻ ♂
21 45 ☽ ☍ ♃
22 02 ☽ ⊻ ♄

08 Friday
08 23 ☽ ⊥ ♂
10 22 ☽ ⋕ ♃
13 14 ☽ ∥ ♀
16 14 ☿ Q ♇
17 57 ☽ □ ☉
19 17 ☽ ⚼ ♀

09 Saturday
00 32 ♂ ⊼ ♃
03 08 ☽ ∠ ♂
03 25 ♂ ⚹ ♄
04 26 ☿ □ ♃
05 46 ☿ △ ♄
07 22 ☽ ☍ ♇
08 10 ☿ ⚹ ♂
10 27 ♄ St D
11 34 ☽ △ ♆
14 54 ☽ ± ♀
19 33 ☽ ☍ ♅

10 Sunday
02 01 ☽ ⋕ ♅
04 10 ☽ ⊼ ♀
05 52 ♀ ⋕ ♃
06 26 ☽ ∥ ♄
07 06 ☽ ⋕ ♄
08 59 ☽ ⊼ ♃
09 03 ☽ ☌ ♄
11 07 ☽ ⚹ ♂
11 31 ☽ ∥ ♅
13 07 ☽ △ ☿
17 45 ☽ ⚼ ♆
19 53 ♂ Q ♆
21 08 ☽ ± ♃

11 Monday
11 34 ☽ △ ☉
12 04 ☽ ± ♆
12 47 ♃ ⊼ ♄
13 09 ☽ Q ♂
15 29 ☽ ⚼ ♃
16 23 ♀ ⊥ ♅
19 51 ☽ ⊼ ♇
22 11 ♂ ⚼ ♇

12 Tuesday
00 11 ☽ ⚼ ☿
00 19 ☽ ⊼ ♆
04 04 ☽ ∥ ♃
06 54 ♀ ⊼ ♄
07 22 ♀ ⚹ ♃
07 43 ☽ ⋕ ♀
08 06 ☽ ± ♇
08 26 ☽ ⊼ ♅
16 29 ☿ Q ♅
20 41 ☽ ± ♅
21 02 ☽ ⚼ ☉
21 18 ☽ ± ☿
22 04 ☽ ⊻ ♄
22 08 ☽ △ ♃
23 42 ☽ ☍ ♀

13 Wednesday
02 27 ☽ ⚼ ♇
04 25 ☽ □ ♂
10 16 ☽ ⊥ ♄
10 45 ☽ ∥ ♇
11 22 ☽ ⊼ ☿
14 31 ☽ ⋕ ♆
15 01 ☽ ⚼ ♅
17 03 ☽ ± ☉

14 Thursday
00 04 ☽ ⋕ ♂
03 48 ♀ ∠ ♇
04 31 ☽ ∠ ♄
06 20 ☽ ⊼ ☉
08 52 ☽ △ ♇
13 26 ☽ ☍ ♆
21 20 ☽ △ ♅
23 33 ☽ ⋕ ☿

15 Friday
01 36 ☽ ⋕ ☉
10 38 ☽ ⚹ ♄
10 48 ☽ □ ♃
12 10 ☉ □ ♇
14 11 ☿ ⊥ ♀
18 38 ☽ ⊼ ♀
20 57 ☽ △ ♂

16 Saturday
06 15 ☉ ∥ ☿
07 44 ☽ ± ♀
08 23 ☽ ☍ ☿
10 23 ☽ Q ♄
20 28 ☽ □ ♇
22 20 ☽ Q ♃
23 21 ☽ ☍ ☉

17 Sunday
01 00 ☽ ⊼ ♆
03 10 ☽ ⚼ ♀
04 21 ☽ ⚼ ♂
08 34 ☽ □ ♅
11 51 ♀ ± ♄
12 40 ☽ ± ♆
19 59 ☉ ⊻ ♆
21 23 ☽ □ ♄
21 34 ☽ ⚹ ♃
22 53 ☽ ± ♂

18 Monday
06 00 ☽ ⚼ ♆
07 13 ☽ Q ♇
11 00 ☽ △ ♀
11 06 ☽ ⊼ ♂
13 58 ♀ □ ♂
19 05 ☽ Q ♅

19 Tuesday
02 08 ☽ ∠ ♃
02 23 ☽ ⊼ ☿
05 57 ☽ ⚹ ♇
10 28 ☽ △ ♆
12 56 ♀ ∠ ♅
13 40 ☽ ⊼ ☉
15 13 ☽ ± ☿
17 40 ☽ ⚹ ♅
19 02 ☽ ⊥ ♃

20 Wednesday
00 48 ♂ ⚼ ♅
01 49 ☽ ± ☉
02 31 ☽ ⋕ ☿
06 03 ☽ △ ♄
06 11 ☽ ⊻ ♃
07 29 ☿ □ ♇
07 40 ☽ ⋕ ☉
09 56 ☽ ∠ ♇
10 16 ☽ ⚼ ☿
19 52 ☽ ⚼ ☉
21 28 ☽ ∠ ♅
22 45 ☽ ☍ ♂

21 Thursday
00 38 ☽ □ ♀
02 29 ☽ ⊥ ♇
09 21 ♂ ∥ ♆
09 38 ☽ ⚼ ♄
13 25 ☽ ⊻ ♇
13 55 ☽ ⊥ ♅
16 45 ☽ ⋕ ♆
16 47 ☉ □ ♅
17 08 ☽ ⋕ ♂
17 26 ☽ △ ☿
17 54 ☽ □ ♆
20 11 ☽ ∥ ♇
21 24 ☿ ⊻ ♆

22 Friday
00 47 ☽ ⊻ ♅
01 27 ☽ △ ☉
01 58 ☽ ± ♄
05 31 ☽ ⋕ ♀
09 59 ♃ St R
12 25 ☽ Q ♀
12 46 ☽ ⊼ ♄
12 50 ☽ ☌ ♃
17 13 ♀ Q ♃
23 39 ☽ ∥ ♃

23 Saturday
02 15 ☽ Q ♆
08 07 ☽ ⊼ ♂
11 46 ☽ ⚹ ♀
13 35 ♂ ± ♇
18 59 ☽ ☌ ♇
19 19 ☽ ± ♂
22 07 ☉ ⊥ ♆
22 13 ♃ ⊼ ♄
23 25 ☽ ⚹ ♆

24 Sunday
05 45 ☽ □ ☿
05 59 ☽ ☌ ♅
07 03 ♂ ⋕ ♇
07 49 ☿ □ ♅
10 48 ☽ □ ☉
11 02 ☽ ∥ ♅
11 56 ☽ ⚼ ♂
14 35 ☽ ∥ ♄
15 01 ☽ ⋕ ♄
16 23 ☽ ∠ ♀
17 32 ☽ ⊻ ♃
17 35 ☽ ☍ ♄
18 33 ☽ ⋕ ♅

25 Monday
01 27 ☽ ∠ ♆
03 51 ☽ ⊥ ♃
06 47 ♀ ⚼ ♄
09 15 ☽ ⊥ ♀
15 10 ☽ △ ♂
16 36 ☽ Q ☿
16 49 ☽ ⊥ ♆
18 35 ☿ ⊥ ♆
19 09 ☽ ∠ ♃
20 02 ☽ Q ☉
20 22 ☽ ⊻ ♀
22 36 ☽ ⊻ ♇

26 Tuesday
02 59 ☽ ⊻ ♆
04 47 ♇ St R
04 53 ☽ ⋕ ♃
08 41 ☽ ⊥ ♇
09 14 ☽ ⊻ ♅
15 19 ☽ ⚹ ☿
17 42 ♀ ∥ ♂
17 43 ☽ ⚹ ☉
19 15 ☽ ⊥ ♅
20 18 ☽ ⚹ ♃
20 29 ☽ ⊼ ♄
22 59 ♀ ⚹ ♇
23 43 ☽ ∠ ♇

27 Wednesday
00 34 ☉ ∠ ♂
03 17 ☽ ∥ ♂
04 17 ☽ ∥ ♀
06 24 ☽ ⋕ ♇
06 28 ☽ ± ♄
09 31 ☽ ∥ ♆
10 11 ☽ ∠ ♅
16 08 ☽ Q ♃
19 13 ☽ ∠ ☿
20 04 ☽ □ ♂
20 25 ☽ ∠ ☉
21 20 ☽ ⚼ ♄

28 Thursday
00 28 ☽ ⚹ ♇
02 43 ☽ ☌ ♀
04 50 ☽ ☌ ♆
05 33 ☉ △ ♃
09 37 ☿ ∠ ♂
09 58 ☉ □ ♄
10 50 ☽ ⚹ ♅
11 47 ☽ ⊥ ☿
11 54 ☿ △ ♃
12 16 ☽ ⊥ ☉
14 06 ☽ ∥ ☉
14 54 ☿ □ ♄
20 05 ☽ Q ♇
20 25 ♀ ⋕ ♇
20 41 ♂ ∠ ♄
21 35 ☽ □ ♃
21 56 ☽ △ ♄
22 46 ☽ ⊻ ☿
22 48 ☽ ⊻ ☉
23 18 ☉ ☌ ☿

29 Friday
00 43 ☽ ∥ ☿
03 10 ☽ Q ♂
05 23 ♀ ☌ ♆
06 26 ☽ Q ♅
23 51 ☽ ⚹ ♂

30 Saturday
01 28 ☽ □ ♇
05 57 ☽ ⊻ ♆
08 04 ☽ ⊻ ♀
11 53 ☽ □ ♅
15 49 ☽ ⊥ ♆
18 48 ☽ ⊥ ♀
22 36 ☽ △ ♃
23 11 ☽ □ ♄

31 Sunday
01 55 ☽ ∠ ♂
03 38 ☽ ☌ ☉
06 01 ☽ ☌ ☿
06 04 ♂ ⊼ ♇
06 43 ☽ ∠ ♆
08 45 ♀ ∥ ♆
11 04 ☽ ∠ ♀
12 24 ☿ ∠ ♆
17 05 ♂ ± ♅
17 45 ☽ ⊥ ♂
23 34 ☽ ⚼ ♃

JANUARY 1968

LONGITUDES

Date	Sidereal time h m s	Sun ☉	Moon ☽	Moon ☽ 24.00	Mercury ☿	Venus ♀	Mars ♂	Jupiter ♃	Saturn ♄	Uranus ♅	Neptune ♆	Pluto ♇
01	18 40 51	10 ♑ 08 18	28 ♑ 01 06	04 ♒ 59 45	12 ♑ 13	29 ♏ 28	23 ♒ 50	05 ♍ 41	06 ♈ 08	29 ♍ 13	25 ♏ 39	22 ♍ 51
02	18 44 48	11 09 29	11 ♒ 52 17	18 ♒ 38 24	13 50	00 ♐ 40	24 36	05 R 38	06 10	29 13	25 41	22 R 51
03	18 48 45	12 10 40	25 ♒ 17 59	01 ♓ 51 04	15 28	01 52	25 23	05 36	06 13	29 13	25 43	22 51
04	18 52 41	13 11 51	08 ♓ 17 51	14 ♓ 38 39	17 05	03 04	26 10	05 34	06 16	29 R 13	25 44	22 50
05	18 56 38	14 13 01	20 ♓ 53 53	27 ♓ 04 05	18 44	04 16	26 57	05 31	06 18	29 13	25 46	22 50
06	19 00 34	15 14 11	03 ♈ 09 50	09 ♈ 11 48	20 22	05 28	27 44	05 28	06 21	29 13	25 48	22 50
07	19 04 31	16 15 20	15 ♈ 10 38	21 ♈ 07 05	22 01	06 40	28 31	05 25	06 24	29 13	25 49	22 49
08	19 08 27	17 16 29	27 ♈ 01 49	02 ♉ 55 34	23 40	07 53	29 ♒ 17	05 22	06 27	29 13	25 51	22 49
09	19 12 24	18 17 38	08 ♉ 49 01	14 ♉ 42 50	25 20	09 05	00 ♓ 04	05 19	06 31	29 12	25 52	22 48
10	19 16 20	19 18 46	20 ♉ 36 39	26 ♉ 34 03	26 59	10 17	00 51	05 15	06 34	29 12	25 54	22 48
11	19 20 17	20 19 54	02 ♊ 32 34	08 ♊ 33 41	28 ♑ 39	11 30	01 38	05 11	06 37	29 12	25 55	22 47
12	19 24 14	21 21 01	14 ♊ 37 49	20 ♊ 45 17	00 ♒ 20	12 42	02 25	05 07	06 41	29 11	25 57	22 47
13	19 28 10	22 22 08	26 ♊ 56 23	03 ♋ 11 15	02 00	13 55	03 11	05 03	06 45	29 11	25 58	22 46
14	19 32 07	23 23 15	09 ♋ 30 01	15 ♋ 52 42	03 40	15 07	03 58	04 59	06 48	29 10	26 00	22 46
15	19 36 03	24 24 21	22 ♋ 19 14	28 ♋ 49 30	05 21	16 20	04 45	04 55	06 52	29 10	26 01	22 45
16	19 40 00	25 25 26	05 ♌ 23 19	12 ♌ 00 28	07 01	17 33	05 32	04 50	06 56	29 09	26 03	22 44
17	19 43 56	26 26 31	18 ♌ 40 43	25 ♌ 23 45	08 41	18 46	06 18	04 45	07 00	29 08	26 04	22 44
18	19 47 53	27 27 36	02 ♍ 09 22	08 ♍ 57 16	10 21	19 58	07 05	04 40	07 04	29 07	26 05	22 43
19	19 51 49	28 28 40	15 ♍ 47 15	22 ♍ 39 09	12 00	21 11	07 52	04 35	07 08	29 07	26 07	22 42
20	19 55 46	29 ♑ 29 43	29 ♍ 32 48	06 ♎ 28 07	13 39	22 24	08 38	04 30	07 12	29 06	26 08	22 41
21	19 59 43	00 ♒ 30 47	13 ♎ 25 01	20 ♎ 23 27	15 16	23 37	09 25	04 25	07 17	29 05	26 09	22 40
22	20 03 39	01 31 49	27 ♎ 23 25	04 ♏ 24 52	16 52	24 50	10 12	04 19	07 21	29 04	26 10	22 40
23	20 07 36	02 32 51	11 ♏ 27 45	18 ♏ 32 00	18 27	26 03	10 58	04 13	07 26	29 03	26 11	22 39
24	20 11 32	03 33 54	25 ♏ 37 27	02 ♐ 43 55	19 59	27 17	11 45	04 08	07 30	29 02	26 13	22 38
25	20 15 29	04 34 56	09 ♐ 51 07	16 ♐ 58 40	21 29	28 30	12 32	04 02	07 35	29 01	26 14	22 37
26	20 19 25	05 35 58	24 ♐ 06 07	01 ♑ 12 57	22 56	29 ♐ 43	13 18	03 55	07 40	28 59	26 15	22 36
27	20 23 22	06 36 58	08 ♑ 18 36	15 ♑ 22 24	24 19	00 ♑ 56	14 05	03 49	07 44	28 58	26 16	22 35
28	20 27 18	07 37 58	22 ♑ 23 43	29 ♑ 21 56	25 38	02 10	14 51	03 43	07 49	28 57	26 17	22 34
29	20 31 15	08 38 57	06 ♒ 16 27	13 ♒ 06 44	26 52	03 23	15 38	03 36	07 54	28 56	26 18	22 33
30	20 35 12	09 39 56	19 ♒ 52 21	26 ♒ 32 58	28 01	04 36	16 24	03 30	07 59	28 54	26 19	22 32
31	20 39 08	10 ♒ 40 53	03 ♓ 08 22	09 ♓ 38 27	29 ♒ 03	05 ♑ 50	17 ♓ 11	03 ♍ 23	08 ♈ 04	28 ♍ 53	26 ♏ 20	22 ♍ 31

Moon True ☊, Mean ☊, Latitude

Date	Moon True ☊	Moon Mean ☊	Moon ☽ Latitude
01	24 ♈ 40	23 ♈ 58	05 S 01
02	24 R 28	23 55	04 48
03	24 19	23 52	04 19
04	24 11	23 49	03 37
05	24 07	23 45	02 46
06	24 05	23 42	01 48
07	24 D 04	23 39	00 S 47
08	24 R 04	23 36	00 N 16
09	24 04	23 33	01 17
10	24 01	23 30	02 16
11	23 56	23 26	03 08
12	23 48	23 23	03 53
13	23 38	23 20	04 28
14	23 25	23 17	04 51
15	23 11	23 14	05 00
16	22 57	23 10	04 53
17	22 45	23 07	04 30
18	22 35	23 04	03 52
19	22 28	23 01	03 00
20	22 24	22 58	01 58
21	22 22	22 55	00 N 47
22	22 D 22	22 51	00 S 27
23	22 R 22	22 48	01 39
24	22 21	22 45	02 46
25	22 17	22 42	03 43
26	22 10	22 39	04 26
27	22 00	22 35	04 53
28	21 49	22 32	05 01
29	21 36	22 29	04 51
30	21 24	22 26	04 25
31	21 ♈ 14	22 ♈ 23	03 S 45

DECLINATIONS

Date	Sun ☉	Moon ☽	Mercury ☿	Venus ♀	Mars ♂	Jupiter ♃	Saturn ♄	Uranus ♅	Neptune ♆	Pluto ♇
01	23 S 03	25 S 28	24 S 46	17 S 51	14 S 38	10 N 26	00 N 12	01 N 01	17 S 32	16 N 49
02	22 59	21 50	24 39	18 08	14 22	10 27	00 13	01 01	17 33	16 50
03	22 53	17 09	24 31	18 24	14 05	10 28	00 14	01 01	17 33	16 51
04	22 47	11 49	24 21	18 39	13 49	10 29	00 16	01 01	17 33	16 51
05	22 41	06 09	24 10	18 54	13 32	10 30	00 17	01 01	17 34	16 52
06	22 35	00 S 24	23 57	19 09	13 15	10 32	00 18	01 01	17 34	16 52
07	22 27	05 N 15	23 43	19 23	12 58	10 33	00 20	01 01	17 34	16 53
08	22 20	10 40	23 27	19 37	12 41	10 35	00 21	01 01	17 35	16 54
09	22 12	15 40	23 10	19 50	12 24	10 36	00 23	01 01	17 35	16 54
10	22 03	20 06	22 51	20 03	12 07	10 38	00 24	01 01	17 35	16 55
11	21 54	23 45	22 30	20 15	11 50	10 39	00 26	01 02	17 36	16 56
12	21 45	26 25	22 08	20 27	11 32	10 41	00 27	01 02	17 36	16 56
13	21 35	27 53	21 44	20 38	11 15	10 43	00 29	01 02	17 36	16 57
14	21 25	27 57	21 19	20 49	10 57	10 44	00 31	01 02	17 37	16 58
15	21 15	26 31	20 52	20 59	10 40	10 46	00 32	01 03	17 37	16 58
16	21 04	23 39	20 24	21 08	10 22	10 48	00 34	01 03	17 37	16 59
17	20 52	19 30	19 55	21 17	10 04	10 50	00 36	01 03	17 37	17 00
18	20 40	14 19	19 24	21 26	09 46	10 52	00 38	01 04	17 38	17 00
19	20 28	08 22	18 51	21 33	09 28	10 54	00 40	01 04	17 38	17 01
20	20 16	01 N 59	18 18	21 41	09 10	10 56	00 41	01 04	17 38	17 02
21	20 03	04 S 34	17 43	21 47	08 52	10 58	00 43	01 05	17 38	17 03
22	19 49	10 58	17 07	21 53	08 33	11 01	00 45	01 05	17 39	17 03
23	19 36	16 51	16 30	21 58	08 15	11 03	00 47	01 06	17 39	17 04
24	19 22	21 52	15 53	22 03	07 57	11 05	00 49	01 06	17 39	17 05
25	19 07	25 37	15 15	22 07	07 38	11 08	00 51	01 06	17 39	17 06
26	18 53	27 45	14 37	22 10	07 20	11 10	00 53	01 07	17 39	17 06
27	18 37	28 03	13 58	22 13	07 01	11 13	00 55	01 08	17 40	17 07
28	18 22	26 32	13 20	22 15	06 42	11 15	00 57	01 08	17 40	17 08
29	18 06	23 12	12 42	22 17	06 24	11 18	01 00	01 09	17 40	17 09
30	17 50	19 03	12 06	22 18	06 05	11 20	01 02	01 09	17 40	17 10
31	17 S 34	13 S 51	11 S 31	22 S 18	05 S 46	11 N 23	01 N 04	01 N 10	17 S 40	17 N 10

ZODIAC SIGN ENTRIES

Date	h	m	Planets
01	15	23	☽ ♒
01	22	37	♀ ♐
03	20	35	☽ ♓
06	05	45	☽ ♈
08	18	02	☽ ♉
09	09	49	♂ ♓
11	06	54	☽ ♊
12	07	19	☿ ♒
13	17	54	☽ ♋
16	02	09	☽ ♌
18	08	11	☽ ♍
20	12	47	☽ ♎
20	23	54	☉ ♒
22	16	28	☽ ♏
24	19	23	☽ ♐
26	17	35	♀ ♑
26	21	57	☽ ♑
29	01	06	☽ ♒
31	06	16	☽ ♓

LATITUDES

Date	Mercury ☿	Venus ♀	Mars ♂	Jupiter ♃	Saturn ♄	Uranus ♅	Neptune ♆	Pluto ♇
01	01 S 53	02 N 15	01 S 07	01 N 05	02 S 26	00 N 46	01 N 41	15 N 16
04	02 01	02 10	01 05	01 06	02 25	00 46	01 41	15 18
07	02 06	02 04	01 03	01 06	02 25	00 46	01 42	15 19
10	02 07	01 58	01 01	01 07	02 24	00 46	01 42	15 21
13	02 04	01 51	00 59	01 08	02 23	00 46	01 42	15 22
16	01 57	01 44	00 57	01 08	02 23	00 46	01 42	15 24
19	01 44	01 36	00 54	01 09	02 22	00 47	01 42	15 25
22	01 24	01 28	00 52	01 10	02 21	00 47	01 42	15 27
25	00 57	01 19	00 50	01 10	02 21	00 47	01 42	15 28
28	00 S 23	01 10	00 48	01 11	02 20	00 47	01 43	15 29
31	00 N 19	01 N 01	00 S 46	01 N 12	02 S 20	00 N 47	01 N 43	15 N 30

DATA

Julian Date	2439857
Delta T	+38 seconds
Ayanamsa	23° 24' 30"
Synetic vernal point	05° ♓ 42' 29"
True obliquity of ecliptic	23° 26' 44"

MOON'S PHASES, APSIDES AND POSITIONS ☽

Date	h	m	Phase	Longitude	Eclipse Indicator
07	14	23	☽	16 ♈ 21	
15	16	11	○	24 ♋ 35	
22	19	38	☾	01 ♏ 51	
29	16	29	●	08 ♒ 50	

Day	h	m		
09	12	44	Apogee	
24	23	31	Perigee	
06	13	41	ON	
14	01	03	Max dec	28° N 06'
20	19	15	OS	
27	03	55	Max dec	28° S 10'

LONGITUDES

Date	Chiron ⚷	Ceres ⚳	Pallas ⚴	Juno ⚵	Vesta ⚶	Black Moon Lilith ⚸
01	25 ♓ 35	24 ♎ 57	00 ♎ 57	04 ♏ 51	29 ♑ 19	11 ♉ 12
11	25 ♓ 51	27 ♎ 59	02 ♎ 28	07 ♏ 15	04 ♒ 29	12 ♉ 31
21	26 ♓ 12	00 ♏ 38	03 ♎ 10	09 ♏ 30	09 ♒ 40	13 ♉ 25
31	26 ♓ 37	02 ♏ 51	02 ♎ 59	11 ♏ 02	14 ♒ 50	14 ♉ 32

ASPECTARIAN

h m	Aspects
01 Monday	
03 13	☽ △ ♇
04 27	☽ ⊻ ♂
05 22	☽ Q ♄
06 52	♀ ⚹ ♅
07 58	☽ ⚹ ♆
14 03	☽ △ ♅
14 43	☽ ⚹ ♀
14 50	☽ ± ♃
17 35	☽ ∥ ☿
02 Tuesday	
00 31	☉ ∠ ♆
01 09	☽ ⊼ ♃
02 00	☽ ⚹ ♄
04 39	☽ Q ♆
04 57	☽ ⅊ ♇
05 08	☽ ∥ ☉
10 39	☽ ⊻ ☉
13 32	☽ Q ♀
15 56	☽ ⊻ ☿
16 08	☽ ⅊ ♅
20 48	☽ ± ♇
22 07	☽ ⊥ ☉
03 Wednesday	
04 05	☽ ⊥ ☿
04 35	☽ ∠ ♄
06 20	☽ ∥ ♀
07 33	☽ ⊼ ♇
08 14	☽ ± ♅
10 06	☽ ∥ ♆
12 10	☽ ☌ ♂
12 45	☽ □ ♆
13 27	☽ # ♇
15 42	☽ ∠ ☉
19 09	☽ ⊼ ♅
21 01	☽ ⊥ ♄
22 19	♂ □ ♆
22 47	☽ ∠ ☿
04 Thursday	
01 15	☽ □ ♀
02 48	☽ ∥ ♂
06 15	♅ St R
06 54	☽ ☍ ♃
08 10	☽ ⊻ ♄
17 42	☽ # ♃
22 04	☽ ⚹ ☉
05 Friday	
07 11	☽ ⚹ ☿
15 45	☽ ☍ ♇
21 29	☽ △ ♆
23 15	♀ Q ♇
23 16	☽ Q ☉
06 Saturday	
00 34	☽ ⊻ ♂
04 13	☽ ☍ ♅
09 26	☽ # ♅
10 11	☽ Q ☿
12 04	♀ □ ♃
12 24	☽ # ♄
13 12	☽ ⊥ ♂
13 28	☿ ⅊ ♃
14 58	☽ ∥ ♄
16 33	☽ ⊼ ♃
17 05	☽ △ ♀
17 26	☿ ∠ ♀
17 57	☽ ∥ ♅
18 22	☽ ☌ ♄
07 Sunday	
03 14	☽ ⅊ ♆
04 29	☽ ± ♃
06 28	♀ △ ♄
08 25	☽ ∠ ♂
14 23	☽ □ ☉
21 24	☽ ± ♆
22 33	☽ ⅊ ♃
23 38	☿ △ ♇
08 Monday	
01 38	☿ ⊥ ♂
02 36	☽ ⅊ ♀
03 26	☽ ⊼ ♇
04 04	☽ □ ☿
09 32	♂ ⊼ ♅
09 36	☽ ⊼ ♆
11 36	☽ ∥ ♃
15 37	☽ ± ♇
16 26	☽ ⊼ ♅
16 55	☽ ⚹ ♂
20 58	☽ # ♂
22 59	☽ ± ♀
23 49	☿ Q ♄
09 Tuesday	
04 39	☽ ± ♅
04 54	☽ △ ♃
07 17	☽ ⊻ ♄
09 56	☽ ⅊ ♇
12 36	☽ ⊼ ♀
18 24	☽ ∥ ♇
19 06	☽ Q ♂
19 33	☽ ⊥ ♄
20 02	☿ ⚹ ♆
21 59	☽ # ♆
22 58	☽ ⅊ ♅
10 Wednesday	
02 36	♂ ⊥ ♄
09 05	☽ △ ☉
11 45	☽ # ♀
13 55	☽ ∠ ♄
16 23	☽ △ ♇
22 41	☽ ☍ ♆
23 41	☽ # ☉
11 Thursday	
02 56	☽ △ ☿
03 56	☽ # ☿
05 17	☽ △ ♅
06 03	♀ Q ♅
08 51	☉ ⅊ ♃
10 03	☽ □ ♂
17 16	☽ □ ♃
18 05	☽ ⅊ ☉
19 23	☿ ± ♃
19 42	☿ △ ♅
20 11	☽ ⚹ ♄
12 Friday	
07 47	☽ ☍ ♀
13 33	☽ ± ☉
13 35	☽ ⅊ ☿
20 00	☽ Q ♄
13 Saturday	
02 21	☽ ⊼ ☉
03 56	☽ □ ♇
04 31	☽ Q ♃
09 54	☽ ± ☿
10 08	☽ ⊼ ♆
16 19	☽ □ ♅
21 21	☉ △ ♇
21 42	☽ ± ♆
23 13	☽ ⊼ ☿
14 Sunday	
00 48	☽ △ ♂
02 45	☉ ∥ ☿
03 29	☽ ⚹ ♃
06 52	☽ □ ♄
14 23	☽ Q ♇
14 50	☽ ⅊ ♆
19 56	☿ ⊻ ♂
23 41	☽ ⊼ ♀
15 Monday	
02 24	☽ Q ♅
03 55	♂ # ♃
06 01	☿ ⊼ ♃
06 55	☽ ⅊ ♂
07 33	☽ ∠ ♃
07 53	☿ ∥ ♀
12 02	☽ ± ♀
12 48	☽ ⚹ ♇
16 11	☽ ☍ ☉
16 34	♂ ☍ ♃
18 51	☽ △ ♆
23 38	☉ Q ♄
16 Tuesday	
00 05	☽ ⊥ ♃
00 36	☽ ⚹ ♅
00 37	☽ ± ♂
06 17	☽ ⅊ ♀
06 25	☉ ∥ ♀
10 42	☿ ⚹ ♄
11 00	☽ ⊻ ♃
12 16	☽ ⊼ ♂
14 49	☽ △ ♄
15 24	☽ ☍ ☿
16 16	☽ ∠ ♇
22 15	☿ ⅊ ♇
17 Wednesday	
02 49	☽ # ♀
02 55	☿ Q ♆
02 56	☉ ⚹ ♆
03 51	☽ ∠ ♅
04 29	☽ # ☉
08 30	☽ ⊥ ♇
09 40	☽ # ☿
12 10	☽ △ ♀
17 58	☽ ⅊ ♄
19 14	☽ ⊻ ♇
19 58	☽ ⊥ ♅
21 12	☽ # ♆
18 Thursday	
00 07	☽ ∥ ♇
01 13	☽ □ ♆
03 00	☽ ⊼ ☉
06 38	☽ ⊻ ♅
10 04	☽ ± ♄
11 24	♂ ⊻ ♄
14 30	☽ ± ☉
16 25	☽ ☌ ♃
20 43	☽ ⊼ ♄
21 14	☽ ☍ ♂
19 Friday	
02 08	☽ ∥ ♃
04 27	☽ ⊼ ☿
07 33	☽ # ♂
07 37	☽ ⅊ ☉
09 03	☽ Q ♆
14 24	☉ ± ♃
16 25	☽ ± ☿
22 22	☽ □ ♀
20 Saturday	
00 04	☽ ☌ ♇
02 43	☉ △ ♅
06 03	☽ ⚹ ♆
10 13	☽ ⅊ ☿
11 13	☽ ☌ ♅
11 54	☽ △ ☉
15 19	☽ ∥ ♅
16 42	☽ ∥ ♄
17 31	♀ □ ♇
18 36	☿ ⅊ ♅
20 32	☽ ⊻ ♃
21 49	☽ # ♄
23 11	☽ # ♅
21 Sunday	
01 21	☽ ☍ ♄
04 41	☽ ⊼ ♂
06 51	☽ ⊥ ♃
08 05	☽ ∠ ♆
08 36	☽ Q ♀
15 04	☿ ∥ ♆
15 36	☽ △ ☿
15 39	☽ ± ♂
22 15	☽ ∠ ♃
23 36	☽ ⊥ ♆
22 Monday	
03 14	☽ ∥ ♂
03 54	☽ ⊻ ♇
07 13	☽ ⚹ ♀
08 01	☽ ⅊ ♂
08 53	☿ ± ♇
09 55	☽ ⊻ ♆
12 12	☽ # ♃
14 10	☽ ⊥ ♇
14 25	☿ # ♇
14 52	☽ ⊻ ♅
19 38	☽ □ ☉
23 45	☽ ⚹ ♃
23 Tuesday	
01 06	☽ ⊥ ♅
05 06	☽ ⊼ ♄
05 31	☽ ∠ ♇
10 41	☽ ∥ ☿
11 07	☽ △ ♂
11 15	☽ ∠ ♀
12 58	☽ # ♇
14 39	♀ ⊻ ♆
15 21	☽ ± ♄
15 31	☽ ∥ ♆
16 23	☽ ∠ ♅
20 02	☽ Q ♃
23 58	☽ ∥ ☉
24 Wednesday	
01 18	☽ □ ☿
03 57	☽ ⊥ ♀
04 36	☽ Q ☉
06 42	☽ ⅊ ♄
06 57	☽ ⚹ ♇
12 59	☽ ☌ ♆
13 03	☽ ∥ ♀
15 03	☽ ⊻ ♀
17 45	☽ ⚹ ♅
25 Thursday	
00 03	☉ ⊼ ♃
02 15	☽ □ ♃
02 26	☽ ⚹ ☉
03 11	☽ Q ♇
08 09	☽ △ ♄
11 18	☽ Q ☿
13 57	☽ Q ♅
16 46	☽ □ ♂
21 58	♀ □ ♅
22 12	☉ ⊥ ♀
26 Friday	
05 39	☽ ∠ ☉
06 30	☿ ⊼ ♇
07 11	☿ ∠ ♄
09 28	☽ □ ♇
09 48	☽ ⚹ ☿
13 02	☿ ± ♅
15 37	☽ ⊻ ♆
20 14	☽ □ ♅
21 59	☽ ⊥ ☉
22 21	☽ ☌ ♀
27 Saturday	
00 51	☽ Q ♂
01 45	☽ ⊥ ♆
04 27	☽ △ ♃
08 55	☽ ⊻ ☉
11 02	☽ □ ♄
13 53	☽ ∠ ☿
17 01	☽ ∠ ♆
22 22	☽ ⚹ ♂
28 Sunday	
05 45	☽ ⅊ ♃
06 48	☽ ⊥ ☿
10 28	☉ ⅊ ♇
12 18	☽ △ ♇
14 24	♀ ⊥ ♆
16 51	☉ ⚹ ♄
17 55	☽ Q ♄
18 08	☽ ⊻ ☿
18 41	☽ ⚹ ♆
21 05	☽ ± ♃
23 16	☽ △ ♅
29 Monday	
00 25	☿ □ ♆
01 36	☽ ∠ ♂
03 31	☉ Q ♆
06 28	☽ ⊻ ♀
07 23	☽ ⊼ ♃
14 14	☽ ⅊ ♇
14 52	☽ ⚹ ♄
15 32	☽ Q ♆
16 03	♀ △ ♃
16 29	☽ ☌ ☉
17 58	☽ ⊥ ♀
18 14	☽ ⊥ ♂
18 47	☽ ∥ ♀
30 Tuesday	
01 24	♀ Q ♂
01 25	☽ ⅊ ♅
05 27	☽ ⊻ ♂
06 04	☽ ± ♇
11 28	☽ ∠ ♀
16 45	☽ ⊼ ♇
17 25	☽ ± ♅
17 37	☽ ∠ ♄
18 14	☽ ∥ ☉
18 41	☽ ∥ ♆
21 04	☽ # ♇
23 35	☽ □ ♆
31 Wednesday	
02 32	☉ ∥ ♆
03 56	☽ ☌ ☿
04 15	☽ ⊼ ♅
07 57	☿ ⊼ ♅
10 02	☽ ⊥ ♄
12 27	☽ ☍ ♃
17 01	☿ # ♃
17 27	☽ ⚹ ♀
21 09	☽ ⊻ ♄
22 33	☽ # ♃
23 12	☽ ∥ ☿

All ephemeris data is given at 12.00 UT and the Moon's longitude is additionally given for 24.00 UT

FEBRUARY 1968

LONGITUDES

Date	Sidereal time h m s	Sun ☉ ° ' "	Moon ☽ ° ' "	Moon ☽ 24.00 ° ' "	Mercury ☿ ° '	Venus ♀ ° '	Mars ♂ ° '	Jupiter ♃ ° '	Saturn ♄ ° '	Uranus ♅ ° '	Neptune ♆ ° '	Pluto ♇ ° '
01	20 43 05	11 ♒ 41 48	16 ♓ 03 15	22 ♓ 22 54	29 ♒ 58	07 ♑ 03	17 ♓ 57	03 ♍ 16	08 ♈ 10	28 ♍ 51	26 ♏ 21	22 ♍ 30
02	20 47 01	12 42 43	28 ♓ 37 40	04 ♈ 47 54	00 ♓ 45	08 16	18 44	03 R 09	08 15	28 R 50	26 21	22 R 29
03	20 50 58	13 43 36	10 ♈ 54 03	16 ♈ 56 36	01 23	09 30	19 30	03 02	08 20	28 48	26 22	22 27
04	20 54 54	14 44 28	22 ♈ 56 08	28 ♈ 53 17	01 51	10 43	20 17	02 55	08 26	28 47	26 23	22 26
05	20 58 51	15 45 19	04 ♉ 48 42	10 ♉ 43 02	02 09	11 57	21 03	02 48	08 31	28 45	26 24	22 25
06	21 02 47	16 46 08	16 ♉ 37 01	22 ♉ 31 18	02 17	13 10	21 49	02 40	08 37	28 43	26 24	22 24
07	21 06 44	17 46 56	28 ♉ 26 36	04 ♊ 23 35	02 R 14	14 24	22 35	02 33	08 42	28 42	26 25	22 23
08	21 10 41	18 47 42	10 ♊ 22 52	16 ♊ 25 03	01 59	15 37	23 22	02 26	08 48	28 40	26 26	22 21
09	21 14 37	19 48 27	22 ♊ 30 42	28 ♊ 40 18	01 34	16 51	24 08	02 18	08 54	28 38	26 26	22 20
10	21 18 34	20 49 11	04 ♋ 54 14	11 ♋ 12 51	00 59	18 05	24 54	02 10	09 00	28 36	26 27	22 19
11	21 22 30	21 49 52	17 ♋ 36 23	24 ♋ 04 57	00 ♓ 14	19 18	25 40	02 03	09 06	28 34	26 28	22 17
12	21 26 27	22 50 33	00 ♌ 38 35	07 ♌ 17 10	29 ♒ 22	20 32	26 27	01 55	09 12	28 32	26 28	22 16
13	21 30 23	23 51 12	14 ♌ 00 33	20 ♌ 48 25	28 22	21 46	27 13	01 47	09 18	28 30	26 29	22 15
14	21 34 20	24 51 49	27 ♌ 40 24	04 ♍ 36 02	27 18	22 59	27 59	01 40	09 24	28 28	26 29	22 13
15	21 38 16	25 52 25	11 ♍ 34 50	18 ♍ 36 15	26 10	24 13	28 45	01 32	09 30	28 26	26 29	22 12
16	21 42 13	26 52 59	25 ♍ 40 11	02 ♎ 44 48	25 01	25 27	29 ♓ 31	01 24	09 36	28 24	26 30	22 11
17	21 46 10	27 53 32	09 ♎ 50 54	16 ♎ 57 37	23 52	26 40	00 ♈ 17	01 16	09 42	28 22	26 30	22 09
18	21 50 06	28 54 04	24 ♎ 04 32	01 ♏ 11 18	22 46	27 54	01 03	01 08	09 48	28 20	26 31	22 08
19	21 54 03	29 ♒ 54 34	08 ♏ 17 39	15 ♏ 23 20	21 43	29 ♑ 08	01 48	01 00	09 55	28 18	26 31	22 06
20	21 57 59	00 ♓ 55 04	22 ♏ 28 09	29 ♏ 31 57	20 45	00 ♒ 22	02 34	00 52	10 01	28 16	26 31	22 05
21	22 01 56	01 55 32	06 ♐ 34 35	13 ♐ 35 55	19 53	01 36	03 20	00 44	10 08	28 14	26 31	22 03
22	22 05 52	02 55 59	20 ♐ 35 49	27 ♐ 34 06	19 07	02 49	04 06	00 37	10 14	28 11	26 31	22 02
23	22 09 49	03 56 24	04 ♑ 30 38	11 ♑ 25 11	18 29	04 03	04 52	00 29	10 21	28 09	26 32	22 00
24	22 13 45	04 56 48	18 ♑ 17 32	25 ♑ 07 28	17 58	05 17	05 37	00 21	10 27	28 07	26 32	21 59
25	22 17 42	05 57 11	01 ♒ 54 42	08 ♒ 39 00	17 34	06 31	06 23	00 13	10 34	28 04	26 32	21 57
26	22 21 39	06 57 32	15 ♒ 20 05	21 ♒ 57 45	17 17	07 45	07 09	00 ♍ 05	10 41	28 02	26 32	21 56
27	22 25 35	07 57 51	28 ♒ 31 46	05 ♓ 01 59	17 08	08 59	07 54	29 ♌ 57	10 47	28 00	26 R 32	21 54
28	22 29 32	08 58 09	11 ♓ 28 18	17 ♓ 50 39	17 D 06	10 13	08 40	29 50	10 54	27 57	26 32	21 53
29	22 33 28	09 ♓ 58 25	24 ♓ 09 03	00 ♈ 23 35	17 ♒ 10	11 ♒ 27	09 ♈ 26	29 ♌ 42	11 ♈ 01	27 ♍ 55	26 ♏ 32	21 ♍ 51

Date	Moon True ☊ ° '	Moon Mean ☊ ° '	Moon ☽ Latitude ° '
01	21 ♈ 06	22 ♈ 20	02 S 54
02	21 R 01	22 16	01 56
03	20 59	22 13	00 S 53
04	20 D 58	22 10	00 N 10
05	20 59	22 07	01 13
06	20 59	22 04	02 12
07	20 R 59	22 01	03 06
08	20 56	21 57	03 52
09	20 51	21 54	04 29
10	20 44	21 51	04 53
11	20 35	21 48	05 04
12	20 24	21 45	05 00
13	20 14	21 41	04 39
14	20 05	21 38	04 02
15	19 57	21 35	03 10
16	19 52	21 32	02 06
17	19 50	21 29	00 N 53
18	19 D 49	21 26	00 S 23
19	19 50	21 22	01 38
20	19 51	21 19	02 47
21	19 R 51	21 16	03 45
22	19 50	21 13	04 29
23	19 46	21 10	04 58
24	19 41	21 07	05 08
25	19 34	21 03	05 01
26	19 26	21 00	04 38
27	19 19	20 57	04 00
28	19 13	20 54	03 10
29	19 ♈ 08	20 ♈ 51	02 S 11

DECLINATIONS

Date	Sun ☉ ° '	Moon ☽ ° '	Mercury ☿ ° '	Venus ♀ ° '	Mars ♂ ° '	Jupiter ♃ ° '	Saturn ♄ ° '	Uranus ♅ ° '	Neptune ♆ ° '	Pluto ♇ ° '
01	17 S 17	08 S 11	10 S 57	22 S 17	05 S 27	11 N 26	01 N 06	01 N 10	17 S 40	17 N 11
02	17 00	02 S 19	10 25	22 16	05 08	11 28	01 08	01 11	17 41	17 12
03	16 43	03 N 30	09 57	22 14	04 50	11 31	01 11	01 12	17 41	17 13
04	16 25	09 05	09 31	22 12	04 31	11 34	01 13	01 12	17 41	17 14
05	16 07	14 17	09 08	22 09	04 12	11 36	01 15	01 13	17 41	17 14
06	15 49	18 55	08 50	22 05	03 53	11 39	01 18	01 14	17 41	17 15
07	15 30	22 51	08 35	22 01	03 34	11 42	01 20	01 14	17 41	17 16
08	15 12	25 50	08 26	21 56	03 15	11 45	01 22	01 15	17 41	17 17
09	14 53	27 42	08 20	21 50	02 56	11 48	01 25	01 16	17 41	17 18
10	14 33	28 14	08 20	21 43	02 37	11 51	01 27	01 17	17 41	17 18
11	14 14	27 19	08 24	21 36	02 18	11 53	01 30	01 17	17 42	17 19
12	13 54	24 54	08 32	21 29	01 59	11 56	01 32	01 18	17 42	17 20
13	13 34	21 05	08 44	21 20	01 40	11 59	01 35	01 19	17 42	17 21
14	13 14	16 04	09 00	21 12	01 21	12 02	01 37	01 20	17 42	17 22
15	12 54	10 09	09 19	21 02	01 02	12 05	01 40	01 21	17 42	17 22
16	12 33	03 N 39	09 41	20 52	00 43	12 08	01 42	01 22	17 42	17 23
17	12 13	03 S 05	10 04	20 41	00 24	12 11	01 45	01 22	17 42	17 24
18	11 52	09 42	10 28	20 30	00 S 05	12 14	01 47	01 23	17 42	17 25
19	11 30	15 49	10 53	20 18	00 N 14	12 17	01 50	01 24	17 42	17 26
20	11 09	21 05	11 18	20 05	00 33	12 20	01 53	01 25	17 42	17 27
21	10 48	25 06	11 42	19 52	00 52	12 22	01 55	01 26	17 42	17 27
22	10 26	27 35	12 05	19 38	01 11	12 25	01 58	01 27	17 42	17 28
23	10 04	28 20	12 27	19 24	01 29	12 28	02 01	01 28	17 42	17 29
24	09 42	27 17	12 48	19 09	01 48	12 31	02 03	01 29	17 42	17 30
25	09 20	24 38	13 07	18 53	02 07	12 34	02 06	01 30	17 42	17 31
26	08 58	20 40	13 25	18 37	02 26	12 37	02 09	01 31	17 42	17 31
27	08 35	15 44	13 40	18 21	02 44	12 39	02 12	01 32	17 42	17 32
28	08 13	10 11	13 54	18 04	03 03	12 42	02 14	01 33	17 42	17 33
29	07 S 50	04 S 20	14 S 05	17 S 46	03 N 22	12 N 45	02 N 17	01 N 33	17 S 42	17 N 34

ZODIAC SIGN ENTRIES

Date	h	m	Planets
01	12	57	☿ ♓
02	14	39	☽ ♈
05	02	15	☽ ♉
07	15	09	☽ ♊
10	02	34	☽ ♋
11	18	54	☿ ♒
12	10	50	☽ ♌
14	16	02	☽ ♍
16	19	21	☽ ♎
17	03	18	♂ ♈
18	22	00	☽ ♏
19	14	09	☉ ♓
20	04	55	♀ ♒
21	00	48	☽ ♐
23	04	12	☽ ♑
25	08	37	☽ ♒
27	03	33	♃ ♌
27	14	42	☽ ♓
29	23	14	☽ ♈

LATITUDES

Date	Mercury ☿ ° '	Venus ♀ ° '	Mars ♂ ° '	Jupiter ♃ ° '	Saturn ♄ ° '	Uranus ♅ ° '	Neptune ♆ ° '	Pluto ♇ ° '
01	00 N 34	00 N 58	00 S 45	01 N 12	02 S 19	00 N 47	01 N 43	15 N 31
04	01 24	00 49	00 43	01 12	02 19	00 47	01 43	15 32
07	02 14	00 40	00 41	01 13	02 18	00 47	01 43	15 33
10	03 00	00 30	00 39	01 13	02 18	00 47	01 43	15 34
13	03 31	00 21	00 36	01 14	02 17	00 47	01 43	15 35
16	03 43	00 12	00 34	01 14	02 17	00 47	01 44	15 36
19	03 35	00 N 03	00 32	01 14	02 17	00 48	01 44	15 37
22	03 10	00 S 06	00 30	01 14	02 16	00 48	01 44	15 37
25	02 34	00 15	00 28	01 15	02 16	00 48	01 44	15 38
28	01 55	00 23	00 25	01 15	02 16	00 48	01 44	15 39
31	01 N 14	00 S 31	00 S 23	01 N 15	02 S 15	00 N 48	01 N 44	15 N 39

DATA

Julian Date	2439888
Delta T	+38 seconds
Ayanamsa	23° 24' 36"
Synetic vernal point	05° ♓ 42' 24"
True obliquity of ecliptic	23° 26' 45"

MOON'S PHASES, APSIDES AND POSITIONS ☽

Date	h	m	Phase	Longitude ° '	Eclipse Indicator
06	12	20	☽	16 ♉ 47	
14	06	43	○	24 ♌ 38	
21	03	28	☾	01 ♐ 34	
28	06	56	●	08 ♓ 45	

Day	h	m	
06	09	59	Apogee
18	16	32	Perigee

02	21	30	ON	
10	09	01	Max dec	28° N 15'
17	01	02	0S	
23	09	48	Max dec	28° S 20'

LONGITUDES

Date	Chiron ⚷ ° '	Ceres ⚳ ° '	Pallas ⚴ ° '	Juno ⚵ ° '	Vesta ⚶ ° '	Black Moon Lilith ⚸ ° '
01	26 ♓ 40	03 ♏ 03	02 ♎ 55	11 ♏ 11	15 ♒ 21	14 ♉ 38
11	27 ♓ 09	04 ♏ 42	01 ♎ 41	12 ♏ 25	20 ♒ 29	15 ♉ 45
21	27 ♓ 41	05 ♏ 45	29 ♍ 34	13 ♏ 09	25 ♒ 35	16 ♉ 51
31	28 ♓ 15	06 ♏ 09	26 ♍ 44	13 ♏ 20	00 ♓ 39	17 ♉ 58

All ephemeris data is given at 12.00 UT and the Moon's longitude is additionally given for 24.00 UT

ASPECTARIAN

h m Aspects

01 Thursday
03 08 ☽ ⊻ ☉
15 22 ☽ ⊥ ☉
15 49 ☽ ☌ ♂
18 16 ☽ Q ♀
19 56 ☉ ♯ ♇
23 49 ☽ ∥ ♂

02 Friday
00 12 ☽ ☍ ♇
07 37 ☽ △ ♆
10 04 ☽ ∠ ☉
11 29 ♀ □ ♄
12 24 ☽ ☍ ♅
13 40 ☉ ⊥ ♂
16 21 ☽ ⊻ ☿
16 38 ☽ ♯ ♅
16 48 ☽ ♯ ♄
20 42 ☽ ⊼ ♃

03 Saturday
02 18 ☽ ∥ ♄
02 25 ☽ ∥ ♅
04 43 ☽ ⊥ ☿
06 54 ☽ ☌ ♄
08 21 ☽ ± ♃
08 55 ☽ □ ♀
12 56 ☽ ⚻ ♆
13 50 ☉ ⚻ ♅
17 19 ☽ ♯ ♂
18 07 ☽ ⚹ ☉
23 22 ☽ ∠ ☿

04 Sunday
02 02 ☽ ⚻ ♃
02 41 ♄ ∥ ♅
06 18 ☽ ⊻ ♂
06 52 ☽ ± ♆
11 00 ☽ ⊼ ♇
13 47 ☽ ♯ ☿
18 57 ☽ ⊼ ♆
19 11 ☽ ⊥ ♂
20 23 ☽ Q ☉
23 04 ☽ ± ♇
23 17 ☽ ∥ ♃
23 45 ☽ ⊼ ♅

05 Monday
01 05 ♀ ∠ ♆
06 30 ☽ ⚹ ☿
07 57 ☽ △ ♃
11 53 ☽ ± ♅
14 41 ☽ ∠ ♂
17 17 ☽ ⚻ ♇
19 36 ☽ ⊻ ♄
20 35 ☽ ♯ ☉

06 Tuesday
02 57 ☽ ∥ ♇
03 22 ☉ ± ♇
04 11 ☽ △ ♀
05 16 ☽ ♯ ♆
06 08 ☽ ⚻ ♅
07 14 ☽ Q ☿
07 53 ☽ ⊥ ♄
12 20 ☽ □ ☉
16 40 ☿ St R
18 45 ♃ ± ♄
23 19 ☽ ⚹ ♂
23 44 ☽ △ ♇

07 Wednesday
02 20 ☽ ∠ ♄
05 29 ♂ ☍ ♇
06 34 ☽ ♯ ♀
07 54 ☽ ☍ ♆
12 30 ☽ △ ♅
14 09 ☽ ⚻ ♀
19 32 ☽ □ ☿
20 12 ☽ □ ♃

08 Thursday
01 15 ☽ Q ♂
08 49 ☽ ⚹ ♄
10 19 ☽ ± ♀
23 36 ☽ ⊼ ♀

09 Friday
06 13 ☽ △ ☉
07 42 ☽ Q ♃
08 04 ☿ ∠ ♀
08 48 ☽ Q ♄
11 39 ☽ □ ♇
15 23 ☽ □ ♂
19 40 ☽ ⊼ ♆
19 58 ♀ ⚻ ♃
23 54 ☽ □ ♅

10 Saturday
04 50 ☽ △ ☿
06 49 ☽ ⚹ ♃
07 17 ☽ ± ♆
13 54 ☽ ⚻ ☉
19 51 ☽ □ ♄
22 17 ☽ Q ♇

11 Sunday
00 27 ☽ ⚻ ♆
07 50 ☽ ⚻ ☿
08 24 ☽ ± ☉
10 05 ☽ Q ♅
10 58 ☽ ∠ ♃
15 30 ☽ ☍ ♀
20 31 ☽ ⊼ ☉
20 41 ☽ ⚹ ♇
22 40 ☉ ⊼ ♇
23 32 ☽ ± ☿

12 Monday
03 27 ☽ ⊥ ♃
03 51 ☽ △ ♂
04 23 ☽ △ ♆
05 04 ☉ ± ♅
08 11 ☽ ⚹ ♅
09 49 ☽ ⊼ ☿
12 50 ♂ △ ♆
14 18 ☽ ⊻ ♃
23 57 ☽ ∠ ♇

13 Tuesday
03 32 ☽ △ ♄
08 37 ☽ ⚻ ♂
08 41 ☿ ⊼ ♅
10 31 ☽ ♯ ♀
11 07 ☽ ∠ ♅
15 57 ☽ ⊥ ♇
17 52 ♂ ♯ ♄
18 26 ☿ ⊥ ♀
21 19 ♀ △ ♇
23 35 ☉ ∠ ♄

14 Wednesday
01 28 ☽ ± ♂
02 30 ☽ ⊻ ♇
02 57 ☽ ⊥ ♅
03 01 ☽ ⊼ ♀
03 12 ☿ ⊻ ♂
04 44 ☽ ♯ ♆
06 14 ☽ ⚻ ♄
06 17 ☽ ∥ ♇
06 43 ☽ ☍ ☉
09 56 ☽ □ ♆
11 23 ☽ ☍ ☿
12 34 ☽ ⊼ ♂
13 20 ♂ ♯ ♅
13 23 ☽ ⊻ ♅
14 30 ☽ ± ♀
18 51 ☽ ☌ ♃
21 59 ☽ ± ♄

15 Thursday
00 35 ☽ ♯ ☉
02 53 ♂ ☍ ♅
04 33 ☽ ∥ ♃
05 13 ☿ □ ♆
07 33 ☽ ⚻ ♀
08 24 ☽ ⊼ ♄
15 00 ☽ ♯ ☿
15 16 ☉ ☌ ☿
16 59 ☽ Q ♆

16 Friday
02 46 ☉ □ ♆
06 05 ☽ ☌ ♇
07 40 ☿ ⊻ ♀
10 59 ☽ ⊼ ☿
11 10 ♀ ± ♃
11 36 ☽ △ ♀
13 25 ☽ ⚹ ♆
14 14 ☽ ⊼ ☉
16 38 ☽ ☌ ♅
18 54 ☽ ☍ ♂
18 56 ☽ ∥ ♄
19 58 ☿ ∠ ♄
20 11 ☽ ∥ ♅
20 23 ☽ ± ☿
21 38 ☽ ⊻ ♃
23 00 ☽ ♯ ♂

17 Saturday
01 10 ☽ ± ☉
02 53 ☽ ∥ ♂
05 54 ☽ ♯ ♅
06 51 ☿ ⊥ ♂
07 13 ☽ ♯ ♄
07 41 ☽ ⊥ ♃
08 39 ♀ ⚹ ♆
10 28 ☽ ⚻ ☿
11 45 ☽ ☍ ♄
13 44 ☉ ♯ ♃
14 48 ☽ ∠ ♆
17 32 ☽ ⚻ ☉
22 44 ☽ ∠ ♃
23 01 ☉ ⊼ ♅

18 Sunday
05 59 ☽ ⊥ ♆
08 43 ☽ ⊻ ♇
09 57 ☽ △ ☿
09 57 ♀ Q ♄
14 28 ♂ ⊼ ♃
15 06 ☽ ∥ ☿
16 06 ☽ ⊻ ♆
18 49 ☽ ⊥ ♇
19 04 ☽ □ ♀
19 10 ☽ ⊻ ♅
19 45 ☽ ∥ ☉
20 12 ♀ △ ♅
20 46 ☽ △ ☉
21 42 ☽ ♯ ♃
21 54 ☿ ± ♅
23 48 ☽ ⚹ ♃

19 Monday
00 25 ☽ ⊼ ♂
02 39 ☿ ⊼ ♇
05 16 ☽ ⊥ ♅
10 00 ☽ ∠ ♇
11 08 ☽ ± ♂
14 45 ☽ ⊼ ♄
18 53 ☽ ♯ ♇
19 53 ☽ Q ♃
20 03 ☽ ∥ ♆
20 27 ☽ ∠ ♅

20 Tuesday
00 59 ☽ ± ♄
03 14 ☽ ⚻ ♂
04 23 ☽ Q ♀
07 16 ☽ ∥ ♀
07 31 ☉ ∥ ☿
09 16 ☽ □ ☿
11 02 ☉ ☍ ♃
11 20 ☽ ⚹ ♇
16 22 ☽ ⚻ ♄
18 53 ☽ ☌ ♆
20 58 ♀ ⊼ ♃
21 49 ☽ ⚹ ♅

21 Wednesday
02 09 ☽ □ ♃
02 42 ☽ ⚹ ♀
03 28 ☽ □ ☉
06 10 ☽ △ ♂
07 42 ☽ Q ♇
14 06 ☽ Q ☿
18 07 ☽ △ ♄
18 13 ☽ Q ♅

22 Thursday
06 47 ☽ ∠ ♀
09 35 ☽ ⚹ ☿
12 20 ☿ ∠ ♂
12 37 ☽ Q ☉
14 28 ☽ □ ♇
22 12 ☽ ⊻ ♆
23 45 ☉ ⊻ ♀
23 45 ☽ ⊥ ♀

23 Friday
01 02 ☽ □ ♅
05 05 ☽ △ ♃
08 34 ☽ ⊥ ♆
09 52 ♂ ∥ ♅
10 17 ☽ ∠ ☿
10 56 ☽ ⚹ ☉
11 08 ☽ ⊻ ♀
12 39 ☽ □ ♂
12 52 ☿ ♯ ♃
22 13 ☽ □ ♄
22 50 ☉ ⊥ ♄

24 Saturday
00 11 ☽ ∠ ♆
01 19 ☽ ⊥ ☿
06 54 ☽ ⚻ ♃
11 26 ☽ ⊻ ☿
15 08 ☽ ∠ ☉
18 27 ☽ △ ♇
21 55 ☽ Q ♂
22 32 ☽ ± ♃

25 Sunday
02 29 ☽ ⚹ ♆
05 13 ☽ △ ♅
05 21 ♀ ⚹ ♂
06 01 ☽ Q ♄
07 23 ♂ ± ♃
08 15 ☽ ⊥ ☉
09 01 ☽ ⊼ ♃
10 47 ♂ ∥ ♄
19 46 ☽ ⊻ ☉
20 22 ♀ ⚻ ♇
20 26 ☽ ⚹ ♂
20 57 ☽ ⚻ ♇
21 01 ☽ ☌ ♀
23 47 ☽ Q ♆

26 Monday
03 33 ☽ ⚹ ♄
07 52 ☽ ⚻ ♅
13 04 ☽ ± ♇
15 29 ☽ ☌ ☿
23 01 ☽ ∥ ♀
23 55 ☽ ⊼ ♇

27 Tuesday
00 06 ☽ ± ♅
01 05 ☽ ∠ ♂
02 53 ☽ ∥ ♆
03 17 ♀ Q ♆
03 40 ☽ ♯ ♇
06 24 ☉ ⊻ ♂
06 56 ☽ ∠ ♄
08 20 ☽ □ ♆
08 53 ♆ St R
11 01 ☽ ⊼ ♅
14 36 ☽ ☍ ♃
18 36 ☽ ⊥ ♂
20 48 ☽ ∥ ☿
23 39 ☽ ⊥ ♄

28 Wednesday
01 26 ☽ ♯ ♃
06 26 ☽ ⊻ ♂
06 56 ☽ ☌ ☉
08 37 ☿ St D
09 23 ☽ ⊻ ♀
10 55 ☽ ⊻ ♄
20 46 ☽ ∥ ☉
21 52 ☽ ⊥ ♀
22 37 ☽ ⊻ ☿

29 Thursday
02 52 ♀ ⚹ ♄
07 37 ☽ ☍ ♇
10 06 ☽ ⊥ ☿
15 45 ☽ ♯ ♂
16 34 ☽ △ ♆
16 52 ☽ ∠ ♀
18 15 ♀ ∥ ♆
19 12 ☽ ☍ ♅
20 14 ☽ ♯ ♄
22 33 ☽ ⊼ ♃
23 13 ☽ ♯ ♅

MARCH 1968

LONGITUDES

Date	Sidereal time h m s	Sun ☉ ° ' "	Moon ☽ ° ' "	Moon ☽ 24.00 ° ' "	Mercury ☿ ° '	Venus ♀ ° '	Mars ♂ ° '	Jupiter ♃ ° '	Saturn ♄ ° '	Uranus ♅ ° '	Neptune ♆ ° '	Pluto ♇ ° '
01	22 37 25	10 ♓ 58 39	06 ♈ 34 25	12 ♈ 41 45	17 ♒ 20	12 ♒ 40	10 ♈ 11	29 ♌ 34	11 ♈ 08	27 ♍ 52	26 ♏ 32	21 ♍ 49
02	22 41 21	11 58 51	18 ♈ 45 52	24 ♈ 47 09	17 37	13 54	10 56	29 R 27	11 15	27 R 50	26 R 32	21 R 48
03	22 45 18	12 59 01	00 ♉ 46 00	06 ♉ 42 52	17 58	15 08	11 42	29 19	11 21	27 47	26 31	21 46
04	22 49 14	13 59 09	12 ♉ 38 16	18 ♉ 32 44	18 25	16 22	12 27	29 12	11 29	27 45	26 31	21 45
05	22 53 11	14 59 15	24 ♉ 26 53	00 ♊ 21 19	18 57	17 36	13 13	29 04	11 36	27 42	26 31	21 43
06	22 57 08	15 59 20	06 ♊ 16 39	12 ♊ 13 32	19 33	18 50	13 58	28 57	11 43	27 40	26 31	21 42
07	23 01 04	16 59 21	18 ♊ 12 35	24 ♊ 14 27	20 13	20 04	14 43	28 50	11 50	27 37	26 30	21 40
08	23 05 01	17 59 21	00 ♋ 19 45	06 ♋ 29 03	20 57	21 18	15 28	28 42	11 57	27 35	26 30	21 38
09	23 08 57	18 59 19	12 ♋ 42 52	19 ♋ 01 43	21 45	22 32	16 13	28 35	12 04	27 32	26 30	21 37
10	23 12 54	19 59 14	25 ♋ 25 58	01 ♌ 55 56	22 36	23 46	16 59	28 28	12 11	27 30	26 29	21 35
11	23 16 50	20 59 08	08 ♌ 31 51	15 ♌ 13 48	23 30	25 00	17 44	28 22	12 18	27 27	26 29	21 33
12	23 20 47	21 58 59	22 ♌ 01 44	28 ♌ 55 31	24 27	26 14	18 29	28 15	12 26	27 24	26 29	21 32
13	23 24 43	22 58 48	05 ♍ 54 50	12 ♍ 59 15	25 27	27 27	19 14	28 08	12 33	27 22	26 28	21 30
14	23 28 40	23 58 35	20 ♍ 08 14	27 ♍ 21 07	26 30	28 41	19 59	28 02	12 40	27 19	26 28	21 29
15	23 32 37	24 58 19	04 ♎ 37 10	11 ♎ 55 34	27 35	29 ♒ 55	20 43	27 55	12 48	27 17	26 27	21 27
16	23 36 33	25 58 02	19 ♎ 15 31	26 ♎ 36 11	28 42	01 ♓ 09	21 28	27 49	12 55	27 14	26 26	21 25
17	23 40 30	26 57 43	03 ♏ 56 45	11 ♏ 16 28	29 ♒ 52	02 23	22 13	27 43	13 02	27 11	26 26	21 24
18	23 44 26	27 57 23	18 ♏ 34 40	25 ♏ 50 44	01 ♓ 04	03 37	22 58	27 37	13 10	27 09	26 25	21 22
19	23 48 23	28 57 00	03 ♐ 04 10	10 ♐ 14 32	02 17	04 51	23 42	27 31	13 17	27 06	26 25	21 21
20	23 52 19	29 ♓ 56 36	17 ♐ 21 32	24 ♐ 24 55	03 33	06 05	24 27	27 25	13 24	27 04	26 24	21 19
21	23 56 16	00 ♈ 56 11	01 ♑ 24 31	08 ♑ 20 14	04 50	07 19	25 12	27 20	13 32	27 01	26 23	21 17
22	00 00 12	01 55 43	15 ♑ 12 02	21 ♑ 59 54	06 09	08 33	25 56	27 14	13 39	26 58	26 22	21 16
23	00 04 09	02 55 14	28 ♑ 43 54	05 ♒ 24 03	07 30	09 47	26 41	27 09	13 47	26 56	26 22	21 14
24	00 08 06	03 54 43	12 ♒ 00 28	18 ♒ 33 12	08 53	11 01	27 25	27 04	13 54	26 53	26 21	21 13
25	00 12 02	04 54 10	25 ♒ 02 23	01 ♓ 28 05	10 17	12 15	28 10	26 59	14 02	26 51	26 20	21 11
26	00 15 59	05 53 35	07 ♓ 50 25	14 ♓ 09 29	11 43	13 29	28 54	26 54	14 09	26 48	26 19	21 10
27	00 19 55	06 52 59	20 ♓ 25 24	26 ♓ 38 19	13 10	14 43	29 ♈ 38	26 49	14 17	26 45	26 18	21 08
28	00 23 52	07 52 20	02 ♈ 48 21	08 ♈ 55 39	14 39	15 57	00 ♉ 23	26 45	14 24	26 43	26 17	21 06
29	00 27 48	08 51 39	15 ♈ 00 25	21 ♈ 02 50	16 09	17 11	01 07	26 41	14 32	26 40	26 16	21 05
30	00 31 45	09 50 56	27 ♈ 03 08	03 ♉ 01 35	17 41	18 25	01 51	26 36	14 39	26 38	26 15	21 03
31	00 35 41	10 ♈ 50 11	08 ♉ 58 27	14 ♉ 54 06	19 ♓ 14	19 ♓ 39	02 ♉ 35	26 ♌ 32	14 ♈ 47	26 ♍ 35	26 ♏ 14	21 ♍ 02

Date	Moon True ☊ ° '	Moon Mean ☊ ° '	Moon ☽ Latitude ° '
01	19 ♈ 05	20 ♈ 47	01 S 08
02	19 R 04	20 44	00 S 02
03	19 D 05	20 41	01 N 03
04	19 06	20 38	02 05
05	19 08	20 35	03 01
06	19 10	20 32	03 50
07	19 R 10	20 28	04 29
08	19 09	20 25	04 57
09	19 07	20 22	05 11
10	19 04	20 19	05 11
11	19 00	20 16	04 55
12	18 55	20 13	04 23
13	18 51	20 09	03 34
14	18 48	20 06	02 31
15	18 46	20 03	01 N 17
16	18 D 46	20 00	00 S 03
17	18 46	19 57	01 23
18	18 47	19 53	02 37
19	18 49	19 50	03 41
20	18 50	19 47	04 30
21	18 R 50	19 44	05 01
22	18 50	19 41	05 15
23	18 49	19 38	05 11
24	18 47	19 34	04 50
25	18 45	19 31	04 15
26	18 43	19 28	03 27
27	18 41	19 25	02 30
28	18 40	19 22	01 27
29	18 40	19 19	00 S 20
30	18 D 40	19 15	00 N 46
31	18 ♈ 40	19 ♈ 12	01 N 50

DECLINATIONS

Date	Sun ☉ ° '	Moon ☽ ° '	Mercury ☿ ° '	Venus ♀ ° '	Mars ♂ ° '	Jupiter ♃ ° '	Saturn ♄ ° '	Uranus ♅ ° '	Neptune ♆ ° '	Pluto ♇ ° '
01	07 S 27	01 N 35	14 S 15	17 S 28	03 N 40	12 N 48	02 N 20	01 N 34	17 S 41	17 N 34
02	07 04	07 20	14 14	17 09	03 59	12 50	02 23	01 35	17 41	17 35
03	06 41	12 44	14 29	16 50	04 17	12 53	02 25	01 36	17 41	17 36
04	06 18	17 37	14 33	16 31	04 35	12 56	02 28	01 37	17 41	17 37
05	05 55	21 49	14 35	16 11	04 54	12 58	02 31	01 38	17 41	17 37
06	05 32	25 08	14 35	15 50	05 12	13 01	02 34	01 40	17 41	17 38
07	05 08	27 23	14 34	15 29	05 30	13 03	02 37	01 41	17 41	17 39
08	04 45	28 23	14 31	15 08	05 48	13 06	02 40	01 42	17 41	17 40
09	04 22	28 00	14 26	14 46	06 06	13 08	02 42	01 43	17 41	17 40
10	03 58	26 09	14 19	14 24	06 24	13 11	02 45	01 44	17 40	17 41
11	03 34	22 53	14 11	14 02	06 42	13 13	02 48	01 45	17 40	17 42
12	03 11	18 19	14 02	13 39	07 00	13 15	02 51	01 46	17 40	17 43
13	02 47	12 40	13 50	13 15	07 18	13 18	02 54	01 47	17 40	17 43
14	02 24	06 N 13	13 38	12 52	07 35	13 20	02 57	01 48	17 40	17 44
15	02 00	00 S 39	13 23	12 28	07 53	13 22	03 00	01 49	17 40	17 45
16	01 36	07 35	13 08	12 03	08 10	13 24	03 03	01 50	17 39	17 45
17	01 13	14 08	12 50	11 38	08 28	13 26	03 05	01 51	17 39	17 46
18	00 49	19 52	12 32	11 13	08 45	13 28	03 08	01 52	17 39	17 47
19	00 25	24 23	12 12	10 48	09 02	13 30	03 11	01 53	17 39	17 47
20	00 S 01	27 19	11 50	10 22	09 19	13 32	03 14	01 54	17 39	17 48
21	00 N 22	28 28	11 27	09 56	09 36	13 34	03 17	01 55	17 38	17 48
22	00 46	27 48	11 03	09 30	09 53	13 36	03 20	01 56	17 38	17 49
23	01 10	25 30	10 38	09 04	10 10	13 38	03 23	01 57	17 38	17 50
24	01 33	21 50	10 11	08 37	10 27	13 39	03 26	01 58	17 38	17 50
25	01 57	17 11	09 43	08 10	10 43	13 41	03 29	01 59	17 38	17 51
26	02 20	11 50	09 13	07 43	11 00	13 43	03 32	02 00	17 37	17 51
27	02 44	06 06	08 43	07 16	11 16	13 44	03 35	02 01	17 37	17 52
28	03 07	00 S 13	08 11	06 48	11 32	13 46	03 38	02 02	17 37	17 52
29	03 31	05 N 36	07 38	06 20	11 48	13 47	03 41	02 03	17 37	17 53
30	03 54	11 09	07 05	05 52	12 04	13 48	03 44	02 04	17 36	17 53
31	04 N 17	16 N 14	06 S 28	05 S 24	12 N 20	13 N 50	03 N 46	02 N 05	17 S 36	17 N 54

ZODIAC SIGN ENTRIES

Date	h	m	Planets
03	10	27	☽ ♉
05	23	17	☽ ♊
08	11	21	☽ ♋
10	20	27	☽ ♌
13	01	51	☽ ♍
15	04	23	☽ ♎
15	13	32	♀ ♓
17	05	33	☽ ♏
17	14	45	☿ ♓
19	06	53	☽ ♐
20	13	22	☉ ♈
21	09	34	☽ ♑
23	14	16	☽ ♒
25	21	15	☽ ♓
27	23	43	♂ ♉
28	06	32	☽ ♈
30	17	55	☽ ♉

LATITUDES

Date	Mercury ☿ ° '	Venus ♀ ° '	Mars ♂ ° '	Jupiter ♃ ° '	Saturn ♄ ° '	Uranus ♅ ° '	Neptune ♆ ° '	Pluto ♇ ° '
01	01 N 28	00 S 29	00 S 24	01 N 15	02 S 15	00 N 48	01 N 44	15 N 39
04	00 48	00 36	00 22	01 15	02 15	00 48	01 45	15 40
07	00 N 11	00 44	00 19	01 15	02 15	00 48	01 45	15 40
10	00 S 22	00 51	00 17	01 15	02 14	00 48	01 45	15 40
13	00 51	00 57	00 15	01 15	02 14	00 48	01 45	15 40
16	01 17	01 03	00 13	01 15	02 14	00 48	01 45	15 40
19	01 38	01 09	00 11	01 15	02 14	00 48	01 45	15 40
22	01 56	01 14	00 09	01 14	02 14	00 48	01 46	15 40
25	02 09	01 18	00 07	01 14	02 14	00 48	01 46	15 40
28	02 19	01 22	00 05	01 14	02 14	00 48	01 46	15 40
31	02 S 24	01 S 25	00 S 03	01 N 14	02 S 14	00 N 48	01 N 46	15 N 40

DATA

Julian Date	2439917
Delta T	+39 seconds
Ayanamsa	23° 24' 40"
Synetic vernal point	05° ♓ 42' 20"
True obliquity of ecliptic	23° 26' 45"

MOON'S PHASES, APSIDES AND POSITIONS ☽

Date	h	m	Phase	Longitude ° '	Eclipse Indicator
07	09	20	☽	16 ♊ 53	
14	18	52	○	24 ♍ 16	
21	11	07	☾	00 ♑ 54	
28	22	48	●	08 ♈ 19	Partial

Day	h	m		
05	06	34	Apogee	
17	01	42	Perigee	
01	05	34	0N	
08	17	33	Max dec	28° N 26'
15	09	44	0S	
21	15	01	Max dec	28° S 29'
28	12	52	0N	

LONGITUDES

Date	Chiron ⚷ ° '	Ceres ⚳ ° '	Pallas ⚴ ° '	Juno ⚵ ° '	Vesta ⚶ ° '	Black Moon Lilith ° '
01	28 ♓ 11	06 ♏ 09	27 ♍ 02	13 ♏ 21	00 ♓ 08	17 ♉ 51
11	28 ♓ 46	05 ♏ 55	23 ♍ 51	13 ♏ 01	05 ♓ 09	18 ♉ 58
21	29 ♓ 22	04 ♏ 59	20 ♍ 40	12 ♏ 07	10 ♓ 06	20 ♉ 05
31	29 ♓ 57	03 ♏ 26	17 ♍ 51	10 ♏ 41	14 ♓ 58	21 ♉ 11

All ephemeris data is given at 12.00 UT and the Moon's longitude is additionally given for 24.00 UT

ASPECTARIAN

h m	Aspects
01 Friday	
03 38	☽ ∠ ☿
03 57	♀ # ♇
10 04	☽ ± ♃
12 00	☽ ∥ ♅
15 07	☽ ∥ ♄
15 45	♀ bQ ♅
16 08	☉ ⊻ ♄
19 32	☽ ☌ ♂
21 00	☽ ☌ ♄
21 06	☽ ∥ ♂
21 24	☽ ⊻ ☉
21 42	☽ bQ ♆
02 Saturday	
01 18	☽ ✶ ♀
03 32	☽ bQ ♃
09 39	☽ ✶ ☿
10 18	☽ ⊥ ☉
10 58	☽ # ☉
15 30	☽ ± ♆
18 01	☽ ⊼ ♇
23 23	♂ ☌ ♄
03 Sunday	
03 29	☽ ⊼ ♆
03 52	☽ Q ♀
05 54	☽ ∠ ☉
06 00	☽ ± ♇
06 02	☽ ⊼ ♅
06 29	♂ bQ ♆
09 07	☽ △ ♃
10 21	☽ Q ☿
12 42	☽ ∥ ♃
18 04	☽ ± ♅
20 22	☽ # ☿
04 Monday	
00 05	☽ bQ ♇
00 07	♀ ± ♇
06 40	☽ # ♀
09 37	☽ ⊻ ♄
11 36	☽ ⊻ ♂
11 57	☽ ∥ ♇
12 13	☽ bQ ♅
12 20	☽ # ♆
14 59	☽ ✶ ☉
20 27	☽ □ ♀
20 31	♄ bQ ♆
21 56	☽ ⊥ ♄
05 Tuesday	
00 16	☽ □ ☿
00 37	☽ ⊥ ♂
06 28	☽ △ ♇
16 12	☽ ☍ ♆
16 24	☽ ∠ ♄
17 38	☽ Q ☉
18 36	☽ △ ♅
20 10	☽ ∠ ♂
21 18	☽ □ ♃
06 Wednesday	
11 31	♂ bQ ♃
23 05	☽ ✶ ♄
23 24	☉ # ♂
07 Thursday	
04 32	☽ ✶ ♂
09 16	☽ Q ♃
09 20	☽ □ ☉
16 07	☽ △ ♀
16 15	☽ △ ☿
18 52	☽ □ ♇
19 04	☿ ☌ ♀
23 18	☽ Q ♄
08 Friday	
04 28	☽ ⊼ ♆
06 01	☽ Q ♂
06 37	☽ □ ♅
08 51	☽ ✶ ♃
16 15	☽ ± ♆
17 19	♀ ± ♅
18 31	♀ ⊼ ♇
23 42	☽ bQ ☿
09 Saturday	
00 55	☽ bQ ♀
06 03	☽ Q ♇
06 05	☿ ± ♅
08 08	☿ ⊼ ♇
09 40	☽ bQ ♆
10 45	☽ □ ♄
13 40	☽ ∠ ♃
16 41	♆ # ♇
17 22	☽ Q ♅
18 12	☽ ± ☿
19 07	☽ □ ♂
20 03	☽ ± ♀
10 Sunday	
00 56	☽ △ ☉
04 50	☽ ✶ ♇
06 20	☽ ⊼ ☿
06 32	☽ ⊥ ♃
08 33	☽ ⊼ ♀
13 58	☽ △ ♆
15 49	☽ ✶ ♅
17 35	☽ ⊻ ♃
19 33	☿ ∥ ♀
11 Monday	
07 01	☽ bQ ☉
08 26	☽ ∠ ♇
18 51	☽ △ ♄
19 01	☽ ∠ ♅
12 Tuesday	
00 29	☽ ± ☉
00 34	☽ ⊥ ♇
01 25	☉ ☍ ♇
05 23	☽ △ ♂
10 55	☽ ⊥ ♅
11 08	☽ ⊻ ♇
11 55	☽ ⊼ ☉
14 46	☽ ∥ ♇
14 57	☽ # ♆
16 34	☽ ☍ ☿
16 51	♀ □ ♆
19 45	☽ □ ♆
20 02	☽ ☍ ♀
21 21	☽ ⊻ ♅
21 30	☽ bQ ♄
22 45	☽ ☌ ♃
13 Wednesday	
05 57	☉ # ♄
07 08	☽ # ☿
08 57	☽ bQ ♂
09 29	☽ # ♀
09 30	☽ ∥ ♃
09 46	♀ # ♃
10 14	♀ ⊼ ♅
13 06	☽ ± ♄
14 00	♀ ∠ ♄
23 21	☽ ⊼ ♄
14 Thursday	
00 11	♀ ☍ ♃
01 06	☽ ± ♂
02 29	☽ Q ♆
07 17	☽ ∥ ♂
11 06	☿ □ ♆
11 43	☽ ⊼ ♂
14 14	☽ ☌ ♇
18 52	☽ ☍ ☉
22 31	☽ ✶ ♆
23 26	☽ ⊼ ☿
23 28	☽ ∥ ♄
23 55	☽ ☌ ♅
15 Friday	
01 02	☽ ⊻ ♃
02 17	☽ # ☉
03 20	♂ ± ♆
03 30	☽ ∥ ♅
03 32	☽ ⊼ ♀
05 34	☿ ⊼ ♅
10 09	☽ ± ☿
10 52	☽ ⊥ ♃
13 41	☿ # ♃
14 20	☽ ± ♀
15 59	☽ # ♅
16 22	☽ ∥ ☉
17 06	☿ ∠ ♄
18 44	☿ ☍ ♃
20 06	☽ # ♄
22 48	☉ # ♅
23 13	☽ ∠ ♆
16 Saturday	
01 32	☽ ☍ ♄
01 32	☽ ∠ ♃
01 49	♃ bQ ♄
02 09	☽ bQ ☿
06 27	☽ bQ ♀
10 33	♂ ⊼ ♇
13 56	☽ ⊥ ♆
14 11	☽ # ♂
15 32	☽ ⊻ ♇
15 49	☽ ☍ ♂
23 18	☉ △ ♆
23 44	☽ ⊻ ♆
23 45	☽ ⊼ ☉
17 Sunday	
00 59	☽ ⊻ ♅
01 19	☽ ⊥ ♇
01 53	☽ ✶ ♃
03 10	☽ ∥ ♀
04 45	☽ △ ☿
07 17	☽ ∥ ☿
09 13	☽ △ ♀
09 19	☽ # ♃
10 16	☽ ± ☉
10 46	☽ ⊥ ♅
16 00	☽ ∠ ♇
17 16	☉ ☍ ♅
21 23	☽ Q ♃
18 Monday	
01 28	☽ ∠ ♅
02 05	☽ bQ ☉
02 11	☽ ∥ ♆
02 41	☽ # ♇
03 01	☽ ⊼ ♄
04 33	☉ ⊼ ♃
12 58	☽ ± ♄
16 36	☽ ✶ ♇
19 37	☽ ⊼ ♂
19 Tuesday	
00 56	☽ ☌ ♆
02 07	☽ ✶ ♅
02 50	☽ □ ♃
03 58	☽ bQ ♄
04 39	☽ △ ☉
06 06	☽ ± ♂
10 34	☽ □ ☿
12 27	☽ Q ♇
15 15	☽ □ ♀
21 56	☽ bQ ♂
22 03	☽ Q ♅
20 Wednesday	
05 16	☽ △ ♄
18 42	☽ □ ♇
19 49	☽ Q ☿
21 Thursday	
00 32	☽ Q ♀
00 44	☽ △ ♂
03 23	☽ ⊻ ♆
04 28	☽ □ ♅
05 02	☽ △ ♃
11 07	☽ □ ☉
13 41	☽ ⊥ ♆
16 39	♀ ⊥ ♄
18 32	☽ ✶ ☿
23 14	☽ ✶ ♀
23 16	♀ # ♂
22 Friday	
05 18	☽ ∠ ♆
06 51	☽ bQ ♃
09 16	☽ □ ♄
21 00	☽ Q ☉
22 35	☽ ± ♃
22 41	☽ △ ♇
23 40	☽ ∠ ☿
23 Saturday	
01 50	♂ ⊼ ♆
04 14	☽ ∠ ♀
07 46	☽ ✶ ♆
08 07	☽ □ ♂
08 47	☽ △ ♅
09 11	☽ ⊼ ♃
17 08	☉ ± ♃
17 20	☿ ⊥ ♄
17 31	☽ Q ♄
17 33	☽ ⊥ ☿
19 38	♂ ⊼ ♅
20 08	☽ ✶ ☉
22 00	☽ ⊥ ♀
24 Sunday	
01 29	☽ bQ ♇
01 40	♂ △ ♃
03 18	☿ # ♂
05 21	☽ Q ♆
05 28	♂ ± ♇
05 38	☽ ⊻ ☿
10 00	☽ ⊻ ♀
11 47	☽ bQ ♅
15 30	☽ ✶ ♄
17 51	☽ ± ♇
18 37	☽ Q ♂
25 Monday	
01 42	☽ ∠ ☉
04 15	☽ ± ♅
04 52	☽ ⊼ ♇
08 48	☽ # ♇
09 52	☽ ∥ ♆
14 19	☉ ∥ ♅
14 24	☽ □ ♆
15 21	☽ ⊼ ♅
15 35	☽ ☍ ♃
18 10	☽ ✶ ♂
19 31	☽ ∠ ♄
19 48	☽ ⊥ ☉
26 Tuesday	
03 53	☽ # ♃
08 01	☽ ⊻ ☉
12 36	☽ ⊥ ♄
15 26	☽ # ♂
20 18	☽ ☌ ☿
23 52	☽ ☌ ♀
27 Wednesday	
00 07	☽ ⊻ ♄
00 08	☽ ∥ ☿
00 14	☽ ∠ ♂
02 38	♀ ⊻ ♄
06 48	☽ ∥ ♀
08 29	♀ ∠ ♂
13 22	☽ ☍ ♇
18 35	☽ ⊥ ♂
22 12	☽ # ♄
23 20	☽ △ ♆
28 Thursday	
00 11	☽ ☍ ♅
00 17	☽ ⊼ ♃
00 53	☽ # ☉
04 35	☽ # ♅
06 58	☽ ⊻ ♂
07 45	☿ ⊻ ♄
11 53	☽ ± ♃
21 13	☽ ∥ ♅
22 48	☽ ☌ ☉
29 Friday	
02 40	☽ ∥ ☉
03 54	☽ ∥ ♄
04 38	☽ bQ ♆
05 27	☽ bQ ♃
10 51	☿ ∠ ♂
11 03	☽ ☌ ♄
14 36	☽ ⊻ ☿
14 47	♃ ⊻ ♅
14 52	☽ # ♀
16 47	☽ ⊻ ♀
19 49	☽ # ☿
22 26	☽ ± ♆
23 24	☉ ∥ ♄
30 Saturday	
00 03	☽ ⊼ ♇
04 16	☽ ⊥ ☿
06 06	☽ ⊥ ♀
10 24	☽ ⊼ ♆
11 07	☽ △ ♃
11 09	☽ ⊼ ♅
12 01	☽ ± ♇
16 24	☽ ∥ ♂
22 16	☽ ☌ ♂
23 10	☽ ± ♅
31 Sunday	
00 16	☽ ∥ ♃
01 00	☽ ∠ ☿
02 15	☽ ∠ ♀
06 04	☽ bQ ♇
12 02	♂ ± ♅
16 06	☽ ⊻ ☉
17 16	☽ bQ ♅
18 58	☽ # ♆
20 33	☽ ∥ ♇
21 37	☉ bQ ♆
23 53	☽ ⊻ ♄

APRIL 1968

LONGITUDES

Date	Sidereal time h m s	Sun ☉ ° ' "	Moon ☽ ° ' "	Moon ☽ 24.00 ° ' "	Mercury ☿ ° '	Venus ♀ ° '	Mars ♂ ° '	Jupiter ♃ ° '	Saturn ♄ ° '	Uranus ♅ ° '	Neptune ♆ ° '	Pluto ♇ ° '
01	00 39 38	11 ♈ 49 24	20 ♉ 48 51	26 ♉ 43 07	20 ♓ 49	20 ♓ 52	03 ♉ 19	26 ♌ 28	14 ♈ 54	26 ♍ 33	26 ♏ 13	21 ♍ 00
02	00 43 35	12 48 35	02 ♊ 37 20	08 ♊ 31 57	22 25	22 06	04 03	26 R 25	15 02	26 R 30	26 R 12	20 R 59
03	00 47 31	13 47 44	14 ♊ 27 27	20 ♊ 24 22	24 03	23 20	04 47	26 21	15 10	26 28	26 11	20 58
04	00 51 28	14 46 50	26 ♊ 23 14	02 ♋ 24 36	25 42	24 34	05 31	26 18	15 17	26 25	26 10	20 56
05	00 55 24	15 45 54	08 ♋ 29 03	14 ♋ 37 07	27 22	25 48	06 15	26 15	15 25	26 23	26 09	20 55
06	00 59 21	16 44 56	20 ♋ 49 23	27 ♋ 06 23	29 ♓ 04	27 02	06 59	26 12	15 32	26 21	26 08	20 53
07	01 03 17	17 43 55	03 ♌ 28 37	09 ♌ 56 32	00 ♈ 47	28 16	07 43	26 09	15 40	26 18	26 07	20 52
08	01 07 14	18 42 52	16 ♌ 30 31	23 ♌ 10 52	02 32	29 ♓ 30	08 27	26 07	15 47	26 16	26 05	20 50
09	01 11 10	19 41 47	29 ♌ 57 47	06 ♍ 51 21	04 19	00 ♈ 44	09 10	26 04	15 55	26 13	26 04	20 49
10	01 15 07	20 40 39	13 ♍ 51 29	20 ♍ 58 00	06 07	01 58	09 54	26 02	16 03	26 11	26 03	20 48
11	01 19 04	21 39 30	28 ♍ 10 30	05 ♎ 28 27	07 56	03 11	10 37	26 00	16 10	26 09	26 02	20 46
12	01 23 00	22 38 18	12 ♎ 51 09	20 ♎ 17 46	09 47	04 25	11 21	25 58	16 18	26 07	26 00	20 45
13	01 26 57	23 37 04	27 ♎ 47 20	05 ♏ 18 48	11 39	05 39	12 05	25 56	16 25	26 04	25 59	20 44
14	01 30 53	24 35 48	12 ♏ 51 03	20 ♏ 22 58	13 33	06 53	12 48	25 55	16 33	26 02	25 58	20 42
15	01 34 50	25 34 30	27 ♏ 53 27	05 ♐ 21 30	15 29	08 07	13 31	25 54	16 40	26 00	25 57	20 41
16	01 38 46	26 33 11	12 ♐ 46 11	20 ♐ 06 42	17 26	09 21	14 15	25 53	16 48	25 58	25 55	20 40
17	01 42 43	27 31 50	27 ♐ 22 24	04 ♑ 32 49	19 24	10 35	14 58	25 52	16 55	25 56	25 54	20 39
18	01 46 39	28 30 27	11 ♑ 37 36	18 ♑ 36 34	21 24	11 48	15 41	25 51	17 03	25 54	25 52	20 37
19	01 50 36	29 ♈ 29 02	25 ♑ 29 38	02 ♒ 16 53	23 25	13 02	16 24	25 50	17 10	25 52	25 51	20 36
20	01 54 33	00 ♉ 27 36	08 ♒ 58 28	15 ♒ 34 37	25 28	14 16	17 08	25 50	17 18	25 50	25 50	20 35
21	01 58 29	01 26 09	22 ♒ 05 38	28 ♒ 31 51	27 32	15 30	17 51	25 50	17 25	25 48	25 48	20 34
22	02 02 26	02 24 39	04 ♓ 53 37	11 ♓ 11 18	29 ♈ 37	16 44	18 34	25 D 50	17 32	25 46	25 47	20 33
23	02 06 22	03 23 08	17 ♓ 25 18	23 ♓ 35 59	01 ♉ 44	17 58	19 17	25 50	17 40	25 44	25 45	20 32
24	02 10 19	04 21 35	29 ♓ 43 41	05 ♈ 48 46	03 51	19 11	20 00	25 50	17 47	25 42	25 44	20 31
25	02 14 15	05 20 01	11 ♈ 51 33	17 ♈ 52 21	05 59	20 25	20 43	25 51	17 54	25 40	25 42	20 30
26	02 18 12	06 18 24	23 ♈ 51 26	29 ♈ 49 06	08 07	21 39	21 25	25 52	18 02	25 38	25 41	20 29
27	02 22 08	07 16 46	05 ♉ 45 35	11 ♉ 41 11	10 16	22 53	22 08	25 53	18 09	25 37	25 39	20 28
28	02 26 05	08 15 06	17 ♉ 36 07	23 ♉ 30 39	12 25	24 07	22 51	25 54	18 16	25 35	25 38	20 27
29	02 30 02	09 13 25	29 ♉ 25 04	05 ♊ 19 36	14 34	25 20	23 34	25 55	18 24	25 33	25 36	20 26
30	02 33 58	10 ♉ 11 41	11 ♊ 14 35	17 ♊ 10 17	16 ♉ 43	26 ♈ 34	24 ♉ 16	25 ♌ 57	18 ♈ 31	25 ♍ 32	25 ♏ 35	20 ♍ 25

Date	Moon True ☊ ° '	Moon Mean ☊ ° '	Moon ☽ Latitude ° '
01	18 ♈ 41	19 ♈ 09	02 N 49
02	18 D 42	19 06	03 41
03	18 42	19 03	04 23
04	18 43	18 59	04 54
05	18 43	18 56	05 13
06	18 R 43	18 53	05 17
07	18 43	18 50	05 07
08	18 D 43	18 47	04 41
09	18 43	18 44	03 59
10	18 43	18 40	03 02
11	18 43	18 37	01 52
12	18 43	18 34	00 N 33
13	18 R 43	18 31	00 S 50
14	18 43	18 28	02 10
15	18 43	18 24	03 21
16	18 42	18 21	04 17
17	18 41	18 18	04 56
18	18 40	18 15	05 15
19	18 40	18 12	05 15
20	18 D 40	18 09	04 57
21	18 40	18 05	04 25
22	18 41	18 02	03 40
23	18 43	17 59	02 45
24	18 44	17 56	01 43
25	18 45	17 53	00 S 38
26	18 R 45	17 50	00 N 28
27	18 44	17 46	01 33
28	18 42	17 43	02 33
29	18 39	17 40	03 26
30	18 ♈ 36	17 ♈ 37	04 N 10

DECLINATIONS

Date	Sun ☉ ° '	Moon ☽ ° '	Mercury ☿ ° '	Venus ♀ ° '	Mars ♂ ° '	Jupiter ♃ ° '	Saturn ♄ ° '	Uranus ♅ ° '	Neptune ♆ ° '	Pluto ♇ ° '
01	04 N 41	20 N 41	05 S 51	04 S 56	12 N 36	13 N 51	03 N 49	02 N 06	17 S 36	17 N 54
02	05 04	24 18	05 14	04 28	12 51	13 52	03 52	02 07	17 35	17 55
03	05 27	26 53	04 35	03 59	13 07	13 53	03 55	02 08	17 35	17 55
04	05 50	28 18	03 55	03 30	13 22	13 54	03 58	02 09	17 35	17 56
05	06 12	28 22	03 14	03 02	13 37	13 55	04 01	02 10	17 35	17 56
06	06 35	27 03	02 32	02 33	13 52	13 56	04 04	02 11	17 34	17 57
07	06 58	24 21	01 49	02 04	14 07	13 57	04 07	02 12	17 34	17 57
08	07 20	20 21	01 05	01 35	14 22	13 58	04 10	02 13	17 34	17 57
09	07 42	15 13	00 S 20	01 06	14 37	13 58	04 12	02 14	17 33	17 58
10	08 05	09 09	00 N 27	00 37	14 51	13 59	04 15	02 15	17 33	17 58
11	08 27	02 N 26	01 14	00 S 08	15 05	13 59	04 18	02 15	17 33	17 59
12	08 49	04 S 35	02 01	00 N 22	15 20	14 00	04 21	02 16	17 32	17 59
13	09 10	11 28	02 50	00 51	15 34	14 00	04 24	02 17	17 32	17 59
14	09 32	17 46	03 40	01 20	15 48	14 01	04 27	02 18	17 32	18 00
15	09 53	22 57	04 30	01 49	16 01	14 01	04 30	02 19	17 31	18 00
16	10 15	26 35	05 21	02 18	16 15	14 01	04 32	02 20	17 31	18 00
17	10 36	28 21	06 12	02 47	16 28	14 02	04 35	02 20	17 31	18 00
18	10 57	28 10	07 05	03 17	16 41	14 02	04 38	02 21	17 30	18 01
19	11 18	26 12	07 57	03 46	16 54	14 02	04 41	02 22	17 30	18 01
20	11 38	22 48	08 50	04 15	17 07	14 02	04 44	02 23	17 30	18 01
21	11 59	18 19	09 43	04 43	17 20	14 02	04 46	02 24	17 29	18 01
22	12 19	13 07	10 36	05 12	17 33	14 02	04 49	02 24	17 29	18 02
23	12 39	07 30	11 29	05 41	17 45	14 01	04 52	02 25	17 28	18 02
24	12 59	01 S 41	12 22	06 09	17 57	14 01	04 55	02 26	17 28	18 02
25	13 18	04 N 06	13 15	06 38	18 09	14 01	04 57	02 26	17 28	18 02
26	13 38	09 42	14 06	07 06	18 21	14 00	05 00	02 27	17 27	18 02
27	13 57	14 54	14 58	07 34	18 33	14 00	05 03	02 28	17 27	18 02
28	14 16	19 32	15 48	08 02	18 44	13 59	05 05	02 28	17 27	18 02
29	14 34	23 23	16 37	08 30	18 56	13 59	05 08	02 29	17 26	18 02
30	14 N 53	26 N 16	17 N 24	08 N 58	19 N 07	13 N 58	05 N 11	02 N 30	17 S 26	18 N 03

ZODIAC SIGN ENTRIES

Date	h m	Planets
02	06 40	☽ ♊
04	19 13	☽ ♋
07	01 01	☿ ♈
07	05 28	☽ ♌
08	21 48	♀ ♈
09	12 04	☽ ♍
11	15 01	☽ ♎
13	15 32	☽ ♏
15	15 23	☽ ♐
17	16 23	☽ ♑
19	19 57	☽ ♒
20	00 41	☉ ♉
22	02 46	☽ ♓
22	16 18	☿ ♉
24	12 32	☽ ♈
27	00 22	☽ ♉
29	13 11	☽ ♊

LATITUDES

Date	Mercury ☿ ° '	Venus ♀ ° '	Mars ♂ ° '	Jupiter ♃ ° '	Saturn ♄ ° '	Uranus ♅ ° '	Neptune ♆ ° '	Pluto ♇ ° '
01	02 S 25	01 S 26	00 S 02	01 N 14	02 S 14	00 N 48	01 N 46	15 N 39
04	02 24	01 28	00 00	01 13	02 14	00 48	01 46	15 39
07	02 19	01 30	00 N 02	01 13	02 14	00 47	01 46	15 39
10	02 10	01 31	00 04	01 13	02 14	00 47	01 46	15 38
13	01 56	01 31	00 06	01 12	02 14	00 47	01 46	15 37
16	01 37	01 31	00 08	01 12	02 14	00 47	01 46	15 37
19	01 14	01 31	00 10	01 11	02 14	00 47	01 47	15 36
22	00 47	01 29	00 12	01 11	02 14	00 47	01 47	15 35
25	00 S 18	01 27	00 14	01 10	02 14	00 47	01 47	15 34
28	00 N 14	01 25	00 16	01 10	02 15	00 47	01 47	15 33
31	00 N 46	01 S 22	00 N 17	01 N 10	02 S 15	00 N 47	01 N 47	15 N 32

DATA

Julian Date	2439948
Delta T	+39 seconds
Ayanamsa	23° 24' 43"
Synetic vernal point	05° ♓ 42' 17"
True obliquity of ecliptic	23° 26' 45"

MOON'S PHASES, APSIDES AND POSITIONS ☽

Date	h m	Phase	Longitude ° '	Eclipse Indicator
06	03 27	☽	16 ♋ 24	
13	04 52	○	23 ♎ 20	total
19	19 35	☾	29 ♑ 48	
27	15 21	●	07 ♉ 25	

Day	h m		
01	23 34	Apogee	
14	06 56	Perigee	
29	09 20	Apogee	
05	01 26	Max dec	28° N 30'
11	20 23	0S	
17	21 38	Max dec	28° S 30'
24	18 57	0N	

LONGITUDES

Date	Chiron ⚷ ° '	Ceres ⚳ ° '	Pallas ⚴ ° '	Juno ⚵ ° '	Vesta ⚶ ° '	Black Moon Lilith ⚸ ° '
01	00 ♈ 01	03 ♏ 15	17 ♍ 36	10 ♏ 30	15 ♓ 27	21 ♉ 18
11	00 ♈ 35	01 ♏ 12	15 ♍ 33	08 ♏ 35	20 ♓ 14	22 ♉ 25
21	01 ♈ 08	28 ♎ 56	14 ♍ 20	06 ♏ 23	24 ♓ 54	23 ♉ 31
31	01 ♈ 38	26 ♎ 45	13 ♍ 57	04 ♏ 06	29 ♓ 28	24 ♉ 38

ASPECTARIAN

h m Aspects

01 Monday
04 02 ☉ ⚼ ♃
05 22 ☽ ⊥ ☉
12 00 ☽ ✶ ☿
12 08 ☽ ✶ ♀
12 12 ☽ ⊥ ♄
12 24 ☽ △ ♇
14 33 ♀ ☍ ♇
14 52 ☿ ☍ ♇
16 00 ☿ ☌ ♀
19 09 ☉ # ♀
22 58 ☽ ☍ ♆
23 27 ☽ □ ♃
23 37 ☽ △ ♅

02 Tuesday
01 20 ☽ ∠ ☉
06 41 ☽ ∠ ♄
15 06 ☽ ⚺ ♂
15 22 ☽ Q ♀
15 57 ☉ # ☿
16 13 ☽ Q ☿

03 Wednesday
04 05 ☽ ⊥ ♂
10 32 ☽ ✶ ☉
11 48 ☽ Q ♃
13 26 ☽ ✶ ♄
14 58 ♀ # ♄
23 28 ☽ ∠ ♂

04 Thursday
01 05 ☽ □ ♇
07 57 ☽ □ ♀
10 20 ☿ # ♄
10 23 ☽ □ ☿
11 34 ☽ ⚻ ♆
11 50 ☽ ✶ ♃
12 04 ☽ □ ♅
12 51 ☽ Q ☉
13 49 ☽ Q ♄
18 45 ☿ △ ♆
20 28 ☿ ⚻ ♃
22 15 ☿ ☍ ♅
23 30 ☽ ± ♆

05 Friday
01 06 ♂ ⚼ ♇
02 08 ☉ ☌ ♄
07 19 ☽ ✶ ♂
12 50 ☽ Q ♇
17 13 ☽ ⚼ ♆
17 24 ☽ ∠ ♃
18 40 ♀ △ ♆
20 22 ♀ ⚻ ♃
22 58 ♀ ☍ ♅
23 30 ☽ Q ♅

06 Saturday
01 41 ☽ □ ♄
03 27 ☽ □ ☉
08 14 ☽ Q ♂
10 30 ☿ ∥ ♀
10 48 ☽ ⊥ ♃
12 07 ☽ ✶ ♇
18 06 ♂ ∥ ♃
22 08 ☽ △ ♆
22 14 ☽ ⚺ ♃
22 31 ☽ ✶ ♅
23 31 ☿ # ♅

07 Sunday
01 08 ☽ △ ♀
05 35 ♀ # ♅
06 10 ☽ △ ☿
16 27 ☽ ∠ ♇
20 21 ☽ □ ♂

08 Monday
02 28 ☽ ∠ ♅
06 16 ☿ ± ♃
07 58 ☽ ⚼ ♀
08 58 ☽ ⊥ ♇
09 45 ☿ ⊥ ♂
10 41 ☽ △ ♄
14 09 ☽ ⚼ ☿
16 18 ☽ △ ☉
18 45 ☽ ⊥ ♅
19 48 ☽ ⚺ ♇
23 47 ☽ ∥ ♇

09 Tuesday
01 40 ☽ # ♆
01 50 ☽ ± ♀
05 09 ☽ □ ♆
05 09 ☽ ☌ ♃
05 26 ☽ ⚺ ♅
08 40 ☽ ± ☿
11 31 ♃ □ ♆
13 28 ☽ ⚻ ♀
13 41 ☽ ⚼ ♄
14 26 ☽ ∥ ♂
17 11 ☽ ∥ ♃
20 43 ☽ ⚻ ☿
20 54 ☽ ⚼ ☉
20 58 ☉ ± ♆

10 Wednesday
04 52 ☽ △ ♂
05 25 ☽ ± ♄
12 20 ☽ Q ♆
13 24 ♀ ± ♃
13 30 ☽ ± ☉
14 48 ☉ ⚻ ♇
15 12 ☿ # ♀
15 44 ☽ ⚻ ♄
15 45 ☽ ∥ ☉
23 42 ☽ ☌ ♇

11 Thursday
00 22 ☽ ⚻ ☉
05 32 ☽ ∥ ♄
07 33 ☽ ⚼ ♂
08 24 ☽ ⚺ ♃
08 27 ☽ ✶ ♆
08 39 ☽ ☌ ♅
12 37 ☽ ∥ ♅
15 45 ☽ ∥ ☿
18 17 ☽ ⊥ ♃
20 14 ☽ ∥ ♀
20 33 ☽ # ♀
21 01 ☽ ☍ ♀
23 10 ☽ ± ♂

12 Friday
02 10 ☽ # ☿
04 07 ☽ # ♅
04 27 ♂ ⚼ ♅
06 18 ☽ ☍ ☿
08 57 ☽ ∠ ♃
09 01 ☽ ∠ ♆
09 26 ☽ ⚻ ♂
11 13 ☽ # ♄
17 36 ☽ ☍ ♄
19 29 ☿ ∥ ♅
23 31 ☽ ⊥ ♆

13 Saturday
00 43 ☽ ⚺ ♇
03 01 ☿ ⚼ ♃
03 24 ☽ # ☉
03 34 ☿ ⚼ ♆
04 52 ☽ ☍ ☉
09 03 ☽ ✶ ♃
09 07 ☽ ⚺ ♆
09 16 ☽ ⚺ ♅
10 18 ☽ ⊥ ♇
18 49 ☽ ⊥ ♅
20 38 ☿ ⚺ ♂
21 18 ☽ # ♃

14 Sunday
00 39 ☽ ∠ ♇
01 39 ☽ ⚻ ♀
03 48 ☽ # ♂
04 09 ☽ Q ♃
08 01 ♀ ⊥ ♂
09 07 ☽ ∠ ♅
11 02 ☽ ∥ ♆
11 55 ☽ ☍ ♂
12 03 ☽ ± ♀
12 57 ☽ # ♇
13 17 ☽ ⚻ ☿
17 56 ☽ ⚻ ♄

15 Monday
00 14 ☽ ± ☿
00 30 ☽ ✶ ♇
03 35 ☽ ± ♄
03 41 ☽ ⚼ ♀
08 02 ☽ ⚻ ☉
08 48 ☽ □ ♃
08 53 ☽ ☌ ♆
08 59 ☽ ✶ ♅
11 48 ☿ ∥ ♄
16 46 ☽ ⚼ ☿
18 07 ☽ ⚼ ♄
18 19 ☽ ± ☉
19 41 ☉ △ ♃
19 41 ☽ Q ♇
20 48 ☉ ⚻ ♆
22 04 ☉ ⚻ ♅

16 Tuesday
03 42 ☿ ☌ ♄
04 14 ☽ Q ♅
05 56 ☽ △ ♀
09 53 ☽ ⚼ ☉
13 07 ♀ ∥ ♅
14 31 ☽ ⚻ ♂
14 41 ☉ ± ♇
18 37 ☽ △ ♄
20 47 ☽ △ ☿

17 Wednesday
00 51 ☽ ± ♂
00 53 ☽ □ ♇
09 29 ☽ △ ♃
09 33 ☽ ⚺ ♆
09 37 ☽ □ ♅
12 17 ☽ △ ☉
16 33 ☽ ⚼ ♂
17 29 ♀ ⚼ ♃
17 53 ☿ ± ♆
18 09 ♀ ⚼ ♆
19 32 ☽ ⊥ ♆

18 Thursday
02 47 ☿ ⚻ ♇
10 40 ☽ ⚼ ♃
10 43 ☽ ∠ ♆
12 20 ☽ □ ♀
19 20 ☽ △ ♂
21 23 ☽ □ ♄

19 Friday
02 08 ☽ ± ♃
03 28 ☽ △ ♇
07 45 ☽ □ ☿
12 36 ☽ ⚻ ♃
12 38 ☽ ✶ ♆
12 39 ☽ △ ♅
19 35 ☽ □ ☉
22 46 ☽ Q ♀

20 Saturday
04 21 ♃ □ ♆
05 19 ☽ Q ♄
05 55 ☽ ⚼ ♇
07 35 ♃ ⚺ ♅
09 56 ☽ Q ♆
13 49 ♅ ✶ ♆
15 21 ☽ ⚼ ♅
16 06 ☿ ⚻ ♅
16 07 ☿ ⚻ ♆
16 13 ☿ △ ♃
18 42 ♂ ⚺ ♄
21 40 ☽ Q ☿
22 10 ☽ ± ♇
22 36 ☽ ✶ ♀

21 Sunday
00 52 ☿ ± ♇
03 17 ☽ ✶ ♄
03 42 ☽ □ ♂
06 41 ☽ Q ☉
07 46 ☽ ± ♅
09 11 ☽ ⚻ ♇
13 27 ☽ # ♇
14 44 ♀ ∥ ♄
16 03 ☽ ∥ ♆
16 35 ☽ # ♂
18 52 ☽ ⚻ ♅
18 53 ☽ □ ♆
18 57 ☽ ☍ ♃
20 33 ☉ ± ♅
23 30 ♃ St D

22 Monday
00 06 ☽ ✶ ☿
04 50 ♂ # ♆
05 23 ☽ ∠ ♀
06 55 ☽ ✶ ☉
07 30 ☽ ∠ ♄
08 00 ☽ # ♃
15 21 ☽ # ☉
15 22 ☽ Q ♂
21 26 ☽ # ☿

23 Tuesday
00 20 ☽ ⊥ ♀
00 48 ☽ ⊥ ♄
05 34 ♀ ☌ ♄
10 23 ☽ ∠ ☿
12 01 ☿ ± ♅
12 28 ☽ ⚺ ♄
13 09 ☽ ⚺ ♀
14 01 ☽ ∠ ☉
15 49 ☽ ✶ ♂
18 01 ☽ ☍ ♇
18 59 ☽ # ♀
22 50 ☽ # ♄

24 Wednesday
04 07 ☽ ☍ ♅
04 11 ☽ △ ♆
04 22 ☽ ⚻ ♃
07 32 ☽ ⊥ ☿
08 57 ☽ # ♅
09 05 ☽ ⊥ ☉
16 09 ☽ ± ♃
21 08 ♂ ∥ ♇
21 51 ☽ ⚺ ☿
21 55 ☽ ⚺ ☉
22 19 ♀ ± ♆
22 36 ☉ ☌ ☿
23 02 ☽ ∠ ♂

25 Thursday
04 55 ♂ △ ♇
05 02 ☽ ∥ ♅
06 34 ☿ ⚼ ♇
09 43 ☽ ⚼ ♆
09 59 ☽ ⚼ ♃
13 25 ♀ ⚻ ♇
14 40 ☉ ∥ ☿
15 36 ☽ ∥ ♄
15 53 ☉ ⚼ ♇
18 02 ☽ ⊥ ♂
23 40 ☽ ∥ ♀

26 Friday
00 12 ☽ ☌ ♄
01 29 ♀ ⚺ ♂
03 38 ☽ ± ♆
05 13 ☽ ⚻ ♇
06 48 ☽ ⚺ ♂
07 04 ☽ ☌ ♀
09 10 ☿ ∥ ♃
15 34 ☽ ⚻ ♅
15 39 ☽ ⚻ ♆
16 02 ☽ △ ♃
17 16 ☽ ± ♇

27 Saturday
03 38 ☽ ± ♅
07 06 ☽ ∥ ☉
07 40 ☽ ∥ ♃
11 24 ☽ ⚼ ♇
12 20 ☽ ∥ ☿
15 21 ☽ ☌ ☉
15 43 ☿ ⚼ ♅
15 53 ☉ ∥ ♃
21 48 ☽ ⚼ ♅
23 10 ☽ ☌ ☿

28 Sunday
00 46 ☽ # ♆
03 53 ☽ ∥ ♇
07 28 ☽ ∥ ♂
13 23 ☽ ⚺ ♄
17 46 ☽ △ ♇
23 21 ☽ ☌ ♂

29 Monday
01 41 ☽ ⊥ ♄
02 45 ☽ ⚺ ♀
04 10 ☽ △ ♅
04 16 ☽ ☍ ♆
04 52 ☽ □ ♃
16 04 ♀ ⚻ ♅
16 22 ☽ ⊥ ♀
17 02 ♀ ⚻ ♆
20 10 ☽ ∠ ♄
23 28 ♀ △ ♃

30 Tuesday
08 57 ♀ ± ♇
09 41 ☽ ⚺ ☉
12 44 ☽ ∠ ♀
12 54 ☿ # ♆
17 29 ☽ Q ♃
19 59 ☉ ⚼ ♅
21 46 ♂ ⊥ ♄
22 55 ☽ ⊥ ☉

All ephemeris data is given at 12.00 UT and the Moon's longitude is additionally given for 24.00 UT

MAY 1968

LONGITUDES

Date	Sidereal time h m s	Sun ☉ ° ' "	Moon ☽ ° ' "	Moon ☽ 24.00 ° ' "	Mercury ☿ ° '	Venus ♀ ° '	Mars ♂ ° '	Jupiter ♃ ° '	Saturn ♄ ° '	Uranus ♅ ° '	Neptune ♆ ° '	Pluto ♇ ° '
01	02 37 55	11 ♉ 09 56	23 ♊ 07 04	29 ♊ 05 17	18 ♉ 50	27 ♈ 48	24 ♉ 59	25 ♌ 58	18 ♈ 38	25 ♍ 30	25 ♏ 33	20 ♍ 24
02	02 41 51	12 08 09	05 ♋ 05 19	11 ♋ 07 33	20 57	29 ♈ 02	25 42	26 00	18 45	25 R 28	25 R 32	20 R 23
03	02 45 48	13 06 20	17 ♋ 12 28	23 ♋ 20 29	23 02	00 ♉ 16	26 24	26 03	18 52	25 27	25 30	20 22
04	02 49 44	14 04 28	29 ♋ 32 06	05 ♌ 47 47	25 05	01 29	27 07	26 04	19 00	25 26	25 28	20 21
05	02 53 41	15 02 35	12 ♌ 08 01	18 ♌ 33 18	27 07	02 43	27 49	26 07	19 07	25 24	25 27	20 21
06	02 57 37	16 00 40	25 ♌ 04 04	01 ♍ 40 44	29 ♉ 06	03 57	28 31	26 09	19 14	25 23	25 25	20 20
07	03 01 34	16 58 43	08 ♍ 23 40	15 ♍ 13 08	01 ♊ 03	05 11	29 14	26 12	19 21	25 21	25 24	20 19
08	03 05 31	17 56 44	22 ♍ 09 18	29 ♍ 12 12	02 58	06 24	29 ♉ 56	26 15	19 28	25 20	25 22	20 18
09	03 09 27	18 54 43	06 ♎ 21 45	13 ♎ 37 39	04 49	07 38	00 ♊ 38	26 18	19 35	25 19	25 20	20 18
10	03 13 24	19 52 40	20 ♎ 59 27	28 ♎ 26 30	06 38	08 52	01 20	26 21	19 41	25 18	25 19	20 17
11	03 17 20	20 50 36	05 ♏ 57 58	13 ♏ 32 49	08 23	10 05	02 03	26 25	19 48	25 17	25 17	20 16
12	03 21 17	21 48 30	21 ♏ 09 55	28 ♏ 47 58	10 05	11 19	02 45	26 28	19 55	25 16	25 16	20 16
13	03 25 13	22 46 22	06 ♐ 25 41	14 ♐ 01 43	11 44	12 33	03 27	26 32	20 02	25 14	25 14	20 15
14	03 29 10	23 44 13	21 ♐ 34 50	29 ♐ 03 53	13 20	13 47	04 09	26 36	20 09	25 14	25 12	20 15
15	03 33 06	24 42 03	06 ♑ 27 51	13 ♑ 45 57	14 52	15 00	04 51	26 40	20 15	25 13	25 11	20 14
16	03 37 03	25 39 52	20 ♑ 57 33	28 ♑ 02 16	16 20	16 14	05 33	26 44	20 22	25 12	25 09	20 14
17	03 41 00	26 37 39	04 ♒ 59 52	11 ♒ 50 21	17 45	17 28	06 14	26 48	20 29	25 11	25 07	20 13
18	03 44 56	27 35 25	18 ♒ 33 51	25 ♒ 10 38	19 07	18 41	06 56	26 53	20 35	25 10	25 06	20 13
19	03 48 53	28 33 10	01 ♓ 41 05	08 ♓ 05 37	20 24	19 55	07 38	26 58	20 42	25 09	25 04	20 12
20	03 52 49	29 ♉ 30 54	14 ♓ 24 48	20 ♓ 39 08	21 38	21 09	08 20	27 03	20 48	25 09	25 03	20 12
21	03 56 46	00 ♊ 28 37	26 ♓ 49 11	02 ♈ 55 33	22 48	22 22	09 01	27 08	20 55	25 08	25 01	20 12
22	04 00 42	01 26 19	08 ♈ 58 44	14 ♈ 59 17	23 54	23 36	09 43	27 13	21 01	25 07	24 59	20 12
23	04 04 39	02 23 59	20 ♈ 57 43	26 ♈ 54 29	24 57	24 50	10 25	27 18	21 08	25 07	24 58	20 11
24	04 08 35	03 21 39	02 ♉ 50 02	08 ♉ 44 44	25 55	26 03	11 06	27 24	21 14	25 06	24 56	20 11
25	04 12 32	04 19 18	14 ♉ 38 59	20 ♉ 33 04	26 49	27 17	11 48	27 29	21 20	25 06	24 54	20 11
26	04 16 29	05 16 55	26 ♉ 27 17	02 ♊ 21 55	27 39	28 31	12 29	27 35	21 26	25 06	24 53	20 11
27	04 20 25	06 14 31	08 ♊ 17 11	14 ♊ 13 18	28 25	29 ♉ 45	13 10	27 41	21 32	25 05	24 51	20 11
28	04 24 22	07 12 07	20 ♊ 10 29	26 ♊ 08 56	29 07	00 ♊ 58	13 52	27 47	21 38	25 05	24 50	20 10
29	04 28 18	08 09 41	02 ♋ 08 50	08 ♋ 10 25	29 ♊ 45	02 12	14 33	27 53	21 44	25 05	24 48	20 10
30	04 32 15	09 07 13	14 ♋ 13 54	20 ♋ 19 32	00 ♋ 18	03 26	15 14	28 00	21 50	25 05	24 47	20 10
31	04 36 11	10 ♊ 04 45	26 ♋ 27 33	02 ♌ 38 17	00 ♋ 46	04 ♊ 39	15 ♊ 56	28 ♌ 06	21 ♈ 56	25 ♍ 05	24 ♏ 45	20 ♍ 10

Moon nodes and latitude

Date	Moon True ☊ ° '	Moon Mean ☊ ° '	Moon ☽ Latitude ° '
01	18 ♈ 32	17 ♈ 34	04 N 44
02	18 R 29	17 31	05 06
03	18 26	17 27	05 14
04	18 24	17 24	05 08
05	18 23	17 21	04 48
06	18 D 23	17 18	04 12
07	18 24	17 15	03 22
08	18 26	17 11	02 22
09	18 27	17 08	01 N 06
10	18 R 27	17 05	00 S 14
11	18 26	17 02	01 34
12	18 24	16 59	02 49
13	18 20	16 56	03 52
14	18 15	16 52	04 38
15	18 10	16 49	05 05
16	18 05	16 46	05 11
17	18 02	16 43	04 58
18	18 00	16 40	04 28
19	18 D 00	16 36	03 45
20	18 00	16 33	02 52
21	18 02	16 30	01 53
22	18 03	16 27	00 S 49
23	18 R 03	16 24	00 N 16
24	18 02	16 21	01 19
25	17 59	16 17	02 19
26	17 53	16 14	03 12
27	17 46	16 11	03 57
28	17 37	16 08	04 32
29	17 28	16 05	04 55
30	17 18	16 02	05 04
31	17 ♈ 10	15 ♈ 58	05 N 02

DECLINATIONS

Date	Sun ☉ ° '	Moon ☽ ° '	Mercury ☿ ° '	Venus ♀ ° '	Mars ♂ ° '	Jupiter ♃ ° '	Saturn ♄ ° '	Uranus ♅ ° '	Neptune ♆ ° '	Pluto ♇ ° '
01	15 N 11	28 N 00	18 N 10	09 N 25	19 N 18	13 N 57	05 N 13	02 N 30	17 S 25	18 N 03
02	15 29	28 26	18 54	09 52	19 29	13 57	05 13	02 31	17 25	18 03
03	15 47	27 32	19 36	10 19	19 39	13 56	05 19	02 31	17 25	18 03
04	16 04	25 16	20 16	10 46	19 50	13 55	05 21	02 32	17 24	18 03
05	16 21	21 46	20 54	11 13	20 00	13 54	05 24	02 33	17 24	18 03
06	16 38	17 07	21 29	11 39	20 10	13 53	05 26	02 33	17 24	18 03
07	16 55	11 33	22 02	12 05	20 20	13 52	05 29	02 34	17 23	18 03
08	17 11	05 N 15	22 33	12 31	20 30	13 51	05 32	02 34	17 23	18 03
09	17 27	01 S 31	23 01	12 56	20 39	13 50	05 34	02 34	17 22	18 03
10	17 43	08 25	23 26	13 21	20 48	13 48	05 37	02 35	17 22	18 02
11	17 58	15 00	23 49	13 46	20 57	13 47	05 39	02 35	17 22	18 02
12	18 13	20 47	24 09	14 11	21 06	13 46	05 42	02 36	17 21	18 02
13	18 28	25 12	24 27	14 35	21 15	13 44	05 44	02 36	17 21	18 02
14	18 43	27 48	24 42	14 58	21 23	13 43	05 46	02 36	17 20	18 02
15	18 57	28 22	24 56	15 22	21 32	13 41	05 49	02 37	17 20	18 02
16	19 11	26 56	25 06	15 45	21 40	13 40	05 51	02 37	17 20	18 02
17	19 24	23 50	25 15	16 07	21 48	13 38	05 53	02 37	17 19	18 01
18	19 38	19 31	25 22	16 30	21 55	13 36	05 56	02 38	17 19	18 01
19	19 51	14 23	25 27	16 52	22 03	13 35	05 58	02 38	17 19	18 01
20	20 03	08 47	25 29	17 13	22 10	13 33	06 00	02 38	17 18	18 01
21	20 15	02 S 59	25 31	17 34	22 17	13 31	06 03	02 38	17 18	18 01
22	20 27	02 N 54	25 30	17 54	22 24	13 29	06 05	02 39	17 17	18 00
23	20 39	08 26	25 28	18 14	22 31	13 27	06 07	02 39	17 17	18 00
24	20 50	13 42	25 24	18 34	22 37	13 25	06 09	02 39	17 17	18 00
25	21 01	18 27	25 19	18 53	22 43	13 23	06 11	02 39	17 16	17 59
26	21 11	22 29	25 12	19 12	22 49	13 21	06 14	02 39	17 16	17 59
27	21 21	25 36	25 04	19 30	22 55	13 19	06 16	02 39	17 16	17 59
28	21 31	27 36	24 55	19 47	23 01	13 17	06 18	02 39	17 15	17 58
29	21 40	28 21	24 45	20 04	23 06	13 14	06 20	02 39	17 15	17 58
30	21 49	27 45	24 34	20 21	23 11	13 12	06 22	02 39	17 15	17 58
31	21 N 58	25 N 48	24 N 22	20 N 37	23 N 16	13 N 10	06 N 24	02 N 39	17 S 14	17 N 57

ZODIAC SIGN ENTRIES

Date	h	m	Planets
02	01	50	☽ ♋
03	06	56	♀ ♉
04	12	54	☽ ♌
06	20	58	☽ ♍
06	22	56	☿ ♊
08	14	14	♂ ♊
09	01	21	☽ ♎
11	02	30	☽ ♏
13	01	53	☽ ♐
15	01	31	☽ ♑
17	03	22	☽ ♒
19	08	53	☽ ♓
21	00	06	☉ ♊
21	18	14	☽ ♈
24	06	15	☽ ♉
26	19	12	☽ ♊
27	17	02	♀ ♊
29	07	43	☽ ♋
29	22	44	☿ ♋
31	18	53	☽ ♌

LATITUDES

Date	Mercury ☿ ° '	Venus ♀ ° '	Mars ♂ ° '	Jupiter ♃ ° '	Saturn ♄ ° '	Uranus ♅ ° '	Neptune ♆ ° '	Pluto ♇ ° '
01	00 N 46	01 S 22	00 N 17	01 N 10	02 S 15	00 N 47	01 N 47	15 N 32
04	01 16	01 18	00 19	01 09	02 15	00 47	01 47	15 31
07	01 42	01 14	00 21	01 09	02 15	00 47	01 47	15 30
10	02 02	01 10	00 23	01 08	02 16	00 47	01 47	15 29
13	02 16	01 05	00 24	01 08	02 16	00 46	01 47	15 28
16	02 23	01 00	00 26	01 07	02 17	00 46	01 47	15 27
19	02 21	00 54	00 28	01 07	02 17	00 46	01 47	15 26
22	02 12	00 48	00 29	01 07	02 17	00 46	01 47	15 25
25	01 54	00 42	00 31	01 06	02 18	00 46	01 47	15 23
28	01 29	00 35	00 33	01 06	02 18	00 46	01 47	15 22
31	00 N 55	00 S 28	00 N 34	01 N 05	02 S 19	00 N 46	01 N 47	15 N 21

DATA

Julian Date	2439978
Delta T	+39 seconds
Ayanamsa	23° 24' 47"
Synetic vernal point	05° ♓ 42' 13"
True obliquity of ecliptic	23° 26' 45"

MOON'S PHASES, APSIDES AND POSITIONS ☽

Date	h	m	Phase	Longitude ° '	Eclipse Indicator
05	17	54	☽	15 ♌ 17	
12	13	05	○	21 ♏ 51	
19	05	44	☾	28 ♒ 18	
27	07	30	●	06 ♊ 04	

Day	h	m		
12	16	54	Perigee	
26	12	05	Apogee	
02	07	58	Max dec	28° N 27'
09	06	42	0S	
15	06	23	Max dec	28° S 25'
22	00	20	0N	
29	13	23	Max dec	28° N 21'

LONGITUDES

Date	Chiron ⚷ ° '	Ceres ⚳ ° '	Pallas ⚴ ° '	Juno ⚵ ° '	Vesta ⚶ ° '	Black Moon Lilith ⚸ ° '
01	01 ♈ 38	26 ♎ 45	13 ♍ 57	04 ♏ 06	29 ♓ 28	24 ♉ 38
11	02 ♈ 06	24 ♎ 54	14 ♍ 22	01 ♏ 57	03 ♈ 54	25 ♉ 45
21	02 ♈ 30	23 ♎ 34	15 ♍ 27	00 ♏ 07	08 ♈ 11	26 ♉ 52
31	02 ♈ 50	22 ♎ 52	17 ♍ 08	28 ♎ 44	12 ♈ 17	27 ♉ 58

All ephemeris data is given at 12.00 UT and the Moon's longitude is additionally given for 24.00 UT

ASPECTARIAN

h m Aspects

01 Wednesday
01 29 ☽ ⚺ ☿
02 52 ☽ ✶ ♄
06 31 ☽ □ ♇
08 04 ☿ ∥ ♇
09 35 ☿ ⚺ ♄
16 00 ☽ ⚺ ♂
16 12 ☽ ⊥ ☿
16 47 ☽ □ ♅
16 53 ☽ ⚻ ♆
17 45 ☽ ✶ ♃
18 40 ☽ ∠ ☉
22 30 ☽ ✶ ♀

02 Thursday
03 15 ☽ Q ♄
04 48 ☽ ⊥ ♂
04 49 ♂ △ ♅
04 54 ☽ ± ♆
05 38 ☿ △ ♇
06 30 ♂ ☍ ♆
14 04 ☽ ∠ ☿
18 33 ☽ Q ♇
22 47 ☽ bQ ♆
22 52 ♂ □ ♃
23 47 ☽ ∠ ♃
23 50 ☽ ∠ ♂

03 Friday
01 09 ☽ Q ♀
03 13 ☽ ✶ ☉
04 37 ☽ Q ♅
14 29 ☿ ∥ ♂
15 18 ☽ □ ♄
17 34 ☽ ⊥ ♃
18 11 ☽ ✶ ♇

04 Saturday
01 41 ☽ ✶ ☿
04 04 ☽ ✶ ♅
04 10 ☽ △ ♆
04 45 ☽ Q ☉
05 17 ☽ ⚺ ♃
07 02 ☽ ✶ ♂
10 46 ☿ ⊥ ♄
10 47 ♀ ± ♅
15 54 ☿ △ ♅
16 10 ☽ □ ♀
16 27 ☿ ☍ ♆
23 09 ☽ ∠ ♇
23 47 ☿ □ ♃

05 Sunday
05 14 ☽ Q ☿
07 23 ☽ Q ♂
08 44 ☽ ∠ ♅
16 09 ☽ ⊥ ♇
16 17 ☽ ∥ ☿
17 54 ☽ □ ☉
21 24 ☽ ∥ ♂

06 Monday
01 01 ☿ ☌ ♂
01 09 ☽ △ ♄
01 33 ☽ ⊥ ♅
03 18 ☽ ⚺ ♇
07 38 ☽ ∥ ♇
10 44 ☽ # ♆
12 34 ☽ ⚺ ♅
12 39 ☽ □ ♆
13 59 ☽ ☌ ♃
14 09 ☽ ∥ ☉
18 39 ☽ □ ♂
20 38 ☽ □ ☿

07 Tuesday
02 25 ☽ ∥ ♃
04 43 ☽ bQ ♄
05 42 ☽ △ ♀
09 59 ☽ ∥ ♀
14 44 ♀ bQ ♇
20 11 ♄ ± ♆
20 48 ☽ Q ♆
20 48 ☽ ± ♄

08 Wednesday
04 12 ☽ △ ☉
07 19 ☽ ⚻ ♄
08 49 ☽ ☌ ♇
10 35 ☽ bQ ♀
10 59 ☽ ∥ ♄
17 26 ☽ ☌ ♅
17 29 ☽ ✶ ♆
19 01 ☽ ⚺ ♃
21 37 ☽ ∥ ♅

09 Thursday
01 55 ☽ △ ♂
03 21 ☽ ± ♀
05 10 ☉ # ♆
05 11 ☽ ⊥ ♃
07 37 ☽ bQ ☉
08 35 ☿ ∠ ♄
09 03 ☽ △ ☿
14 18 ☽ ⚻ ♀
15 40 ☽ # ♅
18 35 ☽ ∠ ♆
20 12 ☽ ∠ ♃
23 35 ☽ ± ☉

10 Friday
02 09 ☽ # ♄
04 04 ☽ bQ ♂
06 45 ☉ ⚺ ♄
09 17 ☽ ⊥ ♆
09 53 ☽ ☍ ♄
10 04 ☽ ⚻ ☉
10 51 ☽ ⚺ ♇
13 11 ☽ bQ ☿
18 56 ☽ ⚺ ♅
18 58 ☽ ⚺ ♆
19 22 ☽ ± ♂
20 32 ☽ ⊥ ♇
20 41 ☽ ✶ ♃
21 58 ☉ △ ♇

11 Saturday
04 33 ☽ ⊥ ♅
05 27 ☽ ⚻ ♂
05 34 ☽ ± ☿
07 02 ☽ # ♀
07 26 ☽ # ♃
10 54 ☽ ∠ ♇
12 57 ♀ ∥ ♃
15 36 ♀ bQ ♅
15 53 ☽ Q ♃
16 20 ☽ ⚻ ☿
18 27 ☉ ∥ ♇
18 49 ☽ ∠ ♅
19 07 ☽ ☍ ♀
21 17 ☽ ∥ ♆

12 Sunday
00 04 ☽ # ♇
00 20 ☽ # ☉
10 02 ☽ ⚻ ♄
10 35 ☽ ✶ ♇
11 52 ♅ ✶ ♆
13 05 ☽ ☍ ☉
13 35 ☽ # ♂
18 25 ☽ ☌ ♆
18 26 ☽ ✶ ♅
19 32 ☽ ± ♄
20 22 ☽ □ ♃

13 Monday
05 26 ☽ Q ♇
06 50 ☽ # ☿
07 05 ☽ ☍ ♂
09 47 ☽ bQ ♄
13 17 ☽ Q ♅
21 22 ☽ ☍ ☿
22 30 ☽ ⚻ ♀

14 Tuesday
08 52 ☽ ± ♀
09 42 ☽ △ ♄
09 52 ☽ □ ♇
15 41 ☽ ⚻ ☉
17 47 ☽ ⚺ ♆
17 49 ☽ □ ♅
20 04 ☽ △ ♃

15 Wednesday
00 34 ☽ bQ ♀
01 59 ☽ ± ☉
03 26 ☽ ⊥ ♆
07 48 ♄ ⚻ ♇
08 40 ☿ Q ♃
09 14 ☽ ⚻ ♂
17 40 ☽ bQ ☉
18 04 ☽ ∠ ♆
19 32 ☽ ± ♂
20 34 ☽ bQ ♃
23 33 ☉ ☍ ♆

16 Thursday
00 29 ☉ △ ♅
00 51 ☿ ⚺ ♀
03 21 ☽ △ ♀
03 24 ☽ ⚻ ☿
04 55 ♂ ∠ ♄
10 46 ☽ △ ♇
11 00 ☽ □ ♄
11 16 ☽ bQ ♂
11 37 ☽ ± ♃
14 35 ☽ ± ☿
19 04 ☽ ✶ ♆
19 09 ☽ △ ♅
20 32 ☽ △ ☉
21 50 ☽ ⚻ ♃

17 Friday
02 42 ☽ # ☿
07 41 ☽ bQ ☿
07 50 ☉ ⊥ ♄
12 23 ☽ bQ ♇
14 17 ☽ △ ♂
15 42 ☽ Q ♆
16 53 ☉ □ ♃
18 08 ☽ Q ♄
21 04 ☽ bQ ♅
23 41 ☽ # ♂

18 Saturday
04 13 ☽ ± ♇
11 27 ☽ # ☉
12 15 ☽ □ ♀
13 05 ☽ ± ♅
13 05 ☽ △ ☿
14 58 ☽ ⚻ ♇
15 41 ☽ ✶ ♄
19 18 ☽ # ♇
22 39 ☽ ∥ ♆
23 50 ☽ □ ♆
23 58 ☽ ⚻ ♅

19 Sunday
01 31 ☽ # ♀
03 13 ☽ ☍ ♃
05 44 ☽ □ ☉
08 21 ☿ □ ♇
15 34 ☽ # ♃
17 40 ♀ △ ♇
18 12 ☿ ✶ ♄
19 33 ☽ ∠ ♄
23 46 ☽ □ ♂

20 Monday
00 55 ☽ Q ♀
04 44 ♀ ⚺ ♄
12 45 ☽ ⊥ ♄
17 52 ♀ # ♆
18 27 ☽ Q ☉
23 07 ☽ ☍ ♇
23 29 ☽ # ♄

21 Tuesday
00 24 ☽ ⚺ ♄
02 23 ☽ ✶ ♀
03 22 ☽ □ ☿
08 29 ☽ △ ♆
08 43 ☽ ☍ ♅
12 25 ☽ Q ♂
12 36 ☽ ⚻ ♃
13 26 ☽ # ♅
19 47 ☽ ✶ ☉

22 Wednesday
00 29 ☽ ± ♃
11 10 ☽ ∠ ♀
11 18 ☽ ∥ ♅
13 33 ☽ ✶ ♂
14 00 ☽ bQ ♆
18 24 ☽ Q ☿
18 30 ☽ bQ ♃
19 01 ♀ ∥ ♇

23 Thursday
01 56 ☽ ∥ ♄
04 13 ☽ ∠ ☉
07 13 ☽ ⊥ ♀
07 59 ☽ ± ♆
10 27 ☽ ⚻ ♇
12 20 ☽ ☌ ♄
12 26 ☿ ⚻ ♆
14 31 ♀ ☍ ♆
16 06 ☿ □ ♅
17 32 ♀ △ ♅
20 03 ☽ ⚻ ♆
20 22 ☽ ⚻ ♅
20 42 ☽ ⚺ ♀
20 46 ☽ ✶ ☿
21 32 ☽ ∠ ♂
22 32 ☽ ± ♇
23 24 ☿ ⚺ ♀
23 56 ☽ ⊥ ☉

24 Friday
00 54 ☽ △ ♃
08 30 ☽ ± ♅
10 41 ☽ ∥ ♃
13 10 ☽ ⚺ ☉
16 46 ☽ bQ ♇
16 53 ☽ ⊥ ♂

25 Saturday
02 45 ☽ bQ ♅
05 46 ☽ # ♆
05 47 ☽ ∠ ☿
05 50 ☽ ⚺ ♂
09 34 ☽ ∥ ♇
13 02 ♀ ⊥ ♄
14 35 ☽ ∥ ♀
16 17 ♀ □ ♃
23 15 ☽ △ ♇

26 Sunday
01 31 ☽ ⊥ ☿
01 43 ☽ ⚺ ♄
03 23 ☽ ∥ ☉
08 49 ☽ ☍ ♆
09 14 ☽ △ ♅
09 31 ☿ ✶ ♃
14 01 ☽ ⊥ ♄
14 19 ☽ □ ♃
14 22 ☽ ∥ ♂
14 37 ☽ ⚺ ☿
16 40 ☽ ☌ ♀

27 Monday
07 30 ☽ ☌ ☉
07 32 ☽ ∥ ☿
08 26 ☽ ∠ ♄
20 20 ☉ ∠ ♄
22 30 ☽ ☌ ♂

28 Tuesday
03 05 ☽ Q ♃
12 00 ☽ □ ♇
14 59 ☽ ✶ ♄
21 20 ☽ ⚻ ♆
21 52 ☽ □ ♅

29 Wednesday
03 25 ☽ ✶ ♃
06 57 ☽ ☌ ☿
09 19 ☽ ± ♆
12 07 ☽ ⚺ ♀
15 12 ☽ Q ♄

30 Thursday
00 00 ☽ Q ♇
01 00 ☽ ⚺ ☉
01 25 ☽ ⊥ ♀
03 12 ☽ bQ ♆
09 32 ☽ ∠ ♃
09 43 ☽ Q ♅
13 54 ☽ ⊥ ☉
14 07 ☽ ⚺ ♂
21 12 ☽ ∠ ♀
23 42 ☽ ✶ ♇

31 Friday
02 37 ☽ ⊥ ♂
03 06 ☽ □ ♄
03 24 ☽ ⊥ ♃
04 04 ♇ St D
08 40 ☽ △ ♆
09 05 ☽ ∠ ☉
09 18 ☽ ✶ ♅
10 52 ☿ ± ♆
15 14 ☽ ⚺ ♃
19 14 ♂ Q ♃
20 41 ☽ ⚺ ☿
21 12 ☽ ∠ ♂

JUNE 1968

Date	Sidereal time h m s	Sun ☉ ° ′ ″	Moon ☽ ° ′ ″	Moon ☽ 24.00 ° ′ ″	Mercury ☿ ° ′	Venus ♀ ° ′	Mars ♂ ° ′	Jupiter ♃ ° ′	Saturn ♄ ° ′	Uranus ♅ ° ′	Neptune ♆ ° ′	Pluto ♇ ° ′
01	04 40 08	11 ♊ 02 15	08 ♌ 52 02	15 ♌ 09 09	01 ♋ 10	05 ♊ 53	16 ♊ 37	28 ♌ 13	22 ♈ 02	25 ♍ 04	24 ♏ 43	20 ♍ 10
02	04 44 04	11 59 44	21 ♌ 30 00	27 ♌ 54 59	01 30	07 07	17 18	28 20	22 08	25 D 04	24 R 42	20 D 10
03	04 48 01	12 57 12	04 ♍ 24 29	10 ♍ 58 53	01 45	08 20	17 59	28 26	22 14	25 05	24 40	20 10
04	04 51 57	13 54 39	17 ♍ 38 36	24 ♍ 23 56	01 55	09 34	18 40	28 33	22 19	25 05	24 39	20 11
05	04 55 54	14 52 04	01 ♎ 15 10	08 ♎ 12 32	02 00	10 48	19 21	28 41	22 25	25 05	24 37	20 11
06	04 59 51	15 49 28	15 ♎ 16 06	22 ♎ 25 50	02 R 01	12 01	20 02	28 48	22 31	25 05	24 36	20 11
07	05 03 47	16 46 51	29 ♎ 41 32	07 ♏ 02 48	01 58	13 15	20 43	28 55	22 36	25 05	24 34	20 11
08	05 07 44	17 44 12	14 ♏ 29 04	21 ♏ 59 34	01 50	14 29	21 24	29 03	22 41	25 06	24 33	20 11
09	05 11 40	18 41 33	29 ♏ 33 19	07 ♐ 09 11	01 38	15 42	22 05	29 11	22 47	25 06	24 31	20 12
10	05 15 37	19 38 54	14 ♐ 45 56	22 ♐ 22 13	01 22	16 56	22 46	29 18	22 52	25 06	24 30	20 12
11	05 19 33	20 36 13	29 ♐ 56 41	07 ♑ 28 03	01 02	18 10	23 26	29 26	22 57	25 07	24 29	20 12
12	05 23 30	21 33 32	14 ♑ 55 05	22 ♑ 16 47	00 39	19 23	24 07	29 34	23 02	25 07	24 27	20 13
13	05 27 26	22 30 50	29 ♑ 32 17	06 ♒ 40 57	00 ♋ 12	20 37	24 48	29 42	23 07	25 08	24 26	20 13
14	05 31 23	23 28 07	13 ♒ 42 22	20 ♒ 36 20	29 ♊ 43	21 51	25 28	29 51	23 12	25 09	24 24	20 14
15	05 35 20	24 25 24	27 ♒ 22 51	04 ♓ 02 06	29 12	23 04	26 09	29 ♌ 59	23 17	25 09	24 23	20 14
16	05 39 16	25 22 41	10 ♓ 34 23	17 ♓ 00 09	28 40	24 18	26 49	00 ♍ 08	23 22	25 10	24 22	20 15
17	05 43 13	26 19 58	23 ♓ 19 56	29 ♓ 34 18	28 06	25 32	27 30	00 16	23 27	25 11	24 20	20 15
18	05 47 09	27 17 14	05 ♈ 43 54	11 ♈ 49 24	27 32	26 45	28 10	00 25	23 32	25 12	24 19	20 16
19	05 51 06	28 14 30	17 ♈ 51 25	23 ♈ 50 38	26 58	27 59	28 51	00 34	23 36	25 13	24 18	20 16
20	05 55 02	29 ♊ 11 45	29 ♈ 47 40	05 ♉ 43 07	26 24	29 ♊ 13	29 ♊ 31	00 43	23 41	25 14	24 16	20 17
21	05 58 59	00 ♋ 09 01	11 ♉ 37 33	17 ♉ 31 28	25 52	00 ♋ 27	00 ♋ 12	00 52	23 45	25 15	24 15	20 17
22	06 02 55	01 06 16	23 ♉ 25 21	29 ♉ 19 38	25 21	01 40	00 52	01 01	23 50	25 16	24 14	20 18
23	06 06 52	02 03 31	05 ♊ 14 40	11 ♊ 10 48	24 53	02 54	01 32	01 10	23 54	25 17	24 13	20 19
24	06 10 49	03 00 46	17 ♊ 08 18	23 ♊ 07 24	24 28	04 08	02 13	01 19	23 58	25 18	24 12	20 20
25	06 14 45	03 58 01	29 ♊ 08 17	05 ♋ 11 07	24 06	05 22	02 53	01 29	24 02	25 19	24 10	20 20
26	06 18 42	04 55 16	11 ♋ 16 02	17 ♋ 23 07	23 47	06 35	03 33	01 38	24 06	25 20	24 09	20 21
27	06 22 38	05 52 30	23 ♋ 32 30	29 ♋ 44 15	23 33	07 49	04 13	01 48	24 10	25 22	24 08	20 22
28	06 26 35	06 49 44	05 ♌ 58 29	12 ♌ 15 19	23 22	09 03	04 53	01 58	24 14	25 23	24 07	20 23
29	06 30 31	07 46 57	18 ♌ 34 52	24 ♌ 57 18	23 16	10 17	05 33	02 07	24 18	25 24	24 06	20 24
30	06 34 28	08 ♋ 44 10	01 ♍ 22 48	07 ♍ 51 36	23 ♊ 15	11 ♋ 30	06 ♋ 13	02 ♍ 17	24 ♈ 22	25 ♍ 26	24 ♏ 05	20 ♍ 25

Date	Moon True ☊ ° ′	Moon Mean ☊ ° ′	Moon ☽ Latitude ° ′
01	17 ♈ 04	15 ♈ 55	04 N 44
02	16 R 59	15 52	04 13
03	16 57	15 49	03 28
04	16 D 57	15 46	02 30
05	16 58	15 42	01 23
06	16 58	15 39	00 N 09
07	16 R 58	15 36	01 S 07
08	16 55	15 33	02 21
09	16 50	15 30	03 27
10	16 43	15 27	04 18
11	16 34	15 23	04 51
12	16 25	15 20	05 03
13	16 16	15 17	04 55
14	16 08	15 14	04 29
15	16 03	15 11	03 48
16	16 00	15 08	02 57
17	16 00	15 04	01 58
18	16 D 00	15 01	00 S 54
19	16 R 00	14 58	00 N 10
20	15 59	14 55	01 13
21	15 56	14 52	02 12
22	15 51	14 48	03 05
23	15 43	14 45	03 50
24	15 33	14 42	04 25
25	15 20	14 39	04 48
26	15 07	14 36	04 59
27	14 54	14 33	04 57
28	14 42	14 29	04 40
29	14 33	14 26	04 10
30	14 ♈ 26	14 ♈ 23	03 N 26

DECLINATIONS

Date	Sun ☉ ° ′	Moon ☽ ° ′	Mercury ☿ ° ′	Venus ♀ ° ′	Mars ♂ ° ′	Jupiter ♃ ° ′	Saturn ♄ ° ′	Uranus ♅ ° ′	Neptune ♆ ° ′	Pluto ♇ ° ′
01	22 N 06	22 N 37	24 N 09	20 N 52	23 N 21	13 N 07	06 N 26	02 N 39	17 S 14	17 N 57
02	22 14	18 19	23 55	21 07	23 25	13 05	06 28	02 39	17 13	17 57
03	22 22	13 07	23 41	21 21	23 30	13 02	06 30	02 39	17 13	17 56
04	22 29	07 11	23 26	21 35	23 34	13 00	06 32	02 39	17 13	17 56
05	22 35	00 N 46	23 10	21 48	23 38	12 57	06 34	02 39	17 12	17 55
06	22 41	05 S 52	22 54	22 00	23 41	12 54	06 36	02 39	17 12	17 55
07	22 47	12 25	22 38	22 12	23 45	12 52	06 38	02 39	17 12	17 54
08	22 53	18 26	22 21	22 23	23 48	12 49	06 39	02 39	17 11	17 54
09	22 58	23 25	22 04	22 34	23 51	12 46	06 41	02 38	17 11	17 53
10	23 03	26 51	21 47	22 43	23 54	12 43	06 43	02 38	17 11	17 53
11	23 07	28 18	21 30	22 53	23 56	12 40	06 45	02 38	17 10	17 52
12	23 11	27 38	21 14	23 01	23 59	12 37	06 46	02 38	17 10	17 52
13	23 14	25 04	20 57	23 09	24 01	12 34	06 48	02 37	17 10	17 51
14	23 17	21 01	20 41	23 16	24 03	12 31	06 50	02 37	17 09	17 51
15	23 20	15 57	20 25	23 23	24 05	12 28	06 51	02 37	17 09	17 50
16	23 22	10 20	20 10	23 29	24 06	12 25	06 53	02 37	17 09	17 49
17	23 24	04 S 27	19 56	23 34	24 08	12 22	06 54	02 36	17 09	17 49
18	23 25	01 N 27	19 43	23 38	24 09	12 19	06 56	02 36	17 08	17 48
19	23 26	07 10	19 30	23 42	24 10	12 16	06 57	02 35	17 08	17 48
20	23 27	12 32	19 19	23 45	24 10	12 13	06 59	02 35	17 08	17 47
21	23 27	17 25	19 09	23 48	24 11	12 09	07 00	02 35	17 07	17 46
22	23 26	21 37	19 01	23 50	24 11	12 06	07 02	02 34	17 07	17 46
23	23 26	24 57	18 54	23 51	24 11	12 03	07 03	02 34	17 07	17 45
24	23 25	27 13	18 48	23 51	24 11	11 59	07 04	02 33	17 07	17 44
25	23 23	28 15	18 44	23 50	24 11	11 56	07 06	02 33	17 06	17 44
26	23 21	27 56	18 41	23 49	24 10	11 52	07 07	02 32	17 06	17 43
27	23 19	26 16	18 40	23 47	24 09	11 49	07 08	02 31	17 06	17 42
28	23 16	23 18	18 41	23 45	24 09	11 45	07 09	02 31	17 06	17 42
29	23 13	19 13	18 43	23 42	24 07	11 41	07 11	02 30	17 06	17 41
30	23 N 09	14 N 12	18 N 47	23 N 38	24 N 06	11 N 38	07 N 12	02 N 30	17 S 05	17 N 40

ZODIAC SIGN ENTRIES

Date	h	m	Planets
03	03	52	☽ ♍
05	09	49	☽ ♎
07	12	30	☽ ♏
09	12	42	☽ ♐
11	12	05	☽ ♑
13	12	46	☽ ♒
13	22	32	☿ ♊
15	14	44	♃ ♍
15	16	42	☽ ♓
18	00	50	☽ ♈
20	12	25	☽ ♉
21	03	20	♀ ♋
21	05	03	♂ ♋
21	08	13	☉ ♋
23	01	22	☽ ♊
25	13	43	☽ ♋
28	00	30	☽ ♌
30	09	26	☽ ♍

LATITUDES

Date	Mercury ☿ ° ′	Venus ♀ ° ′	Mars ♂ ° ′	Jupiter ♃ ° ′	Saturn ♄ ° ′	Uranus ♅ ° ′	Neptune ♆ ° ′	Pluto ♇ ° ′
01	00 N 42	00 S 26	00 N 35	01 N 05	02 S 19	00 N 46	01 N 47	15 N 20
04	00 00	00 19	00 36	01 05	02 20	00 45	01 47	15 19
07	00 S 48	00 12	00 38	01 05	02 20	00 45	01 47	15 18
10	01 39	00 S 05	00 39	01 04	02 21	00 45	01 47	15 16
13	02 30	00 N 02	00 40	01 04	02 21	00 45	01 46	15 15
16	03 16	00 09	00 42	01 04	02 22	00 45	01 46	15 14
19	03 54	00 16	00 43	01 03	02 23	00 45	01 46	15 13
22	04 21	00 23	00 45	01 03	02 23	00 45	01 46	15 11
25	04 35	00 30	00 46	01 03	02 24	00 45	01 46	15 10
28	04 36	00 37	00 47	01 02	02 25	00 44	01 46	15 09
31	04 S 25	00 N 43	00 N 48	01 N 02	02 S 26	00 N 44	01 N 46	15 N 08

DATA

Julian Date	2440009
Delta T	+39 seconds
Ayanamsa	23° 24′ 52″
Synetic vernal point	05° ♓ 42′ 08″
True obliquity of ecliptic	23° 26′ 45″

MOON'S PHASES, APSIDES AND POSITIONS ☽

Date	h	m	Phase	Longitude ° ′	Eclipse Indicator
04	04	47	☽	13 ♍ 37	
10	20	13	○	19 ♐ 59	
17	18	14	☾	26 ♓ 35	
25	22	24	●	04 ♋ 23	

Day	h	m		
10	02	42	Perigee	
22	19	14	Apogee	
05	14	49	OS	
11	16	22	Max dec	28° S 20′
18	06	04	ON	
25	18	35	Max dec	28° N 18′

LONGITUDES

Date	Chiron ⚷ ° ′	Ceres ⚳ ° ′	Pallas ⚴ ° ′	Juno ⚵ ° ′	Vesta ⚶ ° ′	Black Moon Lilith ⚸ ° ′
01	02 ♈ 52	22 ♎ 50	17 ♍ 19	28 ♎ 37	12 ♈ 41	28 ♉ 05
11	03 ♈ 07	22 ♎ 51	19 ♍ 32	27 ♎ 49	16 ♈ 35	29 ♉ 12
21	03 ♈ 17	23 ♎ 29	22 ♍ 08	27 ♎ 34	20 ♈ 15	00 ♊ 19
31	03 ♈ 22	24 ♎ 41	25 ♍ 04	27 ♎ 50	23 ♈ 39	01 ♊ 26

All ephemeris data is given at 12.00 UT and the Moon's longitude is additionally given for 24.00 UT

ASPECTARIAN

h m	Aspects
01 Saturday	
00 49	☽ ∥ ☿
04 54	☽ ∠ ♇
05 38	☽ ⚹ ♀
07 18	☽ ∥ ♂
08 39	☽ ⊥ ☿
14 19	☽ ∠ ♅
15 05	☽ ∥ ☉
16 30	☽ ⚹ ☉
21 53	☽ ∥ ♀
22 08	☽ ⊥ ♇
02 Sunday	
00 36	♅ St D
02 19	☽ ∠ ☿
03 37	☽ ⚹ ♂
07 01	☽ Q ♀
07 26	☽ ⊥ ♅
09 30	☽ ⊻ ♇
12 29	♀ ∠ ♄
13 12	☽ △ ♄
13 54	☽ ∥ ♇
17 04	☽ Q ☉
17 25	☽ # ♆
17 59	☽ □ ♆
18 42	☽ ⊻ ♅
03 Monday	
00 52	☽ ☌ ♃
03 23	☽ Q ♂
07 01	☽ ⚹ ☿
12 21	☽ ∥ ♃
17 12	☽ ⚼ ♄
19 56	☽ □ ♀
04 Tuesday	
01 59	☿ ∥ ♂
03 02	☽ Q ♆
04 47	☽ □ ☉
05 14	☽ Q ☿
09 37	☽ ± ♄
13 56	☽ □ ♂
14 31	☽ ∥ ♄
16 31	☽ ☌ ♇
20 23	☽ ⊼ ♄
05 Wednesday	
00 25	☽ ⚹ ♆
01 12	☽ ☌ ♅
05 07	☽ ∥ ♅
07 28	☽ ⊻ ♃
13 19	☽ □ ☿
17 59	☽ ⊥ ♃
06 Thursday	
00 24	☽ # ♅
02 23	☽ ∠ ♆
05 17	☿ St R
05 59	☽ △ ♀
09 30	☽ ∠ ♃
13 00	☽ △ ☉
14 37	☽ # ♄
17 11	♂ □ ♇
17 35	☽ ⊥ ♆
20 15	☽ ⊻ ♇
20 24	☽ △ ♂
07 Friday	
00 12	☽ ☍ ♄
01 37	☉ ∥ ☿
03 34	☽ ⊻ ♆
04 24	☽ ⊻ ♅
06 13	☽ ⊥ ♇
09 25	☽ ⚼ ♀
10 43	☽ ⚹ ♃
13 40	☽ # ♃
14 17	☽ ⊥ ♅
15 40	☽ ⚼ ☉
15 42	☽ △ ☿
16 05	☉ Q ♃
20 59	☽ ∠ ♇
22 19	☽ ⚼ ♂
08 Saturday	
01 27	☽ ± ♀
04 56	☽ ∠ ♅
06 25	☽ Q ♃
06 45	☽ ∥ ♆
07 17	☽ ± ☉
09 42	☽ # ♇
10 18	☿ ∥ ♀
11 59	☽ ⊼ ♀
13 32	☽ ± ♂
15 43	☽ ⚼ ☿
17 34	☽ ⊼ ☉
21 08	☽ ⚹ ♇
23 35	☽ ⊼ ♂
09 Sunday	
01 11	☽ ⊼ ♄
04 02	☽ ☌ ♆
04 56	☽ ⚹ ♅
05 12	☽ # ☿
05 52	☽ ± ☿
07 08	☽ # ♀
09 26	☽ # ☉
10 46	☽ ± ♄
11 24	☽ □ ♃
14 28	☽ # ♂
15 14	☽ ⊼ ☿
16 11	☽ Q ♇
23 55	☽ Q ♅
10 Monday	
01 04	☽ ⚼ ♄
15 43	☽ ☍ ♀
16 21	♂ ⚹ ♄
20 09	♀ Q ♃
20 13	☽ ☍ ☉
20 35	☽ □ ♇
11 Tuesday	
00 52	☽ △ ♄
01 12	☽ ☍ ♂
01 57	☉ □ ♇
03 21	☽ ⊻ ♆
04 20	☽ □ ♅
11 11	☽ △ ♃
12 51	☽ ⊥ ♆
13 41	☽ ☍ ☿
12 Wednesday	
03 12	☽ ∠ ♆
11 26	☽ ⚼ ♃
19 56	☽ ⊼ ♀
20 37	☽ △ ♇
23 28	♂ ⊼ ♆
23 28	☿ ± ♆
23 34	☽ ⊼ ☉
13 Thursday	
01 19	☽ □ ♄
02 15	☽ ± ♃
03 33	☽ ⚹ ♆
03 45	☽ ⊼ ♂
04 11	♀ □ ♇
04 41	☽ △ ♅
06 42	☽ ± ♀
10 10	☽ ± ☉
12 17	☽ ⊼ ♃
13 05	☽ ⊼ ☿
14 12	☽ ± ♂
18 54	☽ # ♂
21 32	☽ ⚼ ♇
22 50	☽ ± ☿
23 10	☽ ⚼ ♀
23 33	☽ Q ♆
23 36	☽ # ☉
23 52	☽ # ♀
14 Friday	
00 10	♂ □ ♅
02 23	☽ ⚼ ☉
04 47	☉ ⚹ ♄
05 53	☽ ⚼ ♅
06 10	☽ ⚼ ♂
07 29	☿ ⚹ ♃
07 41	☽ Q ♄
12 54	☽ ± ♇
13 42	☽ ⚼ ☿
13 48	☽ # ☿
16 16	☉ ∥ ♀
21 27	☽ ± ♅
23 20	☽ ⊼ ♇
15 Saturday	
03 29	☽ # ♇
03 35	☽ △ ♀
04 41	☽ ⚹ ♄
06 20	☽ △ ☉
06 37	☽ ∥ ♆
06 40	☽ □ ♆
08 02	☽ ⊼ ♅
09 41	☽ △ ♂
11 01	☉ ⊼ ♆
15 09	☽ △ ☿
16 31	♀ ⚹ ♄
16 43	☽ ☍ ♃
16 Sunday	
03 11	☽ # ♃
06 38	☉ □ ♅
07 54	☽ ∠ ♄
13 09	♀ ⊼ ♆
17 Monday	
00 46	☽ ⊥ ♄
02 04	☽ # ♄
05 07	♀ □ ♅
06 08	☽ ☍ ♇
12 13	☽ ⊻ ♄
13 55	☽ △ ♆
15 33	☽ ☍ ♅
16 40	☽ □ ♀
18 14	☽ □ ☉
19 29	☽ # ♅
20 27	☽ □ ♂
20 45	☽ □ ☿
23 34	☿ ☌ ♂
18 Tuesday	
01 30	☽ ⊼ ♃
13 21	☽ ± ♃
15 48	☉ ☌ ☿
16 45	☽ ∥ ♅
19 02	☽ ⚼ ♆
22 17	☿ ☌ ♀
19 Wednesday	
06 29	☽ Q ☿
07 22	☽ ⚼ ♃
07 51	☽ Q ♀
08 30	☽ Q ☉
09 52	☽ Q ♂
11 07	☽ ∥ ♄
12 52	☽ ± ♆
16 50	☽ ⊼ ♇
23 36	☽ ☌ ♄
20 Thursday	
00 53	☽ ⊼ ♆
02 46	☽ ⊼ ♅
04 54	☽ ± ♇
05 27	☽ ⚹ ☿
10 22	☉ ☌ ♀
10 30	☽ ∥ ♃
10 41	☽ ⚹ ☉
10 42	☽ ⚹ ♀
11 25	☽ ⚹ ♂
13 52	☽ △ ♃
14 54	☽ ± ♅
23 07	☽ ⚼ ♇
21 Friday	
01 16	♀ ☌ ♂
08 21	♀ ± ♆
09 11	☽ ⚼ ♅
10 31	☽ # ♆
10 31	☽ ∠ ☿
13 55	☽ ∥ ♇
14 02	♂ ± ♆
14 33	☉ ± ♆
15 47	☉ ☌ ♂
19 42	☽ ∠ ♂
19 48	☽ ∠ ☉
20 40	☽ ∠ ♀
21 08	☽ ∥ ☿
21 17	♀ ⚹ ♃
22 Saturday	
04 04	☽ ⊥ ☿
05 39	☽ △ ♇
09 14	☉ ⚹ ♃
12 50	☽ ⊻ ♄
13 39	☽ ☍ ♆
15 07	☽ ⊥ ♂
15 43	☽ ⊥ ☉
15 45	☽ △ ♅
15 46	☽ ⊻ ☿
16 32	☿ □ ♅
17 06	☽ ⊥ ♀
18 44	♂ ⚹ ♃
23 Sunday	
00 17	☽ ∥ ☉
01 06	☽ ⊥ ♄
03 14	☽ ∥ ♀
03 37	☽ □ ♃
04 02	☽ ⊻ ♂
04 59	☽ ⊻ ☉
05 53	☽ ∥ ♂
06 42	☽ ⊻ ♀
19 26	☽ ∠ ♄
24 Monday	
16 27	☽ Q ♃
18 24	☽ □ ♇
25 Tuesday	
01 46	☽ ⚹ ♄
02 07	☽ ⊼ ♆
02 14	☽ ☌ ☿
04 22	☽ □ ♅
06 12	☿ ⊼ ♆
14 03	☽ ± ♆
15 17	☿ ⚹ ♄
16 43	☽ ⚹ ♃
19 52	☽ ☌ ♂
22 24	☽ ☌ ☉
26 Wednesday	
01 44	☽ ☌ ♀
01 46	☽ Q ♄
02 04	♀ Q ♄
06 15	☽ Q ♇
07 51	☽ ⚼ ♆
16 05	☽ Q ♅
22 41	☽ ∠ ♃
27 Thursday	
01 26	♄ ⊼ ♆
05 49	☽ ⚹ ♇
12 00	☽ ⊻ ☿
13 09	☽ △ ♆
13 14	☽ □ ♄
15 32	☽ ⚹ ♅
16 26	☽ ⊥ ♃
20 03	☉ Q ♄
22 53	♀ Q ♇
23 27	☽ ⊥ ☿
28 Friday	
04 11	☽ ⊻ ♃
06 04	☽ ∥ ♂
08 53	☽ ∥ ♀
09 48	☽ ⊻ ♂
10 52	☽ ∠ ♇
12 13	☽ ∥ ☉
13 23	♀ ⚼ ♆
13 46	☽ ⊻ ☉
16 33	☽ ∠ ☿
18 31	☽ ⊻ ♀
20 27	☽ ∠ ♅
21 55	☽ ⊥ ♂
29 Saturday	
02 10	☽ ⊥ ☉
04 04	☽ ⊥ ♇
07 10	☽ ⊥ ♀
13 34	☽ ⊥ ♅
14 32	☽ ∥ ☿
15 26	☽ ⊻ ♇
15 56	☽ ∠ ♂
19 46	☽ ∥ ♇
20 34	☽ ∠ ☉
20 49	☽ ⚹ ☿
22 23	☽ □ ♆
22 38	☽ # ♆
22 50	☽ △ ♄
30 Sunday	
00 52	☽ ⊻ ♅
01 56	☽ ∠ ♀
03 43	☉ Q ♇
06 09	☿ St D
13 43	☽ ☌ ♃
17 38	♂ Q ♄
19 12	☽ Q ☿
20 33	☉ ⚼ ♆
21 28	☽ ⚹ ♂
23 10	☽ ∥ ♃

JULY 1968

Date	Sidereal time h m s	Sun ☉ ° ' "	Moon ☽ ° ' "	Moon ☽ 24.00 ° ' "	Mercury ☿ ° '	Venus ♀ ° '	Mars ♂ ° '	Jupiter ♃ ° '	Saturn ♄ ° '	Uranus ♅ ° '	Neptune ♆ ° '	Pluto ♇ ° '
01	06 38 24	09 ♋ 41 23	14 ♍ 23 55	21 ♍ 00 01	23 ♊ 19	12 ♋ 44	06 ♋ 53	02 ♍ 27	24 ♈ 25	25 ♍ 27	24 ♏ 04	20 ♍ 26
02	06 42 21	10 38 36	27 ♍ 40 10	04 ♎ 24 38	23 D 27	13 58	07 33	02 37	24 29	25 29	24 R 03	20 27
03	06 46 18	11 35 48	11 ♎ 13 40	18 ♎ 07 29	23 41	15 12	08 13	02 48	24 32	25 31	24 02	20 28
04	06 50 14	12 32 59	25 ♎ 06 12	02 ♏ 09 53	23 59	16 25	08 53	02 58	24 36	25 32	24 01	20 29
05	06 54 11	13 30 11	09 ♏ 18 28	16 ♏ 31 47	24 23	17 39	09 33	03 08	24 39	25 34	24 00	20 30
06	06 58 07	14 27 22	23 ♏ 49 28	01 ♐ 11 01	24 51	18 53	10 13	03 19	24 42	25 36	23 59	20 31
07	07 02 04	15 24 33	08 ♐ 35 46	16 ♐ 02 50	25 25	20 07	10 52	03 29	24 45	25 37	23 58	20 32
08	07 06 00	16 21 45	23 ♐ 31 17	00 ♑ 59 59	26 03	21 20	11 32	03 40	24 48	25 39	23 57	20 33
09	07 09 57	17 18 56	08 ♑ 27 47	15 ♑ 53 32	26 46	22 34	12 12	03 50	24 51	25 41	23 57	20 34
10	07 13 53	18 16 07	23 ♑ 16 06	00 ♒ 34 27	27 35	23 48	12 52	04 01	24 54	25 43	23 56	20 36
11	07 17 50	19 13 18	07 ♒ 47 40	14 ♒ 55 00	28 28	25 02	13 31	04 12	24 57	25 45	23 55	20 37
12	07 21 47	20 10 30	21 ♒ 55 56	28 ♒ 50 05	29 ♊ 25	26 16	14 11	04 23	25 00	25 47	23 54	20 38
13	07 25 43	21 07 42	05 ♓ 37 17	12 ♓ 17 34	00 ♋ 28	27 29	14 50	04 34	25 02	25 49	23 54	20 39
14	07 29 40	22 04 55	18 ♓ 51 06	25 ♓ 18 11	01 35	28 43	15 30	04 45	25 05	25 51	23 53	20 41
15	07 33 36	23 02 08	01 ♈ 39 15	07 ♈ 54 49	02 47	29 ♋ 57	16 09	04 56	25 07	25 53	23 52	20 42
16	07 37 33	23 59 22	14 ♈ 05 29	20 ♈ 11 51	04 03	01 ♌ 11	16 49	05 07	25 09	25 56	23 52	20 43
17	07 41 29	24 56 36	26 ♈ 14 36	02 ♉ 14 23	05 24	02 25	17 28	05 18	25 11	25 58	23 51	20 45
18	07 45 26	25 53 51	08 ♉ 11 54	14 ♉ 07 48	06 49	03 38	18 08	05 30	25 13	26 00	23 50	20 46
19	07 49 22	26 51 07	20 ♉ 02 44	25 ♉ 57 18	08 18	04 52	18 47	05 41	25 15	26 02	23 50	20 48
20	07 53 19	27 48 24	01 ♊ 52 03	07 ♊ 47 33	09 52	06 06	19 26	05 52	25 17	26 05	23 49	20 49
21	07 57 16	28 45 41	13 ♊ 44 15	19 ♊ 42 36	11 29	07 20	20 06	06 04	25 19	26 07	23 49	20 51
22	08 01 12	29 ♋ 42 59	25 ♊ 42 56	01 ♋ 45 34	13 10	08 34	20 45	06 15	25 21	26 10	23 49	20 52
23	08 05 09	00 ♌ 40 18	07 ♋ 50 46	13 ♋ 58 43	14 55	09 48	21 24	06 27	25 22	26 12	23 48	20 54
24	08 09 05	01 37 38	20 ♋ 09 32	26 ♋ 23 19	16 43	11 02	22 03	06 39	25 24	26 15	23 48	20 55
25	08 13 02	02 34 58	02 ♌ 40 07	08 ♌ 59 55	18 34	12 15	22 43	06 50	25 25	26 17	23 47	20 57
26	08 16 58	03 32 19	15 ♌ 22 43	21 ♌ 48 28	20 28	13 29	23 22	07 02	25 26	26 20	23 47	20 58
27	08 20 55	04 29 41	28 ♌ 17 07	04 ♍ 48 38	22 25	14 43	24 01	07 14	25 27	26 23	23 47	21 00
28	08 24 51	05 27 03	11 ♍ 22 58	18 ♍ 00 08	24 23	15 57	24 40	07 26	25 28	26 25	23 46	21 02
29	08 28 48	06 24 26	24 ♍ 40 08	01 ♎ 22 59	26 24	17 11	25 19	07 38	25 29	26 28	23 46	21 03
30	08 32 45	07 21 49	08 ♎ 08 45	14 ♎ 57 29	28 ♋ 27	18 25	25 58	07 50	25 30	26 31	23 46	21 05
31	08 36 41	08 ♌ 19 13	21 ♎ 49 16	28 ♎ 44 10	00 ♌ 30	19 ♌ 39	26 ♋ 37	08 ♍ 02	25 ♈ 31	26 ♍ 34	23 ♏ 46	21 ♍ 07

Date	Moon True ☊ ° '	Moon Mean ☊ ° '	Moon ☽ Latitude ° '
01	14 ♈ 22	14 ♈ 20	02 N 31
02	14 R 20	14 17	01 27
03	14 D 20	14 13	00 N 16
04	14 R 20	14 10	00 S 57
05	14 19	14 07	02 08
06	14 15	14 04	03 12
07	14 09	14 01	04 05
08	14 01	13 58	04 42
09	13 51	13 54	04 59
10	13 39	13 51	04 56
11	13 29	13 48	04 34
12	13 20	13 45	03 56
13	13 13	13 42	03 04
14	13 09	13 39	02 05
15	13 07	13 35	01 S 00
16	13 D 07	13 32	00 N 05
17	13 R 07	13 29	01 09
18	13 07	13 26	02 09
19	13 04	13 23	03 02
20	13 00	13 19	03 48
21	12 53	13 16	04 24
22	12 43	13 13	04 48
23	12 32	13 10	05 00
24	12 20	13 07	04 59
25	12 08	13 04	04 43
26	11 57	13 00	04 12
27	11 48	12 57	03 29
28	11 42	12 54	02 33
29	11 38	12 51	01 29
30	11 37	12 48	00 19
31	11 ♈ 37	12 ♈ 45	00 S 54

DECLINATIONS

Date	Sun ☉ ° '	Moon ☽ ° '	Mercury ☿ ° '	Venus ♀ ° '	Mars ♂ ° '	Jupiter ♃ ° '	Saturn ♄ ° '	Uranus ♅ ° '	Neptune ♆ ° '	Pluto ♇ ° '
01	23 N 06	08 N 28	18 N 52	23 N 33	24 N 04	11 N 34	07 N 13	02 N 29	17 S 05	17 N 39
02	23 01	02 N 15	18 58	23 28	24 03	11 30	07 14	02 28	17 05	17 39
03	22 56	04 S 11	19 05	23 22	24 01	11 27	07 15	02 28	17 05	17 38
04	22 51	10 36	19 13	23 15	23 59	11 23	07 16	02 27	17 05	17 37
05	22 46	16 37	19 24	23 08	23 56	11 19	07 17	02 26	17 04	17 36
06	22 40	21 50	19 35	22 59	23 54	11 15	07 18	02 26	17 04	17 36
07	22 33	25 46	19 46	22 51	23 51	11 11	07 19	02 25	17 04	17 35
08	22 27	27 58	19 58	22 41	23 48	11 07	07 19	02 24	17 04	17 34
09	22 19	28 09	20 11	22 31	23 45	11 03	07 20	02 23	17 04	17 33
10	22 12	26 18	20 24	22 20	23 41	11 00	07 21	02 22	17 04	17 32
11	22 04	22 44	20 38	22 09	23 38	10 56	07 22	02 22	17 03	17 32
12	21 56	17 55	20 51	21 57	23 34	10 51	07 23	02 21	17 03	17 31
13	21 47	12 19	21 04	21 44	23 30	10 47	07 23	02 20	17 03	17 30
14	21 38	06 19	21 17	21 31	23 26	10 43	07 24	02 19	17 03	17 29
15	21 29	00 S 16	21 30	21 17	23 22	10 39	07 24	02 18	17 03	17 28
16	21 19	05 N 38	21 41	21 02	23 17	10 35	07 25	02 17	17 03	17 27
17	21 09	11 12	21 52	20 47	23 13	10 31	07 26	02 16	17 03	17 26
18	20 58	16 16	22 02	20 31	23 08	10 27	07 26	02 15	17 03	17 26
19	20 48	20 41	22 11	20 15	23 03	10 22	07 27	02 14	17 03	17 25
20	20 36	24 16	22 18	19 58	22 58	10 18	07 27	02 13	17 03	17 24
21	20 25	26 49	22 23	19 41	22 52	10 14	07 27	02 12	17 02	17 23
22	20 13	28 11	22 27	19 22	22 47	10 10	07 28	02 11	17 02	17 22
23	20 01	28 13	22 29	19 04	22 41	10 05	07 28	02 10	17 02	17 21
24	19 48	26 51	22 28	18 45	22 35	10 01	07 28	02 09	17 02	17 20
25	19 35	24 09	22 26	18 25	22 29	09 56	07 28	02 08	17 02	17 19
26	19 22	20 15	22 20	18 05	22 23	09 52	07 29	02 07	17 02	17 19
27	19 09	15 20	22 12	17 44	22 16	09 48	07 29	02 06	17 02	17 18
28	18 55	09 40	22 02	17 23	22 10	09 43	07 29	02 05	17 02	17 17
29	18 41	03 N 29	21 49	17 01	22 03	09 39	07 29	02 04	17 02	17 16
30	18 26	02 S 57	21 33	16 39	21 56	09 34	07 29	02 03	17 02	17 15
31	18 N 11	09 S 21	21 N 14	16 N 17	21 N 49	09 N 30	07 N 29	02 N 02	17 S 02	17 N 14

ZODIAC SIGN ENTRIES

Date	h	m	Planets
02	16	10	☽ ♎
04	20	20	☽ ♏
06	22	05	☽ ♐
08	22	24	☽ ♑
10	23	03	☽ ♒
13	01	30	☿ ♋
13	02	03	☽ ♓
15	08	51	☽ ♈
15	12	59	♀ ♌
17	19	30	☽ ♉
20	08	13	☽ ♊
22	19	07	☉ ♌
22	20	31	☽ ♋
25	06	55	☽ ♌
27	15	10	☽ ♍
29	21	32	☽ ♎
31	06	11	☿ ♌

LATITUDES

Date	Mercury ☿ ° '	Venus ♀ ° '	Mars ♂ ° '	Jupiter ♃ ° '	Saturn ♄ ° '	Uranus ♅ ° '	Neptune ♆ ° '	Pluto ♇ ° '
01	04 S 25	00 N 43	00 N 48	01 N 02	02 S 26	00 N 44	01 N 46	15 N 08
04	04 05	00 49	00 50	01 02	02 26	00 44	01 46	15 07
07	03 36	00 55	00 51	01 02	02 27	00 44	01 45	15 05
10	03 01	01 00	00 52	01 01	02 28	00 44	01 45	15 04
13	02 23	01 05	00 53	01 01	02 29	00 44	01 45	15 03
16	01 42	01 10	00 54	01 01	02 29	00 44	01 45	15 02
19	01 00	01 14	00 56	01 01	02 30	00 44	01 45	15 01
22	00 S 21	01 18	00 57	01 01	02 31	00 44	01 45	15 00
25	00 N 16	01 21	00 58	01 01	02 32	00 43	01 44	14 59
28	00 48	01 23	00 59	01 02	02 33	00 43	01 44	14 58
31	01 N 13	01 N 25	01 N 00	01 N 00	02 S 33	00 N 43	01 N 44	14 N 58

DATA

Julian Date	2440039
Delta T	+39 seconds
Ayanamsa	23° 24' 58"
Synetic vernal point	05° ♓ 42' 02"
True obliquity of ecliptic	23° 26' 45"

MOON'S PHASES, APSIDES AND POSITIONS ☽

Date	h	m	Phase	Longitude ° '	Eclipse Indicator
03	12	42	☽	11 ♎ 37	
10	03	18	○	17 ♑ 55	
17	09	11	☾	24 ♈ 50	
25	11	49	●	02 ♌ 35	

Day	h	m		
08	08	40	Perigee	
20	09	15	Apogee	
02	20	27	0S	
09	02	00	Max dec	28° S 20'
15	13	03	0N	
23	00	33	Max dec	28° N 22'
30	01	03	0S	

LONGITUDES

Date	Chiron ⚷ ° '	Ceres ⚳ ° '	Pallas ⚴ ° '	Juno ⚵ ° '	Vesta ⚶ ° '	Black Moon Lilith ⚸ ° '
01	03 ♈ 22	24 ♎ 41	25 ♍ 04	27 ♎ 50	23 ♈ 39	01 ♊ 26
11	03 ♈ 22	26 ♎ 23	28 ♍ 17	28 ♎ 35	26 ♈ 45	02 ♊ 33
21	03 ♈ 16	28 ♎ 31	01 ♎ 44	29 ♎ 47	29 ♈ 28	03 ♊ 39
31	03 ♈ 06	01 ♏ 00	05 ♎ 21	01 ♏ 21	01 ♉ 45	04 ♊ 46

All ephemeris data is given at 12.00 UT and the Moon's longitude is additionally given for 24.00 UT

ASPECTARIAN

h m	Aspects
01 Monday	
02 41	☽ ⚹ ☉
02 50	☽ bQ ♄
07 44	☽ Q ♆
08 38	☽ ⚹ ♀
16 57	☽ ∥ ♄
19 22	☽ ± ♄
20 36	☽ Q ♂
22 59	☽ ☌ ♇
02 Tuesday	
02 16	☽ Q ☉
02 24	♀ Q ♅
04 20	☽ □ ☿
05 30	☽ ⚹ ♆
06 15	☽ ⚻ ♄
08 04	☽ ☌ ♅
08 38	☽ Q ♀
11 11	☽ ∥ ♅
20 57	☽ ⊻ ♃
03 Wednesday	
05 36	☽ # ♅
06 27	☽ □ ♂
07 41	☽ ⊥ ♃
08 09	☽ ∠ ♆
12 42	☽ □ ☉
19 35	☽ □ ♀
20 59	♂ Q ♇
23 23	☽ # ♄
23 34	☽ ∠ ♃
23 50	☽ ⊥ ♆
04 Thursday	
04 03	☽ ⊻ ♇
10 02	☽ △ ☿
10 09	☽ ⊻ ♆
11 08	☽ ☍ ♄
12 45	☽ ⊻ ♅
14 00	☿ ⚻ ♆
14 21	☽ ⊥ ♇
14 59	☽ # ♃
16 41	♂ bQ ♆
22 58	☽ ⊥ ♅
05 Friday	
01 31	☽ ⚹ ♃
05 36	☽ ∠ ♇
12 07	☽ bQ ☿
12 25	☽ △ ♂
13 36	☉ Q ♅
13 56	☽ ∥ ♆
14 06	☽ ∠ ♅
16 12	☽ # ♇
19 29	☽ △ ☉
21 49	☽ Q ♃
23 02	♀ ∠ ♃
06 Saturday	
00 38	☽ # ☿
03 08	☽ △ ♀
03 32	☽ ± ☿
04 14	☿ ⚹ ♄
06 34	☽ ⚹ ♇
12 16	☽ ☌ ♆
13 27	☽ ⚻ ♄
13 44	☽ ⚻ ☿
14 23	☽ bQ ♂
14 54	☽ ⚹ ♅
16 16	☽ # ☉
17 58	☽ # ♀
21 50	☽ bQ ☉
23 15	☽ # ♂
23 16	☽ ± ♄
07 Sunday	
02 11	☽ Q ♇
03 38	☽ □ ♃
05 42	☽ ± ♂
05 51	☽ bQ ♀
10 26	☽ Q ♅
13 24	☽ ± ☉
13 53	☽ bQ ♄
15 51	☽ ⚻ ♂
20 23	♀ ⚹ ♇
20 51	☿ □ ♅
21 41	☽ ± ♀
23 43	☽ ⚻ ☉
08 Monday	
07 14	☽ □ ♇
08 11	☽ ⚻ ♀
12 42	☽ ⊻ ♆
14 04	☽ △ ♄
15 26	☽ □ ♅
16 15	☽ ☍ ☿
22 19	☽ ⊥ ♆
09 Tuesday	
04 28	☽ △ ♃
12 46	☽ ∠ ♆
18 18	☽ ☍ ♂
10 Wednesday	
03 18	☽ ☍ ☉
04 59	☽ bQ ♃
07 38	☽ △ ♇
12 57	☽ ☍ ♀
13 05	☽ ⚹ ♆
14 31	♀ △ ♆
14 41	☽ □ ♄
16 01	☽ △ ♅
19 30	☽ ⚻ ☿
19 53	☽ ± ♃
11 Thursday	
05 56	☽ ⚻ ♃
06 04	☽ ± ☿
06 44	☽ # ♂
08 21	☽ bQ ♇
08 52	☽ Q ♆
10 21	♀ □ ♄
11 18	☉ ∠ ♃
15 23	☽ # ♀
15 47	☽ # ☉
16 58	☽ bQ ♅
20 42	☽ Q ♄
20 54	♂ Q ♅
22 06	☽ ⚻ ♂
22 13	☽ bQ ☿
22 38	☽ # ☿
23 30	☽ ± ♇
12 Friday	
02 29	♀ ⚹ ♅
08 18	☽ ± ♅
08 45	☽ ⚻ ☉
08 50	☽ ± ♂
09 46	☽ ⚻ ♇
13 49	☽ # ♇
15 24	☽ □ ♆
15 51	☽ ∥ ♆
17 19	☽ ⚹ ♄
18 31	☉ ∥ ♀
18 41	☽ ⚻ ♅
19 54	☽ ± ☉
20 14	☽ ⚻ ♀
23 11	☿ ± ♆
23 50	☉ ⚹ ♇
13 Saturday	
01 15	☽ bQ ♂
02 06	☽ △ ☿
07 50	☽ ± ♀
10 05	☽ ☍ ♃
12 58	☽ bQ ☉
18 15	☽ # ♃
19 56	☽ ∠ ♄
14 Sunday	
01 38	☽ bQ ♀
05 31	☽ △ ♂
07 45	☽ # ♄
12 25	☽ ⊥ ♄
12 38	♀ ⊥ ♃
15 23	☽ ☍ ♇
18 28	☽ △ ☉
21 20	☽ △ ♆
23 37	☽ ⊻ ♄
15 Monday	
00 28	☿ ∥ ♀
01 04	☽ ☍ ♅
03 53	☽ # ♅
08 25	☽ △ ♀
11 05	☉ ∥ ☿
14 23	☽ □ ☿
18 21	☽ ⚻ ♃
22 17	☽ ∥ ♅
16 Tuesday	
01 50	☽ bQ ♆
06 07	☽ ± ♃
08 48	☉ △ ♆
17 38	☽ □ ♂
19 23	☽ ± ♆
19 30	☽ ∥ ♄
17 Wednesday	
00 02	☽ bQ ♃
01 04	☽ ⚻ ♇
05 37	☽ Q ☿
07 15	☽ ⚻ ♆
08 58	☽ ∥ ♃
09 11	☽ □ ☉
09 53	☽ ☌ ♄
10 08	☿ ⚹ ♃
11 27	☽ ⚻ ♅
13 00	☽ ± ♇
18 23	☉ □ ♄
23 29	☽ ± ♅
18 Thursday	
01 46	☽ □ ♀
06 27	☽ △ ♃
07 05	☽ bQ ♇
07 35	☽ Q ♂
08 49	☽ ⚹ ☿
14 46	☉ ⚹ ♅
15 55	☽ # ♆
17 41	☽ bQ ♅
17 53	☽ ∥ ♇
18 50	☿ Q ♄
19 Friday	
00 32	☽ Q ☉
09 17	☽ ⚹ ♂
09 36	☽ ∥ ♀
12 37	☽ ∥ ☉
13 31	☽ △ ♇
18 24	☽ Q ♀
19 36	☽ ∠ ☿
19 41	☽ ☍ ♆
19 48	☿ Q ♇
20 15	☿ bQ ♆
21 38	☽ ∥ ☿
22 36	☽ ⊻ ♄
20 Saturday	
00 13	☽ △ ♅
02 45	☽ ∥ ♂
03 02	☽ ⚹ ☉
06 20	♀ ∠ ♇
06 43	♀ ⊻ ♃
10 49	☽ ⊥ ♄
16 40	☽ ⊥ ☿
17 31	☽ ∠ ♂
20 15	☽ □ ♃
21 34	☽ ⚹ ♀
21 Sunday	
05 05	☽ ∠ ♄
06 43	☽ ⊻ ☿
12 03	☽ ∠ ☉
12 46	☽ ⊥ ♂
22 Monday	
01 31	☽ ⊻ ♂
02 18	☽ □ ♇
07 13	☽ ∠ ♀
07 40	☽ ⊥ ☉
08 12	☽ ⚻ ♆
09 03	☽ Q ♃
11 15	☽ ⚹ ♄
12 54	☽ □ ♅
16 30	♂ ⚹ ♇
20 08	☽ ± ♆
20 38	☽ ⊻ ☉
23 Tuesday	
02 09	☿ Q ♅
03 07	☽ ⊥ ♀
05 02	☉ ⊥ ♃
09 13	☽ ⚹ ♃
11 04	☽ Q ♄
13 53	☽ bQ ♆
14 04	☽ Q ♇
14 27	♂ ∠ ♃
16 15	☽ ⊻ ♀
24 Wednesday	
00 29	☽ Q ♅
04 10	☽ ☌ ☿
13 28	☽ ⚹ ♇
14 55	☽ ∠ ♃
15 52	☽ ☌ ♂
16 26	♀ ∠ ♅
19 01	☽ △ ♆
22 06	☽ □ ♄
23 46	☽ ⚹ ♅
25 Thursday	
08 28	☽ ⊥ ♃
11 49	☽ ☌ ☉
18 14	☽ ∠ ♇
20 03	☽ ⊻ ♃
23 24	☽ ∥ ♂
23 41	☽ ∥ ☿
26 Friday	
04 22	☽ ∠ ♅
08 05	☽ ☌ ♀
11 14	☽ ⊥ ♇
16 51	☽ ∥ ☉
18 23	☿ ⚹ ♇
21 17	☽ ⊥ ♅
22 28	☽ ⊻ ♇
23 11	☽ ⊻ ☿
23 57	☽ ∥ ♀
27 Saturday	
02 54	☽ ∥ ♇
03 22	♂ △ ♆
03 40	☽ □ ♆
03 41	☽ ⊻ ♂
04 09	☽ # ♆
06 46	☽ △ ♄
08 28	☽ ⊻ ♅
09 37	☿ ∠ ♃
12 16	☽ ⊥ ☿
15 22	☽ ⊥ ♂
17 33	♀ ⊥ ♇
28 Sunday	
00 19	☽ ⊻ ☉
04 36	☿ △ ♆
04 41	☽ ☌ ♃
07 43	☽ ∠ ☿
08 43	☽ ∠ ♂
10 21	☽ bQ ♄
11 46	☽ ∥ ♃
12 08	☽ ⊥ ☉
12 43	☽ Q ♆
16 55	☿ ☌ ♂
19 27	♀ ∥ ♇
20 38	☽ ∥ ♄
21 08	☽ ⊻ ♀
29 Monday	
01 03	☿ □ ♄
02 40	☽ ± ♄
02 52	☉ ∠ ♇
05 29	☽ ☌ ♇
05 41	☽ ∠ ☉
09 04	☽ ⊥ ♀
10 23	☽ ⚹ ♆
11 08	♀ # ♆
12 46	☿ ⚹ ♅
13 13	☽ ⚹ ♂
13 28	☽ ⚻ ♄
15 14	☽ ☌ ♅
15 40	☽ ⚹ ☿
17 20	☽ ∥ ♅
18 26	♂ □ ♄
30 Tuesday	
02 46	☽ ∠ ♀
08 40	☽ # ♅
10 31	☽ ⚹ ☉
11 26	☽ ⊻ ♃
11 40	☽ Q ♂
13 06	☽ ∠ ♆
16 46	☽ Q ☿
22 10	☽ ⊥ ♃
31 Wednesday	
02 50	☉ ⊻ ♃
04 55	☽ ⊥ ♆
04 57	☽ # ♄
07 50	☽ ⚹ ♀
09 11	☽ Q ☉
09 36	♂ ⚹ ♅
10 46	☽ ⊻ ♇
12 34	☽ # ♃
14 08	☽ ∠ ♃
15 23	☽ ⊻ ♆
18 26	☽ ☍ ♄
20 16	☽ ⊻ ♅
20 45	☽ □ ♂
21 12	☽ ⊥ ♇

AUGUST 1968

LONGITUDES

Date	Sidereal time h m s	Sun ☉	Moon ☽	Moon ☽ 24.00	Mercury ☿	Venus ♀	Mars ♂	Jupiter ♃	Saturn ♄	Uranus ♅	Neptune ♆	Pluto ♇
		° ' "	° ' "	° ' "	° '	° '	° '	° '	° '	° '	° '	° '
01	08 40 38	09 ♌ 16 37	05 ♏ 42 12	12 ♏ 43 23	02 ♌ 34	20 ♌ 53	27 ♋ 16	08 ♍ 14	25 ♈ 32	26 ♍ 36	23 ♏ 46	21 ♍ 08
02	08 44 34	10 14 02	19 ♏ 47 39	26 ♏ 54 52	04 39	22 07	27 55	08 26	25 32	26 39	23 R 46	21 10
03	08 48 31	11 11 28	04 ♐ 04 48	11 ♐ 17 08	06 45	23 20	28 34	08 38	25 32	26 42	23 46	21 12
04	08 52 27	12 08 54	18 ♐ 31 25	25 ♐ 47 06	08 50	24 34	29 13	08 51	25 33	26 45	23 46	21 14
05	08 56 24	13 06 22	03 ♑ 03 32	10 ♑ 19 59	10 55	25 48	29 ♋ 52	09 03	25 33	26 48	23 D 46	21 15
06	09 00 20	14 03 50	17 ♑ 35 38	24 ♑ 49 41	12 59	27 02	00 ♌ 31	09 15	25 33	26 51	23 46	21 17
07	09 04 17	15 01 19	02 ♒ 01 16	09 ♒ 09 36	15 03	28 16	01 09	09 27	25 R 33	26 54	23 46	21 19
08	09 08 14	15 58 48	16 ♒ 13 58	23 ♒ 13 43	17 06	29 ♌ 30	01 48	09 40	25 33	26 57	23 46	21 21
09	09 12 10	16 56 19	00 ♓ 08 21	06 ♓ 57 28	19 08	00 ♍ 44	02 27	09 52	25 33	27 00	23 46	21 23
10	09 16 07	17 53 51	13 ♓ 40 51	20 ♓ 18 23	21 08	01 58	03 06	10 05	25 33	27 03	23 46	21 25
11	09 20 03	18 51 24	26 ♓ 50 06	03 ♈ 16 11	23 08	03 11	03 44	10 17	25 32	27 06	23 46	21 27
12	09 24 00	19 48 59	09 ♈ 36 52	15 ♈ 52 32	25 06	04 25	04 23	10 30	25 32	27 10	23 46	21 29
13	09 27 56	20 46 35	22 ♈ 03 39	28 ♈ 10 42	27 03	05 39	05 02	10 42	25 31	27 13	23 47	21 30
14	09 31 53	21 44 12	04 ♉ 14 15	10 ♉ 14 55	28 ♌ 59	06 53	05 40	10 55	25 30	27 16	23 47	21 32
15	09 35 49	22 41 51	16 ♉ 13 20	22 ♉ 10 07	00 ♍ 53	08 07	06 19	11 07	25 29	27 19	23 47	21 34
16	09 39 46	23 39 31	28 ♉ 05 55	04 ♊ 01 23	02 45	09 21	06 57	11 20	25 29	27 23	23 48	21 36
17	09 43 43	24 37 14	09 ♊ 57 08	15 ♊ 53 45	04 37	10 35	07 36	11 33	25 28	27 26	23 48	21 38
18	09 47 39	25 34 57	21 ♊ 51 48	27 ♊ 51 49	06 27	11 49	08 14	11 45	25 26	27 29	23 48	21 40
19	09 51 36	26 32 43	03 ♋ 54 17	09 ♋ 59 37	08 15	13 02	08 53	11 58	25 25	27 33	23 49	21 42
20	09 55 32	27 30 30	16 ♋ 08 10	22 ♋ 20 15	10 02	14 16	09 31	12 11	25 24	27 36	23 49	21 44
21	09 59 29	28 28 18	28 ♋ 36 05	04 ♌ 55 51	11 48	15 30	10 10	12 24	25 22	27 39	23 50	21 47
22	10 03 25	29 ♌ 26 08	11 ♌ 19 36	17 ♌ 47 22	13 32	16 44	10 48	12 37	25 21	27 43	23 50	21 49
23	10 07 22	00 ♍ 24 00	24 ♌ 19 07	00 ♍ 54 42	15 15	17 58	11 27	12 49	25 19	27 46	23 51	21 51
24	10 11 18	01 21 53	07 ♍ 34 00	14 ♍ 16 46	16 57	19 12	12 05	13 02	25 18	27 50	23 52	21 53
25	10 15 15	02 19 47	21 ♍ 02 49	27 ♍ 51 51	18 37	20 26	12 43	13 15	25 16	27 53	23 52	21 55
26	10 19 12	03 17 43	04 ♎ 43 38	11 ♎ 37 53	20 16	21 40	13 22	13 28	25 14	27 57	23 53	21 57
27	10 23 08	04 15 40	18 ♎ 34 21	25 ♎ 32 48	21 53	22 53	14 00	13 41	25 12	28 00	23 54	21 59
28	10 27 05	05 13 39	02 ♏ 33 00	09 ♏ 34 43	23 29	24 07	14 38	13 54	25 10	28 04	23 54	22 01
29	10 31 01	06 11 38	16 ♏ 37 46	23 ♏ 41 56	25 04	25 21	15 16	14 07	25 07	28 07	23 55	22 03
30	10 34 58	07 09 40	00 ♐ 47 00	07 ♐ 52 46	26 38	26 35	15 55	14 20	25 05	28 11	23 56	22 05
31	10 38 54	08 ♍ 07 42	14 ♐ 58 59	22 ♐ 05 23	28 ♍ 10	27 ♍ 49	16 ♌ 33	14 ♍ 33	25 ♈ 03	28 ♍ 15	23 ♏ 57	22 ♍ 08

Date	Moon True ☊	Moon Mean ☊	Moon ☽ Latitude
	° '	° '	° '
01	11 ♈ 38	12 ♈ 41	02 S 05
02	11 R 38	12 38	03 09
03	11 36	12 35	04 02
04	11 31	12 32	04 41
05	11 25	12 29	05 01
06	11 17	12 25	05 03
07	11 08	12 22	04 45
08	11 00	12 19	04 10
09	10 52	12 16	03 20
10	10 47	12 13	02 20
11	10 44	12 10	01 14
12	10 43	12 06	00 S 06
13	10 D 44	12 03	01 N 01
14	10 45	12 00	02 03
15	10 46	11 57	03 00
16	10 R 46	11 54	03 47
17	10 45	11 51	04 25
18	10 41	11 47	04 52
19	10 36	11 44	05 07
20	10 30	11 41	05 07
21	10 23	11 38	04 53
22	10 15	11 35	04 25
23	10 09	11 31	03 42
24	10 04	11 28	02 47
25	10 00	11 25	01 41
26	09 59	11 22	00 N 29
27	09 D 59	11 19	00 S 47
28	10 00	11 16	02 00
29	10 02	11 12	03 07
30	10 03	11 09	04 02
31	10 ♈ 03	11 ♈ 06	04 S 43

DECLINATIONS

Date	Sun ☉	Moon ☽	Mercury ☿	Venus ♀	Mars ♂	Jupiter ♃	Saturn ♄	Uranus ♅	Neptune ♆	Pluto ♇
	° '	° '	° '	° '	° '	° '	° '	° '	° '	° '
01	17 N 56	15 S 23	20 N 53	15 N 54	21 N 42	09 N 25	07 N 29	02 N 01	17 S 02	17 N 13
02	17 41	20 43	20 30	15 30	21 34	09 20	07 29	01 59	17 02	17 12
03	17 25	24 56	20 04	15 07	21 27	09 16	07 29	01 58	17 02	17 11
04	17 09	27 37	19 36	14 42	21 19	09 11	07 29	01 57	17 02	17 10
05	16 53	28 26	19 05	14 18	21 11	09 07	07 29	01 56	17 02	17 09
06	16 37	27 17	18 33	13 53	21 03	09 02	07 28	01 55	17 02	17 08
07	16 20	24 20	17 59	13 27	20 55	08 57	07 28	01 53	17 02	17 07
08	16 03	19 56	17 23	13 01	20 47	08 53	07 28	01 52	17 03	17 07
09	15 46	14 32	16 46	12 35	20 38	08 48	07 28	01 51	17 03	17 06
10	15 28	08 34	16 08	12 09	20 30	08 43	07 27	01 50	17 03	17 05
11	15 11	02 S 23	15 28	11 42	20 21	08 38	07 27	01 48	17 03	17 04
12	14 53	03 N 43	14 47	11 15	20 12	08 34	07 26	01 47	17 03	17 03
13	14 34	09 32	14 05	10 48	20 03	08 29	07 26	01 46	17 03	17 02
14	14 16	14 52	13 23	10 20	19 54	08 24	07 25	01 45	17 03	17 01
15	13 57	19 34	12 40	09 52	19 45	08 19	07 25	01 43	17 03	17 00
16	13 38	23 26	11 56	09 24	19 35	08 15	07 24	01 42	17 04	16 59
17	13 19	26 19	11 12	08 55	19 26	08 10	07 24	01 41	17 04	16 58
18	13 00	28 03	10 27	08 27	19 16	08 05	07 23	01 39	17 04	16 57
19	12 40	28 30	09 42	07 58	19 06	08 00	07 22	01 38	17 04	16 56
20	12 20	27 33	08 57	07 29	18 56	07 55	07 22	01 36	17 04	16 55
21	12 01	25 14	08 12	06 59	18 46	07 50	07 21	01 35	17 04	16 54
22	11 40	21 37	07 27	06 30	18 36	07 45	07 20	01 34	17 05	16 53
23	11 20	16 55	06 41	06 00	18 25	07 40	07 19	01 32	17 05	16 53
24	11 00	11 19	05 56	05 30	18 15	07 36	07 18	01 31	17 05	16 52
25	10 39	05 N 06	05 11	05 00	18 04	07 31	07 17	01 30	17 05	16 51
26	10 18	01 S 27	04 26	04 30	17 54	07 26	07 17	01 28	17 06	16 50
27	09 57	08 00	03 41	04 00	17 43	07 21	07 16	01 27	17 06	16 49
28	09 36	14 14	02 56	03 30	17 32	07 16	07 15	01 25	17 06	16 48
29	09 15	19 47	02 11	02 59	17 21	07 11	07 14	01 24	17 06	16 47
30	08 53	24 16	01 27	02 28	17 10	07 06	07 12	01 22	17 06	16 46
31	08 N 31	27 S 17	00 43	01 N 58	16 N 59	07 N 01	07 N 11	01 N 21	17 S 07	16 N 45

ZODIAC SIGN ENTRIES

Date	h	m	Planets	
01	02	11	☽	♏
03	05	11	☽	♐
05	06	57	☽	♑
05	17	07	♂	♌
07	08	37	☽	♒
08	21	49	♀	♍
09	11	45	☽	♓
11	17	53	☽	♈
14	03	36	☽	♉
15	00	53	☿	♍
16	15	51	☽	♊
19	04	15	☽	♋
21	14	40	☽	♌
23	02	21	☉	♍
23	22	21	☽	♍
26	03	45	☽	♎
28	07	38	☽	♏
30	10	40	☽	♐

LATITUDES

Date	Mercury ☿	Venus ♀	Mars ♂	Jupiter ♃	Saturn ♄	Uranus ♅	Neptune ♆	Pluto ♇
	° '	° '	° '	° '	° '	° '	° '	° '
01	01 N 20	01 N 26	01 N 00	01 N 00	02 S 34	00 N 43	01 N 44	14 N 57
04	01 36	01 27	01 01	01 00	02 35	00 43	01 44	14 57
07	01 44	01 28	01 02	01 00	02 35	00 43	01 44	14 56
10	01 46	01 28	01 03	01 00	02 36	00 43	01 44	14 55
13	01 42	01 27	01 04	01 00	02 37	00 43	01 43	14 55
16	01 33	01 26	01 05	01 00	02 38	00 43	01 43	14 54
19	01 19	01 24	01 06	01 00	02 39	00 43	01 43	14 54
22	01 03	01 22	01 07	01 00	02 39	00 43	01 43	14 53
25	00 44	01 19	01 07	01 00	02 40	00 43	01 43	14 53
28	00 23	01 16	01 08	01 00	02 41	00 43	01 42	14 53
31	00 00	01 N 11	01 N 09	01 N 00	02 S 42	00 N 42	01 N 42	14 N 53

DATA

Julian Date	2440070
Delta T	+39 seconds
Ayanamsa	23° 25' 03"
Synetic vernal point	05° ♓ 41' 57"
True obliquity of ecliptic	23° 26' 45"

MOON'S PHASES, APSIDES AND POSITIONS ☽

Date	h	m	Phase	Longitude ° '	Eclipse Indicator
01	18	34	☽	09 ♏ 32	
08	11	32	○	15 ♒ 58	
16	02	13	☾	23 ♉ 16	
23	23	57	●	00 ♍ 53	
30	23	34	☽	07 ♐ 38	

Day	h	m		
05	03	21	Perigee	
17	02	47	Apogee	
31	02	20	Perigee	
05	10	02	Max dec	28° S 26'
11	21	18	ON	
19	07	43	Max dec	28° N 31'
26	06	45	OS	

LONGITUDES

Date	Chiron ⚷	Ceres ⚳	Pallas ⚴	Juno ⚵	Vesta ⚶	Black Moon Lilith ⚸
	° '	° '	° '	° '	° '	° '
01	03 ♈ 04	01 ♏ 16	05 ♎ 44	01 ♏ 32	01 ♉ 57	04 ♊ 53
11	02 ♈ 49	04 ♏ 06	09 ♎ 32	03 ♏ 28	03 ♉ 41	06 ♊ 00
21	02 ♈ 29	07 ♏ 11	13 ♎ 29	05 ♏ 42	04 ♉ 50	07 ♊ 07
31	02 ♈ 07	10 ♏ 28	17 ♎ 32	08 ♏ 10	05 ♉ 17	08 ♊ 14

ASPECTARIAN

h m	Aspects
01 Thursday	
05 41	☽ □ ☿
06 30	♀ ⊥ ♅
06 40	☽ ⊥ ♅
06 40	☽ Q ♀
07 38	☿ ⊥ ♃
12 45	☽ ∠ ♇
14 00	☽ # ♀
16 24	☽ ⚹ ♃
17 14	♀ ⊻ ♇
18 34	☽ □ ☉
19 01	☽ ∥ ♆
19 47	☽ # ♇
22 08	☽ ∠ ♅
22 29	☽ # ☉
02 Friday	
10 59	☽ # ☿
13 06	☽ Q ♃
14 20	☽ ⚹ ♇
16 12	☽ # ♂
16 17	☽ □ ♀
18 42	☽ ☌ ♆
21 41	☽ ⊼ ♄
23 36	☽ ⚹ ♅
03 Saturday	
02 20	☽ △ ♂
05 37	☿ ∠ ♇
07 45	☽ ± ♄
10 31	☽ Q ♇
17 12	☽ △ ☿
19 42	☽ □ ♃
19 44	☽ Q ♅
20 11	♀ □ ♆
22 46	☽ ⚼ ♄
04 Sunday	
00 41	☽ △ ☉
01 30	☉ ∠ ♅
04 32	☽ ⚼ ♂
10 41	☉ ∥ ♇
12 08	☿ ⊻ ♃
16 29	☽ □ ♇
20 07	☽ ± ♂
20 39	☽ ⊻ ♆
22 15	☽ ⚼ ☿
22 24	☉ # ♆
22 55	☽ △ ♀
23 37	☽ △ ♄
05 Monday	
01 17	♆ St D
01 38	☽ □ ♅
03 15	☽ ⚼ ☉
06 29	☽ ⊼ ♂
06 33	☽ ⊥ ♆
07 04	♀ △ ♄
15 34	☽ ± ☿
19 09	☽ ± ☉
21 24	☽ ∠ ♆
22 01	☽ △ ♃
22 30	☿ ∠ ♅
06 Tuesday	
01 57	☽ ⚼ ♀
03 07	☽ ⊼ ☿
05 45	☽ ⊼ ☉
08 16	♀ ⊻ ♅
18 08	☽ △ ♇
18 14	☽ ± ♀
22 14	☽ ⚹ ♆
23 12	☽ ⚼ ♃
07 Wednesday	
01 12	☽ □ ♄
02 22	♄ St R
03 25	☽ △ ♅
05 08	☽ ⊼ ♀
10 29	☽ ☍ ♂
11 25	☉ ☌ ☿
14 26	☽ ± ♃
15 12	☿ ⊥ ♇
18 16	☽ Q ♆
19 14	☽ ⚼ ♇
19 41	☉ ⊥ ♇
08 Thursday	
00 41	☽ ⊼ ♃
04 42	☽ ⚼ ♅
07 26	☽ Q ♄
07 46	☽ # ♂
10 29	☽ ± ♇
11 32	☽ ☍ ☉
13 43	☽ ☍ ☿
20 07	☽ ± ♅
20 47	☽ ⊼ ♇
23 12	☿ ∥ ♇
09 Friday	
00 55	☽ □ ♆
01 03	☽ # ♇
01 17	☽ # ☿
01 17	☽ ∥ ♆
01 21	☿ # ♆
04 01	☽ ⚹ ♄
06 31	☽ ⊼ ♅
06 34	☽ # ☉
13 08	☽ ☍ ♀
16 15	☽ ⊼ ♂
20 40	☽ # ♀
10 Saturday	
03 23	☽ ± ♂
05 27	☽ ☍ ♃
06 23	☽ ∠ ♄
10 58	☿ ⊥ ♅
11 25	☽ # ♃
15 20	☿ ⊻ ♇
16 22	☽ # ♄
20 13	☽ ⊼ ☉
20 23	☽ ⚼ ♂
22 36	☽ ⊥ ♄
11 Sunday	
02 03	☽ ☍ ♇
03 57	☽ ⊼ ☿
06 21	☽ △ ♆
08 03	☽ ± ☉
09 36	☽ ⊻ ♄
12 30	☽ ☍ ♅
14 16	☽ # ♅
17 02	☽ ± ☿
19 46	☿ □ ♆
12 Monday	
01 07	☽ ⊼ ♀
01 34	☽ △ ♂
02 10	☽ ⚼ ☉
04 21	☽ ∥ ♅
04 21	♆ # ♇
06 17	☉ ∥ ☿
10 24	♀ ⊻ ♂
10 24	☽ ⚼ ♆
13 06	☽ ⚼ ☿
13 42	☽ ± ♀
13 42	☽ ⊼ ♃
15 35	♃ ⚼ ♄
17 12	☿ △ ♄
18 11	♂ ⊥ ♃
13 Tuesday	
01 25	☽ ± ♃
03 08	☽ ∥ ♄
03 40	☽ ± ♆
07 36	☽ ∥ ♃
08 57	☽ ⚼ ♀
09 17	☽ △ ☉
10 55	☽ ⊼ ♇
14 05	☿ ⊻ ♅
15 21	☽ ⊼ ♆
17 02	☽ ∥ ♀
18 45	☽ ☌ ♄
19 15	☽ ⚼ ♃
22 09	☽ ⊼ ♅
22 43	☽ ± ♇
23 35	☉ ⊥ ♅
23 37	☽ △ ☿
14 Wednesday	
05 53	☽ ∥ ☿
06 56	☉ ⊻ ♇
09 18	☽ ∥ ☉
10 04	☽ ± ♅
15 01	☽ □ ♂
16 36	☽ ⚼ ♇
17 52	☽ △ ♀
22 29	☽ ∥ ♇
22 44	☽ # ♆
15 Thursday	
01 34	☽ △ ♃
04 07	☽ ⚼ ♅
12 59	☽ ∥ ♂
22 15	♂ ∠ ♇
22 50	☽ △ ♇
16 Friday	
02 13	☽ □ ☉
03 17	☽ ☍ ♆
05 16	☽ Q ♂
06 42	☽ ⊻ ♄
10 32	☽ △ ♅
15 24	☉ □ ♆
18 50	☽ ⊥ ♄
23 12	☽ □ ☿
17 Saturday	
06 58	☽ ⚹ ♂
09 42	♀ ⚼ ♄
13 01	☽ ∠ ♄
13 25	☽ □ ♀
15 17	☽ □ ♃
17 52	☽ Q ☉
18 Sunday	
08 31	☉ △ ♄
10 47	♀ ☌ ♃
11 37	☽ □ ♇
11 57	♀ Q ♆
14 55	☽ ∠ ♂
15 54	☽ ⊼ ♆
17 50	♃ Q ♆
18 05	☽ Q ☿
19 09	☽ ⚹ ♄
20 06	☽ ⚹ ☉
23 18	☽ □ ♅
19 Monday	
03 53	☽ ± ♆
04 03	☽ Q ♃
05 41	☽ Q ♀
09 52	☽ ⊥ ♂
09 54	♀ ∥ ♃
18 56	☽ Q ♄
21 41	☽ ⚼ ♆
22 03	☽ ⚹ ☿
22 22	☽ ⊻ ♂
23 28	☽ Q ♇
20 Tuesday	
01 11	☿ ⊻ ♂
04 09	☽ ⚹ ♃
04 19	☽ ∠ ☉
07 58	☽ ⚹ ♀
10 57	☽ Q ♅
14 25	☉ ⊻ ♅
16 50	☿ ⚼ ♄
18 04	♀ ∥ ♄
22 53	☽ ⚹ ♇
23 16	☽ ⊥ ☉
21 Wednesday	
02 52	☽ △ ♆
05 51	☽ □ ♄
08 00	☽ ∠ ☿
09 40	☽ ∠ ♃
10 11	☽ ⚹ ♅
11 44	☽ ⊻ ☉
12 29	☿ Q ♆
16 01	☽ ∠ ♀
21 23	☿ ☌ ♃
22 Thursday	
00 56	☿ ∥ ♃
03 01	☽ ⊥ ♃
03 31	☽ ∠ ♇
03 47	☽ ⊥ ☿
10 47	☽ ⊥ ♀
10 58	☽ ☌ ♂
14 26	☽ ⊻ ♃
14 36	☽ ∠ ♅
14 49	♀ ⊥ ♂
15 36	☿ ∥ ♄
16 45	☽ ⊻ ☿
20 22	☽ ⊥ ♇
23 06	☽ ⊻ ♀
23 Friday	
04 33	☽ ∥ ♂
07 19	☽ ⊥ ♅
07 27	☽ ⊻ ♇
11 09	☽ □ ♆
11 13	☽ # ♆
12 10	☽ ∥ ♇
13 50	☽ △ ♄
18 20	☽ ⊻ ♅
23 57	☽ ☌ ☉
24 Saturday	
13 22	☽ ∥ ☉
13 48	♀ ± ♄
16 52	☽ ⚼ ♄
19 42	☽ Q ♆
20 29	☽ ⊻ ♂
21 57	☽ ☌ ♃
25 Sunday	
02 47	☽ ∥ ♃
03 43	☽ ∥ ♄
07 05	☽ ☌ ☿
07 41	☽ ⊥ ♂
08 51	☽ ± ♄
10 48	☽ ☌ ♀
11 40	☽ ∥ ☿
12 23	☽ ∥ ♀
13 32	☽ ☌ ♇
14 36	☿ ⊥ ♂
16 59	☽ ⚹ ♆
18 49	♂ ∠ ♅
19 25	☽ ⊼ ♄
21 15	☿ ± ♄
26 Monday	
00 06	☽ ☌ ♅
00 20	☽ ∠ ♂
01 22	☽ ∥ ♅
04 29	☿ ∥ ♀
09 19	☽ ⊻ ☉
12 06	☽ # ♅
17 48	♀ ☌ ♇
17 55	♂ ⊻ ♃
19 14	☽ ∠ ♆
20 33	☽ ⊥ ☉
21 45	☽ # ☿
22 21	☽ # ♀
27 Tuesday	
03 25	☽ ⊻ ♃
03 43	☽ ⚹ ♂
09 16	☽ # ♄
09 36	☽ # ♃
10 50	☽ ⊥ ♆
13 17	☽ ∠ ☉
13 30	☿ ☌ ♇
13 56	☽ ⊥ ♃
17 53	☽ ⊻ ♇
18 27	☽ ⊻ ☿
18 56	☽ # ☉
20 09	☽ ⊻ ♀
21 10	☽ ⊻ ♆
23 22	☽ ☍ ♄
28 Wednesday	
01 23	☽ Q ♂
04 13	☽ ⊥ ♇
04 17	☽ ⊻ ♅
05 39	☽ ∠ ♃
06 05	☽ ⊥ ☿
07 27	☽ ⊥ ♀
07 45	♀ ⚹ ♆
14 36	☽ ⊥ ♅
16 55	☽ ⚹ ☉
18 20	☿ ⚹ ♆
19 07	♃ ∥ ♄
19 39	☽ ∠ ♇
22 34	☽ # ♇
23 26	☽ ∠ ☿
23 54	☽ ∥ ♆
29 Thursday	
00 18	☽ ∠ ♀
01 20	☽ # ♂
06 01	☽ ∠ ♅
07 39	☽ ⚹ ♃
07 40	♀ ⊼ ♄
09 35	☽ □ ♂
12 46	☿ ⊼ ♄
14 51	☽ Q ☉
21 14	☽ ⚹ ♇
30 Friday	
00 23	☽ ☌ ♆
02 23	☽ ⊼ ♄
04 06	☽ ⚹ ☿
04 13	☽ ⚹ ♀
04 20	☽ Q ♃
07 35	☽ ⚹ ♅
08 20	☿ ☌ ♀
12 31	☽ ± ♄
14 44	☿ ∥ ♅
17 36	☽ Q ♇
19 08	♂ ⊥ ♇
19 19	♂ # ♆
23 34	☽ □ ☉
31 Saturday	
02 26	☽ Q ♀
02 53	☽ Q ☿
03 41	☽ ⚼ ♄
03 58	☽ Q ♅
11 15	☽ □ ♃
13 11	☿ ☌ ♅
14 46	☽ △ ♂
20 49	♀ ☌ ♅

All ephemeris data is given at 12.00 UT and the Moon's longitude is additionally given for 24.00 UT

SEPTEMBER 1968

LONGITUDES

Date	Sidereal time h m s	Sun ☉ ° ′ ″	Moon ☽ ° ′ ″	Moon ☽ 24.00 ° ′ ″	Mercury ☿ ° ′	Venus ♀ ° ′	Mars ♂ ° ′	Jupiter ♃ ° ′	Saturn ♄ ° ′	Uranus ♅ ° ′	Neptune ♆ ° ′	Pluto ♇ ° ′
01	10 42 51	09 ♍ 05 47	29 ♐ 11 40	06 ♑ 17 31	29 ♍ 41	29 ♍ 03	17 ♌ 11	14 ♍ 46	25 ♈ 00	28 ♍ 18	23 ♏ 58	22 ♍ 10
02	10 46 47	10 03 52	13 ♑ 22 33	20 ♑ 26 22	01 ♎ 11	00 ♎ 16	17 49	14 58	24 R 58	28 22	23 58	22 12
03	10 50 44	11 01 59	27 ♑ 28 34	04 ♒ 28 41	02 39	01 30	18 27	15 11	24 55	28 26	23 59	22 14
04	10 54 41	12 00 07	11 ♒ 26 19	18 ♒ 21 01	04 07	02 44	19 05	15 24	24 52	28 29	24 00	22 16
05	10 58 37	12 58 17	25 ♒ 12 24	02 ♓ 00 06	05 32	03 58	19 43	15 37	24 49	28 33	24 01	22 18
06	11 02 34	13 56 28	08 ♓ 43 49	15 ♓ 23 18	06 57	05 12	20 21	15 50	24 46	28 37	24 02	22 21
07	11 06 30	14 54 41	21 ♓ 58 24	28 ♓ 29 01	08 20	06 25	21 00	16 03	24 43	28 40	24 03	22 23
08	11 10 27	15 52 56	04 ♈ 55 08	11 ♈ 16 49	09 42	07 39	21 38	16 16	24 40	28 44	24 04	22 25
09	11 14 23	16 51 13	17 ♈ 34 14	23 ♈ 47 35	11 02	08 53	22 15	16 29	24 37	28 48	24 06	22 27
10	11 18 20	17 49 31	29 ♈ 57 10	06 ♉ 03 21	12 21	10 07	22 53	16 42	24 34	28 51	24 07	22 29
11	11 22 16	18 47 52	12 ♉ 06 31	18 ♉ 07 09	13 38	11 20	23 31	16 55	24 31	28 55	24 08	22 31
12	11 26 13	19 46 15	24 ♉ 05 44	00 ♊ 02 49	14 54	12 34	24 09	17 08	24 27	28 59	24 09	22 34
13	11 30 10	20 44 40	05 ♊ 58 57	11 ♊ 54 44	16 08	13 48	24 47	17 21	24 24	29 03	24 10	22 36
14	11 34 06	21 43 07	17 ♊ 50 45	23 ♊ 47 36	17 20	15 01	25 25	17 34	24 20	29 06	24 11	22 38
15	11 38 03	22 41 36	29 ♊ 45 53	05 ♋ 46 10	18 30	16 15	26 03	17 47	24 17	29 10	24 13	22 40
16	11 41 59	23 40 07	11 ♋ 49 02	17 ♋ 55 00	19 39	17 29	26 41	18 00	24 13	29 14	24 14	22 43
17	11 45 56	24 38 41	24 ♋ 04 34	00 ♌ 18 12	20 46	18 43	27 19	18 13	24 09	29 18	24 15	22 45
18	11 49 52	25 37 17	06 ♌ 36 15	12 ♌ 59 04	21 50	19 56	27 56	18 26	24 05	29 21	24 17	22 47
19	11 53 49	26 35 54	19 ♌ 26 52	25 ♌ 59 49	22 52	21 10	28 34	18 39	24 01	29 25	24 18	22 49
20	11 57 45	27 34 34	02 ♍ 37 57	09 ♍ 21 14	23 52	22 24	29 12	18 52	23 57	29 29	24 19	22 51
21	12 01 42	28 33 16	16 ♍ 09 33	23 ♍ 02 37	24 49	23 37	29 ♌ 50	19 05	23 53	29 33	24 21	22 54
22	12 05 39	29 ♍ 32 00	00 ♎ 00 07	07 ♎ 01 38	25 44	24 51	00 ♍ 27	19 18	23 49	29 37	24 22	22 56
23	12 09 35	00 ♎ 30 46	14 ♎ 06 39	21 ♎ 14 38	26 35	26 05	01 05	19 31	23 45	29 40	24 24	22 58
24	12 13 32	01 29 34	28 ♎ 24 58	05 ♏ 37 02	27 23	27 18	01 43	19 44	23 41	29 44	24 25	23 00
25	12 17 28	02 28 23	12 ♏ 50 12	20 ♏ 03 52	28 07	28 32	02 20	19 57	23 37	29 48	24 27	23 02
26	12 21 25	03 27 15	27 ♏ 17 25	04 ♐ 30 21	28 48	29 ♎ 45	02 58	20 09	23 32	29 52	24 28	23 04
27	12 25 21	04 26 08	11 ♐ 42 07	18 ♐ 52 19	29 25	00 ♏ 59	03 35	20 22	23 28	29 56	24 30	23 07
28	12 29 18	05 25 03	26 ♐ 00 33	03 ♑ 06 30	29 ♎ 57	02 13	04 13	20 35	23 24	29 ♍ 59	24 32	23 09
29	12 33 14	06 24 00	10 ♑ 09 54	17 ♑ 10 31	00 ♏ 24	03 26	04 50	20 48	23 19	00 ♎ 03	24 33	23 11
30	12 37 11	07 ♎ 22 59	24 ♑ 08 12	01 ♒ 02 48	00 ♏ 46	04 ♏ 40	05 ♍ 28	21 ♍ 01	23 ♈ 15	00 ♎ 07	24 ♏ 35	23 ♍ 13

Date	Moon True ☊ ° ′	Moon Mean ☊ ° ′	Moon ☽ Latitude ° ′
01	10 ♈ 01	11 ♈ 03	05 S 07
02	09 R 59	11 00	05 12
03	09 55	10 57	04 58
04	09 51	10 53	04 27
05	09 47	10 50	03 40
06	09 43	10 47	02 42
07	09 41	10 44	01 36
08	09 40	10 41	00 S 26
09	09 D 40	10 37	00 N 43
10	09 41	10 34	01 49
11	09 43	10 31	02 49
12	09 44	10 28	03 41
13	09 46	10 25	04 23
14	09 46	10 22	04 53
15	09 R 46	10 18	05 11
16	09 45	10 15	05 15
17	09 44	10 12	05 06
18	09 42	10 09	04 42
19	09 40	10 06	04 03
20	09 38	10 02	03 11
21	09 37	09 59	02 06
22	09 36	09 56	00 N 53
23	09 D 36	09 53	00 S 25
24	09 37	09 50	01 42
25	09 37	09 47	02 54
26	09 38	09 43	03 55
27	09 38	09 40	04 40
28	09 39	09 37	05 08
29	09 R 39	09 34	05 17
30	09 ♈ 39	09 ♈ 31	05 S 07

DECLINATIONS

Date	Sun ☉ ° ′	Moon ☽ ° ′	Mercury ☿ ° ′	Venus ♀ ° ′	Mars ♂ ° ′	Jupiter ♃ ° ′	Saturn ♄ ° ′	Uranus ♅ ° ′	Neptune ♆ ° ′	Pluto ♇ ° ′
01	08 N 10	28 S 34	00 00	01 N 27	16 N 47	06 N 56	07 N 10	01 N 19	17 S 07	16 N 44
02	07 48	27 57	00 S 43	00 56	16 36	06 51	07 09	01 18	17 07	16 43
03	07 26	25 32	01 26	00 N 25	16 24	06 46	07 08	01 17	17 08	16 42
04	07 04	21 37	02 08	00 S 05	16 13	06 41	07 07	01 15	17 08	16 41
05	06 41	16 34	02 50	00 36	16 01	06 36	07 06	01 14	17 08	16 41
06	06 19	10 48	03 31	01 07	15 49	06 31	07 04	01 12	17 08	16 40
07	05 57	04 S 39	04 11	01 38	15 37	06 26	07 03	01 11	17 09	16 39
08	05 34	01 N 33	04 51	02 09	15 25	06 21	07 02	01 09	17 09	16 38
09	05 12	07 34	05 30	02 40	15 13	06 16	07 00	01 08	17 09	16 37
10	04 49	13 10	06 09	03 11	15 01	06 11	06 59	01 06	17 10	16 36
11	04 26	18 09	06 47	03 41	14 49	06 06	06 58	01 05	17 10	16 35
12	04 03	22 22	07 24	04 12	14 36	06 01	06 56	01 03	17 10	16 35
13	03 40	25 37	08 00	04 43	14 24	05 56	06 55	01 02	17 11	16 34
14	03 17	27 45	08 35	05 13	14 11	05 51	06 53	01 00	17 11	16 33
15	02 54	28 38	09 10	05 44	13 59	05 46	06 52	00 59	17 11	16 32
16	02 31	28 09	09 43	06 14	13 46	05 41	06 50	00 57	17 12	16 31
17	02 08	26 19	10 16	06 44	13 33	05 36	06 49	00 56	17 12	16 30
18	01 44	23 10	10 47	07 14	13 20	05 31	06 47	00 54	17 13	16 30
19	01 21	18 50	11 17	07 44	13 08	05 26	06 46	00 53	17 13	16 29
20	00 58	13 30	11 46	08 14	12 55	05 21	06 44	00 51	17 13	16 28
21	00 35	07 24	12 14	08 44	12 42	05 15	06 42	00 50	17 14	16 27
22	00 N 11	00 N 49	12 40	09 13	12 28	05 10	06 41	00 48	17 14	16 26
23	00 S 12	05 S 57	13 05	09 42	12 15	05 05	06 39	00 47	17 15	16 26
24	00 36	12 30	13 28	10 11	12 02	05 00	06 38	00 45	17 15	16 25
25	00 59	18 27	13 50	10 40	11 49	04 55	06 36	00 44	17 15	16 24
26	01 22	23 22	14 09	11 09	11 35	04 51	06 34	00 42	17 16	16 23
27	01 46	26 49	14 26	11 37	11 22	04 46	06 33	00 41	17 16	16 23
28	02 09	28 31	14 42	12 05	11 08	04 41	06 31	00 39	17 17	16 22
29	02 33	28 19	14 55	12 33	10 55	04 36	06 29	00 38	17 17	16 21
30	02 S 56	26 S 19	15 S 05	13 S 00	10 N 41	04 N 31	06 N 27	00 N 36	17 S 17	16 N 20

ZODIAC SIGN ENTRIES

Date	h	m	Planets
01	13	22	☽ ♑
01	16	59	☿ ♎
02	06	39	♀ ♎
03	16	19	☽ ♒
05	20	27	☽ ♓
08	02	49	☽ ♈
10	12	06	☽ ♉
12	23	54	☽ ♊
15	12	28	☽ ♋
17	23	25	☽ ♌
20	07	15	☽ ♍
21	18	39	♂ ♍
22	12	00	☽ ♎
22	23	26	☉ ♎
24	14	39	☽ ♏
26	16	30	☽ ♐
26	16	45	♀ ♏
28	14	40	☿ ♏
28	16	09	♅ ♎
28	18	44	☽ ♑
30	22	11	☽ ♒

LATITUDES

Date	Mercury ☿ ° ′	Venus ♀ ° ′	Mars ♂ ° ′	Jupiter ♃ ° ′	Saturn ♄ ° ′	Uranus ♅ ° ′	Neptune ♆ ° ′	Pluto ♇ ° ′
01	00 S 08	01 N 10	01 N 09	01 N 00	02 S 42	00 N 42	01 N 42	14 N 53
04	00 33	01 05	01 10	01 00	02 42	00 42	01 42	14 52
07	00 57	01 00	01 11	01 00	02 43	00 42	01 42	14 52
10	01 23	00 54	01 12	01 01	02 44	00 42	01 42	14 52
13	01 47	00 48	01 12	01 01	02 44	00 42	01 42	14 52
16	02 12	00 41	01 13	01 01	02 45	00 42	01 41	14 53
19	02 35	00 34	01 14	01 01	02 45	00 42	01 41	14 53
22	02 56	00 26	01 14	01 01	02 46	00 42	01 41	14 53
25	03 14	00 18	01 15	01 01	02 46	00 42	01 41	14 53
28	03 28	00 10	01 16	01 02	02 46	00 42	01 41	14 54
31	03 S 36	00 N 02	01 N 16	01 N 02	02 S 47	00 N 42	01 N 41	14 N 54

DATA

Julian Date	2440101
Delta T	+39 seconds
Ayanamsa	23° 25′ 07″
Synetic vernal point	05° ♓ 41′ 52″
True obliquity of ecliptic	23° 26′ 45″

MOON'S PHASES, APSIDES AND POSITIONS ☽

Date	h	m	Phase	Longitude ° ′	Eclipse Indicator
06	22	07	○	14 ♓ 21	
14	20	31	☾	22 ♊ 04	
22	11	08	●	29 ♍ 30	Total
29	05	07	☽	06 ♑ 07	

Day	h	m		
13	21	53	Apogee	
25	20	16	Perigee	
01	16	11	Max dec	28° S 35′
08	05	57	0N	
15	15	45	Max dec	28° N 39′
22	14	53	0S	
28	21	30	Max dec	28° S 40′

LONGITUDES

Date	Chiron ⚷ ° ′	Ceres ⚳ ° ′	Pallas ⚴ ° ′	Juno ⚵ ° ′	Vesta ⚶ ° ′	Black Moon Lilith ⚸ ° ′
01	02 ♈ 04	10 ♏ 49	17 ♎ 57	08 ♏ 25	05 ♉ 18	08 ♊ 21
11	01 ♈ 39	14 ♏ 18	22 ♎ 06	11 ♏ 06	04 ♉ 57	09 ♊ 28
21	01 ♈ 12	17 ♏ 57	26 ♎ 20	13 ♏ 58	03 ♉ 51	10 ♊ 35
31	00 ♈ 45	21 ♏ 43	00 ♏ 38	16 ♏ 59	02 ♉ 04	11 ♊ 42

All ephemeris data is given at 12.00 UT and the Moon's longitude is additionally given for 24.00 UT

ASPECTARIAN

01 Sunday
00 05 ☽ □ ♇
03 09 ☽ ⊻ ♆
04 56 ☽ △ ♄
10 29 ☽ □ ♅
11 43 ☽ □ ♀
12 56 ☽ □ ☿
13 18 ☽ ⊥ ♆
17 17 ☽ ⚼ ♂
18 12 ♀ ∥ ♅
19 24 ♂ ∥ ♇

02 Monday
04 32 ☽ ∠ ♆
05 59 ☽ △ ☉
09 14 ☽ ± ♂
09 32 ☉ ⚼ ♄
14 45 ☽ △ ♃
16 16 ☿ ⋕ ♀
19 54 ☽ ⊼ ♂

03 Tuesday
03 02 ☽ △ ♇
06 03 ☽ ⚹ ♆
06 57 ☿ ⋕ ♅
07 39 ☽ □ ♄
09 21 ☽ ⚼ ☉
13 38 ☽ △ ♅
16 43 ☽ ⚼ ♃
19 34 ☽ △ ♀
21 54 ☽ △ ☿

04 Wednesday
01 55 ☽ ± ☉
02 37 ☽ Q ♆
04 47 ☽ ⚼ ♇
08 26 ☽ ± ♃
08 29 ☉ ∥ ♄
11 25 ☿ ∠ ♂
12 06 ☉ Q ♆
13 03 ☽ ⊼ ☉
14 28 ☽ Q ♄
15 34 ☽ ⚼ ♅
18 59 ☽ ⊼ ♃
20 24 ☽ ± ♇

05 Thursday
00 00 ☽ ⚼ ♀
01 56 ☽ ☍ ♂
02 53 ☽ ⚼ ☿
06 54 ☽ ⊼ ♇
07 19 ☽ ± ♅
09 31 ☽ ∥ ♆
09 55 ☽ □ ♆
11 20 ☽ ⚹ ♄
11 33 ☽ ⋕ ♇
14 30 ☽ ⋕ ♂
17 20 ☽ ± ♀
17 55 ☽ ⊼ ♅
19 51 ☉ ∥ ♃
20 32 ☽ ± ☿

06 Friday
05 02 ☽ ⊼ ♀
08 26 ☽ ⊼ ☿
13 52 ☽ ∠ ♄
15 36 ♀ ⋕ ♅
18 39 ♀ ∠ ♂
22 07 ☽ ☍ ☉

07 Saturday
01 02 ☽ ☍ ♃
02 41 ☽ ⋕ ♄
05 02 ☽ ⋕ ♃
06 05 ☽ ⊥ ♄
06 41 ☽ ⋕ ☉
10 07 ☽ ⊼ ♂
12 45 ☽ ☍ ♇
13 37 ☽ ∥ ☿
15 50 ☽ △ ♆
17 02 ☽ ⊻ ♄
21 43 ☽ ± ♂
22 44 ☽ ∥ ♀

08 Sunday
00 24 ☽ ☍ ♅
00 54 ☿ ∠ ♆
01 27 ☽ ⋕ ♅
10 26 ☽ ∥ ♅
14 32 ☽ ⋕ ♀
15 22 ☽ ⚼ ♂
17 41 ☽ ☍ ♀
19 50 ☽ ⚼ ♆
22 04 ☽ ☍ ☿

09 Monday
00 26 ☉ ☌ ♃
02 36 ☽ ⋕ ☿
02 56 ☽ ∥ ☉
04 43 ☉ ⋕ ☿
06 45 ☽ ∥ ♃
09 43 ☽ ∥ ♄
09 54 ☽ ⊼ ♃
10 31 ☽ ⊼ ☉
13 00 ☽ ± ♆
16 12 ♀ ∠ ♆
19 48 ♂ ⊻ ♇
21 26 ☽ ⊼ ♇
21 31 ☽ △ ♂
21 39 ☽ ± ♃
23 03 ☽ ± ☉

10 Tuesday
00 36 ☽ ⊼ ♆
01 33 ☽ ☌ ♄
09 08 ☽ ± ♇
09 51 ☽ ⊼ ♅
10 33 ♂ ⊥ ♅
13 05 ☿ ⋕ ♃
15 30 ☽ ⚼ ♃
18 07 ☽ ⚼ ☉
20 12 ☽ ∥ ♂
21 41 ☽ ± ♅

11 Wednesday
02 53 ☽ ⚼ ♇
04 06 ☽ ∥ ♇
05 17 ☉ ± ♄
06 57 ☽ ⋕ ♆
10 18 ☽ ⊼ ♀
15 23 ☽ ⊼ ☿
15 37 ☽ ⚼ ♅
18 49 ☿ ⋕ ♄
21 47 ☽ △ ♃
23 38 ☽ ± ♀

12 Thursday
02 32 ☽ △ ☉
04 48 ☽ ± ☿
07 56 ☉ ⋕ ♀
08 54 ☽ △ ♇
11 50 ♂ □ ♆
12 07 ☽ ☍ ♆
12 08 ☽ □ ♂
12 43 ☽ ⊻ ♄
19 48 ☽ ⚼ ♀
21 54 ☽ △ ♅
22 23 ♂ △ ♄

13 Friday
00 45 ☽ ⊥ ♄
01 03 ☽ ⚼ ☿
18 52 ☽ ∠ ♄

14 Saturday
02 33 ☽ Q ♂
05 38 ☽ △ ♀
10 51 ☽ △ ☿
11 26 ☽ □ ♃
17 58 ☿ ⊻ ♃
20 31 ☽ □ ☉
21 42 ☽ □ ♇

15 Sunday
00 49 ☽ ⊼ ♆
01 02 ☽ ⚹ ♄
04 07 ☽ ⚹ ♂
05 49 ☿ ⊥ ♆
10 48 ☽ □ ♅
11 27 ☉ ☌ ♇
12 54 ☽ ± ♆
13 12 ♀ ⋕ ♃

16 Monday
00 16 ☽ Q ♃
00 56 ☽ Q ♄
06 08 ♄ ⊼ ♆
06 53 ☽ ⚼ ♆
09 48 ☽ Q ♇
11 41 ☽ Q ☉
11 43 ☽ ∠ ♂
22 43 ☽ Q ♅

17 Tuesday
00 23 ☽ □ ♀
00 24 ☽ ⚹ ♃
00 25 ♀ ⊻ ♃
00 35 ☉ ⊼ ♄
02 13 ☉ ⚹ ♆
02 58 ♀ ⊥ ♆
04 55 ☽ □ ☿
05 55 ♃ ± ♄
06 20 ☽ ⊥ ♂
09 25 ☽ ⚹ ♇
12 09 ☽ □ ♄
12 21 ☽ △ ♆
13 12 ☽ ⚹ ☉
15 19 ♀ ⋕ ♄
18 35 ☽ ⊻ ♂
22 07 ☽ ⚹ ♅

18 Wednesday
05 53 ☽ ∠ ♃
14 14 ☽ ∠ ♇
14 47 ☽ Q ♀
18 38 ☽ Q ☿
20 12 ☽ ∠ ☉
23 10 ☽ ⊥ ♃

19 Thursday
02 38 ☽ ∠ ♅
07 07 ☽ ⊥ ♇
10 31 ☽ ⊻ ♃
10 43 ☿ ⊻ ♇
14 18 ☽ ⊥ ☉
15 29 ☽ ⚹ ♀
18 13 ☽ ⊻ ♇
18 49 ☽ ⚹ ☿
19 20 ☽ ⊥ ♅
19 43 ☽ ⋕ ♆
20 21 ☽ △ ♄
20 55 ☽ □ ♆
23 08 ☽ ∥ ♇

20 Friday
02 08 ☽ ⊻ ☉
05 30 ☽ ☌ ♂
06 18 ☽ ⊻ ♅
14 02 ☿ ☍ ♄
14 32 ☽ ∥ ♂
18 33 ☽ ⋕ ☿
19 24 ☉ ∥ ♅
21 18 ♀ ⊻ ♇
21 22 ☽ ∠ ♀
23 14 ☽ ⚼ ♄
23 40 ☿ ⊻ ♆
23 59 ☽ ∠ ☿

21 Saturday
00 10 ♂ ⊻ ♅
05 17 ☽ Q ♆
07 20 ☽ ⋕ ♀
14 36 ☽ ∥ ♄
14 49 ☽ ⊥ ♀
15 01 ☽ ± ♄
16 57 ♀ ☍ ♄
16 59 ☽ ⊥ ☿
17 12 ☽ ☌ ♃
20 03 ☽ ∥ ♃
20 59 ☿ ⊥ ♃
23 46 ☽ ☌ ♇

22 Sunday
01 24 ☽ ⊼ ♄
02 16 ☽ ⊻ ♀
02 17 ☽ ⚹ ♆
02 31 ♀ ⊻ ♆
04 08 ☽ ⊻ ☿
04 40 ☿ ⋕ ♂
11 08 ☽ ☌ ☉
11 19 ☽ ☌ ♅
12 01 ☽ ∥ ♅
12 49 ☽ ⊻ ♂
14 01 ☉ ☌ ♅
14 21 ☽ ∥ ☉
15 21 ☽ ⋕ ☉
17 42 ☽ ⋕ ♅
22 41 ♀ ⊥ ♃
23 32 ☽ ⊥ ♂

23 Monday
04 01 ☽ ∠ ♆
08 59 ☽ ⋕ ♃
14 30 ☽ ⋕ ♄
15 29 ☽ ∠ ♂
19 14 ☽ ⊥ ♆
21 14 ☽ ⊻ ♃

24 Tuesday
02 39 ☽ ∥ ♀
02 55 ☽ ⊻ ♇
04 07 ☽ ☍ ♄
05 19 ☽ ⊻ ♆
07 27 ☽ ⊥ ♃
09 58 ☽ ☌ ♀
10 11 ☽ ☌ ☿
10 16 ☽ ⋕ ♂
12 59 ☽ ⊥ ♇
14 13 ☽ ⊻ ♅
15 55 ☽ ∥ ☿
16 09 ☿ ☌ ♀
17 30 ☽ ⊻ ☉
17 45 ☽ ⚹ ♂
21 07 ☉ ⋕ ♅
22 41 ☽ ∠ ♃

25 Wednesday
00 15 ☽ ⊥ ♅
02 45 ☉ ⊻ ♂
03 20 ☽ ⋕ ♇
04 00 ☽ ∠ ♇
04 13 ☽ ⊥ ☉
06 51 ☽ ∥ ♆
14 36 ☽ Q ♂
15 16 ☽ ∠ ♅
20 16 ☽ ∠ ☉
22 14 ♀ ⊥ ♇
23 59 ☽ ⚹ ♃

26 Thursday
04 59 ☽ ⚹ ♇
05 48 ☽ ⊼ ♄
07 19 ☽ ☌ ♆
14 11 ♀ ⊻ ♅
14 38 ☽ ⊻ ☿
15 43 ☽ ± ♄
16 18 ☽ ⚹ ♅
16 29 ☽ ⊻ ♀
20 13 ☽ Q ♃
21 52 ☽ □ ♂
23 00 ☽ ⚹ ☉
23 00 ☿ ⊥ ♇

27 Friday
00 59 ☽ Q ♇
01 04 ☽ ⊥ ☿
03 10 ♀ ⋕ ♂
03 24 ☽ ⊥ ♀
06 38 ☽ ⚼ ♄
12 23 ☽ Q ♅
16 43 ☽ ∠ ☿
19 50 ☽ ∠ ♀
20 30 ☽ Q ☉

28 Saturday
02 44 ☽ □ ♃
07 10 ☽ □ ♇
07 37 ☽ △ ♄
09 30 ☽ ⊻ ♆
14 27 ☿ ⊻ ♅
18 45 ☽ □ ♅
18 53 ☽ ⚹ ☿
19 39 ☽ ⊥ ♆
23 28 ☽ ⚹ ♀

29 Sunday
02 31 ☽ △ ♂
05 07 ☽ □ ☉
10 57 ☽ ∠ ♆
15 56 ☽ Q ☿
21 53 ☽ Q ♀

30 Monday
05 22 ☽ ⚼ ♂
06 31 ☽ △ ♃
10 25 ☽ △ ♇
10 28 ☽ □ ♄
12 46 ☽ ⚹ ♆
17 52 ♄ ⊼ ♇
21 41 ☽ ± ♂
22 25 ☽ △ ♅
23 45 ☽ □ ☿

OCTOBER 1968

LONGITUDES

Date	Sidereal time h m s	Sun ☉ ° ' "	Moon ☽ ° ' "	Moon ☽ 24.00 ° ' "	Mercury ☿ ° '	Venus ♀ ° '	Mars ♂ ° '	Jupiter ♃ ° '	Saturn ♄ ° '	Uranus ♅ ° '	Neptune ♆ ° '	Pluto ♇ ° '
01	12 41 08	08 ♎ 21 59	07 ♒ 54 13	14 ♒ 42 22	01 ♏ 02	05 ♏ 53	06 ♍ 05	21 ♍ 13	23 ♈ 10	00 ♎ 11	24 ♏ 37	23 ♍ 15
02	12 45 04	09 21 01	21 ♒ 27 13	28 ♒ 08 41	01 12	07 07	06 43	21 26	23 R 06	00 14	24 38	23 17
03	12 49 01	10 20 05	04 ♓ 46 47	11 ♓ 21 28	01 R 15	08 20	07 20	21 39	23 01	00 18	24 40	23 20
04	12 52 57	11 19 10	17 ♓ 52 44	24 ♓ 20 36	01 11	09 34	07 58	21 51	22 57	00 22	24 42	23 22
05	12 56 54	12 18 17	00 ♈ 45 06	07 ♈ 06 15	01 00	10 47	08 35	22 04	22 52	00 26	24 44	23 24
06	13 00 50	13 17 27	13 ♈ 24 09	19 ♈ 38 52	00 41	12 01	09 13	22 16	22 47	00 29	24 45	23 26
07	13 04 47	14 16 38	25 ♈ 50 31	01 ♉ 59 16	00 ♏ 15	13 14	09 50	22 29	22 43	00 33	24 47	23 28
08	13 08 43	15 15 52	08 ♉ 05 18	14 ♉ 08 51	29 ♎ 40	14 27	10 27	22 41	22 38	00 37	24 49	23 30
09	13 12 40	16 15 07	20 ♉ 10 10	26 ♉ 09 33	28 57	15 41	11 05	22 54	22 33	00 40	24 51	23 32
10	13 16 37	17 14 25	02 ♊ 07 22	08 ♊ 03 58	28 06	16 54	11 42	23 06	22 28	00 44	24 53	23 34
11	13 20 33	18 13 46	13 ♊ 59 49	19 ♊ 55 20	27 09	18 08	12 19	23 19	22 24	00 48	24 54	23 36
12	13 24 30	19 13 08	25 ♊ 51 01	01 ♋ 47 23	26 05	19 21	12 56	23 31	22 19	00 51	24 56	23 38
13	13 28 26	20 12 33	07 ♋ 44 59	13 ♋ 44 22	24 57	20 34	13 34	23 43	22 14	00 55	24 58	23 40
14	13 32 23	21 12 00	19 ♋ 46 07	25 ♋ 50 48	23 45	21 48	14 11	23 55	22 09	00 59	25 00	23 42
15	13 36 19	22 11 30	01 ♌ 59 01	08 ♌ 11 17	22 32	23 01	14 48	24 08	22 05	01 02	25 02	23 44
16	13 40 16	23 11 01	14 ♌ 28 10	20 ♌ 50 08	21 19	24 14	15 25	24 20	22 00	01 06	25 04	23 46
17	13 44 12	24 10 35	27 ♌ 17 38	03 ♍ 51 02	20 10	25 28	16 02	24 32	21 55	01 10	25 06	23 48
18	13 48 09	25 10 11	10 ♍ 30 35	17 ♍ 16 28	19 05	26 41	16 39	24 44	21 50	01 13	25 08	23 50
19	13 52 06	26 09 50	24 ♍ 08 42	01 ♎ 07 13	18 06	27 54	17 16	24 56	21 46	01 17	25 10	23 52
20	13 56 02	27 09 30	08 ♎ 11 45	15 ♎ 21 54	17 17	29 ♏ 07	17 53	25 08	21 41	01 20	25 12	23 54
21	13 59 59	28 09 13	22 ♎ 37 06	29 ♎ 56 38	16 37	00 ♐ 21	18 30	25 20	21 36	01 24	25 14	23 56
22	14 03 55	29 ♎ 08 57	07 ♏ 19 41	14 ♏ 45 16	16 07	01 34	19 07	25 32	21 31	01 27	25 16	23 58
23	14 07 52	00 ♏ 08 44	22 ♏ 12 24	29 ♏ 40 13	15 49	02 47	19 44	25 44	21 27	01 31	25 18	24 00
24	14 11 48	01 08 33	07 ♐ 07 05	14 ♐ 32 35	15 43	04 00	20 21	25 55	21 22	01 34	25 20	24 02
25	14 15 45	02 08 23	21 ♐ 55 38	29 ♐ 15 24	15 D 47	05 13	20 58	26 07	21 17	01 37	25 23	24 03
26	14 19 41	03 08 15	06 ♑ 31 15	13 ♑ 42 38	16 02	06 26	21 35	26 19	21 13	01 41	25 25	24 05
27	14 23 38	04 08 09	20 ♑ 49 11	27 ♑ 50 40	16 28	07 39	22 12	26 30	21 08	01 44	25 27	24 07
28	14 27 35	05 08 05	04 ♒ 46 59	11 ♒ 38 08	17 03	08 53	22 49	26 42	21 03	01 47	25 29	24 09
29	14 31 31	06 08 02	18 ♒ 24 13	25 ♒ 05 26	17 47	10 06	23 26	26 53	20 59	01 51	25 31	24 11
30	14 35 28	07 08 00	01 ♓ 41 59	08 ♓ 14 08	18 38	11 19	24 03	27 04	20 54	01 54	25 33	24 12
31	14 39 24	08 ♏ 08 00	14 ♓ 42 13	21 ♓ 06 29	19 ♎ 36	12 ♐ 32	24 ♍ 39	27 ♍ 16	20 ♈ 50	01 ♎ 57	25 ♏ 35	24 ♍ 14

Date	Moon True ☊ ° '	Moon Mean ☊ ° '	Moon ☽ Latitude ° '
01	09 ♈ 38	09 ♈ 28	04 S 39
02	09 R 38	09 24	03 57
03	09 D 38	09 21	03 02
04	09 38	09 18	01 58
05	09 39	09 15	00 S 49
06	09 R 39	09 12	00 N 21
07	09 38	09 08	01 29
08	09 38	09 05	02 31
09	09 37	09 02	03 26
10	09 36	08 59	04 11
11	09 35	08 56	04 46
12	09 34	08 53	05 07
13	09 33	08 49	05 16
14	09 33	08 46	05 11
15	09 D 33	08 43	04 53
16	09 34	08 40	04 20
17	09 35	08 37	03 33
18	09 36	08 34	02 34
19	09 37	08 30	01 24
20	09 38	08 27	00 N 08
21	09 R 37	08 24	01 S 11
22	09 36	08 21	02 27
23	09 34	08 18	03 33
24	09 31	08 14	04 26
25	09 28	08 11	05 00
26	09 25	08 08	05 14
27	09 24	08 05	05 08
28	09 23	08 02	04 44
29	09 D 23	07 59	04 04
30	09 25	07 55	03 12
31	09 ♈ 26	07 ♈ 52	02 S 12

DECLINATIONS

Date	Sun ☉ ° '	Moon ☽ ° '	Mercury ☿ ° '	Venus ♀ ° '	Mars ♂ ° '	Jupiter ♃ ° '	Saturn ♄ ° '	Uranus ♅ ° '	Neptune ♆ ° '	Pluto ♇ ° '
01	03 S 19	22 S 48	15 S 12	13 S 28	10 N 28	04 N 26	06 N 26	00 N 35	17 S 18	16 N 20
02	03 42	18 05	15 16	13 55	10 14	04 21	06 24	00 33	17 18	16 19
03	04 06	12 35	15 17	14 21	10 00	04 16	06 22	00 32	17 19	16 18
04	04 29	06 36	15 15	14 47	09 46	04 11	06 20	00 30	17 19	16 18
05	04 52	00 S 27	15 08	15 13	09 32	04 06	06 19	00 29	17 20	16 17
06	05 15	05 N 37	14 58	15 39	09 19	04 01	06 17	00 27	17 20	16 16
07	05 38	11 22	14 43	16 04	09 05	03 56	06 15	00 26	17 21	16 16
08	06 01	16 36	14 24	16 29	08 51	03 51	06 13	00 24	17 21	16 15
09	06 24	21 06	14 00	16 53	08 37	03 47	06 12	00 23	17 22	16 14
10	06 46	24 42	13 31	17 17	08 23	03 42	06 10	00 21	17 22	16 14
11	07 09	27 13	12 58	17 40	08 09	03 37	06 08	00 20	17 23	16 13
12	07 32	28 30	12 22	18 03	07 54	03 32	06 06	00 18	17 23	16 13
13	07 54	28 29	11 41	18 26	07 40	03 27	06 04	00 17	17 24	16 12
14	08 16	27 07	10 58	18 48	07 26	03 23	06 03	00 16	17 24	16 12
15	08 39	24 28	10 13	19 10	07 12	03 18	06 01	00 14	17 25	16 11
16	09 01	20 38	09 27	19 31	06 58	03 13	05 59	00 13	17 25	16 10
17	09 23	15 45	08 42	19 52	06 43	03 09	05 57	00 11	17 26	16 10
18	09 45	10 00	07 58	20 12	06 29	03 04	05 56	00 10	17 26	16 09
19	10 06	03 N 37	07 17	20 31	06 15	02 59	05 54	00 08	17 27	16 09
20	10 28	03 S 08	06 40	20 50	06 00	02 55	05 52	00 07	17 27	16 08
21	10 49	09 54	06 08	21 09	05 46	02 50	05 50	00 06	17 28	16 08
22	11 10	16 16	05 40	21 27	05 32	02 45	05 49	00 04	17 28	16 07
23	11 32	21 46	05 19	21 44	05 17	02 41	05 47	00 03	17 29	16 07
24	11 52	25 52	05 03	22 01	05 03	02 36	05 45	00 N 02	17 29	16 07
25	12 13	28 11	04 53	22 17	04 48	02 32	05 44	00 00	17 30	16 06
26	12 34	28 30	04 49	22 33	04 34	02 27	05 42	00 S 01	17 30	16 06
27	12 54	26 54	04 50	22 48	04 20	02 23	05 40	00 02	17 31	16 05
28	13 14	23 40	04 56	23 02	04 05	02 18	05 39	00 04	17 31	16 05
29	13 34	19 11	05 07	23 16	03 51	02 14	05 37	00 05	17 32	16 05
30	13 54	13 52	05 23	23 29	03 36	02 10	05 36	00 06	17 32	16 04
31	14 S 13	08 S 03	05 S 41	23 S 41	03 N 22	02 N 05	05 N 34	00 S 08	17 S 33	16 N 04

ZODIAC SIGN ENTRIES

Date	h	m	Planets
03	03	21	☽ ♓
05	10	35	☽ ♈
07	20	07	☽ ♉
07	22	46	☿ ♎
10	07	43	☽ ♊
12	20	23	☽ ♋
15	08	08	☽ ♌
17	16	59	☽ ♍
19	22	05	☽ ♎
21	05	16	♀ ♐
22	00	05	☽ ♏
23	08	30	☉ ♏
24	00	32	☽ ♐
26	01	13	☽ ♑
28	03	43	☽ ♒
30	08	54	☽ ♓

LATITUDES

Date	Mercury ☿ ° '	Venus ♀ ° '	Mars ♂ ° '	Jupiter ♃ ° '	Saturn ♄ ° '	Uranus ♅ ° '	Neptune ♆ ° '	Pluto ♇ ° '
01	03 S 36	00 N 02	01 N 16	01 N 02	02 S 47	00 N 42	01 N 41	14 N 54
04	03 35	00 S 07	01 17	01 02	02 47	00 42	01 41	14 55
07	03 22	00 16	01 17	01 02	02 47	00 42	01 40	14 55
10	02 55	00 24	01 18	01 03	02 47	00 42	01 40	14 56
13	02 11	00 33	01 18	01 03	02 47	00 42	01 40	14 57
16	01 14	00 42	01 19	01 03	02 47	00 42	01 40	14 58
19	00 S 12	00 51	01 19	01 04	02 47	00 42	01 40	14 59
22	00 N 44	00 59	01 20	01 04	02 47	00 43	01 40	14 59
25	01 27	01 08	01 20	01 04	02 47	00 43	01 40	15 00
28	01 54	01 16	01 21	01 05	02 47	00 43	01 40	15 02
31	02 N 08	01 S 24	01 N 21	01 N 05	02 S 46	00 N 43	01 N 40	15 N 03

DATA

Julian Date	2440131
Delta T	+39 seconds
Ayanamsa	23° 25' 11"
Synetic vernal point	05° ♓ 41' 49"
True obliquity of ecliptic	23° 26' 46"

MOON'S PHASES, APSIDES AND POSITIONS ☽

Date	h	m	Phase	Longitude ° '	Eclipse Indicator
06	11	46	○	13 ♈ 17	total
14	15	05	☾	21 ♋ 20	
21	21	44	●	28 ♎ 33	
28	12	40	☽	05 ♒ 10	

Day	h	m		
11	16	58	Apogee	
23	15	09	Perigee	
05	13	46	ON	
12	23	38	Max dec	28° N 39'
20	00	57	OS	
26	03	50	Max dec	28° S 37'

LONGITUDES

Date	Chiron ⚷ ° '	Ceres ⚳ ° '	Pallas ⚴ ° '	Juno ⚵ ° '	Vesta ⚶ ° '	Black Moon Lilith ⚸ ° '
01	00 ♈ 45	21 ♏ 43	00 ♏ 38	16 ♏ 59	02 ♉ 04	11 ♊ 42
11	00 ♈ 18	25 ♏ 35	04 ♏ 58	20 ♏ 07	29 ♈ 45	12 ♊ 49
21	29 ♓ 54	29 ♏ 32	09 ♏ 21	23 ♏ 21	27 ♈ 11	13 ♊ 56
31	29 ♓ 33	03 ♐ 33	13 ♏ 45	26 ♏ 40	24 ♈ 40	15 ♊ 03

All ephemeris data is given at 12.00 UT and the Moon's longitude is additionally given for 24.00 UT

ASPECTARIAN

h m Aspects

01 Tuesday
08 07 ☽ □ ♀
08 40 ☽ ⊼ ♂
09 00 ☽ ⚼ ♃
09 43 ☽ Q ♆
12 37 ☽ ⚼ ♇
12 53 ☽ △ ☉
17 43 ☽ Q ♄
18 00 ♀ ⊥ ♅
19 54 ♀ ∠ ♃
20 09 ♀ ⚹ ♂

02 Wednesday
00 54 ☽ ⚼ ♅
01 07 ☽ ± ♃
04 34 ☽ ± ♇
11 58 ☽ ⊼ ♃
14 55 ☽ ⚹ ♄
15 18 ☽ ⊼ ♇
15 34 ☽ ∥ ♆
17 01 ☽ ± ♅
17 36 ☽ ⚼ ☉
17 43 ☽ □ ♆
19 14 ☉ ∠ ♆
20 02 ☽ # ♇

03 Thursday
00 31 ☽ ∥ ☿
03 51 ☽ ⊼ ♅
05 05 ☽ ∥ ♀
05 36 ☽ △ ☿
11 08 ☽ ± ☉
11 39 ☿ St R
11 46 ♀ ∠ ♇
16 53 ☽ ☍ ♂
17 52 ☽ ∠ ♄
19 09 ☽ △ ♀
20 48 ☉ # ♃
22 55 ☽ # ♂
22 57 ☽ ⊼ ☉

04 Friday
08 55 ☽ ⚼ ☿
10 17 ☽ ⊥ ♄
11 17 ♂ ⚼ ♄
13 02 ☽ # ♄
19 29 ☽ ☍ ♃
19 49 ☽ ∥ ☉
21 20 ☽ ⊻ ♄
21 35 ☽ # ♃
22 12 ☽ ☍ ♇

05 Saturday
00 41 ☽ △ ♆
01 25 ☽ ± ☿
01 43 ☽ ⚼ ♀
08 34 ☿ ∥ ♀
11 23 ☽ ☍ ♅
11 54 ☽ # ♅
12 28 ☽ ⊼ ☿
15 37 ☽ ∥ ♅
20 25 ☽ ± ♀

06 Sunday
03 35 ☽ ⊼ ♂
05 01 ☽ ⚼ ♆
05 42 ☽ ∥ ♃
09 03 ☽ ⊼ ♀
10 26 ☽ # ☉
11 46 ☽ ☍ ☉
14 42 ☽ ∥ ♄
15 39 ☽ ± ♂
22 18 ☿ ⊻ ♅
22 18 ☽ ± ♆

07 Monday
02 38 ☽ ∥ ♂
05 22 ☽ ⊼ ♃
05 57 ☽ ☌ ♄
07 22 ☽ ⊼ ♇
09 56 ☽ ⚼ ♂
09 56 ☽ ⊼ ♆
17 14 ☽ ± ♃
19 05 ☽ ± ♇
20 14 ☽ ☍ ☿
21 14 ☽ ⊼ ♅
23 12 ♀ # ♇

08 Tuesday
02 12 ☽ # ☿
07 07 ♃ ⊼ ♄
09 05 ☽ ± ♅
10 21 ☽ ∥ ♇
11 12 ☽ ⚼ ♃
11 23 ☽ # ♀
12 49 ☽ ⚼ ♇
15 47 ☽ # ♆
16 56 ☽ △ ♂
17 33 ☿ ⊥ ♇

09 Wednesday
00 14 ☉ # ♄
02 02 ☽ ☍ ♀
02 59 ☽ ⚼ ♅
03 29 ☽ ⊼ ☉
11 52 ♀ ∠ ♅
13 14 ☿ ⊥ ♃
16 32 ☽ ± ☉
16 44 ☽ ⊻ ♄
17 33 ☽ △ ♃
18 45 ☽ △ ♇
21 23 ☽ ☍ ♆

10 Thursday
04 28 ☽ ⊼ ☿
04 42 ☽ ⊥ ♄
09 11 ☽ △ ♅
12 16 ☽ ⚼ ☉
15 43 ☽ ± ☿
17 32 ♀ ∥ ♆
22 44 ☽ ∠ ♄

11 Friday
08 25 ☽ □ ♂
08 33 ☽ ⚼ ☿
09 29 ☿ ∠ ♂
17 44 ☉ ⊥ ♂
21 20 ☽ ⊼ ♀
21 21 ☽ △ ☉
22 32 ☉ ⊻ ♀

12 Saturday
04 54 ☽ ⚹ ♄
05 00 ☉ ⊥ ♆
07 12 ☽ □ ♃
07 31 ☽ □ ♇
10 09 ☽ ⊼ ♆
10 52 ☽ ± ♀
12 04 ♂ Q ♆
12 26 ☽ △ ☿
22 10 ☽ □ ♅
22 19 ☽ ± ♆
22 51 ☽ Q ♂

13 Sunday
02 59 ☉ # ♂
04 59 ☽ Q ♄
05 12 ♃ ☌ ♇
07 07 ☽ ⚼ ♀
11 28 ☿ ⊻ ♆
16 28 ☽ ⚼ ♆
19 53 ☽ Q ♇
20 06 ☽ Q ♃

14 Monday
00 16 ☽ ⚹ ♂
09 01 ☿ ⊻ ♃
10 26 ☽ Q ♅
12 50 ☿ ⊻ ♇
15 05 ☽ □ ☉
16 28 ☽ △ ♀
16 42 ☽ □ ♄
18 41 ♀ ⊼ ♄
19 09 ☽ □ ☿
19 48 ☽ ⚹ ♇
20 21 ☽ ⚹ ♃
22 22 ☽ △ ♆

15 Tuesday
07 13 ☿ ⊻ ♀
07 31 ☽ ∠ ♂
09 25 ☉ ☍ ♄
10 09 ☽ ⚹ ♅
15 40 ☉ ☌ ☿
21 34 ☿ ☍ ♄

16 Wednesday
01 06 ☽ ∠ ♇
01 52 ☽ ⊥ ♂
02 02 ☽ ∠ ♃
02 36 ♀ ⚹ ♇
03 02 ☽ Q ☿
05 12 ☽ Q ☉
10 44 ☿ ⊥ ♂
13 53 ☽ ⊻ ♂
14 12 ♀ ⚹ ♃
15 06 ☽ ∠ ♅
17 31 ☽ # ♀
18 16 ☽ ⊥ ♇
19 25 ☽ ⊥ ♃
21 28 ☉ ∥ ☿
23 49 ☽ ⚹ ☿

17 Thursday
02 05 ☽ △ ♄
02 43 ☉ ⊻ ♇
04 16 ☽ # ♆
04 48 ♀ ☌ ♆
05 31 ☽ ⊻ ♇
05 45 ☽ ⚹ ☉
06 49 ☽ ⊻ ♃
07 55 ♂ ± ♄
07 56 ☽ □ ♆
08 02 ☽ ⊥ ♅
08 15 ☽ □ ♀
10 07 ☽ ∥ ♇
19 08 ☽ ⊻ ♅
19 11 ☿ ⊥ ♀
22 48 ☉ ⊻ ♃

18 Friday
01 16 ☽ ∠ ☿
05 27 ☽ ⚼ ♄
10 40 ☿ ⊥ ♆
11 09 ☉ ⊻ ♆
11 21 ☽ ∠ ☉
12 58 ☽ # ☉
16 15 ☽ ⊥ ☿
16 42 ☽ Q ♆
20 09 ☽ Q ♀
20 50 ☽ # ☿
21 25 ☽ ± ♄
23 26 ☽ ☌ ♂

19 Saturday
01 59 ☽ ∥ ♂
02 08 ☽ ⊻ ☿
03 37 ☽ ∥ ♄
04 32 ☽ ⊥ ☉
07 53 ☽ ⊼ ♄
09 22 ♀ ± ♄
11 31 ☽ ☌ ♇
13 23 ☽ ☌ ♃
13 47 ☽ ⚹ ♆
14 19 ☽ ∥ ♃
15 46 ☽ ⊻ ☉
19 06 ☽ ⚹ ♀

20 Sunday
00 19 ☽ ☌ ♅
00 30 ☽ ∥ ♅
01 25 ☽ # ♅
01 33 ☿ ⊻ ♂
11 14 ☽ # ♃
15 23 ☽ ∠ ♆
21 36 ☽ # ♄
21 47 ☽ # ♂
22 06 ♃ ⚹ ♆
22 51 ☽ ∠ ♀
23 30 ☽ ∥ ☿

21 Monday
02 29 ☽ ☌ ☿
03 43 ♂ ∥ ♄
04 55 ☽ ⊻ ♂
06 25 ☽ ⊥ ♆
10 20 ☽ ☍ ♄
14 10 ☽ ⊻ ♇
15 05 ☽ ⊥ ♀
15 15 ☽ ⊥ ♂
15 31 ☽ ∥ ☉
16 19 ☽ ⊻ ♆
16 31 ☽ ⊻ ♃
18 35 ♀ Q ♂
21 44 ☽ ☌ ☉

22 Tuesday
00 01 ☽ ⊥ ♇
01 48 ☽ ⊻ ♀
02 25 ☽ ⊻ ♅
02 27 ☽ ⊥ ♃
03 29 ☿ # ♄
05 36 ☿ ∠ ♀
06 35 ☽ ∠ ♂
09 43 ♀ ⚹ ♅
11 24 ☽ # ♇
12 12 ☽ ⊥ ♅
14 39 ☽ ∠ ♇
16 51 ☽ ∥ ♆
17 15 ☽ ∠ ♃

23 Wednesday
01 53 ☽ ⊻ ☿
02 48 ☽ ∠ ♅
07 52 ☽ ⚹ ♂
08 18 ☉ ⊥ ♇
10 47 ☽ ⊼ ♄
11 23 ☽ ⊥ ☿
11 53 ☽ ∥ ♀
14 53 ☽ ⚹ ♇
17 00 ☽ ☌ ♆
17 44 ☽ ⚹ ♃
20 23 ☽ ± ♄
21 50 ☿ # ♂

24 Thursday
01 41 ☽ ⊻ ☉
01 43 ☽ ∠ ☿
03 01 ☽ ⚹ ♅
04 00 ☽ Q ♂
06 32 ☽ ☌ ♀
10 14 ☽ Q ♇
10 48 ☽ ⚼ ♄
12 03 ☽ ⊥ ☉
13 19 ☽ Q ♃
13 31 ☿ # ♂
14 17 ☿ St D
22 28 ☽ Q ♅
22 49 ☉ ⊻ ♅

25 Friday
01 56 ☽ ⚹ ☿
03 39 ☽ ∠ ☉
10 23 ☽ □ ♂
10 58 ☽ △ ♄
11 18 ☉ ⊥ ♃
15 29 ☽ □ ♇
17 39 ☽ ⊻ ♆
18 56 ☽ □ ♃
21 43 ☽ Q ☿
22 57 ♂ ⊼ ♄

26 Saturday
03 32 ☽ ⊥ ♆
03 58 ☽ □ ♅
04 55 ♀ Q ♇
05 59 ☽ ⚹ ☉
07 46 ♀ ⚼ ♄
11 51 ☽ ⊻ ♀
18 30 ☽ ∠ ♆
22 47 ☽ ⊥ ♀

27 Sunday
03 29 ☽ Q ☉
04 23 ☽ □ ☿
12 32 ☽ □ ♄
14 27 ☽ △ ♂
15 25 ☽ ∠ ♀
17 38 ☽ △ ♇
19 54 ☽ ⚹ ♆
21 50 ☽ △ ♃

28 Monday
06 47 ☽ △ ♅
07 45 ♀ Q ♃
12 40 ☽ □ ☉
15 32 ☽ ∥ ♀
16 43 ☽ Q ♆
17 32 ☽ ⚼ ♂
19 26 ☽ Q ♄
19 39 ☽ ⚼ ♇
19 51 ☽ ⚹ ♀

29 Tuesday
00 16 ☽ ⚼ ♃
09 13 ☽ ⚼ ♅
10 11 ☽ ± ♂
10 49 ☽ △ ☿
11 36 ☽ ± ♇
16 30 ☽ ± ♃
16 35 ☽ ⚹ ♄
19 16 ☽ Q ♀
19 48 ☽ ∥ ♆
21 26 ☽ ⊼ ♂
22 23 ☽ ⊼ ♇

30 Wednesday
00 48 ☽ □ ♆
01 25 ☽ ± ♅
02 24 ☽ # ♇
03 28 ☽ ⊼ ♃
11 53 ☽ ∥ ☉
12 22 ☽ ⊼ ♅
15 48 ☽ ⚼ ☿
18 46 ♂ ☌ ♇
19 40 ☽ ∠ ♄
22 48 ☽ △ ☉

31 Thursday
03 39 ☿ # ♄
07 29 ☉ ⊥ ♅
07 32 ☽ □ ♀
09 46 ☽ ± ☿
11 40 ☿ ⊥ ♆
12 14 ☽ ⊥ ♄
20 54 ☽ ∥ ☿
21 58 ☽ # ♄
21 59 ☽ ⊼ ☿
23 25 ☽ ⊻ ♄

NOVEMBER 1968

LONGITUDES

Date	Sidereal time h m s	Sun ☉ ° ' "	Moon ☽ ° ' "	Moon ☽ 24.00 ° ' "	Mercury ☿ ° '	Venus ♀ ° '	Mars ♂ ° '	Jupiter ♃ ° '	Saturn ♄ ° '	Uranus ♅ ° '	Neptune ♆ ° '	Pluto ♇ ° '
01	14 43 21	09 ♏ 08 02	27 ♓ 27 15	03 ♈ 44 48	20 ♎ 41	13 ♐ 45	25 ♍ 16	27 ♍ 27	20 ♈ 46	02 ♎ 01	25 ♏ 38	24 ♍ 16
02	14 47 17	10 08 05	09 ♈ 59 24	16 ♈ 11 19	21 50	14 58	25 53	27 38	20 R 41	02 04	25 40	24 17
03	14 51 14	11 08 10	22 ♈ 20 45	28 ♈ 27 56	23 05	16 10	26 29	27 49	20 37	02 07	25 42	24 19
04	14 55 10	12 08 17	04 ♉ 33 04	10 ♉ 36 20	24 23	17 23	27 06	28 00	20 33	02 10	25 44	24 21
05	14 59 07	13 08 26	16 ♉ 37 54	22 ♉ 37 57	25 44	18 36	27 43	28 11	20 28	02 13	25 46	24 22
06	15 03 04	14 08 36	28 ♉ 36 41	04 ♊ 34 16	27 09	19 49	28 19	28 22	20 24	02 16	25 49	24 24
07	15 07 00	15 08 49	10 ♊ 30 57	16 ♊ 26 58	28 ♎ 35	21 02	28 56	28 33	20 20	02 19	25 51	24 26
08	15 10 57	16 09 03	22 ♊ 22 34	28 ♊ 18 03	00 ♏ 04	22 15	29 ♍ 32	28 43	20 16	02 22	25 53	24 27
09	15 14 53	17 09 19	04 ♋ 13 46	10 ♋ 10 05	01 34	23 27	00 ♎ 09	28 54	20 12	02 25	25 55	24 29
10	15 18 50	18 09 37	16 ♋ 07 26	22 ♋ 06 14	03 05	24 40	00 45	29 04	20 08	02 28	25 57	24 30
11	15 22 46	19 09 57	28 ♋ 07 00	04 ♌ 10 15	04 38	25 53	01 22	29 15	20 04	02 31	26 00	24 32
12	15 26 43	20 10 19	10 ♌ 16 31	16 ♌ 26 22	06 11	27 05	01 58	29 25	20 00	02 34	26 02	24 33
13	15 30 39	21 10 43	22 ♌ 40 24	28 ♌ 59 11	07 45	28 18	02 35	29 35	19 57	02 37	26 04	24 34
14	15 34 36	22 11 09	05 ♍ 23 17	11 ♍ 53 14	09 20	29 ♐ 30	03 11	29 45	19 53	02 40	26 06	24 36
15	15 38 33	23 11 37	18 ♍ 29 29	25 ♍ 12 28	10 54	00 ♑ 43	03 47	29 ♍ 56	19 50	02 42	26 09	24 37
16	15 42 29	24 12 06	02 ♎ 02 26	08 ♎ 59 35	12 30	01 55	04 24	00 ♎ 05	19 46	02 45	26 11	24 38
17	15 46 26	25 12 37	16 ♎ 03 55	23 ♎ 15 14	14 05	03 08	05 00	00 15	19 43	02 48	26 13	24 40
18	15 50 22	26 13 10	00 ♏ 33 09	07 ♏ 57 04	15 40	04 20	05 36	00 25	19 39	02 50	26 15	24 41
19	15 54 19	27 13 45	15 ♏ 26 10	22 ♏ 59 26	17 16	05 33	06 12	00 35	19 36	02 53	26 18	24 42
20	15 58 15	28 14 22	00 ♐ 35 41	08 ♐ 13 34	18 51	06 45	06 49	00 44	19 33	02 56	26 20	24 44
21	16 02 12	29 ♏ 15 00	15 ♐ 51 43	23 ♐ 28 44	20 26	07 57	07 25	00 54	19 30	02 58	26 22	24 45
22	16 06 08	00 ♐ 15 39	01 ♑ 03 16	08 ♑ 34 05	22 02	09 10	08 01	01 03	19 27	03 01	26 24	24 46
23	16 10 05	01 16 20	16 ♑ 00 09	23 ♑ 20 35	23 37	10 22	08 37	01 12	19 24	03 03	26 27	24 47
24	16 14 02	02 17 01	00 ♒ 34 46	07 ♒ 42 16	25 12	11 34	09 13	01 22	19 21	03 05	26 29	24 48
25	16 17 58	03 17 44	14 ♒ 42 53	21 ♒ 36 35	26 47	12 46	09 49	01 31	19 18	03 08	26 31	24 49
26	16 21 55	04 18 28	28 ♒ 23 32	05 ♓ 03 59	28 22	13 58	10 25	01 39	19 15	03 10	26 33	24 50
27	16 25 51	05 19 12	11 ♓ 38 19	18 ♓ 07 00	29 ♏ 57	15 10	11 01	01 48	19 13	03 12	26 36	24 51
28	16 29 48	06 19 58	24 ♓ 30 30	00 ♈ 49 22	01 ♐ 32	16 22	11 37	01 57	19 10	03 14	26 38	24 52
29	16 33 44	07 20 45	07 ♈ 04 06	13 ♈ 15 15	03 06	17 34	12 12	02 05	19 08	03 17	26 40	24 53
30	16 37 41	08 ♐ 21 32	19 ♈ 23 16	25 ♈ 28 39	04 ♐ 41	18 ♑ 46	12 ♎ 48	02 ♎ 14	19 ♈ 06	03 ♎ 19	26 ♏ 42	24 ♍ 54

Date	Moon True ☊ ° '	Moon Mean ☊ ° '	Moon ☽ Latitude ° '
01	09 ♈ 28	07 ♈ 49	01 S 05
02	09 R 28	07 46	00 N 03
03	09 28	07 43	01 10
04	09 25	07 40	02 13
05	09 21	07 36	03 09
06	09 15	07 33	03 56
07	09 09	07 30	04 32
08	09 02	07 27	04 57
09	08 55	07 24	05 09
10	08 49	07 20	05 07
11	08 45	07 17	04 52
12	08 42	07 14	04 24
13	08 D 42	07 11	03 43
14	08 43	07 08	02 51
15	08 44	07 05	01 47
16	08 45	07 01	00 N 36
17	08 R 45	06 58	00 S 39
18	08 43	06 55	01 55
19	08 39	06 52	03 04
20	08 33	06 49	04 02
21	08 25	06 46	04 43
22	08 17	06 42	05 04
23	08 09	06 39	05 03
24	08 03	06 36	04 43
25	07 59	06 33	04 06
26	07 57	06 30	03 16
27	07 D 57	06 26	02 16
28	07 58	06 23	01 11
29	07 59	06 20	00 05
30	07 ♈ 58	06 ♈ 17	01 N 01

DECLINATIONS

Date	Sun ☉ ° '	Moon ☽ ° '	Mercury ☿ ° '	Venus ♀ ° '	Mars ♂ ° '	Jupiter ♃ ° '	Saturn ♄ ° '	Uranus ♅ ° '	Neptune ♆ ° '	Pluto ♇ ° '
01	14 S 33	02 S 01	06 S 04	23 S 53	03 N 07	02 N 01	05 N 32	00 S 09	17 S 33	16 N 04
02	14 52	04 N 00	06 29	24 04	02 53	01 57	05 31	00 10	17 34	16 03
03	15 10	09 47	06 57	24 15	02 38	01 52	05 29	00 11	17 34	16 03
04	15 29	15 07	07 26	24 24	02 24	01 48	05 28	00 13	17 35	16 03
05	15 47	19 49	07 57	24 33	02 10	01 44	05 26	00 14	17 36	16 02
06	16 05	23 41	08 30	24 42	01 55	01 40	05 25	00 15	17 36	16 02
07	16 23	26 31	09 04	24 49	01 41	01 36	05 24	00 16	17 37	16 02
08	16 40	28 10	09 39	24 56	01 26	01 31	05 22	00 17	17 37	16 02
09	16 58	28 31	10 14	25 02	01 12	01 27	05 21	00 18	17 38	16 01
10	17 15	27 33	10 50	25 08	00 57	01 23	05 19	00 20	17 38	16 01
11	17 31	25 19	11 26	25 12	00 43	01 19	05 18	00 21	17 39	16 01
12	17 48	21 54	12 02	25 16	00 29	01 15	05 17	00 22	17 39	16 01
13	18 04	17 29	12 38	25 19	00 N 14	01 12	05 16	00 23	17 40	16 01
14	18 19	12 11	13 14	25 22	00 00	01 08	05 14	00 24	17 40	16 01
15	18 35	06 N 12	13 49	25 24	00 S 15	01 04	05 13	00 25	17 41	16 00
16	18 50	00 S 15	14 24	25 25	00 29	01 00	05 12	00 26	17 41	16 00
17	19 04	06 56	14 59	25 25	00 43	00 56	05 11	00 27	17 42	16 00
18	19 19	13 28	15 33	25 24	00 57	00 53	05 10	00 28	17 42	16 00
19	19 33	19 24	16 07	25 23	01 12	00 49	05 09	00 29	17 43	16 00
20	19 46	24 13	16 40	25 21	01 26	00 45	05 08	00 30	17 43	16 00
21	20 00	27 23	17 12	25 19	01 40	00 42	05 07	00 31	17 44	16 00
22	20 13	28 30	17 44	25 15	01 54	00 38	05 06	00 32	17 44	16 00
23	20 25	27 30	18 15	25 11	02 08	00 35	05 05	00 33	17 45	16 00
24	20 37	24 38	18 45	25 06	02 22	00 31	05 04	00 34	17 46	16 00
25	20 49	20 20	19 14	25 00	02 37	00 28	05 03	00 35	17 46	16 00
26	21 01	15 05	19 42	24 54	02 51	00 24	05 03	00 36	17 47	16 00
27	21 12	09 18	20 09	24 47	03 05	00 21	05 02	00 37	17 47	16 00
28	21 22	03 S 17	20 35	24 39	03 18	00 18	05 01	00 37	17 48	16 00
29	21 33	02 N 44	21 01	24 31	03 32	00 15	05 00	00 38	17 48	16 00
30	21 S 42	08 N 31	21 S 25	24 S 21	03 S 46	00 N 11	05 N 00	00 S 39	17 S 49	16 N 00

ZODIAC SIGN ENTRIES

Date	h	m	Planets
01	16	51	☽ ♈
04	03	01	☽ ♉
06	14	48	☽ ♊
08	11	00	☿ ♏
09	03	26	☽ ♋
09	06	10	♂ ♎
11	15	45	☽ ♌
14	01	55	☽ ♍
14	21	48	♀ ♑
15	22	44	♃ ♎
16	08	26	☽ ♎
18	11	06	☽ ♏
20	11	04	☽ ♐
22	05	49	☉ ♐
22	10	19	☽ ♑
24	11	02	☽ ♒
26	14	52	☽ ♓
27	12	47	☿ ♐
28	22	26	☽ ♈

LATITUDES

Date	Mercury ☿ ° '	Venus ♀ ° '	Mars ♂ ° '	Jupiter ♃ ° '	Saturn ♄ ° '	Uranus ♅ ° '	Neptune ♆ ° '	Pluto ♇ ° '
01	02 N 10	01 S 26	01 N 21	01 N 05	02 S 46	00 N 43	01 N 40	15 N 03
04	02 10	01 34	01 22	01 06	02 46	00 43	01 40	15 04
07	02 02	01 41	01 22	01 06	02 45	00 43	01 40	15 05
10	01 50	01 47	01 22	01 07	02 45	00 43	01 40	15 07
13	01 33	01 53	01 23	01 07	02 44	00 43	01 40	15 08
16	01 15	01 59	01 23	01 08	02 44	00 43	01 40	15 09
19	00 55	02 04	01 23	01 08	02 43	00 43	01 40	15 11
22	00 34	02 08	01 23	01 09	02 43	00 43	01 40	15 12
25	00 N 13	02 11	01 24	01 10	02 42	00 43	01 40	15 14
28	00 S 07	02 14	01 24	01 10	02 41	00 43	01 40	15 15
31	00 S 27	02 S 15	01 N 24	01 N 11	02 S 41	00 N 44	01 N 40	15 N 17

DATA

Julian Date	2440162
Delta T	+39 seconds
Ayanamsa	23° 25' 15"
Synetic vernal point	05° ♓ 41' 45"
True obliquity of ecliptic	23° 26' 45"

MOON'S PHASES, APSIDES AND POSITIONS ☽

Date	h	m	Phase	Longitude ° '	Eclipse Indicator
05	04	25	○	12 ♉ 49	
13	08	53	☾	21 ♌ 03	
20	08	02	●	28 ♏ 04	
26	23	30	☽	04 ♓ 48	

Day	h	m		
08	08	45	Apogee	
20	23	35	Perigee	
01	19	59	ON	
09	06	26	Max dec	28° N 33'
16	11	04	0S	
22	12	29	Max dec	28° S 30'
29	01	02	ON	

LONGITUDES

Date	Chiron ⚷ ° '	Ceres ⚳ ° '	Pallas ⚴ ° '	Juno ⚵ ° '	Vesta ⚶ ° '	Black Moon Lilith ⚸ ° '
01	29 ♓ 31	03 ♐ 58	14 ♏ 11	27 ♏ 00	24 ♈ 26	15 ♊ 10
11	29 ♓ 13	08 ♐ 02	18 ♏ 36	00 ♐ 23	22 ♈ 20	16 ♊ 17
21	29 ♓ 01	12 ♐ 08	23 ♏ 00	03 ♐ 49	20 ♈ 51	17 ♊ 24
31	28 ♓ 53	16 ♐ 15	27 ♏ 22	07 ♐ 15	20 ♈ 07	18 ♊ 31

All ephemeris data is given at 12.00 UT and the Moon's longitude is additionally given for 24.00 UT

ASPECTARIAN

01 Friday
05 10 ☽ ⚼ ☉
05 56 ☽ ☍ ♇
07 25 ☽ # ♂
07 39 ☽ ☍ ♂
08 31 ☽ △ ♆
11 59 ☽ ☍ ♃
12 00 ☽ # ♃
13 40 ☿ ☍ ♄
15 12 ☉ ∠ ♇
17 29 ♀ Q ♅
19 22 ☽ ∥ ♅
20 35 ☽ # ♅
20 43 ☽ ☍ ♅
23 46 ☽ ± ☉

02 Saturday
02 58 ♂ ⚹ ♆
03 49 ☽ ∥ ♃
07 40 ☽ ∥ ♂
12 18 ☽ ⊼ ☉
13 18 ☽ ⚼ ♆
18 08 ☽ ∥ ♄
22 40 ☽ △ ♀
22 59 ☽ # ☿

03 Sunday
06 49 ☽ ± ♆
08 38 ☽ ☌ ♄
13 36 ☽ ☍ ☿
15 52 ☽ ⊼ ♇
18 35 ☽ ⊼ ♆
20 33 ☽ ⊼ ♂
22 53 ☽ ⊼ ♃

04 Monday
03 41 ☽ ± ♇
07 15 ☽ ⚼ ♀
07 16 ☽ ⊼ ♅
08 59 ☽ ± ♂
09 43 ☉ ∠ ♂
10 54 ☽ ± ♃
11 24 ☿ ⊻ ♇
13 50 ☽ # ☉
16 26 ☽ ∥ ♇
19 11 ☽ ± ♅
21 31 ☽ ⚼ ♇

05 Tuesday
00 07 ☽ # ♆
03 04 ☽ ± ♀
03 46 ☽ ⚼ ♂
04 25 ☽ ☍ ☉
05 01 ☽ ⚼ ♃
12 38 ☿ ⊻ ♆
13 11 ☽ ⚼ ♅
13 15 ☉ ∠ ♃
16 23 ☽ ⊼ ♀
19 38 ☽ ⊻ ♄

06 Wednesday
03 31 ☽ △ ♇
06 21 ☽ ☍ ♆
07 35 ☽ ⊥ ♄
07 52 ☉ # ♇
08 39 ☽ ⊼ ☿
11 23 ☽ △ ♂
11 30 ☽ △ ♃
14 23 ♂ ☌ ♃
19 24 ☽ △ ♅
19 51 ☽ # ♀
22 22 ☽ ± ☿
23 00 ♀ △ ♄

07 Thursday
01 36 ☽ ∠ ♄
11 12 ☿ ⊻ ♃
19 05 ☽ ⚼ ☿
21 38 ☿ ⊻ ♂
22 14 ☽ ⊼ ☉
23 44 ♂ ∥ ♃

08 Friday
01 28 ☉ ⊥ ♀
07 45 ☽ ⚹ ♄
11 30 ☽ ± ☉
11 42 ☽ ☍ ♀
16 13 ☽ □ ♇
18 21 ☿ ⊥ ♇
19 08 ☽ ⊼ ♆

09 Saturday
01 03 ☽ □ ♃
03 18 ☽ □ ♂
05 49 ☽ △ ☿
07 19 ☽ ± ♆
07 25 ☽ ⚼ ☉
07 55 ☽ Q ♄
08 19 ☽ □ ♅
18 40 ☉ ∠ ♅

10 Sunday
01 34 ☽ ⚼ ♆
01 57 ☿ ⊻ ♅
04 41 ☽ Q ♇
08 40 ♀ □ ♇
13 56 ☽ Q ♃
16 28 ☽ △ ☉
17 34 ☽ Q ♂
20 01 ☽ □ ♄
20 46 ☽ Q ♅

11 Monday
04 50 ☽ ⚹ ♇
07 02 ☽ ⊼ ♀
07 46 ☽ △ ♆
12 53 ☽ # ♀
14 17 ☽ ⚹ ♃
14 25 ♀ ⊻ ♆
18 47 ☽ ⚹ ♂
20 18 ☽ ± ♀
20 46 ☽ ⚹ ♅
22 42 ☿ ⊥ ♃
23 22 ☉ ∥ ♆

12 Tuesday
02 48 ☽ □ ☿
06 11 ♄ ± ♆
08 19 ☉ ⊼ ♄
10 35 ☽ ∠ ♇
15 56 ☽ ⚼ ♀
20 12 ☽ ∠ ♃
22 26 ♂ # ♅

13 Wednesday
01 42 ☽ ∠ ♂
02 14 ☽ ∠ ♅
04 06 ☽ ⊥ ♇
06 48 ☽ △ ♄
08 53 ☽ □ ☉
09 16 ☽ # ☉
11 05 ☽ # ♆
13 31 ♂ ☌ ♅
13 46 ☽ ⊥ ♃
15 38 ☽ ⊻ ♇
18 30 ☽ □ ♆
18 43 ☽ Q ☿
19 00 ☽ ∥ ♇
19 32 ☽ ⊥ ♅
19 49 ☽ ⊥ ♂
23 49 ☽ △ ♀

14 Thursday
01 19 ☽ ⊻ ♃
01 31 ☿ ⊥ ♅
06 53 ☽ ⊻ ♅
07 41 ☽ ⊻ ♂
07 57 ☽ # ☿
08 27 ☿ ⊥ ♂
11 04 ☽ ⚼ ♄
16 09 ☿ ∠ ♇
17 48 ♀ □ ♃
20 18 ☽ ⚹ ☿
21 37 ☽ Q ☉

15 Friday
03 35 ☽ ± ♄
04 06 ☽ Q ♆
14 23 ☽ ⊼ ♄
15 44 ☽ ∥ ♄
21 06 ☽ ⚹ ☉
22 58 ☽ ☌ ♇

16 Saturday
01 42 ☽ ⚹ ♆
02 59 ☽ ∠ ☿
07 08 ♂ ∥ ♅
07 22 ☽ ∥ ♃
08 33 ☽ ☌ ♃
09 25 ☽ # ♂
09 29 ☽ # ♅
11 47 ☽ □ ♀
12 39 ☽ ∥ ♅
12 50 ☽ ∥ ♂
13 14 ☽ ☌ ♅
14 40 ☽ # ♃
16 16 ☽ ☌ ♂
17 19 ♀ ⊥ ♆
20 42 ☽ ⊥ ☿
22 41 ☉ ⚹ ♇

17 Sunday
01 19 ☽ ∠ ☉
03 47 ☽ ∠ ♆
05 04 ♀ □ ♅
05 45 ☽ # ♄
08 14 ☽ ⊻ ☿
17 40 ☽ ⊥ ☉
18 05 ☽ ☍ ♄
18 59 ☽ ⊥ ♆
21 15 ☽ Q ♀
23 14 ☉ ± ♄

18 Monday
02 21 ☽ ⊻ ♇
04 22 ☽ ⊻ ☉
04 57 ☽ ⊻ ♆
05 31 ♂ # ♃
07 46 ☿ ∠ ♃
11 47 ☽ ⊻ ♃
12 13 ☽ ⊥ ♇
12 57 ☉ ☌ ♆
15 44 ☽ ⊻ ♅
18 42 ☽ ⚹ ♀
20 33 ☽ ⊻ ♂
20 56 ☽ ∥ ☿
21 38 ☽ ⊥ ♃
21 52 ☽ # ♇

19 Tuesday
01 28 ☽ ⊥ ♅
02 49 ☽ ∠ ♇
04 48 ☽ ∥ ♆
06 37 ☽ ⊥ ♂
07 00 ☿ # ♇
12 14 ☽ ∠ ♃
12 39 ☽ ∥ ☉
15 15 ☽ ☌ ☿
15 54 ☽ ∠ ♅
18 36 ☽ ⊼ ♄
20 50 ☽ ∠ ♀
21 33 ☽ ∠ ♂
21 38 ☿ ∠ ♅

20 Wednesday
02 44 ☽ ⚹ ♇
04 04 ☽ ± ♄
05 16 ☽ ☌ ♆
08 02 ☽ ☌ ☉
12 14 ☽ ⚹ ♃
12 16 ☽ ⊥ ♀
14 21 ♀ □ ♂
15 41 ☽ ⚹ ♅
18 12 ☽ ⚼ ♄
19 02 ☽ ∥ ♀
21 39 ☽ Q ♇
22 09 ☿ ⊼ ♄
22 10 ☽ ⚹ ♂
22 31 ☽ ⊻ ♀

21 Thursday
07 17 ☽ Q ♃
10 35 ☽ Q ♅
17 42 ☽ △ ♄
17 49 ☽ Q ♂
20 03 ☽ ⊻ ☿

22 Friday
02 01 ☽ □ ♇
04 37 ☽ ⊻ ♆
06 38 ☽ ⊥ ☿
10 39 ☽ ⊻ ☉
12 00 ☽ □ ♃
12 26 ☿ ∥ ♆
14 10 ☽ ⊥ ♆
15 07 ☽ □ ♅
20 54 ☽ ⊥ ☉
22 40 ☽ ∠ ☿
23 34 ☽ □ ♂

23 Saturday
02 05 ☽ ☌ ♀
04 37 ☽ ∠ ♆
10 10 ☉ ⚹ ♃
11 56 ☿ ∠ ♂
12 28 ☽ ∠ ☉
17 30 ☽ □ ♄
20 30 ♃ # ♅

24 Sunday
01 58 ☽ ⚹ ☿
02 24 ☽ △ ♇
05 10 ☽ ⚹ ♆
05 52 ☿ ⚹ ♇
08 47 ☽ ∥ ♀
10 18 ♀ ∠ ♆
13 19 ☽ △ ♃
14 07 ☿ ± ♄
15 04 ☽ ⚹ ☉
16 13 ☽ △ ♅
23 21 ☽ Q ♄

25 Monday
00 33 ☽ Q ☿
01 22 ☽ Q ♆
03 13 ☽ △ ♂
03 35 ☽ ⚼ ♇
07 52 ☉ ⚹ ♅
07 52 ☿ ☌ ♆
08 20 ☽ ⊻ ♀
09 41 ☽ ∥ ☉
13 05 ☽ Q ☉
15 08 ☽ ⚼ ♃
16 55 ☽ ∥ ☿
17 56 ☽ ⚼ ♅
19 08 ☽ ± ♇
19 42 ☽ ⊥ ♀
19 56 ☽ ⚹ ♄

26 Tuesday
00 09 ☽ ∥ ♆
05 41 ☽ ⊼ ♇
06 28 ☽ ⚼ ♂
07 05 ☽ ± ♃
08 03 ☽ # ♇
08 44 ☽ □ ♆
09 49 ☽ ± ♅
10 50 ☉ ⚼ ♄
11 57 ☽ □ ☿
13 08 ☽ ∠ ♀
17 55 ☽ ⊼ ♃
20 35 ☽ ⊼ ♅
22 30 ☽ ∠ ♄
23 20 ☽ ± ♂
23 30 ☽ □ ☉

27 Wednesday
10 48 ☽ ⊼ ♂
14 54 ☽ ⊥ ♄
19 11 ☽ ⚹ ♀

28 Thursday
01 50 ☿ ∠ ♀
02 00 ☽ ⊻ ♄
05 05 ☽ # ♄
11 53 ☽ ∥ ♂
12 41 ☽ ☍ ♇
13 00 ♂ ∠ ♆
16 02 ☽ △ ♆
19 02 ☿ ⚹ ♃
20 05 ☽ Q ♀
22 31 ☽ ∥ ♅
23 57 ☽ # ♃

29 Friday
00 57 ☉ Q ♇
02 05 ☽ ∥ ♃
02 19 ☽ ☍ ♃
03 16 ☽ △ ☿
03 34 ☽ # ♅
04 41 ☽ ☍ ♅
12 35 ☽ △ ☉
14 42 ☿ ⚹ ♅
15 25 ☽ # ♂
20 57 ☽ ⚼ ♆
21 16 ☽ ∥ ♄
22 28 ☽ ☍ ♂

30 Saturday
03 19 ☿ ⚼ ♄
10 38 ☽ □ ♀
11 26 ☽ ☌ ♄
12 39 ☽ ⚼ ☿
14 36 ☽ ± ♆
18 31 ♀ □ ♄
20 32 ☽ ⚼ ☉
22 53 ☽ ⊼ ♇

DECEMBER 1968

LONGITUDES

Date	Sidereal time h m s	Sun ☉ ° ' "	Moon ☽ ° ' "	Moon ☽ 24.00 ° ' "	Mercury ☿ ° '	Venus ♀ ° '	Mars ♂ ° '	Jupiter ♃ ° '	Saturn ♄ ° '	Uranus ♅ ° '	Neptune ♆ ° '	Pluto ♇ ° '
01	16 41 37	09 ♐ 22 21	01 ♉ 31 49	07 ♉ 33 09	06 ♐ 15	19 ♑ 57	13 ♎ 24	02 ♎ 22	19 ♈ 04	03 ♎ 21	26 ♏ 45	24 ♍ 55
02	16 45 34	10 23 10	13 ♉ 32 59	19 ♉ 31 38	07 49	21 09	14 00	02 30	19 R 01	03 23	26 47	24 56
03	16 49 31	11 24 01	25 ♉ 29 21	01 ♊ 26 23	09 24	22 21	14 35	02 38	18 59	03 25	26 49	24 57
04	16 53 27	12 24 53	07 ♊ 22 56	13 ♊ 19 10	10 58	23 32	15 11	02 46	18 58	03 27	26 51	24 57
05	16 57 24	13 25 45	19 ♊ 15 16	25 ♊ 11 23	12 32	24 44	15 46	02 54	18 56	03 29	26 53	24 58
06	17 01 20	14 26 39	01 ♋ 07 41	07 ♋ 04 21	14 07	25 55	16 22	03 02	18 54	03 30	26 56	24 59
07	17 05 17	15 27 34	13 ♋ 01 34	18 ♋ 59 34	15 41	27 06	16 57	03 09	18 53	03 32	26 58	25 00
08	17 09 13	16 28 30	24 ♋ 58 36	00 ♌ 58 57	17 15	28 17	17 33	03 17	18 51	03 34	27 00	25 00
09	17 13 10	17 29 27	07 ♌ 00 59	13 ♌ 05 02	18 49	29 ♑ 28	18 08	03 24	18 50	03 36	27 02	25 01
10	17 17 06	18 30 26	19 ♌ 11 33	25 ♌ 21 00	20 24	00 ♒ 39	18 44	03 31	18 48	03 37	27 04	25 02
11	17 21 03	19 31 25	01 ♍ 33 52	07 ♍ 50 40	21 58	01 50	19 19	03 38	18 47	03 39	27 06	25 02
12	17 25 00	20 32 25	14 ♍ 11 59	20 ♍ 38 19	23 32	03 01	19 54	03 45	18 46	03 40	27 08	25 03
13	17 28 56	21 33 26	27 ♍ 10 15	03 ♎ 48 16	25 07	04 12	20 30	03 52	18 45	03 42	27 11	25 03
14	17 32 53	22 34 29	10 ♎ 32 49	17 ♎ 24 14	26 42	05 23	21 05	03 58	18 44	03 43	27 13	25 04
15	17 36 49	23 35 32	24 ♎ 22 48	01 ♏ 28 33	28 16	06 33	21 40	04 05	18 44	03 44	27 15	25 04
16	17 40 46	24 36 36	08 ♏ 41 25	16 ♏ 01 03	29 ♐ 51	07 44	22 15	04 11	18 43	03 46	27 17	25 04
17	17 44 42	25 37 42	23 ♏ 26 56	00 ♐ 58 15	01 ♑ 26	08 54	22 50	04 17	18 43	03 47	27 19	25 05
18	17 48 39	26 38 48	08 ♐ 33 58	16 ♐ 12 52	03 02	10 04	23 25	04 23	18 42	03 48	27 21	25 05
19	17 52 35	27 39 55	23 ♐ 53 33	01 ♑ 34 31	04 37	11 14	24 00	04 29	18 42	03 49	27 23	25 05
20	17 56 32	28 41 03	09 ♑ 14 14	16 ♑ 51 14	06 12	12 24	24 35	04 34	18 42	03 50	27 25	25 06
21	18 00 29	29 ♐ 42 11	24 ♑ 24 09	01 ♒ 51 48	07 48	13 34	25 09	04 40	18 D 42	03 52	27 27	25 06
22	18 04 25	00 ♑ 43 19	09 ♒ 13 14	16 ♒ 27 43	09 24	14 44	25 44	04 45	18 42	03 53	27 29	25 06
23	18 08 22	01 44 27	23 ♒ 34 49	00 ♓ 34 16	10 59	15 54	26 19	04 50	18 42	03 53	27 31	25 06
24	18 12 18	02 45 36	07 ♓ 26 05	14 ♓ 10 26	12 35	17 03	26 53	04 55	18 42	03 54	27 33	25 06
25	18 16 15	03 46 44	20 ♓ 47 38	27 ♓ 18 09	14 11	18 13	27 28	05 00	18 43	03 55	27 35	25 07
26	18 20 11	04 47 53	03 ♈ 42 30	10 ♈ 01 18	15 47	19 22	28 02	05 05	18 43	03 56	27 37	25 07
27	18 24 08	05 49 01	16 ♈ 15 09	22 ♈ 24 44	17 23	20 31	28 37	05 09	18 44	03 57	27 38	25 07
28	18 28 04	06 50 10	28 ♈ 30 38	04 ♉ 33 30	18 59	21 40	29 11	05 14	18 44	03 57	27 40	25 R 07
29	18 32 01	07 51 18	10 ♉ 33 55	16 ♉ 32 25	20 35	22 49	29 ♎ 46	05 18	18 45	03 58	27 42	25 07
30	18 35 58	08 52 27	22 ♉ 29 32	28 ♉ 25 41	22 10	23 57	00 ♏ 20	05 22	18 46	03 58	27 44	25 06
31	18 39 54	09 ♑ 53 35	04 ♊ 21 17	10 ♊ 16 42	23 ♑ 45	25 ♒ 05	00 ♏ 54	05 ♎ 26	18 ♈ 47	03 ♎ 59	27 ♏ 46	25 ♍ 06

Date	Moon True ☊ ° '	Moon Mean ☊ ° '	Moon ☽ Latitude ° '
01	07 ♈ 55	06 ♈ 14	02 N 02
02	07 R 50	06 11	02 58
03	07 41	06 07	03 45
04	07 30	06 04	04 21
05	07 18	06 01	04 47
06	07 04	05 58	05 00
07	06 51	05 55	05 00
08	06 39	05 51	04 46
09	06 30	05 48	04 20
10	06 24	05 45	03 42
11	06 20	05 42	02 53
12	06 18	05 39	01 54
13	06 D 18	05 36	00 N 48
14	06 R 18	05 32	00 S 22
15	06 17	05 29	01 34
16	06 14	05 26	02 42
17	06 08	05 23	03 42
18	05 59	05 20	04 27
19	05 48	05 17	04 54
20	05 36	05 13	04 59
21	05 25	05 10	04 44
22	05 16	05 07	04 10
23	05 09	05 04	03 20
24	05 05	05 01	02 20
25	05 03	04 57	01 15
26	05 D 03	04 54	00 S 07
27	05 R 03	04 51	00 N 59
28	05 02	04 48	02 00
29	04 58	04 45	02 56
30	04 52	04 42	03 42
31	04 ♈ 42	04 ♈ 38	04 N 19

DECLINATIONS

Date	Sun ☉ ° '	Moon ☽ ° '	Mercury ☿ ° '	Venus ♀ ° '	Mars ♂ ° '	Jupiter ♃ ° '	Saturn ♄ ° '	Uranus ♅ ° '	Neptune ♆ ° '	Pluto ♇ ° '
01	21 S 52	13 N 55	21 S 48	24 S 11	04 S 00	00 N 08	04 N 59	00 S 40	17 S 49	16 N 01
02	22 01	18 44	22 10	24 01	04 14	00 05	04 59	00 41	17 50	16 01
03	22 09	22 46	22 31	23 50	04 28	00 N 02	04 58	00 41	17 50	16 01
04	22 17	25 51	22 51	23 38	04 41	00 S 01	04 58	00 42	17 51	16 01
05	22 25	27 46	23 10	23 25	04 55	00 04	04 57	00 43	17 51	16 01
06	22 32	28 26	23 28	23 12	05 09	00 06	04 57	00 44	17 52	16 01
07	22 39	27 47	23 44	22 58	05 22	00 09	04 56	00 44	17 52	16 02
08	22 45	25 50	23 59	22 44	05 36	00 12	04 56	00 45	17 52	16 02
09	22 51	22 43	24 13	22 28	05 49	00 15	04 56	00 46	17 53	16 02
10	22 57	18 35	24 26	22 13	06 02	00 17	04 56	00 46	17 53	16 02
11	23 02	13 37	24 38	21 56	06 16	00 20	04 55	00 47	17 54	16 02
12	23 06	07 59	24 48	21 39	06 29	00 22	04 55	00 47	17 54	16 03
13	23 11	01 N 52	24 56	21 22	06 42	00 25	04 55	00 48	17 55	16 03
14	23 14	04 S 31	25 04	21 04	06 55	00 27	04 55	00 48	17 55	16 03
15	23 17	10 55	25 10	20 45	07 09	00 29	04 55	00 49	17 56	16 04
16	23 20	16 58	25 15	20 26	07 22	00 32	04 55	00 49	17 56	16 04
17	23 22	22 13	25 18	20 07	07 35	00 34	04 55	00 50	17 57	16 04
18	23 24	26 08	25 20	19 46	07 48	00 36	04 55	00 50	17 57	16 05
19	23 26	28 12	25 20	19 26	08 00	00 38	04 55	00 51	17 57	16 05
20	23 26	28 06	25 19	19 05	08 13	00 40	04 56	00 51	17 58	16 06
21	23 27	25 54	25 16	18 43	08 26	00 42	04 56	00 51	17 58	16 06
22	23 27	21 58	25 12	18 21	08 38	00 44	04 56	00 52	17 59	16 06
23	23 26	16 49	25 07	17 58	08 51	00 46	04 56	00 52	17 59	16 07
24	23 25	10 57	25 00	17 35	09 04	00 47	04 57	00 52	18 00	16 07
25	23 24	04 S 48	24 51	17 12	09 16	00 49	04 57	00 53	18 00	16 08
26	23 22	01 N 22	24 41	16 48	09 28	00 51	04 58	00 53	18 00	16 08
27	23 19	07 18	24 29	16 24	09 41	00 52	04 58	00 53	18 01	16 09
28	23 16	12 49	24 16	15 59	09 53	00 54	04 59	00 53	18 01	16 09
29	23 13	17 46	24 01	15 34	10 05	00 55	04 59	00 54	18 01	16 10
30	23 09	21 59	23 44	15 09	10 17	00 56	05 00	00 54	18 02	16 10
31	23 S 05	25 N 16	23 S 26	14 S 43	10 S 29	00 S 58	05 N 01	00 S 54	18 S 02	16 N 11

ZODIAC SIGN ENTRIES

Date	h m	Planets
01	08 58	☽ ♉
03	21 06	☽ ♊
06	09 43	☽ ♋
08	22 02	☽ ♌
09	22 40	♀ ♒
11	08 59	☽ ♍
13	17 08	☽ ♎
15	21 31	☽ ♏
16	14 11	☿ ♑
17	22 28	☽ ♐
19	21 32	☽ ♑
21	19 00	☉ ♑
21	20 59	☽ ♒
23	23 01	☽ ♓
26	05 02	☽ ♈
28	14 57	☽ ♉
29	22 07	♂ ♏
31	03 11	☽ ♊

LATITUDES

Date	Mercury ☿ ° '	Venus ♀ ° '	Mars ♂ ° '	Jupiter ♃ ° '	Saturn ♄ ° '	Uranus ♅ ° '	Neptune ♆ ° '	Pluto ♇ ° '
01	00 S 27	02 S 15	01 N 24	01 N 11	02 S 41	00 N 44	01 N 40	15 N 17
04	00 46	02 16	01 24	01 11	02 40	00 44	01 40	15 18
07	01 04	02 16	01 24	01 12	02 39	00 44	01 40	15 20
10	01 20	02 15	01 24	01 13	02 38	00 44	01 40	15 22
13	01 35	02 13	01 24	01 13	02 38	00 44	01 40	15 23
16	01 48	02 10	01 24	01 14	02 37	00 44	01 40	15 25
19	01 58	02 06	01 24	01 15	02 36	00 44	01 40	15 26
22	02 06	02 01	01 24	01 16	02 35	00 44	01 40	15 28
25	02 10	01 55	01 24	01 16	02 34	00 44	01 40	15 30
28	02 11	01 48	01 24	01 17	02 33	00 45	01 40	15 31
31	02 S 07	01 S 39	01 N 24	01 N 18	02 S 33	00 N 45	01 N 40	15 N 33

DATA

Julian Date	2440192
Delta T	+39 seconds
Ayanamsa	23° 25' 19"
Synetic vernal point	05° ♓ 41' 40"
True obliquity of ecliptic	23° 26' 45"

MOON'S PHASES, APSIDES AND POSITIONS ☽

Date	h m	Phase	Longitude ° '	Eclipse Indicator
04	23 07	○	12 ♊ 53	
13	00 49	☾	21 ♍ 05	
19	18 19	●	27 ♐ 56	
26	14 14	☽	04 ♈ 54	

Day	h m		
05	14 30	Apogee	
19	12 16	Perigee	
06	12 02	Max dec	28° N 26'
13	19 04	0S	
19	22 59	Max dec	28° S 26'
26	06 38	0N	

LONGITUDES

Date	Chiron ⚷ ° '	Ceres ⚳ ° '	Pallas ⚴ ° '	Juno ⚵ ° '	Vesta ⚶ ° '	Black Moon Lilith ⚸ ° '
01	28 ♓ 53	16 ♐ 15	27 ♏ 22	07 ♐ 15	20 ♈ 07	18 ♊ 31
11	28 ♓ 51	20 ♐ 23	01 ♐ 43	10 ♐ 43	20 ♈ 07	19 ♊ 38
21	28 ♓ 55	24 ♐ 29	06 ♐ 00	14 ♐ 09	20 ♈ 50	20 ♊ 46
31	29 ♓ 04	28 ♐ 35	10 ♐ 12	17 ♐ 34	22 ♈ 11	21 ♊ 53

ASPECTARIAN

h m Aspects

01 Sunday
02 29 ☽ ⊼ ♆
09 05 ☽ ± ☿
10 47 ☽ ± ♇
13 41 ☽ ⊼ ♃
15 37 ☽ ⊼ ♅
16 00 ☽ ± ☉
18 12 ☉ ∥ ☿
22 01 ☽ ∥ ♇
22 14 ☿ Q ♇
22 49 ☽ ⊼ ☿

02 Monday
01 47 ☽ ± ♃
03 38 ☽ ± ♅
04 45 ☽ bQ ♇
05 05 ☽ ⊼ ☉
07 12 ☽ # ♆
12 56 ☽ ⊼ ♂
20 02 ☽ bQ ♃
21 43 ☽ bQ ♅
22 57 ☽ ⊻ ♄

03 Tuesday
01 37 ☽ ± ♂
04 58 ☽ △ ♀
07 49 ☽ # ☉
10 12 ☽ # ☿
10 54 ☽ △ ♇
11 00 ☽ ⊥ ♄
14 41 ☽ ☍ ♆
18 58 ☽ # ♀
20 42 ☽ bQ ♂

04 Wednesday
02 35 ☽ △ ♃
04 02 ☽ △ ♅
05 06 ☽ ∠ ♄
14 35 ☽ bQ ♀
20 21 ☽ ☍ ☿
23 07 ☽ ☍ ☉

05 Thursday
04 36 ☽ △ ♂
10 49 ☽ ± ♀
11 21 ☽ ⚹ ♄
15 43 ♂ # ♄
17 00 ♀ △ ♇
23 34 ☽ □ ♇
23 35 ☿ ∥ ♀

06 Friday
00 18 ☽ ⊼ ♀
03 29 ☽ ⊼ ♆
11 33 ☽ Q ♄
15 39 ☽ ± ♆
15 53 ☽ □ ♃
16 49 ☽ □ ♅

07 Saturday
02 31 ☉ ☌ ☿
03 18 ☿ Q ♃
03 47 ☉ Q ♃
09 06 ♀ ⚹ ♆
09 46 ☿ Q ♅
09 51 ☽ bQ ♆
11 56 ☽ Q ♇
13 51 ☉ Q ♅
17 21 ☽ ⊼ ☉
18 09 ☽ ⊼ ☿
20 19 ☽ □ ♂
23 44 ☽ □ ♄

08 Sunday
04 31 ☽ Q ♃
05 09 ☽ Q ♅
06 31 ☽ ± ☉
08 01 ☽ ± ☿
09 46 ☉ ∥ ♀
12 03 ☽ ⚹ ♇
16 03 ☽ △ ♆
19 19 ☿ ⚹ ♂
19 21 ☽ ☍ ♀

09 Monday
02 08 ☽ # ☿
02 11 ☽ bQ ☉
04 42 ☽ bQ ☿
04 44 ☽ ⚹ ♃
05 11 ☽ ⚹ ♅
10 10 ☽ Q ♂
11 06 ☽ # ☉
12 07 ☿ △ ♄
13 42 ☽ # ♀
17 57 ☽ ∠ ♇

10 Tuesday
10 32 ☽ △ ☉
10 40 ☽ ∠ ♃
10 53 ☽ ∠ ♅
11 03 ☽ ⚹ ♂
11 15 ☽ △ ♄
11 40 ☽ ⊥ ♇
14 42 ☽ △ ☿
15 06 ♂ ☍ ♄
15 35 ☽ # ♆
18 57 ☉ △ ♄
23 23 ☽ ⊻ ♇

11 Wednesday
00 25 ☉ ⚹ ♂
00 44 ☽ ∥ ♇
03 23 ☽ □ ♆
04 21 ☽ ⊥ ♃
04 26 ☽ ⊥ ♅
12 35 ☽ ⊼ ♀
14 59 ♃ ☌ ♅
16 00 ☽ ⊻ ♅
16 00 ☽ ⊻ ♃
16 15 ☽ bQ ♄
17 32 ☽ ∠ ♂

12 Thursday
01 14 ☽ ± ♀
09 19 ☽ ± ♄
11 25 ☽ ⊥ ♂
13 46 ☽ Q ♆
17 47 ☽ # ♂
19 52 ☽ bQ ♀
20 31 ☽ ⊼ ♄
23 09 ☽ ⊻ ♂

13 Friday
00 11 ☽ ∥ ♄
00 49 ☽ □ ☉
01 31 ♀ △ ♅
04 22 ♀ △ ♃
07 44 ☽ □ ☿
08 08 ☽ ☌ ♇
11 01 ☿ □ ♇
12 01 ☽ ⚹ ♆
16 03 ☽ # ♅
17 29 ☽ # ♃
20 40 ☽ ∥ ♃
22 05 ☽ ∥ ♅
23 50 ☽ ☌ ♅

14 Saturday
00 12 ☽ ☌ ♃
01 56 ☽ △ ♀
12 03 ☽ Q ☉
13 29 ☽ # ♄
14 56 ☽ ∠ ♆
17 47 ♂ ⊥ ♆
20 02 ☿ ⊻ ♆
20 14 ☽ Q ☿
21 17 ☽ ∥ ♂

15 Sunday
02 18 ☽ ☍ ♄
06 37 ☽ ⊥ ♆
07 09 ☽ ☌ ♂
10 33 ☽ ⚹ ☉
13 10 ☽ ⊻ ♇
15 33 ♀ Q ♄
16 53 ☽ ⊻ ♆
19 26 ☽ ⚹ ☿
23 19 ☽ ⊥ ♇

16 Monday
03 49 ☽ ⊻ ♅
04 28 ☽ ⊻ ♃
08 18 ☽ # ♇
10 16 ☽ □ ♀
13 38 ☽ ∠ ☉
13 46 ☽ ⊥ ♅
14 17 ☽ ∠ ♇
14 28 ☽ ⊥ ♃
16 08 ☽ ∥ ♆
23 01 ☉ □ ♇
23 20 ☽ ∠ ☿

17 Tuesday
02 24 ☽ ∥ ♀
04 22 ☽ ⊼ ♄
04 29 ☽ ∠ ♅
05 15 ☽ ∠ ♃
05 24 ☽ ⊥ ☉
10 58 ☽ ⊻ ♂
14 01 ☽ ± ♄
14 37 ☽ ⚹ ♇
15 34 ☽ ⊥ ☿
15 45 ☽ ⊻ ☉
17 59 ☽ Q ♀
18 12 ☽ ☌ ♆
18 15 ☽ ∥ ☉
20 44 ♀ Q ♆
20 57 ☽ ⊥ ♂

18 Wednesday
02 14 ☽ ⊻ ☿
04 20 ☽ bQ ♄
04 29 ☽ ⚹ ♅
05 21 ☽ ⚹ ♃
06 08 ☽ ∥ ☿
09 40 ☽ Q ♇
11 45 ☽ ∠ ♂
12 19 ♀ bQ ♇
14 34 ☽ ⚹ ♀
16 58 ☿ ⊥ ♆
23 22 ☽ Q ♅
23 54 ☿ □ ♅

19 Thursday
00 20 ☽ Q ♃
03 53 ☽ △ ♄
05 06 ☉ ⊻ ♆
09 48 ☿ □ ♃
12 10 ☽ ⚹ ♂
13 52 ☽ □ ♇
15 58 ☽ ∠ ♀
17 28 ☽ ⊻ ♆
18 19 ☽ ☌ ☉

20 Friday
02 51 ☽ ⊥ ♆
03 32 ☽ □ ♅
04 39 ☽ □ ♃
06 42 ☽ ☌ ☿
07 12 ☽ ⊥ ♀
07 40 ☽ Q ♂
17 00 ☽ ∠ ♆
17 24 ☽ ⊻ ♀
20 48 ☿ Q ♂
23 17 ☿ ⊥ ♀

21 Saturday
02 55 ☽ □ ♄
09 35 ♂ ⊻ ♇
11 39 ♄ St D
13 07 ☽ △ ♇
13 15 ☽ □ ♂
16 37 ☽ ∥ ☿
16 53 ☽ ⚹ ♆
21 08 ☽ ⊻ ☉

22 Sunday
03 15 ☽ △ ♅
03 57 ☽ ∥ ☉
04 39 ☽ △ ♃
07 36 ☽ ⊥ ☉
07 52 ☽ Q ♄
12 19 ☽ ⊻ ☿
12 26 ☽ Q ♆
13 27 ☽ bQ ♇
21 55 ☽ ☌ ♀
23 30 ☽ ⊥ ☿
23 35 ☽ ∠ ☉

23 Monday
03 45 ☽ ⚹ ♄
04 04 ☽ bQ ♅
04 26 ☽ ± ♇
05 38 ☽ bQ ♃
06 33 ☽ ∥ ♀
06 53 ☽ ∥ ♆
11 04 ♀ ∥ ♆
14 36 ☽ ⊼ ♇
14 58 ☽ # ♇
16 39 ☽ ∠ ☿
16 52 ☽ △ ♂
18 44 ☽ □ ♆
19 23 ☽ ± ♅
21 04 ☽ ± ♃

24 Tuesday
03 09 ☽ ⚹ ☉
05 27 ☽ ∠ ♄
05 48 ☽ ⊼ ♅
07 33 ☽ ⊼ ♃
11 20 ☿ ∠ ♆
19 12 ☽ ∥ ♂
20 16 ☽ bQ ♂
21 22 ☽ ⊥ ♄
22 24 ☽ ⚹ ☿

25 Wednesday
02 08 ☽ Q ☉
06 50 ☽ ⊻ ♀
07 07 ☉ ⊥ ♆
08 12 ☽ ⊻ ♄
11 23 ☽ # ♄
13 17 ☽ ± ♂
15 20 ☉ □ ♅
16 55 ♂ ⊻ ♆
18 53 ☽ ⊥ ♀
19 56 ☽ ☍ ♇
22 28 ♀ ⚹ ♄
23 20 ☽ Q ☿

26 Thursday
00 33 ☽ △ ♆
00 53 ☽ ⊼ ♂
02 54 ♀ bQ ♅
03 11 ☽ ∥ ♅
03 22 ☽ ∥ ♃
06 43 ♀ ± ♇
09 56 ☽ # ♃
10 06 ☽ # ♅
12 25 ☽ ☍ ♅
13 22 ☽ ∠ ♀
14 14 ☽ □ ☉
14 36 ☽ ☍ ♃
19 09 ☉ □ ♃

27 Friday
02 24 ☽ ∥ ♄
03 58 ♀ bQ ♃
05 01 ☽ bQ ♆
14 31 ☽ □ ☿
16 49 ☽ ☌ ♄
17 08 ♇ St R
21 09 ☽ ⚹ ♀
22 27 ☽ # ♂
22 31 ☽ ± ♆

28 Saturday
02 32 ♀ # ♇
05 18 ☽ ⊼ ♇
06 55 ♃ ∥ ♅
08 20 ☿ □ ♄
10 20 ☽ ⊼ ♆
13 24 ☽ ☍ ♂
17 09 ☽ ± ♇
22 48 ☽ ⊼ ♅
23 17 ☽ Q ♀

29 Sunday
01 24 ☽ ⊼ ♃
01 47 ☽ # ♀
03 48 ☽ ∥ ♇
06 04 ☽ △ ☉
10 48 ☽ ± ♅
11 05 ☽ bQ ♇
13 19 ☽ # ♆
13 28 ☽ ± ♃
20 58 ☉ ∠ ♀

30 Monday
04 29 ☽ ⊻ ♄
04 54 ☽ bQ ♅
07 41 ☽ bQ ♃
11 15 ☽ △ ☿
15 03 ☽ bQ ☉
15 16 ☽ □ ♀
16 37 ☽ ⊥ ♄
17 17 ☽ △ ♇
19 36 ☽ # ☉
22 37 ☽ ☍ ♆
22 59 ☽ # ☿

31 Tuesday
04 39 ☽ ⊼ ♂
10 51 ☽ ∠ ♄
10 59 ☽ ± ☉
11 14 ☽ △ ♅
12 19 ♀ ⊼ ♇
14 11 ☽ △ ♃
17 25 ☽ ± ♂
20 43 ♂ ⊥ ♇
22 17 ☽ bQ ☿

All ephemeris data is given at 12.00 UT and the Moon's longitude is additionally given for 24.00 UT

JANUARY 1969

LONGITUDES

Date	Sidereal time h m s	Sun ☉ ° ' "	Moon ☽ ° ' "	Moon ☽ 24.00 ° ' "	Mercury ☿ ° '	Venus ♀ ° '	Mars ♂ ° '	Jupiter ♃ ° '	Saturn ♄ ° '	Uranus ♅ ° '	Neptune ♆ ° '	Pluto ♇ ° '
01	18 43 51	10 ♑ 54 44	16 ♊ 12 15	22 ♊ 08 10	25 ♑ 20	26 ♒ 14	01 ♏ 28	05 ♎ 29	18 ♈ 48	03 ♎ 59	27 ♏ 48	25 ♍ 06
02	18 47 47	11 55 52	28 ♊ 04 40	04 ♋ 01 58	26 54	27 22	02 02	05 33	18 50	04 00	27 49	25 R 06
03	18 51 44	12 57 01	10 ♋ 00 12	15 ♋ 59 30	28 27	28 30	02 36	05 36	18 51	04 00	27 51	25 06
04	18 55 40	13 58 09	22 ♋ 00 01	28 ♋ 01 50	29 ♑ 59	29 ♒ 37	03 10	05 39	18 53	04 00	27 53	25 06
05	18 59 37	14 59 17	04 ♌ 05 08	10 ♌ 10 01	01 ♒ 30	00 ♓ 45	03 44	05 42	18 54	04 00	27 54	25 05
06	19 03 33	16 00 25	16 ♌ 25 21	22 ♌ 25 21	02 59	01 52	04 17	05 45	18 56	04 00	27 56	25 05
07	19 07 30	17 01 33	28 ♌ 36 13	04 ♍ 49 35	04 26	02 59	04 51	05 47	18 58	04 00	27 58	25 05
08	19 11 27	18 02 42	11 ♍ 05 46	17 ♍ 25 07	05 51	04 06	05 25	05 50	19 00	04 R 01	27 59	25 04
09	19 15 23	19 03 50	23 ♍ 48 01	00 ♎ 14 54	07 13	05 12	05 58	05 52	19 02	04 00	28 01	25 04
10	19 19 20	20 04 58	06 ♎ 46 11	13 ♎ 22 16	08 32	06 18	06 32	05 54	19 04	04 00	28 03	25 03
11	19 23 16	21 06 06	20 ♎ 03 35	26 ♎ 50 29	09 47	07 25	07 05	05 56	19 06	04 00	28 04	25 03
12	19 27 13	22 07 14	03 ♏ 43 15	10 ♏ 42 05	10 58	08 30	07 38	05 57	19 08	04 00	28 06	25 03
13	19 31 09	23 08 22	17 ♏ 47 02	24 ♏ 58 01	12 03	09 36	08 11	05 59	19 11	04 00	28 07	25 02
14	19 35 06	24 09 30	02 ♐ 14 45	09 ♐ 36 46	13 02	10 41	08 45	06 00	19 13	03 59	28 09	25 01
15	19 39 02	25 10 38	17 ♐ 03 23	24 ♐ 33 42	13 55	11 46	09 18	06 01	19 16	03 59	28 10	25 01
16	19 42 59	26 11 46	02 ♑ 06 39	09 ♑ 41 02	14 39	12 51	09 51	06 02	19 19	03 59	28 11	25 00
17	19 46 56	27 12 53	17 ♑ 15 32	24 ♑ 48 48	15 15	13 56	10 23	06 03	19 22	03 58	28 13	25 00
18	19 50 52	28 14 00	02 ♒ 19 31	09 ♒ 46 27	15 41	15 00	10 56	06 03	19 25	03 58	28 14	24 59
19	19 54 49	29 ♑ 15 06	17 ♒ 08 32	24 ♒ 24 51	15 57	16 04	11 29	06 03	19 28	03 57	28 15	24 58
20	19 58 45	00 ♒ 16 12	01 ♓ 34 42	08 ♓ 37 36	16 R 03	17 07	12 01	06 03	19 31	03 56	28 17	24 57
21	20 02 42	01 17 16	15 ♓ 33 18	22 ♓ 21 44	15 56	18 10	12 34	06 R 03	19 34	03 56	28 18	24 57
22	20 06 38	02 18 20	29 ♓ 03 00	05 ♈ 37 22	15 38	19 13	13 06	06 03	19 37	03 55	28 19	24 56
23	20 10 35	03 19 22	12 ♈ 05 15	18 ♈ 27 06	15 09	20 16	13 38	06 03	19 41	03 54	28 20	24 55
24	20 14 31	04 20 24	24 ♈ 43 31	00 ♉ 55 05	14 29	21 18	14 10	06 02	19 44	03 53	28 22	24 54
25	20 18 28	05 21 24	07 ♉ 02 27	13 ♉ 06 16	13 38	22 19	14 42	06 01	19 48	03 52	28 23	24 53
26	20 22 25	06 22 24	19 ♉ 07 11	25 ♉ 05 49	12 39	23 21	15 14	06 00	19 52	03 51	28 24	24 52
27	20 26 21	07 23 22	01 ♊ 02 48	06 ♊ 58 42	11 33	24 22	15 46	05 59	19 56	03 50	28 25	24 51
28	20 30 18	08 24 19	12 ♊ 54 02	18 ♊ 49 20	10 22	25 22	16 18	05 57	20 00	03 49	28 26	24 50
29	20 34 14	09 25 15	24 ♊ 45 02	00 ♋ 41 31	09 07	26 22	16 49	05 56	20 04	03 48	28 27	24 49
30	20 38 11	10 26 10	06 ♋ 39 08	12 ♋ 38 12	07 52	27 22	17 21	05 54	20 08	03 47	28 28	24 48
31	20 42 07	11 ♒ 27 04	18 ♋ 38 58	24 ♋ 41 37	06 ♒ 39	28 ♓ 21	17 ♏ 52	05 ♎ 52	20 ♈ 12	03 ♎ 46	28 ♏ 29	24 ♍ 47

Date	Moon True ☊ ° '	Moon Mean ☊ ° '	Moon ☽ Latitude ° '
01	04 ♈ 30	04 ♈ 35	04 N 44
02	04 R 16	04 32	04 57
03	04 01	04 29	04 58
04	03 47	04 26	04 45
05	03 34	04 23	04 19
06	03 23	04 19	03 41
07	03 15	04 16	02 52
08	03 11	04 13	01 54
09	03 09	04 10	00 N 49
10	03 D 08	04 07	00 S 19
11	03 09	04 03	01 28
12	03 R 08	04 00	02 35
13	03 06	03 57	03 33
14	03 01	03 54	04 20
15	02 53	03 51	04 51
16	02 44	03 48	05 02
17	02 33	03 44	04 53
18	02 23	03 41	04 23
19	02 15	03 38	03 35
20	02 08	03 35	02 35
21	02 05	03 32	01 27
22	02 03	03 29	00 S 16
23	02 D 04	03 25	00 N 53
24	02 05	03 22	01 58
25	02 R 05	03 19	02 56
26	02 04	03 16	03 44
27	02 01	03 13	04 23
28	01 55	03 09	04 49
29	01 47	03 06	05 03
30	01 38	03 03	05 04
31	01 ♈ 28	03 ♈ 00	04 N 52

DECLINATIONS

Date	Sun ☉ ° '	Moon ☽ ° '	Mercury ☿ ° '	Venus ♀ ° '	Mars ♂ ° '	Jupiter ♃ ° '	Saturn ♄ ° '	Uranus ♅ ° '	Neptune ♆ ° '	Pluto ♇ ° '
01	23 S 00	27 N 27	23 S 07	14 S 17	10 S 41	00 S 59	05 N 01	00 S 54	18 S 03	16 N 11
02	22 55	28 23	22 46	13 51	10 52	01 00	05 02	00 54	18 03	16 12
03	22 49	28 01	22 24	13 24	11 04	01 01	05 03	00 54	18 03	16 12
04	22 43	26 20	22 00	12 58	11 16	01 02	05 04	00 54	18 04	16 13
05	22 36	23 25	21 35	12 30	11 27	01 03	05 04	00 54	18 04	16 14
06	22 29	19 28	21 09	12 03	11 39	01 04	05 05	00 54	18 04	16 14
07	22 22	14 39	20 42	11 35	11 50	01 05	05 06	00 54	18 05	16 15
08	22 14	09 10	20 14	11 07	12 01	01 05	05 07	00 54	18 05	16 15
09	22 05	03 N 13	19 45	10 39	12 12	01 06	05 08	00 54	18 05	16 16
10	21 57	02 S 59	19 16	10 11	12 24	01 06	05 09	00 54	18 06	16 17
11	21 47	09 13	18 46	09 43	12 35	01 07	05 10	00 54	18 06	16 17
12	21 38	15 11	18 16	09 14	12 45	01 07	05 12	00 54	18 06	16 18
13	21 28	20 33	17 46	08 45	12 56	01 08	05 13	00 54	18 07	16 19
14	21 17	24 52	17 16	08 16	13 07	01 08	05 14	00 54	18 07	16 19
15	21 06	27 38	16 48	07 47	13 18	01 08	05 15	00 54	18 07	16 20
16	20 55	28 28	16 20	07 18	13 28	01 08	05 17	00 53	18 07	16 21
17	20 43	27 10	15 54	06 48	13 38	01 08	05 18	00 53	18 08	16 21
18	20 31	23 54	15 30	06 19	13 49	01 08	05 19	00 53	18 08	16 22
19	20 19	19 07	15 09	05 49	13 59	01 08	05 21	00 53	18 08	16 23
20	20 06	13 20	14 50	05 19	14 09	01 08	05 22	00 52	18 09	16 24
21	19 53	07 02	14 35	04 50	14 19	01 07	05 23	00 52	18 09	16 24
22	19 39	00 S 37	14 23	04 20	14 29	01 07	05 25	00 52	18 09	16 25
23	19 25	05 N 36	14 14	03 50	14 39	01 07	05 27	00 51	18 09	16 26
24	19 11	11 25	14 09	03 20	14 49	01 06	05 28	00 51	18 09	16 27
25	18 56	16 38	14 08	02 50	14 58	01 06	05 30	00 51	18 09	16 27
26	18 41	21 06	14 11	02 21	15 08	01 05	05 31	00 50	18 10	16 28
27	18 26	24 39	14 17	01 51	15 17	01 04	05 33	00 50	18 10	16 29
28	18 10	27 08	14 26	01 21	15 26	01 03	05 35	00 49	18 10	16 30
29	17 54	28 24	14 38	00 51	15 36	01 02	05 36	00 49	18 10	16 30
30	17 37	28 22	14 52	00 22	15 45	01 01	05 38	00 48	18 10	16 31
31	17 S 21	26 N 58	15 S 07	00 N 08	15 S 54	01 S 00	05 N 40	00 S 48	18 S 10	16 N 32

ZODIAC SIGN ENTRIES

Date	h	m	Planets
02	15	53	☽ ♋
04	12	18	☿ ♒
04	20	07	♀ ♓
05	03	55	☽ ♌
07	14	42	☽ ♍
09	23	32	☽ ♎
12	05	32	☽ ♏
14	08	19	☽ ♐
16	08	39	☽ ♑
18	08	17	☽ ♒
20	05	38	☉ ♒
20	09	20	☽ ♓
22	13	43	☽ ♈
24	22	13	☽ ♉
27	09	53	☽ ♊
29	22	36	☽ ♋

LATITUDES

Date	Mercury ☿ ° '	Venus ♀ ° '	Mars ♂ ° '	Jupiter ♃ ° '	Saturn ♄ ° '	Uranus ♅ ° '	Neptune ♆ ° '	Pluto ♇ ° '
01	02 S 05	01 S 36	01 N 24	01 N 18	02 S 32	00 N 45	01 N 40	15 N 33
04	01 53	01 27	01 24	01 19	02 31	00 45	01 40	15 35
07	01 36	01 16	01 23	01 20	02 31	00 45	01 41	15 37
10	01 10	01 04	01 23	01 21	02 30	00 45	01 41	15 38
13	00 S 36	00 50	01 23	01 22	02 29	00 45	01 41	15 40
16	00 N 07	00 36	01 22	01 22	02 28	00 45	01 41	15 41
19	00 57	00 21	01 22	01 23	02 27	00 45	01 41	15 43
22	01 52	00 S 04	01 21	01 24	02 27	00 45	01 41	15 44
25	02 42	00 N 13	01 21	01 25	02 26	00 46	01 41	15 45
28	03 20	00 32	01 20	01 26	02 25	00 46	01 42	15 47
31	03 N 37	00 N 52	01 N 20	01 N 26	02 S 24	00 N 46	01 N 42	15 N 48

DATA

Julian Date	2440223
Delta T	+39 seconds
Ayanamsa	23° 25' 26"
Synetic vernal point	05° ♓ 41' 34"
True obliquity of ecliptic	23° 26' 44"

MOON'S PHASES, APSIDES AND POSITIONS ☽

Date	h	m	Phase	Longitude ° '	Eclipse Indicator
03	18	28	○	13 ♋ 13	
11	14	00	☾	21 ♎ 11	
18	04	59	●	27 ♑ 56	
25	08	23	☽	05 ♉ 12	

Day	h	m		
01	15	01	Apogee	
16	23	56	Perigee	
29	03	02	Apogee	
02	17	15	Max dec	28° N 25'
10	00	31	0S	
16	09	21	Max dec	28° S 29'
22	14	21	0N	
29	23	10	Max dec	28° N 32'

LONGITUDES

Date	Chiron ⚷ ° '	Ceres ⚳ ° '	Pallas ⚴ ° '	Juno ⚵ ° '	Vesta ⚶ ° '	Black Moon Lilith ⚸ ° '
01	29 ♓ 05	28 ♐ 59	10 ♐ 37	17 ♐ 54	22 ♈ 21	22 ♊ 00
11	29 ♓ 20	03 ♑ 01	14 ♐ 43	21 ♐ 15	24 ♈ 18	23 ♊ 07
21	29 ♓ 39	07 ♑ 00	18 ♐ 42	24 ♐ 31	26 ♈ 42	24 ♊ 14
31	00 ♈ 03	10 ♑ 53	22 ♐ 32	27 ♐ 42	29 ♈ 29	25 ♊ 21

All ephemeris data is given at 12.00 UT and the Moon's longitude is additionally given for 24.00 UT

ASPECTARIAN

h m	Aspects
01 Wednesday	
00 17	☽ ⊼ ☉
08 35	☿ △ ♇
12 34	☽ ♇ ♂
17 17	☽ ⚹ ♄
19 17	☽ ± ☿
23 18	☉ ∥ ☿
02 Thursday	
06 00	☽ □ ♇
09 15	☽ ⊼ ☿
10 24	☽ △ ♀
11 29	☽ ⊼ ♆
17 33	☽ Q ♄
20 22	☽ △ ♂
22 01	♀ □ ♆
23 36	☽ ± ♆
23 55	☽ □ ♅
03 Friday	
01 27	♀ ± ♅
02 36	☿ ⚹ ♆
03 07	☽ □ ♃
09 35	☉ ∠ ♆
14 41	☿ ⊻ ♀
17 43	☽ ♇ ♆
18 12	☽ Q ♇
18 28	☽ ☍ ☉
19 43	☽ ♇ ♀
04 Saturday	
05 45	☽ □ ♄
12 00	☽ Q ♅
12 40	♀ ± ♃
15 18	☽ Q ♃
15 34	☽ ± ♀
18 10	☽ ⚹ ♇
23 44	☽ △ ♆
05 Sunday	
04 42	☽ ⊼ ♀
06 09	☽ ☍ ☿
11 16	☽ □ ♂
11 50	☽ ⚹ ♅
15 12	☽ ⚹ ♃
17 41	☽ # ☉
23 50	♂ ⊻ ♅
23 51	☽ ∠ ♇
06 Monday	
01 16	☽ # ☿
11 25	☽ ⊼ ☉
17 12	☽ △ ♄
17 20	☽ ∠ ♅
17 29	☽ ⊥ ♇
19 20	☽ # ♆
20 46	☽ ∠ ♃
07 Tuesday	
00 12	☽ ± ☉
00 18	☽ Q ♂
02 54	☉ Q ♂
02 58	♀ ∥ ♂
04 27	☽ ∥ ♇
04 53	☿ △ ♅
05 10	☽ ⊻ ♇
10 45	☽ □ ♆
10 51	☽ ⊥ ♅
14 18	☽ ⊥ ♃
19 12	☽ ♇ ☉
21 17	☽ ☍ ♀
22 22	☽ ♇ ♄
22 26	☽ ⊻ ♅
23 31	☿ □ ♂
08 Wednesday	
00 11	☽ # ♂
00 37	☽ ⚹ ♂
00 42	☽ ⊼ ☿
01 53	☽ ⊻ ♃
02 55	☽ # ♀
07 30	♅ St R
09 48	♀ ∠ ♄
10 10	♀ ⊼ ♅
11 32	☿ △ ♃
13 37	☽ ± ☿
15 37	☽ ± ♄
21 19	☽ Q ♆
09 Thursday	
02 20	☽ △ ☉
03 01	☽ ⊼ ♄
04 25	☽ ∥ ♄
06 27	☽ ∠ ♂
07 03	♂ ⊻ ♃
08 26	☿ Q ♄
08 42	☽ ♇ ☿
11 08	☉ □ ♄
14 22	☽ ☌ ♇
19 53	☽ ⚹ ♆
20 16	☽ # ♃
21 01	☽ # ♅
10 Friday	
00 00	☽ ⊥ ♂
02 47	♀ ⊼ ♃
04 01	☽ ∥ ♅
04 47	☽ ∥ ♃
06 56	☽ ☌ ♅
10 24	☽ ☌ ♃
11 05	☽ ⊼ ♀
11 32	☽ ⊻ ♂
15 35	☽ △ ☿
20 22	☽ # ♄
21 40	♀ △ ♂
22 59	☽ ± ♀
23 26	☽ ∠ ♆
11 Saturday	
10 17	☽ ☍ ♄
13 49	☽ ∥ ♀
14 00	☽ □ ☉
15 35	☽ ⊥ ♆
16 33	☽ ♇ ♀
17 10	☿ ♇ ♇
17 40	☿ Q ♆
20 50	☽ ⊻ ♇
12 Sunday	
01 44	☽ ∥ ♂
02 11	☽ ⊻ ♆
07 21	☽ ⊥ ♇
12 29	☽ ⊻ ♅
15 52	☽ ⊻ ♃
16 44	☽ # ♇
19 02	☽ ☌ ♂
19 38	☿ ∥ ♆
20 57	☽ △ ♀
22 48	☽ ⊥ ♅
22 52	☽ ∠ ♇
23 52	☽ Q ☉
13 Monday	
00 11	☽ ∥ ☿
00 37	☽ ∥ ♆
01 31	☽ □ ☿
02 10	☽ ⊥ ♃
14 02	☽ ∠ ♅
14 21	☽ ⊼ ♄
16 23	☽ ∥ ☉
17 22	☽ ∠ ♃
21 39	☽ ⚹ ☉
14 Tuesday	
00 06	☽ ⚹ ♇
00 23	☽ ± ♄
05 15	☽ ☌ ♆
09 53	☽ Q ☿
14 51	☽ ⚹ ♅
15 15	☽ ♇ ♄
18 08	☽ ⚹ ♃
19 48	☽ Q ♇
23 00	☽ ⊻ ♂
15 Wednesday	
00 06	☽ ∠ ☉
02 49	☽ □ ♀
06 39	☽ ⚹ ☿
08 11	☉ △ ♇
09 04	☽ ⊥ ♂
10 17	☽ Q ♅
13 32	☽ Q ♃
15 33	☽ △ ♄
15 39	☽ ⊥ ☉
16 Thursday	
00 01	☽ ∠ ♂
00 43	☽ □ ♇
01 55	☽ ⊻ ☉
05 46	☽ ⊻ ♆
07 55	☽ ∠ ☿
09 51	☽ Q ♀
11 37	☿ # ♇
14 57	☽ □ ♅
15 18	☽ ⊥ ♆
17 50	♂ ⊥ ♅
18 13	☽ □ ♃
18 55	♂ ∠ ♇
22 47	♀ ⊥ ♄
22 49	☽ ⊥ ☿
17 Friday	
00 43	☽ ⚹ ♂
05 35	☽ ∠ ♆
06 19	☽ ⚹ ♀
08 42	☽ ⊻ ☿
15 21	☽ □ ♄
20 27	☽ Q ♂
18 Saturday	
00 17	☽ △ ♇
04 59	☽ ☌ ☉
05 27	☽ ⚹ ♆
07 59	☽ ∠ ♀
12 03	☉ ⚹ ♆
14 37	☽ △ ♅
17 59	☽ △ ♃
20 12	☽ Q ♄
23 34	☽ ⊥ ♀
19 Sunday	
00 19	☽ ♇ ♇
00 46	☽ Q ♆
02 25	☽ □ ♂
06 18	☽ ∥ ☉
09 12	☿ ⊻ ♀
10 02	☽ ☌ ☿
10 05	☽ ⊻ ♀
14 58	☽ ♇ ♅
15 00	☽ ± ♇
15 49	☽ ⚹ ♄
16 18	☽ ∥ ♆
18 26	☽ ♇ ♃
23 42	☽ # ♇
20 Monday	
00 55	☽ ⊼ ♇
05 42	☽ ∥ ☿
05 53	☽ ± ♅
06 27	☽ □ ♆
08 50	☽ ∥ ♂
09 26	☽ ± ♃
09 37	☽ ⊻ ☉
09 59	♀ # ♄
10 56	☿ St R
12 32	♃ St R
13 37	♂ ⊥ ♃
16 00	☽ ⊼ ♅
17 00	☽ ∠ ♄
19 36	☽ ⊼ ♃
20 35	☽ ⊥ ☉
21 Tuesday	
06 35	☽ △ ♂
08 31	☽ ⊥ ♄
12 40	☽ ⊻ ☿
13 23	☽ ∠ ☉
16 58	☽ ☌ ♀
18 06	☽ # ♄
19 05	☽ ⊻ ♄
20 55	☽ ∥ ♀
23 03	☽ ⊥ ☿
22 Wednesday	
04 29	☿ ∥ ♂
04 35	☽ ☍ ♇
10 08	☽ ∥ ♃
10 13	☽ ♇ ♂
10 41	☽ △ ♆
11 06	☽ ∥ ♅
14 48	☽ ∠ ☿
17 37	☽ # ♅
18 25	☽ ⚹ ☉
18 36	☽ # ♃
20 51	☽ ☍ ♅
21 52	♀ ⊻ ♄
23 Thursday	
00 47	☽ ☍ ♃
03 22	☽ ± ♂
05 35	☽ # ♀
11 23	☽ ∥ ♄
14 21	☽ ♇ ♆
15 02	☽ ⊼ ♂
17 30	☽ ⚹ ☿
18 36	☽ Q ☉
24 Friday	
00 49	☿ ⊥ ♀
01 28	☉ △ ♅
02 24	☽ ☌ ♄
04 50	☽ ⊻ ♀
07 27	☽ ± ♆
12 20	☽ ⊼ ♇
15 11	☽ Q ☿
17 25	☽ ⊥ ♀
17 38	☿ □ ♂
19 02	☽ ⊼ ♆
23 57	☽ ± ♇
25 Saturday	
00 09	☽ # ☿
03 45	☽ # ♂
05 47	☽ ⊼ ♅
08 23	☽ □ ☉
09 59	☽ ⊼ ♃
11 07	☽ ∥ ♇
12 37	☽ ∠ ♀
17 35	☽ ± ♅
17 37	☽ ♇ ♇
19 42	☽ # ♆
21 50	☽ ± ♃
23 11	☽ # ☉
26 Sunday	
00 06	☽ □ ☿
03 20	☉ △ ♃
03 53	☽ ☍ ♂
11 28	☽ ♇ ♅
13 30	☽ ⊻ ♄
15 46	☽ ♇ ♃
21 16	☽ ⚹ ♀
23 32	☽ △ ♇
27 Monday	
01 37	☽ ⊥ ♄
06 41	☽ ☍ ♆
17 38	☽ △ ♅
19 54	☽ ∠ ♄
21 57	☽ △ ♃
23 33	♀ ☍ ♇
23 45	☽ Q ♀
28 Tuesday	
01 36	☉ Q ♄
02 02	☽ △ ☉
07 20	☽ △ ☿
10 32	☿ Q ♆
11 53	☿ ∠ ♀
12 07	☉ ∥ ♆
19 12	☽ ⊼ ♂
22 17	☿ ♇ ♇
29 Wednesday	
02 27	☽ ⚹ ♄
02 46	♀ ∥ ♃
07 55	☽ ± ♂
08 49	☉ ☌ ☿
10 51	☽ ♇ ☿
11 16	☽ ♇ ☉
12 09	☽ □ ♇
14 06	♀ ∥ ♅
15 34	☽ □ ♀
19 29	☽ ⊼ ♆
21 21	☉ ♇ ♇
30 Thursday	
02 51	☽ Q ♄
02 56	☽ ♇ ♂
03 17	☽ ± ☿
06 14	☽ □ ♅
07 08	☽ ± ☉
07 17	☿ Q ♄
07 36	☽ ± ♆
10 29	☽ □ ♃
12 46	☉ Q ♆
14 13	☽ ⊼ ☿
20 18	☽ ⊼ ☉
31 Friday	
00 19	☽ Q ♇
01 41	☽ ♇ ♆
10 22	☽ △ ♂
15 06	☽ □ ♄
15 23	♀ △ ♆
18 11	☽ Q ♅
22 20	☽ Q ♃

FEBRUARY 1969

LONGITUDES

Date	Sidereal time h m s	Sun ☉ ° ' "	Moon ☽ ° ' "	Moon ☽ 24.00 ° ' "	Mercury ☿ ° '	Venus ♀ ° '	Mars ♂ ° '	Jupiter ♃ ° '	Saturn ♄ ° '	Uranus ♅ ° '	Neptune ♆ ° '	Pluto ♇ ° '
01	20 46 04	12 ♒ 27 56	00 ♌ 46 20	06 ♌ 53 15	05 ♒ 28	29 ♓ 19	18 ♏ 23	05 ♎ 50	20 ♈ 16	03 ♎ 44	28 ♏ 30	24 ♍ 46
02	20 50 00	13 28 48	13 ♌ 02 27	19 ♌ 14 02	04 R 23	00 ♈ 17	18 54	05 R 47	20 21	03 R 43	28 31	24 R 45
03	20 53 57	14 29 38	25 ♌ 28 04	01 ♍ 44 37	03 24	01 15	19 25	05 45	20 25	03 42	28 32	24 44
04	20 57 54	15 30 27	08 ♍ 03 45	14 ♍ 25 34	02 33	02 12	19 56	05 42	20 30	03 40	28 33	24 43
05	21 01 50	16 31 15	20 ♍ 50 08	27 ♍ 17 36	01 49	03 08	20 26	05 39	20 34	03 39	28 33	24 42
06	21 05 47	17 32 02	03 ♎ 48 05	10 ♎ 21 45	01 14	04 04	20 57	05 36	20 39	03 37	28 34	24 40
07	21 09 43	18 32 48	16 ♎ 58 46	23 ♎ 39 20	00 47	04 59	21 27	05 33	20 44	03 36	28 35	24 39
08	21 13 40	19 33 33	00 ♏ 23 37	07 ♏ 11 46	00 28	05 54	21 58	05 29	20 49	03 34	28 36	24 38
09	21 17 36	20 34 17	14 ♏ 03 57	21 ♏ 00 14	00 18	06 48	22 28	05 26	20 53	03 33	28 36	24 37
10	21 21 33	21 35 00	28 ♏ 00 39	05 ♐ 05 08	00 D 15	07 41	22 58	05 22	20 58	03 31	28 37	24 36
11	21 25 29	22 35 43	12 ♐ 13 30	19 ♐ 25 29	00 19	08 34	23 28	05 18	21 04	03 29	28 37	24 34
12	21 29 26	23 36 24	26 ♐ 40 40	03 ♑ 58 30	00 30	09 26	23 57	05 14	21 09	03 27	28 38	24 33
13	21 33 23	24 37 04	11 ♑ 18 20	18 ♑ 39 23	00 48	10 17	24 27	05 09	21 14	03 26	28 39	24 32
14	21 37 19	25 37 43	26 ♑ 00 48	03 ♒ 21 38	01 11	11 07	24 56	05 05	21 19	03 24	28 39	24 30
15	21 41 16	26 38 20	10 ♒ 40 59	17 ♒ 57 53	01 39	11 56	25 25	05 00	21 25	03 22	28 40	24 29
16	21 45 12	27 38 56	25 ♒ 11 28	02 ♓ 20 58	02 13	12 45	25 54	04 55	21 30	03 20	28 40	24 27
17	21 49 09	28 39 30	09 ♓ 25 41	16 ♓ 25 05	02 51	13 33	26 23	04 50	21 36	03 18	28 40	24 26
18	21 53 05	29 ♒ 40 03	23 ♓ 18 46	00 ♈ 06 29	03 34	14 20	26 52	04 45	21 41	03 16	28 41	24 25
19	21 57 02	00 ♓ 40 34	06 ♈ 48 07	13 ♈ 23 44	04 20	15 06	27 20	04 40	21 47	03 14	28 41	24 23
20	22 00 58	01 41 04	19 ♈ 53 27	26 ♈ 17 34	05 10	15 51	27 48	04 35	21 53	03 12	28 41	24 22
21	22 04 55	02 41 31	02 ♉ 36 26	08 ♉ 50 30	06 03	16 34	28 16	04 29	21 59	03 10	28 42	24 20
22	22 08 52	03 41 57	15 ♉ 00 15	21 ♉ 06 14	07 00	17 17	28 44	04 23	22 04	03 08	28 42	24 19
23	22 12 48	04 42 21	27 ♉ 09 02	03 ♊ 09 16	07 59	17 59	29 12	04 18	22 10	03 06	28 42	24 17
24	22 16 45	05 42 43	09 ♊ 07 31	15 ♊ 04 24	09 01	18 40	29 ♏ 39	04 12	22 16	03 03	28 42	24 16
25	22 20 41	06 43 03	21 ♊ 00 31	26 ♊ 56 27	10 05	19 19	00 ♐ 06	04 05	22 22	03 01	28 42	24 14
26	22 24 38	07 43 21	02 ♋ 52 46	08 ♋ 49 59	11 12	19 57	00 33	03 59	22 29	02 59	28 43	24 13
27	22 28 34	08 43 37	14 ♋ 48 35	20 ♋ 49 03	12 20	20 34	01 00	03 53	22 35	02 57	28 43	24 11
28	22 32 31	09 ♓ 43 51	26 ♋ 51 45	02 ♌ 57 03	13 ♒ 31	21 ♈ 09	01 ♐ 27	03 ♎ 46	22 ♈ 41	02 ♎ 54	28 ♏ 43	24 ♍ 10

Date	Moon True ☊ ° '	Moon Mean ☊ ° '	Moon ☽ Latitude ° '
01	01 ♈ 18	02 ♈ 57	04 N 27
02	01 R 09	02 54	03 49
03	01 02	02 50	02 59
04	00 57	02 47	02 00
05	00 55	02 44	00 N 54
06	00 D 54	02 41	00 S 16
07	00 55	02 38	01 26
08	00 57	02 35	02 33
09	00 58	02 31	03 32
10	00 R 58	02 28	04 21
11	00 56	02 25	04 54
12	00 53	02 22	05 10
13	00 48	02 19	05 05
14	00 43	02 15	04 41
15	00 37	02 12	03 58
16	00 33	02 09	03 00
17	00 29	02 06	01 52
18	00 28	02 03	00 S 39
19	00 D 28	02 00	00 N 35
20	00 29	01 56	01 44
21	00 31	01 53	02 47
22	00 32	01 50	03 40
23	00 33	01 47	04 22
24	00 R 34	01 44	04 52
25	00 33	01 40	05 10
26	00 31	01 37	05 14
27	00 28	01 34	05 04
28	00 ♈ 24	01 ♈ 31	04 N 41

DECLINATIONS

Date	Sun ☉ ° '	Moon ☽ ° '	Mercury ☿ ° '	Venus ♀ ° '	Mars ♂ ° '	Jupiter ♃ ° '	Saturn ♄ ° '	Uranus ♅ ° '	Neptune ♆ ° '	Pluto ♇ ° '
01	17 S 04	24 N 20	15 S 23	00 N 37	16 S 02	00 S 59	05 N 42	00 S 47	18 S 11	16 N 33
02	16 47	20 33	15 40	01 07	16 11	00 58	05 44	00 47	18 11	16 34
03	16 29	15 51	15 57	01 36	16 20	00 57	05 46	00 46	18 11	16 34
04	16 11	10 24	16 14	02 05	16 28	00 56	05 47	00 45	18 11	16 35
05	15 53	04 N 28	16 30	02 34	16 37	00 54	05 49	00 45	18 11	16 36
06	15 35	01 S 45	16 46	03 03	16 45	00 53	05 51	00 44	18 11	16 37
07	15 16	08 00	17 01	03 32	16 53	00 51	05 53	00 44	18 11	16 38
08	14 57	14 00	17 16	04 01	17 02	00 50	05 55	00 43	18 11	16 38
09	14 38	19 27	17 29	04 29	17 10	00 48	05 57	00 42	18 12	16 39
10	14 19	23 57	17 41	04 57	17 17	00 46	06 00	00 42	18 12	16 40
11	13 59	27 07	17 51	05 25	17 25	00 44	06 02	00 41	18 12	16 41
12	13 39	28 34	18 01	05 53	17 33	00 43	06 04	00 40	18 12	16 42
13	13 19	28 02	18 09	06 20	17 41	00 41	06 06	00 39	18 12	16 43
14	12 59	25 33	18 16	06 47	17 48	00 39	06 08	00 39	18 12	16 43
15	12 38	21 23	18 22	07 14	17 55	00 37	06 10	00 38	18 12	16 44
16	12 18	15 58	18 26	07 41	18 03	00 34	06 13	00 37	18 12	16 45
17	11 57	09 46	18 29	08 07	18 10	00 32	06 15	00 36	18 12	16 46
18	11 36	03 S 15	18 30	08 33	18 17	00 30	06 17	00 35	18 12	16 47
19	11 14	03 N 14	18 30	08 59	18 24	00 28	06 19	00 35	18 12	16 48
20	10 53	09 23	18 29	09 24	18 30	00 25	06 22	00 34	18 12	16 48
21	10 31	15 00	18 27	09 49	18 37	00 23	06 24	00 33	18 12	16 49
22	10 09	19 51	18 23	10 14	18 44	00 21	06 26	00 32	18 12	16 50
23	09 47	23 47	18 18	10 38	18 50	00 18	06 29	00 31	18 12	16 51
24	09 25	26 38	18 11	11 02	18 57	00 16	06 31	00 30	18 12	16 52
25	09 03	28 17	18 03	11 25	19 03	00 13	06 33	00 29	18 12	16 52
26	08 41	28 39	17 54	11 48	19 09	00 10	06 36	00 29	18 12	16 53
27	08 18	27 40	17 43	12 10	19 15	00 08	06 38	00 28	18 12	16 54
28	07 S 55	25 N 23	17 S 31	12 N 32	19 S 21	00 S 05	06 N 41	00 S 27	18 S 12	16 N 55

ZODIAC SIGN ENTRIES

Date	h	m	Planets
01	10	29	☽ ♌
02	04	45	♀ ♈
03	20	40	☽ ♍
06	05	00	☽ ♎
08	11	18	☽ ♏
10	15	23	☽ ♐
12	17	28	☽ ♑
14	18	30	☽ ♒
16	20	03	☽ ♓
18	19	55	☉ ♓
18	23	48	☽ ♈
21	07	02	☽ ♉
23	17	41	☽ ♊
25	06	21	♂ ♐
26	06	11	☽ ♋
28	18	12	☽ ♌

LATITUDES

Date	Mercury ☿ ° '	Venus ♀ ° '	Mars ♂ ° '	Jupiter ♃ ° '	Saturn ♄ ° '	Uranus ♅ ° '	Neptune ♆ ° '	Pluto ♇ ° '
01	03 N 38	00 N 58	01 N 19	01 N 27	02 S 24	00 N 46	01 N 42	15 N 48
04	03 27	01 19	01 18	01 28	02 23	00 46	01 42	15 50
07	03 02	01 41	01 17	01 28	02 23	00 46	01 42	15 51
10	02 29	02 04	01 16	01 29	02 22	00 46	01 42	15 52
13	01 53	02 28	01 15	01 30	02 21	00 46	01 42	15 53
16	01 16	02 52	01 14	01 30	02 21	00 46	01 43	15 54
19	00 42	03 18	01 12	01 31	02 20	00 46	01 43	15 54
22	00 N 09	03 44	01 11	01 32	02 20	00 46	01 43	15 55
25	00 S 20	04 10	01 09	01 32	02 19	00 46	01 43	15 56
28	00 47	04 37	01 08	01 33	02 19	00 46	01 43	15 57
31	01 S 10	05 N 04	01 N 06	01 N 33	02 S 18	00 N 47	01 N 43	15 N 57

DATA

Julian Date	2440254
Delta T	+39 seconds
Ayanamsa	23° 25' 32"
Synetic vernal point	05° ♓ 41' 28"
True obliquity of ecliptic	23° 26' 45"

MOON'S PHASES, APSIDES AND POSITIONS ☽

Date	h	m	Phase	Longitude ° '	Eclipse Indicator
02	12	56	○	13 ♌ 31	
10	00	08	☾	21 ♏ 05	
16	16	25	●	27 ♒ 50	
24	04	30	☽	05 ♊ 24	

Day	h	m		
14	03	41	Perigee	
25	22	01	Apogee	
06	05	17	0S	
12	17	42	Max dec	28° S 37'
18	23	57	0N	
26	06	21	Max dec	28° N 41'

LONGITUDES

Date	Chiron ⚷ ° '	Ceres ⚳ ° '	Pallas ⚴ ° '	Juno ⚵ ° '	Vesta ⚶ ° '	Black Moon Lilith ⚸ ° '
01	00 ♈ 06	11 ♑ 16	22 ♐ 54	28 ♐ 01	29 ♈ 47	25 ♊ 28
11	00 ♈ 33	15 ♑ 04	26 ♐ 32	01 ♑ 03	02 ♉ 55	26 ♊ 35
21	01 ♈ 04	18 ♑ 44	29 ♐ 56	03 ♑ 55	06 ♉ 18	27 ♊ 42
31	01 ♈ 37	22 ♑ 15	03 ♑ 05	06 ♑ 36	09 ♉ 53	28 ♊ 50

All ephemeris data is given at 12.00 UT and the Moon's longitude is additionally given for 24.00 UT

ASPECTARIAN

h m Aspects

01 Saturday
00 10 ☽ ⚹ ♇
04 20 ☿ △ ♃
07 31 ☽ △ ♆
08 54 ☽ △ ♀
17 49 ☽ ⚹ ♅
19 47 ♀ # ♅
20 28 ☽ ☍ ☿
21 54 ☽ ⚹ ♃

02 Sunday
03 53 ♂ ∠ ♅
05 06 ♀ # ♃
05 36 ☽ ∠ ♇
12 56 ☽ ☍ ☉
16 44 ☽ bQ ♀
22 59 ☽ ∠ ♅
23 03 ☽ ⊥ ♇
23 51 ☽ □ ♂

03 Monday
00 38 ☽ # ♆
02 13 ☽ △ ♄
02 57 ☽ ∠ ♃
04 27 ☿ △ ♅
05 20 ☉ # ♇
08 33 ☽ ∥ ♇
08 45 ☽ # ☉
09 46 ☽ # ♂
10 36 ☽ ⊻ ♇
11 32 ☽ # ☿
11 33 ☽ ± ♀
16 16 ☽ ⊥ ♅
17 52 ☽ □ ♆
20 10 ☽ ⊥ ♃
20 37 ☉ ∥ ♂
23 58 ☽ ⊼ ♀

04 Tuesday
02 10 ☽ ⊼ ☿
03 41 ☽ ⊻ ♅
07 06 ☽ bQ ♄
07 32 ☽ ⊻ ♃
10 17 ☉ ∥ ☿
11 44 ☽ Q ♂
12 52 ☽ ± ☿
16 51 ☿ ⚹ ♀
23 40 ☿ Q ♂

05 Wednesday
00 12 ☽ ± ♄
03 14 ☽ ⊼ ☉
03 59 ☽ Q ♆
04 52 ☽ bQ ☿
06 40 ☽ ∥ ♄
09 26 ♂ # ♇
11 14 ☽ ⚹ ♂
11 30 ☽ ⊼ ♄
15 24 ☽ ± ☉
18 50 ☽ ∥ ♀
19 08 ♂ ⊼ ♄
19 10 ☽ ☌ ♇
20 41 ☿ # ♇
21 02 ♂ ∠ ♃

06 Thursday
00 44 ♀ ☍ ♅
01 52 ☽ # ♃
02 21 ☽ ⚹ ♆
02 26 ☽ # ♅
07 28 ☽ △ ☿
08 00 ☿ ∥ ♂
08 08 ☽ ∥ ♅
08 40 ☽ ∥ ♃
09 28 ☽ bQ ☉
11 40 ☽ ☌ ♅
12 32 ☽ ☍ ♀
15 17 ☽ ☌ ♃
16 06 ☽ ∠ ♂
17 24 ☽ # ♀

07 Friday
03 49 ☽ # ♄
05 51 ☽ ∠ ♆
09 08 ☽ ⊥ ♂
13 10 ☉ bQ ♅
14 30 ☉ ± ♇
15 04 ☽ △ ☉
18 47 ☽ ☍ ♄
20 23 ☽ ⊻ ♂
22 05 ☽ ⊥ ♆

08 Saturday
01 45 ♀ ☍ ♃
01 46 ☽ ⊻ ♇
08 48 ☽ ⊻ ♆
12 08 ☽ □ ☿
12 26 ☽ ⊥ ♇
15 48 ☽ ∥ ☉
17 36 ☽ ⊻ ♅
20 58 ☽ ⊻ ♃
22 25 ☽ ⊼ ♀
23 17 ☽ # ♇

09 Sunday
01 16 ☽ ∥ ♂
02 35 ☽ ∥ ☿
04 08 ☽ ⊥ ♅
04 15 ☽ ∠ ♇
06 11 ☽ ∥ ♆
07 26 ☽ ⊥ ♃
08 47 ☉ bQ ♃
09 39 ☽ ± ♀
19 17 ☽ Q ☿
19 45 ☽ ∠ ♅
20 15 ☉ ⚹ ♄
22 58 ☽ ∠ ♃
23 53 ☽ ⊼ ♄

10 Monday
00 08 ☽ □ ☉
02 17 ☽ bQ ♀
03 03 ☽ ☌ ♂
06 10 ☽ ⚹ ♇
09 39 ☿ St D
10 13 ☽ ± ♄
13 02 ☽ ☌ ♆
15 49 ☽ ⚹ ☿
21 20 ☽ ⚹ ♅

11 Tuesday
00 25 ☽ ⚹ ♃
01 35 ☽ bQ ♄
02 31 ☽ Q ♇
05 22 ♀ bQ ♂
05 27 ☽ △ ♀
09 04 ☽ Q ☉
17 13 ☽ ∠ ☿
17 26 ☽ Q ♅
20 25 ☽ Q ♃

12 Wednesday
02 48 ☽ △ ♄
06 33 ☽ ⚹ ☉
07 21 ☽ ⊻ ♂
08 21 ☽ ⊥ ☿
08 29 ☽ □ ♇
15 14 ☽ ⊻ ♆
17 35 ☽ ⊥ ♂
18 25 ☽ ⊻ ☿
22 24 ♀ ∥ ♄
23 08 ☽ □ ♅

13 Thursday
01 05 ☽ ⊥ ♆
01 59 ☽ □ ♃
04 00 ☉ □ ♂
08 51 ☽ ∠ ♂
09 02 ☽ ∠ ☉
09 52 ☉ ⊼ ♇
10 13 ☽ □ ♀
15 49 ☽ ∠ ♆
15 49 ♂ ⚹ ♇
21 25 ☿ ∥ ♆

14 Friday
00 49 ☽ ⊥ ☉
04 18 ☽ □ ♄
09 32 ☽ △ ♇
10 11 ☽ ⚹ ♂
11 17 ♃ ∥ ♅
11 20 ☽ ⊻ ☉
16 19 ☽ ⚹ ♆
17 22 ☽ Q ♀
20 42 ☽ ☌ ☿

15 Saturday
00 02 ☽ △ ♅
02 44 ☽ △ ♃
06 28 ☽ Q ♂
09 54 ☽ Q ♄
10 02 ☽ bQ ♇
11 58 ☽ Q ♆
14 11 ☽ ⚹ ♀

16 Sunday
00 38 ☽ bQ ♅
00 50 ☽ ± ♇
01 43 ☽ ∥ ☿
02 36 ☽ ∥ ♆
03 18 ☽ bQ ♃
03 28 ☽ ∥ ♂
04 44 ☉ ± ♅
05 49 ☽ ⚹ ♄
08 45 ☽ # ♇
10 47 ☽ ⊼ ♇
13 14 ☽ □ ♂
15 34 ☽ ± ♅
16 25 ☽ ☌ ☉
16 32 ☽ ∠ ♀
17 49 ☽ □ ♆
18 12 ☽ ± ♃
23 47 ☉ ∠ ♀

17 Monday
00 19 ☽ ⊻ ☿
01 38 ☽ ⊼ ♅
03 19 ☽ ∥ ☉
04 15 ☽ ⊼ ♃
07 09 ☽ ∠ ♄
08 37 ☽ ⊥ ♀
10 58 ☽ ⊥ ☿
12 22 ☉ □ ♆
15 52 ♀ bQ ♆
16 01 ☉ ± ♃
17 44 ☽ # ♀
19 28 ☽ ⊻ ♀
19 58 ♂ ∥ ♆
22 39 ☽ ⊥ ♄

18 Tuesday
00 58 ☽ # ♄
02 41 ☿ △ ♅
03 15 ☽ ∠ ☿
09 08 ☽ ⊻ ♄
13 55 ☽ ☍ ♇
18 28 ☽ △ ♂
21 28 ☽ △ ♆
21 47 ☽ ∥ ♅
22 10 ☽ ∥ ♃

19 Wednesday
00 07 ☽ ⊻ ☉
01 43 ☽ # ♃
02 07 ☽ # ♅
05 36 ☽ ☍ ♅
07 17 ☽ ⚹ ☿
08 11 ☽ ☍ ♃
11 45 ☽ ⊥ ☉
20 50 ☿ △ ♃
22 26 ☽ bQ ♂
23 54 ☽ ∥ ♄

20 Thursday
00 32 ☽ bQ ♆
04 03 ☽ ☌ ♀
05 33 ☽ ∠ ☉
06 36 ☽ Q ☿
08 34 ☿ ∥ ♂
12 04 ☽ ∥ ♀
15 42 ☽ ± ♂
15 44 ☽ ☌ ♄
16 55 ♂ ± ♄
17 14 ☽ ± ♆
17 45 ☽ # ☉
20 21 ☽ ⊼ ♇

21 Friday
03 25 ☽ ⊼ ♂
04 33 ☽ ⊼ ♆
07 41 ☽ ± ♇
12 11 ☽ ⚹ ☉
13 04 ☽ ⊼ ♅
15 34 ☽ ⊼ ♃
19 09 ☽ □ ☿
20 35 ☽ ∥ ♇
22 52 ☉ ⊼ ♅

22 Saturday
00 35 ☽ ± ♅
00 56 ☽ bQ ♇
03 04 ☽ ± ♃
03 23 ☽ # ♆
04 25 ☽ # ☿
05 56 ☽ # ♂
09 37 ☉ # ♀
10 08 ♂ ☌ ♆
13 29 ☽ Q ☉
16 45 ☽ ⊻ ♀
18 07 ☽ bQ ♅
20 33 ☽ bQ ♃

23 Sunday
02 02 ☽ ⊻ ♄
03 01 ☉ ⊼ ♃
05 19 ☽ ⊥ ♀
06 19 ☽ △ ♇
14 03 ☽ ⊥ ♄
15 06 ☽ ☍ ♆
16 14 ☽ ☍ ♂
23 50 ☽ △ ♅

24 Monday
00 22 ☽ ∠ ♀
02 10 ☽ △ ♃
04 30 ☽ □ ☉
08 14 ☽ ∠ ♄
08 40 ☿ ∥ ♆
11 45 ☽ △ ☿
17 34 ☿ bQ ♇

25 Tuesday
08 23 ☽ ⚹ ♀
14 47 ☽ ⚹ ♄
18 31 ☽ □ ♇
19 01 ☿ Q ♄
21 05 ☽ bQ ☿

26 Wednesday
01 37 ☿ Q ♆
03 34 ☽ ⊼ ♆
05 27 ☉ ∠ ♄
07 08 ☽ ⊼ ♂
10 01 ☽ Q ♀
12 12 ☽ □ ♅
14 13 ☽ □ ♃
15 15 ☽ Q ♄
15 41 ☽ ± ♆
17 09 ☽ ± ☿
19 42 ☽ ± ♂
22 40 ☽ △ ☉

27 Thursday
06 31 ☽ ⊼ ☿
06 45 ☽ Q ♇
09 48 ☽ bQ ♆
14 29 ☽ bQ ♂

28 Friday
00 06 ☽ □ ♀
00 13 ☽ Q ♅
02 00 ☽ Q ♃
03 38 ☽ □ ♄
06 39 ☽ ⚹ ♇
07 24 ☽ bQ ☉
09 46 ☿ Q ♂
15 39 ☽ △ ♆
18 11 ♄ ± ♆
20 21 ♆ St R
21 23 ☽ △ ♂
23 52 ☽ ⚹ ♅

MARCH 1969

LONGITUDES

Date	Sidereal time h m s	Sun ☉	Moon ☽	Moon ☽ 24.00	Mercury ☿	Venus ♀	Mars ♂	Jupiter ♃	Saturn ♄	Uranus ♅	Neptune ♆	Pluto ♇
01	22 36 27	10 ♓ 44 03	09 ♌ 05 16	15 ♌ 16 39	14 ♒ 44	21 ♈ 43	01 ♐ 53	03 ♎ 40	22 ♈ 47	02 ♎ 52	28 ♏ 43	24 ♍ 08
02	22 40 24	11 44 14	21 ♌ 31 22	27 ♌ 49 35	15 58	22 16	02 19	03 R 33	22 54	02 R 50	28 R 43	24 R 07
03	22 44 21	12 44 22	04 ♍ 11 23	10 ♍ 36 48	17 14	22 47	02 45	03 26	23 00	02 47	28 42	24 05
04	22 48 17	13 44 28	17 ♍ 05 49	23 ♍ 38 24	18 32	23 16	03 11	03 20	23 06	02 45	28 42	24 03
05	22 52 14	14 44 33	00 ♎ 14 27	06 ♎ 53 53	19 52	23 44	03 36	03 13	23 13	02 42	28 42	24 02
06	22 56 10	15 44 35	13 ♎ 36 34	20 ♎ 22 20	21 13	24 10	04 02	03 06	23 20	02 40	28 42	24 00
07	23 00 07	16 44 36	27 ♎ 11 02	04 ♏ 02 30	22 35	24 34	04 26	02 58	23 26	02 38	28 42	23 59
08	23 04 03	17 44 36	10 ♏ 56 34	17 ♏ 53 03	23 59	24 57	04 51	02 51	23 33	02 35	28 42	23 57
09	23 08 00	18 44 34	24 ♏ 51 46	01 ♐ 52 32	25 24	25 18	05 15	02 44	23 39	02 33	28 41	23 55
10	23 11 56	19 44 30	08 ♐ 55 09	15 ♐ 59 24	26 51	25 36	05 40	02 37	23 46	02 30	28 41	23 54
11	23 15 53	20 44 24	23 ♐ 05 02	00 ♑ 11 47	28 18	25 53	06 03	02 29	23 53	02 28	28 41	23 52
12	23 19 50	21 44 17	07 ♑ 19 20	14 ♑ 27 23	29 ♒ 48	26 08	06 27	02 22	24 00	02 25	28 40	23 50
13	23 23 46	22 44 09	21 ♑ 35 31	28 ♑ 43 22	01 ♓ 18	26 20	06 50	02 14	24 07	02 22	28 40	23 49
14	23 27 43	23 43 58	05 ♒ 50 29	12 ♒ 56 24	02 50	26 31	07 13	02 06	24 14	02 20	28 40	23 47
15	23 31 39	24 43 46	20 ♒ 00 39	27 ♒ 02 45	04 23	26 39	07 36	01 59	24 21	02 17	28 39	23 46
16	23 35 36	25 43 32	04 ♓ 02 15	10 ♓ 58 41	05 57	26 45	07 58	01 51	24 28	02 15	28 39	23 44
17	23 39 32	26 43 16	17 ♓ 51 39	24 ♓ 40 48	07 32	26 49	08 20	01 43	24 35	02 12	28 38	23 42
18	23 43 29	27 42 59	01 ♈ 25 50	08 ♈ 06 30	09 09	26 R 50	08 42	01 36	24 42	02 10	28 37	23 41
19	23 47 25	28 42 39	14 ♈ 42 40	21 ♈ 14 14	10 47	26 48	09 03	01 28	24 49	02 07	28 37	23 39
20	23 51 22	29 ♓ 42 17	27 ♈ 41 13	04 ♉ 03 42	12 26	26 45	09 24	01 20	24 56	02 04	28 36	23 38
21	23 55 19	00 ♈ 41 53	10 ♉ 21 50	16 ♉ 35 52	14 07	26 39	09 44	01 12	25 03	02 02	28 36	23 36
22	23 59 15	01 41 26	22 ♉ 46 07	28 ♉ 52 55	15 49	26 30	10 05	01 05	25 10	01 59	28 35	23 34
23	00 03 12	02 40 58	04 ♊ 56 41	10 ♊ 57 55	17 32	26 19	10 24	00 57	25 18	01 56	28 34	23 33
24	00 07 08	03 40 27	16 ♊ 57 06	22 ♊ 54 46	19 16	26 05	10 44	00 49	25 25	01 54	28 33	23 31
25	00 11 05	04 39 54	28 ♊ 51 29	04 ♋ 47 50	21 02	25 49	11 03	00 41	25 32	01 51	28 33	23 30
26	00 15 01	05 39 19	10 ♋ 44 24	16 ♋ 41 45	22 48	25 30	11 21	00 34	25 40	01 49	28 32	23 28
27	00 18 58	06 38 41	22 ♋ 40 30	28 ♋ 41 11	24 37	25 10	11 40	00 26	25 47	01 46	28 31	23 26
28	00 22 54	07 38 01	04 ♌ 44 22	10 ♌ 50 34	26 26	24 46	11 58	00 18	25 54	01 43	28 30	23 25
29	00 26 51	08 37 19	17 ♌ 00 14	23 ♌ 13 49	28 ♓ 17	24 21	12 15	00 11	26 02	01 41	28 29	23 23
30	00 30 48	09 36 35	29 ♌ 31 40	05 ♍ 54 06	00 ♈ 10	23 53	12 32	00 ♎ 03	26 09	01 38	28 28	23 22
31	00 34 44	10 ♈ 35 48	12 ♍ 21 21	18 ♍ 53 32	02 ♈ 03	23 ♈ 24	12 ♐ 48	29 ♍ 55	26 ♈ 17	01 ♎ 36	28 ♏ 27	23 ♍ 20

MOON NODES AND LATITUDE / DECLINATIONS

Date	Moon True ☊	Moon Mean ☊	Moon Latitude	Sun ☉	Moon ☽	Mercury ☿	Venus ♀	Mars ♂	Jupiter ♃	Saturn ♄	Uranus ♅	Neptune ♆	Pluto ♇
01	00 ♈ 21	01 ♈ 28	04 N 05	07 S 33	21 N 56	17 S 18	12 N 53	19 S 27	00 S 02	06 N 43	00 S 26	18 S 12	16 N 56
02	00 R 18	01 25	03 17	07 10	17 27	17 03	13 10	19 33	00 N 01	06 46	00 25	18 12	16 56
03	00 16	01 21	02 19	06 47	12 08	16 47	13 33	19 39	00 03	06 48	00 24	18 12	16 57
04	00 14	01 18	01 N 12	06 24	06 N 12	16 30	13 53	19 44	00 06	06 51	00 23	18 12	16 58
05	00 D 14	01 15	00 00	06 01	00 S 06	16 12	14 11	19 50	00 09	06 53	00 22	18 12	16 58
06	00 14	01 12	01 S 13	05 37	06 30	15 52	14 29	19 55	00 12	06 56	00 21	18 11	16 59
07	00 15	01 09	02 24	05 14	12 42	15 31	14 47	20 00	00 15	06 58	00 20	18 11	17 00
08	00 16	01 06	03 26	04 51	18 23	15 08	15 03	20 06	00 18	07 01	00 19	18 11	17 01
09	00 17	01 02	04 18	04 27	23 09	14 44	15 19	20 11	00 21	07 04	00 18	18 11	17 02
10	00 18	00 59	04 55	04 04	26 39	14 19	15 34	20 16	00 24	07 06	00 17	18 11	17 02
11	00 R 18	00 56	05 14	03 40	28 29	13 53	15 48	20 21	00 27	07 09	00 16	18 11	17 03
12	00 18	00 53	05 14	03 17	28 28	13 25	16 01	20 25	00 30	07 12	00 15	18 11	17 04
13	00 17	00 50	04 55	02 53	26 34	12 57	16 13	20 30	00 33	07 14	00 14	18 11	17 05
14	00 16	00 46	04 18	02 29	22 59	12 26	16 25	20 35	00 36	07 17	00 13	18 11	17 05
15	00 16	00 43	03 25	02 06	18 03	11 55	16 35	20 39	00 39	07 19	00 12	18 10	17 06
16	00 15	00 40	02 20	01 42	12 12	11 23	16 44	20 44	00 42	07 22	00 11	18 10	17 07
17	00 15	00 37	01 S 08	01 18	05 S 51	10 49	16 52	20 48	00 46	07 24	00 10	18 10	17 07
18	00 D 15	00 34	00 N 07	00 55	00 N 40	10 14	16 59	20 53	00 49	07 27	00 09	18 10	17 08
19	00 15	00 31	01 19	00 31	07 01	09 38	17 05	20 57	00 52	07 30	00 08	18 10	17 09
20	00 15	00 27	02 26	00 S 07	12 56	09 00	17 10	21 00	00 55	07 33	00 07	18 10	17 09
21	00 R 15	00 24	03 24	00 N 17	18 10	08 22	17 13	21 05	00 58	07 36	00 06	18 09	17 10
22	00 15	00 21	04 12	00 40	22 31	07 42	17 15	21 09	01 01	07 38	00 05	18 09	17 11
23	00 15	00 18	04 47	01 04	25 49	07 01	17 15	21 13	01 04	07 41	00 04	18 09	17 11
24	00 14	00 15	05 08	01 28	27 55	06 19	17 14	21 17	01 07	07 44	00 03	18 09	17 12
25	00 14	00 12	05 17	01 51	28 43	05 36	17 12	21 21	01 10	07 46	00 S 02	18 09	17 12
26	00 D 14	00 08	05 12	02 15	28 11	04 51	17 08	21 25	01 14	07 49	00 00	18 08	17 13
27	00 14	00 05	04 53	02 38	26 21	04 06	17 03	21 28	01 17	07 52	00 N 01	18 08	17 14
28	00 15	00 ♈ 02	04 21	03 02	23 19	03 19	16 56	21 32	01 20	07 55	00 02	18 08	17 14
29	00 16	29 ♓ 59	03 37	03 25	19 12	02 32	16 47	21 35	01 23	07 57	00 03	18 08	17 15
30	00 17	29 56	02 42	03 49	14 10	01 43	16 37	21 39	01 26	08 00	00 04	18 07	17 15
31	00 ♈ 18	29 ♓ 52	01 N 38	04 N 12	08 N 26	00 S 54	16 N 26	21 S 42	01 N 29	08 N 03	00 N 05	18 S 07	17 N 16

ZODIAC SIGN ENTRIES

Date	h	m	Planets
03	04	07	☽ ♍
05	11	34	☽ ♎
07	16	56	☽ ♏
09	20	48	☽ ♐
11	23	40	☽ ♑
12	15	19	☿ ♓
14	02	09	☽ ♒
16	05	04	☽ ♓
18	09	27	☽ ♈
20	16	20	☽ ♉
20	19	08	☉ ♈
23	02	12	☽ ♊
25	14	18	☽ ♋
28	02	37	☽ ♌
30	09	59	☿ ♈
30	12	54	☽ ♍
30	21	36	♃ ♍

LATITUDES

Date	Mercury ☿	Venus ♀	Mars ♂	Jupiter ♃	Saturn ♄	Uranus ♅	Neptune ♆	Pluto ♇
01	00 S 55	04 N 46	01 N 07	01 N 33	02 S 18	00 N 46	01 N 43	15 N 57
04	01 18	05 13	01 05	01 33	02 18	00 47	01 43	15 57
07	01 36	05 39	01 03	01 34	02 18	00 47	01 44	15 58
10	01 52	06 05	01 01	01 34	02 17	00 47	01 44	15 58
13	02 04	06 31	00 58	01 34	02 17	00 47	01 44	15 58
16	02 12	06 54	00 55	01 34	02 16	00 47	01 44	15 58
19	02 16	07 15	00 53	01 35	02 16	00 47	01 44	15 58
22	02 17	07 33	00 49	01 35	02 16	00 47	01 44	15 58
25	02 13	07 46	00 46	01 35	02 15	00 47	01 45	15 58
28	02 05	07 54	00 42	01 35	02 15	00 47	01 45	15 58
31	01 S 52	07 N 55	00 N 38	01 N 35	02 S 15	00 N 47	01 N 45	15 N 58

DATA

Julian Date	2440282
Delta T	+39 seconds
Ayanamsa	23° 25' 36"
Synetic vernal point	05° ♓ 41' 24"
True obliquity of ecliptic	23° 26' 45"

MOON'S PHASES, APSIDES AND POSITIONS ☽

Date	h	m	Phase	Longitude	Eclipse Indicator
04	05	17	○	13 ♍ 28	
11	07	44	☾	20 ♐ 34	
18	04	51	●	27 ♓ 25	Annular
26	00	48	☽	05 ♋ 12	

Day	h	m		
13	01	34	Perigee	
25	18	29	Apogee	
05	11	38	0S	
11	23	45	Max dec	28° S 43'
18	09	31	0N	
25	14	23	Max dec	28° N 44'

LONGITUDES

Date	Chiron ⚷	Ceres ⚳	Pallas ⚴	Juno ⚵	Vesta ⚶	Black Moon Lilith ⚸
01	01 ♈ 31	21 ♑ 33	02 ♑ 29	06 ♑ 05	09 ♉ 09	28 ♊ 36
11	02 ♈ 05	24 ♑ 57	05 ♑ 23	08 ♑ 36	12 ♉ 53	29 ♊ 43
21	02 ♈ 40	28 ♑ 08	07 ♑ 55	10 ♑ 50	16 ♉ 46	00 ♋ 51
31	03 ♈ 15	01 ♒ 06	10 ♑ 03	12 ♑ 46	20 ♉ 46	01 ♋ 58

ASPECTARIAN

h m Aspects

01 Saturday
01 30 ☽ ⚹ ♃
02 44 ☽ ± ☉
12 05 ☽ ∠ ♇
15 29 ☽ ⊼ ☉

02 Sunday
00 09 ☽ ☍ ☿
01 35 ☽ # ♂
04 56 ☽ ∠ ♅
05 28 ☽ ⊥ ♇
06 22 ☽ ∠ ♃
08 18 ☽ # ♆
13 29 ☽ △ ♀
13 59 ☽ # ☿
14 26 ☽ ∥ ♇
14 39 ☽ △ ♄
16 56 ☽ ⊻ ♇
22 04 ☽ ⊥ ♅
22 22 ☿ # ♇
23 23 ☽ ⊥ ♃

03 Monday
01 40 ☽ □ ♆
06 13 ☽ ∥ ♀
08 36 ♀ ± ♆
09 13 ☽ □ ♂
09 22 ☽ ⊻ ♅
10 36 ☉ # ♄
10 36 ☽ ⊻ ♃
13 41 ♂ ⚹ ♅
19 00 ☽ Ᵽ ♀
19 12 ☽ Ᵽ ♄
21 53 ☿ Ᵽ ♅

04 Tuesday
01 36 ♀ ☌ ♄
03 20 ☿ ± ♇
05 17 ☽ ☍ ☉
08 27 ☿ Ᵽ ♃
09 29 ☽ ∥ ♄
11 11 ☽ # ☉
11 17 ☽ Q ♆
12 01 ☽ ± ♄
12 20 ☽ ± ♀
14 57 ☽ ⊼ ☿
18 24 ♂ ⚹ ♃
19 45 ☽ Q ♂
23 07 ☽ ⊼ ♄
23 45 ☽ ⊼ ♀

05 Wednesday
00 44 ☽ ☌ ♇
03 09 ☽ ± ☿
09 13 ☽ ⚹ ♆
10 16 ☽ # ♅
11 04 ☽ ∥ ♃
12 13 ☽ # ♃
13 00 ☽ ∥ ♅
16 27 ☽ ☌ ♅
17 19 ☽ ☌ ♃
18 17 ☽ ⚹ ♂
21 16 ☽ Ᵽ ☿

06 Thursday
03 10 ♀ ⊼ ♇
08 54 ☽ ∥ ☉
12 10 ☽ ∠ ♆
13 39 ☽ # ♄
16 06 ☽ ⊼ ☉
21 55 ☽ ∠ ♂

07 Friday
02 59 ☽ △ ☿
03 34 ☽ ± ☉
04 06 ☽ ⊥ ♆
05 21 ☽ ☍ ♄
06 22 ☽ ⊻ ♇
07 17 ☽ ☍ ♀
14 16 ☽ ⊥ ♂
14 39 ☽ ⊻ ♆
16 53 ☽ ⊥ ♇
20 37 ☽ Ᵽ ☉
20 52 ☽ # ♀
21 30 ☽ ⊻ ♅
22 03 ☽ ⊻ ♃
22 48 ☽ ∥ ☿

08 Saturday
01 05 ☽ ⊻ ♂
03 58 ☿ ⚹ ♄
05 55 ☽ # ♇
06 41 ☉ ⊥ ♄
07 55 ☽ ⊥ ♅
08 24 ☽ ⊥ ♃
08 33 ☽ ∠ ♇
11 08 ☽ ∥ ♆
11 29 ☿ ⊼ ♇
14 59 ☿ # ♀
17 54 ♃ # ♅
20 08 ☽ ∥ ♂
23 27 ☽ ∠ ♅
23 51 ☽ ∠ ♃

09 Sunday
00 40 ☽ △ ☉
09 43 ☿ ⚹ ♀
09 55 ☽ ⊼ ♄
10 23 ☽ ⚹ ♇
12 46 ☽ ⊼ ♀
13 02 ☽ □ ☿
18 33 ☽ ☌ ♆
20 17 ☽ ± ♄
23 16 ☽ ± ♀

10 Monday
01 06 ☽ ⚹ ♅
01 21 ☽ ⚹ ♃
06 17 ☽ ☌ ♂
06 30 ☿ ± ♅
06 52 ☽ Q ♇
07 24 ☉ ⊥ ♀
08 26 ☿ ± ♃
11 45 ☽ Ᵽ ♄
14 56 ☽ Ᵽ ♀
21 27 ☽ Q ♅
21 35 ☽ Q ♃
23 13 ☽ Q ☿

11 Tuesday
01 16 ♂ Q ♇
07 44 ☽ □ ☉
09 19 ♄ ⊼ ♇
13 19 ☽ □ ♇
13 22 ☽ △ ♄
16 49 ☽ △ ♀
18 01 ☿ □ ♆
19 40 ♃ ☌ ♅
21 26 ☽ ⊻ ♆
21 50 ☽ ⚹ ☿

12 Wednesday
03 43 ☽ □ ♃
03 46 ☽ □ ♅
07 32 ☽ ⊥ ♆
10 29 ☽ ⊻ ♂
16 22 ☽ Q ☉
20 52 ☽ ⊥ ♂
22 41 ☽ ∠ ♆

13 Thursday
02 03 ☽ ∠ ☿
12 25 ☽ ∠ ♂
14 04 ☽ ⚹ ☉
15 44 ☽ △ ♇
16 16 ☽ □ ♄
18 59 ☽ ⊥ ☿
20 06 ☽ □ ♀
23 54 ☽ ⚹ ♆

14 Friday
01 35 ☿ ⊼ ♃
04 26 ☿ ⊼ ♅
05 46 ☽ △ ♃
06 06 ☽ △ ♅
06 18 ☽ ⊻ ☿
13 17 ☉ ☍ ♇
14 23 ☽ ⚹ ♂
16 58 ☽ Ᵽ ♇
17 15 ☽ ∠ ☉
20 08 ☽ Q ♆
22 53 ☽ Q ♄

15 Saturday
00 10 ☽ ∥ ♂
01 29 ☉ ⊻ ♄
02 49 ☽ Q ♀
06 54 ☽ Ᵽ ♃
07 23 ☽ Ᵽ ♅
08 11 ☽ ± ♇
09 39 ☽ ⊥ ☉
11 17 ☽ Q ♂
11 28 ☽ ∥ ♆
16 08 ☽ # ♇
18 09 ☽ # ♀
18 22 ☽ ⊼ ♇
19 27 ☽ ⚹ ♄
20 39 ☽ ⊻ ☉
22 05 ☽ ± ♃
22 40 ☽ ± ♅
23 25 ☽ ⚹ ♀

16 Sunday
02 44 ☽ □ ♆
08 17 ☽ ⊼ ♃
08 56 ☽ ⊼ ♅
15 32 ☽ ∥ ☿
15 43 ☽ ☌ ☿
18 58 ☽ □ ♂
21 27 ☽ ∠ ♄

17 Monday
01 24 ☽ ∠ ♀
06 14 ☽ # ♄
13 16 ☽ ⊥ ♄
14 11 ☉ ⊻ ♀
17 11 ☽ ⊥ ♀
22 16 ☽ ☍ ♇
23 55 ☽ ⊻ ♄

18 Tuesday
03 16 ☿ □ ♂
03 48 ☽ ⊻ ♀
04 51 ☽ ☌ ☉
05 48 ☽ ∥ ☉
06 34 ☽ # ♃
07 00 ☽ △ ♆
08 59 ☽ ∥ ♅
10 04 ☽ # ♅
11 49 ♀ St R
12 18 ☽ ☍ ♃
12 32 ☽ ∥ ♃
12 50 ☽ # ☉
13 18 ☽ ☍ ♅
17 11 ☉ # ♃
20 39 ☿ ∠ ♄

19 Wednesday
01 25 ☽ △ ♂
03 50 ☽ ⊻ ☿
09 42 ☉ △ ♆
10 00 ☽ Ᵽ ♆
13 54 ☽ ∥ ♄
16 21 ☽ ⊥ ☿
21 21 ☽ # ☿

20 Thursday
02 25 ☿ ∠ ♀
02 32 ☽ ± ♆
04 27 ☽ ⊼ ♇
05 42 ☽ Ᵽ ♂
06 49 ☽ ☌ ♄
09 50 ♀ ∥ ♇
10 15 ☽ ☌ ♀
11 28 ☽ ∠ ☿
12 23 ☉ ∥ ♅
13 43 ☽ ⊼ ♆
15 38 ☽ ± ♇
16 06 ☽ ⊻ ☉
18 47 ☽ ⊼ ♃
20 13 ☽ ⊼ ♅
23 03 ☽ ± ♂

21 Friday
01 19 ☉ # ♅
04 24 ☽ ⊥ ☉
06 02 ☽ ± ♃
07 07 ☽ ∥ ♇
07 18 ☽ ∥ ♀
07 33 ☽ ± ♅
08 38 ☽ Ᵽ ♇
10 46 ☽ ⊼ ♂
11 57 ☽ # ♆
20 20 ☽ ⚹ ☿
22 53 ☉ ☍ ♃
23 08 ☽ Ᵽ ♃
23 09 ☽ ∠ ☉

22 Saturday
00 47 ☽ Ᵽ ♅
03 43 ☽ # ♂
13 34 ☽ △ ♇
13 58 ☿ # ♄
16 45 ☽ ⊻ ♄
18 49 ☉ ☍ ♅
19 13 ☽ ⊻ ♀
23 03 ♂ Ᵽ ♄
23 24 ☽ ☍ ♆
23 30 ☽ Q ☿

23 Sunday
04 10 ☽ △ ♃
04 41 ☽ ⊥ ♄
06 04 ☽ △ ♅
06 52 ☽ ⊥ ♀
07 07 ☽ ⚹ ☉
12 19 ☉ ∥ ♃
22 46 ☽ ∠ ♄
23 11 ☽ ☍ ♂

24 Monday
00 28 ☽ ∠ ♀
09 12 ☽ Q ☉
14 13 ☿ ⊥ ♄
17 27 ☽ □ ☿
21 48 ☿ ⊥ ♀

25 Tuesday
01 11 ☽ □ ♇
02 43 ♀ Ᵽ ♂
05 13 ☽ ⚹ ♄
06 00 ☽ ⚹ ♀
08 25 ♀ ∥ ♇
11 22 ☽ ⊼ ♆
15 40 ☽ □ ♃
18 02 ☽ □ ♅
23 29 ☽ ± ♆

26 Wednesday
00 48 ☽ □ ☉
03 50 ♀ ☌ ♄
05 39 ☽ Q ♀
05 43 ☽ Q ♄
13 17 ☽ ⊼ ♂
13 28 ☽ Q ♇
17 37 ☽ Ᵽ ♆
20 40 ☿ ☍ ♇

27 Thursday
01 41 ☽ ± ♂
03 35 ☽ Q ♃
06 12 ☽ Q ♅
13 32 ☽ ⚹ ♇
16 34 ☽ △ ☿
16 49 ☽ □ ♀
18 00 ☿ ⊻ ♀
18 16 ☽ □ ♄
20 10 ☽ Ᵽ ♂
23 39 ☽ △ ♆

28 Friday
03 18 ☽ ⚹ ♃
04 31 ☿ ⊻ ♄
06 03 ☽ ⚹ ♅
17 57 ☉ # ☿
18 12 ☽ △ ☉
19 13 ☽ ∠ ♇
22 56 ☽ # ♂

29 Saturday
02 32 ☽ △ ♂
03 30 ☽ Ᵽ ☿
07 52 ♂ Q ♃
08 29 ☽ ∠ ♃
11 23 ☽ ∠ ♅
12 44 ☽ ⊥ ♇
14 35 ☿ △ ♆
17 27 ☽ # ♆
21 46 ☽ ∥ ♇
23 59 ☽ ± ☿

30 Sunday
00 17 ☽ ⊻ ♇
00 23 ☽ ∥ ♀
01 39 ☽ △ ♀
01 40 ☽ ⊥ ♃
01 50 ☽ Ᵽ ☉
04 38 ☽ ⊥ ♅
05 31 ☽ △ ♄
10 00 ☽ □ ♆
10 43 ☿ ☍ ♃
12 59 ☽ ⊻ ♃
13 24 ☽ ⊼ ☿
15 59 ☽ ⊻ ♅
20 02 ☿ # ♃
20 21 ☽ ± ☉

31 Monday
04 56 ☽ Ᵽ ♀
06 22 ☿ ☍ ♅
08 29 ☽ ⊼ ☉
09 59 ☽ Ᵽ ♄
12 51 ☽ □ ♂
13 31 ☽ ∥ ♄
15 11 ♀ ⊼ ♇
19 32 ☽ Q ♆
20 56 ☽ ± ♀

All ephemeris data is given at 12.00 UT and the Moon's longitude is additionally given for 24.00 UT

APRIL 1969

LONGITUDES

Date	Sidereal time h m s	Sun ☉	Moon ☽	Moon ☽ 24.00	Mercury ☿	Venus ♀	Mars ♂	Jupiter ♃	Saturn ♄	Uranus ♅	Neptune ♆	Pluto ♇
01	00 38 41	11 ♈ 34 59	25 ♍ 30 44	02 ♎ 12 53	03 ♈ 58	22 ♈ 53	13 ♐ 04	29 ♍ 48	26 ♈ 24	01 ♎ 33	28 ♏ 27	23 ♍ 19
02	00 42 37	12 34 08	08 ♎ 59 52	15 ♎ 51 26	05 54	22 R 20	13 20	29 R 41	26 31	01 R 31	28 R 26	23 R 17
03	00 46 34	13 33 14	22 ♎ 47 16	29 ♎ 46 58	07 52	21 46	13 35	29 33	26 39	01 28	28 25	23 16
04	00 50 30	14 32 19	06 ♏ 50 02	13 ♏ 55 57	09 51	21 10	13 50	29 26	26 46	01 26	28 23	23 14
05	00 54 27	15 31 22	21 ♏ 04 08	28 ♏ 13 58	11 51	20 34	14 04	29 19	26 54	01 23	28 22	23 13
06	00 58 23	16 30 24	05 ♐ 24 52	12 ♐ 36 14	13 53	19 57	14 18	29 11	27 02	01 21	28 21	23 11
07	01 02 20	17 29 23	19 ♐ 47 31	26 ♐ 58 11	15 55	19 19	14 31	29 04	27 09	01 18	28 20	23 10
08	01 06 17	18 28 21	04 ♑ 07 49	11 ♑ 15 59	17 58	18 41	14 43	28 57	27 17	01 16	28 19	23 08
09	01 10 13	19 27 17	18 ♑ 22 22	25 ♑ 26 40	20 03	18 03	14 55	28 50	27 24	01 13	28 18	23 07
10	01 14 10	20 26 11	02 ♒ 28 41	09 ♒ 28 14	22 08	17 26	15 07	28 44	27 32	01 11	28 17	23 05
11	01 18 06	21 25 04	16 ♒ 25 11	23 ♒ 19 26	24 14	16 49	15 18	28 37	27 39	01 08	28 16	23 04
12	01 22 03	22 23 54	00 ♓ 10 52	06 ♓ 59 26	26 20	16 12	15 28	28 30	27 47	01 06	28 14	23 03
13	01 25 59	23 22 43	13 ♓ 45 04	20 ♓ 27 42	28 ♈ 26	15 37	15 38	28 24	27 55	01 04	28 13	23 01
14	01 29 56	24 21 30	27 ♓ 07 16	03 ♈ 43 42	00 ♉ 32	15 03	15 47	28 17	28 02	01 01	28 12	23 00
15	01 33 52	25 20 16	10 ♈ 16 58	16 ♈ 47 00	02 38	14 30	15 55	28 11	28 10	00 59	28 11	22 59
16	01 37 49	26 18 59	23 ♈ 13 47	29 ♈ 37 15	04 43	13 59	16 03	28 05	28 18	00 57	28 09	22 57
17	01 41 46	27 17 40	05 ♉ 57 27	12 ♉ 14 24	06 47	13 30	16 10	27 59	28 25	00 54	28 08	22 56
18	01 45 42	28 16 20	18 ♉ 28 11	24 ♉ 38 54	08 49	13 02	16 17	27 53	28 33	00 52	28 07	22 55
19	01 49 39	29 ♈ 14 58	00 ♊ 48 21	06 ♊ 51 48	10 50	12 37	16 23	27 47	28 41	00 50	28 05	22 54
20	01 53 35	00 ♉ 13 33	12 ♊ 54 28	18 ♊ 55 00	12 49	12 14	16 28	27 42	28 48	00 48	28 04	22 52
21	01 57 32	01 12 06	24 ♊ 53 45	00 ♋ 51 08	14 46	11 53	16 33	27 36	28 56	00 46	28 02	22 51
22	02 01 28	02 10 38	06 ♋ 47 36	12 ♋ 43 37	16 40	11 34	16 37	27 31	29 03	00 44	28 01	22 50
23	02 05 25	03 09 07	18 ♋ 39 44	24 ♋ 36 30	18 31	11 18	16 40	27 25	29 11	00 42	28 00	22 49
24	02 09 21	04 07 34	00 ♌ 34 31	06 ♌ 34 21	20 19	11 03	16 42	27 20	29 19	00 40	27 58	22 48
25	02 13 18	05 05 59	12 ♌ 36 38	18 ♌ 41 59	22 03	10 52	16 44	27 16	29 26	00 37	27 57	22 47
26	02 17 15	06 04 22	24 ♌ 50 59	01 ♍ 04 12	23 44	10 43	16 45	27 11	29 34	00 36	27 55	22 45
27	02 21 11	07 02 42	07 ♍ 22 12	13 ♍ 45 28	25 21	10 36	16 R 46	27 06	29 42	00 34	27 54	22 44
28	02 25 08	08 01 01	20 ♍ 14 24	26 ♍ 49 21	26 55	10 32	16 45	27 02	29 49	00 32	27 52	22 43
29	02 29 04	08 59 17	03 ♎ 30 33	10 ♎ 18 04	28 24	10 30	16 44	26 57	29 ♈ 57	00 30	27 51	22 42
30	02 33 01	09 ♉ 57 32	17 ♎ 11 54	24 ♎ 11 50	29 ♉ 49	10 ♈ 30	16 ♐ 42	26 ♍ 53	00 ♉ 04	00 ♎ 28	27 ♏ 49	22 ♍ 41

MOON

Date	Moon True ☊	Moon Mean ☊	Moon ☽ Latitude
01	00 ♈ 18	29 ♓ 49	00 N 26
02	00 R 18	29 46	00 S 48
03	00 17	29 43	02 01
04	00 16	29 40	03 08
05	00 14	29 37	04 05
06	00 12	29 33	04 46
07	00 10	29 30	05 10
08	00 09	29 27	05 14
09	00 08	29 24	04 59
10	00 D 08	29 21	04 26
11	00 10	29 18	03 38
12	00 11	29 14	02 38
13	00 12	29 11	01 29
14	00 13	29 08	00 S 17
15	00 R 13	29 05	00 N 55
16	00 11	29 02	02 03
17	00 08	28 58	03 03
18	00 04	28 55	03 54
19	29 ♓ 59	28 52	04 33
20	29 54	28 49	04 58
21	29 49	28 46	05 11
22	29 45	28 43	05 09
23	29 42	28 39	04 55
24	29 41	28 36	04 28
25	29 D 42	28 33	03 49
26	29 43	28 30	02 58
27	29 44	28 27	01 59
28	29 46	28 23	00 N 52
29	29 R 46	28 20	00 S 20
30	29 ♓ 45	28 ♓ 17	01 S 33

DECLINATIONS

Date	Sun ☉	Moon ☽	Mercury ☿	Venus ♀	Mars ♂	Jupiter ♃	Saturn ♄	Uranus ♅	Neptune ♆	Pluto ♇
01	04 N 35	02 N 11	00 S 03	16 N 13	21 S 46	01 N 32	08 N 06	00 N 06	18 S 07	17 N 16
02	04 58	04 S 18	00 N 48	15 59	21 49	01 34	08 08	00 07	18 07	17 17
03	05 21	10 44	01 41	15 43	21 52	01 37	08 11	00 08	18 06	17 17
04	05 44	16 46	02 34	15 26	21 56	01 40	08 14	00 09	18 06	17 18
05	06 07	21 58	03 27	15 08	21 59	01 43	08 17	00 10	18 06	17 18
06	06 29	25 54	04 22	14 48	22 02	01 46	08 19	00 11	18 06	17 19
07	06 52	28 12	05 17	14 28	22 05	01 49	08 22	00 12	18 05	17 19
08	07 15	28 37	06 12	14 07	22 08	01 51	08 25	00 13	18 05	17 19
09	07 37	27 08	07 08	13 44	22 11	01 54	08 28	00 13	18 05	17 20
10	07 59	23 56	08 03	13 22	22 14	01 57	08 30	00 14	18 04	17 20
11	08 21	19 23	08 59	12 58	22 17	01 59	08 33	00 15	18 04	17 21
12	08 43	13 52	09 55	12 35	22 20	02 02	08 36	00 16	18 04	17 21
13	09 05	07 46	10 50	12 11	22 23	02 04	08 39	00 17	18 03	17 21
14	09 27	01 S 24	11 44	11 47	22 26	02 07	08 41	00 18	18 03	17 22
15	09 48	04 N 55	12 38	11 23	22 29	02 09	08 44	00 19	18 03	17 22
16	10 10	10 56	13 31	11 00	22 31	02 11	08 47	00 20	18 03	17 22
17	10 31	16 24	14 23	10 36	22 34	02 14	08 50	00 21	18 02	17 23
18	10 52	21 04	15 13	10 13	22 37	02 16	08 52	00 22	18 02	17 23
19	11 13	24 46	16 01	09 51	22 40	02 18	08 55	00 23	18 02	17 23
20	11 33	27 17	16 48	09 29	22 42	02 20	08 58	00 23	18 01	17 24
21	11 54	28 31	17 33	09 08	22 45	02 22	09 00	00 24	18 01	17 24
22	12 14	28 25	18 16	08 47	22 48	02 24	09 03	00 25	18 01	17 24
23	12 34	27 01	18 56	08 28	22 50	02 26	09 06	00 26	18 00	17 24
24	12 54	24 23	19 35	08 09	22 53	02 28	09 08	00 27	18 00	17 24
25	13 14	20 41	20 10	07 52	22 55	02 30	09 11	00 27	18 00	17 25
26	13 33	16 03	20 43	07 35	22 58	02 31	09 14	00 28	17 59	17 25
27	13 52	10 39	21 14	07 20	23 01	02 33	09 16	00 29	17 59	17 25
28	14 11	04 N 39	21 42	07 05	23 03	02 35	09 19	00 30	17 58	17 25
29	14 30	01 S 42	22 08	06 52	23 06	02 36	09 22	00 30	17 58	17 25
30	14 N 48	08 S 11	22 N 31	06 N 40	23 S 08	02 N 38	09 N 24	00 N 31	17 S 58	17 N 25

ZODIAC SIGN ENTRIES

Date	h	m	Planets
01	20	03	☽ ♎
04	00	22	☽ ♏
06	02	57	☽ ♐
08	05	04	☽ ♑
10	07	46	☽ ♒
12	11	41	☽ ♓
14	05	55	☿ ♉
14	17	13	☽ ♈
17	00	43	☽ ♉
19	10	28	☽ ♊
20	06	27	☉ ♉
21	22	17	☽ ♋
24	10	51	☽ ♌
26	21	57	☽ ♍
29	05	44	☽ ♎
29	22	23	♄ ♉
30	15	18	☿ ♊

LATITUDES

Date	Mercury ☿	Venus ♀	Mars ♂	Jupiter ♃	Saturn ♄	Uranus ♅	Neptune ♆	Pluto ♇
01	01 S 46	07 N 54	00 N 37	01 N 35	02 S 15	00 N 47	01 N 45	15 N 58
04	01 28	07 45	00 33	01 34	02 15	00 47	01 45	15 57
07	01 04	07 28	00 28	01 34	02 15	00 46	01 45	15 57
10	00 37	07 04	00 23	01 34	02 14	00 46	01 45	15 57
13	00 S 06	06 33	00 17	01 34	02 14	00 46	01 45	15 56
16	00 N 27	05 57	00 11	01 33	02 14	00 46	01 45	15 55
19	01 00	05 17	00 N 05	01 33	02 14	00 46	01 45	15 55
22	01 30	04 35	00 S 01	01 32	02 14	00 46	01 46	15 54
25	01 57	03 52	00 08	01 32	02 14	00 46	01 46	15 53
28	02 18	03 10	00 16	01 31	02 14	00 46	01 46	15 52
31	02 N 31	02 N 30	00 S 24	01 N 31	02 S 14	00 N 46	01 N 46	15 N 51

DATA

Julian Date	2440313
Delta T	+40 seconds
Ayanamsa	23° 25' 39"
Synetic vernal point	05° ♓ 41' 21"
True obliquity of ecliptic	23° 26' 46"

MOON'S PHASES, APSIDES AND POSITIONS ☽

Date	h	m	Phase	Longitude	Eclipse Indicator
02	18	45	○	12 ♎ 51	
09	13	58	☾	19 ♑ 32	
16	18	16	●	26 ♈ 34	
24	19	45	☽	04 ♌ 26	

Day	h	m		
07	00	13	Perigee	
22	13	40	Apogee	
01	20	09	0S	
08	05	07	Max dec	28° S 42'
14	17	17	0N	
21	13	28	Max dec	28° N 38'
29	05	39	0S	

LONGITUDES

Date	Chiron ⚷	Ceres ⚳	Pallas ⚴	Juno ⚵	Vesta ⚶	Black Moon Lilith ⚸
01	03 ♈ 19	01 ♒ 23	10 ♑ 14	12 ♑ 56	21 ♉ 10	02 ♋ 05
11	03 ♈ 54	04 ♒ 03	11 ♑ 49	14 ♑ 28	25 ♉ 16	03 ♋ 12
21	04 ♈ 27	06 ♒ 25	12 ♑ 49	15 ♑ 34	29 ♉ 27	04 ♋ 19
31	04 ♈ 58	08 ♒ 24	13 ♑ 10	16 ♑ 12	03 ♊ 40	05 ♋ 26

All ephemeris data is given at 12.00 UT and the Moon's longitude is additionally given for 24.00 UT

ASPECTARIAN

(h m Aspects)

01 Tuesday
02 39 ☽ ± ♄
03 29 ☽ ∥ ☉
07 26 ☽ ⊼ ♀
08 02 ☽ ☌ ♇
10 49 ☿ # ♅
13 37 ☽ ⊼ ♄
14 27 ☽ ∥ ♃
16 10 ☿ ∥ ♅
17 15 ☽ ✶ ♆
19 22 ☽ ∥ ☿
19 37 ☽ ☌ ♃
19 47 ☽ ∥ ♅
20 31 ☽ # ♅
21 10 ☽ # ☿
22 10 ☽ Q ♂
22 47 ☽ ☌ ♅

02 Wednesday
01 53 ☽ # ♃
05 38 ☽ ☍ ☿
07 51 ♀ ± ♆
14 36 ☽ # ☉
18 45 ☽ ☍ ☉
19 45 ☽ ✶ ♂
19 45 ☽ ∠ ♆

03 Thursday
02 18 ☽ # ♄
02 23 ♂ Q ♅
08 31 ☉ bQ ♆
10 18 ☽ ☍ ♀
10 27 ☿ ∥ ♃
11 21 ☽ ⊥ ♆
12 49 ☽ ⊻ ♇
13 02 ☉ △ ♂
18 42 ☽ ☍ ♄
21 38 ☽ ⊻ ♆
22 08 ☽ ∠ ♂
23 05 ☽ ⊥ ♇
23 30 ☽ ⊻ ♃

04 Friday
02 50 ☽ ⊻ ♅
06 44 ☽ # ♀
09 38 ☽ ⊥ ♃
13 00 ☽ ⊥ ♅
13 43 ☽ ⊥ ♂
14 16 ☽ # ♇
14 22 ☽ ∠ ♇
17 45 ☽ ∥ ♆
17 57 ☽ ⊼ ☿

05 Saturday
00 02 ☽ ⊻ ♂
00 44 ☽ ∠ ♃
01 59 ☽ ⊼ ☉
04 09 ☽ ∠ ♅
05 43 ☽ ± ☿
11 11 ☽ ⊼ ♀
12 06 ☽ ∥ ♂
12 49 ☽ ± ☉
15 35 ☽ ✶ ♇
20 50 ☽ ± ♀
21 51 ☽ ⊼ ♄
23 17 ☽ bQ ☿

06 Sunday
00 13 ☽ ☌ ♆
01 41 ☽ ✶ ♃
05 00 ☽ bQ ☉
05 13 ☽ ✶ ♅
05 55 ☿ bQ ♆
07 59 ☽ ± ♄
11 15 ☽ bQ ♀
11 37 ☽ Q ♇
17 32 ☿ △ ♂
21 34 ☽ Q ♃
23 08 ☽ bQ ♄

07 Monday
01 12 ☽ Q ♅
03 03 ☽ ☌ ♂
04 27 ☽ △ ☿
07 52 ☽ △ ☉
11 14 ☽ △ ♀
17 37 ☽ □ ♇

08 Tuesday
00 25 ☽ △ ♄
02 16 ☽ ⊻ ♆
03 24 ☽ □ ♃
07 12 ☽ □ ♅
12 19 ☽ ⊥ ♆
15 10 ☉ ☌ ♀
18 20 ☿ ☌ ♀
23 00 ☉ ☌ ☿

09 Wednesday
03 26 ☽ ∠ ♆
06 05 ☽ ⊻ ♂
11 29 ☽ □ ♀
13 58 ☽ □ ☉
15 19 ☽ □ ☿
16 23 ☽ ⊥ ♂
20 02 ☽ △ ♇

10 Thursday
03 29 ☽ □ ♄
04 50 ☽ ✶ ♆
05 39 ☽ △ ♃
07 54 ☽ ∠ ♂
09 04 ☉ ∥ ☿
09 47 ☽ △ ♅
13 41 ☿ ± ♆
16 50 ☽ Q ♀
21 36 ☽ ∥ ♂
21 37 ☽ bQ ♇
22 52 ☿ ⊼ ♇
23 00 ☽ Q ☉

11 Friday
00 17 ☿ ∥ ♄
01 22 ☽ Q ♆
03 28 ☽ Q ☿
07 11 ☽ bQ ♃
10 02 ☽ ✶ ♂
10 40 ☽ Q ♄
11 31 ☽ bQ ♅
12 39 ☽ ✶ ♀
13 07 ☽ ± ♇
18 04 ☽ ∥ ♆
21 16 ☽ # ♇
21 21 ☽ ✶ ☉
22 41 ☽ ± ♃
23 32 ☽ ⊼ ♇

12 Saturday
02 48 ☉ ∥ ♄
03 08 ☽ ± ♅
04 02 ☽ ✶ ☿
07 11 ☽ Q ♂
07 45 ☽ ✶ ♄
08 11 ☉ ± ♆
08 36 ☽ □ ♆
09 05 ☽ ⊼ ♃
13 37 ☽ ⊼ ♅
13 44 ☽ ∠ ♀
17 35 ☽ # ♀

13 Sunday
01 42 ☽ # ☿
01 43 ☽ ∠ ☉
03 27 ☉ ⊼ ♇
04 57 ☽ ⊥ ♀
05 42 ☿ ☌ ♄
07 13 ☽ # ☉
08 40 ☽ # ♄
09 36 ☿ ⊼ ♆
10 29 ☽ ∠ ♄
11 20 ☽ ∠ ☿
11 37 ☿ ⊼ ♃
11 39 ♀ △ ♂
15 12 ☽ ⊻ ♀
15 23 ☽ □ ♂
18 40 ☿ ± ♇
18 59 ☽ ⊥ ☉

14 Monday
02 45 ☽ ⊥ ♄
04 34 ☽ ☍ ♇
06 27 ☽ ⊥ ☿
06 37 ☽ ⊻ ☉
09 22 ☽ # ♃
12 52 ☿ ∥ ♀
13 41 ☽ ⊻ ♄
13 57 ☽ △ ♆
14 06 ☽ ☍ ♃
15 03 ☿ bQ ♂
16 08 ☽ # ♅
17 30 ☿ ⊼ ♅
18 26 ☽ ∥ ♅
19 03 ☽ ☍ ♅
19 21 ☽ ⊻ ☿

15 Tuesday
01 21 ☽ ∥ ♃
13 33 ♄ ⊼ ♆
13 55 ♃ ⊼ ♄
14 34 ♃ ✶ ♆
17 19 ☽ bQ ♆
19 29 ☽ ☌ ♀
22 31 ☽ △ ♂

16 Wednesday
03 09 ☽ ∥ ♄
05 04 ☿ ± ♃
08 37 ☽ ∥ ☉
10 00 ☽ ± ♆
11 29 ☽ ⊼ ♇
12 14 ☽ ∥ ♀
18 16 ☽ ☌ ☉
21 02 ☽ ⊼ ♃
21 13 ☽ ⊼ ♆
21 36 ☽ ☌ ♄
22 44 ☽ ± ♇

17 Thursday
01 04 ☽ ∥ ☿
02 27 ☽ ⊼ ♅
02 51 ☽ bQ ♂
08 17 ☽ ± ♃
13 29 ☿ ± ♅
13 48 ☽ ± ♅
13 52 ☽ ☌ ☿
14 53 ☉ ∥ ♀
15 45 ☽ bQ ♇
16 43 ☽ ∥ ♇
19 56 ☽ # ♆
20 07 ☽ ± ♂

18 Friday
01 19 ☽ bQ ♃
01 24 ☿ bQ ♇
01 54 ☽ ⊻ ♀
03 19 ☉ ⊼ ♃
07 00 ☽ bQ ♅
07 45 ☽ ⊼ ♂
08 05 ♀ bQ ♆
08 06 ☉ ⊼ ♆
13 04 ☽ ⊥ ♀
19 48 ☉ ☌ ♄
20 36 ☽ △ ♇
21 19 ☽ # ♂

19 Saturday
03 25 ☉ ± ♇
06 00 ☽ ∠ ♀
06 11 ☽ △ ♃
06 17 ☿ ± ♂
06 44 ☽ ☍ ♆
07 50 ☽ ⊻ ♄
08 44 ☽ ⊻ ☉
12 07 ☽ △ ♅
19 45 ☽ ⊥ ♄
21 35 ☽ ⊥ ☉

20 Sunday
05 57 ☿ ⊻ ♀
10 30 ☿ bQ ♃
10 41 ☽ ✶ ♀
11 48 ☽ ⊻ ☿
13 48 ☽ ∠ ♄
17 02 ☽ ∠ ☉
19 09 ☽ ☍ ♂
23 14 ♄ ± ♇

21 Monday
01 34 ☉ ⊼ ♅
02 06 ☽ ⊥ ☿
06 50 ☿ ∥ ♇
07 54 ☽ □ ♇
10 01 ☽ Q ♀
17 24 ☽ □ ♃
18 19 ☽ ⊼ ♆
19 32 ♀ ∥ ♄
20 13 ☽ ✶ ♄
21 08 ☉ bQ ♂
23 41 ☽ ∠ ☿
23 47 ☽ □ ♅

22 Tuesday
00 18 ☿ bQ ♅
01 50 ☽ ✶ ☉
03 18 ☿ # ♆
06 24 ☽ ± ♆
11 18 ☿ ⊼ ♂
20 09 ☽ Q ♇
20 43 ☽ Q ♄
21 25 ☽ □ ♀
22 03 ☿ ⊥ ♀

23 Wednesday
00 34 ☽ bQ ♆
04 16 ☽ Q ☉
05 30 ☽ Q ♃
07 57 ☽ ⊼ ♂
11 39 ☽ ✶ ☿
12 04 ☽ Q ♅
18 11 ☉ ± ♃
20 07 ☽ ± ♂
20 22 ☽ ✶ ♇

24 Thursday
05 33 ☽ ✶ ♃
06 47 ☽ △ ♆
09 26 ☽ □ ♄
12 10 ☽ ✶ ♅
14 17 ☽ bQ ♂
16 05 ☽ Q ☿
19 45 ☽ □ ☉
22 26 ☽ # ♂

25 Friday
02 25 ☽ ∠ ♇
08 35 ☽ △ ♀
11 18 ☽ ∠ ♃
14 32 ☽ ∥ ☿
17 56 ☽ ∠ ♅
20 09 ☽ △ ♂
20 12 ☽ ⊥ ♇
22 03 ☿ △ ♇

26 Saturday
02 27 ☽ # ♆
04 54 ☽ ⊥ ♃
05 21 ☽ ∥ ♇
07 56 ☽ ⊻ ♇
09 30 ☽ □ ☿
11 30 ☽ ⊥ ♅
13 39 ☽ bQ ♀
16 29 ☽ ⊻ ♃
17 56 ☽ □ ♆
21 12 ☽ △ ♄
22 49 ☽ ∥ ☉
23 03 ☽ ⊻ ♅

27 Sunday
00 24 ☉ ± ♅
06 47 ☽ ± ♀
11 20 ☽ △ ☉
11 24 ♂ St R
15 30 ☿ ∠ ♀
17 38 ☽ ∥ ♄
18 03 ☽ ⊼ ♀

28 Monday
01 53 ☽ bQ ♄
02 06 ☽ ∥ ♀
03 57 ☽ Q ♆
04 51 ☉ bQ ♇
05 34 ☽ □ ♂
13 47 ☿ △ ♃
16 32 ☽ ☌ ♇
17 29 ☽ bQ ☉
18 37 ☽ ± ♄
19 55 ☽ ∥ ♃

29 Tuesday
00 18 ☽ ☌ ♃
01 42 ☽ △ ☿
01 52 ☽ ✶ ♆
03 09 ☿ ☍ ♆
03 47 ☽ ∥ ♅
05 34 ☽ ⊼ ♄
06 38 ☽ ☌ ♅
07 32 ☽ # ♅
11 00 ☽ ± ☉
14 11 ☽ Q ♂
15 20 ☽ # ♃
19 20 ♀ St D
22 26 ☽ ⊼ ☉

30 Wednesday
00 20 ☽ ☍ ♀
04 25 ☽ ∠ ♆
06 29 ☽ # ♀
07 24 ☽ bQ ☿
11 09 ☽ ✶ ♂
16 34 ☽ # ♄
17 01 ☿ ⊻ ♄
19 56 ☽ ⊥ ♆
21 25 ☽ ⊻ ♇
23 17 ☿ △ ♅

MAY 1969

Date	Sidereal time h m s	Sun ☉ ° ′ ″	Moon ☽ ° ′ ″	Moon ☽ 24.00 ° ′ ″	Mercury ☿ ° ′	Venus ♀ ° ′	Mars ♂ ° ′	Jupiter ♃ ° ′	Saturn ♄ ° ′	Uranus ♅ ° ′	Neptune ♆ ° ′	Pluto ♇ ° ′
01	02 36 57	10 ♉ 55 44	01 ♏ 17 32	08 ♏ 28 29	01 ♊ 09	10 ♈ 33	16 ♐ 40	26 ♍ 49	00 ♉ 12	00 ♎ 26	27 ♏ 48	22 ♍ 40
02	02 40 54	11 53 55	15 ♏ 43 59	23 ♏ 03 16	02 26	10 D 38	16 R 36	26 R 45	00 19	00 R 24	27 R 46	22 R 39
03	02 44 50	12 52 05	00 ♐ 25 23	07 ♐ 49 21	03 37	10 45	16 32	26 42	00 27	00 23	27 45	22 39
04	02 48 47	13 50 12	15 ♐ 14 07	22 ♐ 38 42	04 45	10 55	16 27	26 38	00 34	00 21	27 43	22 38
05	02 52 44	14 48 19	00 ♑ 02 06	07 ♑ 23 25	05 47	11 07	16 22	26 35	00 42	00 19	27 41	22 37
06	02 56 40	15 46 23	14 ♑ 41 53	21 ♑ 56 52	06 45	11 20	16 15	26 32	00 49	00 18	27 40	22 36
07	03 00 37	16 44 27	29 ♑ 07 53	06 ♒ 14 33	07 39	11 36	16 08	26 29	00 57	00 16	27 38	22 35
08	03 04 33	17 42 29	13 ♒ 16 42	20 ♒ 14 13	08 27	11 54	16 00	26 26	01 04	00 15	27 37	22 34
09	03 08 30	18 40 29	27 ♒ 07 07	03 ♓ 55 31	09 11	12 14	15 52	26 24	01 12	00 13	27 35	22 34
10	03 12 26	19 38 29	10 ♓ 39 34	17 ♓ 19 28	09 49	12 35	15 42	26 21	01 19	00 12	27 34	22 33
11	03 16 23	20 36 26	23 ♓ 55 28	00 ♈ 27 48	10 23	12 58	15 32	26 19	01 27	00 11	27 32	22 32
12	03 20 19	21 34 23	06 ♈ 56 42	13 ♈ 22 22	10 52	13 23	15 21	26 17	01 34	00 09	27 30	22 32
13	03 24 16	22 32 18	19 ♈ 45 03	26 ♈ 04 53	11 16	13 50	15 09	26 15	01 41	00 08	27 29	22 31
14	03 28 13	23 30 12	02 ♉ 22 03	08 ♉ 36 40	11 34	14 18	14 57	26 13	01 49	00 07	27 27	22 30
15	03 32 09	24 28 05	14 ♉ 48 53	20 ♉ 58 46	11 48	14 47	14 44	26 12	01 56	00 06	27 25	22 30
16	03 36 06	25 25 56	27 ♉ 06 35	03 ♊ 12 01	11 57	15 18	14 30	26 10	02 03	00 04	27 24	22 29
17	03 40 02	26 23 46	09 ♊ 15 36	15 ♊ 17 19	12 01	15 51	14 16	26 09	02 10	00 03	27 22	22 29
18	03 43 59	27 21 34	21 ♊ 17 22	27 ♊ 15 54	12 R 00	16 24	14 01	26 08	02 17	00 02	27 21	22 28
19	03 47 55	28 19 21	03 ♋ 13 12	09 ♋ 09 30	11 54	16 59	13 45	26 08	02 25	00 01	27 19	22 28
20	03 51 52	29 ♉ 17 07	15 ♋ 05 10	21 ♋ 00 32	11 44	17 36	13 29	26 07	02 32	00 ♎ 00	27 17	22 27
21	03 55 48	00 ♊ 14 50	26 ♋ 56 02	02 ♌ 52 09	11 30	18 13	13 13	26 07	02 39	29 ♍ 59	27 16	22 27
22	03 59 45	01 12 33	08 ♌ 49 22	14 ♌ 48 15	11 11	18 52	12 55	26 06	02 46	29 59	27 14	22 26
23	04 03 42	02 10 13	20 ♌ 49 23	26 ♌ 53 22	10 49	19 31	12 38	26 D 06	02 53	29 58	27 12	22 26
24	04 07 38	03 07 52	03 ♍ 00 50	09 ♍ 12 25	10 24	20 12	12 19	26 06	03 00	29 57	27 11	22 26
25	04 11 35	04 05 30	15 ♍ 28 45	21 ♍ 50 27	09 56	20 54	12 01	26 07	03 07	29 56	27 09	22 26
26	04 15 31	05 03 06	28 ♍ 18 04	04 ♎ 52 07	09 26	21 37	11 42	26 07	03 14	29 56	27 08	22 25
27	04 19 28	06 00 41	11 ♎ 32 59	18 ♎ 20 59	08 54	22 20	11 23	26 08	03 21	29 55	27 06	22 25
28	04 23 24	06 58 14	25 ♎ 16 14	02 ♏ 18 45	08 21	23 05	11 03	26 09	03 27	29 55	27 04	22 25
29	04 27 21	07 55 46	09 ♏ 28 16	16 ♏ 44 23	07 48	23 50	10 43	26 10	03 34	29 54	27 03	22 25
30	04 31 17	08 53 16	24 ♏ 06 27	01 ♐ 33 34	07 14	24 37	10 23	26 11	03 41	29 54	27 01	22 25
31	04 35 14	09 ♊ 50 46	09 ♐ 04 43	16 ♐ 38 40	06 ♊ 41	25 ♈ 24	10 ♐ 03	26 ♍ 12	03 ♉ 48	29 ♍ 53	27 ♏ 00	22 ♍ 25

Date	Moon True ☊ ° ′	Moon Mean ☊ ° ′	Moon ☽ Latitude ° ′
01	29 ♓ 41	28 ♓ 14	02 S 42
02	29 R 37	28 11	03 43
03	29 30	28 08	04 30
04	29 24	28 04	04 59
05	29 17	28 01	05 08
06	29 12	27 58	04 57
07	29 09	27 55	04 27
08	29 08	27 52	03 41
09	29 D 08	27 49	02 44
10	29 09	27 45	01 38
11	29 10	27 42	00 S 28
12	29 R 10	27 39	00 N 42
13	29 08	27 36	01 48
14	29 03	27 33	02 48
15	28 56	27 29	03 39
16	28 47	27 26	04 19
17	28 37	27 23	04 47
18	28 26	27 20	05 01
19	28 15	27 17	05 02
20	28 06	27 14	04 51
21	27 59	27 10	04 26
22	27 54	27 07	03 50
23	27 51	27 04	03 04
24	27 51	27 01	02 08
25	27 D 51	26 58	01 N 05
26	27 R 51	26 55	00 S 02
27	27 51	26 51	01 12
28	27 48	26 48	02 20
29	27 43	26 45	03 22
30	27 36	26 42	04 13
31	27 ♓ 26	26 ♓ 39	04 S 46

DECLINATIONS

Date	Sun ☉ ° ′	Moon ☽ ° ′	Mercury ☿ ° ′	Venus ♀ ° ′	Mars ♂ ° ′	Jupiter ♃ ° ′	Saturn ♄ ° ′	Uranus ♅ ° ′	Neptune ♆ ° ′	Pluto ♇ ° ′
01	15 N 07	14 S 28	22 N 52	06 N 29	23 S 10	02 N 39	09 N 27	00 N 32	17 S 57	17 N 25
02	15 25	20 06	23 10	06 19	23 13	02 40	09 29	00 32	17 57	17 25
03	15 42	24 38	23 26	06 10	23 15	02 42	09 32	00 33	17 57	17 25
04	16 00	27 35	23 40	06 02	23 18	02 43	09 35	00 34	17 56	17 25
05	16 17	28 35	23 51	05 55	23 20	02 44	09 37	00 34	17 56	17 26
06	16 34	27 33	24 00	05 50	23 22	02 45	09 40	00 35	17 56	17 26
07	16 51	24 41	24 07	05 45	23 24	02 46	09 42	00 35	17 55	17 26
08	17 07	20 22	24 11	05 42	23 27	02 47	09 45	00 36	17 55	17 25
09	17 23	15 02	24 14	05 39	23 29	02 48	09 47	00 37	17 55	17 25
10	17 39	09 05	24 15	05 38	23 31	02 49	09 50	00 37	17 54	17 25
11	17 54	02 S 51	24 13	05 37	23 33	02 49	09 52	00 38	17 54	17 25
12	18 10	03 N 24	24 10	05 37	23 35	02 50	09 55	00 38	17 53	17 25
13	18 25	09 24	24 05	05 39	23 37	02 50	09 57	00 39	17 53	17 25
14	18 39	14 56	23 58	05 41	23 39	02 51	10 00	00 39	17 53	17 25
15	18 54	19 46	23 49	05 44	23 40	02 51	10 02	00 39	17 52	17 25
16	19 08	23 46	23 39	05 47	23 42	02 52	10 05	00 40	17 52	17 25
17	19 21	26 34	23 27	05 52	23 44	02 52	10 07	00 40	17 52	17 25
18	19 35	28 10	23 13	05 57	23 45	02 52	10 09	00 41	17 51	17 24
19	19 48	28 27	22 58	06 03	23 47	02 52	10 12	00 41	17 51	17 24
20	20 00	27 24	22 42	06 10	23 48	02 52	10 14	00 41	17 50	17 24
21	20 12	25 08	22 24	06 17	23 50	02 52	10 16	00 42	17 50	17 24
22	20 24	21 45	22 05	06 25	23 51	02 52	10 19	00 42	17 50	17 24
23	20 36	17 27	21 45	06 34	23 52	02 52	10 21	00 42	17 49	17 23
24	20 47	12 24	21 25	06 43	23 53	02 51	10 23	00 42	17 49	17 23
25	20 58	06 44	21 03	06 52	23 54	02 51	10 26	00 43	17 49	17 23
26	21 09	00 N 38	20 41	07 03	23 55	02 51	10 28	00 43	17 48	17 23
27	21 19	05 S 41	20 19	07 13	23 55	02 50	10 30	00 43	17 48	17 22
28	21 29	11 57	19 57	07 24	23 56	02 50	10 32	00 43	17 48	17 22
29	21 38	17 50	19 35	07 36	23 56	02 49	10 34	00 43	17 47	17 22
30	21 47	22 52	19 14	07 48	23 57	02 48	10 37	00 44	17 47	17 21
31	21 N 56	26 S 32	18 N 53	08 N 01	23 S 57	02 N 48	10 N 39	00 N 44	17 S 47	17 N 21

ZODIAC SIGN ENTRIES

Date	h	m	Planets
01	09	50	☽ ♏
03	11	19	☽ ♐
05	11	57	☽ ♑
07	13	28	☽ ♒
09	17	04	☽ ♓
11	23	09	☽ ♈
14	07	28	☽ ♉
16	17	41	☽ ♊
19	05	30	☽ ♋
20	20	51	♅ ♍
21	05	50	☉ ♊
21	18	12	☽ ♌
24	06	07	☽ ♍
26	15	07	☽ ♎
28	20	05	☽ ♏
30	21	30	☽ ♐

LATITUDES

Date	Mercury ☿ ° ′	Venus ♀ ° ′	Mars ♂ ° ′	Jupiter ♃ ° ′	Saturn ♄ ° ′	Uranus ♅ ° ′	Neptune ♆ ° ′	Pluto ♇ ° ′
01	02 N 31	02 N 30	00 S 24	01 N 31	02 S 14	00 N 46	01 N 46	15 N 51
04	02 37	01 52	00 32	01 30	02 14	00 46	01 46	15 50
07	02 33	01 16	00 41	01 30	02 15	00 46	01 46	15 49
10	02 21	00 43	00 50	01 29	02 15	00 46	01 46	15 48
13	01 58	00 N 12	01 00	01 28	02 15	00 45	01 46	15 47
16	01 26	00 S 16	01 10	01 28	02 15	00 45	01 46	15 46
19	00 N 45	00 41	01 20	01 27	02 15	00 45	01 46	15 44
22	00 S 02	01 03	01 30	01 26	02 15	00 45	01 46	15 43
25	00 54	01 23	01 40	01 25	02 16	00 45	01 46	15 42
28	01 46	01 41	01 51	01 25	02 16	00 45	01 46	15 41
31	02 S 35	01 S 57	02 S 01	01 N 24	02 S 17	00 N 45	01 N 46	15 N 39

DATA

Julian Date	2440343
Delta T	+40 seconds
Ayanamsa	23° 25′ 42″
Synetic vernal point	05° ♓ 41′ 17″
True obliquity of ecliptic	23° 26′ 45″

MOON'S PHASES, APSIDES AND POSITIONS ☽

Date	h	m	Phase	Longitude ° ′	Eclipse Indicator
02	05	13	○	11 ♏ 37	
08	20	12	☾	18 ♒ 02	
16	08	26	●	25 ♉ 17	
24	12	15	☽	03 ♍ 08	
31	13	18	○	09 ♐ 54	

Day	h	m		
04	10	45	Perigee	
20	05	26	Apogee	
05	11	46	Max dec	28° S 35′
11	22	53	ON	
19	04	57	Max dec	28° N 30′
26	14	27	OS	

LONGITUDES

Date	Chiron ⚷ ° ′	Ceres ⚳ ° ′	Pallas ⚴ ° ′	Juno ⚵ ° ′	Vesta ⚶ ° ′	Black Moon Lilith ⚸ ° ′
01	04 ♈ 58	08 ♒ 24	13 ♑ 10	16 ♑ 12	03 ♊ 40	05 ♋ 26
11	05 ♈ 27	09 ♒ 59	12 ♑ 49	16 ♑ 17	07 ♊ 57	06 ♋ 34
21	05 ♈ 52	11 ♒ 05	11 ♑ 43	15 ♑ 49	12 ♊ 15	07 ♋ 41
31	06 ♈ 13	11 ♒ 39	09 ♑ 54	14 ♑ 45	16 ♊ 35	08 ♋ 48

All ephemeris data is given at 12.00 UT and the Moon's longitude is additionally given for 24.00 UT

ASPECTARIAN

h m	Aspects
01 Thursday	
00 33	☽ ± ☿
02 02	☉ ⊻ ♀
04 29	☽ ⊻ ♃
05 40	☉ ± ♂
06 07	☽ ⊻ ♆
07 36	☽ ⊥ ♇
10 09	☽ ☍ ♄
10 34	☽ ⊻ ♅
11 45	☽ ⊼ ☿
12 37	☽ ∠ ♂
14 33	☽ ⊥ ♃
14 44	☽ # ☉
20 35	☽ ⊥ ♅
22 39	☽ ∠ ♇
02 Friday	
00 08	☽ # ♇
02 24	☽ ∥ ♆
03 32	☽ ⊼ ♀
03 34	☽ ⊥ ♂
05 13	☽ ☍ ☉
05 28	☽ ∠ ♃
08 44	☉ bQ ♃
11 28	☽ ∠ ♅
13 26	☽ ⊻ ♂
13 30	☽ ± ♀
16 30	☿ # ♂
23 20	☽ ✱ ♇
03 Saturday	
01 05	♄ ⊼ ♅
03 47	☽ ∥ ♂
04 20	☽ bQ ♀
04 25	☽ # ☿
05 58	☽ ✱ ♃
07 39	☽ ☌ ♆
11 56	☽ ✱ ♅
12 03	☽ ⊼ ♄
17 38	☽ ☍ ☿
18 50	☽ Q ♇
21 51	☽ ± ♄
04 Sunday	
01 22	☽ Q ♃
04 56	☽ △ ♀
07 20	☽ Q ♅
09 35	☽ ⊼ ☉
12 33	☽ bQ ♄
13 58	☽ ☌ ♂
19 58	☽ ± ☉
23 58	☽ □ ♇
05 Monday	
06 25	☽ □ ♃
08 12	☽ ⊻ ♆
11 36	☽ bQ ☉
12 28	☽ □ ♅
13 05	☽ △ ♄
17 57	☽ ⊥ ♆
22 04	☽ ⊼ ☿
06 Tuesday	
00 31	☉ bQ ♅
06 23	☽ □ ♀
08 35	☽ ± ☿
08 40	☽ ∠ ♆
13 54	☽ △ ☉
14 02	☿ ⊥ ♄
14 33	☽ ⊻ ♂
22 41	☉ ⊼ ♂
07 Wednesday	
00 25	☽ ⊥ ♂
00 28	☽ bQ ☿
01 04	☽ △ ♇
07 35	☽ △ ♃
09 30	☽ ✱ ♆
12 48	☽ Q ♀
13 55	☽ △ ♅
15 05	☽ □ ♄
15 20	☽ ∠ ♂
15 36	☽ # ☿
19 50	☽ ∥ ♂
08 Thursday	
02 16	☽ bQ ♇
03 16	☽ △ ☿
05 42	♂ bQ ♄
05 45	☽ Q ♆
08 52	☽ bQ ♃
09 35	☽ ✱ ♀
15 23	☽ bQ ♅
16 38	☽ ✱ ♂
17 40	☽ ± ♇
19 02	☉ ⊥ ♀
20 12	☽ □ ☉
22 04	☽ Q ♄
23 31	☽ ∥ ♆
09 Friday	
00 19	☽ ± ♃
01 42	☽ # ♇
02 21	☽ # ☉
04 03	☽ ⊼ ♇
06 57	☽ ± ♅
10 44	☽ ⊼ ♃
12 12	☽ ∠ ♀
12 49	☽ □ ♆
13 17	☽ Q ♂
15 31	☉ ∥ ♇
17 27	☽ ⊼ ♅
19 14	☽ ✱ ♄
10 Saturday	
04 31	☽ ⊥ ♀
06 11	☽ Q ☉
09 05	☽ # ♄
10 26	☽ □ ☿
10 35	♀ bQ ♆
15 33	☽ ⊻ ♀
20 57	☽ □ ♂
22 17	☽ ∠ ♄
11 Sunday	
01 23	☽ # ♀
05 29	☽ ✱ ☉
09 28	☽ ☍ ♇
10 57	☉ # ♆
12 06	☽ # ♃
14 48	☽ ⊥ ♄
16 22	☽ ☍ ♃
18 36	☽ △ ♆
20 28	☽ # ♅
20 30	☽ Q ☿
23 27	☽ ☍ ♅
12 Monday	
01 18	☽ ∥ ♅
01 56	☽ ⊻ ♄
09 48	☽ ∥ ♃
11 15	☽ ∠ ☉
19 33	☽ ✱ ☿
20 46	☽ ∥ ♀
22 21	☽ bQ ♆
13 Tuesday	
00 27	☽ ☌ ♀
03 29	☽ △ ♂
05 27	☽ ⊥ ☉
11 24	☉ △ ♇
14 20	☽ ∥ ♄
15 16	☽ ± ♆
17 13	☽ ⊼ ♇
17 43	☽ ⊻ ☉
14 Wednesday	
00 18	☽ ⊼ ♃
00 41	☽ ∠ ☿
02 38	☽ ⊼ ♆
04 37	☽ ± ♇
07 27	☽ bQ ♂
07 41	☽ ⊼ ♅
10 55	☽ ☌ ♄
11 43	☽ ± ♃
18 17	☽ ⊥ ☿
19 11	☽ ± ♅
21 52	☽ bQ ♇
23 50	☽ ∥ ♇
15 Thursday	
00 26	☽ ± ♂
02 07	☽ # ♆
05 00	☽ bQ ♃
06 04	☽ ⊻ ☿
07 04	☽ ∥ ☉
10 15	♀ △ ♂
11 51	☽ ⊼ ♂
11 57	☽ ⊻ ♀
12 32	☽ bQ ♅
16 Friday	
00 08	☽ ⊥ ♀
02 57	☽ △ ♇
05 54	☿ # ♂
08 26	☽ ☌ ☉
10 10	☽ △ ♃
11 33	☽ ∥ ☿
11 54	☽ # ♂
12 34	☽ ☍ ♆
17 49	☽ △ ♅
18 34	☽ ∠ ♀
21 50	☽ ⊻ ♄
17 Saturday	
06 07	☉ △ ♃
09 12	♃ ± ♄
09 49	☽ ⊥ ♄
17 29	☽ ☌ ☿
19 06	☿ St R
21 46	☽ ☍ ♂
18 Sunday	
01 45	☽ ✱ ♀
03 55	☽ ∠ ♄
11 36	☉ ☍ ♆
14 22	☽ □ ♇
21 43	☽ □ ♃
19 Monday	
00 08	☽ ⊼ ♆
01 16	☽ ⊻ ☉
03 02	☽ Q ♀
05 33	☽ □ ♅
10 21	☽ ✱ ♄
12 12	☽ ± ♆
14 25	☽ ⊥ ☉
20 Tuesday	
02 38	☽ Q ♇
05 19	☽ ⊻ ☿
06 21	☽ bQ ♆
08 50	☽ ⊼ ♂
10 02	☽ Q ♃
10 14	☽ ∠ ☉
10 52	☽ Q ♄
17 16	☽ ⊥ ☿
17 22	☽ □ ♀
17 54	☽ Q ♅
20 43	☽ ± ♂
21 Wednesday	
02 55	☽ ✱ ♇
05 42	☉ △ ♅
10 20	☽ ✱ ♃
11 08	☽ ∠ ☿
12 40	☽ △ ♆
14 31	☽ bQ ♂
18 11	☽ ✱ ♅
19 18	☽ ✱ ☉
22 05	☽ # ♂
23 40	☽ □ ♄
22 Thursday	
09 13	☽ ∠ ♇
09 44	☽ ∥ ☿
16 35	☽ ∠ ♃
16 37	☽ ✱ ☿
19 42	☽ ∥ ☉
20 02	☽ △ ♂
21 34	☽ Q ☉
23 Friday	
00 20	☽ ∠ ♅
03 16	☽ ⊥ ♇
08 20	♃ St D
09 16	☽ △ ♀
10 07	☽ # ♆
10 34	☽ ⊥ ♃
12 21	☽ ∥ ♇
15 12	☽ ⊻ ♇
15 51	☽ Q ☿
18 13	☽ ⊥ ♅
22 27	☽ ⊻ ♃
24 Saturday	
00 36	☽ □ ♆
06 01	☽ ⊻ ♅
08 14	☉ ⊻ ♄
11 58	☽ △ ♄
12 15	☽ □ ☉
16 31	☽ bQ ♀
20 42	☽ ∥ ♄
25 Sunday	
01 48	☽ □ ☿
05 33	☽ □ ♂
10 50	☽ ± ♀
11 23	☽ Q ♆
11 26	☽ ∥ ♀
15 41	☉ ∥ ☿
17 02	☽ bQ ♄
17 59	♂ Q ♅
20 19	♀ ± ♆
22 50	☽ ⊼ ♀
26 Monday	
01 05	☽ ☌ ♇
03 27	☽ ∥ ♃
07 58	☽ ☌ ♃
09 50	☽ ✱ ♆
10 00	☽ ± ♄
11 42	☽ ∥ ♅
14 31	☽ Q ♂
14 59	☽ ☌ ♅
17 12	☽ # ♅
19 52	☿ ⊥ ♄
21 06	☽ ⊼ ♄
27 Tuesday	
01 17	☽ # ♃
01 18	☽ △ ☉
07 28	☽ △ ☿
11 42	☽ ✱ ♂
12 59	☽ ∠ ♆
14 33	♀ ⊼ ♇
18 02	☽ # ♀
28 Wednesday	
04 46	☽ ⊥ ♆
05 53	☽ bQ ☉
06 29	☽ # ♄
07 05	☽ ⊻ ♇
08 01	☽ ☍ ♀
08 50	☽ bQ ☿
13 19	☽ ∠ ♂
13 30	☽ ⊻ ♃
15 05	☽ ⊻ ♆
16 58	☿ ∠ ♀
17 23	☽ ⊥ ♇
19 55	☽ ⊻ ♅
22 26	☽ ± ☉
23 37	☽ ± ☿
23 44	☽ ⊥ ♃
29 Thursday	
02 03	☽ ☍ ♄
04 15	☽ ⊥ ♂
06 03	☽ ⊥ ♅
08 34	☽ ∠ ♇
09 15	☽ ⊼ ☉
09 18	☽ ⊼ ☿
09 50	☉ ☌ ☿
09 57	☽ # ♇
11 48	☽ ∥ ♆
14 02	☽ ⊻ ♂
14 48	☽ ∠ ♃
19 18	☽ # ☿
20 59	☽ ∠ ♅
30 Friday	
06 10	☽ # ☉
09 15	☽ ✱ ♇
12 52	☽ ⊼ ♀
15 21	☽ ✱ ♃
16 42	☽ ☌ ♆
18 05	☽ ∥ ♂
21 20	☽ ✱ ♅
23 03	☽ ± ♀
31 Saturday	
03 31	☽ ⊼ ♄
04 34	☽ Q ♇
04 41	♀ bQ ♂
08 19	☽ ☍ ☿
10 32	☉ ⊥ ♄
10 36	☽ Q ♃
13 09	☽ ± ♄
13 18	☽ ☍ ☉
13 31	☽ ☌ ♂
14 13	☽ bQ ♀
15 50	☉ ☍ ♂
16 28	☽ Q ♅

JUNE 1969

LONGITUDES

Date	Sidereal time h m s	Sun ☉ ° ' "	Moon ☽ ° ' "	Moon ☽ 24.00 ° ' "	Mercury ☿ ° '	Venus ♀ ° '	Mars ♂ ° '	Jupiter ♃ ° '	Saturn ♄ ° '	Uranus ♅ ° '	Neptune ♆ ° '	Pluto ♇ ° '
01	04 39 11	10 ♊ 48 15	24 ♐ 14 07	01 ♑ 49 43	06 ♊ 08	26 ♈ 12	09 ♐ 43	26 ♍ 14	03 ♉ 54	29 ♍ 53	26 ♏ 58	22 ♍ 25
02	04 43 07	11 45 42	09 ♑ 24 07	16 ♑ 56 05	05 R 38	27 00	09 R 23	26 15	04 01	29 R 53	26 R 57	22 R 24
03	04 47 04	12 43 09	24 ♑ 24 30	01 ♒ 48 24	05 10	27 50	09 02	26 17	04 08	29 52	26 55	22 D 25
04	04 51 00	13 40 35	09 ♒ 07 04	16 ♒ 19 56	04 44	28 40	08 42	26 19	04 14	29 52	26 53	22 25
05	04 54 57	14 38 01	23 ♒ 26 42	00 ♓ 27 13	04 22	29 ♈ 31	08 21	26 22	04 20	29 52	26 52	22 25
06	04 58 53	15 35 25	07 ♓ 21 29	14 ♓ 09 41	04 02	00 ♉ 22	08 01	26 24	04 27	29 52	26 50	22 25
07	05 02 50	16 32 49	20 ♓ 52 04	27 ♓ 28 58	03 47	01 14	07 41	26 27	04 33	29 D 52	26 49	22 25
08	05 06 46	17 30 13	04 ♈ 00 47	10 ♈ 27 58	03 36	02 07	07 21	26 29	04 40	29 52	26 47	22 25
09	05 10 43	18 27 36	16 ♈ 50 54	23 ♈ 10 04	03 28	03 00	07 01	26 32	04 46	29 52	26 46	22 25
10	05 14 40	19 24 58	29 ♈ 25 49	05 ♉ 38 34	03 26	03 53	06 42	26 35	04 52	29 52	26 44	22 25
11	05 18 36	20 22 20	11 ♉ 48 39	17 ♉ 56 23	03 D 27	04 48	06 22	26 39	04 58	29 53	26 43	22 26
12	05 22 33	21 19 41	24 ♉ 02 03	00 ♊ 05 51	03 33	05 42	06 04	26 42	05 04	29 53	26 41	22 26
13	05 26 29	22 17 02	06 ♊ 08 02	12 ♊ 08 46	03 44	06 38	05 45	26 46	05 10	29 53	26 40	22 26
14	05 30 26	23 14 22	18 ♊ 08 14	24 ♊ 06 34	03 59	07 34	05 27	26 50	05 16	29 54	26 39	22 27
15	05 34 22	24 11 42	00 ♋ 03 56	06 ♋ 00 29	04 19	08 30	05 10	26 53	05 22	29 54	26 37	22 27
16	05 38 19	25 09 01	11 ♋ 56 26	17 ♋ 51 56	04 43	09 26	04 53	26 58	05 28	29 54	26 36	22 28
17	05 42 15	26 06 19	23 ♋ 47 16	29 ♋ 42 40	05 11	10 23	04 36	27 02	05 34	29 55	26 34	22 28
18	05 46 12	27 03 37	05 ♌ 38 26	11 ♌ 34 57	05 44	11 21	04 21	27 06	05 40	29 56	26 33	22 29
19	05 50 09	28 00 54	17 ♌ 32 36	23 ♌ 31 50	06 22	12 19	04 06	27 11	05 45	29 56	26 32	22 29
20	05 54 05	28 58 10	29 ♌ 33 07	05 ♍ 37 00	07 04	13 17	03 51	27 15	05 51	29 57	26 30	22 30
21	05 58 02	29 ♊ 55 26	11 ♍ 44 03	17 ♍ 54 50	07 49	14 16	03 37	27 20	05 56	29 58	26 29	22 30
22	06 01 58	00 ♋ 52 40	24 ♍ 09 58	00 ♎ 30 04	08 40	15 15	03 24	27 25	06 02	29 58	26 28	22 31
23	06 05 55	01 49 54	06 ♎ 55 43	13 ♎ 27 29	09 34	16 14	03 12	27 31	06 07	29 ♍ 59	26 27	22 31
24	06 09 51	02 47 08	20 ♎ 05 51	26 ♎ 51 16	10 32	17 14	03 00	27 36	06 13	00 ♎ 00	26 25	22 32
25	06 13 48	03 44 21	03 ♏ 43 59	10 ♏ 44 09	11 34	18 14	02 49	27 41	06 18	00 01	26 24	22 33
26	06 17 44	04 41 33	17 ♏ 51 45	25 ♏ 06 32	12 40	19 14	02 39	27 47	06 23	00 02	26 23	22 34
27	06 21 41	05 38 45	02 ♐ 28 00	09 ♐ 55 27	13 50	20 15	02 30	27 53	06 28	00 03	26 22	22 34
28	06 25 38	06 35 57	17 ♐ 27 57	25 ♐ 04 19	15 03	21 16	02 21	27 59	06 33	00 04	26 21	22 35
29	06 29 34	07 33 08	02 ♑ 43 17	10 ♑ 23 23	16 20	22 17	02 14	28 05	06 38	00 05	26 20	22 36
30	06 33 31	08 ♋ 30 20	18 ♑ 03 09	25 ♑ 41 10	17 ♊ 41	23 ♉ 19	02 ♐ 07	28 ♍ 11	06 ♉ 43	00 ♎ 06	26 ♏ 18	22 ♍ 37

MOON NODES AND LATITUDE

Date	Moon True ☊ ° '	Moon Mean ☊ ° '	Moon ☽ Latitude ° '
01	27 ♓ 16	26 ♓ 35	05 S 01
02	27 R 06	26 32	04 54
03	26 57	26 29	04 28
04	26 52	26 26	03 44
05	26 48	26 23	02 47
06	26 47	26 20	01 41
07	26 D 47	26 16	00 S 31
08	26 R 46	26 13	00 N 38
09	26 45	26 10	01 44
10	26 42	26 07	02 43
11	26 36	26 04	03 34
12	26 27	26 01	04 14
13	26 15	25 57	04 41
14	26 02	25 54	04 57
15	25 48	25 51	04 58
16	25 35	25 48	04 47
17	25 23	25 45	04 24
18	25 13	25 41	03 49
19	25 06	25 38	03 04
20	25 02	25 35	02 10
21	25 00	25 32	01 10
22	24 D 59	25 29	00 N 04
23	24 R 59	25 26	01 S 03
24	24 59	25 22	02 09
25	24 56	25 19	03 09
26	24 51	25 16	04 01
27	24 44	25 13	04 39
28	24 34	25 10	04 59
29	24 24	25 07	04 58
30	24 ♓ 14	25 ♓ 03	04 S 36

DECLINATIONS

Date	Sun ☉ ° '	Moon ☽ ° '	Mercury ☿ ° '	Venus ♀ ° '	Mars ♂ ° '	Jupiter ♃ ° '	Saturn ♄ ° '	Uranus ♅ ° '	Neptune ♆ ° '	Pluto ♇ ° '
01	22 N 04	28 S 20	18 N 34	08 N 14	23 S 57	02 N 47	10 N 41	00 N 44	17 S 46	17 N 20
02	22 12	28 00	18 15	08 27	23 57	02 46	10 43	00 44	17 46	17 20
03	22 20	25 38	17 58	08 41	23 57	02 45	10 45	00 44	17 46	17 20
04	22 27	21 34	17 42	08 54	23 57	02 44	10 47	00 44	17 45	17 19
05	22 34	16 20	17 28	09 09	23 57	02 43	10 49	00 44	17 45	17 19
06	22 40	10 22	17 16	09 23	23 57	02 42	10 51	00 44	17 45	17 18
07	22 46	04 S 06	17 06	09 38	23 56	02 40	10 53	00 44	17 44	17 18
08	22 51	02 N 11	16 57	09 53	23 56	02 39	10 55	00 44	17 44	17 17
09	22 57	08 13	16 51	10 08	23 55	02 38	10 57	00 44	17 44	17 17
10	23 01	13 49	16 47	10 24	23 55	02 36	10 59	00 44	17 43	17 16
11	23 06	18 46	16 44	10 39	23 54	02 35	11 01	00 44	17 43	17 16
12	23 10	22 53	16 44	10 55	23 53	02 33	11 03	00 43	17 43	17 15
13	23 13	25 57	16 46	11 11	23 53	02 31	11 05	00 43	17 42	17 15
14	23 16	27 50	16 49	11 27	23 52	02 30	11 06	00 43	17 42	17 14
15	23 19	28 25	16 55	11 43	23 51	02 28	11 08	00 43	17 42	17 14
16	23 21	27 41	17 02	11 59	23 50	02 26	11 10	00 43	17 42	17 13
17	23 23	25 41	17 11	12 15	23 50	02 24	11 12	00 42	17 41	17 13
18	23 25	22 34	17 21	12 32	23 49	02 22	11 13	00 42	17 41	17 12
19	23 26	18 30	17 33	12 48	23 48	02 20	11 15	00 42	17 41	17 11
20	23 26	13 40	17 46	13 04	23 47	02 18	11 17	00 42	17 40	17 11
21	23 27	08 14	18 01	13 21	23 47	02 16	11 19	00 41	17 40	17 10
22	23 27	02 N 23	18 16	13 37	23 46	02 14	11 20	00 41	17 40	17 10
23	23 26	03 S 43	18 33	13 53	23 45	02 12	11 22	00 40	17 40	17 09
24	23 25	09 51	18 50	14 10	23 45	02 09	11 23	00 40	17 39	17 08
25	23 24	15 44	19 08	14 26	23 44	02 07	11 25	00 40	17 39	17 08
26	23 22	21 01	19 27	14 42	23 44	02 04	11 26	00 39	17 39	17 07
27	23 19	25 13	19 46	14 58	23 43	02 02	11 28	00 39	17 39	17 06
28	23 17	27 49	20 05	15 14	23 43	01 59	11 29	00 38	17 38	17 05
29	23 14	28 23	20 24	15 30	23 43	01 57	11 31	00 38	17 38	17 05
30	23 N 10	26 S 47	20 N 44	15 N 45	23 S 43	01 N 54	11 N 32	00 N 37	17 S 38	17 N 04

ZODIAC SIGN ENTRIES

Date	h	m	Planets
01	21	07	☽ ♑
03	21	03	☽ ♒
05	23	13	☽ ♓
06	01	48	♀ ♉
08	04	36	☽ ♈
10	13	06	☽ ♉
12	23	48	☽ ♊
15	11	52	☽ ♋
18	00	35	☽ ♌
20	12	53	☽ ♍
21	13	55	☉ ♋
22	23	03	☽ ♎
24	10	36	♅ ♎
25	05	31	☽ ♏
27	08	00	☽ ♐
29	07	44	☽ ♑

LATITUDES

Date	Mercury ☿ ° '	Venus ♀ ° '	Mars ♂ ° '	Jupiter ♃ ° '	Saturn ♄ ° '	Uranus ♅ ° '	Neptune ♆ ° '	Pluto ♇ ° '
01	02 S 49	02 S 02	02 S 04	01 N 24	02 S 17	00 N 45	01 N 46	15 N 39
04	03 27	02 14	02 13	01 23	02 17	00 45	01 46	15 38
07	03 53	02 25	02 22	01 22	02 17	00 44	01 46	15 36
10	04 09	02 35	02 31	01 22	02 18	00 44	01 45	15 35
13	04 13	02 42	02 39	01 21	02 18	00 44	01 45	15 34
16	04 07	02 48	02 46	01 20	02 19	00 44	01 45	15 32
19	03 53	02 52	02 53	01 19	02 19	00 44	01 45	15 31
22	03 31	02 55	02 59	01 19	02 20	00 44	01 45	15 30
25	03 04	02 57	03 04	01 18	02 20	00 44	01 45	15 28
28	02 32	02 57	03 08	01 17	02 21	00 44	01 45	15 27
31	01 S 57	02 S 56	03 S 12	01 N 17	02 S 21	00 N 43	01 N 45	15 N 26

DATA

Julian Date	2440374
Delta T	+40 seconds
Ayanamsa	23° 25' 48"
Synetic vernal point	05° ♓ 41' 12"
True obliquity of ecliptic	23° 26' 44"

MOON'S PHASES, APSIDES AND POSITIONS ☽

Date	h	m	Phase	Longitude ° '	Eclipse Indicator
07	03	39	☾	16 ♓ 13	
14	23	09	●	23 ♊ 41	
23	01	44	☽	01 ♎ 25	
29	20	04	○	07 ♑ 52	

Day	h	m		
01	14	48	Perigee	
16	14	55	Apogee	
29	23	40	Perigee	
01	20	24	Max dec	28° S 28'
08	03	37	0N	
15	10	31	Max dec	28° N 25'
22	21	28	0S	
29	06	19	Max dec	28° S 26'

LONGITUDES

Date	Chiron ⚷ ° '	Ceres ⚳ ° '	Pallas ⚴ ° '	Juno ⚵ ° '	Vesta ⚶ ° '	Black Moon Lilith ⚸ ° '
01	06 ♈ 15	11 ♒ 41	09 ♑ 41	14 ♑ 37	17 ♊ 01	08 ♋ 55
11	06 ♈ 32	11 ♒ 37	07 ♑ 14	12 ♑ 59	21 ♊ 21	10 ♋ 02
21	06 ♈ 44	10 ♒ 58	04 ♑ 26	10 ♑ 57	25 ♊ 41	11 ♋ 09
31	06 ♈ 51	09 ♒ 44	01 ♑ 34	08 ♑ 40	00 ♋ 01	12 ♋ 16

All ephemeris data is given at 12.00 UT and the Moon's longitude is additionally given for 24.00 UT

ASPECTARIAN

01 Sunday
01 50 ♂ ± ♄
03 31 ☽ ⊑ ♄
09 07 ☽ □ ♇
13 03 ♀ ⊼ ♃
15 09 ☽ □ ♃
15 16 ☽ △ ♀
16 19 ☽ ⊻ ♆
20 55 ☽ □ ♅

02 Monday
01 47 ☽ ⊥ ♆
03 24 ☽ △ ♄
06 13 ☽ ⊼ ☿
10 11 ♀ ⊼ ♆
11 58 ☽ ⊻ ♂
14 01 ♇ St D
15 26 ☽ ± ☿
16 00 ☽ ⊼ ☉
16 02 ☽ ∠ ♆
21 18 ☽ ⊥ ♂

03 Tuesday
02 14 ☽ ± ☉
05 22 ☽ ⊑ ☿
08 47 ☽ △ ♇
11 25 ☽ ∠ ♂
15 03 ☽ △ ♃
16 03 ☽ ✱ ♆
17 43 ☽ ⊑ ☉
17 52 ☽ □ ♀
20 51 ☽ △ ♅
23 00 ☽ ∥ ♂

04 Wednesday
03 54 ☽ □ ♄
04 44 ♀ ± ♇
04 59 ☽ △ ☿
06 51 ☿ # ♆
07 35 ☽ # ☉
09 11 ☽ ⊑ ♇
09 15 ☉ ∠ ♀
11 19 ☽ ✱ ♂
11 38 ☽ Q ♆
13 25 ☿ ⊥ ♀
15 40 ☽ ⊑ ♃
20 06 ☽ △ ☉
21 33 ☽ ⊑ ♅

05 Thursday
00 08 ☽ ± ♇
01 20 ☽ Q ♀
05 53 ☽ ∥ ♆
06 46 ☽ ± ♃
06 53 ☽ Q ♂
06 54 ☽ # ☿
07 46 ☽ # ♇
10 07 ☽ Q ♄
10 15 ☽ ⊼ ♇
11 33 ♂ Q ♃
12 43 ☽ ± ♅
13 00 ☿ ⊻ ♄
16 59 ☽ ⊼ ♃
17 49 ☽ □ ♆
22 09 ♀ ⊼ ♅
23 00 ☽ ⊼ ♅
23 03 ☽ ✱ ♀

06 Friday
06 20 ☽ □ ☿
06 28 ☿ ∥ ♇
06 53 ☽ ✱ ♄
10 08 ☽ # ♄
13 08 ☽ □ ♂
15 40 ☽ # ♀

07 Saturday
03 07 ☽ ∠ ♀
03 39 ☽ □ ☉
06 33 ♅ St D
09 37 ☽ ∠ ♄
13 38 ☽ Q ☿
14 48 ☽ ☍ ♇
17 26 ☽ # ♃
20 28 ☽ ⊥ ♀
20 54 ♀ ± ♂
22 09 ☽ ☍ ♃
22 46 ☽ △ ♆

08 Sunday
00 49 ☽ # ♅
02 04 ☽ ⊥ ♄
04 22 ☽ ☍ ♅
06 25 ☽ ∥ ♅
08 14 ☽ ⊻ ♀
11 14 ☽ ✱ ☿
13 12 ☽ ⊻ ♄
13 49 ☽ ∥ ♃
14 59 ☽ Q ☉
18 02 ☽ △ ♂
22 59 ♀ ± ♃

09 Monday
02 27 ☽ ⊑ ♆
15 03 ☽ ∠ ☿
15 18 ☽ ✱ ☉
19 25 ☽ ± ♆
20 20 ☽ ∥ ♀
21 34 ☽ ⊑ ♂
22 35 ☽ ⊼ ♇
23 28 ☽ ∥ ♄
23 58 ☿ ⊻ ♀

10 Tuesday
06 31 ☽ ⊼ ♃
06 51 ☽ ⊼ ♆
08 09 ☽ ⊥ ☿
10 04 ☽ ± ♇
12 51 ☽ ⊼ ♅
14 22 ☽ ± ♂
15 47 ☿ St D
18 07 ☽ ± ♃
19 42 ☽ ⊻ ☿
21 17 ☽ ☌ ♀
22 26 ☽ ∠ ☉
22 35 ☽ ☌ ♄

11 Wednesday
00 27 ☽ ± ♅
00 42 ☉ ∠ ♄
01 41 ☽ ⊼ ♂
01 47 ☽ ∥ ☿
03 28 ☽ ⊑ ♇
04 20 ☽ ∥ ♇
06 35 ☽ # ♆
11 41 ☽ ⊑ ♃
17 12 ♀ ☌ ♄
17 26 ☽ ⊥ ☉
18 00 ☽ ⊑ ♅

12 Thursday
06 13 ☽ ⊻ ☉
08 35 ♃ ✱ ♆
08 50 ☽ △ ♇
13 56 ☽ ∥ ☉
16 32 ♀ ± ♅
17 14 ☽ ☍ ♆
17 18 ☽ △ ♃
18 52 ♀ ⊼ ♂
18 59 ☽ # ♂
23 34 ☽ △ ♅

13 Friday
01 21 ♀ ∥ ♄
07 08 ☽ ☌ ☿
10 04 ☽ ⊻ ♄
11 16 ☽ ☍ ♂
13 04 ☽ ⊻ ♀
15 56 ☉ □ ♇
22 08 ☽ ⊥ ♄

14 Saturday
02 03 ☽ ⊥ ♀
09 04 ♀ ⊑ ♇
16 19 ☽ ∠ ♄
20 39 ☽ □ ♇
21 38 ☽ ∠ ♀
23 06 ♂ ⊼ ♄
23 09 ☽ ☌ ☉

15 Sunday
05 04 ☽ ⊼ ♆
05 34 ☽ □ ♃
11 40 ☽ □ ♅
17 09 ☽ ± ♆
20 51 ☽ ⊻ ☿
22 03 ☽ ⊼ ♂
22 48 ☽ ✱ ♄

16 Monday
06 30 ☽ ✱ ♀
09 00 ☽ Q ♇
09 25 ☽ ⊥ ☿
09 54 ☽ ± ♂
11 18 ☽ ⊑ ♆
17 33 ☿ ☍ ♂
18 09 ☽ Q ♃
23 17 ☽ Q ♄

17 Tuesday
00 06 ☽ Q ♅
03 43 ☽ ⊑ ♂
04 23 ☽ ∠ ☿
08 55 ☽ Q ♀
09 19 ☽ ✱ ♇
16 00 ☿ ∥ ♇
17 06 ☽ ⊻ ☉
17 38 ☽ △ ♆
18 37 ☽ ✱ ♃
23 31 ☉ ⊼ ♆

18 Wednesday
00 19 ♂ Q ♇
00 25 ☽ ✱ ♅
03 16 ☽ # ♂
06 13 ☽ ∥ ☉
06 19 ☽ ⊥ ☉
08 02 ☿ ⊻ ♄
09 26 ☽ △ ♂
12 02 ☽ □ ♄
12 13 ☽ ✱ ☿
13 11 ☉ □ ♃
15 43 ☽ ∠ ♇

19 Thursday
00 33 ☽ □ ♀
01 08 ☽ ∠ ♃
02 06 ☽ ∠ ☉
06 45 ☽ ∠ ♅
08 23 ♀ ⊑ ♃
09 52 ☽ ⊥ ♇
13 45 ☽ Q ☿
16 19 ☽ # ♆
16 45 ☽ ∥ ☿
18 51 ☽ ∥ ♇
19 21 ☽ ⊥ ♃
21 55 ☽ ⊻ ♇

20 Friday
00 49 ☽ ⊥ ♅
01 39 ☿ # ♆
05 57 ☽ □ ♆
07 24 ☽ ⊻ ♃
10 45 ☽ ✱ ☉
12 47 ☽ ⊻ ♅
14 35 ☽ ∥ ♀
20 21 ☽ □ ♂
22 45 ☽ ∥ ♄

21 Saturday
00 33 ☽ △ ♄
03 49 ☽ □ ☿
12 24 ☽ Q ☉
12 54 ☉ □ ♅
17 21 ☽ △ ♀
17 21 ☽ Q ♆
17 23 ♄ ± ♅

22 Sunday
05 15 ♀ ⊑ ♅
05 58 ☽ ⊑ ♄
06 48 ☽ Q ♂
08 50 ☽ ☌ ♇
12 37 ☽ ∥ ♃
16 22 ☽ ✱ ♆
18 14 ☽ ☌ ♃
18 47 ☽ ∥ ♅
23 01 ☽ ☌ ♅
23 12 ☽ ± ♄

23 Monday
00 08 ☽ # ♅
00 29 ☽ ⊑ ♀
01 44 ☽ □ ☉
05 10 ☽ ✱ ♂
06 06 ☽ # ♃
10 30 ☽ ⊼ ♄
17 14 ☽ △ ☿
18 36 ☽ ± ♀
20 18 ☽ ∠ ♆

24 Tuesday
03 04 ☉ ± ♆
06 25 ☽ ⊼ ♀
08 17 ☽ ∠ ♂
12 35 ☽ ⊥ ♆
16 21 ☽ ⊻ ♇
16 33 ☉ ⊼ ♂
18 11 ☽ # ♄
22 27 ☽ ⊑ ☿
23 13 ☽ ⊻ ♆

25 Wednesday
00 06 ☽ ⊥ ♂
01 24 ☽ ⊻ ♃
02 58 ☽ ⊥ ♇
05 32 ☽ ⊻ ♅
06 16 ☽ # ♀
10 26 ☽ ⊻ ♂
11 55 ☽ ⊥ ♃
12 01 ☽ △ ☉
15 25 ☽ ± ☿
15 56 ☽ ⊥ ♅
16 27 ☽ ☍ ♄
17 59 ☽ # ♇
18 34 ☽ ∠ ♇
20 18 ☽ ∥ ♆

26 Thursday
02 31 ☽ ⊼ ☿
03 25 ☽ ∠ ♃
03 57 ☽ # ☿
05 37 ☿ ⊥ ♄
07 15 ☽ ∠ ♅
14 28 ☽ ☍ ♀
15 16 ☽ ⊑ ☉
19 48 ☽ ✱ ♇

27 Friday
00 21 ☽ # ☉
02 04 ☽ ☌ ♆
02 34 ☽ ∥ ♂
04 29 ☽ ✱ ♃
07 06 ☽ ± ☉
08 04 ☽ ✱ ♅
12 03 ☽ ☌ ♂
15 24 ☽ Q ♇
17 29 ☽ ⊼ ☉
18 30 ☽ ⊼ ♄

28 Saturday
00 00 ☽ Q ♃
03 25 ☽ Q ♅
04 10 ☽ ± ♄
07 50 ☽ ☍ ☿
10 49 ☉ ✱ ♄
18 26 ☽ ⊼ ♀
18 30 ☽ ⊑ ♄
20 06 ☽ □ ♇

29 Sunday
01 59 ☽ ⊻ ♆
04 33 ☽ ± ♀
04 40 ☽ □ ♃
07 52 ☽ □ ♅
11 14 ☽ ⊻ ♂
11 23 ☽ ⊥ ♆
18 10 ☽ △ ♄
19 31 ♀ △ ♇
19 39 ☽ ⊑ ♀
20 04 ☽ ☍ ☉
20 33 ☽ ⊥ ♂

30 Monday
01 27 ☽ ∠ ♆
03 12 ☉ ± ♂
10 33 ☽ ∠ ♂
11 23 ☽ ⊼ ☿
19 10 ☽ △ ♇
20 51 ☽ △ ♀
21 45 ☽ ± ☿

JULY 1969

LONGITUDES

Date	Sidereal time h m s	Sun ☉ ° ' "	Moon ☽ ° ' "	Moon ☽ 24.00 ° ' "	Mercury ☿ ° '	Venus ♀ ° '	Mars ♂ ° '	Jupiter ♃ ° '	Saturn ♄ ° '	Uranus ♅ ° '	Neptune ♆ ° '	Pluto ♇ ° '
01	06 37 27	09 ♋ 27 31	03 ♒ 16 04	10 ♒ 46 39	19 ♊ 06	24 ♉ 20	02 ♐ 01	28 ♍ 17	06 ♉ 48	00 ♎ 08	26 ♏ 17	22 ♍ 38
02	06 41 24	10 24 42	18 ♒ 11 57	25 ♒ 31 11	20 35	25 22	01 R 56	28 24	06 53	00 09	26 R 16	22 39
03	06 45 20	11 21 53	02 ♓ 43 47	09 ♓ 49 26	22 06	26 25	01 51	28 30	06 58	00 10	26 15	22 40
04	06 49 17	12 19 04	16 ♓ 48 02	23 ♓ 39 37	23 42	27 27	01 48	28 37	07 02	00 12	26 14	22 41
05	06 53 13	13 16 16	00 ♈ 24 24	07 ♈ 02 43	25 21	28 30	01 45	28 44	07 07	00 13	26 13	22 42
06	06 57 10	14 13 28	13 ♈ 34 59	20 ♈ 01 40	27 03	29 ♉ 33	01 43	28 51	07 11	00 15	26 12	22 43
07	07 01 07	15 10 40	26 ♈ 23 16	02 ♉ 40 20	28 ♊ 48	00 ♊ 37	01 42	28 58	07 16	00 16	26 11	22 44
08	07 05 03	16 07 53	08 ♉ 53 22	15 ♉ 02 55	00 ♋ 37	01 40	01 D 42	29 05	07 20	00 18	26 10	22 45
09	07 09 00	17 05 06	21 ♉ 09 26	27 ♉ 13 23	02 28	02 44	01 43	29 13	07 24	00 19	26 10	22 46
10	07 12 56	18 02 19	03 ♊ 15 13	09 ♊ 15 17	04 23	03 48	01 44	29 20	07 28	00 21	26 09	22 47
11	07 16 53	18 59 33	15 ♊ 13 57	21 ♊ 11 31	06 20	04 53	01 46	29 28	07 32	00 23	26 08	22 48
12	07 20 49	19 56 47	27 ♊ 08 16	03 ♋ 04 26	08 19	05 57	01 49	29 36	07 36	00 25	26 07	22 49
13	07 24 46	20 54 02	09 ♋ 00 15	14 ♋ 55 54	10 20	07 02	01 53	29 43	07 40	00 27	26 06	22 51
14	07 28 42	21 51 17	20 ♋ 51 36	26 ♋ 47 32	12 24	08 07	01 58	29 51	07 44	00 28	26 05	22 52
15	07 32 39	22 48 32	02 ♌ 43 53	08 ♌ 40 53	14 29	09 12	02 04	29 ♍ 59	07 48	00 30	26 05	22 53
16	07 36 36	23 45 48	14 ♌ 38 44	20 ♌ 37 43	16 35	10 17	02 10	00 ♎ 08	07 51	00 32	26 04	22 54
17	07 40 32	24 43 04	26 ♌ 38 05	02 ♍ 40 10	18 42	11 22	02 17	00 16	07 55	00 34	26 03	22 56
18	07 44 29	25 40 20	08 ♍ 44 20	14 ♍ 50 57	20 50	12 28	02 25	00 25	07 58	00 36	26 03	22 57
19	07 48 25	26 37 36	21 ♍ 00 27	27 ♍ 13 17	22 58	13 34	02 34	00 33	08 02	00 38	26 02	22 59
20	07 52 22	27 34 52	03 ♎ 29 56	09 ♎ 50 55	25 06	14 39	02 43	00 42	08 05	00 41	26 02	23 00
21	07 56 18	28 32 09	16 ♎ 16 41	22 ♎ 47 45	27 14	15 46	02 53	00 51	08 08	00 43	26 01	23 01
22	08 00 15	29 ♋ 29 26	29 ♎ 24 33	06 ♏ 07 30	29 ♋ 22	16 52	03 04	00 59	08 11	00 45	26 00	23 03
23	08 04 11	00 ♌ 26 44	12 ♏ 56 53	19 ♏ 52 55	01 ♌ 29	17 58	03 16	01 08	08 14	00 47	26 00	23 04
24	08 08 08	01 24 02	26 ♏ 55 41	04 ♐ 05 05	03 35	19 05	03 28	01 18	08 17	00 50	26 00	23 06
25	08 12 05	02 21 20	11 ♐ 20 50	18 ♐ 42 26	05 40	20 11	03 41	01 27	08 20	00 52	25 59	23 07
26	08 16 01	03 18 38	26 ♐ 09 13	03 ♑ 40 17	07 44	21 18	03 55	01 36	08 22	00 54	25 59	23 09
27	08 19 58	04 15 58	11 ♑ 14 34	18 ♑ 50 51	09 46	22 25	04 09	01 45	08 25	00 57	25 58	23 11
28	08 23 54	05 13 17	26 ♑ 27 50	04 ♒ 04 11	11 47	23 32	04 24	01 55	08 27	00 59	25 58	23 12
29	08 27 51	06 10 38	11 ♒ 38 34	19 ♒ 09 45	13 47	24 40	04 40	02 05	08 30	01 02	25 58	23 14
30	08 31 47	07 07 59	26 ♒ 36 39	03 ♓ 58 18	15 45	25 47	04 56	02 14	08 32	01 04	25 57	23 15
31	08 35 44	08 ♌ 05 21	11 ♓ 13 59	18 ♓ 23 10	17 ♌ 42	26 ♊ 55	05 ♐ 13	02 ♎ 24	08 ♉ 34	01 ♎ 07	25 ♏ 57	23 ♍ 17

Date	Moon True ☊ ° '	Moon Mean ☊ ° '	Moon ☽ Latitude ° '
01	24 ♓ 05	25 ♓ 00	03 S 54
02	23 R 59	24 57	02 57
03	23 55	24 54	01 50
04	23 54	24 51	00 S 38
05	23 D 54	24 47	00 N 35
06	23 54	24 44	01 43
07	23 R 54	24 41	02 43
08	23 51	24 38	03 35
09	23 47	24 35	04 15
10	23 40	24 32	04 44
11	23 30	24 28	04 59
12	23 19	24 25	05 02
13	23 07	24 22	04 51
14	22 56	24 19	04 28
15	22 46	24 16	03 53
16	22 38	24 12	03 08
17	22 32	24 09	02 14
18	22 29	24 06	01 13
19	22 28	24 03	00 N 08
20	22 D 28	24 00	00 S 59
21	22 29	23 57	02 04
22	22 30	23 53	03 05
23	22 R 29	23 50	03 57
24	22 26	23 47	04 37
25	22 22	23 44	05 02
26	22 16	23 41	05 06
27	22 09	23 38	04 50
28	22 02	23 34	04 14
29	21 56	23 31	03 20
30	21 52	23 28	02 12
31	21 ♓ 50	23 ♓ 25	00 S 57

DECLINATIONS

Date	Sun ☉ ° '	Moon ☽ ° '	Mercury ☿ ° '	Venus ♀ ° '	Mars ♂ ° '	Jupiter ♃ ° '	Saturn ♄ ° '	Uranus ♅ ° '	Neptune ♆ ° '	Pluto ♇ ° '
01	23 N 06	23 S 13	21 N 03	16 N 01	23 S 43	01 N 51	11 N 34	00 N 37	17 S 38	17 N 03
02	23 02	18 11	21 22	16 16	23 43	01 49	11 35	00 36	17 38	17 03
03	22 58	12 13	21 40	16 31	23 43	01 46	11 36	00 36	17 37	17 02
04	22 52	05 S 47	21 58	16 46	23 43	01 43	11 38	00 35	17 37	17 01
05	22 47	00 N 41	22 14	17 01	23 43	01 40	11 39	00 34	17 37	17 00
06	22 41	06 56	22 30	17 15	23 44	01 37	11 40	00 34	17 37	17 00
07	22 35	12 43	22 44	17 29	23 44	01 34	11 41	00 33	17 37	16 59
08	22 28	17 51	22 56	17 43	23 45	01 31	11 43	00 32	17 37	16 58
09	22 21	22 09	23 07	17 57	23 46	01 28	11 44	00 32	17 36	16 57
10	22 14	25 27	23 16	18 11	23 47	01 25	11 45	00 31	17 36	16 56
11	22 06	27 35	23 23	18 24	23 48	01 22	11 46	00 30	17 36	16 56
12	21 58	28 27	23 28	18 37	23 49	01 18	11 47	00 30	17 36	16 55
13	21 49	27 59	23 30	18 49	23 50	01 15	11 48	00 29	17 36	16 54
14	21 40	26 14	23 30	19 01	23 52	01 12	11 49	00 28	17 36	16 53
15	21 31	23 20	23 27	19 13	23 53	01 08	11 50	00 27	17 36	16 52
16	21 21	19 26	23 21	19 25	23 55	01 05	11 51	00 26	17 35	16 51
17	21 11	14 44	23 13	19 36	23 56	01 01	11 52	00 26	17 35	16 50
18	21 01	09 25	23 02	19 46	23 58	00 58	11 53	00 25	17 35	16 50
19	20 50	03 N 41	22 48	19 57	24 00	00 54	11 54	00 24	17 35	16 49
20	20 39	02 S 18	22 31	20 07	24 02	00 51	11 54	00 23	17 35	16 48
21	20 28	08 19	22 12	20 16	24 04	00 47	11 55	00 22	17 35	16 47
22	20 16	14 09	21 51	20 25	24 06	00 43	11 56	00 21	17 35	16 46
23	20 04	19 30	21 27	20 34	24 09	00 39	11 57	00 20	17 35	16 45
24	19 51	23 58	21 01	20 42	24 11	00 36	11 57	00 19	17 35	16 44
25	19 38	27 07	20 33	20 50	24 14	00 32	11 58	00 18	17 35	16 43
26	19 25	28 30	20 03	20 57	24 16	00 28	11 59	00 17	17 35	16 42
27	19 12	27 48	19 31	21 04	24 19	00 24	11 59	00 16	17 35	16 42
28	18 58	25 01	18 58	21 10	24 21	00 20	12 00	00 15	17 35	16 41
29	18 44	20 29	18 24	21 16	24 24	00 16	12 00	00 14	17 35	16 40
30	18 30	14 43	17 48	21 21	24 27	00 12	12 01	00 13	17 35	16 39
31	18 N 15	08 S 14	17 N 11	21 N 26	24 S 30	00 N 08	12 N 01	00 N 12	17 S 35	16 N 38

ZODIAC SIGN ENTRIES

Date	h	m	Planets
01	06	49	☽ ♒
03	07	26	☽ ♓
05	11	16	☽ ♈
06	22	04	♀ ♊
07	18	53	☽ ♉
08	03	58	☿ ♋
10	05	31	☽ ♊
12	17	47	☽ ♋
15	06	29	☽ ♌
15	13	30	♃ ♎
17	18	42	☽ ♍
20	05	20	☽ ♎
22	13	04	☽ ♏
22	19	11	☿ ♌
23	00	48	☉ ♌
24	17	10	☽ ♐
26	18	09	☽ ♑
28	17	34	☽ ♒
30	17	30	☽ ♓

LATITUDES

Date	Mercury ☿ ° '	Venus ♀ ° '	Mars ♂ ° '	Jupiter ♃ ° '	Saturn ♄ ° '	Uranus ♅ ° '	Neptune ♆ ° '	Pluto ♇ ° '
01	01 S 57	02 S 56	03 S 12	01 N 17	02 S 21	00 N 43	01 N 45	15 N 26
04	01 20	02 54	03 15	01 16	02 22	00 43	01 45	15 25
07	00 43	02 51	03 18	01 16	02 23	00 43	01 44	15 23
10	00 S 06	02 47	03 20	01 15	02 23	00 43	01 44	15 22
13	00 N 28	02 42	03 22	01 14	02 24	00 43	01 44	15 21
16	00 57	02 37	03 23	01 14	02 25	00 43	01 44	15 20
19	01 19	02 30	03 23	01 13	02 25	00 43	01 44	15 19
22	01 36	02 23	03 24	01 13	02 26	00 43	01 44	15 18
25	01 45	02 16	03 24	01 12	02 27	00 42	01 44	15 17
28	01 47	02 07	03 23	01 12	02 28	00 42	01 43	15 16
31	01 N 44	01 S 58	03 S 23	01 N 11	02 S 28	00 N 42	01 N 43	15 N 15

DATA

Julian Date	2440404
Delta T	+40 seconds
Ayanamsa	23° 25' 54"
Synetic vernal point	05° ♓ 41' 06"
True obliquity of ecliptic	23° 26' 44"

MOON'S PHASES, APSIDES AND POSITIONS ☽

Date	h	m	Phase	Longitude ° '	Eclipse Indicator
06	13	17	☾	14 ♈ 17	
14	14	11	●	21 ♋ 57	
22	12	09	☽	29 ♎ 30	
29	02	45	○	05 ♒ 49	

Day	h	m		
13	18	04	Apogee	
28	09	07	Perigee	
05	09	25	0N	
12	15	35	Max dec	28° N 28'
20	02	52	0S	
26	16	06	Max dec	28° S 32'

LONGITUDES

Date	Chiron ⚷ ° '	Ceres ⚳ ° '	Pallas ⚴ ° '	Juno ⚵ ° '	Vesta ⚶ ° '	Black Moon Lilith ⚸ ° '
01	06 ♈ 51	09 ♒ 44	01 ♑ 34	08 ♑ 40	00 ♋ 01	12 ♋ 16
11	06 ♈ 53	08 ♒ 02	28 ♐ 57	06 ♑ 21	04 ♋ 20	13 ♋ 23
21	06 ♈ 49	05 ♒ 58	26 ♐ 48	04 ♑ 15	08 ♋ 37	14 ♋ 31
31	06 ♈ 41	03 ♒ 47	25 ♐ 18	02 ♑ 32	12 ♋ 52	15 ♋ 38

ASPECTARIAN

h m	Aspects
01 Tuesday	
00 58	☽ ⚹ ♆
04 03	☽ △ ♃
07 01	☽ △ ♅
09 17	☽ ∥ ♂
10 01	☽ ⚹ ♂
12 38	☽ # ☉
13 28	☽ Ꝑ ☿
17 40	☽ □ ♄
18 58	☽ Ꝑ ♇
20 00	☽ Q ♆
22 25	☽ # ☿
22 33	☽ ⊼ ☉
02 Wednesday	
04 09	☽ Ꝑ ♃
05 07	☽ Q ♂
07 03	☽ Ꝑ ♅
08 54	☽ ± ☉
09 28	☽ ± ♇
14 22	☽ ∥ ♆
16 19	☽ △ ☿
16 50	☽ # ♇
17 56	☉ Q ♇
18 54	☽ ± ♃
19 16	☽ ⊼ ♇
19 41	☽ # ♀
21 45	☽ ± ♅
22 32	☉ ∠ ♀
23 00	☽ Q ♄
03 Thursday	
00 39	☽ Ꝑ ☉
00 40	☽ □ ♀
01 14	☽ □ ♆
04 53	☽ ⊼ ♃
07 43	☽ ⊼ ♅
08 23	♀ ☍ ♆
09 15	☉ Ꝑ ♆
09 38	☿ ∠ ♄
10 33	☽ □ ♂
14 18	☽ # ♄
19 10	☽ ⚹ ♄
20 32	☿ □ ♇
04 Friday	
03 42	☽ △ ☉
09 29	☽ Q ♀
21 12	☽ ∠ ♄
22 17	☽ ☍ ♇
05 Saturday	
01 43	☽ □ ☿
03 09	☽ # ♃
04 32	☽ △ ♆
07 17	☽ # ♅
08 19	☽ ⚹ ♀
08 59	☽ ☍ ♃
11 21	♀ ∥ ♇
11 34	☽ ∥ ♅
11 40	☽ ☍ ♅
13 16	☽ ⊥ ♄
14 25	☽ △ ♂
15 39	☽ ∥ ♃
17 47	♀ △ ♃
06 Sunday	
00 11	☽ ⊻ ♄
00 21	☿ ⊼ ♆
07 37	☽ Ꝑ ♆
13 17	☽ □ ☉
13 58	☽ ∠ ♀
15 08	☽ Q ☿
17 49	☽ Ꝑ ♂
07 Monday	
00 19	☽ ± ♆
01 21	☉ ∥ ☿
04 01	♀ △ ♅
05 04	☽ ⊼ ♇
07 31	☽ ∥ ♄
08 20	☽ ⊥ ♀
10 42	☽ ± ♂
11 37	☽ ⊼ ♆
14 22	☿ □ ♃
16 27	☽ ± ♇
16 57	☽ ⊼ ♃
17 21	☽ ⚹ ☿
19 25	☽ ⊼ ♅
20 48	☽ ⊻ ♀
22 08	☽ ⊼ ♂
08 Tuesday	
00 10	♀ # ♆
02 03	☽ Q ☉
04 34	☽ ± ♃
06 07	♂ St D
06 58	☽ ± ♅
07 35	☽ ∥ ♇
07 49	☿ □ ♅
08 58	☽ ☌ ♄
09 47	☽ Ꝑ ♇
10 45	☽ # ♆
11 18	☽ ∥ ♀
12 34	♀ ☍ ♂
22 13	☽ Ꝑ ♃
09 Wednesday	
00 31	☽ Ꝑ ♅
02 11	☿ ⊼ ♂
02 24	☉ Ꝑ ♂
03 19	☽ ⚹ ☉
03 26	☽ ∠ ☿
08 03	☿ ± ♆
13 12	☽ ∥ ☉
15 11	☽ △ ♇
15 38	☉ Q ♃
18 34	☽ ∥ ☿
19 48	☿ ⊻ ♀
21 53	☽ ☍ ♆
22 46	☽ # ♂
10 Thursday	
00 27	☽ ⊥ ☿
04 07	☽ △ ♃
06 12	☽ △ ♅
08 58	☽ ☍ ♂
11 32	☽ ∠ ☉
13 12	☽ ☌ ♀
14 40	☽ ⊻ ☿
20 09	☉ Q ♅
20 28	☽ ⊻ ♄
11 Friday	
07 06	☽ ⊥ ☉
08 34	☽ ⊥ ♄
20 14	☽ ⊻ ☉
12 Saturday	
02 45	☉ Q ♄
02 48	☽ ∠ ♄
03 11	☿ ⚹ ♄
03 16	☽ □ ♇
05 57	☿ ± ♂
09 56	☽ ⊼ ♆
17 01	☽ □ ♃
18 38	☽ □ ♅
21 31	☽ ⊼ ♂
22 03	☽ ± ♆
13 Sunday	
07 36	☽ ⊻ ♀
09 17	☽ ⚹ ♄
09 44	☽ ± ♂
15 16	☽ ☌ ☿
15 44	☽ Q ♇
16 15	☽ Ꝑ ♆
17 58	☿ Q ♇
20 54	☿ Ꝑ ♆
20 58	☽ ⊥ ♀
14 Monday	
03 07	♀ ⊻ ♄
04 04	☽ Ꝑ ♂
05 51	☽ Q ♃
07 09	☽ Q ♅
09 42	☽ Q ♄
14 11	☽ ☌ ☉
16 04	☽ ⚹ ♇
17 00	☽ ∠ ♀
22 34	☽ △ ♆
15 Tuesday	
06 24	☽ ⚹ ♃
07 30	☽ ⚹ ♅
08 05	☽ # ♂
10 38	☽ △ ♂
11 11	☽ ∥ ☿
13 58	☉ ⚹ ♇
22 16	☽ □ ♄
22 25	☽ ∠ ♇
16 Wednesday	
00 23	☽ ∥ ☉
02 20	☽ ⚹ ♀
04 55	☿ ⊥ ♀
12 08	☽ ∥ ♀
12 59	☽ ∠ ♃
13 48	☽ ∠ ♅
16 33	☽ ⊥ ♇
16 43	☽ ⊻ ☿
19 00	☿ Ꝑ ♂
21 51	☽ # ♆
17 Thursday	
01 39	☽ ∥ ♇
04 35	☽ ⊻ ♇
04 50	☽ Q ♀
06 45	☿ Q ♃
07 13	☽ ⊥ ♃
07 19	☽ ⊥ ☿
07 51	☽ ⊻ ☉
07 52	☽ ⊥ ♅
10 31	☿ Q ♅
10 51	☽ □ ♆
19 19	☽ ⊻ ♃
19 52	☽ ⊻ ♅
20 49	☽ ⊥ ☉
23 21	☽ □ ♂
23 49	♄ Ꝑ ♇
18 Friday	
01 13	☽ ∥ ♄
02 02	☿ Q ♄
05 02	☽ ∠ ☿
10 29	☽ △ ♄
16 08	☽ ∠ ☉
20 03	☽ □ ♀
21 18	☉ △ ♆
22 25	☽ Q ♆
19 Saturday	
11 07	☽ Q ♂
12 05	☿ ⚹ ♇
15 49	☽ ☌ ♇
15 56	☽ Ꝑ ♄
16 35	☽ ⚹ ☿
21 43	☽ ⚹ ♆
22 44	♀ ⊥ ♄
23 22	☽ ∥ ♃
23 45	☽ ⚹ ☉
20 Sunday	
01 18	☽ ∥ ♅
04 25	☽ # ♅
06 17	☽ # ♃
06 36	☽ ☌ ♃
06 36	☽ ☌ ♅
07 57	♃ ☌ ♅
09 17	☽ ± ♄
10 29	☽ ⚹ ♂
20 13	☽ Q ☿
20 42	☽ ⊼ ♄
22 18	☿ △ ♆
21 Monday	
00 26	☽ Q ☉
02 12	☽ ∠ ♆
10 57	☽ △ ♀
15 01	☽ ∠ ♂
18 54	☽ ⊥ ♆
22 Tuesday	
00 26	☽ ⊻ ♇
01 05	☉ ∥ ♀
02 42	☽ # ♄
05 51	☽ ⊻ ♆
07 43	☽ ⊥ ♂
11 21	☽ ⊥ ♇
11 54	☽ □ ☿
12 09	☽ □ ☉
14 25	☽ ⊻ ♅
14 34	☉ ☌ ☿
14 52	☽ ⊻ ♃
16 48	☽ Ꝑ ♀
18 39	☽ ⊻ ♂
23 21	☽ # ♇
23 Wednesday	
01 09	☽ ⊥ ♅
01 41	☽ ⊥ ♃
03 03	☽ ∥ ♆
03 26	☽ ∠ ♇
03 42	☽ ☍ ♄
04 00	☿ ⚹ ♅
07 50	☿ ⚹ ♃
10 08	☽ ± ♀
14 38	☽ # ☉
16 57	☽ ∠ ♅
17 24	☽ # ♀
17 37	☽ ∠ ♃
20 57	☽ # ☿
21 00	☉ ⚹ ♅
21 28	☽ ⊼ ♀
24 Thursday	
05 29	☽ ⚹ ♇
08 45	☉ ⚹ ♃
10 25	☽ ☌ ♆
10 32	☿ △ ♂
13 20	☽ ∥ ♂
18 35	☽ ⚹ ♅
19 25	☽ ⚹ ♃
20 03	☽ △ ☉
23 08	☽ ☌ ♂
25 Friday	
00 15	☿ ∠ ♀
00 49	☿ ∥ ♀
01 03	☽ △ ☿
01 42	☽ Q ♇
07 01	☽ ⊼ ♄
14 30	☽ Q ♅
15 28	☽ Q ♃
16 54	☽ ± ♄
22 29	☽ Ꝑ ☉
26 Saturday	
03 34	☽ ☍ ♀
05 37	☽ Ꝑ ☿
07 10	☽ □ ♇
07 31	☽ Ꝑ ♄
11 43	☽ ⊻ ♆
13 59	☽ ± ☉
17 00	☿ ∠ ♇
19 37	☽ □ ♅
19 43	☿ □ ♄
20 48	☽ □ ♃
21 18	☽ ⊥ ♆
22 18	☽ ± ☿
27 Sunday	
00 12	☽ ⊼ ☉
00 35	☽ ⊻ ♂
07 31	☽ △ ♄
08 06	☉ △ ♂
09 19	☽ ⊼ ☿
10 15	☽ ⊥ ♂
11 34	☽ ∠ ♆
28 Monday	
00 41	☽ ∠ ♂
04 33	♀ □ ♇
06 51	☽ △ ♇
07 02	☽ ⊼ ♀
10 08	♀ ∠ ♄
11 13	☽ ⚹ ♆
12 17	☉ ∥ ☿
16 03	☽ ∥ ♂
17 14	☽ ± ♀
19 09	☽ △ ♅
20 41	☽ △ ♃
29 Tuesday	
00 44	☽ ⚹ ♂
02 45	☽ ☍ ☉
06 10	☽ Q ♆
06 34	☽ Ꝑ ♇
07 00	☽ □ ♄
08 27	☽ # ♀
08 36	☽ Ꝑ ♀
15 55	☽ ☍ ☿
19 01	☽ Ꝑ ♅
20 06	☽ # ☉
20 09	☽ Q ♂
20 45	☽ Ꝑ ♃
20 55	☽ ± ♇
22 16	☽ # ☿
30 Wednesday	
00 36	☽ ∥ ♆
04 23	☽ # ♇
04 51	♃ ∥ ♅
06 34	☽ ⊼ ♇
09 30	☽ ± ♅
10 33	☽ △ ♀
10 57	☽ □ ♆
11 23	☽ ± ♃
11 53	☽ Q ♄
15 37	♀ ⊼ ♆
16 03	☿ ∠ ♅
19 17	☽ ⊼ ♅
20 34	☿ # ♆
21 16	☽ ⊼ ♃
22 12	☽ # ♄
31 Thursday	
01 51	☽ □ ♂
06 25	☽ ⊼ ☉
06 50	☿ ⊥ ♇
07 34	☽ ⚹ ♄
07 59	☿ ∠ ♃
17 03	☉ ∠ ♇
17 07	☽ ± ☉
18 32	♂ Q ♇

All ephemeris data is given at 12.00 UT and the Moon's longitude is additionally given for 24.00 UT

AUGUST 1969

Date	Sidereal time h m s	Sun ☉ ° ' "	Moon ☽ ° ' "	Moon ☽ 24.00 ° ' "	Mercury ☿ ° '	Venus ♀ ° '	Mars ♂ ° '	Jupiter ♃ ° '	Saturn ♄ ° '	Uranus ♅ ° '	Neptune ♆ ° '	Pluto ♇ ° '
01	08 39 40	09 ♌ 02 44	25 ♓ 25 30	02 ♈ 20 51	19 ♌ 37	28 ♊ 03	05 ♐ 30	02 ♎ 34	08 ♉ 36	01 ♎ 10	25 ♏ 57	23 ♍ 19
02	08 43 37	10 00 08	09 ♈ 09 14	15 ♈ 50 50	21 30	29 ♊ 10	05 48	02 44	08 38	01 13	25 R 57	23 20
03	08 47 34	10 57 33	22 ♈ 25 55	28 ♈ 54 53	23 22	00 ♋ 18	06 07	02 54	08 40	01 15	25 57	23 22
04	08 51 30	11 55 00	05 ♉ 18 12	11 ♉ 36 21	25 12	01 27	06 26	03 04	08 42	01 18	25 57	23 24
05	08 55 27	12 52 28	17 ♉ 49 54	23 ♉ 59 24	27 01	02 35	06 46	03 14	08 44	01 21	25 56	23 26
06	08 59 23	13 49 57	00 ♊ 05 25	06 ♊ 08 28	28 ♌ 48	03 43	07 06	03 24	08 45	01 23	25 56	23 27
07	09 03 20	14 47 27	12 ♊ 09 06	18 ♊ 07 50	00 ♍ 33	04 52	07 27	03 35	08 47	01 26	25 56	23 29
08	09 07 16	15 44 59	24 ♊ 05 07	00 ♋ 01 25	02 17	06 00	07 49	03 45	08 48	01 29	25 D 56	23 31
09	09 11 13	16 42 32	05 ♋ 57 07	11 ♋ 52 37	04 00	07 09	08 11	03 56	08 50	01 32	25 56	23 33
10	09 15 09	17 40 06	17 ♋ 48 14	23 ♋ 44 17	05 41	08 18	08 33	04 07	08 51	01 35	25 57	23 35
11	09 19 06	18 37 42	29 ♋ 41 02	05 ♌ 38 45	07 20	09 27	08 56	04 17	08 52	01 38	25 57	23 37
12	09 23 03	19 35 18	11 ♌ 37 39	17 ♌ 37 57	08 58	10 36	09 20	04 28	08 53	01 41	25 57	23 39
13	09 26 59	20 32 56	23 ♌ 39 51	29 ♌ 43 33	10 34	11 45	09 44	04 39	08 54	01 44	25 57	23 41
14	09 30 56	21 30 35	05 ♍ 49 15	11 ♍ 57 09	12 08	12 55	10 08	04 50	08 55	01 47	25 57	23 42
15	09 34 52	22 28 16	18 ♍ 07 28	24 ♍ 20 26	13 42	14 04	10 33	05 01	08 55	01 50	25 57	23 44
16	09 38 49	23 25 57	00 ♎ 36 17	06 ♎ 55 17	15 13	15 14	10 58	05 12	08 56	01 53	25 58	23 46
17	09 42 45	24 23 39	13 ♎ 17 42	19 ♎ 43 48	16 43	16 23	11 24	05 23	08 56	01 57	25 58	23 48
18	09 46 42	25 21 23	26 ♎ 13 54	02 ♏ 48 15	18 12	17 33	11 50	05 34	08 57	02 00	25 58	23 50
19	09 50 38	26 19 08	09 ♏ 27 08	16 ♏ 10 47	19 39	18 43	12 17	05 46	08 57	02 03	25 59	23 52
20	09 54 35	27 16 54	22 ♏ 59 24	29 ♏ 53 05	21 04	19 53	12 44	05 57	08 57	02 06	25 59	23 54
21	09 58 32	28 14 41	06 ♐ 51 53	13 ♐ 55 45	22 28	21 03	13 12	06 08	08 R 57	02 10	25 59	23 56
22	10 02 28	29 ♌ 12 29	21 ♐ 04 31	28 ♐ 17 51	23 50	22 13	13 40	06 20	08 57	02 13	26 00	23 58
23	10 06 25	00 ♍ 10 18	05 ♑ 35 21	12 ♑ 56 24	25 11	23 24	14 08	06 31	08 57	02 16	26 00	24 01
24	10 10 21	01 08 09	20 ♑ 20 17	27 ♑ 46 11	26 30	24 34	14 37	06 43	08 56	02 20	26 01	24 03
25	10 14 18	02 06 01	05 ♒ 13 10	12 ♒ 40 14	27 47	25 44	15 06	06 55	08 56	02 23	26 02	24 05
26	10 18 14	03 03 54	20 ♒ 06 21	27 ♒ 30 32	29 ♍ 02	26 55	15 35	07 06	08 55	02 26	26 02	24 07
27	10 22 11	04 01 48	04 ♓ 51 47	12 ♓ 09 16	00 ♎ 16	28 06	16 05	07 18	08 55	02 30	26 03	24 09
28	10 26 07	04 59 44	19 ♓ 22 10	26 ♓ 29 53	01 27	29 ♋ 16	16 35	07 30	08 54	02 33	26 03	24 11
29	10 30 04	05 57 42	03 ♈ 31 54	10 ♈ 27 53	02 37	00 ♌ 27	17 06	07 42	08 53	02 37	26 04	24 13
30	10 34 01	06 55 41	17 ♈ 17 40	24 ♈ 01 09	03 44	01 38	17 36	07 54	08 52	02 40	26 05	24 15
31	10 37 57	07 ♍ 53 42	00 ♉ 38 28	07 ♉ 09 46	04 ♎ 49	02 ♌ 49	18 ♐ 08	08 ♎ 06	08 ♉ 51	02 ♎ 44	26 ♏ 06	24 ♍ 17

Date	Moon True ☊ ° '	Moon Mean ☊ ° '	Moon ☽ Latitude ° '
01	21 ♓ 49	23 ♓ 22	00 N 20
02	21 D 50	23 18	01 33
03	21 52	23 15	02 38
04	21 53	23 12	03 34
05	21 R 53	23 09	04 18
06	21 51	23 06	04 49
07	21 48	23 03	05 06
08	21 44	22 59	05 10
09	21 38	22 56	05 01
10	21 32	22 53	04 39
11	21 26	22 50	04 05
12	21 21	22 47	03 19
13	21 17	22 44	02 25
14	21 14	22 40	01 23
15	21 13	22 37	00 N 17
16	21 D 14	22 34	00 S 51
17	21 15	22 31	01 58
18	21 16	22 28	03 00
19	21 18	22 24	03 55
20	21 19	22 21	04 37
21	21 R 18	22 18	05 05
22	21 17	22 15	05 15
23	21 15	22 12	05 05
24	21 13	22 09	04 35
25	21 10	22 05	03 47
26	21 08	22 02	02 43
27	21 07	21 59	01 28
28	21 06	21 56	00 S 10
29	21 D 07	21 53	01 N 08
30	21 07	21 50	02 20
31	21 ♓ 08	21 ♓ 46	03 N 22

DECLINATIONS

Date	Sun ☉	Moon ☽	Mercury ☿	Venus ♀	Mars ♂	Jupiter ♃	Saturn ♄	Uranus ♅	Neptune ♆	Pluto ♇
01	18 N 00	01 S 31	16 N 33	21 N 30	24 S 33	00 N 04	12 N 02	00 N 11	17 S 35	16 N 37
02	17 45	05 N 03	15 54	21 34	24 36	00 00	12 02	00 10	17 35	16 36
03	17 29	11 11	15 14	21 38	24 39	00 S 04	12 03	00 09	17 35	16 35
04	17 13	16 39	14 33	21 40	24 42	00 08	12 03	00 08	17 35	16 34
05	16 57	21 16	13 52	21 42	24 45	00 12	12 03	00 07	17 35	16 33
06	16 41	24 52	13 11	21 44	24 48	00 17	12 03	00 05	17 35	16 32
07	16 24	27 19	12 28	21 45	24 51	00 21	12 04	00 04	17 35	16 31
08	16 07	28 29	11 46	21 46	24 54	00 25	12 04	00 03	17 35	16 30
09	15 50	28 19	11 04	21 45	24 57	00 29	12 04	00 02	17 35	16 29
10	15 33	26 52	10 21	21 45	25 01	00 34	12 04	00 N 01	17 35	16 28
11	15 15	24 12	09 38	21 44	25 04	00 38	12 04	00 S 01	17 35	16 27
12	14 57	20 30	08 55	21 42	25 07	00 43	12 04	00 02	17 35	16 27
13	14 39	15 55	08 12	21 39	25 10	00 47	12 04	00 03	17 35	16 26
14	14 20	10 40	07 29	21 37	25 13	00 51	12 04	00 04	17 35	16 25
15	14 02	04 N 57	06 46	21 33	25 16	00 56	12 04	00 06	17 36	16 24
16	13 43	01 S 06	06 03	21 29	25 19	01 00	12 04	00 07	17 36	16 23
17	13 24	07 04	05 21	21 24	25 22	01 05	12 04	00 08	17 36	16 22
18	13 04	12 56	04 39	21 19	25 25	01 09	12 04	00 09	17 36	16 21
19	12 45	18 21	03 57	21 13	25 28	01 14	12 04	00 11	17 36	16 20
20	12 25	22 59	03 15	21 07	25 31	01 19	12 04	00 12	17 36	16 19
21	12 05	26 28	02 34	21 00	25 34	01 23	12 03	00 13	17 36	16 18
22	11 45	28 23	01 54	20 52	25 37	01 28	12 03	00 15	17 37	16 17
23	11 25	28 24	01 13	20 44	25 40	01 33	12 03	00 16	17 37	16 16
24	11 04	26 26	00 N 34	20 35	25 43	01 37	12 03	00 17	17 37	16 15
25	10 44	22 38	00 S 05	20 26	25 45	01 42	12 02	00 19	17 37	16 14
26	10 23	17 21	00 44	20 16	25 48	01 47	12 02	00 20	17 37	16 13
27	10 02	11 06	01 22	20 05	25 50	01 51	12 01	00 21	17 38	16 12
28	09 41	04 S 21	01 59	19 54	25 53	01 56	12 01	00 23	17 38	16 11
29	09 20	02 N 27	02 35	19 43	25 55	02 01	12 00	00 24	17 38	16 10
30	08 58	08 57	03 10	19 31	25 57	02 06	12 00	00 26	17 38	16 09
31	08 N 37	14 N 51	03 S 45	19 N 18	25 S 59	02 S 10	11 N 59	00 S 27	17 S 38	16 N 08

ZODIAC SIGN ENTRIES

Date	h	m	Planets
01	19	54	☽ ♈
03	05	30	♀ ♋
04	02	02	☽ ♉
06	11	49	☽ ♊
07	04	21	☿ ♍
08	23	57	☽ ♋
11	12	38	☽ ♌
14	00	32	☽ ♍
16	10	51	☽ ♎
18	18	54	☽ ♏
21	00	12	☽ ♐
23	02	49	☽ ♑
23	07	43	☉ ♍
25	03	36	☽ ♒
27	04	03	☽ ♓
27	06	50	☿ ♎
29	02	48	♀ ♌
29	05	57	☽ ♈
31	10	50	☽ ♉

LATITUDES

Date	Mercury ☿	Venus ♀	Mars ♂	Jupiter ♃	Saturn ♄	Uranus ♅	Neptune ♆	Pluto ♇
01	01 N 41	01 S 55	03 S 22	01 N 11	02 S 29	00 N 42	01 N 43	15 N 15
04	01 31	01 46	03 21	01 11	02 29	00 42	01 43	15 14
07	01 17	01 36	03 20	01 10	02 30	00 42	01 43	15 13
10	00 59	01 26	03 19	01 10	02 31	00 42	01 43	15 12
13	00 38	01 16	03 17	01 10	02 32	00 42	01 42	15 12
16	00 N 15	01 06	03 16	01 09	02 32	00 42	01 42	15 11
19	00 S 10	00 56	03 14	01 09	02 33	00 42	01 42	15 11
22	00 36	00 45	03 12	01 09	02 34	00 42	01 42	15 10
25	01 04	00 35	03 10	01 08	02 35	00 42	01 42	15 10
28	01 31	00 25	03 07	01 08	02 35	00 42	01 42	15 09
31	01 S 59	00 S 15	03 S 05	01 N 08	02 S 36	00 N 41	01 N 41	15 N 09

DATA

Julian Date	2440435
Delta T	+40 seconds
Ayanamsa	23° 25' 59"
Synetic vernal point	05° ♓ 41' 00"
True obliquity of ecliptic	23° 26' 45"

MOON'S PHASES, APSIDES AND POSITIONS ☽

Date	h	m	Phase	Longitude ° '	Eclipse Indicator
05	01	38	☾	12 ♉ 28	
13	05	16	●	20 ♌ 17	
20	20	03	☽	27 ♏ 36	
27	10	32	○	03 ♓ 58	

Day	h	m		
10	00	57	Apogee	
25	15	33	Perigee	
01	17	28	ON	
08	21	10	Max dec	28° N 35'
16	07	56	0S	
23	00	23	Max dec	28° S 38'
29	03	18	ON	

LONGITUDES

Date	Chiron ⚷ ° '	Ceres ⚳ ° '	Pallas ⚴ ° '	Juno ⚵ ° '	Vesta ⚶ ° '	Black Moon Lilith ⚸ ° '
01	06 ♈ 40	03 ♒ 34	25 ♐ 11	02 ♑ 23	13 ♋ 18	15 ♋ 44
11	06 ♈ 26	01 ♒ 29	24 ♐ 26	01 ♑ 15	17 ♋ 30	16 ♋ 52
21	06 ♈ 08	29 ♑ 43	24 ♐ 22	00 ♑ 44	21 ♋ 36	17 ♋ 59
31	05 ♈ 46	28 ♑ 26	24 ♐ 56	00 ♑ 46	25 ♋ 43	19 ♋ 06

All ephemeris data is given at 12.00 UT and the Moon's longitude is additionally given for 24.00 UT

ASPECTARIAN

h m	Aspects
01 Friday	
00 32	☽ ⊼ ☿
00 35	☉ □ ♄
08 22	☽ ☍ ♇
08 52	☽ ∠ ♄
09 13	☿ ∥ ♇
09 28	☽ ⚼ ☉
12 22	☽ ± ☿
12 54	☽ △ ♆
16 49	☽ # ♅
16 55	☽ □ ♀
17 16	☽ # ♃
17 40	☽ ∥ ♃
18 07	☽ ∥ ♅
19 54	♃ ± ♄
21 58	☽ ☍ ♅
02 Saturday	
00 29	☽ ⊥ ♄
00 32	☽ ☍ ♃
05 56	☽ △ ♂
06 33	☽ ⚼ ☿
11 05	☽ ⊻ ♄
13 38	☽ △ ☉
15 12	☽ ⚼ ♆
03 Sunday	
03 32	☉ # ♆
03 45	☽ Q ♀
07 27	☽ ± ♆
09 32	☽ ⚼ ♂
12 02	☿ ⊻ ♇
13 44	☽ ⊼ ♇
14 00	☽ △ ☿
15 35	☽ ∥ ♄
18 29	☽ ⊼ ♆
04 Monday	
00 53	☽ ± ♇
02 37	☽ ± ♂
03 28	☽ ∥ ☿
04 01	☽ ⚹ ♀
04 26	☽ ⊼ ♅
07 43	☽ ⊼ ♃
08 47	♀ □ ♅
09 52	♃ # ♅
11 37	☽ ∥ ♇
13 15	☿ ⊥ ♅
14 12	☽ ⊼ ♂
14 36	☽ ∥ ☉
15 48	☽ ± ♅
16 28	☽ # ♆
17 53	☽ ⚼ ♇
18 28	☽ ☌ ♄
19 15	☽ ± ♃
21 44	☿ □ ♆
22 31	♀ ± ♆
05 Tuesday	
01 38	☽ □ ☉
09 06	☽ ⚼ ♅
11 28	☽ ∠ ♀
12 48	☽ ⚼ ♃
14 35	☽ ∥ ♀
15 16	☿ ⊥ ♃
22 56	☽ △ ♇
06 Wednesday	
03 49	☽ ☍ ♆
04 15	♀ □ ♃
06 51	☽ ⊥ ♀
09 01	☽ □ ☿
11 25	☽ # ♂
14 35	☽ △ ♅
15 45	☽ Q ☉
18 40	☽ △ ♃
19 56	☽ ⊻ ♀
07 Thursday	
01 06	☉ ∥ ♇
02 20	☽ ☍ ♂
05 15	☽ ⊻ ♄
14 54	♆ St D
17 17	☽ ⊥ ♄
17 45	☽ ⚹ ☉
08 Friday	
00 30	☿ ⊻ ♅
02 01	☿ ∥ ♄
03 03	☽ Q ☿
10 51	☽ □ ♇
11 26	☽ ∠ ♄
15 45	☽ ⊼ ♆
09 Saturday	
02 39	☽ ∠ ☉
03 01	☽ □ ♅
03 53	☽ ± ♆
07 23	☽ ⚹ ☿
07 25	☉ ∠ ♅
07 51	☽ □ ♃
11 00	☿ ⊻ ♃
14 42	☽ ☌ ♀
16 39	☽ ⊼ ♂
17 50	☽ ⚹ ♄
22 06	☽ ⚼ ♆
22 29	☽ ⊥ ☉
23 22	☽ Q ♇
10 Sunday	
05 12	☽ ± ♂
09 44	☉ ⊥ ♇
11 42	☽ ⊻ ☉
15 37	☽ Q ♅
18 10	☽ Q ♄
18 45	☽ ∠ ☿
19 48	♀ ⊼ ♂
20 50	☽ Q ♃
23 34	♀ ⚹ ♄
23 43	☽ ⚹ ♇
11 Monday	
00 01	☽ ⚼ ♂
04 27	☽ △ ♆
05 25	☽ # ♂
07 26	♂ ⊼ ♄
15 51	☽ ⊥ ☿
15 57	☽ ⚹ ♅
21 25	☽ ⚹ ♃
12 Tuesday	
04 48	☽ ∥ ♀
05 49	☽ ⊻ ☿
06 00	☽ ∠ ♇
06 29	☽ □ ♄
07 14	☽ △ ♂
08 17	☉ ∠ ♃
09 44	☽ ⊻ ♀
10 50	☿ △ ♄
19 08	♀ ⚼ ♆
19 16	☿ □ ♂
22 09	☽ ∠ ♅
23 00	☽ ⊥ ♀
13 Wednesday	
00 03	☽ ⊥ ♇
03 41	☽ # ♆
03 54	☽ ∠ ♃
05 16	☽ ☌ ☉
09 30	☽ ∥ ♇
10 14	♀ Q ♇
12 01	☽ ⊻ ♇
16 07	☽ ⊥ ♅
16 32	☽ □ ♆
18 27	☽ ∥ ☉
18 47	☽ ∠ ♀
22 01	☽ ⊥ ♃
14 Thursday	
04 02	☽ ⊻ ♅
05 50	☽ ∥ ♄
10 02	☽ ⊻ ♃
18 03	☽ △ ♄
20 44	☽ □ ♂
15 Friday	
02 09	☽ ☌ ☿
03 19	☽ ⚹ ♀
03 30	☽ ∥ ☿
03 54	☽ Q ♆
16 08	☿ Q ♆
21 06	☽ ⊻ ☉
22 52	☽ ☌ ♇
23 12	☽ ⚼ ♄
16 Saturday	
03 07	☽ ⚹ ♆
04 01	☽ # ♃
04 53	☽ Q ♀
07 30	☽ # ♅
08 22	☽ ∥ ♅
08 46	☽ Q ♂
09 35	☽ ⊥ ☉
11 56	☽ ∥ ♃
12 45	☿ ⚹ ♀
14 28	☽ ☌ ♅
16 26	☽ ± ♄
20 48	☉ ⊻ ♇
20 52	☽ ☌ ♃
17 Sunday	
03 48	☽ ⊼ ♄
04 04	☽ ∠ ☉
05 53	☽ # ☿
07 38	☽ ∠ ♆
08 19	☽ ⚹ ♂
18 22	☽ □ ♀
19 14	☽ ⊻ ☿
18 Monday	
00 27	☽ ⊥ ♆
07 35	☽ ⊻ ♇
07 47	☽ ⊥ ☿
08 23	☽ # ♄
10 16	☽ ⚹ ☉
11 31	☽ ⊻ ♆
12 34	☽ # ☉
13 09	☽ ∠ ♂
18 37	☽ ⊥ ♇
21 29	♀ ± ♂
22 35	☽ ⊻ ♅
19 Tuesday	
02 17	☽ ∠ ☿
02 46	☽ # ♇
03 26	☉ □ ♆
04 55	☉ ⊥ ♅
05 15	☽ ⊻ ♃
06 06	☽ ⊥ ♂
08 31	☽ ∥ ♆
09 28	☽ ⊥ ♅
09 48	☽ Q ☉
10 57	☽ ∠ ♇
11 05	☽ ☍ ♄
16 11	☽ ⊥ ♃
17 15	☽ ⊻ ♂
20 Wednesday	
01 36	☽ ∠ ♅
01 55	☽ # ♀
06 02	☽ △ ♀
08 15	☽ ⚹ ☿
08 22	☽ ∠ ♃
13 36	☽ ⚹ ♇
16 45	♀ Q ♅
17 14	☽ ☌ ♆
20 03	☽ □ ☉
21 Thursday	
03 54	☽ ⚹ ♅
04 42	☽ ∥ ♂
05 44	♄ St R
07 27	☽ Q ☿
09 52	♀ Q ♄
10 25	☽ Q ♇
10 29	☽ ⚼ ♀
10 45	☽ ⚹ ♃
14 11	☉ ∥ ♄
15 33	☽ ⊼ ♄
17 44	♄ ⚼ ♇
23 07	☽ ☌ ♂
22 Friday	
00 26	☽ Q ♅
01 43	☽ ± ♄
03 08	☽ ± ♀
07 21	☽ Q ♃
13 55	☿ ⚼ ♄
14 05	☽ ⊼ ♀
14 28	☿ ☌ ♇
16 47	☽ ⚼ ♄
16 51	☽ □ ♇
17 05	☽ □ ☿
20 12	☽ ⊻ ♆
23 Saturday	
01 42	☿ # ♃
02 28	☽ △ ☉
06 07	☽ ⊥ ♆
06 32	☽ □ ♅
13 33	☽ □ ♃
17 29	☽ △ ♄
19 52	♂ Q ♅
20 52	☽ ∠ ♆
22 56	☉ ⊥ ♃
24 Sunday	
00 59	♀ ⚹ ♇
02 24	☽ ⊻ ♂
03 08	☿ ⚹ ♆
04 43	☽ ⚼ ☉
12 27	☽ ⊥ ♂
15 41	♀ Q ♃
17 28	☽ ∥ ♂
18 00	☽ △ ♇
19 25	☽ ☍ ♀
20 17	☽ ± ☉
21 11	☽ ⚹ ♆
21 39	☿ # ♅
22 54	☽ △ ☿
25 Monday	
03 28	☽ ∠ ♂
04 03	♂ ± ♄
06 38	☽ ⊼ ☉
07 25	☽ △ ♅
14 46	☽ △ ♃
16 31	☽ Q ♆
17 53	♀ △ ♆
17 59	☽ □ ♄
18 14	☽ ⚼ ♇
19 28	☉ ⊻ ♅
20 36	☿ ∥ ♅
23 03	☽ # ♀
26 Tuesday	
01 18	☽ ⚼ ☿
04 27	☽ ⚹ ♂
07 41	☽ ⚼ ♅
08 46	☽ ± ♇
10 55	☽ ∥ ♆
15 17	☽ ⚼ ♃
16 37	☽ # ♇
17 10	☽ ± ☿
18 30	☽ ⊼ ♇
21 37	☽ □ ♆
22 18	☽ ± ♅
23 03	☽ Q ♄
27 Wednesday	
00 00	☽ ⊼ ♀
00 33	☽ Q ♂
03 48	☽ ⊼ ☿
04 53	☉ ⊥ ♀
06 06	☽ ± ♃
08 07	☽ ⊼ ♅
08 37	☽ # ♄
10 32	☽ ☍ ☉
10 38	☽ ± ♀
16 03	☽ ⊼ ♃
16 06	☽ # ☉
18 39	☽ ⚹ ♄
28 Thursday	
02 45	☽ ⚼ ♀
07 11	☽ □ ♂
10 10	☿ ∥ ♃
19 36	☽ ∠ ♄
19 40	☽ ∥ ☿
20 06	☽ ☍ ♇
20 24	☽ ∥ ♃
23 16	☽ △ ♆
29 Friday	
01 55	☽ ∥ ♅
04 43	☽ # ♅
06 15	☽ △ ♀
10 17	☽ ☍ ☿
10 25	☽ ☍ ♅
10 26	☽ # ♃
10 54	☽ ⊥ ♄
12 00	☿ ☌ ♅
12 31	☽ # ☿
16 30	☽ ⊼ ☉
17 46	☿ ± ♄
19 18	☽ ☍ ♃
21 15	☽ ⊻ ♄
30 Saturday	
01 04	☽ ⚼ ♆
03 43	☽ ± ☉
12 06	☽ ∥ ☉
12 35	☽ △ ♂
16 57	☽ ± ♆
20 53	☽ ⚼ ☉
31 Sunday	
00 02	☽ ∥ ♄
00 27	☽ ⊼ ♇
03 44	☽ ⊼ ♆
10 04	♀ ⚹ ♅
11 21	☽ ± ♇
15 50	☽ ⊼ ♅
16 23	☽ □ ♀
16 45	☽ ⚼ ♂
17 44	☽ ∥ ♇
18 15	☉ ⊻ ♃
20 21	☽ ⊼ ☿
23 09	♀ ⚼ ♂

SEPTEMBER 1969

LONGITUDES

Date	Sidereal time h m s	Sun ☉ ° ' "	Moon ☽ ° ' "	Moon ☽ 24.00 ° ' "	Mercury ☿ ° '	Venus ♀ ° '	Mars ♂ ° '	Jupiter ♃ ° '	Saturn ♄ ° '	Uranus ♅ ° '	Neptune ♆ ° '	Pluto ⯓ ° '
01	10 41 54	08 ♍ 51 46	13 ♉ 35 22	19 ♉ 55 39	05 ♎ 52	04 ♌ 00	18 ♐ 39	08 ♎ 18	08 ♉ 50	02 ♎ 47	26 ♏ 06	24 ♍ 20
02	10 45 50	09 49 51	26 ♉ 11 02	02 ♊ 22 02	06 53	05 12	19 11	08 30	08 R 49	02 51	26 07	24 22
03	10 49 47	10 47 58	08 ♊ 29 10	14 ♊ 32 59	07 51	06 23	19 43	08 42	08 48	02 54	26 08	24 24
04	10 53 43	11 46 07	20 ♊ 34 04	26 ♊ 32 58	08 46	07 34	20 15	08 54	08 46	02 58	26 09	24 26
05	10 57 40	12 44 18	02 ♋ 30 15	08 ♋ 26 29	09 38	08 46	20 48	09 06	08 45	03 02	26 10	24 28
06	11 01 36	13 42 31	14 ♋ 22 11	20 ♋ 17 52	10 27	09 57	21 21	09 19	08 43	03 05	26 11	24 30
07	11 05 33	14 40 46	26 ♋ 14 00	02 ♌ 11 03	11 13	11 09	21 54	09 31	08 41	03 09	26 12	24 33
08	11 09 30	15 39 03	08 ♌ 09 24	14 ♌ 09 27	11 55	12 21	22 28	09 43	08 39	03 12	26 13	24 35
09	11 13 26	16 37 21	20 ♌ 11 31	26 ♌ 15 55	12 34	13 33	23 02	09 56	08 38	03 16	26 14	24 37
10	11 17 23	17 35 42	02 ♍ 22 53	08 ♍ 32 38	13 08	14 44	23 36	10 08	08 35	03 20	26 15	24 39
11	11 21 19	18 34 05	14 ♍ 45 22	21 ♍ 01 13	13 38	15 56	24 10	10 21	08 33	03 23	26 16	24 41
12	11 25 16	19 32 29	27 ♍ 20 17	03 ♎ 42 41	14 04	17 09	24 45	10 33	08 31	03 27	26 17	24 44
13	11 29 12	20 30 55	10 ♎ 08 27	16 ♎ 37 37	14 24	18 21	25 20	10 46	08 29	03 31	26 18	24 46
14	11 33 09	21 29 23	23 ♎ 10 14	29 ♎ 46 16	14 39	19 33	25 55	10 58	08 26	03 35	26 19	24 48
15	11 37 05	22 27 53	06 ♏ 25 44	13 ♏ 08 36	14 48	20 45	26 31	11 11	08 24	03 38	26 20	24 50
16	11 41 02	23 26 24	19 ♏ 54 49	26 ♏ 44 21	14 52	21 58	27 06	11 23	08 21	03 42	26 22	24 52
17	11 44 59	24 24 57	03 ♐ 37 06	10 ♐ 33 01	14 R 49	23 10	27 42	11 36	08 19	03 46	26 23	24 55
18	11 48 55	25 23 32	17 ♐ 31 57	24 ♐ 33 46	14 39	24 22	28 18	11 49	08 16	03 50	26 24	24 57
19	11 52 52	26 22 09	01 ♑ 38 15	08 ♑ 45 11	14 22	25 35	28 55	12 01	08 13	03 53	26 26	24 59
20	11 56 48	27 20 47	15 ♑ 54 16	23 ♑ 05 10	13 59	26 48	29 ♐ 31	12 14	08 10	03 57	26 27	25 01
21	12 00 45	28 19 26	00 ♒ 17 27	07 ♒ 30 41	13 28	28 00	00 ♑ 08	12 27	08 07	04 01	26 28	25 04
22	12 04 41	29 ♍ 18 08	14 ♒ 44 21	21 ♒ 57 52	12 50	29 ♌ 13	00 45	12 40	08 04	04 05	26 30	25 06
23	12 08 38	00 ♎ 16 51	29 ♒ 10 40	06 ♓ 22 07	12 05	00 ♍ 26	01 23	12 53	08 00	04 08	26 31	25 08
24	12 12 34	01 15 36	13 ♓ 31 06	20 ♓ 38 29	11 14	01 39	02 00	13 05	07 57	04 12	26 33	25 10
25	12 16 31	02 14 22	27 ♓ 42 13	04 ♈ 42 15	10 18	02 52	02 38	13 18	07 54	04 16	26 34	25 12
26	12 20 28	03 13 11	11 ♈ 38 07	18 ♈ 29 25	09 16	04 05	03 16	13 31	07 50	04 20	26 35	25 15
27	12 24 24	04 12 01	25 ♈ 15 52	01 ♉ 57 13	08 11	05 18	03 54	13 44	07 47	04 24	26 37	25 17
28	12 28 21	05 10 54	08 ♉ 33 24	15 ♉ 04 23	07 04	06 31	04 32	13 57	07 43	04 27	26 39	25 19
29	12 32 17	06 09 49	21 ♉ 30 16	27 ♉ 51 14	05 57	07 45	05 10	14 10	07 40	04 31	26 40	25 21
30	12 36 14	07 ♎ 08 47	04 ♊ 07 32	10 ♊ 19 32	04 ♎ 50	08 ♍ 58	05 ♑ 49	14 ♎ 23	07 ♉ 36	04 ♎ 35	26 ♏ 42	25 ♍ 23

DECLINATIONS

Date	Moon True ☊ ° '	Moon Mean ☊ ° '	Moon ☽ Latitude ° '	Sun ☉ ° '	Moon ☽ ° '	Mercury ☿ ° '	Venus ♀ ° '	Mars ♂ ° '	Jupiter ♃ ° '	Saturn ♄ ° '	Uranus ♅ ° '	Neptune ♆ ° '	Pluto ⯓ ° '
01	21 ♓ 10	21 ♓ 43	04 N 11	08 N 15	19 N 55	04 S 18	19 N 05	26 S 01	02 S 15	11 N 59	00 S 28	17 S 39	16 N 08
02	21 D 10	21 40	04 47	07 53	23 57	04 50	18 51	26 03	02 20	11 58	00 30	17 39	16 07
03	21 11	21 37	05 09	07 31	26 48	05 22	18 36	26 05	02 25	11 58	00 31	17 39	16 06
04	21 R 11	21 34	05 16	07 09	28 22	05 52	18 22	26 07	02 30	11 57	00 33	17 39	16 05
05	21 10	21 30	05 10	06 47	28 36	06 20	18 06	26 08	02 35	11 56	00 34	17 40	16 04
06	21 09	21 27	04 51	06 25	27 30	06 47	17 50	26 09	02 40	11 55	00 36	17 40	16 03
07	21 08	21 24	04 19	06 02	25 09	07 13	17 34	26 11	02 45	11 55	00 37	17 40	16 02
08	21 08	21 21	03 36	05 40	21 43	07 37	17 17	26 12	02 49	11 54	00 39	17 40	16 01
09	21 07	21 18	02 43	05 17	17 20	07 59	17 00	26 13	02 54	11 53	00 40	17 41	16 00
10	21 06	21 15	01 42	04 54	12 13	08 19	16 42	26 14	02 59	11 52	00 42	17 41	15 59
11	21 06	21 11	00 N 35	04 31	06 33	08 37	16 23	26 14	03 04	11 51	00 43	17 41	15 59
12	21 D 06	21 08	00 S 35	04 09	00 N 32	08 53	16 04	26 15	03 09	11 50	00 44	17 42	15 58
13	21 06	21 05	01 44	03 46	05 S 36	09 06	15 45	26 15	03 14	11 49	00 46	17 42	15 57
14	21 R 06	21 02	02 49	03 23	11 37	09 17	15 25	26 16	03 19	11 48	00 47	17 42	15 56
15	21 06	20 59	03 46	02 59	17 13	09 24	15 05	26 16	03 24	11 47	00 49	17 43	15 55
16	21 06	20 56	04 31	02 36	22 04	09 29	14 45	26 16	03 29	11 46	00 50	17 43	15 54
17	21 06	20 52	05 02	02 13	25 50	09 30	14 23	26 15	03 34	11 45	00 52	17 43	15 53
18	21 06	20 49	05 16	01 50	28 07	09 27	14 02	26 15	03 39	11 44	00 53	17 44	15 52
19	21 D 06	20 46	05 12	01 27	28 38	09 21	13 40	26 14	03 44	11 43	00 55	17 44	15 52
20	21 06	20 43	04 48	01 03	27 16	09 10	13 18	26 14	03 49	11 42	00 56	17 44	15 51
21	21 06	20 40	04 06	00 40	24 06	08 56	12 55	26 13	03 54	11 41	00 58	17 45	15 50
22	21 07	20 36	03 08	00 N 17	19 25	08 37	12 32	26 12	03 59	11 40	00 59	17 45	15 49
23	21 08	20 33	01 59	00 S 07	13 37	08 13	12 09	26 10	04 04	11 39	01 01	17 46	15 48
24	21 08	20 30	00 S 42	00 30	07 08	07 46	11 45	26 09	04 09	11 37	01 02	17 46	15 48
25	21 R 08	20 27	00 N 36	00 53	00 S 22	07 14	11 21	26 07	04 14	11 36	01 04	17 46	15 47
26	21 08	20 24	01 51	01 17	06 N 18	06 39	10 57	26 05	04 19	11 35	01 05	17 47	15 46
27	21 06	20 21	02 58	01 40	12 32	06 00	10 32	26 03	04 24	11 34	01 07	17 47	15 45
28	21 05	20 17	03 53	02 04	18 02	05 19	10 07	26 01	04 29	11 32	01 08	17 47	15 45
29	21 03	20 14	04 35	02 27	22 34	04 37	09 41	25 58	04 34	11 31	01 10	17 48	15 44
30	21 ♓ 00	20 ♓ 11	05 N 02	02 S 50	25 N 55	03 S 54	09 N 16	25 S 56	04 S 39	11 N 30	01 S 12	17 S 48	15 N 43

ZODIAC SIGN ENTRIES

Date	h	m	Planets
02	19	23	☽ ♊
05	06	57	☽ ♋
07	19	36	☽ ♌
10	07	20	☽ ♍
12	17	01	☽ ♎
15	00	25	☽ ♏
17	05	42	☽ ♐
19	09	14	☽ ♑
21	06	35	♂ ♑
21	11	31	☽ ♒
23	03	06	♀ ♍
23	05	07	☉ ♎
23	13	22	☽ ♓
25	15	55	☽ ♈
27	20	29	☽ ♉
30	04	05	☽ ♊

LATITUDES

Date	Mercury ☿ ° '	Venus ♀ ° '	Mars ♂ ° '	Jupiter ♃ ° '	Saturn ♄ ° '	Uranus ♅ ° '	Neptune ♆ ° '	Pluto ⯓ ° '
01	02 S 09	00 S 11	03 S 04	01 N 08	02 S 36	00 N 41	01 N 41	15 N 09
04	02 36	00 S 01	03 02	01 07	02 37	00 41	01 41	15 09
07	03 01	00 N 08	02 59	01 07	02 38	00 41	01 41	15 09
10	03 24	00 17	02 56	01 07	02 38	00 41	01 41	15 09
13	03 43	00 26	02 54	01 07	02 39	00 41	01 41	15 09
16	03 56	00 35	02 51	01 07	02 40	00 41	01 40	15 09
19	04 00	00 43	02 48	01 06	02 40	00 41	01 40	15 09
22	03 51	00 50	02 45	01 06	02 41	00 41	01 40	15 09
25	03 26	00 57	02 42	01 06	02 42	00 41	01 40	15 10
28	02 44	01 04	02 39	01 06	02 42	00 41	01 40	15 10
31	01 S 49	01 N 10	02 S 36	01 N 06	02 S 42	00 N 41	01 N 40	15 N 10

DATA

Julian Date	2440466
Delta T	+40 seconds
Ayanamsa	23° 26' 03"
Synetic vernal point	05° ♓ 40' 57"
True obliquity of ecliptic	23° 26' 45"

MOON'S PHASES, APSIDES AND POSITIONS ☽

Date	h	m	Phase	Longitude ° '	Eclipse Indicator
03	16	58	☾	11 ♊ 00	
11	19	56	●	18 ♍ 53	Annular
19	02	24	☽	25 ♐ 59	
25	20	21	○	02 ♈ 35	

Day	h	m		
06	14	59	Apogee	
22	10	53	Perigee	
05	03	58	Max dec	28° N 40'
12	14	05	0S	
19	06	41	Max dec	28° S 40'
25	13	16	0N	

LONGITUDES

Date	Chiron ⚷ ° '	Ceres ⚳ ° '	Pallas ⚴ ° '	Juno ⚵ ° '	Vesta ⚶ ° '	Black Moon Lilith ⚸ ° '
01	05 ♈ 44	28 ♑ 20	25 ♐ 01	00 ♑ 51	26 ♋ 07	19 ♋ 12
11	05 ♈ 19	27 ♑ 41	26 ♐ 10	01 ♑ 35	00 ♌ 05	20 ♋ 20
21	04 ♈ 53	27 ♑ 38	27 ♐ 48	02 ♑ 50	03 ♌ 55	21 ♋ 27
31	04 ♈ 26	28 ♑ 09	29 ♐ 49	04 ♑ 35	06 ♌ 37	22 ♋ 34

All ephemeris data is given at 12.00 UT and the Moon's longitude is additionally given for 24.00 UT

ASPECTARIAN

h m	Aspects
01 Monday	
00 42	☽ # ♆
01 57	☽ ⊼ ♃
02 26	☽ △ ☉
02 59	☽ ± ♅
03 07	☽ ☌ ♄
04 00	☽ ♇ ⯓
07 55	☽ ∥ ♀
08 30	☽ ± ☿
10 10	☽ ± ♂
11 23	☉ △ ♄
13 21	☽ ± ♃
19 58	☽ ♇ ♅
21 59	☽ ⊼ ♂
02 Tuesday	
03 01	☽ ♇ ☿
03 04	♄ ± ♅
05 38	☽ Q ♀
06 45	☽ ♇ ♃
08 29	☽ △ ⯓
11 53	☽ ☍ ♆
03 Wednesday	
01 00	☽ △ ♅
04 35	☿ Q ♂
04 44	☽ # ♂
07 25	☽ ✶ ♀
10 38	☽ △ ☿
12 26	☽ △ ♃
12 36	☽ ⊻ ♄
16 58	☽ □ ☉
22 08	♃ ⊼ ♄
04 Thursday	
00 28	☽ ⊥ ♄
11 21	☽ ☍ ♂
12 12	☿ ⊼ ♄
16 27	☽ ∠ ♀
16 51	☿ ☌ ♃
18 24	☽ ∠ ♄
19 46	☽ □ ⯓
23 12	☽ ⊼ ♆
05 Friday	
08 07	☽ Q ☉
11 19	☽ ± ♆
11 40	♀ □ ♄
12 35	☽ ⊥ ♀
13 03	☽ □ ♅
20 20	♀ ✶ ♃
06 Saturday	
00 35	☽ ✶ ♄
00 50	☉ # ☿
01 35	☽ □ ♃
02 04	☽ ⊻ ♀
02 42	♀ ∠ ⯓
03 30	☽ □ ☿
05 32	☽ ♇ ♆
08 13	☽ Q ⯓
09 07	♂ Q ♃
10 33	☽ ✶ ☉
23 49	☉ Q ♆
07 Sunday	
00 49	☽ Q ♄
01 40	☽ Q ♅
02 49	☽ ⊼ ♂
02 57	☽ # ♂
03 01	♀ # ♆
08 34	☽ ✶ ⯓
11 17	☿ ∠ ♆
11 55	☽ △ ♆
14 38	☽ Q ♃
15 28	☿ ✶ ♀
15 33	☽ ± ♂
18 25	☽ Q ☿
19 34	☽ ∠ ☉
08 Monday	
02 00	☽ ✶ ♅
10 33	☽ ♇ ♂
13 00	☽ □ ♄
14 52	☽ ∠ ⯓
15 11	☽ ✶ ♃
15 15	☽ ⊥ ☉
19 58	☽ ✶ ☿
21 19	☽ ☌ ♀
09 Tuesday	
04 17	☽ ⊻ ☉
08 10	☽ ∠ ♅
08 52	☽ ⊥ ⯓
10 16	☽ # ♆
13 49	☽ ∥ ♀
17 54	☽ △ ♂
18 34	☽ ∥ ⯓
20 47	☽ ⊻ ⯓
21 31	☽ ∠ ♃
23 57	☽ □ ♆
10 Wednesday	
02 03	☽ ⊥ ♅
03 17	☽ ∠ ☿
11 40	♂ ♇ ♄
13 32	☽ ∥ ♄
13 52	☽ ⊻ ♅
15 29	☽ ⊥ ♃
21 41	☽ ⊥ ☿
11 Thursday	
00 03	☽ △ ♄
03 20	☽ ⊻ ♃
03 50	☽ # ☿
09 46	☽ ⊻ ☿
11 03	☽ Q ♆
14 31	☽ ⊻ ♀
19 56	☽ ☌ ☉
20 44	☽ ∥ ☉
12 Friday	
01 47	☽ # ♃
03 12	☽ ⊥ ♀
04 47	☽ ♇ ♄
06 52	☽ □ ♂
07 02	☽ ☌ ⯓
10 00	☽ ✶ ♆
10 55	♂ □ ⯓
11 11	☽ # ♅
17 00	☽ ∥ ♅
20 56	♀ ∥ ⯓
21 44	☽ ± ♄
22 00	☽ ∠ ♀
23 34	☽ ☌ ♅
13 Saturday	
02 36	☽ ∥ ♃
05 12	☽ # ☉
08 55	☽ ⊼ ♄
13 10	☽ ☌ ♃
14 10	☽ ∠ ♆
15 35	♀ ∠ ♅
18 12	☽ Q ♂
20 04	☽ ☌ ☿
20 38	♀ ⊥ ⯓
14 Sunday	
02 17	☽ ∥ ☿
04 42	☽ ✶ ♀
06 47	☽ ⊥ ♆
08 41	☽ ⊻ ☉
12 47	☽ # ♄
14 53	☉ # ♃
14 59	☽ ⊻ ⯓
17 15	☽ ✶ ♂
17 45	☽ ⊻ ♆
20 29	☽ ⊥ ☉
15 Monday	
01 54	☽ ⊥ ⯓
03 07	☽ # ♀
04 44	☽ Q ♀
04 53	♂ ⊻ ♆
06 13	☽ # ⯓
06 58	☽ ⊻ ♅
14 00	☽ ∠ ☉
14 16	☽ ∥ ♆
15 31	☽ ☍ ♄
17 47	☽ ⊥ ♅
18 07	☽ ∠ ⯓
20 38	☽ ⊻ ♃
21 31	☽ ∠ ♂
16 Tuesday	
01 38	☿ Q ♂
03 02	☽ ⊻ ☿
07 28	☽ ⊥ ♃
08 33	♃ ∠ ♆
09 51	☽ ∠ ♅
09 58	☉ ♇ ♄
12 41	☿ St R
13 40	☽ ⊥ ☿
14 12	☽ ⊥ ♂
15 57	☽ □ ♀
18 41	☽ ✶ ☉
20 45	☽ ✶ ⯓
23 21	☽ ☌ ♆
23 34	☽ ∠ ♃
17 Wednesday	
01 13	☽ ⊻ ♂
05 25	☽ ∠ ☿
12 15	☽ ✶ ♅
15 29	☽ ∥ ♂
17 13	☽ Q ☉
17 44	☽ Q ⯓
20 06	☽ ⊼ ♄
18 Thursday	
00 40	☉ ☌ ⯓
02 01	☽ ✶ ♃
06 24	☽ ± ♄
07 07	☽ ✶ ☿
09 04	☽ Q ♅
21 45	☽ ♇ ♄
22 53	☽ Q ♃
23 46	♀ ⊻ ⯓
19 Friday	
00 41	☽ □ ⯓
00 46	☽ △ ♀
02 24	☽ □ ☉
03 09	☽ ⊻ ♆
03 16	☽ Q ☿
07 11	☽ ☌ ♂
13 20	☽ ⊥ ♆
13 26	☉ ✶ ♆
15 49	☽ □ ♅
23 03	☽ △ ♄
20 Saturday	
04 28	☽ ♇ ♀
04 31	☽ ∠ ♆
05 01	♀ □ ♆
05 46	☽ □ ♃
08 52	☽ □ ☿
18 39	☉ # ♅
20 56	☽ ± ♀
21 24	☽ ∥ ♂
22 37	♀ ∠ ♃
21 Sunday	
03 16	☽ △ ⯓
05 38	☽ ✶ ♆
07 51	☽ ⊼ ♀
08 29	☽ △ ☉
11 44	☽ ⊻ ♂
12 10	♀ ⊥ ♅
18 05	☿ ∠ ♀
18 13	☽ △ ♅
22 09	☽ ⊥ ♂
22 Monday	
00 57	☽ □ ♄
01 37	☽ Q ♆
04 16	☽ ♇ ⯓
08 30	☽ △ ♃
08 59	☽ △ ☿
11 13	☽ ♇ ☉
13 46	☽ ∠ ♂
16 28	☿ ☌ ♃
19 14	☽ ♇ ♅
19 15	☽ ± ⯓
19 17	☽ ∥ ♆
20 26	☉ ⊻ ♀
23 Tuesday	
03 15	☽ ± ☉
03 20	☽ # ⯓
05 15	☽ ⊼ ⯓
06 45	☽ Q ♄
07 34	☽ □ ♆
08 42	☽ ♇ ☿
09 48	☽ ♇ ♃
10 16	☽ ± ♅
13 58	☽ ⊼ ☉
14 17	☽ ☍ ♀
15 50	☽ ✶ ♂
17 58	☽ # ♀
19 31	☽ # ♄
20 19	☽ ⊼ ♅
22 54	☽ ± ☿
24 Wednesday	
01 03	☽ ± ♃
02 41	☽ ✶ ♄
03 56	☿ ∠ ♆
08 23	☽ ⊼ ☿
09 32	☽ ∥ ☿
11 15	☽ ⊼ ♃
12 50	☽ Q ♂
20 02	♀ ∥ ♄
22 27	☽ ∥ ♃
25 Thursday	
02 21	♀ △ ♂
03 51	☽ ∠ ♄
04 03	☉ ± ♄
07 44	☽ ☍ ⯓
09 30	☽ ∥ ♅
09 33	♂ ⊥ ♆
10 03	☽ △ ♆
10 13	☽ ∥ ☉
16 42	☽ # ☉
17 05	☽ # ♅
19 09	☽ ⊥ ♄
20 21	☽ ☍ ☉
20 50	☽ □ ♂
21 41	☽ ⊼ ♀
23 18	☽ ☍ ♅
23 33	☉ ∥ ♅
26 Friday	
03 23	☿ ⊥ ♀
04 39	☽ # ♃
05 26	☽ ⊻ ♄
08 11	☽ ☍ ☿
09 02	☽ ± ♀
11 55	☽ ♇ ♆
13 09	☽ # ☿
14 43	☉ □ ♂
15 20	☽ ☍ ♃
17 08	♀ ⊻ ♅
27 Saturday	
02 20	☽ ♇ ♀
03 44	☽ ± ♆
04 32	☽ ∥ ♀
08 06	☽ ∥ ♄
12 02	☽ ⊼ ⯓
14 25	☽ ⊼ ♆
17 02	☉ ☌ ♅
21 18	☿ ⊼ ♄
22 49	☽ ± ⯓
28 Sunday	
01 34	☽ ∥ ⯓
04 17	☽ △ ♂
04 30	☽ ⊼ ♅
05 21	☽ ⊼ ☉
07 54	☽ △ ♀
08 56	♂ □ ♅
09 30	☽ ⊼ ☿
10 29	☽ ☌ ♄
10 51	☽ # ♆
15 14	☽ ♇ ⯓
15 30	☽ ± ♅
17 12	☽ ± ☉
17 36	☿ ⊻ ♀
19 38	☽ ± ☿
22 05	☽ ⊼ ♃
29 Monday	
08 16	☽ ♇ ♅
09 22	☽ ♇ ♂
09 27	☽ ± ♃
09 29	☉ ☌ ☿
10 27	♀ △ ♄
11 02	☽ ♇ ☿
11 18	☽ ♇ ☉
13 14	☿ ∥ ♃
19 17	☽ △ ⯓
21 46	☽ ☍ ♆
22 04	♀ ⊥ ♃
22 35	☿ □ ♂
30 Tuesday	
02 45	☽ ♇ ♃
03 17	☽ ± ♂
12 04	☽ # ♂
12 53	☽ △ ♅
13 16	☽ △ ☿
15 26	☽ ⊼ ♂
17 22	☿ ☌ ♅
18 20	☽ △ ☉
18 40	☽ ⊻ ♄
22 19	☉ ⊼ ♄
22 23	☽ □ ♀

OCTOBER 1969

LONGITUDES

Date	Sidereal time h m s	Sun ☉ ° ′ ″	Moon ☽ ° ′ ″	Moon ☽ 24.00 ° ′ ″	Mercury ☿ ° ′	Venus ♀ ° ′	Mars ♂ ° ′	Jupiter ♃ ° ′	Saturn ♄ ° ′	Uranus ♅ ° ′	Neptune ♆ ° ′	Pluto ♇ ° ′
01	12 40 10	08 ♎ 07 46	16 ♊ 27 37	22 ♊ 32 17	03 ♎ 47	10 ♍ 11	06 ♑ 27	14 ♎ 36	07 ♉ 32	04 ♎ 39	26 ♏ 43	25 ♍ 26
02	12 44 07	09 06 48	28 ♊ 34 00	04 ♋ 33 21	02 R 48	11 25	07 06	14 49	07 R 28	04 42	26 45	25 28
03	12 48 03	10 05 52	10 ♋ 30 54	16 ♋ 27 14	01 56	12 38	07 45	15 02	07 24	04 46	26 47	25 30
04	12 52 00	11 04 59	22 ♋ 22 58	28 ♋ 18 42	01 12	13 52	08 25	15 15	07 20	04 50	26 48	25 32
05	12 55 57	12 04 07	04 ♌ 15 02	10 ♌ 12 32	00 36	15 05	09 04	15 28	07 16	04 54	26 50	25 34
06	12 59 53	13 03 18	16 ♌ 11 48	22 ♌ 13 19	00 ♎ 11	16 19	09 44	15 41	07 12	04 58	26 52	25 36
07	13 03 50	14 02 32	28 ♌ 17 37	04 ♍ 25 07	29 ♍ 56	17 33	10 23	15 54	07 08	05 01	26 54	25 39
08	13 07 46	15 01 47	10 ♍ 36 13	16 ♍ 51 15	29 D 51	18 47	11 03	16 07	07 03	05 05	26 55	25 41
09	13 11 43	16 01 05	23 ♍ 10 29	29 ♍ 34 07	29 ♍ 57	20 00	11 43	16 20	06 59	05 09	26 57	25 43
10	13 15 39	17 00 24	06 ♎ 02 14	12 ♎ 34 53	00 ♎ 14	21 14	12 24	16 33	06 55	05 13	26 59	25 45
11	13 19 36	17 59 46	19 ♎ 11 59	25 ♎ 53 25	00 41	22 28	13 04	16 46	06 50	05 16	27 01	25 47
12	13 23 32	18 59 10	02 ♏ 38 59	09 ♏ 28 21	01 18	23 42	13 44	16 59	06 46	05 20	27 03	25 49
13	13 27 29	19 58 36	16 ♏ 21 13	23 ♏ 17 11	02 03	24 56	14 25	17 12	06 41	05 24	27 04	25 51
14	13 31 26	20 58 04	00 ♐ 15 48	07 ♐ 16 40	02 56	26 10	15 06	17 25	06 37	05 27	27 06	25 53
15	13 35 22	21 57 34	14 ♐ 19 19	21 ♐ 23 19	03 57	27 25	15 47	17 38	06 32	05 31	27 08	25 55
16	13 39 19	22 57 05	28 ♐ 28 16	05 ♑ 33 47	05 04	28 39	16 28	17 51	06 28	05 35	27 10	25 57
17	13 43 15	23 56 39	12 ♑ 39 32	19 ♑ 45 14	06 17	29 ♍ 53	17 09	18 04	06 23	05 38	27 12	25 59
18	13 47 12	24 56 14	26 ♑ 50 36	03 ♒ 55 24	07 35	01 ♎ 07	17 50	18 17	06 18	05 42	27 14	26 01
19	13 51 08	25 55 50	10 ♒ 59 26	18 ♒ 02 31	08 57	02 21	18 32	18 30	06 14	05 46	27 16	26 03
20	13 55 05	26 55 29	25 ♒ 04 28	02 ♓ 05 06	10 22	03 36	19 13	18 43	06 09	05 49	27 18	26 05
21	13 59 01	27 55 09	09 ♓ 04 12	16 ♓ 01 34	11 51	04 50	19 55	18 56	06 04	05 53	27 20	26 07
22	14 02 58	28 54 50	22 ♓ 57 00	29 ♓ 50 13	13 23	06 05	20 36	19 09	05 59	05 57	27 22	26 09
23	14 06 55	29 ♎ 54 34	06 ♈ 40 58	13 ♈ 28 59	14 56	07 19	21 18	19 21	05 55	06 00	27 24	26 11
24	14 10 51	00 ♏ 54 19	20 ♈ 13 59	26 ♈ 55 43	16 31	08 34	22 00	19 34	05 50	06 04	27 26	26 13
25	14 14 48	01 54 06	03 ♉ 33 56	10 ♉ 08 25	18 07	09 48	22 42	19 47	05 45	06 07	27 28	26 15
26	14 18 44	02 53 56	16 ♉ 39 02	23 ♉ 05 39	19 45	11 03	23 24	20 00	05 40	06 11	27 30	26 17
27	14 22 41	03 53 47	29 ♉ 28 13	05 ♊ 46 46	21 23	12 17	24 07	20 13	05 35	06 14	27 32	26 19
28	14 26 37	04 53 41	12 ♊ 01 24	18 ♊ 12 17	23 02	13 32	24 49	20 26	05 31	06 18	27 35	26 21
29	14 30 34	05 53 36	24 ♊ 19 40	00 ♋ 23 52	24 41	14 47	25 31	20 39	05 26	06 21	27 37	26 23
30	14 34 30	06 53 34	06 ♋ 25 16	12 ♋ 24 19	26 21	16 01	26 14	20 51	05 21	06 24	27 39	26 24
31	14 38 27	07 ♏ 53 34	18 ♋ 21 31	24 ♋ 17 24	28 ♎ 00	17 ♎ 16	26 ♑ 56	21 ♎ 04	05 ♉ 16	06 ♎ 28	27 ♏ 41	26 ♍ 26

Date	Moon True ☊ ° ′	Moon Mean ☊ ° ′	Moon ☽ Latitude ° ′
01	20 ♓ 59	20 ♓ 08	05 N 14
02	20 R 57	20 05	05 13
03	20 D 57	20 01	04 57
04	20 57	19 58	04 29
05	20 59	19 55	03 50
06	21 00	19 52	03 00
07	21 02	19 49	02 02
08	21 03	19 46	00 N 57
09	21 R 03	19 42	00 S 12
10	21 03	19 39	01 21
11	21 00	19 36	02 28
12	20 57	19 33	03 28
13	20 53	19 30	04 17
14	20 48	19 27	04 52
15	20 44	19 23	05 10
16	20 40	19 20	05 09
17	20 38	19 17	04 49
18	20 D 38	19 14	04 12
19	20 39	19 11	03 19
20	20 40	19 07	02 15
21	20 41	19 04	01 S 03
22	20 R 42	19 01	00 N 12
23	20 41	18 58	01 26
24	20 38	18 55	02 33
25	20 32	18 52	03 31
26	20 26	18 48	04 17
27	20 18	18 45	04 48
28	20 10	18 42	05 05
29	20 02	18 39	05 07
30	19 56	18 36	04 55
31	19 ♓ 52	18 ♓ 33	04 N 31

DECLINATIONS

Date	Sun ☉ ° ′	Moon ☽ ° ′	Mercury ☿ ° ′	Venus ♀ ° ′	Mars ♂ ° ′	Jupiter ♃ ° ′	Saturn ♄ ° ′	Uranus ♅ ° ′	Neptune ♆ ° ′	Pluto ♇ ° ′
01	03 S 14	27 N 58	03 S 10	08 N 50	25 S 53	04 S 44	11 N 28	01 S 13	17 S 49	15 N 42
02	03 37	28 39	02 29	08 24	25 50	04 49	11 27	01 15	17 49	15 42
03	04 00	27 58	01 49	07 57	25 46	04 54	11 26	01 16	17 50	15 41
04	04 23	26 01	01 13	07 30	25 43	04 59	11 24	01 18	17 50	15 40
05	04 46	22 55	00 40	07 03	25 39	05 05	11 23	01 19	17 50	15 39
06	05 09	18 51	00 S 12	06 36	25 35	05 10	11 21	01 20	17 51	15 39
07	05 32	13 59	00 N 10	06 09	25 31	05 15	11 20	01 22	17 51	15 38
08	05 55	08 29	00 28	05 41	25 27	05 20	11 18	01 23	17 52	15 37
09	06 18	02 N 32	00 40	05 14	25 22	05 25	11 17	01 25	17 52	15 37
10	06 41	03 S 39	00 46	04 46	25 17	05 30	11 15	01 26	17 53	15 36
11	07 04	09 48	00 48	04 18	25 12	05 35	11 14	01 28	17 53	15 35
12	07 26	15 39	00 44	03 49	25 07	05 40	11 12	01 29	17 54	15 35
13	07 49	20 50	00 35	03 21	25 01	05 44	11 11	01 31	17 54	15 34
14	08 11	24 58	00 21	02 52	24 56	05 49	11 09	01 32	17 54	15 34
15	08 33	27 39	00 N 04	02 24	24 50	05 54	11 08	01 34	17 55	15 33
16	08 56	28 35	00 S 18	01 55	24 44	05 59	11 06	01 35	17 55	15 33
17	09 18	27 39	00 43	01 26	24 37	06 04	11 05	01 37	17 56	15 32
18	09 39	24 55	01 11	00 57	24 31	06 09	11 03	01 38	17 56	15 31
19	10 01	20 40	01 41	00 N 28	24 24	06 14	11 02	01 39	17 57	15 31
20	10 23	15 17	02 14	00 S 01	24 17	06 19	11 00	01 41	17 57	15 30
21	10 44	09 09	02 48	00 30	24 10	06 24	10 58	01 42	17 58	15 30
22	11 05	02 S 37	03 24	00 59	24 02	06 29	10 57	01 44	17 58	15 29
23	11 27	03 N 58	04 02	01 28	23 55	06 34	10 55	01 45	17 59	15 29
24	11 48	10 16	04 40	01 57	23 47	06 39	10 54	01 47	17 59	15 28
25	12 08	16 01	05 20	02 26	23 39	06 43	10 52	01 48	18 00	15 28
26	12 29	20 55	06 00	02 55	23 30	06 48	10 50	01 49	18 00	15 27
27	12 49	24 44	06 40	03 24	23 22	06 53	10 49	01 51	18 01	15 27
28	13 09	27 16	07 21	03 53	23 13	06 58	10 47	01 52	18 01	15 27
29	13 29	28 26	08 01	04 22	23 04	07 03	10 46	01 53	18 02	15 26
30	13 49	28 12	08 42	04 51	22 55	07 07	10 44	01 55	18 02	15 26
31	14 S 09	26 N 40	09 S 23	05 S 20	22 S 46	07 S 12	10 N 43	01 S 56	18 S 03	15 N 25

ZODIAC SIGN ENTRIES

Date	h m	Planets
02	14 52	☽ ♋
05	03 25	☽ ♌
07	02 57	☿ ♍
07	15 21	☽ ♍
09	16 56	☿ ♎
10	00 48	☽ ♎
12	07 19	☽ ♏
14	11 33	☽ ♐
16	14 35	☽ ♑
17	14 17	♀ ♎
18	17 21	☽ ♒
20	20 26	☽ ♓
23	00 17	☽ ♈
23	14 11	☉ ♏
25	05 32	☽ ♉
27	13 00	☽ ♊
29	23 13	☽ ♋

LATITUDES

Date	Mercury ☿ ° ′	Venus ♀ ° ′	Mars ♂ ° ′	Jupiter ♃ ° ′	Saturn ♄ ° ′	Uranus ♅ ° ′	Neptune ♆ ° ′	Pluto ♇ ° ′
01	01 S 49	01 N 10	02 S 36	01 N 06	02 S 42	00 N 41	01 N 40	15 N 10
04	00 S 48	01 15	02 32	01 06	02 42	00 41	01 40	15 11
07	00 N 09	01 20	02 29	01 06	02 43	00 41	01 39	15 11
10	00 57	01 24	02 26	01 06	02 43	00 41	01 39	15 12
13	01 31	01 27	02 22	01 06	02 43	00 41	01 39	15 13
16	01 52	01 30	02 19	01 06	02 43	00 41	01 39	15 13
19	02 02	01 32	02 15	01 06	02 44	00 41	01 39	15 14
22	02 02	01 34	02 12	01 06	02 44	00 41	01 39	15 15
25	01 56	01 34	02 08	01 06	02 44	00 41	01 39	15 16
28	01 44	01 35	02 05	01 06	02 44	00 41	01 39	15 17
31	01 N 29	01 N 34	02 S 01	01 N 06	02 S 44	00 N 41	01 N 39	15 N 18

DATA

Julian Date	2440496
Delta T	+40 seconds
Ayanamsa	23° 26′ 07″
Synetic vernal point	05° ♓ 40′ 53″
True obliquity of ecliptic	23° 26′ 45″

MOON'S PHASES, APSIDES AND POSITIONS ☽

Date	h m	Phase	Longitude ° ′	Eclipse Indicator
03	11 05	☾	10 ♋ 04	
11	09 39	●	17 ♎ 54	
18	08 32	☽	24 ♑ 48	
25	08 44	○	01 ♉ 46	

Day	h m		
04	09 16	Apogee	
18	04 20	Perigee	
02	11 51	Max dec	28° N 39′
09	21 54	0S	
16	11 58	Max dec	28° S 35′
22	21 29	0N	
29	19 58	Max dec	28° N 31′

LONGITUDES

Date	Chiron ⚷ ° ′	Ceres ⚳ ° ′	Pallas ⚴ ° ′	Juno ⚵ ° ′	Vesta ⚶ ° ′	Black Moon Lilith ⚸ ° ′
01	04 ♈ 26	28 ♑ 09	29 ♐ 49	04 ♑ 35	07 ♌ 37	22 ♋ 34
11	03 ♈ 59	29 ♑ 13	02 ♑ 09	06 ♑ 44	11 ♌ 09	23 ♋ 41
21	03 ♈ 34	00 ♒ 46	04 ♑ 47	09 ♑ 16	14 ♌ 27	24 ♋ 48
31	03 ♈ 11	02 ♒ 44	07 ♑ 38	12 ♑ 08	17 ♌ 29	25 ♋ 55

ASPECTARIAN

h m	Aspects
01 Wednesday	
06 17	☽ ⊥ ♄
08 16	☽ △ ♃
10 53	☉ ∥ ☿
23 56	☽ ∠ ♄
02 Thursday	
05 48	☽ □ ♇
08 22	☽ ⊼ ♆
13 53	☽ Q ♀
19 53	☽ □ ☿
20 24	☽ ± ♆
03 Friday	
00 07	♂ △ ♄
00 22	☽ □ ♅
05 46	☽ ✱ ♄
06 07	☽ ☍ ♂
11 05	☽ □ ☉
14 33	☽ Ꝑ ♆
16 46	☽ ✱ ♀
18 03	☽ Q ♇
21 17	☽ □ ♃
04 Saturday	
05 52	☽ Q ♄
05 54	☽ Q ☿
06 34	☿ ± ♄
08 53	☿ ∥ ♅
12 55	☽ Q ♅
14 50	☽ # ♂
18 24	☽ ✱ ♇
20 58	☽ △ ♆
05 Sunday	
02 38	☽ ∠ ♀
02 47	☽ Q ☉
04 57	☽ ✱ ☿
06 07	☉ ∠ ♆
06 54	♀ Q ♆
10 23	☽ Q ♃
13 19	☽ ✱ ♅
18 03	☽ □ ♄
20 49	♀ ⊻ ♃
22 16	☽ ⊼ ♂
22 52	☽ ⊥ ♀
06 Monday	
00 46	☽ ∠ ♇
05 09	☽ ✱ ☉
10 01	☽ ∠ ☿
10 57	☽ ✱ ♃
11 00	☽ ± ♂
12 12	☉ ∥ ♃
12 16	☽ ⊻ ♀
17 15	☽ # ♆
18 49	☽ ⊥ ♇
19 33	☽ ∠ ♅
07 Tuesday	
03 31	☽ ⊥ ☿
04 12	☽ ∥ ♇
05 57	☽ Ꝑ ♂
06 46	☽ ⊻ ♇
09 14	☽ □ ♆
13 27	☽ ⊥ ♅
13 36	☽ ∠ ☉
15 10	☽ ⊻ ☿
17 12	☽ ∠ ♃
23 53	☽ ∥ ♄
08 Wednesday	
01 15	☽ ⊻ ♅
05 10	☽ △ ♄
05 22	☉ # ♀
08 42	☽ ⊥ ☉
09 53	☿ St D
11 02	☽ ⊥ ♃
12 55	☽ △ ♂
20 19	☽ Q ♆
21 14	☽ ⊻ ☉
21 51	☽ # ☉
22 46	☽ ⊻ ♃
09 Thursday	
00 23	☽ ∥ ♀
00 43	☽ # ♃
04 00	♀ # ♃
05 21	☽ ☌ ♀
09 46	☽ Ꝑ ♄
16 22	☽ # ♅
16 48	☽ ☌ ♇
19 07	☽ ✱ ♆
19 09	☽ ∥ ☿
20 37	♂ ∠ ♆
21 38	☉ ☌ ♃
10 Friday	
00 45	☽ # ☿
00 58	☽ ☌ ☿
02 34	☽ ± ♄
03 27	☽ ∥ ♅
10 28	☽ ☌ ♅
13 36	☽ ⊼ ♄
16 01	☽ # ♀
19 15	☽ ∥ ♃
22 56	☽ ∠ ♆
11 Saturday	
00 17	☽ □ ♂
00 24	♀ Ꝑ ♄
00 32	☽ ∥ ☉
07 31	☽ ☌ ♃
09 39	☽ ☌ ☉
15 16	☽ ⊥ ♆
17 40	☽ # ♄
17 58	☿ ± ♄
18 29	☽ ⊻ ♀
23 51	☽ ⊻ ♇
12 Sunday	
02 02	☽ ⊻ ♆
06 16	☽ ⊥ ♀
09 28	☽ ⊻ ☿
10 18	☽ Q ♂
10 32	☽ ⊥ ♇
11 41	☽ # ♇
16 45	☽ ⊻ ♅
19 13	☽ ☍ ♄
20 37	☽ ⊥ ☿
21 54	☽ ∥ ♆
23 42	☽ ∠ ♀
13 Monday	
02 24	☽ ∠ ♇
03 19	☽ ⊥ ♅
08 27	☽ ✱ ♂
13 17	☽ ∠ ☿
13 29	☽ ⊻ ♃
18 46	☽ ⊻ ☉
19 02	☽ ∠ ♅
14 Tuesday	
00 02	☽ ⊥ ♃
04 17	☽ ✱ ♀
04 28	☽ ✱ ♇
05 54	☽ ⊥ ☉
06 19	♀ ☌ ♇
06 34	☽ ☌ ♆
11 42	☽ ∠ ♂
11 43	☽ ∥ ♂
15 27	☉ ⊥ ♆
15 44	☽ ∠ ♃
16 55	☽ ✱ ☿
20 56	☽ ✱ ♅
22 30	☽ ∠ ☉
22 49	☽ ⊼ ♄
15 Wednesday	
01 04	☽ Q ♇
02 50	☽ Q ♀
03 52	☽ ⊥ ♂
06 36	♀ ✱ ♆
08 59	☽ ± ♄
14 36	☽ ⊻ ♂
14 59	☽ Q ☿
17 27	☽ Q ♅
17 42	☽ ✱ ♃
16 Thursday	
00 11	☽ Ꝑ ♄
01 57	☽ ✱ ☉
07 44	☽ □ ♇
09 48	☽ ⊻ ♆
12 19	☽ □ ♀
14 22	☽ Q ♃
19 58	☽ ⊥ ♆
22 58	☿ ☌ ♅
23 47	☽ Q ☉
17 Friday	
00 05	☽ □ ♅
00 11	☽ □ ☿
01 27	☽ △ ♄
03 46	♀ # ♅
11 14	☽ ∠ ♆
13 54	☿ ⊼ ♄
19 59	☽ ☌ ♂
21 10	♀ ± ♄
21 17	☽ □ ♃
18 Saturday	
06 31	☿ # ♀
08 32	☽ □ ☉
10 37	☽ △ ♇
12 40	☽ ✱ ♆
14 43	☽ ∥ ♂
19 56	☽ △ ♀
19 Sunday	
03 05	☽ △ ♅
03 57	☽ □ ♄
08 08	☽ △ ☿
09 04	☽ Q ♆
10 28	♂ □ ♃
10 50	☿ ∥ ♅
12 07	☽ Ꝑ ♇
15 10	☉ ⊻ ♇
23 53	☽ Ꝑ ♀
20 Monday	
00 39	☽ ∥ ♆
00 58	☽ △ ♃
01 29	☽ ⊻ ♂
03 28	☽ ± ♇
04 43	☽ Ꝑ ♅
10 26	☽ Q ♄
11 05	☽ # ♇
12 15	☽ ⊥ ♂
12 34	☽ Ꝑ ☿
13 44	☽ ⊼ ♇
15 24	☽ △ ☉
15 49	☽ □ ♆
16 44	☽ ± ♀
20 09	☽ ± ♅
21 25	☉ ⊻ ♆
21 Tuesday	
03 01	☽ Ꝑ ♃
04 01	☽ ⊼ ♀
04 29	☽ ∠ ♂
05 02	☽ # ♄
05 48	☽ ± ☿
06 17	☽ ∥ ☉
06 30	☽ ⊼ ♅
06 52	☽ ✱ ♄
17 23	☽ ⊼ ☿
18 45	☽ ± ♃
19 09	☽ Ꝑ ☉
19 50	☿ ∠ ♆
22 03	☽ ∥ ♃
22 Wednesday	
02 51	☉ # ♄
05 17	☽ ⊼ ♃
07 43	☽ ✱ ♂
08 37	☽ ∠ ♄
09 15	♀ ☌ ♅
09 22	☽ ∥ ☿
10 27	♀ ⊼ ♄
11 56	☽ ± ☉
15 12	☽ ∥ ♅
17 31	☽ ∥ ♀
17 35	☽ ☍ ♇
19 43	☽ △ ♆
20 35	♄ ⊼ ♅
23 12	☽ ⊼ ☉
23 Thursday	
00 12	☽ ⊥ ♄
02 05	☽ # ♀
03 50	☽ # ♅
05 45	☽ Q ♂
10 39	☽ ⊻ ♄
10 48	☽ ☍ ♅
12 16	☽ # ☿
13 14	☽ ☍ ♀
21 49	☽ # ♃
22 07	☽ Ꝑ ♆
24 Friday	
02 54	♀ ∥ ♅
04 30	☽ ☍ ☿
10 48	☽ ☍ ♃
14 09	☽ ± ♆
14 27	☽ ∥ ♄
15 20	☽ □ ♂
18 27	☽ # ☉
22 45	☽ ⊼ ♇
25 Saturday	
00 57	☽ ⊼ ♆
08 44	☽ ☍ ☉
09 34	☽ ∥ ♇
09 37	☽ ± ♇
15 57	☽ ☌ ♄
16 40	☽ ⊼ ♅
20 42	☉ ⊥ ♇
21 10	☽ # ♆
26 Sunday	
00 34	☽ ⊼ ♀
02 04	☽ Ꝑ ♇
03 42	☽ ± ♅
12 49	☽ ± ♀
16 17	☿ ☌ ♃
18 20	☽ ⊼ ♃
18 35	☽ ⊼ ☿
20 27	☽ Ꝑ ♅
27 Monday	
01 19	☽ △ ♂
02 53	☽ # ♂
05 45	☽ ± ♃
06 02	☽ △ ♇
07 26	☽ Ꝑ ♀
07 29	☽ ± ☿
08 21	☽ ☍ ♆
14 18	☿ ⊥ ♆
17 00	♀ ∠ ♆
20 42	☿ ∥ ♃
21 08	☽ ⊼ ☉
23 07	☽ Ꝑ ♃
23 34	☽ ⊻ ♄
28 Tuesday	
00 56	☽ △ ♅
03 09	☽ Ꝑ ☿
07 29	☽ Ꝑ ♂
09 38	☽ ± ☉
11 01	☽ ⊥ ♄
15 15	☽ △ ♀
29 Wednesday	
01 42	☉ ☍ ♄
02 00	☽ ± ♂
04 24	☽ ∠ ♄
04 38	☽ △ ♃
04 39	☽ Ꝑ ☉
12 49	☽ △ ☿
14 30	☽ ⊼ ♂
16 03	☽ □ ♇
18 30	☽ ⊼ ♆
23 37	☉ ⊻ ♅
30 Thursday	
06 27	☽ ± ♆
09 09	☿ □ ♂
09 52	☽ ✱ ♄
11 58	☽ □ ♅
12 54	☿ ⊻ ♇
13 02	☽ △ ☉
18 09	♂ △ ♇
31 Friday	
00 31	☽ Ꝑ ♆
04 04	☽ Q ♇
07 16	☿ ⊻ ♆
09 33	☽ □ ♀
09 49	☽ Q ♄
17 35	☽ □ ♃

All ephemeris data is given at 12.00 UT and the Moon's longitude is additionally given for 24.00 UT

NOVEMBER 1969

Date	Sidereal time h m s	Sun ☉ ° ' "	Moon ☽ ° ' "	Moon ☽ 24.00 ° ' "	Mercury ☿ ° '	Venus ♀ ° '	Mars ♂ ° '	Jupiter ♃ ° '	Saturn ♄ ° '	Uranus ♅ ° '	Neptune ♆ ° '	Pluto ♇ ° '
01	14 42 24	08 ♏ 53 36	00 ♌ 12 34	06 ♌ 07 38	29 ♎ 40	18 ♎ 31	27 ♑ 39	21 ♎ 17	05 ♉ 11	06 ♎ 31	27 ♏ 43	26 ♍ 28
02	14 46 20	09 53 39	12 ♌ 03 15	18 ♌ 00 03	01 ♏ 19	19 46	28 22	21 30	05 R 07	06 34	27 45	26 30
03	14 50 17	10 53 46	23 ♌ 58 44	29 ♌ 59 56	02 59	21 01	29 05	21 42	05 02	06 38	27 47	26 31
04	14 54 13	11 53 54	06 ♍ 04 17	12 ♍ 12 25	04 38	22 16	29 ♑ 48	21 55	04 57	06 41	27 50	26 33
05	14 58 10	12 54 04	18 ♍ 24 53	24 ♍ 42 11	06 17	23 30	00 ♒ 31	22 07	04 52	06 44	27 52	26 35
06	15 02 06	13 54 16	01 ♎ 04 47	07 ♎ 33 01	07 55	24 45	01 14	22 20	04 47	06 48	27 54	26 36
07	15 06 03	14 54 30	14 ♎ 07 07	20 ♎ 47 11	09 33	26 00	01 57	22 33	04 43	06 51	27 56	26 38
08	15 09 59	15 54 46	27 ♎ 33 14	04 ♏ 25 03	11 11	27 15	02 40	22 45	04 38	06 54	27 58	26 40
09	15 13 56	16 55 04	11 ♏ 22 22	18 ♏ 24 41	12 49	28 30	03 24	22 58	04 33	06 57	28 01	26 41
10	15 17 53	17 55 24	25 ♏ 31 26	02 ♐ 41 53	14 26	29 ♎ 45	04 07	23 10	04 29	07 00	28 03	26 43
11	15 21 49	18 55 46	09 ♐ 55 14	17 ♐ 10 39	16 04	01 ♏ 00	04 51	23 22	04 24	07 03	28 05	26 44
12	15 25 46	19 56 09	24 ♐ 27 13	01 ♑ 44 07	17 40	02 16	05 34	23 35	04 19	07 06	28 07	26 46
13	15 29 42	20 56 34	09 ♑ 00 32	16 ♑ 15 44	19 17	03 31	06 18	23 47	04 15	07 09	28 10	26 47
14	15 33 39	21 57 00	23 ♑ 29 07	00 ♒ 40 10	20 53	04 46	07 01	23 59	04 10	07 12	28 12	26 49
15	15 37 35	22 57 27	07 ♒ 48 31	14 ♒ 53 53	22 29	06 01	07 45	24 12	04 06	07 15	28 14	26 50
16	15 41 32	23 57 56	21 ♒ 56 08	28 ♒ 55 10	24 04	07 16	08 29	24 24	04 02	07 18	28 16	26 52
17	15 45 28	24 58 25	05 ♓ 51 00	12 ♓ 43 41	25 39	08 31	09 13	24 36	03 57	07 21	28 19	26 53
18	15 49 25	25 58 56	19 ♓ 33 18	26 ♓ 19 58	27 14	09 47	09 57	24 48	03 53	07 24	28 21	26 54
19	15 53 22	26 59 29	03 ♈ 03 46	09 ♈ 44 47	28 ♏ 49	11 02	10 41	25 00	03 49	07 27	28 23	26 56
20	15 57 18	28 00 02	16 ♈ 23 05	22 ♈ 58 44	00 ♐ 24	12 17	11 25	25 12	03 45	07 29	28 25	26 57
21	16 01 15	29 ♏ 00 37	29 ♈ 31 41	06 ♉ 01 58	01 58	13 32	12 09	25 24	03 40	07 32	28 28	26 58
22	16 05 11	00 ♐ 01 13	12 ♉ 29 31	18 ♉ 54 16	03 32	14 47	12 53	25 35	03 36	07 35	28 30	27 00
23	16 09 08	01 01 51	25 ♉ 16 11	01 ♊ 35 11	05 06	16 03	13 37	25 47	03 32	07 37	28 32	27 01
24	16 13 04	02 02 29	07 ♊ 51 15	14 ♊ 04 21	06 40	17 18	14 21	25 59	03 28	07 40	28 34	27 02
25	16 17 01	03 03 10	20 ♊ 14 32	26 ♊ 21 51	08 14	18 33	15 05	26 11	03 24	07 43	28 37	27 03
26	16 20 57	04 03 52	02 ♋ 26 26	08 ♋ 28 28	09 47	19 49	15 49	26 22	03 21	07 45	28 39	27 04
27	16 24 54	05 04 35	14 ♋ 28 12	20 ♋ 25 57	11 20	21 04	16 34	26 34	03 17	07 48	28 41	27 05
28	16 28 51	06 05 19	26 ♋ 22 04	02 ♌ 17 00	12 54	22 19	17 18	26 45	03 13	07 50	28 43	27 06
29	16 32 47	07 06 06	08 ♌ 11 13	14 ♌ 05 17	14 27	23 35	18 02	26 57	03 10	07 52	28 46	27 07
30	16 36 44	08 ♐ 06 53	19 ♌ 59 46	25 ♌ 55 18	16 ♐ 00	24 ♏ 50	18 ♒ 47	27 ♎ 08	03 ♉ 06	07 ♎ 55	28 ♏ 48	27 ♍ 08

Date	Moon True ☊ ° '	Moon Mean ☊ ° '	Moon ☽ Latitude ° '
01	19 ♓ 49	18 ♓ 29	03 N 55
02	19 D 49	18 26	03 09
03	19 50	18 23	02 15
04	19 51	18 20	01 14
05	19 52	18 17	00 N 08
06	19 R 51	18 13	01 S 00
07	19 48	18 10	02 06
08	19 43	18 07	03 08
09	19 35	18 04	03 59
10	19 26	18 01	04 38
11	19 16	17 58	05 00
12	19 07	17 54	05 02
13	18 59	17 51	04 46
14	18 53	17 48	04 11
15	18 50	17 45	03 20
16	18 49	17 42	02 18
17	18 D 49	17 39	01 S 09
18	18 R 50	17 35	00 N 04
19	18 49	17 32	01 15
20	18 46	17 29	02 21
21	18 40	17 26	03 18
22	18 31	17 23	04 04
23	18 19	17 19	04 37
24	18 06	17 16	04 56
25	17 53	17 13	05 00
26	17 40	17 10	04 51
27	17 28	17 07	04 28
28	17 19	17 04	03 54
29	17 13	17 00	03 10
30	17 ♓ 10	16 ♓ 57	02 N 18

DECLINATIONS

Date	Sun ☉ ° '	Moon ☽ ° '	Mercury ☿ ° '	Venus ♀ ° '	Mars ♂ ° '	Jupiter ♃ ° '	Saturn ♄ ° '	Uranus ♅ ° '	Neptune ♆ ° '	Pluto ♇ ° '
01	14 S 28	23 N 56	10 S 03	05 S 49	22 S 36	07 S 17	10 N 41	01 S 57	18 S 03	15 N 25
02	14 47	20 12	10 43	06 17	22 26	07 21	10 40	01 59	18 04	15 25
03	15 06	15 39	11 23	06 46	22 16	07 26	10 38	02 00	18 04	15 24
04	15 25	10 26	12 03	07 14	22 06	07 31	10 36	02 01	18 05	15 24
05	15 43	04 N 42	12 41	07 43	21 55	07 35	10 35	02 02	18 05	15 24
06	16 01	01 S 21	13 20	08 11	21 45	07 40	10 33	02 04	18 06	15 23
07	16 19	07 31	13 57	08 39	21 34	07 45	10 32	02 05	18 06	15 23
08	16 36	13 31	14 34	09 06	21 23	07 49	10 30	02 06	18 07	15 23
09	16 54	19 02	15 10	09 34	21 12	07 54	10 29	02 07	18 07	15 23
10	17 11	23 39	15 46	10 01	21 00	07 58	10 28	02 09	18 08	15 22
11	17 27	26 53	16 21	10 28	20 48	08 03	10 26	02 10	18 08	15 22
12	17 44	28 22	16 55	10 55	20 37	08 07	10 25	02 11	18 09	15 22
13	18 00	27 53	17 28	11 22	20 25	08 12	10 23	02 12	18 09	15 22
14	18 16	25 31	18 00	11 48	20 12	08 16	10 22	02 13	18 10	15 22
15	18 31	21 33	18 31	12 14	20 00	08 21	10 21	02 14	18 10	15 21
16	18 46	16 23	19 02	12 40	19 47	08 25	10 19	02 16	18 11	15 21
17	19 01	10 26	19 31	13 06	19 35	08 29	10 18	02 17	18 11	15 21
18	19 15	04 S 05	20 00	13 31	19 22	08 34	10 17	02 18	18 12	15 21
19	19 29	02 N 22	20 28	13 56	19 08	08 38	10 15	02 19	18 12	15 21
20	19 43	08 37	20 54	14 20	18 55	08 42	10 14	02 20	18 13	15 21
21	19 57	14 24	21 20	14 44	18 42	08 47	10 13	02 21	18 13	15 21
22	20 10	19 28	21 44	15 08	18 28	08 51	10 12	02 22	18 14	15 21
23	20 22	23 34	22 08	15 31	18 14	08 55	10 10	02 23	18 14	15 21
24	20 35	26 29	22 30	15 54	18 00	08 59	10 09	02 24	18 15	15 21
25	20 46	28 05	22 51	16 17	17 46	09 03	10 08	02 25	18 15	15 21
26	20 58	28 16	23 11	16 39	17 32	09 07	10 07	02 26	18 16	15 21
27	21 09	27 06	23 30	17 01	17 17	09 12	10 06	02 27	18 16	15 21
28	21 20	24 43	23 48	17 22	17 03	09 16	10 05	02 28	18 17	15 21
29	21 30	21 17	24 04	17 43	16 48	09 20	10 04	02 29	18 17	15 21
30	21 S 40	17 N 00	24 S 19	18 S 03	16 S 33	09 S 24	10 N 03	02 S 30	18 S 18	15 N 21

ZODIAC SIGN ENTRIES

Date	h	m	Planets
01	11	35	☽ ♌
01	16	53	☿ ♏
04	00	00	☽ ♍
04	18	51	♂ ♒
06	09	59	☽ ♎
08	16	18	☽ ♏
10	16	40	♀ ♏
10	19	30	☽ ♐
12	21	08	☽ ♑
14	22	53	☽ ♒
17	01	52	☽ ♓
19	06	32	☽ ♈
20	06	00	☿ ♐
21	12	52	☽ ♉
22	11	31	☉ ♐
23	20	59	☽ ♊
26	07	10	☽ ♋
28	19	22	☽ ♌

LATITUDES

Date	Mercury ☿ ° '	Venus ♀ ° '	Mars ♂ ° '	Jupiter ♃ ° '	Saturn ♄ ° '	Uranus ♅ ° '	Neptune ♆ ° '	Pluto ♇ ° '
01	01 N 23	01 N 34	02 S 00	01 N 06	02 S 44	00 N 41	01 N 39	15 N 18
04	01 05	01 32	01 57	01 06	02 43	00 41	01 39	15 20
07	00 46	01 30	01 53	01 06	02 43	00 42	01 39	15 21
10	00 26	01 28	01 49	01 07	02 43	00 42	01 39	15 22
13	00 N 05	01 25	01 46	01 07	02 43	00 42	01 39	15 23
16	00 S 15	01 21	01 42	01 07	02 42	00 42	01 39	15 25
19	00 34	01 17	01 38	01 07	02 42	00 42	01 39	15 26
22	00 53	01 12	01 35	01 08	02 41	00 42	01 39	15 27
25	01 11	01 07	01 31	01 08	02 41	00 42	01 39	15 29
28	01 27	01 01	01 28	01 08	02 40	00 42	01 39	15 30
31	01 S 42	00 N 55	01 S 24	01 N 08	02 S 39	00 N 42	01 N 39	15 N 32

DATA

Julian Date	2440527
Delta T	+40 seconds
Ayanamsa	23° 26' 11"
Synetic vernal point	05° ♓ 40' 49"
True obliquity of ecliptic	23° 26' 45"

MOON'S PHASES, APSIDES AND POSITIONS ☽

Date	h	m	Phase	Longitude ° '	Eclipse Indicator
02	07	14	☾	09 ♌ 42	
09	22	11	●	17 ♏ 21	
16	15	45	☽	24 ♒ 07	
23	23	54	○	01 ♊ 32	

Day	h	m		
01	05	29	Apogee	
13	01	27	Perigee	
29	01	17	Apogee	
06	06	44	0S	
12	18	13	Max dec	28° S 26'
19	03	10	0N	
26	03	17	Max dec	28° N 22'

LONGITUDES

Date	Chiron ⚷ ° '	Ceres ⚳ ° '	Pallas ⚴ ° '	Juno ⚵ ° '	Vesta ⚶ ° '	Black Moon Lilith ⚸ ° '
01	03 ♈ 09	02 ♒ 57	07 ♑ 56	12 ♑ 26	17 ♌ 47	26 ♋ 01
11	02 ♈ 51	05 ♒ 19	10 ♑ 59	15 ♑ 36	20 ♌ 27	27 ♋ 08
21	02 ♈ 36	08 ♒ 00	14 ♑ 11	19 ♑ 00	22 ♌ 44	28 ♋ 15
31	02 ♈ 27	10 ♒ 56	17 ♑ 30	22 ♑ 37	24 ♌ 32	29 ♋ 22

All ephemeris data is given at 12.00 UT and the Moon's longitude is additionally given for 24.00 UT

ASPECTARIAN

h m	Aspects
01 Saturday	
00 25	☽ Q ♅
04 23	☽ ✶ ♇
06 29	☽ ☍ ♂
06 56	☽ △ ♆
10 43	☽ □ ☿
14 20	♂ ✶ ♆
21 45	☽ # ♂
22 02	☽ □ ♄
02 Sunday	
00 51	☽ ✶ ♅
02 18	☽ Q ♀
06 43	☽ Q ♃
07 14	☽ □ ☉
09 45	☿ # ♄
10 52	☽ ∠ ♇
23 46	☽ # ♆
03 Monday	
05 01	☽ Q ☿
05 03	☽ ⊥ ♇
05 19	☿ ⊥ ♇
05 21	☽ ✶ ♀
07 16	☽ ∠ ♅
07 22	☽ ✶ ♃
13 11	☽ ∥ ♇
14 30	☽ # ☉
17 06	☽ ⊻ ♇
19 38	☽ □ ♆
22 42	☽ Q ☉
22 49	☽ ⊼ ♂
23 57	♃ ⊥ ♆
04 Tuesday	
01 19	☽ ⊥ ♅
03 27	☉ ∠ ♇
03 27	♀ ⊥ ♆
04 03	♀ ☌ ♃
05 37	☽ # ☿
08 43	☽ ✶ ☿
09 48	☽ △ ♄
11 12	☽ ∥ ♄
11 22	☉ # ♇
11 25	☽ ± ♂
13 13	☽ ⊻ ♅
13 41	☽ ∠ ♃
14 36	☽ ∠ ♀
16 28	☿ ☍ ♄
05 Wednesday	
00 16	☽ # ♃
00 24	☽ ✶ ☉
00 33	☽ # ♀
04 42	♀ ∥ ♃
06 04	☽ ꝗ ♂
07 04	☽ Q ♆
07 31	☽ ⊥ ♃
07 54	☉ ⊥ ♅
10 04	☽ ⊥ ♀
14 46	☽ ꝗ ♄
18 18	☽ ∠ ☿
19 00	☿ ⊻ ♅
19 13	☽ ⊻ ♃
22 38	☽ # ♅
22 48	☽ ⊻ ♀
06 Thursday	
03 35	☽ ☌ ♇
06 01	☽ ✶ ♆
07 35	☽ ∠ ☉
07 44	☽ ± ♄
12 18	☽ △ ♂
13 48	☽ ⊥ ☿
14 48	☽ ∥ ♅
18 52	☽ ⊼ ♄
22 39	☽ ☌ ♅
07 Friday	
01 42	☽ ⊥ ☉
02 30	☽ ⊻ ☿
09 51	☽ ∠ ♆
12 56	☽ ∥ ♃
13 33	☽ ⊻ ☉
16 49	☽ ∥ ♀
23 52	☽ # ♄
08 Saturday	
00 20	♀ ⊻ ♇
02 05	☽ ⊥ ♆
03 22	☽ ☌ ♃
10 25	☽ ⊻ ♇
11 25	☽ ☌ ♀
12 44	☽ ⊻ ♆
16 52	☽ ∥ ☿
19 02	☿ ∠ ♇
19 46	☽ # ♇
20 58	☽ ⊥ ♇
21 28	☽ □ ♂
09 Sunday	
00 18	☽ ☍ ♄
01 46	☽ ∥ ☉
02 12	♀ ⊻ ♆
04 22	☽ ⊻ ♅
07 46	☽ ∥ ♆
12 32	☽ ∠ ♇
14 01	☿ ⊥ ♅
14 43	☽ ⊥ ♅
14 48	☽ ☌ ☿
20 09	☿ # ♇
22 06	☽ ∥ ♂
22 11	☽ ☌ ☉
10 Monday	
05 58	☽ Q ♂
06 03	☽ ∠ ♅
07 59	☽ ⊻ ♃
14 00	☽ ✶ ♇
16 15	☽ ☌ ♆
18 12	☽ ⊥ ♃
19 46	☽ ⊻ ♀
22 47	♂ □ ♄
11 Tuesday	
02 53	☽ ⊼ ♄
03 07	☽ ✶ ♂
06 43	☽ ⊥ ♀
07 14	☽ ✶ ♅
09 24	☽ ∠ ♃
10 02	☽ Q ♇
10 07	♀ # ♄
12 47	☽ ± ♄
23 01	☽ ∠ ♀
23 25	☽ ⊻ ☿
12 Wednesday	
03 09	☽ Q ♅
03 35	☽ ꝗ ♄
04 00	☽ ⊻ ☉
05 15	☽ ∠ ♂
10 32	☽ ✶ ♃
10 33	☽ ⊥ ☿
14 37	☽ ⊥ ☉
15 49	☽ □ ♇
18 04	☽ ⊻ ♆
20 52	☽ ⊥ ♂
21 52	♀ ⊥ ♇
13 Thursday	
02 04	☽ ✶ ♀
03 13	☽ ∠ ☿
03 59	☽ ⊥ ♆
04 11	☽ △ ♄
06 34	☽ ∠ ☉
06 36	☽ Q ♃
07 17	☽ ⊻ ♂
08 56	☽ □ ♅
18 53	☽ ∠ ♆
23 46	☽ Q ♀
14 Friday	
01 20	♀ ☍ ♄
02 47	☉ ∥ ♆
07 07	☽ ✶ ☿
09 15	☽ ✶ ☉
12 51	☽ □ ♃
17 34	☽ △ ♇
18 23	☉ ∠ ♅
18 25	♂ △ ♅
19 27	☿ ∥ ♆
19 53	☽ ✶ ♆
15 Saturday	
05 41	☽ Q ☿
05 47	☽ □ ♄
06 50	☽ Q ☉
08 33	☿ ∠ ♅
08 41	☽ □ ♀
11 04	☽ △ ♅
11 21	☉ ∥ ☿
11 54	☽ ☌ ♂
16 06	☽ Q ♆
18 49	☽ ꝗ ♇
20 01	☽ ∥ ♂
16 Sunday	
01 16	☽ ∥ ☿
01 55	☽ ∥ ☉
04 07	☽ ∥ ♆
07 47	☉ ☌ ☿
10 10	☽ ± ♇
12 09	☽ Q ♄
12 38	☽ ꝗ ♅
12 39	♀ ⊻ ♅
15 45	☽ □ ☉
16 07	☽ □ ☿
16 17	☽ △ ♃
16 18	☽ # ♇
17 39	☿ ⊻ ♃
20 28	☽ ⊼ ♇
22 55	☽ □ ♆
17 Monday	
00 46	☉ ⊻ ♃
02 13	☽ ∥ ♀
04 10	☽ ± ♅
08 44	☽ ✶ ♄
12 31	☽ # ♄
13 47	☿ ∥ ♂
14 37	☽ ⊼ ♅
17 07	☽ △ ♀
18 11	☽ ⊻ ♂
18 37	☽ ꝗ ♃
19 21	☽ ∥ ♃
18 Tuesday	
05 17	☽ ⊥ ♂
06 54	☿ ✶ ♇
10 39	☽ ± ♃
10 49	☽ ∠ ♄
17 28	☉ ∥ ♂
18 36	☽ ∥ ♅
19 46	♀ □ ♂
21 25	☽ ⊼ ♃
22 05	☽ ∠ ♂
22 11	☽ ꝗ ♀
19 Wednesday	
00 18	☽ △ ☉
01 03	☽ ☍ ♇
01 56	♂ Q ♆
02 41	☽ ⊥ ♄
03 25	☽ △ ☿
03 38	☽ △ ♆
05 15	☿ ☌ ♆
07 59	☿ Q ♂
10 29	☉ ✶ ♇
11 49	☽ # ♅
13 20	☽ ⊻ ♄
15 53	☽ ± ♀
19 53	☽ ☍ ♅
20 Thursday	
02 28	☽ ✶ ♂
03 48	☽ ⊼ ♀
05 22	☽ ꝗ ☉
05 32	♀ ∠ ♇
06 37	☽ ꝗ ♆
09 58	☽ ꝗ ☿
12 24	☽ # ♃
18 29	☽ ∥ ♄
22 25	☉ ☌ ♆
23 01	☽ ± ♆
23 04	☽ ± ☉
21 Friday	
01 33	☽ Q ♂
02 01	☿ ⊥ ♃
04 18	☽ ☍ ♃
04 34	☽ ± ☿
06 14	♂ ꝗ ♇
07 18	☽ ⊼ ♇
10 02	☽ ⊼ ♆
10 58	☽ ⊼ ☉
11 58	♀ ⊥ ♅
13 38	☽ # ♀
16 14	☽ ∥ ♇
17 06	☽ ⊼ ☿
18 21	☽ ± ♇
19 36	☽ ☌ ♄
22 Saturday	
02 50	☽ ⊼ ♅
05 46	☽ # ♆
07 11	☽ # ♂
11 04	☽ ꝗ ♇
12 46	☽ □ ♂
13 02	☿ ⊼ ♄
14 02	☽ ± ♅
15 53	☽ # ☉
16 45	☽ ☍ ♀
23 Sunday	
01 03	♀ # ♇
01 52	☽ # ☿
06 59	☽ ꝗ ♅
11 50	♂ ∥ ♆
13 00	☽ ⊼ ♃
15 18	☽ △ ♇
18 13	☽ ☍ ♆
23 54	☽ ☍ ☉
24 Monday	
00 35	☽ ± ♃
03 39	☽ ⊻ ♄
09 23	☽ ☍ ☿
10 17	☉ ⊥ ♃
11 38	☽ △ ♅
15 06	☽ ⊥ ♄
18 07	☽ ꝗ ♃
25 Tuesday	
01 20	☽ △ ♂
03 50	☿ ✶ ♅
08 20	☽ ⊼ ♀
08 26	☽ ∠ ♄
14 56	☉ Q ♂
19 56	☉ ⊼ ♄
21 25	☽ ± ♀
23 49	☽ △ ♃
26 Wednesday	
00 53	☿ Q ♇
01 23	☽ □ ♇
04 29	☽ ⊼ ♆
05 30	☿ ± ♄
08 35	☽ ꝗ ♂
13 47	☽ ✶ ♄
15 31	☽ ⊼ ☉
16 24	☽ ± ♆
17 15	☽ ꝗ ♀
22 36	☽ □ ♅
27 Thursday	
03 39	☽ ± ♂
04 34	☽ ± ☉
04 48	☽ ⊼ ☿
10 25	☽ ꝗ ♆
13 15	☽ Q ♇
13 37	☽ Q ♄
15 54	☿ ∠ ♃
16 29	☽ ⊼ ♂
18 38	☽ ± ☿
22 57	♀ ∥ ♂
28 Friday	
00 20	☽ ꝗ ☉
02 51	☽ △ ♀
10 55	☽ Q ♅
12 48	☽ □ ♃
13 30	☽ ✶ ♇
15 34	☽ ꝗ ☿
16 47	☽ △ ♆
18 36	☽ # ☿
22 06	♀ ∠ ♅
29 Saturday	
01 50	☽ □ ♄
09 35	☽ △ ☉
10 44	☽ # ☉
11 22	☽ ✶ ♅
20 01	☽ ∠ ♇
23 39	♄ ± ♇
30 Sunday	
01 58	☽ Q ♃
02 40	☽ △ ☿
05 13	☽ # ♆
06 53	☽ # ♀
07 00	☉ ✶ ♅
09 22	☽ ☍ ♂
13 15	♃ ⊻ ♇
14 20	☽ ⊥ ♇
14 29	☽ # ♂
17 56	☽ ∠ ♅
20 22	☽ ∥ ♇
22 58	☽ □ ♀

DECEMBER 1969

LONGITUDES

Date	Sidereal time h m s	Sun ☉ ° ' "	Moon ☽ ° ' "	Moon ☽ 24.00 ° ' "	Mercury ☿ ° '	Venus ♀ ° '	Mars ♂ ° '	Jupiter ♃ ° '	Saturn ♄ ° '	Uranus ♅ ° '	Neptune ♆ ° '	Pluto ♇ ° '
01	16 40 40	09 ♐ 07 42	01 ♍ 52 33	07 ♍ 52 11	17 ♐ 33	26 ♏ 05	19 ♒ 31	27 ♎ 19	03 ♉ 03	07 ♎ 57	28 ♏ 50	27 ♍ 09
02	16 44 37	10 08 32	13 ♍ 54 55	20 ♍ 01 27	19 06	27 21	20 16	27 30	02 R 59	07 59	28 52	27 10
03	16 48 33	11 09 24	26 ♍ 12 27	02 ♎ 28 35	20 39	28 36	21 00	27 42	02 56	08 01	28 55	27 11
04	16 52 30	12 10 17	08 ♎ 50 27	15 ♎ 18 36	22 12	29 ♏ 52	21 44	27 53	02 53	08 04	28 57	27 12
05	16 56 26	13 11 11	21 ♎ 53 26	28 ♎ 35 17	23 45	01 ♐ 07	22 29	28 03	02 50	08 06	28 59	27 13
06	17 00 23	14 12 07	05 ♏ 24 19	12 ♏ 20 31	25 17	02 22	23 14	28 14	02 47	08 08	29 01	27 14
07	17 04 20	15 13 04	19 ♏ 23 41	26 ♏ 33 24	26 50	03 38	23 58	28 25	02 44	08 10	29 03	27 15
08	17 08 16	16 14 02	03 ♐ 49 03	11 ♐ 09 50	28 23	04 53	24 43	28 36	02 41	08 12	29 06	27 15
09	17 12 13	17 15 01	18 ♐ 34 45	26 ♐ 02 41	29 ♐ 55	06 09	25 27	28 47	02 38	08 14	29 08	27 16
10	17 16 09	18 16 02	03 ♑ 32 26	11 ♑ 02 45	01 ♑ 27	07 24	26 12	28 57	02 36	08 16	29 10	27 17
11	17 20 06	19 17 02	18 ♑ 32 25	26 ♑ 00 19	02 59	08 40	26 57	29 08	02 33	08 17	29 12	27 18
12	17 24 02	20 18 04	03 ♒ 25 27	10 ♒ 46 59	04 30	09 55	27 41	29 18	02 31	08 19	29 14	27 18
13	17 27 59	21 19 06	18 ♒ 04 13	25 ♒ 16 43	06 02	11 11	28 26	29 28	02 28	08 21	29 16	27 19
14	17 31 55	22 20 09	02 ♓ 24 09	09 ♓ 26 24	07 33	12 26	29 11	29 38	02 26	08 23	29 18	27 19
15	17 35 52	23 21 11	16 ♓ 23 26	23 ♓ 15 23	09 03	13 42	29 ♒ 56	29 48	02 24	08 24	29 21	27 20
16	17 39 49	24 22 15	00 ♈ 02 27	06 ♈ 44 52	10 33	14 57	00 ♓ 40	29 ♎ 58	02 22	08 26	29 23	27 20
17	17 43 45	25 23 18	13 ♈ 22 57	19 ♈ 57 00	12 01	16 13	01 25	00 ♏ 08	02 20	08 27	29 25	27 21
18	17 47 42	26 24 22	26 ♈ 27 18	02 ♉ 54 11	13 29	17 28	02 10	00 18	02 18	08 29	29 27	27 21
19	17 51 38	27 25 26	09 ♉ 17 53	15 ♉ 38 39	14 56	18 44	02 55	00 28	02 16	08 30	29 29	27 22
20	17 55 35	28 26 31	21 ♉ 56 41	28 ♉ 12 09	16 22	19 59	03 39	00 37	02 15	08 32	29 31	27 22
21	17 59 31	29 ♐ 27 36	04 ♊ 25 11	10 ♊ 35 55	17 46	21 15	04 24	00 47	02 13	08 33	29 33	27 22
22	18 03 28	00 ♑ 28 41	16 ♊ 44 26	22 ♊ 50 50	19 08	22 30	05 09	00 56	02 12	08 34	29 35	27 23
23	18 07 24	01 29 47	28 ♊ 55 11	04 ♋ 57 35	20 28	23 46	05 54	01 05	02 10	08 35	29 37	27 23
24	18 11 21	02 30 54	10 ♋ 58 09	16 ♋ 57 01	21 45	25 01	06 38	01 15	02 09	08 36	29 39	27 23
25	18 15 18	03 32 00	22 ♋ 54 22	28 ♋ 50 23	22 59	26 16	07 23	01 24	02 08	08 38	29 41	27 23
26	18 19 14	04 33 07	04 ♌ 45 20	10 ♌ 39 33	24 09	27 32	08 08	01 33	02 07	08 39	29 43	27 23
27	18 23 11	05 34 15	16 ♌ 33 21	22 ♌ 27 09	25 15	28 ♐ 47	08 53	01 41	02 06	08 40	29 45	27 23
28	18 27 07	06 35 22	28 ♌ 21 25	04 ♍ 16 40	26 17	00 ♑ 03	09 38	01 50	02 06	08 40	29 47	27 24
29	18 31 04	07 36 30	10 ♍ 13 26	16 ♍ 12 19	27 12	01 18	10 22	01 59	02 05	08 41	29 49	27 24
30	18 35 00	08 37 39	22 ♍ 13 57	28 ♍ 19 00	28 02	02 34	11 07	02 07	02 04	08 42	29 50	27 R 24
31	18 38 57	09 ♑ 38 48	04 ♎ 28 06	10 ♎ 41 56	28 ♑ 43	03 ♑ 49	11 ♓ 52	02 ♏ 15	02 ♉ 04	08 ♎ 43	29 ♏ 52	27 ♍ 24

Date	Moon True ☊ ° '	Moon Mean ☊ ° '	Moon ☽ Latitude ° '
01	17 ♓ 09	16 ♓ 54	01 N 20
02	17 D 09	16 51	00 N 17
03	17 R 08	16 48	00 S 48
04	17 07	16 45	01 52
05	17 03	16 41	02 53
06	16 57	16 38	03 45
07	16 48	16 35	04 27
08	16 37	16 32	04 53
09	16 24	16 29	05 00
10	16 13	16 25	04 47
11	16 02	16 22	04 14
12	15 55	16 19	03 24
13	15 50	16 16	02 21
14	15 48	16 13	01 S 10
15	15 D 48	16 10	00 N 03
16	15 R 48	16 06	01 15
17	15 47	16 03	02 20
18	15 43	16 00	03 17
19	15 37	15 57	04 03
20	15 28	15 54	04 36
21	15 17	15 50	04 55
22	15 03	15 47	05 00
23	14 49	15 44	04 52
24	14 36	15 41	04 30
25	14 24	15 38	03 56
26	14 14	15 35	03 13
27	14 08	15 31	02 21
28	14 04	15 28	01 23
29	14 03	15 25	00 N 20
30	14 D 03	15 22	00 S 44
31	14 ♓ 03	15 ♓ 19	01 S 47

DECLINATIONS

Date	Sun ☉ ° '	Moon ☽ ° '	Mercury ☿ ° '	Venus ♀ ° '	Mars ♂ ° '	Jupiter ♃ ° '	Saturn ♄ ° '	Uranus ♅ ° '	Neptune ♆ ° '	Pluto ♇ ° '
01	21 S 49	12 N 03	24 S 33	18 S 23	16 S 18	09 S 27	10 N 02	02 S 31	18 S 18	15 N 21
02	21 59	06 36	24 46	18 42	16 03	09 31	10 01	02 31	18 19	15 21
03	22 07	00 N 47	24 57	19 01	15 47	09 35	10 00	02 32	18 19	15 21
04	22 15	05 S 13	25 07	19 20	15 32	09 39	09 59	02 33	18 20	15 21
05	22 23	11 12	25 16	19 37	15 16	09 43	09 58	02 34	18 20	15 21
06	22 31	16 52	25 22	19 55	15 00	09 47	09 58	02 35	18 20	15 22
07	22 38	21 51	25 29	20 11	14 45	09 50	09 57	02 35	18 21	15 22
08	22 44	25 42	25 34	20 27	14 29	09 54	09 56	02 36	18 21	15 22
09	22 50	27 56	25 37	20 43	14 12	09 58	09 55	02 37	18 22	15 22
10	22 56	28 10	25 38	20 58	13 56	10 01	09 55	02 38	18 22	15 22
11	23 01	26 21	25 39	21 12	13 40	10 05	09 54	02 38	18 23	15 23
12	23 05	22 42	25 37	21 26	13 23	10 08	09 53	02 39	18 23	15 23
13	23 10	17 39	25 34	21 39	13 07	10 12	09 53	02 40	18 24	15 23
14	23 13	11 43	25 30	21 51	12 50	10 15	09 52	02 40	18 24	15 23
15	23 17	05 S 19	25 24	22 03	12 33	10 19	09 52	02 41	18 24	15 24
16	23 20	01 N 09	25 17	22 14	12 16	10 22	09 51	02 41	18 25	15 24
17	23 22	07 26	25 08	22 25	11 59	10 26	09 51	02 42	18 25	15 24
18	23 24	13 16	24 58	22 34	11 42	10 29	09 51	02 43	18 26	15 25
19	23 25	18 26	24 47	22 44	11 25	10 32	09 50	02 43	18 26	15 25
20	23 26	22 42	24 34	22 52	11 08	10 35	09 50	02 44	18 27	15 25
21	23 27	25 52	24 19	23 00	10 50	10 38	09 50	02 44	18 27	15 26
22	23 27	27 46	24 03	23 07	10 33	10 42	09 50	02 45	18 27	15 26
23	23 26	28 18	23 46	23 13	10 15	10 45	09 49	02 45	18 28	15 27
24	23 25	27 28	23 28	23 19	09 58	10 48	09 49	02 45	18 28	15 27
25	23 24	25 23	23 09	23 24	09 40	10 51	09 49	02 46	18 29	15 27
26	23 22	22 12	22 49	23 28	09 22	10 54	09 49	02 46	18 29	15 28
27	23 20	18 07	22 28	23 31	09 04	10 56	09 49	02 47	18 29	15 28
28	23 17	13 20	22 07	23 34	08 46	10 59	09 49	02 47	18 30	15 29
29	23 14	08 03	21 46	23 36	08 28	11 02	09 49	02 47	18 30	15 29
30	23 10	02 N 25	21 24	23 37	08 10	11 05	09 49	02 47	18 30	15 30
31	23 S 06	03 S 25	21 S 02	23 S 38	07 S 52	11 S 08	09 N 49	02 S 48	18 S 31	15 N 30

ZODIAC SIGN ENTRIES

Date	h	m	Planets	
01	08	14	☽	♍
03	19	17	☽	♎
04	14	41	♀	♐
06	02	30	☽	♏
08	05	43	☽	♐
09	13	21	☿	♑
10	06	20	☽	♑
12	06	27	☽	♒
14	07	56	☽	♓
15	14	22	♂	♓
16	11	56	☽	♈
16	15	56	♃	♏
18	18	35	☽	♉
21	03	28	☽	♊
22	00	44	☉	♑
23	14	08	☽	♋
26	02	21	☽	♌
28	11	04	♀	♑
28	15	20	☽	♍
31	03	18	☽	♎

LATITUDES

Date	Mercury ☿ ° '	Venus ♀ ° '	Mars ♂ ° '	Jupiter ♃ ° '	Saturn ♄ ° '	Uranus ♅ ° '	Neptune ♆ ° '	Pluto ♇ ° '
01	01 S 42	00 N 55	01 S 24	01 N 08	02 S 39	00 N 42	01 N 39	15 N 32
04	01 55	00 49	01 20	01 09	02 39	00 42	01 39	15 34
07	02 05	00 43	01 17	01 09	02 38	00 42	01 39	15 35
10	02 12	00 36	01 13	01 09	02 37	00 42	01 39	15 37
13	02 16	00 29	01 10	01 10	02 37	00 43	01 39	15 38
16	02 16	00 22	01 06	01 10	02 36	00 43	01 39	15 40
19	02 11	00 15	01 03	01 10	02 35	00 43	01 39	15 42
22	02 00	00 N 07	00 59	01 11	02 34	00 43	01 39	15 43
25	01 41	00 00	00 56	01 12	02 33	00 43	01 39	15 45
28	01 14	00 S 07	00 53	01 12	02 32	00 43	01 39	15 47
31	00 S 38	00 S 14	00 S 49	01 N 12	02 S 32	00 N 43	01 N 39	15 N 48

DATA

Julian Date	2440557
Delta T	+40 seconds
Ayanamsa	23° 26' 16"
Synetic vernal point	05° ♓ 40' 44"
True obliquity of ecliptic	23° 26' 44"

MOON'S PHASES, APSIDES AND POSITIONS ☽

Date	h	m	Phase	Longitude ° '	Eclipse Indicator
02	03	50	☾	09 ♍ 48	
09	09	42	●	17 ♐ 09	
16	01	09	☽	23 ♓ 55	
23	17	35	○	01 ♋ 44	
31	22	52	☾	10 ♎ 06	

Day	h	m		
10	23	38	Perigee	
26	16	17	Apogee	
03	15	08	OS	
10	02	48	Max dec	28° S 20'
16	07	41	ON	
23	09	19	Max dec	28° N 19'
30	22	00	OS	

LONGITUDES

Date	Chiron ⚷ ° '	Ceres ⚳ ° '	Pallas ⚴ ° '	Juno ⚵ ° '	Vesta ⚶ ° '	Black Moon Lilith ⚸ ° '
01	02 ♈ 27	10 ♒ 56	17 ♑ 30	22 ♑ 37	24 ♌ 32	29 ♋ 22
11	02 ♈ 22	14 ♒ 06	20 ♑ 54	26 ♑ 25	25 ♌ 45	00 ♌ 29
21	02 ♈ 24	17 ♒ 26	24 ♑ 03	00 ♒ 22	26 ♌ 19	01 ♌ 36
31	02 ♈ 31	20 ♒ 56	27 ♑ 54	04 ♒ 28	26 ♌ 09	02 ♌ 43

ASPECTARIAN

h m Aspects

01 Monday
02 29 ☽ ⊻ ♇
02 41 ☽ ✶ ♃
05 48 ♀ ‖ ♆
05 52 ☽ □ ♆
10 09 ☉ ± ♄
12 09 ☽ ⊥ ♅
12 43 ☉ Q ♇
14 20 ☽ △ ♄
19 21 ☿ bQ ♄
21 08 ☽ ‖ ♄
23 30 ☽ ǂ ♃

02 Tuesday
00 12 ☽ ⊻ ♅
03 50 ☽ □ ☉
08 40 ♀ ✶ ♇
09 10 ☽ ∠ ♃
15 09 ☽ Q ♀
15 36 ♀ ⊻ ♃
17 51 ☽ Q ♆
19 59 ☽ bQ ♄
23 40 ☽ □ ☿

03 Wednesday
01 15 ☽ ⊼ ♂
02 02 ☿ Q ♅
03 07 ☽ ⊥ ♃
04 52 ☽ ǂ ♅
10 03 ♂ Q ♄
13 24 ☽ ± ♄
13 37 ☽ ± ♂
13 54 ☽ ☌ ♇
14 54 ☽ ⊻ ♃
17 07 ☽ ✶ ♀
17 12 ☽ ✶ ♆
18 01 ♀ ☌ ♆
18 10 ☽ Q ☉
18 14 ♂ ± ♇
22 21 ☿ ✶ ♂

04 Thursday
00 49 ☽ ⊼ ♄
01 20 ☽ ‖ ♅
07 49 ☽ bQ ♂
10 32 ☽ ☌ ♅
14 53 ☽ Q ☿
18 44 ☽ ✶ ☉
21 31 ☽ ∠ ♆

05 Friday
00 22 ☽ ∠ ♀
04 02 ♂ ǂ ♇
05 55 ☽ ‖ ♃
07 02 ☽ ǂ ♄
08 18 ☉ ∠ ♃
13 08 ☽ △ ♂
13 59 ☽ ⊥ ♆
15 47 ☽ ✶ ☿
18 24 ☽ ⊥ ♀
21 34 ☽ ⊻ ♇
23 13 ☽ ☌ ♃

06 Saturday
00 12 ☽ ∠ ☉
00 44 ☽ ⊻ ♆
04 15 ☽ ‖ ♂
05 24 ☽ ǂ ♇
06 09 ☽ ⊻ ♀
07 25 ☽ ☍ ♄
08 12 ☽ ⊥ ♇
08 45 ♂ bQ ♅
16 45 ☽ ⊻ ♅
17 15 ☽ ⊥ ☉
18 41 ☽ ‖ ♆
19 28 ♀ ⊼ ♄
21 32 ☽ ∠ ☿
23 49 ☽ ∠ ♇

07 Sunday
02 57 ☽ ‖ ♀
03 06 ☽ ⊥ ♅
04 22 ☽ ⊻ ☉
14 43 ☽ ⊥ ☿
16 18 ☽ ‖ ☉
18 21 ☽ ∠ ♅
18 27 ☿ □ ♇
20 06 ☽ □ ♂

08 Monday
01 09 ☽ ✶ ♇
01 57 ☽ ⊻ ☿
03 17 ☽ ⊻ ♃
04 11 ☽ ☌ ♆
05 32 ♀ ⊥ ♃
10 09 ☽ ⊼ ♄
10 53 ☽ ‖ ☿
13 18 ☽ ⊥ ♃
13 55 ☽ ☌ ♀
15 56 ☿ ✶ ♃
19 11 ☽ ✶ ♅
19 56 ☽ ± ♄
20 54 ☽ Q ♇
22 47 ♃ ǂ ♄
23 27 ☿ ⊻ ♆

09 Tuesday
03 17 ☽ Q ♂
04 09 ☽ ∠ ♃
09 42 ☽ ☌ ☉
10 29 ☽ bQ ♄
14 40 ☽ Q ♅
20 47 ☉ bQ ♄
23 38 ☽ ✶ ♂

10 Wednesday
01 58 ☽ □ ♇
04 34 ☽ ✶ ♃
04 59 ☽ ⊻ ♆
08 16 ☽ ☌ ☿
10 30 ☽ △ ♄
14 36 ☽ ⊥ ♆
18 45 ☽ ⊻ ♀
19 34 ☽ □ ♅
23 59 ☽ Q ♃

11 Thursday
00 53 ☽ ∠ ♂
04 44 ♀ ✶ ♅
05 02 ☽ ∠ ♆
05 13 ☽ ⊥ ♀
05 28 ☿ △ ♄
09 58 ♀ ± ♄
13 17 ☽ ⊻ ☉
16 04 ☽ ⊥ ♂
17 38 ☽ ‖ ☿
20 59 ☽ ∠ ♀
23 20 ♂ ⊼ ♇
23 38 ☽ ⊥ ☉

12 Friday
00 07 ♀ Q ♇
01 10 ♃ ⊻ ♆
02 05 ☽ △ ♇
02 14 ☽ ⊻ ♂
05 12 ☽ ✶ ♆
05 14 ☽ □ ♃
09 51 ☽ ‖ ☉
10 31 ☽ □ ♄
12 28 ☉ Q ♅
13 58 ☽ ⊻ ☿
15 17 ☽ ∠ ☉
18 17 ☽ ‖ ♀
19 59 ☽ △ ♅
23 35 ☽ ✶ ♀
23 46 ☿ ⊥ ♆

13 Saturday
00 46 ☽ Q ♆
00 53 ☽ ⊥ ☿
02 30 ☽ bQ ♇
08 46 ☽ ‖ ♆
15 58 ☽ Q ♄
17 23 ☽ ± ♇
17 29 ☽ ∠ ☿
17 48 ☽ ✶ ☉
20 47 ☽ bQ ♅
21 18 ☽ Q ♀
21 28 ☽ ǂ ♇

14 Sunday
03 25 ☽ ⊼ ♇
06 15 ☽ ☌ ♂
06 45 ☽ □ ♆
07 16 ☽ △ ♃
07 27 ☽ ‖ ♂
11 57 ☽ ± ♅
12 03 ☽ ✶ ♄
15 32 ☽ Q ☉
16 16 ♂ □ ♆
17 31 ☽ ‖ ♃
19 02 ☽ ǂ ♄
21 48 ☽ ✶ ☿
22 12 ☽ ⊼ ♅

15 Monday
01 32 ☿ □ ♅
06 52 ☽ □ ♀
07 02 ♂ △ ♃
09 13 ☽ bQ ♃
13 45 ☽ ∠ ♄
21 07 ☽ Q ☿
21 45 ☽ ‖ ♅

16 Tuesday
01 08 ☽ ± ♃
01 09 ☽ □ ☉
05 30 ☽ ⊥ ♄
07 12 ☽ ☍ ♇
10 49 ☽ △ ♆
11 53 ☽ ⊼ ♃
12 27 ♀ ∠ ♃
13 11 ☽ ⊻ ♂
16 08 ☽ ⊻ ♄
17 46 ☽ ǂ ♅

17 Wednesday
00 34 ☽ ⊥ ♂
03 04 ☽ ☍ ♅
09 13 ☽ □ ☿
13 53 ☽ bQ ♆
14 05 ☿ Q ♃
17 42 ☽ △ ♀
17 52 ☽ ∠ ♂
21 39 ☽ ‖ ♄

18 Thursday
00 07 ☽ ǂ ♃
05 37 ☽ ǂ ♂
06 25 ☽ ± ♆
08 54 ♀ bQ ♄
11 54 ☽ △ ☉
13 40 ☽ ⊼ ♇
16 16 ♂ ✶ ♄
17 34 ☽ ⊼ ♆
19 14 ☽ ☍ ♃
21 32 ☽ ‖ ♇
22 29 ♂ ± ♅
22 51 ☽ ☌ ♄
23 16 ☽ ✶ ♂

19 Friday
00 24 ☽ bQ ♀
00 51 ☽ ± ♇
04 12 ☿ ∠ ♆
10 28 ☉ □ ♇
10 30 ☽ ⊼ ♅
12 02 ☽ ǂ ♆
17 47 ☽ bQ ♇
18 25 ☽ bQ ☉
19 11 ☽ ± ♀
21 51 ☽ ± ♅
23 16 ☽ Q ♂

20 Saturday
00 01 ☽ △ ☿
07 50 ☽ ⊼ ♀
13 02 ☽ ± ☉
13 08 ☽ ǂ ♀
15 02 ☽ bQ ♅
16 54 ☽ ǂ ☉
22 24 ☽ △ ♇
22 31 ♀ Q ♅

21 Sunday
00 05 ☽ ǂ ☿
01 34 ☽ ⊼ ☉
02 34 ☽ ☍ ♆
04 52 ☽ ⊼ ♃
07 45 ☽ ⊻ ♄
08 23 ☽ bQ ☿
11 58 ☽ □ ♂
14 09 ☉ ⊻ ♆
16 38 ☽ ± ♃
19 21 ☽ ⊥ ♄
20 01 ☽ △ ♅

22 Monday
01 42 ♂ ‖ ♃
04 04 ☽ ± ☿
10 24 ☽ bQ ♃
12 54 ☽ ∠ ♄
17 16 ☽ ⊼ ☿

23 Tuesday
00 37 ☽ ☍ ♀
00 43 ☉ ✶ ♃
08 57 ☽ □ ♇
13 23 ☽ ⊼ ♆
16 22 ☽ △ ♃
17 35 ☽ ☍ ☉
18 24 ♀ Q ♂
18 27 ☽ ✶ ♄

24 Wednesday
01 21 ☽ ± ♆
02 47 ☽ △ ♂
03 40 ☉ △ ♄
07 14 ☿ ∠ ♂
07 16 ☽ □ ♅
16 13 ☉ ‖ ☿
18 22 ☽ Q ♄
19 24 ☽ bQ ♆
20 51 ☽ Q ♇
21 41 ☿ ‖ ♀
23 25 ♂ ǂ ♄

25 Thursday
10 53 ☽ bQ ♂
12 10 ☽ ☍ ☿
13 06 ☉ ‖ ♀
19 32 ☽ Q ♅
19 37 ☽ ⊼ ♀
21 04 ☽ ✶ ♇

26 Friday
01 45 ☽ △ ♆
03 28 ☽ ǂ ♀
03 55 ☽ ǂ ☉
05 24 ☽ □ ♃
06 19 ☽ ± ♂
06 39 ☽ □ ♄
07 22 ☽ ǂ ☿
09 13 ☽ ± ♀
09 15 ♀ □ ♇
11 33 ☽ ⊼ ☉
19 20 ☽ ⊼ ♂
19 55 ☽ ✶ ♅

27 Saturday
00 54 ☽ ± ☉
03 31 ☽ ∠ ♇
04 45 ♂ ⊼ ♅
05 42 ☽ bQ ♀
09 58 ☽ ǂ ♆
16 15 ☉ ⊥ ♆
18 27 ☽ Q ♃
20 57 ☽ bQ ☉
21 50 ☽ ⊥ ♇

28 Sunday
01 38 ☽ ‖ ♇
02 28 ☽ ∠ ♅
06 41 ♀ ⊻ ♆
07 23 ☽ ⊼ ☿
10 02 ☽ ⊻ ♇
14 53 ☽ □ ♆
15 51 ☽ △ ♀
19 08 ☽ ✶ ♃
19 34 ☽ △ ♄
20 39 ☽ ± ☿
20 46 ☽ ⊥ ♅
22 49 ☽ ǂ ♃

29 Monday
04 11 ☽ ‖ ♄
06 14 ☽ △ ☉
08 54 ☽ ⊻ ♅
10 02 ☽ ǂ ♂
12 19 ☽ ☍ ♂
16 18 ☽ bQ ☿
17 15 ☿ △ ♇

30 Tuesday
01 42 ☽ ∠ ♃
01 44 ☽ bQ ♄
02 23 ♀ ✶ ♃
02 39 ♀ △ ♄
03 14 ☽ Q ♆
04 44 ♃ ☍ ♄
08 06 ♇ St R
10 26 ☽ ǂ ♅
13 47 ☉ □ ♅
19 35 ☽ ± ♄
19 46 ☽ ⊥ ♃
22 11 ☽ ☌ ♇

31 Wednesday
00 09 ☽ △ ☿
03 01 ☽ ✶ ♆
07 20 ☽ ⊼ ♄
07 39 ☽ ⊻ ♃
09 29 ☽ ‖ ♅
10 37 ☽ □ ♀
20 12 ☽ ☌ ♅
22 52 ☽ □ ☉

All ephemeris data is given at 12.00 UT and the Moon's longitude is additionally given for 24.00 UT

JANUARY 1970

LONGITUDES

Date	Sidereal time h m s	Sun ☉ ° ' "	Moon ☽ ° ' "	Moon ☽ 24.00 ° ' "	Mercury ☿ ° '	Venus ♀ ° '	Mars ♂ ° '	Jupiter ♃ ° '	Saturn ♄ ° '	Uranus ♅ ° '	Neptune ♆ ° '	Pluto ♇ ° '
01	18 42 53	10 ♑ 39 57	17 ♎ 01 10	23 ♎ 26 24	29 ♑ 17	05 ♑ 05	12 ♓ 37	02 ♏ 24	02 ♉ 04	08 ♎ 44	29 ♏ 54	27 ♍ 24
02	18 46 50	11 41 07	29 ♎ 58 11	06 ♏ 36 59	29 42	06 20	13 21	02 32	02 R 03	08 44	29 56	27 R 23
03	18 50 47	12 42 17	13 ♏ 23 10	20 ♏ 16 56	29 ♑ 56	07 36	14 06	02 40	02 03	08 45	29 58	27 23
04	18 54 43	13 43 27	27 ♏ 18 19	04 ♐ 27 09	00 ♒ 00	08 51	14 51	02 47	02 D 03	08 45	29 ♏ 59	27 23
05	18 58 40	14 44 38	11 ♐ 43 01	19 ♐ 05 20	29 ♑ 52	10 07	15 35	02 55	02 03	08 46	00 ♐ 01	27 23
06	19 02 36	15 45 49	26 ♐ 33 13	04 ♑ 05 37	29 R 33	11 22	16 20	03 03	02 04	08 46	00 03	27 23
07	19 06 33	16 46 59	11 ♑ 41 20	19 ♑ 19 01	29 02	12 38	17 05	03 10	02 04	08 47	00 05	27 22
08	19 10 29	17 48 10	26 ♑ 57 16	04 ♒ 34 41	28 19	13 53	17 50	03 17	02 04	08 47	00 06	27 22
09	19 14 26	18 49 20	12 ♒ 09 57	19 ♒ 41 53	27 26	15 09	18 34	03 25	02 05	08 47	00 08	27 22
10	19 18 22	19 50 30	27 ♒ 09 27	04 ♓ 31 51	26 23	16 24	19 19	03 32	02 06	08 47	00 10	27 21
11	19 22 19	20 51 39	11 ♓ 48 27	18 ♓ 58 50	25 12	17 40	20 04	03 38	02 07	08 47	00 11	27 21
12	19 26 15	21 52 48	26 ♓ 02 47	03 ♈ 00 15	23 57	18 55	20 48	03 45	02 07	08 48	00 13	27 21
13	19 30 12	22 53 56	09 ♈ 51 22	16 ♈ 36 18	22 38	20 11	21 33	03 52	02 08	08 R 48	00 14	27 20
14	19 34 09	23 55 03	23 ♈ 15 23	29 ♈ 48 58	21 19	21 26	22 18	03 58	02 10	08 47	00 16	27 20
15	19 38 05	24 56 10	06 ♉ 17 29	12 ♉ 41 20	20 01	22 42	23 02	04 04	02 11	08 47	00 17	27 19
16	19 42 02	25 57 16	19 ♉ 00 58	25 ♉ 16 49	18 48	23 57	23 47	04 10	02 12	08 47	00 19	27 19
17	19 45 58	26 58 21	01 ♊ 29 17	07 ♊ 38 45	17 41	25 13	24 31	04 16	02 14	08 47	00 20	27 18
18	19 49 55	27 59 25	13 ♊ 45 35	19 ♊ 50 06	16 41	26 28	25 16	04 22	02 15	08 47	00 22	27 17
19	19 53 51	29 ♑ 00 29	25 ♊ 52 35	01 ♋ 53 20	15 50	27 44	26 00	04 28	02 17	08 46	00 23	27 17
20	19 57 48	00 ♒ 01 32	07 ♋ 52 33	13 ♋ 50 30	15 08	28 ♑ 59	26 45	04 32	02 19	08 46	00 24	27 16
21	20 01 44	01 02 34	19 ♋ 47 22	25 ♋ 43 21	14 36	00 ♒ 14	27 29	04 39	02 21	08 46	00 26	27 15
22	20 05 41	02 03 36	01 ♌ 38 39	07 ♌ 33 29	14 12	01 30	28 14	04 44	02 23	08 45	00 27	27 15
23	20 09 38	03 04 36	13 ♌ 28 04	19 ♌ 22 37	13 58	02 45	28 58	04 49	02 25	08 45	00 28	27 14
24	20 13 34	04 05 36	25 ♌ 17 25	01 ♍ 12 45	13 52	04 01	29 ♓ 42	04 54	02 27	08 44	00 30	27 13
25	20 17 31	05 06 35	07 ♍ 08 55	13 ♍ 06 18	13 D 55	05 16	00 ♈ 27	04 58	02 29	08 43	00 31	27 12
26	20 21 27	06 07 34	19 ♍ 05 17	25 ♍ 06 16	14 05	06 31	01 11	05 03	02 32	08 43	00 32	27 11
27	20 25 24	07 08 31	01 ♎ 09 46	07 ♎ 16 13	14 22	07 47	01 55	05 07	02 34	08 42	00 33	27 11
28	20 29 20	08 09 28	13 ♎ 25 26	19 ♎ 40 06	14 45	09 02	02 40	05 12	02 37	08 41	00 34	27 10
29	20 33 17	09 10 25	25 ♎ 58 42	02 ♏ 22 21	15 14	10 17	03 24	05 16	02 40	08 40	00 35	27 09
30	20 37 13	10 11 21	08 ♏ 51 35	15 ♏ 26 53	15 49	11 33	04 08	05 19	02 43	08 39	00 36	27 08
31	20 41 10	11 ♒ 12 16	22 ♏ 08 38	28 ♏ 57 09	16 ♑ 29	12 ♒ 48	04 ♈ 52	05 ♏ 23	02 ♉ 46	08 ♎ 38	00 ♐ 37	27 ♍ 07

Date	Moon True ☊ ° '	Moon Mean ☊ ° '	Moon ☽ Latitude ° '
01	14 ♓ 03	15 ♓ 16	02 S 47
02	14 R 02	15 12	03 40
03	13 58	15 09	04 23
04	13 51	15 06	04 52
05	13 43	15 03	05 04
06	13 34	15 00	04 57
07	13 24	14 56	04 29
08	13 16	14 53	03 41
09	13 10	14 50	02 38
10	13 07	14 47	01 25
11	13 06	14 44	00 S 07
12	13 D 06	14 41	01 N 09
13	13 07	14 37	02 19
14	13 08	14 34	03 19
15	13 R 07	14 31	04 07
16	13 05	14 28	04 42
17	13 00	14 25	05 02
18	12 53	14 22	05 08
19	12 45	14 18	05 00
20	12 36	14 15	04 39
21	12 28	14 12	04 06
22	12 20	14 09	03 22
23	12 15	14 06	02 30
24	12 11	14 02	01 31
25	12 10	13 59	00 N 27
26	12 D 10	13 56	00 S 38
27	12 11	13 53	01 42
28	12 13	13 50	02 43
29	12 14	13 47	03 37
30	12 15	13 43	04 22
31	12 ♓ 15	13 ♓ 40	04 S 54

DECLINATIONS

Date	Sun ☉ ° '	Moon ☽ ° '	Mercury ☿ ° '	Venus ♀ ° '	Mars ♂ ° '	Jupiter ♃ ° '	Saturn ♄ ° '	Uranus ♅ ° '	Neptune ♆ ° '	Pluto ♇ ° '
01	23 S 01	09 S 15	20 S 41	23 S 38	07 S 34	11 S 10	09 N 49	02 S 48	18 S 31	15 N 31
02	22 56	14 53	20 20	23 37	07 16	11 13	09 50	02 48	18 32	15 31
03	22 50	20 02	20 01	23 35	06 58	11 15	09 50	02 48	18 32	15 32
04	22 44	24 18	19 43	23 33	06 40	11 18	09 50	02 49	18 32	15 32
05	22 38	27 13	19 27	23 30	06 21	11 20	09 50	02 49	18 33	15 33
06	22 31	28 21	19 12	23 26	06 03	11 23	09 51	02 49	18 33	15 34
07	22 23	27 24	19 00	23 21	05 44	11 25	09 51	02 49	18 33	15 34
08	22 16	24 23	18 50	23 16	05 26	11 27	09 52	02 49	18 33	15 35
09	22 07	19 41	18 42	23 10	05 08	11 30	09 52	02 49	18 34	15 35
10	21 59	13 47	18 37	23 03	04 49	11 32	09 53	02 49	18 34	15 36
11	21 50	07 15	18 33	22 56	04 31	11 34	09 53	02 49	18 35	15 37
12	21 40	00 S 31	18 32	22 48	04 12	11 36	09 54	02 49	18 35	15 37
13	21 30	06 N 18	18 33	22 39	03 54	11 38	09 54	02 49	18 35	15 38
14	21 20	12 07	18 36	22 29	03 35	11 40	09 55	02 49	18 35	15 39
15	21 09	17 30	18 40	22 19	03 16	11 42	09 56	02 49	18 36	15 39
16	20 58	21 59	18 45	22 08	02 58	11 44	09 57	02 49	18 36	15 40
17	20 46	25 23	18 51	21 57	02 39	11 46	09 57	02 49	18 36	15 41
18	20 34	27 33	18 59	21 44	02 21	11 47	09 58	02 49	18 36	15 41
19	20 22	28 23	19 07	21 31	02 02	11 49	09 59	02 48	18 37	15 42
20	20 09	27 51	19 15	21 18	01 43	11 51	10 00	02 48	18 37	15 43
21	19 56	26 02	19 24	21 04	01 25	11 52	10 01	02 48	18 37	15 43
22	19 42	23 05	19 33	20 49	01 06	11 54	10 02	02 48	18 37	15 44
23	19 29	19 10	19 42	20 34	00 48	11 55	10 03	02 48	18 37	15 45
24	19 14	14 31	19 52	20 18	00 29	11 57	10 04	02 47	18 38	15 46
25	19 00	09 19	20 01	20 01	00 S 11	11 58	10 05	02 47	18 38	15 46
26	18 45	03 N 44	20 10	19 44	00 N 08	12 00	10 06	02 47	18 38	15 47
27	18 29	02 S 01	20 18	19 26	00 26	12 01	10 07	02 46	18 38	15 48
28	18 14	07 48	20 26	19 08	00 45	12 02	10 08	02 46	18 38	15 49
29	17 58	13 24	20 34	18 49	01 03	12 03	10 09	02 46	18 39	15 50
30	17 42	18 35	20 41	18 29	01 22	12 04	10 11	02 45	18 39	15 50
31	17 S 25	23 S 03	20 S 47	18 S 09	01 N 40	12 S 05	10 N 12	02 S 45	18 S 39	15 N 51

ZODIAC SIGN ENTRIES

Date	h m	Planets
02	12 03	☽ ♏
04	04 22	☿ ♒
04	11 58	☿ ♑
04	16 33	☽ ♐
04	19 55	♆ ♐
06	17 30	☽ ♑
08	16 47	☽ ♒
10	16 37	☽ ♓
12	18 48	☽ ♈
15	00 20	☽ ♉
17	09 07	☽ ♊
19	20 13	☽ ♋
20	11 24	☉ ♒
21	07 26	♀ ♒
22	08 40	☽ ♌
24	21 29	♂ ♈
24	21 33	☽ ♍
27	09 42	☽ ♎
29	19 34	☽ ♏

LATITUDES

Date	Mercury ☿ ° '	Venus ♀ ° '	Mars ♂ ° '	Jupiter ♃ ° '	Saturn ♄ ° '	Uranus ♅ ° '	Neptune ♆ ° '	Pluto ♇ ° '
01	00 S 23	00 S 17	00 S 48	01 N 13	02 S 31	00 N 43	01 N 39	15 N 49
04	00 N 27	00 24	00 45	01 13	02 30	00 43	01 39	15 51
07	01 23	00 31	00 42	01 14	02 30	00 43	01 39	15 52
10	02 19	00 37	00 38	01 14	02 29	00 44	01 39	15 54
13	03 02	00 44	00 35	01 15	02 28	00 44	01 40	15 55
16	03 24	00 50	00 32	01 15	02 27	00 44	01 40	15 57
19	03 25	00 55	00 29	01 16	02 26	00 44	01 40	15 58
22	03 09	01 01	00 26	01 17	02 25	00 44	01 40	16 00
25	02 43	01 05	00 23	01 17	02 24	00 44	01 40	16 01
28	02 12	01 10	00 20	01 18	02 23	00 44	01 40	16 03
31	01 N 40	01 S 14	00 S 17	01 N 18	02 S 23	00 N 44	01 N 40	16 N 04

DATA

Julian Date	2440588
Delta T	+40 seconds
Ayanamsa	23° 26' 21"
Synetic vernal point	05° ♓ 40' 38"
True obliquity of ecliptic	23° 26' 44"

MOON'S PHASES, APSIDES AND POSITIONS ☽

Date	h m	Phase	Longitude ° '	Eclipse Indicator
07	20 35	●	17 ♑ 09	
14	13 18	☽	23 ♈ 58	
22	12 55	○	02 ♌ 06	
30	14 38	☾	10 ♏ 18	

Day	h m		
08	09 47	Perigee	
22	19 46	Apogee	
06	13 13	Max dec	28° S 21'
12	13 51	0N	
19	14 35	Max dec	28° N 23'
27	03 37	0S	

LONGITUDES

Date	Chiron ⚷ ° '	Ceres ⚳ ° '	Pallas ⚴ ° '	Juno ⚵ ° '	Vesta ⚶ ° '	Black Moon Lilith ⚸ ° '
01	02 ♈ 32	21 ♒ 18	28 ♑ 15	04 ♒ 53	26 ♌ 05	02 ♌ 49
11	02 ♈ 44	24 ♒ 56	01 ♒ 47	09 ♒ 07	25 ♌ 04	03 ♌ 56
21	03 ♈ 02	28 ♒ 40	05 ♒ 20	13 ♒ 27	23 ♌ 19	05 ♌ 03
31	03 ♈ 24	02 ♓ 29	08 ♒ 52	17 ♒ 52	21 ♌ 00	06 ♌ 10

All ephemeris data is given at 12.00 UT and the Moon's longitude is additionally given for 24.00 UT

ASPECTARIAN

h m	Aspects
01 Thursday	
03 07	☽ ⊼ ♂
05 24	☽ ∥ ♂
07 59	☽ ∠ ♆
14 23	☽ # ♄
15 10	☽ ± ♂
20 05	☽ ∥ ♃
02 Friday	
00 33	☽ Q ♀
00 53	☽ ⊥ ♆
04 00	♀ ⊥ ♆
07 17	☽ ⊻ ♇
08 52	☽ bQ ♂
11 26	☽ Q ☉
11 29	☽ □ ☿
11 56	☽ ⊻ ♆
14 49	☽ # ♇
15 47	☽ ☍ ♄
16 41	☽ ☌ ♃
18 12	☽ ⊥ ♇
03 Saturday	
00 42	☽ ⚹ ♀
03 48	☽ ⊻ ♅
04 38	☽ ∥ ♆
10 15	☽ ∠ ♇
10 42	☽ ⚹ ☉
11 56	☽ ∥ ☿
13 20	☽ △ ♂
14 23	☽ ⊥ ♅
17 34	☿ ⚹ ♆
20 01	☽ Q ☿
21 06	♄ St D
04 Sunday	
02 41	☽ ∥ ☉
05 34	☽ ∠ ♀
05 57	☽ ∠ ♅
07 18	☽ ∥ ♀
08 09	☿ St R
10 03	♀ □ ♅
12 08	☽ ⚹ ♇
14 35	☽ ∠ ☉
15 11	☿ ⚹ ♆
16 32	☽ ⚹ ☿
16 32	☽ ☌ ♆
16 39	☿ ∠ ♂
20 00	☽ ⊼ ♄
21 19	☽ ⊻ ♃
22 14	☽ ⊥ ♀
05 Monday	
05 59	☽ ± ♄
06 45	☽ ⊥ ☉
07 08	☽ ⚹ ♅
07 22	☽ ⊥ ♃
08 10	☽ Q ♇
09 07	☽ ⊻ ♀
16 42	☉ Q ♃
17 04	☽ ∠ ☿
17 19	☽ ⊻ ☉
18 40	☽ □ ♂
18 41	☉ ∠ ♆
20 43	☽ bQ ♄
22 12	☽ ∠ ♃
06 Tuesday	
02 43	☽ Q ♅
07 19	☽ ⊥ ☿
13 19	☽ □ ♇
16 39	☽ ⊻ ☿
17 35	☽ ⊻ ♆
20 47	☽ △ ♄
22 25	☽ ⚹ ♃
07 Wednesday	
01 02	☽ Q ♂
03 07	☽ ⊥ ♆
07 24	☽ □ ♅
11 31	♂ ∠ ♄
13 37	☽ ☌ ♀
17 21	☽ ∠ ♆
17 31	☽ Q ♃
20 35	☽ ☌ ☉
20 55	☽ ⚹ ♂
08 Thursday	
12 39	☽ △ ♇
14 02	☽ ☌ ☿
14 07	☉ ⚹ ♂
16 58	☽ ⚹ ♆
18 38	☽ ∥ ♀
20 04	☽ □ ♄
21 43	☽ ∠ ♂
22 03	☽ □ ♃
09 Friday	
00 08	☽ ∥ ☉
03 20	♃ ⊥ ♇
05 46	♂ bQ ♃
06 39	☽ △ ♅
11 40	♀ ∠ ♆
11 57	☽ Q ♆
12 19	☽ bQ ♇
12 41	☽ ⊥ ♂
13 32	☿ △ ♇
16 21	☽ ∥ ☿
16 52	☽ ∥ ♆
17 10	☽ ⊻ ♀
17 28	♀ Q ♃
22 44	☽ ⊻ ♂
23 22	☽ ⊻ ☉
10 Saturday	
00 38	☽ Q ♄
02 40	☽ ± ♇
03 38	☽ ⊥ ♀
04 59	☽ # ♇
06 34	☽ bQ ♅
09 43	☽ ⊥ ☉
10 50	☽ ⊻ ☿
12 19	☽ ⊼ ♇
16 53	☽ □ ♆
19 33	☽ ∠ ♀
19 53	☽ ⊥ ☿
20 02	☽ ⚹ ♄
20 24	☽ ∥ ♃
21 09	☽ ± ♅
22 26	☽ △ ♃
11 Sunday	
01 27	☽ ∠ ☉
02 29	☽ # ♄
03 51	☿ ∥ ♆
07 00	☽ ⊼ ♅
09 33	☽ ∠ ☿
20 52	☽ ∠ ♄
22 13	☽ ∥ ♂
22 44	☽ ⚹ ♀
23 31	☽ bQ ♃
12 Monday	
02 35	☽ ☌ ♂
03 46	☽ ∥ ♅
04 21	☽ ⚹ ☉
08 43	☽ ⚹ ☿
12 08	☽ ⊥ ♄
14 13	☽ ☍ ♇
14 57	☽ ± ♃
19 11	☽ △ ♆
21 14	☽ Q ♀
22 29	☽ ⊻ ♄
13 Tuesday	
00 04	☽ # ♅
01 24	☽ ⊼ ♃
02 36	☽ Q ☉
03 38	☽ Q ☿
04 22	☽ # ♂
06 09	♅ St R
09 15	☉ ☌ ☿
10 08	☽ ☍ ♅
21 34	☽ bQ ♆
14 Wednesday	
00 33	☿ ⚹ ♂
03 02	☽ ∥ ♄
07 53	☿ ∥ ♆
08 22	☽ □ ♀
08 47	☽ □ ☿
10 08	☽ # ♃
10 09	☽ ⊻ ♂
10 47	☿ ☌ ♀
13 18	☽ □ ☉
13 50	☽ ± ♆
19 26	☽ ⊼ ♇
21 45	☽ ⊥ ♂
15 Thursday	
00 51	☽ ⊼ ♆
03 19	☽ ∥ ♇
04 21	☽ ☌ ♄
06 29	☽ ± ♇
07 50	☽ ☍ ♃
15 27	☽ ∠ ♂
16 40	☽ ⊼ ♅
17 22	☽ # ♆
17 49	☽ # ☿
23 18	☽ bQ ♇
16 Friday	
03 49	♀ ⚹ ♂
03 58	☽ ± ♅
06 15	☽ # ☉
11 38	☽ △ ☿
12 51	☽ # ♀
21 08	☽ bQ ♅
21 41	☽ ⚹ ♂
22 30	☽ △ ♀
23 36	♂ ∥ ♅
17 Saturday	
02 29	☽ △ ☉
03 54	☽ △ ♇
09 46	☽ ☍ ♆
13 26	☽ ⊻ ♄
14 08	☽ bQ ☿
17 28	☽ ⊼ ♃
19 41	☉ △ ♇
22 26	☽ Q ♂
18 Sunday	
01 10	☽ ⊥ ♄
02 13	☽ △ ♅
05 17	☽ ± ♃
06 24	☽ ± ☿
06 59	☽ bQ ♀
10 21	☽ bQ ☉
17 23	☽ ⊼ ☿
18 55	☽ ∠ ♄
19 41	☿ Q ♃
23 10	☽ bQ ♃
19 Monday	
02 47	☽ ± ♀
03 35	♀ △ ♇
05 46	☽ ± ☉
12 16	☽ □ ♂
14 48	☽ □ ♇
16 07	☽ ⊼ ♀
18 49	☽ ⊼ ☉
21 01	☽ ⊼ ♆
21 22	♂ ⊥ ♄
20 Tuesday	
00 49	☽ ⚹ ♄
02 27	☿ ∠ ♆
05 17	☽ △ ♃
09 03	☽ ± ♆
13 47	☽ □ ♅
19 00	☿ Q ♂
21 12	☉ ⚹ ♆
21 Wednesday	
00 59	☽ Q ♄
01 56	☽ ☍ ☿
02 52	☽ Q ♇
03 11	☽ bQ ♆
04 40	♂ ☍ ♇
15 41	♀ ⚹ ♆
22 Thursday	
02 06	☽ Q ♅
03 06	☽ ⚹ ♇
04 37	☽ △ ♂
09 35	☽ △ ♆
11 40	☽ ☍ ♀
12 55	☽ ☍ ☉
13 30	☽ □ ♄
18 19	☽ □ ♃
19 47	☉ □ ♄
21 40	☉ ∥ ☿
23 Friday	
02 25	☽ ⚹ ♅
03 31	☽ # ♀
05 20	♀ □ ♄
06 26	♂ ± ♃
09 06	☽ # ☿
09 30	☽ ∠ ♇
10 13	☽ # ☉
13 00	☽ ⊼ ☿
13 05	☽ bQ ♂
15 01	☽ # ♆
24 Saturday	
01 03	☽ ± ☿
03 45	☽ ⊥ ♇
05 54	☽ ∥ ♇
07 07	☽ Q ♃
08 35	☽ ± ♂
08 51	☽ ∠ ♅
15 54	☽ ⊻ ♇
16 39	☿ St D
19 15	☽ bQ ☿
20 27	☉ ☌ ♀
21 33	☽ ⊼ ♂
22 34	☽ □ ♆
25 Sunday	
00 03	☽ # ♃
02 33	☽ △ ♄
03 04	☽ ⊥ ♅
06 05	♀ □ ♃
07 30	☽ ⊼ ☉
07 35	☽ ⚹ ♃
07 45	☽ ⊼ ♀
08 32	☉ □ ♃
08 35	☽ ∥ ♄
12 05	☿ ∥ ♀
14 13	♂ △ ♆
15 10	☽ ⊻ ♅
20 44	☽ ± ☉
21 17	☽ ± ♀
26 Monday	
01 47	☽ △ ☿
08 52	☽ bQ ♄
10 53	☽ Q ♆
13 56	☽ ∠ ♃
16 03	☽ # ♅
16 27	☽ bQ ☉
17 26	☽ bQ ♀
22 28	☉ ∥ ♆
27 Tuesday	
02 19	☽ ∥ ♂
02 53	☽ ± ♄
04 07	☽ ☌ ♇
05 05	☽ # ♂
07 57	☽ ⊥ ♃
10 48	☽ ⚹ ♆
13 36	☽ ☍ ♂
14 47	☽ ⊼ ♄
15 06	☽ ∥ ♅
19 50	☽ ⊻ ♃
28 Wednesday	
00 49	☽ △ ☉
02 28	☽ △ ♀
02 46	☽ ☌ ♅
05 25	♀ △ ♅
10 25	♂ ⊻ ♄
14 38	☽ □ ☿
16 08	☽ ∠ ♆
21 54	☽ # ♄
29 Thursday	
00 17	☉ △ ♅
06 04	☽ ∥
09 22	☽ ⊥
14 12	☽ ⊻
20 41	☽ ⊻ ♆
22 56	☽ # ♇
30 Friday	
00 25	♀ ∥ ♆
00 35	☽ ☍ ♄
01 25	☽ ⊥ ♇
02 14	☽ Q ☿
02 45	☽ ⊼ ♂
03 25	☿ ∠ ♆
05 27	☽ ☌ ♃
07 54	☽ ∥ ☉
11 33	☽ ∥ ♀
11 38	☽ ⊻ ♅
12 18	☽ ∥ ♆
14 29	☽ ± ♂
14 38	☽ □ ☉
17 26	☽ □ ♀
17 58	☽ ∠ ♇
22 33	☽ ⊥ ♅
22 55	☽ ∥ ☿
23 02	♀ bQ ♇
31 Saturday	
01 19	☽ ⚹ ☿
07 43	☽ bQ ♂
08 35	♀ Q ♆
14 39	☽ ∠ ♅
20 46	☽ ⚹ ♇

FEBRUARY 1970

Date	Sidereal time h m s	LONGITUDES Sun ☉ ° ' "	Moon ☽ ° ' "	Moon ☽ 24.00 ° ' "	Mercury ☿ ° '	Venus ♀ ° '	Mars ♂ ° '	Jupiter ♃ ° '	Saturn ♄ ° '	Uranus ♅ ° '	Neptune ♆ ° '	Pluto ♇ ° '
01	20 45 07	12 ♒ 13 10	05 ♐ 52 36	12 ♐ 55 02	17 ♑ 13	14 ♒ 03	05 ♈ 37	05 ♏ 27	02 ♉ 49	08 ♎ 37	00 ♐ 38	27 ♍ 06
02	20 49 03	13 14 04	20 ♐ 04 19	27 ♐ 20 09	18 01	15 19	06 21	05 30	02 52	08 R 36	00 39	27 R 05
03	20 53 00	14 14 57	04 ♑ 41 59	12 ♑ 09 08	18 53	16 34	07 05	05 33	02 55	08 35	00 40	27 04
04	20 56 56	15 15 49	19 ♑ 40 41	27 ♑ 15 32	19 48	17 49	07 49	05 36	02 58	08 34	00 41	27 03
05	21 00 53	16 16 40	04 ♒ 52 30	12 ♒ 30 18	20 46	19 05	08 33	05 39	03 02	08 33	00 42	27 01
06	21 04 49	17 17 30	20 ♒ 07 37	27 ♒ 43 08	21 47	20 20	09 17	05 41	03 05	08 32	00 43	27 00
07	21 08 46	18 18 18	05 ♓ 15 40	12 ♓ 44 08	22 51	21 35	10 01	05 44	03 09	08 30	00 44	26 59
08	21 12 42	19 19 05	20 ♓ 07 35	27 ♓ 25 17	23 57	22 50	10 45	05 46	03 13	08 29	00 45	26 58
09	21 16 39	20 19 51	04 ♈ 36 42	11 ♈ 41 27	25 05	24 06	11 29	05 48	03 17	08 27	00 45	26 57
10	21 20 36	21 20 35	18 ♈ 39 21	25 ♈ 30 23	26 15	25 21	12 13	05 50	03 20	08 26	00 46	26 56
11	21 24 32	22 21 17	02 ♉ 14 40	08 ♉ 52 27	27 27	26 36	12 57	05 52	03 24	08 24	00 47	26 54
12	21 28 29	23 21 58	15 ♉ 24 05	21 ♉ 49 56	28 41	27 51	13 41	05 53	03 29	08 23	00 47	26 53
13	21 32 25	24 22 38	28 ♉ 10 30	04 ♊ 26 16	29 ♑ 56	29 ♒ 07	14 24	05 54	03 33	08 21	00 48	26 52
14	21 36 22	25 23 15	10 ♊ 37 44	16 ♊ 45 27	01 ♒ 13	00 ♓ 22	15 08	05 55	03 37	08 20	00 49	26 50
15	21 40 18	26 23 51	22 ♊ 49 54	28 ♊ 51 35	02 32	01 37	15 52	05 56	03 41	08 18	00 49	26 49
16	21 44 15	27 24 25	04 ♋ 51 01	10 ♋ 48 38	03 52	02 52	16 36	05 57	03 46	08 16	00 50	26 48
17	21 48 11	28 24 58	16 ♋ 44 52	22 ♋ 40 07	05 13	04 07	17 19	05 58	03 50	08 15	00 50	26 46
18	21 52 08	29 ♒ 25 29	28 ♋ 34 46	04 ♌ 29 09	06 35	05 22	18 03	05 58	03 55	08 13	00 51	26 45
19	21 56 05	00 ♓ 25 58	10 ♌ 23 36	16 ♌ 18 23	07 59	06 37	18 47	05 58	04 00	08 11	00 51	26 44
20	22 00 01	01 26 25	22 ♌ 13 49	28 ♌ 10 07	09 24	07 52	19 30	05 R 58	04 05	08 09	00 51	26 42
21	22 03 58	02 26 51	04 ♍ 07 34	10 ♍ 06 23	10 50	09 07	20 14	05 58	04 09	08 07	00 52	26 41
22	22 07 54	03 27 15	16 ♍ 06 49	22 ♍ 09 06	12 17	10 22	20 57	05 58	04 14	08 05	00 52	26 39
23	22 11 51	04 27 37	28 ♍ 13 29	04 ♎ 20 12	13 45	11 37	21 41	05 57	04 19	08 03	00 52	26 38
24	22 15 47	05 27 58	10 ♎ 29 31	16 ♎ 41 44	15 15	12 52	22 24	05 56	04 25	08 01	00 53	26 36
25	22 19 44	06 28 18	22 ♎ 57 07	29 ♎ 15 58	16 45	14 07	23 07	05 55	04 30	07 59	00 53	26 35
26	22 23 40	07 28 36	05 ♏ 38 36	12 ♏ 05 18	18 16	15 22	23 51	05 54	04 35	07 57	00 53	26 33
27	22 27 37	08 28 52	18 ♏ 36 24	25 ♏ 12 09	19 49	16 37	24 34	05 53	04 40	07 55	00 53	26 32
28	22 31 34	09 ♓ 29 07	01 ♐ 52 50	08 ♐ 38 39	21 ♒ 22	17 ♓ 52	25 ♈ 17	05 ♏ 51	04 ♉ 46	07 ♎ 53	00 ♐ 53	26 ♍ 30

Date	Moon True ☊ ° '	Moon Mean ☊ ° '	Moon ☽ Latitude ° '	DECLINATIONS Sun ☉ ° '	Moon ☽ ° '	Mercury ☿ ° '	Venus ♀ ° '	Mars ♂ ° '	Jupiter ♃ ° '	Saturn ♄ ° '	Uranus ♅ ° '	Neptune ♆ ° '	Pluto ♇ ° '
01	12 ♓ 13	13 ♓ 37	05 S 11	17 S 08	26 S 24	20 S 52	17 S 49	01 N 59	12 S 06	10 N 13	02 S 44	18 S 39	15 N 52
02	12 R 09	13 34	05 10	16 51	28 14	20 57	17 28	02 17	12 07	10 15	02 44	18 39	15 53
03	12 06	13 31	04 50	16 34	28 11	21 00	17 06	02 35	12 08	10 16	02 44	18 39	15 54
04	12 01	13 28	04 09	16 16	26 06	21 03	16 44	02 53	12 09	10 17	02 43	18 40	15 54
05	11 58	13 24	03 10	15 58	22 07	21 05	16 22	03 12	12 09	10 19	02 43	18 40	15 55
06	11 55	13 21	01 57	15 39	16 38	21 05	15 59	03 30	12 10	10 20	02 42	18 40	15 56
07	11 54	13 18	00 S 37	15 21	10 09	21 05	15 36	03 48	12 11	10 22	02 41	18 40	15 57
08	11 D 54	13 15	00 N 45	15 02	03 S 13	21 03	15 12	04 06	12 11	10 23	02 41	18 40	15 58
09	11 55	13 12	02 02	14 43	03 N 42	21 00	14 48	04 24	12 12	10 25	02 40	18 40	15 58
10	11 57	13 08	03 09	14 23	10 14	20 56	14 23	04 42	12 12	10 26	02 40	18 40	15 59
11	11 58	13 05	04 03	14 04	16 04	20 51	13 58	05 00	12 12	10 28	02 39	18 40	16 00
12	11 59	13 02	04 43	13 44	20 58	20 45	13 33	05 18	12 13	10 30	02 38	18 40	16 01
13	11 R 59	12 59	05 07	13 24	24 44	20 38	13 07	05 36	12 13	10 31	02 38	18 40	16 02
14	11 58	12 56	05 16	13 04	27 15	20 29	12 41	05 53	12 13	10 33	02 37	18 41	16 03
15	11 57	12 53	05 10	12 43	28 25	20 19	12 15	06 11	12 13	10 35	02 36	18 41	16 03
16	11 55	12 49	04 51	12 23	28 12	20 08	11 48	06 29	12 13	10 36	02 36	18 41	16 04
17	11 53	12 46	04 20	12 02	26 41	19 55	11 21	06 46	12 13	10 38	02 35	18 41	16 05
18	11 51	12 43	03 37	11 41	24 00	19 41	10 53	07 04	12 13	10 40	02 34	18 41	16 06
19	11 49	12 40	02 45	11 19	20 18	19 26	10 26	07 21	12 13	10 42	02 34	18 41	16 07
20	11 48	12 37	01 46	10 58	15 47	19 10	09 58	07 38	12 13	10 44	02 33	18 41	16 07
21	11 47	12 34	00 N 42	10 36	10 39	18 53	09 30	07 55	12 13	10 45	02 32	18 41	16 08
22	11 D 47	12 30	00 S 24	10 15	05 N 07	18 34	09 01	08 12	12 13	10 47	02 31	18 41	16 09
23	11 48	12 27	01 30	09 53	00 S 40	18 14	08 33	08 29	12 12	10 49	02 30	18 41	16 10
24	11 48	12 24	02 33	09 31	06 30	17 52	08 04	08 46	12 11	10 51	02 30	18 41	16 11
25	11 49	12 21	03 29	09 08	12 10	17 29	07 35	09 03	12 11	10 53	02 29	18 41	16 12
26	11 49	12 18	04 17	08 46	17 26	17 05	07 05	09 20	12 10	10 55	02 28	18 41	16 12
27	11 50	12 14	04 52	08 24	22 02	16 40	06 36	09 37	12 10	10 57	02 27	18 41	16 13
28	11 ♓ 50	12 ♓ 11	05 S 13	08 S 01	25 S 39	16 S 14	06 S 06	09 N 53	12 S 09	10 N 59	02 S 26	18 S 41	16 N 14

ZODIAC SIGN ENTRIES

Date	h	m	Planets
01	01	50	☽ ♐
03	04	22	☽ ♑
05	04	19	☽ ♒
07	03	37	☽ ♓
09	04	17	☽ ♈
11	07	59	☽ ♉
13	13	08	☿ ♒
13	15	29	☽ ♊
14	05	04	♀ ♓
16	02	17	☽ ♋
18	14	53	☽ ♌
19	01	42	☉ ♓
21	03	42	☽ ♍
23	15	30	☽ ♎
26	01	23	☽ ♏
28	08	38	☽ ♐

LATITUDES

Date	Mercury ☿ ° '	Venus ♀ ° '	Mars ♂ ° '	Jupiter ♃ ° '	Saturn ♄ ° '	Uranus ♅ ° '	Neptune ♆ ° '	Pluto ♇ ° '
01	01 N 29	01 S 15	00 S 16	01 N 19	02 S 22	00 N 44	01 N 40	16 N 04
04	00 57	01 18	00 14	01 19	02 21	00 44	01 41	16 06
07	00 N 26	01 21	00 11	01 20	02 21	00 44	01 41	16 07
10	00 S 02	01 24	00 08	01 20	02 20	00 45	01 41	16 08
13	00 28	01 25	00 06	01 21	02 19	00 45	01 41	16 09
16	00 52	01 26	00 03	01 22	02 18	00 45	01 41	16 10
19	01 12	01 27	00 S 01	01 23	02 18	00 45	01 41	16 11
22	01 30	01 27	00 N 02	01 23	02 17	00 45	01 42	16 12
25	01 45	01 26	00 04	01 24	02 16	00 45	01 42	16 12
28	01 57	01 25	00 07	01 24	02 16	00 45	01 42	16 13
31	02 S 05	01 S 23	00 N 09	01 N 25	02 S 15	00 N 45	01 N 42	16 N 14

DATA

Julian Date	2440619
Delta T	+40 seconds
Ayanamsa	23° 26' 27"
Synetic vernal point	05° ♓ 40' 33"
True obliquity of ecliptic	23° 26' 44"

MOON'S PHASES, APSIDES AND POSITIONS ☽

Date	h	m	Phase	Longitude ° '	Eclipse Indicator
06	07	13	●	17 ♒ 05	
13	04	10	☽	24 ♉ 05	
21	08	19	○	02 ♍ 18	partial

Day	h	m		
05	22	49	Perigee	
18	21	55	Apogee	
02	23	29	Max dec	28° S 28'
08	23	04	0N	
15	20	16	Max dec	28° N 30'
23	09	15	0S	

LONGITUDES

Date	Chiron ⚷ ° '	Ceres ⚳ ° '	Pallas ⚴ ° '	Juno ⚵ ° '	Vesta ⚶ ° '	Black Moon Lilith ⚸ ° '
01	03 ♈ 27	02 ♓ 52	09 ♒ 13	18 ♒ 19	20 ♌ 45	06 ♌ 16
11	03 ♈ 53	06 ♓ 45	12 ♒ 42	22 ♒ 49	18 ♌ 08	07 ♌ 23
21	04 ♈ 23	10 ♓ 40	16 ♒ 08	27 ♒ 22	15 ♌ 35	08 ♌ 30
31	04 ♈ 55	14 ♓ 36	19 ♒ 30	01 ♓ 59	13 ♌ 25	09 ♌ 36

All ephemeris data is given at 12.00 UT and the Moon's longitude is additionally given for 24.00 UT

ASPECTARIAN

h m Aspects

01 Sunday
01 26 ☽ Q ☉
02 56 ☽ ☌ ♆
04 45 ☽ Q ♀
05 18 ☽ ∠ ☿
06 06 ♂ ⊼ ♃
06 41 ☽ ⊼ ♄
09 08 ☉ bQ ♇
11 15 ☽ ⊻ ♃
11 31 ☽ △ ♂
16 42 ☽ ⚹ ♅
17 03 ☽ ± ♄
17 30 ☽ Q ♇
19 37 ☿ Q ♃
21 32 ☽ ⊥ ♃
21 39 ☽ ⊥ ☿
22 09 ☉ Q ♆
23 39 ☽ ⚹ ☉

02 Monday
03 16 ☽ ⚹ ♀
08 18 ☽ bQ ♄
08 22 ☽ ⊻ ☿
12 43 ☽ ∠ ♃
12 53 ☽ Q ♅
23 34 ☽ □ ♇

03 Tuesday
02 28 ☽ ∠ ☉
05 27 ☽ ⊻ ♆
06 26 ☽ ∠ ♀
09 06 ☽ △ ♄
13 23 ☽ ⚹ ♃
15 12 ☽ ⊥ ♆
16 03 ☽ □ ♂
18 09 ☽ ⊥ ☉
18 16 ☽ □ ♅
22 19 ☽ ⊥ ♀
22 39 ♂ # ♅

04 Wednesday
04 28 ☽ ⊻ ☉
05 39 ☽ ∠ ♆
08 41 ☽ Q ♃
08 47 ☽ ⊻ ♀
12 12 ☽ ☌ ☿
22 13 ☽ Q ♂
23 39 ☽ △ ♇

05 Thursday
05 26 ☽ ⚹ ♆
09 05 ☽ □ ♄
10 24 ♄ ± ♇
11 53 ♂ ☍ ♅
13 13 ☽ □ ♃
15 17 ☉ # ♇
17 01 ☽ ∥ ☿
17 46 ☽ △ ♅
18 04 ☽ ⚹ ♂
23 14 ☽ bQ ♇

06 Friday
00 19 ☽ Q ♆
03 43 ☽ ∥ ♆
07 13 ☽ ☌ ☉
12 21 ☽ ☌ ♀
13 23 ☽ ± ♇
13 31 ☽ Q ♄
14 42 ☽ ∥ ♀
14 43 ☽ # ♇
14 49 ☽ ⊻ ☿
15 06 ♀ # ♇
15 59 ☽ ∥ ☉
17 21 ☽ bQ ♅
18 54 ☽ ∠ ♂
22 51 ☽ ⊼ ♇

07 Saturday
00 41 ♀ ± ♇
01 01 ☽ ⊥ ☿
03 13 ♀ Q ♄
04 46 ☽ ∥ ♃
04 47 ☽ □ ♆
07 36 ☽ ± ♅
08 37 ☽ ⚹ ♄
09 55 ☽ ⊥ ♂
11 15 ☽ # ♄
12 45 ☽ △ ♃
16 28 ☽ ∠ ☿
17 11 ☽ ⊼ ♅
20 01 ☽ ⊻ ♂

08 Sunday
08 52 ☽ ∠ ♄
09 06 ☽ # ♂
10 35 ☽ ⊻ ☉
13 03 ☽ bQ ♃
13 51 ☽ ∥ ♅
16 52 ☽ ⊻ ♀
18 47 ☽ ⚹ ☿
21 10 ☽ ⊥ ☉
23 14 ☽ ☍ ♇
23 42 ☽ ⊥ ♄

09 Monday
00 01 ♀ bQ ♅
03 43 ☽ ⊥ ♀
03 56 ☽ ± ♃
05 32 ☽ △ ♆
08 23 ☽ # ♅
09 45 ☽ ⊻ ♄
13 18 ☽ ∠ ☉
14 00 ☽ ⊼ ♃
14 37 ☽ ∥ ♂
16 32 ☽ Q ☿
18 29 ☽ ☍ ♅
20 19 ☽ ∠ ♀

10 Tuesday
00 17 ☽ ☌ ♂
02 18 ☉ ± ♇
07 00 ☽ bQ ♆
10 37 ☉ ∥ ♀
11 57 ☉ Q ♄
12 50 ☽ ∥ ♄
17 03 ☽ ⚹ ☉
19 46 ☽ # ♃
22 42 ☽ ± ♆

11 Wednesday
00 55 ☽ ⚹ ♀
01 19 ☿ △ ♇
02 29 ☽ ⊼ ♇
02 36 ☽ □ ☿
03 37 ☽ # ♀
03 52 ☽ # ☉
09 22 ☽ ⊼ ♆
11 44 ☽ ∥ ♇
13 11 ☽ ± ♇
14 06 ☽ ☌ ♄
16 07 ☽ Q ☉
17 41 ♀ ⊼ ♇
18 32 ☽ ☍ ♃
23 08 ☽ ⊼ ♅

12 Thursday
00 10 ☽ # ♆
00 43 ☽ Q ♀
05 31 ☽ bQ ♇
08 37 ☽ ⊻ ♂
10 07 ☽ ± ♅
10 53 ☽ # ☿
12 22 ☉ bQ ♅
20 26 ☽ ⊥ ♂

13 Friday
02 53 ☽ bQ ♅
04 10 ☽ □ ☉
09 30 ☽ △ ♇
13 59 ☽ □ ♀
14 30 ☽ ∠ ♂
15 45 ☽ △ ☿
17 01 ☽ ☍ ♆
22 21 ☽ ⊻ ♄

14 Saturday
01 40 ♀ ∠ ♂
02 51 ☽ ⊼ ♃
04 18 ☿ ⚹ ♆
07 32 ☽ △ ♅
10 01 ☽ ⊥ ♄
14 32 ☽ ± ♃
20 41 ♀ □ ♆
21 22 ☽ ⚹ ♂

15 Sunday
00 15 ☽ bQ ☿
03 46 ☽ ∠ ♄
08 15 ☽ bQ ♃
10 30 ♂ bQ ♆
13 16 ♀ ∥ ♃
19 44 ☽ △ ☉
19 55 ☽ □ ♇
20 16 ☽ ± ☿
21 47 ☉ ⊼ ♇
22 40 ☽ Q ♂

16 Monday
00 51 ♀ ± ♅
03 56 ☽ ⊼ ♆
07 33 ☽ △ ♀
09 46 ☽ ⊼ ☿
09 48 ☽ ⚹ ♄
10 10 ☿ □ ♄
14 13 ☽ △ ♃
15 59 ☽ ± ♆
18 52 ☽ □ ♅
22 44 ☉ ∥ ♃

17 Tuesday
04 38 ☽ bQ ☉
06 19 ♀ ⚹ ♄
08 01 ☽ Q ♇
10 09 ☽ Q ♄
10 09 ☽ bQ ♆
13 14 ☽ □ ♂
16 03 ☿ Q ♂
17 22 ☽ bQ ♀

18 Wednesday
00 34 ☽ ± ☉
01 09 ☿ □ ♃
07 12 ☽ Q ♅
08 17 ☽ ⚹ ♇
13 48 ☽ ± ♀
13 53 ☽ ⊼ ☉
16 36 ☽ △ ♆
22 55 ☽ □ ♄
23 03 ♀ # ♄
23 28 ♀ △ ♃

19 Thursday
03 01 ☽ □ ♃
03 26 ☽ ⊼ ♀
06 27 ☽ ☍ ☿
07 31 ☽ ⚹ ♅
14 42 ☽ ∠ ♇
15 19 ☿ △ ♅
17 09 ☽ # ☿
21 02 ☽ # ♆
21 58 ♃ St R
22 03 ☉ □ ♆

20 Friday
06 07 ☽ △ ♂
08 55 ☽ ⊥ ♇
10 18 ☽ ∥ ♇
13 51 ☽ ∠ ♅
15 31 ☽ Q ♃
17 11 ♀ ⊼ ♅
21 01 ☽ ⊻ ♇

21 Saturday
02 42 ☉ # ♄
04 24 ☉ ± ♅
04 58 ☽ # ♃
05 26 ☽ □ ♆
07 58 ☽ ⊥ ♅
08 19 ☽ ☍ ☉
11 33 ☽ ∥ ♄
12 04 ☽ △ ♄
12 15 ☽ # ☉
14 21 ☽ bQ ♂
15 42 ☽ ⚹ ♃
17 38 ☽ # ♀
20 00 ☽ ⊻ ♅
23 12 ☽ ☍ ♀
23 25 ☽ ∥ ♂

22 Sunday
01 49 ☿ bQ ♇
03 13 ☿ ∥ ♆
03 18 ☽ ⊼ ☿
09 32 ☽ ± ♂
16 55 ☽ ± ☿
17 29 ☽ Q ♆
18 16 ☽ bQ ♄
21 38 ☽ ∠ ♃
21 38 ☿ Q ♆
22 14 ☽ ⊼ ♂
22 51 ☽ # ♅

23 Monday
08 26 ☉ ⚹ ♄
08 52 ☽ ☌ ♇
12 12 ☽ ± ♄
13 11 ☽ bQ ☿
13 39 ♀ # ♂
15 24 ☽ ⊥ ♃
17 13 ☽ ⚹ ♆
19 32 ☽ ∥ ♅

24 Tuesday
00 03 ☽ ⊼ ♄
01 20 ☽ ⊼ ☉
03 08 ☽ ⊻ ♃
07 12 ☽ ☌ ♅
14 03 ☽ ± ☉
17 08 ☽ ⊼ ♀
18 01 ☽ ∥ ♀
22 00 ☽ # ♂
22 26 ☽ ∠ ♆
22 28 ☽ △ ☿
23 04 ☉ △ ♃
23 51 ☽ ∥ ☉

25 Wednesday
05 59 ☽ ± ♀
06 27 ☽ # ♄
08 55 ☽ bQ ☉
12 05 ☽ ∥ ♃
12 21 ☽ ☍ ♂
15 08 ☉ # ♂
15 41 ☽ ⊥ ♆
18 54 ☽ ⊻ ♇

26 Thursday
01 01 ☽ bQ ♀
03 03 ☽ ⊻ ♆
06 10 ☽ # ♇
06 13 ☽ ⊥ ♇
10 00 ☽ ☍ ♄
10 28 ☽ ∥ ☿
12 29 ☽ ☌ ♃
15 43 ☽ △ ☉
16 18 ☽ ⊻ ♅
18 05 ☽ ∥ ♆
22 55 ☉ ⊼ ♅
22 59 ☽ ∠ ♇

27 Friday
03 24 ☽ ⊥ ♅
07 59 ☽ △ ♀
14 30 ☽ □ ☿
19 50 ☽ ∠ ♅
22 46 ♂ ± ♆
22 52 ☿ ± ♇
23 28 ☽ ⊼ ♂

28 Saturday
02 22 ☽ ⚹ ♇
10 14 ☽ ☌ ♆
10 53 ☽ ± ♂
11 37 ☿ # ♇
17 10 ☽ ⊼ ♄
19 03 ☽ ⊻ ♃
22 37 ☽ ⚹ ♅
23 44 ☽ Q ♇

MARCH 1970

LONGITUDES

Date	Sidereal time h m s	Sun ☉ ° ' "	Moon ☽ ° ' "	Moon ☽ 24.00 ° ' "	Mercury ☿ ° '	Venus ♀ ° '	Mars ♂ ° '	Jupiter ♃ ° '	Saturn ♄ ° '	Uranus ♅ ° '	Neptune ♆ ° '	Pluto ♇ ° '
01	22 35 30	10 ♓ 29 21	15 ♐ 29 46	22 ♐ 26 15	22 ♒ 57	19 ♓ 07	26 ♈ 00	05 ♏ 50	04 ♉ 51	07 ♎ 51	00 ♐ 53	26 ♍ 29
02	22 39 27	11 29 33	29 ♐ 28 07	06 ♑ 35 14	24 33	20 22	26 44	05 R 48	04 57	07 R 48	00 53	26 R 27
03	22 43 23	12 29 44	13 ♑ 47 21	21 ♑ 04 06	26 09	21 37	27 27	05 46	05 02	07 46	00 R 53	26 26
04	22 47 20	13 29 53	28 ♑ 24 57	05 ♒ 49 17	27 47	22 52	28 10	05 43	05 08	07 44	00 53	26 24
05	22 51 16	14 30 01	13 ♒ 16 17	20 ♒ 45 04	29 ♒ 26	24 07	28 53	05 41	05 14	07 41	00 53	26 22
06	22 55 13	15 30 06	28 ♒ 14 40	05 ♓ 44 01	01 ♓ 06	25 21	29 ♈ 36	05 38	05 20	07 39	00 53	26 21
07	22 59 09	16 30 10	13 ♓ 12 06	20 ♓ 37 51	02 47	26 36	00 ♉ 19	05 35	05 26	07 37	00 53	26 19
08	23 03 06	17 30 13	28 ♓ 00 17	05 ♈ 18 32	04 29	27 51	01 02	05 32	05 32	07 34	00 53	26 18
09	23 07 03	18 30 13	12 ♈ 31 50	19 ♈ 39 32	06 12	29 ♓ 06	01 45	05 29	05 38	07 32	00 53	26 16
10	23 10 59	19 30 11	26 ♈ 41 11	03 ♉ 36 27	07 57	00 ♈ 21	02 28	05 26	05 44	07 30	00 53	26 14
11	23 14 56	20 30 07	10 ♉ 25 10	17 ♉ 07 20	09 42	01 35	03 10	05 22	05 50	07 27	00 52	26 13
12	23 18 52	21 30 01	23 ♉ 43 04	00 ♊ 12 36	11 29	02 50	03 53	05 19	05 56	07 25	00 52	26 11
13	23 22 49	22 29 53	06 ♊ 36 17	12 ♊ 54 31	13 17	04 05	04 36	05 15	06 02	07 22	00 52	26 10
14	23 26 45	23 29 42	19 ♊ 07 47	25 ♊ 16 38	15 06	05 19	05 19	05 11	06 09	07 20	00 51	26 08
15	23 30 42	24 29 30	01 ♋ 21 37	07 ♋ 23 19	16 56	06 34	06 01	05 07	06 15	07 17	00 51	26 06
16	23 34 38	25 29 15	13 ♋ 22 20	19 ♋ 19 15	18 47	07 48	06 44	05 02	06 21	07 15	00 51	26 05
17	23 38 35	26 28 58	25 ♋ 14 40	01 ♌ 09 07	20 40	09 03	07 27	04 58	06 28	07 12	00 50	26 03
18	23 42 32	27 28 38	07 ♌ 03 09	12 ♌ 57 18	22 34	10 17	08 09	04 53	06 34	07 10	00 50	26 01
19	23 46 28	28 28 17	18 ♌ 52 01	24 ♌ 47 45	24 29	11 32	08 52	04 48	06 41	07 07	00 49	26 00
20	23 50 25	29 ♓ 27 53	00 ♍ 44 55	06 ♍ 43 53	26 25	12 46	09 34	04 43	06 48	07 05	00 49	25 58
21	23 54 21	00 ♈ 27 27	12 ♍ 44 57	18 ♍ 48 26	28 ♓ 22	14 01	10 17	04 38	06 54	07 02	00 48	25 57
22	23 58 18	01 26 59	24 ♍ 54 32	01 ♎ 03 30	00 ♈ 20	15 15	10 59	04 33	07 01	06 59	00 47	25 55
23	00 02 14	02 26 29	07 ♎ 15 28	13 ♎ 30 34	02 19	16 30	11 41	04 27	07 08	06 57	00 47	25 53
24	00 06 11	03 25 56	19 ♎ 48 54	26 ♎ 10 33	04 19	17 44	12 24	04 22	07 15	06 54	00 46	25 52
25	00 10 07	04 25 22	02 ♏ 35 34	09 ♏ 03 58	06 20	18 58	13 06	04 16	07 21	06 52	00 45	25 50
26	00 14 04	05 24 46	15 ♏ 35 45	22 ♏ 10 58	08 22	20 13	13 48	04 10	07 28	06 49	00 45	25 48
27	00 18 01	06 24 09	28 ♏ 49 36	05 ♐ 31 37	10 24	21 27	14 30	04 04	07 35	06 46	00 44	25 47
28	00 21 57	07 23 29	12 ♐ 17 03	19 ♐ 05 51	12 26	22 41	15 12	03 58	07 42	06 44	00 43	25 45
29	00 25 54	08 22 48	25 ♐ 58 00	02 ♑ 53 28	14 29	23 55	15 54	03 52	07 49	06 41	00 42	25 44
30	00 29 50	09 22 05	09 ♑ 52 11	16 ♑ 54 03	16 31	25 09	16 36	03 46	07 56	06 39	00 41	25 42
31	00 33 47	10 ♈ 21 20	23 ♑ 58 55	01 ♒ 06 38	18 ♈ 34	26 ♈ 24	17 ♉ 18	03 ♏ 39	08 ♉ 04	06 ♎ 36	00 ♐ 41	25 ♍ 41

Date	Moon True ☊ ° '	Moon Mean ☊ ° '	Moon ☽ Latitude ° '
01	11 ♓ 50	12 ♓ 08	05 S 17
02	11 R 50	12 05	05 03
03	11 D 50	12 02	04 30
04	11 50	11 59	03 39
05	11 50	11 55	02 32
06	11 50	11 52	01 S 15
07	11 R 50	11 49	00 N 08
08	11 50	11 46	01 29
09	11 50	11 43	02 42
10	11 49	11 39	03 44
11	11 48	11 36	04 31
12	11 47	11 33	05 02
13	11 46	11 30	05 16
14	11 45	11 27	05 14
15	11 D 46	11 24	04 59
16	11 46	11 20	04 30
17	11 47	11 17	03 50
18	11 49	11 14	03 01
19	11 50	11 11	02 04
20	11 51	11 08	01 N 01
21	11 R 52	11 05	00 S 05
22	11 51	11 01	01 11
23	11 49	10 58	02 16
24	11 47	10 55	03 14
25	11 43	10 52	04 04
26	11 40	10 49	04 42
27	11 36	10 45	05 06
28	11 33	10 42	05 14
29	11 32	10 39	05 04
30	11 D 31	10 36	04 37
31	11 ♓ 32	10 ♓ 33	03 S 52

DECLINATIONS

Date	Sun ☉ ° '	Moon ☽ ° '	Mercury ☿ ° '	Venus ♀ ° '	Mars ♂ ° '	Jupiter ♃ ° '	Saturn ♄ ° '	Uranus ♅ ° '	Neptune ♆ ° '	Pluto ♇ ° '
01	07 S 38	27 S 54	15 S 46	05 S 36	10 N 10	12 S 08	11 N 01	02 S 25	18 S 41	16 N 15
02	07 15	28 30	15 16	05 06	10 26	12 07	11 03	02 24	18 41	16 16
03	06 52	27 12	14 46	04 36	10 42	12 07	11 05	02 24	18 41	16 16
04	06 29	24 03	14 14	04 06	10 58	12 06	11 07	02 23	18 41	16 17
05	06 06	19 16	13 41	03 36	11 14	12 05	11 09	02 22	18 40	16 18
06	05 43	13 15	13 07	03 05	11 30	12 04	11 11	02 21	18 40	16 19
07	05 20	06 S 29	12 31	02 35	11 46	12 02	11 14	02 20	18 40	16 20
08	04 56	00 N 34	11 54	02 04	12 02	12 01	11 16	02 19	18 40	16 20
09	04 33	07 27	11 16	01 33	12 18	12 00	11 18	02 18	18 40	16 21
10	04 09	13 46	10 37	01 03	12 33	11 59	11 20	02 17	18 40	16 22
11	03 46	19 14	09 56	00 32	12 48	11 57	11 22	02 16	18 40	16 23
12	03 22	23 34	09 14	00 S 01	13 04	11 56	11 25	02 15	18 40	16 23
13	02 59	26 36	08 31	00 N 30	13 19	11 55	11 27	02 14	18 40	16 24
14	02 35	28 13	07 47	01 00	13 34	11 53	11 29	02 13	18 40	16 25
15	02 11	28 25	07 01	01 31	13 49	11 52	11 31	02 12	18 39	16 25
16	01 48	27 15	06 14	02 02	14 03	11 50	11 34	02 11	18 39	16 26
17	01 24	24 52	05 27	02 33	14 18	11 48	11 36	02 10	18 39	16 27
18	01 00	21 26	04 38	03 03	14 32	11 47	11 38	02 09	18 39	16 28
19	00 36	17 08	03 47	03 34	14 47	11 45	11 40	02 08	18 39	16 28
20	00 S 13	12 10	02 56	04 05	15 01	11 43	11 43	02 07	18 39	16 29
21	00 N 11	06 42	02 04	04 35	15 15	11 41	11 45	02 06	18 39	16 30
22	00 35	00 N 56	01 11	05 05	15 29	11 39	11 47	02 05	18 38	16 30
23	00 58	04 S 57	00 S 17	05 36	15 42	11 37	11 50	02 04	18 38	16 31
24	01 22	10 45	00 N 37	06 06	15 56	11 35	11 52	02 03	18 38	16 31
25	01 45	16 12	01 33	06 36	16 09	11 33	11 54	02 02	18 38	16 32
26	02 09	21 01	02 29	07 06	16 23	11 31	11 57	02 01	18 38	16 33
27	02 33	24 53	03 25	07 35	16 36	11 29	11 59	02 00	18 37	16 33
28	02 56	27 27	04 22	08 05	16 49	11 27	12 02	01 59	18 37	16 34
29	03 19	28 27	05 19	08 34	17 02	11 25	12 04	01 58	18 37	16 34
30	03 43	27 40	06 16	09 03	17 14	11 23	12 06	01 57	18 37	16 35
31	04 N 06	25 S 07	07 N 12	09 N 32	17 N 27	11 S 20	12 N 09	01 S 56	18 S 37	16 N 36

ZODIAC SIGN ENTRIES

Date	h	m	Planets
02	12	54	☽ ♑
04	14	34	☽ ♒
05	20	10	☿ ♓
06	14	49	☽ ♓
07	01	28	♂ ♉
08	15	16	☽ ♈
10	05	25	♀ ♈
10	17	43	☽ ♉
12	23	37	☽ ♊
15	09	18	☽ ♋
17	21	40	☽ ♌
20	10	30	☽ ♍
21	00	56	☉ ♈
22	07	59	☿ ♈
22	21	56	☽ ♎
25	07	10	☽ ♏
27	14	07	☽ ♐
29	19	00	☽ ♑
31	22	08	☽ ♒

LATITUDES

Date	Mercury ☿ ° '	Venus ♀ ° '	Mars ♂ ° '	Jupiter ♃ ° '	Saturn ♄ ° '	Uranus ♅ ° '	Neptune ♆ ° '	Pluto ♇ ° '
01	02 S 00	01 S 25	00 N 07	01 N 25	02 S 15	00 N 45	01 N 42	16 N 13
04	02 07	01 23	00 10	01 25	02 15	00 45	01 42	16 14
07	02 11	01 20	00 12	01 26	02 14	00 45	01 42	16 14
10	02 11	01 17	00 14	01 26	02 13	00 45	01 42	16 15
13	02 06	01 14	00 16	01 27	02 13	00 45	01 43	16 15
16	01 58	01 09	00 18	01 27	02 12	00 45	01 43	16 15
19	01 44	01 05	00 20	01 28	02 12	00 45	01 43	16 15
22	01 26	01 00	00 22	01 28	02 11	00 45	01 43	16 16
25	01 03	00 54	00 24	01 29	02 11	00 45	01 43	16 15
28	00 36	00 48	00 26	01 29	02 10	00 45	01 43	16 15
31	00 S 05	00 S 42	00 N 28	01 N 29	02 S 10	00 N 45	01 N 43	16 N 15

DATA

Julian Date	2440647
Delta T	+40 seconds
Ayanamsa	23° 26' 31"
Synetic vernal point	05° ♓ 40' 29"
True obliquity of ecliptic	23° 26' 44"

MOON'S PHASES, APSIDES AND POSITIONS ☽

Date	h	m	Phase	Longitude ° '	Eclipse Indicator
01	02	33	☾	10 ♐ 06	
07	17	42	●	16 ♓ 44	Total
14	21	16	☽	23 ♊ 53	
23	01	52	○	02 ♎ 01	
30	11	05	☾	09 ♑ 20	

Day	h	m		
06	09	25	Perigee	
18	11	34	Apogee	
02	07	42	Max dec	28° S 31'
08	10	05	0N	
15	03	18	Max dec	28° N 31'
22	15	48	0S	
29	13	40	Max dec	28° S 27'

LONGITUDES

Date	Chiron ⚷ ° '	Ceres ⚳ ° '	Pallas ⚴ ° '	Juno ⚵ ° '	Vesta ⚶ ° '	Black Moon Lilith ⚸ ° '
01	04 ♈ 48	13 ♓ 48	18 ♒ 50	01 ♓ 04	13 ♌ 49	09 ♌ 23
11	05 ♈ 22	17 ♓ 45	22 ♒ 07	05 ♓ 42	12 ♌ 09	10 ♌ 30
21	05 ♈ 57	21 ♓ 41	25 ♒ 18	10 ♓ 23	11 ♌ 15	11 ♌ 36
31	06 ♈ 32	25 ♓ 35	28 ♒ 21	15 ♓ 06	11 ♌ 08	12 ♌ 43

All ephemeris data is given at 12.00 UT and the Moon's longitude is additionally given for 24.00 UT

ASPECTARIAN

h m	Aspects
01 Sunday	
02 33	☽ □ ☉
03 01	☽ Q ☿
03 43	☽ ꝑ ♂
03 50	☽ ± ♄
05 36	☽ ⊥ ♃
10 23	☿ ꝑ ♅
10 26	☿ Q ♄
18 54	☽ □ ♀
19 31	☽ Q ♅
19 36	☽ ꝑ ♄
21 12	☽ ∠ ♃
02 Monday	
02 32	☽ ✶ ☿
03 13	♂ ⊼ ♇
03 14	♀ ∠ ♄
06 53	☽ □ ♇
07 05	☽ △ ♂
12 03	☽ Q ☉
14 25	☽ ⊻ ♆
19 59	♀ ꝑ ♃
21 18	☽ △ ♄
22 39	☽ ✶ ♃
03 Tuesday	
00 31	☽ ⊥ ♆
02 00	☽ □ ♅
04 14	♀ ⊥ ♂
04 24	☽ Q ♀
07 05	☽ ∠ ☿
07 40	☉ ∠ ♂
09 01	♆ St R
09 42	☽ ✶ ☉
15 29	☽ ∠ ♆
15 55	☿ ⊼ ♇
18 33	☽ Q ♃
23 49	☽ ⊥ ☿
04 Wednesday	
02 06	☽ ✶ ♀
08 43	☽ △ ♇
10 51	☽ ⊻ ☿
11 34	☽ □ ♂
12 09	☽ ∠ ☉
16 01	☽ ✶ ♆
21 45	☿ ✶ ♂
22 58	☽ □ ♄
23 48	☽ □ ♃
05 Thursday	
03 02	☽ △ ♅
03 10	♂ ∥ ♄
03 46	☽ ⊥ ☉
04 42	☽ ∠ ♀
08 57	☽ ꝑ ♇
11 23	☽ Q ♆
14 07	☽ ⊻ ☉
14 35	☽ ∥ ♆
18 05	☽ Q ♂
20 28	☽ ⊥ ♀
23 23	☽ ± ♇
06 Friday	
00 20	☽ # ♇
03 04	☽ ꝑ ♅
04 05	☽ Q ♄
06 57	☽ ⊻ ♀
08 58	☿ □ ♆
08 58	☽ ⊼ ♇
12 35	☽ ∥ ☿
14 17	☽ ✶ ♂
16 14	☽ □ ♆
16 25	☽ ∥ ♃
17 09	☽ ☌ ☿
17 27	☽ ± ♅
18 09	☽ # ♂
19 29	☽ # ♄
19 43	☿ ± ♅
23 25	☽ ✶ ♄
23 49	☽ △ ♃
07 Saturday	
03 02	☽ ⊼ ♅
06 40	♀ ☍ ♇
15 35	☽ ∠ ♂
16 12	☽ ∥ ☉
17 42	☽ ☌ ☉
23 45	☽ ∠ ♄
23 54	☽ ꝑ ♃
23 59	♀ ∥ ♅
08 Sunday	
02 12	☽ ∥ ♅
02 22	☽ ∥ ♀
06 54	☽ ⊥ ♂
07 06	♂ ⊼ ♆
07 23	☿ ∥ ♃
08 32	☿ # ♂
09 13	☽ ☍ ♇
11 00	♂ # ♃
11 43	☽ ☌ ♀
14 26	♃ ☍ ♄
14 30	☽ ± ♃
14 30	☽ ⊥ ♄
16 43	☽ △ ♆
16 48	☽ # ♀
17 13	☽ ⊻ ♂
18 00	☽ # ♅
09 Monday	
00 04	☽ ⊻ ☿
00 20	☽ ⊼ ♃
00 27	☽ ⊻ ♄
02 18	☽ # ☉
02 18	☿ △ ♃
03 25	☿ ✶ ♄
03 42	☽ ☍ ♅
10 56	☿ # ♄
11 23	☽ ⊥ ☿
17 37	☽ ꝑ ♆
19 13	♀ ± ♃
22 48	☽ ⊻ ☉
23 06	♀ ⊥ ♄
10 Tuesday	
00 50	☽ # ☿
02 22	☽ ∥ ♄
04 41	☽ ∠ ☿
04 53	♂ ± ♇
04 56	☽ # ♃
05 55	☿ ⊼ ♅
06 55	☽ ∥ ♂
08 54	☽ ± ♆
09 49	☽ ⊥ ☉
11 14	☽ ⊼ ♇
18 56	☽ ⊻ ♀
19 14	☽ ⊼ ♆
21 36	☽ ± ♇
22 16	♀ △ ♆
22 33	☽ ☌ ♂
22 52	☽ ∥ ♇
11 Wednesday	
02 38	☽ ∠ ☉
03 08	☽ ☍ ♃
03 50	☽ ☌ ♄
06 29	☽ ⊥ ♀
06 46	☽ ⊼ ♅
09 04	☉ ꝑ ♃
09 17	☽ # ♆
10 33	☽ ✶ ☿
13 25	☽ ꝑ ♇
17 23	☽ ± ♅
20 46	☉ ∠ ♄
12 Thursday	
00 10	☽ ∠ ♀
07 37	☽ ✶ ☉
09 37	☽ ꝑ ♅
11 30	☽ Q ☿
16 32	☽ △ ♇
13 Friday	
01 13	☽ ☍ ♆
06 43	☽ ✶ ♀
07 41	☽ Q ☉
08 00	☽ ⊻ ♂
09 27	☽ ⊼ ♃
10 55	☽ ⊻ ♄
13 27	☽ △ ♅
20 03	☽ ⊥ ♂
20 46	☽ ± ♃
22 25	☽ ⊥ ♄
14 Saturday	
02 52	☽ □ ☿
07 56	♂ ☍ ♃
08 06	☽ Q ♀
09 26	♀ ⊼ ♃
11 40	♀ ⊻ ♂
14 02	☽ ꝑ ♃
14 26	☽ ∠ ♂
15 57	☽ ∠ ♄
21 16	☽ □ ☉
15 Sunday	
01 39	☽ □ ♇
05 21	♀ ⊻ ♄
10 59	☽ ⊼ ♆
11 11	☉ ∥ ♅
19 25	☽ △ ♃
20 57	♂ ☌ ♄
21 49	☽ ✶ ♄
21 51	☽ ✶ ♂
22 55	☽ ± ♆
23 33	☽ □ ♀
23 45	☽ □ ♅
16 Monday	
01 32	♀ ☍ ♅
05 39	☉ # ♀
13 25	☽ Q ♇
16 58	☽ ꝑ ♆
18 50	♀ # ♅
22 08	☽ Q ♄
23 30	☽ Q ♂
17 Tuesday	
00 58	☽ △ ☿
01 52	☉ ☍ ♇
03 23	☿ ꝑ ♃
04 20	♂ ⊼ ♅
11 55	☽ Q ♅
13 38	☽ ✶ ♇
14 45	☽ △ ☉
22 46	☿ ∠ ♄
23 21	☽ △ ♆
18 Wednesday	
07 37	☽ □ ♃
11 01	☽ □ ♄
12 13	☽ ✶ ♅
13 14	☽ ꝑ ☿
14 23	☽ □ ♂
19 21	☽ △ ♀
20 03	☽ ∠ ♇
23 50	☿ ∠ ♂
19 Thursday	
00 03	☽ ꝑ ☉
04 00	☽ # ♆
11 03	☽ ± ☿
14 17	☽ ⊥ ♇
15 22	☽ ∥ ♇
15 58	☿ # ♀
18 34	☽ ∠ ♅
19 25	☉ ± ♃
19 55	☽ Q ♃
19 58	☽ ± ☉
23 12	☽ ∥ ♂
20 Friday	
01 33	☽ ⊼ ☿
02 23	☽ ⊻ ♇
05 19	☽ ꝑ ♀
06 38	☿ ☍ ♇
09 11	☽ ⊼ ☉
12 07	☽ □ ♆
12 39	☽ ⊥ ♅
13 49	♃ # ♄
14 02	☽ ∥ ♄
14 02	☽ # ♃
19 55	☽ ✶ ♃
21 Saturday	
00 14	☽ △ ♄
00 39	☽ ⊻ ♅
01 29	☽ ± ♀
06 46	☽ △ ♂
11 10	☿ ∥ ♅
14 48	☽ ⊼ ♀
15 13	☿ ± ♃
20 11	☽ ∥ ♀
20 12	☉ △ ♆
23 59	☽ Q ♆
22 Sunday	
00 11	☉ ⊥ ♄
01 32	☽ ∠ ♃
06 16	☽ ꝑ ♄
07 15	☽ # ♅
07 57	♄ ⊼ ♅
09 48	♂ ꝑ ♇
10 46	☽ # ☿
13 21	☽ ∥ ☉
13 58	☽ ☌ ♇
14 14	☽ ꝑ ♂
17 32	☿ △ ♆
18 36	☽ # ☉
19 04	☽ ⊥ ♃
19 30	☽ ∥ ☿
20 48	☿ ⊥ ♄
22 18	♀ ꝑ ♆
23 21	☉ # ☿
23 28	☽ ✶ ♆
23 Monday	
00 02	☽ ± ♄
00 15	☽ ∥ ♅
00 37	☽ ☍ ☿
01 52	☽ ☍ ☉
06 38	☽ ⊻ ♃
08 47	☽ ± ♂
11 24	☽ ☌ ♅
11 45	☽ ⊼ ♄
14 51	☽ # ♀
14 58	☉ ☌ ☿
21 01	☽ ⊼ ♂
24 Tuesday	
04 19	☽ ∠ ♆
07 37	☽ ☍ ♀
12 32	☿ ⊼ ♃
15 34	☽ ∥ ♃
16 49	☽ # ♄
21 21	☽ ⊥ ♆
23 23	☽ ⊻ ♇
25 Wednesday	
04 24	♂ ± ♅
08 35	☽ ⊻ ♆
08 37	☉ ⊼ ♃
10 35	☽ ⊥ ♇
11 50	☽ # ♂
13 36	☽ # ♇
15 06	☽ ☌ ♃
15 41	☽ ⊼ ☉
18 07	☿ ☍ ♅
19 54	☽ ⊻ ♅
20 14	☽ ⊼ ☿
20 55	☽ ☍ ♄
21 28	☉ ∥ ☿
23 39	☽ ∥ ♆
26 Thursday	
00 19	☿ # ♅
00 52	☿ ⊻ ♄
01 52	☿ ⊥ ♂
03 14	☽ ∠ ♇
03 42	☽ ± ☉
04 03	☉ # ♅
06 55	☽ ⊥ ♅
08 31	☽ ☍ ♂
09 19	☽ ± ☿
21 17	☽ ⊼ ♀
21 30	☽ ꝑ ☉
23 18	☽ ∠ ♅
27 Friday	
04 42	☽ ꝑ ☿
06 31	☽ ✶ ♇
07 00	♂ ∥ ♇
09 16	☽ ± ♀
15 25	☽ ☌ ♆
20 38	☉ ☍ ♅
21 20	☽ ⊻ ♃
28 Saturday	
02 11	☽ ✶ ♅
02 38	☽ △ ☉
03 01	☽ ꝑ ♀
03 48	☽ ⊼ ♄
03 59	☽ Q ♇
07 56	☽ ⊥ ♃
12 19	☽ △ ☿
14 32	☽ ± ♄
17 26	☽ ⊼ ♂
20 38	☉ ⊻ ♄
23 19	☽ Q ♅
23 42	☽ ∠ ♃
29 Sunday	
04 32	☽ ± ♂
06 28	☽ ꝑ ♄
08 05	☽ △ ♀
11 35	☽ □ ♇
20 13	☽ ⊻ ♆
21 02	☽ ꝑ ♂
30 Monday	
01 35	☽ ✶ ♃
02 17	☿ ꝑ ♆
03 03	♀ ± ♆
06 29	☽ □ ♅
06 33	☽ ⊥ ♆
08 40	☽ △ ♄
11 05	☽ □ ☉
13 32	☿ ⊻ ♂
21 56	☽ ∠ ♆
21 59	☽ Q ♃
22 21	♀ ⊼ ♇
31 Tuesday	
00 06	☽ △ ♂
01 16	☽ □ ☿
14 51	☽ △ ♇
16 27	☽ □ ♀
19 55	☽ Q ☉
23 16	☽ ✶ ♆

APRIL 1970

LONGITUDES

Date	Sidereal time h m s	Sun ☉ ° ' "	Moon ☽ ° ' "	Moon ☽ 24.00 ° ' "	Mercury ☿ ° '	Venus ♀ ° '	Mars ♂ ° '	Jupiter ♃ ° '	Saturn ♄ ° '	Uranus ♅ ° '	Neptune ♆ ° '	Pluto ♇ ° '
01	00 37 43	11 ♈ 20 34	08 ♒ 16 55	15 ♒ 29 27	20 ♈ 35	27 ♈ 38	18 ♉ 00	03 ♏ 33	08 ♉ 11	06 ♎ 33	00 ♐ 40	25 ♍ 39
02	00 41 40	12 19 46	22 ♒ 43 52	29 ♒ 59 40	22 35	28 ♈ 52	18 42	03 R 26	08 18	06 R 31	00 R 39	25 R 37
03	00 45 36	13 18 56	07 ♓ 16 19	14 ♓ 33 10	24 35	00 ♉ 06	19 24	03 20	08 25	06 28	00 38	25 36
04	00 49 33	14 18 04	21 ♓ 49 33	28 ♓ 04 45	26 32	01 20	20 06	03 13	08 32	06 26	00 37	25 34
05	00 53 30	15 17 10	06 ♈ 18 00	13 ♈ 28 34	28 ♈ 27	02 34	20 48	03 06	08 40	06 23	00 36	25 33
06	00 57 26	16 16 14	20 ♈ 35 45	27 ♈ 38 53	00 ♉ 20	03 48	21 30	02 59	08 47	06 21	00 35	25 31
07	01 01 23	17 15 16	04 ♉ 37 24	11 ♉ 30 50	02 10	05 02	22 11	02 52	08 54	06 18	00 34	25 30
08	01 05 19	18 14 16	18 ♉ 18 49	25 ♉ 01 08	03 57	06 16	22 53	02 44	09 02	06 15	00 33	25 28
09	01 09 16	19 13 14	01 ♊ 37 40	08 ♊ 08 27	05 40	07 30	23 35	02 37	09 09	06 13	00 32	25 27
10	01 13 12	20 12 10	14 ♊ 33 36	20 ♊ 53 24	07 20	08 44	24 16	02 30	09 16	06 10	00 30	25 25
11	01 17 09	21 11 03	27 ♊ 08 11	03 ♋ 18 23	08 55	09 57	24 58	02 23	09 24	06 08	00 29	25 24
12	01 21 05	22 09 54	09 ♋ 24 29	15 ♋ 27 03	10 25	11 11	25 40	02 15	09 31	06 06	00 28	25 23
13	01 25 02	23 08 43	21 ♋ 26 40	27 ♋ 23 59	11 52	12 25	26 21	02 08	09 39	06 03	00 27	25 21
14	01 28 59	24 07 30	03 ♌ 19 37	09 ♌ 14 14	13 13	13 39	27 03	02 00	09 46	06 01	00 26	25 20
15	01 32 55	25 06 14	15 ♌ 08 29	21 ♌ 02 59	14 29	14 52	27 44	01 53	09 54	05 58	00 25	25 18
16	01 36 52	26 04 57	26 ♌ 58 22	02 ♍ 55 13	15 40	16 06	28 25	01 45	10 01	05 56	00 23	25 17
17	01 40 48	27 03 37	08 ♍ 54 06	14 ♍ 55 31	16 45	17 20	29 07	01 37	10 09	05 53	00 22	25 16
18	01 44 45	28 02 14	20 ♍ 59 55	27 ♍ 07 42	17 45	18 33	29 ♉ 48	01 30	10 17	05 51	00 21	25 14
19	01 48 41	29 00 50	03 ♎ 19 13	09 ♎ 34 44	18 40	19 47	00 ♊ 29	01 22	10 24	05 49	00 19	25 13
20	01 52 38	29 ♈ 59 24	15 ♎ 54 24	22 ♎ 18 21	19 28	21 00	01 10	01 15	10 32	05 46	00 18	25 12
21	01 56 34	00 ♉ 57 55	28 ♎ 46 35	05 ♏ 19 04	20 11	22 14	01 52	01 07	10 39	05 44	00 17	25 11
22	02 00 31	01 56 25	11 ♏ 55 40	18 ♏ 36 11	20 48	23 27	02 33	00 59	10 47	05 42	00 15	25 09
23	02 04 28	02 54 53	25 ♏ 20 21	02 ♐ 07 53	21 20	24 41	03 14	00 52	10 55	05 40	00 14	25 08
24	02 08 24	03 53 19	08 ♐ 58 25	15 ♐ 51 39	21 45	25 54	03 55	00 44	11 02	05 37	00 13	25 07
25	02 12 21	04 51 44	22 ♐ 47 12	29 ♐ 44 44	22 05	27 07	04 36	00 36	11 10	05 35	00 11	25 06
26	02 16 17	05 50 07	06 ♑ 43 57	13 ♑ 44 33	22 19	28 21	05 17	00 29	11 18	05 33	00 10	25 05
27	02 20 14	06 48 29	20 ♑ 46 17	27 ♑ 48 57	22 27	29 ♉ 34	05 58	00 21	11 25	05 31	00 08	25 03
28	02 24 10	07 46 48	04 ♒ 52 22	11 ♒ 56 22	22 R 29	00 ♊ 47	06 39	00 13	11 33	05 29	00 07	25 02
29	02 28 07	08 45 07	19 ♒ 00 48	26 ♒ 05 31	22 26	02 00	07 20	00 ♏ 06	11 41	05 27	00 05	25 01
30	02 32 03	09 ♉ 43 24	03 ♓ 10 21	10 ♓ 15 08	22 ♉ 18	03 ♊ 13	08 ♊ 00	29 ♎ 58	11 ♉ 48	05 ♎ 25	00 ♐ 04	25 ♍ 00

Date	Moon True ☊ ° '	Moon Mean ☊ ° '	Moon ☽ Latitude ° '
01	11 ♓ 33	10 ♓ 30	02 S 52
02	11 D 34	10 26	01 42
03	11 35	10 23	00 S 24
04	11 R 35	10 20	00 N 56
05	11 33	10 17	02 11
06	11 29	10 14	03 17
07	11 24	10 11	04 09
08	11 18	10 07	04 46
09	11 12	10 04	05 06
10	11 07	10 01	05 10
11	11 02	09 58	04 59
12	11 00	09 55	04 34
13	10 59	09 51	03 57
14	10 D 59	09 48	03 10
15	11 00	09 45	02 16
16	11 02	09 42	01 16
17	11 03	09 39	00 N 12
18	11 R 02	09 36	00 S 54
19	11 00	09 32	01 57
20	10 55	09 29	02 57
21	10 48	09 26	03 48
22	10 40	09 23	04 29
23	10 31	09 20	04 55
24	10 23	09 17	05 06
25	10 15	09 13	04 59
26	10 10	09 10	04 34
27	10 07	09 07	03 53
28	10 05	09 04	02 57
29	10 D 06	09 01	01 50
30	10 ♓ 06	08 ♓ 57	00 S 37

DECLINATIONS

Date	Sun ☉ ° '	Moon ☽ ° '	Mercury ☿ ° '	Venus ♀ ° '	Mars ♂ ° '	Jupiter ♃ ° '	Saturn ♄ ° '	Uranus ♅ ° '	Neptune ♆ ° '	Pluto ♇ ° '
01	04 N 29	20 S 58	08 N 08	10 N 01	17 N 39	11 S 18	12 N 11	01 S 55	18 S 36	16 N 36
02	04 52	15 32	09 04	10 29	17 52	11 16	12 14	01 54	18 36	16 37
03	05 15	09 13	09 59	10 58	18 04	11 13	12 16	01 53	18 36	16 37
04	05 38	02 S 23	10 52	11 26	18 15	11 11	12 18	01 52	18 36	16 38
05	06 01	04 N 30	11 45	11 53	18 27	11 09	12 21	01 51	18 35	16 38
06	06 24	11 05	12 35	12 21	18 39	11 06	12 23	01 50	18 35	16 39
07	06 47	16 59	13 24	12 48	18 50	11 04	12 26	01 49	18 35	16 39
08	07 09	21 52	14 11	13 15	19 01	11 01	12 28	01 48	18 35	16 40
09	07 32	25 29	14 56	13 41	19 12	10 59	12 31	01 47	18 34	16 40
10	07 54	27 41	15 39	14 07	19 23	10 56	12 33	01 46	18 34	16 40
11	08 16	28 23	16 19	14 33	19 34	10 54	12 35	01 45	18 34	16 41
12	08 38	27 40	16 57	14 59	19 44	10 51	12 38	01 44	18 34	16 41
13	09 00	25 38	17 32	15 24	19 55	10 49	12 40	01 43	18 33	16 42
14	09 22	22 30	18 05	15 49	20 05	10 46	12 43	01 42	18 33	16 42
15	09 43	18 28	18 35	16 13	20 15	10 43	12 45	01 41	18 33	16 42
16	10 04	13 42	19 02	16 37	20 25	10 41	12 48	01 40	18 32	16 43
17	10 26	08 25	19 26	17 00	20 34	10 38	12 50	01 39	18 32	16 43
18	10 47	02 N 45	19 47	17 23	20 44	10 36	12 53	01 38	18 32	16 43
19	11 08	03 S 07	20 06	17 46	20 53	10 33	12 55	01 37	18 32	16 44
20	11 28	08 59	20 22	18 08	21 02	10 30	12 57	01 36	18 31	16 44
21	11 49	14 36	20 34	18 30	21 11	10 28	13 00	01 36	18 31	16 44
22	12 09	19 41	20 45	18 51	21 19	10 25	13 02	01 35	18 31	16 44
23	12 29	23 53	20 52	19 12	21 28	10 22	13 05	01 34	18 30	16 45
24	12 49	26 50	20 56	19 32	21 36	10 20	13 07	01 33	18 30	16 45
25	13 09	28 13	20 58	19 52	21 44	10 17	13 09	01 32	18 30	16 45
26	13 28	27 50	20 57	20 11	21 52	10 15	13 12	01 31	18 29	16 45
27	13 47	25 40	20 53	20 29	22 00	10 12	13 14	01 30	18 29	16 46
28	14 06	21 55	20 47	20 47	22 07	10 10	13 17	01 30	18 29	16 46
29	14 25	16 53	20 37	21 05	22 15	10 07	13 19	01 29	18 28	16 46
30	14 N 44	10 S 55	20 N 26	21 N 22	22 N 22	10 S 04	13 N 21	01 S 28	18 S 28	16 N 46

ZODIAC SIGN ENTRIES

Date	h	m	Planets
03	00	01	☽ ♓
03	10	05	♀ ♉
05	01	32	☽ ♈
06	07	40	☿ ♉
07	04	02	☽ ♉
09	09	02	☽ ♊
11	17	33	☽ ♋
14	05	16	☽ ♌
16	18	07	☽ ♍
18	18	59	♂ ♊
19	05	35	☽ ♎
20	12	15	☉ ♉
21	14	15	☽ ♏
23	20	15	☽ ♐
26	00	26	☽ ♑
27	20	33	♀ ♊
28	03	43	☽ ♒
30	06	37	☽ ♓
30	06	43	♃ ♎

LATITUDES

Date	Mercury ☿ ° '	Venus ♀ ° '	Mars ♂ ° '	Jupiter ♃ ° '	Saturn ♄ ° '	Uranus ♅ ° '	Neptune ♆ ° '	Pluto ♇ ° '
01	00 N 06	00 S 40	00 N 28	01 N 29	02 S 10	00 N 45	01 N 44	16 N 15
04	00 41	00 33	00 30	01 30	02 10	00 45	01 44	16 15
07	01 15	00 26	00 32	01 30	02 10	00 45	01 44	16 14
10	01 47	00 18	00 34	01 30	02 09	00 45	01 44	16 14
13	02 15	00 11	00 35	01 30	02 09	00 45	01 44	16 13
16	02 36	00 S 03	00 37	01 30	02 08	00 45	01 44	16 13
19	02 50	00 N 05	00 38	01 30	02 08	00 45	01 44	16 12
22	02 53	00 13	00 40	01 30	02 08	00 45	01 44	16 11
25	02 46	00 21	00 41	01 30	02 08	00 45	01 44	16 11
28	02 28	00 29	00 43	01 29	02 08	00 45	01 44	16 10
31	01 N 58	00 N 37	00 N 44	01 N 29	02 S 08	00 N 45	01 N 44	16 N 09

DATA

Julian Date	2440678
Delta T	+41 seconds
Ayanamsa	23° 26' 35"
Synetic vernal point	05° ♓ 40' 25"
True obliquity of ecliptic	23° 26' 44"

MOON'S PHASES, APSIDES AND POSITIONS ☽

Date	h	m	Phase	Longitude ° '	Eclipse Indicator
06	04	09	●	15 ♈ 57	
13	15	44	☽	23 ♋ 18	
21	16	21	○	01 ♏ 09	
28	17	18	☾	08 ♒ 00	

Day	h	m		
03	10	38	Perigee	
15	06	17	Apogee	
30	04	04	Perigee	
04	20	16	ON	
11	11	31	Max dec	28° N 23'
18	23	18	OS	
25	18	54	Max dec	28° S 18'

LONGITUDES

Date	Chiron ⚷ ° '	Ceres ⚳ ° '	Pallas ⚴ ° '	Juno ⚵ ° '	Vesta ⚶ ° '	Black Moon Lilith ⚸ ° '
01	06 ♈ 36	25 ♓ 59	28 ♒ 39	15 ♓ 34	11 ♌ 10	12 ♌ 50
11	07 ♈ 11	29 ♓ 51	01 ♓ 32	20 ♓ 18	11 ♌ 53	13 ♌ 56
21	07 ♈ 44	03 ♈ 40	04 ♓ 13	25 ♓ 02	13 ♌ 17	15 ♌ 03
31	08 ♈ 16	07 ♈ 25	06 ♓ 41	29 ♓ 45	15 ♌ 17	16 ♌ 09

All ephemeris data is given at 12.00 UT and the Moon's longitude is additionally given for 24.00 UT

ASPECTARIAN

h m	Aspects
01 Wednesday	
04 09	☽ □ ♃
09 08	☽ △ ♅
11 49	☽ □ ♄
12 35	☽ Q ☿
15 56	☽ ⚼ ♇
17 29	☽ ✶ ☉
19 17	☽ Q ♆
23 04	☽ ∥ ♆
02 Thursday	
01 22	☽ Q ♀
02 41	☽ # ♂
05 00	☽ □ ♂
06 52	☽ ± ♇
07 38	☽ # ♇
10 00	☽ ⚼ ♅
11 44	☽ ✶ ☿
16 46	☽ ⚻ ♇
17 57	☽ Q ♄
20 09	☽ ∠ ☉
23 04	☽ ✶ ♀
03 Friday	
00 48	☽ # ♄
00 49	☽ ± ♅
01 04	☽ □ ♆
04 35	☽ ∥ ♃
05 33	☽ △ ♃
06 02	☽ # ♀
09 33	☽ # ☿
10 41	☽ ⚻ ♅
12 05	☽ ⊥ ☉
12 14	☽ Q ♂
12 39	☿ ± ♆
13 54	☽ ✶ ♄
16 24	☽ ∠ ☿
19 28	☉ ⊥ ♂
22 13	♀ ⚻ ♆
22 41	☽ ⊻ ☉
04 Saturday	
00 19	☿ ⚻ ♇
00 27	♀ # ♃
01 17	☽ # ☉
02 06	☽ ∠ ♀
06 05	☽ ⚼ ♃
09 01	☽ ✶ ♂
09 32	☽ ⊥ ☿
13 50	☽ ∥ ♅
14 51	☽ ∠ ♄
16 34	♀ ± ♇
18 11	☽ ☍ ♇
18 20	☽ ⊥ ♀
20 10	☿ # ♃
20 50	☽ ± ♃
20 59	☽ ⊻ ☿
05 Sunday	
02 32	☽ △ ♆
02 42	☽ # ♅
05 12	☽ ⊻ ♀
05 53	☽ ⊥ ♄
06 43	☽ ⚻ ♃
11 07	☽ ∠ ♂
12 09	☽ ☍ ♅
15 58	☽ ⊻ ♄
17 42	☽ ∥ ☉
19 28	☉ ⚼ ♆
20 52	☿ ∥ ♀
21 25	♀ ☍ ♃
06 Monday	
02 57	☽ ⊥ ♂
03 33	☽ ⚼ ♆
04 09	☽ ☌ ☉
04 45	♂ # ♆
05 55	☿ ∥ ♄
07 02	♂ ⚼ ♅
12 05	☽ # ♃
13 36	☽ ⊻ ♂
14 23	♀ ∥ ♄
15 08	☿ ⚻ ♆
17 04	☽ ∥ ♄
17 16	☽ ∥ ♀
18 44	☽ ∥ ☿
18 45	☽ ± ♆
20 21	☽ ⚻ ♇
07 Tuesday	
03 14	☿ ± ♇
05 00	☽ ⚻ ♆
06 37	☽ ± ♇
07 08	☽ ☌ ☿
08 59	☽ ☍ ♃
10 34	☽ ∥ ♇
12 47	☽ ☌ ♀
14 54	☽ ⚻ ♅
19 17	☽ # ♆
19 30	☽ ☌ ♄
20 39	☿ ☍ ♃
20 49	☽ ∥ ♂
22 12	☽ ⚼ ♇
08 Wednesday	
01 20	☽ ± ♅
11 51	☽ ⊻ ☉
11 53	♀ ⚻ ♅
17 14	☽ ⚼ ♅
20 37	☽ ☌ ♂
23 26	☽ ⊥ ☉
09 Thursday	
00 48	☽ △ ♇
10 00	☽ ☍ ♆
13 48	☽ ⚻ ♃
17 09	☽ ∠ ☉
19 38	☿ ⚻ ♅
20 25	☽ △ ♅
20 32	☽ ⊻ ☿
23 56	☽ ⊻ ♀
10 Friday	
00 46	☽ ± ♃
02 01	☽ ⊻ ♄
09 20	☽ ⊥ ☿
12 21	☽ ⊥ ♀
13 21	☽ ⊥ ♄
17 30	☽ ⚼ ♃
23 36	☽ ✶ ☉
23 51	♀ ☌ ♄
11 Saturday	
04 54	☽ ∠ ☿
06 40	☽ ∠ ♄
07 20	☽ ∠ ♀
07 34	☽ ⊻ ♂
08 39	☽ □ ♇
18 30	☽ ⚻ ♆
19 53	☽ ⊥ ♂
20 15	☿ ☌ ♄
20 27	♀ ⚼ ♇
22 05	☽ △ ♃
12 Sunday	
00 47	☽ Q ☉
01 35	☿ ∥ ♇
02 27	♂ △ ♇
05 29	☽ □ ♅
06 13	☽ ± ♆
11 13	☿ ⚼ ♇
12 14	☽ ✶ ♄
14 17	☽ ✶ ☿
14 38	☽ ∠ ♂
15 55	☽ ✶ ♀
19 51	☽ Q ♇
13 Monday	
00 01	☽ ⚼ ♆
05 05	♀ ± ♅
12 25	☽ Q ♄
15 14	☿ ± ♅
15 44	☽ □ ☉
17 14	☽ Q ♅
17 30	☽ Q ☿
18 40	☽ Q ♀
19 51	☽ ✶ ♇
22 30	☽ ✶ ♂
14 Tuesday	
06 08	☽ △ ♆
09 21	☽ □ ♃
17 26	☽ ✶ ♅
19 19	☉ ± ♆
15 Wednesday	
00 20	☽ Q ♂
01 14	☽ □ ♄
02 11	☽ ∠ ♇
02 21	☽ ∥ ♂
10 26	☿ # ♆
10 30	☽ □ ☿
11 24	☽ □ ♀
11 26	☽ ∥ ☿
11 32	☽ # ♆
16 51	☉ ⚻ ♇
20 27	☽ ⊥ ♇
21 14	☽ ∥ ♇
21 31	☽ Q ♃
22 49	☽ ∥ ♀
23 48	☽ ∠ ♅
16 Thursday	
08 35	☽ ⊻ ♇
10 02	☽ △ ☉
15 07	☽ □ ♂
16 16	☽ ∥ ♄
17 57	☽ ⊥ ♅
18 07	♀ ∥ ♇
18 53	☽ □ ♆
21 33	☽ ✶ ♃
17 Friday	
02 05	☽ # ♃
03 39	☽ ∥ ☉
05 59	☽ ⊻ ♅
14 31	☽ △ ♄
18 52	☽ ⚼ ☉
18 Saturday	
00 39	☉ # ♃
03 13	☽ ∠ ♃
05 03	☽ △ ☿
06 03	♄ ⚼ ♇
06 39	☽ △ ♀
06 47	☽ Q ♆
14 13	☽ ± ☉
16 36	☽ # ♅
20 18	☽ ☌ ♇
20 28	☽ ⚼ ♄
20 44	☽ ⊥ ♃
19 Sunday	
02 57	☽ ⚻ ☉
05 56	☽ ∥ ♅
06 12	☽ △ ♂
06 13	☽ ✶ ♆
06 27	♂ ☍ ♆
08 16	☽ ⊻ ♃
12 42	☽ ⚼ ☿
14 06	☽ ± ♄
15 07	☽ ⚼ ♀
16 47	☽ ☌ ♅
20 Monday	
01 42	☽ ⚻ ♄
07 06	☽ ± ☿
07 36	♀ ⚼ ♅
10 07	☽ ± ♀
10 52	☽ ∠ ♆
12 32	☽ ⚼ ♂
13 59	♂ ⚻ ♃
18 20	☽ ∥ ♃
19 07	☽ ⚻ ☿
19 29	☉ ⚻ ♆
22 18	♃ ⊥ ♇
22 35	☽ ⚻ ♀
23 08	☽ # ☉
21 Tuesday	
03 41	☽ ⊥ ♆
04 57	☽ # ♄
05 21	☽ ⊻ ♇
06 18	☽ ± ♂
13 17	♀ # ♆
14 46	☽ ⊻ ♆
15 14	☉ ☍ ♃
16 16	☽ ☌ ♃
16 21	☽ ☍ ☉
16 24	☽ ⊥ ♇
17 03	☉ ± ♇
17 59	☽ ⚻ ♂
21 44	☽ # ♇
22 Wednesday	
00 44	☽ ⊻ ♅
06 11	☽ ∥ ♆
07 32	☽ # ♀
07 45	☿ ⚼ ♅
08 48	☽ ∠ ♇
09 55	☽ ☍ ♄
11 35	☽ ⊥ ♅
17 45	☽ # ☿
21 01	☽ # ♂
23 Thursday	
03 42	☽ ∠ ♅
04 36	☽ ☍ ☿
10 43	☽ ☍ ♀
11 38	☽ ✶ ♇
20 38	☽ ☌ ♆
20 49	♀ △ ♇
21 40	☽ ⊻ ♃
24 Friday	
02 24	☽ ⚻ ☉
02 40	☽ ☍ ♂
06 09	☽ ✶ ♅
08 07	☽ ⊥ ♃
08 45	☽ Q ♇
13 43	☽ ± ☉
14 12	☉ ⊻ ♂
15 38	☽ ⚻ ♄
23 40	☽ ∠ ♃
25 Saturday	
02 11	☽ ± ♄
03 01	☽ Q ♅
06 34	☽ ⚼ ☉
10 45	☽ ⚻ ☿
13 01	☉ ∥ ♄
15 59	☽ □ ♇
17 53	☽ ⚼ ♄
20 12	☽ ⚻ ♀
21 18	☽ ± ☿
26 Sunday	
00 44	☽ ⊻ ♆
01 21	☽ ✶ ♃
05 17	☉ ⚻ ♅
07 31	☽ ± ♀
09 23	☽ ⚻ ♂
09 59	☽ □ ♅
10 21	☽ △ ☉
11 01	☽ ⊥ ♆
13 00	☽ ⚼ ☿
19 53	☽ △ ♄
20 11	☽ ± ♂
21 04	♂ △ ♅
21 45	☽ Q ♃
27 Monday	
00 24	☽ ⚼ ♀
02 24	☽ ∠ ♆
12 21	☽ ⚼ ♂
14 52	☽ △ ☿
19 18	☽ △ ♇
23 03	♀ ☍ ♆
23 26	♂ ± ♃
28 Tuesday	
01 56	♄ ± ♅
01 59	♀ ⚻ ♃
03 55	☽ ✶ ♆
04 10	☽ □ ♃
04 24	☽ △ ♀
10 52	☿ St R
10 54	☽ # ♂
11 06	☿ ∥ ♀
13 02	☽ △ ♅
15 10	☽ △ ♂
17 18	☽ □ ☉
17 30	☽ # ♀
18 06	☽ # ☿
20 46	☽ ⚼ ♇
23 27	☽ □ ♄
29 Wednesday	
00 16	☽ Q ♆
04 55	☽ ∥ ♆
12 01	☽ ± ♇
12 29	☽ # ♇
13 57	♃ ⊻ ♆
14 26	☽ ⚼ ♅
17 46	☽ □ ☿
21 45	☽ # ☉
22 10	☽ ⚻ ♇
30 Thursday	
02 05	☽ Q ☉
02 35	☽ # ♄
05 39	☽ ± ♅
06 15	☽ Q ♄
06 37	☽ △ ♃
06 45	☽ □ ♆
12 06	☽ □ ♀
15 14	☽ ∥ ♃
15 48	☽ ⚻ ♅
18 47	☉ ⚼ ♇
20 37	☽ □ ♂
23 55	☽ Q ☿
23 55	☽ ✶ ☉

MAY 1970

LONGITUDES

Date	Sidereal time h m s	Sun ☉ ° ' "	Moon ☽ ° ' "	Moon ☽ 24.00 ° ' "	Mercury ☿ ° '	Venus ♀ ° '	Mars ♂ ° '	Jupiter ♃ ° '	Saturn ♄ ° '	Uranus ♅ ° '	Neptune ♆ ° '	Pluto ♇ ° '
01	02 36 00	10 ♉ 41 39	17 ♓ 19 39	24 ♓ 23 36	22 ♉ 05	04 ♊ 27	08 ♊ 41	29 ♎ 51	11 ♉ 56	05 ♎ 23	00 ♐ 02	24 ♍ 59
02	02 39 57	11 39 53	01 ♈ 26 42	08 ♈ 28 35	21 R 47	05 40	09 22	29 R 43	12 04	05 R 21	00 ♐ 01	24 R 58
03	02 43 53	12 38 05	15 ♈ 28 51	22 ♈ 27 02	21 25	06 53	10 03	29 36	12 12	05 19	29 ♏ 59	24 57
04	02 47 50	13 36 16	29 ♈ 22 43	06 ♉ 15 25	20 59	08 06	10 43	29 29	12 19	05 17	29 R 58	24 56
05	02 51 46	14 34 25	13 ♉ 04 43	19 ♉ 50 10	20 30	09 19	11 24	29 22	12 27	05 15	29 56	24 55
06	02 55 43	15 32 32	26 ♉ 31 28	03 ♊ 08 18	19 58	10 32	12 05	29 14	12 35	05 14	29 55	24 54
07	02 59 39	16 30 38	09 ♊ 40 30	16 ♊ 07 57	19 24	11 45	12 45	29 07	12 42	05 12	29 53	24 53
08	03 03 36	17 28 42	22 ♊ 30 40	28 ♊ 48 44	18 49	12 57	13 26	29 00	12 50	05 10	29 52	24 53
09	03 07 32	18 26 45	05 ♋ 02 21	11 ♋ 11 49	18 12	14 10	14 06	28 53	12 58	05 08	29 50	24 52
10	03 11 29	19 24 45	17 ♋ 17 30	23 ♋ 19 50	17 36	15 23	14 47	28 47	13 06	05 07	29 48	24 51
11	03 15 26	20 22 44	29 ♋ 19 21	05 ♌ 16 36	16 59	16 36	15 27	28 40	13 13	05 05	29 47	24 50
12	03 19 22	21 20 41	11 ♌ 12 12	17 ♌ 06 48	16 24	17 49	16 08	28 33	13 21	05 04	29 45	24 49
13	03 23 19	22 18 36	23 ♌ 01 03	28 ♌ 55 39	15 50	19 01	16 48	28 27	13 29	05 02	29 44	24 49
14	03 27 15	23 16 29	04 ♍ 51 16	10 ♍ 48 36	15 19	20 14	17 28	28 20	13 36	05 01	29 42	24 48
15	03 31 12	24 14 20	16 ♍ 48 18	22 ♍ 51 00	14 50	21 26	18 09	28 14	13 44	04 59	29 40	24 47
16	03 35 08	25 12 10	28 ♍ 57 17	05 ♎ 07 42	14 24	22 39	18 49	28 08	13 51	04 58	29 39	24 47
17	03 39 05	26 09 58	11 ♎ 22 42	17 ♎ 42 41	14 02	23 51	19 29	28 02	13 59	04 57	29 37	24 46
18	03 43 01	27 07 44	24 ♎ 07 56	00 ♏ 38 38	13 44	25 04	20 09	27 56	14 07	04 55	29 36	24 46
19	03 46 58	28 05 29	07 ♏ 14 52	13 ♏ 56 33	13 29	26 16	20 50	27 50	14 14	04 54	29 34	24 45
20	03 50 55	29 ♉ 03 12	20 ♏ 43 29	27 ♏ 35 24	13 19	27 28	21 30	27 44	14 22	04 53	29 32	24 45
21	03 54 51	00 ♊ 00 54	04 ♐ 31 49	11 ♐ 32 14	13 13	28 41	22 10	27 39	14 29	04 52	29 31	24 44
22	03 58 48	00 58 35	18 ♐ 36 02	25 ♐ 42 33	13 D 12	29 ♊ 53	22 50	27 33	14 37	04 51	29 29	24 44
23	04 02 44	01 56 15	02 ♑ 51 05	10 ♑ 00 57	13 15	01 ♋ 05	23 30	27 28	14 44	04 50	29 27	24 43
24	04 06 41	02 53 54	17 ♑ 11 29	24 ♑ 22 07	13 23	02 17	24 10	27 23	14 52	04 49	29 26	24 43
25	04 10 37	03 51 31	01 ♒ 32 17	08 ♒ 41 33	13 36	03 29	24 50	27 18	14 59	04 48	29 24	24 42
26	04 14 34	04 49 08	15 ♒ 49 35	22 ♒ 56 05	13 52	04 42	25 30	27 13	15 07	04 47	29 23	24 42
27	04 18 30	05 46 43	00 ♓ 00 51	07 ♓ 03 47	14 13	05 54	26 10	27 08	15 14	04 46	29 21	24 42
28	04 22 27	06 44 18	14 ♓ 04 46	21 ♓ 03 46	14 39	07 05	26 50	27 04	15 21	04 45	29 19	24 42
29	04 26 24	07 41 51	28 ♓ 00 44	04 ♈ 55 39	15 09	08 17	27 29	26 59	15 29	04 44	29 18	24 41
30	04 30 20	08 39 24	11 ♈ 48 27	18 ♈ 39 03	15 43	09 29	28 09	26 55	15 36	04 44	29 16	24 41
31	04 34 17	09 ♊ 36 56	25 ♈ 27 23	02 ♉ 13 18	16 ♉ 20	10 ♋ 41	28 ♊ 49	26 ♎ 51	15 ♉ 43	04 ♎ 43	29 ♏ 14	24 ♍ 41

Date	Moon True ☊ ° '	Moon Mean ☊ ° '	Moon ☽ Latitude ° '
01	10 ♓ 06	08 ♓ 54	00 N 39
02	10 R 04	08 51	01 51
03	10 00	08 48	02 57
04	09 53	08 45	03 51
05	09 44	08 42	04 31
06	09 33	08 38	04 55
07	09 22	08 35	05 03
08	09 11	08 32	04 55
09	09 02	08 29	04 33
10	08 55	08 26	03 58
11	08 51	08 23	03 14
12	08 49	08 19	02 21
13	08 D 49	08 16	01 23
14	08 49	08 13	00 N 21
15	08 R 49	08 10	00 S 42
16	08 47	08 07	01 45
17	08 43	08 03	02 44
18	08 37	08 00	03 36
19	08 28	07 57	04 18
20	08 17	07 54	04 47
21	08 06	07 51	05 00
22	07 54	07 48	04 55
23	07 44	07 44	04 33
24	07 36	07 41	03 53
25	07 31	07 38	02 58
26	07 29	07 35	01 52
27	07 D 28	07 32	00 S 39
28	07 R 28	07 28	00 N 35
29	07 27	07 25	01 46
30	07 25	07 22	02 51
31	07 ♓ 20	07 ♓ 19	03 N 45

DECLINATIONS

Date	Sun ☉ ° '	Moon ☽ ° '	Mercury ☿ ° '	Venus ♀ ° '	Mars ♂ ° '	Jupiter ♃ ° '	Saturn ♄ ° '	Uranus ♅ ° '	Neptune ♆ ° '	Pluto ♇ ° '
01	15 N 02	04 S 25	20 N 11	21 N 38	22 N 29	10 S 02	13 N 24	01 S 27	18 S 28	16 N 46
02	15 20	02 N 17	19 55	21 54	22 35	09 59	13 26	01 27	18 27	16 46
03	15 38	08 49	19 36	22 09	22 42	09 57	13 29	01 26	18 27	16 46
04	15 56	14 51	19 16	22 24	22 48	09 54	13 31	01 25	18 27	16 46
05	16 13	20 04	18 53	22 38	22 54	09 52	13 33	01 25	18 26	16 46
06	16 30	24 10	18 29	22 51	23 00	09 50	13 36	01 24	18 26	16 46
07	16 47	26 54	18 04	23 04	23 06	09 47	13 38	01 23	18 26	16 46
08	17 03	28 08	17 38	23 16	23 12	09 45	13 40	01 23	18 25	16 47
09	17 19	27 53	17 11	23 27	23 17	09 43	13 43	01 22	18 25	16 46
10	17 35	26 16	16 44	23 38	23 22	09 40	13 45	01 21	18 25	16 46
11	17 51	23 27	16 17	23 48	23 27	09 38	13 47	01 21	18 24	16 46
12	18 06	19 41	15 50	23 58	23 32	09 36	13 50	01 20	18 24	16 46
13	18 21	15 09	15 25	24 06	23 36	09 34	13 52	01 20	18 24	16 46
14	18 36	10 04	15 00	24 14	23 40	09 32	13 54	01 19	18 23	16 46
15	18 50	04 N 34	14 36	24 22	23 44	09 30	13 56	01 18	18 23	16 46
16	19 04	01 S 11	14 15	24 28	23 48	09 28	13 59	01 18	18 23	16 46
17	19 18	07 01	13 54	24 34	23 52	09 26	14 01	01 17	18 22	16 46
18	19 31	12 42	13 36	24 39	23 55	09 24	14 03	01 17	18 22	16 46
19	19 44	17 59	13 20	24 44	23 59	09 22	14 05	01 17	18 22	16 46
20	19 57	22 33	13 07	24 48	24 02	09 20	14 07	01 16	18 21	16 45
21	20 10	25 58	12 55	24 51	24 05	09 18	14 10	01 16	18 21	16 45
22	20 22	27 52	12 46	24 53	24 07	09 16	14 12	01 15	18 21	16 45
23	20 33	27 57	12 40	24 55	24 10	09 15	14 14	01 15	18 20	16 45
24	20 45	26 11	12 35	24 56	24 12	09 13	14 16	01 15	18 20	16 45
25	20 56	22 43	12 34	24 56	24 14	09 11	14 18	01 14	18 20	16 44
26	21 06	17 52	12 34	24 55	24 16	09 10	14 20	01 14	18 19	16 44
27	21 17	12 05	12 36	24 54	24 17	09 08	14 22	01 14	18 19	16 44
28	21 26	05 S 44	12 41	24 52	24 19	09 07	14 25	01 13	18 19	16 43
29	21 36	00 N 50	12 48	24 50	24 20	09 06	14 27	01 13	18 18	16 43
30	21 45	07 17	12 57	24 46	24 21	09 04	14 29	01 13	18 18	16 43
31	21 N 54	13 N 20	13 N 08	24 N 42	24 N 22	09 S 03	14 N 31	01 S 13	18 S 18	16 N 42

ZODIAC SIGN ENTRIES

Date	h	m	Planets
02	09	32	☽ ♈
03	01	31	♆ ♏
04	13	05	☽ ♉
06	18	17	☽ ♊
09	02	17	☽ ♋
11	13	22	☽ ♌
14	02	10	☽ ♍
16	14	02	☽ ♎
18	22	49	☽ ♏
21	04	11	☽ ♐
21	11	37	☉ ♊
22	14	19	♀ ♋
23	07	13	☽ ♑
25	09	25	☽ ♒
27	11	59	☽ ♓
29	15	27	☽ ♈
31	20	03	☽ ♉

LATITUDES

Date	Mercury ☿ ° '	Venus ♀ ° '	Mars ♂ ° '	Jupiter ♃ ° '	Saturn ♄ ° '	Uranus ♅ ° '	Neptune ♆ ° '	Pluto ♇ ° '
01	01 N 58	00 N 37	00 N 44	01 N 29	02 S 08	00 N 45	01 N 44	16 N 09
04	01 18	00 45	00 45	01 29	02 07	00 45	01 45	16 08
07	00 N 30	00 52	00 46	01 28	02 07	00 44	01 45	16 07
10	00 S 22	01 00	00 48	01 28	02 07	00 44	01 45	16 05
13	01 14	01 07	00 49	01 27	02 07	00 44	01 45	16 04
16	02 01	01 14	00 50	01 27	02 07	00 44	01 45	16 03
19	02 40	01 20	00 51	01 26	02 07	00 44	01 45	16 02
22	03 11	01 26	00 52	01 26	02 07	00 44	01 45	16 01
25	03 31	01 32	00 53	01 25	02 08	00 44	01 45	15 59
28	03 43	01 37	00 54	01 24	02 08	00 44	01 45	15 58
31	03 S 45	01 N 42	00 N 55	01 N 24	02 S 08	00 N 44	01 N 45	15 N 57

DATA

Julian Date	2440708
Delta T	+41 seconds
Ayanamsa	23° 26' 38"
Synetic vernal point	05° ♓ 40' 21"
True obliquity of ecliptic	23° 26' 44"

MOON'S PHASES, APSIDES AND POSITIONS ☽

Date	h	m	Phase	Longitude ° '	Eclipse Indicator
05	14	51	●	14 ♉ 41	
13	10	26	☽	22 ♌ 15	
21	03	38	○	29 ♏ 41	
27	22	32	☾	06 ♓ 12	

Day	h	m		
13	01	46	Apogee	
25	07	45	Perigee	
02	03	50	0N	
08	19	52	Max dec	28° N 13'
16	07	06	0S	
23	01	14	Max dec	28° S 09'
29	08	57	0N	

LONGITUDES

Date	Chiron ⚷ ° '	Ceres ⚳ ° '	Pallas ⚴ ° '	Juno ⚵ ° '	Vesta ⚶ ° '	Black Moon Lilith ⚸ ° '
01	08 ♈ 16	07 ♈ 25	06 ♓ 41	29 ♓ 45	15 ♌ 17	16 ♌ 09
11	08 ♈ 46	11 ♈ 04	08 ♓ 55	04 ♈ 29	17 ♌ 46	17 ♌ 16
21	09 ♈ 12	14 ♈ 38	10 ♓ 50	09 ♈ 11	20 ♌ 42	18 ♌ 22
31	09 ♈ 35	18 ♈ 04	12 ♓ 26	13 ♈ 50	23 ♌ 59	19 ♌ 29

ASPECTARIAN

01 Friday
- 02 46 ☽ ⚹ ♄
- 07 50 ☽ bQ ♃
- 19 55 ☽ ⚹ ☿
- 21 30 ☽ Q ♀
- 22 38 ☽ ∥ ♅
- 22 59 ☽ ± ♃

02 Saturday
- 00 59 ☽ ☍ ♇
- 03 16 ☽ ∠ ☉
- 04 28 ☽ ∠ ♄
- 04 29 ☉ ± ♅
- 04 42 ☽ Q ♂
- 06 02 ♀ △ ♅
- 09 00 ☽ # ♅
- 09 06 ☽ ⊼ ♃
- 09 34 ☽ △ ♆
- 13 08 ♀ ± ♃
- 18 39 ☽ ☍ ♅
- 19 44 ☽ ⊥ ☉
- 19 52 ☽ ⚹ ♀
- 19 57 ☽ ⊥ ♄
- 20 54 ☽ ∠ ☿
- 23 25 ☉ ☌ ♄

03 Sunday
- 02 13 ☽ ⚹ ♂
- 06 19 ☽ ⊻ ♄
- 06 45 ☽ ⊻ ☉
- 11 09 ☽ bQ ♆
- 11 54 ☽ ⊥ ☿
- 16 17 ☽ # ♃
- 21 56 ☽ ⊻ ☿

04 Monday
- 00 04 ☽ ∠ ♀
- 02 38 ☽ ± ♆
- 04 18 ☽ ⊼ ♇
- 05 20 ☽ ∠ ♂
- 06 23 ☽ ∥ ♄
- 12 11 ☽ ☍ ♃
- 13 01 ☽ ⊼ ♆
- 14 42 ☽ ± ♇
- 16 51 ☽ ∥ ☉
- 17 11 ☽ ⊥ ♀
- 20 19 ☽ ∥ ♇
- 21 48 ☽ ⊥ ♂
- 22 17 ☽ ⊼ ♅

05 Tuesday
- 03 59 ☽ # ♆
- 04 43 ☽ ⊻ ♀
- 06 26 ☽ bQ ♇
- 06 32 ☽ ∥ ☿
- 08 48 ☽ ± ♅
- 08 53 ☽ ⊻ ♂
- 10 53 ☽ ☌ ♄
- 14 51 ☽ ☌ ☉

06 Wednesday
- 00 07 ☿ bQ ♅
- 00 42 ☽ ☌ ☿
- 00 43 ☽ bQ ♅
- 02 58 ☽ ∥ ♀
- 04 13 ☽ ∥ ♂
- 09 05 ☽ △ ♇
- 15 00 ☿ # ♆
- 16 52 ☽ ⊼ ♃
- 18 07 ☽ ☍ ♆

07 Thursday
- 03 42 ☽ ± ♃
- 03 47 ☽ △ ♅
- 09 53 ♂ ⊻ ♄
- 11 45 ☉ ∥ ♇
- 16 13 ☽ ☌ ♀
- 17 40 ☽ ⊻ ♄
- 18 01 ☽ ☌ ♂
- 19 45 ♀ ∥ ♂
- 20 11 ☽ bQ ♃

08 Friday
- 01 45 ☽ ⊻ ☉
- 05 00 ☽ ⊥ ♄
- 05 20 ☽ ⊻ ☿
- 09 18 ♀ ⊻ ♄
- 13 59 ☽ ⊥ ☉
- 16 10 ☽ ⊥ ☿
- 16 29 ☽ □ ♇
- 22 14 ☽ ∠ ♄

09 Saturday
- 00 16 ☽ △ ♃
- 01 59 ☽ ⊼ ♆
- 05 26 ♂ bQ ♃
- 06 56 ♀ bQ ♃
- 07 19 ☉ ∥ ☿
- 08 21 ☉ ☌ ☿
- 08 37 ☽ ∠ ☿
- 08 39 ☽ ∠ ☉
- 09 08 ♀ ☌ ♂
- 12 12 ☽ □ ♅
- 13 32 ☽ ± ♆

10 Sunday
- 03 15 ☽ Q ♇
- 03 38 ☽ ⚹ ♄
- 06 45 ☽ ⊻ ♂
- 07 06 ☽ bQ ♆
- 07 49 ☽ ⊻ ♀
- 09 43 ☿ ∥ ♇
- 12 34 ☽ ⚹ ☿
- 16 34 ☽ ⚹ ☉
- 19 20 ☽ ⊥ ♂
- 21 02 ☽ ⊥ ♀
- 23 33 ☽ Q ♅

11 Monday
- 03 01 ☽ ⚹ ♇
- 03 41 ☽ Q ♄
- 04 58 ☉ bQ ♅
- 09 34 ☽ ∥ ♀
- 10 42 ☽ □ ♃
- 11 22 ☽ Q ☿
- 12 03 ☽ ∥ ♂
- 12 55 ☽ △ ♆
- 14 25 ☽ ∠ ♂
- 17 06 ☽ ∠ ♀
- 17 12 ☿ ⊻ ♀
- 18 42 ☽ Q ☉
- 23 36 ☽ ⚹ ♅

12 Tuesday
- 09 12 ☽ ∠ ♇
- 16 24 ☽ □ ♄
- 17 14 ☿ ⊻ ♂
- 19 10 ☽ # ♆
- 20 19 ☽ ∥ ☉
- 22 04 ☽ □ ☿
- 22 36 ☽ ⚹ ♂
- 22 46 ☽ Q ♃

13 Wednesday
- 02 57 ☽ ⚹ ♀
- 03 28 ☽ ⊥ ♇
- 03 48 ☽ ∥ ♇
- 05 57 ☽ ∠ ♅
- 10 26 ☽ □ ☉
- 10 37 ☽ ∥ ☿
- 15 39 ☽ ⊻ ♇
- 16 01 ☉ # ♆
- 18 16 ☽ ∥ ♄
- 22 06 ♀ ⊥ ♄
- 22 55 ☽ ⚹ ♃

14 Thursday
- 00 12 ☽ ⊥ ♅
- 00 27 ☽ Q ♂
- 01 35 ☽ □ ♆
- 06 05 ☽ Q ♀
- 12 19 ☽ ⊻ ♅
- 14 25 ☽ # ♃

15 Friday
- 03 20 ☿ ⊥ ♀
- 04 55 ☽ ∠ ♃
- 05 47 ☽ △ ♄
- 08 13 ☽ △ ☿
- 13 44 ☽ Q ♆
- 14 50 ☽ □ ♂
- 22 14 ☽ □ ♀
- 22 41 ☽ ⊥ ♃

16 Saturday
- 01 34 ☉ △ ♇
- 01 41 ☽ # ♅
- 03 49 ☽ ☌ ♇
- 04 00 ☽ △ ☉
- 10 24 ☽ ⊻ ♃
- 11 48 ☽ bQ ♄
- 12 28 ☽ ∥ ♅
- 12 51 ☽ bQ ☿
- 13 21 ☽ ⚹ ♆
- 23 40 ☽ ☌ ♅

17 Sunday
- 05 01 ☿ ∥ ♄
- 05 26 ☽ ± ♄
- 05 47 ☽ ± ☿
- 11 34 ☽ bQ ☉
- 14 43 ☿ ☌ ♄
- 16 56 ☽ ⊼ ☿
- 17 00 ☽ ⊼ ♄
- 18 09 ☽ ∠ ♆
- 22 00 ☽ ∥ ♃

18 Monday
- 01 16 ☿ ⊥ ♂
- 04 11 ☽ △ ♂
- 05 57 ☽ ± ☉
- 06 01 ♀ □ ♇
- 09 53 ♂ ⊥ ♄
- 11 00 ☽ ⊥ ♆
- 13 10 ☽ ⊻ ♇
- 13 54 ☽ △ ♀
- 15 46 ☽ # ☿
- 17 56 ☽ # ♄
- 17 59 ☽ ⊼ ☉
- 18 58 ☽ ☌ ♃
- 22 03 ☽ ⊻ ♆

19 Tuesday
- 00 12 ☽ ⊥ ♇
- 06 08 ☉ ⊼ ♃
- 06 10 ☽ # ♇
- 07 46 ☽ ⊻ ♅
- 09 18 ☽ bQ ♂
- 13 47 ☽ ∥ ♆
- 16 30 ☽ ∠ ♇
- 18 34 ☽ ⊥ ♅
- 19 56 ☽ bQ ♀
- 21 05 ☽ # ☉
- 23 02 ☽ ☍ ☿

20 Wednesday
- 00 38 ☽ ☍ ♄
- 02 17 ☽ ± ♂
- 10 31 ☽ ∠ ♅
- 13 25 ☽ ⊼ ♂
- 13 27 ☽ ± ♀
- 16 52 ♀ △ ♃
- 19 02 ☽ ⚹ ♇
- 21 26 ☽ # ♂
- 23 45 ☉ ☍ ♆

21 Thursday
- 00 10 ☽ ⊻ ♃
- 00 56 ☽ ⊼ ♀
- 02 55 ☽ # ♀
- 03 21 ☽ ☌ ♆
- 03 26 ☿ ∠ ♀
- 03 38 ☽ ☍ ☉
- 10 29 ☽ ⊥ ♃
- 12 34 ☽ ⚹ ♅
- 15 47 ☽ Q ♇

22 Friday
- 01 48 ☽ ∠ ♃
- 02 50 ☽ ⊼ ☿
- 04 12 ♀ ⊼ ♆
- 05 11 ☽ ⊼ ♄
- 05 58 ♀ ∠ ♄
- 06 47 ☿ St D
- 09 02 ☽ Q ♅
- 13 01 ☽ ± ☿
- 15 26 ☽ ± ♄
- 19 30 ☽ ☍ ♂
- 22 20 ☽ □ ♇

23 Saturday
- 03 01 ☽ ⚹ ♃
- 04 15 ☽ bQ ☿
- 06 19 ☽ ⊻ ♆
- 06 44 ☽ bQ ♄
- 08 46 ☽ ☍ ♀
- 10 21 ☽ ⊼ ☉
- 15 19 ☽ □ ♅
- 16 22 ☽ ⊥ ♆
- 21 08 ☽ ± ☉
- 23 01 ☽ Q ♃

24 Sunday
- 05 34 ☽ △ ☿
- 07 23 ☽ ∠ ♆
- 08 04 ☽ △ ♄
- 13 16 ☽ bQ ☉
- 21 57 ☽ # ♀
- 23 05 ☉ ± ♃

25 Monday
- 00 14 ☽ ⊼ ♂
- 00 34 ☽ △ ♇
- 02 50 ☽ # ♂
- 04 56 ☽ □ ♃
- 07 35 ♂ □ ♇
- 08 26 ☽ ⚹ ♆
- 10 46 ☽ ± ♂
- 15 34 ☽ ⊼ ♀
- 16 10 ☽ △ ☉
- 17 27 ☽ △ ♅
- 21 12 ☽ # ☉

26 Tuesday
- 01 42 ☽ bQ ♇
- 02 34 ☽ ± ♀
- 02 36 ☽ bQ ♂
- 04 31 ☽ Q ♆
- 08 38 ☽ □ ☿
- 10 00 ☽ ∥ ♆
- 10 47 ☽ □ ♄
- 11 03 ☉ △ ♅
- 13 45 ♀ □ ♅
- 16 51 ☽ ± ♇
- 17 00 ☽ # ♇
- 18 40 ☽ bQ ♅
- 19 07 ☽ bQ ♀

27 Wednesday
- 00 39 ☉ ⊻ ♀
- 01 22 ♀ ± ♆
- 02 55 ☽ # ♄
- 02 59 ☽ ⊼ ♇
- 05 09 ☽ △ ♂
- 07 09 ☽ △ ♃
- 09 53 ☽ ± ♅
- 09 58 ☽ # ☿
- 10 52 ☽ □ ♆
- 15 52 ☽ Q ☿
- 17 31 ☽ Q ♄
- 20 05 ☽ ⊼ ♅
- 22 32 ☽ □ ☉
- 22 56 ☽ △ ♀
- 23 21 ☽ ∥ ♃

28 Thursday
- 08 34 ☽ bQ ♃
- 13 01 ☽ ⚹ ☿
- 14 13 ☽ ⚹ ♄
- 19 37 ♂ △ ♃
- 23 56 ☽ ± ♃

29 Friday
- 04 30 ☽ ∥ ♅
- 06 15 ☽ ☍ ♇
- 07 42 ☽ Q ☉
- 10 14 ☽ ⊼ ♃
- 11 03 ☽ □ ♂
- 13 24 ☽ # ♅
- 14 13 ☽ △ ♆
- 15 51 ☽ ∠ ☿
- 16 19 ☽ ∠ ♄
- 23 40 ☽ ☍ ♅

30 Saturday
- 06 05 ☽ ⚹ ☉
- 06 31 ☿ ☌ ♄
- 07 34 ☽ □ ♀
- 08 07 ☽ ⊥ ♄
- 08 10 ☽ ⊥ ☿
- 16 18 ☽ bQ ♆
- 18 42 ☽ ⊻ ♄
- 18 50 ☽ # ♃
- 19 09 ☽ ⊻ ☿
- 20 00 ☽ Q ♂

31 Sunday
- 08 06 ☽ ± ♆
- 10 24 ☽ ∠ ☉
- 10 38 ☽ ⊼ ♇
- 11 08 ☽ ∥ ☿
- 14 27 ☽ ☍ ♃
- 17 03 ☽ ∥ ♄
- 18 16 ☽ ⚹ ♂
- 18 16 ☽ Q ♀
- 18 42 ☽ ⊼ ♆
- 21 16 ☽ ± ♇

All ephemeris data is given at 12.00 UT and the Moon's longitude is additionally given for 24.00 UT

JUNE 1970

LONGITUDES

Date	Sidereal time h m s	Sun ☉ ° ' "	Moon ☽ ° ' "	Moon ☽ 24.00 ° ' "	Mercury ☿ ° '	Venus ♀ ° '	Mars ♂ ° '	Jupiter ♃ ° '	Saturn ♄ ° '	Uranus ♅ ° '	Neptune ♆ ° '	Pluto ♇ ° '
01	04 38 13	10 ♊ 34 28	08 ♉ 56 39	15 ♉ 37 14	17 ♉ 02	11 ♋ 53	29 ♊ 29	26 ♎ 47	15 ♉ 51	04 ♎ 42	29 ♏ 13	24 ♍ 41
02	04 42 10	11 31 58	22 ♉ 14 52	28 ♉ 49 22	17 48	13 05	00 ♋ 09	26 R 43	15 58	04 R 42	29 R 11	24 R 41
03	04 46 06	12 29 28	05 ♊ 20 32	11 ♊ 48 13	18 37	14 16	00 48	26 40	16 05	04 41	29 10	24 41
04	04 50 03	13 26 56	18 ♊ 12 18	24 ♊ 32 42	19 30	15 28	01 28	26 36	16 12	04 41	29 08	24 41
05	04 53 59	14 24 24	00 ♋ 49 23	07 ♋ 02 26	20 27	16 39	02 08	26 33	16 19	04 41	29 07	24 D 41
06	04 57 56	15 21 51	13 ♋ 11 57	19 ♋ 18 07	21 27	17 51	02 47	26 30	16 27	04 40	29 05	24 41
07	05 01 53	16 19 17	25 ♋ 21 13	01 ♌ 21 34	22 30	19 02	03 27	26 27	16 34	04 40	29 04	24 41
08	05 05 49	17 16 41	07 ♌ 19 34	13 ♌ 15 41	23 36	20 14	04 06	26 24	16 41	04 40	29 02	24 41
09	05 09 46	18 14 05	19 ♌ 10 25	25 ♌ 04 21	24 46	21 25	04 46	26 22	16 48	04 40	29 00	24 41
10	05 13 42	19 11 27	00 ♍ 58 05	06 ♍ 52 15	25 59	22 36	05 25	26 19	16 55	04 39	28 59	24 41
11	05 17 39	20 08 49	12 ♍ 47 31	18 ♍ 44 34	27 16	23 48	06 05	26 17	17 01	04 39	28 57	24 41
12	05 21 35	21 06 09	24 ♍ 44 05	00 ♎ 46 45	28 35	24 59	06 44	26 15	17 08	04 D 39	28 56	24 42
13	05 25 32	22 03 29	06 ♎ 53 14	13 ♎ 04 09	29 ♉ 57	26 10	07 24	26 13	17 15	04 39	28 55	24 42
14	05 29 28	23 00 47	19 ♎ 20 06	25 ♎ 41 35	01 ♊ 23	27 21	08 03	26 11	17 22	04 39	28 53	24 42
15	05 33 25	23 58 05	02 ♏ 09 01	08 ♏ 42 44	02 51	28 32	08 42	26 10	17 29	04 40	28 52	24 42
16	05 37 22	24 55 22	15 ♏ 22 55	22 ♏ 09 38	04 23	29 ♋ 43	09 22	26 09	17 35	04 40	28 50	24 43
17	05 41 18	25 52 38	29 ♏ 02 46	06 ♐ 02 03	05 57	00 ♌ 54	10 01	26 07	17 42	04 40	28 49	24 43
18	05 45 15	26 49 53	13 ♐ 07 02	20 ♐ 17 08	07 34	02 04	10 40	26 06	17 49	04 40	28 47	24 44
19	05 49 11	27 47 08	27 ♐ 31 37	04 ♑ 49 37	09 15	03 15	11 20	26 06	17 55	04 41	28 46	24 44
20	05 53 08	28 44 23	12 ♑ 10 12	19 ♑ 32 25	10 58	04 26	11 59	26 05	18 02	04 41	28 45	24 44
21	05 57 04	29 ♊ 41 37	26 ♑ 55 15	04 ♒ 17 49	12 44	05 36	12 38	26 05	18 08	04 42	28 43	24 45
22	06 01 01	00 ♋ 38 50	11 ♒ 39 14	18 ♒ 58 45	14 32	06 47	13 17	26 04	18 14	04 42	28 42	24 46
23	06 04 57	01 36 04	26 ♒ 15 44	03 ♓ 29 41	16 24	07 57	13 56	26 D 04	18 21	04 43	28 41	24 46
24	06 08 54	02 33 17	10 ♓ 40 12	17 ♓ 47 03	18 18	09 07	14 36	26 04	18 27	04 43	28 39	24 47
25	06 12 51	03 30 30	24 ♓ 50 03	01 ♈ 49 09	20 15	10 18	15 15	26 05	18 33	04 44	28 38	24 47
26	06 16 47	04 27 44	08 ♈ 44 21	15 ♈ 35 42	22 14	11 28	15 54	26 05	18 40	04 45	28 37	24 48
27	06 20 44	05 24 57	22 ♈ 23 17	29 ♈ 07 13	24 15	12 38	16 33	26 06	18 46	04 45	28 36	24 49
28	06 24 40	06 22 10	05 ♉ 47 37	12 ♉ 24 35	26 18	13 48	17 12	26 07	18 52	04 46	28 34	24 49
29	06 28 37	07 19 24	18 ♉ 58 13	25 ♉ 28 36	28 ♊ 23	14 58	17 51	26 08	18 58	04 47	28 33	24 50
30	06 32 33	08 ♋ 16 37	01 ♊ 55 49	08 ♊ 19 54	00 ♋ 30	16 ♌ 08	18 ♋ 30	26 ♎ 09	19 ♉ 04	04 ♎ 48	28 ♏ 32	24 ♍ 51

DECLINATIONS

Date	Moon True ☊ ° '	Moon Mean ☊ ° '	Moon ☽ Latitude ° '	Sun ☉ ° '	Moon ☽ ° '	Mercury ☿ ° '	Venus ♀ ° '	Mars ♂ ° '	Jupiter ♃ ° '	Saturn ♄ ° '	Uranus ♅ ° '	Neptune ♆ ° '	Pluto ♇ ° '
01	07 ♓ 12	07 ♓ 16	04 N 25	22 N 02	18 N 40	13 N 20	24 N 38	24 N 22	09 S 02	14 N 33	01 S 12	18 S 17	16 N 42
02	07 R 01	07 13	04 51	22 10	23 01	13 34	24 32	24 22	09 01	14 35	01 12	18 17	16 42
03	06 49	07 09	05 00	22 18	26 07	13 50	24 26	24 23	09 00	14 37	01 12	18 17	16 41
04	06 36	07 06	04 54	22 25	27 48	14 08	24 19	24 23	08 59	14 39	01 12	18 16	16 41
05	06 23	07 03	04 34	22 32	28 00	14 27	24 12	24 22	08 58	14 41	01 12	18 16	16 41
06	06 13	07 00	04 01	22 38	26 47	14 47	24 03	24 22	08 57	14 43	01 12	18 16	16 40
07	06 05	06 57	03 17	22 45	24 18	15 08	23 55	24 21	08 56	14 45	01 12	18 15	16 40
08	05 59	06 54	02 25	22 50	20 47	15 30	23 45	24 20	08 55	14 46	01 12	18 15	16 39
09	05 56	06 50	01 28	22 55	16 28	15 54	23 35	24 19	08 55	14 48	01 12	18 15	16 39
10	05 55	06 47	00 N 26	23 00	11 33	16 18	23 24	24 18	08 54	14 50	01 11	18 14	16 38
11	05 D 55	06 44	00 S 36	23 05	06 12	16 43	23 13	24 17	08 53	14 52	01 12	18 14	16 38
12	05 R 55	06 41	01 38	23 09	00 N 35	17 08	23 00	24 15	08 53	14 54	01 12	18 14	16 37
13	05 54	06 38	02 37	23 12	05 S 08	17 34	22 48	24 13	08 53	14 56	01 12	18 14	16 37
14	05 51	06 34	03 29	23 16	10 48	18 01	22 34	24 11	08 52	14 58	01 12	18 13	16 36
15	05 46	06 31	04 13	23 18	16 10	18 28	22 20	24 09	08 52	14 59	01 12	18 13	16 36
16	05 39	06 28	04 44	23 21	20 59	18 55	22 06	24 06	08 52	15 01	01 12	18 13	16 35
17	05 30	06 25	05 01	23 23	24 50	19 22	21 51	24 04	08 51	15 03	01 12	18 12	16 35
18	05 19	06 22	05 00	23 24	27 20	19 48	21 35	24 01	08 51	15 05	01 12	18 12	16 34
19	05 09	06 19	04 40	23 26	28 06	20 15	21 19	23 58	08 51	15 06	01 12	18 12	16 33
20	05 00	06 15	04 02	23 26	26 54	20 41	21 02	23 55	08 51	15 08	01 13	18 12	16 33
21	04 53	06 12	03 07	23 27	23 50	21 06	20 44	23 51	08 51	15 10	01 13	18 11	16 32
22	04 49	06 09	02 00	23 27	19 13	21 31	20 26	23 48	08 52	15 11	01 13	18 11	16 32
23	04 47	06 06	00 S 45	23 26	13 29	21 55	20 08	23 44	08 52	15 13	01 13	18 11	16 31
24	04 D 47	06 03	00 N 31	23 25	07 05	22 17	19 49	23 40	08 52	15 15	01 14	18 11	16 30
25	04 48	06 00	01 45	23 24	00 S 27	22 38	19 29	23 36	08 52	15 16	01 14	18 10	16 30
26	04 R 48	05 56	02 51	23 22	06 N 05	22 57	19 10	23 31	08 53	15 18	01 14	18 10	16 29
27	04 46	05 53	03 46	23 20	12 13	23 15	18 49	23 27	08 53	15 19	01 14	18 10	16 28
28	04 43	05 50	04 28	23 18	17 40	23 30	18 28	23 22	08 54	15 21	01 15	18 10	16 28
29	04 37	05 47	04 54	23 15	22 10	23 44	18 07	23 17	08 54	15 22	01 15	18 09	16 27
30	04 ♓ 29	05 ♓ 44	05 N 05	23 N 11	25 N 32	23 N 55	17 N 45	23 N 12	08 S 55	15 N 24	01 S 16	18 S 09	16 N 26

ZODIAC SIGN ENTRIES

Date	h	m	Planets
02	06	51	♂ ♋
03	02	10	☽ ♊
05	10	25	☽ ♋
07	21	17	☽ ♌
10	10	02	☽ ♍
12	22	28	☽ ♎
13	12	46	☿ ♊
15	08	02	☽ ♏
16	17	49	♀ ♌
17	13	39	☽ ♐
19	16	04	☽ ♑
21	17	00	☽ ♒
21	19	43	☉ ♋
23	18	11	☽ ♓
25	20	52	☽ ♈
28	01	35	☽ ♉
30	06	22	☿ ♋
30	08	24	☽ ♊

LATITUDES

Date	Mercury ☿ ° '	Venus ♀ ° '	Mars ♂ ° '	Jupiter ♃ ° '	Saturn ♄ ° '	Uranus ♅ ° '	Neptune ♆ ° '	Pluto ♇ ° '
01	03 S 44	01 N 43	00 N 55	01 N 24	02 S 08	00 N 43	01 N 45	15 N 56
04	03 37	01 47	00 56	01 23	02 08	00 43	01 44	15 55
07	03 23	01 50	00 57	01 22	02 08	00 43	01 44	15 54
10	03 03	01 53	00 58	01 21	02 08	00 43	01 44	15 52
13	02 38	01 54	00 59	01 20	02 09	00 43	01 44	15 51
16	02 09	01 55	01 00	01 20	02 09	00 43	01 44	15 49
19	01 37	01 56	01 00	01 19	02 09	00 43	01 44	15 48
22	01 02	01 55	01 01	01 18	02 10	00 43	01 44	15 47
25	00 S 27	01 54	01 02	01 17	02 10	00 42	01 44	15 45
28	00 N 07	01 52	01 02	01 17	02 10	00 42	01 44	15 44
31	00 N 38	01 N 49	01 N 03	01 N 16	02 S 11	00 N 42	01 N 44	15 N 43

DATA

Julian Date	2440739
Delta T	+41 seconds
Ayanamsa	23° 26' 43"
Synetic vernal point	05° ♓ 40' 16"
True obliquity of ecliptic	23° 26' 43"

MOON'S PHASES, APSIDES AND POSITIONS ☽

Date	h	m	Phase	Longitude ° '	Eclipse Indicator
04	02	21	●	13 ♊ 04	
12	04	06	☽	20 ♍ 47	
19	12	27	○	27 ♐ 48	
26	04	01	☾	04 ♈ 09	

Day	h	m		
09	20	10	Apogee	
21	18	08	Perigee	
05	03	15	Max dec	28° N 06'
12	14	29	0S	
19	09	30	Max dec	28° S 06'
25	13	37	0N	

LONGITUDES

Date	Chiron ⚷ ° '	Ceres ⚳ ° '	Pallas ⚴ ° '	Juno ⚵ ° '	Vesta ⚶ ° '	Black Moon Lilith ⚸ ° '
01	09 ♈ 37	18 ♈ 24	12 ♓ 34	14 ♈ 18	24 ♌ 20	19 ♌ 36
11	09 ♈ 56	21 ♈ 41	13 ♓ 43	18 ♈ 55	27 ♌ 57	20 ♌ 42
21	10 ♈ 09	24 ♈ 46	14 ♓ 25	23 ♈ 27	01 ♍ 50	21 ♌ 49
31	10 ♈ 18	27 ♈ 39	14 ♓ 37	27 ♈ 54	05 ♍ 55	22 ♌ 55

All ephemeris data is given at 12.00 UT and the Moon's longitude is additionally given for 24.00 UT

ASPECTARIAN

h m	Aspects
01 Monday	
02 43	☽ ∥ ♇
02 46	♂ ⊼ ♆
03 36	☽ ⊥ ☉
04 26	☽ ⊼ ♅
10 10	☽ ⋕ ♆
13 19	☽ ⚼ ♇
15 09	☽ ⊻ ☉
15 10	☽ ± ♅
17 47	☽ ⚹ ♀
22 28	☽ ∠ ♂
02 Tuesday	
00 31	☽ ☌ ♄
03 26	☽ ☌ ☿
04 02	♀ Q ♇
06 40	☽ ∥ ☉
07 23	☽ ⚼ ♅
15 38	☽ ⊥ ♂
16 25	☉ ⚼ ♃
16 26	☽ △ ♇
20 07	☽ ⊼ ♃
21 19	☽ ∥ ♂
22 11	☽ ∥ ♀
23 42	☽ ∠ ♀
03 Wednesday	
00 39	☽ ☍ ♆
03 11	☽ ⊻ ♂
07 04	☽ ± ♃
09 52	♀ ⚼ ♆
10 48	☽ △ ♅
17 59	☽ ⊥ ♀
23 41	☽ ⚼ ♃
04 Thursday	
00 22	♀ ∥ ♂
00 26	♂ ∠ ♄
02 21	☽ ☌ ☉
06 19	☽ ⊻ ♀
08 12	☽ ⊻ ♄
14 38	☽ ⊻ ☿
16 43	☿ ⚼ ♅
19 38	☽ ⊥ ♄
05 Friday	
00 15	☽ □ ♇
02 25	♇ St D
02 55	☽ ⊥ ☿
03 51	☽ △ ♃
04 32	♀ ⚹ ♄
08 43	☽ ⊼ ♆
12 58	☽ ∠ ♄
14 39	☽ ☌ ♂
19 25	☽ □ ♅
20 15	☽ ± ♆
21 40	☽ ∠ ☿
06 Saturday	
06 54	☿ ∥ ♄
10 59	☽ Q ♇
13 44	☽ ⚼ ♆
16 36	☽ ⊻ ☉
18 25	☽ ⚹ ♄
22 08	☽ ☌ ♀
07 Sunday	
05 27	☽ ⊥ ☉
05 46	☽ ⚹ ☿
06 40	☽ Q ♅
10 39	☽ ⚹ ♇
11 36	☽ ∥ ♂
14 10	☽ □ ♃
15 12	☽ ∥ ♀
18 27	☽ Q ♄
18 49	☉ ⊻ ♄
19 23	☽ △ ♆
23 08	☽ ∥ ☉
08 Monday	
00 58	☽ ∠ ☉
05 08	☽ ⊻ ♂
06 38	☽ ⚹ ♅
08 10	☽ Q ☿
16 45	☽ ∠ ♇
17 57	☽ ⊥ ♂
09 Tuesday	
02 16	☽ Q ♃
02 34	☽ ⋕ ♆
07 07	☽ □ ♄
08 11	♂ □ ♅
09 55	☽ ⚹ ☉
10 11	☿ △ ♇
11 00	☽ ⊥ ♇
11 03	☽ ∥ ♇
12 59	☽ ∠ ♅
13 16	☽ ∠ ♂
14 42	☽ ∥ ☿
17 05	☽ ⊻ ♀
20 19	☽ ∥ ♄
20 33	♂ ± ♆
23 13	☽ ⊻ ♇
10 Wednesday	
00 41	☽ □ ☿
02 34	☽ ⚹ ♃
06 39	☽ ⊥ ♀
07 18	☽ ⊥ ♅
07 58	☽ □ ♆
12 30	☽ Q ☉
13 00	♀ Q ♅
18 08	☿ ⊼ ♃
19 30	☽ ⊻ ♅
21 36	☽ ⚹ ♂
11 Thursday	
00 05	☽ ⋕ ♃
03 00	☽ ∠ ♀
07 27	☿ ∥ ♇
08 57	☽ ∠ ♃
20 23	☽ Q ♆
20 37	☽ △ ♄
23 17	☽ Q ♂
23 37	☉ ∥ ♀
12 Friday	
03 03	☽ ⊥ ♃
04 06	☽ □ ☉
06 08	♀ ⚹ ♇
09 27	☽ ⋕ ♅
09 39	♅ St D
11 55	☽ ☌ ♇
12 33	☽ ⚹ ♀
15 00	☽ ⊻ ♃
18 07	☿ ☍ ♆
19 29	☽ ∥ ♅
20 20	☽ ⚹ ♆
20 37	☽ △ ☿
13 Saturday	
02 49	☽ ⚼ ♄
07 38	☽ ☌ ♅
13 01	♀ □ ♃
13 03	☽ □ ♂
14 46	☽ Q ♀
20 34	☽ ± ♄
14 Sunday	
01 35	☽ ∠ ♆
03 45	☽ ∥ ♃
05 38	☽ ⚼ ☿
08 13	☽ ⊼ ♄
18 43	☽ ⊥ ♆
19 32	☽ △ ☉
22 02	☉ ⊥ ♄
22 09	☽ ⊻ ♇
22 51	☿ ⋕ ♆
15 Monday	
00 53	☽ ± ☿
00 54	☽ ☌ ♃
01 04	☿ ± ♃
04 37	☽ □ ♀
05 55	☽ ⊻ ♆
06 31	☽ ⋕ ♄
07 50	☿ ⊥ ♂
09 20	☽ ⊥ ♇
13 28	☽ ⊼ ☿
13 59	☽ ⋕ ♇
16 37	☽ ⊻ ♅
18 32	♀ △ ♆
21 45	☽ ∥ ♆
16 Tuesday	
00 06	☽ ⋕ ☿
00 37	☽ △ ♂
01 26	☽ ⚼ ☉
01 48	☽ ∠ ♇
03 32	☽ ⊥ ♅
06 41	☉ □ ♇
09 12	♀ Q ♄
15 58	☽ ☍ ♄
16 26	☿ △ ♅
17 56	☽ ⋕ ♀
18 46	☽ ± ☉
19 36	☽ ∠ ♅
17 Wednesday	
01 59	☽ ⋕ ☉
04 29	☽ ⚹ ♇
04 39	☽ ⚼ ♂
06 05	☽ ⊼ ☉
06 35	☽ ⋕ ♂
06 56	☽ ⊻ ♃
11 36	☽ ☌ ♆
15 29	☽ △ ♀
17 18	☽ ⊥ ♃
18 04	☉ △ ♃
20 58	☽ ± ♂
21 40	☽ ⚹ ♅
18 Thursday	
01 10	☽ Q ♇
01 23	☽ ☍ ☿
07 41	☽ ⊼ ♂
08 37	☽ ∠ ♃
17 58	☽ Q ♅
19 14	☽ ⚼ ♀
19 56	☽ ⊼ ♄
19 Friday	
05 59	☽ ± ♄
07 23	☽ □ ♇
09 38	☽ ⚹ ♃
11 30	☽ ± ♀
12 27	☽ ☍ ☉
14 03	☽ ⊻ ♆
20 56	☽ ⚼ ♄
22 14	☽ ⊼ ♀
23 46	☽ □ ♅
23 53	☽ ⊥ ♆
20 Saturday	
05 20	☽ Q ♃
09 46	☽ ⊼ ☿
11 41	☽ ☍ ♂
12 08	☉ ⊼ ♆
13 39	☿ ⚼ ♃
14 34	☽ ∠ ♆
17 16	♀ ⚹ ♅
20 52	☽ ± ☿
21 37	☽ △ ♄
23 36	☿ ∥ ♀
21 Sunday	
08 28	☽ △ ♇
10 00	☿ ⊻ ♂
10 38	☽ □ ♃
11 55	☽ ⋕ ♂
13 30	☽ ⚼ ☿
14 23	☽ ⋕ ☉
14 55	☽ ⚹ ♆
16 18	♂ Q ♇
16 49	☽ ⊼ ☉
21 43	☉ ⊥ ♀
22 Monday	
00 39	☽ △ ♅
01 48	☽ ⋕ ☿
03 16	☽ ± ☉
03 21	☽ ☍ ♀
05 54	☽ ⋕ ♀
08 54	☽ ⚼ ♇
10 27	☽ Q ♆
14 48	☽ ⊼ ♂
16 37	☽ ∥ ♆
17 24	☽ △ ☿
18 59	☽ ⚼ ☉
22 52	☽ □ ♄
23 39	☽ ± ♇
23 44	☽ ⋕ ♇
23 Tuesday	
01 05	☽ ± ♂
01 12	☽ ⚼ ♅
02 43	♂ ⚼ ♆
05 07	☽ ⋕ ♄
09 32	☽ ⊼ ♇
09 42	♃ St D
11 41	☽ △ ♃
16 00	☽ □ ♆
16 03	☽ ± ♅
16 38	☽ ⚼ ♂
21 29	☽ △ ☉
24 Wednesday	
02 02	☽ ⊼ ♅
04 53	☽ Q ♄
05 28	☽ ∥ ♃
09 10	☽ ⊼ ♀
12 41	☽ ⚼ ♃
13 59	☿ ⊻ ♄
18 55	☽ △ ♂
20 10	☽ ± ♀
25 Thursday	
01 14	☽ ⚹ ♄
01 32	♀ ∠ ♇
02 55	☽ □ ☿
03 53	☽ ± ♃
09 10	☽ ∥ ♅
11 55	☽ ☍ ♇
12 52	☽ ⚼ ♀
13 20	☉ ∠ ♄
14 08	☽ ⊼ ♃
18 05	☽ ⋕ ♅
18 30	☽ △ ♆
26 Friday	
03 07	☽ ∠ ♄
04 01	☽ □ ☉
05 03	☽ ☍ ♅
15 02	☽ Q ☿
15 46	☉ ± ♆
17 12	☽ △ ♀
18 54	☽ ⊥ ♄
19 11	☉ □ ♅
20 31	☽ ⚼ ♆
22 42	☽ ⋕ ♃
27 Saturday	
01 09	☽ □ ♂
05 32	☽ ⊻ ♄
12 22	☽ ± ♆
13 58	☽ Q ☉
15 54	☽ ⚹ ☿
16 19	☽ ⊼ ♇
18 20	☿ ⊥ ♄
18 36	☽ ☍ ♃
18 40	☿ □ ♇
18 46	☉ ∥ ☿
23 03	☽ ⊼ ♆
28 Sunday	
01 19	☽ ∥ ♄
02 05	☿ ∥ ♂
03 03	☽ ± ♇
06 24	☽ ∥ ♇
09 45	☿ △ ♃
10 09	☽ ⊼ ♅
10 52	☽ Q ♂
13 07	☽ ⚹ ☉
14 25	☽ ⋕ ♆
15 40	☽ ∥ ♀
18 28	♀ Q ♃
19 18	☽ ⚼ ♇
21 02	☽ ± ♅
23 51	☽ ∠ ☿
29 Monday	
03 57	☽ □ ♀
09 05	♀ ⋕ ♆
09 50	☽ ⚹ ♂
11 59	☽ ☌ ♄
13 30	☽ ⚼ ♅
13 54	☿ ⊼ ♆
18 38	☽ ∥ ☉
18 40	☽ ∠ ☉
18 53	☽ ∥ ♂
19 30	☽ ⊥ ☿
22 37	☽ ∥ ☿
22 49	☽ △ ♇
30 Tuesday	
01 13	☽ ⊼ ♃
05 41	☽ ☍ ♆
08 48	☽ ⊻ ☿
12 24	☽ ± ♃
12 42	☽ ⊥ ☉
15 05	☽ ∠ ♂
16 31	☽ Q ♀
17 23	☽ △ ♅

JULY 1970

LONGITUDES

Date	Sidereal time h m s	Sun ☉ ° ' "	Moon ☽ ° ' "	Moon ☽ 24.00 ° ' "	Mercury ☿ ° '	Venus ♀ ° '	Mars ♂ ° '	Jupiter ♃ ° '	Saturn ♄ ° '	Uranus ♅ ° '	Neptune ♆ ° '	Pluto ♇ ° '
01	06 36 30	09 ♋ 13 51	14 ♊ 40 54	20 ♊ 58 53	02 ♋ 38	17 ♌ 18	19 ♋ 09	26 ♎ 10	19 ♉ 10	04 ♎ 49	28 ♏ 31	24 ♍ 52
02	06 40 26	10 11 05	27 ♊ 13 53	03 ♋ 25 58	04 47	18 27	19 48	26 12	19 15	04 50	28 R 30	24 53
03	06 44 23	11 08 18	09 ♋ 35 13	15 ♋ 41 43	06 57	19 37	20 27	26 13	19 21	04 51	28 29	24 54
04	06 48 20	12 05 32	21 ♋ 45 38	27 ♋ 47 07	09 07	20 46	21 06	26 15	19 27	04 52	28 28	24 54
05	06 52 16	13 02 45	03 ♌ 46 25	09 ♌ 43 46	11 17	21 56	21 45	26 17	19 32	04 53	28 27	24 55
06	06 56 13	13 59 59	15 ♌ 39 28	21 ♌ 33 55	13 28	23 05	22 23	26 20	19 38	04 55	28 26	24 56
07	07 00 09	14 57 12	27 ♌ 27 29	03 ♍ 20 38	15 38	24 14	23 02	26 22	19 44	04 56	28 25	24 57
08	07 04 06	15 54 25	09 ♍ 13 52	15 ♍ 07 42	17 47	25 24	23 41	26 25	19 49	04 57	28 24	24 58
09	07 08 02	16 51 38	21 ♍ 02 43	26 ♍ 59 31	19 55	26 33	24 20	26 27	19 54	04 59	28 23	24 59
10	07 11 59	17 48 51	02 ♎ 58 44	09 ♎ 00 58	22 03	27 42	24 59	26 30	20 00	05 00	28 22	25 01
11	07 15 55	18 46 04	15 ♎ 06 54	21 ♎ 17 09	24 09	28 50	25 38	26 33	20 05	05 02	28 21	25 02
12	07 19 52	19 43 16	27 ♎ 32 19	03 ♏ 52 59	26 14	29 ♌ 59	26 16	26 37	20 10	05 03	28 20	25 03
13	07 23 49	20 40 29	10 ♏ 19 40	16 ♏ 52 47	28 ♋ 18	01 ♍ 08	26 55	26 40	20 15	05 05	28 19	25 04
14	07 27 45	21 37 42	23 ♏ 32 41	00 ♐ 19 34	00 ♌ 20	02 16	27 34	26 44	20 20	05 06	28 18	25 05
15	07 31 42	22 34 55	07 ♐ 13 29	14 ♐ 14 21	02 20	03 25	28 13	26 47	20 25	05 08	28 18	25 06
16	07 35 38	23 32 08	21 ♐ 21 52	28 ♐ 35 35	04 19	04 33	28 51	26 51	20 30	05 10	28 17	25 08
17	07 39 35	24 29 22	05 ♑ 54 49	13 ♑ 18 48	06 16	05 41	29 ♋ 30	26 55	20 35	05 12	28 16	25 09
18	07 43 31	25 26 35	20 ♑ 46 31	28 ♑ 16 56	08 11	06 49	00 ♌ 09	27 00	20 39	05 13	28 15	25 10
19	07 47 28	26 23 49	05 ♒ 48 53	13 ♒ 21 14	10 04	07 57	00 47	27 04	20 44	05 15	28 15	25 12
20	07 51 24	27 21 04	20 ♒ 52 51	28 ♒ 22 40	11 56	09 05	01 26	27 09	20 49	05 17	28 14	25 13
21	07 55 21	28 18 19	05 ♓ 49 44	13 ♓ 13 12	13 45	10 13	02 04	27 13	20 53	05 19	28 13	25 14
22	07 59 18	29 ♋ 15 35	20 ♓ 32 26	27 ♓ 46 52	15 33	11 20	02 43	27 18	20 57	05 21	28 13	25 16
23	08 03 14	00 ♌ 12 51	04 ♈ 56 10	12 ♈ 00 05	17 20	12 28	03 22	27 23	21 02	05 23	28 12	25 17
24	08 07 11	01 10 09	18 ♈ 58 31	25 ♈ 51 29	19 04	13 35	04 00	27 28	21 06	05 25	28 12	25 19
25	08 11 07	02 07 27	02 ♉ 39 04	09 ♉ 21 26	20 47	14 42	04 39	27 34	21 10	05 27	28 11	25 20
26	08 15 04	03 04 46	15 ♉ 58 48	22 ♉ 31 26	22 27	15 49	05 17	27 39	21 14	05 30	28 11	25 22
27	08 19 00	04 02 07	28 ♉ 59 37	05 ♊ 23 37	24 07	16 56	05 56	27 45	21 18	05 32	28 10	25 23
28	08 22 57	04 59 28	11 ♊ 43 44	18 ♊ 00 17	25 44	18 02	06 34	27 51	21 22	05 34	28 10	25 25
29	08 26 53	05 56 51	24 ♊ 13 31	00 ♋ 23 42	27 19	19 09	07 13	27 56	21 26	05 36	28 09	25 26
30	08 30 50	06 54 14	06 ♋ 31 07	12 ♋ 36 00	28 ♌ 53	20 15	07 51	28 03	21 29	05 39	28 09	25 28
31	08 34 47	07 ♌ 51 38	18 ♋ 38 35	24 ♋ 39 07	00 ♍ 25	21 ♍ 22	08 ♌ 30	28 ♎ 09	21 ♉ 33	05 ♎ 41	28 ♏ 09	25 ♍ 29

MOON NODES AND LATITUDE

Date	Moon True ☊ ° '	Moon Mean ☊ ° '	Moon ☽ Latitude ° '
01	04 ♓ 20	05 ♓ 40	05 N 00
02	04 R 10	05 37	04 41
03	04 01	05 34	04 09
04	03 53	05 31	03 26
05	03 47	05 28	02 34
06	03 43	05 25	01 36
07	03 41	05 21	00 N 34
08	03 D 41	05 18	00 S 30
09	03 42	05 15	01 33
10	03 43	05 12	02 32
11	03 44	05 09	03 26
12	03 R 44	05 06	04 11
13	03 42	05 02	04 45
14	03 39	04 59	05 05
15	03 34	04 56	05 10
16	03 28	04 53	04 55
17	03 22	04 50	04 22
18	03 17	04 46	03 30
19	03 13	04 43	02 24
20	03 11	04 40	01 S 07
21	03 D 10	04 37	00 N 15
22	03 11	04 34	01 34
23	03 13	04 31	02 45
24	03 14	04 27	03 45
25	03 R 14	04 24	04 30
26	03 14	04 21	04 59
27	03 12	04 18	05 12
28	03 08	04 15	05 10
29	03 04	04 12	04 52
30	03 00	04 08	04 22
31	02 ♓ 56	04 ♓ 05	03 N 39

DECLINATIONS

Date	Sun ☉ ° '	Moon ☽ ° '	Mercury ☿ ° '	Venus ♀ ° '	Mars ♂ ° '	Jupiter ♃ ° '	Saturn ♄ ° '	Uranus ♅ ° '	Neptune ♆ ° '	Pluto ♇ ° '
01	23 N 07	27 N 32	24 N 03	17 N 23	23 N 07	08 S 56	15 N 25	01 S 16	18 S 09	16 N 25
02	23 03	28 06	24 09	17 00	23 02	08 57	15 27	01 17	18 09	16 25
03	22 59	27 14	24 12	16 37	22 56	08 58	15 28	01 17	18 09	16 24
04	22 54	25 04	24 13	16 13	22 50	08 59	15 30	01 18	18 08	16 23
05	22 48	21 48	24 10	15 50	22 44	09 00	15 31	01 18	18 08	16 22
06	22 43	17 40	24 05	15 25	22 38	09 01	15 32	01 19	18 08	16 22
07	22 36	12 53	23 57	15 01	22 32	09 02	15 34	01 19	18 08	16 21
08	22 30	07 39	23 47	14 36	22 25	09 03	15 35	01 20	18 08	16 20
09	22 23	02 N 08	23 33	14 11	22 19	09 04	15 36	01 20	18 07	16 19
10	22 16	03 S 31	23 18	13 45	22 12	09 05	15 37	01 21	18 07	16 19
11	22 08	09 07	22 59	13 19	22 05	09 07	15 39	01 22	18 07	16 18
12	22 00	14 30	22 39	12 53	21 58	09 08	15 40	01 22	18 07	16 17
13	21 51	19 25	22 16	12 26	21 51	09 10	15 41	01 23	18 07	16 16
14	21 42	23 35	21 52	11 59	21 43	09 11	15 42	01 24	18 07	16 15
15	21 33	26 36	21 25	11 32	21 36	09 13	15 43	01 24	18 07	16 14
16	21 24	28 05	20 57	11 05	21 28	09 15	15 44	01 25	18 07	16 14
17	21 14	27 41	20 28	10 37	21 20	09 16	15 46	01 26	18 06	16 13
18	21 03	25 18	19 57	10 10	21 12	09 18	15 47	01 26	18 06	16 12
19	20 53	21 09	19 24	09 42	21 04	09 20	15 48	01 27	18 06	16 11
20	20 42	15 35	18 51	09 13	20 55	09 22	15 49	01 28	18 06	16 10
21	20 30	09 09	18 16	08 45	20 47	09 24	15 50	01 29	18 06	16 09
22	20 19	02 S 19	17 41	08 16	20 38	09 26	15 51	01 30	18 06	16 08
23	20 07	04 N 29	17 05	07 48	20 29	09 28	15 52	01 30	18 06	16 07
24	19 54	10 54	16 28	07 19	20 20	09 30	15 52	01 31	18 06	16 06
25	19 42	16 37	15 50	06 50	20 11	09 32	15 53	01 32	18 06	16 06
26	19 28	21 24	15 12	06 20	20 02	09 34	15 54	01 33	18 06	16 05
27	19 15	25 01	14 34	05 51	19 53	09 36	15 55	01 34	18 06	16 04
28	19 01	27 19	13 55	05 22	19 43	09 39	15 56	01 35	18 06	16 03
29	18 47	28 11	13 16	04 52	19 34	09 41	15 57	01 36	18 06	16 02
30	18 33	27 38	12 36	04 22	19 24	09 43	15 57	01 37	18 06	16 01
31	18 N 18	25 N 46	11 N 57	03 N 52	19 N 14	09 S 46	15 N 58	01 S 38	18 S 06	16 N 00

ZODIAC SIGN ENTRIES

Date	h	m	Planets
02	17	21	☽ ♋
05	04	26	☽ ♌
07	17	11	☽ ♍
10	06	02	☽ ♎
12	12	16	♀ ♍
12	16	41	☽ ♏
14	08	06	☿ ♌
14	23	26	☽ ♐
17	02	19	☽ ♑
18	06	43	♂ ♌
19	02	44	☽ ♒
21	02	36	☽ ♓
23	03	42	☽ ♈
23	06	37	☉ ♌
25	07	18	☽ ♉
27	13	53	☽ ♊
29	23	14	☽ ♋
31	05	21	☿ ♍

LATITUDES

Date	Mercury ☿ ° '	Venus ♀ ° '	Mars ♂ ° '	Jupiter ♃ ° '	Saturn ♄ ° '	Uranus ♅ ° '	Neptune ♆ ° '	Pluto ♇ ° '
01	00 N 38	01 N 49	01 N 03	01 N 16	02 S 11	00 N 42	01 N 44	15 N 43
04	01 05	01 45	01 04	01 15	02 11	00 42	01 44	15 41
07	01 26	01 40	01 04	01 14	02 12	00 42	01 43	15 40
10	01 40	01 34	01 05	01 13	02 12	00 42	01 43	15 39
13	01 48	01 27	01 05	01 12	02 13	00 42	01 43	15 38
16	01 49	01 20	01 06	01 12	02 13	00 42	01 43	15 37
19	01 45	01 12	01 06	01 11	02 14	00 41	01 43	15 35
22	01 35	01 02	01 06	01 10	02 14	00 41	01 43	15 34
25	01 20	00 52	01 07	01 09	02 15	00 41	01 42	15 33
28	01 02	00 41	01 07	01 09	02 15	00 41	01 42	15 32
31	00 N 40	00 N 29	01 N 08	01 N 08	02 S 16	00 N 41	01 N 42	15 N 31

DATA

Julian Date	2440769
Delta T	+41 seconds
Ayanamsa	23° 26' 49"
Synetic vernal point	05° ♓ 40' 10"
True obliquity of ecliptic	23° 26' 43"

MOON'S PHASES, APSIDES AND POSITIONS ☽

Date	h	m	Phase	Longitude ° '	Eclipse Indicator
03	15	18	●	11 ♋ 16	
11	19	43	☽	19 ♎ 04	
18	19	58	○	25 ♑ 46	
25	11	00	☾	02 ♉ 05	

Day	h	m		
07	11	59	Apogee	
19	21	39	Perigee	
02	09	20	Max dec	28° N 07'
09	21	06	0S	
16	19	08	Max dec	28° S 10'
22	20	05	0N	
29	14	37	Max dec	28° N 12'

LONGITUDES

Date	Chiron ⚷ ° '	Ceres ⚳ ° '	Pallas ⚴ ° '	Juno ⚵ ° '	Vesta ⚶ ° '	Black Moon Lilith ⚸ ° '
01	10 ♈ 18	27 ♈ 39	14 ♓ 37	27 ♈ 54	05 ♍ 55	22 ♌ 55
11	10 ♈ 22	00 ♉ 16	14 ♓ 15	02 ♉ 14	10 ♍ 12	24 ♌ 02
21	10 ♈ 21	02 ♉ 36	13 ♓ 19	06 ♉ 24	14 ♍ 39	25 ♌ 08
31	10 ♈ 14	04 ♉ 35	11 ♓ 50	10 ♉ 23	19 ♍ 15	26 ♌ 15

ASPECTARIAN

01 Wednesday
00 52 ☽ ⚺ ☉
03 42 ☿ ∠ ♀
04 45 ☉ ∥ ♂
05 20 ☽ ⚼ ♃
08 56 ☽ ⊥ ♂
12 31 ♂ ✱ ♄
17 28 ☽ ✱ ♀
20 35 ☽ ⚺ ♄
20 58 ☽ ⚺ ♂

02 Thursday
05 53 ☿ ∠ ♄
07 28 ☽ □ ♇
08 10 ☽ ⊥ ♄
08 52 ☿ ± ♆
10 00 ☽ △ ♃
12 36 ☿ □ ♅
14 26 ☽ ⚻ ♆
20 51 ♀ ⊥ ♇

03 Friday
01 17 ☽ ∠ ♀
01 42 ☽ ∠ ♄
02 03 ☽ ± ♆
02 45 ☽ □ ♅
05 44 ☽ ☌ ☿
06 05 ♀ □ ♄
15 18 ☽ ☌ ☉
17 00 ♀ ∠ ♅
18 29 ☽ Q ♇
19 38 ☽ ⚼ ♆
20 44 ☽ ⊥ ♀

04 Saturday
01 45 ♀ ∥ ♇
07 23 ☽ ✱ ♄
09 50 ☽ ⚺ ♀
10 36 ☽ ☌ ♂
14 13 ☽ Q ♅
18 16 ☽ ✱ ♇
19 09 ☽ ∥ ☿
20 58 ☽ □ ♃

05 Sunday
01 20 ☽ △ ♆
03 09 ♀ ⚺ ♂
05 12 ☽ ∥ ☉
05 40 ☽ ∥ ♂
07 29 ☽ Q ♄
08 51 ☉ Q ♇
14 15 ☽ ✱ ♅
21 50 ☉ ⚼ ♆

06 Monday
00 24 ☽ ∠ ♇
05 35 ♀ ∥ ♄
06 11 ☿ Q ♇
06 33 ☽ ⚺ ☿
08 21 ☽ ⚺ ☉
09 17 ☽ Q ♃
09 30 ☽ # ♆
11 38 ☿ ⚼ ♆
18 40 ☽ ⊥ ♇
18 52 ☽ ∥ ♇
20 08 ☽ □ ♄
20 39 ☽ ∠ ♅
21 28 ☽ ⊥ ☿
21 35 ☽ ⊥ ☉
22 39 ☉ ☌ ☿
22 59 ☽ ∥ ♄

07 Tuesday
00 42 ☽ ∥ ♀
02 29 ☽ ⚺ ♂
04 44 ☽ ☌ ♀
06 54 ☽ ⚺ ♇
07 56 ♂ Q ♅
09 46 ☽ ✱ ♃
13 56 ☽ □ ♆
15 01 ☽ ⊥ ♅
15 25 ☽ ⊥ ♂
17 32 ☽ ∠ ☉
19 55 ☽ ∠ ☿

08 Wednesday
03 06 ♀ ⚺ ♇
03 16 ☽ ⚺ ♅
05 45 ☽ # ♃
10 50 ☽ ∠ ♂
16 27 ☽ ∠ ♃

09 Thursday
02 33 ☽ Q ♆
02 46 ☽ ✱ ☉
09 14 ☽ ✱ ☿
09 41 ☽ △ ♄
10 03 ♀ ✱ ♃
10 48 ☽ ⊥ ♃
11 47 ☿ ✱ ♄
15 23 ☽ # ♅
19 02 ☽ ✱ ♂
19 59 ☽ ☌ ♇
22 58 ☽ ⚺ ♃

10 Friday
00 17 ☽ ⚺ ♀
02 46 ☽ ✱ ♆
02 49 ☽ ∥ ♅
03 12 ☿ ⊥ ♀
05 07 ☽ Q ☉
13 05 ♂ ✱ ♇
13 35 ☽ ⊥ ♀
14 36 ☽ Q ☿
15 10 ♄ ⚼ ♅
16 03 ☽ ☌ ♅
16 03 ☽ ⚼ ♄
20 25 ☽ Q ♂
22 58 ☿ Q ♅

11 Saturday
01 48 ♀ □ ♆
08 32 ☽ ∠ ♆
09 15 ☽ ∠ ♀
09 58 ☽ ± ♄
12 00 ☽ ∥ ♃
15 59 ♀ ⊥ ♅
19 43 ☽ □ ☉
21 44 ☽ ⚻ ♄
22 07 ☿ ✱ ♇

12 Sunday
02 02 ☽ ⊥ ♆
05 11 ☽ # ♀
07 14 ☽ ⚺ ♇
09 02 ☽ □ ☿
09 27 ☽ □ ♂
10 13 ☽ ☌ ♃
12 35 ☿ ☌ ♂
13 31 ☽ ⚺ ♆
16 27 ☿ □ ♃
17 07 ☽ ✱ ♀
17 28 ☽ # ♄
18 40 ☽ ⊥ ♇
20 21 ☽ # ♇

13 Monday
00 19 ☉ ✱ ♄
01 45 ♂ □ ♃
02 13 ☽ ⚺ ♅
05 19 ☽ ∥ ♆
11 31 ☽ ∠ ♇
12 16 ☿ △ ♆
13 23 ☽ ⊥ ♅
17 39 ☽ Q ♀
22 29 ☉ ∥ ♂

14 Tuesday
00 52 ☽ # ☉
00 53 ☽ # ♂
02 22 ☽ # ☿
05 49 ☽ ∠ ♅
06 13 ☽ ☍ ♄
08 18 ☽ △ ☉
14 45 ☽ ✱ ♇
17 41 ☽ ⚺ ♃
19 30 ☽ △ ♂
20 26 ☽ ☌ ♆
23 14 ☿ ∥ ♂

15 Wednesday
01 24 ☉ ∥ ☿
02 04 ☽ △ ☿
04 16 ☽ ⊥ ♃
04 48 ☽ □ ♀
08 23 ☽ ✱ ♅
11 48 ☽ Q ♇
12 40 ☽ ⚼ ☉
13 03 ☿ Q ♄
15 03 ♂ △ ♆
19 52 ☽ ∠ ♃
22 44 ☽ ⚼ ♂

16 Thursday
02 20 ☉ Q ♅
04 56 ☽ Q ♅
05 07 ☽ ± ☉
08 01 ☽ ⚼ ☿
10 33 ☽ ⚻ ♄
14 36 ☽ ± ♂
15 53 ☽ ⚻ ☉
18 17 ☽ □ ♇
19 02 ☿ ⚺ ♀
20 35 ☽ ± ♄
21 10 ☽ ✱ ♃
22 37 ☿ ✱ ♅
23 28 ☽ ⚺ ♆

17 Friday
01 00 ☽ ⚻ ♂
01 16 ♀ ⚺ ♅
01 19 ☽ ± ☿
02 44 ♀ ⊥ ♂
09 19 ☽ ⊥ ♆
10 49 ☽ □ ♅
11 27 ☽ ⚼ ♄
11 36 ☽ △ ♀
12 39 ☽ ⚻ ☿
16 55 ☽ Q ♃
23 55 ☽ ∠ ♆

18 Saturday
04 58 ☉ ✱ ♇
11 48 ☽ △ ♄
13 49 ☽ ⚼ ♀
19 03 ☽ △ ♇
19 58 ☽ ☍ ☉
22 00 ☽ □ ♃
23 57 ☽ ✱ ♆

19 Sunday
03 38 ☽ ☍ ♂
05 21 ☽ ± ♀
11 06 ☽ △ ♅
12 24 ☽ # ♂
13 19 ☽ # ☉
13 38 ☿ ∠ ♇
15 41 ☽ ⚻ ♀
18 59 ☽ ⚼ ♇
19 03 ☽ Q ♆
19 44 ☽ ☍ ☿
20 59 ☽ # ☿

20 Monday
01 44 ☽ ∥ ♆
05 29 ♀ # ♃
06 19 ☉ □ ♃
09 20 ☽ ± ♇
09 42 ☽ # ♇
11 03 ☽ ⚼ ♅
11 08 ☽ # ♄
11 53 ☽ □ ♄
18 56 ☽ ⚻ ♇
22 04 ☽ △ ♃
23 03 ☽ ⚻ ☉
23 46 ☽ □ ♆

21 Tuesday
01 29 ☽ ± ♅
05 40 ☽ ⚻ ♂
09 22 ☽ ± ☉
09 57 ☉ △ ♆
11 08 ☽ ∥ ♃
11 10 ☽ ⚻ ♅
13 32 ☽ # ♀
15 48 ☽ ± ♂
16 58 ☽ Q ♄
19 09 ☿ # ♆
19 41 ☽ ☍ ♀
22 26 ☽ ⚼ ♃

22 Wednesday
00 59 ☽ ⚼ ☉
02 41 ☽ ⚻ ☿
07 08 ☽ ⚼ ♂
08 27 ☿ Q ♃
12 41 ☽ ✱ ♄
13 16 ☽ ± ♃
13 54 ☽ ± ☿
14 52 ☽ ∥ ♅
19 49 ☽ ☍ ♇
22 07 ♂ Q ♄
23 16 ☽ ⚻ ♃

23 Thursday
00 43 ☽ △ ♆
01 22 ☽ # ♅
03 30 ☽ △ ☉
06 59 ☽ ⚼ ☿
09 13 ☽ △ ♂
10 20 ♀ ∠ ♃
12 46 ☽ ☍ ♅
13 51 ☽ ∠ ♄
23 17 ☽ ∥ ♀

24 Friday
01 54 ☽ ⚻ ♀
02 03 ☽ ⚼ ♆
05 17 ☽ ⊥ ♄
06 32 ☽ # ♃
12 11 ☽ △ ☿
13 08 ☽ ± ♀
15 27 ☿ ⊥ ♇
15 42 ☽ ⚺ ♄
17 35 ☽ ± ♆
23 03 ☽ ⚻ ♇

25 Saturday
02 04 ☿ ∥ ♇
02 56 ☽ ☍ ♃
04 06 ☽ ⚻ ♆
06 18 ☽ ⚼ ♀
07 24 ☿ ∠ ♅
08 44 ☽ ∥ ♄
08 53 ☽ ∥ ☿
09 39 ☽ ∥ ♇
09 39 ☽ ± ♇
10 07 ☿ ∥ ♄
11 00 ☽ □ ☉
15 44 ☽ □ ♂
17 01 ☽ ⚻ ♅
17 47 ☿ □ ♄
18 54 ☽ # ♆

26 Sunday
01 47 ☽ ⚼ ♇
02 10 ☽ ∥ ☉
03 50 ☽ ± ♅
04 50 ☽ ∥ ♂
11 40 ☽ △ ♀
16 11 ☉ Q ♄
19 45 ♀ Q ♆
20 15 ♂ ✱ ♅
20 17 ☽ ⚼ ♅
21 40 ☽ ☌ ♄
22 05 ☽ Q ☉

27 Monday
01 37 ☽ □ ☿
02 06 ☽ Q ♂
05 17 ☽ △ ♇
09 40 ☽ ⚻ ♃
10 28 ☽ ☍ ♆
20 58 ☽ ± ♃
22 12 ☽ ✱ ☉

28 Tuesday
00 18 ☽ △ ♅
01 42 ☽ ✱ ♂
07 09 ☿ ⚺ ♇
14 08 ☽ ⚼ ♃
16 22 ☽ Q ☿

29 Wednesday
01 15 ☽ □ ♀
03 05 ☉ ✱ ♅
05 08 ☽ ∠ ☉
06 34 ☽ ⚺ ♄
07 53 ☽ ∠ ♂
14 21 ☽ □ ♇
18 15 ☽ ⊥ ♄
18 54 ☽ ✱ ☿
19 17 ☽ △ ♃
19 38 ☽ ⚻ ♆
22 04 ☿ ✱ ♃

30 Thursday
00 04 ☽ ⊥ ☉
00 41 ☿ □ ♆
02 21 ☽ ⊥ ♂
07 21 ☽ ± ♆
10 17 ☽ □ ♅
11 57 ☽ ∠ ♄
12 49 ☽ ⚺ ☉
14 46 ☽ ⚺ ♂
15 46 ☽ Q ♀

31 Friday
00 07 ☿ ⊥ ♅
01 05 ☽ ⚼ ♆
01 45 ☽ Q ♇
04 40 ☽ ∠ ☿
12 01 ♃ ⚺ ♆
16 21 ♀ △ ♄
17 50 ☽ ✱ ♄
17 58 ☽ ✱ ♀
22 06 ☽ Q ♅

All ephemeris data is given at 12.00 UT and the Moon's longitude is additionally given for 24.00 UT

AUGUST 1970

LONGITUDES

Date	Sidereal time h m s	Sun ☉ ° ' "	Moon ☽ ° ' "	Moon ☽ 24.00 ° ' "	Mercury ☿ ° '	Venus ♀ ° '	Mars ♂ ° '	Jupiter ♃ ° '	Saturn ♄ ° '	Uranus ♅ ° '	Neptune ♆ ° '	Pluto ♇ ° '
01	08 38 43	08 ♌ 49 04	00 ♌ 37 49	06 ♌ 34 57	01 ♍ 56	22 ♍ 28	09 ♌ 08	28 ♎ 15	21 ♉ 37	05 ♎ 44	28 ♏ 08	25 ♍ 31
02	08 42 40	09 46 30	12 ♌ 30 44	18 ♌ 25 25	03 24	23 34	09 47	28 22	21 40	05 46	28 R 08	25 33
03	08 46 36	10 43 57	24 ♌ 19 18	00 ♍ 12 41	04 51	24 39	10 25	28 28	21 43	05 49	28 08	25 34
04	08 50 33	11 41 24	06 ♍ 05 51	11 ♍ 59 10	06 16	25 45	11 03	28 35	21 47	05 51	28 08	25 36
05	08 54 29	12 38 53	17 ♍ 53 00	23 ♍ 47 46	07 39	26 50	11 42	28 42	21 50	05 54	28 08	25 38
06	08 58 26	13 36 22	29 ♍ 43 53	05 ♎ 41 50	09 00	27 56	12 20	28 49	21 53	05 56	28 07	25 40
07	09 02 22	14 33 52	11 ♎ 42 04	17 ♎ 45 07	10 19	29 ♍ 01	12 59	28 56	21 56	05 59	28 07	25 41
08	09 06 19	15 31 23	23 ♎ 51 31	00 ♏ 01 47	11 36	00 ♎ 05	13 37	29 03	21 59	06 02	28 07	25 43
09	09 10 16	16 28 55	06 ♏ 16 28	12 ♏ 36 04	12 51	01 10	14 15	29 11	22 01	06 04	28 07	25 45
10	09 14 12	17 26 28	19 ♏ 01 06	25 ♏ 32 01	14 04	02 14	14 54	29 18	22 04	06 07	28 D 07	25 47
11	09 18 09	18 24 02	02 ♐ 09 13	08 ♐ 53 00	15 15	03 19	15 32	29 26	22 07	06 10	28 07	25 49
12	09 22 05	19 21 36	15 ♐ 43 34	22 ♐ 41 01	16 24	04 23	16 10	29 34	22 09	06 13	28 07	25 51
13	09 26 02	20 19 12	29 ♐ 45 17	06 ♑ 56 07	17 30	05 26	16 49	29 42	22 11	06 16	28 07	25 53
14	09 29 58	21 16 48	14 ♑ 13 09	21 ♑ 35 46	18 34	06 30	17 27	29 50	22 14	06 19	28 08	25 54
15	09 33 55	22 14 26	29 ♑ 03 13	06 ♒ 34 37	19 36	07 33	18 05	29 ♎ 58	22 16	06 22	28 08	25 56
16	09 37 51	23 12 05	14 ♒ 08 53	21 ♒ 44 52	20 34	08 36	18 43	00 ♏ 06	22 18	06 25	28 08	25 58
17	09 41 48	24 09 44	29 ♒ 21 23	06 ♓ 57 13	21 30	09 39	19 22	00 15	22 20	06 28	28 08	26 00
18	09 45 45	25 07 25	14 ♓ 31 10	22 ♓ 02 09	22 23	10 41	20 00	00 23	22 22	06 31	28 08	26 02
19	09 49 41	26 05 08	29 ♓ 29 10	06 ♈ 51 23	23 13	11 44	20 38	00 32	22 23	06 34	28 09	26 04
20	09 53 38	27 02 52	14 ♈ 08 07	21 ♈ 18 51	24 00	12 46	21 16	00 40	22 25	06 37	28 09	26 06
21	09 57 34	28 00 37	28 ♈ 23 14	05 ♉ 21 06	24 43	13 47	21 55	00 49	22 27	06 40	28 09	26 08
22	10 01 31	28 58 25	12 ♉ 12 25	18 ♉ 57 15	25 23	14 49	22 33	00 58	22 28	06 43	28 10	26 10
23	10 05 27	29 ♌ 56 14	25 ♉ 35 50	02 ♊ 08 26	25 59	15 50	23 11	01 07	22 30	06 46	28 10	26 12
24	10 09 24	00 ♍ 54 05	08 ♊ 35 26	14 ♊ 57 13	26 30	16 50	23 49	01 17	22 31	06 50	28 11	26 14
25	10 13 20	01 51 57	21 ♊ 14 15	27 ♊ 26 58	26 58	17 51	24 28	01 26	22 32	06 53	28 11	26 16
26	10 17 17	02 49 52	03 ♋ 35 51	09 ♋ 41 22	27 21	18 51	25 06	01 35	22 33	06 56	28 12	26 18
27	10 21 14	03 47 48	15 ♋ 43 59	21 ♋ 44 07	27 38	19 51	25 44	01 45	22 34	06 59	28 12	26 21
28	10 25 10	04 45 46	27 ♋ 42 13	03 ♌ 38 40	27 51	20 50	26 22	01 54	22 35	07 03	28 13	26 23
29	10 29 07	05 43 46	09 ♌ 33 50	15 ♌ 28 06	27 59	21 49	27 00	02 04	22 35	07 06	28 13	26 25
30	10 33 03	06 41 47	21 ♌ 21 47	27 ♌ 15 13	28 R 00	22 48	27 39	02 14	22 36	07 09	28 14	26 27
31	10 37 00	07 ♍ 39 50	03 ♍ 08 41	09 ♍ 02 29	27 ♍ 56	23 ♎ 47	28 ♌ 17	02 ♏ 24	22 ♉ 36	07 ♎ 13	28 ♏ 15	26 ♍ 29

MOON

Date	Moon True ☊ ° '	Moon Mean ☊ ° '	Moon ☽ Latitude ° '
01	02 ♓ 52	04 ♓ 02	02 N 48
02	02 R 50	03 59	01 50
03	02 48	03 56	00 N 47
04	02 D 48	03 52	00 S 18
05	02 49	03 49	01 22
06	02 50	03 46	02 23
07	02 52	03 43	03 19
08	02 53	03 40	04 06
09	02 54	03 37	04 43
10	02 R 54	03 33	05 08
11	02 54	03 30	05 17
12	02 53	03 27	05 09
13	02 51	03 24	04 44
14	02 50	03 21	03 58
15	02 49	03 17	02 57
16	02 48	03 14	01 42
17	02 48	03 11	00 S 19
18	02 D 48	03 08	01 N 05
19	02 48	03 05	02 23
20	02 49	03 02	03 30
21	02 49	02 58	04 22
22	02 49	02 55	04 58
23	02 50	02 52	05 15
24	02 R 50	02 49	05 16
25	02 49	02 46	05 02
26	02 D 49	02 43	04 33
27	02 50	02 39	03 53
28	02 50	02 36	03 03
29	02 50	02 33	02 06
30	02 50	02 30	01 03
31	02 ♓ 50	02 ♓ 27	00 S 02

DECLINATIONS

Date	Sun ☉ ° '	Moon ☽ ° '	Mercury ☿ ° '	Venus ♀ ° '	Mars ♂ ° '	Jupiter ♃ ° '	Saturn ♄ ° '	Uranus ♅ ° '	Neptune ♆ ° '	Pluto ♇ ° '
01	18 N 04	22 N 45	11 N 17	03 N 23	19 N 04	09 S 48	15 N 59	01 S 39	18 S 06	15 N 59
02	17 48	18 48	10 38	02 53	18 54	09 51	16 00	01 40	18 06	15 58
03	17 33	14 09	09 58	02 23	18 43	09 53	16 00	01 41	18 06	15 57
04	17 17	09 00	09 19	01 53	18 33	09 56	16 01	01 42	18 06	15 56
05	17 01	03 N 32	08 40	01 23	18 22	09 59	16 02	01 43	18 06	15 55
06	16 45	02 S 05	08 01	00 52	18 12	10 01	16 02	01 44	18 06	15 54
07	16 28	07 41	07 22	00 N 22	18 01	10 04	16 03	01 45	18 06	15 53
08	16 11	13 04	06 44	00 S 08	17 50	10 07	16 03	01 46	18 06	15 52
09	15 54	18 04	06 06	00 38	17 39	10 10	16 04	01 47	18 06	15 51
10	15 37	22 24	05 29	01 08	17 28	10 13	16 04	01 49	18 06	15 51
11	15 19	25 46	04 52	01 38	17 16	10 15	16 05	01 50	18 06	15 50
12	15 01	27 48	04 16	02 08	17 05	10 18	16 05	01 51	18 06	15 49
13	14 43	28 09	03 41	02 38	16 54	10 21	16 05	01 52	18 06	15 48
14	14 25	26 38	03 06	03 08	16 42	10 24	16 06	01 53	18 06	15 47
15	14 06	23 14	02 33	03 38	16 30	10 27	16 06	01 54	18 06	15 46
16	13 47	18 13	02 00	04 08	16 19	10 31	16 06	01 56	18 06	15 45
17	13 28	12 00	01 28	04 38	16 07	10 34	16 07	01 57	18 06	15 44
18	13 09	05 S 06	00 58	05 07	15 55	10 37	16 07	01 58	18 07	15 43
19	12 50	01 N 59	00 N 28	05 37	15 42	10 40	16 07	01 59	18 07	15 42
20	12 30	08 48	00 00	06 06	15 30	10 43	16 07	02 00	18 07	15 41
21	12 10	14 59	00 S 26	06 35	15 18	10 47	16 08	02 02	18 07	15 40
22	11 50	20 13	00 51	07 04	15 06	10 50	16 08	02 03	18 07	15 39
23	11 30	24 16	01 15	07 33	14 53	10 53	16 08	02 04	18 07	15 38
24	11 09	26 57	01 36	08 02	14 40	10 57	16 08	02 06	18 07	15 37
25	10 49	28 10	01 56	08 31	14 28	11 00	16 08	02 07	18 07	15 36
26	10 28	27 57	02 13	08 59	14 15	11 03	16 08	02 08	18 08	15 35
27	10 07	26 22	02 28	09 28	14 02	11 07	16 08	02 10	18 08	15 34
28	09 46	23 37	02 40	09 56	13 49	11 10	16 08	02 11	18 08	15 33
29	09 25	19 53	02 50	10 24	13 36	11 14	16 08	02 12	18 08	15 32
30	09 03	15 23	02 57	10 51	13 23	11 17	16 08	02 14	18 08	15 31
31	08 N 42	10 N 20	03 S 01	11 S 19	13 N 10	11 S 21	16 N 08	02 S 15	18 S 09	15 N 30

ZODIAC SIGN ENTRIES

Date	h	m	Planets
01	10	44	☽ ♌
03	23	34	☽ ♍
06	12	32	☽ ♎
08	09	59	♀ ♎
08	23	57	☽ ♏
11	08	07	☽ ♐
13	12	25	☽ ♑
15	13	31	☽ ♒
15	17	58	♃ ♏
17	13	01	☽ ♓
19	12	50	☽ ♈
21	14	46	☽ ♉
23	13	34	☉ ♍
23	20	03	☽ ♊
26	04	58	☽ ♋
28	16	38	☽ ♌
31	05	36	☽ ♍

LATITUDES

Date	Mercury ☿ ° '	Venus ♀ ° '	Mars ♂ ° '	Jupiter ♃ ° '	Saturn ♄ ° '	Uranus ♅ ° '	Neptune ♆ ° '	Pluto ♇ ° '
01	00 N 32	00 N 25	01 N 08	01 N 08	02 S 16	00 N 41	01 N 42	15 N 31
04	00 N 06	00 N 12	01 08	01 07	02 17	00 41	01 42	15 30
07	00 S 21	00 S 01	01 08	01 06	02 17	00 41	01 42	15 29
10	00 51	00 16	01 08	01 06	02 18	00 41	01 42	15 28
13	01 22	00 31	01 09	01 05	02 18	00 41	01 41	15 28
16	01 53	00 47	01 09	01 04	02 19	00 41	01 41	15 27
19	02 25	01 03	01 09	01 04	02 20	00 40	01 41	15 26
22	02 56	01 20	01 09	01 03	02 20	00 40	01 41	15 26
25	03 25	01 38	01 09	01 02	02 21	00 40	01 41	15 25
28	03 51	01 56	01 09	01 02	02 22	00 40	01 40	15 25
31	04 S 10	02 S 15	01 N 10	01 N 01	02 S 22	00 N 40	01 N 40	15 N 25

DATA

Julian Date	2440800
Delta T	+41 seconds
Ayanamsa	23° 26' 55"
Synetic vernal point	05° ♓ 40' 05"
True obliquity of ecliptic	23° 26' 43"

MOON'S PHASES, APSIDES AND POSITIONS ☽

Date	h	m	Phase	Longitude ° '	Eclipse Indicator
02	05	58	●	09 ♌ 32	
10	08	50	☽	17 ♏ 19	
17	03	15	○	23 ♒ 49	partial
23	20	34	☾	00 ♊ 17	
31	22	01	●	08 ♍ 04	Annular

Day	h	m		
03	22	11	Apogee	
17	06	43	Perigee	
31	01	16	Apogee	
06	03	06	OS	
13	04	47	Max dec	28° S 15'
19	05	15	ON	
25	20	12	Max dec	28° N 15'

LONGITUDES

Date	Chiron ⚷ ° '	Ceres ⚳ ° '	Pallas ⚴ ° '	Juno ⚵ ° '	Vesta ⚶ ° '	Black Moon Lilith ⚸ ° '
01	10 ♈ 13	04 ♉ 45	11 ♓ 39	10 ♉ 46	19 ♍ 43	26 ♌ 21
11	10 ♈ 01	06 ♉ 17	09 ♓ 37	14 ♉ 27	24 ♍ 26	27 ♌ 28
21	09 ♈ 45	07 ♉ 19	07 ♓ 13	17 ♉ 49	29 ♍ 16	28 ♌ 34
31	09 ♈ 24	07 ♉ 49	04 ♓ 40	20 ♉ 47	04 ♎ 12	29 ♌ 41

ASPECTARIAN

h m	Aspects
01 Saturday	
01 12	☽ ⊥ ☿
01 43	☽ ✱ ♇
06 55	♀ ⊥ ♃
07 00	☽ △ ♆
07 11	☽ □ ♃
08 49	☉ # ♆
14 25	♄ ∥ ♇
14 59	☽ ⊻ ☿
18 02	☽ Q ♄
22 18	☽ ✱ ♅
02 Sunday	
03 11	☽ ∠ ♀
05 58	☽ ☌ ☉
06 08	☽ ☌ ♂
08 00	☽ ∠ ♇
11 30	☽ ∥ ♂
12 01	☉ ☌ ♂
15 54	☽ # ♆
17 46	☽ ∥ ☉
19 53	☽ Q ♃
23 18	☽ ⊥ ♀
03 Monday	
02 19	☽ ⊥ ♇
02 50	☽ ∥ ♄
03 02	☽ ∥ ♇
04 50	☽ ∠ ♅
06 41	☽ □ ♄
07 56	☉ ∠ ♇
12 45	☽ ⊻ ♀
14 33	☽ ⊻ ♇
14 52	☿ # ♃
18 14	♂ ∠ ♇
19 46	☽ □ ♆
20 32	☽ ✱ ♃
23 13	☽ ⊥ ♅
04 Tuesday	
04 48	☿ ⊻ ♅
07 48	☽ # ♃
08 41	♀ ☌ ♇
10 21	☽ ∥ ☿
11 30	☽ ⊻ ♅
12 23	☽ ☌ ☿
20 13	♀ # ♅
22 41	☽ ⊻ ♂
05 Wednesday	
00 24	☽ ⊻ ☉
03 24	☽ ∠ ♃
04 19	♀ ∠ ♂
08 26	☽ Q ♆
11 36	☽ ⊥ ♂
13 41	☽ ⊥ ☉
19 45	☽ # ♅
20 03	☽ △ ♄
21 52	☽ ⊥ ♃
22 08	☽ ∥ ♀
06 Thursday	
03 45	☽ ☌ ♇
06 53	☽ ∠ ♂
07 15	☽ # ♀
07 59	☽ ☌ ♀
08 45	☽ ✱ ♆
09 32	☽ ∠ ☉
10 08	☽ ⊻ ♃
10 30	☽ ∥ ♅
16 23	♀ ✱ ♆
07 Friday	
00 32	☽ ☌ ♅
01 23	♂ # ♆
02 26	☽ bQ ♄
08 54	☽ ⊻ ☿
10 07	♀ ⊻ ♃
10 49	☽ # ☿
14 41	☽ ✱ ♂
14 50	☽ ∠ ♆
18 11	☽ ✱ ☉
20 26	☽ ± ♄
22 15	☽ ⊥ ☿
22 34	☽ ∥ ♃
08 Saturday	
08 18	☽ ⊼ ♄
08 36	☽ ⊥ ♆
15 37	☽ Q ♂
15 39	☽ ⊻ ♇
17 58	☽ ∠ ☿
19 45	☽ Q ☉
20 18	☽ ⊻ ♆
22 13	☽ ☌ ♃
22 54	☉ ∥ ♄
09 Sunday	
01 06	☽ # ♇
01 16	☽ ⊻ ♀
01 52	☽ # ☉
02 02	☽ # ♄
03 18	☽ ⊥ ♇
09 57	☽ # ♂
11 37	☽ ⊻ ♅
12 09	☽ ∥ ♆
13 52	☽ ⊥ ♀
15 46	☉ ∥ ♇
20 32	☽ ∠ ♇
23 03	☽ ⊥ ♅
10 Monday	
01 48	☽ ✱ ☿
02 21	♆ St D
03 54	☽ □ ♂
08 06	☉ Q ♃
08 24	☽ ∠ ♀
08 50	☽ □ ☉
15 54	☽ ∠ ♅
17 14	☿ ∠ ♃
17 39	☽ ☍ ♄
11 Tuesday	
00 29	☽ ✱ ♇
02 16	☽ Q ☿
04 43	☽ ☌ ♆
07 03	☽ ⊻ ♃
14 15	☽ ✱ ♀
17 56	☽ ⊥ ♃
19 12	☽ ✱ ♅
21 30	♀ ∥ ♅
22 08	☽ Q ♇
12 Wednesday	
00 55	☿ ⊻ ♂
06 07	☿ Q ♆
08 08	☉ ∠ ♀
09 57	☽ ∠ ♃
12 49	☽ △ ♂
13 14	☽ Q ♀
13 16	☽ □ ☿
16 20	☽ Q ♅
18 45	☽ △ ☉
23 07	☽ ⊼ ♄
13 Thursday	
00 30	☉ ⊥ ♇
05 25	☽ □ ♇
09 15	☽ ⊻ ♆
09 21	☽ ± ♄
11 54	☽ ✱ ♃
15 37	☽ bQ ♂
19 19	☽ ⊥ ♆
21 59	☽ bQ ☉
22 16	☽ □ ♀
22 55	☽ □ ♅
14 Friday	
00 27	☽ bQ ♄
07 15	☽ ± ♂
07 34	♀ ☌ ♅
08 03	☽ Q ♃
10 13	☽ ∠ ♆
11 20	☿ # ♀
12 50	☉ ∠ ♅
13 51	☽ ± ☉
17 30	☽ ⊼ ♂
19 38	☽ △ ☿
15 Saturday	
00 17	☽ ⊼ ☉
01 03	☽ △ ♄
05 11	♀ bQ ♄
06 17	♂ Q ♃
06 59	☽ △ ♇
10 31	☽ ✱ ♆
12 36	☉ □ ♄
13 28	☽ □ ♃
21 28	☽ bQ ☿
23 42	☽ △ ♅
16 Sunday	
02 34	☽ △ ♀
05 39	☽ Q ♆
06 58	☽ bQ ♇
12 27	☽ ∥ ♆
12 43	☽ ± ☿
15 11	☿ # ♅
19 33	☽ ☍ ♂
20 01	☽ # ♂
20 33	☽ # ♄
21 13	☽ ± ♇
22 00	☽ # ♇
22 49	☽ ⊼ ☿
23 30	☽ bQ ♅
17 Monday	
00 54	☽ □ ♄
03 15	☽ ☍ ☉
04 02	☽ bQ ♀
06 20	☽ # ☉
06 42	☽ ⊼ ♇
10 04	☽ □ ♆
11 41	♂ ∥ ♄
13 25	☽ △ ♃
13 45	☽ ± ♅
17 06	☽ ∥ ♃
19 16	☽ ± ♀
23 15	☽ ⊼ ♅
18 Tuesday	
05 23	☽ Q ♄
05 28	☽ ⊼ ♀
11 09	☿ △ ♄
11 56	☽ ∥ ♀
13 23	☽ bQ ♃
13 30	♂ ⊥ ♇
21 08	☽ ⊼ ♂
22 34	☽ ∥ ♅
19 Wednesday	
00 33	☽ ✱ ♄
01 19	☽ ☍ ☿
03 03	☽ # ☿
03 56	☽ ± ♃
06 08	☽ ⊼ ☉
06 28	☽ ☍ ♇
07 11	☽ ∥ ☿
07 11	☽ ± ♂
09 50	☽ △ ♆
11 35	☉ ⊻ ♇
12 01	☽ # ♅
13 27	♂ ∥ ♇
13 42	☽ ⊼ ♃
16 31	☽ ± ☉
22 27	☽ bQ ♂
23 34	☽ ☍ ♅
20 Thursday	
00 54	☽ ∠ ♄
01 32	☽ # ♀
08 18	☽ bQ ☉
09 33	☽ ☍ ♀
10 22	☽ bQ ♆
15 48	☽ ⊥ ♄
19 12	☽ # ♃
21 10	♀ ∠ ♆
21 Friday	
00 29	☽ △ ♂
01 15	☽ ∥ ☉
01 25	☽ ± ♆
01 53	☽ ⊻ ♄
01 59	♂ ∠ ♅
05 26	☽ ⊼ ☿
08 09	☽ ⊼ ♇
11 19	☽ △ ☉
11 36	☽ ⊼ ♆
13 16	☽ ∥ ♂
14 52	☽ ∥ ♇
15 40	☉ □ ♆
16 13	☽ ± ☿
16 13	☽ ☍ ♃
16 24	☿ ⊥ ♃
16 51	☽ ∥ ♄
18 27	☽ ± ♇
22 Saturday	
01 45	☽ # ♆
02 21	☽ ⊼ ♅
08 38	☽ bQ ☿
08 55	♂ □ ♄
10 10	☽ bQ ♇
12 55	☽ ± ♅
16 59	☽ ⊼ ♀
23 Sunday	
04 36	☽ ± ♀
05 02	☽ bQ ♅
06 21	☽ ☌ ♄
07 24	☽ □ ♂
12 44	☽ △ ☿
13 07	☽ △ ♇
16 42	☽ ☍ ♆
20 34	☽ □ ☉
22 14	☽ ⊼ ♃
22 23	☽ bQ ♀
22 29	☿ ☌ ♇
24 Monday	
04 02	♀ ± ♄
08 41	☽ △ ♅
09 31	☽ ± ♃
10 02	☉ ⊥ ♅
18 24	☽ Q ♂
23 05	☉ ✱ ♃
25 Tuesday	
00 56	☉ # ♃
02 41	☽ bQ ♃
04 57	☽ △ ♀
09 09	☽ Q ☉
14 30	☽ ⊻ ♄
18 33	☽ ✱ ♂
21 45	☽ □ ♇
23 26	☽ □ ☿
26 Wednesday	
01 26	☽ ⊼ ♆
02 07	☽ ⊥ ♄
04 35	☿ ∥ ♅
08 01	☽ △ ♃
10 22	☽ ✱ ☉
13 10	☽ ± ♆
18 35	☽ □ ♅
19 47	☽ ∠ ♄
27 Thursday	
01 31	☽ ∠ ♂
06 58	☽ bQ ♆
09 13	☽ Q ♇
11 49	☽ Q ☿
18 39	☽ ∠ ☉
20 26	☽ ⊥ ♂
20 58	☽ □ ♀
28 Friday	
01 41	☽ ✱ ♄
06 37	☽ Q ♅
07 13	☉ # ♀
09 10	☽ ⊻ ♂
09 19	☽ ✱ ♇
12 19	☽ ✱ ☿
12 21	♂ ⊻ ♇
13 02	☽ △ ♆
14 19	☽ ⊥ ☉
20 36	☽ □ ♃
29 Saturday	
01 54	☽ Q ♄
03 32	☽ ⊻ ☉
06 59	☽ ✱ ♅
12 35	☽ Q ♀
15 46	☽ ∠ ♇
18 58	☽ ∠ ☿
21 44	☽ # ♆
21 51	♀ ⊥ ♆
30 Sunday	
06 57	♀ ⊼ ♄
07 27	☿ St R
08 13	☽ ∥ ♄
09 39	☽ Q ♃
10 08	☽ ⊥ ♇
11 19	☽ ∥ ♇
13 19	☽ ⊥ ☿
13 38	☽ ∠ ♅
14 31	☽ □ ♄
15 12	☽ ✱ ♀
22 13	☽ ∥ ♂
22 23	☽ ⊻ ♇
31 Monday	
00 10	☉ ⊻ ♅
00 54	☿ ⊻ ♂
01 29	☽ ⊻ ☿
01 31	☽ ☌ ♂
02 01	☽ □ ♆
07 23	☽ # ♃
07 50	☽ # ♀
08 03	☽ ⊥ ♅
10 27	☽ ✱ ♃
10 42	♂ □ ♆
13 56	♀ ∥ ♃
19 53	☽ ∥ ☉
20 19	☽ ⊻ ♅
22 01	☽ ☌ ☉

All ephemeris data is given at 12.00 UT and the Moon's longitude is additionally given for 24.00 UT

SEPTEMBER 1970

LONGITUDES

Date	Sidereal time h m s	Sun ☉ ° ' "	Moon ☽ ° ' "	Moon ☽ 24.00 ° ' "	Mercury ☿ ° '	Venus ♀ ° '	Mars ♂ ° '	Jupiter ♃ ° '	Saturn ♄ ° '	Uranus ♅ ° '	Neptune ♆ ° '	Pluto ♇ ° '
01	10 40 56	08 ♍ 37 54	14 ♍ 56 53	20 ♍ 52 11	27 ♍ 46	24 ♎ 45	28 ♌ 55	02 ♏ 34	22 ♉ 37	07 ♎ 16	28 ♏ 15	26 ♍ 31
02	10 44 53	09 36 00	26 ♍ 48 38	02 ♎ 46 32	27 R 30	25 42	29 ♌ 33	02 44	22 37	07 20	28 16	26 33
03	10 48 49	10 34 08	08 ♎ 46 10	14 ♎ 47 50	27 07	26 39	00 ♍ 11	02 54	22 37	07 23	28 17	26 36
04	10 52 46	11 32 17	20 ♎ 51 50	26 ♎ 58 30	26 39	27 36	00 49	03 04	22 37	07 27	28 18	26 38
05	10 56 43	12 30 28	03 ♏ 08 10	09 ♏ 21 12	26 04	28 32	01 28	03 14	22 R 37	07 30	28 19	26 40
06	11 00 39	13 28 41	15 ♏ 37 56	21 ♏ 58 46	25 23	29 ♎ 28	02 06	03 25	22 37	07 34	28 19	26 42
07	11 04 36	14 26 55	28 ♏ 24 02	04 ♐ 54 07	24 36	00 ♏ 23	02 44	03 35	22 37	07 37	28 20	26 44
08	11 08 32	15 25 10	11 ♐ 29 21	18 ♐ 10 00	23 45	01 18	03 22	03 46	22 37	07 41	28 21	26 46
09	11 12 29	16 23 27	24 ♐ 56 21	01 ♑ 48 33	22 50	02 12	04 00	03 57	22 36	07 44	28 22	26 49
10	11 16 25	17 21 46	08 ♑ 46 42	15 ♑ 50 48	21 51	03 06	04 38	04 07	22 36	07 48	28 23	26 51
11	11 20 22	18 20 06	23 ♑ 00 42	00 ♒ 16 06	20 51	03 59	05 16	04 18	22 35	07 52	28 24	26 53
12	11 24 18	19 18 28	07 ♒ 36 36	15 ♒ 01 36	19 50	04 52	05 54	04 29	22 34	07 55	28 25	26 55
13	11 28 15	20 16 51	22 ♒ 30 21	00 ♓ 01 58	18 49	05 43	06 33	04 40	22 33	07 59	28 26	26 58
14	11 32 12	21 15 16	07 ♓ 35 27	15 ♓ 09 40	17 51	06 35	07 11	04 51	22 32	08 03	28 27	27 00
15	11 36 08	22 13 42	22 ♓ 43 30	00 ♈ 15 44	16 56	07 25	07 49	05 02	22 31	08 06	28 28	27 02
16	11 40 05	23 12 10	07 ♈ 45 16	15 ♈ 11 01	16 06	08 15	08 27	05 13	22 30	08 10	28 30	27 04
17	11 44 01	24 10 41	22 ♈ 32 02	29 ♈ 47 32	15 22	09 04	09 05	05 25	22 28	08 14	28 31	27 06
18	11 47 58	25 09 13	06 ♉ 56 53	13 ♉ 59 36	14 45	09 53	09 43	05 36	22 27	08 17	28 32	27 09
19	11 51 54	26 07 48	20 ♉ 55 26	27 ♉ 44 17	14 17	10 41	10 21	05 48	22 26	08 21	28 33	27 11
20	11 55 51	27 06 25	04 ♊ 26 12	11 ♊ 01 22	13 58	11 27	10 59	05 59	22 24	08 25	28 35	27 13
21	11 59 47	28 05 04	17 ♊ 30 06	23 ♊ 52 49	13 48	12 14	11 37	06 11	22 22	08 29	28 36	27 15
22	12 03 44	29 ♍ 03 46	00 ♋ 10 00	06 ♋ 22 11	13 D 48	12 59	12 15	06 22	22 20	08 32	28 37	27 18
23	12 07 41	00 ♎ 02 29	12 ♋ 29 56	18 ♋ 33 52	13 57	13 43	12 53	06 34	22 18	08 36	28 39	27 20
24	12 11 37	01 01 15	24 ♋ 34 34	00 ♌ 32 39	14 17	14 26	13 32	06 46	22 16	08 40	28 40	27 22
25	12 15 34	02 00 03	06 ♌ 28 41	12 ♌ 23 14	14 46	15 09	14 10	06 58	22 14	08 44	28 41	27 24
26	12 19 30	02 58 54	18 ♌ 16 51	24 ♌ 10 02	15 24	15 50	14 48	07 09	22 12	08 47	28 43	27 27
27	12 23 27	03 57 46	00 ♍ 03 15	05 ♍ 56 56	16 11	16 31	15 26	07 21	22 10	08 51	28 44	27 29
28	12 27 23	04 56 40	11 ♍ 51 27	17 ♍ 47 12	17 06	17 10	16 04	07 33	22 07	08 55	28 46	27 31
29	12 31 20	05 55 37	23 ♍ 44 27	29 ♍ 43 29	18 09	17 48	16 42	07 45	22 05	08 59	28 47	27 33
30	12 35 16	06 ♎ 54 36	05 ♎ 44 33	11 ♎ 47 51	19 ♍ 18	18 ♏ 25	17 ♍ 20	07 ♏ 57	22 ♉ 02	09 ♎ 02	28 ♏ 49	27 ♍ 36

Date	Moon True ☊ ° '	Moon Mean ☊ ° '	Moon ☽ Latitude ° '
01	02 ♓ 50	02 ♓ 23	01 S 07
02	02 R 50	02 20	02 09
03	02 49	02 17	03 07
04	02 48	02 14	03 56
05	02 46	02 11	04 36
06	02 45	02 08	05 03
07	02 44	02 04	05 16
08	02 44	02 01	05 13
09	02 D 44	01 58	04 53
10	02 45	01 55	04 16
11	02 46	01 52	03 23
12	02 47	01 49	02 15
13	02 48	01 45	00 S 57
14	02 R 48	01 42	00 N 26
15	02 47	01 39	01 48
16	02 45	01 36	03 01
17	02 43	01 33	04 01
18	02 40	01 29	04 44
19	02 37	01 26	05 08
20	02 34	01 23	05 15
21	02 32	01 20	05 04
22	02 D 32	01 17	04 39
23	02 33	01 14	04 02
24	02 34	01 10	03 14
25	02 36	01 07	02 19
26	02 37	01 04	01 18
27	02 38	01 01	00 N 14
28	02 R 38	00 58	00 S 50
29	02 36	00 55	01 53
30	02 ♓ 32	00 ♓ 51	02 S 51

DECLINATIONS

Date	Sun ☉ ° '	Moon ☽ ° '	Mercury ☿ ° '	Venus ♀ ° '	Mars ♂ ° '	Jupiter ♃ ° '	Saturn ♄ ° '	Uranus ♅ ° '	Neptune ♆ ° '	Pluto ♇ ° '
01	08 N 20	04 N 54	03 S 01	11 S 46	12 N 56	11 S 24	16 N 08	02 S 16	18 S 09	15 N 29
02	07 58	00 S 43	02 58	12 13	12 43	11 28	16 08	02 18	18 09	15 28
03	07 36	06 20	02 51	12 40	12 30	11 32	16 08	02 19	18 09	15 27
04	07 14	11 47	02 41	13 06	12 16	11 35	16 07	02 20	18 10	15 26
05	06 52	16 52	02 27	13 33	12 03	11 39	16 07	02 22	18 10	15 26
06	06 30	21 21	02 09	13 59	11 49	11 43	16 07	02 23	18 10	15 25
07	06 07	24 56	01 47	14 24	11 35	11 46	16 07	02 25	18 10	15 24
08	05 45	27 20	01 21	14 50	11 21	11 50	16 06	02 26	18 11	15 23
09	05 22	28 14	00 53	15 15	11 07	11 54	16 06	02 28	18 11	15 22
10	05 00	27 25	00 S 21	15 39	10 54	11 58	16 06	02 29	18 11	15 21
11	04 37	24 49	00 N 13	16 04	10 40	12 01	16 05	02 30	18 11	15 20
12	04 14	20 33	00 50	16 28	10 26	12 05	16 05	02 32	18 12	15 19
13	03 51	14 54	01 27	16 51	10 11	12 09	16 05	02 33	18 12	15 18
14	03 28	08 19	02 05	17 15	09 57	12 13	16 04	02 35	18 12	15 17
15	03 05	01 S 14	02 43	17 38	09 43	12 17	16 04	02 36	18 13	15 16
16	02 42	05 N 51	03 19	18 00	09 29	12 21	16 03	02 38	18 13	15 16
17	02 19	12 30	03 54	18 22	09 14	12 25	16 03	02 39	18 13	15 15
18	01 56	18 18	04 27	18 44	09 00	12 28	16 02	02 41	18 13	15 14
19	01 32	22 57	04 56	19 05	08 46	12 32	16 01	02 42	18 14	15 13
20	01 09	26 11	05 22	19 26	08 31	12 36	16 01	02 44	18 14	15 12
21	00 46	27 54	05 43	19 47	08 17	12 40	16 00	02 45	18 14	15 11
22	00 N 22	28 06	06 00	20 07	08 02	12 44	16 00	02 47	18 15	15 10
23	00 S 01	26 52	06 13	20 26	07 47	12 48	15 59	02 48	18 15	15 10
24	00 24	24 24	06 21	20 45	07 33	12 52	15 58	02 50	18 15	15 09
25	00 48	20 54	06 24	21 04	07 18	12 56	15 58	02 51	18 16	15 08
26	01 11	16 35	06 23	21 22	07 03	13 00	15 57	02 53	18 16	15 07
27	01 35	11 41	06 17	21 40	06 48	13 04	15 56	02 54	18 17	15 06
28	01 58	06 20	06 06	21 57	06 33	13 08	15 55	02 56	18 17	15 06
29	02 21	00 45	05 51	22 14	06 19	13 12	15 54	02 57	18 17	15 05
30	02 S 45	04 S 54	05 N 33	22 S 30	06 N 04	13 S 16	15 N 54	02 S 59	18 S 18	15 N 04

ZODIAC SIGN ENTRIES

Date	h m	Planets
02	18 25	☽ ♎
03	04 57	♂ ♍
05	05 54	☽ ♏
07	01 54	♀ ♏
07	14 58	☽ ♐
09	20 51	☽ ♑
11	23 34	☽ ♒
13	23 57	☽ ♓
15	23 35	☽ ♈
18	00 21	☽ ♉
20	04 02	☽ ♊
22	11 41	☽ ♋
23	10 59	☉ ♎
24	22 54	☽ ♌
27	11 53	☽ ♍
30	00 33	☽ ♎

LATITUDES

Date	Mercury ☿ ° '	Venus ♀ ° '	Mars ♂ ° '	Jupiter ♃ ° '	Saturn ♄ ° '	Uranus ♅ ° '	Neptune ♆ ° '	Pluto ♇ ° '
01	04 S 15	02 S 21	01 N 10	01 N 01	02 S 22	00 N 40	01 N 40	15 N 24
04	04 23	02 40	01 10	01 01	02 23	00 40	01 40	15 24
07	04 16	02 59	01 10	01 00	02 24	00 40	01 40	15 24
10	03 54	03 18	01 10	01 00	02 24	00 40	01 40	15 24
13	03 14	03 38	01 10	00 59	02 25	00 40	01 40	15 24
16	02 21	03 57	01 09	00 59	02 26	00 40	01 39	15 24
19	01 22	04 17	01 09	00 58	02 26	00 40	01 39	15 24
22	00 S 24	04 36	01 09	00 58	02 27	00 40	01 39	15 24
25	00 N 26	04 54	01 09	00 57	02 27	00 40	01 39	15 24
28	01 06	05 12	01 09	00 57	02 28	00 40	01 39	15 25
31	01 N 33	05 S 29	01 N 09	00 N 57	02 S 28	00 N 40	01 N 39	15 N 25

DATA

Julian Date	2440831
Delta T	+41 seconds
Ayanamsa	23° 26' 59"
Synetic vernal point	05° ♓ 40' 01"
True obliquity of ecliptic	23° 26' 44"

MOON'S PHASES, APSIDES AND POSITIONS ☽

Date	h m	Phase	Longitude ° '	Eclipse Indicator
08	19 38	☽	15 ♐ 44	
15	11 09	○	22 ♓ 12	
22	09 42	☾	28 ♊ 58	
30	14 31	●	07 ♎ 01	

Day	h m		
14	17 07	Perigee	
27	07 44	Apogee	
02	08 59	0S	
09	12 56	Max dec	28° S 14'
15	16 08	0N	
22	03 02	Max dec	28° N 12'
29	15 13	0S	

LONGITUDES

Date	Chiron ⚷ ° '	Ceres ⚳ ° '	Pallas ⚴ ° '	Juno ⚵ ° '	Vesta ⚶ ° '	Black Moon Lilith ⚸ ° '
01	09 ♈ 22	07 ♉ 50	04 ♓ 24	21 ♉ 03	04 ♎ 42	29 ♌ 47
11	08 ♈ 59	07 ♉ 40	01 ♓ 55	23 ♉ 26	09 ♎ 44	00 ♍ 54
21	08 ♈ 33	06 ♉ 53	29 ♒ 43	25 ♉ 11	14 ♎ 49	02 ♍ 00
31	08 ♈ 06	05 ♉ 29	27 ♒ 58	26 ♉ 12	19 ♎ 59	03 ♍ 07

ASPECTARIAN

h m	Aspects
01 Tuesday	
00 29	☽ ∠ ♀
04 48	♃ ⊥ ♇
14 39	☽ Q ♆
17 22	☽ ∠ ♃
20 10	☽ # ☿
20 22	☽ ⊥ ♀
23 16	☽ # ♅
02 Wednesday	
03 32	☽ △ ♄
09 34	☽ ⊻ ♀
11 29	☽ ☌ ♇
11 50	☽ ⊥ ♃
13 21	☽ ☌ ☿
14 56	☽ ✱ ♆
17 50	☽ ⊻ ♂
18 46	☽ ∥ ♅
21 27	☽ ∥ ☿
03 Thursday	
00 04	☽ ⊻ ♃
05 50	♀ # ♂
06 33	☽ ⊥ ♂
09 13	☽ ☌ ♅
09 42	☽ ⚼ ♄
10 22	♀ ⊻ ♇
15 54	☽ ⊻ ☉
17 09	☽ # ☉
20 05	☿ ⊻ ♀
20 48	☿ ⊥ ♃
21 00	☽ ∠ ♆
04 Friday	
01 29	☽ ∠ ♂
03 37	☽ ± ♄
04 52	☽ ⊥ ☉
11 05	☽ ∥ ♃
12 35	☿ ☌ ♇
13 56	♄ St R
14 05	☽ # ♂
14 49	☽ ⊥ ♆
15 28	☽ ⊼ ♄
18 34	☽ ∥ ♀
22 51	☽ ⊻ ☿
23 21	☽ ⊻ ♇
05 Saturday	
00 06	☽ ∠ ☉
02 19	☽ ☌ ♀
02 36	☽ ⊻ ♆
04 56	☽ # ♇
06 04	♀ ⊻ ♆
08 17	☽ # ♄
08 34	☽ ✱ ♂
10 01	☽ ⊥ ☿
11 05	☽ ⊥ ♇
12 12	☽ ☌ ♃
13 52	♂ ⊥ ♅
18 27	☿ ∥ ♅
18 32	☽ ∥ ♆
20 29	☽ ⊻ ♅
06 Sunday	
02 30	☽ ∠ ☿
04 29	☽ ∠ ♇
07 33	☽ ✱ ☉
08 02	☽ ⊥ ♅
08 55	☽ Q ♂
20 30	♂ # ♃
07 Monday	
01 09	☽ ∠ ♅
01 12	☽ ☍ ♄
05 21	☽ ✱ ☿
08 04	☽ Q ☉
08 54	☽ ✱ ♇
09 50	♄ ⚼ ♅
11 53	☽ ☌ ♆
15 02	☿ ⊥ ♀
15 58	☽ ⊻ ♀
20 25	☽ □ ♂
21 43	☽ ⊻ ♃
08 Tuesday	
02 13	☽ Q ☿
03 49	☽ ⊥ ♀
05 03	☽ ✱ ♅
07 04	☽ Q ♇
08 50	☽ ⊥ ♃
19 38	☽ □ ☉
21 17	☽ ∠ ♀
09 Wednesday	
01 14	☽ ∠ ♃
02 46	☽ Q ♅
07 53	☽ ⊼ ♄
08 32	☽ □ ☿
08 57	♂ ✱ ♃
11 27	☉ Q ♆
15 18	☽ □ ♇
17 49	☿ △ ♄
18 01	☽ ⊻ ♆
18 25	☽ ± ♄
18 45	♀ # ♇
10 Thursday	
01 34	☽ ✱ ♀
03 54	☽ ✱ ♃
04 27	☽ ⊥ ♆
04 33	☽ △ ♂
04 58	♀ ⊥ ♇
09 58	☽ ⚼ ♄
10 19	☽ □ ♅
19 51	☽ ∠ ♆
23 27	☽ Q ♀
11 Friday	
00 37	☽ Q ♃
03 37	☽ △ ☉
07 13	☽ ⚼ ♂
08 38	☽ △ ☿
11 17	☽ △ ♄
13 39	♀ # ♄
18 26	☽ △ ♇
20 56	☽ ✱ ♆
22 50	☽ ± ♂
22 59	♀ ☌ ♃
12 Saturday	
06 14	☽ ⚼ ☉
06 51	☽ □ ♃
07 14	☽ □ ♀
07 46	☽ ⚼ ☿
09 06	☽ ⊼ ♂
11 38	☿ ∠ ♀
12 30	☽ △ ♅
16 34	☽ Q ♆
17 23	☉ ∠ ♃
18 18	☉ ☌ ☿
18 55	☿ ∠ ♃
19 00	☽ ⚼ ♇
21 26	☽ ± ☿
21 52	☽ ± ☉
22 37	☽ ∥ ♆
13 Sunday	
04 44	☽ ∥ ♀
06 28	☽ ⊼ ☿
07 24	☽ # ♄
08 12	☽ ⊼ ☉
09 31	☽ ± ♇
10 27	☽ # ♇
12 04	☽ □ ♄
12 46	☽ ⚼ ♅
19 07	☽ ⊼ ♇
21 28	☽ □ ♆
22 16	☽ ∥ ♃
14 Monday	
03 10	☽ ± ♅
06 03	☽ # ♂
07 36	☽ △ ♃
10 18	☽ △ ♀
11 19	☽ ☍ ♂
12 43	☽ ⊼ ♅
16 40	☽ Q ♄
15 Tuesday	
03 19	☽ ☍ ☿
05 27	☽ # ☉
07 25	☽ ∥ ♅
07 27	☽ # ☿
07 41	☽ ⚼ ♃
07 45	☿ # ♅
11 09	☽ ☍ ☉
11 29	☽ ⚼ ♀
11 40	☽ ✱ ♄
18 52	☽ ☍ ♇
18 58	☉ △ ♄
20 57	☉ ∥ ☿
21 10	☽ △ ♆
22 10	☽ ± ♃
16 Wednesday	
00 13	♂ ⊻ ♅
00 29	☿ Q ♆
00 57	☽ # ♅
01 47	☽ ∥ ☉
02 31	☽ ∥ ☿
02 40	☽ ± ♀
07 53	☽ ⊼ ♃
09 14	♀ ⊻ ♅
11 35	☽ ∠ ♄
12 40	☽ ☍ ♅
12 51	☽ ⊼ ♀
13 10	☽ ⊼ ♂
16 08	☉ # ♅
20 04	☉ ∠ ♀
21 16	☽ ⚼ ♆
23 17	☽ ± ♂
17 Thursday	
00 22	☽ ∥ ♂
00 49	☽ ⊼ ☿
01 55	♀ ∥ ♆
02 07	☽ ⊥ ♄
10 10	☽ ± ☿
11 40	☽ # ♃
11 54	☽ ⊻ ♄
11 58	☽ ± ♆
13 06	♀ ✱ ♂
14 40	☽ ⚼ ♂
14 54	☽ ⊼ ☉
19 34	☽ ⊼ ♇
21 53	☽ ⊼ ♆
22 49	☽ ∥ ♇
18 Friday	
00 24	☽ ⚼ ☿
01 34	☽ ± ☉
02 07	☽ ∥ ♄
05 35	☽ ± ♇
09 42	☽ ☍ ♃
11 39	☽ # ♆
14 08	☽ # ♀
14 17	☽ ⊼ ♅
16 55	☽ △ ♂
17 16	☽ ☍ ♀
17 50	☽ ⚼ ☉
20 51	☽ ⚼ ♇
19 Saturday	
00 34	☽ ± ♅
00 51	☽ △ ☿
14 37	☽ ☌ ♄
16 16	☽ ⚼ ♅
21 52	☽ △ ☉
23 03	☽ △ ♇
20 Sunday	
01 28	☽ ☍ ♆
14 51	☽ ⊼ ♃
14 53	☉ ☌ ♇
19 15	☽ △ ♅
21 Monday	
00 33	☽ □ ♂
01 36	☽ ⊼ ♀
01 58	☽ ± ♃
05 10	☽ □ ☿
13 04	♀ ∠ ♇
13 26	☽ ± ♀
19 00	☽ ⚼ ♃
21 07	☽ ⊻ ♄
22 Tuesday	
00 17	☿ St D
00 53	☉ ✱ ♆
06 29	☽ □ ♇
07 32	☽ ⚼ ♀
08 30	☽ ⊥ ♄
09 02	☽ ⊼ ♆
09 42	☽ □ ☉
12 11	☽ Q ♂
15 09	☽ Q ☿
20 37	☽ ± ♆
23 Wednesday	
00 12	☽ △ ♃
01 51	☽ ∠ ♄
04 19	☽ □ ♅
12 49	☽ ✱ ♂
14 15	☽ ⚼ ♆
14 33	☽ △ ♀
14 56	☽ ✱ ☿
17 36	☽ Q ♇
23 56	☽ Q ☉
24 Thursday	
01 02	☿ ✱ ♀
04 03	☉ ⊥ ♃
07 24	☽ ✱ ♄
16 13	☽ Q ♅
17 37	☽ ✱ ♇
20 12	♀ ⊥ ♅
20 14	☽ △ ♆
20 23	☽ ∠ ♂
21 49	☽ ∠ ☿
25 Friday	
02 08	☽ ✱ ☉
07 29	☽ Q ♄
11 02	☽ # ♀
12 59	☽ □ ♃
15 36	☽ ⊥ ♂
16 35	☽ ✱ ♅
16 53	☽ ⊥ ☿
26 Saturday	
00 05	☽ ∠ ♇
03 06	☽ # ♆
04 30	☽ ⊻ ♂
05 46	☽ ⊻ ☿
06 43	☽ □ ♀
11 20	☽ ∠ ☉
15 18	☽ ∥ ♄
18 28	☽ ⊥ ♇
19 29	☽ ∥ ♇
19 58	☽ □ ♄
23 17	☽ ∠ ♅
27 Sunday	
02 15	☽ Q ♃
05 32	☽ # ♃
06 44	☽ ⊻ ♇
07 21	☽ ⊥ ☉
09 19	☽ □ ♆
17 44	☽ ⊥ ♅
20 41	☽ ⊻ ☉
21 37	☽ Q ♀
28 Monday	
03 07	☽ ✱ ♃
03 12	☿ Q ♆
06 00	☽ ⊻ ♅
11 00	☽ ∥ ♂
13 05	☽ ∥ ☿
16 24	☿ ✱ ♀
21 00	☽ ☌ ♂
21 57	☽ Q ♆
23 22	☽ ✱ ♀
23 37	☽ ☌ ☿
29 Tuesday	
02 42	☽ # ♅
05 39	☽ # ☉
08 40	☽ △ ♄
09 59	☽ ∠ ♃
15 25	♂ Q ♆
19 41	☽ ☌ ♇
22 09	☽ ✱ ♆
30 Wednesday	
02 10	☽ ∥ ☉
03 48	☽ ∥ ♅
04 20	☽ ⊥ ♃
07 08	☽ ∠ ♀
14 31	☽ ☌ ☉
14 33	☽ ⚼ ♄
14 36	☽ # ☿
14 55	☉ ⚼ ♄
16 28	☽ ⊻ ♃
16 46	☽ # ♂
18 35	☽ ☌ ♅

All ephemeris data is given at 12.00 UT and the Moon's longitude is additionally given for 24.00 UT

OCTOBER 1970

Date	Sidereal time h m s	Sun ☉ ° ' "	Moon ☽ ° ' "	Moon ☽ 24.00 ° ' "	Mercury ☿ ° '	Venus ♀ ° '	Mars ♂ ° '	Jupiter ♃ ° '	Saturn ♄ ° '	Uranus ♅ ° '	Neptune ♆ ° '	Pluto ⯓ ° '
01	12 39 13	07 ♎ 53 36	17 ♎ 53 34	24 ♎ 01 50	20 ♍ 33	19 ♏ 01	17 ♍ 58	08 ♏ 10	21 ♉ 59	09 ♎ 06	28 ♏ 50	27 ♍ 38
02	12 43 10	08 52 39	00 ♏ 12 48	06 ♏ 26 36	21 54	19 36	18 36	08 22	21 R 57	09 10	28 52	27 40
03	12 47 06	09 51 44	12 ♏ 43 20	19 ♏ 03 07	23 19	20 09	19 14	08 34	21 54	09 14	28 54	27 42
04	12 51 03	10 50 50	25 ♏ 26 06	01 ♐ 52 23	24 49	20 41	19 52	08 46	21 51	09 18	28 55	27 44
05	12 54 59	11 49 59	08 ♐ 22 06	14 ♐ 55 26	26 21	21 11	20 31	08 59	21 48	09 21	28 57	27 47
06	12 58 56	12 49 09	21 ♐ 32 29	28 ♐ 13 27	27 57	21 40	21 09	09 11	21 44	09 25	28 58	27 49
07	13 02 52	13 48 22	04 ♑ 58 28	11 ♑ 47 41	29 ♍ 35	22 07	21 47	09 24	21 41	09 29	29 00	27 51
08	13 06 49	14 47 36	18 ♑ 41 12	25 ♑ 39 05	01 ♎ 15	22 33	22 25	09 36	21 38	09 33	29 02	27 53
09	13 10 45	15 46 51	02 ♒ 41 22	09 ♒ 47 58	02 56	22 57	23 03	09 49	21 34	09 37	29 04	27 55
10	13 14 42	16 46 08	16 ♒ 58 43	24 ♒ 13 23	04 39	23 19	23 41	10 01	21 31	09 40	29 05	27 57
11	13 18 39	17 45 27	01 ♓ 31 32	08 ♓ 52 41	06 22	23 40	24 19	10 14	21 27	09 44	29 07	28 00
12	13 22 35	18 44 48	16 ♓ 16 10	23 ♓ 41 13	08 06	23 59	24 57	10 26	21 24	09 48	29 09	28 02
13	13 26 32	19 44 11	01 ♈ 06 55	08 ♈ 32 20	09 50	24 15	25 35	10 39	21 20	09 52	29 11	28 04
14	13 30 28	20 43 35	15 ♈ 56 26	23 ♈ 18 12	11 34	24 30	26 13	10 52	21 16	09 55	29 13	28 06
15	13 34 25	21 43 02	00 ♉ 36 39	07 ♉ 50 52	13 19	24 42	26 51	11 05	21 12	09 59	29 15	28 08
16	13 38 21	22 42 31	15 ♉ 00 02	22 ♉ 03 30	15 03	24 53	27 29	11 17	21 09	10 03	29 16	28 10
17	13 42 18	23 42 01	29 ♉ 00 47	05 ♊ 51 31	16 47	25 01	28 07	11 30	21 05	10 06	29 18	28 12
18	13 46 14	24 41 35	12 ♊ 35 33	19 ♊ 12 53	18 31	25 07	28 46	11 43	21 00	10 10	29 20	28 14
19	13 50 11	25 41 10	25 ♊ 43 41	02 ♋ 08 14	20 14	25 11	29 ♍ 24	11 56	20 56	10 14	29 22	28 16
20	13 54 08	26 40 48	08 ♋ 26 56	14 ♋ 40 15	21 57	25 13	00 ♎ 02	12 09	20 52	10 18	29 24	28 18
21	13 58 04	27 40 28	20 ♋ 48 47	26 ♋ 53 07	23 39	25 R 12	00 40	12 22	20 48	10 21	29 26	28 20
22	14 02 01	28 40 10	02 ♌ 53 56	08 ♌ 51 53	25 21	25 09	01 18	12 35	20 44	10 25	29 28	28 22
23	14 05 57	29 ♎ 39 54	14 ♌ 47 41	20 ♌ 41 58	27 03	25 03	01 56	12 48	20 39	10 29	29 30	28 24
24	14 09 54	00 ♏ 39 41	26 ♌ 35 26	02 ♍ 28 43	28 ♎ 44	24 55	02 34	13 01	20 35	10 32	29 32	28 26
25	14 13 50	01 39 29	08 ♍ 22 25	14 ♍ 17 07	00 ♏ 24	24 45	03 12	13 14	20 30	10 36	29 34	28 28
26	14 17 47	02 39 20	20 ♍ 13 19	26 ♍ 11 30	02 04	24 32	03 50	13 27	20 26	10 39	29 36	28 30
27	14 21 43	03 39 13	02 ♎ 12 06	08 ♎ 15 26	03 43	24 17	04 28	13 40	20 21	10 43	29 38	28 32
28	14 25 40	04 39 08	14 ♎ 21 48	20 ♎ 31 25	05 22	24 00	05 06	13 53	20 17	10 47	29 40	28 34
29	14 29 37	05 39 06	26 ♎ 44 26	03 ♏ 00 57	07 00	23 40	05 45	14 06	20 12	10 50	29 42	28 36
30	14 33 33	06 39 05	09 ♏ 20 58	15 ♏ 44 29	08 38	23 18	06 23	14 19	20 08	10 54	29 45	28 38
31	14 37 30	07 ♏ 39 06	22 ♏ 11 23	28 ♏ 41 36	10 ♏ 15	22 ♏ 54	07 ♎ 01	14 ♏ 33	20 ♉ 03	10 ♎ 57	29 ♏ 47	28 ♍ 40

Date	Moon True ☊ ° '	Moon Mean ☊ ° '	Moon ☽ Latitude ° '
01	02 ♓ 27	00 ♓ 48	03 S 42
02	02 R 21	00 45	04 23
03	02 14	00 42	04 53
04	02 08	00 39	05 08
05	02 02	00 35	05 08
06	01 58	00 32	04 52
07	01 56	00 29	04 20
08	01 D 56	00 26	03 33
09	01 57	00 23	02 32
10	01 58	00 20	01 20
11	01 59	00 16	00 S 02
12	01 R 58	00 13	01 N 17
13	01 55	00 10	02 31
14	01 50	00 07	03 34
15	01 43	00 04	04 23
16	01 35	00 ♓ 01	04 54
17	01 26	29 ♒ 57	05 06
18	01 18	29 54	05 01
19	01 12	29 51	04 39
20	01 08	29 48	04 05
21	01 06	29 45	03 19
22	01 D 06	29 41	02 26
23	01 07	29 38	01 27
24	01 08	29 35	00 N 24
25	01 R 08	29 32	00 S 39
26	01 06	29 29	01 40
27	01 01	29 26	02 38
28	00 54	29 22	03 29
29	00 44	29 19	04 11
30	00 33	29 16	04 42
31	00 ♓ 21	29 ♒ 13	04 S 59

DECLINATIONS

Date	Sun ☉ ° '	Moon ☽ ° '	Mercury ☿ ° '	Venus ♀ ° '	Mars ♂ ° '	Jupiter ♃ ° '	Saturn ♄ ° '	Uranus ♅ ° '	Neptune ♆ ° '	Pluto ⯓ ° '
01	03 S 08	10 S 26	05 N 10	22 S 45	05 N 49	13 S 20	15 N 53	03 S 00	18 S 18	15 N 03
02	03 31	15 39	04 45	23 00	05 34	13 24	15 52	03 01	18 18	15 02
03	03 54	20 18	04 16	23 15	05 19	13 28	15 51	03 03	18 19	15 02
04	04 18	24 06	03 44	23 28	05 04	13 32	15 50	03 04	18 19	15 01
05	04 41	26 47	03 10	23 41	04 49	13 36	15 49	03 06	18 20	15 00
06	05 04	28 02	02 34	23 54	04 34	13 40	15 48	03 07	18 20	15 00
07	05 27	27 41	01 56	24 06	04 19	13 44	15 47	03 09	18 20	14 59
08	05 50	25 39	01 17	24 17	04 03	13 49	15 46	03 10	18 21	14 58
09	06 13	22 02	00 N 36	24 27	03 48	13 53	15 45	03 12	18 21	14 57
10	06 36	17 02	00 S 05	24 37	03 33	13 57	15 44	03 13	18 22	14 57
11	06 58	10 58	00 48	24 46	03 18	14 01	15 43	03 15	18 22	14 56
12	07 21	04 S 15	01 31	24 54	03 03	14 05	15 42	03 16	18 23	14 55
13	07 43	02 N 45	02 15	25 02	02 48	14 09	15 41	03 18	18 23	14 55
14	08 06	09 34	02 59	25 08	02 32	14 13	15 40	03 19	18 23	14 54
15	08 28	15 48	03 43	25 14	02 17	14 17	15 39	03 21	18 24	14 54
16	08 50	21 01	04 27	25 19	02 02	14 21	15 38	03 22	18 24	14 53
17	09 12	24 55	05 12	25 23	01 47	14 25	15 37	03 24	18 25	14 52
18	09 34	27 17	05 56	25 26	01 31	14 29	15 36	03 25	18 25	14 52
19	09 56	28 02	06 40	25 28	01 16	14 33	15 34	03 27	18 26	14 51
20	10 17	27 15	07 23	25 29	01 01	14 37	15 33	03 28	18 26	14 51
21	10 39	25 07	08 07	25 29	00 46	14 41	15 32	03 29	18 26	14 50
22	11 00	21 53	08 50	25 28	00 30	14 45	15 31	03 31	18 27	14 50
23	11 21	17 47	09 32	25 26	00 N 15	14 49	15 30	03 32	18 27	14 49
24	11 42	13 02	10 14	25 22	00 00	14 53	15 29	03 34	18 28	14 49
25	12 03	07 50	10 55	25 17	00 S 15	14 57	15 27	03 35	18 28	14 48
26	12 24	02 N 20	11 36	25 11	00 31	15 01	15 26	03 37	18 29	14 48
27	12 44	03 S 17	12 16	25 04	00 46	15 05	15 25	03 38	18 29	14 47
28	13 05	08 53	12 55	24 56	01 01	15 09	15 24	03 39	18 30	14 47
29	13 25	14 13	13 34	24 46	01 16	15 13	15 22	03 41	18 30	14 46
30	13 44	19 04	14 12	24 35	01 32	15 17	15 21	03 42	18 31	14 46
31	14 S 04	23 S 08	14 S 49	24 S 23	01 S 47	15 S 21	15 N 20	03 S 43	18 S 31	14 N 45

ZODIAC SIGN ENTRIES

Date	h	m	Planets
02	11	35	☽ ♏
04	20	31	☽ ♐
07	03	10	☽ ♑
07	18	04	☿ ♎
09	07	26	☽ ♒
11	09	30	☽ ♓
13	10	12	☽ ♈
15	11	00	☽ ♉
17	13	43	☽ ♊
19	19	59	☽ ♋
20	10	57	♂ ♎
22	06	12	☽ ♌
23	20	04	☉ ♏
24	18	57	☽ ♍
25	06	16	☿ ♏
27	07	37	☽ ♎
29	18	15	☽ ♏

LATITUDES

Date	Mercury ☿ ° '	Venus ♀ ° '	Mars ♂ ° '	Jupiter ♃ ° '	Saturn ♄ ° '	Uranus ♅ ° '	Neptune ♆ ° '	Pluto ⯓ ° '
01	01 N 33	05 S 29	01 N 09	00 N 57	02 S 28	00 N 40	01 N 39	15 N 25
04	01 49	05 46	01 09	00 56	02 29	00 40	01 39	15 25
07	01 56	06 00	01 08	00 56	02 29	00 40	01 38	15 26
10	01 55	06 13	01 08	00 56	02 30	00 40	01 38	15 26
13	01 48	06 24	01 08	00 55	02 30	00 40	01 38	15 27
16	01 36	06 32	01 08	00 55	02 30	00 40	01 38	15 28
19	01 21	06 36	01 07	00 55	02 31	00 40	01 38	15 29
22	01 03	06 36	01 07	00 55	02 31	00 40	01 38	15 29
25	00 45	06 32	01 06	00 54	02 31	00 40	01 38	15 30
28	00 25	06 22	01 06	00 54	02 31	00 40	01 38	15 31
31	00 N 05	06 S 05	01 N 06	00 N 54	02 S 31	00 N 40	01 N 38	15 N 32

DATA

Julian Date	2440861
Delta T	+41 seconds
Ayanamsa	23° 27' 02"
Synetic vernal point	05° ♓ 39' 58"
True obliquity of ecliptic	23° 26' 44"

MOON'S PHASES, APSIDES AND POSITIONS ☽

Date	h	m	Phase	Longitude ° '	Eclipse Indicator
08	04	43	☽	14 ♑ 30	
14	20	21	○	21 ♈ 04	
22	02	47	☾	28 ♋ 17	
30	06	28	●	06 ♏ 25	

Day	h	m		
13	00	50	Perigee	
24	22	31	Apogee	
06	18	59	Max dec	28° S 07'
13	02	35	0N	
19	11	20	Max dec	28° N 02'
26	22	00	0S	

LONGITUDES

Date	Chiron ⚷ ° '	Ceres ⚳ ° '	Pallas ⚴ ° '	Juno ⚵ ° '	Vesta ⚶ ° '	Black Moon Lilith ⚸ ° '
01	08 ♈ 06	05 ♉ 29	27 ♒ 58	26 ♉ 12	19 ♎ 59	03 ♍ 07
11	07 ♈ 39	03 ♉ 36	26 ♒ 47	26 ♉ 21	25 ♎ 11	04 ♍ 13
21	07 ♈ 13	01 ♉ 24	26 ♒ 11	25 ♉ 37	00 ♏ 27	05 ♍ 19
31	06 ♈ 49	29 ♈ 07	26 ♒ 10	24 ♉ 06	05 ♏ 44	06 ♍ 26

ASPECTARIAN

h m	Aspects
01 Thursday	
01 55	☽ ⊥ ♀
03 16	☉ ∥ ♅
04 01	☽ ∠ ♆
08 17	☽ ± ♄
12 10	☽ ⚻ ♂
14 19	☽ ⚻ ♀
17 50	☽ ⚻ ☿
19 59	☽ ⊼ ♄
20 12	☉ ⚻ ♃
21 42	☽ ⊥ ♆
02 Friday	
00 32	☽ ⊥ ♂
01 16	☽ ∥ ♃
06 57	☽ ⊥ ☿
07 03	☽ ⚻ ⯓
09 05	☽ # ⯓
09 23	☽ ⚻ ♆
12 44	☿ △ ♄
13 02	☽ # ♄
18 41	☽ ⊥ ⯓
18 54	☽ ∠ ♂
19 32	☉ ☌ ♅
03 Saturday	
01 17	☽ ∥ ♆
02 30	☽ ∠ ☿
03 56	☽ ☌ ♃
05 18	☽ ⚻ ♅
06 05	☽ ⚻ ☉
11 58	☽ ∠ ⯓
16 43	☿ ∠ ♃
16 47	☽ ⊥ ♅
18 28	☽ ⊥ ☉
21 28	☉ # ☿
04 Sunday	
01 00	☽ ⚹ ♂
02 41	☽ ☌ ♀
05 17	☽ ☍ ♄
07 12	☽ ∥ ♀
09 51	☽ ∠ ♅
10 40	☽ ⚹ ☿
12 50	☽ ∠ ☉
16 19	☽ ⚹ ⯓
18 31	☽ ☌ ♆
05 Monday	
00 37	☽ Q ♂
11 58	☽ Q ☿
13 08	☽ ⚻ ♃
13 49	☽ ⚹ ♅
14 36	☽ Q ⯓
14 50	☿ # ♅
16 59	☉ # ♂
18 52	☽ ⚹ ☉
06 Tuesday	
00 18	☽ ⊥ ♃
09 55	☿ ☌ ⯓
11 15	☽ □ ♂
11 47	☽ Q ♅
12 14	☽ ⚻ ♀
12 21	☽ ⊼ ♄
15 25	♀ ☍ ♄
16 50	☽ ∠ ♃
18 22	☽ Q ☉
23 05	☽ ± ♄
23 18	☽ □ ⯓
23 24	☽ ⊥ ♀
07 Wednesday	
01 06	☽ □ ☿
01 22	☽ ⚻ ♆
03 23	☿ ⚹ ♆
08 49	♂ △ ♄
12 03	☽ ⊥ ♆
15 01	☽ ♇ ♄
15 55	☽ ∠ ♀
16 56	☉ ∠ ♆
19 55	☽ ⚹ ♃
19 59	☽ □ ♅
08 Thursday	
02 46	♃ ⚻ ♅
03 54	☽ ∠ ♆
04 43	☽ □ ☉
17 04	☽ △ ♄
17 07	☽ Q ♃
18 45	☽ △ ♂
18 53	☽ ⚹ ♀
21 49	☽ ∥ ♀
09 Friday	
02 32	♀ ⚹ ♂
03 52	☽ △ ⯓
05 49	☽ ⚹ ♆
07 16	☉ ± ♄
12 28	☽ △ ☿
15 56	☽ Q ♀
21 29	☽ ♇ ♂
23 44	☽ △ ♅
10 Saturday	
00 12	☽ □ ♃
02 03	☿ ⊥ ♃
02 09	☽ Q ♆
05 17	☽ ♇ ⯓
06 08	☽ ∥ ♆
11 38	☽ △ ☉
13 13	☽ ± ♂
17 01	☽ ♇ ☿
17 26	☽ # ♄
19 30	☽ □ ♄
20 17	☽ ± ⯓
20 40	☽ # ⯓
22 46	☽ □ ♀
23 37	☽ ⊼ ♂
11 Sunday	
00 31	☽ ∥ ♃
00 48	☽ ♇ ♅
06 12	☽ ⊼ ⯓
08 03	☽ □ ♆
08 38	☉ ⊥ ♀
09 51	☽ ± ☿
13 14	☿ ♇ ♄
14 10	☽ ♇ ☉
15 38	☽ ± ♅
16 25	♂ # ♅
20 58	☽ ⊼ ☿
12 Monday	
00 53	☽ Q ♄
01 27	☽ ⊼ ♅
01 43	☽ ∥ ☉
02 24	☽ △ ♃
05 53	☽ ± ☉
15 20	☽ ∥ ♅
16 17	☽ # ♂
16 18	☽ ⊼ ☉
20 16	☽ ⚹ ♄
20 30	☽ ∥ ☿
13 Tuesday	
00 43	☽ △ ♀
02 32	☿ ∠ ♀
02 40	☽ ☍ ♂
03 03	☽ ♇ ♃
07 04	☽ ☍ ⯓
08 52	☽ △ ♆
10 05	☽ # ☿
12 09	☽ ∥ ♂
12 23	☿ ☌ ♅
13 54	☽ # ♅
15 49	♂ ∠ ♃
17 48	☽ ± ♃
20 24	☽ ∠ ♄
14 Wednesday	
00 53	☿ ⚻ ♃
01 16	☿ # ♂
01 23	☽ ♇ ♀
02 12	☽ ☍ ♅
03 39	☽ ⊼ ♃
03 58	☽ ☍ ☿
06 22	☽ # ☉
09 11	☽ ♇ ♆
10 55	☽ ⊥ ♄
16 13	☽ ± ♀
20 21	☽ ☍ ☉
20 39	☽ ⚻ ♄
23 27	☿ ∥ ♅
23 52	☽ ± ♆
15 Thursday	
00 24	☉ ⊼ ♄
02 10	☽ ⊼ ♀
05 32	☽ ⊼ ♂
05 48	☽ # ♃
07 55	☽ ⊼ ⯓
08 19	☽ ∥ ⯓
09 44	☽ ⊼ ♆
11 25	☽ ∥ ♄
15 53	☽ ± ♂
17 50	☽ ± ⯓
23 18	☽ # ♆
16 Friday	
01 05	☿ ∠ ♆
03 38	☽ ⊼ ♅
05 40	☽ ☍ ♃
07 34	☽ ♇ ♂
08 54	☽ ♇ ⯓
12 06	☽ ⊼ ☿
13 14	☿ ± ♄
13 46	☽ ± ♅
22 23	☽ ☌ ♄
23 43	☽ ± ☿
17 Saturday	
02 07	☽ ⊼ ☉
02 08	☉ ⊥ ♆
05 02	☽ ☍ ♀
05 13	☽ ♇ ♅
10 23	☽ △ ♂
10 35	☽ △ ⯓
12 31	☽ ☍ ♆
13 17	☽ ± ☉
15 13	♂ ☌ ⯓
15 42	☽ # ♀
17 32	☽ ♇ ☿
18 Sunday	
06 24	☽ ♇ ☉
07 38	☽ △ ♅
10 24	☽ ⊼ ♃
20 55	☿ ⊥ ♀
21 25	☽ ± ♃
23 16	☉ ⚻ ♀
19 Monday	
00 20	☽ △ ☿
03 12	☽ ⚻ ♄
11 00	☽ ⊼ ♀
11 04	♂ ⚹ ♆
11 55	☽ △ ☉
14 14	☽ ⊥ ♄
14 17	☽ ♇ ♃
16 45	☽ □ ⯓
18 49	☽ ⊼ ♆
19 12	☽ □ ♂
21 29	☿ ⊼ ♄
22 14	☽ ± ♀
20 Tuesday	
06 10	☽ ± ♆
07 06	☽ ∠ ♄
15 23	☽ ♇ ♀
15 33	☽ □ ♅
15 56	♀ St R
19 15	☽ △ ♃
23 31	☽ ♇ ♆
21 Wednesday	
03 13	☽ Q ⯓
07 33	☽ Q ♂
08 38	☽ # ♀
08 50	☿ ⊥ ♆
11 58	☽ ⚹ ♄
18 31	☽ □ ☿
20 38	☽ △ ♀
22 Thursday	
02 47	☽ □ ☉
02 56	☽ ⚹ ⯓
03 00	☽ Q ♅
04 38	☉ ⚻ ⯓
05 07	☽ △ ♆
08 37	☽ ⚹ ♂
09 12	☿ ⚻ ♀
11 40	☽ Q ♄
23 Friday	
03 13	☽ ⚹ ♅
07 53	☽ □ ♃
07 56	☉ ⚻ ♆
08 19	☽ # ♆
09 11	☽ ∠ ⯓
11 52	♃ # ⯓
12 36	☽ Q ☿
16 35	☽ ∠ ♂
18 22	☽ Q ☉
23 50	☽ □ ♄
23 57	☽ ∥ ♄
24 Saturday	
03 05	☽ # ♃
03 20	☽ ∥ ⯓
03 31	☽ ⊥ ⯓
07 50	☿ ⚻ ⯓
08 39	☽ □ ♀
09 50	☽ ∠ ♅
11 57	☽ ⊥ ♂
15 47	☽ ⚻ ⯓
17 05	☽ ⚹ ☿
17 54	☽ # ☉
18 01	☽ □ ♆
21 04	☽ ⚹ ☉
21 11	☽ Q ♃
23 37	☽ # ☿
23 50	☿ ⚻ ♆
25 Sunday	
00 52	☽ ⚻ ♂
04 16	☽ ⊥ ♅
16 32	☽ ⚻ ♅
20 44	☽ Q ♀
22 03	☽ ⚹ ♃
26 Monday	
04 35	☽ ∠ ☿
06 21	☽ ∠ ☉
06 33	☽ # ♅
06 42	☽ Q ♆
12 25	☽ △ ♄
19 13	♃ ∠ ⯓
19 28	☽ # ♂
20 31	☽ ⚹ ♀
27 Tuesday	
00 45	☽ ∥ ♂
01 37	☽ ⊥ ☿
02 06	☽ ⊥ ☉
04 40	☽ ☌ ⯓
04 49	☽ ∠ ♃
06 53	☽ ⚹ ♆
09 40	☉ ☌ ☿
13 27	☽ ∥ ♅
15 09	☽ ⚻ ☉
15 29	☽ ⚻ ☿
16 46	☽ ☌ ♂
18 14	☽ ♇ ♄
23 02	☽ ⊥ ♃
28 Wednesday	
00 11	☿ ⊥ ⯓
01 42	☽ ∠ ♀
04 56	☽ ☌ ♅
05 54	☿ ⚻ ♂
09 56	☉ ⊥ ⯓
11 03	☽ ⚻ ♃
11 50	☽ ± ♄
12 36	☽ ∠ ♆
17 51	♂ ♇ ♄
18 54	☽ ⊥ ♀
23 27	☽ ⊼ ♄
23 34	☉ ∥ ☿
29 Thursday	
06 09	☽ ⊥ ♆
06 14	☽ ⚻ ♀
08 00	☽ ∥ ☉
08 35	☽ ∥ ☿
14 36	☽ # ⯓
15 35	☽ ⚻ ⯓
16 47	☽ ∥ ♃
17 27	☽ # ♄
17 42	☽ ⚻ ♆
17 58	☉ ⚻ ♂
30 Friday	
03 03	☽ ⊥ ⯓
06 05	☽ ⚻ ♂
06 28	☽ ☌ ☉
09 05	☽ ∥ ♆
10 27	☽ ☌ ☿
14 55	☽ ⚻ ♅
18 00	☽ ⊥ ♂
20 04	☽ ∠ ⯓
21 31	☽ ☌ ♃
31 Saturday	
02 13	☽ ⊥ ♅
07 23	♃ # ♄
08 03	☽ ☍ ♄
09 23	☿ # ⯓
11 39	☽ ∠ ♂
13 16	☽ ☌ ♀
18 59	☽ ∠ ♅
20 18	☽ ∥ ♀
22 49	☿ ⚻ ♅
23 58	☽ ⚹ ⯓

All ephemeris data is given at 12.00 UT and the Moon's longitude is additionally given for 24.00 UT

NOVEMBER 1970

Date	Sidereal time h m s	Sun ☉ ° ' "	Moon ☽ ° ' "	Moon ☽ 24.00 ° ' "	Mercury ☿ ° '	Venus ♀ ° '	Mars ♂ ° '	Jupiter ♃ ° '	Saturn ♄ ° '	Uranus ♅ ° '	Neptune ♆ ° '	Pluto ♇ ° '
01	14 41 26	08 ♏ 39 09	05 ♐ 14 57	11 ♐ 51 19	11 ♏ 52	22 ♏ 28	07 ♎ 39	14 ♏ 46	19 ♉ 58	11 ♎ 01	29 ♏ 49	28 ♍ 42
02	14 45 23	09 39 14	18 ♐ 30 31	25 ♐ 12 26	13 28	21 R 59	08 17	14 59	19 R 53	11 04	29 51	28 43
03	14 49 19	10 39 20	01 ♑ 56 55	08 ♑ 43 54	15 04	21 30	08 55	15 12	19 49	11 07	29 53	28 45
04	14 53 16	11 39 28	15 ♑ 33 17	22 ♑ 25 02	16 39	20 58	09 33	15 25	19 44	11 11	29 55	28 47
05	14 57 12	12 39 38	29 ♑ 19 09	06 ♒ 15 36	18 14	20 26	10 11	15 39	19 39	11 14	29 ♏ 57	28 49
06	15 01 09	13 39 49	13 ♒ 14 24	20 ♒ 15 33	19 49	19 52	10 49	15 52	19 34	11 18	00 ♐ 00	28 50
07	15 05 06	14 40 01	27 ♒ 19 00	04 ♓ 24 39	21 23	19 17	11 27	16 05	19 29	11 21	00 02	28 52
08	15 09 02	15 40 15	11 ♓ 32 23	18 ♓ 41 57	22 57	18 41	12 06	16 18	19 24	11 24	00 04	28 54
09	15 12 59	16 40 30	25 ♓ 53 02	03 ♈ 05 14	24 30	18 04	12 44	16 31	19 20	11 28	00 06	28 56
10	15 16 55	17 40 47	10 ♈ 18 01	17 ♈ 30 48	26 04	17 28	13 22	16 45	19 15	11 31	00 08	28 57
11	15 20 52	18 41 05	24 ♈ 42 53	01 ♉ 53 32	27 36	16 51	14 00	16 58	19 10	11 34	00 11	28 59
12	15 24 48	19 41 25	09 ♉ 01 59	16 ♉ 07 29	29 ♏ 09	16 15	14 38	17 11	19 05	11 37	00 13	29 00
13	15 28 45	20 41 47	23 ♉ 09 19	00 ♊ 06 49	00 ♐ 41	15 39	15 16	17 24	19 00	11 40	00 15	29 02
14	15 32 41	21 42 10	06 ♊ 59 27	13 ♊ 46 45	02 13	15 04	15 54	17 37	18 55	11 44	00 17	29 03
15	15 36 38	22 42 35	20 ♊ 28 27	27 ♊ 04 21	03 45	14 30	16 32	17 51	18 50	11 47	00 20	29 05
16	15 40 35	23 43 02	03 ♋ 34 27	09 ♋ 58 51	05 16	13 57	17 10	18 04	18 45	11 50	00 22	29 07
17	15 44 31	24 43 31	16 ♋ 17 48	22 ♋ 31 37	06 47	13 26	17 48	18 17	18 41	11 53	00 24	29 08
18	15 48 28	25 44 01	28 ♋ 40 46	04 ♌ 45 45	08 18	12 56	18 27	18 30	18 36	11 56	00 26	29 09
19	15 52 24	26 44 33	10 ♌ 47 10	16 ♌ 45 39	09 48	12 28	19 05	18 43	18 31	11 59	00 29	29 11
20	15 56 21	27 45 07	22 ♌ 41 52	28 ♌ 36 31	11 18	12 02	19 43	18 57	18 26	12 02	00 31	29 12
21	16 00 17	28 45 43	04 ♍ 30 19	10 ♍ 23 57	12 48	11 38	20 21	19 10	18 21	12 05	00 33	29 14
22	16 04 14	29 ♏ 46 20	16 ♍ 18 09	22 ♍ 13 33	14 18	11 16	20 59	19 23	18 17	12 07	00 35	29 15
23	16 08 10	00 ♐ 46 59	28 ♍ 10 50	04 ♎ 10 35	15 47	10 56	21 37	19 36	18 12	12 10	00 38	29 16
24	16 12 07	01 47 39	10 ♎ 13 21	16 ♎ 19 37	17 16	10 39	22 15	19 49	18 07	12 13	00 40	29 17
25	16 16 04	02 48 22	22 ♎ 29 48	28 ♎ 44 15	18 44	10 24	22 53	20 02	18 03	12 16	00 42	29 19
26	16 20 00	03 49 05	05 ♏ 03 12	11 ♏ 26 47	20 12	10 12	23 32	20 15	17 58	12 19	00 44	29 20
27	16 23 57	04 49 51	17 ♏ 55 03	24 ♏ 27 57	21 39	10 02	24 10	20 28	17 53	12 21	00 47	29 21
28	16 27 53	05 50 37	01 ♐ 05 18	07 ♐ 46 53	23 06	09 54	24 48	20 41	17 49	12 24	00 49	29 22
29	16 31 50	06 51 25	14 ♐ 32 21	21 ♐ 21 20	24 32	09 50	25 26	20 54	17 44	12 26	00 51	29 23
30	16 35 46	07 ♐ 52 15	28 ♐ 13 24	05 ♑ 08 05	25 ♐ 57	09 ♏ 47	26 ♎ 04	21 ♏ 07	17 ♉ 40	12 ♎ 29	00 ♐ 53	29 ♍ 25

Date	Moon True ☊ ° '	Moon Mean ☊ ° '	Moon ☽ Latitude ° '
01	00 ♓ 09	29 ♒ 10	05 S 01
02	29 ♒ 58	29 06	04 47
03	29 50	29 03	04 17
04	29 45	29 00	03 32
05	29 43	28 57	02 34
06	29 42	28 54	01 26
07	29 D 42	28 51	00 S 13
08	29 R 42	28 47	01 N 02
09	29 40	28 44	02 14
10	29 35	28 41	03 17
11	29 27	28 38	04 08
12	29 17	28 35	04 42
13	29 04	28 32	04 59
14	28 52	28 28	04 57
15	28 40	28 25	04 39
16	28 30	28 22	04 07
17	28 23	28 19	03 23
18	28 18	28 16	02 30
19	28 16	28 12	01 31
20	28 D 15	28 09	00 N 29
21	28 R 15	28 06	00 S 33
22	28 14	28 03	01 34
23	28 12	28 00	02 31
24	28 07	27 57	03 23
25	28 00	27 53	04 05
26	27 49	27 50	04 37
27	27 37	27 47	04 56
28	27 23	27 44	05 00
29	27 10	27 41	04 47
30	26 ♒ 58	27 ♒ 38	04 S 17

DECLINATIONS

Date	Sun ☉	Moon ☽	Mercury ☿	Venus ♀	Mars ♂	Jupiter ♃	Saturn ♄	Uranus ♅	Neptune ♆	Pluto ♇
01	14 S 23	26 S 07	15 S 26	24 S 09	02 S 02	15 S 25	15 N 19	03 S 45	18 S 32	14 N 45
02	14 43	27 43	16 02	23 54	02 17	15 29	15 17	03 46	18 32	14 45
03	15 01	27 42	16 36	23 38	02 32	15 33	15 16	03 47	18 33	14 44
04	15 20	26 02	17 10	23 21	02 47	15 36	15 15	03 49	18 33	14 44
05	15 38	22 48	17 44	23 02	03 03	15 40	15 14	03 50	18 33	14 43
06	15 57	18 14	18 16	22 43	03 18	15 44	15 12	03 51	18 34	14 43
07	16 15	12 36	18 47	22 22	03 33	15 48	15 11	03 53	18 34	14 43
08	16 32	06 S 17	19 18	22 01	03 48	15 52	15 10	03 54	18 35	14 43
09	16 49	00 N 25	19 47	21 38	04 03	15 56	15 08	03 55	18 35	14 42
10	17 07	07 06	20 16	21 16	04 18	16 00	15 07	03 56	18 36	14 42
11	17 23	13 25	20 43	20 52	04 33	16 03	15 06	03 58	18 36	14 42
12	17 40	18 58	21 10	20 28	04 48	16 07	15 05	03 59	18 37	14 41
13	17 56	23 23	21 35	20 03	05 03	16 11	15 03	04 00	18 37	14 41
14	18 12	26 22	21 59	19 39	05 17	16 15	15 02	04 01	18 38	14 41
15	18 27	27 45	22 23	19 14	05 32	16 18	15 01	04 03	18 38	14 41
16	18 42	27 30	22 45	18 50	05 47	16 22	15 00	04 04	18 39	14 41
17	18 57	25 48	23 06	18 25	06 02	16 26	14 58	04 05	18 39	14 40
18	19 12	22 52	23 26	18 02	06 16	16 29	14 57	04 06	18 40	14 40
19	19 26	19 00	23 45	17 38	06 31	16 33	14 56	04 07	18 40	14 40
20	19 40	14 25	24 02	17 15	06 46	16 37	14 55	04 08	18 41	14 40
21	19 53	09 21	24 18	16 53	07 00	16 40	14 55	04 09	18 41	14 40
22	20 06	03 N 58	24 33	16 32	07 15	16 44	14 52	04 11	18 42	14 40
23	20 19	01 S 35	24 47	16 12	07 29	16 47	14 51	04 12	18 42	14 40
24	20 32	07 09	25 00	15 52	07 44	16 51	14 50	04 13	18 42	14 40
25	20 44	12 33	25 11	15 34	07 58	16 54	14 49	04 14	18 43	14 40
26	20 55	17 34	25 21	15 16	08 12	16 58	14 48	04 15	18 43	14 40
27	21 06	21 54	25 29	15 00	08 26	17 01	14 46	04 16	18 44	14 40
28	21 17	25 16	25 36	14 45	08 41	17 05	14 45	04 17	18 44	14 40
29	21 28	27 18	25 42	14 31	08 55	17 08	14 44	04 18	18 45	14 40
30	21 S 38	27 S 43	25 S 46	14 S 18	09 S 09	17 S 12	14 N 43	04 S 19	18 S 45	14 N 40

ZODIAC SIGN ENTRIES

Date	h	m	Planets
01	02	24	☽ ♐
03	08	32	☽ ♑
05	13	11	☽ ♒
06	16	32	♆ ♐
07	16	33	☽ ♓
09	18	52	☽ ♈
11	20	50	☽ ♉
13	01	16	☿ ♐
13	23	48	☽ ♊
16	05	23	☽ ♋
18	14	36	☽ ♌
21	02	50	☽ ♍
22	17	25	☉ ♐
23	15	39	☽ ♎
26	02	25	☽ ♏
28	10	02	☽ ♐
30	15	05	☽ ♑

LATITUDES

Date	Mercury ☿	Venus ♀	Mars ♂	Jupiter ♃	Saturn ♄	Uranus ♅	Neptune ♆	Pluto ♇
01	00 S 02	05 S 58	01 N 05	00 N 54	02 S 31	00 N 40	01 N 38	15 N 33
04	00 22	05 33	01 05	00 54	02 31	00 40	01 38	15 34
07	00 42	05 01	01 04	00 53	02 31	00 40	01 37	15 35
10	01 01	04 23	01 04	00 53	02 31	00 40	01 37	15 36
13	01 19	03 41	01 03	00 53	02 31	00 40	01 37	15 37
16	01 35	02 56	01 03	00 53	02 31	00 40	01 37	15 39
19	01 50	02 09	01 02	00 53	02 30	00 40	01 37	15 40
22	02 03	01 23	01 01	00 53	02 30	00 40	01 37	15 42
25	02 13	00 S 39	01 00	00 53	02 30	00 40	01 37	15 43
28	02 20	00 N 02	01 00	00 53	02 29	00 40	01 37	15 44
31	02 S 24	00 N 40	00 N 59	00 N 53	02 S 29	00 N 40	01 N 37	15 N 46

DATA

Julian Date	2440892
Delta T	+41 seconds
Ayanamsa	23° 27' 06"
Synetic vernal point	05° ♓ 39' 54"
True obliquity of ecliptic	23° 26' 43"

MOON'S PHASES, APSIDES AND POSITIONS ☽

Date	h	m	Phase	Longitude ° '	Eclipse Indicator
06	12	47	☽	13 ♒ 42	
13	07	28	○	20 ♉ 30	
20	23	13	☾	28 ♌ 13	
28	21	14	●	06 ♐ 14	

Day	h	m		
09	20	30	Perigee	
21	18	06	Apogee	
03	00	00	Max dec	27° S 55'
09	10	32	0N	
15	20	18	Max dec	27° N 51'
23	05	09	0S	
30	06	01	Max dec	27° S 47'

LONGITUDES

Date	Chiron ⚷ ° '	Ceres ⚳ ° '	Pallas ⚴ ° '	Juno ⚵ ° '	Vesta ⚶ ° '	Black Moon Lilith ⚸ ° '
01	06 ♈ 47	28 ♈ 53	26 ♒ 12	23 ♉ 55	06 ♏ 16	06 ♍ 33
11	06 ♈ 27	26 ♈ 48	26 ♒ 48	21 ♉ 48	11 ♏ 35	07 ♍ 39
21	06 ♈ 11	25 ♈ 08	27 ♒ 53	19 ♉ 34	16 ♏ 54	08 ♍ 46
31	05 ♈ 59	24 ♈ 03	29 ♒ 24	17 ♉ 38	22 ♏ 14	09 ♍ 52

All ephemeris data is given at 12.00 UT and the Moon's longitude is additionally given for 24.00 UT

ASPECTARIAN

h m	Aspects
01 Sunday	
02 02	☽ ☌ ♆
07 18	☿ # ♄
07 49	♀ ∠ ♂
11 09	☿ ∥ ♃
16 35	☽ ✶ ♂
18 42	☽ ⊻ ☉
21 55	☽ Q ♇
22 31	☽ ✶ ♅
02 Monday	
01 40	☽ ⊻ ☿
05 33	☽ ⊻ ♃
06 27	☽ ⊥ ☉
13 57	☽ ⊥ ☿
14 28	☽ ⊼ ♄
14 33	☉ # ♇
15 21	☽ Q ♂
15 55	☿ ∠ ♇
16 31	☽ ⊥ ♃
18 02	☽ ⊻ ♀
20 13	☽ Q ♅
23 54	☽ ∠ ☉
03 Tuesday	
01 08	☽ ± ♄
04 22	☽ ⊥ ♀
06 19	☽ □ ♇
08 12	☽ ∠ ☿
08 19	☿ ⊥ ♂
08 19	☽ ⊻ ♆
08 51	☽ ∠ ♃
14 23	☿ ☌ ♃
17 03	☽ ⚼ ♄
18 59	☽ ⊥ ♆
19 45	☽ ∠ ♀
23 54	☉ ⊻ ♅
04 Wednesday	
00 56	☽ □ ♂
04 17	☽ □ ♅
04 26	♂ ⊥ ♃
04 37	☽ ✶ ☉
05 35	☉ # ♄
10 53	☽ ∠ ♆
11 46	☽ ✶ ♃
14 11	☽ ✶ ☿
19 16	☽ △ ♄
20 15	☿ ⊥ ♅
21 07	☽ ✶ ♀
05 Thursday	
03 16	☽ Q ☉
09 03	☽ Q ♃
10 25	☽ ∥ ♀
11 07	☽ △ ♇
13 06	☽ ✶ ♆
13 48	☽ Q ☿
15 07	☉ ∥ ♃
17 10	☽ Q ♀
06 Friday	
07 39	☽ △ ♂
08 26	☿ ☍ ♄
08 39	☽ △ ♅
09 51	☽ Q ♆
10 24	☽ ∥ ♆
11 50	☽ ∥ ☿
12 29	☿ ☌ ♀
12 47	☽ □ ☉
13 02	☽ ⚼ ♇
16 23	☉ ∠ ♇
16 34	☽ □ ♃
21 41	☽ ∥ ☉
22 46	☽ □ ♄
22 52	☽ □ ♀
22 58	☽ ∥ ♃
07 Saturday	
00 39	☽ □ ☿
01 25	☽ # ♄
01 54	♀ ☍ ♄
01 57	☿ ∥ ♆
03 24	☽ # ♇
04 26	☽ ± ♇
07 30	♂ ☌ ♅
10 21	☽ ⚼ ♅
10 28	☽ ⚼ ♂
14 38	☽ ⊼ ♇
16 36	☽ □ ♆
08 Sunday	
01 38	☽ ± ♅
02 24	☽ ± ♂
05 05	☽ Q ♄
11 46	☽ ⊼ ♅
12 58	☽ ⊼ ♂
19 27	☽ △ ☉
20 07	☽ △ ♃
20 35	☽ ∥ ♅
20 40	☽ ∥ ♂
22 51	♂ ∥ ♅
23 22	♀ ⊥ ♂
23 29	☽ △ ♀
09 Monday	
01 07	☽ ✶ ♄
07 21	☉ ☌ ♃
09 26	☽ △ ☿
17 05	☽ ☍ ♇
19 03	☽ △ ♆
21 33	☽ ⚼ ♃
22 22	☽ ⚼ ☉
23 30	☽ ⚼ ♀
10 Tuesday	
00 32	☽ # ♅
01 27	☽ # ♂
01 59	☽ ∠ ♄
07 48	☉ ⊥ ♅
08 04	♂ ± ♄
08 49	☉ ☌ ♀
10 16	♀ ⊥ ♅
12 45	☽ ± ♃
13 25	☽ ⚼ ☿
13 52	☽ ± ♀
14 01	☽ ☍ ♅
14 28	☽ ± ☉
16 52	☽ ⊥ ♄
17 20	☽ ☍ ♂
19 16	☿ ∠ ♅
20 04	☽ ⚼ ♆
22 53	☽ ⊼ ♃
23 26	☽ ⊼ ♀
11 Wednesday	
01 12	☽ ⊼ ☉
02 48	☽ ⊻ ♄
06 12	☽ ± ☿
08 53	♀ ☌ ♃
11 06	☽ ± ♆
16 03	☿ ∥ ♀
17 12	☽ ∥ ♇
17 25	☽ ⊼ ☿
18 51	☽ ∥ ♄
19 08	☽ ⊼ ♇
21 09	☽ ⊼ ♆
22 34	☉ ☍ ♄
23 03	☽ # ♃
12 Thursday	
05 13	☽ ± ♇
05 39	☽ # ☉
09 44	☿ ✶ ♇
10 21	☽ # ♆
16 23	☽ ⊼ ♅
18 48	☽ # ♀
20 25	☽ ⚼ ♇
21 55	☽ ⊼ ♂
23 43	☽ ☍ ♀
13 Friday	
00 14	☽ # ☿
00 51	☿ ∠ ♂
02 01	☽ ☍ ♃
02 36	☽ ± ♅
04 56	☽ ☌ ♄
05 01	☿ ☌ ♆
07 28	☽ ☍ ☉
08 37	☽ ± ♂
11 20	♂ ∠ ♆
18 04	☽ ⚼ ♅
19 33	♀ ⊻ ♂
22 09	☽ △ ♇
14 Saturday	
00 16	☽ ☍ ♆
00 51	☽ ⚼ ♂
02 37	☽ ☍ ☿
20 23	☽ △ ♅
15 Sunday	
00 54	☉ ⊥ ♂
01 43	☽ ⊼ ♀
02 22	♃ ⊥ ♅
04 34	☽ △ ♂
07 11	☽ ⊼ ♃
09 04	☽ ⊻ ♄
12 03	☽ ± ♀
16 23	☽ ⊼ ☉
18 13	☽ ± ♃
19 52	☽ ⊥ ♄
16 Monday	
03 43	☽ □ ♇
03 47	☽ ⚼ ♀
04 16	☽ ± ☉
05 21	♀ ∠ ♇
05 57	☉ ∥ ♆
06 02	☽ ⊼ ♆
11 02	☽ ⚼ ♃
12 20	☽ ∠ ♄
15 35	☽ ⊼ ☿
16 24	☉ ∥ ♀
17 13	☽ ± ♆
22 27	☽ ⚼ ☉
22 33	♀ ∥ ♆
17 Tuesday	
03 33	☽ □ ♅
04 24	☽ ± ☿
06 45	☽ △ ♀
10 17	☽ ⚼ ♆
13 36	☽ Q ♇
15 03	☽ □ ♂
15 53	☽ △ ♃
16 32	☽ ✶ ♄
18 Wednesday	
00 02	☽ ⚼ ☿
05 43	☽ △ ☉
08 23	☽ # ☿
12 56	☽ ✶ ♇
14 28	☽ Q ♅
15 28	☽ △ ♆
15 32	♂ ⊻ ♃
15 45	☽ Q ♄
17 08	♂ ⊼ ♄
19 20	♃ ☍ ♄
19 Thursday	
04 11	☽ Q ♂
09 38	☽ # ☉
09 45	☽ △ ☿
13 49	☽ # ♆
14 24	☽ ✶ ♅
15 14	☽ □ ♀
17 54	☉ ∠ ♅
18 49	☽ ∠ ♇
20 09	☽ # ♀
20 Friday	
01 01	☽ # ♃
03 26	☽ □ ♄
04 16	☽ □ ♃
05 37	☽ ✶ ♂
09 32	☽ ∥ ♄
10 18	☿ Q ♇
10 45	☽ ∥ ♇
11 57	♀ ⊻ ♅
13 02	☽ ⊥ ♇
20 49	☽ ∠ ♅
21 01	☿ ⊻ ♀
23 13	☽ □ ☉
23 55	☿ ✶ ♅
21 Saturday	
01 14	☽ ⊻ ♇
02 23	☽ Q ♀
03 55	☽ □ ♆
13 49	☽ ∠ ♂
15 13	☽ ⊥ ♅
17 31	☽ Q ♃
22 08	☽ # ♂
23 17	☉ ✶ ♇
22 Sunday	
00 29	♀ ∥ ♃
02 03	☽ ✶ ♀
03 28	☽ ⊻ ♅
07 20	☽ □ ☿
09 10	☽ ⊥ ♂
11 04	☽ # ♅
15 16	☽ Q ☉
15 59	☽ △ ♄
16 39	☽ Q ♆
18 22	☽ ✶ ♃
22 02	☽ ⊻ ♂
23 Monday	
07 36	☽ ∠ ♀
08 07	☉ ☌ ♆
14 12	☽ ☌ ♇
16 55	☽ ✶ ♆
17 20	♄ ± ♅
17 42	☽ ✶ ☉
21 59	☽ ⚼ ♄
23 14	☽ ∥ ♅
24 Tuesday	
00 47	☽ Q ☿
01 05	☽ ∠ ♃
01 11	☽ ⊥ ♀
03 35	☿ ⊥ ♀
12 49	☽ ⊻ ♀
14 36	☽ ∥ ♂
15 43	☽ ± ♄
15 57	☽ ☌ ♅
19 13	☽ ⊥ ♃
22 44	☽ ∠ ♆
25 Wednesday	
01 17	☿ ⊼ ♄
02 04	☽ ∠ ☉
03 24	☽ ⊼ ♄
03 43	☽ ✶ ☿
07 09	☽ ⊻ ♃
12 48	☽ ☌ ♂
16 16	☽ ⊥ ♆
21 02	☽ ⊥ ☉
21 49	☽ # ♇
22 30	☽ # ♄
26 Thursday	
01 07	☽ ⊻ ♇
01 21	☽ ∥ ♀
03 48	☽ ⊻ ♆
08 59	☽ ∥ ♃
09 28	☽ ⊻ ☉
12 18	☽ ∠ ☿
12 32	☽ ⊥ ♇
13 07	☿ ⊻ ♃
18 01	☽ ∥ ♆
21 32	☽ ☌ ♀
27 Friday	
01 39	☽ ⊻ ♅
05 24	☽ ∠ ♇
06 58	☽ ∥ ☉
07 17	☽ ⊥ ☿
11 57	☽ ☍ ♄
12 48	☽ ⊥ ♅
16 47	☽ ☌ ♃
19 43	☽ ⊻ ☿
28 Saturday	
00 02	☽ ⊻ ♂
05 18	☽ ∠ ♅
08 54	☽ ✶ ♇
11 27	☽ ⊥ ♂
11 30	☽ ☌ ♆
11 34	♀ # ♄
12 40	♂ ⊥ ♆
15 11	☽ ∥ ☿
21 08	♀ # ♇
21 14	☽ ☌ ☉
23 25	☿ ± ♄
29 Sunday	
03 41	☽ ⊻ ♀
04 22	☽ ∠ ♂
06 25	☽ Q ♇
08 16	☽ ✶ ♅
10 28	☿ Q ♅
14 16	☽ ⊥ ♀
16 49	☿ ∠ ♀
17 37	☽ ⊼ ♄
23 24	☽ ⊻ ♃
30 Monday	
04 05	☽ ± ♄
05 27	☽ Q ♅
06 01	☽ ∠ ♀
07 35	☽ ☌ ☿
08 04	☽ ✶ ♂
10 03	☽ ⊥ ♃
14 04	☽ □ ♇
15 44	☿ ✶ ♂
16 39	☽ ⊻ ♆
19 41	☽ ⚼ ♄

DECEMBER 1970

LONGITUDES

Date	Sidereal time h m s	Sun ☉ ° ′ ″	Moon ☽ ° ′ ″	Moon ☽ 24.00 ° ′ ″	Mercury ☿ ° ′	Venus ♀ ° ′	Mars ♂ ° ′	Jupiter ♃ ° ′	Saturn ♄ ° ′	Uranus ♅ ° ′	Neptune ♆ ° ′	Pluto ⯓ ° ′
01	16 39 43	08 ♐ 53 05	12 ♑ 04 57	19 ♑ 03 34	27 ♐ 21	09 ♏ 47	26 ♎ 42	21 ♏ 20	17 ♉ 36	12 ♎ 32	00 ♐ 56	29 ♍ 26
02	16 43 39	09 53 57	26 ♑ 03 31	03 ♒ 04 29	28 ♐ 44	09 D 50	27 20	21 33	17 R 31	12 34	00 58	29 27
03	16 47 36	10 54 49	10 ♒ 06 09	17 ♒ 08 16	00 ♑ 06	09 54	27 58	21 46	17 27	12 36	01 00	29 28
04	16 51 33	11 55 42	24 ♒ 10 41	01 ♓ 13 14	01 26	10 01	28 37	21 59	17 23	12 39	01 02	29 29
05	16 55 29	12 56 36	08 ♓ 15 49	15 ♓ 18 21	02 45	10 11	29 15	22 12	17 19	12 41	01 05	29 30
06	16 59 26	13 57 30	22 ♓ 20 46	29 ♓ 22 57	04 02	10 22	29 ♎ 53	22 25	17 14	12 43	01 07	29 30
07	17 03 22	14 58 25	06 ♈ 24 47	13 ♈ 26 07	05 16	10 36	00 ♏ 31	22 37	17 10	12 46	01 09	29 31
08	17 07 19	15 59 21	20 ♈ 26 43	27 ♈ 26 21	06 28	10 52	01 09	22 50	17 06	12 48	01 11	29 32
09	17 11 15	17 00 18	04 ♉ 24 41	11 ♉ 21 20	07 37	11 10	01 47	23 03	17 03	12 50	01 13	29 33
10	17 15 12	18 01 15	18 ♉ 15 56	25 ♉ 08 04	08 43	11 30	02 25	23 15	16 59	12 52	01 16	29 34
11	17 19 08	19 02 13	01 ♊ 57 16	08 ♊ 43 10	09 44	11 51	03 03	23 28	16 55	12 54	01 18	29 35
12	17 23 05	20 03 12	15 ♊ 25 21	22 ♊ 03 32	10 41	12 15	03 41	23 41	16 51	12 56	01 20	29 35
13	17 27 02	21 04 12	28 ♊ 37 26	05 ♋ 06 52	11 32	12 41	04 19	23 53	16 48	12 58	01 22	29 36
14	17 30 58	22 05 12	11 ♋ 31 47	17 ♋ 52 10	12 18	13 08	04 57	24 06	16 44	13 00	01 24	29 37
15	17 34 55	23 06 14	24 ♋ 08 06	00 ♌ 19 49	12 56	13 37	05 36	24 18	16 41	13 02	01 26	29 37
16	17 38 51	24 07 16	06 ♌ 27 35	12 ♌ 31 46	13 27	14 07	06 14	24 30	16 37	13 04	01 28	29 38
17	17 42 48	25 08 19	18 ♌ 32 47	24 ♌ 31 09	13 50	14 40	06 52	24 43	16 34	13 06	01 31	29 38
18	17 46 44	26 09 23	00 ♍ 27 26	06 ♍ 22 12	14 03	15 13	07 30	24 55	16 31	13 07	01 33	29 39
19	17 50 41	27 10 27	12 ♍ 16 08	18 ♍ 09 51	14 R 05	15 48	08 08	25 07	16 28	13 09	01 35	29 39
20	17 54 37	28 11 33	24 ♍ 04 04	29 ♍ 59 27	13 57	16 25	08 46	25 19	16 25	13 11	01 37	29 40
21	17 58 34	29 ♐ 12 39	05 ♎ 56 41	11 ♎ 56 27	13 37	17 02	09 24	25 32	16 22	13 12	01 39	29 40
22	18 02 31	00 ♑ 13 46	17 ♎ 59 24	24 ♎ 06 07	13 05	17 42	10 02	25 44	16 19	13 14	01 41	29 41
23	18 06 27	01 14 54	00 ♏ 17 10	06 ♏ 33 03	12 22	18 22	10 40	25 56	16 16	13 15	01 43	29 41
24	18 10 24	02 16 02	12 ♏ 54 09	19 ♏ 20 47	11 28	19 03	11 18	26 08	16 14	13 17	01 45	29 41
25	18 14 20	03 17 11	25 ♏ 53 10	02 ♐ 31 22	10 24	19 46	11 56	26 19	16 11	13 18	01 47	29 41
26	18 18 17	04 18 21	09 ♐ 15 19	16 ♐ 04 52	09 12	20 30	12 34	26 31	16 09	13 19	01 49	29 42
27	18 22 13	05 19 31	22 ♐ 59 39	29 ♐ 59 15	07 54	21 14	13 12	26 43	16 06	13 20	01 51	29 42
28	18 26 10	06 20 41	07 ♑ 03 07	14 ♑ 10 36	06 33	22 00	13 50	26 55	16 04	13 22	01 53	29 42
29	18 30 06	07 21 52	21 ♑ 20 59	28 ♑ 33 32	05 11	22 47	14 29	27 06	16 02	13 23	01 55	29 42
30	18 34 03	08 23 03	05 ♒ 47 30	13 ♒ 02 10	03 51	23 35	15 07	27 18	16 00	13 24	01 57	29 42
31	18 38 00	09 ♑ 24 14	20 ♒ 16 51	27 ♒ 30 55	02 ♑ 35	24 ♏ 23	15 ♏ 45	27 ♏ 29	15 ♉ 58	13 ♎ 25	01 ♐ 59	29 ♍ 42

Date	Moon True ☊ ° ′	Moon Mean ☊ ° ′	Moon ☽ Latitude ° ′
01	26 ♒ 49	27 ♒ 34	03 S 33
02	26 R 43	27 31	02 34
03	26 40	27 28	01 27
04	26 39	27 25	00 S 13
05	26 D 40	27 22	01 N 01
06	26 R 39	27 18	02 12
07	26 38	27 15	03 14
08	26 33	27 12	04 05
09	26 26	27 09	04 41
10	26 17	27 06	04 59
11	26 06	27 03	05 01
12	25 54	26 59	04 45
13	25 42	26 56	04 15
14	25 33	26 53	03 31
15	25 26	26 50	02 38
16	25 21	26 47	01 39
17	25 19	26 44	00 N 36
18	25 D 19	26 40	00 S 28
19	25 20	26 37	01 30
20	25 21	26 34	02 28
21	25 R 21	26 31	03 20
22	25 19	26 28	04 04
23	25 15	26 24	04 38
24	25 08	26 21	05 00
25	25 00	26 18	05 07
26	24 51	26 15	04 57
27	24 42	26 12	04 31
28	24 34	26 09	03 47
29	24 28	26 05	02 49
30	24 24	26 02	01 39
31	24 ♒ 23	25 ♒ 59	00 S 22

DECLINATIONS

Date	Sun ☉ ° ′	Moon ☽ ° ′	Mercury ☿ ° ′	Venus ♀ ° ′	Mars ♂ ° ′	Jupiter ♃ ° ′	Saturn ♄ ° ′	Uranus ♅ ° ′	Neptune ♆ ° ′	Pluto ⯓ ° ′
01	21 S 47	26 S 25	25 S 49	14 S 07	09 S 23	17 S 15	14 N 42	04 S 20	18 S 46	14 N 40
02	21 56	23 28	25 51	13 56	09 37	17 19	14 41	04 21	18 46	14 40
03	22 05	19 06	25 51	13 47	09 51	17 22	14 40	04 22	18 47	14 40
04	22 13	13 40	25 49	13 39	10 04	17 25	14 39	04 22	18 47	14 40
05	22 21	07 32	25 46	13 31	10 18	17 29	14 38	04 23	18 48	14 40
06	22 29	01 S 01	25 42	13 25	10 32	17 32	14 37	04 24	18 48	14 40
07	22 36	05 N 31	25 37	13 20	10 46	17 35	14 36	04 25	18 48	14 40
08	22 42	11 46	25 30	13 17	10 59	17 38	14 35	04 26	18 49	14 41
09	22 49	17 23	25 21	13 14	11 13	17 41	14 34	04 27	18 49	14 41
10	22 54	22 04	25 12	13 12	11 26	17 45	14 33	04 28	18 50	14 41
11	23 00	25 28	25 01	13 11	11 39	17 48	14 33	04 28	18 50	14 41
12	23 04	27 22	24 49	13 11	11 52	17 51	14 32	04 29	18 51	14 41
13	23 09	27 41	24 36	13 11	12 06	17 54	14 31	04 30	18 51	14 42
14	23 13	26 27	24 23	13 13	12 19	17 57	14 30	04 31	18 51	14 42
15	23 16	23 53	24 08	13 15	12 32	18 00	14 29	04 31	18 52	14 42
16	23 19	20 16	23 53	13 19	12 45	18 03	14 29	04 32	18 52	14 42
17	23 21	15 51	23 37	13 23	12 57	18 06	14 28	04 33	18 53	14 43
18	23 23	10 53	23 21	13 27	13 10	18 09	14 27	04 33	18 53	14 43
19	23 25	05 35	23 05	13 32	13 23	18 12	14 27	04 34	18 53	14 43
20	23 26	00 N 06	22 49	13 38	13 35	18 15	14 26	04 34	18 54	14 44
21	23 27	05 S 26	22 33	13 45	13 48	18 18	14 25	04 35	18 54	14 44
22	23 27	10 49	22 17	13 51	14 00	18 20	14 25	04 36	18 55	14 44
23	23 26	15 55	22 01	13 59	14 12	18 23	14 24	04 36	18 55	14 45
24	23 26	20 28	21 46	14 07	14 24	18 26	14 24	04 37	18 55	14 45
25	23 24	24 12	21 32	14 15	14 36	18 29	14 23	04 37	18 56	14 45
26	23 22	26 44	21 18	14 24	14 48	18 32	14 23	04 38	18 56	14 46
27	23 20	27 46	21 05	14 33	15 00	18 34	14 22	04 38	18 56	14 46
28	23 18	27 02	20 54	14 42	15 12	18 37	14 22	04 38	18 57	14 47
29	23 14	24 32	20 43	14 52	15 24	18 40	14 22	04 39	18 57	14 47
30	23 11	20 25	20 34	15 02	15 35	18 42	14 21	04 39	18 58	14 48
31	23 S 07	15 S 05	20 S 26	15 S 12	15 S 47	18 S 45	14 N 21	04 S 40	18 S 58	14 N 48

ZODIAC SIGN ENTRIES

Date	h m	Planets
02	18 45	☽ ♒
03	10 14	☿ ♑
04	21 55	☽ ♓
06	16 34	♂ ♏
07	01 03	☽ ♈
09	04 24	☽ ♉
11	08 33	☽ ♊
13	14 32	☽ ♋
15	23 21	☽ ♌
18	11 04	☽ ♍
21	00 01	☽ ♎
22	06 36	☉ ♑
23	11 27	☽ ♏
25	19 28	☽ ♐
28	00 01	☽ ♑
30	02 24	☽ ♒

LATITUDES

Date	Mercury ☿ ° ′	Venus ♀ ° ′	Mars ♂ ° ′	Jupiter ♃ ° ′	Saturn ♄ ° ′	Uranus ♅ ° ′	Neptune ♆ ° ′	Pluto ⯓ ° ′
01	02 S 24	00 N 40	00 N 59	00 N 53	02 S 29	00 N 40	01 N 37	15 N 46
04	02 23	01 15	00 58	00 53	02 28	00 41	01 37	15 48
07	02 16	01 45	00 57	00 53	02 28	00 41	01 37	15 49
10	02 03	02 12	00 56	00 53	02 27	00 41	01 37	15 51
13	01 40	02 35	00 56	00 53	02 27	00 41	01 37	15 52
16	01 07	02 54	00 55	00 53	02 26	00 41	01 37	15 54
19	00 S 23	03 10	00 53	00 53	02 25	00 41	01 38	15 56
22	00 N 31	03 23	00 52	00 53	02 24	00 41	01 38	15 57
25	01 31	03 34	00 51	00 53	02 24	00 41	01 38	15 59
28	02 24	03 42	00 50	00 53	02 23	00 41	01 38	16 01
31	02 N 59	03 N 47	00 N 49	00 N 53	02 S 22	00 N 41	01 N 38	16 N 03

DATA

Julian Date	2440922
Delta T	+41 seconds
Ayanamsa	23° 27' 11"
Synetic vernal point	05° ♓ 39' 49"
True obliquity of ecliptic	23° 26' 42"

MOON'S PHASES, APSIDES AND POSITIONS ☽

Date	h m	Phase	Longitude ° ′	Eclipse Indicator
05	20 36	☽	13 ♓ 18	
12	21 03	○	20 ♊ 26	
20	21 09	☾	28 ♍ 35	
28	10 43	●	06 ♑ 17	

Day	h m		
05	05 25	Perigee	
19	15 21	Apogee	
31	09 21	Perigee	
06	15 43	ON	
13	04 37	Max dec	27° N 45'
20	12 24	OS	
27	14 20	Max dec	27° S 46'

LONGITUDES

Date	Chiron ⚷ ° ′	Ceres ⚳ ° ′	Pallas ⚴ ° ′	Juno ⚵ ° ′	Vesta ⚶ ° ′	Black Moon Lilith ⚸ ° ′
01	05 ♈ 59	24 ♈ 03	29 ♒ 24	17 ♉ 38	22 ♏ 14	09 ♍ 52
11	05 ♈ 53	23 ♈ 36	01 ♓ 18	16 ♉ 23	27 ♏ 33	10 ♍ 59
21	05 ♈ 52	23 ♈ 49	03 ♓ 32	16 ♉ 01	02 ♐ 51	12 ♍ 05
31	05 ♈ 57	24 ♈ 39	06 ♓ 03	16 ♉ 35	08 ♐ 07	13 ♍ 12

All ephemeris data is given at 12.00 UT and the Moon's longitude is additionally given for 24.00 UT

ASPECTARIAN

h m	Aspects
01 Tuesday	
00 03	♀ St D
01 56	☽ ∠ ♃
03 05	☽ ⊥ ♆
05 53	☽ Q ♂
06 03	☽ ⊻ ☉
08 02	☽ ⚹ ♀
11 46	☿ ⊥ ♃
12 46	☽ □ ♅
17 12	☽ ⊥ ☉
18 02	☽ ∥ ☿
18 38	☽ ∠ ♆
21 26	☽ △ ♄
02 Wednesday	
04 10	☽ ⚹ ♃
04 43	☽ Q ♀
09 52	☽ ∠ ☉
10 09	☉ ⊻ ♀
14 18	☽ □ ♂
17 05	☽ ⊻ ☿
17 48	☽ △ ⯓
20 25	☽ ⚹ ♆
20 56	☽ ∥ ☉
03 Thursday	
00 34	☿ □ ⯓
01 01	☽ Q ♃
04 26	☽ ⊥ ☿
11 40	☽ □ ♀
13 30	☽ ⚹ ☉
13 35	☽ ∥ ♆
15 18	♄ ∥ ⯓
16 17	☽ △ ♅
16 58	☽ Q ♆
19 27	☽ ♇ ⯓
20 04	☽ ∥ ♃
21 26	☽ ∠ ☿
04 Friday	
00 28	☽ □ ♄
01 09	☉ Q ⯓
04 33	☿ ⊻ ♆
07 53	☽ # ⯓
07 56	☽ # ♄
08 12	☽ □ ♃
10 48	☽ ± ⯓
11 32	☽ Q ☉
12 08	☽ ∥ ♀
17 56	☽ ♇ ♅
19 54	☽ △ ♂
21 02	☽ ⊼ ⯓
23 43	☽ □ ♆
05 Saturday	
01 39	☽ ⚹ ☿
01 47	☽ ∥ ♂
04 16	☿ ♇ ♄
05 40	☉ ⚹ ♅
06 59	☽ Q ♄
09 18	☽ ± ♅
15 18	☽ △ ♀
19 33	☽ ⊼ ♅
20 36	☽ □ ☉
21 38	♂ ⊻ ⯓
22 40	☽ ♇ ♂
23 38	☽ ∥ ♅
06 Sunday	
00 10	☽ Q ☿
03 20	☽ ⚹ ♄
12 07	☽ △ ♃
14 44	☽ ± ♂
17 14	☽ ♇ ♀
07 Monday	
00 14	☽ ☍ ⯓
01 27	☽ ⊼ ♂
02 59	☽ △ ♆
04 48	☽ ∠ ♄
07 54	☽ # ♅
08 51	☽ ± ♀
09 52	☽ □ ☿
14 06	☽ ♇ ♃
19 17	☽ ⊼ ♀
20 06	☽ ⊥ ♄
22 53	☽ ☍ ♅
08 Tuesday	
03 46	☽ △ ☉
04 41	☽ ♇ ♆
05 43	☽ ± ♃
06 19	☽ ⊻ ♄
08 46	☽ # ♂
13 28	♂ ⊻ ♆
16 10	☽ ⊼ ♃
18 04	☽ # ♀
20 09	☽ ± ♆
22 04	☉ ∠ ♂
23 35	☽ ∥ ♄
09 Wednesday	
00 00	☽ ∥ ⯓
03 17	☿ ⊥ ♆
03 37	☽ ⊼ ⯓
06 29	☽ ⊼ ♆
07 15	☽ ☍ ♂
07 32	☽ ♇ ☉
12 51	☉ ⊼ ♄
13 24	☽ # ♃
13 58	☽ ± ⯓
17 27	☉ ⊥ ♀
18 01	☽ △ ☿
18 47	☽ # ♆
23 28	☿ ∠ ♃
23 57	☽ ☍ ♀
10 Thursday	
00 18	☽ ± ☉
02 36	☽ ⊼ ♅
05 33	☽ ♇ ⯓
09 46	☽ ☌ ♄
11 32	☽ ⊼ ☉
13 03	☽ ± ♅
17 16	☽ # ☉
20 51	☽ ☍ ♃
22 17	☽ ♇ ☿
11 Friday	
04 51	☽ ♇ ♅
07 48	☽ △ ⯓
08 24	☽ # ☿
10 50	☽ ☍ ♆
14 02	☽ ⊼ ♂
15 23	☽ ± ☿
12 Saturday	
01 13	☽ ± ♂
02 52	☽ ⊼ ☿
06 08	☽ ⊼ ♀
07 32	☽ △ ♅
14 34	☽ ⊻ ♄
17 16	☽ ± ♀
18 11	☽ ♇ ♂
21 03	☽ ☍ ☉
13 Sunday	
01 23	☽ ⊥ ♄
03 11	☽ ⊼ ♃
10 12	☽ ♇ ♀
13 48	☽ □ ⯓
14 22	☽ ± ♃
17 04	☽ ⊼ ♆
17 49	☽ ∠ ♄
23 04	☽ △ ♂
14 Monday	
04 15	☽ ± ♆
04 48	♀ ⊻ ♅
07 21	☽ ♇ ♃
13 32	☽ ☍ ☿
14 47	☽ □ ♅
15 08	☽ △ ♀
21 14	☽ ♇ ♆
21 48	☽ ⚹ ♄
23 31	☽ Q ⯓
15 Tuesday	
02 31	☉ ± ♄
09 50	☽ ⊼ ☉
09 56	☽ # ☿
12 19	☽ △ ♃
13 03	♂ ⊥ ⯓
16 12	☿ □ ♅
16 33	☽ # ☉
20 45	☽ Q ♄
22 29	☽ ± ☉
22 38	☽ ⚹ ⯓
16 Wednesday	
01 24	☽ Q ♅
02 12	☽ △ ♆
11 31	☽ □ ♂
17 44	☽ ♇ ☉
19 57	☽ # ♆
23 24	☉ ⊻ ♃
17 Thursday	
00 19	☽ # ♃
01 06	☽ ⚹ ♅
02 19	☽ ⊼ ☿
03 52	☽ □ ♀
04 11	☽ ∠ ⯓
08 04	☽ □ ♄
10 55	☉ Q ♅
11 07	♀ ∠ ⯓
14 38	☽ ± ☿
17 41	☽ ∥ ⯓
18 55	☽ ∥ ♄
22 14	☽ ⊥ ⯓
18 Friday	
00 02	☽ # ♀
00 36	☽ □ ♃
01 24	☽ Q ♂
01 39	☽ # ♂
02 29	☽ △ ☉
07 16	☽ ∠ ♅
09 06	☉ ∥ ☿
09 06	☽ ♇ ☿
10 22	☽ ⊻ ⯓
14 13	☽ □ ♆
17 53	☽ Q ♀
19 Saturday	
01 34	☽ ⊥ ♅
03 06	☽ ⚹ ♂
05 58	☿ St R
13 46	☽ Q ♃
13 48	☽ ⊻ ♅
15 41	☽ △ ☿
16 28	☽ # ♅
19 35	☽ ⚹ ♀
20 30	☽ △ ♄
20 Sunday	
02 55	☽ Q ♆
11 21	☽ ∠ ♂
12 05	♀ ☍ ♄
14 36	☽ ⚹ ♃
21 09	☽ □ ☉
23 18	♀ ∥ ♂
23 21	☽ ☌ ⯓
21 Monday	
02 49	☽ ♇ ♄
03 19	☽ ⚹ ♆
03 42	☽ ∠ ♀
06 36	☽ ⊥ ♂
08 19	☽ ∥ ♅
19 19	☽ ⊻ ♂
20 49	☽ ± ♄
21 20	☽ ∠ ♃
22 47	☽ ⊥ ♀
22 53	☉ □ ⯓
22 Tuesday	
02 33	☽ ☌ ♅
02 44	☽ □ ☿
06 45	☿ □ ♅
08 42	☽ ⊼ ♄
09 25	☽ ∠ ♆
11 23	☽ ⊻ ♀
12 31	☽ Q ☉
15 29	☽ ⊥ ♃
23 Wednesday	
02 25	☽ ∥ ♀
03 07	☽ ⊥ ♆
03 22	☽ ∥ ♂
03 25	☽ ⊻ ♃
04 41	☽ # ♄
06 17	☽ # ⯓
10 50	☽ ⊻ ⯓
12 09	☽ Q ☿
12 33	☉ ♇ ♄
14 01	☽ ⚹ ☉
14 46	☽ ⊻ ♆
22 21	☽ ⊥ ⯓
23 24	☉ ⊻ ♆
24 Thursday	
00 43	☽ ∥ ♃
03 25	☽ ∥ ♆
07 51	☉ ⊥ ♃
08 50	☽ ☌ ♂
09 30	☽ ⚹ ☿
10 48	♂ # ♄
12 42	☽ ⊻ ♅
14 25	☿ ⚹ ♂
15 21	☽ ∠ ⯓
18 12	☽ ☍ ♄
18 48	☿ ∠ ♃
19 15	☽ ∥ ☿
19 45	♀ ⊥ ♅
20 51	☽ ∠ ☉
23 53	☽ ⊥ ♅
25 Friday	
00 07	☽ ☌ ♀
06 21	☽ ∥ ☉
11 11	☽ ∠ ☿
12 48	☽ ☌ ♃
14 46	☽ ⊥ ☉
16 23	☽ ∠ ♅
18 54	☽ ⚹ ⯓
22 42	☽ ☌ ♆
26 Saturday	
02 08	☽ ⊥ ☿
02 28	☽ ⊻ ☉
06 34	♂ # ⯓
09 52	♀ # ♄
11 55	☽ ⊻ ☿
16 19	☽ Q ⯓
18 08	☽ ⊻ ♂
19 11	☽ ⚹ ♅
27 Sunday	
00 05	☽ ⊼ ♄
05 08	☽ ⊥ ♂
08 48	☽ ⊻ ♀
10 28	☽ ± ♄
12 55	☿ ⊥ ♆
16 03	☽ Q ♅
17 15	♂ ⊻ ♅
18 30	☽ ⊻ ♃
19 43	☽ ⊥ ♀
21 23	☽ ∠ ♂
23 31	☽ □ ⯓
28 Monday	
01 52	☽ ♇ ♄
03 13	☽ ⊻ ♆
04 53	☽ ⊥ ♃
06 51	☿ ∠ ♀
10 43	☽ ☌ ☉
11 13	☽ ☌ ☿
11 55	☽ ∠ ♀
13 24	☽ ⊥ ♆
14 01	☉ ☌ ☿
20 18	☽ ∠ ♃
22 39	☽ □ ♅
23 58	☽ ⚹ ♂
29 Tuesday	
00 05	♀ # ⯓
03 08	☽ △ ♄
04 34	☽ ∠ ♆
14 32	☽ ⚹ ♀
20 38	☽ ∥ ☉
20 42	♂ ∠ ⯓
20 56	☽ Q ♂
21 43	☽ ⚹ ♃
30 Wednesday	
01 21	☉ ⊥ ♆
01 54	☽ △ ⯓
05 37	☽ ⚹ ♆
09 03	☽ ⊻ ☿
11 18	☽ ∥ ☿
11 37	☽ Q ♀
16 37	☽ ⊻ ☉
17 53	☽ Q ♃
18 10	☽ ⊥ ☿
19 04	☽ ∥ ♆
20 12	☽ ∥ ♃
20 56	☿ ⊥ ♃
31 Thursday	
00 37	☽ △ ♅
01 32	☽ Q ♆
02 46	☽ ♇ ⯓
03 19	☽ ⊥ ☉
04 08	☽ □ ♂
04 52	☽ □ ♄
07 53	☽ ∠ ☿
09 12	☽ ∥ ♂
09 53	☉ ∠ ♀
11 32	☽ ∥ ♀
13 08	☽ # ⯓
14 59	☽ # ♄
17 41	☽ ± ⯓
19 13	☽ □ ♀
19 21	☽ ∠ ☉
20 06	♂ ☍ ♄

JANUARY 1971

LONGITUDES

Date	Sidereal time h m s	Sun ☉ ° ' "	Moon ☽ ° ' "	Moon ☽ 24.00 ° ' "	Mercury ☿ ° '	Venus ♀ ° '	Mars ♂ ° '	Jupiter ♃ ° '	Saturn ♄ ° '	Uranus ♅ ° '	Neptune ♆ ° '	Pluto ♇ ° '
01	18 41 56	10 ♑ 25 24	04 ♓ 43 51	11 ♓ 55 13	01 ♑ 27	25 ♏ 13	16 ♏ 23	27 ♏ 41	15 ♉ 56	13 ♎ 26	02 ♐ 01	29 ♍ 42
02	18 45 53	11 26 34	19 ♓ 04 37	26 ♓ 11 48	00 ♑ 26	26 03	17 01	27 52	15 R 54	13 27	02 02	29 R 42
03	18 49 49	12 27 44	03 ♈ 16 33	10 ♈ 18 42	29 ♐ 35	26 54	17 39	28 03	15 53	13 28	02 04	29 42
04	18 53 46	13 28 54	17 ♈ 18 10	24 ♈ 14 51	28 R 54	27 45	18 17	28 14	15 51	13 29	02 06	29 42
05	18 57 42	14 30 03	01 ♉ 08 44	07 ♉ 59 46	28 23	28 38	18 54	28 25	15 50	13 29	02 08	29 42
06	19 01 39	15 31 12	14 ♉ 47 55	21 ♉ 33 07	28 02	29 ♏ 31	19 32	28 36	15 49	13 30	02 10	29 42
07	19 05 35	16 32 21	28 ♉ 15 20	04 ♊ 54 31	27 51	00 ♐ 25	20 10	28 47	15 48	13 31	02 11	29 42
08	19 09 32	17 33 30	11 ♊ 30 36	18 ♊ 03 30	27 D 49	01 19	20 48	28 58	15 46	13 31	02 13	29 42
09	19 13 29	18 34 38	24 ♊ 33 09	00 ♋ 59 30	27 56	02 14	21 26	29 09	15 45	13 32	02 15	29 41
10	19 17 25	19 35 45	07 ♋ 22 31	13 ♋ 42 12	28 11	03 10	22 04	29 20	15 45	13 32	02 16	29 41
11	19 21 22	20 36 53	19 ♋ 58 32	26 ♋ 11 35	28 33	04 06	22 42	29 30	15 44	13 33	02 18	29 41
12	19 25 18	21 38 00	02 ♌ 21 27	08 ♌ 28 16	29 02	05 03	23 20	29 41	15 43	13 33	02 20	29 41
13	19 29 15	22 39 06	14 ♌ 32 16	20 ♌ 33 39	29 ♐ 37	06 01	23 58	29 ♏ 51	15 43	13 33	02 21	29 40
14	19 33 11	23 40 13	26 ♌ 32 44	02 ♍ 29 51	00 ♑ 17	06 58	24 36	00 ♐ 01	15 42	13 33	02 23	29 40
15	19 37 08	24 41 19	08 ♍ 25 25	14 ♍ 19 53	01 02	07 57	25 14	00 12	15 42	13 34	02 25	29 39
16	19 41 04	25 42 24	20 ♍ 13 42	26 ♍ 07 25	01 51	08 56	25 52	00 22	15 42	13 34	02 26	29 39
17	19 45 01	26 43 30	02 ♎ 01 35	07 ♎ 56 48	02 45	09 55	26 29	00 32	15 42	13 34	02 28	29 38
18	19 48 58	27 44 35	13 ♎ 53 39	19 ♎ 52 47	03 42	10 55	27 07	00 42	15 D 42	13 R 34	02 29	29 38
19	19 52 54	28 45 40	25 ♎ 54 48	02 ♏ 00 20	04 42	11 55	27 45	00 51	15 42	13 34	02 31	29 37
20	19 56 51	29 ♑ 46 44	08 ♏ 10 00	14 ♏ 24 22	05 45	12 56	28 23	01 01	15 43	13 34	02 32	29 37
21	20 00 47	00 ♒ 47 49	20 ♏ 43 57	27 ♏ 09 15	06 50	13 57	29 01	01 11	15 43	13 34	02 33	29 36
22	20 04 44	01 48 52	03 ♐ 40 38	10 ♐ 18 23	07 58	14 58	29 ♏ 39	01 20	15 43	13 33	02 35	29 35
23	20 08 40	02 49 56	17 ♐ 02 43	23 ♐ 53 37	09 08	16 00	00 ♐ 16	01 29	15 44	13 33	02 36	29 35
24	20 12 37	03 50 59	00 ♑ 51 02	07 ♑ 54 40	10 20	17 02	00 54	01 39	15 45	13 33	02 37	29 34
25	20 16 33	04 52 01	15 ♑ 04 07	22 ♑ 18 48	11 33	18 05	01 32	01 48	15 45	13 32	02 39	29 33
26	20 20 30	05 53 03	29 ♑ 37 59	07 ♒ 00 50	12 49	19 08	02 10	01 57	15 46	13 32	02 40	29 32
27	20 24 27	06 54 04	14 ♒ 26 25	21 ♒ 53 44	14 06	20 11	02 47	02 06	15 48	13 32	02 41	29 32
28	20 28 23	07 55 03	29 ♒ 21 45	06 ♓ 49 27	15 24	21 14	03 25	02 15	15 49	13 31	02 42	29 31
29	20 32 20	08 56 02	14 ♓ 15 53	21 ♓ 40 10	16 43	22 18	04 03	02 23	15 50	13 30	02 44	29 30
30	20 36 16	09 57 00	29 ♓ 01 31	06 ♈ 19 17	18 04	23 22	04 41	02 32	15 51	13 30	02 45	29 29
31	20 40 13	10 ♒ 57 56	13 ♈ 32 55	20 ♈ 42 03	19 ♑ 26	24 ♐ 27	05 ♐ 18	02 ♐ 40	15 ♉ 53	13 ♎ 29	02 ♐ 46	29 ♍ 28

Date	Moon True ☊ ° '	Moon Mean ☊ ° '	Moon ☽ Latitude ° '
01	24 ♒ 23	25 ♒ 56	00 N 56
02	24 D 24	25 53	02 10
03	24 26	25 50	03 15
04	24 R 26	25 46	04 08
05	24 25	25 43	04 46
06	24 21	25 40	05 06
07	24 16	25 37	05 10
08	24 10	25 34	04 57
09	24 04	25 30	04 28
10	23 58	25 27	03 46
11	23 53	25 24	02 54
12	23 49	25 21	01 55
13	23 47	25 18	00 N 50
14	23 D 46	25 15	00 S 15
15	23 47	25 11	01 20
16	23 49	25 08	02 20
17	23 51	25 05	03 15
18	23 53	25 02	04 02
19	23 53	24 59	04 39
20	23 R 53	24 55	05 03
21	23 52	24 52	05 15
22	23 49	24 49	05 11
23	23 46	24 46	04 50
24	23 43	24 43	04 12
25	23 41	24 40	03 17
26	23 39	24 36	02 09
27	23 38	24 33	00 S 51
28	23 D 37	24 30	00 N 32
29	23 38	24 27	01 52
30	23 39	24 24	03 04
31	23 ♒ 40	24 ♒ 21	04 N 03

DECLINATIONS

Date	Sun ☉ ° '	Moon ☽ ° '	Mercury ☿ ° '	Venus ♀ ° '	Mars ♂ ° '	Jupiter ♃ ° '	Saturn ♄ ° '	Uranus ♅ ° '	Neptune ♆ ° '	Pluto ♇ ° '
01	23 S 02	08 S 55	20 S 20	15 S 22	15 S 58	18 S 47	14 N 21	04 S 40	18 S 58	14 N 49
02	22 57	02 S 20	20 16	15 33	16 09	18 50	14 21	04 40	18 59	14 49
03	22 52	04 N 17	20 14	15 43	16 20	18 52	14 20	04 41	18 59	14 50
04	22 46	10 37	20 13	15 54	16 31	18 55	14 20	04 41	18 59	14 50
05	22 39	16 20	20 15	16 05	16 42	18 57	14 20	04 41	19 00	14 51
06	22 33	21 09	20 18	16 16	16 53	18 59	14 20	04 41	19 00	14 51
07	22 25	24 48	20 22	16 27	17 03	19 02	14 20	04 42	19 00	14 52
08	22 18	27 04	20 28	16 38	17 14	19 04	14 20	04 42	19 00	14 52
09	22 09	27 48	20 35	16 49	17 24	19 06	14 20	04 42	19 01	14 53
10	22 01	27 00	20 43	17 01	17 35	19 09	14 20	04 42	19 01	14 54
11	21 52	24 50	20 52	17 12	17 45	19 11	14 20	04 42	19 01	14 54
12	21 42	21 30	21 01	17 22	17 55	19 13	14 20	04 42	19 02	14 55
13	21 32	17 17	21 10	17 33	18 05	19 15	14 20	04 42	19 02	14 56
14	21 22	12 26	21 19	17 44	18 14	19 17	14 20	04 43	19 02	14 56
15	21 12	07 11	21 28	17 55	18 24	19 19	14 20	04 43	19 02	14 57
16	21 00	01 N 43	21 38	18 05	18 34	19 21	14 21	04 43	19 03	14 57
17	20 49	03 S 47	21 46	18 15	18 43	19 23	14 21	04 43	19 03	14 58
18	20 37	09 12	21 55	18 26	18 52	19 25	14 21	04 43	19 03	14 59
19	20 25	14 20	22 03	18 35	19 01	19 27	14 22	04 42	19 03	15 00
20	20 12	19 01	22 10	18 45	19 10	19 29	14 22	04 42	19 04	15 00
21	19 59	23 00	22 16	18 54	19 19	19 31	14 23	04 42	19 04	15 01
22	19 46	25 58	22 22	19 04	19 28	19 33	14 23	04 42	19 04	15 02
23	19 32	27 37	22 27	19 12	19 36	19 35	14 23	04 42	19 04	15 02
24	19 18	27 38	22 31	19 21	19 45	19 37	14 24	04 42	19 05	15 03
25	19 03	25 52	22 34	19 29	19 53	19 38	14 24	04 42	19 05	15 04
26	18 48	22 20	22 36	19 37	20 01	19 40	14 25	04 42	19 05	15 04
27	18 33	17 19	22 36	19 45	20 09	19 42	14 26	04 41	19 05	15 05
28	18 18	11 12	22 36	19 52	20 17	19 44	14 26	04 41	19 05	15 06
29	18 02	04 S 28	22 35	19 59	20 25	19 45	14 27	04 41	19 06	15 07
30	17 46	02 N 25	22 33	20 05	20 32	19 47	14 28	04 41	19 06	15 08
31	17 S 29	09 N 04	22 S 29	20 S 11	20 S 40	19 S 48	14 N 28	04 S 40	19 S 06	15 N 08

ZODIAC SIGN ENTRIES

Date	h	m	Planets
01	04	08	☽ ♓
02	23	36	☿ ♐
03	06	26	☽ ♈
05	10	00	☽ ♉
07	01	00	♀ ♐
07	15	08	☽ ♊
09	22	09	☽ ♋
12	07	24	☽ ♌
14	02	16	☿ ♑
14	08	49	♃ ♐
14	18	57	☽ ♍
17	07	53	☽ ♎
19	20	04	☽ ♏
20	17	13	☉ ♒
22	05	16	☽ ♐
23	01	34	♂ ♐
24	10	33	☽ ♑
26	12	36	☽ ♒
28	13	01	☽ ♓
30	13	36	☽ ♈

LATITUDES

Date	Mercury ☿ ° '	Venus ♀ ° '	Mars ♂ ° '	Jupiter ♃ ° '	Saturn ♄ ° '	Uranus ♅ ° '	Neptune ♆ ° '	Pluto ♇ ° '
01	03 N 06	03 N 49	00 N 48	00 N 53	02 S 22	00 N 41	01 N 38	16 N 03
04	03 13	03 51	00 47	00 53	02 21	00 42	01 38	16 05
07	03 03	03 52	00 46	00 53	02 20	00 42	01 38	16 06
10	02 43	03 51	00 44	00 53	02 19	00 42	01 38	16 08
13	02 17	03 49	00 43	00 53	02 18	00 42	01 38	16 10
16	01 48	03 45	00 41	00 53	02 18	00 42	01 38	16 11
19	01 19	03 40	00 40	00 54	02 17	00 42	01 39	16 13
22	00 50	03 34	00 38	00 54	02 16	00 42	01 39	16 14
25	00 N 23	03 26	00 36	00 54	02 15	00 42	01 39	16 16
28	00 S 03	03 18	00 34	00 54	02 14	00 42	01 39	16 17
31	00 S 27	03 N 09	00 N 32	00 N 55	02 S 13	00 N 42	01 N 39	16 N 19

DATA

Julian Date	2440953
Delta T	+41 seconds
Ayanamsa	23° 27' 17"
Synetic vernal point	05° ♓ 39' 42"
True obliquity of ecliptic	23° 26' 42"

MOON'S PHASES, APSIDES AND POSITIONS ☽

Date	h	m	Phase	Longitude ° '	Eclipse Indicator
04	04	55	☽	13 ♈ 11	
11	13	20	○	20 ♋ 40	
19	18	08	☾	29 ♎ 01	
26	22	55	●	06 ♒ 21	

Day	h	m		
16	11	13	Apogee	
28	10	20	Perigee	
02	20	24	0N	
09	11	23	Max dec	27° N 48'
16	19	29	0S	
24	00	20	Max dec	27° S 51'
30	03	32	0N	

LONGITUDES

Date	Chiron ⚷ ° '	Ceres ⚳ ° '	Pallas ⚴ ° '	Juno ⚵ ° '	Vesta ⚶ ° '	Black Moon Lilith ⚸ ° '
01	05 ♈ 58	24 ♈ 45	06 ♓ 19	16 ♉ 41	08 ♐ 38	13 ♍ 18
11	06 ♈ 09	26 ♈ 12	09 ♓ 05	18 ♉ 13	13 ♐ 50	14 ♍ 25
21	06 ♈ 25	28 ♈ 08	12 ♓ 03	20 ♉ 30	18 ♐ 58	15 ♍ 31
31	06 ♈ 45	00 ♉ 28	15 ♓ 11	23 ♉ 25	24 ♐ 01	16 ♍ 38

All ephemeris data is given at 12.00 UT and the Moon's longitude is additionally given for 24.00 UT

ASPECTARIAN

h m Aspects

01 Friday
00 07 ☽ □ ♃
00 08 ☿ ⊻ ♆
01 31 ☽ ⚼ ♅
03 39 ☽ ⚻ ♇
06 55 ☽ ⚹ ☿
07 28 ☽ □ ♆
10 41 ☽ Q ♄
12 56 ☿ ∠ ♂
14 57 ☿ ⊥ ♀
16 30 ☽ ± ♅
22 01 ♇ St R
22 13 ☽ ⚹ ☉
23 55 ☿ ⚼ ♄

02 Saturday
01 34 ☽ Q ☿
02 33 ☽ ⚻ ♅
03 33 ☽ ∥ ♅
06 41 ☽ ⚹ ♄
08 22 ☽ △ ♂
19 55 ☽ Q ☉

03 Sunday
00 29 ☽ △ ♀
03 01 ☽ △ ♃
05 57 ☽ ☍ ♇
06 03 ☽ □ ☿
07 56 ☽ ∠ ♄
08 05 ☿ □ ♇
09 57 ☽ △ ♆
10 52 ☽ ⚼ ♂
13 26 ☽ # ♅
23 15 ☽ ⊥ ♄

04 Monday
02 57 ☽ ± ♂
03 40 ☽ ⚼ ♀
04 55 ☽ □ ☉
04 55 ☽ ⚼ ♃
05 02 ☉ ∠ ♃
05 25 ☽ ☍ ♅
09 31 ☽ ⊻ ♄
11 39 ☽ ⚼ ♆
11 51 ☉ □ ♅
13 45 ☽ ⚻ ♂
20 12 ☽ ± ♀
20 38 ☽ ± ♃

05 Tuesday
03 15 ☽ ± ♆
03 16 ☽ ∥ ♄
04 54 ♀ ☌ ♃
05 26 ☽ ∥ ♇
07 12 ☽ ⚻ ♃
07 19 ☽ ⚻ ♀
07 20 ☽ △ ☿
07 25 ☿ ⊻ ♀
08 04 ♀ ∠ ♅
09 29 ☽ ⚻ ♇
10 10 ☿ ⊻ ♃
10 52 ☽ # ♀
13 43 ☽ ⚻ ♆
13 45 ☽ # ♂
19 58 ☽ ± ♇
20 49 ♃ ∠ ♅

06 Wednesday
00 30 ☽ # ♃
00 37 ☽ # ♆
07 15 ☽ # ☿
08 56 ☽ ⚼ ☿
09 42 ☽ ⚻ ♅
10 23 ♂ ⊥ ♅
11 50 ☽ ⚼ ♇
13 23 ☽ △ ☉
13 47 ☽ ☌ ♄
17 00 ♀ ⚹ ♇
17 15 ♃ ∥ ♆
18 44 ☉ △ ♄
19 56 ☽ # ☉
20 21 ☽ ± ♅
20 50 ☽ ☍ ♂

07 Thursday
00 39 ☽ ± ☿
11 16 ☽ ⚻ ☿
12 27 ☽ ⚼ ♅
12 58 ☽ ☍ ♃
14 36 ☽ △ ♇
16 10 ☽ ☍ ♀
18 24 ☽ ⚼ ☉
19 06 ☽ ☍ ♆

08 Friday
03 46 ☉ ∠ ♆
04 37 ☿ St D
12 06 ☽ ± ☉
15 41 ☽ △ ♅
19 48 ☽ ⊻ ♄

09 Saturday
00 01 ☽ ⚻ ☉
05 56 ☽ ⚻ ♂
06 50 ☽ ⊥ ♄
12 13 ♀ ☌ ♆
17 39 ☽ ± ♂
18 23 ☽ ☍ ☿
20 41 ☽ ⚻ ♃
21 34 ☽ □ ♇
23 33 ☽ ∠ ♄

10 Sunday
02 23 ☽ ⚻ ♆
03 27 ☽ ⚻ ♀
08 05 ☽ ± ♃
11 24 ☽ ⚼ ♂
13 42 ☽ ± ♆
15 39 ☽ ± ♀
20 57 ☿ ⊥ ♂
23 41 ☽ □ ♅

11 Monday
01 23 ☽ ⚼ ♃
03 53 ☽ ⚹ ♄
06 52 ☽ ⚼ ♆
07 36 ☽ Q ♇
10 12 ☽ ⚼ ♀
13 20 ☽ ☍ ☉
17 32 ☽ △ ♂

12 Tuesday
02 58 ☽ Q ♄
05 14 ☽ ⚻ ☿
06 42 ☽ △ ♃
06 46 ☽ ⚹ ♇
10 25 ☽ Q ♅
10 38 ☽ # ☉
11 47 ♃ ⚹ ♇
11 57 ☽ △ ♆
14 55 ☽ # ☿
17 29 ☽ ± ☿
17 43 ☽ △ ♀

13 Wednesday
01 23 ☽ # ♃
02 33 ☽ # ♆
07 54 ☽ # ♂
10 03 ☽ ⚹ ♅
10 36 ☽ # ♀
12 09 ☽ ⚼ ☿
12 16 ☽ ∠ ♇
14 12 ☿ □ ♇
14 20 ☽ □ ♄
23 56 ☽ ∥ ♇

14 Thursday
00 03 ☿ ⊻ ♃
02 51 ☽ ∥ ♄
05 41 ☽ ⚻ ☉
06 13 ☽ ⊥ ♇
07 52 ☽ □ ♂
15 42 ☉ ∥ ☿
16 03 ☽ ∠ ♅
18 17 ☽ ⊻ ♇
18 53 ☽ ± ☉
19 06 ☽ □ ♃
20 01 ☽ △ ☿
23 48 ☽ □ ♆

15 Friday
01 46 ☿ ⚼ ♄
10 15 ☽ ⊥ ♅
10 57 ☽ □ ♀
14 49 ☽ ⚼ ☉
22 19 ☽ Q ♂
22 26 ☽ ⊻ ♅
22 56 ☽ # ♅

16 Saturday
02 47 ☽ △ ♄
08 09 ☽ Q ♃
12 21 ☿ ⊥ ♂
12 25 ☽ Q ♆
21 30 ☉ ⚹ ♂

17 Sunday
00 07 ☽ ⚹ ♂
00 12 ☽ △ ☉
02 53 ☽ Q ♀
04 15 ☿ ⊻ ♆
07 09 ☽ ☌ ♇
08 55 ☽ ⚹ ♃
09 18 ☽ ⚼ ♄
12 53 ☽ ⚹ ♆
13 02 ♄ St D
13 35 ☽ □ ☿
16 02 ☽ ∥ ♅

18 Monday
03 32 ☽ ± ♄
05 27 ☽ ⚹ ♀
06 52 ♅ St R
08 14 ☽ ∠ ♂
11 20 ☽ ☌ ♅
15 38 ☽ ⚻ ♄
15 40 ☽ ∠ ♃
19 13 ☽ ∠ ♆

19 Tuesday
03 17 ☽ ⊥ ♂
04 56 ♀ Q ♇
05 01 ☽ Q ☿
09 53 ☽ ⊥ ♃
12 09 ☽ # ♄
13 11 ☽ ⊥ ♆
14 10 ☽ ∠ ♀
15 14 ☽ # ♇
15 50 ☽ ⊻ ♂
17 47 ♂ ∥ ♆
18 08 ☽ □ ☉
19 19 ☽ ⊻ ♇
21 53 ☽ ⊻ ♃

20 Wednesday
01 01 ☽ ⊻ ♆
06 51 ☽ ⚹ ☿
07 03 ☽ ⊥ ♇
08 06 ☉ △ ♇
09 24 ☽ ⊥ ♀
10 31 ☽ ∥ ♀
12 16 ☽ ∥ ♆
12 54 ☽ ∥ ♂
14 39 ☽ ∥ ♃
18 21 ☽ ∥ ☉
18 48 ♂ ∠ ♅
21 59 ☽ ⊻ ♀
22 23 ☽ ⊻ ♅

21 Thursday
00 23 ☽ ∠ ♇
02 29 ☽ ☍ ♄
02 52 ♀ ⚹ ♅
07 06 ☽ ∥ ☿
08 02 ☽ Q ☉
09 47 ☽ ⊥ ♅
14 16 ☽ ∠ ☿
20 33 ☿ ⊥ ♃
22 37 ☉ ⚹ ♃

22 Friday
02 36 ☽ ∠ ♅
04 14 ☽ ☌ ♂
04 31 ☽ ⚹ ♇
07 40 ☽ ☌ ♃
08 18 ☽ ⚹ ☉
08 34 ☽ ⊥ ☿
09 59 ♂ ⚹ ♇
09 59 ☽ ☌ ♆
13 22 ♀ ∥ ♆
20 31 ☽ ⊻ ☿

23 Saturday
00 59 ☿ ⊥ ♆
02 17 ☽ Q ♇
05 35 ♀ ⚻ ♄
05 48 ☽ ⚹ ♅
06 26 ☉ ⚹ ♆
06 39 ♂ ∥ ♃
07 09 ☉ ∥ ♂
07 22 ☉ ∥ ♃
09 40 ☽ ⚻ ♄
10 00 ☽ ☌ ♀
13 30 ☽ ∠ ☉
20 14 ☽ ± ♄

24 Sunday
02 52 ☽ Q ♅
06 26 ☽ ⊥ ☉
08 28 ☉ ∥ ♀
09 48 ☽ □ ♇
11 49 ☽ ⚼ ♄
12 06 ☽ ⊻ ♂
13 22 ☽ ⊻ ♃
15 02 ☽ ⊻ ♆
17 31 ☽ ⊻ ☉
22 47 ☽ ⊥ ♂
23 41 ☽ ⊥ ♃

25 Monday
01 13 ☽ ⊥ ♆
05 35 ☽ ☌ ☿
09 27 ☉ ∥ ♆
09 27 ☽ □ ♅
13 09 ☽ △ ♄
14 33 ☽ ∠ ♂
14 54 ☽ ∠ ♃
16 17 ☽ ∠ ♆
17 24 ☽ ⊻ ♀

26 Tuesday
01 17 ♂ ☌ ♃
04 04 ☽ ⊥ ♀
10 33 ☽ ∥ ☿
11 51 ☽ △ ♇
15 49 ☽ ⚹ ♃
16 18 ☽ ⚹ ♂
16 57 ☽ ⚹ ♆
19 53 ☽ ∠ ♀
22 55 ☽ ☌ ☉
23 31 ☽ ∥ ♂

27 Wednesday
00 01 ♀ ∥ ♃
01 23 ☽ ∥ ♀
01 25 ☽ ∥ ♃
01 29 ☿ □ ♅
04 12 ☽ ∥ ♆
06 20 ☽ ∥ ☉
07 54 ♂ ☌ ♆
10 32 ☽ △ ♅
11 23 ☽ ⊻ ☿
11 26 ☽ Q ♃
12 08 ☽ ⚼ ♇
12 24 ☽ Q ♆
12 35 ☽ Q ♂
14 11 ☽ □ ♄
21 07 ☽ # ♇
21 57 ☽ ⚹ ♀
21 58 ☽ ⊥ ☿
23 44 ☽ # ♄

28 Thursday
02 37 ☽ ± ♇
10 38 ☽ ⚼ ♅
12 15 ☽ ⚻ ♇
13 49 ☽ ∠ ☿
16 41 ☽ □ ♃
17 23 ☽ □ ♆
18 43 ☽ Q ♀
18 49 ☽ □ ♂
19 10 ☽ Q ♄
19 41 ☿ △ ♄

29 Friday
01 06 ☽ ± ♅
01 09 ♀ ± ♄
02 46 ☽ ⊻ ☉
10 47 ☽ ⚻ ♅
11 16 ☽ ∥ ♅
13 10 ☽ ⊥ ☉
14 32 ☽ ⚹ ♄
16 22 ☽ ⚹ ☿

30 Saturday
01 21 ☿ ∠ ♃
02 03 ☽ □ ♀
04 51 ☽ ∠ ☉
06 14 ☿ ∠ ♆
12 45 ☽ ☍ ♇
13 53 ☽ Q ☿
15 00 ☽ ∠ ♄
17 48 ☽ △ ♃
18 07 ☽ △ ♆
19 57 ☽ # ♅
21 42 ☽ △ ♂

31 Sunday
05 53 ☽ ⊥ ♄
07 22 ☽ ⚹ ☉
11 54 ☽ ☍ ♅
15 54 ☽ ⊻ ♄
18 58 ☽ ⚼ ♃
19 04 ☽ ⚼ ♆
22 55 ☽ □ ☿
23 51 ☽ ⚼ ♂

FEBRUARY 1971

Date	Sidereal time h m s	Sun ☉ ° ' "	Moon ☽ ° ' "	Moon ☽ 24.00 ° ' "	Mercury ☿ ° '	Venus ♀ ° '	Mars ♂ ° '	Jupiter ♃ ° '	Saturn ♄ ° '	Uranus ♅ ° '	Neptune ♆ ° '	Pluto ♇ ° '
01	20 44 09	11 ♒ 58 51	27 ♈ 46 22	04 ♉ 45 44	20 ♑ 49	25 ♐ 31	05 ♐ 56	02 ♐ 48	15 ♉ 55	13 ♎ 28	02 ♐ 47	29 ♍ 27
02	20 48 06	12 59 44	11 ♉ 40 05	18 ♉ 29 27	22 13	26 36	06 33	02 57	15 56	13 R 27	02 48	29 R 26
03	20 52 02	14 00 37	25 ♉ 13 56	01 ♊ 53 40	23 38	27 41	07 11	03 05	15 58	13 27	02 49	29 25
04	20 55 59	15 01 28	08 ♊ 28 53	14 ♊ 59 47	25 04	28 46	07 49	03 13	16 00	13 26	02 50	29 24
05	20 59 56	16 02 17	21 ♊ 26 37	27 ♊ 49 37	26 31	29 ♐ 52	08 26	03 20	16 02	13 25	02 51	29 23
06	21 03 52	17 03 05	04 ♋ 09 02	10 ♋ 25 08	27 58	00 ♑ 58	09 04	03 28	16 04	13 24	02 52	29 22
07	21 07 49	18 03 52	16 ♋ 38 09	22 ♋ 48 18	29 ♑ 27	02 04	09 41	03 35	16 07	13 23	02 53	29 21
08	21 11 45	19 04 37	28 ♋ 55 50	05 ♌ 00 56	00 ♒ 57	03 10	10 19	03 43	16 09	13 22	02 54	29 20
09	21 15 42	20 05 21	11 ♌ 03 52	17 ♌ 04 48	02 27	04 16	10 56	03 50	16 11	13 20	02 54	29 19
10	21 19 38	21 06 04	23 ♌ 04 00	29 ♌ 01 40	03 58	05 23	11 34	03 57	16 14	13 19	02 55	29 17
11	21 23 35	22 06 45	04 ♍ 58 03	10 ♍ 53 26	05 30	06 30	12 11	04 04	16 17	13 18	02 56	29 16
12	21 27 31	23 07 25	16 ♍ 48 04	22 ♍ 42 16	07 03	07 37	12 48	04 11	16 20	13 17	02 57	29 15
13	21 31 28	24 08 03	28 ♍ 36 22	04 ♎ 30 43	08 37	08 44	13 26	04 18	16 22	13 15	02 57	29 14
14	21 35 25	25 08 40	10 ♎ 25 43	16 ♎ 21 47	10 12	09 51	14 03	04 24	16 25	13 14	02 58	29 12
15	21 39 21	26 09 16	22 ♎ 19 21	28 ♎ 18 55	11 47	10 59	14 40	04 30	16 29	13 12	02 59	29 11
16	21 43 18	27 09 51	04 ♏ 20 57	10 ♏ 25 59	13 24	12 06	15 18	04 37	16 32	13 11	02 59	29 10
17	21 47 14	28 10 25	16 ♏ 34 32	22 ♏ 47 09	15 01	13 14	15 55	04 43	16 35	13 09	03 00	29 09
18	21 51 11	29 ♒ 10 57	29 ♏ 04 21	05 ♐ 26 39	16 39	14 22	16 32	04 49	16 39	13 08	03 00	29 07
19	21 55 07	00 ♓ 11 28	11 ♐ 54 31	18 ♐ 28 24	18 18	15 31	17 09	04 54	16 42	13 06	03 01	29 06
20	21 59 04	01 11 58	25 ♐ 08 38	01 ♑ 55 30	19 58	16 39	17 47	05 00	16 46	13 04	03 01	29 04
21	22 03 00	02 12 27	08 ♑ 49 10	15 ♑ 49 39	21 39	17 47	18 24	05 05	16 49	13 03	03 02	29 03
22	22 06 57	03 12 54	22 ♑ 56 49	00 ♒ 10 24	23 21	18 56	19 01	05 11	16 53	13 01	03 02	29 02
23	22 10 54	04 13 20	07 ♒ 29 55	14 ♒ 54 43	25 04	20 05	19 38	05 16	16 57	12 59	03 03	29 00
24	22 14 50	05 13 44	22 ♒ 23 58	29 ♒ 56 43	26 47	21 13	20 15	05 21	17 01	12 57	03 03	28 59
25	22 18 47	06 14 07	07 ♓ 31 49	15 ♓ 08 06	28 ♒ 32	22 22	20 52	05 25	17 05	12 55	03 03	28 57
26	22 22 43	07 14 28	22 ♓ 44 18	00 ♈ 19 12	00 ♓ 18	23 31	21 29	05 30	17 09	12 54	03 04	28 56
27	22 26 40	08 14 47	07 ♈ 51 36	15 ♈ 20 25	02 05	24 41	22 06	05 34	17 13	12 52	03 04	28 54
28	22 30 36	09 ♓ 15 04	22 ♈ 44 42	00 ♉ 03 40	03 ♓ 52	25 ♑ 50	22 ♐ 43	05 ♐ 39	17 ♉ 18	12 ♎ 50	03 ♐ 04	28 ♍ 53

Date	Moon True ☊ ° '	Moon Mean ☊ ° '	Moon ☽ Latitude ° '
01	23 ♒ 41	24 ♒ 17	04 N 45
02	23 D 41	24 14	05 10
03	23 R 41	24 11	05 17
04	23 40	24 08	05 07
05	23 40	24 05	04 41
06	23 39	24 01	04 02
07	23 38	23 58	03 11
08	23 37	23 55	02 13
09	23 37	23 52	01 09
10	23 37	23 49	00 N 03
11	23 D 37	23 46	01 S 03
12	23 37	23 42	02 05
13	23 R 37	23 39	03 03
14	23 37	23 36	03 52
15	23 36	23 33	04 32
16	23 36	23 30	05 00
17	23 36	23 27	05 15
18	23 36	23 23	05 16
19	23 D 36	23 20	05 02
20	23 36	23 17	04 31
21	23 37	23 14	03 44
22	23 38	23 11	02 42
23	23 39	23 07	01 28
24	23 39	23 04	00 S 07
25	23 R 39	23 01	01 N 16
26	23 38	22 58	02 34
27	23 36	22 55	03 41
28	23 ♒ 34	22 ♒ 52	04 N 32

DECLINATIONS

Date	Sun ☉ ° '	Moon ☽ ° '	Mercury ☿ ° '	Venus ♀ ° '	Mars ♂ ° '	Jupiter ♃ ° '	Saturn ♄ ° '	Uranus ♅ ° '	Neptune ♆ ° '	Pluto ♇ ° '
01	17 S 12	15 N 07	22 S 24	20 S 17	20 S 47	19 S 50	14 N 29	04 S 40	19 S 06	15 N 09
02	16 55	20 15	22 18	20 22	20 54	19 51	14 30	04 40	19 06	15 10
03	16 38	24 12	22 11	20 27	21 01	19 53	14 31	04 39	19 06	15 11
04	16 20	26 46	22 02	20 31	21 08	19 54	14 31	04 39	19 06	15 12
05	16 02	27 50	21 52	20 35	21 15	19 56	14 32	04 38	19 07	15 12
06	15 44	27 24	21 41	20 39	21 21	19 57	14 33	04 38	19 07	15 13
07	15 25	25 34	21 29	20 41	21 27	19 58	14 34	04 38	19 07	15 14
08	15 06	22 33	21 15	20 44	21 34	20 00	14 35	04 37	19 07	15 15
09	14 47	18 34	21 01	20 46	21 40	20 01	14 36	04 37	19 07	15 16
10	14 28	13 53	20 44	20 47	21 46	20 02	14 37	04 36	19 07	15 17
11	14 09	08 43	20 27	20 48	21 51	20 03	14 38	04 36	19 07	15 17
12	13 49	03 N 17	20 08	20 49	21 57	20 05	14 39	04 35	19 07	15 18
13	13 29	02 S 14	19 47	20 49	22 02	20 06	14 40	04 35	19 07	15 19
14	13 09	07 41	19 26	20 48	22 08	20 07	14 41	04 34	19 08	15 20
15	12 48	12 53	19 03	20 47	22 13	20 08	14 43	04 33	19 08	15 21
16	12 28	17 40	18 39	20 45	22 18	20 09	14 44	04 33	19 08	15 21
17	12 07	21 50	18 13	20 43	22 23	20 10	14 45	04 32	19 08	15 22
18	11 46	25 06	17 46	20 40	22 27	20 11	14 46	04 32	19 08	15 23
19	11 24	27 12	17 17	20 37	22 32	20 12	14 47	04 31	19 08	15 24
20	11 03	27 52	16 48	20 33	22 36	20 13	14 49	04 30	19 08	15 25
21	10 41	26 53	16 17	20 28	22 40	20 14	14 50	04 29	19 08	15 26
22	10 20	24 10	15 44	20 23	22 44	20 14	14 51	04 29	19 08	15 26
23	09 58	19 50	15 10	20 18	22 48	20 15	14 53	04 28	19 08	15 27
24	09 36	14 10	14 35	20 12	22 52	20 16	14 54	04 27	19 08	15 28
25	09 14	07 34	13 58	20 05	22 55	20 17	14 55	04 27	19 08	15 29
26	08 51	00 S 31	13 21	19 58	22 59	20 18	14 57	04 26	19 08	15 30
27	08 29	06 N 30	12 41	19 50	23 02	20 18	14 58	04 25	19 08	15 31
28	08 S 06	13 N 03	12 S 01	19 S 42	23 S 05	20 S 19	15 N 00	04 S 24	19 S 08	15 N 31

ZODIAC SIGN ENTRIES

Date	h	m	Planets
01	15	49	☽ ♉
03	20	34	☽ ♊
05	14	57	♀ ♑
06	04	07	☽ ♋
07	20	51	☿ ♒
08	14	06	☽ ♌
11	01	58	☽ ♍
13	14	50	☽ ♎
16	03	22	☽ ♏
18	13	45	☽ ♐
19	07	27	☉ ♓
20	20	37	☽ ♑
22	23	43	☽ ♒
25	00	05	☽ ♓
26	07	57	☿ ♓
26	23	30	☽ ♈
28	23	54	☽ ♉

LATITUDES

Date	Mercury ☿ ° '	Venus ♀ ° '	Mars ♂ ° '	Jupiter ♃ ° '	Saturn ♄ ° '	Uranus ♅ ° '	Neptune ♆ ° '	Pluto ♇ ° '
01	00 S 34	03 N 05	00 N 32	00 N 55	02 S 13	00 N 42	01 N 39	16 N 19
04	00 56	02 55	00 30	00 55	02 12	00 43	01 39	16 20
07	01 15	02 44	00 27	00 55	02 11	00 43	01 39	16 22
10	01 31	02 33	00 25	00 55	02 10	00 43	01 40	16 23
13	01 44	02 21	00 23	00 56	02 09	00 43	01 40	16 24
16	01 55	02 09	00 20	00 56	02 09	00 43	01 40	16 25
19	02 03	01 57	00 18	00 56	02 08	00 43	01 40	16 26
22	02 07	01 44	00 15	00 57	02 07	00 43	01 40	16 27
25	02 07	01 31	00 13	00 57	02 06	00 43	01 40	16 28
28	02 04	01 19	00 10	00 57	02 05	00 43	01 40	16 28
31	01 S 56	01 N 06	00 N 07	00 N 57	02 S 05	00 N 43	01 N 41	16 N 29

DATA

Julian Date	2440984
Delta T	+41 seconds
Ayanamsa	23° 27' 22"
Synetic vernal point	05° ♓ 39' 37"
True obliquity of ecliptic	23° 26' 42"

MOON'S PHASES, APSIDES AND POSITIONS ☽

Date	h	m	Phase	Longitude ° '	Eclipse Indicator
02	14	31	☽	13 ♉ 06	
10	07	41	○	20 ♌ 55	total
18	12	14	☾	29 ♏ 12	
25	09	49	●	06 ♓ 09	Partial

Day	h	m		
13	00	58	Apogee	
25	20	57	Perigee	
05	16	54	Max dec	27° N 52'
13	02	17	0S	
20	10	00	Max dec	27° S 52'
26	13	45	0N	

LONGITUDES

Date	Chiron ⚷ ° '	Ceres ⚳ ° '	Palias ⚴ ° '	Juno ⚵ ° '	Vesta ⚶ ° '	Black Moon Lilith ⚸ ° '
01	06 ♈ 47	00 ♉ 44	15 ♓ 30	23 ♉ 44	24 ♐ 31	16 ♍ 44
11	07 ♈ 12	03 ♉ 27	18 ♓ 48	27 ♉ 13	29 ♐ 27	17 ♍ 51
21	07 ♈ 41	06 ♉ 29	22 ♓ 12	01 ♊ 05	04 ♑ 14	18 ♍ 58
31	08 ♈ 12	09 ♉ 45	25 ♓ 42	05 ♊ 17	08 ♑ 52	20 ♍ 04

ASPECTARIAN

h m Aspects

01 Monday
05 03 ☽ Q ☉
06 48 ♃ ☌ ♆
07 51 ☽ △ ♀
09 19 ☽ ∥ ♄
10 19 ☽ ± ♆
10 20 ☽ ± ♃
10 56 ♀ Q ♅
12 09 ☽ ∥ ♇
14 52 ☽ ⊼ ♇
15 38 ☿ ∠ ♂
15 52 ☽ ± ♂
20 36 ☽ ⊼ ♆
20 42 ☽ # ☉
20 43 ☽ ⊼ ♃

02 Tuesday
01 11 ☽ ± ♇
02 41 ☽ ⊼ ♂
06 10 ☽ # ♆
09 57 ☽ # ♃
11 52 ☽ Ҏ ♀
12 38 ☽ # ♀
14 31 ☽ □ ☉
15 08 ☽ ⊼ ♅
15 34 ☽ # ♂
16 51 ☽ Ҏ ♇
19 31 ☽ ☌ ♄
22 47 ☉ △ ♅
23 09 ☽ # ☿

03 Wednesday
01 42 ☽ ± ♅
05 07 ☽ ± ♀
08 48 ☽ △ ☿
16 47 ☽ ⊼ ♀
17 46 ☽ Ҏ ♅
19 31 ☽ △ ♇
21 32 ☉ Ҏ ♇

04 Thursday
01 41 ☽ ☍ ♆
02 17 ☽ ☍ ♃
07 24 ☉ Q ♆
10 42 ☽ ☍ ♂
15 16 ☽ Ҏ ☿
17 02 ☉ Q ♃
21 05 ☽ △ ♅

05 Friday
01 05 ☽ △ ☉
01 37 ♀ □ ♇
01 54 ☽ ⊻ ♄
10 02 ☽ ± ☿
11 57 ☉ □ ♄
13 07 ☽ ⊥ ♄
22 45 ☽ ⊼ ☿

06 Saturday
02 56 ☽ □ ♇
05 22 ☽ ☍ ♀
06 08 ☽ ∠ ♄
07 39 ☽ Ҏ ☉
09 33 ☽ ⊼ ♆
10 41 ☽ ⊼ ♃
14 30 ♀ Ҏ ♄
21 02 ☽ ± ♆
21 53 ☽ ⊼ ♂
22 16 ☽ ± ♃

07 Sunday
02 23 ☽ ± ☉
05 42 ☽ □ ♅
10 04 ☽ ± ♂
10 22 ☿ △ ♇
10 59 ☽ ⚹ ♄
13 23 ☽ Q ♇
14 07 ☿ ∥ ♂
14 25 ☽ Ҏ ♆
15 01 ☽ ⊼ ☉
15 50 ☽ Ҏ ♃

08 Monday
01 40 ☉ # ♇
04 31 ☽ Ҏ ♂
06 02 ♀ ⊻ ♆
10 28 ☽ Q ♄
12 47 ☽ ⚹ ♇
16 31 ☽ ☍ ☿
16 46 ☽ Q ♅
18 17 ☽ # ♂
19 49 ☽ △ ♆
20 54 ☽ # ☿
21 11 ☽ ⊼ ♀
21 31 ☽ △ ♃
23 28 ☽ # ♀

09 Tuesday
01 22 ♀ ⊻ ♃
03 51 ☽ # ♃
08 56 ☽ # ♆
10 16 ☽ ± ♀
11 44 ☽ △ ♂
16 31 ☽ ⚹ ♅
18 27 ☽ ∠ ♇
19 19 ☿ ⚹ ♆
22 16 ☽ □ ♄

10 Wednesday
01 21 ☉ # ♄
02 01 ♂ Q ♇
05 10 ☽ ∥ ♇
06 04 ☽ Ҏ ♀
07 41 ☽ ☍ ☉
07 52 ☿ ∥ ♀
08 25 ☽ ∥ ♄
08 56 ☽ # ☉
11 41 ☿ ⚹ ♃
12 27 ☽ ⊥ ♇
22 33 ☽ ∠ ♅

11 Thursday
00 31 ☽ ⊻ ♇
07 53 ☽ □ ♆
10 10 ☽ □ ♃
13 15 ☽ ⊼ ☿
15 25 ☽ △ ♀
16 43 ☽ ⊥ ♅

12 Friday
03 14 ☽ ± ☿
03 26 ☽ □ ♂
04 51 ☽ ⊻ ♅
06 19 ☽ # ♅
11 02 ☽ △ ♄
14 57 ☉ ± ♇
15 40 ☿ ∥ ♃
20 26 ☽ Q ♆
23 02 ☽ Q ♃

13 Saturday
00 18 ☽ Ҏ ☿
02 03 ☽ ⊼ ☉
05 30 ♂ ⚹ ♅
13 16 ☽ ☌ ♇
15 24 ☽ ± ☉
16 53 ♀ ⊥ ♆
17 39 ☽ Ҏ ♄
18 03 ☿ ⊻ ♀
18 03 ☽ Q ♂
20 51 ☽ ⚹ ♆
22 12 ☽ ∥ ♅
23 40 ☽ ⚹ ♃

14 Sunday
10 43 ☽ □ ♀
11 22 ☽ Ҏ ☉
11 27 ☽ △ ☿
12 00 ☽ ± ♄
17 40 ☽ ☌ ♅
19 44 ☽ ⚹ ♂

15 Monday
00 11 ☽ ⊼ ♄
00 52 ♀ ⊥ ♃
03 15 ☽ ∠ ♆
06 17 ☽ ∠ ♃
07 11 ☿ ∥ ♆
11 37 ☽ ∥ ☉
20 23 ☽ △ ☉
20 52 ☽ # ♄
21 20 ☽ ⊥ ♆

16 Tuesday
00 01 ☽ # ♇
00 29 ☽ ⊥ ♃
01 43 ☽ ⊻ ♇
02 42 ☽ Q ♀
03 31 ☽ ∠ ♂
08 53 ☿ △ ♅
09 18 ☽ ⊻ ♆
12 31 ☽ ⊻ ♃
13 37 ☽ ⊥ ♇
16 47 ☽ ∥ ☿
19 56 ☽ ∥ ♆
20 05 ☉ Q ♂
22 17 ☽ ⊥ ♂
23 10 ☉ ∠ ♀
23 17 ☿ Ҏ ♇

17 Wednesday
01 50 ☽ ∥ ♃
04 50 ☽ ⚹ ♀
05 12 ☽ ∥ ♀
05 21 ☽ ⊻ ♅
07 16 ☽ ∠ ♇
08 30 ☽ □ ☿
10 16 ♀ □ ♅
10 39 ☽ ⊻ ♂
11 34 ☉ Ҏ ♅
11 46 ☿ Q ♆
12 01 ☽ ☍ ♄
15 39 ☽ ∥ ♂
16 59 ☽ ⊥ ♅

18 Thursday
09 20 ☿ ⚹ ♂
10 13 ☽ ∠ ♅
10 33 ☉ ⊼ ♇
11 52 ☿ □ ♄
12 05 ☽ ⚹ ♇
12 14 ☽ □ ☉
12 37 ☽ ∠ ♀
14 27 ☿ Q ♃
16 27 ♂ ⊼ ♄
19 26 ☽ ☌ ♆
22 54 ☽ ☌ ♃

19 Friday
00 04 ☽ Q ☿
07 08 ☽ ⊥ ♀
10 30 ☽ Q ♇
14 11 ☽ ⚹ ♅
19 14 ☽ ⊻ ♀
20 49 ☽ ⊼ ♄
22 05 ☽ ☌ ♂

20 Saturday
00 26 ☽ Q ☉
01 22 ☽ ⚹ ☿
07 43 ☽ ± ♄
11 52 ☽ Q ♅
14 31 ♀ △ ♄
18 58 ☽ □ ♇
23 35 ☽ ⚹ ☉
23 46 ☽ Ҏ ♄

21 Sunday
01 56 ☽ ⊻ ♆
05 29 ☽ ⊻ ♃
07 43 ☽ ∠ ☿
12 22 ☽ ⊥ ♆
15 56 ☽ ⊥ ♃
17 09 ♀ ∠ ♆
19 15 ☽ □ ♅

22 Monday
01 17 ☽ ⊥ ☿
01 45 ☽ △ ♄
03 26 ☽ ∠ ☉
03 44 ☽ ∠ ♆
04 40 ☽ ☌ ♀
05 06 ☽ ⊻ ♂
07 19 ☽ ∠ ♃
07 33 ☿ ± ♇
07 45 ☉ □ ♆
12 46 ☽ ⊻ ☿
15 37 ☽ ⊥ ♂
15 54 ♀ ⊻ ♂
19 38 ☽ ⊥ ☉
20 43 ☽ ∥ ♂
22 06 ☽ △ ♇

23 Tuesday
00 19 ☿ # ♇
04 43 ☽ ⚹ ♆
06 15 ☽ ⊻ ☉
07 07 ☽ ∠ ♂
08 20 ☽ ⚹ ♃
09 41 ☽ ∥ ♀
09 57 ☽ ∥ ♃
15 14 ☽ ∥ ♆
16 10 ♀ ∠ ♃
20 53 ☽ △ ♅
21 34 ♀ ∥ ♃
22 31 ☽ Ҏ ♇
23 39 ☿ # ♄

24 Wednesday
00 13 ☽ Q ♆
03 21 ☽ □ ♄
03 52 ☽ Q ♃
06 33 ☉ Q ♄
06 52 ☽ # ♇
08 25 ☽ ⚹ ♂
09 07 ☽ # ♄
09 58 ☽ ⊻ ♀
10 10 ☽ ∥ ☿
12 55 ☽ ± ♇
14 57 ☉ □ ♃
19 54 ☽ ☌ ☿
20 19 ☽ ⊥ ♀
20 49 ☽ Ҏ ♅
22 27 ☽ ⊼ ♇

25 Thursday
03 45 ☿ Ҏ ♅
04 19 ☽ Q ♂
04 55 ☽ □ ♆
05 27 ☿ ⊥ ♀
05 50 ☽ ∥ ☉
08 07 ☽ Q ♄
08 39 ☽ □ ♃
09 49 ☽ ☌ ☉
11 03 ☽ ± ♅
11 44 ☽ ∠ ♀
17 39 ☿ ⊼ ♇
20 30 ☽ ⊼ ♅
22 45 ☽ ∥ ♅

26 Friday
03 09 ☽ ⚹ ♄
03 56 ☉ ± ♅
09 56 ☽ □ ♂
13 20 ☽ ⚹ ♀
21 47 ☽ ☍ ♇

27 Saturday
01 33 ☽ ⊻ ☿
02 59 ☽ ∠ ♄
04 21 ☽ △ ♆
04 47 ☽ # ♅
08 20 ☽ △ ♃
09 57 ☽ Q ♀
12 24 ☽ ⊥ ☿
12 40 ☽ ⊻ ☉
17 24 ☽ ⊥ ♄
18 37 ☽ # ☉
19 59 ☽ ☍ ♅
22 59 ☽ ⊥ ☉

28 Sunday
01 12 ☿ □ ♆
03 07 ☽ ⊻ ♄
04 24 ☽ Ҏ ♆
04 50 ☽ ∠ ☿
08 23 ☽ # ☿
08 34 ☽ Ҏ ♃
11 57 ☽ △ ♂
14 38 ☽ ∠ ☉
17 28 ☽ □ ♀
19 04 ☽ ± ♆
19 47 ☽ ∥ ♄
21 57 ☽ ∥ ♇
22 02 ☽ ⊼ ♇
23 22 ☽ ± ♃

All ephemeris data is given at 12.00 UT and the Moon's longitude is additionally given for 24.00 UT

MARCH 1971

LONGITUDES

Date	Sidereal time h m s	Sun ☉ ° ′ ″	Moon ☽ ° ′ ″	Moon ☽ 24.00 ° ′ ″	Mercury ☿ ° ′	Venus ♀ ° ′	Mars ♂ ° ′	Jupiter ♃ ° ′	Saturn ♄ ° ′	Uranus ♅ ° ′	Neptune ♆ ° ′	Pluto ♇ ° ′
01	22 34 33	10 ♓ 15 19	07 ♉ 16 42	14 ♉ 23 23	05 ♓ 41	26 ♑ 59	23 ♐ 20	05 ♐ 43	17 ♉ 22	12 ♎ 48	03 ♐ 04	28 ♍ 51
02	22 38 29	11 15 32	21 ♉ 23 29	28 ♉ 16 55	07 31	28 09	23 57	05 47	17 27	12 R 46	03 04	28 R 50
03	22 42 26	12 15 43	05 ♊ 03 44	11 ♊ 44 09	09 22	29 ♑ 18	24 34	05 50	17 31	12 43	03 04	28 48
04	22 46 23	13 15 52	18 ♊ 18 28	24 ♊ 47 01	11 14	00 ♒ 28	25 10	05 54	17 36	12 41	03 04	28 47
05	22 50 19	14 15 59	01 ♋ 10 15	07 ♋ 28 39	13 06	01 38	25 47	05 57	17 41	12 39	03 04	28 45
06	22 54 16	15 16 04	13 ♋ 42 40	19 ♋ 52 48	15 00	02 47	26 24	06 01	17 46	12 37	03 R 04	28 44
07	22 58 12	16 16 07	25 ♋ 59 32	02 ♌ 03 21	16 55	03 57	27 00	06 04	17 50	12 35	03 04	28 42
08	23 02 09	17 16 07	08 ♌ 04 42	14 ♌ 03 58	18 50	05 07	27 37	06 06	17 56	12 33	03 04	28 40
09	23 06 05	18 16 06	20 ♌ 01 36	25 ♌ 57 56	20 47	06 17	28 13	06 09	18 01	12 30	03 04	28 39
10	23 10 02	19 16 02	01 ♍ 53 18	07 ♍ 48 02	22 44	07 28	28 50	06 12	18 06	12 28	03 04	28 37
11	23 13 58	20 15 57	13 ♍ 42 24	19 ♍ 36 41	24 41	08 38	29 ♐ 26	06 14	18 11	12 26	03 04	28 36
12	23 17 55	21 15 49	25 ♍ 31 08	01 ♎ 25 59	26 39	09 48	00 ♑ 03	06 16	18 16	12 24	03 04	28 34
13	23 21 52	22 15 40	07 ♎ 21 29	13 ♎ 17 51	28 ♓ 37	10 59	00 39	06 18	18 22	12 21	03 03	28 32
14	23 25 48	23 15 28	19 ♎ 15 20	25 ♎ 14 12	00 ♈ 36	12 09	01 15	06 20	18 27	12 19	03 03	28 31
15	23 29 45	24 15 15	01 ♏ 14 42	07 ♏ 17 08	02 34	13 20	01 52	06 21	18 33	12 16	03 03	28 29
16	23 33 41	25 15 00	13 ♏ 21 48	19 ♏ 29 03	04 32	14 30	02 28	06 23	18 38	12 14	03 02	28 27
17	23 37 38	26 14 43	25 ♏ 39 13	01 ♐ 52 43	06 29	15 41	03 04	06 24	18 44	12 11	03 02	28 26
18	23 41 34	27 14 25	08 ♐ 09 56	14 ♐ 31 16	08 25	16 52	03 40	06 25	18 50	12 09	03 02	28 24
19	23 45 31	28 14 05	20 ♐ 57 10	27 ♐ 28 02	10 20	18 03	04 16	06 26	18 55	12 06	03 01	28 22
20	23 49 27	29 ♓ 13 43	04 ♑ 04 16	10 ♑ 46 13	12 13	19 14	04 52	06 27	19 01	12 04	03 01	28 21
21	23 53 24	00 ♈ 13 19	17 ♑ 34 11	24 ♑ 28 24	14 03	20 24	05 28	06 27	19 07	12 01	03 00	28 19
22	23 57 21	01 12 54	01 ♒ 29 00	08 ♒ 35 58	15 52	21 35	06 04	06 27	19 13	11 59	03 00	28 18
23	00 01 17	02 12 27	15 ♒ 49 08	23 ♒ 08 12	17 37	22 47	06 40	06 R 27	19 19	11 56	02 59	28 16
24	00 05 14	03 11 58	00 ♓ 32 39	08 ♓ 01 46	19 18	23 58	07 16	06 27	19 25	11 54	02 59	28 14
25	00 09 10	04 11 27	15 ♓ 34 40	23 ♓ 10 17	20 56	25 09	07 51	06 27	19 32	11 51	02 58	28 13
26	00 13 07	05 10 54	00 ♈ 47 26	08 ♈ 24 49	22 29	26 20	08 27	06 27	19 38	11 49	02 57	28 11
27	00 17 03	06 10 19	16 ♈ 01 05	23 ♈ 34 55	23 57	27 31	09 02	06 26	19 44	11 46	02 57	28 09
28	00 21 00	07 09 42	01 ♉ 05 07	08 ♉ 30 32	25 20	28 43	09 38	06 25	19 51	11 43	02 56	28 08
29	00 24 56	08 09 03	15 ♉ 50 14	23 ♉ 03 29	26 38	29 ♒ 54	10 13	06 24	19 57	11 41	02 55	28 06
30	00 28 53	09 08 22	00 ♊ 09 45	07 ♊ 08 44	27 50	01 ♓ 05	10 49	06 23	20 03	11 38	02 54	28 05
31	00 32 50	10 ♈ 07 39	14 ♊ 00 18	20 ♊ 44 33	28 ♈ 56	02 ♓ 17	11 ♑ 24	06 ♐ 21	20 ♉ 10	11 ♎ 36	02 ♐ 54	28 ♍ 03

Date	Moon True ☊ ° ′	Moon Mean ☊ ° ′	Moon ☽ Latitude ° ′
01	23 ♒ 32	22 ♒ 48	05 N 04
02	23 R 30	22 45	05 16
03	23 29	22 42	05 10
04	23 D 29	22 39	04 47
05	23 30	22 36	04 11
06	23 32	22 33	03 23
07	23 33	22 29	02 26
08	23 35	22 26	01 25
09	23 36	22 23	00 N 20
10	23 R 35	22 20	00 S 46
11	23 34	22 17	01 48
12	23 31	22 13	02 46
13	23 26	22 10	03 37
14	23 21	22 07	04 19
15	23 16	22 04	04 50
16	23 10	22 01	05 08
17	23 06	21 58	05 12
18	23 02	21 54	05 01
19	23 01	21 51	04 36
20	23 D 01	21 48	03 56
21	23 02	21 45	03 01
22	23 03	21 42	01 55
23	23 04	21 39	00 S 39
24	23 R 04	21 35	00 N 41
25	23 02	21 32	01 59
26	22 59	21 29	03 10
27	22 53	21 26	04 08
28	22 46	21 23	04 47
29	22 39	21 19	05 07
30	22 33	21 16	05 06
31	22 ♒ 28	21 ♒ 13	04 N 47

DECLINATIONS

Date	Sun ☉ ° ′	Moon ☽ ° ′	Mercury ☿ ° ′	Venus ♀ ° ′	Mars ♂ ° ′	Jupiter ♃ ° ′	Saturn ♄ ° ′	Uranus ♅ ° ′	Neptune ♆ ° ′	Pluto ♇ ° ′
01	07 S 43	18 N 43	11 S 19	19 S 33	23 S 08	20 S 19	15 N 01	04 S 23	19 S 08	15 N 32
02	07 21	23 12	10 35	19 24	23 11	20 20	15 03	04 23	19 08	15 33
03	06 58	26 13	09 51	19 14	23 13	20 21	15 04	04 22	19 08	15 34
04	06 35	27 42	09 05	19 03	23 16	20 21	15 06	04 21	19 08	15 35
05	06 12	27 37	08 18	18 52	23 18	20 22	15 07	04 20	19 08	15 35
06	05 48	26 06	07 30	18 41	23 20	20 22	15 09	04 19	19 08	15 36
07	05 25	23 21	06 40	18 28	23 22	20 22	15 10	04 18	19 08	15 37
08	05 02	19 37	05 50	18 16	23 24	20 23	15 12	04 17	19 08	15 38
09	04 38	15 07	04 58	18 03	23 26	20 23	15 14	04 17	19 07	15 39
10	04 15	10 06	04 06	17 49	23 27	20 24	15 15	04 16	19 07	15 39
11	03 51	04 N 44	03 12	17 35	23 28	20 24	15 17	04 15	19 07	15 40
12	03 28	00 S 46	02 18	17 20	23 30	20 24	15 19	04 14	19 07	15 41
13	03 04	06 15	01 23	17 05	23 31	20 24	15 20	04 13	19 07	15 42
14	02 41	11 32	00 S 27	16 49	23 32	20 24	15 22	04 12	19 07	15 42
15	02 17	16 26	00 N 29	16 33	23 32	20 25	15 24	04 11	19 07	15 43
16	01 53	20 44	01 25	16 17	23 33	20 25	15 26	04 10	19 07	15 44
17	01 30	24 13	02 21	16 00	23 33	20 25	15 27	04 09	19 07	15 45
18	01 06	26 38	03 17	15 42	23 34	20 25	15 29	04 08	19 07	15 45
19	00 42	27 43	04 13	15 24	23 34	20 25	15 31	04 07	19 06	15 46
20	00 S 18	27 18	05 08	15 06	23 34	20 25	15 33	04 06	19 06	15 47
21	00 N 05	25 17	06 02	14 47	23 34	20 25	15 34	04 05	19 06	15 47
22	00 29	21 42	06 55	14 28	23 34	20 25	15 36	04 04	19 06	15 48
23	00 53	16 44	07 47	14 08	23 33	20 25	15 38	04 03	19 06	15 49
24	01 16	10 39	08 37	13 48	23 33	20 25	15 40	04 02	19 06	15 49
25	01 40	03 S 51	09 26	13 27	23 32	20 25	15 42	04 01	19 06	15 50
26	02 04	03 N 13	10 12	13 07	23 31	20 25	15 44	04 00	19 05	15 51
27	02 27	10 07	10 56	12 45	23 30	20 24	15 45	03 59	19 05	15 51
28	02 51	16 20	11 37	12 24	23 29	20 24	15 47	03 58	19 05	15 52
29	03 14	21 28	12 15	12 02	23 28	20 24	15 49	03 57	19 05	15 52
30	03 37	25 10	12 51	11 40	23 26	20 24	15 51	03 56	19 05	15 53
31	04 N 01	27 N 14	13 N 24	11 S 17	23 S 25	20 S 23	15 N 53	03 S 55	19 S 04	15 N 54

ZODIAC SIGN ENTRIES

Date	h m	Planets
03	03 01	☽ ♊
04	02 24	♀ ♒
05	09 47	☽ ♋
07	19 55	☽ ♌
10	08 10	☽ ♍
12	10 11	♂ ♑
12	21 06	☽ ♎
14	04 46	☿ ♈
15	09 31	☽ ♏
17	20 23	☽ ♐
20	04 37	☽ ♑
21	06 38	☉ ♈
22	09 29	☽ ♒
24	11 07	☽ ♓
26	10 45	☽ ♈
28	10 15	☽ ♉
29	14 02	♀ ♓
30	11 43	☽ ♊

LATITUDES

Date	Mercury ☿ ° ′	Venus ♀ ° ′	Mars ♂ ° ′	Jupiter ♃ ° ′	Saturn ♄ ° ′	Uranus ♅ ° ′	Neptune ♆ ° ′	Pluto ♇ ° ′
01	02 S 02	01 N 14	00 N 09	00 N 57	02 S 05	00 N 43	01 N 41	16 N 29
04	01 52	01 02	00 06	00 58	02 04	00 43	01 41	16 29
07	01 38	00 49	00 N 03	00 58	02 04	00 43	01 41	16 30
10	01 19	00 37	00 S 01	00 58	02 03	00 43	01 41	16 30
13	00 55	00 25	00 04	00 58	02 02	00 43	01 41	16 31
16	00 S 25	00 13	00 08	00 59	02 02	00 43	01 41	16 31
19	00 N 08	00 N 01	00 11	00 59	02 01	00 43	01 41	16 31
22	00 44	00 S 10	00 15	00 59	02 00	00 43	01 42	16 31
25	01 21	00 20	00 19	01 00	02 00	00 43	01 42	16 31
28	01 57	00 30	00 23	01 00	01 59	00 43	01 42	16 31
31	02 N 28	00 S 40	00 S 28	01 N 00	01 S 59	00 N 43	01 N 42	16 N 31

DATA

Julian Date	2441012
Delta T	+41 seconds
Ayanamsa	23° 27′ 26″
Synetic vernal point	05° ♓ 39′ 33″
True obliquity of ecliptic	23° 26′ 43″

MOON'S PHASES, APSIDES AND POSITIONS ☽

Date	h m	Phase	Longitude ° ′	Eclipse Indicator
04	02 01	☽	12 ♊ 51	
12	02 34	○	20 ♍ 52	
20	02 30	☾	28 ♐ 50	
26	19 23	●	05 ♈ 29	

Day	h m		
12	03 41	Apogee	
26	08 59	Perigee	
04	22 38	Max dec	27° N 51′
12	08 41	0S	
19	17 40	Max dec	27° S 46′
26	01 05	0N	

LONGITUDES

Date	Chiron ⚷ ° ′	Ceres ⚳ ° ′	Pallas ⚴ ° ′	Juno ⚵ ° ′	Vesta ⚶ ° ′	Black Moon Lilith ⚸ ° ′
01	08 ♈ 06	09 ♉ 04	25 ♓ 00	04 ♊ 25	07 ♑ 58	19 ♍ 51
11	08 ♈ 39	12 ♉ 30	28 ♓ 34	08 ♊ 48	12 ♑ 26	20 ♍ 58
21	09 ♈ 13	16 ♉ 07	02 ♈ 12	13 ♊ 21	16 ♑ 42	22 ♍ 04
31	09 ♈ 48	19 ♉ 52	05 ♈ 53	18 ♊ 06	20 ♑ 43	23 ♍ 11

All ephemeris data is given at 12.00 UT and the Moon's longitude is additionally given for 24.00 UT

ASPECTARIAN

h m	Aspects
01 Monday	
04 55	☿ Q ♂
04 59	☽ ⚻ ♆
07 38	☿ Q ♄
07 57	☽ ± ♇
08 57	☽ ✶ ☿
09 22	☽ ⚻ ♃
12 20	☿ □ ♃
13 41	♂ ± ♄
13 51	☽ ⚼ ♂
13 56	☽ # ♆
15 50	☽ # ♀
17 23	☽ ✶ ☉
19 50	☽ # ♃
21 16	☽ ⚻ ♅
23 04	☽ ⚼ ♇
02 Tuesday	
02 16	☿ ± ♅
05 11	☽ ☌ ♄
05 48	☽ ± ♂
07 29	☽ ± ♅
08 17	☽ Q ☿
11 54	☽ # ♂
15 29	☽ Q ☉
16 38	☽ ⚻ ♂
23 03	☽ ⚼ ♅
03 Wednesday	
00 51	☽ △ ♀
00 56	☽ △ ♇
01 53	♀ △ ♇
08 27	☽ ☍ ♆
13 24	☽ ☍ ♃
18 07	♂ Q ♅
20 58	☽ □ ☿
22 40	☉ ⚻ ♅
04 Thursday	
01 30	♀ ∥ ♆
01 45	☽ △ ♅
02 01	☽ □ ☉
06 17	☽ ⚼ ♀
10 41	☽ ⊻ ♄
21 51	☽ ⊥ ♄
05 Friday	
00 32	☽ ± ♀
01 22	☽ ☍ ♂
06 20	☿ ⚻ ♅
07 27	☽ □ ♇
12 57	☽ ⚻ ♀
14 52	☽ ∠ ♄
15 36	☽ ⚻ ♆
18 10	♆ St R
18 47	♀ ⊥ ♂
21 08	☽ ⚻ ♃
06 Saturday	
03 04	☽ ± ♆
08 42	☽ ± ♃
09 53	☽ □ ♅
14 57	☽ △ ☿
15 17	☽ △ ☉
17 49	♀ ✶ ♆
17 50	☽ Q ♇
19 03	☉ ☌ ☿
19 55	☽ ✶ ♄
20 28	☽ ⚼ ♆
07 Sunday	
02 16	☽ ⚼ ♃
11 53	☽ # ♂
14 06	☽ ⚻ ♂
17 20	☽ ✶ ♇
19 40	☽ Q ♄
21 03	☽ Q ♅
23 23	☽ ⚼ ☉
08 Monday	
00 07	☿ ✶ ♄
01 56	☽ ⚼ ☿
02 01	☽ △ ♆
02 37	☽ ± ♂
05 28	☽ ☍ ♀
07 29	☽ # ♃
08 03	☽ △ ♃
14 47	☽ # ♆
18 58	☽ ± ☉
19 58	☽ # ♀
20 55	☽ ✶ ♅
21 34	☽ ⚼ ♂
23 11	☽ ∠ ♇
23 22	☽ ± ☿
09 Tuesday	
05 13	☉ ✶ ♄
07 54	☽ □ ♄
08 08	☽ ⚻ ☉
09 02	♀ ✶ ♃
09 23	☽ ∥ ♇
11 28	☽ ∥ ♄
13 48	☽ ⚻ ☿
17 17	☽ ⊥ ♇
10 Wednesday	
03 04	☽ ∠ ♅
03 54	☿ ∠ ♀
03 57	♂ □ ♇
04 26	☉ ∥ ☿
05 23	☽ ⊻ ♇
05 28	☽ △ ♂
07 23	☿ ∥ ♅
11 13	☉ ∥ ♅
14 23	☽ □ ♆
20 46	☽ □ ♃
21 16	☽ ⊥ ♅
11 Thursday	
00 33	☽ ⚻ ♀
09 25	☽ ⊻ ♅
14 05	☽ ± ♀
14 11	☽ # ♅
16 10	☽ # ☉
20 03	☽ # ☿
21 10	☽ △ ♄
12 Friday	
02 34	☽ ☍ ☉
02 57	☽ Q ♆
09 27	☽ Q ♃
10 23	☽ ⚼ ♀
14 45	☽ ☍ ☿
17 43	☽ ∥ ☿
18 10	☽ ☌ ♇
21 41	☽ □ ♂
22 58	☽ ∥ ☉
13 Saturday	
03 04	☽ ∥ ♅
03 18	☽ ✶ ♆
03 51	☽ ⚼ ♄
09 45	♄ ± ♅
09 51	☽ ✶ ♃
10 59	☿ ☍ ♇
20 07	☽ △ ♀
22 03	☽ ☌ ♅
22 11	☽ ± ♄
14 Sunday	
09 35	☽ ∠ ♆
10 22	☽ ⚻ ♄
12 00	☽ Q ♂
15 06	♀ △ ♅
16 11	☽ ∠ ♃
20 46	☽ ⚻ ☉
23 37	☿ □ ♂
15 Monday	
03 37	☽ ⊥ ♆
06 30	☽ ⊻ ♇
06 42	☽ # ♄
08 21	☽ # ♇
09 51	☽ ± ☉
10 14	☽ ⊥ ♃
12 37	☽ ∥ ♀
13 18	☽ ✶ ♂
15 06	♀ ⚼ ♇
15 09	☽ ⚻ ☿
15 35	☽ ⊻ ♆
17 51	☿ △ ♆
18 26	☽ ⊥ ♇
22 11	☽ ⊻ ♃
16 Tuesday	
00 32	☿ ∠ ♄
02 29	☽ ∥ ♆
05 19	☽ ⚼ ☉
05 20	☽ ± ☿
09 46	☽ ⊻ ♅
10 04	☽ ∥ ♃
12 11	☽ ∠ ♇
14 29	☽ □ ♀
20 28	☽ ∠ ♂
20 33	☉ # ☿
21 31	☽ ⊥ ♅
22 26	☽ ☍ ♄
22 51	♀ Q ♆
17 Wednesday	
02 22	☽ ⚼ ☿
06 52	☽ ∥ ♂
10 42	♂ ⊻ ♆
10 59	☿ △ ♃
13 15	☽ △ ☉
14 53	☽ ⊥ ♂
14 57	☽ ∠ ♅
17 21	☽ ✶ ♇
18 Thursday	
02 13	☽ ☌ ♆
03 00	☽ ⊻ ♂
05 04	☽ Q ♀
08 02	♀ # ♇
08 40	☽ ☌ ♃
12 34	☽ △ ☿
16 14	☽ Q ♇
19 31	♂ ⚼ ♄
19 31	☽ ✶ ♅
19 Friday	
04 02	♀ # ♄
06 03	☽ ✶ ♀
08 12	☽ ⚻ ♄
09 34	☿ # ♅
15 17	☉ ☍ ♇
17 49	☽ Q ♅
19 23	☽ ± ♄
19 58	♀ Q ♃
20 Saturday	
01 38	☽ □ ♇
02 30	☽ □ ☉
07 30	♀ □ ♄
10 05	☽ ⊻ ♆
10 09	☿ ☍ ♅
11 55	☽ ⚼ ♄
12 18	☽ ∠ ♀
13 31	☽ ☌ ♂
16 16	☽ ⊻ ♃
20 52	☽ ⊥ ♆
23 04	☿ ⊥ ♄
21 Sunday	
02 15	☽ □ ♅
02 59	☽ ⊥ ♃
04 52	☽ □ ☿
05 55	☽ ⊥ ♀
12 46	☽ ∠ ♆
13 14	☽ Q ☉
14 33	♀ ∠ ♂
14 44	☽ △ ♄
17 25	☽ ⊻ ♀
18 46	☽ ∠ ♃
22 Monday	
00 48	☽ ∥ ♂
06 34	☽ △ ♇
11 31	☽ ✶ ☉
14 34	☽ ✶ ♆
16 36	☽ Q ☿
18 49	☽ ∥ ♃
20 05	☽ ⊻ ♂
20 24	☽ ✶ ♃
23 Tuesday	
01 18	☽ ∥ ♆
01 53	♀ ± ♇
03 38	♂ ⊻ ♃
05 35	☽ △ ♅
06 33	☽ ⊥ ♂
07 47	☽ ⚼ ♇
10 38	☽ Q ♆
11 30	♃ St R
14 27	☽ ∠ ☉
15 21	☽ ✶ ☿
15 52	☽ # ♇
16 21	☽ Q ♃
16 35	☽ # ♄
17 15	☿ ⚼ ♆
17 48	☽ □ ♄
22 00	☽ ∠ ♂
22 34	☽ ± ♇
23 19	☽ ∥ ♀
24 Wednesday	
00 25	☽ ☌ ♀
06 07	☽ ⚼ ♅
06 13	☽ ⊥ ☉
06 40	☉ △ ♆
08 17	☽ ⚻ ♇
13 55	☿ ⊻ ♄
15 55	☽ □ ♆
16 34	☽ ⊻ ☉
18 34	☽ # ☿
18 47	☽ ∠ ☿
20 34	☽ ± ♅
21 29	☽ □ ♃
23 07	☽ Q ♄
23 13	☽ ✶ ♂
25 Thursday	
06 06	☽ ⚻ ♅
10 51	☽ ⊥ ☿
11 26	☽ ∥ ♅
18 18	☽ ✶ ♄
19 02	☽ Q ♂
19 03	☽ # ☉
19 56	☿ ⚼ ♃
21 07	☉ ∠ ♄
21 26	☽ ⊻ ☿
26 Friday	
04 23	☽ ⊻ ♀
07 49	☽ ∥ ☉
07 54	☽ ☍ ♇
14 38	☽ ⊥ ♀
14 39	☽ # ♅
15 24	☽ △ ♆
18 05	☽ ∠ ♄
19 23	☽ ☌ ☉
20 53	☽ △ ♃
21 16	♀ ⚼ ♅
27 Saturday	
00 33	☽ □ ♂
05 18	☽ ☍ ♅
06 01	☽ ∠ ♀
08 09	♂ ⊥ ♆
08 22	☽ ⊥ ♄
15 03	☽ ⚼ ♆
15 20	☽ ∥ ☿
17 56	☽ ⊻ ♄
18 12	☉ △ ♃
20 34	☽ ⚼ ♃
21 16	☽ # ♀
28 Sunday	
00 32	♀ ⚻ ♇
01 53	☽ ☌ ☿
05 21	☽ ± ♆
07 16	☽ ⚻ ♇
07 52	☽ ✶ ♀
09 44	☽ ∥ ♄
10 03	☽ ∥ ♇
10 56	☽ ± ♃
14 58	☽ ⚻ ♆
16 54	☽ ± ♇
20 36	☽ ⚻ ♃
22 31	☽ ⊻ ☉
29 Monday	
00 06	☽ # ♆
02 24	☽ △ ♂
04 58	☽ Q ♀
05 12	☽ ⚻ ♅
06 28	☽ # ♃
06 31	☿ # ♀
07 31	☽ ⚼ ♇
09 01	☽ ⊥ ☉
15 02	☽ ± ♅
17 32	☿ ± ♆
18 52	☽ ☌ ♄
23 36	☽ # ♂
30 Tuesday	
01 03	☽ ∠ ☉
04 18	☽ ⚼ ♂
06 02	☽ ⚼ ♅
07 41	☽ ⊻ ☿
08 28	☽ △ ♇
13 44	☽ □ ♀
16 41	☽ ☍ ♆
17 05	☿ ⚻ ♇
18 50	☽ ⊥ ☿
20 19	☽ ± ♂
22 39	☽ ☍ ♃
31 Wednesday	
04 40	☽ ✶ ☉
06 33	☉ # ♅
07 12	☽ ⚻ ♂
07 46	☽ △ ♅
11 51	☽ ∠ ☿
19 32	♂ □ ♅
22 19	♄ ∥ ♇
23 03	☽ ⊻ ♄

APRIL 1971

LONGITUDES

Date	Sidereal time h m s	Sun ☉ ° ' "	Moon ☽ ° ' "	Moon ☽ 24.00 ° ' "	Mercury ☿ ° '	Venus ♀ ° '	Mars ♂ ° '	Jupiter ♃ ° '	Saturn ♄ ° '	Uranus ♅ ° '	Neptune ♆ ° '	Pluto ♇ ° '
01	00 36 46	11 ♈ 06 53	27 ♊ 21 40	03 ♋ 52 03	29 ♈ 55	03 ♓ 28	11 ♑ 28	06 ♐ 20	20 ♉ 17	11 ♎ 33	02 ♐ 53	28 ♍ 01
02	00 40 43	12 06 05	10 ♋ 16 08	16 ♋ 34 27	00 ♉ 48	04 40	12 34	06 R 18	20 23	11 R 30	02 R 52	28 R 00
03	00 44 39	13 05 15	22 ♋ 47 37	28 ♋ 56 14	01 34	05 51	13 09	06 16	20 30	11 28	02 51	27 58
04	00 48 36	14 04 22	05 ♌ 00 57	11 ♌ 02 22	02 13	07 03	13 44	06 14	20 37	11 25	02 50	27 57
05	00 52 32	15 03 27	17 ♌ 01 07	22 ♌ 57 48	02 46	08 14	14 19	06 11	20 43	11 23	02 49	27 55
06	00 56 29	16 02 29	28 ♌ 52 58	04 ♍ 47 09	03 11	09 26	14 53	06 09	20 50	11 20	02 48	27 54
07	01 00 25	17 01 30	10 ♍ 40 49	16 ♍ 34 25	03 29	10 38	15 28	06 06	20 57	11 18	02 47	27 52
08	01 04 22	18 00 28	22 ♍ 28 21	28 ♍ 22 57	03 41	11 49	16 03	06 03	21 04	11 15	02 46	27 50
09	01 08 19	18 59 24	04 ♎ 18 31	10 ♎ 15 19	03 46	13 01	16 37	06 00	21 11	11 12	02 45	27 49
10	01 12 15	19 58 18	16 ♎ 13 36	22 ♎ 13 31	03 R 44	14 13	17 11	05 57	21 18	11 10	02 44	27 47
11	01 16 12	20 57 09	28 ♎ 15 16	04 ♏ 18 59	03 36	15 25	17 46	05 54	21 25	11 07	02 43	27 46
12	01 20 08	21 55 59	10 ♏ 24 47	16 ♏ 32 50	03 21	16 37	18 20	05 50	21 32	11 05	02 42	27 45
13	01 24 05	22 54 47	22 ♏ 43 13	28 ♏ 56 07	03 02	17 48	18 54	05 46	21 39	11 02	02 41	27 43
14	01 28 01	23 53 34	05 ♐ 11 40	11 ♐ 30 03	02 37	19 00	19 28	05 43	21 46	11 00	02 40	27 42
15	01 31 58	24 52 18	17 ♐ 51 29	24 ♐ 16 10	02 07	20 12	20 02	05 39	21 54	10 57	02 38	27 40
16	01 35 54	25 51 01	00 ♑ 44 22	07 ♑ 16 22	01 33	21 24	20 35	05 34	22 01	10 55	02 37	27 39
17	01 39 51	26 49 42	13 ♑ 52 26	20 ♑ 32 53	00 56	22 36	21 09	05 30	22 08	10 52	02 36	27 37
18	01 43 48	27 48 22	27 ♑ 17 58	04 ♒ 07 57	00 ♉ 17	23 48	21 43	05 25	22 15	10 50	02 35	27 36
19	01 47 44	28 46 59	11 ♒ 03 02	18 ♒ 03 21	29 ♈ 36	25 00	22 16	05 21	22 23	10 47	02 34	27 35
20	01 51 41	29 ♈ 45 35	25 ♒ 08 54	02 ♓ 19 38	28 53	26 12	22 49	05 16	22 30	10 45	02 32	27 33
21	01 55 37	00 ♉ 44 10	09 ♓ 35 17	16 ♓ 55 28	28 11	27 25	23 22	05 11	22 37	10 43	02 31	27 32
22	01 59 34	01 42 42	24 ♓ 19 36	01 ♈ 46 57	27 29	28 37	23 55	05 06	22 45	10 40	02 30	27 31
23	02 03 30	02 41 13	09 ♈ 16 35	16 ♈ 47 28	26 48	29 ♓ 49	24 28	05 00	22 52	10 38	02 28	27 29
24	02 07 27	03 39 43	24 ♈ 18 25	01 ♉ 48 14	26 10	01 ♈ 01	25 01	04 55	23 00	10 35	02 27	27 28
25	02 11 23	04 38 10	09 ♉ 15 42	16 ♉ 39 40	25 34	02 13	25 34	04 49	23 07	10 33	02 26	27 27
26	02 15 20	05 36 36	23 ♉ 59 03	01 ♊ 12 57	25 01	03 25	26 06	04 44	23 15	10 31	02 24	27 26
27	02 19 17	06 34 59	08 ♊ 20 39	15 ♊ 21 38	24 32	04 38	26 38	04 38	23 22	10 29	02 23	27 25
28	02 23 13	07 33 21	22 ♊ 15 32	29 ♊ 02 16	24 07	05 50	27 10	04 32	23 30	10 26	02 21	27 23
29	02 27 10	08 31 41	05 ♋ 41 52	12 ♋ 14 34	23 46	07 02	27 42	04 26	23 37	10 24	02 20	27 22
30	02 31 06	09 ♉ 29 59	18 ♋ 40 43	25 ♋ 00 47	23 ♈ 30	08 ♈ 14	28 ♑ 14	04 ♐ 19	23 ♉ 45	10 ♎ 22	02 ♐ 18	27 ♍ 21

Date	Moon True ☊ ° '	Moon Mean ☊ ° '	Moon ☽ Latitude ° '
01	22 ♒ 25	21 ♒ 10	04 N 13
02	22 R 24	21 07	03 28
03	22 D 25	21 04	02 33
04	22 26	21 00	01 33
05	22 27	20 57	00 N 29
06	22 R 27	20 54	00 S 35
07	22 25	20 51	01 36
08	22 20	20 48	02 34
09	22 14	20 44	03 25
10	22 05	20 41	04 07
11	21 54	20 38	04 39
12	21 43	20 35	04 58
13	21 32	20 32	05 04
14	21 22	20 29	04 55
15	21 14	20 25	04 32
16	21 09	20 22	03 55
17	21 06	20 19	03 05
18	21 05	20 16	02 03
19	21 D 06	20 13	00 S 53
20	21 R 06	20 10	00 N 22
21	21 05	20 06	01 37
22	21 01	20 03	02 47
23	20 55	20 00	03 46
24	20 46	19 57	04 30
25	20 36	19 54	04 56
26	20 25	19 50	05 01
27	20 15	19 47	04 47
28	20 06	19 44	04 16
29	20 00	19 41	03 32
30	19 ♒ 56	19 ♒ 38	02 N 37

DECLINATIONS

Date	Sun ☉ ° '	Moon ☽ ° '	Mercury ☿ ° '	Venus ♀ ° '	Mars ♂ ° '	Jupiter ♃ ° '	Saturn ♄ ° '	Uranus ♅ ° '	Neptune ♆ ° '	Pluto ♇ ° '
01	04 N 24	27 N 39	13 N 53	10 S 54	23 S 23	20 S 23	15 N 55	03 S 54	19 S 04	15 N 54
02	04 47	26 30	14 20	10 31	23 22	20 23	15 57	03 53	19 04	15 55
03	05 10	24 02	14 43	10 07	23 20	20 22	15 59	03 52	19 04	15 55
04	05 33	20 31	15 02	09 43	23 18	20 22	16 01	03 51	19 04	15 56
05	05 56	16 12	15 18	09 19	23 16	20 21	16 03	03 50	19 03	15 56
06	06 19	11 19	15 30	08 55	23 14	20 21	16 05	03 49	19 03	15 57
07	06 41	06 05	15 39	08 30	23 11	20 20	16 07	03 48	19 03	15 57
08	07 04	00 N 38	15 44	08 05	23 09	20 20	16 08	03 47	19 03	15 58
09	07 26	04 S 51	15 46	07 40	23 06	20 19	16 10	03 46	19 02	15 58
10	07 49	10 11	15 43	07 15	23 04	20 18	16 12	03 45	19 02	15 59
11	08 11	15 11	15 38	06 49	23 01	20 18	16 14	03 44	19 02	15 59
12	08 33	19 39	15 28	06 23	22 58	20 17	16 16	03 43	19 02	16 00
13	08 55	23 21	15 16	05 57	22 55	20 16	16 18	03 42	19 01	16 00
14	09 16	26 00	15 00	05 31	22 52	20 16	16 20	03 41	19 01	16 00
15	09 38	27 24	14 41	05 05	22 49	20 15	16 22	03 40	19 01	16 01
16	09 59	27 21	14 19	04 38	22 46	20 14	16 24	03 39	19 01	16 01
17	10 21	25 47	13 55	04 11	22 42	20 13	16 26	03 38	19 00	16 02
18	10 42	22 43	13 29	03 44	22 39	20 13	16 28	03 37	19 00	16 02
19	11 03	18 19	13 01	03 17	22 35	20 12	16 30	03 36	19 00	16 02
20	11 23	12 48	12 32	02 50	22 32	20 11	16 32	03 36	19 00	16 03
21	11 44	06 S 29	12 02	02 23	22 28	20 10	16 34	03 35	18 59	16 03
22	12 04	00 N 18	11 32	01 56	22 24	20 09	16 36	03 34	18 59	16 03
23	12 24	07 09	11 02	01 28	22 20	20 08	16 38	03 33	18 59	16 03
24	12 44	13 37	10 32	01 01	22 17	20 07	16 40	03 32	18 58	16 04
25	13 04	19 15	10 03	00 33	22 13	20 06	16 42	03 31	18 58	16 04
26	13 24	23 38	09 35	00 S 06	22 09	20 05	16 44	03 30	18 58	16 04
27	13 43	26 25	09 09	00 N 22	22 05	20 04	16 46	03 29	18 58	16 04
28	14 02	27 29	08 45	00 50	22 00	20 03	16 48	03 28	18 57	16 05
29	14 21	26 51	08 22	01 17	21 56	20 02	16 50	03 28	18 57	16 05
30	14 N 40	24 N 45	08 N 02	01 N 45	21 S 52	20 S 00	16 N 52	03 S 27	18 S 57	16 N 05

ZODIAC SIGN ENTRIES

Date	h	m	Planets
01	14	11	☿ ♉
01	16	51	☽ ♋
04	02	05	☽ ♌
06	14	16	☽ ♍
09	03	17	☽ ♎
11	15	28	☽ ♏
14	02	03	☽ ♐
16	10	38	☽ ♑
18	16	46	☽ ♒
18	21	52	☿ ♈
20	17	54	☉ ♉
20	20	08	☽ ♓
22	21	08	☽ ♈
23	15	44	♀ ♈
24	21	06	☽ ♉
26	21	58	☽ ♊
29	01	43	☽ ♋

LATITUDES

Date	Mercury ☿ ° '	Venus ♀ ° '	Mars ♂ ° '	Jupiter ♃ ° '	Saturn ♄ ° '	Uranus ♅ ° '	Neptune ♆ ° '	Pluto ♇ ° '
01	02 N 37	00 S 43	00 S 29	01 N 00	01 S 58	00 N 43	01 N 42	16 N 31
04	02 58	00 51	00 34	01 00	01 58	00 43	01 42	16 31
07	03 10	01 00	00 39	01 01	01 57	00 43	01 42	16 31
10	03 09	01 07	00 44	01 01	01 57	00 43	01 42	16 30
13	02 55	01 14	00 49	01 01	01 57	00 43	01 43	16 30
16	02 28	01 20	00 54	01 01	01 56	00 43	01 43	16 29
19	01 49	01 25	01 00	01 01	01 56	00 43	01 43	16 28
22	01 01	01 30	01 06	01 01	01 55	00 43	01 43	16 28
25	00 N 10	01 34	01 12	01 01	01 55	00 43	01 43	16 27
28	00 S 39	01 37	01 18	01 01	01 55	00 43	01 43	16 26
31	01 S 25	01 S 40	01 S 25	01 N 01	01 S 55	00 N 43	01 N 43	16 N 25

DATA

Julian Date	2441043
Delta T	+42 seconds
Ayanamsa	23° 27' 30"
Synetic vernal point	05° ♓ 39' 30"
True obliquity of ecliptic	23° 26' 42"

MOON'S PHASES, APSIDES AND POSITIONS ☽

Date	h	m	Phase	Longitude ° '	Eclipse Indicator
02	15	46	☽	12 ♋ 15	
10	20	10	○	20 ♎ 18	
18	12	58	☾	27 ♑ 51	
25	04	02	●	04 ♉ 19	

Day	h	m		
08	07	22	Apogee	
23	17	46	Perigee	
01	05	54	Max dec	27° N 42'
08	14	45	0S	
15	23	17	Max dec	27° S 34'
22	10	58	0N	
28	14	41	Max dec	27° N 29'

LONGITUDES

Date	Chiron ⚷ ° '	Ceres ⚳ ° '	Pallas ⚴ ° '	Juno ⚵ ° '	Vesta ⚶ ° '	Black Moon Lilith ⚸ ° '
01	09 ♈ 52	20 ♉ 15	06 ♈ 15	18 ♊ 35	21 ♑ 06	23 ♍ 17
11	10 ♈ 27	24 ♉ 08	09 ♈ 59	23 ♊ 25	24 ♑ 45	24 ♍ 24
21	11 ♈ 01	28 ♉ 07	13 ♈ 44	28 ♊ 19	28 ♑ 03	25 ♍ 31
31	11 ♈ 34	02 ♊ 10	17 ♈ 31	03 ♋ 15	00 ♒ 54	26 ♍ 38

ASPECTARIAN

h m	Aspects
01 Thursday	
00 14	♀ □ ♆
03 39	☽ Q ☉
10 00	☽ ⊥ ♄
13 13	☽ □ ♇
17 03	☽ ⚹ ☿
22 09	☽ ⊼ ♆
22 10	☉ ☍ ♅
22 33	☿ ± ♃
02 Friday	
00 24	☽ △ ♀
01 33	♂ ⊥ ♃
02 45	☽ ∠ ♄
04 33	☽ ⊼ ♃
09 21	☽ ± ♆
14 20	☽ □ ♅
15 46	☽ □ ☉
15 50	☽ ± ♃
16 34	☽ ☍ ♂
17 07	☽ Q ☿
22 52	☽ Q ♇
03 Saturday	
02 28	☽ ⚼ ♆
04 25	♀ ± ♅
07 31	☽ ⚹ ♄
07 51	☽ ⚼ ♀
09 03	☽ ⚼ ♃
15 43	☉ □ ♂
17 24	☽ # ♂
20 03	♀ □ ♃
22 05	☽ ⚹ ♇
04 Sunday	
00 59	☽ Q ♅
03 18	☽ ± ♀
06 10	☽ □ ☿
07 12	☽ Q ♄
07 41	☽ △ ♆
12 55	☽ # ♃
14 24	☽ △ ♃
16 29	☽ ⊼ ♀
20 33	☽ # ♆
05 Monday	
00 43	☽ ⚹ ♅
02 48	☉ ⊥ ♄
03 47	☽ ∠ ♇
06 17	☽ ⊼ ♂
07 42	☽ △ ☉
12 49	☽ ∥ ♄
13 22	☽ ∥ ♇
14 57	☿ ⊼ ♆
16 25	☽ ∥ ☿
18 59	☽ ± ♂
19 33	☽ □ ♄
21 52	☽ ⊥ ♇
22 43	♀ Q ♄
06 Tuesday	
06 51	☽ ∠ ♅
10 00	☽ ⊻ ♇
14 09	☽ ⚼ ♂
16 47	☽ ⚼ ☉
19 57	☽ □ ♆
21 00	☽ △ ☿
07 Wednesday	
00 08	☽ # ♀
01 04	☽ ⊥ ♅
02 43	☽ □ ♃
09 26	☽ ∥ ☉
11 53	☽ ☍ ♀
12 46	☽ ± ☉
13 14	☽ ⊻ ♅
22 07	☽ # ♅
22 15	☽ △ ♂
08 Thursday	
00 51	♀ ⊼ ♅
02 06	☽ ⊼ ☉
04 12	☽ ⚼ ☿
06 19	☉ ⚼ ♆
08 33	☽ Q ♆
09 07	☽ △ ♄
15 12	☽ Q ♃
22 41	☽ ± ☿
22 53	☽ ☌ ♇
09 Friday	
07 16	☽ ∥ ♅
08 36	☉ ⊥ ♀
08 52	☽ ⚹ ♆
10 54	☽ ⊼ ☿
15 25	☽ ⚹ ♃
15 49	☽ ⚼ ♄
17 11	☿ St R
18 53	☉ # ♀
23 38	☽ ∥ ♀
10 Saturday	
00 24	☽ # ☉
01 52	☽ ☌ ♅
07 31	☽ ⊼ ♀
10 07	☽ ± ♄
14 02	☽ □ ♂
15 01	☽ ∠ ♆
20 10	☽ ☍ ☉
20 52	☽ ± ♀
21 25	☽ ∠ ♃
22 15	☽ ⊼ ♄
11 Sunday	
08 57	☽ ⊥ ♆
10 13	♂ ∠ ♆
10 42	☉ ⚼ ♃
11 02	☽ ⊻ ♇
14 10	☽ # ☿
15 14	☽ ⊥ ♃
16 03	☽ # ♇
16 45	☽ ⚼ ♀
17 23	☽ # ♄
20 50	☽ ⊻ ♆
22 24	☽ ☍ ☿
22 54	☽ ⊥ ♇
12 Monday	
00 54	☉ ⊻ ♄
03 03	☽ ⊻ ♃
03 35	☽ Q ♂
08 24	☽ ∥ ♆
13 18	☽ ⊻ ♅
15 44	☽ ∥ ♃
16 33	☽ ∠ ♇
13 Tuesday	
01 00	☽ ⊥ ♅
01 26	☽ △ ♀
04 13	☽ ⚹ ♂
08 56	☽ ∥ ♂
09 55	☽ ☍ ♄
12 24	☽ ⊼ ☉
15 19	☿ ∠ ♀
18 24	☽ ∠ ♅
21 38	☽ ⚹ ♇
14 Wednesday	
00 59	☽ ± ☉
07 10	☽ ☌ ♆
07 13	☽ ⊼ ☿
09 07	☿ ⊼ ♆
10 32	☽ ∠ ♂
12 59	☽ ☌ ♃
18 17	☽ ± ☿
19 38	☽ ⚼ ☉
20 33	☽ Q ♇
23 00	☽ ⚹ ♅
15 Thursday	
04 27	☽ ⊥ ♂
05 20	♀ ⚹ ♂
10 39	☽ ⚼ ☿
16 16	☽ ⊻ ♂
16 52	☽ □ ♀
19 38	☽ ⊼ ♄
21 31	☽ Q ♅
16 Friday	
02 12	☽ △ ☉
06 17	☽ □ ♇
06 55	☽ ± ♄
11 15	♂ ∠ ♃
13 26	☽ △ ☿
15 28	☽ ⊻ ♆
20 50	☽ ⊻ ♃
23 38	☽ ⚼ ♄
17 Saturday	
01 31	♀ ⚹ ♄
02 26	☽ ⊥ ♆
05 29	☽ Q ♀
06 32	☉ ± ♆
06 34	☽ □ ♅
07 43	☽ ⊥ ♃
18 42	☽ ∠ ♆
23 51	☽ ∠ ♃
18 Sunday	
01 39	☽ ☌ ♂
02 58	☽ △ ♄
05 12	☽ ⚹ ♀
07 04	☉ ⊼ ♇
12 27	☽ ∥ ♂
12 32	☽ △ ♇
12 58	☽ □ ☉
17 00	☽ □ ☿
18 04	☿ ⊥ ♀
18 25	♀ ∥ ♅
21 16	☽ ⚹ ♆
19 Monday	
02 10	☽ ⚹ ♃
02 31	☽ ∥ ♃
08 40	☽ ∥ ♆
10 02	☽ ∠ ♀
11 33	☽ △ ♅
14 38	☽ ⚼ ♇
18 01	☽ Q ♆
18 06	♂ △ ♄
20 19	☽ # ♄
21 34	☿ ± ♃
22 26	☽ # ♇
22 34	☽ Q ☉
22 41	☽ Q ☿
22 44	☽ Q ♃
23 34	☉ ☌ ☿
20 Tuesday	
00 44	☉ ± ♃
02 53	☽ ⊥ ♀
05 57	☽ ± ♇
07 30	☽ □ ♄
07 55	☽ ⊻ ♂
13 00	☽ ⚼ ♅
13 10	☽ # ☿
13 57	☽ ⊻ ♀
16 02	☽ ⊼ ♇
17 18	☽ # ☉
17 58	☽ ⚹ ☿
18 24	☽ ⊥ ♂
20 17	☽ ⚹ ☉
21 Wednesday	
00 20	☽ □ ♆
03 58	☽ ± ♅
04 46	☽ □ ♃
09 55	☽ ∠ ♂
13 43	☽ Q ♄
13 50	☽ ⊼ ♅
14 27	♀ ☍ ♇
17 37	☽ ∠ ☿
20 41	☉ ∥ ☿
21 44	☿ ⊻ ♀
22 26	☽ ∥ ♅
22 47	☽ ∠ ☉
22 Thursday	
03 40	☽ ∥ ♀
07 36	☽ ⊥ ☿
09 25	☽ ⚹ ♄
10 57	☿ ⊼ ♇
11 19	☽ ⚹ ♂
14 24	☽ ⊥ ☉
16 52	☽ ⊻ ☿
17 08	☽ ☍ ♇
17 20	☽ # ♀
19 31	☽ ☌ ♀
23 21	☽ # ♅
23 Friday	
00 43	☽ ⊻ ☉
01 07	☽ △ ♆
05 12	☽ △ ♃
06 49	☉ ⊼ ♆
07 21	☽ Q ♂
09 44	☽ ∠ ♄
14 09	☽ ☍ ♅
24 Saturday	
00 14	☽ ⊥ ♄
00 48	☿ ± ♆
01 04	☽ ⚼ ♆
01 07	☽ ∥ ☿
05 02	☽ ⚼ ♃
07 22	☉ ± ♇
08 26	☽ ∥ ☉
09 53	☽ ⊻ ♄
13 11	☽ □ ♂
14 51	☽ ☌ ☿
15 25	☽ ± ♆
17 03	☽ ⊼ ♇
19 19	☽ ± ♃
21 53	☽ ∥ ♇
23 40	☽ ⊻ ♀
25 Sunday	
00 31	☽ ∥ ♄
01 01	☽ ⊼ ♆
02 39	☽ ± ♇
04 02	☽ ☌ ☉
04 53	☽ ⊼ ♃
10 10	☽ ⊥ ♀
10 41	☽ # ♆
12 09	☿ □ ♂
14 05	☽ ⊼ ♅
16 04	♀ △ ♆
16 05	☽ # ♃
16 10	☉ ⊼ ♃
17 09	☽ ⚼ ♇
23 48	☽ ± ♅
26 Monday	
02 04	☽ ∠ ♀
03 04	☽ # ♂
10 46	☽ ☌ ♄
13 39	☽ ⊻ ☿
14 31	☽ ⚼ ♅
15 38	☽ △ ♂
17 41	☽ △ ♇
23 17	☽ ⊥ ☿
27 Tuesday	
01 58	☽ ☍ ♆
05 08	☽ ⚹ ♀
05 46	☽ ☍ ♃
08 48	☽ ⊻ ☉
12 04	♀ △ ♃
13 58	☽ ∠ ☿
15 37	☽ △ ♅
17 50	☽ ⚼ ♂
19 46	☽ ⊥ ☉
28 Wednesday	
03 32	☽ Q ♀
10 01	☽ ± ♂
12 34	☽ ∠ ☉
14 12	☽ ⊻ ♄
15 11	☽ ⚹ ☿
21 02	☽ ⊼ ♂
21 03	☽ □ ♇
21 26	♂ △ ♇
29 Thursday	
00 57	☽ ⊥ ♄
05 55	☽ ⊼ ♆
09 43	☽ ⊼ ♃
12 07	☽ Q ☿
14 41	☽ □ ♀
16 48	☽ ± ♆
17 23	☽ ∠ ♄
17 35	☽ ⚹ ☉
20 10	☿ ⊻ ♄
20 35	☽ □ ♅
20 35	☽ ± ♃
30 Friday	
05 47	☽ Q ♇
09 26	☽ ⚼ ♆
13 12	☽ ⚼ ♃
17 46	☽ Q ☉
20 57	☽ □ ☿
21 41	☽ ⚹ ♄
23 24	♀ ∠ ♄

All ephemeris data is given at 12.00 UT and the Moon's longitude is additionally given for 24.00 UT

MAY 1971

LONGITUDES

Date	Sidereal time h m s	Sun ☉ ° ' "	Moon ☽ ° ' "	Moon ☽ 24.00 ° ' "	Mercury ☿ ° '	Venus ♀ ° '	Mars ♂ ° '	Jupiter ♃ ° '	Saturn ♄ ° '	Uranus ♅ ° '	Neptune ♆ ° '	Pluto ♇ ° '
01	02 35 03	10 ♉ 28 14	01 ♌ 15 19	07 ♌ 24 56	23 ♈ 18	09 ♈ 27	28 ♑ 46	04 ♐ 13	23 ♉ 53	10 ♎ 20	02 ♐ 17	27 ♍ 20
02	02 38 59	11 26 28	13 ♌ 30 17	19 ♌ 32 03	23 R 11	10 39	29 17	04 R 07	24 00	10 R 18	02 R 16	27 R 19
03	02 42 56	12 24 39	25 ♌ 30 56	01 ♍ 27 35	23 D 09	11 51	29 ♑ 48	04 00	24 08	10 16	02 14	27 18
04	02 46 52	13 22 49	07 ♍ 22 41	13 ♍ 16 51	23 12	13 03	00 ♒ 19	03 54	24 16	10 14	02 13	27 17
05	02 50 49	14 20 56	19 ♍ 10 43	25 ♍ 04 48	23 19	14 16	00 50	03 47	24 23	10 12	02 11	27 16
06	02 54 46	15 19 02	00 ♎ 59 39	06 ♎ 55 42	23 31	15 28	01 21	03 40	24 31	10 10	02 09	27 15
07	02 58 42	16 17 06	12 ♎ 53 22	18 ♎ 53 00	23 48	16 41	01 52	03 33	24 39	10 08	02 08	27 14
08	03 02 39	17 15 08	24 ♎ 54 52	00 ♏ 59 12	24 09	17 53	02 22	03 26	24 46	10 06	02 06	27 13
09	03 06 35	18 13 08	07 ♏ 06 12	13 ♏ 15 58	24 35	19 05	02 52	03 19	24 54	10 04	02 05	27 12
10	03 10 32	19 11 07	19 ♏ 28 35	25 ♏ 44 05	25 05	20 18	03 22	03 12	25 02	10 02	02 03	27 11
11	03 14 28	20 09 04	02 ♐ 02 29	08 ♐ 23 45	25 39	21 30	03 52	03 05	25 10	10 00	02 02	27 10
12	03 18 25	21 06 59	14 ♐ 47 53	21 ♐ 14 50	26 16	22 43	04 21	02 57	25 17	09 59	02 00	27 10
13	03 22 21	22 04 53	27 ♐ 44 37	04 ♑ 17 13	26 58	23 55	04 51	02 50	25 25	09 57	01 59	27 09
14	03 26 18	23 02 46	10 ♑ 52 39	17 ♑ 30 59	27 43	25 08	05 20	02 43	25 33	09 55	01 57	27 08
15	03 30 15	24 00 38	24 ♑ 12 18	00 ♒ 56 40	28 32	26 20	05 49	02 35	25 41	09 54	01 55	27 07
16	03 34 11	24 58 28	07 ♒ 44 14	14 ♒ 35 06	29 ♈ 24	27 33	06 17	02 28	25 48	09 52	01 54	27 07
17	03 38 08	25 56 17	21 ♒ 29 23	28 ♒ 27 09	00 ♉ 20	28 45	06 45	02 20	25 56	09 50	01 52	27 06
18	03 42 04	26 54 05	05 ♓ 28 28	12 ♓ 33 17	01 19	29 ♈ 58	07 14	02 13	26 04	09 49	01 50	27 05
19	03 46 01	27 51 52	19 ♓ 41 28	26 ♓ 52 50	02 20	01 ♉ 10	07 41	02 05	26 12	09 48	01 49	27 05
20	03 49 57	28 49 38	04 ♈ 07 00	11 ♈ 23 32	03 25	02 23	08 09	01 57	26 19	09 46	01 47	27 04
21	03 53 54	29 ♉ 47 22	18 ♈ 41 49	26 ♈ 01 09	04 33	03 35	08 36	01 50	26 27	09 45	01 46	27 03
22	03 57 50	00 ♊ 45 06	03 ♉ 20 41	10 ♉ 39 33	05 44	04 48	09 03	01 42	26 35	09 43	01 44	27 03
23	04 01 47	01 42 48	17 ♉ 56 48	25 ♉ 11 32	06 57	06 01	09 30	01 35	26 43	09 42	01 42	27 02
24	04 05 44	02 40 29	02 ♊ 22 49	09 ♊ 29 52	08 13	07 13	09 56	01 27	26 50	09 41	01 41	27 02
25	04 09 40	03 38 09	16 ♊ 32 00	23 ♊ 28 37	09 32	08 26	10 22	01 19	26 58	09 40	01 39	27 02
26	04 13 37	04 35 48	00 ♋ 19 20	07 ♋ 03 54	10 53	09 38	10 48	01 12	27 06	09 39	01 37	27 01
27	04 17 33	05 33 26	13 ♋ 42 13	20 ♋ 14 22	12 17	10 51	11 13	01 04	27 14	09 37	01 36	27 01
28	04 21 30	06 31 02	26 ♋ 40 31	03 ♌ 01 01	13 43	12 04	11 38	00 56	27 21	09 36	01 34	27 00
29	04 25 26	07 28 36	09 ♌ 16 15	15 ♌ 26 45	15 12	13 16	12 03	00 49	27 29	09 35	01 33	27 00
30	04 29 23	08 26 09	21 ♌ 33 04	27 ♌ 35 49	16 44	14 29	12 27	00 41	27 37	09 34	01 31	27 00
31	04 33 19	09 ♊ 23 41	03 ♍ 35 38	09 ♍ 33 13	18 ♉ 18	15 ♉ 42	12 ♒ 51	00 ♐ 34	27 ♉ 44	09 ♎ 34	01 ♐ 29	27 ♍ 00

Date	Moon True ☊ ° '	Moon Mean ☊ ° '	Moon ☽ Latitude ° '
01	19 ♒ 55	19 ♒ 35	01 N 37
02	19 D 55	19 31	00 N 34
03	19 R 55	19 28	00 S 30
04	19 54	19 25	01 31
05	19 51	19 22	02 28
06	19 46	19 19	03 19
07	19 37	19 16	04 01
08	19 27	19 12	04 33
09	19 14	19 09	04 53
10	19 01	19 06	05 00
11	18 48	19 03	04 52
12	18 36	19 00	04 30
13	18 26	18 56	03 53
14	18 20	18 53	03 04
15	18 16	18 50	02 03
16	18 14	18 47	00 S 55
17	18 D 14	18 44	00 N 17
18	18 R 14	18 41	01 30
19	18 13	18 37	02 38
20	18 09	18 34	03 37
21	18 03	18 31	04 23
22	17 54	18 28	04 52
23	17 44	18 25	05 01
24	17 33	18 22	04 51
25	17 22	18 18	04 24
26	17 13	18 15	03 41
27	17 07	18 12	02 47
28	17 03	18 09	01 46
29	17 01	18 06	00 N 41
30	17 D 01	18 02	00 S 24
31	17 ♒ 01	17 ♒ 59	01 S 27

DECLINATIONS

Date	Sun ☉ ° '	Moon ☽ ° '	Mercury ☿ ° '	Venus ♀ ° '	Mars ♂ ° '	Jupiter ♃ ° '	Saturn ♄ ° '	Uranus ♅ ° '	Neptune ♆ ° '	Pluto ♇ ° '
01	14 N 58	21 N 28	07 N 44	02 N 13	21 S 48	19 S 59	16 N 54	03 S 26	18 S 56	16 N 05
02	15 16	17 19	07 29	02 41	21 43	19 58	16 56	03 25	18 56	16 05
03	15 34	12 33	07 17	03 08	21 39	19 57	16 58	03 24	18 56	16 05
04	15 52	07 24	07 06	03 36	21 35	19 56	17 00	03 24	18 55	16 05
05	16 09	02 N 01	06 59	04 04	21 30	19 54	17 02	03 23	18 55	16 05
06	16 26	03 S 26	06 54	04 31	21 26	19 53	17 04	03 22	18 55	16 06
07	16 43	08 48	06 51	04 59	21 21	19 52	17 05	03 21	18 54	16 06
08	16 59	13 53	06 51	05 26	21 17	19 51	17 07	03 21	18 54	16 06
09	17 16	18 30	06 53	05 53	21 12	19 49	17 09	03 20	18 54	16 06
10	17 31	22 24	06 58	06 20	21 08	19 48	17 11	03 19	18 54	16 06
11	17 47	25 20	07 05	06 48	21 03	19 47	17 13	03 19	18 53	16 06
12	18 02	27 02	07 14	07 15	20 59	19 45	17 15	03 18	18 53	16 06
13	18 18	27 18	07 25	07 41	20 54	19 44	17 17	03 17	18 53	16 06
14	18 32	26 03	07 38	08 08	20 50	19 43	17 19	03 17	18 52	16 06
15	18 47	23 18	07 53	08 35	20 45	19 41	17 21	03 16	18 52	16 05
16	19 01	19 14	08 09	09 01	20 41	19 40	17 23	03 15	18 52	16 05
17	19 15	14 04	08 28	09 27	20 36	19 38	17 24	03 15	18 51	16 05
18	19 28	08 07	08 48	09 53	20 32	19 37	17 26	03 14	18 51	16 05
19	19 41	01 S 39	09 10	10 19	20 27	19 36	17 28	03 14	18 51	16 05
20	19 54	04 N 58	09 33	10 45	20 23	19 34	17 30	03 13	18 50	16 05
21	20 07	11 23	09 57	11 10	20 18	19 33	17 32	03 13	18 50	16 05
22	20 19	17 12	10 23	11 35	20 14	19 31	17 34	03 12	18 50	16 05
23	20 31	22 00	10 50	12 00	20 10	19 30	17 35	03 12	18 49	16 04
24	20 42	25 24	11 18	12 25	20 06	19 28	17 37	03 11	18 49	16 04
25	20 53	27 08	11 47	12 49	20 01	19 27	17 39	03 11	18 49	16 04
26	21 04	27 08	12 17	13 13	19 57	19 26	17 41	03 10	18 48	16 04
27	21 14	25 31	12 47	13 37	19 53	19 24	17 43	03 10	18 48	16 03
28	21 24	22 33	13 19	14 00	19 49	19 23	17 44	03 10	18 48	16 03
29	21 34	18 36	13 51	14 23	19 46	19 21	17 46	03 09	18 47	16 03
30	21 43	13 57	14 24	14 46	19 42	19 20	17 48	03 09	18 47	16 03
31	21 N 52	08 N 50	14 N 58	15 N 09	19 S 38	19 S 18	17 N 50	03 S 09	18 S 47	16 N 02

ZODIAC SIGN ENTRIES

Date	h	m	Planets
01	09	34	☽ ♌
03	20	57	♂ ♒
03	21	03	☽ ♍
06	09	59	☽ ♎
08	22	03	☽ ♏
11	08	08	☽ ♐
13	16	09	☽ ♑
15	22	19	☽ ♒
17	03	32	☿ ♉
18	02	39	☽ ♓
18	12	48	♀ ♉
20	05	11	☽ ♈
21	17	15	☉ ♊
22	06	31	☽ ♉
24	08	01	☽ ♊
26	11	26	☽ ♋
28	18	16	☽ ♌
31	04	48	☽ ♍

LATITUDES

Date	Mercury ☿ ° '	Venus ♀ ° '	Mars ♂ ° '	Jupiter ♃ ° '	Saturn ♄ ° '	Uranus ♅ ° '	Neptune ♆ ° '	Pluto ♇ ° '
01	01 S 25	01 S 40	01 S 25	01 N 01	01 S 55	00 N 43	01 N 43	16 N 25
04	02 03	01 42	01 32	01 01	01 54	00 43	01 43	16 24
07	02 34	01 43	01 39	01 01	01 54	00 43	01 43	16 23
10	02 57	01 43	01 46	01 01	01 54	00 43	01 43	16 22
13	03 12	01 43	01 54	01 01	01 54	00 43	01 43	16 21
16	03 19	01 42	02 02	01 01	01 53	00 43	01 43	16 20
19	03 20	01 40	02 11	01 01	01 53	00 42	01 43	16 18
22	03 14	01 38	02 19	01 00	01 53	00 42	01 43	16 17
25	03 03	01 35	02 28	01 00	01 53	00 42	01 43	16 16
28	02 46	01 32	02 38	01 00	01 53	00 42	01 43	16 14
31	02 S 25	01 S 28	02 S 48	00 N 59	01 S 53	00 N 42	01 N 43	16 N 13

DATA

Julian Date	2441073
Delta T	+42 seconds
Ayanamsa	23° 27' 34"
Synetic vernal point	05° ♓ 39' 26"
True obliquity of ecliptic	23° 26' 42"

MOON'S PHASES, APSIDES AND POSITIONS ☽

Date	h	m	Phase	Longitude ° '	Eclipse Indicator
02	07	34	☽	11 ♌ 16	
10	11	24	○	19 ♏ 10	
17	20	15	☾	26 ♒ 16	
24	12	32	●	02 ♊ 42	

Day	h	m		
05	20	44	Apogee	
21	17	03	Perigee	
05	20	51	0S	
13	04	17	Max dec	27° S 23'
19	18	01	0N	
25	23	51	Max dec	27° N 21'

LONGITUDES

Date	Chiron ⚷ ° '	Ceres ⚳ ° '	Pallas ⚴ ° '	Juno ⚵ ° '	Vesta ⚶ ° '	Black Moon Lilith ⚸ ° '
01	11 ♈ 34	02 ♊ 10	17 ♈ 31	03 ♋ 15	00 ♒ 54	26 ♍ 38
11	12 ♈ 04	06 ♊ 18	21 ♈ 18	08 ♋ 13	03 ♒ 14	27 ♍ 44
21	12 ♈ 32	10 ♊ 28	25 ♈ 05	13 ♋ 11	04 ♒ 58	28 ♍ 51
31	12 ♈ 56	14 ♊ 40	28 ♈ 51	18 ♋ 08	06 ♒ 00	29 ♍ 58

All ephemeris data is given at 12.00 UT and the Moon's longitude is additionally given for 24.00 UT

ASPECTARIAN

h m	Aspects
01 Saturday	
04 27	☽ ⚹ ♇
06 23	☽ Q ♅
06 58	☽ ☍ ♂
08 42	☉ ⊼ ♅
09 49	☽ # ♂
13 59	☽ △ ♆
17 42	☽ △ ♃
21 05	☽ Q ♄
21 05	☽ # ♃
02 Sunday	
03 07	☽ # ♆
05 13	♀ ☍ ♅
05 41	☽ ⚹ ♅
05 44	☽ △ ♀
07 34	☽ □ ☉
09 39	☽ ∠ ♇
14 02	☽ ∥ ♄
18 27	☽ ∥ ♇
22 00	☽ ∥ ☉
03 Monday	
03 32	☽ ⊥ ♇
07 15	☽ △ ☿
09 11	☽ □ ♄
09 15	☉ ⚼ ♇
10 23	♀ Q ♂
10 25	☿ St D
11 30	☽ ∠ ♅
15 00	☽ ⚼ ♀
15 35	☽ ⊻ ♇
21 03	☽ ⊼ ♂
04 Tuesday	
01 32	☽ □ ♆
01 37	♀ # ♅
05 00	☽ □ ♃
05 38	☽ ⊥ ♅
09 46	☽ ± ♂
11 17	☽ ± ♀
13 20	☽ ∥ ☿
13 40	☽ ⚼ ☿
17 47	☽ ⊻ ♅
05 Wednesday	
00 52	☽ ⊼ ♀
01 18	☽ △ ☉
03 39	☽ ∥ ♀
04 54	☽ ⚼ ♂
05 56	☽ # ♅
07 14	☉ ∥ ♇
08 10	☽ ± ☿
14 02	☽ Q ♆
17 15	☽ Q ♃
20 33	☽ ⊼ ☿
20 39	☉ ⊻ ♀
22 42	☽ △ ♄
06 Thursday	
04 25	☽ ☌ ♇
10 30	☽ ⚼ ☉
11 42	☽ ∥ ♅
12 45	☽ △ ♂
14 21	☽ ⚹ ♆
17 13	☽ # ♀
17 22	☽ ⚹ ♃
07 Friday	
03 15	☽ # ☿
05 24	☽ ⚼ ♄
06 18	☽ ± ☉
06 28	☽ ☌ ♅
08 17	☉ ± ♅
19 24	☽ ⊼ ☉
20 26	☽ ☍ ♀
20 29	☽ ∠ ♆
20 54	♀ ⚼ ♆
23 14	☽ ∠ ♃
23 39	☽ ± ♄
08 Saturday	
00 15	♂ ⚹ ♆
10 27	☽ ☍ ☿
11 43	☽ ⊼ ♄
14 21	☽ ⊥ ♆
16 33	☽ ⊻ ♇
16 57	☽ ⊥ ♃
22 02	♀ ⚼ ♃
23 07	☽ # ♇
09 Sunday	
01 33	☉ ∥ ♄
02 10	☽ ⊻ ♆
03 21	☽ □ ♂
04 22	☽ ⊥ ♇
04 38	☽ # ♄
04 39	☽ ⊻ ♃
04 48	☽ # ☉
07 50	♀ ⊥ ♄
14 14	☽ ∥ ♆
17 46	☽ ⊻ ♅
19 32	☽ ∥ ♃
21 55	☽ ∠ ♇
10 Monday	
03 44	☽ ∥ ♂
05 23	☽ ⊥ ♅
05 25	♂ ⚹ ♃
09 00	☿ ⊻ ♄
11 24	☽ ☍ ☉
13 02	♄ ⚼ ♅
13 45	☽ ⊼ ♀
15 47	☽ Q ♂
22 38	☽ ∠ ♅
22 46	☽ ☍ ♄
23 14	☽ ⊼ ☿
11 Tuesday	
02 27	☽ ± ♀
02 45	☽ ⚹ ♇
11 13	☽ ± ☿
11 58	☽ ☌ ♆
13 57	☽ ☌ ♃
15 35	☽ ⚹ ♂
21 19	☽ ⚼ ♀
12 Wednesday	
01 27	☽ Q ♇
02 19	☿ ± ♆
03 00	☽ ⚹ ♅
05 03	☽ ⚼ ☿
10 48	☿ ∥ ♀
20 49	☽ ∠ ♂
13 Thursday	
00 42	☽ ⊼ ☉
01 19	☽ Q ♅
04 13	☽ △ ♀
07 40	☽ ⊼ ♄
08 12	☿ ± ♃
10 29	☽ △ ☿
10 54	☽ □ ♇
12 40	☽ ± ☉
14 06	☽ ⊥ ♂
17 44	☿ ⊼ ♇
18 49	☽ ± ♄
19 45	☽ ⊻ ♆
21 15	☽ ⊻ ♃
14 Friday	
01 31	☽ ⊻ ♂
06 27	☽ ⚼ ☉
06 41	☽ ⊥ ♆
08 06	☽ ⊥ ♃
10 16	☽ □ ♅
11 24	☽ ⚼ ♄
21 22	♀ ⊻ ♄
22 57	☽ ∠ ♆
15 Saturday	
00 14	☽ ∠ ♃
04 00	♀ ± ♆
11 38	☽ △ ☉
14 39	☽ △ ♄
16 10	☽ □ ♀
16 35	♀ ± ♃
17 12	☽ △ ♇
20 15	☽ □ ☿
20 29	☉ # ♆
16 Sunday	
01 42	☽ ⚹ ♆
02 47	☽ ⚹ ♃
03 30	♀ ⊼ ♇
04 06	☽ ∥ ♂
09 21	☽ ☌ ♂
09 24	☉ ⚼ ♅
09 43	☽ ∥ ♃
13 03	☽ # ☉
13 53	☽ ∥ ♆
15 44	☽ △ ♅
19 40	☽ ⚼ ♇
21 06	☽ # ♄
22 46	☽ Q ♆
23 41	☽ Q ♃
17 Monday	
02 59	☽ Q ♀
03 07	☽ # ♇
06 07	☽ Q ☿
11 19	☽ ± ♇
11 53	☉ ☌ ♄
17 47	☽ ⚼ ♅
19 45	☽ □ ♄
20 15	☽ □ ☉
21 40	☽ ⊼ ♇
18 Tuesday	
01 42	☽ ⚹ ♀
04 21	☽ ⚹ ☿
05 34	☽ # ♀
05 49	☽ □ ♆
06 29	☽ □ ♃
09 11	☽ ± ♅
09 30	☽ # ☿
15 05	☽ ⊻ ♂
16 35	☉ △ ♇
19 21	☽ ⊼ ♅
19 Wednesday	
00 11	☿ ⊼ ♆
01 35	☽ ⊥ ♂
02 34	☉ # ♃
02 41	☽ Q ♄
05 07	☽ Q ☉
05 33	☽ ∠ ♀
06 15	☽ ∥ ♅
06 46	☿ ⊼ ♃
07 45	☽ ∠ ☿
17 11	☽ ∠ ♂
21 59	☽ ⊥ ♀
22 57	☽ ⚹ ♄
23 59	☽ ⊥ ☿
20 Thursday	
00 19	☽ ☍ ♇
00 31	♀ ⊼ ♆
02 37	☽ ⚹ ☉
04 19	☿ ± ♇
04 27	♀ ⊼ ♃
05 41	☽ # ♅
08 09	☽ △ ♆
08 28	☽ △ ♃
08 52	☽ ⊻ ♀
10 45	☽ ⊻ ☿
18 52	☽ ⚹ ♂
21 19	☽ ☍ ♅
23 59	☽ ∠ ♄
21 Friday	
01 33	♀ ± ♇
05 08	☽ ∠ ☉
06 08	☽ ∥ ☿
08 50	☽ ⚼ ♆
08 58	☽ ⚼ ♃
11 08	☽ ∥ ♀
14 54	☽ ⊥ ♄
15 13	☽ Q ♂
20 56	☽ ⊥ ☉
23 33	☽ ± ♆
23 35	☽ ± ♃
22 Saturday	
00 49	☽ ⊻ ♄
01 42	☽ ⊼ ♇
04 57	♃ ☌ ♆
05 13	☉ # ♂
07 06	☽ ∥ ♇
07 27	☽ ⊻ ☉
09 20	☽ ⊼ ♃
09 22	☽ ⊼ ♆
11 31	☽ ± ♇
13 39	☽ ∥ ♄
14 36	☽ ☌ ♀
16 15	☽ ☌ ☿
19 30	☽ # ♆
21 39	☽ □ ♂
22 27	☽ ⊼ ♅
22 50	☽ # ♃
23 Sunday	
02 15	☽ # ♂
02 17	☽ ⚼ ♇
03 31	☽ ∥ ☉
08 18	☽ ± ♅
08 59	☉ ☍ ♃
11 49	☉ ☍ ♆
22 39	♂ △ ♅
23 10	☽ ⚼ ♅
24 Monday	
02 39	☽ ☌ ♄
03 04	☽ △ ♇
10 27	☽ ☍ ♃
10 50	☽ ☍ ♆
12 32	☽ ☌ ☉
20 54	☽ ⊻ ♀
22 49	☽ ⊻ ☿
25 Tuesday	
00 18	☽ △ ♅
01 09	☽ △ ♂
08 03	☽ ⊥ ♀
10 06	☽ ⊥ ☿
14 21	☿ ⊼ ♅
22 15	♄ △ ♇
26 Wednesday	
01 04	☽ ∠ ♀
03 20	☽ ∠ ☿
03 48	☽ ⚼ ♂
06 11	☽ □ ♇
06 16	☽ ⊻ ♄
09 52	☿ □ ♂
12 01	♀ ⊼ ♅
13 32	☽ ⊼ ♃
14 18	☽ ⊼ ♆
16 58	☽ ⊥ ♄
20 10	☽ ⊻ ☉
20 12	☽ ± ♂
27 Thursday	
00 07	☽ ± ♃
00 59	☽ ± ♆
04 37	☽ □ ♅
06 18	☽ ⚹ ♀
07 21	☽ ⊼ ♂
07 28	☿ ⚼ ♇
07 48	☽ ⊥ ☉
09 06	☽ ⚹ ☿
09 17	☽ ∠ ♄
14 23	☽ Q ♇
16 17	☽ ⚼ ♃
17 17	☽ ⚼ ♆
23 13	♀ □ ♂
28 Friday	
01 36	☽ ∠ ☉
06 36	☽ Q ♀
09 59	☽ Q ☿
10 54	♀ ⚼ ♇
12 37	☽ ⚹ ♇
13 17	☽ ⚹ ♄
13 45	☽ Q ♅
19 16	☽ ∥ ☉
19 58	☽ △ ♃
21 14	☽ △ ♆
29 Saturday	
05 22	☽ # ♂
07 45	☽ # ♃
08 15	☽ ⚹ ☉
09 08	♂ ⚼ ♇
10 57	☽ # ♆
12 25	☽ Q ♄
12 37	☽ ⚹ ♅
16 29	☽ ∥ ♄
17 17	☽ ∠ ♇
17 34	☽ ☍ ♂
18 02	☿ ± ♅
20 37	☽ □ ♀
30 Sunday	
01 09	☽ □ ☿
01 31	☽ ∥ ♇
08 13	☽ ∥ ♀
09 37	☽ Q ☉
09 58	☽ ∥ ☿
10 54	☽ ⊥ ♇
17 59	☽ ∠ ♅
22 32	♂ Q ♃
22 48	☽ ⊻ ♇
31 Monday	
00 09	☽ □ ♄
05 59	☽ □ ♃
07 47	☽ □ ♆
09 16	♀ ± ♅
11 56	☽ ⊥ ♅
16 02	☉ △ ♅

JUNE 1971

Date	Sidereal time h m s	Sun ☉	Moon ☽	Moon ☽ 24.00	Mercury ☿	Venus ♀	Mars ♂	Jupiter ♃	Saturn ♄	Uranus ♅	Neptune ♆	Pluto ♇
01	04 37 16	10 ♊ 21 12	15 ♍ 29 12	21 ♍ 24 17	19 ♉ 54	16 ♉ 55	13 ♒ 15	00 ♐ 26	27 ♉ 52	09 ♎ 33	01 ♐ 28	26 ♍ 59
02	04 41 13	11 18 41	27 ♍ 19 06	03 ♎ 14 17	21 33	18 07	13 38	00 R 19	28 00	09 R 32	01 R 26	26 R 59
03	04 45 09	12 16 08	09 ♎ 10 26	15 ♎ 08 07	23 15	19 20	14 01	00 12	28 07	09 31	01 25	26 59
04	04 49 06	13 13 35	21 ♎ 07 50	27 ♎ 10 01	24 58	20 33	14 24	00 ♐ 04	28 15	09 30	01 23	26 59
05	04 53 02	14 11 00	03 ♏ 15 06	09 ♏ 23 23	26 45	21 46	14 46	29 ♏ 57	28 22	09 30	01 21	26 59
06	04 56 59	15 08 25	15 ♏ 35 09	21 ♏ 50 34	28 ♉ 33	22 58	15 08	29 50	28 30	09 29	01 20	26 59
07	05 00 55	16 05 48	28 ♏ 09 45	04 ♐ 32 45	00 ♊ 25	24 11	15 29	29 43	28 37	09 29	01 18	26 59
08	05 04 52	17 03 11	10 ♐ 59 32	17 ♐ 30 03	02 18	25 24	15 50	29 36	28 45	09 28	01 17	26 D 59
09	05 08 48	18 00 33	24 ♐ 04 08	00 ♑ 41 38	04 14	26 37	16 10	29 29	28 52	09 28	01 15	26 59
10	05 12 45	18 57 53	07 ♑ 22 21	14 ♑ 06 03	06 12	27 50	16 30	29 22	29 00	09 27	01 14	26 59
11	05 16 41	19 55 14	20 ♑ 52 33	27 ♑ 41 36	08 12	29 ♉ 02	16 49	29 15	29 07	09 27	01 12	26 59
12	05 20 38	20 52 33	04 ♒ 33 02	11 ♒ 26 39	10 14	00 ♊ 15	17 08	29 09	29 15	09 27	01 11	26 59
13	05 24 35	21 49 52	18 ♒ 22 18	25 ♒ 19 52	12 18	01 28	17 27	29 02	29 22	09 26	01 09	26 59
14	05 28 31	22 47 11	02 ♓ 19 13	09 ♓ 20 15	14 24	02 41	17 45	28 55	29 30	09 26	01 08	27 00
15	05 32 28	23 44 29	16 ♓ 22 50	23 ♓ 26 51	16 32	03 54	18 02	28 49	29 37	09 26	01 06	27 00
16	05 36 24	24 41 47	00 ♈ 32 09	07 ♈ 38 32	18 40	05 07	18 19	28 43	29 44	09 26	01 05	27 00
17	05 40 21	25 39 04	14 ♈ 45 45	21 ♈ 53 29	20 50	06 20	18 36	28 37	29 52	09 26	01 03	27 00
18	05 44 17	26 36 21	29 ♈ 01 24	06 ♉ 09 02	23 01	07 33	18 52	28 31	29 ♉ 59	09 D 26	01 02	27 01
19	05 48 14	27 33 38	13 ♉ 15 57	20 ♉ 21 35	25 12	08 46	19 07	28 25	00 ♊ 06	09 26	01 00	27 01
20	05 52 11	28 30 55	27 ♉ 25 23	04 ♊ 26 49	27 24	09 59	19 22	28 19	00 13	09 26	00 59	27 01
21	05 56 07	29 ♊ 28 12	11 ♊ 25 18	18 ♊ 20 19	29 ♊ 36	11 12	19 36	28 13	00 20	09 26	00 58	27 02
22	06 00 04	00 ♋ 25 28	25 ♊ 11 26	01 ♋ 58 13	01 ♋ 47	12 25	19 49	28 08	00 27	09 27	00 56	27 02
23	06 04 00	01 22 44	08 ♋ 40 23	15 ♋ 17 43	03 58	13 38	20 02	28 02	00 34	09 27	00 55	27 03
24	06 07 57	02 19 59	21 ♋ 50 06	28 ♋ 17 31	06 09	14 51	20 14	27 57	00 42	09 27	00 54	27 03
25	06 11 53	03 17 14	04 ♌ 40 04	10 ♌ 57 57	08 18	16 04	20 26	27 52	00 48	09 28	00 52	27 04
26	06 15 50	04 14 29	17 ♌ 11 25	23 ♌ 20 50	10 26	17 17	20 37	27 47	00 55	09 28	00 51	27 05
27	06 19 46	05 11 43	29 ♌ 26 37	05 ♍ 29 16	12 33	18 30	20 47	27 42	01 02	09 29	00 50	27 05
28	06 23 43	06 08 56	11 ♍ 29 17	17 ♍ 27 14	14 38	19 44	20 57	27 38	01 09	09 29	00 48	27 06
29	06 27 40	07 06 09	23 ♍ 23 45	29 ♍ 19 24	16 41	20 57	21 06	27 33	01 16	09 30	00 47	27 06
30	06 31 36	08 ♋ 03 21	05 ♎ 14 50	11 ♎ 10 41	18 ♋ 43	22 ♊ 10	21 ♒ 14	27 ♏ 29	01 ♊ 23	09 ♎ 30	00 ♐ 46	27 ♍ 07

Date	Moon True ☊	Moon Mean ☊	Moon ☽ Latitude
01	17 ♒ 01	17 ♒ 56	02 S 26
02	17 R 00	17 53	03 17
03	16 57	17 50	04 01
04	16 51	17 47	04 34
05	16 43	17 43	04 55
06	16 33	17 40	05 04
07	16 23	17 37	04 57
08	16 12	17 34	04 36
09	16 03	17 31	04 00
10	15 56	17 28	03 11
11	15 51	17 24	02 09
12	15 49	17 21	01 S 00
13	15 D 48	17 18	00 N 14
14	15 49	17 15	01 28
15	15 50	17 12	02 37
16	15 R 50	17 08	03 37
17	15 48	17 05	04 24
18	15 45	17 02	04 55
19	15 39	16 59	05 07
20	15 32	16 56	05 01
21	15 25	16 53	04 37
22	15 18	16 49	03 57
23	15 12	16 46	03 04
24	15 08	16 43	02 03
25	15 05	16 40	00 N 56
26	15 D 05	16 37	00 S 11
27	15 06	16 33	01 17
28	15 07	16 30	02 19
29	15 09	16 27	03 13
30	15 ♒ 10	16 ♒ 24	03 S 59

DECLINATIONS

Date	Sun ☉	Moon ☽	Mercury ☿	Venus ♀	Mars ♂	Jupiter ♃	Saturn ♄	Uranus ♅	Neptune ♆	Pluto ♇
01	22 N 00	03 N 29	15 N 32	15 N 31	19 S 34	19 S 17	17 N 51	03 S 08	18 S 46	16 N 02
02	22 08	01 S 57	16 06	15 53	19 31	19 16	17 53	03 08	18 46	16 02
03	22 16	07 19	16 40	16 14	19 27	19 14	17 55	03 08	18 46	16 01
04	22 23	12 29	17 14	16 35	19 24	19 13	17 56	03 08	18 46	16 01
05	22 30	17 13	17 49	16 55	19 21	19 11	17 58	03 07	18 45	16 00
06	22 37	21 21	18 23	17 16	19 18	19 10	18 00	03 07	18 45	16 00
07	22 43	24 35	18 57	17 35	19 15	19 09	18 01	03 07	18 45	16 00
08	22 49	26 39	19 30	17 55	19 12	19 07	18 03	03 07	18 44	15 59
09	22 54	27 19	20 03	18 13	19 09	19 06	18 05	03 07	18 44	15 59
10	22 59	26 25	20 34	18 32	19 07	19 05	18 06	03 07	18 44	15 58
11	23 04	23 57	21 05	18 50	19 04	19 03	18 08	03 06	18 43	15 58
12	23 08	20 06	21 35	19 07	19 02	19 02	18 09	03 06	18 43	15 57
13	23 12	15 06	22 03	19 24	19 00	19 01	18 11	03 06	18 43	15 57
14	23 15	09 17	22 29	19 40	18 58	19 00	18 12	03 06	18 43	15 56
15	23 18	02 S 58	22 54	19 56	18 56	18 59	18 14	03 06	18 42	15 56
16	23 20	03 N 32	23 16	20 12	18 55	18 57	18 15	03 06	18 42	15 55
17	23 22	09 52	23 36	20 27	18 53	18 56	18 17	03 06	18 42	15 55
18	23 24	15 43	23 54	20 41	18 52	18 55	18 18	03 06	18 42	15 54
19	23 25	20 42	24 10	20 55	18 51	18 54	18 20	03 06	18 41	15 54
20	23 26	24 28	24 22	21 08	18 50	18 53	18 21	03 07	18 41	15 53
21	23 27	26 43	24 32	21 20	18 50	18 52	18 23	03 07	18 41	15 52
22	23 27	27 18	24 40	21 32	18 49	18 51	18 24	03 07	18 41	15 52
23	23 26	26 13	24 44	21 44	18 49	18 50	18 26	03 07	18 40	15 51
24	23 25	23 41	24 45	21 55	18 49	18 49	18 27	03 07	18 40	15 51
25	23 24	20 01	24 44	22 05	18 49	18 48	18 28	03 07	18 40	15 50
26	23 23	15 30	24 40	22 15	18 50	18 47	18 30	03 07	18 40	15 49
27	23 21	10 28	24 33	22 24	18 50	18 46	18 31	03 08	18 39	15 49
28	23 18	05 N 07	24 24	22 32	18 51	18 45	18 33	03 08	18 39	15 48
29	23 15	00 S 20	24 13	22 40	18 52	18 44	18 34	03 08	18 39	15 47
30	23 N 12	05 S 45	23 N 58	22 N 47	18 S 54	18 S 44	18 N 35	03 S 09	18 S 39	15 N 47

ZODIAC SIGN ENTRIES

Date	h	m	Planets
02	17	26	☽ ♎
05	02	12	♃ ♏
05	05	36	☽ ♏
07	06	45	☿ ♊
07	15	28	☽ ♐
09	22	45	☽ ♑
12	04	03	☽ ♒
12	06	58	♀ ♊
14	08	01	☽ ♓
16	11	06	☽ ♈
18	13	39	☽ ♉
18	16	09	♄ ♊
20	16	24	☽ ♊
21	16	25	☿ ♋
22	01	20	☉ ♋
22	20	30	☽ ♋
25	03	12	☽ ♌
27	13	06	☽ ♍
30	01	22	☽ ♎

LATITUDES

Date	Mercury ☿	Venus ♀	Mars ♂	Jupiter ♃	Saturn ♄	Uranus ♅	Neptune ♆	Pluto ♇
01	02 S 17	01 S 26	02 S 51	00 N 59	01 S 53	00 N 42	01 N 43	16 N 13
04	01 50	01 22	03 01	00 59	01 53	00 42	01 43	16 11
07	01 19	01 16	03 12	00 58	01 53	00 42	01 43	16 10
10	00 47	01 11	03 23	00 58	01 53	00 42	01 43	16 08
13	00 S 14	01 05	03 34	00 57	01 53	00 41	01 43	16 07
16	00 N 18	00 59	03 46	00 57	01 53	00 41	01 43	16 06
19	00 48	00 52	03 57	00 56	01 53	00 41	01 43	16 04
22	01 14	00 45	04 10	00 56	01 53	00 41	01 43	16 03
25	01 33	00 38	04 22	00 55	01 53	00 41	01 43	16 01
28	01 46	00 31	04 35	00 54	01 54	00 41	01 42	16 00
31	01 N 53	00 S 24	04 S 48	00 N 54	01 S 54	00 N 41	01 N 42	15 N 59

DATA

Julian Date	2441104
Delta T	+42 seconds
Ayanamsa	23° 27' 39"
Synetic vernal point	05° ♓ 39' 21"
True obliquity of ecliptic	23° 26' 41"

MOON'S PHASES, APSIDES AND POSITIONS ☽

Date	h	m	Phase	Longitude	Eclipse Indicator
01	00	42	☽	09 ♍ 54	
09	00	04	○	17 ♐ 32	
16	01	24	☾	24 ♓ 16	
22	21	57	●	00 ♋ 49	
30	18	11	☽	08 ♎ 18	

Day	h	m		
02	14	19	Apogee	
17	09	38	Perigee	
30	08	58	Apogee	
02	03	23	0S	
09	10	20	Max dec	27° S 19'
15	22	58	0N	
22	08	08	Max dec	27° N 19'
29	10	31	0S	

LONGITUDES

Date	Chiron ⚷	Ceres ⚳	Pallas ⚴	Juno ⚵	Vesta ⚶	Black Moon Lilith ⚸
01	12 ♈ 59	15 ♊ 06	29 ♈ 13	18 ♋ 38	06 ♒ 04	00 ♎ 05
11	13 ♈ 18	19 ♊ 20	02 ♉ 57	23 ♋ 34	06 ♒ 15	01 ♎ 11
21	13 ♈ 34	23 ♊ 34	06 ♉ 39	28 ♋ 28	05 ♒ 38	02 ♎ 18
31	13 ♈ 45	27 ♊ 50	10 ♉ 17	03 ♌ 20	04 ♒ 14	03 ♎ 25

ASPECTARIAN

h m Aspects

01 Tuesday
00 00 ☽ ⊻ ♅
00 42 ☽ □ ☉
07 19 ☽ ⊼ ♂
10 30 ☿ ∥ ♀
13 32 ☽ ⋕ ♅
15 13 ☽ △ ♀
17 55 ☽ Q ♃
19 54 ☽ ± ♂
20 03 ☽ Q ♆
22 24 ☽ △ ☿

02 Wednesday
00 13 ♂ Q ♆
09 08 ☿ ∥ ♇
11 20 ☽ ☌ ♇
13 23 ☽ △ ♄
14 46 ☽ ⚼ ♂
17 14 ☽ ∥ ♅
18 01 ☽ ⚹ ♃
20 20 ☽ ⚹ ♆
21 56 ♀ ∥ ♇

03 Thursday
01 06 ☽ ⚼ ♀
09 48 ☽ ⚼ ☿
12 42 ☽ ☌ ♅
18 47 ☽ △ ☉
20 02 ☽ ⚼ ♄
21 20 ☽ ± ♀
22 05 ☽ △ ♂

04 Friday
00 00 ☽ ∠ ♃
02 32 ☽ ∠ ♆
05 37 ☿ ⚼ ♅
06 57 ☽ ± ☿
10 42 ☽ ⊼ ♀
14 15 ☽ ± ♄
17 48 ☽ ⊥ ♃
20 27 ☽ ⊥ ♆
20 57 ☽ ⊼ ☿
23 38 ☽ ⊻ ♇

05 Saturday
02 17 ☽ ⊼ ♄
03 18 ☽ ⚼ ☉
05 34 ☽ ⊻ ♃
05 36 ☽ ⋕ ♇
08 17 ☽ ⊻ ♆
10 16 ☽ ⋕ ♀
11 28 ☽ ⊥ ♇
15 09 ☿ △ ♇
15 40 ☽ ⋕ ☿
16 03 ☽ ⋕ ♄
18 46 ☿ ∥ ♄
20 25 ☽ ∥ ♆
22 28 ☽ ± ☉
22 51 ☽ ∥ ♃
23 41 ☽ ∥ ♂

06 Sunday
00 12 ☽ ⊻ ♅
05 02 ☽ ∠ ♇
11 04 ☽ ⊼ ☉
11 05 ☽ □ ♂
11 11 ☿ ☌ ♄
11 24 ☉ △ ♂
11 48 ☽ ⊥ ♅
20 49 ☽ ⋕ ☉

07 Monday
03 27 ☿ ⋕ ♆
03 35 ☿ ☍ ♃
03 40 ☽ ☍ ♀
05 01 ☽ ∠ ♅
09 46 ☽ ⚹ ♇
12 53 ☽ ☍ ♄
14 54 ☽ ☌ ♃
15 17 ♇ St D
16 58 ☽ ☍ ☿
17 43 ♀ ⚼ ♅
17 55 ☽ ☌ ♆
20 16 ☿ ⋕ ♃
22 17 ☽ Q ♂
23 17 ☿ ☍ ♆

08 Tuesday
00 02 ☿ ⋕ ♂
08 16 ☽ Q ♇
09 10 ☽ ⚹ ♅
21 10 ☽ ⚹ ♂
23 35 ♀ ∥ ♄

09 Wednesday
00 04 ☽ ☍ ☉
07 15 ☽ Q ♅
17 05 ☽ ⊼ ♀
17 17 ☽ □ ♇
19 17 ♀ △ ♇
20 48 ☽ ⊼ ♄
21 44 ☽ ⊻ ♃

10 Thursday
00 59 ☽ ⊻ ♆
01 11 ☽ ∠ ♂
05 00 ☽ ± ♀
07 42 ☽ ± ♄
08 26 ☽ ⊥ ♃
09 32 ☽ ⊼ ☿
11 45 ☽ ⊥ ♆
15 43 ☽ □ ♅
17 44 ☽ ⊥ ♂
22 06 ☽ ± ☿
22 42 ☽ ⚼ ♀
23 56 ☽ ⚼ ♄

11 Friday
00 22 ☽ ∠ ♃
03 45 ☽ ∠ ♆
03 49 ♀ ⋕ ♆
04 40 ☽ ⊻ ♂
10 11 ☽ ⊼ ☉
13 49 ♀ ☌ ♄
15 51 ♀ ☍ ♃
16 49 ☽ ⚼ ☿
18 14 ☽ ⋕ ☉
21 33 ☽ ± ☉
22 45 ☽ △ ♇

12 Saturday
01 14 ♃ ☍ ♄
02 37 ☽ ⚹ ♃
02 38 ☽ △ ♄
02 44 ☿ △ ♅
03 46 ☽ △ ♀
04 35 ☽ ⋕ ☿
05 42 ♀ ⋕ ♃
05 58 ♀ ⋕ ♂
06 07 ☽ ⚹ ♆
11 23 ♂ ∥ ♃
14 29 ☽ ⚼ ☉
16 50 ☽ ⋕ ♀
17 34 ☽ ∥ ♃
17 35 ☽ ∥ ♂
19 09 ☽ ∥ ♆
20 31 ☽ △ ♅
21 53 ☽ ⋕ ♄
23 23 ☽ Q ♃
23 38 ☽ △ ☿

13 Sunday
00 57 ☽ ⚼ ♇
02 59 ☽ Q ♆
05 52 ♀ ☍ ♆
08 15 ☽ ⋕ ♇
10 22 ☽ ☌ ♂
16 31 ☽ ± ♇
18 25 ☽ △ ☉
22 28 ☽ ⚼ ♅

14 Monday
02 51 ☽ ⊼ ♇
06 13 ☽ □ ♃
07 07 ☽ □ ♄
09 58 ☽ □ ♆
12 41 ☽ □ ♀
13 55 ☽ ± ♅

15 Tuesday
00 10 ☽ ⊼ ♅
11 30 ☽ ∥ ♅
12 18 ☽ □ ☿
14 07 ☽ Q ♄
14 53 ☽ ⊻ ♂
22 15 ☽ Q ♀

16 Wednesday
01 16 ☽ ⊥ ♂
01 24 ☽ □ ☉
06 01 ☽ ☍ ♇
07 32 ☿ △ ♂
08 56 ☽ △ ♃
10 26 ☽ ⋕ ♅
10 38 ☽ ⚹ ♄
12 55 ☽ △ ♆
16 48 ☽ ∠ ♂
17 19 ☉ ∥ ☿
20 28 ☽ ⚹ ♀

17 Thursday
00 13 ☽ Q ☿
03 01 ☽ ☍ ♅
10 00 ☽ Q ☉
10 04 ☽ ⚼ ♃
12 10 ☽ ∠ ♄
14 10 ☽ ⚼ ♆
14 51 ♅ St D
18 35 ☽ ⚹ ♂

18 Friday
00 04 ☽ ⚹ ☿
00 05 ☽ ∠ ♀
01 07 ☽ ± ♃
03 26 ☽ ⊥ ♄
05 18 ☽ ± ♆
07 38 ☽ ⚹ ☉
08 37 ☽ ⊼ ♇
11 09 ☽ ⊼ ♃
12 50 ☽ ∥ ♇
13 37 ☽ ⊻ ♄
15 09 ☽ Q ♂
15 22 ☽ ⊼ ♆
16 39 ☽ ⊥ ♀
18 43 ☽ ± ♇
22 15 ☉ □ ♇
23 55 ☽ ∥ ♄

19 Saturday
01 43 ☽ ⋕ ♆
02 32 ☽ ⋕ ♂
02 46 ☽ ⋕ ♃
03 42 ☽ ⊻ ♀
05 32 ☽ ⊼ ♅
05 54 ☽ ∠ ☿
09 53 ☽ ⚼ ♇
10 43 ☽ ∠ ☉
13 13 ☽ ∥ ♀
15 40 ☽ ± ♅
22 04 ☽ □ ♂
23 53 ☽ ⊥ ☿

20 Sunday
01 12 ♀ △ ♅
03 03 ☽ ⊥ ☉
04 27 ☽ ∥ ☉
06 55 ☽ ⚼ ♅
07 26 ☉ ⊼ ♃
07 51 ☿ □ ♇
11 13 ☽ ∥ ☿
11 19 ☽ △ ♇
11 57 ☽ ⊻ ☿
13 31 ☽ ☍ ♃
14 00 ☽ ⊻ ☉
16 49 ☽ ☌ ♄
18 04 ☽ ☍ ♆
21 35 ☿ ⊼ ♃

21 Monday
08 35 ☽ △ ♅
09 32 ☉ ☌ ☿
11 35 ☽ ☌ ♀
20 34 ☿ ⊻ ♄

22 Tuesday
02 26 ☽ △ ♂
02 47 ☿ ⊼ ♆
12 55 ☉ ⊻ ♄
15 16 ☽ □ ♇
17 09 ☽ ⊼ ♃
21 23 ☽ ⊻ ♄
21 57 ☽ ☌ ☉
22 09 ☽ ⊼ ♆

23 Wednesday
00 37 ☉ ⊼ ♆
01 56 ☽ ☌ ☿
03 45 ☽ ± ♃
05 22 ☽ ⚼ ♂
08 12 ☽ ⊥ ♄
08 51 ☽ ± ♆
12 43 ☿ ± ♃
13 24 ☽ □ ♅
19 51 ☽ ⚼ ♃
21 52 ☽ ± ♂
21 53 ☽ ⊻ ♀
23 33 ☽ Q ♇

24 Thursday
00 37 ☽ ∠ ♄
00 59 ☿ ⚼ ♂
01 07 ☽ ⚼ ♆
03 20 ☽ ∥ ☿
08 00 ♂ ∥ ♃
09 01 ☽ ⊼ ♂
10 00 ☽ ⊥ ♀
14 00 ☽ ∥ ☉
18 27 ☿ ⊥ ♄
20 16 ☿ ± ♆
21 42 ☽ ⚹ ♇
22 26 ☽ Q ♅
23 17 ☽ △ ♃
23 50 ☽ ∥ ♀

25 Friday
04 30 ☽ ∠ ♀
04 39 ☽ ⚹ ♄
04 51 ☽ △ ♆
09 11 ☽ ⊻ ☉
18 42 ☽ ⋕ ♂
18 51 ☽ ⋕ ♃
19 34 ☽ ⋕ ♆
20 19 ☽ ⊻ ☿
20 33 ☽ ∥ ♄
21 07 ☽ ⚹ ♅
21 31 ☽ ⊥ ☉
23 04 ♄ ☍ ♆

26 Saturday
01 06 ☿ □ ♅
01 25 ☉ ± ♃
02 07 ☽ ∠ ♇
03 41 ☽ Q ♄
10 13 ☽ ⊥ ☿
10 25 ☽ ∥ ♇
12 13 ☽ ⚹ ♀
16 19 ☽ ∠ ☉
18 45 ☽ ☍ ♂
19 34 ☽ ⊥ ♇

27 Sunday
02 12 ☽ ∠ ♅
07 20 ☽ ⊻ ♇
07 28 ☽ ∠ ☿
08 35 ☽ □ ♃
13 47 ☿ ⚼ ♃
14 20 ☽ Q ♀
14 44 ☽ □ ♆
15 11 ☽ □ ♄
20 00 ☽ ⊥ ♅

28 Monday
00 24 ☽ ⚹ ☉
05 57 ☉ ⚼ ♂
07 59 ☽ ⊻ ♅
15 59 ☿ ± ♂
17 27 ☿ Q ♇
19 38 ☽ ⚹ ☿
20 16 ☽ Q ♃
20 45 ☽ ⋕ ♅

29 Tuesday
01 34 ☿ ⚼ ♆
02 34 ☽ Q ☉
02 42 ☽ Q ♆
04 14 ☉ ± ♆
06 29 ☽ □ ♀
06 46 ☿ ∠ ♄
07 18 ☽ ⊼ ♂
13 59 ♂ ± ♇
15 22 ♀ △ ♂
16 42 ☉ ⊥ ♄
19 31 ☽ ☌ ♇
19 35 ☽ ± ♂
20 22 ☽ ⚹ ♃

30 Wednesday
00 22 ☽ ∥ ♅
00 56 ☽ Q ☿
02 56 ☽ ⚹ ♆
04 05 ☽ △ ♄
14 01 ☽ ⚼ ♂
18 11 ☽ □ ☉
20 38 ☽ ☌ ♅

All ephemeris data is given at 12.00 UT and the Moon's longitude is additionally given for 24.00 UT

JULY 1971

LONGITUDES

Date	Sidereal time h m s	Sun ☉ ° ' "	Moon ☽ ° ' "	Moon ☽ 24.00 ° ' "	Mercury ☿ ° '	Venus ♀ ° '	Mars ♂ ° '	Jupiter ♃ ° '	Saturn ♄ ° '	Uranus ♅ ° '	Neptune ♆ ° '	Pluto ♇ ° '
01	06 35 33	09 ♋ 00 33	17 ♎ 07 33	23 ♎ 06 03	20 ♋ 43	23 ♊ 23	21 ♒ 22	27 ♏ 25	01 ♊ 30	09 ♎ 31	00 ♐ 45	27 ♍ 08
02	06 39 29	09 57 45	29 ♎ 06 45	05 ♏ 10 12	22 41	24 36	21 29	27 R 21	01 36	09 32	00 R 44	27 09
03	06 43 26	10 54 57	11 ♏ 16 55	17 ♏ 27 19	24 37	25 50	21 35	27 17	01 43	09 33	00 42	27 10
04	06 47 22	11 52 08	23 ♏ 41 49	00 ♐ 00 44	26 32	27 03	21 40	27 13	01 49	09 33	00 41	27 10
05	06 51 19	12 49 19	06 ♐ 24 17	12 ♐ 52 38	28 ♋ 24	28 16	21 45	27 10	01 56	09 34	00 40	27 11
06	06 55 15	13 46 30	19 ♐ 25 50	26 ♐ 03 53	00 ♌ 14	29 ♊ 29	21 49	27 06	02 02	09 35	00 39	27 12
07	06 59 12	14 43 40	02 ♑ 46 39	09 ♑ 33 55	02 02	00 ♋ 43	21 52	27 03	02 09	09 36	00 38	27 13
08	07 03 09	15 40 51	16 ♑ 25 25	23 ♑ 20 47	03 49	01 56	21 54	27 00	02 15	09 37	00 37	27 14
09	07 07 05	16 38 02	00 ♒ 19 37	07 ♒ 21 27	05 33	03 09	21 56	26 57	02 22	09 39	00 36	27 15
10	07 11 02	17 35 13	14 ♒ 25 49	21 ♒ 32 12	07 15	04 23	21 57	26 54	02 28	09 40	00 35	27 16
11	07 14 58	18 32 24	28 ♒ 40 06	05 ♓ 49 04	08 55	05 36	21 R 57	26 52	02 34	09 41	00 34	27 17
12	07 18 55	19 29 36	12 ♓ 58 37	20 ♓ 08 20	10 33	06 49	21 57	26 50	02 40	09 42	00 33	27 18
13	07 22 51	20 26 48	27 ♓ 17 49	04 ♈ 26 43	12 09	08 03	21 56	26 47	02 46	09 44	00 32	27 19
14	07 26 48	21 24 01	11 ♈ 34 41	18 ♈ 41 26	13 43	09 16	21 53	26 46	02 52	09 45	00 31	27 20
15	07 30 44	22 21 14	25 ♈ 46 41	02 ♉ 50 11	15 15	10 30	21 51	26 44	02 58	09 46	00 30	27 22
16	07 34 41	23 18 28	09 ♉ 51 42	16 ♉ 51 00	16 45	11 43	21 47	26 42	03 04	09 48	00 30	27 23
17	07 38 38	24 15 43	23 ♉ 47 53	00 ♊ 42 07	18 13	12 57	21 42	26 41	03 10	09 49	00 29	27 24
18	07 42 34	25 12 58	07 ♊ 33 31	14 ♊ 21 53	19 39	14 10	21 37	26 39	03 16	09 51	00 28	27 25
19	07 46 31	26 10 14	21 ♊ 07 02	27 ♊ 48 48	21 03	15 24	21 31	26 38	03 22	09 53	00 27	27 26
20	07 50 27	27 07 31	04 ♋ 26 04	11 ♋ 01 41	22 24	16 38	21 25	26 37	03 27	09 54	00 27	27 28
21	07 54 24	28 04 49	17 ♋ 32 35	23 ♋ 59 43	23 44	17 51	21 17	26 37	03 33	09 56	00 26	27 29
22	07 58 20	29 02 06	00 ♌ 23 05	06 ♌ 42 43	25 01	19 05	21 09	26 36	03 38	09 58	00 25	27 30
23	08 02 17	29 ♋ 59 25	12 ♌ 58 44	19 ♌ 11 15	26 16	20 18	21 00	26 36	03 44	10 00	00 25	27 32
24	08 06 13	00 ♌ 56 44	25 ♌ 20 27	01 ♍ 26 36	27 28	21 32	20 51	26 36	03 49	10 01	00 24	27 33
25	08 10 10	01 54 03	07 ♍ 29 59	13 ♍ 30 56	28 38	22 46	20 41	26 D 36	03 55	10 03	00 23	27 35
26	08 14 07	02 51 23	19 ♍ 29 50	25 ♍ 27 07	29 ♌ 46	24 00	20 30	26 36	04 00	10 05	00 23	27 36
27	08 18 03	03 48 43	01 ♎ 23 15	07 ♎ 18 44	00 ♍ 51	25 13	20 18	26 36	04 05	10 07	00 22	27 38
28	08 22 00	04 46 04	13 ♎ 14 04	19 ♎ 09 49	01 54	26 27	20 07	26 37	04 10	10 09	00 22	27 39
29	08 25 56	05 43 25	25 ♎ 06 33	01 ♏ 04 50	02 53	27 41	19 54	26 38	04 15	10 11	00 21	27 41
30	08 29 53	06 40 47	07 ♏ 05 16	13 ♏ 08 24	03 50	28 ♋ 55	19 41	26 39	04 20	10 14	00 21	27 42
31	08 33 49	07 ♌ 38 09	19 ♏ 14 49	25 ♏ 25 02	04 ♍ 44	00 ♌ 08	19 ♒ 28	26 ♏ 40	04 ♊ 25	10 ♎ 16	00 ♐ 21	27 ♍ 44

DECLINATIONS

Date	Moon True ☊ ° '	Moon Mean ☊ ° '	Moon ☽ Latitude ° '	Sun ☉ ° '	Moon ☽ ° '	Mercury ☿ ° '	Venus ♀ ° '	Mars ♂ ° '	Jupiter ♃ ° '	Saturn ♄ ° '	Uranus ♅ ° '	Neptune ♆ ° '	Pluto ♇ ° '
01	15 ♒ 09	16 ♒ 21	04 S 35	23 N 08	10 S 58	23 N 42	22 N 53	18 S 55	18 S 43	18 N 36	03 S 09	18 S 39	15 N 46
02	15 R 07	16 18	05 00	23 04	15 50	23 24	22 59	18 57	18 42	18 38	03 09	18 38	15 45
03	15 04	16 14	05 11	23 00	20 08	23 04	23 04	18 59	18 42	18 39	03 10	18 38	15 44
04	15 00	16 11	05 08	22 55	23 40	22 42	23 08	19 01	18 41	18 40	03 10	18 38	15 44
05	14 55	16 08	04 50	22 50	26 09	22 18	23 12	19 04	18 40	18 41	03 10	18 38	15 43
06	14 50	16 05	04 17	22 44	27 18	21 53	23 15	19 06	18 40	18 43	03 11	18 38	15 42
07	14 45	16 02	03 29	22 38	26 54	21 27	23 18	19 09	18 39	18 44	03 11	18 37	15 41
08	14 42	15 59	02 29	22 31	24 53	20 59	23 20	19 13	18 39	18 45	03 12	18 37	15 41
09	14 40	15 55	01 18	22 25	21 20	20 30	23 21	19 16	18 39	18 46	03 12	18 37	15 40
10	14 39	15 52	00 S 01	22 17	16 32	20 00	23 21	19 20	18 38	18 47	03 13	18 37	15 39
11	14 D 40	15 49	01 N 16	22 10	10 45	19 29	23 21	19 24	18 38	18 48	03 13	18 37	15 38
12	14 41	15 46	02 29	22 02	04 S 23	18 57	23 20	19 28	18 37	18 49	03 14	18 37	15 37
13	14 42	15 43	03 33	21 53	02 N 11	18 25	23 18	19 32	18 37	18 51	03 14	18 36	15 36
14	14 43	15 39	04 24	21 45	08 37	17 52	23 16	19 36	18 37	18 52	03 15	18 36	15 36
15	14 R 44	15 36	04 58	21 35	14 35	17 19	23 13	19 41	18 37	18 53	03 15	18 36	15 35
16	14 43	15 33	05 14	21 26	19 44	16 45	23 09	19 46	18 37	18 54	03 16	18 36	15 34
17	14 41	15 30	05 11	21 16	23 46	16 11	23 05	19 51	18 37	18 55	03 17	18 36	15 33
18	14 39	15 27	04 50	21 06	26 21	15 36	22 59	19 57	18 37	18 56	03 17	18 36	15 32
19	14 36	15 24	04 13	20 55	27 22	15 01	22 54	20 02	18 37	18 57	03 18	18 36	15 31
20	14 34	15 20	03 23	20 44	26 45	14 26	22 47	20 08	18 37	18 58	03 19	18 36	15 31
21	14 32	15 17	02 23	20 33	24 40	13 51	22 40	20 13	18 37	18 59	03 19	18 36	15 30
22	14 31	15 14	01 17	20 21	21 20	13 17	22 32	20 19	18 37	19 00	03 20	18 35	15 29
23	14 30	15 11	00 N 08	20 09	17 03	12 42	22 24	20 25	18 37	19 00	03 21	18 35	15 28
24	14 D 30	15 08	01 S 00	19 57	12 08	12 08	22 15	20 32	18 37	19 01	03 22	18 35	15 27
25	14 31	15 05	02 04	19 44	06 50	11 34	22 05	20 38	18 37	19 02	03 22	18 35	15 26
26	14 32	15 01	03 02	19 31	01 N 22	11 00	21 55	20 44	18 38	19 03	03 23	18 35	15 25
27	14 33	14 58	03 51	19 18	04 S 05	10 27	21 44	20 51	18 38	19 04	03 24	18 35	15 24
28	14 34	14 55	04 31	19 05	09 23	09 54	21 32	20 57	18 38	19 05	03 25	18 35	15 23
29	14 35	14 52	04 59	18 51	14 21	09 22	21 20	21 03	18 38	19 06	03 26	18 35	15 22
30	14 R 35	14 49	05 14	18 36	18 49	08 51	21 07	21 10	18 39	19 06	03 27	18 35	15 21
31	14 ♒ 35	14 ♒ 45	05 S 16	18 N 22	22 S 36	08 N 20	20 N 53	21 S 16	18 S 40	19 N 07	03 S 28	18 S 35	15 N 20

ZODIAC SIGN ENTRIES

Date	h	m	Planets
02	13	46	☽ ♏
04	23	59	☽ ♐
06	08	53	☿ ♌
06	22	02	♀ ♋
07	07	03	☽ ♑
09	11	26	☽ ♒
11	14	14	☽ ♓
13	16	32	☽ ♈
15	19	10	☽ ♉
17	22	47	☽ ♊
20	03	56	☽ ♋
22	11	16	☽ ♌
23	12	15	☉ ♌
24	21	09	☽ ♍
26	17	03	☿ ♍
27	09	12	☽ ♎
29	21	50	☽ ♏
31	09	15	♀ ♌

LATITUDES

Date	Mercury ☿ ° '	Venus ♀ ° '	Mars ♂ ° '	Jupiter ♃ ° '	Saturn ♄ ° '	Uranus ♅ ° '	Neptune ♆ ° '	Pluto ♇ ° '
01	01 N 53	00 S 24	04 S 48	00 N 54	01 S 54	00 N 41	01 N 42	15 N 59
04	01 53	00 16	05 01	00 53	01 54	00 41	01 42	15 57
07	01 47	00 09	05 13	00 52	01 54	00 40	01 42	15 56
10	01 35	00 S 01	05 26	00 51	01 54	00 40	01 42	15 55
13	01 19	00 N 06	05 38	00 51	01 55	00 40	01 42	15 53
16	00 58	00 13	05 50	00 50	01 55	00 40	01 42	15 52
19	00 34	00 20	06 02	00 49	01 55	00 40	01 42	15 51
22	00 N 06	00 27	06 12	00 49	01 56	00 40	01 41	15 50
25	00 S 25	00 34	06 22	00 48	01 56	00 40	01 41	15 49
28	00 58	00 41	06 31	00 47	01 56	00 40	01 41	15 48
31	01 S 33	00 N 47	06 S 38	00 N 46	01 S 57	00 N 40	01 N 41	15 N 46

DATA

Julian Date	2441134
Delta T	+42 seconds
Ayanamsa	23° 27' 44"
Synetic vernal point	05° ♓ 39' 16"
True obliquity of ecliptic	23° 26' 41"

MOON'S PHASES, APSIDES AND POSITIONS ☽

Date	h	m	Phase	Longitude ° '	Eclipse Indicator
08	10	37	○	15 ♑ 38	
15	05	47	☾	22 ♈ 06	
22	09	15	●	28 ♋ 56	Partial
30	11	07	☽	06 ♏ 39	

Day	h	m	
12	15	21	Perigee
28	03	20	Apogee

Day	h	m		
06	18	10	Max dec	27° S 21'
13	04	02	ON	
19	14	56	Max dec	27° N 22'
26	18	00	OS	

LONGITUDES

Date	Chiron ⚷ ° '	Ceres ⚳ ° '	Pallas ⚴ ° '	Juno ⚵ ° '	Vesta ⚶ ° '	Black Moon Lilith ⚸ ° '
01	13 ♈ 45	27 ♊ 50	10 ♉ 17	03 ♌ 20	04 ♒ 14	03 ♎ 25
11	13 ♈ 51	02 ♋ 04	13 ♉ 03	08 ♌ 10	02 ♒ 13	04 ♎ 32
21	13 ♈ 51	06 ♋ 18	17 ♉ 17	12 ♌ 56	29 ♑ 50	05 ♎ 39
31	13 ♈ 47	10 ♋ 30	20 ♉ 35	17 ♌ 39	27 ♑ 26	06 ♎ 46

All ephemeris data is given at 12.00 UT and the Moon's longitude is additionally given for 24.00 UT

ASPECTARIAN

h m	Aspects
01 Thursday	
02 33	☽ ∠ ♃
09 14	☽ ∠ ♆
10 43	☽ bQ ♄
20 16	☿ ⊼ ♂
20 34	☽ ⊥ ♃
20 36	☽ △ ♂
20 39	☽ □ ☿
02 Friday	
00 58	☉ □ ♅
02 00	☽ △ ♀
03 16	☽ ⊥ ♆
04 56	☽ ± ♄
08 05	☽ ⊻ ♇
08 30	☽ ⊻ ♃
11 37	☽ # ♇
15 12	☽ ⊻ ♆
17 00	☽ ⊼ ♄
20 01	☽ ⊥ ♇
22 38	♄ # ♆
03 Saturday	
01 09	☉ ∥ ♀
03 13	☽ ∥ ♆
03 14	☽ # ♄
03 33	☽ ∥ ♃
05 08	☽ ∥ ♂
08 36	☽ ⊻ ♅
11 01	☽ bQ ♀
11 13	☽ △ ☉
11 49	☿ ∥ ♀
13 43	☽ ∠ ♇
17 56	☉ ∥ ☿
20 18	☽ ⊥ ♅
04 Sunday	
05 28	☽ # ☿
06 22	☽ ± ♀
06 26	☽ # ☉
07 52	☽ # ♀
08 05	☽ □ ♂
13 39	☽ ∠ ♅
14 30	♀ □ ♇
15 12	♀ ⊼ ♃
18 21	☽ △ ☿
18 32	☽ bQ ☉
18 38	☽ ⚹ ♇
18 41	☽ ☌ ♃
19 04	☽ ⊼ ♀
20 15	☉ bQ ♃
20 17	☿ ⚹ ♇
20 31	☿ △ ♃
23 18	♃ # ♄
05 Monday	
01 16	☽ ☌ ♆
01 16	☿ Q ♅
02 31	♃ ⚹ ♇
03 33	☽ ☍ ♄
07 03	☿ ⊻ ♀
12 50	☽ ± ☉
17 11	☽ Q ♇
17 54	☽ ⚹ ♅
18 15	☽ Q ♂
06 Tuesday	
00 50	☽ ⊼ ☉
03 06	☽ bQ ☿
15 56	☽ Q ♅
16 21	☽ ⚹ ♂
17 27	☿ △ ♆
22 05	☽ ± ☿
07 Wednesday	
01 49	☽ ⊻ ♃
02 04	☽ □ ♇
07 57	☽ ☍ ♀
08 11	☽ ⊻ ♆
10 29	☽ ⊼ ☿
10 32	♀ ⊼ ♆
10 52	☽ ⊼ ♄
12 29	☽ ⊥ ♃
13 32	☿ ⚹ ♄
18 50	☽ ⊥ ♆
19 16	☽ ∠ ♂
21 35	☽ ± ♄
08 Thursday	
00 05	☽ □ ♅
00 33	☉ Q ♇
04 18	☽ ∠ ♃
10 26	☉ bQ ♆
10 36	☽ ∠ ♆
10 37	☽ ☍ ☉
11 06	☽ ⊥ ♂
13 27	☽ bQ ♄
17 56	☉ ± ♂
18 54	♀ ⊻ ♄
21 32	☽ ⊻ ♂
23 45	☽ # ♀
09 Friday	
05 34	☽ # ☉
06 14	☽ ⚹ ♃
06 43	☽ △ ♇
08 09	♀ ± ♃
12 28	☽ ⚹ ♆
15 30	☽ △ ♄
17 17	☽ # ☿
17 18	☽ ⊼ ♀
22 09	☽ ☍ ☿
22 56	☽ ∥ ♂
10 Saturday	
01 29	☽ # ♄
02 10	☽ ∥ ♃
02 16	☽ ∥ ♆
02 40	☽ Q ♃
03 54	☽ △ ♅
04 29	☽ ± ♀
08 20	☽ bQ ♇
08 32	☉ ∠ ♄
08 53	☽ Q ♆
15 53	☽ # ♇
17 43	☽ ⊼ ☉
21 09	☽ bQ ♀
23 34	☽ ± ♇
11 Sunday	
00 42	☽ ☌ ♂
04 34	☽ ± ☉
05 17	☽ bQ ♅
06 30	♂ St R
08 59	☽ □ ♃
09 40	☽ ⊼ ♇
15 11	☽ □ ♆
15 47	☿ # ♂
18 36	☽ □ ♄
20 26	☽ ± ♅
20 46	☽ bQ ☉
23 20	☿ ⚹ ♅
12 Monday	
00 43	☽ △ ♀
06 31	☽ ⊼ ♅
06 44	♀ ± ♆
07 25	☽ ⊼ ☿
14 23	♀ bQ ♂
16 15	☽ ∥ ♅
17 44	☿ ∥ ♄
18 45	☽ ± ☿
23 42	☽ △ ☉
13 Tuesday	
00 59	☽ Q ♄
02 57	☿ # ♃
03 01	☽ ⊻ ♂
03 33	☿ # ♆
11 09	☽ △ ♃
11 44	☽ bQ ☿
12 02	☽ ☍ ♇
13 03	☽ ⊥ ♂
14 33	☿ ∠ ♇
15 52	☽ # ♅
17 26	☽ △ ♆
21 15	☽ ⚹ ♄
14 Wednesday	
03 28	♀ ⊥ ♄
04 08	☽ ∠ ♂
07 45	☽ □ ♀
08 55	☽ ☍ ♅
12 18	☽ bQ ♃
16 03	☽ △ ☿
18 39	☽ bQ ♆
21 33	♀ □ ♅
22 42	☽ ∠ ♄
23 46	☉ ⊼ ♂
15 Thursday	
03 28	☽ ± ♃
05 21	☽ ⚹ ♂
05 47	☽ □ ☉
07 12	☿ Q ♄
09 51	☽ ± ♆
13 37	☽ ⊼ ♃
14 02	☽ ⊥ ♄
14 41	☽ ⊼ ♇
16 20	☽ ∥ ♇
17 03	☽ Q ♀
20 02	☽ ⊼ ♆
22 58	☽ ∥ ☿
16 Friday	
00 19	☽ ⊻ ♄
00 55	☽ ± ♇
01 39	☽ Q ♂
06 20	☽ # ♆
06 23	☽ # ♃
07 46	☽ ∥ ♄
11 37	♀ bQ ♃
11 53	☽ ⊼ ♅
12 14	☽ # ♂
14 40	☽ Q ☉
15 30	☽ ⚹ ♀
16 19	☽ bQ ♇
20 57	☽ ∥ ☉
22 12	☽ ± ♅
17 Saturday	
01 14	☽ □ ☿
07 32	☽ ∥ ♀
08 24	☽ □ ♂
12 52	☽ ⚹ ☉
13 47	☽ bQ ♅
16 59	☽ ☍ ♃
18 16	☽ △ ♇
19 54	☽ ∠ ♀
23 36	☽ ☍ ♆
18 Sunday	
04 25	☽ ☌ ♄
12 11	☽ Q ☿
13 11	☽ ⊥ ♀
14 28	☿ ∥ ♇
16 02	☽ △ ♅
17 02	☽ ∠ ☉
19 Monday	
00 49	☽ ⊻ ♀
10 11	☽ ⊥ ☉
11 52	☽ ⚹ ☿
12 43	☽ △ ♂
12 50	♀ Q ♇
13 05	♀ bQ ♆
14 11	♀ ± ♂
19 01	☿ ⊥ ♇
19 42	☿ ☍ ♂
21 45	☽ ⊻ ☉
21 53	☽ ⊼ ♃
23 21	☽ □ ♇
23 36	☉ △ ♃
20 Tuesday	
02 34	♂ ± ♇
04 45	☽ ⊼ ♆
08 41	☽ ± ♃
10 11	☽ ⊻ ♄
15 32	☽ bQ ♂
15 37	☽ ± ♆
17 59	☽ ∠ ☿
20 42	☉ ⚹ ♇
21 11	☽ ⊥ ♄
21 58	☽ □ ♅
21 Wednesday	
01 05	☽ bQ ♃
07 52	☽ ± ♂
08 06	☽ bQ ♆
08 11	☽ Q ♇
08 12	☉ Q ♅
12 23	☽ ⊥ ☿
12 38	☽ ☌ ♀
13 53	☽ ∠ ♄
18 53	☽ ⊼ ♂
22 Thursday	
00 48	☽ ⊻ ☿
02 42	♀ ∠ ♄
03 50	☽ ∥ ♀
04 53	☽ △ ♃
06 34	☽ ⚹ ♇
07 26	☽ Q ♅
09 15	☽ ☌ ☉
10 59	☿ ∠ ♅
12 04	☽ △ ♆
14 43	☉ # ♂
17 57	☽ # ♂
18 11	☽ ∥ ☉
18 12	☽ ⚹ ♄
23 Friday	
01 37	☽ ∥ ♄
03 45	☽ # ♃
03 52	☽ # ♆
06 15	☽ ⚹ ♅
11 08	☽ ∠ ♇
17 21	☽ Q ♄
18 34	☿ □ ♃
20 05	☽ ∥ ♇
22 26	☉ △ ♆
24 Saturday	
00 04	♀ ⊼ ♂
03 20	☽ ☍ ♂
03 45	☽ ⊻ ♀
04 35	☽ ⊥ ♇
11 23	☽ ∠ ♅
12 04	☽ ∥ ☿
13 42	☿ ⊻ ♇
14 28	☽ □ ♃
16 21	☽ ⊻ ♇
16 38	☽ ☌ ☿
16 47	☽ ⊥ ♀
19 13	♃ St D
21 56	☽ □ ♆
23 57	☽ ⊻ ☉
25 Sunday	
04 49	☽ □ ♄
05 09	☽ ⊥ ♅
12 35	☽ ∠ ♀
12 52	☽ ⊥ ☉
17 06	☽ ⊻ ♅
26 Monday	
02 10	☽ Q ♃
03 12	☽ # ♅
08 25	☽ ∠ ☉
09 45	☽ Q ♆
13 59	☽ ⊼ ♂
22 06	☽ ⚹ ♀
27 Tuesday	
01 21	☿ □ ♆
01 53	☽ ± ♂
02 20	☽ ⚹ ♃
04 23	☽ ☌ ♇
08 56	☽ ∥ ♅
09 57	☽ ⚹ ♆
10 49	☽ ⊻ ☿
17 20	☽ ⚹ ☉
17 30	☽ △ ♄
19 29	☉ ⚹ ♄
19 49	☽ bQ ♂
28 Wednesday	
00 09	☽ ⊥ ☿
01 11	☽ Q ♀
05 45	☽ ☌ ♅
08 43	☽ ∠ ♃
11 45	☉ ∥ ♄
14 11	☽ # ☿
15 17	♀ △ ♃
16 19	☽ ∠ ♆
19 47	☽ Q ☉
20 06	☽ ∠ ☿
29 Thursday	
00 06	☽ bQ ♄
01 41	☽ △ ♂
02 57	☽ ⊥ ♃
10 29	☽ ⊥ ♆
11 57	♀ ⚹ ♇
15 04	☽ ⊻ ♃
17 11	☽ ⊻ ♇
17 12	☽ # ♇
17 46	☽ □ ♀
18 22	☽ ± ♄
22 14	♀ Q ♅
22 33	☽ ⊻ ♆
30 Friday	
04 58	☽ ⚹ ☿
05 14	☽ ⊥ ♇
06 28	☽ ⊼ ♄
07 34	☉ # ♃
08 05	♀ # ♂
10 37	☽ ∥ ♆
10 49	☽ # ☉
11 01	☽ ∥ ♃
11 07	☽ □ ☉
13 38	☽ # ♄
14 23	☉ # ♆
18 15	☽ ⊻ ♅
22 52	☿ ⊥ ♅
23 10	☽ ∠ ♇
31 Saturday	
01 07	☽ # ♀
02 40	☽ ∥ ♂
02 42	☿ □ ♄
06 08	☽ ⊥ ♅
06 41	☽ Q ☿
12 25	☽ □ ♂
15 55	♀ △ ♆
23 44	☽ ∠ ♅

AUGUST 1971

Date	Sidereal time h m s	Sun ☉ ° ' "	Moon ☽ ° ' "	Moon ☽ 24.00 ° ' "	Mercury ☿ ° '	Venus ♀ ° '	Mars ♂ ° '	Jupiter ♃ ° '	Saturn ♄ ° '	Uranus ♅ ° '	Neptune ♆ ° '	Pluto ♇ ° '
01	08 37 46	08 ♌ 35 32	01 ♐ 39 34	07 ♐ 58 53	05 ♍ 34	01 ♌ 22	19 ♒ 14	26 ♏ 41	04 ♊ 30	10 ♎ 18	00 ♐ 20	27 ♍ 45
02	08 41 42	09 32 55	14 ♐ 23 22	20 ♐ 53 21	06 21	02 36	19 R 00	26 43	04 34	10 20	00 R 20	27 47
03	08 45 39	10 30 20	27 ♐ 29 04	04 ♑ 10 39	07 05	03 50	18 45	26 44	04 39	10 22	00 20	27 49
04	08 49 36	11 27 45	10 ♑ 58 08	17 ♑ 51 26	07 45	05 04	18 30	26 46	04 44	10 25	00 19	27 50
05	08 53 32	12 25 11	24 ♑ 50 20	01 ♒ 54 28	08 22	06 18	18 15	26 48	04 48	10 27	00 19	27 52
06	08 57 29	13 22 37	09 ♒ 03 23	16 ♒ 16 30	08 54	07 32	18 00	26 51	04 53	10 30	00 19	27 54
07	09 01 25	14 20 05	23 ♒ 33 07	00 ♓ 52 29	09 22	08 46	17 44	26 53	04 57	10 32	00 19	27 56
08	09 05 22	15 17 33	08 ♓ 13 44	15 ♓ 36 03	09 46	10 00	17 29	26 55	05 01	10 35	00 19	27 57
09	09 09 18	16 15 03	22 ♓ 58 34	00 ♈ 20 26	10 05	11 14	17 13	26 58	05 05	10 37	00 18	27 59
10	09 13 15	17 12 34	07 ♈ 40 52	14 ♈ 59 10	10 19	12 28	16 57	27 01	05 09	10 40	00 18	28 01
11	09 17 11	18 10 06	22 ♈ 14 42	29 ♈ 26 58	10 29	13 42	16 41	27 04	05 13	10 42	00 18	28 03
12	09 21 08	19 07 40	06 ♉ 35 31	13 ♉ 40 03	10 33	14 56	16 25	27 07	05 17	10 45	00 18	28 04
13	09 25 05	20 05 16	20 ♉ 40 22	27 ♉ 36 19	10 R 32	16 10	16 09	27 11	05 21	10 48	00 D 18	28 06
14	09 29 01	21 02 52	04 ♊ 27 53	11 ♊ 15 03	10 25	17 24	15 54	27 14	05 25	10 50	00 18	28 08
15	09 32 58	22 00 31	17 ♊ 57 55	24 ♊ 36 35	10 13	18 38	15 38	27 18	05 28	10 53	00 18	28 10
16	09 36 54	22 58 11	01 ♋ 11 11	07 ♋ 41 55	09 56	19 52	15 23	27 22	05 32	10 56	00 18	28 12
17	09 40 51	23 55 52	14 ♋ 08 55	20 ♋ 32 23	09 33	21 07	15 08	27 26	05 35	10 59	00 18	28 14
18	09 44 47	24 53 35	26 ♋ 52 30	03 ♌ 09 26	09 04	22 21	14 53	27 30	05 38	11 02	00 19	28 16
19	09 48 44	25 51 20	09 ♌ 23 23	15 ♌ 34 32	08 30	23 35	14 38	27 35	05 42	11 05	00 19	28 18
20	09 52 40	26 49 06	21 ♌ 43 02	27 ♌ 49 06	07 51	24 49	14 24	27 40	05 45	11 08	00 19	28 20
21	09 56 37	27 46 53	03 ♍ 52 55	09 ♍ 54 42	07 08	26 03	14 11	27 44	05 48	11 11	00 20	28 22
22	10 00 33	28 44 41	15 ♍ 54 40	21 ♍ 53 03	06 21	27 18	13 57	27 49	05 51	11 14	00 20	28 24
23	10 04 30	29 ♌ 42 31	27 ♍ 50 08	03 ♎ 46 12	05 31	28 32	13 45	27 54	05 54	11 17	00 20	28 26
24	10 08 27	00 ♍ 40 22	09 ♎ 41 35	15 ♎ 36 37	04 38	29 ♌ 46	13 33	27 59	05 56	11 20	00 21	28 28
25	10 12 23	01 38 14	21 ♎ 31 42	27 ♎ 27 15	03 44	01 ♍ 01	13 21	28 05	05 59	11 23	00 21	28 30
26	10 16 20	02 36 08	03 ♏ 23 43	09 ♏ 21 35	02 50	02 15	13 10	28 10	06 02	11 26	00 21	28 32
27	10 20 16	03 34 03	15 ♏ 21 20	21 ♏ 23 30	01 57	03 29	13 00	28 16	06 04	11 29	00 22	28 34
28	10 24 13	04 32 00	27 ♏ 28 39	03 ♐ 37 18	01 05	04 44	12 50	28 22	06 06	11 32	00 22	28 36
29	10 28 09	05 29 58	09 ♐ 50 02	16 ♐ 07 22	00 ♍ 17	05 58	12 42	28 28	06 09	11 35	00 23	28 38
30	10 32 06	06 27 57	22 ♐ 29 49	28 ♐ 57 52	29 ♌ 32	07 12	12 33	28 34	06 11	11 39	00 23	28 40
31	10 36 02	07 ♍ 25 57	05 ♑ 31 57	12 ♑ 12 23	28 ♌ 53	08 ♍ 27	12 ♒ 26	28 ♏ 40	06 ♊ 13	11 ♎ 42	00 ♐ 24	28 ♍ 43

Date	Moon True ☊ ° '	Moon Mean ☊ ° '	Moon ☽ Latitude ° '
01	14 ♒ 34	14 ♒ 42	05 S 03
02	14 R 34	14 39	04 35
03	14 33	14 36	03 52
04	14 33	14 33	02 56
05	14 33	14 30	01 47
06	14 33	14 26	00 S 30
07	14 D 33	14 23	00 N 50
08	14 R 33	14 20	02 08
09	14 32	14 17	03 17
10	14 32	14 14	04 14
11	14 32	14 11	04 53
12	14 32	14 07	05 14
13	14 D 32	14 04	05 15
14	14 32	14 01	04 58
15	14 32	13 58	04 25
16	14 33	13 55	03 38
17	14 34	13 51	02 41
18	14 35	13 48	01 37
19	14 35	13 45	00 N 29
20	14 R 35	13 42	00 S 39
21	14 34	13 39	01 45
22	14 33	13 36	02 45
23	14 31	13 32	03 37
24	14 28	13 29	04 19
25	14 26	13 26	04 50
26	14 23	13 23	05 09
27	14 22	13 20	05 15
28	14 21	13 17	05 07
29	14 D 21	13 13	04 45
30	14 21	13 10	04 08
31	14 ♒ 23	13 ♒ 07	03 S 18

DECLINATIONS

Date	Sun ☉ ° '	Moon ☽ ° '	Mercury ☿ ° '	Venus ♀ ° '	Mars ♂ ° '	Jupiter ♃ ° '	Saturn ♄ ° '	Uranus ♅ ° '	Neptune ♆ ° '	Pluto ♇ ° '
01	18 N 07	25 S 26	07 N 51	20 N 39	21 S 23	18 S 40	19 N 08	03 S 28	18 S 35	15 N 20
02	17 52	27 05	07 22	20 24	21 29	18 41	19 09	03 29	18 35	15 19
03	17 37	27 18	06 55	20 09	21 35	18 41	19 09	03 30	18 35	15 18
04	17 21	25 55	06 29	19 53	21 42	18 42	19 10	03 31	18 35	15 17
05	17 05	22 55	06 04	19 37	21 48	18 43	19 11	03 32	18 35	15 16
06	16 49	18 29	05 41	19 19	21 54	18 43	19 11	03 33	18 35	15 16
07	16 32	12 53	05 19	19 02	21 59	18 44	19 12	03 34	18 35	15 14
08	16 15	06 S 31	05 00	18 44	22 05	18 45	19 13	03 36	18 35	15 13
09	15 58	00 N 14	04 42	18 25	22 11	18 46	19 13	03 37	18 35	15 12
10	15 41	06 56	04 26	18 06	22 16	18 47	19 14	03 37	18 35	15 11
11	15 23	13 12	04 13	17 46	22 21	18 48	19 14	03 38	18 35	15 10
12	15 05	18 39	04 02	17 26	22 26	18 49	19 15	03 39	18 35	15 09
13	14 47	22 59	03 53	17 05	22 31	18 50	19 15	03 40	18 35	15 08
14	14 29	25 55	03 47	16 44	22 35	18 51	19 16	03 42	18 35	15 07
15	14 11	27 18	03 44	16 23	22 39	18 52	19 16	03 43	18 35	15 06
16	13 52	27 04	03 44	16 00	22 43	18 53	19 17	03 44	18 35	15 05
17	13 33	25 21	03 47	15 38	22 47	18 54	19 17	03 45	18 36	15 04
18	13 14	22 22	03 54	15 15	22 50	18 55	19 18	03 46	18 36	15 03
19	12 54	18 22	04 03	14 52	22 53	18 56	19 18	03 47	18 36	15 02
20	12 35	13 39	04 16	14 28	22 56	18 58	19 18	03 48	18 36	15 01
21	12 15	08 28	04 31	14 04	22 58	18 59	19 19	03 50	18 36	15 00
22	11 55	03 N 02	04 50	13 39	23 01	19 00	19 19	03 51	18 36	14 59
23	11 35	02 S 27	05 11	13 14	23 03	19 02	19 20	03 52	18 36	14 58
24	11 14	07 49	05 35	12 49	23 04	19 03	19 20	03 53	18 36	14 57
25	10 54	12 53	06 02	12 23	23 05	19 04	19 20	03 54	18 36	14 56
26	10 33	17 29	06 29	11 57	23 06	19 06	19 20	03 56	18 37	14 55
27	10 12	21 28	06 59	11 31	23 06	19 07	19 21	03 57	18 37	14 54
28	09 51	24 35	07 29	11 04	23 06	19 09	19 21	03 58	18 37	14 53
29	09 30	26 37	07 59	10 37	23 06	19 10	19 21	04 00	18 37	14 52
30	09 08	27 22	08 29	10 10	23 06	19 12	19 21	04 01	18 37	14 51
31	08 N 47	26 S 37	08 N 58	09 N 42	23 S 05	19 S 13	19 N 22	04 S 02	18 S 37	14 N 50

ZODIAC SIGN ENTRIES

Date	h	m	Planets
01	08	49	☽ ♐
03	16	32	☽ ♑
05	20	47	☽ ♒
07	22	34	☽ ♓
09	23	27	☽ ♈
12	00	55	☽ ♉
14	04	10	☽ ♊
16	09	50	☽ ♋
18	17	57	☽ ♌
21	04	19	☽ ♍
23	16	22	☽ ♎
23	19	15	☉ ♍
24	16	25	♀ ♍
26	05	09	☽ ♏
28	16	56	☽ ♐
29	20	41	☿ ♌
31	01	54	☽ ♑

LATITUDES

Date	Mercury ☿ ° '	Venus ♀ ° '	Mars ♂ ° '	Jupiter ♃ ° '	Saturn ♄ ° '	Uranus ♅ ° '	Neptune ♆ ° '	Pluto ♇ ° '
01	01 S 45	00 N 49	06 S 40	00 N 46	01 S 57	00 N 39	01 N 41	15 N 46
04	02 21	00 54	06 45	00 46	01 57	00 39	01 41	15 45
07	02 57	01 00	06 49	00 45	01 58	00 39	01 40	15 44
10	03 31	01 04	06 50	00 44	01 58	00 39	01 40	15 43
13	04 02	01 09	06 50	00 43	01 58	00 39	01 40	15 43
16	04 26	01 13	06 49	00 43	01 59	00 39	01 40	15 42
19	04 40	01 16	06 45	00 42	01 59	00 39	01 40	15 41
22	04 41	01 19	06 40	00 41	02 00	00 39	01 40	15 40
25	04 25	01 21	06 33	00 41	02 00	00 39	01 39	15 40
28	03 52	01 23	06 25	00 40	02 01	00 39	01 39	15 39
31	03 S 05	01 N 24	06 S 16	00 N 40	02 S 01	00 N 39	01 N 39	15 N 39

DATA

Julian Date	2441165
Delta T	+42 seconds
Ayanamsa	23° 27' 49"
Synetic vernal point	05° ♓ 39' 10"
True obliquity of ecliptic	23° 26' 41"

MOON'S PHASES, APSIDES AND POSITIONS ☽

Date	h	m	Phase	Longitude ° '	Eclipse Indicator
06	19	42	○	13 ♒ 41	total
13	10	55	☾	20 ♉ 03	
20	22	53	●	27 ♌ 15	Partial
29	02	56	☽	05 ♐ 08	

Day	h	m	
09	01	10	Perigee
24	19	48	Apogee

Day	h	m		
03	03	18	Max dec	27° S 24'
09	11	10	0N	
15	20	34	Max dec	27° N 24'
23	01	14	0S	
30	12	23	Max dec	27° S 22'

LONGITUDES

Date	Chiron ⚷ ° '	Ceres ⚳ ° '	Pallas ⚴ ° '	Juno ⚵ ° '	Vesta ⚶ ° '	Black Moon Lilith ⚸ ° '
01	13 ♈ 46	10 ♋ 55	20 ♉ 54	18 ♌ 07	27 ♑ 12	06 ♎ 53
11	13 ♈ 36	15 ♋ 04	24 ♉ 01	22 ♌ 46	25 ♑ 11	07 ♎ 59
21	13 ♈ 21	19 ♋ 09	26 ♉ 53	27 ♌ 21	23 ♑ 45	09 ♎ 06
31	13 ♈ 02	23 ♋ 10	29 ♉ 27	01 ♍ 52	23 ♑ 05	10 ♎ 13

All ephemeris data is given at 12.00 UT and the Moon's longitude is additionally given for 24.00 UT

ASPECTARIAN

h m	Aspects
01 Sunday	
02 26	☽ ☌ ♃
04 30	☽ ⚹ ♇
09 28	☽ ☌ ♆
11 23	☽ △ ♀
17 26	☽ ☍ ♄
19 57	☽ □ ☿
22 24	☽ Q ♂
02 Monday	
02 13	☽ △ ☉
03 22	☽ Q ♇
04 24	☽ ⚹ ♅
18 35	☽ ⚼ ♀
20 22	☽ ⚹ ♂
03 Tuesday	
02 41	☽ Q ♅
08 08	☽ ⚼ ☉
08 35	☉ ⚹ ♅
10 39	☽ ⊻ ♃
12 36	☽ □ ♇
12 42	☽ ± ♀
17 07	☽ ⊻ ♆
21 28	☽ ⊥ ♃
23 03	☽ ∠ ♂
04 Wednesday	
00 32	☽ ⊼ ♀
00 55	☽ ⊼ ♄
01 33	☽ ± ☉
03 49	☽ ⊥ ♆
04 59	♀ ⚹ ♄
06 03	☽ △ ☿
11 01	☽ □ ♅
11 34	☽ ± ♄
12 56	☽ ⊼ ☉
13 25	☽ ∠ ♃
14 39	☽ ⊥ ♂
19 36	☽ ∠ ♆
05 Thursday	
00 53	☽ ⊻ ♂
03 19	☽ ⚼ ♄
09 22	☽ ⚼ ☿
15 22	☽ ⚹ ♃
17 10	☽ △ ♇
18 41	☽ ∥ ♂
21 19	☽ ⚹ ♆
23 35	☉ ∠ ♇
06 Friday	
01 17	☽ ± ☿
04 58	☽ △ ♄
07 42	☽ # ♀
08 38	☽ # ♄
09 12	☽ ☍ ♀
10 52	☽ ∥ ♃
11 32	☽ ∥ ♆
11 38	☽ Q ♃
11 44	☽ ⊼ ☿
14 24	☽ △ ♅
17 26	☽ Q ♆
18 25	☽ ⚼ ♇
19 42	☽ ☍ ☉
20 04	☽ # ☉
22 53	♀ ∥ ♄
07 Saturday	
02 26	☽ # ♇
02 35	☽ ☌ ♂
09 19	☽ ± ♇
15 16	☽ ⚼ ♅
17 29	☽ □ ♃
19 11	☽ ⊼ ♇
23 05	☽ □ ♆
08 Sunday	
05 35	☿ ⊻ ♀
06 02	☽ ± ♅
06 44	☽ □ ♄
10 23	♀ # ♃
14 34	☽ ☍ ☿
15 08	☽ ⊼ ♀
15 50	☽ ⊼ ♅
17 44	☽ # ☿
22 27	☽ ∥ ♅
23 20	♀ # ♆
23 43	♀ ⚹ ♅
09 Monday	
00 18	☽ ⊼ ☉
01 48	☽ ± ♀
02 47	☽ ⊻ ♂
10 44	☽ ± ☉
12 11	☽ Q ♄
12 23	☽ ⊥ ♂
17 47	☽ ⚼ ♀
18 32	☽ △ ♃
20 10	☽ ☍ ♇
23 57	☽ △ ♆
23 59	☽ # ♅
10 Tuesday	
02 26	☽ ⚼ ☉
02 47	☽ ∠ ♂
03 16	☽ ∥ ☿
06 53	☉ ☍ ♂
07 51	☽ ⚹ ♄
10 30	☉ Q ♄
16 23	☽ ⊼ ☿
16 54	☽ ☍ ♅
19 08	☽ ⚼ ♃
20 34	☽ △ ♀
23 00	♀ ∠ ♇
11 Wednesday	
00 31	☽ ⚼ ♆
02 23	☽ ± ☿
02 58	☽ ⚹ ♂
04 46	☽ △ ☉
08 38	☽ ∠ ♄
10 03	☽ ± ♃
15 25	☽ ± ♆
17 25	☽ ⚼ ☿
20 03	☽ ⊼ ♃
20 10	☽ ∥ ♇
20 41	☽ ∥ ☉
21 40	☽ ⊼ ♇
22 31	☽ Q ♂
23 40	☽ ⊥ ♄
12 Thursday	
01 26	☽ ⊼ ♆
06 35	☽ ∥ ♀
07 02	☉ ∥ ♇
07 45	☽ ± ♇
09 47	☽ ⊻ ♄
11 40	☽ # ♆
12 46	☽ # ♃
14 55	☽ ∥ ♄
15 13	♆ St D
18 42	☽ △ ☿
19 04	☽ ⊼ ♅
19 13	☿ St R
23 01	☽ ⚼ ♇
13 Friday	
03 31	☽ □ ♀
04 23	☽ □ ♂
05 19	☽ ± ♅
08 56	☽ # ♂
10 55	☽ □ ☉
11 49	♀ ☍ ♂
20 53	☽ ⚼ ♅
23 19	☽ ☍ ♃
14 Saturday	
00 54	☽ △ ♇
04 42	☽ ☍ ♆
12 11	♀ Q ♄
13 40	☽ ☌ ♄
13 49	☽ Q ♀
20 42	☽ Q ☉
22 24	☽ □ ☿
23 19	☽ △ ♅
15 Sunday	
07 54	☽ △ ♂
13 20	☽ ⚹ ♀
16 08	☉ ⊥ ♇
19 51	☽ ⚹ ☉
16 Monday	
04 59	☽ ⊼ ♃
06 11	☽ Q ☿
06 31	☽ □ ♇
10 23	☽ ⊼ ♆
10 33	☽ ⚼ ♂
16 02	☽ ± ♃
19 29	☽ ∠ ♀
20 01	☽ ⊻ ♄
21 26	☽ ± ♆
17 Tuesday	
01 30	☽ ∠ ☉
02 49	☽ ± ♂
03 41	☽ ⚹ ☿
06 04	☽ □ ♅
07 12	☽ ⊥ ♄
08 47	☽ ⚼ ♃
13 48	☽ ⊼ ♂
13 59	☽ ⊥ ♀
14 10	☽ ⚼ ♆
15 55	☽ Q ♇
19 40	☽ ⊥ ☉
18 Wednesday	
00 08	☽ ∠ ♄
02 29	☽ ⊻ ♀
06 52	☽ ∠ ☿
07 56	☽ ⊻ ☉
08 46	☽ # ♂
10 23	♀ ⊥ ♇
13 13	☽ △ ♃
14 39	☽ ⚹ ♇
16 07	☽ Q ♅
18 33	☽ △ ♆
23 20	☽ ⊥ ☿
19 Thursday	
00 44	♀ ∥ ♇
04 51	☽ ⚹ ♄
06 50	☽ ∥ ♄
08 52	☽ # ♃
10 22	☽ ⊻ ☿
10 46	☽ # ♆
15 17	☽ ⚹ ♅
17 50	☉ ∠ ♅
19 35	☽ ∠ ♇
21 59	☽ ☍ ♂
20 Friday	
04 12	☽ Q ♄
05 18	☽ ∥ ♇
07 43	☽ ∥ ♀
13 12	☽ ⊥ ♇
17 27	☽ ∥ ☉
18 47	☽ ☌ ♀
20 42	☽ ∠ ♅
22 53	☽ ☌ ☉
23 46	☽ □ ♃
21 Saturday	
01 03	☽ ⊻ ♇
04 57	☽ □ ♆
10 49	☉ □ ♃
14 24	♀ ∠ ♅
14 35	☽ ⊥ ♅
15 49	☽ □ ♄
18 05	☽ ☌ ☿
22 Sunday	
02 35	☽ ⊻ ♅
03 02	☉ ⊻ ♇
04 32	☽ ∥ ☿
08 09	☽ ⊼ ♂
08 26	☽ # ♅
11 49	☽ Q ♃
16 51	☽ Q ♆
19 59	☽ ± ♂
22 54	♀ □ ♃
23 Monday	
01 51	☿ □ ♄
09 57	♀ ⊻ ♇
12 08	☽ ⚹ ♃
13 12	☽ ☌ ♇
13 34	☽ ⊻ ♀
13 49	☽ ⚼ ♂
16 07	☽ ⊻ ☉
17 03	☽ ⚹ ♆
18 16	☿ ⊥ ♅
18 16	☽ ∥ ♅
24 Tuesday	
01 06	☽ # ☿
02 28	☽ ⊻ ☿
03 08	☽ ⊥ ♀
03 43	☉ □ ♆
04 22	☽ △ ♄
05 20	☽ ⊥ ☉
13 47	☽ ⊥ ☿
15 20	☽ ☌ ♅
18 44	☽ ∠ ♃
19 41	☽ △ ♂
20 24	♂ ⚼ ♇
23 06	♀ □ ♆
23 28	☽ ∠ ♆
23 30	☽ ∠ ♀
25 Wednesday	
01 12	☽ ∠ ☉
03 00	☽ # ☉
06 45	☽ ∠ ☿
09 46	☽ # ♀
10 54	☽ ⚼ ♄
13 08	☽ ⊥ ♃
17 43	☽ ⊥ ♆
22 20	☽ # ♇
26 Thursday	
01 22	☽ ⊻ ♃
02 09	☽ ⊻ ♇
05 11	☽ ± ♄
05 52	☽ ⊻ ♆
09 25	☽ ⚹ ♀
10 16	☽ ⚹ ☉
10 57	☽ ⚹ ☿
14 18	☽ ⊥ ♇
15 01	☉ ☌ ☿
17 19	☽ ⊼ ♄
18 21	☽ ∥ ♆
18 36	☿ ☌ ♀
21 15	☽ ∥ ♃
22 39	☽ # ♄
27 Friday	
04 14	☽ ⊻ ♅
07 22	☽ □ ♂
08 25	☽ ∠ ♇
09 23	☽ Q ☿
12 18	☽ Q ♀
12 28	☽ Q ☉
16 16	☽ ⊥ ♅
18 54	☉ ☌ ♀
23 46	☽ ∥ ♂
28 Saturday	
10 09	☽ ∠ ♅
13 45	☽ ☌ ♃
14 13	☽ ⚹ ♇
17 40	☽ ☌ ♆
18 30	☽ Q ♂
18 37	☽ □ ☿
29 Sunday	
02 56	☽ □ ☉
03 43	☽ □ ♀
04 22	♀ ⊥ ♅
04 52	☽ ☍ ♄
08 51	☿ □ ♆
13 33	☽ Q ♇
14 25	☉ ⊥ ♅
15 23	☽ ⚹ ♅
15 32	♀ □ ♄
17 25	☽ ⚹ ♂
30 Monday	
04 40	☉ □ ♄
14 09	☽ Q ♅
21 19	☽ ∠ ♂
23 22	☽ ⊻ ♃
23 30	☽ □ ♇
31 Tuesday	
00 25	☽ △ ☿
02 38	☽ ⊻ ♆
06 36	☉ ∥ ☿
10 26	☽ ⊥ ♃
13 14	☽ ⊼ ♄
13 34	☽ ⊥ ♆
13 37	☽ ⊥ ♂
15 42	☽ △ ☉
17 48	☽ △ ♀
17 56	♂ Q ♆
18 40	☿ ⊻ ♇
19 14	☿ □ ♃
23 08	☽ □ ♅

SEPTEMBER 1971

LONGITUDES

Date	Sidereal time h m s	Sun ☉ ° ' "	Moon ☽ ° ' "	Moon ☽ 24.00 ° ' "	Mercury ☿ ° '	Venus ♀ ° '	Mars ♂ ° '	Jupiter ♃ ° '	Saturn ♄ ° '	Uranus ♅ ° '	Neptune ♆ ° '	Pluto ♇ ° '
01	10 39 59	08 ♍ 23 59	18 ♑ 59 26	25 ♑ 53 14	28 ♌ 20	09 ♍ 41	12 ♒ 19	28 ♏ 47	06 ♊ 15	11 ♎ 45	00 ♐ 25	28 ♍ 45
02	10 43 56	09 22 02	02 ♒ 53 45	10 ♒ 00 50	27 R 54	10 56	12 R 13	28 53	06 17	11 49	00 25	28 47
03	10 47 52	10 20 07	17 ♒ 14 09	24 ♒ 33 11	27 35	12 10	12 08	29 00	06 18	11 52	00 26	28 49
04	10 51 49	11 18 13	01 ♓ 57 13	09 ♓ 25 25	27 24	13 25	12 04	29 07	06 20	11 55	00 27	28 51
05	10 55 45	12 16 21	16 ♓ 56 44	24 ♓ 30 04	27 D 22	14 39	12 00	29 14	06 22	11 59	00 27	28 53
06	10 59 42	13 14 30	02 ♈ 04 12	09 ♈ 37 55	27 29	15 54	11 57	29 21	06 23	12 02	00 28	28 55
07	11 03 38	14 12 41	17 ♈ 10 01	24 ♈ 39 23	27 44	17 08	11 55	29 28	06 24	12 06	00 29	28 58
08	11 07 35	15 10 55	02 ♉ 05 01	09 ♉ 26 04	28 08	18 23	11 54	29 36	06 26	12 09	00 30	29 00
09	11 11 31	16 09 10	16 ♉ 41 50	23 ♉ 51 50	28 41	19 37	11 53	29 43	06 27	12 13	00 31	29 02
10	11 15 28	17 07 28	00 ♊ 55 45	07 ♊ 53 25	29 ♌ 22	20 52	11 D 54	29 51	06 28	12 16	00 32	29 04
11	11 19 25	18 05 47	14 ♊ 44 52	21 ♊ 30 13	00 ♍ 11	22 06	11 55	29 ♏ 59	06 29	12 20	00 33	29 07
12	11 23 21	19 04 09	28 ♊ 09 41	04 ♋ 43 36	01 09	23 21	11 57	00 ♐ 07	06 29	12 23	00 34	29 09
13	11 27 18	20 02 33	11 ♋ 12 20	17 ♋ 36 18	02 13	24 35	12 00	00 15	06 30	12 27	00 35	29 11
14	11 31 14	21 00 59	23 ♋ 55 56	00 ♌ 11 39	03 24	25 50	12 03	00 23	06 31	12 30	00 36	29 13
15	11 35 11	21 59 27	06 ♌ 23 53	12 ♌ 33 03	04 41	27 05	12 07	00 31	06 31	12 34	00 37	29 15
16	11 39 07	22 57 57	18 ♌ 39 31	24 ♌ 43 39	06 04	28 19	12 13	00 40	06 31	12 37	00 38	29 18
17	11 43 04	23 56 29	00 ♍ 45 48	06 ♍ 46 14	07 32	29 ♍ 34	12 18	00 48	06 32	12 41	00 39	29 20
18	11 47 00	24 55 03	12 ♍ 45 15	18 ♍ 43 05	09 04	00 ♎ 48	12 25	00 57	06 32	12 45	00 40	29 22
19	11 50 57	25 53 39	24 ♍ 39 58	00 ♎ 36 07	10 40	02 03	12 33	01 06	06 R 32	12 48	00 41	29 24
20	11 54 54	26 52 17	06 ♎ 31 45	12 ♎ 27 05	12 19	03 18	12 41	01 15	06 32	12 52	00 42	29 27
21	11 58 50	27 50 57	18 ♎ 22 19	24 ♎ 17 41	14 01	04 32	12 50	01 24	06 31	12 56	00 44	29 29
22	12 02 47	28 49 39	00 ♏ 13 26	06 ♏ 09 50	15 45	05 47	13 00	01 33	06 31	12 59	00 45	29 31
23	12 06 43	29 ♍ 48 23	12 ♏ 07 12	18 ♏ 05 51	17 31	07 02	13 10	01 42	06 31	13 03	00 46	29 34
24	12 10 40	00 ♎ 47 08	24 ♏ 06 09	00 ♐ 08 31	19 18	08 16	13 21	01 52	06 30	13 07	00 48	29 36
25	12 14 36	01 45 55	06 ♐ 13 23	12 ♐ 21 12	21 06	09 31	13 33	02 01	06 29	13 11	00 49	29 38
26	12 18 33	02 44 45	18 ♐ 32 31	24 ♐ 47 48	22 55	10 46	13 46	02 11	06 29	13 14	00 50	29 40
27	12 22 29	03 43 35	01 ♑ 07 36	07 ♑ 32 28	24 44	12 01	13 59	02 20	06 28	13 18	00 52	29 43
28	12 26 26	04 42 28	14 ♑ 02 53	20 ♑ 39 20	26 34	13 15	14 13	02 30	06 27	13 22	00 53	29 45
29	12 30 23	05 41 22	27 ♑ 22 14	04 ♒ 11 57	28 ♍ 23	14 30	14 28	02 40	06 26	13 26	00 55	29 47
30	12 34 19	06 ♎ 40 18	11 ♒ 08 40	18 ♒ 12 28	00 ♎ 12	15 ♎ 45	14 ♒ 43	02 ♐ 50	06 ♊ 25	13 ♎ 29	00 ♐ 56	29 ♍ 49

Moon nodes and latitude

Date	Moon True ☊ ° '	Moon Mean ☊ ° '	Moon ☽ Latitude ° '
01	14 ♒ 24	13 ♒ 04	02 S 15
02	14 D 25	13 01	01 S 03
03	14 R 26	12 57	00 N 15
04	14 25	12 54	01 35
05	14 22	12 51	02 49
06	14 19	12 48	03 52
07	14 15	12 45	04 38
08	14 11	12 42	05 05
09	14 07	12 38	05 12
10	14 05	12 35	04 59
11	14 04	12 32	04 29
12	14 D 04	12 29	03 45
13	14 05	12 26	02 50
14	14 07	12 22	01 48
15	14 08	12 19	00 N 42
16	14 R 08	12 16	00 S 25
17	14 07	12 13	01 29
18	14 03	12 10	02 29
19	13 58	12 07	03 22
20	13 51	12 03	04 05
21	13 43	12 00	04 38
22	13 34	11 57	04 59
23	13 26	11 54	05 07
24	13 18	11 51	05 02
25	13 13	11 48	04 43
26	13 10	11 44	04 11
27	13 09	11 41	03 26
28	13 D 09	11 38	02 30
29	13 10	11 35	01 24
30	13 ♒ 11	11 ♒ 32	00 S 11

DECLINATIONS

Date	Sun ☉ ° '	Moon ☽ ° '	Mercury ☿ ° '	Venus ♀ ° '	Mars ♂ ° '	Jupiter ♃ ° '	Saturn ♄ ° '	Uranus ♅ ° '	Neptune ♆ ° '	Pluto ♇ ° '
01	08 N 25	24 S 20	09 N 26	09 N 15	23 S 03	19 S 15	19 N 22	04 S 03	18 S 38	14 N 49
02	08 04	20 32	09 53	08 46	23 03	19 17	19 22	04 05	18 38	14 48
03	07 42	15 26	10 17	08 18	23 00	19 18	19 22	04 06	18 38	14 47
04	07 20	09 18	10 38	07 50	22 58	19 20	19 22	04 07	18 38	14 47
05	06 57	02 S 34	10 57	07 21	22 55	19 22	19 22	04 09	18 38	14 46
06	06 35	04 N 22	11 12	06 52	22 53	19 24	19 22	04 10	18 39	14 45
07	06 13	11 02	11 24	06 23	22 50	19 25	19 22	04 11	18 39	14 44
08	05 50	16 58	11 33	05 54	22 46	19 27	19 22	04 13	18 39	14 43
09	05 28	21 48	11 37	05 24	22 43	19 29	19 22	04 14	18 39	14 42
10	05 05	25 13	11 38	04 55	22 39	19 31	19 22	04 16	18 40	14 41
11	04 42	27 01	11 35	04 25	22 34	19 32	19 22	04 17	18 40	14 40
12	04 20	27 11	11 28	03 55	22 30	19 34	19 22	04 18	18 40	14 39
13	03 57	25 48	11 18	03 25	22 25	19 36	19 22	04 20	18 40	14 38
14	03 34	23 06	11 04	02 55	22 20	19 38	19 22	04 21	18 41	14 37
15	03 11	19 21	10 46	02 24	22 15	19 40	19 22	04 23	18 41	14 36
16	02 48	14 51	10 24	01 54	22 10	19 42	19 22	04 24	18 41	14 35
17	02 24	09 49	10 01	01 24	22 04	19 44	19 22	04 25	18 41	14 34
18	02 01	04 N 29	09 32	00 53	21 58	19 46	19 22	04 27	18 42	14 34
19	01 38	00 S 58	09 02	00 N 23	21 52	19 48	19 22	04 28	18 42	14 33
20	01 15	06 21	08 29	00 S 08	21 45	19 50	19 22	04 30	18 42	14 32
21	00 51	11 29	07 54	00 38	21 38	19 52	19 21	04 31	18 43	14 31
22	00 28	16 13	07 16	01 09	21 32	19 54	19 21	04 33	18 43	14 30
23	00 N 05	20 21	06 37	01 39	21 24	19 56	19 21	04 34	18 43	14 29
24	00 S 19	23 40	05 56	02 10	21 17	19 58	19 21	04 36	18 44	14 29
25	00 42	26 00	05 14	02 40	21 10	20 00	19 21	04 37	18 44	14 28
26	01 06	27 07	04 31	03 11	21 02	20 02	19 20	04 38	18 44	14 27
27	01 29	26 53	03 47	03 41	20 54	20 04	19 20	04 40	18 45	14 26
28	01 52	25 11	03 02	04 12	20 46	20 06	19 19	04 41	18 45	14 25
29	02 16	22 04	02 17	04 42	20 37	20 08	19 19	04 43	18 45	14 24
30	02 S 39	17 S 37	01 N 31	05 S 12	20 S 29	20 S 10	19 N 19	04 S 44	18 S 46	14 N 23

ZODIAC SIGN ENTRIES

Date	h	m	Planets
02	07	04	☽ ♒
04	08	51	☽ ♓
06	08	43	☽ ♈
08	08	37	☽ ♉
10	10	25	☽ ♊
11	06	45	☿ ♍
11	15	33	♃ ♐
12	15	21	☽ ♋
14	23	38	☽ ♌
17	10	29	☽ ♍
17	20	25	♀ ♎
19	22	47	☽ ♎
22	11	33	☽ ♏
23	16	45	☉ ♎
24	23	43	☽ ♐
27	09	53	☽ ♑
29	16	39	☽ ♒
30	09	19	☿ ♎

LATITUDES

Date	Mercury ☿ ° '	Venus ♀ ° '	Mars ♂ ° '	Jupiter ♃ ° '	Saturn ♄ ° '	Uranus ♅ ° '	Neptune ♆ ° '	Pluto ♇ ° '
01	02 S 47	01 N 24	06 S 12	00 N 39	02 S 01	00 N 39	01 N 39	15 N 39
04	01 51	01 25	06 00	00 39	02 02	00 39	01 39	15 38
07	00 55	01 25	05 51	00 38	02 02	00 38	01 39	15 38
10	00 S 04	01 24	05 39	00 38	02 03	00 38	01 38	15 38
13	00 N 39	01 23	05 27	00 37	02 03	00 38	01 38	15 38
16	01 12	01 21	05 14	00 36	02 04	00 38	01 38	15 38
19	01 35	01 18	05 01	00 36	02 04	00 38	01 38	15 38
22	01 47	01 15	04 49	00 35	02 05	00 38	01 38	15 38
25	01 52	01 12	04 36	00 35	02 05	00 38	01 38	15 38
28	01 49	01 08	04 23	00 34	02 06	00 38	01 38	15 38
31	01 N 41	01 N 03	04 S 11	00 N 34	02 S 06	00 N 38	01 N 37	15 N 39

DATA

Julian Date	2441196
Delta T	+42 seconds
Ayanamsa	23° 27' 54"
Synetic vernal point	05° ♓ 39' 06"
True obliquity of ecliptic	23° 26' 41"

LONGITUDES

Date	Chiron ⚷ ° '	Ceres ⚳ ° '	Pallas ⚴ ° '	Juno ⚵ ° '	Vesta ⚶ ° '	Black Moon Lilith ⚸ ° '
01	13 ♈ 00	23 ♋ 34	29 ♉ 41	02 ♍ 19	23 ♑ 03	10 ♎ 20
11	12 ♈ 37	27 ♋ 29	01 ♊ 48	06 ♍ 44	23 ♑ 15	11 ♎ 27
21	12 ♈ 12	01 ♌ 16	03 ♊ 25	11 ♍ 04	24 ♑ 14	12 ♎ 34
31	11 ♈ 45	04 ♌ 55	04 ♊ 23	15 ♍ 17	25 ♑ 46	13 ♎ 41

MOON'S PHASES, APSIDES AND POSITIONS ☽

Date	h	m	Phase	Longitude ° '	Eclipse Indicator
05	04	02	○	11 ♓ 57	
11	18	23	☾	18 ♊ 21	
19	14	42	●	26 ♍ 00	
27	17	17	☽	03 ♑ 57	

Day	h	m		
06	04	56	Perigee	
21	06	30	Apogee	
05	20	52	0N	
12	02	13	Max dec	27° N 19'
19	07	45	0S	
26	19	58	Max dec	27° S 12'

ASPECTARIAN

h m	Aspects
01 Wednesday	
00 03	☽ ± ♄
00 08	♃ ✶ ♇
00 18	☽ ⊻ ♂
02 22	☽ bQ ☿
02 44	☽ ∠ ♃
05 41	☽ ∠ ♆
06 53	☿ ∥ ♀
15 57	☽ bQ ♄
17 38	☽ ± ☿
20 16	☽ bQ ☉
21 12	☽ ∥ ♂
22 54	☽ bQ ♀
02 Thursday	
03 41	☽ ⊼ ☿
04 58	☽ △ ♇
05 07	☽ ✶ ♃
07 47	☽ ✶ ♆
12 52	☽ ± ☉
15 47	☽ ± ♀
17 44	☽ △ ♄
18 02	☽ # ♄
18 26	☽ ∥ ♃
21 37	☽ ∥ ♆
23 42	☽ ⊼ ☉
03 Friday	
01 34	☽ Q ♃
02 48	☽ ⊼ ♀
03 04	☽ △ ♅
03 35	☽ ☌ ♂
04 02	☽ Q ♆
05 51	♀ ⊻ ♅
06 19	☽ bQ ♇
11 21	♀ ⊼ ♂
14 40	☽ # ♇
21 11	☽ ± ♇
04 Saturday	
03 49	☽ bQ ♅
04 42	☽ ☍ ☿
06 58	☽ ⊼ ♇
07 18	☽ # ☿
07 23	☽ □ ♃
09 34	☽ □ ♆
17 49	☽ # ♀
18 25	☽ ± ♅
19 04	☽ □ ♄
19 38	☽ # ☉
05 Sunday	
04 02	☽ ☍ ☉
04 03	☽ ⊼ ♅
04 09	☽ ⊻ ♂
04 16	☉ ⊻ ♅
05 38	☉ ⊼ ♂
06 02	☿ St D
06 30	☽ ∥ ♅
08 01	☽ ☍ ♀
13 40	☽ ⊥ ♂
16 48	♂ △ ♅
18 41	♃ # ♄
23 48	☽ Q ♄
06 Monday	
03 55	☽ ∠ ♂
04 38	☽ ⊼ ☿
07 00	☽ ☍ ♇
07 39	☽ △ ♃
09 28	☽ △ ♆
11 18	☽ # ♅
14 16	☽ ± ☿
18 51	☽ ✶ ♄
19 24	☽ ∥ ☉
20 13	☽ ∥ ♀
07 Tuesday	
03 39	☽ ✶ ♂
03 53	☽ ☍ ♅
04 47	☽ bQ ☿
06 58	☽ ⊼ ☉
07 40	☽ bQ ♃
09 19	☽ bQ ♆
11 57	☽ ⊼ ♀
13 28	☽ ∥ ☿
17 12	☽ ± ☉
18 47	☽ ∠ ♄
19 16	♀ Q ♃
22 11	☽ ± ♃
22 25	☽ ± ♀
22 48	☽ Q ♂
23 44	☽ ± ♆
08 Wednesday	
02 27	☽ ∥ ♇
02 53	♀ ± ♂
05 24	☽ △ ☿
06 59	☽ ⊼ ♇
07 56	☽ ⊼ ♃
08 42	☽ bQ ☉
09 18	☽ ⊥ ♄
09 26	☽ ⊼ ♆
14 18	☽ bQ ♀
14 23	♀ Q ♆
16 45	☽ ± ♇
19 05	☽ ⊻ ♄
19 40	☽ # ♆
23 08	☽ ∥ ♄
23 16	☉ ∥ ♀
23 35	☽ # ♃
09 Thursday	
04 03	☽ □ ♂
04 32	☽ ⊼ ♅
07 34	☽ bQ ♇
11 02	☽ △ ☉
13 51	♂ St D
14 32	☽ ± ♅
17 19	☽ # ♂
17 20	☽ △ ♀
10 Friday	
01 40	☿ ⊻ ♇
05 44	☽ bQ ♅
08 49	☽ △ ♇
09 11	☽ □ ☿
10 08	☽ ☍ ♃
11 19	☽ ☍ ♆
21 32	♂ Q ♃
21 32	☽ ☌ ♄
11 Saturday	
05 09	☿ □ ♃
07 01	☽ △ ♂
07 24	☉ ± ♂
07 43	☽ △ ♅
08 42	☉ Q ♃
17 58	♀ # ♅
18 23	☽ □ ☉
18 32	☽ Q ☿
21 26	☿ □ ♆
23 14	☉ Q ♆
12 Sunday	
02 25	☽ □ ♀
09 48	☽ bQ ♂
13 12	☉ # ♅
13 48	☽ □ ♇
15 35	☽ ⊼ ♃
16 22	☽ ⊼ ♆
17 53	☽ ✶ ☿
13 Monday	
02 18	☽ ± ♂
02 42	☽ ± ♃
03 16	☽ ⊻ ♄
03 24	☽ ± ♆
05 39	☽ Q ☉
13 28	☽ ⊼ ♂
14 19	☽ □ ♅
14 25	☽ ⊥ ♄
14 52	☽ Q ♀
19 39	☽ bQ ♃
20 12	☽ bQ ♆
23 14	☽ Q ♇
14 Tuesday	
00 23	☽ ∠ ☿
05 59	☽ ✶ ☉
07 23	☽ ∠ ♄
16 02	☽ ✶ ♀
17 31	☽ # ♂
19 22	☽ ⊥ ☿
22 09	☽ ✶ ♇
15 Wednesday	
00 30	☽ △ ♃
00 39	☽ Q ♅
00 47	☽ △ ♆
08 16	☽ ⊻ ☿
10 13	☽ # ♃
11 55	☽ ∥ ♄
12 14	☽ ✶ ♄
12 59	♀ bQ ♂
13 15	☽ ∠ ☉
15 49	☽ # ♆
23 14	☽ ☍ ♂
16 Thursday	
00 05	☽ ✶ ♅
00 19	☽ ∠ ♀
03 24	☽ ∠ ♇
05 29	♃ ☌ ♆
08 23	☽ ⊥ ☉
11 44	☽ Q ♄
13 17	☽ ∥ ♇
19 42	☿ □ ♄
20 03	☽ ⊥ ♀
21 11	☽ ⊥ ♇
21 15	☽ ⊻ ☉
21 45	☿ ⊥ ♅
17 Friday	
05 50	☽ ∠ ♅
07 25	♀ ☌ ♇
09 09	☽ ⊻ ♇
09 20	☽ ⊻ ♀
11 06	☽ ∥ ☿
11 46	☽ □ ♆
12 05	☽ □ ♃
23 31	☽ □ ♄
23 53	☽ ⊥ ♅
18 Saturday	
03 29	☽ ☌ ☿
09 16	♀ ✶ ♆
11 19	☽ ⊼ ♂
11 59	☽ ⊻ ♅
12 09	☽ # ♅
15 08	♀ ✶ ♃
23 31	☽ ± ♂
23 41	☽ ∥ ☉
23 55	☽ Q ♆
19 Sunday	
00 37	☽ Q ♃
02 17	♄ St R
05 28	☽ ∥ ♀
09 38	☽ # ♀
14 42	☽ ☌ ☉
14 45	☽ # ☉
17 53	☽ bQ ♂
21 37	☽ ☌ ♇
20 Monday	
00 12	☽ ✶ ♆
01 10	☽ ✶ ♃
03 38	☽ ∥ ♅
04 41	☽ ☌ ♀
12 00	☽ △ ♄
17 00	♂ Q ♆
17 45	☿ ⊼ ♂
20 12	☿ ⊻ ♅
20 48	☽ # ☿
21 Tuesday	
00 37	☽ △ ♂
00 55	☽ ☌ ♅
01 40	☽ ⊻ ☿
06 38	☽ ∠ ♆
07 57	☽ ∠ ♃
11 28	☉ bQ ♂
15 53	☽ ⊥ ☿
17 49	☉ # ♀
18 23	☽ bQ ♄
22 Wednesday	
00 54	☽ ⊥ ♆
02 25	☽ ⊥ ♃
03 01	☽ # ♇
08 55	☽ ⊻ ☉
10 34	☽ ⊻ ♇
11 08	♂ △ ♅
12 36	☽ ± ♄
13 04	☽ ⊻ ♆
13 14	☽ ∠ ☿
14 43	☽ ⊻ ♃
22 08	☽ ⊥ ☉
22 44	☽ ⊥ ♇
23 Thursday	
00 33	☽ ⊻ ♀
00 42	☽ ⊼ ♄
02 03	☽ ∥ ♆
02 04	♀ △ ♄
05 42	☉ ☌ ♇
05 49	☽ # ♄
09 22	☽ ∥ ♃
13 53	☽ ⊻ ♅
14 02	☽ ⊥ ♀
14 08	☽ □ ♂
16 55	☽ ∠ ♇
17 53	☽ ∠ ☉
18 46	☽ ∥ ♂
24 Friday	
00 43	☽ ✶ ☿
01 59	☽ ⊥ ♅
05 10	☿ Q ♆
10 09	☽ ∠ ♀
12 09	☉ ✶ ♆
12 54	☿ ± ♂
20 01	☽ ∠ ♅
20 15	☿ Q ♃
22 57	☽ ✶ ♇
25 Saturday	
01 19	☽ ☌ ♆
02 26	☽ ✶ ☉
02 38	☽ Q ♂
03 36	☽ ☌ ♃
04 46	☽ Q ☿
12 32	☽ ☍ ♄
19 12	☽ ✶ ♀
19 23	☉ ✶ ♃
22 38	☽ Q ♇
26 Sunday	
01 40	☽ ✶ ♅
02 35	☽ ✶ ♂
04 01	☽ Q ☉
08 05	☿ # ♅
21 01	☽ Q ♀
21 50	☽ □ ☿
27 Monday	
00 54	☽ Q ♅
07 53	☽ ∠ ♂
09 19	☽ □ ♇
11 30	☽ ⊻ ♆
13 49	☿ # ♀
14 19	☽ ⊻ ♃
17 17	☽ □ ☉
21 59	☽ ⊼ ♄
22 45	☽ ⊥ ♆
28 Tuesday	
01 04	☽ ⊥ ♂
01 39	☽ ⊥ ♃
09 04	☽ ± ♄
10 24	☽ □ ♀
10 44	☽ □ ♅
12 20	☽ ⊻ ♂
14 13	♀ ☌ ♅
15 22	☽ ∠ ♆
18 23	☽ ∠ ♃
29 Wednesday	
01 25	☽ bQ ♄
11 14	♀ △ ♂
12 21	☉ # ☿
12 44	♀ ∥ ♅
14 04	☽ △ ☿
16 17	☽ △ ♇
18 15	☽ ✶ ♆
20 49	☽ ∥ ♂
21 26	☽ ✶ ♃
23 07	☽ ∥ ♃
30 Thursday	
03 34	☽ # ♄
03 42	☽ △ ☉
03 51	☽ △ ♄
04 36	☿ bQ ♂
05 45	☉ △ ♄
06 24	☽ ∥ ♆
06 52	☿ ☌ ♇
15 04	☽ Q ♆
15 43	♀ ∠ ♆
16 01	☽ △ ♅
18 13	☽ ☌ ♂
18 17	☽ bQ ♇
18 22	☽ Q ♃
19 56	☽ bQ ☿
20 36	☽ △ ♀
21 46	☿ ✶ ♆
22 31	♂ bQ ♇

All ephemeris data is given at 12.00 UT and the Moon's longitude is additionally given for 24.00 UT

OCTOBER 1971

LONGITUDES

Date	Sidereal time h m s	Sun ☉ ° ' "	Moon ☽ ° ' "	Moon ☽ 24.00 ° ' "	Mercury ☿ ° '	Venus ♀ ° '	Mars ♂ ° '	Jupiter ♃ ° '	Saturn ♄ ° '	Uranus ♅ ° '	Neptune ♆ ° '	Pluto ♇ ° '
01	12 38 16	07 ♎ 39 16	25 ♒ 23 18	02 ♓ 40 52	02 ♎ 01	16 ♎ 59	14 ♒ 59	03 ♐ 00	06 ♊ 23	13 ♎ 33	00 ♐ 57	29 ♍ 52
02	12 42 12	08 38 15	10 ♓ 04 39	17 ♓ 33 58	03 50	18 14	15 16	03 10	06 R 22	13 37	00 59	29 54
03	12 46 09	09 37 17	25 ♓ 07 52	02 ♈ 45 13	05 38	19 29	15 33	03 21	06 20	13 41	01 01	29 56
04	12 50 05	10 36 20	10 ♈ 24 42	18 ♈ 04 57	07 25	20 43	15 51	03 31	06 19	13 44	01 02	29 ♍ 58
05	12 54 02	11 35 25	25 ♈ 44 29	03 ♉ 21 52	09 12	21 58	16 09	03 42	06 17	13 48	01 04	00 ♎ 01
06	12 57 58	12 34 33	10 ♉ 55 46	18 ♉ 24 59	10 58	23 13	16 28	03 52	06 15	13 52	01 05	00 03
07	13 01 55	13 33 43	25 ♉ 48 31	03 ♊ 05 35	12 43	24 28	16 48	04 03	06 13	13 56	01 07	00 05
08	13 05 52	14 32 55	10 ♊ 15 39	17 ♊ 18 22	14 28	25 42	17 08	04 14	06 11	14 00	01 09	00 07
09	13 09 48	15 32 09	24 ♊ 13 40	01 ♋ 01 36	16 12	26 57	17 29	04 24	06 09	14 03	01 10	00 09
10	13 13 45	16 31 26	07 ♋ 42 26	14 ♋ 16 33	17 55	28 12	17 50	04 35	06 07	14 07	01 12	00 12
11	13 17 41	17 30 45	20 ♋ 44 23	27 ♋ 06 30	19 37	29 ♎ 27	18 11	04 46	06 05	14 11	01 14	00 14
12	13 21 38	18 30 06	03 ♌ 23 28	09 ♌ 35 52	21 19	00 ♏ 41	18 33	04 57	06 02	14 15	01 16	00 16
13	13 25 34	19 29 30	15 ♌ 44 20	21 ♌ 49 26	23 00	01 56	18 56	05 09	06 00	14 18	01 17	00 18
14	13 29 31	20 28 56	27 ♌ 51 44	03 ♍ 51 45	24 40	03 11	19 19	05 20	05 57	14 22	01 19	00 20
15	13 33 27	21 28 24	09 ♍ 50 00	15 ♍ 46 54	26 19	04 26	19 43	05 31	05 54	14 26	01 21	00 22
16	13 37 24	22 27 55	21 ♍ 42 53	27 ♍ 38 17	27 58	05 40	20 07	05 43	05 52	14 30	01 23	00 25
17	13 41 21	23 27 27	03 ♎ 33 24	09 ♎ 28 31	29 ♎ 36	06 55	20 31	05 54	05 49	14 34	01 25	00 27
18	13 45 17	24 27 02	15 ♎ 23 53	21 ♎ 19 40	01 ♏ 14	08 10	20 56	06 06	05 46	14 37	01 27	00 29
19	13 49 14	25 26 38	27 ♎ 16 04	03 ♏ 13 15	02 51	09 25	21 21	06 17	05 43	14 41	01 28	00 31
20	13 53 10	26 26 17	09 ♏ 11 21	15 ♏ 10 33	04 27	10 40	21 47	06 29	05 40	14 45	01 30	00 33
21	13 57 07	27 25 58	21 ♏ 11 00	27 ♏ 12 53	06 02	11 54	22 13	06 41	05 36	14 48	01 32	00 35
22	14 01 03	28 25 41	03 ♐ 16 25	09 ♐ 21 50	07 37	13 09	22 40	06 53	05 33	14 52	01 34	00 37
23	14 05 00	29 ♎ 25 25	15 ♐ 29 25	21 ♐ 39 28	09 12	14 24	23 07	07 05	05 30	14 56	01 36	00 39
24	14 08 56	00 ♏ 25 12	27 ♐ 52 21	04 ♑ 08 28	10 46	15 39	23 34	07 17	05 26	15 00	01 38	00 41
25	14 12 53	01 25 00	10 ♑ 28 13	16 ♑ 52 04	12 19	16 53	24 02	07 29	05 23	15 03	01 40	00 43
26	14 16 50	02 24 50	23 ♑ 20 29	29 ♑ 53 55	13 52	18 08	24 30	07 41	05 19	15 07	01 42	00 45
27	14 20 46	03 24 41	06 ♒ 32 50	13 ♒ 17 37	15 24	19 23	24 58	07 53	05 15	15 11	01 44	00 47
28	14 24 43	04 24 34	20 ♒ 08 38	27 ♒ 06 06	16 56	20 38	25 27	08 05	05 11	15 14	01 46	00 49
29	14 28 39	05 24 29	04 ♓ 10 09	11 ♓ 20 44	18 27	21 53	25 56	08 18	05 07	15 18	01 48	00 51
30	14 32 36	06 24 25	18 ♓ 37 39	26 ♓ 00 25	19 58	23 07	26 25	08 30	05 03	15 22	01 50	00 53
31	14 36 32	07 ♏ 24 23	03 ♈ 28 26	11 ♈ 00 46	21 ♏ 28	24 ♏ 22	26 ♒ 55	08 ♐ 43	04 ♊ 59	15 ♎ 25	01 ♐ 52	00 ♎ 55

Date	Moon True ☊ ° '	Moon Mean ☊ ° '	Moon ☽ Latitude ° '
01	13 ♒ 10	11 ♒ 28	01 N 05
02	13 R 08	11 25	02 19
03	13 03	11 22	03 25
04	12 55	11 19	04 17
05	12 47	11 16	04 51
06	12 38	11 13	05 04
07	12 30	11 09	04 56
08	12 23	11 06	04 29
09	12 19	11 03	03 47
10	12 17	11 00	02 53
11	12 D 16	10 57	01 52
12	12 17	10 54	00 N 47
13	12 R 17	10 50	00 S 18
14	12 16	10 47	01 22
15	12 12	10 44	02 21
16	12 06	10 41	03 13
17	11 57	10 38	03 56
18	11 45	10 34	04 30
19	11 32	10 31	04 51
20	11 18	10 28	05 00
21	11 04	10 25	04 56
22	10 52	10 22	04 38
23	10 43	10 19	04 07
24	10 36	10 15	03 25
25	10 32	10 12	02 32
26	10 31	10 09	01 29
27	10 D 31	10 06	00 S 21
28	10 R 31	10 03	00 N 51
29	10 29	10 00	02 01
30	10 26	09 56	03 07
31	10 ♒ 19	09 ♒ 53	04 N 01

DECLINATIONS

Date	Sun ☉ ° '	Moon ☽ ° '	Mercury ☿ ° '	Venus ♀ ° '	Mars ♂ ° '	Jupiter ♃ ° '	Saturn ♄ ° '	Uranus ♅ ° '	Neptune ♆ ° '	Pluto ♇ ° '
01	03 S 02	12 S 02	00 N 44	05 S 42	20 S 20	20 S 13	19 N 18	04 S 46	18 S 46	14 N 23
02	03 26	05 S 39	00 S 02	06 12	20 11	20 15	19 18	04 47	18 46	14 22
03	03 49	01 N 12	00 48	06 42	20 02	20 17	19 18	04 49	18 47	14 21
04	04 12	08 04	01 35	07 12	19 53	20 19	19 17	04 50	18 47	14 20
05	04 35	14 28	02 21	07 41	19 43	20 21	19 17	04 52	18 47	14 20
06	04 58	19 55	03 08	08 11	19 34	20 23	19 16	04 53	18 48	14 19
07	05 21	24 00	03 54	08 40	19 24	20 25	19 16	04 55	18 48	14 18
08	05 44	26 26	04 39	09 09	19 14	20 27	19 15	04 56	18 49	14 17
09	06 07	27 06	05 25	09 38	19 04	20 30	19 15	04 58	18 49	14 17
10	06 30	26 06	06 10	10 07	18 54	20 32	19 14	04 59	18 49	14 16
11	06 53	23 42	06 54	10 35	18 44	20 34	19 14	05 01	18 50	14 15
12	07 15	20 10	07 38	11 03	18 33	20 36	19 13	05 02	18 50	14 15
13	07 38	15 50	08 22	11 31	18 22	20 38	19 13	05 03	18 51	14 14
14	08 00	10 56	09 05	11 59	18 12	20 40	19 12	05 05	18 51	14 13
15	08 22	05 43	09 47	12 26	18 01	20 43	19 12	05 06	18 51	14 13
16	08 45	00 N 20	10 29	12 53	17 50	20 45	19 11	05 08	18 52	14 12
17	09 07	05 S 02	11 10	13 20	17 38	20 47	19 11	05 09	18 52	14 11
18	09 29	10 12	11 51	13 46	17 27	20 49	19 10	05 11	18 53	14 11
19	09 51	15 01	12 31	14 12	17 15	20 51	19 09	05 12	18 53	14 10
20	10 12	19 18	13 10	14 38	17 04	20 53	19 09	05 14	18 53	14 09
21	10 34	22 48	13 48	15 03	16 52	20 55	19 08	05 15	18 54	14 09
22	10 55	25 22	14 26	15 28	16 40	20 57	19 07	05 17	18 54	14 08
23	11 16	26 45	15 02	15 53	16 28	21 00	19 07	05 18	18 55	14 08
24	11 37	26 50	15 39	16 17	16 15	21 02	19 06	05 19	18 55	14 07
25	11 58	25 33	16 14	16 41	16 03	21 04	19 05	05 21	18 55	14 07
26	12 19	22 54	16 48	17 04	15 50	21 06	19 05	05 22	18 56	14 06
27	12 39	18 59	17 22	17 27	15 38	21 08	19 04	05 24	18 56	14 06
28	13 00	13 58	17 55	17 50	15 25	21 10	19 03	05 25	18 57	14 05
29	13 20	08 06	18 26	18 12	15 12	21 12	19 02	05 26	18 57	14 05
30	13 40	01 38	18 57	18 33	14 59	21 14	19 02	05 28	18 58	14 04
31	13 S 59	05 N 04	19 S 27	18 S 54	14 S 46	21 S 16	19 N 01	05 S 29	18 S 58	14 N 04

ZODIAC SIGN ENTRIES

Date	h m	Planets
01	19 37	☽ ♓
03	19 40	☽ ♈
05	06 14	♇ ♎
05	18 42	☽ ♉
07	18 53	☽ ♊
09	22 10	☽ ♋
11	22 43	♀ ♏
12	05 30	☽ ♌
14	16 16	☽ ♍
17	04 47	☽ ♎
17	17 49	☿ ♏
19	17 31	☽ ♏
22	05 31	☽ ♐
24	01 53	☉ ♏
24	16 05	☽ ♑
27	00 11	☽ ♒
29	04 57	☽ ♓
31	06 26	☽ ♈

LATITUDES

Date	Mercury ☿ ° '	Venus ♀ ° '	Mars ♂ ° '	Jupiter ♃ ° '	Saturn ♄ ° '	Uranus ♅ ° '	Neptune ♆ ° '	Pluto ♇ ° '
01	01 N 41	01 N 03	04 S 11	00 N 34	02 S 06	00 N 38	01 N 37	15 N 39
04	01 29	00 58	03 59	00 33	02 07	00 38	01 37	15 39
07	01 14	00 53	03 47	00 33	02 07	00 38	01 37	15 39
10	00 56	00 47	03 35	00 32	02 07	00 38	01 37	15 40
13	00 37	00 41	03 24	00 32	02 08	00 38	01 37	15 40
16	00 N 17	00 34	03 13	00 32	02 08	00 38	01 37	15 41
19	00 S 03	00 27	03 02	00 31	02 08	00 38	01 37	15 42
22	00 24	00 20	02 51	00 31	02 09	00 38	01 37	15 43
25	00 44	00 13	02 41	00 30	02 09	00 38	01 36	15 43
28	01 03	00 N 05	02 31	00 30	02 09	00 38	01 36	15 44
31	01 S 22	00 S 02	02 S 22	00 N 30	02 S 09	00 N 38	01 N 36	15 N 45

DATA

Julian Date	2441226
Delta T	+42 seconds
Ayanamsa	23° 27' 57"
Synetic vernal point	05° ♓ 39' 02"
True obliquity of ecliptic	23° 26' 41"

MOON'S PHASES, APSIDES AND POSITIONS ☽

Date	h m	Phase	Longitude ° '	Eclipse Indicator
04	12 19	○	10 ♈ 37	
11	05 29	☾	17 ♋ 15	
19	07 59	●	25 ♎ 17	
27	05 54	☽	03 ♒ 09	

Day	h m		
04	14 57	Perigee	
18	08 33	Apogee	
03	07 51	0N	
09	09 15	Max dec	27° N 07'
16	13 28	0S	
24	01 35	Max dec	26° S 58'
30	17 54	0N	

LONGITUDES

Date	Chiron ⚷ ° '	Ceres ⚳ ° '	Pallas ⚴ ° '	Juno ⚵ ° '	Vesta ⚶ ° '	Black Moon Lilith ⚸ ° '
01	11 ♈ 45	04 ♌ 55	04 ♊ 23	15 ♍ 17	25 ♑ 46	13 ♎ 41
11	11 ♈ 18	08 ♌ 22	04 ♊ 35	19 ♍ 24	27 ♑ 57	14 ♎ 48
21	10 ♈ 52	11 ♌ 36	03 ♊ 51	23 ♍ 23	00 ♒ 37	15 ♎ 55
31	10 ♈ 27	14 ♌ 32	02 ♊ 10	27 ♍ 13	03 ♒ 41	17 ♎ 03

All ephemeris data is given at 12.00 UT and the Moon's longitude is additionally given for 24.00 UT

ASPECTARIAN

h m Aspects

01 Friday
02 25 ☽ # ♇
07 07 ☽ ᖘ ☉
09 27 ☽ ± ♇
13 12 ☽ ± ☿
15 05 ♂ Q ♃
17 15 ☽ ᖘ ♅
19 24 ☽ ⊼ ♇
21 12 ☽ □ ♆
23 04 ☽ ± ☉
23 53 ☽ ᖘ ♀

02 Saturday
00 27 ☽ ⊼ ☿
00 40 ☽ □ ♃
02 24 ☿ ⚹ ♃
04 32 ♂ ∥ ♃
06 00 ☽ □ ♄
08 00 ☽ ± ♅
09 31 ☽ ⊼ ☉
10 08 ☽ ∥ ♀
10 38 ♀ ∠ ♃
15 03 ☽ ∥ ♅
15 47 ☽ ± ♀
17 42 ☽ ⊼ ♅
19 28 ☽ ∥ ☉
20 29 ☽ ⊻ ♂

03 Sunday
02 14 ☽ ⊼ ♀
05 45 ☽ ∥ ☿
06 14 ☽ ⊥ ♂
10 28 ☽ # ☿
10 45 ☽ Q ♄
19 35 ☽ ☍ ♇
20 43 ☽ ∠ ♂
21 17 ☽ △ ♆
21 25 ☿ △ ♄
21 36 ☽ # ☉

04 Monday
00 35 ☽ # ♅
01 05 ☽ △ ♃
05 36 ☽ ⚹ ♄
06 42 ☽ ☍ ☿
08 41 ☽ # ♀
12 19 ☽ ☍ ☉
17 14 ☽ ☍ ♅
20 41 ☽ ⚹ ♂
20 49 ☽ ᖘ ♆
23 06 ♀ ᖘ ♄

05 Tuesday
00 50 ☽ ᖘ ♃
05 02 ☽ ∠ ♄
05 34 ☽ ☍ ♀
10 56 ☽ ± ♆
11 27 ☽ ∥ ♇
15 06 ☽ ± ♃
15 52 ☽ Q ♂
18 44 ☽ ⊼ ♇
19 08 ☽ ⊥ ♄
20 23 ☽ ⊼ ♆

06 Wednesday
00 40 ☽ ⊼ ♃
04 13 ☽ ± ♇
04 35 ☽ ⊻ ♄
06 28 ☉ ∥ ♅
06 36 ☽ # ♆
08 52 ☽ ∥ ♄
10 19 ☽ # ♂
12 04 ☽ ⊼ ☿
14 23 ☽ # ♃
14 49 ☽ ⊼ ☉
16 43 ☽ ⊼ ♅
18 36 ☽ ᖘ ♇
21 04 ☽ □ ♂
22 57 ☽ ± ☿

07 Thursday
01 08 ☽ ± ☉
02 24 ☽ ± ♅
09 36 ☽ ⊼ ♀
15 33 ☽ ᖘ ☿
16 50 ☽ ᖘ ☉
17 08 ☽ ᖘ ♅
19 02 ☽ △ ♇
20 22 ☽ ± ♀
20 44 ☽ ☍ ♆
21 33 ☉ ☌ ♅

08 Friday
00 55 ♀ ⊥ ♆
01 45 ☽ ☍ ♃
05 11 ☽ ☌ ♄
05 16 ☿ ☌ ♅
09 00 ♂ # ♄
12 50 ☽ ᖘ ♀
14 46 ☉ ☌ ☿
18 21 ☽ △ ♅
19 50 ☽ △ ☉
20 08 ☽ △ ☿
21 07 ☿ ∥ ♅

09 Saturday
00 00 ☽ △ ♂
11 42 ☿ ∠ ♆
17 16 ☽ △ ♀
22 29 ☽ □ ♇

10 Sunday
00 17 ☽ ⊼ ♆
02 58 ☽ ᖘ ♂
03 55 ☉ ∠ ♆
06 18 ☽ ⊼ ♃
09 08 ☽ ⊻ ♄
10 29 ☿ △ ♂
11 05 ☽ ± ♆
17 19 ☽ ± ♃
19 43 ☽ ± ♂
20 01 ☽ ⊥ ♄
20 50 ♀ ⊥ ♃
22 29 ♂ ∥ ♆
23 46 ☽ □ ♅

11 Monday
03 35 ☽ ᖘ ♆
05 29 ☽ □ ☉
07 06 ☽ ⊼ ♂
07 18 ☽ Q ♇
09 35 ☽ □ ☿
10 04 ☉ ∥ ☿
10 10 ☽ ᖘ ♃
12 38 ☽ ∠ ♄
14 27 ☿ ∠ ♃
23 51 ♀ ± ♄

12 Tuesday
03 35 ♀ ⊻ ♇
06 00 ☽ ⚹ ♇
06 15 ☽ □ ♀
07 54 ☽ △ ♆
08 12 ☿ ᖘ ♄
09 23 ☽ # ♃
09 47 ☽ Q ♅
14 05 ☉ △ ♂
15 04 ☽ △ ♃
17 05 ☽ ⚹ ♄
17 35 ☽ ∥ ♄
18 31 ☽ Q ☉
19 47 ☽ # ♆
21 48 ☽ # ♂
23 14 ♀ ⊻ ♆

13 Wednesday
01 15 ☽ Q ☿
09 11 ☽ ⚹ ♅
11 08 ☽ ∠ ♇
16 26 ☽ Q ♄
18 29 ☽ ☍ ♂
20 03 ☽ ⚹ ☉
20 06 ☽ ∥ ♇
21 13 ☽ Q ♀

14 Thursday
04 37 ☽ ⚹ ☿
04 58 ☽ ⊥ ♇
07 29 ☽ # ♀
07 31 ☉ ∠ ♃
15 02 ☽ ∠ ♅
16 57 ☽ ⊻ ♇
18 55 ☽ □ ♆
19 37 ☽ # ☿
21 37 ☿ ⊥ ♆
22 55 ☉ ᖘ ♄
23 52 ☽ ⚹ ♀

15 Friday
00 42 ☽ # ☉
03 11 ☽ □ ♃
04 08 ☽ □ ♄
04 38 ☽ ∠ ☉
09 10 ☽ ⊥ ♅
14 42 ☽ # ♅
15 29 ☽ ∠ ☿
21 20 ☽ ⊻ ♅

16 Saturday
00 25 ☽ ⊥ ☉
07 16 ☽ Q ♆
08 38 ☽ ⊼ ♂
09 39 ☽ ∠ ♀
12 36 ☽ ⊥ ☿
12 52 ♀ ⊻ ♃
13 39 ☽ ⊻ ☉
15 28 ♀ ⊼ ♄
16 07 ☽ Q ♃
21 13 ☽ ± ♂

17 Sunday
02 35 ♀ ⊥ ♇
02 42 ☽ ⊻ ☿
02 59 ♃ ☍ ♄
05 40 ☽ ☌ ♇
06 02 ☽ ⊥ ♀
07 38 ☽ ⚹ ♆
12 34 ☽ ∥ ♅
14 59 ☿ ± ♄
16 07 ☽ ᖘ ♂
16 33 ☽ △ ♄
16 50 ☽ ⚹ ♃
17 00 ☿ ⊥ ♃
19 37 ☽ ⊻ ♀

18 Monday
00 40 ☿ ⊻ ♇
08 16 ☽ ∥ ☉
10 25 ☽ ☌ ♅
14 07 ☽ ∠ ♆
15 13 ☿ ⊻ ♆
21 10 ☽ ∥ ☿
22 49 ☽ ᖘ ♄
23 37 ☽ △ ♂
23 43 ☽ ∠ ♃

19 Tuesday
07 18 ☽ ∥ ♀
07 34 ☽ # ♇
07 59 ☽ ☌ ☉
08 22 ☽ ⊥ ♆
10 03 ♀ # ♇
12 44 ☉ ⊥ ♆
16 55 ☽ ± ♄
18 12 ☽ ⊥ ♃
18 34 ☽ ⊻ ♇
20 30 ☽ ⊻ ♆
23 33 ☽ ∥ ♂

20 Wednesday
00 59 ☽ ☌ ☿
04 56 ☽ ⊼ ♄
06 29 ☽ ⊻ ♃
06 41 ☽ ⊥ ♇
09 33 ☽ ∥ ♆
11 06 ☽ # ♄
15 18 ☽ ☌ ♀
22 16 ☽ ∥ ♃
23 12 ☽ ⊻ ♅

21 Thursday
00 47 ☽ ∠ ♇
05 40 ☿ ⊼ ♄
11 15 ☽ ⊥ ♅
14 08 ☽ □ ♂
20 26 ☿ ⊥ ♇
23 08 ☿ ⊻ ♃

22 Friday
01 03 ☿ # ♇
01 33 ☽ ⊻ ☉
05 14 ☽ ∠ ♅
06 44 ☽ ⚹ ♇
08 37 ☽ ☌ ♆
14 29 ☽ ⊥ ☉
16 28 ☽ ☍ ♄
19 14 ☽ ☌ ♃
21 51 ☽ ⊻ ☿

23 Saturday
03 06 ☽ Q ♂
06 26 ☽ Q ♇
09 38 ☽ ⊻ ♀
09 44 ☽ ∠ ☉
10 54 ☽ ⚹ ♅
11 21 ☽ ⊥ ☿
13 35 ☉ ± ♄
22 38 ☽ ⊥ ♀
22 49 ♀ ⊻ ♅

24 Sunday
03 23 ☽ ⚹ ♂
07 21 ☽ ∠ ☿
10 18 ☽ Q ♅
10 49 ♀ ∥ ♂
12 50 ♀ ∠ ♇
17 19 ☽ ⚹ ☉
17 25 ☽ □ ♇
17 54 ☽ ∠ ♀
18 40 ☉ ⊻ ♇
19 14 ☽ ⊻ ♆

25 Monday
02 24 ☽ ⊼ ♄
06 15 ☽ ⊻ ♃
06 29 ☿ ∥ ♂
06 41 ☽ ⊥ ♆
09 10 ☽ ∠ ♂
13 42 ☽ ± ♄
13 57 ☉ ⊥ ♃
15 57 ☽ ⚹ ☿
17 45 ☽ ⊥ ♃
18 01 ☽ Q ☉
18 18 ☉ ⊻ ♆
20 39 ☽ □ ♅
23 40 ☽ ∠ ♆

26 Tuesday
01 20 ☽ ⚹ ♀
02 42 ☽ ⊥ ♂
06 26 ☽ ᖘ ♄
10 46 ☽ ∠ ♃
14 12 ☽ ⊻ ♂
17 15 ☽ Q ☿
23 49 ☽ ∥ ♃

27 Wednesday
01 35 ☽ △ ♇
01 44 ☽ Q ♀
02 09 ♂ ± ♇
03 19 ☽ ⚹ ♆
05 54 ☽ □ ☉
08 23 ☿ ⊻ ♅
09 41 ☽ △ ♄
11 33 ☽ # ♄
12 13 ☽ ∥ ♆
14 26 ☽ ⚹ ♃
18 13 ☿ ∠ ♇
19 16 ☽ ∥ ♀
19 26 ☽ ∥ ☿

28 Thursday
00 31 ☿ ∥ ♀
00 49 ☽ Q ♆
03 23 ☽ △ ♅
04 26 ☽ ᖘ ♇
05 15 ☽ ∥ ♂
05 41 ☽ □ ☿
11 30 ☽ # ♇
11 54 ☽ Q ♃
12 56 ☽ □ ♀
15 59 ☽ ∥ ☉
20 06 ☽ ± ♇
21 30 ☽ ☌ ♂

29 Friday
00 19 ♀ ⊥ ♅
05 25 ☽ ᖘ ♅
05 36 ☉ ⊼ ♄
06 23 ☽ ⊼ ♇
08 00 ☽ □ ♆
13 36 ☽ □ ♄
14 14 ☽ △ ☉
19 01 ☽ □ ♃
20 38 ☽ ± ♅
22 02 ☽ ∥ ♅

30 Saturday
06 37 ☽ ⊼ ♅
12 19 ☿ ∥ ♆
14 26 ☽ △ ☿
15 19 ☿ # ♄
16 52 ☽ ᖘ ☉
19 11 ☽ Q ♄
20 00 ☽ △ ♀
23 52 ☉ ⊥ ♇

31 Sunday
01 06 ☽ ⊻ ♂
07 54 ☽ ☍ ♇
08 27 ☽ ± ☉
09 26 ☽ △ ♆
11 05 ☽ ⊥ ♂
11 15 ☿ ⊥ ♅
13 31 ☽ # ♅
14 25 ☽ ⚹ ♄
16 33 ♀ ∥ ♆
17 18 ☽ ᖘ ☿
17 32 ☉ # ♇
18 43 ☽ ⊼ ☉
19 20 ♀ # ♄
20 28 ☽ △ ♃
22 14 ☽ ᖘ ♀

NOVEMBER 1971

Date	Sidereal time h m s	Sun ☉ ° ' "	Moon ☽ ° ' "	Moon ☽ 24.00 ° ' "	Mercury ☿ ° '	Venus ♀ ° '	Mars ♂ ° '	Jupiter ♃ ° '	Saturn ♄ ° '	Uranus ♅ ° '	Neptune ♆ ° '	Pluto ♇ ° '
01	14 40 29	08 ♏ 24 23	18 ♈ 36 23	26 ♈ 13 59	22 ♏ 57	25 ♏ 37	27 ♒ 25	08 ♐ 55	04 ♊ 55	15 ♎ 29	01 ♐ 55	00 ♎ 57
02	14 44 25	09 24 24	03 ♉ 52 13	11 ♉ 29 38	24 27	26 52	27 55	09 08	04 R 51	15 32	01 57	00 59
03	14 48 22	10 24 27	19 ♉ 04 49	26 ♉ 36 25	25 55	28 06	28 26	09 20	04 47	15 36	01 59	01 01
04	14 52 19	11 24 33	04 ♊ 03 12	11 ♊ 24 11	27 23	29 ♏ 21	28 57	09 33	04 43	15 39	02 01	01 03
05	14 56 15	12 24 40	18 ♊ 38 33	25 ♊ 45 44	28 ♏ 51	00 ♐ 36	29 28	09 46	04 38	15 43	02 03	01 04
06	15 00 12	13 24 50	02 ♋ 45 25	09 ♋ 37 29	00 ♐ 18	01 51	29 ♒ 59	09 58	04 34	15 46	02 05	01 06
07	15 04 08	14 25 01	16 ♋ 22 01	22 ♋ 59 18	01 45	03 05	00 ♓ 31	10 11	04 30	15 50	02 07	01 08
08	15 08 05	15 25 14	29 ♋ 29 42	05 ♌ 53 44	03 10	04 20	01 03	10 24	04 25	15 53	02 10	01 10
09	15 12 01	16 25 30	12 ♌ 11 58	18 ♌ 25 03	04 36	05 35	01 35	10 37	04 21	15 57	02 12	01 11
10	15 15 58	17 25 47	24 ♌ 33 37	00 ♍ 38 21	06 00	06 50	02 07	10 50	04 16	16 00	02 14	01 13
11	15 19 54	18 26 06	06 ♍ 39 55	12 ♍ 38 58	07 24	08 04	02 40	11 03	04 12	16 03	02 16	01 15
12	15 23 51	19 26 28	18 ♍ 36 06	24 ♍ 31 54	08 47	09 19	03 13	11 16	04 07	16 07	02 18	01 16
13	15 27 48	20 26 51	00 ♎ 26 54	06 ♎ 21 36	10 09	10 34	03 46	11 29	04 02	16 10	02 21	01 18
14	15 31 44	21 27 16	12 ♎ 16 26	18 ♎ 11 47	11 30	11 48	04 19	11 42	03 57	16 13	02 23	01 20
15	15 35 41	22 27 43	24 ♎ 07 58	00 ♏ 05 17	12 49	13 03	04 53	11 55	03 53	16 16	02 25	01 21
16	15 39 37	23 28 12	06 ♏ 03 58	12 ♏ 04 11	14 08	14 18	05 26	12 08	03 48	16 20	02 27	01 23
17	15 43 34	24 28 42	18 ♏ 06 05	24 ♏ 09 49	15 25	15 33	06 00	12 21	03 43	16 23	02 29	01 24
18	15 47 30	25 29 14	00 ♐ 15 27	06 ♐ 23 05	16 40	16 47	06 34	12 34	03 38	16 26	02 32	01 26
19	15 51 27	26 29 48	12 ♐ 32 48	18 ♐ 44 41	17 54	18 02	07 08	12 48	03 33	16 29	02 34	01 27
20	15 55 23	27 30 23	24 ♐ 58 50	01 ♑ 15 23	19 06	19 17	07 43	13 01	03 28	16 32	02 36	01 29
21	15 59 20	28 30 59	07 ♑ 34 28	13 ♑ 56 17	20 15	20 32	08 18	13 14	03 24	16 35	02 38	01 30
22	16 03 17	29 ♏ 31 37	20 ♑ 21 04	26 ♑ 49 02	21 22	21 46	08 52	13 28	03 19	16 38	02 41	01 32
23	16 07 13	00 ♐ 32 16	03 ♒ 20 29	09 ♒ 55 43	22 25	23 01	09 27	13 41	03 14	16 41	02 43	01 33
24	16 11 10	01 32 56	16 ♒ 35 01	23 ♒ 18 43	23 25	24 16	10 03	13 54	03 09	16 44	02 45	01 35
25	16 15 06	02 33 37	00 ♓ 07 03	07 ♓ 00 15	24 22	25 30	10 38	14 08	03 04	16 47	02 48	01 36
26	16 19 03	03 34 19	13 ♓ 58 28	21 ♓ 01 45	25 14	26 45	11 13	14 21	02 59	16 50	02 50	01 37
27	16 22 59	04 35 02	28 ♓ 10 02	05 ♈ 23 05	26 00	28 00	11 49	14 35	02 54	16 53	02 52	01 38
28	16 26 56	05 35 47	12 ♈ 40 33	20 ♈ 01 52	26 41	29 ♐ 14	12 25	14 48	02 49	16 56	02 54	01 40
29	16 30 52	06 36 32	27 ♈ 26 18	04 ♉ 52 59	27 16	00 ♑ 29	13 01	15 02	02 44	16 59	02 57	01 41
30	16 34 49	07 ♐ 37 18	12 ♉ 20 54	19 ♉ 48 56	27 ♐ 44	01 ♑ 44	13 ♓ 37	15 ♐ 15	02 ♊ 39	17 ♎ 01	02 ♐ 59	01 ♎ 42

Date	Moon True ☊ ° '	Moon Mean ☊ ° '	Moon ☽ Latitude ° '
01	10 ♒ 10	09 ♒ 50	04 N 39
02	09 R 59	09 47	04 58
03	09 47	09 44	04 56
04	09 36	09 40	04 33
05	09 27	09 37	03 53
06	09 21	09 34	03 00
07	09 17	09 31	01 58
08	09 16	09 28	00 N 51
09	09 D 16	09 25	00 S 16
10	09 R 16	09 21	01 20
11	09 14	09 18	02 19
12	09 11	09 15	03 12
13	09 05	09 12	03 56
14	08 56	09 09	04 29
15	08 44	09 05	04 51
16	08 31	09 02	05 00
17	08 17	08 59	04 56
18	08 03	08 56	04 38
19	07 51	08 53	04 08
20	07 41	08 50	03 25
21	07 35	08 46	02 32
22	07 31	08 43	01 30
23	07 30	08 40	00 S 22
24	07 D 30	08 37	00 N 48
25	07 30	08 34	01 57
26	07 R 30	08 31	03 01
27	07 27	08 27	03 56
28	07 23	08 24	04 36
29	07 15	08 21	05 00
30	07 ♒ 07	08 ♒ 18	05 N 03

DECLINATIONS

Date	Sun ☉ ° '	Moon ☽ ° '	Mercury ☿ ° '	Venus ♀ ° '	Mars ♂ ° '	Jupiter ♃ ° '	Saturn ♄ ° '	Uranus ♅ ° '	Neptune ♆ ° '	Pluto ♇ ° '
01	14 S 19	11 N 36	19 S 56	19 S 15	14 S 33	21 S 18	19 N 00	05 S 31	18 S 59	14 N 03
02	14 38	17 29	20 24	19 35	14 19	21 20	18 59	05 32	18 59	14 03
03	14 57	22 14	20 51	19 54	14 06	21 22	18 58	05 33	18 59	14 02
04	15 15	25 26	21 17	20 13	13 52	21 24	18 58	05 35	19 00	14 02
05	15 34	26 50	21 42	20 32	13 39	21 26	18 57	05 36	19 00	14 02
06	15 52	26 25	22 06	20 49	13 25	21 28	18 56	05 37	19 01	14 01
07	16 10	24 23	22 29	21 07	13 11	21 30	18 55	05 39	19 01	14 01
08	16 28	21 06	22 51	21 23	12 57	21 32	18 54	05 40	19 02	14 01
09	16 45	16 54	23 11	21 39	12 43	21 34	18 54	05 41	19 02	14 00
10	17 02	12 05	23 31	21 55	12 29	21 36	18 53	05 42	19 03	14 00
11	17 19	06 54	23 49	22 09	12 14	21 38	18 52	05 44	19 03	14 00
12	17 36	01 N 34	24 06	22 24	12 00	21 40	18 51	05 45	19 03	13 59
13	17 52	03 S 47	24 21	22 37	11 46	21 42	18 50	05 46	19 04	13 59
14	18 08	08 59	24 36	22 50	11 31	21 44	18 49	05 48	19 04	13 59
15	18 23	13 52	24 49	23 02	11 16	21 46	18 48	05 49	19 05	13 59
16	18 39	18 15	25 01	23 14	11 01	21 48	18 48	05 50	19 05	13 58
17	18 54	21 57	25 11	23 25	10 47	21 49	18 47	05 51	19 06	13 58
18	19 08	24 45	25 20	23 35	10 32	21 51	18 46	05 52	19 06	13 58
19	19 23	26 24	25 28	23 44	10 17	21 53	18 45	05 54	19 07	13 58
20	19 36	26 46	25 34	23 53	10 02	21 55	18 44	05 55	19 07	13 58
21	19 50	25 45	25 38	24 01	09 46	21 57	18 43	05 56	19 07	13 58
22	20 03	23 23	25 42	24 08	09 31	21 58	18 42	05 57	19 08	13 57
23	20 16	19 46	25 44	24 15	09 16	22 00	18 41	05 58	19 08	13 57
24	20 29	15 06	25 44	24 21	09 00	22 02	18 41	05 59	19 09	13 57
25	20 41	09 36	25 43	24 26	08 45	22 03	18 40	06 00	19 09	13 57
26	20 52	03 S 31	25 40	24 30	08 29	22 05	18 39	06 02	19 10	13 57
27	21 04	02 N 53	25 36	24 34	08 14	22 07	18 38	06 03	19 10	13 57
28	21 15	09 15	25 30	24 37	07 58	22 08	18 37	06 04	19 11	13 57
29	21 25	15 13	25 22	24 39	07 42	22 10	18 36	06 05	19 11	13 57
30	21 S 35	20 N 21	25 S 14	24 S 40	07 S 26	22 S 11	18 N 35	06 S 06	19 S 11	13 N 57

ZODIAC SIGN ENTRIES

Date	h	m	Planets
02	05	55	☽ ♉
04	05	27	☽ ♊
05	00	30	♀ ♐
06	06	59	☿ ♐
06	07	15	☽ ♋
06	12	31	♂ ♓
08	12	56	☽ ♌
10	22	44	☽ ♍
13	11	05	☽ ♎
15	23	49	☽ ♏
18	11	30	☽ ♐
20	21	36	☽ ♑
22	23	14	☉ ♐
23	05	52	☽ ♒
25	11	48	☽ ♓
27	15	04	☽ ♈
29	02	41	♀ ♑
29	16	08	☽ ♉

LATITUDES

Date	Mercury ☿ ° '	Venus ♀ ° '	Mars ♂ ° '	Jupiter ♃ ° '	Saturn ♄ ° '	Uranus ♅ ° '	Neptune ♆ ° '	Pluto ♇ ° '
01	01 S 28	00 S 05	02 S 19	00 N 30	02 S 09	00 N 38	01 N 36	15 N 46
04	01 45	00 13	02 10	00 29	02 10	00 38	01 36	15 47
07	02 01	00 20	02 01	00 29	02 10	00 38	01 36	15 48
10	02 14	00 28	01 53	00 28	02 10	00 38	01 36	15 49
13	02 24	00 35	01 45	00 28	02 10	00 38	01 36	15 50
16	02 32	00 43	01 37	00 28	02 10	00 38	01 36	15 52
19	02 35	00 50	01 30	00 27	02 10	00 38	01 36	15 53
22	02 32	00 57	01 22	00 27	02 09	00 38	01 36	15 54
25	02 23	01 04	01 15	00 27	02 09	00 38	01 36	15 56
28	02 06	01 10	01 09	00 27	02 09	00 39	01 36	15 57
31	01 S 37	01 S 16	01 S 02	00 N 26	02 S 09	00 N 39	01 N 36	15 N 59

DATA

Julian Date	2441257
Delta T	+42 seconds
Ayanamsa	23° 28' 00"
Synetic vernal point	05° ♓ 38' 59"
True obliquity of ecliptic	23° 26' 41"

LONGITUDES

Date	Chiron ⚷ ° '	Ceres ⚳ ° '	Pallas ⚴ ° '	Juno ⚵ ° '	Vesta ⚶ ° '	Black Moon Lilith ⚸ ° '
01	10 ♈ 25	14 ♌ 49	01 ♊ 56	27 ♍ 35	04 ♒ 01	17 ♎ 09
11	10 ♈ 03	17 ♌ 23	29 ♉ 16	01 ♎ 14	07 ♒ 27	18 ♎ 16
21	09 ♈ 46	19 ♌ 32	25 ♉ 59	04 ♎ 40	11 ♒ 11	19 ♎ 24
31	09 ♈ 32	21 ♌ 11	22 ♉ 36	07 ♎ 52	15 ♒ 08	20 ♎ 31

MOON'S PHASES, APSIDES AND POSITIONS ☽

Date	h	m	Phase	Longitude ° '	Eclipse Indicator
02	21	19	○	09 ♉ 48	
09	20	51	☾	16 ♌ 48	
18	01	46	●	25 ♏ 03	
25	16	37	☽	02 ♓ 45	

Day	h	m		
02	02	31	Perigee	
14	14	59	Apogee	
30	10	55	Perigee	
05	18	13	Max dec	26° N 54'
12	19	00	0S	
20	06	27	Max dec	26° S 48'
27	01	16	0N	

All ephemeris data is given at 12.00 UT and the Moon's longitude is additionally given for 24.00 UT

ASPECTARIAN

01 Monday
01 54 ☽ ∠ ♂
07 03 ☽ ☍ ♅
09 07 ☽ ± ☿
09 19 ☽ bQ ♆
13 44 ☽ ± ♀
14 04 ☽ ∠ ♄
19 36 ☽ ⊼ ☿
20 29 ☽ bQ ♃
21 36 ☽ ∥ ♇
22 26 ☉ ∥ ♂
23 10 ☽ # ♂
23 14 ☽ # ☉
23 31 ☽ ± ♆

02 Tuesday
00 00 ☽ ⊼ ♀
02 20 ☽ ✶ ♂
03 30 ☉ ⊻ ♃
04 09 ☽ ⊥ ♄
07 27 ☽ ⊼ ♇
08 58 ☽ ⊼ ♆
10 49 ☽ ± ♃
13 32 ☽ ⊻ ♄
16 54 ☽ ± ♇
17 11 ♄ # ♆
18 54 ☽ ∥ ♄
18 54 ☽ # ♆
20 23 ☽ ⊼ ♃
21 19 ☽ ☍ ☉
21 51 ☽ Q ♂
22 28 ☽ # ♀

03 Wednesday
03 25 ☽ # ☿
06 27 ☽ ⊼ ♅
07 01 ☽ # ♃
07 08 ☽ bQ ♇
16 01 ☽ ± ♅
18 17 ♂ # ♇
22 44 ♀ □ ♂

04 Thursday
00 05 ☽ ☍ ☿
03 28 ☽ □ ♂
03 43 ☽ ☍ ♀
06 29 ☽ bQ ♅
07 07 ☽ △ ♇
08 42 ☽ ☍ ♆
13 04 ☽ ☌ ♄
19 17 ☿ ∥ ♃
21 05 ☽ ☍ ♃

05 Friday
00 54 ☽ ⊼ ☉
07 06 ☽ △ ♅
11 35 ☽ ± ☉
14 21 ♀ ∠ ♅
21 23 ♀ ✶ ♇

06 Saturday
03 50 ☿ □ ♂
03 57 ☽ bQ ☉
07 02 ☽ △ ♂
07 16 ☽ ⊼ ☿
09 08 ☽ □ ♇
10 16 ☽ ⊼ ♀
10 50 ☽ ⊼ ♆
15 08 ☽ ⊻ ♄
16 51 ♀ ☌ ♆
18 54 ☽ ± ☿
20 06 ☿ ∠ ♅
21 19 ☽ ± ♆
21 45 ☽ ± ♀

07 Sunday
00 49 ☽ ⊼ ♃
01 35 ☿ ✶ ♇
01 36 ☽ ⊥ ♄
08 13 ☽ △ ☉
10 25 ☽ bQ ♂
11 02 ☽ □ ♅
11 40 ☽ ± ♃
12 46 ☽ bQ ☿
13 22 ☽ bQ ♆
15 25 ☽ bQ ♀
17 00 ☽ Q ♇
17 37 ☽ ∠ ♄
18 31 ☿ ☌ ♆

08 Monday
01 24 ☽ # ☿
03 26 ☽ ± ♂
03 51 ♂ bQ ♅
04 18 ☽ bQ ♃
09 13 ☽ # ♃
10 17 ☽ # ♀
13 34 ♀ ☍ ♄
15 01 ☽ ⊼ ♂
15 07 ☽ ✶ ♇
16 59 ☽ △ ♆
17 26 ♂ ⊼ ♇
19 44 ☽ △ ☿
20 15 ☽ Q ♅
21 10 ☽ ✶ ♄
22 02 ☽ △ ♀
23 47 ☉ ⊻ ♅

09 Tuesday
00 20 ☽ # ♆
01 05 ☽ ∥ ♄
03 25 ♀ ∥ ♃
06 14 ☉ ∠ ♇
07 59 ☿ ☍ ♄
08 55 ☽ △ ♃
12 42 ☽ # ☉
19 14 ☽ ✶ ♅
19 42 ☽ ∠ ♇
19 56 ☽ Q ♄
20 51 ☽ □ ☉

10 Wednesday
02 41 ☽ ∥ ♇
10 00 ☽ # ♂
13 18 ☽ ⊥ ♇
17 15 ♂ □ ♆

11 Thursday
00 46 ☽ ∠ ♅
01 11 ☽ ⊻ ♇
03 13 ☽ □ ♆
03 39 ☽ ☍ ♂
07 06 ☽ □ ♄
11 30 ☽ Q ☉
13 39 ☽ □ ☿
15 08 ☽ □ ♀
17 18 ☽ # ♅
18 49 ☽ ⊥ ♅
20 56 ☽ □ ♃

12 Friday
06 57 ☽ ⊻ ♅
13 51 ☽ ✶ ☉
15 27 ☽ Q ♆

13 Saturday
06 43 ☽ Q ☿
07 43 ☽ Q ♀
10 00 ☽ Q ♃
13 44 ☽ ☌ ♇
15 51 ☽ ✶ ♆
19 03 ☽ ⊼ ♂
19 14 ☽ △ ♄
21 07 ☽ ∥ ♅
22 22 ♂ □ ♄
23 06 ☽ ∠ ☉

14 Sunday
07 50 ☽ ± ♂
09 23 ♀ ☌ ♃
10 13 ☽ ✶ ☿
10 48 ☽ ✶ ♃
10 57 ☽ ✶ ♀
16 21 ☿ ☌ ♃
19 03 ☽ ⊥ ☉
20 02 ☽ ☌ ♅
22 23 ☽ ∠ ♆
23 39 ☽ ∥ ♂

15 Monday
01 27 ☽ bQ ♄
02 58 ☽ bQ ♂
07 17 ☉ ⊥ ♅
08 19 ☽ ⊻ ☉
12 36 ☽ # ♇
16 37 ☽ ⊥ ♆
17 43 ☽ ∠ ♃
17 58 ♀ Q ♇
19 30 ☽ ± ♄
20 22 ☽ ∠ ☿
20 50 ☽ ∠ ♀
21 56 ☿ Q ♇

16 Tuesday
02 35 ☽ ⊻ ♇
04 44 ☽ ⊻ ♆
07 29 ☽ ⊼ ♄
10 41 ☽ △ ♂
12 08 ☽ ⊥ ♃
14 27 ☽ ∥ ☉
14 38 ☽ ⊥ ♇
15 10 ☽ # ♄
16 38 ☽ ⊥ ☿
16 58 ☽ ∥ ♆
16 59 ☽ ⊥ ♀

17 Wednesday
00 21 ☽ ⊻ ♃
01 22 ☉ # ♄
06 02 ☽ ⊻ ☿
06 20 ☽ ⊻ ♀
08 34 ☽ ⊻ ♅
08 38 ☽ ∠ ♇
11 01 ☽ ∥ ♃
20 31 ☽ ⊥ ♅

18 Thursday
00 10 ☽ ∥ ♀
01 46 ☽ ☌ ☉
04 51 ♀ ✶ ♅
07 11 ☿ ✶ ♅
08 19 ☉ ∥ ♆
14 19 ☽ ✶ ♇
14 19 ☽ ∠ ♅
16 28 ☽ ☌ ♆
18 35 ☽ ☍ ♄
19 16 ☽ ∥ ☿

19 Friday
00 58 ☽ □ ♂
12 29 ☽ ☌ ♃
13 46 ☽ Q ♇
19 40 ☽ ✶ ♅
23 29 ☽ ☌ ☿
23 49 ☽ ☌ ♀

20 Saturday
13 29 ☽ Q ♂
17 16 ☽ ⊻ ☉
18 50 ☽ Q ♅

21 Sunday
00 27 ☽ □ ♇
02 37 ☽ ⊻ ♆
04 07 ☽ ⊼ ♄
05 42 ☽ ⊥ ☉
13 25 ☽ ✶ ♂
13 28 ☽ ∥ ☿
14 01 ☽ ⊥ ♆
15 25 ☽ ± ♄
22 52 ☽ ⊻ ♃

22 Monday
00 10 ☽ ∠ ☉
05 02 ☽ □ ♅
05 54 ☽ ∥ ♀
07 00 ☽ ∠ ♆
08 13 ☽ bQ ♄
10 19 ☽ ⊥ ♃
14 03 ☽ ⊻ ☿
14 56 ☽ ⊻ ♀
18 51 ☽ ∠ ♂
20 29 ♃ Q ♇
22 14 ☽ ∥ ♃

23 Tuesday
02 10 ☽ ⊥ ☿
03 13 ☽ ⊥ ♀
03 18 ☽ ∠ ♃
06 26 ☽ ✶ ☉
08 43 ☽ △ ♇
09 14 ☽ ∥ ☉
10 51 ☽ ✶ ♆
11 48 ☽ △ ♄
12 13 ☽ ⊥ ♂
15 34 ☽ ∥ ♆
18 02 ☽ # ♄
20 04 ☽ ∠ ☿
21 25 ☽ ∠ ♀
23 40 ☽ ⊻ ♂

24 Wednesday
06 06 ☽ Q ☉
07 06 ☽ ✶ ♃
08 42 ☽ Q ♆
11 59 ☽ bQ ♇
12 17 ☽ △ ♅
12 40 ☉ ✶ ♇
16 46 ☉ ∠ ♅
17 17 ☽ # ♇

25 Thursday
01 07 ☽ ✶ ☿
03 04 ☽ ✶ ♀
04 02 ☽ ± ♇
04 52 ☽ Q ♃
14 36 ☽ ⊼ ♇
14 56 ☽ bQ ♅
15 39 ☽ ∥ ♂
16 37 ☽ □ ☉
16 41 ☽ □ ♆
17 07 ☽ □ ♄
17 43 ☉ ☌ ♆
19 00 ♂ ± ♅
23 05 ☉ ☍ ♄
23 38 ☽ Q ☿

26 Friday
02 08 ☽ Q ♀
02 20 ☽ ∥ ♅
06 36 ☽ ± ♅
07 04 ☽ ☌ ♂
12 39 ☽ □ ♃
16 54 ☽ ⊼ ♅
23 51 ☽ Q ♄

27 Saturday
08 11 ☽ □ ☿
11 41 ☽ □ ♀
17 48 ☽ ☍ ♇
18 50 ♄ ☍ ♆
19 50 ☽ ✶ ♄
19 51 ☽ △ ♆
23 29 ☽ △ ☉
23 52 ☽ # ♅

28 Sunday
05 51 ♀ Q ♅
07 17 ☽ # ♂
11 33 ☽ ⊻ ♂
15 32 ☽ △ ♃
18 59 ☽ ☍ ♅
20 21 ☽ ∠ ♄
20 34 ☽ bQ ♆
21 46 ☽ ⊥ ♂

29 Monday
01 52 ☽ bQ ☉
06 41 ☽ ∥ ♇
10 52 ☽ ⊥ ♄
11 12 ☽ ± ♆
11 43 ☽ △ ☿
12 58 ☽ ∠ ♂
16 14 ☽ bQ ♃
17 22 ☽ △ ♀
17 29 ☽ ± ☉
18 51 ☽ ⊼ ♇
20 30 ☽ ⊻ ♄
20 54 ☽ ⊼ ♆

30 Tuesday
03 13 ☽ ∥ ♄
03 51 ☽ ⊼ ☉
04 31 ☽ ± ♇
06 07 ☽ # ♆
06 57 ☽ ± ♃
07 41 ♀ Q ♂
11 31 ♀ □ ♇
12 37 ☽ bQ ☿
14 07 ☽ ✶ ♂
16 44 ☽ ⊼ ♃
19 00 ☽ bQ ♇
19 07 ☽ # ☉
19 32 ☽ ⊼ ♅
19 40 ☽ bQ ♀
22 29 ☽ # ♃

DECEMBER 1971

LONGITUDES

Date	Sidereal time h m s	Sun ☉ ° ' "	Moon ☽ ° ' "	Moon ☽ 24.00 ° ' "	Mercury ☿ ° '	Venus ♀ ° '	Mars ♂ ° '	Jupiter ♃ ° '	Saturn ♄ ° '	Uranus ♅ ° '	Neptune ♆ ° '	Pluto ♇ ° '
01	16 38 46	08 ♐ 38 05	27 ♉ 15 57	04 ♊ 40 46	28 ♐ 03	02 ♑ 58	14 ♓ 13	15 ♐ 29	02 ♊ 34	17 ♎ 04	03 ♐ 01	01 ♎ 43
02	16 42 42	09 38 54	12 ♊ 02 17	19 ♊ 19 31	28 14	04 13	14 49	15 42	02 R 30	17 07	03 03	01 44
03	16 46 39	10 39 44	26 ♊ 31 35	03 ♋ 37 49	28 R 15	05 27	15 26	15 56	02 25	17 09	03 06	01 45
04	16 50 35	11 40 35	10 ♋ 37 41	17 ♋ 30 54	28 05	06 42	16 02	16 09	02 20	17 12	03 08	01 47
05	16 54 32	12 41 27	24 ♋ 17 17	00 ♌ 56 55	27 45	07 57	16 39	16 23	02 15	17 15	03 10	01 48
06	16 58 28	13 42 21	07 ♌ 29 57	13 ♌ 56 43	27 14	09 11	17 15	16 36	02 10	17 17	03 12	01 49
07	17 02 25	14 43 15	20 ♌ 17 38	26 ♌ 33 13	26 31	10 26	17 52	16 50	02 06	17 20	03 15	01 50
08	17 06 21	15 44 11	02 ♍ 44 02	08 ♍ 50 41	25 38	11 40	18 29	17 03	02 01	17 22	03 17	01 51
09	17 10 18	16 45 08	14 ♍ 53 50	20 ♍ 54 07	24 34	12 55	19 06	17 17	01 56	17 24	03 19	01 51
10	17 14 15	17 46 07	26 ♍ 52 11	02 ♎ 48 41	23 22	14 09	19 44	17 31	01 52	17 27	03 21	01 52
11	17 18 11	18 47 06	08 ♎ 44 15	14 ♎ 39 27	22 04	15 24	20 21	17 44	01 47	17 29	03 23	01 53
12	17 22 08	19 48 07	20 ♎ 34 52	26 ♎ 31 00	20 42	16 38	20 58	17 58	01 42	17 31	03 26	01 54
13	17 26 04	20 49 08	02 ♏ 28 20	08 ♏ 27 16	19 19	17 53	21 36	18 11	01 38	17 34	03 28	01 55
14	17 30 01	21 50 11	14 ♏ 28 10	20 ♏ 31 21	17 58	19 07	22 13	18 25	01 33	17 36	03 30	01 55
15	17 33 57	22 51 15	26 ♏ 37 04	02 ♐ 45 30	16 41	20 22	22 51	18 39	01 29	17 38	03 32	01 56
16	17 37 54	23 52 19	08 ♐ 56 49	15 ♐ 11 05	15 31	21 36	23 29	18 52	01 25	17 40	03 34	01 57
17	17 41 50	24 53 25	21 ♐ 28 23	27 ♐ 48 44	14 29	22 51	24 06	19 06	01 20	17 42	03 36	01 57
18	17 45 47	25 54 31	04 ♑ 12 06	10 ♑ 38 27	13 37	24 05	24 44	19 19	01 16	17 44	03 39	01 58
19	17 49 44	26 55 37	17 ♑ 07 47	23 ♑ 40 02	12 56	25 20	25 22	19 33	01 12	17 46	03 41	01 59
20	17 53 40	27 56 44	00 ♒ 15 10	06 ♒ 53 09	12 26	26 34	26 00	19 46	01 08	17 48	03 43	01 59
21	17 57 37	28 57 51	13 ♒ 33 59	20 ♒ 17 40	12 07	27 48	26 39	20 00	01 04	17 49	03 45	02 00
22	18 01 33	29 ♐ 58 59	27 ♒ 04 12	03 ♓ 53 37	11 58	29 ♑ 03	27 17	20 13	01 00	17 51	03 47	02 00
23	18 05 30	01 ♑ 00 07	10 ♓ 45 55	17 ♓ 41 08	12 D 00	00 ♒ 17	27 55	20 27	00 56	17 53	03 49	02 01
24	18 09 26	02 01 15	24 ♓ 39 14	01 ♈ 40 10	12 10	01 31	28 33	20 40	00 52	17 55	03 51	02 01
25	18 13 23	03 02 23	08 ♈ 43 49	15 ♈ 50 01	12 29	02 45	29 12	20 54	00 48	17 56	03 53	02 01
26	18 17 19	04 03 31	22 ♈ 58 31	00 ♉ 09 00	12 56	04 00	29 ♓ 50	21 07	00 44	17 58	03 55	02 02
27	18 21 16	05 04 39	07 ♉ 21 01	14 ♉ 34 06	13 29	05 14	00 ♈ 29	21 21	00 41	17 59	03 57	02 02
28	18 25 13	06 05 47	21 ♉ 47 37	29 ♉ 00 58	14 09	06 28	01 07	21 34	00 37	18 01	03 59	02 02
29	18 29 09	07 06 55	06 ♊ 13 25	13 ♊ 24 16	14 54	07 42	01 46	21 48	00 34	18 02	04 01	02 03
30	18 33 06	08 08 03	20 ♊ 32 47	27 ♊ 38 17	15 44	08 56	02 25	22 01	00 30	18 03	04 03	02 03
31	18 37 02	09 ♑ 09 11	04 ♋ 40 09	11 ♋ 37 49	16 ♐ 39	10 ♒ 10	03 ♈ 03	22 ♐ 14	00 ♊ 27	18 ♎ 05	04 ♐ 05	02 ♎ 03

Date	Moon True ☊ ° '	Moon Mean ☊ ° '	Moon ☽ Latitude ° '
01	06 ♒ 57	08 ♒ 15	04 N 45
02	06 R 48	08 11	04 09
03	06 40	08 08	03 17
04	06 35	08 05	02 14
05	06 32	08 02	01 N 05
06	06 D 31	07 59	00 S 05
07	06 32	07 56	01 13
08	06 33	07 52	02 16
09	06 34	07 49	03 11
10	06 R 34	07 46	03 57
11	06 31	07 43	04 33
12	06 27	07 40	04 56
13	06 20	07 37	05 07
14	06 12	07 33	05 05
15	06 04	07 30	04 48
16	05 55	07 27	04 19
17	05 48	07 24	03 36
18	05 42	07 21	02 42
19	05 38	07 17	01 39
20	05 37	07 14	00 S 29
21	05 D 37	07 11	00 N 43
22	05 38	07 08	01 54
23	05 40	07 05	03 00
24	05 41	07 02	03 56
25	05 R 41	06 58	04 39
26	05 40	06 55	05 05
27	05 37	06 52	05 12
28	05 33	06 49	05 00
29	05 28	06 46	04 29
30	05 24	06 43	03 41
31	05 ♒ 21	06 ♒ 39	02 N 40

DECLINATIONS

Date	Sun ☉ ° '	Moon ☽ ° '	Mercury ☿ ° '	Venus ♀ ° '	Mars ♂ ° '	Jupiter ♃ ° '	Saturn ♄ ° '	Uranus ♅ ° '	Neptune ♆ ° '	Pluto ♇ ° '
01	21 S 45	24 N 11	25 S 03	24 S 41	07 S 10	22 S 13	18 N 35	06 S 07	19 S 12	13 N 57
02	21 54	26 21	24 51	24 41	06 54	22 15	18 34	06 08	19 12	13 57
03	22 03	26 41	24 38	24 40	06 38	22 16	18 33	06 09	19 13	13 57
04	22 11	25 15	24 23	24 38	06 22	22 18	18 32	06 10	19 13	13 57
05	22 19	22 20	24 06	24 36	06 06	22 19	18 31	06 11	19 14	13 57
06	22 27	18 19	23 48	24 33	05 50	22 20	18 30	06 12	19 14	13 57
07	22 34	13 34	23 28	24 29	05 34	22 22	18 30	06 13	19 14	13 57
08	22 41	08 23	23 07	24 24	05 18	22 23	18 29	06 13	19 15	13 58
09	22 47	03 N 00	22 44	24 19	05 01	22 25	18 28	06 14	19 15	13 58
10	22 53	02 S 23	22 21	24 13	04 45	22 26	18 27	06 15	19 16	13 58
11	22 58	07 38	21 57	24 06	04 29	22 27	18 26	06 16	19 16	13 58
12	23 03	12 37	21 32	23 58	04 12	22 29	18 26	06 17	19 16	13 58
13	23 08	17 08	21 08	23 50	03 56	22 30	18 25	06 18	19 17	13 58
14	23 12	21 02	20 46	23 41	03 39	22 31	18 24	06 18	19 17	13 59
15	23 15	24 05	20 25	23 31	03 23	22 33	18 23	06 19	19 18	13 59
16	23 18	26 03	20 06	23 21	03 06	22 34	18 23	06 20	19 18	13 59
17	23 21	26 46	19 50	23 10	02 50	22 35	18 22	06 21	19 18	13 59
18	23 23	26 05	19 37	22 58	02 33	22 36	18 21	06 21	19 19	14 00
19	23 25	23 59	19 28	22 45	02 17	22 37	18 21	06 22	19 19	14 00
20	23 26	20 35	19 22	22 32	02 00	22 38	18 20	06 23	19 19	14 00
21	23 26	16 04	19 19	22 18	01 43	22 39	18 19	06 24	19 20	14 00
22	23 27	10 42	19 18	22 04	01 27	22 41	18 19	06 24	19 20	14 01
23	23 26	04 S 45	19 21	21 48	01 10	22 42	18 18	06 25	19 21	14 01
24	23 26	01 N 29	19 26	21 33	00 53	22 43	18 18	06 25	19 21	14 02
25	23 25	07 44	19 33	21 16	00 37	22 44	18 17	06 26	19 21	14 02
26	23 23	13 39	19 42	20 59	00 20	22 45	18 16	06 27	19 22	14 02
27	23 21	18 53	19 53	20 42	00 S 03	22 46	18 16	06 27	19 22	14 03
28	23 18	23 02	20 04	20 23	00 N 13	22 47	18 15	06 28	19 22	14 03
29	23 15	25 46	20 17	20 05	00 30	22 47	18 15	06 28	19 23	14 04
30	23 12	26 47	20 30	19 45	00 47	22 48	18 14	06 29	19 23	14 04
31	23 S 08	26 N 02	20 S 44	19 S 25	01 N 04	22 S 49	18 N 14	06 S 29	19 S 23	14 N 04

ZODIAC SIGN ENTRIES

Date	h	m	Planets
01	16	25	☽ ♊
03	17	51	☽ ♋
05	22	17	☽ ♌
08	06	40	☽ ♍
10	18	19	☽ ♎
13	07	01	☽ ♏
15	18	37	☽ ♐
18	04	07	☽ ♑
20	11	32	☽ ♒
22	12	24	☉ ♑
22	17	10	☽ ♓
23	06	32	♀ ♒
24	21	09	☽ ♈
26	18	04	♂ ♈
26	23	45	☽ ♉
29	01	38	☽ ♊
31	04	01	☽ ♋

LATITUDES

Date	Mercury ☿ ° '	Venus ♀ ° '	Mars ♂ ° '	Jupiter ♃ ° '	Saturn ♄ ° '	Uranus ♅ ° '	Neptune ♆ ° '	Pluto ♇ ° '
01	01 S 37	01 S 16	01 S 02	00 N 26	02 S 09	00 N 39	01 N 36	15 N 59
04	00 57	01 22	00 56	00 26	02 08	00 39	01 36	16 00
07	00 S 04	01 27	00 50	00 26	02 08	00 39	01 36	16 02
10	00 N 56	01 32	00 45	00 26	02 08	00 39	01 36	16 04
13	01 53	01 36	00 39	00 25	02 07	00 39	01 36	16 05
16	02 34	01 39	00 34	00 25	02 07	00 39	01 36	16 07
19	02 55	01 42	00 29	00 25	02 06	00 39	01 36	16 09
22	02 57	01 45	00 24	00 25	02 06	00 39	01 36	16 10
25	02 46	01 46	00 19	00 24	02 05	00 39	01 36	16 12
28	02 27	01 47	00 15	00 24	02 04	00 39	01 36	16 14
31	02 N 03	01 S 48	00 S 10	00 N 24	02 S 04	00 N 39	01 N 36	16 N 15

DATA

Julian Date	2441287
Delta T	+42 seconds
Ayanamsa	23° 28' 05"
Synetic vernal point	05° ♓ 38' 54"
True obliquity of ecliptic	23° 26' 40"

MOON'S PHASES, APSIDES AND POSITIONS ☽

Date	h	m	Phase	Longitude ° '	Eclipse Indicator
02	07	48	○	09 ♊ 28	
09	16	02	☾	16 ♍ 55	
17	19	03	●	25 ♐ 11	
25	01	35	☽	02 ♈ 36	
31	20	20	○	09 ♋ 30	

Day	h	m		
12	07	00	Apogee	
28	04	47	Perigee	
03	04	17	Max dec	26° N 47'
10	01	21	0S	
17	12	23	Max dec	26° S 46'
24	06	20	0N	
30	13	41	Max dec	26° N 47'

LONGITUDES

Date	Chiron ⚷ ° '	Ceres ⚳ ° '	Pallas ⚴ ° '	Juno ⚵ ° '	Vesta ⚶ ° '	Black Moon Lilith ° '
01	09 ♈ 32	21 ♌ 11	22 ♉ 36	07 ♎ 52	15 ♒ 08	20 ♎ 31
11	09 ♈ 24	22 ♌ 16	19 ♉ 40	10 ♎ 48	19 ♒ 17	21 ♎ 38
21	09 ♈ 21	22 ♌ 41	17 ♉ 38	13 ♎ 25	23 ♒ 35	22 ♎ 45
31	09 ♈ 24	22 ♌ 24	16 ♉ 45	15 ♎ 40	28 ♒ 00	23 ♎ 52

ASPECTARIAN

h m	Aspects
01 Wednesday	
03 26	☽ ± ☿
04 50	♀ ⊼ ♄
05 13	☽ ± ♅
10 14	☽ Q ♂
11 29	☽ ± ♀
12 57	♀ ⊻ ♆
13 17	☽ ⊼ ☿
16 09	☽ # ♀
19 01	☽ # ☿
19 13	☽ △ ♇
19 47	☽ bQ ♅
20 32	☽ ☌ ♄
21 20	☽ ☍ ♆
22 04	☽ ⊼ ♀
02 Thursday	
07 48	☽ ☍ ☉
16 46	☽ □ ♂
18 07	☽ ☍ ♃
20 22	☽ △ ♅
03 Friday	
02 32	☿ St R
08 07	☿ ∥ ♀
14 52	☽ ☍ ☿
20 50	☽ □ ♇
21 52	☽ ⊻ ♄
23 07	☽ ⊼ ♆
04 Saturday	
04 35	☽ ☍ ♀
08 04	☽ ⊥ ♄
09 24	☽ ± ♆
13 57	☽ ⊼ ☉
18 09	☽ # ♀
19 25	♂ □ ♃
21 19	☽ # ☿
21 46	☽ ⊼ ♃
21 51	☽ △ ♂
23 29	☽ □ ♅
23 37	☽ ∠ ♄
05 Sunday	
01 07	☽ bQ ♆
01 16	☽ ± ☉
04 00	☽ Q ♇
05 48	♂ ∥ ♅
08 32	☽ ± ♃
10 32	☉ ∥ ♃
12 03	☽ # ☉
12 06	☽ # ♃
13 32	♄ bQ ♅
17 36	♀ ± ♄
18 01	☽ ⊼ ☿
18 37	☽ bQ ☉
06 Monday	
01 00	☽ bQ ♃
01 33	☽ ⚹ ♇
01 55	☽ bQ ♂
02 17	☽ ⚹ ♄
04 05	☽ △ ♆
04 30	☽ ± ☿
06 58	☽ # ♆
07 55	☽ Q ♅
10 58	☽ ∥ ♄
12 22	♀ ⊥ ♆
13 11	♂ ⊼ ♅
14 30	☉ Q ♇
15 27	☽ ⊼ ♀
19 19	☽ ± ♂
20 22	☽ bQ ☿
07 Tuesday	
00 21	☽ Q ♄
00 33	☽ △ ☉
03 53	☽ ± ♀
05 19	☽ △ ♃
05 25	☽ ∠ ♇
06 21	☽ ⚹ ♅
07 10	☽ ⊼ ♂
10 07	☽ ∥ ♇
22 37	☽ ⊥ ♇
22 55	☽ bQ ♀
23 10	☽ △ ☿
08 Wednesday	
10 15	☽ ⊻ ♇
10 36	☽ □ ♄
11 17	☽ ∠ ♅
13 04	☽ □ ♆
21 40	☽ # ♅
09 Thursday	
02 34	☽ # ♂
05 03	☽ ⊥ ♅
07 36	☽ △ ♀
09 37	☉ ∥ ☿
16 02	☽ □ ☉
16 51	☽ □ ♃
17 01	☽ ⊻ ♅
20 52	☽ ☍ ♂
10 Friday	
00 52	☽ Q ♆
03 45	♃ ⚹ ♅
04 05	☉ ⚹ ♅
04 10	☉ ☌ ♃
05 37	☽ □ ☿
06 54	☿ ∥ ♃
08 37	♄ △ ♇
16 35	♂ Q ♄
22 00	☽ △ ♄
22 07	☽ ☌ ♇
22 10	☽ ∥ ♂
11 Saturday	
01 08	☽ ⚹ ♆
05 37	☽ ∥ ♅
05 48	☽ Q ♃
07 40	☽ Q ☉
14 26	☽ Q ☿
12 Sunday	
03 05	☽ □ ♀
04 12	☽ bQ ♄
05 47	☽ ☌ ♅
06 36	☽ ⚹ ♃
07 37	☽ ∠ ♆
08 51	☿ □ ♂
10 16	☽ ⚹ ☉
12 14	☽ ⚹ ☿
12 50	☽ ⊼ ♂
13 13	♀ bQ ♄
18 56	☽ # ♇
21 02	☉ ☌ ☿
22 18	☽ ± ♄
13 Monday	
01 38	☽ ± ♂
01 53	☽ ⊥ ♆
05 35	♀ □ ♅
10 19	☽ ⊼ ♄
10 52	☽ ⊻ ♇
13 28	☽ ∠ ♃
14 00	☽ ⊻ ♆
15 20	☽ ∠ ☿
19 19	♀ ⊻ ♃
19 21	☽ ∠ ☉
19 25	☽ # ♄
19 38	☽ Q ♀
20 44	☽ bQ ♂
22 56	☽ ⊥ ♇
23 35	♀ ∠ ♆
14 Tuesday	
00 41	☽ ∥ ♆
01 16	☿ ⊻ ♀
05 05	☿ ☌ ♃
07 31	☽ ⊥ ☿
07 50	☽ ⊥ ♃
10 23	☽ ∥ ☿
14 58	☽ ⊥ ☉
16 53	☽ ∠ ♇
18 13	☽ ⊻ ♅
18 16	☽ ⊻ ☿
18 36	☿ ⚹ ♅
19 59	☽ ⊻ ♃
22 17	☽ ⚹ ♀
22 55	☽ ∥ ♃
15 Wednesday	
03 56	☽ ⊻ ☉
04 11	☽ △ ♂
04 33	☽ ∥ ☉
06 07	☽ ⊥ ♅
07 16	☽ ∥ ♀
11 34	☉ □ ♂
21 27	☽ ☍ ♄
22 24	☽ ⚹ ♇
23 47	☽ ∠ ♅
16 Thursday	
01 33	☽ ☌ ♆
06 58	☽ ∠ ♀
11 04	☿ ⊥ ♀
16 27	☉ ∥ ♀
21 38	☽ Q ♇
23 38	☽ ☌ ☿
17 Friday	
02 13	☽ ⊥ ♀
04 47	☽ ⚹ ♅
07 24	☽ ☌ ♃
14 53	☽ ⊻ ♀
17 16	☽ □ ♂
19 03	☽ ☌ ☉
18 Saturday	
01 56	☿ Q ♇
03 35	☽ Q ♅
06 32	☽ ⊼ ♄
07 49	☽ □ ♇
10 57	☽ ⊻ ♆
17 42	☽ ± ♄
22 10	☽ ⊥ ♆
19 Sunday	
04 37	☽ ⊻ ☿
04 42	☽ Q ♂
10 17	☽ bQ ♄
13 10	☽ □ ♅
13 51	♀ ⚹ ♂
14 51	☽ ∠ ♆
15 11	☽ ⊥ ☿
16 32	☽ ⊻ ♃
16 41	☽ ∥ ☉
17 17	♀ ⊥ ♃
22 22	☽ ∥ ♀
22 35	☽ ∥ ♃
20 Monday	
01 24	♀ ∥ ♃
03 42	☽ ⊥ ♃
03 53	☽ ⚹ ♂
04 36	☽ ☌ ♀
07 02	☽ ∠ ☿
07 27	☽ ⊻ ☉
13 35	☽ △ ♄
15 09	☽ △ ♇
18 17	☽ ⚹ ♆
19 08	☽ ∥ ☿
19 12	☽ ∥ ♆
19 15	☽ ⊥ ☉
20 20	☽ ∠ ♃
21 Tuesday	
00 05	☿ ∥ ♆
00 36	☽ # ♄
01 05	☿ ∠ ♀
08 23	☽ ∠ ♂
09 27	☽ ⚹ ☿
12 46	☽ ∠ ☉
15 55	☽ Q ♆
18 08	☽ bQ ♇
19 37	☽ △ ♅
21 35	☽ # ♇
23 40	☽ ⚹ ♃
22 Wednesday	
01 14	☽ ⊥ ♂
06 33	☽ Q ☿
08 51	☉ Q ♅
10 07	☽ ± ♇
12 23	☽ ⊻ ♂
15 50	☽ ⊻ ♀
17 33	☽ ⚹ ☉
18 53	☽ □ ♄
20 41	☽ ⊼ ♇
20 48	☿ St D
21 14	☽ Q ♃
22 12	☽ bQ ♅
23 50	☽ □ ♆
23 Thursday	
03 24	☽ ⊥ ♀
05 29	☽ ∥ ♅
08 26	☿ ∥ ♆
10 22	☉ ⊼ ♄
13 57	☽ ± ♅
14 09	☽ □ ☿
16 12	☽ Q ☉
20 37	☽ ∠ ♀
23 54	♀ △ ♄
24 Friday	
00 22	☽ ⊼ ♅
02 05	☽ Q ♄
02 30	☽ ∥ ♂
05 03	☽ □ ♃
09 49	☽ # ♂
11 55	☉ □ ♇
19 00	☽ ☌ ♂
21 41	♀ △ ♇
22 35	☽ ⚹ ♄
25 Saturday	
00 36	☽ ☍ ♇
00 53	☽ ⚹ ♀
01 35	☽ □ ☉
03 45	☽ △ ♆
06 57	☽ # ♅
18 32	☽ △ ☿
23 09	☽ Q ♀
23 54	☽ ∠ ♄
26 Sunday	
03 34	☽ ☍ ♅
05 10	☽ bQ ♆
08 35	☉ ⊻ ♆
08 51	☽ △ ♃
10 28	♀ ⚹ ♆
13 41	☽ ∥ ♇
14 57	☽ ⊥ ♄
19 00	☉ ⊻ ♀
20 18	☽ ± ♆
20 36	☽ bQ ☿
27 Monday	
00 01	☽ ⊻ ♂
00 56	☽ ⊻ ♄
03 08	☽ ⊼ ♇
06 20	☽ ⊼ ♆
07 56	☽ △ ☉
08 08	☽ □ ♀
08 57	☽ ∥ ♄
10 18	☽ bQ ♃
10 29	☽ ⊥ ♂
12 14	☽ ± ☿
13 08	☽ ± ♇
14 29	☽ # ♆
17 23	☽ # ☿
18 48	♂ ⚹ ♄
20 58	☽ # ♀
22 40	☽ ⊼ ☿
28 Tuesday	
01 30	☽ ± ♃
02 09	☽ ∠ ♂
04 06	☽ bQ ♇
05 43	☽ ⊼ ♅
08 12	☿ Q ♇
10 12	☽ # ♃
10 45	☽ bQ ☉
11 37	☽ ⊼ ♃
13 48	☽ # ☉
14 26	♀ ∠ ♃
15 41	☽ ± ♅
23 40	☉ ± ♄
29 Wednesday	
02 37	☽ ☌ ♄
02 44	☿ ∥ ♀
02 51	☽ ± ☉
04 14	☽ ⚹ ♂
05 02	☽ △ ♇
06 41	☽ bQ ♅
08 19	☽ ☍ ♆
13 36	☽ ⊼ ☉
14 42	☽ △ ♀
22 18	♂ ☍ ♇
30 Thursday	
01 12	☽ Q ♂
03 24	☽ ☍ ☿
07 48	☽ △ ♅
14 31	☽ ☍ ♃
18 17	☽ bQ ♀
31 Friday	
04 49	☽ ⊻ ♄
07 31	☽ □ ♇
09 06	☽ □ ♂
11 00	☽ ⊼ ♆
11 04	☽ ± ♀
14 22	♀ ∥ ♆
15 03	☽ ⊥ ♄
20 20	☽ ☍ ☉
21 21	☽ ± ♆
22 24	☽ ⊼ ♀

All ephemeris data is given at 12.00 UT and the Moon's longitude is additionally given for 24.00 UT

JANUARY 1972

Date	Sidereal time h m s	Sun ☉ ° ' "	Moon ☽ ° ' "	Moon ☽ 24.00 ° ' "	Mercury ☿ ° '	Venus ♀ ° '	Mars ♂ ° '	Jupiter ♃ ° '	Saturn ♄ ° '	Uranus ♅ ° '	Neptune ♆ ° '	Pluto ♇ ° '
01	18 40 59	10 ♑ 10 20	18 ♋ 30 49	25 ♋ 18 48	17 ♐ 37	11 ♒ 24	03 ♈ 42	22 ♐ 28	00 ♊ 24	18 ♎ 06	04 ♐ 07	02 ♎ 03
02	18 44 55	11 11 28	02 ♌ 01 31	08 ♌ 38 52	18 39	12 38	04 21	22 41	00 R 21	18 07	04 09	02 03
03	18 48 52	12 12 37	15 ♌ 10 49	21 ♌ 37 30	19 44	13 52	05 00	22 54	00 18	18 08	04 11	02 03
04	18 52 48	13 13 46	27 ♌ 59 05	04 ♍ 15 52	20 51	15 06	05 39	23 07	00 15	18 09	04 13	02 03
05	18 56 45	14 14 54	10 ♍ 28 14	16 ♍ 36 38	22 01	16 20	06 18	23 21	00 12	18 10	04 14	02 R 03
06	19 00 42	15 16 03	22 ♍ 41 31	28 ♍ 43 28	23 13	17 34	06 57	23 34	00 09	18 11	04 16	02 03
07	19 04 38	16 17 13	04 ♎ 43 01	10 ♎ 40 46	24 27	18 48	07 36	23 47	00 07	18 12	04 18	02 03
08	19 08 35	17 18 22	16 ♎ 37 20	22 ♎ 33 19	25 43	20 02	08 15	24 00	00 04	18 13	04 20	02 03
09	19 12 31	18 19 31	28 ♎ 29 20	04 ♏ 25 58	27 00	21 16	08 54	24 13	00 ♊ 02	18 14	04 22	02 03
10	19 16 28	19 20 41	10 ♏ 23 48	16 ♏ 23 23	28 19	22 29	09 33	24 26	29 ♉ 59	18 15	04 23	02 03
11	19 20 24	20 21 50	22 ♏ 25 13	28 ♏ 29 46	29 ♐ 39	23 43	10 12	24 39	29 57	18 15	04 25	02 02
12	19 24 21	21 22 59	04 ♐ 37 29	10 ♐ 48 42	01 ♑ 00	24 57	10 51	24 52	29 55	18 16	04 27	02 02
13	19 28 17	22 24 09	17 ♐ 03 45	23 ♐ 22 51	02 22	26 11	11 31	25 05	29 53	18 17	04 28	02 02
14	19 32 14	23 25 18	29 ♐ 46 10	06 ♑ 13 49	03 45	27 24	12 10	25 18	29 51	18 17	04 30	02 02
15	19 36 11	24 26 27	12 ♑ 45 48	19 ♑ 22 04	05 09	28 37	12 49	25 30	29 49	18 18	04 32	02 01
16	19 40 07	25 27 35	26 ♑ 02 30	02 ♒ 46 54	06 34	29 ♒ 51	13 28	25 43	29 47	18 18	04 33	02 01
17	19 44 04	26 28 43	09 ♒ 35 02	16 ♒ 26 36	08 00	01 ♓ 04	14 08	25 56	29 46	18 18	04 35	02 00
18	19 48 00	27 29 50	23 ♒ 21 17	00 ♓ 18 43	09 27	02 18	14 47	26 08	29 44	18 19	04 36	02 00
19	19 51 57	28 30 57	07 ♓ 18 31	14 ♓ 20 19	10 54	03 31	15 27	26 21	29 43	18 19	04 38	01 59
20	19 55 53	29 ♑ 32 03	21 ♓ 25 03	28 ♓ 28 22	12 22	04 44	16 06	26 33	29 41	18 19	04 39	01 59
21	19 59 50	00 ♒ 33 07	05 ♈ 33 54	12 ♈ 39 57	13 51	05 57	16 45	26 46	29 40	18 19	04 41	01 58
22	20 03 46	01 34 11	19 ♈ 46 15	26 ♈ 52 26	15 20	07 10	17 25	26 58	29 39	18 19	04 42	01 58
23	20 07 43	02 35 14	03 ♉ 58 16	11 ♉ 03 26	16 50	08 23	18 04	27 11	29 38	18 R 19	04 44	01 57
24	20 11 40	03 36 16	18 ♉ 07 40	25 ♉ 10 44	18 20	09 36	18 44	27 23	29 37	18 19	04 45	01 56
25	20 15 36	04 37 17	02 ♊ 12 21	09 ♊ 12 16	19 52	10 49	19 23	27 35	29 37	18 19	04 46	01 56
26	20 19 33	05 38 17	16 ♊ 10 13	23 ♊ 05 56	21 24	12 02	20 03	27 47	29 36	18 19	04 48	01 55
27	20 23 29	06 39 16	29 ♊ 59 10	06 ♋ 49 39	22 56	13 15	20 42	27 59	29 36	18 19	04 49	01 54
28	20 27 26	07 40 13	13 ♋ 37 09	20 ♋ 21 26	24 29	14 28	21 22	28 11	29 35	18 19	04 50	01 54
29	20 31 22	08 41 10	27 ♋ 02 18	03 ♌ 39 34	26 03	15 40	22 02	28 23	29 35	18 18	04 52	01 53
30	20 35 19	09 42 05	10 ♌ 13 07	16 ♌ 42 49	27 38	16 53	22 41	28 35	29 35	18 18	04 53	01 52
31	20 39 15	10 ♒ 43 00	23 ♌ 08 40	29 ♌ 30 39	29 ♑ 13	18 ♓ 05	23 ♈ 21	28 ♐ 47	29 ♉ 35	18 ♎ 17	04 ♐ 54	01 ♎ 51

DECLINATIONS

Date	Moon True ☊ ° '	Moon Mean ☊ ° '	Moon ☽ Latitude ° '	Sun ☉ ° '	Moon ☽ ° '	Mercury ☿ ° '	Venus ♀ ° '	Mars ♂ ° '	Jupiter ♃ ° '	Saturn ♄ ° '	Uranus ♅ ° '	Neptune ♆ ° '	Pluto ♇ ° '
01	05 ♒ 18	06 ♒ 36	01 N 31	23 S 03	23 N 40	20 S 58	19 S 05	01 N 20	22 S 50	18 N 14	06 S 30	19 S 24	14 N 05
02	05 R 17	06 33	00 N 18	22 58	20 00	21 12	18 44	01 37	22 51	18 13	06 30	19 24	14 05
03	05 D 18	06 30	00 S 54	22 53	15 25	21 25	18 22	01 54	22 52	18 13	06 30	19 24	14 06
04	05 19	06 27	02 02	22 47	10 16	21 39	18 01	02 10	22 52	18 12	06 31	19 25	14 06
05	05 21	06 23	03 02	22 41	04 N 50	21 52	17 38	02 27	22 53	18 12	06 31	19 25	14 07
06	05 22	06 20	03 52	22 34	00 S 39	22 05	17 15	02 43	22 54	18 12	06 31	19 25	14 07
07	05 23	06 17	04 32	22 27	06 02	22 17	16 52	03 00	22 54	18 11	06 32	19 26	14 08
08	05 24	06 14	04 59	22 19	11 08	22 28	16 28	03 17	22 55	18 11	06 32	19 26	14 09
09	05 R 24	06 11	05 13	22 11	15 49	22 39	16 03	03 33	22 56	18 11	06 32	19 26	14 09
10	05 22	06 08	05 14	22 03	19 55	22 49	15 39	03 50	22 56	18 11	06 33	19 26	14 10
11	05 21	06 04	05 02	21 54	23 14	22 58	15 14	04 06	22 57	18 10	06 33	19 27	14 10
12	05 18	06 01	04 35	21 45	25 35	23 06	14 48	04 23	22 58	18 10	06 33	19 27	14 11
13	05 16	05 58	03 56	21 35	26 44	23 13	14 22	04 39	22 58	18 10	06 33	19 27	14 11
14	05 15	05 55	03 04	21 25	26 30	23 19	13 56	04 56	22 59	18 10	06 33	19 28	14 12
15	05 13	05 52	02 01	21 14	24 50	23 24	13 29	05 12	22 59	18 10	06 34	19 28	14 13
16	05 13	05 49	00 S 51	21 03	21 46	23 28	13 02	05 28	23 00	18 10	06 34	19 28	14 13
17	05 D 13	05 45	00 N 24	20 52	17 28	23 31	12 35	05 45	23 00	18 09	06 34	19 28	14 14
18	05 13	05 42	01 39	20 40	12 11	23 33	12 07	06 01	23 01	18 09	06 34	19 29	14 15
19	05 13	05 39	02 49	20 28	06 S 13	23 33	11 39	06 17	23 01	18 09	06 34	19 29	14 15
20	05 14	05 36	03 49	20 15	00 N 06	23 32	11 11	06 33	23 01	18 09	06 34	19 29	14 16
21	05 15	05 33	04 36	20 02	06 26	23 30	10 42	06 49	23 02	18 09	06 34	19 29	14 17
22	05 15	05 29	05 06	19 49	12 27	23 27	10 13	07 05	23 02	18 09	06 34	19 29	14 17
23	05 R 15	05 26	05 17	19 35	17 49	23 22	09 44	07 22	23 02	18 10	06 34	19 30	14 18
24	05 15	05 23	05 09	19 21	22 11	23 16	09 15	07 37	23 03	18 10	06 34	19 30	14 19
25	05 15	05 20	04 43	19 07	25 13	23 09	08 45	07 53	23 03	18 10	06 34	19 30	14 20
26	05 D 15	05 17	04 00	18 52	26 42	23 01	08 15	08 09	23 03	18 10	06 34	19 30	14 20
27	05 15	05 14	03 03	18 37	26 30	22 51	07 46	08 25	23 04	18 10	06 34	19 30	14 21
28	05 15	05 10	01 57	18 21	24 41	22 40	07 15	08 41	23 04	18 10	06 34	19 31	14 22
29	05 15	05 07	00 N 45	18 06	21 30	22 27	06 45	08 56	23 04	18 10	06 33	19 31	14 23
30	05 R 15	05 04	00 S 27	17 49	17 15	22 13	06 15	09 12	23 04	18 11	06 33	19 31	14 23
31	05 ♒ 15	05 ♒ 01	01 S 38	17 S 33	12 N 16	21 S 58	05 S 44	09 N 28	23 S 04	18 N 11	06 S 33	19 S 31	14 N 24

ZODIAC SIGN ENTRIES

Date	h	m	Planets
02	08	22	☽ ♌
04	15	50	☽ ♍
07	02	33	☽ ♎
09	15	03	☽ ♏
10	03	43	♄ ♉
11	18	18	☿ ♑
12	02	57	☽ ♐
14	12	26	☽ ♑
16	15	01	♀ ♓
16	19	04	☽ ♒
18	23	28	☽ ♓
20	22	59	☉ ♒
21	02	35	☽ ♈
23	05	17	☽ ♉
25	08	14	☽ ♊
27	12	01	☽ ♋
29	17	21	☽ ♌
31	23	46	☿ ♒

LATITUDES

Date	Mercury ☿ ° '	Venus ♀ ° '	Mars ♂ ° '	Jupiter ♃ ° '	Saturn ♄ ° '	Uranus ♅ ° '	Neptune ♆ ° '	Pluto ♇ ° '
01	01 N 55	01 S 47	00 S 09	00 N 24	02 S 03	00 N 39	01 N 36	16 N 16
04	01 29	01 47	00 05	00 24	02 03	00 40	01 36	16 18
07	01 03	01 45	00 S 01	00 23	02 02	00 40	01 37	16 19
10	00 37	01 43	00 N 03	00 23	02 01	00 40	01 37	16 21
13	00 N 12	01 40	00 06	00 23	02 00	00 40	01 37	16 23
16	00 S 11	01 36	00 10	00 23	02 00	00 40	01 37	16 24
19	00 33	01 32	00 13	00 23	01 59	00 40	01 37	16 26
22	00 53	01 27	00 16	00 23	01 58	00 40	01 37	16 27
25	01 12	01 21	00 19	00 22	01 57	00 40	01 37	16 29
28	01 27	01 14	00 22	00 22	01 56	00 40	01 37	16 30
31	01 S 41	01 S 07	00 N 25	00 N 22	01 S 56	00 N 40	01 N 38	16 N 32

DATA

Julian Date	2441318
Delta T	+42 seconds
Ayanamsa	23° 28' 12"
Synetic vernal point	05° ♓ 38' 48"
True obliquity of ecliptic	23° 26' 39"

MOON'S PHASES, APSIDES AND POSITIONS ☽

Date	h	m	Phase	Longitude ° '	Eclipse Indicator
08	13	31	☾	17 ♎ 22	
16	10	52	●	25 ♑ 25	Annular
23	09	29	☽	02 ♉ 29	
30	10	58	○	09 ♌ 39	total

Day	h	m		
09	03	33	Apogee	
22	05	21	Perigee	
06	09	07	0S	
13	20	20	Max dec	26° S 49'
20	11	38	0N	
26	21	06	Max dec	26° N 49'

LONGITUDES

Date	Chiron ⚷ ° '	Ceres ⚳ ° '	Pallas ⚴ ° '	Juno ⚵ ° '	Vesta ⚶ ° '	Black Moon Lilith ⚸ ° '
01	09 ♈ 25	22 ♌ 20	16 ♉ 43	15 ♎ 52	28 ♒ 27	23 ♎ 59
11	09 ♈ 33	21 ♌ 15	17 ♉ 07	17 ♎ 38	02 ♓ 59	25 ♎ 06
21	09 ♈ 47	19 ♌ 32	18 ♉ 36	18 ♎ 55	07 ♓ 35	26 ♎ 13
31	10 ♈ 06	17 ♌ 22	21 ♉ 00	19 ♎ 38	12 ♓ 15	27 ♎ 21

All ephemeris data is given at 12.00 UT and the Moon's longitude is additionally given for 24.00 UT

ASPECTARIAN

h m Aspects

01 Saturday
06 34 ☽ ∠ ♄
10 19 ☽ ⊼ ☿
10 37 ☉ ⊥ ♆
11 16 ☽ □ ♅
13 03 ☽ ⚼ ♆
14 42 ☽ Q ♇
16 42 ☽ # ☉
18 09 ☽ # ♃
19 04 ☽ ⊼ ♃
21 43 ☽ ± ☿
23 36 ☿ ⚹ ♅

02 Sunday
04 03 ♂ △ ♆
05 21 ☽ # ☿
05 54 ☽ ± ♃
09 00 ☽ ⚹ ♄
12 03 ☽ ⚹ ♇
15 11 ☽ ⚼ ☿
15 25 ☽ # ♆
15 50 ☽ △ ♆
16 25 ☽ △ ♂
19 25 ☽ Q ♅
19 41 ☽ # ♀
21 50 ☽ ∥ ♄
22 25 ☽ ⚼ ♃

03 Monday
06 04 ☽ ⊼ ☉
06 42 ☽ Q ♄
09 20 ☽ ☍ ♀
15 28 ☽ ∠ ♇
17 19 ☉ ∥ ♃
17 30 ☽ ⚹ ♅
18 06 ☽ ± ☉
18 22 ☽ ∥ ♇
21 15 ☽ △ ☿
21 26 ☽ ⚼ ♂
22 53 ♀ # ♄

04 Tuesday
02 39 ☽ △ ♃
08 20 ☽ ⊥ ♇
12 30 ☽ ⚼ ☉
14 43 ♇ St R
15 20 ☽ ± ♂
16 17 ☽ □ ♄
19 45 ☽ ⊻ ♇
21 53 ☽ ∠ ♅
23 55 ☽ □ ♆

05 Wednesday
03 28 ☽ ⊼ ♂
04 39 ☽ # ♅
10 02 ♀ Q ♆
15 19 ☽ ⊥ ♅
20 02 ☽ △ ☉
21 56 ☽ ∥ ♂

06 Thursday
00 45 ☽ ⊼ ♀
01 55 ♀ ⚼ ♇
03 06 ☽ ⊻ ♅
09 25 ☉ ⚼ ♄
11 10 ☽ Q ♆
13 10 ☽ □ ☿
13 45 ☽ □ ♃
13 56 ☽ ± ♀
20 08 ☿ ☌ ♃
21 38 ☽ # ♂

07 Friday
00 16 ♀ △ ♅
02 48 ☽ △ ♄
06 39 ☽ ☌ ♇
09 57 ☽ ⚼ ♀
11 10 ☽ ⚹ ♆
14 18 ☽ ∥ ♅
18 07 ☽ ☍ ♂

08 Saturday
00 40 ☉ ∥ ☿
02 29 ☽ Q ♃
05 26 ☽ Q ☿
08 52 ☽ ⚼ ♄
13 31 ☽ □ ☉
15 14 ☽ ☌ ♅
17 29 ☽ ∠ ♆
19 41 ☽ △ ♀

09 Sunday
03 00 ☽ ± ♄
03 10 ☽ # ♇
03 12 ☽ ⚹ ♃
08 38 ☽ ⚹ ☿
09 49 ☉ □ ♅
11 44 ☽ ⊥ ♆
13 13 ☽ ∥ ♀
15 06 ☽ ⊼ ♄
19 11 ☽ ⊻ ♇
23 53 ☽ ⊻ ♆

10 Monday
01 22 ☽ # ♄
05 17 ☽ Q ☉
07 16 ☽ ⊥ ♇
09 01 ☽ ∥ ♆
10 02 ☽ ∠ ♃
10 12 ☽ ⊼ ♂
13 04 ☉ ∠ ♆
18 35 ☽ ∠ ☿
22 55 ☽ ± ♂

11 Tuesday
01 18 ☽ ∠ ♇
02 02 ☽ ∥ ☉
03 43 ☽ ⊻ ♅
04 22 ☽ ⊥ ♃
07 32 ☽ ⚹ ☉
08 51 ☿ ∥ ♃
09 38 ☽ ∥ ♃
09 41 ☽ ∥ ☿
14 44 ☽ ⊥ ☿
14 51 ☽ □ ♀
15 39 ☽ ⊥ ♅
16 30 ☽ ⊻ ♃
17 15 ☿ ⊼ ♄
17 49 ☽ ⚼ ♂
22 58 ☿ Q ♅

12 Wednesday
02 49 ☽ ☍ ♄
04 02 ☽ ⊻ ☿
06 57 ☽ ⚹ ♇
09 21 ☽ ∠ ♅
10 06 ♀ ⚹ ♃
11 39 ☽ ☌ ♆
15 44 ☽ ∠ ☉

13 Thursday
00 45 ☽ △ ♂
05 52 ☽ Q ♀
06 07 ☿ □ ♇
06 12 ☽ Q ♇
09 16 ♀ ± ♇
10 38 ☽ ⊥ ☉
14 19 ☽ ⚹ ♅
21 25 ♀ # ♇
23 02 ☽ ⊻ ☉

14 Friday
02 15 ♀ ∠ ♂
03 28 ☽ ☌ ♃
07 05 ☽ ⚹ ♀
12 09 ☽ ⊼ ♄
12 58 ☽ Q ♅
16 12 ☽ □ ♇
20 19 ☽ ☌ ☿
20 49 ☽ ⊻ ♆
23 16 ☽ ± ♄

15 Saturday
01 04 ☿ ⊻ ♆
07 54 ☽ ⊥ ♆
12 06 ☽ □ ♂
13 44 ☽ ∠ ♀
15 44 ☽ ⚼ ♄
22 04 ☽ □ ♅
23 00 ☿ ± ♄

16 Sunday
00 15 ☽ ∥ ☿
00 19 ☽ ∠ ♆
03 39 ☽ ∥ ♃
07 40 ☽ ⊥ ♀
10 52 ☽ ☌ ☉
10 53 ♀ □ ♄
11 25 ☽ ⊻ ♃
16 44 ☽ ∥ ☉
18 40 ☽ △ ♄
19 28 ☽ ⊻ ♀
19 40 ☉ ⊻ ♃
22 10 ☽ Q ♂
22 16 ☽ ⊥ ♃
22 38 ☽ △ ♇

17 Monday
01 36 ☽ ∥ ♆
03 10 ☽ ⚹ ♆
08 31 ☽ # ♄
08 53 ☽ ⊻ ☿
14 24 ☽ ∠ ♃
20 22 ☽ ⚹ ♂
20 39 ☽ ⊥ ☿

18 Tuesday
00 16 ☽ Q ♆
00 58 ☽ ⚼ ♇
03 05 ☽ # ♇
03 15 ☽ △ ♅
06 16 ♀ ⊼ ♇
10 13 ♂ ∠ ♄
12 18 ☽ ∥ ♀
14 06 ☽ ∠ ☿
16 34 ☽ ± ♇
16 53 ☽ ⚹ ♃
19 43 ☽ ⊻ ☉
22 59 ☽ □ ♄
23 39 ☽ ∠ ♂

19 Wednesday
02 53 ☽ ⊼ ♇
04 53 ☽ ☌ ♀
05 10 ☽ ⚼ ♅
06 51 ☽ ⊥ ☉
07 25 ☽ □ ♆
07 33 ☿ ⊥ ♆
08 07 ♀ ⚼ ♅
10 39 ☽ ∥ ♅
11 45 ☽ # ♂
13 48 ☽ Q ♃
15 50 ☽ ⊥ ♂
18 51 ☽ ⚹ ☿
20 33 ☽ ± ♅
23 25 ☽ ∠ ☉

20 Thursday
02 34 ☽ ⊻ ♂
05 43 ☽ Q ♄
06 46 ☽ ⊼ ♅
10 27 ♀ □ ♆
13 05 ♂ # ♅
15 38 ☉ △ ♄
17 37 ☽ Q ☿
20 53 ☽ □ ♃

21 Friday
02 03 ☽ ⚹ ♄
02 52 ☽ ⚹ ☉
05 56 ☽ ☍ ♇
10 30 ☽ △ ♆
12 32 ☽ # ♅
12 43 ☽ ⊻ ♀
13 36 ☽ ∥ ♂
23 49 ☽ ⊥ ♀

22 Saturday
00 43 ☽ Q ☉
01 14 ☿ ⚼ ♄
03 22 ☽ ∠ ♄
03 35 ☽ # ♀
03 37 ☽ □ ☿
07 50 ☽ ☌ ♂
09 33 ☽ ☍ ♅
11 53 ☽ ⚼ ♆
16 26 ☽ ∠ ♀
18 33 ☽ ⊥ ♄
19 53 ☽ ∥ ♇
21 10 ☉ △ ♇

23 Sunday
00 21 ☽ △ ♃
03 07 ☽ ± ♆
04 41 ☽ ⊻ ♄
05 29 ♅ St R
08 35 ☽ ⊼ ♇
09 29 ☽ □ ☉
13 17 ☽ ⊼ ♆
13 43 ☽ ∥ ♄
18 44 ☽ ± ♇
20 11 ☽ ⚹ ♀
20 33 ☽ # ♆
20 35 ☽ # ☉
21 06 ♂ ☍ ♅
21 20 ☉ ∥ ♆

24 Monday
02 06 ☽ ⚼ ♃
05 27 ☉ ⊥ ♃
06 40 ♀ Q ♃
09 59 ☽ ⚼ ♇
11 39 ☉ ⊥ ♀
11 42 ☿ □ ♅
12 20 ☽ ⊼ ♅
12 24 ☽ △ ☿
13 05 ☽ ⊻ ♂
17 37 ☽ ± ♃
17 49 ☽ # ♃
18 29 ☽ Q ♀
19 11 ☽ # ☿
22 32 ☽ ± ♅
22 54 ☿ □ ♂
23 47 ☽ ⊥ ♂

25 Tuesday
03 59 ☽ ⊼ ♃
07 34 ☽ ☌ ♄
10 36 ☿ ∠ ♆
11 32 ☽ △ ♇
13 54 ☽ ⚼ ♅
15 42 ☉ ⚹ ♆
15 55 ☽ ∠ ♂
16 24 ☽ ☍ ♆
16 28 ☽ △ ☉
17 06 ☽ ⚼ ☿

26 Wednesday
02 30 ♂ ⚼ ♆
04 12 ☽ □ ♀
05 31 ☿ ∥ ♃
10 30 ☽ ± ☿
15 16 ♀ # ♂
15 43 ☽ △ ♅
17 32 ♀ ± ♅
19 03 ☽ ⚹ ♂
20 21 ☽ ⚼ ☉
22 10 ☽ ⊼ ☿

27 Thursday
08 28 ☽ ☍ ♃
11 19 ☽ ⊻ ♄
13 16 ☽ ± ☉
15 21 ☽ □ ♇
17 00 ☽ Q ♂
20 29 ☽ ⊼ ♆
21 49 ☽ ⊥ ♄

28 Friday
00 38 ☽ ⊼ ☉
07 04 ☽ ± ♆
13 39 ☽ △ ♀
13 43 ☽ ∠ ♄
20 20 ☽ □ ♅
23 06 ☽ ⚼ ♆
23 10 ☽ Q ♇

29 Saturday
01 21 ☽ # ♃
02 31 ☽ □ ♂
04 46 ☉ # ♄
05 17 ☽ # ☿
09 59 ☽ ☍ ☿
14 29 ☽ ⊼ ♃
16 36 ☽ ⚹ ♄
19 14 ☽ ⚼ ♀
20 45 ☽ ⚹ ♇
21 13 ♀ ∥ ♅
23 49 ☽ # ♆

30 Sunday
01 32 ☽ ± ♃
02 13 ☽ △ ♆
03 25 ♀ ⊥ ♂
04 49 ☽ Q ♅
07 08 ☽ ∥ ♄
08 48 ☽ # ☉
10 58 ☽ ☍ ☉
13 21 ☽ ± ♀
14 30 ☽ Q ♄
18 18 ☽ ⚼ ♃

31 Monday
00 16 ☽ ∠ ♇
01 35 ☽ ⊼ ♀
01 51 ♀ Q ♄
02 03 ☽ ∥ ♇
02 56 ☽ ⚹ ♅
04 33 ☿ ⊻ ♃
10 23 ♄ St D
12 24 ☽ △ ♂
15 59 ♀ ⊼ ♅
17 05 ☽ ⊥ ♇
17 27 ☿ △ ♄
20 32 ♂ ⊥ ♄
22 48 ☽ △ ♃

FEBRUARY 1972

LONGITUDES

Date	Sidereal time h m s	Sun ☉ ° ' "	Moon ☽ ° ' "	Moon ☽ 24.00 ° ' "	Mercury ☿ ° '	Venus ♀ ° '	Mars ♂ ° '	Jupiter ♃ ° '	Saturn ♄ ° '	Uranus ♅ ° '	Neptune ♆ ° '	Pluto ♇ ° '
01	20 43 12	11 ♒ 43 53	05 ♍ 48 51	12 ♍ 03 22	00 ♒ 49	19 ♓ 18	24 ♈ 00	28 ♐ 59	29 ♉ 35	18 ♎ 17	04 ♐ 55	01 ♎ 50
02	20 47 09	12 44 46	18 ♍ 14 24	24 ♍ 22 11	02 26	20 30	24 40	29 11	29 D 35	18 R 16	04 56	01 R 49
03	20 51 05	13 45 38	00 ♎ 27 02	06 ♎ 29 16	04 03	21 42	25 19	29 22	29 35	18 16	04 57	01 48
04	20 55 02	14 46 28	12 ♎ 29 19	18 ♎ 27 37	05 41	22 54	25 59	29 34	29 36	18 15	04 58	01 47
05	20 58 58	15 47 18	24 ♎ 24 40	00 ♏ 20 57	07 20	24 07	26 39	29 45	29 36	18 15	05 00	01 46
06	21 02 55	16 48 07	06 ♏ 17 03	12 ♏ 13 32	08 59	25 19	27 18	29 ♐ 56	29 37	18 14	05 01	01 45
07	21 06 51	17 48 55	18 ♏ 10 59	24 ♏ 10 00	10 39	26 30	27 58	00 ♑ 08	29 38	18 13	05 02	01 44
08	21 10 48	18 49 42	00 ♐ 11 10	06 ♐ 15 06	12 20	27 42	28 37	00 19	29 38	18 12	05 02	01 43
09	21 14 44	19 50 28	12 ♐ 22 20	18 ♐ 33 25	14 02	28 ♓ 54	29 17	00 30	29 39	18 11	05 03	01 42
10	21 18 41	20 51 13	24 ♐ 48 51	01 ♑ 09 05	15 45	00 ♈ 06	29 ♈ 57	00 41	29 40	18 10	05 04	01 41
11	21 22 38	21 51 57	07 ♑ 34 29	14 ♑ 05 21	17 28	01 17	00 ♉ 36	00 52	29 42	18 09	05 05	01 40
12	21 26 34	22 52 40	20 ♑ 41 53	27 ♑ 24 10	19 13	02 28	01 16	01 03	29 43	18 08	05 06	01 39
13	21 30 31	23 53 22	04 ♒ 12 12	11 ♒ 05 49	20 58	03 40	01 55	01 14	29 44	18 07	05 07	01 38
14	21 34 27	24 54 02	18 ♒ 04 44	25 ♒ 08 33	22 44	04 51	02 35	01 24	29 46	18 06	05 07	01 36
15	21 38 24	25 54 41	02 ♓ 16 43	09 ♓ 28 35	24 31	06 02	03 15	01 35	29 48	18 05	05 08	01 35
16	21 42 20	26 55 18	16 ♓ 43 27	24 ♓ 00 29	26 18	07 13	03 54	01 45	29 49	18 03	05 09	01 34
17	21 46 17	27 55 54	01 ♈ 18 51	08 ♈ 37 43	28 07	08 24	04 34	01 56	29 51	18 02	05 09	01 33
18	21 50 13	28 56 28	15 ♈ 56 16	23 ♈ 13 44	29 ♒ 56	09 35	05 14	02 06	29 53	18 01	05 10	01 31
19	21 54 10	29 ♒ 57 00	00 ♉ 29 25	07 ♉ 42 44	01 ♓ 46	10 45	05 53	02 16	29 55	17 59	05 11	01 30
20	21 58 07	00 ♓ 57 30	14 ♉ 53 12	22 ♉ 00 26	03 37	11 56	06 33	02 26	29 ♉ 57	17 58	05 11	01 29
21	22 02 03	01 57 59	29 ♉ 04 08	06 ♊ 04 10	05 28	13 06	07 12	02 36	00 ♊ 00	17 56	05 12	01 27
22	22 06 00	02 58 26	13 ♊ 00 24	19 ♊ 52 51	07 20	14 17	07 52	02 46	00 02	17 55	05 12	01 26
23	22 09 56	03 58 50	26 ♊ 41 31	03 ♋ 26 31	09 13	15 27	08 32	02 56	00 05	17 53	05 13	01 24
24	22 13 53	04 59 13	10 ♋ 07 57	16 ♋ 45 55	11 06	16 37	09 11	03 05	00 07	17 52	05 13	01 23
25	22 17 49	05 59 34	23 ♋ 20 34	29 ♋ 52 01	13 00	17 47	09 51	03 15	00 10	17 50	05 13	01 22
26	22 21 46	06 59 53	06 ♌ 20 23	12 ♌ 45 46	14 53	18 56	10 30	03 24	00 13	17 48	05 14	01 20
27	22 25 42	08 00 10	19 ♌ 08 16	25 ♌ 27 57	16 47	20 06	11 10	03 33	00 16	17 46	05 14	01 19
28	22 29 39	09 00 25	01 ♍ 44 55	07 ♍ 59 14	18 40	21 15	11 49	03 42	00 19	17 45	05 14	01 17
29	22 33 36	10 ♓ 00 39	14 ♍ 10 58	20 ♍ 20 13	20 ♓ 33	22 ♈ 24	12 ♉ 29	03 ♑ 51	00 ♊ 22	17 ♎ 43	05 ♐ 15	01 ♎ 16

Date	Moon True ☊ ° '	Moon Mean ☊ ° '	Moon ☽ Latitude ° '
01	05 ♒ 14	04 ♒ 58	02 S 41
02	05 R 14	04 54	03 36
03	05 12	04 51	04 20
04	05 11	04 48	04 52
05	05 10	04 45	05 11
06	05 09	04 42	05 16
07	05 D 09	04 39	05 08
08	05 09	04 35	04 47
09	05 10	04 32	04 12
10	05 11	04 29	03 25
11	05 13	04 26	02 27
12	05 14	04 23	01 19
13	05 14	04 20	00 S 06
14	05 R 14	04 16	01 N 10
15	05 12	04 13	02 23
16	05 10	04 10	03 29
17	05 07	04 07	04 21
18	05 03	04 04	04 57
19	05 01	04 00	05 13
20	04 58	03 57	05 09
21	04 58	03 54	04 47
22	04 D 58	03 51	04 07
23	04 59	03 48	03 15
24	05 01	03 45	02 12
25	05 02	03 41	01 N 04
26	05 R 03	03 38	00 S 07
27	05 02	03 35	01 16
28	04 59	03 32	02 20
29	04 ♒ 55	03 ♒ 29	03 S 17

DECLINATIONS

Date	Sun ☉ ° '	Moon ☽ ° '	Mercury ☿ ° '	Venus ♀ ° '	Mars ♂ ° '	Jupiter ♃ ° '	Saturn ♄ ° '	Uranus ♅ ° '	Neptune ♆ ° '	Pluto ♇ ° '
01	17 S 16	06 N 53	21 S 41	05 S 13	09 N 43	23 S 04	18 N 11	06 S 33	19 S 31	14 N 25
02	16 59	01 N 20	21 23	04 42	09 58	23 05	18 12	06 33	19 31	14 26
03	16 42	04 S 09	21 03	04 11	10 14	23 05	18 12	06 32	19 32	14 26
04	16 24	09 25	20 42	03 40	10 29	23 05	18 12	06 32	19 32	14 27
05	16 06	14 17	20 20	03 09	10 44	23 05	18 13	06 32	19 32	14 28
06	15 48	18 35	19 56	02 38	10 59	23 05	18 13	06 31	19 32	14 29
07	15 30	22 11	19 31	02 06	11 14	23 05	18 13	06 31	19 32	14 30
08	15 11	24 52	19 04	01 35	11 29	23 05	18 14	06 31	19 32	14 30
09	14 52	26 27	18 36	01 03	11 44	23 05	18 14	06 30	19 32	14 31
10	14 33	26 46	18 07	00 S 32	11 59	23 05	18 15	06 30	19 32	14 32
11	14 13	25 40	17 35	00 00	12 13	23 05	18 15	06 30	19 33	14 33
12	13 54	23 10	17 03	00 N 31	12 28	23 05	18 16	06 29	19 33	14 34
13	13 34	19 18	16 29	01 03	12 42	23 05	18 17	06 29	19 33	14 35
14	13 13	14 18	15 54	01 34	12 57	23 05	18 17	06 28	19 33	14 35
15	12 53	08 26	15 17	02 06	13 11	23 05	18 18	06 28	19 33	14 36
16	12 32	02 S 02	14 39	02 37	13 25	23 05	18 18	06 27	19 33	14 37
17	12 12	04 N 31	13 59	03 09	13 39	23 05	18 19	06 27	19 33	14 38
18	11 51	10 50	13 18	03 40	13 53	23 05	18 20	06 26	19 33	14 39
19	11 30	16 31	12 35	04 11	14 07	23 04	18 20	06 26	19 33	14 40
20	11 08	21 13	11 52	04 42	14 21	23 04	18 21	06 25	19 33	14 40
21	10 47	24 37	11 07	05 13	14 35	23 04	18 22	06 24	19 33	14 41
22	10 25	26 27	10 20	05 44	14 48	23 04	18 23	06 24	19 33	14 42
23	10 03	26 39	09 33	06 15	15 01	23 04	18 23	06 23	19 33	14 43
24	09 41	25 15	08 44	06 46	15 15	23 04	18 24	06 23	19 33	14 44
25	09 19	22 28	07 54	07 16	15 28	23 03	18 25	06 22	19 33	14 45
26	08 57	18 35	07 04	07 47	15 41	23 03	18 26	06 21	19 33	14 45
27	08 34	13 53	06 12	08 17	15 54	23 03	18 27	06 21	19 33	14 46
28	08 12	08 40	05 20	08 47	16 07	23 03	18 28	06 20	19 33	14 47
29	07 S 49	03 N 12	04 S 27	09 N 17	16 N 20	23 S 03	18 N 29	06 S 19	19 S 33	14 N 48

ZODIAC SIGN ENTRIES

Date	h	m	Planets
01	00	56	☽ ♍
03	11	06	☽ ♎
05	23	18	☽ ♏
06	19	37	♃ ♑
08	11	38	☽ ♐
10	10	08	♀ ♈
10	14	04	♂ ♉
10	21	50	☽ ♑
13	04	36	☽ ♒
15	08	11	☽ ♓
17	09	51	☽ ♈
18	12	53	☿ ♓
19	11	11	☽ ♉
19	13	11	☉ ♓
21	13	35	☽ ♊
21	14	52	♄ ♊
23	17	52	☽ ♋
26	00	15	☽ ♌
28	08	39	☽ ♍

LATITUDES

Date	Mercury ☿ ° '	Venus ♀ ° '	Mars ♂ ° '	Jupiter ♃ ° '	Saturn ♄ ° '	Uranus ♅ ° '	Neptune ♆ ° '	Pluto ♇ ° '
01	01 S 45	01 S 04	00 N 26	00 N 22	01 S 55	00 N 40	01 N 38	16 N 32
04	01 55	00 56	00 29	00 22	01 55	00 41	01 38	16 34
07	02 02	00 47	00 31	00 22	01 54	00 41	01 38	16 35
10	02 05	00 37	00 34	00 22	01 53	00 41	01 38	16 36
13	02 05	00 27	00 36	00 21	01 52	00 41	01 38	16 38
16	02 01	00 16	00 38	00 21	01 51	00 41	01 38	16 39
19	01 52	00 S 05	00 40	00 21	01 50	00 41	01 38	16 40
22	01 38	00 N 07	00 42	00 21	01 50	00 41	01 39	16 41
25	01 19	00 19	00 44	00 21	01 49	00 41	01 39	16 42
28	00 55	00 32	00 46	00 21	01 48	00 41	01 39	16 42
31	00 S 25	00 N 45	00 N 48	00 N 21	01 S 47	00 N 41	01 N 39	16 N 43

DATA

Julian Date	2441349
Delta T	+42 seconds
Ayanamsa	23° 28' 17"
Synetic vernal point	05° ♓ 38' 42"
True obliquity of ecliptic	23° 26' 40"

MOON'S PHASES, APSIDES AND POSITIONS ☽

Date	h	m	Phase	Longitude ° '	Eclipse Indicator
07	11	12	☾	17 ♏ 47	
15	00	29	●	25 ♒ 26	
21	17	20	☽	02 ♊ 11	
29	03	12	○	09 ♍ 39	

Day	h	m		
06	00	42	Apogee	
17	18	48	Perigee	
02	17	46	0S	
10	05	32	Max dec	26° S 49'
16	19	27	0N	
23	02	48	Max dec	26° N 46'

LONGITUDES

Date	Chiron ⚷ ° '	Ceres ⚳ ° '	Pallas ⚴ ° '	Juno ⚵ ° '	Vesta ⚶ ° '	Black Moon Lilith ⚸ ° '
01	10 ♈ 08	17 ♌ 08	21 ♉ 17	19 ♎ 40	12 ♓ 43	27 ♎ 27
11	10 ♈ 32	14 ♌ 48	24 ♉ 31	19 ♎ 43	17 ♓ 26	28 ♎ 35
21	10 ♈ 59	12 ♌ 36	28 ♉ 22	19 ♎ 08	22 ♓ 09	29 ♎ 42
31	11 ♈ 30	10 ♌ 51	02 ♊ 41	17 ♎ 54	26 ♓ 54	00 ♏ 49

ASPECTARIAN

h m	Aspects
01 Tuesday	
00 05	☽ ∥ ♂
00 08	☽ □ ♄
01 06	☽ ⊼ ☿
04 25	☽ ⊻ ♇
07 10	☽ ∠ ♅
10 17	☽ □ ♆
13 27	☽ ⋕ ♅
14 12	☽ ± ☿
18 28	☽ ꝗ ♂
19 57	☽ ⋕ ♀
02 Wednesday	
00 23	☽ ⊼ ☉
00 26	☽ ⊥ ♅
03 07	☿ △ ♇
06 26	☉ Q ♂
10 10	☽ ꝗ ☿
12 04	☽ ⊻ ♅
12 52	☽ ± ♂
13 05	☽ ± ☉
16 54	☽ ☍ ♀
21 12	☽ Q ♆
03 Thursday	
01 18	☽ ⊼ ♂
08 21	☽ ꝗ ☉
09 49	☽ □ ♃
10 17	☽ △ ♄
12 07	☽ ∥ ♀
14 41	☽ ☌ ♇
20 15	☽ △ ☿
20 58	☽ ⚹ ♆
22 41	☽ ∥ ♅
04 Friday	
01 33	☿ ⚹ ♆
05 45	☉ ∠ ♃
10 01	☿ ⊥ ♃
16 14	☽ ꝗ ♄
16 28	♃ ⊼ ♄
17 01	☽ △ ☉
17 22	☽ ⋕ ♂
22 21	☽ Q ♃
23 34	☽ ☌ ♅
05 Saturday	
03 04	☽ ∠ ♆
10 22	☽ ± ♄
11 19	☽ ⊼ ♀
13 00	☽ ⋕ ♇
16 46	☽ ☍ ♂
21 09	☽ ∥ ☉
21 16	☽ ⊥ ♆
22 30	☽ ⊼ ♄
22 58	☽ ⚹ ♃
06 Sunday	
00 49	☽ ± ♀
02 52	☽ ⊻ ♇
09 25	☽ ⊻ ♆
09 47	☽ ⋕ ♄
10 57	☉ ꝗ ♇
14 58	☽ ⊥ ♇
16 59	☉ Q ♆
17 51	☽ ∥ ♆
18 21	☽ □ ☿
19 35	☽ ∥ ☿
21 03	☽ ꝗ ♀
07 Monday	
05 45	☽ ∠ ♃
09 06	☽ ∠ ♇
10 56	☿ ∥ ♆
11 12	☽ □ ☉
12 04	☽ ⊻ ♅
19 10	☽ ∥ ♃
21 21	☉ △ ♅
22 27	♃ Q ♅
08 Tuesday	
00 05	☽ ⊥ ♅
00 07	☽ ⊥ ♃
06 31	☽ △ ♀
08 43	☽ ⊼ ♂
10 55	☽ ☍ ♄
12 16	☽ ⊻ ♃
12 21	☽ Q ☿
15 03	☽ ⚹ ♇
17 58	☽ ∠ ♅
21 18	☽ ± ♂
21 37	☽ ☌ ♆
09 Wednesday	
02 20	☽ Q ☉
03 33	♂ ± ♆
05 29	☿ ∠ ♀
14 35	☽ Q ♇
15 46	☽ ⚹ ☿
15 56	☽ ꝗ ♂
23 16	☽ ⚹ ♅
10 Thursday	
01 57	♂ ⊻ ♄
03 28	♀ ⚹ ♄
03 45	☽ ⚹ ☉
05 16	♀ ⊻ ♂
05 28	☿ ⋕ ♄
11 00	☿ ∠ ♃
12 49	☉ ⋕ ♇
21 14	☽ ⊼ ♄
22 08	☽ Q ♅
22 15	☽ △ ♂
23 02	☽ □ ♀
23 17	☽ ☌ ♃
11 Friday	
00 55	☿ ꝗ ♇
00 59	☽ □ ♇
01 00	☽ ∠ ☿
02 03	♀ □ ♃
06 35	☿ Q ♆
07 22	☽ ⊻ ♆
08 30	☽ ± ♄
10 34	☽ ∠ ☉
18 30	☽ ⊥ ♆
19 33	♀ ☍ ♇
20 18	☽ ⊥ ☿
21 20	☿ △ ♅
12 Saturday	
01 08	☽ ꝗ ♄
01 08	♂ △ ♃
04 31	☽ ⊥ ☉
07 22	☽ □ ♅
08 54	☽ ⊻ ☿
10 55	☽ ∠ ♆
11 33	☽ Q ♀
12 34	☽ ∥ ♃
13 11	☿ Q ♂
16 14	☽ ⊻ ☉
13 Sunday	
01 30	♂ ⊼ ♇
04 08	☽ △ ♄
06 42	☽ ⊻ ♃
07 29	☽ △ ♇
07 48	☽ □ ♂
10 41	☽ ∥ ♆
10 58	☽ ⚹ ♀
13 35	☽ ⚹ ♆
17 21	☽ ⊥ ♃
17 22	☽ ⋕ ♄
14 Monday	
03 56	☽ ∥ ☿
09 06	☽ ∠ ♃
09 29	☽ ꝗ ♇
10 22	☽ Q ♆
10 44	☽ ⋕ ♇
12 02	☽ △ ♅
15 18	☽ ∠ ♀
16 29	☽ Q ♂
16 56	☽ ∥ ☉
17 35	☽ ⋕ ♂
17 35	♀ △ ♆
21 03	☽ ☌ ☿
23 36	☉ ⋕ ♂
15 Tuesday	
00 29	☽ ☌ ☉
00 46	☽ ± ♇
04 25	☉ ± ♇
07 49	☽ □ ♄
07 54	☽ ⊥ ♀
10 49	☽ ⚹ ♃
10 50	☽ ⊼ ♇
12 47	♃ □ ♇
13 20	☽ ꝗ ♅
13 42	☽ ⚹ ♂
16 47	☽ □ ♆
18 50	☽ ⊻ ♀
19 33	☽ ∥ ♅
16 Wednesday	
02 15	☿ ± ♇
04 17	☽ ± ♅
07 02	☽ Q ♃
10 00	☽ ⋕ ♀
12 54	☿ ⋕ ♇
13 49	☽ Q ♄
14 12	☽ ⊼ ♅
15 46	☽ ∠ ♂
17 Thursday	
06 00	☽ ⊻ ☿
06 02	☽ ⊻ ☉
06 31	☽ ∥ ♀
06 38	☉ ☌ ☿
07 17	☽ ⊥ ♂
09 36	☽ ⚹ ♄
12 22	☽ ☍ ♇
13 01	☽ □ ♃
16 37	☽ ⊥ ☉
17 14	☽ ⊥ ☿
17 35	☽ ⊻ ♂
18 19	☽ △ ♆
19 10	☽ ⋕ ♅
20 38	☿ ⋕ ♂
18 Friday	
00 39	☽ ☌ ♀
08 29	☽ ∠ ☉
09 51	♂ ⊼ ♆
10 07	☽ ∠ ☿
10 16	☽ ∠ ♄
11 22	☿ □ ♄
15 24	☽ ☍ ♅
15 49	☽ ⋕ ☉
18 58	☽ ꝗ ♆
20 55	☽ ⋕ ☿
19 Saturday	
00 59	☽ ∥ ♂
01 07	☽ ⊥ ♄
03 44	☽ ∥ ♇
08 33	☿ ⊼ ♇
09 49	☽ ± ♆
11 02	☽ ⚹ ☉
11 03	☽ ⊻ ♄
11 16	☉ □ ♄
13 40	☽ ⊼ ♇
14 25	☽ ⚹ ☿
14 59	☽ △ ♃
19 09	☿ ⚹ ♃
19 47	☽ ⊼ ♆
20 37	☽ ∥ ♄
21 23	☽ ☌ ♂
23 38	☽ ± ♇
20 Sunday	
02 45	☽ ⋕ ♆
03 41	☿ ꝗ ♅
06 36	☽ ⊻ ♀
08 31	☽ Q ☉
13 24	☽ Q ☿
14 40	☽ ꝗ ♇
16 20	☽ ꝗ ♃
17 10	☽ ⊼ ♅
17 35	☽ ⊥ ♀
23 47	☽ ⋕ ♃
21 Monday	
00 04	☉ ⊼ ♇
03 17	☽ ± ♅
07 44	☽ ± ♃
08 25	☿ □ ♆
10 12	☽ ∠ ♀
13 35	☽ ☌ ♄
16 04	☽ △ ♇
17 20	☽ □ ☉
18 07	☽ ⊼ ♃
18 37	☽ ꝗ ♅
20 40	♂ ± ♇
22 30	☽ ☍ ♆
22 Tuesday	
00 40	☽ □ ☿
00 42	♂ ∥ ♇
02 39	☽ ⊻ ♂
06 00	☉ ⚹ ♃
07 32	☉ ∥ ☿
10 36	☉ ꝗ ♅
13 34	☽ ⊥ ♂
14 25	☽ ⚹ ♀
20 32	☽ △ ♅
22 24	☿ ⚹ ♂
23 Wednesday	
04 09	♀ ∠ ♄
06 08	☽ ∠ ♂
13 28	☽ Q ♀
18 01	☽ ⊻ ♄
18 13	♀ ⋕ ♅
19 36	☿ ⊥ ♀
20 21	☽ □ ♇
23 13	☽ ☍ ♃
24 Thursday	
02 01	☽ △ ☉
03 10	☽ ⊼ ♆
04 46	☽ ⊥ ♄
10 13	☽ ⚹ ♂
13 57	☽ ± ♆
14 02	☽ △ ☿
17 32	☉ □ ♆
21 03	☽ ∠ ♄
21 28	☿ ± ♅
25 Friday	
00 51	☽ □ ♀
01 58	☽ □ ♅
04 44	☽ Q ♇
06 18	☽ ꝗ ♆
07 21	☽ ꝗ ☉
07 42	☽ ⋕ ♃
09 07	☽ Q ♂
13 08	♀ ☍ ♅
22 00	☽ ꝗ ☿
23 15	☿ ⋕ ♀
26 Saturday	
00 36	☽ ⚹ ♄
01 16	☽ ± ☉
02 44	☽ ⚹ ♇
06 28	☽ ⊼ ♃
06 28	☽ ⋕ ♆
09 56	☽ △ ♆
11 00	☽ Q ♅
12 48	☽ ∥ ♄
13 20	☽ ⊼ ☉
17 34	☽ ± ☿
17 47	☽ ± ♃
19 05	☿ Q ♃
20 12	☽ □ ♂
23 01	☽ Q ♄
27 Sunday	
02 31	☽ ∥ ♂
06 41	☽ ∠ ♇
06 47	☽ ⊼ ☿
07 42	☽ ∥ ♇
07 57	☿ ∥ ♅
09 26	☽ ⚹ ♅
10 53	☽ ꝗ ♃
14 00	☽ △ ♀
14 53	♀ ꝗ ♆
19 55	☉ ⋕ ♀
23 41	☽ ⊥ ♇
28 Monday	
00 26	☿ ⊼ ♅
07 20	☿ Q ♄
09 14	☽ □ ♄
11 07	☽ ⊻ ♇
11 32	☽ ∥ ♀
13 54	☽ ∠ ♅
14 15	☽ ⋕ ☉
15 48	☽ △ ♃
18 42	☽ □ ♆
21 32	☽ ꝗ ♀
22 21	☽ ⋕ ♅
29 Tuesday	
03 12	☽ ☍ ☉
05 32	☽ ⋕ ☿
07 13	☽ ⊥ ♅
08 31	☽ △ ♂
16 46	☽ ± ♀
18 52	☽ ⊻ ♅

All ephemeris data is given at 12.00 UT and the Moon's longitude is additionally given for 24.00 UT

MARCH 1972

LONGITUDES

Date	Sidereal time h m s	Sun ☉ ° ' "	Moon ☽ ° ' "	Moon ☽ 24.00 ° ' "	Mercury ☿ ° '	Venus ♀ ° '	Mars ♂ ° '	Jupiter ♃ ° '	Saturn ♄ ° '	Uranus ♅ ° '	Neptune ♆ ° '	Pluto ♇ ° '
01	22 37 32	11 ♓ 00 50	26 ♍ 27 06	02 ♎ 31 44	22 ♓ 25	23 ♈ 33	13 ♉ 09	04 ♑ 00	00 ♊ 25	17 ♎ 41	05 ♐ 15	01 ♎ 14
02	22 41 29	12 01 00	08 ♎ 34 18	14 ♎ 34 59	24 16	24 42	13 48	04 09	00 28	17 R 39	05 15	01 R 13
03	22 45 25	13 01 08	20 ♎ 34 03	26 ♎ 31 46	26 06	25 51	14 28	04 18	00 32	17 37	05 15	01 11
04	22 49 22	14 01 15	02 ♏ 28 27	08 ♏ 24 30	27 53	27 00	15 07	04 26	00 35	17 35	05 15	01 10
05	22 53 18	15 01 20	14 ♏ 20 19	20 ♏ 16 23	29 ♓ 38	28 08	15 47	04 35	00 39	17 33	05 15	01 08
06	22 57 15	16 01 23	26 ♏ 13 11	02 ♐ 11 16	01 ♈ 21	29 ♈ 16	16 26	04 43	00 43	17 31	05 15	01 06
07	23 01 11	17 01 25	08 ♐ 11 13	14 ♐ 13 36	03 00	00 ♉ 24	17 05	04 51	00 46	17 29	05 R 15	01 05
08	23 05 08	18 01 25	20 ♐ 19 04	26 ♐ 28 13	04 35	01 32	17 45	04 59	00 50	17 27	05 15	01 03
09	23 09 05	19 01 24	02 ♑ 41 40	09 ♑ 00 00	06 06	02 40	18 24	05 07	00 54	17 25	05 15	01 02
10	23 13 01	20 01 20	15 ♑ 23 47	21 ♑ 53 32	07 32	03 47	19 04	05 15	00 58	17 23	05 15	01 00
11	23 16 58	21 01 16	28 ♑ 29 39	05 ♒ 12 28	08 53	04 55	19 43	05 23	01 03	17 20	05 15	00 59
12	23 20 54	22 01 09	12 ♒ 02 10	18 ♒ 58 50	10 08	06 02	20 23	05 30	01 07	17 18	05 15	00 57
13	23 24 51	23 01 01	26 ♒ 02 19	03 ♓ 12 18	11 16	07 09	21 02	05 37	01 11	17 16	05 15	00 55
14	23 28 47	24 00 51	10 ♓ 28 18	17 ♓ 49 36	12 17	08 15	21 41	05 45	01 16	17 14	05 14	00 54
15	23 32 44	25 00 38	25 ♓ 15 19	02 ♈ 44 24	13 11	09 22	22 21	05 52	01 20	17 11	05 14	00 52
16	23 36 40	26 00 24	10 ♈ 15 41	17 ♈ 47 56	13 58	10 28	23 00	05 59	01 25	17 09	05 14	00 50
17	23 40 37	27 00 08	25 ♈ 19 52	02 ♉ 50 19	14 37	11 34	23 39	06 05	01 29	17 07	05 14	00 49
18	23 44 34	27 59 50	10 ♉ 18 08	17 ♉ 42 21	15 07	12 40	24 19	06 12	01 34	17 04	05 13	00 47
19	23 48 30	28 59 29	25 ♉ 02 08	02 ♊ 16 53	15 30	13 45	24 58	06 18	01 39	17 02	05 13	00 45
20	23 52 27	29 ♓ 59 07	09 ♊ 26 08	16 ♊ 29 38	15 44	14 50	25 37	06 25	01 44	16 59	05 12	00 44
21	23 56 23	00 ♈ 58 42	23 ♊ 27 19	00 ♋ 19 12	15 50	15 55	26 17	06 31	01 49	16 57	05 12	00 42
22	00 00 20	01 58 15	07 ♋ 05 29	13 ♋ 46 25	15 R 49	17 00	26 56	06 37	01 54	16 54	05 11	00 40
23	00 04 16	02 57 45	20 ♋ 22 20	26 ♋ 53 35	15 39	18 05	27 35	06 43	01 59	16 52	05 11	00 39
24	00 08 13	03 57 13	03 ♌ 20 34	09 ♌ 43 41	15 22	19 09	28 15	06 48	02 05	16 49	05 10	00 37
25	00 12 09	04 56 38	16 ♌ 03 17	22 ♌ 19 46	14 59	20 13	28 54	06 54	02 10	16 47	05 10	00 36
26	00 16 06	05 56 02	28 ♌ 33 25	04 ♍ 44 34	14 29	21 16	29 ♉ 33	06 59	02 15	16 44	05 09	00 34
27	00 20 03	06 55 23	10 ♍ 53 28	17 ♍ 00 20	13 53	22 19	00 ♊ 12	07 05	02 21	16 42	05 09	00 32
28	00 23 59	07 54 42	23 ♍ 05 23	29 ♍ 08 47	13 13	23 22	00 51	07 10	02 26	16 39	05 08	00 31
29	00 27 56	08 53 59	05 ♎ 10 42	11 ♎ 11 15	12 28	24 25	01 31	07 14	02 32	16 37	05 07	00 29
30	00 31 52	09 53 13	17 ♎ 10 36	23 ♎ 08 54	11 41	25 27	02 10	07 19	02 37	16 35	05 07	00 27
31	00 35 49	10 ♈ 52 26	29 ♎ 06 18	05 ♏ 02 58	10 ♈ 52	26 ♉ 29	02 ♊ 49	07 ♑ 24	02 ♊ 43	16 ♎ 32	05 ♐ 06	00 ♎ 26

DECLINATIONS

Date	Moon True ☊ ° '	Moon Mean ☊ ° '	Moon ☽ Latitude ° '	Sun ☉ ° '	Moon ☽ ° '	Mercury ☿ ° '	Venus ♀ ° '	Mars ♂ ° '	Jupiter ♃ ° '	Saturn ♄ ° '	Uranus ♅ ° '	Neptune ♆ ° '	Pluto ♇ ° '
01	04 ♒ 49	03 ♒ 26	04 S 03	07 S 26	02 S 19	03 S 33	09 N 46	16 N 32	23 S 02	18 N 29	06 S 18	19 S 33	14 N 49
02	04 R 42	03 22	04 38	07 03	07 40	02 40	10 16	16 45	23 02	18 30	06 18	19 33	14 49
03	04 35	03 19	05 00	06 40	12 40	01 46	10 45	16 57	23 02	18 31	06 17	19 33	14 50
04	04 28	03 16	05 09	06 17	17 10	00 S 52	11 14	17 09	23 02	18 32	06 16	19 33	14 51
05	04 22	03 13	05 05	05 54	20 59	00 N 01	11 43	17 21	23 02	18 33	06 15	19 33	14 52
06	04 18	03 10	04 47	05 31	23 58	00 53	12 11	17 33	23 01	18 34	06 15	19 33	14 53
07	04 16	03 06	04 17	05 08	25 54	01 44	12 39	17 45	23 01	18 35	06 14	19 33	14 54
08	04 D 15	03 03	03 35	04 44	26 40	02 34	13 07	17 56	23 01	18 36	06 13	19 33	14 54
09	04 16	03 00	02 42	04 21	26 07	03 22	13 35	18 08	23 01	18 37	06 12	19 33	14 55
10	04 17	02 57	01 40	03 57	24 13	04 09	14 02	18 19	23 00	18 39	06 11	19 33	14 56
11	04 19	02 54	00 S 31	03 34	20 59	04 53	14 29	18 31	23 00	18 40	06 10	19 33	14 57
12	04 R 18	02 51	00 N 42	03 10	16 31	05 35	14 56	18 42	23 00	18 41	06 10	19 33	14 57
13	04 16	02 47	01 54	02 46	11 03	06 14	15 22	18 53	23 00	18 42	06 09	19 33	14 58
14	04 12	02 44	03 02	02 23	04 S 50	06 49	15 48	19 03	22 59	18 43	06 08	19 33	14 59
15	04 06	02 41	03 59	01 59	01 N 48	07 22	16 14	19 14	22 59	18 44	06 07	19 33	15 00
16	03 58	02 38	04 40	01 35	08 21	07 50	16 39	19 25	22 59	18 45	06 06	19 32	15 00
17	03 50	02 35	05 02	01 12	14 29	08 16	17 04	19 35	22 58	18 46	06 05	19 32	15 01
18	03 42	02 32	05 04	00 48	19 43	08 37	17 29	19 45	22 58	18 48	06 04	19 32	15 02
19	03 36	02 28	04 45	00 S 24	23 38	08 54	17 53	19 55	22 58	18 49	06 03	19 32	15 03
20	03 31	02 25	04 08	00 00	25 57	09 06	18 17	20 05	22 58	18 50	06 02	19 32	15 03
21	03 30	02 22	03 17	00 N 23	26 34	09 15	18 40	20 15	22 57	18 51	06 01	19 32	15 04
22	03 D 29	02 19	02 16	00 47	25 31	09 19	19 03	20 24	22 57	18 52	06 00	19 32	15 05
23	03 30	02 16	01 10	01 11	23 03	09 19	19 25	20 34	22 57	18 54	05 59	19 32	15 05
24	03 31	02 12	00 N 01	01 34	19 26	09 14	19 47	20 43	22 57	18 55	05 59	19 31	15 06
25	03 R 30	02 09	01 S 06	01 58	14 58	09 05	20 09	20 52	22 56	18 56	05 58	19 31	15 07
26	03 27	02 06	02 09	02 21	09 57	08 52	20 30	21 01	22 56	18 57	05 57	19 31	15 07
27	03 21	02 03	03 05	02 45	04 N 38	08 35	20 51	21 10	22 56	18 59	05 56	19 31	15 08
28	03 13	02 00	03 52	03 08	00 S 48	08 15	21 11	21 18	22 56	19 00	05 55	19 31	15 09
29	03 02	01 57	04 27	03 32	06 09	07 51	21 31	21 27	22 56	19 01	05 54	19 31	15 09
30	02 50	01 53	04 50	03 55	11 13	07 25	21 50	21 35	22 55	19 03	05 53	19 30	15 10
31	02 ♒ 37	01 ♒ 50	05 S 01	04 N 18	15 S 50	06 N 57	22 N 09	21 N 43	22 S 55	19 N 04	05 S 52	19 S 30	15 N 10

ZODIAC SIGN ENTRIES

Date	h	m	Planets
01	19	00	☽ ♎
04	07	00	☽ ♏
05	16	59	☿ ♈
06	19	36	☽ ♐
07	03	25	♀ ♉
09	06	49	☽ ♑
11	14	43	☽ ♒
13	18	39	☽ ♓
15	19	37	☽ ♈
17	19	27	☽ ♉
19	20	12	☽ ♊
20	12	21	☉ ♈
21	23	26	☽ ♋
24	05	46	☽ ♌
26	14	48	☽ ♍
27	04	30	♂ ♊
29	01	42	☽ ♎
31	13	48	☽ ♏

LATITUDES

Date	Mercury ☿ ° '	Venus ♀ ° '	Mars ♂ ° '	Jupiter ♃ ° '	Saturn ♄ ° '	Uranus ♅ ° '	Neptune ♆ ° '	Pluto ♇ ° '
01	00 S 36	00 N 40	00 N 47	00 N 21	01 S 47	00 N 41	01 N 39	16 N 43
04	00 S 02	00 53	00 49	00 20	01 47	00 41	01 39	16 44
07	00 N 35	01 06	00 50	00 20	01 46	00 41	01 39	16 44
10	01 15	01 20	00 52	00 20	01 45	00 41	01 40	16 45
13	01 55	01 33	00 53	00 20	01 44	00 41	01 40	16 45
16	02 32	01 47	00 55	00 20	01 44	00 41	01 40	16 45
19	03 02	02 00	00 56	00 20	01 43	00 41	01 40	16 46
22	03 21	02 13	00 57	00 20	01 42	00 41	01 40	16 46
25	03 27	02 26	00 58	00 19	01 42	00 41	01 40	16 46
28	03 17	02 39	01 00	00 19	01 42	00 41	01 40	16 46
31	02 N 52	02 N 52	01 N 01	00 N 19	01 S 41	00 N 41	01 N 41	16 N 46

DATA

Julian Date	2441378
Delta T	+42 seconds
Ayanamsa	23° 28' 21"
Synetic vernal point	05° ♓ 38' 39"
True obliquity of ecliptic	23° 26' 40"

MOON'S PHASES, APSIDES AND POSITIONS ☽

Date	h	m	Phase	Longitude ° '	Eclipse Indicator
08	07	05	☾	17 ♐ 49	
15	11	35	●	25 ♓ 00	
22	02	12	☽	01 ♋ 34	
29	20	05	○	09 ♎ 14	

Day	h	m	
04	19	13	Apogee
16	20	44	Perigee

01	01	53	0S	
08	14	15	Max dec	26° S 40'
15	05	38	0N	
21	08	34	Max dec	26° N 35'
28	08	27	0S	

LONGITUDES

Date	Chiron ⚷ ° '	Ceres ⚳ ° '	Pallas ⚴ ° '	Juno ⚵ ° '	Vesta ⚶ ° '	Black Moon Lilith ⚸ ° '
01	11 ♈ 27	11 ♌ 00	02 ♊ 14	18 ♎ 03	26 ♓ 25	00 ♏ 42
11	11 ♈ 59	09 ♌ 48	06 ♊ 54	16 ♎ 20	01 ♈ 09	01 ♏ 50
21	12 ♈ 33	09 ♌ 19	11 ♊ 52	14 ♎ 12	05 ♈ 53	02 ♏ 57
31	13 ♈ 08	09 ♌ 33	17 ♊ 03	11 ♎ 50	10 ♈ 35	04 ♏ 04

All ephemeris data is given at 12.00 UT and the Moon's longitude is additionally given for 24.00 UT

ASPECTARIAN

h m Aspects

01 Wednesday
02 39 ☽ ☍ ☿
05 42 ☽ Q ♆
05 43 ☽ ⊼ ♀
15 31 ☽ ⚼ ♂
16 42 ☽ ∥ ☿
19 52 ☽ △ ♄
21 25 ☽ ☌ ♇

02 Thursday
03 07 ☽ □ ♃
03 31 ☉ ± ♅
05 24 ☽ ⚹ ♆
05 46 ☽ ∥ ♅
06 53 ♀ ⊥ ♄
09 25 ☽ ∥ ☉
10 23 ☽ ± ♂
19 30 ☽ ⊼ ☉
23 02 ☽ ⊼ ♂

03 Friday
01 32 ☽ # ♀
01 51 ☽ ⚼ ♄
03 23 ☿ ⊻ ♀
06 06 ☽ ☌ ♅
08 36 ☽ ± ☉
11 22 ☽ ∠ ♆
15 31 ☽ Q ♃
20 01 ☽ ± ♄
23 13 ☽ # ♇
23 46 ☽ ☍ ♀

04 Saturday
01 06 ☽ ⊼ ☿
04 23 ☽ ⚼ ☉
05 30 ☽ ⊥ ♆
08 10 ☽ ⊼ ♄
09 21 ☽ ⊻ ♇
11 54 ☽ # ♂
13 13 ☉ ∥ ♅
15 21 ☽ ± ☿
16 01 ☽ ⚹ ♃
17 37 ☽ ⊻ ♆
20 07 ☽ # ♄
21 27 ☽ ⊥ ♇

05 Sunday
02 24 ☽ ∥ ♆
12 43 ☽ ⚼ ☿
13 31 ☽ △ ☉
15 05 ☽ ☍ ♂
15 37 ☽ ∠ ♇
18 29 ☽ ⊻ ♅
22 44 ☽ ∠ ♃

06 Monday
00 35 ♂ ⚼ ♇
02 36 ☿ ⚹ ♄
03 33 ☽ ∥ ♃
06 34 ☽ ⊥ ♅
08 37 ☿ ☍ ♇
11 41 ♀ ± ♆
13 59 ☿ ∠ ♂
17 05 ☽ ⊥ ♃
18 47 ☽ ⊼ ♀
21 05 ☽ ☍ ♄
21 49 ☽ ⚹ ♇
23 59 ☽ △ ☿

07 Tuesday
00 38 ☽ ∠ ♅
05 16 ☽ ⊻ ♃
05 18 ♆ St R
06 09 ☽ ☌ ♆
07 19 ☉ Q ♃
08 04 ☽ ± ♀
16 42 ☉ ⚹ ♂
20 17 ♀ ⊻ ♄
21 43 ☽ Q ♇
22 39 ☉ ⊼ ♅

08 Wednesday
01 36 ♂ ⊼ ♅
02 02 ♀ ⊼ ♇
03 48 ☽ ⚼ ♀
06 23 ☽ ⚹ ♅
06 40 ☽ ⊼ ♂
07 05 ☽ □ ☉
18 47 ☿ □ ♃
19 05 ☽ ± ♂
22 23 ☿ △ ♆

09 Thursday
05 43 ☽ Q ♅
08 33 ☽ ⊼ ♄
08 49 ☽ □ ♇
08 57 ☉ Q ♄
11 56 ☽ △ ♀
13 26 ☽ ⚼ ♂
16 41 ☽ ☌ ♃
16 53 ☽ ⊻ ♆
19 22 ☽ □ ☿
20 04 ☽ ± ♄
20 58 ☽ Q ☉

10 Friday
04 15 ☽ ⊥ ♆
07 56 ☉ # ☿
12 31 ♃ ⊻ ♆
13 05 ☽ ⚼ ♄
15 40 ☽ □ ♅
19 09 ☽ △ ♂
19 16 ♄ △ ♇
20 59 ☽ ∠ ♆
21 17 ☽ ⚹ ☉
22 10 ☽ ∥ ♃

11 Saturday
08 46 ☽ Q ☿
16 27 ☽ △ ♇
16 36 ☽ △ ♄
19 18 ♀ ⊼ ♆
20 24 ☽ ∥ ♆
23 18 ♀ △ ♃

12 Sunday
00 04 ☽ ⚹ ♆
00 25 ☽ ⊻ ♃
00 30 ☽ □ ♀
01 11 ☽ # ♄
01 29 ☽ # ♂
02 30 ☽ ∠ ☉
08 21 ☽ ⚹ ☿
09 35 ♂ ∥ ♄
11 04 ☽ ⊥ ♃
13 32 ♀ ∥ ♇
17 38 ♂ ⚼ ♃
18 46 ☽ ⚼ ♇
18 50 ☽ # ♀
19 16 ☽ # ♇
19 26 ☽ ⊥ ☉
21 01 ☽ Q ♆
21 06 ☽ △ ♅

13 Monday
02 44 ☽ ∠ ♃
03 06 ☽ □ ♂
06 30 ☽ ⊻ ☉
07 21 ♀ ± ♇
08 59 ☿ # ♅
10 07 ☽ ± ♇
10 21 ☽ Q ♀
12 24 ☽ ∠ ☿
20 11 ☽ ⊼ ♇
20 41 ☽ □ ♄
22 24 ☽ ⚼ ♅

14 Tuesday
03 23 ☽ □ ♆
04 09 ☽ ⚹ ♃
04 36 ☽ ⊥ ☿
05 11 ☽ # ☿
07 09 ☽ ∥ ♅
08 03 ☽ ⚹ ♀
10 39 ☽ Q ♂
13 14 ☽ ± ♅
15 11 ☽ ⊻ ☿
21 36 ☽ ∥ ☉
23 00 ☽ ⊼ ♅
23 58 ☽ Q ♃

15 Wednesday
02 24 ☽ Q ♄
07 06 ☽ ⚹ ♂
10 27 ☽ ∠ ♀
11 35 ☽ ☌ ☉
12 44 ☽ # ☉
20 59 ☽ ☍ ♇
21 48 ☽ ⚹ ♄

16 Thursday
02 01 ☽ ⊥ ♀
03 41 ☽ # ♅
03 59 ☽ △ ♆
05 07 ☽ □ ♃
08 14 ☽ ∠ ♂
09 56 ☽ ∥ ☿
12 21 ☽ ⊻ ♀
17 03 ♂ ± ♅
18 11 ☽ ☌ ☿
21 51 ☽ ∠ ♄
22 56 ☽ ☍ ♅
23 13 ☽ ⊥ ♂

17 Friday
03 52 ☽ ⚼ ♆
05 59 ♂ # ♆
09 13 ☽ ⊻ ♂
12 15 ☽ ⊥ ♄
14 14 ☽ ∥ ♇
14 51 ☽ ⊻ ☉
18 13 ☽ ± ♆
20 44 ☽ ⊼ ♇
21 54 ☽ ⊻ ♄

18 Saturday
00 10 ☽ ∥ ♀
01 08 ☽ ⊥ ☉
03 49 ☽ ⊼ ♆
05 21 ☽ △ ♃
06 21 ☽ ± ♇
07 21 ☽ ∥ ♄
11 06 ☽ # ♆
12 13 ☽ ∥ ♂
16 07 ☽ ☌ ♀
16 40 ☽ ∠ ☉
20 02 ☽ ⊻ ☿
20 52 ☽ ⚼ ♇
22 56 ☽ ⊼ ♅

19 Sunday
05 50 ☽ ⚼ ♃
06 03 ☽ ⊥ ☿
07 12 ☽ # ♃
08 43 ☽ ± ♅
11 53 ☽ ☌ ♂
19 01 ☽ ⚹ ☉
20 47 ☽ ± ♃
21 13 ☽ ∠ ☿
21 27 ☽ △ ♇
23 01 ☽ ☌ ♄
23 33 ☽ ⚼ ♅

20 Monday
04 53 ☽ ☍ ♆
06 52 ☽ ⊼ ♃
16 39 ☽ Q ☉
21 56 ☽ ⊻ ♀
22 49 ☽ ⚹ ☿

21 Tuesday
00 49 ☽ △ ♅
05 30 ☉ ☍ ♇
07 13 ♀ ⚼ ♇
09 07 ☽ ⊥ ♀
10 04 ☿ ⊻ ♀
17 10 ☽ ⊻ ♂
18 39 ☿ St R
19 39 ☽ Q ☿

22 Wednesday
00 25 ♀ ∥ ♄
00 39 ☽ □ ♇
02 11 ☽ ∠ ♀
02 12 ☽ □ ☉
02 44 ☽ ⊻ ♄
03 05 ☉ ∠ ♀
04 14 ☽ ⊥ ♂
08 37 ☽ ⊼ ♆
09 58 ♀ ⊼ ♅
10 10 ☉ ⚹ ♄
11 09 ☽ ☍ ♃
13 09 ♄ ⚼ ♅
13 27 ☽ ⊥ ♄
19 20 ☽ ± ♆
21 08 ☽ ∠ ♂

23 Thursday
03 32 ☽ □ ☿
05 38 ☽ □ ♅
05 47 ☽ ∠ ♄
07 26 ☽ ⚹ ♀
08 51 ☽ Q ♇
11 39 ☽ ⚼ ♆
12 45 ☽ # ♃
18 41 ♀ # ♆

24 Friday
02 00 ☽ ⚹ ♂
04 28 ☽ ∥ ♂
06 56 ☽ ⚹ ♇
07 32 ☽ Q ♀
09 37 ☽ ⚹ ♄
10 02 ☽ ∥ ♀
11 26 ☽ # ♆
13 14 ☽ △ ☉
14 46 ☽ Q ♅
14 57 ☽ ∥ ♄
15 26 ☽ △ ♆
18 33 ☽ ⊼ ♃

25 Saturday
01 41 ☽ Q ♂
05 58 ☽ ± ♃
08 23 ☽ Q ♄
10 01 ☽ △ ☿
11 07 ☽ ∠ ♇
11 18 ☽ ∥ ♇
13 23 ☽ ⚹ ♅
17 17 ☉ △ ♆
20 04 ☽ ⚼ ☉
20 40 ☽ □ ♀
23 15 ☽ ⚼ ♃
23 58 ☿ ⊥ ♀

26 Sunday
04 19 ☽ ⊥ ♇
10 30 ☿ ∠ ♂
13 42 ☽ ⚼ ☿
14 02 ☽ □ ♂
14 54 ☽ ± ☉
15 53 ☽ ⊻ ♇
17 16 ☽ ∥ ☿
18 09 ☽ ∠ ♅
19 13 ☽ □ ♄

27 Monday
00 47 ☽ □ ♆
03 34 ☽ ⊼ ☉
04 30 ☽ △ ♃
05 50 ♀ ⚼ ♃
06 12 ☽ # ♅
06 24 ☽ ± ☿
11 38 ☽ ⊥ ♅
16 04 ☉ □ ♃
17 34 ☽ ⊼ ☿
19 45 ☽ ∥ ☉
20 14 ♀ ± ♅
23 21 ☽ ⊻ ♅
23 46 ♂ △ ♇

28 Tuesday
12 05 ☽ Q ♆
12 37 ☽ △ ♀
23 13 ☽ # ☉

29 Wednesday
00 44 ♂ ± ♃
02 41 ☽ ☌ ♇
03 01 ♀ ∥ ♂
04 17 ☽ △ ♂
06 41 ☽ △ ♄
10 52 ☽ ∥ ♅
11 53 ☽ ⚹ ♆
15 36 ♂ ⚼ ♅
16 09 ☽ □ ♃
19 18 ☽ # ☿
20 05 ☽ ☍ ☉
21 16 ☽ ⚼ ♀

30 Thursday
01 41 ☽ ☍ ☿
10 47 ☽ ☌ ♅
11 58 ☽ ⚼ ♂
12 54 ☽ ⚼ ♄
17 00 ☽ ± ♀
17 53 ☽ ∠ ♆

31 Friday
04 04 ☿ ∠ ♀
04 29 ☽ Q ♃
06 13 ☽ ⊼ ♀
07 07 ☽ ± ♂
07 09 ☽ ± ♄
07 59 ♂ ☌ ♄
08 20 ☽ # ♇
11 59 ☽ ⊥ ♆
12 00 ☉ ☌ ☿
14 40 ☽ ⊻ ♇
19 21 ☽ ⊼ ♄
19 55 ☽ ⊼ ♂

APRIL 1972

LONGITUDES

Date	Sidereal time h m s	Sun ☉ ° ' "	Moon ☽ ° ' "	Moon ☽ 24.00 ° ' "	Mercury ☿ ° '	Venus ♀ ° '	Mars ♂ ° '	Jupiter ♃ ° '	Saturn ♄ ° '	Uranus ♅ ° '	Neptune ♆ ° '	Pluto ♇ ° '
01	00 39 45	11 ♈ 51 37	10 ♏ 59 08	16 ♏ 55 01	10 ♈ 03	27 ♉ 31	03 ♊ 28	07 ♑ 28	02 ♊ 49	16 ♎ 29	05 ♐ 05	00 ♎ 24
02	00 43 42	12 50 46	22 ♏ 50 57	28 ♏ 47 13	09 R 13	28 32	04 07	07 32	02 55	16 R 27	05 R 04	00 R 23
03	00 47 38	13 49 53	04 ♐ 44 12	10 ♐ 42 21	08 24	29 ♉ 33	04 46	07 36	03 01	16 24	05 03	00 21
04	00 51 35	14 48 58	16 ♐ 42 06	22 ♐ 43 58	07 38	00 ♊ 33	05 25	07 40	03 07	16 21	05 02	00 19
05	00 55 32	15 48 02	28 ♐ 48 31	04 ♑ 56 19	06 54	01 33	06 04	07 44	03 13	16 19	05 02	00 18
06	00 59 28	16 47 03	11 ♑ 07 59	17 ♑ 24 07	06 13	02 33	06 43	07 47	03 19	16 16	05 01	00 16
07	01 03 25	17 46 03	23 ♑ 45 21	00 ♒ 12 15	05 37	03 32	07 22	07 51	03 25	16 14	05 00	00 15
08	01 07 21	18 45 02	06 ♒ 45 24	13 ♒ 25 15	05 05	04 31	08 01	07 54	03 32	16 11	04 59	00 13
09	01 11 18	19 43 58	20 ♒ 12 12	27 ♒ 06 31	04 38	05 29	08 40	07 57	03 38	16 08	04 58	00 12
10	01 15 14	20 42 53	04 ♓ 08 18	11 ♓ 17 26	04 16	06 27	09 19	08 00	03 44	16 06	04 57	00 10
11	01 19 11	21 41 45	18 ♓ 33 38	25 ♓ 56 22	03 59	07 24	09 58	08 02	03 51	16 03	04 56	00 09
12	01 23 07	22 40 36	03 ♈ 24 50	10 ♈ 58 03	03 48	08 21	10 37	08 05	03 57	16 01	04 55	00 07
13	01 27 04	23 39 25	18 ♈ 34 48	26 ♈ 13 44	03 42	09 18	11 16	08 07	04 04	15 58	04 53	00 06
14	01 31 01	24 38 13	03 ♉ 53 23	11 ♉ 32 17	03 D 41	10 14	11 55	08 09	04 10	15 56	04 52	00 04
15	01 34 57	25 36 58	19 ♉ 09 01	26 ♉ 42 15	03 46	11 09	12 34	08 11	04 17	15 53	04 51	00 03
16	01 38 54	26 35 41	04 ♊ 10 51	11 ♊ 33 53	03 55	12 04	13 13	08 13	04 24	15 51	04 50	00 01
17	01 42 50	27 34 22	18 ♊ 50 39	26 ♊ 00 40	04 10	12 58	13 52	08 14	04 30	15 48	04 49	00 ♎ 00
18	01 46 47	28 33 01	03 ♋ 03 42	09 ♋ 59 42	04 29	13 51	14 31	08 15	04 37	15 46	04 48	29 ♍ 58
19	01 50 43	29 ♈ 31 37	16 ♋ 48 47	23 ♋ 31 13	04 53	14 44	15 09	08 17	04 44	15 43	04 46	29 57
20	01 54 40	00 ♉ 30 12	00 ♌ 07 23	06 ♌ 37 43	05 21	15 36	15 48	08 17	04 51	15 41	04 45	29 56
21	01 58 36	01 28 44	13 ♌ 02 45	19 ♌ 22 59	05 53	16 28	16 27	08 18	04 58	15 38	04 44	29 54
22	02 02 33	02 27 14	25 ♌ 38 57	01 ♍ 51 12	06 29	17 19	17 06	08 19	05 05	15 36	04 43	29 53
23	02 06 30	03 25 41	08 ♍ 00 13	14 ♍ 06 28	07 10	18 09	17 45	08 19	05 12	15 33	04 41	29 52
24	02 10 26	04 24 07	20 ♍ 10 25	26 ♍ 12 25	07 54	18 58	18 23	08 19	05 19	15 31	04 40	29 50
25	02 14 23	05 22 30	02 ♎ 12 51	08 ♎ 12 02	08 41	19 47	19 02	08 R 19	05 26	15 28	04 39	29 49
26	02 18 19	06 20 52	14 ♎ 10 12	20 ♎ 07 37	09 32	20 35	19 41	08 19	05 33	15 26	04 37	29 48
27	02 22 16	07 19 12	26 ♎ 04 29	02 ♏ 00 59	10 26	21 22	20 19	08 19	05 40	15 24	04 36	29 47
28	02 26 12	08 17 29	07 ♏ 57 17	13 ♏ 53 34	11 23	22 08	20 58	08 18	05 48	15 21	04 34	29 45
29	02 30 09	09 15 45	19 ♏ 50 00	25 ♏ 46 44	12 23	22 53	21 37	08 18	05 55	15 19	04 33	29 44
30	02 34 05	10 ♉ 14 00	01 ♐ 44 00	07 ♐ 42 00	13 ♈ 26	23 ♊ 37	22 ♊ 15	08 ♑ 17	06 ♊ 02	15 ♎ 17	04 ♐ 32	29 ♍ 43

Date	Moon True ☊ ° '	Moon Mean ☊ ° '	Moon ☽ Latitude ° '
01	02 ♒ 24	01 ♒ 47	04 S 58
02	02 R 13	01 44	04 42
03	02 05	01 41	04 15
04	01 59	01 38	03 35
05	01 55	01 34	02 46
06	01 54	01 31	01 48
07	01 D 54	01 28	00 S 43
08	01 R 54	01 25	00 N 26
09	01 53	01 22	01 35
10	01 50	01 18	02 42
11	01 44	01 15	03 40
12	01 36	01 12	04 25
13	01 25	01 09	04 53
14	01 14	01 06	05 00
15	01 04	01 03	04 46
16	00 55	00 59	04 12
17	00 48	00 56	03 22
18	00 45	00 53	02 20
19	00 43	00 50	01 13
20	00 D 43	00 47	00 N 03
21	00 R 43	00 43	01 S 05
22	00 41	00 40	02 08
23	00 38	00 37	03 03
24	00 32	00 34	03 49
25	00 23	00 31	04 25
26	00 11	00 28	04 48
27	29 ♑ 58	00 24	04 59
28	29 43	00 21	04 56
29	29 30	00 18	04 41
30	29 ♑ 18	00 ♒ 15	04 S 14

DECLINATIONS

Date	Sun ☉ ° '	Moon ☽ ° '	Mercury ☿ ° '	Venus ♀ ° '	Mars ♂ ° '	Jupiter ♃ ° '	Saturn ♄ ° '	Uranus ♅ ° '	Neptune ♆ ° '	Pluto ♇ ° '
01	04 N 41	19 S 50	06 N 26	22 N 27	21 N 51	22 S 55	19 N 05	05 S 51	19 S 30	15 N 11
02	05 04	23 02	05 55	22 45	21 59	22 55	19 07	05 50	19 30	15 12
03	05 27	25 15	05 23	23 03	22 06	22 55	19 08	05 49	19 30	15 12
04	05 50	26 21	04 52	23 19	22 14	22 54	19 09	05 48	19 30	15 13
05	06 13	26 12	04 20	23 36	22 21	22 54	19 11	05 47	19 29	15 13
06	06 36	24 46	03 50	23 51	22 28	22 54	19 12	05 46	19 29	15 14
07	06 58	22 04	03 20	24 07	22 35	22 54	19 13	05 45	19 29	15 14
08	07 21	18 10	02 53	24 21	22 41	22 54	19 15	05 44	19 29	15 15
09	07 43	13 15	02 27	24 35	22 48	22 54	19 16	05 43	19 29	15 15
10	08 05	07 29	02 03	24 49	22 54	22 54	19 17	05 42	19 28	15 16
11	08 27	01 S 09	01 42	25 02	23 00	22 54	19 19	05 41	19 28	15 16
12	08 49	05 N 25	01 24	25 15	23 06	22 53	19 20	05 40	19 28	15 17
13	09 11	11 48	01 07	25 27	23 12	22 53	19 21	05 39	19 28	15 17
14	09 33	17 31	00 54	25 38	23 18	22 53	19 23	05 38	19 27	15 17
15	09 54	22 05	00 43	25 49	23 23	22 53	19 24	05 37	19 27	15 18
16	10 16	25 06	00 35	26 00	23 28	22 53	19 26	05 36	19 27	15 18
17	10 37	26 20	00 29	26 09	23 33	22 53	19 27	05 35	19 27	15 19
18	10 58	25 45	00 26	26 19	23 38	22 53	19 28	05 34	19 26	15 19
19	11 18	23 35	00 25	26 28	23 43	22 53	19 30	05 33	19 26	15 19
20	11 39	20 11	00 26	26 36	23 47	22 53	19 31	05 32	19 26	15 20
21	11 59	15 52	00 30	26 43	23 51	22 53	19 33	05 31	19 26	15 20
22	12 20	10 58	00 37	26 51	23 56	22 53	19 34	05 30	19 25	15 20
23	12 40	05 44	00 45	26 57	23 59	22 53	19 35	05 29	19 25	15 21
24	12 59	00 N 23	00 56	27 03	24 03	22 53	19 37	05 28	19 25	15 21
25	13 19	04 S 56	01 08	27 09	24 07	22 53	19 38	05 28	19 25	15 21
26	13 38	10 01	01 23	27 14	24 10	22 53	19 40	05 27	19 24	15 21
27	13 57	14 42	01 39	27 18	24 13	22 54	19 41	05 26	19 24	15 22
28	14 16	18 50	01 57	27 23	24 16	22 54	19 42	05 25	19 24	15 22
29	14 35	22 13	02 17	27 26	24 19	22 54	19 44	05 24	19 23	15 22
30	14 N 53	24 S 39	02 N 38	27 N 29	24 N 21	22 S 54	19 N 45	05 S 23	19 S 23	15 N 22

ZODIAC SIGN ENTRIES

Date	h	m	Planets
03	02	27	☽ ♐
03	22	48	♀ ♊
05	14	20	☽ ♑
07	23	37	☽ ♒
10	04	58	☽ ♓
12	06	32	☽ ♈
14	05	54	☽ ♉
16	05	16	☽ ♊
17	07	49	♇ ♍
18	06	46	☽ ♋
19	23	37	☉ ♉
20	11	46	☽ ♌
22	20	24	☽ ♍
25	07	34	☽ ♎
27	19	56	☽ ♏
30	08	31	☽ ♐

LATITUDES

Date	Mercury ☿ ° '	Venus ♀ ° '	Mars ♂ ° '	Jupiter ♃ ° '	Saturn ♄ ° '	Uranus ♅ ° '	Neptune ♆ ° '	Pluto ♇ ° '
01	02 N 41	02 N 56	01 N 01	00 N 19	01 S 40	00 N 41	01 N 41	16 N 46
04	02 00	03 07	01 02	00 19	01 40	00 41	01 41	16 45
07	01 12	03 18	01 03	00 19	01 39	00 41	01 41	16 45
10	00 N 24	03 29	01 04	00 19	01 39	00 41	01 41	16 45
13	00 S 23	03 38	01 05	00 18	01 38	00 41	01 41	16 44
16	01 04	03 47	01 05	00 18	01 38	00 41	01 41	16 44
19	01 40	03 55	01 06	00 18	01 37	00 41	01 41	16 43
22	02 08	04 01	01 07	00 18	01 37	00 41	01 41	16 42
25	02 31	04 07	01 07	00 18	01 36	00 41	01 41	16 42
28	02 46	04 10	01 08	00 17	01 36	00 41	01 41	16 41
31	02 S 56	04 N 12	01 N 09	00 N 17	01 S 35	00 N 41	01 N 42	16 N 40

DATA

Julian Date	2441409
Delta T	+42 seconds
Ayanamsa	23° 28' 24"
Synetic vernal point	05° ♓ 38' 36"
True obliquity of ecliptic	23° 26' 40"

MOON'S PHASES, APSIDES AND POSITIONS ☽

Date	h	m	Phase	Longitude ° '	Eclipse Indicator
06	23	44	☾	17 ♑ 16	
13	20	31	●	24 ♈ 00	
20	12	45	☽	00 ♌ 32	
28	12	44	○	08 ♏ 19	

Day	h	m		
01	07	03	Apogee	
14	06	16	Perigee	
28	09	47	Apogee	
04	21	16	Max dec	26° S 27'
11	16	14	0N	
17	16	02	Max dec	26° N 21'
24	13	41	0S	

LONGITUDES

Date	Chiron ⚷ ° '	Ceres ⚳ ° '	Pallas ⚴ ° '	Juno ⚵ ° '	Vesta ⚶ ° '	Black Moon Lilith ⚸ ° '
01	13 ♈ 12	09 ♌ 37	17 ♊ 35	11 ♎ 36	11 ♈ 03	04 ♏ 11
11	13 ♈ 47	10 ♌ 35	22 ♊ 56	09 ♎ 16	15 ♈ 43	05 ♏ 18
21	14 ♈ 21	12 ♌ 09	28 ♊ 24	07 ♎ 12	20 ♈ 21	06 ♏ 25
31	14 ♈ 55	14 ♌ 14	03 ♋ 56	05 ♎ 34	24 ♈ 55	07 ♏ 32

ASPECTARIAN

h m Aspects

01 Saturday
00 05 ☽ ⊻ ♆
02 45 ☽ ⊥ ♇
04 51 ☽ ⚹ ♃
07 05 ☽ # ♄
09 47 ☽ ∥ ♆
10 13 ☽ ⊼ ☿
13 56 ☽ ⊼ ☉
20 55 ☽ ∠ ♇
21 34 ☽ ± ☿
23 05 ☽ ⊻ ♅

02 Sunday
02 53 ☽ # ♂
03 10 ☽ ± ☉
09 16 ☽ # ♀
10 55 ☽ ∥ ♃
11 11 ☽ ⊥ ♅
11 22 ☽ ∠ ♃
14 35 ☽ ⚼ ☿
16 21 ☿ # ♅
23 01 ☽ ⚼ ☉

03 Monday
00 34 ☽ ☍ ♀
01 06 ♀ # ♃
03 10 ☽ ⚹ ♇
05 18 ☽ ∠ ♅
05 39 ☽ ⊥ ♃
08 30 ☽ ☍ ♄
10 15 ☉ ∥ ☿
12 04 ☽ ☍ ♂
12 38 ☽ ☌ ♆
17 48 ☽ ⊻ ♃
18 55 ☽ △ ☿
22 18 ♂ ☍ ♆

04 Tuesday
03 16 ☽ Q ♇
06 38 ♀ △ ♇
07 54 ☽ △ ☉
09 22 ☉ # ♅
10 47 ☿ □ ♃
11 19 ☽ ⚹ ♅

05 Wednesday
06 28 ♀ ⚼ ♅
11 02 ☽ Q ♅
14 55 ☽ □ ♇
16 33 ♀ ± ♃
17 52 ☽ ⊼ ♀
20 43 ☽ ⊼ ♄

06 Thursday
00 00 ☉ ☍ ♅
00 09 ☽ ⊻ ♆
02 51 ☿ ⚹ ♂
02 59 ☽ □ ☿
03 00 ☽ ⊼ ♂
05 31 ☽ ☌ ♃
06 34 ☽ ± ♀
08 28 ☽ ± ♄
11 46 ☽ ⊥ ♆
15 13 ☽ ± ♂
20 32 ☽ # ♀
21 49 ☽ □ ♅
23 44 ☽ □ ☉

07 Friday
01 19 ☽ ⚼ ♀
01 51 ☽ ⚼ ♄
04 56 ☽ ∠ ♆
05 37 ☽ ∥ ♃
08 17 ☽ # ♂
08 58 ♀ ☌ ♄
09 16 ☽ ⚼ ♂
11 45 ☽ Q ☿

08 Saturday
00 03 ☽ △ ♇
04 40 ☽ ∥ ♆
05 54 ☉ ∠ ♄
06 04 ☽ △ ♄
06 04 ☽ # ♄
07 03 ♂ ⊼ ♃
07 35 ☽ △ ♀
08 46 ☽ ⚹ ♆
09 04 ☽ ⚹ ☿
11 59 ☽ Q ☉
14 05 ☽ ⊻ ♃
14 25 ☽ △ ♂
17 36 ☿ △ ♆
21 29 ☿ ⚹ ♀
23 15 ♀ ☍ ♆

09 Sunday
00 54 ☽ ⊥ ♃
02 47 ☽ # ♇
03 10 ☽ ⚼ ♇
04 52 ☽ △ ♅
06 18 ☽ Q ♆
11 02 ☽ ∠ ☿
11 07 ☽ ⚹ ☉
16 49 ☽ ∠ ♃
17 30 ☉ ⚼ ♆
18 57 ☽ ± ♇

10 Monday
02 14 ☽ ⊥ ☿
05 16 ☽ ⊼ ♇
06 51 ☽ ⚼ ♅
09 44 ☽ # ☉
09 49 ♂ # ♃
11 19 ☽ □ ♄
12 13 ☽ ⊻ ☿
13 22 ☽ □ ♆
14 52 ☽ ∠ ☉
16 11 ☽ □ ♀
18 31 ☽ ⚹ ♃
18 58 ☽ ∥ ♅
21 08 ☽ □ ♂
21 59 ☽ ± ♅

11 Tuesday
06 57 ☽ ⊥ ☉
07 54 ☽ ⊼ ♅
09 51 ☽ # ☿
14 25 ☽ Q ♃
17 24 ☽ Q ♄
17 29 ☽ ⊻ ☉
21 58 ☽ ∥ ☿
22 53 ☿ ⚹ ♄
23 54 ☽ Q ♀

12 Wednesday
03 58 ☽ Q ♂
04 41 ♀ ⊼ ♃
06 44 ☽ ☍ ♇
12 37 ☽ ☌ ☿
12 52 ☽ ⚹ ♄
12 55 ☽ # ♅
14 23 ☽ △ ♆
19 27 ☽ □ ♃
20 23 ☽ ⚹ ♀
23 58 ☽ ⚹ ♂

13 Thursday
01 24 ☽ ∥ ☉
07 54 ☽ ☍ ♅
12 46 ☽ ∠ ♄
14 03 ☽ ⚼ ♆
20 31 ☽ ☌ ☉
21 33 ☽ ∠ ♀

14 Friday
00 36 ☽ ∠ ♂
02 12 ☽ ∥ ♇
02 59 ☽ ⊥ ♄
03 29 ☿ St D
04 09 ☽ ± ♆
06 01 ☽ ⊼ ♇
11 41 ☽ ⊻ ☿
12 27 ☽ ⊻ ♄
12 34 ☽ ⊥ ♀
13 32 ☽ ⊼ ♆
15 19 ☽ ⊥ ♂
15 24 ☽ ± ♇
18 42 ☽ △ ♃
21 00 ☽ ∥ ♄
21 07 ☽ ⊥ ☿
21 20 ☽ # ♆
22 35 ☽ ⊻ ♀

15 Saturday
01 09 ☽ ⊻ ♂
05 32 ☽ ⚼ ♇
06 51 ☽ ⊼ ♅
11 23 ☽ ∠ ☿
16 19 ☽ ± ♅
17 12 ☽ # ♃
18 24 ☽ ⚼ ♃
20 58 ☽ ∥ ♂
22 59 ☽ ⊻ ☉

16 Sunday
05 19 ☽ △ ♇
06 38 ☽ ⚼ ♅
08 49 ☽ ± ♃
09 16 ☽ ⊥ ☉
11 34 ☽ ⚹ ☿
12 21 ☽ ☌ ♄
13 03 ☽ ☍ ♆
18 32 ☽ ⊼ ♃

17 Monday
00 55 ☽ ∠ ☉
01 39 ☽ ☌ ♀
03 23 ☽ ☌ ♂
03 23 ☽ ∥ ♀
06 42 ♄ # ♆
06 59 ☽ △ ♅
07 28 ☽ Q ☿

18 Tuesday
00 31 ☽ ∥ ♀
03 44 ☽ ⚹ ☉
06 44 ☽ □ ♇
13 57 ☉ ⊥ ♄
14 30 ☽ □ ☿
14 42 ☽ ⊻ ♄
14 58 ☽ ⊼ ♆
17 51 ☉ ± ♆
20 59 ☽ ☍ ♃

19 Wednesday
00 44 ☿ ⚹ ♄
01 12 ☽ ⊥ ♄
01 22 ☽ ± ♆
01 58 ☽ Q ☉
06 26 ☿ △ ♆
08 04 ☽ ⊻ ♀
08 55 ☽ ⊻ ♂
10 04 ☽ □ ♅
10 59 ☽ ∥ ♂
14 01 ☽ Q ♇
17 15 ☽ ∠ ♄
17 16 ☽ ⚼ ♆
17 39 ☽ # ♃
18 47 ♄ ☍ ♆
19 29 ☽ ⊥ ♀
20 09 ☽ ⊥ ♂
22 08 ☉ ⊼ ♇

20 Thursday
07 33 ♂ △ ♅
11 38 ☽ ⚹ ♇
12 45 ☽ □ ☉
12 57 ☽ ∠ ♀
13 19 ☽ ∠ ♂
13 48 ♀ △ ♅
15 57 ☽ ∥ ♄
16 30 ☽ # ♆
18 31 ☽ Q ♅
20 30 ☽ △ ♆
20 47 ☽ ⚹ ♄
22 01 ☽ △ ☿

21 Friday
03 07 ☽ ⊼ ♃
09 40 ☉ ∠ ♀
09 55 ☉ ∠ ♂
10 03 ♀ ☌ ♂
14 22 ☽ ± ♃
14 45 ☽ ∥ ♇
15 30 ☽ ∠ ♇
16 52 ☽ ⚹ ♅
18 46 ☽ ⚹ ♂
18 56 ☽ ⚹ ♀
19 28 ☽ Q ♄

22 Saturday
03 36 ☽ ⚼ ☿
05 58 ☽ ∥ ☉
07 31 ☽ ⚼ ♃
08 37 ☽ ⊥ ♇
19 01 ☽ Q ♂
19 35 ☽ Q ♀
20 10 ☽ ⊻ ♇
21 31 ☽ ∠ ♅
21 52 ☽ ± ☿

23 Sunday
02 18 ☽ △ ☉
05 32 ☽ □ ♆
06 28 ☽ □ ♄
10 15 ☽ ⊼ ☿
12 37 ☽ △ ♃
13 07 ☽ # ♅
15 02 ☽ ⊥ ♅

24 Monday
02 48 ☽ ⊻ ♅
08 16 ☽ □ ♂
09 27 ☽ □ ♀
09 37 ☽ ∥ ☿
10 20 ☽ ⚼ ☉
16 56 ☽ Q ♆
18 04 ☽ # ☿
18 20 ☉ ⊼ ♆

25 Tuesday
00 21 ♃ St R
01 17 ☿ □ ♃
05 49 ☽ ± ☉
07 13 ☽ ☌ ♇
13 38 ☉ ⊻ ♄
14 26 ☽ ∥ ♅
16 51 ☽ ⚹ ♆
18 31 ☽ △ ♄
18 54 ☽ ⊼ ☉
22 40 ☉ ± ♇

26 Wednesday
00 15 ☽ □ ♃
01 57 ☽ ☍ ☿
14 32 ☽ ☌ ♅
22 57 ☽ ∠ ♆
23 44 ☽ △ ♂

27 Thursday
00 59 ☽ ⚼ ♄
01 49 ☽ △ ♀
07 42 ☽ # ☉
12 29 ☽ Q ♃
15 34 ☽ # ♇
17 05 ☽ ⊥ ♆
19 20 ☽ ± ♄
19 28 ☽ ⊻ ♇

28 Friday
05 11 ☽ ⊻ ♆
07 34 ☽ ⊥ ♇
07 35 ☽ ⊼ ♄
07 45 ☽ ⚼ ♂
10 13 ☽ ⚼ ♀
12 19 ☉ △ ♃
12 42 ☽ ⚹ ♃
12 44 ☽ ☍ ☉
15 37 ☽ ∥ ♆
17 42 ☽ # ♄
19 33 ☽ ⊼ ☿

29 Saturday
01 43 ☽ ∠ ♇
02 55 ☽ ⊻ ♅
02 59 ☽ ± ♂
05 38 ☽ ± ♀
08 48 ☽ ± ☿
14 59 ☽ ⊥ ♅
15 48 ☽ ⊼ ♂
17 51 ☽ ∥ ♃
18 33 ☽ ⊼ ♀
18 58 ☽ ∠ ♃

30 Sunday
04 42 ☽ ⚼ ☿
07 57 ☽ ⚹ ♇
08 22 ☽ # ♂
09 05 ☽ ∠ ♅
13 06 ☽ ⊥ ♃
17 37 ☽ ☌ ♆
20 44 ☽ ☍ ♄

All ephemeris data is given at 12.00 UT and the Moon's longitude is additionally given for 24.00 UT

MAY 1972

LONGITUDES

Date	Sidereal time h m s	Sun ☉ ° ' "	Moon ☽ ° ' "	Moon ☽ 24.00 ° ' "	Mercury ☿ ° '	Venus ♀ ° '	Mars ♂ ° '	Jupiter ♃ ° '	Saturn ♄ ° '	Uranus ♅ ° '	Neptune ♆ ° '	Pluto ♇ ° '
01	02 38 02	11 ♉ 12 12	13 ♐ 40 59	19 ♐ 41 17	14 ♈ 32	24 ♊ 20	22 ♊ 54	08 ♑ 15	06 ♊ 09	15 ♎ 15	04 ♐ 30	29 ♍ 42
02	02 41 59	12 10 23	25 ♐ 43 11	01 ♑ 47 05	15 40	25 02	23 33	08 R 14	06 17	15 R 12	04 R 29	29 R 41
03	02 45 55	13 08 33	07 ♑ 53 25	14 ♑ 02 36	16 51	25 43	24 11	08 13	06 24	15 10	04 27	29 40
04	02 49 52	14 06 41	20 ♑ 15 09	26 ♑ 31 35	18 04	26 23	24 50	08 11	06 31	15 08	04 26	29 39
05	02 53 48	15 04 48	02 ♒ 52 26	09 ♒ 18 14	19 20	27 02	25 28	08 09	06 39	15 06	04 24	29 37
06	02 57 45	16 02 53	15 ♒ 49 31	22 ♒ 26 46	20 38	27 40	26 07	08 07	06 46	15 04	04 23	29 36
07	03 01 41	17 00 56	29 ♒ 10 24	06 ♓ 00 45	21 59	28 16	26 46	08 05	06 54	15 02	04 21	29 35
08	03 05 38	17 58 59	12 ♓ 58 04	20 ♓ 02 24	23 22	28 51	27 24	08 02	07 01	15 00	04 20	29 34
09	03 09 34	18 57 00	27 ♓ 13 39	04 ♈ 31 30	24 47	29 25	28 03	08 00	07 09	14 58	04 18	29 34
10	03 13 31	19 54 59	11 ♈ 55 23	19 ♈ 24 35	26 14	29 ♊ 58	28 41	07 57	07 16	14 56	04 17	29 33
11	03 17 28	20 52 58	26 ♈ 58 04	04 ♉ 34 42	27 43	00 ♋ 29	29 20	07 54	07 24	14 54	04 15	29 32
12	03 21 24	21 50 55	12 ♉ 13 08	19 ♉ 51 59	29 ♈ 14	00 58	29 ♊ 58	07 51	07 32	14 52	04 13	29 31
13	03 25 21	22 48 50	27 ♉ 29 47	05 ♊ 05 10	00 ♉ 48	01 26	00 ♋ 36	07 48	07 39	14 50	04 12	29 30
14	03 29 17	23 46 44	12 ♊ 36 50	20 ♊ 03 41	02 24	01 53	01 15	07 44	07 47	14 48	04 10	29 29
15	03 33 14	24 44 37	27 ♊ 24 47	04 ♋ 39 26	04 02	02 18	01 53	07 41	07 55	14 46	04 09	29 28
16	03 37 10	25 42 28	11 ♋ 47 10	18 ♋ 47 43	05 41	02 41	02 32	07 37	08 02	14 45	04 07	29 28
17	03 41 07	26 40 17	25 ♋ 41 03	02 ♌ 27 15	07 23	03 02	03 10	07 33	08 10	14 43	04 05	29 27
18	03 45 03	27 38 04	09 ♌ 06 36	15 ♌ 39 28	09 08	03 22	03 49	07 29	08 18	14 41	04 04	29 26
19	03 49 00	28 35 50	22 ♌ 06 18	28 ♌ 27 39	10 54	03 39	04 27	07 24	08 25	14 40	04 02	29 26
20	03 52 57	29 ♉ 33 34	04 ♍ 44 03	10 ♍ 56 05	12 42	03 55	05 05	07 20	08 33	14 38	04 01	29 25
21	03 56 53	00 ♊ 31 16	17 ♍ 04 20	23 ♍ 09 22	14 32	04 08	05 44	07 15	08 41	14 36	03 59	29 24
22	04 00 50	01 28 57	29 ♍ 11 43	05 ♎ 11 56	16 25	04 20	06 22	07 10	08 49	14 35	03 57	29 24
23	04 04 46	02 26 36	11 ♎ 10 28	17 ♎ 07 47	18 19	04 29	07 00	07 05	08 56	14 33	03 56	29 23
24	04 08 43	03 24 14	23 ♎ 04 16	29 ♎ 00 18	20 16	04 37	07 39	07 00	09 04	14 32	03 54	29 23
25	04 12 39	04 21 51	04 ♏ 56 12	10 ♏ 52 15	22 15	04 42	08 17	06 55	09 12	14 31	03 52	29 22
26	04 16 36	05 19 26	16 ♏ 48 43	22 ♏ 45 49	24 15	04 44	08 55	06 50	09 20	14 29	03 51	29 22
27	04 20 32	06 17 00	28 ♏ 43 46	04 ♐ 42 44	26 18	04 R 44	09 34	06 44	09 27	14 28	03 49	29 21
28	04 24 29	07 14 32	10 ♐ 42 56	16 ♐ 44 31	28 ♉ 22	04 42	10 12	06 38	09 35	14 27	03 48	29 21
29	04 28 26	08 12 04	22 ♐ 47 40	28 ♐ 52 36	00 ♊ 28	04 38	10 50	06 33	09 43	14 26	03 46	29 21
30	04 32 22	09 09 35	04 ♑ 59 31	11 ♑ 08 39	02 35	04 31	11 28	06 27	09 51	14 24	03 45	29 20
31	04 36 19	10 ♊ 07 04	17 ♑ 20 15	23 ♑ 34 38	04 ♊ 44	04 ♋ 22	12 ♋ 07	06 ♑ 21	09 ♊ 59	14 ♎ 23	03 ♐ 43	29 ♍ 20

Date	Moon True ☊ ° '	Moon Mean ☊ ° '	Moon ☽ Latitude ° '
01	29 ♑ 08	00 ♒ 12	03 S 35
02	29 R 01	00 09	02 46
03	28 57	00 05	01 49
04	28 55	00 ♒ 02	00 S 46
05	28 D 55	29 ♑ 59	00 N 21
06	28 55	29 56	01 28
07	28 R 55	29 53	02 33
08	28 53	29 49	03 31
09	28 48	29 46	04 18
10	28 41	29 43	04 50
11	28 32	29 40	05 02
12	28 22	29 37	04 54
13	28 13	29 34	04 24
14	28 05	29 30	03 37
15	27 59	29 27	02 35
16	27 56	29 24	01 25
17	27 55	29 21	00 N 12
18	27 D 55	29 18	00 S 59
19	27 56	29 15	02 05
20	27 R 56	29 11	03 03
21	27 55	29 08	03 52
22	27 51	29 05	04 28
23	27 45	29 02	04 53
24	27 37	28 59	05 04
25	27 28	28 55	05 03
26	27 18	28 52	04 48
27	27 08	28 49	04 21
28	27 00	28 46	03 42
29	26 53	28 43	02 52
30	26 49	28 40	01 55
31	26 ♑ 47	28 ♑ 36	00 S 51

DECLINATIONS

Date	Sun ☉ ° '	Moon ☽ ° '	Mercury ☿ ° '	Venus ♀ ° '	Mars ♂ ° '	Jupiter ♃ ° '	Saturn ♄ ° '	Uranus ♅ ° '	Neptune ♆ ° '	Pluto ♇ ° '
01	15 N 12	26 S 00	03 N 01	27 N 32	24 N 24	22 S 54	19 N 46	05 S 22	19 S 23	15 N 22
02	15 30	26 08	03 26	27 34	24 26	22 54	19 48	05 22	19 23	15 22
03	15 47	25 01	03 52	27 35	24 28	22 54	19 49	05 21	19 22	15 23
04	16 05	22 40	04 19	27 36	24 30	22 55	19 51	05 20	19 22	15 23
05	16 22	19 11	04 48	27 37	24 31	22 55	19 52	05 19	19 22	15 23
06	16 39	14 41	05 18	27 37	24 33	22 55	19 53	05 18	19 21	15 23
07	16 55	09 22	05 49	27 37	24 34	22 55	19 55	05 18	19 21	15 23
08	17 12	03 S 26	06 21	27 36	24 35	22 55	19 56	05 17	19 21	15 23
09	17 28	02 N 51	06 54	27 35	24 36	22 56	19 57	05 16	19 21	15 23
10	17 43	09 09	07 29	27 34	24 36	22 56	19 59	05 15	19 20	15 23
11	17 59	15 05	08 04	27 32	24 37	22 56	20 00	05 15	19 20	15 23
12	18 14	20 10	08 40	27 29	24 37	22 56	20 01	05 14	19 20	15 23
13	18 29	23 54	09 17	27 27	24 37	22 57	20 03	05 13	19 19	15 23
14	18 43	25 54	09 55	27 24	24 37	22 57	20 04	05 12	19 19	15 23
15	18 57	26 00	10 33	27 20	24 36	22 57	20 05	05 12	19 19	15 23
16	19 11	24 20	11 12	27 16	24 36	22 58	20 07	05 11	19 18	15 23
17	19 25	21 12	11 52	27 12	24 35	22 58	20 08	05 11	19 18	15 23
18	19 38	17 02	12 32	27 07	24 34	22 58	20 09	05 10	19 18	15 23
19	19 51	12 10	13 13	27 02	24 33	22 59	20 11	05 09	19 18	15 23
20	20 04	06 56	13 53	26 57	24 32	22 59	20 12	05 09	19 17	15 22
21	20 16	01 N 33	14 34	26 51	24 30	22 59	20 13	05 08	19 17	15 22
22	20 28	03 S 47	15 15	26 45	24 29	23 00	20 14	05 08	19 17	15 22
23	20 39	08 54	15 56	26 38	24 27	23 00	20 16	05 07	19 16	15 22
24	20 50	13 40	16 37	26 32	24 25	23 00	20 17	05 07	19 16	15 22
25	21 01	17 55	17 17	26 24	24 22	23 01	20 18	05 06	19 16	15 22
26	21 12	21 27	17 57	26 17	24 20	23 01	20 19	05 06	19 15	15 21
27	21 22	24 07	18 37	26 09	24 17	23 01	20 20	05 05	19 15	15 21
28	21 31	25 43	19 15	26 01	24 15	23 02	20 22	05 05	19 15	15 21
29	21 41	26 07	19 53	25 52	24 12	23 02	20 23	05 04	19 15	15 21
30	21 50	25 16	20 29	25 43	24 08	23 03	20 24	05 04	19 14	15 20
31	21 N 58	23 S 09	21 N 04	25 N 34	24 N 05	23 S 03	20 N 26	05 S 03	19 S 14	15 N 20

ZODIAC SIGN ENTRIES

Date	h	m	Planets
02	20	29	☽ ♑
05	06	35	☽ ♒
07	13	28	☽ ♓
09	16	35	☽ ♈
10	13	51	♀ ♋
11	16	47	☽ ♉
12	13	14	♂ ♋
12	23	45	☿ ♉
13	15	57	☽ ♊
15	16	16	☽ ♋
17	19	38	☽ ♌
20	02	56	☽ ♍
20	23	00	☉ ♊
22	13	36	☽ ♎
25	02	01	☽ ♏
27	14	33	☽ ♐
29	06	46	☿ ♊
30	02	13	☽ ♑

LATITUDES

Date	Mercury ☿ ° '	Venus ♀ ° '	Mars ♂ ° '	Jupiter ♃ ° '	Saturn ♄ ° '	Uranus ♅ ° '	Neptune ♆ ° '	Pluto ♇ ° '
01	02 S 56	04 N 12	01 N 09	00 N 17	01 S 35	00 N 41	01 N 42	16 N 40
04	03 00	04 13	01 09	00 17	01 35	00 41	01 42	16 39
07	02 58	04 11	01 09	00 16	01 35	00 41	01 42	16 38
10	02 50	04 07	01 10	00 16	01 34	00 41	01 42	16 37
13	02 38	04 00	01 10	00 16	01 34	00 41	01 42	16 35
16	02 21	03 51	01 11	00 16	01 34	00 41	01 42	16 34
19	01 59	03 39	01 11	00 16	01 33	00 41	01 42	16 33
22	01 34	03 23	01 11	00 15	01 33	00 40	01 42	16 32
25	01 05	03 03	01 11	00 15	01 33	00 40	01 42	16 30
28	00 34	02 39	01 12	00 15	01 33	00 40	01 42	16 29
31	00 S 02	02 N 11	01 N 12	00 N 14	01 S 32	00 N 40	01 N 42	16 N 28

DATA

Julian Date	2441439
Delta T	+42 seconds
Ayanamsa	23° 28' 27"
Synetic vernal point	05° ♓ 38' 32"
True obliquity of ecliptic	23° 26' 39"

MOON'S PHASES, APSIDES AND POSITIONS ☽

Date	h	m	Phase	Longitude ° '	Eclipse Indicator
06	12	26	☾	16 ♒ 04	
13	04	08	●	22 ♉ 30	
20	01	16	☽	29 ♌ 08	
28	04	28	○	06 ♐ 56	

Day	h	m	
12	16	42	Perigee
25	14	30	Apogee

02	02	42	Max dec	26° S 14'
09	01	14	0N	
15	01	20	Max dec	26° N 11'
21	18	56	0S	
29	07	46	Max dec	26° S 08'

LONGITUDES

Date	Chiron ⚷ ° '	Ceres ⚳ ° '	Pallas ⚴ ° '	Juno ⚵ ° '	Vesta ⚶ ° '	Black Moon Lilith ⚸ ° '
01	14 ♈ 55	14 ♌ 14	03 ♋ 56	05 ♎ 34	24 ♈ 55	07 ♏ 32
11	15 ♈ 26	16 ♌ 45	09 ♋ 29	04 ♎ 29	29 ♈ 26	08 ♏ 40
21	15 ♈ 55	19 ♌ 48	15 ♋ 36	03 ♎ 58	03 ♉ 08	09 ♏ 47
31	16 ♈ 20	22 ♌ 50	20 ♋ 36	04 ♎ 02	08 ♉ 52	10 ♏ 54

ASPECTARIAN

h m Aspects

01 Monday
01 08 ☽ ⚺ ♃
06 36 ☽ ⚻ ☉
08 02 ☽ Q ♇
13 52 ☽ △ ☿
15 07 ☽ ⚹ ♅
19 40 ☽ ± ☉

02 Tuesday
02 22 ☉ ∥ ♇
02 42 ☿ ☍ ♅
07 26 ☽ ☍ ♂
10 33 ☽ ☍ ♀
14 56 ☽ Q ♅
15 08 ☽ P ☉
19 50 ☽ □ ♇

03 Wednesday
05 16 ☽ ⚺ ♆
09 03 ☽ ⚻ ♄
12 38 ☽ ☌ ♃
17 00 ☽ ⊥ ♆
18 52 ☽ # ♂
20 54 ☽ ± ♄
23 07 ☽ △ ☉

04 Thursday
02 09 ☽ □ ♅
07 20 ☽ □ ☿
10 02 ☽ ∥ ♃
10 25 ☽ ∠ ♆
14 28 ☽ P ♄
21 15 ☽ ⚻ ♂

05 Friday
00 22 ☽ ⚻ ♀
00 55 ☉ P ♇
05 53 ☽ △ ♇
07 50 ☽ # ♄
09 14 ☽ ± ♂
10 54 ☽ ∥ ♆
12 19 ☽ ± ♀
12 25 ☉ ⚻ ♅
13 15 ☿ P ♆
14 52 ☽ ⚹ ♆
19 08 ☽ △ ♄
21 16 ☽ Q ☿
21 50 ☽ ⚺ ♃

06 Saturday
02 42 ☽ # ☉
02 54 ☽ P ♂
05 55 ☽ P ♀
08 35 ☽ # ♇
08 53 ☽ ⊥ ♃
09 47 ☽ P ♇
10 37 ☽ △ ♅
12 22 ☿ # ♅
12 26 ☽ □ ☉
13 00 ☽ Q ♆
21 43 ☽ ⚹ ☿

07 Sunday
01 10 ☽ ∠ ♃
02 04 ☽ ± ♇
07 30 ☽ △ ♂
10 19 ☽ △ ♀
10 23 ☿ ∠ ♄
12 44 ☽ ⚻ ♇
13 30 ☽ P ♅
21 05 ☽ □ ♆
23 02 ☽ Q ☉

08 Monday
01 25 ☽ # ☿
01 40 ☽ □ ♄
03 11 ☽ ∠ ☿
03 33 ☽ ⚹ ♃
04 44 ☽ ∥ ♅
05 11 ☽ ± ♅
15 27 ☽ ⚻ ♅
20 18 ☽ ⊥ ☿
21 09 ☽ ⚹ ☉
23 58 ☽ Q ♃

09 Tuesday
07 29 ☽ ⚺ ☿
08 31 ☽ Q ♄
13 25 ☽ □ ♂
15 46 ☽ □ ♀
15 51 ☽ ☍ ♇
17 59 ♀ □ ♇
21 07 ☽ # ♅
23 37 ☽ △ ♆
23 50 ☽ ∠ ☉

10 Wednesday
04 25 ☽ ⚹ ♄
04 54 ☽ ∥ ☿
05 36 ☽ □ ♃
15 26 ☽ ⊥ ☉
16 49 ☽ ☍ ♅
19 59 ☽ Q ♂
22 02 ☽ Q ♀
23 46 ☽ P ♆

11 Thursday
01 41 ☽ ⚺ ☉
04 42 ☽ ∠ ♄
12 17 ☉ ± ♅
13 17 ☽ ∥ ♇
13 19 ☽ ☌ ☿
14 01 ☽ ± ♆
15 53 ☽ ⚹ ♂
16 02 ☽ ⚻ ♇
17 44 ☽ ⚹ ♀
19 03 ☽ ⊥ ♄
19 25 ♂ □ ♇
20 19 ☿ ± ♆
23 28 ☽ ⚻ ♆

12 Friday
01 29 ☽ ± ♇
01 43 ☽ ∥ ☉
04 34 ☽ ⚺ ♄
05 10 ☽ △ ♃
07 39 ☽ # ♆
11 16 ☽ ∥ ♄
15 36 ☽ P ♇
16 08 ☽ ⚻ ♅
16 12 ☿ ⚻ ♇
16 30 ☽ ∠ ♂
18 04 ☽ ∠ ♀

13 Saturday
01 32 ☽ ± ♅
04 08 ☽ ☌ ☉
04 38 ☽ P ♃
04 49 ☽ # ♃
07 03 ☿ ⚹ ♂
07 15 ☽ ⊥ ♂
08 39 ☽ ⊥ ♀
11 32 ☉ P ♃
15 09 ☽ △ ♇
15 41 ☽ P ♅
17 08 ☽ ⚺ ♂
17 49 ☽ ⚺ ☿
18 25 ☽ ⚺ ♀
18 28 ☽ ∥ ♂
18 46 ☽ ± ♃
22 34 ☽ ☍ ♆

14 Sunday
01 25 ☿ ⚹ ♀
02 01 ☿ ⊥ ♄
04 13 ☽ ☌ ♄
04 15 ☽ ⚻ ♃
04 27 ☽ ⊥ ☿
06 12 ♃ ⚻ ♄
15 30 ☽ △ ♅
20 38 ☽ ∠ ☿

15 Monday
07 19 ☽ ⚺ ☉
13 42 ☿ ⚻ ♆
15 24 ☽ □ ♇
17 53 ☽ ⊥ ☉
19 44 ☽ ☌ ♂
20 17 ☽ ☌ ♀
23 07 ☽ ⚻ ♆

16 Tuesday
00 22 ☽ ⚹ ☿
04 59 ☽ ☍ ♃
05 36 ☽ ⚺ ♄
08 45 ☿ ± ♇
09 11 ☽ ± ♆
09 19 ☽ ∥ ♂
10 02 ☽ ∠ ☉
15 52 ☽ ⊥ ♄
17 02 ☽ □ ♅
21 41 ☽ Q ♇
23 29 ☽ Q ☿
23 43 ☽ # ♃

17 Wednesday
00 14 ☉ # ♆
00 32 ☽ P ♆
00 49 ♀ ☌ ♂
07 33 ☽ ∠ ♄
13 52 ☽ ⚹ ☉
14 04 ☿ △ ♃
18 38 ☽ ∥ ♄
18 39 ☽ ⚹ ♇
22 20 ☽ ∥ ☉
23 33 ☽ # ♆
23 38 ☿ ⚺ ♄

18 Thursday
00 26 ☽ Q ♅
01 22 ☽ ⚺ ♀
01 48 ☉ ⊥ ♀
01 57 ☽ ⚺ ♂
02 54 ☽ △ ♆
09 03 ☽ ⚻ ♃
10 30 ☽ ⚹ ♄
12 02 ☽ □ ☿
12 28 ☽ ⊥ ♀
13 02 ☽ Q ☉
13 20 ☽ ⊥ ♂
19 56 ☽ ± ♃
20 26 ☽ ∥ ♇
21 07 ♂ ⚻ ♆
21 44 ☽ ∠ ♇
22 11 ☽ ⚹ ♅

19 Friday
01 04 ☉ ⊥ ♂
05 25 ☽ ∠ ♀
06 47 ☽ ∠ ♂
07 38 ☽ ∥ ☿
08 49 ☽ Q ♄
12 33 ☽ P ♃
14 29 ☽ ⊥ ♇

20 Saturday
01 16 ☽ □ ☉
01 50 ☽ ⚺ ♇
02 15 ☽ ∠ ♅
08 27 ☉ △ ♇
10 23 ☽ ⚹ ♀
10 37 ☽ □ ♆
12 43 ☽ ⚹ ♂
13 45 ☉ P ♅
16 58 ☽ △ ♃
19 27 ☽ □ ♄
19 31 ☽ ⊥ ♅
20 00 ☽ # ♅
20 44 ♀ ⚻ ♆

21 Sunday
06 08 ☉ ∥ ♄
06 09 ☽ △ ☿
07 10 ☽ ⚺ ♅
10 08 ☽ Q ♀
10 16 ☿ P ♇
12 50 ☿ ⚻ ♅
13 22 ☽ Q ♂
21 39 ☽ Q ♆

22 Monday
04 50 ☉ ± ♃
12 24 ☽ ☌ ♇
16 02 ☿ ∥ ♇
16 58 ☽ △ ☉
17 16 ☽ P ☿
18 10 ☽ ∥ ♅
21 29 ☽ ⚹ ♆
22 25 ☽ □ ♀

23 Tuesday
03 09 ☽ □ ♂
03 51 ☽ □ ♃
07 27 ☽ △ ♄
14 44 ♂ ☍ ♃
14 46 ☽ ± ☿
18 48 ☽ ☌ ♅

24 Wednesday
01 45 ☽ P ☉
03 26 ☿ ∠ ♀
03 36 ☽ ∠ ♆
05 13 ☽ ⚻ ☿
14 02 ☽ P ♄
15 12 ☿ ± ♅
15 53 ☽ Q ♃
21 09 ☽ # ♇
21 32 ☽ ± ☉
21 45 ☽ ⊥ ♆

25 Thursday
00 06 ☉ ☍ ♆
00 45 ☽ ⚺ ♇
07 28 ☽ # ☿
08 13 ☿ P ♃
08 27 ☽ ± ♄
09 51 ☽ ⚺ ♆
10 44 ☽ ⚻ ☉
11 30 ☽ △ ♀
12 53 ☽ ⊥ ♇
15 59 ☽ ⚹ ♃
19 09 ☽ △ ♂
20 31 ☽ ∥ ♆
20 42 ☉ ⚺ ♀
20 43 ☽ ⚻ ♄

26 Friday
03 40 ☽ # ♄
06 15 ☿ ∠ ♂
07 03 ☽ ∠ ♇
07 19 ☽ ⚺ ♅
09 53 ☽ # ☉
17 55 ☽ P ♀
19 24 ☽ ⊥ ♅
22 02 ☽ ∠ ♃

27 Saturday
01 02 ☽ ∥ ♃
03 08 ☽ P ♂
03 14 ♀ St R
06 05 ☽ ☍ ☿
07 10 ♂ ⚺ ♄
12 01 ☽ ± ♀
13 15 ☽ ⚹ ♇
13 29 ☽ ∠ ♅
13 56 ☽ # ♂
16 00 ☽ ⊥ ♃
21 26 ♂ ± ♆
22 11 ☽ ☌ ♆
22 14 ☽ ± ♂
22 19 ☉ ⚻ ♃

28 Sunday
00 02 ☽ ⚻ ♀
03 55 ☽ ⚺ ♃
04 28 ☽ ☍ ☉
09 43 ☽ ☍ ♄
10 54 ☽ ⚻ ♂
11 53 ☿ # ♆
13 16 ☽ Q ♇
15 51 ☿ ⊥ ♀
19 19 ☽ # ♀
19 25 ☽ ⚹ ♅
23 18 ☿ △ ♇

29 Monday
00 19 ☿ P ♅
12 56 ☿ ± ♃
19 10 ☽ Q ♅

30 Tuesday
00 55 ☽ □ ♇
01 44 ☽ # ♀
06 17 ☽ ⚻ ☿
08 55 ☿ ∥ ♄
09 33 ☽ ⚺ ♆
11 05 ☽ ☍ ♀
14 49 ☽ ☌ ♃
20 29 ☽ ± ☿
20 49 ☽ ⚻ ☉
21 15 ☽ ⊥ ♆
21 35 ☽ ⚻ ♄

31 Wednesday
00 49 ☿ ☍ ♆
01 19 ☽ ☍ ♂
02 48 ☽ # ♂
06 18 ☽ □ ♅
07 55 ☉ ☌ ♄
08 15 ☿ ⚺ ♀
09 21 ☽ ± ♄
09 27 ☽ ± ☉
12 57 ☽ ∥ ♃
14 39 ☽ ∠ ♆
17 34 ☽ P ☿
21 19 ☽ # ☉

All ephemeris data is given at 12.00 UT and the Moon's longitude is additionally given for 24.00 UT

JUNE 1972

LONGITUDES

Date	Sidereal time h	m	s	Sun ☉ °	'	"	Moon ☽ °	'	"	Moon ☽ 24.00 °	'	Mercury ☿ °	'	Venus ♀ °	'	Mars ♂ °	'	Jupiter ♃ °	'	Saturn ♄ °	'	Uranus ♅ °	'	Neptune ♆ °	'	Pluto ⯓ °	'
01	04	40	15	11 ♊ 04	33		29 ♑ 52	06		06 ♒ 12	58	06 ♊ 54		04 ♋ 10		12 ♋ 45		06 ♑ 15		10 ♊ 06		14 ♎ 22		03 ♐ 41		29 ♍ 20	
02	04	44	12	12 02	01		12 ♒ 37	37		19 ♒ 06	24	09 04		03 R 56		13 23		06 R 08		10 14		14 R 21		03 R 40		29 R 19	
03	04	48	08	12 59	28		25 ♒ 39	41		02 ♓ 17	47	11 16		03 40		14 01		06 02		10 22		14 20		03 38		29 19	
04	04	52	05	13 56	55		09 ♓ 01	03		15 ♓ 49	43	13 28		03 21		14 39		05 55		10 30		14 19		03 36		29 19	
05	04	56	01	14 54	20		22 ♓ 43	59		29 ♓ 43	54	15 40		03 00		15 17		05 49		10 38		14 18		03 35		29 19	
06	04	59	58	15 51	45		06 ♈ 49	27		14 ♈ 00	27	17 52		02 36		15 56		05 42		10 45		14 18		03 33		29 19	
07	05	03	55	16 49	10		21 ♈ 16	32		28 ♈ 37	12	20 04		02 11		16 34		05 35		10 53		14 17		03 32		29 19	
08	05	07	51	17 46	34		06 ♉ 01	46		13 ♉ 29	23	22 15		01 44		17 12		05 28		11 01		14 16		03 30		29 19	
09	05	11	48	18 43	57		20 ♉ 59	04		28 ♉ 29	44	24 25		01 15		17 50		05 21		11 09		14 16		03 28		29 D 19	
10	05	15	44	19 41	20		06 ♊ 00	13		13 ♊ 29	23	26 34		00 44		18 28		05 14		11 16		14 15		03 27		29 19	
11	05	19	41	20 38	43		20 ♊ 56	06		28 ♊ 19	18	28 ♊ 41		00 ♋ 11		19 06		05 07		11 24		14 14		03 25		29 19	
12	05	23	37	21 36	04		05 ♋ 38	04		12 ♋ 51	37	00 ♋ 47		29 ♊ 38		19 45		05 00		11 32		14 14		03 24		29 19	
13	05	27	34	22 33	25		19 ♋ 59	43		27 ♋ 00	48	02 52		29 03		20 23		04 53		11 40		14 13		03 22		29 19	
14	05	31	30	23 30	45		03 ♌ 55	45		10 ♌ 44	05	04 54		28 27		21 01		04 45		11 47		14 13		03 21		29 19	
15	05	35	27	24 28	04		17 ♌ 25	51		24 ♌ 01	16	06 55		27 50		21 39		04 38		11 55		14 13		03 19		29 19	
16	05	39	24	25 25	22		00 ♍ 30	37		06 ♍ 54	17	08 54		27 13		22 17		04 30		12 03		14 12		03 18		29 20	
17	05	43	20	26 22	39		13 ♍ 12	44		19 ♍ 26	28	10 50		26 35		22 55		04 23		12 10		14 12		03 16		29 20	
18	05	47	17	27 19	56		25 ♍ 36	02		01 ♎ 42	00	12 44		25 57		23 33		04 15		12 18		14 12		03 15		29 20	
19	05	51	13	28 17	11		07 ♎ 44	57		13 ♎ 45	26	14 36		25 20		24 11		04 08		12 26		14 12		03 14		29 20	
20	05	55	10	29 ♊ 14	26		19 ♎ 44	02		25 ♎ 41	16	16 26		24 43		24 49		04 00		12 33		14 12		03 12		29 21	
21	05	59	06	00 ♋ 11	41		01 ♏ 37	39		07 ♏ 33	41	18 14		24 07		25 27		03 53		12 41		14 12		03 11		29 21	
22	06	03	03	01 08	54		13 ♏ 29	48		19 ♏ 26	26	19 59		23 31		26 05		03 45		12 48		14 D 12		03 09		29 22	
23	06	06	59	02 06	07		25 ♏ 23	58		01 ♐ 22	44	21 42		22 57		26 43		03 37		12 56		14 12		03 08		29 22	
24	06	10	56	03 03	20		07 ♐ 23	02		13 ♐ 25	10	23 22		22 24		27 21		03 30		13 03		14 12		03 07		29 23	
25	06	14	53	04 00	32		19 ♐ 29	20		25 ♐ 35	46	25 01		21 53		27 59		03 22		13 11		14 12		03 05		29 23	
26	06	18	49	04 57	44		01 ♑ 44	39		07 ♑ 56	07	26 37		21 23		28 37		03 14		13 18		14 12		03 04		29 24	
27	06	22	46	05 54	56		14 ♑ 10	20		20 ♑ 27	26	28 11		20 55		29 15		03 07		13 26		14 12		03 03		29 24	
28	06	26	42	06 52	07		26 ♑ 47	30		03 ♒ 10	41	29 ♋ 42		20 29		29 ♋ 53		02 59		13 33		14 13		03 01		29 25	
29	06	30	39	07 49	19		09 ♒ 37	04		16 ♒ 06	47	01 ♌ 11		20 05		00 ♌ 31		02 51		13 41		14 13		03 00		29 25	
30	06	34	35	08 ♋ 46	30		22 ♒ 39	54		29 ♒ 16	34	02 ♌ 37		19 ♊ 43		01 ♌ 09		02 ♑ 44		13 ♊ 48		14 ♎ 14		02 ♐ 59		29 ♍ 26	

Moon / DECLINATIONS

Date	Moon True ☊ ° '	Moon Mean ☊ ° '	Moon ☽ Latitude ° '	Sun ☉ ° '	Moon ☽ ° '	Mercury ☿ ° '	Venus ♀ ° '	Mars ♂ ° '	Jupiter ♃ ° '	Saturn ♄ ° '	Uranus ♅ ° '	Neptune ♆ ° '	Pluto ⯓ ° '
01	26 ♑ 46	28 ♑ 33	00 N 17	22 N 06	19 S 55	21 N 37	25 N 24	24 N 01	23 S 03	20 N 27	05 S 03	19 S 14	15 N 20
02	26 D 47	28 30	01 25	22 14	15 40	22 08	25 14	23 58	23 04	20 28	05 03	19 13	15 19
03	26 48	28 27	02 30	22 22	10 37	22 38	25 03	23 54	23 04	20 29	05 02	19 13	15 19
04	26 49	28 24	03 29	22 29	04 S 58	23 05	24 52	23 50	23 05	20 30	05 02	19 13	15 19
05	26 R 49	28 21	04 17	22 35	01 N 03	23 29	24 40	23 46	23 05	20 31	05 02	19 13	15 18
06	26 47	28 17	04 51	22 42	07 10	23 52	24 29	23 41	23 06	20 33	05 01	19 12	15 18
07	26 44	28 14	05 09	22 47	13 04	24 11	24 16	23 37	23 06	20 34	05 01	19 12	15 17
08	26 39	28 11	05 06	22 53	18 20	24 28	24 04	23 32	23 06	20 35	05 01	19 12	15 17
09	26 34	28 08	04 42	22 58	22 33	24 42	23 51	23 27	23 07	20 36	05 01	19 11	15 17
10	26 29	28 05	04 00	23 03	25 15	24 54	23 37	23 22	23 07	20 37	05 01	19 11	15 16
11	26 24	28 01	03 01	23 07	26 08	25 02	23 24	23 16	23 08	20 38	05 00	19 11	15 16
12	26 21	27 58	01 50	23 11	25 10	25 08	23 10	23 11	23 08	20 39	05 00	19 10	15 15
13	26 20	27 55	00 N 35	23 14	22 31	25 11	22 55	23 05	23 08	20 40	05 00	19 10	15 15
14	26 D 20	27 52	00 S 41	23 17	18 36	25 11	22 41	22 59	23 09	20 41	05 00	19 10	15 14
15	26 21	27 49	01 53	23 20	13 49	25 09	22 26	22 53	23 09	20 42	05 00	19 10	15 14
16	26 22	27 46	02 56	23 22	08 33	25 04	22 11	22 47	23 10	20 43	05 00	19 10	15 13
17	26 24	27 42	03 49	23 24	03 N 05	24 57	21 56	22 41	23 10	20 45	05 00	19 09	15 13
18	26 24	27 39	04 29	23 25	02 S 22	24 48	21 41	22 34	23 10	20 46	05 00	19 09	15 12
19	26 R 24	27 36	04 57	23 26	07 37	24 37	21 26	22 28	23 11	20 47	05 00	19 09	15 11
20	26 22	27 33	05 11	23 26	12 31	24 24	21 12	22 21	23 11	20 48	05 00	19 09	15 11
21	26 20	27 30	05 12	23 27	16 55	24 09	20 57	22 14	23 11	20 49	05 00	19 08	15 10
22	26 16	27 27	04 59	23 26	20 39	23 52	20 43	22 07	23 12	20 49	05 00	19 08	15 10
23	26 12	27 23	04 34	23 26	23 33	23 33	20 29	22 00	23 12	20 50	05 00	19 08	15 09
24	26 08	27 20	03 56	23 25	25 26	23 14	20 15	21 52	23 12	20 51	05 00	19 08	15 08
25	26 05	27 17	03 07	23 23	26 09	22 52	20 02	21 45	23 13	20 52	05 00	19 07	15 08
26	26 02	27 14	02 10	23 21	25 36	22 30	19 49	21 37	23 13	20 53	05 00	19 07	15 07
27	26 01	27 11	01 S 05	23 19	23 46	22 07	19 37	21 29	23 13	20 54	05 00	19 07	15 06
28	26 D 01	27 07	00 N 04	23 16	20 44	21 42	19 26	21 21	23 14	20 55	05 00	19 07	15 06
29	26 01	27 04	01 14	23 13	16 39	21 17	19 15	21 13	23 14	20 56	05 01	19 07	15 05
30	26 ♑ 02	27 ♑ 01	02 N 22	23 N 09	11 S 44	20 N 51	19 N 04	21 N 04	23 S 14	20 N 57	05 S 01	19 S 06	15 N 04

ZODIAC SIGN ENTRIES

Date	h	m	Planets
01	12	15	☽ ♒
03	19	52	☽ ♓
06	00	27	☽ ♈
08	02	15	☽ ♉
10	02	24	☽ ♊
11	20	08	♀ ♊
12	02	45	☽ ♋
12	02	56	☿ ♋
14	05	10	☽ ♌
16	11	03	☽ ♍
18	20	39	☽ ♎
21	07	06	☉ ♋
21	08	43	☽ ♏
23	21	14	☽ ♐
26	08	36	☽ ♑
28	16	09	♂ ♌
28	16	52	☿ ♌
28	18	02	☽ ♒

LATITUDES

Date	Mercury ☿ ° '	Venus ♀ ° '	Mars ♂ ° '	Jupiter ♃ ° '	Saturn ♄ ° '	Uranus ♅ ° '	Neptune ♆ ° '	Pluto ⯓ ° '
01	00 N 09	02 N 01	01 N 12	00 N 14	01 S 32	00 N 40	01 N 42	16 N 27
04	00 40	01 28	01 12	00 14	01 32	00 40	01 42	16 26
07	01 07	00 51	01 12	00 14	01 32	00 40	01 42	16 24
10	01 30	00 N 11	01 12	00 13	01 32	00 40	01 42	16 23
13	01 46	00 S 31	01 12	00 13	01 32	00 40	01 41	16 21
16	01 56	01 14	01 12	00 13	01 32	00 40	01 41	16 20
19	01 59	01 55	01 12	00 12	01 32	00 39	01 41	16 19
22	01 56	02 35	01 12	00 12	01 32	00 39	01 41	16 17
25	01 46	03 10	01 11	00 11	01 31	00 39	01 41	16 16
28	01 31	03 41	01 12	00 11	01 31	00 39	01 41	16 14
31	01 N 11	04 S 07	01 N 12	00 N 10	01 S 31	00 N 39	01 N 41	16 N 13

DATA

Julian Date	2441470
Delta T	+42 seconds
Ayanamsa	23° 28' 33"
Synetic vernal point	05° ♓ 38' 27"
True obliquity of ecliptic	23° 26' 38"

MOON'S PHASES, APSIDES AND POSITIONS ☽

Date	h	m	Phase	Longitude ° '	Eclipse Indicator
04	21	22	☾	14 ♓ 19	
11	11	30	●	20 ♊ 38	
18	15	41	☽	27 ♍ 29	
26	18	46	○	05 ♑ 14	

Day	h	m		
10	00	09	Perigee	
22	03	21	Apogee	
05	07	52	ON	
11	11	19	Max dec	26° N 08'
18	01	30	0S	
25	13	45	Max dec	26° S 09'

LONGITUDES

Date	Chiron ⚷ ° '	Ceres ⚳ ° '	Pallas ⚴ ° '	Juno ⚵ ° '	Vesta ⚶ ° '	Black Moon Lilith ⚸ ° '
01	16 ♈ 22	23 ♌ 10	21 ♋ 09	04 ♎ 04	08 ♉ 39	11 ♏ 01
11	16 ♈ 43	26 ♌ 38	26 ♋ 40	04 ♎ 42	12 ♉ 54	12 ♏ 08
21	17 ♈ 00	00 ♍ 20	02 ♌ 08	05 ♎ 48	17 ♉ 02	13 ♏ 15
31	17 ♈ 13	04 ♍ 11	07 ♌ 33	07 ♎ 18	21 ♉ 03	14 ♏ 22

All ephemeris data is given at 12.00 UT and the Moon's longitude is additionally given for 24.00 UT

ASPECTARIAN

h m	Aspects
01 Thursday	
02 03	☽ # ☿
02 50	☽ ♇ ♄
04 11	☽ ♇ ☉
05 09	☿ ⊼ ♃
08 32	☽ # ♄
09 43	☿ ⊥ ♂
10 58	☽ △ ⯓
16 14	☽ ∥ ♆
19 13	☽ ⚹ ♆
20 00	☽ ⊼ ♀
23 57	☽ ⊻ ♃
02 Friday	
04 00	☽ △ ☿
07 04	☽ ± ♀
07 30	☽ △ ♄
10 48	☽ △ ☉
11 06	☽ ⊥ ♃
13 29	☽ ⊼ ♂
13 46	☽ # ⯓
15 09	☽ ♇ ⯓
15 12	☽ △ ♅
17 37	☽ Q ♆
18 30	☉ ∥ ☿
23 27	☽ ♇ ♀
03 Saturday	
01 09	☽ ± ♂
01 34	☿ ☌ ♄
03 37	☽ ∠ ♃
07 44	☽ ± ⯓
14 21	♀ ⊼ ♆
18 24	☽ ♇ ♂
18 38	☽ ⊼ ⯓
18 40	☽ ♇ ♅
23 43	♂ □ ♅
04 Sunday	
02 07	☽ △ ♀
02 22	☽ □ ♆
06 32	☽ ⚹ ♃
10 46	☽ ± ♅
11 44	☽ ∥ ♅
12 03	☿ # ♃
14 39	☽ □ ♄
21 14	☉ △ ♅
21 18	☿ △ ♅
21 21	☽ ⊼ ♅
21 21	☉ ☌ ☿
21 22	☽ □ ☉
21 22	☽ □ ☿
22 26	☽ △ ♂
05 Monday	
03 32	☽ Q ♃
06 16	☿ ⊻ ♂
22 13	☽ Q ♄
23 17	☽ ☍ ⯓
06 Tuesday	
02 17	☿ ∥ ♂
03 34	☽ # ♅
05 05	☽ □ ♀
06 30	☽ △ ♆
06 39	☽ Q ☉
10 06	☽ Q ☿
10 08	☽ □ ♃
16 52	☉ ⊻ ♂
18 39	☽ ⚹ ♄
07 Wednesday	
00 28	☽ ☍ ♅
03 53	☽ □ ♂
04 08	☽ ⚹ ☉
07 29	☽ ♇ ♆
09 39	☽ ⚹ ☿
10 16	☽ Q ♀
16 00	☿ ∥ ♀
19 37	☽ ∠ ♄
21 41	☽ ∥ ⯓
22 12	☽ ± ♆
08 Thursday	
01 08	☽ ⊼ ⯓
03 16	♂ ⊥ ♄
05 16	☽ ⚹ ♀
06 23	☽ ∠ ☉
07 55	☽ ⊼ ♆
10 21	☽ ⊥ ♄
10 36	☽ Q ♂
10 51	☽ ± ⯓
11 07	☽ △ ♃
14 18	☽ ∠ ☿
16 14	♂ Q ⯓
16 22	☽ # ♆
20 06	☽ ⊻ ♄
21 53	☽ ⊥ ☉
23 56	☽ ∥ ♄
09 Friday	
01 14	☽ ⊼ ♅
01 19	☽ ♇ ⯓
04 39	☽ ∠ ♀
05 45	⯓ St D
06 44	☽ ⚹ ♂
07 12	☽ ⊥ ☿
08 09	☽ ⊻ ☉
10 50	☽ ± ♅
11 00	☽ ♇ ♃
15 02	☽ ∥ ☉
16 02	☽ # ♃
18 23	☽ ∥ ♂
18 24	☽ ⊻ ☿
18 35	☽ ⊥ ♀
21 04	☽ ∥ ♀
10 Saturday	
01 13	☽ ♇ ♅
01 16	☽ ± ♃
01 18	☽ △ ⯓
03 51	☽ ⊻ ♀
07 24	☽ ∥ ☿
07 46	☽ ∠ ♂
07 55	☽ ☍ ♆
10 47	☽ ⊼ ♃
11 12	♂ ♇ ♆
20 31	☽ ☌ ♄
22 49	☽ ⊥ ♂
11 Sunday	
01 13	☽ △ ♅
08 55	☽ ⊻ ♂
11 30	☽ ☌ ☉
16 38	☉ # ♃
19 06	☿ □ ⯓
12 Monday	
01 30	☿ ☌ ♀
01 37	☽ □ ⯓
02 30	☽ ☌ ♀
02 43	☽ ☌ ☿
08 19	☽ ⊼ ♆
08 22	♀ ∥ ♂
10 30	☉ ∥ ♀
10 58	☽ ☍ ♃
12 23	☽ ∥ ☿
12 30	☉ ∥ ♂
14 41	♀ # ♃
18 13	☽ ± ♆
21 52	☽ ⊻ ♄
23 42	♂ # ♃
13 Tuesday	
00 50	♀ □ ⯓
02 17	☽ □ ♅
06 46	☽ ∥ ☉
07 26	☽ # ♃
07 29	☽ Q ⯓
07 42	☽ ∥ ♂
08 02	☽ ⊥ ♄
08 51	☽ ∥ ♀
09 16	☽ ♇ ♆
12 41	☽ ☌ ♂
16 41	☽ ⊻ ☉
17 52	☿ ⊼ ♆
23 30	☽ ∠ ♄
14 Wednesday	
00 06	☽ ∥ ♄
02 52	☽ ⊻ ♀
03 45	☽ ⊥ ☉
03 59	☽ ⚹ ⯓
08 53	☽ # ♆
09 01	☽ Q ♅
10 18	☿ ☍ ♃
10 59	☽ △ ♆
12 52	☽ ⊥ ♀
13 26	☽ ⊼ ♃
14 01	☽ ⊻ ☿
20 40	☽ ∠ ☉
23 56	☽ ± ♃
15 Thursday	
02 01	☽ ⚹ ♄
02 28	☽ ⊥ ☿
04 06	☽ ∠ ♀
05 15	☽ ∥ ⯓
06 12	☽ ⚹ ♅
06 24	☽ ∠ ⯓
15 57	☽ ♇ ♃
20 03	☽ ⊻ ♂
21 36	☽ ∠ ☿
22 43	☽ ⊥ ⯓
23 56	☽ Q ♄
16 Friday	
01 50	☽ ⚹ ☉
06 09	☽ ⚹ ♀
07 39	☽ ⊥ ♂
09 34	☽ ∠ ♅
09 48	☽ ⊻ ⯓
16 54	☿ ± ♆
17 12	☽ □ ♆
19 25	☽ △ ♃
17 Saturday	
01 23	☽ ∠ ♂
02 02	☽ Q ☉
02 27	☽ ⊥ ♅
03 36	☽ Q ♀
03 37	☽ # ♅
06 38	☽ ⚹ ☿
09 59	☽ □ ♄
13 54	☽ ⊻ ♅
15 09	☉ ☌ ♀
18 Sunday	
03 32	☽ Q ♆
06 01	☿ ⊻ ♄
07 47	☽ ⚹ ♂
10 01	☽ Q ☿
12 40	☽ □ ♀
15 41	☽ □ ☉
19 20	☽ ☌ ⯓
23 51	☽ ∥ ♅
19 Monday	
03 02	☽ ⚹ ♆
04 53	☽ □ ♃
06 40	☿ □ ♅
08 43	☽ Q ♂
21 26	☽ △ ♄
20 Tuesday	
00 53	☽ ☌ ♅
04 11	☽ □ ☿
08 55	☽ ∠ ♆
10 02	♀ ⊻ ♂
14 41	☉ □ ⯓
16 31	☽ Q ♃
21 33	☽ △ ♀
22 50	☽ □ ♂
21 Wednesday	
00 09	☿ Q ⯓
02 08	☽ # ⯓
03 02	☽ ⊥ ♆
03 56	☽ ♇ ♄
07 24	☽ ⊻ ⯓
08 51	☽ △ ☉
11 19	☿ ♇ ♆
15 08	☽ ⊻ ♆
16 30	☽ ⚹ ♃
17 28	♅ St D
18 35	☿ ⊥ ♄
19 32	☽ ⊥ ⯓
22 20	☽ ± ♄
22 Thursday	
01 27	♀ ∥ ♄
01 43	☽ ∥ ♆
02 25	☽ ♇ ♀
10 35	☽ ⊼ ♄
12 28	☽ # ♀
13 18	☽ # ♄
13 24	☽ ⊻ ♅
13 45	☽ ∠ ⯓
17 49	☽ ♇ ☉
19 45	☽ ± ♀
22 30	☽ ∠ ♃
22 47	☽ # ♂
23 Friday	
01 31	☽ ⊥ ♅
03 19	☽ △ ☿
07 18	☽ ⊼ ♀
08 42	☽ ∥ ♃
10 53	☽ # ☉
12 08	☽ # ☿
13 32	☽ ± ☉
14 49	☽ △ ♂
16 25	☽ ⊥ ♃
19 37	☽ ∠ ♅
19 58	☽ ⚹ ⯓
22 09	☉ ∥ ☿
24 Saturday	
01 25	☿ ⊻ ♀
02 37	☽ ⊼ ☉
03 29	☽ ☌ ♆
04 19	☽ ⊻ ♃
13 19	☉ ⊼ ♆
13 21	☿ # ♃
14 18	☽ ♇ ☿
19 57	☽ Q ⯓
21 43	☉ ☍ ♃
22 26	☽ ♇ ♂
23 24	☽ ☍ ♄
25 Sunday	
01 33	☽ ⚹ ♅
09 38	♀ ⊥ ♂
10 55	☽ ± ☿
16 31	☽ ☍ ♀
17 12	☽ ± ♂
20 59	♂ ∠ ♄
26 Monday	
00 30	☽ ⊼ ☿
01 11	☽ Q ♅
05 35	☽ ⊼ ♂
07 25	☽ □ ⯓
14 34	☽ ⊻ ♆
14 52	☽ ☌ ♃
18 46	☽ ☍ ☉
20 58	☿ ⊥ ♀
27 Tuesday	
02 09	☽ ⊥ ♆
10 34	☽ ⊼ ♄
12 04	☽ □ ♅
16 18	☿ ∠ ♄
16 19	☽ # ☉
17 01	☽ ∥ ♃
17 35	♂ ⚹ ⯓
19 23	☽ ∠ ♆
22 09	☽ ± ♄
28 Wednesday	
00 26	☽ ⊼ ♀
03 05	♃ ⊻ ♆
04 15	☽ # ☿
07 27	☿ ⚹ ⯓
07 35	☽ # ♂
10 45	☽ # ♄
11 26	☽ ± ♀
15 21	☽ ♇ ♄
16 56	☽ △ ⯓
17 24	☿ ☌ ♂
18 08	☽ ☍ ♂
18 12	☽ ☍ ☿
20 44	☽ # ♀
22 13	☽ ∥ ♆
23 31	☽ ⊻ ♃
23 41	☽ ⚹ ♆
29 Thursday	
03 43	☉ # ♃
03 48	☽ ♇ ♀
08 24	☽ ⊼ ☉
10 36	☽ ⊥ ♃
17 29	☿ ∥ ♂
19 35	☽ △ ♄
20 03	☽ # ⯓
20 24	☽ ± ☉
20 31	☽ △ ♅
20 54	☽ ♇ ⯓
21 56	☽ Q ♆
30 Friday	
03 03	☽ ∠ ♃
05 22	☿ Q ♅
06 30	☿ ∥ ♄
06 41	♀ # ♆
06 45	☽ △ ♀
13 24	☽ ± ⯓
13 41	☿ ⊼ ♃
14 11	☽ ♇ ☉
17 03	☉ ± ♆
18 01	☿ △ ♆
23 55	☽ ♇ ♅

JULY 1972

Date	Sidereal time h m s	Sun ☉ ° ' "	Moon ☽ ° ' "	Moon ☽ 24.00 ° ' "	Mercury ☿ ° '	Venus ♀ ° '	Mars ♂ ° '	Jupiter ♃ ° '	Saturn ♄ ° '	Uranus ♅ ° '	Neptune ♆ ° '	Pluto ♇ ° '
01	06 38 32	09 ♋ 43 41	05 ♓ 56 50	12 ♓ 40 49	04 ♌ 01	19 ♊ 24	01 ♌ 47	02 ♑ 36	13 ♊ 55	14 ♎ 14	02 ♐ 58	29 ♍ 27
02	06 42 28	10 40 53	19 ♓ 28 33	26 ♓ 20 07	05 23	19 R 07	02 25	02 R 29	14 02	14 15	02 R 56	29 28
03	06 46 25	11 38 04	03 ♈ 15 28	10 ♈ 14 34	06 42	18 52	03 03	02 21	14 10	14 15	02 55	29 28
04	06 50 22	12 35 16	17 ♈ 17 19	24 ♈ 23 30	07 59	18 39	03 41	02 14	14 17	14 16	02 54	29 29
05	06 54 18	13 32 28	01 ♉ 32 52	08 ♉ 45 04	09 13	18 29	04 19	02 06	14 24	14 17	02 53	29 30
06	06 58 15	14 29 41	15 ♉ 59 39	23 ♉ 16 04	10 24	18 22	04 57	01 59	14 31	14 17	02 52	29 31
07	07 02 11	15 26 54	00 ♊ 33 44	07 ♊ 51 56	11 33	18 17	05 35	01 51	14 38	14 18	02 51	29 32
08	07 06 08	16 24 08	15 ♊ 09 58	22 ♊ 27 02	12 39	18 14	06 13	01 44	14 45	14 19	02 50	29 33
09	07 10 04	17 21 21	29 ♊ 42 23	06 ♋ 55 16	13 42	18 D 13	06 51	01 37	14 52	14 20	02 49	29 34
10	07 14 01	18 18 35	14 ♋ 04 59	21 ♋ 10 52	14 42	18 15	07 29	01 30	14 59	14 21	02 47	29 34
11	07 17 57	19 15 50	28 ♋ 12 23	05 ♌ 09 04	15 39	18 19	08 07	01 23	15 06	14 22	02 46	29 36
12	07 21 54	20 13 04	12 ♌ 00 35	18 ♌ 46 42	16 33	18 25	08 45	01 16	15 13	14 23	02 46	29 37
13	07 25 51	21 10 18	25 ♌ 27 17	02 ♍ 02 22	17 23	18 34	09 23	01 09	15 20	14 24	02 45	29 38
14	07 29 47	22 07 33	08 ♍ 32 01	14 ♍ 56 27	18 10	18 44	10 01	01 02	15 27	14 25	02 44	29 39
15	07 33 44	23 04 47	21 ♍ 15 58	27 ♍ 30 53	18 54	18 57	10 39	00 56	15 34	14 26	02 43	29 40
16	07 37 40	24 02 02	03 ♎ 41 39	09 ♎ 48 44	19 34	19 12	11 17	00 49	15 40	14 28	02 42	29 41
17	07 41 37	24 59 17	15 ♎ 52 38	21 ♎ 53 53	20 10	19 28	11 55	00 43	15 47	14 29	02 41	29 42
18	07 45 33	25 56 32	27 ♎ 53 04	03 ♏ 50 45	20 42	19 46	12 33	00 36	15 54	14 30	02 40	29 43
19	07 49 30	26 53 47	09 ♏ 47 29	15 ♏ 43 51	21 09	20 07	13 11	00 30	16 00	14 32	02 39	29 45
20	07 53 26	27 51 03	21 ♏ 40 24	27 ♏ 37 39	21 33	20 29	13 49	00 24	16 07	14 33	02 39	29 46
21	07 57 23	28 48 19	03 ♐ 36 09	09 ♐ 36 22	21 52	20 52	14 27	00 18	16 13	14 35	02 38	29 47
22	08 01 20	29 ♋ 45 35	15 ♐ 38 44	21 ♐ 43 40	22 06	21 18	15 05	00 12	16 20	14 36	02 37	29 49
23	08 05 16	00 ♌ 42 51	27 ♐ 51 32	04 ♑ 02 39	22 16	21 44	15 43	00 07	16 26	14 38	02 37	29 50
24	08 09 13	01 40 08	10 ♑ 17 15	16 ♑ 35 35	22 21	22 13	16 21	00 ♑ 01	16 32	14 40	02 36	29 51
25	08 13 09	02 37 26	22 ♑ 57 45	29 ♑ 23 53	22 R 21	22 43	16 59	29 ♐ 56	16 38	14 41	02 35	29 53
26	08 17 06	03 34 44	05 ♒ 54 00	12 ♒ 28 04	22 15	23 14	17 37	29 50	16 45	14 43	02 35	29 54
27	08 21 02	04 32 03	19 ♒ 06 02	25 ♒ 47 45	22 05	23 47	18 15	29 45	16 51	14 45	02 34	29 56
28	08 24 59	05 29 23	02 ♓ 33 04	09 ♓ 21 47	21 50	24 21	18 53	29 40	16 57	14 47	02 34	29 57
29	08 28 55	06 26 43	16 ♓ 13 38	23 ♓ 08 22	21 29	24 56	19 31	29 35	17 03	14 49	02 33	29 ♍ 58
30	08 32 52	07 24 05	00 ♈ 05 43	07 ♈ 05 22	21 04	25 33	20 09	29 31	17 09	14 51	02 33	00 ♎ 00
31	08 36 49	08 ♌ 21 27	14 ♈ 07 02	21 ♈ 10 24	20 ♌ 35	26 ♊ 10	20 ♌ 47	29 ♐ 26	17 ♊ 15	14 ♎ 53	02 ♐ 32	00 ♎ 02

Date	Moon True ☊ ° '	Moon Mean ☊ ° '	Moon ☽ Latitude ° '
01	26 ♑ 03	26 ♑ 58	03 N 23
02	26 D 04	26 55	04 14
03	26 05	26 52	04 51
04	26 R 05	26 48	05 12
05	26 05	26 45	05 15
06	26 04	26 42	04 57
07	26 03	26 39	04 21
08	26 01	26 36	03 27
09	26 00	26 32	02 20
10	26 00	26 29	01 N 06
11	26 D 00	26 26	00 S 12
12	26 00	26 23	01 28
13	26 00	26 20	02 36
14	26 01	26 17	03 34
15	26 01	26 13	04 21
16	26 01	26 10	04 53
17	26 01	26 07	05 12
18	26 R 01	26 04	05 17
19	26 D 02	26 01	05 08
20	26 02	25 58	04 46
21	26 02	25 54	04 11
22	26 02	25 51	03 26
23	26 03	25 48	02 30
24	26 03	25 45	01 26
25	26 03	25 42	00 S 17
26	26 R 03	25 38	00 N 54
27	26 03	25 35	02 04
28	26 02	25 32	03 08
29	26 01	25 29	04 03
30	25 59	25 26	04 44
31	25 ♑ 58	25 ♑ 23	05 N 09

DECLINATIONS

Date	Sun ☉ ° '	Moon ☽ ° '	Mercury ☿ ° '	Venus ♀ ° '	Mars ♂ ° '	Jupiter ♃ ° '	Saturn ♄ ° '	Uranus ♅ ° '	Neptune ♆ ° '	Pluto ♇ ° '
01	23 N 05	06 S 11	20 N 24	18 N 55	20 N 56	23 S 15	20 N 58	05 S 01	19 S 06	15 N 04
02	23 01	00 S 16	19 56	18 46	20 47	23 15	20 59	05 01	19 06	15 03
03	22 56	05 N 45	19 29	18 38	20 38	23 15	20 59	05 02	19 06	15 02
04	22 51	11 36	19 00	18 30	20 30	23 15	21 00	05 02	19 06	15 02
05	22 45	16 56	18 32	18 23	20 20	23 16	21 01	05 02	19 05	15 01
06	22 39	21 22	18 03	18 17	20 11	23 16	21 02	05 02	19 05	15 00
07	22 33	24 31	17 35	18 12	20 02	23 16	21 03	05 03	19 05	14 59
08	22 26	26 03	17 06	18 07	19 52	23 16	21 03	05 03	19 05	14 58
09	22 19	25 47	16 37	18 03	19 43	23 17	21 04	05 04	19 05	14 58
10	22 12	23 47	16 09	18 00	19 33	23 17	21 05	05 04	19 05	14 57
11	22 04	20 19	15 41	17 57	19 23	23 17	21 06	05 04	19 05	14 56
12	21 55	15 48	15 13	17 55	19 13	23 17	21 06	05 05	19 04	14 55
13	21 47	10 35	14 46	17 54	19 03	23 18	21 07	05 05	19 04	14 54
14	21 38	05 N 03	14 19	17 53	18 53	23 18	21 08	05 06	19 04	14 54
15	21 28	00 S 32	13 53	17 53	18 42	23 18	21 09	05 06	19 04	14 53
16	21 18	05 57	13 28	17 53	18 32	23 18	21 09	05 07	19 04	14 52
17	21 08	11 03	13 04	17 53	18 21	23 18	21 10	05 07	19 04	14 51
18	20 58	15 39	12 41	17 54	18 10	23 19	21 11	05 08	19 04	14 50
19	20 47	19 37	12 20	17 56	17 59	23 19	21 11	05 09	19 04	14 49
20	20 36	22 47	11 59	17 57	17 48	23 19	21 12	05 09	19 03	14 48
21	20 24	24 59	11 40	18 00	17 37	23 19	21 12	05 10	19 03	14 48
22	20 12	26 05	11 22	18 02	17 26	23 19	21 13	05 10	19 03	14 47
23	20 00	25 56	11 07	18 05	17 15	23 19	21 14	05 11	19 03	14 46
24	19 48	24 29	10 53	18 08	17 03	23 20	21 14	05 12	19 03	14 45
25	19 35	21 46	10 41	18 11	16 51	23 20	21 15	05 12	19 03	14 44
26	19 21	17 55	10 31	18 14	16 40	23 20	21 15	05 13	19 03	14 43
27	19 08	13 08	10 23	18 18	16 28	23 20	21 16	05 14	19 03	14 42
28	18 54	07 38	10 17	18 22	16 16	23 20	21 17	05 15	19 03	14 41
29	18 40	01 S 42	10 14	18 25	16 04	23 20	21 17	05 15	19 03	14 40
30	18 25	04 N 23	10 13	18 29	15 52	23 20	21 18	05 16	19 03	14 39
31	18 N 11	10 N 19	10 N 15	18 N 33	15 N 40	23 S 21	21 N 18	05 S 17	19 S 03	14 N 38

ZODIAC SIGN ENTRIES

Date	h	m	Planets
01	01	18	☽ ♓
03	06	22	☽ ♈
05	09	25	☽ ♉
07	11	05	☽ ♊
09	12	29	☽ ♋
11	15	05	☽ ♌
13	20	16	☽ ♍
16	04	49	☽ ♎
18	16	15	☽ ♏
21	04	46	☽ ♐
22	18	03	☉ ♌
23	16	10	☽ ♑
24	16	42	♃ ♐
26	01	07	☽ ♒
28	07	29	☽ ♓
30	11	39	♇ ♎
30	11	50	☽ ♈

LATITUDES

Date	Mercury ☿ ° '	Venus ♀ ° '	Mars ♂ ° '	Jupiter ♃ ° '	Saturn ♄ ° '	Uranus ♅ ° '	Neptune ♆ ° '	Pluto ♇ ° '
01	01 N 11	04 S 07	01 N 12	00 N 10	01 S 31	00 N 39	01 N 41	16 N 13
04	00 45	04 28	01 12	00 10	01 31	00 39	01 41	16 11
07	00 N 16	04 45	01 11	00 10	01 31	00 39	01 41	16 10
10	00 S 18	04 57	01 11	00 09	01 32	00 39	01 41	16 09
13	00 54	05 04	01 11	00 09	01 32	00 38	01 40	16 07
16	01 34	05 09	01 11	00 08	01 32	00 38	01 40	16 06
19	02 14	05 10	01 10	00 08	01 32	00 38	01 40	16 05
22	02 56	05 08	01 10	00 07	01 32	00 38	01 40	16 04
25	03 35	05 04	01 10	00 07	01 32	00 38	01 40	16 02
28	04 10	04 58	01 09	00 07	01 32	00 38	01 40	16 01
31	04 S 37	04 S 50	01 N 09	00 N 06	01 S 32	00 N 38	01 N 39	16 N 00

DATA

Julian Date	2441500
Delta T	+42 seconds
Ayanamsa	23° 28' 38"
Synetic vernal point	05° ♓ 38' 21"
True obliquity of ecliptic	23° 26' 38"

MOON'S PHASES, APSIDES AND POSITIONS ☽

Date	h	m	Phase	Longitude ° '	Eclipse Indicator
04	03	25	☾	12 ♈ 15	
10	19	39	●	18 ♋ 37	Total
18	07	46	☽	25 ♎ 46	
26	07	24	○	03 ♒ 24	partial

Day	h	m		
07	22	44	Perigee	
19	20	20	Apogee	
02	13	05	0N	
08	20	29	Max dec	26° N 10'
15	09	42	0S	
22	21	11	Max dec	26° S 10'
29	18	43	0N	

LONGITUDES

Date	Chiron ⚷ ° '	Ceres ⚳ ° '	Pallas ⚴ ° '	Juno ⚵ ° '	Vesta ⚶ ° '	Black Moon Lilith ⚸ ° '
01	17 ♈ 13	04 ♍ 11	07 ♌ 33	07 ♎ 18	21 ♉ 03	14 ♏ 22
11	17 ♈ 20	08 ♍ 11	12 ♌ 55	09 ♎ 09	24 ♉ 55	15 ♏ 30
21	17 ♈ 22	12 ♍ 19	18 ♌ 13	11 ♎ 19	28 ♉ 36	16 ♏ 37
31	17 ♈ 19	16 ♍ 32	23 ♌ 28	13 ♎ 43	02 ♊ 05	17 ♏ 44

All ephemeris data is given at 12.00 UT and the Moon's longitude is additionally given for 24.00 UT

ASPECTARIAN

h m	Aspects
01 Saturday	
00 18	☽ ⊼ ♇
04 10	☽ ⊼ ♂
06 03	☽ ✶ ♃
06 39	☽ □ ♆
07 18	♂ ∥ ♄
08 09	☽ ⊼ ☿
15 28	☽ ± ♂
16 06	☽ ± ♅
16 50	☽ ∥ ♅
17 20	☿ ∠ ♀
19 16	☽ △ ☉
20 06	☽ ± ☿
02 Sunday	
02 19	☽ □ ♄
02 46	☽ ⊼ ♅
03 16	☽ Q ♃
05 06	♂ Q ♅
08 13	☽ ⚼ ♂
11 22	☽ □ ♀
13 39	♂ ⊼ ♃
13 46	☽ ⚼ ☿
03 Monday	
05 27	☽ ☍ ♇
06 58	♂ △ ♆
09 06	☽ ⋕ ♅
10 05	☽ Q ♄
10 27	☽ □ ♃
11 25	☽ △ ♆
11 38	☽ △ ♂
18 07	☽ Q ♀
18 32	☽ △ ☿
04 Tuesday	
03 25	☽ □ ☉
06 51	☽ ✶ ♄
06 52	☽ ☍ ♅
07 31	☿ ⋕ ♆
08 22	♄ △ ♅
11 00	♀ ∠ ♂
13 02	☽ ⚼ ♆
14 17	☽ ✶ ♀
16 19	☿ ± ♃
05 Wednesday	
03 02	☽ ∥ ♇
04 12	☽ ± ♆
08 23	☽ ∠ ♄
08 34	☽ ⊼ ♇
11 59	☽ Q ☉
12 55	☽ △ ♃
14 13	☽ ⊼ ♆
15 13	☽ ∠ ♀
16 51	☽ □ ♂
18 36	☽ ± ♇
19 08	☽ ∥ ♀
19 18	☽ ∥ ☿
21 18	☿ ∥ ♀
22 58	☽ ⋕ ♆
23 31	☽ ⊥ ♄
06 Thursday	
01 55	☽ □ ☿
05 15	☽ ∥ ♂
06 02	☽ ⊥ ♀
06 45	☉ □ ♅
09 11	☽ ⊼ ♅
09 21	☽ ✶ ☉
09 33	☽ ⚺ ♄
09 33	☽ ⚼ ♇
09 57	☽ ∥ ♄
12 44	☉ ⚺ ♄
13 37	☽ ⚼ ♃
15 53	☽ ⚺ ♀
19 06	☽ ± ♅
20 17	☽ ∥ ☉
07 Friday	
00 00	☽ Q ♂
01 05	☽ ⋕ ♃
04 20	☽ ± ♃
09 56	☽ ⚼ ♅
10 12	☽ Q ☿
10 18	☽ △ ♇
11 48	☽ ∠ ☉
14 07	☽ ⊼ ♃
15 45	☽ ☍ ♆
20 38	☽ ✶ ♂
08 Saturday	
03 37	☽ ⊥ ☉
07 31	☽ ✶ ☿
10 36	☽ △ ♅
11 19	☽ ☌ ♄
14 11	☽ ⚺ ☉
17 02	☽ ☌ ♀
22 26	☽ ∠ ♂
09 Sunday	
04 55	♀ St D
10 12	☽ ∠ ☿
11 45	☽ □ ♇
14 00	☽ ⊥ ♂
15 09	☽ ☍ ♃
17 09	☽ ⊼ ♆
17 11	☉ Q ♇
23 11	☉ ⚼ ♆
10 Monday	
00 26	☽ ⚺ ♂
02 18	☽ ⊥ ☿
03 08	☽ ± ♆
03 21	☿ ✶ ♅
08 57	☿ ∠ ♇
10 28	☉ ⚺ ♀
12 21	♂ ± ♃
12 27	☽ □ ♅
13 07	☽ ⚺ ☿
13 32	☽ ⚺ ♄
16 08	☽ ⋕ ♃
17 54	☽ Q ♇
18 15	☽ ⚼ ♆
19 04	☽ ⚺ ♀
19 39	☽ ☌ ☉
20 15	☿ ✶ ♄
23 46	☽ ⊥ ♄
11 Tuesday	
00 38	☽ ∥ ☉
05 18	☽ ⊥ ♀
07 20	☽ ∥ ♄
14 23	☽ ✶ ♇
15 18	☽ ∠ ♄
17 25	☽ ⊼ ♃
17 37	☽ ∥ ♂
19 07	☽ ⋕ ♆
19 11	☽ Q ♅
19 52	☽ △ ♆
20 53	☽ ∠ ♀
12 Wednesday	
01 13	☽ ∥ ♀
03 45	☽ ± ♃
05 17	☿ ⚼ ♃
06 00	☽ ☌ ♂
15 03	☽ ∥ ☿
16 12	☽ ✶ ♅
16 12	☽ ∥ ♇
16 36	☽ ∠ ♇
17 43	☽ ✶ ♄
19 28	☽ ⚼ ♃
20 34	☽ ☌ ☿
23 29	☽ ✶ ♀
13 Thursday	
03 41	☽ ⚺ ☉
04 09	☿ ∥ ♇
08 42	☽ ⊥ ♇
08 48	♂ ⋕ ♆
15 21	☽ ⊥ ☉
15 26	☽ Q ♄
16 39	☉ ⊥ ♄
19 11	☽ ∠ ♅
19 36	☽ ⚺ ♇
21 25	☽ Q ♀
22 17	☽ △ ♃
14 Friday	
01 17	☽ □ ♆
09 11	☽ ∠ ☉
11 47	☽ ⊥ ♅
11 49	☽ ⋕ ♅
14 55	☽ ⚺ ♂
23 02	☽ ⚺ ♅
15 Saturday	
01 04	☽ □ ♄
02 46	☽ ⊥ ♂
07 13	☽ ⚺ ☿
07 31	☽ □ ♀
10 57	☽ Q ♆
14 47	☿ ✶ ♀
15 45	☽ ✶ ☉
19 22	☽ ⊥ ☿
20 52	☽ ∠ ♂
16 Sunday	
04 11	☽ ☌ ♇
06 27	☽ □ ♃
08 12	☽ ∥ ♅
10 04	☽ ✶ ♆
13 47	☽ ∠ ☿
16 58	☽ Q ☉
17 Monday	
03 44	☽ ✶ ♂
08 19	☉ ∥ ♄
09 14	☽ ☌ ♅
11 49	☽ △ ♄
15 35	☽ ∠ ♆
17 36	☽ Q ♃
19 20	☽ △ ♀
20 57	☽ ✶ ☿
21 26	☽ ⋕ ☿
18 Tuesday	
04 56	☽ Q ♂
05 41	☉ ⊥ ♀
07 34	☽ ⋕ ♇
07 46	☽ □ ☉
09 34	☽ ⊥ ♆
15 42	☽ ⚺ ♇
17 26	☽ ✶ ♃
18 07	☽ ⚼ ♄
21 37	☽ ⚺ ♆
22 05	☽ Q ☿
19 Wednesday	
01 12	☽ ⋕ ♀
02 06	☽ ⋕ ♂
02 16	☽ ⚼ ♀
03 49	☽ ⊥ ♇
08 22	☽ ∥ ♆
12 26	☽ ± ♄
18 54	♀ ∥ ♂
19 14	☽ □ ♂
19 41	☽ ⋕ ☉
21 00	☽ ± ♀
21 36	☽ ⚺ ♅
22 01	☽ ∠ ♇
23 08	☽ ⋕ ♄
23 27	☽ ∠ ♃
20 Thursday	
00 40	☽ ⊼ ♄
09 30	☽ ⊼ ♀
09 44	☽ ⊥ ♅
11 45	☽ □ ☿
16 54	☽ ∥ ♃
17 27	☽ ⊥ ♃
21 Friday	
01 32	☽ △ ☉
03 55	☽ ∠ ♅
04 20	☽ ✶ ♇
05 26	☽ ⚺ ♃
10 03	☽ ☌ ♆
17 07	♂ ✶ ♅
22 Saturday	
01 12	♂ ∠ ♇
04 22	☽ Q ♇
09 56	☽ ✶ ♅
10 06	☽ ⚼ ☉
10 49	☽ △ ♂
13 15	☉ ✶ ♇
13 22	☽ ☍ ♄
16 02	♂ ⚼ ♃
22 12	☉ ⊼ ♃
23 34	☽ ☍ ♀
23 Sunday	
00 56	☽ △ ☿
05 20	☽ ± ☉
09 36	☽ Q ♅
15 51	☽ □ ♇
16 21	☽ ☌ ♃
17 51	☽ ⚼ ♂
18 01	☽ ⊼ ☉
21 13	☽ ⚺ ♆
24 Monday	
06 20	☽ ⚼ ☿
08 17	☉ ∠ ♄
08 46	☽ ⊥ ♆
12 07	☽ ± ♂
18 52	☿ ✶ ♀
20 21	☽ □ ♅
20 32	♂ ✶ ♄
23 02	☿ St R
23 33	☽ ± ☿
23 33	☽ ∥ ♃
25 Tuesday	
00 00	☽ ⊼ ♄
00 09	☽ ⊼ ♂
01 53	☽ ∠ ♆
10 51	☽ ⊼ ☿
11 06	☉ △ ♆
11 24	☽ ± ♄
11 31	☽ ⊼ ♀
13 41	☉ Q ♅
15 41	☽ ⋕ ♄
22 42	♃ □ ♇
23 10	☽ ± ♀
26 Wednesday	
00 54	☽ ⚺ ♃
00 55	☽ △ ♇
03 18	☽ ⋕ ☉
04 17	☽ ⚼ ♄
05 36	☽ ∥ ♆
05 53	☽ ✶ ♆
07 24	☽ ☍ ☉
10 16	☽ ⋕ ♀
11 53	☽ ⊥ ♃
16 28	☽ ⚼ ♀
19 03	☽ ⋕ ♂
27 Thursday	
03 49	☽ Q ♆
04 07	☽ △ ♅
04 12	☽ ∠ ♃
04 27	☽ ⚼ ♇
04 33	☽ ⋕ ♇
07 54	☽ △ ♄
10 23	☽ ☍ ♂
17 16	☽ ☍ ☿
20 40	☽ ± ♇
20 46	☉ ⋕ ♆
20 46	☽ △ ♀
28 Friday	
00 33	☽ ⋕ ☿
06 56	☽ ✶ ♃
07 05	☽ ⚼ ♅
07 23	☽ ⊼ ♇
12 01	☽ □ ♆
16 11	☉ ± ♃
17 35	☽ ⊼ ☉
21 48	☽ ∥ ♅
23 00	☽ ± ♅
29 Saturday	
03 57	☽ Q ♃
04 54	☽ ± ☉
09 31	☽ ⊼ ♅
13 26	☽ □ ♄
17 59	☽ ⊼ ♂
20 54	☽ ⊼ ☿
21 44	☽ ⚼ ☉
30 Sunday	
03 48	☽ □ ♀
04 52	☽ ± ♂
06 57	☽ ± ☿
07 07	☉ ∥ ♀
11 00	☽ □ ♃
11 50	☽ ☍ ♇
15 31	☽ ⋕ ♅
16 12	☽ △ ♆
20 44	☽ Q ♄
21 05	☽ ⚼ ♂
21 56	☽ ⚼ ☿
31 Monday	
01 27	☽ △ ☉
07 51	☿ ☌ ♂
11 44	☽ ∥ ☿
12 06	☽ Q ♀
13 18	☽ ☍ ♅
17 22	☽ ✶ ♄
17 49	☽ ⚼ ♆
22 35	☽ △ ☿
23 52	☽ △ ♂

AUGUST 1972

LONGITUDES

Date	Sidereal time h m s	Sun ☉ ° ' "	Moon ☽ ° ' "	Moon ☽ 24.00 ° ' "	Mercury ☿ ° '	Venus ♀ ° '	Mars ♂ ° '	Jupiter ♃ ° '	Saturn ♄ ° '	Uranus ♅ ° '	Neptune ♆ ° '	Pluto ♇ ° '
01	08 40 45	09 ♌ 18 51	28 ♈ 15 10	05 ♉ 21 02	20 ♌ 01	26 ♊ 49	21 ♌ 25	29 ♐ 22	17 ♊ 20	14 ♎ 55	02 ♐ 32	00 ♎ 03
02	08 44 42	10 16 16	12 ♉ 27 43	19 ♉ 34 53	19 R 23	27 29	22 03	29 R 18	17 26	14 57	02 R 31	00 05
03	08 48 38	11 13 42	26 ♉ 42 14	03 ♊ 49 28	18 42	28 10	22 41	29 13	17 32	14 59	02 31	00 06
04	08 52 35	12 11 09	10 ♊ 56 15	18 ♊ 02 16	17 59	28 52	23 19	29 10	17 38	15 01	02 31	00 08
05	08 56 31	13 08 38	25 ♊ 07 10	02 ♋ 10 35	17 13	29 ♊ 35	23 57	29 06	17 43	15 03	02 30	00 10
06	09 00 28	14 06 08	09 ♋ 12 10	16 ♋ 11 34	16 26	00 ♋ 19	24 35	29 02	17 49	15 05	02 30	00 11
07	09 04 24	15 03 39	23 ♋ 08 23	00 ♌ 02 17	15 39	01 04	25 13	28 59	17 54	15 08	02 30	00 13
08	09 08 21	16 01 12	06 ♌ 52 56	13 ♌ 40 01	14 52	01 50	25 51	28 56	17 59	15 10	02 30	00 15
09	09 12 18	16 58 45	20 ♌ 23 16	27 ♌ 02 29	14 06	02 37	26 29	28 53	18 05	15 12	02 30	00 17
10	09 16 14	17 56 19	03 ♍ 37 28	10 ♍ 08 10	13 23	03 24	27 07	28 50	18 10	15 15	02 29	00 18
11	09 20 11	18 53 55	16 ♍ 34 31	22 ♍ 56 34	12 43	04 12	27 45	28 47	18 15	15 17	02 29	00 20
12	09 24 07	19 51 31	29 ♍ 14 28	05 ♎ 28 22	12 07	05 01	28 23	28 45	18 20	15 20	02 29	00 22
13	09 28 04	20 49 09	11 ♎ 38 33	17 ♎ 45 19	11 35	05 51	29 01	28 42	18 25	15 22	02 29	00 24
14	09 32 00	21 46 47	23 ♎ 49 05	29 ♎ 50 17	11 09	06 41	29 ♌ 39	28 40	18 30	15 25	02 D 29	00 26
15	09 35 57	22 44 27	05 ♏ 49 23	11 ♏ 46 55	10 49	07 32	00 ♍ 17	28 38	18 35	15 27	02 29	00 28
16	09 39 53	23 42 07	17 ♏ 43 26	23 ♏ 39 32	10 35	08 24	00 56	28 37	18 39	15 30	02 29	00 29
17	09 43 50	24 39 49	29 ♏ 35 48	05 ♐ 32 50	10 28	09 16	01 34	28 35	18 44	15 33	02 29	00 31
18	09 47 47	25 37 32	11 ♐ 31 16	17 ♐ 31 41	10 D 29	10 09	02 12	28 34	18 49	15 36	02 29	00 33
19	09 51 43	26 35 15	23 ♐ 34 39	29 ♐ 40 45	10 37	11 02	02 50	28 32	18 53	15 38	02 30	00 35
20	09 55 40	27 33 00	05 ♑ 50 29	12 ♑ 04 21	10 52	11 57	03 28	28 31	18 57	15 41	02 30	00 37
21	09 59 36	28 30 47	18 ♑ 22 43	24 ♑ 45 57	11 15	12 51	04 06	28 31	19 02	15 44	02 30	00 39
22	10 03 33	29 ♌ 28 34	01 ♒ 14 19	07 ♒ 47 57	11 46	13 46	04 44	28 30	19 06	15 47	02 30	00 41
23	10 07 29	00 ♍ 26 23	14 ♒ 26 56	21 ♒ 11 11	12 24	14 42	05 22	28 30	19 10	15 50	02 31	00 43
24	10 11 26	01 24 13	28 ♒ 00 34	04 ♓ 54 46	13 10	15 38	06 01	28 29	19 14	15 53	02 31	00 45
25	10 15 22	02 22 04	11 ♓ 53 24	18 ♓ 55 59	14 03	16 35	06 39	28 D 29	19 18	15 56	02 31	00 47
26	10 19 19	03 19 57	26 ♓ 01 54	03 ♈ 10 33	15 03	17 32	07 17	28 29	19 22	15 59	02 32	00 49
27	10 23 16	04 17 52	10 ♈ 21 13	17 ♈ 33 13	16 09	18 30	07 55	28 30	19 26	16 02	02 32	00 51
28	10 27 12	05 15 48	24 ♈ 45 52	01 ♉ 58 30	17 22	19 28	08 33	28 30	19 30	16 05	02 33	00 54
29	10 31 09	06 13 46	09 ♉ 10 32	16 ♉ 21 26	18 41	20 26	09 12	28 31	19 33	16 08	02 33	00 56
30	10 35 05	07 11 46	23 ♉ 30 46	00 ♊ 38 11	20 06	21 25	09 50	28 32	19 37	16 11	02 34	00 58
31	10 39 02	08 ♍ 09 48	07 ♊ 43 23	14 ♊ 46 11	21 ♌ 35	22 ♋ 24	10 ♍ 28	28 ♐ 33	19 ♊ 40	16 ♎ 14	02 ♐ 34	01 ♎ 00

Date	Moon True ☊ ° '	Moon Mean ☊ ° '	Moon ☽ Latitude ° '
01	25 ♑ 57	25 ♑ 19	05 N 16
02	25 D 57	25 16	05 03
03	25 58	25 13	04 32
04	25 58	25 10	03 44
05	26 00	25 07	02 42
06	26 01	25 04	01 31
07	26 01	25 00	00 N 16
08	26 R 01	24 57	01 S 00
09	26 00	24 54	02 10
10	25 57	24 51	03 12
11	25 54	24 48	04 03
12	25 50	24 44	04 41
13	25 47	24 41	05 04
14	25 43	24 38	05 13
15	25 41	24 35	05 09
16	25 40	24 32	04 51
17	25 D 40	24 29	04 20
18	25 41	24 25	03 39
19	25 42	24 22	02 47
20	25 44	24 19	01 47
21	25 45	24 16	00 S 40
22	25 R 45	24 13	00 N 30
23	25 44	24 10	01 40
24	25 41	24 06	02 47
25	25 36	24 03	03 45
26	25 31	24 00	04 30
27	25 25	23 57	04 59
28	25 20	23 54	05 10
29	25 15	23 50	05 01
30	25 13	23 47	04 33
31	25 ♑ 12	23 ♑ 44	03 N 49

DECLINATIONS

Date	Sun ☉ ° '	Moon ☽ ° '	Mercury ☿ ° '	Venus ♀ ° '	Mars ♂ ° '	Jupiter ♃ ° '	Saturn ♄ ° '	Uranus ♅ ° '	Neptune ♆ ° '	Pluto ♇ ° '
01	17 N 56	15 N 46	10 N 19	18 N 37	15 N 27	23 S 21	21 N 19	05 S 18	19 S 03	14 N 37
02	17 40	20 23	10 26	18 41	15 15	23 21	21 19	05 19	19 03	14 36
03	17 25	23 49	10 35	18 45	15 02	23 21	21 20	05 19	19 03	14 36
04	17 09	25 47	10 46	18 49	14 50	23 21	21 20	05 20	19 03	14 35
05	16 53	26 03	10 59	18 52	14 37	23 21	21 20	05 21	19 03	14 34
06	16 36	24 39	11 15	18 56	14 24	23 21	21 21	05 22	19 03	14 33
07	16 19	21 43	11 32	18 59	14 11	23 21	21 21	05 23	19 03	14 32
08	16 02	17 36	11 50	19 03	13 58	23 21	21 22	05 24	19 03	14 31
09	15 45	12 38	12 10	19 06	13 45	23 22	21 22	05 25	19 03	14 30
10	15 27	07 11	12 30	19 09	13 32	23 22	21 22	05 26	19 03	14 29
11	15 10	01 N 34	12 51	19 12	13 19	23 22	21 23	05 27	19 03	14 28
12	14 52	03 S 59	13 13	19 14	13 05	23 22	21 23	05 28	19 03	14 27
13	14 33	09 16	13 34	19 17	12 52	23 22	21 24	05 29	19 03	14 26
14	14 15	14 06	13 55	19 19	12 38	23 22	21 24	05 30	19 03	14 25
15	13 56	18 19	14 15	19 20	12 25	23 22	21 24	05 31	19 03	14 24
16	13 37	21 46	14 35	19 22	12 11	23 22	21 25	05 32	19 03	14 23
17	13 18	24 18	14 53	19 23	11 57	23 23	21 25	05 33	19 03	14 22
18	12 59	25 47	15 09	19 24	11 44	23 23	21 25	05 34	19 03	14 21
19	12 39	26 04	15 24	19 25	11 30	23 23	21 25	05 35	19 03	14 20
20	12 20	25 06	15 37	19 25	11 16	23 23	21 26	05 36	19 03	14 19
21	12 00	22 51	15 48	19 25	11 02	23 23	21 26	05 37	19 04	14 18
22	11 39	19 24	15 56	19 24	10 48	23 23	21 26	05 39	19 04	14 17
23	11 19	14 54	16 02	19 23	10 33	23 23	21 27	05 40	19 04	14 16
24	10 59	09 34	16 06	19 22	10 19	23 23	21 27	05 41	19 04	14 15
25	10 38	03 S 39	16 06	19 20	10 05	23 24	21 27	05 42	19 04	14 14
26	10 17	02 N 33	16 04	19 18	09 51	23 24	21 27	05 43	19 04	14 13
27	09 56	08 41	15 58	19 16	09 36	23 24	21 27	05 44	19 04	14 12
28	09 35	14 23	15 50	19 13	09 22	23 24	21 28	05 46	19 04	14 11
29	09 14	19 18	15 38	19 10	09 07	23 24	21 28	05 47	19 05	14 10
30	08 52	23 03	15 23	19 06	08 52	23 24	21 28	05 48	19 05	14 09
31	08 N 31	25 N 22	15 N 05	19 N 01	08 N 38	23 S 24	21 N 28	05 S 49	19 S 05	14 N 08

ZODIAC SIGN ENTRIES

Date	h	m	Planets
01	14	57	☽ ♉
03	17	33	☽ ♊
05	20	18	☽ ♋
06	01	26	♀ ♋
07	23	56	☽ ♌
10	05	23	☽ ♍
12	13	27	☽ ♎
15	00	19	☽ ♏
15	00	59	♂ ♍
17	12	49	☽ ♐
20	00	38	☽ ♑
22	09	43	☽ ♒
23	01	03	☉ ♍
24	15	28	☽ ♓
26	18	40	☽ ♈
28	20	43	☽ ♉
30	22	56	☽ ♊

LATITUDES

Date	Mercury ☿ ° '	Venus ♀ ° '	Mars ♂ ° '	Jupiter ♃ ° '	Saturn ♄ ° '	Uranus ♅ ° '	Neptune ♆ ° '	Pluto ♇ ° '
01	04 S 44	04 S 47	01 N 09	00 N 06	01 S 32	00 N 38	01 N 39	16 N 00
04	04 55	04 38	01 09	00 05	01 32	00 38	01 39	15 59
07	04 50	04 27	01 08	00 05	01 33	00 37	01 39	15 58
10	04 29	04 15	01 08	00 05	01 33	00 37	01 39	15 57
13	03 54	04 03	01 07	00 04	01 33	00 37	01 39	15 56
16	03 08	03 49	01 07	00 04	01 33	00 37	01 39	15 55
19	02 16	03 35	01 06	00 03	01 34	00 37	01 38	15 54
22	01 23	03 21	01 06	00 03	01 34	00 37	01 38	15 54
25	00 S 32	03 06	01 05	00 03	01 34	00 37	01 38	15 53
28	00 N 12	02 51	01 04	00 02	01 34	00 37	01 38	15 52
31	00 N 49	02 S 35	01 N 04	00 N 02	01 S 35	00 N 37	01 N 38	15 N 52

DATA

Julian Date	2441531
Delta T	+42 seconds
Ayanamsa	23° 28' 43"
Synetic vernal point	05° ♓ 38' 17"
True obliquity of ecliptic	23° 26' 38"

MOON'S PHASES, APSIDES AND POSITIONS ☽

Date	h	m	Phase	Longitude ° '	Eclipse Indicator
02	08	02	☾	10 ♉ 07	
09	05	26	●	16 ♌ 43	
17	01	09	☽	24 ♏ 14	
24	18	22	○	01 ♓ 40	
31	12	48	☾	08 ♊ 12	

Day	h	m		
03	14	52	Perigee	
16	15	00	Apogee	
28	20	10	Perigee	
05	03	52	Max dec	26° N 09'
11	18	42	0S	
19	05	37	Max dec	26° S 07'
26	02	10	0N	

LONGITUDES

Date	Chiron ⚷ ° '	Ceres ⚳ ° '	Pallas ⚴ ° '	Juno ⚵ ° '	Vesta ⚶ ° '	Black Moon Lilith ⚸ ° '
01	17 ♈ 18	16 ♍ 58	23 ♌ 59	13 ♎ 58	02 ♊ 25	17 ♏ 51
11	17 ♈ 09	21 ♍ 17	29 ♌ 09	16 ♎ 37	05 ♊ 38	18 ♏ 58
21	16 ♈ 56	25 ♍ 39	04 ♍ 15	19 ♎ 26	08 ♊ 33	20 ♏ 05
31	16 ♈ 38	00 ♎ 05	09 ♍ 17	22 ♎ 24	11 ♊ 08	21 ♏ 12

All ephemeris data is given at 12.00 UT and the Moon's longitude is additionally given for 24.00 UT

ASPECTARIAN

h m Aspects

01 Tuesday
06 44 ☽ ∥ ♇
09 05 ☽ ± ♆
09 28 ☽ ✶ ♀
10 37 ☽ ∥ ♂
13 52 ☽ △ ♃
15 03 ☽ ⊼ ♇
18 58 ☽ ∠ ♄
19 14 ☽ ⊼ ♆
22 04 ☽ ∥ ☉

02 Wednesday
01 13 ☽ ± ♇
02 23 ☽ ∥ ♀
04 29 ☽ # ♆
08 02 ☽ □ ☉
10 16 ☽ ⊥ ♄
12 03 ☽ ∠ ♀
15 04 ☽ ⚼ ♃
16 12 ☽ ⊼ ♅
16 25 ☽ ⚼ ♇
17 42 ☽ ∥ ♄
20 27 ☽ ⊻ ♄
23 09 ☽ □ ☿

03 Thursday
02 20 ☽ ± ♅
03 59 ☽ ⊥ ♀
04 54 ☽ □ ♂
06 10 ☽ ± ♃
08 00 ☽ # ♃
14 36 ☽ ⊻ ♀
16 14 ☽ ⊼ ♃
16 34 ☽ Q ☉
17 32 ☽ ⚼ ♅
17 45 ☽ △ ♇
21 47 ☽ ☍ ♆

04 Friday
04 03 ☽ Q ☿
12 40 ☽ Q ♂
14 16 ☽ ✶ ☉
18 54 ☽ △ ♅
20 51 ♀ ☍ ♃
22 00 ☿ ✶ ♄
23 18 ☽ ✶ ☿
23 23 ☽ ☌ ♄

05 Saturday
09 55 ☽ ✶ ♂
17 31 ☽ ∠ ☉
18 44 ☽ ☍ ♃
18 58 ♂ ∥ ♇
20 01 ☽ ☌ ♀
20 28 ♂ ⊥ ♇
20 35 ☽ □ ♇
23 26 ☽ ∠ ☿

06 Sunday
00 33 ☽ ⊼ ♆
07 27 ♀ □ ♇
09 59 ☽ ⊥ ☉
10 30 ☉ ⚼ ♃
10 48 ☽ ± ♆
12 41 ☽ ∠ ♂
14 00 ☽ ⊥ ☿
21 01 ☽ ⊻ ☉
22 08 ☽ □ ♅
23 45 ☽ ⊻ ☿

07 Monday
00 00 ☽ # ♃
02 15 ☽ ⚼ ♆
02 53 ☽ ⊻ ♄
03 28 ☽ Q ♇
04 53 ☽ ⊥ ♂
05 22 ☿ ∠ ♀
13 20 ☽ ⊥ ♄
13 45 ☉ ✶ ♅
14 24 ☽ ∥ ♄
15 47 ☽ ⊻ ♂
16 04 ☉ ∠ ♇
20 03 ☉ ☌ ☿
22 07 ☽ ⊼ ♃

08 Tuesday
00 20 ☽ ✶ ♇
00 37 ☿ ∠ ♇
02 37 ☽ ⊻ ♀
03 06 ☿ ✶ ♅
04 12 ☽ # ♆
04 18 ☽ △ ♆
04 18 ☽ ∥ ♀
05 07 ☽ ∠ ♄
05 27 ☽ Q ♅
08 35 ☽ ± ♃
12 21 ♀ # ♆
13 47 ☽ ⊥ ♀
20 24 ☽ ∥ ☉

09 Wednesday
00 25 ☽ ⚼ ♃
01 22 ☽ ☌ ☿
02 43 ☽ ✶ ♅
02 51 ☽ ∠ ♇
03 20 ☽ ∥ ♇
05 26 ☽ ☌ ☉
06 37 ☽ ∥ ♂
06 43 ☽ ∠ ♀
07 50 ☽ ✶ ♄
08 20 ♀ ⊼ ♆
14 02 ☽ ∥ ☿
19 01 ☽ ⊥ ♇
19 56 ☿ ⚼ ♃
23 33 ☽ ☌ ♂

10 Thursday
03 17 ☽ △ ♃
05 38 ☽ Q ♄
05 48 ☽ ∠ ♅
05 55 ☽ ⊻ ♇
09 55 ☽ □ ♆
11 34 ☽ ✶ ♀
18 10 ☉ ✶ ♄
19 31 ☽ # ♅
22 23 ☽ ⊥ ♅

11 Friday
05 08 ☽ ⊻ ☿
09 35 ☽ ⊻ ♅
11 16 ☽ Q ♀
15 10 ☽ □ ♄
15 50 ☽ ⊥ ☿
16 43 ☽ ⊻ ☉
19 21 ☽ Q ♆

12 Saturday
05 00 ☽ ⊥ ☉
06 58 ☿ ∥ ♂
08 06 ☽ ∠ ☿
10 17 ☽ ⊻ ♂
11 03 ☽ □ ♃
14 10 ☽ ☌ ♇
18 14 ☽ ✶ ♆
18 35 ☽ ∥ ♅
22 26 ☽ ⊥ ♂
23 43 ☽ ∠ ☉
23 55 ☽ □ ♀

13 Sunday
00 44 ♂ △ ♃
07 14 ☿ ⊥ ♀
11 53 ☽ ✶ ☿
16 55 ☽ ∠ ♂
17 18 ☉ ∠ ♀
19 20 ☽ ☌ ♅
21 54 ☽ Q ♃
22 35 ☉ ∥ ♇
23 28 ☽ ∠ ♆

14 Monday
01 24 ☽ △ ♄
03 08 ♆ St D
04 50 ☽ # ♂
07 36 ☽ ✶ ☉
10 43 ☽ Q ☿
11 00 ☽ # ☿
12 47 ☽ # ☉
13 42 ☽ # ♇
17 19 ☽ ⊥ ♆
21 38 ☽ ✶ ♃

15 Tuesday
00 15 ☉ ∥ ☿
00 17 ☽ ✶ ♂
01 13 ☽ ⊻ ♇
05 18 ☽ ⊻ ♆
07 28 ☽ ⚼ ♄
09 38 ☽ Q ☉
13 17 ☽ ⊥ ♇
15 43 ☽ △ ♀
16 42 ☽ ∥ ♆
18 38 ☽ # ♀
18 40 ♂ ⊻ ♇
18 46 ♂ ∠ ♅
21 50 ☽ □ ☿
21 54 ☿ ∥ ♇

16 Wednesday
00 20 ♂ Q ♄
01 42 ☽ ± ♄
01 46 ☽ Q ♂
03 42 ☽ ∠ ♃
07 29 ☽ ∠ ♇
07 30 ☽ ⊻ ♅
09 14 ☽ # ♄
13 54 ☽ ⊼ ♄
14 31 ♀ ± ♆
19 40 ☽ ⊥ ♅
21 51 ☽ ⊥ ♃

17 Thursday
00 23 ☽ ⚼ ♀
01 09 ☽ □ ☉
02 05 ☽ ∥ ♃
08 22 ☉ ⊥ ♇
09 58 ☽ ⊻ ♃
13 53 ☽ ✶ ♇
13 56 ☽ ∠ ♅
16 11 ☽ □ ♂
17 50 ☽ ☌ ♆
20 00 ☽ ± ♀
22 38 ☿ St D

18 Friday
09 02 ☽ ⊼ ♀
09 54 ☽ △ ☿
14 05 ☽ Q ♇
20 10 ☽ ✶ ♅
21 55 ☿ ⊻ ♀
23 11 ♂ □ ♆

19 Saturday
02 38 ☽ ☍ ♄
16 05 ☽ ⚼ ☿
18 27 ☽ △ ☉
20 02 ☽ Q ♅
21 45 ☽ ☌ ♃

20 Sunday
01 49 ☽ □ ♇
05 30 ☽ ⊻ ♆
07 08 ☽ △ ♂
10 04 ☽ ± ☿
17 08 ☽ ⊥ ♆
21 58 ☽ ⊼ ☿

21 Monday
00 40 ☽ ☍ ♀
01 59 ☽ ⚼ ☉
06 58 ☽ □ ♅
07 18 ☽ ∥ ♃
10 20 ☽ ∠ ♆
11 57 ☉ △ ♃
13 14 ☽ ⊼ ♄
13 26 ☽ ⚼ ♂
20 25 ☽ ± ☉
22 49 ☽ # ♄

22 Tuesday
00 34 ☽ ± ♄
06 57 ☽ ⊻ ♃
07 09 ☽ ± ♂
08 29 ☽ ⊼ ☉
10 59 ☽ △ ♇
11 59 ☽ # ♀
14 01 ☽ ∥ ♆
14 20 ☽ ✶ ♆
17 17 ☽ ⚼ ♄
17 59 ☽ ⊥ ♃
18 45 ☽ ⊼ ♂

23 Wednesday
06 29 ☽ # ☿
08 07 ☽ ☍ ☿
10 17 ☽ ∠ ♃
12 07 ☽ Q ♆
12 29 ☽ ⊼ ♀
14 17 ☽ ⚼ ♇
14 29 ☽ △ ♅
15 04 ☽ # ♇
19 15 ☉ ⊻ ♇
20 28 ☽ △ ♄
22 12 ☉ ∠ ♅
23 58 ☽ ± ♀

24 Thursday
05 32 ☽ # ☉
06 17 ☽ ± ♇
07 32 ☉ Q ♄
08 38 ☽ # ♂
12 50 ☽ ✶ ♃
16 48 ☽ ⊼ ♇
16 55 ☽ ⚼ ♀
17 01 ☽ ⚼ ♅
18 22 ☽ ☍ ☉
18 30 ♀ □ ♅
19 51 ☽ □ ♆
21 10 ☿ ⚼ ♃

25 Friday
02 34 ☽ ☍ ♂
03 52 ☽ ∥ ♅
07 57 ♃ St D
08 37 ☽ ± ♅
09 36 ☽ Q ♃
15 51 ☉ □ ♆
15 57 ☽ ⊼ ☿
18 55 ☽ ⊼ ♅
20 35 ☽ △ ♀

26 Saturday
00 41 ☽ □ ♄
02 55 ☽ ± ☿
11 52 ♀ ⚼ ♆
16 08 ☽ □ ♃
19 18 ☽ ⚼ ☿
20 04 ☽ ☍ ♇
22 55 ☽ △ ♆

27 Sunday
00 19 ☽ # ♅
01 09 ☽ ⊼ ☉
05 31 ☿ ∠ ♇
07 06 ☽ Q ♄
07 45 ☽ ⊼ ♂
09 10 ☿ ✶ ♅
11 54 ☽ ± ☉
15 34 ☽ ∥ ♂
16 47 ☽ ∥ ☉
18 13 ☽ ± ♂
21 23 ♀ Q ♇
21 29 ☽ ☍ ♅
22 33 ☽ △ ☿
23 58 ☽ ⚼ ♆

28 Monday
02 32 ☽ □ ♀
03 11 ☽ ✶ ♄
03 58 ☽ ⚼ ☉
09 54 ☽ ⚼ ♂
11 05 ☽ ∥ ♇
12 50 ♀ ⊻ ♄
14 58 ☽ ± ♆
18 14 ☽ △ ♃
18 21 ☽ ∥ ☿
22 13 ☽ ⊼ ♇

29 Tuesday
00 57 ☽ ⊼ ♆
04 16 ☽ ∠ ♄
06 44 ☽ △ ☉
08 14 ☽ ± ♇
10 41 ☽ Q ♀
10 48 ☽ # ♆
11 15 ☽ ∥ ♀
12 02 ☽ △ ♂
19 15 ☽ ⚼ ♃
19 20 ☽ ⊥ ♄
23 18 ☽ ⚼ ♇
23 40 ☽ ⊼ ♅

30 Wednesday
00 48 ☽ ∥ ♄
03 31 ☿ ✶ ♄
05 25 ☽ ⊻ ♄
05 38 ☽ □ ☿
08 13 ☽ ✶ ♀
09 45 ☽ ± ♅
10 21 ☽ ± ♃
11 16 ☉ ∥ ♂
14 49 ☽ # ♃
17 28 ♀ # ♆
20 27 ☽ ⊼ ♃

31 Thursday
00 35 ☽ △ ♇
00 58 ☽ ⚼ ♅
02 27 ♂ ⊥ ♅
03 16 ☽ ☍ ♆
11 25 ☽ ∠ ♀
12 48 ☽ □ ☉
15 34 ☽ Q ☿
16 53 ☽ □ ♂

SEPTEMBER 1972

Date	Sidereal time h m s	Sun ☉ ° ' "	Moon ☽ ° ' "	Moon ☽ 24.00 ° ' "	Mercury ☿ ° '	Venus ♀ ° '	Mars ♂ ° '	Jupiter ♃ ° '	Saturn ♄ ° '	Uranus ♅ ° '	Neptune ♆ ° '	Pluto ♇ ° '
01	10 42 58	09 ♍ 07 53	21 ♊ 46 27	28 ♊ 44 04	23 ♌ 10	23 ♋ 24	11 ♍ 06	28 ♐ 34	19 ♊ 43	16 ♎ 17	02 ♐ 35	01 ♎ 02
02	10 46 55	10 05 59	05 ♋ 39 01	12 ♋ 31 16	24 48	24 24	11 45	28 36	19 47	16 20	02 35	01 04
03	10 50 51	11 04 07	19 ♋ 20 47	26 ♋ 07 34	26 30	25 25	12 23	28 37	19 50	16 24	02 36	01 06
04	10 54 48	12 02 17	02 ♌ 51 36	09 ♌ 32 51	28 ♌ 14	26 26	13 01	28 39	19 53	16 27	02 37	01 08
05	10 58 45	13 00 29	16 ♌ 11 15	22 ♌ 46 45	00 ♍ 02	27 27	13 39	28 41	19 56	16 30	02 37	01 11
06	11 02 41	13 58 42	29 ♌ 19 16	05 ♍ 48 44	01 51	28 28	14 18	28 43	19 59	16 34	02 38	01 13
07	11 06 38	14 56 58	12 ♍ 15 04	18 ♍ 38 13	03 43	29 ♋ 30	14 56	28 46	20 01	16 37	02 39	01 15
08	11 10 34	15 55 15	24 ♍ 58 09	01 ♎ 14 51	05 35	00 ♌ 33	15 34	28 48	20 04	16 40	02 40	01 17
09	11 14 31	16 53 34	07 ♎ 28 20	13 ♎ 38 43	07 29	01 35	16 13	28 51	20 07	16 44	02 40	01 19
10	11 18 27	17 51 54	19 ♎ 46 06	25 ♎ 50 42	09 23	02 38	16 51	28 54	20 09	16 47	02 41	01 22
11	11 22 24	18 50 17	01 ♏ 52 45	07 ♏ 52 34	11 17	03 41	17 30	28 57	20 11	16 51	02 42	01 24
12	11 26 20	19 48 41	13 ♏ 50 31	19 ♏ 47 00	13 12	04 44	18 08	29 00	20 14	16 54	02 43	01 26
13	11 30 17	20 47 06	25 ♏ 42 32	01 ♐ 37 36	15 06	05 48	18 46	29 04	20 16	16 57	02 44	01 28
14	11 34 14	21 45 34	07 ♐ 32 47	13 ♐ 28 40	17 00	06 52	19 25	29 07	20 18	17 01	02 45	01 31
15	11 38 10	22 44 03	19 ♐ 25 55	25 ♐ 25 09	18 53	07 56	20 03	29 11	20 20	17 04	02 46	01 33
16	11 42 07	23 42 33	01 ♑ 27 02	07 ♑ 32 13	20 46	09 01	20 42	29 15	20 21	17 08	02 47	01 35
17	11 46 03	24 41 06	13 ♑ 41 22	19 ♑ 55 06	22 39	10 06	21 20	29 19	20 23	17 12	02 48	01 37
18	11 50 00	25 39 40	26 ♑ 13 59	02 ♒ 38 32	24 30	11 11	21 59	29 23	20 25	17 15	02 49	01 40
19	11 53 56	26 38 15	09 ♒ 09 11	15 ♒ 46 15	26 21	12 16	22 37	29 28	20 26	17 19	02 50	01 42
20	11 57 53	27 36 52	22 ♒ 29 56	29 ♒ 20 19	28 10	13 21	23 16	29 33	20 28	17 22	02 52	01 44
21	12 01 49	28 35 31	06 ♓ 17 16	13 ♓ 20 30	29 ♍ 59	14 27	23 54	29 38	20 29	17 26	02 53	01 46
22	12 05 46	29 ♍ 34 12	20 ♓ 29 32	27 ♓ 43 45	01 ♎ 47	15 33	24 33	29 43	20 30	17 30	02 54	01 49
23	12 09 43	00 ♎ 32 55	05 ♈ 02 20	12 ♈ 24 20	03 34	16 39	25 11	29 48	20 31	17 33	02 55	01 51
24	12 13 39	01 31 40	19 ♈ 48 43	27 ♈ 14 22	05 20	17 46	25 50	29 53	20 32	17 37	02 57	01 53
25	12 17 36	02 30 27	04 ♉ 40 13	12 ♉ 05 11	07 05	18 52	26 28	29 ♐ 59	20 33	17 41	02 58	01 56
26	12 21 32	03 29 16	19 ♉ 28 20	26 ♉ 48 47	08 49	19 59	27 07	00 ♑ 04	20 34	17 44	02 59	01 58
27	12 25 29	04 28 07	04 ♊ 05 51	11 ♊ 19 00	10 32	21 06	27 46	00 10	20 34	17 48	03 01	02 00
28	12 29 25	05 27 01	18 ♊ 27 51	25 ♊ 32 11	12 15	22 14	28 24	00 16	20 35	17 52	03 02	02 02
29	12 33 22	06 25 57	02 ♋ 31 53	09 ♋ 26 58	13 56	23 21	29 03	00 22	20 35	17 55	03 03	02 05
30	12 37 18	07 ♎ 24 55	16 ♋ 17 34	23 ♋ 03 51	15 ♎ 36	24 ♌ 29	29 ♍ 42	00 ♑ 28	20 ♊ 36	17 ♎ 59	03 ♐ 05	02 ♎ 07

Date	Moon True ☊ ° '	Moon Mean ☊ ° '	Moon ☽ Latitude ° '
01	25 ♑ 12	23 ♑ 41	02 N 51
02	25 D 14	23 38	01 44
03	25 15	23 35	00 N 32
04	25 R 15	23 31	00 S 41
05	25 13	23 28	01 50
06	25 08	23 25	02 53
07	25 02	23 22	03 45
08	24 54	23 19	04 25
09	24 44	23 16	04 52
10	24 34	23 12	05 04
11	24 25	23 09	05 03
12	24 18	23 06	04 48
13	24 12	23 03	04 21
14	24 09	23 00	03 43
15	24 08	22 56	02 55
16	24 D 08	22 53	01 58
17	24 09	22 50	00 S 56
18	24 R 09	22 47	00 N 11
19	24 08	22 44	01 19
20	24 05	22 41	02 25
21	23 59	22 37	03 24
22	23 51	22 34	04 13
23	23 41	22 31	04 47
24	23 31	22 28	05 02
25	23 21	22 25	04 56
26	23 13	22 21	04 31
27	23 08	22 18	03 49
28	23 05	22 15	02 52
29	23 03	22 12	01 47
30	23 ♑ 04	22 ♑ 09	00 N 36

DECLINATIONS

Date	Sun ☉ ° '	Moon ☽ ° '	Mercury ☿ ° '	Venus ♀ ° '	Mars ♂ ° '	Jupiter ♃ ° '	Saturn ♄ ° '	Uranus ♅ ° '	Neptune ♆ ° '	Pluto ♇ ° '
01	08 N 09	26 N 02	14 N 44	18 N 57	08 N 23	23 S 25	21 N 28	05 S 50	19 S 05	14 N 07
02	07 47	25 03	14 20	18 52	08 08	23 25	21 28	05 52	19 05	14 06
03	07 25	22 34	13 54	18 46	07 53	23 25	21 29	05 53	19 05	14 05
04	07 03	18 51	13 24	18 40	07 39	23 25	21 29	05 54	19 06	14 04
05	06 41	14 14	12 53	18 33	07 24	23 25	21 29	05 56	19 06	14 03
06	06 18	09 01	12 18	18 26	07 09	23 25	21 29	05 57	19 06	14 02
07	05 56	03 N 30	11 42	18 19	06 54	23 25	21 29	05 58	19 06	14 01
08	05 33	02 S 04	11 04	18 11	06 39	23 26	21 29	05 59	19 06	14 00
09	05 11	07 26	10 24	18 02	06 23	23 26	21 29	06 01	19 07	13 59
10	04 48	12 25	09 43	17 53	06 08	23 26	21 29	06 02	19 07	13 58
11	04 25	16 52	09 00	17 44	05 53	23 26	21 29	06 03	19 07	13 57
12	04 02	20 34	08 16	17 34	05 38	23 26	21 30	06 05	19 07	13 56
13	03 39	23 25	07 32	17 23	05 23	23 26	21 30	06 06	19 08	13 55
14	03 16	25 14	06 46	17 12	05 07	23 26	21 30	06 07	19 08	13 55
15	02 53	25 55	06 00	17 01	04 52	23 27	21 30	06 09	19 08	13 54
16	02 30	25 24	05 13	16 49	04 37	23 27	21 30	06 10	19 08	13 53
17	02 07	23 40	04 26	16 37	04 21	23 27	21 30	06 12	19 09	13 52
18	01 43	20 44	03 39	16 24	04 06	23 27	21 30	06 13	19 09	13 51
19	01 20	16 42	02 52	16 11	03 50	23 27	21 30	06 14	19 09	13 50
20	00 57	11 44	02 04	15 57	03 35	23 27	21 30	06 16	19 09	13 49
21	00 34	06 S 02	01 17	15 43	03 19	23 27	21 30	06 17	19 10	13 48
22	00 N 10	00 N 07	00 N 29	15 28	03 04	23 28	21 30	06 19	19 10	13 47
23	00 S 13	06 23	00 S 18	15 13	02 48	23 28	21 30	06 20	19 10	13 46
24	00 36	12 24	01 05	14 57	02 33	23 28	21 30	06 21	19 11	13 46
25	01 00	17 44	01 52	14 41	02 17	23 28	21 30	06 23	19 11	13 45
26	01 23	21 57	02 38	14 25	02 02	23 28	21 30	06 24	19 11	13 44
27	01 47	24 43	03 25	14 08	01 46	23 28	21 29	06 26	19 11	13 43
28	02 10	25 48	04 10	13 50	01 30	23 28	21 29	06 27	19 12	13 42
29	02 33	25 12	04 56	13 32	01 15	23 28	21 29	06 28	19 12	13 41
30	02 S 57	23 N 03	05 S 41	13 N 14	00 N 59	23 S 28	21 N 29	06 S 30	19 S 12	13 N 40

ZODIAC SIGN ENTRIES

Date	h	m	Planets
02	02	11	☽ ♋
04	06	54	☽ ♌
05	11	36	☿ ♍
06	13	15	☽ ♍
07	23	27	♀ ♌
08	21	36	☽ ♎
11	08	15	☽ ♏
13	20	42	☽ ♐
16	09	07	☽ ♑
18	19	05	☽ ♒
21	01	09	☽ ♓
21	12	12	☿ ♎
22	22	33	☉ ♎
23	03	44	☽ ♈
25	04	27	☽ ♉
25	18	20	♃ ♑
27	05	14	☽ ♊
29	07	39	☽ ♋
30	23	23	♂ ♎

LATITUDES

Date	Mercury ☿ ° '	Venus ♀ ° '	Mars ♂ ° '	Jupiter ♃ ° '	Saturn ♄ ° '	Uranus ♅ ° '	Neptune ♆ ° '	Pluto ♇ ° '
01	01 N 00	02 S 30	01 N 04	00 N 02	01 S 35	00 N 37	01 N 38	15 N 52
04	01 24	02 15	01 03	00 01	01 35	00 37	01 37	15 51
07	01 40	01 59	01 02	00 N 01	01 35	00 37	01 37	15 51
10	01 47	01 44	01 02	00 00	01 36	00 37	01 37	15 51
13	01 48	01 29	01 01	00 00	01 36	00 36	01 37	15 51
16	01 42	01 14	01 00	00 00	01 36	00 36	01 37	15 50
19	01 32	00 59	00 59	00 S 01	01 36	00 36	01 37	15 50
22	01 18	00 45	00 59	00 01	01 37	00 36	01 36	15 50
25	01 02	00 31	00 58	00 01	01 37	00 36	01 36	15 51
28	00 43	00 17	00 57	00 02	01 37	00 36	01 36	15 51
31	00 N 24	00 S 04	00 N 56	00 S 02	01 S 38	00 N 36	01 N 36	15 N 51

DATA

Julian Date	2441562
Delta T	+42 seconds
Ayanamsa	23° 28' 47"
Synetic vernal point	05° ♓ 38' 12"
True obliquity of ecliptic	23° 26' 38"

MOON'S PHASES, APSIDES AND POSITIONS ☽

Date	h	m	Phase	Longitude ° '	Eclipse Indicator
07	17	28	●	15 ♍ 10	
15	19	13	☽	23 ♐ 02	
23	04	07	○	00 ♈ 14	
29	19	16	☾	06 ♋ 44	

Day	h	m		
13	09	58	Apogee	
25	07	10	Perigee	
01	09	39	Max dec	26° N 02'
08	03	03	0S	
15	13	54	Max dec	25° S 56'
22	11	34	0N	
28	15	12	Max dec	25° N 49'

LONGITUDES

Date	Chiron ⚷ ° '	Ceres ⚳ ° '	Pallas ⚴ ° '	Juno ⚵ ° '	Vesta ⚶ ° '	Black Moon Lilith ⚸ ° '
01	16 ♈ 36	00 ♎ 31	09 ♍ 47	22 ♎ 43	11 ♊ 22	21 ♏ 19
11	16 ♈ 14	05 ♎ 00	14 ♍ 44	25 ♎ 49	13 ♊ 30	22 ♏ 26
21	15 ♈ 50	09 ♎ 29	19 ♍ 37	29 ♎ 01	15 ♊ 09	23 ♏ 33
31	15 ♈ 23	14 ♎ 00	24 ♍ 25	02 ♏ 18	16 ♊ 14	24 ♏ 40

ASPECTARIAN

h m	Aspects
01 Friday	
02 33	☽ △ ♅
03 56	☽ ⊥ ♀
08 28	☽ ☌ ♄
14 42	☽ ⚹ ☿
15 01	☽ ⊻ ♀
21 30	☿ ⊻ ♀
21 55	☽ Q ☉
23 44	☽ ☍ ♃
02 Saturday	
01 15	☽ Q ♂
04 01	☽ □ ♇
06 40	☽ ⊼ ♆
16 00	☿ ⊥ ♇
17 07	☽ ± ♆
18 20	☉ ⊥ ♅
20 14	☽ ∠ ☿
20 21	☽ ⚹ ☉
23 09	☽ ⚹ ♂
03 Sunday	
01 48	☿ ∥ ♇
05 13	☽ ♯ ♃
06 47	☽ □ ♅
08 55	☽ ⊡ ♆
11 34	☽ Q ♇
12 51	☽ ⊻ ♄
14 19	☽ ⊥ ☿
19 51	☽ ∥ ♄
22 24	♀ ⊥ ♄
23 31	☽ ⊥ ♄
23 36	☽ ☌ ♀
04 Monday	
00 49	☽ ∠ ☉
02 31	☽ ⊻ ☿
02 56	☽ ∠ ♂
04 28	☽ ⊼ ♃
08 55	☽ ⚹ ♇
10 39	☽ ♯ ♆
11 33	☽ △ ♆
13 08	☽ ∥ ♀
14 51	☽ Q ♅
15 13	☽ ± ♃
15 38	☽ ∠ ♄
17 39	☿ △ ♃
18 08	☽ ⊥ ☉
19 50	☽ ⊥ ♂
05 Tuesday	
05 47	☽ ⊻ ☉
07 11	☽ ⊻ ♂
07 27	☽ ⊡ ♃
11 59	☽ ∠ ♇
12 35	☽ ⚹ ♅
12 52	☽ ∥ ♇
18 50	☽ ⚹ ♄
19 13	☽ ∥ ☿
06 Wednesday	
03 26	☿ ⊻ ♇
04 26	☽ ⊥ ♇
08 00	☿ ∠ ♅
10 19	☽ ⊻ ♀
10 53	☽ △ ♃
13 37	☿ Q ♄
15 30	☽ ⊻ ♇
16 09	☽ ∠ ♅
16 55	☽ Q ♄
17 27	☽ ☌ ☿
17 56	♀ ⊼ ♃
18 07	☽ □ ♆
20 36	☽ ∥ ♂
22 10	☿ □ ♆
22 21	☽ ⊥ ♀
07 Thursday	
00 44	☽ ∥ ☉
01 22	☽ ♯ ♅
08 56	☽ ⊥ ♅
09 37	☉ ♯ ♅
10 57	☉ ☌ ♂
16 36	☽ ∠ ♀
17 18	☽ ☌ ♂
17 28	☽ ☌ ☉
20 14	☽ ⊻ ♅
08 Friday	
02 40	☽ □ ♄
03 49	☽ Q ♆
13 51	♀ ∠ ♂
19 20	☽ □ ♃
23 37	☽ ⚹ ♀
09 Saturday	
00 07	☽ ☌ ♇
02 26	☽ ♯ ☉
02 44	☽ ⚹ ♆
05 30	☽ ∥ ♅
05 47	♀ ⚹ ♇
07 28	☽ ♯ ♂
07 40	☉ ⊻ ♅
12 01	☽ ⊻ ☿
15 01	☿ ⊥ ♀
10 Sunday	
00 19	☽ ♯ ☿
00 59	☽ Q ♀
01 49	☽ ⊥ ☿
05 57	☽ ⊻ ♂
06 07	☽ ☌ ♅
06 20	☽ Q ♃
07 54	☽ ∠ ♆
07 56	☽ ⊻ ☉
09 12	♂ ⊻ ♅
12 45	☽ △ ♄
13 18	♀ △ ♆
18 25	☽ ⊥ ♂
19 58	☽ ♯ ♇
20 47	☽ ⊥ ☉
21 03	♂ ♯ ♅
22 48	☽ ∠ ☿
11 Monday	
01 41	☽ ⊥ ♆
06 08	☽ ⚹ ♃
06 14	☿ ⊥ ♅
11 02	☽ ⊻ ♇
13 18	☽ ∠ ♂
13 39	☽ ⊻ ♆
15 57	☽ □ ♀
16 15	☽ ∠ ☉
16 59	☽ ♯ ♀
18 38	☽ ⊡ ♄
23 05	☽ ⊥ ♇
12 Tuesday	
01 59	☽ ∥ ♆
10 27	☽ ⚹ ☿
12 19	☽ ∠ ♃
12 47	☽ ± ♄
15 48	♀ Q ♅
17 15	☽ ∠ ♇
18 12	☽ ⊻ ♅
18 14	♀ ± ♃
18 58	☽ ♯ ♄
21 09	☽ ⚹ ♂
22 38	☉ □ ♄
23 23	♀ ∠ ♄
13 Wednesday	
00 56	☽ ⊼ ♄
01 08	☽ ⚹ ☉
06 24	☽ ⊥ ♅
06 36	☽ ⊥ ♃
07 17	☉ ∠ ♀
10 44	☉ Q ♆
12 17	☽ ∥ ♃
15 21	☽ Q ☿
18 50	☽ ⊻ ♃
22 51	☽ Q ♂
23 44	☽ ⚹ ♇
14 Thursday	
00 44	☽ ∠ ♅
02 16	☽ ☌ ♆
03 38	☽ Q ☉
10 29	☽ △ ♀
12 13	☿ ⊻ ♅
15 Friday	
00 06	☽ Q ♇
07 14	☽ ⚹ ♅
07 32	☿ ♯ ♅
10 42	☽ □ ☿
13 19	☽ □ ♂
13 48	☽ ☍ ♄
19 13	☽ □ ☉
19 44	☽ ⊡ ♀
22 47	♂ □ ♄
16 Saturday	
06 37	☿ □ ♄
07 23	☽ Q ♅
07 37	☽ ☌ ♃
10 29	☿ ☌ ♂
12 10	☿ Q ♆
12 16	☽ □ ♇
14 39	☽ ⊻ ♆
15 24	☽ ± ♀
15 30	♂ Q ♆
17 Sunday	
02 28	☽ ⊥ ♆
04 20	☽ ⊼ ♀
14 08	☽ ∥ ♃
15 49	☿ ∥ ♂
18 48	☽ □ ♅
19 57	☽ ∠ ♆
18 Monday	
00 55	☽ ⊼ ♄
03 30	☽ △ ♂
06 32	☽ ♯ ♄
08 10	☽ △ ☿
10 50	☽ △ ☉
12 20	☽ ± ♄
17 58	☽ ⊻ ♃
22 08	☽ ∥ ♆
22 12	☽ △ ♇
19 Tuesday	
00 21	☽ ⚹ ♆
05 10	☽ ⊡ ♄
05 11	☽ ⊥ ♃
09 03	☽ ⊡ ♂
14 51	☽ ♯ ♀
16 38	☽ ⊡ ☿
16 53	☽ ⊡ ☉
18 11	☽ ☍ ♀
20 15	☉ ☌ ☿
21 42	☽ ∠ ♃
22 20	☽ Q ♆
20 Wednesday	
01 42	☽ ⊡ ♇
02 12	☽ ± ♂
02 22	☽ ♯ ♇
02 50	☽ △ ♅
08 23	☽ △ ♄
10 19	☽ ± ☉
11 20	☽ ± ☿
13 25	☽ ⊼ ♂
17 43	☽ ± ♇
18 06	☿ ∠ ♀
21 41	☽ ⊼ ☉
23 29	☽ ⊼ ☿
21 Thursday	
00 26	☽ ⚹ ♃
04 12	☽ ⊼ ♇
05 20	☽ ⊡ ♅
06 08	☽ □ ♆
06 59	☿ □ ♃
11 00	☽ ∥ ♅
16 06	♀ ⊡ ♃
20 48	☽ ± ♅
21 09	☽ Q ♃
23 13	☽ ♯ ♂
22 Friday	
03 02	☽ ⊼ ♀
06 58	☽ ⊼ ♅
09 23	☽ ♯ ☿
10 51	☽ ♯ ☉
12 01	☽ □ ♄
12 13	☽ ∥ ☉
12 24	☿ ☌ ♇
13 16	☽ ∥ ☿
13 55	☽ ± ♀
15 43	☉ □ ♃
19 03	☽ ☍ ♂
22 50	☽ ∥ ♂
23 Saturday	
01 24	☉ ♯ ☿
03 11	☿ ⚹ ♆
03 21	☽ □ ♃
04 07	☽ ☍ ☉
06 00	☽ ⊡ ♀
06 46	☽ ☍ ♇
06 58	☉ ∥ ☿
08 32	☽ △ ♆
09 15	☽ ☍ ☿
11 48	☽ ♯ ♅
16 24	♀ ∠ ♇
17 41	☽ Q ♄
24 Sunday	
08 25	☽ △ ♀
08 26	☽ ☍ ♅
08 38	♀ ⚹ ♅
08 58	☽ ⊡ ♆
13 10	☽ ⚹ ♄
17 46	☽ ∥ ♇
21 12	☉ ☌ ♇
22 10	☽ ⊼ ♂
22 29	☽ ∥ ♀
23 32	☽ ± ♆
25 Monday	
04 22	☽ △ ♃
07 33	☽ ⊼ ♇
08 16	☽ ⊼ ☉
08 17	☽ ± ♂
09 14	☽ ⊼ ♆
13 25	☽ ∠ ♄
16 25	☽ ⊼ ☿
17 17	☽ ± ♇
18 39	☽ ± ☉
19 29	☽ ♯ ♆
21 46	☿ ♯ ♂
23 26	☉ ⚹ ♆
23 30	☽ ⊡ ♂
26 Tuesday	
03 26	☽ ± ☿
04 01	☽ ⊥ ♄
04 48	☽ ⊡ ♃
07 54	☽ ⊡ ♇
08 58	☽ ∥ ♄
09 10	☽ ⊼ ♅
10 17	☽ ⊡ ☉
12 54	☽ □ ♀
13 47	☽ ⊻ ♄
19 00	☽ ± ♅
19 33	☽ ± ♃
20 02	☽ ⊡ ☿
23 26	☽ ♯ ♃
27 Wednesday	
00 30	♀ ⚹ ♄
01 04	☽ △ ♂
05 28	☽ ⊼ ♃
08 32	☽ △ ♇
09 51	☽ ⊡ ♅
10 12	☽ ☍ ♆
11 41	☉ ♯ ♂
12 40	☽ △ ☉
21 01	☽ Q ♀
28 Thursday	
00 08	☽ △ ☿
10 59	☽ △ ♅
15 35	☽ ☌ ♄
18 55	☽ ⚹ ♀
20 58	♀ ⊥ ♂
23 29	♀ ∥ ♇
29 Friday	
05 43	☽ □ ♂
08 15	☽ ☍ ♃
11 13	☽ □ ♇
12 54	☽ ⊼ ♆
19 16	☽ □ ☉
22 59	☽ ∠ ♀
23 20	☽ ± ♆
30 Saturday	
08 14	☽ ♯ ♃
10 37	☽ □ ☿
14 36	☽ Q ♂
15 00	☽ □ ♅
15 10	☽ ⊡ ♆
16 13	☽ ⊥ ♀
18 47	☽ Q ♇
19 37	☽ ⊻ ♄
23 54	☽ ∥ ♄

All ephemeris data is given at 12.00 UT and the Moon's longitude is additionally given for 24.00 UT

OCTOBER 1972

Date	Sidereal time h m s	Sun ☉ ° ' "	Moon ☽ ° ' "	Moon ☽ 24.00 ° ' "	Mercury ☿ ° '	Venus ♀ ° '	Mars ♂ ° '	Jupiter ♃ ° '	Saturn ♄ ° '	Uranus ♅ ° '	Neptune ♆ ° '	Pluto ♇ ° '
01	12 41 15	08 ♎ 23 56	29 ♋ 46 01	06 ♌ 24 20	17 ♎ 16	25 ♌ 37	00 ♎ 20	00 ♑ 35	20 ♊ 36	18 ♎ 03	03 ♐ 06	02 ♎ 09
02	12 45 12	09 22 59	12 ♌ 59 03	19 ♌ 30 22	18 55	26 45	00 59	00 41	20 36	18 07	03 08	02 12
03	12 49 08	10 22 04	25 ♌ 58 33	02 ♍ 23 45	20 33	27 53	01 38	00 48	20 R 36	18 10	03 09	02 14
04	12 53 05	11 21 11	08 ♍ 46 10	15 ♍ 05 54	22 10	29 ♌ 01	02 17	00 55	20 36	18 14	03 11	02 16
05	12 57 01	12 20 21	21 ♍ 23 03	27 ♍ 37 44	23 46	00 ♍ 10	02 55	01 02	20 35	18 18	03 12	02 18
06	13 00 58	13 19 33	03 ♎ 49 59	09 ♎ 59 52	25 21	01 19	03 34	01 09	20 35	18 22	03 14	02 21
07	13 04 54	14 18 46	16 ♎ 07 27	22 ♎ 12 47	26 56	02 28	04 13	01 17	20 35	18 25	03 16	02 23
08	13 08 51	15 18 02	28 ♎ 15 59	04 ♏ 17 10	28 ♎ 30	03 37	04 52	01 24	20 34	18 29	03 17	02 25
09	13 12 47	16 17 20	10 ♏ 16 29	16 ♏ 14 09	00 ♏ 03	04 46	05 31	01 32	20 33	18 33	03 19	02 27
10	13 16 44	17 16 40	22 ♏ 10 24	28 ♏ 05 33	01 36	05 55	06 10	01 40	20 32	18 37	03 20	02 30
11	13 20 41	18 16 02	03 ♐ 59 57	09 ♐ 54 01	03 07	07 05	06 48	01 47	20 32	18 41	03 22	02 32
12	13 24 37	19 15 26	15 ♐ 48 13	21 ♐ 43 04	04 38	08 14	07 27	01 56	20 31	18 44	03 24	02 34
13	13 28 34	20 14 51	27 ♐ 39 07	03 ♑ 36 59	06 09	09 24	08 06	02 04	20 29	18 48	03 26	02 36
14	13 32 30	21 14 18	09 ♑ 37 18	15 ♑ 40 43	07 38	10 34	08 45	02 12	20 28	18 52	03 27	02 38
15	13 36 27	22 13 47	21 ♑ 47 57	27 ♑ 59 38	09 07	11 44	09 24	02 20	20 27	18 56	03 29	02 41
16	13 40 23	23 13 18	04 ♒ 16 28	10 ♒ 39 04	10 35	12 55	10 03	02 29	20 25	19 00	03 31	02 43
17	13 44 20	24 12 51	17 ♒ 08 02	23 ♒ 43 50	12 03	14 05	10 42	02 38	20 24	19 03	03 33	02 45
18	13 48 16	25 12 25	00 ♓ 26 51	07 ♓ 17 22	13 29	15 15	11 21	02 47	20 22	19 07	03 35	02 47
19	13 52 13	26 12 01	14 ♓ 15 27	21 ♓ 20 58	14 55	16 26	12 00	02 56	20 20	19 11	03 37	02 49
20	13 56 10	27 11 39	28 ♓ 33 36	05 ♈ 52 47	16 20	17 37	12 39	03 05	20 18	19 15	03 38	02 51
21	14 00 06	28 11 18	13 ♈ 17 45	20 ♈ 47 28	17 44	18 48	13 18	03 14	20 16	19 18	03 40	02 53
22	14 04 03	29 ♎ 11 00	28 ♈ 20 48	05 ♉ 56 25	19 08	19 59	13 58	03 23	20 14	19 22	03 42	02 55
23	14 07 59	00 ♏ 10 43	13 ♉ 32 57	21 ♉ 09 01	20 30	21 10	14 37	03 33	20 12	19 26	03 44	02 58
24	14 11 56	01 10 29	28 ♉ 43 17	06 ♊ 14 35	21 52	22 21	15 16	03 42	20 10	19 30	03 46	03 00
25	14 15 52	02 10 17	13 ♊ 41 50	21 ♊ 04 14	23 12	23 32	15 55	03 52	20 07	19 33	03 48	03 02
26	14 19 49	03 10 07	28 ♊ 21 07	05 ♋ 32 05	24 32	24 44	16 34	04 02	20 05	19 37	03 50	03 04
27	14 23 45	04 10 00	12 ♋ 36 54	19 ♋ 35 29	25 50	25 56	17 13	04 11	20 02	19 41	03 52	03 06
28	14 27 42	05 09 54	26 ♋ 27 58	03 ♌ 14 32	27 07	27 07	17 53	04 21	20 00	19 44	03 54	03 08
29	14 31 39	06 09 51	09 ♌ 55 30	16 ♌ 31 12	28 23	28 19	18 32	04 32	19 57	19 48	03 56	03 10
30	14 35 35	07 09 50	23 ♌ 02 03	29 ♌ 28 28	29 ♏ 37	29 ♍ 31	19 11	04 42	19 54	19 52	03 58	03 12
31	14 39 32	08 ♏ 09 51	05 ♍ 50 52	12 ♍ 09 37	00 ♐ 50	00 ♎ 43	19 ♎ 51	04 ♑ 52	19 ♊ 51	19 ♎ 55	04 ♐ 00	03 ♎ 14

Date	Moon True ☊ ° '	Moon Mean ☊ ° '	Moon ☽ Latitude ° '
01	23 ♑ 03	22 ♑ 06	00 S 35
02	23 R 02	22 02	01 43
03	22 58	21 59	02 45
04	22 52	21 56	03 36
05	22 42	21 53	04 17
06	22 30	21 50	04 44
07	22 16	21 47	04 58
08	22 02	21 43	04 58
09	21 48	21 40	04 45
10	21 37	21 37	04 19
11	21 28	21 34	03 42
12	21 21	21 31	02 56
13	21 18	21 27	02 01
14	21 17	21 24	01 S 01
15	21 D 16	21 21	00 N 03
16	21 R 16	21 18	01 08
17	21 15	21 15	02 12
18	21 11	21 12	03 11
19	21 05	21 08	04 01
20	20 57	21 05	04 38
21	20 46	21 02	04 58
22	20 35	20 59	04 58
23	20 24	20 56	04 37
24	20 14	20 53	03 56
25	20 08	20 49	03 00
26	20 04	20 46	01 52
27	20 02	20 43	00 N 39
28	20 D 02	20 40	00 S 34
29	20 R 02	20 37	01 43
30	20 01	20 33	02 45
31	19 ♑ 58	20 ♑ 30	03 S 38

DECLINATIONS

Date	Sun ☉ ° '	Moon ☽ ° '	Mercury ☿ ° '	Venus ♀ ° '	Mars ♂ ° '	Jupiter ♃ ° '	Saturn ♄ ° '	Uranus ♅ ° '	Neptune ♆ ° '	Pluto ♇ ° '
01	03 S 20	19 N 38	06 S 25	12 N 55	00 N 43	23 S 28	21 N 29	06 S 31	19 S 13	13 N 40
02	03 43	15 16	07 09	12 36	00 28	23 29	21 29	06 33	19 13	13 39
03	04 06	10 17	07 52	12 17	00 N 12	23 29	21 29	06 34	19 13	13 38
04	04 30	04 N 56	08 35	11 57	00 S 04	23 29	21 29	06 36	19 14	13 37
05	04 53	00 S 31	09 17	11 37	00 19	23 29	21 29	06 37	19 14	13 37
06	05 16	05 52	09 59	11 16	00 35	23 29	21 29	06 39	19 14	13 36
07	05 39	10 55	10 40	10 55	00 51	23 29	21 29	06 40	19 15	13 35
08	06 02	15 30	11 20	10 34	01 06	23 29	21 28	06 41	19 15	13 34
09	06 24	19 24	12 00	10 12	01 22	23 29	21 28	06 43	19 15	13 34
10	06 47	22 29	12 39	09 50	01 38	23 29	21 28	06 44	19 16	13 33
11	07 10	24 35	13 17	09 28	01 53	23 29	21 28	06 46	19 16	13 32
12	07 32	25 36	13 54	09 05	02 09	23 29	21 28	06 47	19 17	13 31
13	07 55	25 27	14 31	08 42	02 25	23 29	21 28	06 49	19 17	13 31
14	08 17	24 07	15 07	08 19	02 40	23 29	21 27	06 50	19 17	13 30
15	08 39	21 38	15 42	07 55	02 56	23 29	21 27	06 52	19 18	13 29
16	09 02	18 05	16 16	07 31	03 12	23 29	21 27	06 53	19 18	13 29
17	09 23	13 36	16 50	07 07	03 27	23 29	21 27	06 54	19 18	13 28
18	09 45	08 20	17 22	06 43	03 43	23 29	21 27	06 56	19 19	13 27
19	10 07	02 S 29	17 54	06 18	03 58	23 28	21 27	06 57	19 19	13 27
20	10 29	03 N 41	18 25	05 53	04 14	23 28	21 26	06 59	19 20	13 26
21	10 50	09 49	18 55	05 28	04 29	23 28	21 26	07 00	19 20	13 26
22	11 11	15 31	19 23	05 03	04 45	23 28	21 26	07 02	19 20	13 25
23	11 32	20 18	19 51	04 37	05 00	23 28	21 26	07 03	19 21	13 24
24	11 53	23 43	20 18	04 11	05 16	23 28	21 25	07 04	19 21	13 24
25	12 14	25 25	20 44	03 46	05 31	23 28	21 25	07 06	19 22	13 23
26	12 34	25 18	21 09	03 19	05 46	23 27	21 25	07 07	19 22	13 23
27	12 55	23 30	21 32	02 53	06 02	23 27	21 25	07 09	19 22	13 22
28	13 15	20 18	21 55	02 27	06 17	23 27	21 24	07 10	19 23	13 22
29	13 35	16 06	22 16	02 00	06 32	23 27	21 24	07 11	19 23	13 21
30	13 54	11 14	22 36	01 33	06 48	23 26	21 24	07 13	19 24	13 21
31	14 S 14	06 N 00	22 S 55	01 N 06	07 S 03	23 S 26	21 N 24	07 S 14	19 S 24	13 N 20

ZODIAC SIGN ENTRIES

Date	h	m	Planets
01	12	25	☽ ♌
03	19	31	☽ ♍
05	08	33	♀ ♍
06	04	35	☽ ♎
08	15	27	☽ ♏
09	11	11	☿ ♏
11	03	52	☽ ♐
13	16	44	☽ ♑
16	03	51	☽ ♒
18	11	12	☽ ♓
20	14	22	☽ ♈
22	14	37	☽ ♉
23	07	41	☉ ♏
24	14	02	☽ ♊
26	14	44	☽ ♋
28	18	14	☽ ♌
30	19	27	☿ ♐
30	21	40	♀ ♎
31	00	59	☽ ♍

LATITUDES

Date	Mercury ☿ ° '	Venus ♀ ° '	Mars ♂ ° '	Jupiter ♃ ° '	Saturn ♄ ° '	Uranus ♅ ° '	Neptune ♆ ° '	Pluto ♇ ° '
01	00 N 24	00 S 04	00 N 56	00 S 02	01 S 38	00 N 36	01 N 36	15 N 51
04	00 N 03	00 N 09	00 55	00 02	01 38	00 36	01 36	15 51
07	00 S 18	00 21	00 54	00 03	01 38	00 36	01 36	15 52
10	00 39	00 32	00 54	00 03	01 39	00 36	01 36	15 52
13	01 00	00 43	00 53	00 03	01 39	00 36	01 35	15 53
16	01 20	00 53	00 52	00 03	01 39	00 36	01 35	15 53
19	01 39	01 02	00 51	00 04	01 39	00 36	01 35	15 54
22	01 57	01 10	00 50	00 04	01 39	00 36	01 35	15 55
25	02 14	01 18	00 48	00 04	01 40	00 36	01 35	15 56
28	02 27	01 25	00 47	00 05	01 40	00 36	01 35	15 56
31	02 S 38	01 N 31	00 N 46	00 S 05	01 S 40	00 N 36	01 N 35	15 N 57

DATA

Julian Date	2441592
Delta T	+42 seconds
Ayanamsa	23° 28' 51"
Synetic vernal point	05° ♓ 38' 09"
True obliquity of ecliptic	23° 26' 38"

MOON'S PHASES, APSIDES AND POSITIONS ☽

Date	h	m	Phase	Longitude ° '	Eclipse Indicator
07	08	08	●	14 ♎ 09	
15	12	55	☽	22 ♑ 16	
22	13	25	○	29 ♈ 15	
29	04	41	☾	05 ♌ 52	

Day	h	m		
11	03	04	Apogee	
23	12	30	Perigee	
05	09	43	0S	
12	20	56	Max dec	25° S 41'
19	21	45	0N	
25	22	24	Max dec	25° N 35'

LONGITUDES

Date	Chiron ⚷ ° '	Ceres ⚳ ° '	Pallas ⚴ ° '	Juno ⚵ ° '	Vesta ⚶ ° '	Black Moon Lilith ⚸ ° '
01	15 ♈ 23	14 ♎ 00	24 ♍ 25	02 ♏ 18	16 ♊ 14	24 ♏ 40
11	14 ♈ 56	18 ♎ 30	29 ♍ 09	05 ♏ 38	16 ♊ 40	25 ♏ 47
21	14 ♈ 29	23 ♎ 00	03 ♎ 47	09 ♏ 05	16 ♊ 22	26 ♏ 54
31	14 ♈ 04	27 ♎ 28	08 ♎ 19	12 ♏ 26	15 ♊ 22	28 ♏ 01

ASPECTARIAN

h m	Aspects
01 Sunday	
03 51	☽ ⊻ ♀
05 29	☽ Q ☉
06 19	☽ ⊥ ♄
13 05	☽ ⚹ ♂
13 29	☽ ⊼ ♃
14 29	☽ ♯ ♆
15 33	☿ ∥ ♅
16 19	☽ ⚹ ♇
18 02	☽ △ ♆
22 32	☽ ∠ ♄
22 49	♂ □ ♃
23 21	☽ Q ☿
23 24	☽ Q ♅
23 50	☿ ☌ ♅
23 56	♀ ⊥ ♇
02 Monday	
00 24	☿ ∠ ♆
00 25	☽ ± ♃
04 53	☽ ⚹ ☉
08 33	☿ Q ♃
16 25	♄ St R
17 01	☽ ⚼ ♃
17 48	☽ ∠ ♂
19 45	☽ ∠ ♇
20 05	☽ ∥ ♇
21 28	☽ ⚹ ♅
23 40	♅ ∠ ♆
03 Tuesday	
00 28	☽ ⚹ ☿
01 59	☽ ∥ ♀
02 01	☽ ⚹ ♄
10 46	☽ ∠ ☉
11 19	☽ ⊥ ♂
12 28	☽ ⊥ ♇
12 49	☿ △ ♄
15 54	☽ ☌ ♀
21 06	☽ △ ♃
21 39	☽ ♯ ☿
23 08	☽ ⊻ ♂
23 43	☽ ⊻ ♇
04 Wednesday	
00 23	☽ Q ♄
01 27	☽ □ ♆
01 31	☽ ∠ ♅
04 41	☽ ♯ ♅
05 01	☽ ⊥ ☉
08 31	☽ ∠ ☿
11 39	♂ ☌ ♇
13 50	☽ ♯ ☉
17 18	☽ ⊻ ☉
18 36	☽ ⊥ ♅
05 Thursday	
04 04	☽ ⊥ ☿
06 04	☽ ⊻ ♅
08 28	☽ ♯ ♂
10 29	☽ □ ♄
11 05	☽ ∥ ♂
11 39	☽ Q ♆
17 14	☽ ⊻ ☿
22 56	♂ ⚹ ♆
06 Friday	
06 37	☽ ⊻ ♀
06 46	☽ □ ♃
08 23	♀ △ ♃
09 00	☽ ∥ ☉
09 06	☽ ☌ ♇
10 50	☽ ⚹ ♆
11 28	☽ ☌ ♂
15 34	☽ ∥ ♅
19 27	☽ ⊥ ♀
07 Saturday	
08 08	☽ ☌ ☉
10 18	♀ ⊻ ♇
10 30	☽ ∥ ☿
11 59	☽ ♯ ♀
14 27	♀ Q ♄
14 54	☽ ∠ ♀
16 12	☽ ∠ ♆
16 33	☽ ☌ ♅
17 03	☿ ⊥ ♆
18 01	☿ ♯ ♀
18 16	☽ Q ♃
20 46	☽ △ ♄
08 Sunday	
01 34	☽ ♯ ♇
05 05	♀ □ ♆
09 18	♀ ∠ ♅
10 03	☽ ⊥ ♆
12 32	☽ ☌ ☿
18 19	☽ ⚹ ♃
20 18	☽ ⊻ ♇
22 01	☽ ⊻ ♆
23 47	☽ ⚹ ♀
09 Monday	
01 54	☽ ⊻ ♂
02 33	☽ ⚼ ♄
08 20	☽ ⊥ ♇
11 02	☽ ∥ ♆
14 38	☽ ⊥ ♂
20 36	☽ ± ♄
10 Tuesday	
00 44	☽ ∠ ♃
01 12	☽ ⊻ ☉
02 28	☽ Q ♀
02 30	☽ ∠ ♇
03 18	☽ ♯ ♄
04 46	☽ ⊻ ♅
08 42	☽ ⊼ ♄
08 46	☉ ∥ ♅
09 50	☽ ∠ ♂
13 07	☿ ⚹ ♃
14 26	☽ ⊥ ☉
16 58	☽ ⊥ ♅
19 09	☽ ⊥ ♃
21 58	☽ ∥ ♃
23 14	♀ ⊻ ♂
11 Wednesday	
02 25	☿ ⊻ ♇
07 28	☽ ⊻ ♃
09 00	☽ ⚹ ♇
09 57	☽ ⊻ ☿
10 23	☽ ∠ ☉
10 43	☽ ☌ ♆
11 20	☽ ∠ ♅
14 34	☉ ∠ ♆
15 58	☿ ⊻ ♆
18 03	☽ ⚹ ♂
18 57	☽ □ ♀
21 34	☿ ♯ ♇
22 36	☉ ☌ ♅
23 58	☽ ⊥ ☿
12 Thursday	
09 29	☽ Q ♇
18 00	☽ ⚹ ♅
19 39	☽ ⚹ ☉
19 51	☽ Q ♂
20 55	☽ ∠ ☿
21 32	☽ ☍ ♄
13 Friday	
01 40	☿ ⚼ ♄
06 46	☉ Q ♃
17 45	☉ △ ♄
18 23	☽ Q ♅
20 59	☽ ☌ ♃
22 00	☽ □ ♇
22 05	☽ Q ☉
23 39	☽ ⊻ ♆
14 Saturday	
07 29	☽ ⚹ ☿
10 10	☽ □ ♂
11 40	☽ ⊥ ♆
12 52	☉ ♯ ♀
14 05	☽ △ ♀
19 17	☽ ∥ ♃
15 Sunday	
04 38	☿ ⊥ ♇
05 30	☽ ∠ ♆
06 22	☽ □ ♅
09 22	☽ ⊼ ♄
10 29	☽ Q ☿
12 55	☽ □ ☉
13 24	☽ ♯ ♄
20 21	☿ ⊻ ♂
21 00	☽ ± ♄
22 35	☽ ⚼ ♀
16 Monday	
04 32	☽ ∥ ♆
08 33	☽ ⊻ ♃
09 01	☽ △ ♇
10 34	☽ ⚹ ♆
13 46	♀ ⊥ ♅
14 10	☽ ⚼ ♄
17 29	☽ ± ♀
20 02	☽ ⊥ ♃
21 08	☽ ∥ ☿
23 28	☽ △ ♂
17 Tuesday	
01 24	☽ □ ☿
05 49	☽ ⊼ ♀
09 05	☽ Q ♆
12 40	☽ ♯ ♇
12 55	☽ ∠ ♃
13 08	☽ ⚼ ♇
15 32	☽ △ ♅
17 57	☽ △ ♄
23 59	♀ ♯ ♅
18 Wednesday	
01 54	☽ △ ☉
04 20	☽ ⚼ ♂
05 28	☽ ± ♇
06 12	☽ ∥ ☉
13 32	♃ □ ♇
16 08	☽ ⊼ ♇
16 09	☽ ⚹ ♃
17 32	☽ □ ♆
17 55	☽ ∥ ♅
18 30	☽ ⚼ ♅
19 22	☽ ♯ ♀
21 03	☽ ± ♂
19 Thursday	
02 28	☿ ± ♄
06 19	☽ ∥ ♂
06 22	☽ ⚼ ☉
07 57	☽ ⊼ ♂
10 09	☽ ± ♅
13 09	☽ Q ♃
13 15	☽ △ ☿
16 03	☽ ☍ ♀
20 23	☽ ⊼ ♅
22 17	☽ □ ♄
22 49	☽ ± ☉
20 Friday	
09 35	☽ ⊼ ☉
14 13	☽ ♯ ♂
17 03	☽ ⚼ ☿
19 05	☽ ☍ ♇
19 30	☽ □ ♃
20 00	☽ ∥ ♀
20 22	☽ △ ♆
23 09	☉ ⊥ ♆
21 Saturday	
00 50	☽ ♯ ♅
03 54	☽ Q ♄
09 14	☽ ± ☿
12 01	☽ ☍ ♂
14 39	☿ ∠ ♇
16 20	☽ ♯ ☉
19 51	☽ ⊼ ☿
20 38	☽ ⚼ ♆
21 30	☿ ∠ ♃
21 34	☽ ⊼ ♀
21 40	☽ ☍ ♅
22 56	♀ ⊻ ♅
23 09	☽ ⚹ ♄
22 Sunday	
02 49	☽ ∥ ♇
07 56	☽ ± ♀
09 23	☿ ∥ ♆
10 59	☽ ± ♆
13 25	☽ ☍ ☉
16 21	☿ ⊻ ♅
17 07	♀ □ ♄
19 15	☽ ⊼ ♇
20 03	☽ △ ♃
20 29	☽ ⊼ ♆
22 35	♀ ♯ ♂
22 52	☽ ∠ ♄
23 22	☽ ⚼ ♀
23 Monday	
04 45	☽ ± ♇
06 41	☽ ♯ ♆
06 51	☿ ⊼ ♄
09 10	☽ ♯ ☿
13 02	☽ ⊥ ♄
13 45	☽ ⊼ ♂
15 35	☿ ⊥ ♂
18 48	☽ ∥ ♄
18 58	☽ ⚼ ♇
19 58	☽ ⚼ ♃
21 19	☽ ⊼ ♅
22 29	☽ ⊻ ♄
23 39	☽ ± ♂
23 55	♀ Q ♆
24 Tuesday	
00 04	☽ ☍ ☿
01 03	☽ △ ♀
06 51	☽ ± ♅
09 45	☽ ♯ ♃
10 22	☽ ± ♃
14 34	☽ ⚼ ♂
16 11	☽ ⊼ ☉
18 49	☽ △ ♇
20 01	☽ ⊼ ♃
20 04	☽ ☍ ♆
21 14	☽ ⚼ ♅
25 Wednesday	
00 50	♃ ⊻ ♆
02 27	☽ ± ☉
15 46	☽ △ ♂
18 03	☽ ⚼ ☉
21 34	☽ △ ♅
22 25	☽ ☌ ♄
26 Thursday	
05 03	☽ ⊼ ☿
05 29	☽ □ ♀
09 21	☉ ⊻ ♇
15 59	☽ ± ☿
19 52	☽ □ ♇
20 38	☽ △ ☉
21 10	☽ ⊼ ♆
21 35	☽ ☍ ♃
27 Friday	
04 19	☿ ♯ ♄
04 37	☉ ⊻ ♆
07 19	☽ ± ♆
08 40	☽ ⚼ ☿
08 55	☿ ⊥ ♅
12 25	☽ ♯ ♃
12 42	☉ ⚹ ♃
14 27	☽ Q ♀
20 18	☽ □ ♂
22 47	☽ ⚼ ♆
28 Saturday	
00 12	☽ □ ♅
00 44	☽ ⊻ ♄
02 09	☽ ♯ ☿
02 39	☽ Q ♇
04 39	☽ ∥ ♄
08 06	☉ ⚼ ♄
11 11	☽ ⊥ ♄
11 55	☿ ⚹ ♀
13 16	☽ ⚹ ♀
13 16	☽ △ ☿
17 44	☽ ♯ ♆
20 02	☉ ♯ ♇
23 50	☽ ⚹ ♇
29 Sunday	
01 13	☽ △ ♆
02 10	☽ ⊼ ♃
03 05	☽ ∠ ♄
04 41	☽ □ ☉
05 34	☽ Q ♂
08 09	☽ Q ♅
13 06	☽ ± ♃
15 11	☿ ⊥ ♃
18 46	☽ ∠ ♀
23 57	☽ ♯ ☉
30 Monday	
01 53	☽ ∥ ♇
03 03	☽ ∠ ♇
03 41	♂ ∠ ♆
04 31	☽ ⚹ ♂
05 45	☽ ⚼ ♃
06 07	☽ ⚹ ♅
06 14	☽ ⚹ ♄
12 59	☽ ⊥ ♀
16 17	☽ Q ☉
19 46	☽ ⊥ ♇
20 26	♄ △ ♅
31 Tuesday	
01 20	☽ ⊻ ♀
01 34	☽ □ ☿
04 29	☽ Q ♄
06 26	☽ ♯ ♅
07 03	☽ ⊻ ♇
07 29	☽ ♯ ♂
08 31	☽ □ ♆
10 00	☽ ∠ ♂
10 07	☽ △ ♃
10 15	☽ ∠ ♅
12 19	♂ △ ♄
15 16	♂ ☌ ♅
16 46	☽ ⚹ ☉

All ephemeris data is given at 12.00 UT and the Moon's longitude is additionally given for 24.00 UT

NOVEMBER 1972

LONGITUDES

Date	Sidereal time h m s	Sun ☉ ° ' "	Moon ☽ ° ' "	Moon ☽ 24.00 ° ' "	Mercury ☿ ° '	Venus ♀ ° '	Mars ♂ ° '	Jupiter ♃ ° '	Saturn ♄ ° '	Uranus ♅ ° '	Neptune ♆ ° '	Pluto ♇ ° '
01	14 43 28	09 ♏ 09 54	18 ♍ 25 07	24 ♍ 37 42	02 ♐ 01	01 ♎ 55	20 ♎ 30	05 ♑ 03	19 ♊ 48	19 ♎ 59	04 ♐ 02	03 ♎ 16
02	14 47 25	10 09 59	00 ♎ 47 41	06 ♎ 55 20	03 09	03 07	21 09	05 13	19 R 45	20 03	04 05	03 18
03	14 51 21	11 10 06	13 ♎ 00 52	19 ♎ 04 31	04 16	04 20	21 49	05 24	19 42	20 06	04 07	03 20
04	14 55 18	12 10 15	25 ♎ 06 28	01 ♏ 06 51	05 20	05 32	22 28	05 35	19 38	20 10	04 09	03 21
05	14 59 14	13 10 26	07 ♏ 05 52	13 ♏ 03 37	06 22	06 45	23 07	05 45	19 35	20 13	04 11	03 23
06	15 03 11	14 10 39	19 ♏ 00 17	24 ♏ 56 02	07 20	07 57	23 47	05 56	19 31	20 17	04 13	03 25
07	15 07 08	15 10 54	00 ♐ 51 04	06 ♐ 45 35	08 15	09 10	24 26	06 07	19 28	20 21	04 15	03 27
08	15 11 04	16 11 11	12 ♐ 39 50	18 ♐ 34 09	09 06	10 23	25 06	06 18	19 24	20 24	04 17	03 29
09	15 15 01	17 11 29	24 ♐ 28 49	00 ♑ 24 16	09 52	11 35	25 45	06 30	19 20	20 28	04 20	03 31
10	15 18 57	18 11 49	06 ♑ 20 54	12 ♑ 19 13	10 34	12 48	26 25	06 41	19 17	20 31	04 22	03 32
11	15 22 54	19 12 10	18 ♑ 19 42	24 ♑ 22 56	11 10	14 01	27 05	06 52	19 13	20 34	04 24	03 34
12	15 26 50	20 12 33	00 ♒ 29 29	06 ♒ 39 59	11 40	15 14	27 44	07 04	19 09	20 38	04 26	03 36
13	15 30 47	21 12 57	12 ♒ 55 03	19 ♒ 15 17	12 03	16 27	28 24	07 15	19 05	20 41	04 28	03 37
14	15 34 43	22 13 22	25 ♒ 41 17	02 ♓ 13 36	12 19	17 41	29 03	07 27	19 01	20 45	04 31	03 39
15	15 38 40	23 13 49	08 ♓ 52 43	15 ♓ 39 01	12 27	18 54	29 ♎ 43	07 39	18 57	20 48	04 33	03 41
16	15 42 37	24 14 17	22 ♓ 32 45	29 ♓ 34 01	12 R 25	20 07	00 ♏ 23	07 51	18 52	20 51	04 35	03 42
17	15 46 33	25 14 46	06 ♈ 42 44	13 ♈ 58 34	12 14	21 21	01 02	08 03	18 48	20 55	04 37	03 44
18	15 50 30	26 15 17	21 ♈ 21 01	28 ♈ 49 18	11 53	22 34	01 42	08 15	18 44	20 58	04 39	03 46
19	15 54 26	27 15 49	06 ♉ 22 27	13 ♉ 59 16	11 21	23 47	02 22	08 27	18 39	21 01	04 42	03 47
20	15 58 23	28 16 22	21 ♉ 38 27	29 ♉ 18 35	10 39	25 01	03 02	08 39	18 35	21 04	04 44	03 49
21	16 02 19	29 ♏ 16 57	06 ♊ 58 14	14 ♊ 36 00	09 47	26 15	03 41	08 51	18 30	21 08	04 46	03 50
22	16 06 16	00 ♐ 17 34	22 ♊ 10 35	29 ♊ 40 51	08 45	27 28	04 21	09 03	18 26	21 11	04 48	03 52
23	16 10 12	01 18 12	07 ♋ 05 52	14 ♋ 24 52	07 35	28 42	05 01	09 15	18 21	21 14	04 51	03 53
24	16 14 09	02 18 52	21 ♋ 37 21	28 ♋ 42 59	06 18	29 ♎ 56	05 41	09 28	18 17	21 17	04 53	03 55
25	16 18 06	03 19 33	05 ♌ 41 40	12 ♌ 33 26	04 58	01 ♏ 10	06 21	09 40	18 12	21 20	04 55	03 56
26	16 22 02	04 20 16	19 ♌ 18 29	25 ♌ 57 05	03 35	02 23	07 01	09 53	18 07	21 23	04 57	03 57
27	16 25 59	05 21 00	02 ♍ 29 39	08 ♍ 55 37	02 14	03 37	07 41	10 05	18 03	21 26	05 00	03 59
28	16 29 55	06 21 46	15 ♍ 18 12	21 ♍ 35 42	00 ♐ 57	04 51	08 21	10 18	17 58	21 29	05 02	04 00
29	16 33 52	07 22 33	27 ♍ 48 49	03 ♎ 58 19	29 ♏ 46	06 06	09 01	10 31	17 53	21 32	05 04	04 01
30	16 37 48	08 ♐ 23 22	10 ♎ 04 41	16 ♎ 08 22	28 ♏ 44	07 ♏ 20	09 ♏ 41	10 ♑ 44	17 ♊ 48	21 ♎ 35	05 ♐ 07	04 ♎ 02

Date	Moon True ☊ ° '	Moon Mean ☊ ° '	Moon ☽ Latitude ° '
01	19 ♑ 52	20 ♑ 27	04 S 18
02	19 R 44	20 24	04 46
03	19 33	20 21	05 00
04	19 20	20 18	05 00
05	19 07	20 14	04 47
06	18 54	20 11	04 22
07	18 43	20 08	03 45
08	18 35	20 05	02 59
09	18 29	20 02	02 04
10	18 26	19 59	01 S 04
11	18 25	19 55	00 00
12	18 D 26	19 52	01 N 04
13	18 27	19 49	02 08
14	18 R 27	19 46	03 06
15	18 26	19 43	03 57
16	18 22	19 39	04 36
17	18 17	19 36	05 01
18	18 09	19 33	05 06
19	18 01	19 30	04 51
20	17 53	19 27	04 16
21	17 47	19 24	03 21
22	17 42	19 20	02 13
23	17 40	19 17	00 N 57
24	17 D 39	19 14	00 S 21
25	17 40	19 11	01 36
26	17 42	19 08	02 43
27	17 43	19 05	03 39
28	17 R 42	19 01	04 22
29	17 40	18 58	04 49
30	17 ♑ 36	18 ♑ 55	05 S 07

DECLINATIONS

Date	Sun ☉ ° '	Moon ☽ ° '	Mercury ☿ ° '	Venus ♀ ° '	Mars ♂ ° '	Jupiter ♃ ° '	Saturn ♄ ° '	Uranus ♅ ° '	Neptune ♆ ° '	Pluto ♇ ° '
01	14 S 33	00 N 37	23 S 12	00 N 39	07 S 18	23 S 26	21 N 23	07 S 15	19 S 24	13 N 20
02	14 52	04 S 41	23 28	00 N 12	07 33	23 26	21 23	07 17	19 25	13 19
03	15 11	09 44	23 43	00 S 15	07 48	23 25	21 23	07 18	19 25	13 19
04	15 29	14 22	23 56	00 42	08 03	23 25	21 22	07 19	19 26	13 18
05	15 48	18 24	24 08	01 09	08 18	23 24	21 22	07 21	19 26	13 18
06	16 06	21 40	24 18	01 37	08 33	23 24	21 22	07 22	19 26	13 18
07	16 24	24 00	24 26	02 04	08 48	23 24	21 22	07 23	19 27	13 17
08	16 41	25 16	24 33	02 31	09 03	23 23	21 21	07 25	19 27	13 17
09	16 58	25 24	24 38	02 59	09 17	23 23	21 21	07 26	19 28	13 17
10	17 15	24 22	24 41	03 26	09 32	23 22	21 21	07 27	19 28	13 16
11	17 32	22 12	24 42	03 54	09 47	23 22	21 20	07 29	19 29	13 16
12	17 48	19 00	24 41	04 21	10 01	23 21	21 20	07 30	19 29	13 16
13	18 04	14 54	24 38	04 48	10 16	23 21	21 20	07 31	19 29	13 15
14	18 20	10 02	24 32	05 16	10 30	23 20	21 19	07 32	19 30	13 15
15	18 35	04 S 35	24 24	05 43	10 45	23 20	21 19	07 34	19 30	13 15
16	18 50	01 N 16	24 13	06 10	10 59	23 19	21 19	07 35	19 31	13 15
17	19 05	07 16	24 00	06 37	11 13	23 18	21 18	07 36	19 31	13 14
18	19 19	13 03	23 43	07 04	11 27	23 18	21 18	07 37	19 31	13 14
19	19 33	18 13	23 24	07 31	11 42	23 17	21 18	07 39	19 32	13 14
20	19 47	22 17	23 01	07 58	11 56	23 16	21 17	07 40	19 32	13 14
21	20 00	24 47	22 36	08 24	12 10	23 16	21 17	07 41	19 33	13 14
22	20 13	25 26	22 07	08 51	12 23	23 15	21 16	07 42	19 33	13 13
23	20 25	24 21	21 36	09 17	12 37	23 14	21 16	07 43	19 34	13 13
24	20 38	21 21	21 03	09 43	12 51	23 13	21 16	07 44	19 34	13 13
25	20 49	17 18	20 29	10 09	13 04	23 13	21 15	07 46	19 34	13 13
26	21 01	12 27	19 54	10 35	13 18	23 12	21 15	07 47	19 35	13 13
27	21 12	07 11	19 20	11 01	13 31	23 11	21 15	07 48	19 35	13 13
28	21 23	01 N 46	18 48	11 26	13 45	23 10	21 14	07 49	19 36	13 13
29	21 33	03 S 36	18 18	11 51	13 58	23 09	21 14	07 50	19 36	13 13
30	21 S 43	08 S 42	17 S 51	12 S 16	14 S 11	23 S 08	21 N 13	07 S 51	19 S 36	13 N 13

ZODIAC SIGN ENTRIES

Date	h	m	Planets
02	10	27	☽ ♎
04	21	46	☽ ♏
07	10	16	☽ ♐
09	23	11	☽ ♑
12	11	02	☽ ♒
14	19	56	☽ ♓
15	22	17	♂ ♏
17	00	44	☽ ♈
19	01	53	☽ ♉
21	01	05	☽ ♊
22	05	03	☉ ♐
23	00	31	☽ ♋
24	13	23	♀ ♏
25	02	12	☽ ♌
27	07	24	☽ ♍
29	07	08	☿ ♏
29	16	15	☽ ♎

LATITUDES

Date	Mercury ☿ ° '	Venus ♀ ° '	Mars ♂ ° '	Jupiter ♃ ° '	Saturn ♄ ° '	Uranus ♅ ° '	Neptune ♆ ° '	Pluto ♇ ° '
01	02 S 41	01 N 33	00 N 46	00 S 05	01 S 40	00 N 36	01 N 35	15 N 58
04	02 47	01 38	00 45	00 05	01 40	00 36	01 35	15 59
07	02 47	01 42	00 44	00 06	01 40	00 36	01 35	16 00
10	02 40	01 46	00 42	00 06	01 40	00 36	01 35	16 01
13	02 24	01 48	00 41	00 06	01 40	00 36	01 35	16 02
16	01 57	01 50	00 40	00 06	01 40	00 36	01 35	16 04
19	01 16	01 51	00 39	00 07	01 40	00 36	01 35	16 05
22	00 S 22	01 51	00 37	00 07	01 40	00 36	01 34	16 06
25	00 N 39	01 50	00 36	00 07	01 40	00 36	01 34	16 08
28	01 36	01 49	00 34	00 07	01 40	00 36	01 34	16 09
31	02 N 16	01 N 47	00 N 33	00 S 08	01 S 40	00 N 36	01 N 34	16 N 11

DATA

Julian Date	2441623
Delta T	+42 seconds
Ayanamsa	23° 28' 54"
Synetic vernal point	05° ♓ 38' 05"
True obliquity of ecliptic	23° 26' 38"

MOON'S PHASES, APSIDES AND POSITIONS ☽

Date	h	m	Phase	Longitude ° '	Eclipse Indicator
06	01	21	●	13 ♏ 44	
14	05	01	☽	21 ♒ 56	
20	23	07	○	28 ♉ 44	
27	17	45	☾	05 ♍ 36	

Day	h	m		
07	13	06	Apogee	
21	00	09	Perigee	
01	14	47	0S	
09	02	36	Max dec	25° S 29'
16	06	52	0N	
22	08	02	Max dec	25° N 27'
28	19	49	0S	

LONGITUDES

Date	Chiron ⚷ ° '	Ceres ⚳ ° '	Pallas ⚴ ° '	Juno ⚵ ° '	Vesta ⚶ ° '	Black Moon Lilith ⚸ ° '
01	14 ♈ 01	27 ♎ 55	08 ♎ 46	12 ♏ 47	15 ♊ 13	28 ♏ 07
11	13 ♈ 39	02 ♏ 21	13 ♎ 11	16 ♏ 12	13 ♊ 26	29 ♏ 14
21	13 ♈ 20	06 ♏ 44	17 ♎ 28	19 ♏ 37	11 ♊ 07	00 ♐ 21
31	13 ♈ 05	11 ♏ 02	21 ♎ 37	23 ♏ 00	08 ♊ 32	01 ♐ 28

All ephemeris data is given at 12.00 UT and the Moon's longitude is additionally given for 24.00 UT

ASPECTARIAN

h m	Aspects
01 Wednesday	
03 27	☽ ⊥ ♅
04 03	☽ ⊥ ♂
07 37	♂ ∥ ♅
11 49	☽ ∥ ♀
14 23	☉ ⊥ ♇
14 39	☽ □ ♄
15 02	☽ ⊻ ♅
15 23	☽ Q ☿
16 14	☽ ⊻ ♂
17 16	☽ # ♀
19 01	☽ Q ♆
02 Thursday	
00 05	☽ ∠ ☉
08 08	☿ ∥ ♃
14 58	☿ ✶ ♇
15 30	♀ ☌ ♇
16 54	☽ ☌ ♇
17 03	☽ ☌ ♀
17 05	☽ ✶ ☿
18 26	☽ ✶ ♆
19 11	☽ ⊥ ☉
20 47	☽ □ ♃
22 11	☿ ✶ ♀
03 Friday	
00 11	☽ ∥ ♅
02 07	☽ ∥ ♂
07 33	♀ ✶ ♆
08 02	☽ ⊻ ☉
08 26	☿ ☌ ♆
04 Saturday	
00 06	☽ ∠ ♆
01 10	☽ △ ♄
01 36	☽ ∠ ☿
02 07	☽ ☌ ♅
06 15	☽ # ♇
06 26	☽ ☌ ♂
07 49	☿ ∠ ♅
08 54	☽ Q ♃
12 57	♀ □ ♃
18 05	☽ ⊥ ♆
18 39	☿ ⊻ ♃
18 47	☽ ∥ ☉
21 15	☽ ⊥ ☿
05 Sunday	
04 32	☽ ⊻ ♇
06 08	☽ ⊻ ♆
06 58	☽ bQ ♄
09 16	☽ ✶ ♃
10 23	☽ ⊻ ☿
11 12	☽ ⊻ ♀
16 37	☽ ⊥ ♇
18 57	☽ ∥ ♆
21 13	☉ ± ♄
06 Monday	
00 39	☽ ⊥ ♀
00 59	☽ ± ♄
01 21	☽ ☌ ☉
09 28	☽ # ♄
10 49	☽ ∠ ♇
13 03	☽ ⊼ ♄
14 36	☽ ⊻ ♅
15 58	☽ ∠ ♃
19 54	♂ Q ♃
20 54	☽ ∠ ♀
22 14	☽ ⊻ ♂
07 Tuesday	
02 48	☽ ⊥ ♅
04 40	☽ ∥ ♃
10 30	☽ ⊥ ♃
11 07	☽ ⊥ ♂
14 08	☉ ⊥ ♀
17 17	☽ ✶ ♇
18 44	☽ ∥ ☿
18 56	☽ ☌ ♆
21 10	☽ ∠ ♅
22 52	☽ ⊻ ♃
08 Wednesday	
04 12	☽ ☌ ☿
06 29	☽ ∠ ♂
06 49	☽ ✶ ♀
17 44	☽ Q ♇
19 49	☽ ⊻ ☉
09 Thursday	
01 37	☽ ☍ ♄
03 48	☽ ✶ ♅
09 09	☽ ⊥ ☉
09 59	☽ Q ♀
14 44	☽ ✶ ♂
10 Friday	
04 14	☽ Q ♅
05 03	☽ ∠ ☉
06 19	☽ □ ♇
06 59	☽ ∥ ☿
07 59	☽ ⊻ ♆
12 41	☽ ☌ ♃
16 24	☽ Q ♂
20 06	☽ ⊥ ♆
20 24	☉ ∠ ♇
20 57	☽ ⊻ ☿
11 Saturday	
00 31	☽ ∥ ♃
02 26	☽ □ ♀
09 34	☽ ⊥ ☿
12 14	☉ ⊼ ♄
13 45	☽ ⊼ ♄
13 54	☽ ✶ ☉
14 08	☽ ∠ ♆
16 29	☽ □ ♅
19 15	☽ # ♄
12 Sunday	
01 34	☽ ± ♄
04 12	☽ ∠ ☿
06 18	☽ □ ♂
08 49	☽ ∥ ♆
15 39	☽ Q ☉
18 04	☽ △ ♇
19 05	☽ bQ ♄
19 05	☽ ∥ ☉
19 42	☽ ✶ ♆
22 41	☉ ⊻ ♅
13 Monday	
00 58	☽ ⊻ ♃
10 19	☽ ✶ ☿
12 39	☽ ⊥ ♃
14 57	♂ ⊥ ♆
18 46	☽ Q ♆
19 26	☽ △ ♀
20 31	☽ # ♇
22 50	☽ bQ ♇
23 37	☽ △ ♄
14 Tuesday	
02 45	☽ △ ♅
05 01	☽ □ ☉
05 54	☽ ∠ ♃
09 25	☽ Q ☿
09 55	☽ ∥ ♂
15 38	☽ ± ♇
18 32	☽ △ ♂
18 46	☉ ∠ ♃
23 12	☽ ∥ ♅
15 Wednesday	
02 07	☽ bQ ♀
02 37	☽ ⊼ ♇
04 11	☽ □ ♆
06 27	☽ bQ ♅
07 32	☽ ∥ ♀
09 46	☽ ✶ ♃
12 52	♀ △ ♄
18 22	☽ □ ☿
19 51	☽ ± ♀
20 26	☿ St R
22 33	☽ ± ♅
22 53	☽ bQ ♂
16 Thursday	
01 10	♀ ∠ ♆
05 40	☽ □ ♄
07 15	☽ Q ♃
07 24	☽ ⊼ ♀
09 04	☽ ⊼ ♅
15 08	☽ △ ☉
15 19	☽ ± ♂
17 Friday	
02 02	☽ ⊼ ♂
03 11	♀ ☌ ♅
07 01	☽ ☍ ♇
08 30	☽ △ ♆
09 12	☽ # ♀
12 09	☽ Q ♄
13 23	☽ # ♅
14 15	☽ □ ♃
18 18	☽ bQ ☉
20 58	☽ △ ☿
18 Saturday	
04 55	☽ # ♂
07 46	☽ ✶ ♄
09 15	☽ bQ ♆
10 06	☽ ± ☉
11 23	☽ ☍ ♅
12 47	☽ ∥ ♇
14 08	☽ ☍ ♀
20 28	☽ ⊼ ☉
20 38	☽ bQ ☿
23 46	☽ ± ♆
19 Sunday	
05 21	☽ ☍ ♂
05 53	☉ ⊥ ♅
07 43	☽ ∠ ♄
07 54	☽ ⊼ ♇
09 20	☽ ⊼ ♆
09 48	☉ ∥ ♆
10 27	☽ ± ☿
15 19	☽ △ ♃
17 24	☽ ± ♇
18 56	☽ # ♆
19 05	♀ ∥ ♅
19 26	☽ # ☉
19 33	☽ ⊼ ☿
19 56	☿ ∥ ♃
21 52	☽ ⊥ ♄
20 Monday	
05 16	☽ ∥ ♄
07 14	☽ ⊻ ♄
07 34	☽ bQ ♇
11 07	☽ ⊼ ♅
15 11	☽ bQ ♃
16 55	☽ # ☿
17 44	☽ ⊼ ♀
19 31	☿ ∠ ♀
19 37	☽ # ♃
20 32	☽ ± ♅
23 07	☽ ☍ ☉
21 Tuesday	
03 57	☽ ± ♀
05 27	☽ ± ♃
06 04	♂ bQ ♄
06 37	☽ ⊼ ♂
07 05	☽ △ ♇
08 32	☽ ☍ ♆
10 40	☽ bQ ♅
13 21	☿ ⊥ ♂
14 59	☽ ⊼ ♃
16 09	☽ ☍ ☿
16 28	☽ ± ♂
17 30	♂ ⊻ ♇
19 18	☽ bQ ♀
22 Wednesday	
02 12	♀ Q ♃
06 05	☽ ☌ ♄
06 23	☿ ⊻ ♃
07 19	☽ bQ ♂
10 25	☽ △ ♅
21 12	☽ △ ♀
23 Thursday	
01 56	☽ ⊼ ☉
05 23	♂ ⊻ ♆
06 46	☽ □ ♇
08 20	☽ ⊼ ♆
08 28	☽ △ ♂
12 22	☽ ± ☉
12 44	☽ ⊼ ☿
14 56	♀ ⊥ ♆
15 35	☽ ☍ ♃
18 08	☽ ± ♆
21 37	☽ # ♃
21 47	☽ ± ☿
24 Friday	
03 06	☿ # ♄
04 16	☽ bQ ☉
06 26	☽ ⊻ ♄
09 05	☽ bQ ♆
11 26	☽ □ ♅
11 31	☽ bQ ☿
12 23	☿ ∠ ♅
12 29	☽ Q ♇
12 37	☽ ∥ ♄
14 21	☽ # ☿
15 33	☿ ⊥ ♀
16 27	☽ ⊥ ♄
16 32	☽ # ☉
19 31	☿ ⊻ ♂
23 18	☽ # ♆
25 Saturday	
01 25	☉ ∥ ☿
03 25	☽ □ ♀
07 35	☽ △ ☉
07 43	☽ ∠ ♄
08 57	☽ ✶ ♇
10 39	☽ △ ♆
10 51	☽ △ ☿
12 41	☿ ☌ ♆
13 11	☽ □ ♂
18 22	☽ Q ♅
19 02	☽ ⊼ ♃
22 22	☉ ⊥ ♃
26 Sunday	
02 42	☉ ✶ ♇
03 07	♂ # ♇
04 28	☉ ☌ ☿
05 42	☿ ✶ ♇
05 47	☽ ± ♃
07 33	☿ ⊥ ♃
08 10	☽ # ♂
08 23	☽ ∥ ♇
09 53	☽ ✶ ♄
11 22	☽ ∠ ♇
14 09	☽ Q ♀
15 45	☽ ✶ ♅
20 00	☽ # ♀
22 13	☽ bQ ♃
22 50	☽ Q ♂
23 04	☿ ⊻ ♀
27 Monday	
01 22	♀ bQ ♄
01 30	☿ ∥ ♆
03 16	☉ ☌ ♆
03 41	☽ ⊥ ♇
07 31	☽ Q ♄
09 15	☽ # ♅
11 34	☽ □ ☿
14 19	☽ ✶ ♀
14 45	☽ ⊻ ♇
16 39	☽ □ ♆
17 45	☽ □ ☉
17 46	☉ # ♄
18 59	♀ ⊻ ♇
19 20	☽ ∠ ♅
22 10	☽ ✶ ♂
28 Tuesday	
02 23	☽ △ ♃
12 20	☽ ⊥ ♅
15 02	☉ ∠ ♅
15 31	♀ ⊻ ♆
17 01	☽ □ ♄
18 19	☽ Q ☿
21 37	☽ ∠ ♀
23 50	☽ ⊻ ♅
29 Wednesday	
02 49	☽ Q ♆
04 14	☽ ∠ ♂
06 52	☽ Q ☉
15 29	☽ ✶ ☿
16 55	☽ ⊥ ♀
22 42	☽ ⊥ ♂
30 Thursday	
00 07	☽ ☌ ♇
02 12	☽ ✶ ♆
05 58	☽ ⊻ ♀
07 52	☽ ∥ ♅
08 22	☽ ✶ ☉
11 10	☽ ⊻ ♂
13 18	☽ □ ♃
18 42	☽ ∠ ☿

DECEMBER 1972

LONGITUDES

Date	Sidereal time h m s	Sun ☉ ° ' "	Moon ☽ ° ' "	Moon ☽ 24.00 ° ' "	Mercury ☿ ° '	Venus ♀ ° '	Mars ♂ ° '	Jupiter ♃ ° '	Saturn ♄ ° '	Uranus ♅ ° '	Neptune ♆ ° '	Pluto ♇ ° '
01	16 41 45	09 ♐ 24 13	22 ♎ 09 47	28 ♎ 09 21	27 ♏ 52	08 ♏ 34	10 ♏ 21	10 ♑ 57	17 ♊ 43	21 ♎ 38	05 ♐ 09	04 ♎ 04
02	16 45 41	10 25 04	04 ♏ 07 24	10 ♏ 04 16	27 R 40	09 48	11 01	11 09	17 R 38	21 41	05 11	04 05
03	16 49 38	11 25 57	16 ♏ 00 16	21 ♏ 55 38	26 40	11 02	11 41	11 22	17 33	21 43	05 13	04 06
04	16 53 35	12 26 52	27 ♏ 50 38	03 ♐ 45 30	26 22	12 16	12 21	11 35	17 29	21 46	05 16	04 07
05	16 57 31	13 27 47	09 ♐ 40 27	15 ♐ 35 42	26 14	13 31	13 01	11 48	17 24	21 49	05 18	04 08
06	17 01 28	14 28 44	21 ♐ 31 28	27 ♐ 27 59	26 D 18	14 45	13 41	12 02	17 19	21 52	05 20	04 09
07	17 05 24	15 29 41	03 ♑ 25 28	09 ♑ 24 12	26 31	16 00	14 22	12 15	17 14	21 54	05 22	04 10
08	17 09 21	16 30 40	15 ♑ 24 28	21 ♑ 26 34	26 53	17 14	15 02	12 28	17 09	21 57	05 24	04 11
09	17 13 17	17 31 39	27 ♑ 30 50	03 ♒ 37 40	27 23	18 28	15 42	12 41	17 04	21 59	05 27	04 12
10	17 17 14	18 32 39	09 ♒ 47 26	16 ♒ 00 33	28 01	19 43	16 22	12 55	16 59	22 02	05 29	04 13
11	17 21 10	19 33 39	22 ♒ 17 29	28 ♒ 38 40	28 45	20 57	17 03	13 08	16 54	22 04	05 31	04 14
12	17 25 07	20 34 40	05 ♓ 04 32	11 ♓ 35 33	29 ♏ 35	22 12	17 43	13 21	16 49	22 07	05 33	04 15
13	17 29 04	21 35 42	18 ♓ 11 05	24 ♓ 54 31	00 ♐ 30	23 26	18 23	13 35	16 44	22 09	05 36	04 16
14	17 33 00	22 36 44	01 ♈ 43 05	08 ♈ 38 01	01 29	24 41	19 04	13 48	16 39	22 11	05 38	04 17
15	17 36 57	23 37 46	15 ♈ 39 19	22 ♈ 46 57	02 32	25 56	19 44	14 02	16 34	22 14	05 40	04 17
16	17 40 53	24 38 48	00 ♉ 00 37	07 ♉ 19 56	03 39	27 10	20 24	14 15	16 29	22 16	05 42	04 18
17	17 44 50	25 39 53	14 ♉ 44 17	22 ♉ 12 51	04 48	28 25	21 05	14 29	16 24	22 18	05 44	04 19
18	17 48 46	26 40 56	29 ♉ 44 43	07 ♊ 18 46	06 01	29 ♏ 40	21 45	14 42	16 20	22 20	05 46	04 19
19	17 52 43	27 42 01	14 ♊ 53 51	22 ♊ 28 44	07 15	00 ♐ 54	22 26	14 56	16 15	22 22	05 49	04 20
20	17 56 39	28 43 05	00 ♋ 02 10	07 ♋ 32 59	08 32	02 09	23 06	15 10	16 10	22 24	05 51	04 21
21	18 00 36	29 ♐ 44 11	15 ♋ 00 05	22 ♋ 22 31	09 50	03 24	23 47	15 24	16 05	22 26	05 53	04 21
22	18 04 33	00 ♑ 45 16	29 ♋ 39 30	06 ♌ 50 25	11 09	04 39	24 27	15 37	16 01	22 28	05 55	04 22
23	18 08 29	01 46 23	13 ♌ 54 48	20 ♌ 52 26	12 31	05 53	25 08	15 51	15 56	22 30	05 57	04 22
24	18 12 26	02 47 30	27 ♌ 43 11	04 ♍ 27 08	13 53	07 08	25 49	16 05	15 51	22 32	05 59	04 23
25	18 16 22	03 48 37	11 ♍ 04 28	17 ♍ 35 29	15 16	08 23	26 29	16 19	15 47	22 34	06 01	04 23
26	18 20 19	04 49 45	24 ♍ 00 35	00 ♎ 20 13	16 41	09 38	27 10	16 33	15 42	22 36	06 03	04 24
27	18 24 15	05 50 54	06 ♎ 34 53	12 ♎ 45 07	18 06	10 53	27 51	16 46	15 38	22 37	06 05	04 24
28	18 28 12	06 52 03	18 ♎ 51 30	24 ♎ 54 34	19 32	12 08	28 31	17 00	15 33	22 39	06 07	04 24
29	18 32 08	07 53 12	00 ♏ 54 54	06 ♏ 53 03	20 58	13 22	29 12	17 14	15 29	22 41	06 09	04 25
30	18 36 05	08 54 23	12 ♏ 49 31	18 ♏ 44 49	22 25	14 37	29 ♏ 53	17 28	15 24	22 42	06 11	04 25
31	18 40 02	09 ♑ 55 33	24 ♏ 39 26	00 ♐ 33 48	23 ♐ 53	15 ♐ 52	00 ♐ 34	17 ♑ 42	15 ♊ 20	22 ♎ 44	06 ♐ 13	04 ♎ 25

MOON / DECLINATIONS

Date	Moon True ☊ ° '	Moon Mean ☊ ° '	Moon ☽ Latitude ° '	Sun ☉	Moon ☽	Mercury ☿	Venus ♀	Mars ♂	Jupiter ♃	Saturn ♄	Uranus ♅	Neptune ♆	Pluto ♇
01	17 ♑ 30	18 ♑ 52	05 S 09	21 S 52	13 S 24	17 S 28	12 S 40	14 S 24	23 S 07	21 N 13	07 S 52	19 S 37	13 N 13
02	17 R 23	18 49	04 57	22 01	17 33	17 10	13 04	14 37	23 06	21 13	07 53	19 37	13 13
03	17 16	18 45	04 32	22 09	20 58	16 56	13 28	14 50	23 05	21 12	07 54	19 38	13 13
04	17 09	18 42	03 56	22 18	23 31	16 47	13 52	15 03	23 04	21 12	07 55	19 38	13 13
05	17 03	18 39	03 09	22 25	25 02	16 42	14 15	15 15	23 03	21 11	07 56	19 38	13 13
06	16 59	18 36	02 14	22 32	25 25	16 42	14 38	15 28	23 02	21 11	07 57	19 39	13 13
07	16 56	18 33	01 13	22 39	24 37	16 45	15 01	15 40	23 01	21 11	07 58	19 39	13 13
08	16 55	18 30	00 S 08	22 46	22 41	16 51	15 23	15 53	23 00	21 10	07 59	19 40	13 13
09	16 D 56	18 26	00 N 58	22 52	19 43	17 00	15 45	16 05	22 59	21 10	08 00	19 40	13 13
10	16 57	18 23	02 02	22 57	15 50	17 12	16 07	16 17	22 58	21 10	08 01	19 40	13 13
11	16 59	18 20	03 02	23 02	11 12	17 26	16 28	16 29	22 56	21 09	08 02	19 41	13 14
12	17 01	18 17	03 55	23 07	06 01	17 42	16 48	16 41	22 55	21 09	08 03	19 41	13 14
13	17 01	18 14	04 36	23 11	00 S 26	17 59	17 09	16 52	22 54	21 08	08 04	19 41	13 14
14	17 R 01	18 10	05 04	23 14	05 N 20	18 17	17 29	17 04	22 52	21 08	08 04	19 42	13 14
15	17 00	18 07	05 15	23 17	11 00	18 36	17 48	17 15	22 51	21 08	08 05	19 42	13 14
16	16 57	18 04	05 07	23 20	16 16	18 55	18 07	17 27	22 50	21 07	08 06	19 43	13 14
17	16 55	18 01	04 39	23 22	20 42	19 15	18 25	17 38	22 48	21 07	08 07	19 43	13 15
18	16 52	17 58	03 51	23 24	23 52	19 35	18 43	17 49	22 47	21 07	08 08	19 43	13 15
19	16 49	17 55	02 47	23 25	25 21	19 56	19 01	18 00	22 46	21 06	08 08	19 44	13 15
20	16 48	17 51	01 31	23 26	24 58	20 15	19 18	18 11	22 44	21 06	08 09	19 44	13 15
21	16 47	17 48	00 N 10	23 27	22 46	20 35	19 34	18 22	22 43	21 06	08 10	19 44	13 15
22	16 D 47	17 45	01 S 11	23 26	19 04	20 54	19 50	18 32	22 41	21 05	08 11	19 45	13 16
23	16 48	17 42	02 25	23 26	14 21	21 13	20 05	18 42	22 40	21 05	08 11	19 45	13 16
24	16 50	17 39	03 28	23 25	09 01	21 31	20 20	18 53	22 38	21 05	08 12	19 45	13 17
25	16 51	17 36	04 17	23 23	03 N 26	21 49	20 34	19 03	22 37	21 04	08 13	19 46	13 17
26	16 51	17 32	04 52	23 21	02 S 06	22 05	20 48	19 13	22 35	21 04	08 13	19 46	13 18
27	16 R 52	17 29	05 12	23 19	07 23	22 21	21 01	19 22	22 33	21 04	08 14	19 46	13 18
28	16 51	17 26	05 17	23 16	12 16	22 36	21 13	19 32	22 32	21 03	08 14	19 47	13 18
29	16 51	17 23	05 08	23 13	16 36	22 50	21 25	19 42	22 30	21 03	08 15	19 47	13 19
30	16 50	17 20	04 45	23 09	20 13	23 03	21 36	19 51	22 28	21 03	08 15	19 47	13 19
31	16 ♑ 49	17 ♑ 16	04 S 11	23 S 04	22 S 59	23 S 15	21 S 47	20 S 00	22 S 27	21 N 02	08 S 16	19 S 48	13 N 20

ZODIAC SIGN ENTRIES

Date	h	m	Planets
02	03	42	☽ ♏
04	16	22	☽ ♐
07	05	06	☽ ♑
09	16	53	☽ ♒
12	02	33	☽ ♓
12	23	20	☿ ♐
14	08	59	☽ ♈
16	11	59	☽ ♉
18	12	24	☽ ♊
18	18	34	♀ ♐
20	11	57	☽ ♋
21	18	13	☉ ♑
22	12	34	☽ ♌
24	16	03	☽ ♍
26	23	21	☽ ♎
29	10	10	☽ ♏
30	16	12	♂ ♐
31	22	51	☽ ♐

LATITUDES

Date	Mercury ☿ ° '	Venus ♀ ° '	Mars ♂ ° '	Jupiter ♃ ° '	Saturn ♄ ° '	Uranus ♅ ° '	Neptune ♆ ° '	Pluto ♇ ° '
01	02 N 16	01 N 47	00 N 33	00 S 08	01 S 40	00 N 36	01 N 34	16 N 11
04	02 38	01 44	00 32	00 08	01 40	00 37	01 34	16 12
07	02 42	01 41	00 30	00 08	01 40	00 37	01 34	16 14
10	02 35	01 37	00 28	00 08	01 39	00 37	01 34	16 15
13	02 20	01 32	00 27	00 09	01 39	00 37	01 35	16 17
16	02 00	01 27	00 25	00 09	01 39	00 37	01 35	16 19
19	01 37	01 22	00 24	00 09	01 38	00 37	01 35	16 20
22	01 14	01 16	00 22	00 10	01 38	00 37	01 35	16 22
25	00 50	01 09	00 20	00 10	01 37	00 37	01 35	16 24
28	00 26	01 02	00 18	00 10	01 37	00 37	01 35	16 26
31	00 N 03	00 N 55	00 N 17	00 S 10	01 S 36	00 N 37	01 N 35	16 N 27

DATA

Julian Date	2441653
Delta T	+42 seconds
Ayanamsa	23° 28' 59"
Synetic vernal point	05° ♓ 38' 01"
True obliquity of ecliptic	23° 26' 37"

MOON'S PHASES, APSIDES AND POSITIONS ☽

Date	h	m	Phase	Longitude ° '	Eclipse Indicator
05	20	24	●	13 ♐ 49	
13	18	36	☽	21 ♓ 52	
20	09	45	○	28 ♊ 37	
27	10	27	☾	05 ♎ 47	

Day	h	m		
04	13	42	Apogee	
19	12	51	Perigee	
31	22	02	Apogee	
06	07	52	Max dec	25° S 26'
13	13	48	0N	
19	19	03	Max dec	25° N 26'
26	02	50	0S	

LONGITUDES

Date	Chiron ⚷ ° '	Ceres ⚳ ° '	Pallas ⚴ ° '	Juno ⚵ ° '	Vesta ⚶ ° '	Black Moon Lilith ☾ ° '
01	13 ♈ 05	11 ♏ 02	21 ♎ 37	23 ♏ 00	08 ♊ 32	01 ♐ 28
11	12 ♈ 56	15 ♏ 15	25 ♎ 35	26 ♏ 20	05 ♊ 58	02 ♐ 35
21	12 ♈ 52	19 ♏ 22	29 ♎ 21	29 ♏ 36	03 ♊ 45	03 ♐ 42
31	12 ♈ 53	23 ♏ 21	02 ♏ 52	02 ♐ 46	02 ♊ 06	04 ♐ 49

ASPECTARIAN

h m	Aspects
01 Friday	
01 25	♂ ⊥ ♇
03 12	☽ △ ♄
07 41	☽ ∥ ♀
07 58	☽ ∠ ♆
10 56	☽ ☌ ♅
10 57	☽ # ♇
11 26	☽ ⊥ ☿
16 54	☽ ∠ ☉
17 44	☽ ∥ ♂
18 51	☿ ⊥ ♅
22 44	☽ ⊻ ☿
02 Saturday	
01 50	☽ Q ♃
02 02	☽ ⊥ ♆
09 02	☽ ⚼ ♄
09 45	☽ ∥ ☿
11 55	☽ ⊻ ♇
12 39	☽ ⊥ ☉
14 09	☽ ⊻ ♆
15 36	♃ ⊥ ♆
17 32	♀ ⊥ ♇
19 36	♂ ⚹ ♃
20 18	♀ # ♇
03 Sunday	
00 02	☽ ⊥ ♇
00 47	☽ ☌ ♀
01 50	☽ ∥ ♆
01 53	☽ ⊻ ☉
02 27	☽ ⚹ ♃
02 44	☽ ☌ ♂
03 04	☽ ± ♄
08 02	♂ ± ♄
10 13	☉ ⊻ ♃
13 51	☽ # ♄
15 07	☽ ⊼ ♄
18 17	☽ ∠ ♇
19 55	♀ ⚹ ♃
21 29	♀ ± ♄
22 38	☽ ∥ ☉
23 38	☽ ⊻ ♅
04 Monday	
00 28	☿ ∠ ♃
05 17	☉ ⊻ ♂
07 05	☽ ∥ ♃
09 03	☽ ☌ ☿
09 25	☽ ∠ ♃
11 51	☽ ⊥ ♅
15 13	♀ ☌ ♂
05 Tuesday	
00 45	☽ ⚹ ♇
03 06	☽ ☌ ♆
04 01	☽ ⊥ ♃
06 11	☽ ∠ ♅
06 38	☉ ⊻ ♀
16 23	☿ St D
16 24	☽ ⊻ ♃
19 11	☽ ⊻ ♂
20 24	☽ ☌ ☉
20 41	☽ ⊻ ♀
06 Wednesday	
01 07	☽ Q ♇
03 32	☽ ☍ ♄
08 04	☽ ⊥ ♂
10 15	☽ ⊥ ♀
12 41	☽ ⚹ ♅
21 47	☽ ⊻ ☿
07 Thursday	
03 20	☽ ∠ ♂
06 32	☽ ∠ ♀
10 07	☽ ⊥ ☿
12 58	☽ Q ♅
13 30	☽ □ ♇
15 55	☽ ⊻ ♆
08 Friday	
03 59	☽ ⊥ ♆
04 15	☉ Q ♇
04 42	☽ ∠ ☿
06 01	☽ ☌ ♃
08 56	☽ ∥ ♃
10 27	♀ ⊼ ♄
11 12	☽ ⚹ ♂
11 21	☽ ∥ ☉
14 24	☽ ⊻ ☉
15 26	☽ ⊼ ♄
16 03	☽ ⚹ ♀
21 59	☽ ∠ ♆
09 Saturday	
01 03	☽ □ ♅
01 15	☽ # ♄
01 52	☉ ☍ ♄
03 16	☽ ± ♄
03 25	☽ ⊥ ☉
11 44	☽ ⚹ ☿
12 21	☽ ∥ ♆
12 23	☽ Q ♂
18 29	☽ Q ♀
20 53	☽ ⚼ ♄
22 44	☽ ∠ ☉
10 Sunday	
01 09	☽ △ ♇
02 19	♀ ∠ ♇
03 36	☽ ⚹ ♆
04 30	☽ ∥ ☿
06 31	☿ ∠ ♃
09 35	☽ ∥ ♂
10 34	☽ ∥ ♀
12 27	☽ Q ☿
12 40	☿ ⊥ ♅
13 41	☉ ∥ ♃
18 09	☽ ⊻ ♃
11 Monday	
01 25	☽ □ ♂
01 46	☽ △ ♄
01 57	☽ # ♇
02 52	☽ Q ♆
05 52	☽ ⊥ ♃
06 10	☽ ⚼ ♇
06 21	☽ ⚹ ☉
07 23	♂ ⊼ ♄
09 11	☽ □ ♀
11 35	☽ △ ♅
14 56	♀ ∥ ♂
23 14	☽ ∠ ♃
23 15	☽ ± ♇
12 Tuesday	
01 01	☽ □ ☿
02 53	☽ ∥ ♅
06 58	☽ Q ☉
10 16	♀ ⊻ ♅
10 28	☽ ⊼ ♇
12 54	☽ □ ♆
15 47	☽ ⚼ ♅
13 Wednesday	
03 29	☽ ⚹ ♃
08 17	☽ ± ♅
09 22	☽ □ ♄
12 21	☽ △ ♂
18 36	☽ □ ☉
19 06	☽ ⊼ ♅
22 21	☽ △ ♀
14 Thursday	
01 25	☽ Q ♃
01 38	☉ ⚹ ♅
11 33	☽ △ ☿
16 18	☽ ⚼ ♂
16 28	☽ ☍ ♇
17 05	☽ Q ♄
18 50	☽ △ ♆
19 51	♂ ∠ ♇
23 31	☽ # ♅
15 Friday	
03 09	☽ ⚼ ♀
08 34	☽ ± ♂
09 12	☽ □ ♃
13 32	☽ ⚹ ♄
15 27	☽ ⚼ ☿
19 14	☽ ⊼ ♂
19 54	☽ ± ♀
20 28	☽ ⚼ ♆
21 53	☽ ∥ ♇
23 06	☽ ☍ ♅
16 Saturday	
02 26	☽ △ ☉
06 52	☽ ⊼ ♀
07 46	☽ ± ☿
11 29	☽ ± ♆
14 25	☽ ∠ ♄
18 10	☽ # ♂
18 29	☽ ⊼ ☿
19 03	☽ ⊼ ♇
21 22	☽ ⊼ ♆
22 03	☽ # ♀
17 Sunday	
01 49	☿ ⚹ ♇
02 50	☽ # ☿
04 50	☽ ± ♇
04 56	☽ ⚼ ☉
05 02	☽ ⊥ ♄
06 08	☽ # ♆
09 45	♀ ⊥ ♅
11 35	☽ △ ♃
14 38	☽ ∥ ♄
14 40	☽ ⊻ ♄
19 22	☽ ⚼ ♇
20 30	☽ ± ☉
22 40	☽ ☍ ♂
18 Monday	
00 10	☽ ⊼ ♅
02 31	☽ # ♃
06 47	☽ ⊼ ☉
07 12	☿ ☌ ♆
07 35	☽ # ☉
09 45	☽ ± ♅
11 51	☽ ☍ ♀
11 56	☽ ⚼ ♃
13 09	♀ ∠ ♃
19 16	☽ △ ♇
21 27	☿ ∥ ♆
21 35	☽ ☍ ♆
22 49	☽ ☍ ☿
19 Tuesday	
00 04	☽ ⚼ ♅
02 26	☽ ± ♃
09 51	♂ ⊻ ♅
12 04	☽ ⊼ ♃
14 07	☽ ☌ ♄
14 23	☿ ∠ ♅
23 52	☽ △ ♅
20 Wednesday	
00 29	☽ ⊼ ♂
09 45	☽ ☍ ☉
10 27	☽ ± ♂
15 40	☽ ⊼ ♀
18 53	☽ □ ♇
21 17	☽ ⊼ ♆
21 Thursday	
01 30	☽ ⚼ ♂
02 08	☽ ± ♀
02 19	☿ ⊥ ♃
02 51	☽ ⊼ ☿
06 14	☽ # ☉
06 57	☽ ± ♆
12 25	☽ # ♃
12 39	☽ ☍ ♃
13 28	☽ ± ☿
13 45	☽ ⊻ ♄
15 06	☉ ⊥ ♂
18 01	☽ ⚼ ♀
21 35	☽ ⚼ ♆
23 28	☽ ⊥ ♄
23 55	☽ ∥ ♄
23 58	☽ Q ♇
22 Friday	
00 08	☽ □ ♅
01 57	☽ # ☿
03 00	☽ △ ♂
03 51	♀ ∥ ♆
05 38	☽ ⚼ ☿
06 36	♀ ⚹ ♇
07 55	☽ # ♀
08 11	☽ # ♆
13 58	☽ ⊼ ☉
14 14	☽ ∠ ♄
14 52	☽ # ♂
19 51	☽ ⚹ ♇
21 06	☽ △ ♀
22 28	☽ △ ♆
23 Saturday	
00 46	☽ ± ☉
01 39	☿ # ♄
06 11	☽ Q ♅
09 21	☽ △ ☿
13 12	♀ ☌ ♆
15 23	☽ ⊼ ♃
15 26	☽ ⚹ ♄
16 59	☽ ∥ ♇
17 18	☽ ⚼ ☉
18 20	♃ ⊼ ♄
21 24	☽ ∠ ♇
24 Sunday	
01 56	☽ ± ♃
02 52	☽ ⚹ ♅
08 27	☽ □ ♂
12 14	☽ Q ♄
13 10	☽ ⊥ ♇
15 32	☽ # ♅
18 04	☽ ⚼ ♃
19 54	♀ ∠ ♅
21 46	☽ △ ☉
23 53	☽ ⊻ ♇
25 Monday	
02 48	☽ □ ♆
05 36	☽ ∠ ♅
06 35	☽ □ ♀
18 36	☽ Q ♂
20 15	☿ ☍ ♄
20 36	☽ □ ♄
20 38	☽ □ ☿
21 48	☽ △ ♃
22 07	☽ ⊥ ♅
26 Tuesday	
01 40	☉ □ ♇
06 20	☉ Q ♅
07 11	☿ Q ♇
09 17	☿ ⊻ ♃
09 20	☽ ⊻ ♅
12 05	☽ Q ♆
18 18	☽ ⚹ ♂
19 36	☽ Q ♀
27 Wednesday	
07 47	☽ ☌ ♇
09 31	♀ ⊥ ♃
10 27	☽ □ ☉
10 56	☽ Q ☿
11 02	☽ ⚹ ♆
16 00	☽ ∥ ♅
17 12	♀ # ♄
17 48	☉ ⊻ ♆
21 17	☽ ⚹ ♀
28 Thursday	
00 54	☽ ∠ ♂
05 32	☽ △ ♄
05 36	☿ ∥ ♃
08 16	☽ □ ♃
13 30	☽ ⚹ ☿
16 29	☽ ∠ ♆
16 46	♂ ⊥ ♅
17 28	☽ # ♇
19 32	☽ ☌ ♅
19 41	☽ ⊥ ♂
29 Friday	
01 01	☽ Q ☉
06 19	☽ ∠ ♀
08 22	☽ ⊻ ♂
10 28	☽ ⊥ ♆
11 08	☽ ⚼ ♄
19 01	☽ ⊻ ♇
20 51	☽ Q ♃
22 33	☽ ⊻ ♆
23 34	☽ ∠ ☿
30 Saturday	
02 07	♂ ∥ ♆
02 31	☽ ⊥ ♀
03 20	☽ ⚹ ☉
05 08	☽ ± ♄
07 07	☽ ⊥ ♇
08 52	☽ ∥ ♆
09 11	☽ ∥ ♂
16 04	☽ ⊻ ♀
16 41	☿ ⚹ ♅
17 12	☽ ⊼ ♄
18 24	☽ # ♄
20 02	☉ ∥ ☿
20 19	☽ ⊥ ☿
21 36	☽ ⚹ ♃
23 47	☽ ∥ ♀
31 Sunday	
01 21	☽ ∠ ♇
02 15	♀ ☍ ♄
06 39	☽ ∥ ♃
08 05	☽ ⊻ ♅
10 13	☽ ⊻ ☿
12 36	☽ ∠ ☉
12 52	☽ ∥ ☉
15 08	☽ ∥ ☿
20 18	☽ ⊥ ♅
22 29	♀ Q ♇

All ephemeris data is given at 12.00 UT and the Moon's longitude is additionally given for 24.00 UT

JANUARY 1973

LONGITUDES

Date	Sidereal time h m s	Sun ☉ ° ' "	Moon ☽ ° ' "	Moon ☽ 24.00 ° ' "	Mercury ☿ ° '	Venus ♀ ° '	Mars ♂ ° '	Jupiter ♃ ° '	Saturn ♄ ° '	Uranus ♅ ° '	Neptune ♆ ° '	Pluto ♇ ° '
01	18 43 58	10 ♑ 56 44	06 ♐ 28 18	12 ♐ 23 21	25 ♐ 22	17 ♐ 07	01 ♐ 14	17 ♑ 56	15 ♊ 16	22 ♎ 45	06 ♐ 15	04 ♎ 25
02	18 47 55	11 57 55	18 ♐ 19 16	24 ♐ 16 22	26 51	18 22	01 55	18 10	15 R 12	22 47	06 17	04 25
03	18 51 51	12 59 06	00 ♑ 14 55	06 ♑ 15 11	28 20	19 37	02 36	18 24	15 08	22 48	06 19	04 25
04	18 55 48	14 00 17	12 ♑ 17 24	18 ♑ 21 44	29 ♐ 50	20 52	03 17	18 38	15 04	22 49	06 21	04 26
05	18 59 44	15 01 28	24 ♑ 28 25	00 ♒ 37 37	01 ♑ 20	22 07	03 58	18 52	15 00	22 51	06 23	04 26
06	19 03 41	16 02 39	06 ♒ 49 29	13 ♒ 04 12	02 51	23 22	04 39	19 06	14 56	22 52	06 25	04 R 26
07	19 07 37	17 03 50	19 ♒ 21 56	25 ♒ 42 50	04 22	24 37	05 20	19 20	14 52	22 53	06 26	04 26
08	19 11 34	18 05 00	02 ♓ 07 06	08 ♓ 34 52	05 54	25 52	06 01	19 34	14 48	22 54	06 28	04 26
09	19 15 31	19 06 10	15 ♓ 06 20	21 ♓ 41 38	07 26	27 07	06 42	19 48	14 45	22 55	06 30	04 25
10	19 19 27	20 07 19	28 ♓ 20 58	05 ♈ 04 27	08 59	28 22	07 23	20 02	14 41	22 56	06 32	04 25
11	19 23 24	21 08 28	11 ♈ 52 12	18 ♈ 44 18	10 32	29 ♐ 37	08 04	20 16	14 38	22 57	06 34	04 25
12	19 27 20	22 09 36	25 ♈ 40 47	02 ♉ 41 35	12 05	00 ♑ 52	08 45	20 30	14 34	22 58	06 35	04 25
13	19 31 17	23 10 44	09 ♉ 46 37	16 ♉ 55 40	13 39	02 07	09 26	20 44	14 31	22 58	06 37	04 25
14	19 35 13	24 11 51	24 ♉ 08 26	01 ♊ 24 31	15 13	03 22	10 07	20 58	14 28	22 59	06 39	04 24
15	19 39 10	25 12 57	08 ♊ 43 24	16 ♊ 04 27	16 48	04 38	10 49	21 12	14 25	23 00	06 40	04 24
16	19 43 06	26 14 03	23 ♊ 26 57	00 ♋ 50 08	18 23	05 53	11 30	21 26	14 21	23 01	06 42	04 24
17	19 47 03	27 15 08	08 ♋ 13 07	15 ♋ 35 01	19 59	07 08	12 11	21 40	14 19	23 01	06 44	04 23
18	19 51 00	28 16 12	22 ♋ 54 57	00 ♌ 12 03	21 36	08 23	12 52	21 54	14 16	23 02	06 45	04 23
19	19 54 56	29 ♑ 17 15	07 ♌ 25 30	14 ♌ 34 35	23 13	09 38	13 33	22 08	14 13	23 02	06 47	04 23
20	19 58 53	00 ♒ 18 18	21 ♌ 38 40	28 ♌ 37 16	24 50	10 53	14 15	22 22	14 10	23 02	06 48	04 22
21	20 02 49	01 19 20	05 ♍ 30 02	12 ♍ 16 44	26 28	12 08	14 56	22 36	14 08	23 03	06 50	04 22
22	20 06 46	02 20 22	18 ♍ 57 16	25 ♍ 31 42	28 07	13 23	15 37	22 50	14 05	23 03	06 51	04 21
23	20 10 42	03 21 23	02 ♎ 00 12	08 ♎ 23 01	29 ♑ 46	14 38	16 19	23 04	14 03	23 03	06 53	04 21
24	20 14 39	04 22 24	14 ♎ 40 31	20 ♎ 53 10	01 ♒ 26	15 53	17 00	23 18	14 01	23 04	06 54	04 20
25	20 18 35	05 23 24	27 ♎ 01 27	03 ♏ 05 55	03 06	17 08	17 41	23 32	13 58	23 04	06 55	04 19
26	20 22 32	06 24 24	09 ♏ 07 10	15 ♏ 05 47	04 47	18 23	18 23	23 46	13 56	23 04	06 57	04 19
27	20 26 29	07 25 23	21 ♏ 02 24	26 ♏ 57 39	06 29	19 38	19 04	24 00	13 54	23 R 04	06 58	04 18
28	20 30 25	08 26 21	02 ♐ 52 07	08 ♐ 46 24	08 12	20 53	19 46	24 13	13 53	23 04	06 59	04 17
29	20 34 22	09 27 19	14 ♐ 41 06	20 ♐ 36 45	09 55	22 08	20 27	24 27	13 51	23 04	07 01	04 17
30	20 38 18	10 28 16	26 ♐ 33 51	02 ♑ 32 52	11 38	23 24	21 09	24 41	13 49	23 03	07 02	04 16
31	20 42 15	11 ♒ 29 12	08 ♑ 34 15	14 ♑ 38 21	13 ♒ 23	24 ♑ 39	21 ♐ 50	24 ♑ 55	13 ♊ 48	23 ♎ 03	07 ♐ 03	04 ♎ 15

Moon nodes and latitude

Date	Moon True ☊ ° '	Moon Mean ☊ ° '	Moon ☽ Latitude ° '
01	16 ♑ 48	17 ♑ 13	03 S 26
02	16 R 47	17 10	02 32
03	16 47	17 07	01 31
04	16 46	17 04	00 S 25
05	16 D 46	17 01	00 N 43
06	16 46	16 57	01 49
07	16 R 47	16 54	02 51
08	16 46	16 51	03 46
09	16 46	16 48	04 31
10	16 46	16 45	05 01
11	16 46	16 42	05 16
12	16 D 46	16 38	05 14
13	16 46	16 35	04 52
14	16 46	16 32	04 13
15	16 47	16 29	03 16
16	16 48	16 26	02 06
17	16 48	16 22	00 N 47
18	16 R 48	16 19	00 S 34
19	16 48	16 16	01 52
20	16 47	16 13	03 01
21	16 45	16 10	03 58
22	16 42	16 07	04 40
23	16 40	16 03	05 06
24	16 38	16 00	05 16
25	16 37	15 57	05 11
26	16 D 37	15 54	04 52
27	16 38	15 51	04 21
28	16 39	15 48	03 39
29	16 41	15 44	02 48
30	16 43	15 41	01 49
31	16 ♑ 44	15 ♑ 38	00 S 45

DECLINATIONS

Date	Sun ☉ ° '	Moon ☽ ° '	Mercury ☿ ° '	Venus ♀ ° '	Mars ♂ ° '	Jupiter ♃ ° '	Saturn ♄ ° '	Uranus ♅ ° '	Neptune ♆ ° '	Pluto ♇ ° '
01	23 S 00	24 S 46	23 S 26	21 S 57	20 S 09	22 S 25	21 N 02	08 S 17	19 S 48	13 N 20
02	22 54	25 27	23 36	22 06	20 18	22 23	21 02	08 17	19 48	13 21
03	22 49	24 57	23 45	22 14	20 27	22 21	21 01	08 17	19 48	13 21
04	22 42	23 17	23 52	22 22	20 35	22 19	21 01	08 18	19 49	13 21
05	22 36	20 32	23 59	22 30	20 44	22 18	21 01	08 18	19 49	13 22
06	22 29	16 49	24 04	22 36	20 52	22 16	21 01	08 19	19 49	13 23
07	22 21	12 18	24 08	22 42	21 00	22 14	21 00	08 19	19 50	13 23
08	22 13	07 12	24 11	22 48	21 08	22 12	21 00	08 20	19 50	13 24
09	22 05	01 S 43	24 12	22 52	21 15	22 10	21 00	08 20	19 50	13 24
10	21 56	03 N 57	24 12	22 56	21 23	22 08	21 00	08 20	19 50	13 25
11	21 47	09 33	24 11	22 59	21 30	22 06	21 00	08 21	19 51	13 25
12	21 37	14 47	24 08	23 02	21 37	22 04	20 59	08 21	19 51	13 26
13	21 27	19 22	24 05	23 03	21 44	22 02	20 59	08 21	19 51	13 26
14	21 17	22 53	23 59	23 04	21 51	22 00	20 59	08 21	19 51	13 27
15	21 06	24 59	23 53	23 05	21 58	21 58	20 59	08 22	19 52	13 28
16	20 54	25 23	23 45	23 04	22 04	21 56	20 59	08 22	19 52	13 28
17	20 43	23 59	23 35	23 03	22 10	21 54	20 59	08 22	19 52	13 29
18	20 31	20 56	23 24	23 02	22 16	21 52	20 59	08 22	19 52	13 30
19	20 18	16 37	23 12	22 59	22 22	21 49	20 59	08 22	19 53	13 30
20	20 05	11 26	22 58	22 56	22 28	21 47	20 58	08 22	19 53	13 31
21	19 52	05 48	22 43	22 52	22 33	21 45	20 58	08 23	19 53	13 32
22	19 39	00 N 05	22 26	22 48	22 38	21 43	20 58	08 23	19 53	13 32
23	19 25	05 S 28	22 08	22 42	22 43	21 41	20 58	08 23	19 53	13 33
24	19 10	10 38	21 48	22 36	22 48	21 38	20 58	08 23	19 54	13 34
25	18 56	15 15	21 27	22 30	22 53	21 36	20 58	08 23	19 54	13 34
26	18 40	19 09	21 04	22 22	22 57	21 34	20 58	08 23	19 54	13 35
27	18 25	22 13	20 40	22 14	23 02	21 32	20 58	08 23	19 54	13 36
28	18 09	24 19	20 14	22 06	23 06	21 29	20 58	08 23	19 54	13 37
29	17 53	25 20	19 46	21 56	23 09	21 27	20 58	08 23	19 55	13 37
30	17 37	25 13	19 18	21 46	23 13	21 25	20 58	08 23	19 55	13 38
31	17 S 20	23 S 55	18 S 47	21 S 35	23 S 17	21 S 22	20 N 58	08 S 22	19 S 55	13 N 39

ZODIAC SIGN ENTRIES

Date	h	m	Planets
03	11	30	☽ ♑
04	14	41	☿ ♑
05	22	47	☽ ♒
08	08	03	☽ ♓
10	14	57	☽ ♈
11	19	15	♀ ♑
12	19	24	☽ ♉
14	21	41	☽ ♊
16	22	39	☽ ♋
18	23	40	☽ ♌
20	04	48	☉ ♒
21	02	23	☽ ♍
23	08	16	☽ ♎
23	15	23	☿ ♒
25	17	52	☽ ♏
28	06	10	☽ ♐
30	18	54	☽ ♑

LATITUDES

Date	Mercury ☿ ° '	Venus ♀ ° '	Mars ♂ ° '	Jupiter ♃ ° '	Saturn ♄ ° '	Uranus ♅ ° '	Neptune ♆ ° '	Pluto ♇ ° '
01	00 S 04	00 N 53	00 N 16	00 S 10	01 S 36	00 N 37	01 N 35	16 N 28
04	00 26	00 45	00 14	00 11	01 36	00 37	01 35	16 30
07	00 46	00 38	00 12	00 11	01 35	00 37	01 35	16 31
10	01 04	00 30	00 10	00 11	01 35	00 37	01 35	16 33
13	01 20	00 22	00 08	00 12	01 34	00 38	01 35	16 35
16	01 35	00 14	00 06	00 12	01 33	00 38	01 35	16 36
19	01 46	00 N 06	00 04	00 12	01 33	00 38	01 35	16 38
22	01 56	00 S 01	00 N 02	00 12	01 32	00 38	01 35	16 40
25	02 02	00 09	00 00	00 13	01 31	00 38	01 36	16 41
28	02 05	00 17	00 S 03	00 13	01 31	00 38	01 36	16 43
31	02 S 04	00 S 24	00 S 05	00 S 13	01 S 30	00 N 38	01 N 36	16 N 44

DATA

Julian Date	2441684
Delta T	+43 seconds
Ayanamsa	23° 29' 04"
Synetic vernal point	05° ♓ 37' 55"
True obliquity of ecliptic	23° 26' 36"

MOON'S PHASES, APSIDES AND POSITIONS ☽

Date	h	m	Phase	Longitude ° '	Eclipse Indicator
04	15	42	●	14 ♑ 10	Annular
12	05	27	☽	21 ♈ 53	
18	21	28	○	28 ♋ 40	
26	06	05	☾	06 ♏ 09	

Day	h	m		
16	20	56	Perigee	
28	15	47	Apogee	
02	13	59	Max dec	25° S 27'
09	19	17	0N	
16	05	15	Max dec	25° N 27'
22	12	19	0S	
29	21	26	Max dec	25° S 26'

LONGITUDES

Date	Chiron ⚷ ° '	Ceres ⚳ ° '	Pallas ⚴ ° '	Juno ⚵ ° '	Vesta ⚶ ° '	Black Moon Lilith ⚸ ° '
01	12 ♈ 53	23 ♏ 44	03 ♏ 13	03 ♐ 05	01 ♊ 58	04 ♐ 55
11	13 ♈ 01	27 ♏ 32	06 ♏ 26	06 ♐ 08	01 ♊ 07	06 ♐ 02
21	13 ♈ 13	01 ♐ 09	09 ♏ 18	09 ♐ 02	01 ♊ 00	07 ♐ 09
31	13 ♈ 31	04 ♐ 32	11 ♏ 46	11 ♐ 45	01 ♊ 36	08 ♐ 16

All ephemeris data is given at 12.00 UT and the Moon's longitude is additionally given for 24.00 UT

ASPECTARIAN

h m	Aspects
01 Monday	
00 44	☽ ☌ ♂
04 40	☽ ∠ ♃
07 50	☽ ⚹ ♇
08 37	☽ ⊥ ☉
11 33	☽ ☌ ♆
14 36	☽ ∠ ♅
21 56	☽ ⊻ ☉
23 18	☽ ⊥ ♃
02 Tuesday	
05 43	☽ ☍ ♄
07 09	♀ ⊻ ♃
08 10	☽ Q ♇
11 41	☽ ⊻ ♃
12 07	☽ ☌ ♀
19 44	☉ ⊥ ♆
21 00	☽ ⚹ ♅
03 Wednesday	
07 37	☽ ☌ ☿
17 00	☽ ⊻ ♂
20 21	☽ □ ♇
21 07	☽ Q ♅
04 Thursday	
00 09	☽ ⊻ ♆
04 31	♀ ∥ ♃
05 37	☽ ∥ ☿
05 41	☽ ⊥ ♂
12 07	☽ ⊥ ♆
15 42	☽ ☌ ☉
17 27	☽ ⊼ ♄
18 10	☽ ∥ ☉
20 37	☽ ∥ ♀
21 35	☽ ∥ ♃
05 Friday	
00 33	☽ ∠ ♂
00 47	☽ ☌ ♃
05 13	☽ ± ♄
05 55	☽ ∠ ♆
06 34	♂ ∠ ♃
06 52	☽ ⊻ ♀
08 20	☽ # ♄
08 48	☽ □ ♅
10 36	☽ ∥ ♂
11 21	☉ ⊼ ♄
17 04	☽ ∥ ♆
19 56	☽ ⊥ ♀
22 40	☉ ∥ ♀
22 43	☽ ⚼ ♄
06 Saturday	
02 04	♀ ⚹ ♅
03 15	☽ ⊻ ☿
04 10	♂ ⚹ ♇
06 54	♇ St R
07 22	☽ △ ♇
07 33	☽ ⚹ ♂
11 12	☽ ⚹ ♆
15 19	☽ ∠ ♀
16 27	☽ ⊥ ☿
07 Sunday	
03 29	☽ △ ♄
06 33	☽ # ♇
07 14	☽ ⊻ ☉
07 55	☽ Q ♂
10 14	☽ Q ♆
11 56	☽ ⊻ ♃
12 01	☽ ∠ ☿
12 07	☽ ⚼ ♇
12 52	☿ □ ♇
14 07	♂ # ♄
18 40	☽ △ ♅
19 37	☽ ⊥ ☉
20 06	☿ Q ♅
23 01	☽ ⚹ ♀
23 30	☽ ⊥ ♃
08 Monday	
05 06	☽ ± ♇
06 53	☽ ∥ ♅
13 57	☽ ∠ ☉
15 18	☿ ⊻ ♂
16 18	☽ ⊼ ♇
16 39	☽ ∠ ♃
17 16	☉ ∥ ♃
19 39	☽ □ ♂
19 59	☽ ⚹ ☿
20 07	☽ □ ♆
21 07	☿ ⊻ ♆
22 45	☽ ⚼ ♅
23 50	☽ Q ♀
09 Tuesday	
04 41	♂ ☌ ♆
11 21	☽ □ ♄
15 19	☽ ± ♅
19 54	☽ ⚹ ☉
20 43	☽ ⚹ ♃
20 56	☽ Q ☿
10 Wednesday	
02 14	☽ ⊼ ♅
09 19	☉ ☌ ♃
12 03	☽ □ ♀
18 42	☽ Q ♃
19 18	☽ Q ☉
19 43	☽ Q ♄
22 50	☽ ☍ ♇
11 Thursday	
00 32	☉ ± ♄
02 37	☽ △ ♆
04 57	☽ △ ♂
06 45	☽ # ♅
07 40	♂ ∠ ♅
09 20	☽ □ ☿
16 49	☽ ⚹ ♄
22 07	☉ ∠ ♆
12 Friday	
02 54	☽ □ ♃
04 56	☽ ⚼ ♆
05 27	☽ □ ☉
05 32	☽ ∥ ♇
07 19	☽ ☍ ♅
08 31	☽ ⚼ ♂
12 00	☉ ∥ ♂
17 40	♃ ± ♄
18 39	☽ ∠ ♄
19 53	☿ ⊥ ♆
20 26	☽ ± ♆
21 46	☽ △ ♀
13 Saturday	
00 43	☽ ± ♂
02 55	☽ ⊼ ♇
06 39	☽ ⊼ ♆
07 07	☉ □ ♅
09 53	☽ ⊥ ♄
11 24	☽ ⊼ ♂
13 04	☽ ± ♇
14 54	☽ # ♆
19 19	☽ △ ☿
19 56	☽ ⊻ ♄
22 02	☽ ∥ ♄
14 Sunday	
00 33	☽ # ☉
00 47	☿ ⊼ ♄
01 30	☽ ⚼ ♀
03 41	☽ # ♂
04 08	☽ ⚼ ♇
05 08	☽ # ♃
06 39	☽ △ ♃
10 05	☽ ⊼ ♅
12 06	☽ △ ☉
13 35	☽ # ♀
17 51	☽ ± ♀
20 01	☽ ± ♅
21 54	☽ # ☿
23 16	☽ ⚼ ☿
15 Monday	
04 39	☽ ⊼ ♀
04 55	☽ △ ♇
07 45	♀ □ ♇
07 48	☽ ⚼ ♃
08 38	☽ ☍ ♆
10 49	☽ ⚼ ♅
12 10	☿ ⊥ ♂
13 04	♂ ∥ ♃
14 37	☽ ⚼ ☉
15 35	☽ ☍ ♂
15 49	☽ ± ☿
19 13	♀ Q ♅
21 15	☽ ☌ ♄
22 45	☽ ± ♃
16 Tuesday	
02 39	☉ # ♄
02 47	☽ ⊼ ☿
06 23	☽ ± ☉
08 41	☽ ⊼ ♃
11 17	☽ △ ♅
16 52	☽ ⊼ ☉
17 Wednesday	
04 08	♀ ⊻ ♆
05 47	☽ □ ♇
06 54	☉ ∠ ♂
09 34	☽ ⊼ ♆
10 04	☽ ☍ ♀
16 14	☽ # ☿
16 39	☿ ± ♄
18 27	♃ ∠ ♆
18 46	☽ ⊼ ♂
19 21	☽ ± ♆
20 44	☽ # ♀
21 53	☽ ⊻ ♄
18 Thursday	
03 09	☽ # ♂
05 03	☽ ± ♂
05 46	☽ # ♃
07 40	☽ ⊥ ♄
09 34	☽ ☍ ☿
10 05	☽ ⚼ ♆
10 19	☽ ☍ ♃
11 08	☽ Q ♇
11 46	☽ ∥ ♄
12 11	☽ □ ♅
14 22	☿ ∠ ♆
14 51	☽ # ☉
17 21	☿ ☌ ♃
18 31	☽ # ♆
20 33	☽ ⚼ ♂
21 28	☽ ☍ ☉
22 25	☽ ∠ ♄
19 Friday	
06 56	☽ ⚹ ♇
09 23	☿ □ ♅
10 22	☉ ⚼ ♄
10 55	☽ △ ♆
16 02	☽ ⊼ ♀
18 03	☽ Q ♅
22 48	☽ △ ♂
23 21	☽ ⚹ ♄
20 Saturday	
02 42	☽ ∥ ♇
03 07	☽ ± ♀
08 07	☽ ∠ ♇
09 35	♂ ☍ ♄
13 16	☽ ⊼ ♃
14 23	☽ ⚹ ♅
16 10	☿ ∥ ♀
18 11	☽ ⊼ ☿
19 45	☽ Q ♄
19 59	☽ ⚼ ♀
23 33	☽ ⊥ ♇
23 46	☽ ± ♃
21 Sunday	
01 10	☽ # ♅
04 07	☽ ⊼ ☉
05 58	☽ ± ☿
10 00	☽ ⊻ ♇
10 19	☉ ∥ ♆
14 20	☽ □ ♆
15 28	☽ ± ☉
15 46	☽ ⚼ ♃
16 29	☽ ∠ ♅
22 37	☿ ∥ ♂
22 Monday	
00 01	☽ ⚼ ☿
00 56	☽ △ ♀
01 39	♀ ⊥ ♆
03 16	☽ □ ♄
05 40	☽ □ ♂
08 34	☽ ⊥ ♅
08 50	☽ ⚼ ☉
19 11	☽ △ ♃
19 28	☽ ⊻ ♅
22 47	☽ Q ♆
23 Tuesday	
01 06	♀ ⊼ ♄
01 50	☿ ⚼ ♄
07 13	☽ △ ☿
09 34	♀ ∥ ♂
10 53	♃ □ ♅
13 05	♂ Q ♇
14 45	☽ △ ☉
16 23	☽ ☌ ♇
16 34	☽ Q ♂
21 10	☽ ⚹ ♆
24 Wednesday	
01 14	☽ ∥ ♅
10 44	☽ △ ♄
11 03	☉ △ ♇
14 35	☽ □ ♀
16 44	☽ ⚹ ♂
25 Thursday	
00 14	☿ ∥ ♃
01 56	☿ ∠ ♂
02 00	☽ ∠ ♆
02 54	☽ # ♇
03 37	♂ ⊥ ♃
04 14	☽ ☌ ♅
05 01	☽ □ ♃
15 50	☽ ⚼ ♄
19 42	☽ ⊥ ♆
23 52	☽ ∠ ♂
26 Friday	
01 57	☽ □ ☿
02 25	☽ ⊻ ♇
05 15	☿ △ ♇
05 55	☽ Q ♀
06 05	☽ □ ☉
07 39	☽ ⊻ ♆
09 02	☽ ∥ ☉
09 39	☽ ± ♄
11 45	♀ ⊻ ♂
14 23	☽ ⊥ ♇
17 19	☽ ∥ ♆
17 24	☽ Q ♃
17 47	☿ # ♄
18 56	☽ ⊥ ♂
21 39	☽ ⊼ ♄
27 Saturday	
00 30	☽ ∥ ☿
01 03	☉ ⚹ ♆
01 23	☽ # ♄
05 29	♅ St R
06 02	☽ ∥ ♃
07 47	☽ ⊻ ♂
08 29	☽ ∠ ♇
08 50	☽ ⚹ ♀
12 13	☽ ∥ ♀
16 06	☽ ⊻ ♅
17 01	♀ ± ♄
18 06	☽ ⚹ ♃
18 55	☿ ⚹ ♆
20 09	☽ Q ☿
20 15	☽ ∥ ♂
21 43	☽ Q ☉
28 Sunday	
04 16	☽ ⊥ ♅
14 53	☽ ⚹ ♇
18 52	☽ ∠ ♀
20 24	☽ ☌ ♆
20 31	☉ ☌ ☿
22 33	☽ ∠ ♅
29 Monday	
00 23	☽ ⚹ ☉
00 39	☽ ⚹ ☿
01 10	☽ ∠ ♃
05 03	☿ ∥ ♆
09 30	♀ ∠ ♆
10 18	☽ ☍ ♄
15 13	☽ Q ♇
15 18	☽ ⊥ ♀
19 47	☽ ⊥ ♃
30 Tuesday	
00 24	☽ ☌ ♂
04 52	☽ ⊻ ♀
04 56	☽ ⚹ ♅
05 35	♀ □ ♅
08 09	☽ ⊻ ♃
09 36	☽ ∠ ☉
12 10	☽ ∠ ☿
31 Wednesday	
03 24	☽ □ ♇
05 00	☽ Q ♅
05 18	☽ ⊥ ☉
08 59	☽ ⊻ ♆
09 14	☽ ⊥ ☿
17 39	☿ △ ♄
18 18	☽ ⊻ ☉
18 20	♀ ☌ ♃
19 14	☽ ∥ ♂
20 53	☽ ⊥ ♆
22 19	☽ ⊼ ♄
23 06	☽ ⊻ ☿

FEBRUARY 1973

LONGITUDES

Date	Sidereal time h m s	Sun ☉ ° ′ ″	Moon ☽ ° ′ ″	Moon ☽ 24.00 ° ′ ″	Mercury ☿ ° ′	Venus ♀ ° ′	Mars ♂ ° ′	Jupiter ♃ ° ′	Saturn ♄ ° ′	Uranus ♅ ° ′	Neptune ♆ ° ′	Pluto ♇ ° ′
01	20 46 11	12 ♒ 30 08	20 ♑ 45 29	26 ♑ 55 56	15 ♒ 08	25 ♑ 54	22 ♐ 32	25 ♑ 09	13 ♊ 46	23 ♎ 03	07 ♐ 04	04 ♎ 14
02	20 50 08	13 31 02	03 ♒ 09 53	09 ♒ 27 30	16 53	27 09	23 14	25 22	13 R 45	23 R 03	07 06	04 R 13
03	20 54 04	14 31 55	15 ♒ 48 51	22 ♒ 13 58	18 39	28 24	23 55	25 36	13 44	23 02	07 07	04 12
04	20 58 01	15 32 47	28 ♒ 42 49	05 ♓ 15 20	20 26	29 ♑ 39	24 37	25 50	13 43	23 02	07 08	04 11
05	21 01 58	16 33 38	11 ♓ 51 24	18 ♓ 30 53	22 13	00 ♒ 54	25 19	26 03	13 42	23 01	07 09	04 10
06	21 05 54	17 34 28	25 ♓ 13 36	01 ♈ 59 22	24 01	02 09	26 00	26 17	13 41	23 01	07 10	04 09
07	21 09 51	18 35 16	08 ♈ 47 59	15 ♈ 39 15	25 49	03 24	26 42	26 30	13 40	23 00	07 11	04 08
08	21 13 47	19 36 02	22 ♈ 33 00	29 ♈ 29 01	27 38	04 39	27 24	26 44	13 39	23 00	07 12	04 07
09	21 17 44	20 36 47	06 ♉ 27 10	13 ♉ 27 16	29 ♒ 26	05 54	28 05	26 57	13 39	22 59	07 13	04 06
10	21 21 40	21 37 31	20 ♉ 29 09	27 ♉ 32 41	01 ♓ 15	07 09	28 47	27 11	13 39	22 58	07 14	04 05
11	21 25 37	22 38 12	04 ♊ 37 41	11 ♊ 43 57	03 03	08 24	29 ♐ 29	27 24	13 38	22 57	07 15	04 04
12	21 29 33	23 38 53	18 ♊ 51 17	25 ♊ 59 24	04 51	09 39	00 ♑ 11	27 38	13 38	22 56	07 16	04 03
13	21 33 30	24 39 31	03 ♋ 08 00	10 ♋ 16 45	06 38	10 54	00 53	27 51	13 38	22 56	07 16	04 02
14	21 37 27	25 40 08	17 ♋ 25 13	24 ♋ 32 57	08 25	12 09	01 34	28 04	13 D 38	22 55	07 17	04 01
15	21 41 23	26 40 43	01 ♌ 39 26	08 ♌ 44 09	10 10	13 24	02 16	28 17	13 38	22 54	07 18	03 59
16	21 45 20	27 41 16	15 ♌ 46 32	22 ♌ 46 02	11 53	14 39	02 58	28 31	13 38	22 52	07 19	03 58
17	21 49 16	28 41 48	29 ♌ 42 08	06 ♍ 34 21	13 35	15 54	03 40	28 44	13 39	22 51	07 19	03 57
18	21 53 13	29 ♒ 42 19	13 ♍ 22 15	20 ♍ 05 29	15 14	17 09	04 22	28 57	13 39	22 50	07 20	03 56
19	21 57 09	00 ♓ 42 47	26 ♍ 43 48	03 ♎ 17 04	16 50	18 24	05 04	29 10	13 40	22 49	07 21	03 54
20	22 01 06	01 43 15	09 ♎ 45 13	16 ♎ 08 20	18 23	19 39	05 46	29 23	13 41	22 48	07 21	03 53
21	22 05 02	02 43 40	22 ♎ 26 34	28 ♎ 40 10	19 51	20 54	06 28	29 36	13 42	22 46	07 22	03 52
22	22 08 59	03 44 05	04 ♏ 49 30	10 ♏ 54 59	21 14	22 09	07 10	29 ♑ 49	13 43	22 45	07 22	03 50
23	22 12 56	04 44 28	16 ♏ 57 08	22 ♏ 56 28	22 33	23 24	07 52	00 ♒ 01	13 44	22 44	07 23	03 49
24	22 16 52	05 44 50	28 ♏ 53 37	04 ♐ 49 11	23 44	24 39	08 34	00 14	13 45	22 42	07 23	03 48
25	22 20 49	06 45 10	10 ♐ 43 50	16 ♐ 38 14	24 50	25 54	09 16	00 27	13 46	22 41	07 24	03 46
26	22 24 45	07 45 29	22 ♐ 33 03	28 ♐ 28 57	25 48	27 09	09 58	00 39	13 47	22 39	07 24	03 45
27	22 28 42	08 45 46	04 ♑ 26 36	10 ♑ 26 36	26 38	28 24	10 40	00 52	13 49	22 38	07 25	03 43
28	22 32 38	09 ♓ 46 02	16 ♑ 29 33	22 ♑ 35 58	27 ♓ 19	29 ♒ 39	11 ♑ 23	01 ♒ 04	13 ♊ 51	22 ♎ 36	07 ♐ 25	03 ♎ 42

Date	Moon True ☊ ° ′	Moon Mean ☊ ° ′	Moon ☽ Latitude ° ′
01	16 ♑ 44	15 ♑ 35	00 N 22
02	16 R 43	15 32	01 29
03	16 40	15 28	02 33
04	16 36	15 25	03 30
05	16 32	15 22	04 17
06	16 26	15 19	04 51
07	16 21	15 16	05 09
08	16 18	15 13	05 10
09	16 15	15 09	04 53
10	16 14	15 06	04 18
11	16 D 15	15 03	03 28
12	16 16	15 00	02 24
13	16 17	14 57	01 N 11
14	16 R 18	14 54	00 S 06
15	16 17	14 50	01 23
16	16 14	14 47	02 33
17	16 09	14 44	03 33
18	16 02	14 41	04 20
19	15 54	14 38	04 51
20	15 46	14 34	05 07
21	15 39	14 31	05 06
22	15 33	14 28	04 51
23	15 29	14 25	04 23
24	15 27	14 22	03 44
25	15 D 26	14 19	02 56
26	15 27	14 15	02 00
27	15 28	14 12	00 59
28	15 ♑ 29	14 ♑ 09	00 N 05

DECLINATIONS

Date	Sun ☉ ° ′	Moon ☽ ° ′	Mercury ☿ ° ′	Venus ♀ ° ′	Mars ♂ ° ′	Jupiter ♃ ° ′	Saturn ♄ ° ′	Uranus ♅ ° ′	Neptune ♆ ° ′	Pluto ♇ ° ′
01	17 S 03	21 S 29	18 S 15	21 S 24	23 S 20	21 S 20	20 N 58	08 S 22	19 S 55	13 N 40
02	16 46	18 00	17 42	21 12	23 23	21 17	20 58	08 22	19 55	13 40
03	16 29	13 40	17 07	21 00	23 26	21 15	20 59	08 22	19 55	13 41
04	16 11	08 38	16 31	20 46	23 28	21 13	20 59	08 22	19 55	13 42
05	15 53	03 S 09	15 53	20 32	23 31	21 10	20 59	08 22	19 56	13 43
06	15 34	02 N 33	15 14	20 18	23 33	21 08	20 59	08 21	19 56	13 44
07	15 15	08 14	14 33	20 03	23 35	21 05	20 59	08 21	19 56	13 44
08	14 57	13 34	13 52	19 47	23 37	21 03	20 59	08 21	19 56	13 45
09	14 37	18 17	13 08	19 31	23 38	21 00	20 59	08 21	19 56	13 46
10	14 18	22 01	12 24	19 14	23 39	20 58	21 00	08 20	19 56	13 47
11	13 58	24 28	11 39	18 57	23 41	20 55	21 00	08 20	19 56	13 48
12	13 38	25 22	10 52	18 39	23 41	20 53	21 00	08 20	19 56	13 49
13	13 18	24 35	10 05	18 20	23 42	20 50	21 00	08 19	19 56	13 49
14	12 58	22 12	09 17	18 01	23 43	20 48	21 00	08 19	19 57	13 50
15	12 37	18 27	08 28	17 42	23 43	20 45	21 01	08 18	19 57	13 51
16	12 17	13 40	07 39	17 22	23 43	20 43	21 01	08 18	19 57	13 52
17	11 56	08 15	06 50	17 01	23 43	20 40	21 01	08 18	19 57	13 53
18	11 35	02 N 32	06 01	16 40	23 42	20 37	21 02	08 17	19 57	13 53
19	11 13	03 S 09	05 13	16 19	23 42	20 35	21 02	08 17	19 57	13 54
20	10 52	08 34	04 25	15 57	23 41	20 32	21 02	08 16	19 57	13 55
21	10 30	13 28	03 38	15 35	23 40	20 30	21 03	08 16	19 57	13 56
22	10 08	17 42	02 52	15 12	23 39	20 27	21 03	08 15	19 57	13 57
23	09 46	21 06	02 08	14 49	23 37	20 25	21 03	08 15	19 57	13 58
24	09 24	23 34	01 26	14 25	23 35	20 22	21 04	08 14	19 57	13 58
25	09 02	24 58	00 47	14 01	23 33	20 19	21 04	08 13	19 57	13 59
26	08 40	25 14	00 S 10	13 37	23 31	20 17	21 04	08 13	19 57	14 00
27	08 17	24 21	00 N 23	13 12	23 29	20 14	21 05	08 12	19 57	14 01
28	07 S 54	22 S 20	00 N 53	12 S 47	23 S 26	20 S 12	21 N 05	08 S 12	19 S 57	14 N 02

ZODIAC SIGN ENTRIES

Date	h m	Planets
02	05 55	☽ ♒
04	14 22	☽ ♓
04	18 43	♀ ♒
06	20 29	☽ ♈
09	00 53	☽ ♉
09	19 30	☿ ♓
11	04 10	☽ ♊
12	05 51	♂ ♑
13	06 44	☽ ♋
15	09 12	☽ ♌
17	12 31	☽ ♍
18	19 01	☉ ♓
19	17 58	☽ ♎
22	02 35	☽ ♏
23	09 28	♃ ♒
24	14 14	☽ ♐
27	03 04	☽ ♑
28	18 45	♀ ♓

LATITUDES

Date	Mercury ☿ ° ′	Venus ♀ ° ′	Mars ♂ ° ′	Jupiter ♃ ° ′	Saturn ♄ ° ′	Uranus ♅ ° ′	Neptune ♆ ° ′	Pluto ♇ ° ′
01	02 S 03	00 S 26	00 S 06	00 S 13	01 S 30	00 N 38	01 N 36	16 N 45
04	01 56	00 33	00 08	00 14	01 29	00 38	01 36	16 46
07	01 45	00 40	00 11	00 14	01 28	00 38	01 36	16 47
10	01 28	00 46	00 13	00 14	01 28	00 38	01 36	16 49
13	01 05	00 52	00 16	00 15	01 27	00 38	01 37	16 50
16	00 36	00 58	00 18	00 15	01 26	00 39	01 37	16 51
19	00 S 01	01 03	00 21	00 15	01 25	00 39	01 37	16 52
22	00 N 40	01 08	00 24	00 16	01 25	00 39	01 37	16 53
25	01 23	01 12	00 26	00 16	01 24	00 39	01 37	16 54
28	02 07	01 16	00 29	00 17	01 23	00 39	01 37	16 55
31	02 N 47	01 S 19	00 S 32	00 S 17	01 S 23	00 N 39	01 N 37	16 N 56

DATA

Julian Date	2441715
Delta T	+43 seconds
Ayanamsa	23° 29′ 10″
Synetic vernal point	05° ♓ 37′ 50″
True obliquity of ecliptic	23° 26′ 36″

MOON'S PHASES, APSIDES AND POSITIONS ☽

Date	h m	Phase	Longitude ° ′	Eclipse Indicator
03	09 23	●	14 ♒ 25	
10	14 05	☽	21 ♉ 43	
17	10 07	○	28 ♌ 37	
25	03 10	☾	06 ♐ 23	

Day	h m		
13	10 51	Perigee	
25	12 36	Apogee	
06	01 18	ON	
12	12 57	Max dec	25° N 22′
18	22 36	OS	
26	05 42	Max dec	25° S 17′

LONGITUDES

Date	Chiron ⚷ ° ′	Ceres ⚳ ° ′	Pallas ⚴ ° ′	Juno ⚵ ° ′	Vesta ⚶ ° ′	Black Moon Lilith ⚸ ° ′
01	13 ♈ 33	04 ♐ 52	11 ♏ 59	12 ♐ 01	01 ♊ 42	08 ♐ 22
11	13 ♈ 55	07 ♐ 57	13 ♏ 54	14 ♐ 30	02 ♊ 59	09 ♐ 29
21	14 ♈ 22	10 ♐ 44	15 ♏ 15	16 ♐ 43	04 ♊ 50	10 ♐ 36
31	14 ♈ 51	13 ♐ 08	15 ♏ 55	18 ♐ 38	07 ♊ 08	11 ♐ 42

All ephemeris data is given at 12.00 UT and the Moon's longitude is additionally given for 24.00 UT

ASPECTARIAN

h m	Aspects
01 Thursday	
10 04	☽ ± ♄
12 38	☽ ∥ ♀
13 11	☽ ∥ ♃
14 34	☽ ∠ ♆
15 40	☽ ⊻ ♂
15 57	☽ # ♄
16 28	☽ □ ♅
20 42	☽ ☌ ♃
23 04	♀ ∥ ♃
23 07	☽ ☌ ♀
23 36	☽ ∥ ♆
02 Friday	
03 31	☽ P ♄
03 59	☽ ⊥ ♂
05 44	♂ ⚹ ♅
14 01	☽ △ ♇
14 10	☽ ∥ ☿
17 22	☉ △ ♄
19 31	☽ ⚹ ♆
19 51	☽ ∥ ☉
22 13	☽ ∠ ♂
03 Saturday	
08 05	☽ △ ♄
09 23	☽ ☌ ☉
11 53	☽ # ♇
13 51	♀ # ♄
18 11	☽ ☌ ☿
18 11	☽ Q ♆
18 14	☿ Q ♆
18 14	♀ P ♄
18 21	☽ P ♇
19 21	☿ P ♇
04 Sunday	
01 29	☽ △ ♅
04 00	☽ ⚹ ♂
06 35	☽ ⊻ ♃
11 02	☽ ± ♇
13 15	☽ ∥ ♅
13 54	☽ ⊻ ♀
17 50	☽ ⊥ ♃
22 02	☽ ⊼ ♇
05 Monday	
02 03	☽ ⊥ ♀
03 17	☽ Q ♂
03 27	☽ □ ♆
05 03	☽ P ♅
10 31	☽ ∠ ♃
12 43	☉ ∥ ☿
15 19	☽ □ ♄
20 03	☽ ∠ ♀
21 11	☽ ⊻ ☉
21 19	☽ ± ♅
22 38	☿ △ ♅
06 Tuesday	
05 35	♀ ⊥ ♂
08 03	☽ ⊼ ♅
08 49	☽ ⊥ ☉
09 31	☽ ⊻ ☿
13 27	☽ □ ♂
13 55	☽ ⚹ ♃
21 49	☽ ⊥ ☿
23 27	☽ Q ♄
07 Wednesday	
01 32	☽ ⚹ ♀
02 05	☽ ∠ ☉
02 14	♂ ⊻ ♃
03 49	☽ ☍ ♇
09 09	☽ △ ♆
11 29	☽ Q ♃
12 33	☽ # ♅
16 05	☽ ∠ ☿
20 31	☽ ⚹ ♄
22 26	☿ ⊻ ♃
22 42	♀ ∥ ♆
08 Thursday	
00 43	☽ Q ♀
00 53	☉ P ♇
01 57	♀ △ ♇
02 22	☉ Q ♆
06 28	☽ ⚹ ☉
06 58	☿ ⚹ ♂
11 24	☽ P ♆
12 46	☽ ☍ ♅
12 53	☽ ∥ ♇
13 12	☽ # ☿
15 29	☿ # ♇
18 12	☽ # ☉
18 32	☿ ± ♇
19 22	☽ □ ♃
20 50	☽ △ ♂
22 06	☽ ⚹ ☿
22 34	☽ ∠ ♄
09 Friday	
02 59	☽ ± ♆
04 53	☽ Q ☉
07 58	☽ ⊼ ♇
10 58	☽ □ ♀
13 19	☽ ⊼ ♆
14 03	☽ ⊥ ♄
18 16	☽ ± ♇
18 43	☽ # ♀
19 54	♃ # ♄
21 46	☽ # ♆
21 49	☽ Q ☿
23 58	☽ P ♂
10 Saturday	
00 20	☽ ⊻ ♄
04 30	☽ # ♃
04 37	☽ ∥ ♄
09 37	☽ P ♇
13 28	♀ ⚹ ♆
14 05	☽ □ ☉
16 07	☽ ± ♂
16 13	☽ ⊼ ♅
23 34	☽ △ ♃
11 Sunday	
02 24	☽ ± ♅
02 28	☽ # ♂
02 50	☽ ⊼ ♂
08 56	☽ □ ☿
11 03	☽ △ ♇
16 26	☽ ☍ ♆
17 24	☿ ⊥ ♃
17 37	☽ P ♅
19 00	☽ △ ♀
19 27	☉ △ ♅
12 Monday	
00 18	☉ # ♇
01 21	☽ P ♃
01 27	☿ ⊼ ♇
03 12	☽ ☌ ♄
16 44	☽ ± ♃
18 52	☽ △ ♅
20 41	☽ △ ☉
22 42	☽ P ♀
13 Tuesday	
02 59	☽ ⊼ ♃
08 01	☽ ☍ ♂
12 49	♄ St D
13 30	☽ □ ♇
15 16	☽ ± ♀
18 43	☽ △ ☿
18 58	☽ ⊼ ♆
20 39	☿ □ ♆
22 49	☽ # ♂
23 47	☽ P ☉
14 Wednesday	
02 18	☽ ⊼ ♀
05 03	☽ ± ♆
05 15	☿ P ♅
05 38	☽ ⊻ ♄
15 43	☽ ⊥ ♄
16 04	☽ ± ☉
19 43	☽ Q ♇
20 12	☽ P ♆
20 34	☽ ∥ ♄
21 14	☽ □ ♅
22 06	☽ # ♃
23 30	☽ P ☿
15 Thursday	
02 57	☽ ⊼ ☉
03 21	☽ # ♆
06 13	☽ ☍ ♃
06 54	☽ ∠ ♄
13 06	☽ ⊼ ♂
15 57	☽ ⚹ ♇
16 25	♀ △ ♄
16 26	☽ # ♀
16 50	☽ ± ☿
16 54	☿ ∥ ♅
21 34	☽ △ ♆
23 48	☽ ± ♂
16 Friday	
03 39	☽ Q ♅
04 27	☽ ⊼ ☿
08 21	☽ ⚹ ♄
09 54	☽ ☍ ♀
11 07	☽ ∥ ♇
15 57	☽ P ♂
17 28	☽ ∠ ♇
18 35	☉ ± ♇
18 49	☽ # ☉
17 Saturday	
00 10	☽ ⚹ ♅
02 49	♃ P ♄
04 58	☽ Q ♄
08 58	☽ ⊥ ♇
10 07	☽ ☍ ☉
10 17	☽ ⊼ ♃
11 48	☽ # ♅
12 55	☿ □ ♄
12 58	☉ ⊻ ♃
14 24	☿ ∠ ♃
18 59	☽ # ☿
19 17	☽ △ ♂
19 24	☽ ⊻ ♇
20 55	☽ ± ♃
21 23	♂ □ ♇
18 Sunday	
01 20	☽ □ ♆
02 14	☽ ∠ ♅
12 30	☽ □ ♄
13 02	☽ P ♃
15 46	☽ ☍ ☿
18 10	☽ ⊥ ♅
19 26	☽ ⊼ ♀
19 Monday	
03 40	♂ Q ♅
04 55	☽ ⊻ ♅
07 21	☽ ± ♀
09 29	☽ Q ♆
11 42	☿ ± ♅
16 31	☽ △ ♃
18 24	☿ Q ♂
19 47	☽ ∥ ☿
19 53	☽ ⊼ ☉
21 26	♀ P ♇
20 Tuesday	
01 07	☽ ☌ ♇
01 31	☽ P ♀
04 10	☽ □ ♂
06 11	♀ Q ♆
07 32	☽ ⚹ ♆
07 53	☽ ± ☉
10 40	☽ ∥ ♅
19 22	☽ △ ♄
22 11	☽ ∥ ☉
21 Wednesday	
02 14	☽ P ☉
06 24	☽ ⊼ ☿
08 44	☽ △ ♀
11 51	☽ ∠ ♆
12 38	☽ ☌ ♅
14 28	☽ # ♇
16 07	☽ Q ♂
19 23	☽ ± ☿
22 32	☽ ∥ ♀
22 Thursday	
00 04	☽ P ♄
02 02	☽ □ ♃
05 15	☽ ⊥ ♆
09 40	☽ △ ☉
10 04	☽ ⊻ ♇
12 31	♀ ∠ ♂
14 27	☉ ⊼ ♇
15 07	☽ P ☿
16 53	☽ ⚹ ♂
17 01	☽ ⊻ ♆
17 40	☽ ± ♄
19 10	♂ ⊻ ♆
21 51	☽ ⊥ ♇
23 14	♀ △ ♅
23 Friday	
03 08	☽ ∥ ♆
05 34	☽ ⊼ ♄
06 37	☽ ∥ ♃
11 35	☽ # ♄
14 11	☽ Q ♃
15 31	☿ ⊼ ♅
15 43	☽ ∠ ♇
23 33	☽ ⊻ ♅
24 Saturday	
00 29	☽ △ ☿
00 35	☽ ∠ ♂
02 27	☽ □ ♀
11 37	☽ ⊥ ♅
12 20	☽ ∥ ♂
14 46	☽ ⚹ ♃
19 54	☽ ⊥ ♂
21 54	☽ ⚹ ♇
25 Sunday	
02 44	☉ ⊥ ♃
03 10	☽ □ ☉
05 13	☽ ☌ ♆
05 49	☽ ∠ ♅
08 51	☽ ⊻ ♂
13 45	♀ # ♇
18 11	☽ ☍ ♄
19 12	☽ Q ♀
21 45	☽ ∠ ♃
22 13	☽ Q ♇
26 Monday	
03 28	☉ □ ♆
09 33	☉ P ♅
12 12	☽ ⚹ ♅
16 21	☽ ⊥ ♃
19 06	☽ Q ☉
19 06	☽ □ ☿
22 24	☽ ⚹ ♀
23 13	♀ ± ♇
27 Tuesday	
04 41	☽ ⊻ ♃
10 33	☽ □ ♇
12 22	☽ Q ♅
17 19	☉ ∥ ♅
17 57	☽ ⊻ ♆
21 26	☽ ⚹ ☉
28 Wednesday	
00 15	☽ ∥ ♂
01 14	☽ ☌ ♂
05 54	☽ ⊥ ♆
06 45	☽ ⊼ ♄
07 56	☽ ∠ ♀
09 34	☽ Q ☿
18 37	☽ ± ♄
22 40	☽ # ♄
23 39	☽ ∠ ♆
23 58	☽ □ ♅

MARCH 1973

LONGITUDES

Date	Sidereal time h m s	Sun ☉ ° ' "	Moon ☽ ° ' "	Moon ☽ 24.00 ° ' "	Mercury ☿ ° '	Venus ♀ ° '	Mars ♂ ° '	Jupiter ♃ ° '	Saturn ♄ ° '	Uranus ♅ ° '	Neptune ♆ ° '	Pluto ⯓ ° '
01	22 36 35	10 ♓ 46 17	28 ♑ 46 22	05 ♒ 01 08	27 ♓ 52	00 ♓ 54	12 ♑ 05	01 ♒ 17	13 ♊ 52	22 ♎ 34	07 ♐ 25	03 ♎ 40
02	22 40 31	11 46 29	11 ♒ 20 37	17 ♒ 45 01	28 16	02 09	12 47	01 29	13 54	22 R 33	07 25	03 R 39
03	22 44 28	12 46 40	24 ♒ 14 30	00 ♓ 49 04	28 31	03 24	13 29	01 42	13 56	22 31	07 26	03 37
04	22 48 25	13 46 50	07 ♓ 28 38	14 ♓ 13 01	28 36	04 39	14 11	01 54	13 58	22 29	07 26	03 36
05	22 52 21	14 46 57	21 ♓ 01 54	27 ♓ 54 53	28 R 31	05 53	14 54	02 06	14 00	22 27	07 26	03 34
06	22 56 18	15 47 03	04 ♈ 51 30	11 ♈ 51 11	28 18	07 08	15 36	02 18	14 03	22 25	07 26	03 33
07	23 00 14	16 47 06	18 ♈ 53 23	25 ♈ 57 30	27 56	08 23	16 18	02 30	14 05	22 23	07 26	03 31
08	23 04 11	17 47 08	03 ♉ 02 58	10 ♉ 09 14	27 26	09 38	17 00	02 42	14 07	22 22	07 26	03 30
09	23 08 07	18 47 07	17 ♉ 15 50	24 ♉ 22 18	26 49	10 53	17 43	02 54	14 10	22 20	07 26	03 28
10	23 12 04	19 47 04	01 ♊ 28 19	08 ♊ 33 35	26 05	12 08	18 25	03 05	14 13	22 18	07 R 26	03 26
11	23 16 00	20 47 00	15 ♊ 37 54	22 ♊ 41 06	25 16	13 22	19 07	03 17	14 16	22 15	07 26	03 25
12	23 19 57	21 46 53	29 ♊ 43 05	06 ♋ 43 46	24 23	14 37	19 50	03 29	14 18	22 13	07 26	03 23
13	23 23 54	22 46 43	13 ♋ 43 05	20 ♋ 40 58	23 27	15 52	20 32	03 40	14 21	22 11	07 26	03 22
14	23 27 50	23 46 32	27 ♋ 37 20	04 ♌ 32 05	22 30	17 07	21 14	03 52	14 25	22 09	07 26	03 20
15	23 31 47	24 46 18	11 ♌ 25 05	18 ♌ 16 09	21 32	18 21	21 57	04 03	14 28	22 07	07 26	03 18
16	23 35 43	25 46 01	25 ♌ 05 04	01 ♍ 51 36	20 35	19 36	22 39	04 14	14 31	22 05	07 25	03 17
17	23 39 40	26 45 43	08 ♍ 35 30	15 ♍ 16 28	19 41	20 51	23 22	04 25	14 34	22 03	07 25	03 15
18	23 43 36	27 45 23	21 ♍ 54 15	28 ♍ 28 36	18 50	22 06	24 04	04 36	14 38	22 00	07 25	03 14
19	23 47 33	28 45 00	04 ♎ 59 17	11 ♎ 26 10	18 02	23 20	24 47	04 47	14 42	21 58	07 25	03 12
20	23 51 29	29 ♓ 44 35	17 ♎ 49 07	24 ♎ 08 06	17 20	24 35	25 29	04 58	14 45	21 56	07 24	03 10
21	23 55 26	00 ♈ 44 09	00 ♏ 23 10	06 ♏ 34 26	16 42	25 49	26 11	05 09	14 49	21 53	07 24	03 09
22	23 59 23	01 43 41	12 ♏ 42 06	18 ♏ 46 28	16 10	27 04	26 54	05 20	14 53	21 51	07 24	03 07
23	00 03 19	02 43 11	24 ♏ 47 51	00 ♐ 46 43	15 44	28 19	27 36	05 30	14 57	21 49	07 23	03 05
24	00 07 16	03 42 39	06 ♐ 43 31	12 ♐ 38 50	15 25	29 ♓ 33	28 19	05 41	15 01	21 46	07 23	03 04
25	00 11 12	04 42 05	18 ♐ 33 14	24 ♐ 27 21	15 11	00 ♈ 48	29 02	05 51	15 05	21 44	07 22	03 02
26	00 15 09	05 41 30	00 ♑ 21 52	06 ♑ 17 28	15 03	02 03	29 ♑ 44	06 01	15 09	21 41	07 22	03 00
27	00 19 05	06 40 53	12 ♑ 14 49	18 ♑ 14 39	15 D 01	03 17	00 ♒ 27	06 12	15 14	21 39	07 21	02 59
28	00 23 02	07 40 14	24 ♑ 17 37	00 ♒ 24 24	15 05	04 32	01 09	06 22	15 18	21 37	07 20	02 57
29	00 26 58	08 39 33	06 ♒ 35 37	12 ♒ 51 49	15 14	05 46	01 52	06 32	15 22	21 34	07 20	02 55
30	00 30 55	09 38 50	19 ♒ 33 10	25 ♒ 41 04	15 28	07 01	02 34	06 41	15 27	21 32	07 19	02 54
31	00 34 51	10 ♈ 38 06	02 ♓ 14 46	08 ♓ 54 47	15 ♓ 48	08 ♈ 15	03 ♒ 17	06 ♒ 51	15 ♊ 32	21 ♎ 29	07 ♐ 18	02 ♎ 52

DECLINATIONS

Date	Moon True ☊ ° '	Moon Mean ☊ ° '	Moon ☽ Latitude ° '	Sun ☉ ° '	Moon ☽ ° '	Mercury ☿ ° '	Venus ♀ ° '	Mars ♂ ° '	Jupiter ♃ ° '	Saturn ♄ ° '	Uranus ♅ ° '	Neptune ♆ ° '	Pluto ⯓ ° '
01	15 ♑ 28	14 ♑ 06	01 N 11	07 S 32	19 S 15	01 N 19	12 S 21	23 S 24	20 S 09	21 N 06	08 S 11	19 S 57	14 N 03
02	15 R 25	14 03	02 14	07 09	15 14	01 41	11 56	23 21	20 06	21 06	08 10	19 57	14 03
03	15 19	13 59	03 12	06 46	10 25	01 58	11 29	23 17	20 04	21 07	08 10	19 57	14 04
04	15 11	13 56	04 01	06 23	05 S 02	02 11	11 03	23 14	20 01	21 07	08 09	19 57	14 05
05	15 01	13 53	04 38	06 00	00 N 42	02 18	10 36	23 10	19 59	21 08	08 08	19 57	14 06
06	14 51	13 50	04 59	05 36	06 30	02 22	10 09	23 07	19 56	21 08	08 08	19 57	14 07
07	14 41	13 47	05 03	05 13	12 04	02 20	09 42	23 03	19 53	21 09	08 07	19 57	14 08
08	14 33	13 44	04 48	04 50	17 02	02 13	09 15	22 58	19 51	21 09	08 06	19 57	14 08
09	14 26	13 40	04 16	04 26	21 05	02 02	08 47	22 54	19 48	21 10	08 05	19 57	14 09
10	14 22	13 37	03 28	04 03	23 51	01 47	08 19	22 49	19 46	21 10	08 05	19 57	14 10
11	14 21	13 34	02 27	03 39	25 06	01 27	07 50	22 44	19 43	21 11	08 04	19 57	14 11
12	14 D 21	13 31	01 17	03 16	24 44	01 05	07 22	22 39	19 40	21 11	08 03	19 57	14 11
13	14 21	13 28	00 N 03	02 52	22 48	00 39	06 53	22 34	19 38	21 12	08 02	19 57	14 12
14	14 R 21	13 25	01 S 10	02 28	19 30	00 N 11	06 25	22 28	19 35	21 12	08 01	19 57	14 13
15	14 18	13 21	02 19	02 05	15 08	00 S 19	05 56	22 23	19 33	21 13	08 01	19 56	14 14
16	14 12	13 18	03 18	01 41	10 28	00 49	05 26	22 17	19 30	21 14	08 00	19 56	14 15
17	14 04	13 15	04 06	01 17	04 N 33	01 21	04 57	22 11	19 28	21 14	07 59	19 56	14 15
18	13 53	13 12	04 40	00 54	01 S 04	01 52	04 28	22 04	19 25	21 15	07 58	19 56	14 16
19	13 41	13 09	04 58	00 30	06 32	02 22	03 58	21 58	19 23	21 16	07 57	19 56	14 17
20	13 28	13 05	05 00	00 S 06	11 37	02 52	03 29	21 51	19 20	21 16	07 56	19 56	14 17
21	13 15	13 02	04 48	00 N 18	16 06	03 20	02 59	21 44	19 18	21 17	07 56	19 56	14 18
22	13 05	12 59	04 22	00 41	19 49	03 46	02 29	21 37	19 15	21 17	07 55	19 56	14 19
23	12 57	12 56	03 45	01 05	22 37	04 10	01 59	21 30	19 13	21 18	07 54	19 56	14 20
24	12 52	12 53	02 59	01 29	24 22	04 32	01 29	21 22	19 11	21 19	07 53	19 56	14 20
25	12 49	12 50	02 05	01 52	25 01	04 52	00 59	21 15	19 08	21 19	07 52	19 55	14 21
26	12 48	12 46	01 05	02 16	24 32	05 09	00 S 29	21 07	19 06	21 20	07 51	19 55	14 22
27	12 D 48	12 43	00 S 03	02 39	22 56	05 23	00 N 01	20 59	19 03	21 21	07 50	19 55	14 22
28	12 R 48	12 40	01 N 01	03 03	20 16	05 35	00 31	20 51	19 01	21 21	07 49	19 55	14 23
29	12 46	12 37	02 02	03 26	16 39	05 45	01 01	20 42	18 59	21 22	07 48	19 55	14 24
30	12 42	12 34	03 00	03 49	12 12	05 52	01 32	20 34	18 56	21 23	07 47	19 55	14 24
31	12 ♑ 36	12 ♑ 31	03 N 50	04 N 13	07 S 06	05 S 56	02 N 02	20 S 25	18 S 54	21 N 24	07 S 46	19 S 55	14 N 25

ZODIAC SIGN ENTRIES

Date	h	m	Planets
01	14	22	☽ ♒
03	22	31	☽ ♓
06	03	37	☽ ♈
08	06	51	☽ ♉
10	09	31	☽ ♊
12	12	29	☽ ♋
14	16	07	☽ ♌
16	20	42	☽ ♍
19	02	48	☽ ♎
20	18	12	☉ ♈
21	11	15	☽ ♏
23	22	26	☽ ♐
24	20	34	♀ ♈
26	11	16	☽ ♑
26	20	59	♂ ♒
28	23	12	☽ ♒
31	07	55	☽ ♓

LATITUDES

Date	Mercury ☿ ° '	Venus ♀ ° '	Mars ♂ ° '	Jupiter ♃ ° '	Saturn ♄ ° '	Uranus ♅ ° '	Neptune ♆ ° '	Pluto ⯓ ° '
01	02 N 21	01 S 17	00 S 30	00 S 17	01 S 23	00 N 39	01 N 37	16 N 55
04	02 59	01 20	00 33	00 17	01 22	00 39	01 37	16 56
07	03 26	01 22	00 36	00 17	01 22	00 39	01 38	16 57
10	03 38	01 24	00 39	00 18	01 21	00 39	01 38	16 57
13	03 32	01 26	00 42	00 18	01 20	00 39	01 38	16 58
16	03 10	01 26	00 45	00 19	01 20	00 39	01 38	16 58
19	02 33	01 26	00 48	00 19	01 20	00 39	01 38	16 59
22	01 50	01 26	00 51	00 20	01 18	00 39	01 38	16 59
25	01 04	01 25	00 55	00 20	01 18	00 39	01 39	16 59
28	00 N 19	01 23	00 58	00 20	01 17	00 39	01 39	16 59
31	00 S 22	01 S 21	01 S 01	00 S 21	01 S 16	00 N 39	01 N 39	16 N 59

DATA

Julian Date	2441743
Delta T	+43 seconds
Ayanamsa	23° 29' 14"
Synetic vernal point	05° ♓ 37' 46"
True obliquity of ecliptic	23° 26' 37"

LONGITUDES

Date	Chiron ⚷ ° '	Ceres ⚳ ° '	Pallas ⚴ ° '	Juno ⚵ ° '	Vesta ⚶ ° '	Black Moon Lilith ⚸ ° '
01	14 ♈ 45	12 ♐ 41	15 ♏ 50	18 ♐ 17	06 ♊ 38	11 ♐ 29
11	15 ♈ 17	14 ♐ 44	15 ♏ 54	19 ♐ 55	09 ♊ 16	12 ♐ 36
21	15 ♈ 51	16 ♐ 19	15 ♏ 09	21 ♐ 09	12 ♊ 13	13 ♐ 42
31	16 ♈ 26	17 ♐ 21	13 ♏ 36	21 ♐ 57	15 ♊ 28	14 ♐ 49

MOON'S PHASES, APSIDES AND POSITIONS ☽

Date	h	m	Phase	Longitude ° '	Eclipse Indicator
05	00	07	●	14 ♓ 17	
11	21	26	☽	21 ♊ 11	
18	23	33	○	28 ♍ 14	
26	23	46	☾	06 ♑ 11	

Day	h	m		
10	08	09	Perigee	
25	08	50	Apogee	
05	09	06	0N	
11	18	32	Max dec	25° N 10'
18	07	25	0S	
25	13	43	Max dec	25° S 01'

ASPECTARIAN

h m	Aspects
01 Thursday	
03 38	☽ ⊥ ♀
05 40	☽ ∠ ☉
05 42	☽ ∥ ♃
07 11	☽ ∥ ♆
10 11	☽ ⚹ ☿
12 11	☽ ♇ ♄
16 33	☽ ⊻ ♀
16 55	☽ ☌ ♃
20 50	♀ ⊻ ♃
21 24	☽ △ ⯓
02 Friday	
00 32	☽ ⊥ ☉
04 35	☽ ⚹ ♆
12 53	☽ ⊻ ☉
14 52	☽ ⊻ ♂
15 43	☽ ∠ ☿
16 49	☽ △ ♄
18 10	☽ # ⯓
03 Saturday	
01 39	☽ ♇ ⯓
02 43	☽ ⊥ ♂
03 07	☽ Q ♆
06 27	☽ ∥ ♀
08 46	☽ ⊥ ☿
08 50	☽ △ ♅
09 58	♂ ⊥ ♆
16 19	♀ ⊼ ⯓
18 11	☽ ± ⯓
19 53	☽ ⊻ ☿
20 12	☽ ∠ ♂
22 21	☽ ∥ ♅
04 Sunday	
01 48	☽ ⊻ ♃
04 04	♂ ⊼ ♄
05 02	☽ ⊼ ⯓
05 44	☽ ∥ ☉
06 23	☽ ☌ ♀
11 55	☽ □ ♆
12 01	☽ ♇ ♅
12 46	☽ ⊥ ♃
12 58	☿ St R
16 41	☉ □ ♄
23 36	☽ □ ♄
23 45	☽ # ☿
05 Monday	
00 07	☽ ☌ ☉
00 36	☽ ⚹ ♂
03 58	☽ ± ♅
04 59	☽ ∠ ♃
14 29	☽ ⊼ ♅
18 42	☽ ∥ ☿
21 02	☉ ⚹ ♂
22 47	☽ Q ♂
06 Tuesday	
00 53	☽ ☌ ☿
02 00	♃ ∥ ♆
02 57	♅ ∠ ♆
07 08	☽ Q ♄
07 31	☽ ⚹ ♃
08 29	☽ # ☉
09 45	☽ ☍ ⯓
16 18	☽ ⊻ ♀
16 26	☽ △ ♆
17 21	♀ ♇ ♅
17 44	♀ □ ♆
18 49	☽ # ♅
07 Wednesday	
02 22	☽ # ♀
02 50	☉ ± ♅
03 35	☽ ⊥ ♀
03 47	☽ ⚹ ♄
04 06	☿ Q ♂
04 25	☽ Q ♃
07 22	☽ □ ♂
08 09	☽ ⊻ ☉
14 36	♀ ⊥ ♃
17 56	☽ ☍ ♅
18 02	☽ ♇ ♆
19 07	☽ ⊥ ☉
20 23	☽ ∠ ♀
21 32	☽ ∥ ⯓
08 Thursday	
02 51	☽ ⊻ ☿
05 20	☽ ∠ ♄
09 16	☽ ± ♆
09 22	☉ ∠ ♃
11 24	☽ □ ♃
11 31	☽ ∠ ☉
12 38	☽ ⊥ ☿
12 45	☽ ⊼ ⯓
19 25	☽ ⊼ ♆
20 36	☽ ⊥ ♄
22 52	☽ ± ⯓
09 Friday	
00 11	☽ ⚹ ♀
03 12	☽ ∠ ☿
03 46	☽ # ♃
04 35	☽ # ♆
06 45	☽ ⊻ ♄
12 36	☽ ∥ ♄
12 48	☽ △ ♂
14 02	☽ ♇ ⯓
14 34	♆ St R
14 46	☽ ⚹ ☉
20 31	☽ ⊼ ♅
22 24	☽ Q ♀
10 Saturday	
01 54	☽ # ♂
03 21	☽ ⚹ ☿
06 38	☽ ± ♅
12 34	☽ Q ☉
14 47	☽ △ ♃
15 20	☽ △ ⯓
15 28	☽ ♇ ♂
21 50	☽ ♇ ♅
22 06	☽ ☍ ♆
22 36	☽ Q ☿
11 Sunday	
00 20	♀ ∥ ♅
07 31	☽ ± ♂
07 48	☽ □ ♀
09 40	☽ ☌ ♄
16 34	☽ ♇ ♃
18 15	☽ ⊼ ♂
21 26	☽ □ ☉
23 15	☽ △ ♅
12 Monday	
02 05	♃ △ ⯓
03 26	☽ □ ☿
05 45	♀ □ ♄
08 07	☽ ± ♃
18 16	☽ □ ⯓
18 31	☽ ⊼ ♃
22 17	☉ ⊼ ♅
13 Tuesday	
01 13	☽ ⊼ ♆
05 30	♂ ± ♄
11 31	☽ ± ♆
13 06	☽ ⊻ ♄
14 05	☽ # ♂
16 03	☽ △ ♀
18 03	♀ ± ♅
20 16	☉ ☌ ☿
23 29	☽ ⊥ ♄
14 Wednesday	
00 22	☽ ☍ ♂
00 41	☽ ∥ ♄
01 09	☽ Q ⯓
02 34	☽ □ ♅
03 01	☽ ♇ ♆
03 42	☽ △ ☿
04 50	☽ △ ☉
09 12	☽ # ♆
11 25	☽ # ♃
15 07	☽ ∠ ♄
20 34	☽ ♇ ♀
20 50	☿ ⊼ ♅
21 54	☽ ⚹ ⯓
22 59	☽ ☍ ♃
15 Thursday	
04 02	☽ ♇ ☿
05 02	☽ △ ♆
06 01	☿ ⚹ ♂
08 54	☽ ♇ ☉
09 44	☽ Q ♅
13 48	☽ ± ♀
16 29	☽ ∥ ⯓
17 21	☽ ⚹ ♄
17 27	♂ □ ♅
18 44	☽ ± ☿
16 Friday	
00 03	☽ ∠ ⯓
01 22	☽ ⊼ ♀
01 54	☽ ± ☉
03 43	♀ ∠ ♃
04 15	♂ ∠ ♆
04 35	☽ ⊼ ☿
06 43	☽ ⚹ ♅
07 29	☽ ⊼ ♂
13 18	☽ ⊼ ☉
14 33	☽ Q ♄
15 52	☽ ⊥ ⯓
18 40	☽ ± ♂
21 06	☽ # ♅
22 57	☿ ☌ ♀
17 Saturday	
02 30	☽ ⊻ ⯓
04 27	☽ ⊼ ♃
09 14	☽ ∠ ♅
09 54	☽ □ ♆
10 05	☽ # ♀
10 28	☉ ∥ ☿
11 34	☽ ♇ ♂
15 20	☽ ± ♃
17 50	☿ ∠ ♃
22 47	☽ □ ♄
18 Sunday	
00 30	☽ # ☿
01 21	☽ ⊥ ♅
02 58	☽ # ☉
06 45	☽ ☍ ☿
07 46	☽ ♇ ♃
10 23	♀ ⊼ ♅
11 17	☽ ∥ ☉
12 11	☽ ⊻ ♅
12 23	☽ ☍ ♀
15 47	☽ ∥ ☿
16 10	☽ △ ♂
18 24	☽ Q ♆
23 33	☽ ☍ ☉
19 Monday	
01 32	☽ ∥ ♀
08 42	☽ ☌ ⯓
11 38	☽ △ ♃
16 30	☽ ⚹ ♆
18 29	☽ ∥ ♅
20 Tuesday	
06 12	☽ △ ♄
11 07	☽ ⊼ ☿
19 46	☽ ☌ ♅
20 42	☽ ∠ ♆
21 57	☽ ± ☿
21 Wednesday	
01 55	☽ # ⯓
02 16	☽ ⊼ ♀
03 08	☿ ∥ ♀
03 27	☽ □ ♂
10 54	☽ ♇ ♄
12 44	☽ ⊼ ☉
13 57	☽ ⊥ ♆
14 26	☽ ♇ ☿
15 05	☽ ± ♀
17 19	☽ ⊻ ⯓
21 22	☽ □ ♃
22 Thursday	
01 24	☽ ± ☉
01 36	☽ ⊻ ♆
04 25	♀ ⚹ ♂
04 28	☽ ± ♄
04 59	☽ ⊥ ⯓
08 04	☽ ∥ ♃
10 37	☽ ♇ ♀
12 51	☽ ∥ ♆
16 19	☽ ⊼ ♄
16 36	☽ Q ♂
18 36	☽ △ ☿
20 39	☽ ♇ ☉
22 40	☽ ∠ ⯓
23 40	☽ # ♄
23 Friday	
01 52	☽ ∥ ♂
06 04	☽ ⊻ ♅
07 14	☿ ± ♅
09 23	☽ Q ♃
17 54	☉ Q ♄
17 59	☽ ⚹ ♂
18 01	☽ ⊥ ♅
19 52	☽ △ ♀
20 40	☉ ☍ ⯓
24 Saturday	
04 37	☽ ⚹ ⯓
05 21	☽ △ ☉
09 51	☽ ⚹ ♃
12 06	☽ ∠ ♅
12 13	☉ # ♀
13 19	☽ ☌ ♆
22 48	♂ # ♄
25 Sunday	
02 13	☽ ∠ ♂
04 52	☽ Q ⯓
04 54	☽ ☍ ♄
05 15	☽ □ ☿
16 45	☽ ∠ ♃
18 26	☽ ⚹ ♅
21 40	☽ ⊥ ♂
21 56	☿ □ ♄
26 Monday	
10 38	☽ ⊻ ♂
11 18	☽ ⊥ ♃
15 48	☽ □ ♀
17 20	☽ □ ⯓
17 24	☽ Q ☿
18 43	☽ Q ♅
21 43	☉ ⚹ ♃
21 45	☿ ∠ ♂
23 38	☽ ⊻ ♃
23 46	☽ □ ☉
27 Tuesday	
02 09	☽ ⊻ ♆
03 48	♂ ♇ ♄
06 11	♀ ☍ ⯓
08 19	☿ St D
10 47	♀ Q ♄
14 13	☽ ⊥ ♆
17 34	☽ ⚹ ☿
18 01	☽ ⊼ ♄
28 Wednesday	
03 15	☽ # ♄
04 04	☉ △ ♆
06 02	☽ ± ♄
06 42	☽ □ ♅
07 15	☽ ∥ ♂
08 07	☽ Q ♀
08 08	☽ ∠ ♆
14 37	☽ ∥ ♆
14 57	☽ Q ☉
21 06	☽ ∥ ♃
23 29	☽ ∠ ☿
23 52	☽ ♇ ♄
29 Thursday	
02 17	☽ ☌ ♂
04 55	☽ △ ⯓
10 14	☽ ⚹ ♀
11 52	☽ ☌ ♃
13 25	☽ ⚹ ♆
16 19	☽ ⚹ ☉
17 09	☽ ⊥ ☿
30 Friday	
00 41	☽ # ⯓
04 47	☽ ⊻ ☿
04 49	♀ ⚹ ♃
04 52	☽ △ ♄
09 31	☽ ♇ ⯓
09 41	☿ □ ♄
12 11	☽ Q ♆
15 59	☿ ± ♅
16 17	☽ △ ♅
17 45	☽ ∠ ♀
17 52	♀ △ ♆
22 29	♂ △ ⯓
22 55	☽ ∠ ☉
31 Saturday	
02 12	☽ ± ⯓
08 57	☽ ∥ ♅
12 01	☽ ⊥ ♀
13 07	☽ ⊼ ⯓
13 59	☽ ⊻ ♂
16 40	☽ ⊥ ☉
17 05	☽ ∥ ☿
19 38	☽ ♇ ♅
20 25	☽ ⊻ ♃
21 07	☽ □ ♆
23 47	☽ # ☉
23 56	☽ ⊻ ♀

All ephemeris data is given at 12.00 UT and the Moon's longitude is additionally given for 24.00 UT

APRIL 1973

LONGITUDES

Date	Sidereal time h m s	Sun ☉ ° ' "	Moon ☽ ° ' "	Moon ☽ 24.00 ° ' "	Mercury ☿ ° '	Venus ♀ ° '	Mars ♂ ° '	Jupiter ♃ ° '	Saturn ♄ ° '	Uranus ♅ ° '	Neptune ♆ ° '	Pluto ♇ ° '
01	00 38 48	11 ♈ 37 19	15 ♓ 41 06	22 ♓ 33 35	16 ♓ 12	09 ♈ 30	04 ♒ 00	07 ♒ 01	15 ♊ 36	21 ♎ 27	07 ♐ 18	02 ♎ 50
02	00 42 45	12 36 31	29 ♓ 31 54	06 ♈ 35 35	16 41	10 44	04 42	07 10	15 41	21 R 24	07 R 17	02 R 49
03	00 46 41	13 35 41	13 ♈ 44 00	20 ♈ 56 23	17 14	11 59	05 25	07 20	15 46	21 21	07 16	02 47
04	00 50 38	14 34 48	28 ♈ 11 54	05 ♉ 29 38	17 51	13 13	06 07	07 29	15 51	21 19	07 15	02 46
05	00 54 34	15 33 54	12 ♉ 48 38	20 ♉ 07 59	18 33	14 28	06 50	07 38	15 56	21 16	07 15	02 44
06	00 58 31	16 32 57	27 ♉ 26 50	04 ♊ 44 25	19 17	15 42	07 33	07 47	16 01	21 14	07 14	02 42
07	01 02 27	17 31 59	12 ♊ 00 03	19 ♊ 13 13	20 06	16 56	08 15	07 56	16 07	21 11	07 13	02 41
08	01 06 24	18 30 58	26 ♊ 23 29	03 ♋ 30 36	20 57	18 11	08 58	08 05	16 12	21 09	07 12	02 39
09	01 10 20	19 29 54	10 ♋ 34 23	17 ♋ 34 44	21 52	19 25	09 40	08 13	16 17	21 06	07 11	02 38
10	01 14 17	20 28 49	24 ♋ 31 41	01 ♌ 25 14	22 50	20 40	10 23	08 22	16 23	21 03	07 10	02 36
11	01 18 14	21 27 41	08 ♌ 15 31	15 ♌ 02 37	23 50	21 54	11 06	08 30	16 28	21 01	07 09	02 34
12	01 22 10	22 26 30	21 ♌ 46 38	28 ♌ 27 40	24 53	23 08	11 48	08 38	16 34	20 58	07 08	02 33
13	01 26 07	23 25 18	05 ♍ 05 47	11 ♍ 41 02	25 59	24 23	12 31	08 46	16 40	20 56	07 07	02 31
14	01 30 03	24 24 03	18 ♍ 13 27	24 ♍ 43 01	27 08	25 37	13 13	08 54	16 45	20 53	07 06	02 30
15	01 34 00	25 22 45	01 ♎ 09 43	07 ♎ 33 32	28 18	26 51	13 56	09 02	16 51	20 51	07 05	02 28
16	01 37 56	26 21 26	13 ♎ 54 25	20 ♎ 12 20	29 ♓ 31	28 05	14 39	09 10	16 57	20 48	07 04	02 27
17	01 41 53	27 20 05	26 ♎ 27 18	02 ♏ 39 18	00 ♈ 46	29 ♈ 20	15 21	09 17	17 03	20 45	07 03	02 25
18	01 45 49	28 18 42	08 ♏ 48 24	14 ♏ 54 41	02 04	00 ♉ 34	16 04	09 25	17 09	20 43	07 01	02 24
19	01 49 46	29 ♈ 17 17	20 ♏ 58 19	26 ♏ 59 28	03 23	01 48	16 46	09 32	17 15	20 40	07 00	02 23
20	01 53 43	00 ♉ 15 50	02 ♐ 58 25	08 ♐ 55 27	04 44	03 02	17 29	09 39	17 21	20 38	06 59	02 21
21	01 57 39	01 14 22	14 ♐ 50 58	20 ♐ 45 22	06 08	04 16	18 12	09 46	17 27	20 35	06 58	02 20
22	02 01 36	02 12 52	26 ♐ 39 09	02 ♑ 32 50	07 33	05 30	18 54	09 53	17 33	20 33	06 57	02 18
23	02 05 32	03 11 20	08 ♑ 26 59	14 ♑ 22 13	09 01	06 45	19 37	10 00	17 40	20 30	06 55	02 17
24	02 09 29	04 09 46	20 ♑ 19 10	26 ♑ 18 30	10 30	07 59	20 19	10 06	17 46	20 28	06 54	02 16
25	02 13 25	05 08 11	02 ♒ 20 54	08 ♒ 27 02	12 01	09 13	21 02	10 13	17 52	20 25	06 53	02 14
26	02 17 22	06 06 34	14 ♒ 37 33	20 ♒ 53 06	13 34	10 27	21 45	10 19	17 59	20 23	06 51	02 13
27	02 21 18	07 04 56	27 ♒ 14 16	03 ♓ 41 34	15 08	11 41	22 27	10 25	18 05	20 20	06 50	02 12
28	02 25 15	08 03 16	10 ♓ 15 26	16 ♓ 56 09	16 45	12 55	23 10	10 31	18 12	20 18	06 49	02 10
29	02 29 12	09 01 35	23 ♓ 43 56	00 ♈ 38 45	18 23	14 09	23 52	10 36	18 19	20 16	06 47	02 09
30	02 33 08	09 ♉ 59 52	07 ♈ 40 27	14 ♈ 48 41	20 ♈ 03	15 ♉ 23	24 ♒ 35	10 ♒ 42	18 ♊ 25	20 ♎ 13	06 ♐ 46	02 ♎ 08

Date	Moon True ☊ ° '	Moon Mean ☊ ° '	Moon ☽ Latitude ° '
01	12 ♑ 26	12 ♑ 27	04 N 29
02	12 R 15	12 24	04 53
03	12 03	12 21	05 00
04	11 51	12 18	04 48
05	11 41	12 15	04 18
06	11 33	12 11	03 30
07	11 28	12 08	02 29
08	11 26	12 05	01 19
09	11 D 26	12 02	00 N 05
10	11 R 26	11 59	01 S 09
11	11 25	11 56	02 17
12	11 22	11 52	03 16
13	11 17	11 49	04 04
14	11 08	11 46	04 38
15	10 58	11 43	04 57
16	10 45	11 40	05 01
17	10 33	11 37	04 50
18	10 20	11 33	04 25
19	10 10	11 30	03 49
20	10 02	11 27	03 03
21	09 56	11 24	02 09
22	09 53	11 21	01 10
23	09 52	11 17	00 S 08
24	09 D 53	11 14	00 N 56
25	09 53	11 11	01 57
26	09 R 53	11 08	02 55
27	09 51	11 05	03 45
28	09 47	11 02	04 26
29	09 40	10 58	04 53
30	09 ♑ 32	10 ♑ 55	05 N 05

DECLINATIONS

Date	Sun ☉ ° '	Moon ☽ ° '	Mercury ☿ ° '	Venus ♀ ° '	Mars ♂ ° '	Jupiter ♃ ° '	Saturn ♄ ° '	Uranus ♅ ° '	Neptune ♆ ° '	Pluto ♇ ° '
01	04 N 36	01 S 31	05 S 58	02 N 32	20 S 16	18 S 52	21 N 24	07 S 45	19 S 54	14 N 25
02	04 59	04 N 18	05 58	03 02	20 07	18 50	21 25	07 45	19 54	14 26
03	05 22	10 02	05 56	03 32	19 58	18 47	21 26	07 44	19 54	14 27
04	05 45	15 19	05 51	04 02	19 48	18 45	21 26	07 43	19 54	14 27
05	06 08	19 46	05 45	04 31	19 39	18 43	21 27	07 42	19 54	14 28
06	06 30	23 00	05 36	05 01	19 29	18 41	21 28	07 41	19 54	14 28
07	06 53	24 42	05 25	05 31	19 19	18 39	21 29	07 40	19 53	14 29
08	07 16	24 42	05 13	06 00	19 09	18 37	21 29	07 39	19 53	14 29
09	07 38	23 06	04 58	06 30	18 59	18 34	21 30	07 38	19 53	14 30
10	08 00	20 05	04 42	06 59	18 48	18 32	21 31	07 37	19 53	14 30
11	08 22	16 00	04 24	07 28	18 38	18 30	21 32	07 36	19 53	14 31
12	08 44	11 09	04 05	07 57	18 27	18 28	21 32	07 35	19 52	14 31
13	09 06	05 52	03 44	08 25	18 16	18 26	21 33	07 34	19 52	14 32
14	09 28	00 N 24	03 21	08 54	18 05	18 25	21 34	07 33	19 52	14 32
15	09 49	05 S 00	02 57	09 22	17 54	18 23	21 34	07 32	19 52	14 33
16	10 10	10 06	02 32	09 50	17 43	18 21	21 35	07 31	19 52	14 33
17	10 32	14 42	02 05	10 18	17 31	18 19	21 36	07 30	19 51	14 33
18	10 53	18 37	01 36	10 46	17 20	18 17	21 37	07 29	19 51	14 34
19	11 13	21 41	01 07	11 13	17 08	18 15	21 37	07 28	19 51	14 34
20	11 34	23 45	00 36	11 40	16 56	18 14	21 38	07 27	19 51	14 35
21	11 54	24 43	00 S 04	12 07	16 44	18 12	21 39	07 26	19 50	14 35
22	12 15	24 34	00 N 30	12 34	16 32	18 10	21 40	07 25	19 50	14 35
23	12 35	23 18	01 04	13 00	16 20	18 09	21 40	07 25	19 50	14 36
24	12 55	20 59	01 40	13 26	16 07	18 07	21 41	07 24	19 50	14 36
25	13 14	17 44	02 17	13 52	15 55	18 06	21 42	07 23	19 49	14 36
26	13 34	13 40	02 54	14 17	15 42	18 04	21 43	07 22	19 49	14 36
27	13 53	08 54	03 33	14 42	15 30	18 03	21 43	07 21	19 49	14 37
28	14 12	03 S 37	04 13	15 06	15 17	18 01	21 44	07 20	19 49	14 37
29	14 30	02 N 00	04 53	15 31	15 04	18 00	21 45	07 19	19 48	14 37
30	14 N 49	07 N 43	05 N 35	15 N 54	14 S 51	17 S 58	21 N 46	07 S 18	19 S 48	14 N 37

ZODIAC SIGN ENTRIES

Date	h	m	Planets
02	12	48	☽ ♈
04	14	58	☽ ♉
06	16	12	☽ ♊
08	18	04	☽ ♋
10	21	31	☽ ♌
13	02	47	☽ ♍
15	09	50	☽ ♎
16	21	17	☿ ♈
17	18	51	☽ ♏
18	01	05	♀ ♉
20	05	30	☉ ♉
20	06	02	☽ ♐
22	18	49	☽ ♑
25	07	21	☽ ♒
27	17	10	☽ ♓
29	22	53	☽ ♈

LATITUDES

Date	Mercury ☿ ° '	Venus ♀ ° '	Mars ♂ ° '	Jupiter ♃ ° '	Saturn ♄ ° '	Uranus ♅ ° '	Neptune ♆ ° '	Pluto ♇ ° '
01	00 S 34	01 S 21	01 S 02	00 S 21	01 S 16	00 N 39	01 N 39	16 N 59
04	01 09	01 18	01 06	00 22	01 16	00 39	01 39	16 59
07	01 38	01 14	01 09	00 22	01 15	00 39	01 39	16 58
10	02 01	01 11	01 13	00 23	01 14	00 39	01 39	16 58
13	02 20	01 06	01 16	00 23	01 14	00 39	01 39	16 58
16	02 33	01 01	01 20	00 24	01 13	00 39	01 39	16 57
19	02 41	00 56	01 23	00 24	01 13	00 39	01 40	16 57
22	02 43	00 51	01 27	00 25	01 12	00 39	01 40	16 56
25	02 41	00 45	01 30	00 25	01 12	00 39	01 40	16 55
28	02 34	00 39	01 34	00 26	01 11	00 39	01 40	16 54
31	02 S 22	00 S 32	01 S 38	00 S 26	01 S 11	00 N 39	01 N 40	16 N 53

DATA

Julian Date	2441774
Delta T	+43 seconds
Ayanamsa	23° 29' 17"
Synetic vernal point	05° ♓ 37' 43"
True obliquity of ecliptic	23° 26' 37"

MOON'S PHASES, APSIDES AND POSITIONS ☽

Date	h	m	Phase	Longitude ° '	Eclipse Indicator
03	11	45	●	13 ♈ 35	
10	04	28	☽	20 ♋ 10	
17	13	51	○	27 ♎ 25	
25	17	59	☾	05 ♒ 23	

Day	h	m		
06	04	03	Perigee	
22	02	03	Apogee	
01	18	18	0N	
08	00	10	Max dec	24° N 55'
14	13	44	0S	
21	20	42	Max dec	24° S 48'
29	03	34	0N	

LONGITUDES

Date	Chiron ⚷ ° '	Ceres ⚳ ° '	Pallas ⚴ ° '	Juno ⚵ ° '	Vesta ⚶ ° '	Black Moon Lilith ⚸ ° '
01	16 ♈ 29	17 ♐ 26	13 ♏ 24	22 ♐ 00	15 ♊ 48	14 ♐ 56
11	17 ♈ 05	17 ♐ 48	11 ♏ 01	22 ♐ 14	19 ♊ 18	16 ♐ 02
21	17 ♈ 40	17 ♐ 32	08 ♏ 09	21 ♐ 56	22 ♊ 59	17 ♐ 09
31	18 ♈ 13	16 ♐ 37	05 ♏ 05	21 ♐ 04	26 ♊ 50	18 ♐ 15

All ephemeris data is given at 12.00 UT and the Moon's longitude is additionally given for 24.00 UT

ASPECTARIAN

h m	Aspects
01 Sunday	
01 22	☽ ⊥ ♂
04 15	☽ ⊻ ☉
07 14	☽ ⊥ ♃
08 05	☽ # ♀
11 34	☽ ± ♅
11 52	☽ □ ♄
12 56	☽ ☌ ☿
18 07	☽ ∠ ♂
22 02	☽ ⚻ ♅
23 11	☽ ∠ ♃
02 Monday	
06 18	☽ ∥ ♀
15 02	☽ ∥ ☉
17 35	☽ ☍ ♇
18 53	☽ # ☿
19 07	☽ Q ♄
21 16	☽ ✱ ♂
03 Tuesday	
01 07	☽ ✱ ♃
01 09	☽ △ ♆
02 16	☽ # ♅
04 01	♃ ✱ ♆
08 47	☽ ☌ ♀
11 45	☽ ☌ ☉
15 25	☽ ✱ ♄
18 06	☽ ⊻ ☿
18 27	☽ Q ♂
21 25	☽ Q ♃
21 34	♂ ∥ ♆
04 Wednesday	
00 39	☽ ☍ ♅
02 12	☽ ⅊ ♆
04 30	☽ ⊥ ☿
07 50	☽ ∥ ♇
16 24	☽ ∠ ♄
17 02	☽ ± ♆
17 24	☉ # ☿
19 30	☽ ⚻ ♇
20 02	☽ ∠ ☿
05 Thursday	
01 42	☽ □ ♂
02 53	☽ ⚻ ♆
03 25	☽ □ ♃
05 20	☽ ± ♇
05 47	☽ # ♃
07 16	☽ ⊥ ♄
11 14	☽ # ♂
12 45	☽ # ♆
14 57	☽ ⊻ ♀
16 50	☽ ⊻ ☉
17 09	☽ ⊻ ♄
20 03	☽ ⅊ ♇
21 53	☽ ✱ ☿
21 57	☉ ✱ ♄
23 13	☽ ∥ ♄
06 Friday	
01 36	♂ ✱ ♆
01 42	☽ ⊥ ♀
01 50	☽ ⚻ ♅
03 23	☽ ⊥ ☉
11 39	☽ ± ♅
17 51	☽ ∠ ♀
18 41	☽ Q ☿
18 42	♀ ✱ ♄
19 14	☽ ∠ ☉
20 38	☽ △ ♇
22 15	♂ ☌ ♃
07 Saturday	
02 25	☽ ⅊ ♅
04 05	☽ ☍ ♆
05 12	☽ △ ♃
05 29	☽ △ ♂
08 52	☿ # ♀
18 52	☽ ☌ ♄
20 59	☽ ✱ ♀
21 52	☽ ✱ ☉
08 Sunday	
02 18	☽ □ ☿
03 14	☽ △ ♅
06 23	☽ ⅊ ♃
07 43	☽ ⅊ ♂
16 49	☿ ⚻ ♅
18 59	☽ Q ♀
19 27	☽ Q ☉
21 40	☽ ± ♃
22 32	☽ □ ♇
23 40	☽ ± ♂
09 Monday	
06 14	☽ ⚻ ♆
07 57	☽ ⚻ ♃
10 23	☽ ⚻ ♂
11 57	☉ # ♅
16 27	☽ ± ♆
19 13	☉ ☌ ♀
21 51	☽ ⊻ ♄
10 Tuesday	
01 54	☽ ∥ ♄
04 28	☽ □ ☉
04 39	☽ □ ♀
05 13	☽ Q ♇
05 29	♀ Q ♃
06 01	☽ □ ♅
07 55	☽ ⅊ ♆
08 15	☽ ⊥ ♄
08 38	☉ Q ♃
08 50	☽ △ ☿
13 24	☽ # ♆
19 26	♀ ☍ ♅
20 32	☽ # ♂
21 50	☽ # ♃
11 Wednesday	
00 01	☽ ∠ ♄
01 31	☉ ☍ ♅
02 02	☽ ✱ ♇
02 52	☿ ∠ ♃
10 03	☽ △ ♆
12 26	☽ ☍ ♃
13 06	☽ ⅊ ☿
13 20	☽ Q ♅
16 49	♀ ⅊ ♆
17 17	☽ ☍ ♂
18 27	♀ # ♅
19 40	☽ ∥ ♇
12 Thursday	
02 38	☽ ✱ ♄
04 28	☽ ∠ ♇
04 36	☉ ⅊ ♆
06 24	☽ ± ☿
08 14	♂ ∥ ♃
10 34	☽ ✱ ♅
13 17	☽ △ ☉
14 41	☽ △ ♀
18 04	☽ ⚻ ☿
20 32	☽ ⊥ ♇
22 25	☽ ∥ ☉
13 Friday	
00 16	☽ Q ♄
01 29	☽ ∥ ♀
04 23	☽ # ♅
07 21	☽ ⊻ ♇
13 30	☽ ∠ ♅
15 40	☽ □ ♆
18 17	♀ Q ♂
18 32	☽ ⅊ ☉
18 45	☽ ⚻ ♃
20 36	☽ ⅊ ♀
22 03	☽ # ☿
14 Saturday	
02 18	☽ ⚻ ♂
05 50	☽ ± ♃
05 53	☽ ⊥ ♅
09 17	☽ □ ♄
12 21	☽ ± ☉
13 57	☽ ± ♂
14 50	☽ ± ♀
16 54	☽ ⊻ ♅
22 36	☽ ⅊ ♃
15 Sunday	
00 21	☽ ⚻ ☉
00 41	☽ Q ♆
03 07	☽ ⚻ ♀
03 27	☽ ∥ ☿
06 07	☽ ☍ ☿
07 36	☽ ⅊ ♂
14 27	☽ ☌ ♇
23 05	☽ ✱ ♆
23 40	☽ ∥ ♅
16 Monday	
02 56	☽ △ ♃
10 35	☽ # ♀
12 23	☽ # ☉
13 29	☽ △ ♂
17 40	☿ ∠ ♂
17 50	☽ △ ♄
17 Tuesday	
01 06	☽ ☌ ♅
03 32	☽ ∠ ♆
11 11	☽ # ♇
13 46	☉ Q ♂
13 51	☽ ☍ ☉
18 10	☽ ☍ ♀
20 51	☽ ⊥ ♆
21 18	☽ ⚻ ☿
22 54	☽ ⅊ ♄
23 32	☽ ⊻ ♇
18 Wednesday	
03 56	☽ ∥ ♂
08 31	☽ ⊻ ♆
09 45	☽ ∥ ♃
10 22	☽ ± ☿
11 12	☽ ⊥ ♇
13 12	☽ □ ♃
16 38	☽ ± ♄
18 05	☿ ☍ ♇
20 47	☽ ∥ ♆
20 49	♀ ± ♆
19 Thursday	
03 10	☽ □ ♂
04 33	☽ ⚻ ♄
04 53	☽ ∠ ♇
06 14	☽ ⅊ ☿
11 24	☽ ⊻ ♅
11 27	☽ # ♄
12 17	☉ ∥ ♀
21 30	♀ ∠ ♄
22 59	♀ ⚻ ♇
23 19	☽ ⊥ ♅
20 Friday	
01 13	☽ Q ♃
06 04	☽ ⚻ ☉
06 41	♂ △ ♄
07 39	♂ ⅊ ♇
10 45	☽ ✱ ♇
12 08	☽ ⚻ ♀
16 01	☽ △ ☿
17 20	☽ ∠ ♅
17 23	☽ Q ♂
19 13	☽ ± ☉
20 04	☽ ☌ ♆
23 26	☿ Q ♄
21 Saturday	
01 36	☽ ✱ ♃
01 39	☽ ± ♀
05 22	☉ ± ♆
10 57	☽ Q ♇
15 04	☽ ⅊ ☉
17 20	☽ ☍ ♄
19 14	☽ ✱ ♂
22 02	☽ ⅊ ♀
23 37	☽ ✱ ♅
22 Sunday	
01 55	☿ △ ♆
08 22	☽ ∠ ♃
13 17	♂ Q ♆
14 11	☉ ⚻ ♇
21 27	☉ ∠ ♄
23 29	☽ □ ♇
23 57	☽ Q ♅
23 Monday	
00 21	☽ △ ☉
02 52	☽ ⊥ ♃
03 42	☽ ∠ ♂
08 08	☽ △ ♀
08 54	☽ ⊻ ♆
13 18	☽ □ ☿
15 10	☽ ⊻ ♃
15 26	♀ ⚻ ♆
21 03	☽ ⊥ ♆
23 08	☽ ⊥ ♂
24 Tuesday	
05 14	☿ ✱ ♃
05 48	☽ # ♄
06 49	☽ ⚻ ♄
12 01	☽ ⊻ ♂
12 17	☽ □ ♅
15 10	☽ ∠ ♆
16 26	♂ △ ♅
17 23	♀ ± ♇
18 59	☽ ± ♄
21 22	☽ ∥ ♆
25 Wednesday	
06 42	☽ Q ☿
09 39	☽ ∥ ♃
11 47	☽ △ ♇
13 03	☽ ⅊ ♄
17 59	☽ □ ☉
20 54	☽ ✱ ♆
23 54	☽ ∥ ♂
26 Thursday	
03 00	☽ □ ♀
03 34	☽ ☌ ♃
06 49	☽ # ♇
08 55	☽ # ♀
09 09	♀ □ ♃
09 38	☽ ✱ ☿
12 31	☽ # ☉
16 59	☽ ⅊ ♇
18 31	☽ △ ♄
20 07	☽ Q ♆
23 00	☽ △ ♅
27 Friday	
02 26	☽ ☌ ♂
06 02	☉ ⚻ ♆
07 01	♀ ∥ ♇
07 37	☽ Q ☉
10 03	☽ ± ♇
17 03	☽ Q ♀
18 11	☽ ∠ ☿
19 19	☽ ∥ ♅
20 43	♀ ⊥ ♄
21 13	☽ ⚻ ♇
28 Saturday	
02 59	☽ ⅊ ♅
05 44	☽ □ ♆
07 41	☽ ✱ ☉
09 42	☽ # ☿
12 28	☽ ⊻ ♃
13 00	☽ ⊥ ☿
14 51	☉ ± ♇
17 18	☽ ✱ ♀
18 46	♀ # ♂
19 16	☽ ± ♅
23 20	☽ ⊥ ♃
29 Sunday	
01 16	☽ ⊻ ☿
02 22	☽ □ ♄
05 55	☽ ⚻ ♅
10 52	☿ ✱ ♄
12 15	☽ ⊻ ♂
12 33	☽ ∠ ☉
15 18	☽ ∠ ♃
20 47	☉ ∥ ♇
22 20	☽ ∠ ♀
23 14	☽ ⊥ ♂
30 Monday	
01 45	☽ ∥ ☿
02 34	☽ ☍ ♇
05 17	☽ ⊥ ☉
09 52	☽ Q ♄
10 17	☽ # ♅
10 28	☽ △ ♆
13 28	☉ # ♂
14 19	☿ ☍ ♅
15 10	☽ ⊥ ♀
15 23	☽ ∠ ♂
16 13	☽ ⊻ ☉
17 09	☽ ✱ ♃

MAY 1973

LONGITUDES

Date	Sidereal time h m s	Sun ☉ ° ' "	Moon ☽ ° ' "	Moon ☽ 24.00 ° ' "	Mercury ☿ ° '	Venus ♀ ° '	Mars ♂ ° '	Jupiter ♃ ° '	Saturn ♄ ° '	Uranus ♅ ° '	Neptune ♆ ° '	Pluto ♇ ° '
01	02 37 05	10 ♉ 58 07	22 ♈ 02 53	29 ♈ 22 19	21 ♈ 45	16 ♉ 37	25 ♒ 17	10 ♒ 47	18 ♊ 32	20 ♎ 11	06 ♐ 45	02 ♎ 07
02	02 41 01	11 56 21	06 ♉ 46 06	14 ♉ 13 10	23 29	17 51	26 00	10 53	18 39	20 R 08	06 R 43	02 R 05
03	02 44 58	12 54 33	21 ♉ 42 24	29 ♉ 12 39	25 15	19 05	26 42	10 58	18 46	20 06	06 42	02 04
04	02 48 54	13 52 43	06 ♊ 42 46	14 ♊ 11 39	27 02	20 19	27 24	11 03	18 52	20 04	06 40	02 03
05	02 52 51	14 50 52	21 ♊ 38 17	29 ♊ 01 50	28 ♈ 51	21 33	28 07	11 07	18 59	20 02	06 39	02 02
06	02 56 47	15 48 58	06 ♋ 21 35	13 ♋ 36 58	00 ♉ 42	22 47	28 49	11 12	19 06	19 59	06 37	02 01
07	03 00 44	16 47 03	20 ♋ 47 36	27 ♋ 53 14	02 35	24 01	29 ♒ 32	11 16	19 13	19 57	06 36	02 00
08	03 04 41	17 45 06	04 ♌ 53 46	11 ♌ 49 12	04 30	25 15	00 ♓ 14	11 21	19 20	19 55	06 34	01 59
09	03 08 37	18 43 07	18 ♌ 39 37	25 ♌ 25 10	06 27	26 29	00 56	11 25	19 27	19 53	06 33	01 58
10	03 12 34	19 41 06	02 ♍ 06 04	08 ♍ 42 33	08 25	27 42	01 38	11 29	19 35	19 51	06 31	01 57
11	03 16 30	20 39 03	15 ♍ 14 53	21 ♍ 43 18	10 25	28 ♉ 56	02 21	11 32	19 42	19 49	06 30	01 56
12	03 20 27	21 36 58	28 ♍ 08 03	04 ♎ 29 24	12 27	00 ♊ 10	03 03	11 36	19 49	19 46	06 28	01 55
13	03 24 23	22 34 51	10 ♎ 47 32	17 ♎ 02 40	14 30	01 24	03 45	11 39	19 56	19 44	06 27	01 54
14	03 28 20	23 32 43	23 ♎ 14 59	29 ♎ 24 41	16 35	02 38	04 27	11 42	20 03	19 42	06 25	01 53
15	03 32 16	24 30 33	05 ♏ 31 54	11 ♏ 36 49	18 42	03 52	05 09	11 45	20 10	19 40	06 23	01 52
16	03 36 13	25 28 22	17 ♏ 39 36	23 ♏ 40 24	20 50	05 05	05 51	11 48	20 18	19 38	06 22	01 51
17	03 40 10	26 26 09	29 ♏ 39 25	05 ♐ 36 51	22 59	06 19	06 33	11 51	20 25	19 37	06 20	01 51
18	03 44 06	27 23 55	11 ♐ 32 56	17 ♐ 27 55	25 08	07 33	07 15	11 53	20 32	19 35	06 19	01 50
19	03 48 03	28 21 39	23 ♐ 22 07	29 ♐ 15 50	27 19	08 46	07 57	11 55	20 40	19 33	06 17	01 49
20	03 51 59	29 ♉ 19 22	05 ♑ 09 28	11 ♑ 03 24	29 ♉ 30	10 00	08 39	11 57	20 48	19 31	06 16	01 48
21	03 55 56	00 ♊ 17 04	16 ♑ 58 05	22 ♑ 53 59	01 ♊ 42	11 14	09 21	11 59	20 55	19 29	06 14	01 48
22	03 59 52	01 14 45	28 ♑ 51 38	04 ♒ 51 35	03 54	12 27	10 03	12 01	21 03	19 28	06 12	01 47
23	04 03 49	02 12 25	10 ♒ 54 23	17 ♒ 00 38	06 05	13 41	10 44	12 03	21 10	19 26	06 11	01 46
24	04 07 45	03 10 04	23 ♒ 10 54	29 ♒ 25 49	08 16	14 55	11 26	12 04	21 18	19 24	06 09	01 46
25	04 11 42	04 07 41	05 ♓ 45 54	12 ♓ 11 43	10 26	16 08	12 08	12 05	21 25	19 23	06 07	01 45
26	04 15 39	05 05 18	18 ♓ 43 43	25 ♓ 22 19	12 35	17 22	12 49	12 06	21 33	19 21	06 06	01 45
27	04 19 35	06 02 54	02 ♈ 07 47	09 ♈ 00 19	14 42	18 36	13 31	12 07	21 41	19 20	06 04	01 44
28	04 23 32	07 00 29	15 ♈ 59 56	23 ♈ 06 28	16 48	19 49	14 13	12 07	21 48	19 18	06 03	01 44
29	04 27 28	07 58 03	00 ♉ 19 35	07 ♉ 38 46	18 53	21 03	14 54	12 08	21 56	19 17	06 01	01 43
30	04 31 25	08 55 36	15 ♉ 03 17	22 ♉ 32 16	20 55	22 16	15 35	12 08	22 04	19 15	05 59	01 43
31	04 35 21	09 ♊ 53 09	00 ♊ 04 40	07 ♊ 39 20	22 ♊ 55	23 ♊ 30	16 ♓ 17	12 ♒ 08	22 ♊ 11	19 ♎ 14	05 ♐ 58	01 ♎ 42

DECLINATIONS

Date	Moon True ☊ ° '	Moon Mean ☊ ° '	Moon ☽ Latitude ° '	Sun ☉ ° '	Moon ☽ ° '	Mercury ☿ ° '	Venus ♀ ° '	Mars ♂ ° '	Jupiter ♃ ° '	Saturn ♄ ° '	Uranus ♅ ° '	Neptune ♆ ° '	Pluto ♇ ° '
01	09 ♑ 23	10 ♑ 52	04 N 58	15 N 07	13 N 11	06 N 17	16 N 18	14 S 38	17 S 57	21 N 46	07 S 17	19 S 48	14 N 38
02	09 R 14	10 49	04 31	15 25	18 02	07 00	16 41	14 25	17 56	21 47	07 16	19 48	14 38
03	09 06	10 46	03 45	15 43	21 49	07 43	17 03	14 11	17 55	21 48	07 16	19 47	14 38
04	09 01	10 43	02 44	16 00	24 08	08 28	17 25	13 58	17 54	21 48	07 15	19 47	14 38
05	08 57	10 39	01 32	16 18	24 42	09 13	17 47	13 44	17 52	21 49	07 14	19 47	14 38
06	08 56	10 36	00 N 14	16 35	23 31	09 58	18 08	13 31	17 51	21 50	07 13	19 47	14 38
07	08 D 56	10 33	01 S 03	16 51	20 47	10 44	18 29	13 17	17 50	21 51	07 12	19 46	14 38
08	08 57	10 30	02 15	17 08	16 51	11 30	18 49	13 03	17 49	21 51	07 11	19 46	14 39
09	08 58	10 27	03 17	17 24	12 06	12 16	19 09	12 49	17 48	21 52	07 11	19 46	14 39
10	08 R 57	10 23	04 07	17 40	06 53	13 03	19 28	12 35	17 48	21 53	07 10	19 46	14 39
11	08 54	10 20	04 42	17 55	01 N 28	13 49	19 47	12 21	17 47	21 53	07 09	19 45	14 39
12	08 50	10 17	05 03	18 10	03 S 53	14 36	20 05	12 07	17 46	21 54	07 08	19 45	14 39
13	08 43	10 14	05 08	18 25	08 59	15 22	20 23	11 53	17 45	21 55	07 08	19 45	14 39
14	08 35	10 11	04 58	18 40	13 39	16 08	20 40	11 39	17 45	21 55	07 07	19 44	14 39
15	08 27	10 08	04 35	18 54	17 41	16 53	20 56	11 25	17 44	21 56	07 06	19 44	14 39
16	08 20	10 04	03 59	19 08	20 55	17 37	21 12	11 10	17 43	21 57	07 05	19 44	14 39
17	08 13	10 01	03 13	19 22	23 13	18 21	21 28	10 56	17 43	21 57	07 05	19 44	14 39
18	08 08	09 58	02 19	19 35	24 28	19 03	21 41	10 41	17 42	21 58	07 04	19 43	14 39
19	08 05	09 55	01 19	19 48	24 36	19 44	21 57	10 27	17 42	21 59	07 03	19 43	14 38
20	08 04	09 52	00 S 16	20 01	23 36	20 23	22 10	10 12	17 42	21 59	07 03	19 43	14 38
21	08 D 04	09 48	00 N 48	20 13	21 34	21 01	22 23	09 58	17 41	22 00	07 02	19 43	14 38
22	08 05	09 45	01 51	20 25	18 35	21 36	22 35	09 43	17 41	22 01	07 02	19 42	14 38
23	08 07	09 42	02 50	20 36	14 46	22 10	22 47	09 28	17 41	22 01	07 01	19 42	14 38
24	08 08	09 39	03 42	20 48	10 18	22 41	22 58	09 14	17 41	22 02	07 00	19 42	14 38
25	08 R 09	09 36	04 25	20 59	05 S 18	23 09	23 08	08 59	17 41	22 02	07 00	19 41	14 38
26	08 08	09 33	04 55	21 09	00 N 04	23 36	23 18	08 44	17 40	22 03	06 59	19 41	14 37
27	08 06	09 29	05 11	21 19	05 36	23 59	23 27	08 29	17 40	22 04	06 59	19 41	14 37
28	08 02	09 26	05 10	21 29	11 04	24 20	23 35	08 14	17 41	22 04	06 58	19 41	14 37
29	07 58	09 23	04 50	21 38	16 06	24 38	23 43	07 59	17 41	22 05	06 58	19 40	14 37
30	07 54	09 20	04 10	21 48	20 20	24 53	23 50	07 44	17 41	22 05	06 57	19 40	14 36
31	07 ♑ 51	09 ♑ 17	03 N 12	21 N 56	23 N 18	25 N 06	23 N 56	07 S 29	17 S 41	22 N 06	06 S 57	19 S 40	14 N 36

ZODIAC SIGN ENTRIES

Date	h	m	Planets
02	01	01	☽ ♉
04	01	16	☽ ♊
06	01	35	☽ ♋
06	02	55	☿ ♉
08	03	36	☽ ♌
08	04	09	♂ ♓
10	08	13	☽ ♍
12	08	42	♀ ♊
12	15	31	☽ ♎
15	01	09	☽ ♏
17	12	41	☽ ♐
20	01	30	☽ ♑
20	17	24	☿ ♊
21	04	54	☉ ♊
22	14	17	☽ ♒
25	01	05	☽ ♓
27	08	14	☽ ♈
29	11	28	☽ ♉
31	11	53	☽ ♊

LATITUDES

Date	Mercury ☿ ° '	Venus ♀ ° '	Mars ♂ ° '	Jupiter ♃ ° '	Saturn ♄ ° '	Uranus ♅ ° '	Neptune ♆ ° '	Pluto ♇ ° '
01	02 S 22	00 S 32	01 S 38	00 S 26	01 S 11	00 N 39	01 N 40	16 N 53
04	02 06	00 25	01 41	00 27	01 10	00 39	01 40	16 52
07	01 45	00 18	01 45	00 28	01 10	00 39	01 40	16 51
10	01 20	00 11	01 49	00 28	01 10	00 39	01 40	16 50
13	00 52	00 S 04	01 53	00 29	01 09	00 39	01 40	16 49
16	00 S 21	00 N 03	01 56	00 29	01 09	00 38	01 40	16 48
19	00 N 10	00 11	02 00	00 30	01 08	00 38	01 40	16 47
22	00 41	00 18	02 04	00 31	01 08	00 38	01 40	16 45
25	01 09	00 25	02 08	00 31	01 08	00 38	01 40	16 44
28	01 33	00 32	02 11	00 32	01 07	00 38	01 40	16 43
31	01 N 51	00 N 39	02 S 15	00 S 33	01 S 07	00 N 38	01 N 40	16 N 41

DATA

Julian Date	2441804
Delta T	+43 seconds
Ayanamsa	23° 29' 20"
Synetic vernal point	05° ♓ 37' 40"
True obliquity of ecliptic	23° 26' 36"

MOON'S PHASES, APSIDES AND POSITIONS ☽

Date	h	m	Phase	Longitude ° '	Eclipse Indicator
02	20	55	●	12 ♉ 18	
09	12	07	☽	18 ♌ 43	
17	04	58	○	26 ♏ 09	
25	08	40	☾	04 ♓ 00	

Day	h	m		
04	05	35	Perigee	
19	13	18	Apogee	
05	07	42	Max dec	24° N 44'
11	18	32	0S	
19	02	41	Max dec	24° S 41'
26	11	41	0N	

LONGITUDES

Date	Chiron ⚷ ° '	Ceres ⚳ ° '	Pallas ⚴ ° '	Juno ⚵ ° '	Vesta ⚶ ° '	Black Moon Lilith ⚸ ° '
01	18 ♈ 13	16 ♐ 37	05 ♏ 05	21 ♐ 04	26 ♊ 50	18 ♐ 15
11	18 ♈ 46	15 ♐ 08	02 ♏ 11	19 ♐ 41	00 ♋ 49	19 ♐ 22
21	19 ♈ 16	13 ♐ 12	29 ♎ 46	17 ♐ 54	04 ♋ 54	20 ♐ 29
31	19 ♈ 42	11 ♐ 01	28 ♎ 04	15 ♐ 42	09 ♋ 06	21 ♐ 35

All ephemeris data is given at 12.00 UT and the Moon's longitude is additionally given for 24.00 UT

ASPECTARIAN

h m Aspects

01 Tuesday
02 10 ☽ ⊻ ♀
06 09 ☽ ⚹ ♄
07 09 ☉ □ ♃
08 55 ☽ ☍ ♅
11 27 ☽ ☌ ☿
11 30 ☽ bQ ♆
11 52 ☿ bQ ♆
12 17 ♂ # ♇
13 14 ☽ Q ♃
17 35 ☽ ⚹ ♂
18 27 ☽ # ♂
18 44 ☽ ∥ ♇
21 27 ♀ bQ ♇
21 41 ☽ ∥ ☉

02 Wednesday
02 13 ☽ ± ♆
03 14 ☿ Q ♃
04 12 ☽ ∥ ♀
04 26 ☽ ⊼ ♇
06 55 ☽ ∠ ♄
11 26 ☽ # ♃
11 55 ☽ ⊼ ♆
14 05 ☽ Q ♂
14 08 ☽ ± ♇
15 14 ♂ ± ♇
18 40 ☽ □ ♃
20 55 ☽ ☌ ☉
21 06 ☿ # ♅
21 33 ☽ ⊥ ♄
22 10 ☽ # ♆

03 Thursday
04 35 ☽ bQ ♇
05 04 ♀ ⊻ ♄
07 15 ☽ ⊻ ♄
07 25 ☽ ☌ ♀
07 50 ☉ ⊥ ♄
09 26 ☽ ⊼ ♅
11 49 ☽ ∥ ♄
18 25 ☽ ⊻ ☿
19 01 ☽ ± ♅
20 23 ☽ □ ♂

04 Friday
04 33 ☽ △ ♇
05 18 ☽ ⊥ ☿
07 13 ♀ ⊼ ♅
09 22 ☽ bQ ♅
11 56 ☽ ☍ ♆
18 59 ☽ △ ♃
20 09 ☿ ⚹ ♂
21 42 ☽ ∠ ☿

05 Saturday
00 17 ☽ ⊻ ☉
07 41 ☽ ☌ ♄
09 24 ☽ △ ♅
10 38 ☽ ⊥ ☉
11 51 ☽ ⊻ ♀
17 46 ♀ # ♃
19 18 ☽ bQ ♃
22 28 ☽ ⊥ ♀
23 02 ☽ △ ♂

06 Sunday
01 23 ☽ ⚹ ☿
02 16 ☽ ∠ ☉
04 53 ☽ □ ♇
10 05 ☽ ± ♃
10 57 ☿ ± ♆
12 26 ☽ ⊼ ♆
14 34 ☽ ∠ ♀
20 02 ☽ ⊼ ♃
22 20 ☽ ± ♆

07 Monday
00 03 ☽ Q ☿
00 58 ☽ bQ ♂
04 06 ☽ ∥ ♄
04 36 ☿ ⊼ ♇
04 48 ☽ ⚹ ☉
09 20 ☽ ⊻ ♄
10 35 ☽ □ ♅
10 40 ☽ Q ♇
13 21 ☽ bQ ♆
16 51 ☽ ± ♂
17 13 ☉ bQ ♇
17 57 ☽ ⚹ ♀
18 51 ☽ # ♆
19 32 ☽ ⊥ ♄

08 Tuesday
01 39 ☽ ∥ ♀
02 32 ☽ Q ☉
03 34 ☽ ⊼ ♂
06 36 ☽ # ♃
07 00 ☽ ⚹ ♇
09 53 ☿ ∠ ♄
10 36 ☽ ∥ ☉
11 02 ☽ ∠ ♄
11 12 ☽ □ ☿
14 53 ☽ △ ♆
16 27 ☽ Q ♀
17 12 ☽ Q ♅
23 13 ☽ ☍ ♃
23 35 ☽ ∥ ♇

09 Wednesday
00 41 ♀ ± ♅
08 24 ☽ # ♂
09 01 ☽ ∠ ♇
11 19 ☽ ∥ ☿
12 07 ☽ □ ☉
13 16 ☿ ⊼ ♆
13 25 ☽ ⚹ ♄
14 09 ☽ ⚹ ♅

10 Thursday
00 57 ☽ ⊥ ♇
01 07 ☿ # ♂
03 18 ☽ □ ♀
06 24 ☿ ± ♇
08 56 ☉ ⊻ ♄
10 44 ☽ # ♅
11 03 ☽ Q ♄
11 07 ☽ ☍ ♂
11 43 ☽ ⊻ ♇
15 48 ☉ ⊼ ♅
16 57 ☽ ∠ ♅
20 00 ☽ □ ♆
22 17 ♂ ⊼ ♇
23 48 ☉ # ♃

11 Friday
01 31 ☽ △ ☿
05 09 ☽ ⊼ ♃
09 21 ☽ ⊥ ♅
09 52 ♀ # ♆
16 15 ☽ ± ♃
20 19 ☽ □ ♄
20 25 ☽ ⊻ ♅
22 49 ☽ △ ☉

12 Saturday
01 41 ☿ □ ♃
05 09 ☽ Q ♆
05 28 ♄ △ ♅
09 06 ☽ bQ ♃
10 28 ☽ bQ ☿
13 32 ☿ ∥ ♇
16 14 ☽ △ ♀
19 07 ☽ ☌ ♇
21 48 ☽ ⊼ ♂

13 Sunday
03 04 ☽ ∥ ♅
03 44 ☽ ⚹ ♆
05 00 ☿ ⊥ ♄
05 22 ☽ bQ ☉
06 46 ☽ ± ☿
09 53 ☽ ± ♂
13 39 ☽ △ ♃
20 32 ☽ ⊼ ☿
21 40 ♀ △ ♇
23 56 ☽ bQ ♀

14 Monday
00 02 ☽ ± ☉
01 53 ☽ ∥ ♂
04 12 ☽ bQ ♂
05 09 ☽ ☌ ♅
05 45 ☽ △ ♄
08 27 ☽ ∠ ♆
09 37 ☿ Q ♂
12 37 ☽ ⊼ ☉
15 23 ☿ bQ ♇
17 35 ☽ # ♇
19 18 ☽ ± ♀
20 22 ♂ bQ ♅

15 Tuesday
01 56 ☽ ⊥ ♆
04 49 ☽ ⊻ ♇
05 39 ☽ # ☿
08 21 ☽ ⊼ ♀
11 12 ☽ △ ♂
11 18 ☽ bQ ♄
12 20 ☽ ∥ ♃
13 41 ☽ ⊻ ♆
16 36 ☽ ⊥ ♇
20 55 ☽ # ☉
22 53 ☿ ⊼ ♅

16 Wednesday
00 20 ☽ □ ♃
02 25 ☽ ∥ ♆
03 30 ♀ bQ ♅
05 15 ☽ ± ♄
05 45 ☿ ⊻ ♄
10 24 ☽ ∠ ♇
14 43 ☽ # ♀
15 19 ☿ # ♃
15 56 ☽ ⊻ ♅
16 04 ☉ ± ♅
17 19 ☽ ⊼ ♄
19 41 ☽ ☍ ☿
21 27 ☽ # ♄

17 Thursday
03 54 ☽ ⊥ ♅
04 55 ♂ □ ♆
04 58 ☽ ☍ ☉
12 23 ☽ Q ♃
12 25 ♀ ☍ ♆
16 24 ☽ ⚹ ♇
21 57 ☽ ∠ ♅
22 42 ♀ □ ♂

18 Friday
01 26 ☽ ☌ ♆
02 46 ☽ □ ♂
02 58 ☽ ☍ ♀
12 41 ☽ ⚹ ♃
16 37 ☽ Q ♇
16 46 ☿ ± ♅

19 Saturday
03 13 ☉ # ♆
04 15 ☽ ⚹ ♅
06 27 ☽ ☍ ♄
11 33 ☿ # ♆
15 34 ☉ ∥ ☿
15 42 ♀ ∥ ♄
17 35 ☽ Q ♂
19 15 ☽ ∠ ♃
21 52 ☽ ⊼ ☿
23 04 ☽ ⊼ ☉

20 Sunday
04 36 ☽ Q ♅
05 11 ☽ □ ♇
08 24 ☉ ☌ ☿
12 22 ☽ ± ☉
12 52 ☽ ± ☿
13 38 ☽ ⊥ ♃
14 14 ☽ ⊻ ♆
19 33 ☽ ⚹ ♂
23 00 ☽ ⊼ ♀

21 Monday
01 52 ☽ ⊻ ♃
02 25 ☽ ⊥ ♆
04 25 ☽ # ♀
07 47 ☽ # ♄
08 17 ☽ bQ ☉
11 20 ☽ bQ ☿
12 36 ☽ ± ♀
13 02 ☿ △ ♇
16 07 ☽ # ☿
17 06 ☽ □ ♅
20 05 ☽ ⊼ ♄
20 37 ☽ ∠ ♆
23 00 ☽ # ☉

22 Tuesday
03 10 ♀ △ ♃
03 44 ☽ ∥ ♆
03 51 ☽ ∠ ♂
08 19 ☽ ± ♄
08 52 ☽ bQ ♀
17 12 ☽ △ ☉
17 51 ☽ △ ♇
18 06 ☽ ∥ ♃
18 08 ☿ bQ ♅
23 01 ☽ ⊥ ♂

23 Wednesday
00 19 ☽ △ ☿
01 18 ☉ △ ♇
02 31 ☽ bQ ♄
02 39 ☽ ⚹ ♆
05 34 ☿ ∥ ♄
11 39 ☽ ⊻ ♂
12 49 ☽ # ♇
13 02 ☿ ☍ ♆
14 15 ☽ ☌ ♃
18 06 ☽ △ ♀
23 32 ☽ bQ ♇

24 Thursday
02 15 ☽ Q ♆
04 41 ☽ △ ♅
08 19 ☽ △ ♄
16 58 ☽ ± ♇
17 37 ☽ ∥ ♂

25 Friday
04 03 ☽ ∥ ♅
04 26 ☽ ⊼ ♇
08 40 ☽ □ ☉
09 24 ☽ bQ ♅
10 17 ♂ ⊻ ♃
10 25 ☿ ∥ ♀
12 40 ☽ □ ♆
18 05 ☉ bQ ♅
22 28 ☽ □ ☿
23 48 ☽ ⊻ ♃

26 Saturday
00 33 ☽ ☌ ♂
02 10 ☽ ± ♅
06 35 ☿ △ ♃
09 16 ☽ □ ♀
10 51 ☽ ⊥ ♃
13 08 ☽ ⊼ ♅
16 06 ☿ □ ♂
17 10 ☽ □ ♄
20 31 ☽ Q ☉

27 Sunday
03 06 ☽ ∠ ♃
05 38 ♂ ± ♅
11 19 ☽ ☍ ♇
12 31 ☉ ☍ ♆
13 12 ☽ Q ☿
17 56 ☽ # ♅
18 53 ☽ △ ♆
19 23 ☽ ⚹ ☉
20 35 ☽ Q ♀

28 Monday
00 00 ☽ # ♂
01 17 ☽ Q ♄
02 03 ♀ △ ♅
05 22 ☽ ⚹ ♃
08 47 ☽ ⊻ ♂
13 37 ☽ ⚹ ☿
17 35 ☽ ☍ ♅
19 05 ☽ ⚹ ♀
19 30 ☽ ⊥ ♂
20 31 ☽ bQ ♆
21 54 ☽ ⚹ ♄
22 53 ☽ ∠ ☉

29 Tuesday
01 42 ☽ Q ♃
04 35 ☽ ∥ ♇
11 16 ☽ ∠ ♂
11 29 ☽ ± ♆
14 18 ☽ ⊼ ♇
14 54 ☽ ⊥ ☉
16 37 ☿ △ ♅
18 48 ☽ ∠ ☿
20 17 ☽ # ♃
21 19 ☽ ⊼ ♆
22 15 ☽ ∠ ♀
22 56 ☽ ∠ ♄

30 Wednesday
00 07 ☽ ± ♇
01 24 ☽ ⊻ ☉
07 17 ☽ □ ♃
07 18 ♀ ☌ ♄
07 46 ☽ # ♆
11 45 ☽ ⊥ ☿
12 54 ☽ ⚹ ♂
13 38 ☽ ⊥ ♄
14 08 ☽ ⊥ ♀
14 40 ☽ bQ ♇
18 44 ☽ ⊼ ♅
22 12 ♃ St R
22 51 ☽ ∥ ☉
22 52 ☽ ⊻ ☿
23 20 ☽ ⊻ ♄

31 Thursday
00 36 ☽ ⊻ ♀
00 43 ☽ ∥ ♄
02 31 ☿ ☌ ♄
04 19 ☽ ± ♅
09 00 ☽ Q ♂
14 35 ☽ △ ♇
18 35 ☽ bQ ♅
20 26 ☽ ∥ ♀
21 18 ☽ ☍ ♆

JUNE 1973

LONGITUDES

Date	Sidereal time h m s	Sun ☉ ° ' "	Moon ☽ ° ' "	Moon ☽ 24.00 ° ' "	Mercury ☿ ° '	Venus ♀ ° '	Mars ♂ ° '	Jupiter ♃ ° '	Saturn ♄ ° '	Uranus ♅ ° '	Neptune ♆ ° '	Pluto ♇ ° '
01	04 39 18	10 ♊ 50 40	15 ♊ 15 03	22 ♊ 50 34	24 ♊ 54	24 ♊ 44	16 ♓ 58	12 ♒ 08	22 ♊ 19	19 ♎ 13	05 ♐ 56	01 ♎ 42
02	04 43 14	11 48 11	00 ♋ 24 41	07 ♋ 56 18	26 49	25 57	17 39	12 R 07	22 27	19 R 11	05 R 54	01 R 42
03	04 47 11	12 45 40	15 ♋ 24 23	22 ♋ 48 05	28 ♊ 43	27 11	18 20	12 07	22 34	19 10	05 53	01 42
04	04 51 08	13 43 08	00 ♌ 06 42	07 ♌ 19 43	00 ♋ 33	28 24	19 01	12 06	22 42	19 09	05 51	01 41
05	04 55 04	14 40 35	14 ♌ 26 46	21 ♌ 27 39	02 22	29 ♊ 38	19 42	12 05	22 50	19 08	05 50	01 41
06	04 59 01	15 38 00	28 ♌ 22 19	05 ♍ 10 49	04 07	00 ♋ 51	20 23	12 04	22 58	19 07	05 48	01 41
07	05 02 57	16 35 25	11 ♍ 53 20	18 ♍ 30 07	05 50	02 04	21 04	12 02	23 05	19 06	05 46	01 41
08	05 06 54	17 32 48	25 ♍ 01 28	01 ♎ 27 44	07 31	03 18	21 44	12 01	23 13	19 05	05 45	01 41
09	05 10 50	18 30 10	07 ♎ 49 19	14 ♎ 06 36	09 09	04 31	22 25	11 59	23 21	19 04	05 43	01 41
10	05 14 47	19 27 31	20 ♎ 20 01	26 ♎ 29 56	10 44	05 45	23 06	11 57	23 29	19 03	05 42	01 40
11	05 18 43	20 24 52	02 ♏ 36 45	08 ♏ 40 52	12 16	06 58	23 46	11 55	23 37	19 02	05 40	01 40
12	05 22 40	21 22 11	14 ♏ 42 37	20 ♏ 42 20	13 45	08 11	24 26	11 53	23 44	19 02	05 39	01 D 40
13	05 26 37	22 19 29	26 ♏ 40 22	02 ♐ 37 01	15 12	09 25	25 07	11 50	23 52	19 01	05 37	01 40
14	05 30 33	23 16 47	08 ♐ 32 34	14 ♐ 27 19	16 36	10 38	25 47	11 48	24 00	19 00	05 36	01 41
15	05 34 30	24 14 04	20 ♐ 21 33	26 ♐ 15 31	17 57	11 51	26 27	11 45	24 08	19 00	05 34	01 41
16	05 38 26	25 11 20	02 ♑ 09 31	08 ♑ 03 50	19 16	13 05	27 07	11 42	24 16	18 59	05 32	01 41
17	05 42 23	26 08 36	13 ♑ 58 45	19 ♑ 54 36	20 31	14 18	27 46	11 39	24 24	18 59	05 31	01 41
18	05 46 19	27 05 52	25 ♑ 51 40	01 ♒ 50 20	21 43	15 31	28 26	11 35	24 31	18 58	05 30	01 41
19	05 50 16	28 03 07	07 ♒ 50 57	13 ♒ 53 53	22 53	16 45	29 06	11 32	24 39	18 58	05 28	01 41
20	05 54 12	29 00 21	19 ♒ 59 33	26 ♒ 08 22	23 59	17 58	29 ♓ 45	11 28	24 47	18 57	05 27	01 42
21	05 58 09	29 ♊ 57 35	02 ♓ 20 47	08 ♓ 37 12	25 02	19 11	00 ♈ 25	11 24	24 55	18 57	05 25	01 42
22	06 02 06	00 ♋ 54 50	14 ♓ 58 05	21 ♓ 23 51	26 02	20 24	01 04	11 20	25 03	18 57	05 24	01 42
23	06 06 02	01 52 04	27 ♓ 54 53	04 ♈ 31 34	26 58	21 37	01 43	11 16	25 10	18 56	05 22	01 43
24	06 09 59	02 49 17	11 ♈ 14 11	18 ♈ 02 58	27 51	22 51	02 22	11 11	25 18	18 56	05 21	01 43
25	06 13 55	03 46 31	24 ♈ 58 02	01 ♉ 59 25	28 41	24 04	03 01	11 07	25 26	18 56	05 19	01 43
26	06 17 52	04 43 45	09 ♉ 06 59	16 ♉ 20 29	29 ♋ 27	25 17	03 40	11 02	25 34	18 56	05 18	01 44
27	06 21 48	05 40 59	23 ♉ 39 27	01 ♊ 03 20	00 ♌ 09	26 30	04 19	10 57	25 41	18 D 56	05 17	01 44
28	06 25 45	06 38 13	08 ♊ 31 20	16 ♊ 02 36	00 47	27 43	04 57	10 52	25 49	18 56	05 15	01 45
29	06 29 41	07 35 27	23 ♊ 36 05	01 ♋ 10 41	01 22	28 ♋ 56	05 35	10 47	25 57	18 56	05 14	01 46
30	06 33 38	08 ♋ 32 42	08 ♋ 45 15	16 ♋ 18 36	01 ♌ 52	00 ♌ 09	06 ♈ 13	10 ♒ 42	26 ♊ 05	18 ♎ 57	05 ♐ 13	01 ♎ 46

Date	Moon True ☊ ° '	Moon Mean ☊ ° '	Moon ☽ Latitude ° '
01	07 ♑ 48	09 ♑ 14	02 N 01
02	07 R 47	09 10	00 N 40
03	07 D 47	09 07	00 S 42
04	07 48	09 04	02 00
05	07 49	09 01	03 09
06	07 50	08 58	04 04
07	07 51	08 54	04 44
08	07 R 51	08 51	05 08
09	07 50	08 48	05 16
10	07 48	08 45	05 08
11	07 46	08 42	04 47
12	07 43	08 39	04 13
13	07 41	08 35	03 28
14	07 39	08 32	02 34
15	07 38	08 29	01 34
16	07 37	08 26	00 S 30
17	07 D 37	08 23	00 N 35
18	07 38	08 20	01 40
19	07 38	08 16	02 40
20	07 39	08 13	03 34
21	07 40	08 10	04 19
22	07 41	08 07	04 53
23	07 41	08 04	05 13
24	07 R 41	08 00	05 17
25	07 41	07 57	05 03
26	07 40	07 54	04 31
27	07 40	07 51	03 41
28	07 40	07 48	02 35
29	07 D 40	07 45	01 17
30	07 ♑ 40	07 ♑ 41	00 S 06

DECLINATIONS

Date	Sun ☉ ° '	Moon ☽ ° '	Mercury ☿ ° '	Venus ♀ ° '	Mars ♂ ° '	Jupiter ♃ ° '	Saturn ♄ ° '	Uranus ♅ ° '	Neptune ♆ ° '	Pluto ♇ ° '
01	22 N 04	24 N 38	25 N 16	24 N 02	07 S 15	17 S 41	22 N 06	06 S 56	19 S 40	14 N 36
02	22 12	24 07	25 23	24 07	07 00	17 42	22 07	06 56	19 39	14 36
03	22 20	21 52	25 28	24 11	06 45	17 42	22 07	06 55	19 39	14 35
04	22 27	18 11	25 31	24 14	06 30	17 42	22 08	06 55	19 39	14 35
05	22 34	13 29	25 31	24 17	06 15	17 43	22 08	06 55	19 38	14 35
06	22 40	08 14	25 29	24 19	06 00	17 43	22 09	06 54	19 38	14 34
07	22 46	02 N 44	25 25	24 20	05 45	17 44	22 09	06 54	19 38	14 34
08	22 52	02 S 44	25 19	24 21	05 30	17 45	22 10	06 53	19 38	14 33
09	22 57	07 56	25 11	24 21	05 15	17 45	22 10	06 53	19 37	14 33
10	23 01	12 42	25 02	24 20	05 00	17 46	22 11	06 53	19 37	14 33
11	23 06	16 52	24 51	24 18	04 45	17 47	22 11	06 53	19 37	14 32
12	23 10	20 16	24 38	24 16	04 30	17 48	22 12	06 52	19 37	14 32
13	23 13	22 47	24 24	24 13	04 15	17 49	22 12	06 52	19 36	14 31
14	23 16	24 16	24 09	24 09	04 00	17 50	22 13	06 52	19 36	14 31
15	23 19	24 43	23 52	24 05	03 46	17 51	22 13	06 52	19 36	14 30
16	23 21	23 56	23 35	24 00	03 31	17 52	22 13	06 52	19 36	14 30
17	23 23	22 08	23 17	23 54	03 16	17 53	22 14	06 51	19 35	14 29
18	23 25	19 21	22 57	23 47	03 01	17 54	22 14	06 51	19 35	14 29
19	23 26	15 44	22 38	23 40	02 47	17 55	22 15	06 51	19 35	14 28
20	23 26	11 26	22 17	23 32	02 32	17 56	22 15	06 51	19 35	14 28
21	23 27	06 36	21 56	23 23	02 17	17 58	22 15	06 51	19 34	14 27
22	23 26	01 S 25	21 35	23 14	02 03	17 59	22 16	06 51	19 34	14 26
23	23 26	03 N 58	21 13	23 04	01 48	18 00	22 16	06 51	19 34	14 26
24	23 25	09 18	20 52	22 53	01 34	18 02	22 16	06 51	19 34	14 25
25	23 23	14 22	20 30	22 42	01 19	18 03	22 17	06 51	19 34	14 25
26	23 22	18 49	20 08	22 30	01 05	18 05	22 17	06 51	19 33	14 24
27	23 19	22 15	19 46	22 18	00 51	18 06	22 17	06 51	19 33	14 23
28	23 17	24 17	19 25	22 04	00 37	18 08	22 18	06 51	19 33	14 23
29	23 14	24 35	19 04	21 50	00 22	18 09	22 18	06 51	19 33	14 22
30	23 N 10	23 N 03	18 N 43	21 N 36	00 S 08	18 S 11	22 N 18	06 S 51	19 S 33	14 N 21

ZODIAC SIGN ENTRIES

Date	h	m	Planets	
02	11	21	☽	♋
04	04	42	☿	♋
04	11	49	☽	♌
05	19	20	♀	♋
06	14	51	☽	♍
08	21	16	☽	♎
11	06	52	☽	♏
13	18	43	☽	♐
16	07	37	☽	♑
18	20	19	☽	♒
20	20	54	♂	♈
21	07	29	☽	♓
21	13	01	☉	♋
23	15	48	☽	♈
25	20	37	☽	♉
27	06	42	☿	♌
27	22	18	☽	♊
29	22	08	☽	♋
30	08	55	♀	♌

LATITUDES

Date	Mercury ☿ ° '	Venus ♀ ° '	Mars ♂ ° '	Jupiter ♃ ° '	Saturn ♄ ° '	Uranus ♅ ° '	Neptune ♆ ° '	Pluto ♇ ° '
01	01 N 55	00 N 42	02 S 16	00 S 33	01 S 07	00 N 38	01 N 40	16 N 41
04	02 04	00 48	02 20	00 34	01 07	00 38	01 40	16 39
07	02 06	00 55	02 24	00 34	01 06	00 38	01 40	16 38
10	02 02	01 01	02 27	00 35	01 06	00 38	01 40	16 37
13	01 50	01 06	02 31	00 36	01 06	00 38	01 40	16 35
16	01 32	01 12	02 35	00 36	01 06	00 37	01 40	16 34
19	01 08	01 17	02 38	00 37	01 05	00 37	01 40	16 32
22	00 39	01 21	02 42	00 38	01 05	00 37	01 40	16 31
25	00 N 04	01 25	02 45	00 38	01 05	00 37	01 40	16 29
28	00 S 35	01 29	02 48	00 39	01 05	00 37	01 40	16 28
31	01 S 18	01 N 32	02 S 52	00 S 39	01 S 05	00 N 37	01 N 39	16 N 26

DATA

Julian Date	2441835
Delta T	+43 seconds
Ayanamsa	23° 29' 25"
Synetic vernal point	05° ♓ 37' 35"
True obliquity of ecliptic	23° 26' 35"

MOON'S PHASES, APSIDES AND POSITIONS ☽

Date	h	m	Phase	Longitude ° '	Eclipse Indicator
01	04	34	●	10 ♊ 33	
07	21	11	◐	16 ♍ 57	
15	20	35	○	24 ♐ 35	
23	19	45	◑	02 ♈ 11	
30	11	39	●	08 ♋ 32	Total

Day	h	m		
01	14	03	Perigee	
15	16	48	Apogee	
29	23	47	Perigee	
01	17	22	Max dec	24° N 40'
07	23	54	0S	
15	08	20	Max dec	24° S 40'
22	18	21	0N	
29	03	55	Max dec	24° N 41'

LONGITUDES

Date	Chiron ⚷ ° '	Ceres ⚳ ° '	Pallas ⚴ ° '	Juno ⚵ ° '	Vesta ⚶ ° '	Black Moon Lilith ⚸ ° '
01	19 ♈ 45	10 ♐ 47	27 ♎ 56	15 ♐ 29	09 ♋ 31	21 ♐ 42
11	20 ♈ 08	08 ♐ 36	27 ♎ 06	13 ♐ 13	13 ♋ 47	22 ♐ 48
21	20 ♈ 27	06 ♐ 41	27 ♎ 02	11 ♐ 04	18 ♋ 07	23 ♐ 55
31	20 ♈ 41	05 ♐ 12	27 ♎ 41	09 ♐ 12	22 ♋ 31	25 ♐ 01

All ephemeris data is given at 12.00 UT and the Moon's longitude is additionally given for 24.00 UT

ASPECTARIAN

h m Aspects

01 Friday
04 34 ☽ ☌ ☉
06 29 ☿ ☌ ♀
07 04 ☽ △ ♃
14 50 ☽ □ ♂
17 54 ☉ ∥ ♄
18 15 ☽ △ ♅
23 16 ☽ ☌ ♄

02 Saturday
04 18 ☽ ☌ ♀
05 29 ☽ ☌ ☿
06 47 ☽ bQ ♃
12 06 ☽ ∥ ♀
14 03 ☽ □ ♇
15 44 ☿ bQ ♃
18 18 ♂ ∥ ♅
19 51 ☉ △ ♃
20 44 ☽ ⊼ ♆
21 05 ☽ ± ♃

03 Sunday
04 10 ♂ ⊥ ♃
06 20 ☽ ± ♆
06 42 ☽ ⊼ ♃
07 27 ☽ ⊻ ☉
08 17 ☽ ∥ ☉
09 53 ☽ ∥ ♄
10 42 ♀ bQ ♃
16 58 ☽ △ ♂
17 48 ☽ ⊥ ☉
18 05 ☽ □ ♅
18 56 ☽ Q ♇
20 51 ☽ bQ ♆
23 44 ☽ ⊻ ♄

04 Monday
03 19 ☽ # ♆
08 55 ☽ ⊻ ♀
09 32 ☽ ∠ ☉
09 39 ☽ ⊥ ♄
12 51 ☽ ⊻ ☿
14 36 ☽ # ♃
14 36 ☽ ✶ ♇
16 31 ♂ ⊼ ♅
18 48 ☽ bQ ♂
19 46 ☽ ⊥ ♀
21 31 ☽ △ ♆
23 41 ☽ Q ♅

05 Tuesday
00 16 ☽ ⊥ ☿
00 44 ☽ ∠ ♄
02 56 ☿ □ ♇
06 46 ☽ ∥ ♇
08 00 ☽ ☍ ♃
10 40 ☽ ± ♂
12 20 ☽ ∠ ♀
12 25 ☽ ✶ ☉
15 49 ☽ ∠ ♇
17 41 ☽ ∠ ☿
19 59 ☽ ✶ ♅
21 26 ☽ ⊼ ♂

06 Wednesday
02 30 ☽ ✶ ♄
07 18 ☽ ⊥ ♇
10 37 ☽ Q ☉
16 47 ☽ ✶ ♀
17 49 ☽ ⊻ ♇
17 49 ☽ # ♅
22 06 ☽ ∠ ♅
22 15 ☽ # ♂
23 37 ☽ ✶ ☿
23 43 ☽ Q ♄

07 Thursday
01 05 ☽ □ ♆
04 15 ♀ □ ♇
11 04 ☿ ⊼ ♆
12 16 ☽ ⊼ ♃
14 11 ☽ ⊥ ♅
14 45 ☿ ± ♃
16 21 ☽ Q ♀
21 11 ☽ □ ☉
23 08 ☽ ± ♃

08 Friday
00 23 ☽ Q ☿
01 05 ☽ ⊻ ♅
05 36 ☽ ☍ ♂
08 38 ☽ □ ♄
09 39 ☽ Q ♆
15 41 ☽ bQ ♃
23 59 ☽ ∥ ♂

09 Saturday
00 24 ☽ ☌ ♇
05 06 ☽ □ ♀
07 02 ☽ ∥ ♅
08 02 ☽ ✶ ♆
14 53 ☽ □ ☿
19 54 ☽ △ ♃

10 Sunday
01 58 ☉ △ ♅
09 31 ☽ ☌ ♅
10 10 ☽ △ ☉
11 02 ♀ ⊼ ♆
12 42 ☽ ∠ ♆
15 56 ♀ ± ♃
17 40 ☽ ⊼ ♂
18 11 ☽ △ ♄
22 08 ☽ # ♇

11 Monday
02 47 ☿ ± ♆
05 10 ♂ □ ♄
06 04 ☽ ± ♂
06 13 ☽ ⊥ ♆
06 38 ☿ ⊼ ♃
10 09 ☽ ⊻ ♇
17 58 ☽ ∥ ♃
18 00 ☽ bQ ☉
18 01 ☽ ⊻ ♆
20 14 ♇ St D
21 34 ☽ △ ♀
22 00 ☽ ⊥ ♇
23 59 ☽ bQ ♄

12 Tuesday
00 53 ☽ bQ ♂
06 22 ☽ □ ♃
06 51 ☽ ∥ ♆
09 50 ☽ △ ☿
13 26 ☽ ± ☉
15 55 ☽ ∠ ♇
18 07 ☽ ± ♄
20 37 ☽ ⊻ ♅

13 Wednesday
02 29 ☽ ⊼ ☉
05 33 ☽ # ♄
06 18 ☽ ⊼ ♄
06 56 ☽ bQ ♀
08 40 ☽ △ ♂
08 40 ☽ ⊥ ♅
17 45 ☽ # ☉
18 22 ☽ Q ♃
20 05 ☽ bQ ☿
22 06 ☽ ✶ ♇

14 Thursday
02 49 ☽ ∠ ♅
03 11 ☽ ± ♀
06 02 ☽ ☌ ♆
09 23 ☽ # ♀
09 38 ☽ # ☿
11 06 ☿ ∥ ♀
16 44 ☽ ± ☿
16 44 ☽ ⊼ ♀
18 34 ☽ ✶ ♃
22 25 ☽ Q ♇

15 Friday
06 24 ♀ ± ♆
06 29 ☽ ⊼ ☿
09 00 ☉ ☌ ♄
09 13 ☽ ✶ ♅
09 53 ♀ ⊼ ♃
19 46 ☽ ☍ ♄
20 35 ☽ ☍ ☉
22 04 ♂ ∠ ♃

16 Saturday
00 56 ☽ ∠ ♃
01 07 ☽ □ ♂
06 54 ☿ □ ♅
09 37 ☽ Q ♅
10 41 ☽ # ♀
11 02 ☽ □ ♇
18 52 ☽ ⊻ ♆
19 09 ☽ ⊥ ♃
19 18 ☽ # ☿
19 57 ☿ Q ♇
21 05 ☽ # ☉

17 Sunday
04 10 ☉ ∥ ☿
07 01 ☽ ⊥ ♆
07 17 ☽ ⊻ ♃
10 53 ☽ # ♄
12 02 ☿ bQ ♆
12 44 ☽ ☍ ♀
15 51 ☽ Q ♂
22 06 ☽ □ ♅
23 51 ☉ bQ ♃

18 Monday
01 12 ☽ ∠ ♆
02 44 ☽ ☍ ☿
09 16 ☽ ⊼ ♄
10 12 ☽ ∥ ♆
14 42 ☽ ⊼ ☉
17 29 ☽ ✶ ♂
21 28 ☽ ± ♄
22 10 ☽ ∥ ♃
23 42 ☽ △ ♇

19 Tuesday
03 46 ☽ ± ☉
07 16 ☽ ✶ ♆
15 38 ☽ bQ ♄
19 16 ☽ ☌ ♃
19 25 ☽ # ♇
23 12 ☽ bQ ☉

20 Wednesday
01 06 ☽ ∠ ♂
05 31 ☽ bQ ♇
07 00 ☽ Q ♆
07 34 ☽ ⊼ ♀
09 58 ☽ △ ♅
14 26 ☿ ∥ ♄
19 47 ☽ ⊥ ♂
20 32 ☽ ⊼ ☿
20 37 ☽ ± ♀
21 28 ☽ △ ♄
23 08 ☽ ± ♇

21 Thursday
03 11 ☉ ∥ ♀
07 01 ☽ △ ☉
07 24 ♀ □ ♅
08 04 ☽ ⊻ ♂
08 49 ☿ ⊻ ♄
09 14 ☽ ± ☿
10 45 ☽ ⊼ ♇
10 49 ☽ ∥ ♅
15 05 ☽ bQ ♅
15 55 ☽ bQ ♀
17 53 ☽ □ ♆
22 11 ♀ Q ♇

22 Friday
03 57 ☽ bQ ☿
05 11 ☽ ⊻ ♃
08 12 ☽ ± ♅
09 00 ☽ ∥ ♂
11 50 ♀ bQ ♆
16 24 ☽ ⊥ ♃
19 26 ☽ ⊼ ♅
23 13 ☽ △ ♀

23 Saturday
00 17 ☉ □ ♂
02 50 ☽ # ♂
06 55 ☽ □ ♄
08 01 ☉ □ ♇
08 59 ☽ ∠ ♃
10 09 ☽ △ ☿
11 38 ♂ ☍ ♇
18 55 ☽ ☍ ♇
19 17 ☽ ☌ ♂
19 45 ☽ □ ☉

24 Sunday
00 53 ☽ # ♅
01 30 ☽ △ ♆
11 55 ☽ ✶ ♃
15 42 ☽ Q ♄

25 Monday
01 33 ☽ ☍ ♅
03 59 ☽ bQ ♆
06 05 ☽ Q ☉
08 49 ☽ Q ♃
10 17 ☽ □ ♀
12 13 ☽ ∥ ♇
12 48 ☽ ✶ ♄
18 45 ☽ □ ☿
19 27 ☽ ± ♆
23 33 ☽ ⊼ ♇

26 Tuesday
02 24 ☽ ⊻ ♂
04 06 ☽ ✶ ☉
05 36 ☽ ⊼ ♆
07 40 ☽ # ♃
09 41 ☽ ± ♇
12 58 ☽ ⊥ ♂
14 26 ☽ ∠ ♄
15 11 ☽ □ ♃
16 32 ☽ # ♆
18 08 ♀ ⊻ ♄
19 03 ☉ ± ♃
19 30 ☽ ∥ ☿
19 35 ☽ Q ♀
22 00 ♅ St D

27 Wednesday
00 39 ☽ bQ ♇
02 05 ☉ ⊼ ♆
02 32 ☽ Q ☿
04 16 ☽ ⊼ ♅
04 33 ☽ ∠ ♂
05 27 ☽ ⊥ ♄
06 48 ☽ ∠ ☉
12 08 ♀ ∥ ♄
12 17 ☽ ∥ ♀
12 18 ☽ ∥ ♄
14 05 ☽ ± ♅
15 20 ☽ ⊻ ♄
17 03 ☽ ✶ ♀
22 06 ☽ ∥ ☉
22 27 ☽ ⊥ ☉
23 01 ☽ ✶ ☿

28 Thursday
01 07 ☽ △ ♇
02 44 ☿ # ♆
04 39 ☽ bQ ♅
06 01 ☽ ✶ ♂
06 46 ☽ ☍ ♆
08 46 ☽ ⊻ ☉
15 44 ☽ △ ♃
17 30 ♃ bQ ♄
19 18 ☽ ∠ ♀
23 10 ♂ △ ♆

29 Friday
00 04 ☽ ∠ ☿
02 02 ☽ Q ♂
04 36 ☽ △ ♅
10 51 ☽ ⊥ ♀
14 53 ☽ ⊥ ☿
15 26 ☽ bQ ♃
15 45 ☽ ☌ ♄
21 12 ☽ ⊻ ♀

30 Saturday
00 44 ☽ ⊻ ☿
00 56 ☽ □ ♇
05 36 ☽ ± ♃
06 24 ☽ ⊼ ♆
07 05 ☿ ✶ ♇
07 49 ☽ □ ♂
10 48 ☽ ∥ ☉
11 39 ☽ ☌ ☉
15 03 ☽ ⊼ ♃
15 54 ☽ ± ♆
18 49 ☽ ∥ ♄

JULY 1973

LONGITUDES

Date	Sidereal time h m s	Sun ☉ ° ' "	Moon ☽ ° ' "	Moon ☽ 24.00 ° ' "	Mercury ☿ ° '	Venus ♀ ° '	Mars ♂ ° '	Jupiter ♃ ° '	Saturn ♄ ° '	Uranus ♅ ° '	Neptune ♆ ° '	Pluto ♇ ° '
01	06 37 35	09 ♋ 29 55	23 ♋ 49 37	01 ♌ 17 14	02 ♌ 18	01 ♌ 22	06 ♈ 51	10 ♒ 36	26 ♊ 12	18 ♎ 57	05 ♐ 12	01 ♎ 47
02	06 41 31	10 27 09	08 ♌ 40 32	15 ♌ 58 41	02 39	02 36	07 29	10 R 30	26 20	18 57	05 R 10	01 47
03	06 45 28	11 24 22	23 ♌ 11 03	00 ♍ 17 10	02 57	03 49	08 07	10 25	26 28	18 57	05 09	01 48
04	06 49 24	12 21 35	07 ♍ 16 44	14 ♍ 09 36	03 09	05 02	08 44	10 19	26 35	18 58	05 08	01 49
05	06 53 21	13 18 48	20 ♍ 55 46	27 ♍ 35 23	03 17	06 15	09 22	10 13	26 43	18 58	05 07	01 50
06	06 57 17	14 16 00	04 ♎ 08 41	10 ♎ 36 01	03 21	07 28	09 59	10 07	26 50	18 59	05 05	01 50
07	07 01 14	15 13 13	16 ♎ 57 47	23 ♎ 14 27	03 R 19	08 41	10 36	10 00	26 58	18 59	05 04	01 51
08	07 05 10	16 10 25	29 ♎ 26 31	05 ♏ 34 32	03 13	09 53	11 12	09 54	27 06	19 00	05 03	01 52
09	07 09 07	17 07 37	11 ♏ 39 00	17 ♏ 40 27	03 03	11 06	11 49	09 47	27 13	19 00	05 02	01 53
10	07 13 04	18 04 49	23 ♏ 39 27	29 ♏ 36 28	02 47	12 19	12 25	09 41	27 21	19 01	05 01	01 54
11	07 17 00	19 02 01	05 ♐ 31 59	11 ♐ 26 30	02 28	13 32	13 02	09 34	27 28	19 02	05 00	01 55
12	07 20 57	19 59 13	17 ♐ 20 26	23 ♐ 14 12	02 04	14 45	13 38	09 27	27 35	19 03	04 59	01 56
13	07 24 53	20 56 25	29 ♐ 08 10	05 ♑ 02 41	01 36	15 58	14 13	09 20	27 43	19 03	04 58	01 57
14	07 28 50	21 53 37	10 ♑ 58 05	16 ♑ 54 40	01 05	17 11	14 49	09 13	27 50	19 04	04 57	01 58
15	07 32 46	22 50 50	22 ♑ 52 43	28 ♑ 52 29	00 ♌ 31	18 23	15 24	09 06	27 58	19 05	04 56	01 59
16	07 36 43	23 48 03	04 ♒ 54 12	10 ♒ 58 06	29 ♋ 54	19 36	15 59	08 59	28 05	19 06	04 55	02 00
17	07 40 39	24 45 16	17 ♒ 04 25	23 ♒ 13 21	29 15	20 49	16 34	08 52	28 12	19 07	04 54	02 01
18	07 44 36	25 42 30	29 ♒ 25 08	05 ♓ 39 57	28 35	22 01	17 09	08 44	28 20	19 08	04 53	02 02
19	07 48 33	26 39 45	11 ♓ 58 03	18 ♓ 19 37	27 54	23 14	17 43	08 37	28 27	19 10	04 52	02 03
20	07 52 29	27 37 00	24 ♓ 44 54	01 ♈ 14 07	27 13	24 27	18 17	08 29	28 34	19 11	04 51	02 04
21	07 56 26	28 34 15	07 ♈ 47 28	14 ♈ 25 10	26 33	25 39	18 51	08 22	28 41	19 12	04 51	02 06
22	08 00 22	29 ♋ 31 32	21 ♈ 07 24	27 ♈ 54 19	25 54	26 52	19 25	08 14	28 48	19 13	04 50	02 07
23	08 04 19	00 ♌ 28 49	04 ♉ 46 03	11 ♉ 42 40	25 17	28 04	19 58	08 07	28 55	19 15	04 49	02 08
24	08 08 15	01 26 08	18 ♉ 44 08	25 ♉ 50 22	24 43	29 ♌ 17	20 31	07 59	29 02	19 16	04 48	02 10
25	08 12 12	02 23 27	03 ♊ 01 11	10 ♊ 16 16	24 13	00 ♍ 30	21 04	07 51	29 09	19 18	04 48	02 11
26	08 16 08	03 20 48	17 ♊ 35 12	24 ♊ 57 26	23 47	01 42	21 37	07 44	29 16	19 19	04 47	02 12
27	08 20 05	04 18 09	02 ♋ 22 17	09 ♋ 48 58	23 25	02 55	22 09	07 36	29 23	19 21	04 46	02 14
28	08 24 02	05 15 32	17 ♋ 16 35	24 ♋ 44 10	23 09	04 07	22 41	07 28	29 30	19 22	04 46	02 15
29	08 27 58	06 12 55	02 ♌ 10 44	09 ♌ 35 15	22 58	05 19	23 12	07 20	29 37	19 24	04 45	02 17
30	08 31 55	07 10 18	16 ♌ 56 44	24 ♌ 14 17	22 52	06 32	23 44	07 13	29 44	19 26	04 45	02 18
31	08 35 51	08 ♌ 07 43	01 ♍ 27 05	08 ♍ 34 27	22 ♋ 53	07 ♍ 44	24 ♈ 15	07 ♒ 05	29 ♊ 50	19 ♎ 27	04 ♐ 44	02 ♎ 20

Date	Moon True ☊ ° '	Moon Mean ☊ ° '	Moon ☽ Latitude ° '
01	07 ♑ 40	07 ♑ 38	01 S 29
02	07 R 40	07 35	02 44
03	07 39	07 32	03 47
04	07 39	07 29	04 34
05	07 38	07 26	05 04
06	07 38	07 22	05 17
07	07 D 38	07 19	05 13
08	07 38	07 16	04 55
09	07 39	07 13	04 23
10	07 40	07 10	03 41
11	07 41	07 06	02 49
12	07 42	07 03	01 51
13	07 42	07 00	00 S 47
14	07 R 43	06 57	00 N 18
15	07 42	06 54	01 23
16	07 41	06 51	02 25
17	07 39	06 47	03 21
18	07 36	06 44	04 08
19	07 33	06 41	04 45
20	07 31	06 38	05 08
21	07 29	06 35	05 15
22	07 28	06 32	05 06
23	07 D 27	06 28	04 40
24	07 28	06 25	03 57
25	07 29	06 22	02 59
26	07 31	06 19	01 48
27	07 32	06 16	00 N 28
28	07 R 31	06 12	00 S 53
29	07 30	06 09	02 11
30	07 27	06 06	03 20
31	07 ♑ 23	06 ♑ 03	04 S 14

DECLINATIONS

Date	Sun ☉ ° '	Moon ☽ ° '	Mercury ☿ ° '	Venus ♀ ° '	Mars ♂ ° '	Jupiter ♃ ° '	Saturn ♄ ° '	Uranus ♅ ° '	Neptune ♆ ° '	Pluto ♇ ° '
01	23 N 06	19 N 53	18 N 23	21 N 21	00 N 06	18 S 13	22 N 19	06 S 51	19 S 32	14 N 21
02	23 02	15 27	18 04	21 05	00 20	18 15	22 19	06 51	19 32	14 20
03	22 57	10 13	17 45	20 49	00 33	18 16	22 19	06 51	19 32	14 19
04	22 52	04 N 36	17 27	20 32	00 47	18 18	22 19	06 52	19 32	14 18
05	22 47	01 S 04	17 10	20 15	01 01	18 20	22 20	06 52	19 32	14 18
06	22 41	06 30	16 55	19 57	01 14	18 22	22 20	06 52	19 32	14 17
07	22 34	11 29	16 40	19 38	01 28	18 24	22 20	06 52	19 31	14 16
08	22 28	15 52	16 27	19 19	01 41	18 26	22 20	06 53	19 31	14 15
09	22 21	19 30	16 15	18 59	01 55	18 28	22 20	06 53	19 31	14 15
10	22 13	22 15	16 05	18 39	02 08	18 29	22 21	06 53	19 31	14 14
11	22 05	24 00	15 56	18 19	02 21	18 31	22 21	06 53	19 31	14 13
12	21 57	24 41	15 49	17 58	02 34	18 33	22 21	06 54	19 31	14 12
13	21 49	24 14	15 43	17 36	02 47	18 35	22 21	06 54	19 30	14 11
14	21 40	22 41	15 40	17 14	03 00	18 38	22 21	06 55	19 30	14 11
15	21 30	20 08	15 37	16 52	03 12	18 40	22 22	06 55	19 30	14 10
16	21 21	16 42	15 37	16 29	03 25	18 42	22 22	06 55	19 30	14 09
17	21 11	12 32	15 38	16 05	03 37	18 44	22 22	06 56	19 30	14 08
18	21 00	07 48	15 41	15 41	03 50	18 46	22 22	06 56	19 30	14 07
19	20 50	02 S 41	15 45	15 17	04 02	18 48	22 22	06 57	19 30	14 06
20	20 38	02 N 37	15 51	14 53	04 14	18 50	22 22	06 57	19 30	14 06
21	20 27	07 55	15 59	14 28	04 26	18 52	22 22	06 58	19 30	14 05
22	20 15	12 58	16 07	14 02	04 38	18 54	22 22	06 58	19 29	14 04
23	20 03	17 30	16 17	13 37	04 50	18 56	22 22	06 59	19 29	14 03
24	19 51	21 12	16 28	13 11	05 01	18 59	22 22	06 59	19 29	14 02
25	19 38	23 41	16 39	12 44	05 13	19 01	22 23	07 00	19 29	14 01
26	19 25	24 39	16 51	12 18	05 24	19 03	22 23	07 01	19 29	14 00
27	19 11	23 54	17 04	11 51	05 35	19 05	22 23	07 01	19 29	13 59
28	18 57	21 27	17 17	11 23	05 46	19 07	22 23	07 02	19 29	13 58
29	18 43	17 33	17 31	10 56	05 57	19 09	22 23	07 02	19 29	13 57
30	18 29	12 35	17 44	10 28	06 08	19 11	22 23	07 03	19 29	13 56
31	18 N 14	07 N 00	17 N 58	10 N 00	06 N 18	19 S 13	22 N 23	07 S 04	19 S 29	13 N 56

ZODIAC SIGN ENTRIES

Date	h	m	Planets
01	21	55	☽ ♌
03	23	31	☽ ♍
06	04	23	☽ ♎
08	13	05	☽ ♏
11	00	48	☽ ♐
13	13	45	☽ ♑
16	02	15	☽ ♒
16	08	03	☿ ♋
18	13	07	☽ ♓
20	21	43	☽ ♈
22	23	56	☉ ♌
23	03	41	☽ ♉
25	02	13	♀ ♍
25	06	58	☽ ♊
27	08	10	☽ ♋
29	08	29	☽ ♌
31	09	34	☽ ♍

LATITUDES

Date	Mercury ☿ ° '	Venus ♀ ° '	Mars ♂ ° '	Jupiter ♃ ° '	Saturn ♄ ° '	Uranus ♅ ° '	Neptune ♆ ° '	Pluto ♇ ° '
01	01 S 18	01 N 32	02 S 52	00 S 39	01 S 05	00 N 37	01 N 39	16 N 26
04	02 03	01 34	02 55	00 40	01 05	00 37	01 39	16 25
07	02 49	01 36	02 58	00 41	01 04	00 37	01 39	16 23
10	03 33	01 37	03 01	00 41	01 04	00 36	01 39	16 22
13	04 11	01 37	03 04	00 42	01 04	00 36	01 39	16 20
16	04 40	01 37	03 07	00 42	01 04	00 36	01 39	16 19
19	04 55	01 36	03 10	00 43	01 04	00 36	01 39	16 18
22	04 56	01 35	03 12	00 43	01 04	00 36	01 38	16 16
25	04 41	01 32	03 15	00 44	01 04	00 36	01 38	16 15
28	04 14	01 29	03 17	00 44	01 04	00 36	01 38	16 14
31	03 S 35	01 N 26	03 S 20	00 S 45	01 S 04	00 N 36	01 N 38	16 N 13

DATA

Julian Date	2441865
Delta T	+43 seconds
Ayanamsa	23° 29' 30"
Synetic vernal point	05° ♓ 37' 29"
True obliquity of ecliptic	23° 26' 35"

MOON'S PHASES, APSIDES AND POSITIONS ☽

Date	h	m	Phase	Longitude ° '	Eclipse Indicator
07	08	26	☽	15 ♎ 05	
15	11	56	○	22 ♑ 51	
23	03	58	☾	00 ♉ 10	
29	18	59	●	06 ♌ 30	

Day	h	m		
12	21	38	Apogee	
28	07	14	Perigee	
05	07	26	0S	
12	14	26	Max dec	24° S 41'
20	00	12	0N	
26	13	39	Max dec	24° N 39'

LONGITUDES

Date	Chiron ⚷ ° '	Ceres ⚳ ° '	Pallas ⚴ ° '	Juno ⚵ ° '	Vesta ⚶ ° '	Black Moon Lilith ⚸ ° '
01	20 ♈ 41	05 ♐ 12	27 ♎ 41	09 ♐ 12	22 ♋ 31	25 ♐ 01
11	20 ♈ 50	04 ♐ 17	28 ♎ 59	07 ♐ 47	26 ♋ 57	26 ♐ 08
21	20 ♈ 54	03 ♐ 58	00 ♏ 48	06 ♐ 53	01 ♌ 25	27 ♐ 14
31	20 ♈ 53	04 ♐ 15	03 ♏ 04	06 ♐ 33	05 ♌ 56	28 ♐ 21

All ephemeris data is given at 12.00 UT and the Moon's longitude is additionally given for 24.00 UT

ASPECTARIAN

h m	Aspects
01 Sunday	
01 27	☽ ∥ ♀
04 03	☿ ⊥ ♄
04 12	☽ □ ♅
05 31	☽ Q ♇
06 12	☽ ⚼ ♆
14 07	☽ # ♆
15 51	☽ ⊻ ♄
20 03	♀ ✱ ♇
21 32	☽ ∥ ☿
21 41	☽ # ♃
23 25	☿ # ♃
02 Monday	
00 48	☽ ✱ ♇
01 14	☽ ☌ ♀
01 36	☽ ⊥ ♄
01 59	☽ ☌ ☿
06 16	♀ ⊥ ♄
06 18	☽ △ ♆
09 11	☽ Q ♅
09 58	☽ △ ♂
13 15	☉ ⊼ ♃
13 46	☿ ☌ ♀
14 59	☽ ☍ ♃
15 07	☽ ⊻ ☉
16 23	☽ ∠ ♄
17 24	☽ ∥ ♇
03 Tuesday	
01 21	☽ ∠ ♇
01 41	☽ ⊥ ☉
04 56	☽ ✱ ♅
05 43	☉ ± ♆
11 53	☽ ⚼ ♂
16 24	☽ ⊥ ♇
17 34	☽ ✱ ♄
17 49	☽ ∠ ☉
04 Wednesday	
02 25	☽ # ♅
02 36	☽ ⊻ ♇
03 49	☽ ± ♂
04 48	☽ ⊻ ☿
06 17	☽ ∠ ♅
07 45	☽ ⊻ ♀
08 18	☽ □ ♆
14 01	♀ △ ♆
14 17	☽ Q ♄
14 39	☽ ⊼ ♂
15 18	☽ ⊥ ☿
17 14	☽ ⊼ ♃
19 09	☽ ⊥ ♀
21 30	☽ ✱ ☉
21 54	☽ ⊥ ♅
05 Thursday	
03 28	☽ ∥ ♂
03 41	☽ ± ♃
07 16	☽ ∠ ☿
08 30	☽ ⊻ ♅
11 46	☽ # ♂
12 37	☽ ∠ ♀
15 54	☽ Q ♆
19 38	☽ ⚼ ♃
20 29	☽ Q ☉
22 31	☽ □ ♄
06 Friday	
02 25	♀ Q ♅
07 45	☽ ☌ ♇
10 31	☽ ✱ ☿
13 43	☽ ∥ ♅
13 45	☽ ✱ ♆
16 18	♂ ✱ ♃
17 00	☿ St R
18 47	☽ ✱ ♀
22 59	☽ △ ♃
23 23	☽ ☍ ♂
07 Saturday	
08 26	☽ □ ☉
08 54	☽ Q ☿
15 51	☽ ☌ ♅
17 55	☽ ∠ ♆
19 50	☽ Q ♀
20 36	♀ # ♆
08 Sunday	
02 45	☽ # ♇
07 23	☽ △ ♄
11 15	☽ ⊥ ♆
12 08	♀ ☍ ♃
15 20	☽ # ☿
16 44	☽ ⊻ ♇
19 17	☽ □ ☿
22 51	☿ ⊥ ♄
22 57	☽ ⊻ ♆
09 Monday	
04 26	☽ ∥ ♃
04 32	☽ ⊥ ♇
08 21	☽ □ ♃
08 35	☽ # ♀
10 48	☽ □ ♀
12 06	☽ ∥ ♆
12 21	☽ ⊼ ♂
12 50	☉ ∥ ♄
13 08	☽ ⚼ ♄
22 26	☽ ∠ ♇
23 51	☽ △ ☉
10 Tuesday	
00 56	☽ ± ♂
02 41	☽ ⊻ ♅
07 18	☽ ± ♄
11 41	☽ # ☉
12 30	♀ ∠ ♄
12 59	☽ # ♄
14 44	☽ ⊥ ♅
16 03	♀ △ ♂
19 30	☽ ⊼ ♄
20 00	☽ ⚼ ♂
20 02	☽ Q ♃
22 31	♀ # ♃
11 Wednesday	
04 39	☽ ✱ ♇
05 57	☽ △ ☿
08 42	☽ ⚼ ☉
08 57	☽ ∠ ♅
10 55	☽ ☌ ♆
11 55	☉ □ ♅
20 07	☽ ✱ ♃
12 Thursday	
04 02	☽ △ ♂
04 35	☽ ± ☉
05 03	☽ Q ♇
06 07	☽ △ ♀
10 30	☉ Q ♇
11 27	☽ ⚼ ☿
11 52	☉ ⚼ ♆
15 28	☽ ✱ ♅
17 52	☽ ⊼ ☉
19 06	☿ ✱ ♇
13 Friday	
02 20	☽ ∠ ♃
05 06	☽ ± ☿
09 05	☽ ☍ ♄
15 55	☽ Q ♅
16 08	☽ ⚼ ♀
16 48	☽ ⊼ ☿
17 43	☽ □ ♇
20 27	☽ ⊥ ♃
23 49	☽ ⊻ ♆
14 Saturday	
07 43	♀ ∠ ♇
08 30	☽ ⊻ ♃
11 58	☽ ⊥ ♆
12 28	☽ ± ♀
15 44	☽ # ♄
20 11	☽ □ ♂
23 28	☽ # ☉
15 Sunday	
01 57	☽ ⊼ ♀
04 22	☽ □ ♅
06 05	☽ ∠ ♆
11 56	☽ ☍ ☉
16 54	☽ ∥ ♆
22 17	☽ ⊼ ♄
22 54	☽ ∥ ♃
16 Monday	
02 03	♀ ✱ ♅
02 32	☽ ☍ ☿
06 13	☽ △ ♇
10 05	☽ Q ♂
10 21	☽ ± ♄
12 02	☽ ✱ ♆
13 32	☽ # ♀
16 55	♂ Q ♄
18 35	☽ # ☿
20 00	☽ ☌ ♃
17 Tuesday	
03 09	☽ # ♇
04 20	☽ ⚼ ♄
10 58	☽ ✱ ♂
11 40	☽ Q ♆
11 53	☽ ⚼ ♇
16 01	☽ △ ♅
20 06	☽ ☍ ♀
18 Wednesday	
04 14	☽ ⊼ ☉
05 27	☽ ± ♇
09 52	☽ △ ♄
10 28	☽ ⊼ ☿
12 22	☿ ∥ ♀
16 08	☽ ∥ ♅
16 46	☽ ± ☉
17 03	☽ ⊼ ♇
17 30	☽ ∠ ♂
19 34	☿ ⊻ ♄
21 06	☽ ⚼ ♅
21 24	☽ ± ☿
22 30	☽ □ ♆
19 Thursday	
05 42	☽ ⊻ ♃
06 02	☽ # ♂
11 22	☽ ⚼ ☉
11 31	☽ ⊥ ♂
13 40	☽ ⚼ ☿
14 16	☽ ± ♅
16 58	☽ ⊥ ♃
23 22	☽ ⊻ ♂
20 Friday	
01 35	☽ ⊼ ♅
06 05	☉ ☌ ☿
09 41	☽ ∠ ♃
11 23	☽ ⊼ ♀
16 21	☽ △ ☿
17 45	☽ △ ☉
19 09	☽ □ ♄
19 33	☽ ∥ ♂
23 38	☽ ± ♀
21 Saturday	
01 34	☽ ☍ ♇
06 38	☽ △ ♆
07 37	☽ # ♅
13 02	☽ ✱ ♃
15 17	☉ ⊻ ♄
17 44	☽ ⚼ ♀
20 52	♀ ⊥ ♇
23 26	☿ ⊻ ♀
22 Sunday	
03 22	♂ ☍ ♅
04 13	☽ Q ♄
08 36	☽ ☍ ♅
08 50	☽ ☌ ♂
09 42	☽ ⚼ ♆
10 26	☽ Q ♃
10 39	♀ ∥ ♇
16 55	☽ ∥ ♀
17 28	☽ ∥ ♇
20 05	☽ □ ☿
23 10	☽ △ ♀
23 Monday	
01 37	☽ ± ♆
01 42	☽ ✱ ♄
03 58	☽ □ ☉
04 52	☽ ∥ ☿
05 30	♂ ⚼ ♆
07 25	☽ ⊼ ♇
12 05	☽ ⊼ ♆
16 56	♂ Q ♃
17 45	☽ □ ♃
17 51	☽ ± ♇
20 39	☽ # ♃
24 Tuesday	
00 03	☽ # ♆
02 08	☽ Q ☿
02 55	☽ ∥ ☉
03 56	☽ ∠ ♄
06 37	♀ ✱ ♄
09 19	☽ ⚼ ♇
12 54	☽ ⊼ ♅
13 16	☽ Q ☉
15 09	☽ ⊻ ♂
19 21	☽ ⊥ ♄
21 46	☽ ∥ ♄
21 46	☽ ✱ ☿
23 04	☽ ± ♅
25 Wednesday	
01 40	☽ ⊥ ♂
05 30	☽ ⊻ ♄
06 38	☉ ✱ ♇
07 24	☽ □ ♀
08 07	☿ ⊥ ♀
10 36	☽ △ ♇
10 53	☽ ✱ ☉
14 07	☽ ⚼ ♅
14 57	☽ ☍ ♆
17 16	☽ ∠ ♂
19 57	☽ △ ♃
21 57	☽ ∠ ☿
26 Thursday	
03 47	☉ # ♆
12 19	☽ ⊥ ☿
13 20	☽ ∠ ☉
14 50	☽ △ ♅
15 46	☽ Q ♀
18 49	☽ ✱ ♂
20 18	☽ ⚼ ♃
21 50	☽ ⊻ ☿
22 12	♀ ⊻ ♇
27 Friday	
04 58	☽ ⊥ ☉
07 08	☽ ☌ ♄
10 46	☽ ± ♃
11 46	☽ □ ♇
12 57	☽ ✱ ♀
14 59	☽ Q ♂
15 20	☽ ⊻ ☉
15 52	☽ ⊼ ♆
20 21	☽ ⊼ ♃
21 26	☉ # ♃
23 42	☉ △ ♆
28 Saturday	
01 32	☽ ± ♆
04 32	☽ ∥ ♄
15 13	☽ ∠ ♀
15 22	☽ □ ♅
16 00	☽ ⚼ ♆
16 48	☽ Q ♇
17 10	♀ ∠ ♅
18 56	☉ ⊥ ♄
21 01	☽ □ ♂
21 18	☽ ☌ ☿
29 Sunday	
00 45	♀ □ ♆
01 02	☽ # ♆
03 05	☽ # ♃
03 13	☿ □ ♂
05 07	☽ ∥ ☉
06 59	☽ ⊥ ♀
07 50	☽ ⊻ ♄
12 09	☽ ∥ ☿
12 09	☽ ✱ ♇
16 10	☽ △ ♆
17 32	☽ ⊻ ♀
17 36	☽ ⊥ ♄
18 59	☽ ☌ ☉
20 17	☽ ☍ ♃
20 28	☽ Q ♅
30 Monday	
05 48	☽ ∥ ♇
08 21	☽ ∠ ♄
12 35	☽ ∠ ♇
12 49	☉ ☍ ♃
16 05	☽ ✱ ♅
18 35	☉ Q ♅
21 43	☽ ⊻ ☿
21 47	☿ St D
22 13	☽ ∥ ♀
23 34	☽ △ ♂
31 Tuesday	
00 11	♀ ⊼ ♃
03 27	☽ ⊥ ♇
07 41	☽ ⊥ ☿
09 17	☽ ✱ ♄
11 46	☽ # ♅
13 28	☽ ⊻ ♇
14 50	☽ ∥ ♂
14 59	☿ ∠ ♀
17 03	☽ ∠ ♅
17 31	☽ □ ♆
21 23	☽ ⊼ ♃
22 53	☽ ∠ ☿
23 34	☽ ☌ ♀

AUGUST 1973

Date	Sidereal time h m s	Sun ☉ ° ′ ″	Moon ☽ ° ′ ″	Moon ☽ 24.00 ° ′ ″	Mercury ☿ ° ′	Venus ♀ ° ′	Mars ♂ ° ′	Jupiter ♃ ° ′	Saturn ♄ ° ′	Uranus ♅ ° ′	Neptune ♆ ° ′	Pluto ♇ ° ′
01	08 39 48	09 ♌ 05 08	15 ♍ 35 52	22 ♍ 30 57	22 ♋ 59	08 ♍ 57	24 ♈ 45	06 ♒ 57	29 ♊ 57	19 ♎ 29	04 ♐ 44	02 ♎ 21
02	08 43 44	10 02 34	29 ♍ 19 29	06 ♎ 01 25	23 D 13	10 09	25 15	06 R 49	00 ♋ 04	19 31	04 R 43	02 23
03	08 47 41	11 00 01	12 ♎ 36 50	19 ♎ 05 57	23 32	11 21	25 45	06 41	00 10	19 33	04 43	02 24
04	08 51 37	11 57 28	25 ♎ 29 07	01 ♏ 46 44	23 58	12 33	26 15	06 34	00 17	19 35	04 42	02 26
05	08 55 34	12 54 56	07 ♏ 59 19	14 ♏ 07 24	24 31	13 46	26 44	06 26	00 23	19 37	04 42	02 27
06	08 59 31	13 52 25	20 ♏ 11 35	26 ♏ 12 29	25 10	14 58	27 13	06 18	00 30	19 39	04 42	02 29
07	09 03 27	14 49 54	02 ♐ 10 43	08 ♐ 06 56	25 55	16 10	27 41	06 11	00 36	19 41	04 41	02 31
08	09 07 24	15 47 25	14 ♐ 01 44	19 ♐ 55 43	26 47	17 22	28 09	06 03	00 42	19 43	04 41	02 32
09	09 11 20	16 44 56	25 ♐ 49 29	01 ♑ 43 33	27 45	18 34	28 37	05 55	00 49	19 45	04 41	02 34
10	09 15 17	17 42 28	07 ♑ 38 26	13 ♑ 34 36	28 49	19 46	29 04	05 48	00 55	19 47	04 41	02 36
11	09 19 13	18 40 01	19 ♑ 32 28	25 ♑ 32 25	29 ♋ 59	20 58	29 30	05 41	01 01	19 50	04 40	02 38
12	09 23 10	19 37 35	01 ♒ 34 47	07 ♒ 39 49	01 ♌ 15	22 10	29 ♈ 57	05 33	01 07	19 52	04 40	02 39
13	09 27 06	20 35 10	13 ♒ 47 44	19 ♒ 58 44	02 36	23 22	00 ♉ 23	05 26	01 13	19 54	04 40	02 41
14	09 31 03	21 32 47	26 ♒ 12 56	02 ♓ 30 23	04 02	24 34	00 48	05 19	01 19	19 57	04 40	02 43
15	09 35 00	22 30 24	08 ♓ 51 10	15 ♓ 15 16	05 33	25 45	01 13	05 12	01 25	19 59	04 40	02 45
16	09 38 56	23 28 03	21 ♓ 42 40	28 ♓ 13 20	07 08	26 57	01 38	05 05	01 31	20 01	04 40	02 47
17	09 42 53	24 25 43	04 ♈ 47 14	11 ♈ 24 18	08 48	28 09	02 02	04 58	01 37	20 04	04 D 40	02 49
18	09 46 49	25 23 25	18 ♈ 04 30	24 ♈ 47 47	10 32	29 ♍ 21	02 25	04 51	01 42	20 06	04 40	02 50
19	09 50 46	26 21 08	01 ♉ 34 08	08 ♉ 23 31	12 18	00 ♎ 32	02 48	04 44	01 48	20 09	04 40	02 52
20	09 54 42	27 18 53	15 ♉ 15 56	22 ♉ 11 21	14 08	01 44	03 11	04 37	01 53	20 12	04 40	02 54
21	09 58 39	28 16 40	29 ♉ 09 46	06 ♊ 11 07	16 01	02 56	03 32	04 31	01 59	20 14	04 40	02 56
22	10 02 35	29 ♌ 14 28	13 ♊ 15 19	20 ♊ 22 16	17 55	04 07	03 54	04 24	02 04	20 17	04 41	02 58
23	10 06 32	00 ♍ 12 19	27 ♊ 31 45	04 ♋ 43 32	19 51	05 19	04 15	04 18	02 10	20 20	04 41	03 00
24	10 10 29	01 10 11	11 ♋ 57 14	19 ♋ 12 25	21 48	06 30	04 35	04 12	02 15	20 22	04 41	03 02
25	10 14 25	02 08 05	26 ♋ 28 33	03 ♌ 45 00	23 46	07 41	04 54	04 06	02 20	20 25	04 41	03 04
26	10 18 22	03 06 00	11 ♌ 01 03	18 ♌ 15 57	25 45	08 53	05 14	04 00	02 25	20 28	04 42	03 06
27	10 22 18	04 03 57	25 ♌ 28 54	02 ♍ 39 08	27 44	10 04	05 32	03 54	02 30	20 31	04 42	03 08
28	10 26 15	05 01 56	09 ♍ 45 52	16 ♍ 48 24	29 ♌ 43	11 15	05 50	03 48	02 35	20 34	04 42	03 10
29	10 30 11	05 59 56	23 ♍ 46 08	00 ♎ 38 35	01 ♍ 42	12 27	06 07	03 43	02 40	20 37	04 43	03 12
30	10 34 08	06 57 57	07 ♎ 25 23	14 ♎ 06 18	03 41	13 38	06 23	03 37	02 45	20 40	04 43	03 15
31	10 38 04	07 ♍ 56 00	20 ♎ 41 16	27 ♎ 10 20	05 ♍ 39	14 ♎ 49	06 ♉ 39	03 ♒ 32	02 ♋ 50	20 ♎ 43	04 ♐ 44	03 ♎ 17

Date	Moon True ☊ ° ′	Moon Mean ☊ ° ′	Moon ☽ Latitude ° ′
01	07 ♑ 18	06 ♑ 00	04 S 51
02	07 R 14	05 57	05 10
03	07 10	05 53	05 11
04	07 07	05 50	04 57
05	07 06	05 47	04 28
06	07 D 06	05 44	03 49
07	07 07	05 41	02 59
08	07 09	05 37	02 03
09	07 10	05 34	01 S 01
10	07 R 11	05 31	00 N 03
11	07 10	05 28	01 07
12	07 07	05 25	02 09
13	07 02	05 22	03 05
14	06 56	05 18	03 54
15	06 48	05 15	04 32
16	06 40	05 12	04 57
17	06 33	05 09	05 07
18	06 26	05 06	05 01
19	06 22	05 03	04 39
20	06 19	04 59	04 00
21	06 D 18	04 56	03 06
22	06 19	04 53	02 01
23	06 20	04 50	00 N 47
24	06 R 20	04 47	00 S 30
25	06 18	04 43	01 46
26	06 14	04 40	02 55
27	06 08	04 37	03 52
28	06 00	04 34	04 34
29	05 50	04 31	04 58
30	05 40	04 28	05 04
31	05 ♑ 31	04 ♑ 24	04 S 54

DECLINATIONS

Date	Sun ☉ ° ′	Moon ☽ ° ′	Mercury ☿ ° ′	Venus ♀ ° ′	Mars ♂ ° ′	Jupiter ♃ ° ′	Saturn ♄ ° ′	Uranus ♅ ° ′	Neptune ♆ ° ′	Pluto ♇ ° ′
01	17 N 59	01 N 13	18 N 11	09 N 31	06 N 29	19 S 16	22 N 23	07 S 05	19 S 29	13 N 55
02	17 44	04 S 28	18 24	09 02	06 39	19 18	22 23	07 05	19 29	13 54
03	17 28	09 45	18 36	08 34	06 49	19 20	22 23	07 06	19 29	13 53
04	17 12	14 27	18 47	08 05	06 59	19 22	22 23	07 07	19 29	13 52
05	16 56	18 24	18 57	07 35	07 09	19 24	22 23	07 08	19 29	13 51
06	16 40	21 28	19 06	07 06	07 19	19 26	22 23	07 08	19 29	13 50
07	16 23	23 32	19 14	06 36	07 28	19 28	22 23	07 09	19 29	13 49
08	16 06	24 31	19 20	06 06	07 38	19 30	22 23	07 10	19 29	13 48
09	15 49	24 24	19 25	05 36	07 47	19 32	22 23	07 11	19 29	13 47
10	15 32	23 11	19 27	05 06	07 56	19 34	22 23	07 12	19 29	13 46
11	15 14	20 55	19 28	04 36	08 05	19 36	22 23	07 13	19 29	13 45
12	14 56	17 43	19 26	04 05	08 13	19 38	22 23	07 13	19 29	13 44
13	14 38	13 44	19 22	03 35	08 22	19 39	22 23	07 14	19 29	13 43
14	14 19	09 06	19 15	03 04	08 30	19 41	22 23	07 15	19 29	13 42
15	14 01	04 S 03	19 06	02 34	08 38	19 43	22 22	07 16	19 29	13 41
16	13 42	01 N 16	18 54	02 03	08 46	19 45	22 22	07 17	19 29	13 40
17	13 23	06 36	18 39	01 32	08 54	19 47	22 22	07 18	19 29	13 39
18	13 04	11 44	18 22	01 01	09 02	19 48	22 22	07 19	19 29	13 38
19	12 44	16 22	18 02	00 N 30	09 09	19 50	22 22	07 20	19 29	13 37
20	12 24	20 14	17 39	00 S 01	09 17	19 52	22 22	07 21	19 29	13 36
21	12 04	23 00	17 13	00 32	09 24	19 53	22 22	07 22	19 29	13 35
22	11 44	24 24	16 45	01 03	09 31	19 55	22 22	07 23	19 29	13 34
23	11 24	24 12	16 15	01 34	09 37	19 56	22 22	07 24	19 30	13 33
24	11 04	22 24	15 42	02 05	09 44	19 58	22 22	07 25	19 30	13 32
25	10 43	19 08	15 08	02 36	09 50	20 00	22 22	07 26	19 30	13 31
26	10 22	14 40	14 31	03 07	09 56	20 01	22 21	07 27	19 30	13 30
27	10 01	09 23	13 53	03 37	10 02	20 02	22 21	07 29	19 30	13 29
28	09 40	03 N 41	13 13	04 08	10 08	20 04	22 21	07 30	19 30	13 28
29	09 19	02 S 05	12 31	04 39	10 14	20 05	22 21	07 31	19 30	13 27
30	08 57	07 36	11 49	05 10	10 19	20 06	22 21	07 32	19 30	13 26
31	08 N 36	12 S 37	11 N 05	05 S 40	10 N 24	20 S 08	22 N 21	07 S 33	19 S 31	13 N 25

ZODIAC SIGN ENTRIES

Date	h	m	Planets
01	22	20	♄ ♋
02	13	12	☽ ♎
04	20	35	☽ ♏
07	07	37	☽ ♐
09	20	30	☽ ♑
11	12	21	☿ ♌
12	08	52	☽ ♒
12	14	56	♂ ♉
14	19	14	☽ ♓
17	03	16	☽ ♈
19	01	10	♀ ♎
19	09	14	☽ ♉
21	13	26	☽ ♊
23	06	54	☉ ♍
23	16	08	☽ ♋
25	17	49	☽ ♌
27	19	33	☽ ♍
28	15	22	☿ ♍
29	22	52	☽ ♎

LATITUDES

Date	Mercury ☿ ° ′	Venus ♀ ° ′	Mars ♂ ° ′	Jupiter ♃ ° ′	Saturn ♄ ° ′	Uranus ♅ ° ′	Neptune ♆ ° ′	Pluto ♇ ° ′
01	03 S 21	01 N 24	03 S 20	00 S 45	01 S 04	00 N 36	01 N 38	16 N 12
04	02 34	01 20	03 22	00 45	01 04	00 36	01 38	16 11
07	01 46	01 14	03 24	00 45	01 04	00 35	01 38	16 10
10	00 58	01 08	03 26	00 46	01 04	00 35	01 37	16 09
13	00 S 14	01 02	03 28	00 46	01 04	00 35	01 37	16 08
16	00 N 25	00 55	03 29	00 46	01 04	00 35	01 37	16 07
19	00 58	00 47	03 30	00 46	01 04	00 35	01 37	16 07
22	01 22	00 39	03 31	00 47	01 04	00 35	01 37	16 06
25	01 37	00 30	03 32	00 47	01 04	00 35	01 37	16 05
28	01 45	00 21	03 32	00 47	01 04	00 35	01 36	16 04
31	01 N 46	00 N 11	03 S 32	00 S 47	01 S 04	00 N 35	01 N 36	16 N 04

DATA

Julian Date	2441896
Delta T	+43 seconds
Ayanamsa	23° 29′ 35″
Synetic vernal point	05° ♓ 37′ 24″
True obliquity of ecliptic	23° 26′ 35″

MOON'S PHASES, APSIDES AND POSITIONS ☽

Date	h	m	Phase	Longitude ° ′	Eclipse Indicator
05	22	27	☽	13 ♏ 20	
14	02	17	○	21 ♒ 09	
21	10	22	☾	28 ♉ 13	
28	03	25	●	04 ♍ 41	

Day	h	m		
09	10	04	Apogee	
25	06	34	Perigee	
01	17	02	0S	
08	21	26	Max dec	24° S 36′
16	06	19	0N	
22	21	16	Max dec	24° N 31′
29	03	16	0S	

LONGITUDES

Date	Chiron ⚷ ° ′	Ceres ⚳ ° ′	Pallas ⚴ ° ′	Juno ⚵ ° ′	Vesta ⚶ ° ′	Black Moon Lilith ⚸ ° ′
01	20 ♈ 53	04 ♐ 19	03 ♏ 19	06 ♐ 33	06 ♌ 23	28 ♐ 27
11	20 ♈ 46	05 ♐ 13	06 ♏ 00	06 ♐ 49	10 ♌ 54	29 ♐ 34
21	20 ♈ 34	06 ♐ 38	09 ♏ 00	07 ♐ 36	15 ♌ 26	00 ♑ 40
31	20 ♈ 18	08 ♐ 28	12 ♏ 15	08 ♐ 52	19 ♌ 58	01 ♑ 47

All ephemeris data is given at 12.00 UT and the Moon's longitude is additionally given for 24.00 UT

ASPECTARIAN

01 Wednesday
00 04 ☽ ⊻ ☉
01 38 ☽ bQ ♂
01 59 ☉ ∥ ☿
05 42 ☽ Q ♄
07 30 ☽ ± ♃
08 22 ☽ ⊥ ♅
11 03 ☽ ⊥ ☉
17 39 ☽ ± ♂
18 44 ☽ ⊻ ♅
22 54 ☽ bQ ♃

02 Thursday
00 22 ☽ Q ♆
01 01 ☽ ⚹ ☿
01 55 ☉ ⊻ ♀
03 52 ☽ ∠ ☉
04 32 ☽ ⊼ ♂
13 19 ☽ □ ♄
15 45 ♀ bQ ♂
17 27 ☽ ☌ ♇
21 39 ☽ ⚹ ♆
21 58 ☽ # ♂
22 46 ☽ Q ☿
23 39 ☽ ∥ ♅

03 Friday
01 18 ☽ △ ♃
06 51 ☽ # ♀
08 49 ☽ ⚹ ☉
09 27 ☽ ⊻ ♀
21 39 ☽ ⊥ ♀

04 Saturday
00 52 ☽ ☌ ♅
01 08 ☽ ∠ ♆
06 00 ♀ Q ♄
08 46 ☽ # ♇
08 53 ☽ Q ☉
09 01 ☽ □ ☿
12 06 ♀ ± ♃
13 30 ☽ ☍ ♂
16 21 ☽ ∠ ♀
18 07 ☽ ⊥ ♆
21 13 ☽ △ ♄

05 Sunday
01 17 ☽ ⊻ ♇
03 07 ☽ # ☉
05 38 ☽ ⊻ ♆
08 05 ♂ # ♅
09 00 ♀ ⊥ ♅
09 01 ☽ □ ♃
12 55 ☽ ⊥ ♇
16 00 ☽ # ☿
16 21 ♃ ± ♄
19 07 ☽ ∥ ♃
19 39 ☽ ∥ ♆
22 27 ☽ □ ☉

06 Monday
00 31 ☽ ⚹ ♀
02 37 ☽ bQ ♄
04 04 ♀ ∥ ♂
06 37 ☽ ∠ ♇
09 56 ♀ # ♅
10 55 ☽ ⊻ ♅
20 06 ☽ Q ♃
20 39 ☽ ± ♄
21 14 ☽ # ♄
22 33 ☽ △ ☿
22 54 ☽ ⊥ ♅

07 Tuesday
02 35 ☽ ⊼ ♂
03 01 ☽ Q ♀
08 48 ☽ ⊼ ♄
12 40 ☽ ⚹ ♇
15 10 ☽ ± ♂
17 04 ☽ ∠ ♅
17 04 ☽ ☌ ♆
17 14 ♅ ∠ ♆
19 59 ☽ ⚹ ♃
23 09 ♃ ∥ ♆

08 Wednesday
07 04 ☽ bQ ☿
09 41 ☉ ∠ ♄
10 09 ☽ bQ ♂
13 02 ☽ Q ♇
15 54 ☽ △ ☉
19 33 ☽ □ ♀
23 36 ☽ ⚹ ♅

09 Thursday
02 08 ☽ ∠ ♃
02 56 ☽ ± ☿
15 42 ♂ ± ♆
16 17 ☽ ⊼ ☿
17 54 ☽ △ ♂
20 15 ☽ ⊥ ♃
22 14 ☽ ☍ ♄

10 Friday
00 05 ☽ Q ♅
01 07 ☽ bQ ☉
01 45 ☽ □ ♇
06 00 ☽ ⊻ ♆
08 18 ☽ ⊻ ♃
09 09 ☉ ∠ ♇
12 27 ♀ ⊻ ♅
18 09 ☽ ⊥ ♆
20 35 ☿ □ ♂
20 57 ☽ ± ☉
21 48 ☽ # ♄

11 Saturday
06 44 ♀ bQ ♃
10 06 ☽ ⊼ ☉
12 16 ☽ ∠ ♆
12 34 ☽ □ ♅
15 10 ☽ △ ♀
22 38 ☽ ∥ ♃
23 36 ☽ ∥ ♆
23 50 ☽ # ☿

12 Sunday
08 39 ☽ □ ♂
09 32 ☿ ⊻ ♄
11 05 ☽ ⊼ ♄
11 15 ☽ ☍ ☿
14 08 ☽ △ ♇
18 07 ☽ ⚹ ♆
18 11 ☉ ⚹ ♅
19 46 ☽ ☌ ♃
22 08 ♀ Q ♆
23 01 ☽ ± ♄

13 Monday
00 13 ☽ bQ ♀
06 32 ☽ # ☉
12 04 ☽ # ♇
13 38 ☿ ⚹ ♇
16 45 ☽ bQ ♄
17 35 ☽ Q ♆
19 35 ☽ bQ ♇
19 41 ☽ ± ♀
21 13 ☽ Q ♂
23 53 ☽ △ ♅

14 Tuesday
02 17 ☽ ☍ ☉
08 29 ☽ ⊼ ♀
12 58 ☽ ± ♇
14 53 ☽ # ♂
19 25 ♀ ± ♂
20 57 ☽ ∥ ♅
21 04 ☽ ⚹ ♂
21 49 ☽ △ ♄
22 14 ☿ △ ♆

15 Wednesday
00 26 ☽ ⊼ ♇
04 06 ☽ □ ♆
04 40 ☽ bQ ♅
04 54 ☽ ⊼ ☿
05 09 ☽ ⊻ ♃
06 53 ☿ ☍ ♃
16 21 ☽ ⊥ ♃
17 46 ☽ ± ☿
19 29 ☽ # ♀
21 39 ☽ ± ♅

16 Thursday
02 15 ☽ ∠ ♂
03 17 ♂ ⚹ ♄
08 52 ☽ ⊼ ♅
09 00 ☽ ∠ ♃
12 55 ☽ bQ ☿
14 43 ☉ ∥ ♇
15 11 ☽ ∥ ♀
15 30 ☽ ⊼ ☉
16 07 ♆ St D
17 49 ☿ ⊥ ♄
19 27 ☽ ⊥ ♂
22 39 ☽ ☍ ♀

17 Friday
01 11 ☿ Q ♅
03 25 ☽ ± ☉
06 10 ☽ □ ♄
06 48 ☽ ⊻ ♂
08 23 ☽ ☍ ♇
11 47 ☽ △ ♆
12 19 ☽ ⚹ ♃
15 12 ☽ # ♅
20 22 ☽ △ ☿
21 05 ☽ bQ ☉
22 50 ☽ ∥ ♂

18 Saturday
09 49 ☽ Q ♃
14 51 ☽ bQ ♆
14 56 ☽ Q ♄
15 39 ☽ ☍ ♅
18 09 ☽ ∥ ☉
21 25 ☽ ∥ ♇

19 Sunday
02 03 ☽ △ ☉
06 52 ☽ ± ♆
10 00 ☽ ⊼ ♀
12 24 ☽ ⚹ ♄
14 14 ☽ ☌ ♂
14 18 ☽ ⊼ ♇
17 01 ♂ ⊼ ♇
17 28 ☽ ⊼ ♆
17 32 ☽ □ ♃
20 48 ☽ ∥ ☿
21 35 ☽ ± ♀

20 Monday
00 52 ☽ ± ♇
01 20 ♃ ⚹ ♆
01 26 ☉ ⊥ ♇
06 51 ☽ # ♆
09 22 ☽ # ♃
09 44 ☽ □ ☿
14 47 ☽ bQ ♀
14 51 ☽ ∠ ♄
15 28 ♀ □ ♄
16 36 ☽ bQ ♇
20 34 ☽ ⊼ ♅

21 Tuesday
05 23 ☽ ∥ ♄
06 30 ☽ ⊥ ♄
06 57 ☽ ± ♅
10 22 ☽ □ ☉
12 15 ♀ ☌ ♇
16 52 ☽ ⊻ ♄
18 29 ☽ △ ♇
19 02 ☽ △ ♀
19 41 ☽ ⊻ ♂
21 05 ☽ △ ♃
21 26 ☽ ☍ ♆
21 35 ☽ Q ☿
22 25 ☽ bQ ♅

22 Wednesday
00 56 ☿ ∠ ♄
05 41 ♀ ⊼ ♂
06 10 ☽ ⊥ ♂
12 43 ☿ ∠ ♇
17 19 ♀ △ ♃
19 13 ☽ Q ☉
21 06 ☽ ⚹ ☿
21 46 ☽ ∠ ♂
22 18 ☽ bQ ♃
23 17 ♀ ⚹ ♆
23 53 ☽ △ ♅

23 Thursday
13 17 ☽ ± ♃
15 01 ♂ □ ♃
16 47 ☽ ⚹ ☉
18 02 ☿ ⚹ ♅
19 47 ☽ ☌ ♄
21 09 ☽ □ ♇
23 13 ☽ ⊼ ♃
23 28 ☽ ⚹ ♂
23 56 ☽ ⊼ ♆

24 Friday
02 07 ☽ ∠ ☿
02 08 ☽ □ ♀
02 34 ☿ ∠ ♀
09 54 ☽ ± ♆
12 24 ☽ ∥ ♄
19 22 ☽ ⊥ ☿
19 29 ☽ ∠ ☉
19 36 ♂ ⊼ ♆
19 50 ☽ Q ♂

25 Saturday
00 48 ☽ bQ ♆
01 58 ☽ □ ♅
03 03 ☽ Q ♇
06 34 ☽ # ♃
06 50 ☽ ⊻ ☿
09 43 ☽ # ♆
10 35 ☽ Q ♀
11 24 ☽ ⊥ ☉
17 31 ☉ ⚹ ♄
21 44 ☽ ⊻ ♄
22 00 ☽ ⊻ ☉
22 54 ☽ ⚹ ♇

26 Sunday
00 29 ☽ ☍ ♃
01 33 ☽ △ ♆
02 13 ☽ □ ♂
07 41 ☽ ⊥ ♄
07 46 ☽ Q ♅
08 09 ☽ ⚹ ♀
12 07 ☉ ⊻ ♇
12 49 ☽ ∥ ☿
17 35 ☽ ∥ ♇
22 40 ☽ ∠ ♄
23 46 ☽ ∠ ♇

27 Monday
03 42 ☽ ⚹ ♅
04 38 ☿ ⊥ ♇
08 15 ☉ ⊼ ♃
09 04 ☽ ∥ ☉
09 14 ☽ ∥ ♂
11 04 ☉ ∥ ♂
11 15 ☽ ∠ ♀
11 42 ☉ ⊥ ♀
14 46 ☽ ⊥ ♇
16 22 ☽ ☌ ☿
20 08 ☽ # ♅
23 49 ☽ ⚹ ♄

28 Tuesday
00 51 ☽ ⊻ ♇
02 00 ☽ ⊼ ♃
02 42 ☿ ∥ ♇
03 25 ☽ ☌ ☉
03 27 ☽ □ ♆
03 41 ☽ ⊥ ♀
03 50 ☉ □ ♆
04 52 ☽ ∠ ♅
05 12 ☽ △ ♂
10 16 ☽ # ♀
12 04 ☽ ± ♃
14 46 ☽ ⊻ ♀
20 11 ☽ ⊥ ♅
20 15 ☽ Q ♄

29 Wednesday
01 51 ☉ ∠ ♅
03 20 ☽ bQ ♃
06 31 ☽ ⊻ ♅
07 19 ☽ bQ ♂
10 10 ☽ Q ♆
15 59 ☉ △ ♂
23 18 ☽ ± ♂

30 Thursday
00 05 ☽ ∥ ♀
00 15 ☿ ⚹ ♄
03 40 ☽ □ ♄
04 13 ☽ ⊻ ☿
04 34 ☽ ☌ ♇
05 18 ☽ △ ♃
06 37 ☿ ⊻ ♇
07 12 ☽ ⚹ ♆
10 07 ☽ ⊼ ♂
11 07 ☽ ⊻ ☉
11 23 ☿ ⊼ ♃
11 41 ☽ ∥ ♅
16 43 ☽ ⊥ ☿
17 48 ☽ # ☉
22 43 ☽ ⊥ ☉

31 Friday
00 14 ☽ ☌ ♀
00 46 ☿ □ ♆
00 51 ☽ # ♂
05 21 ☽ # ☿
10 14 ☽ ∠ ♆
11 54 ☽ ∠ ☿
12 02 ☽ ☌ ♅
12 51 ☿ ∠ ♅
16 08 ☽ # ♇
16 28 ☽ ∠ ☉

SEPTEMBER 1973

Date	Sidereal time h m s	Sun ☉ ° ' "	Moon ☽ ° ' "	Moon ☽ 24.00 ° ' "	Mercury ☿ ° '	Venus ♀ ° '	Mars ♂ ° '	Jupiter ♃ ° '	Saturn ♄ ° '	Uranus ♅ ° '	Neptune ♆ ° '	Pluto ♇ ° '
01	10 42 01	08 ♍ 54 05	03 ♏ 33 41	09 ♏ 51 36	07 ♍ 36	16 ♎ 00	06 ♉ 54	03 ♒ 27	02 ♋ 54	20 ♎ 46	04 ♐ 44	03 ♎ 19
02	10 45 58	09 52 11	16 ♏ 04 31	22 ♏ 12 53	09 32	17 11	07 09	03 R 22	02 59	20 49	04 45	03 21
03	10 49 54	10 50 18	28 ♏ 17 16	04 ♐ 18 16	11 27	18 22	07 22	03 18	03 04	20 52	04 45	03 23
04	10 53 51	11 48 27	10 ♐ 16 32	16 ♐ 12 43	13 22	19 33	07 35	03 13	03 08	20 55	04 46	03 25
05	10 57 47	12 46 37	22 ♐ 07 30	28 ♐ 01 35	15 15	20 44	07 47	03 09	03 12	20 58	04 46	03 27
06	11 01 44	13 44 49	03 ♑ 55 37	09 ♑ 50 16	17 07	21 55	07 59	03 04	03 16	21 01	04 47	03 30
07	11 05 40	14 43 02	15 ♑ 46 11	21 ♑ 43 55	18 59	23 05	08 10	03 00	03 21	21 04	04 48	03 32
08	11 09 37	15 41 17	27 ♑ 44 03	03 ♒ 47 03	20 49	24 16	08 20	02 56	03 25	21 08	04 49	03 34
09	11 13 33	16 39 33	09 ♒ 53 22	16 ♒ 03 21	22 37	25 27	08 29	02 53	03 29	21 11	04 49	03 36
10	11 17 30	17 37 51	22 ♒ 17 17	28 ♒ 35 22	24 25	26 37	08 37	02 49	03 32	21 14	04 50	03 38
11	11 21 27	18 36 11	04 ♓ 57 42	11 ♓ 24 21	26 12	27 48	08 45	02 46	03 36	21 17	04 51	03 41
12	11 25 23	19 34 32	17 ♓ 55 13	24 ♓ 30 12	27 57	28 ♎ 58	08 52	02 43	03 40	21 21	04 52	03 43
13	11 29 20	20 32 55	01 ♈ 09 03	07 ♈ 51 32	29 ♍ 42	00 ♏ 09	08 58	02 40	03 44	21 24	04 53	03 45
14	11 33 16	21 31 20	14 ♈ 37 19	21 ♈ 26 05	01 ♎ 25	01 19	09 03	02 37	03 47	21 28	04 54	03 47
15	11 37 13	22 29 48	28 ♈ 17 27	05 ♉ 21 04	03 07	02 29	09 07	02 34	03 50	21 31	04 54	03 50
16	11 41 09	23 28 17	12 ♉ 06 38	19 ♉ 03 51	04 48	03 39	09 11	02 32	03 54	21 34	04 55	03 52
17	11 45 06	24 26 48	26 ♉ 02 27	03 ♊ 02 13	06 28	04 49	09 14	02 29	03 57	21 38	04 56	03 54
18	11 49 02	25 25 22	10 ♊ 02 59	17 ♊ 04 37	08 07	05 59	09 15	02 27	04 00	21 41	04 57	03 56
19	11 52 59	26 23 58	24 ♊ 07 01	01 ♋ 10 05	09 45	07 09	09 16	02 25	04 03	21 45	04 59	03 59
20	11 56 56	27 22 36	08 ♋ 13 43	15 ♋ 17 50	11 22	08 19	09 R 16	02 24	04 06	21 48	05 00	04 01
21	12 00 52	28 21 16	22 ♋ 22 15	29 ♋ 26 49	12 58	09 29	09 15	02 22	04 09	21 52	05 01	04 03
22	12 04 49	29 ♍ 19 59	06 ♌ 31 17	13 ♌ 35 19	14 33	10 39	09 13	02 21	04 12	21 55	05 02	04 06
23	12 08 45	00 ♎ 18 43	20 ♌ 38 35	27 ♌ 40 39	16 07	11 48	09 11	02 20	04 14	21 59	05 03	04 08
24	12 12 42	01 17 30	04 ♍ 41 02	11 ♍ 39 13	17 40	12 58	09 07	02 19	04 17	22 03	05 04	04 10
25	12 16 38	02 16 19	18 ♍ 34 43	25 ♍ 27 00	19 12	14 08	09 03	02 18	04 19	22 06	05 06	04 12
26	12 20 35	03 15 10	02 ♎ 15 35	09 ♎ 00 03	20 44	15 17	08 57	02 18	04 22	22 10	05 07	04 15
27	12 24 31	04 14 03	15 ♎ 40 04	22 ♎ 15 22	22 14	16 26	08 51	02 17	04 24	22 13	05 08	04 17
28	12 28 28	05 12 58	28 ♎ 45 46	05 ♏ 11 15	23 43	17 36	08 44	02 17	04 26	22 17	05 09	04 19
29	12 32 25	06 11 55	11 ♏ 31 51	17 ♏ 47 43	25 12	18 45	08 36	02 D 17	04 28	22 21	05 11	04 22
30	12 36 21	07 ♎ 10 54	23 ♏ 59 09	00 ♐ 06 28	26 ♎ 39	19 ♏ 54	08 ♉ 27	02 ♒ 18	04 ♋ 30	22 ♎ 24	05 ♐ 12	04 ♎ 24

Date	Moon True ☊ ° '	Moon Mean ☊ ° '	Moon ☽ Latitude ° '
01	05 ♑ 24	04 ♑ 21	04 S 29
02	05 R 19	04 18	03 51
03	05 16	04 15	03 04
04	05 15	04 12	02 09
05	05 D 16	04 09	01 10
06	05 16	04 05	00 S 07
07	05 R 15	04 02	00 N 56
08	05 13	03 59	01 57
09	05 08	03 56	02 53
10	05 01	03 53	03 42
11	04 51	03 49	04 22
12	04 40	03 46	04 49
13	04 28	03 43	05 01
14	04 16	03 40	04 56
15	04 06	03 37	04 35
16	03 58	03 34	03 58
17	03 53	03 30	03 06
18	03 51	03 27	02 02
19	03 D 50	03 24	00 N 51
20	03 R 50	03 21	00 S 23
21	03 49	03 18	01 36
22	03 47	03 15	02 44
23	03 41	03 11	03 41
24	03 33	03 08	04 24
25	03 22	03 05	04 51
26	03 10	03 02	05 00
27	02 57	02 59	04 53
28	02 45	02 55	04 30
29	02 35	02 52	03 55
30	02 ♑ 27	02 ♑ 49	03 S 08

DECLINATIONS

Date	Sun ☉	Moon ☽	Mercury ☿	Venus ♀	Mars ♂	Jupiter ♃	Saturn ♄	Uranus ♅	Neptune ♆	Pluto ♇
01	08 N 14	16 S 55	10 N 21	06 S 11	10 N 29	20 S 09	22 N 21	07 S 34	19 S 31	13 N 24
02	07 52	20 20	09 36	06 41	10 34	20 10	22 21	07 35	19 31	13 23
03	07 30	22 46	08 50	07 12	10 38	20 11	22 21	07 37	19 31	13 22
04	07 08	24 07	08 03	07 42	10 43	20 12	22 20	07 38	19 31	13 21
05	06 46	24 22	07 17	08 12	10 47	20 13	22 20	07 39	19 31	13 20
06	06 24	23 30	06 30	08 41	10 51	20 14	22 20	07 40	19 32	13 19
07	06 01	21 35	05 42	09 11	10 55	20 15	22 20	07 41	19 32	13 18
08	05 39	18 43	04 55	09 41	10 58	20 16	22 20	07 43	19 32	13 17
09	05 16	15 00	04 08	10 10	11 02	20 17	22 20	07 44	19 32	13 16
10	04 53	10 35	03 20	10 39	11 05	20 18	22 20	07 45	19 32	13 15
11	04 31	05 38	02 33	11 08	11 08	20 18	22 20	07 46	19 32	13 14
12	04 08	00 S 21	01 46	11 36	11 11	20 19	22 19	07 48	19 33	13 13
13	03 45	05 N 03	00 59	12 05	11 13	20 20	22 19	07 49	19 33	13 12
14	03 22	10 19	00 N 12	12 33	11 15	20 21	22 19	07 50	19 33	13 11
15	02 59	15 09	00 S 34	13 01	11 18	20 21	22 19	07 52	19 33	13 10
16	02 36	19 14	01 21	13 28	11 19	20 22	22 19	07 53	19 34	13 09
17	02 12	22 16	02 06	13 56	11 21	20 22	22 19	07 54	19 34	13 08
18	01 49	23 59	02 52	14 23	11 23	20 23	22 19	07 55	19 34	13 07
19	01 26	24 10	03 37	14 49	11 24	20 23	22 19	07 57	19 34	13 07
20	01 03	22 48	04 21	15 16	11 25	20 23	22 18	07 58	19 35	13 06
21	00 39	20 00	05 05	15 42	11 26	20 24	22 18	07 59	19 35	13 05
22	00 N 16	16 00	05 49	16 08	11 27	20 24	22 18	08 01	19 35	13 04
23	00 S 07	11 08	06 32	16 33	11 27	20 24	22 18	08 02	19 35	13 03
24	00 31	05 42	07 14	16 58	11 27	20 24	22 18	08 03	19 36	13 02
25	00 54	00 N 04	07 56	17 23	11 27	20 25	22 18	08 05	19 36	13 01
26	01 18	05 S 29	08 38	17 47	11 27	20 25	22 18	08 06	19 36	13 00
27	01 41	10 40	09 18	18 11	11 27	20 25	22 18	08 08	19 36	12 59
28	02 04	15 15	09 58	18 34	11 26	20 25	22 18	08 09	19 37	12 59
29	02 28	19 01	10 38	18 57	11 25	20 25	22 17	08 10	19 37	12 58
30	02 S 51	21 S 49	11 S 16	19 S 20	11 N 24	20 S 25	22 N 17	08 S 12	19 S 37	12 N 57

ZODIAC SIGN ENTRIES

Date	h	m	Planets
01	05	17	☽ ♏
03	15	24	☽ ♐
06	04	01	☽ ♑
08	16	30	☽ ♒
11	02	40	☽ ♓
13	09	05	♀ ♏
13	09	56	☽ ♈
13	16	16	☿ ♎
15	14	59	☽ ♉
17	18	48	☽ ♊
19	22	01	☽ ♋
22	00	56	☽ ♌
23	04	21	☉ ♎
24	03	58	☽ ♍
26	08	00	☽ ♎
28	14	18	☽ ♏
30	23	47	☽ ♐

LATITUDES

Date	Mercury ☿	Venus ♀	Mars ♂	Jupiter ♃	Saturn ♄	Uranus ♅	Neptune ♆	Pluto ♇
01	01 N 45	00 N 07	03 S 32	00 S 47	01 S 04	00 N 35	01 N 36	16 N 04
04	01 38	00 S 03	03 31	00 47	01 04	00 35	01 36	16 03
07	01 28	00 13	03 30	00 47	01 04	00 35	01 36	16 03
10	01 13	00 24	03 29	00 47	01 04	00 34	01 36	16 02
13	00 56	00 36	03 27	00 47	01 04	00 34	01 35	16 02
16	00 37	00 47	03 24	00 47	01 04	00 34	01 35	16 02
19	00 N 16	00 58	03 20	00 47	01 04	00 34	01 35	16 02
22	00 S 05	01 10	03 18	00 47	01 04	00 34	01 35	16 02
25	00 27	01 21	03 13	00 47	01 05	00 34	01 35	16 02
28	00 49	01 33	03 08	00 47	01 05	00 34	01 35	16 02
31	01 S 11	01 S 44	03 S 02	00 S 47	01 S 05	00 N 34	01 N 34	16 N 02

DATA

Julian Date	2441927
Delta T	+43 seconds
Ayanamsa	23° 29' 39"
Synetic vernal point	05° ♓ 37' 21"
True obliquity of ecliptic	23° 26' 35"

MOON'S PHASES, APSIDES AND POSITIONS ☽

Date	h	m	Phase	Longitude ° '	Eclipse Indicator
04	15	22	☽	11 ♐ 57	
12	15	16	○	19 ♓ 42	
19	16	11	☾	26 ♊ 34	
26	13	54	●	03 ♎ 20	

Day	h	m	
06	03	14	Apogee
20	22	07	Perigee

05	05	12	Max dec	24° S 25'
12	13	32	0N	
19	02	54	Max dec	24° N 17'
25	12	15	0S	

LONGITUDES

Date	Chiron ⚷ ° '	Ceres ⚳ ° '	Pallas ⚴ ° '	Juno ⚵ ° '	Vesta ⚶ ° '	Black Moon Lilith ⚸ ° '
01	20 ♈ 16	08 ♐ 41	12 ♏ 35	09 ♐ 00	20 ♌ 25	01 ♑ 53
11	19 ♈ 56	10 ♐ 56	16 ♏ 04	10 ♐ 43	24 ♌ 57	03 ♑ 00
21	19 ♈ 32	13 ♐ 31	19 ♏ 43	12 ♐ 48	29 ♌ 28	04 ♑ 06
31	19 ♈ 06	16 ♐ 22	23 ♏ 31	15 ♐ 31	03 ♍ 57	05 ♑ 13

ASPECTARIAN

h m	Aspects
01 Saturday	
02 13	☿ △ ♂
02 55	☽ ⊥ ♆
08 00	☿ ∥ ♂
10 45	☽ △ ♄
11 32	☽ ⊻ ♇
11 48	☽ □ ♃
14 14	☽ ⊻ ♆
18 28	☽ ☍ ♂
21 04	☽ ✱ ☿
22 59	☽ ⊥ ♇
23 01	☽ ✱ ☉
02 Sunday	
00 36	☉ ± ♃
05 34	☽ ∥ ♆
10 04	☿ ± ♃
10 35	☽ ∥ ♃
14 23	☽ ⊻ ♀
15 44	☽ bQ ♄
16 26	☽ ∠ ♇
16 39	♃ △ ♇
20 24	☉ ☌ ☿
21 17	☽ ⊻ ♅
22 17	☽ Q ♃
03 Monday	
00 18	☽ Q ☉
00 40	☽ Q ☿
03 24	☽ ⊥ ♀
05 21	☉ # ♅
06 55	☽ # ♄
09 10	☽ ⊥ ♅
09 33	☽ ± ♄
20 34	☉ # ♀
21 34	☽ ⊼ ♄
21 55	☽ ✱ ♃
22 11	☽ ✱ ♇
23 14	☽ ∠ ♀
04 Tuesday	
00 55	☽ ☌ ♆
03 11	☽ ∠ ♅
06 29	☽ ⊼ ♂
08 48	♀ ∥ ♅
15 22	☽ □ ☉
16 23	♀ ∠ ♆
18 15	☿ ⊥ ♀
18 47	☿ # ♀
18 48	☽ ± ♂
19 25	☽ □ ☿
22 26	☽ Q ♇
05 Wednesday	
00 48	☿ # ♅
01 47	♃ ⊼ ♄
03 58	☽ ∠ ♃
08 15	☿ ⊥ ♅
08 51	☽ ✱ ♀
09 38	☽ ✱ ♅
11 21	☿ Q ♄
13 23	☽ bQ ♂
17 01	♀ ☌ ♅
22 08	☽ ⊥ ♃
06 Thursday	
10 09	☽ Q ♅
10 16	☽ ⊻ ♃
10 40	☽ ☍ ♄
11 07	☽ □ ♇
11 58	☽ Q ♀
13 45	☽ ⊻ ♆
17 48	☉ ∥ ☿
20 22	☽ △ ♂
23 49	☿ bQ ♃
07 Friday	
01 56	☽ ⊥ ♆
04 03	☽ # ♄
09 41	☽ △ ☉
19 38	☽ △ ☿
20 07	☽ ∠ ♆
21 19	☉ ⊥ ♅
22 44	☽ □ ♅
08 Saturday	
00 02	☽ ∥ ♃
04 20	☽ □ ♀
04 39	☉ Q ♄
05 53	☽ ∥ ♆
16 18	☿ ⊻ ♅
18 23	☽ bQ ☉
22 17	☽ ☌ ♃
23 19	☽ ⊼ ♄
23 36	☽ △ ♇
09 Sunday	
02 02	☽ ✱ ♆
06 47	☽ bQ ☿
09 13	☽ □ ♂
11 11	☽ ± ♄
13 38	☽ ± ☉
14 39	☿ Q ♆
21 49	☽ # ♇
10 Monday	
00 27	☿ bQ ♂
01 30	☽ Q ♆
02 17	☽ ⊼ ☉
03 19	☽ ± ☿
04 46	☽ bQ ♄
04 58	☽ bQ ♇
09 27	☽ # ♂
09 58	☽ △ ♅
11 40	☽ ∥ ♀
16 22	☉ bQ ♃
16 45	☽ ⊼ ☿
20 22	☽ Q ♂
21 07	☽ △ ♀
22 14	☽ ± ♇
11 Tuesday	
01 54	☽ ∥ ♅
07 53	☽ ⊻ ♃
09 27	☽ △ ♄
09 35	☽ ⊼ ♇
11 47	☽ □ ♆
12 02	♀ # ♂
14 30	☽ bQ ♅
17 36	☽ # ☉
19 04	☽ ⊥ ♃
19 08	☽ ✱ ♂
12 Wednesday	
04 01	☽ bQ ♀
04 34	☽ # ☿
07 15	☽ ± ♅
09 46	♀ ⊥ ♆
11 37	☽ ∠ ♃
15 16	☽ ☍ ☉
18 17	☽ ⊼ ♅
20 11	☽ ∥ ☿
22 07	☽ ± ♀
22 56	☽ ∠ ♂
13 Thursday	
06 34	☽ ∥ ☉
08 59	☽ ☍ ☿
10 01	☽ ⊼ ♀
14 42	☽ ✱ ♃
15 17	☽ ⊥ ♂
16 38	☽ □ ♄
16 41	☽ ☍ ♇
18 41	☽ △ ♆
14 Friday	
00 30	☽ # ♅
02 04	☽ ⊻ ♂
07 36	☿ ⊻ ♀
10 20	☉ ⊻ ♅
11 59	☽ Q ♃
16 31	☽ ∥ ♂
17 25	♄ □ ♇
21 18	☽ bQ ♆
23 53	☽ # ♀
15 Saturday	
00 06	☽ ☍ ♅
00 40	☽ Q ♄
01 05	☽ ⊼ ☉
01 52	☽ ∥ ♇
04 26	☿ △ ♃
12 08	☿ ± ♂
12 23	☽ ± ☉
13 05	☽ ± ♆
13 39	♀ □ ♃
19 26	☽ □ ♃
19 59	☽ ☍ ♀
20 01	♀ # ♇
21 35	☽ ⊼ ☿
21 40	☽ ⊼ ♇
21 42	☽ ✱ ♄
22 18	☉ Q ♆
22 19	☿ ☌ ♇
22 39	☿ □ ♄
23 32	☽ ⊼ ♆
16 Sunday	
05 13	☽ bQ ☉
06 55	☽ ☌ ♂
08 06	☽ ± ♇
09 25	☽ ± ☿
13 47	☿ ✱ ♆
14 09	☽ # ♆
16 28	♀ ⊻ ♇
17 14	♀ △ ♄
19 51	☽ # ♃
23 41	☽ bQ ♇
23 45	☽ ∠ ♄
17 Monday	
03 05	☽ bQ ☿
04 23	☽ ⊼ ♅
06 21	☉ bQ ♂
09 03	☽ △ ☉
12 23	☽ ∥ ♄
14 08	☉ # ☿
14 28	♀ ⊻ ♆
14 44	☽ ± ♅
15 17	☽ ⊥ ♄
23 02	☽ △ ♃
18 Tuesday	
01 31	☽ △ ♇
01 37	☽ ⊻ ♄
03 17	☽ ☍ ♆
04 25	☽ ⊼ ♀
06 14	☽ bQ ♅
08 16	☽ △ ☿
10 38	☽ ⊻ ♂
15 37	☽ ± ♀
20 54	☽ ⊥ ♂
19 Wednesday	
00 37	☽ bQ ♃
04 52	☿ ⊼ ♂
07 57	☽ △ ♅
08 21	☽ bQ ♀
12 16	☽ ∠ ♂
15 55	☽ ± ♃
16 11	☽ □ ☉
23 19	♂ St R
20 Thursday	
02 06	☽ ⊼ ♃
04 49	☽ □ ♇
04 58	☽ ☌ ♄
06 30	☽ ⊼ ♆
12 10	☽ △ ♀
13 46	☽ ✱ ♂
16 42	☽ ± ♆
17 13	☽ ∥ ♄
18 01	☽ □ ☿
21 Friday	
01 02	☽ Q ☉
07 21	♀ ☍ ♂
08 00	☽ bQ ♆
09 12	☽ # ♃
10 07	☽ Q ♂
11 08	☽ □ ♅
11 28	☽ Q ♇
14 52	☽ # ♆
22 54	☽ ✱ ☉
22 Saturday	
00 11	♀ ⊥ ♇
04 25	☽ Q ☿
04 56	☽ ☍ ♃
07 52	☽ ✱ ♇
08 02	☽ ⊻ ♄
09 28	☽ △ ♆
11 24	☽ # ♀
16 34	☽ □ ♂
17 48	☽ Q ♅
18 15	☽ ⊥ ♄
19 38	☽ □ ♀
23 Sunday	
02 15	☽ ∠ ☉
02 55	☽ ∥ ♇
03 20	☽ ✱ ☿
09 25	☽ ∠ ♇
09 36	☽ ∠ ♄
10 30	☽ ∥ ♂
14 17	☽ ✱ ♅
18 43	☽ ⊥ ☉
24 Monday	
00 49	☽ ⊥ ♇
01 49	☽ # ♅
05 03	☽ Q ♀
05 45	☽ ⊻ ☉
06 06	☽ # ☿
07 57	☽ ⊼ ♃
08 07	☽ ∠ ☿
11 07	☽ ⊻ ♇
11 18	☽ ✱ ♄
12 40	☽ □ ♆
16 04	☽ ∠ ♅
18 14	☽ ± ♃
19 35	☽ △ ♂
25 Tuesday	
01 32	☽ ⊥ ☿
03 34	☽ ✱ ♀
07 41	☽ ⊥ ♅
08 04	☽ Q ♄
08 39	☽ # ☉
09 47	☽ bQ ♃
12 48	☉ △ ♃
13 14	☽ ⊻ ☿
16 25	☽ ∥ ☉
17 07	☿ ∥ ♅
18 10	☽ ⊻ ♅
19 53	☽ Q ♆
21 28	☽ bQ ♂
26 Wednesday	
02 09	☿ ∠ ♆
05 17	☉ ± ♂
08 11	☽ ∠ ♀
12 04	☽ △ ♃
13 13	☽ ± ♂
13 54	☽ ☌ ☉
15 32	☽ ☌ ♇
15 44	☽ □ ♄
17 04	☽ ✱ ♆
23 49	☽ ⊼ ♂
23 52	☽ ∥ ♅
27 Thursday	
01 41	☽ ⊥ ♀
04 26	☽ ∥ ☿
11 53	☿ ☌ ♅
13 15	☉ ☌ ♇
13 32	☽ ⊻ ♀
15 49	☽ # ♂
16 06	☉ □ ♄
20 08	☽ ∠ ♆
23 40	☽ # ♇
28 Friday	
00 00	☽ ☌ ♅
01 30	☽ ☌ ☿
10 30	☉ ✱ ♆
12 44	☽ ⊥ ♆
13 30	♃ St D
18 34	☽ □ ♃
22 24	☽ ⊻ ♇
22 36	☽ △ ♄
23 58	☽ ⊻ ♆
29 Saturday	
01 04	☽ ⊻ ☉
06 29	☽ ☍ ♂
09 46	☽ ⊥ ♇
11 32	☽ ∥ ♀
13 23	☽ ⊥ ☉
16 33	☽ ∥ ♆
22 55	☽ ∥ ♃
30 Sunday	
01 14	♀ ∠ ♇
03 04	☽ ∠ ♇
03 15	☽ ☌ ♀
03 16	☽ bQ ♄
03 24	♀ bQ ♄
04 50	☽ Q ♃
08 11	☽ ∠ ☉
08 55	☽ ⊻ ♅
16 51	☿ # ♂
17 21	☽ # ♄
17 54	☽ ⊻ ☿
20 22	♀ Q ♃
20 42	☽ ⊥ ♅
20 51	☽ ± ♄

All ephemeris data is given at 12.00 UT and the Moon's longitude is additionally given for 24.00 UT

OCTOBER 1973

Date	Sidereal time h m s	Sun ☉ ° ′ ″	Moon ☽ ° ′ ″	Moon ☽ 24.00 ° ′ ″	Mercury ☿ ° ′	Venus ♀ ° ′	Mars ♂ ° ′	Jupiter ♃ ° ′	Saturn ♄ ° ′	Uranus ♅ ° ′	Neptune ♆ ° ′	Pluto ♇ ° ′
01	12 40 18	08 ♎ 09 54	06 ♐ 10 07	12 ♐ 10 37	28 ♎ 06	21 ♏ 03	08 ♉ 17	02 ♒ 18	04 ♋ 32	22 ♎ 28	05 ♐ 14	04 ♎ 26
02	12 44 14	09 08 57	18 ♐ 08 30	24 ♐ 04 24	29 ♎ 31	22 12	08 R 06	02 19	04 33	22 32	05 15	04 28
03	12 48 11	10 08 01	29 ♐ 58 58	05 ♑ 52 52	00 ♏ 56	23 21	07 55	02 20	04 35	22 35	05 16	04 31
04	12 52 07	11 07 07	11 ♑ 46 48	17 ♑ 41 28	02 19	24 29	07 43	02 21	04 36	22 39	05 18	04 33
05	12 56 04	12 06 15	23 ♑ 37 33	29 ♑ 35 45	03 41	25 38	07 30	02 22	04 38	22 43	05 19	04 35
06	13 00 00	13 05 24	05 ♒ 36 41	11 ♒ 41 00	05 03	26 46	07 16	02 24	04 39	22 47	05 21	04 38
07	13 03 57	14 04 36	17 ♒ 49 14	24 ♒ 01 53	06 23	27 55	07 02	02 25	04 40	22 50	05 22	04 40
08	13 07 54	15 03 49	00 ♓ 19 21	06 ♓ 41 59	07 42	29 ♏ 03	06 47	02 27	04 41	22 54	05 24	04 42
09	13 11 50	16 03 04	13 ♓ 09 59	19 ♓ 43 28	09 00	00 ♐ 11	06 31	02 29	04 42	22 58	05 26	04 44
10	13 15 47	17 02 20	26 ♓ 22 23	03 ♈ 06 37	10 16	01 19	06 15	02 31	04 43	23 02	05 27	04 47
11	13 19 43	18 01 39	09 ♈ 55 52	16 ♈ 49 46	11 31	02 27	05 58	02 34	04 43	23 05	05 29	04 49
12	13 23 40	19 01 00	23 ♈ 47 50	00 ♉ 49 28	12 45	03 35	05 41	02 36	04 44	23 09	05 31	04 51
13	13 27 36	20 00 23	07 ♉ 54 04	15 ♉ 00 57	13 57	04 42	05 23	02 39	04 44	23 13	05 32	04 53
14	13 31 33	20 59 48	22 ♉ 09 29	29 ♉ 19 00	15 07	05 50	05 05	02 42	04 45	23 17	05 34	04 56
15	13 35 29	21 59 15	06 ♊ 28 55	13 ♊ 38 43	16 16	06 57	04 46	02 46	04 45	23 21	05 36	04 58
16	13 39 26	22 58 45	20 ♊ 47 57	27 ♊ 56 15	17 22	08 04	04 27	02 49	04 45	23 24	05 37	05 00
17	13 43 23	23 58 16	05 ♋ 03 18	12 ♋ 08 54	18 26	09 11	04 08	02 52	04 R 45	23 28	05 39	05 02
18	13 47 19	24 57 51	19 ♋ 12 53	26 ♋ 15 08	19 28	10 18	03 48	02 56	04 45	23 32	05 41	05 05
19	13 51 16	25 57 27	03 ♌ 15 35	10 ♌ 14 10	20 27	11 25	03 29	03 00	04 45	23 36	05 43	05 07
20	13 55 12	26 57 06	17 ♌ 10 48	24 ♌ 05 26	21 23	12 31	03 09	03 04	04 45	23 39	05 45	05 09
21	13 59 09	27 56 47	00 ♍ 57 58	07 ♍ 48 15	22 16	13 38	02 48	03 09	04 44	23 43	05 47	05 11
22	14 03 05	28 56 30	14 ♍ 36 11	21 ♍ 21 33	23 05	14 44	02 28	03 13	04 44	23 47	05 48	05 13
23	14 07 02	29 ♎ 56 15	28 ♍ 04 11	04 ♎ 43 52	23 50	15 50	02 08	03 18	04 43	23 51	05 50	05 15
24	14 10 58	00 ♏ 56 03	11 ♎ 20 24	17 ♎ 53 35	24 31	16 56	01 48	03 23	04 42	23 55	05 52	05 17
25	14 14 55	01 55 52	24 ♎ 23 15	00 ♏ 49 16	25 07	18 02	01 27	03 28	04 41	23 58	05 54	05 20
26	14 18 52	02 55 44	07 ♏ 11 32	13 ♏ 30 01	25 38	19 07	01 07	03 33	04 40	24 02	05 56	05 22
27	14 22 48	03 55 38	19 ♏ 44 45	25 ♏ 55 51	26 03	20 13	00 48	03 38	04 39	24 06	05 58	05 24
28	14 26 45	04 55 33	02 ♐ 03 27	08 ♐ 07 48	26 21	21 18	00 28	03 44	04 38	24 10	06 00	05 26
29	14 30 41	05 55 31	14 ♐ 09 12	20 ♐ 08 01	26 33	22 23	00 ♉ 09	03 49	04 37	24 13	06 02	05 28
30	14 34 38	06 55 30	26 ♐ 04 42	01 ♑ 59 43	26 R 36	23 27	29 ♈ 50	03 55	04 35	24 17	06 04	05 30
31	14 38 34	07 ♏ 55 30	07 ♑ 53 36	13 ♑ 46 57	26 ♏ 31	24 ♐ 32	29 ♈ 31	04 ♒ 01	04 ♋ 34	24 ♎ 21	06 ♐ 06	05 ♎ 32

Date	Moon True ☊ ° ′	Moon Mean ☊ ° ′	Moon ☽ Latitude ° ′
01	02 ♑ 23	02 ♑ 46	02 S 14
02	02 R 20	02 43	01 15
03	02 D 20	02 40	00 S 12
04	02 R 20	02 36	00 N 50
05	02 19	02 33	01 50
06	02 17	02 30	02 47
07	02 13	02 27	03 37
08	02 06	02 24	04 17
09	01 57	02 21	04 46
10	01 46	02 17	05 00
11	01 34	02 14	04 58
12	01 22	02 11	04 39
13	01 12	02 08	04 03
14	01 05	02 05	03 11
15	01 00	02 01	02 06
16	00 58	01 58	00 N 54
17	00 D 58	01 55	00 S 22
18	00 58	01 52	01 36
19	00 R 58	01 49	02 43
20	00 56	01 46	03 40
21	00 52	01 42	04 24
22	00 45	01 39	04 52
23	00 36	01 36	05 04
24	00 26	01 33	04 59
25	00 15	01 30	04 38
26	00 05	01 26	04 04
27	29 ♐ 56	01 23	03 18
28	29 50	01 20	02 23
29	29 46	01 17	01 23
30	29 45	01 14	00 20
31	29 ♐ 45	01 ♑ 11	00 N 44

DECLINATIONS

Date	Sun ☉ ° ′	Moon ☽ ° ′	Mercury ☿ ° ′	Venus ♀ ° ′	Mars ♂ ° ′	Jupiter ♃ ° ′	Saturn ♄ ° ′	Uranus ♅ ° ′	Neptune ♆ ° ′	Pluto ♇ ° ′
01	03 S 14	23 S 32	11 S 54	19 S 42	11 N 23	20 S 24	22 N 17	08 S 13	19 S 38	12 N 56
02	03 38	24 09	12 32	20 04	11 22	20 24	22 17	08 14	19 38	12 55
03	04 01	23 39	13 08	20 25	11 20	20 24	22 17	08 16	19 38	12 54
04	04 24	22 05	13 44	20 45	11 19	20 24	22 17	08 17	19 38	12 54
05	04 47	19 34	14 19	21 06	11 17	20 23	22 17	08 19	19 39	12 53
06	05 10	16 10	14 53	21 25	11 15	20 23	22 17	08 20	19 39	12 52
07	05 33	12 03	15 26	21 45	11 13	20 23	22 17	08 21	19 39	12 51
08	05 56	07 21	15 58	22 03	11 10	20 22	22 17	08 23	19 40	12 50
09	06 19	02 S 13	16 30	22 22	11 08	20 22	22 17	08 24	19 40	12 50
10	06 42	03 N 09	17 00	22 39	11 05	20 21	22 17	08 26	19 40	12 49
11	07 04	08 30	17 29	22 56	11 03	20 20	22 17	08 27	19 41	12 48
12	07 27	13 34	17 58	23 13	11 00	20 20	22 17	08 28	19 41	12 47
13	07 49	17 58	18 25	23 29	10 57	20 19	22 16	08 30	19 41	12 47
14	08 12	21 23	18 51	23 44	10 54	20 18	22 16	08 31	19 42	12 46
15	08 34	23 28	19 15	23 59	10 51	20 18	22 16	08 33	19 42	12 45
16	08 56	24 01	19 39	24 14	10 47	20 17	22 16	08 34	19 42	12 45
17	09 18	22 59	20 01	24 27	10 44	20 16	22 16	08 35	19 43	12 44
18	09 40	20 29	20 22	24 40	10 41	20 15	22 16	08 37	19 43	12 43
19	10 02	16 47	20 41	24 53	10 37	20 14	22 16	08 38	19 43	12 43
20	10 23	12 11	20 59	25 05	10 34	20 13	22 16	08 40	19 44	12 42
21	10 45	07 01	21 15	25 16	10 30	20 12	22 16	08 41	19 44	12 41
22	11 06	01 N 34	21 30	25 26	10 27	20 11	22 16	08 42	19 45	12 41
23	11 27	03 S 53	21 42	25 36	10 24	20 10	22 16	08 44	19 45	12 40
24	11 48	09 04	21 53	25 46	10 20	20 09	22 16	08 45	19 45	12 39
25	12 09	13 46	22 01	25 54	10 17	20 07	22 16	08 47	19 46	12 39
26	12 29	17 45	22 07	26 02	10 13	20 06	22 16	08 48	19 46	12 38
27	12 50	20 51	22 11	26 10	10 10	20 05	22 16	08 49	19 46	12 38
28	13 10	22 55	22 12	26 17	10 07	20 04	22 16	08 51	19 47	12 37
29	13 30	23 53	22 10	26 23	10 04	20 02	22 16	08 52	19 47	12 37
30	13 50	23 43	22 05	26 28	10 01	20 01	22 16	08 53	19 47	12 36
31	14 S 09	22 S 29	21 S 56	26 S 33	09 N 58	19 S 59	22 N 16	08 S 55	19 S 48	12 N 36

ZODIAC SIGN ENTRIES

Date	h	m	Planets
02	20	12	☿ ♏
03	12	02	☽ ♑
06	00	49	☽ ♒
08	11	23	☽ ♓
09	08	08	♀ ♐
10	18	29	☽ ♈
12	22	36	☽ ♉
15	01	09	☽ ♊
17	03	28	☽ ♋
19	06	25	☽ ♌
21	10	19	☽ ♍
23	13	30	☉ ♏
23	15	28	☽ ♎
25	22	28	☽ ♏
28	07	57	☽ ♐
29	22	56	♂ ♈
30	19	57	☽ ♑

LATITUDES

Date	Mercury ☿ ° ′	Venus ♀ ° ′	Mars ♂ ° ′	Jupiter ♃ ° ′	Saturn ♄ ° ′	Uranus ♅ ° ′	Neptune ♆ ° ′	Pluto ♇ ° ′
01	01 S 11	01 S 44	03 S 02	00 S 47	01 S 05	00 N 34	01 N 34	16 N 02
04	01 33	01 55	02 56	00 47	01 05	00 34	01 34	16 02
07	01 53	02 06	02 48	00 46	01 05	00 34	01 34	16 03
10	02 13	02 17	02 40	00 46	01 05	00 34	01 34	16 03
13	02 30	02 27	02 31	00 46	01 05	00 34	01 34	16 03
16	02 45	02 36	02 21	00 46	01 05	00 34	01 34	16 04
19	02 56	02 45	02 11	00 46	01 05	00 34	01 34	16 05
22	03 03	02 53	02 00	00 46	01 05	00 34	01 34	16 05
25	03 04	03 01	01 49	00 46	01 05	00 34	01 33	16 06
28	02 56	03 08	01 37	00 45	01 05	00 34	01 33	16 07
31	02 S 38	03 S 13	01 S 26	00 S 45	01 S 05	00 N 34	01 N 33	16 N 08

DATA

Julian Date	2441957
Delta T	+43 seconds
Ayanamsa	23° 29′ 42″
Synetic vernal point	05° ♓ 37′ 18″
True obliquity of ecliptic	23° 26′ 35″

MOON'S PHASES, APSIDES AND POSITIONS ☽

Date	h	m	Phase	Longitude ° ′	Eclipse Indicator
04	10	32	☽	11 ♑ 04	
12	03	09	○	18 ♈ 39	
18	22	33	☾	25 ♋ 24	
26	03	17	●	02 ♏ 34	

Day	h	m		
03	22	47	Apogee	
16	01	02	Perigee	
31	18	59	Apogee	
02	13	06	Max dec	24° S 09′
09	21	58	0N	
16	08	16	Max dec	24° N 02′
22	18	51	0S	
29	20	30	Max dec	23° S 57′

LONGITUDES

Date	Chiron ⚷ ° ′	Ceres ⚳ ° ′	Pallas ⚴ ° ′	Juno ⚵ ° ′	Vesta ⚶ ° ′	Black Moon Lilith ⚸ ° ′
01	19 ♈ 06	16 ♐ 22	23 ♏ 31	15 ♐ 11	03 ♍ 57	05 ♑ 13
11	18 ♈ 39	19 ♐ 28	27 ♏ 25	17 ♐ 52	08 ♍ 23	06 ♑ 19
21	18 ♈ 11	22 ♐ 45	01 ♐ 25	20 ♐ 46	12 ♍ 46	07 ♑ 26
31	17 ♈ 45	26 ♐ 12	05 ♐ 30	23 ♐ 53	17 ♍ 04	08 ♑ 32

All ephemeris data is given at 12.00 UT and the Moon's longitude is additionally given for 24.00 UT

ASPECTARIAN

h m Aspects

01 Monday
04 20 ☽ ⚹ ♃
07 14 ♀ ∥ ♆
07 19 ☽ ⊥ ☿
08 33 ☽ ⚹ ♇
08 44 ☽ ⊼ ♄
10 07 ☽ ☌ ♆
14 25 ☉ ⊼ ♂
14 36 ☽ ∠ ♅
16 09 ☽ ⊼ ♂
16 20 ☽ ⚹ ☉
02 Tuesday
03 43 ☽ ∠ ☿
04 00 ☽ ± ♂
07 23 ☿ ⊥ ♆
08 38 ☽ Q ♇
10 20 ☽ ∠ ♃
18 38 ☽ Q ☉
19 23 ♀ ⊻ ♅
20 55 ☽ ⚹ ♅
21 04 ☽ ⊻ ♀
21 53 ☽ ꝑ ♂
03 Wednesday
03 01 ☿ # ♇
04 34 ☽ ⊥ ♃
10 33 ☽ ⊥ ♀
11 05 ♀ ∥ ♃
14 11 ☽ ⚹ ☿
16 47 ☽ ⊻ ♃
21 15 ☽ □ ♇
21 22 ☽ ☍ ♄
21 25 ☽ Q ♅
22 47 ☽ ⊻ ♆
04 Thursday
03 52 ☽ △ ♂
06 50 ☽ ∠ ♀
09 42 ☽ # ♄
10 32 ☽ □ ☉
11 01 ☽ ⊥ ♆
12 32 ☿ □ ♃
17 50 ☽ Q ☿
05 Friday
00 07 ☽ ∥ ♀
05 01 ☽ ∥ ♃
05 19 ☽ ∠ ♆
10 09 ☽ □ ♅
11 19 ☽ ∥ ♆
16 28 ☽ ⚹ ♀
06 Saturday
04 19 ☿ ⊻ ♇
04 46 ☿ △ ♄
05 35 ☽ ☌ ♃
10 02 ☽ △ ♇
10 05 ☽ ⊼ ♄
10 44 ☽ □ ☿
11 29 ☽ ⚹ ♆
15 14 ☽ □ ♂
17 29 ☿ ⊻ ♆
18 54 ☽ Q ♀
18 59 ☽ ∥ ☿
21 59 ☽ ± ♄
07 Sunday
04 03 ☽ △ ☉
07 37 ☽ # ♇
11 08 ☽ Q ♆
12 49 ♄ □ ♇
15 35 ☽ ꝑ ♄
15 35 ☽ ꝑ ♇
16 32 ☽ # ♂
21 46 ☽ △ ♅
21 56 ☿ ☍ ♂
08 Monday
01 39 ☽ Q ♂
04 11 ♀ ± ♄
06 58 ☽ ∥ ♅
08 45 ♀ ⊥ ♅
08 55 ☽ ± ♇
09 21 ☽ □ ♀
11 28 ☽ ꝑ ☉
16 02 ☽ ⊻ ♃
18 18 ☽ ∥ ☉
20 14 ☽ △ ♄
20 17 ☽ ⊼ ♇
21 35 ☽ □ ♆
23 55 ☽ ⚹ ♂
09 Tuesday
02 19 ☽ ꝑ ♅
03 19 ☽ ⊥ ♃
03 26 ☽ △ ☿
05 22 ♀ # ♄
05 46 ☽ ± ☉
17 44 ☽ ⊼ ☉
19 00 ☽ ± ♅
19 57 ☽ ∠ ♃
10 Wednesday
02 58 ☽ ∠ ♂
05 57 ☽ ⊼ ♅
09 49 ☽ ꝑ ☿
18 48 ☽ ⊥ ♂
21 38 ☽ △ ♀
21 57 ☿ ⊥ ♇
23 00 ☽ ⚹ ♃
11 Thursday
02 51 ☽ □ ♄
02 59 ☽ ☍ ♇
03 29 ☽ ± ☿
04 10 ☽ △ ♆
05 00 ☽ # ☉
05 12 ☽ ⊻ ♂
11 45 ☽ # ♅
14 35 ♀ ⚹ ♃
15 03 ☽ ⊼ ☿
20 06 ☽ Q ♃
23 40 ☽ ∥ ♂
12 Friday
02 14 ☽ ꝑ ♀
03 09 ☽ ☍ ☉
06 20 ☽ ꝑ ♆
08 12 ☽ ∥ ♇
10 10 ☽ Q ♄
10 53 ☽ ☍ ♅
19 02 ☽ ± ♀
21 47 ☽ ± ♆
13 Saturday
01 08 ♂ ⊼ ♆
03 05 ☽ □ ♃
06 07 ☽ ⊼ ♀
06 39 ☽ ⚹ ♄
06 53 ☽ ⊼ ♇
07 50 ☽ ☌ ♂
08 00 ☽ ⊼ ♆
12 48 ♀ ⊼ ♄
15 04 ☽ # ☿
16 09 ♀ ⚹ ♇
17 04 ☽ ± ♇
23 06 ☽ # ♆
23 07 ☽ ☍ ☿
23 33 ♀ ⊼ ♂
14 Sunday
01 16 ☉ ∠ ♆
03 31 ☽ # ♃
06 17 ♀ ☌ ♆
07 57 ☽ ∠ ♄
08 15 ☽ ꝑ ♇
09 54 ☽ ⊼ ☉
13 53 ☽ ⊼ ♅
20 23 ☽ ∥ ♄
20 43 ☽ ± ☉
22 55 ♂ ⊼ ♇
23 03 ☽ ⊥ ♄
23 59 ☽ ± ♅
15 Monday
05 44 ☽ △ ♃
09 06 ☽ ⊻ ♄
09 12 ☽ ⊻ ♂
09 27 ☽ △ ♇
10 27 ☉ ∥ ♅
10 31 ☽ ☍ ♆
12 51 ☽ ☍ ♀
12 55 ☽ ꝑ ☉
13 47 ♂ ⚹ ♄
15 08 ☽ ꝑ ♅
19 02 ☽ ⊥ ♂
19 04 ☉ ∠ ♀
16 Tuesday
05 46 ☽ ⊼ ☿
06 58 ☽ ꝑ ♃
09 48 ☽ ∠ ♂
15 37 ☿ ∥ ♆
15 56 ☽ △ ☉
16 24 ☽ △ ♅
16 40 ☽ ± ☿
19 38 ♀ ∠ ♅
22 09 ☽ ± ♃
23 00 ☉ ☌ ♅
17 Wednesday
05 51 ♄ St R
08 18 ☽ ⊼ ♃
09 03 ☽ ꝑ ☿
10 29 ☽ ⚹ ♂
11 29 ☽ ☌ ♄
11 58 ☽ □ ♇
13 01 ☽ ⊼ ♆
19 35 ☽ ⊼ ♀
20 19 ☽ ∥ ♄
23 11 ☽ ± ♆
18 Thursday
03 47 ♀ ± ♂
04 03 ☿ ∥ ♃
06 20 ☽ Q ♂
06 37 ☽ ± ♀
12 27 ☽ △ ☿
12 49 ☽ # ☿
13 47 ☽ # ♃
14 30 ☽ ꝑ ♆
17 36 ☽ # ♆
18 36 ☽ Q ♇
18 57 ☿ ꝑ ♄
19 23 ☽ □ ♅
22 33 ☽ □ ☉
23 16 ☽ ꝑ ♀
19 Friday
03 25 ☿ ∠ ♇
11 33 ☽ ☍ ♃
12 22 ☽ □ ♂
14 33 ☽ ⊻ ♄
15 11 ☽ ⚹ ♇
16 14 ☽ △ ♆
20 Saturday
00 53 ☽ ⊥ ♄
02 24 ☽ Q ♅
03 15 ☽ △ ♀
03 15 ☿ Q ♃
07 51 ☽ Q ☉
09 29 ☽ ∥ ♇
16 12 ♂ □ ♃
16 27 ☽ ∠ ♄
17 10 ☽ ∠ ♇
19 48 ☽ □ ☿
19 50 ☽ ∥ ♂
20 00 ☽ # ☉
22 12 ☉ # ♂
23 18 ☽ ⚹ ♅
21 Sunday
04 29 ☽ # ♅
06 19 ☽ ⚹ ☉
08 53 ☽ ⊥ ♇
15 09 ☽ △ ♂
15 50 ☽ ⊼ ♃
18 36 ☽ ⚹ ♄
19 25 ☽ ⊻ ♇
20 27 ☽ □ ♆
22 Monday
01 41 ☽ ∠ ♅
02 26 ☽ ± ♃
05 23 ☽ Q ☿
10 44 ☽ ∠ ☉
12 15 ☽ □ ♀
15 46 ☽ Q ♄
16 57 ☽ ꝑ ♂
17 40 ☽ ⊥ ♅
18 27 ☽ ꝑ ♃
23 Tuesday
03 59 ☽ ⚹ ☿
04 01 ☽ ⊥ ☉
04 24 ☽ ⊻ ♅
04 24 ☽ Q ♆
08 36 ☽ ± ♂
09 33 ☉ ⊥ ♆
12 17 ☿ ⊻ ♅
15 38 ☽ ⊻ ☉
19 07 ☽ ⊼ ♂
21 28 ☽ △ ♃
23 19 ☽ Q ♀
23 58 ☽ □ ♄
24 Wednesday
00 59 ☽ ☌ ♇
02 02 ☽ ⚹ ♆
08 31 ☽ ∠ ☿
09 38 ♀ ꝑ ♂
10 29 ☽ ∥ ♅
18 08 ☽ # ♂
20 05 ♀ Q ♇
23 11 ☽ ⚹ ♀
25 Thursday
01 49 ☽ ⊥ ☿
02 42 ☽ ∥ ☉
03 27 ☉ ☍ ♂
05 32 ☽ ∠ ♆
06 02 ☽ # ♇
11 13 ☽ ☌ ♅
13 26 ☽ ⊻ ☿
22 14 ♀ ∠ ♃
22 18 ☽ ⊥ ♆
26 Friday
00 51 ☽ ☍ ♂
03 17 ☽ ☌ ☉
05 04 ☽ □ ♃
05 40 ☽ ∠ ♀
07 15 ☽ △ ♄
08 32 ☽ ⊻ ♇
09 37 ☽ ⊻ ♆
19 56 ☽ ⊥ ♇
22 08 ☉ # ♇
27 Saturday
00 21 ☽ ⊥ ♀
02 47 ☽ ∥ ♆
04 16 ☉ □ ♃
05 22 ☽ ∥ ♃
11 49 ☽ ꝑ ♄
12 59 ☽ ⊻ ♀
13 16 ☽ ∠ ♇
15 41 ☽ Q ♃
20 28 ☽ ⊻ ♅
28 Sunday
00 34 ☽ ☌ ☿
02 07 ☽ ∥ ☿
03 05 ☽ # ♄
05 09 ☉ △ ♄
05 18 ☽ ± ♄
08 15 ☽ ⊥ ♅
08 57 ☽ ⊼ ♂
15 19 ☽ ⚹ ♃
17 04 ☽ ⊼ ♄
18 10 ☽ ⊻ ☉
18 40 ☽ ⚹ ♇
19 48 ☽ ☌ ♆
20 29 ☽ ± ♂
29 Monday
00 34 ☉ ⊻ ♇
02 07 ☽ ∠ ♅
07 09 ☽ ⊥ ☉
13 56 ☽ ꝑ ♂
14 45 ☉ ⊻ ♆
18 39 ☽ Q ♇
19 27 ♂ ± ♆
21 26 ☽ ∠ ♃
30 Tuesday
02 50 ☽ ∠ ☉
06 10 ☽ ☌ ♀
08 21 ☽ ⚹ ♅
10 28 ☿ St R
13 03 ☽ ⊻ ☿
15 46 ☽ ⊥ ♃
19 24 ☽ △ ♂
31 Wednesday
01 11 ☽ ⊥ ☿
04 03 ☽ ⊻ ♃
05 14 ☽ ☍ ♄
07 11 ☽ □ ♇
07 35 ♀ ⚹ ♅
08 21 ☽ ⊻ ♆
08 50 ☽ Q ♅
12 04 ☽ ⚹ ☉
14 43 ☽ # ♄
19 17 ☽ ∠ ☿
19 35 ☽ ∥ ☿
20 36 ☽ ⊥ ♆

NOVEMBER 1973

LONGITUDES

Date	Sidereal time h m s	Sun ☉ ° ' "	Moon ☽ ° ' "	Moon ☽ 24.00 ° ' "	Mercury ☿ ° '	Venus ♀ ° '	Mars ♂ ° '	Jupiter ♃ ° '	Saturn ♄ ° '	Uranus ♅ ° '	Neptune ♆ ° '	Pluto ♇ ° '
01	14 42 31	08 ♏ 55 33	19 ♑ 40 21	25 ♑ 34 28	26 ♏ 18	25 ♐ 36	29 ♈ 13	04 ♒ 07	04 ♋ 32	24 ♎ 24	06 ♐ 08	05 ♎ 34
02	14 46 27	09 55 37	01 ♒ 29 57	07 ♒ 27 29	25 R 55	26 40	28 R 55	04 14	04 R 30	24 28	06 10	05 36
03	14 50 24	10 55 42	13 ♒ 27 44	19 ♒ 31 21	25 22	27 44	28 38	04 20	04 29	24 32	06 12	05 38
04	14 54 21	11 55 50	25 ♒ 39 00	01 ♓ 51 15	24 40	28 47	28 22	04 27	04 27	24 35	06 14	05 40
05	14 58 17	12 55 58	08 ♓ 08 40	14 ♓ 31 43	23 48	29 ♐ 50	28 06	04 34	04 25	24 39	06 17	05 42
06	15 02 14	13 56 08	21 ♓ 00 47	27 ♓ 36 09	22 48	00 ♑ 53	27 50	04 41	04 22	24 43	06 19	05 44
07	15 06 10	14 56 20	04 ♈ 17 58	11 ♈ 06 15	21 40	01 56	27 36	04 48	04 20	24 46	06 21	05 46
08	15 10 07	15 56 33	18 ♈ 00 50	25 ♈ 01 26	20 26	02 58	27 22	04 55	04 18	24 50	06 23	05 48
09	15 14 03	16 56 48	02 ♉ 07 36	09 ♉ 18 43	19 08	04 00	27 08	05 03	04 15	24 53	06 25	05 49
10	15 18 00	17 57 05	16 ♉ 34 04	23 ♉ 52 48	17 48	05 02	26 55	05 11	04 13	24 57	06 27	05 51
11	15 21 56	18 57 23	01 ♊ 14 02	08 ♊ 36 47	16 29	06 03	26 43	05 18	04 10	25 01	06 29	05 53
12	15 25 53	19 57 44	16 ♊ 00 09	23 ♊ 23 13	15 14	07 04	26 32	05 26	04 07	25 04	06 32	05 55
13	15 29 50	20 58 06	00 ♋ 45 09	08 ♋ 05 12	14 04	08 04	26 22	05 34	04 05	25 08	06 34	05 57
14	15 33 46	21 58 30	15 ♋ 22 44	22 ♋ 37 13	13 02	09 05	26 12	05 42	04 02	25 11	06 36	05 58
15	15 37 43	22 58 55	29 ♋ 48 16	06 ♌ 55 33	12 10	10 04	26 03	05 51	03 59	25 15	06 38	06 00
16	15 41 39	23 59 23	13 ♌ 58 55	20 ♌ 58 14	11 28	11 04	25 55	05 59	03 55	25 18	06 40	06 02
17	15 45 36	24 59 53	27 ♌ 53 27	04 ♍ 44 36	10 58	12 03	25 48	06 08	03 52	25 21	06 43	06 03
18	15 49 32	26 00 24	11 ♍ 31 45	18 ♍ 14 59	10 40	13 01	25 41	06 17	03 49	25 25	06 45	06 05
19	15 53 29	27 00 57	24 ♍ 54 25	01 ♎ 30 09	10 33	13 59	25 36	06 26	03 46	25 28	06 47	06 07
20	15 57 25	28 01 32	08 ♎ 02 17	14 ♎ 30 58	10 D 38	14 57	25 31	06 35	03 42	25 32	06 49	06 08
21	16 01 22	29 ♏ 02 08	20 ♎ 56 17	27 ♎ 18 20	10 53	15 54	25 27	06 44	03 38	25 35	06 52	06 10
22	16 05 19	00 ♐ 02 47	03 ♏ 37 13	09 ♏ 53 02	11 17	16 51	25 23	06 53	03 35	25 38	06 54	06 11
23	16 09 15	01 03 26	16 ♏ 05 54	22 ♏ 15 55	11 50	17 47	25 21	07 02	03 31	25 41	06 56	06 13
24	16 13 12	02 04 08	28 ♏ 23 13	04 ♐ 27 56	12 31	18 42	25 19	07 12	03 27	25 45	06 58	06 14
25	16 17 08	03 04 51	10 ♐ 30 15	16 ♐ 30 21	13 19	19 37	25 19	07 22	03 23	25 48	07 01	06 16
26	16 21 05	04 05 35	22 ♐ 28 29	28 ♐ 24 55	14 13	20 32	25 D 18	07 31	03 20	25 51	07 03	06 17
27	16 25 01	05 06 20	04 ♑ 19 57	10 ♑ 13 57	15 12	21 25	25 19	07 41	03 15	25 54	07 05	06 19
28	16 28 58	06 07 07	16 ♑ 07 16	22 ♑ 00 22	16 16	22 18	25 21	07 51	03 11	25 57	07 07	06 20
29	16 32 54	07 07 54	27 ♑ 53 41	03 ♒ 47 44	17 24	23 11	25 23	08 02	03 07	26 01	07 10	06 22
30	16 36 51	08 ♐ 08 43	09 ♒ 43 02	15 ♒ 40 09	18 ♏ 35	24 ♑ 02	25 ♈ 26	08 ♒ 12	03 ♋ 03	26 ♎ 04	07 ♐ 12	06 ♎ 23

DECLINATIONS

Date	Moon True ☊ ° '	Moon Mean ☊ ° '	Moon ☽ Latitude ° '	Sun ☉	Moon ☽	Mercury ☿	Venus ♀	Mars ♂	Jupiter ♃	Saturn ♄	Uranus ♅	Neptune ♆	Pluto ♇
01	29 ♐ 46	01 ♑ 07	01 N 46	14 S 29	20 S 16	21 S 44	26 S 37	09 N 55	19 S 58	22 N 17	08 S 56	19 S 48	12 N 35
02	29 D 48	01 04	02 43	14 48	17 11	21 28	26 41	09 53	19 56	22 17	08 58	19 49	12 35
03	29 R 48	01 01	03 34	15 06	13 22	21 08	26 43	09 50	19 55	22 17	08 59	19 49	12 34
04	29 47	00 58	04 16	15 25	08 57	20 44	26 46	09 48	19 53	22 17	09 00	19 49	12 34
05	29 44	00 55	04 48	15 43	04 S 04	20 16	26 47	09 46	19 51	22 17	09 02	19 50	12 33
06	29 39	00 52	05 06	16 01	01 N 07	19 44	26 48	09 44	19 50	22 17	09 03	19 50	12 33
07	29 32	00 48	05 08	16 19	06 25	19 09	26 48	09 42	19 48	22 17	09 04	19 51	12 32
08	29 25	00 45	04 53	16 37	11 35	18 30	26 48	09 40	19 46	22 17	09 05	19 51	12 32
09	29 18	00 42	04 20	16 54	16 17	17 50	26 47	09 39	19 44	22 17	09 07	19 51	12 32
10	29 11	00 39	03 30	17 11	20 09	17 08	26 45	09 38	19 42	22 17	09 08	19 52	12 31
11	29 07	00 36	02 26	17 28	22 47	16 26	26 43	09 37	19 40	22 17	09 09	19 52	12 31
12	29 04	00 32	01 N 11	17 44	23 53	15 45	26 40	09 36	19 38	22 17	09 11	19 52	12 31
13	29 D 04	00 29	00 S 09	18 00	23 17	15 07	26 37	09 36	19 36	22 17	09 12	19 53	12 30
14	29 04	00 26	01 28	18 16	21 06	14 32	26 33	09 36	19 34	22 17	09 13	19 53	12 30
15	29 06	00 23	02 40	18 31	17 35	14 01	26 28	09 36	19 32	22 18	09 14	19 54	12 30
16	29 07	00 20	03 41	18 46	13 06	13 35	26 23	09 36	19 30	22 18	09 16	19 54	12 29
17	29 R 07	00 17	04 28	19 01	08 01	13 15	26 17	09 36	19 28	22 18	09 17	19 54	12 29
18	29 06	00 13	04 58	19 16	02 N 38	13 00	26 10	09 37	19 26	22 18	09 18	19 55	12 29
19	29 04	00 10	05 12	19 30	02 S 45	12 51	26 03	09 38	19 23	22 18	09 19	19 55	12 29
20	29 00	00 07	05 09	19 43	07 55	12 47	25 56	09 39	19 21	22 18	09 21	19 56	12 28
21	28 55	00 04	04 50	19 57	12 39	12 47	25 48	09 40	19 19	22 18	09 22	19 56	12 28
22	28 50	00 ♑ 01	04 17	20 10	16 45	12 53	25 39	09 42	19 16	22 18	09 23	19 56	12 28
23	28 45	29 ♐ 58	03 33	20 22	20 03	13 02	25 30	09 44	19 14	22 18	09 24	19 57	12 28
24	28 41	29 54	02 39	20 35	22 23	13 14	25 21	09 46	19 11	22 19	09 25	19 57	12 28
25	28 38	29 51	01 38	20 47	23 39	13 30	25 11	09 48	19 09	22 19	09 27	19 57	12 28
26	28 37	29 48	00 S 34	20 58	23 47	13 48	25 00	09 51	19 06	22 19	09 28	19 58	12 27
27	28 D 37	29 45	00 N 32	21 09	22 51	14 08	24 49	09 54	19 04	22 19	09 29	19 58	12 27
28	28 38	29 42	01 35	21 20	20 54	14 30	24 38	09 57	19 01	22 19	09 30	19 59	12 27
29	28 39	29 38	02 35	21 30	18 03	14 54	24 26	10 00	18 59	22 19	09 31	19 59	12 27
30	28 ♐ 41	29 ♐ 35	03 N 28	21 S 40	14 S 28	15 S 18	24 S 14	10 N 04	18 S 56	22 N 19	09 S 32	19 S 59	12 N 27

ZODIAC SIGN ENTRIES

Date	h	m	Planets
02	08	58	☽ ♒
04	20	26	☽ ♓
05	15	39	♀ ♑
07	04	19	☽ ♈
09	08	25	☽ ♉
11	09	59	☽ ♊
13	10	46	☽ ♋
15	12	20	☽ ♌
17	15	41	☽ ♍
19	21	15	☽ ♎
22	05	07	☽ ♏
22	10	54	☉ ♐
24	15	11	☽ ♐
27	03	13	☽ ♑
29	16	17	☽ ♒

LATITUDES

Date	Mercury ☿ ° '	Venus ♀ ° '	Mars ♂ ° '	Jupiter ♃ ° '	Saturn ♄ ° '	Uranus ♅ ° '	Neptune ♆ ° '	Pluto ♇ ° '
01	02 S 29	03 S 15	01 S 22	00 S 45	01 S 05	00 N 34	01 N 33	16 N 08
04	01 51	03 19	01 10	00 45	01 05	00 34	01 33	16 09
07	01 S 00	03 23	00 59	00 45	01 05	00 34	01 33	16 10
10	00 N 01	03 25	00 48	00 45	01 05	00 34	01 33	16 11
13	01 00	03 25	00 37	00 45	01 05	00 34	01 33	16 12
16	01 46	03 24	00 27	00 45	01 05	00 34	01 33	16 14
19	02 15	03 22	00 17	00 45	01 05	00 34	01 33	16 15
22	02 28	03 18	00 S 08	00 45	01 05	00 34	01 33	16 16
25	02 27	03 12	00 N 01	00 44	01 05	00 34	01 33	16 18
28	02 18	03 04	00 09	00 44	01 05	00 34	01 33	16 19
31	02 N 03	02 S 55	00 N 17	00 S 44	01 S 05	00 N 34	01 N 33	16 N 21

DATA

Julian Date	2441988
Delta T	+43 seconds
Ayanamsa	23° 29' 46"
Synetic vernal point	05° ♓ 37' 14"
True obliquity of ecliptic	23° 26' 34"

MOON'S PHASES, APSIDES AND POSITIONS ☽

Date	h	m	Phase	Longitude ° '	Eclipse Indicator
03	06	29	☽	10 ♒ 42	
10	14	27	○	18 ♉ 03	
17	06	34	☾	24 ♌ 46	
24	19	55	●	02 ♐ 24	

Day	h	m		
12	14	53	Perigee	
28	12	48	Apogee	
06	06	53	0N	
12	15	36	Max dec	23° N 54'
18	23	42	0S	
26	03	09	Max dec	23° S 52'

LONGITUDES

Date	Chiron ⚷ ° '	Ceres ⚳ ° '	Pallas ⚴ ° '	Juno ⚵ ° '	Vesta ⚶ ° '	Black Moon Lilith ⚸ ° '
01	17 ♈ 43	26 ♐ 33	05 ♐ 54	24 ♐ 12	17 ♍ 29	08 ♑ 39
11	17 ♈ 19	00 ♑ 09	10 ♐ 02	27 ♐ 30	21 ♍ 40	09 ♑ 45
21	16 ♈ 58	03 ♑ 51	14 ♐ 47	00 ♑ 57	25 ♍ 09	10 ♑ 52
31	16 ♈ 42	07 ♑ 40	18 ♐ 21	04 ♑ 32	29 ♍ 43	11 ♑ 58

All ephemeris data is given at 12.00 UT and the Moon's longitude is additionally given for 24.00 UT

ASPECTARIAN

h m	Aspects
01 Thursday	
14 40	☽ ∥ ♃
14 47	☽ Q ☉
14 59	☽ ∠ ♆
15 59	☽ ∥ ♆
21 41	☽ □ ♅
23 48	☿ ⊻ ♀
02 Friday	
01 05	☽ ⚹ ☿
01 15	☽ ⊻ ♀
06 55	☽ □ ♂
14 35	☽ ⊥ ♀
17 34	☽ ☌ ♃
18 03	☽ ⊼ ♄
20 17	☽ △ ♇
21 27	☽ ⚹ ♆
03 Saturday	
00 23	☽ Q ☿
02 20	☽ ∥ ☉
06 04	☽ ± ♄
06 29	☽ □ ☉
10 24	☽ ∠ ♀
16 33	☽ # ♇
18 10	☽ Q ♂
21 26	☽ Q ♆
23 53	☽ ⚼ ♄
04 Sunday	
02 13	☽ ⚼ ♇
03 27	♀ ⊥ ♃
04 19	♀ △ ♂
05 26	☉ ⊥ ♇
07 33	☽ # ♂
09 55	☽ △ ♅
10 12	☽ □ ☿
10 50	♃ ⊼ ♄
11 42	☽ ∥ ♅
14 10	☿ ⊻ ♅
17 09	☽ ⚹ ♂
18 39	☽ ⚹ ♀
19 48	☽ ± ♇
05 Monday	
04 55	☽ △ ♄
05 08	☽ ⊻ ♃
07 20	☽ ⊼ ♇
08 27	☽ □ ♆
11 37	☿ ⊥ ♀
14 52	☽ ⚼ ♅
16 37	☽ ⊥ ♃
19 36	☽ Q ♀
21 08	☽ ∠ ♂
21 47	☽ △ ☉
06 Tuesday	
04 54	♃ ∥ ♆
07 45	☽ ± ♅
08 00	☿ ∥ ♆
08 12	☿ ∥ ♃
09 32	☽ ∠ ♃
13 29	☽ ⊥ ♂
14 26	☿ Q ♃
15 02	☽ △ ☿
18 47	☽ ⊼ ♅
07 Wednesday	
00 12	☽ ⊻ ♂
03 35	☽ ⚼ ☉
07 25	☽ □ ♀
12 04	☽ □ ♄
12 54	☽ ⚹ ♃
14 36	☽ ☍ ♇
15 38	☽ △ ♆
15 52	☽ ⚼ ☿
20 51	☽ ± ☉
08 Thursday	
00 13	☽ # ♅
03 01	☽ ∥ ♂
05 28	☿ ∠ ♇
06 20	☽ ± ☿
08 09	☽ ⊼ ☉
10 06	☽ Q ♃
15 49	☽ ⊼ ☿
16 37	☽ ∥ ♇
17 48	☽ ⚼ ♆
19 20	☽ Q ♄
23 43	☽ ☍ ♅
09 Friday	
03 43	☽ ☌ ♂
09 07	☽ ± ♆
09 48	☿ ⚼ ♄
13 24	☿ ∠ ♀
15 23	☽ △ ♀
15 34	☽ ⚹ ♄
15 42	☽ # ☉
16 56	☽ □ ♃
17 41	♀ ☍ ♄
18 12	☽ ⊼ ♇
19 12	☽ ⊼ ♆
19 34	☽ # ☿
10 Saturday	
04 12	☽ ± ♇
08 53	☽ # ♃
09 57	☽ # ♆
10 32	☉ ☌ ☿
10 42	☉ ∥ ☿
13 52	☽ ☍ ☿
14 27	☽ ☍ ☉
15 56	♀ ⊻ ♃
16 20	☽ ∠ ♄
18 07	☽ ⚼ ♀
19 03	☽ ⚼ ♇
11 Sunday	
01 48	☽ ⊼ ♅
04 45	☽ ⊻ ♂
06 17	☽ ∥ ♄
07 01	☽ ⊥ ♄
07 58	♀ □ ♇
09 56	☽ ± ♀
11 38	☽ ± ♅
14 24	☽ ⊥ ♂
16 46	☽ ⊻ ♄
16 51	☉ ⚼ ♄
18 41	☽ △ ♃
19 35	☽ △ ♇
20 25	☽ ⊼ ♀
20 34	☽ ☍ ♆
22 47	♀ ⊻ ♆
12 Monday	
02 20	☽ ⚼ ♅
04 50	☽ ∠ ♂
10 50	☽ ⊼ ☿
12 05	♀ Q ♅
18 54	☽ ⊼ ☉
19 16	☽ ⚼ ♃
19 51	☽ ± ☿
13 Tuesday	
02 48	☽ △ ♅
04 56	☽ ⚹ ♂
05 23	☽ ± ☉
09 26	☽ ⚼ ☿
10 03	☽ ± ♃
11 23	☉ ∠ ♇
17 25	☽ ☌ ♄
19 57	☽ ⊼ ♃
20 30	☽ □ ♇
21 09	☽ ⚼ ☉
21 32	☽ ⊼ ♆
14 Wednesday	
00 19	☽ Q ♂
00 52	☽ ☍ ♀
00 58	☽ ∥ ♄
07 24	☽ ± ♆
08 22	☽ △ ☿
21 13	☽ # ♆
22 20	☽ ⚼ ♆
23 33	☽ # ♃
23 45	☽ △ ☉
15 Thursday	
02 17	☽ Q ♇
04 20	☽ □ ♅
05 47	☽ □ ♂
06 40	☽ # ☉
16 50	☿ ⊥ ♇
18 59	☽ ⊻ ♄
22 17	☽ ☍ ♃
22 27	☽ ⚹ ♇
23 32	☽ △ ♆
16 Friday	
05 07	☽ ⊥ ♄
06 39	☽ ⊼ ♀
07 54	☽ □ ☿
09 23	☽ # ☿
10 50	☽ Q ♅
11 58	☉ Q ♃
15 02	☽ ∥ ♇
17 41	☽ ± ♀
18 17	☿ ⚹ ♀
20 26	☽ ∠ ♄
20 35	♃ △ ♇
17 Saturday	
00 08	☽ ∠ ♇
04 44	☽ ∥ ♂
06 14	☽ # ♅
06 34	☽ □ ☉
07 35	☽ ⚹ ♅
08 23	☽ △ ♂
10 25	☽ ⚼ ♀
13 50	☽ Q ☿
15 47	☽ ⊥ ♇
21 05	☉ ⊻ ♅
22 25	☽ ⚹ ♄
18 Sunday	
02 21	☽ ⊻ ♇
02 36	☽ ⊼ ♃
03 31	☽ □ ♆
04 59	♀ ⊥ ♆
05 07	☉ ⊼ ♂
10 01	☽ ∠ ♅
10 30	☽ ⚹ ☿
10 31	☽ ⚼ ♂
13 21	☽ ± ♃
14 52	☽ △ ♀
16 46	☽ Q ☉
19 36	☽ Q ♄
19 Monday	
02 09	☽ ⊥ ♅
02 29	☽ ± ♂
02 40	☉ ∥ ♃
05 38	☽ ⚼ ♃
11 47	☽ Q ♆
13 02	☽ ⊻ ♅
13 11	☽ ∠ ☿
13 14	☽ ⊼ ♂
14 15	☿ St D
16 08	☽ ⚹ ☉
20 Tuesday	
04 04	☽ □ ♄
04 42	☉ ± ♄
05 40	☽ ⊥ ☿
08 30	☽ ☌ ♇
09 00	♂ ☍ ♅
09 17	☽ △ ♃
09 45	☽ ⚹ ♆
16 51	☽ ⊻ ☿
19 00	☽ ∥ ♅
20 31	☽ # ♂
22 01	☽ ∠ ☉
21 Wednesday	
01 50	☽ □ ♀
10 22	☉ ∥ ♆
11 02	☽ # ♇
12 46	☽ ∥ ☿
13 44	☽ ∠ ♆
16 17	☽ ⊥ ☉
20 26	☽ ☍ ♂
20 47	☽ ☌ ♅
22 Thursday	
04 36	☽ ⊻ ☉
06 48	☽ ⊥ ♆
11 55	☽ △ ♄
14 30	♃ ⚹ ♆
14 31	☽ Q ♀
16 55	☽ ⊻ ♇
18 17	☽ ⊻ ♆
18 19	☽ □ ♃
23 Friday	
03 21	☽ ☌ ☿
04 29	☽ ⊥ ♇
05 28	☽ ∥ ♃
11 07	☽ ∥ ♆
15 02	☽ ∥ ☉
15 31	☽ ⚹ ♀
16 40	☽ ⚼ ♄
18 40	☉ ± ♂
21 58	☽ ∠ ♇
24 Saturday	
02 29	☿ ⊥ ♇
03 54	☉ ⊥ ♅
05 40	☽ Q ♃
05 59	☽ ⊼ ♂
06 47	☽ ⊻ ♅
10 11	☽ ± ♄
11 04	☽ # ♄
17 46	☽ ± ♂
18 39	☽ ⊥ ♅
19 55	☽ ☌ ☉
21 57	☽ ⊼ ♄
23 21	☽ ∠ ♀
25 Sunday	
03 33	☽ ⚹ ♇
05 01	☽ ☌ ♆
05 40	☽ ⚹ ♃
11 36	☽ ⚼ ♂
12 35	☽ ∠ ♅
18 03	☽ ⊻ ☿
18 44	☽ ⊥ ♀
18 55	☉ ⊼ ♄
26 Monday	
00 06	♂ St D
03 34	☽ Q ♇
07 03	☽ ⊥ ☿
07 45	☽ ⊻ ♀
12 06	☽ ∠ ♃
17 43	☽ △ ♂
18 51	☽ ⚹ ♅
27 Tuesday	
02 50	☽ ∠ ☿
06 34	☽ ⊥ ♃
09 50	☽ ☍ ♄
13 43	☽ ⊻ ☉
16 02	☽ □ ♇
17 37	☽ ⊻ ♆
18 55	☽ ⊻ ♃
19 18	☽ Q ♅
19 47	☽ # ♄
28 Wednesday	
03 04	☽ ⊥ ☉
05 52	☽ ⊥ ♆
06 47	♀ ∠ ♆
07 49	☽ ∥ ☉
12 20	☽ ⚹ ☿
17 18	☉ ⚹ ♇
20 32	☽ ∥ ♆
23 09	☽ ∠ ☉
29 Thursday	
00 17	☽ ∠ ♆
01 37	☽ ☌ ♀
04 49	☽ ∥ ♃
06 52	☽ □ ♂
08 09	☽ □ ♅
12 42	☉ ☌ ♆
15 24	☽ Q ☿
22 34	☽ ⊼ ♄
30 Friday	
01 51	☿ ⚼ ♄
05 14	☽ △ ♇
06 53	☽ ⚹ ♆
07 19	☽ ∥ ☿
08 31	☽ ⚹ ☉
08 53	☽ ☌ ♃
10 40	☽ ± ♄
13 30	☉ ⚹ ♃
19 33	☽ Q ♂
23 59	☽ # ♇

LONGITUDES

Date	Sidereal time h m s	Sun ☉ ° ' "	Moon ☽ ° ' "	Moon ☽ 24.00 ° ' "	Mercury ☿ ° '	Venus ♀ ° '	Mars ♂ ° '	Jupiter ♃ ° '	Saturn ♄ ° '	Uranus ♅ ° '	Neptune ♆ ° '	Pluto ♇ ° '
01	16 40 48	09 ♐ 09 32	21 ♒ 39 41	27 ♒ 42 12	19 ♏ 50	24 ♑ 53	25 ♈ 30	08 ♒ 22	02 ♋ 59	26 ♎ 07	07 ♐ 14	06 ♎ 24
02	16 44 44	10 10 23	03 ♓ 48 20	09 ♓ 58 39	21 07	25 44	25 35	08 33	02 R 54	26 10	07 16	06 25
03	16 48 41	11 11 14	16 ♓ 13 45	22 ♓ 34 11	22 26	26 33	25 40	08 43	02 50	26 13	07 19	06 27
04	16 52 37	12 12 06	29 ♓ 00 26	05 ♈ 32 58	23 47	27 22	25 47	08 54	02 46	26 15	07 21	06 28
05	16 56 34	13 12 59	12 ♈ 12 06	18 ♈ 58 05	25 10	28 09	25 55	09 05	02 41	26 18	07 23	06 29
06	17 00 30	14 13 53	25 ♈ 51 02	02 ♉ 50 53	26 34	28 56	26 01	09 16	02 37	26 21	07 25	06 30
07	17 04 27	15 14 47	09 ♉ 57 27	17 ♉ 10 21	27 59	29 ♑ 42	26 09	09 27	02 32	26 24	07 28	06 31
08	17 08 23	16 15 43	24 ♉ 29 01	01 ♊ 52 45	29 ♏ 26	00 ♒ 27	26 18	09 38	02 27	26 27	07 30	06 32
09	17 12 20	17 16 39	09 ♊ 20 38	16 ♊ 51 41	00 ♐ 53	01 11	26 28	09 49	02 23	26 30	07 32	06 34
10	17 16 17	18 17 36	24 ♊ 24 46	01 ♋ 58 45	02 21	01 53	26 38	10 01	02 18	26 32	07 34	06 35
11	17 20 13	19 18 34	09 ♋ 32 28	17 ♋ 04 46	03 49	02 35	26 49	10 12	02 13	26 35	07 37	06 36
12	17 24 10	20 19 33	24 ♋ 34 36	02 ♌ 01 01	05 18	03 15	27 01	10 23	02 08	26 38	07 39	06 37
13	17 28 06	21 20 33	09 ♌ 23 12	16 ♌ 40 30	06 48	03 55	27 13	10 35	02 04	26 40	07 41	06 37
14	17 32 03	22 21 34	23 ♌ 52 24	00 ♍ 58 35	08 18	04 33	27 25	10 47	01 59	26 43	07 43	06 38
15	17 35 59	23 22 36	07 ♍ 58 50	14 ♍ 53 05	09 48	05 09	27 39	10 59	01 54	26 45	07 45	06 39
16	17 39 56	24 23 39	21 ♍ 41 24	28 ♍ 23 57	11 19	05 45	27 53	11 10	01 49	26 48	07 48	06 40
17	17 43 52	25 24 42	05 ♎ 00 57	11 ♎ 32 42	12 50	06 19	28 07	11 22	01 44	26 50	07 50	06 41
18	17 47 49	26 25 47	17 ♎ 59 32	24 ♎ 21 49	14 21	06 51	28 22	11 34	01 39	26 52	07 52	06 42
19	17 51 46	27 26 53	00 ♏ 39 54	06 ♏ 54 12	15 53	07 22	28 38	11 47	01 34	26 55	07 54	06 42
20	17 55 42	28 27 59	13 ♏ 05 03	19 ♏ 12 51	17 25	07 52	28 54	11 59	01 29	26 57	07 56	06 43
21	17 59 39	29 ♐ 29 06	25 ♏ 17 54	01 ♐ 20 35	18 57	08 20	29 10	12 11	01 24	26 59	07 58	06 44
22	18 03 35	00 ♑ 30 14	07 ♐ 21 10	13 ♐ 19 58	20 29	08 46	29 27	12 23	01 19	27 01	08 01	06 44
23	18 07 32	01 31 23	19 ♐ 17 15	25 ♐ 13 18	22 01	09 10	29 ♈ 45	12 36	01 14	27 04	08 03	06 45
24	18 11 28	02 32 32	01 ♑ 08 22	07 ♑ 02 42	23 34	09 33	00 ♉ 03	12 48	01 09	27 06	08 05	06 46
25	18 15 25	03 33 41	12 ♑ 56 34	18 ♑ 50 14	25 07	09 53	00 21	13 01	01 04	27 08	08 07	06 46
26	18 19 21	04 34 51	24 ♑ 43 58	00 ♒ 38 03	26 40	10 12	00 40	13 14	01 00	27 10	08 09	06 47
27	18 23 18	05 36 00	06 ♒ 32 48	12 ♒ 28 31	28 14	10 28	01 00	13 27	00 55	27 12	08 11	06 47
28	18 27 15	06 37 10	18 ♒ 25 35	24 ♒ 24 22	29 ♐ 47	10 43	01 20	13 39	00 50	27 14	08 13	06 47
29	18 31 11	07 38 20	00 ♓ 25 16	06 ♓ 28 43	01 ♑ 21	10 55	01 40	13 52	00 45	27 15	08 15	06 48
30	18 35 08	08 39 30	12 ♓ 35 09	18 ♓ 45 03	02 56	11 05	02 01	14 05	00 40	27 17	08 17	06 48
31	18 39 04	09 ♑ 40 39	24 ♓ 58 55	01 ♈ 17 12	04 ♑ 30	11 ♒ 13	02 ♉ 22	14 ♒ 18	00 ♋ 35	27 ♎ 19	08 ♐ 19	06 ♎ 48

Date	Moon True ☊ ° '	Moon Mean ☊ ° '	Moon ☽ Latitude ° '
01	28 ♐ 43	29 ♐ 32	04 N 13
02	28 D 43	29 29	04 47
03	28 R 43	29 26	05 09
04	28 43	29 23	05 17
05	28 41	29 19	05 08
06	28 39	29 16	04 42
07	28 37	29 13	03 58
08	28 36	29 10	02 58
09	28 35	29 07	01 45
10	28 34	29 04	00 N 23
11	28 D 34	29 00	01 S 01
12	28 35	28 57	02 20
13	28 36	28 54	03 28
14	28 36	28 51	04 22
15	28 37	28 48	04 58
16	28 37	28 44	05 16
17	28 R 37	28 41	05 16
18	28 37	28 38	05 00
19	28 37	28 35	04 30
20	28 36	28 32	03 47
21	28 D 36	28 29	02 55
22	28 37	28 25	01 56
23	28 37	28 22	00 S 52
24	28 R 37	28 19	00 N 14
25	28 37	28 16	01 19
26	28 36	28 13	02 20
27	28 35	28 10	03 16
28	28 34	28 06	04 03
29	28 33	28 03	04 40
30	28 32	28 00	05 05
31	28 ♐ 31	27 ♐ 57	05 N 17

DECLINATIONS

Date	Sun ☉ ° '	Moon ☽ ° '	Mercury ☿ ° '	Venus ♀ ° '	Mars ♂ ° '	Jupiter ♃ ° '	Saturn ♄ ° '	Uranus ♅ ° '	Neptune ♆ ° '	Pluto ♇ ° '
01	21 S 50	10 S 17	15 S 44	24 S 01	10 N 07	18 S 53	22 N 20	09 S 33	20 00	12 N 27
02	21 59	05 39	16 10	23 48	10 11	18 50	22 20	09 34	20 00	12 27
03	22 07	00 S 41	16 36	23 35	10 16	18 48	22 20	09 35	20 00	12 27
04	22 16	04 N 27	17 03	23 21	10 20	18 45	22 20	09 36	20 01	12 27
05	22 23	09 33	17 29	23 07	10 25	18 42	22 20	09 37	20 01	12 27
06	22 31	14 22	17 56	22 53	10 29	18 39	22 20	09 38	20 02	12 27
07	22 38	18 34	18 22	22 38	10 34	18 36	22 20	09 39	20 02	12 27
08	22 44	21 46	18 48	22 23	10 40	18 33	22 21	09 40	20 02	12 27
09	22 50	23 35	19 13	22 08	10 45	18 30	22 21	09 41	20 03	12 27
10	22 56	23 43	19 38	21 53	10 51	18 27	22 21	09 42	20 03	12 27
11	23 01	22 05	20 03	21 37	10 56	18 24	22 21	09 43	20 03	12 27
12	23 05	18 55	20 26	21 21	11 02	18 21	22 21	09 44	20 04	12 28
13	23 10	14 34	20 49	21 05	11 08	18 18	22 21	09 45	20 04	12 28
14	23 13	09 27	21 11	20 49	11 15	18 14	22 22	09 46	20 04	12 28
15	23 17	03 N 58	21 32	20 33	11 21	18 11	22 22	09 47	20 05	12 28
16	23 19	01 S 32	21 52	20 17	11 28	18 08	22 22	09 48	20 05	12 28
17	23 22	06 50	22 11	20 01	11 34	18 05	22 22	09 49	20 05	12 28
18	23 24	11 41	22 30	19 44	11 41	18 01	22 22	09 49	20 06	12 29
19	23 25	15 55	22 47	19 28	11 48	17 58	22 22	09 50	20 06	12 29
20	23 26	19 23	23 03	19 11	11 55	17 54	22 22	09 51	20 06	12 29
21	23 26	21 55	23 18	18 55	12 02	17 51	22 22	09 52	20 07	12 29
22	23 26	23 26	23 32	18 39	12 10	17 47	22 23	09 52	20 07	12 30
23	23 26	23 52	23 44	18 22	12 17	17 44	22 23	09 53	20 07	12 30
24	23 25	23 12	23 56	18 06	12 25	17 40	22 23	09 54	20 08	12 30
25	23 24	21 30	24 06	17 50	12 33	17 37	22 23	09 55	20 08	12 31
26	23 22	18 53	24 15	17 34	12 41	17 33	22 23	09 56	20 08	12 31
27	23 19	15 29	24 23	17 18	12 49	17 30	22 23	09 56	20 09	12 31
28	23 17	11 27	24 30	17 03	12 57	17 26	22 24	09 57	20 09	12 32
29	23 13	06 57	24 35	16 47	13 05	17 22	22 24	09 58	20 09	12 32
30	23 10	02 08	24 39	16 32	13 13	17 18	22 24	09 58	20 10	12 32
31	23 S 05	02 N 51	24 S 41	16 S 18	13 N 21	17 S 15	22 N 24	09 S 59	20 S 10	12 N 33

ZODIAC SIGN ENTRIES

Date	h	m	Planets
02	04	32	☽ ♓
04	13	50	☽ ♈
06	19	08	☽ ♉
07	21	37	♀ ♒
08	20	58	☽ ♊
08	21	29	☿ ♐
10	20	52	☽ ♋
12	20	44	☽ ♌
14	22	20	☽ ♍
17	02	53	☽ ♎
19	10	44	☽ ♏
21	21	20	☽ ♐
22	00	08	☉ ♑
24	08	09	☽ ♑
24	09	41	♂ ♉
26	22	43	☽ ♒
28	15	14	☿ ♑
29	11	10	☽ ♓
31	21	34	☽ ♈

LATITUDES

Date	Mercury ☿ ° '	Venus ♀ ° '	Mars ♂ ° '	Jupiter ♃ ° '	Saturn ♄ ° '	Uranus ♅ ° '	Neptune ♆ ° '	Pluto ♇ ° '
01	02 N 03	02 S 55	00 N 17	00 S 44	01 S 05	00 N 34	01 N 33	16 N 21
04	01 44	02 43	00 24	00 44	01 05	00 34	01 33	16 22
07	01 23	02 28	00 31	00 44	01 05	00 34	01 33	16 24
10	01 01	02 11	00 37	00 44	01 04	00 34	01 33	16 25
13	00 39	01 52	00 42	00 44	01 04	00 34	01 33	16 27
16	00 N 16	01 29	00 47	00 44	01 04	00 34	01 33	16 29
19	00 S 05	01 04	00 52	00 44	01 04	00 34	01 33	16 30
22	00 26	00 35	00 57	00 44	01 03	00 35	01 33	16 32
25	00 45	00 S 04	01 01	00 44	01 03	00 35	01 33	16 34
28	01 03	00 N 32	01 04	00 44	01 03	00 35	01 33	16 35
31	01 S 19	01 N 10	01 N 08	00 S 44	01 S 02	00 N 35	01 N 33	16 N 37

DATA

Julian Date	2442018
Delta T	+43 seconds
Ayanamsa	23° 29' 50"
Synetic vernal point	05° ♓ 37' 09"
True obliquity of ecliptic	23° 26' 33"

MOON'S PHASES, APSIDES AND POSITIONS ☽

Date	h	m	Phase	Longitude ° '	Eclipse Indicator
03	01	29	☽	10 ♓ 45	
10	01	35	○	17 ♊ 51	partial
16	17	13	☾	24 ♍ 37	
24	15	07	●	02 ♑ 40	Annular

Day	h	m		
10	22	30	Perigee	
25	21	45	Apogee	
03	15	13	0N	
10	01	46	Max dec	23° N 52'
16	05	14	0S	
23	09	19	Max dec	23° S 53'
30	22	20	0N	

LONGITUDES

Date	Chiron ⚷ ° '	Ceres ⚳ ° '	Pallas ⚴ ° '	Juno ⚵ ° '	Vesta ⚶ ° '	Black Moon Lilith ⚸ ° '
01	16 ♈ 42	07 ♑ 40	18 ♐ 21	04 ♑ 32	29 ♍ 36	11 ♑ 58
11	16 ♈ 30	11 ♑ 32	22 ♐ 31	08 ♑ 14	03 ♎ 15	13 ♑ 05
21	16 ♈ 24	15 ♑ 27	26 ♐ 40	12 ♑ 00	06 ♎ 39	14 ♑ 11
31	16 ♈ 23	19 ♑ 25	00 ♑ 47	15 ♑ 51	09 ♎ 44	15 ♑ 18

ASPECTARIAN

01 Saturday
04 41 ☽ ⚼ ♄
07 08 ☽ Q ♆
07 55 ☽ □ ☿
10 54 ☽ Q ☉
11 29 ☽ ⚼ ♇
12 53 ☽ ⋕ ♂
15 56 ☽ ∥ ♅
18 54 ☽ ⊻ ♀
19 42 ☽ ⚹ ♂
20 53 ☽ △ ♅

02 Sunday
05 21 ☽ ± ♇
07 22 ♀ □ ♂
07 38 ☽ ⊥ ♀
10 15 ☽ △ ♄
17 07 ☽ ⊼ ♇
17 46 ☿ ∠ ♇
18 47 ☽ □ ♆
21 22 ☽ ⊻ ♃
22 36 ☉ ⚼ ♂

03 Monday
01 16 ☽ ∠ ♂
01 23 ♀ □ ♅
01 29 ☽ □ ☉
02 20 ☽ ⚼ ♅
02 24 ☽ ∠ ♀
09 05 ☽ ⊥ ♃
12 33 ☉ ∠ ♅
18 35 ☽ ⊥ ♂
19 35 ☽ ± ♅
22 38 ♃ ± ♄

04 Tuesday
01 08 ☽ △ ☿
02 22 ☽ ∠ ♃
05 57 ☽ ⊻ ♂
06 53 ☽ ⊼ ♅
08 45 ☽ ⚹ ♀
18 52 ☽ □ ♄

05 Wednesday
01 41 ☽ ☍ ♇
01 58 ☉ ⋕ ♄
03 19 ☽ △ ♆
05 25 ☉ ∠ ♀
06 19 ☽ ⚹ ♃
07 56 ☽ ⚼ ☿
08 06 ☽ Q ♀
12 22 ☽ ⋕ ♅
13 58 ☽ △ ☉
16 14 ☽ ∥ ♂

06 Thursday
01 39 ☿ ⊼ ♂
01 45 ☽ ± ☿
02 12 ☽ ∥ ♇
02 56 ☽ Q ♄
03 55 ☽ Q ♃
06 02 ☽ ⚼ ♆
08 14 ☿ ⊻ ♅
12 17 ☽ ☌ ♂
12 40 ☿ ± ♄
12 52 ☽ ☍ ♅
13 23 ☽ ⊼ ☿
17 37 ☽ □ ♀
18 17 ☽ ⚼ ☉
21 36 ☽ ± ♆
23 32 ☽ ⚹ ♄

07 Friday
01 32 ☿ Q ♃
06 13 ☽ ⊼ ♇
07 48 ☽ ⊼ ♆
10 34 ☽ ⋕ ☿
10 43 ☽ ± ☉
11 08 ☽ □ ♃
12 12 ☽ ⋕ ♃
12 23 ☉ ∥ ♀
16 18 ☽ ± ♇
21 29 ☽ ⊼ ☉
21 54 ☽ ⋕ ♆
23 33 ☿ ∥ ♃

08 Saturday
00 32 ☽ ∠ ♄
07 11 ☽ ⚼ ♇
15 00 ☽ ⊻ ♂
15 12 ☽ ⊥ ♄
15 12 ☽ ⊼ ♅
15 49 ♀ ⋕ ♄
17 32 ☽ ⋕ ♀
17 43 ☽ ∥ ♄
20 54 ☽ ☍ ☿
22 11 ☽ △ ♀
22 48 ☽ ⋕ ☉

09 Sunday
00 49 ☽ ⊥ ♂
00 52 ☽ ⊻ ♄
00 57 ☽ ± ♅
07 32 ☽ △ ♇
09 06 ☽ ☍ ♆
12 46 ☽ △ ♃
15 26 ☽ ∠ ♂
15 27 ☽ ⚼ ♅
17 26 ♂ ☍ ♅
21 34 ☿ ⚹ ♀
23 27 ☽ ⚼ ♀

10 Monday
01 35 ☽ ☍ ☉
11 18 ☿ ⊼ ♄
12 57 ☽ ⚼ ♃
14 27 ☽ ± ♀
15 15 ☿ ⊥ ♅
15 23 ☽ △ ♅
15 34 ☽ ⚹ ♂
17 25 ☿ ± ♂
18 48 ☉ Q ♇

11 Tuesday
00 26 ☽ ⊼ ♀
00 27 ☽ ☌ ♄
00 41 ♀ ⊼ ♄
01 55 ☽ ⋕ ☉
01 56 ☽ ⊼ ☿
03 25 ☽ ± ♃
07 19 ☽ □ ♇
08 56 ☽ ⊼ ♆
09 22 ☽ ∥ ♄
10 50 ☽ Q ♂
12 30 ☽ ± ☿
12 49 ☿ ∥ ♆
13 04 ☽ ⊼ ♃
16 52 ☽ ⋕ ♀
18 29 ☽ ± ♆

12 Wednesday
02 53 ☽ ⋕ ☿
04 24 ☽ ⚼ ☿
04 29 ☽ ⋕ ♆
04 41 ☽ ⊼ ☉
08 54 ☽ ⚼ ♆
12 03 ☽ Q ♇
15 01 ☽ ± ☉
15 18 ☽ □ ♅
15 34 ☽ ⋕ ♃
15 58 ☽ □ ♂

13 Thursday
00 08 ☽ ⊻ ♄
02 40 ☽ ☍ ♀
06 40 ☽ ⚼ ☉
07 18 ☽ △ ☿
07 29 ☽ ⚹ ♇
09 11 ☿ ⚹ ♇
09 13 ☽ △ ♆
09 51 ☽ ⊥ ♄
13 59 ☽ ☍ ♃
20 43 ☽ Q ♅
22 10 ☽ ∥ ♇
22 22 ☿ ∥ ♀

14 Friday
00 34 ☽ ∠ ♄
02 33 ☿ ☌ ♆
04 01 ☽ ∥ ♂
08 15 ☽ ∠ ♇
09 16 ☽ △ ☉
10 34 ☽ ⋕ ♅
16 47 ☽ ⚹ ♅
18 04 ☽ △ ♂
23 26 ☽ ⊥ ♇

15 Saturday
01 38 ☽ ⚹ ♄
06 55 ☽ ⊼ ♀
09 43 ☽ ⊻ ♇
11 37 ☽ □ ♆
15 32 ☽ □ ☿
17 16 ☽ ⊼ ♃
17 45 ☽ ± ♀
18 33 ☽ ∠ ♅
20 14 ☽ ⚼ ♂
22 13 ☽ Q ♄

16 Sunday
03 54 ☽ ± ♃
09 25 ☿ ⚹ ♃
10 15 ☽ ⚼ ♀
10 24 ☽ ⊥ ♅
12 20 ☽ ± ♂
17 13 ☽ □ ☉
19 20 ☽ Q ♆
19 47 ☿ ∠ ♅
20 07 ☽ ⚼ ♃
21 08 ☽ ⊻ ♅
23 16 ☽ ⊼ ♂

17 Monday
03 25 ☽ Q ☿
05 05 ♀ ∥ ♆
06 04 ☽ □ ♄
14 29 ☽ △ ♀
15 03 ☽ ☌ ♇
17 10 ☽ ⚹ ♆
17 23 ☿ ⚼ ♂
23 52 ☽ △ ♃

18 Tuesday
01 57 ☿ ⋕ ♄
02 29 ☽ ∥ ♅
04 18 ☽ ⚹ ☿
04 32 ♀ △ ♇
04 47 ☽ Q ☉
12 02 ☽ ⋕ ♂
16 15 ☉ ∠ ♃
16 15 ☽ ⋕ ♇
21 12 ☽ ∠ ♆
22 53 ☉ ⚹ ♅

19 Wednesday
04 49 ☽ ☌ ♅
05 19 ☽ ⚹ ☉
08 01 ☽ ☍ ♂
12 28 ☽ ∠ ☿
13 43 ☽ △ ♄
14 23 ☽ ⊥ ♆
20 12 ♀ ± ♄
23 38 ☽ ⊻ ♇

20 Thursday
01 15 ☽ ∥ ♃
01 27 ☽ □ ♀
01 58 ☽ ⊻ ♆
08 17 ☽ ⊥ ☿
09 49 ☽ □ ♃
10 39 ☽ ∥ ♀
11 17 ☽ ⊥ ♇
12 49 ☽ ∠ ☉
16 08 ♀ ⚹ ♆
18 03 ☽ ∥ ♆
18 36 ☽ ⚼ ♄
21 40 ☽ ⊻ ☿

21 Friday
01 48 ☉ △ ♂
04 57 ☽ ∠ ♇
08 05 ☽ ⊥ ☉
08 37 ☿ Q ♇
12 13 ☽ ± ♄
14 07 ☽ Q ♀
15 21 ☽ ⊻ ♅
17 41 ☽ ⋕ ♄
19 52 ☽ ⊼ ♂
21 04 ☽ ⊻ ☉
21 52 ☽ Q ♃

22 Saturday
00 03 ☽ ⊼ ♄
02 52 ☉ ∥ ☿
03 19 ☽ ⊥ ♅
08 07 ☽ ± ♂
10 46 ☽ ⚹ ♇
12 02 ☽ ∥ ☉
13 19 ☽ ☌ ♆
14 55 ☽ ⚹ ♀
15 05 ☽ ∥ ☿
21 24 ☽ ∠ ♅
22 17 ☽ ⚹ ♃

23 Sunday
02 37 ☽ ⚼ ♂
05 51 ☉ ☍ ♄
10 55 ☽ Q ♇
18 21 ☽ ☌ ☿
18 32 ☽ ∥ ☿
22 12 ☽ ∠ ♀

24 Monday
03 46 ☽ ⚹ ♅
05 07 ☽ ∠ ♃
07 12 ☽ ∥ ☉
09 44 ☽ △ ♂
12 02 ☽ ☍ ♄
15 07 ☽ ☌ ☉
17 02 ☽ ⊥ ♀
23 26 ☽ □ ♇
23 44 ☽ ⊥ ♃

25 Tuesday
01 17 ☽ ⋕ ♄
02 09 ☽ ⊻ ♆
04 13 ☽ Q ♅
05 06 ♂ ∥ ♇
05 36 ☽ ⊻ ♀
07 27 ☿ ∠ ♀
12 09 ☽ ⊻ ♃
14 24 ☽ ⊥ ♆
23 06 ☉ ⊥ ♀

26 Wednesday
01 29 ☽ ∥ ♆
08 46 ☽ ∠ ♆
13 33 ♀ ∥ ♃
16 32 ☽ ⊻ ☿
16 57 ☽ □ ♅
19 46 ☿ ⚹ ♅
22 10 ☽ ∥ ♃
22 44 ☽ ∥ ♀

27 Thursday
00 25 ☽ □ ♂
00 38 ☽ ⊼ ♄
06 35 ☽ ⊥ ☿
06 49 ♂ ⚹ ♄
09 54 ☽ ⊻ ☉
12 29 ☽ △ ♇
12 44 ☽ ± ♄
15 20 ☽ ⚹ ♆
15 49 ☿ ∠ ♃
20 07 ☽ ☌ ♀
23 12 ☽ ⊥ ☉

28 Friday
02 12 ☽ ☌ ♃
03 34 ☽ ∠ ☿
03 45 ☽ ⋕ ♂
05 54 ☽ ⋕ ♇
06 48 ☽ ⚼ ♄
13 52 ☽ Q ♂
15 37 ☽ Q ♆
16 02 ☉ □ ♇
18 46 ☽ ⚼ ♇
19 01 ☽ ∠ ☉
20 15 ☽ ∥ ♅

29 Saturday
03 08 ☿ ☍ ♄
05 41 ☽ △ ♅
12 38 ☽ △ ♄
12 45 ☽ ± ♇
14 08 ☽ ⚹ ☿
14 33 ☽ ⚹ ♂
18 07 ☿ △ ♂
18 54 ☉ ⊥ ♃

30 Sunday
00 38 ☽ ⊼ ♇
02 56 ☉ ⊻ ♆
03 32 ☽ □ ♆
03 36 ☽ ⚹ ☉
09 02 ☽ ⊻ ♀
11 25 ☽ ⚼ ♅
14 59 ☽ ⊻ ♃
17 14 ☽ Q ☿
20 53 ☽ ⊥ ♀
20 53 ☽ ∠ ♂

31 Monday
02 50 ☽ ⊥ ♃
03 16 ☉ Q ♅
04 57 ☽ ± ♅
05 05 ☽ Q ☉
08 36 ♂ ± ♆
14 23 ☽ ∠ ♀
14 43 ☽ ⊥ ♂
16 28 ☽ ⊼ ♅
20 23 ☽ ∠ ♃
22 36 ☽ □ ♄
23 40 ☿ ⊥ ♀

All ephemeris data is given at 12.00 UT and the Moon's longitude is additionally given for 24.00 UT

JANUARY 1974

Date	Sidereal time h m s	Sun ☉ ° ' "	Moon ☽ ° ' "	Moon ☽ 24.00 ° ' "	Mercury ☿ ° '	Venus ♀ ° '	Mars ♂ ° '	Jupiter ♃ ° '	Saturn ♄ ° '	Uranus ♅ ° '	Neptune ♆ ° '	Pluto ♇ ° '
01	18 43 01	10 ♑ 41 49	07 ♈ 40 23	14 ♈ 08 56	06 ♑ 05	11 ♒ 19	02 ♉ 43	14 ♒ 31	00 ♋ 30	27 ♎ 21	08 ♐ 21	06 ♎ 49
02	18 46 57	11 42 58	20 ♈ 43 17	27 ♈ 23 47	07 40	11 22	03 05	14 44	00 R 25	27 22	08 23	06 49
03	18 50 54	12 44 07	04 ♉ 10 45	11 ♉ 04 23	09 16	11 R 22	03 28	14 58	00 20	27 24	08 25	06 49
04	18 54 50	13 45 16	18 ♉ 04 45	25 ♉ 11 49	10 51	11 20	03 50	15 11	00 16	27 26	08 27	06 49
05	18 58 47	14 46 25	02 ♊ 25 22	09 ♊ 45 00	12 28	11 16	04 13	15 24	00 11	27 27	08 29	06 49
06	19 02 44	15 47 33	17 ♊ 10 09	24 ♊ 40 04	14 04	11 09	04 37	15 37	00 06	27 28	08 31	06 50
07	19 06 40	16 48 41	02 ♋ 13 47	09 ♋ 50 15	15 41	11 00	05 00	15 51	00 ♋ 02	27 30	08 33	06 50
08	19 10 37	17 49 49	17 ♋ 28 14	25 ♋ 06 29	17 19	10 48	05 24	16 04	29 ♊ 57	27 31	08 35	06 50
09	19 14 33	18 50 57	02 ♌ 43 40	10 ♌ 18 31	18 57	10 33	05 49	16 18	29 52	27 32	08 36	06 R 50
10	19 18 30	19 52 04	17 ♌ 49 50	25 ♌ 16 34	20 35	10 16	06 13	16 31	29 48	27 34	08 38	06 50
11	19 22 26	20 53 11	02 ♍ 37 49	09 ♍ 52 50	22 14	09 57	06 38	16 45	29 43	27 35	08 40	06 50
12	19 26 23	21 54 18	17 ♍ 01 08	24 ♍ 02 23	23 53	09 35	07 03	16 59	29 39	27 36	08 42	06 49
13	19 30 19	22 55 25	00 ♎ 56 27	07 ♎ 43 22	25 32	09 11	07 29	17 12	29 35	27 37	08 44	06 49
14	19 34 16	23 56 32	14 ♎ 23 20	20 ♎ 56 39	27 12	08 45	07 55	17 26	29 30	27 38	08 45	06 49
15	19 38 13	24 57 39	27 ♎ 23 43	03 ♏ 45 02	28 ♑ 53	08 17	08 21	17 40	29 26	27 39	08 47	06 49
16	19 42 09	25 58 45	10 ♏ 01 06	16 ♏ 12 31	00 ♒ 34	07 48	08 47	17 54	29 22	27 40	08 49	06 49
17	19 46 06	26 59 52	22 ♏ 19 49	28 ♏ 23 35	02 15	07 16	09 13	18 08	29 18	27 41	08 50	06 48
18	19 50 02	28 00 58	04 ♐ 24 24	10 ♐ 22 47	03 57	06 43	09 40	18 22	29 14	27 42	08 52	06 48
19	19 53 59	29 ♑ 02 04	16 ♐ 19 36	22 ♐ 14 19	05 39	06 09	10 07	18 36	29 10	27 42	08 54	06 48
20	19 57 55	00 ♒ 03 09	28 ♐ 08 23	04 ♑ 01 53	07 22	05 34	10 35	18 50	29 06	27 43	08 55	06 47
21	20 01 52	01 04 14	09 ♑ 55 11	15 ♑ 48 37	09 05	04 58	11 02	19 04	29 02	27 44	08 57	06 47
22	20 05 48	02 05 18	21 ♑ 42 30	27 ♑ 37 06	10 48	04 21	11 30	19 18	28 58	27 44	08 58	06 47
23	20 09 45	03 06 22	03 ♒ 32 38	09 ♒ 29 22	12 31	03 44	11 58	19 32	28 55	27 45	09 00	06 46
24	20 13 42	04 07 25	15 ♒ 27 28	21 ♒ 27 09	14 15	03 08	12 26	19 46	28 51	27 45	09 01	06 46
25	20 17 38	05 08 27	27 ♒ 28 36	03 ♓ 32 01	15 58	02 31	12 55	20 00	28 48	27 46	09 03	06 45
26	20 21 35	06 09 28	09 ♓ 37 36	15 ♓ 45 34	17 41	01 54	13 23	20 14	28 44	27 46	09 04	06 44
27	20 25 31	07 10 28	21 ♓ 56 09	28 ♓ 09 40	19 24	01 19	13 52	20 28	28 41	27 46	09 06	06 44
28	20 29 28	08 11 26	04 ♈ 26 12	10 ♈ 46 14	21 06	00 44	14 21	20 42	28 38	27 46	09 07	06 43
29	20 33 24	09 12 24	17 ♈ 10 02	23 ♈ 37 55	22 47	00 ♒ 11	14 50	20 57	28 34	27 47	09 08	06 43
30	20 37 21	10 13 21	00 ♉ 10 14	06 ♉ 47 20	24 28	29 ♑ 39	15 20	21 11	28 31	27 47	09 10	06 42
31	20 41 17	11 ♒ 14 16	13 ♉ 29 30	20 ♉ 17 02	26 ♒ 07	29 ♑ 08	15 ♉ 50	21 ♒ 25	28 ♊ 28	27 ♎ 47	09 ♐ 11	06 ♎ 41

Date	Moon True ☊ ° '	Moon Mean ☊ ° '	Moon ☽ Latitude ° '
01	28 ♐ 30	27 ♐ 54	05 N 13
02	28 D 30	27 50	04 54
03	28 31	27 47	04 18
04	28 32	27 44	03 26
05	28 34	27 41	02 20
06	28 35	27 38	01 N 03
07	28 R 35	27 35	00 S 20
08	28 34	27 31	01 43
09	28 32	27 28	02 58
10	28 30	27 25	04 00
11	28 26	27 22	04 44
12	28 23	27 19	05 09
13	28 21	27 15	05 15
14	28 19	27 12	05 03
15	28 D 19	27 09	04 36
16	28 20	27 06	03 56
17	28 21	27 03	03 06
18	28 23	27 00	02 08
19	28 25	26 56	01 06
20	28 25	26 53	00 S 02
21	28 R 25	26 50	01 N 03
22	28 22	26 47	02 04
23	28 18	26 44	03 00
24	28 12	26 41	03 49
25	28 05	26 37	04 27
26	27 58	26 34	04 55
27	27 51	26 31	05 08
28	27 45	26 28	05 08
29	27 40	26 25	04 52
30	27 38	26 22	04 24
31	27 ♐ 37	26 ♐ 18	03 N 36

DECLINATIONS

Date	Sun ☉ ° '	Moon ☽ ° '	Mercury ☿ ° '	Venus ♀ ° '	Mars ♂ ° '	Jupiter ♃ ° '	Saturn ♄ ° '	Uranus ♅ ° '	Neptune ♆ ° '	Pluto ♇ ° '
01	23 S 01	07 N 50	24 S 42	16 S 03	13 N 30	17 S 11	22 N 24	09 S 59	20 S 10	12 N 33
02	22 56	12 38	24 42	15 49	13 38	17 07	22 24	10 00	20 10	12 34
03	22 50	16 57	24 41	15 35	13 47	17 03	22 24	10 00	20 10	12 34
04	22 44	20 31	24 38	15 22	13 55	16 59	22 25	10 01	20 11	12 35
05	22 37	22 56	24 33	15 09	14 04	16 55	22 25	10 01	20 11	12 35
06	22 30	23 52	24 27	14 57	14 13	16 51	22 25	10 02	20 11	12 36
07	22 23	23 05	24 20	14 45	14 22	16 47	22 25	10 02	20 12	12 36
08	22 15	20 36	24 11	14 33	14 31	16 43	22 25	10 03	20 12	12 37
09	22 07	16 40	24 00	14 24	14 39	16 39	22 25	10 03	20 12	12 37
10	21 58	11 41	23 49	14 12	14 48	16 35	22 25	10 04	20 13	12 38
11	21 49	06 07	23 35	14 02	14 57	16 31	22 26	10 04	20 13	12 38
12	21 40	00 N 23	23 20	13 53	15 06	16 27	22 26	10 05	20 13	12 39
13	21 30	05 S 11	23 04	13 45	15 15	16 23	22 26	10 05	20 13	12 39
14	21 19	10 19	22 46	13 37	15 24	16 19	22 26	10 05	20 13	12 40
15	21 08	14 50	22 26	13 29	15 34	16 15	22 26	10 06	20 14	12 41
16	20 57	18 33	22 05	13 23	15 43	16 10	22 26	10 06	20 14	12 41
17	20 46	21 21	21 42	13 16	15 52	16 06	22 26	10 06	20 14	12 42
18	20 34	23 08	21 18	13 11	16 01	16 02	22 27	10 07	20 14	12 42
19	20 21	23 50	20 52	13 06	16 10	15 58	22 27	10 07	20 15	12 43
20	20 09	23 27	20 25	13 02	16 19	15 53	22 27	10 07	20 15	12 44
21	19 55	22 02	19 56	12 59	16 28	15 49	22 27	10 07	20 15	12 44
22	19 42	19 39	19 25	12 56	16 38	15 45	22 27	10 07	20 15	12 45
23	19 28	16 27	18 54	12 54	16 47	15 40	22 27	10 07	20 15	12 46
24	19 14	12 34	18 22	12 52	16 56	15 36	22 27	10 08	20 16	12 46
25	18 59	08 17	17 46	12 51	17 05	15 31	22 27	10 08	20 16	12 47
26	18 44	03 S 25	17 09	12 51	17 14	15 27	22 28	10 08	20 16	12 48
27	18 29	01 N 32	16 32	12 51	17 23	15 23	22 28	10 08	20 16	12 48
28	18 13	06 28	15 54	12 51	17 33	15 18	22 28	10 08	20 16	12 49
29	17 57	11 15	15 14	12 52	17 42	15 14	22 28	10 08	20 16	12 50
30	17 41	15 37	14 33	12 54	17 51	15 09	22 28	10 08	20 17	12 51
31	17 S 24	19 N 20	13 S 52	12 S 56	18 N 00	15 S 04	22 N 28	10 S 08	20 S 17	12 N 51

ZODIAC SIGN ENTRIES

Date	h	m	Planets
03	04	38	☽ ♉
05	08	00	☽ ♊
07	08	28	☽ ♋
07	20	26	♄ ♊
09	07	42	☽ ♌
11	07	41	☽ ♍
13	10	21	☽ ♎
15	16	54	☽ ♏
16	03	56	☿ ♒
18	03	12	☽ ♐
20	10	46	☉ ♒
20	15	47	☽ ♑
23	04	50	☽ ♒
25	17	00	☽ ♓
28	03	32	☽ ♈
29	19	51	♀ ♑
30	11	41	☽ ♉

LATITUDES

Date	Mercury ☿ ° '	Venus ♀ ° '	Mars ♂ ° '	Jupiter ♃ ° '	Saturn ♄ ° '	Uranus ♅ ° '	Neptune ♆ ° '	Pluto ♇ ° '
01	01 S 24	01 N 23	01 N 09	00 S 44	01 S 02	00 N 35	01 N 33	16 N 38
04	01 38	02 06	01 12	00 44	01 02	00 35	01 33	16 40
07	01 49	02 50	01 14	00 44	01 02	00 35	01 33	16 41
10	01 58	03 36	01 17	00 44	01 01	00 35	01 33	16 43
13	02 04	04 22	01 19	00 44	01 01	00 35	01 33	16 45
16	02 06	05 07	01 21	00 44	01 00	00 35	01 34	16 46
19	02 06	05 50	01 23	00 44	01 00	00 35	01 34	16 48
22	01 58	06 25	01 25	00 45	00 59	00 35	01 34	16 50
25	01 47	06 55	01 26	00 45	00 59	00 35	01 34	16 51
28	01 30	07 18	01 28	00 45	00 58	00 35	01 34	16 53
31	01 S 07	07 N 33	01 N 29	00 S 45	00 S 58	00 N 36	01 N 34	16 N 54

DATA

Julian Date	2442049
Delta T	+45 seconds
Ayanamsa	23° 29' 56"
Synetic vernal point	05° ♓ 37' 04"
True obliquity of ecliptic	23° 26' 33"

LONGITUDES

Date	Chiron ⚷ ° '	Ceres ⚳ ° '	Pallas ⚴ ° '	Juno ⚵ ° '	Vesta ⚶ ° '	Black Moon Lilith ☾ ° '
01	16 ♈ 23	19 ♑ 48	01 ♑ 11	16 ♑ 14	10 ♎ 01	15 ♑ 24
11	16 ♈ 29	23 ♑ 47	05 ♑ 14	20 ♑ 09	12 ♎ 39	16 ♑ 31
21	16 ♈ 40	27 ♑ 45	09 ♑ 12	24 ♑ 05	14 ♎ 48	17 ♑ 37
31	16 ♈ 55	01 ♒ 43	13 ♑ 05	28 ♑ 03	16 ♎ 22	18 ♑ 44

MOON'S PHASES, APSIDES AND POSITIONS ☽

Date	h	m	Phase	Longitude ° '	Eclipse Indicator
01	18	06	☽	10 ♈ 57	
08	12	36	○	17 ♋ 51	
15	07	04	☾	24 ♎ 45	
23	11	02	●	03 ♒ 04	
31	07	39	☽	11 ♉ 03	

Day	h	m		
08	11	27	Perigee	
21	21	55	Apogee	
06	13	19	Max dec	23° N 52'
12	13	36	0S	
19	15	35	Max dec	23° S 51'
27	04	38	0N	

ASPECTARIAN

h m	Aspects
01 Tuesday	
02 27	☽ ⊻ ♂
08 36	☽ □ ☿
10 23	☽ ☍ ♇
13 16	☽ △ ♆
18 06	☽ □ ☉
18 48	☽ ⚹ ♀
22 37	☽ # ♅
23 05	☿ □ ♇
02 Wednesday	
00 54	☽ ⚹ ♃
03 20	☉ ⊻ ♀
07 51	☽ Q ♄
11 40	☽ ∥ ♇
16 47	☽ Q ♀
16 49	☽ bQ ♆
17 30	☽ ∥ ♂
23 00	☽ Q ♃
23 03	☿ ⊻ ♆
23 59	☽ ☍ ♅
03 Thursday	
04 26	☽ # ♀
05 16	☽ ⚹ ♄
06 06	♀ St R
06 45	☿ ⊥ ♃
08 54	☽ ± ♆
10 42	☽ ☌ ♂
12 34	☽ # ♃
14 08	☿ Q ♅
16 37	☽ ⊼ ♇
19 25	☽ ⊼ ♆
22 01	☽ △ ☿
04 Friday	
00 30	☽ □ ♀
03 01	☽ ± ♇
04 02	☽ △ ☉
06 59	☽ □ ♃
07 13	☽ ∠ ♄
09 26	☽ # ♆
18 20	☽ bQ ♇
18 29	♃ bQ ♄
19 00	☿ ⊻ ♀
22 23	☽ ⊥ ♄
05 Saturday	
02 45	☽ bQ ☿
03 45	☽ ⊼ ♅
04 55	☉ ⊥ ♆
05 29	☽ ∥ ♄
07 17	☽ bQ ☉
08 11	☽ # ☉
08 19	☽ ⊻ ♄
13 42	☽ ± ♅
15 02	☽ ⊻ ♂
19 14	☽ △ ♇
19 27	☽ ± ☿
21 57	☽ ☍ ♆
23 11	☽ ± ☉
06 Sunday	
01 07	☽ ⊥ ♂
02 22	☽ △ ♀
04 25	☽ bQ ♅
06 24	☽ ⊼ ☿
06 57	☉ ⊻ ♃
09 29	☽ △ ♃
09 37	☽ ⊼ ☉
16 01	☽ ∠ ♂
18 43	☿ ⊥ ♆
07 Monday	
02 13	☽ bQ ♀
04 29	☽ △ ♅
06 01	☉ # ♄
08 32	☽ ☌ ♄
09 47	☽ bQ ♃
14 45	☿ ⊻ ♃
16 19	☽ ± ♀
16 30	☽ ⚹ ♂
19 16	☽ □ ♇
20 15	☽ ∥ ♄
21 07	☽ # ☉
21 59	☽ ⊼ ♆
08 Tuesday	
00 12	☽ ± ♃
01 39	☽ ⊼ ♀
07 27	☽ ± ♆
09 46	☽ ⊼ ♃
11 43	☽ ☍ ☿
11 54	☽ Q ♂
12 36	☽ ☍ ☉
14 53	☽ # ♆
15 18	♀ # ♂
21 37	☽ bQ ♆
23 34	☽ Q ♇
09 Wednesday	
00 54	♇ St R
03 49	☽ □ ♅
07 31	☽ ⊻ ♄
08 22	☉ ☌ ☿
12 03	☽ # ♃
16 57	☽ ⊥ ♄
17 00	☽ □ ♂
18 29	☽ ⚹ ♇
21 19	☽ △ ♆
21 52	☽ ∥ ♂
23 58	☽ # ♀
10 Thursday	
00 10	☽ ☍ ♀
07 10	☽ ∠ ♄
07 44	☽ ∥ ♇
08 22	☽ Q ♅
09 53	☽ ☍ ♃
15 31	☽ ⊼ ☉
16 58	☽ ⊼ ☿
18 25	☽ ∠ ♇
19 10	☽ # ♅
11 Friday	
01 55	☽ ± ☉
03 44	☽ ⚹ ♅
03 53	☽ ± ☿
07 16	☽ ⚹ ♄
09 03	☽ ⊥ ♇
17 46	☽ bQ ☉
18 48	☽ △ ♂
18 55	☽ ⊻ ♇
20 34	☽ bQ ☿
22 00	☽ □ ♆
22 55	♂ ⊼ ♇
23 50	☽ ⊼ ♀
12 Saturday	
03 00	☽ Q ♄
04 32	☽ ∠ ♅
09 20	☿ ∠ ♆
09 39	☽ ± ♀
11 56	☽ ⊼ ♃
19 49	☽ ⊥ ♅
20 51	☽ bQ ♂
20 59	☽ △ ☉
22 21	☽ ± ♃
13 Sunday	
00 36	☽ bQ ♀
01 19	☽ △ ☿
04 38	☽ Q ♆
06 11	☽ ⊻ ♅
09 38	☽ □ ♄
12 59	☽ ± ♂
14 16	☽ bQ ♃
22 24	☽ ☌ ♇
23 57	☽ ⊼ ♂
14 Monday	
01 49	☽ ⚹ ♆
02 10	☽ △ ♀
07 29	☉ ∠ ♆
10 50	☽ ∥ ♅
12 07	♀ ⚹ ♆
17 39	☽ △ ♃
18 12	☿ □ ♅
15 Tuesday	
00 01	☽ # ♇
04 38	☽ ∥ ♀
05 15	☽ ∠ ♆
07 04	☽ □ ☉
10 39	♀ □ ♂
12 00	☿ # ♄
12 29	☽ ☌ ♅
15 13	☽ □ ☿
15 49	☽ △ ♄
16 29	☽ # ♂
19 35	☿ ⊼ ♄
20 21	☽ ∥ ♃
22 11	☽ ⊥ ♆
16 Wednesday	
05 51	☽ ⊻ ♇
07 54	☽ □ ♀
09 32	☽ ☍ ♂
09 40	☽ ⊻ ♆
13 47	♂ ⊼ ♆
17 24	☽ ⊥ ♇
20 21	☽ Q ☉
20 22	☽ bQ ♄
17 Thursday	
01 28	☽ ∥ ♆
03 35	☽ □ ♃
06 40	☽ ∥ ☉
07 16	☽ Q ☿
10 58	☽ ∠ ♇
13 54	☽ ± ♄
15 18	☽ ∥ ☿
17 34	☽ Q ♀
22 05	☽ ⚹ ☉
22 36	☽ ⊻ ♅
18 Friday	
01 01	☽ # ♄
01 43	☽ ⊼ ♄
04 18	☉ □ ♅
08 30	♀ △ ♇
10 34	☽ ⊥ ♅
10 57	☽ ⚹ ☿
13 43	♂ # ♃
16 00	☽ Q ♃
16 26	☽ ⚹ ♀
16 48	☽ ⚹ ♇
20 59	☽ ☌ ♆
22 59	☽ ⊼ ♂
19 Saturday	
04 41	☽ ∠ ♅
05 19	☿ ± ♄
06 56	☽ ∠ ☉
11 35	☽ ± ♂
14 53	☉ ⊼ ♄
16 42	☽ ⚹ ♃
17 01	☽ Q ♇
17 11	☿ ☌ ♀
21 20	☽ ∠ ♀
22 17	☽ ∠ ☿
20 Sunday	
00 29	☉ ∥ ♆
02 54	☽ ⊥ ☉
03 56	☿ △ ♇
06 35	☽ bQ ♂
11 08	☽ ⚹ ♅
13 57	☽ ☍ ♄
14 46	☽ ⊥ ♀
16 16	☽ ⊻ ☉
19 42	☽ ⊥ ☿
20 22	☿ ∥ ♆
23 49	☽ ∠ ♃
21 Monday	
02 24	☽ ⊻ ♀
05 37	☽ □ ♇
06 28	☽ # ♄
08 53	♀ ± ♄
10 00	☽ ⊻ ☿
10 01	☽ ⊻ ♆
10 04	☿ ⚹ ♆
11 37	☽ Q ♅
12 45	☉ ∥ ☿
14 22	☽ △ ♂
18 32	☽ ⊥ ♃
22 16	☽ ⊥ ♆
22 Tuesday	
06 42	☽ ∥ ♆
06 59	☽ ⊻ ♃
11 34	☽ ∥ ☉
14 19	☽ ∥ ☿
16 37	☽ ∠ ♆
23 Wednesday	
00 15	☽ □ ♅
01 20	☿ □ ♂
02 40	☽ ⊼ ♄
09 48	☽ # ♂
11 02	☽ ☌ ☉
12 23	☽ ☌ ♀
14 45	☽ ± ♄
17 11	☽ ∥ ♃
18 30	☽ △ ♇
21 19	☉ ☌ ♀
23 02	☽ ⚹ ♆
24 Thursday	
05 41	☽ □ ♂
06 41	☿ bQ ♄
08 48	☽ bQ ♄
09 10	☽ ☌ ☿
10 12	☽ ∥ ♀
10 46	☽ # ♇
20 48	☽ ☌ ♃
23 10	☽ Q ♆
25 Friday	
00 36	☽ bQ ♇
01 33	☽ ∥ ♅
04 03	♂ ± ♇
04 15	☉ ± ♄
12 34	☽ △ ♅
14 36	☽ △ ♄
18 29	☽ ± ♇
19 05	☽ Q ♂
21 30	☽ ⊻ ♀
26 Saturday	
04 33	☽ ⊻ ☉
06 20	☽ ⊼ ♇
08 47	☽ ⊥ ♀
09 26	☿ # ♂
10 54	☽ □ ♆
17 25	☽ ⊥ ☉
18 09	☽ bQ ♅
19 40	☽ ⚹ ♂
27 Sunday	
01 35	☽ ∠ ♀
01 38	☉ △ ♇
03 34	♂ ∠ ♄
06 18	☽ ⊻ ☿
09 06	☽ ⊻ ♃
11 41	☽ ± ♅
12 30	☽ ∠ ☉
19 45	☽ ⊥ ☿
20 55	☽ ⊥ ♃
23 15	☽ ⊼ ♅
28 Monday	
00 57	☽ □ ♄
01 54	☽ ∠ ♂
05 15	☽ ⚹ ♀
05 31	☿ ☌ ♃
12 14	☿ Q ♆
14 28	☽ ∠ ♃
15 39	☽ ∠ ☿
16 20	☽ ☍ ♇
19 44	☽ ⊥ ♂
19 45	☽ ⚹ ☉
20 42	☿ bQ ♇
20 54	☽ △ ♆
29 Tuesday	
03 02	☽ Q ♀
06 17	☽ # ♅
07 29	☽ ⊻ ♂
10 24	☉ ⚹ ♆
10 54	☽ Q ♄
12 15	☿ ∥ ♃
19 10	☽ ⚹ ♃
20 09	☽ Q ☉
20 26	☽ ∥ ♇
20 41	☽ # ♀
30 Wednesday	
00 00	☽ ⚹ ☿
00 58	☽ bQ ♆
02 49	☉ # ♂
06 45	☽ # ☿
07 38	☽ ☍ ♅
09 00	☽ ⚹ ♄
09 21	☽ # ♃
09 51	♃ Q ♆
11 04	☽ □ ♀
17 27	☽ ± ♆
17 35	☽ Q ♃
23 49	☽ ⊼ ♇
23 51	☽ # ☉
31 Thursday	
01 02	☽ Q ☿
02 22	☽ ∥ ♂
04 18	☽ ⊼ ♆
07 39	☽ □ ☉
10 34	☽ ± ♇
11 58	☽ ∠ ♄
16 18	☽ ☌ ♂
19 17	☽ # ♆

All ephemeris data is given at 12.00 UT and the Moon's longitude is additionally given for 24.00 UT

FEBRUARY 1974

Date	Sidereal time h m s	Sun ☉	Moon ☽	Moon ☽ 24.00	Mercury ☿	Venus ♀	Mars ♂	Jupiter ♃	Saturn ♄	Uranus ♅	Neptune ♆	Pluto ♇
01	20 45 14	12 ♒ 15 10	27 ♉ 10 10	04 ♊ 09 02	27 ♒ 44	28 ♑ 39	16 ♉ 19	21 ♒ 39	28 ♊ 25	27 ♎ 47	09 ♐ 12	06 ♎ 40
02	20 49 11	13 16 03	11 ♊ 13 42	18 ♊ 24 05	29 ♒ 19	28 R 12	16 49	21 54	28 R 23	27 R 47	09 14	06 R 40
03	20 53 07	14 16 54	25 ♊ 39 57	03 ♋ 00 57	00 ♓ 51	27 47	17 19	22 08	28 20	27 47	09 15	06 39
04	20 57 04	15 17 44	10 ♋ 26 29	17 ♋ 55 49	02 20	27 25	17 50	22 22	28 17	27 47	09 16	06 38
05	21 01 00	16 18 33	25 ♋ 28 02	03 ♌ 02 04	03 45	27 04	18 20	22 37	28 15	27 46	09 17	06 37
06	21 04 57	17 19 20	10 ♌ 36 43	18 ♌ 10 45	05 05	26 46	18 51	22 51	28 13	27 46	09 18	06 36
07	21 08 53	18 20 06	25 ♌ 42 51	03 ♍ 11 48	06 21	26 30	19 21	23 05	28 10	27 46	09 19	06 35
08	21 12 50	19 20 51	10 ♍ 36 28	17 ♍ 55 49	07 30	26 17	19 52	23 20	28 08	27 45	09 20	06 34
09	21 16 46	20 21 34	25 ♍ 09 02	02 ♎ 15 30	08 33	26 06	20 23	23 34	28 06	27 45	09 21	06 33
10	21 20 43	21 22 17	09 ♎ 14 48	16 ♎ 06 42	09 28	25 58	20 54	23 49	28 04	27 44	09 22	06 32
11	21 24 40	22 22 58	22 ♎ 51 13	29 ♎ 28 28	10 15	25 52	21 26	24 03	28 02	27 44	09 23	06 31
12	21 28 36	23 23 38	05 ♏ 58 48	12 ♏ 22 37	10 53	25 49	21 57	24 17	28 00	27 43	09 24	06 30
13	21 32 33	24 24 17	18 ♏ 40 27	24 ♏ 52 53	11 22	25 D 48	22 29	24 32	27 59	27 43	09 25	06 29
14	21 36 29	25 24 55	01 ♐ 00 33	07 ♐ 04 09	11 41	25 50	23 00	24 46	27 57	27 42	09 26	06 28
15	21 40 26	26 25 32	13 ♐ 04 21	19 ♐ 01 50	11 50	25 54	23 32	25 01	27 56	27 41	09 27	06 27
16	21 44 22	27 26 08	24 ♐ 57 14	00 ♑ 51 14	11 R 48	26 00	24 04	25 15	27 54	27 40	09 28	06 25
17	21 48 19	28 26 43	06 ♑ 44 24	12 ♑ 37 18	11 36	26 09	24 36	25 29	27 53	27 39	09 28	06 24
18	21 52 15	29 ♒ 27 16	18 ♑ 30 29	24 ♑ 24 24	11 13	26 20	25 08	25 44	27 52	27 39	09 29	06 23
19	21 56 12	00 ♓ 27 48	00 ♒ 19 28	06 ♒ 16 02	10 41	26 33	25 41	25 58	27 51	27 38	09 30	06 22
20	22 00 09	01 28 18	12 ♒ 14 26	18 ♒ 14 55	10 01	26 48	26 13	26 13	27 50	27 37	09 31	06 21
21	22 04 05	02 28 47	24 ♒ 17 41	00 ♓ 22 53	09 12	27 05	26 45	26 27	27 49	27 36	09 31	06 19
22	22 08 02	03 29 14	06 ♓ 30 38	12 ♓ 41 01	08 17	27 25	27 18	26 41	27 49	27 34	09 32	06 18
23	22 11 58	04 29 39	18 ♓ 54 05	25 ♓ 09 52	07 17	27 46	27 51	26 56	27 48	27 33	09 32	06 17
24	22 15 55	05 30 03	01 ♈ 28 23	07 ♈ 49 41	06 14	28 08	28 24	27 10	27 48	27 32	09 33	06 15
25	22 19 51	06 30 25	14 ♈ 13 47	20 ♈ 40 43	05 09	28 33	28 56	27 24	27 47	27 31	09 34	06 14
26	22 23 48	07 30 45	27 ♈ 10 36	03 ♉ 43 30	04 04	28 59	29 ♉ 29	27 39	27 47	27 29	09 34	06 12
27	22 27 44	08 31 03	10 ♉ 19 33	16 ♉ 58 55	03 01	29 27	00 ♊ 03	27 53	27 47	27 28	09 34	06 11
28	22 31 41	09 ♓ 31 20	23 ♉ 41 46	00 ♊ 28 16	02 ♓ 00	29 ♑ 57	00 ♊ 36	28 ♒ 07	27 ♊ 47	27 ♎ 27	09 ♐ 35	06 ♎ 10

Date	Moon True ☊	Moon Mean ☊	Moon ☽ Latitude
01	27 ♐ 38	26 ♐ 15	02 N 38
02	27 D 40	26 12	01 28
03	27 40	26 09	00 N 11
04	27 R 40	26 06	01 S 08
05	27 37	26 02	02 24
06	27 32	25 59	03 30
07	27 25	25 56	04 21
08	27 16	25 53	04 54
09	27 08	25 50	05 06
10	27 00	25 47	05 00
11	26 54	25 43	04 36
12	26 50	25 40	03 58
13	26 48	25 37	03 10
14	26 D 48	25 34	02 14
15	26 49	25 31	01 13
16	26 50	25 27	00 S 10
17	26 R 49	25 24	00 N 53
18	26 47	25 21	01 53
19	26 42	25 18	02 49
20	26 34	25 15	03 37
21	26 23	25 12	04 17
22	26 11	25 08	04 45
23	25 58	25 05	05 00
24	25 45	25 02	05 00
25	25 34	24 59	04 46
26	25 26	24 56	04 17
27	25 20	24 53	03 34
28	25 ♐ 17	24 ♐ 49	02 N 39

DECLINATIONS

Date	Sun ☉	Moon ☽	Mercury ☿	Venus ♀	Mars ♂	Jupiter ♃	Saturn ♄	Uranus ♅	Neptune ♆	Pluto ♇
01	17 S 08	22 N 05	13 S 10	12 S 59	18 N 09	15 S 00	22 N 28	10 S 08	20 S 17	12 N 52
02	16 50	23 35	12 27	13 01	18 18	14 55	22 29	10 08	20 17	12 53
03	16 33	23 33	11 45	13 05	18 27	14 51	22 29	10 08	20 17	12 54
04	16 15	21 54	11 02	13 08	18 36	14 46	22 29	10 08	20 17	12 54
05	15 57	18 41	10 20	13 12	18 44	14 42	22 29	10 08	20 17	12 55
06	15 39	14 12	09 38	13 16	18 53	14 37	22 29	10 08	20 18	12 56
07	15 20	08 51	08 57	13 20	19 02	14 32	22 29	10 07	20 18	12 57
08	15 01	03 N 03	08 18	13 24	19 11	14 28	22 30	10 07	20 18	12 58
09	14 42	02 S 46	07 41	13 29	19 19	14 23	22 30	10 07	20 18	12 58
10	14 23	08 15	07 05	13 33	19 28	14 18	22 30	10 07	20 18	12 59
11	14 03	13 09	06 33	13 38	19 36	14 13	22 30	10 07	20 18	13 00
12	13 43	17 16	06 03	13 43	19 45	14 09	22 30	10 06	20 18	13 01
13	13 23	20 26	05 38	13 47	19 53	14 04	22 30	10 06	20 18	13 02
14	13 03	22 33	05 16	13 52	20 02	13 59	22 30	10 06	20 18	13 02
15	12 43	23 35	04 58	13 57	20 10	13 55	22 31	10 06	20 18	13 03
16	12 22	23 31	04 45	14 01	20 18	13 50	22 31	10 05	20 19	13 04
17	12 01	22 23	04 36	14 06	20 26	13 45	22 31	10 05	20 19	13 05
18	11 40	20 18	04 33	14 10	20 34	13 40	22 31	10 05	20 19	13 06
19	11 19	17 20	04 34	14 14	20 42	13 35	22 31	10 04	20 19	13 07
20	10 57	13 39	04 40	14 18	20 50	13 31	22 31	10 04	20 19	13 07
21	10 36	09 23	04 51	14 22	20 58	13 26	22 32	10 04	20 19	13 08
22	10 14	04 S 43	05 05	14 25	21 05	13 21	22 32	10 03	20 19	13 09
23	09 52	00 N 12	05 24	14 28	21 13	13 16	22 32	10 03	20 19	13 10
24	09 30	05 11	05 46	14 31	21 20	13 11	22 32	10 02	20 19	13 11
25	09 07	10 00	06 10	14 34	21 28	13 06	22 32	10 02	20 19	13 12
26	08 45	14 28	06 36	14 36	21 35	13 02	22 33	10 01	20 19	13 12
27	08 23	18 19	07 03	14 39	21 42	12 57	22 33	10 01	20 19	13 13
28	08 S 00	21 N 16	07 S 31	14 S 40	21 N 50	12 S 52	22 N 33	10 S 00	20 S 19	13 N 14

ZODIAC SIGN ENTRIES

Date	h	m	Planets
01	16	53	☽ ♊
02	22	42	☿ ♓
03	19	06	☽ ♋
05	19	12	☽ ♌
07	18	52	☽ ♍
09	20	10	☽ ♎
12	00	58	☽ ♏
14	10	01	☽ ♐
16	22	16	☽ ♑
19	00	59	☉ ♓
19	11	21	☽ ♒
21	23	15	☽ ♓
24	09	12	☽ ♈
26	17	11	☽ ♉
27	10	11	♂ ♊
28	14	25	♀ ♒
28	23	10	☽ ♊

LATITUDES

Date	Mercury ☿	Venus ♀	Mars ♂	Jupiter ♃	Saturn ♄	Uranus ♅	Neptune ♆	Pluto ♇
01	00 S 58	07 N 37	01 N 29	00 S 45	00 S 58	00 N 36	01 N 34	16 N 55
04	00 S 25	07 41	01 30	00 45	00 57	00 36	01 34	16 56
07	00 N 15	07 39	01 31	00 45	00 56	00 36	01 34	16 58
10	01 00	07 32	01 32	00 46	00 56	00 36	01 35	16 59
13	01 49	07 19	01 33	00 46	00 55	00 36	01 35	17 00
16	02 35	07 03	01 33	00 46	00 55	00 36	01 35	17 02
19	03 14	06 44	01 34	00 46	00 54	00 36	01 35	17 03
22	03 38	06 23	01 34	00 46	00 54	00 36	01 35	17 04
25	03 43	06 00	01 35	00 47	00 53	00 36	01 35	17 05
28	03 28	05 37	01 35	00 47	00 52	00 36	01 35	17 06
31	02 N 58	05 N 13	01 N 35	00 S 47	00 S 52	00 N 36	01 N 36	17 N 07

DATA

Julian Date	2442080
Delta T	+45 seconds
Ayanamsa	23° 30' 01"
Synetic vernal point	05° ♓ 36' 59"
True obliquity of ecliptic	23° 26' 33"

MOON'S PHASES, APSIDES AND POSITIONS ☽

Date	h	m	Phase	Longitude	Eclipse Indicator
06	23	24	○	17 ♌ 48	
14	00	04	☾	24 ♏ 55	
22	05	34	●	03 ♓ 13	

Day	h	m		
06	00	00	Perigee	
18	07	55	Apogee	
02	23	41	Max dec	23° N 46'
09	00	31	0S	
15	22	31	Max dec	23° S 41'
23	11	01	0N	

LONGITUDES

Date	Chiron ⚷	Ceres ⚳	Pallas ⚴	Juno ⚵	Vesta ⚶	Black Moon Lilith ⚸
01	16 ♈ 57	02 ♒ 07	13 ♑ 28	28 ♑ 27	16 ♎ 30	18 ♑ 50
11	17 ♈ 18	06 ♒ 02	17 ♑ 13	02 ♒ 25	17 ♎ 29	19 ♑ 57
21	17 ♈ 44	09 ♒ 55	20 ♑ 49	06 ♒ 23	17 ♎ 21	21 ♑ 04
31	18 ♈ 42	13 ♒ 43	24 ♑ 16	10 ♒ 19	16 ♎ 35	22 ♑ 10

ASPECTARIAN

h m Aspects

01 Friday
02 14 ☽ □ ♃
02 26 ☽ Ᵽ ♇
02 58 ♅ St R
03 47 ☽ ⊥ ♄
12 47 ☿ △ ♅
13 03 ☽ ⊼ ♅
13 06 ☽ □ ☿
13 20 ♃ Ᵽ ♇
14 10 ☽ ⊻ ♄
14 29 ☽ △ ♀
16 34 ☽ ∥ ♄
18 01 ☿ ∥ ♀
21 51 ☿ ⧺ ♇
22 11 ☿ △ ♄
22 48 ☿ ⊻ ♀
23 22 ☽ ± ♅

02 Saturday
01 17 ♀ ⊼ ♄
04 17 ☽ △ ♇
08 37 ☽ ☍ ♆
14 31 ☉ Ᵽ ♄
14 37 ☽ Ᵽ ♅
15 13 ☽ Ᵽ ♀
15 41 ☽ △ ☉
21 42 ☽ ⊻ ♂

03 Sunday
05 47 ☽ ± ♀
06 05 ☽ △ ♃
08 00 ☽ ⊥ ♂
08 51 ☿ ± ♇
12 35 ♀ □ ♅
15 23 ☽ ⊼ ♀
15 28 ☽ △ ♅
16 21 ☽ ☌ ♄
18 22 ☽ Ᵽ ☉
21 26 ☽ △ ☿
23 16 ☽ ∠ ♂

04 Monday
05 45 ☽ ∥ ♄
05 52 ☽ □ ♇
06 59 ☽ Ᵽ ♃
10 01 ☽ ± ☉
10 06 ☽ ⊼ ♆
19 45 ☽ ± ♆
20 21 ☽ ⊼ ☉
21 40 ☽ ± ♃

05 Tuesday
00 12 ☽ Ᵽ ☿
00 15 ☽ ✶ ♂
01 19 ☽ ⧺ ♆
02 35 ☿ ⊥ ♀
07 23 ☽ ⊼ ♃
10 07 ☽ Ᵽ ♆
10 39 ☽ Q ♇
11 40 ☽ ∥ ♂
14 29 ☽ ☍ ♀
15 39 ☽ □ ♅
15 59 ☽ ± ☿
16 24 ☽ ⊻ ♄
18 53 ☿ ∥ ♅
20 00 ☽ Q ♂

06 Wednesday
01 53 ☽ ⊥ ♄
02 25 ☽ ⊼ ☿
04 22 ☽ ⧺ ☉
05 39 ☽ ✶ ♇
09 56 ☽ △ ♆
09 58 ☽ ⧺ ♃
16 06 ☽ ∠ ♄
16 23 ☽ ⧺ ♀
17 57 ☽ ∥ ♇
20 10 ☽ Q ♅
23 24 ☽ ☍ ☉

07 Thursday
01 31 ☽ □ ♂
05 25 ☽ ∠ ♇
06 31 ☽ ⧺ ♅
07 45 ☽ ☍ ♃
11 28 ☽ ⧺ ☿
13 15 ☽ ⊼ ♀
15 16 ☽ ✶ ♅
15 55 ☽ ✶ ♄
16 47 ☿ ⊼ ♇
19 48 ☽ ⊥ ♇
22 43 ☽ ± ♀

08 Friday
05 27 ☽ ⊻ ♇
06 32 ☽ ☍ ☿
09 56 ☽ □ ♆
11 14 ☽ Q ♄
13 05 ☽ Ᵽ ♀
15 30 ☽ ∠ ♅

09 Saturday
03 25 ☽ ⊼ ☉
03 46 ☽ △ ♂
04 18 ☿ Q ♂
06 20 ☽ ⊥ ♅
09 19 ☽ ⊼ ♃
13 26 ☉ □ ♂
13 35 ☽ △ ♀
14 11 ☽ ± ☉
15 43 ☽ Q ♆
16 22 ☽ ⊻ ♅
16 57 ☽ □ ♄
19 34 ☽ ± ♃

10 Sunday
06 01 ☽ Ᵽ ♂
06 39 ☽ Ᵽ ☉
07 14 ☽ ∥ ☿
07 20 ☽ ☌ ♇
09 20 ☿ □ ♆
11 14 ☽ Ᵽ ♃
12 03 ☉ Q ♆
12 13 ☽ ✶ ♆
12 24 ☽ ⊼ ☿
15 50 ☉ Ᵽ ♇
19 30 ☉ ∥ ♃
20 42 ☽ ∥ ♅
22 16 ☽ ± ♂
23 34 ☽ ± ☿

11 Monday
09 21 ☽ ⊼ ♂
11 05 ☽ △ ☉
11 11 ☽ ⧺ ♇
14 12 ☽ △ ♃
14 36 ☽ ∥ ♀
14 46 ☽ ∠ ♆
15 55 ♂ Ᵽ ♇
16 30 ☽ ∥ ☉
16 34 ☽ Ᵽ ☿
17 25 ☽ □ ♀
17 41 ☽ ∥ ♃
20 49 ☽ ☌ ♅
21 21 ☽ △ ♄

12 Tuesday
07 13 ☽ ⊥ ♆
09 06 ☿ ∠ ♀
12 40 ☉ ∥ ♀
12 58 ☽ ⊻ ♇
14 16 ♂ ⊥ ♄
18 25 ☽ ⊻ ♆
21 36 ☽ △ ☿

13 Wednesday
00 13 ☽ ⊥ ♇
01 10 ☽ Ᵽ ♄
02 42 ☽ Q ♀
07 04 ☽ ⧺ ♂
07 28 ♀ St D
10 56 ☽ ∥ ♆
15 53 ☉ ☌ ♃
17 24 ☽ ∠ ♇
18 21 ☽ ± ♄
19 40 ☽ ☍ ♂
23 32 ☽ □ ♃

14 Thursday
00 04 ☽ □ ☉
01 49 ☽ ✶ ♀
05 30 ☽ ⊻ ♅
06 00 ☽ ⊼ ♄
11 24 ☽ ⧺ ♄
12 40 ☉ ⧺ ♇
17 18 ☽ ⊥ ♅
22 31 ☉ ⊻ ♀
22 47 ☽ ✶ ♇

15 Friday
04 44 ☽ ☌ ♆
06 32 ♀ ∥ ♃
07 37 ☽ ∠ ♀
09 30 ☽ □ ☿
11 13 ☽ ∠ ♅
11 52 ☽ Q ♃
14 58 ☽ Q ☉
19 45 ☿ St R
22 48 ☽ Q ♇

16 Saturday
01 25 ☿ Q ♂
01 52 ☽ ⊥ ♀
10 07 ☽ ⊼ ♂
12 37 ☽ ✶ ♃
13 47 ♂ ⧺ ♆
14 10 ☽ ⊻ ♀
17 31 ☽ ✶ ☉
17 31 ☽ ✶ ♅
17 33 ☉ △ ♅
17 59 ☽ ☍ ♄
21 44 ☽ Q ☿
22 54 ☽ ± ♂
22 57 ☉ △ ♄

17 Sunday
10 07 ☽ ⧺ ♄
11 19 ☽ □ ♇
17 35 ☽ ⊻ ♆
17 57 ☽ Q ♅
18 07 ☽ Ᵽ ♂
19 49 ☽ ∠ ♃
21 38 ☽ ✶ ☿

18 Monday
02 58 ☽ ∠ ☉
05 50 ☽ ⊥ ♆
07 47 ☿ ∠ ♀
09 30 ☽ ⧺ ♂
11 50 ☽ ∥ ♆
14 32 ☽ ⊥ ♃
23 00 ☽ ⊥ ☉

19 Tuesday
00 11 ☽ ∠ ♆
02 08 ☽ △ ♂
02 59 ☽ ⊻ ♃
03 03 ☽ ∠ ☿
04 12 ☽ ☌ ♀
06 33 ☽ □ ♅
07 00 ☽ ⊼ ♄
09 40 ☉ ± ♇
12 18 ☽ ⊻ ☉
19 07 ☽ ± ♄
20 22 ☽ ⊥ ☿

20 Wednesday
00 10 ☽ △ ♇
06 31 ☽ ✶ ♆
07 47 ☽ ⊻ ☿
08 07 ☽ ∥ ♀
11 24 ♂ □ ♃
12 51 ☽ ∥ ♃
13 11 ☽ Ᵽ ♄
15 07 ☽ ⧺ ♇

21 Thursday
03 03 ☿ □ ♆
04 54 ☽ ∥ ☉
06 07 ☽ Ᵽ ♇
06 30 ☽ Q ♆
08 24 ☽ ∥ ♅
16 20 ☽ ☌ ♃
17 06 ☽ □ ♂
17 40 ☽ ⊻ ♀
18 30 ☽ △ ♅
18 57 ☽ △ ♄
23 52 ☽ ± ♇

22 Friday
05 34 ☽ ☌ ☉
05 46 ☽ ⊥ ♀
09 14 ☉ ⊥ ♀
10 13 ☽ ∥ ☿
11 35 ☽ ⊼ ♇
15 13 ☽ ☌ ☿
17 53 ☽ □ ♆
22 54 ♀ □ ♅
23 36 ♂ ⊼ ♅
23 46 ☽ Ᵽ ♅
23 48 ☽ ∠ ♀

23 Saturday
00 01 ☉ ∥ ♅
00 56 ♀ △ ♂
05 51 ☽ Q ♂
10 05 ♂ ⊻ ♄
14 39 ♀ ⊼ ♄
17 05 ☽ ± ♅

24 Sunday
03 39 ☽ ⊻ ♃
04 32 ☽ ⊼ ♅
05 01 ☽ □ ♄
05 28 ☽ ✶ ♀
05 53 ☽ ✶ ♂
11 39 ☿ ⊼ ♇
14 08 ♃ ⧺ ♇
15 05 ☽ ⧺ ☿
15 16 ☽ ⊥ ♃
20 16 ☽ ⊻ ☉
20 18 ☽ ⊻ ☿
20 30 ☉ ☌ ☿
21 01 ☽ ☍ ♇

25 Monday
03 15 ☽ △ ♆
04 53 ☽ Q ♀
05 34 ☉ ⊼ ♇
06 42 ☽ ⊥ ☿
07 50 ☽ ⧺ ☉
08 30 ☽ ⊥ ☉
08 31 ☽ ∠ ♃
11 26 ☽ ∠ ♂
12 06 ☽ ⧺ ♅
14 54 ☽ Q ♄
21 35 ☿ ⊥ ♀
22 00 ♃ △ ♅
22 11 ☽ ∠ ☿

26 Tuesday
02 40 ☽ ∠ ☉
04 06 ☽ ⧺ ♃
04 54 ☽ ⊥ ♂
04 55 ☽ ∥ ♇
07 11 ☽ Ᵽ ♆
12 35 ☽ ☍ ♅
12 49 ☽ ⧺ ♀
12 52 ☽ ✶ ♃
13 07 ☽ ✶ ♄
15 27 ☽ □ ♀
16 26 ☽ ⊻ ♂
23 41 ☽ ✶ ☿
23 43 ☽ ± ♆

27 Wednesday
02 13 ♃ △ ♄
04 30 ☽ ⊼ ♇
08 27 ☽ ✶ ☉
10 38 ☽ ⊼ ♆
11 11 ☽ Q ♃
15 21 ☽ ± ♇
16 27 ☽ ∠ ♄
19 51 ☽ Q ☿
21 13 ♄ St D

28 Thursday
03 20 ☽ ⧺ ♆
07 29 ☽ Ᵽ ♇
07 49 ☽ Q ☉
08 36 ☽ ⊥ ♄
13 24 ☉ □ ♆
18 14 ☽ ∥ ♂
18 39 ☽ ⊼ ♅
19 15 ☽ ⊻ ♄
19 59 ☽ □ ♃
23 31 ☽ △ ♀

All ephemeris data is given at 12.00 UT and the Moon's longitude is additionally given for 24.00 UT

MARCH 1974

LONGITUDES

Date	Sidereal time h m s	Sun ☉ ° ' "	Moon ☽ ° ' "	Moon ☽ 24.00 ° ' "	Mercury ☿ ° '	Venus ♀ ° '	Mars ♂ ° '	Jupiter ♃ ° '	Saturn ♄ ° '	Uranus ♅ ° '	Neptune ♆ ° '	Pluto ♇ ° '
01	22 35 38	10 ♓ 31 34	07 ♊ 18 37	14 ♊ 12 57	01 ♓ 03	00 ♒ 28	01 ♊ 09	28 ♒ 21	27 ♊ 47	27 ♎ 25	09 ♐ 35	06 ♎ 08
02	22 39 34	11 31 46	21 ♊ 11 23	28 ♊ 13 59	00 ♓ 12	01 01	01 42	28 36	27 D 47	27 R 24	09 36	06 R 07
03	22 43 31	12 31 56	05 ♋ 20 43	12 ♋ 31 26	29 ♒ 26	01 34	02 16	28 50	27 48	27 22	09 36	06 05
04	22 47 27	13 32 04	19 ♋ 45 53	27 ♋ 03 39	28 R 46	02 10	02 49	29 04	27 48	27 21	09 36	06 04
05	22 51 24	14 32 10	04 ♌ 24 11	11 ♌ 46 47	28 13	02 46	03 23	29 18	27 49	27 19	09 36	06 02
06	22 55 20	15 32 14	19 ♌ 10 38	26 ♌ 34 45	27 47	03 24	03 56	29 32	27 49	27 17	09 37	06 01
07	22 59 17	16 32 16	03 ♍ 58 08	11 ♍ 19 44	27 27	04 03	04 30	29 ♒ 46	27 50	27 16	09 37	05 59
08	23 03 13	17 32 15	18 ♍ 38 30	25 ♍ 53 26	27 15	04 43	05 04	00 ♓ 00	27 51	27 14	09 37	05 58
09	23 07 10	18 32 13	03 ♎ 03 39	10 ♎ 08 25	27 09	05 25	05 38	00 15	27 52	27 12	09 37	05 56
10	23 11 07	19 32 09	17 ♎ 07 08	23 ♎ 59 24	27 D 09	06 07	06 12	00 29	27 53	27 10	09 37	05 55
11	23 15 03	20 32 03	00 ♏ 44 59	07 ♏ 23 51	27 16	06 50	06 46	00 43	27 55	27 08	09 37	05 53
12	23 19 00	21 31 56	13 ♏ 56 05	20 ♏ 21 58	27 28	07 35	07 20	00 56	27 56	27 07	09 R 37	05 51
13	23 22 56	22 31 47	26 ♏ 41 53	02 ♐ 56 17	27 46	08 20	07 54	01 10	27 57	27 05	09 37	05 50
14	23 26 53	23 31 36	09 ♐ 05 45	15 ♐ 10 54	28 09	09 07	08 28	01 24	27 59	27 03	09 37	05 48
15	23 30 49	24 31 23	21 ♐ 12 25	27 ♐ 10 58	28 37	09 54	09 02	01 38	28 01	27 01	09 37	05 47
16	23 34 46	25 31 09	03 ♑ 07 16	09 ♑ 02 02	29 10	10 42	09 37	01 52	28 02	26 59	09 37	05 45
17	23 38 42	26 30 53	14 ♑ 55 55	20 ♑ 49 36	29 ♒ 47	11 31	10 11	02 06	28 04	26 57	09 37	05 43
18	23 42 39	27 30 36	26 ♑ 43 43	02 ♒ 38 52	00 ♓ 27	12 20	10 45	02 19	28 06	26 54	09 36	05 42
19	23 46 36	28 30 16	08 ♒ 35 35	14 ♒ 34 22	01 12	13 10	11 20	02 33	28 09	26 52	09 36	05 40
20	23 50 32	29 ♓ 29 55	20 ♒ 35 38	26 ♒ 39 45	02 00	14 01	11 54	02 47	28 11	26 50	09 36	05 38
21	23 54 29	00 ♈ 29 32	02 ♓ 47 00	08 ♓ 57 38	02 52	14 53	12 29	03 00	28 13	26 48	09 36	05 37
22	23 58 25	01 29 07	15 ♓ 11 46	21 ♓ 29 30	03 46	15 45	13 03	03 14	28 15	26 46	09 35	05 35
23	00 02 22	02 28 39	27 ♓ 50 49	04 ♈ 15 41	04 44	16 38	13 38	03 28	28 18	26 43	09 35	05 33
24	00 06 18	03 28 10	10 ♈ 43 59	17 ♈ 15 36	05 44	17 32	14 13	03 41	28 21	26 41	09 35	05 32
25	00 10 15	04 27 39	23 ♈ 50 20	00 ♉ 28 02	06 47	18 26	14 48	03 54	28 23	26 39	09 34	05 30
26	00 14 11	05 27 06	07 ♉ 08 30	13 ♉ 51 33	07 52	19 21	15 23	04 08	28 26	26 37	09 34	05 28
27	00 18 08	06 26 31	20 ♉ 37 03	27 ♉ 24 53	09 00	20 16	15 57	04 21	28 29	26 34	09 33	05 27
28	00 22 05	07 25 53	04 ♊ 14 57	11 ♊ 07 11	10 10	21 12	16 32	04 34	28 32	26 32	09 33	05 25
29	00 26 01	08 25 13	18 ♊ 01 34	24 ♊ 58 05	11 22	22 08	17 07	04 47	28 35	26 29	09 32	05 23
30	00 29 58	09 24 31	01 ♋ 56 43	08 ♋ 57 27	12 36	23 05	17 42	05 01	28 39	26 27	09 31	05 22
31	00 33 54	10 ♈ 23 47	16 ♋ 00 14	23 ♋ 04 59	13 ♓ 52	24 ♒ 02	18 ♊ 17	05 ♓ 14	28 ♊ 42	26 ♎ 25	09 ♐ 31	05 ♎ 20

DECLINATIONS

Date	Moon True ☊ ° '	Moon Mean ☊ ° '	Moon ☽ Latitude ° '	Sun ☉ ° '	Moon ☽ ° '	Mercury ☿ ° '	Venus ♀ ° '	Mars ♂ ° '	Jupiter ♃ ° '	Saturn ♄ ° '	Uranus ♅ ° '	Neptune ♆ ° '	Pluto ♇ ° '
01	25 ♐ 16	24 ♐ 46	01 N 34	07 S 37	23 N 04	07 S 59	14 S 42	21 N 57	12 S 47	22 N 33	10 00	20 S 19	13 N 15
02	25 D 16	24 43	00 N 22	07 14	23 31	08 27	14 43	22 03	12 42	22 33	09 59	20 19	13 16
03	25 R 16	24 40	00 S 53	06 52	22 27	08 53	14 44	22 10	12 37	22 34	09 58	20 19	13 17
04	25 14	24 37	02 06	06 28	19 55	09 19	14 44	22 17	12 32	22 34	09 58	20 19	13 17
05	25 10	24 33	03 11	06 05	16 04	09 42	14 44	22 24	12 28	22 34	09 57	20 19	13 18
06	25 02	24 30	04 04	05 42	11 12	10 04	14 43	22 30	12 23	22 34	09 57	20 19	13 19
07	24 53	24 27	04 41	05 19	05 N 41	10 24	14 42	22 36	12 18	22 34	09 56	20 19	13 20
08	24 41	24 24	04 59	04 56	00 S 06	10 41	14 41	22 43	12 13	22 35	09 55	20 19	13 21
09	24 29	24 21	04 57	04 32	05 46	10 57	14 39	22 49	12 08	22 35	09 55	20 19	13 21
10	24 18	24 18	04 37	04 09	11 00	11 10	14 37	22 55	12 03	22 35	09 54	20 19	13 22
11	24 08	24 14	04 02	03 45	15 31	11 21	14 34	23 01	11 58	22 35	09 53	20 19	13 23
12	24 02	24 11	03 15	03 22	19 07	11 29	14 31	23 06	11 54	22 35	09 53	20 19	13 24
13	23 58	24 08	02 19	02 58	21 40	11 36	14 27	23 12	11 49	22 36	09 52	20 19	13 25
14	23 56	24 05	01 18	02 34	23 06	11 41	14 23	23 17	11 44	22 36	09 51	20 19	13 25
15	23 D 55	24 02	00 S 14	02 11	23 19	11 43	14 19	23 23	11 39	22 36	09 51	20 19	13 26
16	23 R 55	23 59	00 N 49	01 47	22 36	11 43	14 14	23 28	11 34	22 36	09 50	20 19	13 27
17	23 54	23 55	01 49	01 23	20 48	11 42	14 08	23 33	11 29	22 36	09 49	20 18	13 28
18	23 52	23 52	02 44	00 59	18 07	11 38	14 02	23 38	11 25	22 37	09 48	20 18	13 29
19	23 46	23 49	03 33	00 36	14 41	11 33	13 55	23 43	11 20	22 37	09 48	20 18	13 29
20	23 38	23 46	04 13	00 S 12	10 38	11 25	13 48	23 48	11 15	22 37	09 47	20 18	13 30
21	23 27	23 43	04 41	00 N 12	06 06	11 16	13 41	23 52	11 10	22 37	09 46	20 18	13 31
22	23 15	23 39	04 57	00 35	01 S 16	11 05	13 33	23 57	11 05	22 38	09 45	20 18	13 31
23	23 01	23 36	04 59	00 59	03 N 43	10 53	13 24	24 01	11 01	22 38	09 44	20 18	13 32
24	22 48	23 33	04 46	01 23	08 39	10 39	13 15	24 05	10 56	22 38	09 43	20 18	13 33
25	22 36	23 30	04 18	01 46	13 14	10 23	13 06	24 09	10 51	22 38	09 43	20 18	13 34
26	22 27	23 27	03 35	02 10	17 17	10 06	12 56	24 13	10 46	22 38	09 42	20 18	13 34
27	22 21	23 24	02 39	02 33	20 28	09 47	12 46	24 16	10 42	22 39	09 41	20 17	13 35
28	22 17	23 20	01 34	02 57	22 32	09 27	12 35	24 20	10 37	22 39	09 40	20 17	13 36
29	22 16	23 17	00 N 23	03 20	23 17	09 05	12 24	24 23	10 32	22 39	09 39	20 17	13 36
30	22 D 16	23 14	00 S 51	03 44	22 35	08 42	12 12	24 27	10 28	22 39	09 38	20 17	13 37
31	22 ♐ 16	23 ♐ 11	02 S 02	04 N 07	20 N 27	08 S 17	12 S 00	24 N 30	10 S 23	22 N 39	09 S 37	20 S 17	13 N 37

ZODIAC SIGN ENTRIES

Date	h	m	Planets
02	17	49	☿ ♒
03	03	00	☽ ♋
05	04	49	☽ ♌
07	05	33	☽ ♍
08	11	11	♃ ♓
09	06	52	☽ ♎
11	10	40	☽ ♏
13	18	20	☽ ♐
16	05	41	☽ ♑
17	20	11	☿ ♓
18	18	38	☽ ♒
21	00	07	☉ ♈
21	06	33	☽ ♓
23	16	02	☽ ♈
25	23	09	☽ ♉
28	04	33	☽ ♊
30	08	40	☽ ♋

LATITUDES

Date	Mercury ☿ ° '	Venus ♀ ° '	Mars ♂ ° '	Jupiter ♃ ° '	Saturn ♄ ° '	Uranus ♅ ° '	Neptune ♆ ° '	Pluto ♇ ° '
01	03 N 20	05 N 29	01 N 35	00 S 47	00 S 52	00 N 36	01 N 36	17 N 06
04	02 46	05 05	01 35	00 47	00 52	00 36	01 36	17 07
07	02 02	05 04	01 35	00 48	00 51	00 36	01 36	17 08
10	01 23	04 16	01 35	00 48	00 51	00 36	01 36	17 08
13	00 41	03 51	01 35	00 48	00 50	00 36	01 36	17 09
16	00 N 03	03 28	01 35	00 49	00 49	00 36	01 36	17 09
19	00 S 32	03 04	01 35	00 49	00 49	00 36	01 36	17 10
22	01 02	02 41	01 35	00 49	00 48	00 37	01 37	17 10
25	01 28	02 19	01 35	00 50	00 48	00 37	01 37	17 10
28	01 49	01 57	01 35	00 50	00 47	00 37	01 37	17 10
31	02 S 06	01 N 36	01 N 35	00 S 51	00 S 47	00 N 37	01 N 37	17 N 10

DATA

Julian Date	2442108
Delta T	+45 seconds
Ayanamsa	23° 30' 04"
Synetic vernal point	05° ♓ 36' 55"
True obliquity of ecliptic	23° 26' 33"

MOON'S PHASES, APSIDES AND POSITIONS ☽

Date	h	m	Phase	Longitude ° '	Eclipse Indicator
01	18	03	☽	10 ♊ 47	
08	10	03	○	17 ♍ 27	
15	19	15	☾	24 ♐ 49	
23	21	24	●	02 ♈ 52	
31	01	44	☽	09 ♋ 58	

Day	h	m		
06	06	16	Perigee	
18	01	51	Apogee	
02	07	09	Max dec	23° N 32'
08	11	37	0S	
15	16	58	Max dec	23° S 25'
22	18	06	0N	
29	12	27	Max dec	23° N 17'

LONGITUDES

Date	Chiron ⚷ ° '	Ceres ⚳ ° '	Pallas ⚴ ° '	Juno ⚵ ° '	Vesta ⚶ ° '	Black Moon Lilith ° '
01	18 ♈ 06	12 ♒ 58	23 ♑ 35	09 ♒ 32	16 ♎ 48	21 ♑ 57
11	18 ♈ 37	16 ♒ 42	26 ♑ 52	13 ♒ 26	15 ♎ 25	23 ♑ 04
21	19 ♈ 11	20 ♒ 21	29 ♑ 55	17 ♒ 17	13 ♎ 21	24 ♑ 10
31	19 ♈ 45	23 ♒ 52	02 ♒ 43	21 ♒ 04	10 ♎ 53	25 ♑ 17

ASPECTARIAN

h m Aspects

01 Friday
00 44 ☽ ☌ ♂
01 33 ☉ ∥ ☿
01 44 ☽ □ ☿
03 05 ☽ ∥ ♄
05 12 ☽ ± ♅
09 57 ☽ △ ♇
10 29 ☿ □ ♂
15 58 ☽ ☍ ♆
18 03 ☽ □ ☉
20 53 ☽ ⚼ ♅
21 57 ☿ ⊻ ♀

02 Saturday
02 44 ☽ ⚼ ♀
14 31 ☿ ± ♇
18 47 ☽ ± ♀
22 34 ☽ △ ♅
23 15 ☽ ☌ ♄

03 Sunday
00 50 ☽ △ ♃
02 31 ☽ △ ☿
05 23 ☽ ⊼ ♀
06 36 ☽ ⊻ ♂
08 13 ☉ ⚼ ♅
10 30 ☽ ∥ ♄
13 15 ☽ □ ♇
15 17 ☽ ∥ ♂
17 05 ☽ ⊥ ♂
19 07 ☽ ⊼ ♆

04 Monday
00 55 ☽ △ ☉
02 24 ☽ ⚼ ♃
02 29 ☽ ⚼ ☿
03 40 ☿ ☌ ♃
05 07 ☽ ± ♆
08 39 ☽ ∠ ♂
08 54 ☽ # ♆
16 45 ☽ ± ☿
17 32 ☽ ± ♃
19 04 ☽ Q ♇
19 58 ☽ ⚼ ♆

05 Tuesday
00 26 ☽ □ ♅
01 14 ☽ ⊻ ♄
02 14 ☽ ⊼ ☿
03 28 ☽ ⚼ ☉
03 32 ☽ ⊼ ♃
09 13 ☽ ☍ ♀
09 26 ♂ ± ♅
10 16 ☽ ⚹ ♂
11 02 ☽ ⊥ ♄
14 40 ☽ ⚹ ♇
19 03 ☽ # ♀
19 13 ☽ ± ☉
20 28 ☽ △ ♆

06 Wednesday
01 41 ☽ ∠ ♄
02 05 ☽ ∥ ♇
04 08 ☿ ∥ ♅
05 40 ☽ ⊼ ☉
05 42 ☽ Q ♅
06 27 ☽ # ♃
06 33 ☽ Q ♂
09 20 ☿ △ ♄
14 58 ☽ ∠ ♇
16 49 ☽ # ☿
17 40 ☽ # ♅

07 Thursday
01 08 ☽ ⚹ ♅
01 38 ☽ ☍ ☿
02 02 ☽ ⚹ ♄
03 59 ♀ ± ♄
04 21 ♂ ∥ ♄
05 04 ☽ ☍ ♃
05 32 ☽ ⊥ ♇
12 09 ☽ ⊼ ♀
12 54 ☽ □ ♂
13 40 ☽ # ☉
15 17 ☽ ⊻ ♇
21 12 ☽ □ ♆
21 34 ☽ Q ♄
22 23 ☽ ± ♀

08 Friday
01 30 ☽ ∠ ♅
07 42 ♃ ± ♇
10 03 ☽ ☍ ☉
13 52 ☽ ⚼ ♀
15 33 ☿ △ ♅
16 16 ☽ ⊥ ♅

09 Saturday
02 09 ☽ ⊼ ☿
02 12 ☽ ⊻ ♅
02 52 ☽ Q ♆
03 17 ☽ □ ♄
07 02 ☽ ∥ ☉
07 11 ☽ ⊼ ♃
12 09 ☽ ± ☿
16 10 ☽ △ ♀
16 31 ☽ △ ♂
16 51 ☽ ☌ ♇
17 28 ☽ ± ♃
22 17 ☿ St D
23 06 ☽ ⚹ ♆

10 Sunday
00 24 ♂ △ ♇
03 25 ☽ ⚼ ☿
05 13 ♀ △ ♇
06 46 ☽ ∥ ♅
09 07 ☽ ⚼ ♃
12 52 ☽ ∥ ☿
15 34 ☿ △ ♅
16 32 ☽ ⊼ ☉
17 11 ☽ ∥ ♃
19 24 ☽ ⚼ ♂

11 Monday
00 06 ♀ △ ♂
00 07 ☽ # ♇
01 06 ☽ ∠ ♆
03 54 ☽ ± ☉
05 35 ☽ ☌ ♅
05 43 ☽ △ ☿
06 39 ☽ ∥ ♀
06 56 ☽ △ ♄
11 56 ☽ △ ♃
12 01 ☽ ± ♂
17 09 ☽ ⊥ ♆
21 14 ☽ ⊻ ♇
21 19 ☽ ⚼ ☉
23 20 ☽ ⊼ ♂
23 38 ☽ □ ♀

12 Tuesday
01 17 ♆ St R
02 08 ☉ ± ♅
04 03 ☽ ⊻ ♆
08 10 ☽ ⊥ ♇
10 09 ☽ ⚼ ♄
21 58 ☽ ∥ ♆

13 Wednesday
00 54 ☽ ∠ ♇
02 59 ☽ ± ♄
03 24 ☽ △ ☉
11 16 ☽ Q ♀
12 43 ☽ ⊻ ♅
14 07 ☽ □ ☿
14 25 ☽ ⊼ ♄
20 45 ☽ □ ♃

14 Thursday
00 14 ☽ ⊥ ♅
01 12 ☿ △ ♄
01 12 ☽ # ♄
05 35 ☽ ⚹ ♇
10 42 ☽ ☍ ♂
12 02 ☽ ⚹ ♀
13 01 ☽ ☌ ♆
17 47 ☽ ∠ ♅
20 03 ☽ # ♂
22 23 ☿ ∥ ♃

15 Friday
02 28 ☽ Q ☿
03 30 ♀ ⚹ ♆
05 10 ☽ Q ♇
08 48 ☽ Q ♃
12 42 ☽ # ♂
19 15 ☽ □ ☉
19 55 ☽ ∠ ♀
23 37 ☽ ⚹ ♅

16 Saturday
01 42 ☽ ☍ ♄
03 36 ☽ ⚹ ☿
09 25 ☽ ⚹ ♃
11 51 ☽ # ♄
12 09 ♂ ☍ ♆
15 25 ☽ ⊥ ♀
17 19 ☽ □ ♇
23 51 ☽ Q ♅

17 Sunday
01 10 ☽ ⊻ ♆
01 50 ☽ ⊼ ♂
04 31 ☽ ⊻ ♀
10 05 ☿ ± ♇
11 04 ☽ Q ☉
11 08 ☉ ∠ ♀
11 40 ☽ ∠ ☿
13 23 ☽ ⊥ ♆
14 40 ☽ ± ♂
16 30 ☽ ∠ ♃
17 04 ☽ ∥ ♆
21 57 ☉ ⊼ ♅

18 Monday
07 05 ☽ ⊥ ☿
07 41 ☽ ∠ ♆
09 55 ☽ ⚼ ♂
11 10 ☽ ⊥ ♃
12 22 ☽ □ ♅
13 44 ☽ ⚹ ☉
14 48 ☽ ⊼ ♄
20 03 ☽ ⊻ ☿
23 34 ☽ ⊻ ♃

19 Tuesday
02 55 ☉ □ ♄
03 00 ☽ ± ♄
06 07 ☽ △ ♇
11 04 ♀ ⚼ ♄
14 02 ☽ ⚹ ♆
16 57 ☽ ∥ ♀
17 47 ☽ △ ♂
19 27 ☽ # ♇
21 10 ☽ ⚼ ♄
21 54 ☽ ☌ ♀
22 45 ☽ ∠ ☉

20 Wednesday
07 26 ☽ ∥ ☿
08 29 ☽ ∥ ♃
09 17 ♂ ⚼ ♅
12 05 ☽ ⚼ ♇
14 00 ☽ Q ♆
16 43 ☽ ∥ ♅
18 16 ☽ ⊥ ☉

21 Thursday
00 18 ☽ △ ♅
03 02 ☽ △ ♄
05 49 ☽ ± ♇
07 08 ☽ ⊻ ☉
11 22 ☉ Q ♂
12 10 ☽ ☌ ☿
12 27 ☽ ☌ ♃
17 22 ☿ ☌ ♃
17 30 ☽ ⊼ ♇

22 Friday
01 13 ☽ □ ♆
05 25 ☽ ⚼ ♅
07 42 ☽ □ ♂
12 01 ☿ ∥ ♃
13 09 ☽ ⊻ ♀
15 00 ☽ # ☉
15 56 ♀ # ♇
21 43 ☽ ∥ ☉
22 35 ☽ ± ♅

23 Saturday
01 26 ☽ ⊥ ♀
09 54 ☽ ⊼ ♅
12 51 ☽ □ ♄
19 26 ☽ Q ♂
19 38 ☽ ∠ ♀
21 24 ☽ ☌ ☉
22 41 ☽ ⊻ ♃

24 Sunday
01 57 ☽ ⊻ ☿
02 23 ☽ ☍ ♇
07 23 ☿ ⊼ ♇
09 52 ☽ △ ♆
10 02 ☽ ⊥ ♃
14 00 ☽ ⊥ ☿
17 30 ☽ # ♅
18 40 ☉ ⊻ ♃
18 43 ☽ ⚹ ♂
21 42 ☽ # ☿
22 22 ☽ Q ♄
23 30 ☽ # ♃

25 Monday
01 25 ☽ ⚹ ♀
02 51 ☽ ∠ ♃
07 55 ☽ ∠ ☿
11 16 ☽ # ♀
13 19 ☽ ⚼ ♆
13 46 ☽ ∥ ♇
17 05 ☽ ☍ ♅
20 17 ☽ ⚹ ♄
23 17 ☽ ∠ ♂

26 Tuesday
00 49 ☽ Q ♀
05 35 ☽ ± ♆
06 30 ☽ ⚹ ♃
08 44 ☽ ⊻ ☉
09 01 ☽ ⊼ ♇
12 31 ☉ ☍ ♇
13 25 ☽ ⚹ ☿
16 11 ☽ ⊥ ♂
16 20 ☽ ⊼ ♆
19 44 ☽ ± ♇
20 19 ☽ ⊥ ☉
23 17 ☽ ∠ ♄

27 Wednesday
03 22 ☽ ⊻ ♂
04 18 ☽ Q ♃
10 27 ☽ # ♆
11 20 ☽ □ ♀
11 42 ☽ ⚼ ♇
12 44 ☽ Q ☿
13 34 ☽ ∠ ☉
15 19 ☽ ⊥ ♄
16 25 ♀ ⚼ ♇
19 41 ☿ ∥ ♅
22 29 ☽ ⊼ ♅
23 24 ☿ □ ♆

28 Thursday
01 56 ☽ ⊻ ♄
09 00 ☽ ± ♅
12 34 ☽ □ ♃
13 55 ☽ ∥ ♄
14 03 ☽ △ ♇
18 00 ☽ ⚹ ☉
20 44 ♀ Q ♆
21 15 ☽ ☍ ♆
23 19 ☽ □ ☿

29 Friday
00 41 ☽ ⚼ ♅
10 22 ☽ ☌ ♂
14 21 ☿ ⚼ ♅
16 28 ☽ Q ☉
19 38 ☽ △ ♀

30 Saturday
02 35 ☽ △ ♅
06 19 ☽ ☌ ♄
10 39 ☽ ∥ ♄
14 48 ☉ △ ♆
17 20 ☽ △ ♃
17 51 ☽ □ ♇
23 16 ☽ ⚼ ♀

31 Sunday
00 58 ☽ ⊼ ♆
01 44 ☽ □ ☉
08 01 ☽ △ ☿
11 10 ☽ ± ♆
13 28 ☽ # ♆
15 42 ☽ ± ♀
16 03 ☽ ⊻ ♂
19 17 ☽ ⚼ ♃
22 28 ♃ ⊼ ♇

All ephemeris data is given at 12.00 UT and the Moon's longitude is additionally given for 24.00 UT

APRIL 1974

LONGITUDES

Date	Sidereal time h m s	Sun ☉ ° ' "	Moon ☽ ° ' "	Moon ☽ 24.00 ° ' "	Mercury ☿ ° '	Venus ♀ ° '	Mars ♂ ° '	Jupiter ♃ ° '	Saturn ♄ ° '	Uranus ♅ ° '	Neptune ♆ ° '	Pluto ♇ ° '
01	00 37 51	11 ♈ 23 00	00 ♌ 11 34	07 ♌ 19 44	15 ♓ 11	25 ♒ 00	18 ♊ 52	05 ♓ 27	28 ♊ 45	26 ♎ 22	09 ♐ 30	05 ♎ 18
02	00 41 47	12 22 11	14 ♌ 29 13	21 ♌ 39 35	16 31	25 58	19 28	05 40	28 49	26 R 20	09 R 30	05 R 17
03	00 45 44	13 21 19	28 ♌ 50 22	06 ♍ 01 00	17 52	26 56	20 03	05 53	28 53	26 17	09 29	05 15
04	00 49 40	14 20 25	13 ♍ 10 50	20 ♍ 19 11	19 16	27 55	20 38	06 05	28 56	26 15	09 28	05 14
05	00 53 37	15 19 29	27 ♍ 25 20	04 ♎ 28 37	20 41	28 55	21 13	06 18	29 00	26 12	09 27	05 12
06	00 57 34	16 18 30	11 ♎ 28 21	18 ♎ 23 57	22 08	29 ♒ 54	21 48	06 31	29 04	26 10	09 27	05 10
07	01 01 30	17 17 30	25 ♎ 14 54	02 ♏ 00 48	23 36	00 ♓ 54	22 24	06 44	29 08	26 07	09 26	05 09
08	01 05 27	18 16 28	08 ♏ 41 24	15 ♏ 16 32	25 07	01 55	22 59	06 56	29 12	26 05	09 25	05 07
09	01 09 23	19 15 23	21 ♏ 46 11	28 ♏ 10 27	26 38	02 55	23 34	07 09	29 16	26 02	09 24	05 05
10	01 13 20	20 14 17	04 ♐ 29 33	10 ♐ 43 47	28 12	03 57	24 10	07 21	29 21	26 00	09 23	05 04
11	01 17 16	21 13 10	16 ♐ 53 35	22 ♐ 59 24	29 ♓ 47	04 58	24 45	07 33	29 25	25 57	09 22	05 02
12	01 21 13	22 12 00	29 ♐ 01 48	05 ♑ 01 21	01 ♈ 23	06 00	25 21	07 46	29 29	25 54	09 21	05 01
13	01 25 09	23 10 49	10 ♑ 58 43	16 ♑ 54 31	03 01	07 02	25 56	07 58	29 34	25 52	09 20	04 59
14	01 29 06	24 09 36	22 ♑ 49 26	28 ♑ 44 08	04 41	08 04	26 32	08 10	29 39	25 49	09 19	04 58
15	01 33 03	25 08 21	04 ♒ 39 16	10 ♒ 35 30	06 22	09 07	27 07	08 22	29 43	25 47	09 18	04 56
16	01 36 59	26 07 04	16 ♒ 33 26	22 ♒ 33 40	08 05	10 09	27 43	08 34	29 48	25 44	09 17	04 54
17	01 40 56	27 05 46	28 ♒ 36 45	04 ♓ 43 09	09 50	11 12	28 18	08 46	29 53	25 42	09 16	04 53
18	01 44 52	28 04 26	10 ♓ 53 17	17 ♓ 07 32	11 36	12 16	28 54	08 58	29 ♊ 58	25 39	09 15	04 51
19	01 48 49	29 ♈ 03 04	23 ♓ 26 08	29 ♓ 49 16	13 23	13 19	29 ♊ 30	09 09	00 ♋ 03	25 36	09 14	04 50
20	01 52 45	00 ♉ 01 40	06 ♈ 17 03	12 ♈ 49 27	15 13	14 23	00 ♋ 06	09 21	00 08	25 34	09 13	04 48
21	01 56 42	01 00 15	19 ♈ 26 24	26 ♈ 07 42	17 04	15 27	00 41	09 32	00 13	25 31	09 12	04 47
22	02 00 38	01 58 48	02 ♉ 53 07	09 ♉ 42 18	18 56	16 31	01 17	09 44	00 18	25 29	09 10	04 46
23	02 04 35	02 57 19	16 ♉ 34 54	23 ♉ 30 30	20 50	17 36	01 53	09 55	00 24	25 26	09 09	04 44
24	02 08 32	03 55 48	00 ♊ 28 41	07 ♊ 29 00	22 46	18 41	02 29	10 07	00 29	25 24	09 08	04 43
25	02 12 28	04 54 15	14 ♊ 31 05	21 ♊ 34 31	24 43	19 45	03 04	10 18	00 35	25 21	09 07	04 41
26	02 16 25	05 52 40	28 ♊ 38 57	05 ♋ 44 03	26 42	20 50	03 40	10 29	00 40	25 19	09 05	04 40
27	02 20 21	06 51 03	12 ♋ 49 34	19 ♋ 55 13	28 ♈ 43	21 56	04 16	10 40	00 46	25 16	09 04	04 39
28	02 24 18	07 49 24	27 ♋ 00 48	04 ♌ 06 06	00 ♉ 45	23 01	04 52	10 51	00 51	25 14	09 03	04 37
29	02 28 14	08 47 42	11 ♌ 10 55	18 ♌ 15 02	02 49	24 07	05 28	11 01	00 57	25 11	09 02	04 36
30	02 32 11	09 ♉ 45 59	25 ♌ 18 16	02 ♍ 20 23	04 ♉ 53	25 ♓ 12	06 ♋ 04	11 ♓ 12	01 ♋ 03	25 ♎ 09	09 ♐ 00	04 ♎ 35

Date	Moon True ☊ ° '	Moon Mean ☊ ° '	Moon ☽ Latitude ° '
01	22 ♐ 15	23 ♐ 08	03 S 07
02	22 R 11	23 04	04 00
03	22 05	23 01	04 38
04	21 57	22 58	04 59
05	21 47	22 55	05 01
06	21 36	22 52	04 45
07	21 26	22 49	04 12
08	21 17	22 45	03 26
09	21 11	22 42	02 30
10	21 07	22 39	01 28
11	21 06	22 36	00 S 23
12	21 D 06	22 33	00 N 42
13	21 07	22 30	01 45
14	21 08	22 26	02 42
15	21 R 07	22 23	03 32
16	21 05	22 20	04 13
17	21 00	22 17	04 44
18	20 54	22 14	05 02
19	20 46	22 10	05 06
20	20 37	22 07	04 55
21	20 28	22 04	04 29
22	20 20	22 01	03 47
23	20 14	21 58	02 51
24	20 10	21 55	01 44
25	20 09	21 51	00 N 30
26	20 D 09	21 48	00 S 46
27	20 10	21 45	02 00
28	20 11	21 42	03 06
29	20 R 11	21 39	04 01
30	20 ♐ 10	21 ♐ 36	04 S 42

DECLINATIONS

Date	Sun ☉ ° '	Moon ☽ ° '	Mercury ☿ ° '	Venus ♀ ° '	Mars ♂ ° '	Jupiter ♃ ° '	Saturn ♄ ° '	Uranus ♅ ° '	Neptune ♆ ° '	Pluto ♇ ° '
01	04 N 30	17 N 04	07 S 51	11 S 47	24 N 33	10 S 18	22 N 40	09 S 37	20 S 17	13 N 38
02	04 53	12 39	07 24	11 34	24 35	10 14	22 40	09 36	20 17	13 39
03	05 16	07 32	06 56	11 20	24 38	10 09	22 40	09 35	20 17	13 39
04	05 39	02 N 00	06 26	11 06	24 41	10 04	22 40	09 34	20 16	13 40
05	06 02	03 S 35	05 55	10 52	24 43	10 00	22 40	09 33	20 16	13 41
06	06 25	08 54	05 22	10 37	24 45	09 55	22 41	09 32	20 16	13 41
07	06 47	13 41	04 49	10 22	24 47	09 51	22 41	09 31	20 16	13 42
08	07 10	17 39	04 14	10 06	24 49	09 46	22 41	09 30	20 16	13 42
09	07 32	20 37	03 38	09 50	24 51	09 42	22 41	09 29	20 16	13 43
10	07 55	22 29	03 01	09 33	24 52	09 37	22 41	09 28	20 15	13 43
11	08 17	23 10	02 23	09 16	24 54	09 33	22 42	09 27	20 15	13 44
12	08 39	22 44	01 44	08 59	24 55	09 28	22 42	09 27	20 15	13 44
13	09 01	21 15	01 04	08 41	24 56	09 24	22 42	09 26	20 15	13 45
14	09 22	18 51	00 S 23	08 23	24 57	09 20	22 42	09 25	20 15	13 45
15	09 44	15 40	00 N 19	08 05	24 57	09 15	22 42	09 24	20 15	13 46
16	10 05	11 51	01 03	07 46	24 58	09 11	22 42	09 23	20 14	13 46
17	10 26	07 31	01 47	07 27	24 58	09 07	22 43	09 22	20 14	13 47
18	10 47	02 S 50	02 32	07 08	24 59	09 02	22 43	09 21	20 14	13 47
19	11 08	02 N 05	03 18	06 48	24 59	08 58	22 43	09 20	20 14	13 47
20	11 29	07 01	04 04	06 28	24 59	08 54	22 43	09 19	20 14	13 48
21	11 49	11 45	04 52	06 08	24 58	08 50	22 43	09 18	20 13	13 48
22	12 10	16 01	05 40	05 47	24 58	08 46	22 43	09 17	20 13	13 49
23	12 30	19 32	06 29	05 26	24 57	08 42	22 44	09 16	20 13	13 49
24	12 50	21 57	07 18	05 05	24 56	08 37	22 44	09 15	20 13	13 49
25	13 09	23 03	08 08	04 43	24 56	08 33	22 44	09 15	20 12	13 50
26	13 29	22 40	08 59	04 22	24 54	08 29	22 44	09 14	20 12	13 50
27	13 48	20 50	09 49	04 00	24 53	08 25	22 44	09 13	20 12	13 50
28	14 07	17 43	10 40	03 38	24 52	08 21	22 44	09 12	20 12	13 50
29	14 26	13 33	11 31	03 15	24 50	08 18	22 44	09 11	20 12	13 51
30	14 N 45	08 N 40	12 N 22	02 S 53	24 N 48	08 S 14	22 N 44	09 S 10	20 S 11	13 N 51

ZODIAC SIGN ENTRIES

Date	h	m	Planets
01	11	41	☽ ♌
03	13	56	☽ ♍
05	16	22	☽ ♎
06	14	17	♀ ♓
07	20	25	☽ ♏
10	03	27	☽ ♐
11	15	20	☿ ♈
12	13	56	☽ ♑
15	02	34	☽ ♒
17	14	44	☽ ♓
18	22	34	♄ ♋
20	00	20	☽ ♈
20	08	18	♂ ♋
20	11	19	☉ ♉
22	06	53	☽ ♉
24	11	11	☽ ♊
26	14	17	☽ ♋
28	03	10	☿ ♉
28	17	03	☽ ♌
30	20	00	☽ ♍

LATITUDES

Date	Mercury ☿ ° '	Venus ♀ ° '	Mars ♂ ° '	Jupiter ♃ ° '	Saturn ♄ ° '	Uranus ♅ ° '	Neptune ♆ ° '	Pluto ♇ ° '
01	02 S 11	01 N 30	01 N 34	00 S 51	00 S 47	00 N 37	01 N 37	17 N 10
04	02 22	01 10	01 34	00 51	00 46	00 37	01 37	17 10
07	02 29	00 51	01 34	00 52	00 46	00 37	01 37	17 10
10	02 31	00 33	01 33	00 52	00 45	00 37	01 37	17 10
13	02 28	00 N 16	01 33	00 53	00 45	00 37	01 37	17 09
16	02 21	00 S 01	01 33	00 53	00 44	00 37	01 38	17 09
19	02 10	00 16	01 32	00 54	00 44	00 37	01 38	17 08
22	01 54	00 30	01 32	00 54	00 43	00 36	01 38	17 08
25	01 33	00 44	01 31	00 55	00 43	00 36	01 38	17 07
28	01 08	00 56	01 31	00 56	00 42	00 36	01 38	17 06
31	00 S 40	01 S 07	01 N 30	00 S 56	00 S 42	00 N 36	01 N 38	17 N 05

DATA

Julian Date	2442139
Delta T	+45 seconds
Ayanamsa	23° 30' 07"
Synetic vernal point	05° ♓ 36' 52"
True obliquity of ecliptic	23° 26' 33"

MOON'S PHASES, APSIDES AND POSITIONS ☽

Date	h	m	Phase	Longitude ° '	Eclipse Indicator
06	21	00	○	16 ♎ 41	
14	14	57	☾	24 ♑ 17	
22	10	16	●	01 ♉ 55	
29	07	39	☽	08 ♌ 37	

Day	h	m		
02	16	11	Perigee	
14	21	42	Apogee	
27	15	48	Perigee	
04	20	34	0S	
11	14	29	Max dec	23° S 11'
19	01	54	0N	
25	17	56	Max dec	23° N 06'

LONGITUDES

Date	Chiron ⚷ ° '	Ceres ⚳ ° '	Pallas ⚴ ° '	Juno ⚵ ° '	Vesta ⚶ ° '	Black Moon Lilith ⚸ ° '
01	19 ♈ 49	24 ♒ 13	02 ♒ 59	21 ♒ 26	10 ♎ 38	25 ♑ 23
11	20 ♈ 24	27 ♒ 34	05 ♒ 26	25 ♒ 07	08 ♎ 08	26 ♑ 30
21	21 ♈ 00	00 ♓ 45	07 ♒ 32	28 ♒ 41	05 ♎ 57	27 ♑ 37
31	21 ♈ 34	03 ♓ 45	09 ♒ 13	02 ♓ 05	04 ♎ 23	28 ♑ 44

All ephemeris data is given at 12.00 UT and the Moon's longitude is additionally given for 24.00 UT

ASPECTARIAN

h m Aspects

01 Monday
00 24 ☽ Q ♇
02 25 ☽ bQ ♆
02 36 ☽ ⊼ ♀
02 39 ☽ ⊥ ♂
05 34 ☽ □ ♅
09 34 ☽ ⊻ ♄
10 43 ☽ ± ♃
11 58 ☽ bQ ☿
13 57 ☉ ⊥ ♃
18 28 ☽ ∠ ♂
19 43 ☽ ⊥ ♄
20 35 ☽ ⚹ ♇
20 58 ☽ ⊼ ♃

02 Tuesday
03 38 ☽ △ ♆
04 39 ☽ ± ☿
07 00 ☽ ∥ ♇
08 11 ☽ △ ☉
10 52 ☽ ∠ ♄
11 44 ☽ Q ♅
15 44 ☽ ⊼ ☿
17 36 ☽ # ♀
20 35 ♀ △ ♅
20 41 ☽ ⚹ ♂
21 41 ☽ ∠ ♇
23 52 ☽ # ♃

03 Wednesday
02 40 ☽ # ♅
07 45 ☽ ⚹ ♅
08 36 ☽ ☍ ♀
11 08 ☽ bQ ☉
12 04 ☽ ⚹ ♄
12 41 ☽ ⊥ ♇
14 55 ☽ # ☿
17 35 ☽ Q ♂
21 16 ☽ ∥ ☉
22 42 ☽ ⊻ ♇
23 57 ☽ ☍ ♃

04 Thursday
03 17 ☽ ± ☉
05 47 ☽ □ ♆
08 14 ☽ Q ♄
08 46 ☽ ∠ ♅
14 05 ☽ ⊼ ☉
23 21 ☽ ☍ ☿
23 50 ☽ ⊥ ♅

05 Friday
01 04 ☽ □ ♂
04 12 ☿ ± ♅
08 43 ☉ # ☿
09 57 ☽ ⊻ ♅
12 03 ☽ Q ♆
14 23 ♀ △ ♄
14 42 ☽ □ ♄
14 43 ☽ ⊼ ♀
18 46 ♀ ± ♇
21 21 ☽ ∥ ☿
23 40 ☽ # ☉

06 Saturday
01 12 ☽ ☌ ♇
01 43 ☽ ± ♀
03 05 ☿ □ ♂
03 21 ☽ ⊼ ♃
08 31 ☽ ⚹ ♆
13 50 ☽ ± ♃
14 59 ☽ ∥ ♅
16 46 ☽ ∥ ♃
18 24 ☽ bQ ♀
19 49 ☽ ∥ ♀
21 00 ☽ ☍ ☉

07 Sunday
05 43 ☽ bQ ♃
06 46 ☽ △ ♂
07 53 ☉ Q ♄
08 45 ☽ ⊼ ☿
10 34 ☽ ∠ ♆
12 05 ☽ # ♇
13 32 ☽ ☌ ♅
18 55 ☽ △ ♄
20 40 ☽ ± ☿
22 50 ☽ △ ♀

08 Monday
02 31 ☽ ⊥ ♆
05 34 ☽ ⊻ ♇
08 47 ☽ △ ♃
10 40 ☽ bQ ♂
13 19 ☽ ⊻ ♆
14 54 ☽ bQ ☿
16 24 ☽ ⊥ ♇
22 05 ☽ bQ ♄

09 Tuesday
02 50 ☿ ⊼ ♅
03 52 ☽ ± ♂
06 57 ☽ ⊼ ☉
08 32 ☽ ∥ ♆
08 53 ☽ ∠ ♇
14 49 ☽ ± ♄
15 32 ☽ ⊼ ♂
19 03 ☽ ± ☉
19 57 ☽ ⊻ ♅
22 22 ☽ △ ☿

10 Wednesday
02 09 ☽ ⊼ ♄
04 11 ♀ ∥ ♃
07 15 ☽ ⊥ ♅
10 51 ☽ □ ♀
13 05 ☽ ⚹ ♇
13 33 ☽ bQ ☉
16 24 ☽ # ♄
17 34 ☽ □ ♃
19 28 ♀ ∥ ♅
21 23 ☽ ☌ ♆

11 Thursday
00 28 ☽ ∠ ♅
06 17 ☿ □ ♄
12 17 ☽ Q ♇
13 38 ♀ ⊼ ♇
21 15 ☽ △ ☉

12 Friday
01 03 ☽ Q ♀
04 17 ☽ ☍ ♂
05 23 ☽ Q ♃
05 48 ☽ ⚹ ♅
12 51 ☽ # ♄
12 55 ☽ ☍ ♄
17 27 ☽ □ ☿
23 57 ☽ □ ♇

13 Saturday
00 25 ☉ # ♀
00 56 ♃ ∥ ♅
03 17 ☽ ⚹ ♀
05 18 ☉ ∠ ♃
05 44 ☽ Q ♅
05 49 ☽ ⚹ ♃
08 42 ☽ ⊻ ♆
09 16 ♂ △ ♅
20 48 ☽ ⊥ ♆
23 07 ☽ ∥ ♆

14 Sunday
09 35 ☉ # ♃
11 40 ☽ Q ☿
12 32 ☽ ∠ ♀
12 42 ☽ ∠ ♃
14 35 ☉ # ♅
14 51 ♀ ☌ ♃
14 57 ☽ □ ☉
15 02 ☽ ∠ ♆
15 52 ☿ ☍ ♇
15 54 ☉ bQ ♆
18 04 ☽ □ ♅
19 55 ☽ ⊼ ♂

15 Monday
01 56 ☽ ⊼ ♄
07 17 ☽ ⊥ ♃
08 34 ☽ ⊥ ♀
08 44 ☽ ± ♂
12 34 ☽ △ ♇
14 10 ☽ ± ♄
16 04 ☽ ⚹ ☿
16 26 ♀ □ ♆
19 38 ☽ ⊻ ♃
21 23 ☽ ⚹ ♆
21 53 ☽ ⊻ ♀

16 Tuesday
00 25 ☽ # ♇
03 02 ☉ ☍ ♅
03 52 ☽ bQ ♂
06 40 ☽ Q ☉
08 27 ☽ bQ ♄
18 42 ☽ bQ ♇
19 30 ☿ ⊻ ♃
21 18 ☽ # ☉
21 27 ☽ Q ♆

17 Wednesday
00 45 ♀ bQ ♅
02 02 ☽ ∥ ♅
03 14 ☽ ∠ ☿
03 19 ☽ ∥ ♃
04 25 ☿ △ ♆
06 15 ☽ △ ♅
08 44 ☽ ⚹ ☉
11 22 ☽ △ ♂
12 23 ☽ ∥ ♀
12 32 ☽ ± ♇
14 31 ☽ △ ♄
23 59 ☽ ⊥ ☿

18 Thursday
00 18 ☽ ⊼ ♇
08 12 ☽ ☌ ♃
08 50 ☽ □ ♆
11 33 ☽ bQ ♅
13 16 ☽ # ☿
13 36 ☽ ⊻ ☿
14 54 ☽ ☌ ♀
16 35 ☽ ∠ ☉

19 Friday
04 46 ☽ ± ♅
09 50 ☿ ⊻ ♀
11 13 ☽ ⊥ ☉
16 05 ☽ ⊼ ♅
18 58 ☽ ∥ ☿
20 40 ♃ □ ♆
23 26 ☽ ⊻ ☉
23 57 ☽ □ ♂

20 Saturday
00 30 ☽ □ ♄
09 17 ☽ ☍ ♇
09 30 ☽ # ♀
13 53 ♂ ☌ ♄
14 01 ☿ ⊥ ♃
14 48 ☉ ⚹ ♄
16 01 ☉ ⚹ ♂
17 23 ☽ △ ♆
17 44 ☽ ⊻ ♃
21 14 ☽ # ♃
23 27 ☽ # ♅

21 Sunday
04 09 ☽ ⊻ ♀
04 50 ☽ ⊥ ♃
07 00 ☽ ☌ ☿
09 47 ☽ Q ♄
10 35 ☽ Q ♂
12 26 ☽ ∥ ☉
15 56 ☽ ⊥ ♀
20 32 ☽ bQ ♆
21 18 ☽ ∠ ♃
22 53 ☽ ☍ ♅
23 08 ☽ ∥ ♇

22 Monday
03 35 ☿ Q ♄
07 24 ☽ ⚹ ♄
09 02 ☽ ⚹ ♂
09 23 ☽ ∠ ♀
10 16 ☽ ☌ ☉
12 31 ☽ ± ♆
14 26 ☿ # ♀
15 18 ☽ ⊼ ♇
18 26 ☿ Q ♂
23 03 ☽ ⊼ ♆

23 Tuesday
00 13 ☽ ⚹ ♃
01 49 ☽ ± ♇
09 55 ☽ ∠ ♄
12 32 ☽ ∠ ♂
13 55 ☽ ⚹ ♀
16 47 ☉ ± ♆
17 28 ☽ bQ ♇
17 44 ☽ # ♆
20 34 ☽ ⊻ ☿
21 23 ☽ Q ♃

24 Wednesday
01 37 ☽ ⊥ ♄
03 17 ☽ ⊼ ♅
04 49 ☽ ⊥ ♂
08 35 ☽ ⊥ ☿
12 01 ☽ ⊻ ♄
12 22 ☽ Q ♀
13 34 ☽ ± ♅
15 35 ☽ ⊻ ♂
18 22 ☽ ⊻ ☉
19 15 ☽ △ ♇

25 Thursday
01 08 ☽ ∥ ♄
02 30 ☽ ∠ ☿
02 48 ☽ ☍ ♆
03 22 ♀ ± ♅
04 36 ☿ bQ ♆
04 42 ☽ □ ♃
04 55 ☽ bQ ♅
05 23 ☽ ⊥ ☉
06 50 ☉ ⊼ ♇
18 03 ♃ bQ ♅
19 28 ☿ ☍ ♅
19 39 ☿ ∠ ♃
21 39 ☽ □ ♀
21 50 ☽ ∠ ☉
23 09 ☿ # ♃

26 Friday
06 21 ☽ △ ♅
08 10 ☽ ⚹ ☿
10 37 ☽ ∥ ♄
15 27 ☽ ☌ ♄
15 30 ☿ ⊥ ♀
19 02 ☿ # ♅
19 49 ☉ ∠ ♀
20 53 ☽ ☌ ♂
22 11 ☽ □ ♇

27 Saturday
01 09 ☽ ⚹ ☉
05 39 ☽ ⊼ ♆
07 50 ☽ Q ☿
08 18 ☽ △ ♃
14 29 ☉ ∥ ♇
15 47 ☽ ± ♆
17 43 ☽ # ♆
22 56 ☽ Q ☉

28 Sunday
02 29 ♂ □ ♇
04 35 ☽ Q ♇
04 41 ☽ △ ♀
06 59 ☽ bQ ♆
08 59 ☽ □ ♅
10 00 ☽ bQ ♃
13 17 ☿ ⚹ ♄
18 33 ☽ ⊻ ♄
19 24 ☽ □ ☿

29 Monday
00 52 ☽ ⚹ ♇
01 26 ☽ ± ♃
01 53 ☽ ⊻ ♂
04 47 ☽ ⊥ ♄
07 36 ☽ ∥ ☉
07 39 ☽ □ ☉
08 12 ☽ bQ ♀
08 21 ☽ △ ♆
10 27 ☽ ∥ ♇
11 44 ☽ ⊼ ♃
12 30 ☽ ⊥ ♂
14 29 ☿ ± ♆
15 23 ☽ Q ♅
17 34 ☉ ⊼ ♆
20 09 ☽ ∠ ♄
20 46 ☽ ∥ ☿

30 Tuesday
00 45 ☽ ± ♀
02 16 ☽ ∠ ♇
04 28 ☽ ∠ ♂
08 27 ☿ ⊼ ♇
09 37 ☽ # ♅
10 42 ♀ ⊼ ♅
11 44 ☽ ⚹ ♅
11 49 ☽ ⊼ ♀
14 02 ☽ # ♃
17 34 ☽ ⊥ ♇
21 52 ☽ ⚹ ♄

MAY 1974

LONGITUDES

Date	Sidereal time h m s	Sun ☉ ° ' "	Moon ☽ ° ' "	Moon ☽ 24.00 ° ' "	Mercury ☿ ° '	Venus ♀ ° '	Mars ♂ ° '	Jupiter ♃ ° '	Saturn ♄ ° '	Uranus ♅ ° '	Neptune ♆ ° '	Pluto ♇ ° '
01	02 36 07	10 ♉ 44 13	09 ♍ 21 06	16 ♍ 20 11	07 ♉ 00	26 ♓ 18	06 ♋ 40	11 ♓ 23	01 ♋ 09	25 ♎ 06	08 ♐ 59	04 ♎ 33
02	02 40 04	11 42 25	23 ♍ 17 19	00 ♎ 12 11	09 07	27 24	07 16	11 33	01 15	25 R 04	08 R 57	04 R 32
03	02 44 01	12 40 36	07 ♎ 04 28	13 ♎ 53 52	11 15	28 31	07 52	11 43	01 21	25 01	08 56	04 31
04	02 47 57	13 38 44	20 ♎ 40 03	27 ♎ 22 46	13 24	29 ♓ 37	08 28	11 54	01 27	24 59	08 55	04 30
05	02 51 54	14 36 51	04 ♏ 01 44	10 ♏ 36 47	15 34	00 ♈ 43	09 04	12 04	01 33	24 57	08 53	04 29
06	02 55 50	15 34 56	17 ♏ 07 46	23 ♏ 34 36	17 44	01 50	09 40	12 14	01 39	24 54	08 52	04 27
07	02 59 47	16 32 59	29 ♏ 57 17	06 ♐ 15 52	19 54	02 57	10 16	12 24	01 45	24 52	08 50	04 26
08	03 03 43	17 31 01	12 ♐ 30 30	18 ♐ 41 22	22 04	04 04	10 52	12 34	01 52	24 50	08 49	04 25
09	03 07 40	18 29 01	24 ♐ 48 46	00 ♑ 53 01	24 14	05 11	11 28	12 43	01 58	24 47	08 47	04 24
10	03 11 36	19 27 00	06 ♑ 54 30	12 ♑ 53 40	26 23	06 18	12 04	12 53	02 04	24 45	08 46	04 23
11	03 15 33	20 24 58	18 ♑ 51 01	24 ♑ 47 04	28 ♉ 31	07 25	12 41	13 02	02 11	24 43	08 44	04 22
12	03 19 30	21 22 54	00 ♒ 42 23	06 ♒ 37 32	00 ♊ 37	08 33	13 17	13 12	02 17	24 41	08 43	04 21
13	03 23 26	22 20 48	12 ♒ 33 08	18 ♒ 29 47	02 42	09 40	13 53	13 21	02 24	24 38	08 41	04 20
14	03 27 23	23 18 42	24 ♒ 28 06	00 ♓ 28 41	04 46	10 48	14 29	13 30	02 31	24 36	08 40	04 19
15	03 31 19	24 16 34	06 ♓ 32 08	12 ♓ 39 00	06 47	11 56	15 05	13 39	02 37	24 34	08 38	04 18
16	03 35 16	25 14 25	18 ♓ 49 49	25 ♓ 05 03	08 46	13 04	15 42	13 48	02 44	24 32	08 37	04 17
17	03 39 12	26 12 15	01 ♈ 25 07	07 ♈ 50 22	10 42	14 12	16 18	13 56	02 51	24 30	08 35	04 16
18	03 43 09	27 10 03	14 ♈ 21 01	20 ♈ 57 15	12 36	15 20	16 54	14 05	02 58	24 28	08 33	04 15
19	03 47 05	28 07 51	27 ♈ 39 06	04 ♉ 26 31	14 28	16 28	17 31	14 13	03 04	24 26	08 32	04 15
20	03 51 02	29 ♉ 05 37	11 ♉ 19 17	18 ♉ 17 08	16 17	17 36	18 07	14 22	03 11	24 24	08 30	04 14
21	03 54 59	00 ♊ 03 22	25 ♉ 19 38	02 ♊ 26 17	18 02	18 45	18 43	14 30	03 18	24 22	08 29	04 13
22	03 58 55	01 01 06	09 ♊ 36 29	16 ♊ 49 35	19 44	19 53	19 20	14 38	03 25	24 20	08 27	04 12
23	04 02 52	01 58 48	24 ♊ 04 52	01 ♋ 21 36	21 23	21 02	19 56	14 46	03 32	24 18	08 25	04 12
24	04 06 48	02 56 29	08 ♋ 39 05	15 ♋ 56 35	23 00	22 11	20 32	14 54	03 39	24 16	08 24	04 11
25	04 10 45	03 54 09	23 ♋ 13 28	00 ♌ 29 07	24 33	23 19	21 09	15 01	03 47	24 15	08 22	04 10
26	04 14 41	04 51 47	07 ♌ 42 59	14 ♌ 54 37	26 03	24 28	21 45	15 09	03 54	24 13	08 21	04 10
27	04 18 38	05 49 23	22 ♌ 03 38	29 ♌ 09 41	27 29	25 37	22 22	15 16	04 01	24 11	08 19	04 09
28	04 22 34	06 46 58	06 ♍ 12 33	13 ♍ 12 01	28 ♊ 53	26 46	22 58	15 23	04 08	24 09	08 17	04 08
29	04 26 31	07 44 32	20 ♍ 07 58	27 ♍ 00 19	00 ♋ 13	27 55	23 35	15 30	04 15	24 08	08 16	04 08
30	04 30 28	08 42 04	03 ♎ 49 00	10 ♎ 34 00	01 30	29 ♈ 04	24 11	15 37	04 23	24 06	08 14	04 07
31	04 34 24	09 ♊ 39 34	17 ♎ 15 21	23 ♎ 53 02	02 ♋ 43	00 ♉ 14	24 ♋ 48	15 ♓ 44	04 ♋ 30	24 ♎ 05	08 ♐ 12	04 ♎ 07

Date	Moon True ☊ ° '	Moon Mean ☊ ° '	Moon ☽ Latitude ° '
01	20 ♐ 07	21 ♐ 32	05 S 05
02	20 R 03	21 29	05 10
03	19 58	21 26	04 57
04	19 52	21 23	04 27
05	19 46	21 20	03 43
06	19 41	21 16	02 48
07	19 38	21 13	01 45
08	19 37	21 10	00 S 39
09	19 D 37	21 07	00 N 28
10	19 38	21 04	01 33
11	19 39	21 01	02 34
12	19 41	20 57	03 27
13	19 42	20 54	04 11
14	19 R 43	20 51	04 45
15	19 42	20 48	05 06
16	19 40	20 45	05 14
17	19 38	20 42	05 08
18	19 34	20 38	04 46
19	19 31	20 35	04 08
20	19 28	20 32	03 15
21	19 26	20 29	02 09
22	19 25	20 26	00 N 54
23	19 D 25	20 22	00 S 26
24	19 26	20 19	01 44
25	19 27	20 16	02 56
26	19 28	20 13	03 56
27	19 29	20 10	04 41
28	19 29	20 07	05 08
29	19 R 29	20 03	05 17
30	19 28	20 00	05 07
31	19 ♐ 27	19 ♐ 57	04 S 40

DECLINATIONS

Date	Sun ☉	Moon ☽	Mercury ☿	Venus ♀	Mars ♂	Jupiter ♃	Saturn ♄	Uranus ♅	Neptune ♆	Pluto ♇
01	15 N 03	03 N 21	13 N 13	02 S 30	24 N 46	08 S 10	22 N 44	09 S 09	20 S 11	13 N 51
02	15 21	02 S 05	14 04	02 07	24 44	08 06	22 45	09 08	20 11	13 51
03	15 39	07 21	14 53	01 44	24 42	08 02	22 45	09 07	20 11	13 52
04	15 56	12 11	15 43	01 20	24 40	07 59	22 45	09 07	20 10	13 52
05	16 13	16 21	16 31	00 57	24 37	07 55	22 45	09 06	20 10	13 52
06	16 30	19 38	17 18	00 33	24 34	07 51	22 45	09 05	20 10	13 52
07	16 47	21 52	18 04	00 S 09	24 31	07 48	22 45	09 04	20 10	13 52
08	17 04	22 56	18 49	00 N 15	24 28	07 44	22 45	09 03	20 09	13 52
09	17 20	22 52	19 32	00 39	24 25	07 41	22 45	09 02	20 09	13 52
10	17 36	21 42	20 12	01 03	24 21	07 37	22 45	09 02	20 09	13 52
11	17 51	19 35	20 51	01 27	24 18	07 34	22 45	09 01	20 09	13 53
12	18 07	16 38	21 28	01 52	24 14	07 30	22 45	09 00	20 08	13 53
13	18 22	13 02	22 02	02 16	24 10	07 27	22 45	08 59	20 08	13 53
14	18 36	08 53	22 34	02 40	24 06	07 24	22 45	08 58	20 08	13 53
15	18 51	04 S 22	23 03	03 05	24 02	07 21	22 45	08 58	20 08	13 53
16	19 05	00 N 24	23 29	03 30	23 57	07 17	22 45	08 57	20 07	13 53
17	19 18	05 16	23 53	03 54	23 53	07 14	22 45	08 56	20 07	13 53
18	19 32	10 03	24 15	04 19	23 48	07 11	22 45	08 55	20 07	13 53
19	19 45	14 29	24 34	04 44	23 43	07 08	22 45	08 55	20 07	13 53
20	19 58	18 19	24 50	05 08	23 38	07 05	22 45	08 54	20 06	13 52
21	20 10	21 11	25 04	05 33	23 33	07 02	22 45	08 53	20 06	13 52
22	20 22	22 47	25 15	05 57	23 28	06 59	22 45	08 53	20 06	13 52
23	20 34	22 53	25 24	06 22	23 22	06 57	22 45	08 52	20 06	13 52
24	20 45	21 25	25 31	06 46	23 16	06 54	22 45	08 51	20 05	13 52
25	20 56	18 33	25 36	07 11	23 11	06 51	22 45	08 51	20 05	13 52
26	21 07	14 32	25 39	07 35	23 05	06 48	22 45	08 50	20 05	13 52
27	21 17	09 43	25 39	08 00	22 58	06 46	22 44	08 50	20 05	13 52
28	21 27	04 N 27	25 38	08 24	22 52	06 43	22 44	08 49	20 04	13 51
29	21 36	00 S 57	25 36	08 48	22 46	06 41	22 44	08 48	20 04	13 51
30	21 45	06 12	25 31	09 12	22 39	06 38	22 44	08 48	20 04	13 51
31	21 N 54	11 S 05	25 N 25	09 N 36	22 N 32	06 S 36	22 N 44	08 S 47	20 S 04	13 N 51

ZODIAC SIGN ENTRIES

Date	h	m	Planets
02	23	39	☽ ♎
04	20	21	♀ ♈
05	04	43	☽ ♏
07	12	05	☽ ♐
09	22	15	☽ ♑
11	04	55	☿ ♊
12	10	34	☽ ♒
14	23	03	☽ ♓
17	09	20	☽ ♈
19	16	10	☽ ♉
21	10	36	☉ ♊
21	19	54	☽ ♊
23	21	46	☽ ♋
25	23	12	☽ ♌
28	01	25	☽ ♍
29	08	03	☿ ♋
30	05	16	☽ ♎
31	07	19	♀ ♉

LATITUDES

Date	Mercury ☿	Venus ♀	Mars ♂	Jupiter ♃	Saturn ♄	Uranus ♅	Neptune ♆	Pluto ♇
01	00 S 40	01 S 07	01 N 30	00 S 56	00 S 42	00 N 36	01 N 38	17 N 05
04	00 S 10	01 17	01 29	00 57	00 41	00 36	01 38	17 05
07	00 N 22	01 27	01 29	00 57	00 41	00 36	01 38	17 04
10	00 53	01 35	01 28	00 58	00 41	00 36	01 38	17 02
13	01 21	01 42	01 28	00 59	00 40	00 36	01 38	17 01
16	01 45	01 48	01 27	00 59	00 40	00 36	01 38	17 00
19	02 02	01 54	01 26	01 00	00 39	00 36	01 38	16 59
22	02 13	01 58	01 26	01 00	00 39	00 36	01 38	16 58
25	02 16	02 01	01 25	01 02	00 39	00 36	01 38	16 56
28	02 12	02 04	01 24	01 02	00 38	00 36	01 38	16 55
31	02 N 01	02 S 05	01 N 23	01 S 03	00 S 38	00 N 36	01 N 38	16 N 54

DATA

Julian Date	2442169
Delta T	+45 seconds
Ayanamsa	23° 30' 11"
Synetic vernal point	05° ♓ 36' 49"
True obliquity of ecliptic	23° 26' 32"

LONGITUDES

Date	Chiron ⚷ ° '	Ceres ⚳ ° '	Pallas ⚴ ° '	Juno ⚵ ° '	Vesta ⚶ ° '	Black Moon Lilith ⚸ ° '
01	21 ♈ 34	03 ♓ 45	09 ♒ 13	02 ♓ 05	04 ♎ 23	28 ♑ 44
11	22 ♈ 08	06 ♓ 29	10 ♒ 25	05 ♓ 19	03 ♎ 37	29 ♑ 50
21	22 ♈ 57	08 ♓ 57	11 ♒ 04	08 ♓ 21	03 ♎ 39	00 ♒ 57
31	23 ♈ 07	11 ♓ 06	11 ♒ 08	11 ♓ 06	04 ♎ 30	02 ♒ 04

MOON'S PHASES, APSIDES AND POSITIONS ☽

Date	h	m	Phase	Longitude ° '	Eclipse Indicator
06	08	55	○	15 ♏ 27	
14	09	29	☾	23 ♒ 13	
21	20	34	●	00 ♊ 24	
28	13	03	☽	06 ♍ 49	

Day	h	m		
12	16	48	Apogee	
24	13	15	Perigee	
02	02	48	0S	
08	22	25	Max dec	23° S 03'
16	10	00	0N	
23	01	32	Max dec	23° N 02'
29	07	48	0S	

ASPECTARIAN

h m Aspects

01 Wednesday
03 48 ☽ ⊻ ♇
06 48 ☿ ✶ ♂
07 12 ☽ ✶ ♂
07 14 ☽ △ ☿
07 39 ☉ ± ♇
11 22 ☽ □ ♆
13 17 ☽ ∠ ♅
14 33 ☽ △ ☉
15 31 ☽ ☍ ♃
16 04 ☽ # ♀
18 34 ☽ Q ♄

02 Thursday
04 43 ☽ ⊥ ♅
04 44 ☽ Q ♂
06 11 ☿ ∥ ♇
07 19 ☉ ✶ ♃
10 15 ☿ ⊼ ♆
12 09 ☽ ∥ ♀
13 41 ☽ ⚼ ☿
15 04 ☽ ⊻ ♅
18 21 ☽ Q ♆
18 22 ☽ ⚼ ☉
19 45 ☽ ☍ ♀

03 Friday
01 55 ☽ □ ♄
03 51 ☿ ± ♇
07 32 ☽ ☌ ♇
08 13 ☽ ± ☿
11 15 ☽ ± ☉
13 27 ☽ □ ♂
15 14 ☽ ∥ ♃
15 15 ☽ ✶ ♆
17 45 ☿ ✶ ♃
20 16 ☽ ⊼ ♃
20 28 ☽ ∥ ♅
20 42 ☽ ⊼ ☿
22 36 ☽ ⊼ ☉

04 Saturday
07 01 ☽ ± ♃
16 55 ☉ ☌ ☿
17 46 ☽ ∠ ♆
19 41 ☽ ☌ ♅
21 06 ☽ # ♇
22 18 ☉ ∥ ☿
23 16 ☽ ⚼ ♃

05 Sunday
05 05 ♂ ⊼ ♆
05 28 ☽ ⊼ ♀
07 28 ☽ △ ♄
09 56 ☽ ⊥ ♆
11 05 ☽ # ☉
12 49 ☽ ⊻ ♇
13 19 ☽ # ☿
15 37 ☿ ∠ ♀
17 21 ☽ ± ♀
20 49 ☽ ⊻ ♆
21 37 ☽ △ ♂
23 27 ☿ ∠ ♄
23 44 ☽ ⊥ ♇

06 Monday
02 51 ☽ △ ♃
07 39 ♀ □ ♄
08 55 ☽ ☍ ☉
11 06 ☽ ⚼ ♄
11 24 ☽ ⚼ ♀
13 20 ☽ ☍ ☿
16 19 ☽ ∠ ♇
16 46 ☽ ∥ ♆

07 Tuesday
02 27 ☽ ⊻ ♅
02 44 ☽ ⚼ ♂
04 01 ☽ ± ♄
06 55 ☿ ⚼ ♇
13 43 ☽ ⊥ ♅
15 26 ☽ ⊼ ♄
17 43 ☉ ∠ ♄
18 13 ☽ △ ♀
20 30 ☽ ✶ ♇
20 36 ☽ ± ♂

08 Wednesday
04 54 ☽ ☌ ♆
05 07 ☽ # ♄
06 51 ☽ ∠ ♅
08 41 ☽ ⊼ ♂
12 06 ☽ □ ♃
19 33 ♀ ☍ ♇
19 34 ☽ Q ♇
22 33 ☽ ⊼ ☉

09 Thursday
10 36 ☽ ⊼ ☿
11 18 ☽ ± ☉
11 57 ☽ ✶ ♅
15 53 ☽ # ♄
17 55 ☿ Q ♃
18 07 ☿ ⊼ ♅
23 50 ☽ Q ♃

10 Friday
01 01 ☽ ± ☿
02 17 ☽ ☍ ♄
06 40 ☽ ⚼ ☉
06 58 ☽ □ ♇
08 24 ☿ ⊥ ♄
09 57 ☿ # ♆
10 22 ☉ ⚼ ♇
10 40 ☽ □ ♀
11 41 ☽ Q ♅
15 42 ☽ ⊻ ♆
22 54 ☿ ∠ ♂
22 54 ☽ ☍ ♂
22 54 ☽ ⚼ ☿

11 Saturday
00 08 ☽ ✶ ♃
01 54 ☽ # ☿
03 44 ☽ ⊥ ♆
06 30 ☽ ∥ ♆
15 27 ☽ △ ☉
21 52 ☽ ∠ ♆
23 49 ☽ □ ♅

12 Sunday
01 38 ☽ # ☉
02 41 ☽ Q ♀
06 50 ☽ ∠ ♃
07 21 ♂ △ ♃
11 47 ☽ △ ☿
12 37 ☿ ± ♅
15 14 ☽ ⊼ ♄
15 29 ♀ △ ♆
19 23 ☽ △ ♇

13 Monday
01 19 ☽ ⊥ ♃
03 31 ☽ ± ♄
04 12 ☽ ✶ ♆
05 34 ☽ ✶ ♀
06 40 ☽ # ♇
08 15 ☿ ⊻ ♄
13 38 ☽ ⊻ ♃
14 50 ☽ ⊼ ♂
21 53 ☽ ⚼ ♄

14 Tuesday
01 40 ☽ ⚼ ♇
03 35 ☽ ± ♂
04 23 ☽ Q ♆
06 49 ☿ △ ♇
09 29 ☽ □ ☉
11 32 ☽ ∥ ♅
12 16 ☽ △ ♅
14 57 ☽ ∠ ♀
18 40 ♂ ± ♆
19 41 ☽ ± ♇
20 13 ☽ ∥ ♃
21 14 ☿ ∥ ♄
22 33 ☽ ⚼ ♂

15 Wednesday
04 11 ☽ △ ♄
07 35 ☽ ⊼ ♇
10 41 ☽ ⊥ ♀
12 35 ☽ □ ☿
16 08 ☽ □ ♆
17 57 ☽ ⚼ ♅
18 02 ☽ # ♀
19 01 ☉ ⊼ ♅
23 40 ☽ ⊻ ♀

16 Thursday
00 14 ☽ Q ☉
02 07 ☽ ☌ ♃
05 37 ☽ △ ♂
10 08 ☿ ☍ ♆
11 26 ☽ ± ♅
21 16 ☿ ⚼ ♅
22 55 ☽ ⊼ ♅

17 Friday
01 19 ☽ ✶ ☉
04 15 ☉ Q ♃
04 37 ☿ ⊥ ♂
04 38 ☽ ∥ ♀
05 47 ♀ ⊻ ♃
05 58 ☽ Q ☿
11 36 ☿ ∥ ♂
14 42 ☽ □ ♄
17 21 ☽ ☍ ♇
21 40 ☽ # ♃

18 Saturday
01 21 ☽ △ ♆
06 06 ☉ ⊥ ♄
06 17 ☽ # ♅
07 41 ☽ ∠ ☉
08 16 ☽ ✶ ☿
11 30 ☽ ⊻ ♃
13 58 ☽ ☌ ♀
16 53 ☽ □ ♂
22 32 ☽ ⊥ ♃

19 Sunday
00 07 ☽ Q ♄
01 21 ☽ ⊥ ☉
04 39 ☽ ⚼ ♆
06 16 ☽ ☍ ♅
08 31 ☽ ∥ ♇
08 38 ☿ □ ♃
12 55 ☽ ⊻ ☉
14 49 ☽ ∠ ♃
15 43 ☽ ∠ ☿
20 37 ☽ ± ♆
21 41 ☽ ✶ ♄
23 38 ☽ ⊼ ♇

20 Monday
02 31 ☽ Q ♂
07 07 ☽ ⊼ ♆
09 54 ☽ ⊥ ☿
10 06 ☽ ± ♇
17 18 ☽ ✶ ♃
21 47 ☽ ⊻ ☿
23 48 ☽ ⊻ ♀
23 56 ☽ ∠ ♄

21 Tuesday
00 14 ☽ ✶ ♂
01 34 ☽ ∥ ☉
01 36 ☽ ⚼ ♇
01 49 ☽ # ♆
04 56 ☉ # ♆
10 22 ☽ ⊼ ♅
10 47 ♀ □ ♂
10 56 ☽ ⊥ ♀
14 00 ☽ Q ♃
15 23 ☽ ⊥ ♄
19 28 ☉ ± ♅
20 30 ☽ ± ♅
20 34 ☽ ☌ ☉

22 Wednesday
01 34 ☽ ⊻ ♄
02 47 ☽ ∠ ♂
02 58 ☽ △ ♇
03 01 ☿ ⊻ ♂
03 26 ☽ ∠ ♀
10 04 ☽ ☍ ♆
11 04 ☽ ∥ ♄
11 33 ☽ ⚼ ♅
18 28 ☽ ⊥ ♂
19 01 ☿ ✶ ♀
20 26 ☽ □ ♃

23 Thursday
04 51 ☽ ⊻ ♂
05 37 ♀ ⊥ ♃
06 32 ☽ ✶ ♀
06 59 ☽ ☌ ☿
12 22 ☽ △ ♅
15 57 ☽ ∥ ♄
23 50 ♀ Q ♄

24 Friday
01 56 ☽ ⊻ ☉
03 43 ☽ ☌ ♄
04 01 ☽ Q ♀
04 39 ☽ □ ♇
11 35 ☽ ⊼ ♆
12 31 ☽ ⊥ ☉
18 17 ☽ ∥ ☉
18 27 ♀ # ♃
21 26 ☽ ± ♆
22 22 ☽ △ ♃

25 Saturday
00 30 ☽ # ♆
04 22 ☽ ∠ ☉
07 18 ☿ △ ♅
08 22 ☉ ⊻ ♄
08 25 ☽ ☌ ♂
10 16 ☽ Q ♇
12 11 ☽ □ ♀
12 14 ☽ ⚼ ♆
12 58 ♀ ⚼ ♆
13 41 ☽ □ ♅
14 26 ☽ ⊻ ☿
18 38 ☉ △ ♇
23 20 ☽ ⚼ ♃

26 Sunday
01 31 ☽ ⊥ ☿
05 36 ☽ ⊻ ♄
06 06 ☽ ✶ ♇
06 46 ♀ ☍ ♅
06 55 ☽ ✶ ☉
13 02 ☽ △ ♆
14 24 ☽ ± ♃
15 33 ☽ ∥ ♇
15 39 ☽ ⊥ ♄
18 11 ☽ ∠ ☿
19 29 ☽ Q ♅

27 Monday
00 30 ☽ ⊼ ♃
03 47 ♂ Q ♇
04 22 ☽ Q ☉
06 50 ☽ ∠ ♄
07 07 ☽ ∠ ♇
12 32 ☽ ⊻ ♂
15 34 ☽ ✶ ♅
16 11 ☽ # ♅
18 32 ☽ △ ♀
19 27 ☽ ∥ ♀
22 11 ☽ ✶ ☿
22 17 ☽ ⊥ ♇
23 07 ☽ ⊥ ♂

28 Tuesday
01 45 ☽ # ♃
08 26 ☽ ✶ ♄
08 28 ☽ ⊻ ♇
13 03 ☽ □ ☉
13 16 ♄ □ ♇
15 09 ☽ ∠ ♂
15 33 ☽ □ ♆
17 02 ☽ ∠ ♅
20 52 ☽ Q ☿
22 24 ☽ ⚼ ♀

29 Wednesday
00 05 ♂ ⚼ ♆
03 54 ☽ ☍ ♃
05 13 ☽ Q ♄
08 32 ☽ ⊥ ♅
12 30 ♀ # ♅
15 24 ☽ ± ♀
17 02 ♂ ∥ ♄
18 17 ☽ ✶ ♂
18 57 ☽ ⊻ ♅
22 41 ☽ Q ♆

30 Thursday
00 40 ☉ ☍ ♆
02 51 ☽ ⊼ ♀
07 29 ☽ □ ☿
08 51 ♂ □ ♅
12 33 ☽ ☌ ♇
13 00 ☽ □ ♄
14 01 ☽ ∥ ♃
16 24 ☽ Q ♂
19 50 ☽ ✶ ♆
21 20 ☽ △ ☉
21 47 ☉ ⚼ ♅

31 Friday
00 24 ☽ ∥ ♅
03 42 ☽ # ♀
09 14 ☽ ⊼ ♃
20 09 ☽ ± ♃
21 17 ☉ ∠ ♂
22 45 ☽ ∠ ♆
23 33 ♀ ∠ ♃

All ephemeris data is given at 12.00 UT and the Moon's longitude is additionally given for 24.00 UT

JUNE 1974

LONGITUDES

Date	Sidereal time h m s	Sun ☉ ° ' "	Moon ☽ ° ' "	Moon ☽ 24.00 ° ' "	Mercury ☿ ° '	Venus ♀ ° '	Mars ♂ ° '	Jupiter ♃ ° '	Saturn ♄ ° '	Uranus ♅ ° '	Neptune ♆ ° '	Pluto ♇ ° '
01	04 38 21	10 ♊ 37 04	00 ♏ 27 07	06 ♏ 57 38	03 ♋ 53	01 ♉ 23	25 ♋ 24	15 ♓ 50	04 ♋ 37	24 ♎ 03	08 ♐ 11	04 ♎ 07
02	04 42 17	11 34 32	13 ♏ 24 39	19 ♏ 48 15	04 59	02 32	26 01	15 57	04 45	24 R 02	08 R 09	04 R 06
03	04 46 14	12 31 59	26 ♏ 08 31	02 ♐ 25 32	06 02	03 42	26 37	16 03	04 52	24 00	08 08	04 06
04	04 50 10	13 29 25	08 ♐ 39 28	14 ♐ 50 24	07 01	04 51	27 14	16 09	05 00	23 59	08 06	04 05
05	04 54 07	14 26 51	20 ♐ 58 33	27 ♐ 04 04	07 56	06 01	27 50	16 15	05 07	23 57	08 04	04 05
06	04 58 03	15 24 15	03 ♑ 08 10	09 ♑ 08 10	08 48	07 10	28 27	16 21	05 15	23 56	08 03	04 05
07	05 02 00	16 21 38	15 ♑ 07 17	21 ♑ 04 51	09 35	08 20	29 04	16 26	05 22	23 55	08 01	04 05
08	05 05 57	17 19 01	27 ♑ 01 12	02 ♒ 56 45	10 19	09 30	29 ♋ 40	16 32	05 30	23 54	08 00	04 05
09	05 09 53	18 16 23	08 ♒ 51 53	14 ♒ 47 05	10 58	10 40	00 ♌ 17	16 37	05 37	23 53	07 58	04 04
10	05 13 50	19 13 44	20 ♒ 42 47	26 ♒ 39 32	11 34	11 50	00 54	16 42	05 45	23 51	07 56	04 04
11	05 17 46	20 11 05	02 ♓ 37 49	08 ♓ 38 12	12 05	13 00	01 30	16 47	05 52	23 50	07 55	04 04
12	05 21 43	21 08 26	14 ♓ 41 13	20 ♓ 47 27	12 32	14 10	02 07	16 52	06 00	23 49	07 53	04 04
13	05 25 39	22 05 45	26 ♓ 57 25	03 ♈ 11 41	12 54	15 20	02 44	16 56	06 08	23 48	07 52	04 04
14	05 29 36	23 03 05	09 ♈ 30 44	15 ♈ 55 02	13 12	16 30	03 20	17 01	06 15	23 47	07 50	04 04
15	05 33 32	24 00 24	22 ♈ 24 59	29 ♈ 00 56	13 25	17 40	03 57	17 05	06 23	23 47	07 49	04 D 04
16	05 37 29	24 57 43	05 ♉ 43 08	12 ♉ 31 42	13 34	18 50	04 34	17 09	06 31	23 46	07 47	04 04
17	05 41 26	25 55 01	19 ♉ 26 39	26 ♉ 27 51	13 38	20 01	05 11	17 13	06 38	23 45	07 46	04 04
18	05 45 22	26 52 19	03 ♊ 35 03	10 ♊ 47 47	13 R 38	21 11	05 48	17 17	06 46	23 44	07 44	04 04
19	05 49 19	27 49 37	18 ♊ 05 31	25 ♊ 27 29	13 33	22 22	06 24	17 20	06 54	23 44	07 42	04 04
20	05 53 15	28 46 54	02 ♋ 52 52	10 ♋ 20 41	13 24	23 32	07 01	17 24	07 02	23 43	07 41	04 04
21	05 57 12	29 ♊ 44 11	17 ♋ 49 56	25 ♋ 19 33	13 11	24 43	07 38	17 27	07 09	23 42	07 40	04 05
22	06 01 08	00 ♋ 41 27	02 ♌ 48 28	10 ♌ 15 42	12 53	25 53	08 15	17 30	07 17	23 42	07 38	04 05
23	06 05 05	01 38 43	17 ♌ 40 18	25 ♌ 01 27	12 32	27 04	08 52	17 33	07 25	23 41	07 37	04 05
24	06 09 01	02 35 58	02 ♍ 18 28	09 ♍ 30 46	12 08	28 15	09 29	17 35	07 33	23 41	07 35	04 05
25	06 12 58	03 33 12	16 ♍ 37 59	23 ♍ 39 50	11 40	29 ♉ 25	10 06	17 38	07 41	23 41	07 34	04 06
26	06 16 55	04 30 26	00 ♎ 36 11	07 ♎ 27 01	11 09	00 ♊ 36	10 43	17 40	07 48	23 40	07 32	04 06
27	06 20 51	05 27 39	14 ♎ 12 25	20 ♎ 52 33	10 37	01 47	11 20	17 42	07 56	23 40	07 31	04 07
28	06 24 48	06 24 52	27 ♎ 27 40	03 ♏ 58 02	10 02	02 58	11 57	17 44	08 04	23 40	07 30	04 07
29	06 28 44	07 22 04	10 ♏ 23 57	16 ♏ 45 45	09 27	04 09	12 34	17 46	08 12	23 40	07 28	04 08
30	06 32 41	08 ♋ 19 16	23 ♏ 03 47	29 ♏ 18 22	08 ♋ 51	05 ♊ 20	13 ♌ 11	17 ♓ 47	08 ♋ 20	23 ♎ 40	07 ♐ 27	04 ♎ 08

Date	Moon True ☊ ° '	Moon Mean ☊ ° '	Moon ☽ Latitude ° '
01	19 ♐ 26	19 ♐ 54	03 S 59
02	19 R 25	19 51	03 07
03	19 24	19 48	02 05
04	19 23	19 44	00 S 59
05	19 D 23	19 41	00 N 09
06	19 23	19 38	01 15
07	19 24	19 35	02 18
08	19 24	19 32	03 14
09	19 24	19 28	04 01
10	19 24	19 25	04 39
11	19 R 24	19 22	05 04
12	19 24	19 19	05 16
13	19 D 24	19 16	05 15
14	19 24	19 13	04 58
15	19 25	19 09	04 26
16	19 25	19 06	03 40
17	19 26	19 03	02 39
18	19 26	19 00	01 27
19	19 26	18 57	00 N 07
20	19 R 26	18 53	01 S 14
21	19 26	18 50	02 31
22	19 24	18 47	03 38
23	19 23	18 44	04 29
24	19 22	18 41	05 03
25	19 21	18 38	05 16
26	19 20	18 34	05 10
27	19 D 20	18 31	04 47
28	19 21	18 28	04 09
29	19 22	18 25	03 19
30	19 ♐ 24	18 ♐ 22	02 S 20

DECLINATIONS

Date	Sun ☉ ° '	Moon ☽ ° '	Mercury ☿ ° '	Venus ♀ ° '	Mars ♂ ° '	Jupiter ♃ ° '	Saturn ♄ ° '	Uranus ♅ ° '	Neptune ♆ ° '	Pluto ♇ ° '
01	22 N 02	15 S 22	25 N 18	09 N 59	22 N 25	06 S 34	22 N 44	08 S 47	20 S 03	13 N 50
02	22 10	18 50	25 09	10 23	22 18	06 31	22 44	08 46	20 03	13 50
03	22 18	21 19	25 00	10 46	22 11	06 29	22 44	08 46	20 03	13 50
04	22 25	22 43	24 49	11 10	22 04	06 27	22 43	08 45	20 03	13 49
05	22 32	22 59	24 37	11 33	21 56	06 25	22 43	08 45	20 02	13 49
06	22 39	22 09	24 24	11 56	21 48	06 23	22 43	08 44	20 02	13 49
07	22 45	20 18	24 11	12 18	21 41	06 21	22 43	08 44	20 02	13 48
08	22 50	17 35	23 56	12 41	21 33	06 19	22 43	08 44	20 02	13 48
09	22 55	14 09	23 42	13 03	21 25	06 18	22 42	08 43	20 01	13 48
10	23 00	10 11	23 26	13 25	21 16	06 16	22 42	08 43	20 01	13 47
11	23 05	05 48	23 10	13 47	21 08	06 14	22 42	08 42	20 01	13 47
12	23 09	01 S 10	22 54	14 08	20 59	06 13	22 42	08 42	20 01	13 46
13	23 12	03 N 36	22 38	14 29	20 51	06 11	22 41	08 42	20 01	13 46
14	23 16	08 20	22 21	14 50	20 42	06 10	22 41	08 41	20 00	13 46
15	23 18	12 51	22 04	15 11	20 33	06 08	22 41	08 41	20 00	13 45
16	23 21	16 53	21 48	15 31	20 24	06 07	22 41	08 41	20 00	13 45
17	23 23	20 08	21 31	15 51	20 15	06 06	22 40	08 41	20 00	13 44
18	23 24	22 18	21 15	16 10	20 06	06 04	22 40	08 40	19 59	13 44
19	23 25	23 02	20 59	16 30	19 56	06 03	22 40	08 40	19 59	13 43
20	23 26	22 11	20 43	16 49	19 46	06 02	22 39	08 40	19 59	13 43
21	23 27	19 46	20 28	17 07	19 37	06 01	22 39	08 40	19 59	13 42
22	23 26	16 00	20 14	17 25	19 27	06 00	22 39	08 40	19 59	13 41
23	23 26	11 15	20 00	17 43	19 17	06 00	22 38	08 39	19 58	13 41
24	23 25	05 57	19 46	18 00	19 07	05 59	22 38	08 39	19 58	13 40
25	23 24	00 N 25	19 34	18 17	18 57	05 58	22 38	08 39	19 58	13 40
26	23 22	04 S 59	19 23	18 34	18 46	05 58	22 37	08 39	19 58	13 40
27	23 20	10 01	19 12	18 50	18 36	05 57	22 37	08 39	19 58	13 39
28	23 17	14 27	19 03	19 05	18 25	05 57	22 37	08 39	19 57	13 38
29	23 14	18 05	18 54	19 21	18 14	05 56	22 36	08 39	19 57	13 37
30	23 N 11	20 S 48	18 N 47	19 N 35	18 N 04	05 S 56	22 N 36	08 S 39	19 S 57	13 N 37

ZODIAC SIGN ENTRIES

Date	h	m	Planets
01	11	10	☽ ♏
03	19	21	☽ ♐
06	05	48	☽ ♑
08	18	02	☽ ♒
09	00	54	♂ ♌
11	06	43	☽ ♓
13	17	52	☽ ♈
16	01	46	☽ ♉
18	05	59	☽ ♊
20	07	21	☽ ♋
21	18	38	☉ ♋
22	07	30	☽ ♌
24	08	11	☽ ♍
25	23	44	♀ ♊
26	10	57	☽ ♎
28	16	40	☽ ♏

LATITUDES

Date	Mercury ☿ ° '	Venus ♀ ° '	Mars ♂ ° '	Jupiter ♃ ° '	Saturn ♄ ° '	Uranus ♅ ° '	Neptune ♆ ° '	Pluto ♇ ° '
01	01 N 55	02 S 06	01 N 23	01 S 04	00 S 38	00 N 36	01 N 38	16 N 53
04	01 34	02 06	01 22	01 04	00 38	00 36	01 38	16 52
07	01 05	02 06	01 21	01 05	00 37	00 35	01 38	16 50
10	00 N 30	02 04	01 21	01 06	00 37	00 35	01 38	16 49
13	00 S 11	02 02	01 20	01 07	00 37	00 35	01 38	16 47
16	00 57	01 59	01 19	01 08	00 36	00 35	01 38	16 46
19	01 47	01 56	01 18	01 09	00 36	00 35	01 38	16 44
22	02 36	01 52	01 17	01 09	00 36	00 35	01 38	16 43
25	03 22	01 47	01 16	01 10	00 35	00 35	01 38	16 41
28	04 02	01 42	01 15	01 11	00 35	00 35	01 38	16 40
31	04 S 30	01 S 36	01 N 15	01 S 12	00 S 35	00 N 35	01 N 38	16 N 38

DATA

Julian Date	2442200
Delta T	+45 seconds
Ayanamsa	23° 30' 15"
Synetic vernal point	05° ♓ 36' 45"
True obliquity of ecliptic	23° 26' 31"

MOON'S PHASES, APSIDES AND POSITIONS ☽

Date	h	m	Phase	Longitude ° '	Eclipse Indicator
04	22	10	○	13 ♐ 54	partial
13	01	45	☾	21 ♓ 41	
20	04	56	●	28 ♊ 30	Total
26	19	20	☽	04 ♎ 48	

Day	h	m		
09	09	26	Apogee	
21	13	43	Perigee	
05	05	36	Max dec	23° S 02'
12	17	54	0N	
19	11	21	Max dec	23° N 02'
25	13	50	0S	

LONGITUDES

Date	Chiron ⚷ ° '	Ceres ⚳ ° '	Pallas ⚴ ° '	Juno ⚵ ° '	Vesta ⚶ ° '	Black Moon Lilith ⚸ ° '
01	23 ♈ 10	11 ♓ 18	11 ♒ 06	11 ♓ 22	04 ♎ 37	02 ♒ 11
11	23 ♈ 34	13 ♓ 01	10 ♒ 28	13 ♓ 46	06 ♎ 14	03 ♒ 17
21	23 ♈ 55	14 ♓ 19	09 ♒ 10	15 ♓ 48	08 ♎ 28	04 ♒ 24
31	24 ♈ 11	15 ♓ 06	07 ♒ 17	17 ♓ 22	11 ♎ 13	05 ♒ 31

All ephemeris data is given at 12.00 UT and the Moon's longitude is additionally given for 24.00 UT

ASPECTARIAN

h m Aspects

01 Saturday
00 20 ☽ ☌ ♅
02 19 ☽ □ ♂
02 28 ☽ ꝑ ☉
02 59 ☽ # ♇
12 43 ☽ ꝑ ♃
13 52 ☽ ☍ ♀
15 10 ☽ ⊥ ♆
16 52 ☿ □ ♇
18 44 ☽ ⊻ ♇
18 54 ☽ △ ☿
19 45 ☽ △ ♄
20 17 ☽ ± ☉

02 Sunday
02 14 ☽ ⊻ ♆
04 15 ♀ ± ♆
05 50 ☽ ⊥ ♇
06 02 ☿ ☌ ♄
08 18 ☽ ⊼ ☉
16 47 ☽ △ ♃
22 32 ☽ ∥ ♆
22 40 ☽ ∠ ♇

03 Monday
00 00 ☽ ꝑ ♄
00 27 ☉ ∥ ♂
01 27 ☽ ꝑ ☿
07 57 ☽ ⊻ ♅
12 58 ☽ △ ♂
17 15 ☽ ± ♄
19 21 ☽ ⊥ ♅
20 03 ☽ ± ☿
20 19 ♀ ⊼ ♇
23 21 ☽ # ♂

04 Tuesday
03 12 ☽ ⚹ ♇
03 55 ☽ ⊼ ♀
04 01 ☽ # ☉
04 52 ☽ ⊼ ♄
08 33 ☽ ⊼ ☿
10 55 ☽ ☌ ♆
11 58 ☽ # ♄
12 37 ☽ ∠ ♅
15 16 ♀ ⚹ ♄
16 41 ☽ ± ♀
19 17 ☽ ꝑ ♂
22 10 ☽ ☍ ☉

05 Wednesday
02 26 ☽ Q ♇
02 40 ☽ □ ♃
12 05 ☽ ꝑ ♀
13 47 ☽ ± ♂
15 40 ☿ ⊼ ♆
17 51 ☽ ⚹ ♅
23 31 ☽ # ♄

06 Thursday
02 15 ☽ ⊼ ♂
02 39 ☽ # ☉
13 55 ☽ □ ♇
14 27 ☽ Q ♃
16 16 ☽ ☍ ♄
17 36 ☽ Q ♅
17 59 ☽ # ♂
20 56 ☽ △ ♀
21 48 ☽ ⊻ ♆

07 Friday
00 08 ☽ ☍ ☿
04 54 ☉ ∥ ♄
05 40 ♀ ⊼ ♆
09 48 ☽ ⊥ ♆
14 09 ☉ □ ♃
14 40 ☽ ⚹ ♃
14 42 ☽ ∥ ♆
14 43 ☽ ⊼ ☉

08 Saturday
03 52 ☽ ± ☉
03 53 ☽ ∠ ♆
05 42 ☽ □ ♅
17 40 ☽ ☍ ♂
21 12 ☽ ∠ ♃
23 40 ☽ ꝑ ☉
23 51 ♀ ± ♇

09 Sunday
02 17 ☽ △ ♇
05 21 ☽ ⊼ ♄
10 11 ☽ ⚹ ♆
14 19 ☽ # ♇
15 35 ☽ ⊥ ♃
16 02 ☽ □ ♀
16 30 ☽ ⊼ ☿
17 39 ☽ ± ♄
18 24 ☽ # ♀

10 Monday
01 16 ☿ ⚹ ♀
03 50 ☽ ⊻ ♃
05 18 ☽ ± ☿
08 40 ☽ ꝑ ♇
08 44 ☽ △ ☉
10 26 ☽ Q ♆
12 04 ☽ ꝑ ♄
18 21 ☽ △ ♅
20 17 ☽ ∥ ♅

11 Tuesday
00 21 ☽ ꝑ ☿
02 50 ☽ ± ♇
08 22 ☽ Q ♀
09 37 ☽ ⊼ ♂
09 42 ☽ ∥ ♃
12 25 ♀ ∥ ♇
14 53 ☽ ⊼ ♇
18 33 ☽ △ ♄
18 35 ☉ ∥ ☿
22 16 ☽ ± ♂
22 32 ☽ □ ♆

12 Wednesday
00 23 ☽ ꝑ ♅
00 39 ♂ ꝑ ♃
07 34 ☽ △ ☿
10 51 ☽ ⚹ ♀
16 19 ☽ ☌ ♃
17 03 ☽ ꝑ ♂
18 10 ☽ ± ♅

13 Thursday
01 45 ☽ □ ☉
05 54 ☽ ⊼ ♅
06 29 ☿ ∥ ♄
19 11 ☽ ∠ ♀
23 41 ☽ △ ♂

14 Friday
00 57 ☽ # ♃
01 40 ☽ ☍ ♇
05 46 ☽ □ ♄
08 50 ☽ △ ♆
13 20 ♇ St D
13 50 ☽ # ♅
14 03 ☽ ⊥ ♀
15 08 ☽ Q ☉
19 04 ☽ □ ☿
23 17 ♀ ⚹ ♃

15 Saturday
02 07 ☽ ⊻ ♃
02 23 ☽ ⊻ ♀
06 19 ☉ △ ♅
12 43 ☽ ꝑ ♆
13 14 ☽ ⊥ ♃
14 29 ☽ ☍ ♅
15 08 ☽ ⚹ ☉
15 38 ☽ Q ♄
16 21 ♂ ⚹ ♇
17 07 ☽ ∥ ♇

16 Sunday
02 40 ☽ ∥ ♀
04 31 ☽ Q ☿
05 00 ☽ ± ♆
05 37 ☽ ∠ ♃
09 03 ☽ ⊼ ♇
09 51 ☽ □ ♂
13 25 ☽ ⚹ ♄
15 39 ☽ ⊼ ♆
16 36 ♀ ꝑ ♇
19 41 ☽ ± ♇
20 04 ☽ ∠ ☉

17 Monday
01 28 ☉ ⊥ ♀
01 54 ☽ ⚹ ☿
08 08 ☽ ⚹ ♃
10 45 ☽ # ♆
11 21 ☽ ꝑ ♇
12 52 ☽ ⊥ ☉
12 53 ☽ ∥ ♂
13 04 ☽ ☌ ♀
15 49 ☽ ∠ ♄
18 42 ☽ Q ♂
19 22 ☽ ⊼ ♅
22 38 ☿ St R
23 52 ☽ ⊻ ☉

18 Tuesday
00 06 ☽ ∥ ☿
03 42 ☽ ∠ ☿
04 44 ☽ Q ♃
05 33 ☽ ± ♅
07 14 ☽ ⊥ ♄
12 49 ☽ △ ♇
15 51 ☽ ⚹ ♂
17 22 ☽ ⊻ ♄
18 44 ☽ ⊥ ☿
18 55 ☽ ☍ ♆
19 00 ☽ ∥ ♄
20 35 ☽ ꝑ ♅

19 Wednesday
01 24 ♀ ∠ ♄
03 54 ♂ # ♆
04 36 ☽ ⊻ ☿
10 46 ☽ □ ♃
17 39 ☽ ∠ ♂
19 34 ☽ ⊻ ♀
21 11 ☽ △ ♅

20 Thursday
03 40 ☽ ∥ ♄
04 56 ☽ ☌ ☉
06 08 ☽ ⊥ ♀
08 52 ☽ ⊥ ♂
12 23 ♂ ⊻ ♄
13 55 ☽ □ ♇
15 40 ♀ ⊼ ♅
18 44 ☽ ☌ ♄
18 57 ☽ ⊻ ♂
19 43 ☽ ⊼ ♆
21 52 ☽ ∠ ♀

21 Friday
04 40 ☽ ☌ ☿
05 19 ☽ ± ♆
05 42 ☽ ∥ ☿
10 17 ☽ # ♆
11 23 ☽ △ ♃
12 54 ♂ △ ♆
13 10 ☽ ∥ ♂
18 48 ☽ Q ♇
19 43 ☽ ꝑ ♆
21 24 ☽ □ ♅
23 57 ☽ ⚹ ♀

22 Saturday
04 20 ☽ ∥ ♀
08 22 ☽ ⊻ ☉
11 30 ☽ ꝑ ♃
14 03 ☽ ⚹ ♇
18 40 ☽ ⊥ ☉
19 16 ☽ ⊻ ♄
19 45 ☽ △ ♆
20 52 ☽ Q ♀
21 08 ☽ ☌ ♂

23 Sunday
00 12 ☽ ∥ ♇
02 03 ☽ ± ♃
02 19 ☽ Q ♅
03 24 ♀ Q ♂
03 53 ☽ ⊻ ☿
05 02 ☽ ⊥ ♄
10 13 ☽ ∠ ☉
11 48 ☽ ⊼ ♃
13 22 ☽ ⊥ ☿
14 18 ☽ ∠ ♇
14 24 ☿ # ♆
19 13 ☿ ∠ ♀
19 48 ☽ ∠ ♄
21 49 ☽ ⚹ ♅
23 57 ☽ # ♅

24 Monday
03 42 ☽ ∠ ☿
04 42 ☽ □ ♀
05 02 ☽ ⊥ ♇
11 50 ☽ # ♃
12 31 ☽ ⚹ ☉
14 58 ☽ ⊻ ♇
18 07 ♄ ⊼ ♆
20 46 ☽ □ ♆
20 47 ☽ ⚹ ♄
22 36 ☽ ∠ ♅

25 Tuesday
00 29 ☽ ⊻ ♂
03 53 ☽ ⚹ ☿
10 02 ☽ Q ☉
11 03 ☽ ⊥ ♂
13 42 ☽ ☍ ♃
13 46 ☽ ⊥ ♅
16 19 ♀ Q ♃
17 08 ♀ ± ♅
17 13 ☽ Q ♄
23 35 ☽ Q ☿

26 Wednesday
00 01 ☽ ⊻ ♅
01 46 ☉ □ ♇
03 08 ☽ ∠ ♂
03 14 ☽ Q ♆
12 00 ☽ △ ♀
15 29 ♃ ± ♅
16 28 ☽ ∥ ♃
18 07 ☽ ☌ ♇
19 20 ☽ □ ☉
21 18 ☿ ⊻ ♂
23 11 ♀ ∥ ♂

27 Thursday
00 08 ☽ ⚹ ♆
00 45 ☽ □ ♄
02 26 ☉ ⊥ ♂
05 16 ☽ ∥ ♅
05 51 ☽ □ ☿
06 37 ☽ ⚹ ♂
15 29 ♀ ⊥ ♄
17 04 ☽ ꝑ ♀
18 17 ☽ ⊼ ♃

28 Friday
01 13 ♂ Q ♅
02 57 ☽ ∠ ♆
03 21 ♂ ± ♃
05 04 ☽ ☌ ♅
05 10 ☽ ± ♃
05 15 ☽ Q ♂
07 18 ☽ # ♇
09 09 ☿ ∥ ♀
11 00 ☽ ± ♀
19 24 ☽ ⊥ ♆
21 44 ☽ ꝑ ♃
23 10 ☽ ⊼ ♀

29 Saturday
00 17 ☽ ⊻ ♇
02 31 ☿ ⊥ ♀
05 52 ☽ △ ☉
06 32 ☽ ⊻ ♆
07 50 ☽ △ ♄
10 18 ☽ △ ☿
11 29 ☽ ⊥ ♇
11 32 ♀ △ ♇
13 04 ☽ # ♂
14 31 ☉ ⊼ ♆
16 16 ☽ □ ♂
18 05 ☽ # ☿
22 59 ☽ # ♀

30 Sunday
01 55 ☽ △ ♃
03 25 ☽ ∥ ♆
04 30 ☽ ∠ ♇
12 11 ☉ ☌ ♄
12 31 ☽ ꝑ ♄
12 32 ☽ ꝑ ☉
13 09 ☽ ⊻ ♅
13 26 ☽ ꝑ ☿
20 06 ☉ ☌ ☿

JULY 1974

Date	Sidereal time h m s	LONGITUDES Sun ☉ ° ' "	Moon ☽ ° ' "	Moon ☽ 24.00 ° ' "	Mercury ☿ ° '	Venus ♀ ° '	Mars ♂ ° '	Jupiter ♃ ° '	Saturn ♄ ° '	Uranus ♅ ° '	Neptune ♆ ° '	Pluto ♇ ° '
01	06 36 37	09 ♋ 16 27	05 ♐ 29 50	11 ♐ 38 29	08 ♋ 15	06 ♊ 31	13 ♌ 48	17 ♓ 48	08 ♋ 27	23 ♎ 40	07 ♐ 26	04 ♎ 09
02	06 40 34	10 13 38	17 ♐ 44 38	23 ♐ 48 33	07 R 40	07 42	14 25	17 49	08 35	23 D 40	07 R 24	04 09
03	06 44 30	11 10 49	29 ♐ 50 31	05 ♑ 50 46	07 06	08 53	15 02	17 50	08 43	23 40	07 23	04 10
04	06 48 27	12 08 00	11 ♑ 49 34	17 ♑ 47 09	06 34	10 04	15 39	17 51	08 51	23 40	07 22	04 10
05	06 52 24	13 05 11	23 ♑ 43 45	29 ♑ 39 38	06 04	11 16	16 16	17 52	08 59	23 40	07 20	04 11
06	06 56 20	14 02 22	05 ♒ 35 03	11 ♒ 30 15	05 37	12 27	16 53	17 52	09 06	23 40	07 19	04 12
07	07 00 17	14 59 33	17 ♒ 25 33	23 ♒ 21 14	05 14	13 38	17 30	17 52	09 14	23 40	07 18	04 13
08	07 04 13	15 56 45	29 ♒ 17 38	05 ♓ 15 08	04 55	14 50	18 07	17 R 52	09 22	23 41	07 17	04 13
09	07 08 10	16 53 56	11 ♓ 14 06	17 ♓ 14 58	04 40	16 01	18 44	17 52	09 30	23 41	07 16	04 14
10	07 12 06	17 51 08	23 ♓ 18 09	29 ♓ 24 09	04 30	17 13	19 22	17 51	09 38	23 41	07 14	04 15
11	07 16 03	18 48 20	05 ♈ 33 27	11 ♈ 46 32	04 25	18 24	19 59	17 51	09 45	23 42	07 13	04 16
12	07 19 59	19 45 33	18 ♈ 03 56	24 ♈ 26 08	04 D 24	19 36	20 36	17 50	09 53	23 42	07 12	04 17
13	07 23 56	20 42 47	00 ♉ 53 37	07 ♉ 26 50	04 29	20 47	21 13	17 49	10 01	23 43	07 11	04 18
14	07 27 53	21 40 01	14 ♉ 06 11	20 ♉ 52 00	04 39	21 59	21 50	17 47	10 09	23 44	07 10	04 19
15	07 31 49	22 37 15	27 ♉ 44 29	04 ♊ 43 45	04 55	23 11	22 28	17 46	10 16	23 44	07 09	04 20
16	07 35 46	23 34 30	11 ♊ 49 34	19 ♊ 02 16	05 16	24 23	23 05	17 44	10 24	23 45	07 08	04 21
17	07 39 42	24 31 46	26 ♊ 20 57	03 ♋ 45 10	05 42	25 34	23 42	17 43	10 32	23 46	07 07	04 22
18	07 43 39	25 29 03	11 ♋ 14 09	18 ♋ 46 58	06 14	26 46	24 20	17 41	10 39	23 47	07 06	04 23
19	07 47 35	26 26 19	26 ♋ 22 28	03 ♌ 59 25	06 52	27 58	24 57	17 38	10 47	23 48	07 05	04 24
20	07 51 32	27 23 37	11 ♌ 36 33	19 ♌ 12 30	07 35	29 ♊ 10	25 34	17 36	10 54	23 48	07 04	04 25
21	07 55 28	28 20 54	26 ♌ 46 01	04 ♍ 15 56	08 23	00 ♋ 22	26 12	17 33	11 02	23 49	07 03	04 26
22	07 59 25	29 ♋ 18 12	11 ♍ 41 12	19 ♍ 00 58	09 16	01 34	26 49	17 31	11 10	23 50	07 03	04 27
23	08 03 22	00 ♌ 15 31	26 ♍ 14 35	03 ♎ 21 35	10 15	02 46	27 27	17 28	11 17	23 52	07 02	04 28
24	08 07 18	01 12 49	10 ♎ 21 44	17 ♎ 14 57	11 19	03 59	28 04	17 25	11 25	23 53	07 01	04 30
25	08 11 15	02 10 08	24 ♎ 01 20	00 ♏ 41 07	12 28	05 11	28 42	17 21	11 32	23 54	07 00	04 31
26	08 15 11	03 07 28	07 ♏ 14 38	13 ♏ 42 18	13 42	06 23	29 19	17 18	11 40	23 55	06 59	04 32
27	08 19 08	04 04 47	20 ♏ 04 37	26 ♏ 22 04	15 01	07 35	29 ♌ 57	17 14	11 47	23 56	06 59	04 34
28	08 23 04	05 02 08	02 ♐ 35 12	08 ♐ 44 32	16 25	08 48	00 ♍ 34	17 10	11 55	23 58	06 58	04 35
29	08 27 01	05 59 29	14 ♐ 50 37	20 ♐ 53 55	17 53	10 00	01 12	17 06	12 02	23 59	06 57	04 36
30	08 30 57	06 56 50	26 ♐ 54 56	02 ♑ 54 05	19 26	11 12	01 49	17 02	12 09	24 01	06 57	04 38
31	08 34 54	07 ♌ 54 12	08 ♑ 51 47	14 ♑ 48 24	21 ♋ 03	12 ♋ 25	02 ♍ 27	16 ♓ 57	12 ♋ 17	24 ♎ 02	06 ♐ 56	04 ♎ 39

Date	Moon True ☊ ° '	Moon Mean ☊ ° '	Moon ☽ Latitude ° '
01	19 ♐ 25	18 ♐ 19	01 S 16
02	19 D 25	18 15	00 S 09
03	19 R 25	18 12	00 N 57
04	19 24	18 09	02 00
05	19 21	18 06	02 58
06	19 17	18 03	03 47
07	19 13	17 59	04 27
08	19 09	17 56	04 55
09	19 05	17 53	05 10
10	19 02	17 50	05 12
11	18 59	17 47	05 00
12	18 59	17 44	04 34
13	18 D 59	17 40	03 53
14	19 00	17 37	02 59
15	19 02	17 34	01 54
16	19 03	17 31	00 N 39
17	19 R 03	17 28	00 S 40
18	19 01	17 25	01 58
19	18 58	17 21	03 09
20	18 53	17 18	04 07
21	18 48	17 15	04 48
22	18 42	17 12	05 08
23	18 37	17 09	05 07
24	18 34	17 05	04 48
25	18 32	17 02	04 13
26	18 D 31	16 59	03 25
27	18 32	16 56	02 28
28	18 34	16 53	01 25
29	18 35	16 50	00 S 20
30	18 R 34	16 46	00 N 45
31	18 ♐ 32	16 ♐ 43	01 N 47

Date	DECLINATIONS Sun ☉ ° '	Moon ☽ ° '	Mercury ☿ ° '	Venus ♀ ° '	Mars ♂ ° '	Jupiter ♃ ° '	Saturn ♄ ° '	Uranus ♅ ° '	Neptune ♆ ° '	Pluto ♇ ° '
01	23 N 07	22 S 28	18 N 41	19 N 50	17 N 53	05 S 56	22 N 35	08 S 39	19 S 57	13 N 36
02	23 03	23 02	18 37	20 03	17 42	05 55	22 35	08 39	19 57	13 35
03	22 58	22 29	18 33	20 17	17 30	05 55	22 35	08 39	19 56	13 34
04	22 53	20 55	18 31	20 29	17 19	05 55	22 34	08 39	19 56	13 34
05	22 48	18 26	18 31	20 41	17 08	05 55	22 34	08 39	19 56	13 33
06	22 42	15 12	18 31	20 53	16 56	05 56	22 33	08 39	19 56	13 32
07	22 36	11 23	18 34	21 04	16 45	05 56	22 33	08 40	19 56	13 32
08	22 29	07 07	18 37	21 15	16 33	05 56	22 32	08 40	19 56	13 31
09	22 22	02 S 34	18 41	21 25	16 21	05 57	22 32	08 40	19 55	13 30
10	22 15	02 N 07	18 47	21 34	16 09	05 57	22 31	08 40	19 55	13 29
11	22 07	06 48	18 54	21 43	15 57	05 58	22 31	08 40	19 55	13 28
12	21 59	11 18	19 02	21 52	15 45	05 58	22 30	08 41	19 55	13 28
13	21 51	15 26	19 10	21 59	15 33	05 59	22 30	08 41	19 55	13 27
14	21 42	18 56	19 20	22 07	15 20	06 00	22 29	08 41	19 55	13 26
15	21 33	21 30	19 30	22 13	15 08	06 00	22 29	08 41	19 55	13 25
16	21 23	22 52	19 41	22 19	14 55	06 01	22 28	08 42	19 55	13 24
17	21 13	22 44	19 52	22 24	14 43	06 02	22 28	08 42	19 54	13 24
18	21 03	21 00	20 03	22 29	14 30	06 03	22 27	08 42	19 54	13 23
19	20 52	17 47	20 15	22 33	14 17	06 04	22 27	08 43	19 54	13 22
20	20 41	13 21	20 26	22 37	14 04	06 06	22 26	08 43	19 54	13 21
21	20 30	08 05	20 37	22 39	13 51	06 07	22 26	08 43	19 54	13 20
22	20 18	02 N 26	20 48	22 42	13 38	06 08	22 25	08 44	19 54	13 19
23	20 06	03 S 12	20 58	22 43	13 25	06 10	22 25	08 44	19 54	13 18
24	19 53	08 31	21 08	22 44	13 12	06 11	22 24	08 45	19 54	13 18
25	19 41	13 14	21 16	22 44	12 58	06 13	22 23	08 45	19 54	13 17
26	19 28	17 09	21 24	22 44	12 45	06 14	22 23	08 46	19 54	13 16
27	19 14	20 05	21 31	22 43	12 31	06 16	22 22	08 46	19 54	13 15
28	19 01	22 05	21 34	22 41	12 18	06 18	22 22	08 47	19 53	13 14
29	18 47	22 55	21 37	22 39	12 04	06 20	22 21	08 47	19 53	13 13
30	18 32	22 39	21 38	22 36	11 50	06 22	22 20	08 48	19 53	13 12
31	18 N 18	21 S 22	21 N 37	22 N 32	11 N 36	06 S 24	22 N 20	08 S 48	19 S 53	13 N 11

ZODIAC SIGN ENTRIES

Date	h	m	Planets
01	01	20	☽ ♐
03	12	19	☽ ♑
06	00	41	☽ ♒
08	13	25	☽ ♓
11	01	10	☽ ♈
13	10	21	☽ ♉
15	15	54	☽ ♊
17	17	56	☽ ♋
19	17	43	☽ ♌
21	04	34	♀ ♋
21	17	10	☽ ♍
23	05	30	☉ ♌
23	18	19	☽ ♎
25	22	45	☽ ♏
27	14	04	♂ ♍
28	07	00	☽ ♐
30	18	11	☽ ♑

LATITUDES

Date	Mercury ☿ ° '	Venus ♀ ° '	Mars ♂ ° '	Jupiter ♃ ° '	Saturn ♄ ° '	Uranus ♅ ° '	Neptune ♆ ° '	Pluto ♇ ° '
01	04 S 30	01 S 36	01 N 15	01 S 12	00 S 35	00 N 35	01 N 38	16 N 38
04	04 46	01 29	01 14	01 13	00 35	00 34	01 38	16 37
07	04 47	01 23	01 13	01 14	00 34	00 34	01 37	16 35
10	04 35	01 16	01 12	01 15	00 34	00 34	01 37	16 34
13	04 12	01 08	01 11	01 16	00 34	00 34	01 37	16 32
16	03 40	01 00	01 10	01 17	00 34	00 34	01 37	16 31
19	03 01	00 52	01 09	01 18	00 34	00 34	01 37	16 29
22	02 19	00 44	01 08	01 18	00 33	00 34	01 37	16 28
25	01 35	00 36	01 07	01 19	00 33	00 34	01 37	16 27
28	00 52	00 28	01 06	01 20	00 33	00 34	01 36	16 25
31	00 S 11	00 S 19	01 N 04	01 S 21	00 S 33	00 N 33	01 N 36	16 N 24

DATA

Julian Date	2442230
Delta T	+45 seconds
Ayanamsa	23° 30' 20"
Synetic vernal point	05° ♓ 36' 39"
True obliquity of ecliptic	23° 26' 31"

MOON'S PHASES, APSIDES AND POSITIONS ☽

Date	h	m	Phase	Longitude ° '	Eclipse Indicator
04	12	40	○	12 ♑ 10	
12	15	28	☾	19 ♈ 54	
19	12	07	●	26 ♋ 27	
26	03	51	☽	02 ♏ 48	

Day	h	m		
06	21	05	Apogee	
19	21	29	Perigee	
02	12	04	Max dec	23° S 02'
10	01	11	ON	
16	22	05	Max dec	23° N 00'
22	22	16	OS	
29	18	16	Max dec	22° S 57'

LONGITUDES

Date	Chiron ⚷ ° '	Ceres ⚳ ° '	Pallas ⚴ ° '	Juno ⚵ ° '	Vesta ⚶ ° '	Black Moon Lilith ☾ ° '
01	24 ♈ 11	15 ♓ 06	07 ♒ 17	17 ♓ 22	11 ♎ 13	05 ♒ 31
11	24 ♈ 22	15 ♓ 21	04 ♒ 56	18 ♓ 23	14 ♎ 25	06 ♒ 38
21	24 ♈ 29	15 ♓ 00	02 ♒ 17	18 ♓ 47	17 ♎ 58	07 ♒ 45
31	24 ♈ 30	14 ♓ 04	29 ♑ 37	18 ♓ 30	21 ♎ 51	08 ♒ 52

ASPECTARIAN

01 Monday
00 41 ☽ ⊥ ♅
05 02 ☿ ☌ ♄
05 59 ☽ ± ☿
06 02 ☽ ± ♄
07 19 ☽ ± ☉
09 22 ☽ ✶ ♇
14 12 ☽ ☍ ♀
14 42 ☽ # ♄
15 45 ☽ ☌ ♆
17 07 ☽ ⊼ ☿
17 50 ☽ ⊼ ♄
18 10 ☽ ∠ ♅
19 59 ☽ ⊼ ☉
20 59 ☿ ⊥ ♂

02 Tuesday
00 17 ♅ St D
00 21 ♀ # ♆
05 05 ☽ △ ♂
06 05 ♀ ☍ ♆
08 52 ☽ Q ♇
11 25 ☿ ⊻ ♀
12 09 ☽ □ ♃
20 44 ♂ ⊥ ♄
23 06 ☿ ⊼ ♆
23 42 ☽ ✶ ♅

03 Wednesday
07 24 ♀ bQ ♅
08 10 ♀ ⊻ ♄
09 57 ☽ # ♄
12 23 ☽ bQ ♂
20 38 ☽ □ ♇
23 38 ☽ Q ♅

04 Thursday
00 00 ☽ Q ♃
01 52 ☽ ☍ ☿
03 03 ☽ ⊻ ♆
05 57 ☽ ☍ ♄
07 23 ☽ ± ♂
08 05 ☽ ⊼ ♀
12 40 ☽ ☍ ☉
15 05 ☽ ⊥ ♆
16 25 ☽ # ♀
20 06 ☽ ⊼ ♂
21 30 ☽ ± ♀
22 30 ☽ ∥ ♆

05 Friday
00 08 ☽ ✶ ♃
09 12 ☽ ∠ ♆
11 24 ☽ # ☿
11 52 ☽ □ ♅
17 41 ☽ bQ ♀
18 15 ☉ ± ♆
23 00 ☽ # ♂

06 Saturday
06 29 ☽ ∠ ♃
09 11 ☽ △ ♇
12 05 ☽ ⊼ ☿
15 31 ☽ ✶ ♆
19 13 ☽ ⊼ ♄
22 53 ☽ # ♇
23 51 ☽ ± ☿

07 Sunday
00 44 ☽ ⊥ ♃
03 28 ☽ △ ♀
06 38 ☽ ⊼ ☉
07 31 ☽ ± ♄
12 10 ☽ ☍ ♂
12 54 ☽ ⊻ ♃
15 37 ☽ bQ ♇
15 47 ☽ Q ♆
16 18 ♃ St R
17 32 ☽ bQ ☿
19 51 ☽ ± ☉

08 Monday
00 29 ☉ ∥ ♄
00 39 ☽ △ ♅
01 56 ☽ bQ ♄
02 12 ♂ ⊼ ♃
03 31 ☽ ∥ ♅
09 50 ☽ ± ♇
15 37 ☽ bQ ☉
18 20 ☽ ∥ ♃
21 56 ☽ ⊼ ♇
23 05 ☽ △ ☿

09 Tuesday
04 03 ☽ □ ♆
06 53 ☽ bQ ♅
08 29 ☽ △ ♄
22 36 ☽ □ ♀

10 Wednesday
00 16 ☽ △ ☉
00 52 ☽ ± ♅
01 13 ☽ ☌ ♃
03 46 ☽ ⊼ ♂
07 39 ♂ ∠ ♇
12 03 ☉ △ ♃
12 46 ☽ ⊼ ♅
16 16 ☽ ± ♂
16 36 ☿ ∠ ♂

11 Thursday
00 51 ♀ □ ♃
07 38 ☽ # ♃
09 29 ☽ ☍ ♇
09 47 ☽ □ ☿
10 49 ☽ bQ ♂
13 49 ☽ Q ♀
15 13 ☽ △ ♆
20 12 ☽ □ ♄
21 49 ☽ # ♅

12 Friday
01 56 ☿ St D
11 33 ☽ ⊻ ♃
15 12 ☽ ✶ ♀
15 28 ☽ □ ☉
17 02 ☽ △ ♂
19 48 ☽ bQ ♆
20 13 ☽ Q ☿
22 39 ☽ ☍ ♅
22 51 ☽ ⊥ ♃
23 12 ☉ ∥ ♀

13 Saturday
00 10 ☽ ∥ ♇
04 18 ☉ ⊻ ♀
06 37 ☽ Q ♄
12 32 ☽ ± ♆
12 42 ☽ ∥ ♂
15 32 ☽ ∠ ♃
18 15 ☽ ⊼ ♇
18 40 ☽ ✶ ☿
21 53 ☽ ∠ ♀
23 31 ☽ ⊼ ♆

14 Sunday
03 24 ☽ Q ☉
04 49 ☽ ✶ ♄
05 10 ☽ ± ♇
05 57 ♀ ✶ ♂
15 25 ☽ ∥ ☿
15 41 ☽ ⊥ ♀
18 33 ☽ ✶ ♃
20 05 ☽ # ♆
21 16 ☽ bQ ♇
22 02 ☽ ∠ ☿

15 Monday
00 24 ☉ bQ ♆
00 31 ☉ ⊻ ♂
02 22 ☽ □ ♂
02 25 ☽ ✶ ☉
03 18 ☽ ⊻ ♀
04 26 ☉ Q ♇
05 02 ☽ ⊼ ♅
07 40 ☽ ∠ ♄
12 26 ☽ ∥ ☉
14 04 ☽ ⊥ ☿
15 27 ☽ ± ♅
15 30 ☽ Q ♃
22 27 ☽ ∥ ♀
23 17 ♀ △ ♅
23 19 ☽ ⊥ ♄
23 20 ☽ △ ♇

16 Tuesday
00 36 ☽ ⊻ ☿
02 16 ☽ ∥ ♄
04 06 ☽ ☍ ♆
06 08 ☽ ∠ ☉
06 20 ♃ ± ♅
06 49 ☽ bQ ♅
09 35 ☽ ⊻ ♄
10 42 ☽ Q ♂
16 28 ☉ □ ♅
21 50 ☽ □ ♃
22 15 ☽ ⊥ ☉

17 Wednesday
07 29 ☽ ✶ ♂
07 46 ☽ △ ♅
08 49 ☽ ⊻ ☉
10 37 ☽ ☌ ♀
14 16 ♂ ✶ ♅
17 09 ☿ # ♆
17 45 ☽ ∥ ♄
18 22 ☽ ∥ ♀

18 Thursday
01 00 ☽ □ ♇
03 39 ♀ ∥ ♄
03 41 ☽ ☌ ☿
05 24 ☽ ⊼ ♆
08 49 ☽ ∠ ♂
11 04 ☽ ☌ ♄
11 36 ☽ ∥ ☉
14 58 ☽ ± ♆
19 46 ☽ ∥ ☿
21 27 ☽ # ♆
22 13 ☽ △ ♃
23 46 ☽ ⊥ ♂

19 Friday
05 14 ☽ bQ ♆
05 43 ☽ Q ♇
07 55 ☽ □ ♅
09 39 ☽ ⊻ ♂
12 07 ☽ ☌ ☉
14 44 ☽ ⊻ ♀
19 47 ☿ ⊼ ♆
21 51 ☽ bQ ♃

20 Saturday
00 39 ☽ ✶ ♇
01 00 ☽ ⊥ ♀
04 52 ☽ △ ♆
05 19 ☽ ⊻ ☿
08 14 ☽ ∥ ♂
10 53 ☽ ⊻ ♄
11 57 ☽ ∥ ♇
11 59 ☽ ± ♃
12 19 ☽ Q ♅
15 16 ☽ ⊥ ☿
16 23 ☽ ∠ ♀
20 26 ☽ ⊥ ♄
21 26 ☽ ⊼ ♃

21 Sunday
00 21 ☽ ∠ ♇
03 44 ☉ ∥ ☿
04 08 ♂ ∠ ♄
06 18 ☽ ∠ ☿
07 19 ☽ ✶ ♅
09 12 ☽ # ♅
10 49 ☽ ∠ ♄
11 03 ☽ ☌ ♂
14 40 ☽ ⊥ ♇
14 42 ☽ ⊻ ☉
18 15 ☽ ✶ ♀
20 25 ☽ # ♃

22 Monday
00 17 ☽ ⊻ ♇
00 58 ☽ ⊥ ☉
04 29 ☽ □ ♆
07 23 ☽ ∠ ♅
07 49 ☽ ✶ ☿
11 08 ☽ ✶ ♄
15 21 ☽ Q ♀
16 34 ☽ ∠ ☉
21 29 ☽ ☍ ♃
22 05 ☽ ⊥ ♅

23 Tuesday
04 51 ☽ Q ☿
07 02 ☽ Q ♄
08 01 ☽ ⊻ ♅
09 59 ☽ Q ♆
14 06 ☽ ⊻ ♂
19 14 ☽ ✶ ☉

24 Wednesday
00 02 ☽ □ ♀
00 30 ♂ ∥ ♇
00 43 ☽ ⊥ ♂
01 09 ☽ ∥ ♃
01 55 ☽ ☌ ♇
06 15 ☽ ✶ ♆
11 27 ☉ # ♆
13 06 ☽ ∥ ♅
13 48 ☽ □ ☿
13 50 ☽ □ ♄
14 14 ☿ ☌ ♄
16 55 ☽ ∠ ♂
17 19 ☽ Q ☉
22 31 ♀ □ ♇

25 Thursday
00 14 ☽ ⊼ ♃
04 54 ♂ ⊥ ♇
08 24 ☽ ∠ ♆
10 38 ☽ # ♂
10 49 ☽ ± ♃
11 47 ☽ ☌ ♅
12 15 ☽ # ♇
16 22 ☉ bQ ♃
20 49 ☽ ✶ ♂
22 25 ☿ ± ♆

26 Friday
00 34 ☽ ⊥ ♆
02 58 ☽ bQ ♃
03 51 ☽ □ ☉
07 01 ☽ ⊻ ♇
10 15 ☽ △ ♀
11 32 ☽ ⊻ ♆
18 06 ☽ ⊥ ♇
19 56 ☽ Q ♂
20 16 ☽ △ ♄
23 59 ♀ ⊼ ♆

27 Saturday
01 21 ☽ △ ☿
04 26 ☽ # ☉
06 39 ☽ △ ♃
09 26 ☿ ∠ ♂
09 39 ☽ ∥ ♆
11 01 ☽ ∠ ♇
17 17 ☽ bQ ♀
19 22 ☽ ⊻ ♅

28 Sunday
00 22 ☉ ✶ ♇
00 56 ☽ bQ ♄
03 58 ☽ # ☿
06 55 ☽ ⊥ ♅
07 53 ☽ □ ♂
09 27 ☽ bQ ☿
12 27 ☽ ± ♀
15 53 ☽ ✶ ♇
17 10 ☽ △ ☉
17 20 ☽ # ♄
18 32 ☽ ± ♄
20 31 ☽ ☌ ♆
23 51 ☿ △ ♃

29 Monday
00 27 ☽ ∠ ♅
01 03 ☽ # ♀
01 25 ☽ ⊼ ♀
05 22 ☽ ± ☿
06 24 ☽ ⊼ ♄
15 30 ☽ Q ♇
16 26 ☽ □ ♃
18 54 ☽ ⊼ ☿

30 Tuesday
01 14 ☽ bQ ☉
06 11 ☽ ✶ ♅
12 00 ☉ △ ♆
13 45 ☽ # ♀
19 59 ☽ # ♄
20 47 ☽ ± ☉
22 23 ☽ △ ♂

31 Wednesday
03 30 ☽ □ ♇
04 11 ☽ Q ♃
06 18 ☽ Q ♅
08 07 ☽ ⊻ ♆
08 27 ☽ # ☿
08 58 ♀ ☌ ♄
09 54 ☽ ⊼ ☉
18 58 ☽ ☍ ♄
19 59 ☽ ☍ ♀
20 13 ☽ ⊥ ♆
22 18 ♀ ± ♆

All ephemeris data is given at 12.00 UT and the Moon's longitude is additionally given for 24.00 UT

AUGUST 1974

LONGITUDES

Date	Sidereal time h m s	Sun ☉ ° ' "	Moon ☽ ° ' "	Moon ☽ 24.00 ° ' "	Mercury ☿ ° '	Venus ♀ ° '	Mars ♂ ° '	Jupiter ♃ ° '	Saturn ♄ ° '	Uranus ♅ ° '	Neptune ♆ ° '	Pluto ♇ ° '
01	08 38 51	08 ♌ 51 35	20 ♑ 44 15	26 ♑ 39 39	22 ♋ 44	13 ♋ 37	03 ♍ 05	16 ♓ 53	12 ♋ 53	24 ♎ 04	06 ♐ 56	04 ♎ 41
02	08 42 47	09 48 58	02 ♒ 34 52	08 ♒ 30 07	24 29	14 50	03 42	16 R 48	12 31	24 05	06 R 55	04 42
03	08 46 44	10 46 22	14 ♒ 25 39	20 ♒ 21 40	26 17	16 03	04 20	16 43	12 38	24 07	06 55	04 44
04	08 50 40	11 43 48	26 ♒ 18 22	02 ♓ 15 57	28 ♋ 08	17 15	04 58	16 38	12 46	24 09	06 54	04 45
05	08 54 37	12 41 14	08 ♓ 14 38	14 ♓ 14 37	00 ♌ 01	18 28	05 35	16 33	12 53	24 10	06 54	04 47
06	08 58 33	13 38 41	20 ♓ 16 09	26 ♓ 19 30	01 58	19 41	06 13	16 28	13 00	24 12	06 53	04 48
07	09 02 30	14 36 09	02 ♈ 24 59	08 ♈ 32 53	03 56	20 53	06 51	16 22	13 07	24 14	06 53	04 50
08	09 06 26	15 33 39	14 ♈ 43 35	20 ♈ 57 27	05 56	22 06	07 29	16 17	13 14	24 16	06 53	04 52
09	09 10 23	16 31 10	27 ♈ 14 56	03 ♉ 36 26	07 57	23 19	08 07	16 11	13 21	24 18	06 52	04 53
10	09 14 20	17 28 42	10 ♉ 02 25	16 ♉ 33 20	09 59	24 32	08 44	16 05	13 28	24 20	06 52	04 55
11	09 18 16	18 26 16	23 ♉ 09 36	29 ♉ 51 39	12 02	25 45	09 22	15 59	13 35	24 22	06 52	04 57
12	09 22 13	19 23 52	06 ♊ 39 47	13 ♊ 34 18	14 05	26 58	10 00	15 53	13 42	24 24	06 51	04 58
13	09 26 09	20 21 28	20 ♊ 35 19	27 ♊ 42 53	16 09	28 11	10 38	15 46	13 49	24 26	06 51	05 00
14	09 30 06	21 19 07	04 ♋ 56 49	12 ♋ 16 48	18 12	29 ♋ 24	11 16	15 40	13 55	24 28	06 51	05 02
15	09 34 02	22 16 47	19 ♋ 42 16	27 ♋ 12 27	20 14	00 ♌ 37	11 54	15 33	14 02	24 30	06 51	05 04
16	09 37 59	23 14 28	04 ♌ 46 25	12 ♌ 22 59	22 16	01 50	12 32	15 26	14 09	24 32	06 51	05 06
17	09 41 55	24 12 10	20 ♌ 00 53	27 ♌ 38 45	24 17	03 04	13 10	15 20	14 15	24 35	06 51	05 07
18	09 45 52	25 09 54	05 ♍ 15 09	12 ♍ 48 44	26 17	04 17	13 48	15 13	14 22	24 37	06 51	05 09
19	09 49 49	26 07 39	20 ♍ 18 15	27 ♍ 42 35	28 ♌ 16	05 30	14 26	15 06	14 28	24 39	06 D 51	05 11
20	09 53 45	27 05 25	05 ♎ 00 51	12 ♎ 22 19	00 ♍ 14	06 44	15 04	14 58	14 35	24 42	06 51	05 13
21	09 57 42	28 03 12	19 ♎ 16 37	26 ♎ 13 26	02 11	07 57	15 43	14 51	14 41	24 44	06 51	05 15
22	10 01 38	29 01 01	03 ♏ 02 47	09 ♏ 44 49	04 07	09 10	16 21	14 44	14 48	24 47	06 51	05 17
23	10 05 35	29 ♌ 58 51	16 ♏ 19 49	22 ♏ 48 15	06 01	10 24	16 59	14 37	14 54	24 49	06 51	05 19
24	10 09 31	00 ♍ 56 42	29 ♏ 10 35	05 ♐ 27 27	07 54	11 37	17 37	14 29	15 00	24 52	06 51	05 21
25	10 13 28	01 54 34	11 ♐ 39 25	17 ♐ 47 10	09 46	12 51	18 15	14 22	15 06	24 54	06 51	05 23
26	10 17 24	02 52 27	23 ♐ 51 20	29 ♐ 52 32	11 36	14 05	18 54	14 14	15 13	24 57	06 52	05 25
27	10 21 21	03 50 22	05 ♑ 51 23	11 ♑ 48 27	13 25	15 18	19 32	14 06	15 19	25 00	06 52	05 27
28	10 25 18	04 48 18	17 ♑ 44 16	23 ♑ 39 21	15 13	16 32	20 10	13 59	15 25	25 02	06 52	05 29
29	10 29 14	05 46 16	29 ♑ 34 07	05 ♒ 28 59	16 59	17 46	20 49	13 51	15 31	25 05	06 53	05 31
30	10 33 11	06 44 14	11 ♒ 24 17	17 ♒ 20 20	18 44	18 59	21 27	13 43	15 37	25 08	06 53	05 33
31	10 37 07	07 ♍ 42 15	23 ♒ 17 22	29 ♒ 15 36	20 ♍ 28	20 ♌ 13	22 ♍ 05	13 ♓ 35	15 ♋ 42	25 ♎ 11	06 ♐ 53	05 ♎ 35

Date	Moon True ☊ ° '	Moon Mean ☊ ° '	Moon ☽ Latitude ° '
01	18 ♐ 28	16 ♐ 40	02 N 44
02	18 R 21	16 37	03 34
03	18 13	16 34	04 14
04	18 03	16 31	04 44
05	17 53	16 27	05 01
06	17 43	16 24	05 05
07	17 35	16 21	04 55
08	17 28	16 18	04 32
09	17 24	16 15	03 55
10	17 22	16 11	03 06
11	17 D 22	16 08	02 06
12	17 23	16 05	00 N 57
13	17 R 23	16 02	00 S 17
14	17 22	15 59	01 32
15	17 18	15 56	02 43
16	17 13	15 52	03 44
17	17 04	15 49	04 30
18	16 55	15 46	04 56
19	16 45	15 43	05 02
20	16 36	15 40	04 47
21	16 29	15 36	04 15
22	16 24	15 33	03 28
23	16 21	15 30	02 32
24	16 21	15 27	01 30
25	16 D 21	15 24	00 S 25
26	16 R 21	15 21	00 N 40
27	16 19	15 17	01 41
28	16 15	15 14	02 38
29	16 09	15 11	03 27
30	16 00	15 08	04 08
31	15 ♐ 48	15 ♐ 05	04 N 37

DECLINATIONS

Date	Sun ☉ ° '	Moon ☽ ° '	Mercury ☿ ° '	Venus ♀ ° '	Mars ♂ ° '	Jupiter ♃ ° '	Saturn ♄ ° '	Uranus ♅ ° '	Neptune ♆ ° '	Pluto ♇ ° '
01	18 N 03	19 S 08	21 N 33	22 N 28	11 N 22	06 S 26	22 N 19	08 S 49	19 S 53	13 N 10
02	17 48	16 07	21 27	22 23	11 08	06 28	22 19	08 50	19 53	13 09
03	17 32	12 27	21 19	22 18	10 54	06 30	22 18	08 50	19 53	13 08
04	17 16	08 18	21 08	22 12	10 40	06 32	22 17	08 51	19 53	13 07
05	17 00	03 S 50	20 54	22 05	10 26	06 34	22 17	08 52	19 53	13 06
06	16 44	00 N 49	20 38	21 57	10 12	06 36	22 16	08 52	19 53	13 05
07	16 27	05 28	20 19	21 49	09 57	06 39	22 16	08 53	19 53	13 04
08	16 10	09 58	19 58	21 40	09 43	06 41	22 15	08 54	19 53	13 03
09	15 53	14 09	19 34	21 31	09 29	06 44	22 14	08 54	19 53	13 02
10	15 36	17 46	19 07	21 21	09 14	06 46	22 14	08 55	19 53	13 01
11	15 18	20 35	18 39	21 10	08 59	06 49	22 13	08 56	19 53	13 00
12	15 00	22 22	18 08	20 59	08 45	06 51	22 12	08 57	19 53	12 59
13	14 42	22 49	17 35	20 47	08 30	06 54	22 12	08 58	19 53	12 58
14	14 24	21 49	17 00	20 35	08 15	06 57	22 11	08 58	19 53	12 57
15	14 05	19 18	16 23	20 22	08 00	07 00	22 10	08 59	19 53	12 56
16	13 46	15 27	15 45	20 08	07 46	07 02	22 10	09 00	19 53	12 55
17	13 27	10 35	15 05	19 54	07 31	07 05	22 09	09 01	19 53	12 54
18	13 08	04 N 59	14 25	19 39	07 16	07 08	22 08	09 02	19 53	12 53
19	12 49	00 S 47	13 43	19 24	07 01	07 11	22 08	09 03	19 53	12 52
20	12 29	06 23	13 00	19 08	06 45	07 14	22 07	09 04	19 54	12 51
21	12 09	11 28	12 16	18 51	06 30	07 17	22 07	09 05	19 54	12 50
22	11 49	15 48	11 32	18 34	06 15	07 20	22 06	09 05	19 54	12 49
23	11 29	19 09	10 47	18 17	06 00	07 23	22 05	09 06	19 54	12 48
24	11 08	21 26	10 02	17 59	05 45	07 26	22 05	09 07	19 54	12 47
25	10 48	22 36	09 16	17 40	05 29	07 29	22 04	09 08	19 54	12 46
26	10 27	22 38	08 30	17 21	05 14	07 32	22 03	09 09	19 54	12 45
27	10 06	21 37	07 44	17 02	04 59	07 35	22 03	09 10	19 54	12 44
28	09 45	19 39	06 58	16 42	04 43	07 38	22 02	09 11	19 54	12 43
29	09 24	16 52	06 12	16 21	04 28	07 41	22 01	09 12	19 54	12 42
30	09 02	13 23	05 25	16 00	04 12	07 44	22 01	09 13	19 54	12 41
31	08 N 41	09 S 23	04 N 39	15 N 39	03 N 57	07 S 47	22 N 00	09 S 14	19 S 55	12 N 40

ZODIAC SIGN ENTRIES

Date	h	m	Planets
02	06	46	☽ ♒
04	19	26	☽ ♓
05	11	42	☿ ♌
07	07	15	☽ ♈
09	17	13	☽ ♉
12	00	15	☽ ♊
14	03	49	☽ ♋
14	23	47	♀ ♌
16	04	26	☽ ♌
18	03	42	☽ ♍
20	03	45	☽ ♎
20	09	04	☿ ♍
22	06	37	☽ ♏
23	12	29	☉ ♍
24	13	34	☽ ♐
27	00	15	☽ ♑
29	12	53	☽ ♒

LATITUDES

Date	Mercury ☿ ° '	Venus ♀ ° '	Mars ♂ ° '	Jupiter ♃ ° '	Saturn ♄ ° '	Uranus ♅ ° '	Neptune ♆ ° '	Pluto ♇ ° '
01	00 N 02	00 S 17	01 N 04	01 S 21	00 S 33	00 N 33	01 N 36	16 N 24
04	00 36	00 S 08	01 03	01 22	00 32	00 33	01 36	16 22
07	01 05	00 00	01 02	01 23	00 32	00 33	01 36	16 21
10	01 26	00 N 08	01 01	01 23	00 32	00 33	01 36	16 20
13	01 39	00 16	01 00	01 24	00 32	00 33	01 36	16 19
16	01 45	00 24	00 59	01 25	00 32	00 33	01 35	16 18
19	01 45	00 31	00 57	01 25	00 32	00 33	01 35	16 17
22	01 39	00 38	00 56	01 26	00 32	00 33	01 35	16 16
25	01 28	00 45	00 55	01 26	00 31	00 33	01 35	16 16
28	01 14	00 51	00 54	01 26	00 31	00 32	01 35	16 15
31	00 N 57	00 N 57	00 N 53	01 S 27	00 S 31	00 N 32	01 N 35	16 N 14

DATA

Julian Date	2442261
Delta T	+45 seconds
Ayanamsa	23° 30' 25"
Synetic vernal point	05° ♓ 36' 34"
True obliquity of ecliptic	23° 26' 31"

MOON'S PHASES, APSIDES AND POSITIONS ☽

Date	h	m	Phase	Longitude ° '	Eclipse Indicator
03	03	57	○	10 ♒ 27	
11	02	46	☾	18 ♉ 04	
17	19	02	●	24 ♌ 29	
24	15	38	☽	01 ♐ 05	

Day	h	m		
03	01	08	Apogee	
17	07	13	Perigee	
30	05	09	Apogee	
06	07	49	0N	
13	07	49	Max dec	22° N 51'
19	08	43	0S	
26	00	51	Max dec	22° S 45'

LONGITUDES

Date	Chiron ⚷ ° '	Ceres ⚳ ° '	Pallas ⚴ ° '	Juno ⚵ ° '	Vesta ⚶ ° '	Black Moon Lilith ⚸ ° '
01	24 ♈ 30	13 ♓ 56	29 ♑ 21	18 ♓ 26	22 ♎ 15	08 ♒ 59
11	24 ♈ 25	12 ♓ 25	26 ♑ 56	17 ♓ 22	26 ♎ 25	10 ♒ 05
21	24 ♈ 15	10 ♓ 28	24 ♑ 56	15 ♓ 38	00 ♏ 47	11 ♒ 12
31	24 ♈ 00	08 ♓ 17	23 ♑ 30	13 ♓ 25	05 ♏ 21	12 ♒ 19

All ephemeris data is given at 12.00 UT and the Moon's longitude is additionally given for 24.00 UT

ASPECTARIAN

h m	Aspects
01 Thursday	
00 41	☿ ⚼ ♆
04 15	☽ ⚹ ♃
04 51	☽ ∥ ♆
06 19	☽ ⚼ ♂
11 14	☿ Q ♇
14 25	☽ ∠ ♆
16 44	☽ ☍ ☿
18 45	☽ □ ♅
22 12	☽ # ☉
02 Friday	
01 34	☽ ± ♂
06 39	☿ □ ♅
10 26	☽ ∠ ♃
14 25	☽ ⊼ ♂
16 19	☽ △ ♇
20 47	☽ ⚹ ♆
03 Saturday	
03 57	☽ ☍ ☉
04 33	☽ ⊥ ♃
07 43	☽ # ♇
08 21	☽ ⊼ ♄
10 43	♀ ∥ ♄
10 50	☉ ± ♃
15 38	☽ ⊼ ♀
16 37	☽ ⊻ ♃
20 37	☽ ± ♄
21 04	☽ Q ♆
21 45	☽ # ♂
22 45	☽ ⚼ ♇
04 Sunday	
00 37	♀ △ ♃
03 42	♂ ⊻ ♇
05 08	☽ ± ♀
07 38	☽ △ ♅
08 57	☽ ∥ ♅
14 58	☽ ⚼ ♄
16 21	☽ ⊼ ☿
16 57	☽ ± ♇
21 34	☽ ∥ ♃
22 42	☉ Q ♅
05 Monday	
01 20	☽ ⚼ ♀
03 54	☿ ⊥ ♂
05 02	☽ ⊼ ♇
06 23	☽ ☍ ♂
06 42	☽ ± ☿
09 18	☽ □ ♆
13 52	☽ ⚼ ♅
15 02	♄ ± ♆
17 30	☉ ⊻ ♄
21 22	☽ △ ♄
21 40	☽ ⊼ ☉
06 Tuesday	
04 09	☽ ⚼ ☿
04 29	☽ ☌ ♃
06 07	☿ ⚼ ♃
07 53	☽ ± ♅
10 39	☽ ± ☉
10 41	☽ △ ♀
19 49	☽ ⊼ ♅
07 Wednesday	
06 00	☽ ⚼ ☉
13 12	♂ □ ♆
15 33	☽ △ ☿
16 45	☽ ☍ ♇
18 13	☽ # ♃
20 45	☽ △ ♆
21 09	☽ ⊼ ♂
23 00	☿ ⚹ ♇
08 Thursday	
06 07	☽ # ♅
07 32	♀ ⚼ ♆
09 05	☽ □ ♄
09 27	☽ ± ♂
10 40	☽ ∥ ♂
13 45	☽ △ ☉
14 58	☽ ⊻ ♃
16 56	☿ # ♆
23 12	☿ △ ♆
09 Friday	
01 45	☽ ⚼ ♆
02 25	☽ ⊥ ♃
03 20	♀ Q ♇
03 29	♀ ∠ ♂
03 42	☽ ⚼ ♂
03 43	☽ □ ♀
04 16	☉ ⊼ ♃
05 25	☽ ∥ ♇
06 22	☽ ☍ ♅
14 41	☿ ⊻ ♂
18 51	☽ ± ♆
19 23	☽ ∠ ♃
19 50	☽ Q ♄
22 13	☽ ∥ ☉
10 Saturday	
02 26	☽ ⊼ ♇
06 06	☽ ⊼ ♆
07 51	♀ □ ♅
09 28	☽ △ ♂
11 53	☽ □ ☿
13 00	☿ ± ♃
13 38	☽ ± ♇
17 05	☽ Q ♀
18 23	☽ ⚹ ♄
21 07	☽ ∥ ☿
23 03	☽ ⚹ ♃
11 Sunday	
02 46	☽ □ ☉
05 12	☽ # ♆
06 10	☽ ⚼ ♇
11 36	♂ ∠ ♅
14 10	☽ ⊼ ♅
15 51	☿ Q ♅
17 07	☽ ⚹ ♀
17 20	♂ # ♅
17 49	☽ ∥ ♀
20 35	☽ Q ♃
21 48	☽ ∠ ♄
12 Monday	
00 55	☽ ± ♅
02 31	☽ Q ☿
07 06	☿ ⊻ ♄
09 01	☽ ∥ ♄
09 02	☽ △ ♇
12 20	☽ ☍ ♆
13 23	☽ Q ☉
13 49	☽ ⊥ ♄
16 47	☽ ⚼ ♅
18 06	☽ □ ♂
20 27	☉ ⊥ ♄
22 07	☽ ∠ ♀
13 Tuesday	
00 19	☽ ⊻ ♄
02 52	☉ ∠ ♇
03 07	☽ ⚹ ☿
03 50	☽ □ ♃
07 50	☿ ⊼ ♃
11 35	☽ ⚹ ☉
14 57	☽ ⊥ ♀
18 31	☽ △ ♅
14 Wednesday	
01 58	☽ ⊻ ♀
02 10	☽ Q ♂
06 22	☽ ∥ ♄
08 38	☽ ∠ ☿
12 09	☽ □ ♇
14 25	☽ ∠ ☉
15 08	☽ ⊼ ♆
22 49	☽ ⚹ ♂
15 Thursday	
00 56	☽ ± ♆
01 46	☽ ⊥ ☿
02 43	☽ ∥ ♀
02 47	☽ ☌ ♄
05 21	☽ △ ♃
06 06	☽ ⊥ ☉
07 28	☽ # ♆
09 30	☿ ⊥ ♄
09 57	☿ ∠ ♇
10 45	♀ ⚼ ♃
12 59	☽ ⊻ ☿
15 27	☽ ⚼ ♆
16 25	☽ ⊻ ☉
17 24	☽ Q ♇
19 42	☽ □ ♅
16 Friday	
00 01	☽ ∠ ♂
05 11	☽ ⚼ ♃
06 57	☽ ☌ ♀
10 03	☽ ∥ ☿
12 30	☽ ⚹ ♇
14 54	☽ ⊥ ♂
15 17	☽ △ ♆
19 19	☽ ± ♃
21 17	☽ ∥ ☉
17 Saturday	
00 17	☽ Q ♅
00 46	☽ ⊻ ♂
00 54	☽ ∥ ♇
02 53	☽ ⊻ ♄
04 41	☽ ⊼ ♃
10 07	☉ ☌ ☿
12 10	☽ ∠ ♇
12 23	☽ ⊥ ♄
12 51	♀ # ♆
15 34	☿ ⚹ ♅
18 46	☽ # ♅
19 02	☽ ☌ ☉
19 12	☽ ⚹ ♅
19 44	☽ ☌ ☿
21 46	☉ ⚹ ♅
18 Sunday	
01 57	☽ ∥ ♂
02 21	☽ ⊥ ♇
02 39	☽ ∠ ♄
03 01	☽ # ♃
10 20	☽ ⊻ ♀
11 51	☽ ⊻ ♇
14 31	☽ □ ♆
18 56	☽ ∠ ♅
20 41	☽ ⊥ ♀
22 10	♂ # ♃
19 Monday	
02 11	☽ ☌ ♂
02 35	☽ ⚹ ♄
03 35	♆ St D
03 42	☽ ☍ ♃
05 35	♀ ⚹ ♇
07 02	☉ ∥ ♇
09 21	☽ ⊥ ♅
12 21	☽ ∠ ♀
13 40	♂ ⚹ ♄
19 03	☽ ⊻ ♅
19 20	☽ Q ♆
22 04	☽ Q ♄
22 05	☽ ⊻ ☉
23 18	☿ ⊥ ♇
20 Tuesday	
02 55	☽ ⊻ ☿
03 29	☿ ∠ ♄
08 36	☽ ⊥ ☉
08 52	♂ ☍ ♃
12 20	☽ ☌ ♇
13 35	☽ # ♂
14 20	♀ △ ♆
14 21	☽ ⊥ ☿
15 02	☽ ⚹ ♆
15 06	☽ ⚹ ♀
15 50	☽ ∥ ♃
16 53	☿ ∥ ♇
21 Wednesday	
00 16	☽ ∥ ♅
00 40	☽ ∠ ☉
04 08	☽ □ ♄
04 32	☽ ⊼ ♃
05 38	☽ ⊻ ♂
07 52	☽ ∠ ☿
13 16	☽ Q ♀
14 41	☽ ± ♃
15 14	☽ # ☉
15 33	☽ # ☿
16 23	☽ ⊥ ♂
16 25	☽ ∠ ♆
19 04	☽ # ♇
19 22	☉ ∥ ☿
21 26	☽ ☌ ♅
22 Thursday	
04 07	♀ ± ♃
04 21	☽ ⚹ ☉
05 19	♃ △ ♄
06 12	☽ ⚼ ♃
08 07	☽ ⊥ ♆
08 50	☽ ∠ ♂
14 13	☽ ⚹ ☿
15 59	☽ ⊻ ♇
18 47	☽ ⊻ ♆
18 50	☉ ⊥ ♇
23 Friday	
00 05	☽ □ ♀
02 49	☽ ⊥ ♇
02 58	☿ ⊻ ♇
03 25	☽ Q ☉
05 37	☽ # ♀
08 52	☽ △ ♃
09 21	☽ △ ♄
09 46	☉ ∠ ♄
13 16	☽ ⚹ ♂
15 38	☽ Q ☿
18 39	☽ ∥ ♆
19 23	☽ ∠ ♇
22 36	☿ □ ♆
24 Saturday	
03 50	☽ ⊻ ♅
11 45	♀ ⊥ ♂
12 53	☽ Q ♂
13 35	☽ ⚼ ♄
15 13	☽ ⊥ ♅
15 38	☽ □ ☉
22 16	☽ # ♄
23 49	☽ ⚹ ♇
25 Sunday	
02 42	☽ ☌ ♆
07 00	☽ ± ♄
07 40	☽ □ ☿
08 35	☽ ∠ ♅
13 08	♀ Q ♅
13 55	☿ ∠ ♅
14 35	☽ △ ♀
16 09	☿ # ♅
17 13	☽ □ ♃
18 48	☽ ⊼ ♄
23 14	☽ Q ♇
26 Monday	
01 38	☽ □ ♂
14 11	☽ ⚹ ♅
14 17	♂ ⊥ ♅
14 45	♀ ⊼ ♃
23 35	☽ ⚼ ♀
27 Tuesday	
04 30	☽ # ♄
04 33	☽ Q ♃
07 35	☽ △ ☉
11 11	☽ □ ♇
12 10	♀ ⊻ ♄
14 02	☽ ⊻ ♆
14 18	☽ Q ♅
16 33	☿ # ♃
19 45	☽ ± ♀
20 32	☿ ☍ ♃
28 Wednesday	
02 09	☽ ⊥ ♆
04 28	☽ ⚹ ♃
05 59	☽ △ ☿
07 15	☽ ☍ ♄
09 16	☽ ⊼ ♀
09 28	☽ ∥ ♆
14 50	☿ ⚹ ♄
16 34	☽ ⚼ ☉
17 13	☽ △ ♂
20 23	☽ ∠ ♆
29 Thursday	
02 52	☽ □ ♅
05 27	☉ ⊻ ♇
10 33	☽ ∠ ♃
12 27	☽ ± ☉
16 15	☽ # ♀
17 46	☽ ⚼ ☿
30 Friday	
00 06	☽ △ ♇
00 13	☉ # ♅
01 23	☽ ⚼ ♂
01 42	☽ ⊼ ☉
02 50	☽ ⚹ ♆
04 37	☽ ⊥ ♃
15 10	☽ ± ☿
15 36	☉ □ ♆
16 25	☽ # ♇
16 38	☽ ⊻ ♃
17 36	☿ ⊥ ♅
20 34	☽ ⊼ ♄
20 39	☽ ± ♂
23 57	☿ ⊻ ♀
31 Saturday	
03 07	☽ Q ♆
05 06	☽ ☍ ♀
05 21	☽ ⊼ ☿
06 32	☽ ⚼ ♇
08 47	☽ ± ♄
09 27	☽ ⊼ ♂
12 49	☽ ∥ ♅
15 49	☽ △ ♅
16 21	☽ # ☉
19 22	♀ ∠ ♇
20 50	☽ ∥ ♃

SEPTEMBER 1974

LONGITUDES

Date	Sidereal time h m s	Sun ☉ ° ' "	Moon ☽ ° ' "	Moon ☽ 24.00 ° ' "	Mercury ☿ ° '	Venus ♀ ° '	Mars ♂ ° '	Jupiter ♃ ° '	Saturn ♄ ° '	Uranus ♅ ° '	Neptune ♆ ° '	Pluto ♇ ° '
01	10 41 04	08 ♍ 40 16	05 ♓ 15 15	11 ♓ 16 25	22 ♍ 11	21 ♌ 27	22 ♍ 44	13 ♓ 27	15 ♋ 48	25 ♎ 14	06 ♐ 54	05 ♎ 37
02	10 45 00	09 38 20	17 ♓ 19 17	23 ♓ 23 56	23 52	22 41	23 22	13 R 19	15 54	25 16	06 54	05 39
03	10 48 57	10 36 25	29 ♓ 30 29	05 ♈ 39 05	25 32	23 55	24 01	13 11	15 59	25 19	06 55	05 41
04	10 52 53	11 34 32	11 ♈ 49 51	18 ♈ 02 56	27 11	25 09	24 39	13 03	16 05	25 22	06 55	05 44
05	10 56 50	12 32 40	24 ♈ 18 32	00 ♉ 36 50	28 ♍ 48	26 23	25 18	12 56	16 10	25 25	06 56	05 46
06	11 00 47	13 30 51	06 ♉ 58 07	13 ♉ 22 38	00 ♎ 25	27 37	25 56	12 48	16 16	25 28	06 56	05 48
07	11 04 43	14 29 03	19 ♉ 50 42	26 ♉ 22 38	02 00	28 ♌ 51	26 35	12 40	16 21	25 31	06 57	05 50
08	11 08 40	15 27 18	02 ♊ 58 48	09 ♊ 39 31	03 34	00 ♍ 05	27 14	12 32	16 26	25 34	06 58	05 52
09	11 12 36	16 25 35	16 ♊ 25 05	23 ♊ 15 49	05 07	01 19	27 52	12 24	16 32	25 38	06 58	05 54
10	11 16 33	17 23 54	00 ♋ 11 54	07 ♋ 13 28	06 39	02 33	28 31	12 16	16 37	25 41	06 59	05 57
11	11 20 29	18 22 15	14 ♋ 20 30	21 ♋ 32 52	08 09	03 47	29 10	12 08	16 42	25 44	07 00	05 59
12	11 24 26	19 20 38	28 ♋ 50 15	06 ♌ 12 09	09 39	05 02	29 ♍ 48	12 00	16 47	25 47	07 00	06 01
13	11 28 22	20 19 03	13 ♌ 37 54	21 ♌ 06 38	11 07	06 16	00 ♎ 27	11 52	16 52	25 50	07 01	06 03
14	11 32 19	21 17 30	28 ♌ 37 19	06 ♍ 08 47	12 34	07 30	01 06	11 45	16 56	25 54	07 02	06 06
15	11 36 16	22 15 59	13 ♍ 39 50	21 ♍ 09 12	14 00	08 44	01 45	11 37	17 01	25 57	07 03	06 08
16	11 40 12	23 14 30	28 ♍ 35 38	05 ♎ 58 02	15 25	09 59	02 24	11 29	17 06	26 00	07 04	06 10
17	11 44 09	24 13 03	13 ♎ 15 23	20 ♎ 26 53	16 48	11 13	03 03	11 22	17 10	26 03	07 05	06 12
18	11 48 05	25 11 38	27 ♎ 31 53	04 ♏ 30 00	18 10	12 28	03 42	11 14	17 15	26 07	07 06	06 15
19	11 52 02	26 10 14	11 ♏ 21 00	18 ♏ 04 53	19 31	13 42	04 21	11 07	17 19	26 10	07 07	06 17
20	11 55 58	27 08 53	24 ♏ 41 48	01 ♐ 12 02	20 51	14 57	05 00	10 59	17 23	26 14	07 08	06 19
21	11 59 55	28 07 32	07 ♐ 36 02	13 ♐ 54 16	22 09	16 11	05 39	10 52	17 28	26 17	07 09	06 22
22	12 03 51	29 ♍ 06 14	20 ♐ 07 21	26 ♐ 15 55	23 26	17 26	06 18	10 45	17 32	26 20	07 10	06 24
23	12 07 48	00 ♎ 04 57	02 ♑ 20 36	08 ♑ 22 04	24 41	18 40	06 57	10 38	17 36	26 24	07 11	06 26
24	12 11 45	01 03 42	14 ♑ 21 01	20 ♑ 18 05	25 55	19 55	07 36	10 31	17 40	26 27	07 12	06 28
25	12 15 41	02 02 29	26 ♑ 13 54	02 ♒ 09 03	27 07	21 09	08 15	10 24	17 44	26 31	07 13	06 31
26	12 19 38	03 01 18	08 ♒ 04 06	13 ♒ 59 33	28 17	22 24	08 55	10 17	17 47	26 34	07 14	06 33
27	12 23 34	04 00 08	19 ♒ 55 52	25 ♒ 53 28	29 ♎ 26	23 39	09 34	10 11	17 51	26 38	07 16	06 35
28	12 27 31	04 59 00	01 ♓ 52 42	07 ♓ 53 51	00 ♏ 32	24 53	10 13	10 04	17 55	26 42	07 17	06 38
29	12 31 27	05 57 54	13 ♓ 57 10	20 ♓ 02 49	01 36	26 08	10 52	09 58	17 58	26 45	07 18	06 40
30	12 35 24	06 ♎ 56 49	26 ♓ 10 59	02 ♈ 21 43	02 ♏ 38	27 ♍ 23	11 ♎ 32	09 ♓ 51	18 ♋ 01	26 ♎ 49	07 ♐ 19	06 ♎ 42

Date	Moon True ☊ ° '	Moon Mean ☊ ° '	Moon ☽ Latitude ° '
01	15 ♐ 35	15 ♐ 02	04 N 55
02	15 R 21	14 58	05 00
03	15 08	14 55	04 51
04	14 56	14 52	04 28
05	14 47	14 49	03 52
06	14 41	14 46	03 05
07	14 37	14 42	02 07
08	14 36	14 39	01 N 01
09	14 D 36	14 36	00 S 10
10	14 R 35	14 33	01 22
11	14 34	14 30	02 30
12	14 30	14 27	03 31
13	14 23	14 23	04 19
14	14 14	14 20	04 50
15	14 04	14 17	05 01
16	13 52	14 14	04 51
17	13 42	14 11	04 22
18	13 33	14 08	03 37
19	13 27	14 04	02 41
20	13 24	14 01	01 37
21	13 23	13 58	00 S 31
22	13 D 23	13 55	00 N 36
23	13 R 23	13 52	01 39
24	13 22	13 48	02 37
25	13 19	13 45	03 27
26	13 14	13 42	04 08
27	13 06	13 39	04 38
28	12 56	13 36	04 57
29	12 45	13 33	05 02
30	12 ♐ 32	13 ♐ 29	04 N 54

DECLINATIONS

Date	Sun ☉ ° '	Moon ☽ ° '	Mercury ☿ ° '	Venus ♀ ° '	Mars ♂ ° '	Jupiter ♃ ° '	Saturn ♄ ° '	Uranus ♅ ° '	Neptune ♆ ° '	Pluto ♇ ° '
01	08 N 19	05 S 00	03 N 53	15 N 17	03 N 41	07 S 51	22 N 00	09 S 16	19 S 55	12 N 39
02	07 57	00 S 25	03 07	14 55	03 26	07 54	21 59	09 17	19 55	12 38
03	07 35	04 N 15	02 21	14 32	03 10	07 57	21 58	09 18	19 55	12 37
04	07 13	08 47	01 35	14 09	02 54	08 00	21 58	09 19	19 55	12 36
05	06 51	13 01	00 50	13 45	02 39	08 03	21 57	09 20	19 55	12 35
06	06 29	16 45	00 N 05	13 21	02 23	08 06	21 56	09 21	19 55	12 34
07	06 07	19 44	00 S 40	12 57	02 07	08 09	21 56	09 22	19 56	12 33
08	05 44	21 45	01 24	12 32	01 51	08 12	21 55	09 23	19 56	12 32
09	05 21	22 35	02 08	12 07	01 36	08 15	21 55	09 24	19 56	12 31
10	04 59	22 05	02 51	11 42	01 20	08 18	21 54	09 26	19 56	12 30
11	04 36	20 11	03 34	11 16	01 04	08 21	21 54	09 27	19 56	12 29
12	04 13	16 57	04 17	10 50	00 48	08 25	21 53	09 28	19 56	12 28
13	03 50	12 36	04 58	10 24	00 32	08 28	21 52	09 29	19 57	12 27
14	03 27	07 25	05 40	09 58	00 17	08 30	21 52	09 30	19 57	12 26
15	03 04	01 N 48	06 21	09 31	00 N 01	08 33	21 51	09 32	19 57	12 26
16	02 41	03 S 53	07 01	09 04	00 S 15	08 36	21 51	09 33	19 57	12 25
17	02 18	09 15	07 40	08 36	00 31	08 39	21 50	09 34	19 57	12 24
18	01 55	13 58	08 19	08 09	00 47	08 42	21 50	09 35	19 58	12 23
19	01 31	17 47	08 57	07 41	01 03	08 45	21 49	09 37	19 58	12 22
20	01 08	20 31	09 34	07 13	01 19	08 48	21 49	09 38	19 58	12 21
21	00 45	22 05	10 11	06 44	01 35	08 50	21 48	09 39	19 58	12 20
22	00 N 21	22 29	10 47	06 16	01 50	08 53	21 48	09 40	19 58	12 19
23	00 S 02	21 46	11 22	05 47	02 06	08 56	21 47	09 42	19 59	12 18
24	00 25	20 04	11 56	05 18	02 22	08 58	21 47	09 43	19 59	12 17
25	00 49	17 31	12 29	04 49	02 38	09 01	21 46	09 44	19 59	12 16
26	01 12	14 16	13 01	04 20	02 54	09 03	21 46	09 45	19 59	12 15
27	01 35	10 26	13 32	03 51	03 10	09 06	21 45	09 47	20 00	12 14
28	01 59	06 11	14 02	03 21	03 26	09 08	21 45	09 48	20 00	12 13
29	02 22	01 40	14 30	02 52	03 41	09 11	21 45	09 49	20 00	12 13
30	02 S 45	02 N 59	14 S 58	02 N 22	03 S 57	09 S 13	21 N 44	09 S 51	20 S 00	12 N 12

ZODIAC SIGN ENTRIES

Date	h m	Planets
01	01 29	☽ ♓
03	12 58	☽ ♈
05	22 50	☽ ♉
06	05 48	☿ ♎
08	06 36	☽ ♊
08	10 28	♀ ♍
10	11 40	☽ ♋
12	13 54	☽ ♌
12	19 08	♂ ♎
14	14 12	☽ ♍
16	14 17	☽ ♎
18	16 14	☽ ♏
20	21 46	☽ ♐
23	07 22	☽ ♑
23	09 58	☉ ♎
25	19 38	☽ ♒
28	00 20	☿ ♏
28	08 15	☽ ♓
30	19 25	☽ ♈

LATITUDES

Date	Mercury ☿ ° '	Venus ♀ ° '	Mars ♂ ° '	Jupiter ♃ ° '	Saturn ♄ ° '	Uranus ♅ ° '	Neptune ♆ ° '	Pluto ♇ ° '
01	00 N 51	00 N 59	00 N 52	01 S 27	00 S 31	00 N 32	01 N 34	16 N 14
04	00 31	01 04	00 51	01 27	00 31	00 32	01 34	16 14
07	00 N 09	01 09	00 50	01 27	00 31	00 32	01 34	16 13
10	00 S 14	01 13	00 48	01 27	00 31	00 32	01 34	16 13
13	00 37	01 17	00 47	01 28	00 31	00 32	01 34	16 12
16	01 01	01 20	00 46	01 28	00 30	00 32	01 34	16 12
19	01 25	01 22	00 45	01 28	00 30	00 32	01 33	16 12
22	01 48	01 24	00 43	01 27	00 30	00 32	01 33	16 12
25	02 11	01 26	00 42	01 27	00 30	00 32	01 33	16 12
28	02 32	01 27	00 40	01 27	00 30	00 32	01 33	16 12
31	02 S 51	01 N 27	00 N 39	01 S 27	00 S 30	00 N 32	01 N 33	16 N 12

DATA

Julian Date	2442292
Delta T	+45 seconds
Ayanamsa	23° 30' 29"
Synetic vernal point	05° ♓ 36' 30"
True obliquity of ecliptic	23° 26' 31"

MOON'S PHASES, APSIDES AND POSITIONS ☽

Date	h m	Phase	Longitude ° '	Eclipse Indicator
01	19 25	○	08 ♓ 58	
09	12 01	☾	16 ♊ 26	
16	02 45	●	22 ♍ 52	
23	07 08	☽	29 ♐ 53	

Day	h m	
14	15 22	Perigee
26	17 16	Apogee

Day	h m		
02	14 06	0N	
09	15 14	Max dec	22° N 36'
15	19 32	0S	
22	08 20	Max dec	22° S 30'
29	20 38	0N	

LONGITUDES

Date	Chiron ⚷ ° '	Ceres ⚳ ° '	Pallas ⚴ ° '	Juno ⚵ ° '	Vesta ⚶ ° '	Black Moon Lilith ⚸ ° '
01	23 ♈ 59	08 ♓ 04	23 ♑ 24	13 ♓ 11	05 ♏ 49	12 ♒ 26
11	23 ♈ 39	05 ♓ 53	22 ♑ 38	10 ♓ 43	10 ♏ 33	13 ♒ 33
21	23 ♈ 17	03 ♓ 57	22 ♑ 30	08 ♓ 23	15 ♏ 26	14 ♒ 40
31	22 ♈ 51	02 ♓ 26	22 ♑ 56	06 ♓ 29	20 ♏ 25	15 ♒ 47

All ephemeris data is given at 12.00 UT and the Moon's longitude is additionally given for 24.00 UT

ASPECTARIAN

h m	Aspects
01 Sunday	
00 41	☽ ± ♇
03 01	☽ bQ ♄
12 44	☽ ⊼ ♇
15 17	☽ □ ♆
19 09	☽ # ☿
19 24	☽ # ♂
19 25	☽ ☍ ☉
19 27	♀ ⊥ ♄
21 16	☿ ∥ ♂
21 57	☽ bQ ♅
02 Monday	
00 37	☿ ☌ ♂
04 09	☽ ☌ ♃
09 09	☽ △ ♄
15 36	☉ # ♃
15 53	☽ ± ♅
23 46	☽ ⊼ ♀
03 Tuesday	
00 36	☽ ☍ ♂
02 58	☽ ☍ ☿
02 59	☿ Q ♆
03 35	☽ ∥ ☿
03 45	☽ ⊼ ♅
04 36	☉ ∠ ♅
06 42	☽ ∥ ♂
08 53	☿ ⊻ ♅
12 53	☽ ± ♀
16 04	♀ ⊻ ♂
04 Wednesday	
00 07	☽ ☍ ♇
02 28	☽ △ ♆
04 14	☽ ∥ ☉
07 42	☽ # ♃
08 22	☽ bQ ♀
11 28	☽ ⊼ ☉
14 21	☽ ⊻ ♃
14 53	☽ # ♅
16 38	♀ ✶ ♅
20 16	☽ □ ♄
22 03	♂ Q ♆
05 Thursday	
00 01	☽ ± ☉
01 48	☽ ⊥ ♃
02 05	☿ Q ♄
07 27	☽ bQ ♆
09 26	☽ ∥ ♇
13 59	☽ ⊼ ♂
14 08	☽ ☍ ♅
16 01	☽ ∥ ♀
16 22	☽ △ ♀
17 06	♂ ⊻ ♅
18 41	☽ bQ ☉
18 49	☽ ∠ ♃
20 18	☉ ☍ ♃
21 49	☽ ⊼ ☿
06 Friday	
00 36	☽ ± ♆
02 00	☽ ± ♂
06 52	☽ Q ♄
09 48	☽ ⊼ ♇
10 48	☽ ± ☿
11 57	☽ ⊼ ♆
19 50	☽ bQ ♂
21 05	☽ ± ♇
22 48	☽ ✶ ♃
07 Saturday	
01 15	☽ △ ☉
05 29	☽ ✶ ♄
06 00	☽ bQ ☿
13 50	☽ # ♆
13 50	☽ bQ ♇
20 46	☽ Q ♃
22 29	☽ ⊼ ♅
08 Sunday	
01 01	☽ △ ♂
06 12	☽ □ ♀
07 50	♀ ⊥ ♇
09 12	☽ ∠ ♄
09 27	☽ ± ♅
12 04	♀ ∥ ♇
13 12	☽ △ ☿
14 52	☽ ∥ ♄
17 14	☽ △ ♇
19 10	☽ ☍ ♆
23 04	☿ # ♂
09 Monday	
01 29	☽ ⊥ ♄
01 41	☽ bQ ♅
04 57	☽ □ ♃
12 01	☽ □ ☉
12 12	☽ ⊻ ♄
14 43	☉ ✶ ♄
16 25	♀ ∠ ♄
17 36	☽ Q ♀
10 Tuesday	
00 41	☿ ☌ ♇
04 10	☽ △ ♅
08 58	☽ □ ♂
15 22	☽ ∥ ♄
16 07	♂ Q ♄
16 26	☽ ✶ ♀
17 21	☿ ✶ ♆
21 33	☽ Q ☉
21 51	☽ □ ♇
23 36	☽ ⊼ ♆
11 Wednesday	
00 20	☽ □ ☿
08 20	☽ △ ♃
09 44	☽ ± ♆
14 12	☽ # ♆
15 58	☽ ☌ ♄
16 56	☽ Q ♂
19 13	☽ ✶ ☉
20 07	☽ ∠ ♀
12 Thursday	
00 45	☽ bQ ♆
04 04	☽ Q ♇
06 58	☽ □ ♅
09 01	☽ bQ ♃
09 50	☽ Q ☿
10 45	☉ # ☿
12 20	☽ ⊥ ♀
13 40	☽ ✶ ♂
21 37	☽ ∠ ☉
23 01	☽ ⊻ ♀
23 30	☉ ⊥ ♅
23 35	☽ ± ♃
23 44	☽ ✶ ♇
13 Friday	
01 19	☽ △ ♆
07 30	☽ ✶ ☿
07 52	♀ ⊻ ♇
09 12	☽ ⊼ ♃
12 20	☽ Q ♅
12 42	☽ ∥ ♇
13 11	☽ ⊥ ☉
15 04	☽ ∠ ♂
17 13	☽ ⊻ ♄
23 26	☿ ⊼ ♃
23 29	☽ ⊻ ☉
23 33	☽ ∥ ♀
23 57	☽ ∠ ♇
14 Saturday	
02 43	☽ # ♅
02 51	♀ □ ♆
02 53	☽ ⊥ ♄
06 08	☽ ⊥ ♂
07 14	☽ # ♃
07 38	☽ ✶ ♅
10 09	☽ ∠ ☿
14 21	☽ ⊥ ♇
16 08	☽ ⊻ ♂
17 19	☽ ∠ ♄
18 48	☽ # ☿
23 57	☽ ⊻ ♇
15 Sunday	
01 26	☽ □ ♆
02 01	☽ ⊥ ☿
03 26	☽ ☌ ♀
06 16	☽ ∥ ☉
07 39	☽ ∠ ♅
08 45	☽ ☍ ♃
11 14	♀ # ♅
12 36	☽ ⊻ ☿
17 24	☽ ✶ ♄
19 15	☽ # ♂
19 51	☽ ∥ ♂
22 06	☽ ⊥ ♅
16 Monday	
02 45	☽ ☌ ☉
06 17	☽ Q ♆
07 11	☽ # ☉
07 48	☽ ⊻ ♅
12 49	☽ Q ♄
18 28	☽ ☌ ♂
17 Tuesday	
00 22	☽ ☌ ♇
01 49	☽ ✶ ♆
03 38	☽ ∥ ☿
08 19	☽ ⊻ ♀
08 43	♀ ∠ ♅
08 54	☽ ⊼ ♃
09 10	☽ ∥ ♃
09 12	☽ # ♀
09 34	♀ # ♃
13 29	☽ ∥ ♅
14 30	♀ ☍ ♃
18 31	☽ ☌ ☿
18 33	☽ □ ♄
18 46	☽ ± ♃
18 47	☿ □ ♄
19 13	☽ ⊥ ♀
20 55	☿ ± ♃
18 Wednesday	
02 46	☽ ∠ ♆
03 26	☽ # ♇
07 43	☽ ⊻ ☉
08 12	☿ # ♀
09 32	☉ Q ♆
09 34	☽ ☌ ♅
09 49	☽ bQ ♃
11 52	☽ ∠ ♀
18 07	☽ ⊥ ♆
18 45	☽ ⊥ ☉
23 08	☽ ⊻ ♂
19 Thursday	
03 05	☽ ⊻ ♇
03 37	☿ ∥ ♃
04 22	♃ bQ ♅
04 32	☽ ⊻ ♆
10 08	☽ ⊥ ♂
11 35	☽ △ ♃
11 40	☽ ∠ ☉
11 59	☉ ⊻ ♅
13 39	☽ ⊥ ♇
16 35	☽ ✶ ♀
22 41	☽ △ ♄
20 Friday	
03 01	☽ ∠ ♂
04 13	☽ ⊻ ☿
05 30	☉ # ♂
05 50	☽ ∠ ♇
06 14	☽ ∥ ♆
14 12	☿ ∥ ♅
14 49	☽ ⊻ ♅
16 23	☽ ⊥ ☿
16 33	☽ Q ♀
16 52	☽ ✶ ☉
21 Saturday	
01 58	☽ ⊥ ♅
02 18	☽ bQ ♄
06 02	☽ # ♄
08 07	☽ ✶ ♂
09 39	☽ ✶ ♇
11 04	☽ ∠ ☿
11 08	☽ ☌ ♆
11 52	☿ ∠ ♆
17 11	☽ Q ☉
18 08	☽ □ ♃
19 01	☽ ∠ ♅
19 22	☽ ± ♄
22 Sunday	
06 12	☽ □ ♀
06 57	☽ ⊼ ♄
07 29	☿ ⊥ ♀
08 16	☽ Q ♂
08 39	☽ Q ♇
14 05	♀ ✶ ♄
15 51	♂ ☌ ♇
19 11	☽ ✶ ☿
23 12	☉ Q ♄
23 Monday	
00 12	☽ ✶ ♅
04 44	☽ Q ♃
07 08	☽ □ ☉
11 43	☽ # ♄
20 10	☽ □ ♇
20 46	♂ ✶ ♆
21 38	☽ Q ☿
21 39	☽ ⊻ ♆
21 42	☽ □ ♂
24 Tuesday	
00 07	☽ Q ♅
04 22	☽ ✶ ♃
04 48	☿ bQ ♃
09 41	☽ ⊥ ♆
13 01	☽ ∥ ♆
13 36	☿ ⊥ ♀
18 42	☽ ☍ ♄
23 01	♀ ⊥ ♅
23 19	☿ ☌ ♅
25 Wednesday	
00 32	☽ △ ♀
03 01	☿ # ♇
03 52	☽ ∠ ♆
10 20	☽ ∠ ♃
12 35	☽ □ ♅
13 59	☽ □ ☿
26 Thursday	
00 51	☽ △ ☉
04 24	☽ ⊥ ♃
08 55	☽ △ ♇
10 19	☽ ✶ ♆
10 29	☽ bQ ♀
13 48	☽ △ ♂
16 27	☽ ⊻ ♃
19 10	☽ ∥ ☿
27 Friday	
01 00	☽ # ♇
06 51	☽ ± ♀
07 47	☽ ⊼ ♄
09 57	☽ bQ ☉
10 39	☽ Q ♆
15 21	☽ bQ ♇
15 49	☽ ∥ ♅
19 42	☽ ∥ ♃
19 56	☽ ± ♄
20 21	☽ ⊼ ♀
21 52	☽ bQ ♂
28 Saturday	
01 33	☽ △ ♅
05 41	☽ ± ☉
07 16	♂ ⊼ ♃
09 02	☽ △ ☿
09 29	☽ ± ♇
09 43	♀ # ♂
14 04	☽ bQ ♄
16 57	☽ ± ♂
18 45	☽ ⊼ ☉
19 41	♀ Q ♆
21 30	☽ ⊼ ♇
22 48	☽ □ ♆
29 Sunday	
01 57	☽ ∥ ♂
04 10	☽ ☌ ♃
04 58	☽ # ♀
05 01	☿ ⊥ ♆
05 34	☽ ⊼ ♂
07 38	☽ bQ ♅
08 36	☽ ∥ ☉
17 43	☽ bQ ☿
19 57	☽ △ ♄
30 Monday	
00 30	♀ ⊻ ♅
01 25	☉ # ♀
01 27	☽ ± ♅
05 51	☉ ☌ ♇
09 10	☽ ∥ ♀
10 46	☽ # ☉
12 58	☽ ± ☿
13 14	☽ ⊼ ♅
14 36	☽ ☍ ♀
17 22	☽ # ♂
21 27	☉ ✶ ♆

OCTOBER 1974

Date	Sidereal time h m s	LONGITUDES Sun ☉ ° ' "	Moon ☽ ° ' "	Moon ☽ 24.00 ° ' "	Mercury ☿ ° '	Venus ♀ ° '	Mars ♂ ° '	Jupiter ♃ ° '	Saturn ♄ ° '	Uranus ♅ ° '	Neptune ♆ ° '	Pluto ♇ ° '
01	12 39 20	07 ♎ 55 47	08 ♈ 35 06	14 ♈ 51 10	03 ♏ 38	28 ♍ 38	12 ♎ 11	09 ♓ 45	18 ♋ 05	26 ♎ 52	07 ♐ 21	06 ♎ 45
02	12 43 17	08 54 47	21 ♈ 09 55	27 ♈ 31 23	04 35	29 ♍ 52	12 51	09 R 39	18 08	26 56	07 22	06 47
03	12 47 14	09 53 48	03 ♉ 55 32	10 ♉ 22 25	05 29	01 ♎ 07	13 30	09 33	18 11	27 00	07 24	06 49
04	12 51 10	10 52 52	16 ♉ 52 02	23 ♉ 24 27	06 19	02 22	14 10	09 28	18 14	27 03	07 25	06 52
05	12 55 07	11 51 59	29 ♉ 59 44	06 ♊ 38 00	07 07	03 37	14 49	09 22	18 17	27 07	07 26	06 54
06	12 59 03	12 51 07	13 ♊ 19 22	20 ♊ 03 56	07 50	04 52	15 29	09 17	18 20	27 11	07 28	06 56
07	13 03 00	13 50 18	26 ♊ 51 53	03 ♋ 43 20	08 30	06 07	16 08	09 12	18 22	27 14	07 29	06 58
08	13 06 56	14 49 31	10 ♋ 38 22	17 ♋ 37 03	09 05	07 22	16 48	09 06	18 25	27 18	07 31	07 01
09	13 10 53	15 48 47	24 ♋ 39 23	01 ♌ 45 17	09 35	08 37	17 27	09 02	18 27	27 22	07 32	07 03
10	13 14 49	16 48 05	08 ♌ 54 34	16 ♌ 06 55	10 00	09 51	18 07	08 57	18 30	27 25	07 34	07 05
11	13 18 46	17 47 25	23 ♌ 21 54	00 ♍ 39 00	10 19	11 06	18 47	08 52	18 32	27 29	07 36	07 08
12	13 22 43	18 46 47	07 ♍ 57 31	15 ♍ 16 41	10 31	12 21	19 27	08 48	18 34	27 33	07 37	07 10
13	13 26 39	19 46 12	22 ♍ 35 39	29 ♍ 53 30	10 37	13 36	20 07	08 44	18 36	27 37	07 39	07 12
14	13 30 36	20 45 39	07 ♎ 09 20	14 ♎ 22 15	10 R 36	14 52	20 46	08 40	18 38	27 40	07 41	07 14
15	13 34 32	21 45 08	21 ♎ 31 27	28 ♎ 36 10	10 27	16 07	21 26	08 36	18 40	27 44	07 42	07 17
16	13 38 29	22 44 39	05 ♏ 35 48	12 ♏ 29 54	10 10	17 22	22 06	08 32	18 42	27 48	07 44	07 19
17	13 42 25	23 44 12	19 ♏ 18 09	26 ♏ 00 21	09 44	18 37	22 46	08 28	18 43	27 52	07 46	07 21
18	13 46 22	24 43 47	02 ♐ 36 31	09 ♐ 06 44	09 10	19 52	23 26	08 25	18 45	27 55	07 47	07 23
19	13 50 18	25 43 23	15 ♐ 31 15	21 ♐ 50 25	08 27	21 07	24 06	08 22	18 46	27 59	07 49	07 26
20	13 54 15	26 43 02	28 ♐ 04 38	04 ♑ 14 25	07 36	22 22	24 46	08 19	18 48	28 03	07 51	07 28
21	13 58 12	27 42 42	10 ♑ 20 20	16 ♑ 22 58	06 37	23 37	25 26	08 16	18 49	28 07	07 53	07 30
22	14 02 08	28 42 24	22 ♑ 22 56	28 ♑ 20 53	05 31	24 52	26 07	08 14	18 50	28 11	07 55	07 32
23	14 06 05	29 ♎ 42 08	04 ♒ 17 27	10 ♒ 13 16	04 20	26 08	26 47	08 12	18 51	28 14	07 56	07 35
24	14 10 01	00 ♏ 41 53	16 ♒ 08 57	22 ♒ 05 06	03 06	27 23	27 27	08 09	18 52	28 18	07 58	07 37
25	14 13 58	01 41 40	28 ♒ 02 17	04 ♓ 01 01	01 50	28 38	28 07	08 07	18 52	28 22	08 00	07 39
26	14 17 54	02 41 29	10 ♓ 01 47	16 ♓ 05 01	00 ♏ 35	29 ♎ 53	28 48	08 06	18 53	28 26	08 02	07 41
27	14 21 51	03 41 20	22 ♓ 11 06	28 ♓ 20 19	29 ♎ 22	01 ♏ 08	29 ♎ 28	08 04	18 54	28 29	08 04	07 43
28	14 25 47	04 41 12	04 ♈ 32 56	10 ♈ 49 08	28 16	02 23	00 ♏ 08	08 03	18 54	28 33	08 06	07 45
29	14 29 44	05 41 06	17 ♈ 09 02	23 ♈ 32 41	27 16	03 39	00 49	08 02	18 54	28 37	08 08	07 47
30	14 33 41	06 41 02	00 ♉ 00 05	06 ♉ 31 10	26 25	04 54	01 29	08 01	18 54	28 41	08 10	07 49
31	14 37 37	07 ♏ 41 00	13 ♉ 05 50	19 ♉ 43 57	25 ♎ 45	06 ♏ 09	02 ♏ 10	08 ♓ 00	18 ♋ 54	28 ♎ 44	08 ♐ 12	07 ♎ 52

Date	Moon True ☊ ° '	Moon Mean ☊ ° '	Moon ☽ Latitude ° '
01	12 ♐ 21	13 ♐ 26	04 N 32
02	12 R 10	13 23	03 56
03	12 02	13 20	03 08
04	11 57	13 17	02 09
05	11 54	13 14	01 N 03
06	11 D 53	13 10	00 S 08
07	11 54	13 07	01 19
08	11 55	13 04	02 28
09	11 R 54	13 01	03 28
10	11 52	12 58	04 17
11	11 48	12 54	04 51
12	11 42	12 51	05 05
13	11 34	12 48	05 01
14	11 25	12 45	04 36
15	11 18	12 42	03 55
16	11 11	12 39	02 59
17	11 07	12 35	01 55
18	11 05	12 32	00 S 46
19	11 D 04	12 29	00 N 24
20	11 05	12 26	01 31
21	11 07	12 23	02 32
22	11 08	12 20	03 25
23	11 R 08	12 16	04 09
24	11 07	12 13	04 42
25	11 03	12 10	05 02
26	10 59	12 07	05 10
27	10 53	12 04	05 04
28	10 46	12 00	04 44
29	10 40	11 57	04 10
30	10 34	11 54	03 23
31	10 ♐ 30	11 ♐ 51	02 N 24

DECLINATIONS

Date	Sun ☉ ° '	Moon ☽ ° '	Mercury ☿ ° '	Venus ♀ ° '	Mars ♂ ° '	Jupiter ♃ ° '	Saturn ♄ ° '	Uranus ♅ ° '	Neptune ♆ ° '	Pluto ♇ ° '
01	03 S 09	07 N 34	15 S 24	01 N 52	04 S 13	09 S 15	21 N 44	09 S 52	20 S 01	12 N 11
02	03 32	11 54	15 49	01 23	04 29	09 17	21 43	09 53	20 01	12 10
03	03 55	15 46	16 12	00 53	04 45	09 19	21 43	09 55	20 01	12 09
04	04 18	18 57	16 34	00 N 23	05 00	09 21	21 43	09 56	20 02	12 08
05	04 42	21 11	16 53	00 S 07	05 16	09 23	21 42	09 57	20 02	12 07
06	05 05	22 16	17 11	00 37	05 32	09 25	21 42	09 59	20 02	12 07
07	05 28	22 05	17 28	01 07	05 47	09 27	21 42	10 00	20 02	12 06
08	05 51	20 34	17 41	01 37	06 03	09 29	21 41	10 01	20 03	12 05
09	06 13	17 47	17 53	02 07	06 19	09 31	21 41	10 03	20 03	12 04
10	06 36	13 54	18 02	02 37	06 34	09 32	21 41	10 04	20 03	12 03
11	06 59	09 09	18 08	03 07	06 50	09 34	21 40	10 05	20 04	12 03
12	07 21	03 N 52	18 12	03 37	07 05	09 35	21 40	10 07	20 04	12 02
13	07 44	01 S 40	18 12	04 07	07 21	09 37	21 40	10 08	20 04	12 01
14	08 06	07 04	18 08	04 37	07 36	09 38	21 40	10 09	20 05	12 00
15	08 29	12 01	18 01	05 07	07 52	09 40	21 40	10 11	20 05	12 00
16	08 51	16 12	17 50	05 36	08 07	09 41	21 39	10 12	20 05	11 59
17	09 13	19 24	17 34	06 06	08 22	09 42	21 39	10 13	20 05	11 58
18	09 35	21 26	17 14	06 35	08 38	09 43	21 39	10 15	20 06	11 58
19	09 57	22 15	16 50	07 04	08 53	09 44	21 39	10 16	20 06	11 57
20	10 18	21 55	16 21	07 33	09 08	09 45	21 39	10 17	20 06	11 56
21	10 40	20 31	15 48	08 02	09 23	09 46	21 39	10 19	20 07	11 56
22	11 01	18 13	15 11	08 31	09 38	09 47	21 38	10 20	20 07	11 55
23	11 22	15 10	14 30	09 00	09 53	09 47	21 38	10 22	20 07	11 54
24	11 43	11 31	13 47	09 28	10 08	09 48	21 38	10 23	20 08	11 54
25	12 04	07 25	13 01	09 56	10 23	09 49	21 38	10 24	20 08	11 53
26	12 24	03 S 01	12 16	10 24	10 38	09 49	21 38	10 26	20 08	11 52
27	12 45	01 N 34	11 31	10 52	10 52	09 49	21 38	10 27	20 09	11 52
28	13 05	06 09	10 48	11 19	11 07	09 50	21 38	10 28	20 09	11 51
29	13 25	10 35	10 09	11 46	11 22	09 50	21 38	10 30	20 09	11 51
30	13 45	14 38	09 33	12 13	11 36	09 50	21 38	10 31	20 10	11 50
31	14 S 04	18 N 03	09 S 02	12 S 40	11 S 51	09 S 50	21 N 38	10 S 32	20 S 10	11 N 50

ZODIAC SIGN ENTRIES

Date	h	m	Planets
02	14	27	♀ ♎
03	04	39	☽ ♉
05	12	00	☽ ♊
07	17	30	☽ ♋
09	21	03	☽ ♌
11	22	56	☽ ♍
14	00	11	☽ ♎
16	02	23	☽ ♏
18	07	14	☽ ♐
20	15	44	☽ ♑
23	03	20	☽ ♒
23	19	11	☉ ♏
25	15	57	☽ ♓
26	14	12	♀ ♏
26	23	21	☿ ♎
28	03	13	☽ ♈
28	07	05	♂ ♏
30	12	00	☽ ♉

LATITUDES

Date	Mercury ☿ ° '	Venus ♀ ° '	Mars ♂ ° '	Jupiter ♃ ° '	Saturn ♄ ° '	Uranus ♅ ° '	Neptune ♆ ° '	Pluto ♇ ° '
01	02 S 51	01 N 27	00 N 39	01 S 27	00 S 30	00 N 32	01 N 33	16 N 12
04	03 06	01 26	00 38	01 26	00 30	00 32	01 33	16 12
07	03 18	01 25	00 36	01 26	00 30	00 32	01 33	16 12
10	03 24	01 24	00 35	01 26	00 29	00 32	01 32	16 12
13	03 21	01 22	00 33	01 25	00 29	00 32	01 32	16 13
16	03 07	01 19	00 32	01 25	00 29	00 32	01 32	16 13
19	02 39	01 16	00 30	01 24	00 29	00 32	01 32	16 14
22	01 55	01 12	00 29	01 24	00 29	00 32	01 32	16 14
25	00 S 58	01 08	00 27	01 23	00 29	00 31	01 32	16 15
28	00 N 03	01 03	00 26	01 23	00 29	00 31	01 32	16 16
31	00 N 59	00 N 58	00 N 24	01 S 22	00 S 28	00 N 31	01 N 32	16 N 17

DATA

Julian Date	2442322
Delta T	+45 seconds
Ayanamsa	23° 30' 32"
Synetic vernal point	05° ♓ 36' 28"
True obliquity of ecliptic	23° 26' 31"

MOON'S PHASES, APSIDES AND POSITIONS ☽

Date	h	m	Phase	Longitude ° '	Eclipse Indicator
01	10	38	○	07 ♈ 52	
08	19	46	☾	15 ♋ 09	
15	12	25	●	21 ♎ 46	
23	01	53	☽	29 ♑ 17	
31	01	19	○	07 ♉ 14	

Day	h	m		
12	15	34	Perigee	
24	11	08	Apogee	
06	20	34	Max dec	22° N 21'
13	04	47	0S	
19	16	44	Max dec	22° S 17'
27	03	52	0N	

LONGITUDES

Date	Chiron ⚷ ° '	Ceres ⚳ ° '	Pallas ⚴ ° '	Juno ⚵ ° '	Vesta ⚶ ° '	Black Moon Lilith ⚸ ° '
01	22 ♈ 51	02 ♓ 26	22 ♑ 56	06 ♓ 29	20 ♏ 25	15 ♒ 47
11	22 ♈ 24	01 ♓ 29	23 ♑ 52	05 ♓ 16	25 ♏ 29	16 ♒ 54
21	21 ♈ 57	01 ♓ 08	25 ♑ 15	04 ♓ 52	00 ♐ 39	18 ♒ 01
31	21 ♈ 30	01 ♓ 24	27 ♑ 02	05 ♓ 18	05 ♐ 52	19 ♒ 09

ASPECTARIAN

01 Tuesday
01 38 ☽ ⊼ ☿
08 27 ☽ ☍ ♇
09 37 ☽ △ ♆
10 38 ☽ ☍ ☉
14 14 ☽ ⊻ ♃
19 17 ☽ ☍ ♂
21 12 ☽ # ♃

02 Wednesday
00 34 ☽ # ♅
01 37 ☽ ⊥ ♃
06 13 ☽ □ ♄
13 33 ☽ ∥ ♇
14 17 ☽ bQ ♆
17 11 ♀ Q ♄
18 33 ☽ ∠ ♃
22 56 ☽ ☍ ♅

03 Thursday
04 28 ☉ ⊼ ♃
06 11 ☽ ⊼ ♀
07 15 ☽ ± ♆
15 06 ☽ ☍ ☿
15 15 ☽ # ☿
16 14 ☽ Q ♄
17 25 ☽ ⊼ ♇
18 28 ☽ ⊼ ♆
18 36 ☽ ± ♀
22 25 ☽ ✶ ♃

04 Friday
00 02 ☽ ⊼ ☉
04 35 ☽ ± ♇
06 44 ☽ ⊼ ♂
12 02 ☽ ± ☉
13 01 ☽ bQ ♀
14 31 ☽ ✶ ♄
18 22 ☽ ± ♂
20 23 ☽ Q ♃
21 12 ☽ bQ ♇
22 16 ☽ # ♆

05 Saturday
05 00 ☿ ⊻ ♇
05 51 ☽ bQ ☉
06 44 ☽ ⊼ ♅
11 40 ☽ bQ ♂
17 41 ☽ ± ♅
17 58 ☽ ∠ ♄
19 14 ☽ △ ♀
20 25 ☽ ∥ ♄
23 01 ☿ ⊻ ♆

06 Sunday
00 31 ☽ △ ♇
01 29 ☽ ☍ ♆
01 37 ☽ ⊼ ☿
04 48 ☽ □ ♃
05 43 ♂ ± ♃
09 56 ☽ bQ ♅
10 13 ☽ ⊥ ♄
11 06 ☽ △ ☉
12 58 ☽ ± ☿
16 02 ☽ △ ♂
20 57 ☽ ⊻ ♄

07 Monday
05 47 ☽ bQ ☿
12 40 ☽ △ ♅
20 34 ☽ ∥ ♄

08 Tuesday
05 08 ♀ ☌ ♇
05 42 ☽ □ ♇
05 46 ☽ □ ♀
06 35 ☽ ⊼ ♆
09 11 ☽ △ ☿
09 22 ☽ △ ♃
13 08 ☿ △ ♃
15 03 ♀ ✶ ♆
16 58 ☽ ± ♆
17 21 ☽ # ♆
18 20 ☉ ± ♃
19 46 ☽ □ ☉
23 07 ☽ □ ♂

09 Wednesday
01 24 ☽ ☌ ♄
08 24 ☽ bQ ♆
10 56 ☽ bQ ♃
11 17 ☽ # ☿
12 40 ☽ Q ♇
15 38 ☽ Q ♀
16 36 ☽ □ ♅
19 32 ♀ ⊼ ♃

10 Thursday
02 04 ☽ ± ♃
04 37 ☽ Q ☉
05 32 ☉ ∥ ♂
07 06 ☽ Q ♂
08 57 ☽ ✶ ♇
09 45 ☽ △ ♆
12 04 ☽ ⊼ ♃
13 44 ☽ ✶ ♀
13 52 ☽ □ ☿
15 41 ☿ ⊻ ♀
21 46 ☽ ∥ ♇
22 54 ☽ Q ♅

11 Friday
02 06 ☽ ✶ ☉
02 26 ♂ □ ♄
03 59 ☽ ⊻ ♄
04 04 ☽ ✶ ♂
07 33 ☽ # ♅
09 57 ☽ ∠ ♇
10 03 ☽ # ♃
13 56 ☽ ⊥ ♄
16 57 ☽ ∠ ♀
18 49 ☽ ✶ ♅
20 18 ☽ Q ☿
21 24 ☽ # ☉
22 15 ☽ # ♂

12 Saturday
00 49 ☽ ⊥ ♇
04 39 ☽ ∠ ☉
04 47 ☽ ∠ ♄
05 58 ☽ ∠ ♂
06 42 ☉ □ ♄
09 08 ☽ ⊥ ♀
10 42 ☽ ⊻ ♇
11 27 ☽ □ ♆
12 58 ☽ # ♀
13 22 ☽ ☍ ♃
16 15 ☽ ✶ ☿
19 33 ☽ ∠ ♅
19 53 ☽ ⊻ ♀
20 29 ☽ ⊥ ☉
21 25 ☽ ⊥ ♂

13 Sunday
05 26 ☽ ✶ ♄
07 02 ☽ ⊻ ☉
07 44 ☽ ⊻ ♂
10 23 ☽ ⊥ ♅
16 59 ☽ ∠ ☿
17 02 ☽ Q ♆
19 47 ☿ St R
20 17 ☽ ⊻ ♅
23 50 ☽ ∥ ♀

14 Monday
01 12 ☽ Q ♄
07 48 ☽ ⊥ ☿
08 21 ♀ ± ♃
12 09 ☽ ☌ ♇
12 52 ☽ ✶ ♆
12 56 ☉ ☌ ♂
14 29 ☽ ⊼ ♃
14 37 ☽ ∥ ♂
17 12 ☽ ∥ ☉
17 41 ☽ ⊻ ☿

15 Tuesday
00 10 ☽ ∥ ♃
00 25 ☽ ± ♃
02 02 ☽ ☌ ♀
02 41 ☽ ∥ ♅
07 11 ☽ □ ♄
11 51 ☽ ☌ ♂
11 54 ☽ # ♇
12 25 ☽ ☌ ☉
14 00 ☽ ∠ ♆
15 29 ☽ bQ ♃
22 34 ☽ ☌ ♅

16 Wednesday
05 20 ☽ ⊥ ♆
11 42 ☉ ∠ ♆
14 59 ☽ ⊻ ♇
15 42 ☽ ⊻ ♆
17 04 ☽ △ ♃
19 42 ☽ ☌ ☿
22 27 ☽ ∥ ☿

17 Thursday
01 28 ☽ ⊥ ♇
06 01 ☉ bQ ♃
10 39 ☽ ⊻ ♀
10 58 ☽ △ ♄
11 39 ♂ ∠ ♆
14 09 ♀ □ ♄
17 27 ☽ ∠ ♇
18 31 ☽ ⊻ ♂
18 48 ☽ ∥ ♆
20 33 ☽ ⊻ ☉
22 28 ☽ ⊥ ♀

18 Friday
03 26 ☽ ⊻ ♅
05 54 ☽ ⊥ ♂
08 17 ☽ ⊥ ☉
11 25 ♂ bQ ♃
14 06 ☽ bQ ♄
14 25 ☽ ⊥ ♅
15 58 ☽ # ♄
16 35 ☽ ∠ ♀
20 50 ☽ ✶ ♇
21 34 ☽ ☌ ♆
21 36 ☉ ∥ ♃
22 40 ☽ □ ♃
23 20 ☽ ∠ ♂
23 29 ☽ ⊻ ☿

19 Saturday
02 15 ☽ ∠ ☉
06 49 ☽ ± ♄
07 13 ☽ ∠ ♅
10 06 ☽ ⊥ ☿
14 32 ☿ △ ♃
18 10 ☽ ⊼ ♄
19 25 ☽ Q ♇
23 48 ☽ ✶ ♀

20 Sunday
02 09 ☽ ∠ ☿
05 16 ☽ ✶ ♂
05 23 ☿ ⊻ ♆
08 37 ☽ Q ♃
09 09 ☽ ✶ ☉
11 10 ☉ ∥ ♅
11 57 ☽ ✶ ♅
15 10 ☿ ⊻ ♇
18 18 ☽ # ♄
21 27 ♀ ∠ ♆

21 Monday
01 39 ☽ Q ♀
05 13 ☽ ✶ ☿
05 34 ♀ bQ ♃
05 57 ☽ Q ♂
06 23 ☽ □ ♇
07 08 ☽ ⊻ ♆
07 56 ☽ ✶ ♃
10 39 ☽ Q ☉
11 33 ☽ Q ♅
16 54 ☽ ∥ ♆
19 02 ☽ ⊥ ♆
22 20 ☉ ☌ ♅

22 Tuesday
03 06 ☽ Q ☿
04 53 ☽ ☍ ♄
13 04 ☽ ∠ ♆
13 42 ☽ ∠ ♃
17 35 ☽ □ ♀
19 56 ☽ □ ♂
23 43 ☽ □ ♅

23 Wednesday
01 53 ☽ □ ☉
02 24 ♂ ∥ ♃
07 46 ☽ ⊥ ♃
12 05 ☽ □ ☿
17 48 ☽ ∥ ☿
18 40 ☽ △ ♇
19 24 ☽ ✶ ♆
19 52 ☽ ⊻ ♃

24 Thursday
09 39 ☽ # ♇
10 51 ☽ ∥ ☉
14 57 ♀ ☌ ♂
17 29 ☽ ⊼ ♄
18 51 ☽ ∥ ♅
19 45 ☽ Q ♆
19 52 ☽ ∥ ♂
22 18 ☽ ∥ ♃
23 02 ☽ ∥ ♀
23 48 ☉ # ♇

25 Friday
01 06 ☽ bQ ♇
05 28 ♀ ∥ ♃
05 37 ☽ ± ♄
06 37 ♀ ☌ ♅
08 50 ☿ ⊥ ♆
12 11 ☽ △ ♂
12 40 ☽ △ ♅
13 20 ☽ △ ♀
13 28 ☉ ☌ ☿
14 19 ♂ ∥ ♅
18 54 ☽ △ ☿
19 16 ☽ ± ♇
19 42 ☉ ⊥ ♆
20 01 ☽ △ ☉
21 36 ♂ ☌ ♅
23 43 ☽ bQ ♄

26 Saturday
07 19 ☽ ⊼ ♇
08 01 ☽ □ ♆
08 09 ☽ ☌ ♃
08 54 ☉ ∥ ☿
13 27 ♀ ∥ ♅
18 42 ☿ ☌ ♀
18 47 ☽ bQ ♅
19 55 ☽ bQ ♂
22 00 ☽ bQ ☿
22 44 ☽ bQ ♀

27 Sunday
00 41 ☿ # ♇
04 31 ☽ bQ ☉
05 32 ☽ △ ♄
10 49 ☿ ☌ ♂
12 36 ☽ ± ♅
13 10 ♃ □ ♆
13 33 ♀ ∥ ♂
14 08 ☽ ± ☿
14 39 ☽ ± ♂
18 26 ☽ ± ♀
23 41 ☽ ± ☉

28 Monday
00 21 ☽ ⊼ ♅
00 50 ☽ ⊼ ☿
01 22 ☿ ∥ ♀
03 00 ☽ ⊼ ♂
04 00 ☿ ∥ ♂
05 47 ☿ ☌ ♅
06 16 ♀ ⊥ ♆
07 23 ☽ ⊼ ♀
12 17 ☽ ⊼ ☉
18 10 ☽ ☍ ♇
18 42 ☽ ⊻ ♃
18 50 ☽ △ ♆
23 27 ☿ ∥ ♅

29 Tuesday
06 06 ☽ ⊥ ♃
07 48 ☽ # ♃
09 51 ☽ # ☿
11 29 ☽ # ♅
15 18 ☽ □ ♄
15 50 ♀ # ♇
16 40 ☽ # ♂
19 09 ☽ ∥ ♇
19 33 ☽ # ♀
23 01 ☽ ∠ ♃
23 16 ☽ bQ ♆

30 Wednesday
00 08 ☿ ∥ ♃
05 44 ☽ ☍ ☿
05 58 ☽ # ☉
09 32 ☽ ☍ ♅
14 54 ☽ ☍ ♂
16 00 ☽ ± ♆
21 59 ☽ ☍ ♀

31 Thursday
00 43 ☽ Q ♄
01 19 ☽ ☍ ☉
02 26 ☽ ⊼ ♇
02 43 ☽ ✶ ♃
03 03 ☽ ⊼ ♆
10 01 ♂ # ♇
13 23 ☽ ± ♇
13 27 ♂ ⊥ ♆
14 56 ♄ St R
16 23 ☉ ⊻ ♇
19 35 ☉ △ ♃
22 31 ☽ ✶ ♄

All ephemeris data is given at 12.00 UT and the Moon's longitude is additionally given for 24.00 UT

NOVEMBER 1974

LONGITUDES

Date	Sidereal time h m s	Sun ☉ ° ' "	Moon ☽ ° ' "	Moon ☽ 24.00 ° ' "	Mercury ☿ ° '	Venus ♀ ° '	Mars ♂ ° '	Jupiter ♃ ° '	Saturn ♄ ° '	Uranus ♅ ° '	Neptune ♆ ° '	Pluto ♇ ° '
01	14 41 34	08 ♏ 40 59	26 ♉ 25 19	03 ♊ 09 45	25 ♎ 16	07 ♏ 24	02 ♏ 50	08 ♓ 00	18 ♋ 54	28 ♎ 48	08 ♐ 14	07 ♎ 54
02	14 45 30	09 41 01	09 ♊ 57 03	16 ♊ 47 01	24 R 58	08 40	03 31	07 R 59	18 R 54	28 52	08 16	07 56
03	14 49 27	10 41 05	23 ♊ 39 27	00 ♋ 34 08	24 52	09 55	04 11	07 59	18 54	28 56	08 18	07 58
04	14 53 23	11 41 11	07 ♋ 30 54	14 ♋ 29 34	24 D 57	11 10	04 52	07 D 59	18 54	28 59	08 20	08 00
05	14 57 20	12 41 19	21 ♋ 29 58	28 ♋ 31 56	25 13	12 25	05 33	08 00	18 53	29 03	08 22	08 02
06	15 01 16	13 41 29	05 ♌ 35 17	12 ♌ 39 50	25 39	13 41	06 13	08 00	18 53	29 07	08 24	08 04
07	15 05 13	14 41 41	19 ♌ 45 21	26 ♌ 51 36	26 14	14 56	06 54	08 01	18 52	29 10	08 26	08 06
08	15 09 10	15 41 55	03 ♍ 58 18	11 ♍ 05 06	26 57	16 11	07 35	08 02	18 51	29 14	08 28	08 08
09	15 13 06	16 42 11	18 ♍ 11 39	25 ♍ 17 33	27 48	17 27	08 16	08 03	18 50	29 18	08 31	08 10
10	15 17 03	17 42 29	02 ♎ 22 21	09 ♎ 25 35	28 46	18 42	08 57	08 04	18 49	29 21	08 33	08 11
11	15 20 59	18 42 49	16 ♎ 26 46	23 ♎ 25 26	29 ♎ 49	19 57	09 38	08 06	18 48	29 25	08 35	08 13
12	15 24 56	19 43 11	00 ♏ 21 08	07 ♏ 13 25	00 ♏ 57	21 13	10 19	08 07	18 47	29 29	08 37	08 15
13	15 28 52	20 43 35	14 ♏ 01 54	20 ♏ 46 18	02 09	22 28	11 00	08 09	18 45	29 32	08 39	08 17
14	15 32 49	21 44 01	27 ♏ 26 21	04 ♐ 01 52	03 25	23 43	11 41	08 12	18 44	29 36	08 41	08 19
15	15 36 45	22 44 28	10 ♐ 32 47	16 ♐ 59 05	04 45	24 59	12 22	08 14	18 42	29 39	08 44	08 21
16	15 40 42	23 44 56	23 ♐ 20 52	29 ♐ 38 17	06 06	26 14	13 03	08 16	18 40	29 43	08 46	08 22
17	15 44 38	24 45 27	05 ♑ 51 35	12 ♑ 01 03	07 31	27 29	13 44	08 19	18 39	29 46	08 48	08 24
18	15 48 35	25 45 58	18 ♑ 07 06	24 ♑ 10 09	08 57	28 ♏ 45	14 25	08 22	18 37	29 50	08 50	08 26
19	15 52 32	26 46 31	00 ♒ 10 40	06 ♒ 09 10	10 24	00 ♐ 00	15 07	08 25	18 35	29 53	08 52	08 28
20	15 56 28	27 47 05	12 ♒ 06 13	18 ♒ 02 23	11 53	01 16	15 48	08 29	18 33	29 ♎ 57	08 55	08 29
21	16 00 25	28 47 40	23 ♒ 58 15	29 ♒ 54 26	13 23	02 31	16 29	08 32	18 30	00 ♏ 00	08 57	08 31
22	16 04 21	29 ♏ 48 17	05 ♓ 51 32	11 ♓ 50 08	14 53	03 46	17 11	08 36	18 28	00 04	08 59	08 33
23	16 08 18	00 ♐ 48 54	17 ♓ 50 49	23 ♓ 54 08	16 25	05 02	17 52	08 40	18 25	00 07	09 01	08 34
24	16 12 14	01 49 33	00 ♈ 00 36	06 ♈ 10 42	17 57	06 17	18 33	08 44	18 23	00 11	09 04	08 36
25	16 16 11	02 50 13	12 ♈ 24 52	18 ♈ 43 28	19 29	07 32	19 15	08 48	18 20	00 14	09 06	08 37
26	16 20 07	03 50 54	25 ♈ 06 46	01 ♉ 35 01	21 02	08 48	19 56	08 53	18 18	00 17	09 08	08 39
27	16 24 04	04 51 36	08 ♉ 08 19	14 ♉ 46 43	22 35	10 03	20 38	08 57	18 15	00 20	09 10	08 40
28	16 28 01	05 52 19	21 ♉ 30 09	28 ♉ 18 27	24 08	11 18	21 20	09 02	18 12	00 24	09 13	08 42
29	16 31 57	06 53 04	05 ♊ 11 22	12 ♊ 08 32	25 42	12 34	22 01	09 07	18 09	00 27	09 15	08 43
30	16 35 54	07 ♐ 53 50	19 ♊ 09 32	26 ♊ 13 52	27 ♏ 15	13 ♐ 49	22 ♏ 43	09 ♓ 12	18 ♋ 06	00 ♏ 30	09 ♐ 17	08 ♎ 45

DECLINATIONS

Date	Moon True ☊ ° '	Moon Mean ☊ ° '	Moon ☽ Latitude ° '	Sun ☉ ° '	Moon ☽ ° '	Mercury ☿ ° '	Venus ♀ ° '	Mars ♂ ° '	Jupiter ♃ ° '	Saturn ♄ ° '	Uranus ♅ ° '	Neptune ♆ ° '	Pluto ♇ ° '
01	10 ♐ 27	11 ♐ 48	01 N 16	14 S 24	20 N 35	08 S 37	13 S 06	12 S 05	09 S 50	21 N 38	10 S 34	20 S 10	11 N 49
02	10 R 27	11 45	00 N 03	14 43	21 59	08 18	13 32	12 20	09 50	21 38	10 35	20 11	11 49
03	10 D 27	11 41	01 S 12	15 02	22 06	08 05	13 58	12 34	09 50	21 38	10 36	20 11	11 48
04	10 29	11 38	02 23	15 20	20 51	07 57	14 23	12 48	09 50	21 39	10 38	20 12	11 48
05	10 30	11 35	03 26	15 39	18 20	07 55	14 48	13 02	09 49	21 39	10 39	20 12	11 47
06	10 31	11 32	04 18	15 57	14 42	07 58	15 12	13 16	09 49	21 39	10 40	20 12	11 47
07	10 R 31	11 29	04 54	16 15	10 14	08 06	15 37	13 30	09 49	21 39	10 41	20 13	11 46
08	10 30	11 25	05 12	16 32	05 N 12	08 19	16 00	13 44	09 48	21 39	10 43	20 13	11 46
09	10 28	11 22	05 11	16 50	00 S 07	08 35	16 24	13 57	09 47	21 39	10 44	20 13	11 45
10	10 25	11 19	04 52	17 07	05 24	08 54	16 47	14 11	09 47	21 40	10 45	20 14	11 45
11	10 22	11 16	04 14	17 24	10 23	09 16	17 09	14 25	09 46	21 40	10 47	20 14	11 44
12	10 19	11 13	03 22	17 40	14 45	09 41	17 31	14 38	09 45	21 40	10 48	20 14	11 44
13	10 16	11 10	02 20	17 56	18 16	10 08	17 53	14 51	09 44	21 40	10 49	20 15	11 44
14	10 15	11 06	01 S 10	18 12	20 44	10 36	18 14	15 05	09 43	21 40	10 50	20 15	11 43
15	10 D 14	11 03	00 N 02	18 28	22 00	11 06	18 34	15 18	09 42	21 41	10 52	20 15	11 43
16	10 15	11 00	01 12	18 43	22 02	11 37	18 54	15 31	09 41	21 41	10 53	20 16	11 43
17	10 16	10 57	02 17	18 58	21 12	12 08	19 13	15 44	09 40	21 41	10 54	20 16	11 42
18	10 17	10 54	03 14	19 12	19 00	12 40	19 33	15 56	09 38	21 42	10 55	20 16	11 42
19	10 19	10 51	04 02	19 26	16 10	13 13	19 52	16 09	09 37	21 42	10 57	20 17	11 42
20	10 20	10 47	04 39	19 40	12 42	13 46	20 09	16 22	09 36	21 42	10 58	20 17	11 42
21	10 20	10 44	05 04	19 54	08 45	14 19	20 27	16 34	09 34	21 43	10 59	20 18	11 41
22	10 R 20	10 41	05 16	20 07	04 S 28	14 51	20 44	16 46	09 33	21 43	11 00	20 18	11 41
23	10 20	10 38	05 16	20 19	00 N 01	15 24	21 00	16 58	09 31	21 43	11 01	20 18	11 41
24	10 19	10 35	04 58	20 32	04 34	15 56	21 15	17 10	09 29	21 44	11 03	20 19	11 41
25	10 18	10 31	04 29	20 44	09 01	16 28	21 30	17 22	09 27	21 44	11 04	20 19	11 41
26	10 17	10 28	03 45	20 55	13 12	16 59	21 45	17 34	09 25	21 45	11 05	20 19	11 40
27	10 16	10 25	02 49	21 06	16 53	17 30	21 59	17 46	09 24	21 45	11 06	20 20	11 40
28	10 16	10 22	01 42	21 17	19 47	18 00	22 12	17 57	09 22	21 46	11 07	20 20	11 40
29	10 15	10 19	00 N 28	21 28	21 38	18 29	22 24	18 09	09 19	21 46	11 08	20 20	11 40
30	10 ♐ 15	10 ♐ 16	00 S 49	21 S 38	22 N 11	18 S 58	22 S 36	18 S 20	09 S 17	21 N 47	11 S 09	20 S 21	11 N 40

ZODIAC SIGN ENTRIES

Date	h	m	Planets
01	18	23	☽ ♊
03	23	01	☽ ♋
06	02	30	☽ ♌
08	05	18	☽ ♍
10	07	58	☽ ♎
11	16	05	☿ ♏
12	11	23	☽ ♏
14	16	39	☽ ♐
17	00	42	☽ ♑
19	11	39	☽ ♒
19	11	56	♀ ♐
21	09	32	♅ ♏
22	00	11	☽ ♓
22	16	38	☉ ♐
24	11	59	☽ ♈
26	21	05	☽ ♉
29	02	58	☽ ♊

LATITUDES

Date	Mercury ☿ ° '	Venus ♀ ° '	Mars ♂ ° '	Jupiter ♃ ° '	Saturn ♄ ° '	Uranus ♅ ° '	Neptune ♆ ° '	Pluto ♇ ° '
01	01 N 14	00 N 56	00 N 24	01 S 22	00 S 28	00 N 31	01 N 32	16 N 17
04	01 50	00 51	00 22	01 21	00 28	00 31	01 32	16 18
07	02 10	00 45	00 21	01 21	00 28	00 31	01 31	16 19
10	02 17	00 38	00 19	01 20	00 28	00 31	01 31	16 20
13	02 14	00 32	00 17	01 19	00 28	00 31	01 31	16 21
16	02 04	00 25	00 16	01 19	00 28	00 31	01 31	16 22
19	01 49	00 18	00 14	01 18	00 27	00 32	01 31	16 23
22	01 31	00 11	00 12	01 17	00 27	00 32	01 31	16 25
25	01 11	00 N 04	00 10	01 17	00 27	00 32	01 31	16 26
28	00 50	00 S 03	00 09	01 16	00 27	00 32	01 31	16 28
31	00 N 29	00 S 11	00 N 07	01 S 15	00 S 27	00 N 32	01 N 31	16 N 29

DATA

Julian Date	2442353
Delta T	+45 seconds
Ayanamsa	23° 30' 35"
Synetic vernal point	05° ♓ 36' 25"
True obliquity of ecliptic	23° 26' 31"

MOON'S PHASES, APSIDES AND POSITIONS ☽

Date	h	m	Phase	Longitude ° '	Eclipse Indicator
07	02	47	☾	14 ♌ 19	
14	00	53	●	21 ♏ 16	
21	22	39	☽	29 ♒ 15	
29	15	10	○	07 ♊ 01	total

Day	h	m		
08	03	32	Perigee	
21	07	46	Apogee	
03	01	54	Max dec	22° N 13'
09	11	31	0S	
16	01	30	Max dec	22° S 11'
23	11	55	0N	
30	09	32	Max dec	22° N 11'

LONGITUDES

Date	Chiron ⚷ ° '	Ceres ⚳ ° '	Pallas ⚴ ° '	Juno ⚵ ° '	Vesta ⚶ ° '	Black Moon Lilith ⚸ ° '
01	21 ♈ 27	01 ♓ 28	27 ♑ 13	05 ♓ 23	06 ♐ 24	19 ♒ 15
11	21 ♈ 02	02 ♓ 22	29 ♑ 21	06 ♓ 41	11 ♐ 41	20 ♒ 22
21	20 ♈ 40	03 ♓ 46	01 ♒ 45	08 ♓ 42	17 ♐ 01	21 ♒ 30
31	20 ♈ 22	05 ♓ 39	04 ♒ 23	11 ♓ 21	22 ♐ 22	22 ♒ 37

ASPECTARIAN

h m Aspects

01 Friday
00 29 ☽ Q ♃
00 49 ☉ ⊻ ♆
05 40 ☽ ⚼ ♇
07 18 ☽ # ♆
09 59 ☽ ⊼ ☿
16 16 ☽ ⊼ ♅
20 24 ☽ ± ☿
21 35 ♀ ⊻ ♇
23 11 ♀ △ ♃

02 Saturday
00 01 ☽ ⊼ ♂
01 19 ☽ ∠ ♄
02 59 ☽ ± ♅
03 36 ☽ ∥ ♄
04 15 ♀ ⊻ ♆
08 25 ☽ △ ♇
08 32 ☽ □ ♃
09 01 ☽ ☍ ♆
09 30 ☽ ⊼ ♀
11 11 ☽ ± ♂
11 30 ☽ ⊼ ☉
12 02 ☽ ⚼ ☿
17 12 ☽ ⊥ ♄
18 55 ☽ ⚼ ♅
21 07 ☽ ± ♀
22 52 ☽ ± ☉

03 Sunday
03 42 ☽ ⊻ ♄
03 48 ☽ ⚼ ♂
12 11 ♃ St D
12 52 ☿ St D
14 06 ☽ △ ☿
14 24 ☽ ⚼ ♀
15 48 ☽ ⚼ ☉
21 12 ☽ △ ♅
23 57 ☽ ∥ ♄

04 Monday
06 43 ♃ ⊼ ♇
07 12 ☽ △ ♂
12 49 ☽ △ ♃
12 50 ☽ □ ♇
13 25 ☽ ⊼ ♆
18 55 ☽ △ ♀
19 29 ☽ # ♆
19 44 ☽ △ ☉
23 45 ☽ ± ♆

05 Tuesday
07 32 ☽ ☌ ♄
14 33 ☽ ⚼ ♃
15 12 ☽ ⚼ ♆
18 31 ☽ □ ☿
19 45 ☽ Q ♇

06 Wednesday
00 56 ☽ □ ♅
05 00 ☽ # ☉
05 54 ☽ ± ♃
09 18 ☽ # ♀
13 08 ☽ □ ♂
13 09 ☉ ☌ ♀
16 06 ☽ ⊼ ♃
16 13 ☽ ⚹ ♇
16 47 ☽ △ ♆
19 31 ♀ ⊥ ♇
19 46 ☽ # ♂
21 10 ☉ ⊥ ♇

07 Thursday
02 13 ☽ Q ☿
02 47 ☽ □ ☉
03 03 ☽ □ ♀
04 11 ☽ ∥ ♇
07 37 ☽ Q ♅
09 44 ☽ # ♅
10 30 ☽ ⊻ ♄
14 08 ☽ # ♃
17 39 ☽ ∠ ♇
20 37 ☽ ⊥ ♄
21 08 ☽ Q ♂
22 03 ☽ # ☿
23 30 ☽ ⚹ ☿

08 Friday
03 58 ☽ ⚹ ♅
08 53 ☽ ⊥ ♇
11 30 ☽ Q ☉
11 48 ☽ ∠ ♄
12 24 ☽ Q ♀
18 24 ☽ ⚹ ♂
18 51 ☽ ☍ ♃
19 02 ☽ ⊻ ♇
19 37 ☽ □ ♆

09 Saturday
02 18 ☽ ∠ ☿
04 10 ♂ △ ♃
05 23 ☽ ∠ ♅
08 07 ♂ ⊻ ♇
09 17 ☽ ⚹ ☉
10 37 ☽ ⚹ ♀
13 05 ☽ ⚹ ♄
18 31 ☽ ⊥ ☿
20 40 ☽ ⊥ ♅
21 00 ☽ ∠ ♂
21 08 ♂ ⊻ ♆

10 Sunday
02 06 ☽ Q ♆
05 25 ☽ ⊻ ☿
06 52 ☽ ⊻ ♅
09 22 ☽ Q ♄
12 37 ☽ ∠ ☉
13 01 ☽ ⊥ ♂
14 12 ♀ △ ♄
14 28 ☽ ∠ ♀
21 42 ☽ ⊼ ♃
21 55 ☽ ☌ ♇
22 31 ☽ ⚹ ♆
23 45 ☽ ⊻ ♂

11 Monday
02 42 ☿ ☌ ♅
05 07 ☽ ⊥ ☉
05 59 ☽ ∥ ☿
07 19 ☽ ⊥ ♀
07 58 ☽ ± ♃
08 56 ☽ ∥ ♃
13 58 ☉ △ ♄
14 03 ☽ ∥ ♅
16 02 ☽ □ ♄
16 12 ☽ ⊻ ☉
18 37 ☽ ⊻ ♀
19 04 ☽ # ♇
23 27 ☽ ⚼ ♃

12 Tuesday
00 18 ☽ ∠ ♆
10 28 ☽ ☌ ♅
11 14 ☽ ∥ ♂
13 08 ☽ ☌ ☿
15 46 ☿ ∥ ♃
15 57 ☽ ⊥ ♆

13 Wednesday
01 37 ☽ △ ♃
01 50 ☽ ⊻ ♇
02 29 ☽ ⊻ ♆
06 21 ☽ ☌ ♂
08 33 ☽ ∥ ♀
09 09 ☽ ∥ ☉
12 27 ☽ ⊥ ♇
20 23 ☽ △ ♄
21 50 ☿ ⊥ ♆

14 Thursday
00 53 ☽ ☌ ☉
04 00 ♀ ∠ ♇
04 05 ☉ ∥ ♀
04 33 ☽ ∠ ♇
04 36 ☽ ☌ ♀
06 19 ☽ ∥ ♆
15 56 ☽ ⊻ ♅
23 25 ☽ ⚼ ♄

15 Friday
00 06 ☽ ⊻ ☿
00 11 ☿ ∥ ♅
02 56 ☽ ⊥ ♅
03 14 ☽ # ♄
07 42 ☽ □ ♃
07 55 ☽ ⚹ ♇
08 37 ☽ ☌ ♆
12 24 ☽ ⊥ ☿
15 34 ☽ ⊻ ♂
16 00 ☽ ± ♄
19 41 ☽ ∠ ♅

16 Saturday
02 48 ☉ ∠ ♇
03 11 ☽ ⊼ ♄
03 25 ☽ ⊥ ♂
06 22 ☽ Q ♇
07 15 ☽ ∠ ☿
12 50 ☽ ⊻ ☉
16 37 ☿ # ♇
17 35 ☽ Q ♃
18 06 ☽ ⊻ ♀
21 28 ☽ ∠ ♂

17 Sunday
00 02 ☽ # ♄
00 12 ☽ ⚹ ♅
01 17 ☽ ⊥ ☉
06 54 ☽ ⊥ ♀
15 37 ☽ ⚹ ☿
16 48 ☽ ⚹ ♃
16 57 ☽ □ ♇
17 44 ☽ ⊻ ♆
20 16 ☽ ∠ ☉
22 18 ☽ ∥ ♆
23 35 ☽ Q ♅

18 Monday
02 05 ☿ △ ♃
02 24 ☽ ∠ ♀
03 20 ☿ ⊻ ♇
04 17 ☽ ⚹ ♂
05 31 ☽ ⊥ ♆
07 07 ☽ ∥ ♀
10 11 ☿ ⊻ ♆
10 13 ☽ ∥ ☉
12 20 ♂ ⊥ ♇
12 58 ☽ ☍ ♄
18 21 ☽ Q ☿
22 27 ☽ ∠ ♃
23 22 ☽ ∠ ♆

19 Tuesday
04 34 ☽ ⚹ ☉
05 30 ☽ Q ♂
09 45 ♀ ⊻ ♅
11 25 ☽ □ ♅
11 37 ☽ ⚹ ♀
12 07 ☽ ∥ ♂
16 31 ☽ ⊥ ♃

20 Wednesday
04 39 ☽ ⊻ ♃
04 41 ☽ △ ♇
05 32 ☽ ⚹ ♆
05 55 ☽ ∥ ☿
06 53 ☽ Q ☉
11 29 ☽ □ ☿
14 37 ☽ Q ♀
18 20 ☽ # ♇
19 56 ☽ □ ♂
22 20 ♃ ⊼ ♇
22 44 ♀ ∥ ♆
22 46 ☽ ∥ ♅

21 Thursday
00 59 ☽ ⊼ ♄
05 52 ☽ Q ♆
07 09 ☽ ∥ ♃
11 05 ☽ ⚼ ♇
13 05 ☽ ± ♄
22 39 ☽ □ ☉

22 Friday
00 15 ☽ △ ♅
05 18 ☽ ± ♇
06 20 ♀ ⚼ ♄
06 25 ☿ ⊥ ♇
07 12 ☽ ⚼ ♄
07 18 ☽ □ ♀
17 25 ☽ ⊼ ♇
17 32 ☽ ☌ ♃
18 18 ☽ □ ♆
18 30 ☉ ⊻ ♅

23 Saturday
06 32 ☽ ⚼ ♅
08 44 ☽ △ ☿
09 43 ☉ ∥ ♆
12 03 ☽ △ ♂
13 09 ☽ △ ♄

24 Sunday
00 29 ☽ ± ♅
06 16 ♂ △ ♄
09 50 ♀ ⊥ ♅
12 19 ☽ ⊼ ♅
15 52 ☽ △ ☉
18 32 ☽ ⚼ ☿
18 38 ☿ △ ♄
19 20 ☽ ⚼ ♂

25 Monday
01 35 ☽ △ ♀
04 42 ☽ ☍ ♇
05 01 ☽ ⊻ ♃
05 20 ☿ ☌ ♂
05 37 ☽ △ ♆
13 41 ☽ ± ♂
14 20 ☽ ± ☿
14 22 ☽ # ♃
16 35 ☽ ⊥ ♃
23 13 ☽ ⚼ ☉
23 14 ☽ □ ♄
23 24 ☉ ⚼ ♄
23 28 ☽ # ♅

26 Tuesday
01 44 ☽ ⊼ ♂
02 56 ☽ ∥ ♇
03 18 ☽ ⊼ ☿
09 07 ♀ ⚹ ♇
09 16 ☽ ⚼ ♀
09 40 ☽ ∠ ♃
10 10 ☽ ⚼ ♆
11 53 ♀ # ♄
13 37 ♀ □ ♃
17 31 ☽ ± ☉
18 40 ♀ ☌ ♆
21 39 ☽ ☍ ♅

27 Wednesday
02 54 ☽ ± ♆
03 45 ☽ ± ♀
05 31 ☽ ⊼ ☉
08 34 ☽ Q ♄
12 58 ☽ ⊼ ♇
13 29 ☽ ⚹ ♃
13 53 ☽ ⊼ ♆
15 50 ☽ ⊼ ♀
17 20 ☽ # ☿
18 58 ☽ # ♂
23 50 ☽ ± ♇

28 Thursday
05 08 ☿ ∠ ♇
06 08 ☽ ⚹ ♄
08 56 ☿ ∥ ♂
11 10 ☽ Q ♃
11 40 ☽ ☍ ♂
15 54 ☽ ⚼ ♇
17 16 ☽ ☍ ☿
17 46 ☽ # ♆

29 Friday
01 07 ☉ ⊥ ♅
03 43 ☽ ⊼ ♅
04 21 ♀ ± ♄
08 28 ☽ ∠ ♄
08 32 ☽ # ☉
14 12 ☽ ± ♅
14 57 ☽ ∥ ♄
15 10 ☽ ☍ ☉
18 07 ☽ △ ♇
18 50 ☽ □ ♃
19 02 ☽ ☍ ♆
23 58 ☽ ⊥ ♄

30 Saturday
01 59 ☽ ☍ ♀
05 44 ☽ ⚼ ♅
10 12 ☽ ⊻ ♄
18 22 ☽ ⊼ ♂

All ephemeris data is given at 12.00 UT and the Moon's longitude is additionally given for 24.00 UT

DECEMBER 1974

Date	Sidereal time h m s	Sun ☉ ° ' "	Moon ☽ ° ' "	Moon ☽ 24.00 ° ' "	Mercury ☿ ° '	Venus ♀ ° '	Mars ♂ ° '	Jupiter ♃ ° '	Saturn ♄ ° '	Uranus ♅ ° '	Neptune ♆ ° '	Pluto ♇ ° '
01	16 39 50	08 ♐ 54 37	03 ♋ 20 59	10 ♋ 30 18	28 ♏ 49	15 ♐ 05	23 ♏ 25	09 ♓ 18	18 ♋ 03	00 ♏ 33	09 ♐ 19	08 ♎ 46
02	16 43 47	09 55 26	17 ♋ 41 12	24 ♋ 53 04	00 ♐ 22	16 20	24 07	09 23	17 R 59	00 37	09 22	08 47
03	16 47 43	10 56 16	02 ♌ 05 19	09 ♌ 17 24	01 56	17 35	24 48	09 29	17 56	00 40	09 24	08 49
04	16 51 40	11 57 07	16 ♌ 28 47	23 ♌ 39 00	03 30	18 51	25 30	09 35	17 52	00 43	09 26	08 50
05	16 55 36	12 57 59	00 ♍ 47 40	07 ♍ 54 25	05 03	20 06	26 12	09 41	17 49	00 46	09 28	08 51
06	16 59 33	13 58 53	14 ♍ 58 58	22 ♍ 01 04	06 37	21 21	26 54	09 47	17 45	00 49	09 31	08 53
07	17 03 30	14 59 48	29 ♍ 00 34	05 ♎ 57 16	08 11	22 37	27 36	09 54	17 42	00 52	09 33	08 54
08	17 07 26	16 00 45	12 ♎ 51 06	19 ♎ 41 58	09 45	23 52	28 18	10 00	17 38	00 55	09 35	08 55
09	17 11 23	17 01 42	26 ♎ 29 46	03 ♏ 14 29	11 19	25 08	29 00	10 07	17 34	00 58	09 37	08 56
10	17 15 19	18 02 41	09 ♏ 56 02	16 ♏ 34 24	12 53	26 23	29 ♏ 42	10 14	17 30	01 01	09 40	08 57
11	17 19 16	19 03 41	23 ♏ 09 31	29 ♏ 41 23	14 26	27 38	00 ♐ 24	10 21	17 26	01 03	09 42	08 58
12	17 23 12	20 04 42	06 ♐ 09 59	12 ♐ 35 17	16 00	28 ♐ 54	01 07	10 28	17 22	01 06	09 44	08 59
13	17 27 09	21 05 44	18 ♐ 57 18	25 ♐ 16 05	17 34	00 ♑ 09	01 49	10 36	17 18	01 09	09 46	09 00
14	17 31 05	22 06 47	01 ♑ 31 41	07 ♑ 44 11	19 09	01 25	02 31	10 43	17 14	01 12	09 49	09 01
15	17 35 02	23 07 50	13 ♑ 53 43	20 ♑ 00 28	20 43	02 40	03 14	10 51	17 09	01 14	09 51	09 02
16	17 38 59	24 08 54	26 ♑ 04 38	02 ♒ 06 27	22 17	03 55	03 56	10 59	17 05	01 17	09 53	09 03
17	17 42 55	25 09 59	08 ♒ 06 15	14 ♒ 04 22	23 52	05 11	04 38	11 07	17 01	01 20	09 55	09 04
18	17 46 52	26 11 04	20 ♒ 01 11	25 ♒ 57 08	25 26	06 26	05 21	11 15	16 56	01 22	09 57	09 05
19	17 50 48	27 12 09	01 ♓ 52 42	07 ♓ 48 22	27 01	07 41	06 03	11 23	16 52	01 25	10 00	09 06
20	17 54 45	28 13 15	13 ♓ 44 41	19 ♓ 42 13	28 ♐ 36	08 57	06 46	11 32	16 47	01 27	10 02	09 07
21	17 58 41	29 ♐ 14 21	25 ♓ 41 32	01 ♈ 43 15	00 ♑ 11	10 12	07 28	11 40	16 43	01 30	10 04	09 07
22	18 02 38	00 ♑ 15 27	07 ♈ 47 57	13 ♈ 56 14	01 46	11 27	08 11	11 49	16 38	01 32	10 06	09 08
23	18 06 34	01 16 33	20 ♈ 08 40	26 ♈ 25 49	03 22	12 43	08 54	11 58	16 33	01 34	10 08	09 09
24	18 10 31	02 17 40	02 ♉ 48 09	09 ♉ 16 08	04 57	13 58	09 36	12 07	16 29	01 37	10 10	09 10
25	18 14 28	03 18 46	15 ♉ 50 07	22 ♉ 30 21	06 33	15 13	10 19	12 16	16 24	01 39	10 13	09 10
26	18 18 24	04 19 53	29 ♉ 17 01	06 ♊ 10 06	08 09	16 29	11 02	12 25	16 19	01 41	10 15	09 11
27	18 22 21	05 21 00	13 ♊ 09 30	20 ♊ 14 54	09 46	17 44	11 44	12 35	16 14	01 43	10 17	09 11
28	18 26 17	06 22 07	27 ♊ 25 52	04 ♋ 41 47	11 22	18 59	12 27	12 44	16 10	01 46	10 19	09 12
29	18 30 14	07 23 15	12 ♋ 01 53	19 ♋ 25 17	12 59	20 15	13 10	12 54	16 05	01 48	10 21	09 12
30	18 34 10	08 24 22	26 ♋ 50 58	04 ♌ 17 55	14 36	21 30	13 53	13 04	16 00	01 50	10 23	09 13
31	18 38 07	09 ♑ 25 30	11 ♌ 45 02	19 ♌ 11 18	16 ♑ 14	22 ♑ 45	14 ♐ 36	13 ♓ 14	15 ♋ 55	01 ♏ 52	10 ♐ 25	09 ♎ 13

Date	Moon True ☊ ° '	Moon Mean ☊ ° '	Moon ☽ Latitude ° '
01	10 ♐ 16	10 ♐ 12	02 S 05
02	10 D 16	10 09	03 13
03	10 R 16	10 06	04 10
04	10 16	10 03	04 50
05	10 15	10 00	05 13
06	10 D 15	09 57	05 16
07	10 15	09 53	05 00
08	10 16	09 50	04 28
09	10 16	09 47	03 40
10	10 17	09 44	02 41
11	10 18	09 41	01 34
12	10 18	09 37	00 S 23
13	10 R 18	09 34	00 N 48
14	10 17	09 31	01 55
15	10 16	09 28	02 55
16	10 14	09 25	03 47
17	10 12	09 22	04 28
18	10 09	09 18	04 56
19	10 07	09 15	05 12
20	10 05	09 12	05 15
21	10 05	09 09	05 04
22	10 D 05	09 06	04 39
23	10 06	09 03	04 02
24	10 07	08 59	03 11
25	10 09	08 56	02 10
26	10 10	08 53	01 N 00
27	10 R 11	08 50	00 S 16
28	10 10	08 47	01 33
29	10 07	08 43	02 46
30	10 04	08 40	03 48
31	09 ♐ 59	08 ♐ 37	04 S 36

DECLINATIONS

Date	Sun ☉ ° '	Moon ☽ ° '	Mercury ☿ ° '	Venus ♀ ° '	Mars ♂ ° '	Jupiter ♃ ° '	Saturn ♄ ° '	Uranus ♅ ° '	Neptune ♆ ° '	Pluto ♇ ° '
01	21 S 47	21 N 19	19 S 25	22 S 47	18 S 31	09 S 15	21 N 47	11 S 10	20 S 21	11 N 40
02	21 56	19 05	19 52	22 57	18 42	09 13	21 48	11 12	20 21	11 40
03	22 05	15 38	20 18	23 07	18 53	09 10	21 48	11 13	20 22	11 40
04	22 13	11 17	20 43	23 16	19 03	09 08	21 49	11 14	20 22	11 40
05	22 21	06 19	21 08	23 24	19 14	09 06	21 49	11 15	20 22	11 40
06	22 29	01 N 03	21 31	23 32	19 24	09 03	21 50	11 16	20 23	11 40
07	22 36	04 S 12	21 53	23 39	19 34	09 00	21 50	11 17	20 23	11 40
08	22 42	09 11	22 14	23 45	19 44	08 58	21 51	11 18	20 23	11 40
09	22 49	13 38	22 34	23 51	19 54	08 55	21 51	11 19	20 24	11 40
10	22 54	17 20	22 53	23 55	20 04	08 52	21 52	11 20	20 24	11 40
11	22 59	20 04	23 11	23 59	20 13	08 49	21 53	11 21	20 24	11 40
12	23 04	21 43	23 27	24 03	20 23	08 46	21 53	11 22	20 25	11 40
13	23 09	22 11	23 43	24 05	20 32	08 43	21 54	11 23	20 25	11 40
14	23 12	21 31	23 57	24 07	20 41	08 40	21 54	11 24	20 25	11 40
15	23 16	19 49	24 10	24 08	20 50	08 37	21 55	11 25	20 26	11 40
16	23 19	17 13	24 21	24 08	20 58	08 34	21 56	11 25	20 26	11 40
17	23 21	13 56	24 32	24 08	21 07	08 31	21 56	11 26	20 26	11 41
18	23 23	10 07	24 41	24 06	21 15	08 28	21 57	11 27	20 27	11 41
19	23 25	05 57	24 49	24 04	21 23	08 24	21 58	11 28	20 27	11 41
20	23 26	01 S 33	24 55	24 02	21 31	08 21	21 58	11 29	20 27	11 41
21	23 26	02 N 56	25 00	23 58	21 39	08 17	21 59	11 30	20 27	11 41
22	23 26	07 22	25 04	23 54	21 46	08 14	22 00	11 31	20 28	11 42
23	23 26	11 36	25 06	23 49	21 54	08 10	22 00	11 31	20 28	11 42
24	23 25	15 26	25 07	23 43	22 01	08 07	22 01	11 32	20 28	11 42
25	23 24	18 39	25 07	23 37	22 08	08 03	22 02	11 33	20 29	11 43
26	23 22	20 58	25 05	23 30	22 14	07 59	22 02	11 34	20 29	11 43
27	23 20	22 07	25 01	23 22	22 21	07 56	22 03	11 34	20 29	11 43
28	23 17	21 52	24 56	23 13	22 27	07 52	22 04	11 35	20 30	11 43
29	23 14	20 09	24 50	23 04	22 33	07 48	22 04	11 36	20 30	11 44
30	23 10	17 03	24 42	22 54	22 39	07 44	22 05	11 37	20 30	11 44
31	23 S 06	12 N 51	24 S 32	22 S 44	22 S 45	07 S 40	22 N 06	11 S 37	20 S 30	11 N 45

ZODIAC SIGN ENTRIES

Date	h	m	Planets
01	06	22	☽ ♋
02	06	17	☿ ♐
03	08	31	☽ ♌
05	10	40	☽ ♍
07	13	42	☽ ♎
09	18	13	☽ ♏
10	22	05	♂ ♐
12	00	34	☽ ♐
13	09	06	♀ ♑
14	09	04	☽ ♑
16	19	48	☽ ♒
19	08	12	☽ ♓
21	09	16	☿ ♑
21	20	35	☽ ♈
22	05	56	☉ ♑
24	06	45	☽ ♉
26	13	15	☽ ♊
28	16	15	☽ ♋
30	17	05	☽ ♌

LATITUDES

Date	Mercury ☿ ° '	Venus ♀ ° '	Mars ♂ ° '	Jupiter ♃ ° '	Saturn ♄ ° '	Uranus ♅ ° '	Neptune ♆ ° '	Pluto ♇ ° '
01	00 N 29	00 S 11	00 N 07	01 S 15	00 S 27	00 N 32	01 N 31	16 N 29
04	00 N 08	00 18	00 05	01 15	00 26	00 32	01 31	16 31
07	00 S 13	00 25	00 03	01 14	00 26	00 32	01 31	16 32
10	00 32	00 32	00 N 01	01 14	00 26	00 32	01 31	16 34
13	00 51	00 39	00 00	01 13	00 26	00 32	01 31	16 35
16	01 09	00 45	00 S 02	01 13	00 25	00 32	01 31	16 37
19	01 24	00 51	00 04	01 12	00 25	00 32	01 31	16 39
22	01 38	00 57	00 06	01 11	00 25	00 32	01 31	16 40
25	01 50	01 03	00 08	01 11	00 25	00 32	01 31	16 42
28	01 59	01 08	00 10	01 10	00 24	00 32	01 31	16 44
31	02 S 06	01 S 13	00 S 12	01 S 10	00 S 24	00 N 32	01 N 31	16 N 46

DATA

Julian Date	2442383
Delta T	+45 seconds
Ayanamsa	23° 30' 39"
Synetic vernal point	05° ♓ 36' 20"
True obliquity of ecliptic	23° 26' 30"

MOON'S PHASES, APSIDES AND POSITIONS ☽

Date	h	m	Phase	Longitude ° '	Eclipse Indicator
06	10	10	☾	13 ♍ 54	
13	16	25	●	21 ♐ 17	Partial
21	19	43	☽	29 ♓ 34	
29	03	51	○	07 ♋ 02	

Day	h	m		
03	06	50	Perigee	
19	04	31	Apogee	
31	00	31	Perigee	
06	16	46	0S	
13	09	44	Max dec	22° S 12'
20	20	18	0N	
27	19	57	Max dec	22° N 11'

LONGITUDES

Date	Chiron ⚷ ° '	Ceres ⚳ ° '	Pallas ⚴ ° '	Juno ⚵ ° '	Vesta ⚶ ° '	Black Moon Lilith ⚸ ° '
01	20 ♈ 22	05 ♓ 39	04 ♒ 23	11 ♓ 21	22 ♐ 22	22 ♒ 37
11	20 ♈ 08	07 ♓ 54	07 ♒ 11	14 ♓ 33	27 ♐ 44	23 ♒ 44
21	20 ♈ 00	10 ♓ 30	10 ♒ 09	18 ♓ 16	03 ♑ 07	24 ♒ 51
31	19 ♈ 57	13 ♓ 23	13 ♒ 14	22 ♓ 16	08 ♑ 29	25 ♒ 58

ASPECTARIAN

h m Aspects

01 Sunday
03 25 ☽ ⊼ ☿
03 39 ☽ ∥ ♄
04 35 ☽ # ☉
05 02 ☽ ± ♂
07 17 ☽ △ ♅
08 33 ☉ ⚹ ♇
11 31 ☉ # ♄
14 45 ☽ ± ☿
20 56 ☽ ⚻ ♂
21 06 ☽ □ ♇
21 34 ♀ ∠ ♅
21 59 ☉ □ ♃
22 02 ☽ △ ♃
22 02 ☽ ⊼ ☉
22 03 ☽ ⊼ ♆
22 08 ☉ ☌ ♆

02 Monday
00 08 ☽ # ♆
00 34 ♃ □ ♆
00 41 ♂ ∠ ♇
06 01 ☽ # ☿
07 40 ☽ ⚻ ☿
08 06 ☽ ± ♆
08 50 ☽ ± ☉
09 31 ☽ ⊼ ♀
12 30 ☽ ☌ ♄
14 55 ☽ # ♂
15 46 ☿ ⊻ ♅
20 29 ☽ ± ♀
23 09 ☽ ⚻ ♆
23 15 ☽ ⚻ ♃
23 15 ☽ △ ♂

03 Tuesday
00 59 ☽ ⚻ ☉
03 12 ☽ Q ♇
09 37 ☽ □ ♅
11 43 ☽ △ ☿
12 55 ☽ ⚻ ♀
14 20 ☽ ± ♃
15 11 ☿ ∥ ♆
18 16 ♀ ⊼ ♄
23 13 ☽ ⚹ ♇

04 Wednesday
00 13 ☽ △ ♆
00 24 ☽ ⊼ ♃
02 48 ☿ ⚻ ♄
03 52 ☽ △ ☉
10 02 ☽ ∥ ♇
10 16 ☉ ± ♄
12 15 ☽ # ♅
14 19 ☽ ⊻ ♄
15 45 ☽ Q ♅
16 20 ☽ △ ♀
22 43 ☽ # ♃

05 Thursday
00 20 ☽ ⊥ ♄
00 20 ☽ ∠ ♇
03 53 ☽ □ ♂
11 57 ☽ ⚹ ♅
15 23 ☽ ∠ ♄
15 29 ☽ ⊥ ♇
20 04 ☽ □ ☿

06 Friday
01 38 ☽ ⊻ ♇
02 40 ♀ Q ♇
02 41 ☽ □ ♆
03 07 ☽ ☍ ♃
10 10 ☽ □ ☉
11 51 ☽ Q ♂
13 25 ☽ ∠ ♅
15 03 ☿ ⊥ ♅
16 42 ☽ ⚹ ♄
23 56 ☽ □ ♀

07 Saturday
04 51 ☽ ⊥ ♅
06 32 ☽ Q ☿
09 02 ☿ # ♄
09 27 ☽ ⚹ ♂
09 29 ☽ Q ♆
13 10 ☽ Q ♄
15 12 ☽ ⊻ ♅
19 25 ☽ Q ☉
23 06 ☿ ⚹ ♇

08 Sunday
05 08 ☽ ☌ ♇
05 54 ☽ ⚹ ☿
06 17 ☽ ⚹ ♆
07 00 ☽ ⊼ ♃
09 28 ☿ ☌ ♆
09 30 ☉ ∠ ♅
10 07 ☽ Q ♀
10 55 ☽ ∥ ♃
12 50 ☽ ∠ ♂
16 13 ☿ □ ♃
17 33 ☽ ± ♃
17 58 ☽ ⚹ ☉
20 20 ☽ □ ♄
23 01 ☽ ∥ ♅

09 Monday
00 56 ☽ # ♇
05 29 ☽ ⊥ ♂
08 40 ☽ ∠ ♆
09 20 ☽ ⚹ ♀
09 32 ☽ ⚻ ♃
11 38 ☽ ∠ ☿
15 46 ☿ ± ♄
16 42 ☽ ⊻ ♂
19 58 ☽ ☌ ♅
22 38 ☽ ∠ ☉
23 56 ☉ ⊼ ♄

10 Tuesday
00 43 ☽ ⊥ ♆
05 47 ☽ ⊥ ☿
10 14 ☽ ⊻ ♇
11 30 ☽ ⊻ ♆
12 32 ☽ △ ♃
14 36 ☉ ∥ ☿
14 53 ☽ ∠ ♀
16 07 ☽ ⊥ ☉
18 01 ☽ ⊻ ☿
21 05 ☽ ⊥ ♇

11 Wednesday
01 37 ☽ △ ♄
03 54 ☽ ⊻ ☉
08 56 ☽ ⊥ ♀
13 30 ☽ ∠ ♇
13 48 ☽ ∥ ♂
15 45 ☽ ∥ ♆
21 06 ☽ ⊻ ♀

12 Thursday
02 05 ☽ ☌ ♂
02 35 ☽ ⊻ ♅
02 59 ♀ Q ♃
04 59 ☽ ⚻ ♄
11 43 ♂ ⊻ ♅
13 31 ☿ ∠ ♅
13 45 ☽ ⊥ ♅
16 21 ☽ # ♄
16 55 ♂ ∥ ♆
17 16 ☽ ⚹ ♇
18 41 ☽ ☌ ♆
20 06 ☽ □ ♃
21 39 ☽ ± ♄

13 Friday
06 41 ☽ ∠ ♅
07 57 ☿ ⊼ ♄
08 53 ☽ ⊼ ♄
09 01 ☽ ☌ ☿
09 53 ☉ Q ♇
15 54 ☽ Q ♇
16 25 ☽ ☌ ☉

14 Saturday
02 53 ☽ # ♄
02 58 ♂ ⚻ ♄
06 33 ☽ Q ♃
07 46 ♀ ⚹ ♅
11 21 ☽ ⚹ ♅
11 45 ☽ ☌ ♀
14 01 ☽ ⊻ ♂

15 Sunday
00 21 ☽ ∥ ♂
02 20 ☽ ⊥ ♂
02 31 ☽ □ ♇
04 05 ☽ ⊻ ♆
04 46 ☽ ∥ ♆
05 59 ☽ ⚹ ♃
08 14 ☿ ∥ ♀
10 43 ☽ Q ♅
15 50 ☽ ⊥ ♆
17 02 ☿ Q ♇
18 21 ☽ ☍ ♄
21 01 ☽ ∠ ♂

16 Monday
03 23 ☽ ⊻ ☿
07 50 ☽ ⊻ ☉
09 38 ☽ ∠ ♆
11 48 ☽ ∠ ♃
12 26 ♀ ⊻ ♂
17 03 ☽ ⊥ ☿
20 50 ☽ ⊥ ☉
22 24 ☽ □ ♅

17 Tuesday
04 37 ☽ ⚹ ♂
05 27 ☽ ⊻ ♀
05 56 ☽ ⊥ ♃
13 45 ☽ ∠ ☿
13 57 ☽ △ ♇
15 39 ☽ ⚹ ♆
16 32 ☽ ∠ ☉
18 07 ☽ ⊻ ♃
18 54 ☽ ⊥ ♀

18 Wednesday
02 30 ☽ # ♇
03 55 ☽ ∥ ♅
05 49 ☽ ⊼ ♄
06 15 ☽ Q ♂
15 12 ☽ ∠ ♀
15 56 ☽ Q ♆
17 52 ☽ ± ♄
20 14 ☽ ⚻ ♇
21 52 ☽ ∥ ♃

19 Thursday
00 38 ☽ ⚹ ☿
01 38 ☽ ⚹ ☉
11 03 ☽ △ ♅
11 58 ☽ ⚻ ♄
14 28 ☽ ± ♇
20 02 ☉ ☌ ☿
20 59 ☽ □ ♂

20 Friday
01 09 ☽ ⚹ ♀
02 38 ☽ ⊼ ♇
04 13 ☽ Q ☉
04 28 ☽ □ ♆
04 40 ☽ Q ☿
07 28 ☽ ☌ ♃
15 13 ♀ □ ♇
17 29 ☽ ⚻ ♅
18 06 ☽ △ ♄

21 Saturday
03 32 ☿ Q ♃
04 12 ☽ Q ♀
09 20 ♀ ⊻ ♆
11 36 ☽ ± ♅
12 50 ♂ ⊥ ♅
19 43 ☽ □ ☉
22 18 ☽ □ ☿
23 35 ☽ ⊼ ♅
23 54 ☉ Q ♃

22 Sunday
08 23 ☿ ⚹ ♅
12 48 ☽ △ ♂
14 38 ☽ ☍ ♇
16 32 ☽ △ ♆
16 42 ☽ # ♃
19 49 ♀ ⚹ ♃
19 58 ☽ ⊻ ♃
19 59 ☽ □ ♀

23 Monday
05 07 ☽ □ ♄
07 45 ☽ ⊥ ♃
11 32 ☽ # ♅
12 34 ☽ ∥ ♇
19 18 ☉ ⚹ ♅
19 36 ☽ ⚻ ♂
20 47 ♂ ⚹ ♇
21 34 ☽ ⚻ ♆

24 Tuesday
01 10 ☽ ∠ ♃
04 59 ♀ Q ♅
09 46 ☽ ☍ ♅
10 58 ☽ △ ☉
12 14 ♂ # ♄
13 35 ☽ ± ♂
14 34 ☽ ± ♆
15 06 ☽ Q ♄
16 35 ☽ △ ☿
23 48 ☽ ⊼ ♇

25 Wednesday
01 21 ☽ ⊼ ♂
01 42 ☽ ⊼ ♆
05 26 ☽ ⚹ ♃
08 13 ♂ ☌ ♆
10 46 ☽ △ ♀
10 47 ☽ ± ♇
13 01 ☽ ⚹ ♄
14 33 ♂ ± ♄
16 51 ☽ ⚻ ☉
23 41 ☽ ⚻ ☿

26 Thursday
02 58 ☽ ⚻ ♇
03 19 ☽ Q ♃
05 54 ☽ # ♆
07 22 ♀ ⊥ ♆
09 08 ♀ ☍ ♄
10 11 ☽ ± ☉
15 33 ☽ ∠ ♄
16 13 ☽ ⊼ ♅
16 14 ☽ ⚻ ♀
17 41 ☽ ± ☿
21 31 ☽ ⊼ ☉

27 Friday
02 40 ☽ ± ♅
03 24 ☿ □ ♇
03 53 ♄ ± ♆
05 12 ☽ △ ♇
05 26 ☽ ⊼ ☿
07 02 ☽ ⊥ ♄
07 04 ☽ ☍ ♆
09 19 ☽ ∥ ♄
09 20 ☽ ± ♀
09 27 ☽ ☍ ♂
11 00 ☽ □ ♃
12 18 ♀ ⊥ ♂
17 12 ☽ ⊻ ♄
18 04 ☽ ⚻ ♅
19 53 ☿ ⊻ ♆
20 30 ☉ ∥ ♀
20 31 ☽ ⊼ ♀

28 Saturday
06 07 ☽ ∥ ♄
19 11 ☽ △ ♅

29 Sunday
00 23 ♂ □ ♃
03 51 ☽ ☍ ☉
07 23 ☽ □ ♇
08 21 ☽ # ♆
09 15 ☽ ⊼ ♆
10 36 ☿ ⚹ ♃
13 26 ☽ △ ♃
13 45 ☽ ☍ ☿
13 57 ☽ ⊼ ♂
16 51 ☿ ⊻ ♂
18 33 ☽ ☌ ♄
19 02 ☽ ± ♆

30 Monday
00 11 ☽ ± ♂
00 13 ☿ Q ♅
02 34 ☽ ☍ ♀
09 38 ☽ ⚻ ♆
12 35 ☽ Q ♇
13 59 ☽ ⚻ ♃
15 27 ☽ ⚻ ♂
20 02 ☽ □ ♅

31 Tuesday
04 38 ☽ ± ♃
07 08 ☉ □ ♇
07 34 ☿ ☍ ♄
07 55 ☽ ⚹ ♇
07 59 ☽ ⊼ ☉
09 51 ☽ △ ♆
09 51 ♀ ∥ ♂
14 25 ☽ ⊼ ♃
14 48 ☿ ⊥ ♆
16 50 ☽ △ ♂
17 36 ☽ ∥ ♇
18 12 ☽ # ♅
18 22 ☽ ± ☉
18 41 ☽ ⊻ ♄
20 07 ☽ ⊼ ☿

All ephemeris data is given at 12.00 UT and the Moon's longitude is additionally given for 24.00 UT

JANUARY 1975

LONGITUDES

Date	Sidereal time h m s	Sun ☉ ° ' "	Moon ☽ ° ' "	Moon ☽ 24.00 ° ' "	Mercury ☿ ° '	Venus ♀ ° '	Mars ♂ ° '	Jupiter ♃ ° '	Saturn ♄ ° '	Uranus ♅ ° '	Neptune ♆ ° '	Pluto ♇ ° '
01	18 42 03	10 ♑ 26 38	26 ♌ 35 42	03 ♍ 57 22	17 ♑ 51	24 ♑ 00	15 ♐ 19	13 ♓ 24	15 ♋ 50	01 ♏ 54	10 ♐ 27	09 ♎ 14
02	18 46 00	11 27 47	11 ♍ 15 32	18 ♍ 29 36	19 29	25 16	16 02	13 34	15 R 45	01 56	10 28	09 14
03	18 49 57	12 28 55	25 ♍ 39 06	02 ♎ 43 44	21 07	26 31	16 45	13 44	15 40	01 57	10 31	09 14
04	18 53 53	13 30 04	09 ♎ 43 20	16 ♎ 37 51	22 46	27 46	17 28	13 55	15 35	01 59	10 33	09 14
05	18 57 50	14 31 13	23 ♎ 27 22	00 ♏ 12 00	24 24	29 ♑ 02	18 11	14 05	15 30	02 01	10 35	09 15
06	19 01 46	15 32 23	06 ♏ 51 59	13 ♏ 27 34	26 02	00 ♒ 17	18 55	14 16	15 25	02 03	10 37	09 15
07	19 05 43	16 33 33	19 ♏ 59 03	26 ♏ 26 43	27 41	01 32	19 38	14 27	15 20	02 04	10 39	09 15
08	19 09 39	17 34 43	02 ♐ 50 50	09 ♐ 11 43	29 ♑ 19	02 47	20 21	14 38	15 15	02 06	10 41	09 15
09	19 13 36	18 35 53	15 ♐ 29 36	21 ♐ 44 44	00 ♒ 57	04 02	21 05	14 49	15 10	02 07	10 43	09 15
10	19 17 32	19 37 02	27 ♐ 57 19	04 ♑ 07 33	02 35	05 18	21 48	15 00	15 05	02 09	10 45	09 15
11	19 21 29	20 38 12	10 ♑ 15 36	16 ♑ 21 38	04 13	06 33	22 31	15 11	15 00	02 10	10 46	09 15
12	19 25 26	21 39 22	22 ♑ 25 46	28 ♑ 28 09	05 50	07 48	23 15	15 22	14 56	02 12	10 48	09 R 15
13	19 29 22	22 40 31	04 ♒ 28 57	10 ♒ 28 18	07 26	09 03	23 58	15 34	14 51	02 13	10 50	09 15
14	19 33 19	23 41 39	16 ♒ 26 33	22 ♒ 23 24	09 01	10 18	24 42	15 45	14 46	02 14	10 52	09 15
15	19 37 15	24 42 47	28 ♒ 19 35	04 ♓ 15 13	10 35	11 34	25 25	15 57	14 41	02 16	10 54	09 15
16	19 41 12	25 43 55	10 ♓ 10 36	16 ♓ 06 05	12 07	12 49	26 09	16 08	14 36	02 17	10 55	09 15
17	19 45 08	26 45 01	22 ♓ 02 04	27 ♓ 59 00	13 37	14 04	26 53	16 20	14 31	02 18	10 57	09 15
18	19 49 05	27 46 07	03 ♈ 57 22	09 ♈ 57 40	15 04	15 19	27 36	16 32	14 27	02 19	10 59	09 15
19	19 53 01	28 47 13	16 ♈ 00 30	22 ♈ 06 25	16 29	16 34	28 20	16 44	14 22	02 20	11 00	09 14
20	19 56 58	29 ♑ 48 17	28 ♈ 16 03	04 ♉ 30 00	17 50	17 49	29 04	16 56	14 17	02 21	11 02	09 14
21	20 00 55	00 ♒ 49 21	10 ♉ 48 52	17 ♉ 13 16	19 07	19 04	29 ♐ 48	17 08	14 12	02 22	11 04	09 14
22	20 04 51	01 50 23	23 ♉ 43 43	00 ♊ 20 44	20 19	20 19	00 ♑ 31	17 20	14 08	02 23	11 05	09 13
23	20 08 48	02 51 25	07 ♊ 04 42	13 ♊ 55 53	21 25	21 34	01 15	17 33	14 03	02 23	11 07	09 13
24	20 12 44	03 52 26	20 ♊ 54 27	28 ♊ 00 19	22 26	22 49	01 59	17 45	13 59	02 24	11 08	09 13
25	20 16 41	04 53 25	05 ♋ 13 16	12 ♋ 32 50	23 19	24 04	02 43	17 57	13 54	02 25	11 10	09 12
26	20 20 37	05 54 24	19 ♋ 58 20	27 ♋ 28 52	24 03	25 19	03 27	18 10	13 50	02 25	11 11	09 12
27	20 24 34	06 55 22	05 ♌ 03 19	12 ♌ 40 25	24 39	26 34	04 11	18 23	13 46	02 26	11 13	09 11
28	20 28 30	07 56 19	20 ♌ 18 47	27 ♌ 57 00	25 06	27 49	04 55	18 35	13 41	02 26	11 14	09 11
29	20 32 27	08 57 15	05 ♍ 33 38	13 ♍ 07 23	25 21	29 ♒ 04	05 39	18 48	13 37	02 27	11 16	09 10
30	20 36 24	09 58 10	20 ♍ 37 03	28 ♍ 01 40	25 R 26	00 ♓ 19	06 23	19 01	13 33	02 27	11 17	09 09
31	20 40 20	10 ♒ 59 04	05 ♎ 20 25	12 ♎ 32 47	25 ♒ 20	01 ♓ 33	07 ♑ 07	19 ♓ 14	13 ♋ 29	02 ♏ 28	11 ♐ 19	09 ♎ 09

Date	Moon True ☊ ° '	Moon Mean ☊ ° '	Moon ☽ Latitude ° '
01	09 ♐ 55	08 ♐ 34	05 S 04
02	09 R 51	08 31	05 12
03	09 49	08 28	05 01
04	09 47	08 24	04 31
05	09 D 48	08 21	03 46
06	09 49	08 18	02 50
07	09 51	08 15	01 46
08	09 52	08 12	00 S 38
09	09 R 52	08 09	00 N 31
10	09 50	08 05	01 37
11	09 46	08 02	02 37
12	09 40	07 59	03 30
13	09 33	07 56	04 12
14	09 24	07 53	04 43
15	09 15	07 49	05 02
16	09 06	07 46	05 07
17	08 59	07 43	04 59
18	08 53	07 40	04 39
19	08 50	07 37	04 05
20	08 49	07 34	03 20
21	08 D 50	07 30	02 25
22	08 51	07 27	01 20
23	08 51	07 24	00 N 10
24	08 R 51	07 21	01 S 04
25	08 48	07 18	02 16
26	08 43	07 14	03 21
27	08 35	07 11	04 14
28	08 26	07 08	04 49
29	08 16	07 05	05 04
30	08 07	07 02	04 57
31	08 ♐ 01	06 ♐ 59	04 S 30

DECLINATIONS

Date	Sun ☉ ° '	Moon ☽ ° '	Mercury ☿ ° '	Venus ♀ ° '	Mars ♂ ° '	Jupiter ♃ ° '	Saturn ♄ ° '	Uranus ♅ ° '	Neptune ♆ ° '	Pluto ♇ ° '
01	23 S 02	07 N 53	24 S 21	22 S 32	22 S 50	07 S 36	22 N 06	11 S 38	20 S 31	11 N 45
02	22 57	02 N 31	24 08	22 20	22 56	07 32	22 07	11 38	20 31	11 45
03	22 51	02 S 52	23 54	22 07	23 01	07 28	22 08	11 39	20 31	11 46
04	22 45	08 00	23 38	21 54	23 06	07 24	22 08	11 40	20 31	11 46
05	22 39	12 37	23 21	21 40	23 10	07 20	22 09	11 40	20 32	11 47
06	22 32	16 29	23 02	21 25	23 15	07 15	22 10	11 41	20 32	11 47
07	22 25	19 27	22 41	21 10	23 19	07 11	22 10	11 41	20 32	11 48
08	22 17	21 21	22 19	20 54	23 23	07 07	22 11	11 42	20 32	11 48
09	22 09	22 09	21 56	20 37	23 26	07 02	22 12	11 42	20 33	11 49
10	22 00	21 49	21 31	20 20	23 30	06 58	22 12	11 43	20 33	11 49
11	21 51	20 26	21 05	20 02	23 33	06 53	22 13	11 43	20 33	11 50
12	21 42	18 07	20 37	19 44	23 36	06 49	22 14	11 44	20 33	11 50
13	21 32	15 03	20 08	19 25	23 39	06 44	22 14	11 44	20 34	11 51
14	21 22	11 24	19 37	19 06	23 41	06 40	22 15	11 45	20 34	11 51
15	21 11	07 20	19 06	18 46	23 44	06 35	22 16	11 45	20 34	11 52
16	21 00	03 S 00	18 33	18 25	23 46	06 31	22 16	11 45	20 34	11 52
17	20 48	01 N 26	18 00	18 04	23 48	06 26	22 17	11 46	20 34	11 53
18	20 37	05 50	17 26	17 42	23 49	06 21	22 18	11 46	20 35	11 54
19	20 24	10 04	16 51	17 20	23 51	06 16	22 18	11 47	20 35	11 54
20	20 12	13 59	16 16	16 58	23 52	06 12	22 19	11 47	20 35	11 55
21	19 59	17 22	15 41	16 35	23 53	06 07	22 19	11 47	20 35	11 55
22	19 45	20 00	15 07	16 11	23 53	06 02	22 20	11 47	20 35	11 56
23	19 31	21 39	14 33	15 47	23 54	05 57	22 21	11 48	20 36	11 57
24	19 17	22 04	14 00	15 23	23 54	05 52	22 21	11 48	20 36	11 57
25	19 03	21 04	13 28	14 58	23 54	05 47	22 22	11 48	20 36	11 58
26	18 48	18 38	12 58	14 33	23 53	05 42	22 22	11 48	20 36	11 59
27	18 33	14 54	12 31	14 07	23 53	05 37	22 23	11 48	20 36	11 59
28	18 17	10 09	12 06	13 41	23 52	05 32	22 24	11 49	20 36	12 00
29	18 01	04 N 46	11 44	13 15	23 51	05 27	22 24	11 49	20 37	12 01
30	17 45	00 S 58	11 26	12 48	23 50	05 22	22 25	11 49	20 37	12 02
31	17 S 29	06 S 15	11 S 11	12 S 21	23 S 48	05 S 17	22 N 25	11 S 49	20 S 37	12 N 02

ZODIAC SIGN ENTRIES

Date	h	m	Planets
01	17	32	☽ ♍
03	19	21	☽ ♎
05	23	39	☽ ♏
06	06	39	♀ ♒
08	06	39	☽ ♐
08	21	58	☿ ♒
10	15	58	☽ ♑
13	03	03	☽ ♒
15	15	23	☽ ♓
18	04	03	☽ ♈
20	15	21	☽ ♉
20	16	36	☉ ♒
21	18	49	♂ ♑
22	23	43	☽ ♊
25	03	20	☽ ♋
27	04	00	☽ ♌
29	03	30	☽ ♍
30	06	05	♀ ♓
31	03	13	☽ ♎

LATITUDES

Date	Mercury ☿ ° '	Venus ♀ ° '	Mars ♂ ° '	Jupiter ♃ ° '	Saturn ♄ ° '	Uranus ♅ ° '	Neptune ♆ ° '	Pluto ♇ ° '
01	02 S 07	01 S 15	00 S 13	01 S 10	00 S 24	00 N 32	01 N 31	16 N 46
04	02 09	01 19	00 15	01 09	00 24	00 32	01 31	16 48
07	02 06	01 23	00 17	01 09	00 23	00 32	01 31	16 50
10	01 59	01 26	00 19	01 09	00 23	00 32	01 32	16 51
13	01 46	01 29	00 21	01 08	00 23	00 32	01 32	16 53
16	01 27	01 31	00 23	01 08	00 22	00 32	01 32	16 55
19	01 00	01 32	00 25	01 07	00 22	00 33	01 32	16 57
22	00 S 25	01 33	00 27	01 07	00 22	00 33	01 32	16 58
25	00 N 18	01 33	00 29	01 07	00 21	00 33	01 32	17 00
28	01 08	01 33	00 31	01 07	00 21	00 33	01 32	17 01
31	02 N 00	01 S 32	00 S 33	01 S 06	00 S 21	00 N 33	01 N 32	17 N 03

DATA

Julian Date	2442414
Delta T	+46 seconds
Ayanamsa	23° 30' 45"
Synetic vernal point	05° ♓ 36' 14"
True obliquity of ecliptic	23° 26' 29"

MOON'S PHASES, APSIDES AND POSITIONS ☽

Date	h	m	Phase	Longitude ° '	Eclipse Indicator
04	19	04	☾	13 ♎ 48	
12	10	20	●	21 ♑ 35	
20	15	14	☽	29 ♈ 57	
27	15	09	○	07 ♌ 03	

Day	h	m		
15	21	25	Apogee	
28	09	26	Perigee	
02	23	10	0S	
09	16	48	Max dec	22° S 10'
17	04	18	0N	
24	07	19	Max dec	22° N 05'
30	08	26	0S	

LONGITUDES

Date	Chiron ⚷ ° '	Ceres ⚳ ° '	Pallas ⚴ ° '	Juno ⚵ ° '	Vesta ⚶ ° '	Black Moon Lilith ⚸ ° '
01	19 ♈ 57	13 ♓ 41	13 ♒ 33	22 ♓ 42	09 ♑ 02	26 ♒ 05
11	20 ♈ 00	16 ♓ 50	16 ♒ 44	27 ♓ 08	14 ♑ 23	27 ♒ 12
21	20 ♈ 09	20 ♓ 11	20 ♒ 00	01 ♈ 51	19 ♑ 43	28 ♒ 19
31	20 ♈ 23	23 ♓ 41	23 ♒ 18	06 ♈ 49	25 ♑ 00	29 ♒ 27

All ephemeris data is given at 12.00 UT and the Moon's longitude is additionally given for 24.00 UT

ASPECTARIAN

h m	Aspects
01 Wednesday	
01 07	☽ Q ♅
04 19	☽ ⊥ ♄
07 01	☽ ± ☿
07 25	☽ ⊼ ♀
08 09	☽ ∠ ♇
10 00	☽ bQ ☉
12 09	☉ ⊻ ♆
13 18	☽ # ♃
18 04	☽ ± ♀
18 52	☽ ∠ ♄
20 39	☽ ⚹ ♅
22 49	☽ ⊥ ♇
23 29	☽ bQ ☿
02 Thursday	
03 29	♂ ⊼ ♄
08 39	☽ ⊻ ♇
10 12	☽ bQ ♀
10 43	☽ □ ♆
12 22	☽ △ ☉
14 34	☉ ∥ ♂
15 52	☽ ☍ ♃
16 21	♀ ∠ ♆
19 24	☽ ⚹ ♄
20 19	☽ □ ♂
21 25	☽ ∠ ♅
03 Friday	
03 25	☽ △ ☿
11 25	♀ # ♄
12 31	☽ ⊥ ♅
13 36	☽ △ ♀
15 23	☽ Q ♄
16 51	☽ Q ♆
19 05	♂ ∠ ♅
22 42	☽ ⊻ ♅
04 Saturday	
04 18	☽ Q ♂
09 06	☽ ∥ ♃
11 10	☽ ☌ ♇
13 26	☽ ⚹ ♆
19 04	☽ □ ☉
19 21	☽ ⊼ ♃
22 07	☽ □ ♄
23 44	☉ ⚹ ♃
23 45	☉ Q ♅
05 Sunday	
02 13	☽ ⚹ ♂
05 59	☽ ± ♃
06 35	☿ ⊥ ♂
06 48	☽ ∥ ♅
07 23	☽ # ♇
13 25	♀ ∠ ♃
13 54	☽ □ ☿
15 47	☽ ∠ ♆
22 09	☽ bQ ♃
22 55	☽ □ ♀
23 09	☿ ∥ ♂
06 Monday	
03 17	☽ ☌ ♅
05 30	☽ Q ☉
05 40	☿ ∠ ♆
06 22	☽ ∠ ♂
07 55	☽ ⊥ ♆
09 25	☉ ☍ ♄
16 19	☽ ⊻ ♇
18 49	☽ ⊻ ♆
23 38	☽ ⊥ ♂
07 Tuesday	
01 40	☽ △ ♃
02 56	☽ Q ☿
03 17	☽ ⊥ ♇
03 30	☽ △ ♄
05 09	☽ ⚹ ☉
11 05	☽ Q ♀
11 19	☽ ⊻ ♂
14 09	☉ ⊥ ♆
19 55	☽ ∠ ♇
22 31	♀ □ ♅
08 Wednesday	
00 02	☽ ∥ ♆
04 24	☽ ⚹ ☿
05 50	☽ ∥ ♀
07 10	☽ bQ ♄
10 35	☽ ⊻ ♅
11 27	☽ ∠ ☉
11 52	☽ ⚹ ♀
15 38	☉ ∥ ☿
17 04	☿ ∠ ♃
20 30	☿ # ♄
21 56	☽ ⊥ ♅
09 Thursday	
00 02	☽ ± ♄
00 07	☽ ⚹ ♇
02 51	☽ ☌ ♆
04 48	☉ # ♄
05 27	☽ ∥ ☿
05 59	☽ ⊥ ☉
10 40	☽ □ ♃
11 23	☽ ⊼ ♄
12 22	☽ ∥ ☉
13 01	☽ ∠ ☿
15 08	☽ ∠ ♅
17 57	♂ Q ♇
18 29	☽ ⊻ ☉
18 32	♀ ∥ ♆
19 33	☽ ∠ ♀
23 03	☽ Q ♇
23 22	☽ ☌ ♂
10 Friday	
03 22	☽ ∥ ☉
05 25	☿ □ ♅
08 57	☽ ⊥ ☿
14 53	☽ ⊥ ♀
20 10	☽ ⚹ ♅
20 34	♃ △ ♄
21 57	☽ Q ♃
22 22	☽ ⊻ ☿
22 40	☽ ∥ ☿
11 Saturday	
03 54	☽ ⊻ ♀
10 02	☽ □ ♇
10 25	☽ ∥ ♆
13 01	☽ ⊻ ♆
17 36	☽ ∥ ♀
17 41	♇ St R
19 42	☽ Q ♅
21 16	☽ ☍ ♄
21 50	☽ ⚹ ♃
12 Sunday	
00 51	☽ ⊥ ♆
10 20	☽ ☌ ☉
13 43	☽ ⊻ ♂
14 49	☿ ∥ ♆
18 43	☽ ∠ ♆
13 Monday	
02 25	☽ ⊥ ♂
04 02	☽ ∠ ♃
07 28	☽ □ ♅
08 18	♀ ∠ ♂
15 52	♀ △ ♇
18 48	☽ ☌ ☿
21 34	☽ △ ♇
21 34	☽ ∠ ♂
22 14	☽ ☌ ♀
22 20	☽ ⊥ ♃
23 26	♀ ⊥ ♃
14 Tuesday	
00 46	☽ ⚹ ♆
08 39	☽ ⊼ ♄
09 11	☽ # ♇
09 52	☽ ∥ ♅
10 35	☽ ⊻ ♃
15 40	☿ △ ♇
20 39	☽ ± ♄
22 56	♀ ⚹ ♆
15 Wednesday	
00 52	☿ ⊥ ♃
00 59	☽ Q ♆
03 46	☽ bQ ♇
04 00	☽ ⊻ ☉
05 45	☽ ⚹ ♂
07 35	☿ ∠ ♂
14 44	☽ bQ ♄
16 17	☽ ∥ ♃
17 01	☿ ⚹ ♆
17 17	☽ ⊥ ☉
19 58	☽ △ ♅
21 58	☽ ± ♇
16 Thursday	
07 38	☽ Q ♂
10 07	☽ ⊼ ♇
13 14	☽ ∠ ☉
13 31	☽ □ ♆
16 29	☽ ⊻ ☿
16 36	☉ ∠ ♆
17 58	☽ ⊻ ♀
20 54	☽ △ ♄
17 Friday	
00 17	☽ ☌ ♃
02 24	☽ bQ ♅
03 56	☿ ∥ ♀
06 24	☽ ⊥ ☿
07 33	☽ ⊥ ♀
20 16	♀ ⊼ ♄
20 37	☽ ± ♅
22 24	☽ ⚹ ☉
22 24	☽ □ ♂
22 29	☉ ⊻ ♂
18 Saturday	
02 07	☿ ⊼ ♄
03 07	☽ ∠ ☿
03 50	☽ ∠ ♀
08 42	☽ ⊼ ♅
14 49	☽ # ♃
15 48	☉ ∥ ♆
22 34	☽ ☍ ♇
19 Sunday	
00 41	☽ Q ☉
02 03	☽ △ ♆
08 46	☽ □ ♄
13 03	☽ ⚹ ☿
13 14	☽ ⚹ ♀
13 27	☽ ⊻ ♃
15 48	♀ ⊻ ♃
17 09	☿ ⊻ ♃
22 10	☽ # ♅
22 58	☽ ∥ ♇
20 Monday	
01 27	☽ ⊥ ♃
07 11	☿ ☌ ♀
07 39	☽ bQ ♆
13 38	☽ △ ♂
15 14	☽ □ ☉
15 20	☽ Q ♀
15 23	☽ Q ☿
19 12	☽ ∠ ♃
19 42	☽ Q ♄
19 53	☽ ☍ ♅
21 Tuesday	
01 03	☽ ± ♆
01 24	☽ # ☿
06 39	☽ # ♀
09 00	☽ ⊼ ♇
12 28	☽ ⊼ ♆
18 20	☽ ⚹ ♄
19 55	☽ bQ ♂
20 17	☽ ± ♇
22 Wednesday	
00 02	☽ ⚹ ♃
05 04	☽ □ ♀
05 07	☽ □ ☿
08 25	☿ ± ♄
08 37	♀ ± ♄
09 33	☽ # ☉
11 32	☿ ☌ ♀
12 54	☽ bQ ♇
13 32	☽ ± ♂
16 38	♃ bQ ♅
18 54	☽ # ♆
21 45	☽ ∠ ♄
22 21	☽ Q ♃
23 Thursday	
00 49	☉ □ ♅
01 02	☽ ⊼ ♂
02 45	☉ ∠ ♃
03 39	☽ ⊼ ♅
03 53	☽ △ ☉
13 43	☽ ⊥ ♄
14 19	☽ ± ♅
15 46	☽ △ ♇
19 06	☽ ☍ ♆
24 Friday	
00 09	☽ ⊻ ♄
05 59	☽ bQ ♅
06 31	☽ □ ♃
08 15	☽ bQ ☉
14 46	☽ △ ☿
15 34	☽ △ ♀
18 20	♀ Q ♆
25 Saturday	
00 41	☽ ± ☉
01 52	♂ ⚹ ♅
07 21	☽ △ ♅
07 38	☽ ☍ ♂
07 43	☿ Q ♆
11 25	☽ ⊼ ☉
14 34	♀ bQ ♇
17 22	☽ bQ ☿
17 55	☽ # ♆
18 32	☽ □ ♇
18 55	☽ bQ ♀
21 46	☽ ⊼ ♆
26 Sunday	
02 08	☽ ☌ ♄
07 31	☽ ± ♆
08 46	☽ ± ☿
09 03	☽ △ ♃
10 37	☽ # ☉
10 51	☽ ± ♀
16 55	☿ bQ ♇
18 50	☽ ⊼ ☿
21 20	☽ ⊼ ♀
21 58	☽ bQ ♆
23 32	☽ Q ♇
27 Monday	
07 51	☽ □ ♅
09 19	☽ bQ ♃
10 33	☽ ⊼ ♂
15 09	☽ ☍ ☉
16 40	☽ # ♀
18 30	☽ ⚹ ♇
20 29	☽ ± ♂
21 43	☽ △ ♆
23 42	☽ ± ♃
28 Tuesday	
01 38	☽ ⊻ ♄
01 47	☽ # ☿
03 06	☽ ∥ ♇
04 02	☽ # ♅
09 15	☽ ⊼ ♃
11 01	☽ ⊥ ♄
11 21	☽ bQ ♂
12 12	☽ Q ♅
17 40	☿ # ♇
18 04	☽ ∠ ♇
19 40	☽ ☍ ☿
29 Wednesday	
00 50	☽ ☍ ♀
01 06	☽ ∠ ♄
03 54	♀ bQ ♄
06 34	☿ ∥ ♅
07 05	☽ ⚹ ♅
08 13	☽ ⊥ ♇
08 59	☽ # ♃
12 09	☽ △ ♂
16 57	☉ △ ♇
17 42	☽ ⊻ ♇
17 46	☽ ⊼ ☉
21 03	☽ □ ♆
30 Thursday	
00 44	☽ ⚹ ♄
04 00	☽ ± ☉
06 55	☽ ∠ ♅
09 23	☽ ☍ ♃
10 47	☿ St R
19 33	☽ bQ ☉
19 46	☽ ⊼ ☿
19 56	☽ Q ♄
21 27	☽ ⊥ ♅
31 Friday	
02 05	☽ Q ♆
05 11	☽ ⊼ ♀
05 29	☽ ± ☿
07 15	☽ ⊻ ♅
07 37	☽ ∥ ♃
15 07	☽ □ ♂
16 01	☽ ± ♀
16 59	♂ Q ♃
18 18	☽ ☌ ♇
19 51	☉ ⚹ ♆
20 10	☽ bQ ☿
21 57	☽ ⚹ ♆
22 06	☽ △ ☉

FEBRUARY 1975

LONGITUDES

Date	Sidereal time	Sun ☉	Moon ☽	Moon ☽ 24.00	Mercury ☿	Venus ♀	Mars ♂	Jupiter ♃	Saturn ♄	Uranus ♅	Neptune ♆	Pluto ♇
	h m s	° ' "	° ' "	° ' "	° '	° '	° '	° '	° '	° '	° '	° '
01	20 44 17	11 ♒ 59 58	19 ♎ 38 25	26 ♎ 37 12	25 ♒ 03	02 ♓ 48	07 ♑ 52	19 ♓ 27	13 ♋ 25	02 ♏ 28	11 ♐ 20	09 ♎ 08
02	20 48 13	13 00 51	03 ♏ 29 10	10 ♏ 14 32	24 R 34	04 03	08 36	19 40	13 R 21	02 28	11 21	09 R 07
03	20 52 10	14 01 43	16 ♏ 53 38	23 ♏ 26 51	23 55	05 18	09 20	19 53	13 17	02 28	11 22	09 07
04	20 56 06	15 02 34	29 ♏ 54 40	06 ♐ 17 34	23 07	06 32	10 04	20 06	13 13	02 28	11 24	09 06
05	21 00 03	16 03 25	12 ♐ 36 06	18 ♐ 50 44	22 10	07 47	10 49	20 19	13 09	02 28	11 25	09 05
06	21 03 59	17 04 14	25 ♐ 02 00	01 ♑ 10 20	21 07	09 02	11 33	20 33	13 05	02 R 28	11 26	09 04
07	21 07 56	18 05 03	07 ♑ 16 10	13 ♑ 19 53	19 59	10 16	12 17	20 46	13 02	02 28	11 27	09 03
08	21 11 53	19 05 51	19 ♑ 21 50	25 ♑ 22 18	18 48	11 31	13 02	20 59	12 58	02 28	11 28	09 02
09	21 15 49	20 06 37	01 ♒ 21 34	07 ♒ 19 51	17 36	12 46	13 46	21 13	12 55	02 28	11 30	09 01
10	21 19 46	21 07 22	13 ♒ 17 20	19 ♒ 14 13	16 25	14 00	14 31	21 26	12 51	02 28	11 31	09 00
11	21 23 42	22 08 06	25 ♒ 10 37	01 ♓ 06 43	15 18	15 15	15 15	21 40	12 48	02 28	11 32	08 59
12	21 27 39	23 08 49	07 ♓ 02 40	12 ♓ 58 38	14 14	16 29	16 00	21 54	12 45	02 27	11 33	08 58
13	21 31 35	24 09 30	18 ♓ 54 47	24 ♓ 51 22	13 17	17 44	16 45	22 07	12 42	02 27	11 34	08 57
14	21 35 32	25 10 09	00 ♈ 48 37	06 ♈ 46 49	12 26	18 58	17 29	22 21	12 39	02 26	11 35	08 56
15	21 39 28	26 10 47	12 ♈ 46 19	18 ♈ 47 30	11 42	20 13	18 14	22 35	12 36	02 26	11 36	08 55
16	21 43 25	27 11 24	24 ♈ 50 46	00 ♉ 56 36	11 06	21 27	18 58	22 49	12 33	02 25	11 36	08 54
17	21 47 22	28 11 58	07 ♉ 05 32	13 ♉ 18 04	10 37	22 41	19 43	23 02	12 30	02 25	11 37	08 53
18	21 51 18	29 ♒ 12 31	19 ♉ 34 49	25 ♉ 56 21	10 17	23 56	20 28	23 16	12 28	02 24	11 38	08 52
19	21 55 15	00 ♓ 13 02	02 ♊ 23 14	08 ♊ 56 01	10 04	25 10	21 13	23 30	12 25	02 23	11 39	08 51
20	21 59 11	01 13 32	15 ♊ 35 16	22 ♊ 21 19	09 58	26 24	21 58	23 44	12 23	02 23	11 40	08 49
21	22 03 08	02 13 59	29 ♊ 14 34	06 ♋ 15 12	09 D 59	27 38	22 42	23 58	12 21	02 22	11 40	08 48
22	22 07 04	03 14 25	13 ♋ 23 13	20 ♋ 38 26	10 07	28 ♓ 52	23 27	24 12	12 18	02 21	11 41	08 47
23	22 11 01	04 14 48	28 ♋ 00 26	05 ♌ 28 34	10 21	00 ♈ 07	24 12	24 26	12 16	02 20	11 42	08 46
24	22 14 57	05 15 10	13 ♌ 01 54	20 ♌ 39 20	10 41	01 21	24 57	24 41	12 14	02 19	11 42	08 44
25	22 18 54	06 15 30	28 ♌ 19 31	06 ♍ 01 00	11 06	02 35	25 42	24 55	12 12	02 18	11 43	08 43
26	22 22 51	07 15 48	13 ♍ 42 16	21 ♍ 21 46	11 37	03 49	26 27	25 09	12 11	02 17	11 43	08 42
27	22 26 47	08 16 05	28 ♍ 58 06	06 ♎ 29 56	12 12	05 02	27 12	25 23	12 09	02 16	11 44	08 40
28	22 30 44	09 ♓ 16 19	13 ♎ 56 11	21 ♎ 16 00	12 ♒ 51	06 ♈ 16	27 ♑ 57	25 ♓ 37	12 ♋ 07	02 ♏ 15	11 ♐ 44	08 ♎ 39

MOON

Date	Moon True ☊	Moon Mean ☊	Moon ☽ Latitude
	° '	° '	° '
01	07 ♐ 56	06 ♐ 55	03 S 48
02	07 R 54	06 52	02 52
03	07 D 54	06 49	01 49
04	07 54	06 46	00 S 42
05	07 R 54	06 43	00 N 25
06	07 53	06 40	01 30
07	07 49	06 36	02 29
08	07 42	06 33	03 21
09	07 32	06 30	04 03
10	07 20	06 27	04 35
11	07 06	06 24	04 54
12	06 51	06 20	05 00
13	06 37	06 17	04 53
14	06 25	06 14	04 34
15	06 15	06 11	04 02
16	06 09	06 08	03 19
17	06 05	06 05	02 27
18	06 03	06 01	01 26
19	06 D 03	05 58	00 N 19
20	06 R 03	05 55	00 S 50
21	06 02	05 52	01 59
22	05 58	05 49	03 03
23	05 52	05 46	03 58
24	05 43	05 42	04 37
25	05 32	05 39	04 58
26	05 20	05 36	04 57
27	05 09	05 33	04 35
28	05 ♐ 00	05 ♐ 30	03 S 54

DECLINATIONS

Date	Sun ☉	Moon ☽	Mercury ☿	Venus ♀	Mars ♂	Jupiter ♃	Saturn ♄	Uranus ♅	Neptune ♆	Pluto ♇
	° '	° '	° '	° '	° '	° '	° '	° '	° '	° '
01	17 S 12	11 S 12	11 S 01	11 S 54	23 S 46	05 S 12	22 N 26	11 S 49	20 S 37	12 N 03
02	16 55	15 23	10 55	11 26	23 44	05 06	22 26	11 49	20 37	12 04
03	16 37	18 38	10 54	10 58	23 42	05 01	22 27	11 49	20 37	12 05
04	16 20	20 49	10 56	10 30	23 39	04 56	22 27	11 49	20 37	12 05
05	16 02	21 54	11 03	10 02	23 36	04 51	22 28	11 49	20 38	12 06
06	15 43	21 51	11 14	09 33	23 33	04 45	22 28	11 49	20 38	12 07
07	15 25	20 46	11 28	09 04	23 30	04 40	22 29	11 49	20 38	12 08
08	15 06	18 44	11 44	08 35	23 27	04 35	22 29	11 49	20 38	12 08
09	14 47	15 54	12 03	08 05	23 23	04 29	22 30	11 49	20 38	12 09
10	14 28	12 27	12 23	07 35	23 19	04 24	22 30	11 49	20 38	12 10
11	14 08	08 49	12 45	07 06	23 15	04 18	22 31	11 49	20 38	12 11
12	13 48	04 S 17	13 07	06 35	23 10	04 13	22 31	11 49	20 38	12 12
13	13 28	00 N 07	13 28	06 05	23 05	04 07	22 31	11 48	20 38	12 12
14	13 08	04 30	13 50	05 35	23 00	04 02	22 32	11 48	20 38	12 13
15	12 48	08 46	14 10	05 04	22 55	03 57	22 32	11 48	20 39	12 14
16	12 27	12 43	14 30	04 34	22 50	03 51	22 33	11 48	20 39	12 15
17	12 06	16 11	14 48	04 03	22 44	03 45	22 33	11 48	20 39	12 16
18	11 45	19 00	15 05	03 32	22 38	03 40	22 33	11 47	20 39	12 16
19	11 24	20 57	15 20	03 01	22 32	03 34	22 34	11 47	20 39	12 17
20	11 02	21 50	15 34	02 30	22 26	03 29	22 34	11 47	20 39	12 18
21	10 41	21 27	15 46	01 59	22 19	03 23	22 34	11 47	20 39	12 19
22	10 19	19 44	15 56	01 27	22 12	03 18	22 35	11 46	20 39	12 20
23	09 57	16 41	16 05	00 56	22 05	03 12	22 35	11 46	20 39	12 21
24	09 35	12 28	16 12	00 S 25	21 58	03 06	22 35	11 46	20 39	12 21
25	09 13	07 24	16 17	00 N 06	21 50	03 01	22 36	11 45	20 39	12 22
26	08 51	01 N 50	16 21	00 38	21 42	02 55	22 36	11 45	20 39	12 23
27	08 28	03 S 48	16 23	01 09	21 34	02 49	22 36	11 44	20 39	12 24
28	08 S 06	09 S 06	16 S 24	01 N 40	21 S 26	02 S 44	22 N 37	11 S 44	20 S 39	12 N 25

ZODIAC SIGN ENTRIES

Date	h	m	Planets
02	05	53	☽ ♏
04	12	10	☽ ♐
06	21	42	☽ ♑
09	09	16	☽ ♒
11	21	45	☽ ♓
14	10	22	☽ ♈
16	22	09	☽ ♉
19	06	50	☉ ♓
19	07	35	☽ ♊
21	13	18	☽ ♋
23	09	53	♀ ♈
23	15	13	☽ ♌
25	14	37	☽ ♍
27	13	38	☽ ♎

LATITUDES

Date	Mercury ☿	Venus ♀	Mars ♂	Jupiter ♃	Saturn ♄	Uranus ♅	Neptune ♆	Pluto ♇
	° '	° '	° '	° '	° '	° '	° '	° '
01	02 N 17	01 S 32	00 S 34	01 S 06	00 S 20	00 N 33	01 N 32	17 N 04
04	03 02	01 30	00 36	01 06	00 20	00 33	01 32	17 05
07	03 32	01 27	00 38	01 06	00 20	00 33	01 33	17 07
10	03 42	01 24	00 40	01 05	00 19	00 33	01 33	17 08
13	03 30	01 21	00 42	01 05	00 19	00 33	01 33	17 09
16	03 04	01 16	00 44	01 05	00 18	00 33	01 33	17 11
19	02 29	01 11	00 46	01 05	00 18	00 33	01 33	17 12
22	01 50	01 06	00 48	01 05	00 18	00 33	01 33	17 13
25	01 12	01 00	00 51	01 05	00 17	00 33	01 33	17 14
28	00 35	00 53	00 53	01 05	00 17	00 33	01 33	17 15
31	00 N 01	00 S 46	00 S 55	01 S 04	00 S 16	00 N 33	01 N 34	17 N 16

DATA

Julian Date	2442445
Delta T	+46 seconds
Ayanamsa	23° 30' 50"
Synetic vernal point	05° ♓ 36' 10"
True obliquity of ecliptic	23° 26' 30"

MOON'S PHASES, APSIDES AND POSITIONS ☽

Date	h	m	Phase	Longitude ° '	Eclipse Indicator
03	06	23	☾	13 ♏ 47	
11	05	17	●	21 ♒ 51	
19	07	38	☽	00 ♊ 02	
26	01	15	○	06 ♍ 49	

Day	h	m		
12	04	10	Apogee	
25	22	01	Perigee	
05	23	01	Max dec	22° S 01'
13	11	23	ON	
20	17	06	Max dec	21° N 52'
26	19	47	OS	

LONGITUDES

Date	Chiron ⚷	Ceres ⚳	Pallas ⚴	Juno ⚵	Vesta ⚶	Black Moon Lilith ⚸
	° '	° '	° '	° '	° '	° '
01	20 ♈ 25	24 ♓ 03	23 ♒ 38	07 ♈ 19	25 ♑ 32	29 ♒ 33
11	20 ♈ 45	27 ♓ 42	26 ♒ 57	12 ♈ 30	00 ♒ 46	00 ♓ 41
21	21 ♈ 09	01 ♈ 27	00 ♓ 17	17 ♈ 52	05 ♒ 56	01 ♓ 48
31	21 ♈ 36	05 ♈ 17	03 ♓ 37	21 ♈ 22	11 ♒ 02	02 ♓ 55

All ephemeris data is given at 12.00 UT and the Moon's longitude is additionally given for 24.00 UT

ASPECTARIAN

h m	Aspects
01 Saturday	
01 30	☽ □ ♄
04 21	♀ # ♇
05 30	♀ △ ♅
08 34	☽ ⚼ ♀
11 08	☽ ∥ ☿
11 40	☽ ⊼ ♃
15 18	☽ ∥ ♅
15 24	☽ ∥ ♀
16 27	♀ ∥ ♅
16 34	☽ # ♇
18 19	♀ ± ♇
21 00	☽ △ ☿
22 08	☽ ± ♃
23 17	☽ Q ♂
23 31	☽ ∠ ♆
02 Sunday	
10 13	☽ ☌ ♅
13 05	☽ △ ♀
14 07	☽ ⚼ ♃
15 18	☽ ⊥ ♆
19 19	☉ ⊼ ♄
21 33	☽ ∥ ☉
21 35	☽ ⚹ ♂
21 59	☽ ⊻ ♇
03 Monday	
02 01	☽ ⊻ ♆
04 48	♂ □ ♇
05 10	☿ ∠ ♂
05 29	☽ △ ♄
05 35	☿ ⚼ ♇
06 23	☽ □ ☉
07 36	☉ ⊥ ♃
08 46	☽ ⊥ ♇
15 57	☿ ∥ ♀
17 33	☽ △ ♃
04 Tuesday	
00 09	☽ □ ☿
01 12	☽ ∠ ♇
02 27	☽ ∠ ♂
04 18	☿ Q ♆
08 51	☽ ⚼ ♄
09 10	☽ ∥ ♆
16 48	☽ ⊻ ♅
18 23	☽ Q ☉
20 17	☽ ⊥ ♂
05 Wednesday	
01 41	☽ ± ♄
01 49	☽ □ ♀
04 08	☽ ⊥ ♅
05 18	☽ ⚹ ♇
07 42	☽ Q ☿
08 22	☽ ⊻ ♂
09 44	☽ ☌ ♆
13 03	☽ ⊼ ♄
19 13	☽ ⚹ ☉
21 21	☽ ∠ ♅
06 Thursday	
01 49	♅ St R
03 07	☽ □ ♃
04 18	☽ Q ♇
05 00	☽ ⚹ ☿
08 11	♂ ⊻ ♆
12 45	♀ ⊼ ♇
16 20	☽ Q ♀
22 12	☿ ⊻ ♃
07 Friday	
02 33	☽ ⚹ ♅
03 01	☽ ∠ ☉
07 52	☽ ∠ ☿
13 55	☽ ∥ ♆
15 01	☽ Q ♃
15 31	☽ □ ♇
18 37	☽ ⚹ ♀
20 18	☽ ⊻ ♆
22 24	☽ ⊥ ☉
22 35	☽ ☌ ♂
23 21	☽ ☍ ♄
08 Saturday	
00 06	☽ ⊥ ☿
02 16	☽ Q ♅
06 09	☉ ⊥ ♂
08 14	☽ ⊥ ♆
08 17	☿ ± ♄
08 43	☉ ☌ ☿
09 05	☿ ⊥ ♂
09 11	☉ ± ♄
10 13	♂ ☍ ♄
10 58	☽ ⊻ ☿
11 09	♀ □ ♆
11 25	☽ ⊻ ☉
15 18	☽ ⚹ ♃
18 21	☿ ∥ ♅
09 Sunday	
02 14	☽ ∠ ♆
03 57	☽ ∠ ♀
14 14	☽ □ ♅
14 50	♀ △ ♄
19 50	☿ # ♇
21 04	☽ ∥ ☉
21 57	☽ ∠ ♃
10 Monday	
00 07	☽ ⊥ ♀
03 23	☽ △ ♇
08 25	☽ ⚹ ♆
10 27	♂ Q ♅
11 08	☽ ⊼ ♄
12 19	☽ ∥ ☿
13 37	☽ ⊻ ♀
13 46	☽ # ♇
14 38	☽ ⊻ ♂
16 01	☽ ∥ ♅
16 25	☽ ⊥ ♃
17 46	☽ ☌ ☿
21 41	☉ ⊻ ♃
23 11	☽ ± ♄
11 Tuesday	
03 33	☽ ⊥ ♂
04 46	☽ ⊻ ♃
05 17	☽ ☌ ☉
05 19	☿ ⊥ ♃
08 40	☽ Q ♆
09 36	☽ ⚼ ♇
12 29	♀ ⚹ ♂
12 30	☿ ⊻ ♀
12 30	☿ ⊻ ♂
17 17	☽ ⚼ ♄
21 19	☽ ∥ ♀
22 57	☽ ∠ ♂
12 Wednesday	
02 43	☽ △ ♅
03 47	☽ ± ♇
12 21	☽ ∥ ♃
15 54	☽ ⊼ ♇
21 07	☽ □ ♆
21 36	☉ Q ♆
23 29	☽ △ ♄
13 Thursday	
01 26	☽ ⊻ ☿
06 36	♀ ⚼ ♅
07 18	☉ ⚼ ♇
07 19	☽ ⚹ ♂
09 03	☽ ⚼ ♅
09 20	☽ ☌ ♀
11 56	☉ ∥ ☿
12 41	☽ ⊥ ☿
18 36	☽ ☌ ♃
23 35	☽ ⊻ ☉
14 Friday	
03 12	☽ ± ♅
05 02	☿ ⊼ ♄
05 37	☽ ∠ ☿
05 39	☿ ⊥ ♀
09 09	☽ Q ♂
09 26	☽ # ♃
12 47	☽ ⊥ ☉
15 00	♂ ⊥ ♆
15 17	☽ ⊼ ♅
17 19	☽ # ♀
15 Saturday	
04 18	☽ ☍ ♇
08 31	☽ ∠ ☉
09 38	☽ △ ♆
09 58	☽ ⚹ ☿
11 39	☽ □ ♄
15 48	☿ ⚹ ♆
23 36	☽ □ ♂
16 Sunday	
04 31	☽ ⊻ ♀
06 14	☽ # ♅
07 54	☽ ⊻ ♃
08 41	☽ Q ☿
09 03	☽ ∥ ♇
10 28	☽ # ☉
15 29	☽ ⚼ ♆
17 02	☽ ⚹ ☉
17 43	☽ ⊥ ♀
19 58	☽ ⊥ ♃
20 16	☉ ⚼ ♄
23 12	☽ Q ♄
17 Monday	
00 59	☽ # ☿
01 24	☉ # ♇
02 53	☽ ☍ ♅
09 08	☽ ± ♆
13 17	☽ ∠ ♀
13 53	☽ ∠ ♃
15 28	☽ ⊼ ♇
18 33	☽ Q ☉
18 38	☽ □ ☿
20 23	♀ ☌ ♃
20 46	☽ ⊼ ♆
22 26	☽ ⚹ ♄
18 Tuesday	
03 01	☽ ± ♇
04 24	♀ ∥ ♃
09 19	☉ ∥ ♅
13 47	☽ △ ♂
19 07	☽ ⚹ ♃
20 05	☽ ⚼ ♇
21 06	☽ ⚹ ♀
19 Wednesday	
02 48	☽ ∠ ♄
05 32	♂ # ♄
07 12	☽ # ♆
07 38	☽ □ ☉
10 13	☿ ∠ ♀
12 00	☽ ⊼ ♅
17 51	☽ Q ♃
19 24	☽ ⊥ ♄
19 28	☽ ⚼ ♂
21 41	☽ Q ♀
23 00	☽ ± ♅
23 49	☽ △ ♇
20 Thursday	
01 55	☽ △ ☿
04 56	☽ ☍ ♆
06 16	☽ ⊻ ♄
11 33	♀ ± ♅
12 42	☽ ± ♂
15 12	☽ ⚼ ♅
19 28	☿ St D
23 57	☽ ⊼ ♂
21 Friday	
02 41	☽ □ ♃
04 35	☽ ⚼ ☿
08 57	☽ □ ♀
15 06	☉ △ ♅
17 22	☽ △ ♅
17 33	☽ △ ☉
20 12	☽ ± ☿
22 Saturday	
01 19	☉ ± ♇
01 30	☽ # ♆
04 17	☽ □ ♇
06 28	☽ ⊼ ☿
09 09	☽ ⊼ ♆
10 12	☽ ☌ ♄
19 08	☽ ± ♆
20 39	☽ ⚼ ☉
23 Sunday	
05 30	☽ ☍ ♂
06 07	☽ △ ♃
09 52	☽ ⚼ ♆
09 59	☽ Q ♇
12 25	☽ ± ☉
15 41	☽ # ☿
15 42	☽ △ ♀
18 58	☽ □ ♅
22 45	☽ ⊼ ☉
23 11	♂ ⚹ ♃
24 Monday	
05 13	☽ ⚹ ♇
06 36	☽ ⚼ ♃
08 12	☽ ☍ ☿
09 54	☽ △ ♆
10 45	☽ ⊻ ♄
12 35	☽ ∥ ♇
15 36	☽ # ♅
17 41	☽ ⚼ ♀
20 11	☽ ⊥ ♄
21 02	☽ ± ♃
23 28	☽ Q ♅
25 Tuesday	
03 07	☽ # ☉
04 48	☽ ∠ ♇
06 35	☽ ⊼ ♃
06 45	♀ ⊼ ♅
07 41	☽ ⊼ ♂
09 02	☽ ± ♀
10 15	☽ ∠ ♄
17 32	☽ ± ♂
18 12	☽ ⚹ ♅
18 51	☽ ⊥ ♇
19 13	☽ ⊼ ♀
26 Wednesday	
01 15	☽ ☍ ☉
04 11	☽ ⊻ ♇
07 21	☽ # ♃
08 18	☽ ⚼ ♂
08 37	☽ ⊼ ☿
08 54	☽ □ ♆
09 37	☽ ⚹ ♄
16 41	☽ ∥ ♀
16 50	☿ ⚹ ♆
17 36	☽ ∠ ♅
18 21	☽ ± ☿
20 56	♂ ∠ ♆
23 31	☽ # ♀
27 Thursday	
04 24	☽ Q ♄
06 15	☽ ☍ ♃
07 44	☽ ⊥ ♅
07 53	☽ ∥ ♃
09 03	☽ △ ♂
09 05	☽ ⚼ ☿
10 08	☿ ⊼ ♄
13 13	☽ Q ♆
17 14	☽ ⊻ ♅
21 27	☉ ⊼ ♇
22 32	☽ ☍ ♀
28 Friday	
03 28	☽ ☌ ♇
03 55	☽ ⊼ ☉
07 36	☽ ∥ ☉
08 26	☽ ⚹ ♆
09 04	☽ □ ♄
10 10	☽ △ ☿
14 20	☽ ± ☉

MARCH 1975

LONGITUDES

Date	Sidereal time h m s	Sun ☉ ° ' "	Moon ☽ ° ' "	Moon ☽ 24.00 ° ' "	Mercury ☿ ° '	Venus ♀ ° '	Mars ♂ ° '	Jupiter ♃ ° '	Saturn ♄ ° '	Uranus ♅ ° '	Neptune ♆ ° '	Pluto ♇ ° '
01	22 34 40	10 ♓ 16 33	28 ♎ 28 47	05 ♏ 34 08	13 ♒ 35	07 ♈ 30	28 ♑ 42	25 ♓ 52	12 ♋ 06	02 ♏ 14	11 ♐ 45	08 ♎ 38
02	22 38 37	11 16 45	12 ♏ 31 55	19 ♏ 22 12	14 22	08 44	29 ♑ 27	26 06	12 R 04	02 R 12	11 45	08 R 36
03	22 42 33	12 16 55	26 ♏ 05 12	02 ♐ 41 18	15 12	09 58	00 ♒ 12	26 20	12 03	02 11	11 46	08 35
04	22 46 30	13 17 04	09 ♐ 10 58	15 ♐ 34 44	16 06	11 11	00 57	26 35	12 02	02 10	11 46	08 33
05	22 50 26	14 17 11	21 ♐ 53 10	28 ♐ 06 55	17 03	12 25	01 42	26 49	12 01	02 08	11 46	08 32
06	22 54 23	15 17 17	04 ♑ 16 33	10 ♑ 22 42	18 03	13 39	02 28	27 03	12 00	02 07	11 47	08 30
07	22 58 20	16 17 21	16 ♑ 25 56	22 ♑ 26 48	19 05	14 52	03 13	27 18	11 59	02 05	11 47	08 29
08	23 02 16	17 17 24	28 ♑ 25 47	04 ♒ 23 21	20 09	16 06	03 58	27 32	11 59	02 04	11 47	08 27
09	23 06 13	18 17 25	10 ♒ 19 54	16 ♒ 15 49	21 16	17 19	04 43	27 47	11 58	02 02	11 47	08 26
10	23 10 09	19 17 24	22 ♒ 11 25	28 ♒ 06 58	22 26	18 32	05 29	28 01	11 57	02 01	11 47	08 24
11	23 14 06	20 17 21	04 ♓ 02 42	09 ♓ 58 49	23 37	19 46	06 14	28 16	11 57	01 59	11 48	08 23
12	23 18 02	21 17 16	15 ♓ 55 31	21 ♓ 52 56	24 50	20 59	06 59	28 30	11 57	01 57	11 48	08 21
13	23 21 59	22 17 10	27 ♓ 51 15	03 ♈ 50 35	26 05	22 12	07 45	28 45	11 57	01 56	11 48	08 20
14	23 25 55	23 17 01	09 ♈ 51 07	15 ♈ 53 01	27 22	23 26	08 30	28 59	11 D 57	01 54	11 R 48	08 18
15	23 29 52	24 16 50	21 ♈ 56 29	28 ♈ 01 44	28 ♒ 40	24 39	09 15	29 14	11 57	01 52	11 48	08 16
16	23 33 49	25 16 38	04 ♉ 09 03	10 ♉ 18 43	00 ♓ 01	25 52	10 01	29 28	11 57	01 50	11 48	08 15
17	23 37 45	26 16 23	16 ♉ 31 06	22 ♉ 46 33	01 22	27 05	10 46	29 43	11 57	01 48	11 48	08 13
18	23 41 42	27 16 06	29 ♉ 05 56	05 ♊ 28 20	02 46	28 18	11 32	29 ♓ 57	11 58	01 46	11 47	08 11
19	23 45 38	28 15 47	11 ♊ 55 34	18 ♊ 27 37	04 10	29 ♈ 31	12 17	00 ♈ 12	11 58	01 44	11 47	08 10
20	23 49 35	29 ♓ 15 26	25 ♊ 04 55	01 ♋ 47 53	05 37	00 ♉ 43	13 03	00 26	11 59	01 42	11 47	08 08
21	23 53 31	00 ♈ 15 02	08 ♋ 36 49	15 ♋ 32 00	07 05	01 56	13 48	00 41	11 59	01 40	11 47	08 07
22	23 57 28	01 14 36	22 ♋ 33 32	29 ♋ 41 24	08 34	03 09	14 34	00 55	12 00	01 38	11 47	08 05
23	00 01 24	02 14 08	06 ♌ 55 25	14 ♌ 15 10	10 04	04 21	15 19	01 10	12 01	01 36	11 46	08 03
24	00 05 21	03 13 37	21 ♌ 40 04	29 ♌ 09 19	11 36	05 34	16 05	01 24	12 02	01 34	11 46	08 02
25	00 09 18	04 13 04	06 ♍ 41 54	14 ♍ 16 41	13 09	06 46	16 50	01 39	12 04	01 32	11 46	08 00
26	00 13 14	05 12 29	21 ♍ 52 23	29 ♍ 27 39	14 44	07 59	17 36	01 53	12 05	01 30	11 46	07 58
27	00 17 11	06 11 51	07 ♎ 01 09	14 ♎ 31 36	16 20	09 11	18 21	02 08	12 06	01 27	11 45	07 57
28	00 21 07	07 11 12	21 ♎ 57 51	29 ♎ 18 54	17 57	10 23	19 07	02 22	12 08	01 25	11 45	07 55
29	00 25 04	08 10 31	06 ♏ 33 56	13 ♏ 42 23	19 36	11 36	19 52	02 37	12 09	01 23	11 44	07 53
30	00 29 00	09 09 48	20 ♏ 43 50	27 ♏ 38 07	21 16	12 48	20 38	02 51	12 11	01 21	11 44	07 52
31	00 32 57	10 ♈ 09 03	04 ♐ 25 14	11 ♐ 05 22	22 ♓ 58	14 ♉ 00	21 ♒ 24	03 ♈ 06	12 ♋ 13	01 ♏ 18	11 ♐ 43	07 ♎ 50

Date	Moon True ☊ ° '	Moon Mean ☊ ° '	Moon ☽ Latitude ° '
01	04 ♐ 54	05 ♐ 26	02 S 59
02	04 R 50	05 23	01 55
03	04 49	05 20	00 S 46
04	04 D 49	05 17	00 N 23
05	04 R 49	05 14	01 29
06	04 47	05 11	02 29
07	04 44	05 07	03 21
08	04 38	05 04	04 03
09	04 28	05 01	04 35
10	04 17	04 58	04 54
11	04 03	04 55	05 00
12	03 49	04 52	04 54
13	03 36	04 48	04 34
14	03 24	04 45	04 03
15	03 14	04 42	03 20
16	03 08	04 39	02 27
17	03 04	04 36	01 27
18	03 03	04 32	00 N 21
19	03 D 03	04 29	00 S 47
20	03 04	04 26	01 55
21	03 R 03	04 23	02 58
22	03 01	04 20	03 52
23	02 57	04 17	04 34
24	02 51	04 13	04 59
25	02 42	04 10	05 04
26	02 34	04 07	04 47
27	02 25	04 04	04 11
28	02 18	04 01	03 18
29	02 13	03 58	02 13
30	02 10	03 54	01 01
31	02 ♐ 10	03 ♐ 51	00 N 12

DECLINATIONS

Date	Sun ☉ ° '	Moon ☽ ° '	Mercury ☿ ° '	Venus ♀ ° '	Mars ♂ ° '	Jupiter ♃ ° '	Saturn ♄ ° '	Uranus ♅ ° '	Neptune ♆ ° '	Pluto ♇ ° '
01	07 S 43	13 S 43	16 S 22	02 N 12	21 S 18	02 S 38	22 N 37	11 S 44	20 S 39	12 N 26
02	07 20	17 25	16 20	02 43	21 09	02 32	22 37	11 43	20 39	12 26
03	06 57	20 01	16 15	03 14	21 00	02 27	22 37	11 43	20 39	12 27
04	06 34	21 27	16 10	03 45	20 51	02 21	22 38	11 42	20 39	12 28
05	06 11	21 43	16 02	04 16	20 42	02 15	22 38	11 42	20 39	12 29
06	05 48	20 53	15 53	04 47	20 32	02 09	22 38	11 41	20 39	12 30
07	05 25	19 06	15 43	05 18	20 22	02 04	22 38	11 41	20 39	12 31
08	05 01	16 30	15 31	05 49	20 12	01 58	22 38	11 40	20 39	12 31
09	04 38	13 15	15 18	06 19	20 02	01 52	22 39	11 39	20 39	12 32
10	04 14	09 29	15 04	06 50	19 52	01 46	22 39	11 39	20 39	12 33
11	03 51	05 21	14 48	07 20	19 41	01 41	22 39	11 38	20 39	12 34
12	03 27	01 S 02	14 30	07 50	19 31	01 35	22 39	11 38	20 39	12 35
13	03 04	03 N 21	14 11	08 20	19 20	01 29	22 39	11 37	20 39	12 35
14	02 40	07 37	13 51	08 50	19 08	01 23	22 39	11 36	20 39	12 36
15	02 16	11 38	13 30	09 20	18 57	01 17	22 39	11 36	20 39	12 37
16	01 53	15 13	13 07	09 49	18 46	01 12	22 40	11 35	20 39	12 38
17	01 29	18 10	12 43	10 18	18 34	01 06	22 40	11 35	20 39	12 38
18	01 05	20 18	12 18	10 47	18 22	01 00	22 40	11 34	20 39	12 39
19	00 41	21 27	11 51	11 16	18 10	00 54	22 40	11 33	20 39	12 40
20	00 S 18	21 26	11 23	11 44	17 58	00 49	22 40	11 33	20 39	12 41
21	00 N 06	20 12	10 54	12 13	17 45	00 43	22 40	11 32	20 38	12 41
22	00 30	17 44	10 24	12 40	17 33	00 37	22 40	11 32	20 38	12 42
23	00 53	14 07	09 52	13 08	17 20	00 31	22 40	11 31	20 38	12 43
24	01 17	09 34	09 19	13 35	17 07	00 26	22 40	11 30	20 38	12 44
25	01 41	04 N 21	08 45	14 02	16 54	00 20	22 40	11 30	20 38	12 44
26	02 04	01 S 11	08 10	14 29	16 40	00 14	22 40	11 29	20 38	12 45
27	02 28	06 38	07 33	14 55	16 27	00 08	22 40	11 28	20 38	12 46
28	02 51	11 37	06 56	15 21	16 13	00 S 03	22 40	11 27	20 38	12 46
29	03 15	15 48	06 17	15 47	16 00	00 N 03	22 40	11 26	20 38	12 47
30	03 38	18 55	05 37	16 12	15 46	00 09	22 40	11 25	20 38	12 48
31	04 N 01	20 S 50	04 S 56	16 N 36	15 S 32	00 N 15	22 N 40	11 S 24	20 S 37	12 N 48

ZODIAC SIGN ENTRIES

Date	h	m	Planets
01	14	33	☽ ♏
03	05	32	♂ ♒
03	19	05	☽ ♐
06	03	39	☽ ♑
08	15	09	☽ ♒
11	03	49	☽ ♓
13	16	18	☽ ♈
16	03	52	☽ ♉
16	11	50	☿ ♓
18	13	43	☽ ♊
18	16	47	♃ ♈
19	21	42	♀ ♉
20	20	48	☽ ♋
20	15	57	☉ ♈
23	00	31	☽ ♌
25	01	21	☽ ♍
27	00	51	☽ ♎
29	01	08	☽ ♏
31	04	10	☽ ♐

LATITUDES

Date	Mercury ☿ ° '	Venus ♀ ° '	Mars ♂ ° '	Jupiter ♃ ° '	Saturn ♄ ° '	Uranus ♅ ° '	Neptune ♆ ° '	Pluto ♇ ° '
01	00 N 24	00 S 51	00 S 53	01 S 05	00 S 17	00 N 33	01 N 34	17 N 15
04	00 S 09	00 44	00 55	01 04	00 16	00 33	01 34	17 16
07	00 39	00 36	00 58	01 04	00 16	00 34	01 34	17 17
10	01 05	00 28	01 00	01 04	00 16	00 34	01 34	17 18
13	01 27	00 20	01 02	01 03	00 15	00 34	01 34	17 18
16	01 46	00 11	01 04	01 03	00 15	00 34	01 34	17 19
19	02 01	00 S 02	01 06	01 03	00 14	00 34	01 34	17 19
22	02 11	00 N 07	01 08	01 04	00 14	00 34	01 35	17 20
25	02 18	00 16	01 10	01 04	00 14	00 34	01 35	17 20
28	02 20	00 26	01 11	01 04	00 13	00 34	01 35	17 20
31	02 S 20	00 N 36	01 S 13	01 S 05	00 S 13	00 N 34	01 N 35	17 N 20

DATA

Julian Date	2442473
Delta T	+46 seconds
Ayanamsa	23° 30' 53"
Synetic vernal point	05° ♓ 36' 06"
True obliquity of ecliptic	23° 26' 30"

MOON'S PHASES, APSIDES AND POSITIONS ☽

Date	h	m	Phase	Longitude ° '	Eclipse Indicator
04	20	20	☾	13 ♐ 38	
12	23	47	●	21 ♓ 47	
20	20	05	☽	29 ♊ 35	
27	10	36	○	06 ♎ 08	

Day	h	m		
11	05	19	Apogee	
26	08	59	Perigee	
05	05	34	Max dec	21° S 45'
12	17	40	0N	
20	00	01	Max dec	21° N 36'
26	06	54	0S	

LONGITUDES

Date	Chiron ⚷ ° '	Ceres ⚳ ° '	Pallas ⚴ ° '	Juno ⚵ ° '	Vesta ⚶ ° '	Black Moon Lilith ⚸ ° '
01	21 ♈ 30	04 ♈ 31	02 ♓ 57	22 ♈ 16	10 ♒ 01	02 ♓ 42
11	22 ♈ 01	08 ♈ 24	06 ♓ 15	27 ♈ 51	15 ♒ 02	03 ♓ 49
21	22 ♈ 33	12 ♈ 09	09 ♓ 30	03 ♉ 33	19 ♒ 57	04 ♓ 56
31	23 ♈ 08	16 ♈ 19	12 ♓ 42	09 ♉ 20	24 ♒ 46	06 ♓ 03

All ephemeris data is given at 12.00 UT and the Moon's longitude is additionally given for 24.00 UT

ASPECTARIAN

h m	Aspects
01 Saturday	
01 08	☽ ∥ ♅
04 48	☽ # ♇
06 15	☽ bQ ☉
07 33	☽ ⊼ ♃
09 06	☽ ∠ ♆
12 23	☽ □ ♂
17 48	☽ ± ♃
18 18	☽ ☌ ♅
02 Sunday	
00 19	☽ ⊥ ♆
04 16	☽ ∥ ☿
04 47	☽ ⊼ ♀
05 06	♀ # ♃
05 13	☽ ⊻ ♇
09 28	☽ bQ ♃
09 31	♀ ☍ ♇
09 39	☽ △ ☉
10 39	☽ ⊻ ♆
11 12	☽ △ ♄
15 24	☽ □ ☿
15 36	☽ ⊥ ♇
16 13	☽ ± ♀
21 07	☽ Q ♂
21 55	♃ ± ♅
23 29	☉ □ ♆
03 Monday	
06 38	☉ △ ♄
07 30	☽ ∠ ♇
09 46	☽ bQ ♀
12 28	☽ △ ♃
13 44	☽ bQ ♄
19 55	☽ ✶ ♂
20 21	☽ ∥ ♆
23 03	☽ ⊻ ♅
04 Tuesday	
00 38	☽ ∥ ♂
01 54	☽ Q ☿
06 10	☽ ± ♄
10 06	☽ ⊥ ♅
10 50	☽ ✶ ♇
16 09	☽ △ ♀
16 50	☽ ☌ ♆
17 19	☽ ⊼ ♄
20 20	☽ □ ☉
23 22	♀ △ ♆
05 Wednesday	
01 31	☽ ∠ ♂
02 02	☽ ✶ ☿
02 58	☽ ∠ ♅
04 17	♀ □ ♄
09 25	☽ Q ♇
18 15	♂ ∥ ♆
19 49	☽ ⊥ ♂
21 40	☽ □ ♃
06 Thursday	
01 17	♂ □ ♅
07 47	☽ ✶ ♅
08 13	☽ ⊻ ♂
09 22	☽ ∠ ☿
09 54	☽ Q ☉
11 01	☿ ± ♄
16 03	☽ ∥ ♆
18 34	☽ ∥ ♂
20 17	☽ □ ♇
07 Friday	
02 46	☽ ⊻ ♆
03 11	☽ ☍ ♄
04 43	☽ ⊥ ☿
07 21	☽ Q ♅
08 33	☽ □ ♀
09 42	☽ Q ♃
11 41	☽ ✶ ☉
14 41	☽ ⊥ ♆
14 56	☉ # ♀
17 47	☽ ⊻ ☿
08 Saturday	
05 38	♀ Q ♂
06 43	☉ bQ ♅
08 42	☽ ∠ ♆
10 10	☽ ✶ ♃
19 18	☽ □ ♅
20 14	☽ ∥ ☿
20 29	☽ ∠ ☉
23 54	☽ ☌ ♂
09 Sunday	
00 43	☽ Q ♀
08 10	☽ △ ♇
14 57	☽ ✶ ♆
15 18	☽ ⊼ ♄
16 19	☽ ⊥ ☉
16 43	☽ # ♇
17 03	☽ ∠ ♃
22 28	☽ ∥ ♅
10 Monday	
01 20	☿ ⊥ ♃
03 26	☽ ± ♄
03 46	☽ ✶ ♀
05 35	☽ ⊻ ☉
11 39	☽ ⊥ ♃
12 32	☽ ☌ ☿
14 27	☽ bQ ♇
15 15	☽ Q ♆
21 39	☽ bQ ♄
11 Tuesday	
00 03	☽ ⊻ ♃
01 54	☽ # ♀
07 23	☿ bQ ♇
07 50	☽ △ ♅
08 38	☽ ± ♇
13 37	☽ ∠ ♀
15 35	☿ Q ♆
16 44	☽ ⊻ ♂
20 45	☽ ⊼ ♇
21 19	☽ ∥ ☉
12 Wednesday	
03 40	☽ □ ♆
03 58	☽ △ ♄
05 40	☽ ⊥ ♂
08 56	☽ ∥ ♃
09 53	☽ ⊥ ♀
14 04	☽ bQ ♅
23 21	☽ ⊻ ♀
23 47	☽ ☌ ☉
13 Thursday	
01 02	☽ ∠ ♂
02 01	☽ # ♃
08 02	☽ ⊻ ☿
08 08	☽ ± ♅
10 34	☽ # ☉
13 49	☽ ☌ ♃
20 08	☽ ⊼ ♅
20 40	☉ ⊻ ♀
21 29	☽ ⊥ ☿
14 Friday	
04 10	☿ bQ ♄
05 49	♂ △ ♇
08 31	♄ St D
08 55	☽ ☍ ♇
09 07	☽ ✶ ♂
09 59	♆ St R
15 52	☽ △ ♆
16 10	☽ □ ♄
17 36	☽ ∠ ☿
20 01	☽ ∥ ♀
15 Saturday	
09 38	☉ ∠ ♂
10 33	☽ Q ♂
11 46	☽ # ♅
17 02	☽ ⊻ ☉
17 56	☽ ☌ ♀
18 16	☽ ∥ ♇
21 34	☽ bQ ♆
22 55	☽ # ☿
16 Sunday	
00 11	☿ ⊻ ♃
02 38	☽ ⊻ ♃
02 53	☽ ✶ ☿
03 46	☽ Q ♄
05 53	☽ ⊥ ☉
07 29	☽ ☍ ♅
14 37	☽ ⊥ ♃
15 13	☽ ± ♆
19 58	☽ ⊼ ♇
17 Monday	
00 10	☽ □ ♂
00 59	☽ ∠ ☉
01 01	☉ ± ♅
02 52	☽ ⊼ ♆
03 11	☽ ✶ ♄
05 10	☽ Q ☿
06 23	♀ bQ ♆
07 34	☽ ± ♇
08 27	☽ ∠ ♃
15 34	☽ # ♂
16 18	☿ # ♇
19 19	☿ △ ♅
18 Tuesday	
00 49	☽ bQ ♇
02 24	☿ ± ♇
07 58	☽ ∠ ♄
08 15	☽ ✶ ☉
10 20	☽ ⊻ ♀
13 39	☽ ✶ ♃
17 02	☽ ⊼ ♅
17 19	☽ # ♆
18 42	☉ ∥ ♃
19 46	☽ □ ☿
20 19	♂ ✶ ♆
22 49	☽ ⊥ ♀
19 Wednesday	
00 55	☽ ⊥ ♄
01 49	♂ ⊼ ♄
04 15	☽ ± ♅
05 02	☽ △ ♇
08 40	☽ Q ☉
11 45	☽ ☍ ♆
12 05	☽ ⊻ ♄
12 30	☽ Q ♃
12 42	☽ △ ♂
17 15	☽ ∠ ♀
20 50	☽ bQ ♅
21 09	♀ Q ♄
20 Thursday	
02 10	♀ # ♅
02 59	☿ # ♀
03 52	☿ ∥ ♅
04 55	♀ ⊻ ♃
17 38	☽ bQ ♂
20 05	☽ □ ☉
21 45	☽ □ ♃
23 05	☽ ✶ ♀
23 48	☽ △ ♅
21 Friday	
05 51	☽ # ♆
06 54	♀ ☍ ♅
08 59	☽ △ ☿
10 30	☽ ± ♂
11 07	☽ □ ♇
17 31	☽ ⊼ ♆
17 53	☽ ☌ ♄
21 32	☽ ⊼ ♂
22 08	☽ Q ♀
22 Saturday	
01 41	☉ ☌ ♃
03 51	☽ ± ♆
04 27	☿ ⊼ ♇
13 35	☽ # ♂
13 36	♀ ∥ ♇
13 54	☽ bQ ☿
17 56	☽ Q ♇
18 03	☉ # ♃
19 07	☽ bQ ♆
21 11	☉ ⊼ ♅
23 Sunday	
02 18	☽ △ ♃
03 12	☽ □ ♅
03 40	☽ △ ☉
06 44	☽ ± ☿
07 23	☽ □ ♀
13 51	☽ ✶ ♇
17 05	☽ ∥ ♀
17 46	☽ ⊼ ☿
19 53	☽ ∥ ♇
19 57	☽ △ ♆
20 22	☽ ⊻ ♄
24 Monday	
02 20	☽ # ♅
02 29	☽ ☍ ♂
03 21	☽ bQ ♃
06 03	☽ bQ ☉
06 08	☽ ⊥ ♄
08 37	☽ Q ♅
13 22	☽ # ☿
14 11	☽ ∠ ♇
14 34	☿ □ ♆
15 59	♀ ± ♆
18 06	☽ ± ♃
18 53	☿ △ ♄
20 38	☽ ∠ ♄
21 33	☽ ± ☉
25 Tuesday	
01 53	♃ ⊼ ♅
03 14	♂ ∠ ♃
03 48	☽ ✶ ♅
03 50	☽ ⊼ ♃
04 33	☽ ⊥ ♇
07 47	☽ ⊼ ☉
12 08	☽ △ ♀
14 04	☽ ⊻ ♇
20 01	☽ □ ♆
20 30	☽ ✶ ♄
22 54	☽ ∥ ☉
23 25	☽ ☍ ☿
26 Wednesday	
03 31	☽ ∠ ♅
04 53	☽ ⊼ ♂
05 47	☽ # ♃
07 59	☽ ∥ ♃
09 43	♀ ⊥ ♃
11 49	♀ ⊼ ♇
13 54	☽ bQ ♀
14 52	☽ ± ♂
15 30	☽ Q ♄
16 08	☽ # ☉
17 42	☽ ⊥ ♅
27 Thursday	
00 28	☽ Q ♆
03 11	☽ ⊻ ♅
03 51	♂ ± ♄
04 06	☽ ☍ ♃
05 23	☽ ± ♀
05 52	☽ bQ ♂
10 36	☽ ☍ ☉
13 28	☽ ☌ ♇
13 46	☿ bQ ♅
15 45	☽ ⊼ ♀
15 47	☽ ∥ ☿
19 33	☽ ✶ ♆
20 08	☽ □ ♄
28 Friday	
04 43	☽ ⊼ ☿
07 08	☽ △ ♂
11 06	☽ ∥ ♅
15 39	☽ ± ☿
18 09	☽ # ♇
19 47	☽ ∠ ♆
29 Saturday	
03 25	☽ ☌ ♅
05 13	☉ ☍ ♇
05 20	☽ ⊼ ♃
08 19	☽ bQ ☿
10 37	☽ ⊥ ♆
11 50	☽ # ♀
13 13	☽ ∥ ♂
14 12	☽ ⊻ ♇
14 47	♀ ⊼ ♆
14 53	☽ ⊼ ☉
15 29	☽ ± ♃
19 12	☿ ⊻ ♂
20 03	♀ # ♂
20 40	☽ ⊻ ♆
21 12	☽ ☍ ♀
21 24	☽ △ ♄
23 31	♀ ✶ ♄
30 Sunday	
00 17	☽ ⊥ ♇
01 45	☽ ± ☉
06 59	☽ bQ ♃
11 50	☽ □ ♂
13 04	☽ △ ☿
15 40	☽ ∠ ♇
18 24	☽ bQ ☉
23 14	☽ bQ ♄
31 Monday	
06 29	☽ ⊻ ♅
08 27	☽ ∥ ♆
08 48	♀ ± ♇
09 36	☽ △ ♃
15 13	☽ ± ♄
17 09	☽ ⊥ ♅
18 06	☽ ✶ ♇
21 28	☽ Q ♂
23 08	☽ △ ☉

APRIL 1975

LONGITUDES

Date	Sidereal time h m s	Sun ☉ ° ' "	Moon ☽ ° ' "	Moon ☽ 24.00 ° ' "	Mercury ☿ ° '	Venus ♀ ° '	Mars ♂ ° '	Jupiter ♃ ° '	Saturn ♄ ° '	Uranus ♅ ° '	Neptune ♆ ° '	Pluto ♇ ° '
01	00 36 53	11 ♈ 08 16	17 ♐ 38 48	24 ♐ 05 57	24 ♓ 41	15 ♉ 12	22 ♒ 09	03 ♈ 20	12 ♋ 15	01 ♏ 16	11 ♐ 42	07 ♎ 48
02	00 40 50	12 07 28	00 ♑ 27 18	06 ♑ 43 26	26 25	16 24	22 55	03 35	12 17	01 R 14	11 R 42	07 R 47
03	00 44 47	13 06 38	12 ♑ 54 54	19 ♑ 02 20	28 11	17 35	23 41	03 49	12 19	01 11	11 41	07 45
04	00 48 43	14 05 46	25 ♑ 06 21	01 ♒ 07 33	29 ♓ 58	18 47	24 26	04 04	12 21	01 09	11 41	07 43
05	00 52 40	15 04 52	07 ♒ 06 32	13 ♒ 03 51	01 ♈ 46	19 59	25 12	04 18	12 24	01 06	11 40	07 42
06	00 56 36	16 03 56	19 ♒ 00 02	24 ♒ 55 35	03 36	21 10	25 58	04 32	12 26	01 04	11 39	07 40
07	01 00 33	17 02 59	00 ♓ 50 58	06 ♓ 46 35	05 28	22 22	26 43	04 47	12 29	01 02	11 38	07 38
08	01 04 29	18 01 59	12 ♓ 42 48	18 ♓ 39 56	07 21	23 33	27 29	05 01	12 31	00 59	11 38	07 37
09	01 08 26	19 00 58	24 ♓ 38 16	00 ♈ 38 04	09 15	24 45	28 15	05 15	12 34	00 57	11 37	07 35
10	01 12 22	19 59 55	06 ♈ 39 32	12 ♈ 42 50	11 11	25 56	29 00	05 30	12 37	00 54	11 36	07 33
11	01 16 19	20 58 50	18 ♈ 48 07	24 ♈ 55 34	13 08	27 07	29 ♒ 46	05 44	12 40	00 52	11 35	07 32
12	01 20 16	21 57 43	01 ♉ 05 16	07 ♉ 17 22	15 07	28 18	00 ♓ 32	05 58	12 43	00 49	11 34	07 30
13	01 24 12	22 56 34	13 ♉ 31 59	19 ♉ 49 16	17 07	29 ♉ 29	01 18	06 12	12 46	00 46	11 33	07 29
14	01 28 09	23 55 22	26 ♉ 09 22	02 ♊ 32 27	19 08	00 ♊ 40	02 03	06 27	12 50	00 44	11 32	07 27
15	01 32 05	24 54 09	08 ♊ 58 42	15 ♊ 28 19	21 11	01 51	02 49	06 41	12 53	00 41	11 31	07 25
16	01 36 02	25 52 54	22 ♊ 01 30	28 ♊ 38 29	23 14	03 02	03 35	06 55	12 56	00 39	11 30	07 24
17	01 39 58	26 51 36	05 ♋ 19 27	12 ♋ 04 38	25 19	04 12	04 20	07 09	13 00	00 36	11 29	07 22
18	01 43 55	27 50 17	18 ♋ 54 10	25 ♋ 48 09	27 25	05 23	05 06	07 23	13 04	00 34	11 28	07 21
19	01 47 51	28 48 55	02 ♌ 46 40	09 ♌ 49 39	29 ♈ 32	06 33	05 52	07 37	13 07	00 31	11 27	07 19
20	01 51 48	29 ♈ 47 30	16 ♌ 56 57	24 ♌ 08 20	01 ♉ 39	07 43	06 37	07 51	13 11	00 29	11 26	07 18
21	01 55 45	00 ♉ 46 04	01 ♍ 23 25	08 ♍ 41 39	03 47	08 54	07 23	08 05	13 15	00 26	11 25	07 16
22	01 59 41	01 44 35	16 ♍ 02 26	23 ♍ 24 58	05 55	10 04	08 09	08 19	13 19	00 23	11 24	07 15
23	02 03 38	02 43 04	00 ♎ 48 24	08 ♎ 11 50	08 02	11 14	08 54	08 33	13 23	00 21	11 23	07 13
24	02 07 34	03 41 31	15 ♎ 34 17	22 ♎ 54 48	10 10	12 24	09 40	08 47	13 27	00 18	11 22	07 12
25	02 11 31	04 39 56	00 ♏ 12 28	07 ♏ 26 26	12 17	13 33	10 26	09 01	13 32	00 16	11 20	07 10
26	02 15 27	05 38 19	14 ♏ 35 59	21 ♏ 40 28	14 23	14 43	11 11	09 14	13 36	00 13	11 19	07 09
27	02 19 24	06 36 40	28 ♏ 39 25	05 ♐ 32 31	16 27	15 53	11 57	09 28	13 41	00 11	11 18	07 08
28	02 23 20	07 35 00	12 ♐ 19 34	19 ♐ 00 31	18 30	17 02	12 43	09 42	13 45	00 08	11 17	07 06
29	02 27 17	08 33 18	25 ♐ 35 27	02 ♑ 04 35	20 32	18 11	13 28	09 55	13 50	00 06	11 15	07 05
30	02 31 14	09 ♉ 31 35	08 ♑ 28 11	14 ♑ 46 39	22 ♉ 31	19 ♊ 20	14 ♓ 14	10 ♈ 09	13 ♋ 54	00 ♏ 03	11 ♐ 14	07 ♎ 03

Date	Moon True ☊ ° '	Moon Mean ☊ ° '	Moon ☽ Latitude ° '
01	02 ♐ 11	03 ♐ 48	01 N 22
02	02 D 12	03 45	02 26
03	02 12	03 42	03 21
04	02 R 12	03 38	04 06
05	02 09	03 35	04 39
06	02 04	03 32	05 00
07	01 58	03 29	05 08
08	01 50	03 26	05 02
09	01 42	03 23	04 44
10	01 34	03 19	04 13
11	01 27	03 16	03 30
12	01 21	03 13	02 36
13	01 18	03 10	01 35
14	01 16	03 07	00 N 28
15	01 D 17	03 03	00 S 42
16	01 18	03 00	01 51
17	01 19	02 57	02 55
18	01 21	02 54	03 51
19	01 R 21	02 51	04 35
20	01 20	02 48	05 03
21	01 17	02 44	05 12
22	01 14	02 41	05 02
23	01 10	02 38	04 32
24	01 06	02 35	03 44
25	01 03	02 32	02 41
26	01 01	02 29	01 29
27	01 00	02 25	00 S 13
28	01 D 00	02 22	01 N 02
29	01 01	02 19	02 11
30	01 ♐ 03	02 ♐ 16	03 N 12

DECLINATIONS

Date	Sun ☉ ° '	Moon ☽ ° '	Mercury ☿ ° '	Venus ♀ ° '	Mars ♂ ° '	Jupiter ♃ ° '	Saturn ♄ ° '	Uranus ♅ ° '	Neptune ♆ ° '	Pluto ♇ ° '
01	04 N 24	21 S 30	04 S 14	17 N 01	15 S 17	00 N 20	22 N 40	11 S 23	20 S 37	12 N 49
02	04 48	21 00	03 31	17 25	15 03	00 26	22 40	11 22	20 37	12 50
03	05 11	19 28	02 47	17 48	14 49	00 32	22 40	11 22	20 37	12 50
04	05 34	17 05	02 01	18 11	14 34	00 37	22 40	11 21	20 37	12 51
05	05 56	13 59	01 15	18 34	14 19	00 43	22 40	11 20	20 37	12 52
06	06 19	10 22	00 S 28	18 56	14 04	00 49	22 39	11 19	20 37	12 52
07	06 42	06 22	00 N 20	19 18	13 49	00 54	22 39	11 18	20 37	12 53
08	07 04	02 S 08	01 10	19 39	13 34	01 00	22 39	11 17	20 36	12 53
09	07 27	02 N 13	02 00	20 00	13 19	01 06	22 39	11 17	20 36	12 54
10	07 49	06 31	02 50	20 20	13 03	01 11	22 39	11 16	20 36	12 54
11	08 11	10 36	03 42	20 40	12 48	01 17	22 39	11 15	20 36	12 55
12	08 33	14 18	04 34	20 59	12 32	01 23	22 39	11 14	20 36	12 56
13	08 55	17 25	05 27	21 18	12 17	01 28	22 38	11 13	20 36	12 56
14	09 17	19 45	06 20	21 36	12 01	01 34	22 38	11 12	20 36	12 57
15	09 39	21 06	07 14	21 53	11 45	01 39	22 38	11 11	20 35	12 57
16	10 00	21 22	08 08	22 10	11 29	01 45	22 38	11 10	20 35	12 58
17	10 21	20 25	09 03	22 27	11 13	01 50	22 37	11 09	20 35	12 58
18	10 42	18 18	09 57	22 43	10 56	01 56	22 37	11 09	20 35	12 58
19	11 03	15 05	10 51	22 58	10 40	02 01	22 37	11 08	20 35	12 59
20	11 24	10 57	11 45	23 12	10 24	02 07	22 37	11 07	20 34	12 59
21	11 45	06 07	12 39	23 26	10 07	02 12	22 36	11 06	20 34	13 00
22	12 05	00 N 52	13 32	23 40	09 50	02 17	22 36	11 05	20 34	13 00
23	12 25	04 S 29	14 24	23 53	09 34	02 23	22 36	11 04	20 34	13 00
24	12 45	09 34	15 15	24 05	09 17	02 28	22 36	11 03	20 34	13 01
25	13 05	14 03	16 04	24 16	09 00	02 34	22 35	11 02	20 34	13 01
26	13 24	17 38	16 53	24 27	08 43	02 39	22 35	11 01	20 33	13 02
27	13 43	20 04	17 39	24 37	08 26	02 44	22 35	11 01	20 33	13 02
28	14 03	21 15	18 24	24 47	08 09	02 50	22 34	11 00	20 33	13 02
29	14 21	21 11	19 06	24 56	07 52	02 55	22 34	10 59	20 33	13 02
30	14 N 40	19 S 59	19 N 47	25 N 04	07 S 35	03 N 00	22 N 33	10 S 58	20 S 33	13 N 03

ZODIAC SIGN ENTRIES

Date	h	m	Planets
02	11	08	☽ ♑
04	12	28	☿ ♈
04	21	45	☽ ♒
07	10	17	☽ ♓
09	22	44	☽ ♈
11	19	15	♂ ♓
12	09	53	☽ ♉
13	22	26	♀ ♊
14	19	14	☽ ♊
17	02	27	☽ ♋
19	07	14	☽ ♌
19	17	20	☿ ♉
20	17	07	☉ ♉
21	09	42	☽ ♍
23	10	41	☽ ♎
25	11	39	☽ ♏
27	14	20	☽ ♐
29	20	08	☽ ♑

LATITUDES

Date	Mercury ☿ ° '	Venus ♀ ° '	Mars ♂ ° '	Jupiter ♃ ° '	Saturn ♄ ° '	Uranus ♅ ° '	Neptune ♆ ° '	Pluto ♇ ° '
01	02 S 18	00 N 39	01 S 14	01 S 05	00 S 13	00 N 34	01 N 35	17 N 20
04	02 11	00 49	01 16	01 05	00 12	00 34	01 35	17 20
07	02 00	00 58	01 17	01 05	00 12	00 34	01 35	17 20
10	01 43	01 08	01 19	01 05	00 12	00 34	01 35	17 20
13	01 23	01 17	01 21	01 05	00 11	00 34	01 35	17 20
16	00 58	01 26	01 23	01 05	00 11	00 34	01 36	17 19
19	00 S 29	01 35	01 24	01 06	00 11	00 34	01 36	17 19
22	00 N 02	01 43	01 26	01 06	00 10	00 34	01 36	17 18
25	00 35	01 51	01 27	01 06	00 10	00 34	01 36	17 18
28	01 06	01 59	01 29	01 06	00 10	00 34	01 36	17 17
31	01 N 35	02 N 06	01 S 30	01 S 07	00 S 09	00 N 34	01 N 36	17 N 16

DATA

Julian Date	2442504
Delta T	+46 seconds
Ayanamsa	23° 30' 56"
Synetic vernal point	05° ♓ 36' 04"
True obliquity of ecliptic	23° 26' 29"

MOON'S PHASES, APSIDES AND POSITIONS ☽

Date	h	m	Phase	Longitude ° '	Eclipse Indicator
03	12	25	☾	13 ♑ 08	
11	16	39	●	21 ♈ 10	
19	04	41	☽	28 ♋ 31	
25	19	55	○	04 ♏ 59	

Day	h	m		
07	16	24	Apogee	
23	12	55	Perigee	
01	13	26	Max dec	21° S 30'
08	23	47	0N	
16	05	11	Max dec	21° N 24'
22	15	52	0S	
28	22	34	Max dec	21° S 22'

LONGITUDES

Date	Chiron ⚷ ° '	Ceres ⚳ ° '	Pallas ⚴ ° '	Juno ⚵ ° '	Vesta ⚶ ° '	Black Moon Lilith ⚸ ° '
01	23 ♈ 11	16 ♈ 43	13 ♓ 01	09 ♉ 54	25 ♒ 14	06 ♓ 10
11	23 ♈ 47	20 ♈ 42	16 ♓ 07	15 ♉ 45	29 ♒ 53	07 ♓ 17
21	24 ♈ 23	24 ♈ 42	19 ♓ 07	21 ♉ 37	04 ♓ 23	08 ♓ 24
31	24 ♈ 58	28 ♈ 41	21 ♓ 59	27 ♉ 31	08 ♓ 42	09 ♓ 32

All ephemeris data is given at 12.00 UT and the Moon's longitude is additionally given for 24.00 UT

ASPECTARIAN

h m Aspects

01 Tuesday
01 08 ☽ ☌ ♆
02 05 ☽ ⊼ ♄
07 02 ☽ ⊼ ♀
08 10 ☉ # ☿
09 28 ☽ ∠ ♅
15 59 ☽ Q ♇
19 15 ☽ ± ♀
19 57 ☿ ± ♅
20 54 ☽ ⚹ ♂

02 Wednesday
01 43 ☉ △ ♆
03 09 ☽ □ ☿
07 45 ♂ bQ ♇
13 28 ☽ ⚹ ♅
13 58 ☽ bQ ♀
16 00 ☉ □ ♄
18 05 ☽ □ ♃
19 39 ☽ ‖ ♆

03 Thursday
02 00 ☽ □ ♇
03 14 ☽ ∠ ♂
09 37 ☽ ⊻ ♆
10 50 ☽ ☍ ♄
12 17 ♂ Q ♆
12 25 ☽ □ ☉
12 32 ☽ Q ♅
19 28 ☽ Q ☿
21 20 ☽ ⊥ ♆
21 56 ☽ ⊥ ♂
22 09 ☽ △ ♀

04 Friday
03 10 ☽ # ♀
05 50 ☽ Q ♃
10 35 ☽ ⊻ ♂
15 07 ☽ ∠ ♆
18 53 ♀ ∠ ♃
22 54 ☿ ⊥ ♂
23 23 ☽ ⚹ ☿

05 Saturday
00 00 ☽ □ ♅
03 11 ☽ Q ☉
03 22 ☿ ⊼ ♅
06 15 ☽ ⚹ ♃
09 27 ☽ ‖ ♂
13 10 ☽ △ ♇
19 49 ☽ # ♇
21 10 ☽ ⚹ ♆
22 41 ☽ ⊼ ♄

06 Sunday
02 30 ☿ # ♃
05 32 ☽ ⚹ ☉
05 59 ☽ ‖ ♅
10 51 ☽ ± ♄
11 04 ☽ ∠ ☿
13 07 ☽ ∠ ♃
16 53 ☽ □ ♀
19 24 ☽ bQ ♇
21 25 ☽ Q ♆

07 Monday
01 52 ☿ ☌ ♃
03 04 ☽ ☌ ♂
05 09 ☽ bQ ♄
07 43 ☽ ⊥ ♃
08 40 ☽ ⊥ ☿
10 16 ☽ # ☉
12 21 ☽ △ ♅
13 36 ☽ ± ♇
14 39 ☽ ∠ ☉
17 24 ♀ bQ ♇
20 07 ☽ ⊻ ♃
23 06 ☽ ⊻ ☿

08 Tuesday
01 43 ☽ ⊼ ♇
06 46 ☿ ‖ ♃
09 24 ☽ Q ♀
09 49 ☽ □ ♆
10 30 ☽ ⊥ ☉
11 37 ☽ △ ♄
13 21 ♂ bQ ♄
15 19 ☿ ☍ ♇
16 31 ☽ # ☿
18 07 ☽ # ♃
18 35 ☽ bQ ♅
23 41 ☽ ⊻ ☉

09 Wednesday
05 41 ☽ ‖ ♃
10 30 ☽ ‖ ☿
12 14 ☽ ⚹ ♀
12 37 ☽ ± ♅
19 43 ☽ ⊻ ♂

10 Thursday
00 34 ☽ ⊼ ♅
04 02 ☿ ∠ ♀
08 30 ☽ ⊥ ♂
09 38 ☽ ☌ ♃
13 47 ☽ ☍ ♇
17 08 ☿ △ ♆
19 47 ☉ ⊥ ♀
20 15 ☽ ‖ ☉
21 24 ☽ ∠ ♀
21 47 ☽ △ ♆
22 41 ☽ ☌ ☿
23 52 ☽ □ ♄

11 Friday
01 24 ♂ # ♇
03 32 ☽ ∠ ♂
06 09 ☿ □ ♄
07 18 ♀ # ♆
10 20 ♂ ⊥ ♃
15 59 ☽ # ♅
16 39 ☽ ☌ ☉
17 02 ☽ ⊥ ♀
23 38 ♀ ∠ ♄

12 Saturday
00 59 ☽ # ♂
02 43 ☽ ‖ ♇
03 13 ☽ bQ ♆
06 01 ☽ ⊻ ♀
10 51 ☽ ⚹ ♂
11 17 ☽ Q ♄
11 29 ☽ ☍ ♅
20 11 ☿ ∠ ♂
20 31 ♂ △ ♅
20 40 ☽ ± ♆
21 38 ☽ ⊻ ♃

13 Sunday
00 23 ☽ ⊼ ♇
08 13 ☽ ⊼ ♆
09 25 ☽ ⊥ ♃
10 32 ☽ ⚹ ♄
11 31 ☽ Q ♂
11 54 ☽ ± ♇
17 35 ♂ ± ♇
20 09 ☽ ⊻ ☿

14 Monday
02 55 ☽ ∠ ♃
05 00 ☽ bQ ♇
07 26 ☽ ⊻ ☉
09 42 ☽ ⊥ ☿
13 16 ♀ ⊼ ♅
15 10 ☽ ∠ ♄
19 41 ☽ ⊥ ☉
20 35 ☽ ⊼ ♅
21 21 ☽ ☌ ♀
23 47 ☽ □ ♂

15 Tuesday
00 35 ☽ # ♆
05 48 ☽ ∠ ☿
07 39 ☽ ⚹ ♃
07 46 ☽ ± ♅
08 05 ☽ ⊥ ♄
09 07 ☽ △ ♇
13 51 ☽ ∠ ☉
16 43 ☽ ☍ ♆
19 16 ☽ ⊻ ♄

16 Wednesday
00 22 ☽ bQ ♅
06 13 ☽ Q ♃
14 38 ☽ ⚹ ☿
19 34 ☽ ⚹ ☉

17 Thursday
03 05 ☉ bQ ♆
03 34 ☽ △ ♅
09 19 ☽ # ♆
09 48 ☽ ⊻ ♀
10 08 ☽ △ ♂
15 19 ☽ □ ♃
15 39 ☽ □ ♇
16 13 ☽ Q ☿
16 47 ♂ ‖ ♅
18 47 ☽ Q ☉
19 53 ♀ □ ♂
21 31 ☽ ⊥ ♀
22 57 ☽ ⊼ ♆

18 Friday
01 18 ☿ bQ ♆
01 41 ☽ ☌ ♄
04 02 ♀ ‖ ♄
08 20 ♃ ☍ ♇
09 30 ☽ ± ♆
14 13 ☽ bQ ♂
14 49 ☽ ∠ ♀
20 55 ☉ ☌ ☿
20 59 ☉ # ♂
23 11 ☽ Q ♇

19 Saturday
01 09 ☽ bQ ♆
04 41 ☽ □ ☉
05 26 ☽ □ ☿
06 43 ☽ ± ♂
08 07 ☿ # ♂
08 08 ☽ □ ♅
11 21 ♀ ± ♅
16 56 ☉ # ♅
17 34 ☽ ⊼ ♂
19 02 ☽ ⚹ ♀
19 07 ☿ # ♅
19 44 ☽ ⚹ ♇
20 24 ☽ △ ♃
20 33 ☉ ‖ ☿
23 00 ☿ ☍ ♅

20 Sunday
00 21 ♀ ⊥ ♄
00 45 ☽ ‖ ♇
02 44 ☽ △ ♆
03 23 ♀ △ ♇
05 39 ☽ ⊻ ♄
06 37 ☿ Q ♄
08 22 ☽ ‖ ☿
09 44 ☽ ‖ ☉
11 05 ☽ # ♅
13 53 ☿ ⊥ ♀
14 33 ☽ Q ♅
15 04 ☽ # ♂
15 17 ♀ ⚹ ♃
15 46 ☽ ⊥ ♄
17 03 ☽ Q ♀
20 55 ☽ ∠ ♇
22 01 ☽ bQ ♃

21 Monday
04 09 ☉ ☍ ♅
06 48 ☽ ∠ ♄
08 29 ♂ ⊼ ♇
10 26 ☽ ⚹ ♅
10 54 ☽ △ ☉
11 48 ☽ ⊥ ♇
13 10 ☽ ± ♃
16 36 ☽ △ ☿
21 28 ☿ ‖ ♇
21 39 ☽ ⊻ ♇
22 24 ☽ ☍ ♂
23 11 ☽ ⊼ ♃

22 Tuesday
00 49 ☉ Q ♄
01 24 ☽ □ ♀
04 26 ☽ □ ♆
05 41 ☽ ‖ ♃
06 19 ☿ ± ♆
07 33 ☽ ⚹ ♄
10 57 ☽ ∠ ♅
13 14 ☽ bQ ☉
19 45 ♂ ⊻ ♃
21 16 ☽ bQ ☿

23 Wednesday
01 33 ☽ ⊥ ♅
02 22 ☽ # ♃
02 52 ☿ ⊼ ♇
03 10 ☽ Q ♄
04 54 ☽ ± ☉
09 41 ☽ Q ♆
11 16 ☽ ⊻ ♅
14 20 ☽ ± ☿
15 04 ♀ ☍ ♆
15 19 ☽ ⊼ ☉
18 26 ☿ ⊻ ♃
22 24 ☽ ☌ ♇

24 Thursday
00 46 ☽ ☍ ♃
01 43 ☽ ⊼ ☿
01 52 ☽ ⊼ ♂
03 15 ☿ ⚹ ♂
05 09 ☽ ⚹ ♆
06 23 ☽ △ ♀
08 32 ☽ □ ♄
10 40 ☽ ‖ ♂
12 10 ☽ ± ♂
19 30 ☽ ‖ ♅

25 Friday
01 26 ☿ ⊼ ♆
03 42 ☽ bQ ♂
05 38 ☽ ∠ ♆
05 56 ☽ # ☉
06 04 ☽ # ♇
07 41 ☉ ‖ ♇
09 02 ☽ bQ ♀
11 25 ♀ ⊻ ♄
12 05 ☽ ☌ ♅
19 55 ☽ ☍ ☉
20 30 ☽ ⊥ ♆
22 05 ☿ ± ♇
23 32 ☽ ⊻ ♇

26 Saturday
01 16 ☽ ± ♀
02 47 ☿ ⚹ ♄
02 51 ☽ ⊼ ♃
04 19 ☉ ± ♆
04 30 ☽ # ☿
05 57 ☽ △ ♂
06 30 ☽ ⊻ ♆
09 34 ☽ ⊥ ♇
10 19 ☽ △ ♄
11 34 ☽ ☍ ☿
12 13 ☽ ⊼ ♀
13 06 ☽ ± ♃
16 00 ♂ □ ♆
20 47 ☿ ⊻ ♀
22 04 ♀ bQ ♅
23 08 ☿ ⊥ ♃

27 Sunday
00 47 ☽ ∠ ♇
04 40 ☽ bQ ♃
12 02 ☽ bQ ♄
14 38 ☽ ⊻ ♅
19 05 ☽ ‖ ♆

28 Monday
00 24 ☉ ⊼ ♇
01 05 ☽ ⊥ ♅
02 46 ☽ ⚹ ♇
02 57 ☽ ⊼ ☉
03 51 ☽ ± ♄
07 15 ☽ △ ♃
10 08 ☽ ☌ ♆
12 44 ☽ □ ♂
14 25 ☽ ± ☉
14 34 ☽ ⊼ ♄
17 00 ☽ ∠ ♅
21 14 ☽ ☍ ♀

29 Tuesday
00 09 ☽ Q ♇
01 06 ☽ ⊼ ☿
07 58 ☽ bQ ☉
14 02 ☽ ± ☿
20 17 ☽ ⚹ ♅
23 33 ☽ Q ♂

30 Wednesday
00 34 ♂ △ ♄
03 10 ☽ ‖ ♆
06 31 ☿ bQ ♇
09 20 ☽ □ ♇
09 52 ☽ bQ ☿
13 59 ☽ # ☿
14 10 ☽ △ ☉
15 14 ☽ □ ♃
17 14 ☽ ⊻ ♆
18 46 ☽ Q ♅
22 24 ☽ ☍ ♄
23 40 ☽ ⚹ ♂

MAY 1975

LONGITUDES

Date	Sidereal time h m s	Sun ☉ ° ' "	Moon ☽ ° ' "	Moon ☽ 24.00 ° ' "	Mercury ☿ ° '	Venus ♀ ° '	Mars ♂ ° '	Jupiter ♃ ° '	Saturn ♄ ° '	Uranus ♅ ° '	Neptune ♆ ° '	Pluto ♇ ° '
01	02 35 10	10 ♉ 29 50	21 ♑ 00 26	27 ♑ 10 01	24 ♉ 27	20 ♊ 29	14 ♓ 59	10 ♈ 22	13 ♋ 59	00 ♏ 01	11 ♐ 13	07 ♎ 02
02	02 39 07	11 28 03	03 ♒ 15 56	09 ♒ 18 47	26 21	21 38	15 45	10 36	14 04	29 ♎ 58	11 R 11	07 R 01
03	02 43 03	12 26 15	15 ♒ 19 07	21 ♒ 17 31	28 ♉ 12	22 47	16 31	10 49	14 09	29 R 56	11 10	06 59
04	02 47 00	13 24 25	27 ♒ 14 34	03 ♓ 10 51	00 ♊ 00	23 56	17 16	11 03	14 14	29 53	11 09	06 58
05	02 50 56	14 22 34	09 ♓ 06 55	15 ♓ 03 16	01 45	25 04	18 02	11 16	14 19	29 51	11 07	06 57
06	02 54 53	15 20 42	21 ♓ 00 26	26 ♓ 58 51	03 26	26 13	18 47	11 29	14 24	29 48	11 06	06 56
07	02 58 49	16 18 48	02 ♈ 58 56	09 ♈ 01 06	05 04	27 21	19 33	11 42	14 29	29 46	11 05	06 55
08	03 02 46	17 16 52	15 ♈ 05 40	21 ♈ 12 56	06 38	28 29	20 18	11 56	14 35	29 43	11 03	06 53
09	03 06 43	18 14 55	27 ♈ 23 08	03 ♉ 36 28	08 09	29 ♊ 37	21 04	12 09	14 40	29 41	11 02	06 52
10	03 10 39	19 12 57	09 ♉ 53 07	16 ♉ 13 10	09 36	00 ♋ 45	21 49	12 22	14 46	29 39	11 00	06 51
11	03 14 36	20 10 57	22 ♉ 36 41	29 ♉ 03 44	10 58	01 52	22 35	12 35	14 51	29 36	10 59	06 50
12	03 18 32	21 08 55	05 ♊ 34 17	12 ♊ 08 20	12 17	03 00	23 20	12 48	14 57	29 34	10 57	06 49
13	03 22 29	22 06 52	18 ♊ 45 49	25 ♊ 26 40	13 32	04 07	24 05	13 00	15 02	29 32	10 56	06 48
14	03 26 25	23 04 48	02 ♋ 10 47	08 ♋ 58 05	14 43	05 14	24 51	13 13	15 08	29 29	10 54	06 47
15	03 30 22	24 02 41	15 ♋ 48 27	22 ♋ 41 44	15 49	06 21	25 36	13 26	15 14	29 27	10 53	06 46
16	03 34 18	25 00 33	29 ♋ 37 49	06 ♌ 36 31	16 52	07 28	26 21	13 39	15 20	29 25	10 51	06 45
17	03 38 15	25 58 23	13 ♌ 37 39	20 ♌ 41 02	17 50	08 35	27 06	13 51	15 26	29 23	10 50	06 44
18	03 42 12	26 56 12	27 ♌ 46 25	04 ♍ 53 32	18 44	09 41	27 52	14 04	15 32	29 20	10 48	06 43
19	03 46 08	27 53 58	12 ♍ 02 03	19 ♍ 11 38	19 34	10 48	28 37	14 16	15 38	29 18	10 46	06 42
20	03 50 05	28 51 43	26 ♍ 21 53	03 ♎ 32 23	20 19	11 54	29 ♓ 22	14 28	15 44	29 16	10 45	06 41
21	03 54 01	29 ♉ 49 26	10 ♎ 42 38	17 ♎ 52 09	20 59	13 00	00 ♈ 07	14 41	15 50	29 14	10 43	06 40
22	03 57 58	00 ♊ 47 08	25 ♎ 00 24	02 ♏ 06 53	21 35	14 05	00 52	14 53	15 56	29 12	10 42	06 39
23	04 01 54	01 44 48	09 ♏ 11 04	16 ♏ 12 26	22 07	15 11	01 37	15 05	16 03	29 10	10 40	06 39
24	04 05 51	02 42 27	23 ♏ 10 33	00 ♐ 04 57	22 33	16 16	02 22	15 17	16 09	29 08	10 38	06 38
25	04 09 47	03 40 04	06 ♐ 55 18	13 ♐ 41 17	22 55	17 21	03 07	15 29	16 15	29 06	10 37	06 37
26	04 13 44	04 37 40	20 ♐ 22 42	26 ♐ 59 25	23 12	18 26	03 52	15 41	16 22	29 04	10 35	06 36
27	04 17 41	05 35 16	03 ♑ 31 21	09 ♑ 58 33	23 25	19 31	04 37	15 53	16 28	29 02	10 34	06 36
28	04 21 37	06 32 50	16 ♑ 21 09	22 ♑ 39 18	23 33	20 35	05 22	16 04	16 35	29 00	10 32	06 35
29	04 25 34	07 30 23	28 ♑ 53 18	05 ♒ 03 27	23 36	21 39	06 07	16 16	16 41	28 58	10 30	06 35
30	04 29 30	08 27 55	11 ♒ 10 10	17 ♒ 13 52	23 R 34	22 43	06 51	16 27	16 48	28 57	10 29	06 34
31	04 33 27	09 ♊ 25 26	23 ♒ 15 03	29 ♒ 14 14	23 ♊ 28	23 ♋ 47	07 ♈ 36	16 ♈ 39	16 ♋ 55	28 ♎ 55	10 ♐ 27	06 ♎ 33

Date	Moon True ☊ ° '	Moon Mean ☊ ° '	Moon ☽ Latitude ° '
01	01 ♐ 04	02 ♐ 13	04 N 02
02	01 D 05	02 09	04 39
03	01 R 05	02 06	05 04
04	01 05	02 03	05 15
05	01 03	02 00	05 12
06	01 01	01 57	04 56
07	00 59	01 54	04 28
08	00 57	01 50	03 47
09	00 55	01 47	02 55
10	00 54	01 44	01 54
11	00 53	01 41	00 N 46
12	00 D 53	01 38	00 S 26
13	00 54	01 35	01 37
14	00 54	01 31	02 45
15	00 55	01 28	03 44
16	00 56	01 25	04 31
17	00 56	01 22	05 03
18	00 R 56	01 19	05 16
19	00 56	01 15	05 11
20	00 56	01 12	04 46
21	00 56	01 09	04 04
22	00 56	01 06	03 06
23	00 D 56	01 03	01 58
24	00 56	01 00	00 S 43
25	00 R 56	00 56	00 N 33
26	00 55	00 53	01 46
27	00 55	00 50	02 51
28	00 54	00 47	03 46
29	00 54	00 44	04 29
30	00 53	00 41	04 58
31	00 ♐ 52	00 ♐ 37	05 N 14

DECLINATIONS

Date	Sun ☉ ° '	Moon ☽ ° '	Mercury ☿ ° '	Venus ♀ ° '	Mars ♂ ° '	Jupiter ♃ ° '	Saturn ♄ ° '	Uranus ♅ ° '	Neptune ♆ ° '	Pluto ♇ ° '
01	14 N 58	17 S 49	20 N 25	25 N 12	07 S 18	03 N 05	22 N 33	10 S 57	20 S 32	13 N 03
02	15 16	14 54	21 01	25 19	07 00	03 10	22 33	10 56	20 32	13 03
03	15 34	11 24	21 34	25 25	06 43	03 16	22 32	10 55	20 32	13 03
04	15 52	07 30	22 05	25 30	06 26	03 21	22 32	10 55	20 32	13 04
05	16 09	03 S 20	22 33	25 35	06 08	03 26	22 31	10 54	20 32	13 04
06	16 26	00 N 59	22 59	25 39	05 51	03 31	22 31	10 53	20 31	13 04
07	16 43	05 17	23 22	25 43	05 33	03 36	22 31	10 52	20 31	13 04
08	17 00	09 26	23 43	25 46	05 16	03 41	22 30	10 51	20 31	13 04
09	17 16	13 16	24 02	25 48	04 58	03 46	22 30	10 50	20 31	13 04
10	17 32	16 35	24 18	25 49	04 41	03 51	22 29	10 50	20 30	13 05
11	17 47	19 10	24 32	25 50	04 23	03 56	22 29	10 49	20 30	13 05
12	18 03	20 49	24 43	25 50	04 05	04 01	22 28	10 48	20 30	13 05
13	18 18	21 29	24 53	25 50	03 48	04 06	22 28	10 47	20 30	13 05
14	18 33	20 41	25 00	25 49	03 30	04 11	22 27	10 46	20 30	13 05
15	18 47	18 48	25 06	25 47	03 12	04 15	22 26	10 46	20 29	13 05
16	19 01	15 49	25 09	25 44	02 55	04 20	22 26	10 45	20 29	13 05
17	19 15	11 54	25 11	25 41	02 37	04 25	22 25	10 44	20 29	13 05
18	19 28	07 17	25 11	25 37	02 19	04 30	22 25	10 43	20 29	13 05
19	19 42	02 N 15	25 09	25 33	02 02	04 34	22 24	10 43	20 28	13 05
20	19 54	02 S 56	25 05	25 28	01 44	04 39	22 23	10 42	20 28	13 05
21	20 07	07 59	25 00	25 22	01 26	04 43	22 23	10 41	20 28	13 05
22	20 19	12 34	24 54	25 16	01 08	04 48	22 22	10 40	20 28	13 05
23	20 31	16 25	24 46	25 09	00 51	04 53	22 22	10 40	20 27	13 05
24	20 42	19 16	24 37	25 01	00 33	04 57	22 21	10 39	20 27	13 05
25	20 53	20 55	24 27	24 53	00 S 15	05 02	22 20	10 38	20 27	13 04
26	21 04	21 20	24 15	24 44	00 N 02	05 06	22 20	10 38	20 27	13 04
27	21 14	20 33	24 03	24 35	00 20	05 10	22 19	10 37	20 27	13 04
28	21 24	18 42	23 49	24 25	00 37	05 15	22 18	10 36	20 26	13 04
29	21 34	16 00	23 34	24 14	00 55	05 19	22 17	10 36	20 26	13 04
30	21 43	12 39	23 19	24 03	01 12	05 23	22 17	10 35	20 26	13 04
31	21 N 52	08 S 49	23 N 02	23 N 52	01 N 30	05 N 28	22 N 16	10 S 35	20 S 26	13 N 03

ZODIAC SIGN ENTRIES

Date	h m	Planets
01	17 46	♅ ♎
02	05 34	☽ ♒
04	11 55	☿ ♊
04	17 34	☽ ♓
07	06 03	☽ ♈
09	17 03	☽ ♉
09	20 11	♀ ♋
12	01 44	☽ ♊
14	08 08	☽ ♋
16	12 38	☽ ♌
18	15 45	☽ ♍
20	18 05	☽ ♎
21	08 14	♂ ♈
21	16 24	☉ ♊
22	20 25	☽ ♏
24	23 51	☽ ♐
27	05 31	☽ ♑
29	14 09	☽ ♒

LATITUDES

Date	Mercury ☿ ° '	Venus ♀ ° '	Mars ♂ ° '	Jupiter ♃ ° '	Saturn ♄ ° '	Uranus ♅ ° '	Neptune ♆ ° '	Pluto ♇ ° '
01	01 N 35	02 N 06	01 S 30	01 S 07	00 S 09	00 N 34	01 N 36	17 N 16
04	01 59	02 12	01 31	01 07	00 09	00 34	01 36	17 15
07	02 16	02 18	01 32	01 07	00 09	00 34	01 36	17 14
10	02 26	02 23	01 33	01 08	00 09	00 33	01 36	17 13
13	02 28	02 27	01 35	01 08	00 08	00 33	01 36	17 12
16	02 22	02 31	01 36	01 08	00 08	00 33	01 36	17 11
19	02 07	02 33	01 36	01 09	00 08	00 33	01 36	17 10
22	01 44	02 35	01 37	01 09	00 07	00 33	01 36	17 08
25	01 12	02 35	01 38	01 10	00 07	00 33	01 36	17 07
28	00 N 32	02 35	01 39	01 10	00 07	00 33	01 36	17 06
31	00 S 15	02 N 33	01 S 39	01 S 10	00 S 06	00 N 33	01 N 36	17 N 04

DATA

Julian Date	2442534
Delta T	+46 seconds
Ayanamsa	23° 30' 59"
Synetic vernal point	05° ♓ 36' 00"
True obliquity of ecliptic	23° 26' 29"

MOON'S PHASES, APSIDES AND POSITIONS ☽

Date	h m	Phase	Longitude ° '	Eclipse Indicator
03	05 44	☾	12 ♒ 11	
11	07 05	●	19 ♉ 59	Partial
18	10 29	☽	26 ♌ 53	
25	05 51	○	03 ♐ 25	total

Day	h m		
05	09 40	Apogee	
20	19 57	Perigee	
06	06 34	0N	
13	10 51	Max dec	21° N 21'
19	22 26	0S	
26	07 57	Max dec	21° S 21'

LONGITUDES

Date	Chiron ⚷ ° '	Ceres ⚳ ° '	Pallas ⚴ ° '	Juno ⚵ ° '	Vesta ⚶ ° '	Black Moon Lilith ⚸ ° '
01	24 ♈ 58	28 ♈ 41	21 ♓ 59	27 ♉ 31	08 ♓ 42	09 ♓ 32
11	25 ♈ 33	02 ♉ 38	24 ♓ 42	03 ♊ 25	12 ♓ 48	10 ♓ 39
21	26 ♈ 05	06 ♉ 33	27 ♓ 14	09 ♊ 19	16 ♓ 40	11 ♓ 46
31	26 ♈ 35	10 ♉ 26	29 ♓ 34	15 ♊ 11	20 ♓ 15	12 ♓ 53

All ephemeris data is given at 12.00 UT and the Moon's longitude is additionally given for 24.00 UT

ASPECTARIAN

h m Aspects

01 Thursday
04 41 ☽ ⊥ ♆
08 02 ☉ ⊻ ♃
10 54 ☽ ⚻ ♀
12 33 ♂ ⚼ ♅
16 46 ☿ # ♆
19 56 ☽ △ ☿
22 07 ☽ ∠ ♆
23 47 ☽ ± ♀

02 Friday
01 04 ☿ ∠ ♃
02 38 ☽ Q ♃
05 19 ☉ ⚻ ♆
05 31 ☽ □ ♅
06 43 ☽ ∠ ♂
09 27 ☽ # ☉
19 23 ☽ ⚼ ♀
19 25 ☽ △ ♇

03 Saturday
01 05 ☽ # ♇
01 44 ☽ ⊥ ♂
02 50 ☽ ✱ ♃
03 43 ☽ ✱ ♆
05 44 ☽ □ ☉
09 39 ☽ ⚻ ♄
13 00 ♀ Q ♃
14 33 ☽ ⊻ ♂
15 06 ☽ ∥ ♅
18 53 ☿ Q ♂
21 46 ☽ ± ♄

04 Sunday
01 05 ☿ ∠ ♄
01 23 ☽ ⚼ ♇
01 25 ☉ ± ♇
03 45 ☽ Q ♆
04 36 ☽ △ ♀
09 32 ☽ ∠ ♃
10 24 ☿ ⚻ ♅
16 03 ☽ ⚼ ♄
17 19 ☽ △ ♅
18 33 ☽ □ ☿
18 45 ☽ ∥ ♂
19 31 ☽ ± ♇
21 10 ☽ Q ☉
21 51 ♃ △ ♆

05 Monday
04 04 ☽ ⊥ ♃
07 38 ☽ ⚻ ♇
10 21 ☿ ∥ ♄
10 25 ☉ ✱ ♄
11 25 ☽ # ♃
16 03 ☽ □ ♆
16 26 ☽ ⊻ ♃
22 35 ☽ △ ♄
23 32 ☽ ⚼ ♅
23 34 ☽ ✱ ☉

06 Tuesday
07 14 ☽ ☌ ♂
13 01 ☽ Q ☿
17 36 ☽ ± ♅
23 33 ☽ □ ♀

07 Wednesday
02 23 ☽ ∥ ♃
05 36 ☽ ⚻ ♅
08 23 ☽ ∠ ☉
13 27 ☽ # ♂
16 48 ☽ ✱ ☿
19 48 ☽ ☍ ♇
22 12 ☿ ± ♅

08 Thursday
03 50 ☽ ⊥ ☉
04 02 ☽ △ ♆
05 38 ☽ ☌ ♃
10 59 ☽ □ ♄
15 00 ☽ Q ♀
15 50 ☿ △ ♇
16 40 ☽ ⊻ ☉
20 35 ☽ # ♅
22 54 ☽ ⊻ ♂

09 Friday
02 38 ☽ ∠ ☿
08 41 ☉ ⊥ ♃
09 22 ☽ ⚼ ♆
10 46 ☽ ∥ ♇
11 20 ☽ ⊥ ♂
13 24 ♀ △ ♅
16 26 ☽ ☍ ♅
16 44 ☽ ✱ ♀
21 07 ☿ ⊥ ♄
22 16 ☽ Q ♄
22 25 ☽ ⊥ ☿

10 Saturday
02 42 ☽ ± ♆
05 47 ☽ ∠ ♂
06 13 ☽ ⚻ ♇
11 22 ☽ ⊻ ☿
14 07 ☽ ⚻ ♆
16 47 ☽ ⊻ ♃
17 37 ☽ ± ♇
20 52 ☽ ∥ ☉
21 19 ☽ ✱ ♄

11 Sunday
00 11 ☽ ∠ ♀
04 19 ☽ ⊥ ♃
07 05 ☽ ☌ ☉
10 33 ☽ ⚼ ♇
11 56 ☽ ✱ ♂
12 06 ☿ ☍ ♆
18 40 ☽ ⊥ ♀
21 24 ☽ ∠ ♃

12 Monday
00 58 ☽ ⚻ ♅
01 34 ☽ ∠ ♄
06 05 ☽ # ♆
06 49 ☽ ⊻ ♀
11 32 ☽ Q ♂
11 59 ☽ ± ♅
14 17 ☽ △ ♇
16 51 ♂ # ♃
18 13 ☽ ⊥ ♄
19 01 ♂ ± ♅
21 49 ☽ ☍ ♆
23 36 ☿ ✱ ♃

13 Tuesday
01 24 ☽ ✱ ♃
01 34 ☽ ☌ ☿
04 14 ☉ ⚼ ♇
04 22 ☽ ⚼ ♅
05 13 ☽ ⊻ ♄
18 30 ☽ ⊻ ☉
22 09 ☽ □ ♂
23 24 ☽ Q ♃

14 Wednesday
06 04 ☽ ⊥ ☉
07 14 ☽ △ ♅
07 28 ☿ ⚼ ♅
15 15 ☽ # ♆
17 54 ☽ ☌ ♀
20 08 ☽ □ ♇
21 48 ☿ ⊻ ♄
23 14 ☽ ∠ ☉

15 Thursday
03 22 ☽ ⚻ ♆
07 47 ☽ □ ♃
10 59 ☽ ☌ ♄
12 02 ☽ ⊻ ☿
12 07 ☽ ∥ ☉
13 52 ☽ ± ♆
20 37 ♀ □ ♇
23 21 ☽ ⊥ ☿

16 Friday
03 25 ☽ ✱ ☉
03 34 ☽ Q ♇
05 29 ☽ ⚼ ♆
06 01 ☽ △ ♂
11 38 ☽ □ ♅
16 09 ☽ ∠ ☿

17 Saturday
00 13 ☽ ✱ ♇
01 37 ☽ Q ☉
02 38 ☽ ⊻ ♀
05 14 ☽ ∥ ♇
07 14 ☽ △ ♆
09 16 ☽ ⚼ ♂
12 23 ☽ △ ♃
13 46 ☽ ⊥ ♀
15 05 ☽ ⊻ ♄
18 22 ☽ Q ♅
18 23 ☽ # ♅
19 40 ☽ ✱ ☿

18 Sunday
01 21 ☽ ⊥ ♄
01 26 ☽ ± ♂
01 46 ☽ ∠ ♇
06 21 ☽ ∠ ♀
10 29 ☽ □ ☉
12 09 ☽ ⚻ ♂
14 12 ☽ ⚼ ♃
14 38 ☽ ✱ ♅
16 41 ☽ ∠ ♄
16 57 ☽ ⊥ ♇
17 19 ☽ Q ☿

19 Monday
01 18 ☽ ∥ ♃
03 03 ☽ ⊻ ♇
05 35 ☽ ± ♃
09 45 ☽ ✱ ♀
09 53 ☽ □ ♆
11 34 ♀ ⚻ ♆
13 07 ☽ # ♂
15 48 ☽ ∠ ♅
15 48 ☽ ⚻ ♃
18 04 ☽ ✱ ♄

20 Tuesday
01 20 ☽ □ ☿
06 43 ☽ ∥ ♂
06 50 ☽ ⊥ ♅
07 32 ☽ Q ♀
09 02 ♂ ⚻ ♅
14 18 ☽ Q ♄
15 59 ☽ Q ♆
16 29 ☽ △ ☉
16 51 ☽ ⊻ ♅
17 18 ☽ ☍ ♂
20 08 ☽ # ♃
21 48 ☉ ⚻ ♅

21 Wednesday
05 15 ☽ ☌ ♇
07 20 ☉ ∠ ♃
12 01 ☽ ✱ ♆
16 09 ☽ □ ♀
18 44 ☽ ☍ ♃
19 23 ☽ ⚼ ☉
20 39 ☽ □ ♄

22 Thursday
01 44 ☽ ∥ ♅
06 00 ☽ △ ☿
11 36 ☽ ± ☉
13 09 ☽ ∠ ♆
14 54 ☽ # ♇
16 17 ☉ ∠ ♄
19 03 ☽ ☌ ♅
21 34 ☉ ✱ ♂
22 27 ☽ ⚻ ♂
22 28 ☽ ⚻ ☉

23 Friday
04 20 ☽ ⊥ ♆
05 25 ☉ # ♆
07 41 ☽ ⊻ ♇
08 21 ☽ ⚼ ☿
09 11 ☽ ± ♂
09 20 ♀ □ ♃
14 31 ☽ ⊻ ♆
17 54 ☽ ⊥ ♇
22 13 ☽ ⚻ ♃
23 06 ☽ △ ♀
23 48 ☽ △ ♄

24 Saturday
00 14 ☽ ± ☿
01 26 ☽ ⚼ ♂
08 41 ☽ ± ♃
09 04 ♀ ☌ ♄
09 20 ☽ ∠ ♇
10 54 ☽ ⚻ ☿
20 03 ♀ ± ♆
22 19 ☽ ⊻ ♅

25 Sunday
00 32 ☽ ⚼ ♃
01 58 ☽ ⚼ ♄
03 08 ☽ ∥ ♆
03 17 ☽ ⚼ ♀
04 56 ☽ △ ♂
05 51 ☽ ☍ ☉
08 48 ☽ ⊥ ♅
11 03 ☽ # ☉
11 28 ☽ ✱ ♇
17 57 ☽ ± ♄
18 31 ☽ ☌ ♆
20 32 ☽ ± ♀

26 Monday
00 42 ☽ ∠ ♅
03 26 ☽ △ ♃
04 43 ☽ ⚻ ♄
08 11 ☽ ⚻ ♀
08 49 ☽ Q ♇
17 13 ☽ ☍ ☿
22 10 ☽ # ☉
22 38 ☉ ± ♅

27 Tuesday
03 46 ☽ ✱ ♅
13 46 ☽ ∥ ♆
14 09 ☽ □ ♂
16 08 ☽ ⚻ ☉
17 42 ☽ □ ♇

28 Wednesday
01 04 ☽ ⊻ ♆
01 57 ☽ Q ♅
03 53 ♄ ± ♆
04 14 ☽ ± ☉
11 27 ☽ □ ♃
12 21 ☽ ⊥ ♆
12 26 ☽ ☍ ♄
12 57 ☉ △ ♇
20 47 ☽ ☍ ♀
22 42 ☽ ⚼ ☉

29 Thursday
01 47 ☽ ⚻ ☿
02 12 ☽ Q ♂
05 29 ☽ ∠ ♆
12 10 ☽ □ ♅
13 22 ☽ ± ☿
16 01 ☿ St R
22 37 ☽ Q ♃

30 Friday
02 43 ♂ ☍ ♇
02 57 ☽ △ ♇
02 58 ☽ ✱ ♂
06 13 ☽ △ ☉
06 55 ☽ ⚼ ☿
09 12 ☽ # ♇
10 39 ☽ ✱ ♆
22 38 ☽ ✱ ♃
23 15 ☽ ⚻ ♄

31 Saturday
01 15 ☽ ∥ ♅
05 50 ☿ ⊻ ♀
08 37 ☽ ⚼ ♇
10 25 ☽ Q ♆
10 37 ☽ ∠ ♂
11 19 ☽ ± ♄
12 26 ☽ △ ☿
13 09 ☽ ⚻ ♀
23 19 ☽ △ ♅

JUNE 1975

LONGITUDES

Date	Sidereal time h m s	Sun ☉ ° ' "	Moon ☽ ° ' "	Moon ☽ 24.00 ° ' "	Mercury ☿ ° '	Venus ♀ ° '	Mars ♂ ° '	Jupiter ♃ ° '	Saturn ♄ ° '	Uranus ♅ ° '	Neptune ♆ ° '	Pluto ♇ ° '
01	04 37 23	10 ♊ 22 57	05 ♓ 11 57	11 ♓ 08 47	23 ♊ 18	24 ♋ 50	08 ♈ 21	16 ♈ 50	17 ♋ 02	28 ♎ 53	10 ♐ 26	06 ♎ 33
02	04 41 20	11 20 26	17 ♓ 05 18	23 ♓ 02 04	23 R 04	25 53	09 06	17 01	17 08	28 R 51	10 R 24	06 R 32
03	04 45 16	12 17 55	28 ♓ 59 40	04 ♈ 58 41	22 45	26 56	09 50	17 13	17 15	28 50	10 22	06 32
04	04 49 13	13 15 23	10 ♈ 59 40	17 ♈ 03 07	22 24	27 59	10 35	17 24	17 22	28 48	10 21	06 31
05	04 53 10	14 12 51	23 ♈ 09 33	29 ♈ 19 23	21 59	29 ♋ 01	11 19	17 35	17 29	28 47	10 19	06 31
06	04 57 06	15 10 17	05 ♉ 33 02	11 ♉ 50 50	21 31	00 ♌ 03	12 04	17 45	17 36	28 45	10 17	06 31
07	05 01 03	16 07 43	18 ♉ 13 04	24 ♉ 39 55	21 01	01 05	12 48	17 56	17 43	28 44	10 16	06 30
08	05 04 59	17 05 09	01 ♊ 11 30	07 ♊ 47 51	20 30	02 06	13 32	18 07	17 50	28 42	10 14	06 30
09	05 08 56	18 02 33	14 ♊ 28 55	21 ♊ 14 34	19 57	03 07	14 17	18 17	17 57	28 41	10 13	06 30
10	05 12 52	18 59 57	28 ♊ 04 33	04 ♋ 58 34	19 24	04 08	15 01	18 28	18 04	28 40	10 11	06 30
11	05 16 49	19 57 20	11 ♋ 56 14	18 ♋ 57 06	18 50	05 08	15 45	18 38	18 11	28 38	10 09	06 29
12	05 20 45	20 54 42	26 ♋ 00 41	03 ♌ 06 27	18 17	06 08	16 29	18 48	18 19	28 37	10 08	06 29
13	05 24 42	21 52 03	10 ♌ 13 54	17 ♌ 22 28	17 45	07 08	17 13	18 58	18 26	28 36	10 06	06 29
14	05 28 39	22 49 24	24 ♌ 31 39	01 ♍ 40 59	17 14	08 07	17 57	19 08	18 33	28 35	10 05	06 29
15	05 32 35	23 46 43	08 ♍ 49 59	15 ♍ 58 17	16 46	09 06	18 41	19 18	18 40	28 34	10 03	06 29
16	05 36 32	24 44 01	23 ♍ 05 31	00 ♎ 11 23	16 20	10 05	19 25	19 28	18 48	28 32	10 02	06 29
17	05 40 28	25 41 18	07 ♎ 15 37	14 ♎ 18 00	15 58	11 03	20 09	19 38	18 55	28 31	10 00	06 D 29
18	05 44 25	26 38 35	21 ♎ 18 21	28 ♎ 16 29	15 38	12 01	20 53	19 47	19 02	28 30	09 58	06 29
19	05 48 21	27 35 51	05 ♏ 12 16	12 ♏ 05 33	15 23	12 58	21 36	19 57	19 10	28 30	09 57	06 29
20	05 52 18	28 33 06	18 ♏ 56 12	25 ♏ 44 04	15 11	13 55	22 20	20 06	19 17	28 29	09 55	06 29
21	05 56 14	29 ♊ 30 20	02 ♐ 29 02	09 ♐ 10 57	15 04	14 51	23 03	20 15	19 25	28 28	09 54	06 29
22	06 00 11	00 ♋ 27 34	15 ♐ 49 43	22 ♐ 25 11	15 01	15 47	23 47	20 24	19 32	28 27	09 52	06 29
23	06 04 08	01 24 47	28 ♐ 57 16	05 ♑ 25 52	15 D 03	16 42	24 30	20 33	19 40	28 26	09 51	06 30
24	06 08 04	02 22 00	11 ♑ 50 58	18 ♑ 12 31	15 09	17 37	25 13	20 42	19 47	28 26	09 49	06 30
25	06 12 01	03 19 13	24 ♑ 30 35	00 ♒ 45 13	15 20	18 31	25 57	20 51	19 55	28 25	09 48	06 30
26	06 15 57	04 16 25	06 ♒ 56 33	13 ♒ 04 46	15 36	19 25	26 40	20 59	20 03	28 24	09 46	06 30
27	06 19 54	05 13 37	19 ♒ 10 07	25 ♒ 12 54	15 56	20 18	27 23	21 08	20 10	28 24	09 45	06 31
28	06 23 50	06 10 49	01 ♓ 13 27	07 ♓ 12 11	16 22	21 10	28 06	21 16	20 18	28 23	09 44	06 31
29	06 27 47	07 08 01	13 ♓ 09 32	19 ♓ 06 01	16 52	22 02	28 49	21 24	20 25	28 23	09 42	06 31
30	06 31 43	08 ♋ 05 13	25 ♓ 02 08	00 ♈ 58 29	17 ♊ 27	22 ♌ 53	29 ♈ 32	21 ♈ 32	20 ♋ 33	28 ♎ 23	09 ♐ 41	06 ♎ 32

Date	Moon True ☊ ° '	Moon Mean ☊ ° '	Moon ☽ Latitude ° '
01	00 ♐ 52	00 ♐ 34	05 N 16
02	00 D 52	00 31	05 04
03	00 52	00 28	04 39
04	00 53	00 25	04 02
05	00 54	00 21	03 14
06	00 55	00 18	02 16
07	00 56	00 15	01 N 10
08	00 R 57	00 12	00 S 01
09	00 56	00 09	01 14
10	00 55	00 06	02 24
11	00 53	00 ♐ 02	03 27
12	00 50	29 ♏ 59	04 18
13	00 48	29 56	04 54
14	00 45	29 53	05 13
15	00 44	29 50	05 11
16	00 43	29 47	04 51
17	00 D 43	29 43	04 13
18	00 45	29 40	03 20
19	00 46	29 37	02 16
20	00 47	29 34	01 S 05
21	00 R 48	29 31	00 N 09
22	00 47	29 27	01 21
23	00 44	29 24	02 28
24	00 41	29 21	03 26
25	00 36	29 18	04 12
26	00 30	29 15	04 45
27	00 24	29 12	05 05
28	00 19	29 08	05 11
29	00 15	29 05	05 03
30	00 ♐ 13	29 ♏ 02	04 N 42

DECLINATIONS

Date	Sun ☉ ° '	Moon ☽ ° '	Mercury ☿ ° '	Venus ♀ ° '	Mars ♂ ° '	Jupiter ♃ ° '	Saturn ♄ ° '	Uranus ♅ ° '	Neptune ♆ ° '	Pluto ♇ ° '
01	22 N 00	04 S 42	22 N 45	23 N 40	01 N 47	05 N 32	22 N 15	10 S 34	20 S 25	13 N 03
02	22 09	00 S 26	22 27	23 27	02 05	05 36	22 14	10 33	20 25	13 03
03	22 16	03 N 52	22 09	23 14	02 22	05 40	22 14	10 33	20 25	13 03
04	22 24	08 04	21 51	23 00	02 40	05 44	22 13	10 32	20 25	13 02
05	22 30	12 00	21 32	22 46	02 57	05 48	22 12	10 32	20 25	13 02
06	22 37	15 30	21 13	22 31	03 14	05 52	22 11	10 31	20 24	13 02
07	22 43	18 22	20 54	22 16	03 31	05 56	22 10	10 31	20 24	13 01
08	22 49	20 23	20 36	22 01	03 48	06 00	22 10	10 30	20 24	13 01
09	22 54	21 19	20 18	21 45	04 05	06 04	22 09	10 30	20 24	13 01
10	22 59	21 02	20 00	21 28	04 22	06 08	22 08	10 29	20 23	13 00
11	23 04	19 28	19 43	21 11	04 39	06 11	22 07	10 29	20 23	13 00
12	23 08	16 43	19 28	20 54	04 56	06 15	22 06	10 29	20 23	13 00
13	23 11	12 57	19 13	20 37	05 13	06 19	22 05	10 28	20 23	12 59
14	23 15	08 26	19 00	20 19	05 30	06 22	22 04	10 28	20 23	12 59
15	23 18	03 N 27	18 48	20 00	05 46	06 26	22 03	10 27	20 22	12 58
16	23 20	01 S 43	18 37	19 42	06 03	06 29	22 03	10 27	20 22	12 58
17	23 22	06 45	18 28	19 23	06 20	06 33	22 02	10 27	20 22	12 57
18	23 24	11 24	18 21	19 03	06 36	06 36	22 01	10 26	20 22	12 57
19	23 25	15 23	18 16	18 44	06 52	06 40	22 00	10 26	20 22	12 56
20	23 26	18 29	18 12	18 24	07 08	06 43	21 59	10 26	20 21	12 56
21	23 26	20 30	18 10	18 03	07 25	06 46	21 58	10 26	20 21	12 55
22	23 26	21 20	18 10	17 43	07 41	06 49	21 57	10 25	20 21	12 55
23	23 26	20 58	18 12	17 22	07 57	06 52	21 56	10 25	20 21	12 54
24	23 25	19 30	18 15	17 01	08 13	06 56	21 55	10 25	20 21	12 54
25	23 24	17 05	18 20	16 40	08 28	06 59	21 54	10 25	20 20	12 53
26	23 22	13 56	18 26	16 19	08 44	07 02	21 53	10 24	20 20	12 53
27	23 20	10 14	18 34	15 57	09 00	07 05	21 52	10 24	20 20	12 52
28	23 18	06 12	18 43	15 35	09 15	07 07	21 51	10 24	20 20	12 51
29	23 15	01 57	18 54	15 14	09 30	07 10	21 50	10 24	20 20	12 51
30	23 N 12	02 N 21	19 N 05	14 N 52	09 N 46	07 N 13	21 N 49	10 S 24	20 S 20	12 N 50

ZODIAC SIGN ENTRIES

Date	h	m	Planets
01	01	32	☽ ♓
03	14	01	☽ ♈
06	01	19	☽ ♉
06	10	54	♀ ♌
08	09	49	☽ ♊
10	15	21	☽ ♋
12	18	45	☽ ♌
14	21	11	☽ ♍
16	23	41	☽ ♎
19	02	59	☽ ♏
21	07	34	☽ ♐
22	00	26	☉ ♋
23	13	56	☽ ♑
25	22	33	☽ ♒
28	09	33	☽ ♓
30	22	02	☽ ♈

LATITUDES

Date	Mercury ☿ ° '	Venus ♀ ° '	Mars ♂ ° '	Jupiter ♃ ° '	Saturn ♄ ° '	Uranus ♅ ° '	Neptune ♆ ° '	Pluto ♇ ° '
01	00 S 31	02 N 32	01 S 39	01 S 11	00 S 06	00 N 33	01 N 36	17 N 04
04	01 23	02 29	01 40	01 11	00 06	00 33	01 36	17 02
07	02 14	02 25	01 40	01 11	00 06	00 33	01 36	17 01
10	03 01	02 19	01 40	01 12	00 05	00 33	01 36	16 59
13	03 40	02 12	01 40	01 13	00 05	00 33	01 36	16 58
16	04 08	02 03	01 40	01 13	00 05	00 33	01 36	16 56
19	04 24	01 53	01 40	01 14	00 05	00 33	01 36	16 55
22	04 28	01 41	01 40	01 15	00 04	00 32	01 36	16 53
25	04 20	01 28	01 40	01 15	00 04	00 32	01 36	16 52
28	04 02	01 13	01 39	01 16	00 04	00 32	01 36	16 50
31	03 S 37	00 N 56	01 S 39	01 S 16	00 S 04	00 N 32	01 N 36	16 N 48

DATA

Julian Date	2442565
Delta T	+46 seconds
Ayanamsa	23° 31' 04"
Synetic vernal point	05° ♓ 35' 56"
True obliquity of ecliptic	23° 26' 28"

MOON'S PHASES, APSIDES AND POSITIONS ☽

Date	h	m	Phase	Longitude ° '	Eclipse Indicator
01	23	22	☾	10 ♓ 50	
09	18	49	●	18 ♊ 19	
16	14	58	☽	24 ♍ 51	
23	16	54	○	01 ♑ 36	

Day	h	m		
02	04	27	Apogee	
14	22	15	Perigee	
29	22	49	Apogee	
02	14	24	0N	
09	18	34	Max dec	21° N 22'
16	04	03	0S	
22	16	30	Max dec	21° S 22'
29	22	53	0N	

LONGITUDES

Date	Chiron ⚷ ° '	Ceres ⚳ ° '	Pallas ⚴ ° '	Juno ⚵ ° '	Vesta ⚶ ° '	Black Moon Lilith ⚸ ° '
01	26 ♈ 37	10 ♉ 49	29 ♓ 47	15 ♊ 46	20 ♓ 35	13 ♉ 00
11	27 ♈ 04	14 ♉ 37	01 ♈ 50	21 ♊ 36	23 ♓ 48	14 ♉ 07
21	27 ♈ 26	18 ♉ 20	03 ♈ 34	27 ♊ 23	26 ♓ 37	15 ♉ 14
31	27 ♈ 44	21 ♉ 57	04 ♈ 57	03 ♋ 06	28 ♓ 59	16 ♉ 22

All ephemeris data is given at 12.00 UT and the Moon's longitude is additionally given for 24.00 UT

ASPECTARIAN

h m	Aspects
01 Sunday	
02 21	☽ ± ♀
02 38	☽ ± ♇
05 07	☽ ∠ ♃
05 33	☽ ₽ ♄
05 33	♀ Q ♇
05 52	☽ ⊥ ♂
07 23	☽ # ♃
13 03	☉ ☍ ♆
14 43	☽ ⚻ ♇
18 47	☽ ⊻ ♂
22 15	☽ ₽ ♀
22 31	☽ □ ♆
23 22	☽ □ ☉
23 33	☽ ⊥ ♃
02 Monday	
01 10	♀ ₽ ♆
03 22	☽ # ♂
05 30	☽ ₽ ♅
06 15	☉ ⊥ ♄
11 52	☽ ⊻ ♃
12 06	☽ △ ♄
23 37	☽ ± ♅
23 46	☽ □ ☿
03 Tuesday	
03 00	☽ ∥ ♂
04 42	☉ ∥ ♄
05 26	☉ ∥ ☿
05 47	☿ ∥ ♄
07 27	☽ △ ♀
11 40	☽ ⚻ ♅
14 51	☽ Q ☉
22 19	☽ ∥ ♃
04 Wednesday	
03 02	♃ □ ♄
03 06	☽ ☍ ♇
04 42	♂ △ ♆
08 06	☿ Q ♂
10 43	☽ △ ♆
10 51	☽ Q ☿
11 07	☽ ☌ ♂
16 53	☽ ✱ ☉
19 01	☿ ⊥ ♀
05 Thursday	
00 45	☽ □ ♄
00 52	☽ ☌ ♃
01 22	☉ ₽ ♅
02 47	☽ # ♅
06 41	♀ □ ♅
09 46	☽ ✱ ☿
16 12	☽ ₽ ♆
18 42	☽ ∥ ♇
22 55	☽ ☍ ♅
06 Friday	
00 26	☽ □ ♀
00 47	☽ ∠ ☉
05 42	☉ ∥ ♀
09 35	☽ ± ♆
12 06	☽ Q ♄
13 47	☽ ∠ ☿
13 50	☽ ⚻ ♇
19 29	☽ ⊥ ☉
21 02	☽ ⚻ ♆
07 Saturday	
01 11	☽ ⊻ ♂
01 15	☽ ± ♇
06 14	☽ ⊥ ☿
07 46	☽ ⊻ ☉
11 03	☽ ✱ ♄
11 28	☽ ⊻ ♃
13 10	☽ ⊥ ♂
13 45	☽ Q ♀
17 02	☽ ⊻ ☿
18 08	☽ ₽ ♇
21 38	♀ ∥ ♄
22 48	☽ ⊥ ♃
08 Sunday	
06 51	☽ ∠ ♂
07 08	☉ ∠ ♀
07 27	☽ ⚻ ♅
12 19	☽ # ♆
13 48	☽ ✱ ♀
15 00	☽ ∥ ☿
15 01	☽ ∠ ♄
15 33	☽ ∠ ♃
18 23	☽ ± ♅
21 39	☽ △ ♇
09 Monday	
03 46	☿ # ♆
04 22	☽ ☍ ♆
07 26	☽ ⊥ ♄
09 23	☉ ⊻ ♄
10 34	☽ ₽ ♅
11 37	☽ ✱ ♂
18 14	☽ ⊻ ♄
18 49	☽ ☌ ☉
18 52	☽ ✱ ♃
19 00	☽ ∠ ♀
19 34	☉ ✱ ♃
21 20	☽ ☌ ☿
10 Tuesday	
10 03	☽ Q ♂
12 06	☽ ⊥ ♀
13 01	☽ △ ♅
16 02	☿ ∠ ♀
16 13	☽ Q ♃
18 14	☉ ☌ ☿
23 22	☽ ⊻ ♀
11 Wednesday	
00 13	☽ # ♆
02 37	☽ □ ♇
08 38	☽ ∥ ☿
08 57	☽ ⚻ ♆
18 31	☿ ✱ ♃
18 54	☽ □ ♂
19 13	☽ ± ♆
22 48	☽ ☌ ♄
23 21	☽ ⊻ ☿
23 36	☽ □ ♃
12 Thursday	
02 43	☽ ⊻ ☉
09 11	☽ ⊥ ☿
09 25	☽ Q ♇
10 31	☽ ₽ ♆
10 59	☿ ⊻ ♄
13 38	☽ ⊥ ☉
16 24	☽ □ ♅
20 24	♀ ✱ ♇
23 51	☽ ∠ ☿
13 Friday	
05 42	☽ ✱ ♇
05 56	☽ ∠ ☉
06 24	☽ ☌ ♀
11 47	☽ △ ♆
11 47	☽ ∥ ♇
22 06	☿ ✱ ♂
22 41	☽ Q ♅
14 Saturday	
00 11	☽ ✱ ☿
00 23	☽ △ ♂
01 35	☽ # ♅
01 53	☽ ⊻ ♄
02 52	☽ △ ♃
06 42	♀ # ♆
06 54	☽ ∠ ♇
08 56	☽ ✱ ☉
12 02	☽ ⊥ ♄
18 47	☽ ✱ ♅
19 39	☽ Q ☿
21 59	☽ ⊥ ♇
21 59	☽ ∥ ♃
15 Sunday	
01 33	☽ ∥ ♂
02 54	☽ ₽ ♂
03 16	☽ ∠ ♄
04 19	☽ ₽ ♃
06 30	☽ Q ☉
08 03	☽ ⊻ ♇
11 28	♂ □ ♄
12 29	☽ ⊻ ♀
14 03	☽ □ ♆
18 49	☽ ± ♂
19 36	☽ ± ♃
19 56	☽ ∠ ♅
23 19	☽ ⊥ ♀
16 Monday	
00 56	☽ □ ☿
04 42	☽ ✱ ♄
05 28	☽ ⚻ ♂
05 49	☽ ⚻ ♃
10 43	♀ △ ♆
11 04	☽ ⊥ ♅
14 11	♂ ☌ ♃
14 58	☽ □ ☉
15 36	☽ ∠ ♀
20 19	☽ Q ♆
21 12	☽ ⊻ ♅
17 Tuesday	
01 08	☽ Q ♄
04 06	♇ St D
09 48	☽ # ♂
10 41	☽ ☌ ♇
10 59	☽ # ♃
16 39	☽ ✱ ♆
18 55	☽ ✱ ♀
18 Wednesday	
02 30	☽ △ ☿
06 49	☽ ∥ ♅
08 05	☽ □ ♄
09 22	☽ ☍ ♃
11 13	☽ ☍ ♂
12 26	♂ ∥ ♃
16 59	☽ Q ♀
18 18	☽ ∠ ♆
20 46	☽ # ♇
21 52	☽ △ ☉
19 Thursday	
00 23	☽ ☌ ♅
03 47	☽ ₽ ☿
09 49	☽ ⊥ ♆
14 13	☽ ⊻ ♇
19 09	☽ ± ☿
20 14	☽ ⊻ ♆
20 Friday	
00 41	☽ ⊥ ♇
01 51	☽ ₽ ☉
02 31	☽ □ ♀
05 30	☽ ⚻ ☿
09 26	☽ # ☿
10 10	☉ △ ♅
11 13	☽ # ♀
12 38	☽ △ ♄
14 04	☽ ⚻ ♃
16 29	☽ ∠ ♇
18 19	☽ ⚻ ♂
18 51	☽ ± ☉
21 Saturday	
00 48	☽ ± ♃
03 25	☿ ∥ ♀
04 51	☽ ⊻ ♅
05 33	☽ ± ♂
06 17	☽ ⚻ ☉
09 31	☽ ∥ ♆
15 29	☽ ₽ ♄
15 32	☽ ⊥ ♅
17 00	☽ ₽ ♃
17 08	☿ ✱ ♀
19 10	☽ ✱ ♇
22 33	☽ ₽ ♂
22 Sunday	
01 16	☽ ☌ ♆
07 42	☽ ∠ ♅
07 49	☽ ± ♄
10 32	☽ ☍ ☿
11 54	☽ △ ♀
15 18	☿ St D
16 50	☽ Q ♇
18 48	☽ ⚻ ♄
20 25	☽ △ ♃
23 Monday	
03 20	☽ △ ♂
05 18	♀ Q ♅
11 03	☽ ✱ ♅
16 54	☽ ☍ ☉
17 27	☽ ₽ ♀
23 06	♂ ₽ ♆
24 Tuesday	
00 20	☽ ∥ ♆
01 59	☽ □ ♇
08 12	☽ ⊻ ♆
09 20	☽ Q ♅
11 31	☽ ± ♀
18 17	☽ ⚻ ☿
19 28	☽ ⊥ ♆
21 54	☉ Q ♃
23 42	☽ ⚻ ♀
25 Wednesday	
01 05	☽ # ☿
03 09	☽ ☍ ♄
04 56	☽ □ ♃
05 50	☽ ± ☿
12 33	☽ ∠ ♆
14 55	☽ □ ♂
15 59	☽ # ♀
19 29	☽ □ ♅
23 25	☽ ₽ ☿
26 Thursday	
06 23	☽ ⚻ ☉
11 09	☽ △ ♇
16 02	☽ Q ♃
17 31	☽ ✱ ♆
19 03	☽ ± ☉
19 13	☽ # ♇
27 Friday	
04 04	☽ Q ♂
05 26	☽ △ ☿
08 01	♀ ⊻ ♄
10 59	☽ ∥ ♅
14 00	☽ ⚻ ♄
14 16	☽ ₽ ☉
14 24	☽ ☍ ♀
15 55	☽ ✱ ♃
16 38	☽ ₽ ♇
17 06	☽ Q ♆
19 07	☽ # ♂
28 Saturday	
02 03	☽ ± ♄
05 21	☽ ✱ ♂
06 20	☽ △ ♅
06 40	☽ # ♃
08 41	☉ ∠ ♀
10 35	☽ ± ♇
15 09	♀ △ ♃
20 15	☽ ₽ ♄
20 32	☉ □ ♇
21 42	♀ ∠ ♇
21 42	♂ ☍ ♅
22 14	☽ ∠ ♃
22 38	☽ ⚻ ♇
22 48	☽ △ ☉
29 Sunday	
05 03	☽ □ ♆
12 27	☽ ₽ ♅
13 24	☽ ∠ ♂
16 35	☽ ⊥ ♃
19 51	☽ □ ☿
30 Monday	
02 50	☽ △ ♄
04 50	☽ ⊻ ♃
06 38	☽ ± ♅
07 19	☽ ⚻ ♀
08 45	☽ ⊥ ♂
18 45	☽ ⚻ ♅
20 23	☽ ± ♀
21 40	☽ ⊻ ♂

JULY 1975

LONGITUDES

Date	Sidereal time h m s	Sun ☉ ° ' "	Moon ☽ ° ' "	Moon ☽ 24.00 ° ' "	Mercury ☿ ° '	Venus ♀ ° '	Mars ♂ ° '	Jupiter ♃ ° '	Saturn ♄ ° '	Uranus ♅ ° '	Neptune ♆ ° '	Pluto ♇ ° '
01	06 35 40	09 ♋ 02 26	06 ♈ 55 37	12 ♈ 54 10	18 ♊ 06	23 ♌ 44	00 ♉ 14	21 ♈ 40	20 ♋ 41	28 ♎ 22	09 ♐ 39	06 ♎ 32
02	06 39 37	09 59 38	18 ♈ 54 44	24 ♈ 57 56	18 50	24 34	00 57	21 48	20 48	28 R 22	09 R 38	06 33
03	06 43 33	10 56 50	01 ♉ 04 23	07 ♉ 14 40	19 39	25 23	01 40	21 55	20 56	28 22	09 37	06 33
04	06 47 30	11 54 03	13 ♉ 29 20	19 ♉ 48 52	20 32	26 11	02 22	22 03	21 04	28 22	09 35	06 34
05	06 51 26	12 51 16	26 ♉ 13 45	02 ♊ 44 18	21 30	26 59	03 05	22 10	21 12	28 22	09 34	06 34
06	06 55 23	13 48 29	09 ♊ 20 47	16 ♊ 03 22	22 31	27 46	03 47	22 17	21 19	28 21	09 33	06 35
07	06 59 19	14 45 43	22 ♊ 52 02	29 ♊ 46 39	23 38	28 32	04 29	22 24	21 27	28 D 21	09 32	06 36
08	07 03 16	15 42 57	06 ♋ 46 58	13 ♋ 52 32	24 48	29 ♌ 17	05 11	22 31	21 35	28 22	09 30	06 36
09	07 07 12	16 40 10	21 ♋ 02 46	28 ♋ 16 57	26 03	00 ♍ 02	05 53	22 37	21 43	28 22	09 29	06 37
10	07 11 09	17 37 24	05 ♌ 34 16	12 ♌ 53 49	27 22	00 45	06 35	22 44	21 50	28 22	09 28	06 38
11	07 15 06	18 34 38	20 ♌ 14 40	27 ♌ 35 52	28 ♊ 44	01 28	07 17	22 50	21 58	28 22	09 27	06 39
12	07 19 02	19 31 52	04 ♍ 56 31	12 ♍ 15 47	00 ♋ 11	02 09	07 59	22 57	22 06	28 22	09 26	06 39
13	07 22 59	20 29 06	19 ♍ 32 55	26 ♍ 47 17	01 42	02 50	08 40	23 03	22 14	28 23	09 25	06 40
14	07 26 55	21 26 20	03 ♎ 58 25	11 ♎ 05 56	03 17	03 29	09 22	23 08	22 21	28 23	09 23	06 41
15	07 30 52	22 23 34	18 ♎ 09 37	25 ♎ 09 18	04 55	04 07	10 03	23 14	22 29	28 23	09 22	06 42
16	07 34 48	23 20 48	02 ♏ 04 59	08 ♏ 56 41	06 37	04 44	10 45	23 20	22 37	28 24	09 21	06 43
17	07 38 45	24 18 02	15 ♏ 44 32	22 ♏ 28 38	08 22	05 20	11 26	23 25	22 45	28 24	09 20	06 44
18	07 42 41	25 15 16	29 ♏ 09 10	05 ♐ 46 17	10 11	05 55	12 07	23 30	22 53	28 25	09 19	06 45
19	07 46 38	26 12 30	12 ♐ 20 09	18 ♐ 50 55	12 03	06 28	12 48	23 35	23 00	28 25	09 18	06 46
20	07 50 35	27 09 45	25 ♐ 18 42	01 ♑ 43 37	13 57	07 00	13 29	23 40	23 08	28 26	09 17	06 47
21	07 54 31	28 07 00	08 ♑ 05 45	14 ♑ 25 10	15 54	07 31	14 09	23 45	23 16	28 27	09 16	06 48
22	07 58 28	29 ♋ 04 15	20 ♑ 41 56	26 ♑ 56 06	17 54	08 00	14 50	23 49	23 24	28 28	09 15	06 49
23	08 02 24	00 ♌ 01 31	03 ♒ 07 44	09 ♒ 16 52	19 55	08 27	15 31	23 54	23 31	28 28	09 14	06 50
24	08 06 21	00 58 48	15 ♒ 23 38	21 ♒ 28 06	21 58	08 53	16 11	23 58	23 39	28 29	09 14	06 52
25	08 10 17	01 56 05	27 ♒ 30 28	03 ♓ 30 53	24 03	09 18	16 51	24 02	23 47	28 30	09 13	06 53
26	08 14 14	02 53 22	09 ♓ 29 36	15 ♓ 26 54	26 08	09 40	17 31	24 06	23 55	28 31	09 12	06 54
27	08 18 10	03 50 41	21 ♓ 23 07	27 ♓ 18 38	28 ♋ 14	10 01	18 11	24 09	24 02	28 32	09 11	06 55
28	08 22 07	04 48 01	03 ♈ 13 54	09 ♈ 09 24	00 ♌ 21	10 20	18 51	24 13	24 10	28 33	09 10	06 57
29	08 26 04	05 45 21	15 ♈ 05 38	21 ♈ 03 12	02 27	10 38	19 31	24 16	24 18	28 35	09 10	06 58
30	08 30 00	06 42 42	27 ♈ 02 42	03 ♉ 04 45	04 34	10 53	20 10	24 19	24 25	28 36	09 09	06 59
31	08 33 57	07 ♌ 40 05	09 ♉ 09 59	15 ♉ 19 05	06 ♌ 40	11 ♍ 07	20 ♉ 50	24 ♈ 22	24 ♋ 33	28 ♎ 37	09 ♐ 08	07 ♎ 01

Date	Moon True ☊ ° '	Moon Mean ☊ ° '	Moon ☽ Latitude ° '
01	00 ♐ 12	28 ♏ 59	04 N 10
02	00 D 12	28 56	03 26
03	00 13	28 52	02 32
04	00 15	28 49	01 30
05	00 16	28 46	00 N 22
06	00 R 15	28 43	00 S 49
07	00 13	28 40	01 59
08	00 09	28 37	03 04
09	00 04	28 33	03 59
10	29 ♏ 57	28 30	04 40
11	29 49	28 27	05 03
12	29 43	28 24	05 06
13	29 37	28 21	04 49
14	29 34	28 18	04 14
15	29 33	28 14	03 24
16	29 D 33	28 11	02 22
17	29 34	28 08	01 14
18	29 34	28 05	00 S 02
19	29 R 34	28 02	01 N 08
20	29 31	27 58	02 13
21	29 25	27 55	03 11
22	29 18	27 52	03 58
23	29 08	27 49	04 33
24	28 57	27 46	04 55
25	28 46	27 43	05 03
26	28 35	27 39	04 58
27	28 26	27 36	04 39
28	28 19	27 33	04 09
29	28 15	27 30	03 28
30	28 13	27 27	02 38
31	28 ♏ 12	27 ♏ 24	01 N 39

DECLINATIONS

Date	Sun ☉ ° '	Moon ☽ ° '	Mercury ☿ ° '	Venus ♀ ° '	Mars ♂ ° '	Jupiter ♃ ° '	Saturn ♄ ° '	Uranus ♅ ° '	Neptune ♆ ° '	Pluto ♇ ° '
01	23 N 08	06 N 34	19 N 18	14 N 29	10 N 01	07 N 16	21 N 48	10 S 24	20 S 19	12 N 49
02	23 04	10 34	19 31	14 07	10 16	07 18	21 46	10 24	20 19	12 49
03	22 59	14 13	19 45	13 45	10 31	07 21	21 45	10 24	20 19	12 48
04	22 54	17 19	20 00	13 22	10 45	07 24	21 44	10 24	20 19	12 47
05	22 49	19 40	20 15	13 00	11 00	07 26	21 43	10 24	20 19	12 47
06	22 44	21 03	20 31	12 37	11 15	07 29	21 42	10 24	20 19	12 46
07	22 37	21 16	20 46	12 15	11 29	07 31	21 41	10 24	20 18	12 45
08	22 31	20 12	21 02	11 52	11 43	07 33	21 40	10 24	20 18	12 44
09	22 24	17 51	21 17	11 30	11 58	07 35	21 39	10 24	20 18	12 44
10	22 17	14 21	21 32	11 07	12 12	07 38	21 38	10 24	20 18	12 43
11	22 09	09 57	21 46	10 45	12 25	07 40	21 36	10 24	20 18	12 42
12	22 01	04 N 57	22 00	10 22	12 39	07 42	21 35	10 24	20 18	12 41
13	21 53	00 S 17	22 12	10 00	12 53	07 44	21 34	10 24	20 18	12 41
14	21 44	05 28	22 24	09 38	13 06	07 46	21 33	10 25	20 17	12 40
15	21 35	10 15	22 34	09 16	13 20	07 48	21 32	10 25	20 17	12 39
16	21 25	14 25	22 42	08 54	13 33	07 49	21 30	10 25	20 17	12 38
17	21 15	17 44	22 49	08 32	13 46	07 51	21 29	10 25	20 17	12 37
18	21 05	20 00	22 54	08 11	13 59	07 53	21 28	10 25	20 17	12 37
19	20 55	21 09	22 57	07 49	14 12	07 55	21 27	10 26	20 17	12 36
20	20 44	21 08	22 57	07 28	14 24	07 56	21 26	10 26	20 17	12 35
21	20 32	20 01	22 55	07 08	14 37	07 58	21 24	10 26	20 17	12 34
22	20 21	17 56	22 51	06 47	14 49	07 59	21 23	10 26	20 17	12 33
23	20 09	15 02	22 43	06 27	15 01	08 00	21 22	10 27	20 16	12 32
24	19 56	11 31	22 34	06 07	15 13	08 02	21 21	10 27	20 16	12 31
25	19 44	07 36	22 21	05 48	15 25	08 03	21 19	10 28	20 16	12 31
26	19 31	03 S 25	22 06	05 29	15 37	08 04	21 18	10 28	20 16	12 30
27	19 18	00 N 52	21 48	05 10	15 48	08 05	21 17	10 28	20 16	12 29
28	19 04	05 06	21 28	04 52	16 00	08 06	21 16	10 29	20 16	12 28
29	18 50	09 09	21 05	04 35	16 11	08 07	21 14	10 29	20 16	12 27
30	18 36	12 52	20 39	04 18	16 22	08 08	21 13	10 30	20 16	12 26
31	18 N 21	16 N 07	20 N 12	04 N 01	16 N 33	08 N 09	21 N 12	10 S 30	20 S 16	12 N 25

ZODIAC SIGN ENTRIES

Date	h m	Planets
01	03 53	♂ ♉
03	09 54	☽ ♉
05	18 58	☽ ♊
08	00 23	☽ ♋
09	11 06	♀ ♍
10	02 50	☽ ♌
12	03 55	☽ ♍
12	08 56	☿ ♋
14	05 21	☽ ♎
16	08 23	☽ ♏
18	13 32	☽ ♐
20	20 46	☽ ♑
23	05 56	☽ ♒
23	11 22	☉ ♌
25	16 58	☽ ♓
28	05 27	☽ ♈
28	08 05	☿ ♌
30	17 53	☽ ♉

LATITUDES

Date	Mercury ☿ ° '	Venus ♀ ° '	Mars ♂ ° '	Jupiter ♃ ° '	Saturn ♄ ° '	Uranus ♅ ° '	Neptune ♆ ° '	Pluto ♇ ° '
01	03 S 37	00 N 56	01 S 39	01 S 16	00 S 04	00 N 32	01 N 36	16 N 48
04	03 07	00 37	01 38	01 17	00 03	00 32	01 36	16 47
07	02 31	00 N 17	01 38	01 18	00 03	00 32	01 36	16 45
10	01 53	00 S 06	01 37	01 19	00 03	00 32	01 35	16 44
13	01 13	00 30	01 36	01 19	00 02	00 32	01 35	16 42
16	00 S 34	00 56	01 35	01 20	00 02	00 31	01 35	16 41
19	00 N 03	01 25	01 34	01 21	00 02	00 31	01 35	16 39
22	00 36	01 56	01 32	01 22	00 02	00 31	01 35	16 38
25	01 04	02 28	01 31	01 22	00 01	00 31	01 35	16 37
28	01 25	03 03	01 29	01 23	00 01	00 31	01 35	16 35
31	01 N 39	03 S 39	01 S 28	01 S 24	00 S 01	00 N 31	01 N 34	16 N 34

DATA

Julian Date	2442595
Delta T	+46 seconds
Ayanamsa	23° 31' 08"
Synetic vernal point	05° ♓ 35' 51"
True obliquity of ecliptic	23° 26' 28"

MOON'S PHASES, APSIDES AND POSITIONS ☽

Date	h m	Phase	Longitude ° '	Eclipse Indicator
01	16 37	☾	09 ♈ 13	
09	04 10	●	16 ♋ 22	
15	19 47	☽	22 ♎ 42	
23	05 28	○	29 ♑ 46	
31	08 48	☾	07 ♉ 32	

Day	h m		
11	20 06	Perigee	
27	15 22	Apogee	
07	04 17	Max dec	21° N 20'
13	10 40	0S	
19	23 40	Max dec	21° S 17'
27	07 09	0N	

LONGITUDES

Date	Chiron ⚷ ° '	Ceres ⚳ ° '	Pallas ⚴ ° '	Juno ⚵ ° '	Vesta ⚶ ° '	Black Moon Lilith ⚸ ° '
01	27 ♈ 44	21 ♉ 57	04 ♈ 57	03 ♋ 06	28 ♓ 59	16 ♓ 22
11	27 ♈ 58	25 ♉ 27	05 ♈ 55	08 ♋ 46	00 ♈ 48	17 ♓ 29
21	28 ♈ 07	28 ♉ 48	06 ♈ 25	14 ♋ 20	01 ♈ 59	18 ♓ 36
31	28 ♈ 10	01 ♊ 58	06 ♈ 23	19 ♋ 49	02 ♈ 29	19 ♓ 43

ASPECTARIAN

h m	Aspects
01 Tuesday	
10 14	☽ Q ☿
11 13	☽ ☍ ♇
15 54	☽ ⚼ ♀
16 06	☽ ∥ ♃
16 37	☽ □ ☉
17 29	☽ △ ♆
02 Wednesday	
03 11	☉ ⚻ ♆
09 56	☽ ∥ ♂
10 54	☽ # ♅
11 51	☽ ⚹ ☿
15 48	☽ □ ♄
17 47	☽ ☌ ♃
23 20	☽ ⚼ ♆
03 Thursday	
00 01	☽ △ ♀
00 54	♂ # ♅
02 20	☽ ∥ ♇
06 42	☽ ☍ ♅
07 29	☽ Q ☉
09 02	☽ ∥ ♀
13 13	☽ ☌ ♂
16 57	☽ ± ♆
19 29	☽ ∠ ☿
22 40	☽ ⚻ ♇
04 Friday	
03 25	☽ Q ♄
04 32	☽ ⚻ ♆
08 43	☽ ⚹ ☉
10 14	☽ ± ♇
14 09	☽ ⊥ ☿
05 Saturday	
02 25	☽ ⊻ ☿
02 30	☽ ⚹ ♄
03 18	☽ ⚼ ♇
03 35	☿ ⊻ ♄
04 21	☽ ⊻ ♃
13 29	☽ □ ♀
15 15	☽ ∠ ☉
15 37	☽ ⊥ ♃
15 57	☽ ⚻ ♅
17 14	☿ # ♆
19 38	♀ ⊥ ♄
21 03	☽ # ♆
21 50	☽ ∥ ☿
06 Sunday	
01 20	☽ ⊻ ♂
02 38	♀ ∥ ♇
02 58	☽ ± ♅
04 14	♂ ± ♆
05 52	☿ ⚹ ♃
06 29	☽ ∠ ♄
07 00	☽ △ ♇
08 14	☽ ∠ ♃
09 00	☽ ⊥ ☉
12 22	☽ ☍ ♆
12 50	☽ ⊥ ♂
19 12	☽ ⚼ ♅
20 37	☽ ⊻ ☉
22 48	☽ ⊥ ♄
07 Monday	
00 11	☽ Q ♀
03 59	♅ St D
05 45	☽ ∠ ♂
06 28	♀ ⚹ ♅
09 30	☽ ⊻ ♄
11 11	☽ ⚹ ♃
13 27	☽ ☌ ☿
21 33	☽ △ ♅
22 25	☽ ⚹ ♀
23 56	☽ ∥ ☿
08 Tuesday	
06 51	☉ ± ♆
08 06	☽ Q ♃
09 08	☽ ⚹ ♂
10 32	☽ # ♆
11 42	☽ □ ♇
16 37	☽ ⚻ ♆
17 50	♀ ∥ ♂
09 Wednesday	
01 23	☽ ∠ ♀
02 43	☽ ± ♆
04 10	☽ ☌ ☉
06 28	☽ Q ♂
13 07	☽ ☌ ♄
14 39	☽ □ ♃
17 13	☽ ⊥ ♀
17 43	☽ ⚼ ♆
17 57	☽ Q ♇
21 07	☽ ⊻ ☿
10 Thursday	
00 08	☽ □ ♅
03 40	☽ ⊻ ♀
07 55	♀ ⊥ ♇
08 00	☽ ⊥ ☿
13 30	♂ ⚻ ♇
13 44	☽ ⚹ ♇
13 45	☽ □ ♂
18 23	☽ △ ♆
20 26	☿ ∥ ♄
21 26	☽ ∥ ♇
23 40	☽ ∥ ♂
11 Friday	
00 16	☽ ∠ ☿
05 35	☿ △ ♅
05 40	☽ Q ♅
07 34	☽ ∥ ♀
09 05	☽ ⊻ ☉
09 41	☽ # ♅
14 17	☽ ∠ ♇
14 50	☽ ⊻ ♄
16 16	☽ △ ♃
19 34	☽ ⊥ ☉
23 09	☽ ∥ ♃
12 Saturday	
00 43	☽ ⊥ ♄
01 15	☽ ⚹ ♅
03 22	☽ ⚹ ☿
04 59	☽ ⊥ ♇
07 13	☽ ☌ ♀
10 00	♀ # ♅
11 17	☽ ∠ ☉
13 26	☉ ∥ ☿
14 49	☽ ⊻ ♇
15 34	☽ ∠ ♄
15 42	♂ ∥ ♇
16 57	☽ ⚼ ♃
17 13	☽ △ ♂
19 20	☽ □ ♆
13 Sunday	
01 14	☽ Q ☿
01 49	☽ ∠ ♅
07 50	☽ ± ♃
13 39	☽ ⚹ ☉
16 28	☽ ⚹ ♄
16 41	☽ ⊥ ♅
17 49	☽ ⚻ ♃
19 10	☽ ⚼ ♂
14 Monday	
01 01	☽ Q ♆
02 39	☽ ⊻ ♅
05 22	☉ Q ♂
10 41	☽ □ ☿
10 56	☽ ± ♂
11 02	☽ Q ☉
11 08	☽ ⊻ ♀
12 39	☽ Q ♄
12 48	♂ ⚻ ♆
16 34	☽ ☌ ♇
17 02	☿ ⚹ ♀
21 06	☽ ⚹ ♆
21 32	☽ ⚻ ♂
21 42	☽ ⊥ ♀
23 17	☽ # ♃
15 Tuesday	
07 11	☽ # ♀
12 50	☽ ∥ ♅
13 43	☽ ∠ ♀
14 46	☉ ☌ ♄
16 50	☿ Q ♃
19 29	☽ □ ♄
19 47	☽ □ ☉
20 45	☽ ☍ ♃
21 17	☉ ∥ ♄
22 38	☽ ∠ ♆
16 Wednesday	
01 13	☽ # ♇
05 36	☽ ☌ ♅
06 16	☽ # ♂
06 32	♂ Q ♄
11 30	☉ □ ♃
13 25	☿ □ ♇
14 13	☽ ⊥ ♆
16 51	☽ ⚹ ♀
20 06	☽ ⊻ ♇
21 04	☽ △ ☿
17 Thursday	
00 42	☽ ⊻ ♆
03 58	☽ ☍ ♂
06 40	☽ ⊥ ♇
12 54	☉ ⚼ ♆
14 58	☽ Q ♀
22 41	☽ ∠ ♇
23 05	☉ Q ♇
18 Friday	
00 36	☽ △ ♄
00 45	☿ ⚻ ♆
01 46	☽ ⚻ ♃
03 43	☽ ⚼ ☿
04 26	☽ △ ☉
10 40	☽ ⊻ ♅
12 38	☽ ± ♃
16 04	☽ ∥ ♆
21 32	☽ ⊥ ♅
22 36	☽ ± ☿
19 Saturday	
00 48	☽ □ ♀
01 48	☽ ⚹ ♇
04 00	☽ ⚼ ♄
05 06	☽ ⚼ ♃
05 32	☽ # ☉
06 27	☽ ☌ ♆
06 33	♀ ∥ ♃
09 46	☽ ⚼ ☉
10 55	♂ ± ♇
11 23	☽ ⚻ ☿
12 54	☽ ⚻ ♂
14 00	☽ ∠ ♅
20 41	☽ ± ♄
23 52	☽ Q ♇
20 Sunday	
00 34	☽ ± ♂
01 46	♀ ⊻ ♇
02 49	☿ ⚹ ♂
03 40	☽ ± ☉
07 55	☽ ⚻ ♄
08 55	☽ △ ♃
15 44	☽ ⚻ ☉
17 50	☽ ⚹ ♅
18 15	☽ ⚼ ♂
21 Monday	
01 53	☽ # ☉
04 19	☿ ± ♆
08 02	☽ ∥ ♆
09 33	☽ □ ♇
10 51	☽ △ ♀
14 13	☽ ⊻ ♆
16 27	☽ Q ♅
20 26	☉ □ ♅
22 Tuesday	
00 09	☽ △ ♂
01 36	☽ ⊥ ♆
05 37	☽ ☍ ☿
16 35	☽ ⚼ ♀
17 14	☽ ☍ ♄
18 02	☽ □ ♃
18 50	☽ ∠ ♆
20 34	☉ # ♆
23 Wednesday	
02 58	☽ □ ♅
05 28	☽ ☍ ☉
10 38	☽ ± ♀
12 04	☽ # ♂
17 18	♀ ∠ ♄
19 15	☽ △ ♇
22 46	☽ ⚻ ♀
23 55	☽ ⚹ ♆
24 Thursday	
05 13	☽ Q ♃
05 28	☽ # ♇
13 39	☽ □ ♂
17 19	♀ ⚼ ♃
18 45	☽ ∥ ♅
23 31	☽ Q ♆
25 Friday	
00 48	☽ ⚼ ♇
03 41	☽ ⚻ ☿
04 31	☽ ⚻ ♄
05 03	☽ ⚹ ♃
07 15	♀ □ ♆
08 47	☿ ☌ ♄
09 20	☽ # ♃
11 52	☿ □ ♃
13 57	☿ ⚼ ♆
13 59	☽ △ ♅
15 33	☿ ∠ ♀
16 35	☽ ± ♄
18 07	☽ ± ☿
18 45	☽ ± ♇
21 36	☽ ⚻ ☉
21 44	☿ Q ♇
23 18	☽ # ♀
26 Saturday	
03 33	☽ Q ♂
06 47	☽ ⚻ ♇
10 41	☽ ± ☉
10 49	☽ ⚼ ♄
11 12	☽ ∠ ♃
11 25	☽ □ ♆
12 22	☽ ☍ ♀
16 00	☽ ⚼ ☿
20 07	☽ ⚼ ♅
27 Sunday	
05 09	☽ ⚹ ♂
05 26	☽ ⊥ ♃
06 25	☽ ⚼ ☉
14 20	☽ ± ♅
15 28	☿ □ ♅
17 26	☽ △ ♄
17 38	☽ ⊻ ♃
18 47	☉ ⊥ ♀
22 23	♃ ⚼ ♆
28 Monday	
02 31	☽ ⚻ ♅
04 53	☽ △ ☿
10 47	☽ ∥ ♀
13 02	♄ ⚼ ♆
13 20	☽ ∠ ♂
15 27	☽ △ ☉
19 32	☽ ☍ ♇
20 25	☿ Q ♂
29 Tuesday	
00 01	☽ △ ♆
01 28	☿ ∥ ♄
02 46	☽ ⚻ ♀
02 51	♃ □ ♄
05 45	☽ ∥ ♃
08 37	☽ ⊥ ♂
15 10	☽ ± ♀
20 22	☽ # ♅
21 26	☽ ⊻ ♂
30 Wednesday	
06 13	☽ ⚼ ♆
06 32	☽ ☌ ♃
06 42	☽ □ ♄
09 02	☽ ∥ ♇
09 38	☽ ⚼ ♀
15 06	☽ ☍ ♅
16 09	☿ ⊥ ♀
19 07	☉ ⚹ ♇
31 Thursday	
00 08	☽ ± ♆
06 03	☽ □ ☿
07 45	☽ ⚻ ♇
08 37	☿ # ♆
08 48	☽ □ ☉
11 57	☽ ⚻ ♆
15 48	☽ ∥ ♂
15 52	☽ △ ♀
16 03	☿ ⚹ ♇
18 41	☽ Q ♄
19 32	☽ ± ♇

All ephemeris data is given at 12.00 UT and the Moon's longitude is additionally given for 24.00 UT

AUGUST 1975

LONGITUDES

Date	Sidereal time h m s	Sun ☉ ° ' "	Moon ☽ ° ' "	Moon ☽ 24.00 ° ' "	Mercury ☿ ° '	Venus ♀ ° '	Mars ♂ ° '	Jupiter ♃ ° '	Saturn ♄ ° '	Uranus ♅ ° '	Neptune ♆ ° '	Pluto ♇ ° '
01	08 37 53	08 ♌ 37 29	21 ♉ 32 41	27 ♉ 51 24	08 ♌ 45	11 ♍ 18	21 ♉ 29	24 ♈ 25	24 ♋ 41	28 ♎ 38	09 ♐ 08	07 ♎ 02
02	08 41 50	09 34 54	04 ♊ 15 49	10 ♊ 46 27	10 50	11 27	22 08	24 27	24 48	28 40	09 R 07	07 04
03	08 45 46	10 32 20	17 ♊ 23 43	24 ♊ 07 57	12 53	11 34	22 47	24 30	24 56	28 41	09 07	07 05
04	08 49 43	11 29 47	00 ♋ 59 18	07 ♋ 57 48	14 56	11 39	23 26	24 32	25 04	28 42	09 06	07 06
05	08 53 39	12 27 15	15 ♋ 03 14	22 ♋ 15 15	16 57	11 42	24 05	24 34	25 11	28 44	09 05	07 08
06	08 57 36	13 24 45	29 ♋ 33 15	06 ♌ 56 25	18 57	11 R 43	24 44	24 35	25 19	28 45	09 05	07 10
07	09 01 33	14 22 15	14 ♌ 23 47	21 ♌ 54 13	20 56	11 41	25 22	24 37	25 26	28 47	09 05	07 11
08	09 05 29	15 19 47	29 ♌ 26 28	06 ♍ 59 16	22 53	11 37	26 00	24 38	25 34	28 49	09 04	07 13
09	09 09 26	16 17 19	14 ♍ 31 20	22 ♍ 01 29	24 49	11 30	26 38	24 39	25 41	28 50	09 04	07 14
10	09 13 22	17 14 52	29 ♍ 28 38	06 ♎ 51 51	26 43	11 21	27 16	24 40	25 49	28 52	09 03	07 16
11	09 17 19	18 12 27	14 ♎ 10 26	21 ♎ 23 49	28 ♌ 35	11 10	27 54	24 41	25 56	28 54	09 03	07 18
12	09 21 15	19 10 02	28 ♎ 31 39	05 ♏ 33 47	00 ♍ 27	10 56	28 32	24 42	26 04	28 56	09 03	07 19
13	09 25 12	20 07 38	12 ♏ 30 11	19 ♏ 20 57	02 16	10 40	29 09	24 42	26 11	28 57	09 02	07 21
14	09 29 08	21 05 15	26 ♏ 06 19	02 ♐ 46 34	04 05	10 22	29 ♉ 46	24 42	26 18	28 59	09 02	07 23
15	09 33 05	22 02 53	09 ♐ 22 02	15 ♐ 53 04	05 51	10 01	00 ♊ 23	24 R 42	26 26	29 01	09 02	07 24
16	09 37 02	23 00 32	22 ♐ 20 02	28 ♐ 43 20	07 37	09 39	01 00	24 42	26 33	29 03	09 02	07 26
17	09 40 58	23 58 12	05 ♑ 03 16	11 ♑ 20 10	09 21	09 14	01 37	24 42	26 40	29 05	09 02	07 28
18	09 44 55	24 55 53	17 ♑ 34 20	23 ♑ 46 00	11 03	08 47	02 14	24 41	26 47	29 08	09 02	07 30
19	09 48 51	25 53 35	29 ♑ 55 22	06 ♒ 02 40	12 44	08 18	02 50	24 40	26 54	29 10	09 01	07 32
20	09 52 48	26 51 18	12 ♒ 08 02	18 ♒ 11 37	14 23	07 47	03 26	24 39	27 01	29 12	09 01	07 34
21	09 56 44	27 49 03	24 ♒ 13 33	00 ♓ 13 58	16 01	07 15	04 02	24 38	27 09	29 14	09 01	07 35
22	10 00 41	28 46 49	06 ♓ 13 00	12 ♓ 10 49	17 38	06 42	04 38	24 36	27 16	29 16	09 D 01	07 37
23	10 04 37	29 ♌ 44 36	18 ♓ 07 34	24 ♓ 03 29	19 13	06 07	05 13	24 35	27 23	29 19	09 01	07 39
24	10 08 34	00 ♍ 42 25	29 ♓ 58 48	05 ♈ 53 47	20 47	05 31	05 49	24 33	27 30	29 21	09 01	07 41
25	10 12 31	01 40 15	11 ♈ 48 47	17 ♈ 44 10	22 20	04 55	06 24	24 31	27 37	29 23	09 02	07 43
26	10 16 27	02 38 08	23 ♈ 40 21	29 ♈ 37 48	23 51	04 18	06 59	24 29	27 43	29 26	09 02	07 45
27	10 20 24	03 36 02	05 ♉ 37 03	11 ♉ 38 38	25 21	03 41	07 33	24 26	27 50	29 28	09 02	07 47
28	10 24 20	04 33 57	17 ♉ 43 09	23 ♉ 51 14	26 49	03 03	08 08	24 24	27 57	29 31	09 02	07 49
29	10 28 17	05 31 55	00 ♊ 03 30	06 ♊ 20 35	28 16	02 26	08 42	24 21	28 04	29 33	09 02	07 51
30	10 32 13	06 29 54	12 ♊ 43 06	19 ♊ 11 40	29 ♍ 41	01 50	09 16	24 18	28 11	29 36	09 03	07 53
31	10 36 10	07 ♍ 27 56	25 ♊ 46 46	02 ♋ 28 53	01 ♎ 05	01 ♍ 14	09 ♊ 50	24 ♈ 15	28 ♋ 17	29 ♎ 39	09 ♐ 03	07 ♎ 55

Date	Moon True ☊ ° '	Moon Mean ☊ ° '	Moon ☽ Latitude ° '
01	28 ♏ 13	27 ♏ 20	00 N 35
02	28 R 13	27 17	00 S 32
03	28 11	27 14	01 40
04	28 08	27 11	02 44
05	28 02	27 08	03 41
06	27 53	27 04	04 25
07	27 43	27 01	04 53
08	27 32	26 58	05 01
09	27 22	26 55	04 48
10	27 14	26 52	04 15
11	27 09	26 49	03 26
12	27 05	26 45	02 25
13	27 04	26 42	01 16
14	27 D 04	26 39	00 S 05
15	27 R 04	26 36	01 N 05
16	27 02	26 33	02 09
17	26 58	26 30	03 06
18	26 51	26 26	03 53
19	26 42	26 23	04 28
20	26 30	26 20	04 50
21	26 16	26 17	04 59
22	26 02	26 14	04 55
23	25 49	26 10	04 38
24	25 38	26 07	04 08
25	25 29	26 04	03 28
26	25 23	26 01	02 39
27	25 19	25 58	01 42
28	25 18	25 55	00 N 40
29	25 D 18	25 51	00 S 25
30	25 R 18	25 48	01 31
31	25 ♏ 17	25 ♏ 45	02 S 34

DECLINATIONS

Date	Sun ☉ ° '	Moon ☽ ° '	Mercury ☿ ° '	Venus ♀ ° '	Mars ♂ ° '	Jupiter ♃ ° '	Saturn ♄ ° '	Uranus ♅ ° '	Neptune ♆ ° '	Pluto ♇ ° '
01	18 N 06	18 N 43	19 N 42	03 N 45	16 N 44	08 N 09	21 N 11	10 S 31	20 S 16	12 N 24
02	17 51	20 28	19 11	03 30	16 55	08 10	21 09	10 31	20 16	12 23
03	17 36	21 11	18 38	03 16	17 05	08 11	21 08	10 32	20 16	12 22
04	17 20	20 42	18 03	03 02	17 15	08 11	21 07	10 32	20 16	12 21
05	17 04	18 56	17 27	02 49	17 26	08 12	21 06	10 33	20 16	12 20
06	16 48	15 55	16 49	02 37	17 36	08 12	21 04	10 33	20 16	12 19
07	16 31	11 50	16 10	02 25	17 45	08 12	21 03	10 34	20 16	12 18
08	16 14	06 58	15 30	02 15	17 55	08 13	21 02	10 34	20 16	12 17
09	15 57	01 N 40	14 50	02 06	18 05	08 13	21 00	10 35	20 16	12 16
10	15 40	03 S 42	14 08	01 57	18 14	08 13	20 59	10 36	20 16	12 15
11	15 22	08 45	13 26	01 49	18 23	08 13	20 58	10 36	20 16	12 14
12	15 05	13 12	12 43	01 43	18 32	08 13	20 56	10 37	20 16	12 13
13	14 47	16 48	12 00	01 37	18 41	08 13	20 55	10 38	20 16	12 12
14	14 28	19 22	11 16	01 33	18 50	08 13	20 54	10 39	20 16	12 11
15	14 10	20 48	10 33	01 30	18 58	08 13	20 53	10 39	20 16	12 10
16	13 51	21 04	09 48	01 28	19 07	08 12	20 51	10 40	20 16	12 09
17	13 32	20 15	09 04	01 27	19 15	08 12	20 50	10 41	20 16	12 08
18	13 13	18 26	08 20	01 27	19 23	08 11	20 49	10 41	20 16	12 07
19	12 53	15 48	07 35	01 28	19 31	08 11	20 47	10 42	20 16	12 06
20	12 34	12 30	06 51	01 31	19 38	08 10	20 46	10 43	20 16	12 05
21	12 14	08 44	06 06	01 34	19 46	08 09	20 45	10 44	20 16	12 04
22	11 54	04 39	05 22	01 39	19 53	08 09	20 44	10 45	20 16	12 03
23	11 34	00 S 26	04 38	01 45	20 01	08 08	20 42	10 46	20 16	12 02
24	11 13	03 N 47	03 54	01 52	20 08	08 07	20 41	10 46	20 16	12 01
25	10 53	07 52	03 10	02 00	20 15	08 06	20 40	10 47	20 16	12 00
26	10 32	11 39	02 27	02 08	20 22	08 05	20 38	10 48	20 16	11 59
27	10 11	15 00	01 44	02 18	20 28	08 04	20 37	10 49	20 16	11 58
28	09 50	17 45	01 01	02 28	20 35	08 03	20 36	10 50	20 17	11 57
29	09 29	19 45	00 N 19	02 39	20 41	08 01	20 35	10 51	20 17	11 56
30	09 08	20 49	00 S 23	02 51	20 47	08 00	20 33	10 52	20 17	11 55
31	08 N 46	20 N 48	01 S 04	03 N 03	20 N 53	07 N 59	20 N 32	10 S 53	20 S 17	11 N 54

ZODIAC SIGN ENTRIES

Date	h	m	Planets
02	04	02	☽ ♊
04	10	17	☽ ♋
06	12	44	☽ ♌
08	12	53	☽ ♍
10	12	51	☽ ♎
12	06	12	☿ ♍
12	14	30	☽ ♏
14	18	59	☽ ♐
14	20	47	♂ ♊
17	02	25	☽ ♑
19	12	09	☽ ♒
21	23	32	☽ ♓
23	18	24	☉ ♍
24	12	02	☽ ♈
27	00	45	☽ ♉
29	11	53	☽ ♊
30	17	20	☿ ♎
31	19	35	☽ ♋

LATITUDES

Date	Mercury ☿ ° '	Venus ♀ ° '	Mars ♂ ° '	Jupiter ♃ ° '	Saturn ♄ ° '	Uranus ♅ ° '	Neptune ♆ ° '	Pluto ♇ ° '
01	01 N 42	03 S 52	01 S 27	01 S 24	00 S 01	00 N 31	01 N 34	16 N 33
04	01 46	04 30	01 25	01 25	00 S 01	00 31	01 34	16 32
07	01 44	05 08	01 23	01 26	00 00	00 31	01 34	16 31
10	01 37	05 47	01 21	01 27	00 00	00 31	01 34	16 30
13	01 26	06 25	01 19	01 27	00 00	00 30	01 34	16 29
16	01 11	07 00	01 17	01 28	00 N 01	00 30	01 34	16 28
19	00 52	07 31	01 14	01 29	00 01	00 30	01 33	16 27
22	00 31	07 58	01 12	01 30	00 01	00 30	01 33	16 26
25	00 N 08	08 17	01 09	01 30	00 01	00 30	01 33	16 25
28	00 S 16	08 28	01 06	01 31	00 02	00 30	01 33	16 24
31	00 S 41	08 S 32	01 S 03	01 S 32	00 N 02	00 N 30	01 N 33	16 N 23

DATA

Julian Date	2442626
Delta T	+46 seconds
Ayanamsa	23° 31' 13"
Synetic vernal point	05° ♓ 35' 46"
True obliquity of ecliptic	23° 26' 28"

MOON'S PHASES, APSIDES AND POSITIONS ☽

Date	h	m	Phase	Longitude ° '	Eclipse Indicator
07	11	57	●	14 ♌ 22	
14	02	24	◐	20 ♏ 42	
21	19	48	○	28 ♒ 08	
29	23	20	◑	05 ♊ 59	

Day	h	m		
08	20	09	Perigee	
24	03	19	Apogee	
03	14	37	Max dec	21° N 12'
09	19	25	0S	
16	05	48	Max dec	21° S 06'
23	14	27	0N	
30	23	48	Max dec	20° N 57'

LONGITUDES

Date	Chiron ⚷ ° '	Ceres ⚳ ° '	Pallas ⚴ ° '	Juno ⚵ ° '	Vesta ⚶ ° '	Black Moon Lilith ⚸ ° '
01	28 ♈ 10	02 ♊ 16	06 ♈ 21	20 ♋ 22	02 ♈ 29	19 ♓ 50
11	28 ♈ 07	05 ♊ 13	05 ♈ 39	25 ♋ 44	02 ♈ 08	20 ♓ 57
21	27 ♈ 59	07 ♊ 53	04 ♈ 22	01 ♌ 00	01 ♈ 01	22 ♓ 04
31	27 ♈ 47	10 ♊ 16	02 ♈ 30	06 ♌ 07	29 ♓ 12	23 ♓ 11

ASPECTARIAN

h m Aspects

01 Friday
06 14 ☽ ∥ ☉
09 17 ☉ ☌ ☿
11 53 ☽ ☌ ♂
12 56 ☽ ⚼ ♇
16 19 ☿ △ ♆
17 30 ☽ ⊻ ♃
18 03 ☽ ✶ ♄
21 16 ☽ ∥ ☿
22 28 ☽ Q ☉
23 51 ☽ Q ☿

02 Saturday
00 31 ☉ △ ♆
01 30 ☽ ⊼ ♅
04 52 ☽ ⊥ ♃
08 17 ☽ # ♆
08 52 ♂ ⚼ ♇
12 44 ☽ ± ♅
17 11 ☽ △ ♇
19 45 ☿ ⊻ ♀
20 58 ☽ ☍ ♆
21 37 ☽ ∠ ♃
22 20 ☽ ∠ ♄
22 35 ☽ ✶ ☉

03 Sunday
01 22 ☽ □ ♀
02 21 ☽ ✶ ☿
05 17 ☽ ⚼ ♅
07 31 ☽ ∥ ♄
14 47 ☽ ⊥ ♄
22 07 ☽ ⊻ ♂
22 31 ☽ ∥ ♄

04 Monday
00 40 ☽ ✶ ♃
01 32 ☽ ⊻ ♄
03 34 ☽ ∠ ☉
08 01 ☽ △ ♅
09 10 ☽ ⊥ ♂
09 40 ☽ Q ♀
09 51 ☽ ∠ ☿
16 17 ☉ ∥ ♂
16 18 ☉ ⊻ ♀
19 47 ☽ # ♆
20 21 ☽ ⊥ ☉
21 34 ☽ Q ♃
22 33 ☽ □ ♇
23 19 ♄ Q ♇

05 Tuesday
01 26 ☽ ∠ ♂
01 56 ☽ ⊼ ♆
03 56 ☽ ⊥ ☿
06 20 ☽ ✶ ♀
07 18 ☽ ⊻ ☉
09 21 ☿ Q ♅
12 04 ☽ ± ♆
12 32 ☿ ∥ ♂
15 42 ☽ ⊻ ☿

06 Wednesday
00 27 ☽ ∥ ♂
03 02 ☽ ⚼ ♆
03 43 ☽ ✶ ♂
03 50 ☽ □ ♃
04 12 ☽ ∥ ☿
04 47 ☽ Q ♇
04 59 ☽ ☌ ♄
05 19 ☽ ∥ ☉
05 21 ♀ St R
06 35 ♂ ⊻ ♃
07 21 ☽ ∠ ♀
10 42 ☽ □ ♅
13 21 ☉ ∥ ☿
22 00 ☽ ⊥ ♀

07 Thursday
00 11 ☽ Q ♂
00 22 ☽ ✶ ♇
03 27 ☽ △ ♆
07 39 ☽ ⊻ ♀
09 28 ☽ ∥ ♇
11 57 ☽ ☌ ☉
15 13 ♂ ✶ ♄
15 50 ☽ Q ♅
18 34 ☽ # ♅

08 Friday
00 00 ☽ ☌ ☿
00 28 ☽ ∠ ♇
03 38 ☿ ∠ ♇
04 21 ☽ △ ♃
05 47 ☽ ⊻ ♄
06 08 ☽ ∥ ♃
06 18 ☽ □ ♂
11 00 ☽ ✶ ♅
14 49 ☽ ⊥ ♇
15 24 ☽ ⊥ ♄

09 Saturday
00 23 ☽ ⊻ ♇
03 18 ☽ □ ♆
04 14 ☽ ⚼ ♃
05 50 ☽ ∠ ♄
07 14 ☽ ☌ ♀
10 03 ☽ ∥ ♀
10 05 ☿ △ ♃
10 54 ☽ ∠ ♅
15 01 ☽ ⊻ ☉
18 37 ☽ ± ♃
23 49 ☿ ⊻ ♄

10 Sunday
01 17 ☽ ⊥ ☉
01 20 ☽ ⊥ ♅
02 11 ☉ Q ♅
04 15 ☽ ⊼ ♃
04 19 ☽ # ♀
06 02 ☽ ✶ ♄
06 53 ☽ ⊻ ☿
08 05 ☽ Q ♆
08 17 ☽ △ ♂
11 01 ☽ ⊻ ♅
16 48 ☽ ∠ ☉
18 01 ☽ ⊥ ☿
22 44 ☿ □ ♂

11 Monday
00 41 ☽ ☌ ♇
01 40 ☽ Q ♄
03 35 ☽ ✶ ♆
07 07 ☽ ⊻ ♀
09 20 ☽ # ♃
09 49 ☽ ⚼ ♂
10 54 ☽ ∠ ☿
16 00 ☿ ✶ ♅
16 53 ☽ ⊥ ♀
19 10 ☽ ✶ ☉
21 32 ☽ ∥ ♅

12 Tuesday
01 26 ☽ ± ♂
04 14 ♀ ∠ ♄
04 26 ☽ ∠ ♆
05 32 ☽ ☍ ♃
06 21 ☽ # ♇
07 42 ☽ ∠ ♀
07 47 ☽ □ ♄
09 35 ☽ # ☿
12 00 ☽ ⊼ ♂
12 41 ☽ ☌ ♅
15 45 ☽ ✶ ☿
16 49 ☽ Q ☉
19 41 ☽ ⊥ ♆
22 48 ☽ # ☉
23 37 ☿ ⊥ ♇

13 Wednesday
03 03 ☽ ⊻ ♇
04 05 ♂ ⊼ ♅
05 00 ☿ ∥ ♇
06 00 ☽ ⊻ ♆
08 53 ☽ ✶ ♀
10 41 ☿ ⊥ ♄
13 29 ☽ ⊥ ♇
15 34 ☽ Q ☿

14 Thursday
02 24 ☽ □ ☉
05 21 ☽ ∠ ♇
05 30 ☽ Q ♀
05 36 ☽ # ♂
09 30 ☽ ⊼ ♃
12 21 ☽ △ ♄
17 11 ☽ ⊻ ♅
18 54 ☽ ☍ ♂
19 31 ♃ St R
20 15 ☽ ± ♃

15 Friday
00 40 ☽ ∥ ♆
04 03 ☽ ⊥ ♅
04 36 ☽ □ ☿
08 25 ☽ ✶ ♇
08 25 ☿ # ♅
11 23 ☽ ☌ ♆
12 37 ☽ ⚼ ♃
13 10 ☽ □ ♀
14 28 ☽ # ♄
15 49 ☽ ⚼ ♄
20 35 ☽ ∠ ♅
21 16 ☉ ∠ ♇

16 Saturday
00 05 ♂ ⊥ ♃
06 35 ☽ Q ♇
08 37 ♀ ⚼ ♃
08 38 ☽ ± ♄
09 32 ☿ ⊻ ♇
13 22 ☽ △ ☉
16 26 ☽ △ ♃
19 59 ☽ ⊼ ♄
22 14 ☽ # ♄

17 Sunday
00 40 ☽ ✶ ♅
05 09 ☽ ⊼ ♂
07 36 ☿ □ ♆
10 43 ☿ ☌ ♀
11 39 ☽ ∥ ♆
16 37 ☽ □ ♇
16 51 ☿ ⚼ ♃
17 08 ☽ ± ♂
19 35 ☽ ⊻ ♆
19 42 ☽ △ ♀
20 05 ☽ ⚼ ☉
21 29 ☽ △ ☿
23 02 ♀ □ ♆
23 34 ☽ Q ♅

18 Monday
01 42 ☽ # ♂
05 50 ☉ △ ♃
07 05 ☽ ⊥ ♆
11 18 ☽ ⚼ ♂
14 51 ☽ ± ☉
16 34 ☿ ∥ ♃
23 18 ☿ ∠ ♄
23 35 ☽ ⚼ ♀

19 Tuesday
00 30 ☽ ∠ ♆
01 46 ☽ □ ♃
03 28 ☽ ⊼ ☉
06 03 ☽ ☍ ♄
07 03 ☽ ⚼ ☿
10 30 ☽ □ ♅
16 28 ☽ ± ♀
17 59 ☽ △ ♂

20 Wednesday
02 57 ☽ △ ♇
03 27 ☽ ± ☿
03 47 ☽ ⊼ ♀
05 52 ☽ ✶ ♆
09 08 ☿ ∠ ♅
11 34 ☽ # ☉
13 01 ☽ Q ♃
14 47 ☽ # ♇
16 50 ☉ ⊻ ♄
17 10 ☽ ⊼ ☿
21 54 ♀ ⊻ ♇
23 39 ☽ ∥ ♅

21 Thursday
05 37 ☽ Q ♆
08 44 ☽ ⚼ ♇
12 48 ☽ ✶ ♃
14 53 ♆ St D
15 30 ☽ # ♃
17 53 ☽ ⊼ ♄
19 48 ☽ ☍ ☉
22 02 ☽ △ ♅

22 Friday
00 07 ☉ ∥ ♇
02 45 ☽ ± ♇
06 00 ☽ ± ♄
07 02 ☽ # ☿
08 39 ☽ □ ♂
12 55 ☽ ☍ ♀
14 50 ☽ ⊼ ♇
17 39 ☽ □ ♆
18 48 ☽ ∠ ♃

23 Saturday
00 17 ☽ ⚼ ♄
00 45 ☉ ✶ ♅
02 24 ☿ ± ♃
04 16 ☽ ⚼ ♅
04 44 ☽ # ♀
12 55 ☽ ⊥ ♃
14 33 ☽ ☍ ☿
15 51 ♂ ± ♅
22 31 ☽ ± ♅
22 51 ☽ Q ♂

24 Sunday
00 41 ☽ ∥ ♀
01 01 ☽ ⊻ ♃
06 10 ♀ □ ♂
06 55 ☽ △ ♄
10 43 ☽ ⊼ ♅
12 33 ☽ ∥ ☿
13 36 ☽ ⊼ ☉
22 42 ☽ ⊼ ♀

25 Monday
00 27 ☽ ✶ ♂
02 51 ☽ ± ☉
03 40 ☽ ☍ ♇
06 21 ☽ △ ♆
10 16 ☽ ± ♀
13 13 ☉ ⊥ ♇
13 25 ☽ ∥ ♃
17 30 ♂ # ♆
18 15 ☉ # ♅
22 43 ☽ ⚼ ☉

26 Tuesday
03 36 ☽ ⚼ ♀
05 11 ☿ ⊥ ♅
05 18 ☽ ∥ ☉
06 23 ☽ # ♅
08 24 ☽ ∠ ♂
12 24 ☽ ⊼ ☿
12 43 ☽ ⚼ ♆
13 37 ☽ ☌ ♃
14 15 ☽ ∥ ♇
20 15 ☽ □ ♄
20 36 ☿ ∥ ♀
21 50 ☿ ⊼ ♃
23 38 ☽ ☍ ♅

27 Wednesday
02 13 ☽ ± ☿
03 27 ☽ ⊥ ♂
06 48 ♀ ⊥ ♄
06 50 ☽ ± ♆
07 37 ☽ △ ☉
08 19 ☽ △ ♀
13 11 ☉ ☌ ♀
16 04 ☽ ⊻ ♂
16 20 ☽ ⊼ ♇
18 41 ☉ ⊥ ♄
18 49 ☽ ⊼ ♆
22 02 ♂ △ ♇
22 44 ☽ ⚼ ☿

28 Thursday
04 17 ☽ ± ♇
08 29 ☽ Q ♄
15 39 ☿ Q ♆
16 01 ♂ ∥ ♄
22 01 ☽ ⚼ ♇

29 Friday
01 00 ☽ ⊻ ♃
08 05 ☽ △ ☿
08 07 ☽ ✶ ♄
08 24 ☿ ✶ ♄
11 02 ☽ ⊼ ♅
12 33 ☽ ⊥ ♃
16 21 ☽ □ ♀
21 08 ☽ # ♆
22 32 ☽ ± ♅
23 20 ☽ □ ☉

30 Saturday
02 18 ♂ ☍ ♆
02 54 ☽ △ ♇
03 39 ☽ ∥ ♄
05 06 ☽ ☍ ♆
05 14 ☽ ☌ ♂
05 37 ☽ ∠ ♃
09 54 ♀ ⊥ ♇
10 20 ☽ ∥ ♂
10 27 ☿ ⊻ ♅
12 52 ☽ ∠ ♄
13 05 ♂ ∠ ♃
15 31 ☽ ⚼ ♅

31 Sunday
00 35 ☽ Q ♀
05 36 ☽ ⊥ ♄
08 54 ☽ ∥ ♂
09 14 ☽ ✶ ♃
11 23 ☽ Q ☉
13 45 ☿ ⊻ ♀
16 33 ☽ ⊻ ♄
18 58 ☽ △ ♅
20 20 ☽ ∥ ♄
21 22 ☽ ✶ ♀
22 37 ☽ □ ☿
23 41 ☉ ⊻ ♇

All ephemeris data is given at 12.00 UT and the Moon's longitude is additionally given for 24.00 UT

SEPTEMBER 1975

Date	Sidereal time h m s	LONGITUDES Sun ☉ ° ' "	Moon ☽ ° ' "	Moon ☽ 24.00 ° ' "	Mercury ☿ ° '	Venus ♀ ° '	Mars ♂ ° '	Jupiter ♃ ° '	Saturn ♄ ° '	Uranus ♅ ° '	Neptune ♆ ° '	Pluto ♇ ° '
01	10 40 06	08 ♍ 25 59	09 ♋ 18 20	16 ♋ 15 16	02 ♎ 28	00 ♍ 39	10 ♊ 24	24 ♈ 11	28 ♋ 24	29 ♎ 41	09 ♐ 03	07 ♎ 57
02	10 44 03	09 24 04	23 ♋ 19 43	00 ♌ 31 29	03 49	00 ♍ 05	10 57	24 R 08	28 30	29 44	09 04	07 59
03	10 48 00	10 22 11	07 ♌ 50 06	15 ♌ 14 57	05 08	29 ♌ 32	11 30	24 04	28 37	29 47	09 04	08 01
04	10 51 56	11 20 20	22 ♌ 45 06	00 ♍ 19 27	06 26	29 R 01	12 03	24 00	28 43	29 49	09 04	08 04
05	10 55 53	12 18 31	07 ♍ 56 46	15 ♍ 35 37	07 43	28 32	12 35	23 56	28 50	29 52	09 05	08 06
06	10 59 49	13 16 43	23 ♍ 14 34	00 ♎ 52 12	08 57	28 04	13 08	23 52	28 56	29 55	09 05	08 08
07	11 03 46	14 14 57	08 ♎ 27 12	15 ♎ 58 20	10 10	27 39	13 40	23 47	29 02	29 ♎ 58	09 06	08 10
08	11 07 42	15 13 13	23 ♎ 24 36	00 ♏ 45 13	11 21	27 15	14 11	23 43	29 08	00 ♏ 01	09 07	08 12
09	11 11 39	16 11 30	07 ♏ 59 35	15 ♏ 07 21	12 30	26 54	14 43	23 38	29 15	00 04	09 07	08 14
10	11 15 35	17 09 49	22 ♏ 08 21	29 ♏ 02 37	13 37	26 34	15 14	23 33	29 21	00 07	09 08	08 17
11	11 19 32	18 08 10	05 ♐ 50 18	12 ♐ 31 40	14 42	26 17	15 45	23 28	29 27	00 10	09 08	08 19
12	11 23 29	19 06 32	19 ♐ 07 07	25 ♐ 37 03	15 44	26 03	16 15	23 22	29 33	00 13	09 09	08 21
13	11 27 25	20 04 55	02 ♑ 01 58	08 ♑ 22 19	16 44	25 51	16 46	23 17	29 39	00 16	09 10	08 23
14	11 31 22	21 03 21	14 ♑ 38 36	20 ♑ 51 18	17 42	25 41	17 16	23 11	29 45	00 19	09 11	08 25
15	11 35 18	22 01 47	27 ♑ 00 51	03 ♒ 07 40	18 37	25 34	17 45	23 05	29 50	00 22	09 11	08 28
16	11 39 15	23 00 16	09 ♒ 12 08	15 ♒ 14 37	19 29	25 29	18 15	23 00	29 ♋ 56	00 25	09 12	08 30
17	11 43 11	23 58 46	21 ♒ 15 25	27 ♒ 14 49	20 17	25 26	18 44	22 53	00 ♌ 02	00 28	09 13	08 32
18	11 47 08	24 57 17	03 ♓ 13 04	09 ♓ 10 23	21 03	25 D 26	19 12	22 47	00 07	00 32	09 14	08 34
19	11 51 04	25 55 51	15 ♓ 07 00	21 ♓ 03 04	21 44	25 28	19 41	22 41	00 13	00 35	09 15	08 37
20	11 55 01	26 54 26	26 ♓ 58 48	02 ♈ 54 23	22 22	25 33	20 09	22 34	00 18	00 38	09 16	08 39
21	11 58 58	27 53 04	08 ♈ 50 01	14 ♈ 45 55	22 55	25 39	20 36	22 28	00 24	00 41	09 17	08 41
22	12 02 54	28 51 43	20 ♈ 42 20	26 ♈ 39 31	23 24	25 48	21 03	22 21	00 29	00 45	09 18	08 44
23	12 06 51	29 ♍ 50 25	02 ♉ 37 48	08 ♉ 37 30	23 47	25 59	21 30	22 14	00 34	00 48	09 19	08 46
24	12 10 47	00 ♎ 49 08	14 ♉ 38 59	20 ♉ 42 42	24 06	26 13	21 57	22 07	00 39	00 52	09 20	08 48
25	12 14 44	01 47 54	26 ♉ 49 04	02 ♊ 58 36	24 18	26 28	22 23	22 00	00 44	00 55	09 21	08 51
26	12 18 40	02 46 42	09 ♊ 11 47	15 ♊ 29 09	24 25	26 45	22 49	21 53	00 49	00 58	09 22	08 53
27	12 22 37	03 45 32	21 ♊ 51 15	28 ♊ 18 34	24 R 25	27 05	23 14	21 46	00 54	01 02	09 23	08 55
28	12 26 33	04 44 25	04 ♋ 51 38	11 ♋ 30 51	24 18	27 26	23 39	21 38	00 59	01 05	09 24	08 57
29	12 30 30	05 43 20	18 ♋ 16 37	25 ♋ 09 11	24 03	27 49	24 04	21 31	01 04	01 09	09 26	09 00
30	12 34 27	06 ♎ 42 17	02 ♌ 08 41	09 ♌ 15 05	23 ♎ 42	28 ♌ 14	24 ♊ 28	21 ♈ 23	01 ♌ 08	01 ♏ 12	09 ♐ 27	09 ♎ 02

Date	Moon True ☊ ° '	Moon Mean ☊ ° '	Moon Latitude ° '
01	25 ♏ 14	25 ♏ 42	03 S 31
02	25 R 08	25 39	04 17
03	25 00	25 36	04 49
04	24 50	25 32	05 02
05	24 40	25 29	04 54
06	24 30	25 26	04 25
07	24 22	25 23	03 38
08	24 16	25 20	02 36
09	24 13	25 16	01 25
10	24 13	25 13	00 S 11
11	24 D 12	25 10	01 N 02
12	24 R 12	25 07	02 09
13	24 12	25 04	03 07
14	24 09	25 01	03 55
15	24 04	24 57	04 31
16	23 57	24 54	04 54
17	23 47	24 51	05 04
18	23 36	24 48	05 00
19	23 25	24 45	04 43
20	23 15	24 41	04 14
21	23 05	24 38	03 33
22	22 58	24 35	02 44
23	22 54	24 32	01 47
24	22 52	24 29	00 N 44
25	22 D 51	24 26	00 S 21
26	22 52	24 22	01 27
27	22 53	24 19	02 30
28	22 54	24 16	03 27
29	22 R 53	24 13	04 15
30	22 ♏ 51	24 ♏ 10	04 S 49

DECLINATIONS

Date	Sun ☉ ° '	Moon ☽ ° '	Mercury ☿ ° '	Venus ♀ ° '	Mars ♂ ° '	Jupiter ♃ ° '	Saturn ♄ ° '	Uranus ♅ ° '	Neptune ♆ ° '	Pluto ♇ ° '
01	08 N 25	19 N 37	01 S 45	03 N 16	20 N 59	07 N 57	20 N 31	10 S 54	20 S 17	11 N 53
02	08 03	17 12	02 25	03 29	21 05	07 56	20 30	10 55	20 17	11 52
03	07 41	13 40	03 04	03 43	21 10	07 54	20 28	10 56	20 17	11 51
04	07 19	09 11	03 43	03 56	21 16	07 53	20 27	10 57	20 17	11 50
05	06 57	04 N 03	04 21	04 10	21 21	07 51	20 26	10 58	20 17	11 49
06	06 34	01 S 23	04 59	04 24	21 26	07 49	20 25	10 59	20 18	11 48
07	06 12	06 41	05 36	04 37	21 32	07 47	20 24	11 00	20 18	11 47
08	05 49	11 30	06 11	04 51	21 36	07 45	20 22	11 01	20 18	11 46
09	05 27	15 31	06 46	05 05	21 41	07 43	20 21	11 02	20 18	11 45
10	05 04	18 29	07 20	05 18	21 46	07 41	20 20	11 03	20 18	11 44
11	04 41	20 16	07 53	05 31	21 50	07 39	20 19	11 04	20 18	11 43
12	04 19	20 51	08 25	05 44	21 55	07 37	20 18	11 05	20 18	11 42
13	03 56	20 18	08 56	05 56	21 59	07 35	20 17	11 06	20 19	11 41
14	03 33	18 44	09 25	06 08	22 03	07 33	20 15	11 07	20 19	11 40
15	03 10	16 19	09 53	06 19	22 07	07 31	20 14	11 08	20 19	11 39
16	02 47	13 13	10 20	06 30	22 11	07 28	20 13	11 10	20 19	11 38
17	02 23	09 37	10 45	06 40	22 15	07 26	20 12	11 11	20 19	11 37
18	02 00	05 40	11 09	06 50	22 19	07 23	20 11	11 12	20 20	11 36
19	01 37	01 S 31	11 30	06 59	22 23	07 21	20 10	11 13	20 20	11 35
20	01 14	02 N 41	11 50	07 07	22 26	07 18	20 09	11 14	20 20	11 34
21	00 51	06 46	12 08	07 15	22 29	07 16	20 08	11 15	20 20	11 33
22	00 27	10 37	12 23	07 22	22 33	07 13	20 07	11 17	20 20	11 33
23	00 N 04	14 03	12 36	07 29	22 36	07 11	20 06	11 18	20 20	11 32
24	00 S 20	16 56	12 46	07 35	22 39	07 08	20 05	11 19	20 21	11 31
25	00 43	19 06	12 53	07 40	22 42	07 05	20 04	11 20	20 21	11 30
26	01 06	20 24	12 57	07 44	22 45	07 02	20 03	11 21	20 21	11 29
27	01 30	20 42	12 58	07 48	22 48	07 00	20 02	11 23	20 21	11 28
28	01 53	19 54	12 55	07 51	22 51	06 57	20 01	11 24	20 22	11 27
29	02 16	17 59	12 48	07 54	22 54	06 54	20 00	11 25	20 22	11 26
30	02 S 40	14 N 59	12 S 37	07 N 55	22 N 56	06 N 51	19 N 59	11 S 26	20 S 22	11 N 25

ZODIAC SIGN ENTRIES

Date	h	m	Planets
02	15	34	♀ ♌
02	23	08	☽ ♌
04	23	29	☽ ♍
06	22	38	☽ ♎
08	05	16	♅ ♏
08	22	46	☽ ♏
11	01	41	☽ ♐
13	08	11	☽ ♑
15	17	51	☽ ♒
17	04	57	♄ ♌
18	05	32	☽ ♓
20	18	07	☽ ♈
23	06	43	☽ ♉
23	15	55	☉ ♎
25	18	13	☽ ♊
28	03	07	☽ ♋
30	08	20	☽ ♌

LATITUDES

Date	Mercury ☿ ° '	Venus ♀ ° '	Mars ♂ ° '	Jupiter ♃ ° '	Saturn ♄ ° '	Uranus ♅ ° '	Neptune ♆ ° '	Pluto ♇ ° '
01	00 S 50	08 S 31	01 S 02	01 S 32	00 N 02	00 N 30	01 N 33	16 N 23
04	01 16	08 24	00 59	01 33	00 02	00 30	01 33	16 22
07	01 42	08 09	00 55	01 33	00 03	00 30	01 32	16 22
10	02 08	07 49	00 52	01 34	00 03	00 30	01 32	16 21
13	02 33	07 24	00 48	01 34	00 03	00 30	01 32	16 21
16	02 56	06 56	00 44	01 35	00 03	00 30	01 32	16 20
19	03 16	06 25	00 40	01 35	00 04	00 29	01 32	16 20
22	03 33	05 53	00 36	01 36	00 04	00 29	01 32	16 20
25	03 44	05 20	00 31	01 36	00 04	00 29	01 31	16 20
28	03 46	04 48	00 26	01 36	00 05	00 29	01 31	16 20
31	03 S 36	04 S 16	00 S 22	01 S 37	00 N 05	00 N 29	01 N 31	16 N 20

DATA

Julian Date	2442657
Delta T	+46 seconds
Ayanamsa	23° 31' 17"
Synetic vernal point	05° ♓ 35' 43"
True obliquity of ecliptic	23° 26' 28"

LONGITUDES

Date	Chiron ⚷ ° '	Ceres ⚳ ° '	Pallas ⚴ ° '	Juno ⚵ ° '	Vesta ⚶ ° '	Black Moon Lilith ⚸ ° '
01	27 ♈ 45	10 ♊ 29	02 ♈ 17	06 ♌ 38	28 ♓ 59	23 ♓ 18
11	27 ♈ 27	12 ♊ 27	29 ♓ 54	11 ♌ 36	26 ♓ 38	24 ♓ 25
21	27 ♈ 06	13 ♊ 58	27 ♓ 15	16 ♌ 25	24 ♓ 05	25 ♓ 32
31	26 ♈ 41	14 ♊ 59	24 ♓ 33	21 ♌ 03	21 ♓ 42	26 ♓ 39

MOON'S PHASES, APSIDES AND POSITIONS ☽

Date	h	m	Phase	Longitude ° '	Eclipse Indicator
05	19	19	●	12 ♍ 36	
12	11	59	☽	19 ♐ 06	
20	11	50	○	26 ♓ 54	
28	11	46	☾	04 ♋ 44	

Day	h	m		
06	04	16	Perigee	
20	06	51	Apogee	
06	05	56	0S	
12	12	02	Max dec	20° S 51'
19	20	40	0N	
27	06	46	Max dec	20° N 43'

ASPECTARIAN

01 Monday
01 32 ☽ # ♆
06 34 ☽ Q ♃
09 38 ☽ □ ♇
10 22 ☽ ⚹ ☉
11 34 ☽ ⊼ ♆
13 58 ☽ ⊻ ♂
21 57 ☽ ± ♆
22 32 ☽ ∠ ♀

02 Tuesday
00 45 ☽ ⊥ ♂
03 30 ☉ □ ♆
05 40 ☉ bQ ♃
09 11 ☽ Q ☿
13 13 ☽ ⊥ ♀
13 14 ☽ bQ ♆
13 20 ☽ □ ♃
13 56 ☽ ∠ ☉
16 28 ☽ Q ♇
16 34 ☽ ∠ ♂
20 13 ☉ ∥ ♃
20 43 ☽ ☌ ♄
22 43 ☽ □ ♅
22 51 ☽ ⊻ ♀

03 Wednesday
02 12 ♀ ⚹ ♅
05 56 ☽ ⊥ ☉
07 10 ☽ ⚹ ☿
12 03 ♃ bQ ♆
12 19 ☽ ⚹ ♇
14 00 ☽ △ ♆
16 25 ☽ ⊻ ☉
17 14 ☿ ⊥ ♀
18 11 ☽ ⚹ ♂
22 15 ☽ ∥ ♇

04 Thursday
03 04 ☽ # ♅
04 06 ☽ Q ♅
09 43 ☽ ∠ ☿
12 29 ☽ ∠ ♇
13 59 ☽ △ ♃
14 08 ☽ Q ♂
18 21 ☽ ∥ ♃
21 32 ☽ ⊻ ♄
21 38 ☽ ☌ ♀
21 41 ☽ ∥ ☉
23 15 ☽ ⚹ ♅
23 56 ♀ ⊻ ♄

05 Friday
00 37 ☿ # ♀
01 18 ☽ ⊥ ☿
02 46 ☽ ⊥ ♇
07 04 ☽ ⊥ ♄
10 44 ☽ # ☿
11 29 ☽ ∥ ♀
11 36 ☽ ⊻ ☿
12 14 ☽ ⊻ ♇
13 33 ☽ bQ ♃
13 47 ☽ □ ♆
19 19 ☽ ☌ ☉
19 33 ☽ □ ♂
19 37 ☿ ☌ ♇
21 18 ☽ ∠ ♄
22 54 ☽ ∠ ♅

06 Saturday
03 36 ☽ ± ♃
03 36 ☉ □ ♂
12 58 ☽ ⊼ ♃
13 04 ☽ ⊥ ♅
14 44 ☿ ⚹ ♆
18 03 ☽ Q ♆
19 23 ☽ ⊻ ♀
21 01 ☽ ⚹ ♄
22 32 ☽ ⊻ ♅

07 Sunday
02 05 ☽ # ♀
04 35 ☽ ⊥ ♀
06 08 ☉ ∠ ♄
06 16 ☽ ∥ ☿
09 53 ☽ # ☉
11 33 ☽ ☌ ♇
13 02 ☽ ⚹ ♆
14 57 ☽ ☌ ☿
16 08 ☽ Q ♄
17 11 ☽ # ♃
18 30 ☽ ∠ ♀
20 36 ☽ △ ♂
21 53 ☽ ⊻ ☉

08 Monday
02 57 ☉ # ☿
06 37 ☉ ∠ ♅
07 20 ☿ Q ♄
08 12 ☽ ⊥ ☉
09 18 ♂ ∠ ♄
09 22 ☽ ∥ ♅
12 29 ☽ ☍ ♃
13 08 ☽ ∠ ♆
13 25 ☽ # ♇
18 06 ☽ ⚹ ♀
21 25 ☽ □ ♄
21 47 ☽ bQ ♂
22 49 ☽ ☌ ♅
23 55 ☽ ∠ ☉

09 Tuesday
02 13 ☿ ∠ ♀
03 54 ☽ ⊥ ♆
12 25 ☽ ⊻ ♇
13 15 ☽ ± ♂
13 28 ☽ Q ♀
13 53 ☽ ⊻ ♆
17 03 ♀ Q ♂
20 13 ☽ ⊻ ☿
22 32 ☽ ⊥ ♇
23 44 ☽ ⊼ ♂

10 Wednesday
02 50 ☽ ⚹ ☉
02 51 ☉ ∥ ♀
05 51 ♂ bQ ♅
07 17 ☽ ⊥ ☿
13 58 ☽ ∠ ♇
14 25 ☽ ⊼ ♃
19 31 ☽ □ ♀
20 42 ☉ ± ♃

11 Thursday
00 13 ☽ ∠ ☿
00 37 ☽ △ ♄
00 48 ☽ ± ♃
01 09 ☽ Q ☉
01 56 ☽ ⊻ ♅
02 22 ☿ # ♃
12 35 ☽ ⊥ ♅
12 42 ☽ ∥ ♆
12 54 ☽ # ♄
16 26 ☽ ⚹ ♇
16 39 ☽ bQ ♃
17 54 ☽ ☌ ♆
23 27 ♄ # ♆

12 Friday
03 36 ☽ bQ ♄
04 50 ☽ ∠ ♅
05 18 ☽ ⚹ ☿
06 33 ☽ ☍ ♂
11 59 ☽ □ ☉
14 16 ☽ Q ♇
19 47 ☽ △ ♃
20 13 ☽ ± ♄

13 Saturday
00 36 ☽ △ ♀
05 19 ☽ Q ☿
07 29 ☽ ⊼ ♄
08 40 ☽ ⚹ ♅
11 53 ☽ ∥ ♆
12 38 ☽ # ♄
12 56 ☿ △ ♂

14 Sunday
00 04 ☽ □ ♇
01 31 ☽ ⊻ ♆
04 30 ☽ bQ ♀
07 31 ☽ Q ♅
13 02 ☽ ⊥ ♆
17 15 ☽ ⊼ ♂
18 22 ☽ □ ☿
21 37 ☽ ± ♀

15 Monday
01 27 ☽ △ ☉
04 24 ☽ □ ♃
05 22 ☽ ± ♂
06 29 ☽ ∠ ♆
09 11 ☽ ⊼ ♀
17 35 ☽ ☍ ♄
18 36 ☽ □ ♅
23 44 ☽ bQ ♂

16 Tuesday
09 25 ☽ bQ ☉
10 36 ☽ △ ♇
11 44 ☉ ⊼ ♃
12 00 ☽ ⚹ ♆
15 31 ☽ Q ♃
22 58 ☽ # ♇

17 Wednesday
02 00 ☽ ∥ ♅
04 53 ☽ ± ☉
05 27 ☽ ∥ ☿
06 44 ☽ △ ♂
09 56 ☽ △ ☿
11 55 ☽ Q ♆
15 15 ☽ ⚹ ♃
16 35 ☽ bQ ♇
17 56 ☽ ⊼ ☉
20 21 ☽ ☍ ♀

18 Thursday
00 53 ☉ ⊥ ♅
01 37 ☽ # ♃
01 47 ♀ St D
05 20 ☽ # ♀
05 43 ☽ ⊼ ♄
06 34 ☽ △ ♅
10 42 ☽ ± ♇
15 35 ☿ ∥ ♅
17 53 ☽ ± ♄
18 03 ☽ bQ ☿
21 08 ☽ ∠ ♃
22 50 ☽ ⊼ ♇

19 Friday
00 04 ☉ ⊻ ♀
00 08 ☽ □ ♆
11 21 ☽ # ☉
12 12 ☽ bQ ♄
12 57 ☽ bQ ♅
13 19 ☽ ± ☿
15 08 ☽ ⊥ ♃
17 37 ☿ # ♇
21 36 ☽ □ ♂

20 Saturday
02 09 ☽ ⊼ ☿
03 10 ☽ ⊻ ♃
04 24 ☽ ∥ ☉
07 14 ☽ ± ♅
09 04 ☽ ⊼ ♀
11 50 ☽ ☍ ☉
18 47 ☽ △ ♄
19 26 ☿ ☍ ♃
19 26 ☽ ⊼ ♅
20 54 ☉ Q ♆
21 19 ☽ ± ♀

21 Sunday
11 31 ☽ Q ♂
11 42 ☽ ☍ ♇
12 54 ☽ △ ♆
13 09 ♀ ∥ ♃
14 56 ☽ ∥ ♃
15 01 ☽ ∥ ♀
15 44 ☽ bQ ♀

22 Monday
12 44 ☽ ⚹ ♂
15 17 ☽ ☌ ♃
16 24 ☽ # ♅
17 37 ☽ ☍ ☿
18 08 ☽ ∥ ♇
19 15 ☽ bQ ♆
22 26 ☽ △ ♀

23 Tuesday
00 45 ☽ # ☿
05 54 ☽ ⊼ ☉
07 50 ☽ □ ♄
08 19 ☽ ☍ ♅
13 22 ☽ ± ♆
19 00 ☽ ± ☉
20 04 ☽ ∠ ♂

24 Wednesday
00 19 ☽ ⊼ ♇
01 24 ☽ ⊼ ♆
07 33 ☉ ⚹ ♄
12 18 ☽ ± ♇
13 02 ☉ ⊻ ♅
14 31 ☽ bQ ☉
14 40 ☽ ⊥ ♂
19 29 ♂ ⚹ ♃
19 59 ☽ Q ♄

25 Thursday
02 38 ☽ ⊻ ♃
02 59 ☽ ⊻ ♂
06 09 ☽ bQ ♇
07 00 ☽ ⊼ ☿
11 18 ☽ □ ♀
14 18 ☽ ⊥ ♃
18 53 ☽ ± ☿
19 42 ☽ ⚹ ♄
20 02 ☽ ⊼ ♅
21 04 ☿ ∠ ♆
22 33 ☽ △ ☉

26 Friday
03 38 ☽ ∥ ♄
07 36 ☽ ∠ ♃
07 42 ☽ ± ♅
10 43 ☽ # ♆
11 13 ☉ ⊥ ♀
11 24 ☽ △ ♇
12 20 ☽ ☍ ♆
12 25 ☽ bQ ☿
22 53 ☽ Q ♀
23 45 ☿ St R

27 Saturday
00 43 ☽ ∠ ♄
00 59 ☽ bQ ♅
11 50 ☽ ⚹ ♃
14 40 ☽ ☌ ♂
16 45 ☽ △ ☿
17 43 ☽ ⊥ ♄
17 50 ☿ ∠ ♆
21 59 ☽ ⚹ ♀

28 Sunday
02 18 ☽ # ♆
04 53 ☽ ⊻ ♄
05 05 ☽ △ ♅
09 48 ☽ Q ♃
09 56 ☽ ∥ ♄
11 46 ☽ □ ☉
19 26 ☽ □ ♇
20 14 ☽ ⊼ ♆

29 Monday
02 02 ☽ ∠ ♀
06 58 ☽ ± ♆
11 49 ☿ △ ♂
17 38 ☽ □ ♃
18 23 ☽ ⊥ ♀
21 52 ☽ □ ☿
22 15 ☽ Q ☉
22 25 ☽ ⊻ ♂
22 46 ☽ bQ ♆

30 Tuesday
00 18 ♅ # ♇
03 14 ☽ Q ♇
05 05 ☽ ⊻ ♀
09 03 ☽ ⊥ ♂
10 17 ☽ ☌ ♄
10 23 ☽ □ ♅
20 18 ☽ ⚹ ☉
23 40 ☽ ⚹ ♇

All ephemeris data is given at 12.00 UT and the Moon's longitude is additionally given for 24.00 UT

OCTOBER 1975

LONGITUDES

Date	Sidereal time h m s	Sun ☉ ° ' "	Moon ☽ ° ' "	Moon ☽ 24.00 ° ' "	Mercury ☿ ° '	Venus ♀ ° '	Mars ♂ ° '	Jupiter ♃ ° '	Saturn ♄ ° '	Uranus ♅ ° '	Neptune ♆ ° '	Pluto ♇ ° '
01	12 38 23	07 ♎ 41 17	16 ♌ 28 10	23 ♌ 47 32	23 ♎ 12	28 ♌ 40	24 ♊ 52	21 ♈ 16	01 ♌ 13	01 ♏ 16	09 ♐ 28	09 ♎ 04
02	12 42 20	08 40 19	01 ♍ 12 34	08 ♍ 42 24	22 R 35	29 08	25 15	21 R 08	01 17	01 19	09 29	09 07
03	12 46 16	09 39 22	16 ♍ 16 03	23 ♍ 52 18	21 51	29 ♌ 38	25 38	21 00	01 22	01 23	09 31	09 09
04	12 50 13	10 38 29	01 ♎ 29 52	09 ♎ 07 24	21 00	00 ♍ 09	26 00	20 52	01 26	01 26	09 32	09 11
05	12 54 09	11 37 37	16 ♎ 43 33	24 ♎ 17 02	20 02	00 41	26 22	20 45	01 30	01 30	09 33	09 14
06	12 58 06	12 36 47	01 ♏ 46 41	09 ♏ 11 30	18 59	01 15	26 43	20 37	01 35	01 34	09 35	09 16
07	13 02 02	13 35 59	16 ♏ 30 39	23 ♏ 43 32	17 51	01 51	27 04	20 29	01 39	01 37	09 36	09 18
08	13 05 59	14 35 13	00 ♐ 49 42	07 ♐ 48 58	16 41	02 27	27 24	20 21	01 43	01 41	09 38	09 21
09	13 09 56	15 34 29	14 ♐ 41 17	21 ♐ 26 44	15 30	03 05	27 44	20 13	01 47	01 45	09 39	09 23
10	13 13 52	16 33 47	28 ♐ 05 35	04 ♑ 38 10	14 21	03 45	28 03	20 05	01 50	01 48	09 41	09 25
11	13 17 49	17 33 07	11 ♑ 04 55	17 ♑ 26 18	13 14	04 25	28 22	19 56	01 54	01 52	09 42	09 28
12	13 21 45	18 32 28	23 ♑ 42 50	29 ♑ 55 03	12 12	05 06	28 40	19 48	01 58	01 56	09 44	09 30
13	13 25 42	19 31 51	06 ♒ 03 29	12 ♒ 08 42	11 17	05 49	28 57	19 40	02 01	01 59	09 45	09 32
14	13 29 38	20 31 16	18 ♒ 11 10	24 ♒ 11 26	10 30	06 32	29 15	19 32	02 05	02 03	09 47	09 35
15	13 33 35	21 30 42	00 ♓ 09 55	06 ♓ 07 05	09 53	07 17	29 31	19 24	02 08	02 07	09 49	09 37
16	13 37 31	22 30 11	12 ♓ 03 20	17 ♓ 59 02	09 26	08 02	29 ♊ 47	19 16	02 11	02 10	09 50	09 39
17	13 41 28	23 29 41	23 ♓ 54 31	29 ♓ 50 06	09 10	08 49	00 ♋ 02	19 08	02 14	02 14	09 52	09 42
18	13 45 25	24 29 13	05 ♈ 46 03	11 ♈ 42 37	09 D 05	09 36	00 17	19 00	02 17	02 18	09 54	09 44
19	13 49 21	25 28 47	17 ♈ 40 04	23 ♈ 38 36	09 12	10 25	00 31	18 52	02 20	02 22	09 55	09 46
20	13 53 18	26 28 23	29 ♈ 38 25	05 ♉ 39 46	09 29	11 14	00 44	18 44	02 23	02 25	09 57	09 48
21	13 57 14	27 28 01	11 ♉ 42 49	17 ♉ 47 49	09 56	12 04	00 57	18 36	02 26	02 29	09 59	09 51
22	14 01 11	28 27 41	23 ♉ 54 59	00 ♊ 04 34	10 32	12 54	01 09	18 28	02 28	02 33	10 01	09 53
23	14 05 07	29 ♎ 27 24	06 ♊ 16 50	12 ♊ 32 03	11 18	13 46	01 20	18 20	02 31	02 37	10 03	09 55
24	14 09 04	00 ♏ 27 08	18 ♊ 50 31	25 ♊ 12 32	12 11	14 38	01 31	18 13	02 33	02 40	10 04	09 57
25	14 13 00	01 26 55	01 ♋ 38 27	08 ♋ 08 32	13 11	15 31	01 41	18 05	02 35	02 44	10 06	09 59
26	14 16 57	02 26 44	14 ♋ 43 08	21 ♋ 22 30	14 17	16 25	01 50	17 57	02 38	02 48	10 08	10 02
27	14 20 54	03 26 35	28 ♋ 06 54	04 ♌ 56 32	15 29	17 19	01 58	17 50	02 40	02 52	10 10	10 04
28	14 24 50	04 26 28	11 ♌ 51 29	18 ♌ 51 48	16 45	18 14	02 06	17 42	02 42	02 55	10 12	10 06
29	14 28 47	05 26 24	25 ♌ 57 23	03 ♍ 08 02	18 06	19 09	02 13	17 35	02 44	02 59	10 14	10 08
30	14 32 43	06 26 21	10 ♍ 23 23	17 ♍ 42 56	19 29	20 06	02 19	17 28	02 45	03 03	10 16	10 10
31	14 36 40	07 ♏ 26 21	25 ♍ 06 03	02 ♎ 31 57	20 ♎ 56	21 ♍ 02	02 ♋ 25	17 ♈ 20	02 ♌ 47	03 ♏ 07	10 ♐ 18	10 ♎ 12

Date	Moon True ☊ ° '	Moon Mean ☊ ° '	Moon ☽ Latitude ° '
01	22 ♏ 46	24 ♏ 07	05 S 07
02	22 R 40	24 03	05 06
03	22 34	24 00	04 43
04	22 27	23 57	04 01
05	22 22	23 54	03 01
06	22 19	23 51	01 49
07	22 17	23 47	00 S 31
08	22 D 17	23 44	00 N 46
09	22 18	23 41	01 59
10	22 20	23 38	03 03
11	22 21	23 35	03 55
12	22 R 21	23 32	04 34
13	22 20	23 28	05 00
14	22 17	23 25	05 12
15	22 13	23 22	05 09
16	22 08	23 19	04 54
17	22 02	23 16	04 26
18	21 57	23 13	03 46
19	21 53	23 09	02 57
20	21 50	23 06	01 59
21	21 49	23 03	00 N 55
22	21 D 48	23 00	00 S 12
23	21 49	22 57	01 19
24	21 50	22 53	02 24
25	21 52	22 50	03 23
26	21 53	22 47	04 12
27	21 54	22 44	04 49
28	21 R 54	22 41	05 11
29	21 53	22 38	05 15
30	21 51	22 34	05 00
31	21 ♏ 50	22 ♏ 31	04 S 25

DECLINATIONS

Date	Sun ☉ ° '	Moon ☽ ° '	Mercury ☿ ° '	Venus ♀ ° '	Mars ♂ ° '	Jupiter ♃ ° '	Saturn ♄ ° '	Uranus ♅ ° '	Neptune ♆ ° '	Pluto ♇ ° '
01	03 S 03	11 N 01	12 S 22	07 N 56	22 N 59	06 N 48	19 N 58	11 S 27	20 S 22	11 N 24
02	03 26	06 17	12 02	07 57	23 02	06 45	19 57	11 29	20 23	11 23
03	03 50	01 N 04	11 38	07 56	23 04	06 42	19 56	11 30	20 23	11 23
04	04 13	04 S 16	11 09	07 55	23 07	06 39	19 55	11 31	20 23	11 22
05	04 36	09 21	10 36	07 53	23 09	06 36	19 55	11 32	20 23	11 21
06	04 59	13 48	10 00	07 51	23 11	06 33	19 54	11 34	20 24	11 20
07	05 22	17 17	09 19	07 48	23 14	06 30	19 53	11 35	20 24	11 19
08	05 45	19 34	08 37	07 44	23 16	06 27	19 52	11 36	20 24	11 18
09	06 08	20 35	07 52	07 40	23 18	06 24	19 51	11 38	20 24	11 17
10	06 31	20 23	07 07	07 35	23 20	06 21	19 51	11 39	20 25	11 17
11	06 53	19 04	06 23	07 29	23 23	06 18	19 50	11 40	20 25	11 16
12	07 16	16 51	05 39	07 23	23 25	06 15	19 49	11 41	20 25	11 15
13	07 39	13 55	04 59	07 16	23 27	06 12	19 49	11 43	20 25	11 14
14	08 01	10 26	04 22	07 08	23 29	06 09	19 48	11 44	20 26	11 14
15	08 23	06 35	03 50	07 00	23 31	06 06	19 47	11 45	20 26	11 13
16	08 45	02 S 31	03 22	06 51	23 33	06 03	19 47	11 47	20 26	11 12
17	09 07	01 N 39	03 01	06 42	23 36	06 00	19 46	11 48	20 27	11 11
18	09 29	05 45	02 44	06 32	23 38	05 57	19 46	11 49	20 27	11 11
19	09 51	09 39	02 34	06 22	23 40	05 54	19 45	11 50	20 27	11 10
20	10 13	13 12	02 29	06 11	23 42	05 51	19 44	11 52	20 27	11 09
21	10 34	16 14	02 30	05 59	23 44	05 48	19 44	11 53	20 28	11 08
22	10 56	18 34	02 36	05 47	23 46	05 45	19 43	11 54	20 28	11 08
23	11 17	20 04	02 47	05 35	23 49	05 43	19 43	11 56	20 28	11 07
24	11 38	20 35	03 02	05 22	23 51	05 40	19 43	11 57	20 29	11 06
25	11 59	20 03	03 21	05 08	23 53	05 37	19 42	11 58	20 29	11 06
26	12 19	18 27	03 44	04 54	23 56	05 34	19 42	11 59	20 29	11 05
27	12 40	15 49	04 09	04 39	23 58	05 31	19 41	12 01	20 29	11 04
28	13 00	12 15	04 37	04 24	24 00	05 29	19 41	12 02	20 30	11 04
29	13 20	07 55	05 08	04 09	24 03	05 26	19 41	12 03	20 30	11 03
30	13 40	03 03	05 40	03 53	24 05	05 23	19 40	12 05	20 30	11 03
31	14 S 00	02 S 06	06 S 14	03 N 37	24 N 08	05 N 21	19 N 40	12 S 06	20 S 31	11 N 02

ZODIAC SIGN ENTRIES

Date	h	m	Planets
02	10	03	☽ ♍
04	05	19	♀ ♍
04	09	39	☽ ♎
06	09	09	☽ ♏
08	10	35	☽ ♐
10	15	29	☽ ♑
13	00	10	☽ ♒
15	11	40	☽ ♓
17	08	44	♂ ♋
18	00	20	☽ ♈
20	12	43	☽ ♉
22	23	51	☽ ♊
24	01	06	☉ ♏
25	08	57	☽ ♋
27	15	20	☽ ♌
29	18	47	☽ ♍
31	19	55	☽ ♎

LATITUDES

Date	Mercury ☿ ° '	Venus ♀ ° '	Mars ♂ ° '	Jupiter ♃ ° '	Saturn ♄ ° '	Uranus ♅ ° '	Neptune ♆ ° '	Pluto ♇ ° '
01	03 S 36	04 S 16	00 S 22	01 S 37	00 N 05	00 N 29	01 N 31	16 N 20
04	03 12	03 44	00 16	01 37	00 05	00 29	01 31	16 20
07	02 31	03 14	00 11	01 37	00 05	00 29	01 31	16 20
10	01 35	02 45	00 S 05	01 37	00 06	00 29	01 31	16 20
13	00 S 34	02 17	00 N 01	01 37	00 06	00 29	01 31	16 21
16	00 N 24	01 50	00 07	01 37	00 06	00 29	01 30	16 21
19	01 10	01 24	00 14	01 36	00 07	00 29	01 30	16 21
22	01 42	01 00	00 20	01 36	00 07	00 29	01 30	16 22
25	02 01	00 38	00 27	01 36	00 07	00 29	01 30	16 23
28	02 07	00 S 16	00 35	01 35	00 08	00 29	01 30	16 23
31	02 N 05	00 N 04	00 N 43	01 S 35	00 N 08	00 N 29	01 N 30	16 N 24

DATA

Julian Date	2442687
Delta T	+46 seconds
Ayanamsa	23° 31' 20"
Synetic vernal point	05° ♓ 35' 40"
True obliquity of ecliptic	23° 26' 28"

LONGITUDES

Date	Chiron ⚷ ° '	Ceres ⚳ ° '	Pallas ⚴ ° '	Juno ⚵ ° '	Vesta ⚶ ° '	Black Moon Lilith ⚸ ° '
01	26 ♈ 41	14 ♊ 59	24 ♓ 33	21 ♌ 03	21 ♓ 42	26 ♓ 39
11	26 ♈ 14	15 ♊ 25	22 ♓ 04	25 ♌ 28	19 ♓ 44	27 ♓ 46
21	25 ♈ 46	15 ♊ 13	20 ♓ 00	29 ♌ 40	18 ♓ 27	28 ♓ 53
31	25 ♈ 19	14 ♊ 22	18 ♓ 32	03 ♍ 36	17 ♓ 54	00 ♈ 00

MOON'S PHASES, APSIDES AND POSITIONS ☽

Date	h	m	Phase	Longitude ° '	Eclipse Indicator
05	03	23	●	11 ♎ 16	
12	01	15	☽	18 ♑ 06	
20	05	06	○	26 ♈ 11	
27	22	07	☾	03 ♌ 52	

Day	h	m		
04	15	03	Perigee	
17	10	33	Apogee	
03	16	47	0S	
09	19	41	Max dec	20° S 39'
17	02	29	0N	
24	12	08	Max dec	20° N 35'
31	02	15	0S	

ASPECTARIAN

01 Wednesday
00 21 ☽ △ ♆
00 42 ☽ ∠ ♂
03 35 ☽ Q ☿
03 50 ☽ # ☿
09 34 ☽ # ♅
09 50 ☽ ∥ ♇
16 37 ☽ Q ♅
19 48 ☽ △ ♃
22 38 ☽ ✶ ☿
22 56 ☽ ∠ ☉

02 Thursday
00 30 ☽ ∠ ♇
02 06 ☽ ✶ ♂
03 56 ☽ ∥ ♀
08 33 ☽ ☌ ♀
09 43 ☽ ∥ ♃
12 08 ☽ ⊻ ♄
12 11 ☽ ✶ ♅
14 31 ☽ ⊥ ☉
15 04 ☽ ⊥ ♇
15 33 ♂ ⊥ ♄
19 50 ☽ bQ ♃
21 46 ☽ ∠ ☿
21 47 ☽ ⊥ ♄
21 55 ☽ Q ♂
23 13 ☉ ☌ ♇

03 Friday
00 21 ☽ # ☉
00 41 ☽ ⊻ ♇
00 47 ☽ ⊻ ☉
01 16 ☽ □ ♆
08 26 ☉ ✶ ♆
10 01 ☽ ± ♃
11 23 ☽ ⊥ ☿
12 09 ☽ ∠ ♄
12 11 ☽ ∠ ♅
18 51 ☿ ∥ ♅
19 25 ☽ ⊼ ♃
20 22 ☽ ⊻ ☿

04 Saturday
01 50 ☿ # ♇
02 26 ☽ ⊥ ♅
03 08 ☽ □ ♂
05 45 ☽ Q ♆
09 48 ☽ ⊻ ♀
11 42 ☽ ∥ ☉
11 54 ☽ ✶ ♄
11 55 ☽ ⊻ ♅
15 40 ☿ ☍ ♃
18 04 ♄ □ ♅
19 35 ☽ ⊥ ♀
22 52 ☽ # ♃

05 Sunday
00 08 ☽ ☌ ♇
00 40 ☽ ✶ ♆
03 23 ☽ ☌ ☉
04 52 ☽ # ♀
06 53 ☽ Q ♄
10 18 ☽ ∠ ♀
16 54 ☽ ☌ ☿
17 37 ☽ ∥ ☿
18 19 ☽ ☍ ♃
22 11 ☽ # ♇
23 18 ☽ ∥ ♅

06 Monday
00 27 ☽ ∠ ♆
03 41 ☽ △ ♂
11 08 ☽ ✶ ♀
11 39 ☽ ☌ ♅
11 40 ☽ □ ♄
14 55 ☽ ⊥ ♆

07 Tuesday
00 09 ☽ ⊻ ♇
00 39 ☽ ⊻ ♆
01 51 ♀ ✶ ♅
02 50 ♀ ⊻ ♄
04 31 ☽ bQ ♂
06 52 ☽ ⊻ ☉
07 26 ☽ Q ♀
10 01 ☽ ⊥ ♇
13 10 ☉ Q ♄
14 04 ☽ ⊻ ☿
17 29 ☽ ⊥ ☉
18 31 ☽ ⊼ ♃
19 44 ☽ ± ♂
23 18 ☽ ⊥ ☿

08 Wednesday
01 01 ☽ ∠ ♇
01 42 ☿ ∠ ♀
04 28 ☽ ± ♃
06 03 ☽ ⊼ ♂
09 44 ☽ ∠ ☉
13 21 ☽ ∠ ☿
13 28 ☽ ⊻ ♅
13 31 ☽ △ ♄
14 55 ☽ □ ♀
16 41 ☽ # ♄
19 39 ☽ bQ ♃
23 49 ☽ ⊥ ♅

09 Thursday
02 42 ☽ ✶ ♇
03 10 ☽ ☌ ♆
04 24 ☽ ∥ ♆
11 15 ☉ ☌ ☿
13 20 ☽ ✶ ☿
13 41 ☽ ✶ ☉
15 39 ☽ ∠ ♅
15 42 ☽ bQ ♄
19 36 ☿ # ♀
21 42 ☽ △ ♃
23 39 ♀ ⊥ ♇
23 56 ☽ Q ♇

10 Friday
03 11 ☉ # ♃
07 53 ☽ ± ♄
09 05 ☽ Q ☿
11 05 ☽ ∥ ♆
11 55 ☽ ☍ ♂
12 56 ☽ Q ☉
18 49 ☽ ✶ ♅
18 53 ☽ ⊼ ♄
22 08 ☿ Q ♄
22 54 ☽ △ ♀

11 Saturday
00 15 ☽ # ♄
00 58 ☉ ∥ ☿
08 58 ☽ □ ♇
09 25 ☽ ⊻ ♆
14 34 ☿ # ♃
15 43 ☽ □ ☿
17 15 ☽ Q ♅
20 44 ☽ ⊥ ♆

12 Sunday
01 15 ☽ □ ☉
03 22 ♀ bQ ♃
04 35 ☽ □ ♃
04 40 ☽ bQ ♀
13 58 ☽ ∠ ♆
17 25 ☉ # ♀
21 48 ☽ ⊼ ♂
23 03 ☽ ± ♀

13 Monday
03 58 ☿ ⊥ ♀
03 59 ☽ □ ♅
04 03 ☽ ☍ ♄
09 47 ☽ ± ♂
11 29 ☽ ⊼ ♀
14 58 ☉ ☍ ♃
15 08 ☽ Q ♃
18 52 ☽ △ ♇
19 18 ☽ ✶ ♆
21 38 ☽ △ ☿

14 Tuesday
03 27 ☽ ∥ ♅
03 58 ☽ bQ ♂
06 46 ☽ # ♇
14 40 ☽ ✶ ♃
17 05 ☽ △ ☉
19 12 ☽ Q ♆

15 Wednesday
00 49 ☽ bQ ♇
01 52 ☽ bQ ☿
01 57 ☽ ∥ ☉
09 23 ☽ # ♀
10 40 ☽ △ ♂
14 56 ☽ # ♃
14 57 ☿ ✶ ♆
15 56 ☽ △ ♅
15 59 ☽ ⊼ ♄
18 58 ☽ ± ♇
19 11 ☽ ± ☿
20 26 ☽ ∠ ♃
23 54 ☿ ☌ ♇

16 Thursday
01 58 ☽ bQ ☉
03 19 ☽ ☍ ♀
04 08 ☽ ± ♄
06 24 ☽ ∥ ☿
06 51 ☽ ⊼ ☿
07 08 ☽ ⊼ ♇
07 30 ☽ □ ♆
14 25 ☽ ⊥ ♃
16 50 ♀ ⊥ ♄
21 49 ☽ ± ☉
22 25 ☽ bQ ♅
22 26 ☽ bQ ♄

17 Friday
02 26 ☽ ⊻ ♃
11 05 ☽ ⊼ ☉
14 06 ♄ □ ♅
16 44 ☽ ± ♅
19 21 ☽ # ☿
21 10 ☿ ⊻ ♀

18 Saturday
00 40 ☽ □ ♂
04 56 ☽ △ ♄
04 57 ☽ ⊼ ♅
10 15 ☿ St D
13 11 ☽ ∥ ♃
15 57 ♀ ⊻ ♇
16 31 ☽ ∥ ♀
18 44 ☽ ☍ ☿
20 02 ☽ ☍ ♇
20 19 ☽ ⊼ ♀
20 21 ☽ △ ♆
21 02 ♀ □ ♆
22 10 ☉ ∠ ♆

19 Sunday
02 48 ☉ ∠ ♀
09 17 ☽ ± ♀
13 25 ☽ # ☉
13 44 ☽ Q ♂
14 23 ☽ ☌ ♃
21 50 ☽ ∥ ♇

20 Monday
02 32 ☽ # ♅
02 36 ☽ bQ ♆
04 40 ☽ bQ ♀
05 06 ☽ ☍ ☉
11 51 ♂ Q ♃
14 13 ☽ ✶ ♂
17 29 ☽ □ ♄
17 35 ☽ ☍ ♅
20 37 ☽ ± ♆

21 Tuesday
07 47 ☿ ☌ ♇
08 17 ☽ ⊼ ♇
08 18 ☽ ⊼ ☿
08 34 ☽ ⊼ ♆
12 44 ☽ △ ♀
14 34 ☿ ✶ ♆
20 11 ☽ ± ♇
20 30 ☽ ∠ ♂
20 43 ☽ ± ☿

22 Wednesday
01 18 ♀ ± ♃
01 26 ☽ ⊻ ♃
05 14 ☽ Q ♄
13 04 ☽ ⊥ ♃
13 53 ☽ bQ ♇
14 26 ☽ ⊥ ♂
15 21 ☽ bQ ☿
16 35 ♀ ∥ ♃
20 44 ♀ Q ♂
21 38 ☽ ⊼ ☉

23 Thursday
01 12 ☉ # ♇
02 18 ☽ ⊻ ♂
04 42 ☽ ✶ ♄
04 53 ☽ ⊼ ♅
04 58 ☽ ∥ ♄
06 23 ☽ ∠ ♃
10 16 ☽ ± ☉
16 30 ☽ ± ♅
19 01 ☽ △ ♇
19 15 ☽ ☍ ♆
22 19 ☽ △ ☿

24 Friday
01 06 ☽ # ♆
03 25 ☽ □ ♀
05 01 ☽ bQ ☉
09 33 ☽ ∠ ♄
09 46 ☽ bQ ♅
10 49 ☽ ✶ ♃
22 52 ☽ # ♆

25 Saturday
02 34 ☽ ⊥ ♄
09 08 ☽ Q ♃
11 20 ☉ ∥ ♅
11 37 ☽ △ ☉
12 04 ☽ ☌ ♂
13 46 ☽ ⊻ ♄
14 02 ☽ △ ♅
15 44 ☽ Q ♀
18 35 ☉ △ ♂
18 57 ☽ ∥ ♄

26 Sunday
03 26 ☽ □ ♇
03 38 ☽ ⊼ ♆
11 08 ☽ □ ☿
14 34 ☽ ± ♆
15 17 ☽ ✶ ♀
16 33 ☉ □ ♄
17 48 ☽ □ ♃
19 16 ☿ Q ♄
21 04 ☉ ☌ ♅

27 Monday
06 45 ☽ bQ ♆
11 55 ☽ Q ♇
18 52 ☽ ⊻ ♂
19 56 ☽ ∠ ♀
20 02 ☽ ☌ ♄
20 24 ☽ □ ♅
21 29 ♀ ∠ ♄
22 07 ☽ □ ☉
22 24 ☽ Q ☿
23 53 ♀ ⊼ ♃

28 Tuesday
03 25 ♀ ∠ ♅
04 52 ☿ # ♀
05 27 ☽ ⊥ ♂
05 57 ☉ ⊥ ♆
07 43 ☽ # ☉
08 57 ☽ ✶ ♇
09 08 ☽ △ ♆
12 41 ☽ ⊥ ♀
13 15 ☽ # ♅
18 57 ☽ ∥ ♇
21 05 ☽ ∠ ♂
21 16 ☽ ✶ ☿
21 56 ☽ △ ♃
23 41 ☽ ⊻ ♀

29 Wednesday
03 34 ☽ Q ♅
03 43 ☿ ☍ ♃
07 26 ☽ Q ☉
10 37 ☽ ∠ ♇
22 33 ☽ ✶ ♂
22 59 ☽ bQ ♃
23 21 ☽ ⊻ ♄
23 48 ☽ ✶ ♅

30 Thursday
00 31 ☿ # ♃
00 35 ☽ # ☿
00 35 ☽ ∥ ♃
01 12 ☽ ∠ ☿
01 42 ☽ ⊥ ♇
05 00 ☽ ✶ ☉
07 46 ☽ ∥ ♀
09 18 ☽ ⊥ ♄
11 39 ☽ ⊻ ♇
11 47 ☽ □ ♆
13 45 ☽ ± ♃
17 38 ☽ ⊥ ☿
18 30 ☽ Q ♂
23 29 ☽ ⊼ ♃

31 Friday
00 05 ☽ ∠ ♄
00 36 ☽ ∠ ♅
04 30 ☽ ⊻ ☿
04 57 ☽ ☌ ♀
07 22 ☽ ∠ ☉
15 16 ☽ ⊥ ♅
16 54 ☿ ⊻ ♀
17 11 ☽ Q ♆
18 39 ☽ # ♀
22 59 ☽ ⊥ ☉
23 52 ☽ □ ♂

All ephemeris data is given at 12.00 UT and the Moon's longitude is additionally given for 24.00 UT

NOVEMBER 1975

LONGITUDES

Date	Sidereal time h m s	Sun ☉ ° ' "	Moon ☽ ° ' "	Moon ☽ 24.00 ° ' "	Mercury ☿ ° '	Venus ♀ ° '	Mars ♂ ° '	Jupiter ♃ ° '	Saturn ♄ ° '	Uranus ♅ ° '	Neptune ♆ ° '	Pluto ♇ ° '
01	14 40 36	08 ♏ 26 23	09 ♎ 59 44	17 ♎ 28 25	22 ♎ 25	21 ♍ 59	02 ♋ 29	17 ♈ 13	02 ♌ 49	03 ♏ 11	10 ♐ 20	10 ♎ 15
02	14 44 33	09 26 27	24 ♎ 56 57	02 ♏ 24 17	23 56	22 57	02 33	17 R 06	02 50	03 14	10 22	10 17
03	14 48 29	10 26 33	09 ♏ 49 22	17 ♏ 11 13	25 28	23 56	02 36	17 00	02 51	03 18	10 24	10 19
04	14 52 26	11 26 41	24 ♏ 28 58	01 ♐ 41 50	27 02	24 54	02 38	16 53	02 53	03 22	10 26	10 21
05	14 56 23	12 26 51	08 ♐ 49 14	15 ♐ 50 40	28 ♎ 37	25 54	02 39	16 46	02 54	03 26	10 28	10 23
06	15 00 19	13 27 03	22 ♐ 45 51	29 ♐ 34 37	00 ♏ 12	26 53	02 40	16 40	02 55	03 29	10 30	10 25
07	15 04 16	14 27 16	06 ♑ 16 59	12 ♑ 53 02	01 48	27 54	02 R 39	16 34	02 56	03 33	10 32	10 27
08	15 08 12	15 27 30	19 ♑ 23 01	25 ♑ 47 17	03 25	28 54	02 38	16 28	02 56	03 37	10 34	10 29
09	15 12 09	16 27 46	02 ♒ 06 13	08 ♒ 20 18	05 01	29 ♍ 55	02 36	16 22	02 57	03 40	10 36	10 31
10	15 16 05	17 28 04	14 ♒ 30 03	20 ♒ 36 01	06 38	00 ♎ 57	02 33	16 16	02 58	03 44	10 38	10 33
11	15 20 02	18 28 23	26 ♒ 38 46	02 ♓ 38 53	08 15	01 59	02 29	16 10	02 58	03 48	10 40	10 35
12	15 23 58	19 28 43	08 ♓ 36 57	14 ♓ 33 30	09 52	03 01	02 24	16 04	02 58	03 51	10 42	10 37
13	15 27 55	20 29 05	20 ♓ 29 07	26 ♓ 24 19	11 29	04 03	02 19	15 59	02 58	03 55	10 45	10 39
14	15 31 52	21 29 28	02 ♈ 19 35	08 ♈ 15 24	13 06	05 06	02 12	15 54	02 59	03 59	10 47	10 41
15	15 35 48	22 29 52	14 ♈ 12 11	20 ♈ 10 21	14 43	06 10	02 05	15 49	02 R 59	04 02	10 49	10 43
16	15 39 45	23 30 18	26 ♈ 10 14	02 ♉ 12 10	16 20	07 13	01 57	15 44	02 58	04 06	10 51	10 44
17	15 43 41	24 30 46	08 ♉ 16 25	14 ♉ 23 13	17 56	08 17	01 47	15 39	02 58	04 10	10 53	10 46
18	15 47 38	25 31 15	20 ♉ 32 47	26 ♉ 45 17	19 32	09 21	01 37	15 35	02 58	04 13	10 55	10 48
19	15 51 34	26 31 46	03 ♊ 00 50	09 ♊ 19 32	21 08	10 26	01 27	15 30	02 57	04 17	10 58	10 50
20	15 55 31	27 32 18	15 ♊ 41 29	22 ♊ 06 43	22 44	11 31	01 15	15 26	02 57	04 20	11 00	10 52
21	15 59 27	28 32 52	28 ♊ 35 17	05 ♋ 07 12	24 20	12 36	01 02	15 22	02 56	04 24	11 02	10 53
22	16 03 24	29 ♏ 33 27	11 ♋ 42 29	18 ♋ 21 07	25 55	13 42	00 49	15 19	02 55	04 27	11 04	10 55
23	16 07 21	00 ♐ 34 04	25 ♋ 03 07	01 ♌ 48 28	27 31	14 47	00 35	15 15	02 54	04 31	11 06	10 57
24	16 11 17	01 34 43	08 ♌ 37 07	15 ♌ 29 03	29 ♏ 06	15 53	00 20	15 12	02 53	04 34	11 09	10 58
25	16 15 14	02 35 23	22 ♌ 24 12	29 ♌ 22 29	00 ♐ 41	17 00	00 ♋ 04	15 08	02 52	04 38	11 11	11 00
26	16 19 10	03 36 05	06 ♍ 23 46	13 ♍ 27 54	02 15	18 06	29 ♊ 48	15 05	02 51	04 41	11 13	11 02
27	16 23 07	04 36 49	20 ♍ 34 41	27 ♍ 43 51	03 50	19 13	29 31	15 03	02 50	04 45	11 15	11 03
28	16 27 03	05 37 34	04 ♎ 55 04	12 ♎ 07 56	05 24	20 20	29 13	15 00	02 48	04 48	11 18	11 05
29	16 31 00	06 38 21	19 ♎ 22 00	26 ♎ 36 44	06 59	21 27	28 54	14 58	02 47	04 51	11 20	11 06
30	16 34 56	07 ♐ 39 09	03 ♏ 51 33	11 ♏ 05 49	08 ♐ 33	22 ♎ 35	28 ♊ 35	14 ♈ 55	02 ♌ 45	04 ♏ 55	11 ♐ 22	11 ♎ 08

Date	Moon True ☊ ° '	Moon Mean ☊ ° '	Moon ☽ Latitude ° '
01	21 ♏ 48	22 ♏ 28	03 S 31
02	21 R 46	22 25	02 23
03	21 46	22 22	01 S 06
04	21 D 45	22 19	00 N 15
05	21 46	22 15	01 33
06	21 46	22 12	02 44
07	21 47	22 09	03 43
08	21 48	22 06	04 28
09	21 48	22 03	04 59
10	21 49	21 59	05 15
11	21 R 49	21 56	05 16
12	21 48	21 53	05 04
13	21 48	21 50	04 39
14	21 48	21 47	04 02
15	21 D 48	21 44	03 14
16	21 48	21 40	02 18
17	21 48	21 37	01 14
18	21 49	21 34	00 N 07
19	21 R 48	21 31	01 S 02
20	21 48	21 28	02 09
21	21 47	21 25	03 10
22	21 46	21 21	04 03
23	21 45	21 18	04 43
24	21 44	21 15	05 09
25	21 44	21 12	05 17
26	21 D 43	21 09	05 07
27	21 44	21 05	04 38
28	21 45	21 02	03 51
29	21 46	20 59	02 50
30	21 ♏ 47	20 ♏ 56	01 S 38

DECLINATIONS

Date	Sun ☉ ° '	Moon ☽ ° '	Mercury ☿ ° '	Venus ♀ ° '	Mars ♂ ° '	Jupiter ♃ ° '	Saturn ♄ ° '	Uranus ♅ ° '	Neptune ♆ ° '	Pluto ♇ ° '
01	14 S 19	07 S 12	06 S 49	03 N 20	24 N 10	05 N 18	19 N 40	12 S 07	20 S 31	11 N 02
02	14 38	11 53	07 26	03 03	24 13	05 16	19 40	12 08	20 31	11 01
03	14 57	15 48	08 03	02 45	24 15	05 13	19 39	12 10	20 32	11 00
04	15 16	18 39	08 40	02 27	24 18	05 11	19 39	12 11	20 32	11 00
05	15 34	20 14	09 18	02 09	24 21	05 09	19 39	12 12	20 32	10 59
06	15 53	20 31	09 56	01 51	24 24	05 06	19 39	12 14	20 33	10 59
07	16 11	19 35	10 34	01 32	24 26	05 04	19 39	12 15	20 33	10 58
08	16 28	17 37	11 13	01 12	24 29	05 02	19 39	12 16	20 33	10 58
09	16 46	14 50	11 50	00 53	24 32	05 00	19 39	12 17	20 34	10 58
10	17 03	11 27	12 28	00 33	24 35	04 58	19 39	12 19	20 34	10 57
11	17 20	07 40	13 05	00 N 13	24 38	04 56	19 39	12 20	20 34	10 57
12	17 36	03 S 38	13 42	00 S 08	24 41	04 54	19 39	12 21	20 35	10 56
13	17 52	00 N 30	14 19	00 28	24 44	04 52	19 39	12 22	20 35	10 56
14	18 08	04 37	14 54	00 49	24 47	04 50	19 39	12 24	20 35	10 55
15	18 24	08 35	15 29	01 11	24 50	04 49	19 39	12 25	20 36	10 55
16	18 39	12 15	16 04	01 32	24 54	04 47	19 39	12 26	20 36	10 55
17	18 54	15 26	16 38	01 54	24 57	04 45	19 39	12 27	20 36	10 54
18	19 09	18 00	17 11	02 15	25 00	04 44	19 40	12 28	20 36	10 54
19	19 23	19 45	17 43	02 37	25 03	04 42	19 40	12 30	20 37	10 54
20	19 37	20 32	18 14	03 00	25 06	04 41	19 40	12 31	20 37	10 53
21	19 50	20 16	18 45	03 22	25 09	04 40	19 40	12 32	20 37	10 53
22	20 03	18 54	19 14	03 44	25 13	04 39	19 41	12 33	20 38	10 53
23	20 16	16 29	19 43	04 07	25 16	04 38	19 41	12 34	20 38	10 53
24	20 29	13 08	20 10	04 30	25 19	04 36	19 41	12 36	20 38	10 52
25	20 41	09 03	20 37	04 52	25 22	04 36	19 42	12 37	20 39	10 52
26	20 52	04 N 25	21 03	05 15	25 25	04 35	19 42	12 38	20 39	10 52
27	21 04	00 S 31	21 27	05 38	25 28	04 34	19 43	12 39	20 39	10 52
28	21 15	05 30	21 51	06 01	25 31	04 33	19 43	12 40	20 40	10 52
29	21 25	10 12	22 13	06 24	25 33	04 32	19 44	12 41	20 40	10 52
30	21 S 35	14 S 20	22 S 35	06 S 47	25 N 36	04 N 32	19 N 44	12 S 42	20 S 40	10 N 51

ZODIAC SIGN ENTRIES

Date	h	m	Planets
02	20	07	☽ ♏
04	21	10	☽ ♐
06	08	58	☿ ♏
07	00	45	☽ ♑
09	07	59	☽ ♒
09	13	52	♀ ♎
11	18	42	☽ ♓
14	07	17	☽ ♈
16	19	38	☽ ♉
19	06	14	☽ ♊
21	14	36	☽ ♋
22	22	31	☉ ♐
23	20	48	☽ ♌
25	01	44	☿ ♐
25	18	30	♂ ♊
26	01	04	☽ ♍
28	03	48	☽ ♎
30	05	37	☽ ♏

LATITUDES

Date	Mercury ☿ ° '	Venus ♀ ° '	Mars ♂ ° '	Jupiter ♃ ° '	Saturn ♄ ° '	Uranus ♅ ° '	Neptune ♆ ° '	Pluto ♇ ° '
01	02 N 03	00 N 10	00 N 45	01 S 35	00 N 08	00 N 29	01 N 30	16 N 24
04	01 52	00 28	00 53	01 34	00 08	00 29	01 30	16 25
07	01 38	00 45	01 02	01 34	00 09	00 29	01 30	16 26
10	01 20	01 00	01 10	01 33	00 09	00 29	01 30	16 27
13	01 01	01 14	01 19	01 32	00 10	00 29	01 29	16 28
16	00 41	01 27	01 28	01 32	00 10	00 29	01 29	16 29
19	00 N 21	01 38	01 37	01 31	00 10	00 29	01 29	16 30
22	00 00	01 48	01 46	01 30	00 11	00 29	01 29	16 32
25	00 S 20	01 57	01 55	01 29	00 11	00 29	01 29	16 33
28	00 39	02 05	02 04	01 28	00 11	00 29	01 29	16 34
31	00 S 58	02 N 11	02 N 13	01 S 27	00 N 12	00 N 29	01 N 29	16 N 36

DATA

Julian Date	2442718
Delta T	+46 seconds
Ayanamsa	23° 31' 22"
Synetic vernal point	05° ♓ 35' 37"
True obliquity of ecliptic	23° 26' 27"

MOON'S PHASES, APSIDES AND POSITIONS ☽

Date	h	m	Phase	Longitude ° '	Eclipse Indicator
03	13	05	●	10 ♏ 29	Partial
10	18	21	☽	17 ♒ 44	
18	22	28	○	25 ♉ 58	total
26	06	52	☾	03 ♍ 23	

Day	h	m		
02	00	28	Perigee	
13	23	55	Apogee	
30	00	44	Perigee	
06	05	14	Max dec	20° S 34'
13	09	06	0N	
20	17	56	Max dec	20° N 34'
27	09	29	0S	

LONGITUDES

Date	Chiron ⚷ ° '	Ceres ⚳ ° '	Pallas ⚴ ° '	Juno ⚵ ° '	Vesta ⚶ ° '	Black Moon Lilith ⚸ ° '
01	25 ♈ 16	14 ♊ 14	18 ♓ 25	03 ♍ 58	17 ♓ 53	00 ♈ 07
11	24 ♈ 50	12 ♊ 42	17 ♓ 41	07 ♍ 34	18 ♓ 11	01 ♈ 13
21	24 ♈ 27	10 ♊ 40	17 ♓ 37	10 ♍ 48	19 ♓ 10	02 ♈ 20
31	24 ♈ 07	08 ♊ 22	18 ♓ 11	13 ♍ 37	20 ♓ 47	03 ♈ 27

ASPECTARIAN

01 Saturday
00 26 ☽ ✶ ♄
00 59 ☽ ⚺ ♅
03 02 ☽ # ♃
09 19 ☽ ⚺ ☉
09 58 ☽ ∥ ☿
12 24 ☽ ☌ ♇
12 32 ☽ ✶ ♆
19 44 ☽ Q ♄
23 31 ☽ ☍ ♃

02 Sunday
07 20 ☽ # ♇
08 34 ☽ ⚺ ♀
10 10 ☽ ☌ ☿
12 40 ☽ ∠ ♆
13 27 ☽ ∥ ♅
18 53 ☽ ⊥ ♀

03 Monday
00 17 ☽ △ ♂
00 43 ☽ □ ♄
01 24 ☽ ☌ ♅
03 11 ☽ ⊥ ♆
05 46 ☽ ∥ ☉
08 48 ☉ ⚺ ♇
10 27 ☽ ∠ ♀
10 48 ☉ ⚺ ♆
10 48 ☿ ∠ ♆
12 48 ☽ ⚺ ♇
12 56 ☽ ⚺ ♆
13 05 ☽ ☌ ☉
22 36 ☽ ⊥ ♇
23 36 ☽ ⚻ ♃

04 Tuesday
00 43 ☽ bQ ♂
09 23 ☽ ± ♃
12 45 ☽ ✶ ♀
13 26 ☽ ∠ ♇
15 34 ☽ ± ♂
16 44 ☽ ⚺ ☿

05 Wednesday
00 13 ☽ bQ ♃
00 43 ☽ # ♄
01 36 ☽ ⚻ ♂
02 00 ☽ △ ♄
02 51 ☽ ⚺ ♅
04 00 ☽ ⊥ ☿
10 19 ☽ Q ♀
13 02 ☽ ⊥ ♅
14 40 ☽ ✶ ♇
14 48 ☽ ☌ ♆
18 39 ☽ ⚺ ☉
21 13 ☽ ∠ ☿

06 Thursday
00 17 ☽ ∥ ♆
01 30 ☽ △ ♃
03 33 ☽ bQ ♄
04 32 ☽ ∠ ♅
05 47 ☽ ⊥ ☉
09 58 ☽ ∥ ♆
11 23 ☽ Q ♇
12 01 ♂ St R
19 17 ☽ ± ♄
19 49 ☽ □ ♀
22 48 ☽ ∠ ☉

07 Friday
02 53 ☽ ✶ ☿
03 16 ♀ ⊥ ♅
05 30 ☽ ☍ ♂
05 58 ☽ ⚻ ♄
07 04 ☽ ✶ ♅
10 54 ☽ # ♄
19 34 ☽ □ ♇
19 43 ☽ ⚺ ♆

08 Saturday
00 37 ☿ △ ♂
02 57 ☿ # ♇
03 36 ☽ Q ☿
03 42 ♀ Q ♆
04 07 ☽ ✶ ☉
04 54 ☿ □ ♄
04 59 ☽ Q ♅
06 37 ☽ □ ♃
06 46 ☽ ⊥ ♆
15 06 ☿ ☌ ♅
21 35 ☽ ∥ ☉
23 37 ☽ ∠ ♆

09 Sunday
04 28 ☽ Q ☉
05 33 ☿ ⊥ ♆
07 28 ☽ △ ♀
09 44 ☉ ⚻ ♃
12 57 ☽ ⚻ ♂
13 20 ☉ ⊥ ♇
13 37 ☽ ☍ ♄
15 01 ☽ □ ♅
16 17 ☽ Q ♃
18 26 ☽ □ ☿

10 Monday
00 28 ☽ ± ♂
04 16 ☽ △ ♇
04 26 ☽ ✶ ♆
05 45 ☿ ∥ ♅
06 11 ☽ ∥ ☿
06 16 ☽ ∥ ♅
13 51 ☉ bQ ♂
15 05 ☽ bQ ♀
15 20 ☽ # ♇
15 25 ☽ ✶ ♃
17 57 ☽ bQ ♂
18 21 ☽ □ ☉

11 Tuesday
00 27 ☿ ⊥ ♀
04 04 ☽ Q ♆
09 53 ☽ bQ ♇
10 32 ☽ ± ♀
20 58 ☽ ∠ ♃
23 01 ♀ □ ♂
23 36 ☽ △ ♂
23 40 ☽ ⚻ ♀

12 Wednesday
00 39 ☽ ⚻ ♄
02 23 ☽ △ ♅
03 56 ☽ ± ♇
04 32 ☽ # ♃
11 03 ♀ ✶ ♄
12 43 ☽ ± ♄
14 55 ☽ ⊥ ♃
14 56 ☽ △ ☿
16 03 ☽ ⚻ ♇
16 14 ☽ □ ♆
23 13 ☿ ⚺ ♇

13 Thursday
00 38 ☿ ⚺ ♆
02 57 ☽ ⚺ ♃
06 47 ☽ ∥ ♀
06 55 ☽ bQ ♄
08 41 ♀ ⚺ ♅
08 49 ☽ bQ ♅
11 49 ☽ # ♀
12 00 ☽ △ ☉

14 Friday
02 06 ☽ bQ ☿
03 09 ☽ ± ♅
11 45 ☽ □ ♂
13 17 ☽ ∥ ♃
13 19 ☽ △ ♄
15 22 ☽ ⚻ ♅
18 10 ☽ ☍ ♀
19 25 ♄ St R
20 56 ☉ ± ♃
21 13 ☽ bQ ☉
23 12 ☽ ± ☿

15 Saturday
04 56 ☽ ☍ ♇
05 09 ☽ △ ♆
13 12 ☽ ⚻ ☿
15 13 ☽ ☌ ♃
17 03 ☽ ± ☉
23 41 ☽ Q ♂

16 Sunday
03 00 ☽ ∥ ♇
03 32 ☿ ⚻ ♃
06 11 ☽ ⚻ ☉
11 21 ☽ bQ ♆
13 21 ☽ # ♅
18 16 ☿ ⊥ ♇
20 24 ☿ bQ ♂
23 21 ☽ ✶ ♂

17 Monday
01 31 ☽ □ ♄
03 50 ☽ ☍ ♅
05 18 ☽ ± ♆
12 02 ☽ ⚻ ♀
16 55 ☽ ⚻ ♇
17 09 ☽ ⚻ ♆

18 Tuesday
00 56 ☽ ± ♀
01 06 ☽ # ☿
02 23 ☽ ⚺ ♃
04 28 ☽ ∠ ♂
04 42 ☽ ± ♇
09 45 ☽ ☍ ☿
12 48 ☽ Q ♄
13 59 ☽ ⊥ ♃
14 06 ☉ ± ♂
18 51 ☉ ∠ ♇
20 05 ☽ bQ ♀
21 41 ☽ ⊥ ♂
22 11 ☽ bQ ♇
22 28 ☽ ☍ ☉

19 Wednesday
04 36 ☽ # ☉
07 14 ☽ ∠ ♃
09 03 ☽ ⚺ ♂
10 27 ☽ ∥ ♄
11 53 ☽ ✶ ♄
14 26 ☽ ⚻ ♅
17 17 ☿ ± ♃
21 02 ♀ ☌ ♇

20 Thursday
00 04 ♀ ✶ ♆
01 52 ☽ ± ♅
02 53 ☽ △ ♇
03 08 ☽ ☍ ♆
03 25 ☽ △ ♀
11 32 ☽ ✶ ♃
16 13 ☽ ∠ ♄
18 11 ☉ # ♄
18 52 ☽ bQ ♅

21 Friday
03 01 ☽ ⚻ ☿
08 57 ☽ ⊥ ♄
09 46 ☽ Q ♃
11 55 ☽ ⚻ ☉
15 39 ☽ ± ☿
16 27 ☽ ☌ ♂
19 59 ☽ ⚺ ♄
20 15 ☽ # ☉
21 25 ☿ ± ♂
22 44 ☽ △ ♅
23 52 ☽ ± ☉

22 Saturday
00 37 ☽ ∥ ♄
08 27 ☽ # ☿
10 23 ☽ bQ ☿
10 34 ☽ □ ♇
10 50 ☽ ⚻ ♆
11 56 ☿ ∠ ♇
15 55 ☽ □ ♀
17 35 ☽ bQ ☉
18 29 ☽ □ ♃
21 43 ☽ ± ♆

23 Sunday
04 50 ☉ bQ ♃
10 36 ☿ # ♄
12 18 ☉ ⚻ ♂
13 53 ☽ bQ ♆
14 28 ♀ Q ♄
16 58 ☽ △ ☿
18 56 ☽ Q ♇
21 30 ♀ ☍ ♃
21 39 ☽ ⚺ ♂
22 36 ☽ △ ☉

24 Monday
01 55 ☽ ☌ ♄
02 57 ☽ Q ♀
04 51 ☽ □ ♅
08 04 ☽ ⊥ ♂
15 25 ☽ # ♅
16 08 ☽ ✶ ♇
16 26 ☽ △ ♆
18 56 ♀ # ♃
23 27 ☽ △ ♃
23 31 ☽ ∠ ♂

25 Tuesday
01 45 ☽ ∥ ♇
01 49 ☽ ✶ ♀
04 06 ☿ bQ ♃
04 10 ☿ ⚻ ♂
07 47 ☉ ∥ ♆
12 24 ☽ Q ♅
13 34 ☿ ∥ ♆
18 13 ☽ ∠ ♇
18 16 ☉ ∥ ☿
18 30 ☉ △ ♄

26 Wednesday
00 57 ☽ ✶ ♂
01 16 ☽ bQ ♃
04 02 ☽ □ ☿
05 54 ☽ ∠ ♀
05 57 ☽ ⚺ ♄
06 52 ☽ □ ☉
08 06 ☽ # ♀
09 05 ☽ ✶ ♅
09 40 ☽ ⊥ ♇
11 11 ☽ ∥ ♃
16 10 ☽ ⊥ ♄
16 34 ☽ ± ♃
19 53 ☽ ⚺ ♇
20 13 ☽ □ ♆
20 54 ☿ △ ♄
21 00 ☽ Q ♂
22 32 ☽ ⊥ ♀

27 Thursday
02 42 ☽ ⚻ ♃
07 22 ☽ ∠ ♄
09 31 ☽ ⚺ ♀
10 36 ☽ ∠ ♅
14 22 ☽ Q ☿
15 17 ☉ ⚺ ♅
15 41 ☽ Q ☉

28 Friday
01 45 ☽ ⊥ ♅
02 24 ☿ ⚺ ♅
02 35 ☽ Q ♆
02 41 ☽ □ ♂
07 25 ☽ # ♃
08 16 ☿ ∠ ♀
08 29 ☽ ✶ ♄
11 48 ☽ ⚺ ♅
12 55 ☽ ✶ ☿
13 16 ☽ ✶ ☉
14 50 ☽ ∥ ♀
21 21 ☉ ☌ ☿
22 16 ☽ ☌ ♇
22 38 ☽ ✶ ♆

29 Saturday
04 24 ☽ Q ♄
04 43 ☽ ☍ ♃
15 34 ☽ # ♇
15 45 ☽ ☌ ♀
16 03 ☽ ∠ ☉
16 51 ☽ ∠ ☿
23 34 ☽ ∠ ♆

30 Sunday
01 58 ☽ ∥ ♅
03 27 ☽ △ ♂
08 04 ☽ ⊥ ☉
09 34 ☽ ⊥ ☿
10 10 ☽ □ ♄
13 45 ☽ ☌ ♅
14 31 ☽ ⊥ ♆
18 46 ☽ ⚺ ☉
20 44 ☽ ⚺ ☿

All ephemeris data is given at 12.00 UT and the Moon's longitude is additionally given for 24.00 UT

DECEMBER 1975

LONGITUDES

Date	Sidereal time (h m s)	Sun ☉ (° ' ")	Moon ☽ (° ' ")	Moon ☽ 24.00 (° ' ")	Mercury ☿ (° ')	Venus ♀ (° ')	Mars ♂ (° ')	Jupiter ♃ (° ')	Saturn ♄ (° ')	Uranus ♅ (° ')	Neptune ♆ (° ')	Pluto ⯓ (° ')
01	16 38 53	08 ♐ 39 59	18 ♏ 18 51	25 ♏ 30 00	10 ♐ 07	23 ♎ 43	28 ♊ 15	14 ♈ 53	02 ♌ 43	04 ♏ 58	11 ♐ 24	11 ♎ 09
02	16 42 50	09 40 50	02 ♐ 38 34	09 ♐ 43 56	11 42	24 51	27 R 55	14 R 52	02 R 41	05 01	11 27	11 11
03	16 46 46	10 41 42	16 ♐ 45 30	23 ♐ 42 44	13 16	25 59	27 34	14 50	02 39	05 05	11 29	11 12
04	16 50 43	11 42 35	00 ♑ 35 11	07 ♑ 22 33	14 50	27 07	27 13	14 49	02 37	05 08	11 31	11 14
05	16 54 39	12 43 30	14 ♑ 04 34	20 ♑ 41 09	16 24	28 16	26 51	14 47	02 35	05 11	11 33	11 15
06	16 58 36	13 44 25	27 ♑ 12 16	03 ♒ 38 02	17 58	29 ♎ 24	26 29	14 47	02 33	05 14	11 36	11 16
07	17 02 32	14 45 21	09 ♒ 58 39	16 ♒ 14 26	19 32	00 ♏ 33	26 06	14 46	02 30	05 17	11 38	11 18
08	17 06 29	15 46 18	22 ♒ 25 44	28 ♒ 33 01	21 06	01 42	25 44	14 45	02 28	05 20	11 40	11 19
09	17 10 25	16 47 15	04 ♓ 36 47	10 ♓ 37 36	22 40	02 51	25 21	14 45	02 25	05 24	11 43	11 20
10	17 14 22	17 48 13	16 ♓ 36 01	22 ♓ 32 41	24 14	04 01	24 57	14 45	02 22	05 27	11 45	11 21
11	17 18 19	18 49 12	28 ♓ 28 12	04 ♈ 23 13	25 48	05 10	24 34	14 D 45	02 19	05 30	11 47	11 22
12	17 22 15	19 50 11	10 ♈ 18 20	16 ♈ 14 10	27 22	06 20	24 10	14 45	02 17	05 33	11 49	11 24
13	17 26 12	20 51 11	22 ♈ 11 20	28 ♈ 10 23	28 ♐ 57	07 30	23 47	14 46	02 14	05 35	11 52	11 25
14	17 30 08	21 52 11	04 ♉ 11 50	10 ♉ 16 11	00 ♑ 31	08 39	23 24	14 46	02 10	05 38	11 54	11 26
15	17 34 05	22 53 12	16 ♉ 23 51	22 ♉ 35 14	02 05	09 50	23 00	14 47	02 07	05 41	11 56	11 27
16	17 38 01	23 54 14	28 ♉ 50 36	05 ♊ 10 12	03 39	11 00	22 37	14 49	02 04	05 44	11 58	11 28
17	17 41 58	24 55 16	11 ♊ 34 11	18 ♊ 02 36	05 14	12 10	22 14	14 50	02 01	05 47	12 00	11 29
18	17 45 54	25 56 19	24 ♊ 35 28	01 ♋ 12 40	06 48	13 21	21 51	14 51	01 57	05 50	12 03	11 30
19	17 49 51	26 57 23	07 ♋ 54 01	14 ♋ 39 16	08 22	14 31	21 28	14 53	01 54	05 52	12 05	11 31
20	17 53 48	27 58 27	21 ♋ 28 08	28 ♋ 20 13	09 57	15 42	21 06	14 55	01 50	05 55	12 07	11 32
21	17 57 44	28 ♐ 59 31	05 ♌ 15 09	12 ♌ 12 31	11 31	16 53	20 44	14 57	01 46	05 58	12 09	11 32
22	18 01 41	00 ♑ 00 37	19 ♌ 11 53	26 ♌ 12 52	13 05	18 04	20 22	15 00	01 43	06 00	12 11	11 33
23	18 05 37	01 01 43	03 ♍ 15 06	10 ♍ 18 12	14 38	19 15	20 01	15 02	01 39	06 03	12 14	11 34
24	18 09 34	02 02 49	17 ♍ 21 54	24 ♍ 25 54	16 12	20 26	19 40	15 05	01 35	06 05	12 16	11 35
25	18 13 30	03 03 57	01 ♎ 30 00	08 ♎ 34 00	17 45	21 37	19 20	15 08	01 31	06 08	12 18	11 36
26	18 17 27	04 05 05	15 ♎ 37 43	22 ♎ 41 00	19 17	22 49	19 00	15 11	01 27	06 10	12 20	11 36
27	18 21 23	05 06 13	29 ♎ 43 42	06 ♏ 45 40	20 49	24 00	18 41	15 15	01 23	06 13	12 22	11 37
28	18 25 20	06 07 23	13 ♏ 46 41	20 ♏ 46 34	22 20	25 12	18 22	15 18	01 19	06 15	12 24	11 38
29	18 29 17	07 08 32	27 ♏ 45 03	04 ♐ 41 53	23 49	26 24	18 04	15 22	01 14	06 17	12 26	11 38
30	18 33 13	08 09 43	11 ♐ 36 44	18 ♐ 29 18	25 18	27 35	17 47	15 26	01 10	06 20	12 29	11 39
31	18 37 10	09 ♑ 10 53	25 ♐ 19 12	02 ♑ 06 06	26 ♑ 44	28 ♏ 47	17 ♊ 30	15 ♈ 30	01 ♌ 06	06 ♏ 22	12 ♐ 31	11 ♎ 39

Date	Moon True ☊ (° ')	Moon Mean ☊ (° ')	Moon ☽ Latitude (° ')
01	21 ♏ 48	20 ♏ 53	00 S 19
02	21 R 47	20 50	01 N 00
03	21 46	20 46	02 14
04	21 44	20 43	03 18
05	21 41	20 40	04 10
06	21 37	20 37	04 47
07	21 34	20 34	05 08
08	21 31	20 30	05 15
09	21 29	20 27	05 06
10	21 28	20 24	04 45
11	21 D 29	20 21	04 12
12	21 30	20 18	03 27
13	21 32	20 15	02 34
14	21 34	20 11	01 34
15	21 35	20 08	00 N 28
16	21 R 35	20 05	00 S 40
17	21 33	20 02	01 47
18	21 30	19 59	02 50
19	21 25	19 56	03 46
20	21 19	19 52	04 30
21	21 12	19 49	04 58
22	21 06	19 46	05 10
23	21 02	19 43	05 03
24	20 59	19 40	04 38
25	20 58	19 36	03 56
26	20 D 58	19 33	03 00
27	21 00	19 30	01 53
28	21 01	19 27	00 S 39
29	21 R 01	19 24	00 N 36
30	20 59	19 21	01 49
31	20 ♏ 54	19 ♏ 17	02 N 54

DECLINATIONS

Date	Sun ☉	Moon ☽	Mercury ☿	Venus ♀	Mars ♂	Jupiter ♃	Saturn ♄	Uranus ♅	Neptune ♆	Pluto ⯓
01	21 S 45	17 S 35	22 S 55	07 S 11	25 N 39	04 N 31	19 N 45	12 S 43	20 S 41	10 N 51
02	21 54	19 43	23 14	07 34	25 41	04 31	19 45	12 45	20 41	10 51
03	22 03	20 34	23 32	07 57	25 44	04 31	19 46	12 46	20 41	10 51
04	22 11	20 08	23 49	08 20	25 46	04 30	19 46	12 47	20 42	10 51
05	22 19	18 33	24 04	08 43	25 48	04 30	19 47	12 48	20 42	10 51
06	22 27	16 02	24 19	09 06	25 50	04 30	19 48	12 49	20 42	10 51
07	22 34	12 48	24 32	09 29	25 52	04 30	19 48	12 50	20 42	10 51
08	22 41	09 04	24 43	09 52	25 54	04 30	19 49	12 51	20 43	10 51
09	22 47	05 04	24 54	10 14	25 56	04 30	19 50	12 52	20 43	10 51
10	22 53	00 S 55	25 03	10 37	25 57	04 31	19 50	12 53	20 43	10 51
11	22 58	03 N 14	25 11	11 00	25 59	04 31	19 51	12 54	20 44	10 51
12	23 03	07 16	25 17	11 22	26 00	04 31	19 52	12 55	20 44	10 51
13	23 07	11 02	25 22	11 44	26 01	04 32	19 53	12 56	20 44	10 51
14	23 11	14 24	25 25	12 06	26 02	04 32	19 54	12 57	20 45	10 51
15	23 15	17 12	25 27	12 28	26 02	04 33	19 54	12 58	20 45	10 51
16	23 18	19 15	25 28	12 50	26 03	04 34	19 55	12 59	20 45	10 51
17	23 21	20 24	25 27	13 11	26 04	04 35	19 56	13 00	20 45	10 51
18	23 23	20 30	25 25	13 33	26 04	04 35	19 57	13 00	20 46	10 52
19	23 24	19 27	25 21	13 54	26 04	04 36	19 58	13 01	20 46	10 52
20	23 26	17 17	25 16	14 15	26 04	04 37	19 59	13 02	20 46	10 52
21	23 26	14 08	25 09	14 35	26 04	04 39	20 00	13 03	20 47	10 52
22	23 26	10 09	25 00	14 55	26 04	04 40	20 01	13 04	20 47	10 52
23	23 26	05 36	24 50	15 15	26 04	04 41	20 02	13 05	20 47	10 53
24	23 25	00 N 43	24 39	15 35	26 03	04 42	20 03	13 06	20 47	10 53
25	23 24	04 S 12	24 26	15 54	26 03	04 44	20 04	13 06	20 48	10 53
26	23 23	08 55	24 12	16 13	26 02	04 45	20 05	13 07	20 48	10 53
27	23 21	13 08	23 56	16 32	26 01	04 47	20 06	13 08	20 48	10 54
28	23 18	16 36	23 38	16 50	26 01	04 48	20 07	13 09	20 48	10 54
29	23 15	19 04	23 20	17 08	26 00	04 50	20 08	13 09	20 49	10 54
30	23 11	20 23	22 59	17 26	25 59	04 52	20 09	13 10	20 49	10 55
31	23 S 07	20 S 27	22 S 38	17 S 43	25 N 58	04 N 54	20 N 10	13 S 11	20 S 49	10 N 55

ZODIAC SIGN ENTRIES

Date	h	m	Planets
02	07	33	☽ ♐
04	10	58	☽ ♑
06	17	12	☽ ♒
07	00	29	♀ ♏
09	02	52	☽ ♓
11	15	06	☽ ♈
14	03	39	☽ ♉
14	04	10	☿ ♑
16	14	12	☽ ♊
18	21	49	☽ ♋
21	02	54	☽ ♌
22	11	46	☉ ♑
23	06	28	☽ ♍
25	09	27	☽ ♎
27	12	28	☽ ♏
29	15	53	☽ ♐
31	20	16	☽ ♑

LATITUDES

Date	Mercury ☿	Venus ♀	Mars ♂	Jupiter ♃	Saturn ♄	Uranus ♅	Neptune ♆	Pluto ⯓
01	00 S 58	02 N 11	02 N 13	01 S 27	00 N 12	00 N 29	01 N 29	16 N 36
04	01 15	02 16	02 21	01 27	00 12	00 29	01 29	16 37
07	01 30	02 20	02 29	01 26	00 12	00 29	01 29	16 39
10	01 44	02 23	02 37	01 25	00 13	00 29	01 29	16 40
13	01 56	02 24	02 43	01 24	00 13	00 29	01 29	16 42
16	02 05	02 25	02 49	01 23	00 14	00 29	01 29	16 44
19	02 11	02 24	02 54	01 22	00 14	00 29	01 29	16 45
22	02 13	02 23	02 59	01 21	00 14	00 29	01 29	16 47
25	02 11	02 21	03 03	01 20	00 15	00 29	01 29	16 49
28	02 05	02 17	03 05	01 19	00 15	00 29	01 29	16 50
31	01 S 52	02 N 13	03 N 07	01 S 18	00 N 15	00 N 29	01 N 29	16 N 52

DATA

Julian Date	2442748
Delta T	+46 seconds
Ayanamsa	23° 31' 27"
Synetic vernal point	05° ♓ 35' 33"
True obliquity of ecliptic	23° 26' 26"

MOON'S PHASES, APSIDES AND POSITIONS ☽

Date	h	m	Phase	Longitude (° ')	Eclipse Indicator
03	00	50	●	10 ♐ 13	
10	14	39	☽	17 ♓ 55	
18	14	39	○	26 ♊ 03	
25	14	52	☾	03 ♎ 11	

Day	h	m		
11	19	20	Apogee	
26	03	39	Perigee	
03	15	46	Max dec	20° S 35'
10	17	15	0N	
18	01	57	Max dec	20° N 35'
24	15	30	0S	
31	01	28	Max dec	20° S 34'

LONGITUDES

Date	Chiron ⚷	Ceres ⚳	Pallas ⚴	Juno ⚵	Vesta ⚶	Black Moon Lilith ⚸
01	24 ♈ 07	08 ♊ 22	18 ♓ 11	13 ♍ 37	20 ♓ 47	03 ♈ 27
11	23 ♈ 51	06 ♊ 03	19 ♓ 20	15 ♍ 57	22 ♓ 56	04 ♈ 34
21	23 ♈ 41	04 ♊ 01	21 ♓ 00	17 ♍ 43	25 ♓ 31	05 ♈ 41
31	23 ♈ 36	02 ♊ 27	23 ♓ 08	18 ♍ 52	28 ♓ 29	06 ♈ 48

ASPECTARIAN

01 Monday
00 05 ☽ ⊻ ⯓
00 29 ☽ ⊻ ♆
02 33 ☉ ∠ ♀
03 47 ☽ ♇ ♂
06 19 ☽ ⊼ ♃
10 04 ☽ ⊥ ⯓
16 17 ☽ ± ♃
18 26 ☽ ± ♂
21 47 ☽ ⊻ ♀

02 Tuesday
01 07 ☽ ∠ ⯓
01 23 ☿ ⊥ ♅
04 01 ☿ ⚹ ⯓
04 14 ☽ ⊼ ♂
07 19 ☽ ♇ ♃
08 07 ☿ ☌ ♆
08 42 ☽ ⊥ ♀
12 04 ☽ △ ♄
12 37 ☽ # ♄
16 02 ☽ ⊻ ♅

03 Wednesday
00 50 ☽ ☌ ☉
01 15 ☽ ∠ ♀
02 15 ☽ ⊥ ♅
02 29 ☽ ⚹ ⯓
02 57 ☽ ☌ ♆
05 16 ☽ ☌ ☿
08 42 ☽ △ ♃
13 32 ☽ ♇ ♄
17 44 ☽ ∠ ♅
21 32 ☉ ⊥ ♅
22 59 ♀ ∠ ♆
23 08 ☽ Q ⯓

04 Thursday
00 17 ☉ ⚹ ⯓
05 04 ☽ ± ♄
05 23 ☽ ⚹ ♀
06 15 ☽ ☍ ♂
07 20 ☉ ☌ ♆
11 44 ☿ △ ♃
13 32 ♀ △ ♂
15 34 ☽ ⊼ ♄
19 14 ☽ # ♄
20 03 ☽ ⚹ ♅

05 Friday
04 30 ☽ Q ♀
06 54 ☽ □ ⯓
07 27 ☽ ⊻ ♆
09 22 ☽ ⊻ ☉
13 17 ☽ □ ♃
16 06 ♂ Q ♃
16 45 ☽ ⊻ ☿
17 39 ☽ Q ♅
18 19 ☽ ⊥ ♆
21 08 ☽ ⊥ ☉

06 Saturday
05 12 ☽ ⊥ ☿
05 43 ☿ ♇ ♄
07 38 ♂ ⊥ ♄
10 42 ☽ ⊼ ♂
10 52 ☽ ∠ ♆
15 06 ☽ ∠ ☉
16 29 ☽ □ ♀
21 33 ☽ ± ♂
21 55 ☽ ☍ ♄
22 23 ☽ Q ♃

07 Sunday
00 14 ☽ ∠ ☿
03 05 ☽ □ ♅
11 45 ☽ ∥ ♅
12 10 ☉ △ ♃
14 05 ☽ ♇ ♂
14 31 ☽ △ ⯓
15 10 ☽ ⚹ ♆
21 09 ☽ ⚹ ♃
21 57 ☽ ⚹ ☉

08 Monday
00 01 ☿ ∠ ♅
00 52 ☽ # ⯓
07 33 ☽ ∥ ♀
09 02 ☽ ⚹ ☿
14 26 ☽ Q ♆
18 15 ☽ △ ♂
19 37 ☽ ♇ ⯓
23 25 ☽ Q ☉

09 Tuesday
02 22 ☽ ∠ ♃
03 13 ♀ □ ♄
07 39 ☽ ⊼ ♄
08 08 ☽ △ ♀
12 07 ☽ Q ☿
13 26 ☽ ± ⯓
13 33 ☽ △ ♅
15 14 ☽ # ♃
19 33 ☽ ± ♄
20 14 ☽ ⊥ ♃
22 23 ☿ Q ⯓

10 Wednesday
01 26 ☽ ⊼ ⯓
02 13 ☽ □ ♆
02 13 ☉ ♇ ♄
08 16 ☽ ⊻ ♃
12 38 ♃ St D
13 33 ☽ ♇ ♄
14 39 ☽ □ ☉
17 23 ☽ ♇ ♀
19 47 ☽ ♇ ♅
20 51 ☿ ☍ ♂

11 Thursday
02 41 ♀ # ⯓
04 21 ☽ □ ♂
05 46 ☽ □ ☿
13 34 ☽ ± ♀
14 05 ☽ ± ♅
19 01 ♀ ☌ ♅
19 31 ☽ ∥ ♃
19 45 ☿ ± ♄
19 47 ☽ △ ♄

12 Friday
01 09 ♀ ⊥ ♆
02 18 ☽ ⊼ ♅
03 04 ☽ ⊼ ♀
14 12 ☽ ☍ ⯓
15 05 ☽ △ ♆
15 40 ☽ Q ♂
21 01 ☽ ☌ ♃

13 Saturday
05 31 ☉ ∠ ♅
09 04 ☽ △ ☉
10 50 ☽ ∥ ⯓
15 06 ☽ ⚹ ♂
17 23 ☽ # ♀
21 24 ☽ ♇ ♆

14 Sunday
01 13 ☽ # ♅
03 34 ☽ △ ☿
07 54 ♀ ♇ ♂
08 00 ☽ □ ♄
14 52 ☽ ☍ ♅
15 23 ☽ ± ♆
17 47 ☽ ♇ ☉
20 03 ☽ ∠ ♂
21 46 ☽ ☍ ♀

15 Monday
02 18 ☽ ⊼ ⯓
03 15 ☽ ⊼ ♆
08 52 ☽ ⊻ ♃
12 32 ☿ ⊼ ♄
13 02 ☽ ± ☉
13 08 ☽ ⊥ ♂
13 32 ☽ ♇ ☿
13 58 ☉ ☍ ♂
14 03 ☽ ± ⯓
19 12 ☽ Q ♄
20 32 ☽ ⊥ ♃

16 Tuesday
00 25 ☽ ⊻ ♂
01 27 ☉ Q ⯓
01 42 ☽ ⊼ ☉
07 27 ☽ ♇ ⯓
09 25 ☽ ± ☿
13 51 ☽ ∠ ♃
18 06 ☽ ⚹ ♄
21 43 ♀ ⊻ ⯓
22 07 ♀ ∥ ♅
22 26 ☽ ⊼ ☿
23 25 ☽ ∥ ♄

17 Wednesday
01 07 ☽ ⊼ ♅
08 35 ♀ ⊻ ♆
11 50 ☽ △ ⯓
12 24 ☽ ± ♅
12 49 ☽ ☍ ♆
13 14 ☽ ⊼ ♀
18 04 ☽ ⚹ ♃
20 40 ☿ ⚹ ♅
22 03 ☽ ∠ ♄

18 Thursday
01 27 ☽ ± ♀
05 06 ☽ ♇ ♅
07 08 ☽ ☌ ♂
12 21 ☉ ± ♄
14 28 ☽ ⊥ ♄
14 39 ☽ ☍ ☉
16 08 ☽ Q ♃
19 29 ☽ ♇ ♀

19 Friday
01 17 ☽ ⊻ ♄
03 17 ☽ ∥ ♄
08 22 ☽ △ ♅
12 57 ☽ ☍ ☿
18 26 ☽ □ ⯓
19 28 ☽ ⊼ ♆
19 39 ♀ ⊼ ♃

20 Saturday
00 26 ☽ □ ♃
00 53 ☽ △ ♀
02 38 ♀ ± ♂
06 06 ☽ ± ♆
11 22 ☽ ⊻ ♂
21 35 ☽ ⊥ ♂
21 54 ☽ ♇ ♆
22 17 ♂ ♇ ♅

21 Sunday
00 16 ☽ ⊼ ☉
02 05 ☽ Q ⯓
06 00 ☽ ☌ ♄
09 11 ☽ # ♀
11 31 ☽ ± ☉
12 27 ☿ □ ⯓
12 48 ☽ ∠ ♂
13 14 ☽ □ ♅
18 55 ☽ # ♅
22 04 ☿ ⊻ ♆
22 52 ☽ ⚹ ⯓
23 56 ☽ △ ♆

22 Monday
00 10 ☽ ⊼ ☿
01 34 ♀ ⊥ ⯓
04 15 ☽ ♇ ☉
04 46 ☽ △ ♃
07 56 ☽ ∥ ⯓
09 53 ☽ □ ♀
11 46 ☽ ± ☿
13 57 ☽ ⚹ ♂
20 15 ☽ Q ♅

23 Tuesday
00 36 ☽ ∠ ⯓
05 05 ☽ ♇ ☿
06 30 ☽ ♇ ♃
07 55 ☽ △ ☉
09 17 ☽ ⊻ ♄
09 56 ☽ Q ♂
15 57 ☽ ⊥ ⯓
16 33 ☽ ∥ ♃
16 46 ☽ ⚹ ♅
18 19 ☿ □ ♃
19 26 ☽ Q ♀
19 27 ☽ ⊥ ♄
21 53 ☽ ± ♃
23 56 ♀ ⊼ ♂

24 Wednesday
01 44 ☉ ⊼ ♄
02 10 ☽ ⊻ ⯓
03 19 ☽ □ ♆
08 07 ☽ ⊼ ♃
09 46 ☽ △ ☿
10 41 ☽ ∠ ♄
15 49 ☽ □ ♂
17 42 ☽ ⚹ ♀
18 21 ☽ ∠ ♅

25 Thursday
01 40 ♀ ± ♃
09 40 ☽ ⊥ ♅
09 57 ☽ Q ♆
12 02 ☽ ⚹ ♄
14 36 ☽ # ♃
14 52 ☽ □ ☉
18 10 ☿ Q ♅
19 53 ☽ ⊻ ♅
20 49 ☿ ⊥ ♆
21 30 ☽ ∠ ♀

26 Friday
05 09 ☽ ☌ ⯓
06 23 ☽ ⚹ ♆
08 19 ☿ ⊼ ♂
08 19 ☽ Q ♄
11 15 ☽ ☍ ♃
14 12 ☽ ⊥ ♀
17 36 ☽ △ ♂
18 58 ☽ □ ☿
22 48 ☽ # ⯓
23 50 ☽ Q ☉

27 Saturday
01 21 ☽ ⊻ ♀
07 58 ☽ ∠ ♆
11 59 ☽ ∥ ♅
14 48 ☽ □ ♄
18 35 ☽ ♇ ♂
21 53 ☽ ⚹ ☉
23 06 ☽ ☌ ♅
23 22 ☽ ⊥ ♆

28 Sunday
05 23 ☽ Q ☿
08 19 ☽ ⊻ ⯓
09 38 ☽ ± ♂
09 39 ☽ ⊻ ♆
14 12 ☽ ∥ ♀
14 37 ☽ ⊼ ♃
15 07 ☉ ⚹ ♅
18 36 ☽ ⊥ ⯓
19 42 ☽ ⊼ ♂

29 Monday
00 58 ☽ ± ♃
01 35 ☽ ∠ ☉
04 26 ☽ ⚹ ☿
09 27 ☽ ☌ ♀
10 05 ☽ ∠ ⯓
15 22 ☿ ± ♂
16 32 ☽ ♇ ♃
16 53 ♀ ∠ ⯓
17 59 ☽ △ ♄
18 19 ☽ ⊥ ☉
18 51 ☉ ∥ ☿

30 Tuesday
02 48 ☽ ⊻ ♅
05 26 ☽ # ♄
05 32 ☽ ⊻ ☉
09 26 ☽ ∠ ☿
12 03 ☽ ⚹ ⯓
13 15 ☽ ⊥ ♅
13 30 ☽ ☌ ♆
18 41 ☽ △ ♃
19 54 ☽ ♇ ♄
22 33 ☽ ☍ ♂

31 Wednesday
03 00 ☽ ⊥ ☿
05 01 ☽ ∠ ♅
09 04 ☽ Q ⯓
11 37 ☽ ± ♄
14 48 ☽ ⊻ ☿
18 43 ☽ ⊻ ♀
21 02 ☽ # ♄
22 10 ☽ ⊼ ♄

All ephemeris data is given at 12.00 UT and the Moon's longitude is additionally given for 24.00 UT

JANUARY 1976

LONGITUDES

Date	Sidereal time h m s	Sun ☉ ° ' "	Moon ☽ ° ' "	Moon ☽ 24.00 ° ' "	Mercury ☿ ° '	Venus ♀ ° '	Mars ♂ ° '	Jupiter ♃ ° '	Saturn ♄ ° '	Uranus ♅ ° '	Neptune ♆ ° '	Pluto ♇ ° '
01	18 41 06	10 ♑ 12 04	08 ♑ 49 39	15 ♑ 29 34	28 ♑ 09	29 ♏ 59	17 ♊ 14	15 ♈ 34	01 ♌ 01	06 ♏ 24	12 ♐ 33	11 ♎ 40
02	18 45 03	11 13 15	22 ♑ 05 33	28 ♑ 37 25	29 ♑ 32	01 ♐ 11	16 R 59	15 39	00 R 57	06 26	12 35	11 40
03	18 48 59	12 14 26	05 ♒ 05 02	11 ♒ 28 19	00 ♒ 52	02 23	16 45	15 43	00 52	06 28	12 37	11 41
04	18 52 56	13 15 36	17 ♒ 47 18	24 ♒ 02 06	02 08	03 36	16 31	15 48	00 48	06 30	12 39	11 41
05	18 56 52	14 16 47	00 ♓ 12 54	06 ♓ 20 00	03 21	04 48	16 18	15 53	00 43	06 32	12 41	11 41
06	19 00 49	15 17 57	12 ♓ 23 44	18 ♓ 24 33	04 29	06 00	16 06	15 59	00 39	06 34	12 43	11 42
07	19 04 46	16 19 07	24 ♓ 22 55	00 ♈ 19 24	05 32	07 13	15 55	16 04	00 34	06 36	12 45	11 42
08	19 08 42	17 20 16	06 ♈ 14 35	12 ♈ 09 06	06 29	08 25	15 45	16 10	00 29	06 38	12 47	11 42
09	19 12 39	18 21 25	18 ♈ 03 36	23 ♈ 58 46	07 19	09 37	15 35	16 16	00 24	06 40	12 49	11 42
10	19 16 35	19 22 34	29 ♈ 55 16	05 ♉ 53 49	08 02	10 50	15 27	16 22	00 20	06 41	12 51	11 42
11	19 20 32	20 23 42	11 ♉ 55 04	17 ♉ 59 39	08 36	12 03	15 19	16 28	00 15	06 43	12 52	11 42
12	19 24 28	21 24 49	24 ♉ 08 10	00 ♊ 21 12	09 00	13 15	15 12	16 34	00 10	06 45	12 54	11 43
13	19 28 25	22 25 56	06 ♊ 39 13	13 ♊ 02 37	09 15	14 28	15 06	16 41	00 05	06 46	12 56	11 43
14	19 32 21	23 27 03	19 ♊ 31 43	26 ♊ 06 41	09 R 18	15 41	15 00	16 47	00 ♌ 00	06 48	12 58	11 R 43
15	19 36 18	24 28 09	02 ♋ 47 34	09 ♋ 34 17	09 09	16 54	14 55	16 54	29 ♋ 55	06 49	13 00	11 43
16	19 40 15	25 29 14	16 ♋ 26 37	23 ♋ 24 11	08 49	18 07	14 51	17 01	29 50	06 51	13 02	11 43
17	19 44 11	26 30 19	00 ♌ 26 28	07 ♌ 32 49	08 17	19 19	14 48	17 08	29 46	06 52	13 03	11 42
18	19 48 08	27 31 23	14 ♌ 42 31	21 ♌ 54 46	07 34	20 32	14 46	17 15	29 41	06 53	13 05	11 42
19	19 52 04	28 32 27	29 ♌ 08 43	06 ♍ 23 33	06 41	21 45	14 45	17 23	29 36	06 55	13 07	11 42
20	19 56 01	29 ♑ 33 30	13 ♍ 38 28	20 ♍ 52 45	05 39	22 59	14 44	17 30	29 31	06 56	13 09	11 42
21	19 59 57	00 ♒ 34 33	28 ♍ 05 45	05 ♎ 16 57	04 30	24 12	14 D 44	17 38	29 26	06 57	13 10	11 42
22	20 03 54	01 35 35	12 ♎ 25 57	19 ♎ 32 27	03 16	25 25	14 45	17 46	29 21	06 58	13 12	11 41
23	20 07 50	02 36 37	26 ♎ 36 14	03 ♏ 37 13	01 59	26 38	14 46	17 54	29 16	06 59	13 14	11 41
24	20 11 47	03 37 39	10 ♏ 36 01	17 ♏ 30 43	00 ♒ 43	27 51	14 49	18 02	29 11	07 00	13 15	11 41
25	20 15 44	04 38 40	24 ♏ 23 18	01 ♐ 13 13	29 ♑ 28	29 ♐ 05	14 52	18 11	29 06	07 01	13 17	11 40
26	20 19 40	05 39 41	08 ♐ 00 31	14 ♐ 45 16	28 17	00 ♑ 18	14 56	18 19	29 01	07 02	13 18	11 40
27	20 23 37	06 40 41	21 ♐ 27 31	28 ♐ 07 14	27 11	01 31	15 00	18 28	28 56	07 03	13 20	11 40
28	20 27 33	07 41 41	04 ♑ 44 24	11 ♑ 18 58	26 13	02 45	15 05	18 37	28 51	07 04	13 21	11 39
29	20 31 30	08 42 40	17 ♑ 50 50	24 ♑ 19 54	25 23	03 58	15 11	18 46	28 47	07 04	13 23	11 39
30	20 35 26	09 43 38	00 ♒ 46 02	07 ♒ 09 09	24 41	05 12	15 18	18 55	28 42	07 05	13 24	11 39
31	20 39 23	10 ♒ 44 35	13 ♒ 29 10	19 ♒ 45 59	24 ♑ 08	06 ♑ 25	15 ♊ 25	19 ♈ 04	28 ♋ 37	07 ♏ 06	13 ♐ 26	11 ♎ 38

DECLINATIONS

Date	Moon True ☊ ° '	Moon Mean ☊ ° '	Moon ☽ Latitude ° '	Sun ☉ ° '	Moon ☽ ° '	Mercury ☿ ° '	Venus ♀ ° '	Mars ♂ ° '	Jupiter ♃ ° '	Saturn ♄ ° '	Uranus ♅ ° '	Neptune ♆ ° '	Pluto ♇ ° '
01	20 ♏ 48	19 ♏ 14	03 N 49	23 S 03	19 S 21	22 S 15	18 S 00	25 N 57	04 N 56	20 N 11	13 S 12	20 S 49	10 N 55
02	20 R 39	19 11	04 29	22 58	17 12	21 52	18 16	25 56	04 58	20 12	13 12	20 50	10 56
03	20 29	19 08	04 55	22 53	14 13	21 27	18 32	25 55	05 00	20 13	13 13	20 50	10 56
04	20 19	19 05	05 06	22 47	10 39	21 02	18 48	25 54	05 02	20 14	13 14	20 50	10 56
05	20 10	19 02	05 01	22 41	06 41	20 36	19 03	25 53	05 04	20 15	13 14	20 50	10 57
06	20 03	18 58	04 43	22 34	02 S 33	20 10	19 17	25 52	05 07	20 16	13 15	20 51	10 57
07	19 58	18 55	04 13	22 27	01 N 38	19 43	19 31	25 50	05 09	20 18	13 15	20 51	10 58
08	19 55	18 52	03 32	22 19	05 43	19 17	19 45	25 49	05 11	20 19	13 16	20 51	10 58
09	19 54	18 49	02 42	22 11	09 35	18 51	19 58	25 48	05 14	20 20	13 17	20 51	10 59
10	19 D 55	18 46	01 45	22 02	13 05	18 26	20 10	25 47	05 16	20 21	13 17	20 52	10 59
11	19 56	18 42	00 N 43	21 54	16 05	18 02	20 22	25 46	05 19	20 22	13 18	20 52	11 00
12	19 R 56	18 39	00 S 22	21 44	18 27	17 39	20 34	25 45	05 22	20 23	13 18	20 52	11 00
13	19 54	18 36	01 28	21 34	19 58	17 19	20 45	25 44	05 24	20 24	13 19	20 52	11 01
14	19 50	18 33	02 31	21 24	20 31	17 01	20 55	25 44	05 27	20 26	13 19	20 52	11 01
15	19 44	18 30	03 28	21 14	19 57	16 45	21 05	25 43	05 30	20 27	13 20	20 53	11 02
16	19 35	18 27	04 14	21 03	18 15	16 32	21 14	25 42	05 33	20 28	13 20	20 53	11 02
17	19 24	18 23	04 46	20 51	15 24	16 22	21 23	25 41	05 36	20 29	13 21	20 53	11 03
18	19 13	18 20	05 01	20 39	11 37	16 16	21 31	25 40	05 39	20 30	13 21	20 53	11 03
19	19 02	18 17	04 58	20 27	07 07	16 13	21 38	25 40	05 42	20 31	13 21	20 53	11 04
20	18 53	18 14	04 35	20 15	02 N 12	16 13	21 45	25 39	05 45	20 32	13 22	20 53	11 05
21	18 47	18 11	03 55	20 02	02 S 50	16 15	21 52	25 39	05 48	20 34	13 22	20 54	11 05
22	18 43	18 08	03 00	19 48	07 40	16 21	21 57	25 38	05 52	20 35	13 22	20 54	11 06
23	18 42	18 04	01 54	19 35	12 06	16 28	22 02	25 38	05 55	20 36	13 23	20 54	11 06
24	18 D 42	18 01	00 S 43	19 21	15 41	16 38	22 07	25 38	05 58	20 37	13 23	20 54	11 07
25	18 R 42	17 58	00 N 30	19 06	18 23	16 49	22 10	25 37	06 02	20 38	13 23	20 54	11 08
26	18 41	17 55	01 40	18 51	20 00	17 01	22 13	25 37	06 05	20 39	13 24	20 54	11 08
27	18 37	17 52	02 44	18 36	20 26	17 14	22 16	25 37	06 09	20 40	13 24	20 55	11 09
28	18 30	17 48	03 38	18 21	19 44	17 27	22 18	25 37	06 12	20 41	13 24	20 55	11 10
29	18 20	17 45	04 19	18 05	17 58	17 40	22 19	25 37	06 16	20 43	13 24	20 55	11 10
30	18 08	17 42	04 47	17 49	15 19	17 54	22 19	25 37	06 20	20 44	13 25	20 55	11 11
31	17 ♏ 54	17 ♏ 39	04 N 59	17 S 32	12 S 00	18 S 07	22 S 19	25 N 37	06 N 23	20 N 45	13 S 25	20 S 55	11 N 12

ZODIAC SIGN ENTRIES

Date	h	m	Planets
01	12	14	♀ ♐
02	20	22	☿ ♒
03	02	33	☽ ♒
05	11	35	☽ ♓
07	23	21	☽ ♈
10	12	10	☽ ♉
12	23	19	☽ ♊
14	13	16	♄ ♋
15	07	00	☽ ♋
17	11	15	☽ ♌
19	13	25	☽ ♍
20	22	25	☉ ♒
21	15	10	☽ ♎
23	17	48	☽ ♏
25	01	30	☿ ♑
25	21	51	☽ ♐
26	06	09	♀ ♑
28	03	24	☽ ♑
30	10	34	☽ ♒

LATITUDES

Date	Mercury ☿ ° '	Venus ♀ ° '	Mars ♂ ° '	Jupiter ♃ ° '	Saturn ♄ ° '	Uranus ♅ ° '	Neptune ♆ ° '	Pluto ♇ ° '
01	01 S 46	02 N 12	03 N 08	01 S 18	00 N 15	00 N 29	01 N 29	16 N 53
04	01 23	02 07	03 09	01 17	00 16	00 29	01 29	16 55
07	00 51	02 01	03 10	01 16	00 16	00 29	01 30	16 56
10	00 S 11	01 55	03 10	01 15	00 16	00 29	01 30	16 58
13	00 N 39	01 48	03 09	01 15	00 17	00 30	01 30	17 00
16	01 34	01 41	03 08	01 14	00 17	00 30	01 30	17 02
19	02 28	01 33	03 07	01 13	00 17	00 30	01 30	17 03
22	03 10	01 25	03 05	01 12	00 18	00 30	01 30	17 05
25	03 31	01 16	03 04	01 11	00 18	00 30	01 30	17 07
28	03 31	01 07	03 02	01 11	00 18	00 30	01 30	17 08
31	03 N 14	00 N 58	02 N 59	01 S 10	00 N 19	00 N 30	01 N 30	17 N 10

DATA

Julian Date	2442779
Delta T	+47 seconds
Ayanamsa	23° 31' 32"
Synetic vernal point	05° ♓ 35' 27"
True obliquity of ecliptic	23° 26' 26"

MOON'S PHASES, APSIDES AND POSITIONS ☽

Date	h	m	Phase	Longitude ° '	Eclipse Indicator
01	14	40	●	10 ♑ 19	
09	12	40	☽	18 ♈ 23	
17	04	47	○	26 ♋ 12	
23	23	04	☾	03 ♏ 05	
31	06	20	●	10 ♒ 30	

Day	h	m		
08	16	48	Apogee	
20	13	40	Perigee	
07	02	34	ON	
14	12	05	Max dec	20° N 31'
20	22	29	OS	
27	09	04	Max dec	20° S 27'

LONGITUDES

Date	Chiron ⚷ ° '	Ceres ⚳ ° '	Pallas ⚴ ° '	Juno ⚵ ° '	Vesta ⚶ ° '	Black Moon Lilith ⚸ ° '
01	23 ♈ 35	02 ♊ 20	23 ♓ 22	18 ♍ 56	28 ♓ 48	06 ♈ 55
11	23 ♈ 36	01 ♊ 29	25 ♓ 56	19 ♍ 17	02 ♈ 06	08 ♈ 01
21	23 ♈ 43	01 ♊ 19	28 ♓ 50	18 ♍ 53	05 ♈ 40	09 ♈ 08
31	23 ♈ 56	01 ♊ 49	02 ♈ 02	17 ♍ 44	09 ♈ 25	10 ♈ 15

All ephemeris data is given at 12.00 UT and the Moon's longitude is additionally given for 24.00 UT

ASPECTARIAN

h m Aspects

01 Thursday
01 18 ☿ ∠ ♆
06 25 ☽ ⊥ ♀
07 39 ☽ ✱ ♅
14 40 ☽ ☌ ☉
17 05 ☽ □ ♇
18 42 ☽ ⊻ ♆

02 Friday
00 11 ☽ ∠ ♀
00 13 ☽ □ ♃
00 26 ♀ bQ ♃
02 27 ☽ ∥ ♀
02 53 ☽ ⊼ ♂
05 19 ☽ Q ♅
05 35 ☽ ⊥ ♆
07 29 ♀ △ ♄
13 36 ☽ ± ♂
22 06 ☽ ∠ ♆
22 36 ☉ □ ♇

03 Saturday
03 15 ☽ ☌ ☿
04 13 ☽ ☍ ♄
05 54 ☽ bQ ♂
06 28 ☽ ✱ ♀
09 27 ☽ Q ♃
12 13 ☿ ☍ ♄
14 36 ☽ □ ♅
19 04 ☽ ∥ ♅
21 03 ☉ ⊻ ♆

04 Sunday
00 23 ☽ △ ♇
02 03 ☿ bQ ♂
02 12 ☽ ✱ ♆
02 38 ☽ ⊻ ☉
07 22 ☽ Q ♀
08 12 ☽ ✱ ♃
09 37 ☽ △ ♂
10 08 ☽ # ♇
15 04 ☽ ⊥ ☉
22 51 ☿ ∥ ♆

05 Monday
01 13 ☽ Q ♆
05 08 ☽ bQ ♇
10 01 ☽ ∠ ☉
12 59 ☽ ⊼ ♄
13 20 ☽ ∠ ♃
18 47 ☽ ⊻ ☿
21 22 ☽ # ♃
21 58 ☽ □ ♀
22 44 ☽ ± ♇

06 Tuesday
00 11 ☿ Q ♃
00 26 ☽ △ ♅
00 41 ☽ ± ♄
06 02 ☿ # ♄
07 10 ☽ ⊥ ♃
07 50 ☽ ⊥ ☿
10 36 ☽ ⊼ ♇
12 38 ☽ □ ♆
18 19 ☽ ✱ ☉
18 26 ☽ bQ ♄
19 12 ☽ ⊻ ♃
19 17 ☽ □ ♂
23 03 ♂ ✱ ♃
23 35 ♀ ⊻ ♅

07 Wednesday
03 32 ☽ ∠ ☿
04 04 ☉ ⊼ ♂
05 33 ☉ □ ♃
06 23 ☽ bQ ♅
19 07 ☿ ∥ ♀
20 41 ☽ Q ☉

08 Thursday
00 24 ☽ △ ♄
00 36 ☽ ± ♅
07 01 ☽ Q ♂
08 46 ☽ ∥ ♃
12 32 ☽ ✱ ☿
12 47 ☽ ⊼ ♅
16 03 ☿ □ ♅
16 55 ☽ △ ♀
23 05 ☽ ☍ ♇

09 Friday
01 19 ☽ △ ♆
07 03 ☽ ✱ ♂
08 19 ☽ ☌ ♃
12 40 ☽ □ ☉
14 45 ☽ Q ☿
19 21 ☉ Q ♅
21 15 ☽ ∥ ♇
23 01 ☉ ⊥ ♆

10 Saturday
02 49 ☽ bQ ♀
07 48 ☽ bQ ♆
12 49 ☽ □ ♄
13 02 ☽ ∠ ♂
13 29 ☽ # ♅
22 59 ☽ ± ♀

11 Sunday
01 37 ☽ ☍ ♅
01 56 ☽ ± ♆
05 06 ☽ □ ☿
05 20 ♀ ✱ ♇
06 52 ☽ ⊥ ♂
11 24 ♀ # ♄
11 35 ☽ ⊼ ♇
12 17 ☽ ⊼ ♀
13 54 ☽ ⊼ ♆
18 39 ☽ ⊻ ♂
21 04 ☽ ⊻ ♃
23 26 ☽ ± ♇

12 Monday
00 25 ☽ Q ♄
01 40 ♀ ⊥ ♅
04 20 ☽ # ☿
04 53 ♀ ☌ ♆
06 13 ☽ △ ☉
07 18 ☉ ± ♂
08 55 ☽ ⊥ ♃
16 59 ☽ bQ ♇
23 34 ☽ ✱ ♄

13 Tuesday
02 27 ☽ ∠ ♃
12 13 ☽ ⊼ ♅
13 36 ☽ bQ ☉
15 17 ♂ ∠ ♄
16 56 ☽ △ ☿
21 31 ☽ △ ♇
23 24 ♀ ☍ ♂
23 29 ♀ bQ ♄
23 31 ☽ ± ♅
23 50 ☽ ☍ ♆

14 Wednesday
01 55 ☽ ∥ ♄
03 41 ☽ ☌ ♂
03 42 ☽ ∠ ♄
04 10 ☽ ☍ ♀
05 06 ♀ ∥ ♆
06 42 ☿ St R
06 54 ☽ ✱ ♃
07 51 ☽ ± ☉
11 40 ♇ St R
16 10 ☽ bQ ♅
19 46 ☽ ⊼ ☉
20 08 ☽ ⊥ ♄
20 39 ☽ bQ ☿
21 23 ☽ ∥ ♄

15 Thursday
04 58 ☽ Q ♃
06 54 ☽ ⊻ ♄
12 08 ♀ △ ♃
12 38 ☽ ± ☿
17 53 ♂ ∠ ♄
19 10 ☽ △ ♅
22 12 ☉ ∥ ♀
23 02 ☽ ⊼ ☿

16 Friday
03 45 ☽ □ ♇
06 02 ☽ ⊼ ♆
09 15 ☽ ⊻ ♂
13 00 ☽ □ ♃
15 10 ☽ ⊼ ♀
16 29 ☽ ± ♆
19 36 ☽ ⊥ ♂

17 Saturday
02 28 ☽ ± ♀
04 17 ☽ # ☿
04 47 ☽ ☍ ☉
07 56 ☽ bQ ♆
08 33 ☉ ∥ ♆
10 45 ☽ Q ♇
10 51 ☽ ☌ ♄
10 55 ☽ ∠ ♂
19 11 ☽ bQ ♀
22 52 ☽ □ ♅

18 Sunday
00 39 ☽ ☍ ☿
01 40 ☽ # ♅
03 16 ♄ Q ♇
06 59 ☽ ✱ ♇
09 17 ☽ △ ♆
12 06 ☽ ✱ ♂
15 10 ☽ ∥ ♇
16 17 ☽ △ ♃
22 37 ☽ △ ♀

19 Monday
01 39 ☉ ∠ ♆
04 45 ☉ # ♄
04 58 ☽ Q ♅
06 29 ☿ □ ♅
07 57 ☽ ∠ ♇
08 01 ☽ Q ♂
10 55 ☽ ⊼ ☉
11 14 ☿ ∠ ♀
12 44 ☽ ⊻ ♄
15 04 ♀ ∠ ♅
17 24 ☽ bQ ♃
19 00 ☽ ∥ ♃
21 37 ☽ ± ☉
22 37 ☽ ⊥ ♄
22 51 ☽ ⊥ ♇
23 41 ☽ ⊼ ☿

20 Tuesday
00 53 ☽ ✱ ♅
08 26 ☽ ± ♃
08 47 ☽ ⊻ ♇
08 56 ☽ ± ☿
11 00 ☉ ☍ ♄
11 10 ☽ □ ♆
13 26 ☽ ∠ ♄
13 38 ☽ bQ ☉
13 48 ☽ □ ♂
14 51 ☿ Q ♃
16 05 ☉ bQ ♂
18 28 ☽ ⊼ ♃
21 27 ♂ St D
21 54 ♀ ± ♄
22 47 ☽ bQ ☿

21 Wednesday
01 46 ☽ ∠ ♅
02 14 ♀ Q ♇
04 54 ☽ □ ♀
14 13 ☽ ✱ ♄
16 27 ☽ △ ☉
16 46 ☽ ⊥ ♅
17 08 ☽ Q ♆
21 52 ☽ △ ☿

22 Thursday
02 43 ☽ # ♃
02 49 ☽ ⊻ ♅
10 11 ☽ Q ♄
10 45 ☽ ☌ ♇
13 18 ☽ ✱ ♆
13 48 ☽ Q ♀
15 54 ☽ △ ♂
21 05 ☽ ☍ ♃

23 Friday
05 33 ☉ ☌ ☿
05 51 ☿ ⊥ ♀
06 34 ☽ # ♇
09 15 ☉ ⊥ ♀
12 03 ☽ ✱ ♀
14 46 ☽ ∠ ♆
16 31 ☽ □ ♄
17 25 ☽ bQ ♂
20 14 ☽ ∥ ♅
20 26 ☽ □ ☿
23 04 ☽ □ ☉

24 Saturday
05 48 ☽ ☌ ♅
06 14 ☽ ⊥ ♆
08 55 ☽ ± ♂
13 53 ☽ ⊻ ♇
16 18 ☽ ∠ ♀
16 37 ☽ ⊻ ♆
19 20 ☽ ⊼ ♂
19 59 ☽ ∥ ☿

25 Sunday
00 17 ☽ ⊥ ♇
00 54 ☽ Q ☿
01 03 ☽ ⊼ ♃
04 29 ☿ bQ ♂
08 42 ☽ Q ☉
09 29 ☽ ⊥ ♀
11 38 ☽ ± ♃
12 28 ♀ ⊼ ♄
15 47 ☿ ⊻ ♀
16 00 ☽ ∠ ♇
19 39 ☿ ☍ ♄
19 43 ☽ ∥ ☉
20 11 ☽ ✱ ☿
20 13 ☽ △ ♄
21 02 ☽ ⊻ ♀

26 Monday
03 37 ☽ bQ ♃
07 30 ☽ ✱ ☉
10 16 ☽ ⊻ ♅
11 27 ☿ ∠ ♆
18 30 ☽ ✱ ♇
20 39 ☽ ∠ ☿
20 56 ☽ ⊥ ♅
21 26 ☽ ☌ ♆
22 37 ☽ bQ ♄

27 Tuesday
00 22 ☽ ☍ ♂
06 07 ☉ Q ♃
06 34 ☽ △ ♃
11 33 ☽ ⊥ ☿
12 26 ☽ ∠ ☉
13 04 ☽ ∠ ♅
14 39 ☽ ± ♄
15 57 ☽ Q ♇
20 50 ☉ □ ♅
21 36 ☽ ⊻ ☿

28 Wednesday
01 24 ☽ ⊼ ♄
06 01 ☽ ⊥ ☉
08 00 ☽ ☌ ♀
16 14 ☽ ✱ ♅
17 50 ☽ ⊻ ☉

29 Thursday
00 37 ☽ □ ♇
03 46 ☽ ⊻ ♆
07 04 ☽ ⊼ ♂
10 33 ☽ ∥ ☉
13 42 ☽ □ ♃
14 16 ☽ Q ♅
14 49 ☽ ∥ ☿
14 50 ☽ ⊥ ♆
18 13 ☽ ± ♂

30 Friday
01 13 ☽ ☌ ☿
07 35 ☽ ∠ ♆
08 09 ☽ ☍ ♄
08 13 ☉ ∥ ☿
11 06 ☽ bQ ♂
21 11 ☽ ⊻ ♀
23 41 ☽ Q ♃
23 53 ☽ □ ♅

31 Saturday
02 17 ☽ ∥ ♅
06 20 ☽ ☌ ☉
08 28 ☽ △ ♇
09 45 ☽ ⊥ ♀
11 54 ☽ ✱ ♆
15 43 ☽ △ ♂
17 14 ☽ # ♇
22 47 ☽ ✱ ♃

FEBRUARY 1976

LONGITUDES

Date	Sidereal time h m s	Sun ☉ ° ′ ″	Moon ☽ ° ′ ″	Moon ☽ 24.00 ° ′ ″	Mercury ☿ ° ′	Venus ♀ ° ′	Mars ♂ ° ′	Jupiter ♃ ° ′	Saturn ♄ ° ′	Uranus ♅ ° ′	Neptune ♆ ° ′	Pluto ♇ ° ′
01	20 43 19	11 ♒ 45 31	25 ♒ 59 37	02 ♓ 10 04	23 ♑ 44	07 ♑ 39	15 ♊ 33	19 ♈ 13	28 ♋ 32	07 ♏ 06	13 ♐ 27	11 ♎ 37
02	20 47 16	12 46 25	08 ♓ 17 27	14 ♓ 21 54	23 R 28	08 52	15 41	19 23	28 R 28	07 07	13 29	11 R 36
03	20 51 13	13 47 19	20 ♓ 23 38	26 ♓ 22 56	23 21	10 06	15 50	19 32	28 23	07 07	13 30	11 36
04	20 55 09	14 48 11	02 ♈ 20 10	08 ♈ 15 43	23 D 21	11 19	16 00	19 42	28 18	07 07	13 31	11 35
05	20 59 06	15 49 02	14 ♈ 10 06	20 ♈ 03 50	23 29	12 33	16 10	19 52	28 14	07 08	13 33	11 34
06	21 03 02	16 49 52	25 ♈ 57 29	01 ♉ 51 42	23 43	13 46	16 21	20 02	28 09	07 08	13 34	11 34
07	21 06 59	17 50 40	07 ♉ 47 07	13 ♉ 44 26	24 04	15 00	16 33	20 12	28 05	07 08	13 35	11 33
08	21 10 55	18 51 27	19 ♉ 44 21	25 ♉ 47 35	24 30	16 14	16 45	20 22	28 00	07 08	13 36	11 32
09	21 14 52	19 52 12	01 ♊ 54 49	08 ♊ 06 43	25 02	17 27	16 58	20 32	27 56	07 09	13 37	11 31
10	21 18 48	20 52 56	14 ♊ 23 57	20 ♊ 47 03	25 39	18 41	17 11	20 43	27 51	07 09	13 39	11 30
11	21 22 45	21 53 38	27 ♊ 16 31	03 ♋ 52 44	26 20	19 55	17 24	20 53	27 47	07 R 09	13 40	11 29
12	21 26 42	22 54 18	10 ♋ 35 55	17 ♋ 26 12	27 06	21 09	17 39	21 04	27 43	07 09	13 41	11 28
13	21 30 38	23 54 57	24 ♋ 23 27	01 ♌ 27 25	27 55	22 22	17 53	21 15	27 39	07 08	13 42	11 27
14	21 34 35	24 55 35	08 ♌ 37 35	15 ♌ 53 19	28 47	23 36	18 08	21 25	27 34	07 08	13 43	11 26
15	21 38 31	25 56 11	23 ♌ 13 45	00 ♍ 37 52	29 ♑ 43	24 50	18 24	21 36	27 30	07 08	13 44	11 25
16	21 42 28	26 56 45	08 ♍ 04 35	15 ♍ 32 44	00 ♒ 42	26 04	18 40	21 47	27 26	07 08	13 45	11 24
17	21 46 24	27 57 18	23 ♍ 01 08	00 ♎ 28 40	01 43	27 17	18 57	21 59	27 23	07 07	13 46	11 23
18	21 50 21	28 57 49	07 ♎ 54 20	15 ♎ 17 12	02 47	28 31	19 14	22 10	27 19	07 07	13 47	11 22
19	21 54 17	29 ♒ 58 19	22 ♎ 36 34	29 ♎ 51 50	03 53	29 ♑ 45	19 31	22 21	27 15	07 07	13 47	11 21
20	21 58 14	00 ♓ 58 48	07 ♏ 02 38	14 ♏ 08 41	05 02	00 ♒ 59	19 49	22 33	27 11	07 06	13 48	11 20
21	22 02 11	01 59 16	21 ♏ 09 54	28 ♏ 06 17	06 12	02 13	20 07	22 44	27 08	07 06	13 49	11 19
22	22 06 07	02 59 42	04 ♐ 57 58	11 ♐ 45 06	07 24	03 27	20 26	22 56	27 04	07 05	13 50	11 18
23	22 10 04	04 00 08	18 ♐ 27 55	25 ♐ 06 40	08 38	04 41	20 45	23 08	27 01	07 04	13 51	11 16
24	22 14 00	05 00 31	01 ♑ 41 36	08 ♑ 12 57	09 54	05 55	21 04	23 19	26 57	07 04	13 51	11 15
25	22 17 57	06 00 54	14 ♑ 40 58	21 ♑ 05 52	11 12	07 08	21 24	23 31	26 54	07 03	13 52	11 14
26	22 21 53	07 01 15	27 ♑ 27 47	03 ♒ 46 55	12 31	08 22	21 44	23 43	26 51	07 02	13 53	11 12
27	22 25 50	08 01 34	10 ♒ 03 22	16 ♒ 17 14	13 51	09 36	22 05	23 55	26 48	07 01	13 53	11 11
28	22 29 46	09 01 52	22 ♒ 28 36	28 ♒ 37 33	15 13	10 50	22 26	24 08	26 45	07 00	13 54	11 10
29	22 33 43	10 ♓ 02 08	04 ♓ 44 11	10 ♓ 48 33	16 ♒ 36	12 ♒ 04	22 ♊ 47	24 ♈ 20	26 ♋ 42	06 ♏ 59	13 ♐ 54	11 ♎ 08

Moon Nodes and Latitude / DECLINATIONS

Date	Moon True ☊ ° ′	Moon Mean ☊ ° ′	Moon ☽ Latitude ° ′	Sun ☉ ° ′	Moon ☽ ° ′	Mercury ☿ ° ′	Venus ♀ ° ′	Mars ♂ ° ′	Jupiter ♃ ° ′	Saturn ♄ ° ′	Uranus ♅ ° ′	Neptune ♆ ° ′	Pluto ♇ ° ′
01	17 ♏ 40	17 ♏ 36	04 N 57	17 S 16	08 S 12	18 S 19	22 S 18	25 N 37	06 N 27	20 N 46	13 S 25	20 S 55	11 N 13
02	17 R 26	17 33	04 41	16 59	04 S 07	18 31	22 17	25 37	06 31	20 47	13 25	20 55	11 13
03	17 15	17 29	04 12	16 41	00 N 03	18 42	22 15	25 37	06 35	20 48	13 25	20 56	11 14
04	17 06	17 26	03 33	16 24	04 11	18 53	22 12	25 38	06 38	20 49	13 25	20 56	11 15
05	17 00	17 23	02 44	16 06	08 06	19 03	22 08	25 38	06 42	20 50	13 25	20 56	11 15
06	16 57	17 20	01 49	15 48	11 43	19 11	22 04	25 38	06 46	20 51	13 25	20 56	11 16
07	16 D 56	17 17	00 N 48	15 29	14 52	19 19	21 59	25 39	06 50	20 52	13 25	20 56	11 17
08	16 R 56	17 14	00 S 15	15 10	17 26	19 26	21 54	25 39	06 54	20 53	13 25	20 56	11 18
09	16 55	17 10	01 19	14 51	19 16	19 32	21 48	25 39	06 58	20 54	13 25	20 56	11 18
10	16 54	17 07	02 20	14 32	20 13	19 37	21 41	25 40	07 03	20 55	13 25	20 56	11 19
11	16 50	17 04	03 16	14 13	20 09	19 40	21 34	25 40	07 07	20 56	13 25	20 56	11 20
12	16 43	17 01	04 04	13 53	18 58	19 43	21 26	25 41	07 11	20 57	13 25	20 57	11 21
13	16 34	16 58	04 39	13 33	16 40	19 44	21 17	25 41	07 15	20 58	13 25	20 57	11 22
14	16 23	16 54	04 58	13 13	13 18	19 44	21 08	25 42	07 19	20 58	13 25	20 57	11 22
15	16 11	16 51	04 59	12 52	09 04	19 43	20 58	25 42	07 24	20 59	13 25	20 57	11 23
16	15 59	16 48	04 39	12 32	04 N 13	19 41	20 48	25 43	07 28	21 00	13 25	20 57	11 24
17	15 50	16 45	04 00	12 11	00 S 55	19 37	20 36	25 44	07 32	21 01	13 25	20 57	11 25
18	15 43	16 42	03 05	11 50	05 59	19 32	20 25	25 44	07 37	21 02	13 25	20 57	11 26
19	15 38	16 39	01 59	11 29	10 38	19 26	20 12	25 45	07 41	21 03	13 25	20 57	11 26
20	15 37	16 35	00 S 45	11 08	14 35	19 19	19 59	25 45	07 45	21 04	13 24	20 57	11 27
21	15 D 37	16 32	00 N 29	10 46	17 35	19 10	19 46	25 46	07 50	21 04	13 24	20 57	11 28
22	15 R 37	16 29	01 41	10 24	19 28	19 00	19 32	25 46	07 54	21 05	13 24	20 57	11 29
23	15 36	16 26	02 45	10 03	20 12	18 49	19 17	25 47	07 59	21 06	13 24	20 57	11 30
24	15 33	16 23	03 39	09 41	19 47	18 37	19 02	25 47	08 03	21 07	13 24	20 57	11 31
25	15 27	16 19	04 20	09 18	18 19	18 23	18 46	25 48	08 08	21 07	13 23	20 57	11 31
26	15 19	16 16	04 48	08 56	15 58	18 08	18 30	25 48	08 13	21 08	13 23	20 57	11 32
27	15 08	16 13	05 01	08 34	12 53	17 51	18 13	25 49	08 17	21 09	13 23	20 57	11 33
28	14 56	16 10	05 00	08 11	09 18	17 34	17 56	25 49	08 22	21 09	13 22	20 57	11 34
29	14 ♏ 43	16 ♏ 07	04 N 44	07 S 48	05 S 21	17 S 15	17 S 38	25 N 49	08 N 26	21 N 10	13 S 22	20 S 57	11 N 35

ZODIAC SIGN ENTRIES

Date	h	m	Planets
01	19	47	☽ ♓
04	07	17	☽ ♈
06	20	13	☽ ♉
09	08	16	☽ ♊
11	16	59	☽ ♋
13	21	32	☽ ♌
15	19	03	☿ ♒
15	22	59	☽ ♍
17	23	14	☽ ♎
19	12	40	☉ ♓
19	16	50	♀ ♒
20	00	14	☽ ♏
22	03	18	☽ ♐
24	08	54	☽ ♑
26	16	48	☽ ♒
29	02	42	☽ ♓

LATITUDES

Date	Mercury ☿ ° ′	Venus ♀ ° ′	Mars ♂ ° ′	Jupiter ♃ ° ′	Saturn ♄ ° ′	Uranus ♅ ° ′	Neptune ♆ ° ′	Pluto ♇ ° ′
01	03 N 05	00 N 55	02 N 59	01 S 10	00 N 19	00 N 30	01 N 30	17 N 10
04	02 35	00 46	02 56	01 09	00 19	00 30	01 30	17 12
07	02 00	00 37	02 54	01 08	00 19	00 30	01 30	17 14
10	01 26	00 27	02 51	01 08	00 20	00 30	01 31	17 15
13	00 52	00 18	02 49	01 07	00 20	00 30	01 31	17 16
16	00 N 20	00 N 09	02 46	01 06	00 20	00 30	01 31	17 18
19	00 S 10	00 00	02 44	01 06	00 20	00 30	01 31	17 19
22	00 36	00 S 09	02 41	01 05	00 21	00 30	01 31	17 20
25	01 00	00 17	02 38	01 05	00 21	00 30	01 31	17 21
28	01 21	00 26	02 36	01 04	00 21	00 30	01 31	17 22
31	01 S 38	00 S 34	02 N 33	01 S 04	00 N 21	00 N 30	01 N 32	17 N 23

DATA

Julian Date	2442810
Delta T	+47 seconds
Ayanamsa	23° 31' 37"
Synetic vernal point	05° ♓ 35' 22"
True obliquity of ecliptic	23° 26' 26"

MOON'S PHASES, APSIDES AND POSITIONS ☽

Date	h	m	Phase	Longitude ° ′	Eclipse Indicator
08	10	05	☽	18 ♉ 47	
15	16	43	○	26 ♌ 08	
22	08	16	☾	02 ♐ 50	
29	23	25	●	10 ♓ 31	

Day	h	m		
05	13	07	Apogee	
17	10	32	Perigee	
03	11	40	0N	
10	22	33	Max dec	20° N 19'
17	07	46	0S	
23	15	01	Max dec	20° S 13'

LONGITUDES

Date	Chiron ⚷ ° ′	Ceres ⚳ ° ′	Pallas ⚴ ° ′	Juno ⚵ ° ′	Vesta ⚶ ° ′	Black Moon Lilith ⚸ ° ′
01	23 ♈ 57	01 ♊ 54	02 ♈ 23	17 ♍ 35	09 ♈ 48	10 ♈ 22
11	24 ♈ 15	03 ♊ 06	05 ♈ 52	15 ♍ 43	13 ♈ 44	11 ♈ 28
21	24 ♈ 38	04 ♊ 50	09 ♈ 34	13 ♍ 23	17 ♈ 48	12 ♈ 35
31	25 ♈ 04	07 ♊ 02	13 ♈ 29	10 ♍ 50	21 ♈ 59	13 ♈ 42

All ephemeris data is given at 12.00 UT and the Moon's longitude is additionally given for 24.00 UT

ASPECTARIAN

h m	Aspects
01 Sunday	
01 21	♀ ⚹ ♅
04 49	☽ ∠ ♀
07 45	☽ ⚺ ☿
08 41	☉ △ ♇
10 57	☽ Q ♆
13 12	☽ ⚼ ♇
16 54	☽ ⚻ ♄
19 05	☽ ⊥ ☿
22 15	☽ # ♃
02 Monday	
04 13	☽ ∠ ♃
04 32	☽ ± ♄
06 44	☽ ± ♇
07 56	♄ ⚼ ♆
09 41	☽ △ ♅
12 21	☽ ∠ ☿
13 16	☽ ⚹ ♀
18 32	☽ ⚻ ♇
21 39	☽ ⚺ ☉
22 08	☽ ⚼ ♄
22 10	☽ ⊥ ♃
22 15	☽ □ ♆
03 Tuesday	
02 48	☽ □ ♂
04 59	☉ ⚹ ♆
10 16	☽ ⚺ ♃
10 41	☽ ⊥ ☉
15 27	☽ ⚼ ♅
15 47	☽ Q ♀
17 53	☽ ⚹ ☿
22 58	☿ St D
04 Wednesday	
03 55	☽ △ ♄
06 25	☽ ∠ ☉
09 33	☽ ± ♅
15 25	☽ Q ♂
17 06	♀ □ ♇
18 09	☽ Q ☿
21 42	☽ ⚻ ♅
05 Thursday	
03 06	☽ ∥ ♃
06 43	☽ ☍ ♇
08 19	☽ □ ♀
10 43	☽ △ ♆
15 40	☽ ⚹ ☉
16 09	☽ ⚹ ♂
22 14	☉ △ ♂
23 45	☽ ☌ ♃
06 Friday	
07 20	☽ □ ☿
07 48	♀ ⚺ ♆
08 53	☽ ∥ ♇
16 26	☽ □ ♄
17 18	☽ ⚼ ♆
18 24	☽ Q ☉
23 09	☽ ∠ ♂
07 Saturday	
00 31	☽ # ♅
10 42	☽ ☍ ♅
11 35	☽ ± ♆
16 43	☽ # ☉
17 40	☽ ⊥ ♂
19 35	☽ ⚻ ♇
23 42	☽ ⚻ ♆
08 Sunday	
04 11	☽ △ ♀
04 35	☽ Q ♄
05 56	☽ ⚺ ♂
07 36	☽ ± ♇
10 05	☽ □ ☉
13 16	☽ ⚺ ♃
21 52	☽ △ ☿
09 Monday	
00 13	♀ ⚻ ♂
01 19	☽ ⊥ ♃
01 26	☽ ⚼ ♇
04 15	☽ ⚹ ♄
13 11	☽ ⚼ ♀
17 18	☽ # ☿
19 08	☽ ∠ ♃
22 08	☽ ⚻ ♅
10 Tuesday	
04 29	☽ ⚼ ☿
06 30	☽ △ ♇
07 06	☉ ⚹ ♃
08 24	☽ ± ♀
09 05	☽ ∠ ♄
09 37	☽ ± ♅
10 34	☽ ☍ ♆
17 20	☽ ☌ ♂
20 56	☽ ⚻ ♀
20 57	♀ Q ♅
22 10	♅ St R
22 26	☽ ± ☿
11 Wednesday	
00 02	☽ ⚹ ♃
01 13	☽ △ ☉
01 55	☽ ⊥ ♄
02 32	☽ ⚼ ♅
06 59	♀ ⊥ ♆
10 11	☽ ⚻ ☿
12 56	☽ ⚺ ♄
22 21	☽ Q ♃
12 Thursday	
00 01	☽ # ☿
05 51	☽ △ ♅
06 50	☽ ⚼ ☉
08 22	♄ # ♆
10 12	♀ □ ♃
13 33	☽ □ ♇
17 26	☽ ⚻ ♆
23 57	☽ ± ☉
13 Friday	
00 35	☽ ⚺ ♂
03 55	☽ ± ♆
04 54	☿ ☍ ♄
06 31	☽ □ ♃
08 12	☽ ☍ ♀
11 07	☽ ⊥ ♂
11 07	☽ ⚻ ☉
17 31	☽ ☌ ♄
18 23	☽ ☍ ☿
19 21	☽ ⚼ ♆
20 37	☽ Q ♇
21 10	☉ ∥ ♅
14 Saturday	
02 40	☽ ∠ ♂
09 31	☽ □ ♅
09 57	☿ ∠ ♆
11 17	☽ # ♅
12 38	☽ # ☉
16 40	☽ ⚹ ♇
20 26	☽ △ ♆
23 27	☽ ∥ ♇
15 Sunday	
01 19	♀ ± ♂
03 58	☽ ⚹ ♂
07 01	☉ Q ♆
09 05	♀ # ♄
09 19	☽ △ ♃
14 50	☽ ⚻ ♀
14 53	♀ ∥ ♆
15 06	☽ Q ♅
16 43	☽ ☍ ☉
17 11	☽ ∠ ♇
18 55	☽ ⚺ ♄
20 27	☽ ∥ ♃
23 15	☽ ⚻ ☿
23 22	☉ ⚼ ♇
23 50	☽ Q ♂
16 Monday	
01 26	☽ ± ♀
04 34	☽ ⊥ ♄
07 42	☽ ⊥ ♇
09 37	☽ ± ☿
09 54	☽ ⚼ ♃
10 29	☽ ⚹ ♅
17 14	☽ ⚼ ♀
17 21	☽ ⚺ ♇
18 59	☽ ∠ ♄
21 07	☽ □ ♆
23 04	☉ ⚻ ♄
17 Tuesday	
00 33	☽ ± ♃
01 08	☽ ⚼ ☿
05 20	☽ □ ♂
10 18	☽ ⚻ ♃
10 34	☽ ∠ ♅
13 35	♀ ☍ ♄
18 59	☽ ⚹ ♄
19 29	☽ △ ♀
20 31	☽ ⚻ ☉
18 Wednesday	
01 02	☽ ⊥ ♅
02 05	☽ Q ♆
03 04	☽ △ ☿
06 54	☽ ± ☉
10 43	☽ ⚺ ♅
14 16	☽ Q ♄
17 02	♀ ∠ ♆
17 37	☽ ☌ ♇
20 14	☽ # ♃
21 33	☽ ⚹ ♆
22 34	☽ ⚼ ☉
19 Thursday	
06 49	☽ △ ♂
11 34	☽ ☍ ♃
14 44	☉ # ♇
16 24	☽ ∥ ☉
16 33	☽ # ♇
19 38	☽ □ ♄
22 14	☽ ∠ ♆
20 Friday	
00 55	☽ □ ♀
01 06	☽ △ ☉
04 19	☽ ∥ ♅
06 01	☿ ⚼ ♂
08 11	☽ ⚼ ♂
08 19	☽ □ ☿
11 41	☉ ⚺ ♀
12 06	☽ ☌ ♅
13 17	☽ ⊥ ♆
19 13	☽ ⚺ ♇
23 26	☽ ⚺ ♆
23 41	☽ ± ♂
21 Saturday	
05 24	☽ ⊥ ♇
10 10	☽ ⚻ ♂
10 13	☽ Q ♀
14 45	☽ ⚻ ♃
17 43	☽ Q ☿
20 52	☽ ∠ ♇
22 16	☽ △ ♄
22 Sunday	
01 17	☽ ± ♃
05 12	☽ ∥ ☿
05 40	☿ □ ♅
08 16	☽ □ ☉
09 04	☽ ⚹ ♀
12 53	☽ ∥ ♀
13 40	☉ ± ♄
15 44	☽ ⚺ ♅
16 44	☽ ⚹ ☿
17 18	☽ ⚼ ♃
23 10	☽ ⚹ ♇
23 Monday	
00 31	☽ ⚼ ♄
02 21	☽ ⊥ ♅
03 43	☽ ☌ ♆
14 24	☽ ∠ ♀
16 12	☽ ☍ ♂
16 34	☽ ± ♄
18 29	☽ ∠ ♅
18 54	☽ Q ☉
20 32	☽ △ ♃
20 39	☽ Q ♇
22 19	☽ ∠ ☿
24 Tuesday	
03 23	☽ ⚻ ♄
04 50	♂ ⊥ ♄
08 24	☽ ⊥ ♀
16 15	♀ ⚼ ♂
16 30	☽ ⊥ ☿
17 39	☉ ± ♇
18 36	☽ ⚹ ☉
20 33	☽ ⚺ ♀
21 51	☽ ⚹ ♅
25 Wednesday	
04 47	☽ ⚺ ☿
04 58	☽ ∥ ♀
05 35	☽ □ ♇
10 11	♀ □ ♅
10 29	☽ ⚺ ♆
11 08	☽ ∥ ☿
12 39	☿ △ ♇
19 07	☿ Q ♃
20 09	☽ Q ♅
21 42	☽ ⊥ ♆
26 Thursday	
00 51	☽ ∠ ☉
00 54	☽ ⚻ ♂
04 49	☽ □ ♃
10 50	☽ ☍ ♄
12 18	☉ △ ♅
12 32	☽ ± ♂
14 41	☽ ∠ ♆
19 20	☽ ⊥ ☉
27 Friday	
06 08	☽ ⚼ ♂
06 11	☽ □ ♅
07 46	☽ ⚺ ☉
08 06	♂ ⚼ ♅
08 27	☽ ∥ ♅
11 03	☽ ☌ ♀
12 40	☿ ⚹ ♆
14 10	☽ △ ♇
15 39	☽ Q ♃
19 22	☽ ⚹ ♆
20 11	☽ ☌ ☿
21 14	☽ # ♇
28 Saturday	
02 34	☉ # ♃
11 54	☽ △ ♂
14 52	☉ ∠ ♃
15 16	☽ ⚹ ♃
17 42	☽ # ♃
18 13	♀ △ ♇
18 40	☽ Q ♆
19 10	☽ ⚼ ♇
19 40	☽ ∥ ☉
20 17	☽ ⚻ ♄
29 Sunday	
08 00	☽ ± ♄
12 48	☽ ± ♇
16 26	☽ △ ♅
18 03	♀ Q ♃
21 14	☽ ∠ ♃
23 25	☽ ☌ ☉

MARCH 1976

LONGITUDES

Date	Sidereal time h m s	Sun ☉ ° ' "	Moon ☽ ° ' "	Moon ☽ 24.00 ° ' "	Mercury ☿ ° '	Venus ♀ ° '	Mars ♂ ° '	Jupiter ♃ ° '	Saturn ♄ ° '	Uranus ♅ ° '	Neptune ♆ ° '	Pluto ♇ ° '
01	22 37 40	11 ♓ 02 22	16 ♓ 50 48	22 ♓ 51 02	18 ♒ 00	13 ♒ 18	23 ♊ 08	24 ♈ 32	26 ♋ 39	06 ♏ 58	13 ♐ 55	11 ♎ 07
02	22 41 36	12 02 34	28 ♓ 49 27	04 ♈ 46 13	19 26	14 32	23 30	24 45	26 R 36	06 R 57	13 55	11 R 06
03	22 45 33	13 02 45	10 ♈ 41 38	16 ♈ 35 58	20 53	15 46	23 52	24 57	26 34	06 56	13 56	11 04
04	22 49 29	14 02 53	22 ♈ 29 34	28 ♈ 22 51	22 21	17 00	24 15	25 10	26 31	06 55	13 56	11 03
05	22 53 26	15 03 00	04 ♉ 16 16	10 ♉ 10 18	23 51	18 14	24 38	25 22	26 29	06 54	13 56	11 01
06	22 57 22	16 03 05	16 ♉ 05 30	22 ♉ 02 28	25 21	19 28	25 01	25 35	26 26	06 52	13 57	11 00
07	23 01 19	17 03 07	28 ♉ 01 47	04 ♊ 04 07	26 53	20 42	25 24	25 48	26 24	06 51	13 57	10 58
08	23 05 15	18 03 07	10 ♊ 10 07	16 ♊ 20 26	28 ♒ 26	21 56	25 48	26 00	26 22	06 50	13 57	10 57
09	23 09 12	19 03 06	22 ♊ 35 45	28 ♊ 56 39	00 ♓ 00	23 10	26 12	26 13	26 20	06 48	13 58	10 55
10	23 13 09	20 03 02	05 ♋ 23 43	11 ♋ 57 28	01 35	24 24	26 36	26 26	26 18	06 47	13 58	10 54
11	23 17 05	21 02 56	18 ♋ 38 18	25 ♋ 26 28	03 11	25 38	27 01	26 39	26 16	06 45	13 58	10 52
12	23 21 02	22 02 47	02 ♌ 22 05	09 ♌ 25 08	04 49	26 52	27 25	26 52	26 14	06 44	13 58	10 51
13	23 24 58	23 02 37	16 ♌ 35 19	23 ♌ 52 11	06 28	28 06	27 50	27 06	26 13	06 42	13 58	10 49
14	23 28 55	24 02 24	01 ♍ 15 02	08 ♍ 42 59	08 08	29 ♒ 20	28 15	27 19	26 11	06 41	13 58	10 48
15	23 32 51	25 02 09	16 ♍ 14 58	23 ♍ 49 47	09 49	00 ♓ 34	28 41	27 32	26 10	06 39	13 58	10 46
16	23 36 48	26 01 52	01 ♎ 26 08	09 ♎ 02 42	11 31	01 48	29 07	27 45	26 09	06 37	13 R 58	10 45
17	23 40 44	27 01 33	16 ♎ 38 10	24 ♎ 11 22	13 15	03 02	29 33	27 59	26 07	06 36	13 58	10 43
18	23 44 41	28 01 12	01 ♏ 41 11	09 ♏ 06 43	14 59	04 16	29 ♊ 59	28 12	26 06	06 34	13 58	10 41
19	23 48 38	29 ♓ 00 50	16 ♏ 27 14	23 ♏ 42 11	16 45	05 30	00 ♋ 25	28 25	26 05	06 32	13 58	10 40
20	23 52 34	00 ♈ 00 26	00 ♐ 51 12	07 ♐ 54 05	18 32	06 44	00 52	28 39	26 05	06 30	13 58	10 38
21	23 56 31	01 00 00	14 ♐ 50 49	21 ♐ 41 29	20 21	07 58	01 18	28 53	26 04	06 28	13 58	10 36
22	00 00 27	01 59 32	28 ♐ 26 16	05 ♑ 05 26	22 11	09 12	01 45	29 06	26 03	06 26	13 58	10 35
23	00 04 24	02 59 03	11 ♑ 39 19	18 ♑ 08 16	24 02	10 26	02 12	29 20	26 03	06 24	13 57	10 33
24	00 08 20	03 58 32	24 ♑ 32 40	00 ♒ 52 54	25 54	11 40	02 40	29 33	26 02	06 22	13 57	10 31
25	00 12 17	04 57 59	07 ♒ 09 21	13 ♒ 22 22	27 48	12 54	03 07	29 ♈ 47	26 02	06 20	13 57	10 30
26	00 16 13	05 57 24	19 ♒ 32 19	25 ♒ 39 31	29 ♓ 43	14 08	03 35	00 ♉ 01	26 02	06 18	13 56	10 28
27	00 20 10	06 56 47	01 ♓ 44 16	07 ♓ 46 51	01 ♈ 39	15 22	04 03	00 15	26 02	06 16	13 56	10 26
28	00 24 07	07 56 09	13 ♓ 47 31	19 ♓ 46 30	03 36	16 36	04 31	00 29	26 D 02	06 14	13 56	10 25
29	00 28 03	08 55 28	25 ♓ 44 03	01 ♈ 40 23	05 35	17 50	04 59	00 42	26 02	06 12	13 55	10 23
30	00 32 00	09 54 46	07 ♈ 35 43	13 ♈ 30 17	07 35	19 04	05 28	00 56	26 02	06 10	13 55	10 21
31	00 35 56	10 ♈ 54 01	19 ♈ 24 19	25 ♈ 18 05	09 ♈ 35	20 ♓ 18	05 ♋ 56	01 ♉ 10	26 ♋ 02	06 ♏ 08	13 ♐ 54	10 ♎ 20

Date	Moon True ☊ ° '	Moon Mean ☊ ° '	Moon ☽ Latitude ° '
01	14 ♏ 31	16 ♏ 04	04 N 16
02	14 R 21	16 00	03 37
03	14 13	15 57	02 49
04	14 08	15 54	01 53
05	14 05	15 51	00 N 52
06	14 D 05	15 48	00 S 11
07	14 05	15 45	01 14
08	14 06	15 41	02 16
09	14 R 07	15 38	03 12
10	14 05	15 35	04 00
11	14 02	15 32	04 38
12	13 56	15 29	05 01
13	13 49	15 25	05 07
14	13 41	15 22	04 53
15	13 34	15 19	04 19
16	13 27	15 16	03 27
17	13 23	15 13	02 19
18	13 20	15 10	01 S 03
19	13 D 20	15 06	00 N 17
20	13 20	15 03	01 34
21	13 22	15 00	02 43
22	13 23	14 57	03 40
23	13 R 22	14 54	04 25
24	13 20	14 51	04 54
25	13 17	14 47	05 09
26	13 11	14 44	05 09
27	13 05	14 41	04 55
28	12 58	14 38	04 27
29	12 52	14 35	03 49
30	12 47	14 31	03 00
31	12 ♏ 43	14 ♏ 28	02 N 04

DECLINATIONS

Date	Sun ☉ ° '	Moon ☽ ° '	Mercury ☿ ° '	Venus ♀ ° '	Mars ♂ ° '	Jupiter ♃ ° '	Saturn ♄ ° '	Uranus ♅ ° '	Neptune ♆ ° '	Pluto ♇ ° '
01	07 S 26	01 S 15	16 S 54	17 S 19	25 N 50	08 N 31	21 N 11	13 S 22	20 S 57	11 N 35
02	07 03	02 N 51	16 33	17 00	25 50	08 36	21 11	13 21	20 57	11 36
03	06 40	06 49	16 10	16 41	25 50	08 40	21 12	13 21	20 57	11 37
04	06 17	10 30	15 46	16 21	25 50	08 45	21 12	13 21	20 57	11 38
05	05 53	13 46	15 21	16 01	25 50	08 50	21 13	13 20	20 57	11 39
06	05 30	16 29	14 54	15 40	25 50	08 55	21 13	13 20	20 57	11 40
07	05 07	18 31	14 26	15 19	25 50	08 59	21 14	13 19	20 57	11 40
08	04 43	19 44	13 57	14 57	25 50	09 04	21 14	13 19	20 57	11 41
09	04 20	20 02	13 26	14 35	25 50	09 09	21 15	13 18	20 57	11 42
10	03 56	19 20	12 55	14 13	25 50	09 14	21 15	13 18	20 57	11 43
11	03 33	17 33	12 22	13 50	25 50	09 19	21 16	13 17	20 57	11 44
12	03 09	14 44	11 48	13 27	25 50	09 24	21 16	13 17	20 57	11 44
13	02 46	10 59	11 12	13 03	25 49	09 28	21 16	13 16	20 57	11 45
14	02 22	06 28	10 36	12 39	25 49	09 33	21 17	13 16	20 57	11 46
15	01 58	01 N 27	09 58	12 15	25 48	09 38	21 17	13 15	20 57	11 47
16	01 35	03 S 44	09 19	11 51	25 48	09 43	21 17	13 15	20 57	11 48
17	01 11	08 41	08 39	11 26	25 47	09 48	21 18	13 14	20 57	11 48
18	00 47	13 02	07 57	11 00	25 46	09 53	21 18	13 14	20 57	11 49
19	00 S 24	16 29	07 14	10 35	25 45	09 58	21 18	13 13	20 57	11 50
20	00 00	18 48	06 31	10 09	25 45	10 03	21 18	13 12	20 57	11 51
21	00 N 24	19 53	05 46	09 43	25 44	10 07	21 19	13 12	20 57	11 51
22	00 48	19 46	05 00	09 17	25 42	10 12	21 19	13 11	20 57	11 52
23	01 11	18 32	04 12	08 50	25 41	10 17	21 19	13 10	20 57	11 53
24	01 35	16 23	03 24	08 23	25 40	10 22	21 19	13 10	20 57	11 54
25	01 58	13 30	02 35	07 56	25 39	10 27	21 19	13 09	20 57	11 54
26	02 22	10 04	01 44	07 29	25 37	10 32	21 19	13 08	20 56	11 55
27	02 45	06 16	00 53	07 01	25 35	10 37	21 19	13 08	20 56	11 56
28	03 09	02 S 16	00 S 01	06 33	25 34	10 42	21 19	13 07	20 56	11 56
29	03 32	01 N 48	00 N 52	06 05	25 32	10 47	21 19	13 06	20 56	11 57
30	03 56	05 46	01 46	05 37	25 30	10 52	21 19	13 06	20 56	11 58
31	04 N 19	09 N 30	02 N 41	05 S 09	25 N 28	10 N 57	21 N 19	13 S 05	20 S 56	11 N 58

ZODIAC SIGN ENTRIES

Date	h m	Planets
02	14 22	☽ ♈
05	03 18	☽ ♉
07	15 56	☽ ♊
09	12 02	☿ ♓
10	01 59	☽ ♋
12	07 55	☽ ♌
14	09 59	☽ ♍
15	00 59	♀ ♓
16	09 44	☽ ♎
18	09 17	☽ ♏
18	13 15	♂ ♋
20	10 34	☽ ♐
20	11 50	☉ ♈
22	14 48	☽ ♑
24	22 19	☽ ♒
26	10 25	♃ ♉
26	15 36	☿ ♈
27	08 34	☽ ♓
29	20 37	☽ ♈

LATITUDES

Date	Mercury ☿ ° '	Venus ♀ ° '	Mars ♂ ° '	Jupiter ♃ ° '	Saturn ♄ ° '	Uranus ♅ ° '	Neptune ♆ ° '	Pluto ♇ ° '
01	01 S 33	00 S 31	02 N 34	01 S 04	00 N 21	00 N 30	01 N 32	17 N 23
04	01 48	00 39	02 31	01 03	00 22	00 30	01 32	17 24
07	02 00	00 46	02 29	01 03	00 22	00 31	01 32	17 25
10	02 09	00 53	02 26	01 02	00 22	00 31	01 32	17 26
13	02 13	00 59	02 24	01 02	00 22	00 31	01 32	17 26
16	02 14	01 05	02 21	01 02	00 22	00 31	01 32	17 27
19	02 11	01 10	02 19	01 01	00 23	00 31	01 32	17 27
22	02 03	01 15	02 17	01 01	00 23	00 31	01 32	17 28
25	01 51	01 19	02 14	01 01	00 23	00 31	01 33	17 28
28	01 35	01 22	02 12	01 00	00 23	00 31	01 33	17 28
31	01 S 13	01 S 25	02 N 10	01 S 00	00 N 23	00 N 31	01 N 33	17 N 29

DATA

Julian Date	2442839
Delta T	+47 seconds
Ayanamsa	23° 31' 40"
Synetic vernal point	05° ♓ 35' 19"
True obliquity of ecliptic	23° 26' 27"

MOON'S PHASES, APSIDES AND POSITIONS ☽

Date	h m	Phase	Longitude ° '	Eclipse Indicator
09	04 38	☽	18 ♊ 45	
16	02 53	○	25 ♍ 39	
22	18 54	☾	02 ♑ 17	
30	17 08	●	10 ♈ 07	

Day	h m		
04	04 23	Apogee	
16	19 33	Perigee	
31	09 58	Apogee	
01	19 18	ON	
09	07 21	Max dec	20° N 04'
15	18 42	OS	
21	21 20	Max dec	19° S 59'
29	01 20	ON	

LONGITUDES

Date	Chiron ⚷ ° '	Ceres ⚳ ° '	Pallas ⚴ ° '	Juno ⚵ ° '	Vesta ⚶ ° '	Black Moon Lilith ⚸ ° '
01	25 ♈ 01	06 ♊ 47	13 ♈ 05	11 ♍ 06	21 ♈ 33	13 ♈ 35
11	25 ♈ 31	09 ♊ 21	17 ♈ 10	08 ♍ 38	25 ♈ 48	14 ♈ 42
21	26 ♈ 03	12 ♊ 15	21 ♈ 24	06 ♍ 32	00 ♉ 07	15 ♈ 48
31	26 ♈ 38	15 ♊ 27	25 ♈ 47	04 ♍ 58	04 ♉ 28	16 ♈ 55

All ephemeris data is given at 12.00 UT and the Moon's longitude is additionally given for 24.00 UT

ASPECTARIAN

01 Monday
00 38 ☽ ⊼ ♇
01 42 ☽ ⚼ ♄
04 09 ☽ ⊻ ♀
06 09 ☽ □ ♆
13 51 ☉ ⊼ ♇
14 37 ☽ ⊻ ☿
15 26 ☽ ⊥ ♃
17 28 ☽ ⊥ ♀
22 13 ☽ ⚼ ♅
23 56 ♀ ✶ ♆

02 Tuesday
00 58 ☽ □ ♂
01 58 ☉ ⚼ ♄
03 39 ☽ ⊻ ♃
04 15 ☽ ⊥ ☿
07 33 ☽ △ ♄
13 36 ☽ ∠ ♀
16 17 ☽ ± ♅

03 Wednesday
00 53 ☽ ∠ ☿
04 23 ☽ ⊼ ♅
11 07 ☽ ⫲ ☉
12 46 ☽ ☍ ♇
14 28 ☽ Q ♂
17 13 ☽ ⊻ ☉
18 34 ☽ △ ♆
23 31 ☽ ✶ ♀

04 Thursday
00 05 ☽ ∥ ♃
06 33 ☽ ⊥ ☉
09 17 ☉ □ ♆
11 41 ☽ ✶ ☿
15 42 ☽ ✶ ♂
17 32 ☽ ☌ ♃
19 58 ☽ ∥ ♇
20 11 ☽ □ ♄

05 Friday
01 08 ☽ ⚼ ♆
02 36 ☽ ∠ ☉
02 49 ☽ Q ♀
08 38 ☽ ⫲ ♅
15 40 ☽ Q ☿
17 20 ☽ ☍ ♅
19 28 ☽ ± ♆
23 16 ☽ ∠ ♂
23 25 ☽ ⫲ ☿

06 Saturday
01 42 ☽ ⊼ ♇
04 47 ☿ △ ♂
05 08 ☽ ⫲ ♀
07 39 ☽ ⊼ ♆
08 40 ☽ Q ♄
11 55 ☽ ✶ ☉
13 50 ☽ ± ♇
16 12 ☿ ✶ ♃
18 06 ☽ ⊥ ♂
19 36 ☽ □ ♀
21 24 ☿ Q ♆
22 02 ☿ ⚼ ♇

07 Sunday
04 40 ☿ ⊼ ♄
06 34 ☽ ⊻ ♂
07 27 ☽ ⊻ ♃
07 54 ☽ ⚼ ♇
08 45 ☽ ✶ ♄
09 22 ☽ □ ☿
14 13 ☽ Q ☉
19 38 ☽ ⊥ ♃

08 Monday
05 28 ☽ ⊼ ♅
13 31 ☽ △ ♇
13 40 ☽ ∠ ♃
14 20 ☽ ∠ ♄
17 11 ☽ ± ♅
19 23 ☽ ☍ ♆

09 Tuesday
04 38 ☽ □ ☉
07 41 ☽ ⊥ ♄
10 30 ☽ ⚼ ♅
13 12 ☽ △ ♀
15 17 ♂ ✶ ♃
18 18 ☿ ∥ ♅
19 00 ☽ ✶ ♃
19 03 ☽ ☌ ♂
19 04 ☽ ⊻ ♄
19 26 ♂ ⊻ ♄
22 34 ♃ □ ♄

10 Wednesday
03 56 ☽ △ ☿
14 33 ☽ △ ♅
17 41 ☽ Q ♃
20 07 ☽ ⚼ ♀
22 04 ☽ □ ♇
22 31 ☿ ± ♄
23 55 ☉ ⊥ ♃

11 Thursday
03 38 ☽ ⊼ ♆
11 06 ☽ ⚼ ☿
13 57 ☽ ± ♀
14 22 ☽ ± ♆
16 35 ♀ ⚼ ♇
16 37 ☽ △ ☉
18 29 ♀ Q ♆

12 Friday
00 05 ♀ ⊼ ♄
01 25 ☽ ☌ ♄
01 33 ☽ ⊼ ♀
02 21 ☽ □ ♃
03 11 ☽ ⊻ ♂
04 37 ☉ ⚼ ♅
05 03 ☽ ± ☿
05 56 ☽ Q ♇
06 08 ☽ ⚼ ♆
12 09 ♀ ✶ ♃
12 26 ☿ ± ♇
13 52 ☽ ⊥ ♂
14 12 ☿ ⫲ ♇
16 45 ☽ ⊼ ☿
19 26 ☽ □ ♅
20 35 ☽ ⚼ ☉
21 58 ☽ ⫲ ♀
22 01 ☽ ⫲ ♅
22 29 ♀ ∥ ♅

13 Saturday
02 23 ☽ ✶ ♇
04 17 ♀ △ ♂
05 34 ☽ ∠ ♂
07 30 ☽ ∥ ♇
07 38 ☽ △ ♆
10 29 ☽ ⫲ ☿
12 49 ☽ ± ☉
15 28 ☿ △ ♅
20 15 ☽ ∥ ♃
23 25 ☽ ⊼ ☉

14 Sunday
01 21 ☽ Q ♅
03 10 ☽ ∠ ♇
03 48 ☽ ⊻ ♄
05 31 ☽ △ ♃
07 01 ☽ ✶ ♂
08 37 ☽ ☍ ♀
13 31 ☽ ⊥ ♄
17 42 ☽ ⊥ ♇
20 43 ☽ ✶ ♅

15 Monday
00 27 ☽ ☍ ☿
02 53 ☽ Q ♂
03 18 ☽ ⊻ ♇
03 55 ☽ ∠ ♄
06 00 ☽ ⚼ ♃
08 23 ☽ □ ♆
09 19 ☽ ⫲ ☉
20 30 ☽ ± ♃
20 32 ☽ ∠ ♅
20 40 ♆ St R
22 52 ☿ ⫲ ♃

16 Tuesday
01 18 ☿ ⊼ ♇
02 42 ☽ ∥ ☉
02 53 ☽ ☍ ☉
03 40 ☽ ✶ ♄
06 07 ☽ ⊼ ♃
06 50 ☿ ⚼ ♄
08 14 ☽ □ ♂
10 43 ☽ ⊥ ♅
12 37 ☽ ⊼ ♀
12 51 ☽ Q ♆
14 40 ☉ △ ♄
14 47 ♀ ⫲ ♇
18 36 ♀ ± ♄
20 10 ☽ ⊻ ♅
22 34 ☽ Q ♄
22 55 ☽ ± ♀

17 Wednesday
02 39 ☽ ☌ ♇
05 56 ☽ ⊼ ☿
07 47 ☿ ∠ ♃
07 47 ☽ ✶ ♆
11 49 ☽ ∥ ☿
14 24 ☽ ⚼ ♀
16 40 ☽ ± ☿
17 53 ☽ ⫲ ♃
22 03 ☿ □ ♆

18 Thursday
01 19 ☽ ∥ ♀
03 04 ☽ □ ♄
04 48 ☽ ⫲ ♇
05 42 ☽ ⊼ ☉
06 19 ☽ ☍ ♃
07 38 ☽ ∠ ♆
08 55 ☽ ⚼ ☿
09 10 ☽ △ ♂
13 08 ☽ ∥ ♅
16 02 ☽ ± ☉
16 32 ☽ △ ♀
17 36 ☉ ⊻ ♃
19 51 ☽ ☌ ♅
20 04 ♀ ± ♇
22 09 ☽ ⊥ ♆

19 Friday
02 32 ☽ ⊻ ♇
07 42 ☽ ⚼ ☉
07 55 ☽ ⊻ ♆
10 15 ☽ ⚼ ♂
12 20 ☽ ⊥ ♇
12 34 ☽ △ ☿

20 Saturday
01 36 ☽ ± ♂
03 15 ☽ ∠ ♇
03 58 ☽ △ ♄
07 40 ♀ △ ♅
08 13 ☽ ⊼ ♃
10 28 ☽ △ ☉
12 01 ☽ ⊼ ♂
17 00 ♀ ⫲ ♃
18 33 ☽ ± ♃
21 35 ☽ ⊻ ♅
22 58 ☽ □ ♀
23 37 ☉ ± ♅

21 Sunday
04 40 ☽ ✶ ♇
05 27 ☽ ⚼ ♄
07 53 ☽ ⊥ ♅
10 17 ☽ ⚼ ♃
10 28 ☽ ☌ ♆
21 07 ☽ ± ♄
21 12 ♃ ⚼ ♆
23 07 ☽ □ ☿
23 35 ☽ ∠ ♅

22 Monday
01 25 ☉ □ ♂
01 35 ☽ Q ♇
02 30 ☿ ⚼ ♅
07 44 ☽ ⊼ ♄
09 33 ☽ Q ♀
13 13 ☽ △ ♃
18 10 ☽ ☍ ♂
18 54 ☽ □ ☉

23 Tuesday
01 40 ☿ ⊥ ♃
02 25 ☽ ✶ ♅
09 31 ☽ ✶ ♀
09 59 ☽ □ ♇
12 48 ☽ Q ☿
14 18 ♀ ⊼ ♇
16 14 ☽ ⊻ ♆
23 50 ♀ ⚼ ♄

24 Wednesday
00 28 ☽ Q ♅
03 23 ☽ ⊥ ♆
06 46 ☽ Q ☉
13 42 ☿ △ ♄
14 49 ☽ ☍ ♄
15 00 ☽ ✶ ☿
16 26 ☽ ∠ ♀
20 20 ☽ ∠ ♆
21 39 ☽ □ ♃

25 Thursday
03 58 ☽ ⊼ ♂
07 26 ☽ ✶ ☉
10 26 ☽ □ ♅
11 27 ☽ ⊥ ♀
14 36 ☽ ∥ ♅
15 55 ☽ ± ♂
18 25 ☽ △ ♇
23 30 ☽ ⫲ ♇
23 50 ☉ ⫲ ☿

26 Friday
00 18 ☽ ⊻ ♀
00 52 ☽ ∠ ☿
01 06 ☽ ✶ ♆
08 19 ♀ □ ♆
08 58 ☽ ⫲ ♃
08 58 ☽ Q ♃
10 03 ☽ ⚼ ♂
15 01 ☽ ∠ ☉
16 18 ☿ ⊻ ♃
19 16 ☿ ± ♅
20 10 ☉ ⊼ ♅
21 42 ☽ ⊥ ☿
23 36 ☽ ⚼ ♇

27 Saturday
00 33 ☽ Q ♆
00 43 ☽ ⊼ ♄
06 46 ☽ ∥ ♀
08 59 ☽ ✶ ♃
09 09 ♀ ∠ ♃
10 18 ☽ ⊥ ☉
11 47 ☽ ⊻ ☿
12 34 ☽ ± ♄
16 46 ☽ △ ♂
17 21 ☽ ± ♇
19 59 ♄ St D
20 58 ☽ △ ♅
23 16 ☽ ⊻ ☉

28 Sunday
05 16 ☽ ⊼ ♇
06 28 ☽ ⚼ ♄
07 12 ☽ ⫲ ☉
12 16 ☽ □ ♆
15 26 ☽ ∠ ♃
18 16 ☽ ☌ ♀
23 02 ☽ ⫲ ☿

29 Monday
02 33 ☿ □ ♂
02 53 ☽ ⚼ ♅
04 56 ☽ ∥ ☿
09 53 ☽ ⊥ ♃
12 36 ☽ △ ♄
19 20 ☿ ⊼ ♅
20 59 ☽ ± ♅
22 15 ☽ ⊻ ♃
23 30 ☽ ∥ ☉

30 Tuesday
07 29 ☽ □ ♂
09 06 ☽ ⊼ ♅
11 11 ☽ ⫲ ♀
11 57 ☽ ☌ ☿
17 08 ☽ ☌ ☉
17 36 ☽ ☍ ♇
22 31 ☉ ☍ ♇

31 Wednesday
00 49 ☽ △ ♆
14 02 ☽ ⊻ ♀
20 39 ☿ ☍ ♇
20 55 ♂ △ ♅
21 36 ☽ Q ♂
22 09 ☽ ∥ ♃

APRIL 1976

Date	Sidereal time h m s	Sun ☉	Moon ☽	Moon ☽ 24.00	Mercury ☿	Venus ♀	Mars ♂	Jupiter ♃	Saturn ♄	Uranus ♅	Neptune ♆	Pluto ♇
		° ' "	° ' "	° ' "	° '	° '	° '	° '	° '	° '	° '	° '
01	00 39 53	11 ♈ 53 15	01 ♉ 11 51	07 ♉ 05 55	11 ♈ 37	21 ♓ 32	06 ♋ 25	01 ♉ 24	26 ♋ 03	06 ♏ 05	13 ♐ 54	10 ♎ 18
02	00 43 49	12 52 26	13 ♉ 00 37	18 ♉ 56 18	13 40	22 46	06 54	01 38	26 03	06 R 03	13 R 53	10 R 16
03	00 47 46	13 51 35	24 ♉ 53 23	00 ♊ 52 17	15 43	24 00	07 23	01 52	26 04	06 01	13 53	10 15
04	00 51 42	14 50 42	06 ♊ 53 27	12 ♊ 57 23	17 48	25 14	07 52	02 06	26 05	05 58	13 52	10 13
05	00 55 39	15 49 47	19 ♊ 04 35	25 ♊ 15 35	19 52	26 28	08 21	02 20	26 06	05 56	13 51	10 11
06	00 59 36	16 48 50	01 ♋ 30 54	07 ♋ 51 04	21 57	27 42	08 51	02 35	26 07	05 54	13 51	10 10
07	01 03 32	17 47 50	14 ♋ 16 36	20 ♋ 47 58	24 02	28 ♓ 56	09 21	02 49	26 08	05 51	13 50	10 08
08	01 07 29	18 46 48	27 ♋ 25 35	04 ♌ 09 47	26 06	00 ♈ 09	09 50	03 03	26 09	05 49	13 50	10 06
09	01 11 25	19 45 43	11 ♌ 00 48	17 ♌ 58 45	28 ♈ 10	01 23	10 20	03 17	26 10	05 47	13 49	10 05
10	01 15 22	20 44 36	25 ♌ 03 35	02 ♍ 15 04	00 ♉ 13	02 37	10 50	03 31	26 12	05 44	13 48	10 03
11	01 19 18	21 43 27	09 ♍ 32 51	16 ♍ 56 18	02 15	03 51	11 20	03 46	26 13	05 42	13 47	10 02
12	01 23 15	22 42 16	24 ♍ 24 40	01 ♎ 57 01	04 15	05 05	11 51	04 00	26 15	05 39	13 46	10 00
13	01 27 11	23 41 03	09 ♎ 32 14	17 ♎ 09 08	06 13	06 19	12 21	04 14	26 17	05 37	13 45	09 58
14	01 31 08	24 39 47	24 ♎ 46 28	02 ♏ 22 58	08 09	07 33	12 52	04 28	26 19	05 34	13 44	09 57
15	01 35 05	25 38 30	09 ♏ 57 24	17 ♏ 28 38	10 02	08 47	13 22	04 43	26 21	05 32	13 43	09 55
16	01 39 01	26 37 10	24 ♏ 55 41	02 ♐ 17 40	11 52	10 01	13 53	04 57	26 23	05 29	13 42	09 53
17	01 42 58	27 35 49	09 ♐ 33 55	16 ♐ 43 55	13 39	11 14	14 24	05 11	26 25	05 27	13 41	09 52
18	01 46 54	28 34 26	23 ♐ 47 22	00 ♑ 44 06	15 23	12 28	14 55	05 26	26 27	05 24	13 40	09 50
19	01 50 51	29 ♈ 33 02	07 ♑ 34 06	14 ♑ 17 30	17 03	13 42	15 26	05 40	26 30	05 22	13 39	09 49
20	01 54 47	00 ♉ 31 36	20 ♑ 54 33	27 ♑ 25 33	18 39	14 56	15 57	05 54	26 32	05 19	13 38	09 47
21	01 58 44	01 30 08	03 ♒ 50 55	10 ♒ 11 04	20 10	16 10	16 29	06 09	26 35	05 17	13 37	09 46
22	02 02 40	02 28 39	16 ♒ 26 29	22 ♒ 37 39	21 38	17 24	17 00	06 23	26 37	05 14	13 36	09 44
23	02 06 37	03 27 08	28 ♒ 45 06	04 ♓ 49 15	23 00	18 38	17 32	06 37	26 40	05 12	13 35	09 43
24	02 10 34	04 25 35	10 ♓ 50 39	16 ♓ 49 45	24 19	19 52	18 03	06 52	26 43	05 09	13 34	09 41
25	02 14 30	05 24 01	22 ♓ 46 59	28 ♓ 42 46	25 32	21 05	18 35	07 06	26 46	05 06	13 33	09 40
26	02 18 27	06 22 24	04 ♈ 37 30	10 ♈ 31 34	26 41	22 19	19 07	07 20	26 49	05 04	13 32	09 38
27	02 22 23	07 20 47	16 ♈ 25 17	22 ♈ 19 00	27 44	23 33	19 39	07 35	26 52	05 01	13 30	09 37
28	02 26 20	08 19 07	28 ♈ 13 00	04 ♉ 07 36	28 43	24 47	20 11	07 49	26 56	04 59	13 29	09 35
29	02 30 16	09 17 26	10 ♉ 03 03	15 ♉ 59 38	29 ♉ 37	26 01	20 43	08 03	26 59	04 56	13 28	09 34
30	02 34 13	10 ♉ 15 43	21 ♉ 57 37	27 ♉ 57 15	00 ♊ 25	27 ♈ 15	21 ♋ 15	08 ♉ 18	27 ♋ 03	04 ♏ 54	13 ♐ 27	09 ♎ 33

Date	Moon True ☊	Moon Mean ☊	Moon ☽ Latitude
	° '	° '	° '
01	12 ♏ 40	14 ♏ 25	01 N 03
02	12 D 40	14 22	00 S 02
03	12 40	14 19	01 07
04	12 42	14 16	02 09
05	12 44	14 12	03 07
06	12 45	14 09	03 57
07	12 46	14 06	04 37
08	12 R 46	14 03	05 04
09	12 44	14 00	05 15
10	12 42	13 57	05 08
11	12 39	13 53	04 42
12	12 37	13 50	03 56
13	12 34	13 47	02 53
14	12 33	13 44	01 37
15	12 32	13 41	00 S 14
16	12 D 33	13 37	01 N 08
17	12 33	13 34	02 24
18	12 35	13 31	03 29
19	12 36	13 28	04 20
20	12 36	13 25	04 55
21	12 R 36	13 22	05 14
22	12 36	13 18	05 17
23	12 35	13 15	05 05
24	12 34	13 12	04 40
25	12 33	13 09	04 03
26	12 32	13 06	03 15
27	12 32	13 03	02 20
28	12 31	12 59	01 19
29	12 31	12 56	00 14
30	12 ♏ 31	12 ♏ 53	00 S 52

DECLINATIONS

Date	Sun ☉	Moon ☽	Mercury ☿	Venus ♀	Mars ♂	Jupiter ♃	Saturn ♄	Uranus ♅	Neptune ♆	Pluto ♇
	° '	° '	° '	° '	° '	° '	° '	° '	° '	° '
01	04 N 42	12 N 52	03 N 36	04 S 41	25 N 26	11 N 02	21 N 19	13 S 04	20 S 56	11 N 59
02	05 05	15 43	04 32	04 12	25 24	11 07	21 19	13 03	20 56	12 00
03	05 28	17 55	05 28	03 44	25 21	11 12	21 19	13 03	20 56	12 00
04	05 51	19 20	06 24	03 15	25 19	11 17	21 19	13 02	20 56	12 01
05	06 14	19 53	07 20	02 46	25 16	11 21	21 19	13 01	20 55	12 02
06	06 36	19 29	08 17	02 17	25 14	11 26	21 19	13 00	20 55	12 02
07	06 59	18 05	09 13	01 48	25 11	11 31	21 19	12 59	20 55	12 03
08	07 21	15 42	10 08	01 19	25 08	11 36	21 19	12 59	20 55	12 03
09	07 44	12 25	11 03	00 50	25 05	11 41	21 18	12 58	20 55	12 04
10	08 06	08 20	11 57	00 S 21	25 02	11 46	21 18	12 57	20 55	12 04
11	08 28	03 N 38	12 50	00 N 08	24 58	11 51	21 18	12 56	20 55	12 05
12	08 50	01 S 23	13 42	00 38	24 55	11 56	21 18	12 55	20 55	12 06
13	09 12	06 25	14 32	01 07	24 52	12 01	21 17	12 55	20 54	12 06
14	09 33	11 06	15 20	01 36	24 48	12 05	21 17	12 54	20 54	12 07
15	09 55	15 02	16 06	02 05	24 44	12 10	21 17	12 53	20 54	12 07
16	10 16	17 54	16 51	02 34	24 40	12 15	21 16	12 52	20 54	12 08
17	10 37	19 31	17 33	03 03	24 36	12 20	21 16	12 51	20 54	12 08
18	10 58	19 49	18 13	03 33	24 32	12 25	21 16	12 50	20 54	12 09
19	11 19	18 54	18 50	04 02	24 28	12 30	21 15	12 50	20 54	12 09
20	11 39	16 57	19 25	04 30	24 24	12 34	21 15	12 49	20 53	12 09
21	12 00	14 12	19 57	04 59	24 19	12 39	21 14	12 48	20 53	12 10
22	12 20	10 52	20 27	05 28	24 14	12 44	21 14	12 47	20 53	12 10
23	12 40	07 08	20 54	05 57	24 10	12 49	21 13	12 46	20 53	12 11
24	13 00	03 S 11	21 19	06 25	24 05	12 54	21 13	12 45	20 53	12 11
25	13 19	00 N 51	21 41	06 54	24 00	12 58	21 12	12 45	20 53	12 11
26	13 39	04 50	22 00	07 22	23 55	13 03	21 12	12 44	20 52	12 12
27	13 58	08 37	22 17	07 50	23 49	13 08	21 11	12 43	20 52	12 12
28	14 17	12 04	22 32	08 18	23 44	13 12	21 11	12 42	20 52	12 12
29	14 35	15 03	22 44	08 46	23 38	13 17	21 10	12 41	20 52	12 13
30	14 N 54	17 N 25	22 N 53	09 N 13	23 N 33	13 N 22	21 N 09	12 S 40	20 S 52	12 N 13

ZODIAC SIGN ENTRIES

Date	h	m	Planets	
01	09	34	☽	♉
03	22	15	☽	♊
06	09	06	☽	♋
08	08	56	♀	♈
08	16	36	☽	♌
10	09	29	☿	♉
10	20	16	☽	♍
12	20	54	☽	♎
14	20	14	☽	♏
16	20	15	☽	♐
18	22	43	☽	♑
19	23	03	☉	♉
21	04	47	☽	♒
23	14	28	☽	♓
26	02	37	☽	♈
28	15	37	☽	♉
29	23	11	☿	♊

LATITUDES

Date	Mercury ☿	Venus ♀	Mars ♂	Jupiter ♃	Saturn ♄	Uranus ♅	Neptune ♆	Pluto ♇
	° '	° '	° '	° '	° '	° '	° '	° '
01	01 S 05	01 S 26	02 N 09	01 S 00	00 N 23	00 N 31	01 N 33	17 N 29
04	00 38	01 28	02 07	00 59	00 23	00 31	01 33	17 29
07	00 S 07	01 30	02 04	00 59	00 24	00 31	01 33	17 28
10	00 N 26	01 31	02 02	00 59	00 24	00 31	01 33	17 28
13	01 00	01 31	02 00	00 59	00 24	00 31	01 33	17 28
16	01 31	01 31	01 58	00 58	00 24	00 31	01 33	17 28
19	01 59	01 30	01 56	00 58	00 24	00 31	01 34	17 27
22	02 22	01 28	01 54	00 58	00 24	00 31	01 34	17 27
25	02 37	01 26	01 52	00 58	00 25	00 31	01 34	17 26
28	02 43	01 24	01 50	00 58	00 25	00 31	01 34	17 25
31	02 N 41	01 S 21	01 N 48	00 S 58	00 N 25	00 N 31	01 N 34	17 N 24

DATA

Julian Date	2442870
Delta T	+47 seconds
Ayanamsa	23° 31' 43"
Synetic vernal point	05° ♓ 35' 17"
True obliquity of ecliptic	23° 26' 27"

MOON'S PHASES, APSIDES AND POSITIONS ☽

Date	h	m	Phase	Longitude ° '	Eclipse Indicator
07	19	02	☽	18 ♋ 05	
14	11	49	○	24 ♎ 39	
21	07	14	☾	01 ♒ 19	
29	10	19	●	09 ♉ 13	Annular

Day	h	m	
14	07	03	Perigee
27	12	43	Apogee

Day	h	m		
05	13	58	Max dec	19° N 53'
12	05	28	0S	
18	05	38	Max dec	19° S 51'
25	06	56	0N	

LONGITUDES

Date	Chiron ⚷	Ceres ⚳	Pallas ⚴	Juno ⚵	Vesta ⚶	Black Moon Lilith ⚸
	° '	° '	° '	° '	° '	° '
01	26 ♈ 41	15 ♊ 47	26 ♈ 14	04 ♍ 50	04 ♉ 54	17 ♈ 02
11	27 ♈ 17	19 ♊ 14	00 ♉ 46	03 ♍ 59	09 ♉ 17	18 ♈ 08
21	27 ♈ 54	22 ♊ 53	05 ♉ 27	03 ♍ 47	13 ♉ 42	19 ♈ 15
31	28 ♈ 30	26 ♊ 42	10 ♉ 14	04 ♍ 12	18 ♉ 06	20 ♈ 21

All ephemeris data is given at 12.00 UT and the Moon's longitude is additionally given for 24.00 UT

ASPECTARIAN

h m Aspects

01 Thursday
01 31 ☽ □ ♄
03 40 ☽ ⊥ ♀
03 40 ♀ ⚼ ♅
05 21 ☽ ∥ ♇
07 19 ☽ ⚼ ♆
11 27 ☉ ⋕ ♀
12 26 ☽ ☌ ♃
13 32 ☽ ⋕ ♅
18 06 ☉ ☌ ☿
21 55 ☽ ☍ ♅
23 04 ☽ ⚹ ♂

02 Friday
00 07 ☽ ∠ ♀
01 37 ☽ ± ♆
06 28 ☽ ⊼ ♇
06 32 ☿ ⋕ ♀
11 42 ☽ ⊻ ☉
13 36 ☽ ⊻ ☿
13 47 ☽ ⊼ ♆
14 07 ☽ Q ♄
14 35 ☿ △ ♆
18 36 ☽ ± ♇

03 Saturday
00 57 ☽ ⊥ ☉
04 17 ☽ ⊥ ☿
06 44 ☽ ∠ ♂
10 00 ☽ ⚹ ♀
12 21 ☉ ∥ ☿
12 25 ☉ △ ♆
12 43 ☽ ⚼ ♇
14 22 ☽ ⚹ ♄
20 41 ☽ ∠ ☉

04 Sunday
01 34 ☽ ⊥ ♂
02 08 ☽ ∠ ☿
02 17 ☽ ⊻ ♃
10 11 ☽ ⊼ ♅
12 45 ☽ Q ♀
14 01 ☽ ⊻ ♂
14 28 ☽ ⊥ ♃
18 35 ☽ △ ♇
20 19 ☽ ∠ ♄
22 02 ☽ ± ♅

05 Monday
01 47 ☽ ☍ ♆
04 48 ♀ △ ♄
05 06 ☽ ⚹ ☉
08 33 ☽ ∠ ♃
09 08 ♀ ⊥ ♃
13 51 ☽ ⚹ ☿
13 59 ☽ ⊥ ♄
15 37 ☽ ⚼ ♅

06 Tuesday
01 38 ☽ ⊻ ♄
03 53 ☽ □ ♀
06 24 ☽ Q ☉
14 04 ☽ ⚹ ♃
17 32 ☽ Q ☿
20 17 ☽ △ ♅

07 Wednesday
02 26 ☽ ☌ ♂
04 18 ☽ □ ♇
11 11 ☽ ⊼ ♆
13 01 ☽ Q ♃
19 02 ☽ □ ☉
22 14 ☽ ± ♆

08 Thursday
05 35 ♀ ± ♅
09 10 ☽ □ ☿
09 42 ☽ ☌ ♄
12 36 ☿ □ ♄
13 13 ☽ Q ♇
14 30 ☽ ⚼ ♆
17 23 ☽ △ ♀
22 12 ☽ □ ♃

09 Friday
00 17 ♂ □ ♇
02 52 ☽ □ ♅
08 20 ☽ ⋕ ♅
10 23 ☽ ⚹ ♇
10 47 ☽ ⊻ ♂
14 11 ☽ ∥ ♇
14 40 ☿ Q ♂
16 30 ☽ ∥ ♃
16 50 ☽ △ ♆
18 53 ☽ ∥ ☿
19 29 ☿ ⚼ ♆
21 32 ☽ ⊥ ♂
22 10 ☽ ⚼ ♀

10 Saturday
04 10 ☽ △ ☉
06 29 ☿ ∥ ♃
09 47 ☽ Q ♅
11 59 ☽ ∠ ♇
13 08 ☽ ∥ ☉
13 21 ☽ ∠ ♂
13 55 ☽ ⊻ ♄
14 52 ☽ ± ♀
15 21 ☿ ∥ ♇
22 02 ☽ △ ☿
23 56 ☽ ⊥ ♄

11 Sunday
01 47 ☽ ⊼ ♀
02 20 ☽ △ ♃
02 57 ☽ ⊥ ♇
05 42 ☽ ⚹ ♅
07 03 ☽ ⚼ ☉
09 44 ♀ ⊻ ♃
12 47 ☽ ⊻ ♇
14 44 ☽ ∠ ♄
14 47 ☿ ⋕ ♅
15 02 ☽ ⚹ ♂
18 53 ☽ □ ♆
22 45 ☽ ± ☉

12 Monday
02 26 ☽ ⚼ ☿
03 11 ☽ ⚼ ♃
03 20 ☽ ∥ ♀
06 00 ☽ ∠ ♅
08 02 ☽ ⋕ ♀
08 36 ☿ ☌ ♃
09 05 ☽ ⊼ ☉
11 04 ☽ Q ♂
14 57 ☽ ⚹ ♄
17 49 ☽ ± ♃
19 02 ☽ ± ☿
20 20 ☽ ⊥ ♅
22 45 ♀ ⊼ ♅
23 42 ☽ Q ♆

13 Tuesday
03 29 ☽ ⊼ ♃
04 48 ☿ ☍ ♅
05 49 ☽ ⊻ ♅
05 58 ☽ ⊼ ☿
06 28 ☽ ☍ ♀
10 01 ☽ Q ♄
12 41 ☽ ☌ ♇
15 27 ☿ ⊻ ♀
16 36 ☽ □ ♂
18 38 ☽ ⚹ ♆

14 Wednesday
03 03 ☽ ⋕ ☉
06 57 ☿ ± ♆
11 49 ☽ ☍ ☉
14 26 ☽ □ ♄
17 42 ☽ ⋕ ♃
17 43 ☽ ⋕ ♇
18 14 ☽ ∠ ♆
18 41 ♃ ∥ ♇
22 16 ☽ ∥ ♅

15 Thursday
03 33 ☽ ☍ ♃
05 00 ☽ ☌ ♅
08 27 ☽ ⊥ ♆
09 58 ☽ ⊼ ♀
10 33 ☿ ⊼ ♇
11 56 ☽ ⊻ ♇
12 08 ☽ ☍ ☿
17 38 ☽ △ ♂
17 59 ☽ ⊻ ♆
20 22 ☽ ± ♀
21 29 ☽ ⊥ ♇
22 34 ☽ ⋕ ☿

16 Friday
03 55 ♂ ⊼ ♆
05 55 ☉ □ ♄
09 44 ♀ ☍ ♇
11 56 ☽ ∠ ♇
12 09 ☽ ⚼ ♀
14 22 ☽ △ ♄
14 56 ☽ ⊼ ☉
18 40 ☽ ⚼ ♂

17 Saturday
01 26 ☽ ± ☉
04 38 ☽ ⊼ ♃
05 12 ☽ ⊻ ♅
10 00 ☽ ± ♂
12 28 ☿ ⊼ ♆
12 30 ☽ ⚹ ♇
14 45 ☽ ± ♃
15 03 ☽ △ ♀
15 05 ☽ ⚼ ♄
15 07 ☽ ⊥ ♅
17 25 ☽ ⚼ ☉
18 53 ☽ ☌ ♆
19 47 ☽ ⊼ ☿
20 23 ☽ ⊼ ♂
22 42 ☿ Q ♄

18 Sunday
02 36 ☿ ⚹ ♂
06 09 ☽ ⚼ ♃
06 14 ☽ ∠ ♅
06 17 ☽ ± ♄
07 19 ☽ ± ☿
08 40 ☽ Q ♇
10 16 ♃ ☍ ♅
14 25 ☉ ⚼ ♆
16 36 ☽ ⊼ ♄
18 24 ☿ ± ♇
20 52 ☽ △ ☉

19 Monday
00 58 ☽ ⚼ ☿
08 07 ☽ ⚹ ♅
08 35 ☽ △ ♃
11 06 ♀ △ ♆
12 43 ☽ ⋕ ☿
15 59 ☽ □ ♇
22 51 ☽ ⊻ ♆

20 Tuesday
00 03 ☽ □ ♀
02 38 ☽ ☍ ♂
05 29 ☽ Q ♅
07 19 ☽ △ ☿
09 41 ☽ ⊥ ♆
22 23 ☽ ☍ ♄

21 Wednesday
02 14 ☽ ∠ ♆
07 14 ☽ □ ☉
12 40 ☽ Q ♀
14 41 ☽ □ ♅
16 25 ☽ □ ♃
22 31 ♀ □ ♂
22 37 ☽ ∥ ♅
23 10 ☽ △ ♇
23 21 ☽ ⋕ ♃

22 Thursday
00 12 ☉ ∥ ♇
02 49 ☽ ⋕ ☉
03 04 ☽ ⋕ ♇
06 33 ☽ ⚹ ♆
13 08 ☽ ⊼ ♂
14 03 ☽ ⚹ ♀
20 29 ☽ Q ☉
23 20 ☽ □ ☿

23 Friday
01 09 ♃ ⋕ ♅
01 18 ☽ ± ♂
03 44 ☽ Q ♃
04 05 ☽ ⚼ ♇
05 47 ☽ Q ♆
07 54 ☽ ⊼ ♄
10 41 ☿ ⋕ ♆
18 34 ☽ ⋕ ♀
19 13 ☉ ⋕ ♅
19 46 ☽ ± ♄
19 47 ☽ ⚼ ♂
21 47 ☽ ± ♇
22 06 ☽ ⚹ ☉
22 43 ☽ ∠ ♀

24 Saturday
00 42 ☽ △ ♅
01 57 ☉ ∥ ♃
03 54 ☽ ⚹ ♃
05 59 ☿ ∥ ♄
09 41 ☽ ⊼ ♇
13 45 ☽ ⚼ ♄
15 17 ☽ Q ☿
17 26 ☽ □ ♆
18 44 ☽ ⊥ ♀
19 04 ☿ ⚼ ♇

25 Sunday
03 08 ☽ △ ♂
05 05 ☉ ☍ ♅
06 37 ☽ ⚼ ♅
06 46 ☽ ∠ ☉
08 11 ☽ ⊻ ♀
10 35 ☽ ∠ ♃
18 11 ☽ ⚹ ☿
20 06 ☽ △ ♄

26 Monday
00 45 ☽ ± ♅
02 36 ☽ ⊥ ☉
05 11 ☽ ⊥ ♃
12 53 ☽ ⊼ ♅
15 14 ☿ ⚹ ♄
15 52 ☽ ⊻ ☉
17 38 ☽ ⊻ ♃
22 10 ☽ ☍ ♇

27 Tuesday
03 47 ☽ ∠ ☿
05 18 ♃ ± ♆
06 02 ♂ ± ♆
06 04 ☽ △ ♆
06 11 ☽ ∥ ♀
06 40 ♂ Q ♃
15 51 ☉ ± ♆
18 52 ☽ □ ♂
19 35 ☉ ☌ ♃
23 50 ☽ ⊥ ☿

28 Wednesday
04 12 ☽ ☌ ♀
04 13 ☉ Q ♂
09 22 ☽ □ ♄
12 33 ☽ ⚼ ♆
13 03 ☽ ∥ ♇
13 07 ☽ ⊻ ☿
16 45 ☽ ⋕ ♅
20 54 ☽ ∥ ♃

29 Thursday
01 41 ☽ ☍ ♅
06 46 ☽ ± ♆
07 30 ☽ ∥ ☉
07 53 ☽ ☌ ♃
09 10 ☽ Q ♂
10 19 ☽ ☌ ☉
11 01 ☽ ⊼ ♇
18 39 ☉ ⊼ ♇
18 53 ☽ ⊼ ♆
22 01 ☽ Q ♄
23 07 ☽ ± ♇

30 Friday
07 54 ♀ □ ♄
10 30 ☽ ⚹ ♂
17 10 ☽ ⚼ ♇
22 14 ☽ ⚹ ♄
23 47 ☽ ⊻ ♀

MAY 1976

LONGITUDES

Date	Sidereal time h m s	Sun ☉ ° ' "	Moon ☽ ° ' "	Moon ☽ 24.00 ° ' "	Mercury ☿ ° '	Venus ♀ ° '	Mars ♂ ° '	Jupiter ♃ ° '	Saturn ♄ ° '	Uranus ♅ ° '	Neptune ♆ ° '	Pluto ♇ ° '
01	02 38 09	11 ♉ 13 58	03 ♊ 58 49	10 ♊ 02 34	01 ♊ 08	28 ♈ 28	21 ♋ 47	08 ♉ 32	27 ♋ 06	04 ♏ 51	13 ♐ 25	09 ♎ 31
02	02 42 06	12 12 12	16 ♊ 08 49	22 ♊ 17 51	01 46	29 ♈ 42	22 20	08 46	27 10	04 R 49	13 R 24	09 R 30
03	02 46 03	13 10 23	28 ♊ 29 57	04 ♋ 45 27	02 19	00 ♉ 56	22 52	09 01	27 13	04 46	13 23	09 29
04	02 49 59	14 08 33	11 ♋ 04 41	17 ♋ 27 57	02 47	02 10	23 25	09 15	27 17	04 44	13 21	09 27
05	02 53 56	15 06 41	23 ♋ 55 34	00 ♌ 27 52	03 09	03 23	23 57	09 29	27 21	04 41	13 20	09 26
06	02 57 52	16 04 47	07 ♌ 05 07	13 ♌ 47 34	03 25	04 37	24 30	09 44	27 25	04 39	13 19	09 25
07	03 01 49	17 02 50	20 ♌ 35 25	27 ♌ 28 47	03 37	05 51	25 03	09 58	27 29	04 36	13 17	09 23
08	03 05 45	18 00 52	04 ♍ 27 44	11 ♍ 32 11	03 43	07 05	25 36	10 12	27 34	04 34	13 16	09 22
09	03 09 42	18 58 52	18 ♍ 41 59	25 ♍ 56 48	03 R 44	08 19	26 09	10 27	27 38	04 31	13 14	09 21
10	03 13 38	19 56 50	03 ♎ 16 14	10 ♎ 39 40	03 40	09 32	26 42	10 41	27 42	04 29	13 13	09 20
11	03 17 35	20 54 46	18 ♎ 06 24	25 ♎ 35 35	03 31	10 46	27 15	10 55	27 47	04 26	13 11	09 19
12	03 21 32	21 52 41	03 ♏ 06 15	10 ♏ 37 25	03 18	12 00	27 48	11 09	27 51	04 24	13 10	09 18
13	03 25 28	22 50 34	18 ♏ 07 59	25 ♏ 36 52	03 01	13 13	28 21	11 23	27 56	04 22	13 08	09 16
14	03 29 25	23 48 25	03 ♐ 03 04	10 ♐ 25 34	02 39	14 27	28 54	11 38	28 00	04 19	13 07	09 15
15	03 33 21	24 46 15	17 ♐ 43 32	24 ♐ 56 14	02 15	15 41	29 ♋ 28	11 52	28 05	04 17	13 05	09 14
16	03 37 18	25 44 04	02 ♑ 03 04	09 ♑ 03 36	01 47	16 55	00 ♌ 01	12 06	28 10	04 14	13 04	09 13
17	03 41 14	26 41 51	15 ♑ 57 35	22 ♑ 44 54	01 16	18 08	00 35	12 20	28 15	04 12	13 02	09 12
18	03 45 11	27 39 37	29 ♑ 25 35	05 ♒ 59 48	00 44	19 22	01 08	12 34	28 20	04 10	13 01	09 11
19	03 49 07	28 37 22	12 ♒ 27 50	18 ♒ 50 03	00 ♊ 10	20 36	01 42	12 48	28 25	04 08	12 59	09 10
20	03 53 04	29 ♉ 35 06	25 ♒ 06 53	01 ♓ 18 52	29 ♉ 36	21 49	02 16	13 02	28 30	04 05	12 58	09 09
21	03 57 01	00 ♊ 32 49	07 ♓ 26 31	13 ♓ 30 25	29 01	23 03	02 49	13 16	28 35	04 03	12 56	09 08
22	04 00 57	01 30 31	19 ♓ 31 10	25 ♓ 29 20	28 27	24 17	03 23	13 30	28 41	04 01	12 55	09 08
23	04 04 54	02 28 11	01 ♈ 25 31	07 ♈ 20 16	27 54	25 31	03 57	13 44	28 46	03 59	12 53	09 07
24	04 08 50	03 25 51	13 ♈ 14 10	19 ♈ 07 42	27 22	26 44	04 31	13 58	28 51	03 57	12 51	09 06
25	04 12 47	04 23 30	25 ♈ 01 23	00 ♉ 55 40	26 53	27 58	05 05	14 12	28 57	03 55	12 50	09 05
26	04 16 43	05 21 07	06 ♉ 50 59	12 ♉ 47 42	26 26	29 ♉ 12	05 39	14 26	29 03	03 53	12 48	09 04
27	04 20 40	06 18 44	18 ♉ 46 10	24 ♉ 46 41	26 02	00 ♊ 25	06 13	14 40	29 08	03 51	12 47	09 04
28	04 24 36	07 16 19	00 ♊ 49 33	06 ♊ 54 57	25 41	01 39	06 47	14 53	29 14	03 49	12 45	09 03
29	04 28 33	08 13 54	13 ♊ 03 08	19 ♊ 14 14	25 24	02 53	07 22	15 07	29 20	03 47	12 43	09 02
30	04 32 30	09 11 27	25 ♊ 28 23	01 ♋ 45 43	25 11	04 07	07 56	15 21	29 25	03 45	12 42	09 02
31	04 36 26	10 ♊ 08 59	08 ♋ 06 19	14 ♋ 30 16	25 ♉ 02	05 ♊ 20	08 ♌ 30	15 ♉ 35	29 ♋ 31	03 ♏ 43	12 ♐ 40	09 ♎ 01

Date	Moon True ☊ ° '	Moon Mean ☊ ° '	Moon ☽ Latitude ° '
01	12 ♏ 31	12 ♏ 50	01 S 57
02	12 D 31	12 47	02 56
03	12 R 31	12 43	03 49
04	12 31	12 40	04 31
05	12 31	12 37	05 01
06	12 31	12 34	05 16
07	12 D 31	12 31	05 15
08	12 31	12 28	04 55
09	12 31	12 24	04 17
10	12 32	12 21	03 21
11	12 33	12 18	02 12
12	12 33	12 15	00 S 52
13	12 R 33	12 12	00 N 31
14	12 33	12 08	01 51
15	12 32	12 05	03 03
16	12 30	12 02	04 01
17	12 28	11 59	04 44
18	12 27	11 56	05 09
19	12 25	11 53	05 17
20	12 25	11 49	05 09
21	12 D 25	11 46	04 47
22	12 26	11 43	04 13
23	12 27	11 40	03 28
24	12 29	11 37	02 35
25	12 30	11 34	01 35
26	12 31	11 30	00 N 31
27	12 R 31	11 27	00 S 34
28	12 30	11 24	01 39
29	12 27	11 21	02 40
30	12 24	11 18	03 35
31	12 ♏ 20	11 ♏ 14	04 S 19

DECLINATIONS

Date	Sun ☉ ° '	Moon ☽ ° '	Mercury ☿ ° '	Venus ♀ ° '	Mars ♂ ° '	Jupiter ♃ ° '	Saturn ♄ ° '	Uranus ♅ ° '	Neptune ♆ ° '	Pluto ♇ ° '
01	15 N 12	19 N 02	23 N 00	09 N 41	23 N 27	13 N 26	21 N 09	12 S 39	20 S 51	12 N 13
02	15 30	19 48	23 05	10 08	23 21	13 31	21 08	12 39	20 51	12 14
03	15 48	19 37	23 08	10 35	23 15	13 36	21 07	12 38	20 51	12 14
04	16 05	18 28	23 08	11 01	23 09	13 40	21 07	12 37	20 51	12 14
05	16 22	16 22	23 06	11 28	23 02	13 45	21 06	12 36	20 51	12 14
06	16 39	13 24	23 01	11 54	22 56	13 49	21 05	12 35	20 51	12 14
07	16 56	09 39	22 55	12 20	22 49	13 54	21 05	12 35	20 50	12 15
08	17 12	05 18	22 46	12 45	22 43	13 58	21 04	12 34	20 50	12 15
09	17 28	00 N 32	22 35	13 10	22 36	14 03	21 03	12 33	20 50	12 15
10	17 44	04 S 23	22 23	13 35	22 29	14 07	21 02	12 32	20 50	12 15
11	17 59	09 08	22 08	14 00	22 22	14 12	21 01	12 31	20 50	12 15
12	18 14	13 22	21 51	14 24	22 15	14 16	21 00	12 30	20 49	12 15
13	18 29	16 44	21 33	14 48	22 07	14 21	21 00	12 30	20 49	12 15
14	18 44	18 57	21 14	15 12	22 00	14 25	20 59	12 29	20 49	12 15
15	18 58	19 50	20 53	15 35	21 52	14 29	20 58	12 28	20 49	12 16
16	19 12	19 24	20 30	15 58	21 45	14 34	20 57	12 27	20 49	12 16
17	19 25	17 48	20 07	16 20	21 37	14 38	20 56	12 27	20 48	12 16
18	19 38	15 14	19 44	16 42	21 29	14 42	20 55	12 26	20 48	12 16
19	19 51	12 00	19 19	17 04	21 21	14 46	20 54	12 25	20 48	12 16
20	20 04	08 18	18 55	17 25	21 12	14 51	20 53	12 24	20 48	12 16
21	20 16	04 20	18 31	17 46	21 04	14 55	20 52	12 24	20 48	12 15
22	20 28	00 S 16	18 07	18 06	20 56	14 59	20 51	12 23	20 47	12 15
23	20 39	03 N 45	17 44	18 26	20 47	15 03	20 50	12 22	20 47	12 15
24	20 50	07 36	17 22	18 45	20 38	15 07	20 49	12 22	20 47	12 15
25	21 01	11 10	17 01	19 04	20 29	15 12	20 48	12 21	20 47	12 15
26	21 12	14 18	16 41	19 22	20 20	15 16	20 47	12 20	20 47	12 15
27	21 22	16 51	16 23	19 40	20 11	15 20	20 46	12 20	20 46	12 15
28	21 31	18 42	16 07	19 57	20 02	15 24	20 44	12 19	20 46	12 15
29	21 41	19 43	15 53	20 14	19 53	15 28	20 43	12 18	20 46	12 15
30	21 50	19 47	15 41	20 30	19 43	15 32	20 42	12 18	20 46	12 14
31	21 N 58	18 N 53	15 N 31	20 N 46	19 N 34	15 N 36	20 N 41	12 S 17	20 S 46	12 N 14

ZODIAC SIGN ENTRIES

Date	h	m	Planets
01	04	05	☽ ♊
02	17	49	♀ ♉
03	14	53	☽ ♋
05	23	09	☽ ♌
08	04	21	☽ ♍
10	06	39	☽ ♎
12	07	03	☽ ♏
14	07	04	☽ ♐
16	08	31	☽ ♑
16	11	10	♂ ♌
18	13	02	☽ ♒
19	19	21	☿ ♉
20	21	27	☽ ♓
20	22	21	☉ ♊
23	09	07	☽ ♈
25	22	07	☽ ♉
27	03	43	♀ ♊
28	10	22	☽ ♊
30	20	39	☽ ♋

LATITUDES

Date	Mercury ☿ ° '	Venus ♀ ° '	Mars ♂ ° '	Jupiter ♃ ° '	Saturn ♄ ° '	Uranus ♅ ° '	Neptune ♆ ° '	Pluto ♇ ° '
01	02 N 41	01 S 21	01 N 48	00 S 58	00 N 25	00 N 31	01 N 34	17 N 24
04	02 28	01 17	01 46	00 58	00 25	00 31	01 34	17 24
07	02 04	01 13	01 44	00 57	00 25	00 31	01 34	17 23
10	01 31	01 08	01 42	00 57	00 25	00 31	01 34	17 22
13	00 N 49	01 03	01 40	00 57	00 25	00 30	01 34	17 21
16	00 S 01	00 58	01 38	00 57	00 26	00 30	01 34	17 19
19	00 53	00 52	01 36	00 57	00 26	00 30	01 34	17 18
22	01 45	00 46	01 34	00 57	00 26	00 30	01 34	17 17
25	02 31	00 40	01 32	00 57	00 26	00 30	01 34	17 16
28	03 09	00 33	01 30	00 57	00 26	00 30	01 34	17 14
31	03 S 37	00 S 26	01 N 29	00 S 57	00 N 26	00 N 30	01 N 34	17 N 13

DATA

Julian Date	2442900
Delta T	+47 seconds
Ayanamsa	23° 31' 46"
Synetic vernal point	05° ♓ 35' 14"
True obliquity of ecliptic	23° 26' 26"

MOON'S PHASES, APSIDES AND POSITIONS ☽

Date	h	m	Phase	Longitude ° '	Eclipse Indicator
07	05	17	☽	16 ♌ 47	
13	20	04	○	23 ♏ 10	partial
20	21	22	☾	29 ♒ 58	
29	01	47	●	07 ♊ 49	

Day	h	m		
12	16	36	Perigee	
25	00	07	Apogee	
02	19	34	Max dec	19° N 51'
09	14	36	0S	
15	15	53	Max dec	19° S 51'
22	13	37	0N	
30	01	50	Max dec	19° N 53'

LONGITUDES

Date	Chiron ⚷ ° '	Ceres ⚳ ° '	Pallas ⚴ ° '
01	28 ♈ 30	26 ♊ 42	10 ♉ 14
11	29 ♈ 05	00 ♋ 39	15 ♉ 08
21	29 ♈ 38	04 ♋ 44	20 ♉ 09
31	00 ♉ 09	08 ♋ 54	25 ♉ 16

Date	Juno ⚵ ° '	Vesta ⚶ ° '	Black Moon Lilith ⚸ ° '
01	04 ♍ 12	18 ♉ 06	20 ♈ 21
11	05 ♍ 10	22 ♉ 31	21 ♈ 28
21	06 ♍ 37	26 ♉ 54	22 ♈ 34
31	08 ♍ 29	01 ♊ 17	23 ♈ 41

ASPECTARIAN

h m	Aspects
01 Saturday	
06 01	☽ ☌ ☿
11 01	♀ P ♆
13 05	☽ ⊥ ♀
13 43	☽ ⚻ ♅
17 49	☽ ∠ ♂
21 12	☽ ⚺ ♃
22 57	☽ △ ♇
02 Sunday	
01 33	☽ ± ♅
03 35	☽ ⚺ ☉
04 08	☽ ∠ ♄
06 37	☽ ☍ ♆
08 51	☽ ∠ ♀
09 15	☽ ⊥ ♃
12 22	☽ ⊥ ♂
16 22	☽ ⊥ ☉
19 08	☽ P ♅
21 51	☽ ⊥ ♄
03 Monday	
00 36	☽ ⚺ ♂
03 10	☽ ∠ ♃
09 32	☽ ⚺ ♄
11 19	☽ ∠ ☉
16 56	☉ ⚻ ♆
17 11	☽ ⚹ ♀
19 38	☽ ⚺ ☿
23 59	☽ △ ♅
04 Tuesday	
07 30	☽ ⊥ ☿
08 29	☽ ⚹ ♃
08 56	☽ □ ♇
16 17	☽ ⚻ ♆
17 07	☿ ∥ ♂
18 15	☽ ⚹ ☉
18 26	☽ Q ♀
05 Wednesday	
00 58	☽ ∠ ☿
03 30	☽ ± ♆
05 19	☿ ⚺ ♀
06 33	♃ ⚻ ♇
07 25	☽ Q ♃
12 02	☽ ∥ ☉
12 03	☽ ☌ ♂
18 20	☽ Q ☉
18 21	☽ ☌ ♄
18 27	☽ Q ♇
18 28	☉ Q ♄
19 46	☉ ± ♇
20 05	☽ P ♆
06 Thursday	
05 15	☽ ⚹ ☿
07 06	☽ □ ♀
07 36	☽ □ ♅
09 00	☽ ∥ ♃
09 12	♄ Q ♇
12 24	♀ ☍ ♅
16 10	☽ ⚹ ♇
16 50	☽ □ ♃
17 36	☽ # ♅
19 53	☽ ∥ ♇
21 05	☽ ∥ ♀
23 07	☽ △ ♆
07 Friday	
03 07	☽ Q ☿
05 17	☽ □ ☉
07 18	♀ ∥ ♇
15 31	☽ Q ♅
18 38	☽ ∠ ♇
20 06	☽ ⚺ ♂
08 Saturday	
00 05	☽ ⚺ ♄
01 30	♀ # ♅
06 53	☽ ⊥ ♂
10 08	☽ ⊥ ♇
10 27	☽ ⊥ ♄
10 43	☽ □ ☿
12 10	☽ ⚹ ♅
15 29	♀ ± ♆
16 53	☽ △ ♀
20 20	☽ ⚺ ♇
21 55	☽ △ ♃
22 50	☽ ∠ ♂
09 Sunday	
01 47	☽ ∠ ♄
02 53	☽ □ ♆
05 05	☿ St R
08 38	☿ ∥ ♂
12 30	☽ △ ☉
13 22	☽ ∠ ♅
20 21	☽ P ♀
23 21	☽ P ♃
10 Monday	
00 48	☽ ⚹ ♂
02 51	☽ ⚹ ♄
04 11	☽ ⊥ ♅
08 00	♀ ⚻ ♇
08 39	☽ Q ♆
12 29	☽ ± ♀
12 38	☽ △ ☿
13 58	☽ ⚺ ♅
14 20	☽ ± ♃
14 55	☽ P ☉
21 09	☽ Q ♂
21 50	☽ ☌ ♇
22 30	☽ Q ♄
23 06	☽ ⚻ ♀
11 Tuesday	
00 14	☽ ⚻ ♃
04 06	☽ ⚹ ♆
06 30	☽ ± ☉
12 40	☽ P ☿
14 47	♂ Q ♇
15 39	♀ ☌ ♃
16 49	☽ ⚻ ☉
12 Wednesday	
02 02	♀ ∥ ♃
02 53	☽ ± ☿
03 12	☽ □ ♂
03 34	☽ □ ♄
04 08	☽ ∠ ♆
05 17	☽ # ♇
06 48	☽ ∥ ♅
12 19	☽ ⚻ ☿
14 04	☽ ☌ ♅
14 47	♂ ☌ ♄
17 53	☽ # ♃
18 28	☽ ⊥ ♆
19 31	☽ # ♀
21 52	☽ ⚺ ♇
13 Thursday	
01 03	☽ ☍ ♃
03 16	♂ P ♆
03 27	☽ ☍ ♀
04 02	☽ ⚺ ♆
07 26	☽ ⊥ ♇
10 23	♀ ⚻ ♆
20 04	☽ ☍ ☉
21 50	☽ ∠ ♇
14 Friday	
03 49	☽ △ ♄
05 02	☽ △ ♂
08 16	☽ # ☉
11 23	☽ ☍ ☿
14 03	☽ ⚺ ♅
22 04	☽ ⚹ ♇
22 58	☉ P ♇
23 48	☽ ⊥ ♅
15 Saturday	
02 12	☽ ⚻ ♃
03 27	♀ ± ♇
04 19	☽ P ♄
04 23	☽ ☌ ♆
06 00	☿ ∥ ♄
06 24	☽ P ♂
08 19	☽ ⚻ ♀
12 14	☽ ± ♃
12 48	♄ P ♆
14 34	☽ ∠ ♅
16 19	☿ # ♆
17 49	☽ Q ♇
19 11	☽ ± ♀
19 17	☽ ± ♄
20 28	♀ Q ♄
21 55	☽ ± ♂
16 Sunday	
00 34	☽ ⚻ ☉
03 29	☽ P ♃
05 23	☽ ⚻ ♄
08 25	☽ ⚻ ♂
11 25	☽ ± ☉
11 33	☽ ⚻ ☿
11 44	☽ P ♀
15 39	☽ # ☉
15 43	☽ ⚹ ♅
21 28	☽ ± ☿
17 Monday	
00 16	☽ □ ♇
04 00	☽ P ☉
05 34	☽ △ ♃
06 54	☽ ⚺ ♆
12 25	☽ Q ♅
12 32	☽ P ☿
16 13	☽ △ ♀
17 25	☽ ⊥ ♆
18 Tuesday	
00 46	☽ # ♀
03 18	☿ ⚹ ♂
03 44	♀ Q ♂
08 33	☽ △ ☉
09 27	☽ ∠ ♆
10 00	☽ ☍ ♄
14 17	☽ △ ☿
15 15	☽ ☍ ♂
15 25	☉ ∥ ☿
16 11	☽ # ♃
20 37	☽ □ ♅
19 Wednesday	
05 53	☽ △ ♇
06 22	☉ ⚹ ♄
09 04	☽ ∥ ♅
10 11	☽ # ♇
12 39	☽ □ ♃
12 59	☽ ⚹ ♆
20 Thursday	
05 00	♃ ⚻ ♆
05 01	☽ □ ♀
10 10	☽ P ♇
11 42	☽ Q ♆
12 15	☉ ☌ ☿
18 35	☽ ⚻ ♄
20 17	☽ □ ☿
21 22	☽ □ ☉
23 41	☽ Q ♃
21 Friday	
02 30	☽ ⚻ ♂
03 34	☽ ± ♇
05 22	☽ △ ♅
06 21	☽ ± ♄
14 51	☽ ± ♂
15 21	☽ ⚻ ♇
19 56	☽ Q ♀
22 50	☽ □ ♆
23 45	☽ ⚹ ♃
22 Saturday	
00 15	☽ P ♄
03 46	☿ ⚹ ♄
06 09	☽ Q ☿
09 01	♀ P ♇
09 37	☽ P ♂
11 00	☽ P ♅
11 59	☽ Q ☉
12 32	☿ ∥ ♀
22 40	☽ ⚹ ♀
23 Sunday	
02 35	♂ ∥ ♄
05 03	☽ ± ♅
05 11	☽ ⚹ ☿
06 27	☽ ∠ ♃
06 35	☽ △ ♄
11 21	♂ # ♆
13 12	♂ □ ♅
14 18	☽ ⚹ ☉
17 10	☽ ⚻ ♅
17 23	☽ △ ♂
21 08	☉ ∥ ♂
24 Monday	
01 04	☽ ⊥ ♃
03 35	☽ ☍ ♇
04 27	☉ # ♆
08 36	☽ ∠ ♀
08 44	☉ ∥ ♄
10 19	☽ ∠ ☿
11 14	☽ △ ♆
13 31	☽ ⚺ ♃
20 47	☿ ☌ ♀
23 31	☽ ∠ ☉
25 Tuesday	
00 24	☉ ⚻ ♅
03 54	☽ ⊥ ☿
05 03	☽ ⊥ ♀
15 38	☽ ⚺ ☿
17 42	☽ P ♆
18 41	☽ ⚺ ♀
19 27	☽ ⊥ ☉
19 56	☽ ∥ ♇
20 03	☽ □ ♄
20 37	☽ # ♅
26 Wednesday	
06 00	☽ ☍ ♅
08 42	☽ ⚺ ☉
08 46	♀ ⚹ ♄
09 27	☽ □ ♂
11 54	☽ ± ♆
15 12	♄ # ♆
16 29	☽ ⚻ ♇
20 36	☽ ∥ ♃
23 59	☽ ⚻ ♆
27 Thursday	
03 36	☽ ☌ ♃
04 34	☽ ± ♇
06 25	☉ ⚹ ♂
07 39	☽ ∥ ☿
08 42	☽ Q ♄
22 34	☽ P ♇
23 26	☽ Q ♂
28 Friday	
02 05	☽ ☌ ☿
08 49	☽ ⚹ ♄
13 49	☽ ☌ ♀
16 24	♀ ∥ ♂
17 52	☽ ⚻ ♅
29 Saturday	
00 20	☽ ⚹ ♂
01 47	☽ ☌ ☉
03 52	♃ ± ♇
04 10	☽ △ ♇
05 37	☽ ± ♅
11 22	☽ ☍ ♆
13 13	☿ Q ♂
14 30	☽ ∠ ♄
16 06	☽ ⚺ ♃
18 45	☽ ∥ ♂
23 05	☽ P ♅
30 Sunday	
03 56	☽ ⊥ ♃
05 03	♀ ⚻ ♅
06 54	☽ ∠ ♂
07 58	☉ △ ♇
08 02	☽ ⊥ ♄
11 28	☽ ⚺ ☿
16 37	☽ ∥ ♂
19 37	☽ ⚺ ♄
21 29	☽ ∠ ♃
22 46	☽ ⊥ ☿
31 Monday	
00 55	☽ ⊥ ♂
01 24	☉ ± ♅
03 34	☿ ∥ ♃
03 43	☽ △ ♅
04 58	♀ ∥ ♄
06 13	☽ ⚺ ♀
11 36	♀ # ♆
12 48	☽ ⚺ ♂
13 43	☽ □ ♇
15 36	☽ ∠ ☿
16 09	☽ ⚺ ☉
18 43	☽ ⊥ ♀
20 33	☽ ⚻ ♆

All ephemeris data is given at 12.00 UT and the Moon's longitude is additionally given for 24.00 UT

JUNE 1976

LONGITUDES

Date	Sidereal time h m s	Sun ☉ ° ′ ″	Moon ☽ ° ′ ″	Moon ☽ 24.00 ° ′ ″	Mercury ☿ ° ′	Venus ♀ ° ′	Mars ♂ ° ′	Jupiter ♃ ° ′	Saturn ♄ ° ′	Uranus ♅ ° ′	Neptune ♆ ° ′	Pluto ♇ ° ′
01	04 40 23	11 ♊ 06 30	20 ♋ 57 38	27 ♋ 28 28	24 ♉ 58	06 ♊ 34	09 ♌ 05	15 ♉ 48	29 ♋ 37	03 ♏ 41	12 ♐ 38	09 ♎ 00
02	04 44 19	12 04 00	04 ♌ 02 50	10 ♌ 40 47	24 D 57	07 48	09 39	16 02	29 43	03 R 39	12 R 37	09 R 00
03	04 48 16	13 01 28	17 ♌ 22 21	24 ♌ 07 36	25 02	09 01	10 14	16 15	29 49	03 37	12 35	08 59
04	04 52 12	13 58 56	00 ♍ 56 33	07 ♍ 49 13	25 10	10 15	10 49	16 29	29 ♋ 56	03 36	12 34	08 59
05	04 56 09	14 56 22	14 ♍ 45 35	21 ♍ 45 38	25 24	11 29	11 23	16 42	00 ♌ 02	03 34	12 32	08 59
06	05 00 05	15 53 46	28 ♍ 49 14	05 ♎ 56 17	25 41	12 42	11 58	16 56	00 08	03 32	12 30	08 58
07	05 04 02	16 51 10	13 ♎ 06 33	20 ♎ 19 44	26 03	13 56	12 33	17 09	00 14	03 30	12 29	08 58
08	05 07 59	17 48 32	27 ♎ 35 27	04 ♏ 53 13	26 30	15 10	13 08	17 22	00 21	03 29	12 27	08 57
09	05 11 55	18 45 54	12 ♏ 12 29	19 ♏ 32 34	27 00	16 23	13 43	17 35	00 27	03 27	12 25	08 57
10	05 15 52	19 43 14	26 ♏ 52 46	04 ♐ 12 15	27 35	17 37	14 17	17 49	00 34	03 26	12 24	08 57
11	05 19 48	20 40 34	11 ♐ 30 13	18 ♐ 45 50	28 14	18 51	14 52	18 02	00 40	03 24	12 22	08 56
12	05 23 45	21 37 53	25 ♐ 58 17	03 ♑ 06 48	28 58	20 04	15 27	18 15	00 47	03 23	12 21	08 56
13	05 27 41	22 35 11	10 ♑ 10 45	17 ♑ 09 32	29 ♉ 45	21 18	16 03	18 28	00 53	03 22	12 19	08 56
14	05 31 38	23 32 29	24 ♑ 02 44	00 ♒ 50 03	00 ♊ 36	22 32	16 38	18 41	01 00	03 20	12 18	08 56
15	05 35 34	24 29 46	07 ♒ 31 18	14 ♒ 06 27	01 31	23 46	17 13	18 54	01 07	03 19	12 16	08 56
16	05 39 31	25 27 02	20 ♒ 35 38	26 ♒ 59 04	02 29	24 59	17 48	19 07	01 13	03 18	12 14	08 56
17	05 43 28	26 24 19	03 ♓ 17 04	09 ♓ 30 04	03 32	26 13	18 23	19 19	01 20	03 16	12 13	08 56
18	05 47 24	27 21 35	15 ♓ 38 33	21 ♓ 43 04	04 38	27 27	18 59	19 32	01 27	03 15	12 11	08 56
19	05 51 21	28 18 50	27 ♓ 44 13	03 ♈ 42 38	05 47	28 40	19 34	19 45	01 34	03 14	12 10	08 D 56
20	05 55 17	29 ♊ 16 06	09 ♈ 38 57	15 ♈ 33 51	07 00	29 ♊ 54	20 09	19 57	01 41	03 13	12 08	08 56
21	05 59 14	00 ♋ 13 21	21 ♈ 27 57	27 ♈ 21 54	08 17	01 ♋ 08	20 45	20 10	01 48	03 12	12 07	08 56
22	06 03 10	01 10 36	03 ♉ 16 19	09 ♉ 11 47	09 37	02 22	21 20	20 22	01 55	03 11	12 05	08 56
23	06 07 07	02 07 51	15 ♉ 08 51	21 ♉ 08 03	11 00	03 35	21 56	20 35	02 02	03 10	12 04	08 56
24	06 11 03	03 05 06	27 ♉ 09 49	03 ♊ 14 34	12 27	04 49	22 32	20 47	02 09	03 09	12 02	08 56
25	06 15 00	04 02 21	09 ♊ 22 38	15 ♊ 34 18	13 57	06 03	23 07	20 59	02 16	03 08	12 01	08 56
26	06 18 56	04 59 35	21 ♊ 49 45	28 ♊ 09 08	15 31	07 16	23 43	21 12	02 23	03 08	11 59	08 57
27	06 22 53	05 56 50	04 ♋ 32 31	10 ♋ 59 52	17 07	08 30	24 19	21 24	02 30	03 07	11 58	08 57
28	06 26 50	06 54 04	17 ♋ 31 07	24 ♋ 06 08	18 47	09 44	24 55	21 36	02 38	03 06	11 56	08 57
29	06 30 46	07 51 18	00 ♌ 44 46	07 ♌ 26 45	20 30	10 58	25 30	21 48	02 45	03 06	11 55	08 57
30	06 34 43	08 ♋ 48 31	14 ♌ 11 53	20 ♌ 59 54	22 ♊ 16	12 ♋ 11	26 ♌ 06	22 ♉ 00	02 ♌ 52	03 ♏ 05	11 ♐ 53	08 ♎ 58

Date	Moon True ☊ ° ′	Moon Mean ☊ ° ′	Moon ☽ Latitude ° ′
01	12 ♏ 15	11 ♏ 11	04 S 52
02	12 R 11	11 08	05 10
03	12 07	11 05	05 12
04	12 05	11 02	04 56
05	12 05	10 59	04 24
06	12 D 05	10 55	03 35
07	12 07	10 52	02 32
08	12 08	10 49	01 S 19
09	12 R 08	10 46	00 00
10	12 08	10 43	01 N 20
11	12 05	10 40	02 33
12	12 01	10 36	03 36
13	11 55	10 33	04 24
14	11 49	10 30	04 55
15	11 42	10 27	05 09
16	11 36	10 24	05 06
17	11 32	10 20	04 48
18	11 30	10 17	04 17
19	11 29	10 14	03 35
20	11 D 29	10 11	02 44
21	11 30	10 08	01 46
22	11 32	10 05	00 N 45
23	11 R 32	10 01	00 S 20
24	11 31	09 58	01 23
25	11 27	09 55	02 24
26	11 21	09 52	03 19
27	11 14	09 49	04 06
28	11 05	09 46	04 40
29	10 55	09 42	05 00
30	10 ♏ 46	09 ♏ 39	05 S 05

DECLINATIONS

Date	Sun ☉ ° ′	Moon ☽ ° ′	Mercury ☿ ° ′	Venus ♀ ° ′	Mars ♂ ° ′	Jupiter ♃ ° ′	Saturn ♄ ° ′	Uranus ♅ ° ′	Neptune ♆ ° ′	Pluto ♇ ° ′
01	22 N 07	17 N 00	15 N 23	21 N 01	19 N 24	15 N 40	20 N 40	12 S 16	20 S 45	12 N 14
02	22 14	14 13	15 18	21 15	19 14	15 43	20 39	12 16	20 45	12 14
03	22 22	10 40	15 14	21 29	19 04	15 47	20 37	12 15	20 45	12 13
04	22 29	06 31	15 13	21 42	18 54	15 51	20 36	12 15	20 45	12 13
05	22 35	01 N 57	15 14	21 55	18 44	15 55	20 35	12 14	20 45	12 13
06	22 42	02 S 49	15 17	22 07	18 34	15 59	20 34	12 14	20 44	12 13
07	22 47	07 31	15 22	22 19	18 23	16 02	20 32	12 13	20 44	12 12
08	22 53	11 50	15 28	22 29	18 13	16 06	20 31	12 12	20 44	12 12
09	22 58	15 30	15 37	22 40	18 02	16 10	20 30	12 12	20 44	12 12
10	23 03	18 10	15 47	22 49	17 52	16 14	20 28	12 11	20 44	12 11
11	23 07	19 38	15 59	22 58	17 41	16 17	20 27	12 11	20 43	12 11
12	23 11	19 47	16 13	23 06	17 30	16 21	20 26	12 11	20 43	12 11
13	23 14	18 40	16 28	23 14	17 19	16 24	20 24	12 10	20 43	12 10
14	23 17	16 27	16 44	23 21	17 08	16 28	20 23	12 10	20 43	12 10
15	23 20	13 25	17 02	23 27	16 56	16 31	20 21	12 09	20 43	12 09
16	23 22	09 47	17 20	23 32	16 45	16 35	20 20	12 09	20 42	12 09
17	23 23	05 49	17 40	23 37	16 34	16 38	20 18	12 08	20 42	12 08
18	23 25	01 S 43	18 00	23 41	16 22	16 42	20 17	12 08	20 42	12 08
19	23 26	02 N 23	18 21	23 45	16 10	16 45	20 16	12 08	20 42	12 07
20	23 26	06 20	18 43	23 47	15 59	16 48	20 14	12 07	20 42	12 07
21	23 26	10 01	19 05	23 49	15 47	16 52	20 13	12 07	20 41	12 06
22	23 26	13 18	19 27	23 51	15 35	16 55	20 11	12 07	20 41	12 06
23	23 25	16 04	19 50	23 51	15 23	16 58	20 10	12 07	20 41	12 05
24	23 24	18 10	20 12	23 51	15 11	17 01	20 08	12 06	20 41	12 05
25	23 23	19 29	20 35	23 50	14 58	17 05	20 06	12 06	20 41	12 04
26	23 21	19 52	20 57	23 49	14 46	17 08	20 05	12 06	20 41	12 04
27	23 18	19 16	21 18	23 47	14 33	17 11	20 03	12 06	20 40	12 03
28	23 16	17 40	21 39	23 44	14 21	17 14	20 02	12 05	20 40	12 02
29	23 12	15 06	21 59	23 40	14 08	17 17	20 00	12 05	20 40	12 02
30	23 N 09	11 N 42	22 N 18	23 N 36	13 N 56	17 N 20	19 N 59	12 S 05	20 S 40	12 N 01

ZODIAC SIGN ENTRIES

Date	h	m	Planets
02	04	37	☽ ♌
04	10	21	☽ ♍
05	05	09	♄ ♌
06	14	00	☽ ♎
08	15	58	☽ ♏
10	17	07	☽ ♐
12	18	45	☽ ♑
13	19	20	☿ ♊
14	22	31	☽ ♒
17	05	43	☽ ♓
19	16	32	☽ ♈
20	13	56	♀ ♋
21	06	24	☉ ♋
22	05	21	☽ ♉
24	17	37	☽ ♊
27	03	29	☽ ♋
29	10	39	☽ ♌

LATITUDES

Date	Mercury ☿ ° ′	Venus ♀ ° ′	Mars ♂ ° ′	Jupiter ♃ ° ′	Saturn ♄ ° ′	Uranus ♅ ° ′	Neptune ♆ ° ′	Pluto ♇ ° ′
01	03 S 44	00 S 24	01 N 28	00 S 57	00 N 26	00 N 30	01 N 34	17 N 12
04	03 58	00 17	01 26	00 57	00 27	00 30	01 34	17 11
07	04 01	00 10	01 24	00 57	00 27	00 30	01 34	17 09
10	03 56	00 S 03	01 23	00 57	00 27	00 30	01 34	17 08
13	03 43	00 N 04	01 21	00 57	00 27	00 30	01 34	17 06
16	03 23	00 12	01 19	00 57	00 27	00 30	01 34	17 05
19	02 58	00 19	01 17	00 58	00 28	00 30	01 34	17 03
22	02 28	00 26	01 15	00 58	00 28	00 29	01 34	17 02
25	01 55	00 32	01 14	00 58	00 28	00 29	01 34	17 00
28	01 19	00 39	01 12	00 58	00 28	00 29	01 34	16 58
31	00 S 43	00 N 45	01 N 10	00 S 58	00 N 28	00 N 29	01 N 34	16 N 57

DATA

Julian Date	2442931
Delta T	+47 seconds
Ayanamsa	23° 31' 50"
Synetic vernal point	05° ♓ 35' 09"
True obliquity of ecliptic	23° 26' 25"

MOON'S PHASES, APSIDES AND POSITIONS ☽

Date	h	m	Phase	Longitude ° ′	Eclipse Indicator
05	12	20	☽	14 ♍ 57	
12	04	15	○	21 ♐ 19	
19	13	15	☾	28 ♓ 22	
27	14	50	●	06 ♋ 04	

Day	h	m		
09	19	19	Perigee	
21	16	41	Apogee	
05	21	51	0S	
12	02	40	Max dec	19° S 53'
18	21	59	0N	
26	09	40	Max dec	19° N 53'

LONGITUDES

Date	Chiron ⚷ ° ′	Ceres ⚳ ° ′	Pallas ⚴ ° ′	Juno ⚵ ° ′	Vesta ⚶ ° ′	Black Moon Lilith ⚸ ° ′
01	00 ♉ 12	09 ♋ 19	25 ♉ 47	08 ♍ 41	01 ♊ 43	23 ♈ 48
11	00 ♉ 40	13 ♋ 35	01 ♊ 00	10 ♍ 55	06 ♊ 03	24 ♈ 54
21	01 ♉ 04	17 ♋ 54	06 ♊ 19	13 ♍ 25	10 ♊ 21	26 ♈ 01
31	01 ♉ 24	22 ♋ 17	11 ♊ 43	16 ♍ 09	14 ♊ 35	27 ♈ 07

ASPECTARIAN

h m Aspects

01 Tuesday
02 15 ☽ ⚹ ♃
04 17 ☽ ⊥ ☉
07 43 ☽ ± ♆
08 57 ♂ ⚹ ♇
13 14 ☽ ∠ ♀
19 22 ☽ ⚹ ☿
22 15 ☽ ∠ ☉
23 08 ☽ Q ♇

02 Wednesday
00 12 ☽ ∥ ♃
00 17 ☽ ⚼ ♆
00 49 ☽ Q ♃
01 20 ☿ St D
03 19 ☽ ∥ ☿
04 03 ☽ ☌ ♄
11 17 ☽ □ ♅
17 18 ☽ Q ☿
19 29 ☽ ⚹ ♀
20 58 ☽ ⚹ ♇
22 37 ☽ ☌ ♂

03 Thursday
01 20 ☉ ☍ ♆
01 52 ☽ # ♅
02 04 ☽ ∥ ♇
03 27 ☽ △ ♆
03 37 ☽ ⚹ ☉
09 58 ☽ □ ♃
11 23 ♀ △ ♇
19 09 ☽ Q ♀
19 33 ☽ Q ♅
23 26 ♀ ± ♅
23 45 ☽ ∠ ♇

04 Friday
01 43 ☽ □ ☿
02 37 ☽ Q ☉
10 12 ☽ ⊻ ♄
15 34 ☽ ⊥ ♇
16 38 ☽ ⚹ ♅
20 46 ☽ ⊥ ♄

05 Saturday
02 01 ☽ ⊻ ♇
05 47 ☽ □ ♀
05 56 ☽ ⊻ ♂
08 10 ☽ □ ♆
08 40 ♀ ⚹ ♂
12 20 ☽ □ ☉
12 28 ☽ ∠ ♄
14 32 ☉ ∠ ♄
15 24 ☽ △ ♃
16 43 ☽ ⊥ ♂
18 31 ☽ ∠ ♅

06 Sunday
06 34 ☽ △ ☿
08 10 ♀ ☍ ♆
08 44 ☽ ∠ ♂
09 50 ☽ ⊥ ♅
14 14 ☽ ⚹ ♄
14 51 ☽ Q ♆
17 20 ☽ ⚼ ♃
19 57 ☽ ⊻ ♅

07 Monday
05 04 ☽ ☌ ♇
08 28 ☽ ⚼ ☿
08 41 ☽ ± ♃
09 19 ♂ △ ♆
10 32 ☽ Q ♄
10 57 ☽ ⚹ ♆
11 01 ☽ ⚹ ♂
13 30 ☽ △ ♀
18 40 ☽ △ ☉
18 50 ☽ ⊼ ♃
21 41 ☉ ⊻ ♃
23 53 ☽ ± ☿

08 Tuesday
07 46 ☽ Q ♂
10 08 ☽ ⊼ ☿
11 46 ☽ ∠ ♆
14 09 ☽ # ♇
14 12 ☽ ∥ ♅
15 54 ♀ ∠ ♄
16 34 ☽ □ ♄
16 38 ☽ ⚼ ♀
21 11 ☽ ⚼ ☉
21 40 ☽ ☌ ♅

09 Wednesday
02 32 ☽ ⊥ ♆
04 26 ☉ ⚼ ♅
06 40 ☽ ⊻ ♇
08 45 ☽ ± ♀
12 21 ☽ ⊻ ♆
12 58 ☽ # ☿
12 59 ☽ ± ☉
14 33 ☽ □ ♂
16 29 ☽ ⊥ ♇
17 17 ☽ # ♃
19 28 ☽ ⊼ ♀
20 57 ☽ ☍ ♃
23 29 ☽ ⊼ ☉

10 Thursday
07 12 ☽ ∠ ♇
08 51 ☽ # ♂
13 13 ☽ ☍ ☿
16 35 ♀ ⊻ ♃
18 04 ☽ △ ♄
22 43 ☽ ⊻ ♅

11 Friday
03 33 ♀ ⚼ ♅
07 47 ☽ ⚹ ♇
08 33 ☽ ⊥ ♅
13 26 ☽ ☌ ♆
17 48 ☽ △ ♂
18 56 ☽ ⚼ ♄
22 57 ☽ ⊼ ♃
23 23 ☽ ∠ ♅

12 Saturday
01 16 ☽ ☍ ♀
03 37 ☽ Q ♇
04 15 ☽ ☍ ☉
09 05 ☽ ± ♃
09 07 ♅ # ♇
09 59 ☽ ± ♄
17 17 ☽ ⊼ ☿
19 51 ☽ ⚼ ♂
20 07 ☽ ⊼ ♄

13 Sunday
00 25 ☽ ⚼ ♃
00 26 ☽ ⚹ ♅
04 00 ☽ ± ☿
04 30 ☿ ∥ ♃
09 53 ☽ □ ♇
11 45 ☽ ± ♂
13 16 ☉ ∥ ♀
15 39 ☽ ⊻ ♆
20 20 ☽ ⚼ ☿
20 53 ☽ Q ♅
22 31 ☽ ⊼ ♂

14 Monday
01 59 ☽ ⊥ ♆
02 29 ☽ △ ♃
05 10 ☽ # ♂
09 05 ☽ ⊼ ♀
09 40 ☽ # ☿
11 03 ☽ ⊼ ☉
11 55 ☽ # ♃
17 42 ☽ ∠ ♆
20 42 ☽ ± ♀
22 26 ☽ ± ☉

15 Tuesday
00 12 ☿ ⚹ ♄
00 24 ☽ ☍ ♄
00 25 ☽ △ ☿
04 27 ☽ □ ♅
07 35 ☿ ∥ ♂
14 28 ☽ ⚼ ♀
14 33 ☽ △ ♇
15 51 ☽ ⚼ ☉
20 36 ☽ ⚹ ♆
20 40 ☽ # ♇
20 41 ☽ ∥ ♅

16 Wednesday
01 00 ☉ ⊥ ♃
05 31 ☉ ⊥ ♄
06 34 ☽ ☍ ♂
09 12 ☽ □ ♃
12 45 ♅ # ♇
14 56 ♀ ⊥ ♃
17 05 ♀ ⊥ ♄
18 14 ☽ ⚼ ♇
18 49 ☽ Q ♆
21 07 ☽ △ ♀
21 51 ☽ △ ☉

17 Thursday
04 31 ♂ ∥ ♃
06 19 ☿ ⊼ ♅
08 14 ☽ ⊼ ♄
11 19 ☽ ± ♇
11 59 ☽ △ ♅
12 31 ☽ □ ☿
14 55 ♃ Q ♄
19 53 ☽ ± ♄
19 55 ☽ Q ♃
22 53 ☽ ⊼ ♇

18 Friday
04 36 ☉ ☌ ♀
05 15 ☽ □ ♆
13 36 ☽ ⚼ ♄
17 08 ☽ ⚼ ♅
18 54 ☽ ⊼ ♂
19 49 ☽ ⚹ ♃
21 43 ♇ St D

19 Saturday
03 16 ☽ Q ☿
07 26 ☽ ± ♂
11 00 ☽ ± ♅
13 15 ☽ □ ☉
14 05 ☽ □ ♀
19 45 ☽ △ ♄
23 02 ☽ ⊼ ♅
23 28 ♂ □ ♃

20 Sunday
02 21 ☽ ∠ ♃
02 27 ☽ ⚼ ♂
06 02 ☽ ⚹ ☿
10 32 ☽ ☍ ♇
17 02 ☽ △ ♆
20 54 ☽ ⊥ ♃

21 Monday
04 49 ☽ Q ☉
06 42 ☽ Q ♀
09 19 ☽ ⊻ ♃
10 28 ☽ △ ♂
16 10 ☽ ∠ ☿
23 27 ☽ ⚼ ♆
23 45 ☿ △ ♇

22 Tuesday
02 24 ♀ ⊻ ♄
02 50 ☽ ∥ ♇
02 56 ☽ # ♅
03 15 ☿ Q ♂
04 26 ☿ ± ♅
05 53 ♂ Q ♅
07 22 ☽ ⚹ ☉
09 13 ☽ □ ♄
09 56 ☽ ⚹ ♀
11 49 ☽ ☍ ♅
12 47 ☽ ⊥ ☿
17 41 ☽ ± ♆
23 28 ☽ ⊼ ♇

23 Wednesday
02 32 ☽ ⊻ ☿
03 56 ♀ △ ♅
05 48 ☽ ⊼ ♆
05 57 ☽ ∥ ♂
09 08 ☉ ⊻ ♄
11 34 ☽ ± ♇
16 20 ☽ ∠ ☉
19 42 ☽ ∠ ♀
21 31 ☽ ∥ ♃
21 54 ☽ Q ♄
23 05 ☽ ☌ ♃

24 Thursday
02 18 ☽ □ ♂
05 17 ☿ ☍ ♆
05 35 ☽ ⚼ ♇
07 49 ☿ ∥ ♄
11 50 ☽ ⊥ ☉
13 44 ☉ △ ♅
15 38 ☽ ⊥ ♀
21 57 ☽ ⚹ ♄
23 49 ☽ ⊼ ♅

25 Friday
00 41 ☽ ⊻ ☉
04 46 ☽ ⊻ ♀
10 44 ♀ ∠ ♃
11 09 ☽ △ ♇
11 32 ☽ ± ♅
15 34 ☽ Q ♂
17 06 ☽ ☍ ♆
18 43 ☿ # ♆
22 08 ☽ ☌ ☿

26 Saturday
03 25 ☽ ∠ ♄
04 55 ☽ ⚼ ♅
10 46 ☽ ⊻ ♃
15 46 ☽ ⚹ ♂
20 45 ☽ ⊥ ♄
21 11 ♂ ∠ ♇
22 21 ☽ ⊥ ♃

27 Sunday
08 09 ☽ ⊻ ♄
09 20 ☽ △ ♅
14 50 ☽ ☌ ☉
15 31 ☽ ∠ ♃
18 05 ☿ ∠ ♄
20 09 ☽ ☌ ♀
20 12 ☽ □ ♇
20 42 ♀ □ ♇
21 18 ☽ ∠ ♂

28 Monday
01 45 ☽ ⊼ ♆
02 19 ☉ ∠ ♃
02 20 ☿ ⚼ ♅
12 46 ☽ ± ♆
14 40 ☽ ⊻ ☿
14 40 ☽ ⊥ ♂
16 37 ☽ ∥ ♃
18 42 ♀ ∠ ♂
19 34 ☽ ⚹ ♃

29 Tuesday
02 06 ☽ ⊻ ♂
03 12 ☽ ⊥ ☿
05 06 ☽ ⚼ ♆
05 10 ☽ Q ♇
15 38 ☽ ☌ ♄
16 13 ☽ □ ♅
17 33 ☽ Q ♃
19 54 ☽ ∥ ♂
21 48 ☽ ∠ ☿

30 Wednesday
01 42 ☽ ⊻ ☉
02 42 ☽ ⚹ ♇
06 14 ♀ ⊼ ♆
07 52 ☿ ⊻ ♃
07 55 ☽ △ ♆
08 05 ☽ ⊻ ♀
09 34 ☽ # ♅
09 58 ☽ ∥ ♇
13 10 ☽ ⊥ ☉
15 56 ☉ □ ♇
19 45 ☽ ⊥ ♀

All ephemeris data is given at 12.00 UT and the Moon's longitude is additionally given for 24.00 UT

JULY 1976

LONGITUDES

Date	Sidereal time h m s	Sun ☉ ° ' "	Moon ☽ ° ' "	Moon ☽ 24.00 ° ' "	Mercury ☿ ° '	Venus ♀ ° '	Mars ♂ ° '	Jupiter ♃ ° '	Saturn ♄ ° '	Uranus ♅ ° '	Neptune ♆ ° '	Pluto ♇ ° '
01	06 38 39	09 ♋ 45 45	27 ♌ 50 32	04 ♍ 43 33	24 ♊ 05	13 ♋ 25	26 ♌ 42	22 ♉ 11	03 ♌ 00	03 ♏ 05	11 ♐ 52	08 ♎ 58
02	06 42 36	10 42 58	11 ♍ 38 44	18 ♍ 35 53	25 57	14 39	27 18	22 23	03 07	03 R 04	11 R 51	08 59
03	06 46 32	11 40 10	25 ♍ 34 48	02 ♎ 35 22	27 52	15 53	27 54	22 35	03 14	03 04	11 49	08 59
04	06 50 29	12 37 22	09 ♎ 37 25	16 ♎ 40 51	29 ♊ 49	17 07	28 30	22 46	03 22	03 03	11 48	09 00
05	06 54 25	13 34 34	23 ♎ 45 31	00 ♏ 51 17	01 ♋ 48	18 20	29 07	22 58	03 29	03 03	11 47	09 00
06	06 58 22	14 31 46	07 ♏ 57 59	15 ♏ 05 24	03 50	19 34	29 ♌ 43	23 09	03 37	03 03	11 45	09 01
07	07 02 19	15 28 57	22 ♏ 13 14	29 ♏ 21 12	05 53	20 48	00 ♍ 19	23 20	03 44	03 02	11 44	09 01
08	07 06 15	16 26 09	06 ♐ 28 53	13 ♐ 35 50	07 59	22 02	00 55	23 32	03 52	03 02	11 43	09 02
09	07 10 12	17 23 20	20 ♐ 41 33	27 ♐ 45 29	10 05	23 15	01 32	23 43	03 59	03 02	11 41	09 03
10	07 14 08	18 20 32	04 ♑ 47 04	11 ♑ 45 44	12 13	24 29	02 08	23 54	04 07	03 02	11 40	09 03
11	07 18 05	19 17 43	18 ♑ 40 56	25 ♑ 32 10	14 22	25 43	02 44	24 05	04 14	03 D 02	11 39	09 04
12	07 22 01	20 14 55	02 ♒ 19 00	09 ♒ 01 06	16 31	26 57	03 21	24 15	04 22	03 02	11 38	09 05
13	07 25 58	21 12 07	15 ♒ 38 11	22 ♒ 10 09	18 40	28 11	03 57	24 26	04 29	03 02	11 37	09 06
14	07 29 54	22 09 19	28 ♒ 36 57	04 ♓ 58 40	20 49	29 ♋ 24	04 34	24 37	04 37	03 02	11 36	09 07
15	07 33 51	23 06 32	11 ♓ 15 30	17 ♓ 27 44	22 58	00 ♌ 38	05 10	24 47	04 45	03 03	11 34	09 07
16	07 37 48	24 03 45	23 ♓ 35 45	29 ♓ 40 01	25 06	01 52	05 47	24 58	04 52	03 03	11 33	09 08
17	07 41 44	25 00 59	05 ♈ 41 03	11 ♈ 39 25	27 14	03 06	06 24	25 08	05 00	03 03	11 32	09 09
18	07 45 41	25 58 14	17 ♈ 35 46	23 ♈ 30 44	29 ♋ 20	04 20	07 00	25 18	05 08	03 03	11 31	09 10
19	07 49 37	26 55 29	29 ♈ 25 00	05 ♉ 19 14	01 ♌ 26	05 33	07 37	25 28	05 15	03 04	11 30	09 11
20	07 53 34	27 52 45	11 ♉ 14 08	17 ♉ 10 23	03 30	06 47	08 14	25 38	05 23	03 04	11 29	09 12
21	07 57 30	28 50 02	23 ♉ 08 37	29 ♉ 09 28	05 32	08 01	08 51	25 48	05 31	03 05	11 28	09 13
22	08 01 27	29 ♋ 47 20	05 ♊ 13 31	11 ♊ 21 17	07 34	09 15	09 27	25 58	05 38	03 05	11 27	09 14
23	08 05 23	00 ♌ 44 38	17 ♊ 33 13	23 ♊ 49 42	09 33	10 29	10 04	26 08	05 46	03 06	11 26	09 15
24	08 09 20	01 41 57	00 ♋ 11 03	06 ♋ 37 25	11 31	11 43	10 41	26 17	05 54	03 07	11 25	09 16
25	08 13 17	02 39 18	13 ♋ 08 54	19 ♋ 45 29	13 28	12 56	11 18	26 27	06 02	03 07	11 25	09 18
26	08 17 13	03 36 38	26 ♋ 27 00	03 ♌ 13 12	15 22	14 10	11 55	26 36	06 09	03 08	11 24	09 19
27	08 21 10	04 34 00	10 ♌ 03 45	16 ♌ 58 11	17 16	15 24	12 33	26 45	06 17	03 09	11 23	09 20
28	08 25 06	05 31 22	23 ♌ 56 01	00 ♍ 56 41	19 07	16 38	13 10	26 54	06 25	03 10	11 22	09 21
29	08 29 03	06 28 45	07 ♍ 59 37	15 ♍ 04 15	20 57	17 52	13 47	27 03	06 32	03 11	11 21	09 23
30	08 32 59	07 26 09	22 ♍ 10 03	29 ♍ 16 30	22 45	19 06	14 24	27 12	06 40	03 12	11 21	09 24
31	08 36 56	08 ♌ 23 33	06 ♎ 23 10	13 ♎ 29 42	24 ♌ 31	20 ♌ 20	15 ♍ 02	27 ♉ 20	06 ♌ 48	03 ♏ 13	11 ♐ 20	09 ♎ 25

Date	Moon True ☊ ° '	Moon Mean ☊ ° '	Moon ☽ Latitude ° '
01	10 ♏ 38	09 ♏ 36	04 S 52
02	10 R 33	09 33	04 22
03	10 30	09 30	03 36
04	10 29	09 26	02 37
05	10 D 29	09 23	01 28
06	10 29	09 20	00 S 13
07	10 R 29	09 17	01 N 02
08	10 26	09 14	02 14
09	10 21	09 11	03 17
10	10 14	09 07	04 07
11	10 04	09 04	04 42
12	09 53	09 01	05 00
13	09 42	08 58	05 01
14	09 32	08 55	04 46
15	09 24	08 52	04 17
16	09 18	08 48	03 37
17	09 14	08 45	02 48
18	09 13	08 42	01 52
19	09 D 12	08 39	00 N 52
20	09 R 12	08 36	00 S 11
21	09 12	08 32	01 13
22	09 09	08 29	02 13
23	09 05	08 26	03 08
24	08 57	08 23	03 55
25	08 48	08 20	04 32
26	08 36	08 17	04 54
27	08 24	08 13	05 01
28	08 12	08 10	04 50
29	08 02	08 07	04 21
30	07 55	08 04	03 36
31	07 ♏ 50	08 ♏ 01	02 S 38

DECLINATIONS

Date	Sun ☉ ° '	Moon ☽ ° '	Mercury ☿ ° '	Venus ♀ ° '	Mars ♂ ° '	Jupiter ♃ ° '	Saturn ♄ ° '	Uranus ♅ ° '	Neptune ♆ ° '	Pluto ♇ ° '
01	23 N 05	07 N 39	22 N 36	23 N 31	13 N 43	17 N 23	19 N 57	12 S 05	20 S 40	12 N 01
02	23 00	03 N 10	22 52	23 25	13 30	17 26	19 55	12 05	20 40	12 00
03	22 56	01 S 33	23 06	23 18	13 17	17 29	19 54	12 05	20 40	11 59
04	22 50	06 13	23 19	23 11	13 04	17 32	19 52	12 05	20 39	11 59
05	22 45	10 35	23 30	23 04	12 51	17 34	19 50	12 04	20 39	11 58
06	22 39	14 23	23 38	22 55	12 37	17 37	19 49	12 04	20 39	11 57
07	22 33	17 19	23 45	22 46	12 24	17 40	19 47	12 04	20 39	11 56
08	22 26	19 12	23 48	22 36	12 11	17 43	19 45	12 04	20 39	11 56
09	22 19	19 50	23 49	22 26	11 57	17 45	19 44	12 04	20 39	11 55
10	22 11	19 14	23 48	22 14	11 44	17 48	19 42	12 04	20 39	11 54
11	22 03	17 29	23 43	22 03	11 30	17 51	19 40	12 04	20 38	11 54
12	21 55	14 46	23 36	21 50	11 16	17 53	19 38	12 04	20 38	11 53
13	21 46	11 21	23 26	21 37	11 02	17 56	19 37	12 04	20 38	11 52
14	21 37	07 29	23 13	21 23	10 49	17 58	19 35	12 05	20 38	11 51
15	21 28	03 S 22	22 58	21 09	10 35	18 01	19 33	12 05	20 38	11 50
16	21 18	00 N 47	22 40	20 54	10 21	18 03	19 31	12 05	20 38	11 50
17	21 08	04 50	22 20	20 39	10 07	18 05	19 30	12 05	20 38	11 49
18	20 57	08 38	21 58	20 23	09 52	18 08	19 28	12 05	20 38	11 48
19	20 46	12 04	21 33	20 06	09 38	18 10	19 26	12 05	20 38	11 47
20	20 35	15 02	21 06	19 49	09 24	18 12	19 24	12 05	20 37	11 46
21	20 24	17 23	20 38	19 31	09 10	18 15	19 22	12 06	20 37	11 45
22	20 12	18 59	20 07	19 12	08 55	18 17	19 21	12 06	20 37	11 45
23	20 00	19 44	19 35	18 54	08 41	18 19	19 19	12 06	20 37	11 44
24	19 47	19 31	19 02	18 34	08 26	18 21	19 17	12 06	20 37	11 43
25	19 34	18 17	18 28	18 14	08 12	18 23	19 15	12 07	20 37	11 42
26	19 21	16 03	17 52	17 54	07 57	18 25	19 13	12 07	20 37	11 41
27	19 07	12 54	17 15	17 33	07 42	18 27	19 11	12 07	20 37	11 40
28	18 53	08 59	16 38	17 11	07 27	18 29	19 10	12 08	20 37	11 39
29	18 39	04 N 32	15 59	16 49	07 13	18 31	19 08	12 08	20 37	11 38
30	18 25	00 S 12	15 20	16 27	06 58	18 33	19 06	12 08	20 37	11 37
31	18 N 10	04 S 57	14 N 40	16 N 04	06 N 43	18 N 35	19 N 04	12 S 09	20 S 37	11 N 36

ZODIAC SIGN ENTRIES

Date	h	m	Planets
01	15	46	☽ ♍
03	19	34	☽ ♎
04	14	18	☿ ♋
05	22	33	☽ ♏
06	23	27	♂ ♍
08	01	05	☽ ♐
10	03	49	☽ ♑
12	07	53	☽ ♒
14	14	36	☽ ♓
14	23	36	♀ ♌
17	00	40	☽ ♈
18	19	35	☿ ♌
19	13	11	☽ ♉
22	01	40	☽ ♊
22	17	18	☉ ♌
24	11	39	☽ ♋
26	18	19	☽ ♌
28	22	23	☽ ♍
31	01	13	☽ ♎

LATITUDES

Date	Mercury ☿ ° '	Venus ♀ ° '	Mars ♂ ° '	Jupiter ♃ ° '	Saturn ♄ ° '	Uranus ♅ ° '	Neptune ♆ ° '	Pluto ♇ ° '
01	00 S 43	00 N 45	01 N 10	00 S 58	00 N 28	00 N 29	01 N 34	16 N 57
04	00 S 07	00 51	01 08	00 58	00 28	00 29	01 34	16 55
07	00 N 26	00 57	01 07	00 58	00 29	00 29	01 34	16 54
10	00 55	01 02	01 05	00 59	00 29	00 29	01 33	16 52
13	01 18	01 07	01 03	00 59	00 29	00 29	01 33	16 50
16	01 35	01 11	01 01	00 59	00 29	00 29	01 33	16 49
19	01 45	01 15	01 00	00 59	00 30	00 29	01 33	16 47
22	01 48	01 19	00 58	01 00	00 30	00 28	01 33	16 46
25	01 45	01 22	00 56	01 00	00 30	00 28	01 33	16 44
28	01 37	01 24	00 55	01 00	00 30	00 28	01 33	16 43
31	01 N 24	01 N 26	00 N 53	01 S 00	00 N 31	00 N 28	01 N 32	16 N 42

DATA

Julian Date	2442961
Delta T	+47 seconds
Ayanamsa	23° 31' 55"
Synetic vernal point	05° ♓ 35' 04"
True obliquity of ecliptic	23° 26' 25"

MOON'S PHASES, APSIDES AND POSITIONS ☽

Date	h	m	Phase	Longitude ° '	Eclipse Indicator
04	17	28	☽	12 ♎ 50	
11	13	09	○	19 ♑ 20	
19	06	29	☾	26 ♈ 42	
27	01	39	●	04 ♌ 09	

Day	h	m		
07	01	49	Perigee	
19	11	02	Apogee	
03	04	09	0S	
09	12	19	Max dec	19° S 50'
16	07	27	0N	
23	18	51	Max dec	19° N 46'
30	10	58	0S	

LONGITUDES

Date	Chiron ⚷ ° '	Ceres ⚳ ° '	Pallas ⚴ ° '	Juno ⚵ ° '	Vesta ⚶ ° '	Black Moon Lilith ⚸ ° '
01	01 ♉ 24	22 ♋ 17	11 ♊ 43	16 ♍ 09	14 ♊ 35	27 ♈ 07
11	01 ♉ 39	26 ♋ 42	17 ♊ 11	19 ♍ 05	18 ♊ 46	28 ♈ 14
21	01 ♉ 49	01 ♌ 09	22 ♊ 43	22 ♍ 10	22 ♊ 53	29 ♈ 20
31	01 ♉ 54	05 ♌ 37	28 ♊ 18	25 ♍ 23	26 ♊ 54	00 ♉ 26

All ephemeris data is given at 12.00 UT and the Moon's longitude is additionally given for 24.00 UT

ASPECTARIAN

h m	Aspects
01 Thursday	
00 09	☽ Q ♅
01 57	☽ □ ♃
04 24	☽ ⚹ ☿
05 13	☽ ∠ ♇
06 12	☽ ∠ ☉
09 55	☽ ☌ ♂
13 07	☽ ∠ ♀
20 57	☽ ⊥ ♇
21 04	☽ ⊻ ♄
21 07	☽ ⚹ ♅
02 Friday	
03 14	♄ □ ♅
04 35	☽ Q ☿
07 23	☽ ⊻ ♇
07 35	☽ ⊥ ♄
10 16	☽ ⚹ ☉
12 21	☽ □ ♆
17 42	☽ ⚹ ♀
22 26	☉ ∥ ☿
23 05	☽ ∠ ♅
23 16	☽ ∠ ♄
03 Saturday	
03 44	☿ ⊥ ♄
06 47	☽ △ ♃
08 29	☽ Q ☉
12 51	☿ ⚹ ♂
14 32	☽ ⊥ ♅
15 44	☉ ⊼ ♆
16 10	☽ ⊻ ♂
16 19	☽ Q ♀
16 32	☽ □ ☿
19 15	☽ Q ♆
21 55	☿ ⊥ ♃
04 Sunday	
00 48	☽ ⊻ ♅
01 13	☽ ⚹ ♄
02 17	☿ ∥ ♀
02 53	☽ ⊥ ♂
08 48	☽ ⚼ ♃
10 56	☽ ☌ ♇
15 42	☽ ⚹ ♆
17 28	☽ □ ☉
18 54	☽ ∠ ♂
21 51	☽ Q ♄
05 Monday	
00 19	☽ ± ♃
01 14	♀ ± ♆
01 56	☽ □ ♀
10 38	☽ ⊼ ♃
17 06	☽ ∠ ♆
20 12	☽ # ♇
20 53	☽ ∥ ♅
21 27	☽ ⚹ ♂
06 Tuesday	
00 58	☽ # ♂
02 45	☿ △ ♅
03 42	☽ ☌ ♅
03 51	☽ △ ☿
04 35	☽ □ ♄
08 17	☽ ⊥ ♆
09 13	☿ ⊻ ♄
13 46	☽ ⊻ ♇
18 22	☽ ⊻ ♆
18 35	☽ Q ♂
23 51	☽ △ ☉
23 53	☽ ⊥ ♇
07 Wednesday	
00 32	☉ ∠ ♂
09 23	☽ △ ♀
09 23	☽ ⚼ ☿
13 55	☽ ☍ ♃
15 02	☽ ∠ ♇
15 33	☽ # ♃
08 Thursday	
02 13	☽ □ ♂
02 54	☽ ⚼ ☉
03 07	☽ ± ☿
06 12	☽ ⊻ ♅
07 33	☽ △ ♄
13 00	☽ ⚼ ♀
14 58	☽ ⊼ ☿
16 18	☽ ⚹ ♇
16 18	☽ ⊥ ♅
18 51	☿ ∠ ♃
19 09	☽ ± ☉
20 48	☽ ☌ ♆
23 19	♂ # ♅
09 Friday	
00 04	☿ □ ♇
02 38	☽ # ♄
05 38	☽ ± ♀
06 00	☽ ⊼ ☉
07 30	☽ ∠ ♅
09 05	☽ ⚼ ♄
12 36	☽ Q ♇
16 00	♂ ∥ ♇
16 46	☽ ⊼ ♀
17 11	☽ ⊼ ♃
19 28	☉ ± ♆
22 27	♀ ⚹ ♃
23 07	☽ # ♄
10 Saturday	
00 30	☽ ± ♄
03 32	☽ ± ♃
05 53	☿ ⊼ ♆
07 15	☽ △ ♂
09 00	☽ ⚹ ♅
10 50	☽ ⊼ ♄
19 09	☽ ⚼ ♃
19 21	☽ □ ♇
23 50	☽ ⊻ ♆
11 Sunday	
03 07	☽ ☍ ☿
05 40	☽ Q ♅
06 07	♅ St D
08 08	☽ # ♃
08 57	☉ ∥ ♀
10 12	☽ ⊥ ♆
10 17	☽ ⚼ ♂
10 44	♃ ⚼ ♇
13 09	☽ ☍ ☉
21 34	☽ △ ♃
23 44	♂ ⚹ ♅
12 Monday	
01 22	♂ ⊥ ♇
01 32	☽ ☍ ♀
01 57	☽ ∠ ♆
02 47	☽ ± ♂
05 57	♀ ⚼ ♆
13 17	☽ □ ♅
13 55	☽ ⊼ ♂
14 40	♀ Q ♇
15 41	☽ ☍ ♄
13 Tuesday	
00 08	☽ △ ♇
00 20	☿ ± ♆
03 24	♀ ⊥ ♂
04 42	☽ ⚹ ♆
07 16	☽ ∥ ♅
08 38	☽ # ♇
14 11	☽ # ♂
16 25	☿ ∠ ♂
18 39	☽ ⊼ ☿
23 01	☽ ⊼ ☉
14 Wednesday	
02 39	☽ Q ♆
03 35	☽ ⚼ ♇
04 26	☽ □ ♃
07 58	☽ ± ☿
11 04	☽ ± ☉
13 38	☽ ⊼ ♀
14 42	♂ ⊻ ♄
20 20	☽ △ ♅
20 28	☽ ± ♇
23 26	☽ ⊼ ♄
23 47	☽ ☍ ♂
15 Thursday	
02 12	☽ ± ♀
04 24	☽ ⚼ ☿
05 28	☽ ⚼ ☉
07 54	☽ ⊼ ♇
11 00	☽ ± ♄
12 36	☽ □ ♆
14 50	☉ ☌ ☿
14 59	☽ Q ♃
21 23	☽ ⚼ ♀
16 Friday	
01 08	☽ ⚼ ♅
04 37	☽ ⚼ ♄
10 14	☿ ⚹ ♃
13 00	☽ △ ☉
14 44	☽ ⚹ ♃
15 36	☽ △ ☿
18 48	☽ ± ♅
17 Saturday	
04 14	☿ ⚼ ♆
06 14	☽ △ ♀
06 44	☽ ⊼ ♅
10 37	☽ △ ♄
11 07	♀ □ ♅
11 07	☿ Q ♇
13 22	♀ # ♆
13 30	☽ ⊼ ♂
15 36	☉ ⚹ ♃
18 58	☽ ☍ ♇
21 04	☽ ∠ ♃
23 45	☽ △ ♆
18 Sunday	
02 13	☽ ± ♂
15 31	☽ ⊥ ♃
19 47	☽ ∥ ♂
21 26	☽ ⚼ ♂
19 Monday	
01 35	☉ ⚼ ♆
03 52	☽ ⊻ ♃
05 25	♀ ☌ ♄
06 05	☽ ⚼ ♆
06 29	☽ □ ☉
09 52	☽ ∥ ♇
12 05	☽ # ♅
15 06	☿ ⊥ ♂
16 58	☽ □ ☿
18 40	☉ Q ♇
19 25	☽ ☍ ♅
20 Tuesday	
00 00	☽ □ ♄
00 21	☽ ± ♆
01 56	☽ □ ♀
05 34	☽ △ ♂
07 03	☿ □ ♅
07 06	☉ # ♆
07 52	☽ ⊼ ♇
12 30	☽ ⊼ ♆
20 02	☽ ± ♇
22 13	☽ Q ☉
21 Wednesday	
07 11	♀ Q ♃
11 39	☿ ☌ ♄
12 08	☿ # ♆
12 45	☽ Q ♄
12 57	☽ Q ☿
14 09	☽ ⚼ ♇
17 24	☽ ☌ ♃
18 24	☽ Q ♀
23 44	☽ ∥ ♃
22 Thursday	
00 16	♀ ∥ ♄
00 20	☽ ⚹ ☉
03 08	♂ ⊻ ♇
06 20	☉ ∥ ☿
07 47	☽ ⊼ ♅
11 47	♀ ⚹ ♇
12 49	☽ ⚹ ♄
15 43	☽ ∥ ♀
17 18	☿ Q ♃
17 29	☽ ⚹ ☿
19 35	☽ ± ♅
19 53	☽ △ ♇
19 58	☽ ∥ ♄
20 11	♀ ⊻ ♂
20 44	☽ □ ♂
20 46	☽ ⚹ ♀
23 Friday	
00 11	☽ ☍ ♆
08 13	☽ ∠ ☉
08 14	☽ ∥ ☿
08 22	☿ ⚹ ♇
13 03	☽ ⚼ ♅
18 14	☽ ∠ ♄
21 11	☿ ⊻ ♂
24 Saturday	
00 51	☿ ∥ ♄
02 52	☽ ⊥ ☉
03 51	☽ ∠ ☿
04 34	☽ ⊻ ♃
04 45	☽ ∠ ♀
06 29	♀ △ ♆
09 03	☽ Q ♂
10 49	☿ △ ♆
11 27	☽ ⊥ ♄
15 04	☽ ⊻ ☉
15 59	☽ ⊥ ♃
17 29	☽ △ ♅
18 18	☿ ☌ ♀
18 45	☽ ∥ ♄
22 46	☽ ⊻ ♄
23 23	☽ ⊥ ♀
23 43	☽ ⊥ ☿
25 Sunday	
02 07	♀ ∥ ♃
04 55	☽ □ ♇
08 05	☽ ∥ ☿
08 28	☽ ⚹ ♂
08 49	☽ ⊼ ♆
08 50	☽ ∠ ♃
10 35	☽ ∥ ♃
11 35	☽ ⊻ ♀
12 40	☽ ⊻ ☿
12 50	☽ ∥ ♀
14 48	☿ ∥ ♃
15 52	♂ □ ♆
19 45	☽ ± ♆
23 53	☉ □ ♅
26 Monday	
09 16	☿ ∥ ♀
11 54	☽ ⚼ ♆
12 16	☽ ⚹ ♃
12 53	☽ ∠ ♂
13 33	☽ Q ♇
23 16	♂ ⊥ ♄
23 52	☽ □ ♅
27 Tuesday	
01 39	☽ ☌ ☉
03 31	☉ ∥ ♄
05 19	☽ ☌ ♄
05 33	☽ ⊥ ♂
09 41	☽ Q ♃
10 44	☽ ⚹ ♇
14 18	☽ △ ♆
16 32	☽ ⊻ ♂
17 06	☽ # ♅
20 01	☽ ∥ ♇
22 12	☽ ☌ ♀
28 Wednesday	
02 26	☽ ☌ ☿
07 14	☽ Q ♅
12 44	☽ ∠ ♇
17 09	☽ □ ♃
21 00	☽ ∥ ♂
29 Thursday	
03 48	☽ ⚹ ♅
04 08	☽ ⊥ ♇
09 14	☽ ⊻ ☉
09 31	☽ ⊻ ♄
13 49	☉ ☌ ♄
14 21	☽ ⊻ ♇
15 07	☿ Q ♅
17 42	☽ □ ♆
19 47	☽ ⊥ ♄
20 10	☽ ⊥ ☉
22 16	☽ ☌ ♂
23 40	☉ ∥ ♃
30 Friday	
05 17	☽ ∠ ♅
06 19	☽ ⊻ ♀
11 09	☽ ∠ ♄
12 29	☽ ∠ ☉
13 07	☽ ⊻ ☿
17 25	☽ ⊥ ♀
20 30	☽ ⊥ ♅
20 35	☽ △ ♃
31 Saturday	
00 06	☽ Q ♆
00 42	☽ ⊥ ☿
06 38	☽ ⊻ ♅
10 03	☽ ∠ ♀
10 38	☿ ∠ ♇
12 42	☽ ⚹ ♄
15 38	☽ ⚹ ☉
17 08	☽ ☌ ♇
18 02	☽ ∠ ☿
20 20	☽ ⚹ ♆
20 42	☽ # ♂
22 09	☽ ⚼ ♃

AUGUST 1976

LONGITUDES

Date	Sidereal time h m s	Sun ☉ ° ′ ″	Moon ☽ ° ′ ″	Moon ☽ 24.00 ° ′ ″	Mercury ☿ ° ′	Venus ♀ ° ′	Mars ♂ ° ′	Jupiter ♃ ° ′	Saturn ♄ ° ′	Uranus ♅ ° ′	Neptune ♆ ° ′	Pluto ♇ ° ′
01	08 40 52	09 ♌ 20 57	20 ♎ 35 46	27 ♎ 41 10	26 ♌ 16	21 ♌ 33	15 ♍ 39	27 ♉ 29	06 ♌ 56	03 ♏ 14	11 ♐ 19	09 ♎ 27
02	08 44 49	10 18 23	04 ♏ 45 43	11 ♏ 49 16	27 59	22 47	16 16	27 37	07 03	03 15	11 R 19	09 28
03	08 48 46	11 15 48	18 ♏ 51 43	25 ♏ 52 58	29 ♌ 40	24 01	16 54	27 46	07 11	03 16	11 18	09 29
04	08 52 42	12 13 15	02 ♐ 52 57	09 ♐ 51 34	01 ♍ 20	25 15	17 31	27 54	07 19	03 17	11 17	09 31
05	08 56 39	13 10 42	16 ♐ 48 39	23 ♐ 44 04	02 58	26 29	18 09	28 02	07 27	03 19	11 17	09 32
06	09 00 35	14 08 10	00 ♑ 37 37	07 ♑ 29 02	04 35	27 43	18 46	28 10	07 34	03 20	11 16	09 34
07	09 04 32	15 05 39	14 ♑ 18 06	21 ♑ 04 29	06 10	28 ♌ 57	19 24	28 17	07 42	03 21	11 16	09 35
08	09 08 28	16 03 09	27 ♑ 47 55	04 ♒ 28 07	07 43	00 ♍ 10	20 01	28 25	07 50	03 23	11 15	09 37
09	09 12 25	17 00 40	11 ♒ 04 48	17 ♒ 37 44	09 15	01 24	20 39	28 32	07 57	03 24	11 15	09 39
10	09 16 21	17 58 11	24 ♒ 06 44	00 ♓ 31 41	10 45	02 38	21 17	28 40	08 05	03 26	11 14	09 40
11	09 20 18	18 55 44	06 ♓ 52 32	13 ♓ 09 19	12 13	03 52	21 55	28 47	08 13	03 27	11 14	09 42
12	09 24 15	19 53 18	19 ♓ 22 08	25 ♓ 31 11	13 40	05 06	22 32	28 54	08 20	03 29	11 14	09 43
13	09 28 11	20 50 53	01 ♈ 36 43	07 ♈ 39 04	15 05	06 20	23 10	29 01	08 28	03 31	11 13	09 45
14	09 32 08	21 48 30	13 ♈ 38 41	19 ♈ 36 01	16 28	07 34	23 48	29 07	08 36	03 32	11 13	09 47
15	09 36 04	22 46 08	25 ♈ 31 35	01 ♉ 25 59	17 49	08 47	24 26	29 14	08 43	03 34	11 13	09 48
16	09 40 01	23 43 48	07 ♉ 19 50	13 ♉ 13 47	19 09	10 01	25 04	29 20	08 51	03 36	11 13	09 50
17	09 43 57	24 41 29	19 ♉ 08 29	25 ♉ 04 38	20 27	11 15	25 42	29 26	08 58	03 38	11 12	09 52
18	09 47 54	25 39 12	01 ♊ 02 56	07 ♊ 04 01	21 43	12 29	26 20	29 32	09 06	03 40	11 12	09 54
19	09 51 50	26 36 57	13 ♊ 08 35	19 ♊ 17 13	22 58	13 43	26 58	29 38	09 13	03 42	11 12	09 56
20	09 55 47	27 34 43	25 ♊ 30 31	01 ♋ 48 58	24 10	14 57	27 37	29 44	09 21	03 44	11 12	09 57
21	09 59 44	28 32 31	08 ♋ 13 00	14 ♋ 42 56	25 20	16 10	28 15	29 50	09 28	03 46	11 12	09 59
22	10 03 40	29 ♌ 30 21	21 ♋ 18 59	28 ♋ 01 12	26 28	17 24	28 53	29 ♉ 55	09 36	03 48	11 12	10 01
23	10 07 37	00 ♍ 28 12	04 ♌ 49 31	11 ♌ 43 43	27 34	18 38	29 ♍ 31	00 ♊ 00	09 43	03 50	11 D 12	10 03
24	10 11 33	01 26 04	18 ♌ 43 25	25 ♌ 48 07	28 37	19 52	00 ♎ 10	00 05	09 51	03 52	11 12	10 05
25	10 15 30	02 23 59	02 ♍ 57 09	10 ♍ 09 46	29 ♍ 38	21 06	00 48	00 10	09 58	03 54	11 12	10 07
26	10 19 26	03 21 54	17 ♍ 25 11	24 ♍ 42 31	00 ♎ 37	22 20	01 27	00 15	10 05	03 57	11 12	10 09
27	10 23 23	04 19 51	02 ♎ 00 54	09 ♎ 19 30	01 32	23 33	02 05	00 20	10 13	03 59	11 12	10 11
28	10 27 19	05 17 50	16 ♎ 37 34	23 ♎ 54 25	02 25	24 47	02 44	00 24	10 20	04 01	11 12	10 13
29	10 31 16	06 15 50	01 ♏ 09 27	08 ♏ 22 11	03 15	26 01	03 22	00 28	10 27	04 04	11 12	10 15
30	10 35 13	07 13 51	15 ♏ 32 17	22 ♏ 39 26	04 01	27 15	04 01	00 32	10 35	04 06	11 13	10 17
31	10 39 09	08 ♍ 11 54	29 ♏ 43 30	06 ♐ 44 20	04 ♎ 44	28 ♍ 29	04 ♎ 40	00 ♊ 36	10 ♌ 42	04 ♏ 08	11 ♐ 13	10 ♎ 19

Date	Moon True ☊ ° ′	Moon Mean ☊ ° ′	Moon ☽ Latitude ° ′
01	07 ♏ 48	07 ♏ 57	01 S 30
02	07 D 48	07 54	00 S 16
03	07 R 48	07 51	00 N 58
04	07 47	07 48	02 09
05	07 44	07 45	03 11
06	07 38	07 42	04 01
07	07 30	07 38	04 37
08	07 19	07 35	04 57
09	07 07	07 32	05 00
10	06 55	07 29	04 48
11	06 43	07 26	04 21
12	06 34	07 23	03 42
13	06 27	07 19	02 53
14	06 22	07 16	01 58
15	06 20	07 13	00 N 57
16	06 D 20	07 10	00 S 05
17	06 20	07 07	01 08
18	06 R 20	07 03	02 08
19	06 19	07 00	03 03
20	06 15	06 57	03 51
21	06 10	06 54	04 29
22	06 02	06 51	04 54
23	05 52	06 48	05 04
24	05 42	06 44	04 56
25	05 32	06 41	04 30
26	05 23	06 38	03 47
27	05 17	06 35	02 48
28	05 14	06 32	01 38
29	05 12	06 29	00 S 22
30	05 D 12	06 25	00 N 55
31	05 ♏ 13	06 ♏ 22	02 N 08

DECLINATIONS

Date	Sun ☉	Moon ☽	Mercury ☿	Venus ♀	Mars ♂	Jupiter ♃	Saturn ♄	Uranus ♅	Neptune ♆	Pluto ♇
01	17 N 55	09 S 26	14 N 00	15 N 41	06 N 28	18 N 37	19 N 02	12 S 09	20 S 37	11 N 36
02	17 40	13 22	13 19	15 17	06 13	18 39	19 00	12 09	20 37	11 35
03	17 24	16 30	12 38	14 53	05 58	18 40	18 58	12 10	20 37	11 34
04	17 08	18 38	11 57	14 29	05 42	18 42	18 56	12 10	20 37	11 33
05	16 52	19 37	11 16	14 04	05 27	18 44	18 55	12 11	20 37	11 32
06	16 35	19 25	10 34	13 39	05 12	18 46	18 53	12 11	20 37	11 31
07	16 19	18 05	09 52	13 13	04 57	18 47	18 51	12 12	20 37	11 30
08	16 02	15 45	09 11	12 47	04 41	18 49	18 49	12 12	20 37	11 29
09	15 44	12 38	08 29	12 21	04 26	18 50	18 47	12 13	20 37	11 28
10	15 27	08 57	07 47	11 54	04 11	18 52	18 45	12 13	20 37	11 27
11	15 09	04 57	07 06	11 27	03 55	18 53	18 43	12 14	20 37	11 26
12	14 51	00 S 48	06 25	11 00	03 40	18 55	18 41	12 14	20 37	11 25
13	14 33	03 N 17	05 44	10 33	03 24	18 56	18 39	12 15	20 37	11 24
14	14 14	07 11	05 04	10 05	03 09	18 57	18 38	12 16	20 37	11 23
15	13 56	10 45	04 24	09 37	02 53	18 59	18 36	12 16	20 37	11 22
16	13 37	13 52	03 44	09 08	02 38	19 00	18 34	12 17	20 37	11 21
17	13 18	16 25	03 05	08 40	02 22	19 01	18 32	12 18	20 37	11 20
18	12 58	18 17	02 26	08 11	02 06	19 02	18 30	12 18	20 37	11 19
19	12 39	19 21	01 48	07 42	01 51	19 03	18 28	12 19	20 37	11 18
20	12 19	19 31	01 10	07 13	01 35	19 04	18 26	12 20	20 37	11 17
21	11 59	18 43	00 N 34	06 43	01 19	19 06	18 24	12 20	20 37	11 16
22	11 39	16 55	00 S 02	06 14	01 03	19 07	18 22	12 21	20 37	11 15
23	11 18	14 09	00 37	05 44	00 48	19 08	18 21	12 22	20 37	11 14
24	10 58	10 31	01 12	05 14	00 32	19 08	18 19	12 23	20 37	11 13
25	10 37	06 13	01 45	04 44	00 N 16	19 09	18 17	12 23	20 37	11 12
26	10 16	01 N 30	02 17	04 14	00 00	19 10	18 15	12 24	20 37	11 11
27	09 55	03 S 22	02 48	03 43	00 S 16	19 11	18 13	12 25	20 37	11 10
28	09 34	08 02	03 18	03 13	00 31	19 12	18 11	12 26	20 37	11 09
29	09 13	12 13	03 46	02 42	00 47	19 13	18 09	12 27	20 37	11 08
30	08 51	15 37	04 13	02 12	01 03	19 13	18 07	12 28	20 37	11 07
31	08 N 30	18 S 01	04 S 39	01 N 41	01 S 19	19 N 14	18 N 06	12 S 28	20 S 37	11 N 06

ZODIAC SIGN ENTRIES

Date	h	m	Planets
02	03	55	☽ ♏
03	16	41	☿ ♍
04	07	03	☽ ♐
06	10	54	☽ ♑
08	08	36	♀ ♍
08	15	57	☽ ♒
10	23	00	☽ ♓
13	08	49	☽ ♈
15	21	05	☽ ♉
18	09	54	☽ ♊
20	20	34	☽ ♋
23	00	18	☉ ♍
23	03	31	☽ ♌
23	10	24	♃ ♊
24	05	55	♂ ♎
25	07	04	☽ ♍
25	20	52	☿ ♎
27	08	42	☽ ♎
29	10	05	☽ ♏
31	12	28	☽ ♐

LATITUDES

Date	Mercury ☿	Venus ♀	Mars ♂	Jupiter ♃	Saturn ♄	Uranus ♅	Neptune ♆	Pluto ♇
01	01 N 19	01 N 27	00 N 52	01 S 01	00 N 31	00 N 28	01 N 32	16 N 41
04	01 01	01 28	00 51	01 01	00 31	00 28	01 32	16 40
07	00 40	01 28	00 49	01 01	00 31	00 28	01 32	16 39
10	00 N 16	01 28	00 47	01 02	00 31	00 28	01 32	16 37
13	00 S 09	01 27	00 45	01 02	00 32	00 28	01 32	16 36
16	00 37	01 26	00 44	01 02	00 32	00 28	01 32	16 35
19	01 05	01 24	00 42	01 03	00 32	00 28	01 31	16 34
22	01 34	01 21	00 40	01 03	00 33	00 27	01 31	16 33
25	02 04	01 18	00 38	01 03	00 33	00 27	01 31	16 32
28	02 33	01 15	00 37	01 04	00 33	00 27	01 31	16 31
31	03 S 01	01 N 10	00 N 35	01 S 04	00 N 34	00 N 27	01 N 31	16 N 30

DATA

Julian Date	2442992
Delta T	+47 seconds
Ayanamsa	23° 32′ 00″
Synetic vernal point	05° ♓ 35′ 00″
True obliquity of ecliptic	23° 26′ 25″

MOON'S PHASES, APSIDES AND POSITIONS ☽

Date	h	m	Phase	Longitude ° ′	Eclipse Indicator
02	22	07	☽	10 ♏ 43	
09	23	43	○	17 ♒ 29	
18	00	13	☾	25 ♉ 11	
25	11	01	●	02 ♍ 22	

Day	h	m		
01	04	17	Perigee	
16	05	42	Apogee	
28	02	12	Perigee	
05	19	53	Max dec	19° S 41′
12	16	40	0N	
20	04	18	Max dec	19° N 34′
26	19	22	0S	

LONGITUDES

Date	Chiron ⚷ ° ′	Ceres ⚳ ° ′	Pallas ⚴ ° ′	Juno ⚵ ° ′	Vesta ⚶ ° ′	Black Moon Lilith ⚸ ° ′
01	01 ♉ 55	06 ♌ 04	28 ♊ 51	25 ♍ 43	27 ♊ 18	00 ♉ 33
11	01 ♉ 54	10 ♌ 33	04 ♋ 28	29 ♍ 03	01 ♋ 11	01 ♉ 40
21	01 ♉ 48	15 ♌ 02	10 ♋ 05	02 ♎ 27	04 ♋ 58	02 ♉ 46
31	01 ♉ 36	19 ♌ 30	15 ♋ 41	05 ♎ 55	08 ♋ 34	03 ♉ 52

ASPECTARIAN

h m	Aspects
01 Sunday	
03 15	☽ ⊻ ♂
05 30	♀ Q ♅
09 09	☽ Q ♄
13 22	☽ Q ☉
13 31	☽ ± ♃
13 47	☽ ⚹ ♀
13 52	☽ ⊥ ♂
14 26	☉ ⚹ ♇
15 56	☉ Q ♃
21 41	☽ ∠ ♆
22 55	☽ ⚹ ☿
23 46	☽ ⊼ ♃
02 Monday	
00 38	☽ # ♇
04 10	☽ ∥ ♅
05 48	☽ ∠ ♂
06 29	☿ □ ♃
09 26	☽ ☌ ♅
09 26	☉ ⊥ ♂
11 47	☽ # ☿
12 03	☽ Q ♀
12 56	☽ ⊥ ♆
15 56	☽ □ ♄
20 00	☽ ⊻ ♇
22 05	☽ Q ☿
22 07	☽ □ ☉
23 07	☽ ⊻ ♆
03 Tuesday	
00 18	☽ # ♀
06 14	☽ ⊥ ♇
08 29	☽ ⚹ ♂
12 52	☉ △ ♆
19 49	☽ # ☉
21 21	♀ ∠ ♇
21 38	☽ ∠ ♇
21 40	☽ □ ♀
04 Wednesday	
03 22	☽ ☍ ♃
04 22	☿ # ♅
05 58	☽ Q ♂
08 59	☽ □ ☿
12 42	☽ ⊻ ♅
13 07	☽ # ♃
16 59	☽ # ♄
19 41	☽ △ ♄
23 02	☽ ⊥ ♅
23 26	☽ ⚹ ♇
05 Thursday	
02 25	☿ ∥ ♇
02 27	☽ ☌ ♆
05 16	☽ △ ☉
14 25	☽ □ ♂
14 36	☽ ∠ ♅
17 05	☿ ⚹ ♅
18 37	♂ ∠ ♅
20 12	☽ Q ♇
20 33	☿ ⊥ ♇
21 51	☽ bQ ♄
06 Friday	
06 25	☽ △ ♀
07 40	☽ ⊼ ♃
09 12	☽ bQ ☉
13 40	☽ ± ♄
16 44	☽ ⚹ ♅
18 14	☽ ± ♃
19 49	☽ △ ☿
21 45	♀ □ ♃
07 Saturday	
00 16	☽ ⊼ ♄
00 25	☽ # ♄
01 56	☽ # ♃
02 08	☽ ± ☉
03 41	☽ □ ♇
06 39	☽ ⊻ ♆
10 12	☽ bQ ♃
11 18	☽ bQ ♀
13 30	☽ ⊼ ☉
13 52	☽ Q ♅
17 14	☽ ⊥ ♆
21 27	☽ △ ♂
08 Sunday	
01 45	☽ bQ ☿
04 52	☽ ± ♀
09 14	☽ ∠ ♆
09 17	☽ # ☉
13 07	☽ △ ♃
13 51	♃ ∥ ♄
13 55	☿ ⊻ ♄
16 42	☽ ⊼ ♀
19 57	☽ ± ☿
22 03	☽ □ ♅
09 Monday	
01 39	☽ bQ ♂
06 16	☽ ☍ ♄
08 13	☽ ⊼ ☿
09 23	☽ △ ♇
12 18	☽ ⚹ ♆
14 16	☽ # ♀
14 54	☽ ∥ ♅
18 27	☿ ⊻ ♇
18 52	☽ ± ♂
19 08	♀ # ♅
20 00	☽ # ♇
23 43	☽ ☍ ☉
10 Tuesday	
06 29	☽ ⊼ ♂
10 23	☽ Q ♆
13 02	☽ bQ ♇
20 00	☿ □ ♆
20 34	☽ □ ♃
20 37	☽ # ☿
11 Wednesday	
03 48	♀ ⚹ ♅
05 30	☽ △ ♅
05 41	☽ ☍ ♀
05 58	☽ ± ♇
08 35	♀ ⊥ ♇
13 18	♀ ∥ ♇
14 34	☽ ⊼ ♄
17 23	☽ ⊼ ♇
18 22	☽ # ♂
20 19	☽ □ ♆
23 32	☽ ☍ ☿
12 Thursday	
02 11	☽ ± ♄
07 10	☽ Q ♃
10 17	☽ bQ ♅
13 06	☽ ⊼ ☉
18 30	☽ ☍ ♂
19 49	☽ bQ ♄
13 Friday	
00 36	☿ ⊥ ♄
01 48	☽ ± ☉
03 53	☽ ± ♅
06 49	☽ ⚹ ♃
12 38	☽ ∥ ♂
15 46	☽ ⊼ ♅
21 08	☽ bQ ☉
22 26	☽ ⊼ ♀
14 Saturday	
00 39	☽ ∥ ☿
01 46	☽ △ ♄
02 06	♂ ∠ ♄
04 14	☽ ☍ ♇
05 04	☉ Q ♅
07 08	☽ △ ♆
11 48	☽ ± ♀
12 58	☽ ∠ ♃
18 25	☽ ⊼ ☿
15 Sunday	
04 56	☽ ∥ ♀
05 55	☽ △ ☉
07 18	☽ ⊥ ♃
08 04	☽ bQ ♀
08 06	☽ ± ☿
09 40	☽ ⊼ ♂
10 30	♀ ⊻ ♄
13 23	☽ bQ ♆
16 22	☽ ∥ ♇
19 35	☽ ⊻ ♃
22 32	☽ ± ♂
23 13	☽ # ♅
16 Monday	
01 41	☿ ∠ ♅
04 23	☽ ☍ ♅
04 43	☽ bQ ☿
07 41	☽ ± ♆
08 20	♀ ⊻ ♇
10 01	☽ ∥ ☉
15 07	☽ □ ♄
17 07	☽ ⊼ ♇
17 53	☽ bQ ♂
18 07	☽ △ ♀
19 53	☽ ⊼ ♆
17 Tuesday	
05 20	☽ ± ♇
11 07	♀ □ ♆
14 59	☽ △ ☿
16 30	☉ ∠ ♇
23 36	☽ bQ ♇
18 Wednesday	
00 13	☽ □ ☉
02 00	☽ △ ♂
03 59	☽ Q ♄
08 57	☽ ☌ ♃
15 32	☽ ∥ ♄
17 14	☽ ⊼ ♅
19 Thursday	
02 56	☽ ∥ ♃
04 12	☽ ⚹ ♄
05 11	☽ ± ♅
05 39	☽ △ ♇
08 11	☽ ☍ ♆
08 57	☿ ∥ ♂
13 15	☽ □ ♀
15 08	☽ Q ☉
22 53	☽ bQ ♅
20 Friday	
09 09	☽ □ ☿
09 45	☽ ∠ ♄
11 01	☉ # ♅
14 13	☉ ⊻ ♂
16 14	☿ ∠ ♄
16 14	☽ □ ♂
16 17	☽ ⚹ ☉
16 42	♂ ⊥ ♅
20 07	☽ ⊻ ♃
20 49	♀ ⊥ ♄
21 Saturday	
03 02	☽ ⊥ ♄
03 38	☽ Q ♀
03 39	☽ △ ♅
04 10	☽ ∥ ♃
07 31	☽ ⊥ ♃
14 21	☽ ⊻ ♄
15 18	☽ □ ♇
17 22	☽ ∥ ♄
17 32	☽ ⊼ ♆
22 22	☽ Q ☿
22 38	☽ ∠ ☉
22 Sunday	
00 18	☽ ∠ ♃
03 33	☽ Q ♂
04 11	☽ ⚹ ♀
04 32	☽ ± ♆
16 15	☽ ⊥ ☉
20 45	☽ bQ ♆
22 04	☽ ⚹ ☿
23 18	☉ □ ♃
23 46	♂ Q ♆
23 Monday	
00 02	☽ Q ♇
02 03	♆ St D
02 12	☽ ⚹ ♂
03 28	☽ ⚹ ♃
03 45	☽ ⊻ ☉
09 42	☽ ∠ ♀
10 15	☽ □ ♅
15 57	♀ ∠ ♅
16 53	☿ # ♂
17 42	☉ ∥ ♇
18 14	☿ ⊥ ♅
20 36	☽ ☌ ♄
21 07	☽ ⚹ ♇
23 05	☽ △ ♆
24 Tuesday	
00 22	☽ # ♅
00 33	☽ Q ♃
02 33	☽ ∠ ☿
02 53	☽ ⊥ ♀
05 37	☽ ∠ ♂
07 45	☽ ∥ ♇
08 53	♂ △ ♃
09 02	☽ ∥ ☉
14 08	☽ ⊻ ♀
17 22	☽ Q ♅
19 08	☽ ⊥ ☿
21 40	☽ ⊥ ♂
22 49	☽ ∠ ♇
25 Wednesday	
01 33	☿ Q ♆
06 02	☽ ⊻ ☿
07 19	☽ □ ♃
08 14	☽ ⊻ ♂
11 01	☽ ☌ ☉
13 36	☽ ⚹ ♅
13 57	☽ ⊥ ♇
20 39	☽ ∥ ♀
23 47	☽ ⊻ ♄
23 57	☽ ⊻ ♇
26 Thursday	
01 43	☽ □ ♆
02 15	☿ △ ♃
08 27	☽ # ☿
09 47	☽ ⊥ ♄
14 31	☽ ∠ ♅
19 01	☽ # ♂
19 46	☽ ∥ ♂
20 50	☽ ☌ ♀
27 Friday	
00 44	☽ ∠ ♄
02 55	☉ ⚹ ♅
03 20	♄ ⚹ ♇
05 21	☽ ⊥ ♅
07 23	☽ Q ♆
08 08	☉ ⊥ ♇
08 52	☽ ∥ ☿
09 13	☽ △ ♃
11 10	☽ ☌ ☿
12 07	☽ ☌ ♂
13 36	☽ # ♀
15 14	☽ ⊻ ♅
16 04	☽ ⊻ ☉
28 Saturday	
01 26	☽ ☌ ♇
01 34	☽ ⚹ ♄
02 37	☽ ⊥ ☉
03 05	☽ ⚹ ♆
09 58	☽ bQ ♃
09 59	☿ # ♀
18 28	☽ ∠ ☉
19 44	☽ # ☉
21 29	☽ Q ♄
23 51	♀ ∠ ♄
29 Sunday	
00 52	☽ ± ♃
02 42	☽ ⊻ ♀
03 48	☽ ∠ ♆
05 23	☽ # ♇
10 51	☽ ⊼ ♃
13 27	☽ ∥ ♅
13 33	☽ ⊥ ♀
15 40	☽ ⊻ ☿
15 51	☽ ⊻ ♂
16 50	☽ ☌ ♅
18 44	☽ ⊥ ♆
21 06	☽ ⚹ ☉
30 Monday	
02 14	☽ ⊥ ☿
02 19	☽ ⊥ ♂
03 10	☽ ⊻ ♇
03 37	☽ □ ♄
04 45	☽ ⊻ ♆
05 58	☽ ∠ ♀
12 03	☿ ☌ ♂
13 15	☽ ⊥ ♇
14 51	☿ ⊻ ♅
15 18	♂ ⊻ ♅
18 08	☽ ∠ ♂
18 10	☽ ∠ ☿
18 40	☽ Q ☉
31 Tuesday	
04 29	☽ ∠ ♇
05 13	♀ ⊥ ♅
09 40	☽ ⚹ ♀
13 02	☽ # ♄
13 30	☽ ☍ ♃
19 34	☽ ⊻ ♅
20 51	☽ ⚹ ♂
20 59	☽ ⚹ ☿
23 14	♀ # ♂

All ephemeris data is given at 12.00 UT and the Moon's longitude is additionally given for 24.00 UT

SEPTEMBER 1976

LONGITUDES

Date	Sidereal time h m s	Sun ☉ ° ' "	Moon ☽ ° ' "	Moon ☽ 24.00 ° ' "	Mercury ☿ ° '	Venus ♀ ° '	Mars ♂ ° '	Jupiter ♃ ° '	Saturn ♄ ° '	Uranus ♅ ° '	Neptune ♆ ° '	Pluto ♇ ° '
01	10 43 06	09 ♍ 09 58	13 ♐ 41 55	20 ♐ 36 13	05 ♎ 23	29 ♍ 42	05 ♎ 18	00 ♊ 40	10 ♌ 49	04 ♏ 11	11 ♐ 13	10 ♎ 21
02	10 47 02	10 08 03	27 ♐ 27 16	04 ♑ 15 06	05 58	00 ♎ 56	05 57	00 43	10 56	04 13	11 14	10 23
03	10 50 59	11 06 10	10 ♑ 59 45	17 ♑ 41 14	06 29	02 10	06 36	00 46	11 03	04 16	11 14	10 25
04	10 54 55	12 04 18	24 ♑ 19 35	00 ♒ 54 47	06 56	03 24	07 15	00 49	11 10	04 19	11 14	10 27
05	10 58 52	13 02 28	07 ♒ 26 50	13 ♒ 55 44	07 18	04 37	07 54	00 52	11 17	04 21	11 15	10 29
06	11 02 48	14 00 39	20 ♒ 21 26	26 ♒ 43 57	07 34	05 51	08 33	00 55	11 24	04 24	11 15	10 31
07	11 06 45	14 58 51	03 ♓ 03 15	09 ♓ 19 22	07 45	07 05	09 12	00 57	11 31	04 27	11 16	10 34
08	11 10 42	15 57 06	15 ♓ 32 21	21 ♓ 42 15	07 51	08 19	09 51	01 00	11 38	04 30	11 16	10 36
09	11 14 38	16 55 22	27 ♓ 49 13	03 ♈ 53 23	07 R 50	09 32	10 30	01 02	11 45	04 32	11 17	10 38
10	11 18 35	17 53 40	09 ♈ 54 57	15 ♈ 54 13	07 44	10 46	11 09	01 04	11 51	04 35	11 17	10 40
11	11 22 31	18 52 00	21 ♈ 51 27	27 ♈ 47 02	07 30	12 00	11 48	01 06	11 58	04 38	11 18	10 42
12	11 26 28	19 50 22	03 ♉ 41 23	09 ♉ 34 56	07 10	13 13	12 28	01 07	12 05	04 41	11 19	10 45
13	11 30 24	20 48 46	15 ♉ 28 12	21 ♉ 21 43	06 44	14 27	13 07	01 09	12 11	04 44	11 19	10 47
14	11 34 21	21 47 13	27 ♉ 16 04	03 ♊ 11 51	06 10	15 41	13 46	01 10	12 18	04 47	11 20	10 49
15	11 38 17	22 45 41	09 ♊ 09 42	15 ♊ 10 15	05 30	16 54	14 26	01 11	12 25	04 50	11 21	10 51
16	11 42 14	23 44 11	21 ♊ 14 08	27 ♊ 22 00	04 44	18 08	15 05	01 11	12 31	04 53	11 22	10 54
17	11 46 11	24 42 44	03 ♋ 34 27	09 ♋ 52 05	03 53	19 21	15 45	01 12	12 38	04 56	11 22	10 56
18	11 50 07	25 41 19	16 ♋ 15 24	22 ♋ 44 52	02 56	20 35	16 24	01 12	12 44	04 59	11 23	10 58
19	11 54 04	26 39 56	29 ♋ 20 50	06 ♌ 03 32	01 55	21 49	17 04	01 12	12 50	05 02	11 24	11 00
20	11 58 00	27 38 35	12 ♌ 53 06	19 ♌ 49 27	00 ♎ 52	23 02	17 43	01 R 12	12 57	05 05	11 25	11 03
21	12 01 57	28 37 17	26 ♌ 52 24	04 ♍ 01 32	29 ♍ 47	24 16	18 23	01 12	13 03	05 08	11 26	11 05
22	12 05 53	29 ♍ 36 00	11 ♍ 16 19	18 ♍ 36 00	28 43	25 29	19 03	01 12	13 09	05 12	11 27	11 07
23	12 09 50	00 ♎ 34 46	25 ♍ 59 44	03 ♎ 26 32	27 40	26 43	19 43	01 11	13 15	05 15	11 28	11 10
24	12 13 46	01 33 33	10 ♎ 55 21	18 ♎ 25 06	26 40	27 57	20 23	01 10	13 21	05 18	11 29	11 12
25	12 17 43	02 32 23	25 ♎ 54 42	03 ♏ 23 07	25 46	29 ♎ 10	21 03	01 09	13 27	05 21	11 30	11 14
26	12 21 40	03 31 14	10 ♏ 49 25	18 ♏ 12 46	24 58	00 ♏ 24	21 42	01 08	13 33	05 25	11 31	11 16
27	12 25 36	04 30 07	25 ♏ 32 28	02 ♐ 47 58	24 17	01 37	22 22	01 06	13 39	05 28	11 32	11 19
28	12 29 33	05 29 02	09 ♐ 58 50	17 ♐ 04 47	23 45	02 51	23 03	01 05	13 45	05 32	11 33	11 21
29	12 33 29	06 27 59	24 ♐ 05 41	01 ♑ 01 26	23 23	04 04	23 43	01 03	13 50	05 35	11 34	11 23
30	12 37 26	07 ♎ 26 58	07 ♑ 52 06	14 ♑ 37 46	23 ♍ 11	05 ♏ 18	24 ♎ 23	01 ♊ 01	13 ♌ 56	05 ♏ 38	11 ♐ 36	11 ♎ 26

Date	Moon True ☊ ° '	Moon Mean ☊ ° '	Moon ☽ Latitude ° '
01	05 ♏ 13	06 ♏ 19	03 N 11
02	05 R 12	06 16	04 03
03	05 08	06 13	04 40
04	05 03	06 09	05 01
05	04 55	06 06	05 06
06	04 46	06 03	04 56
07	04 37	06 00	04 30
08	04 29	05 57	03 52
09	04 22	05 54	03 04
10	04 17	05 50	02 08
11	04 14	05 47	01 07
12	04 13	05 44	00 N 03
13	04 D 14	05 41	01 S 01
14	04 15	05 38	02 02
15	04 17	05 35	02 59
16	04 17	05 31	03 48
17	04 R 17	05 28	04 28
18	04 15	05 25	04 57
19	04 11	05 22	05 11
20	04 07	05 19	05 08
21	04 01	05 15	04 48
22	03 56	05 12	04 08
23	03 51	05 09	03 12
24	03 48	05 06	02 02
25	03 47	05 03	00 S 43
26	03 D 47	05 00	00 N 39
27	03 48	04 56	01 57
28	03 49	04 53	03 06
29	03 50	04 50	04 02
30	03 ♏ 51	04 ♏ 47	04 N 43

DECLINATIONS

Date	Sun ☉ ° '	Moon ☽ ° '	Mercury ☿ ° '	Venus ♀ ° '	Mars ♂ ° '	Jupiter ♃ ° '	Saturn ♄ ° '	Uranus ♅ ° '	Neptune ♆ ° '	Pluto ♇ ° '
01	08 N 08	19 S 17	05 S 02	01 N 10	01 S 35	19 N 15	18 N 04	12 S 29	20 S 38	11 N 05
02	07 46	19 22	05 24	00 39	01 51	19 15	18 02	12 30	20 38	11 04
03	07 24	18 20	05 44	00 N 08	02 07	19 16	18 00	12 31	20 38	11 03
04	07 02	16 18	06 01	00 S 22	02 23	19 16	17 58	12 32	20 38	11 02
05	06 40	13 28	06 17	00 53	02 39	19 17	17 56	12 33	20 38	11 01
06	06 17	10 01	06 29	01 24	02 54	19 17	17 55	12 34	20 38	11 00
07	05 55	06 11	06 39	01 55	03 10	19 18	17 53	12 35	20 38	10 58
08	05 32	02 S 08	06 46	02 26	03 26	19 18	17 51	12 36	20 38	10 57
09	05 10	01 N 57	06 50	02 57	03 42	19 18	17 49	12 37	20 39	10 56
10	04 47	05 53	06 50	03 28	03 58	19 19	17 47	12 38	20 39	10 55
11	04 24	09 33	06 47	03 58	04 14	19 19	17 46	12 39	20 39	10 54
12	04 01	12 48	06 40	04 29	04 30	19 19	17 44	12 40	20 39	10 53
13	03 38	15 30	06 29	05 00	04 46	19 19	17 42	12 41	20 39	10 52
14	03 15	17 34	06 14	05 30	05 01	19 19	17 40	12 42	20 39	10 51
15	02 52	18 53	05 55	06 00	05 17	19 19	17 39	12 43	20 39	10 50
16	02 29	19 21	05 32	06 31	05 33	19 19	17 37	12 44	20 40	10 50
17	02 06	18 55	05 04	07 01	05 49	19 19	17 35	12 45	20 40	10 49
18	01 43	17 33	04 33	07 31	06 04	19 19	17 34	12 46	20 40	10 48
19	01 20	15 13	03 59	08 01	06 20	19 19	17 32	12 47	20 40	10 47
20	00 56	12 01	03 22	08 30	06 36	19 19	17 30	12 48	20 40	10 46
21	00 33	08 03	02 42	09 00	06 51	19 19	17 29	12 49	20 40	10 45
22	00 N 10	03 N 30	02 02	09 29	07 07	19 19	17 27	12 50	20 41	10 44
23	00 S 14	01 S 21	01 20	09 58	07 23	19 18	17 25	12 51	20 41	10 43
24	00 37	06 12	00 S 39	10 27	07 38	19 18	17 24	12 52	20 41	10 42
25	01 01	10 41	00 N 01	10 56	07 54	19 18	17 22	12 53	20 41	10 41
26	01 24	14 28	00 38	11 24	08 09	19 17	17 21	12 55	20 41	10 40
27	01 47	17 15	01 13	11 53	08 25	19 17	17 19	12 56	20 42	10 39
28	02 11	18 53	01 44	12 20	08 40	19 17	17 18	12 57	20 42	10 38
29	02 34	19 16	02 10	12 48	08 56	19 16	17 16	12 58	20 42	10 37
30	02 S 57	18 S 30	02 N 33	13 S 15	09 S 11	19 N 16	17 N 15	12 S 59	20 S 42	10 N 36

ZODIAC SIGN ENTRIES

Date	h	m	Planets
01	17	44	♀ ♎
02	16	29	☽ ♑
04	22	20	☽ ♒
07	06	11	☽ ♓
09	16	18	☽ ♈
12	04	30	☽ ♉
14	17	32	☽ ♊
17	05	07	☽ ♋
19	13	11	☽ ♌
21	07	15	☿ ♍
21	17	16	☽ ♍
22	21	48	☉ ♎
23	18	28	☽ ♎
25	18	34	☽ ♏
26	04	17	♀ ♏
27	19	21	☽ ♐
29	22	13	☽ ♑

LATITUDES

Date	Mercury ☿ ° '	Venus ♀ ° '	Mars ♂ ° '	Jupiter ♃ ° '	Saturn ♄ ° '	Uranus ♅ ° '	Neptune ♆ ° '	Pluto ♇ ° '
01	03 S 09	01 N 09	00 N 34	01 S 04	00 N 34	00 N 27	01 N 31	16 N 30
04	03 34	01 04	00 33	01 05	00 34	00 27	01 31	16 30
07	03 54	00 58	00 31	01 05	00 34	00 27	01 30	16 29
10	04 07	00 52	00 29	01 05	00 35	00 27	01 30	16 28
13	04 09	00 46	00 27	01 06	00 35	00 27	01 30	16 28
16	03 58	00 39	00 26	01 06	00 35	00 27	01 30	16 27
19	03 30	00 31	00 24	01 06	00 36	00 27	01 30	16 27
22	02 46	00 24	00 22	01 07	00 36	00 27	01 30	16 27
25	01 50	00 16	00 20	01 07	00 37	00 27	01 29	16 26
28	00 S 49	00 N 08	00 19	01 07	00 37	00 27	01 29	16 26
31	00 N 07	00 S 01	00 N 17	01 S 08	00 N 37	00 N 27	01 N 29	16 N 26

DATA

Julian Date	2443023
Delta T	+47 seconds
Ayanamsa	23° 32' 03"
Synetic vernal point	05° ♓ 34' 56"
True obliquity of ecliptic	23° 26' 25"

LONGITUDES

Date	Chiron ⚷ ° '	Ceres ⚳ ° '	Pallas ⚴ ° '	Juno ⚵ ° '	Vesta ⚶ ° '	Black Moon Lilith ⚸ ° '
01	01 ♉ 35	19 ♌ 57	16 ♋ 14	06 ♎ 16	08 ♋ 55	03 ♉ 59
11	01 ♉ 18	24 ♌ 24	21 ♋ 46	09 ♎ 48	12 ♋ 19	05 ♉ 06
21	00 ♉ 57	28 ♌ 49	27 ♋ 13	13 ♎ 21	15 ♋ 30	06 ♉ 12
31	00 ♉ 33	03 ♍ 10	02 ♌ 31	16 ♎ 55	18 ♋ 24	07 ♉ 18

MOON'S PHASES, APSIDES AND POSITIONS ☽

Date	h	m	Phase	Longitude ° '	Eclipse Indicator
01	03	35	☽	08 ♐ 50	
08	12	52	○	15 ♓ 59	
16	17	20	☾	23 ♊ 57	
23	19	55	●	00 ♎ 54	
30	11	12	☽	07 ♑ 25	

Day	h	m		
12	23	12	Apogee	
25	03	13	Perigee	
02	01	47	Max dec	19° S 28'
09	00	30	0N	
16	12	45	Max dec	19° N 21'
23	05	24	0S	
29	07	46	Max dec	19° S 18'

ASPECTARIAN

h m Aspects

01 Wednesday
02 30 ♀ Q ♆
03 35 ☽ □ ☉
05 54 ☽ ⊥ ♅
06 12 ☽ ✶ ♇
06 58 ☽ △ ♄
07 43 ☽ ☌ ♆
08 13 ☽ Q ♀
10 49 ☽ ⫲ ♃
18 34 ☽ Q ♂
18 41 ☽ Q ☿
21 33 ☽ ∠ ♅

02 Thursday
03 05 ☽ Q ♇
07 32 ♀ △ ♃
09 18 ☽ ⚼ ♄
16 21 ☿ ☌ ♂
16 28 ☽ ⫲ ♃
17 46 ☽ ⊼ ♃
18 25 ☉ ⊻ ♇
18 45 ☽ □ ♀
23 59 ☽ ✶ ♅

03 Friday
01 19 ☽ ± ♄
03 40 ☽ □ ☿
03 46 ☽ □ ♂
04 27 ☽ ± ♃
10 32 ☉ ⊻ ♄
10 58 ☽ □ ♇
12 06 ☽ ⊼ ♄
12 12 ☽ △ ☉
12 25 ☽ ⊻ ♆
15 14 ☉ □ ♆
16 53 ☽ ⫲ ♄
20 35 ☽ ⚼ ♃
21 29 ☽ Q ♅
23 11 ☽ ⊥ ♆

04 Saturday
15 29 ☽ ∠ ♆
17 23 ☽ ⚼ ☉
23 53 ☽ △ ♃

05 Sunday
03 32 ♄ △ ♆
06 16 ☽ △ ♀
06 17 ☽ □ ♅
06 35 ♀ ⊻ ♅
11 11 ☽ ± ☉
11 42 ☽ △ ☿
12 52 ☽ △ ♂
17 38 ☽ △ ♇
18 45 ☽ ∥ ♅
19 01 ☽ ✶ ♆
19 09 ☽ ☍ ♄
23 11 ☽ ⊼ ☉

06 Monday
03 40 ☉ ⫲ ☿
05 34 ☽ ⫲ ♇
13 02 ☽ ⚼ ♀
16 14 ☽ ⚼ ☿
17 27 ☽ Q ♆
18 19 ☽ ⚼ ♂
21 45 ☽ ⚼ ♇

07 Tuesday
07 50 ☽ ± ♀
08 00 ☽ □ ♃
09 14 ☽ ∥ ☿
09 30 ☽ ± ☿
12 17 ☽ ± ♂
13 45 ☽ ⫲ ☉
14 40 ☽ △ ♅
14 53 ☽ ± ♇
20 32 ☽ ⊼ ♀
21 05 ☽ ⊼ ☿

08 Wednesday
00 24 ☽ ⊼ ♂
02 25 ☽ ⊼ ♇
02 31 ☿ ☌ ♀
03 45 ☽ □ ♆
04 22 ☽ ⊼ ♄
04 48 ☽ ∥ ♂
10 25 ☽ ∥ ♀
12 52 ☽ ☍ ☉
16 06 ☽ ± ♄
18 44 ☽ Q ♃
19 43 ☽ ⚼ ♅
22 03 ☿ St R

09 Thursday
09 52 ☽ ⚼ ♄
13 25 ☽ ± ♅
17 10 ♂ ☌ ♇
18 21 ☽ ✶ ♃
18 52 ☽ ⫲ ♀
23 18 ☽ ⫲ ♂

10 Friday
01 20 ☽ ⊼ ♅
05 46 ☽ ∥ ☉
07 41 ☽ ☍ ☿
10 05 ♀ ☌ ♇
10 57 ☉ ⊥ ♄
13 31 ☽ ☍ ♇
13 53 ☽ ☍ ♀
14 37 ☽ ☍ ♂
14 45 ☽ △ ♆
15 55 ☽ △ ♄
17 05 ♂ ✶ ♆
18 00 ☽ ⫲ ☿
22 20 ♀ ✶ ♆

11 Saturday
00 21 ☽ ∠ ♃
04 12 ♀ ☌ ♂
05 26 ☽ ⊼ ☉
11 30 ♀ ✶ ♄
18 31 ☉ ⫲ ♂
18 34 ☽ ⊥ ♃
18 38 ☽ ± ☉
19 13 ♂ ✶ ♄
21 00 ☽ ⚼ ♆
21 36 ☽ ∥ ♇
23 38 ☉ ⫲ ♀

12 Sunday
06 46 ☽ ⊻ ♃
07 55 ☉ ∠ ♅
10 56 ☽ ⫲ ♅
13 04 ♀ ∥ ♂
14 02 ☽ ☍ ♅
14 33 ☽ ⚼ ☉
15 18 ☽ ± ♆
18 51 ☽ ⊼ ☿

13 Monday
02 25 ☽ ⊼ ♇
03 32 ☽ ⊼ ♆
05 15 ☽ □ ♄
06 38 ☽ ± ☿
06 55 ☽ ⊼ ♂
09 40 ☽ ⊼ ♀
14 41 ☽ ± ♇
19 52 ☽ ± ♂
23 19 ☽ ± ♀
23 52 ☽ △ ☉

14 Tuesday
00 12 ☽ ⚼ ☿
09 03 ☽ ⚼ ♇
13 29 ☽ ∥ ♄
15 14 ☽ ⚼ ♂
18 12 ☽ Q ♄
19 42 ☽ ⚼ ♀
19 54 ☽ ☌ ♃
21 38 ♀ ⚼ ♃

15 Wednesday
03 15 ☽ ⊼ ♅
05 04 ☽ △ ☿
09 23 ☿ ∥ ♀
15 21 ☽ ± ♅
15 24 ☽ △ ♇
16 23 ☽ ☍ ♆
18 34 ☽ ✶ ♄
23 08 ☽ △ ♂

16 Thursday
05 11 ☽ △ ♀
06 12 ☽ ∥ ♃
08 01 ☿ ⊻ ♅
09 19 ☽ ⚼ ♅
11 12 ☿ ∥ ♂
17 20 ☽ □ ☉
19 17 ☽ ∥ ♃

17 Friday
00 24 ☽ ∠ ♄
07 25 ☽ ⊻ ♃
12 32 ☽ □ ☿
14 37 ☽ △ ♅
17 53 ☽ ⊥ ♄
18 56 ☽ ⊥ ♃

18 Saturday
02 03 ☽ □ ♇
02 51 ☽ ⊼ ♆
04 44 ♂ ⚼ ♃
05 21 ☽ ⊻ ♄
06 48 ☽ Q ☉
11 45 ☽ ∥ ♄
11 54 ☽ ∠ ♃
12 17 ☽ □ ♂
14 06 ☽ ± ♆
20 03 ☽ Q ☿
20 51 ☽ □ ♀

19 Sunday
06 40 ☽ ⚼ ♆
06 46 ☽ ✶ ☉
11 23 ☽ Q ♇
15 21 ☽ ✶ ♃
16 18 ☽ ✶ ☿
18 37 ♃ St R
22 13 ☽ □ ♅
22 46 ☽ Q ♂

20 Monday
04 17 ☿ △ ♃
06 42 ☽ ⫲ ♅
08 27 ☽ Q ♀
08 46 ☽ ✶ ♇
09 26 ☽ △ ♆
11 33 ☽ ∠ ☉
12 06 ☽ ☌ ♄
12 34 ☽ Q ♃
16 48 ☽ ∠ ☿
20 06 ☽ ∥ ♇
20 13 ☉ ∠ ♄
20 48 ☽ ✶ ♂

21 Tuesday
04 15 ☽ ⊥ ☉
05 39 ☽ Q ♅
07 08 ☽ ⊥ ☿
07 10 ☽ ✶ ♀
07 10 ☽ ⫲ ♀
10 39 ☽ ∠ ♇
15 10 ☽ ⊻ ☉
16 34 ☽ ⊻ ☿
18 10 ☽ ⫲ ♂
19 17 ☽ □ ♃
19 46 ☿ Q ♆
23 28 ☽ ∠ ♂

22 Wednesday
01 30 ☉ ⊥ ♅
01 34 ☉ ☌ ☿
01 38 ☿ ⊥ ♅
01 48 ☽ ⊥ ♇
01 55 ☽ ✶ ♅
04 42 ♀ Q ♄
06 16 ♀ ± ♃
08 11 ☉ Q ♆
10 36 ☽ ∠ ♀
11 45 ☽ ⊻ ♇
12 17 ☽ □ ♆
15 04 ☽ ⊥ ♂
15 06 ☽ ⊻ ♄
20 39 ☽ ⫲ ☿
22 05 ♃ Q ♄
23 35 ☿ ∠ ♄

23 Thursday
00 59 ☽ ⊥ ♄
01 20 ☽ ⊻ ♂
02 39 ☽ ∠ ♅
02 40 ☽ ⊥ ♀
04 50 ☽ ⫲ ☉
06 04 ☽ ∥ ☉
06 58 ♀ ∠ ♆
11 57 ☽ ∥ ☿
13 16 ☽ ⊻ ♀
14 31 ☽ ☌ ☿
15 40 ☽ ∠ ♄
17 16 ☽ ⊥ ♅
17 36 ☽ Q ♆
19 55 ☽ ☌ ☉
20 22 ☽ △ ♃
22 08 ☿ ⊻ ♀

24 Friday
02 37 ☉ △ ♃
02 57 ☽ ⊻ ♅
12 26 ☽ ☌ ♇
12 45 ☉ ∥ ☿
12 54 ☽ ✶ ♆
15 55 ☽ ✶ ♄
19 53 ☽ ∥ ♂
20 24 ☽ ⚼ ♃
23 44 ♀ ⫲ ♇

25 Saturday
03 50 ☽ ☌ ♂
10 47 ☽ ± ♃
11 15 ☽ Q ♄
11 46 ☽ ⊻ ☿
12 01 ☽ ⫲ ♇
12 56 ☽ ∠ ♆
13 39 ☽ ∥ ♀
17 41 ☽ ☌ ♀
20 24 ☽ ⊼ ♃
20 53 ☽ ⊥ ☿
23 23 ☽ ⊻ ☉

26 Sunday
01 25 ☽ ∥ ♅
03 14 ☽ ☌ ♅
03 25 ☽ ⊥ ♆
09 45 ☽ ⊥ ☉
10 40 ☽ ∠ ☿
12 44 ☽ ⊻ ♇
13 07 ☽ ⊻ ♆
16 27 ☽ □ ♄
18 57 ☿ ⊥ ♀
22 30 ☽ ⊥ ♇

27 Monday
01 24 ☽ ∠ ☉
02 10 ♀ ⊼ ♃
06 33 ☽ ⊻ ♂
10 01 ☽ ✶ ☿
12 40 ☽ ⫲ ♄
13 16 ☽ ∠ ♇
16 54 ☽ ⊥ ♂
21 10 ☽ ☍ ♃
22 58 ☽ ⊻ ♀

28 Tuesday
03 55 ☽ ✶ ☉
04 31 ☽ ⊻ ♅
05 09 ☽ Q ☿
08 35 ☽ ∠ ♂
09 55 ☽ ⊥ ♀
13 04 ☉ ⊻ ♅
14 19 ☽ ✶ ♇
14 37 ☽ ⊥ ♅
14 39 ☽ ☌ ♆
18 23 ☽ △ ♄

29 Wednesday
01 38 ☽ Q ☉
02 34 ☽ ∠ ♀
03 22 ☽ ⫲ ♃
04 05 ☿ ⊻ ♂
05 57 ☽ ∠ ♅
06 06 ♅ ⊥ ♆
10 47 ☽ Q ♇
10 48 ☽ □ ☿
11 18 ☽ ✶ ♂
12 29 ☽ ⫲ ♃
20 15 ☽ ⚼ ♄
21 00 ♀ ∥ ♅

30 Thursday
00 01 ☽ ⊼ ♃
07 01 ☽ ✶ ♀
08 04 ☽ ✶ ♅
09 15 ☽ Q ♂
10 30 ☽ ± ♃
11 12 ☽ □ ☉
12 07 ☽ ± ♄
17 57 ♀ ⊥ ♆
18 19 ☽ □ ♇
18 36 ☽ ⊻ ♆
19 05 ♀ ☌ ♅
22 50 ☽ ⊼ ♄

All ephemeris data is given at 12.00 UT and the Moon's longitude is additionally given for 24.00 UT

OCTOBER 1976

LONGITUDES

Date	Sidereal time h m s	Sun ☉ ° ' "	Moon ☽ ° ' "	Moon ☽ 24.00 ° ' "	Mercury ☿ ° '	Venus ♀ ° '	Mars ♂ ° '	Jupiter ♃ ° '	Saturn ♄ ° '	Uranus ♅ ° '	Neptune ♆ ° '	Pluto ♇ ° '
01	12 41 22	08 ♎ 25 58	21 ♑ 18 37	27 ♑ 54 50	23 ♍ 09	06 ♏ 31	25 ♎ 03	00 ♊ 59	14 ♌ 01	05 ♏ 42	11 ♐ 37	11 ♎ 28
02	12 45 19	09 25 00	04 ♒ 26 40	10 ♒ 54 20	23 D 18	07 45	25 43	00 R 56	14 07	05 45	11 38	11 31
03	12 49 15	10 24 03	17 ♒ 18 06	23 ♒ 38 14	23 37	08 58	26 24	00 53	14 12	05 49	11 39	11 33
04	12 53 12	11 23 09	29 ♒ 54 56	06 ♓ 08 28	24 06	10 11	27 04	00 51	14 18	05 52	11 41	11 35
05	12 57 09	12 22 16	12 ♓ 19 04	18 ♓ 26 56	24 44	11 25	27 44	00 48	14 23	05 56	11 42	11 38
06	13 01 05	13 21 25	24 ♓ 32 18	00 ♈ 35 22	25 31	12 38	28 25	00 44	14 28	05 59	11 43	11 40
07	13 05 02	14 20 36	06 ♈ 36 22	12 ♈ 35 30	26 25	13 52	29 05	00 41	14 33	06 03	11 45	11 42
08	13 08 58	15 19 49	18 ♈ 33 00	24 ♈ 29 06	27 28	15 05	29 ♎ 46	00 37	14 38	06 06	11 46	11 45
09	13 12 55	16 19 05	00 ♉ 24 05	06 ♉ 18 14	28 36	16 18	00 ♏ 26	00 33	14 43	06 10	11 48	11 47
10	13 16 51	17 18 22	12 ♉ 11 50	18 ♉ 05 14	29 ♍ 51	17 31	01 07	00 29	14 48	06 14	11 49	11 49
11	13 20 48	18 17 41	23 ♉ 58 48	29 ♉ 52 56	01 ♎ 11	18 45	01 48	00 25	14 53	06 17	11 51	11 52
12	13 24 44	19 17 03	05 ♊ 48 03	11 ♊ 44 36	02 35	19 58	02 29	00 21	14 58	06 21	11 52	11 54
13	13 28 41	20 16 27	17 ♊ 43 05	23 ♊ 44 00	04 02	21 11	03 09	00 16	15 03	06 24	11 54	11 56
14	13 32 38	21 15 54	29 ♊ 47 53	05 ♋ 55 17	05 33	22 25	03 50	00 12	15 07	06 28	11 55	11 59
15	13 36 34	22 15 22	12 ♋ 06 43	18 ♋ 22 46	07 07	23 38	04 31	00 07	15 12	06 32	11 57	12 01
16	13 40 31	23 14 53	24 ♋ 43 55	01 ♌ 10 41	08 43	24 51	05 12	00 ♊ 02	15 16	06 35	11 58	12 03
17	13 44 27	24 14 26	07 ♌ 43 29	14 ♌ 22 42	10 20	26 04	05 53	29 ♉ 57	15 20	06 39	12 00	12 06
18	13 48 24	25 14 02	21 ♌ 08 37	28 ♌ 01 23	11 59	27 17	06 34	29 51	15 25	06 43	12 02	12 08
19	13 52 20	26 13 39	05 ♍ 01 02	12 ♍ 07 27	13 38	28 30	07 15	29 46	15 29	06 47	12 03	12 10
20	13 56 17	27 13 19	19 ♍ 20 21	26 ♍ 39 14	15 19	29 ♏ 44	07 56	29 40	15 33	06 50	12 05	12 12
21	14 00 13	28 13 02	04 ♎ 03 29	11 ♎ 32 17	17 00	00 ♐ 57	08 38	29 34	15 37	06 54	12 07	12 15
22	14 04 10	29 ♎ 12 46	19 ♎ 04 37	26 ♎ 39 26	18 41	02 10	09 19	29 28	15 41	06 58	12 09	12 17
23	14 08 07	00 ♏ 12 32	04 ♏ 15 32	11 ♏ 51 42	20 23	03 23	10 00	29 22	15 45	07 01	12 10	12 19
24	14 12 03	01 12 21	19 ♏ 26 44	26 ♏ 59 26	22 04	04 36	10 42	29 16	15 48	07 05	12 12	12 21
25	14 16 00	02 12 11	04 ♐ 28 47	11 ♐ 53 49	23 46	05 49	11 23	29 09	15 52	07 09	12 14	12 24
26	14 19 56	03 12 03	19 ♐ 13 45	26 ♐ 27 58	25 27	07 02	12 05	29 03	15 55	07 13	12 16	12 26
27	14 23 53	04 11 57	03 ♑ 36 03	10 ♑ 37 42	27 08	08 15	12 46	28 56	15 59	07 16	12 18	12 28
28	14 27 49	05 11 52	17 ♑ 32 50	24 ♑ 21 27	28 ♎ 49	09 28	13 28	28 49	16 02	07 20	12 20	12 30
29	14 31 46	06 11 49	01 ♒ 03 43	07 ♒ 39 52	00 ♏ 30	10 41	14 09	28 42	16 05	07 24	12 21	12 33
30	14 35 42	07 11 48	14 ♒ 10 15	20 ♒ 35 14	02 10	11 54	14 51	28 35	16 08	07 28	12 23	12 35
31	14 39 39	08 ♏ 11 48	26 ♒ 55 16	03 ♓ 10 47	03 ♏ 49	13 ♐ 07	15 ♏ 33	28 ♉ 28	16 ♌ 11	07 ♏ 31	12 ♐ 25	12 ♎ 37

Date	Moon True ☊ ° '	Moon Mean ☊ ° '	Moon ☽ Latitude ° '
01	03 ♏ 50	04 ♏ 44	05 N 07
02	03 R 48	04 41	05 15
03	03 46	04 37	05 06
04	03 42	04 34	04 43
05	03 39	04 31	04 06
06	03 35	04 28	03 19
07	03 33	04 25	02 23
08	03 31	04 21	01 22
09	03 30	04 18	00 N 17
10	03 D 31	04 15	00 S 48
11	03 31	04 12	01 51
12	03 33	04 09	02 50
13	03 34	04 06	03 42
14	03 35	04 02	04 24
15	03 36	03 59	04 56
16	03 R 36	03 56	05 14
17	03 36	03 53	05 17
18	03 35	03 50	05 03
19	03 34	03 46	04 31
20	03 33	03 43	03 42
21	03 33	03 40	02 37
22	03 33	03 37	01 S 20
23	03 D 32	03 34	00 N 04
24	03 32	03 31	01 27
25	03 33	03 27	02 44
26	03 33	03 24	03 48
27	03 R 33	03 21	04 36
28	03 33	03 18	05 06
29	03 33	03 15	05 18
30	03 D 33	03 12	05 13
31	03 ♏ 33	03 ♏ 08	04 N 52

DECLINATIONS

Date	Sun ☉ ° '	Moon ☽ ° '	Mercury ☿ ° '	Venus ♀ ° '	Mars ♂ ° '	Jupiter ♃ ° '	Saturn ♄ ° '	Uranus ♅ ° '	Neptune ♆ ° '	Pluto ♇ ° '
01	03 S 21	16 S 41	02 N 50	13 S 42	09 S 26	19 N 15	17 N 13	13 S 00	20 S 42	10 N 35
02	03 44	14 03	03 02	14 09	09 41	19 15	17 12	13 01	20 43	10 34
03	04 07	10 47	03 08	14 36	09 57	19 14	17 10	13 03	20 43	10 34
04	04 30	07 05	03 10	15 02	10 12	19 13	17 09	13 04	20 43	10 33
05	04 53	03 S 09	03 06	15 27	10 27	19 12	17 07	13 05	20 43	10 32
06	05 16	00 N 53	02 58	15 53	10 42	19 12	17 06	13 06	20 43	10 31
07	05 39	04 49	02 45	16 17	10 57	19 11	17 05	13 07	20 44	10 30
08	06 02	08 32	02 28	16 42	11 11	19 10	17 03	13 08	20 44	10 29
09	06 25	11 53	02 07	17 06	11 26	19 09	17 02	13 10	20 44	10 28
10	06 48	14 44	01 43	17 30	11 41	19 08	17 01	13 11	20 44	10 28
11	07 10	16 58	01 15	17 53	11 56	19 07	16 59	13 12	20 45	10 27
12	07 33	18 29	00 45	18 16	12 10	19 06	16 58	13 13	20 45	10 26
13	07 55	19 12	00 N 12	18 38	12 25	19 05	16 57	13 15	20 45	10 25
14	08 18	19 02	00 S 23	19 00	12 39	19 04	16 56	13 16	20 45	10 24
15	08 40	17 59	00 59	19 22	12 54	19 03	16 55	13 17	20 46	10 24
16	09 02	16 02	01 37	19 42	13 08	19 02	16 53	13 18	20 46	10 23
17	09 24	13 14	02 17	20 03	13 22	19 01	16 52	13 19	20 46	10 22
18	09 46	09 40	02 57	20 23	13 36	19 00	16 51	13 21	20 46	10 21
19	10 08	05 28	03 38	20 42	13 50	18 59	16 50	13 22	20 47	10 21
20	10 29	00 N 49	04 20	21 01	14 04	18 57	16 49	13 23	20 47	10 20
21	10 50	04 S 01	05 02	21 19	14 18	18 56	16 48	13 24	20 47	10 19
22	11 12	08 42	05 44	21 37	14 32	18 55	16 47	13 26	20 47	10 18
23	11 33	12 53	06 27	21 54	14 45	18 53	16 46	13 27	20 48	10 18
24	11 54	16 11	07 09	22 10	14 59	18 52	16 45	13 28	20 48	10 17
25	12 14	18 21	07 51	22 26	15 12	18 51	16 44	13 29	20 48	10 16
26	12 35	19 13	08 33	22 42	15 26	18 49	16 43	13 30	20 49	10 16
27	12 55	18 48	09 15	22 56	15 39	18 48	16 42	13 32	20 49	10 15
28	13 15	17 14	09 57	23 10	15 52	18 46	16 42	13 33	20 49	10 15
29	13 35	14 45	10 38	23 24	16 05	18 45	16 41	13 34	20 49	10 14
30	13 55	11 35	11 18	23 37	16 18	18 43	16 40	13 35	20 50	10 13
31	14 S 14	07 S 57	11 S 58	23 S 49	16 S 31	18 N 42	16 N 39	13 S 37	20 S 50	10 N 13

ZODIAC SIGN ENTRIES

Date	h	m	Planets
02	03	49	☽ ♒
04	12	10	☽ ♓
06	22	50	☽ ♈
08	20	23	♂ ♏
09	11	11	☽ ♉
10	14	47	☿ ♎
12	00	14	☽ ♊
14	12	24	☽ ♋
16	20	24	♃ ♉
16	21	49	☽ ♌
19	03	25	☽ ♍
20	17	22	♀ ♐
21	05	26	☽ ♎
23	05	17	☽ ♏
23	06	58	☉ ♏
25	04	49	☽ ♐
27	05	55	☽ ♑
29	04	55	☿ ♏
29	10	05	☽ ♒
31	17	53	☽ ♓

LATITUDES

Date	Mercury ☿ ° '	Venus ♀ ° '	Mars ♂ ° '	Jupiter ♃ ° '	Saturn ♄ ° '	Uranus ♅ ° '	Neptune ♆ ° '	Pluto ♇ ° '
01	00 N 07	00 S 01	00 N 17	01 S 08	00 N 37	00 N 27	01 N 29	16 N 26
04	00 53	00 10	00 15	01 08	00 38	00 26	01 29	16 26
07	01 27	00 18	00 13	01 08	00 38	00 26	01 29	16 26
10	01 48	00 27	00 11	01 08	00 39	00 26	01 29	16 27
13	01 58	00 36	00 10	01 09	00 39	00 26	01 29	16 27
16	01 59	00 45	00 08	01 09	00 40	00 26	01 28	16 27
19	01 54	00 54	00 06	01 09	00 40	00 26	01 28	16 28
22	01 43	01 02	00 04	01 09	00 41	00 26	01 28	16 28
25	01 28	01 11	00 03	01 09	00 41	00 26	01 28	16 29
28	01 11	01 19	00 N 01	01 09	00 42	00 26	01 28	16 29
31	00 N 53	01 S 27	00 S 01	01 S 09	00 N 42	00 N 26	01 N 28	16 N 30

DATA

Julian Date	2443053
Delta T	+47 seconds
Ayanamsa	23° 32' 06"
Synetic vernal point	05° ♓ 34' 54"
True obliquity of ecliptic	23° 26' 25"

MOON'S PHASES, APSIDES AND POSITIONS ☽

Date	h	m	Phase	Longitude ° '	Eclipse Indicator
08	04	55	○	15 ♈ 02	
16	08	59	☾	23 ♋ 07	
23	05	10	●	29 ♎ 56	Total
29	22	05	☽	06 ♒ 37	

Day	h	m		
10	11	55	Apogee	
23	12	55	Perigee	
06	06	45	0N	
13	19	41	Max dec	19° N 14'
20	16	05	0S	
26	15	52	Max dec	19° S 14'

LONGITUDES

Date	Chiron ⚷ ° '	Ceres ⚳ ° '	Pallas ⚴ ° '	Juno ⚵ ° '	Vesta ⚶ ° '	Black Moon Lilith ⚸ ° '
01	00 ♉ 33	03 ♍ 10	02 ♌ 31	16 ♎ 55	18 ♋ 24	07 ♉ 18
11	00 ♉ 07	07 ♍ 28	07 ♌ 37	20 ♎ 30	20 ♋ 57	08 ♉ 25
21	29 ♈ 39	11 ♍ 39	12 ♌ 29	24 ♎ 03	23 ♋ 07	09 ♉ 31
31	29 ♈ 11	15 ♍ 44	17 ♌ 00	27 ♎ 35	24 ♋ 48	10 ♉ 38

ASPECTARIAN

h m	Aspects	h m	Aspects	h m	Aspects
01 Friday		**12 Tuesday**		**22 Friday**	
02 26	☽ ⚼ ♃	01 01	☽ ☌ ♃	00 57	☽ ⚹ ♆
03 59	☿ St D	04 35	☽ △ ☿	01 10	☽ ☌ ♇
05 20	☽ ⊥ ♆	04 51	☽ ⊼ ♂	04 43	☽ ⚼ ♃
05 28	☽ Q ♅	05 44	☽ # ♀	06 35	☽ ⚹ ♄
06 03	☽ # ♄	06 13	☽ Q ♄	08 42	☽ ∠ ♀
06 28	☽ Q ♀	08 39	☽ ⚼ ☉	11 18	☽ ☌ ☿
09 29	♂ ± ♃	08 42	☿ ⊻ ♂	17 34	☉ ⊼ ♃
15 22	☽ △ ☿	13 07	☽ ⊼ ♅	18 55	☽ ± ♃
19 09	☽ □ ♂	17 44	☽ ± ♂	20 46	☽ # ♇
21 38	☽ ∠ ♆	**13 Wednesday**		**23 Saturday**	
02 Saturday		00 17	☽ ☍ ♆	00 12	☽ ⊥ ♀
05 33	☽ △ ♃	00 21	☽ △ ♇	00 48	☽ ∠ ♆
11 17	☽ ∥ ♀	01 17	☽ ± ♅	01 40	☽ Q ♄
14 26	☽ □ ♅	05 53	☽ ∥ ♃	03 08	☽ ∥ ☉
18 45	☽ □ ♀	06 36	☽ ⚹ ♄	04 20	☽ ⊼ ♃
19 18	☽ ⚼ ☿	12 56	☽ ⚼ ♂	05 10	☽ ☌ ☉
19 56	☽ ∥ ♅	17 34	☽ △ ☉	10 30	☽ ⊻ ♀
21 59	☽ △ ☉	19 24	☽ ⚼ ♅	15 02	☽ ⊥ ♆
03 Sunday		19 43	☽ ⊼ ♀	15 41	☽ ∥ ♅
01 10	☽ △ ♇	**14 Thursday**		16 23	☽ ☌ ♅
01 23	☽ ⚹ ♆	08 57	☽ ± ♀	21 30	☽ ☌ ♂
02 12	☿ ∠ ♀	10 21	☽ ∥ ♃	**24 Sunday**	
04 20	♂ Q ♄	12 38	☽ ∠ ♄	00 31	☽ ⊻ ♆
06 08	☽ ☍ ♄	12 47	☽ ⊻ ♃	00 45	☽ ⊻ ♇
12 37	☽ ± ☿	12 47	☽ # ♀	01 38	☽ ∥ ♂
13 32	☽ # ♇	16 29	♀ # ♃	06 13	☽ □ ♄
17 18	☽ ∥ ♂	20 24	☽ △ ♂	10 16	☽ ⊥ ♇
21 41	♂ ∠ ♆	**15 Friday**		16 42	☽ ⊻ ☿
04 Monday		00 27	☽ ⊥ ♃	17 03	☽ # ♄
00 03	☽ Q ♆	00 55	☽ □ ☿	**25 Monday**	
00 23	☽ ⊼ ☿	01 08	☽ △ ♅	00 37	☽ ∠ ♇
04 40	☽ ⚼ ☉	02 41	☿ ⊻ ♅	03 28	☽ ⊥ ☿
05 36	☽ ⚼ ♇	04 32	☽ ⚼ ♀	03 31	☽ ☍ ♃
06 14	☽ △ ♂	06 20	☽ ⊥ ♄	03 50	☿ ± ♃
13 47	☽ □ ♃	11 41	☽ ⊼ ♆	08 05	☽ ⊻ ☉
17 06	☉ ☌ ♇	11 49	☽ □ ♇	14 21	☽ ☌ ♀
19 16	☉ ⚹ ♆	17 44	☽ ∠ ♃	16 19	☽ ⊻ ♅
22 58	☽ ± ♇	17 57	☽ ⊻ ♄	18 26	☽ ⊥ ☉
23 27	☽ ± ☉	23 12	☽ ± ♆	19 48	☽ ∠ ☿
23 32	☽ △ ♅	**16 Saturday**		21 09	☽ # ♃
05 Tuesday		02 36	☽ ∥ ♄	23 43	☽ ⊻ ♂
02 25	☽ ∥ ☉	08 59	☽ □ ☉	**26 Tuesday**	
10 03	☽ △ ♀	12 15	☽ △ ♀	00 34	☽ ☌ ♆
10 39	☽ ⊼ ♇	16 12	☽ ⚼ ♆	00 51	☽ ⚹ ♇
10 48	☽ □ ♆	16 14	☽ Q ☿	02 06	☽ ⊥ ♅
12 07	☽ ⊼ ☉	21 49	☽ ⚹ ♃	06 33	☽ △ ♄
12 14	☽ # ☿	21 57	☽ Q ♇	10 01	☽ ⊥ ♂
12 52	☽ ⚼ ♂	**17 Sunday**		10 11	☽ ∠ ☉
16 04	☽ ⊼ ♄	05 24	☉ ± ♃	15 40	♀ ⊻ ♅
16 18	♀ ⊻ ♇	06 57	♂ ∥ ♅	16 57	☽ ∠ ♅
17 44	♀ ⊻ ♆	08 28	☽ □ ♂	18 46	♂ ⊻ ♆
19 46	♂ # ♇	10 02	☽ □ ♅	20 38	☽ Q ♇
06 Wednesday		11 01	☽ # ♂	23 41	☽ ⚹ ☿
00 37	☽ Q ♃	11 17	☽ # ♅	**27 Wednesday**	
03 55	☽ ± ♄	16 15	♂ ⊥ ♆	01 00	♂ ⊻ ♇
04 57	☽ ⚼ ♅	17 23	☽ ⚹ ☿	01 41	☽ ∠ ♂
07 33	☽ ± ♂	19 35	☽ Q ♃	04 11	☽ ⊼ ♃
14 04	☽ ☍ ☿	19 45	☽ △ ♆	07 33	☽ ⚼ ♄
18 49	☽ ⚼ ♀	19 55	☽ ⚹ ♇	11 55	☽ # ♃
20 08	☽ ⊼ ♂	20 49	☽ Q ☉	13 05	☽ ⚹ ☉
21 51	☽ ⚼ ♄	**18 Monday**		14 14	☽ ± ♃
22 51	☽ ± ♅	01 48	☽ ☌ ♄	14 17	☿ ∠ ♆
07 Thursday		07 39	☽ ∥ ♇	18 17	☽ ⚹ ♅
00 01	☽ ∥ ☿	08 47	♀ ∠ ♇	20 40	☽ ⊻ ♀
00 14	☽ ⚹ ♃	11 25	☽ # ☉	22 44	☽ Q ☿
10 53	☽ ⊼ ♅	12 46	☿ ⚹ ♆	22 56	☽ ± ♄
14 47	☽ ± ♀	14 17	☿ ☌ ♇	**28 Thursday**	
17 41	☉ ⚹ ♄	17 32	♂ ☌ ♅	00 24	☿ Q ♄
17 51	☽ # ☉	18 17	☽ Q ♅	02 54	☽ ⊻ ♆
22 15	☽ ☍ ♇	18 19	☽ Q ♂	03 13	☽ □ ♇
22 19	☽ △ ♆	19 43	☽ ⚹ ☉	04 31	☽ ⚹ ♂
08 Friday		22 29	☽ ∠ ♇	05 34	☽ ⚼ ♃
02 44	♀ □ ♄	23 34	☽ ∠ ☿	08 01	☽ ⊥ ♀
04 04	☽ △ ♄	23 46	☽ □ ♀	09 21	☽ ⊼ ♄
04 13	☽ ⊼ ♀	**19 Tuesday**		11 21	☽ Q ☉
04 55	☽ ☍ ☉	03 04	☽ □ ♃	11 58	☿ ⊼ ♃
06 08	☽ ∠ ♃	05 00	☿ ∠ ♀	13 22	☽ ⊥ ♆
18 39	☉ ⚼ ♃	13 58	☽ ⊥ ♇	15 09	☽ Q ♅
09 Saturday		15 00	☽ ⚹ ♅	17 59	☽ # ♄
00 13	☽ ⊥ ♃	15 59	☽ ⚹ ♂	22 21	☿ # ♇
01 34	☽ ∥ ♇	17 03	☽ ⊥ ☿	**29 Friday**	
04 40	☽ ⚼ ♆	17 51	♀ ∥ ♆	01 00	☽ ∥ ♂
07 57	☽ ⊼ ☿	20 23	☽ # ☿	01 24	☽ ∠ ♀
08 02	♀ # ♄	23 17	☽ ∠ ☉	02 44	☽ Q ♂
08 21	☽ # ♂	23 55	☽ □ ♆	05 20	☽ ∠ ♆
12 05	☽ ☍ ♂	**20 Wednesday**		07 47	☽ △ ♃
12 19	☽ ⊻ ♃	00 06	☽ ⊻ ♇	10 38	☉ ∥ ♅
13 32	☉ ⊻ ♀	02 07	☉ # ♇	10 50	☽ □ ☿
15 49	♂ ⊼ ♃	03 14	☿ ⚼ ♃	16 00	☉ ⊥ ♆
21 32	☽ ± ☿	04 27	☽ ⊻ ☿	20 26	☽ ∥ ☉
22 16	☽ # ♅	05 41	☽ ⊻ ♄	21 24	☽ ∥ ♅
22 59	☽ ± ♆	08 36	☉ ∠ ♆	22 05	☽ □ ☉
23 47	☽ ☍ ♅	09 05	☽ Q ♀	23 34	☽ □ ♅
10 Sunday		10 53	♀ ☍ ♃	**30 Saturday**	
06 49	♆ ⚹ ♇	15 20	☽ ⊥ ☉	07 21	☽ ⚹ ♀
11 07	☿ ∠ ♄	15 29	☿ ⚹ ♄	08 41	☽ ⚹ ♆
11 14	☽ ⊼ ♆	15 39	☽ ⊥ ♄	09 02	☽ △ ♇
11 14	☽ ⊼ ♇	16 08	☽ ∠ ♅	13 20	☽ □ ♂
11 23	☿ Q ♆	18 13	☽ ∠ ♂	13 39	☽ ∥ ☿
17 21	☽ □ ♄	20 26	☉ Q ♄	15 41	☽ ☍ ♄
18 01	♀ ⊥ ♇	**21 Thursday**		18 48	☉ ☌ ♅
18 04	☽ ⚼ ☿	01 52	☽ ⊻ ☉	21 19	☽ # ♇
19 16	☿ ⊥ ♅	04 47	☽ △ ♃	21 57	♀ ☌ ♆
23 12	☿ △ ♃	05 37	☽ Q ♆	**31 Sunday**	
23 22	☽ ⊼ ☉	06 25	☽ ∠ ♄	01 51	♀ ⚹ ♇
23 30	☽ ± ♇	06 31	☽ ⚹ ♀	07 14	☽ Q ♆
11 Monday		06 53	☽ ⊥ ♅	08 12	☽ Q ♀
00 07	☽ ☍ ♀	09 35	☽ ⊥ ♂	13 20	☽ ⚼ ♇
12 13	☽ ∥ ♄	16 35	☽ ⊻ ♅	14 55	☽ □ ♃
12 42	☽ ± ☉	17 58	☽ ∥ ☿	20 33	♀ ⊥ ♅
17 53	☽ ⚼ ♇	19 42	☽ ⊻ ♂		

All ephemeris data is given at 12.00 UT and the Moon's longitude is additionally given for 24.00 UT

NOVEMBER 1976

LONGITUDES

Date	Sidereal time h m s	Sun ☉	Moon ☽	Moon ☽ 24.00	Mercury ☿	Venus ♀	Mars ♂	Jupiter ♃	Saturn ♄	Uranus ♅	Neptune ♆	Pluto ♇
01	14 43 36	09 ♏ 11 50	09 ♓ 22 16	15 ♓ 30 10	05 ♏ 29	14 ♐ 20	16 ♏ 15	28 ♉ 21	16 ♌ 14	07 ♏ 35	12 ♐ 27	12 ♎ 39
02	14 47 32	10 11 54	21 ♓ 34 56	27 ♓ 37 01	07 08	15 33	16 57	28 R 13	16 17	07 39	12 29	12 41
03	14 51 29	11 11 59	03 ♈ 36 51	09 ♈ 34 49	08 46	16 45	17 38	28 06	16 20	07 43	12 31	12 43
04	14 55 25	12 12 05	15 ♈ 31 18	21 ♈ 26 38	10 24	17 58	18 20	27 58	16 22	07 47	12 33	12 45
05	14 59 22	13 12 14	27 ♈ 21 11	03 ♉ 15 13	12 02	19 11	19 02	27 50	16 25	07 50	12 35	12 48
06	15 03 18	14 12 24	09 ♉ 09 02	15 ♉ 02 56	13 39	20 24	19 44	27 43	16 27	07 54	12 37	12 50
07	15 07 15	15 12 36	20 ♉ 57 09	26 ♉ 51 58	15 16	21 36	20 27	27 35	16 30	07 58	12 39	12 52
08	15 11 11	16 12 50	02 ♊ 47 38	08 ♊ 44 25	16 53	22 49	21 09	27 27	16 32	08 02	12 41	12 54
09	15 15 08	17 13 06	14 ♊ 42 51	20 ♊ 42 25	18 29	24 01	21 51	27 19	16 34	08 05	12 43	12 56
10	15 19 05	18 13 23	26 ♊ 44 13	02 ♋ 48 19	20 04	25 14	22 33	27 11	16 36	08 09	12 46	12 58
11	15 23 01	19 13 43	08 ♋ 55 02	15 ♋ 04 42	21 40	26 27	23 16	27 03	16 38	08 13	12 48	13 00
12	15 26 58	20 14 04	21 ♋ 17 44	27 ♋ 34 29	23 15	27 39	23 58	26 55	16 40	08 16	12 50	13 02
13	15 30 54	21 14 27	03 ♌ 55 21	10 ♌ 20 45	24 50	28 ♐ 52	24 40	26 47	16 41	08 20	12 52	13 04
14	15 34 51	22 14 52	16 ♌ 51 02	23 ♌ 26 35	26 24	00 ♑ 04	25 23	26 39	16 43	08 24	12 54	13 06
15	15 38 47	23 15 19	00 ♍ 07 44	06 ♍ 54 44	27 59	01 16	26 05	26 31	16 44	08 27	12 56	13 08
16	15 42 44	24 15 48	13 ♍ 47 48	20 ♍ 47 03	29 ♏ 33	02 29	26 48	26 23	16 45	08 31	12 58	13 10
17	15 46 40	25 16 19	27 ♍ 52 27	05 ♎ 03 53	01 ♐ 06	03 41	27 31	26 14	16 47	08 35	13 01	13 11
18	15 50 37	26 16 52	12 ♎ 21 01	19 ♎ 43 24	02 40	04 53	28 13	26 06	16 48	08 38	13 03	13 13
19	15 54 34	27 17 26	27 ♎ 10 23	04 ♏ 41 11	04 13	06 05	28 56	25 58	16 49	08 42	13 05	13 15
20	15 58 30	28 18 02	12 ♏ 14 48	19 ♏ 50 09	05 46	07 18	29 ♏ 39	25 50	16 50	08 46	13 07	13 17
21	16 02 27	29 ♏ 18 40	27 ♏ 26 03	05 ♐ 01 14	07 19	08 30	00 ♐ 22	25 42	16 50	08 49	13 09	13 19
22	16 06 23	00 ♐ 19 19	12 ♐ 34 28	20 ♐ 04 34	08 52	09 42	01 05	25 33	16 51	08 53	13 11	13 20
23	16 10 20	01 19 59	27 ♐ 30 26	04 ♑ 51 05	10 24	10 54	01 47	25 25	16 52	08 56	13 14	13 22
24	16 14 16	02 20 41	12 ♑ 05 46	19 ♑ 13 53	11 57	12 06	02 31	25 17	16 52	09 00	13 16	13 24
25	16 18 13	03 21 24	26 ♑ 15 01	03 ♒ 08 59	13 29	13 18	03 14	25 09	16 52	09 03	13 18	13 26
26	16 22 09	04 22 08	09 ♒ 55 44	16 ♒ 35 25	15 01	14 30	03 57	25 01	16 53	09 07	13 20	13 27
27	16 26 06	05 22 53	23 ♒ 08 19	29 ♒ 34 47	16 33	15 41	04 40	24 53	16 53	09 10	13 23	13 29
28	16 30 03	06 23 39	05 ♓ 55 19	12 ♓ 10 27	18 05	16 53	05 23	24 45	16 R 53	09 14	13 25	13 31
29	16 33 59	07 24 26	18 ♓ 20 47	24 ♓ 26 53	19 36	18 05	06 06	24 38	16 53	09 17	13 27	13 32
30	16 37 56	08 ♐ 25 14	00 ♈ 29 25	06 ♈ 28 58	21 ♐ 08	19 ♑ 16	06 ♐ 50	24 ♉ 30	16 ♌ 52	09 ♏ 21	13 ♐ 29	13 ♎ 34

Date	Moon True ☊	Moon Mean ☊	Moon ☽ Latitude
01	03 ♏ 33	03 ♏ 05	04 N 18
02	03 D 34	03 02	03 33
03	03 35	02 59	02 39
04	03 35	02 56	01 39
05	03 36	02 52	00 N 35
06	03 R 36	02 49	00 S 31
07	03 35	02 46	01 35
08	03 34	02 43	02 35
09	03 32	02 40	03 29
10	03 30	02 37	04 14
11	03 28	02 33	04 48
12	03 25	02 30	05 09
13	03 24	02 27	05 16
14	03 23	02 24	05 08
15	03 D 23	02 21	04 43
16	03 24	02 18	04 01
17	03 25	02 14	03 04
18	03 27	02 11	01 54
19	03 28	02 08	00 S 35
20	03 R 27	02 05	00 N 48
21	03 26	02 02	02 08
22	03 23	01 58	03 19
23	03 19	01 55	04 15
24	03 15	01 52	04 53
25	03 11	01 49	05 11
26	03 07	01 46	05 11
27	03 05	01 43	04 55
28	03 04	01 39	04 24
29	03 D 05	01 36	03 41
30	03 ♏ 06	01 ♏ 33	02 N 49

DECLINATIONS

Date	Sun ☉	Moon ☽	Mercury ☿	Venus ♀	Mars ♂	Jupiter ♃	Saturn ♄	Uranus ♅	Neptune ♆	Pluto ♇
01	14 S 34	04 S 04	12 S 37	24 S 00	16 S 43	18 N 40	16 N 38	13 S 38	20 S 50	10 N 12
02	14 53	00 S 04	13 16	24 11	16 56	18 39	16 38	13 39	20 51	10 12
03	15 11	03 N 52	13 54	24 21	17 08	18 37	16 37	13 40	20 51	10 11
04	15 30	07 38	14 31	24 30	17 21	18 35	16 37	13 42	20 51	10 10
05	15 48	11 04	15 08	24 39	17 33	18 34	16 36	13 43	20 51	10 10
06	16 06	14 04	15 44	24 47	17 45	18 32	16 35	13 44	20 52	10 09
07	16 24	16 28	16 19	24 54	17 57	18 30	16 35	13 45	20 52	10 09
08	16 41	18 11	16 53	25 00	18 08	18 29	16 34	13 46	20 52	10 08
09	16 59	19 06	17 27	25 06	18 20	18 27	16 34	13 48	20 53	10 08
10	17 15	19 10	17 59	25 11	18 31	18 25	16 33	13 49	20 53	10 07
11	17 32	18 21	18 31	25 16	18 43	18 24	16 33	13 50	20 53	10 07
12	17 48	16 40	19 02	25 19	18 54	18 22	16 33	13 51	20 53	10 07
13	18 04	14 09	19 31	25 22	19 05	18 20	16 32	13 52	20 54	10 06
14	18 20	10 54	20 00	25 24	19 16	18 18	16 32	13 54	20 54	10 06
15	18 35	07 01	20 28	25 26	19 26	18 16	16 32	13 55	20 54	10 05
16	18 50	02 N 40	20 55	25 26	19 37	18 14	16 32	13 56	20 55	10 05
17	19 05	01 S 58	21 21	25 26	19 47	18 13	16 31	13 57	20 55	10 05
18	19 19	06 38	21 45	25 25	19 57	18 11	16 31	13 58	20 55	10 04
19	19 33	11 00	22 09	25 24	20 07	18 09	16 31	13 59	20 55	10 04
20	19 47	14 45	22 31	25 21	20 17	18 07	16 31	14 01	20 56	10 04
21	20 00	17 30	22 53	25 18	20 27	18 05	16 31	14 02	20 56	10 03
22	20 13	19 01	23 13	25 15	20 36	18 04	16 31	14 03	20 56	10 03
23	20 26	19 10	23 32	25 10	20 46	18 02	16 31	14 04	20 57	10 03
24	20 38	18 02	23 50	25 05	20 55	18 00	16 31	14 05	20 57	10 02
25	20 50	15 48	24 07	24 59	21 04	17 58	16 31	14 06	20 57	10 02
26	21 01	12 45	24 22	24 52	21 13	17 57	16 31	14 07	20 57	10 02
27	21 12	09 10	24 37	24 45	21 21	17 55	16 31	14 08	20 58	10 02
28	21 23	05 15	24 49	24 37	21 30	17 53	16 31	14 10	20 58	10 02
29	21 33	01 13	25 01	24 28	21 38	17 51	16 32	14 11	20 58	10 02
30	21 S 43	02 N 47	25 S 11	24 S 18	21 S 46	17 N 50	16 N 32	14 S 12	20 S 59	10 N 01

ZODIAC SIGN ENTRIES

Date	h	m	Planets
03	04	46	☽ ♈
05	17	23	☽ ♉
08	06	21	☽ ♊
10	18	28	☽ ♋
13	04	36	☽ ♌
14	10	42	♀ ♑
15	11	46	☽ ♍
16	19	02	☿ ♐
17	15	34	☽ ♎
19	16	32	☽ ♏
20	23	53	♂ ♐
21	16	03	☽ ♐
22	04	22	☉ ♐
23	16	03	☽ ♑
25	18	30	☽ ♒
28	00	47	☽ ♓
30	11	01	☽ ♈

LATITUDES

Date	Mercury ☿	Venus ♀	Mars ♂	Jupiter ♃	Saturn ♄	Uranus ♅	Neptune ♆	Pluto ♇
01	00 N 46	01 S 29	00 S 02	01 S 09	00 N 42	00 N 26	01 N 28	16 N 30
04	00 26	01 37	00 03	01 09	00 43	00 26	01 28	16 31
07	00 N 06	01 44	00 05	01 09	00 43	00 26	01 28	16 32
10	00 S 14	01 50	00 07	01 09	00 44	00 26	01 28	16 33
13	00 34	01 56	00 09	01 08	00 44	00 26	01 28	16 34
16	00 53	02 01	00 11	01 08	00 45	00 26	01 27	16 35
19	01 11	02 06	00 12	01 08	00 45	00 26	01 27	16 36
22	01 27	02 10	00 14	01 07	00 46	00 26	01 27	16 37
25	01 43	02 13	00 16	01 07	00 46	00 26	01 27	16 39
28	01 56	02 15	00 18	01 06	00 47	00 26	01 27	16 40
31	02 S 06	02 S 17	00 S 20	01 S 06	00 N 47	00 N 26	01 N 27	16 N 41

DATA

Julian Date	2443084
Delta T	+47 seconds
Ayanamsa	23° 32' 09"
Synetic vernal point	05° ♓ 34' 51"
True obliquity of ecliptic	23° 26' 25"

MOON'S PHASES, APSIDES AND POSITIONS ☽

Date	h	m	Phase	Longitude	Eclipse Indicator
06	23	15	○	14 ♉ 41	
14	22	39	☾	22 ♌ 42	
21	15	11	●	29 ♏ 27	
28	12	59	☽	06 ♓ 26	

Day	h	m		
06	14	24	Apogee	
21	01	17	Perigee	
02	12	26	0N	
10	01	49	Max dec	19° N 15'
17	01	54	0S	
23	02	41	Max dec	19° S 16'
29	19	14	0N	

LONGITUDES

Date	Chiron ⚷	Ceres ⚳	Pallas ⚴	Juno ⚵	Vesta ⚶	Black Moon Lilith ⚸
01	29 ♈ 08	16 ♍ 08	17 ♌ 26	27 ♎ 56	24 ♋ 56	10 ♉ 44
11	28 ♈ 41	20 ♍ 48	21 ♌ 30	01 ♏ 24	26 ♋ 00	11 ♉ 51
21	28 ♈ 16	23 ♍ 48	25 ♌ 03	04 ♏ 49	26 ♋ 25	12 ♉ 57
31	27 ♈ 55	27 ♍ 19	27 ♌ 57	08 ♏ 09	26 ♋ 06	14 ♉ 04

ASPECTARIAN

01 Monday
03 05 ♂ # ♄
03 17 ☽ △ ☿
06 42 ☽ ± ♇
08 31 ☽ △ ♅
11 38 ☽ △ ☉
11 45 ♂ □ ♄
18 02 ☽ □ ♆
18 26 ☽ ⊼ ♇
22 46 ☽ □ ♀

02 Tuesday
01 30 ☽ ⊼ ♄
01 31 ☽ Q ♃
02 17 ☽ △ ♂
02 28 ☿ ⊥ ♆
13 15 ☽ bQ ☿
13 24 ☽ ± ♄
14 08 ☽ bQ ♅
19 50 ☽ bQ ☉
19 55 ☿ ☌ ♅

03 Wednesday
01 04 ☽ ⚹ ♃
02 56 ☿ ∥ ♅
03 16 ♀ △ ♄
07 24 ☽ bQ ♄
08 10 ☽ ± ♅
09 56 ☽ bQ ♂
10 02 ☽ ± ☿
15 28 ☽ ± ☉
20 17 ☽ ⊼ ♅

04 Thursday
00 01 ☽ ⊼ ☿
04 40 ☽ ⊼ ☉
05 10 ☽ ± ♂
05 59 ☽ △ ♆
06 24 ☽ ☍ ♇
06 53 ☽ ∠ ♃
13 44 ☽ △ ♄
17 31 ☽ △ ♀
18 04 ☽ ⊼ ♂
20 44 ☉ ⊻ ♆

05 Friday
00 55 ☽ ⊥ ♃
01 49 ☉ ⊻ ♇
03 06 ♂ ⊥ ♇
05 24 ♀ ⊻ ♂
05 26 ☽ ∥ ♇
12 29 ☽ bQ ♆
12 59 ☽ ⊻ ♃
14 38 ☉ ⊥ ♀
18 58 ♃ bQ ♇
20 23 ☿ ⊻ ♆
23 30 ☿ ⊻ ♇

06 Saturday
03 29 ☽ bQ ♀
06 50 ☽ ± ♆
09 08 ☽ # ♅
09 27 ☽ ☍ ♅
19 05 ☽ ⊼ ♆
19 30 ☽ ⊼ ♇
22 37 ☽ ☍ ☿
23 15 ☽ ☍ ☉
23 53 ☽ ± ♀

07 Sunday
02 55 ☽ □ ♄
07 44 ☽ ± ♇
09 38 ☽ # ☿
09 44 ☉ ☌ ☿
10 54 ☽ ☍ ♂
11 03 ☽ # ☉
13 18 ☽ ∥ ♄
13 28 ☽ ⊼ ♀
18 50 ☉ ∥ ☿
22 51 ☿ # ♄

08 Monday
01 18 ☽ ☌ ♃
02 04 ☽ bQ ♇
02 37 ☉ # ♄
06 43 ☿ □ ♄
08 16 ☿ ⊥ ♀
11 06 ☽ # ♂
15 31 ☽ Q ♄
16 24 ♀ ∠ ♅
17 32 ☽ ∥ ♃
19 50 ☉ □ ♄
22 37 ☽ ⊼ ♅

09 Tuesday
08 00 ☽ ☍ ♆
08 25 ☽ △ ♇
10 45 ☽ ± ♅
15 44 ☽ ⚹ ♄
17 29 ☽ ⊼ ☉
18 57 ☿ ⊥ ♇
20 42 ☽ ⊼ ☿

10 Wednesday
00 26 ♂ # ♃
03 10 ☽ ⊼ ♂
04 50 ☽ bQ ♅
06 30 ♀ Q ♇
06 33 ☽ ± ☉
08 41 ☽ ☍ ♀
10 29 ☽ ± ☿
12 53 ☽ ⊻ ♃
15 49 ☽ ± ♂
21 39 ☽ ∠ ♄

11 Thursday
00 37 ☽ ⊥ ♃
01 59 ☽ bQ ☉
05 31 ☽ # ♂
06 18 ☉ ⊥ ♇
06 31 ☿ # ♃
06 56 ☽ bQ ☿
09 46 ☽ # ☿
10 37 ☽ △ ♅
10 38 ☽ bQ ♂
11 18 ☽ ∥ ♃
15 21 ☽ ⊥ ♄
18 03 ☽ ∠ ♃
19 35 ☽ ⊼ ♆
19 59 ☽ □ ♇
22 54 ♀ ⊼ ♃
23 19 ☽ # ☉

12 Friday
02 25 ☿ ∥ ♂
03 02 ☽ ⊻ ♄
07 14 ☽ ± ♆
09 47 ☽ △ ☉
13 18 ☽ ∥ ♄
16 17 ☽ △ ☿
17 25 ☽ △ ♂
22 38 ☽ ⚹ ♃

13 Saturday
00 31 ☽ bQ ♆
01 26 ☽ ⊼ ♀
06 36 ☽ Q ♇
07 39 ☿ ☌ ♂
13 56 ☽ ± ♀
14 14 ☽ # ♅
20 18 ☽ □ ♅
21 00 ☽ Q ♃

14 Sunday
04 43 ☽ △ ♆
05 04 ☽ ⚹ ♇
08 23 ☽ bQ ♀
09 19 ☉ # ♃
11 45 ☽ ☌ ♄
15 25 ☿ ☍ ♃
17 14 ☽ ∥ ♇
22 39 ☽ □ ☉

15 Monday
04 22 ☽ □ ♂
05 24 ☽ Q ♅
05 36 ☽ □ ♃
07 39 ☽ □ ☿
08 25 ☽ ∠ ♇
14 14 ☽ △ ♀
14 22 ☿ ∠ ♇
21 25 ♀ bQ ♄

16 Tuesday
00 01 ♂ ☍ ♃
00 24 ☽ ⊥ ♇
02 47 ☽ ⚹ ♅
09 08 ☽ Q ☉
10 11 ♀ ± ♃
10 34 ☽ □ ♆
10 54 ☽ ⊻ ♇
11 51 ☿ ∥ ♆
13 49 ☽ Q ♂
17 07 ☽ ⊻ ♄
19 16 ☽ Q ☿

17 Wednesday
03 23 ☽ ⊥ ♄
03 34 ♀ ⊥ ♂
04 43 ☽ ∠ ♅
07 17 ☽ ⚹ ☉
09 16 ☽ △ ♃
11 21 ☽ ⚹ ♂
17 16 ☽ Q ♆
18 04 ☽ ⚹ ☿
18 33 ☽ ∠ ♄
19 54 ☽ ⊥ ♅
22 35 ☽ □ ♀

18 Thursday
05 53 ☽ ⊻ ♅
08 16 ☉ ☍ ♃
09 58 ☽ bQ ♃
10 07 ☽ ∠ ☉
12 02 ♂ ∠ ♇
13 08 ☽ ⚹ ♆
13 26 ☽ ☌ ♇
13 30 ☽ ∠ ♂
19 16 ☽ ⚹ ♄
21 41 ☽ ∠ ☿

19 Friday
00 30 ☽ ± ♃
01 51 ☽ ⊥ ☉
04 50 ☽ ⊥ ♂
06 37 ☽ Q ♀
06 38 ☽ # ♇
10 05 ☽ ⊼ ♃
12 12 ☽ ⊻ ☉
13 28 ☽ ∠ ♆
13 52 ☽ ⊥ ☿
14 38 ☽ Q ♄
14 58 ☽ ⊻ ♂

20 Saturday
00 33 ☽ ⊻ ☿
03 28 ☽ ⚹ ♀
03 51 ☽ ⊥ ♆
06 27 ☽ ☌ ♅
06 47 ☽ ∥ ♅
11 33 ☉ ∠ ♇
13 23 ☽ ⊻ ♆
13 39 ☽ ⊻ ♇
19 15 ☽ □ ♄
23 09 ☽ ⊥ ♇

21 Sunday
02 11 ☽ # ♄
05 15 ☽ ∠ ♀
09 17 ☽ ☍ ♃
13 23 ☽ ∠ ♇
15 11 ☽ ☌ ☉
16 51 ☽ ☌ ♂
18 50 ♀ ⚹ ♅
18 59 ☽ # ♃
20 41 ☽ ⊥ ♀

22 Monday
05 26 ☽ ☌ ☿
06 06 ☽ ⊻ ♅
07 02 ☽ ⊻ ♀
12 16 ☿ ⊻ ♅
12 59 ☽ ☌ ♆
13 13 ☽ ⚹ ♇
15 42 ☽ ⊥ ♅
18 50 ☽ △ ♄

23 Tuesday
03 28 ♀ bQ ♃
06 12 ☽ ∠ ♅
08 32 ☽ Q ♇
08 39 ☽ ⊼ ♃
11 15 ♀ ± ♄
18 19 ☽ ± ♃
18 42 ☽ ⊻ ☉
19 06 ☽ bQ ♄
19 20 ☽ ⊻ ♂

24 Wednesday
05 18 ☽ ⊥ ☉
05 44 ☽ ⊥ ♂
06 50 ☽ ⚹ ♅
09 01 ☽ bQ ♃
09 57 ☽ ± ♄
11 43 ☽ ⊻ ☿
12 00 ☽ ☌ ♀
12 26 ☽ # ♃
13 58 ☽ ⊻ ♆
14 11 ☽ □ ♇
17 30 ♂ ∥ ♆
20 00 ☽ ⊼ ♄
21 29 ☽ ∠ ☉
21 34 ☽ ∠ ♂
22 41 ☿ ⊻ ♀
23 01 ☽ ⊥ ☿

25 Thursday
00 05 ☽ ⊥ ♆
01 20 ☉ ☌ ♂
03 04 ☽ Q ♅
05 20 ☽ # ♄
09 06 ☿ ☌ ♆
10 08 ☽ △ ♃
11 06 ☿ ⚹ ♇
12 08 ♀ ⊻ ♆
14 41 ♀ □ ♇
15 33 ☽ ∠ ♆
16 21 ☽ ∠ ☿

26 Friday
00 48 ☽ ⚹ ♂
01 21 ☽ ⚹ ☉
01 57 ☽ ∥ ♅
04 14 ☉ ∥ ♆
10 33 ☽ □ ♅
13 36 ☿ ⊥ ♅
18 08 ☽ ⚹ ♆
18 20 ☽ △ ♇
21 01 ☽ ⊻ ♀
22 21 ☽ ⚹ ☿
23 27 ☽ Q ♂

27 Saturday
00 31 ☽ ☍ ♄
00 34 ☽ Q ☉
06 25 ☽ # ♇
09 04 ☽ ⊥ ♀
15 13 ☽ □ ♃
16 10 ☽ Q ♆
17 09 ☿ △ ♄
18 46 ♄ St R
21 20 ☿ ∥ ♀
21 58 ☽ bQ ♇
23 26 ☽ Q ☿

28 Sunday
03 33 ☽ ∠ ♀
10 55 ☽ □ ♂
11 49 ♀ ⊼ ♄
12 59 ☽ □ ☉
15 02 ☽ ± ♇
18 22 ☽ △ ♅

29 Monday
01 00 ☽ Q ♃
02 27 ☽ □ ♆
02 37 ☽ ⊼ ♇
09 08 ☽ ⊼ ♄
11 26 ☽ ⚹ ♀
14 49 ☽ □ ☿
20 54 ☽ ± ♄
23 45 ☽ bQ ♅

30 Tuesday
00 14 ☽ ⚹ ♃
13 44 ☽ Q ♀
14 45 ☽ bQ ♄
16 29 ♀ ⊥ ♆
17 44 ☽ ± ♅

All ephemeris data is given at 12.00 UT and the Moon's longitude is additionally given for 24.00 UT

DECEMBER 1976

Date	Sidereal time h	m	s	Sun ☉ °	'	"	Moon ☽ °	'	"	Moon ☽ 24.00 °	'	"	Mercury ☿ °	'	Venus ♀ °	'	Mars ♂ °	'	Jupiter ♃ °	'	Saturn ♄ °	'	Uranus ♅ °	'	Neptune ♆ °	'	Pluto ⯓ °	'
01	16	41	52	09 ♐	26	03	12 ♈	26	10	18 ♈	21	34	22 ♐	39	20 ♑	28	07 ♐	33	24 ♉	22	16 ♌	52	09 ♏	24	13 ♐	32	13 ♎	35
02	16	45	49	10	26	52	24 ♈	15	45	00 ♉	09	13	24	10	21	39	08	16	24 R	15	16 R	51	09	28	13	34	13	37
03	16	49	45	11	27	43	06 ♉	02	26	11 ♉	55	52	25	40	22	51	09	00	24	07	16	51	09	31	13	36	13	38
04	16	53	42	12	28	35	17 ♉	49	54	23 ♉	44	53	27	10	24	02	09	43	24	00	16	50	09	34	13	38	13	40
05	16	57	38	13	29	27	29 ♉	41	07	05 ♊	38	53	28 ♐	40	25	13	10	27	23	53	16	49	09	38	13	41	13	41
06	17	01	35	14	30	21	11 ♊	38	24	17 ♊	39	53	00 ♑	10	26	25	11	10	23	45	16	48	09	41	13	43	13	42
07	17	05	32	15	31	16	23 ♊	43	30	29 ♊	49	22	01	39	27	36	11	54	23	38	16	47	09	44	13	45	13	44
08	17	09	28	16	32	11	05 ♋	57	39	12 ♋	08	26	03	07	28	47	12	38	23	31	16	46	09	47	13	47	13	45
09	17	13	25	17	33	08	18 ♋	21	52	24 ♋	38	02	04	34	29 ♑	57	13	22	23	23	16	45	09	50	13	50	13	46
10	17	17	21	18	34	06	00 ♌	57	06	07 ♌	19	11	06	01	01 ♒	08	14	05	23	18	16	44	09	54	13	52	13	48
11	17	21	18	19	35	05	13 ♌	44	27	20 ♌	13	05	07	27	02	19	14	49	23	11	16	42	09	57	13	54	13	49
12	17	25	14	20	36	04	26 ♌	45	17	03 ♍	21	15	08	51	03	29	15	33	23	05	16	41	10	00	13	57	13	50
13	17	29	11	21	37	05	10 ♍	01	12	16 ♍	45	22	10	14	04	40	16	17	22	59	16	39	10	03	13	59	13	51
14	17	33	07	22	38	07	23 ♍	33	56	00 ♎	27	05	11	36	05	50	17	01	22	52	16	37	10	06	14	01	13	53
15	17	37	04	23	39	10	07 ♎	24	56	14 ♎	27	33	12	55	07	01	17	45	22	46	16	35	10	09	14	03	13	54
16	17	41	01	24	40	14	21 ♎	34	53	28 ♎	46	46	14	13	08	11	18	29	22	41	16	33	10	12	14	06	13	55
17	17	44	57	25	41	19	06 ♏	02	56	13 ♏	22	58	15	27	09	21	19	13	22	35	16	31	10	15	14	08	13	56
18	17	48	54	26	42	24	20 ♏	46	16	28 ♏	12	06	16	39	10	31	19	57	22	29	16	29	10	18	14	10	13	57
19	17	52	50	27	43	31	05 ♐	39	36	13 ♐	07	45	17	47	11	41	20	42	22	24	16	27	10	21	14	12	13	58
20	17	56	47	28	44	38	20 ♐	35	30	28 ♐	01	41	18	51	12	50	21	26	22	19	16	24	10	23	14	14	13	59
21	18	00	43	29 ♐	45	46	05 ♑	25	13	12 ♑	45	01	19	49	14	00	22	10	22	14	16	22	10	26	14	17	14	00
22	18	04	40	00 ♑	46	55	20 ♑	00	09	27 ♑	09	45	20	43	15	09	22	55	22	09	16	19	10	29	14	19	14	01
23	18	08	36	01	48	03	04 ♒	13	12	11 ♒	10	01	21	30	16	19	23	39	22	04	16	16	10	32	14	21	14	01
24	18	12	33	02	49	12	17 ♒	59	54	24 ♒	42	48	22	09	17	28	24	24	22	00	16	14	10	34	14	23	14	02
25	18	16	30	03	50	21	01 ♓	18	45	07 ♓	48	01	22	41	18	37	25	08	21	55	16	11	10	37	14	25	14	03
26	18	20	26	04	51	30	14 ♓	10	57	20 ♓	28	01	23	04	19	46	25	53	21	51	16	08	10	40	14	27	14	04
27	18	24	23	05	52	39	26 ♓	39	46	02 ♈	46	50	23	17	20	54	26	37	21	47	16	05	10	42	14	30	14	05
28	18	28	19	06	53	47	08 ♈	49	52	14 ♈	49	34	23 R	19	22	03	27	22	21	44	16	02	10	45	14	32	14	05
29	18	32	16	07	54	56	20 ♈	46	38	26 ♈	41	45	23	09	23	11	28	07	21	40	15	58	10	47	14	34	14	06
30	18	36	12	08	56	05	02 ♉	35	36	08 ♉	28	51	22	48	24	19	28	52	21	37	15	55	10	49	14	36	14	06
31	18	40	09	09 ♑	57	14	14 ♉	22	07	20 ♉	16	00	22 ♑	16	25 ♒	27	29 ♐	36	21 ♉	33	15 ♌	52	10 ♏	52	14 ♐	38	14 ♎	07

Date	Moon True ☊ °	'	Moon Mean ☊ °	'	Moon ☽ Latitude °	'
01	03 ♏	08	01 ♏	30	01 N	51
02	03 D	10	01	27	00 N	48
03	03 R	10	01	24	00 S	16
04	03	09	01	20	01	19
05	03	06	01	17	02	19
06	03	00	01	14	03	14
07	02	53	01	11	04	00
08	02	45	01	08	04	36
09	02	36	01	04	04	59
10	02	28	01	01	05	08
11	02	21	00	58	05	02
12	02	16	00	55	04	41
13	02	14	00	52	04	04
14	02 D	13	00	49	03	13
15	02	13	00	45	02	10
16	02	14	00	42	00 S	57
17	02 R	15	00	39	00 N	20
18	02	13	00	36	01	38
19	02	09	00	33	02	49
20	02	03	00	30	03	49
21	01	54	00	26	04	33
22	01	44	00	23	04	59
23	01	34	00	20	05	04
24	01	26	00	17	04	52
25	01	19	00	14	04	24
26	01	14	00	10	03	44
27	01	12	00	07	02	54
28	01 D	11	00	04	01	57
29	01	12	00 ♏	01	00 N	55
30	01 R	12	29 ♎	58	00 S	07
31	01 ♏	11	29 ♎	55	01 S	10

DECLINATIONS

Date	Sun ☉ °	'	Moon ☽ °	'	Mercury ☿ °	'	Venus ♀ °	'	Mars ♂ °	'	Jupiter ♃ °	'	Saturn ♄ °	'	Uranus ♅ °	'	Neptune ♆ °	'	Pluto ⯓ °	'
01	21 S	52	06 N	37	25 S	20	24 S	08	21 S	54	17 N	48	16 N	32	14 S	13	20 S	59	10 N	01
02	22	01	10	10	25	28	23	57	22	01	17	46	16	32	14	14	20	59	10	01
03	22	09	13	17	25	34	23	46	22	08	17	45	16	33	14	15	20	59	10	01
04	22	18	15	53	25	39	23	34	22	16	17	43	16	33	14	16	21	00	10	01
05	22	25	17	49	25	42	23	21	22	23	17	41	16	34	14	17	21	00	10	01
06	22	32	18	59	25	44	23	07	22	29	17	40	16	34	14	18	21	00	10	01
07	22	39	19	18	25	45	22	53	22	36	17	38	16	35	14	19	21	01	10	01
08	22	46	18	43	25	44	22	38	22	42	17	37	16	35	14	20	21	01	10	01
09	22	51	17	15	25	41	22	23	22	48	17	35	16	36	14	21	21	01	10	01
10	22	57	14	56	25	37	22	07	22	54	17	34	16	36	14	22	21	01	10	01
11	23	02	11	52	25	32	21	50	23	00	17	32	16	37	14	23	21	02	10	01
12	23	06	08	12	25	25	21	33	23	05	17	31	16	37	14	24	21	02	10	01
13	23	10	04 N	03	25	17	21	16	23	10	17	30	16	38	14	25	21	02	10	01
14	23	14	00 S	24	25	07	20	57	23	15	17	28	16	39	14	26	21	02	10	01
15	23	17	04	56	24	56	20	39	23	20	17	27	16	40	14	27	21	03	10	01
16	23	20	09	18	24	44	20	19	23	25	17	26	16	40	14	28	21	03	10	01
17	23	22	13	13	24	31	19	59	23	29	17	25	16	41	14	29	21	03	10	01
18	23	24	16	22	24	16	19	39	23	33	17	23	16	42	14	30	21	03	10	01
19	23	25	18	28	24	00	19	18	23	37	17	22	16	43	14	30	21	04	10	01
20	23	26	19	18	23	43	18	57	23	40	17	21	16	44	14	31	21	04	10	01
21	23	26	18	47	23	26	18	35	23	44	17	20	16	45	14	32	21	04	10	02
22	23	26	17	02	23	08	18	13	23	47	17	19	16	46	14	33	21	04	10	02
23	23	26	14	16	22	49	17	50	23	50	17	18	16	47	14	34	21	05	10	02
24	23	25	10	48	22	31	17	27	23	52	17	17	16	48	14	35	21	05	10	02
25	23	23	06	53	22	12	17	03	23	55	17	16	16	49	14	36	21	05	10	02
26	23	21	02 S	47	21	53	16	40	23	57	17	16	16	50	14	36	21	05	10	03
27	23	19	01 N	20	21	35	16	15	23	58	17	15	16	51	14	37	21	06	10	03
28	23	16	05	17	21	18	15	51	24	00	17	14	16	52	14	38	21	06	10	03
29	23	12	08	58	21	02	15	25	24	01	17	13	16	53	14	39	21	06	10	04
30	23	08	12	15	20	47	15	00	24	02	17	13	16	54	14	39	21	06	10	04
31	23 S	04	15 N	02	20 S	33	14 S	34	24 S	03	17 N	12	16 N	55	14 S	40	21 S	07	10 N	04

ZODIAC SIGN ENTRIES

Date	h	m	Planets
02	23	41	☽ ♉
05	12	38	☽ ♊
06	09	25	☿ ♑
08	00	21	☽ ♋
09	12	53	♀ ♒
10	10	12	☽ ♌
12	17	55	☽ ♍
14	23	13	☽ ♎
17	02	01	☽ ♏
19	02	54	☽ ♐
21	03	12	☽ ♑
21	17	35	☉ ♑
23	04	48	☽ ♒
25	09	36	☽ ♓
27	18	32	☽ ♈
30	06	43	☽ ♉

LATITUDES

Date	Mercury ☿ °	'	Venus ♀ °	'	Mars ♂ °	'	Jupiter ♃ °	'	Saturn ♄ °	'	Uranus ♅ °	'	Neptune ♆ °	'	Pluto ⯓ °	'
01	02 S	06	02 S	17	00 S	20	01 S	06	00 N	47	00 N	26	01 N	27	16 N	41
04	02	14	02	18	00	21	01	05	00	48	00	26	01	27	16	43
07	02	19	02	17	00	23	01	05	00	49	00	26	01	27	16	44
10	02	19	02	16	00	25	01	04	00	49	00	26	01	27	16	46
13	02	15	02	14	00	27	01	03	00	50	00	26	01	27	16	48
16	02	04	02	10	00	28	01	03	00	50	00	26	01	27	16	49
19	01	46	02	06	00	30	01	02	00	51	00	26	01	27	16	51
22	01	18	02	01	00	32	01	01	00	51	00	26	01	27	16	52
25	00 S	41	01	54	00	34	01	00	00	52	00	26	01	27	16	54
28	00 N	08	01	47	00	35	01	00	00	52	00	26	01	27	16	56
31	01 N	04	01 S	38	00 S	37	00 S	59	00 N	53	00 N	26	01 N	27	16 N	58

DATA

Julian Date	2443114
Delta T	+47 seconds
Ayanamsa	23° 32' 13"
Synetic vernal point	05° ♓ 34' 47"
True obliquity of ecliptic	23° 26' 24"

MOON'S PHASES, APSIDES AND POSITIONS ☽

Date	h	m	Phase	Longitude °	'	Eclipse Indicator
06	18	15	○	14 ♊	46	
14	10	14	☾	22 ♍	34	
21	02	08	●	29 ♐	21	
28	07	48	☽	06 ♈	43	

Day	h	m		
03	18	16	Apogee	
19	11	24	Perigee	
31	09	13	Apogee	
07	08	32	Max dec	19° N 18'
14	09	53	0S	
20	14	41	Max dec	19° S 18'
27	04	11	0N	

LONGITUDES

Date	Chiron ⚷ °	'	Ceres ⚳ °	'	Pallas ⚴ °	'	Juno ⚵ °	'	Vesta ⚶ °	'	Black Moon Lilith ⚸ °	'
01	27 ♈	55	27 ♍	19	27 ♌	57	08 ♏	09	26 ♋	06	14 ♉	04
11	27 ♈	38	00 ♎	33	00 ♍	03	11 ♏	22	25 ♋	48	15 ♉	10
21	27 ♈	26	03 ♎	28	01 ♍	15	14 ♏	27	23 ♋	16	16 ♉	17
31	27 ♈	19	06 ♎	00	01 ♍	22	17 ♏	22	20 ♋	57	17 ♉	23

All ephemeris data is given at 12.00 UT and the Moon's longitude is additionally given for 24.00 UT

ASPECTARIAN

h m Aspects

01 Wednesday
01 30 ☽ △ ♂
05 23 ☽ △ ☉
05 51 ☽ ⊼ ♅
05 53 ☽ ∠ ♃
11 14 ☉ ⊻ ♅
14 13 ☽ △ ♆
14 20 ☽ ☍ ⯓
20 58 ☽ △ ♄
23 54 ☽ ⊥ ♃

02 Thursday
06 06 ☽ □ ♀
07 48 ♀ Q ♅
09 51 ☽ ♇ ♂
11 00 ☽ ∥ ⯓
11 45 ☽ △ ☿
11 58 ☽ ⊻ ♃
13 16 ☿ ⊼ ♃
14 38 ☽ ♇ ☉
15 37 ☉ ∥ ♂
16 58 ☿ ∠ ♅
20 48 ☽ ♇ ♆

03 Friday
05 23 ☽ ± ♂
10 43 ☽ ± ☉
11 30 ☿ Q ⯓
15 12 ☽ ± ♆
18 25 ☽ ⊼ ♂
19 07 ☽ ☍ ♅
20 19 ☽ # ♅
22 49 ☽ ♇ ☿

04 Saturday
00 05 ☽ ⊼ ☉
03 27 ☽ ⊼ ♆
03 30 ☽ ⊼ ⯓
06 37 ♂ ⊻ ♅
09 59 ☽ □ ♄
11 18 ♀ △ ♃
15 43 ☽ ± ⯓
19 26 ☽ ∥ ♄
19 46 ☽ ± ☿

05 Sunday
00 23 ☽ ☌ ♃
01 59 ☽ △ ♀
09 39 ☽ ⊼ ☿
09 59 ☽ ♇ ⯓
10 08 ☽ ∥ ♃
16 36 ☉ ☌ ♆
16 41 ☉ ⚹ ⯓
22 20 ☽ Q ♄
22 20 ♆ ⚹ ⯓
23 42 ♀ ∠ ♂

06 Monday
05 58 ☿ ± ♃
08 04 ☽ ⊼ ♅
11 00 ☽ ☍ ♂
11 29 ☽ ♇ ♀
16 08 ☽ △ ⯓
16 09 ☽ ☍ ♆
18 15 ☽ ☍ ☉
20 05 ☽ ± ♅
22 17 ☽ ⚹ ♄

07 Tuesday
07 20 ☽ ± ♀
11 50 ☽ ⊻ ♃
14 00 ☽ ♇ ♅
14 22 ☿ ♇ ♄
17 19 ☉ ⊥ ♅
20 26 ☽ ⊼ ♀
23 32 ☽ ⊥ ♃

08 Wednesday
03 50 ☽ ∠ ♄
03 51 ☉ ∥ ♀
05 41 ☽ ☍ ☿
07 39 ♀ ∥ ♂
12 20 ♀ ∠ ♆
16 56 ☽ ∠ ♃
17 26 ☉ △ ♄
19 28 ☽ △ ♅
21 20 ☽ ⊥ ♄

09 Thursday
01 45 ☽ ⊼ ♂
03 09 ☽ □ ⯓
03 14 ☽ ⊼ ♆
07 25 ☽ ∥ ♃
08 54 ☽ ⊻ ♄
10 18 ☽ ⊼ ☉
14 02 ☽ ± ♂
14 49 ☽ ± ♆
19 38 ☽ ∥ ♄
21 35 ☽ ⚹ ♃
22 49 ☽ ± ☉

10 Friday
02 06 ♂ ⚹ ⯓
04 20 ♂ ☌ ♆
08 02 ☽ ♇ ♆
08 15 ☽ ♇ ♂
12 23 ☽ ☍ ♀
13 36 ☽ Q ⯓
16 51 ☽ # ♅
17 22 ☽ ♇ ☉
20 08 ☽ Q ♃
22 46 ☽ ⊼ ☿

11 Saturday
04 53 ☽ □ ♅
11 23 ☽ ± ☿
12 08 ☽ ⚹ ⯓
12 18 ☽ △ ♆
14 08 ☽ △ ♂
17 29 ☽ ☌ ♄
23 41 ☿ ♇ ♃
23 45 ☽ △ ☉

12 Sunday
00 35 ☽ ∥ ⯓
05 19 ☽ □ ♃
06 03 ☽ ♇ ☿
14 17 ☽ Q ♅
15 48 ☽ ∠ ⯓

13 Monday
01 26 ☽ ⊼ ♀
03 44 ♂ ⊥ ♅
08 07 ☽ ⊥ ⯓
08 30 ☿ ⚹ ♅
11 57 ☉ ∥ ♂
12 03 ☽ ⚹ ♅
12 26 ☽ △ ☿
13 16 ☽ ± ♀
18 52 ☽ ⊻ ⯓
19 00 ☿ ± ♄
19 05 ☽ □ ♆
23 31 ♂ △ ♄
23 47 ☽ ⊻ ♄
23 48 ☽ □ ♂

14 Tuesday
05 34 ♀ ∥ ♆
06 46 ☽ ♇ ♀
10 14 ☽ □ ☉
10 21 ☽ ⊥ ♄
10 48 ☽ △ ♃
14 42 ☽ ∠ ♅
17 06 ☉ ⊼ ♃

15 Wednesday
01 59 ☽ ∠ ♄
02 45 ☽ Q ♆
06 22 ☽ ⊥ ♅
08 59 ☽ Q ♂
11 14 ☽ △ ♀
12 36 ☽ ♇ ♃
16 41 ☽ ⊻ ♅
19 48 ☽ Q ☉
22 21 ☽ □ ☿
23 03 ☽ ☌ ⯓
23 21 ☽ ⚹ ♆

16 Thursday
03 34 ☽ ⚹ ♄
03 48 ☽ ± ♃
06 15 ☿ □ ⯓
06 31 ☽ ⚹ ♂
09 39 ☿ ⊻ ♆
13 49 ☽ ⊼ ♃
16 09 ☽ # ⯓
17 34 ☽ ⚹ ☉
23 36 ☽ Q ♄

17 Friday
00 33 ☽ ∠ ♆
01 04 ☉ ∠ ♅
07 20 ☽ Q ☿
08 50 ☽ ∠ ♂
15 25 ☽ ⊥ ♆
17 47 ☉ Q ⯓
17 52 ☽ □ ♀
18 54 ☽ ☌ ♅
20 10 ☽ ∠ ☉
20 50 ☽ ∥ ♅

18 Saturday
00 21 ☽ ⊥ ♂
00 54 ☽ ⊻ ⯓
01 15 ☽ ⊻ ♆
04 44 ☽ ⚹ ☿
05 04 ☽ □ ♄
07 20 ♀ □ ♅
08 42 ☿ ⊼ ♄
10 37 ☽ ⊻ ♂
10 40 ☽ ⊥ ⯓
11 53 ☽ ⊥ ☉
14 46 ☽ ☍ ♃
15 02 ☽ # ♄
21 54 ☽ # ♃
22 18 ☽ ⊻ ☉

19 Sunday
01 13 ☽ ∠ ⯓
01 34 ☽ Q ♀
07 00 ☽ ∠ ☿
19 33 ☽ ⊻ ♅
22 29 ☽ ⚹ ♀
22 36 ☽ ⊥ ☿

20 Monday
01 21 ☽ ⚹ ⯓
01 39 ☽ ∥ ♀
01 46 ☽ ☌ ♆
02 37 ☉ ± ♃
05 13 ☽ ⊥ ♅
05 17 ☽ △ ♄
08 59 ☽ ⊻ ☿
13 26 ☽ ☌ ♂
14 45 ☽ ⊼ ♃
15 33 ☿ ∥ ♂
19 46 ☽ ∠ ♅
20 42 ☽ Q ⯓

21 Tuesday
00 23 ☽ ± ♃
00 41 ☽ ∠ ♀
02 08 ☽ ☌ ☉
05 25 ☽ ♇ ♄
11 31 ☉ ∥ ☿
11 57 ♀ △ ⯓
13 36 ♂ ⊼ ♃
14 56 ☽ ♇ ♃
16 34 ☽ ⊥ ♀
17 02 ☽ ∥ ♀
17 59 ♀ ⚹ ♆
20 03 ☽ ± ♄
20 13 ☽ ⚹ ♅

22 Wednesday
00 25 ☿ ⊥ ♆
02 04 ☽ □ ⯓
02 33 ☽ ⊻ ♆
03 16 ☽ ⊻ ♀
05 54 ☽ ⊼ ♄
08 50 ☽ # ♃
12 31 ☽ ⊥ ♆
13 15 ☽ ☌ ☿
14 42 ☽ # ♄
15 33 ☽ △ ♃
16 09 ☽ Q ♅
17 07 ☽ ⊻ ♂

23 Thursday
00 07 ☉ ♇ ♄
03 41 ☽ ∠ ♆
03 47 ☽ ⊥ ♂
07 33 ☽ ⊻ ☉
09 45 ☽ ∥ ♅
11 17 ♀ ☍ ♄
18 39 ☽ ⊥ ☉
20 04 ☽ ∠ ♂
22 55 ☽ □ ♅

24 Friday
05 01 ☽ △ ⯓
05 37 ☽ ⚹ ♆
06 16 ☿ △ ♃
08 53 ☽ ☍ ♄
10 58 ☽ ☌ ♀
11 39 ☽ ∠ ☉
16 49 ☽ # ⯓
19 05 ☽ □ ♃
19 45 ☽ ⊻ ☿
22 21 ♀ # ♃

25 Saturday
00 06 ☽ ⚹ ♂
03 04 ☽ Q ♆
07 01 ☽ ⊥ ☿
07 52 ☽ ♇ ⯓
08 00 ☿ Q ♅
17 03 ☽ ⚹ ☉
23 25 ☽ Q ♂

26 Sunday
00 11 ☽ ∠ ☿
00 29 ☽ ± ⯓
02 19 ♀ # ♄
03 53 ☽ Q ♃
04 25 ♂ ∠ ♅
05 20 ☽ △ ♅
11 46 ☽ ⊼ ⯓
12 31 ☽ □ ♆
15 41 ☽ ⊼ ♄
17 32 ☽ Q ☉
18 01 ♂ Q ⯓
23 43 ☽ ⊻ ♀

27 Monday
02 36 ☽ ⚹ ♃
03 08 ☽ ± ♄
05 21 ☽ ⚹ ☿
06 47 ☉ ∠ ♀
10 07 ☽ ♇ ♅
11 55 ☽ □ ♂
12 31 ☽ ⊥ ♀
20 37 ☽ ♇ ♄

28 Tuesday
03 51 ☽ ± ♅
04 33 ☿ St R
05 01 ☽ Q ☿
05 35 ♀ □ ♃
07 48 ☽ □ ☉
07 50 ☽ ∠ ♃
08 05 ☽ ∠ ♀
08 12 ☉ ♇ ♃
15 50 ☽ ⊼ ♅
22 31 ☽ ☍ ⯓
22 40 ♂ ± ♃
23 26 ☽ △ ♆

29 Wednesday
01 44 ☽ ⊥ ♃
02 21 ☽ △ ♄
05 28 ☿ ∥ ♆
11 32 ☿ ⊻ ♀
13 47 ☽ ⊻ ♃
16 43 ☽ □ ☿
17 24 ☽ ⚹ ♀
19 38 ☽ ∥ ⯓

30 Thursday
03 53 ☽ △ ♂
05 53 ☽ ♇ ♆
11 10 ☿ Q ♅
20 24 ☽ Q ♀

31 Friday
00 17 ☽ ± ♆
02 09 ☽ △ ☉
04 50 ☽ ☍ ♅
06 48 ♀ ∥ ♅
08 13 ☽ # ♀
08 27 ☽ # ♅
09 52 ☉ ± ♄
11 29 ☽ ⊼ ⯓
12 31 ☽ ♇ ♂
12 33 ☽ ⊼ ♆
15 01 ☽ □ ♄
23 42 ☽ ± ⯓

JANUARY 1977

LONGITUDES

Date	Sidereal time h m s	Sun ☉ ° ' "	Moon ☽ ° ' "	Moon ☽ 24.00 ° ' "	Mercury ☿ ° '	Venus ♀ ° '	Mars ♂ ° '	Jupiter ♃ ° '	Saturn ♄ ° '	Uranus ♅ ° '	Neptune ♆ ° '	Pluto ♇ ° '
01	18 44 05	10 ♑ 58 22	26 ♉ 11 02	02 ♊ 07 41	21 ♑ 31	26 ♒ 35	00 ♑ 21	21 ♉ 30	15 ♌ 48	10 ♏ 54	14 ♐ 40	14 ♎ 08
02	18 48 02	11 59 31	08 ♊ 06 25	14 ♊ 07 33	20 R 36	27 43	01 06	21 R 28	15 R 44	10 56	14 42	14 08
03	18 51 59	13 00 39	20 ♊ 11 26	26 ♊ 18 16	19 31	28 50	01 51	21 25	15 41	10 59	14 44	14 09
04	18 55 55	14 01 47	02 ♋ 28 14	08 ♋ 41 26	18 19	29 ♒ 57	02 36	21 23	15 37	11 01	14 46	14 09
05	18 59 52	15 02 56	14 ♋ 57 55	21 ♋ 17 41	17 01	01 ♓ 04	03 21	21 21	15 33	11 03	14 48	14 09
06	19 03 48	16 04 04	27 ♋ 40 41	04 ♌ 06 48	15 41	02 11	04 06	21 19	15 29	11 05	14 50	14 10
07	19 07 45	17 05 11	10 ♌ 35 58	17 ♌ 08 03	14 21	03 17	04 51	21 17	15 25	11 07	14 52	14 10
08	19 11 41	18 06 19	23 ♌ 42 55	00 ♍ 20 28	13 02	04 24	05 36	21 15	15 21	11 09	14 54	14 10
09	19 15 38	19 07 27	07 ♍ 00 38	13 ♍ 43 21	11 49	05 30	06 21	21 14	15 17	11 11	14 56	14 10
10	19 19 34	20 08 35	20 ♍ 28 36	27 ♍ 16 22	10 42	06 35	07 07	21 13	15 13	11 13	14 58	14 11
11	19 23 31	21 09 42	04 ♎ 06 43	10 ♎ 59 40	09 43	07 41	07 52	21 12	15 09	11 15	15 00	14 11
12	19 27 28	22 10 50	17 ♎ 55 18	24 ♎ 53 40	08 53	08 46	08 37	21 11	15 05	11 17	15 02	14 11
13	19 31 24	23 11 57	01 ♏ 54 48	08 ♏ 58 39	08 12	09 51	09 22	21 11	15 00	11 19	15 04	14 11
14	19 35 21	24 13 05	16 ♏ 05 08	23 ♏ 14 07	07 42	10 56	10 08	21 10	14 56	11 20	15 06	14 11
15	19 39 17	25 14 12	00 ♐ 25 17	07 ♐ 38 17	07 21	12 00	10 53	21 D 10	14 52	11 22	15 08	14 11
16	19 43 14	26 15 20	14 ♐ 52 36	22 ♐ 07 39	07 09	13 04	11 39	21 11	14 47	11 24	15 10	14 R 11
17	19 47 10	27 16 27	29 ♐ 22 42	06 ♑ 36 59	07 D 06	14 08	12 24	21 11	14 42	11 25	15 11	14 11
18	19 51 07	28 17 33	13 ♑ 49 41	20 ♑ 59 55	07 11	15 11	13 10	21 11	14 38	11 27	15 13	14 11
19	19 55 03	29 ♑ 18 40	28 ♑ 06 54	05 ♒ 09 51	07 25	16 14	13 55	21 12	14 33	11 28	15 15	14 11
20	19 59 00	00 ♒ 19 45	12 ♒ 08 06	19 ♒ 01 06	07 45	17 17	14 41	21 14	14 29	11 30	15 17	14 11
21	20 02 57	01 20 50	25 ♒ 48 26	02 ♓ 29 51	08 11	18 19	15 27	21 14	14 24	11 31	15 18	14 11
22	20 06 53	02 21 53	09 ♓ 05 14	15 ♓ 34 37	08 44	19 21	16 12	21 16	14 19	11 32	15 20	14 11
23	20 10 50	03 22 56	21 ♓ 58 10	28 ♓ 16 11	09 21	20 23	16 58	21 17	14 14	11 34	15 22	14 11
24	20 14 46	04 23 58	04 ♈ 29 05	10 ♈ 37 21	10 04	21 24	17 44	21 19	14 10	11 35	15 23	14 10
25	20 18 43	05 24 59	16 ♈ 41 33	22 ♈ 42 19	10 51	22 25	18 30	21 21	14 05	11 36	15 25	14 10
26	20 22 39	06 25 59	28 ♈ 40 18	04 ♉ 36 11	11 42	23 25	19 15	21 23	14 00	11 37	15 27	14 10
27	20 26 36	07 26 57	10 ♉ 30 41	16 ♉ 24 31	12 36	24 25	20 01	21 25	13 55	11 38	15 28	14 09
28	20 30 32	08 27 55	22 ♉ 18 23	28 ♉ 12 56	13 33	25 25	20 47	21 28	13 50	11 39	15 30	14 09
29	20 34 29	09 28 51	04 ♊ 08 51	10 ♊ 06 43	14 34	26 24	21 33	21 31	13 45	11 40	15 31	14 08
30	20 38 26	10 29 46	16 ♊ 07 08	22 ♊ 10 34	15 37	27 22	22 19	21 34	13 41	11 41	15 33	14 08
31	20 42 22	11 ♒ 30 40	28 ♊ 17 29	04 ♋ 28 14	16 ♑ 43	28 ♓ 20	23 ♑ 05	21 ♉ 37	13 ♌ 36	11 ♏ 42	15 ♐ 34	14 ♎ 07

Date	Moon True ☊ ° '	Moon Mean ☊ ° '	Moon ☽ Latitude ° '
01	01 ♏ 08	29 ♎ 51	02 S 09
02	01 R 02	29 48	03 03
03	00 54	29 45	03 49
04	00 42	29 42	04 26
05	00 30	29 39	04 50
06	00 16	29 35	05 01
07	00 03	29 32	04 56
08	29 ♎ 50	29 29	04 36
09	29 43	29 26	04 00
10	29 37	29 23	03 11
11	29 34	29 20	02 10
12	29 D 34	29 16	01 S 01
13	29 R 34	29 13	00 N 12
14	29 33	29 10	01 26
15	29 30	29 07	02 35
16	29 25	29 04	03 35
17	29 17	29 01	04 21
18	29 06	28 57	04 50
19	28 53	28 54	05 00
20	28 40	28 51	04 52
21	28 29	28 48	04 27
22	28 19	28 45	03 48
23	28 12	28 41	02 59
24	28 07	28 38	02 02
25	28 05	28 35	01 N 00
26	28 D 05	28 32	00 S 03
27	28 R 05	28 29	01 05
28	28 04	28 26	02 05
29	28 02	28 22	02 59
30	27 57	28 19	03 46
31	27 ♎ 49	28 ♎ 16	04 S 23

DECLINATIONS

Date	Sun ☉ ° '	Moon ☽ ° '	Mercury ☿ ° '	Venus ♀ ° '	Mars ♂ ° '	Jupiter ♃ ° '	Saturn ♄ ° '	Uranus ♅ ° '	Neptune ♆ ° '	Pluto ♇ ° '
01	22 S 59	17 N 13	20 S 21	14 S 08	24 S 04	17 N 12	16 N 56	14 S 41	21 S 07	10 N 05
02	22 54	18 39	20 10	13 42	24 04	17 11	16 58	14 42	21 07	10 05
03	22 48	19 16	20 01	13 15	24 04	17 11	16 59	14 42	21 07	10 05
04	22 42	18 59	19 54	12 48	24 04	17 11	17 00	14 43	21 07	10 06
05	22 35	17 48	19 48	12 21	24 04	17 10	17 01	14 44	21 08	10 06
06	22 28	15 43	19 44	11 54	24 03	17 10	17 03	14 44	21 08	10 06
07	22 21	12 50	19 41	11 26	24 02	17 10	17 04	14 45	21 08	10 07
08	22 13	09 17	19 40	10 58	24 01	17 10	17 05	14 45	21 08	10 07
09	22 05	05 13	19 40	10 30	23 59	17 10	17 07	14 46	21 08	10 08
10	21 56	00 N 51	19 41	10 02	23 57	17 10	17 08	14 47	21 09	10 08
11	21 47	03 S 38	19 44	09 33	23 55	17 10	17 10	14 47	21 09	10 09
12	21 37	07 58	19 48	09 05	23 53	17 10	17 11	14 48	21 09	10 09
13	21 27	11 57	19 53	08 36	23 50	17 10	17 12	14 48	21 09	10 10
14	21 16	15 16	19 59	08 07	23 48	17 10	17 14	14 49	21 09	10 10
15	21 05	17 42	20 05	07 38	23 45	17 10	17 15	14 49	21 10	10 11
16	20 54	19 01	20 13	07 09	23 41	17 11	17 17	14 50	21 10	10 11
17	20 42	19 05	20 21	06 39	23 38	17 11	17 18	14 50	21 10	10 12
18	20 30	17 55	20 29	06 10	23 34	17 11	17 20	14 51	21 10	10 12
19	20 18	15 38	20 37	05 41	23 30	17 12	17 21	14 51	21 10	10 13
20	20 05	12 29	20 46	05 11	23 25	17 12	17 23	14 52	21 10	10 13
21	19 52	08 43	20 54	04 41	23 21	17 13	17 24	14 52	21 11	10 14
22	19 38	04 38	21 02	04 12	23 16	17 14	17 26	14 52	21 11	10 15
23	19 24	00 S 27	21 10	03 42	23 11	17 14	17 27	14 53	21 11	10 15
24	19 10	03 N 39	21 17	03 12	23 05	17 15	17 29	14 53	21 11	10 16
25	18 55	07 29	21 24	02 43	23 00	17 16	17 30	14 53	21 11	10 16
26	18 40	10 57	21 30	02 13	22 54	17 17	17 32	14 54	21 11	10 17
27	18 25	13 56	21 35	01 44	22 47	17 17	17 33	14 54	21 11	10 18
28	18 09	16 20	21 40	01 14	22 41	17 18	17 35	14 54	21 12	10 18
29	17 53	18 03	21 44	00 44	22 34	17 19	17 36	14 55	21 12	10 19
30	17 36	18 59	21 46	00 15	22 27	17 20	17 38	14 55	21 12	10 20
31	17 S 20	19 N 03	21 S 48	00 N 14	22 S 20	17 N 21	17 N 39	14 S 55	21 S 12	10 N 20

ZODIAC SIGN ENTRIES

Date	h	m	Planets
01	00	42	♂ ♑
01	19	43	☽ ♊
04	07	12	☽ ♋
04	13	01	♀ ♓
06	16	20	☽ ♌
08	23	23	☽ ♍
11	04	48	☽ ♎
13	08	44	☽ ♏
15	11	18	☽ ♐
17	13	02	☽ ♑
19	15	12	☽ ♒
20	04	14	☉ ♒
21	19	30	☽ ♓
24	03	19	☽ ♈
26	14	41	☽ ♉
29	03	37	☽ ♊
31	15	20	☽ ♋

LATITUDES

Date	Mercury ☿ ° '	Venus ♀ ° '	Mars ♂ ° '	Jupiter ♃ ° '	Saturn ♄ ° '	Uranus ♅ ° '	Neptune ♆ ° '	Pluto ♇ ° '
01	01 N 24	01 S 35	00 S 38	00 S 59	00 N 53	00 N 26	01 N 27	16 N 58
04	02 19	01 24	00 39	00 58	00 53	00 26	01 27	17 00
07	03 00	01 13	00 41	00 57	00 54	00 26	01 27	17 02
10	03 20	01 01	00 42	00 56	00 54	00 26	01 28	17 04
13	03 19	00 47	00 44	00 55	00 55	00 27	01 28	17 05
16	03 02	00 32	00 46	00 54	00 55	00 27	01 28	17 07
19	02 37	00 S 16	00 47	00 54	00 55	00 27	01 28	17 09
22	02 07	00 N 01	00 49	00 53	00 56	00 27	01 28	17 12
25	01 36	00 19	00 50	00 52	00 56	00 27	01 28	17 12
28	01 05	00 39	00 52	00 51	00 57	00 27	01 28	17 14
31	00 N 36	00 N 59	00 S 53	00 S 50	00 N 57	00 N 27	01 N 28	17 N 16

DATA

Julian Date	2443145
Delta T	+48 seconds
Ayanamsa	23° 32' 18"
Synetic vernal point	05° ♓ 34' 42"
True obliquity of ecliptic	23° 26' 24"

MOON'S PHASES, APSIDES AND POSITIONS ☽

Date	h	m	Phase	Longitude ° '	Eclipse Indicator
05	12	10	○	15 ♋ 03	
12	19	55	☾	22 ♎ 31	
19	14	11	●	29 ♑ 24	
27	05	11	☽	07 ♉ 10	

Day	h	m		
16	10	04	Perigee	
28	05	30	Apogee	
03	16	43	Max dec	19° N 17'
10	16	32	0S	
17	01	19	Max dec	19° S 13'
23	14	35	0N	
31	01	58	Max dec	19° N 08'

LONGITUDES

Date	Chiron ⚷ ° '	Ceres ⚳ ° '	Pallas ⚴ ° '	Juno ⚵ ° '	Vesta ⚶ ° '	Black Moon Lilith ⚸ ° '
01	27 ♈ 19	06 ♎ 13	01 ♍ 19	17 ♏ 39	20 ♋ 42	17 ♉ 30
11	27 ♈ 18	08 ♎ 13	00 ♍ 12	20 ♏ 21	18 ♋ 05	18 ♉ 36
21	27 ♈ 24	09 ♎ 40	28 ♌ 01	22 ♏ 49	15 ♋ 31	19 ♉ 43
31	27 ♈ 35	10 ♎ 29	25 ♌ 00	25 ♏ 00	13 ♋ 19	20 ♉ 50

ASPECTARIAN

h m	Aspects
01 Saturday	
02 34	☽ ☌ ♃
03 08	☽ △ ☿
08 02	☽ ± ♂
08 32	☽ ∥ ♄
10 17	☉ ⚹ ♅
11 32	☽ ⚻ ☉
11 50	☽ ∥ ♃
12 18	☿ △ ♃
12 54	☽ □ ♀
13 53	♀ Q ♆
17 57	☽ ⚻ ♇
20 59	☽ ⊼ ♂
23 12	☿ ⊥ ♀
02 Sunday	
01 19	♂ ⚻ ♄
03 17	☽ Q ♄
07 21	☽ ⚻ ☿
07 22	☽ ± ☉
09 32	☿ ⊥ ♆
17 41	☽ ⊼ ♅
20 28	☽ ⊼ ☉
23 55	☽ ± ☿
03 Monday	
00 01	☽ △ ♇
01 11	☽ ☍ ♆
03 08	☽ ⚹ ♄
05 38	☽ ± ♅
10 48	☽ ⊼ ☿
14 25	☽ ⊻ ♃
18 40	♀ ⚻ ♇
23 24	☽ ⚻ ♅
04 Tuesday	
02 08	☽ ⊥ ♃
06 38	☽ △ ♀
08 25	☽ ∠ ♄
12 16	☽ ☍ ♂
14 50	☉ □ ♇
19 32	☽ ∠ ♃
05 Wednesday	
01 43	☽ ⊥ ♄
04 30	☽ △ ♅
06 06	☉ ⊻ ♆
10 27	☽ □ ♇
11 42	☽ ⊼ ♆
12 10	☽ ☍ ☉
13 07	☽ ⊻ ♄
14 18	☽ ⚻ ♀
15 33	☽ ☍ ☿
20 18	☽ ∥ ♃
21 23	☿ ∠ ♀
21 58	☽ ∥ ♄
23 06	☽ ± ♆
23 11	☉ ⊼ ♄
06 Thursday	
00 04	☽ ⚹ ♃
08 07	☉ ☌ ☿
08 56	☽ ± ♀
15 40	☿ ⊼ ♄
16 03	☽ ⚻ ♆
20 23	☽ Q ♇
20 49	☽ # ♅
21 12	☽ ⊼ ♀
22 29	☽ Q ♃
07 Friday	
00 43	☽ ⊼ ♂
02 42	☿ ⊻ ♆
12 29	☽ ± ♂
12 58	☽ □ ♅
15 08	☿ □ ♇
18 15	☽ ⊼ ☿
18 34	☽ ⚹ ♇
19 53	☽ △ ♆
20 49	☽ ☌ ♄
23 24	☽ # ♀
08 Saturday	
00 55	☽ ⊼ ☉
04 14	☽ ± ☿
05 59	☽ ⚻ ♂
06 38	☽ ∥ ♇
07 32	☽ □ ♃
12 46	☽ ± ☉
19 10	☽ ⚻ ☿
21 53	☽ Q ♅
21 54	☽ ∠ ♇
09 Sunday	
06 23	☽ ⚻ ☉
08 17	♂ ⚻ ♃
09 02	☽ ☍ ♀
10 45	☽ △ ♂
14 05	☽ ⊥ ♇
19 30	☽ ⚹ ♅
19 55	☽ △ ☿
10 Monday	
00 43	☿ ⚹ ♅
00 49	☽ ⊻ ♇
02 12	☽ □ ♆
02 43	☽ ⊻ ♄
06 47	♀ # ♇
11 22	☽ △ ☉
13 18	☽ ⊥ ♄
13 18	☽ △ ♃
22 10	☽ ∠ ♅
11 Tuesday	
05 05	☽ ∠ ♄
08 08	☉ ⊥ ♆
10 03	☽ Q ♆
12 54	☉ △ ♃
14 00	☽ ⊥ ♅
14 39	♃ ∥ ♄
15 39	☽ ⚻ ♃
18 46	☽ ⊼ ♀
18 56	☽ □ ♂
21 11	☽ □ ☿
23 10	♀ ± ♇
12 Wednesday	
00 28	☽ ⊻ ♅
01 14	♀ ⚹ ♂
05 13	☿ ± ♄
05 32	☽ ☌ ♇
06 05	☽ ± ♀
07 00	☽ ⚹ ♆
07 06	☽ ⚹ ♄
07 17	☽ ± ♃
13 28	☿ ⚹ ♀
16 11	☿ ☌ ♂
17 38	☽ ⊼ ♃
17 44	☽ ∥ ♀
19 55	☽ □ ☉
21 17	♀ Q ♃
21 54	♄ △ ♆
22 55	☽ ⚻ ♀
13 Thursday	
00 48	☽ # ♇
01 20	♂ ± ♄
02 40	☽ Q ☿
03 40	☽ Q ♄
03 48	☽ Q ♂
08 51	☽ ∠ ♆
14 41	☉ Q ♅
22 17	☽ ⚹ ☿
14 Friday	
00 11	☽ ⊥ ♆
01 23	☽ ⚹ ♂
02 35	☽ △ ♀
03 59	☽ ☌ ♅
04 59	☽ Q ☉
08 17	☽ ∥ ♅
08 48	☽ ⊻ ♇
10 04	☽ □ ♄
10 20	☽ ⊻ ♆
18 54	☽ ⊥ ♇
20 33	☽ ☍ ♃
21 29	♀ △ ♅
22 48	☽ ∠ ☿
15 Saturday	
02 41	☽ ⚹ ☉
03 03	☉ ∥ ♆
04 01	☽ ∠ ♂
05 44	☽ # ♃
06 35	☽ # ♄
09 57	☽ ∠ ♇
10 54	♃ St D
13 30	☽ ⊥ ☿
19 51	☽ ⊥ ♂
23 20	☽ ⊻ ☿
16 Sunday	
03 46	♂ ⚹ ♅
05 33	☽ ∠ ☉
06 13	☽ ⊻ ♅
06 21	☽ ⊻ ♂
07 04	♇ St R
08 46	☽ □ ♀
10 52	☽ ⚹ ♇
11 51	☽ △ ♄
12 28	☽ ☌ ♆
16 10	☽ ⊥ ♅
21 35	☽ ⊥ ☉
22 26	☽ ⊼ ♃
17 Monday	
06 43	☽ Q ♇
07 06	☽ ∠ ♅
08 02	☿ St D
08 15	☽ ⊻ ☉
08 22	☽ ± ♃
12 33	☽ ⚻ ♄
13 23	♀ ⊼ ♇
16 55	☽ Q ♀
23 17	☽ ⚻ ♃
18 Tuesday	
00 15	♀ ⊼ ♄
00 51	☽ ☌ ☿
03 24	☽ ± ♄
08 01	☽ ⚹ ♅
10 50	☽ ☌ ♂
12 36	☽ □ ♇
12 46	♀ □ ♆
13 20	☽ ⊼ ♄
13 40	☉ ∥ ☿
14 20	☽ ⊻ ♆
14 27	☽ ⚹ ♀
19 15	☽ # ♄
20 51	☽ # ♃
19 Wednesday	
00 20	☽ △ ♃
00 24	☽ ⊥ ♆
04 09	☽ Q ♅
14 11	☽ ☌ ☉
15 38	☽ ∠ ♆
17 44	☽ ∠ ♀
18 30	☽ ∥ ♅
20 22	♂ □ ♇
20 Thursday	
04 13	☽ ⊻ ☿
06 08	♂ ⊼ ♄
10 24	☽ ⊥ ♀
10 44	☉ ∠ ♆
10 53	☽ □ ♅
14 52	☽ ⊥ ☿
15 33	☽ △ ♇
16 03	☽ ☍ ♄
16 41	☽ ⊻ ♂
17 28	☽ ⚹ ♆
21 42	☽ ⊻ ♀
21 Friday	
02 42	☽ # ♇
03 49	☽ ⊥ ♂
03 53	☽ □ ♃
07 11	☽ ∠ ☿
07 30	♂ ⊻ ♆
14 41	☽ Q ♆
18 02	☽ ⚻ ♇
20 38	☿ ± ♄
20 48	☽ ∠ ♂
22 45	☽ ⊻ ☉
22 Saturday	
10 20	☽ ± ♇
10 34	☽ ⊥ ☉
11 19	☽ ⚹ ☿
12 19	☽ Q ♃
14 50	☽ ∥ ♀
16 31	☽ △ ♅
21 24	☽ ⊼ ♇
21 36	☽ ⊼ ♄
23 34	☽ □ ♆
23 Sunday	
02 00	☽ ⚹ ♂
04 40	☽ ∠ ☉
08 45	☽ ☌ ♀
08 45	☽ ± ♄
08 54	♀ ± ♄
10 42	☽ ⚹ ♃
10 47	☽ Q ☿
15 30	☿ ∥ ♆
20 44	☽ ⚻ ♅
24 Monday	
01 47	☽ ⚻ ♄
02 12	☽ Q ♂
09 02	♄ ⚹ ♇
09 41	☽ # ♀
09 50	♀ ⚹ ♃
11 49	☽ ⚹ ☉
14 08	☽ ± ♅
15 34	☽ ∠ ♃
23 38	☽ □ ☿
25 Tuesday	
01 54	☽ ⊼ ♅
06 51	☽ △ ♄
06 59	☽ ☍ ♇
09 19	☽ ⊥ ♃
09 28	☽ △ ♆
13 34	☽ Q ☉
15 49	☽ □ ♂
21 18	☽ ⊻ ♃
26 Wednesday	
00 28	☽ ⊻ ♀
07 07	☽ ∥ ♇
09 52	☿ ⚹ ♅
13 39	☽ ⊥ ♀
15 35	☽ ⚻ ♆
27 Thursday	
05 11	☽ □ ☉
09 35	☽ ∠ ♀
09 53	☽ ± ♆
14 17	☽ ☍ ♅
16 37	☽ △ ☿
18 53	☽ □ ♄
19 24	☽ ⊼ ♇
20 54	☽ # ♅
22 07	☽ ⊼ ♆
28 Friday	
07 37	☽ ± ♇
08 41	☽ △ ♂
10 17	☽ ☌ ♃
18 17	☿ ⊼ ♄
18 53	☽ ⚹ ♀
29 Saturday	
00 27	☽ ∥ ♃
01 52	☽ ⚻ ☿
01 53	☽ ⚻ ♇
02 02	☿ □ ♇
04 33	☽ ∥ ♄
07 12	☽ Q ♄
09 35	☽ # ☉
10 43	♂ △ ♃
11 02	♂ ⊥ ♆
17 11	☽ ⚻ ♂
18 43	♀ ⚻ ♅
21 19	☽ Q ♀
21 45	☽ ± ☿
23 44	☽ △ ☉
30 Sunday	
03 08	☽ ⊼ ♅
07 10	☽ ⚹ ♄
08 03	☽ △ ♇
10 17	☿ ⊻ ♆
10 27	☉ # ♄
10 51	☽ ☍ ♆
10 55	☽ ⊼ ☿
12 25	☽ ± ♂
15 06	☽ ± ♅
22 50	☽ ⊻ ♃
31 Monday	
01 06	☽ ⊼ ♂
08 12	☽ ⚻ ☉
08 52	☽ ⚻ ♅
09 48	☉ # ♃
10 40	☽ ⊥ ♃
12 06	☽ □ ♀
12 35	☽ ∠ ♄
16 22	☉ □ ♅
17 57	♀ ⚻ ♄

All ephemeris data is given at 12.00 UT and the Moon's longitude is additionally given for 24.00 UT

FEBRUARY 1977

LONGITUDES

Date	Sidereal time h m s	Sun ☉ ° ' "	Moon ☽ ° ' "	Moon ☽ 24.00 ° ' "	Mercury ☿ ° '	Venus ♀ ° '	Mars ♂ ° '	Jupiter ♃ ° '	Saturn ♄ ° '	Uranus ♅ ° '	Neptune ♆ ° '	Pluto ♇ ° '
01	20 46 19	12 ♒ 31 33	10 ♋ 43 06	17 ♋ 02 16	17 ♑ 51	29 ♓ 18	23 ♑ 51	21 ♉ 40	13 ♌ 31	11 ♏ 42	15 ♐ 36	14 ♎ 07
02	20 50 15	13 32 24	23 ♋ 25 51	29 ♋ 53 51	19 01	00 ♈ 14	24 37	21 44	13 R 26	11 43	15 37	14 R 06
03	20 54 12	14 33 15	06 ♌ 26 11	13 ♌ 02 40	20 13	01 11	25 23	21 47	13 21	11 44	15 38	14 06
04	20 58 08	15 34 04	19 ♌ 43 05	26 ♌ 27 06	21 26	02 06	26 09	21 51	13 16	11 44	15 40	14 05
05	21 02 05	16 34 52	03 ♍ 14 24	10 ♍ 04 35	22 41	03 01	26 55	21 55	13 11	11 45	15 41	14 04
06	21 06 01	17 35 38	16 ♍ 57 16	23 ♍ 52 06	23 58	03 56	27 41	22 00	13 06	11 45	15 42	14 04
07	21 09 58	18 36 24	00 ♎ 48 43	07 ♎ 46 48	25 16	04 49	28 28	22 04	13 01	11 46	15 44	14 03
08	21 13 55	19 37 09	14 ♎ 46 06	21 ♎ 46 22	26 36	05 42	29 ♑ 14	22 09	12 57	11 46	15 45	14 02
09	21 17 51	20 37 52	28 ♎ 47 26	05 ♏ 49 09	27 57	06 35	00 ♒ 00	22 14	12 52	11 46	15 46	14 01
10	21 21 48	21 38 35	12 ♏ 51 25	19 ♏ 54 08	29 ♑ 19	07 26	00 46	22 19	12 47	11 47	15 47	14 00
11	21 25 44	22 39 16	26 ♏ 57 12	04 ♐ 00 31	00 ♒ 42	08 17	01 33	22 24	12 42	11 47	15 49	14 00
12	21 29 41	23 39 57	11 ♐ 03 55	18 ♐ 07 15	02 06	09 07	02 19	22 29	12 37	11 47	15 50	13 59
13	21 33 37	24 40 37	25 ♐ 10 16	02 ♑ 12 41	03 32	09 56	03 05	22 35	12 33	11 47	15 51	13 58
14	21 37 34	25 41 15	09 ♑ 14 09	16 ♑ 14 18	04 58	10 45	03 52	22 40	12 28	11 47	15 52	13 57
15	21 41 30	26 41 52	23 ♑ 12 40	00 ♒ 08 48	06 26	11 32	04 38	22 46	12 23	11 R 47	15 53	13 56
16	21 45 27	27 42 28	07 ♒ 02 15	13 ♒ 52 32	07 54	12 19	05 25	22 52	12 18	11 47	15 54	13 55
17	21 49 24	28 43 02	20 ♒ 39 13	27 ♒ 21 57	09 24	13 04	06 11	22 59	12 14	11 47	15 55	13 54
18	21 53 20	29 ♒ 43 35	04 ♓ 00 24	10 ♓ 34 21	10 55	13 49	06 58	23 05	12 09	11 47	15 56	13 53
19	21 57 17	00 ♓ 44 06	17 ♓ 03 40	23 ♓ 28 17	12 26	14 33	07 44	23 11	12 05	11 47	15 57	13 52
20	22 01 13	01 44 36	29 ♓ 48 15	06 ♈ 03 45	13 58	15 15	08 31	23 18	12 00	11 46	15 58	13 51
21	22 05 10	02 45 04	12 ♈ 15 01	18 ♈ 22 21	15 32	15 56	09 17	23 25	11 56	11 46	15 58	13 49
22	22 09 06	03 45 30	24 ♈ 26 10	00 ♉ 26 57	17 06	16 37	10 04	23 32	11 51	11 46	15 59	13 48
23	22 13 03	04 45 54	06 ♉ 25 13	12 ♉ 21 32	18 42	17 16	10 50	23 39	11 47	11 45	16 00	13 47
24	22 16 59	05 46 17	18 ♉ 16 30	24 ♉ 10 48	20 18	17 53	11 37	23 47	11 43	11 45	16 01	13 46
25	22 20 56	06 46 37	00 ♊ 05 03	05 ♊ 59 56	21 55	18 30	12 23	23 54	11 39	11 44	16 01	13 45
26	22 24 53	07 46 55	11 ♊ 56 06	17 ♊ 54 15	23 33	19 05	13 10	24 02	11 34	11 44	16 02	13 43
27	22 28 49	08 47 12	23 ♊ 54 58	29 ♊ 58 54	25 13	19 38	13 57	24 09	11 30	11 43	16 03	13 42
28	22 32 46	09 ♓ 47 27	06 ♋ 06 35	12 ♋ 18 32	26 ♒ 53	20 ♈ 10	14 ♒ 43	24 ♉ 17	11 ♌ 26	11 ♏ 42	16 ♐ 03	13 ♎ 41

Date	Moon True ☊ ° '	Moon Mean ☊ ° '	Moon ☽ Latitude ° '
01	27 ♎ 39	28 ♎ 13	04 S 49
02	27 R 27	28 10	05 01
03	27 14	28 07	04 58
04	27 01	28 03	04 39
05	26 50	28 00	04 04
06	26 42	27 57	03 14
07	26 37	27 54	02 12
08	26 34	27 51	01 S 02
09	26 D 34	27 47	00 N 12
10	26 34	27 44	01 26
11	26 R 34	27 41	02 35
12	26 33	27 38	03 34
13	26 29	27 35	04 21
14	26 23	27 32	04 51
15	26 15	27 28	05 04
16	26 05	27 25	04 59
17	25 54	27 22	04 37
18	25 45	27 19	04 00
19	25 37	27 16	03 11
20	25 31	27 13	02 13
21	25 28	27 09	01 10
22	25 27	27 06	00 N 05
23	25 D 27	27 03	00 S 59
24	25 28	27 00	02 00
25	25 30	26 57	02 56
26	25 R 30	26 53	03 45
27	25 28	26 50	04 24
28	25 ♎ 25	26 ♎ 47	04 S 52

DECLINATIONS

Date	Sun ☉ ° '	Moon ☽ ° '	Mercury ☿ ° '	Venus ♀ ° '	Mars ♂ ° '	Jupiter ♃ ° '	Saturn ♄ ° '	Uranus ♅ ° '	Neptune ♆ ° '	Pluto ♇ ° '
01	17 S 03	18 N 13	21 S 49	00 N 44	22 S 13	17 N 23	17 N 41	14 S 55	21 S 12	10 N 21
02	16 46	16 28	21 49	01 13	22 05	17 24	17 42	14 55	21 12	10 22
03	16 28	13 51	21 47	01 42	21 57	17 25	17 44	14 56	21 12	10 23
04	16 10	10 30	21 45	02 11	21 49	17 26	17 45	14 56	21 12	10 23
05	15 52	06 32	21 41	02 39	21 41	17 28	17 47	14 56	21 12	10 24
06	15 34	02 N 10	21 37	03 08	21 32	17 29	17 48	14 56	21 13	10 25
07	15 15	02 S 21	21 31	03 36	21 23	17 30	17 50	14 56	21 13	10 26
08	14 56	06 47	21 24	04 05	21 14	17 32	17 51	14 56	21 13	10 26
09	14 37	10 52	21 15	04 33	21 05	17 33	17 53	14 56	21 13	10 27
10	14 17	14 20	21 06	05 00	20 55	17 35	17 54	14 56	21 13	10 28
11	13 58	16 58	20 55	05 28	20 46	17 36	17 55	14 57	21 13	10 29
12	13 38	18 34	20 43	05 55	20 36	17 38	17 57	14 57	21 13	10 29
13	13 18	19 01	20 30	06 22	20 25	17 40	17 58	14 57	21 13	10 30
14	12 57	18 17	20 15	06 49	20 15	17 41	18 00	14 57	21 13	10 31
15	12 37	16 27	19 59	07 15	20 04	17 43	18 01	14 57	21 13	10 32
16	12 16	13 41	19 42	07 41	19 53	17 45	18 02	14 57	21 13	10 32
17	11 55	10 14	19 23	08 07	19 42	17 47	18 04	14 56	21 13	10 33
18	11 34	06 19	19 04	08 33	19 31	17 48	18 05	14 56	21 13	10 34
19	11 13	02 S 11	18 42	08 58	19 19	17 50	18 07	14 56	21 14	10 35
20	10 51	01 N 58	18 20	09 22	19 08	17 52	18 08	14 56	21 14	10 36
21	10 30	05 55	17 56	09 47	18 56	17 54	18 09	14 56	21 14	10 37
22	10 08	09 33	17 31	10 10	18 44	17 56	18 10	14 56	21 14	10 37
23	09 46	12 44	17 05	10 34	18 31	17 58	18 12	14 56	21 14	10 38
24	09 24	15 21	16 37	10 57	18 19	18 00	18 13	14 56	21 14	10 39
25	09 01	17 18	16 08	11 19	18 06	18 02	18 14	14 55	21 14	10 40
26	08 39	18 31	15 38	11 41	17 53	18 04	18 15	14 55	21 14	10 41
27	08 17	18 54	15 06	12 02	17 40	18 06	18 16	14 55	21 14	10 41
28	07 S 54	18 N 26	14 S 33	12 N 23	17 S 27	18 N 09	18 N 18	14 S 55	21 S 14	10 N 42

ZODIAC SIGN ENTRIES

Date	h	m	Planets
02	05	54	♀ ♈
03	00	11	☽ ♌
05	06	17	☽ ♍
07	10	36	☽ ♎
09	11	57	♂ ♒
09	14	04	☽ ♏
10	23	55	☿ ♒
11	17	11	☽ ♐
13	20	14	☽ ♑
15	23	45	☽ ♒
18	04	45	☽ ♓
18	18	30	☉ ♓
20	12	22	☽ ♈
22	23	06	☽ ♉
25	11	50	☽ ♊
28	00	02	☽ ♋

LATITUDES

Date	Mercury ☿ ° '	Venus ♀ ° '	Mars ♂ ° '	Jupiter ♃ ° '	Saturn ♄ ° '	Uranus ♅ ° '	Neptune ♆ ° '	Pluto ♇ ° '
01	00 N 26	01 N 06	00 S 54	00 S 50	00 N 57	00 N 27	01 N 28	17 N 16
04	00 S 01	01 28	00 55	00 49	00 57	00 27	01 28	17 18
07	00 26	01 50	00 56	00 49	00 58	00 27	01 28	17 19
10	00 49	02 14	00 58	00 48	00 58	00 27	01 28	17 21
13	01 10	02 39	00 59	00 47	00 58	00 27	01 29	17 22
16	01 27	03 04	01 00	00 46	00 58	00 27	01 29	17 24
19	01 42	03 30	01 01	00 46	00 58	00 27	01 29	17 25
22	01 54	03 57	01 02	00 45	00 59	00 27	01 29	17 26
25	02 03	04 24	01 04	00 44	00 59	00 27	01 29	17 27
28	02 08	04 52	01 05	00 43	00 59	00 27	01 29	17 28
31	02 S 10	05 N 20	01 S 06	00 S 43	00 N 59	00 N 27	01 N 29	17 N 29

DATA

Julian Date	2443176
Delta T	+48 seconds
Ayanamsa	23° 32' 23"
Synetic vernal point	05° ♓ 34' 37"
True obliquity of ecliptic	23° 26' 24"

MOON'S PHASES, APSIDES AND POSITIONS ☽

Date	h	m	Phase	Longitude ° '	Eclipse Indicator
04	03	56	○	15 ♌ 14	
11	04	07	☾	22 ♏ 19	
18	03	37	●	29 ♒ 22	
26	02	50	☽	07 ♊ 24	

Day	h	m		
11	03	54	Perigee	
25	02	42	Apogee	
06	23	33	0S	
13	08	59	Max dec	19° S 01'
20	00	33	0N	
27	11	10	Max dec	18° N 54'

LONGITUDES

Date	Chiron ⚷ ° '	Ceres ⚳ ° '	Pallas ⚴ ° '	Juno ⚵ ° '	Vesta ⚶ ° '	Black Moon Lilith ⚸ ° '
01	27 ♈ 36	10 ♎ 31	24 ♌ 40	25 ♏ 12	13 ♋ 08	20 ♉ 56
11	27 ♈ 53	10 ♎ 34	21 ♌ 17	27 ♏ 02	11 ♋ 37	22 ♉ 03
21	28 ♈ 14	09 ♎ 53	18 ♌ 06	28 ♏ 28	10 ♋ 51	23 ♉ 09
31	28 ♈ 40	08 ♎ 31	15 ♌ 35	29 ♏ 28	10 ♋ 50	24 ♉ 16

All ephemeris data is given at 12.00 UT and the Moon's longitude is additionally given for 24.00 UT

ASPECTARIAN

h m Aspects

01 Tuesday
03 15 ☽ ± ☉
04 12 ☽ ∠ ♃
05 54 ☽ ⊥ ♄
07 27 ♂ Q ♅
13 53 ☽ △ ♅
15 45 ☽ ⊼ ☉
17 17 ☽ ⊻ ♄
18 28 ☽ □ ♇
20 39 ☽ ∥ ♄
21 17 ☽ ⊼ ♆

02 Wednesday
00 53 ☽ ∥ ♃
02 53 ☽ ☍ ☿
08 09 ☽ # ☉
08 36 ☽ ± ♆
08 48 ☽ ✶ ♃
09 36 ☉ ☍ ♄
14 21 ☽ ☍ ♂

03 Thursday
01 13 ☉ △ ♇
01 21 ☽ bQ ♆
01 37 ☽ △ ♀
03 03 ☽ # ♅
04 03 ☽ Q ♇
07 08 ☽ Q ♃
21 38 ☽ □ ♅

04 Friday
00 28 ☽ ☌ ♄
01 53 ☽ ✶ ♇
03 56 ☽ ☍ ☉
04 43 ☽ △ ♆
06 58 ☽ bQ ♀
12 40 ☽ ∥ ♇
14 19 ☉ ✶ ♆
15 23 ☽ ⊼ ☿
15 50 ☽ □ ♃
16 28 ☿ ⊥ ♆
20 33 ☿ △ ♃

05 Saturday
00 10 ☽ ⊼ ♂
00 13 ☽ ± ♀
03 08 ☽ ± ☿
04 39 ☽ ∠ ♇
05 50 ☽ Q ♅
08 17 ☿ ∥ ♂
11 24 ☽ ± ♂
11 35 ☽ ⊼ ♀
20 29 ☽ ⊥ ♇
20 37 ☽ bQ ☿

06 Sunday
02 56 ☽ ✶ ♅
04 08 ☽ bQ ♂
05 20 ☽ ⊻ ♄
06 58 ☽ ⊻ ♇
07 20 ☽ ∥ ♀
08 01 ☿ Q ♅
09 50 ☽ □ ♆
13 12 ☽ ⊼ ☉
15 43 ☽ ⊥ ♄
20 48 ☽ △ ♃

07 Monday
00 26 ☽ ± ☉
01 26 ☽ △ ☿
05 00 ☽ ∠ ♅
07 13 ☽ ∠ ♄
07 42 ☽ △ ♂
17 02 ☽ Q ♆
17 12 ☽ bQ ☉
19 23 ☽ ☍ ♀
19 30 ☽ # ♀
20 32 ☽ ⊥ ♅
22 50 ☽ bQ ♃

08 Tuesday
06 51 ☽ ⊻ ♅
08 53 ☽ ✶ ♄
10 45 ☽ ☌ ♇
11 42 ☉ ∥ ♅
13 41 ☽ ✶ ♆
13 41 ♀ ± ♅
14 23 ☽ ± ♃
15 53 ♂ ∥ ♆
20 58 ☽ △ ☉

09 Wednesday
00 43 ☽ ⊼ ♃
05 19 ☽ Q ♄
09 27 ☽ # ♇
10 24 ☽ □ ☿
14 11 ☽ □ ♂
15 23 ☽ ∠ ♆
18 42 ☿ ∥ ♆

10 Thursday
02 10 ☽ ⊼ ♀
06 46 ☽ ⊥ ♆
07 58 ♀ ∠ ♃
10 10 ☽ ☌ ♅
11 42 ☽ ∥ ☉
11 52 ☽ □ ♄
12 31 ♂ ∠ ♆
13 03 ☽ ± ♀
13 58 ☽ ⊻ ♇
16 51 ☽ ∥ ♅
17 00 ☽ ⊻ ♆
20 25 ☽ Q ☿
22 40 ☽ Q ♂

11 Friday
00 10 ☽ ⊥ ♇
04 07 ☽ □ ☉
04 12 ☽ ☍ ♃
05 19 ☉ □ ♃
05 22 ☽ bQ ♀
13 52 ☿ ∠ ♆
15 28 ☽ ∠ ♇
19 05 ☽ ✶ ☿
19 52 ☽ # ♃
20 16 ☽ ✶ ♂

12 Saturday
00 27 ☽ # ♄
08 29 ☽ △ ♀
13 06 ☽ Q ☉
13 13 ☽ ⊻ ♅
14 38 ☽ △ ♄
16 57 ☽ ✶ ♇
19 48 ☿ ☌ ♂
20 07 ☽ ☌ ♆
23 15 ☽ ∠ ♂
23 26 ☽ ∠ ☿
23 26 ☽ ⊥ ♅

13 Sunday
07 33 ☽ ⊼ ♃
11 06 ☽ ✶ ☉
13 21 ☽ Q ♇
14 45 ☽ ∠ ♅
15 27 ☽ ⊥ ♂
16 01 ☽ bQ ♄
16 29 ☽ ⊥ ☿
17 51 ☽ ± ♃

14 Monday
02 17 ☽ ⊻ ♂
03 53 ☽ ⊻ ☿
07 17 ☽ ± ♄
09 19 ☽ bQ ♃
12 08 ☿ ∥ ♂
14 41 ☽ ∠ ☉
14 45 ☽ □ ♀
16 22 ☽ ✶ ♅
16 45 ☽ # ♄
17 30 ☽ ⊼ ♄
18 43 ☉ ∠ ♀
19 51 ♅ St R
20 04 ☽ □ ♇
21 13 ☽ # ♃
23 22 ☽ ⊻ ♆

15 Tuesday
07 20 ☽ ⊥ ☉
09 42 ☽ ⊥ ♆
11 14 ☽ △ ♃
13 00 ☽ Q ♅
18 30 ☽ ⊻ ☉
19 39 ♀ ⊼ ♅
23 36 ☽ Q ♀

16 Wednesday
01 17 ☽ ∠ ♆
01 58 ☽ ∥ ♅
08 59 ☽ ☌ ♂
11 48 ♀ △ ♄
13 42 ☽ ☌ ☿
16 35 ☉ Q ♆
20 19 ☽ □ ♅
21 11 ☽ ☍ ♄
21 48 ☽ ✶ ♀
23 30 ☽ ∥ ☉

17 Thursday
00 03 ☽ △ ♇
03 35 ☽ ✶ ♆
09 54 ☽ # ♇
16 10 ☽ □ ♃
16 13 ☉ bQ ♇
23 53 ☽ # ♀

18 Friday
01 00 ☽ Q ♆
02 04 ☽ ∠ ♀
02 44 ☽ bQ ♇
03 37 ☽ ☌ ☉
14 00 ♀ ☍ ♇
17 43 ☽ ⊻ ♂
19 03 ☽ ± ♇
19 22 ☽ ⊥ ♀

19 Saturday
01 03 ☽ Q ♃
01 44 ☿ □ ♅
02 13 ☽ △ ♅
02 17 ☽ ⊻ ☿
02 50 ☽ ⊼ ♄
05 26 ☽ ⊥ ♂
06 04 ☽ ⊼ ♇
06 42 ☿ ☍ ♄
07 03 ☽ ⊻ ♀
09 55 ☽ □ ♆
13 53 ☽ ± ♄
14 54 ☽ ⊥ ☿
23 18 ☽ ∠ ♂
23 34 ☽ ✶ ♃

20 Sunday
06 14 ☽ bQ ♅
06 42 ☽ bQ ♄
10 00 ☿ △ ♇
10 12 ☽ ∠ ☿
16 02 ☽ ⊻ ☉
23 26 ☽ ± ♅
23 50 ☿ # ♄

21 Monday
04 29 ☽ ∠ ♃
04 36 ☽ ⊥ ☉
04 40 ☉ # ♇
05 51 ☽ ✶ ♂
11 04 ☽ ⊼ ♅
11 23 ☽ △ ♄
13 09 ♀ △ ♆
14 01 ☿ # ♃
15 04 ☽ ☍ ♇
18 50 ☿ ✶ ♆
19 17 ☽ △ ♆
19 21 ☽ ✶ ☿
19 39 ☽ ☌ ♀
22 13 ☽ ⊥ ♃
23 02 ☿ ✶ ♀
23 45 ☽ ∠ ☉

22 Tuesday
06 58 ☽ Q ♂
10 11 ☽ ⊻ ♃
10 37 ☉ # ♀
15 40 ☽ # ☉
16 58 ☽ ∥ ♀
19 40 ☽ ∥ ♇
22 43 ☽ Q ☿

23 Wednesday
01 06 ☽ bQ ♆
08 22 ☽ ✶ ☉
16 38 ♀ ∥ ♇
19 14 ☽ ± ♆
21 33 ☽ □ ♂
22 46 ☽ ☍ ♅
22 46 ☽ □ ♄

24 Thursday
00 04 ♄ □ ♅
02 52 ☽ ⊼ ♇
06 30 ♀ ⊥ ♃
07 24 ☽ ⊼ ♆
07 44 ☽ # ♅
10 53 ☽ Q ☉
11 10 ☽ ⊻ ♀
14 49 ♂ ☍ ♄
15 01 ☽ ± ♇
15 59 ♂ □ ♅
16 46 ☽ □ ☿
22 15 ♂ # ♄
23 18 ☽ ☌ ♃
23 42 ☽ # ☿

25 Friday
00 02 ☽ ⊥ ♀
09 17 ☽ bQ ♇
11 07 ☽ Q ♄
18 25 ♂ # ♃
19 17 ☽ ∠ ♀
23 52 ☽ # ♂

26 Saturday
01 02 ☽ ∥ ♃
02 50 ☽ □ ☉
05 15 ☽ ∥ ♄
10 39 ☉ ± ♇
11 17 ☽ ✶ ♄
11 35 ☽ ⊼ ♅
14 40 ☽ △ ♂
15 36 ☽ △ ♇
19 26 ☿ □ ♃
20 16 ☽ ☍ ♆
23 38 ☽ ± ♅

27 Sunday
03 04 ☽ ✶ ♀
04 42 ♂ △ ♇
12 29 ☽ ⊻ ♃
14 59 ☽ △ ☿
17 07 ☽ ∠ ♄
17 33 ☽ bQ ♅
20 24 ☿ ∥ ♅
22 38 ☽ bQ ♂

28 Monday
00 29 ☽ ⊥ ♃
03 58 ☽ Q ♀
10 42 ☽ ⊥ ♄
15 28 ☽ ∥ ♄
17 25 ☽ ± ♂
18 14 ☽ ∠ ♃
18 38 ☽ ∥ ♃
19 46 ☽ △ ☉
22 16 ☽ ⊻ ♄
22 49 ☽ △ ♅

MARCH 1977

Date	Sidereal time h m s	Sun ☉ ° ' "	Moon ☽ ° ' "	Moon ☽ 24.00 ° ' "	Mercury ☿ ° '	Venus ♀ ° '	Mars ♂ ° '	Jupiter ♃ ° '	Saturn ♄ ° '	Uranus ♅ ° '	Neptune ♆ ° '	Pluto ♇ ° '
01	22 36 42	10 ♓ 47 39	18 ♋ 35 10	24 ♋ 56 52	28 ♒ 34	20 ♈ 41	15 ♒ 30	24 ♉ 25	11 ♌ 22	11 ♏ 41	16 ♐ 04	13 ♎ 40
02	22 40 39	11 47 49	01 ♌ 23 52	07 ♌ 56 20	00 ♓ 17	21 10	16 17	24 34	11 R 19	11 R 41	16 05	13 R 38
03	22 44 35	12 47 58	14 ♌ 34 19	21 ♌ 17 44	02 00	21 37	17 03	24 42	11 15	11 40	16 05	13 37
04	22 48 32	13 48 04	28 ♌ 06 23	04 ♍ 59 57	03 44	22 03	17 50	24 50	11 11	11 39	16 06	13 35
05	22 52 28	14 48 09	11 ♍ 58 03	19 ♍ 00 09	05 30	22 27	18 37	24 59	11 07	11 38	16 06	13 34
06	22 56 25	15 48 12	26 ♍ 05 41	03 ♎ 14 00	07 16	22 48	19 24	25 08	11 04	11 37	16 06	13 33
07	23 00 22	16 48 12	10 ♎ 24 28	17 ♎ 36 25	09 04	23 08	20 10	25 16	11 00	11 36	16 07	13 31
08	23 04 18	17 48 11	24 ♎ 49 12	02 ♏ 02 13	10 53	23 27	20 57	25 25	10 57	11 35	16 07	13 30
09	23 08 15	18 48 09	09 ♏ 14 55	16 ♏ 26 49	12 43	23 43	21 44	25 35	10 53	11 33	16 07	13 28
10	23 12 11	19 48 05	23 ♏ 37 29	00 ♐ 46 35	14 34	23 57	22 31	25 44	10 50	11 32	16 08	13 27
11	23 16 08	20 47 59	07 ♐ 53 47	14 ♐ 58 53	16 26	24 08	23 17	25 53	10 47	11 31	16 08	13 25
12	23 20 04	21 47 52	22 ♐ 01 41	29 ♐ 02 02	18 19	24 18	24 04	26 03	10 44	11 30	16 08	13 24
13	23 24 01	22 47 43	05 ♑ 59 47	12 ♑ 54 52	20 14	24 25	24 51	26 12	10 41	11 28	16 08	13 22
14	23 27 57	23 47 32	19 ♑ 47 08	26 ♑ 36 32	22 09	24 30	25 38	26 22	10 38	11 27	16 09	13 21
15	23 31 54	24 47 19	03 ♒ 22 57	10 ♒ 06 18	24 06	24 33	26 25	26 32	10 35	11 25	16 09	13 19
16	23 35 51	25 47 05	16 ♒ 46 29	23 ♒ 23 24	26 03	24 R 33	27 12	26 42	10 32	11 24	16 09	13 18
17	23 39 47	26 46 49	29 ♒ 56 59	06 ♓ 27 10	28 ♓ 01	24 31	27 58	26 52	10 30	11 22	16 09	13 16
18	23 43 44	27 46 31	12 ♓ 53 52	19 ♓ 17 05	00 ♈ 00	24 27	28 45	27 02	10 27	11 21	16 R 09	13 14
19	23 47 40	28 46 12	25 ♓ 36 50	01 ♈ 53 07	02 00	24 20	29 ♒ 32	27 12	10 25	11 19	16 09	13 13
20	23 51 37	29 ♓ 45 50	08 ♈ 06 04	14 ♈ 15 46	04 00	24 10	00 ♓ 19	27 23	10 22	11 18	16 09	13 11
21	23 55 33	00 ♈ 45 26	20 ♈ 22 25	26 ♈ 26 15	06 01	23 58	01 06	27 33	10 20	11 16	16 09	13 10
22	23 59 30	01 45 00	02 ♉ 27 31	08 ♉ 26 34	08 02	23 43	01 53	27 44	10 18	11 14	16 08	13 08
23	00 03 26	02 44 32	14 ♉ 23 46	20 ♉ 19 31	10 03	23 26	02 39	27 54	10 16	11 12	16 08	13 06
24	00 07 23	03 44 02	26 ♉ 14 18	02 ♊ 08 35	12 03	23 07	03 26	28 05	10 14	11 11	16 08	13 05
25	00 11 20	04 43 29	08 ♊ 02 55	13 ♊ 57 51	14 03	22 45	04 13	28 16	10 12	11 09	16 08	13 03
26	00 15 16	05 42 54	19 ♊ 53 58	25 ♊ 51 52	16 03	22 21	05 00	28 27	10 10	11 07	16 08	13 01
27	00 19 13	06 42 17	01 ♋ 52 09	07 ♋ 55 24	18 00	21 55	05 47	28 38	10 09	11 05	16 07	13 00
28	00 23 09	07 41 38	14 ♋ 02 15	20 ♋ 13 14	19 57	21 27	06 34	28 49	10 07	11 03	16 07	12 58
29	00 27 06	08 40 56	26 ♋ 28 54	02 ♌ 49 44	21 51	20 57	07 20	29 01	10 06	11 01	16 07	12 56
30	00 31 02	09 40 12	09 ♌ 16 09	15 ♌ 48 31	23 42	20 25	08 07	29 12	10 05	10 59	16 06	12 55
31	00 34 59	10 ♈ 39 26	22 ♌ 27 04	29 ♌ 11 55	25 ♈ 31	19 ♈ 52	08 ♓ 54	29 ♉ 23	10 ♌ 03	10 ♏ 57	16 ♐ 06	12 ♎ 53

Date	Moon True ☊ ° '	Moon Mean ☊ ° '	Moon ☽ Latitude ° '
01	25 ♎ 20	26 ♎ 44	05 S 07
02	25 R 13	26 41	05 07
03	25 06	26 38	04 52
04	24 58	26 34	04 19
05	24 52	26 31	03 31
06	24 47	26 28	02 29
07	24 45	26 25	01 S 17
08	24 D 44	26 22	00 00
09	24 44	26 19	01 N 18
10	24 46	26 15	02 31
11	24 47	26 12	03 34
12	24 R 48	26 09	04 23
13	24 47	26 06	04 56
14	24 45	26 03	05 12
15	24 41	25 59	05 09
16	24 37	25 56	04 50
17	24 32	25 53	04 15
18	24 28	25 50	03 28
19	24 24	25 47	02 32
20	24 22	25 44	01 28
21	24 21	25 40	00 N 22
22	24 D 21	25 37	00 S 45
23	24 22	25 34	01 48
24	24 24	25 31	02 47
25	24 26	25 28	03 39
26	24 27	25 24	04 21
27	24 28	25 21	04 52
28	24 R 28	25 18	05 11
29	24 27	25 15	05 16
30	24 25	25 12	05 06
31	24 ♎ 24	25 ♎ 09	04 S 40

DECLINATIONS

Date	Sun ☉ ° '	Moon ☽ ° '	Mercury ☿ ° '	Venus ♀ ° '	Mars ♂ ° '	Jupiter ♃ ° '	Saturn ♄ ° '	Uranus ♅ ° '	Neptune ♆ ° '	Pluto ♇ ° '
01	07 S 31	17 N 05	13 S 59	12 N 43	17 S 13	18 N 11	18 N 19	14 S 55	21 S 14	10 N 43
02	07 08	14 51	13 24	13 03	17 00	18 13	18 20	14 54	21 14	10 44
03	06 45	11 48	12 47	13 22	16 46	18 15	18 21	14 54	21 14	10 45
04	06 22	08 04	12 09	13 40	16 32	18 17	18 22	14 54	21 14	10 46
05	05 59	03 N 49	11 29	13 57	16 18	18 20	18 23	14 53	21 14	10 46
06	05 36	00 S 44	10 48	14 14	16 04	18 22	18 24	14 53	21 14	10 47
07	05 13	05 18	10 06	14 30	15 49	18 24	18 25	14 53	21 14	10 48
08	04 49	09 36	09 23	14 45	15 34	18 27	18 26	14 52	21 14	10 49
09	04 26	13 20	08 39	14 59	15 20	18 29	18 27	14 52	21 14	10 50
10	04 02	16 14	07 53	15 13	15 05	18 31	18 28	14 52	21 14	10 50
11	03 39	18 06	07 06	15 25	14 50	18 34	18 29	14 51	21 14	10 51
12	03 15	18 49	06 18	15 36	14 34	18 36	18 30	14 51	21 14	10 52
13	02 52	18 22	05 28	15 47	14 19	18 39	18 31	14 50	21 14	10 53
14	02 28	16 50	04 38	15 56	14 03	18 41	18 31	14 50	21 14	10 54
15	02 04	14 23	03 47	16 04	13 48	18 44	18 32	14 50	21 14	10 54
16	01 41	11 12	02 54	16 11	13 32	18 46	18 33	14 49	21 14	10 55
17	01 17	07 30	02 01	16 17	13 16	18 49	18 34	14 49	21 14	10 56
18	00 53	03 S 30	01 06	16 21	13 00	18 51	18 34	14 48	21 14	10 57
19	00 29	00 N 35	00 S 11	16 24	12 44	18 54	18 35	14 48	21 14	10 57
20	00 S 06	04 34	00 N 44	16 26	12 27	18 56	18 36	14 47	21 13	10 58
21	00 N 18	08 18	01 40	16 26	12 11	18 59	18 36	14 46	21 13	10 59
22	00 42	11 38	02 37	16 25	11 54	19 01	18 37	14 46	21 13	11 00
23	01 05	14 26	03 34	16 23	11 38	19 04	18 37	14 45	21 13	11 01
24	01 29	16 36	04 31	16 19	11 21	19 06	18 38	14 45	21 13	11 01
25	01 53	18 03	05 27	16 13	11 04	19 09	18 38	14 44	21 13	11 02
26	02 16	18 43	06 24	16 06	10 47	19 12	18 39	14 44	21 13	11 03
27	02 40	18 33	07 19	15 58	10 30	19 14	18 39	14 43	21 13	11 03
28	03 03	17 32	08 14	15 48	10 13	19 17	18 40	14 42	21 13	11 04
29	03 27	15 41	09 08	15 36	09 56	19 20	18 40	14 42	21 13	11 05
30	03 50	13 01	10 01	15 23	09 38	19 22	18 41	14 41	21 13	11 06
31	04 N 13	09 N 37	10 N 52	15 N 09	09 S 21	19 N 25	18 N 41	14 S 41	21 S 13	11 N 06

ZODIAC SIGN ENTRIES

Date	h	m	Planets
02	08	09	☿ ♓
02	09	25	☽ ♌
04	15	19	☽ ♍
06	18	34	☽ ♎
08	20	37	☽ ♏
10	22	42	☽ ♐
13	01	40	☽ ♑
15	06	00	☽ ♒
17	12	06	☽ ♓
18	11	56	☿ ♈
19	20	23	☽ ♈
20	02	19	♂ ♓
20	17	42	☉ ♈
22	07	05	☽ ♉
24	19	39	☽ ♊
27	08	16	☽ ♋
29	18	40	☽ ♌

LATITUDES

Date	Mercury ☿ ° '	Venus ♀ ° '	Mars ♂ ° '	Jupiter ♃ ° '	Saturn ♄ ° '	Uranus ♅ ° '	Neptune ♆ ° '	Pluto ♇ ° '
01	02 S 09	05 N 01	01 S 05	00 S 43	00 N 59	00 N 27	01 N 29	17 N 29
04	02 09	05 29	01 06	00 43	00 59	00 27	01 29	17 30
07	02 05	05 56	01 07	00 42	00 59	00 27	01 30	17 30
10	01 57	06 23	01 08	00 41	00 59	00 27	01 30	17 31
13	01 45	06 48	01 09	00 41	00 59	00 27	01 30	17 32
16	01 27	07 11	01 09	00 40	00 59	00 27	01 30	17 33
19	01 05	07 31	01 10	00 40	00 59	00 27	01 30	17 33
22	00 37	07 47	01 11	00 39	00 59	00 28	01 30	17 34
25	00 S 06	07 57	01 11	00 38	00 59	00 28	01 30	17 34
28	00 N 28	08 02	01 12	00 38	00 59	00 28	01 30	17 35
31	01 N 04	07 N 59	01 S 12	00 S 37	00 N 59	00 N 28	01 N 31	17 N 35

DATA

Julian Date	2443204
Delta T	+48 seconds
Ayanamsa	23° 32' 26"
Synetic vernal point	05° ♓ 34' 34"
True obliquity of ecliptic	23° 26' 24"

MOON'S PHASES, APSIDES AND POSITIONS ☽

Date	h	m	Phase	Longitude ° '	Eclipse Indicator
05	17	13	○	15 ♍ 01	
12	11	35	☾	21 ♐ 47	
19	18	33	●	29 ♓ 02	
27	22	27	☽	07 ♋ 08	

Day	h	m		
08	23	16	Perigee	
24	21	54	Apogee	
06	08	10	OS	
12	14	38	Max dec	18° S 50'
19	08	37	ON	
26	19	20	Max dec	18° N 46'

LONGITUDES

Date	Chiron ⚷ ° '	Ceres ⚳ ° '	Pallas ⚴ ° '	Juno ⚵ ° '	Vesta ⚶ ° '	Black Moon Lilith ° '
01	28 ♈ 35	08 ♎ 50	16 ♌ 01	29 ♏ 19	10 ♋ 46	24 ♉ 03
11	29 ♈ 03	07 ♎ 41	14 ♌ 21	29 ♏ 56	11 ♋ 21	25 ♉ 09
21	29 ♈ 35	04 ♎ 50	13 ♌ 22	00 ♐ 51	12 ♋ 34	26 ♉ 16
31	00 ♉ 09	02 ♎ 32	13 ♌ 27	29 ♏ 33	14 ♋ 23	27 ♉ 23

...meris data is given at 12.00 UT and the Moon's longitude is additionally given for 24.00 UT

ASPECTARIAN

01 Tuesday
00 55 ☽ bQ ☿
02 37 ☽ □ ♇
04 50 ☿ Q ♆
05 44 ☽ ⊼ ♂
07 12 ☽ ⊼ ♆
09 45 ☽ # ♂
13 15 ☿ bQ ♇
16 08 ☽ □ ♀
18 35 ☽ ± ♆
20 42 ☽ ± ☿
23 08 ☽ ✶ ♃

02 Wednesday
01 01 ☉ ⊼ ♄
02 44 ☽ bQ ☉
05 39 ♂ ✶ ♆
09 09 ☉ △ ♅
09 36 ☽ ⊼ ☿
11 24 ☽ bQ ♆
11 29 ☽ # ♅
12 26 ☽ Q ♇
20 46 ☽ ± ☉
20 58 ☿ # ♀
21 35 ☽ Q ♃

03 Thursday
01 32 ☽ ∥ ♀
03 14 ☽ # ☿
06 02 ☽ ☌ ♄
06 46 ☽ □ ♅
08 33 ☽ ⊼ ☉
09 10 ☉ Q ♃
10 17 ☽ ✶ ♇
14 43 ☽ △ ♆
16 44 ☽ ☍ ♂
19 13 ☽ ∥ ♇

04 Friday
00 59 ☽ △ ♀
06 12 ☽ □ ♃
07 05 ☉ ⊼ ♇
12 51 ☽ ∠ ♇
14 41 ☽ Q ♅
22 51 ☽ # ☉
23 14 ☽ ☍ ☿

05 Saturday
04 00 ☽ bQ ♀
04 27 ☽ ⊥ ♇
10 33 ☽ ⊻ ♄
11 25 ☽ ✶ ♅
14 44 ☽ ⊻ ♇
17 13 ☽ ☍ ☉
19 04 ☽ □ ♆
19 51 ☽ ± ♀
20 46 ☽ ⊥ ♄

06 Sunday
00 00 ☽ ⊼ ♂
06 18 ☽ ⊼ ♀
10 21 ☽ △ ♃
10 45 ☽ ± ♂
11 57 ☽ ∠ ♄
12 42 ☿ # ♇
12 52 ☽ ∠ ♅
15 35 ☿ ± ♇
19 20 ☉ □ ♆
20 52 ☿ ∠ ♀

07 Monday
01 28 ☽ Q ♆
02 45 ☽ bQ ♂
03 58 ☽ ⊥ ♅
09 27 ☽ ⊼ ☿
11 32 ☽ ∥ ☉
11 47 ☽ bQ ♃
12 59 ☽ ✶ ♄
13 59 ☽ ⊻ ♅
16 32 ☉ ± ♄
17 11 ☽ ☌ ♇
20 53 ☽ ± ☿
21 31 ☽ ✶ ♆
23 27 ☽ ⊼ ☉
23 45 ☉ ⊥ ♀

08 Tuesday
02 36 ♃ ∥ ♄
02 56 ☽ ± ♃
05 12 ☽ △ ♂
08 54 ☽ Q ♄
09 40 ☽ ☍ ♀
10 11 ☽ ± ☉
10 54 ☽ ∥ ☿
12 49 ☿ ⊼ ♄
13 01 ☽ ⊼ ♃
14 01 ☽ bQ ☿
19 19 ☽ # ♇
21 02 ☿ △ ♅
22 29 ☽ ∠ ♆
23 52 ♀ # ♅

09 Wednesday
02 16 ☽ bQ ☉
13 28 ☽ ⊥ ♆
14 43 ☽ □ ♄
15 50 ☽ ☌ ♅
18 37 ☽ △ ☿
19 01 ☽ ⊻ ♇
21 43 ☿ ⊼ ♇
23 28 ☽ ⊻ ♆
23 39 ☽ ∥ ♅

10 Thursday
00 13 ☿ Q ♃
01 46 ☽ # ♀
02 17 ☽ ∥ ♂
05 01 ☽ ⊥ ♇
05 07 ☽ △ ☉
05 14 ♀ # ♂
10 02 ☽ □ ♂
12 32 ☽ ⊼ ♀
15 34 ☽ ☍ ♃
20 04 ☽ ∠ ♇
22 45 ☽ ± ♀

11 Friday
08 08 ☿ □ ♆
09 20 ♂ ∥ ♅
14 08 ☽ bQ ♀
16 19 ☿ ± ♄
16 52 ☽ △ ♄
18 05 ☽ Q ♂
18 07 ☽ ⊻ ♅
20 21 ☽ # ♄
21 20 ☽ ✶ ♇
23 09 ☽ # ♃

12 Saturday
01 58 ☽ ☌ ♆
04 17 ☽ ⊥ ♅
04 42 ☽ □ ☿
11 35 ☽ □ ☉
11 40 ☿ ⊥ ♀
15 42 ☽ ✶ ♂
15 55 ☽ △ ♀
17 45 ☽ Q ♇
18 19 ☽ bQ ♄
18 57 ☽ ⊼ ♃
19 38 ☽ ∠ ♅
20 28 ♀ ✶ ♂

13 Sunday
04 33 ☽ # ♃
05 23 ☽ ± ♃
08 29 ☽ # ♄
09 44 ☽ ± ♄
16 29 ☽ Q ☿
19 05 ☽ ∠ ♂
20 05 ☽ ⊼ ♄
20 58 ☽ Q ☉
21 08 ☽ bQ ♃
21 29 ☽ ✶ ♅

14 Monday
00 46 ☽ □ ♇
05 38 ☽ ⊻ ♆
11 43 ☽ ⊥ ♂
16 08 ☽ ⊥ ♆
16 50 ☽ ✶ ☿
18 25 ☽ Q ♅
19 35 ☽ ✶ ☉
20 20 ☽ □ ♀
21 19 ☽ # ♀
22 54 ☽ ⊻ ♂
23 43 ☽ △ ♃

15 Tuesday
06 04 ☉ ⊻ ♀
08 01 ☽ ∠ ♆
08 12 ☽ ∥ ♅
16 38 ♂ □ ♃
17 16 ☽ ∥ ♂
17 41 ☿ ⊻ ♀
23 56 ☽ ∠ ☿

16 Wednesday
00 21 ☽ ∠ ☉
00 49 ☽ ☍ ♄
02 20 ☽ □ ♅
03 01 ♀ St R
04 24 ☽ Q ♀
05 25 ☉ ☌ ☿
05 44 ☽ △ ♇
05 55 ☿ bQ ♄
06 22 ☉ bQ ♄
10 52 ☽ ✶ ♆
13 54 ☽ # ♇
16 14 ☿ bQ ♅
17 53 ☽ ⊥ ☉
18 58 ☽ ⊥ ☿
20 37 ☿ ✶ ♃

17 Thursday
02 06 ☽ ✶ ♀
02 28 ☉ bQ ♅
05 43 ☽ ⊻ ☉
06 16 ☽ □ ♃
07 50 ☽ ⊻ ☿
08 09 ☽ ☌ ♂
08 41 ☽ Q ♆
08 55 ☽ bQ ♇
11 02 ☿ ⊻ ♂
14 26 ☉ ✶ ♃
17 20 ♂ Q ♆
20 44 ♂ bQ ♇

18 Friday
01 29 ☽ ± ♇
05 36 ☽ ∠ ♀
07 27 ☽ ⊼ ♄
07 36 ♆ St R
09 07 ☽ △ ♅
12 38 ☽ ⊼ ♇
16 03 ☽ Q ♃
18 05 ☽ □ ♆
18 39 ☽ ± ♄
22 18 ☉ ∥ ☿
22 20 ☽ ⊥ ♀

19 Saturday
05 04 ☽ ∥ ☉
06 10 ☽ ∥ ☿
09 34 ☽ ⊻ ♀
10 09 ☽ # ☿
11 32 ☽ # ☉
11 37 ☽ bQ ♄
13 21 ☽ bQ ♅
15 05 ☽ ✶ ♃
18 33 ☽ ☌ ☉
19 59 ☽ ⊻ ♂

20 Sunday
00 21 ☉ # ☿
02 34 ☽ ☌ ☿
06 35 ☽ ± ♅
08 19 ☽ ⊥ ♂
16 24 ☽ △ ♄
18 11 ☽ ⊼ ♅
20 26 ☽ ∠ ♃
21 52 ☽ ☍ ♇

21 Monday
03 01 ☽ ∠ ♂
03 09 ☿ ± ♅
03 41 ☽ △ ♆
14 22 ☽ ⊥ ♃
18 58 ☽ ☌ ♀

22 Tuesday
02 26 ☽ ⊻ ♃
07 08 ☽ ∥ ♇
08 57 ☿ ⊥ ♂
09 22 ☽ bQ ♆
10 27 ☽ ⊻ ☉
10 45 ☽ ✶ ♂
14 00 ☽ # ♂
23 34 ☽ ⊥ ☉

23 Wednesday
01 27 ☽ ⊻ ☿
02 19 ☉ ⊻ ♂
03 25 ☽ ± ♆
03 42 ☽ □ ♄
05 35 ☽ ☍ ♅
09 24 ☽ ⊼ ♇
12 34 ☽ Q ♂
14 34 ☿ △ ♄
15 09 ☽ # ♅
15 31 ☽ ⊼ ♆
16 01 ☽ ⊥ ☿
19 23 ☽ ∠ ☉
21 30 ☽ ± ♇

24 Thursday
01 38 ☿ ⊼ ♅
05 50 ☽ ⊻ ♀
08 29 ☽ ∥ ♀
10 51 ♃ bQ ♇
14 00 ☽ ∠ ☿
15 44 ☽ bQ ♇
15 49 ☽ ☌ ♃
16 03 ☽ Q ♄
17 41 ☽ ⊥ ♀

25 Friday
00 04 ☿ ☍ ♇
01 34 ☿ ∠ ♃
03 40 ☽ □ ♂
04 35 ♃ Q ♄
04 38 ☽ ✶ ☉
11 26 ☽ ∠ ♀
14 53 ♂ # ♇
16 22 ☽ ✶ ♄
18 16 ☽ ⊼ ♅
21 54 ☉ ± ♅
22 07 ☽ △ ♇

26 Saturday
02 39 ☽ ✶ ☿
04 23 ☽ ☍ ♆
06 23 ☽ ± ♅
06 59 ☽ ∥ ♄
07 11 ☽ Q ☉
08 14 ♀ ⊥ ♃
13 03 ☿ △ ♆
16 47 ☽ ✶ ♀
22 35 ☽ ∠ ♄

27 Sunday
00 28 ☽ bQ ♅
05 27 ☽ ⊻ ♃
07 13 ☽ ∥ ♄
07 34 ☽ Q ☿
15 56 ☽ Q ♀
16 31 ☽ ⊥ ♄
17 35 ☽ ⊥ ♃
20 18 ☽ △ ♂
22 27 ☽ □ ☉

28 Monday
04 21 ☽ ⊻ ♄
06 10 ☽ △ ♅
09 55 ☽ □ ♇
10 00 ♀ ∠ ♂
11 34 ☽ ∠ ♃
16 03 ☽ ⊼ ♆

29 Tuesday
00 06 ♂ ± ♇
01 33 ☽ □ ☿
01 50 ☽ □ ♀
03 05 ☿ ☌ ♀
03 33 ☽ bQ ♂
03 39 ☽ ± ♆
12 50 ☽ ∥ ♀
16 52 ☽ ✶ ♃
20 26 ☽ Q ♇
20 46 ☽ bQ ♆
21 43 ☽ # ♅
21 48 ☽ ± ♂
22 51 ☿ ∠ ♂

30 Wednesday
04 16 ☿ # ♂
04 39 ☿ ⊥ ♃
09 44 ☽ ⊼ ♂
12 48 ☽ △ ☉
13 29 ☽ ☌ ♄
15 09 ☽ □ ♅
15 37 ☽ Q ♃
18 41 ☽ ✶ ♇
21 40 ☉ △ ♄

31 Thursday
00 32 ☽ △ ♆
02 07 ☽ ∥ ♇
05 20 ☽ ∥ ☿
07 33 ☽ △ ♀
13 53 ☽ # ♂
18 11 ☽ bQ ☉
18 19 ☽ △ ☿
18 52 ☉ ⊼ ♅
19 02 ☿ ∥ ♇
21 39 ☽ ∠ ♇
23 32 ☽ Q ♅

APRIL 1977

LONGITUDES

Date	Sidereal time h m s	Sun ☉	Moon ☽	Moon ☽ 24.00	Mercury ☿	Venus ♀	Mars ♂	Jupiter ♃	Saturn ♄	Uranus ♅	Neptune ♆	Pluto ♇
01	00 38 55	11 ♈ 38 38	06 ♍ 03 07	13 ♍ 00 30	27 ♈ 17	19 ♈ 17	09 ♓ 41	29 ♉ 35	10 ♌ 02	10 ♏ 55	16 ♐ 05	12 ♎ 51
02	00 42 52	12 37 47	20 ♍ 03 48	27 ♍ 12 37	28 ♈ 59	18 R 42	10 28	29 47	10 R 01	10 R 53	16 R 05	12 R 50
03	00 46 49	13 36 54	04 ♎ 26 22	11 ♎ 44 24	00 ♉ 37	18 05	11 14	29 ♉ 58	10 00	10 51	16 04	12 48
04	00 50 45	14 35 59	19 ♎ 05 52	26 ♎ 29 56	02 11	17 28	12 01	00 ♊ 10	10 00	10 48	16 04	12 46
05	00 54 42	15 35 01	03 ♏ 55 38	11 ♏ 22 00	03 40	16 50	12 48	00 22	09 59	10 46	16 03	12 45
06	00 58 38	16 34 03	18 ♏ 48 06	26 ♏ 13 01	05 04	16 12	13 35	00 34	09 58	10 44	16 03	12 43
07	01 02 35	17 33 02	03 ♐ 35 55	10 ♐ 56 03	06 23	15 34	14 21	00 46	09 58	10 42	16 02	12 41
08	01 06 31	18 31 59	18 ♐ 12 47	25 ♐ 25 34	07 37	14 56	15 08	00 58	09 58	10 39	16 01	12 40
09	01 10 28	19 30 55	02 ♑ 34 03	09 ♑ 37 54	08 45	14 20	15 55	01 10	09 57	10 37	16 01	12 38
10	01 14 24	20 29 49	16 ♑ 36 59	23 ♑ 31 11	09 47	13 44	16 42	01 22	09 57	10 35	16 00	12 36
11	01 18 21	21 28 42	00 ♒ 20 32	07 ♒ 05 06	10 43	13 09	17 28	01 34	09 D 57	10 33	15 59	12 35
12	01 22 18	22 27 32	13 ♒ 45 02	20 ♒ 20 28	11 33	12 35	18 15	01 47	09 57	10 30	15 59	12 33
13	01 26 14	23 26 21	26 ♒ 51 39	03 ♓ 18 46	12 17	12 03	19 02	01 59	09 57	10 28	15 58	12 31
14	01 30 11	24 25 08	09 ♓ 42 04	16 ♓ 01 47	12 55	11 32	19 48	02 12	09 58	10 25	15 57	12 30
15	01 34 07	25 23 53	22 ♓ 18 09	28 ♓ 31 25	13 26	11 04	20 35	02 24	09 58	10 23	15 56	12 28
16	01 38 04	26 22 37	04 ♈ 41 47	10 ♈ 49 29	13 52	10 37	21 21	02 37	09 59	10 21	15 55	12 26
17	01 42 00	27 21 18	16 ♈ 54 45	22 ♈ 57 48	14 11	10 12	22 08	02 49	09 59	10 18	15 54	12 25
18	01 45 57	28 19 58	28 ♈ 58 50	04 ♉ 58 07	14 23	09 50	22 55	03 02	10 00	10 16	15 53	12 23
19	01 49 53	29 ♈ 18 36	10 ♉ 55 52	16 ♉ 52 21	14 30	09 30	23 41	03 15	10 01	10 13	15 52	12 22
20	01 53 50	00 ♉ 17 12	22 ♉ 47 50	28 ♉ 42 37	14 R 30	09 12	24 28	03 28	10 02	10 11	15 51	12 20
21	01 57 47	01 15 46	04 ♊ 37 02	10 ♊ 31 24	14 25	08 57	25 14	03 41	10 03	10 08	15 50	12 18
22	02 01 43	02 14 18	16 ♊ 26 07	22 ♊ 21 34	14 14	08 44	26 01	03 53	10 04	10 06	15 49	12 17
23	02 05 40	03 12 48	28 ♊ 18 11	04 ♋ 16 27	13 58	08 33	26 47	04 06	10 05	10 03	15 48	12 15
24	02 09 36	04 11 15	10 ♋ 16 50	16 ♋ 19 49	13 37	08 25	27 34	04 19	10 07	10 01	15 47	12 14
25	02 13 33	05 09 41	22 ♋ 25 58	28 ♋ 35 46	13 12	08 19	28 20	04 33	10 08	09 58	15 46	12 12
26	02 17 29	06 08 05	04 ♌ 49 46	11 ♌ 08 29	12 43	08 16	29 06	04 46	10 10	09 56	15 45	12 11
27	02 21 26	07 06 26	17 ♌ 32 25	24 ♌ 02 01	12 11	08 D 15	29 ♓ 53	04 59	10 11	09 53	15 44	12 09
28	02 25 22	08 04 45	00 ♍ 37 41	07 ♍ 19 44	11 36	08 16	00 ♈ 39	05 12	10 13	09 50	15 43	12 08
29	02 29 19	09 03 02	14 ♍ 08 25	21 ♍ 03 51	10 59	08 20	01 25	05 25	10 15	09 48	15 41	12 06
30	02 33 16	10 ♉ 01 18	28 ♍ 06 00	05 ♎ 14 42	10 ♉ 20	08 ♈ 26	02 ♈ 12	05 ♊ 39	10 ♌ 17	09 ♏ 45	15 ♐ 40	12 ♎ 05

Date	Moon True ☊	Moon Mean ☊	Moon ☽ Latitude
01	24 ♎ 22	25 ♎ 05	03 S 57
02	24 R 20	25 02	02 59
03	24 19	24 59	01 48
04	24 19	24 56	00 S 29
05	24 D 19	24 53	00 N 53
06	24 19	24 50	02 12
07	24 20	24 46	03 21
08	24 21	24 43	04 17
09	24 21	24 40	04 55
10	24 21	24 37	05 15
11	24 R 21	24 34	05 16
12	24 21	24 30	05 00
13	24 21	24 27	04 28
14	24 D 21	24 24	03 44
15	24 21	24 21	02 49
16	24 21	24 18	01 47
17	24 21	24 15	00 N 41
18	24 R 21	24 11	00 S 26
19	24 21	24 08	01 31
20	24 20	24 05	02 32
21	24 20	24 02	03 25
22	24 19	23 59	04 11
23	24 17	23 56	04 45
24	24 16	23 52	05 08
25	24 16	23 49	05 17
26	24 15	23 46	05 12
27	24 D 15	23 43	04 51
28	24 16	23 40	04 16
29	24 17	23 36	03 25
30	24 ♎ 18	23 ♎ 33	02 S 20

DECLINATIONS

Date	Sun ☉	Moon ☽	Mercury ☿	Venus ♀	Mars ♂	Jupiter ♃	Saturn ♄	Uranus ♅	Neptune ♆	Pluto ♇
01	04 N 36	05 N 37	11 N 41	14 N 53	09 S 03	19 N 27	18 N 41	14 S 40	21 S 13	11 N 07
02	04 59	01 N 12	12 28	14 36	08 46	19 30	18 41	14 39	21 13	11 08
03	05 22	03 S 25	13 14	14 18	08 28	19 33	18 42	14 39	21 12	11 08
04	05 45	07 55	13 57	13 58	08 10	19 35	18 42	14 38	21 12	11 09
05	06 08	12 00	14 37	13 38	07 53	19 38	18 42	14 37	21 12	11 09
06	06 31	15 18	15 15	13 17	07 35	19 41	18 42	14 36	21 12	11 10
07	06 53	17 35	15 50	12 55	07 17	19 43	18 42	14 36	21 12	11 11
08	07 16	18 39	16 22	12 32	06 59	19 46	18 42	14 35	21 12	11 11
09	07 38	18 30	16 52	12 09	06 41	19 48	18 42	14 34	21 12	11 12
10	08 00	17 12	17 18	11 45	06 23	19 51	18 42	14 34	21 12	11 12
11	08 22	14 56	17 42	11 22	06 04	19 54	18 42	14 33	21 12	11 13
12	08 44	11 55	18 02	10 58	05 46	19 56	18 42	14 32	21 11	11 14
13	09 06	08 21	18 19	10 34	05 28	19 59	18 42	14 31	21 11	11 14
14	09 28	04 29	18 33	10 11	05 10	20 02	18 42	14 31	21 11	11 15
15	09 49	00 S 28	18 44	09 48	04 51	20 04	18 42	14 30	21 11	11 15
16	10 11	03 N 30	18 52	09 25	04 33	20 07	18 42	14 29	21 11	11 16
17	10 32	07 17	18 57	09 03	04 15	20 09	18 42	14 28	21 11	11 16
18	10 53	10 43	18 59	08 42	03 56	20 12	18 42	14 28	21 11	11 17
19	11 14	13 40	18 57	08 21	03 38	20 15	18 41	14 27	21 11	11 17
20	11 34	16 02	18 52	08 01	03 19	20 17	18 41	14 26	21 10	11 18
21	11 55	17 42	18 44	07 42	03 01	20 20	18 41	14 25	21 10	11 18
22	12 15	18 36	18 34	07 24	02 42	20 22	18 40	14 24	21 10	11 18
23	12 35	18 41	18 20	07 07	02 24	20 25	18 40	14 24	21 10	11 19
24	12 55	17 56	18 04	06 51	02 05	20 27	18 40	14 23	21 10	11 19
25	13 15	16 22	17 45	06 35	01 47	20 30	18 39	14 22	21 10	11 20
26	13 34	14 01	17 24	06 21	01 28	20 32	18 39	14 21	21 10	11 20
27	13 53	10 57	17 01	06 09	01 10	20 35	18 38	14 21	21 09	11 20
28	14 12	07 16	16 36	05 57	00 51	20 37	18 38	14 20	21 09	11 21
29	14 31	03 05	16 10	05 46	00 33	20 40	18 37	14 19	21 09	11 21
30	14 N 49	01 S 23	15 N 41	05 N 36	00 S 14	20 N 42	18 N 37	14 S 18	21 S 09	11 N 21

ZODIAC SIGN ENTRIES

Date	h m	Planets
01	01 25	☽ ♍
03	02 46	☿ ♉
03	04 39	☽ ♎
03	15 42	♃ ♊
05	05 40	☽ ♏
07	06 08	☽ ♐
09	07 40	☽ ♑
11	11 24	☽ ♒
13	17 49	☽ ♓
16	02 52	☽ ♈
18	14 02	☽ ♉
20	04 57	☉ ♉
21	02 37	☽ ♊
23	15 25	☽ ♋
26	02 43	☽ ♌
27	15 46	♂ ♈
28	10 52	☽ ♍
30	15 13	☽ ♎

LATITUDES

Date	Mercury ☿	Venus ♀	Mars ♂	Jupiter ♃	Saturn ♄	Uranus ♅	Neptune ♆	Pluto ♇
01	01 N 16	07 N 56	01 S 12	00 S 37	00 N 59	00 N 28	01 N 31	17 N 35
04	01 49	07 43	01 13	00 37	00 59	00 28	01 31	17 35
07	02 19	07 22	01 13	00 36	00 59	00 28	01 31	17 35
10	02 42	06 53	01 13	00 36	00 59	00 28	01 31	17 35
13	02 56	06 19	01 13	00 35	00 59	00 28	01 31	17 35
16	03 01	05 41	01 13	00 35	00 59	00 28	01 31	17 34
19	02 54	04 59	01 13	00 34	00 59	00 28	01 31	17 34
22	02 34	04 17	01 13	00 34	00 59	00 28	01 31	17 33
25	02 03	03 35	01 13	00 34	00 59	00 28	01 32	17 33
28	01 21	02 54	01 13	00 33	00 59	00 28	01 32	17 32
31	00 N 32	02 N 15	01 S 12	00 S 33	00 N 59	00 N 28	01 N 32	17 N 31

DATA

Julian Date	2443235
Delta T	+48 seconds
Ayanamsa	23° 32' 28"
Synetic vernal point	05° ♓ 34' 31"
True obliquity of ecliptic	23° 26' 24"

MOON'S PHASES, APSIDES AND POSITIONS ☽

Date	h m	Phase	Longitude	Eclipse Indicator
04	04 09	○	14 ♎ 17	partial
10	19 15	☾	20 ♑ 48	
18	10 35	●	28 ♈ 17	Annular
26	14 42	☽	06 ♌ 15	

Day	h m		
05	20 57	Perigee	
21	12 05	Apogee	
02	18 15	0S	
08	20 52	Max dec	18° S 44'
15	14 47	0N	
23	02 26	Max dec	18° N 45'
30	04 40	0S	

LONGITUDES

Date	Chiron ⚷	Ceres ⚳	Pallas ⚴	Juno ⚵	Vesta ⚶	Black Moon Lilith ⚸
01	00 ♉ 13	02 ♎ 19	13 ♌ 31	29 ♏ 28	14 ♋ 35	27 ♉ 29
11	00 ♉ 49	00 ♎ 16	14 ♌ 29	28 ♏ 24	16 ♋ 56	28 ♉ 36
21	01 ♉ 26	28 ♍ 41	16 ♌ 07	26 ♏ 50	19 ♋ 41	29 ♉ 43
31	02 ♉ 02	27 ♍ 44	18 ♌ 18	24 ♏ 52	22 ♋ 48	00 ♊ 49

All ephemeris data is given at 12.00 UT and the Moon's longitude is additionally given for 24.00 UT

ASPECTARIAN

h m Aspects

01 Friday
00 31 ☽ □ ♃
09 04 ☽ bQ ♀
11 14 ☽ ± ☉
13 24 ☽ ⊥ ♇
17 13 ☽ ∥ ☉
18 39 ☽ ☍ ♂
18 53 ☽ ⊻ ♄
20 23 ☽ ⚹ ♅
22 24 ☽ ⊼ ☉
22 47 ♂ ⊼ ♄
23 43 ☽ ⊻ ♇
23 59 ☽ ± ♀

02 Saturday
00 16 ☽ bQ ☿
05 09 ☽ ⊥ ♄
05 15 ☽ □ ♆
07 31 ♀ # ♅
09 47 ☽ ⊼ ♀
16 41 ☉ ☍ ♇
17 34 ☽ ± ☿
20 20 ☽ ∠ ♄
21 45 ☽ ∠ ♅

03 Sunday
00 22 ♂ △ ♅
01 00 ☿ ⊻ ♃
04 30 ☽ △ ♃
04 53 ☽ ⊼ ☿
11 24 ☽ Q ♆
12 40 ☽ ⊥ ♅
18 49 ☿ bQ ♆
21 09 ☽ ⚹ ♄
22 30 ☽ ⊻ ♅
23 11 ☽ # ☉
23 48 ☽ ⊼ ♂

04 Monday
01 42 ☽ ☌ ♇
02 26 ♀ ⊥ ♂
04 09 ☽ ☍ ☉
05 31 ☽ bQ ♃
07 04 ☽ ⚹ ♆
09 27 ☽ ☍ ♀
10 09 ☽ ± ♂
12 40 ☿ ∥ ♀
13 17 ☽ ∥ ♂
16 42 ☽ Q ♄
20 20 ☽ ± ♃

05 Tuesday
01 33 ☽ bQ ♂
05 16 ☉ ∠ ♃
06 10 ☽ ⊼ ♃
06 44 ☽ # ♇
07 22 ☽ ∠ ♆
10 23 ♂ ⊼ ♇
11 32 ☽ ☍ ☿
12 02 ☿ # ♅
21 46 ☽ □ ♄
21 53 ☽ ⊥ ♆
22 02 ☽ # ♀
23 01 ☽ ☌ ♅
23 25 ☉ △ ♆

06 Wednesday
02 12 ☽ ⊻ ♇
03 06 ☽ △ ♂
06 23 ☽ ∥ ♅
06 29 ☉ ☌ ♀
07 33 ☽ ⊻ ♆
07 58 ☽ ⊼ ♀
08 08 ☽ ⊼ ☉
11 26 ☽ # ☿
11 52 ☽ ⊥ ♇
17 16 ☽ ± ♀
17 47 ♀ △ ♆
18 31 ☽ ± ☉

07 Thursday
02 24 ☽ ∠ ♇
06 20 ♀ ∠ ♃
07 16 ☽ bQ ♀
07 19 ☽ ☍ ♃
10 10 ☽ bQ ☉
16 59 ☽ ⊼ ☿
22 24 ☽ △ ♄
23 35 ☽ ⊻ ♅

08 Friday
01 49 ☉ # ♂
02 51 ☽ ⚹ ♇
03 44 ☽ ± ☿
06 37 ☽ □ ♂
06 49 ☽ △ ♀
08 23 ☽ ☌ ♆
08 41 ♀ ⊻ ♂
09 26 ☽ ⊥ ♅
12 34 ☽ △ ☉
15 33 ☽ # ♄
19 57 ☽ bQ ☿
22 42 ☽ Q ♇
23 13 ☽ bQ ♄

09 Saturday
00 21 ☽ ∠ ♅
02 09 ☽ # ♄
09 36 ☽ ⊼ ♃
13 16 ♂ ± ♄
14 21 ☽ ± ♄
14 24 ☽ Q ♂
15 03 ♂ □ ♆
19 54 ☽ ± ♃
23 21 ☽ △ ☿

10 Sunday
00 33 ☽ ⊼ ♄
01 39 ☽ ⚹ ♅
05 06 ☽ □ ♇
07 13 ☽ □ ♀
10 51 ☽ # ☿
10 56 ☽ ⊻ ♆
11 34 ☽ bQ ♃
12 08 ☽ ⚹ ♂
16 09 ☿ □ ♄
17 19 ☿ ± ♆
19 15 ☽ □ ☉
21 20 ☽ ⊥ ♆
22 20 ☽ Q ♅

11 Monday
05 41 ♄ St D
07 27 ☿ ☍ ♅
13 09 ☽ ∠ ♆
13 22 ☽ Q ♀
14 13 ☽ △ ♃
15 22 ☽ ∥ ♅
16 00 ☽ ∠ ♂
20 36 ♀ ∥ ♇

12 Tuesday
05 09 ☽ ☍ ♄
05 35 ☽ Q ☉
06 09 ☽ □ ♅
07 47 ☽ □ ☿
09 07 ☽ ⊥ ♂
09 50 ☽ △ ♇
09 58 ☽ ⚹ ♀
13 37 ♀ ☍ ♇
16 02 ☽ ⚹ ♆
16 50 ☽ # ♇
19 31 ☽ # ♀
20 41 ☽ ⊻ ♂

13 Wednesday
05 10 ☽ ⚹ ☉
07 16 ☿ ⊻ ♀
07 36 ☽ # ☉
12 20 ☽ ∠ ♀
13 13 ☽ bQ ♇
14 02 ☽ Q ♆
18 42 ☽ Q ☿
20 06 ☿ ⊼ ♇
21 41 ☽ □ ♃

14 Thursday
04 28 ☽ ⊥ ♀
05 58 ☽ ± ♇
07 31 ☽ ∥ ♂
11 25 ☽ ∠ ☉
12 29 ☽ ⊼ ♄
13 22 ☽ △ ♅
15 21 ☽ ⊻ ♀
17 16 ☽ ⊼ ♇
18 22 ☽ ⚹ ☿
23 50 ☽ □ ♆
23 53 ☽ ± ♄

15 Friday
04 26 ♂ Q ♃
05 58 ☽ ⊥ ☉
06 24 ☿ ∥ ♄
08 17 ☽ Q ♃
08 29 ☽ ☌ ♂
11 09 ☉ ∥ ♀
17 08 ☽ bQ ♄
17 55 ☽ bQ ♅
18 28 ☽ ⊻ ☉

16 Saturday
00 17 ☽ ∠ ☿
07 52 ☽ ⚹ ♃
11 19 ☽ ± ♅
17 59 ☽ # ♂
18 22 ☽ ⊥ ☿
19 20 ☉ ⊥ ♃
22 21 ☽ △ ♄
23 01 ☽ ⊼ ♅
23 13 ☽ ☌ ♀

17 Sunday
03 09 ☽ ☍ ♇
05 44 ♀ ⊼ ♅
06 28 ☽ ⊻ ☿
10 01 ☽ △ ♆
13 50 ☽ ∠ ♃
22 53 ☽ ∥ ♀
23 04 ☽ ⊻ ♂

18 Monday
01 29 ♀ △ ♄
08 02 ☽ ⊥ ♃
10 35 ☽ ☌ ☉
11 51 ☽ ⊥ ♂
13 26 ☽ ∥ ☉
15 49 ☽ bQ ♆
16 18 ☽ ∥ ♇
20 16 ☽ ⊻ ♃

19 Tuesday
07 10 ☽ ∠ ♂
09 11 ☽ ⊻ ♀
09 52 ☽ ± ♆
10 09 ☽ □ ♄
10 34 ☽ ☍ ♅
14 53 ☽ ⊼ ♇
16 04 ☉ ∥ ♇
19 12 ☽ # ♅
19 13 ☽ ☌ ☿
20 59 ☽ ⊥ ♀
21 58 ☽ ⊼ ♆

20 Wednesday
02 09 ☿ St R
02 59 ☽ ± ♇
14 47 ☽ ∠ ♀
15 37 ☽ ⚹ ♂
21 11 ☽ bQ ♇
22 38 ☽ Q ♄

21 Thursday
01 46 ☉ bQ ♆
04 34 ☽ ⊻ ☉
05 58 ♂ bQ ♄
08 56 ☉ ⊥ ♂
09 01 ♂ bQ ♅
10 03 ☽ ☌ ♃
17 42 ☽ Q ♂
17 52 ☽ ⊥ ☉
20 37 ☽ ⚹ ♀
21 26 ☿ ∥ ♄
23 03 ☽ ⚹ ♄
23 10 ☽ ⊼ ♅

22 Friday
03 35 ☽ △ ♇
07 37 ☽ ⊻ ☿
10 45 ☽ ☍ ♆
10 59 ☽ ∥ ☿
11 19 ☽ ± ♅
13 46 ☽ ∠ ☉
16 33 ☽ ∥ ♄
19 33 ☽ ⊥ ☿
20 34 ☽ Q ♀
22 25 ♄ □ ♅

23 Saturday
05 28 ☽ bQ ♅
05 30 ☽ ∠ ♄
08 44 ☽ □ ♂
12 40 ☽ ∥ ♄
13 19 ☽ ∠ ☿
22 45 ☽ ⚹ ☉
23 39 ☽ ⊥ ♄
23 53 ☽ ⊻ ♃

24 Sunday
07 57 ☽ ∥ ☿
08 19 ☽ □ ♀
11 28 ☽ △ ♅
11 40 ☽ ⊻ ♄
12 05 ☽ ⊥ ♃
15 52 ☽ □ ♇
16 21 ☉ ⊻ ♃
18 26 ☽ ⚹ ☿
22 54 ☽ ⊼ ♆

25 Monday
00 44 ☽ Q ☉
06 14 ☽ ∠ ♃
09 26 ☿ ∠ ♂
10 42 ☽ ± ♆
17 13 ☽ Q ☿

26 Tuesday
00 15 ☽ △ ♂
03 05 ☽ Q ♇
04 10 ☽ bQ ♆
08 54 ☽ # ♅
11 52 ☽ ⚹ ♃
14 42 ☽ □ ☉
15 27 ☽ ∥ ☉
18 32 ☽ △ ♀
21 40 ☽ □ ♅
22 10 ☽ ☌ ♄

27 Wednesday
01 56 ☽ ⚹ ♇
02 23 ☽ □ ☿
06 43 ☽ bQ ♂
08 38 ☽ △ ♆
09 12 ☽ ∥ ♇
09 49 ♀ St D
10 56 ☽ Q ♃
13 06 ☿ ⊼ ♇
22 34 ☽ bQ ♀

28 Thursday
00 27 ☽ ± ♂
05 40 ☽ ∠ ♇
06 58 ☽ Q ♅
12 03 ☽ ⊼ ♂
14 58 ☽ ± ♀
16 55 ☉ ⊻ ♀
20 13 ☽ ∥ ♀
20 21 ☽ □ ♃
21 21 ☉ # ♅
21 51 ☽ ⊥ ♇

29 Friday
01 43 ☽ ⊼ ♀
02 22 ☽ △ ☉
04 24 ☽ ⚹ ♅
05 09 ☽ ⊻ ♄
06 42 ☽ △ ☿
08 27 ☽ ⊻ ♇
14 42 ☽ □ ♆
15 41 ☽ ⊥ ♄

30 Saturday
02 47 ☽ # ♂
03 28 ☉ ± ♆
05 42 ☉ ☍ ♅
06 18 ☽ ∥ ♂
06 20 ☽ ∠ ♅
06 23 ☽ bQ ☉
07 13 ☽ ∠ ♄
07 31 ☽ bQ ☿
13 55 ☿ □ ♄
16 44 ☉ ☌ ☿
18 47 ☉ □ ♄
19 17 ☽ ☍ ♂
21 21 ☽ Q ♆
21 29 ☽ ⊥ ♅
22 02 ☽ ± ☿
22 41 ☽ ± ☉

MAY 1977

LONGITUDES

Date	Sidereal time h m s	Sun ☉ ° ' "	Moon ☽ ° ' "	Moon ☽ 24.00 ° ' "	Mercury ☿ ° '	Venus ♀ ° '	Mars ♂ ° '	Jupiter ♃ ° '	Saturn ♄ ° '	Uranus ♅ ° '	Neptune ♆ ° '	Pluto ♇ ° '
01	02 37 12	10 ♉ 59 31	12 ♎ 29 36	19 ♎ 50 13	09 ♉ 41	08 ♈ 34	02 ♈ 58	05 ♊ 52	10 ♌ 19	09 ♏ 43	15 ♐ 39	12 ♎ 03
02	02 41 09	11 57 42	27 ♎ 15 49	04 ♏ 45 33	09 R 03	08 44	03 44	06 05	10 21	09 R 40	15 R 38	12 R 02
03	02 45 05	12 55 51	12 ♏ 18 25	19 ♏ 53 18	08 25	08 57	04 30	06 19	10 24	09 38	15 36	12 01
04	02 49 02	13 53 59	27 ♏ 28 58	05 ♐ 04 13	07 48	09 11	05 16	06 32	10 26	09 35	15 35	11 59
05	02 52 58	14 52 06	12 ♐ 37 50	20 ♐ 08 40	07 14	09 28	06 02	06 46	10 29	09 33	15 34	11 58
06	02 56 55	15 50 10	27 ♐ 35 41	04 ♑ 58 00	06 42	09 46	06 49	06 59	10 31	09 30	15 32	11 57
07	03 00 51	16 48 14	12 ♑ 14 53	19 ♑ 25 47	06 13	10 06	07 35	07 13	10 34	09 28	15 31	11 55
08	03 04 48	17 46 16	26 ♑ 30 22	03 ♒ 28 26	05 47	10 29	08 21	07 26	10 37	09 25	15 30	11 54
09	03 08 45	18 44 16	10 ♒ 19 58	17 ♒ 05 03	05 26	10 52	09 07	07 40	10 40	09 23	15 28	11 53
10	03 12 41	19 42 15	23 ♒ 43 55	00 ♓ 16 52	05 09	11 18	09 53	07 54	10 43	09 20	15 27	11 52
11	03 16 38	20 40 13	06 ♓ 44 19	13 ♓ 06 40	04 56	11 45	10 38	08 07	10 46	09 18	15 25	11 50
12	03 20 34	21 38 10	19 ♓ 24 23	25 ♓ 37 56	04 47	12 14	11 24	08 21	10 49	09 15	15 24	11 49
13	03 24 31	22 36 05	01 ♈ 47 48	07 ♈ 54 26	04 43	12 44	12 10	08 35	10 52	09 13	15 23	11 48
14	03 28 27	23 33 59	13 ♈ 58 16	19 ♈ 59 44	04 D 44	13 16	12 56	08 48	10 56	09 10	15 21	11 47
15	03 32 24	24 31 52	25 ♈ 59 14	01 ♉ 57 04	04 49	13 49	13 42	09 02	10 59	09 08	15 20	11 46
16	03 36 20	25 29 43	07 ♉ 53 44	13 ♉ 49 24	04 59	14 23	14 27	09 16	11 03	09 05	15 18	11 45
17	03 40 17	26 27 34	19 ♉ 44 23	25 ♉ 38 59	05 13	14 58	15 13	09 30	11 06	09 03	15 17	11 43
18	03 44 14	27 25 22	01 ♊ 33 25	07 ♊ 27 57	05 32	15 35	15 59	09 44	11 10	09 01	15 15	11 42
19	03 48 10	28 23 10	13 ♊ 22 49	19 ♊ 18 16	05 55	16 13	16 44	09 57	11 14	08 58	15 14	11 41
20	03 52 07	29 ♉ 20 56	25 ♊ 14 32	01 ♋ 11 53	06 22	16 52	17 30	10 11	11 18	08 56	15 12	11 40
21	03 56 03	00 ♊ 18 40	07 ♋ 10 36	13 ♋ 10 58	06 54	17 33	18 15	10 25	11 22	08 54	15 10	11 39
22	04 00 00	01 16 23	19 ♋ 13 19	25 ♋ 17 59	07 29	18 14	19 01	10 39	11 26	08 51	15 09	11 38
23	04 03 56	02 14 05	01 ♌ 25 23	07 ♌ 35 53	08 09	18 56	19 46	10 53	11 30	08 49	15 07	11 38
24	04 07 53	03 11 45	13 ♌ 49 54	20 ♌ 07 54	08 52	19 39	20 32	11 07	11 34	08 47	15 06	11 37
25	04 11 49	04 09 23	26 ♌ 30 20	02 ♍ 57 38	09 39	20 23	21 17	11 21	11 39	08 44	15 04	11 36
26	04 15 46	05 07 00	09 ♍ 30 15	16 ♍ 08 34	10 30	21 08	22 02	11 35	11 43	08 42	15 03	11 35
27	04 19 43	06 04 35	22 ♍ 52 58	29 ♍ 43 42	11 24	21 54	22 47	11 49	11 48	08 40	15 01	11 34
28	04 23 39	07 02 09	06 ♎ 40 59	13 ♎ 44 53	12 21	22 41	23 33	12 03	11 52	08 38	14 59	11 33
29	04 27 36	07 59 42	20 ♎ 55 18	28 ♎ 11 59	13 22	23 28	24 18	12 16	11 57	08 36	14 58	11 33
30	04 31 32	08 57 13	05 ♏ 34 31	13 ♏ 02 14	14 26	24 16	25 03	12 30	12 02	08 34	14 56	11 32
31	04 35 29	09 ♊ 54 43	20 ♏ 34 19	28 ♏ 09 44	15 ♉ 33	25 ♈ 05	25 ♈ 48	12 ♊ 44	12 ♌ 07	08 ♏ 32	14 ♐ 55	11 ♎ 31

Moon nodes, Moon latitude and DECLINATIONS

Date	Moon True ☊ ° '	Moon Mean ☊ ° '	Moon ☽ Latitude ° '	Sun ☉ ° '	Moon ☽ ° '	Mercury ☿ ° '	Venus ♀ ° '	Mars ♂ ° '	Jupiter ♃ ° '	Saturn ♄ ° '	Uranus ♅ ° '	Neptune ♆ ° '	Pluto ♇ ° '
01	24 ♎ 19	23 ♎ 30	01 S 05	15 N 08	05 S 56	15 N 13	05 N 28	00 N 04	20 N 45	18 N 36	14 S 17	21 S 09	11 N 22
02	24 R 20	23 27	00 N 16	15 25	10 15	14 44	05 20	00 23	20 47	18 35	14 17	21 09	11 22
03	24 19	23 24	01 38	15 43	13 59	14 16	05 14	00 41	20 50	18 35	14 16	21 09	11 22
04	24 17	23 21	02 53	16 01	16 47	13 48	05 09	01 00	20 52	18 34	14 15	21 08	11 23
05	24 15	23 17	03 56	16 18	18 25	13 20	05 05	01 18	20 54	18 33	14 14	21 08	11 23
06	24 12	23 14	04 42	16 35	18 43	12 54	05 01	01 36	20 57	18 33	14 13	21 08	11 23
07	24 10	23 11	05 08	16 51	17 45	12 29	04 59	01 55	20 59	18 32	14 13	21 08	11 23
08	24 07	23 08	05 15	17 08	15 42	12 06	04 58	02 13	21 02	18 31	14 12	21 08	11 23
09	24 06	23 05	05 03	17 24	12 47	11 45	04 57	02 31	21 04	18 30	14 11	21 08	11 24
10	24 D 06	23 02	04 34	17 40	09 18	11 26	04 58	02 49	21 06	18 30	14 10	21 07	11 24
11	24 07	22 58	03 52	17 55	05 27	11 09	04 59	03 08	21 09	18 29	14 09	21 07	11 24
12	24 08	22 55	03 00	18 10	01 S 26	10 54	05 02	03 26	21 11	18 28	14 09	21 07	11 24
13	24 10	22 52	02 00	18 25	02 N 33	10 42	05 05	03 44	21 13	18 27	14 08	21 07	11 24
14	24 11	22 49	00 N 56	18 40	06 22	10 32	05 09	04 02	21 15	18 26	14 07	21 07	11 24
15	24 R 12	22 46	00 S 10	18 54	09 53	10 25	05 14	04 20	21 18	18 25	14 06	21 07	11 24
16	24 11	22 42	01 14	19 08	12 58	10 20	05 19	04 38	21 20	18 24	14 06	21 06	11 24
17	24 08	22 39	02 15	19 22	15 30	10 18	05 25	04 56	21 22	18 23	14 05	21 06	11 24
18	24 04	22 36	03 10	19 35	17 22	10 18	05 32	05 13	21 24	18 22	14 04	21 06	11 25
19	23 59	22 33	03 57	19 48	18 29	10 20	05 40	05 31	21 26	18 21	14 03	21 06	11 25
20	23 52	22 30	04 33	20 01	18 48	10 25	05 48	05 49	21 28	18 20	14 03	21 06	11 25
21	23 46	22 27	04 58	20 13	18 17	10 32	05 56	06 06	21 30	18 19	14 02	21 05	11 25
22	23 40	22 23	05 10	20 25	16 57	10 41	06 06	06 24	21 33	18 17	14 01	21 05	11 25
23	23 35	22 20	05 08	20 37	14 50	10 51	06 16	06 41	21 35	18 16	14 01	21 05	11 24
24	23 31	22 17	04 51	20 48	12 01	11 04	06 26	06 59	21 37	18 15	14 00	21 05	11 24
25	23 29	22 14	04 21	20 59	08 35	11 19	06 37	07 16	21 39	18 14	13 59	21 05	11 24
26	23 D 29	22 11	03 36	21 09	04 40	11 35	06 48	07 33	21 41	18 13	13 58	21 05	11 24
27	23 30	22 08	02 39	21 19	00 N 24	11 53	07 00	07 50	21 43	18 11	13 58	21 04	11 24
28	23 31	22 04	01 30	21 29	04 S 02	12 12	07 13	08 07	21 45	18 10	13 57	21 04	11 24
29	23 32	22 01	00 S 14	21 39	08 23	12 33	07 25	08 24	21 46	18 09	13 56	21 04	11 24
30	23 R 32	21 58	01 N 05	21 48	12 22	12 55	07 38	08 41	21 48	18 07	13 56	21 04	11 24
31	23 ♎ 29	21 ♎ 55	02 N 21	21 N 56	15 S 38	13 N 19	07 N 52	08 N 58	21 N 50	18 N 06	13 S 55	21 S 04	11 N 24

ZODIAC SIGN ENTRIES

Date	h	m	Planets
02	16	24	☽ ♏
04	15	59	☽ ♐
06	15	54	☽ ♑
08	18	00	☽ ♒
10	23	29	☽ ♓
13	08	29	☽ ♈
15	20	04	☽ ♉
18	08	50	☽ ♊
20	21	35	☽ ♋
21	04	14	☉ ♊
23	09	13	☽ ♌
25	18	31	☽ ♍
28	00	28	☽ ♎
30	02	57	☽ ♏

LATITUDES

Date	Mercury ☿ ° '	Venus ♀ ° '	Mars ♂ ° '	Jupiter ♃ ° '	Saturn ♄ ° '	Uranus ♅ ° '	Neptune ♆ ° '	Pluto ♇ ° '
01	00 N 32	02 N 15	01 S 12	00 S 33	00 N 59	00 N 28	01 N 32	17 N 31
04	00 S 20	01 38	01 12	00 32	00 59	00 27	01 32	17 31
07	01 10	01 04	01 11	00 32	00 59	00 27	01 32	17 30
10	01 55	00 32	01 11	00 32	00 59	00 27	01 32	17 29
13	02 32	00 N 03	01 10	00 31	00 59	00 27	01 32	17 28
16	03 01	00 S 23	01 10	00 31	00 59	00 27	01 32	17 27
19	03 20	00 47	01 09	00 31	00 59	00 27	01 32	17 25
22	03 31	01 08	01 08	00 30	00 59	00 27	01 32	17 24
25	03 34	01 27	01 07	00 30	00 59	00 27	01 32	17 23
28	03 30	01 44	01 06	00 30	00 59	00 27	01 32	17 21
31	03 S 20	01 S 59	01 S 05	00 S 30	00 N 59	00 N 27	01 N 32	17 N 20

DATA

Julian Date	2443265
Delta T	+48 seconds
Ayanamsa	23° 32' 31"
Synetic vernal point	05° ♓ 34' 28"
True obliquity of ecliptic	23° 26' 24"

MOON'S PHASES, APSIDES AND POSITIONS ☽

Date	h	m	Phase	Longitude ° '	Eclipse Indicator
03	13	03	○	12 ♏ 58	
10	04	08	☾	19 ♒ 23	
18	02	51	●	27 ♉ 03	
26	03	20	☽	04 ♍ 46	

Day	h	m		
04	04	56	Perigee	
18	18	02	Apogee	
06	05	37	Max dec	18° S 46'
12	20	35	0N	
20	09	07	Max dec	18° N 49'
27	14	10	0S	

LONGITUDES

Date	Chiron ⚷ ° '	Ceres ⚳ ° '	Pallas ⚴ ° '	Juno ⚵ ° '	Vesta ⚶ ° '	Black Moon Lilith ° '
01	02 ♉ 02	27 ♍ 44	18 ♌ 18	24 ♏ 52	22 ♋ 48	00 ♊ 49
11	02 ♉ 38	27 ♍ 27	20 ♌ 55	22 ♏ 40	26 ♋ 11	01 ♊ 56
21	03 ♉ 13	27 ♍ 51	23 ♌ 54	20 ♏ 25	29 ♋ 50	03 ♊ 03
31	03 ♉ 46	28 ♍ 52	27 ♌ 09	18 ♏ 21	03 ♌ 41	04 ♊ 10

...eris data is given at 12.00 UT and the Moon's longitude is additionally given for 24.00 UT

ASPECTARIAN

h m	Aspects
01 Sunday	
00 52	☽ △ ♃
05 27	☽ ☍ ♀
07 26	☽ ⊻ ♅
07 35	☽ ⊼ ☿
08 25	☽ ⚹ ♄
09 21	☽ ⊼ ☉
09 34	☽ # ♀
11 09	☿ ☍ ♅
11 17	☽ ☌ ♇
13 39	☿ ± ♆
14 54	☉ ∥ ☿
17 10	☽ ⚹ ♆
02 Monday	
00 18	☿ ⊥ ♂
01 53	☽ ⚼ ♃
04 04	☽ Q ♄
10 08	♂ ± ♅
13 45	☉ ⊼ ♇
16 36	☽ ± ♃
17 23	☽ ∠ ♆
18 45	☽ # ♇
20 48	☿ ⊻ ♀
22 55	☽ ⊼ ♂
03 Tuesday	
02 20	☽ ⊼ ♃
06 04	☽ ☍ ☿
06 35	☽ ⊼ ♀
07 43	☽ ⊥ ♆
07 46	☽ ☌ ♅
08 58	☽ □ ♄
08 59	☽ ± ♂
11 32	☽ ⊻ ♇
12 07	☿ # ♅
13 03	☽ ☍ ☉
13 50	☽ # ☿
14 04	☽ ∥ ♅
16 15	☽ ± ♀
17 13	☽ ⊻ ♆
21 01	☽ ⊥ ♇
04 Wednesday	
00 00	☽ ⚼ ♂
03 25	☽ # ☉
06 42	☽ ⚼ ♀
11 13	☽ ∠ ♇
14 05	♅ ⊥ ♆
05 Thursday	
00 59	☽ △ ♂
02 32	☽ ☍ ♃
03 43	☽ ⊼ ☿
06 52	☽ △ ♀
07 06	☽ ⊻ ♅
08 34	☽ △ ♄
10 57	☽ ⚹ ♇
12 55	☽ ± ☿
15 49	☽ ⊼ ☉
16 04	☽ # ♄
16 38	☽ ⊥ ♅
16 40	☽ ☌ ♆
17 46	♀ ⊼ ♅
06 Friday	
02 04	☽ ± ☉
02 34	☿ ⊻ ♃
02 48	☽ ⚼ ☿
04 49	☉ ⊼ ♆
06 07	☽ Q ♇
07 01	☽ ∠ ♅
08 38	☽ ⚼ ♄
09 32	☉ ⊥ ♀
09 50	☿ ⊻ ♂
17 38	☽ ⚼ ☉
19 46	☽ # ♄
19 50	♂ ⚹ ♃
23 18	☽ ± ♄
07 Saturday	
02 21	☽ △ ☿
03 33	☽ ⊼ ♃
03 51	☽ □ ♂
07 24	☽ ⚹ ♅
08 22	☽ □ ♀
09 12	☽ ⊼ ♄
11 28	☽ □ ♇
13 38	☽ ± ♃
17 26	☽ ⊻ ♆
20 09	☽ △ ☉
22 30	☽ # ☉
08 Sunday	
03 23	☽ Q ♅
03 30	☽ ⊥ ♆
04 58	☽ ⚼ ♃
11 42	☽ Q ♂
15 09	☉ ± ♇
15 28	☽ Q ♀
18 50	☽ ∠ ♆
21 40	♀ △ ♄
09 Monday	
01 11	☽ ∥ ♅
03 37	☽ □ ☿
07 14	☽ △ ♃
09 43	☽ ⚹ ♂
10 19	☽ □ ♅
12 35	☽ ☍ ♄
12 59	☽ ⚹ ♀
14 44	☽ △ ♇
19 54	♂ ⊼ ♅
20 20	☽ # ☿
21 06	☽ ⚹ ♆
21 59	☽ # ♇
10 Tuesday	
04 08	☽ □ ☉
10 57	☽ Q ☿
14 13	☽ ∠ ♂
14 24	☿ ∥ ♇
16 51	☽ ∠ ♀
17 42	☽ ⚼ ♇
18 46	☽ Q ♆
23 08	♃ # ♆
11 Wednesday	
07 50	☽ ⊥ ♂
08 40	☽ ⚹ ☿
10 05	☽ ⊥ ♀
10 19	☽ ± ♇
14 38	☽ □ ♃
14 43	☽ # ♀
15 55	☽ Q ☉
16 08	♂ △ ♄
16 15	♀ ☍ ♇
16 47	☽ △ ♅
19 36	☽ ⊼ ♄
19 48	☽ ⊻ ♂
21 35	☽ ⊼ ♇
21 47	☽ ⊻ ♀
12 Thursday	
00 56	☽ # ♂
04 22	☽ □ ♆
07 02	☽ ± ♄
12 43	☽ ∠ ☿
16 39	☽ ⚹ ☉
21 18	☽ ⚼ ♅
13 Friday	
00 25	☽ ⚼ ♄
00 38	♂ ☍ ♇
01 38	☽ Q ♃
06 00	☽ ⊥ ☿
14 21	☉ ∥ ♄
14 46	☽ ± ♅
17 43	☽ ⊻ ☿
19 53	☽ ∥ ♂
20 51	☿ St D
14 Saturday	
00 23	☽ ∠ ☉
01 35	☽ ⚹ ♃
02 31	☽ ⊼ ♅
04 00	☽ ∥ ♀
05 56	☽ △ ♄
07 40	☽ ☍ ♇
09 48	☽ ☌ ♂
10 31	☽ ☌ ♀
14 44	☽ △ ♆
19 47	☽ ⊥ ☉
15 Sunday	
08 00	☽ ∠ ♃
08 49	☽ ⊻ ☉
15 47	☽ ∥ ☿
20 22	♃ ⊼ ♅
20 42	☽ ⚼ ♆
23 22	☽ ∥ ♇
16 Monday	
02 04	♀ ☌ ♂
02 28	☽ ⊥ ♃
06 00	☽ ☌ ☿
14 24	☽ ☍ ♅
14 49	☽ ⊻ ♃
14 50	☽ ± ♆
18 24	☽ □ ♄
19 46	☽ ⊼ ♇
21 56	☽ # ♅
17 Tuesday	
01 49	☽ ⊻ ♀
02 12	☽ ⊻ ♂
02 58	☽ ⊼ ♆
07 55	☽ ± ♇
13 45	♂ △ ♆
14 39	☽ ⊥ ♀
15 13	☽ ⊥ ♂
18 28	☉ ⚼ ♇
23 26	♀ △ ♆
18 Wednesday	
02 10	☽ ⚼ ♇
02 51	☽ ☌ ☉
07 07	☽ Q ♄
09 55	☽ ∠ ♀
10 45	☽ ∠ ♂
20 19	☽ ⊻ ☿
19 Thursday	
03 05	☽ ⊼ ♅
04 55	☽ ☌ ♃
07 37	☽ ⚹ ♄
07 45	☽ ∥ ♄
08 35	☽ △ ♇
08 55	☽ ⊥ ☿
15 13	☽ ± ♅
15 44	☽ ☍ ♆
18 05	☽ ⚹ ♀
19 16	☽ ⚹ ♂
20 Friday	
03 51	☽ ∠ ☿
09 13	♀ ∥ ♂
09 22	☽ ⚼ ♅
10 38	☉ Q ♄
14 09	☽ ∠ ♄
19 45	☽ Q ♀
21 00	☽ ⊻ ☉
21 10	☽ Q ♂
21 Saturday	
08 21	☽ ⊥ ♄
10 07	☽ ⊥ ☉
11 25	☽ ⚹ ☿
11 26	☽ ∥ ♄
15 25	☽ △ ♅
18 37	☽ ⊻ ♃
20 25	☽ ⊻ ♄
20 57	☽ □ ♇
22 Sunday	
03 56	☽ ⊼ ♆
05 39	☽ ∠ ☉
06 48	☽ ⊥ ♃
09 55	☽ □ ♀
11 34	☽ □ ♂
12 34	☽ Q ☿
15 48	☽ ± ♆
23 Monday	
00 56	☽ ∠ ♃
08 29	☽ Q ♇
09 28	☽ ⚼ ♆
13 43	☽ ⚹ ☉
19 42	☽ # ♅
24 Tuesday	
01 51	☽ □ ☿
02 19	☽ □ ♅
06 41	☽ ⚹ ♃
07 39	☽ ☌ ♄
07 45	☽ ⚹ ♇
09 11	☿ ☍ ♅
14 25	☽ △ ♆
14 50	☽ Q ☉
16 36	☽ ∥ ♇
18 35	☽ ∥ ☿
18 52	☿ ± ♆
22 11	♄ ⚹ ♇
23 46	☽ △ ♀
25 Wednesday	
01 33	☽ △ ♂
05 57	☽ Q ♃
12 10	☽ ∠ ♇
12 26	☽ Q ♅
19 49	☽ ∥ ♂
20 26	☿ ∥ ♇
23 49	☽ ∥ ♀
26 Thursday	
01 36	☉ # ♆
03 20	☽ □ ☉
04 50	☽ ⊥ ♇
05 28	☽ ⚼ ♀
07 13	☽ ⚼ ♂
10 33	☽ ⚹ ♅
12 28	♃ △ ♇
13 56	☽ △ ☿
15 46	☽ ⊻ ♇
15 50	☽ □ ♃
16 03	☽ ⊻ ♄
22 00	☽ □ ♆
22 48	☽ ± ♀
27 Friday	
00 31	☽ ± ♂
02 54	☽ ⊥ ♄
09 46	♃ ⚹ ♄
10 09	☽ ⊼ ♀
11 50	☽ ⊼ ♂
13 23	☽ ∠ ♅
16 23	☿ ⊼ ♇
18 38	☽ ⚼ ☿
18 55	☽ ∠ ♄
23 05	☿ □ ♄
28 Saturday	
01 57	☿ ⊻ ♃
05 03	☽ ⊥ ♅
05 40	☽ Q ♆
11 23	☽ ± ☿
12 39	☽ △ ☉
15 19	☽ ⊻ ♅
20 17	☽ ☌ ♇
20 53	☽ ⚹ ♄
21 16	☽ △ ♃
22 22	☽ ⊼ ☿
29 Sunday	
02 04	☽ ⚹ ♆
06 17	☽ # ♀
12 06	☽ # ♂
15 41	☽ ⚼ ☉
16 28	☽ ☍ ♀
17 02	☽ Q ♄
17 53	☽ ☍ ♂
22 39	☽ ⚼ ♃
30 Monday	
02 33	☉ ⊼ ♅
02 51	☽ ∠ ♆
05 51	☽ # ♇
07 27	☽ ± ☉
13 32	☽ ± ♃
14 47	☉ ∥ ♃
16 08	☽ # ☿
16 49	☽ ☌ ♅
17 25	☽ ⊥ ♆
17 49	☽ ⊼ ☉
21 35	☽ ⊻ ♇
22 26	☽ □ ♄
22 41	☽ ∥ ♅
22 51	☿ ⊼ ♆
23 20	☽ ⊼ ♃
31 Tuesday	
03 00	☽ ⊻ ♆
03 21	☽ ☍ ☿
07 09	☽ ⊥ ♇
19 33	☽ ⊼ ♀
20 42	☽ ⊼ ♂
21 24	☽ ∠ ♇

JUNE 1977

LONGITUDES

Date	Sidereal time h m s	Sun ☉	Moon ☽	Moon ☽ 24.00	Mercury ☿	Venus ♀	Mars ♂	Jupiter ♃	Saturn ♄	Uranus ♅	Neptune ♆	Pluto ♇
01	04 39 25	10 ♊ 52 12	05 ♐ 47 19	13 ♐ 25 45	16 ♉ 43	25 ♈ 55	26 ♈ 33	12 ♊ 58	12 ♌ 11	08 ♏ 30	14 ♐ 53	11 ♎ 31
02	04 43 22	11 49 40	21 ♐ 03 42	28 ♐ 39 48	17 56	26 45	27 18	13 12	12 16	08 R 28	14 R 51	11 R 30
03	04 47 18	12 47 07	06 ♑ 12 45	13 ♑ 41 21	19 12	27 36	28 02	13 26	12 21	08 26	14 50	11 29
04	04 51 15	13 44 33	21 ♑ 04 36	28 ♑ 21 39	20 30	28 28	28 47	13 40	12 27	08 24	14 48	11 29
05	04 55 12	14 41 59	05 ♒ 31 54	12 ♒ 34 57	21 52	29 ♈ 20	29 ♈ 32	13 54	12 32	08 22	14 46	11 28
06	04 59 08	15 39 23	19 ♒ 30 37	26 ♒ 18 54	23 17	00 ♉ 13	00 ♉ 17	14 08	12 37	08 20	14 45	11 28
07	05 03 05	16 36 47	02 ♓ 59 59	09 ♓ 34 11	24 44	01 06	01 01	14 22	12 42	08 18	14 43	11 27
08	05 07 01	17 34 11	16 ♓ 01 55	22 ♓ 23 41	26 14	02 00	01 46	14 36	12 48	08 16	14 42	11 27
09	05 10 58	18 31 33	28 ♓ 40 03	04 ♈ 51 38	27 47	02 55	02 31	14 50	12 53	08 15	14 40	11 26
10	05 14 54	19 28 56	10 ♈ 59 02	17 ♈ 02 51	29 ♉ 23	03 50	03 15	15 04	12 59	08 13	14 38	11 26
11	05 18 51	20 26 18	23 ♈ 03 43	29 ♈ 02 11	01 ♊ 01	04 45	03 59	15 18	13 04	08 11	14 37	11 26
12	05 22 47	21 23 39	04 ♉ 58 49	10 ♉ 54 07	02 43	05 41	04 44	15 32	13 10	08 10	14 35	11 25
13	05 26 44	22 21 00	16 ♉ 48 34	22 ♉ 42 35	04 26	06 38	05 28	15 46	13 16	08 08	14 34	11 25
14	05 30 41	23 18 20	28 ♉ 36 32	04 ♊ 30 47	06 13	07 35	06 12	15 59	13 22	08 07	14 32	11 25
15	05 34 37	24 15 40	10 ♊ 25 38	16 ♊ 21 19	08 02	08 32	06 57	16 13	13 27	08 05	14 30	11 25
16	05 38 34	25 12 59	22 ♊ 18 05	28 ♊ 16 07	09 54	09 29	07 41	16 27	13 33	08 04	14 29	11 24
17	05 42 30	26 10 18	04 ♋ 15 37	10 ♋ 16 44	11 48	10 28	08 25	16 41	13 39	08 02	14 27	11 24
18	05 46 27	27 07 36	16 ♋ 19 37	22 ♋ 24 26	13 44	11 26	09 09	16 55	13 45	08 01	14 26	11 24
19	05 50 23	28 04 54	28 ♋ 31 22	04 ♌ 40 34	15 43	12 25	09 53	17 09	13 51	07 59	14 24	11 24
20	05 54 20	29 02 11	10 ♌ 52 16	17 ♌ 06 41	17 44	13 24	10 37	17 23	13 58	07 58	14 22	11 24
21	05 58 16	29 ♊ 59 27	23 ♌ 24 04	29 ♌ 44 43	19 48	14 24	11 20	17 36	14 04	07 57	14 21	11 24
22	06 02 13	00 ♋ 56 43	06 ♍ 08 56	12 ♍ 37 03	21 53	15 23	12 04	17 50	14 10	07 56	14 19	11 D 24
23	06 06 10	01 53 58	19 ♍ 09 26	25 ♍ 46 25	23 59	16 24	12 48	18 04	14 16	07 54	14 18	11 24
24	06 10 06	02 51 12	02 ♎ 28 21	09 ♎ 15 32	26 07	17 24	13 31	18 17	14 23	07 53	14 16	11 24
25	06 14 03	03 48 26	16 ♎ 08 13	23 ♎ 06 36	28 ♊ 16	18 25	14 15	18 31	14 29	07 52	14 15	11 24
26	06 17 59	04 45 39	00 ♏ 10 45	07 ♏ 20 38	00 ♋ 26	19 26	14 58	18 45	14 36	07 51	14 13	11 24
27	06 21 56	05 42 51	14 ♏ 36 00	21 ♏ 56 31	02 37	20 27	15 42	18 58	14 42	07 50	14 12	11 25
28	06 25 52	06 40 03	29 ♏ 21 35	06 ♐ 50 28	04 48	21 29	16 25	19 12	14 49	07 49	14 10	11 25
29	06 29 49	07 37 15	14 ♐ 22 12	21 ♐ 55 41	06 59	22 31	17 08	19 26	14 55	07 48	14 09	11 25
30	06 33 45	08 ♋ 34 27	29 ♐ 29 43	07 ♑ 03 00	09 ♋ 10	23 ♉ 33	17 ♉ 51	19 ♊ 39	15 ♌ 02	07 ♏ 48	14 ♐ 07	11 ♎ 25

Moon Node and Latitude

Date	Moon True ☊	Moon Mean ☊	Moon ☽ Latitude
01	23 ♎ 25	21 ♎ 52	03 N 28
02	23 R 19	21 48	04 20
03	23 11	21 45	04 54
04	23 04	21 42	05 07
05	22 58	21 39	05 00
06	22 53	21 36	04 35
07	22 50	21 33	03 55
08	22 49	21 29	03 05
09	22 D 50	21 26	02 06
10	22 51	21 23	01 N 03
11	22 R 51	21 20	00 S 01
12	22 50	21 17	01 05
13	22 47	21 13	02 05
14	22 42	21 10	02 59
15	22 34	21 07	03 46
16	22 24	21 04	04 23
17	22 13	21 01	04 48
18	22 01	20 58	05 01
19	21 50	20 54	05 01
20	21 40	20 51	04 46
21	21 32	20 48	04 18
22	21 28	20 45	03 36
23	21 25	20 42	02 42
24	21 24	20 39	01 39
25	21 D 25	20 35	00 S 28
26	21 R 24	20 32	00 N 46
27	21 23	20 29	02 00
28	21 19	20 26	03 07
29	21 12	20 23	04 02
30	21 ♎ 03	20 ♎ 19	04 N 40

DECLINATIONS

Date	Sun ☉	Moon ☽	Mercury ☿	Venus ♀	Mars ♂	Jupiter ♃	Saturn ♄	Uranus ♅	Neptune ♆	Pluto ♇
01	22 N 04	17 S 51	13 N 43	08 N 06	09 N 14	21 N 52	18 N 05	13 S 55	21 S 03	11 N 23
02	22 12	18 48	14 09	08 20	09 31	21 54	18 03	13 54	21 03	11 23
03	22 20	18 24	14 35	08 34	09 47	21 56	18 02	13 53	21 03	11 23
04	22 27	16 44	15 02	08 49	10 03	21 57	18 00	13 53	21 03	11 23
05	22 34	14 02	15 30	09 04	10 20	21 59	17 59	13 52	21 03	11 22
06	22 40	10 37	15 59	09 19	10 36	22 01	17 57	13 52	21 02	11 22
07	22 46	06 44	16 28	09 35	10 52	22 03	17 56	13 51	21 02	11 22
08	22 52	02 S 40	16 58	09 50	11 07	22 04	17 54	13 50	21 02	11 22
09	22 57	01 N 24	17 28	10 06	11 23	22 06	17 53	13 50	21 02	11 21
10	23 01	05 19	17 58	10 22	11 39	22 08	17 51	13 49	21 02	11 21
11	23 06	08 57	18 28	10 38	11 54	22 09	17 50	13 49	21 02	11 21
12	23 10	12 10	18 58	10 55	12 09	22 11	17 48	13 48	21 02	11 20
13	23 13	14 52	19 28	11 11	12 25	22 12	17 46	13 48	21 01	11 20
14	23 16	16 56	19 58	11 28	12 40	22 14	17 45	13 47	21 01	11 20
15	23 19	18 17	20 26	11 44	12 55	22 16	17 43	13 47	21 01	11 19
16	23 21	18 51	20 54	12 01	13 09	22 17	17 41	13 46	21 01	11 19
17	23 23	18 34	21 22	12 18	13 24	22 18	17 40	13 46	21 01	11 18
18	23 24	17 27	21 48	12 34	13 39	22 20	17 38	13 46	21 01	11 18
19	23 26	15 33	22 12	12 51	13 53	22 21	17 36	13 45	21 00	11 17
20	23 26	12 55	22 36	13 08	14 07	22 23	17 34	13 45	21 00	11 17
21	23 26	09 40	22 57	13 24	14 21	22 24	17 33	13 44	21 00	11 16
22	23 26	05 54	23 17	13 41	14 35	22 25	17 31	13 44	21 00	11 16
23	23 26	01 N 48	23 34	13 58	14 49	22 27	17 29	13 44	21 00	11 15
24	23 25	02 S 30	23 50	14 14	15 03	22 28	17 27	13 43	21 00	11 15
25	23 23	06 47	24 02	14 31	15 16	22 29	17 25	13 43	20 59	11 14
26	23 21	10 49	24 13	14 47	15 29	22 30	17 23	13 43	20 59	11 14
27	23 19	14 19	24 20	15 03	15 43	22 32	17 22	13 42	20 59	11 13
28	23 16	16 58	24 25	15 19	15 56	22 33	17 20	13 42	20 59	11 13
29	23 13	18 31	24 27	15 35	16 08	22 34	17 18	13 42	20 59	11 12
30	23 N 10	18 S 46	24 N 26	15 N 51	16 N 21	22 N 35	17 N 16	13 S 42	20 S 59	11 N 11

ZODIAC SIGN ENTRIES

Date	h	m	Planets
01	02	54	☽ ♐
03	02	07	☽ ♑
05	02	44	☽ ♒
06	03	00	♂ ♉
06	06	10	♀ ♉
07	06	35	☽ ♓
09	14	34	☽ ♈
10	21	07	☿ ♊
12	01	56	☽ ♉
14	14	50	☽ ♊
17	03	28	☽ ♋
19	14	53	☽ ♌
21	12	14	☉ ♋
22	00	29	☽ ♍
24	07	35	☽ ♎
26	07	07	☿ ♋
26	11	42	☽ ♏
28	13	02	☽ ♐
30	12	48	☽ ♑

LATITUDES

Date	Mercury ☿	Venus ♀	Mars ♂	Jupiter ♃	Saturn ♄	Uranus ♅	Neptune ♆	Pluto ♇
01	03 S 15	02 S 03	01 S 05	00 S 29	00 N 59	00 N 27	01 N 32	17 N 20
04	02 56	02 15	01 03	00 29	00 59	00 27	01 32	17 18
07	02 33	02 26	01 02	00 29	00 59	00 27	01 32	17 17
10	02 06	02 34	01 01	00 29	00 59	00 27	01 32	17 15
13	01 35	02 41	00 59	00 28	00 59	00 27	01 32	17 14
16	01 02	02 47	00 58	00 28	00 59	00 27	01 32	17 12
19	00 S 28	02 51	00 56	00 28	00 59	00 27	01 32	17 10
22	00 N 05	02 53	00 55	00 28	00 59	00 26	01 32	17 09
25	00 37	02 55	00 53	00 27	00 59	00 26	01 32	17 07
28	01 04	02 55	00 51	00 27	00 59	00 26	01 32	17 06
31	01 N 26	02 S 53	00 S 50	00 S 27	00 N 59	00 N 26	01 N 32	17 N 04

DATA

Julian Date	2443296
Delta T	+48 seconds
Ayanamsa	23° 32' 35"
Synetic vernal point	05° ♓ 34' 24"
True obliquity of ecliptic	23° 26' 23"

MOON'S PHASES, APSIDES AND POSITIONS ☽

Date	h	m	Phase	Longitude	Eclipse Indicator
01	20	31	○	11 ♐ 13	
08	15	07	☾	17 ♓ 42	
16	18	23	●	25 ♊ 28	
24	12	44	☽	02 ♎ 53	

Day	h	m		
01	15	06	Perigee	
14	21	24	Apogee	
29	23	41	Perigee	
02	16	42	Max dec	18° S 50'
09	03	41	0N	
16	16	07	Max dec	18° N 51'
23	22	07	0S	
30	04	22	Max dec	18° S 50'

LONGITUDES

Date	Chiron ⚷	Ceres ⚳	Pallas ⚴	Juno ⚵	Vesta ⚶	Black Moon Lilith ⚸
01	03 ♉ 49	29 ♍ 00	27 ♌ 30	18 ♏ 09	04 ♌ 05	04 ♊ 16
11	04 ♉ 18	00 ♎ 37	01 ♍ 00	16 ♏ 27	08 ♌ 07	05 ♊ 23
21	04 ♉ 45	02 ♎ 42	04 ♍ 42	15 ♏ 13	12 ♌ 19	06 ♊ 30
31	05 ♉ 07	05 ♎ 11	08 ♍ 33	14 ♏ 31	16 ♌ 39	07 ♊ 37

ASPECTARIAN

h m	Aspects
01 Wednesday	
05 33	☽ ± ♀
06 38	☽ ± ♂
15 26	☽ # ♄
16 15	☽ ⚺ ♅
20 31	☽ ☍ ☉
20 31	☽ ⚼ ♀
20 34	☉ ∠ ♀
20 59	☽ ✶ ♇
21 30	☽ ⚼ ♂
22 07	☽ △ ♄
22 36	☿ # ♅
23 27	☽ ☍ ♃
02 Thursday	
01 39	☽ ⊥ ♅
02 15	☽ ☌ ♆
03 44	☿ ± ♇
03 50	☉ △ ♇
06 38	☽ ⚻ ☿
15 46	☽ ∠ ♅
15 50	☽ Q ♇
16 55	☽ ± ☿
21 30	☽ △ ♀
21 51	☽ ⚼ ♄
22 21	☽ △ ♂
03 Friday	
00 14	☉ ✶ ♄
08 29	☽ ⚼ ☿
12 14	☽ ± ♄
15 32	☽ ✶ ♅
19 08	☽ # ♄
20 27	☽ □ ♇
21 55	☽ ⚻ ♄
23 16	☽ ⚻ ☉
23 47	☽ ⚻ ♃
04 Saturday	
01 49	☽ ⚺ ♆
06 28	♂ ∠ ♃
09 35	☉ ☌ ♃
09 40	☽ ± ♃
09 40	☽ ± ☉
10 53	☽ Q ♅
10 59	☽ △ ☿
11 33	☽ ⊥ ♆
17 14	☉ ∠ ♂
19 48	♀ ∠ ♃
05 Sunday	
00 43	☽ ⚼ ♃
00 57	☽ □ ♀
01 24	☽ □ ♂
01 32	☽ ⚼ ☉
01 38	☽ # ☿
02 22	☽ ∠ ♆
03 53	☉ ± ♅
13 16	☽ ∥ ♅
13 50	☉ ☍ ♆
16 47	☽ □ ♅
19 28	♂ ⚼ ♆
22 05	☽ △ ♇
23 41	♀ ⚼ ♆
23 59	☽ ☍ ♄
06 Monday	
02 31	☽ △ ♃
03 45	☽ ✶ ♆
04 48	☽ △ ☉
06 59	☽ # ♇
09 35	☽ Q ♀
09 44	☽ Q ♂
12 06	☽ # ♂
19 24	☽ □ ☿
19 42	☽ # ♀
22 51	♀ ☌ ♂
07 Tuesday	
00 15	☽ ⚼ ♇
00 45	☽ Q ♆
06 13	♃ ± ♅
08 13	☽ ✶ ♂
08 20	☽ ✶ ♀
16 27	☽ ± ♇
21 39	☽ △ ♅
08 Wednesday	
03 28	☽ ⚻ ♇
05 55	☽ ⚻ ♄
08 12	☽ Q ☿
09 16	☽ □ ♃
09 30	☽ □ ♆
13 28	☽ ∠ ♂
13 58	☽ ∠ ♀
15 07	☽ □ ☉
15 14	☿ ⚼ ♇
17 14	☽ ± ♄
20 41	♃ ☍ ♆
09 Thursday	
01 38	☽ ⚼ ♅
07 36	☽ ⊥ ♂
08 22	☽ ⊥ ♀
09 20	♂ ∥ ♇
10 04	☽ ✶ ☿
10 29	☽ ⚼ ♄
18 54	☽ ± ♅
19 54	☽ ⚺ ♂
20 12	☽ Q ♃
20 52	☽ ⚺ ♀
10 Friday	
04 32	☽ Q ☉
06 34	☽ ⚻ ♅
06 50	☿ ∥ ♄
12 53	☽ ☍ ♇
15 58	☽ △ ♄
19 12	☽ △ ♆
19 45	☽ ∠ ☿
20 13	☽ ✶ ♃
11 Saturday	
06 18	☽ ✶ ☉
12 45	☿ Q ♄
16 34	☽ ⊥ ☿
12 Sunday	
01 08	☽ ⚼ ♆
01 19	☽ ∥ ♀
02 50	☽ ∠ ♃
05 29	☽ ∥ ♇
06 38	☽ ⚺ ☿
11 28	☽ ☌ ♂
11 55	☽ ∥ ♂
13 33	☽ ☌ ♀
15 07	☽ ∠ ☉
18 26	☽ ☍ ♅
19 17	☽ ± ♆
21 24	☽ ⊥ ♃
13 Monday	
01 03	☽ ⚻ ♇
01 54	☽ # ♅
04 44	☽ □ ♄
07 26	☽ ⚻ ♆
09 49	☽ ⚺ ♃
10 59	☽ ⊥ ☉
13 14	☽ ± ♇
14 Tuesday	
00 16	☽ ⚺ ☉
00 27	♀ ∥ ♇
07 11	☉ ⚼ ♅
07 32	☽ ⚼ ♇
11 50	☿ ⚺ ♂
17 38	☽ Q ♄
15 Wednesday	
00 25	☽ ∥ ♄
01 06	♀ ☍ ♅
04 28	☽ ⚺ ♂
06 15	☽ ☌ ☿
07 15	☽ ⚻ ♅
07 49	☽ ⚺ ♀
11 25	♀ ± ♆
12 40	☿ ⚻ ♅
13 59	☽ △ ♇
17 26	☽ ⊥ ♂
18 11	☽ ✶ ♄
19 23	☽ ± ♅
20 14	☽ ☍ ♆
21 02	☽ ⊥ ♀
23 58	☽ ☌ ♃
16 Thursday	
01 23	☿ ⚺ ♀
12 49	☽ ∠ ♂
13 31	☽ ⚼ ♅
16 48	☽ ∠ ♀
17 33	☿ # ♆
18 23	☽ ☌ ☉
17 Friday	
00 02	♂ ☍ ♅
00 41	☽ ∠ ♄
07 05	☿ △ ♇
13 15	♂ ± ♆
18 50	☽ ⊥ ♄
19 17	♀ ⊥ ♃
19 31	☽ △ ♅
20 49	☽ ✶ ♂
18 Saturday	
01 27	☽ ✶ ♀
02 14	☽ □ ♇
05 53	☽ ⚺ ☿
06 52	☽ ⚺ ♄
08 15	☽ ⚻ ♆
09 05	☽ ∥ ♄
11 13	♀ ⚻ ♇
12 12	☿ ✶ ♄
13 11	☽ ⚺ ♃
15 17	☿ ± ♅
20 03	☽ ⊥ ☿
20 05	☽ ± ♆
20 15	☿ ☍ ♆
22 08	☽ Q ♂
23 15	♂ # ♅
19 Sunday	
01 15	☽ ⊥ ♃
03 14	☽ Q ♀
11 04	☽ ⚺ ☉
13 43	☽ ⚼ ♆
13 43	☽ Q ♇
14 57	☿ ⊥ ♂
17 08	☽ ∠ ☿
19 13	☽ ∠ ♃
21 28	☿ ∥ ♃
23 45	☽ ⊥ ☉
20 Monday	
02 34	☽ ∥ ♂
05 01	☽ # ♅
06 24	☽ □ ♅
07 08	☿ ☌ ♃
09 52	☉ ∠ ♄
10 23	☽ ∥ ♀
11 28	☽ □ ♂
13 01	☽ ✶ ♇
17 18	☽ □ ♀
18 00	☽ ☌ ♄
18 36	☽ ∠ ☉
18 44	☽ △ ♆
21 Tuesday	
00 34	☽ ∥ ♇
00 44	☽ ✶ ♃
03 08	♀ □ ♄
03 47	☽ ✶ ☿
10 57	♀ ⚻ ♆
13 22	♇ St D
13 57	♂ ⚻ ♇
16 49	☽ Q ♅
17 41	☽ ∠ ♇
23 57	☽ Q ♃
22 Wednesday	
00 43	♂ ⊥ ♃
01 20	☿ ⊥ ♀
01 28	☽ ✶ ☉
06 55	☽ Q ☿
10 36	☽ ⊥ ♇
15 18	☽ ✶ ♅
16 11	♀ # ♅
21 45	☽ ⚺ ♇
23 38	☽ △ ♂
23 53	☿ ⚼ ♅
23 59	☉ ∥ ☿
23 Thursday	
01 36	☽ Q ☉
02 59	☽ ⚺ ♄
03 07	☽ □ ♆
06 31	☽ △ ♀
09 58	☽ □ ♃
14 03	☽ ⊥ ♄
16 13	♄ △ ♆
18 49	☽ ∠ ♅
22 27	☽ □ ☿
24 Friday	
04 32	☽ ⚼ ♂
06 26	☽ ∠ ♄
10 58	☽ ⊥ ♅
11 39	☽ Q ♆
11 52	☽ ⚼ ♀
12 00	♀ ± ♇
12 44	☽ □ ☉
21 27	☽ ± ♂
21 35	☽ ⚺ ♅
25 Saturday	
03 45	☽ ☌ ♇
05 01	☽ ± ♀
08 32	☽ ⚻ ♂
08 43	☽ ✶ ♆
09 07	☽ ✶ ♄
11 56	♂ ⚻ ♆
15 11	♀ ⚺ ♃
16 11	☽ △ ♃
16 15	☽ ⚻ ♀
21 17	♂ □ ♄
26 Sunday	
02 09	☿ ∠ ♄
04 14	☿ ∠ ♂
05 53	☽ Q ♄
10 23	☽ ∠ ♆
12 31	☽ △ ☿
14 39	☽ # ♇
18 05	☽ ⚼ ♃
20 14	☽ △ ☉
27 Monday	
00 50	☽ ☌ ♅
01 26	☽ ⊥ ♆
06 44	☽ ⚺ ♇
07 30	☽ ∥ ♅
09 17	☽ ± ♃
11 20	☽ ⚺ ♆
12 10	☽ □ ♄
13 54	☽ ☍ ♂
16 36	☽ ⊥ ♇
17 49	☽ ⚼ ☿
18 30	☽ # ♀
19 17	☽ ⚻ ♃
22 18	☽ ☍ ♀
22 42	☽ ⚼ ☉
28 Tuesday	
00 31	☽ # ♂
07 15	☽ ∠ ♇
10 57	☽ ± ☿
14 15	☽ ± ☉
16 04	☽ # ♄
22 14	☽ ⚻ ☿
29 Wednesday	
00 31	☽ ⚻ ☉
01 33	☽ ⚺ ♅
07 18	☽ ✶ ♇
11 06	☽ ⊥ ♅
11 39	☽ ☌ ♆
12 53	☽ △ ♄
16 37	☽ ⚻ ♂
16 39	☉ △ ♅
20 09	☽ ☍ ♃
20 58	☿ △ ♅
21 20	♂ ± ♇
23 03	☿ ∠ ♀
30 Thursday	
00 22	☉ ☌ ☿
01 23	☽ ∠ ♅
01 53	☽ ⚻ ♀
02 22	☽ Q ♇
02 37	☽ ± ♂
10 27	☿ ⊥ ♄
12 06	☽ ± ♀
12 52	☽ ⚼ ♄
17 36	☽ ⚼ ♂
18 05	☉ ∠ ♀

All ephemeris data is given at 12.00 UT and the Moon's longitude is additionally given for 24.00 UT

JULY 1977

Date	Sidereal time h m s	Sun ☉ ° ' "	Moon ☽ ° ' "	Moon ☽ 24.00 ° ' "	Mercury ☿ ° '	Venus ♀ ° '	Mars ♂ ° '	Jupiter ♃ ° '	Saturn ♄ ° '	Uranus ♅ ° '	Neptune ♆ ° '	Pluto ♇ ° '
01	06 37 42	09 ♋ 31 38	14 ♑ 34 15	22 ♑ 02 13	11 ♋ 20	24 ♉ 36	18 ♉ 35	19 ♊ 53	15 ♌ 09	07 ♏ 47	14 ♐ 06	11 ♎ 26
02	06 41 39	10 28 49	29 ♑ 25 46	06 ♒ 43 54	13 30	25 38	19 18	20 06	15 16	07 R 46	14 R 05	11 26
03	06 45 35	11 26 00	13 ♒ 55 50	21 ♒ 00 58	15 39	26 41	20 01	20 20	15 22	07 45	14 03	11 26
04	06 49 32	12 23 12	27 ♒ 58 57	04 ♓ 49 35	17 46	27 44	20 44	20 33	15 29	07 45	14 02	11 27
05	06 53 28	13 20 23	11 ♓ 32 56	18 ♓ 09 10	19 52	28 48	21 26	20 46	15 36	07 44	14 00	11 27
06	06 57 25	14 17 35	24 ♓ 38 39	01 ♈ 01 48	21 57	29 ♉ 52	22 09	21 00	15 43	07 44	13 59	11 28
07	07 01 21	15 14 47	07 ♈ 19 10	13 ♈ 31 23	24 00	00 ♊ 56	22 52	21 13	15 50	07 43	13 58	11 28
08	07 05 18	16 11 59	19 ♈ 39 03	25 ♈ 42 52	26 02	02 00	23 34	21 26	15 57	07 43	13 56	11 29
09	07 09 14	17 09 12	01 ♉ 43 30	07 ♉ 41 36	28 ♋ 02	03 04	24 17	21 39	16 04	07 42	13 55	11 29
10	07 13 11	18 06 25	13 ♉ 37 50	19 ♉ 32 48	00 ♌ 00	04 09	24 59	21 53	16 11	07 42	13 54	11 30
11	07 17 08	19 03 39	25 ♉ 27 06	01 ♊ 21 15	01 56	05 13	25 42	22 06	16 18	07 42	13 53	11 31
12	07 21 04	20 00 53	07 ♊ 15 46	13 ♊ 11 04	03 51	06 18	26 24	22 19	16 25	07 42	13 51	11 31
13	07 25 01	20 58 08	19 ♊ 07 32	25 ♊ 05 32	05 43	07 23	27 06	22 32	16 32	07 41	13 50	11 32
14	07 28 57	21 55 23	01 ♋ 05 19	07 ♋ 07 08	07 34	08 29	27 48	22 45	16 40	07 41	13 49	11 33
15	07 32 54	22 52 38	13 ♋ 11 09	19 ♋ 17 30	09 23	09 34	28 31	22 58	16 47	07 41	13 48	11 33
16	07 36 50	23 49 54	25 ♋ 26 18	01 ♌ 37 37	11 10	10 40	29 13	23 11	16 54	07 D 41	13 47	11 34
17	07 40 47	24 47 10	07 ♌ 51 29	14 ♌ 07 59	12 55	11 46	29 ♉ 54	23 23	17 01	07 41	13 45	11 35
18	07 44 43	25 44 27	20 ♌ 27 07	26 ♌ 48 57	14 38	12 52	00 ♊ 36	23 36	17 09	07 41	13 44	11 36
19	07 48 40	26 41 44	03 ♍ 13 33	09 ♍ 40 59	16 19	13 58	01 18	23 49	17 16	07 41	13 43	11 37
20	07 52 37	27 39 01	16 ♍ 11 23	22 ♍ 44 53	17 59	15 04	02 00	24 02	17 24	07 42	13 42	11 38
21	07 56 33	28 36 18	29 ♍ 21 38	06 ♎ 01 50	19 36	16 10	02 41	24 14	17 31	07 42	13 41	11 39
22	08 00 30	29 ♋ 33 36	12 ♎ 45 40	19 ♎ 33 19	21 12	17 17	03 23	24 27	17 38	07 42	13 40	11 40
23	08 04 26	00 ♌ 30 53	26 ♎ 24 59	03 ♏ 20 47	22 46	18 24	04 04	24 39	17 46	07 43	13 39	11 41
24	08 08 23	01 28 12	10 ♏ 20 49	17 ♏ 25 04	24 18	19 31	04 45	24 52	17 53	07 43	13 38	11 42
25	08 12 19	02 25 30	24 ♏ 33 27	01 ♐ 45 46	25 48	20 38	05 26	25 04	18 01	07 43	13 37	11 43
26	08 16 16	03 22 49	09 ♐ 01 37	16 ♐ 20 33	27 16	21 45	06 08	25 16	18 08	07 44	13 36	11 44
27	08 20 12	04 20 09	23 ♐ 41 53	01 ♑ 04 51	28 ♌ 42	22 52	06 49	25 29	18 16	07 44	13 35	11 45
28	08 24 09	05 17 29	08 ♑ 28 32	15 ♑ 51 57	00 ♍ 06	24 00	07 29	25 41	18 23	07 45	13 35	11 46
29	08 28 06	06 14 49	23 ♑ 14 06	00 ♒ 33 54	01 28	25 07	08 10	25 53	18 31	07 46	13 34	11 48
30	08 32 02	07 12 11	07 ♒ 50 23	15 ♒ 02 40	02 49	26 15	08 51	26 05	18 38	07 46	13 33	11 49
31	08 35 59	08 ♌ 09 33	22 ♒ 09 55	29 ♒ 11 33	04 ♍ 07	27 ♊ 23	09 ♊ 32	26 ♊ 17	18 ♌ 46	07 ♏ 47	13 ♐ 32	11 ♎ 50

Date	Moon True ☊ ° '	Moon Mean ☊ ° '	Moon ☽ Latitude ° '
01	20 ♎ 53	20 ♎ 16	04 N 59
02	20 R 42	20 13	04 58
03	20 33	20 10	04 37
04	20 25	20 07	04 00
05	20 20	20 04	03 10
06	20 17	20 00	02 11
07	20 16	19 57	01 08
08	20 D 15	19 54	00 N 03
09	20 R 15	19 51	01 S 00
10	20 14	19 48	02 00
11	20 10	19 45	02 55
12	20 04	19 41	03 41
13	19 55	19 38	04 18
14	19 44	19 35	04 45
15	19 31	19 32	04 58
16	19 17	19 29	04 58
17	19 04	19 25	04 44
18	18 53	19 22	04 16
19	18 44	19 19	03 35
20	18 38	19 16	02 42
21	18 35	19 13	01 40
22	18 34	19 10	00 S 31
23	18 D 34	19 06	00 N 42
24	18 R 34	19 03	01 53
25	18 32	19 00	02 58
26	18 29	18 57	03 54
27	18 22	18 54	04 35
28	18 14	18 51	04 58
29	18 04	18 47	05 01
30	17 53	18 44	04 44
31	17 ♎ 44	18 ♎ 41	04 N 10

DECLINATIONS

Date	Sun ☉ ° '	Moon ☽ ° '	Mercury ☿ ° '	Venus ♀ ° '	Mars ♂ ° '	Jupiter ♃ ° '	Saturn ♄ ° '	Uranus ♅ ° '	Neptune ♆ ° '	Pluto ♇ ° '
01	23 N 06	17 S 41	24 N 23	16 N 07	16 N 34	22 N 36	17 N 14	13 S 41	20 S 59	11 N 11
02	23 02	15 25	24 16	16 22	16 46	22 37	17 12	13 41	20 58	11 10
03	22 57	12 13	24 07	16 37	16 58	22 38	17 10	13 41	20 58	11 10
04	22 52	08 25	23 55	16 52	17 10	22 39	17 08	13 41	20 58	11 09
05	22 46	04 18	23 41	17 07	17 22	22 40	17 06	13 41	20 58	11 08
06	22 40	00 S 07	23 25	17 22	17 34	22 41	17 04	13 41	20 58	11 08
07	22 34	03 N 57	23 06	17 36	17 45	22 42	17 02	13 40	20 58	11 07
08	22 27	07 44	22 45	17 50	17 56	22 43	17 00	13 40	20 58	11 06
09	22 20	11 08	22 22	18 04	18 08	22 44	16 58	13 40	20 58	11 05
10	22 13	14 01	21 57	18 17	18 18	22 45	16 55	13 40	20 57	11 05
11	22 05	16 18	21 31	18 30	18 29	22 46	16 53	13 40	20 57	11 04
12	21 57	17 53	21 03	18 43	18 40	22 47	16 51	13 40	20 57	11 03
13	21 48	18 42	20 34	18 56	18 50	22 47	16 49	13 40	20 57	11 02
14	21 39	18 42	20 03	19 08	19 00	22 48	16 47	13 40	20 57	11 02
15	21 30	17 50	19 32	19 20	19 10	22 49	16 45	13 40	20 57	11 01
16	21 20	16 10	18 59	19 31	19 20	22 50	16 43	13 40	20 57	11 00
17	21 10	13 43	18 25	19 42	19 30	22 50	16 41	13 40	20 57	10 59
18	21 00	10 37	17 51	19 53	19 39	22 51	16 38	13 40	20 57	10 59
19	20 49	06 59	17 16	20 03	19 49	22 52	16 36	13 40	20 57	10 58
20	20 38	02 N 57	16 40	20 13	19 58	22 52	16 34	13 40	20 56	10 57
21	20 26	01 S 16	16 04	20 22	20 07	22 53	16 32	13 40	20 56	10 56
22	20 15	05 31	15 27	20 31	20 15	22 54	16 29	13 41	20 56	10 55
23	20 02	09 33	14 50	20 40	20 24	22 54	16 27	13 41	20 56	10 54
24	19 50	13 08	14 12	20 48	20 32	22 55	16 25	13 41	20 56	10 54
25	19 37	16 01	13 35	20 55	20 41	22 55	16 23	13 41	20 56	10 53
26	19 24	17 57	12 57	21 02	20 48	22 56	16 20	13 41	20 56	10 52
27	19 11	18 43	12 19	21 09	20 56	22 56	16 18	13 41	20 56	10 51
28	18 57	18 13	11 42	21 15	21 04	22 57	16 16	13 42	20 56	10 50
29	18 43	16 30	11 04	21 21	21 11	22 57	16 14	13 42	20 56	10 49
30	18 28	13 44	10 27	21 26	21 18	22 58	16 11	13 42	20 56	10 48
31	18 N 14	10 S 11	09 N 49	21 N 30	21 N 25	22 N 58	16 N 09	13 S 42	20 S 56	10 N 47

ZODIAC SIGN ENTRIES

Date	h	m	Planets
02	12	56	☽ ♒
04	15	31	☽ ♓
06	15	09	♀ ♊
06	22	03	☽ ♈
09	08	33	☽ ♉
10	12	00	☿ ♌
11	21	15	☽ ♊
14	09	50	☽ ♋
16	20	51	☽ ♌
17	15	13	♂ ♊
19	05	58	☽ ♍
21	13	09	☽ ♎
22	23	04	☉ ♌
23	18	13	☽ ♏
25	21	04	☽ ♐
27	22	15	☽ ♑
28	10	15	☿ ♍
29	23	04	☽ ♒

LATITUDES

Date	Mercury ☿ ° '	Venus ♀ ° '	Mars ♂ ° '	Jupiter ♃ ° '	Saturn ♄ ° '	Uranus ♅ ° '	Neptune ♆ ° '	Pluto ♇ ° '
01	01 N 26	02 S 53	00 S 50	00 S 27	00 N 59	00 N 26	01 N 32	17 N 04
04	01 41	02 51	00 48	00 27	00 59	00 26	01 32	17 02
07	01 49	02 48	00 46	00 27	00 59	00 26	01 31	17 01
10	01 51	02 44	00 44	00 26	00 59	00 26	01 31	16 59
13	01 47	02 39	00 42	00 26	00 59	00 26	01 31	16 57
16	01 37	02 33	00 39	00 26	00 59	00 26	01 31	16 56
19	01 23	02 27	00 37	00 26	00 59	00 26	01 31	16 54
22	01 04	02 19	00 35	00 26	01 00	00 26	01 31	16 53
25	00 42	02 12	00 33	00 26	01 00	00 25	01 31	16 51
28	00 N 17	02 03	00 30	00 25	01 00	00 25	01 31	16 50
31	00 S 11	01 S 55	00 S 28	00 S 25	01 N 00	00 N 25	01 N 30	16 N 48

DATA

Julian Date	2443326
Delta T	+48 seconds
Ayanamsa	23° 32' 41"
Synetic vernal point	05° ♓ 34' 19"
True obliquity of ecliptic	23° 26' 23"

MOON'S PHASES, APSIDES AND POSITIONS ☽

Date	h	m	Phase	Longitude ° '	Eclipse Indicator
01	03	24	○	09 ♑ 11	
08	04	39	☾	15 ♈ 54	
16	08	36	●	23 ♋ 42	
23	19	38	☽	00 ♏ 49	
30	10	52	○	07 ♒ 09	

Day	h	m		
12	07	52	Apogee	
28	02	15	Perigee	
06	12	41	ON	
13	23	48	Max dec	18° N 48'
21	04	50	OS	
27	14	37	Max dec	18° S 43'

LONGITUDES

Date	Chiron ⚷ ° '	Ceres ⚳ ° '	Pallas ⚴ ° '	Juno ⚵ ° '	Vesta ⚶ ° '	Black Moon Lilith ⚸ ° '
01	05 ♉ 07	05 ♎ 11	08 ♍ 33	14 ♏ 31	16 ♌ 39	07 ♊ 37
11	05 ♉ 24	08 ♎ 01	12 ♍ 31	14 ♏ 21	21 ♌ 06	08 ♊ 44
21	05 ♉ 37	11 ♎ 07	16 ♍ 35	14 ♏ 42	25 ♌ 38	09 ♊ 51
31	05 ♉ 45	14 ♎ 27	20 ♍ 45	15 ♏ 33	00 ♍ 16	10 ♊ 58

All ephemeris data is given at 12.00 UT and the Moon's longitude is additionally given for 24.00 UT

ASPECTARIAN

01 Friday
01 08 ☉ ⊥ ♄
01 10 ☽ ✱ ♅
03 16 ☽ ± ♄
03 24 ☽ ☍ ☉
03 28 ☽ bQ ♀
05 58 ☽ ☍ ☿
06 58 ☽ □ ♇
11 15 ☽ ⊻ ♆
12 56 ☽ ⊼ ♄
12 57 ☿ □ ♇
17 57 ☽ # ♄
18 45 ☽ △ ♂
20 21 ☽ Q ♅
20 39 ☽ ⊼ ♃
20 52 ☽ ⊥ ♆

02 Saturday
00 10 ☽ # ♂
04 01 ☽ # ♀
05 22 ☽ △ ♀
06 30 ☽ ± ♃
11 25 ☽ ∠ ♆
18 21 ☿ ⊼ ♆
21 27 ☽ bQ ♃

03 Sunday
01 42 ☽ ∥ ♅
01 42 ☽ □ ♅
06 16 ♀ bQ ♇
07 31 ☽ ⊼ ☉
07 49 ☽ △ ♇
08 46 ☿ ⊻ ♄
12 08 ☉ □ ♇
12 12 ☽ ✱ ♆
14 27 ☽ ☍ ♄
15 24 ☽ ⊼ ☿
18 20 ☽ ± ☉
19 04 ☽ # ♇
22 50 ☽ □ ♂
23 00 ☽ △ ♃

04 Monday
03 22 ♂ ⊻ ♃
03 25 ☽ ± ☿
08 02 ♂ ∥ ♄
08 37 ☽ Q ♆
09 20 ☽ bQ ♇
10 53 ☽ bQ ☉
11 33 ☽ □ ♀
21 54 ☽ bQ ☿

05 Tuesday
01 06 ☽ ± ♇
05 10 ☽ △ ♅
08 00 ☽ Q ♂
09 55 ♀ ∥ ♄
11 50 ☽ ⊼ ♇
13 30 ☿ ± ♆
15 29 ☽ △ ☉
16 26 ☽ □ ♆
19 24 ☽ ⊼ ♄
22 22 ☽ Q ♀
23 33 ☿ ⊻ ♃

06 Wednesday
04 24 ☉ ⊼ ♆
05 07 ☽ □ ♃
06 03 ☽ △ ☿
06 31 ☽ ± ♄
07 06 ☽ ✱ ♂
08 26 ☽ bQ ♅
08 39 ☉ ∥ ♃
15 30 ☿ ✱ ♂
22 41 ☽ ✱ ♀
23 31 ☽ bQ ♄

07 Thursday
01 19 ☽ ± ♅
12 46 ☽ ⊼ ♅
13 06 ☽ ∠ ♂
15 43 ☽ Q ♃
20 01 ☽ ☍ ♇

08 Friday
00 50 ☽ △ ♆
04 39 ☽ □ ☉
04 40 ☽ △ ♄
04 48 ☉ ⊻ ♄
06 17 ☽ ∠ ♀
07 40 ☽ ⊥ ♂
13 50 ☿ ∥ ♃
15 35 ☽ ✱ ♃
20 14 ☽ ⊻ ♂

09 Saturday
01 47 ☽ ⊥ ♀
03 09 ☽ □ ☿
06 24 ☽ bQ ♆
06 53 ☿ ⊥ ♃
11 43 ☽ ∥ ♇
14 36 ☉ ∥ ☿
14 57 ☽ ⊻ ♀
19 29 ☽ Q ☉
22 06 ☽ ∠ ♃
22 39 ☿ bQ ♆

10 Sunday
00 01 ☽ ☍ ♅
00 26 ☽ ± ♆
05 49 ☿ Q ♇
07 41 ☽ ⊼ ♇
08 52 ☽ # ♅
12 32 ☽ ⊼ ♆
13 03 ♀ Q ♄
16 38 ☽ ⊥ ♃
17 14 ☽ □ ♄
19 51 ☽ ± ♇
21 52 ☽ ✱ ☉
22 36 ☽ Q ☿
23 22 ♀ ∥ ♂

11 Monday
05 03 ☽ ⊻ ♃
12 32 ☽ ☌ ♂
14 09 ☽ bQ ♇
19 36 ☽ ∥ ♄

12 Tuesday
03 44 ☽ ✱ ☿
06 10 ☽ Q ♄
07 02 ☽ ∠ ☉
08 04 ☉ ± ♆
09 51 ☽ ☌ ♀
12 52 ☽ ⊼ ♅
16 09 ♂ bQ ♇
17 05 ☿ # ♆
20 38 ☽ △ ♇

13 Wednesday
01 02 ☽ ± ♅
01 20 ☽ ☍ ♆
02 53 ☽ ⊥ ☉
06 44 ☽ ✱ ♄
15 48 ☽ ∠ ☿
16 02 ☽ ⊻ ☉
18 39 ♀ ⊼ ♅
18 59 ☽ ☌ ♃
19 10 ☽ bQ ♅

14 Thursday
05 02 ☽ ⊻ ♂
05 36 ♃ bQ ♅
13 07 ☽ ⊥ ☿
13 09 ☽ ∠ ♄
13 36 ☿ □ ♅
14 41 ☿ ∠ ♃
17 45 ☽ ⊥ ♂

15 Friday
01 08 ☽ △ ♅
03 10 ☽ ⊻ ☿
04 09 ☽ ⊻ ♀
07 12 ☽ ⊥ ♄
08 47 ☽ □ ♇
12 41 ☽ ∠ ♂
13 12 ☽ ⊼ ♆
14 45 ☉ ⊻ ♃
17 09 ☽ ⊥ ♀
18 28 ☿ ✱ ♀
18 38 ☿ ∥ ♀
19 09 ☽ ⊻ ♄

16 Saturday
00 04 ☿ ∥ ♂
00 58 ☽ ± ♆
05 09 ☽ ∥ ♄
07 31 ☽ ⊻ ♃
08 36 ☽ ☌ ☉
08 40 ♅ St D
12 29 ☽ ∠ ♀
13 03 ☿ Q ♂
17 38 ☿ ✱ ♇
18 28 ☽ bQ ♆
19 23 ☽ ⊥ ♃
19 45 ☽ ✱ ♂
20 02 ☽ Q ♇

17 Sunday
08 09 ♀ △ ♇
11 40 ☽ □ ♅
12 27 ☽ # ♅
13 02 ☽ ∠ ♃
19 09 ☽ ✱ ♇
20 11 ☽ ✱ ♀
20 12 ☽ Q ♂
23 13 ☽ ☌ ☿
23 16 ☽ △ ♆
23 37 ☿ △ ♆

18 Monday
05 40 ☽ ☌ ♄
09 25 ☽ ∥ ♇
18 03 ☽ ✱ ♃
19 08 ☉ # ♆
21 06 ☽ Q ♀
21 53 ☽ Q ♅
22 47 ☽ ⊻ ☉
23 37 ☽ ∠ ♇

19 Tuesday
06 05 ♀ ± ♅
06 50 ♀ ☍ ♆
08 12 ☽ □ ♂
10 56 ☽ ⊥ ☉
16 27 ☽ ⊥ ♇
16 54 ☽ Q ♃
20 19 ☽ ✱ ♅

20 Wednesday
02 47 ☿ ☌ ♄
03 36 ☽ ⊻ ♇
04 58 ☽ ∠ ☉
07 26 ☽ □ ♆
09 45 ☽ □ ♀
14 14 ☽ ⊻ ♄
15 45 ☽ ⊻ ☿
16 19 ☿ ∥ ♄
21 57 ♅ ⊥ ♆
23 54 ☽ ∠ ♅

21 Thursday
01 18 ☽ ⊥ ♄
02 34 ☽ □ ♃
04 15 ☽ ⊥ ☿
10 32 ☽ ✱ ☉
13 59 ☉ bQ ♆
16 11 ☽ Q ♆
16 13 ☽ ⊥ ♅
16 56 ☉ ∥ ♀
17 45 ☽ ∠ ♄
18 19 ☽ △ ♂
22 44 ☽ ∠ ☿

22 Friday
02 59 ☽ ⊻ ♅
09 42 ☽ Q ☉
10 03 ☽ ☌ ♇
10 57 ☉ ∥ ♂
13 36 ☽ ✱ ♆
14 40 ☉ Q ♇
20 36 ♀ ✱ ♄
20 42 ☽ ✱ ♄
20 43 ☽ △ ♀
22 27 ☽ bQ ♂

23 Saturday
04 48 ☽ ✱ ☿
08 53 ☽ △ ♃
15 01 ☽ ± ♂
15 53 ☽ ∠ ♆
16 30 ☉ ⊥ ♃
17 52 ☽ Q ♄
19 38 ☽ □ ☉
20 37 ☽ # ♇

24 Sunday
01 08 ☽ bQ ♀
01 56 ☽ ⊼ ♂
04 13 ☽ Q ☿
07 22 ☽ ⊥ ♆
07 30 ☽ ☌ ♅
11 10 ☽ bQ ♃
14 18 ☽ ⊻ ♇
16 03 ☽ ∥ ♅
17 35 ☽ ⊻ ♆
17 51 ☽ ± ♀
18 45 ☽ # ☿
22 30 ☿ ✱ ♃

25 Monday
00 29 ☽ ⊥ ♇
00 54 ☽ □ ♄
02 39 ☽ ± ♃
04 51 ☽ ⊼ ♀
08 06 ☿ # ♅
10 52 ☿ Q ♅
12 52 ☽ ⊼ ♃
14 18 ☽ □ ☿
14 56 ♀ # ♆
15 31 ☽ # ♄
15 37 ☽ ∠ ♇

26 Tuesday
02 01 ☽ △ ☉
03 14 ☿ ∠ ♇
06 59 ☽ ☍ ♂
09 52 ☽ ⊻ ♅
12 31 ♂ Q ♄
16 27 ☽ ✱ ♇
19 31 ☽ ☌ ♆
19 44 ☽ ⊥ ♅

27 Wednesday
03 04 ☽ △ ♄
04 24 ☽ bQ ☉
09 10 ♀ bQ ♅
10 26 ☽ ∠ ♅
10 33 ☽ ☍ ♀
11 10 ♂ # ♆
12 06 ☽ Q ♇
14 56 ☽ ☍ ♃
20 04 ☽ ± ☉
20 59 ☽ △ ☿

28 Thursday
03 41 ☽ bQ ♄
06 29 ☽ ⊼ ☉
10 20 ☽ ⊼ ♂
10 49 ☽ ✱ ♅
17 22 ☽ □ ♇
18 24 ☽ ± ♄
20 16 ☽ ⊻ ♆
20 32 ☽ ± ♂
21 17 ♂ ⊼ ♅
23 52 ☽ bQ ☿

29 Friday
04 15 ☽ ⊼ ♄
06 01 ☽ ⊥ ♆
06 20 ☽ Q ♅
11 54 ☽ bQ ♂
14 49 ☽ # ♄
15 21 ☽ ⊼ ♀
16 01 ☽ ± ☿
16 23 ☽ ⊼ ♃
20 42 ☽ ∠ ♆
21 54 ☿ ∥ ♇

30 Saturday
02 00 ☽ ± ♀
02 22 ☽ ± ♃
02 52 ☽ ⊼ ☿
07 38 ♀ ☌ ♃
10 52 ☽ ☍ ☉
11 53 ☽ □ ♅
12 12 ☽ ∥ ♅
13 46 ☽ △ ♂
17 28 ☽ bQ ♃
18 09 ☽ bQ ♀
18 37 ☽ △ ♇
21 29 ☽ ✱ ♆

31 Sunday
02 31 ☉ □ ♅
06 12 ☽ ☍ ♄
08 09 ☽ # ♇
14 38 ☽ # ☿
17 44 ☽ Q ♆
19 07 ☽ △ ♃
19 58 ☽ bQ ♇
21 41 ☽ △ ♀

AUGUST 1977

LONGITUDES

Date	Sidereal time h m s	Sun ☉ ° ' "	Moon ☽ ° ' "	Moon ☽ 24.00 ° ' "	Mercury ☿ ° '	Venus ♀ ° '	Mars ♂ ° '	Jupiter ♃ ° '	Saturn ♄ ° '	Uranus ♅ ° '	Neptune ♆ ° '	Pluto ♇ ° '
01	08 39 55	09 ♌ 06 56	06 ♓ 07 05	12 ♓ 56 13	05 ♍ 23	28 ♊ 31	10 ♊ 12	26 ♊ 29	18 ♌ 54	07 ♏ 48	13 ♐ 31	11 ♎ 51
02	08 43 52	10 04 19	19 ♓ 38 50	26 ♓ 14 59	06 37	29 ♊ 39	10 53	26 41	19 01	07 49	13 R 31	11 53
03	08 47 48	11 01 44	02 ♈ 44 50	09 ♈ 08 42	07 48	00 ♋ 47	11 33	26 53	19 09	07 50	13 30	11 54
04	08 51 45	11 59 10	15 ♈ 27 00	21 ♈ 40 13	08 58	01 56	12 13	27 04	19 17	07 51	13 29	11 56
05	08 55 41	12 56 38	27 ♈ 48 56	03 ♉ 53 46	10 05	03 04	12 54	27 16	19 24	07 52	13 29	11 57
06	08 59 38	13 54 06	09 ♉ 55 20	15 ♉ 54 18	11 09	04 13	13 34	27 27	19 32	07 53	13 28	11 58
07	09 03 35	14 51 36	21 ♉ 51 21	27 ♉ 47 07	12 11	05 22	14 14	27 39	19 39	07 54	13 27	12 00
08	09 07 31	15 49 07	03 ♊ 42 15	09 ♊ 37 21	13 10	06 31	14 54	27 50	19 47	07 55	13 27	12 01
09	09 11 28	16 46 40	15 ♊ 33 01	21 ♊ 29 47	14 07	07 40	15 33	28 01	19 55	07 56	13 26	12 03
10	09 15 24	17 44 14	27 ♊ 28 09	03 ♋ 28 32	15 00	08 49	16 13	28 13	20 02	07 58	13 26	12 04
11	09 19 21	18 41 49	09 ♋ 31 20	15 ♋ 36 51	15 51	09 58	16 53	28 24	20 10	07 59	13 25	12 06
12	09 23 17	19 39 26	21 ♋ 45 22	27 ♋ 57 05	16 38	11 07	17 32	28 35	20 18	08 00	13 25	12 08
13	09 27 14	20 37 03	04 ♌ 12 06	10 ♌ 30 31	17 21	12 17	18 12	28 46	20 26	08 02	13 25	12 09
14	09 31 10	21 34 42	16 ♌ 52 21	23 ♌ 17 34	18 02	13 27	18 51	28 57	20 33	08 03	13 24	12 11
15	09 35 07	22 32 23	29 ♌ 46 05	06 ♍ 17 50	18 38	14 36	19 30	29 07	20 41	08 05	13 24	12 12
16	09 39 04	23 30 04	12 ♍ 52 41	19 ♍ 30 32	19 10	15 46	20 09	29 18	20 49	08 06	13 24	12 14
17	09 43 00	24 27 47	26 ♍ 11 14	02 ♎ 54 41	19 38	16 56	20 48	29 29	20 56	08 08	13 23	12 16
18	09 46 57	25 25 31	09 ♎ 40 47	16 ♎ 29 28	20 02	18 06	21 27	29 39	21 04	08 10	13 23	12 18
19	09 50 53	26 23 16	23 ♎ 20 39	00 ♏ 14 17	20 21	19 16	22 06	29 ♊ 49	21 12	08 11	13 23	12 19
20	09 54 50	27 21 02	07 ♏ 10 20	14 ♏ 08 44	20 35	20 26	22 44	00 ♋ 00	21 19	08 13	13 23	12 21
21	09 58 46	28 18 49	21 ♏ 09 27	28 ♏ 12 22	20 43	21 36	23 23	00 10	21 27	08 15	13 23	12 23
22	10 02 43	29 ♌ 16 37	05 ♐ 17 21	12 ♐ 24 14	20 47	22 47	24 01	00 20	21 35	08 17	13 22	12 25
23	10 06 39	00 ♍ 14 27	19 ♐ 32 44	26 ♐ 42 32	20 R 44	23 57	24 39	00 30	21 42	08 19	13 22	12 27
24	10 10 36	01 12 18	03 ♑ 53 14	11 ♑ 04 20	20 36	25 08	25 17	00 40	21 50	08 21	13 22	12 29
25	10 14 33	02 10 09	18 ♑ 15 16	25 ♑ 25 28	20 22	26 18	25 55	00 49	21 57	08 23	13 22	12 30
26	10 18 29	03 08 03	02 ♒ 34 15	09 ♒ 40 58	20 02	27 29	26 33	00 59	22 05	08 25	13 D 22	12 32
27	10 22 26	04 05 57	16 ♒ 44 58	23 ♒ 45 38	19 36	28 40	27 11	01 09	22 13	08 27	13 22	12 34
28	10 26 22	05 03 53	00 ♓ 42 24	07 ♓ 34 48	19 05	29 ♋ 51	27 49	01 18	22 20	08 29	13 22	12 36
29	10 30 19	06 01 50	14 ♓ 22 27	21 ♓ 05 04	18 27	01 ♌ 02	28 26	01 27	22 28	08 31	13 23	12 38
30	10 34 15	06 59 49	27 ♓ 42 29	04 ♈ 14 39	17 45	02 13	29 04	01 36	22 35	08 33	13 23	12 40
31	10 38 12	07 ♍ 57 50	10 ♈ 41 37	17 ♈ 03 33	16 ♍ 57	03 ♌ 24	29 ♊ 41	01 ♋ 46	22 ♌ 43	08 ♏ 35	13 ♐ 23	12 ♎ 42

Date	Moon True ☊ ° '	Moon Mean ☊ ° '	Moon ☽ Latitude ° '
01	17 ♎ 36	18 ♎ 38	03 N 21
02	17 R 31	18 35	02 22
03	17 28	18 31	01 18
04	17 27	18 28	00 N 11
05	17 D 27	18 25	00 S 55
06	17 27	18 22	01 57
07	17 R 27	18 19	02 53
08	17 26	18 16	03 41
09	17 22	18 12	04 20
10	17 15	18 09	04 47
11	17 07	18 06	05 02
12	16 58	18 03	05 04
13	16 47	18 00	04 51
14	16 37	17 56	04 24
15	16 29	17 53	03 43
16	16 22	17 50	02 49
17	16 18	17 47	01 46
18	16 17	17 44	00 S 35
19	16 D 17	17 41	00 N 38
20	16 17	17 37	01 50
21	16 18	17 34	02 57
22	16 R 18	17 31	03 53
23	16 17	17 28	04 36
24	16 13	17 25	05 02
25	16 08	17 22	05 09
26	16 02	17 18	04 56
27	15 55	17 15	04 26
28	15 49	17 12	03 40
29	15 44	17 09	02 42
30	15 41	17 06	01 36
31	15 ♎ 40	17 ♎ 02	00 N 27

DECLINATIONS

Date	Sun ☉ ° '	Moon ☽ ° '	Mercury ☿ ° '	Venus ♀ ° '	Mars ♂ ° '	Jupiter ♃ ° '	Saturn ♄ ° '	Uranus ♅ ° '	Neptune ♆ ° '	Pluto ♇ ° '
01	17 N 59	06 S 09	09 N 13	21 N 34	21 N 32	22 N 58	16 N 07	13 S 43	20 S 56	10 N 46
02	17 43	01 S 55	08 36	21 38	21 39	22 59	16 04	13 43	20 56	10 45
03	17 28	02 N 17	08 00	21 41	21 45	22 59	16 02	13 43	20 56	10 44
04	17 12	06 15	07 24	21 43	21 51	22 59	16 00	13 44	20 56	10 44
05	16 56	09 50	06 49	21 45	21 57	23 00	15 57	13 44	20 56	10 43
06	16 39	12 56	06 15	21 47	22 03	23 00	15 55	13 44	20 56	10 42
07	16 23	15 27	05 41	21 47	22 09	23 00	15 53	13 45	20 56	10 41
08	16 06	17 16	05 09	21 48	22 14	23 00	15 50	13 45	20 56	10 40
09	15 48	18 21	04 37	21 47	22 20	23 01	15 48	13 46	20 56	10 39
10	15 31	18 38	04 06	21 46	22 25	23 01	15 46	13 46	20 55	10 38
11	15 13	18 05	03 36	21 45	22 29	23 01	15 43	13 47	20 55	10 37
12	14 55	16 41	03 07	21 43	22 34	23 01	15 41	13 47	20 55	10 36
13	14 37	14 30	02 40	21 40	22 39	23 01	15 39	13 48	20 55	10 35
14	14 19	11 35	02 14	21 37	22 43	23 02	15 36	13 48	20 56	10 34
15	14 00	08 04	01 50	21 33	22 47	23 02	15 34	13 49	20 56	10 33
16	13 41	04 N 07	01 28	21 29	22 51	23 02	15 31	13 49	20 56	10 32
17	13 22	00 S 06	01 07	21 24	22 55	23 02	15 29	13 50	20 56	10 31
18	13 03	04 23	00 48	21 18	22 58	23 02	15 27	13 50	20 56	10 30
19	12 43	08 29	00 32	21 12	23 02	23 02	15 24	13 51	20 56	10 29
20	12 24	12 10	00 18	21 05	23 05	23 02	15 22	13 51	20 56	10 28
21	12 04	15 12	00 N 06	20 58	23 08	23 02	15 19	13 52	20 56	10 27
22	11 44	17 21	00 S 03	20 50	23 11	23 02	15 17	13 53	20 56	10 26
23	11 23	18 27	00 09	20 41	23 13	23 02	15 15	13 53	20 56	10 25
24	11 03	18 21	00 12	20 32	23 16	23 02	15 12	13 54	20 56	10 24
25	10 42	17 06	00 12	20 22	23 18	23 02	15 10	13 55	20 56	10 23
26	10 21	14 47	00 08	20 12	23 20	23 02	15 08	13 55	20 56	10 22
27	10 00	11 36	00 S 01	20 01	23 22	23 02	15 05	13 56	20 56	10 21
28	09 39	07 48	00 N 10	19 50	23 24	23 02	15 03	13 57	20 56	10 20
29	09 18	03 S 40	00 24	19 38	23 25	23 02	15 00	13 57	20 56	10 19
30	08 57	00 N 34	00 42	19 26	23 27	23 02	14 58	13 58	20 56	10 18
31	08 N 35	04 N 39	01 N 03	19 N 13	23 N 28	23 N 02	14 N 56	13 S 59	20 S 56	10 N 17

ZODIAC SIGN ENTRIES

Date	h	m	Planets
01	01	23	☽ ♓
02	19	19	♀ ♋
03	06	54	☽ ♈
05	16	18	☽ ♉
08	04	29	☽ ♊
10	17	04	☽ ♋
13	03	57	☽ ♌
15	12	26	☽ ♍
17	18	49	☽ ♎
19	23	35	☽ ♏
20	12	43	♃ ♋
22	03	03	☽ ♐
23	06	00	☉ ♍
24	05	30	☽ ♑
26	07	41	☽ ♒
28	10	46	☽ ♓
28	15	09	♀ ♌
30	16	11	☽ ♈

LATITUDES

Date	Mercury ☿ ° '	Venus ♀ ° '	Mars ♂ ° '	Jupiter ♃ ° '	Saturn ♄ ° '	Uranus ♅ ° '	Neptune ♆ ° '	Pluto ♇ ° '
01	00 S 21	01 S 52	00 S 27	00 S 25	01 N 00	00 N 25	01 N 30	16 N 48
04	00 52	01 42	00 24	00 25	01 00	00 25	01 30	16 47
07	01 25	01 33	00 22	00 25	01 00	00 25	01 30	16 45
10	01 58	01 23	00 19	00 25	01 01	00 25	01 30	16 44
13	02 31	01 13	00 16	00 25	01 01	00 25	01 30	16 43
16	03 04	01 02	00 14	00 25	01 01	00 25	01 30	16 41
19	03 35	00 52	00 11	00 24	01 01	00 25	01 29	16 40
22	04 01	00 42	00 08	00 24	01 02	00 25	01 29	16 39
25	04 21	00 32	00 05	00 24	01 02	00 25	01 29	16 38
28	04 31	00 21	00 S 02	00 24	01 02	00 24	01 29	16 37
31	04 S 27	00 S 11	00 N 01	00 S 24	01 N 02	00 N 24	01 N 29	16 N 36

DATA

Julian Date	2443357
Delta T	+48 seconds
Ayanamsa	23° 32' 45"
Synetic vernal point	05° ♓ 34' 14"
True obliquity of ecliptic	23° 26' 23"

MOON'S PHASES, APSIDES AND POSITIONS ☽

Date	h	m	Phase	Longitude ° '	Eclipse Indicator
06	20	40	☾	14 ♉ 15	
14	21	31	●	21 ♌ 58	
22	01	04	☽	28 ♏ 50	
28	20	10	○	05 ♓ 24	

Day	h	m		
08	23	44	Apogee	
24	09	16	Perigee	
02	22	52	0N	
10	08	04	Max dec	18° N 39'
17	11	25	0S	
23	22	18	Max dec	18° S 33'
30	08	48	0N	

LONGITUDES

Date	Chiron ⚷ ° '	Ceres ⚳ ° '	Pallas ⚴ ° '	Juno ⚵ ° '	Vesta ⚶ ° '	Black Moon Lilith ⚸ ° '
01	05 ♉ 45	14 ♎ 48	21 ♍ 10	15 ♏ 39	00 ♍ 44	11 ♊ 04
11	05 ♉ 47	18 ♎ 21	25 ♍ 24	16 ♏ 58	05 ♍ 27	12 ♊ 11
21	05 ♉ 43	22 ♎ 03	29 ♍ 42	18 ♏ 40	10 ♍ 14	13 ♊ 18
31	05 ♉ 33	25 ♎ 54	04 ♎ 03	20 ♏ 41	15 ♍ 05	14 ♊ 25

ASPECTARIAN

h m	Aspects
01 Monday	
10 35	☽ ☍ ☿
11 33	☽ ± ♇
14 57	☽ △ ♅
17 39	☽ ⚻ ☉
19 33	☽ □ ♂
21 25	☿ ⊥ ♇
22 06	☽ ⚻ ♇
02 Tuesday	
01 02	☽ □ ♆
05 05	☽ ± ☉
05 49	♀ ∥ ♂
10 52	☽ ⚻ ♄
17 44	☽ ⚼ ♅
21 51	☽ ± ♄
22 37	☽ ⚼ ☉
03 Wednesday	
00 59	☽ □ ♃
05 45	☽ Q ♂
08 01	☽ □ ♀
10 18	☽ ± ♅
12 31	☿ ✶ ♅
14 38	☽ ⚼ ♄
21 32	☽ ⚻ ♅
22 26	☽ ⚻ ☿
04 Thursday	
00 56	♂ △ ♇
04 51	☽ △ ☉
05 16	☽ ☍ ♇
05 30	☽ ✶ ♂
08 15	☽ △ ♆
10 26	☉ ✶ ♇
10 58	☽ ± ☿
11 16	☽ Q ♃
14 38	☉ ∠ ♃
14 48	☿ Q ♃
18 25	☽ ∥ ☿
19 26	☽ △ ♄
21 30	☽ Q ♀
05 Friday	
06 06	☽ ⚼ ☿
07 52	☉ ✶ ♂
10 54	☽ ✶ ♃
12 10	☽ ∠ ♂
13 18	☽ ⚼ ♆
18 18	☽ ∥ ♇
23 27	☽ ✶ ♀
06 Saturday	
01 14	☉ △ ♆
07 01	☽ ⊥ ♂
07 06	☽ ± ♆
07 55	☽ ☍ ♅
08 38	♂ ☍ ♆
14 42	☽ △ ☿
16 07	☽ ⚻ ♇
17 09	☽ ∠ ♃
19 05	☽ ⋕ ♅
19 06	☽ ⚻ ♆
19 22	♀ ∠ ♄
19 43	☽ ⚺ ♂
20 40	☽ □ ☉
23 47	♂ ± ♅
07 Sunday	
04 12	☽ ± ♇
07 29	☿ ⚺ ♇
07 31	☽ □ ♄
08 40	☽ ∠ ♀
11 34	☽ ⊥ ♃
16 50	☽ ∥ ♄
21 38	☽ ∥ ☉
22 26	☽ ⚼ ♇
23 55	☽ ⚺ ♃
08 Monday	
04 50	☽ ⊥ ♀
12 15	☽ Q ☉
18 18	☽ ⚺ ♀
18 53	☿ □ ♆
20 22	☽ Q ♄
20 34	☽ ⚻ ♅
09 Tuesday	
04 54	☽ △ ♇
07 44	☽ ☍ ♆
08 44	☽ ± ♅
08 51	☽ □ ☿
12 01	☽ ☌ ♂
12 47	☉ ∥ ♄
14 42	☽ ✶ ☉
17 52	♀ △ ♅
20 54	☽ ✶ ♄
10 Wednesday	
02 56	☽ ⚼ ♅
13 31	☽ ☌ ♃
23 26	☽ ∠ ☉
23 54	☽ Q ☿
11 Thursday	
03 17	☽ ∠ ♄
08 57	☽ △ ♅
12 59	☽ ☌ ♀
17 06	☽ □ ♇
18 48	☽ ⊥ ☉
19 41	☽ ⚻ ♆
21 16	☽ ⊥ ♄
12 Friday	
01 19	☽ ✶ ☿
03 18	☽ ⚺ ♂
07 27	☽ ± ♆
07 34	☽ ⚺ ☉
09 08	☽ ⚺ ♄
15 39	☽ ⊥ ♂
13 Saturday	
00 15	☽ ∥ ♄
00 53	☽ ⚼ ♆
01 25	☽ ⚺ ♃
04 13	☽ Q ♇
06 28	☉ ☌ ♄
08 16	☽ ∠ ☿
09 15	♀ □ ♇
09 58	☽ ∠ ♂
10 39	☽ ∥ ☉
13 05	☽ ⊥ ♃
18 18	☽ ⋕ ♅
19 19	☽ □ ♅
14 Sunday	
02 23	☽ ⊥ ☿
03 09	☽ ✶ ♇
04 54	☽ ⚺ ♀
05 29	☽ △ ♆
06 25	☽ ∠ ♃
11 13	♀ ⚻ ♆
14 17	☽ ⚺ ☿
15 54	☽ ✶ ♂
17 18	☽ ⊥ ♀
18 58	☽ ☌ ♄
19 24	☽ ∥ ♇
21 31	☽ ☌ ☉
15 Monday	
05 10	☽ Q ♅
07 15	☽ ∠ ♇
10 48	☽ ✶ ♃
11 40	☽ ∠ ♀
13 50	♀ ⊥ ♄
15 22	☽ Q ♂
23 52	☽ ⊥ ♇
16 Tuesday	
02 12	☉ ⋕ ♅
03 17	☽ ✶ ♅
09 06	☽ Q ♃
10 50	☽ ⚺ ♇
12 56	☽ □ ♆
17 44	☽ ✶ ♀
23 49	☽ ☌ ☿
17 Wednesday	
01 50	☽ □ ♂
02 29	☽ ⚺ ♄
04 33	☽ ∥ ☿
06 31	☽ ∠ ♅
08 40	☽ ⚺ ☉
13 21	☽ ⊥ ♄
17 16	☽ ⋕ ☿
17 22	☽ Q ♀
17 57	☽ □ ♃
18 19	♂ ✶ ♄
20 14	☽ ⊥ ☉
21 17	☽ Q ♆
22 38	☽ ⊥ ♅
18 Thursday	
05 32	☽ ∠ ♄
09 18	☽ ⚺ ♅
13 25	☽ ∠ ☉
16 37	☽ ☌ ♇
18 32	☽ ✶ ♆
19 Friday	
04 12	☽ □ ♀
06 39	☽ ⚺ ☿
06 53	☉ Q ♅
08 12	☽ ✶ ♄
09 42	☽ △ ♂
14 23	♀ ± ♆
15 19	♂ ∥ ♃
17 20	☽ ⊥ ☿
17 42	☽ ✶ ☉
20 46	☽ ∠ ♆
23 25	☽ △ ♃
20 Saturday	
00 31	☽ ⋕ ♇
05 17	☽ Q ♄
09 12	☽ ∠ ☿
12 03	☉ ∠ ♇
12 21	☽ ⊥ ♆
13 01	☽ ⚼ ♂
13 27	☽ ⋕ ☉
13 48	☽ ☌ ♅
15 28	☿ ✶ ♀
16 02	☽ Q ☉
20 56	☽ ⚺ ♇
22 41	☽ ⚺ ♆
21 Sunday	
00 38	☽ ∥ ♅
01 37	☽ ⚼ ♃
05 14	☽ ± ♂
06 53	♂ ⚼ ♅
07 15	☽ ⊥ ♇
08 25	♀ ⚺ ♄
11 15	☽ ✶ ☿
12 30	☽ □ ♄
12 50	☽ △ ♀
13 06	☽ ⋕ ♄
15 58	☽ ⚻ ♂
17 11	☽ ± ♃
18 10	♀ ⋕ ♆
22 37	☽ ∠ ♇
22 Monday	
01 04	☽ □ ☉
03 30	☽ ⚻ ♃
07 45	☽ Q ☿
14 18	☿ St R
16 35	☽ ⚼ ♀
17 04	☽ ⚺ ♅
23 Tuesday	
00 02	☽ ✶ ♇
01 38	☽ ☌ ♆
03 11	☽ ⊥ ♅
09 05	☽ ± ♀
13 59	☽ □ ☿
15 39	☽ △ ♄
18 19	☽ ∠ ♅
19 43	☉ ✶ ♃
20 03	☽ ⚻ ♀
20 14	☽ Q ♇
20 58	☽ ☍ ♂
24 Wednesday	
06 33	☽ ☍ ♃
07 12	☽ △ ☉
16 58	☽ ⚼ ♄
19 14	♀ ⚺ ♂
19 28	☽ ✶ ♅
20 51	☉ ⊥ ♀
25 Thursday	
02 22	☽ □ ♇
03 50	☽ ⚺ ♆
08 08	☽ ± ♄
10 03	☽ ⚼ ☉
12 08	♆ St D
13 52	☽ ⊥ ♆
15 28	☽ △ ☿
15 33	☽ Q ♅
18 15	☽ ⚻ ♄
26 Friday	
01 26	☽ ⚻ ♂
02 12	☽ ± ☉
02 41	☽ ☍ ♀
04 57	☽ ∠ ♆
08 53	☽ ⋕ ♄
09 18	☽ ⚻ ♃
11 38	☉ ∥ ♇
11 58	☽ ± ♂
13 01	☽ ⚻ ☉
16 03	☽ ⚼ ☿
19 01	☽ ∥ ♅
19 31	☽ ± ♃
21 52	☽ □ ♅
27 Saturday	
03 53	☽ ⚼ ♂
04 53	☽ △ ♇
06 05	♀ ⚼ ♆
06 15	☽ ✶ ♆
06 49	☽ ± ☿
10 57	☽ ⚼ ♃
16 43	☽ ⚻ ☿
20 17	☽ ⋕ ♇
21 25	☽ ☍ ♄
23 28	☽ ⋕ ☉
28 Sunday	
02 46	☽ Q ♆
06 37	☽ ⚼ ♇
06 45	☽ △ ♂
10 22	☽ ⚻ ♀
13 03	☽ △ ♃
20 10	☽ ☍ ☉
21 48	☽ ± ♀
22 19	☽ ± ♇
29 Monday	
01 37	☽ △ ♅
03 50	♀ Q ♇
08 55	☽ ⚻ ♇
10 14	☽ □ ♆
15 14	☽ ⚼ ♀
18 56	☽ ☍ ☿
21 57	♀ ⚺ ♃
30 Tuesday	
02 37	☽ ⚻ ♄
03 34	☉ ⊥ ♇
04 26	☽ ⚼ ♅
05 22	☽ ⋕ ☿
12 50	☽ ∥ ☿
13 37	☽ ± ♄
14 36	☽ □ ♂
18 33	☿ ∠ ♀
19 13	☽ □ ♃
20 54	☽ ± ♅
21 05	☽ △ ♀
31 Wednesday	
06 23	☽ ⚼ ♄
06 29	☽ ⚻ ☉
08 03	☽ ⚻ ♅
15 47	☽ ☍ ♇
17 03	☽ △ ♆
18 39	☽ ± ☉
23 04	☽ ⚻ ☿

All ephemeris data is given at 12.00 UT and the Moon's longitude is additionally given for 24.00 UT

SEPTEMBER 1977

Date	Sidereal time h m s	Sun ☉ ° ' "	Moon ☽ ° ' "	Moon ☽ 24.00 ° ' "	Mercury ☿ ° '	Venus ♀ ° '	Mars ♂ ° '	Jupiter ♃ ° '	Saturn ♄ ° '	Uranus ♅ ° '	Neptune ♆ ° '	Pluto ♇ ° '
01	10 42 08	08 ♍ 55 52	23 ♈ 20 42	29 ♈ 33 25	16 ♍ 06	04 ♌ 35	00 ♋ 18	01 ♋ 54	22 ♌ 50	08 ♏ 38	13 ♐ 23	12 ♎ 44
02	10 46 05	09 53 57	05 ♉ 42 06	11 ♉ 47 14	15 R 11	05 47	00 55	02 03	22 58	08 40	13 23	12 46
03	10 50 02	10 52 03	17 ♉ 49 20	23 ♉ 48 57	14 14	06 58	01 32	02 12	23 05	08 42	13 24	12 48
04	10 53 58	11 50 11	29 ♉ 46 42	05 ♊ 43 11	13 16	08 10	02 09	02 20	23 13	08 45	13 24	12 50
05	10 57 55	12 48 21	11 ♊ 39 00	17 ♊ 34 48	12 18	09 21	02 45	02 29	23 20	08 47	13 24	12 52
06	11 01 51	13 46 34	23 ♊ 31 11	29 ♊ 28 44	11 21	10 33	03 22	02 37	23 27	08 50	13 25	12 55
07	11 05 48	14 44 48	05 ♋ 28 03	11 ♋ 29 39	10 27	11 45	03 58	02 45	23 35	08 52	13 25	12 57
08	11 09 44	15 43 04	17 ♋ 34 03	23 ♋ 41 41	09 37	12 57	04 34	02 53	23 42	08 55	13 25	12 59
09	11 13 41	16 41 22	29 ♋ 52 57	06 ♌ 08 10	08 52	14 09	05 10	03 01	23 49	08 57	13 26	13 01
10	11 17 37	17 39 43	12 ♌ 27 37	18 ♌ 51 28	08 13	15 21	05 46	03 09	23 57	09 00	13 26	13 03
11	11 21 34	18 38 05	25 ♌ 19 49	01 ♍ 52 41	07 41	16 33	06 22	03 17	24 04	09 03	13 27	13 05
12	11 25 31	19 36 29	08 ♍ 30 02	15 ♍ 11 41	07 18	17 45	06 58	03 24	24 11	09 06	13 28	13 07
13	11 29 27	20 34 55	21 ♍ 57 28	28 ♍ 47 06	07 03	18 57	07 33	03 32	24 18	09 08	13 28	13 10
14	11 33 24	21 33 22	05 ♎ 40 16	12 ♎ 36 35	06 57	20 09	08 08	03 39	24 25	09 11	13 29	13 12
15	11 37 20	22 31 52	19 ♎ 35 41	26 ♎ 37 08	07 D 01	21 22	08 44	03 46	24 33	09 14	13 29	13 14
16	11 41 17	23 30 23	03 ♏ 40 32	10 ♏ 45 30	07 13	22 34	09 19	03 53	24 40	09 17	13 30	13 16
17	11 45 13	24 28 56	17 ♏ 51 37	24 ♏ 58 30	07 36	23 47	09 53	03 59	24 47	09 20	13 31	13 19
18	11 49 10	25 27 31	02 ♐ 05 49	09 ♐ 13 14	08 07	24 59	10 28	04 06	24 54	09 23	13 32	13 21
19	11 53 06	26 26 08	16 ♐ 20 26	23 ♐ 27 07	08 48	26 12	11 02	04 12	25 01	09 26	13 32	13 23
20	11 57 03	27 24 46	00 ♑ 33 01	07 ♑ 37 50	09 36	27 25	11 37	04 19	25 07	09 29	13 33	13 25
21	12 01 00	28 23 25	14 ♑ 41 20	21 ♑ 43 14	10 33	28 38	12 11	04 25	25 14	09 32	13 34	13 28
22	12 04 56	29 ♍ 22 07	28 ♑ 43 16	05 ♒ 41 10	11 37	29 ♌ 51	12 45	04 31	25 21	09 35	13 35	13 30
23	12 08 53	00 ♎ 20 50	12 ♒ 36 40	19 ♒ 29 31	12 48	01 ♍ 04	13 19	04 37	25 28	09 38	13 36	13 32
24	12 12 49	01 19 34	26 ♒ 19 28	03 ♓ 06 16	14 04	02 17	13 52	04 42	25 35	09 41	13 37	13 35
25	12 16 46	02 18 21	09 ♓ 49 42	16 ♓ 29 35	15 26	03 30	14 26	04 48	25 41	09 44	13 38	13 37
26	12 20 42	03 17 09	23 ♓ 05 45	29 ♓ 38 06	16 53	04 43	14 59	04 53	25 48	09 47	13 39	13 39
27	12 24 39	04 15 59	06 ♈ 06 34	12 ♈ 31 08	18 24	05 56	15 32	04 58	25 55	09 50	13 40	13 42
28	12 28 35	05 14 52	18 ♈ 51 49	25 ♈ 08 43	19 59	07 09	16 05	05 03	26 01	09 54	13 41	13 44
29	12 32 32	06 13 46	01 ♉ 21 59	07 ♉ 31 50	21 37	08 23	16 38	05 08	26 08	09 57	13 42	13 46
30	12 36 29	07 ♎ 12 42	13 ♉ 38 30	19 ♉ 42 20	23 ♍ 16	09 ♍ 36	17 ♋ 10	05 ♋ 13	26 ♌ 14	10 ♏ 00	13 ♐ 43	13 ♎ 49

Date	Moon True ☊ ° '	Moon Mean ☊ ° '	Moon ☽ Latitude ° '
01	15 ♎ 40	16 ♎ 59	00 S 42
02	15 D 41	16 56	01 47
03	15 43	16 53	02 47
04	15 44	16 50	03 38
05	15 45	16 47	04 19
06	15 R 44	16 43	04 50
07	15 42	16 40	05 08
08	15 39	16 37	05 13
09	15 35	16 34	05 03
10	15 31	16 31	04 39
11	15 26	16 28	04 01
12	15 22	16 24	03 09
13	15 20	16 21	02 05
14	15 18	16 18	00 S 53
15	15 D 18	16 15	00 N 24
16	15 19	16 12	01 40
17	15 20	16 08	02 50
18	15 21	16 05	03 50
19	15 22	16 02	04 36
20	15 R 22	15 59	05 05
21	15 22	15 56	05 16
22	15 20	15 53	05 07
23	15 19	15 49	04 41
24	15 17	15 46	03 59
25	15 15	15 43	03 04
26	15 14	15 40	01 59
27	15 13	15 37	00 N 50
28	15 D 13	15 34	00 S 20
29	15 13	15 30	01 28
30	15 ♎ 14	15 ♎ 27	02 S 31

DECLINATIONS

Date	Sun ☉ ° '	Moon ☽ ° '	Mercury ☿ ° '	Venus ♀ ° '	Mars ♂ ° '	Jupiter ♃ ° '	Saturn ♄ ° '	Uranus ♅ ° '	Neptune ♆ ° '	Pluto ♇ ° '
01	08 N 13	08 N 25	01 N 27	18 N 58	23 N 29	23 N 02	14 N 53	14 S 00	20 S 56	10 N 16
02	07 51	11 44	01 54	18 45	23 30	23 02	14 51	14 00	20 56	10 14
03	07 29	14 29	02 24	18 30	23 31	23 02	14 49	14 01	20 57	10 13
04	07 07	16 33	02 56	18 15	23 31	23 01	14 46	14 02	20 57	10 12
05	06 45	17 54	03 30	17 59	23 32	23 01	14 44	14 03	20 57	10 11
06	06 23	18 27	04 05	17 43	23 32	23 01	14 42	14 04	20 57	10 10
07	06 00	18 12	04 40	17 26	23 32	23 01	14 39	14 04	20 57	10 09
08	05 38	17 07	05 14	17 09	23 32	23 01	14 37	14 05	20 57	10 08
09	05 15	15 14	05 48	16 51	23 32	23 01	14 35	14 06	20 57	10 07
10	04 53	12 36	06 20	16 33	23 31	23 01	14 32	14 07	20 57	10 06
11	04 30	09 18	06 49	16 15	23 31	23 01	14 30	14 08	20 57	10 05
12	04 07	05 28	07 16	15 55	23 30	23 00	14 28	14 09	20 57	10 04
13	03 44	01 N 17	07 40	15 36	23 29	23 00	14 26	14 10	20 58	10 03
14	03 21	03 S 04	08 00	15 16	23 29	23 00	14 23	14 11	20 58	10 02
15	02 58	07 18	08 16	14 55	23 27	23 00	14 21	14 12	20 58	10 01
16	02 35	11 11	08 28	14 34	23 26	23 00	14 19	14 12	20 58	10 00
17	02 12	14 14	08 36	14 13	23 25	23 00	14 17	14 13	20 58	09 59
18	01 48	16 49	08 39	13 51	23 24	22 59	14 14	14 14	20 58	09 58
19	01 25	18 09	08 38	13 29	23 22	22 59	14 12	14 15	20 58	09 57
20	01 02	18 21	08 33	13 07	23 20	22 59	14 10	14 16	20 59	09 56
21	00 38	17 24	08 24	12 44	23 18	22 59	14 08	14 17	20 59	09 55
22	00 N 15	15 24	08 10	12 21	23 17	22 59	14 06	14 18	20 59	09 54
23	00 S 08	12 32	07 53	11 57	23 15	22 59	14 03	14 19	20 59	09 53
24	00 32	09 00	07 32	11 33	23 12	22 59	14 01	14 20	20 59	09 52
25	00 55	05 03	07 07	11 09	23 10	22 58	13 59	14 21	20 59	09 52
26	01 18	00 S 55	06 39	10 44	23 08	22 58	13 57	14 22	21 00	09 51
27	01 42	03 N 12	06 09	10 19	23 05	22 58	13 55	14 23	21 00	09 50
28	02 05	07 05	05 36	09 54	23 03	22 58	13 53	14 24	21 00	09 49
29	02 28	10 34	05 01	09 28	23 00	22 58	13 51	14 25	21 00	09 48
30	02 S 52	13 N 32	04 N 23	09 N 02	22 N 57	22 N 58	13 N 49	14 S 26	21 S 00	09 N 47

ZODIAC SIGN ENTRIES

Date	h	m	Planets
01	00	20	♂ ♋
02	00	52	☽ ♉
04	12	27	☽ ♊
07	01	03	☽ ♋
09	12	14	☽ ♌
11	20	34	☽ ♍
14	02	07	☽ ♎
16	05	45	☽ ♏
18	08	28	☽ ♐
20	11	04	☽ ♑
22	14	12	☽ ♒
22	15	05	♀ ♍
23	03	29	☉ ♎
24	18	30	☽ ♓
27	00	40	☽ ♈
29	09	21	☽ ♉

LATITUDES

Date	Mercury ☿ ° '	Venus ♀ ° '	Mars ♂ ° '	Jupiter ♃ ° '	Saturn ♄ ° '	Uranus ♅ ° '	Neptune ♆ ° '	Pluto ♇ ° '
01	04 S 22	00 S 08	00 N 03	00 S 24	01 N 03	00 N 24	01 N 29	16 N 36
04	03 56	00 N 02	00 06	00 24	01 03	00 24	01 29	16 35
07	03 14	00 11	00 09	00 24	01 03	00 24	01 28	16 35
10	02 20	00 20	00 13	00 23	01 04	00 24	01 28	16 34
13	01 21	00 29	00 16	00 23	01 04	00 24	01 28	16 33
16	00 S 25	00 37	00 20	00 23	01 04	00 24	01 28	16 33
19	00 N 24	00 45	00 23	00 23	01 05	00 24	01 28	16 32
22	01 00	00 52	00 27	00 23	01 05	00 24	01 28	16 32
25	01 30	00 59	00 31	00 23	01 06	00 24	01 27	16 32
28	01 46	01 05	00 35	00 23	01 06	00 24	01 27	16 31
31	01 N 54	01 N 11	00 N 39	00 S 23	01 N 06	00 N 24	01 N 27	16 N 31

DATA

Julian Date	2443388
Delta T	+48 seconds
Ayanamsa	23° 32' 48"
Synetic vernal point	05° ♓ 34' 11"
True obliquity of ecliptic	23° 26' 24"

MOON'S PHASES, APSIDES AND POSITIONS ☽

Date	h	m	Phase	Longitude ° '	Eclipse Indicator
05	14	33	☾	12 ♊ 55	
13	09	23	●	20 ♍ 29	
20	06	18	☽	27 ♐ 11	
27	08	17	○	04 ♈ 07	

Day	h	m		
05	18	13	Apogee	
18	09	21	Perigee	
06	16	30	Max dec	18° N 28'
13	19	06	0S	
20	03	58	Max dec	18° S 25'
26	17	16	0N	

LONGITUDES

Date	Chiron ⚷ ° '	Ceres ⚳ ° '	Pallas ⚴ ° '	Juno ⚵ ° '	Vesta ⚶ ° '	Black Moon Lilith ⚸ ° '
01	05 ♉ 32	26 ♎ 17	04 ♎ 29	20 ♏ 54	15 ♍ 34	14 ♊ 32
11	05 ♉ 17	00 ♏ 15	08 ♎ 52	23 ♏ 14	20 ♍ 28	15 ♊ 39
21	04 ♉ 58	04 ♏ 19	13 ♎ 17	25 ♏ 48	25 ♍ 25	16 ♊ 46
31	04 ♉ 35	08 ♏ 27	17 ♎ 43	28 ♏ 36	00 ♎ 23	17 ♊ 53

All ephemeris data is given at 12.00 UT and the Moon's longitude is additionally given for 24.00 UT

ASPECTARIAN

h m Aspects

01 Thursday
01 52 ☽ Q ♂
04 10 ☉ ⚹ ♅
05 20 ☽ Q ♃
09 46 ☽ ± ☿
10 47 ☽ ∥ ☉
11 01 ☽ △ ♄
13 13 ☽ bQ ☉
21 44 ☽ bQ ♆

02 Friday
00 46 ☽ ∥ ♇
01 58 ☽ bQ ☿
02 09 ☽ ⚹ ♂
04 47 ☽ ⚹ ♃
12 10 ☽ □ ♀
15 19 ☽ ± ♆
17 51 ☽ ☍ ♅
20 59 ☽ △ ☉

03 Saturday
01 59 ☽ ⊼ ♇
03 11 ☽ ⊼ ♆
05 23 ☽ △ ☿
07 33 ☽ # ♅
09 17 ☽ ∠ ♂
10 45 ☽ ∠ ♃
12 47 ☿ Q ♃
13 58 ☽ ± ♇
15 19 ☽ ∥ ♄
22 39 ☽ □ ♄
22 40 ☿ Q ♂

04 Sunday
02 06 ☿ ⊥ ♀
03 54 ☽ Q ♀
04 17 ☽ ⊥ ♂
05 00 ☽ ⊥ ♃
08 05 ☽ bQ ♇
08 46 ☿ □ ♆
11 25 ♀ ⊥ ♂
16 10 ♀ ⊥ ♃
17 02 ☽ ⊻ ♂
17 14 ☽ ⊻ ♃
22 03 ♂ ☌ ♃
22 11 ☿ ⊻ ♇

05 Monday
00 14 ♀ □ ♅
05 41 ☉ ☌ ☿
06 11 ☽ ⊼ ♅
06 50 ☽ ⚹ ♀
11 21 ☽ Q ♄
13 13 ☽ □ ☿
13 45 ☉ ⊻ ♇
13 49 ☽ ∥ ♀
14 29 ☽ △ ♇
14 33 ☽ □ ☉
15 33 ☽ ☍ ♆
18 22 ☽ ± ♅

06 Tuesday
02 53 ☉ □ ♆
11 52 ☽ ⚹ ♄
12 38 ☽ bQ ♅
16 33 ☽ ∠ ♀
21 07 ☿ ⊻ ♀
22 54 ☽ Q ☿

07 Wednesday
06 05 ☽ Q ☉
06 31 ☽ ☌ ♃
08 51 ☽ ☌ ♂
12 18 ☉ Q ♃
12 37 ☽ ⊥ ♀
18 16 ☽ ∠ ♄
18 49 ☽ △ ♅
21 16 ☽ ⚹ ☿

08 Thursday
01 52 ☽ ⊻ ♀
02 55 ☽ □ ♇
03 49 ☽ ⊼ ♆
08 02 ☽ ⚹ ☉
11 17 ☽ ∥ ♀
12 16 ☽ ⊥ ♄
12 46 ♀ ⚹ ♇
15 39 ☽ ± ♆
21 42 ♀ △ ♆
22 06 ☉ ∥ ☿

09 Friday
00 08 ☽ ⊻ ♄
00 59 ☽ ∠ ☿
08 56 ☿ ⚹ ♅
09 12 ☽ bQ ♆
14 11 ☽ Q ♇
15 47 ☽ ∠ ☉
17 26 ☽ ⊥ ☿
18 06 ☽ ⊻ ♃
18 43 ☽ ∥ ♄
22 40 ☽ ⊻ ♂
23 02 ☽ # ♅

10 Saturday
04 20 ☽ ⊻ ☿
05 26 ☽ □ ♅
05 41 ☽ ⊥ ♃
10 22 ☽ ⊥ ☉
10 38 ☽ ⊥ ♂
13 07 ☽ ⚹ ♇
13 51 ☽ △ ♆
17 59 ☽ ☌ ♀
19 02 ☉ Q ♂
22 34 ☽ ⊻ ☉
22 47 ☽ ∠ ♃

11 Sunday
04 19 ☽ ∠ ♂
06 36 ☽ ∥ ♇
09 39 ☽ ☌ ♄
15 10 ☽ Q ♅
17 05 ☽ ∠ ♇

12 Monday
02 09 ☽ ∥ ☿
02 41 ☽ ⚹ ♃
09 06 ☽ ⚹ ♂
09 31 ☽ ⊥ ♇
09 53 ☽ ☌ ☿
13 04 ☽ ⚹ ♅
20 20 ☽ ⊻ ♇
20 43 ☽ ∥ ☉
20 54 ☽ □ ♆
21 05 ☿ ⚹ ♂

13 Tuesday
00 29 ☽ Q ♃
00 54 ☿ ⊥ ♇
02 34 ♀ ∠ ♃
06 10 ☽ ⊻ ♀
07 33 ☽ Q ♂
09 23 ☽ ☌ ☉
15 51 ☽ ∠ ♅
16 10 ☽ ⊻ ♄
17 47 ☽ ⊥ ♀

14 Wednesday
02 47 ☽ ⊥ ♄
04 42 ☽ Q ♆
07 40 ☽ ⊥ ♅
08 27 ☽ □ ♃
11 01 ☽ ∠ ♀
13 28 ☽ # ☉
14 13 ☽ ⊻ ☿
15 04 ☿ St D
16 28 ☽ □ ♂
18 07 ☽ ⊻ ♅
18 33 ☽ ∠ ♄

15 Thursday
00 36 ☽ ⊥ ☿
01 03 ☽ ☌ ♇
01 30 ☽ ⚹ ♆
15 19 ☽ ⚹ ♀
16 11 ☽ ∠ ☿
17 24 ☽ ⊻ ☉
18 04 ☽ # ☿
20 32 ☽ ⚹ ♄

16 Friday
03 12 ☽ ∠ ♆
04 23 ☽ ⊥ ☉
04 25 ☽ # ♇
10 42 ♂ △ ♅
12 21 ☽ △ ♃
13 40 ☽ Q ♀
16 17 ☿ ⊥ ♇
17 06 ☽ Q ♄
18 09 ☽ ⚹ ☿
18 30 ☽ ⊥ ♆
20 48 ☽ ∠ ☉
21 32 ☽ ☌ ♅
21 57 ☽ △ ♂

17 Saturday
04 18 ☽ ⊻ ♇
04 39 ☽ ⊻ ♆
06 10 ♂ ∠ ♄
07 46 ♀ ∥ ♄
08 01 ☉ ∠ ♅
10 15 ☽ ∥ ♅
10 25 ☽ # ♀
10 42 ☽ # ♄
11 45 ♀ # ♅
13 55 ☽ bQ ♃
14 27 ☽ ⊥ ♇
15 02 ☽ Q ☿
20 14 ☉ ⊻ ♄
22 55 ☽ □ ♀
23 46 ☽ □ ♄

18 Sunday
00 00 ☽ ⚹ ☉
00 21 ☽ bQ ♂
05 13 ☽ ± ♃
05 40 ☽ ∠ ♇
09 51 ♀ ☌ ♄
11 51 ♄ # ♅
15 24 ☽ ⊼ ♃
16 10 ☽ ± ♂
21 42 ☽ Q ☉
22 37 ☽ □ ☿

19 Monday
00 18 ☽ ⊻ ♅
02 42 ☽ ⊼ ♂
05 55 ♀ ∠ ♂
07 00 ☽ ⚹ ♇
07 16 ☽ ☌ ♆
10 27 ☽ ⊥ ♅

20 Tuesday
01 42 ☽ ∠ ♅
02 45 ☽ △ ♄
03 18 ☽ Q ♇
06 12 ☽ △ ♀
06 18 ☽ □ ☉
08 14 ☿ ⚹ ♅
11 43 ☉ ⊻ ♀
13 16 ♀ Q ♅
18 25 ☽ ☍ ♃

21 Wednesday
03 11 ☽ ⚹ ♅
04 22 ☽ bQ ♄
04 26 ☽ △ ☿
07 33 ☽ ☍ ♂
08 36 ♀ ∠ ♇
09 54 ☽ □ ♇
10 02 ☽ bQ ♀
10 05 ☽ ⊻ ♆
19 49 ☽ ± ♄
20 20 ☽ ⊥ ♆
23 43 ☽ Q ♅

22 Thursday
02 50 ☽ ± ♀
06 10 ☽ ⊼ ♄
08 04 ☽ bQ ☿
11 46 ☽ ∠ ♆
13 12 ☽ △ ☉
14 07 ☽ ⊼ ♀
21 53 ☽ ∥ ♅
22 03 ☽ ⊼ ♃
23 53 ☽ # ♄

23 Friday
00 57 ☽ ± ☿
06 48 ☽ □ ♅
08 30 ☽ ± ♃
12 21 ☽ ⊼ ☿
13 16 ☽ ⊼ ♂
13 37 ☽ △ ♇
13 43 ☽ ⚹ ♆
16 47 ☽ # ♀
17 08 ☽ bQ ☉
22 29 ♂ □ ♇

24 Saturday
00 11 ☽ ± ♂
00 18 ☽ bQ ♃
00 38 ♂ ⊼ ♆
02 39 ☿ ⊻ ♇
03 29 ☿ □ ♆
05 36 ☿ ⚹ ♂
06 20 ☽ # ♇
10 06 ☽ ± ☉
10 41 ☽ ☍ ♄
10 45 ☽ Q ♆
15 59 ☽ bQ ♇
16 41 ☽ bQ ♂
19 00 ☉ ⊥ ♄
19 09 ☉ Q ♆
21 32 ☽ ⊼ ☉
22 11 ☽ # ☿
23 34 ☽ ☍ ♀

25 Sunday
02 57 ☽ △ ♃
08 02 ☽ ± ♇
11 50 ☽ △ ♅
18 50 ☽ ⊼ ♇
18 50 ☽ □ ♆
20 38 ☽ △ ♂
23 19 ☽ ☍ ☿

26 Monday
02 58 ♆ ⚹ ♇
09 55 ☽ ∥ ☉
11 56 ☿ Q ♃
15 06 ☽ bQ ♅
15 40 ♀ ⚹ ♃
17 00 ☽ ⊼ ♄

27 Tuesday
00 58 ☉ ⊥ ♅
02 13 ☽ # ☉
04 08 ☽ ± ♄
07 46 ☽ ± ♅
08 17 ☽ ☍ ☉
09 52 ☽ □ ♃
11 38 ☽ ⊼ ♀
19 00 ☽ ⊼ ♅
21 03 ☽ bQ ♄

28 Wednesday
00 03 ☽ ± ♀
02 11 ☽ △ ♆
02 15 ☽ ☍ ♇
03 49 ☽ ∥ ☿
06 29 ☽ □ ♂
06 51 ☉ □ ♃
14 26 ☽ ⊼ ☿
17 04 ♀ ∥ ♇
18 57 ☽ bQ ♀
20 03 ☽ Q ♃
23 44 ♀ ⊥ ♇

29 Thursday
01 48 ☽ △ ♄
03 38 ☽ ± ☿
04 59 ☽ ∥ ♀
06 25 ☽ ∥ ♇
06 50 ☽ bQ ♆
18 37 ☽ Q ♂
19 22 ☽ ⚹ ♃
22 17 ☽ ⊼ ☉
23 47 ☽ bQ ☿

30 Friday
00 21 ☽ ± ♆
03 10 ☽ △ ♀
04 49 ☽ ☍ ♅
08 54 ♂ ∥ ♃
11 05 ☽ ± ☉
12 09 ☽ ⊼ ♆
12 20 ☽ ⊼ ♇
14 34 ☽ ∥ ♄
19 17 ☽ ⚹ ♂
20 18 ♀ ⚹ ♅
20 42 ☽ # ♅

OCTOBER 1977

Date	Sidereal time h m s	Sun ☉ ° ' "	Moon ☽ ° ' "	Moon ☽ 24.00 ° ' "	Mercury ☿ ° '	Venus ♀ ° '	Mars ♂ ° '	Jupiter ♃ ° '	Saturn ♄ ° '	Uranus ♅ ° '	Neptune ♆ ° '	Pluto ♇ ° '
01	12 40 25	08 ♎ 11 41	25 ♉ 43 40	01 ♊ 42 55	24 ♍ 58	10 ♍ 49	17 ♋ 42	05 ♋ 17	26 ♌ 21	10 ♏ 04	13 ♐ 44	13 ♎ 51
02	12 44 22	09 10 42	07 ♊ 40 31	13 ♊ 36 59	26 42	12 03	18 14	05 21	26 27	10 07	13 45	13 53
03	12 48 18	10 09 45	19 ♊ 32 49	25 ♊ 28 34	28 ♍ 27	13 16	18 46	05 26	26 33	10 10	13 47	13 56
04	12 52 15	11 08 51	01 ♋ 24 48	07 ♋ 22 05	00 ♎ 12	14 30	19 18	05 30	26 39	10 14	13 48	13 58
05	12 56 11	12 07 59	13 ♋ 21 01	19 ♋ 22 10	01 58	15 44	19 49	05 33	26 46	10 17	13 49	14 00
06	13 00 08	13 07 09	25 ♋ 26 07	01 ♌ 33 25	03 45	16 58	20 21	05 37	26 52	10 20	13 50	14 03
07	13 04 04	14 06 21	07 ♌ 44 36	14 ♌ 00 08	05 31	18 11	20 52	05 40	26 58	10 24	13 52	14 05
08	13 08 01	15 05 36	20 ♌ 20 29	26 ♌ 45 59	07 18	19 25	21 23	05 43	27 04	10 27	13 53	14 07
09	13 11 58	16 04 53	03 ♍ 16 57	09 ♍ 53 33	09 04	20 39	21 53	05 46	27 10	10 31	13 55	14 10
10	13 15 54	17 04 12	16 ♍ 35 55	23 ♍ 24 00	10 50	21 53	22 23	05 49	27 16	10 34	13 56	14 12
11	13 19 51	18 03 33	00 ♎ 17 40	07 ♎ 16 39	12 36	23 07	22 54	05 52	27 21	10 38	13 57	14 15
12	13 23 47	19 02 57	14 ♎ 20 36	21 ♎ 28 58	14 21	24 21	23 23	05 54	27 27	10 41	13 59	14 17
13	13 27 44	20 02 22	28 ♎ 41 10	05 ♏ 56 30	16 06	25 35	23 53	05 57	27 33	10 45	14 00	14 19
14	13 31 40	21 01 50	13 ♏ 14 11	20 ♏ 33 26	17 50	26 49	24 22	05 59	27 38	10 49	14 02	14 22
15	13 35 37	22 01 19	27 ♏ 53 24	05 ♐ 13 18	19 33	28 04	24 51	06 01	27 44	10 52	14 03	14 24
16	13 39 33	23 00 51	12 ♐ 32 20	19 ♐ 49 48	21 16	29 ♍ 18	25 20	06 02	27 49	10 56	14 05	14 26
17	13 43 30	24 00 24	27 ♐ 05 04	04 ♑ 17 36	22 59	00 ♎ 32	25 49	06 04	27 55	10 59	14 07	14 29
18	13 47 27	24 59 59	11 ♑ 26 56	18 ♑ 32 45	24 40	01 46	26 17	06 05	28 00	11 03	14 08	14 31
19	13 51 23	25 59 36	25 ♑ 34 48	02 ♒ 32 55	26 22	03 01	26 45	06 06	28 05	11 07	14 10	14 33
20	13 55 20	26 59 14	09 ♒ 27 02	16 ♒ 17 06	28 02	04 15	27 13	06 07	28 11	11 10	14 11	14 36
21	13 59 16	27 58 54	23 ♒ 03 11	29 ♒ 45 22	29 ♎ 42	05 30	27 40	06 08	28 16	11 14	14 13	14 38
22	14 03 13	28 58 35	06 ♓ 23 44	12 ♓ 58 24	01 ♏ 21	06 44	28 07	06 08	28 21	11 18	14 15	14 40
23	14 07 09	29 ♎ 58 19	19 ♓ 29 31	25 ♓ 57 13	03 00	07 59	28 34	06 08	28 26	11 21	14 17	14 43
24	14 11 06	00 ♏ 58 04	02 ♈ 21 39	08 ♈ 42 55	04 38	09 13	29 00	06 R 09	28 31	11 25	14 18	14 45
25	14 15 02	01 57 51	15 ♈ 01 11	21 ♈ 16 33	06 15	10 28	29 27	06 08	28 35	11 29	14 20	14 47
26	14 18 59	02 57 40	27 ♈ 29 10	03 ♉ 39 11	07 52	11 42	29 ♋ 53	06 08	28 40	11 32	14 22	14 49
27	14 22 56	03 57 30	09 ♉ 46 42	15 ♉ 51 55	09 28	12 57	00 ♌ 18	06 08	28 45	11 36	14 24	14 52
28	14 26 52	04 57 23	21 ♉ 54 59	27 ♉ 56 08	11 04	14 12	00 43	06 07	28 49	11 40	14 25	14 54
29	14 30 49	05 57 18	03 ♊ 55 33	09 ♊ 53 32	12 40	15 26	01 08	06 06	28 54	11 44	14 27	14 56
30	14 34 45	06 57 15	15 ♊ 50 21	21 ♊ 46 20	14 15	16 41	01 33	06 05	28 58	11 47	14 29	14 58
31	14 38 42	07 ♏ 57 14	27 ♊ 41 51	03 ♋ 37 18	15 ♏ 49	17 ♎ 56	01 ♌ 57	06 ♋ 04	29 ♌ 02	11 ♏ 51	14 ♐ 31	15 ♎ 01

Date	Moon True ☊ ° '	Moon Mean ☊ ° '	Moon ☽ Latitude ° '
01	15 ♎ 15	15 ♎ 24	03 S 26
02	15 D 15	15 21	04 12
03	15 16	15 18	04 46
04	15 16	15 14	05 08
05	15 R 16	15 11	05 17
06	15 16	15 08	05 12
07	15 16	15 05	04 53
08	15 16	15 02	04 20
09	15 D 16	14 59	03 33
10	15 16	14 55	02 33
11	15 16	14 52	01 22
12	15 17	14 49	00 S 05
13	15 R 16	14 46	01 N 14
14	15 16	14 43	02 29
15	15 15	14 40	03 35
16	15 15	14 36	04 27
17	15 14	14 33	05 01
18	15 13	14 30	05 16
19	15 13	14 27	05 12
20	15 D 13	14 24	04 49
21	15 14	14 20	04 11
22	15 15	14 17	03 19
23	15 16	14 14	02 18
24	15 17	14 11	01 11
25	15 17	14 08	00 N 02
26	15 R 17	14 05	01 S 07
27	15 16	14 01	02 11
28	15 13	13 58	03 08
29	15 10	13 55	03 57
30	15 06	13 52	04 34
31	15 ♎ 03	13 ♎ 49	05 S 00

DECLINATIONS

Date	Sun ☉ ° '	Moon ☽ ° '	Mercury ☿ ° '	Venus ♀ ° '	Mars ♂ ° '	Jupiter ♃ ° '	Saturn ♄ ° '	Uranus ♅ ° '	Neptune ♆ ° '	Pluto ♇ ° '
01	03 S 15	15 N 51	03 N 44	08 N 36	22 N 54	22 N 58	13 N 47	14 S 28	21 S 01	09 N 46
02	03 38	17 27	03 04	08 10	22 52	22 57	13 45	14 29	21 01	09 45
03	04 01	18 17	02 22	07 43	22 49	22 57	13 43	14 30	21 01	09 44
04	04 25	18 18	01 39	07 17	22 45	22 57	13 41	14 31	21 01	09 43
05	04 48	17 31	00 55	06 49	22 42	22 57	13 39	14 32	21 01	09 42
06	05 11	15 56	00 N 11	06 22	22 39	22 57	13 37	14 33	21 01	09 41
07	05 34	13 36	00 S 34	05 55	22 36	22 57	13 35	14 34	21 02	09 41
08	05 57	10 36	01 19	05 27	22 32	22 57	13 33	14 35	21 02	09 40
09	06 20	07 00	02 04	04 59	22 29	22 57	13 31	14 36	21 02	09 39
10	06 42	02 N 57	02 50	04 31	22 26	22 57	13 29	14 37	21 02	09 38
11	07 05	01 S 22	03 35	04 03	22 22	22 57	13 27	14 38	21 03	09 37
12	07 28	05 44	04 20	03 34	22 19	22 56	13 25	14 40	21 03	09 36
13	07 50	09 52	05 05	03 06	22 15	22 56	13 24	14 41	21 03	09 36
14	08 13	13 27	05 50	02 37	22 11	22 56	13 22	14 42	21 03	09 35
15	08 35	16 12	06 35	02 09	22 08	22 56	13 20	14 43	21 03	09 34
16	08 57	17 53	07 19	01 40	22 04	22 56	13 18	14 44	21 04	09 33
17	09 19	18 23	08 02	01 11	22 00	22 56	13 17	14 45	21 04	09 32
18	09 41	17 42	08 46	00 42	21 56	22 56	13 15	14 46	21 04	09 32
19	10 02	15 55	09 28	00 N 13	21 53	22 56	13 13	14 48	21 04	09 31
20	10 24	13 14	10 10	00 S 16	21 49	22 56	13 12	14 49	21 05	09 30
21	10 45	09 53	10 52	00 45	21 45	22 56	13 10	14 50	21 05	09 29
22	11 07	06 05	11 33	01 14	21 41	22 56	13 09	14 51	21 05	09 29
23	11 28	02 S 03	12 13	01 44	21 37	22 56	13 07	14 52	21 05	09 28
24	11 49	02 N 01	12 52	02 13	21 34	22 57	13 06	14 53	21 06	09 27
25	12 09	05 56	13 31	02 42	21 30	22 57	13 04	14 54	21 06	09 26
26	12 30	09 32	14 09	03 11	21 26	22 57	13 03	14 56	21 06	09 26
27	12 50	12 40	14 47	03 40	21 22	22 57	13 01	14 57	21 06	09 25
28	13 10	15 13	15 23	04 09	21 18	22 57	13 00	14 58	21 06	09 24
29	13 30	17 03	15 59	04 38	21 15	22 57	12 58	14 59	21 07	09 24
30	13 50	18 08	16 34	05 07	21 11	22 57	12 57	15 00	21 07	09 23
31	14 S 10	18 N 25	17 S 08	05 S 36	21 N 07	22 N 57	12 N 56	15 S 01	21 S 07	09 N 23

ZODIAC SIGN ENTRIES

Date	h	m	Planets
01	20	33	☽ ♊
04	09	09	☽ ♋
04	09	16	☿ ♎
06	20	58	☽ ♌
09	05	59	☽ ♍
11	11	29	☽ ♎
13	14	11	☽ ♏
15	15	27	☽ ♐
17	01	37	♀ ♎
17	16	51	☽ ♑
19	19	36	☽ ♒
21	16	23	☿ ♏
22	00	26	☽ ♓
23	12	41	☉ ♏
24	07	34	☽ ♈
26	16	53	☽ ♉
26	18	56	♂ ♌
29	04	08	☽ ♊
31	16	40	☽ ♋

LATITUDES

Date	Mercury ☿ ° '	Venus ♀ ° '	Mars ♂ ° '	Jupiter ♃ ° '	Saturn ♄ ° '	Uranus ♅ ° '	Neptune ♆ ° '	Pluto ♇ ° '
01	01 N 54	01 N 11	00 N 39	00 S 23	01 N 06	00 N 24	01 N 27	16 N 31
04	01 53	01 16	00 43	00 22	01 07	00 24	01 27	16 31
07	01 46	01 21	00 47	00 22	01 07	00 24	01 27	16 31
10	01 35	01 25	00 51	00 22	01 08	00 24	01 27	16 31
13	01 21	01 28	00 56	00 22	01 08	00 23	01 27	16 31
16	01 04	01 30	01 01	00 22	01 09	00 23	01 26	16 32
19	00 45	01 32	01 05	00 22	01 09	00 23	01 26	16 32
22	00 25	01 34	01 10	00 22	01 10	00 23	01 26	16 32
25	00 N 05	01 34	01 15	00 21	01 11	00 23	01 26	16 33
28	00 S 15	01 34	01 21	00 21	01 11	00 23	01 26	16 34
31	00 S 35	01 N 33	01 N 26	00 S 21	01 N 12	00 N 23	01 N 26	16 N 34

DATA

Julian Date	2443418
Delta T	+48 seconds
Ayanamsa	23° 32' 51"
Synetic vernal point	05° ♓ 34' 09"
True obliquity of ecliptic	23° 26' 23"

MOON'S PHASES, APSIDES AND POSITIONS ☽

Date	h	m	Phase	Longitude ° '	Eclipse Indicator
05	09	21	☾	12 ♋ 01	
12	20	31	●	19 ♎ 24	Total
19	12	46	☽	26 ♑ 01	
26	23	35	○	03 ♉ 27	

Day	h	m		
03	13	43	Apogee	
15	08	53	Perigee	
31	08	03	Apogee	
04	00	39	Max dec	18° N 23'
11	04	27	0S	
17	09	54	Max dec	18° S 24'
24	00	01	0N	
31	08	22	Max dec	18° N 26'

LONGITUDES

Date	Chiron ⚷ ° '	Ceres ⚳ ° '	Pallas ⚴ ° '	Juno ⚵ ° '	Vesta ⚶ ° '	Black Moon Lilith ⚸ ° '
01	04 ♉ 35	08 ♏ 27	17 ♎ 43	28 ♏ 36	00 ♎ 23	17 ♊ 53
11	04 ♉ 09	12 ♏ 38	22 ♎ 10	01 ♐ 33	05 ♎ 24	19 ♊ 00
21	03 ♉ 41	16 ♏ 51	26 ♎ 37	04 ♐ 40	10 ♎ 25	20 ♊ 07
31	03 ♉ 12	21 ♏ 07	01 ♏ 04	07 ♐ 55	15 ♎ 26	21 ♊ 14

ASPECTARIAN

h m Aspects

01 Saturday
00 15 ☽ ± ♇
01 05 ☽ ∠ ♃
06 30 ☽ bQ ☉
10 14 ☽ △ ☿
13 14 ☿ ∠ ♅
13 14 ☽ □ ♄
18 16 ☽ bQ ♇
19 10 ☽ ⊥ ♃
23 00 ☉ # ☿

02 Sunday
02 39 ☽ ∠ ♂
07 18 ☽ ⊻ ♃
08 20 ☿ ⊻ ♄
15 18 ☽ △ ☉
16 57 ☽ ⊼ ♅
21 39 ☽ ⊥ ♂
21 51 ☽ □ ♀

03 Monday
00 18 ☽ ☍ ♆
00 36 ☽ △ ♇
01 48 ☽ Q ♄
05 08 ☽ ± ♅
10 21 ☽ ⊻ ♂
12 11 ☉ ⊻ ♅
22 00 ♀ □ ♆
23 26 ☽ bQ ♅

04 Tuesday
01 12 ♀ ⊻ ♇
02 18 ☽ ✶ ♄
09 07 ☽ □ ☿
14 27 ☽ Q ♀
20 16 ☽ ☌ ♃

05 Wednesday
01 53 ☉ ∠ ♄
05 50 ☽ △ ♅
08 47 ☽ ∠ ♄
09 14 ☿ Q ♂
09 21 ☽ □ ☉
09 58 ☿ Q ♆
11 47 ♂ ± ♆
12 56 ☽ ⊼ ♆
13 19 ☽ □ ♇
17 18 ☽ ✶ ♀
23 23 ☿ ⊥ ♄

06 Thursday
00 55 ☽ ± ♆
01 29 ☽ ☌ ♂
02 53 ☽ ⊥ ♄
03 27 ☽ Q ☿
14 50 ☽ ⊻ ♄
18 42 ☽ bQ ♆
20 21 ☿ ⊥ ♅

07 Friday
00 07 ☽ Q ☉
01 00 ☽ Q ♇
01 25 ♀ Q ♃
02 12 ☽ ∠ ♀
03 01 ☽ # ♅
05 58 ☉ ✶ ♆
06 59 ☽ ✶ ☿
07 59 ☽ ⊻ ♃
11 29 ☉ ☌ ♇
12 13 ☽ ∥ ♄
14 08 ☿ □ ♃
17 08 ☽ □ ♅
17 52 ♂ ⊥ ♄
19 35 ☽ ⊥ ♃
21 28 ☽ ⊥ ♀
21 52 ☉ # ♀
23 45 ☽ △ ♆

08 Saturday
00 12 ☽ ✶ ♇
01 14 ☽ ✶ ☉
10 05 ☽ ⊻ ♀
12 43 ☽ ∠ ♃
14 01 ☽ ⊻ ♂
16 15 ☽ ∠ ☿
18 36 ☽ ∥ ♇

09 Sunday
00 39 ☽ ☌ ♄
01 40 ☽ ⊥ ♂
03 12 ☽ Q ♅
04 25 ☽ ∠ ♇
07 38 ☽ ∠ ☉
11 32 ☽ ⊥ ☿
15 46 ☽ # ☉
16 34 ☽ ✶ ♃
18 49 ☽ ∠ ♂
20 54 ☽ ⊥ ♇

10 Monday
00 07 ☽ ⊻ ☿
01 11 ☽ ✶ ♅
01 20 ☽ ⊥ ☉
01 47 ☽ ∥ ♀
07 14 ☽ □ ♆
07 43 ☽ ⊻ ♇
08 20 ☿ ⊻ ♅
12 34 ☽ # ☿
12 54 ☽ ⊻ ☉
14 11 ☽ Q ♃
22 16 ☽ ☌ ♀
22 37 ☽ ✶ ♂

11 Tuesday
03 52 ☽ ∠ ♅
04 35 ♀ ✶ ♂
06 52 ☽ ⊻ ♄
08 34 ☿ ∠ ♄
14 52 ☽ Q ♆
17 19 ☽ ⊥ ♄
19 30 ☽ ⊥ ♅
20 13 ☽ Q ♂
21 03 ☿ # ♀
21 37 ☽ □ ♃

12 Wednesday
01 13 ☽ # ♀
02 38 ☽ ∥ ☿
05 47 ☽ ⊻ ♅
06 53 ☿ ✶ ♆
08 47 ☽ ∠ ♄
11 04 ☿ ☌ ♇
11 23 ☽ ✶ ♆
11 54 ☽ ☌ ♇
12 01 ☽ ☌ ☿
20 31 ☽ ☌ ☉
22 45 ☽ ∥ ☉

13 Thursday
03 44 ☽ □ ♂
06 22 ☽ ⊻ ♀
10 06 ☽ ✶ ♄
10 22 ☽ # ♇
12 32 ☽ ∠ ♆
15 18 ♀ ∠ ♅
17 15 ☽ ⊥ ♀

14 Friday
00 02 ☽ △ ♃
03 26 ☽ ⊥ ♆
06 03 ☽ Q ♄
08 00 ☽ ☌ ♅
09 28 ☽ ∠ ♀
11 26 ☽ # ♄
13 18 ☽ ⊻ ♆
13 51 ☽ ⊻ ♇
20 32 ☽ ⊻ ☿
21 59 ☽ ∥ ♅
23 43 ☽ ⊥ ♇

15 Saturday
00 43 ☽ bQ ♃
01 42 ☽ ⊻ ☉
05 09 ♀ ⊻ ♄
06 52 ☽ △ ♂
07 40 ☽ ⊥ ☿
11 45 ☽ □ ♄
12 14 ☽ ⊥ ☉
12 18 ☽ ✶ ♀
14 29 ☽ ∠ ♇
15 29 ☽ ± ♃

16 Sunday
00 22 ☽ ∠ ☿
01 19 ☽ ⊼ ♃
04 02 ☽ ∠ ☉
08 16 ☽ bQ ♂
09 21 ☽ ⊻ ♅
09 46 ☽ Q ♀
14 32 ☽ ☌ ♆
15 08 ☽ ✶ ♇
19 15 ☽ ⊥ ♅
23 34 ☽ ± ♂

17 Monday
04 18 ☽ ✶ ☿
06 32 ☽ ✶ ☉
09 49 ☽ ⊼ ♂
10 11 ☽ ∠ ♅
10 59 ☽ Q ♇
13 23 ☽ △ ♄
18 16 ☽ □ ♀

18 Tuesday
02 19 ☉ # ♇
02 54 ☽ Q ☿
02 59 ☽ ☍ ♃
03 58 ☽ Q ☉
11 20 ☽ ✶ ♅
14 38 ☽ bQ ♄
16 32 ☽ ⊻ ♆
17 11 ☽ □ ♇
19 09 ♀ Q ♆
23 16 ☉ ☌ ☿

19 Wednesday
02 44 ☽ ⊥ ♆
06 00 ☽ ± ♄
07 45 ☽ Q ♅
12 46 ☽ □ ☉
13 24 ☿ # ♇
13 31 ☽ □ ☿
14 04 ☽ ☍ ♂
16 20 ☽ ⊼ ♄
18 10 ☽ ∠ ♆
19 43 ☿ □ ♂
22 54 ☽ ∥ ♅

20 Thursday
02 04 ☽ △ ♀
06 11 ☽ ⊼ ♃
10 24 ♀ ⊥ ♄
12 21 ☽ # ♄
14 11 ☿ ✶ ♄
15 01 ☽ □ ♅
16 40 ☽ ± ♃
20 20 ☽ ✶ ♆
21 03 ☽ △ ♇
22 00 ☉ □ ♂

21 Friday
04 10 ☉ ∥ ☿
04 58 ☿ ∠ ♆
06 28 ☽ ∥ ☿
06 40 ☽ ∥ ☉
06 42 ♀ ⊥ ♅
06 59 ☽ bQ ♀
08 34 ☽ bQ ♃
14 37 ☽ # ♇
17 40 ☽ Q ♆
19 23 ☉ ✶ ♄
20 32 ☽ ⊼ ♂
21 22 ☽ ☍ ♄
21 31 ☽ △ ☉
23 49 ☽ bQ ♇

22 Saturday
00 21 ♀ □ ♃
00 43 ☽ ± ♀
01 35 ☽ △ ☿
07 44 ☽ ± ♂
11 32 ☽ △ ♃
12 41 ☽ ⊼ ♀
16 09 ☽ ± ♇
18 43 ☉ ∠ ♆
20 58 ☽ △ ♅

23 Sunday
00 42 ☽ bQ ♂
02 22 ☽ □ ♆
02 52 ♂ ⊻ ♄
02 59 ☽ bQ ☉
03 10 ☽ ⊼ ♇
08 50 ☽ bQ ☿
13 40 ☽ ∥ ♀
21 00 ☽ ± ☉

24 Monday
00 49 ☽ bQ ♅
03 59 ☽ ± ☿
04 44 ☽ ⊼ ♄
05 29 ☽ △ ♂
09 10 ☽ ⊼ ☉
10 14 ♃ St R
13 17 ☽ # ♀
16 04 ☽ ± ♄
16 54 ☽ ⊼ ☿
17 47 ☽ ± ♅
19 08 ☽ □ ♃
19 43 ☿ # ♄

25 Tuesday
02 22 ☽ ☍ ♀
05 13 ☽ ⊼ ♅
05 27 ♂ bQ ♆
09 15 ☽ bQ ♄
10 20 ☿ △ ♃
10 41 ☽ △ ♆
11 33 ☽ ☍ ♇

26 Wednesday
05 31 ☽ Q ♃
08 40 ♀ ⊻ ♅
11 14 ☽ ∥ ♇
14 18 ☽ △ ♄
15 39 ☽ bQ ♆
16 48 ☽ □ ♂
17 02 ♀ Q ♂
19 32 ☿ ⊥ ♆
23 35 ☽ ☍ ☉

27 Thursday
04 50 ☽ ✶ ♃
09 16 ☽ ± ♆
11 19 ☽ ☍ ☿
13 37 ☽ # ☉
14 57 ☽ ∥ ♄
15 37 ☽ ☍ ♅
18 44 ☿ ∥ ♅
18 57 ☽ ⊼ ♀
21 07 ☽ ⊼ ♆
22 03 ☽ ⊼ ♇

28 Friday
00 03 ☉ # ♄
04 19 ♀ ∠ ♄
05 26 ☽ Q ♂
08 01 ☿ Q ♄
08 11 ☽ ± ♀
09 21 ☽ # ♅
09 58 ☽ ± ♇
10 24 ☽ ∠ ♃
14 43 ☽ # ☿
16 33 ♀ ✶ ♆
21 19 ☿ ☌ ♅

29 Saturday
01 51 ☽ □ ♄
01 59 ♀ ☌ ♇
03 59 ☽ bQ ♇
04 12 ☽ bQ ♀
04 20 ☽ ⊥ ♃
06 13 ☽ ✶ ♂
15 24 ☉ △ ♃
16 22 ☽ ⊻ ♃
16 27 ☽ ⊼ ☉

30 Sunday
03 47 ☽ ⊼ ♅
05 38 ☽ ± ☉
08 17 ☽ ⊼ ☿
09 16 ☽ ☍ ♆
10 15 ☽ △ ♇
13 29 ☽ ∠ ♂
13 55 ☽ △ ♀
14 18 ☽ Q ♄
15 47 ☿ ⊻ ♆
15 58 ☽ ± ♅
22 16 ☽ ± ☿
23 24 ☿ ⊻ ♇

31 Monday
01 30 ☽ bQ ☉
08 20 ☽ ⊥ ♂
10 17 ☽ bQ ♅
12 16 ♂ # ♆
14 44 ☽ ✶ ♄
19 17 ☽ bQ ☿
20 55 ☽ ⊻ ♂

All ephemeris data is given at 12.00 UT and the Moon's longitude is additionally given for 24.00 UT

NOVEMBER 1977

LONGITUDES

Date	Sidereal time h m s	Sun ☉ ° ' "	Moon ☽ ° ' "	Moon ☽ 24.00 ° ' "	Mercury ☿ ° '	Venus ♀ ° '	Mars ♂ ° '	Jupiter ♃ ° '	Saturn ♄ ° '	Uranus ♅ ° '	Neptune ♆ ° '	Pluto ♇ ° '
01	14 42 38	08 ♏ 57 15	09 ♋ 33 09	15 ♋ 29 50	17 ♏ 23	19 ♎ 11	02 ♌ 21	06 ♋ 02	29 ♌ 07	11 ♏ 55	14 ♐ 33	15 ♎ 03
02	14 46 35	09 57 18	21 ♋ 27 53	27 ♋ 27 50	18 56	20 26	02 44	06 R 00	29 11	11 59	14 35	15 05
03	14 50 31	10 57 23	03 ♌ 30 14	09 ♌ 35 40	20 30	21 40	03 08	05 58	29 15	12 02	14 37	15 07
04	14 54 28	11 57 31	15 ♌ 44 42	21 ♌ 57 56	22 02	22 55	03 30	05 56	29 19	12 06	14 39	15 09
05	14 58 25	12 57 40	28 ♌ 15 55	04 ♍ 39 11	23 35	24 10	03 53	05 54	29 22	12 10	14 41	15 12
06	15 02 21	13 57 51	11 ♍ 08 13	17 ♍ 43 27	25 06	25 25	04 15	05 51	29 26	12 14	14 43	15 14
07	15 06 18	14 58 05	24 ♍ 25 12	01 ♎ 13 42	26 38	26 40	04 36	05 49	29 30	12 17	14 45	15 16
08	15 10 14	15 58 20	08 ♎ 09 02	15 ♎ 11 09	28 09	27 55	04 57	05 46	29 33	12 21	14 47	15 18
09	15 14 11	16 58 38	22 ♎ 19 48	29 ♎ 34 34	29 ♏ 40	29 ♎ 10	05 18	05 43	29 37	12 25	14 49	15 20
10	15 18 07	17 58 57	06 ♏ 54 51	14 ♏ 19 52	01 ♐ 10	00 ♏ 25	05 38	05 39	29 40	12 29	14 51	15 22
11	15 22 04	18 59 18	21 ♏ 48 39	29 ♏ 20 07	02 40	01 41	05 58	05 36	29 43	12 32	14 53	15 24
12	15 26 00	19 59 41	06 ♐ 53 04	14 ♐ 26 16	04 10	02 56	06 17	05 32	29 47	12 36	14 55	15 26
13	15 29 57	21 00 06	21 ♐ 58 31	29 ♐ 28 36	05 39	04 11	06 36	05 28	29 50	12 40	14 57	15 28
14	15 33 54	22 00 32	06 ♑ 55 30	14 ♑ 18 15	07 08	05 26	06 55	05 24	29 53	12 44	14 59	15 30
15	15 37 50	23 00 59	21 ♑ 36 08	28 ♑ 48 34	08 36	06 41	07 13	05 20	29 56	12 47	15 01	15 32
16	15 41 47	24 01 28	05 ♒ 55 11	12 ♒ 55 46	10 04	07 56	07 30	05 16	29 ♌ 58	12 51	15 04	15 34
17	15 45 43	25 01 58	19 ♒ 50 17	26 ♒ 38 49	11 32	09 12	07 47	05 11	00 ♍ 01	12 55	15 06	15 36
18	15 49 40	26 02 29	03 ♓ 21 34	09 ♓ 58 49	12 58	10 27	08 03	05 06	00 04	12 58	15 08	15 38
19	15 53 36	27 03 01	16 ♓ 30 57	22 ♓ 58 20	14 25	11 42	08 19	05 01	00 06	13 02	15 10	15 40
20	15 57 33	28 03 35	29 ♓ 21 23	05 ♈ 40 31	15 50	12 57	08 34	04 56	00 08	13 06	15 12	15 42
21	16 01 29	29 ♏ 04 09	11 ♈ 56 09	18 ♈ 08 40	17 15	14 12	08 49	04 51	00 11	13 09	15 14	15 44
22	16 05 26	00 ♐ 04 45	24 ♈ 18 26	00 ♉ 25 46	18 39	15 28	09 03	04 46	00 13	13 13	15 17	15 46
23	16 09 23	01 05 23	06 ♉ 31 00	12 ♉ 34 22	20 02	16 43	09 17	04 40	00 15	13 17	15 19	15 47
24	16 13 19	02 06 01	18 ♉ 36 08	24 ♉ 36 30	21 25	17 58	09 30	04 34	00 17	13 20	15 21	15 49
25	16 17 16	03 06 41	00 ♊ 35 40	06 ♊ 33 50	22 46	19 14	09 43	04 28	00 19	13 24	15 23	15 51
26	16 21 12	04 07 23	12 ♊ 31 08	18 ♊ 27 47	24 05	20 29	09 55	04 22	00 20	13 28	15 25	15 53
27	16 25 09	05 08 05	24 ♊ 23 56	00 ♋ 19 48	25 23	21 44	10 06	04 16	00 22	13 31	15 28	15 54
28	16 29 05	06 08 49	06 ♋ 15 35	12 ♋ 11 34	26 39	23 00	10 17	04 10	00 23	13 35	15 30	15 56
29	16 33 02	07 09 35	18 ♋ 08 01	24 ♋ 05 15	27 54	24 15	10 27	04 03	00 25	13 38	15 32	15 58
30	16 36 58	08 ♐ 10 22	00 ♌ 03 37	06 ♌ 03 34	29 ♐ 06	25 ♏ 30	10 ♌ 36	03 ♋ 57	00 ♍ 26	13 ♏ 42	15 ♐ 34	15 ♎ 59

Date	Moon True ☊ ° '	Moon Mean ☊ ° '	Moon ☽ Latitude ° '
01	14 ♎ 59	13 ♎ 45	05 S 13
02	14 R 57	13 42	05 12
03	14 55	13 39	04 58
04	14 D 55	13 36	04 30
05	14 56	13 33	03 49
06	14 58	13 30	02 55
07	14 59	13 26	01 50
08	15 00	13 23	00 S 37
09	15 R 00	13 20	00 N 40
10	14 59	13 17	01 57
11	14 55	13 14	03 07
12	14 51	13 11	04 06
13	14 45	13 07	04 47
14	14 40	13 04	05 08
15	14 35	13 01	05 08
16	14 32	12 58	04 49
17	14 30	12 55	04 14
18	14 D 31	12 51	03 25
19	14 32	12 48	02 26
20	14 33	12 45	01 22
21	14 34	12 42	00 N 14
22	14 R 34	12 39	00 S 52
23	14 31	12 36	01 56
24	14 26	12 32	02 53
25	14 19	12 29	03 42
26	14 10	12 26	04 21
27	14 00	12 23	04 48
28	13 49	12 20	05 03
29	13 39	12 17	05 04
30	13 ♎ 31	12 ♎ 13	04 S 53

DECLINATIONS

Date	Sun ☉ ° '	Moon ☽ ° '	Mercury ☿ ° '	Venus ♀ ° '	Mars ♂ ° '	Jupiter ♃ ° '	Saturn ♄ ° '	Uranus ♅ ° '	Neptune ♆ ° '	Pluto ♇ ° '
01	14 S 29	17 N 54	17 S 41	06 S 04	21 N 04	22 N 57	12 N 54	15 S 03	21 S 07	09 N 22
02	14 48	16 35	18 14	06 33	21 00	22 57	12 53	15 04	21 08	09 21
03	15 07	14 32	18 45	07 02	20 57	22 58	12 52	15 05	21 08	09 21
04	15 25	11 49	19 15	07 30	20 53	22 58	12 51	15 06	21 08	09 20
05	15 44	08 30	19 45	07 58	20 50	22 58	12 50	15 07	21 08	09 20
06	16 02	04 41	20 14	08 26	20 46	22 58	12 48	15 08	21 09	09 19
07	16 20	00 N 32	20 41	08 54	20 43	22 58	12 47	15 09	21 09	09 19
08	16 37	03 S 48	21 08	09 22	20 40	22 59	12 46	15 11	21 09	09 18
09	16 54	08 04	21 33	09 49	20 37	22 59	12 45	15 12	21 10	09 17
10	17 11	11 59	21 58	10 16	20 34	22 59	12 44	15 13	21 10	09 17
11	17 28	15 12	22 21	10 43	20 31	22 59	12 43	15 14	21 10	09 17
12	17 44	17 25	22 44	11 10	20 28	22 59	12 42	15 15	21 10	09 16
13	18 00	18 26	23 05	11 36	20 25	23 00	12 42	15 16	21 11	09 16
14	18 16	18 08	23 25	12 03	20 22	23 00	12 41	15 17	21 11	09 15
15	18 32	16 38	23 44	12 29	20 20	23 00	12 40	15 19	21 11	09 15
16	18 47	14 07	24 01	12 54	20 18	23 00	12 39	15 20	21 11	09 14
17	19 01	10 51	24 18	13 20	20 15	23 01	12 38	15 21	21 12	09 14
18	19 16	07 05	24 33	13 45	20 13	23 01	12 38	15 22	21 12	09 14
19	19 30	03 S 05	24 47	14 09	20 11	23 01	12 37	15 23	21 12	09 13
20	19 44	00 N 59	24 59	14 34	20 09	23 02	12 36	15 24	21 12	09 13
21	19 57	04 56	25 11	14 58	20 08	23 02	12 36	15 25	21 13	09 13
22	20 10	08 37	25 21	15 21	20 06	23 02	12 35	15 26	21 13	09 12
23	20 23	11 52	25 29	15 44	20 04	23 02	12 35	15 27	21 13	09 12
24	20 35	14 35	25 36	16 07	20 03	23 03	12 34	15 29	21 13	09 12
25	20 47	16 39	25 42	16 29	20 02	23 03	12 34	15 30	21 14	09 11
26	20 58	17 59	25 47	16 51	20 01	23 03	12 33	15 31	21 14	09 11
27	21 09	18 31	25 50	17 13	20 00	23 04	12 33	15 32	21 14	09 11
28	21 20	18 15	25 51	17 34	20 00	23 04	12 33	15 33	21 14	09 11
29	21 30	17 11	25 51	17 54	19 59	23 04	12 32	15 34	21 15	09 10
30	21 S 40	15 N 22	25 S 50	18 S 15	19 N 59	23 N 05	12 N 32	15 S 35	21 S 15	09 N 10

ZODIAC SIGN ENTRIES

Date	h	m	Planets
03	05	03	☽ ♌
05	15	17	☽ ♍
07	21	51	☽ ♎
09	17	20	☿ ♐
10	00	42	☽ ♏
10	03	52	♀ ♏
12	01	03	☽ ♐
14	00	50	☽ ♑
16	02	00	☽ ♒
17	02	43	♄ ♍
18	05	58	☽ ♓
20	13	13	☽ ♈
22	10	07	☉ ♐
22	23	09	☽ ♉
25	10	48	☽ ♊
27	23	20	☽ ♋
30	11	53	☽ ♌

LATITUDES

Date	Mercury ☿ ° '	Venus ♀ ° '	Mars ♂ ° '	Jupiter ♃ ° '	Saturn ♄ ° '	Uranus ♅ ° '	Neptune ♆ ° '	Pluto ♇ ° '
01	00 S 42	01 N 33	01 N 28	00 S 21	01 N 12	00 N 23	01 N 26	16 N 34
04	01 01	01 31	01 33	00 21	01 12	00 23	01 26	16 35
07	01 19	01 29	01 39	00 20	01 13	00 23	01 26	16 36
10	01 36	01 27	01 45	00 20	01 14	00 23	01 26	16 37
13	01 52	01 23	01 51	00 20	01 14	00 23	01 25	16 38
16	02 05	01 19	01 58	00 20	01 15	00 23	01 25	16 39
19	02 16	01 15	02 04	00 19	01 16	00 23	01 25	16 40
22	02 24	01 10	02 11	00 19	01 16	00 23	01 25	16 41
25	02 28	01 05	02 18	00 19	01 17	00 23	01 25	16 42
28	02 27	00 59	02 25	00 18	01 18	00 23	01 25	16 44
31	02 S 21	00 N 53	02 N 33	00 S 18	01 N 18	00 N 23	01 N 25	16 N 45

DATA

Julian Date	2443449
Delta T	+48 seconds
Ayanamsa	23° 32' 54"
Synetic vernal point	05° ♓ 34' 06"
True obliquity of ecliptic	23° 26' 23"

MOON'S PHASES, APSIDES AND POSITIONS ☽

Date	h	m	Phase	Longitude ° '	Eclipse Indicator
04	03	58	☾	11 ♌ 37	
11	07	09	●	18 ♏ 47	
17	21	52	☽	25 ♒ 27	
25	17	31	○	03 ♊ 21	

Day	h	m		
12	11	54	Perigee	
27	20	14	Apogee	
07	14	57	0S	
13	18	31	Max dec	18° S 28'
20	06	07	0N	
27	15	54	Max dec	18° N 32'

LONGITUDES

Date	Chiron ⚷ ° '	Ceres ⚳ ° '	Pallas ⚴ ° '	Juno ⚵ ° '	Vesta ⚶ ° '	Black Moon Lilith ⚸ ° '
01	03 ♉ 09	21 ♏ 32	01 ♏ 30	08 ♐ 15	15 ♎ 56	21 ♊ 21
11	02 ♉ 42	25 ♏ 48	05 ♏ 55	11 ♐ 36	20 ♎ 58	22 ♊ 28
21	02 ♉ 15	00 ♐ 05	10 ♏ 18	15 ♐ 03	25 ♎ 58	23 ♊ 35
31	01 ♉ 52	04 ♐ 20	14 ♏ 38	18 ♐ 33	00 ♏ 56	24 ♊ 42

All ephemeris data is given at 12.00 UT and the Moon's longitude is additionally given for 24.00 UT

ASPECTARIAN

h m	Aspects
01 Tuesday	
01 59	☉ ⊥ ♆
04 54	☽ ☌ ♃
10 41	☽ △ ☉
15 20	☽ # ☿
16 48	☽ △ ♅
21 15	☽ ∠ ♄
22 07	☽ ⊼ ♆
23 08	☽ □ ♇
02 Wednesday	
06 10	☽ △ ☿
09 40	☽ □ ♀
10 13	☽ ± ♆
15 27	☽ ⊥ ♄
03 Thursday	
03 30	☽ ⊻ ♄
04 16	☽ ⚼ ♆
06 28	☽ # ♅
06 47	☽ # ☉
09 14	☉ ∥ ♅
11 14	☽ ☌ ♂
11 14	☽ Q ♇
11 33	♂ Q ♇
16 52	☽ ⊻ ♃
19 17	☿ ⚼ ♃
19 23	☉ Q ♄
21 58	☿ ⊥ ♇
04 Friday	
01 32	☽ Q ♀
03 31	☽ ∥ ♄
03 58	☽ □ ☉
04 36	☽ ⊥ ♃
04 52	☽ □ ♅
09 52	☽ △ ♆
10 51	☽ ⚹ ♇
15 41	☉ ☌ ♅
22 00	☽ ∠ ♃
05 Saturday	
01 50	☽ □ ☿
03 22	☽ ⚹ ♀
06 21	☽ ∥ ♇
14 06	☽ ☌ ♄
15 07	☽ # ♀
15 36	☽ Q ♅
15 39	☽ ∠ ♇
17 31	☽ Q ☉
22 52	☽ ⊻ ♂
06 Sunday	
02 16	☽ ⚹ ♃
08 29	☽ ⊥ ♇
10 19	☽ ⊥ ♂
10 33	☽ ∠ ♀
14 01	☽ ⚹ ♅
16 05	☽ Q ☿
17 36	☽ ⚹ ☉
18 33	☽ □ ♆
19 30	☽ ⊻ ♇
07 Monday	
00 12	☽ Q ♃
03 09	☽ ∠ ♂
04 37	☽ ⊥ ♀
06 32	☉ ⊻ ♆
13 26	☿ # ♂
15 29	☿ ⊻ ♀
16 24	☽ ⊻ ♀
16 25	☽ ⚹ ☿
17 06	☽ ∠ ♅
19 19	☉ ⊻ ♇
21 00	☽ ⊻ ♄
22 34	☽ ∠ ☉
08 Tuesday	
02 41	☽ Q ♆
06 20	☽ ⚹ ♂
07 30	☽ ⊥ ♄
07 54	☽ □ ♃
08 53	☽ ⊥ ♅
08 59	♀ # ♇
13 22	☿ ∥ ♆
15 22	☽ ⊥ ☉
19 13	☽ ⊻ ♅
21 35	☽ ∠ ☿
22 59	☽ ∠ ♄
23 20	☽ ⚹ ♆
09 Wednesday	
00 13	☽ ☌ ♇
02 21	☽ ⊻ ☉
03 22	☽ Q ♂
11 10	☿ □ ♄
12 44	☿ ± ♃
14 29	☽ ⊥ ☿
19 08	☽ # ♇
20 51	♀ ⚹ ♄
22 54	☿ ∠ ♇
23 39	☽ ∥ ♀
10 Thursday	
00 07	☽ ⚹ ♄
00 24	☽ ☌ ♀
00 25	☽ ∠ ♆
00 39	♀ ∠ ♆
01 33	☽ ⊻ ☿
09 52	☽ □ ♂
09 58	☽ △ ♃
13 14	♂ ⊻ ♃
15 09	☽ ⊥ ♆
17 06	☽ # ♄
19 44	☽ Q ♄
21 03	☽ ☌ ♅
11 Friday	
00 52	☽ ⊻ ♆
01 42	☽ ⊻ ♇
07 09	☽ ☌ ☉
10 04	☽ ⚼ ♃
11 21	☽ ⊥ ♇
12 17	☽ ∥ ♅
12 Saturday	
00 22	☽ ± ♃
00 40	☽ □ ♄
01 44	☽ ∠ ♇
05 09	☽ ⊻ ♀
07 12	☽ ☌ ☿
09 52	☽ ⊼ ♃
11 02	☽ △ ♂
15 32	☽ ⊥ ♀
18 17	☽ ∥ ☉
21 07	☽ ⊻ ♅
13 Sunday	
00 10	☉ ⚼ ♃
00 48	☽ ☌ ♆
01 37	☽ ⚹ ♇
06 08	☿ # ♃
06 42	☽ ⊥ ♅
07 08	☽ ∠ ♀
09 14	☿ ⊼ ♃
10 20	☽ ⊻ ☉
11 24	☽ ⚼ ♂
20 37	☽ ⊥ ☉
20 48	☽ Q ♇
21 08	☽ ∠ ♅
23 33	☉ ⊥ ♇
14 Monday	
00 36	☽ △ ♄
02 06	☽ ± ♂
07 27	☿ △ ♂
09 04	☽ ∥ ☉
09 22	☽ ⚹ ♀
09 33	☽ ☍ ♃
11 30	♀ △ ♃
11 59	☽ ⊼ ♂
12 09	☽ ∠ ☉
12 22	☽ ⊻ ☿
21 28	☽ ⚹ ♅
23 13	☽ ⊥ ☿
15 Tuesday	
00 59	☽ ⚼ ♄
01 09	☽ ⊻ ♆
02 00	☽ □ ♇
06 44	☽ Q ♀
11 02	☽ ⊥ ♆
14 31	☽ ⚹ ☉
15 42	☽ ∠ ☿
15 52	☽ ± ♄
17 19	☽ Q ♅
22 13	♀ # ♄
16 Wednesday	
01 03	♀ □ ♂
01 31	☽ ∥ ♅
01 55	☽ ⊼ ♄
02 04	☽ ∠ ♆
10 53	☽ ⊼ ♃
12 11	☽ Q ☉
14 45	☽ ☍ ♂
15 47	☽ □ ♀
19 55	☽ ⚹ ☿
20 20	☽ ∥ ♀
21 05	☽ ± ♃
23 21	☽ # ♄
23 55	☽ □ ♅
17 Thursday	
03 43	☽ ⚹ ♆
04 36	☽ △ ♇
10 05	♀ ⊥ ♆
12 36	☽ ⚼ ♃
19 15	☽ Q ☿
21 52	☽ □ ☉
22 37	☽ # ♇
18 Friday	
00 50	☽ Q ♆
06 04	☽ ☍ ♄
07 06	☽ ⚼ ♇
11 59	☿ ⊻ ♅
15 08	☽ △ ♃
20 40	☽ ⊼ ♂
23 24	☽ ± ♇
19 Saturday	
02 12	☽ △ ♀
05 34	☽ △ ♅
07 39	☽ □ ☿
07 52	☽ ± ♂
09 30	☽ □ ♆
10 26	☽ ⊼ ♇
19 56	♀ Q ♄
20 Sunday	
00 54	☽ ⚼ ♂
01 00	☿ ☌ ♆
09 04	☽ ⚼ ♀
09 20	☽ △ ☉
09 32	☿ ⚹ ♇
09 37	☽ ⚼ ♅
13 29	☽ ⊼ ♄
14 52	♀ ☌ ♅
22 31	☽ □ ♃
21 Monday	
00 56	☽ ± ♄
02 47	☽ ± ♅
04 03	☽ ± ♀
05 54	☽ △ ♂
07 14	☉ ± ♃
14 22	☽ ⊼ ♅
16 29	☽ ⚼ ☉
16 52	☽ ⊼ ♀
18 16	☽ ⚼ ♄
18 24	☽ △ ♆
19 20	☽ ☍ ♇
23 35	☽ △ ☿
22 Tuesday	
05 07	☉ # ♂
08 20	♀ ⊻ ♆
09 00	☽ Q ♃
11 31	☽ ± ☉
15 18	☉ □ ♄
16 07	☽ ∥ ♇
17 42	♀ ∥ ♅
17 48	♀ ⊻ ♇
22 09	☿ ⊥ ♅
23 36	☽ △ ♄
23 44	☽ ⚼ ♆
23 Wednesday	
00 20	☽ ⊼ ☉
04 38	☉ ∠ ♇
08 23	☽ ⚹ ♃
08 43	☽ ⚼ ☿
17 33	☽ ± ♆
17 35	☽ □ ♂
17 46	☽ ∥ ♄
24 Thursday	
01 28	☽ ☍ ♅
04 50	☽ ± ☿
05 30	☽ ⊼ ♆
06 26	☽ ⊼ ♇
10 36	☽ ☍ ♀
13 55	☽ ∠ ♃
18 19	☽ ⊼ ☿
18 26	☽ ± ♇
21 25	☽ # ♅
25 Friday	
06 07	☽ Q ♂
07 47	☽ ⊥ ♃
09 08	☽ # ♀
11 26	☽ □ ♄
12 31	☽ ⚼ ♇
16 21	♀ ⚼ ♃
17 31	☽ ☍ ☉
19 44	☽ ⊻ ♃
26 Saturday	
06 39	☽ ⚹ ♂
13 54	☽ ⊼ ♅
17 23	☉ ⊼ ♃
17 53	☽ ☍ ♆
18 48	☽ △ ♇
23 47	☽ Q ♄
27 Sunday	
02 05	☽ ± ♅
05 48	☿ ⚼ ♂
05 59	☽ ⊼ ♀
13 26	☽ ∠ ♂
14 14	☽ ☍ ☿
15 16	♀ ⊥ ♇
19 33	☽ ± ♀
20 23	☽ ⚼ ♅
22 49	☉ ∥ ♆
28 Monday	
00 06	☽ ⚹ ♄
02 18	♂ ⊥ ♃
07 48	☽ ☌ ♃
07 56	☽ ⊥ ♂
11 45	☽ ⊼ ☉
15 55	☽ ⚼ ♀
20 15	☽ ⊻ ♂
29 Tuesday	
01 01	☽ ± ☉
01 16	☽ # ♀
02 52	☽ △ ♅
06 30	☽ ∠ ♄
06 44	☽ ⊼ ♆
07 36	☽ □ ♇
13 21	☿ Q ♇
18 53	☽ ± ♆
20 52	☽ ⚼ ☉
30 Wednesday	
00 41	☽ ⊥ ♄
01 47	☽ △ ♀
03 29	☿ ∠ ♅
09 34	☽ # ♅
09 51	☽ ⊼ ☿
12 45	☽ ⊻ ♄
13 02	☽ ⚼ ♆
19 43	☽ ⊻ ♃
19 53	☽ Q ♇
23 10	☽ ± ☿

DECEMBER 1977

LONGITUDES

Date	Sidereal time h m s	Sun ☉ ° ' "	Moon ☽ ° ' "	Moon ☽ 24.00 ° ' "	Mercury ☿ ° '	Venus ♀ ° '	Mars ♂ ° '	Jupiter ♃ ° '	Saturn ♄ ° '	Uranus ♅ ° '	Neptune ♆ ° '	Pluto ♇ ° '
01	16 40 55	09 ♐ 11 10	12 ♌ 05 30	18 ♌ 09 55	00 ♑ 15	26 ♏ 46	10 ♌ 45	03 ♋ 50	00 ♍ 27	13 ♏ 45	15 ♐ 37	16 ♎ 01
02	16 44 52	10 12 00	24 ♌ 17 21	00 ♍ 28 21	01 21	28 01	10 53	03 R 43	00 28	13 49	15 39	16 03
03	16 48 48	11 12 51	06 ♍ 43 28	13 ♍ 03 18	02 24	29 ♏ 17	11 00	03 36	00 29	13 52	15 41	16 04
04	16 52 45	12 13 43	19 ♍ 28 24	25 ♍ 59 21	03 23	00 ♐ 32	11 07	03 29	00 30	13 56	15 43	16 06
05	16 56 41	13 14 37	02 ♎ 36 38	09 ♎ 20 40	04 17	01 47	11 13	03 22	00 31	13 59	15 46	16 07
06	17 00 38	14 15 32	16 ♎ 11 49	23 ♎ 10 17	05 06	03 03	11 18	03 15	00 32	14 02	15 48	16 09
07	17 04 34	15 16 28	00 ♏ 16 05	07 ♏ 29 04	05 49	04 18	11 23	03 07	00 32	14 06	15 50	16 10
08	17 08 31	16 17 26	14 ♏ 48 52	22 ♏ 14 54	06 25	05 34	11 26	03 00	00 32	14 09	15 52	16 12
09	17 12 27	17 18 25	29 ♏ 46 19	07 ♐ 22 03	06 54	06 49	11 29	02 52	00 33	14 12	15 55	16 13
10	17 16 24	18 19 25	15 ♐ 00 50	22 ♐ 41 17	07 14	08 05	11 32	02 45	00 33	14 16	15 57	16 14
11	17 20 21	19 20 25	00 ♑ 21 54	08 ♑ 01 12	07 25	09 20	11 33	02 37	00 33	14 19	15 59	16 16
12	17 24 17	20 21 27	15 ♑ 37 44	23 ♑ 10 12	07 R 26	10 36	11 34	02 29	00 R 33	14 22	16 02	16 17
13	17 28 14	21 22 29	00 ♒ 37 29	07 ♒ 58 39	07 16	11 51	11 R 33	02 21	00 33	14 25	16 04	16 18
14	17 32 10	22 23 32	15 ♒ 13 04	22 ♒ 20 17	06 55	13 07	11 33	02 13	00 33	14 29	16 06	16 19
15	17 36 07	23 24 35	29 ♒ 20 07	06 ♓ 12 34	06 22	14 22	11 31	02 05	00 32	14 32	16 08	16 21
16	17 40 03	24 25 38	12 ♓ 57 50	19 ♓ 36 13	05 38	15 38	11 28	01 57	00 32	14 35	16 11	16 22
17	17 44 00	25 26 42	26 ♓ 08 11	02 ♈ 34 13	04 43	16 53	11 25	01 49	00 31	14 38	16 13	16 23
18	17 47 56	26 27 47	08 ♈ 54 54	15 ♈ 10 48	03 38	18 09	11 21	01 41	00 30	14 41	16 15	16 24
19	17 51 53	27 28 51	21 ♈ 22 31	27 ♈ 30 39	02 25	19 24	11 15	01 33	00 29	14 44	16 17	16 25
20	17 55 50	28 29 56	03 ♉ 35 44	09 ♉ 38 19	01 ♑ 06	20 40	11 10	01 25	00 28	14 47	16 19	16 26
21	17 59 46	29 ♐ 31 01	15 ♉ 38 51	21 ♉ 37 48	29 ♐ 44	21 55	11 03	01 17	00 27	14 50	16 22	16 27
22	18 03 43	00 ♑ 32 07	27 ♉ 35 33	03 ♊ 32 26	28 21	23 11	10 55	01 09	00 26	14 53	16 24	16 28
23	18 07 39	01 33 12	09 ♊ 28 47	15 ♊ 24 49	27 01	24 26	10 47	01 00	00 25	14 56	16 26	16 29
24	18 11 36	02 34 18	21 ♊ 20 47	27 ♊ 16 51	25 45	25 42	10 38	00 52	00 23	14 58	16 28	16 30
25	18 15 32	03 35 25	03 ♋ 13 13	09 ♋ 09 59	24 36	26 57	10 28	00 44	00 22	15 01	16 31	16 31
26	18 19 29	04 36 32	15 ♋ 07 21	21 ♋ 05 25	23 36	28 13	10 17	00 36	00 20	15 04	16 33	16 32
27	18 23 25	05 37 39	27 ♋ 04 22	03 ♌ 04 23	22 46	29 ♐ 28	10 05	00 28	00 19	15 07	16 35	16 33
28	18 27 22	06 38 46	09 ♌ 05 39	15 ♌ 08 26	22 06	00 ♑ 44	09 53	00 20	00 17	15 09	16 37	16 33
29	18 31 19	07 39 54	21 ♌ 13 00	27 ♌ 19 40	21 37	01 59	09 40	00 12	00 15	15 12	16 39	16 34
30	18 35 15	08 41 02	03 ♍ 28 49	09 ♍ 40 50	21 18	03 15	09 26	00 ♋ 04	00 13	15 15	16 41	16 35
31	18 39 12	09 ♑ 42 11	15 ♍ 56 12	22 ♍ 15 22	21 ♐ 09	04 ♑ 30	09 ♌ 11	29 ♊ 56	00 ♍ 11	15 ♏ 17	16 ♐ 43	16 ♎ 36

Date	Moon True ☊ ° '	Moon Mean ☊ ° '	Moon ☽ Latitude ° '
01	13 ♎ 25	12 ♎ 10	04 S 28
02	13 R 21	12 07	03 51
03	13 19	12 04	03 03
04	13 D 20	12 01	02 04
05	13 20	11 57	00 S 57
06	13 R 21	11 54	00 N 15
07	13 19	11 51	01 29
08	13 16	11 48	02 39
09	13 09	11 45	03 41
10	13 01	11 42	04 28
11	12 51	11 38	04 55
12	12 40	11 35	05 02
13	12 31	11 32	04 48
14	12 23	11 29	04 15
15	12 19	11 26	03 27
16	12 16	11 23	02 29
17	12 D 16	11 19	01 25
18	12 16	11 16	00 N 18
19	12 R 16	11 13	00 S 48
20	12 14	11 10	01 50
21	12 09	11 07	02 47
22	12 02	11 03	03 35
23	11 51	11 00	04 14
24	11 38	10 57	04 41
25	11 24	10 54	04 56
26	11 09	10 51	04 59
27	10 54	10 48	04 48
28	10 42	10 44	04 24
29	10 32	10 41	03 48
30	10 25	10 38	03 01
31	10 ♎ 22	10 ♎ 35	02 S 05

DECLINATIONS

Date	Sun ☉	Moon ☽	Mercury ☿	Venus ♀	Mars ♂	Jupiter ♃	Saturn ♄	Uranus ♅	Neptune ♆	Pluto ♇
01	21 S 50	12 N 52	25 S 48	18 S 34	19 N 59	23 N 05	12 N 32	15 S 36	21 S 15	09 N 10
02	21 59	09 47	25 43	18 53	19 59	23 05	12 32	15 37	21 15	09 10
03	22 07	06 13	25 38	19 12	20 00	23 05	12 32	15 38	21 16	09 10
04	22 16	02 N 16	25 31	19 30	20 00	23 06	12 32	15 39	21 16	09 10
05	22 23	01 S 55	25 23	19 47	20 01	23 06	12 32	15 40	21 16	09 09
06	22 31	06 08	25 14	20 04	20 02	23 06	12 32	15 41	21 16	09 09
07	22 38	10 11	25 03	20 21	20 03	23 07	12 32	15 42	21 17	09 09
08	22 44	13 44	24 52	20 36	20 05	23 07	12 32	15 43	21 17	09 09
09	22 50	16 30	24 39	20 51	20 07	23 07	12 32	15 44	21 17	09 09
10	22 56	18 10	24 26	21 06	20 08	23 08	12 32	15 45	21 17	09 09
11	23 01	18 31	24 11	21 20	20 11	23 08	12 32	15 46	21 18	09 09
12	23 05	17 31	23 56	21 33	20 13	23 08	12 32	15 47	21 18	09 09
13	23 10	15 20	23 40	21 46	20 16	23 08	12 33	15 48	21 18	09 09
14	23 13	12 13	23 24	21 58	20 18	23 09	12 33	15 49	21 18	09 09
15	23 17	08 28	23 07	22 10	20 22	23 09	12 33	15 50	21 18	09 09
16	23 19	04 24	22 49	22 20	20 25	23 09	12 34	15 51	21 19	09 09
17	23 22	00 S 15	22 31	22 31	20 28	23 09	12 34	15 52	21 19	09 09
18	23 24	03 N 48	22 13	22 40	20 32	23 10	12 35	15 52	21 19	09 09
19	23 25	07 36	21 55	22 49	20 36	23 10	12 35	15 53	21 19	09 09
20	23 26	10 59	21 37	22 57	20 40	23 10	12 36	15 54	21 20	09 10
21	23 26	13 52	21 20	23 04	20 45	23 10	12 36	15 55	21 20	09 10
22	23 26	16 08	21 04	23 11	20 49	23 11	12 37	15 56	21 20	09 10
23	23 26	17 41	20 49	23 17	20 54	23 11	12 38	15 57	21 20	09 10
24	23 25	18 29	20 36	23 22	20 59	23 11	12 38	15 58	21 20	09 10
25	23 23	18 28	20 25	23 26	21 05	23 11	12 39	15 58	21 21	09 10
26	23 22	17 38	20 16	23 30	21 10	23 11	12 40	15 59	21 21	09 11
27	23 19	16 02	20 10	23 33	21 16	23 12	12 41	16 00	21 21	09 11
28	23 16	13 44	20 06	23 36	21 22	23 12	12 41	16 01	21 21	09 11
29	23 13	10 49	20 04	23 37	21 28	23 12	12 42	16 01	21 22	09 11
30	23 09	07 25	20 05	23 38	21 34	23 12	12 43	16 02	21 22	09 12
31	23 S 05	03 N 38	20 S 08	23 S 38	21 N 40	23 N 12	12 N 44	16 S 03	21 S 22	09 N 12

ZODIAC SIGN ENTRIES

Date	h	m	Planets
01	06	43	☿ ♑
02	23	05	☽ ♍
04	01	49	♀ ♐
05	07	18	☽ ♎
07	11	38	☽ ♏
09	12	22	☽ ♐
11	11	26	☽ ♑
13	10	59	☽ ♒
15	13	09	☽ ♓
17	19	11	☽ ♈
20	04	54	☽ ♉
21	07	18	☿ ♐
21	23	23	☉ ♑
22	16	51	☽ ♊
25	05	30	☽ ♋
27	17	52	☽ ♌
27	22	09	♀ ♑
30	05	13	☽ ♍
30	23	49	♃ ♊

LATITUDES

Date	Mercury ☿	Venus ♀	Mars ♂	Jupiter ♃	Saturn ♄	Uranus ♅	Neptune ♆	Pluto ♇
01	02 S 21	00 N 53	02 N 33	00 S 18	01 N 18	00 N 23	01 N 25	16 N 45
04	02 08	00 47	02 40	00 18	01 19	00 23	01 25	16 46
07	01 45	00 40	02 48	00 17	01 20	00 23	01 25	16 48
10	01 11	00 34	02 56	00 17	01 20	00 23	01 25	16 49
13	00 S 26	00 27	03 04	00 17	01 21	00 23	01 25	16 51
16	00 N 30	00 20	03 12	00 16	01 21	00 23	01 25	16 52
19	01 30	00 12	03 20	00 16	01 22	00 23	01 25	16 54
22	02 22	00 N 05	03 28	00 15	01 22	00 23	01 25	16 56
25	02 55	00 S 02	03 36	00 15	01 23	00 23	01 25	16 57
28	03 07	00 09	03 44	00 15	01 24	00 23	01 25	16 59
31	03 N 01	00 S 17	03 N 51	00 S 14	01 N 25	00 N 23	01 N 25	17 N 01

DATA

Julian Date	2443479
Delta T	+48 seconds
Ayanamsa	23° 32' 58"
Synetic vernal point	05° ♓ 34' 02"
True obliquity of ecliptic	23° 26' 22"

MOON'S PHASES, APSIDES AND POSITIONS ☽

Date	h	m	Phase	Longitude ° '	Eclipse Indicator
03	21	16	☾	11 ♍ 36	
10	17	33	●	18 ♐ 34	
17	10	37	☽	25 ♓ 23	
25	12	49	○	03 ♋ 37	

Day	h	m		
10	23	09	Perigee	
24	20	57	Apogee	
05	01	07	0S	
11	06	11	Max dec	18° S 33'
17	13	25	0N	
24	23	29	Max dec	18° N 34'

LONGITUDES

Date	Chiron ⚷ ° '	Ceres ⚳ ° '	Pallas ⚴ ° '	Juno ⚵ ° '	Vesta ⚶ ° '	Black Moon Lilith ⚸ ° '
01	01 ♉ 52	04 ♐ 20	14 ♏ 38	18 ♐ 33	00 ♏ 56	24 ♊ 42
11	01 ♉ 33	08 ♐ 34	18 ♏ 54	22 ♐ 06	05 ♏ 51	25 ♊ 50
21	01 ♉ 19	12 ♐ 45	23 ♏ 05	25 ♐ 41	10 ♏ 42	26 ♊ 57
31	01 ♉ 10	16 ♐ 52	27 ♏ 10	29 ♐ 17	15 ♏ 27	28 ♊ 04

ASPECTARIAN

h m Aspects

01 Thursday
05 42 ☽ △ ☉
07 34 ☽ ⊥ ♃
09 18 ☽ ☌ ♂
14 50 ☽ ∥ ♄
15 18 ☽ □ ♅
16 27 ☿ △ ♄
18 53 ☽ bQ ☿
18 59 ☽ △ ♆
19 47 ☽ ⚹ ♇

02 Friday
01 12 ☽ ∠ ♃
06 46 ♀ ± ♃
16 24 ☽ ∥ ♇
20 04 ☽ □ ♀

03 Saturday
00 01 ☽ ☌ ♄
01 08 ☽ ∠ ♇
02 39 ☽ Q ♅
02 58 ☽ △ ☿
06 05 ☽ ⚹ ♃
06 24 ☉ △ ♂
18 22 ☽ ⊥ ♇
20 12 ☽ ⊻ ♂
21 16 ☽ □ ☉

04 Sunday
01 35 ☽ ⚹ ♅
04 38 ☽ Q ♃
04 59 ☽ □ ♆
05 42 ☽ ⊻ ♇
07 35 ☽ ⊥ ♂
10 04 ☽ Q ♀
11 24 ♀ □ ♄
14 26 ☿ ☍ ♃
22 56 ♀ ∠ ♇

05 Monday
00 20 ☽ ∠ ♂
05 25 ☽ ∠ ♅
08 13 ☽ ⊻ ♄
09 20 ☽ Q ☉
10 22 ☽ ⚹ ♀
13 21 ☽ □ ♃
14 04 ☽ Q ♆
15 12 ☽ □ ☿
18 59 ☽ ⊥ ♄
21 38 ☽ ⊥ ♅

06 Tuesday
03 24 ☽ ⚹ ♂
06 30 ☉ ⊻ ♅
08 14 ☽ ⊻ ♅
08 21 ☽ ⚹ ☉
08 55 ♀ # ♂
10 50 ☽ ∠ ♄
11 18 ☽ ⚹ ♆
11 54 ☽ ☌ ♇
15 26 ♀ ⊼ ♃
15 31 ☽ ∠ ♀
19 17 ☿ ± ♂

07 Wednesday
00 17 ☽ Q ♂
00 32 ☽ Q ☿
05 44 ☽ # ♇
08 23 ☽ ⊥ ♀
12 01 ☽ ∠ ☉
12 27 ☽ ⚹ ♄
12 57 ☽ ∠ ♆
16 44 ☽ △ ♃
19 23 ☽ ⊻ ♀
21 40 ☽ ⚹ ☿

08 Thursday
01 48 ☉ ☌ ♆
03 20 ☽ # ♄
03 55 ☽ ⊥ ♆
04 04 ☽ ⊥ ☉
06 29 ☽ □ ♂
08 18 ☽ Q ♄
09 38 ☉ ⚹ ♇
10 55 ☽ ☌ ♅
13 44 ☽ ⊻ ♆
14 14 ☽ ⊻ ♇
14 34 ☽ ⊻ ☉
17 07 ☽ bQ ♃
23 03 ☽ ∠ ☿
23 56 ☽ ⊥ ♇

09 Friday
04 20 ☽ ∥ ♅
07 26 ☽ ± ♃
13 14 ☽ □ ♄
13 50 ☽ ⊥ ☿
14 09 ☿ ⊻ ♀
14 18 ☽ ∠ ♇
16 52 ☽ ⊼ ♃
23 33 ☽ ⊻ ☿

10 Saturday
00 09 ☽ ☌ ♀
06 31 ☽ △ ♂
10 49 ☽ ⊻ ♅
13 28 ☽ ☌ ♆
13 55 ☽ ⚹ ♇
17 33 ☽ ☌ ☉
20 14 ☽ ⊥ ♅

11 Sunday
06 02 ☽ bQ ♂
07 31 ♀ ∥ ♆
08 42 ☽ Q ♇
10 21 ☽ ∠ ♅
12 11 ♄ St R
12 17 ☽ △ ♄
15 30 ☽ ☍ ♃
20 08 ☽ ± ♂
23 06 ☽ ☌ ☿

12 Monday
02 10 ☿ St R
03 20 ☽ ⊻ ♀
05 34 ☽ ⊼ ♂
10 00 ☽ ⚹ ♅
11 52 ☽ bQ ♄
12 17 ☉ ⊥ ♅
12 38 ☽ ⊻ ♆
13 02 ☽ □ ♇
13 40 ☽ ⊥ ♀
19 12 ♂ St R
20 03 ☽ ⊻ ☉
22 12 ☽ ⊥ ♆

13 Tuesday
02 12 ☽ ± ♄
04 55 ☉ # ♃
05 12 ☽ Q ♅
05 21 ☽ ∠ ♀
06 22 ☽ ⊥ ☉
06 24 ♀ △ ♂
07 46 ☽ ∥ ♅
11 52 ☽ ⊼ ♄
12 43 ☽ ∠ ♆
14 47 ☽ ⊼ ♃
22 04 ☽ ∠ ☉
22 37 ☽ ⊻ ☿

14 Wednesday
00 30 ☽ ± ♃
05 54 ☽ ☍ ♂
08 09 ☽ ⚹ ♀
08 17 ☽ ⊥ ☿
09 39 ☽ # ♄
10 45 ☽ □ ♅
13 29 ☽ ⚹ ♆
13 51 ☽ △ ♇
15 20 ☽ bQ ♃
22 54 ☽ ∠ ☿

15 Thursday
00 12 ☉ ∥ ☿
01 02 ☽ ⚹ ☉
06 23 ☽ Q ♀
07 50 ☽ # ♇
08 41 ☿ # ♃
09 56 ☽ Q ♆
14 05 ☽ ☍ ♄
15 09 ♀ ⊻ ♅
15 29 ☽ bQ ♇
16 44 ☽ △ ♃
23 27 ☽ Q ☉
23 41 ☽ ⚹ ☿

16 Friday
07 21 ☽ ± ♇
09 20 ☽ ⊼ ♂
14 55 ☽ △ ♅
16 52 ☿ ± ♂
17 17 ☽ □ ♀
17 48 ☽ □ ♆
18 08 ☽ ⊼ ♇
19 54 ☽ Q ☿
20 05 ☽ ± ♂
22 47 ♀ ☌ ♆

17 Saturday
02 16 ♀ ⚹ ♇
10 37 ☽ □ ☉
12 31 ☽ bQ ♂
12 34 ☿ ∥ ♀
18 31 ☽ bQ ♅
20 08 ☽ ⊼ ♄
22 29 ☽ □ ♃

18 Sunday
02 48 ☽ □ ☿
07 25 ☽ ± ♄
09 21 ☉ bQ ♂
11 33 ☽ ± ♅
16 36 ☽ △ ♂
23 05 ☽ ⊼ ♅

19 Monday
00 36 ☽ bQ ♄
02 06 ☽ △ ♆
02 23 ☽ ☍ ♇
07 44 ☽ △ ♀
08 30 ☽ Q ♃
22 41 ☽ ∥ ♇

20 Tuesday
01 02 ☽ △ ☉
05 42 ☿ ☍ ♃
05 50 ☽ △ ♄
07 30 ☽ bQ ♆
07 33 ☽ △ ☿
07 44 ☽ ⚹ ♃
10 32 ☉ Q ♇
14 24 ♀ ⊥ ♅
16 34 ☽ bQ ♀
23 11 ☿ △ ♄

21 Wednesday
00 53 ☽ ∥ ♄
01 25 ☽ ± ♆
02 54 ☽ □ ♂
09 32 ☽ bQ ☉
10 19 ☿ ∠ ♅
10 21 ☽ bQ ☿
10 22 ☽ ☍ ♅
12 36 ☽ ± ♀
12 42 ☿ ∥ ♆
13 15 ☽ ∠ ♃
13 26 ☽ ⊼ ♆
13 37 ☽ ⊼ ♇
14 08 ☉ ☌ ☿
19 46 ☉ ∠ ♅

22 Thursday
01 41 ☽ ± ♇
02 04 ☽ ⊼ ♀
02 32 ☽ ± ☿
05 16 ☽ ± ☉
07 07 ☽ ⊥ ♃
09 35 ☽ # ♅
09 42 ☉ △ ♄
09 57 ☿ Q ♇
11 11 ♀ # ♃
13 22 ☽ ⊼ ☿
14 39 ☽ Q ♂
17 43 ☽ □ ♄
18 29 ☽ ⊼ ☉
19 05 ☽ ⊻ ♃
19 50 ☽ bQ ♇

23 Friday
00 40 ☉ ☍ ♃
05 45 ☿ # ♂
14 36 ☽ ⚹ ♂
23 04 ☽ ⊼ ♅

24 Saturday
02 07 ☽ ☍ ♆
02 11 ☽ △ ♇
06 02 ☽ Q ♄
10 56 ♀ bQ ♂
11 15 ☽ ± ♅
12 35 ☿ ☌ ♀
14 50 ☿ bQ ♂
20 06 ☽ ☍ ☿
20 33 ☽ ∠ ♂
21 50 ☽ ☍ ♀
23 33 ☉ ∥ ♀

25 Sunday
05 31 ☽ bQ ♅
06 15 ☽ ⚹ ♄
07 02 ☽ ☌ ♃
12 49 ☽ ☍ ☉
14 28 ☽ ⊥ ♂
20 43 ♆ ⚹ ♇

26 Monday
02 24 ☽ ⊻ ♂
05 30 ☉ ± ♂
11 53 ☽ △ ♅
12 26 ☽ ∠ ♄
14 50 ☽ □ ♇
14 52 ☽ ⊼ ♆
18 12 ♀ Q ♇

27 Tuesday
02 58 ☽ ± ♆
03 54 ☽ ⊼ ☿
06 29 ☽ ⊥ ♄
12 27 ☽ # ♅
15 12 ☽ ± ☿
17 21 ☽ ⊼ ♀
18 28 ☽ ⊻ ♄
18 43 ☽ ⊻ ♃
21 03 ☽ bQ ♆

28 Wednesday
00 46 ♀ ∠ ♅
02 57 ☽ Q ♇
03 43 ♀ △ ♄
05 13 ♀ ☍ ♃
06 34 ☽ ⊥ ♃
06 41 ☽ ⊼ ☉
06 44 ☽ ± ♀
08 13 ☽ bQ ☿
10 54 ♂ # ♆
13 32 ☽ ☌ ♂
19 42 ☽ ± ☉
21 06 ☽ ∥ ♄
23 53 ♃ ⚹ ♄

29 Thursday
00 05 ☽ □ ♅
00 15 ☽ ∠ ♃
02 41 ☽ bQ ♀
02 49 ☽ ⚹ ♇
02 58 ☽ △ ♆
11 30 ♃ bQ ♅
12 46 ☽ △ ☿
15 07 ☽ bQ ☉
19 26 ☉ # ♃
23 52 ☽ ∥ ♇

30 Friday
05 25 ☽ ⚹ ♃
05 40 ☽ ☌ ♄
08 18 ☽ ∠ ♇
11 29 ☽ △ ♀
11 33 ☽ Q ♅
14 57 ♀ ± ♂
17 18 ☿ ⊥ ♅
22 59 ☽ △ ☉
23 17 ☽ ⊻ ♂

31 Saturday
01 45 ☽ ⊥ ♇
02 07 ☉ ⊼ ♂
04 25 ☽ Q ♃
10 35 ☽ ⊥ ♂
10 46 ☽ ⚹ ♅
13 15 ☽ ⊻ ♇
13 30 ☽ □ ♆
21 54 ☽ □ ☿
22 03 ☿ St D

All ephemeris data is given at 12.00 UT and the Moon's longitude is additionally given for 24.00 UT

JANUARY 1978

Date	Sidereal time h m s	LONGITUDES Sun ☉ ° ' "	Moon ☽ ° ' "	Moon ☽ 24.00 ° ' "	Mercury ☿ ° '	Venus ♀ ° '	Mars ♂ ° '	Jupiter ♃ ° '	Saturn ♄ ° '	Uranus ♅ ° '	Neptune ♆ ° '	Pluto ♇ ° '
01	18 43 08	10 ♑ 43 19	28 ♍ 38 51	05 ♎ 07 09	21 ♐ 10	05 ♑ 46	08 ♌ 55	29 ♊ 48	00 ♍ 09	15 ♏ 20	16 ♐ 46	16 ♎ 36
02	18 47 05	11 44 29	11 ♎ 40 47	18 ♎ 20 13	21 D 20	07 01	08 R 39	29 R 40	00 R 06	15 22	16 48	16 37
03	18 51 01	12 45 38	25 ♎ 05 52	01 ♏ 58 05	21 37	08 17	08 22	29 32	00 04	15 25	16 50	16 37
04	18 54 58	13 46 48	08 ♏ 57 05	16 ♏ 02 58	22 02	09 32	08 04	29 25	00 ♍ 01	15 27	16 52	16 38
05	18 58 54	14 47 58	23 ♏ 15 36	00 ♐ 34 42	22 34	10 48	07 46	29 17	29 ♌ 59	15 30	16 54	16 38
06	19 02 51	15 49 08	07 ♐ 59 43	15 ♐ 29 54	23 12	12 03	07 27	29 10	29 56	15 32	16 56	16 39
07	19 06 48	16 50 19	23 ♐ 04 15	00 ♑ 41 34	23 55	13 19	07 07	29 02	29 53	15 34	16 58	16 39
08	19 10 44	17 51 29	08 ♑ 20 31	15 ♑ 59 37	24 42	14 34	06 47	28 55	29 50	15 36	17 00	16 40
09	19 14 41	18 52 40	23 ♑ 37 26	01 ♒ 12 29	25 35	15 50	06 26	28 48	29 47	15 39	17 02	16 40
10	19 18 37	19 53 50	08 ♒ 43 28	16 ♒ 09 14	26 30	17 05	06 05	28 40	29 44	15 41	17 04	16 40
11	19 22 34	20 54 59	23 ♒ 28 50	00 ♓ 41 33	27 30	18 21	05 43	28 33	29 41	15 43	17 06	16 41
12	19 26 30	21 56 08	07 ♓ 46 57	14 ♓ 44 46	28 32	19 36	05 21	28 26	29 38	15 45	17 08	16 41
13	19 30 27	22 57 17	21 ♓ 34 59	28 ♓ 17 48	29 ♐ 37	20 52	04 58	28 20	29 35	15 47	17 10	16 41
14	19 34 23	23 58 24	04 ♈ 53 30	11 ♈ 22 32	00 ♑ 45	22 07	04 35	28 13	29 31	15 49	17 12	16 41
15	19 38 20	24 59 31	17 ♈ 45 26	24 ♈ 02 48	01 55	23 22	04 12	28 06	29 28	15 51	17 14	16 42
16	19 42 17	26 00 38	00 ♉ 15 16	06 ♉ 23 28	03 07	24 38	03 49	28 00	29 24	15 53	17 16	16 42
17	19 46 13	27 01 43	12 ♉ 28 05	18 ♉ 29 44	04 20	25 53	03 25	27 54	29 20	15 54	17 17	16 42
18	19 50 10	28 02 48	24 ♉ 29 02	00 ♊ 26 33	05 36	27 09	03 01	27 48	29 17	15 56	17 19	16 42
19	19 54 06	29 ♑ 03 52	06 ♊ 22 50	12 ♊ 18 23	06 53	28 24	02 37	27 42	29 13	15 58	17 21	16 R 42
20	19 58 03	00 ♒ 04 55	18 ♊ 13 38	24 ♊ 08 59	08 11	29 ♑ 40	02 13	27 36	29 09	16 00	17 23	16 42
21	20 01 59	01 05 57	00 ♋ 04 45	06 ♋ 01 16	09 30	00 ♒ 55	01 49	27 30	29 05	16 01	17 25	16 42
22	20 05 56	02 06 59	11 ♋ 58 45	17 ♋ 57 25	10 51	02 10	01 25	27 25	29 01	16 03	17 26	16 42
23	20 09 52	03 08 00	23 ♋ 57 27	29 ♋ 58 59	12 13	03 26	01 01	27 19	28 57	16 04	17 28	16 41
24	20 13 49	04 09 00	06 ♌ 02 08	12 ♌ 07 01	13 36	04 41	00 37	27 14	28 53	16 06	17 30	16 41
25	20 17 46	05 09 59	18 ♌ 13 45	24 ♌ 22 25	15 00	05 57	00 ♌ 14	27 09	28 49	16 07	17 31	16 41
26	20 21 42	06 10 57	00 ♍ 33 10	06 ♍ 46 09	16 24	07 12	29 ♋ 50	27 04	28 44	16 08	17 33	16 41
27	20 25 39	07 11 54	13 ♍ 01 32	19 ♍ 19 31	17 50	08 27	29 27	27 00	28 40	16 10	17 35	16 40
28	20 29 35	08 12 51	25 ♍ 40 21	02 ♎ 04 17	19 17	09 43	29 04	26 55	28 36	16 11	17 36	16 40
29	20 33 32	09 13 47	08 ♎ 31 39	15 ♎ 02 44	20 44	10 58	28 42	26 51	28 31	16 12	17 38	16 40
30	20 37 28	10 14 43	21 ♎ 37 52	28 ♎ 17 25	22 12	12 13	28 19	26 47	28 27	16 13	17 40	16 39
31	20 41 25	11 ♒ 15 37	05 ♏ 01 38	11 ♏ 50 50	23 ♑ 41	13 ♒ 29	27 ♋ 58	26 ♊ 43	28 ♌ 22	16 ♏ 14	17 ♐ 41	16 ♎ 39

Date	Moon True ☊ ° '	Moon Mean ☊ ° '	Moon ☽ Latitude ° '
01	10 ♎ 20	10 ♎ 32	01 S 01
02	10 D 20	10 29	00 N 07
03	10 R 20	10 25	01 17
04	10 18	10 22	02 25
05	10 14	10 19	03 26
06	10 07	10 16	04 15
07	09 58	10 13	04 48
08	09 47	10 09	05 00
09	09 35	10 06	04 51
10	09 24	10 03	04 22
11	09 15	10 00	03 36
12	09 09	09 57	02 38
13	09 06	09 54	01 31
14	09 05	09 50	00 N 22
15	09 D 05	09 47	00 S 46
16	09 R 05	09 44	01 50
17	09 04	09 41	02 47
18	09 00	09 38	03 36
19	08 54	09 34	04 15
20	08 46	09 31	04 43
21	08 34	09 28	04 59
22	08 22	09 25	05 01
23	08 09	09 22	04 51
24	07 56	09 19	04 27
25	07 45	09 15	03 51
26	07 37	09 12	03 03
27	07 31	09 09	02 07
28	07 28	09 06	01 S 03
29	07 D 27	09 03	00 N 06
30	07 28	09 00	01 15
31	07 ♎ 29	08 ♎ 56	02 N 22

DECLINATIONS

Date	Sun ☉ ° '	Moon ☽ ° '	Mercury ☿ ° '	Venus ♀ ° '	Mars ♂ ° '	Jupiter ♃ ° '	Saturn ♄ ° '	Uranus ♅ ° '	Neptune ♆ ° '	Pluto ♇ ° '
01	23 S 00	00 S 24	20 S 13	23 S 38	21 N 47	23 N 12	12 N 45	16 S 04	21 S 22	09 N 12
02	22 55	04 31	20 19	23 36	21 53	23 12	12 46	16 04	21 22	09 12
03	22 50	08 31	20 27	23 34	22 00	23 13	12 47	16 05	21 22	09 13
04	22 44	12 11	20 35	23 32	22 07	23 13	12 48	16 06	21 23	09 13
05	22 37	15 16	20 45	23 28	22 14	23 13	12 49	16 07	21 23	09 13
06	22 30	17 27	20 55	23 24	22 20	23 13	12 51	16 07	21 23	09 14
07	22 23	18 28	21 06	23 19	22 28	23 13	12 52	16 08	21 23	09 14
08	22 15	18 11	21 17	23 13	22 35	23 13	12 53	16 08	21 23	09 15
09	22 07	16 35	21 28	23 07	22 42	23 13	12 54	16 09	21 24	09 15
10	21 58	13 51	21 39	23 00	22 49	23 13	12 55	16 10	21 24	09 15
11	21 49	10 17	21 49	22 52	22 56	23 13	12 57	16 10	21 24	09 16
12	21 39	06 13	21 59	22 43	23 03	23 14	12 58	16 11	21 24	09 16
13	21 29	01 S 56	22 09	22 34	23 10	23 14	12 59	16 11	21 24	09 17
14	21 19	02 N 17	22 18	22 24	23 17	23 14	13 01	16 12	21 24	09 17
15	21 08	06 16	22 27	22 14	23 24	23 14	13 02	16 13	21 25	09 18
16	20 57	09 51	22 35	22 02	23 31	23 14	13 04	16 13	21 25	09 18
17	20 45	12 55	22 41	21 50	23 37	23 14	13 05	16 14	21 25	09 19
18	20 33	15 24	22 48	21 38	23 44	23 14	13 07	16 14	21 25	09 19
19	20 21	17 11	22 53	21 25	23 50	23 14	13 08	16 15	21 25	09 20
20	20 08	18 13	22 57	21 11	23 57	23 14	13 10	16 15	21 25	09 20
21	19 55	18 28	23 00	20 56	24 03	23 14	13 11	16 15	21 25	09 21
22	19 41	17 54	23 02	20 41	24 09	23 14	13 13	16 16	21 26	09 21
23	19 27	16 33	23 03	20 25	24 15	23 14	13 14	16 16	21 26	09 22
24	19 13	14 27	23 02	20 09	24 20	23 14	13 16	16 17	21 26	09 23
25	18 59	11 42	23 01	19 52	24 26	23 14	13 17	16 17	21 26	09 23
26	18 44	08 25	22 59	19 34	24 31	23 14	13 19	16 17	21 26	09 24
27	18 28	04 43	22 55	19 16	24 36	23 14	13 21	16 18	21 26	09 24
28	18 13	00 N 46	22 50	18 58	24 41	23 14	13 22	16 18	21 26	09 25
29	17 57	03 S 18	22 43	18 39	24 45	23 15	13 24	16 19	21 26	09 26
30	17 41	07 16	22 36	18 19	24 50	23 15	13 26	16 19	21 27	09 26
31	17 S 24	10 S 58	22 S 27	17 S 59	24 N 54	23 N 15	13 N 28	16 S 19	21 S 27	09 N 27

ZODIAC SIGN ENTRIES

Date	h	m	Planets
01	14	31	☽ ♎
03	20	35	☽ ♏
05	00	44	♄ ♌
05	23	03	☽ ♐
07	22	55	☽ ♑
09	22	05	☽ ♒
11	22	50	☽ ♓
13	20	07	☿ ♑
14	03	05	☽ ♈
16	11	30	☽ ♉
18	23	06	☽ ♊
20	10	04	☉ ♒
20	18	29	♀ ♒
21	11	50	☽ ♋
24	00	02	☽ ♌
26	01	59	♂ ♋
26	10	56	☽ ♍
28	20	08	☽ ♎
31	03	04	☽ ♏

LATITUDES

Date	Mercury ☿ ° '	Venus ♀ ° '	Mars ♂ ° '	Jupiter ♃ ° '	Saturn ♄ ° '	Uranus ♅ ° '	Neptune ♆ ° '	Pluto ♇ ° '
01	02 N 57	00 S 19	03 N 53	00 S 14	01 N 25	00 N 23	01 N 25	17 N 01
04	02 37	00 26	04 00	00 14	01 26	00 23	01 25	17 03
07	02 12	00 33	04 06	00 13	01 27	00 23	01 25	17 05
10	01 45	00 39	04 12	00 13	01 27	00 23	01 25	17 07
13	01 17	00 45	04 16	00 12	01 28	00 23	01 25	17 08
16	00 50	00 51	04 20	00 12	01 28	00 23	01 25	17 10
19	00 N 23	00 57	04 23	00 11	01 29	00 23	01 26	17 12
22	00 S 02	01 02	04 25	00 11	01 29	00 23	01 26	17 14
25	00 25	01 07	04 26	00 10	01 30	00 24	01 26	17 15
28	00 47	01 11	04 26	00 10	01 30	00 24	01 26	17 17
31	01 S 06	01 S 15	04 N 25	00 S 09	01 N 31	00 N 24	01 N 26	17 N 19

DATA

Julian Date	2443510
Delta T	+49 seconds
Ayanamsa	23° 33' 03"
Synetic vernal point	05° ♓ 33' 57"
True obliquity of ecliptic	23° 26' 22"

MOON'S PHASES, APSIDES AND POSITIONS ☽

Date	h	m	Phase	Longitude ° '	Eclipse Indicator
02	12	07	☾	11 ♎ 45	
09	04	00	●	18 ♑ 32	
16	03	03	☽	25 ♈ 38	
24	07	56	○	03 ♌ 59	
31	23	51	☾	11 ♏ 46	

Day	h	m		
08	12	03	Perigee	
21	02	16	Apogee	
01	09	39	OS	
07	18	51	Max dec	18° S 32'
13	22	55	ON	
21	07	16	Max dec	18° N 29'
28	16	31	OS	

LONGITUDES

Date	Chiron ⚷ ° '	Ceres ⚳ ° '	Pallas ⚴ ° '	Juno ⚵ ° '	Vesta ⚶ ° '	Black Moon Lilith ⚸ ° '
01	01 ♉ 09	17 ♐ 17	27 ♏ 34	29 ♐ 39	15 ♏ 55	28 ♊ 11
11	01 ♉ 06	21 ♐ 19	01 ♐ 29	03 ♑ 14	20 ♏ 33	29 ♊ 18
21	01 ♉ 10	25 ♐ 16	05 ♐ 15	06 ♑ 48	25 ♏ 02	00 ♋ 25
31	01 ♉ 18	29 ♐ 05	08 ♐ 48	10 ♑ 19	29 ♏ 20	01 ♋ 32

All ephemeris data is given at 12.00 UT and the Moon's longitude is additionally given for 24.00 UT

ASPECTARIAN

h m	Aspects
01 Sunday	
03 19	☽ ∠ ♂
14 08	☽ □ ♃
14 47	☽ ⚺ ♄
15 09	☽ ∠ ♅
23 22	☽ Q ♆
02 Monday	
01 51	☽ ⊥ ♄
02 35	☽ □ ♀
06 36	☽ ⚹ ♂
07 39	☽ Q ☿
07 47	☽ ⊥ ♅
12 07	☽ □ ☉
17 26	☿ ⊥ ♅
18 11	☽ ∠ ♄
18 42	☽ ⚺ ♅
20 55	☽ ☌ ♇
21 16	☽ ⚹ ♆
03 Tuesday	
03 48	☽ Q ♂
05 42	☽ ⚹ ☿
13 24	♀ ⚻ ♂
14 17	☽ Q ♀
16 21	☽ # ♇
19 43	☽ △ ♃
20 40	☽ ⚹ ♄
22 42	☽ Q ☉
23 47	☽ ∠ ♆
04 Wednesday	
08 37	☽ ∠ ☿
10 32	☽ □ ♂
13 05	☽ ⚹ ♀
15 16	☽ ⊥ ♆
16 24	☽ # ♄
17 12	☽ Q ♄
20 49	☽ ⚹ ☉
21 10	☽ bQ ♃
23 02	☽ ☌ ♅
05 Thursday	
00 25	☽ ⊥ ☿
00 59	☽ ⚺ ♇
01 24	☽ ⚺ ♆
10 48	☽ ⚺ ☿
10 59	☽ ⊥ ♇
12 03	☽ ± ♃
16 03	☉ bQ ♄
16 34	☽ ∠ ♀
17 17	☿ bQ ♂
20 04	☽ ∥ ♅
21 49	☽ ⚻ ♃
22 59	☽ □ ♄
23 32	☽ ∠ ☉
06 Friday	
01 44	☽ ∠ ♇
04 58	☉ ⚹ ♅
08 35	☽ ⊥ ♀
11 08	☽ △ ♂
15 09	☽ ⊥ ☉
19 06	☽ ⚺ ♀
07 Saturday	
00 05	☽ ⚺ ♅
01 25	☽ ⚺ ☉
01 50	☽ ⚹ ♇
02 19	☽ ☌ ♆
04 05	☉ # ♂
07 41	☉ □ ♇
09 37	☽ ⊥ ♅
10 32	☽ bQ ♂
13 24	☽ ☌ ☿
15 07	☉ ⚺ ♆
20 48	☽ Q ♇
21 19	☽ ☍ ♃
22 42	☽ △ ♄
23 50	☽ ∠ ♅
08 Sunday	
00 24	☽ ± ♂
09 36	☽ ⚻ ♂
12 09	♀ # ♃
16 59	♀ bQ ♄
22 09	☽ bQ ♄
22 38	☽ ☌ ♀
23 25	☽ ⚹ ♅
09 Monday	
01 03	☽ □ ♇
01 37	☽ ⚺ ♆
02 33	☿ ∥ ♆
04 00	☽ ☌ ☉
08 25	♀ ⚹ ♅
11 04	☽ ⊥ ♆
12 16	☽ ± ♄
15 16	☽ ⚺ ☿
16 29	☽ ∥ ♅
18 22	☽ Q ♅
20 06	☽ ⚻ ♃
21 43	☽ ⚻ ♄
10 Tuesday	
01 20	☽ ∠ ♆
01 24	☽ ⊥ ☿
04 09	♀ □ ♇
05 34	☽ ± ♃
07 52	☽ ☍ ♂
11 39	♀ ⚺ ♆
16 47	☽ ∠ ☿
18 42	☽ # ♄
19 55	☽ bQ ♃
23 15	☽ □ ♅
11 Wednesday	
00 51	☽ △ ♇
01 31	☽ ⚹ ♆
02 47	☽ ⚺ ♀
05 39	♀ # ♂
07 28	☽ ⚺ ☉
11 21	☉ ∥ ☿
13 34	☽ ⊥ ♀
18 07	☽ ⊥ ☉
18 14	☽ # ♇
19 10	☽ ⚹ ☿
20 21	☽ △ ♃
21 21	☽ Q ♆
22 16	☽ ☍ ♄
12 Thursday	
01 40	☽ bQ ♇
06 04	☽ ∠ ♀
07 58	☽ ⚻ ♂
10 03	☿ ☍ ♃
10 27	☽ ∠ ☉
15 18	☿ Q ♇
16 58	☽ ± ♇
17 07	☽ Q ☿
17 57	☽ ± ♂
13 Friday	
01 24	☿ ± ♂
01 47	☽ △ ♅
03 23	☽ ⚻ ♇
04 12	☽ □ ♆
09 14	☽ bQ ♂
10 35	☽ ⚹ ♀
11 01	☿ △ ♄
14 38	☽ ⚹ ☉
17 06	☉ ⊥ ♆
23 22	☉ ∥ ♆
23 57	☽ □ ♃
14 Saturday	
00 31	♂ ∥ ♃
01 51	☉ ± ♄
02 15	☽ ⚻ ♄
03 44	☽ □ ☿
04 32	☽ bQ ♅
05 46	♂ Q ♇
10 26	☽ Q ♀
11 28	☽ △ ♂
13 09	☽ ± ♄
13 22	☿ ∠ ♅
14 09	☽ Q ☉
19 21	☿ ∥ ♀
21 07	☽ ± ♅
15 Sunday	
05 48	☽ bQ ♄
08 23	☽ ⚻ ♅
08 54	☽ Q ♃
09 08	♀ ⊥ ♆
09 59	☽ ☍ ♇
11 00	☽ △ ♆
13 34	♀ ± ♄
20 02	♂ ⊥ ♃
23 54	☽ □ ♀
16 Monday	
03 03	☽ □ ☉
07 40	☽ ⚹ ♃
08 09	☽ ∥ ♇
10 21	☽ △ ♄
15 55	☽ bQ ♆
18 11	☽ △ ☿
18 43	☽ □ ♂
22 24	☿ ⚻ ♂
17 Tuesday	
09 40	☽ ± ♆
12 51	☽ ∠ ♃
13 25	☽ ∥ ♄
18 51	☽ ☍ ♅
20 24	☽ ⚻ ♇
21 37	☽ ⚻ ♆
18 Wednesday	
03 17	☽ bQ ☿
05 16	☽ Q ♂
06 36	☉ ⚻ ♃
06 39	☽ ⊥ ♃
08 25	☽ ± ♇
09 19	☉ Q ♅
17 59	☽ △ ♀
18 36	☽ ⚺ ♃
19 50	☽ △ ☉
21 36	☽ □ ♄
22 09	☽ # ♅
23 28	♀ ⚻ ♃
23 31	☽ ± ☿
19 Thursday	
00 47	♇ St R
02 32	☽ bQ ♇
03 26	♀ Q ♅
04 39	☽ ⚹ ♂
11 01	♀ ∥ ♆
13 07	☽ ⚻ ☿
15 19	☉ ⚻ ♄
20 Friday	
02 44	♀ ⚻ ♄
03 01	♂ bQ ♆
03 55	☽ bQ ♀
05 02	☽ bQ ☉
07 28	☽ ⚻ ♅
08 54	☽ △ ♇
09 50	☽ Q ♄
10 01	☽ ∠ ♂
10 17	☽ ☍ ♆
19 39	☽ ± ♅
21 Saturday	
00 19	☽ ± ♀
00 59	☽ ± ☉
03 40	☽ ⊥ ♂
06 50	☽ ☌ ♃
10 00	☽ ⚹ ♄
13 54	☽ ⚻ ♀
13 54	☽ bQ ♅
14 15	☽ ⚻ ☉
15 24	☽ ⚺ ♂
22 Sunday	
00 11	☉ ☍ ♂
01 04	♀ ☍ ♂
06 15	☉ ☌ ♀
09 26	☽ ☍ ☿
16 04	☽ ∠ ♄
17 11	♀ ∠ ♆
19 51	☉ ∠ ♆
20 11	☽ △ ♅
21 28	☽ □ ♇
22 59	☽ ⚻ ♆
23 Monday	
10 00	☽ ⊥ ♄
10 05	♀ ± ♃
11 01	☽ ± ♆
15 37	☽ # ♅
16 07	☉ ± ♃
18 40	☽ ⚺ ♃
21 53	☽ ⚺ ♄
24 Tuesday	
01 37	☽ ☌ ♂
04 58	☽ bQ ♆
06 30	☽ ⊥ ♃
07 56	☽ ☍ ☉
09 01	☽ ☍ ♀
09 20	☽ Q ♇
16 42	☿ bQ ♄
22 54	☽ ∥ ♄
25 Wednesday	
00 09	☽ ∠ ♃
04 50	☽ ⚻ ☿
07 51	☽ □ ♅
08 58	☽ ⚹ ♇
10 37	☽ △ ♆
18 06	☽ ± ☿
26 Thursday	
05 12	☽ ∥ ♇
05 17	☽ ⚹ ♃
07 26	☿ ⚹ ♅
08 30	☽ ☌ ♄
10 39	☽ ⚺ ♂
13 52	☽ bQ ☿
14 11	☽ ∠ ♇
16 37	☿ □ ♇
18 57	☽ Q ♅
21 54	☽ ⊥ ♂
23 50	☽ ⚻ ☉
27 Friday	
02 16	☽ ⚻ ♀
04 19	☽ Q ♃
07 30	☽ ⊥ ♇
07 40	☿ ⚺ ♆
12 22	☽ ± ☉
14 38	☽ ∠ ♂
15 02	☽ ± ♀
18 00	☽ ⚹ ♅
18 58	☽ ⚺ ♇
20 42	☽ □ ♆
22 21	☽ △ ☿
28 Saturday	
06 58	☽ bQ ☉
09 59	☽ bQ ♀
14 20	☽ □ ♃
17 28	☽ ⚺ ♄
18 12	☽ ⚹ ♂
22 21	☽ ∠ ♅
29 Sunday	
04 37	☽ ⊥ ♄
06 37	☽ Q ♆
13 25	☽ △ ☉
15 06	☽ ⊥ ♅
15 53	☽ Q ♂
16 59	☽ △ ♀
21 09	☽ ∠ ♄
30 Monday	
01 45	♂ ⚺ ♄
02 08	☽ ⚺ ♅
02 57	☽ ☌ ♇
03 56	♀ bQ ♃
04 46	☽ ⚹ ♆
13 10	☽ □ ☿
15 47	☿ ± ♄
21 14	☽ △ ♃
23 44	☽ □ ♂
31 Tuesday	
00 13	☽ ⚹ ♄
01 50	☽ # ♇
07 50	☽ ∠ ♆
11 58	☿ ⊥ ♆
21 22	☽ Q ♄
22 02	☉ bQ ♃
23 42	☽ bQ ♃
23 44	☽ ⊥ ♆
23 51	☽ □ ☉

FEBRUARY 1978

LONGITUDES

Date	Sidereal time h m s	Sun ☉ ° ' "	Moon ☽ ° ' "	Moon ☽ 24.00 ° ' "	Mercury ☿ ° '	Venus ♀ ° '	Mars ♂ ° '	Jupiter ♃ ° '	Saturn ♄ ° '	Uranus ♅ ° '	Neptune ♆ ° '	Pluto ♇ ° '
01	20 45 21	12 ♒ 16 31	18 ♏ 45 12	25 ♏ 44 51	25 ♑ 11	14 ♒ 44	27 ♋ 36	26 ♊ 39	28 ♌ 18	16 ♏ 15	17 ♐ 43	16 ♎ 39
02	20 49 18	13 17 24	02 ♐ 49 47	09 ♐ 59 53	26 42	15 59	27 R 15	26 R 35	28 R 13	16 16	17 44	16 R 38
03	20 53 14	14 18 17	17 ♐ 14 50	24 ♐ 34 13	28 13	17 15	26 55	26 32	28 09	16 17	17 45	16 38
04	20 57 11	15 19 09	01 ♑ 57 21	09 ♑ 23 28	29 ♑ 45	18 30	26 35	26 29	28 04	16 18	17 47	16 37
05	21 01 08	16 19 59	16 ♑ 51 35	24 ♑ 20 39	01 ♒ 18	19 45	26 16	26 26	27 59	16 19	17 48	16 36
06	21 05 04	17 20 49	01 ♒ 49 31	09 ♒ 17 00	02 51	21 01	25 58	26 23	27 55	16 20	17 50	16 36
07	21 09 01	18 21 38	16 ♒ 41 57	24 ♒ 03 17	04 26	22 16	25 40	26 20	27 50	16 20	17 51	16 35
08	21 12 57	19 22 25	01 ♓ 20 04	08 ♓ 31 28	06 01	23 31	25 22	26 18	27 45	16 21	17 52	16 34
09	21 16 54	20 23 11	15 ♓ 36 52	22 ♓ 35 49	07 37	24 46	25 06	26 16	27 40	16 22	17 54	16 34
10	21 20 50	21 23 55	29 ♓ 28 04	06 ♈ 13 33	09 14	26 02	24 50	26 14	27 35	16 22	17 55	16 33
11	21 24 47	22 24 38	12 ♈ 52 20	19 ♈ 24 39	10 51	27 17	24 35	26 12	27 31	16 23	17 56	16 32
12	21 28 43	23 25 20	25 ♈ 50 50	02 ♉ 11 20	12 30	28 32	24 21	26 10	27 26	16 23	17 57	16 31
13	21 32 40	24 25 59	08 ♉ 26 39	14 ♉ 37 20	14 09	29 ♒ 47	24 07	26 09	27 21	16 23	17 58	16 31
14	21 36 37	25 26 37	20 ♉ 44 01	26 ♉ 47 19	15 49	01 ♓ 02	23 54	26 07	27 16	16 24	18 00	16 30
15	21 40 33	26 27 14	02 ♊ 47 51	08 ♊ 46 16	17 30	02 17	23 42	26 06	27 11	16 24	18 01	16 29
16	21 44 30	27 27 48	14 ♊ 43 09	20 ♊ 39 06	19 12	03 33	23 31	26 06	27 06	16 24	18 02	16 28
17	21 48 26	28 28 21	26 ♊ 34 41	02 ♋ 30 25	20 54	04 48	23 21	26 05	27 01	16 24	18 03	16 27
18	21 52 23	29 ♒ 28 52	08 ♋ 26 47	14 ♋ 24 14	22 38	06 03	23 11	26 05	26 57	16 24	18 04	16 26
19	21 56 19	00 ♓ 29 22	20 ♋ 23 08	26 ♋ 23 50	24 22	07 18	23 02	26 04	26 52	16 24	18 05	16 25
20	22 00 16	01 29 49	02 ♌ 26 39	08 ♌ 31 48	26 08	08 33	22 54	26 D 04	26 47	16 R 24	18 06	16 24
21	22 04 12	02 30 15	14 ♌ 39 29	20 ♌ 49 53	27 54	09 48	22 47	26 05	26 42	16 24	18 07	16 23
22	22 08 09	03 30 39	27 ♌ 03 05	03 ♍ 19 10	29 ♒ 41	11 03	22 40	26 05	26 37	16 24	18 08	16 22
23	22 12 06	04 31 02	09 ♍ 38 12	16 ♍ 00 13	01 ♓ 29	12 18	22 35	26 06	26 33	16 24	18 09	16 21
24	22 16 02	05 31 23	22 ♍ 25 15	28 ♍ 53 16	03 19	13 33	22 30	26 06	26 28	16 24	18 09	16 20
25	22 19 59	06 31 42	05 ♎ 24 20	11 ♎ 58 25	05 09	14 48	22 26	26 07	26 23	16 23	18 10	16 18
26	22 23 55	07 31 59	18 ♎ 35 33	25 ♎ 15 47	07 00	16 03	22 22	26 09	26 18	16 23	18 11	16 17
27	22 27 52	08 32 15	01 ♏ 59 06	08 ♏ 45 35	08 52	17 18	22 20	26 10	26 14	16 23	18 12	16 16
28	22 31 48	09 ♓ 32 30	15 ♏ 35 14	22 ♏ 28 04	10 ♓ 44	18 ♓ 33	22 ♋ 18	26 ♊ 11	26 ♌ 09	16 ♏ 22	18 ♐ 12	16 ♎ 15

Moon / DECLINATIONS

Date	Moon True ☊ ° '	Moon Mean ☊ ° '	Moon ☽ Latitude ° '	Sun ☉ ° '	Moon ☽ ° '	Mercury ☿ ° '	Venus ♀ ° '	Mars ♂ ° '	Jupiter ♃ ° '	Saturn ♄ ° '	Uranus ♅ ° '	Neptune ♆ ° '	Pluto ♇ ° '
01	07 ♎ 29	08 ♎ 53	03 N 23	17 S 07	14 S 09	22 S 17	17 S 38	24 N 57	23 N 15	13 N 29	16 S 19	21 S 27	09 N 28
02	07 R 27	08 50	04 13	16 50	16 36	22 06	17 16	25 01	23 15	13 31	16 20	21 27	09 28
03	07 23	08 47	04 48	16 32	18 03	21 53	16 55	25 04	23 15	13 33	16 20	21 27	09 29
04	07 17	08 44	05 05	16 15	18 21	21 39	16 32	25 07	23 15	13 34	16 20	21 27	09 30
05	07 09	08 40	05 02	15 56	17 23	21 23	16 10	25 10	23 15	13 36	16 20	21 27	09 31
06	07 01	08 37	04 38	15 38	15 14	21 06	15 47	25 13	23 15	13 38	16 21	21 27	09 31
07	06 53	08 34	03 56	15 20	12 05	20 48	15 23	25 15	23 15	13 40	16 21	21 27	09 32
08	06 47	08 31	02 59	15 01	08 13	20 29	14 59	25 17	23 15	13 41	16 21	21 27	09 33
09	06 43	08 28	01 51	14 42	03 S 58	20 08	14 35	25 19	23 15	13 43	16 21	21 27	09 33
10	06 41	08 25	00 N 39	14 22	00 N 28	19 46	14 10	25 21	23 15	13 45	16 21	21 28	09 34
11	06 D 41	08 21	00 S 33	14 03	04 34	19 22	13 45	25 22	23 16	13 47	16 21	21 28	09 35
12	06 42	08 18	01 42	13 43	08 24	18 57	13 19	25 24	23 16	13 49	16 21	21 28	09 36
13	06 43	08 15	02 44	13 23	11 44	18 31	12 53	25 25	23 16	13 50	16 22	21 28	09 36
14	06 45	08 12	03 36	13 02	14 28	18 03	12 27	25 26	23 16	13 52	16 22	21 28	09 37
15	06 R 44	08 09	04 18	12 42	16 30	17 34	12 00	25 26	23 16	13 54	16 22	21 28	09 38
16	06 43	08 06	04 48	12 21	17 48	17 03	11 33	25 27	23 16	13 56	16 22	21 28	09 39
17	06 39	08 02	05 06	12 00	18 18	16 31	11 06	25 27	23 16	13 57	16 22	21 28	09 40
18	06 34	07 59	05 10	11 39	18 01	15 58	10 39	25 27	23 16	13 59	16 22	21 28	09 40
19	06 28	07 56	05 01	11 18	16 56	15 23	10 11	25 27	23 17	14 01	16 22	21 28	09 41
20	06 21	07 53	04 39	10 57	15 05	14 47	09 43	25 26	23 17	14 03	16 22	21 28	09 42
21	06 15	07 50	04 04	10 35	12 33	14 09	09 14	25 26	23 17	14 04	16 22	21 28	09 43
22	06 09	07 46	03 17	10 13	09 25	13 30	08 46	25 25	23 17	14 06	16 22	21 28	09 43
23	06 05	07 43	02 19	09 51	05 48	12 50	08 17	25 24	23 17	14 08	16 22	21 28	09 44
24	06 03	07 40	01 14	09 29	01 N 53	12 08	07 48	25 23	23 17	14 09	16 22	21 28	09 45
25	06 02	07 37	00 S 03	09 07	02 S 12	11 25	07 19	25 22	23 18	14 11	16 21	21 28	09 46
26	06 D 02	07 34	01 N 08	08 45	06 14	10 41	06 49	25 21	23 18	14 13	16 21	21 28	09 47
27	06 04	07 31	02 18	08 22	10 01	09 55	06 20	25 19	23 18	14 14	16 21	21 28	09 48
28	06 ♎ 05	07 ♎ 27	03 N 20	08 S 00	13 S 19	09 S 09	05 S 50	25 N 18	23 N 18	14 N 16	16 S 21	21 S 28	09 N 48

ZODIAC SIGN ENTRIES

Date	h	m	Planets
02	07	13	☽ ♐
04	08	50	☽ ♑
04	15	54	☿ ♒
06	09	04	☽ ♒
08	09	47	☽ ♓
10	12	56	☽ ♈
12	19	50	☽ ♉
13	16	07	♀ ♓
15	06	24	☽ ♊
17	18	56	☽ ♋
19	00	21	☉ ♓
20	07	10	☽ ♌
22	16	11	☿ ♓
22	17	39	☽ ♍
25	02	03	☽ ♎
27	08	28	☽ ♏

LATITUDES

Date	Mercury ☿ ° '	Venus ♀ ° '	Mars ♂ ° '	Jupiter ♃ ° '	Saturn ♄ ° '	Uranus ♅ ° '	Neptune ♆ ° '	Pluto ♇ ° '
01	01 S 12	01 S 16	04 N 24	00 S 09	01 N 31	00 N 24	01 N 26	17 N 19
04	01 28	01 19	04 22	00 09	01 31	00 24	01 26	17 21
07	01 42	01 22	04 19	00 08	01 32	00 24	01 26	17 22
10	01 53	01 24	04 16	00 08	01 32	00 24	01 26	17 24
13	02 01	01 26	04 11	00 07	01 32	00 24	01 26	17 26
16	02 05	01 27	04 07	00 07	01 33	00 24	01 26	17 27
19	02 06	01 27	04 02	00 06	01 33	00 24	01 27	17 28
22	02 03	01 27	03 56	00 06	01 33	00 24	01 27	17 30
25	01 56	01 26	03 51	00 05	01 33	00 24	01 27	17 31
28	01 44	01 25	03 45	00 05	01 33	00 24	01 27	17 32
31	01 S 27	01 S 23	03 N 39	00 S 05	01 N 34	00 N 24	01 N 27	17 N 33

DATA

Julian Date	2443541
Delta T	+49 seconds
Ayanamsa	23° 33' 07"
Synetic vernal point	05° ♓ 33' 52"
True obliquity of ecliptic	23° 26' 22"

MOON'S PHASES, APSIDES AND POSITIONS ☽

Date	h	m	Phase	Longitude ° '	Eclipse Indicator
07	14	54	●	18 ♒ 29	
14	22	11	☽	25 ♉ 52	
23	01	26	○	04 ♍ 04	

Day	h	m		
05	21	12	Perigee	
17	18	24	Apogee	
04	05	38	Max dec	18° S 23'
10	09	52	0N	
17	15	15	Max dec	18° N 19'
24	23	05	0S	

LONGITUDES

Date	Chiron ⚷ ° '	Ceres ⚳ ° '	Pallas ⚴ ° '	Juno ⚵ ° '	Vesta ⚶ ° '	Black Moon Lilith ⚸ ° '
01	01 ♉ 20	29 ♐ 28	09 ♐ 09	10 ♑ 40	29 ♏ 45	01 ♋ 39
11	01 ♉ 35	03 ♑ 08	12 ♐ 26	14 ♑ 06	03 ♐ 48	02 ♋ 46
21	01 ♉ 55	06 ♑ 38	15 ♐ 25	17 ♑ 28	07 ♐ 34	03 ♋ 54
31	02 ♉ 19	09 ♑ 56	18 ♐ 03	20 ♑ 42	11 ♐ 00	05 ♋ 01

All ephemeris data is given at 12.00 UT and the Moon's longitude is additionally given for 24.00 UT

ASPECTARIAN

h m	Aspects
01 Wednesday	
01 09	☽ Q ☿
04 20	☽ □ ♀
06 31	☽ # ♄
07 40	☽ ☌ ♅
08 21	☽ ⊻ ♇
10 12	☽ ⊻ ♆
15 15	☽ ± ♃
18 41	☽ ⊥ ♇
02 Thursday	
00 22	☽ ⚹ ☿
01 29	☽ ⊼ ♃
02 48	☽ △ ♂
04 15	☽ □ ♄
08 53	☽ ∥ ♅
09 12	☽ Q ☉
09 59	☽ ∠ ♇
10 25	☿ ⊼ ♃
14 08	☽ Q ♀
14 33	☽ ∥ ☉
17 29	♀ □ ♅
19 17	☿ ☍ ♂
19 28	☽ ∥ ♀
03 Friday	
00 16	♀ △ ♇
03 23	☽ bQ ♂
04 33	☽ ∠ ☿
06 47	☽ ⚹ ☉
10 25	☽ ⊻ ♅
10 57	☿ ⊼ ♄
10 59	☽ ⚹ ♇
12 00	☽ ⚹ ♀
12 51	☽ ☌ ♆
13 09	☿ Q ♅
17 54	☽ ± ♂
20 17	☽ ⊥ ♅
21 06	☽ ⊥ ☿
22 01	♀ ⚹ ♆
04 Saturday	
03 08	☽ ☍ ♃
03 29	☽ ⊼ ♂
04 33	☉ ∥ ♅
05 43	☽ △ ♄
06 35	☽ Q ♇
08 01	☽ ⊻ ☿
09 09	☽ ∠ ☉
10 56	☽ ∠ ♅
14 44	☽ ∠ ♀
21 42	♂ ⊻ ♃
05 Sunday	
00 45	☽ ⊥ ☉
00 57	♀ ∥ ♅
05 49	☽ bQ ♄
06 11	☿ ∥ ♆
06 33	☽ ⊥ ♀
11 06	☽ ⊻ ☉
11 08	☽ ⚹ ♅
11 34	☉ □ ♅
11 36	☽ □ ♇
13 31	☽ ⊻ ♆
17 04	☽ ⊻ ♀
18 25	☉ △ ♇
20 11	☽ ± ♄
23 09	☽ ⊥ ♆
06 Monday	
01 07	☽ ∥ ♅
02 47	☽ ☍ ♂
03 18	☽ ⊼ ♃
04 55	☿ ± ♃
05 45	☽ ⊼ ♄
05 56	☽ ∥ ♀
06 23	☽ Q ♅
07 44	☽ ∥ ☉
11 33	☿ ∠ ♆
12 53	☽ ± ♃
13 37	☽ ∠ ♆
13 51	☽ ☌ ☿
23 39	☉ ⚹ ♆
07 Tuesday	
00 51	☽ # ♄
03 20	☽ bQ ♃
11 25	☽ □ ♅
11 49	☽ △ ♇
13 52	☽ ⚹ ♆
14 54	☽ ☌ ☉
21 55	☽ ☌ ♀
08 Wednesday	
02 21	☽ ⊼ ♂
03 42	☽ △ ♃
04 07	☽ # ♇
04 34	☉ ∥ ♀
06 06	☽ ☍ ♄
09 35	☽ Q ♆
12 04	☽ ± ♂
12 24	☽ bQ ♇
20 46	☽ ⊻ ☿
09 Thursday	
02 49	☽ bQ ♂
03 26	☽ ± ♇
08 10	☽ ⊥ ☿
13 16	☽ △ ♅
13 37	☽ ⊼ ♇
15 54	☽ □ ♆
17 08	♀ ⊼ ♂
20 49	☽ ⊻ ☉
10 Friday	
01 37	☽ ∠ ☿
04 02	☽ △ ♂
05 21	☽ ⊻ ♀
06 20	☽ □ ♃
08 05	☽ ⊥ ☉
08 43	☽ ⊼ ♄
15 21	☽ bQ ♅
15 45	♀ △ ♃
16 59	☽ ⊥ ♀
19 15	☽ ± ♄
11 Saturday	
01 19	☽ ∠ ☉
07 28	☽ ± ♅
07 49	☽ ⚹ ☿
10 07	♀ # ♄
10 48	☽ ∠ ♀
11 21	☽ bQ ♄
14 25	☽ Q ♃
16 09	♀ ☍ ♄
16 56	☿ bQ ♃
18 25	☽ ⊼ ♅
18 42	☽ ☍ ♇
21 17	☽ △ ♆
12 Sunday	
05 36	☉ # ♄
07 04	☽ ⚹ ☉
09 05	☽ Q ☿
09 14	☽ □ ♂
12 36	☽ ⚹ ♃
14 58	☽ △ ♄
17 37	☽ ⚹ ♀
20 10	☽ ∥ ♇
13 Monday	
01 29	☽ bQ ♆
05 50	☉ ⊼ ♂
07 47	☽ Q ☉
15 41	♀ Q ♆
17 13	☽ ∠ ♃
17 24	♀ ± ♂
18 51	☽ ± ♆
19 00	☽ Q ♂
19 12	☽ Q ♀
20 10	☽ # ♀
14 Tuesday	
00 13	☽ # ☉
00 48	☽ □ ☿
02 30	♂ ± ♆
03 28	☽ ☍ ♅
03 40	☽ ⊼ ♇
06 10	☽ ∥ ♄
06 36	☽ ⊼ ♆
10 48	☽ ⊥ ♃
15 29	☽ ± ♇
18 10	☽ ⚹ ♂
20 20	☿ □ ♅
20 39	♀ bQ ♇
21 40	☿ △ ♇
22 11	☽ □ ☉
22 40	☽ ⊻ ♃
15 Wednesday	
00 52	☽ □ ♄
03 54	☉ △ ♃
09 22	☽ bQ ♇
10 01	☽ # ♅
10 52	☽ □ ♀
19 25	☿ ⚹ ♆
23 41	☽ ∠ ♂
16 Thursday	
00 53	☽ # ☿
04 08	☉ ☍ ♄
12 47	☽ Q ♄
15 24	☽ ⊼ ♅
15 31	☽ △ ♇
17 34	☽ ⊥ ♂
18 42	☽ ☍ ♆
22 34	☽ △ ☿
17 Friday	
03 33	☽ ± ♅
05 32	☽ ⊻ ♂
11 00	☽ ☌ ♃
12 54	☽ ⚹ ♄
16 12	☽ △ ☉
18 52	☿ ∥ ♅
21 46	☽ bQ ♅
18 Saturday	
05 49	☉ ± ♂
06 35	☽ △ ♀
10 04	☽ bQ ☿
19 00	☽ ∠ ♄
19 01	☿ ⊼ ♂
19 Sunday	
01 17	☽ bQ ☉
02 07	☉ Q ♆
04 01	☽ △ ♅
04 03	☽ □ ♇
07 16	☽ ± ☿
07 23	☽ ⊼ ♆
12 57	☽ ⊥ ♄
15 26	♅ St R
16 16	☽ bQ ♀
17 14	☽ ☌ ♂
19 24	☽ ± ♆
20 24	☽ # ♅
20 57	☽ ± ☉
21 19	☽ ⊼ ☿
23 21	☽ ⊻ ♃
20 Monday	
00 26	♅ ⊻ ♇
00 44	♀ bQ ♂
00 50	☽ ⊻ ♄
01 24	♃ St D
09 42	☉ bQ ♇
09 57	☽ ⊼ ☉
11 16	☽ ⊥ ♃
11 16	☿ △ ♃
12 14	☽ ± ♀
12 39	♀ # ♇
13 18	☽ bQ ♆
15 51	☽ Q ♇
16 22	☽ # ☿
20 31	☿ ☍ ♄
22 28	☽ ∥ ♄
21 Tuesday	
01 25	☽ ⊼ ♀
05 00	☽ ∠ ♃
15 02	☿ # ♄
15 21	☽ ⚹ ♇
15 24	☽ □ ♅
18 44	☽ △ ♆
23 00	♀ ± ♇
23 08	☿ ± ♂
22 Wednesday	
03 38	☽ ⊻ ♂
05 27	☽ # ☉
09 46	☽ ∥ ♇
10 08	☽ ⚹ ♃
11 11	☽ ☌ ♄
15 05	☽ ⊥ ♂
17 14	☽ # ♀
17 54	☽ ☍ ☿
17 55	☿ Q ♆
20 15	☽ ∠ ♇
23 Thursday	
01 26	☽ ☍ ☉
02 04	☽ Q ♅
08 07	☽ ∠ ♂
09 04	☽ Q ♃
10 05	☿ bQ ♇
13 20	☽ ⊥ ♇
17 35	☽ ☍ ♀
19 19	☉ # ♇
24 Friday	
00 37	☽ ⊻ ♇
00 44	☽ ⚹ ♅
04 02	☽ □ ♆
12 08	☽ ⚹ ♂
18 51	☽ □ ♃
19 28	☽ ⊻ ♄
25 Saturday	
04 37	☽ ∠ ♅
06 29	☽ ⊥ ♄
10 13	☽ Q ♂
11 27	☽ ⊼ ☿
13 24	☽ Q ♆
14 14	☽ ⊼ ☉
21 07	☽ ⊥ ♅
22 52	☽ ∠ ♄
26 Sunday	
00 12	☽ ± ☿
02 05	☽ ± ☉
06 55	☽ ⊼ ♀
07 50	☽ ☌ ♇
08 01	☽ ⊻ ♅
08 19	☉ bQ ♂
11 15	☽ ⚹ ♆
15 12	☽ ∥ ♀
16 30	♀ ⊼ ♇
16 44	☿ bQ ♂
18 28	♀ △ ♅
18 47	☽ □ ♂
18 52	☽ ± ♀
19 08	☽ bQ ☿
19 41	☽ bQ ☉
27 Monday	
01 36	☽ △ ♃
01 47	☽ ⚹ ♄
02 15	☽ ∥ ☉
03 03	☉ ☌ ☿
10 32	☽ # ♇
11 32	☽ ∥ ☿
12 37	☽ bQ ♀
14 09	☽ ∠ ♆
16 02	☿ # ♇
23 00	☽ Q ♄
28 Tuesday	
00 32	☽ △ ☉
02 07	☽ △ ☿
02 10	♃ ⚹ ♄
04 16	☽ bQ ♃
05 25	♀ □ ♆
05 46	☿ ± ♇
06 04	☽ ⊥ ♆
13 09	☽ ⊻ ♇
13 22	☽ ☌ ♅
16 35	☽ ⊻ ♆
17 41	☽ △ ♀
20 03	☽ ± ♃
20 04	☽ # ♄
23 36	☽ ⊥ ♇
23 41	☽ △ ♂

MARCH 1978

LONGITUDES

Date	Sidereal time h m s	Sun ☉ ° ' "	Moon ☽ ° ' "	Moon ☽ 24.00 ° ' "	Mercury ☿ ° '	Venus ♀ ° '	Mars ♂ ° '	Jupiter ♃ ° '	Saturn ♄ ° '	Uranus ♅ ° '	Neptune ♆ ° '	Pluto ♇ ° '
01	22 35 45	10 ♓ 32 43	29 ♏ 24 05	06 ♐ 23 13	12 ♓ 38	19 ♓ 48	22 ♋ 17	26 ♊ 13	26 ♌ 04	16 ♏ 22	18 ♐ 13	16 ♎ 13
02	22 39 41	11 32 55	13 ♐ 25 24	20 ♐ 30 28	14 32	21 02	22 D 17	26 15	26 R 00	16 R 21	18 14	16 R 12
03	22 43 38	12 33 06	27 ♐ 38 11	04 ♑ 48 16	16 27	22 17	22 17	26 17	25 55	16 21	18 14	16 11
04	22 47 35	13 33 14	12 ♑ 00 19	19 ♑ 13 51	18 23	23 32	22 18	26 20	25 51	16 20	18 15	16 10
05	22 51 31	14 33 22	26 ♑ 28 19	03 ♒ 43 06	20 19	24 47	22 20	26 22	25 46	16 19	18 15	16 08
06	22 55 28	15 33 27	10 ♒ 57 30	18 ♒ 10 49	22 15	26 02	22 23	26 25	25 42	16 18	18 16	16 07
07	22 59 24	16 33 31	25 ♒ 22 19	02 ♓ 31 18	24 12	27 17	22 26	26 28	25 37	16 18	18 16	16 05
08	23 03 21	17 33 33	09 ♓ 37 05	16 ♓ 39 04	26 08	28 31	22 30	26 31	25 33	16 17	18 17	16 04
09	23 07 17	18 33 34	23 ♓ 36 43	00 ♈ 29 36	28 04	29 ♓ 46	22 35	26 34	25 29	16 16	18 17	16 03
10	23 11 14	19 33 32	07 ♈ 17 24	13 ♈ 59 55	29 ♓ 59	01 ♈ 01	22 40	26 38	25 24	16 15	18 17	16 01
11	23 15 10	20 33 28	20 ♈ 37 02	27 ♈ 08 47	01 ♈ 54	02 15	22 46	26 41	25 20	16 14	18 18	16 00
12	23 19 07	21 33 23	03 ♉ 35 18	09 ♉ 56 46	03 47	03 30	22 53	26 45	25 16	16 13	18 18	15 58
13	23 23 04	22 33 15	16 ♉ 13 31	22 ♉ 25 55	05 38	04 45	23 00	26 49	25 12	16 12	18 18	15 57
14	23 27 00	23 33 05	28 ♉ 34 24	04 ♊ 39 28	07 27	05 59	23 08	26 53	25 08	16 11	18 19	15 55
15	23 30 57	24 32 53	10 ♊ 41 38	16 ♊ 41 27	09 14	07 14	23 16	26 58	25 04	16 09	18 19	15 54
16	23 34 53	25 32 38	22 ♊ 39 31	28 ♊ 36 24	10 58	08 29	23 25	27 02	25 00	16 08	18 19	15 52
17	23 38 50	26 32 22	04 ♋ 32 43	10 ♋ 29 02	12 38	09 43	23 35	27 07	24 57	16 07	18 19	15 51
18	23 42 46	27 32 03	16 ♋ 25 56	22 ♋ 23 58	14 14	10 58	23 45	27 12	24 53	16 05	18 19	15 49
19	23 46 43	28 31 42	28 ♋ 23 39	04 ♌ 25 30	15 46	12 12	23 56	27 17	24 49	16 04	18 19	15 47
20	23 50 39	29 ♓ 31 18	10 ♌ 29 56	16 ♌ 37 24	17 12	13 26	24 08	27 22	24 46	16 03	18 19	15 46
21	23 54 36	00 ♈ 30 52	22 ♌ 48 14	29 ♌ 02 45	18 34	14 41	24 20	27 28	24 42	16 01	18 R 19	15 44
22	23 58 33	01 30 25	05 ♍ 21 10	11 ♍ 43 40	19 49	15 55	24 32	27 33	24 39	16 00	18 19	15 43
23	00 02 29	02 29 55	18 ♍ 10 21	24 ♍ 41 17	20 59	17 10	24 45	27 39	24 35	15 58	18 19	15 41
24	00 06 26	03 29 22	01 ♎ 16 25	07 ♎ 55 39	22 02	18 24	24 58	27 45	24 32	15 56	18 19	15 39
25	00 10 22	04 28 48	14 ♎ 38 52	21 ♎ 25 49	22 58	19 38	25 12	27 51	24 29	15 55	18 19	15 38
26	00 14 19	05 28 12	28 ♎ 16 17	05 ♏ 09 57	23 47	20 52	25 27	27 57	24 26	15 53	18 19	15 36
27	00 18 15	06 27 34	12 ♏ 06 29	19 ♏ 05 34	24 29	22 07	25 42	28 03	24 23	15 51	18 18	15 34
28	00 22 12	07 26 54	26 ♏ 06 48	03 ♐ 09 50	25 04	23 21	25 57	28 10	24 20	15 50	18 18	15 33
29	00 26 08	08 26 13	10 ♐ 14 19	17 ♐ 19 51	25 31	24 35	26 13	28 16	24 17	15 48	18 18	15 31
30	00 30 05	09 25 29	24 ♐ 26 08	01 ♑ 32 48	25 50	25 49	26 29	28 23	24 15	15 46	18 18	15 29
31	00 34 02	10 ♈ 24 44	08 ♑ 39 33	15 ♑ 46 03	26 ♈ 03	27 ♈ 03	26 ♋ 46	28 ♊ 30	24 ♌ 12	15 ♏ 44	18 ♐ 17	15 ♎ 28

Date	Moon True ☊ ° '	Moon Mean ☊ ° '	Moon ☽ Latitude ° '
01	06 ♎ 07	07 ♎ 24	04 N 12
02	06 R 07	07 21	04 50
03	06 06	07 18	05 11
04	06 05	07 15	05 13
05	06 02	07 12	04 55
06	05 59	07 08	04 18
07	05 56	07 05	03 25
08	05 54	07 02	02 20
09	05 53	06 59	01 N 07
10	05 D 52	06 56	00 S 08
11	05 53	06 52	01 21
12	05 54	06 49	02 27
13	05 55	06 46	03 25
14	05 56	06 43	04 12
15	05 57	06 40	04 47
16	05 58	06 37	05 08
17	05 R 58	06 33	05 17
18	05 57	06 30	05 11
19	05 56	06 27	04 53
20	05 55	06 24	04 21
21	05 55	06 21	03 37
22	05 54	06 18	02 41
23	05 54	06 14	01 37
24	05 53	06 11	00 S 26
25	05 D 53	06 08	00 N 48
26	05 53	06 05	02 01
27	05 R 53	06 02	03 08
28	05 53	05 58	04 04
29	05 53	05 55	04 46
30	05 53	05 52	05 11
31	05 ♎ 53	05 ♎ 49	05 N 17

DECLINATIONS

Date	Sun ☉ ° '	Moon ☽ ° '	Mercury ☿ ° '	Venus ♀ ° '	Mars ♂ ° '	Jupiter ♃ ° '	Saturn ♄ ° '	Uranus ♅ ° '	Neptune ♆ ° '	Pluto ♇ ° '
01	07 S 37	15 S 55	08 S 20	05 S 20	25 N 16	23 N 18	14 N 18	16 S 21	21 S 28	09 N 49
02	07 14	17 37	07 31	04 50	25 14	23 18	14 19	16 21	21 28	09 50
03	06 51	18 14	06 41	04 20	25 12	23 19	14 21	16 21	21 28	09 51
04	06 28	17 42	05 49	03 49	25 10	23 19	14 22	16 20	21 28	09 52
05	06 05	16 02	04 57	03 19	25 08	23 19	14 24	16 20	21 28	09 52
06	05 42	13 20	04 04	02 49	25 05	23 19	14 25	16 20	21 28	09 53
07	05 18	09 51	03 10	02 18	25 03	23 19	14 27	16 20	21 28	09 54
08	04 55	05 48	02 15	01 47	25 00	23 20	14 28	16 19	21 28	09 55
09	04 32	01 S 30	01 20	01 17	24 57	23 20	14 30	16 19	21 28	09 56
10	04 08	02 N 46	00 S 25	00 46	24 55	23 20	14 31	16 19	21 28	09 56
11	03 44	06 48	00 N 30	00 S 15	24 52	23 20	14 33	16 19	21 28	09 57
12	03 21	10 24	01 26	00 N 16	24 49	23 21	14 34	16 18	21 28	09 58
13	02 57	13 25	02 21	00 47	24 46	23 21	14 36	16 18	21 28	09 59
14	02 34	15 45	03 15	01 17	24 42	23 21	14 37	16 18	21 28	10 00
15	02 10	17 19	04 08	01 48	24 39	23 21	14 38	16 17	21 28	10 00
16	01 46	18 06	05 01	02 19	24 36	23 21	14 39	16 17	21 28	10 01
17	01 23	18 05	05 52	02 50	24 32	23 22	14 41	16 16	21 28	10 02
18	00 59	17 17	06 41	03 20	24 28	23 22	14 42	16 16	21 28	10 03
19	00 35	15 42	07 28	03 51	24 25	23 22	14 43	16 16	21 28	10 04
20	00 S 11	13 25	08 13	04 21	24 21	23 22	14 44	16 15	21 28	10 04
21	00 N 12	10 30	08 56	04 52	24 17	23 23	14 46	16 15	21 28	10 05
22	00 36	07 03	09 36	05 22	24 13	23 23	14 47	16 14	21 28	10 06
23	01 00	03 N 12	10 13	05 52	24 09	23 23	14 48	16 14	21 28	10 07
24	01 23	00 S 54	10 47	06 22	24 04	23 23	14 49	16 13	21 28	10 07
25	01 47	05 02	11 17	06 52	24 00	23 24	14 50	16 13	21 28	10 08
26	02 10	08 58	11 44	07 22	23 56	23 24	14 51	16 12	21 28	10 09
27	02 34	12 29	12 08	07 52	23 51	23 24	14 52	16 12	21 28	10 10
28	02 57	15 20	12 28	08 21	23 46	23 24	14 53	16 11	21 28	10 10
29	03 21	17 16	12 44	08 50	23 42	23 24	14 54	16 11	21 28	10 11
30	03 44	18 09	12 56	09 19	23 37	23 25	14 54	16 10	21 28	10 12
31	04 N 07	17 S 53	13 N 04	09 N 48	23 N 32	23 N 25	14 N 55	16 S 10	21 S 28	10 N 12

ZODIAC SIGN ENTRIES

Date	h	m	Planets
01	13	02	☽ ♐
03	15	58	☽ ♑
05	17	51	☽ ♒
07	19	45	☽ ♓
09	16	29	♀ ♈
09	23	08	☽ ♈
10	12	10	☿ ♈
12	05	18	☽ ♉
14	14	48	☽ ♊
17	02	49	☽ ♋
19	15	12	☽ ♌
20	23	34	☉ ♈
22	01	49	☽ ♍
24	09	41	☽ ♎
26	15	01	☽ ♏
28	18	37	☽ ♐
30	21	23	☽ ♑

LATITUDES

Date	Mercury ☿ ° '	Venus ♀ ° '	Mars ♂ ° '	Jupiter ♃ ° '	Saturn ♄ ° '	Uranus ♅ ° '	Neptune ♆ ° '	Pluto ♇ ° '
01	01 S 39	01 S 24	03 N 43	00 S 05	01 N 33	00 N 24	01 N 27	17 N 32
04	01 20	01 22	03 37	00 04	01 34	00 24	01 27	17 33
07	00 56	01 20	03 31	00 04	01 34	00 24	01 27	17 34
10	00 S 27	01 16	03 25	00 04	01 34	00 24	01 27	17 35
13	00 N 07	01 13	03 20	00 03	01 34	00 24	01 27	17 36
16	00 44	01 08	03 14	00 03	01 34	00 24	01 28	17 37
19	01 22	01 03	03 08	00 03	01 34	00 24	01 28	17 38
22	01 59	00 58	03 03	00 02	01 34	00 24	01 28	17 38
25	02 32	00 53	02 57	00 02	01 34	00 24	01 28	17 39
28	02 58	00 46	02 52	00 01	01 33	00 24	01 28	17 39
31	03 N 14	00 S 40	02 N 47	00 S 01	01 N 33	00 N 24	01 N 28	17 N 39

DATA

Julian Date	2443569
Delta T	+49 seconds
Ayanamsa	23° 33' 10"
Synetic vernal point	05° ♓ 33' 49"
True obliquity of ecliptic	23° 26' 23"

MOON'S PHASES, APSIDES AND POSITIONS ☽

Date	h	m	Phase	Longitude ° '	Eclipse Indicator
02	08	34	☾	11 ♐ 24	
09	02	36	●	18 ♓ 10	
16	18	21	☽	25 ♊ 48	
24	16	20	○	03 ♎ 40	total
31	15	11	☾	10 ♑ 33	

Day	h	m		
05	16	23	Perigee	
17	14	24	Apogee	
31	04	51	Perigee	
03	13	07	Max dec	18° S 14'
09	20	22	0N	
16	23	26	Max dec	18° N 12'
24	06	48	0S	
30	18	36	Max dec	18° S 11'

LONGITUDES

Date	Chiron ⚷ ° '	Ceres ⚳ ° '	Pallas ⚴ ° '	Juno ⚵ ° '	Vesta ⚶ ° '	Black Moon Lilith ⚸ ° '
01	02 ♉ 14	09 ♑ 17	17 ♐ 34	20 ♑ 04	10 ♐ 21	04 ♋ 47
11	02 ♉ 42	12 ♑ 24	19 ♐ 51	23 ♑ 12	13 ♐ 27	05 ♋ 55
21	03 ♉ 13	15 ♑ 15	21 ♐ 40	26 ♑ 10	16 ♐ 05	07 ♋ 02
31	03 ♉ 47	17 ♑ 47	22 ♐ 54	28 ♑ 55	18 ♐ 10	08 ♋ 09

All ephemeris data is given at 12.00 UT and the Moon's longitude is additionally given for 24.00 UT

ASPECTARIAN

01 Wednesday
04 28 ☉ ± ♇
06 17 ☽ □ ♄
06 30 ☽ ⊼ ♃
15 08 ☽ ∠ ♇
17 00 ☽ ∥ ♅

02 Thursday
01 31 ☽ Ꝑ ♂
08 34 ☽ □ ☉
09 56 ♂ St D
14 11 ☽ □ ☿
16 43 ☽ ✶ ♇
16 51 ☽ ± ♂
16 58 ☽ ⊻ ♅
20 09 ☽ ☌ ♆

03 Friday
02 08 ☽ □ ♀
02 59 ☽ ⊼ ♂
03 06 ☽ ⊥ ♅
03 13 ☉ ∥ ☿
08 37 ☿ ⊼ ♇
09 08 ☽ △ ♄
09 44 ☽ ☍ ♃
10 37 ☿ △ ♅
11 54 ♀ △ ♂
12 55 ☽ Q ♇
17 15 ☽ Q ☉
18 13 ☽ ∠ ♅

04 Saturday
01 11 ☽ Q ☿
10 05 ☽ Ꝑ ♄
10 19 ☿ □ ♆
11 09 ☽ Q ♀
14 46 ☽ ✶ ☉
18 53 ☽ □ ♇
19 11 ☽ ✶ ♅
22 22 ☽ ⊻ ♆

05 Sunday
00 13 ☽ ✶ ☿
00 57 ☽ ± ♄
05 08 ☽ ☍ ♂
08 20 ☽ ⊥ ♆
08 36 ☽ ∥ ♅
08 56 ☽ ✶ ♀
10 50 ☽ ⊼ ♄
11 50 ☽ ⊼ ♃
15 04 ☽ Q ♅
17 29 ☽ ∠ ☉
21 48 ☽ ± ♃
23 14 ☽ ∠ ♆

06 Monday
03 22 ☽ # ♄
04 54 ☽ ∠ ☿
05 54 ♀ ⊼ ♄
09 30 ☽ ⊥ ☉
12 08 ☽ ∠ ♀
12 46 ☽ Ꝑ ♃
13 35 ☿ △ ♂
19 43 ♀ □ ♃
20 12 ☽ ⊻ ☉
20 33 ☽ △ ♇
20 53 ☽ □ ♅
22 09 ☽ ⊥ ☿

07 Tuesday
00 09 ☽ ✶ ♆
01 01 ☉ ⊼ ♇
04 31 ☽ ⊥ ♀
05 45 ☉ △ ♅
07 04 ☽ ⊼ ♂
09 43 ☽ ⊻ ☿
11 38 ☽ # ♇
12 25 ☽ ☍ ♄
13 50 ☽ △ ♃
15 30 ☽ ⊻ ♀
17 09 ☽ ± ♂
20 13 ☽ Q ♆
21 35 ☽ Ꝑ ♇

08 Wednesday
05 03 ☿ ⊼ ♄
08 23 ☽ Ꝑ ♂
12 46 ☽ ± ♇
16 54 ☿ □ ♃
17 32 ☽ ∥ ☉
22 59 ☽ ⊼ ♇
23 21 ☽ △ ♅

09 Thursday
02 36 ☽ ☌ ☉
02 48 ☽ □ ♆
05 23 ☉ □ ♆
10 12 ☽ △ ♂
13 12 ☽ ∥ ☿
13 27 ☽ ∥ ♀
15 13 ☽ ⊼ ♄
15 35 ☿ ∥ ♀
17 10 ☽ □ ♃
21 01 ☽ ☌ ☿
23 48 ☽ ☌ ♀

10 Friday
01 03 ☽ # ☿
01 20 ☽ Ꝑ ♅
01 39 ☽ ± ♄
01 51 ☽ # ♀
16 29 ♀ Ꝑ ♅
17 16 ☽ ± ♅
17 32 ☽ Ꝑ ♄
19 11 ☽ # ☉
19 11 ♀ ± ♄

11 Saturday
01 12 ☽ Q ♃
03 37 ☽ ☍ ♇
03 43 ☿ Ꝑ ♅
04 02 ☽ ⊼ ♅
05 13 ☿ ± ♄
07 42 ☿ # ♀
07 46 ☽ △ ♆
11 53 ☽ ⊻ ☉
15 58 ☽ □ ♂
20 37 ☽ △ ♄
23 12 ☽ ✶ ♃
23 49 ☽ ⊥ ☉

12 Sunday
01 33 ☿ ☌ ♀
08 53 ☽ ∥ ♇
11 28 ☽ Ꝑ ♆
11 49 ☽ ⊻ ♀
12 25 ☽ ⊻ ☿
18 03 ☽ ∠ ☉

13 Monday
00 22 ☽ ⊥ ♀
01 42 ☽ ⊥ ☿
01 54 ☽ Q ♂
03 31 ☽ ∠ ♃
04 29 ☽ ± ♆
11 28 ☽ ⊼ ♇
11 56 ☽ ☍ ♅
16 01 ☽ ⊼ ♆
19 33 ☽ ∠ ♀
20 55 ☽ ⊥ ♃
21 59 ☽ ∠ ☿
23 02 ☽ ± ♇
23 14 ☉ # ☿
23 18 ☽ ∥ ♄

14 Tuesday
00 15 ☉ △ ♂
01 14 ☽ ✶ ♂
01 19 ☽ ✶ ☉
05 18 ☽ □ ♄
08 41 ☽ ⊻ ♃
16 36 ☽ Ꝑ ♇
19 05 ☽ # ♅

15 Wednesday
03 00 ☽ Q ☉
04 19 ☽ ✶ ♀
07 07 ☽ ∠ ♂
08 36 ☽ ✶ ☿
16 43 ☽ Q ♄
21 37 ☉ # ♀
22 23 ☽ △ ♇
22 54 ☽ ⊼ ♅
23 05 ☿ Ꝑ ♄
23 48 ☉ ⊼ ♄

16 Thursday
00 32 ☿ ± ♅
01 20 ☽ ⊥ ♂
03 16 ☽ ☍ ♆
07 06 ☽ Q ♀
10 57 ☽ ± ♅
12 43 ☽ Q ☿
13 34 ☽ ⊻ ♂
16 42 ☽ ✶ ♄
18 21 ☽ □ ☉
20 53 ☽ ☌ ♃

17 Friday
05 04 ☽ Ꝑ ♅
16 09 ♀ Ꝑ ♄
19 30 ♀ ± ♅
22 51 ☽ ∠ ♄
23 40 ☽ □ ♀

18 Saturday
03 10 ☉ □ ♃
06 54 ☽ □ ☿
10 46 ☽ □ ♇
11 19 ☽ △ ♅
15 48 ☽ ⊼ ♆
16 54 ☽ ⊥ ♄

19 Sunday
02 57 ☽ ☌ ♂
03 51 ☽ ± ♆
03 53 ☿ Q ♃
04 38 ☽ # ♅
04 53 ☽ ⊻ ♄
09 46 ☽ ⊻ ♃
12 17 ☽ △ ☉
12 27 ☿ ☍ ♇
16 51 ☿ ⊼ ♅
21 48 ☽ ⊥ ♃
21 49 ☽ Ꝑ ♆
22 43 ☽ Q ♇
23 05 ☽ ∥ ♄

20 Monday
15 42 ☽ ∠ ♃
18 26 ☽ △ ♀
18 46 ♆ St R
20 35 ☽ Ꝑ ☉
22 18 ☽ ✶ ♇
22 51 ☽ □ ♅

21 Tuesday
02 47 ☽ △ ☿
03 18 ☽ △ ♆
07 34 ☿ △ ♆
11 26 ♂ ± ♆
14 59 ☽ ⊻ ♂
15 02 ☽ ∥ ♇
15 35 ☽ ± ☉
15 39 ☽ ☌ ♄
16 18 ☉ ± ♄
21 02 ☽ ✶ ♃
21 26 ☽ ∥ ☿
23 52 ☉ Ꝑ ♅

22 Wednesday
02 40 ☽ Ꝑ ♀
02 41 ☽ ⊥ ♂
03 12 ☽ ∠ ♇
04 04 ☽ ⊼ ☉
04 17 ♀ Q ♃
08 01 ♀ ☍ ♇
09 26 ☽ Q ♅
10 54 ☽ Ꝑ ☿
13 21 ♀ ⊼ ♅
19 59 ☽ Q ♃
20 01 ☽ ∠ ♂
20 12 ☽ ⊥ ♇
21 27 ☽ ∥ ♀
21 32 ☽ ± ♀
22 07 ♂ ⊻ ♄

23 Thursday
05 31 ☽ ± ☿
07 23 ☽ ⊻ ♇
07 49 ☿ ∥ ♇
07 55 ☽ ✶ ♅
09 55 ☽ ⊼ ♀
12 16 ☽ □ ♆
17 40 ☽ ⊼ ☿
23 46 ☽ ⊻ ♄
23 52 ☽ ∥ ☉

24 Friday
00 19 ☽ ✶ ♂
05 32 ☽ □ ♃
10 25 ♀ △ ♆
10 40 ☽ ⊥ ♄
11 24 ☽ ∠ ♅
15 08 ☽ # ☉
16 20 ☽ ☍ ☉
21 06 ☽ Q ♆
22 28 ☽ Q ♂

25 Saturday
02 50 ☽ ∠ ♄
03 34 ☽ ⊥ ♅
13 44 ☽ ☌ ♇
14 14 ☽ ⊻ ♅
18 30 ☽ ✶ ♆
21 43 ☽ ☍ ♀

26 Sunday
00 33 ☽ # ♀
03 40 ☽ ☍ ☿
05 18 ☽ ✶ ♄
06 58 ☽ □ ♂
11 26 ☽ △ ♃
19 38 ☽ # ♇
20 47 ☽ ∠ ♆

27 Monday
01 30 ☽ ⊼ ☉
02 09 ☽ Q ♄
08 32 ☿ △ ♄
09 02 ☽ # ☿
12 21 ☽ ⊥ ♆
12 39 ☽ ± ☉
13 39 ☽ Ꝑ ♃
17 57 ☽ ⊻ ♇
18 26 ☽ ☌ ♅
22 39 ☽ ⊻ ♆

28 Tuesday
04 13 ☽ ⊥ ♇
05 12 ☽ ± ♃
05 16 ☽ Ꝑ ☉
06 50 ☽ ⊼ ♀
07 40 ☽ # ♄
08 59 ☽ □ ♄
10 08 ☽ ⊼ ☿
11 43 ☽ △ ♂
15 31 ☽ ⊼ ♃
18 02 ☽ ± ♀
19 32 ☽ ∠ ♇
20 44 ☽ ± ☿
21 12 ☽ ∥ ♅

29 Wednesday
06 28 ♀ △ ♄
08 43 ☽ △ ☉
10 47 ☽ Ꝑ ♀
12 29 ☽ Ꝑ ☿
13 41 ☽ Ꝑ ♂
20 55 ☽ ✶ ♇
21 23 ☽ ⊻ ♅

30 Thursday
01 38 ☽ ☌ ♆
05 12 ☽ ± ♂
07 30 ☽ ⊥ ♅
07 47 ☉ Ꝑ ♄
11 41 ☽ △ ♄
12 29 ☿ ☌ ♀
14 25 ☽ △ ☿
14 34 ☽ △ ♀
15 31 ☽ ⊼ ♂
17 09 ☽ Q ♇
18 43 ☽ ☍ ♃
20 00 ☉ ± ♅
22 39 ☽ ∠ ♅

31 Friday
04 35 ♀ □ ♂
12 55 ☽ Ꝑ ♄
15 11 ☽ □ ☉
23 28 ☽ □ ♇
23 55 ☽ ✶ ♅

APRIL 1978

LONGITUDES

Date	Sidereal time h m s	Sun ☉ ° ' "	Moon ☽ ° ' "	Moon ☽ 24.00 ° ' "	Mercury ☿ ° '	Venus ♀ ° '	Mars ♂ ° '	Jupiter ♃ ° '	Saturn ♄ ° '	Uranus ♅ ° '	Neptune ♆ ° '	Pluto ♇ ° '
01	00 37 58	11 ♈ 23 58	22 ♑ 52 01	29 ♑ 57 10	26 ♈ 08	28 ♈ 17	27 ♋ 03	28 ♊ 37	24 ♌ 09	15 ♏ 42	18 ♐ 17	15 ♎ 26
02	00 41 55	12 23 09	07 ♒ 01 12	14 ♒ 03 50	26 R 05	29 ♈ 32	27 20	28 44	24 R 07	15 R 40	18 R 17	15 R 24
03	00 45 51	13 22 19	21 ♒ 04 48	28 ♒ 03 48	25 56	00 ♉ 46	27 38	28 52	24 05	15 38	18 16	15 23
04	00 49 48	14 21 27	05 ♓ 00 34	11 ♓ 54 49	25 40	02 00	27 56	28 59	24 02	15 36	18 16	15 21
05	00 53 44	15 20 33	18 ♓ 46 17	25 ♓ 34 42	25 19	03 14	28 15	29 07	24 00	15 34	18 15	15 19
06	00 57 41	16 19 37	02 ♈ 19 50	09 ♈ 01 28	24 51	04 28	28 34	29 15	23 58	15 32	18 15	15 18
07	01 01 37	17 18 39	15 ♈ 39 26	22 ♈ 13 34	24 19	05 41	28 54	29 23	23 56	15 30	18 14	15 16
08	01 05 34	18 17 39	28 ♈ 43 48	05 ♉ 10 04	23 42	06 55	29 13	29 31	23 54	15 28	18 14	15 14
09	01 09 31	19 16 37	11 ♉ 32 23	17 ♉ 50 51	23 02	08 09	29 34	29 39	23 53	15 26	18 13	15 13
10	01 13 27	20 15 33	24 ♉ 05 34	00 ♊ 16 44	22 20	09 23	29 ♋ 54	29 47	23 51	15 23	18 12	15 11
11	01 17 24	21 14 27	06 ♊ 24 38	12 ♊ 29 33	21 35	10 37	00 ♌ 15	29 ♊ 56	23 50	15 21	18 12	15 09
12	01 21 20	22 13 18	18 ♊ 31 52	24 ♊ 31 59	20 50	11 51	00 36	00 ♋ 04	23 48	15 19	18 11	15 08
13	01 25 17	23 12 08	00 ♋ 30 22	06 ♋ 27 32	20 05	13 04	00 58	00 13	23 47	15 17	18 10	15 06
14	01 29 13	24 10 55	12 ♋ 23 59	18 ♋ 20 18	19 20	14 18	01 19	00 22	23 46	15 14	18 09	15 04
15	01 33 10	25 09 40	24 ♋ 17 02	00 ♌ 14 48	18 37	15 32	01 42	00 31	23 44	15 12	18 09	15 03
16	01 37 06	26 08 23	06 ♌ 14 11	12 ♌ 15 47	17 56	16 45	02 04	00 40	23 43	15 10	18 08	15 01
17	01 41 03	27 07 04	18 ♌ 20 10	24 ♌ 27 53	17 19	17 59	02 27	00 49	23 43	15 07	18 07	14 59
18	01 45 00	28 05 42	00 ♍ 39 29	06 ♍ 55 25	16 44	19 12	02 50	00 58	23 42	15 05	18 06	14 58
19	01 48 56	29 ♈ 04 18	13 ♍ 16 07	19 ♍ 41 57	16 14	20 26	03 13	01 08	23 41	15 03	18 05	14 56
20	01 52 53	00 ♉ 02 52	26 ♍ 13 10	02 ♎ 49 57	15 48	21 39	03 37	01 17	23 40	15 00	18 04	14 54
21	01 56 49	01 01 24	09 ♎ 32 22	16 ♎ 20 23	15 27	22 53	04 01	01 27	23 40	14 58	18 03	14 53
22	02 00 46	01 59 54	23 ♎ 13 47	00 ♏ 12 19	15 11	24 06	04 25	01 36	23 40	14 55	18 02	14 51
23	02 04 42	02 58 22	07 ♏ 15 32	14 ♏ 22 54	14 59	25 20	04 49	01 46	23 39	14 53	18 01	14 50
24	02 08 39	03 56 48	21 ♏ 33 46	28 ♏ 47 26	14 53	26 33	05 14	01 56	23 39	14 50	18 00	14 48
25	02 12 35	04 55 12	06 ♐ 03 06	13 ♐ 20 00	14 D 51	27 46	05 38	02 06	23 39	14 48	17 59	14 47
26	02 16 32	05 53 35	20 ♐ 37 18	27 ♐ 54 15	14 55	28 ♉ 59	06 04	02 16	23 D 39	14 45	17 58	14 45
27	02 20 29	06 51 56	05 ♑ 10 08	12 ♑ 24 19	15 03	00 ♊ 13	06 29	02 26	23 39	14 43	17 57	14 43
28	02 24 25	07 50 16	19 ♑ 36 16	26 ♑ 45 32	15 17	01 26	06 54	02 37	23 40	14 40	17 56	14 42
29	02 28 22	08 48 34	03 ♒ 51 47	10 ♒ 54 46	15 35	02 39	07 20	02 47	23 40	14 38	17 55	14 40
30	02 32 18	09 ♉ 46 51	17 ♒ 54 21	24 ♒ 50 26	15 ♈ 57	03 ♊ 52	07 ♌ 46	02 ♋ 58	23 ♌ 40	14 ♏ 35	17 ♐ 54	14 ♎ 39

Moon

Date	Moon True ☊ ° '	Moon Mean ☊ ° '	Moon ☽ Latitude ° '
01	05 ♎ 53	05 ♎ 46	05 N 03
02	05 D 53	05 43	04 32
03	05 54	05 39	03 44
04	05 55	05 36	02 43
05	05 55	05 33	01 34
06	05 56	05 30	00 N 20
07	05 R 56	05 27	00 S 54
08	05 55	05 23	02 05
09	05 53	05 20	03 04
10	05 51	05 17	03 56
11	05 49	05 14	04 35
12	05 47	05 11	05 01
13	05 45	05 08	05 14
14	05 44	05 04	05 13
15	05 43	05 01	04 59
16	05 D 43	04 58	04 32
17	05 45	04 55	03 52
18	05 46	04 52	03 02
19	05 48	04 49	02 01
20	05 49	04 45	00 S 53
21	05 R 49	04 42	00 N 20
22	05 48	04 39	01 34
23	05 46	04 36	02 44
24	05 42	04 33	03 45
25	05 38	04 29	04 32
26	05 34	04 26	05 02
27	05 30	04 23	05 12
28	05 28	04 20	05 03
29	05 27	04 17	04 34
30	05 ♎ 27	04 ♎ 14	03 N 51

DECLINATIONS

Date	Sun ☉ ° '	Moon ☽ ° '	Mercury ☿ ° '	Venus ♀ ° '	Mars ♂ ° '	Jupiter ♃ ° '	Saturn ♄ ° '	Uranus ♅ ° '	Neptune ♆ ° '	Pluto ♇ ° '
01	04 N 31	16 S 31	13 N 09	10 N 17	23 N 27	23 N 25	14 N 56	16 S 09	21 S 27	10 N 13
02	04 54	14 08	13 09	10 45	23 22	23 25	14 57	16 09	21 27	10 14
03	05 17	10 56	13 05	11 13	23 17	23 25	14 58	16 08	21 27	10 14
04	05 40	07 09	12 58	11 41	23 11	23 26	14 58	16 07	21 27	10 15
05	06 02	03 S 00	12 46	12 09	23 06	23 26	14 59	16 07	21 27	10 16
06	06 25	01 N 14	12 31	12 36	23 01	23 26	15 00	16 06	21 27	10 16
07	06 48	05 20	12 13	13 03	22 55	23 26	15 00	16 06	21 27	10 17
08	07 10	09 06	11 52	13 30	22 49	23 26	15 01	16 05	21 27	10 18
09	07 33	12 22	11 28	13 56	22 43	23 26	15 01	16 04	21 27	10 18
10	07 55	14 59	11 01	14 22	22 38	23 26	15 02	16 04	21 27	10 19
11	08 17	16 52	10 33	14 48	22 32	23 27	15 02	16 03	21 27	10 19
12	08 39	17 56	10 03	15 13	22 25	23 27	15 03	16 02	21 26	10 20
13	09 01	18 12	09 33	15 38	22 19	23 27	15 03	16 02	21 26	10 21
14	09 23	17 40	09 01	16 02	22 13	23 27	15 03	16 01	21 26	10 21
15	09 44	16 21	08 30	16 26	22 07	23 27	15 04	16 00	21 26	10 22
16	10 06	14 19	08 00	16 50	22 00	23 27	15 04	16 00	21 26	10 22
17	10 27	11 39	07 30	17 13	21 53	23 27	15 04	15 59	21 26	10 23
18	10 48	08 24	07 01	17 36	21 47	23 27	15 04	15 58	21 26	10 23
19	11 09	04 43	06 34	17 59	21 40	23 27	15 04	15 58	21 26	10 24
20	11 29	00 N 42	06 09	18 20	21 33	23 27	15 05	15 57	21 26	10 24
21	11 50	03 S 28	05 46	18 42	21 26	23 27	15 05	15 56	21 25	10 25
22	12 10	07 34	05 25	19 03	21 19	23 27	15 05	15 56	21 25	10 25
23	12 30	11 21	05 07	19 23	21 12	23 27	15 05	15 55	21 25	10 26
24	12 50	14 32	04 51	19 43	21 04	23 27	15 05	15 54	21 25	10 26
25	13 10	16 51	04 38	20 03	20 57	23 27	15 05	15 53	21 25	10 26
26	13 29	18 05	04 27	20 22	20 49	23 27	15 05	15 53	21 25	10 27
27	13 48	18 08	04 19	20 40	20 42	23 27	15 04	15 52	21 25	10 27
28	14 07	17 01	04 14	20 58	20 34	23 27	15 04	15 51	21 25	10 28
29	14 26	14 50	04 11	21 15	20 26	23 27	15 04	15 51	21 25	10 28
30	14 N 45	11 S 48	04 N 11	21 N 32	20 N 18	23 N 26	15 N 04	15 S 50	21 S 24	10 N 28

ZODIAC SIGN ENTRIES

Date	h	m	Planets
02	00	05	☽ ♒
02	21	14	♀ ♉
04	03	20	☽ ♓
06	07	51	☽ ♈
08	14	21	☽ ♉
10	18	50	♂ ♌
10	23	27	☽ ♊
12	00	12	♃ ♋
13	10	59	☽ ♋
15	23	30	☽ ♌
18	10	44	☽ ♍
20	10	50	☉ ♉
20	18	53	☽ ♎
22	23	39	☽ ♏
25	02	00	☽ ♐
27	03	27	☽ ♑
27	07	53	♀ ♊
29	05	28	☽ ♒

LATITUDES

Date	Mercury ☿ ° '	Venus ♀ ° '	Mars ♂ ° '	Jupiter ♃ ° '	Saturn ♄ ° '	Uranus ♅ ° '	Neptune ♆ ° '	Pluto ♇ ° '
01	03 N 17	00 S 38	02 N 45	00 S 01	01 N 33	00 N 24	01 N 28	17 N 39
04	03 16	00 31	02 40	00 S 01	01 33	00 24	01 28	17 39
07	03 00	00 24	02 35	00 00	01 33	00 24	01 29	17 40
10	02 31	00 16	02 31	00 00	01 33	00 24	01 29	17 39
13	01 50	00 08	02 26	00 00	01 32	00 24	01 29	17 39
16	01 02	00 S 01	02 22	00 N 01	01 32	00 24	01 29	17 39
19	00 N 12	00 N 07	02 17	00 01	01 32	00 24	01 29	17 39
22	00 S 36	00 15	02 13	00 01	01 32	00 24	01 29	17 38
25	01 19	00 23	02 09	00 02	01 32	00 24	01 29	17 38
28	01 56	00 31	02 05	00 02	01 31	00 24	01 29	17 37
31	02 S 25	00 N 39	02 N 01	00 N 02	01 N 31	00 N 24	01 N 29	17 N 37

DATA

Julian Date	2443600
Delta T	+49 seconds
Ayanamsa	23° 33' 13"
Synetic vernal point	05° ♓ 33' 46"
True obliquity of ecliptic	23° 26' 23"

MOON'S PHASES, APSIDES AND POSITIONS ☽

Date	h	m	Phase	Longitude ° '	Eclipse Indicator
07	15	15	●	17 ♈ 27	Partial
15	13	56	☽	25 ♋ 14	
23	04	11	○	02 ♏ 39	
29	21	02	☾	09 ♒ 11	

Day	h	m		
14	10	24	Apogee	
26	08	12	Perigee	
06	05	00	0N	
13	07	45	Max dec	18° N 13'
20	16	03	0S	
27	00	57	Max dec	18° S 16'

LONGITUDES

Date	Chiron ⚷ ° '	Ceres ⚳ ° '	Pallas ⚴ ° '	Juno ⚵ ° '	Vesta ⚶ ° '	Black Moon Lilith ⚸ ° '
01	03 ♉ 50	18 ♑ 01	23 ♐ 00	29 ♑ 11	18 ♐ 21	08 ♋ 16
11	04 ♉ 27	20 ♑ 10	23 ♐ 31	01 ♒ 41	19 ♐ 42	09 ♋ 23
21	05 ♉ 04	21 ♑ 53	23 ♐ 17	03 ♒ 54	20 ♐ 18	10 ♋ 30
31	05 ♉ 41	23 ♑ 08	22 ♐ 17	05 ♒ 46	20 ♐ 05	11 ♋ 37

All ephemeris data is given at 12.00 UT and the Moon's longitude is additionally given for 24.00 UT

ASPECTARIAN

h m	Aspects
01 Saturday	
04 04	☽ ± ♄
04 15	☽ ⊻ ♆
08 47	♀ ∥ ♇
14 11	☽ ⊼ ♄
14 24	☽ ⊥ ♆
16 18	☿ St R
16 22	☽ ∥ ♅
17 31	☽ □ ☿
19 05	♀ ⚹ ♃
19 13	☽ ☍ ♂
20 10	☽ Q ♅
21 33	♂ ∥ ♃
21 49	☽ ⊼ ♃
22 04	☽ □ ♀
23 53	☽ Q ☉
02 Sunday	
04 48	☽ # ♄
05 39	☽ ∠ ♆
08 06	☽ ± ♃
20 07	☽ # ☿
21 50	☽ ⚹ ☉
23 33	☽ ⚼ ♃
23 56	☽ Q ☿
03 Monday	
02 16	☽ △ ♇
02 42	☽ □ ♅
07 11	☽ ⚹ ♆
07 39	☽ Q ♀
10 17	☽ # ♀
16 38	☽ # ♇
17 08	☽ ☍ ♄
20 12	☽ ⚹ ☿
23 31	☽ ⊼ ♂
04 Tuesday	
01 29	☽ ∠ ☉
01 30	☽ △ ♃
03 48	☽ Q ♆
03 58	☽ ⚼ ♇
06 16	☽ ⚹ ♀
10 06	☽ ± ♂
18 15	☽ ⊥ ☉
19 31	☽ ± ♇
20 01	☽ # ☉
21 36	☽ ∠ ☿
05 Wednesday	
02 07	☽ ⚼ ♂
05 31	☽ ⊻ ☉
05 58	☽ ⊼ ♇
06 24	☽ △ ♅
10 57	☽ ∠ ♀
11 05	☽ □ ♆
11 33	☉ ☍ ♇
12 32	♀ ⚼ ♆
12 55	☽ ⊥ ☿
17 19	☉ ⊼ ♅
21 12	☽ ⊼ ♄
23 10	☽ ⊻ ☿
06 Thursday	
04 25	☽ ⊥ ♀
05 08	☽ △ ♂
06 27	☽ □ ♃
07 48	☽ ± ♄
08 48	☽ ⚼ ♅
09 31	☿ ∥ ♀
16 11	☽ ⊻ ♀
23 52	☽ ⚼ ♄
07 Friday	
00 53	☽ ± ♅
11 18	☽ ☍ ♇
11 43	☽ ⊼ ♅
13 53	☉ Q ♃
15 10	☽ Q ♃
15 15	☽ ☌ ☉
16 42	☽ △ ♆
21 57	☽ ∥ ☉
08 Saturday	
03 07	☽ △ ♄
03 09	☽ ☌ ☿
03 59	☿ △ ♄
10 21	☉ △ ♆
12 56	☽ □ ♂
13 28	☽ ⚹ ♃
20 16	☽ ∥ ♇
20 22	☽ ⚼ ♆
09 Sunday	
04 55	☽ ☌ ♀
05 43	☽ ∥ ☿
13 17	☽ ± ♆
17 58	☽ ∠ ♃
18 57	☽ ⊼ ♇
19 22	☽ ☍ ♅
22 31	♂ ⊻ ♃
23 46	☽ Q ♂
10 Monday	
00 42	☽ ⊼ ♆
03 59	☽ ⊻ ☉
04 21	☽ ∥ ♀
06 24	☽ ± ♇
08 47	☽ ⊻ ☿
11 24	☽ ⊥ ♃
11 32	☽ □ ♄
12 28	☽ ∥ ♄
16 33	☽ ⊥ ☉
19 45	☽ ⊥ ☿
23 10	☽ ⊻ ♃
23 35	☽ ⚹ ♂
23 47	☽ ⚼ ♇
11 Tuesday	
00 29	☽ # ♅
11 38	☽ ∠ ☉
12 20	☽ ∠ ☿
16 49	☉ ☌ ☿
21 13	☽ ⊻ ♀
22 40	☽ Q ♄
23 00	☿ ∥ ♇
12 Wednesday	
02 04	♀ ∥ ♄
05 14	☽ △ ♇
05 37	☽ ⊼ ♅
06 00	☽ ∠ ♂
10 28	☽ ⊥ ♀
11 18	☽ ☍ ♆
16 19	☽ ⚹ ☿
17 32	☽ ± ♅
18 33	♀ ± ♆
20 02	☽ ⚹ ☉
22 31	☽ ⚹ ♄
13 Thursday	
00 31	☽ ⊥ ♂
06 32	☽ ∠ ♀
11 24	☽ ☌ ♃
11 32	☽ ⚼ ♅
12 57	☽ ⊻ ♂
14 58	☽ Q ☿
20 07	♃ ⚼ ♅
22 19	☽ Q ☉
14 Friday	
01 52	☉ △ ♄
02 24	☉ ∥ ☿
04 39	☽ ∠ ♄
10 58	♀ # ♅
16 17	☽ ⚹ ♀
17 23	☽ □ ♇
17 43	☽ △ ♅
22 49	☽ ⊥ ♄
23 37	☽ ⊼ ♆
15 Saturday	
01 12	☽ □ ☿
02 46	♀ ⊼ ♇
05 49	♀ ☍ ♅
10 54	☽ ⊻ ♄
10 58	☽ ∥ ♀
11 38	♀ ∠ ♃
11 43	☽ ± ♆
13 56	☽ □ ☉
15 01	☿ Q ♃
16 43	☽ # ♅
19 17	☽ Q ♀
16 Sunday	
00 41	☽ ⊻ ♃
03 23	☽ ☌ ♂
04 07	☽ ∥ ♄
04 57	☿ △ ♆
05 34	☽ Q ♇
05 47	☽ ⚼ ♆
12 52	☽ ⊥ ♃
17 Monday	
03 15	☿ ⊻ ♀
05 25	☽ ⚹ ♇
05 41	☽ □ ♅
06 58	☽ ∠ ♃
07 14	☉ ∥ ♇
10 05	☽ △ ☿
11 13	☽ □ ♀
11 34	☽ △ ♆
14 37	♀ ⊼ ♆
20 24	☽ ∥ ☉
21 50	☽ ∥ ♇
22 31	☽ ☌ ♄
18 Tuesday	
06 38	☽ △ ☉
10 40	☽ ∠ ♇
12 36	☽ ⚹ ♃
14 00	☽ ⚼ ☿
16 19	☽ ⊻ ♂
16 39	☽ Q ♅
19 40	♂ Q ♇
22 40	☽ ∥ ☿
19 Wednesday	
03 50	☽ ⊥ ♇
04 07	☽ ⊥ ♂
04 11	♂ ⚼ ♆
06 30	☽ ± ☿
11 44	☽ Q ♃
13 38	☽ ⚼ ☉
15 07	☽ ⊻ ♇
15 19	☽ ⚹ ♅
17 22	☽ ⊼ ☿
21 00	☽ □ ♆
21 32	☽ ∠ ♂
21 40	♀ ± ♇
20 Thursday	
02 45	☽ △ ♀
07 20	☽ ⊻ ♄
07 42	☽ ± ☉
10 43	♀ Q ♂
14 15	☿ ⊥ ♀
18 17	☽ ⊥ ♄
18 52	☽ ∠ ♅
19 31	☽ ⊼ ☉
21 19	☽ □ ♃
21 Friday	
01 49	☽ ⚹ ♂
05 48	☽ Q ♆
08 45	☽ ⚼ ♀
10 27	☽ ∠ ♄
10 59	☽ ⊥ ♅
13 21	♂ # ♆
21 25	☽ ☌ ♇
21 33	☽ ⊻ ♅
22 13	☽ ☍ ☿
23 46	☽ Q ♂
22 Saturday	
00 15	☽ # ☿
00 24	☉ ⚹ ♃
02 13	☽ ± ♀
02 59	☽ ⚹ ♆
03 21	♀ □ ♄
12 45	☽ ⚹ ♄
13 39	☽ ⊼ ♀
23 Sunday	
02 34	☽ △ ♃
04 11	☽ ☍ ☉
04 49	☽ ∠ ♆
05 50	☽ # ♇
07 44	☽ □ ♂
09 17	☽ Q ♄
13 12	☉ ⚼ ♆
20 02	☽ ⊥ ♆
20 59	☽ # ☉
22 05	♀ ⊥ ♃
24 Monday	
00 44	☽ ⊻ ♇
00 48	☽ ☌ ♅
00 54	☽ ⊼ ☿
04 11	☽ ⚼ ♃
06 04	☽ ⊻ ♆
10 44	☽ ⊥ ♇
10 52	☽ ± ☿
15 29	☽ □ ♄
16 49	☽ # ♄
19 21	☽ ± ♃
21 03	☽ ☍ ♀
25 Tuesday	
00 50	☽ ∥ ♅
01 39	☽ ∠ ♇
01 46	☽ ⚼ ☿
05 24	☽ ⊼ ♃
06 48	☿ St D
10 00	☽ ⊼ ☉
11 18	☽ △ ♂
12 16	♄ St D
20 36	☽ ± ☉
26 Wednesday	
02 21	☽ ⚹ ♇
02 22	☽ ⊻ ♅
02 33	☽ △ ☿
07 38	☽ ☌ ♆
12 13	☽ ⊥ ♅
12 29	☽ ⚼ ☉
12 44	☽ ⚼ ♂
17 00	☽ △ ♄
19 12	☉ □ ♂
22 05	☽ Q ♇
22 16	♅ ⊻ ♇
27 Thursday	
02 40	♀ ⚼ ♇
03 01	☽ ∠ ♅
03 03	☽ ⊼ ♀
04 01	☽ ± ♂
07 26	☽ ☍ ♃
08 31	☿ ∠ ♀
13 33	♀ ∥ ♂
13 53	☽ ± ♀
14 14	☽ ⊼ ♂
15 01	☽ △ ☉
17 47	☽ ⚼ ♄
28 Friday	
03 48	☽ ⚹ ♅
03 50	☽ □ ♇
04 39	☽ □ ☿
06 12	☽ ⚼ ♀
08 45	☽ ± ♄
09 13	☽ ⊻ ♆
18 48	☽ ⊼ ♄
19 14	☽ ⊥ ♆
23 49	☽ Q ♅
29 Saturday	
02 05	☽ ∥ ♅
09 45	☽ △ ♀
09 49	☽ # ♄
10 09	☽ ⊼ ♃
10 24	☽ ∠ ♆
11 30	☽ Q ☿
15 09	☽ # ☉
15 11	♀ ⊻ ♃
18 05	☽ ☍ ♂
20 28	☽ ± ♃
21 02	☽ □ ☉
30 Sunday	
01 41	♀ # ♆
06 19	☽ □ ♅
06 25	☽ △ ♇
08 32	☽ ⚹ ☿
11 59	☽ ⚹ ♆
12 06	☽ ⚼ ♃
21 08	☽ # ♇
21 59	☽ ☍ ♄

MAY 1978

LONGITUDES

Date	Sidereal time h m s	Sun ☉	Moon ☽	Moon ☽ 24.00	Mercury ☿	Venus ♀	Mars ♂	Jupiter ♃	Saturn ♄	Uranus ♅	Neptune ♆	Pluto ♇
01	02 36 15	10 ♉ 45 06	01 ♓ 43 00	08 ♓ 32 05	16 ♈ 24	05 ♊ 05	08 ♌ 12	03 ♋ 08	23 ♌ 41	14 ♏ 33	17 ♐ 52	14 ♎ 38
02	02 40 11	11 43 19	15 ♓ 17 44	22 ♓ 00 02	16 55	06 18	08 39	03 19	23 42	14 R 30	17 R 51	14 R 36
03	02 44 08	12 41 32	28 ♓ 39 05	05 ♈ 14 56	17 30	07 31	09 06	03 30	23 43	14 28	17 50	14 35
04	02 48 04	13 39 42	11 ♈ 47 41	18 ♈ 17 24	18 09	08 44	09 32	03 41	23 43	14 25	17 49	14 33
05	02 52 01	14 37 51	24 ♈ 44 08	01 ♉ 07 55	18 52	09 57	09 59	03 51	23 44	14 23	17 47	14 32
06	02 55 58	15 35 59	07 ♉ 28 49	13 ♉ 46 51	19 38	11 10	10 27	04 02	23 46	14 20	17 46	14 30
07	02 59 54	16 34 05	20 ♉ 02 05	26 ♉ 13 54	20 28	12 23	10 54	04 14	23 47	14 18	17 45	14 29
08	03 03 51	17 32 09	02 ♊ 24 23	08 ♊ 31 38	21 21	13 35	11 22	04 25	23 48	14 15	17 44	14 28
09	03 07 47	18 30 12	14 ♊ 36 29	20 ♊ 39 07	22 17	14 48	11 50	04 36	23 50	14 13	17 42	14 26
10	03 11 44	19 28 13	26 ♊ 39 44	02 ♋ 38 38	23 16	16 01	12 18	04 47	23 51	14 10	17 41	14 25
11	03 15 40	20 26 12	08 ♋ 36 08	14 ♋ 32 36	24 19	17 14	12 46	04 59	23 53	14 08	17 39	14 24
12	03 19 37	21 24 09	20 ♋ 28 26	26 ♋ 24 07	25 24	18 26	13 14	05 10	23 54	14 05	17 38	14 22
13	03 23 33	22 22 05	02 ♌ 20 10	08 ♌ 17 06	26 32	19 39	13 43	05 22	23 56	14 03	17 37	14 21
14	03 27 30	23 19 59	14 ♌ 15 30	20 ♌ 15 59	27 43	20 51	14 12	05 34	23 58	14 00	17 35	14 20
15	03 31 27	24 17 51	26 ♌ 19 09	02 ♍ 25 38	28 ♈ 56	22 04	14 41	05 45	24 00	13 58	17 34	14 19
16	03 35 23	25 15 41	08 ♍ 36 03	14 ♍ 51 02	00 ♉ 12	23 16	15 10	05 57	24 02	13 55	17 32	14 18
17	03 39 20	26 13 30	21 ♍ 11 07	27 ♍ 36 52	01 30	24 29	15 39	06 09	24 05	13 53	17 31	14 16
18	03 43 16	27 11 17	04 ♎ 08 44	10 ♎ 47 04	02 51	25 41	16 08	06 21	24 07	13 50	17 29	14 15
19	03 47 13	28 09 02	17 ♎ 32 07	24 ♎ 24 00	04 14	26 53	16 38	06 33	24 09	13 48	17 28	14 14
20	03 51 09	29 ♉ 06 46	01 ♏ 22 40	08 ♏ 27 55	05 40	28 06	17 08	06 45	24 12	13 45	17 26	14 13
21	03 55 06	00 ♊ 04 28	15 ♏ 39 18	22 ♏ 56 14	07 08	29 ♊ 18	17 37	06 57	24 15	13 43	17 25	14 12
22	03 59 02	01 02 09	00 ♐ 17 56	07 ♐ 43 27	08 38	00 ♋ 30	18 07	07 09	24 17	13 40	17 23	14 11
23	04 02 59	01 59 49	15 ♐ 11 52	22 ♐ 41 31	10 11	01 42	18 37	07 21	24 20	13 38	17 22	14 10
24	04 06 56	02 57 27	00 ♑ 11 44	07 ♑ 41 08	11 46	02 54	19 08	07 33	24 23	13 36	17 20	14 09
25	04 10 52	03 55 05	15 ♑ 08 38	22 ♑ 33 15	13 24	04 06	19 38	07 46	24 26	13 33	17 19	14 08
26	04 14 49	04 52 41	29 ♑ 54 07	07 ♒ 10 34	15 03	05 18	20 09	07 58	24 29	13 31	17 17	14 07
27	04 18 45	05 50 17	14 ♒ 22 05	21 ♒ 28 23	16 45	06 30	20 39	08 11	24 32	13 29	17 15	14 06
28	04 22 42	06 47 51	28 ♒ 29 16	05 ♓ 24 44	18 30	07 42	21 10	08 23	24 36	13 27	17 14	14 06
29	04 26 38	07 45 25	12 ♓ 14 53	18 ♓ 59 55	20 16	08 54	21 41	08 36	24 39	13 24	17 12	14 05
30	04 30 35	08 42 57	25 ♓ 40 04	02 ♈ 15 39	22 05	10 05	22 12	08 48	24 43	13 22	17 11	14 04
31	04 34 31	09 ♊ 40 29	08 ♈ 47 01	15 ♈ 14 28	23 ♉ 56	11 ♋ 17	22 ♌ 43	09 ♋ 01	24 ♌ 46	13 ♏ 20	17 ♐ 09	14 ♎ 03

DECLINATIONS

Date	Moon True ☊	Moon Mean ☊	Moon ☽ Latitude	Sun ☉	Moon ☽	Mercury ☿	Venus ♀	Mars ♂	Jupiter ♃	Saturn ♄	Uranus ♅	Neptune ♆	Pluto ♇
01	05 ♎ 28	04 ♎ 10	02 N 54	15 N 03	08 S 09	04 N 13	21 N 48	20 N 10	23 N 26	15 N 04	15 S 49	21 S 24	10 N 29
02	05 D 29	04 07	01 48	15 21	04 S 08	04 17	22 03	20 02	23 26	15 03	15 48	21 24	10 29
03	05 30	04 04	00 N 37	15 39	00 N 02	04 24	22 18	19 53	23 26	15 03	15 48	21 24	10 29
04	05 R 30	04 01	00 S 34	15 56	04 08	04 33	22 32	19 45	23 26	15 02	15 47	21 24	10 29
05	05 29	03 58	01 43	16 14	07 59	04 44	22 46	19 36	23 26	15 02	15 46	21 24	10 30
06	05 25	03 55	02 45	16 31	11 25	04 57	22 59	19 28	23 25	15 02	15 45	21 24	10 30
07	05 20	03 51	03 38	16 47	14 15	05 11	23 11	19 19	23 25	15 01	15 45	21 23	10 30
08	05 13	03 48	04 20	17 04	16 24	05 28	23 23	19 10	23 25	15 01	15 44	21 23	10 30
09	05 05	03 45	04 49	17 20	17 46	05 46	23 34	19 01	23 25	15 00	15 43	21 23	10 31
10	04 57	03 42	05 05	17 36	18 19	06 07	23 44	18 52	23 24	14 59	15 42	21 23	10 31
11	04 49	03 39	05 07	17 51	18 03	06 28	23 54	18 43	23 24	14 59	15 42	21 23	10 31
12	04 43	03 35	04 57	18 07	16 59	06 51	24 03	18 34	23 24	14 58	15 41	21 23	10 31
13	04 39	03 32	04 33	18 22	15 12	07 16	24 11	18 24	23 23	14 57	15 40	21 22	10 31
14	04 37	03 29	03 58	18 36	12 45	07 42	24 19	18 15	23 23	14 57	15 39	21 22	10 31
15	04 D 36	03 26	03 11	18 51	09 44	08 09	24 26	18 05	23 22	14 56	15 39	21 22	10 32
16	04 37	03 23	02 16	19 05	06 15	08 38	24 32	17 56	23 22	14 55	15 38	21 22	10 32
17	04 38	03 20	01 12	19 18	02 N 24	09 07	24 38	17 46	23 22	14 54	15 37	21 22	10 32
18	04 39	03 16	00 S 03	19 32	01 S 41	09 38	24 43	17 36	23 21	14 53	15 37	21 22	10 32
19	04 R 38	03 13	01 N 09	19 45	05 50	10 10	24 47	17 26	23 21	14 53	15 36	21 22	10 32
20	04 35	03 10	02 18	19 58	09 47	10 43	24 50	17 16	23 20	14 52	15 35	21 22	10 32
21	04 30	03 07	03 21	20 10	13 19	11 16	24 53	17 06	23 20	14 51	15 34	21 21	10 32
22	04 23	03 04	04 13	20 22	16 05	11 50	24 55	16 56	23 19	14 50	15 34	21 21	10 32
23	04 15	03 01	04 48	20 34	17 51	12 25	24 56	16 45	23 18	14 49	15 33	21 21	10 32
24	04 06	02 57	05 04	20 45	18 23	13 01	24 57	16 35	23 18	14 48	15 32	21 21	10 32
25	03 58	02 54	04 59	20 56	17 38	13 37	24 57	16 24	23 17	14 47	15 32	21 21	10 32
26	03 52	02 51	04 34	21 07	15 42	14 14	24 56	16 13	23 17	14 45	15 31	21 21	10 32
27	03 48	02 48	03 52	21 17	12 49	14 50	24 54	16 03	23 16	14 44	15 30	21 20	10 32
28	03 46	02 45	02 57	21 27	09 14	15 28	24 52	15 52	23 15	14 43	15 30	21 20	10 32
29	03 D 46	02 41	01 52	21 36	05 14	16 05	24 49	15 41	23 14	14 42	15 29	21 20	10 32
30	03 46	02 38	00 43	21 45	01 04	16 42	24 45	15 30	23 14	14 41	15 28	21 20	10 31
31	03 ♎ 47	02 ♎ 35	00 S 27	21 N 54	03 N 04	17 N 19	24 N 41	15 N 19	23 N 13	14 N 39	15 S 28	21 S 20	10 N 31

ZODIAC SIGN ENTRIES

Date	h m	Planets
01	09 00	☽ ♓
03	14 27	☽ ♈
05	21 52	☽ ♉
08	07 18	☽ ♊
10	18 41	☽ ♋
13	07 17	☽ ♌
15	19 15	☽ ♍
16	08 20	☿ ♉
18	04 24	☽ ♎
20	09 39	☽ ♏
21	10 08	☉ ♊
22	02 03	♀ ♋
22	11 31	☽ ♐
24	11 41	☽ ♑
26	12 10	☽ ♒
28	14 36	☽ ♓
30	19 52	☽ ♈

LATITUDES

Date	Mercury ☿	Venus ♀	Mars ♂	Jupiter ♃	Saturn ♄	Uranus ♅	Neptune ♆	Pluto ♇
01	02 S 25	00 N 39	02 N 01	00 N 02	01 N 31	00 N 24	01 N 29	17 N 37
04	02 47	00 47	01 57	00 02	01 31	00 24	01 29	17 36
07	03 02	00 55	01 54	00 03	01 31	00 24	01 29	17 35
10	03 09	01 02	01 50	00 03	01 30	00 24	01 30	17 34
13	03 11	01 09	01 46	00 03	01 30	00 24	01 30	17 33
16	03 06	01 16	01 43	00 04	01 30	00 24	01 30	17 32
19	02 56	01 23	01 40	00 04	01 30	00 24	01 30	17 31
22	02 41	01 29	01 36	00 04	01 29	00 24	01 30	17 30
25	02 21	01 34	01 33	00 04	01 29	00 24	01 30	17 28
28	01 57	01 39	01 30	00 05	01 29	00 24	01 30	17 27
31	01 S 29	01 N 44	01 N 27	00 N 05	01 N 29	00 N 24	01 N 30	17 N 26

DATA

Julian Date	2443630
Delta T	+49 seconds
Ayanamsa	23° 33' 16"
Synetic vernal point	05° ♓ 33' 43"
True obliquity of ecliptic	23° 26' 22"

MOON'S PHASES, APSIDES AND POSITIONS ☽

Date	h	m	Phase	Longitude	Eclipse Indicator
07	04	47	●	16 ♉ 17	
15	07	39	☽	24 ♌ 07	
22	13	17	○	01 ♐ 05	
29	03	30	☾	07 ♓ 25	

Day	h	m		
12	04	14	Apogee	
24	05	17	Perigee	
03	11	48	0N	
10	16	01	Max dec	18° N 20'
18	02	10	0S	
24	09	56	Max dec	18° S 23'
30	18	06	0N	

LONGITUDES

Date	Chiron ⚷	Ceres ⚳	Pallas ⚴	Juno ⚵	Vesta ⚶	Black Moon Lilith ⚸
01	05 ♉ 41	23 ♑ 08	22 ♐ 17	05 ♒ 46	20 ♐ 05	11 ♋ 37
11	06 ♉ 18	23 ♑ 51	20 ♐ 31	07 ♒ 14	19 ♐ 24	12 ♋ 45
21	06 ♉ 54	23 ♑ 58	18 ♐ 06	08 ♒ 14	17 ♐ 19	13 ♋ 52
31	07 ♉ 28	23 ♑ 30	15 ♐ 16	08 ♒ 42	15 ♐ 05	14 ♋ 59

All ephemeris data is given at 12.00 UT and the Moon's longitude is additionally given for 24.00 UT

ASPECTARIAN

01 Monday
06 25 ☽ Q ☉
08 21 ☽ bQ ♇
08 47 ☽ Q ♆
11 25 ☽ ∠ ☿
12 39 ☉ ∥ ♄
14 31 ☽ △ ♃
18 30 ☽ □ ♀
23 48 ☽ ⚻ ♂

02 Tuesday
00 08 ☽ ± ♇
03 53 ☽ ⊥ ☿
05 09 ☽ ⚹ ☉
10 36 ☽ △ ♅
10 46 ☽ ⚻ ♇
10 48 ☽ ± ♂
11 08 ☽ # ☿
15 01 ☽ ⚺ ☿
15 11 ☉ ± ♆
16 34 ☽ □ ♆

03 Wednesday
03 04 ☽ ⚻ ♄
03 29 ☽ bQ ♂
05 46 ☽ Q ♀
10 08 ☽ ∠ ☉
13 28 ☽ bQ ♅
13 55 ☽ ± ♄
20 55 ☽ □ ♃
23 25 ☉ # ♅

04 Thursday
00 13 ☿ △ ♆
03 48 ☽ ⊥ ☉
05 48 ☽ ⚹ ♀
05 49 ☽ ± ♅
06 21 ☽ bQ ♄
07 42 ☽ △ ♂
14 32 ☽ ∥ ☿
15 43 ☽ ⚺ ☉
16 50 ☽ ⚻ ♅
17 05 ☽ ☍ ♇
23 06 ☽ △ ♆

05 Friday
00 09 ♂ ⊥ ♃
00 24 ☽ ☌ ☿
06 01 ☉ ☍ ♅
06 33 ☽ Q ♃
09 33 ☉ ⚻ ♇
10 08 ☽ △ ♄
12 26 ☽ ∠ ♀
13 21 ♀ ⚹ ♂

06 Saturday
03 06 ☽ bQ ♆
05 14 ☽ ∥ ♇
05 24 ☽ ⚹ ♃
07 09 ☽ ⊥ ♀
17 51 ☽ □ ♂
19 45 ☽ ⚺ ♀
20 09 ☽ ± ♆
23 59 ♀ Q ♄

07 Sunday
01 01 ☽ ☍ ♅
01 22 ☽ ⚻ ♇
04 47 ☽ ☌ ☉
07 37 ☽ ⚻ ♆
10 25 ☽ ∠ ♃
12 52 ☽ ± ♇
12 53 ☽ ⚺ ☿
19 14 ☽ □ ♄
19 38 ☽ ∥ ♄

08 Monday
01 22 ☽ ⊥ ☿
03 38 ☽ # ♅
04 06 ☽ ⊥ ♃
05 50 ☽ Q ♂
06 16 ☽ bQ ♇
15 59 ☽ ⚺ ♃
16 08 ♀ ∥ ♃
16 36 ☉ ⚻ ♆
20 21 ☽ ∠ ☿

09 Tuesday
00 19 ☽ ∥ ☉
00 40 ♀ ⚻ ♅
04 56 ♀ △ ♇
06 17 ☽ ⚹ ♂
06 29 ☽ Q ♄
11 13 ☽ ⚻ ♅
11 40 ☽ △ ♇
12 26 ☽ ☌ ♀
18 07 ☽ ☍ ♆
20 24 ☽ ⚺ ☉
21 40 ☿ Q ♃
23 05 ☽ ± ♅

10 Wednesday
04 37 ☽ ⚹ ☿
06 22 ☽ ⚹ ♄
09 24 ☽ ⊥ ☉
13 19 ☽ ∠ ♂
17 00 ☽ bQ ♅
21 53 ☉ ∠ ♃

11 Thursday
01 50 ☿ △ ♄
04 35 ☽ ☌ ♃
05 03 ☽ ∠ ☉
06 56 ☽ Q ☿
08 09 ☽ ⊥ ♂
11 00 ☉ ± ♇
12 33 ☽ ∠ ♄
16 34 ☽ ∥ ☉
20 24 ♀ ☍ ♆
20 45 ☽ ⚺ ♂
23 07 ☽ △ ♅
23 41 ☽ □ ♇

12 Friday
06 16 ☽ ⚻ ♆
06 48 ☽ ⊥ ♄
07 24 ☽ ⚺ ♀
14 03 ☽ ⚹ ☉
18 23 ☽ ± ♆
18 58 ☽ ⚺ ♄
20 56 ☽ ⊥ ♀
23 01 ☽ □ ☿

13 Saturday
06 30 ☽ # ♅
12 02 ☽ Q ♇
12 33 ☽ bQ ♆
14 43 ☽ ∥ ♄
14 45 ☉ ∥ ♂
16 28 ☽ Q ☉
17 11 ☽ ∠ ♀
18 13 ☽ ⚺ ♃
19 36 ♀ ± ♅

14 Sunday
03 04 ♂ □ ♅
06 30 ☽ ⊥ ♃
11 29 ☽ □ ♅
11 52 ☽ ☌ ♂
12 09 ☽ ⚹ ♇
18 39 ☽ △ ♆
18 39 ♂ ⚹ ♇

15 Monday
00 47 ☽ ∠ ♃
02 38 ☽ ⚹ ♀
04 26 ☉ □ ♄
06 08 ☽ ∥ ♇
07 25 ☽ ☌ ♄
07 39 ☽ □ ☉
17 44 ☽ △ ☿
17 53 ☽ ∠ ♇
21 55 ☽ ∥ ☿
23 03 ☽ Q ♅

16 Tuesday
04 51 ☽ Q ♀
06 47 ☽ ⚹ ♃
11 24 ☽ ⊥ ♇
22 11 ☽ ⚹ ♅
22 55 ☽ ⚺ ♇

17 Wednesday
01 06 ☽ ⚺ ♂
02 07 ☽ bQ ☿
03 48 ♀ ⚹ ♄
05 05 ☽ □ ♆
06 11 ☽ Q ♃
12 54 ☽ ⊥ ♂
17 26 ☽ ⚺ ♄
18 48 ☽ □ ♀
21 01 ☽ ± ☿
22 11 ☽ △ ☉

18 Thursday
02 17 ☽ ∠ ♅
04 36 ☽ ⊥ ♄
05 45 ☿ bQ ♆
06 17 ☽ ∠ ♂
09 22 ☽ ⚻ ☿
14 26 ☽ Q ♆
16 04 ☽ □ ♃
18 40 ☽ ⊥ ♅
21 02 ☽ ∠ ♄

19 Friday
03 38 ☽ bQ ☉
05 24 ☽ ⚺ ♅
06 10 ☽ ☌ ♇
10 21 ☽ ⚹ ♂
11 52 ☽ ⚹ ♆
20 42 ☽ ± ☉
23 37 ☽ ⚹ ♄

20 Saturday
04 18 ☿ ∥ ♇
05 51 ☽ △ ♀
07 50 ☽ ⚻ ☉
08 00 ☽ Q ♂
13 48 ☽ ∠ ♆
14 37 ☉ bQ ♇
16 45 ☽ # ♇
18 57 ☽ # ☿
20 07 ☽ ☍ ☿
20 13 ☽ Q ♄
21 14 ☽ △ ♃

21 Sunday
00 47 ♀ bQ ♅
02 23 ♂ △ ♆
04 57 ☽ ⊥ ♆
08 30 ☿ ⚹ ♃
08 47 ☽ ☌ ♅
09 32 ☽ bQ ♀
09 35 ☽ ⚺ ♇
14 54 ☽ ⚺ ♆
15 22 ☽ □ ♂
19 30 ☽ ⊥ ♇
22 31 ☽ bQ ♃

22 Monday
00 12 ☽ # ♄
01 43 ☽ ± ♀
02 11 ☽ □ ♄
06 47 ☽ ∥ ♅
10 12 ☽ ∠ ♇
12 21 ☽ ⚻ ♀
13 17 ☽ ☍ ☉
13 24 ☽ ± ♃
15 36 ☉ ⊥ ♃
20 46 ☽ # ♂
23 14 ☽ ⚻ ♃

23 Tuesday
03 02 ☽ ⚻ ☿
09 30 ☽ ⚺ ♅
10 21 ☽ ⚹ ♇
13 46 ☽ ± ☿
15 28 ☽ ☌ ♆
17 41 ☽ △ ♂
19 06 ☽ ⊥ ♅

24 Wednesday
02 41 ☽ △ ♄
05 32 ☽ Q ♇
05 34 ☿ ± ♆
05 52 ☽ bQ ☿
09 27 ☽ ∠ ♅
16 42 ☽ ☍ ♀
16 43 ☽ ⚻ ☉
17 41 ☉ ⚺ ♀
18 31 ☽ bQ ♂
23 58 ☽ ☍ ♃

25 Thursday
02 47 ☽ bQ ♄
03 00 ☽ ± ☉
08 50 ☽ △ ☿
09 27 ☽ ⚹ ♅
09 29 ☽ ± ♂
10 23 ☽ □ ♇
14 19 ☿ ☍ ♅
15 30 ☽ ⚺ ♆
17 20 ☽ ± ♄
18 31 ☽ bQ ☉
19 31 ☽ ⚻ ♂
22 41 ☿ ⚻ ♇

26 Friday
01 12 ☽ ⊥ ♆
03 07 ☽ ⚻ ♄
04 51 ☽ Q ♅
06 12 ☽ # ♂
06 35 ♀ ∠ ♂
13 50 ☽ ∥ ♅
15 54 ☽ ∠ ♆
20 45 ☽ # ♄
20 47 ☽ △ ☉
21 42 ☽ ⚻ ♀
22 52 ☽ # ☿

27 Saturday
01 31 ☽ ⚻ ♃
08 09 ☿ ∥ ♄
08 35 ☽ ± ♀
10 31 ☽ □ ♅
11 34 ☽ △ ♇
11 40 ☽ ± ♃
16 34 ☽ □ ☿
16 51 ☽ ⚹ ♆
18 52 ☿ ⚻ ♆
20 33 ☉ # ♆
23 01 ☽ ☍ ♂

28 Sunday
01 10 ☽ bQ ♀
03 07 ☽ bQ ♃
03 45 ☽ # ♇
05 18 ☽ ☍ ♄
13 02 ☽ bQ ♇
13 17 ☽ Q ♆
13 21 ☿ # ♅

29 Monday
00 05 ☿ ∥ ♂
03 30 ☽ □ ☉
03 57 ☽ Q ☿
04 40 ☽ ± ♇
04 42 ♀ ☌ ♃
05 28 ☽ △ ♃
05 31 ☽ △ ♀
09 26 ☿ ± ♇
14 02 ☽ △ ♅
15 14 ☽ ⚻ ♇
20 47 ☽ □ ♆

30 Tuesday
04 00 ♀ ∠ ♄
04 31 ☽ ⚹ ☿
05 30 ☽ ⚻ ♂
10 16 ☽ ⚻ ♄
14 03 ☽ Q ☉
14 08 ☿ □ ♂
14 46 ☉ ⚺ ♃
15 11 ♂ # ♅
16 47 ☽ ± ♂
16 53 ☽ bQ ♅
21 12 ☽ ± ♄

31 Wednesday
09 20 ☽ ± ♅
09 57 ☽ bQ ♂
12 20 ☽ ∠ ☿
12 26 ☽ □ ♃
13 05 ☿ ∠ ♃
13 47 ☽ ⚹ ☉
13 50 ☽ bQ ♄
17 06 ☽ □ ♀
20 25 ☽ ⚻ ♅
21 46 ☽ ☍ ♇
22 58 ☿ □ ♄

JUNE 1978

LONGITUDES

Date	Sidereal time h m s	Sun ☉ ° ' "	Moon ☽ ° ' "	Moon ☽ 24.00 ° ' "	Mercury ☿ ° '	Venus ♀ ° '	Mars ♂ ° '	Jupiter ♃ ° '	Saturn ♄ ° '	Uranus ♅ ° '	Neptune ♆ ° '	Pluto ♇ ° '
01	04 38 28	10 ♊ 38 01	21 ♈ 38 22	27 ♈ 59 00	25 ♉ 50	12 ♋ 29	23 ♌ 15	09 ♋ 13	24 ♌ 50	13 ♏ 18	17 ♐ 07	14 ♎ 02
02	04 42 25	11 35 31	04 ♉ 16 39	10 ♉ 31 35	27 45	13 40	23 46	09 26	24 54	13 R 16	17 R 06	14 R 02
03	04 46 21	12 33 01	16 ♉ 44 02	22 ♉ 54 11	29 ♉ 43	14 52	24 18	09 39	24 57	13 14	17 04	14 01
04	04 50 18	13 30 29	29 ♉ 02 11	05 ♊ 08 13	01 ♊ 43	16 03	24 49	09 52	25 01	13 11	17 03	14 00
05	04 54 14	14 27 57	11 ♊ 12 24	17 ♊ 14 52	03 45	17 15	25 21	10 05	25 05	13 09	17 01	14 00
06	04 58 11	15 25 24	23 ♊ 15 44	29 ♊ 15 09	05 48	18 26	25 53	10 17	25 09	13 07	16 59	13 59
07	05 02 07	16 22 51	05 ♋ 13 17	11 ♋ 10 19	07 54	19 38	26 25	10 30	25 14	13 05	16 58	13 59
08	05 06 04	17 20 16	17 ♋ 06 29	23 ♋ 02 01	10 01	20 49	26 57	10 43	25 18	13 03	16 56	13 58
09	05 10 00	18 17 40	28 ♋ 57 14	04 ♌ 52 29	12 09	22 00	27 30	10 56	25 22	13 01	16 55	13 58
10	05 13 57	19 15 03	10 ♌ 48 10	16 ♌ 44 42	14 19	23 11	28 02	11 09	25 27	13 00	16 53	13 57
11	05 17 54	20 12 25	22 ♌ 42 37	28 ♌ 42 25	16 30	24 22	28 35	11 22	25 31	12 58	16 51	13 57
12	05 21 50	21 09 47	04 ♍ 44 41	10 ♍ 50 01	18 41	25 33	29 07	11 35	25 36	12 56	16 50	13 56
13	05 25 47	22 07 07	16 ♍ 59 02	23 ♍ 12 23	20 53	26 44	29 ♌ 40	11 49	25 40	12 54	16 48	13 56
14	05 29 43	23 04 26	29 ♍ 30 41	05 ♎ 54 33	23 05	27 55	00 ♍ 13	12 02	25 45	12 52	16 46	13 55
15	05 33 40	24 01 44	12 ♎ 24 33	19 ♎ 01 12	25 17	29 ♋ 06	00 46	12 15	25 50	12 51	16 45	13 55
16	05 37 36	24 59 02	25 ♎ 44 53	02 ♏ 35 55	27 28	00 ♌ 17	01 19	12 28	25 55	12 49	16 43	13 55
17	05 41 33	25 56 18	09 ♏ 34 24	16 ♏ 40 20	29 ♊ 39	01 27	01 52	12 41	26 00	12 47	16 42	13 55
18	05 45 29	26 53 34	23 ♏ 53 25	01 ♐ 13 11	01 ♋ 49	02 38	02 25	12 55	26 05	12 46	16 40	13 54
19	05 49 26	27 50 50	08 ♐ 38 55	16 ♐ 09 41	03 58	03 49	02 58	13 08	26 10	12 44	16 38	13 54
20	05 53 23	28 48 04	23 ♐ 44 20	01 ♑ 21 35	06 06	04 59	03 32	13 21	26 15	12 43	16 37	13 54
21	05 57 19	29 ♊ 45 19	09 ♑ 00 02	16 ♑ 38 14	08 12	06 09	04 05	13 34	26 20	12 41	16 35	13 54
22	06 01 16	00 ♋ 42 32	24 ♑ 14 49	01 ♒ 48 26	10 17	07 20	04 39	13 48	26 26	12 40	16 34	13 54
23	06 05 12	01 39 46	09 ♒ 17 59	16 ♒ 42 29	12 19	08 30	05 13	14 01	26 31	12 38	16 32	13 54
24	06 09 09	02 36 59	24 ♒ 01 11	01 ♓ 13 36	14 20	09 40	05 47	14 15	26 36	12 37	16 31	13 D 54
25	06 13 05	03 34 13	08 ♓ 19 24	15 ♓ 18 30	16 19	10 50	06 20	14 28	26 42	12 36	16 29	13 54
26	06 17 02	04 31 26	22 ♓ 10 57	28 ♓ 56 57	18 16	12 00	06 54	14 41	26 47	12 35	16 28	13 54
27	06 20 58	05 28 39	05 ♈ 36 50	12 ♈ 10 59	20 11	13 10	07 28	14 55	26 53	12 33	16 26	13 54
28	06 24 55	06 25 52	18 ♈ 39 49	25 ♈ 03 51	22 04	14 20	08 03	15 08	26 59	12 32	16 25	13 54
29	06 28 52	07 23 05	01 ♉ 23 32	07 ♉ 39 22	23 55	15 30	08 37	15 22	27 05	12 31	16 23	13 54
30	06 32 48	08 ♋ 20 19	13 ♉ 51 47	20 ♉ 01 14	25 ♋ 44	16 ♌ 39	09 ♍ 11	15 ♋ 35	27 ♌ 10	12 ♏ 30	16 ♐ 22	13 ♎ 55

MOON NODES AND LATITUDE

Date	Moon True ☊ ° '	Moon Mean ☊ ° '	Moon ☽ Latitude ° '
01	03 ♎ 45	02 ♎ 32	01 S 34
02	03 R 41	02 29	02 35
03	03 35	02 26	03 27
04	03 26	02 22	04 09
05	03 15	02 19	04 39
06	03 02	02 16	04 57
07	02 49	02 13	05 01
08	02 37	02 10	04 52
09	02 26	02 07	04 30
10	02 18	02 03	03 56
11	02 13	02 00	03 13
12	02 10	01 57	02 20
13	02 09	01 54	01 19
14	02 D 09	01 51	00 S 14
15	02 R 09	01 47	00 N 54
16	02 07	01 44	02 01
17	02 04	01 41	03 04
18	01 57	01 38	03 57
19	01 49	01 35	04 37
20	01 39	01 32	04 58
21	01 28	01 28	04 58
22	01 18	01 25	04 37
23	01 10	01 22	03 57
24	01 04	01 19	03 02
25	01 01	01 16	01 57
26	01 00	01 12	00 N 46
27	01 D 00	01 09	00 S 24
28	00 R 59	01 06	01 32
29	00 58	01 03	02 33
30	00 ♎ 54	01 ♎ 00	03 S 26

DECLINATIONS

Date	Sun ☉ ° '	Moon ☽ ° '	Mercury ☿ ° '	Venus ♀ ° '	Mars ♂ ° '	Jupiter ♃ ° '	Saturn ♄ ° '	Uranus ♅ ° '	Neptune ♆ ° '	Pluto ♇ ° '
01	22 N 02	06 N 59	17 N 56	24 N 36	15 N 07	23 N 12	14 N 38	15 S 27	21 S 20	10 N 31
02	22 10	10 31	18 33	24 30	14 56	23 11	14 37	15 26	21 20	10 31
03	22 18	13 31	19 09	24 24	14 44	23 11	14 35	15 26	21 19	10 31
04	22 25	15 53	19 44	24 16	14 33	23 10	14 34	15 25	21 19	10 31
05	22 32	17 30	20 18	24 09	14 21	23 09	14 33	15 25	21 19	10 30
06	22 39	18 20	20 51	24 00	14 10	23 08	14 31	15 24	21 19	10 30
07	22 45	18 20	21 23	23 51	13 58	23 07	14 30	15 23	21 19	10 30
08	22 50	17 31	21 53	23 41	13 46	23 06	14 28	15 23	21 19	10 30
09	22 55	15 58	22 22	23 30	13 34	23 05	14 27	15 22	21 18	10 29
10	23 00	13 44	22 48	23 19	13 22	23 04	14 25	15 22	21 18	10 29
11	23 05	10 54	23 13	23 07	13 09	23 03	14 24	15 21	21 18	10 29
12	23 09	07 36	23 35	22 55	12 57	23 02	14 22	15 21	21 18	10 28
13	23 12	03 N 55	23 55	22 42	12 45	23 01	14 20	15 20	21 18	10 28
14	23 15	00 S 01	24 12	22 28	12 32	23 00	14 19	15 20	21 18	10 28
15	23 18	04 04	24 26	22 14	12 20	22 59	14 17	15 19	21 18	10 27
16	23 21	08 04	24 38	21 59	12 07	22 58	14 15	15 19	21 17	10 27
17	23 23	11 46	24 47	21 44	11 54	22 57	14 13	15 18	21 17	10 27
18	23 24	14 55	24 53	21 28	11 42	22 55	14 12	15 18	21 17	10 26
19	23 25	17 11	24 56	21 11	11 29	22 54	14 10	15 17	21 17	10 26
20	23 26	18 20	24 57	20 54	11 16	22 53	14 08	15 17	21 17	10 25
21	23 26	18 11	24 54	20 36	11 03	22 52	14 06	15 17	21 17	10 25
22	23 26	16 43	24 49	20 18	10 50	22 50	14 04	15 16	21 17	10 24
23	23 26	14 07	24 42	20 00	10 36	22 49	14 03	15 16	21 16	10 24
24	23 25	10 39	24 32	19 40	10 23	22 48	14 01	15 15	21 16	10 23
25	23 23	06 39	24 20	19 21	10 10	22 47	13 59	15 15	21 16	10 23
26	23 22	02 S 23	24 06	19 00	09 56	22 45	13 57	15 15	21 16	10 22
27	23 20	01 N 51	23 50	18 40	09 43	22 44	13 55	15 14	21 16	10 22
28	23 17	05 54	23 32	18 19	09 29	22 42	13 53	15 14	21 16	10 21
29	23 14	09 34	23 12	17 57	09 16	22 41	13 51	15 14	21 16	10 21
30	23 N 11	12 N 43	22 N 50	17 N 35	09 N 02	22 N 39	13 N 49	15 S 13	21 S 16	10 N 20

ZODIAC SIGN ENTRIES

Date	h	m	Planets
02	03	50	☽ ♉
03	15	26	☿ ♊
04	13	53	☽ ♊
07	01	30	☽ ♋
09	14	07	☽ ♌
12	02	35	☽ ♍
14	02	38	♂ ♍
14	12	55	☽ ♎
16	06	19	♀ ♌
16	19	28	☽ ♏
17	15	49	☿ ♋
18	22	01	☽ ♐
20	21	52	☽ ♑
21	18	10	☉ ♋
22	21	07	☽ ♒
24	21	57	☽ ♓
27	01	53	☽ ♈
29	09	21	☽ ♉

LATITUDES

Date	Mercury ☿ ° '	Venus ♀ ° '	Mars ♂ ° '	Jupiter ♃ ° '	Saturn ♄ ° '	Uranus ♅ ° '	Neptune ♆ ° '	Pluto ♇ ° '
01	01 S 19	01 N 45	01 N 26	00 N 05	01 N 29	00 N 24	01 N 30	17 N 25
04	00 48	01 49	01 23	00 05	01 28	00 24	01 30	17 24
07	00 S 15	01 52	01 20	00 06	01 28	00 24	01 30	17 22
10	00 N 17	01 54	01 17	00 06	01 28	00 24	01 30	17 21
13	00 47	01 56	01 14	00 06	01 28	00 24	01 30	17 19
16	01 13	01 56	01 11	00 06	01 27	00 23	01 30	17 18
19	01 33	01 57	01 09	00 07	01 27	00 23	01 30	17 16
22	01 47	01 56	01 06	00 07	01 27	00 23	01 30	17 14
25	01 55	01 54	01 03	00 07	01 27	00 23	01 29	17 13
28	01 55	01 52	01 01	00 08	01 27	00 23	01 29	17 11
31	01 N 50	01 N 48	00 N 58	00 N 08	01 N 27	00 N 23	01 N 29	17 N 09

DATA

Julian Date	2443661
Delta T	+49 seconds
Ayanamsa	23° 33' 20"
Synetic vernal point	05° ♓ 33' 40"
True obliquity of ecliptic	23° 26' 22"

MOON'S PHASES, APSIDES AND POSITIONS ☽

Date	h	m	Phase	Longitude ° '	Eclipse Indicator
05	19	01	●	14 ♊ 45	
13	22	44	☽	22 ♍ 33	
20	20	31	○	29 ♐ 08	
27	11	44	☾	05 ♈ 28	

Day	h	m		
08	17	39	Apogee	
21	12	04	Perigee	
06	23	58	Max dec	18° N 26'
14	11	53	0S	
20	21	12	Max dec	18° S 26'
27	01	25	0N	

LONGITUDES

Date	Chiron ⚷ ° '	Ceres ⚳ ° '	Pallas ⚴ ° '	Juno ⚵ ° '	Vesta ⚶ ° '	Black Moon Lilith ⚸ ° '
01	07 ♉ 32	23 ♑ 25	14 ♐ 58	08 ♒ 43	14 ♐ 51	15 ♋ 06
11	08 ♉ 03	22 ♑ 17	12 ♐ 01	08 ♒ 33	12 ♐ 27	16 ♋ 13
21	08 ♉ 31	20 ♑ 39	09 ♐ 20	07 ♒ 46	10 ♐ 17	17 ♋ 20
31	08 ♉ 56	18 ♑ 38	07 ♐ 08	06 ♒ 23	08 ♐ 39	18 ♋ 27

All ephemeris data is given at 12.00 UT and the Moon's longitude is additionally given for 24.00 UT

ASPECTARIAN

h m	Aspects
01 Thursday	
03 32	☽ △ ♆
08 00	☽ ⊥ ☿
15 09	☽ △ ♂
18 03	☽ △ ♄
20 10	☽ ∠ ☉
21 19	☽ ⚺ ☿
22 44	☽ Q ♃
02 Friday	
03 57	♀ △ ♅
06 30	☽ Q ♀
07 51	☽ ⚼ ♆
11 59	☽ ∥ ♇
14 44	☽ ⊥ ☉
19 05	♀ □ ♇
21 50	☉ Q ♂
22 04	☽ ⚹ ♃
03 Saturday	
01 04	☽ ± ♆
03 14	☽ ⚺ ☉
03 32	☿ ⚼ ♇
05 14	☽ ☍ ♅
06 44	☽ ⚻ ♇
08 00	☽ ⚹ ♀
12 39	☽ ⚻ ♆
16 30	☿ ∠ ♀
18 22	☽ ± ♇
21 53	☽ ∥ ♄
22 38	☽ ∥ ♂
22 56	☉ Q ♄
04 Sunday	
03 23	☽ □ ♂
03 41	☽ ∠ ♃
04 06	☽ □ ♄
04 19	☉ ⚻ ♅
06 40	☽ # ♅
09 27	♂ ∥ ♄
11 56	☽ ⚼ ♇
14 52	♂ ∠ ♃
16 24	☽ ∠ ♀
18 18	☽ ☌ ☿
21 39	☽ ⊥ ♃
22 17	♂ ☌ ♄
05 Monday	
00 19	☉ △ ♇
07 27	♀ ⚻ ♆
09 43	☽ ⚺ ♃
12 05	☽ ⊥ ♀
14 14	♃ ∠ ♄
15 45	☽ Q ♄
15 51	☽ ⚻ ♅
16 20	☿ ⊥ ♃
16 27	☽ Q ♂
17 32	☽ △ ♇
19 01	☽ ☌ ☉
23 31	☽ ☍ ♆
06 Tuesday	
01 19	☽ ⚺ ♀
03 45	☽ ± ♅
15 49	☽ ⚹ ♄
17 30	☽ ⚹ ♂
21 42	☽ ⚼ ♅
07 Wednesday	
03 26	♀ ⊥ ♄
08 54	☿ # ♆
18 33	☽ ⚺ ☿
22 09	☽ ∠ ♄
22 51	☽ ☌ ♃
08 Thursday	
01 06	☽ ∠ ♂
02 11	☉ ☍ ♆
03 50	☽ △ ♅
05 39	☽ □ ♇
09 18	☽ ⊥ ☿
11 39	☽ ⚻ ♆
12 30	☽ ⚺ ☉
16 28	☽ ⊥ ♄
17 17	♀ ⊥ ♂
20 10	☽ ⊥ ♂
20 20	☽ ☌ ♀
20 50	☿ ⚺ ♃
23 46	☽ ± ♆
09 Friday	
01 44	☽ ⊥ ☉
04 41	☽ ⚺ ♄
07 33	☽ ∠ ☿
08 54	☽ ⚺ ♂
17 59	☽ ⚼ ♆
18 05	☽ Q ♇
19 06	☽ # ♅
21 32	☿ ⚻ ♅
21 34	☽ ∠ ☉
10 Saturday	
01 59	☿ Q ♄
05 12	☽ ∥ ♄
05 45	☉ ± ♅
05 57	♀ ± ♆
07 57	☿ △ ♇
12 44	☽ ⚺ ♃
15 42	☽ ∥ ♂
16 25	☽ □ ♅
18 21	☽ ⚹ ♇
20 41	☽ ⚹ ☿
11 Sunday	
00 15	☽ △ ♆
01 04	☽ ⊥ ♃
02 11	☉ ∥ ☿
02 49	☿ ∥ ♃
05 08	☉ ∥ ♃
06 32	☽ ⚹ ☉
08 29	☿ ∥ ♀
13 12	☿ Q ♂
15 19	☽ ∥ ♇
15 42	☽ ⚺ ♀
15 54	☿ ☍ ♆
16 04	☉ ∥ ♀
17 40	☽ ☌ ♄
19 28	☽ ∠ ♃
21 11	♀ ∥ ♃
12 Monday	
00 18	☽ ☌ ♂
00 28	☽ ∠ ♇
02 09	☽ Q ☿
03 57	♂ ∠ ♇
04 27	☽ Q ♅
04 59	☽ ⊥ ♀
08 36	☽ Q ☉
12 51	♀ ⚺ ♄
14 40	☿ ± ♅
18 18	☽ ⊥ ♇
13 Tuesday	
00 41	☽ ∠ ♀
01 44	☽ ⚹ ♃
04 04	☽ ⚹ ♅
06 04	☽ ⚺ ♇
08 36	☿ ⊥ ♀
11 39	☽ □ ♆
21 09	☽ □ ☿
22 44	☽ □ ☉
14 Wednesday	
01 23	☽ Q ♃
04 49	☽ ⚺ ♄
08 41	☽ ⚹ ♀
08 54	☽ ∠ ♅
11 50	☉ ☌ ☿
13 23	☽ ⚺ ♂
16 15	☽ ⊥ ♄
21 52	☽ Q ♆
15 Thursday	
01 07	☽ ⊥ ♂
01 46	☽ ⊥ ♅
09 05	☽ ∠ ♄
09 22	☽ Q ♀
11 42	☽ □ ♃
12 48	☽ ⚺ ♅
14 46	☽ ☌ ♇
15 25	♂ Q ♅
18 15	☿ ⚹ ♄
18 23	☽ ∠ ♂
19 53	☽ ⚹ ♆
16 Friday	
10 33	☽ △ ☉
12 18	☽ ⚹ ♄
15 37	☽ △ ☿
15 44	☿ ⚼ ♅
20 42	☽ □ ♀
22 10	☽ ⚹ ♂
22 27	☽ ∠ ♆
17 Saturday	
03 06	☽ # ♇
09 17	☽ Q ♄
12 54	☽ # ♂
13 34	☉ ⚹ ♄
13 54	☽ ⊥ ♆
14 30	☽ ⚼ ☉
16 42	♀ ⚼ ♆
17 23	☽ △ ♃
17 27	☽ ☌ ♅
19 21	☽ ⚺ ♇
19 34	☽ Q ♂
21 13	♀ Q ♇
21 49	♃ △ ♅
22 09	☽ ⚼ ☿
18 Sunday	
00 01	☽ ⚺ ♆
03 43	♀ ⚺ ♂
05 24	☽ ⊥ ♇
06 04	☽ # ♄
06 41	☽ ± ☉
15 26	☽ ∥ ♅
15 37	☽ □ ♄
15 44	☽ ± ☿
17 17	☽ ⚻ ☉
18 42	☽ ⚼ ♃
20 14	☽ ∠ ♇
20 57	☿ ⚹ ♂
19 Monday	
02 29	☽ □ ♂
03 11	☽ ⚻ ☿
03 32	☽ △ ♀
03 37	♀ # ♆
07 59	☿ ⚺ ♀
09 19	☉ ⚼ ♅
09 31	☽ ± ♃
16 06	☉ ⊥ ♀
18 32	☽ ⚺ ♅
19 17	☽ ⚻ ♃
20 24	☽ ⚹ ♇
20 Tuesday	
00 44	☽ ☌ ♆
04 04	☽ ⊥ ♅
05 34	☽ ⚼ ♀
15 25	☽ Q ♇
15 59	☽ △ ♄
18 15	☽ ∠ ♅
20 31	☽ ☍ ☉
20 57	☽ ± ♀
21 Wednesday	
04 00	☽ △ ♂
07 10	☽ ⚻ ♀
10 33	☽ ☍ ☿
15 42	☽ ⚼ ♄
17 47	☽ ⚹ ♅
19 17	☽ ☍ ♃
19 42	☽ □ ♇
23 54	☽ ⚺ ♆
22 Thursday	
04 28	☽ ⚼ ♂
05 56	☽ ± ♄
09 21	☽ ⊥ ♆
12 40	☽ Q ♅
15 28	☽ ⚻ ♄
19 15	☽ ± ♂
22 57	☽ ⚻ ☉
22 59	♃ □ ♇
23 35	☽ ∠ ♆
23 Friday	
02 03	☿ ∠ ♄
02 31	☽ ∥ ♅
05 11	☽ ⚻ ♂
09 11	☽ ± ☉
10 36	☽ ☍ ♀
12 35	☽ # ♄
15 43	☿ △ ♅
17 23	☽ □ ♅
17 39	☽ ⚻ ☿
19 26	☽ △ ♇
19 45	☽ ⚻ ♃
23 42	☽ ⚹ ♆
24 Saturday	
00 45	☽ ⚼ ☉
04 59	☽ ± ☿
05 41	☽ ± ♃
06 43	☿ □ ♇
07 41	♇ St D
10 41	☿ ☌ ♃
11 37	♂ ∥ ♇
13 39	☽ # ♇
13 46	☽ # ♂
16 19	☽ ☍ ♄
19 27	☽ Q ♆
20 06	☽ ⚼ ♇
20 49	☽ ⚼ ♃
22 16	☽ ⚼ ☿
25 Sunday	
03 22	☽ △ ☉
08 29	☽ ☍ ♂
11 17	☽ ± ♇
13 56	☿ ⚻ ♆
16 41	☽ ⚻ ♀
19 19	☽ △ ♅
21 34	☽ ⚻ ♇
22 43	☽ △ ♃
26 Monday	
02 01	☽ □ ♆
04 00	☽ ± ♀
04 02	☽ △ ☿
20 13	☽ ⚻ ♄
21 20	☽ ⚼ ♀
21 32	☽ ⚼ ♅
23 38	♀ □ ♅
27 Tuesday	
07 02	☽ ± ♄
11 44	☽ □ ☉
13 43	☽ ± ♅
15 32	☽ ⚻ ♂
21 16	☿ ⊥ ♄
23 32	☽ ⚼ ♄
28 Wednesday	
00 40	☽ ⚻ ♅
03 02	☽ ± ♂
03 10	♀ ⚹ ♇
03 10	☽ ☍ ♇
03 10	☽ △ ♀
05 20	☽ □ ♃
06 03	♂ ⊥ ♇
07 49	☽ △ ♆
16 16	☿ ± ♆
19 28	☽ □ ☿
20 35	☽ ⚼ ♂
23 41	☽ Q ☉
29 Thursday	
03 44	☽ △ ♄
06 11	☿ ∠ ♂
08 37	♀ ⚺ ♃
08 44	☉ ∥ ☿
10 03	☽ ∥ ♂
11 59	☽ ⚼ ♆
15 50	☽ Q ♃
17 33	☽ ∥ ♇
30 Friday	
00 26	☽ ⚹ ☉
02 31	☽ △ ♂
05 14	☽ ± ♆
06 02	♀ △ ♆
09 22	☽ ☍ ♅
11 42	☽ Q ☿
12 05	☽ ⚻ ♇
15 25	☽ ⚹ ♃
16 51	☽ ⚻ ♆
18 00	☽ □ ♀
21 23	☽ ∥ ♄
23 47	☽ ± ♇

JULY 1978

LONGITUDES

Date	Sidereal time h m s	Sun ☉ ° ' "	Moon ☽ ° ' "	Moon ☽ 24.00 ° ' "	Mercury ☿ ° '	Venus ♀ ° '	Mars ♂ ° '	Jupiter ♃ ° '	Saturn ♄ ° '	Uranus ♅ ° '	Neptune ♆ ° '	Pluto ♇ ° '
01	06 36 45	09 ♋ 17 32	26 ♉ 08 05	02 ♊ 12 43	27 ♋ 31	17 ♌ 49	09 ♍ 46	15 ♋ 49	27 ♌ 16	12 ♏ 29	16 ♐ 20	13 ♎ 55
02	06 40 41	10 14 46	08 ♊ 15 26	14 ♊ 16 32	29 ♋ 15	18 58	10 20	16 02	27 22	12 R 28	16 R 19	13 55
03	06 44 38	11 12 00	20 ♊ 16 15	26 ♊ 14 49	00 ♌ 57	20 08	10 55	16 16	27 28	12 27	16 17	13 55
04	06 48 34	12 09 14	02 ♋ 12 24	08 ♋ 09 13	02 38	21 17	11 29	16 29	27 34	12 26	16 16	13 56
05	06 52 31	13 06 27	14 ♋ 05 25	20 ♋ 01 11	04 16	22 26	12 04	16 43	27 40	12 25	16 14	13 56
06	06 56 27	14 03 41	25 ♋ 56 43	01 ♌ 52 11	05 52	23 35	12 39	16 56	27 47	12 25	16 13	13 56
07	07 00 24	15 00 55	07 ♌ 47 49	13 ♌ 43 54	07 25	24 44	13 14	17 10	27 53	12 24	16 12	13 57
08	07 04 21	15 58 09	19 ♌ 40 41	25 ♌ 38 31	08 57	25 53	13 49	17 23	27 59	12 23	16 10	13 57
09	07 08 17	16 55 22	01 ♍ 37 46	07 ♍ 38 50	10 27	27 02	14 24	17 37	28 05	12 23	16 09	13 58
10	07 12 14	17 52 36	13 ♍ 42 12	19 ♍ 48 22	11 54	28 11	14 59	17 50	28 12	12 22	16 08	13 58
11	07 16 10	18 49 49	25 ♍ 57 49	02 ♎ 11 09	13 19	29 ♌ 19	15 35	18 04	28 18	12 21	16 06	13 59
12	07 20 07	19 47 03	08 ♎ 28 55	14 ♎ 51 42	14 42	00 ♍ 28	16 10	18 17	28 25	12 21	16 05	13 59
13	07 24 03	20 44 16	21 ♎ 20 02	27 ♎ 54 27	16 02	01 36	16 46	18 31	28 31	12 21	16 04	14 00
14	07 28 00	21 41 30	04 ♏ 35 25	11 ♏ 23 17	17 20	02 44	17 21	18 44	28 38	12 20	16 02	14 01
15	07 31 56	22 38 44	18 ♏ 18 19	25 ♏ 20 36	18 36	03 53	17 57	18 58	28 44	12 20	16 01	14 01
16	07 35 53	23 35 57	02 ♐ 30 04	09 ♐ 46 26	19 49	05 01	18 32	19 11	28 51	12 20	16 00	14 02
17	07 39 50	24 33 11	17 ♐ 09 10	24 ♐ 37 33	21 00	06 08	19 08	19 25	28 58	12 19	15 59	14 03
18	07 43 46	25 30 25	02 ♑ 10 37	09 ♑ 47 13	22 08	07 16	19 44	19 38	29 05	12 19	15 58	14 04
19	07 47 43	26 27 39	17 ♑ 26 02	25 ♑ 05 41	23 14	08 24	20 20	19 51	29 11	12 19	15 56	14 05
20	07 51 39	27 24 54	02 ♒ 44 41	10 ♒ 21 40	24 16	09 31	20 56	20 05	29 18	12 19	15 55	14 05
21	07 55 36	28 22 09	17 ♒ 55 20	25 ♒ 24 30	25 16	10 39	21 32	20 18	29 25	12 D 19	15 54	14 06
22	07 59 32	29 ♋ 19 25	02 ♓ 48 15	10 ♓ 05 48	26 13	11 46	22 08	20 32	29 32	12 19	15 53	14 07
23	08 03 29	00 ♌ 16 42	17 ♓ 16 39	24 ♓ 20 30	27 07	12 53	22 44	20 45	29 39	12 19	15 52	14 08
24	08 07 25	01 13 59	01 ♈ 17 14	08 ♈ 06 55	27 58	14 00	23 21	20 59	29 46	12 19	15 51	14 09
25	08 11 22	02 11 17	14 ♈ 49 45	21 ♈ 26 05	28 45	15 07	23 57	21 12	29 ♌ 53	12 19	15 50	14 10
26	08 15 19	03 08 36	27 ♈ 56 20	04 ♉ 20 58	29 ♌ 29	16 13	24 33	21 25	00 ♍ 00	12 20	15 49	14 11
27	08 19 15	04 05 57	10 ♉ 40 30	16 ♉ 55 30	00 ♍ 09	17 20	25 10	21 39	00 07	12 20	15 48	14 12
28	08 23 12	05 03 18	23 ♉ 06 28	29 ♉ 13 58	00 46	18 26	25 46	21 52	00 14	12 20	15 47	14 13
29	08 27 08	06 00 40	05 ♊ 18 31	11 ♊ 20 36	01 18	19 32	26 23	22 05	00 21	12 21	15 46	14 14
30	08 31 05	06 58 03	17 ♊ 20 39	23 ♊ 19 08	01 47	20 38	27 00	22 18	00 29	12 21	15 45	14 16
31	08 35 01	07 ♌ 55 27	29 ♊ 16 23	05 ♋ 12 48	02 ♍ 11	21 ♍ 44	27 ♍ 37	22 ♋ 32	00 ♍ 36	12 ♏ 22	15 ♐ 45	14 ♎ 17

Date	Moon True ☊ ° '	Moon Mean ☊ ° '	Moon ☽ Latitude ° '
01	00 ♎ 47	00 ♎ 57	04 S 08
02	00 R 38	00 53	04 38
03	00 26	00 50	04 56
04	00 13	00 47	05 00
05	29 ♍ 59	00 44	04 51
06	29 47	00 41	04 30
07	29 36	00 38	03 57
08	29 27	00 34	03 14
09	29 21	00 31	02 21
10	29 18	00 28	01 22
11	29 17	00 25	00 S 18
12	29 D 17	00 22	00 N 49
13	29 R 17	00 18	01 55
14	29 17	00 15	02 56
15	29 14	00 12	03 50
16	29 10	00 09	04 32
17	29 03	00 06	04 57
18	28 54	00 ♎ 03	05 03
19	28 45	29 ♍ 59	04 48
20	28 37	29 56	04 12
21	28 30	29 53	03 18
22	28 25	29 50	02 12
23	28 23	29 47	00 N 59
24	28 D 22	29 44	00 S 16
25	28 23	29 40	01 27
26	28 24	29 37	02 32
27	28 R 24	29 34	03 27
28	28 22	29 31	04 11
29	28 18	29 28	04 42
30	28 12	29 24	05 01
31	28 ♍ 04	29 ♍ 21	05 S 06

DECLINATIONS

Date	Sun ☉ ° '	Moon ☽ ° '	Mercury ☿ ° '	Venus ♀ ° '	Mars ♂ ° '	Jupiter ♃ ° '	Saturn ♄ ° '	Uranus ♅ ° '	Neptune ♆ ° '	Pluto ♇ ° '
01	23 N 07	15 N 16	22 N 27	17 N 13	08 N 48	22 N 38	13 N 47	15 S 13	21 S 15	10 N 20
02	23 03	17 06	22 03	16 50	08 34	22 36	13 45	15 13	21 15	10 19
03	22 58	18 10	21 37	16 27	08 20	22 35	13 42	15 13	21 15	10 18
04	22 53	18 25	21 10	16 03	08 06	22 33	13 40	15 12	21 15	10 18
05	22 48	17 52	20 43	15 39	07 52	22 32	13 38	15 12	21 15	10 17
06	22 42	16 32	20 14	15 15	07 38	22 30	13 36	15 12	21 15	10 16
07	22 36	14 30	19 44	14 50	07 24	22 29	13 34	15 12	21 15	10 16
08	22 29	11 51	19 14	14 25	07 10	22 27	13 32	15 11	21 15	10 15
09	22 22	08 42	18 43	13 59	06 56	22 25	13 29	15 11	21 15	10 14
10	22 15	05 09	18 12	13 34	06 41	22 24	13 27	15 11	21 14	10 14
11	22 07	01 N 20	17 40	13 08	06 27	22 22	13 25	15 11	21 14	10 13
12	21 59	02 S 37	17 08	12 41	06 12	22 20	13 23	15 11	21 14	10 12
13	21 50	06 33	16 36	12 15	05 58	22 19	13 20	15 11	21 14	10 11
14	21 42	10 17	16 04	11 48	05 43	22 17	13 18	15 11	21 14	10 11
15	21 32	13 35	15 31	11 21	05 29	22 15	13 16	15 11	21 14	10 10
16	21 23	16 13	14 59	10 53	05 14	22 13	13 13	15 11	21 14	10 09
17	21 13	17 53	14 27	10 26	04 59	22 11	13 11	15 11	21 14	10 08
18	21 02	18 22	13 55	09 58	04 44	22 10	13 09	15 11	21 14	10 08
19	20 52	17 33	13 23	09 30	04 29	22 08	13 06	15 10	21 14	10 07
20	20 41	15 27	12 52	09 01	04 15	22 06	13 04	15 10	21 14	10 06
21	20 29	12 19	12 21	08 33	04 00	22 04	13 01	15 11	21 13	10 05
22	20 18	08 25	11 50	08 04	03 45	22 02	12 59	15 11	21 13	10 04
23	20 05	04 S 07	11 21	07 36	03 30	22 00	12 57	15 11	21 13	10 04
24	19 53	00 N 16	10 52	07 07	03 15	21 58	12 54	15 11	21 13	10 03
25	19 40	04 30	10 24	06 38	02 59	21 56	12 52	15 11	21 13	10 02
26	19 27	08 23	09 57	06 08	02 44	21 54	12 49	15 11	21 13	10 01
27	19 14	11 45	09 31	05 39	02 29	21 52	12 47	15 11	21 13	10 00
28	19 00	14 30	09 06	05 10	02 14	21 50	12 44	15 11	21 13	09 59
29	18 46	16 33	08 43	04 40	01 59	21 48	12 42	15 11	21 13	09 58
30	18 32	17 50	08 21	04 10	01 43	21 46	12 39	15 11	21 13	09 57
31	18 N 17	18 N 20	08 N 00	03 N 40	01 N 28	21 N 44	12 N 36	15 S 12	21 S 13	09 N 57

ZODIAC SIGN ENTRIES

Date	h	m	Planets
01	19	37	☽ ♊
02	22	28	☿ ♌
04	07	33	☽ ♋
06	20	13	☽ ♌
09	08	44	☽ ♍
11	19	48	☽ ♎
12	02	14	♀ ♍
14	03	47	☽ ♏
16	07	50	☽ ♐
18	08	33	☽ ♑
20	07	41	☽ ♒
22	07	26	☽ ♓
23	05	00	☉ ♌
24	09	46	☽ ♈
26	12	02	♄ ♍
26	15	50	☽ ♉
27	06	10	☿ ♍
29	01	31	☽ ♊
31	13	28	☽ ♋

LATITUDES

Date	Mercury ☿ ° '	Venus ♀ ° '	Mars ♂ ° '	Jupiter ♃ ° '	Saturn ♄ ° '	Uranus ♅ ° '	Neptune ♆ ° '	Pluto ♇ ° '
01	01 N 50	01 N 48	00 N 58	00 N 08	01 N 27	00 N 23	01 N 29	17 N 09
04	01 39	01 44	00 55	00 08	01 27	00 23	01 29	17 08
07	01 22	01 39	00 53	00 08	01 27	00 23	01 29	17 06
10	01 01	01 33	00 50	00 09	01 26	00 23	01 29	17 04
13	00 36	01 26	00 48	00 09	01 26	00 23	01 29	17 03
16	00 N 07	01 18	00 45	00 09	01 26	00 23	01 29	17 01
19	00 S 25	01 10	00 43	00 10	01 26	00 23	01 29	17 00
22	01 00	01 00	00 41	00 10	01 26	00 22	01 29	16 58
25	01 37	00 49	00 38	00 10	01 26	00 22	01 28	16 56
28	02 15	00 38	00 36	00 10	01 26	00 22	01 28	16 55
31	02 S 53	00 N 26	00 N 34	00 N 11	01 N 26	00 N 22	01 N 28	16 N 53

DATA

Julian Date	2443691
Delta T	+49 seconds
Ayanamsa	23° 33' 25"
Synetic vernal point	05° ♓ 33' 35"
True obliquity of ecliptic	23° 26' 22"

LONGITUDES

Date	Chiron ⚷ ° '	Ceres ⚳ ° '	Pallas ⚴ ° '	Juno ⚵ ° '	Vesta ⚶ ° '	Black Moon Lilith ⚸ ° '
01	08 ♉ 56	18 ♑ 38	07 ♐ 08	06 ♒ 23	08 ♐ 39	18 ♋ 27
11	09 ♉ 16	16 ♑ 26	05 ♐ 38	04 ♒ 29	07 ♐ 45	19 ♋ 35
21	09 ♉ 31	14 ♑ 19	04 ♐ 52	02 ♒ 13	07 ♐ 40	20 ♋ 42
31	09 ♉ 41	12 ♑ 27	04 ♐ 49	29 ♑ 49	08 ♐ 21	21 ♋ 49

MOON'S PHASES, APSIDES AND POSITIONS ☽

Date	h	m	Phase	Longitude ° '	Eclipse Indicator
05	09	50	●	13 ♋ 01	
13	10	49	☽	20 ♎ 41	
20	03	05	○	27 ♑ 04	
26	22	31	☾	03 ♉ 34	

Day	h	m		
05	23	49	Apogee	
19	21	35	Perigee	
04	07	25	Max dec	18° N 26'
11	20	09	0S	
18	09	00	Max dec	18° S 23'
24	10	30	0N	
31	14	32	Max dec	18° N 20'

ASPECTARIAN

01 Saturday
00 04 ☿ ∥ ♃
08 04 ☽ ∠ ☉
08 34 ☿ ⊻ ♄
11 27 ☽ ⋕ ♅
14 15 ☽ □ ♄
15 10 ☽ ⚹ ☿
17 29 ☽ ⊑ ♇
21 24 ☽ ∠ ♃

02 Sunday
03 21 ☽ ⊥ ☉
08 32 ☽ ∥ ♀
09 10 ☽ Q ♀
15 36 ☽ ⊥ ♃
16 18 ☽ ⊻ ☉
16 21 ☽ □ ♂
17 44 ☉ ⚹ ♂
20 23 ☽ ⊼ ♅
23 17 ☽ △ ♇

03 Monday
01 56 ☽ ∠ ☿
02 18 ☽ Q ♄
03 49 ☽ ⊻ ♃
04 02 ☽ ☍ ♆
08 22 ☽ ± ♅
11 41 ☽ ⚹ ♀
14 36 ♃ ⊼ ♆
16 37 ☿ ⊑ ♆
22 57 ☽ ⊥ ☿

04 Tuesday
01 50 ☿ Q ♇
02 24 ☽ ⊑ ♅
02 35 ☽ ⚹ ♄
06 15 ☽ Q ♂
07 46 ☿ ⋕ ♆
12 59 ☽ ⊻ ☿
19 02 ☉ △ ♅
21 06 ☽ ∠ ♀
23 47 ☉ ∠ ♄

05 Wednesday
07 42 ☽ ⚹ ♂
08 38 ☽ △ ♅
09 07 ☽ ∠ ♄
09 50 ☽ ☌ ☉
11 41 ☽ □ ♇
16 20 ☽ ⊼ ♆
17 15 ☽ ⊥ ♀
17 24 ☽ ☌ ♃
19 03 ♀ ⊥ ♃

06 Thursday
02 14 ♂ ⚹ ♅
03 29 ☽ ⊥ ♄
04 28 ☽ ± ♆
06 43 ☽ ⊻ ♀
08 55 ☉ □ ♇
14 41 ♀ ⋕ ♅
15 38 ☽ ∠ ♂
15 45 ☽ ⊻ ♄
22 39 ☽ ⊑ ♆

07 Friday
00 09 ☽ Q ♇
04 36 ☽ ⋕ ♅
07 17 ☿ ⊥ ♂
07 46 ☽ ∥ ♀
10 48 ☽ ⊥ ♂
11 08 ☽ ☌ ☿
21 12 ☽ ∥ ♄
21 18 ☽ □ ♅
23 34 ☽ ⊻ ♂

08 Saturday
00 27 ☽ ⚹ ♇
03 52 ☽ ⊻ ☉
04 56 ☽ △ ♆
07 17 ☽ ⊻ ♃
16 56 ☉ ⊼ ♆
17 01 ☽ ⊥ ☉
17 37 ♂ ⊻ ♇
19 36 ☽ ⊥ ♃
21 29 ☉ ∥ ♃

09 Sunday
00 38 ☽ ∥ ♇
01 49 ☽ ☌ ♀
04 51 ☽ ☌ ♄
06 40 ☽ ∠ ♇
09 30 ☽ Q ♅
12 38 ☽ ∠ ☉
14 00 ☽ ∠ ♃

10 Monday
00 38 ☽ ⊥ ♇
01 08 ☽ ∥ ♂
07 57 ☽ ⊻ ☿
09 22 ☽ ⚹ ♅
10 38 ☉ ☌ ♃
12 25 ♀ ☌ ♄
12 32 ☽ ⊻ ♇
14 40 ☽ ☌ ♂
16 46 ☽ □ ♆
18 34 ♀ ∥ ♄
19 51 ☿ □ ♅
20 17 ☽ ⚹ ♃
20 55 ☽ ⚹ ☉
21 21 ☽ ⊥ ☿

11 Tuesday
04 45 ♀ ∠ ♇
14 42 ☽ ∠ ♅
16 34 ☽ ⊻ ♄
17 07 ☽ ∠ ☿
19 09 ☽ ⊻ ♀
20 04 ☽ Q ♃
22 10 ☽ Q ☉
23 38 ☿ ⚹ ♇

12 Wednesday
03 39 ☽ Q ♆
04 12 ☽ ⊥ ♄
07 47 ☽ ⊥ ♀
07 57 ☽ ⊥ ♅
08 35 ♂ □ ♆
09 36 ♀ Q ♅
19 17 ☽ ⊻ ♅
21 22 ☽ ∠ ♄
22 23 ☽ ☌ ♇

13 Thursday
01 03 ☽ ⚹ ☿
02 15 ☽ ⚹ ♆
02 24 ☽ ∠ ♀
03 08 ☽ ⊻ ♂
06 42 ☽ □ ♃
08 35 ☽ ⋕ ♂
10 49 ☽ □ ☉
12 30 ☿ △ ♆
14 45 ☽ ⊥ ♂

14 Friday
01 13 ☽ ⚹ ♄
01 34 ☽ Q ☿
05 40 ☽ ∠ ♆
07 49 ☽ ∠ ♂
08 24 ☽ ⚹ ♀
11 18 ☽ ⋕ ♇
12 36 ☿ ⊻ ♂
20 35 ☉ ± ♆
21 15 ☽ ⋕ ♀
21 37 ☽ ⊥ ♆
22 46 ☽ Q ♄

15 Saturday
01 39 ☽ ☌ ♅
04 35 ☽ ⊻ ♇
07 26 ☽ Q ♀
08 04 ☽ ⊻ ♆
09 26 ☽ ⋕ ♄
11 21 ☽ ⚹ ♂
12 33 ☽ □ ☿
13 09 ☽ △ ♃
14 12 ♀ ∠ ♃
14 44 ☉ ⊥ ♄
14 57 ☽ ⊥ ♇
19 58 ☽ △ ☉
20 38 ☿ ⊻ ♃

16 Sunday
01 33 ☽ ∥ ♅
01 53 ☽ ⋕ ☿
03 26 ☿ ⋕ ♅
05 52 ☽ □ ♄
06 13 ☽ ∠ ♇
08 35 ☽ Q ♂
14 50 ☽ ⊑ ♃
16 31 ☽ □ ♀
22 47 ☽ ⊑ ☉

17 Monday
04 10 ☽ ⊻ ♅
05 50 ☽ ± ♃
06 58 ☽ ⚹ ♇
09 25 ☉ ⋕ ♆
10 06 ☽ ☌ ♆
13 53 ☽ ⊥ ♅
14 25 ☽ ± ☉
15 20 ☽ □ ♂
15 42 ☽ ⊼ ♃
18 42 ☽ △ ☿

18 Tuesday
00 41 ☽ ⊼ ☉
02 17 ☽ Q ♇
03 19 ♀ ∥ ♇
04 18 ☽ ∠ ♅
05 36 ♂ ⚹ ♃
07 03 ☽ △ ♄
08 22 ♄ ∠ ♇
20 26 ☽ ⊑ ☿
20 41 ☽ △ ♀

19 Wednesday
03 59 ☽ ⚹ ♅
05 03 ♀ ⊥ ♇
06 44 ☽ □ ♇
06 53 ☽ ⊑ ♄
09 40 ☽ ⊻ ♆
11 39 ☽ ± ☿
15 51 ☽ ☍ ♃
16 43 ☽ △ ♂
19 03 ☽ ⊥ ♆
21 05 ☽ ± ♄
21 45 ☽ ⊼ ☿
22 05 ☽ ⊑ ♀
22 47 ☽ Q ♅

20 Thursday
01 49 ☿ ∥ ♄
03 05 ☽ ☍ ☉
06 33 ☽ ⊼ ♄
09 09 ☽ ∠ ♆
13 19 ☽ ± ♀
14 28 ☽ ∥ ♅
17 13 ☽ ⊑ ♂
23 32 ☽ ⊼ ♀

21 Friday
03 06 ☽ □ ♅
05 55 ☽ △ ♇
07 02 ☽ ⋕ ♄
08 02 ☽ ± ♂
08 47 ☽ ⚹ ♆
10 02 ♅ St D
11 43 ☽ ⋕ ☿
15 52 ☽ ⊼ ♃
18 01 ☽ ⊼ ♂

22 Saturday
00 36 ☽ ☍ ☿
01 39 ☽ ± ♃
02 10 ☽ ⋕ ♇
04 01 ☽ Q ♆
05 57 ☽ ⊼ ☉
06 00 ☽ ⊑ ♇
06 38 ☽ ☍ ♄
14 14 ☽ ⋕ ♀
16 25 ☽ ± ☉
16 32 ☽ ⊑ ♃
18 00 ☉ ⊻ ♄
20 44 ☽ ± ♇
22 43 ☿ ⊥ ♃
23 52 ♀ ⚹ ♅

23 Sunday
03 41 ☽ △ ♅
04 01 ☽ ☍ ♀
06 43 ☽ ⊼ ♇
08 24 ☽ ⊑ ☉
09 38 ☽ □ ♆
15 37 ☽ ⋕ ♂
17 58 ☽ △ ♃
21 40 ☽ ☍ ♂

24 Monday
02 35 ☉ ⊑ ♆
05 07 ☽ ⊑ ♅
05 52 ☽ ⊼ ☿
09 20 ☽ ⊼ ♄
11 54 ☽ △ ☉
15 24 ♀ ⊻ ♇
16 59 ☽ ± ☿
19 55 ☽ ± ♄
20 50 ☽ ± ♅

25 Tuesday
03 44 ☽ ∥ ♂
07 30 ☽ ⊼ ♅
09 57 ☽ ⊑ ☿
10 49 ☽ ☍ ♇
11 30 ☉ Q ♇
12 06 ☽ ⊑ ♄
12 33 ☽ ⊼ ♀
13 49 ☽ △ ♆
23 23 ☽ ∥ ♀
23 46 ☽ □ ♃

26 Wednesday
00 28 ☽ ± ♀
01 39 ☿ ∠ ♇
03 28 ♀ □ ♆
05 26 ☽ ⊼ ♂
08 00 ☿ ∥ ♇
15 03 ☽ △ ☿
15 53 ☽ △ ♄
17 07 ☽ ± ♂
17 22 ☽ ⊑ ♆
18 42 ☽ ⊑ ♀
21 27 ☽ ∥ ☿
22 31 ☽ □ ☉
23 07 ☽ ∥ ♇

27 Thursday
10 00 ☽ Q ♃
10 12 ☿ ☌ ♄
10 20 ☽ ± ♆
10 58 ☽ ⊑ ♂
15 10 ☽ ☍ ♅
18 41 ☿ Q ♅
18 46 ☽ ⊼ ♇
20 12 ☽ ∥ ♄
21 49 ☽ ⊼ ♆

28 Friday
02 01 ☽ △ ♀
04 10 ♃ ± ♆
06 23 ☽ ± ♇
09 32 ☽ ⚹ ♃
11 53 ☽ Q ☉
17 29 ☽ △ ♂
19 07 ☽ ⋕ ♅

29 Saturday
00 00 ☽ ⊑ ♇
02 07 ☽ □ ♄
03 44 ☽ □ ☿
09 26 ♄ Q ♅
13 31 ☽ ⚹ ☉
15 36 ☽ ∠ ♃

30 Sunday
02 00 ☽ ⊼ ♅
05 49 ☽ △ ♇
08 50 ☽ ☍ ♆
09 53 ☽ ⊥ ♃
14 01 ☽ ± ♅
14 17 ☽ Q ♄
17 04 ☽ Q ☿
19 16 ☽ □ ♀
22 05 ☽ ∠ ☉
22 09 ☽ ⊻ ♃

31 Monday
02 07 ♂ ∠ ♅
08 08 ☽ ⊑ ♅
08 28 ☽ □ ♂
09 14 ☽ ∥ ☉
14 42 ☽ ⚹ ♄
17 49 ☽ ⊥ ☉
18 03 ☽ ⚹ ☿

All ephemeris data is given at 12.00 UT and the Moon's longitude is additionally given for 24.00 UT

AUGUST 1978

LONGITUDES

Date	Sidereal time h m s	Sun ☉	Moon ☽	Moon ☽ 24.00
01	08 38 58	08 ♌ 52 53	11 ♋ 08 39	17 ♋ 04 15
02	08 42 54	09 50 19	22 ♋ 59 50	28 ♋ 55 39
03	08 46 51	10 47 46	04 ♌ 51 53	10 ♌ 48 47
04	08 50 48	11 45 14	16 ♌ 46 31	22 ♌ 45 18
05	08 54 44	12 42 43	28 ♌ 45 22	04 ♍ 46 56
06	08 58 41	13 40 13	10 ♍ 50 15	16 ♍ 55 36
07	09 02 37	14 37 43	23 ♍ 03 18	29 ♍ 13 41
08	09 06 34	15 35 15	05 ♎ 27 06	11 ♎ 43 57
09	09 10 30	16 32 47	18 ♎ 04 38	24 ♎ 29 33
10	09 14 27	17 30 21	00 ♏ 59 09	07 ♏ 33 47
11	09 18 23	18 27 55	14 ♏ 13 51	20 ♏ 59 39
12	09 22 20	19 25 30	27 ♏ 51 23	04 ♐ 49 13
13	09 26 17	20 23 06	11 ♐ 53 08	19 ♐ 02 59
14	09 30 13	21 20 43	26 ♐ 18 29	03 ♑ 39 07
15	09 34 10	22 18 21	11 ♑ 04 13	18 ♑ 32 57
16	09 38 06	23 15 59	26 ♑ 04 20	03 ♒ 37 15
17	09 42 03	24 13 39	11 ♒ 10 30	18 ♒ 42 53
18	09 45 59	25 11 20	26 ♒ 13 03	03 ♓ 40 22
19	09 49 56	26 09 03	11 ♓ 03 20	18 ♓ 21 16
20	09 53 52	27 06 47	25 ♓ 33 28	02 ♈ 39 26
21	09 57 49	28 04 32	09 ♈ 38 51	16 ♈ 31 33
22	10 01 46	29 ♌ 02 19	23 ♈ 17 32	29 ♈ 56 57
23	10 05 42	00 ♍ 00 08	06 ♉ 30 04	13 ♉ 57 13
24	10 09 39	00 57 58	19 ♉ 18 50	25 ♉ 35 26
25	10 13 35	01 55 50	01 ♊ 47 33	07 ♊ 55 38
26	10 17 32	02 53 44	14 ♊ 00 21	20 ♊ 02 14
27	10 21 28	03 51 40	26 ♊ 01 49	01 ♋ 59 39
28	10 25 25	04 49 38	07 ♋ 56 13	13 ♋ 52 02
29	10 29 21	05 47 37	19 ♋ 47 31	25 ♋ 43 07
30	10 33 18	06 45 38	01 ♌ 39 11	07 ♌ 36 06
31	10 37 15	07 ♍ 43 41	13 ♌ 34 09	19 ♌ 33 39

Date	Mercury ☿	Venus ♀	Mars ♂	Jupiter ♃	Saturn ♄	Uranus ♅	Neptune ♆	Pluto ♇
01	02 ♍ 30	22 ♍ 49	28 ♍ 13	22 ♋ 45	00 ♍ 43	12 ♏ 22	15 ♐ 44	14 ♎ 18
02	02 45	23 55	28 50	22 58	00 50	12 23	15 R 43	14 19
03	02 55	25 00	29 ♍ 27	23 11	00 58	12 23	15 42	14 21
04	03 00	26 05	00 ♎ 04	23 24	01 05	12 24	15 41	14 22
05	03 R 00	27 10	00 42	23 37	01 12	12 25	15 41	14 23
06	02 54	28 15	01 19	23 51	01 20	12 26	15 40	14 25
07	02 44	29 ♍ 19	01 56	24 04	01 27	12 27	15 39	14 26
08	02 28	00 ♎ 24	02 34	24 17	01 34	12 27	15 39	14 27
09	02 06	01 28	03 11	24 30	01 42	12 28	15 38	14 29
10	01 40	02 32	03 49	24 42	01 49	12 29	15 38	14 30
11	01 08	03 35	04 26	24 55	01 57	12 30	15 38	14 32
12	00 ♍ 32	04 39	05 04	25 08	02 04	12 32	15 37	14 33
13	29 ♌ 51	05 42	05 42	25 21	02 12	12 33	15 36	14 35
14	29 07	06 45	06 19	25 34	02 19	12 34	15 36	14 36
15	28 20	07 47	06 57	25 47	02 27	12 35	15 35	14 38
16	27 31	08 50	07 35	25 59	02 34	12 37	15 35	14 40
17	26 40	09 52	08 13	26 12	02 42	12 38	15 35	14 41
18	25 48	10 54	08 51	26 24	02 49	12 39	15 34	14 43
19	24 58	11 55	09 29	26 37	02 57	12 41	15 34	14 45
20	24 08	12 56	10 07	26 49	03 05	12 42	15 34	14 46
21	23 21	13 57	10 46	27 02	03 12	12 44	15 33	14 48
22	22 38	14 58	11 24	27 14	03 20	12 46	15 33	14 50
23	22 00	15 58	12 02	27 27	03 27	12 47	15 33	14 52
24	21 26	16 58	12 41	27 39	03 35	12 49	15 33	14 54
25	20 59	17 58	13 19	27 51	03 42	12 51	15 33	14 55
26	20 39	18 57	13 58	28 03	03 50	12 52	15 33	14 57
27	20 26	19 56	14 37	28 15	03 58	12 54	15 33	14 59
28	20 21	20 54	15 15	28 28	04 05	12 56	15 D 33	15 01
29	20 D 23	21 52	15 54	28 40	04 13	12 58	15 33	15 03
30	20 35	22 50	16 33	28 51	04 20	13 00	15 33	15 05
31	20 ♌ 54	23 ♎ 47	17 ♎ 12	29 ♋ 03	04 ♍ 28	13 ♏ 02	15 ♐ 33	15 ♎ 07

DECLINATIONS

Date	Moon True ☊	Moon Mean ☊	Moon ☽ Latitude	Sun ☉	Moon ☽	Mercury ☿	Venus ♀	Mars ♂	Jupiter ♃	Saturn ♄	Uranus ♅	Neptune ♆	Pluto ♇
01	27 ♍ 55	29 ♍ 18	04 S 59	18 N 02	18 N 01	07 N 42	03 N 11	01 N 13	21 N 42	12 N 34	15 S 12	21 S 13	09 N 56
02	27 R 46	29 15	04 38	17 47	16 55	07 25	02 41	00 57	21 40	12 31	15 12	21 13	09 55
03	27 37	29 12	04 05	17 32	15 05	07 10	02 11	00 42	21 38	12 29	15 12	21 13	09 54
04	27 30	29 09	03 21	17 16	12 36	06 58	01 41	00 26	21 36	12 26	15 12	21 13	09 53
05	27 24	29 05	02 28	17 00	09 35	06 47	01 11	00 N 11	21 33	12 24	15 13	21 13	09 52
06	27 21	29 02	01 28	16 43	06 09	06 39	00 41	00 S 05	21 31	12 21	15 13	21 13	09 51
07	27 20	28 59	00 S 23	16 27	02 N 24	06 34	00 N 11	00 20	21 29	12 18	15 13	21 13	09 50
08	27 D 20	28 56	00 N 44	16 10	01 S 30	06 32	00 S 19	00 36	21 27	12 16	15 14	21 13	09 49
09	27 21	28 53	01 50	15 53	05 23	06 32	00 49	00 51	21 25	12 13	15 14	21 13	09 48
10	27 22	28 50	02 53	15 35	09 07	06 35	01 19	01 07	21 23	12 10	15 14	21 13	09 47
11	27 23	28 46	03 48	15 18	12 29	06 41	01 49	01 22	21 20	12 08	15 15	21 13	09 46
12	27 R 23	28 43	04 31	15 00	15 17	06 50	02 19	01 38	21 18	12 05	15 15	21 13	09 45
13	27 21	28 40	05 00	14 42	17 15	07 01	02 49	01 54	21 16	12 02	15 15	21 13	09 44
14	27 18	28 37	05 11	14 23	18 12	07 16	03 19	02 09	21 14	12 00	15 16	21 13	09 43
15	27 14	28 34	05 02	14 05	17 57	07 33	03 49	02 25	21 11	11 57	15 16	21 13	09 42
16	27 09	28 30	04 33	13 46	16 28	07 52	04 18	02 41	21 09	11 54	15 17	21 13	09 41
17	27 04	28 27	03 44	13 27	13 50	08 14	04 48	02 56	21 07	11 52	15 17	21 13	09 40
18	27 01	28 24	02 40	13 08	10 16	08 37	05 17	03 12	21 05	11 49	15 18	21 13	09 39
19	26 58	28 21	01 26	12 48	06 06	09 02	05 47	03 28	21 02	11 46	15 18	21 13	09 38
20	26 58	28 18	00 N 08	12 28	01 S 39	09 28	06 16	03 43	21 00	11 44	15 18	21 13	09 37
21	26 D 58	28 15	01 S 09	12 09	02 N 46	09 54	06 45	03 59	20 58	11 41	15 19	21 13	09 36
22	26 59	28 11	02 19	11 49	06 53	10 21	07 14	04 15	20 56	11 38	15 19	21 13	09 35
23	27 01	28 08	03 20	11 28	10 32	10 47	07 43	04 30	20 53	11 36	15 20	21 13	09 34
24	27 02	28 05	04 09	11 08	13 34	11 13	08 11	04 46	20 51	11 33	15 21	21 13	09 33
25	27 03	28 02	04 45	10 47	15 52	11 37	08 40	05 02	20 49	11 30	15 21	21 13	09 32
26	27 R 02	27 59	05 06	10 26	17 25	12 00	09 08	05 17	20 46	11 27	15 22	21 13	09 31
27	27 01	27 56	05 14	10 06	18 09	12 22	09 36	05 33	20 44	11 25	15 22	21 13	09 30
28	26 58	27 52	05 09	09 44	18 04	12 41	10 04	05 48	20 42	11 22	15 23	21 13	09 29
29	26 55	27 49	04 50	09 23	17 12	12 58	10 32	06 04	20 39	11 19	15 23	21 13	09 28
30	26 52	27 46	04 19	09 02	15 35	13 10	10 59	06 20	20 37	11 17	15 24	21 13	09 27
31	26 ♍ 48	27 ♍ 43	03 S 36	08 N 40	13 N 18	13 N 23	11 S 27	06 S 35	20 N 35	11 N 14	15 S 25	21 S 13	09 N 26

ZODIAC SIGN ENTRIES

Date	h	m	Planets
03	02	10	☽ ♌
04	09	07	♂ ♎
05	14	29	☽ ♍
08	01	30	☽ ♎
08	03	08	♀ ♎
10	10	11	☽ ♏
12	15	43	☽ ♐
13	07	05	☿ ♌
14	18	03	☽ ♑
16	18	15	☽ ♒
18	18	04	☽ ♓
20	19	29	☽ ♈
23	00	06	☽ ♉
23	11	57	☉ ♍
25	08	31	☽ ♊
27	19	59	☽ ♋
30	08	40	☽ ♌

LATITUDES

Date	Mercury ☿	Venus ♀	Mars ♂	Jupiter ♃	Saturn ♄	Uranus ♅	Neptune ♆	Pluto ♇
01	03 S 06	00 N 22	00 N 33	00 N 11	01 N 26	00 N 22	01 N 28	16 N 53
04	03 41	00 N 08	00 31	00 11	01 26	00 22	01 28	16 51
07	04 13	00 S 06	00 28	00 11	01 26	00 22	01 28	16 50
10	04 36	00 21	00 26	00 12	01 27	00 22	01 28	16 49
13	04 48	00 36	00 24	00 12	01 27	00 22	01 28	16 47
16	04 45	00 53	00 22	00 12	01 27	00 22	01 28	16 46
19	04 25	01 10	00 20	00 13	01 27	00 22	01 27	16 45
22	03 50	01 27	00 17	00 13	01 27	00 22	01 27	16 44
25	03 02	01 45	00 15	00 13	01 27	00 21	01 27	16 43
28	02 08	02 04	00 13	00 14	01 27	00 21	01 27	16 42
31	01 S 13	02 S 23	00 N 11	00 N 14	01 N 28	00 N 21	01 N 27	16 N 41

DATA

Julian Date	2443722
Delta T	+49 seconds
Ayanamsa	23° 33' 29"
Synetic vernal point	05° ♓ 33' 30"
True obliquity of ecliptic	23° 26' 22"

MOON'S PHASES, APSIDES AND POSITIONS ☽

Date	h	m	Phase	Longitude	Eclipse Indicator
04	01	01	●	11 ♌ 19	
11	20	06	☽	18 ♏ 47	
18	10	14	○	25 ♒ 07	
25	12	18	☾	01 ♊ 57	

Day	h	m		
02	03	13	Apogee	
17	06	21	Perigee	
29	13	28	Apogee	
08	02	52	0S	
14	19	15	Max dec	18° S 15'
20	20	52	0N	
27	21	41	Max dec	18° N 13'

LONGITUDES

Date	Chiron ⚷	Ceres ⚳	Pallas ⚴	Juno ⚵	Vesta ⚶	Black Moon Lilith ⚸
01	09 ♉ 42	12 ♑ 18	04 ♐ 51	29 ♑ 35	08 ♐ 28	21 ♋ 56
11	09 ♉ 46	10 ♑ 56	04 ♐ 32	27 ♑ 20	09 ♐ 57	23 ♋ 03
21	09 ♉ 45	10 ♑ 05	06 ♐ 48	25 ♑ 27	12 ♐ 04	24 ♋ 10
31	09 ♉ 38	09 ♑ 57	08 ♐ 32	24 ♑ 08	14 ♐ 42	25 ♋ 17

ASPECTARIAN

h m Aspects

01 Tuesday
07 01 ☽ ⊻ ☉
09 55 ♀ ⚹ ♃
10 40 ☽ ∥ ☉
11 17 ☽ Q ♀
14 29 ☽ △ ♅
18 24 ☽ □ ♇
21 16 ☽ ⊼ ♆
21 21 ☽ ∠ ♄
22 51 ☽ Q ♂

02 Wednesday
01 10 ☽ ∠ ☿
09 25 ☽ ± ♆
11 56 ☽ ☌ ♃
14 03 ☽ ⚹ ♀
15 46 ☽ ⊥ ♄
19 44 ☽ ⊥ ☿

03 Thursday
00 28 ☽ ⚹ ♂
03 36 ☽ bQ ♆
04 02 ☽ ⊻ ♄
06 54 ☽ Q ♇
08 01 ☽ ⊻ ☿
10 40 ☽ # ♅
23 24 ☽ ∠ ♀

04 Friday
01 01 ☽ ☌ ☉
03 11 ☽ □ ♅
07 09 ☽ ⚹ ♇
08 24 ☽ ∠ ♂
09 49 ☽ △ ♆
13 30 ☽ ∥ ♄
19 19 ☽ ⊥ ♀
23 07 ☿ St R

05 Saturday
01 33 ☽ ⊻ ♃
03 26 ☽ ⊥ ♂
04 26 ☉ □ ♅
08 31 ☽ ⊻ ♀
09 56 ☽ ∥ ♇
13 16 ☽ ∠ ♇
13 46 ☽ ⊥ ♃
15 19 ☽ Q ♅
16 04 ☽ ⊻ ♂
16 56 ☽ ☌ ♄
17 30 ♀ ∠ ♅
20 25 ☽ ☌ ☿

06 Sunday
07 11 ☽ ⊥ ♇
07 59 ☽ ∠ ♃
08 27 ☽ ∥ ☿
12 34 ♂ ⊻ ♄
15 09 ☽ ⚹ ♅
18 04 ☽ ⊻ ☉
19 04 ☽ ⊻ ♇
21 31 ☽ □ ♆

07 Monday
06 51 ☽ ⊥ ☉
06 58 ☉ ⚹ ♇
07 03 ♀ # ♂
14 00 ☽ ⚹ ♃
20 33 ☽ ∠ ♅

08 Tuesday
00 00 ☽ # ♂
01 20 ☽ ☌ ♀
01 51 ☽ ∠ ☉
02 09 ☽ # ♀
03 47 ☽ ∥ ♀
04 27 ☽ ⊻ ♄
06 07 ☽ ∥ ♂
06 09 ☽ ☌ ♂
06 23 ☽ ⊻ ☿
08 32 ☽ Q ♆
09 25 ☿ ⊻ ♂
13 28 ☉ △ ♆
13 37 ☽ Q ♃
13 56 ☽ ⊥ ♅
16 06 ☽ ⊥ ♄
17 37 ☽ ⊥ ☿

09 Wednesday
01 24 ☽ ⊻ ♅
05 12 ☽ ☌ ♇
07 24 ☽ ⚹ ♆
08 53 ☽ ⚹ ☉
09 22 ☽ ∠ ♄
10 13 ☽ ∠ ☿
15 12 ♀ ∥ ♂
17 57 ♀ ⊻ ♄
19 15 ☽ # ☿
22 20 ☿ ⊻ ♀

10 Thursday
00 12 ☽ □ ♃
05 09 ♂ Q ♆
05 31 ☿ ☌ ♄
06 57 ☉ ∠ ♀
09 04 ☽ Q ☉
11 21 ☽ ∠ ♆
13 11 ☽ ⚹ ☿
13 33 ☽ ⚹ ♄
15 05 ☽ ⊻ ♀
16 30 ☽ # ♇
17 26 ☽ ⊻ ♂

11 Friday
02 56 ☽ ⊥ ♀
03 43 ☽ ⊥ ♆
04 51 ☽ ⊥ ♂
08 55 ☽ ☌ ♅
09 20 ☽ # ♄
10 07 ☽ Q ☿
11 29 ☽ Q ♄
12 32 ☽ ⊻ ♇
12 41 ♀ Q ♆
14 28 ☽ ⊻ ♆
15 57 ☉ # ♅
18 24 ☿ ⊥ ♃
20 06 ☽ □ ☉
20 24 ☽ ∠ ♀
21 42 ☽ ∠ ♂
23 12 ☽ ⊥ ♇

12 Saturday
07 11 ☽ △ ♃
09 34 ☽ # ☉
11 45 ☽ ∥ ♅
12 08 ☿ Q ♅
14 57 ☽ ∠ ♇
16 25 ☽ □ ☿
19 21 ☽ □ ♄

13 Sunday
00 39 ☽ ⚹ ♀
01 00 ☽ ⚹ ♂
09 22 ☽ bQ ♃
11 42 ♀ ☌ ♂
13 07 ☽ ⊻ ♅
14 11 ☿ ⊥ ♀
14 58 ☿ ⊥ ♂
16 33 ☽ ⚹ ♇
18 15 ☽ ☌ ♆
20 53 ☿ ∠ ♇
22 11 ☽ Q ♂
22 31 ☽ Q ♀
23 11 ☽ ⊥ ♅

14 Monday
00 41 ☽ ± ♃
03 14 ☽ △ ☉
07 49 ♀ ⊥ ♅
10 18 ☉ ∠ ♂
10 45 ☽ ⊼ ♃
12 30 ☽ Q ♇
14 04 ☽ ∠ ♅
16 23 ☽ △ ☿
21 38 ♂ ⊥ ♅
21 55 ☽ △ ♄
23 38 ♃ # ♆

15 Tuesday
05 03 ☽ □ ♂
05 30 ☽ bQ ☉
06 18 ☽ □ ♀
11 37 ♀ Q ♃
14 27 ☽ ⚹ ♅
15 27 ☽ bQ ☿
17 44 ☽ □ ♇
19 15 ☽ ⊻ ♆
20 59 ☽ ± ☉
22 19 ☽ bQ ♄

16 Wednesday
04 51 ☽ ⊥ ♆
05 07 ☽ ± ☿
05 14 ♀ ⊥ ♄
07 14 ☽ ⊼ ☉
09 40 ☽ Q ♅
11 52 ☽ ☍ ♃
12 48 ☽ ± ♄
14 10 ☽ ⊼ ☿
19 10 ☽ ∠ ♆
22 25 ☽ ⊼ ♄
23 58 ☽ ∥ ♅

17 Thursday
07 06 ☽ △ ♂
09 46 ☽ △ ♀
10 50 ♂ Q ♃
14 19 ☽ □ ♅
15 09 ☽ # ☉
17 36 ☽ △ ♇
19 00 ☽ ⚹ ♆
22 29 ☿ ⊻ ♃

18 Friday
02 02 ☽ # ♄
08 02 ☽ bQ ♂
10 14 ☽ ☍ ☉
10 36 ♂ ⊥ ♄
10 54 ☿ ∠ ♀
11 22 ☽ ☍ ☿
11 26 ☽ bQ ♀
12 18 ☽ ⊼ ♃
14 10 ☽ Q ♆
15 42 ☽ # ♇
17 38 ☽ bQ ♇
20 11 ☉ ☌ ☿
20 55 ☽ # ☿
22 06 ☽ ± ♃
22 43 ☽ ☍ ♄
23 09 ☽ ± ♂

19 Saturday
03 01 ☽ ± ♀
08 14 ☽ ± ♇
09 20 ☽ ⊼ ♂
12 56 ☽ bQ ♃
13 31 ☽ ⊼ ♀
13 33 ☽ ∥ ♀
14 40 ☽ △ ♅
18 04 ☽ ⊼ ♇
19 24 ☽ □ ♆
19 42 ☿ ∠ ♂

20 Sunday
01 30 ☽ ∥ ♂
02 50 ☉ ⊻ ♃
06 23 ♀ ⊻ ♅
09 45 ☽ ⊼ ☿
14 10 ☽ △ ♃
14 48 ☽ ⊼ ☉
15 37 ☽ bQ ♅
19 19 ☽ ± ☿
20 28 ☿ ∥ ♇

21 Monday
00 50 ☽ ⊼ ♄
01 43 ☽ ± ☉
06 58 ☽ ± ♅
09 53 ☽ bQ ☿
11 13 ☽ ± ♄
14 01 ☽ ☍ ♂
17 22 ☽ ⊼ ♅
18 24 ☽ bQ ☉
19 19 ☽ # ♂
20 05 ☽ ☍ ♀
21 00 ☽ ☍ ♇
22 18 ☽ △ ♆

22 Tuesday
03 06 ☽ bQ ♄
08 49 ♀ ☌ ♇
10 53 ☽ △ ☿
14 26 ☽ # ♀
19 12 ☽ □ ♃
23 10 ☽ △ ☉

23 Wednesday
01 06 ☽ bQ ♆
01 59 ☉ ∥ ♄
02 04 ♀ ⚹ ♆
05 19 ☽ ∥ ♇
06 21 ☽ △ ♄
08 26 ☉ ∠ ♇
14 04 ☽ ∥ ☿
15 07 ♃ ⊥ ♄
17 39 ☽ ± ♆
18 14 ☽ ∥ ☉
19 43 ☽ ∥ ♄
22 50 ☽ ⊼ ♂
23 43 ☽ ☍ ♅

24 Thursday
03 37 ☽ ⊼ ♇
04 53 ☽ ⊼ ♆
04 57 ☽ Q ♃
07 10 ☽ ⊼ ♀
08 07 ☉ Q ♅
09 27 ☉ ∥ ☿
10 44 ☽ ± ♂
15 01 ☽ ± ♇
15 53 ☽ □ ☿
17 12 ♂ ⊻ ♅
19 34 ☽ ± ♀

25 Friday
04 14 ☽ ⚹ ♃
04 54 ☽ bQ ♂
05 37 ☿ ∥ ♄
05 49 ☽ # ♅
08 22 ☽ bQ ♇
12 18 ☽ □ ☉
14 28 ☽ bQ ♀
15 46 ☽ □ ♄

26 Saturday
01 40 ☽ Q ☿
08 53 ♀ ∠ ♄
09 45 ☽ ⊼ ♅
10 05 ☽ ∠ ♃
11 55 ☽ △ ♂
13 53 ☽ △ ♇
15 03 ☽ ☍ ♆
21 42 ☽ ± ♅
22 42 ☽ △ ♀

27 Sunday
00 57 ☽ ⚹ ☿
02 55 ☽ Q ☉
03 46 ☽ Q ♄
04 19 ☽ ⊥ ♃
06 55 ♀ # ♇
14 53 ☉ ☌ ♄
15 47 ☽ bQ ♅
16 33 ☽ ⊻ ♃
22 59 ☿ ⚹ ♀

28 Monday
00 28 ☉ ⊥ ♃
00 57 ♆ St D
02 20 ☉ # ♀
02 39 ♂ ☌ ♇
04 08 ☽ ⚹ ♄
05 10 ☽ ⚹ ☉
06 46 ☽ ∠ ☿
15 41 ☿ St D
22 09 ☽ △ ♅
22 41 ♂ ⚹ ♆

29 Tuesday
00 59 ☽ ⊥ ☿
02 22 ☽ □ ♇
03 24 ☽ ⊼ ♆
03 40 ☽ □ ♂
06 13 ☉ ∥ ♇
10 49 ☽ ∠ ♄
13 14 ☽ ⊻ ☿
14 13 ☽ ∠ ☉
15 33 ☽ ± ♆
16 35 ☽ □ ♀

30 Wednesday
05 14 ☽ ⊥ ♄
06 15 ☽ ☌ ♃
09 46 ☽ bQ ♆
10 02 ☽ ⊥ ☉
14 13 ☽ # ♅
14 54 ☽ Q ♇
17 29 ☽ ⊻ ♄
18 12 ☽ Q ♂
23 13 ☽ ⊻ ☉

31 Thursday
01 51 ♀ # ♄
08 07 ☽ Q ♀
10 56 ☽ □ ♅
11 17 ☽ ∥ ☿
15 07 ☽ ⚹ ♇
15 58 ☽ △ ♆
19 42 ☽ ⚹ ♂

All ephemeris data is given at 12.00 UT and the Moon's longitude is additionally given for 24.00 UT

SEPTEMBER 1978

LONGITUDES

Date	Sidereal time h m s	Sun ☉ ° ' "	Moon ☽ ° ' "	Moon ☽ 24.00 ° ' "	Mercury ☿ ° '	Venus ♀ ° '	Mars ♂ ° '	Jupiter ♃ ° '	Saturn ♄ ° '	Uranus ♅ ° '	Neptune ♆ ° '	Pluto ♇ ° '
01	10 41 11	08 ♍ 41 45	25 ♌ 34 49	01 ♍ 37 55	21 ♌ 22	24 ♎ 44	17 ♎ 51	29 ♋ 15	04 ♍ 36	13 ♏ 04	15 ♐ 33	15 ♎ 09
02	10 45 08	09 39 51	07 ♍ 43 07	13 ♍ 50 38	21 57	25 41	18 30	29 27	04 43	13 06	15 33	15 11
03	10 49 04	10 37 59	20 ♍ 00 38	26 ♍ 13 17	22 41	26 37	19 09	29 39	04 51	13 08	15 33	15 13
04	10 53 01	11 36 09	02 ♎ 28 44	08 ♎ 47 09	23 33	27 32	19 49	29 ♋ 50	04 58	13 11	15 34	15 15
05	10 56 57	12 34 20	15 ♎ 08 41	21 ♎ 33 30	24 32	28 27	20 28	00 ♌ 02	05 06	13 13	15 34	15 17
06	11 00 54	13 32 32	28 ♎ 01 45	04 ♏ 33 37	25 38	29 ♎ 21	21 07	00 14	05 13	13 15	15 34	15 19
07	11 04 50	14 30 46	11 ♏ 09 14	17 ♏ 48 46	26 50	00 ♏ 15	21 47	00 24	05 21	13 18	15 35	15 21
08	11 08 47	15 29 02	24 ♏ 32 20	01 ♐ 20 03	28 09	01 09	22 26	00 36	05 29	13 20	15 35	15 23
09	11 12 44	16 27 19	08 ♐ 11 58	15 ♐ 08 07	29 ♌ 33	02 02	23 06	00 47	05 36	13 22	15 35	15 25
10	11 16 40	17 25 38	22 ♐ 08 26	29 ♐ 12 49	01 ♍ 02	02 54	23 45	00 58	05 44	13 25	15 36	15 28
11	11 20 37	18 23 59	06 ♑ 21 01	13 ♑ 32 46	02 36	03 45	24 25	01 09	05 51	13 27	15 36	15 30
12	11 24 33	19 22 20	20 ♑ 47 38	28 ♑ 05 06	04 13	04 36	25 05	01 20	05 59	13 30	15 37	15 32
13	11 28 30	20 20 44	05 ♒ 24 33	12 ♒ 45 18	05 54	05 27	25 45	01 31	06 06	13 32	15 37	15 34
14	11 32 26	21 19 09	20 ♒ 06 34	27 ♒ 27 33	07 37	06 16	26 25	01 42	06 14	13 35	15 38	15 36
15	11 36 23	22 17 35	04 ♓ 47 23	12 ♓ 05 14	09 23	07 05	27 05	01 53	06 21	13 38	15 38	15 39
16	11 40 19	23 16 04	19 ♓ 20 18	26 ♓ 31 50	11 11	07 53	27 45	02 03	06 28	13 40	15 39	15 41
17	11 44 16	24 14 34	03 ♈ 39 10	10 ♈ 41 44	13 00	08 40	28 25	02 14	06 36	13 43	15 40	15 43
18	11 48 13	25 13 06	17 ♈ 39 05	24 ♈ 30 53	14 50	09 27	29 05	02 24	06 43	13 46	15 40	15 45
19	11 52 09	26 11 40	01 ♉ 16 57	07 ♉ 57 11	16 41	10 12	29 ♎ 45	02 35	06 51	13 49	15 41	15 47
20	11 56 06	27 10 16	14 ♉ 31 38	21 ♉ 00 28	18 33	10 57	00 ♏ 25	02 45	06 58	13 51	15 42	15 50
21	12 00 02	28 08 55	27 ♉ 23 54	03 ♊ 42 17	20 24	11 41	01 06	02 55	07 05	13 54	15 42	15 52
22	12 03 59	29 ♍ 07 35	09 ♊ 56 01	16 ♊ 05 33	22 16	12 24	01 46	03 05	07 12	13 57	15 43	15 54
23	12 07 55	00 ♎ 06 18	22 ♊ 11 25	28 ♊ 14 08	24 07	13 05	02 27	03 15	07 20	14 00	15 44	15 57
24	12 11 52	01 05 03	04 ♋ 14 16	10 ♋ 12 22	25 58	13 46	03 07	03 25	07 27	14 03	15 45	15 59
25	12 15 48	02 03 51	16 ♋ 09 08	22 ♋ 05 02	27 49	14 26	03 48	03 35	07 34	14 06	15 46	16 01
26	12 19 45	03 02 40	28 ♋ 00 40	03 ♌ 56 35	29 ♍ 39	15 05	04 28	03 44	07 41	14 09	15 47	16 04
27	12 23 42	04 01 32	09 ♌ 53 18	15 ♌ 51 21	01 ♎ 28	15 42	05 09	03 54	07 48	14 12	15 48	16 06
28	12 27 38	05 00 26	21 ♌ 51 10	27 ♌ 53 11	03 17	16 19	05 50	04 03	07 55	14 15	15 49	16 08
29	12 31 35	05 59 22	03 ♍ 57 47	10 ♍ 05 18	05 05	16 54	06 31	04 13	08 02	14 18	15 50	16 10
30	12 35 31	06 ♎ 58 21	16 ♍ 16 01	22 ♍ 30 10	06 ♎ 52	17 ♏ 28	07 ♏ 12	04 ♌ 22	08 ♍ 09	14 ♏ 21	15 ♐ 51	16 ♎ 13

Date	Moon True ☊ ° '	Moon Mean ☊ ° '	Moon ☽ Latitude ° '
01	26 ♍ 46	27 ♍ 40	02 S 44
02	26 R 44	27 36	01 43
03	26 43	27 33	00 S 37
04	26 D 43	27 30	00 N 32
05	26 44	27 27	01 40
06	26 45	27 24	02 45
07	26 46	27 21	03 42
08	26 47	27 17	04 28
09	26 47	27 14	05 01
10	26 R 47	27 11	05 16
11	26 47	27 08	05 13
12	26 46	27 05	04 50
13	26 46	27 01	04 08
14	26 45	26 58	03 10
15	26 45	26 55	01 59
16	26 44	26 52	00 N 41
17	26 D 44	26 49	00 S 38
18	26 44	26 46	01 54
19	26 45	26 42	03 00
20	26 R 45	26 39	03 55
21	26 44	26 36	04 37
22	26 44	26 33	05 04
23	26 44	26 30	05 16
24	26 D 44	26 27	05 15
25	26 44	26 23	04 59
26	26 45	26 20	04 31
27	26 45	26 17	03 52
28	26 46	26 14	03 02
29	26 47	26 11	02 03
30	26 ♍ 48	26 ♍ 07	00 S 58

DECLINATIONS

Date	Sun ☉ ° '	Moon ☽ ° '	Mercury ☿ ° '	Venus ♀ ° '	Mars ♂ ° '	Jupiter ♃ ° '	Saturn ♄ ° '	Uranus ♅ ° '	Neptune ♆ ° '	Pluto ♇ ° '
01	08 N 19	10 N 26	13 N 31	11 S 54	06 S 51	20 N 32	11 N 11	15 S 25	21 S 13	09 N 25
02	07 57	07 57	13 36	12 20	07 06	20 30	11 08	15 26	21 13	09 24
03	07 35	03 N 23	13 37	12 47	07 22	20 28	11 06	15 27	21 13	09 23
04	07 13	00 S 30	13 35	13 13	07 37	20 25	11 03	15 27	21 14	09 22
05	06 51	04 25	13 30	13 39	07 53	20 23	11 00	15 28	21 14	09 21
06	06 28	08 12	13 21	14 05	08 08	20 21	10 58	15 29	21 14	09 20
07	06 06	11 39	13 08	14 30	08 23	20 18	10 55	15 30	21 14	09 19
08	05 43	14 34	12 53	14 55	08 39	20 16	10 52	15 30	21 14	09 18
09	05 21	16 44	12 33	15 20	08 54	20 14	10 50	15 31	21 14	09 17
10	04 58	17 57	12 11	15 44	09 09	20 12	10 47	15 32	21 14	09 16
11	04 35	18 05	11 46	16 08	09 25	20 09	10 44	15 33	21 14	09 15
12	04 12	17 03	11 17	16 32	09 40	20 07	10 42	15 33	21 14	09 14
13	03 50	14 54	10 46	16 55	09 55	20 05	10 39	15 34	21 14	09 13
14	03 27	11 47	10 13	17 18	10 10	20 02	10 36	15 35	21 15	09 12
15	03 03	07 55	09 37	17 41	10 25	20 00	10 34	15 36	21 15	09 11
16	02 40	03 S 36	08 59	18 03	10 40	19 58	10 31	15 37	21 15	09 10
17	02 17	00 N 52	08 20	18 25	10 55	19 56	10 28	15 37	21 15	09 09
18	01 54	05 11	07 39	18 46	11 10	19 53	10 26	15 38	21 15	09 08
19	01 31	09 06	06 57	19 07	11 25	19 51	10 23	15 39	21 15	09 07
20	01 07	12 27	06 13	19 27	11 40	19 49	10 20	15 40	21 15	09 06
21	00 44	15 05	05 29	19 47	11 54	19 47	10 18	15 41	21 15	09 05
22	00 N 21	16 56	04 43	20 07	12 09	19 45	10 15	15 42	21 16	09 04
23	00 S 03	17 57	03 58	20 26	12 23	19 42	10 13	15 43	21 16	09 03
24	00 26	18 08	03 11	20 45	12 38	19 40	10 10	15 43	21 16	09 02
25	00 49	17 30	02 25	21 03	12 52	19 38	10 07	15 44	21 16	09 01
26	01 13	16 07	01 38	21 20	13 07	19 36	10 05	15 45	21 16	09 00
27	01 36	14 03	00 51	21 37	13 21	19 34	10 02	15 46	21 16	08 59
28	01 59	11 21	00 N 03	21 54	13 35	19 32	10 00	15 47	21 16	08 58
29	02 23	08 08	00 S 44	22 10	13 49	19 30	09 57	15 48	21 17	08 57
30	02 S 46	04 N 32	01 S 31	22 S 25	14 S 03	19 N 28	09 N 55	15 S 49	21 S 17	08 N 56

ZODIAC SIGN ENTRIES

Date	h	m	Planets
01	20	46	☽ ♍
04	07	15	☽ ♎
05	08	31	♃ ♌
06	15	38	☽ ♏
07	05	07	♀ ♏
08	21	39	☽ ♐
09	19	23	☿ ♍
11	01	20	☽ ♑
13	03	09	☽ ♒
15	04	09	☽ ♓
17	05	50	☽ ♈
19	09	43	☽ ♉
19	20	57	♂ ♏
21	16	36	☽ ♊
23	09	25	☉ ♎
24	03	31	☽ ♋
26	16	02	☽ ♌
26	16	40	☿ ♎
29	04	11	☽ ♍

LATITUDES

Date	Mercury ☿ ° '	Venus ♀ ° '	Mars ♂ ° '	Jupiter ♃ ° '	Saturn ♄ ° '	Uranus ♅ ° '	Neptune ♆ ° '	Pluto ♇ ° '
01	00 S 55	02 S 29	00 N 10	00 N 14	01 N 28	00 N 21	01 N 27	16 N 40
04	00 S 05	02 49	00 08	00 15	01 28	00 21	01 26	16 39
07	00 N 37	03 08	00 06	00 15	01 28	00 21	01 26	16 39
10	01 09	03 28	00 04	00 15	01 28	00 21	01 26	16 38
13	01 32	03 48	00 N 02	00 16	01 29	00 21	01 26	16 37
16	01 45	04 08	00 00	00 16	01 29	00 21	01 26	16 37
19	01 50	04 28	00 S 02	00 17	01 29	00 21	01 26	16 36
22	01 48	04 47	00 04	00 17	01 30	00 21	01 25	16 36
25	01 41	05 07	00 06	00 17	01 30	00 21	01 25	16 35
28	01 29	05 25	00 08	00 18	01 31	00 21	01 25	16 35
31	01 N 14	05 S 42	00 S 10	00 N 18	01 N 31	00 N 21	01 N 25	16 N 35

DATA

Julian Date	2443753
Delta T	+49 seconds
Ayanamsa	23° 33' 33"
Synetic vernal point	05° ♓ 33' 27"
True obliquity of ecliptic	23° 26' 23"

MOON'S PHASES, APSIDES AND POSITIONS ☽

Date	h	m	Phase	Longitude ° '	Eclipse Indicator
02	16	09	●	09 ♍ 50	
10	03	20	☽	17 ♐ 05	
16	19	01	○	23 ♓ 33	total
24	05	08	☾	00 ♋ 48	

Day	h	m	
14	09	44	Perigee
26	05	49	Apogee

04	08	57	0S	
11	02	50	Max dec	18° S 10'
17	07	19	0N	
24	05	22	Max dec	18° N 10'

LONGITUDES

Date	Chiron ⚷ ° '	Ceres ⚳ ° '	Pallas ⚴ ° '	Juno ⚵ ° '	Vesta ⚶ ° '	Black Moon Lilith ⚸ ° '
01	09 ♉ 37	09 ♑ 58	08 ♐ 44	24 ♑ 02	15 ♐ 00	25 ♋ 24
11	09 ♉ 24	10 ♑ 25	10 ♐ 56	23 ♑ 26	18 ♐ 08	26 ♋ 31
21	09 ♉ 06	11 ♑ 24	13 ♐ 28	23 ♑ 31	21 ♐ 38	27 ♋ 38
31	08 ♉ 44	12 ♑ 53	16 ♐ 17	24 ♑ 16	25 ♐ 26	28 ♋ 45

All ephemeris data is given at 12.00 UT and the Moon's longitude is additionally given for 24.00 UT

ASPECTARIAN

h m Aspects

01 Friday
01 52 ☽ # ♀
03 13 ☽ ☌ ☿
05 57 ☽ ∥ ♄
10 11 ☽ ⚹ ♀
19 25 ☽ ⊻ ♃
19 36 ☽ ∥ ♇
21 05 ☽ ∠ ♇
22 55 ☽ Q ♅
23 38 ☉ ⊥ ♇

02 Saturday
03 14 ☽ ∠ ♂
05 21 ☽ ∥ ☉
06 02 ☽ ☌ ♄
07 28 ☽ ⊥ ♃
11 51 ☽ # ♂
14 53 ☽ ⊥ ♇
16 09 ☽ ☌ ☉
18 17 ☽ ∠ ♀
21 55 ☽ ⊥ ♂
22 35 ☽ ⚹ ♅

03 Sunday
01 23 ☽ ∠ ♃
02 39 ☽ ⊻ ♇
03 20 ☽ □ ♆
10 15 ☽ ⊻ ♂
13 15 ☽ ⊥ ♀
17 33 ☽ ⊻ ☿
20 22 ☉ # ♂

04 Monday
01 46 ☽ ⊻ ♀
03 44 ☽ ∠ ♅
05 57 ☽ ⊥ ☿
06 52 ☽ ⚹ ♃
14 04 ☽ Q ♆
16 49 ☽ ⊻ ♄
19 27 ♂ ∠ ♄
20 59 ☽ ⊥ ♅

05 Tuesday
00 30 ☽ ∠ ☿
04 18 ☽ ⊥ ♄
05 15 ☿ # ♀
06 02 ☽ Q ♃
06 45 ☽ ⊻ ☉
08 22 ☽ ⊻ ♅
12 16 ☽ ☌ ♇
12 47 ☽ ⚹ ♆
18 57 ☽ ⊥ ☉
21 22 ☽ ∠ ♄
22 30 ☽ ☌ ♂

06 Wednesday
01 48 ☽ # ☉
04 33 ☉ ⚹ ♅
07 08 ☽ ⚹ ☿
11 29 ☽ ∥ ♂
13 01 ☽ ∠ ☉
14 38 ☽ ☌ ♀
16 06 ☽ □ ♃
16 41 ☽ ∠ ♆
19 28 ☽ # ♇

07 Thursday
01 21 ☽ ⚹ ♄
06 39 ☽ # ♄
07 22 ☽ Q ☿
09 08 ☽ ⊥ ♆
15 53 ☽ ☌ ♅
17 10 ♀ □ ♃
18 33 ☽ ⚹ ☉
19 36 ☽ ⊻ ♇
19 59 ☽ ⊻ ♆
20 38 ♀ ∠ ♆
22 41 ☽ # ☿
23 17 ☽ Q ♄

08 Friday
06 23 ☽ ⊥ ♇
08 04 ☽ ⊻ ♂
09 34 ☉ ⊻ ♇
10 01 ♃ ⚻ ♆
14 26 ☉ □ ♆
15 27 ☉ ∠ ♃
16 01 ☽ ∥ ♀
17 37 ☽ Q ☉
19 07 ☽ □ ☿
19 15 ☽ ⊥ ♂
21 22 ☽ ∥ ♅
22 22 ☽ ∠ ♇
22 51 ☽ △ ♃

09 Saturday
00 29 ☽ ⊻ ♀
07 26 ☽ □ ♄
11 41 ☽ ⊥ ♀
11 49 ☽ ∠ ♂
20 59 ☽ ⊻ ♅
23 27 ♀ ∥ ♅

10 Sunday
00 32 ☽ ⚹ ♇
00 47 ☽ ☌ ♆
01 18 ☽ ⚻ ♃
02 37 ☿ ∠ ♇
03 20 ☽ □ ☉
04 16 ☽ ∠ ♀
07 20 ☽ ⊥ ♅
10 48 ☿ ⊻ ♃
14 53 ☽ ⚹ ♂
16 53 ☽ ± ♃
18 04 ☿ Q ♅
21 04 ☽ Q ♇
21 27 ♂ # ♇
22 41 ☽ ∠ ♅

11 Monday
03 09 ☽ ⊼ ♃
04 55 ☽ △ ☿
07 22 ☽ ⚹ ♀
11 10 ☽ △ ♄
12 07 ☽ Q ♂
13 30 ☉ ⊥ ♂
23 53 ☽ ⚹ ♅

12 Tuesday
03 16 ☽ □ ♇
03 25 ☽ ⊻ ♆
04 39 ☽ Q ♀
09 04 ☽ ⚻ ☿
09 29 ☽ △ ☉
12 19 ☽ ⚻ ♄
13 21 ☽ ⊥ ♆
17 56 ☽ ∥ ♀
19 24 ☽ □ ♂
19 46 ☽ Q ♅
23 17 ☿ ⚹ ♀

13 Wednesday
01 47 ☽ ± ☿
03 14 ☽ ± ♄
04 09 ☽ ∠ ♆
05 33 ☽ ☍ ♃
05 48 ☽ ∥ ♅
11 53 ☽ ⚻ ☉
12 03 ☽ □ ♀
12 54 ☽ ⊼ ☿
13 09 ☽ ⊼ ♄
15 09 ☿ ☌ ♄
17 56 ☿ ∥ ♄

14 Thursday
01 19 ☽ □ ♅
03 38 ☽ ± ☉
03 57 ☉ ∠ ♀
04 38 ☽ △ ♇
04 41 ☽ ⚹ ♆
10 36 ♀ ⚹ ♄
13 11 ☿ ⊥ ♃
13 20 ☿ # ♂
14 07 ☽ ⊼ ☉
19 48 ☽ # ♄
21 49 ☽ ∥ ♂
22 46 ☽ △ ♂

15 Friday
00 00 ☽ # ☿
00 17 ☽ Q ♆
04 29 ☽ # ♇
05 11 ☽ ⚻ ♇
07 10 ☽ ⊼ ♃
09 09 ♆ ⚹ ♇
14 35 ☽ ☍ ♄
15 32 ☿ ⊥ ♇
15 59 ☽ △ ♀
17 08 ☽ ± ♃
20 00 ☽ ± ♇
20 36 ☽ ☍ ☿
23 29 ♂ # ♄

16 Saturday
00 33 ☽ ⚻ ♂
02 35 ☽ △ ♅
05 25 ☿ ∥ ♇
05 53 ☽ □ ♆
05 55 ☽ ⊼ ♇
08 10 ☽ ⚻ ♃
16 12 ☽ ± ♂
17 26 ☽ # ☉
18 15 ☽ ⚻ ♀
19 01 ☽ ☍ ☉

17 Sunday
02 43 ☽ ⊼ ♂
03 39 ☽ ⚻ ♅
09 34 ☽ △ ♃
10 14 ☽ ± ♀
17 03 ☽ ⊼ ♄
18 56 ☽ ± ♅
19 07 ☽ ∥ ☉
20 30 ☿ ∠ ♂
21 02 ☽ ⊼ ♀
21 39 ☿ ⚹ ♅

18 Monday
03 24 ☽ ± ♄
05 15 ☽ ⊼ ♅
06 23 ☽ ⊼ ☿
08 34 ☽ △ ♆
08 42 ☽ ☍ ♇
18 25 ☽ ± ☿
19 10 ☽ ⚻ ♄
19 14 ♀ ⊥ ♆
22 56 ☿ □ ♆

19 Tuesday
00 10 ☿ ⊻ ♇
00 31 ☽ ∥ ☿
02 16 ☽ ⊼ ☉
09 07 ☽ ☍ ♂
10 56 ☽ ⚻ ♆
12 04 ☽ ∥ ♇
12 50 ☽ ⚻ ☿
13 46 ☽ ± ☉
14 21 ☽ □ ♃
20 33 ☽ ∥ ♄
22 05 ☽ △ ♄

20 Wednesday
00 42 ☿ ∠ ♃
03 09 ☽ ± ♆
05 03 ☽ ☍ ♀
05 23 ☽ # ♂
07 20 ☽ ⚻ ☉
10 46 ☽ ☍ ♅
14 09 ☽ ⊼ ♆
14 24 ☽ ⊼ ♇
20 40 ☽ △ ☿
21 59 ♂ ∠ ♆
23 40 ☽ Q ♃

21 Thursday
01 34 ☽ ± ♇
11 27 ♀ # ♃
13 32 ☽ △ ☉
18 36 ☽ ⚻ ♇
18 44 ☽ # ♅
19 25 ☽ ⊼ ♂
22 38 ☽ ⚹ ♃

22 Friday
06 41 ☽ □ ♄
07 30 ☉ ∠ ♅
07 34 ☽ ± ♂
17 04 ☽ ⊼ ♀
19 51 ☽ ⊼ ♅
23 17 ☽ ☍ ♆
23 40 ☽ △ ♇

23 Saturday
02 06 ☽ ⚻ ♂
04 08 ☽ ∠ ♃
05 31 ☽ ± ♀
07 40 ☽ ± ♅
16 30 ☽ □ ☿
18 17 ☽ Q ♄
22 11 ☽ ⊥ ♃

24 Sunday
00 25 ☽ ⚻ ♀
01 35 ☽ ⚻ ♅
05 08 ☽ □ ☉
09 37 ☽ △ ♂
10 20 ☽ ⊻ ♃
18 31 ☽ ⚹ ♄
22 46 ♀ ☌ ♅

25 Monday
01 55 ♂ □ ♃
07 50 ☽ △ ♅
08 20 ☽ △ ♀
11 11 ☽ Q ☿
11 13 ☽ ⊼ ♆
11 39 ☿ ⊥ ♂
11 44 ☽ □ ♇
20 37 ☽ Q ☉
23 22 ☽ ± ♆

26 Tuesday
01 07 ☽ ∠ ♄
05 20 ☿ ∠ ♅
06 19 ♀ ∥ ♆
15 55 ☽ ⚹ ☿
16 51 ☽ # ♅
17 37 ☽ ⚻ ♆
19 31 ☽ ⊥ ♄
20 30 ☉ # ☿
20 45 ☿ ∠ ♀
23 06 ☽ ⚹ ☉
23 45 ☽ ☌ ♃

27 Wednesday
00 16 ☽ Q ♇
01 52 ☽ □ ♂
06 19 ☉ Q ♆
07 45 ☽ ⊻ ♄
08 17 ☉ ⚹ ♃
15 39 ♀ ⊻ ♆
18 09 ☽ # ♂
20 43 ☽ □ ♅
23 54 ☽ △ ♆

28 Thursday
00 20 ☽ □ ♀
00 31 ☽ ⚹ ♇
03 35 ☽ ∠ ☿
04 27 ♀ ⊻ ♇
07 59 ☽ ∠ ☉
16 11 ☽ Q ♂
19 12 ☿ Q ♆
22 41 ☽ ∥ ♄
23 20 ☿ ⚹ ♃

29 Friday
00 41 ☽ ⊥ ☿
03 28 ☽ ⊥ ☉
04 29 ♃ Q ♇
06 14 ☽ ∥ ♇
06 29 ☽ ∠ ♇
08 43 ☽ Q ♅
12 30 ☽ ⊻ ♃
13 56 ☽ Q ♀
14 34 ☽ ⊻ ☿
16 20 ☽ ⊻ ☉
17 18 ☽ ⚹ ♂
20 05 ☽ ☌ ♄

30 Saturday
00 12 ☽ ⊥ ♇
00 24 ☽ ⊥ ♃
08 17 ☽ ⚹ ♅
11 11 ☽ □ ♆
11 54 ☽ ⊻ ♇
14 25 ☽ ⚹ ♀
15 22 ☉ ☌ ☿
18 03 ☽ ∠ ♃
19 20 ☿ ⊻ ♂
22 01 ☽ # ☉

OCTOBER 1978

LONGITUDES

Date	Sidereal time h m s	Sun ☉ ° ' "	Moon ☽ ° ' "	Moon ☽ 24.00 ° ' "	Mercury ☿ ° '	Venus ♀ ° '	Mars ♂ ° '	Jupiter ♃ ° '	Saturn ♄ ° '	Uranus ♅ ° '	Neptune ♆ ° '	Pluto ♇ ° '
01	12 39 28	07 ♎ 57 21	28 ♍ 47 55	05 ♎ 09 23	08 ♎ 38	18 ♏ 00	07 ♏ 53	04 ♌ 31	08 ♍ 16	14 ♏ 25	15 ♐ 52	16 ♎ 15
02	12 43 24	08 56 24	11 ♎ 34 39	18 ♎ 03 43	10 23	18 31	08 34	04 40	08 23	14 28	15 53	16 18
03	12 47 21	09 55 28	24 ♎ 36 33	01 ♏ 13 06	12 08	19 01	09 15	04 49	08 30	14 31	15 54	16 20
04	12 51 17	10 54 34	07 ♏ 53 12	14 ♏ 36 44	13 52	19 29	09 56	04 58	08 37	14 34	15 55	16 22
05	12 55 14	11 53 43	21 ♏ 23 31	28 ♏ 13 21	15 35	19 55	10 37	05 06	08 44	14 38	15 57	16 25
06	12 59 11	12 52 53	05 ♐ 06 01	12 ♐ 01 17	17 17	20 20	11 19	05 15	08 51	14 41	15 58	16 27
07	13 03 07	13 52 05	18 ♐ 58 56	25 ♐ 58 43	18 58	20 44	12 00	05 23	08 57	14 44	15 59	16 29
08	13 07 04	14 51 19	03 ♑ 00 26	10 ♑ 03 50	20 39	21 05	12 42	05 31	09 04	14 48	16 00	16 32
09	13 11 00	15 50 35	17 ♑ 08 40	24 ♑ 14 41	22 19	21 25	13 23	05 39	09 11	14 51	16 02	16 34
10	13 14 57	16 49 52	01 ♒ 21 38	08 ♒ 29 13	23 57	21 42	14 05	05 47	09 17	14 54	16 03	16 36
11	13 18 53	17 49 11	15 ♒ 37 09	22 ♒ 45 06	25 36	21 58	14 47	05 55	09 24	14 58	16 04	16 39
12	13 22 50	18 48 32	29 ♒ 52 41	06 ♓ 59 33	27 13	22 12	15 28	06 03	09 30	15 01	16 06	16 41
13	13 26 46	19 47 55	14 ♓ 05 15	21 ♓ 09 23	28 ♎ 50	22 23	16 10	06 11	09 37	15 05	16 07	16 44
14	13 30 43	20 47 19	28 ♓ 11 28	05 ♈ 11 04	00 ♏ 26	22 33	16 52	06 18	09 43	15 08	16 09	16 46
15	13 34 40	21 46 45	12 ♈ 07 44	19 ♈ 01 03	02 01	22 40	17 34	06 25	09 49	15 12	16 10	16 48
16	13 38 36	22 46 13	25 ♈ 50 37	02 ♉ 36 07	03 36	22 45	18 16	06 33	09 56	15 15	16 12	16 51
17	13 42 33	23 45 44	09 ♉ 17 17	15 ♉ 53 54	05 10	22 48	18 58	06 40	10 02	15 19	16 13	16 53
18	13 46 29	24 45 16	22 ♉ 25 52	28 ♉ 53 08	06 44	22 R 48	19 40	06 46	10 08	15 22	16 15	16 55
19	13 50 26	25 44 51	05 ♊ 15 45	11 ♊ 33 53	08 16	22 46	20 22	06 53	10 14	15 26	16 16	16 58
20	13 54 22	26 44 27	17 ♊ 47 44	23 ♊ 57 36	09 49	22 42	21 04	07 00	10 20	15 29	16 18	17 00
21	13 58 19	27 44 06	00 ♋ 03 52	06 ♋ 06 58	11 20	22 35	21 47	07 06	10 26	15 33	16 20	17 02
22	14 02 15	28 43 48	12 ♋ 07 23	18 ♋ 05 39	12 51	22 26	22 29	07 13	10 32	15 37	16 21	17 05
23	14 06 12	29 ♎ 43 31	24 ♋ 02 21	29 ♋ 58 05	14 21	22 14	23 11	07 19	10 38	15 40	16 23	17 07
24	14 10 09	00 ♏ 43 17	05 ♌ 53 28	11 ♌ 49 07	15 51	22 00	23 54	07 25	10 44	15 44	16 25	17 09
25	14 14 05	01 43 05	17 ♌ 45 41	23 ♌ 43 48	17 20	21 44	24 36	07 30	10 50	15 48	16 26	17 12
26	14 18 02	02 42 55	29 ♌ 44 04	05 ♍ 47 04	18 49	21 25	25 19	07 36	10 55	15 51	16 28	17 14
27	14 21 58	03 42 47	11 ♍ 53 21	18 ♍ 03 25	20 17	21 04	26 02	07 42	11 01	15 55	16 30	17 16
28	14 25 55	04 42 42	24 ♍ 17 43	00 ♎ 36 38	21 44	20 41	26 44	07 47	11 06	15 59	16 32	17 19
29	14 29 51	05 42 38	07 ♎ 00 25	13 ♎ 29 19	23 11	20 16	27 27	07 52	11 12	16 02	16 33	17 21
30	14 33 48	06 42 37	20 ♎ 03 23	26 ♎ 42 39	24 37	19 49	28 10	07 57	11 17	16 06	16 35	17 23
31	14 37 44	07 ♏ 42 37	03 ♏ 26 57	10 ♏ 16 04	26 ♏ 03	19 ♏ 20	28 ♏ 53	08 ♌ 02	11 ♍ 23	16 ♏ 10	16 ♐ 37	17 ♎ 26

Date	Moon True ☊ ° '	Moon Mean ☊ ° '	Moon ☽ Latitude ° '
01	26 ♍ 48	26 ♍ 04	00 N 11
02	26 R 47	26 01	01 21
03	26 46	25 58	02 28
04	26 45	25 55	03 28
05	26 43	25 52	04 18
06	26 41	25 48	04 54
07	26 39	25 45	05 13
08	26 38	25 42	05 14
09	26 D 37	25 39	04 56
10	26 38	25 36	04 20
11	26 39	25 33	03 27
12	26 40	25 29	02 22
13	26 42	25 26	01 N 09
14	26 R 42	25 23	00 S 08
15	26 41	25 20	01 24
16	26 39	25 17	02 33
17	26 36	25 13	03 33
18	26 31	25 10	04 19
19	26 27	25 07	04 52
20	26 22	25 04	05 09
21	26 18	25 01	05 12
22	26 15	24 58	05 01
23	26 13	24 54	04 37
24	26 D 13	24 51	04 01
25	26 14	24 48	03 15
26	26 16	24 45	02 20
27	26 18	24 42	01 18
28	26 19	24 39	00 S 11
29	26 R 18	24 35	00 N 58
30	26 16	24 32	02 05
31	26 ♍ 11	24 ♍ 29	03 N 08

DECLINATIONS

Date	Sun ☉ ° '	Moon ☽ ° '	Mercury ☿ ° '	Venus ♀ ° '	Mars ♂ ° '	Jupiter ♃ ° '	Saturn ♄ ° '	Uranus ♅ ° '	Neptune ♆ ° '	Pluto ♇ ° '
01	03 S 09	00 N 39	02 S 17	22 S 40	14 S 17	19 N 26	09 N 52	15 S 50	21 S 17	08 N 55
02	03 33	03 S 20	03 04	22 54	14 31	19 24	09 50	15 51	21 17	08 54
03	03 56	07 14	03 50	23 08	14 45	19 22	09 47	15 52	21 17	08 53
04	04 19	10 51	04 36	23 21	14 59	19 20	09 45	15 53	21 17	08 52
05	04 42	13 57	05 21	23 33	15 12	19 18	09 43	15 54	21 18	08 51
06	05 05	16 20	06 06	23 44	15 26	19 16	09 40	15 55	21 18	08 51
07	05 28	17 47	06 51	23 55	15 39	19 14	09 38	15 56	21 18	08 50
08	05 51	18 11	07 35	24 05	15 52	19 12	09 35	15 57	21 18	08 49
09	06 14	17 27	08 18	24 15	16 05	19 10	09 33	15 58	21 18	08 48
10	06 37	15 38	09 01	24 23	16 18	19 08	09 31	15 59	21 18	08 47
11	07 00	12 51	09 43	24 31	16 31	19 07	09 28	16 00	21 19	08 46
12	07 22	09 17	10 25	24 38	16 44	19 05	09 26	16 01	21 19	08 45
13	07 45	05 12	11 06	24 44	16 57	19 03	09 24	16 02	21 19	08 45
14	08 07	00 S 51	11 46	24 49	17 09	19 02	09 22	16 03	21 19	08 44
15	08 29	03 N 30	12 26	24 53	17 22	19 00	09 19	16 04	21 19	08 43
16	08 51	07 36	13 05	24 56	17 34	18 58	09 17	16 05	21 20	08 42
17	09 13	11 14	13 43	24 58	17 46	18 57	09 15	16 06	21 20	08 41
18	09 35	14 12	14 20	25 00	17 58	18 55	09 13	16 07	21 20	08 40
19	09 57	16 24	14 57	25 00	18 10	18 53	09 11	16 08	21 20	08 40
20	10 19	17 45	15 33	24 58	18 22	18 52	09 08	16 09	21 20	08 39
21	10 40	18 15	16 08	24 56	18 34	18 51	09 06	16 10	21 21	08 38
22	11 01	17 54	16 42	24 52	18 45	18 49	09 04	16 11	21 21	08 37
23	11 23	16 46	17 16	24 48	18 57	18 48	09 02	16 12	21 21	08 37
24	11 43	14 54	17 48	24 42	19 08	18 46	09 00	16 14	21 21	08 36
25	12 04	12 25	18 20	24 34	19 19	18 45	08 58	16 15	21 21	08 35
26	12 25	09 23	18 51	24 26	19 30	18 44	08 56	16 16	21 22	08 34
27	12 45	05 54	19 21	24 16	19 40	18 42	08 54	16 17	21 22	08 34
28	13 06	02 N 06	19 50	24 04	19 51	18 41	08 52	16 18	21 22	08 33
29	13 26	01 S 54	20 18	23 52	20 01	18 40	08 50	16 19	21 22	08 32
30	13 45	05 54	20 44	23 38	20 12	18 39	08 49	16 20	21 23	08 32
31	14 S 05	09 S 43	21 S 10	23 S 22	20 S 22	18 N 38	08 N 47	16 S 21	21 S 23	08 N 31

ZODIAC SIGN ENTRIES

Date	h	m	Planets
01	14	17	☽ ♎
03	21	48	☽ ♏
06	03	07	☽ ♐
08	06	52	☽ ♑
10	09	42	☽ ♒
12	12	12	☽ ♓
14	05	30	☿ ♏
14	15	06	☽ ♈
16	19	22	☽ ♉
19	02	05	☽ ♊
21	11	52	☽ ♋
23	18	37	☉ ♏
24	00	04	☽ ♌
26	12	32	☽ ♍
28	22	51	☽ ♎
31	05	53	☽ ♏

LATITUDES

Date	Mercury ☿ ° '	Venus ♀ ° '	Mars ♂ ° '	Jupiter ♃ ° '	Saturn ♄ ° '	Uranus ♅ ° '	Neptune ♆ ° '	Pluto ♇ ° '
01	01 N 14	05 S 42	00 S 10	00 N 18	01 N 31	00 N 21	01 N 25	16 N 35
04	00 57	05 58	00 12	00 19	01 31	00 21	01 25	16 34
07	00 38	06 13	00 13	00 19	01 32	00 21	01 25	16 34
10	00 N 18	06 26	00 15	00 20	01 32	00 21	01 25	16 34
13	00 S 02	06 36	00 17	00 20	01 33	00 20	01 24	16 34
16	00 23	06 43	00 19	00 21	01 33	00 20	01 24	16 35
19	00 44	06 46	00 21	00 21	01 34	00 20	01 24	16 35
22	01 04	06 44	00 23	00 22	01 34	00 20	01 24	16 35
25	01 23	06 36	00 24	00 22	01 35	00 20	01 24	16 36
28	01 41	06 23	00 26	00 23	01 35	00 20	01 24	16 36
31	01 S 58	06 S 03	00 S 28	00 N 23	01 N 36	00 N 20	01 N 24	16 N 37

DATA

Julian Date	2443783
Delta T	+49 seconds
Ayanamsa	23° 33' 35"
Synetic vernal point	05° ♓ 33' 24"
True obliquity of ecliptic	23° 26' 23"

MOON'S PHASES, APSIDES AND POSITIONS ☽

Date	h	m	Phase	Longitude ° '	Eclipse Indicator
02	06	41	●	08 ♎ 43	Partial
09	09	38	☽	15 ♑ 45	
16	06	10	○	22 ♈ 32	
24	00	34	☾	00 ♌ 15	
31	20	07	●	08 ♏ 03	

Day	h	m		
11	16	12	Perigee	
24	01	23	Apogee	
01	15	55	0S	
08	08	29	Max dec	18° S 12'
14	16	37	0N	
21	13	55	Max dec	18° N 15'
29	00	41	0S	

LONGITUDES

Date	Chiron ⚷ ° '	Ceres ⚳ ° '	Pallas ⚴ ° '	Juno ⚵ ° '	Vesta ⚶ ° '	Black Moon Lilith ⚸ ° '
01	08 ♉ 44	12 ♑ 53	16 ♐ 17	24 ♑ 16	25 ♐ 26	28 ♋ 45
11	08 ♉ 19	14 ♑ 47	19 ♐ 20	25 ♑ 37	29 ♐ 30	29 ♋ 52
21	07 ♉ 51	17 ♑ 04	22 ♐ 34	27 ♑ 31	03 ♑ 46	00 ♌ 59
31	07 ♉ 22	19 ♑ 40	25 ♐ 57	29 ♑ 53	08 ♑ 13	02 ♌ 06

All ephemeris data is given at 12.00 UT and the Moon's longitude is additionally given for 24.00 UT

ASPECTARIAN

01 Sunday
00 04 ☽ ∠ ♂
03 39 ☽ ⫲ ☿
05 51 ☉ ⊻ ♂
06 46 ☿ ⊻ ♄
08 53 ☿ ⊥ ♅
13 10 ☽ ∠ ♅
18 10 ☽ ⊥ ♂
20 17 ☽ ∠ ♀
20 47 ☉ ⊻ ♄
21 35 ☽ Q ♆
22 56 ☽ ⚹ ♃
23 42 ☉ ⊥ ♅

02 Monday
04 38 ♂ ⚹ ♄
06 00 ☽ ⊻ ♄
06 04 ☽ ⊻ ♂
06 10 ☽ ⊥ ♅
06 41 ☽ ☌ ☉
09 27 ☽ ☌ ☿
09 57 ☽ ∥ ☿
13 23 ☽ ∥ ☉
13 49 ☽ ⊥ ♀
17 16 ☽ ⊥ ♄
17 23 ☽ ⊻ ♅
20 00 ☽ ⚹ ♆
20 46 ☽ ☌ ♇
21 32 ☽ Q ♃

03 Tuesday
01 21 ☽ ⊻ ♀
09 58 ☽ ∠ ♄
18 17 ☉ ∥ ☿
22 35 ☽ ⫲ ♇
23 27 ☽ ∠ ♆

04 Wednesday
04 27 ☽ ⫲ ♄
04 45 ☿ ⊥ ♀
06 41 ☽ □ ♃
11 32 ♂ ⊥ ♆
13 19 ☽ ⚹ ♄
15 39 ☽ ⊥ ♆
15 52 ☽ ☌ ♂
17 50 ☽ ⊻ ☉
22 11 ☿ ⊻ ♅
23 16 ☿ ⊥ ♄
23 58 ☽ ☌ ♅

05 Thursday
00 14 ☽ ⊻ ☿
02 21 ☽ ⊻ ♆
03 10 ☽ ⊻ ♇
05 20 ☽ ⊥ ☉
09 20 ☽ ☌ ♀
10 50 ☽ Q ♄
12 23 ☽ ⊥ ☿
13 48 ☽ ⊥ ♇
17 10 ☿ ⚹ ♆
22 26 ☽ ∠ ☉
23 58 ☿ ☌ ♇

06 Friday
00 44 ☽ ∥ ♂
05 37 ☽ ∠ ♇
06 24 ☽ ∠ ☿
07 02 ☽ ∥ ♅
11 26 ☿ Q ♃
12 15 ☽ △ ♃
18 33 ☽ □ ♄
23 20 ☽ ⊻ ♂

07 Saturday
02 31 ☽ ⚹ ☉
04 40 ☽ ⊻ ♅
06 50 ☽ ☌ ♆
07 42 ☽ ⚹ ♇
10 14 ☽ ⊥ ♂
11 59 ☽ ⚹ ☿
14 26 ☽ ⚼ ♃
15 02 ☽ ⊥ ♅
15 05 ☽ ⊻ ♀

08 Sunday
00 42 ☽ Q ☉
01 38 ☽ ⊥ ♀
02 28 ☽ ∠ ♂
04 20 ☽ Q ♇
06 00 ☽ ± ♃
06 30 ☽ ∠ ♅
10 22 ☉ ⊻ ♅
10 26 ♀ Q ♄
11 18 ☽ Q ☿
16 19 ☽ ⊼ ♃
17 22 ☽ ∠ ♀
17 49 ☉ ⊥ ♄
19 53 ☿ ⊻ ♀
20 23 ☉ ⊥ ♀
21 06 ♂ ∥ ♅
22 23 ☽ △ ♄

09 Monday
05 19 ☽ ⚹ ♂
08 06 ☽ ⚹ ♅
09 38 ☽ □ ☉
10 07 ☽ ⊻ ♆
11 01 ☽ □ ♇
16 37 ☉ ⚹ ♆
19 22 ☽ ⚹ ♀
20 16 ☽ ⊥ ♆
21 53 ☽ □ ☿
23 59 ☽ ⚼ ♄

10 Tuesday
02 39 ☽ Q ♂
04 25 ☿ ⫲ ♇
04 28 ☽ Q ♅
05 17 ☽ ∥ ♂
06 22 ☉ ☌ ♇
08 20 ☽ ∥ ♅
11 29 ☽ ∠ ♆
15 16 ☽ ± ♄
16 02 ☽ Q ♀
17 10 ☿ ∠ ♄
19 32 ☽ ☍ ♃

11 Wednesday
01 27 ☽ ⊼ ♄
04 03 ☿ ⫲ ♄
10 30 ☽ □ ♂
10 53 ☽ □ ♅
12 46 ☽ ⚹ ♆
13 44 ☽ △ ♇
14 51 ☉ Q ♃
15 59 ☽ △ ☉
19 02 ♂ ☌ ♅
22 52 ☽ □ ♀

12 Thursday
06 00 ☽ ∥ ☿
06 57 ☽ △ ☿
09 00 ☽ Q ♆
11 06 ☽ ⫲ ♄
15 03 ☽ ⚼ ♇
15 18 ☽ ⫲ ♇
19 07 ☽ ⚼ ☉
22 30 ☽ ⊼ ♃
22 36 ☽ ∥ ☉

13 Friday
04 22 ☽ ☍ ♄
06 18 ☽ ± ♇
08 44 ☽ ± ♃
10 21 ♂ ⊻ ♆
11 28 ☽ ± ☉
11 31 ☽ ⚼ ☿
13 41 ☽ △ ♅
15 27 ☽ □ ♆
15 42 ☽ △ ♂
16 29 ☽ ⊼ ♇
22 25 ☽ ⊼ ☉

14 Saturday
00 09 ☽ ⚼ ♃
02 16 ☽ △ ♀
04 46 ☽ ± ☿
08 25 ♂ ⊻ ♇
15 21 ☽ ⚼ ♅
16 20 ☽ ⊼ ☿
18 37 ☽ ⚼ ♂
22 54 ☿ ∠ ♆

15 Sunday
02 03 ☽ △ ♃
04 13 ☽ ⚼ ♀
06 54 ☽ ± ♅
07 59 ☽ ⊼ ♄
10 58 ☽ ± ♂
17 21 ☽ ⊼ ♅
18 28 ☽ ± ♄
19 02 ☽ △ ♆
19 57 ☽ ± ♀
20 10 ☽ ☍ ♇
21 58 ☽ ⊼ ♂

16 Monday
02 14 ☉ ⫲ ♇
06 10 ☽ ☍ ☉
06 31 ☽ ⊼ ♀
10 22 ☽ ⚼ ♄
11 31 ☉ ⊻ ♀
18 52 ☽ ∥ ♇
20 45 ☽ ⫲ ☉
21 30 ☽ ⚼ ♆
22 34 ☽ ∥ ♄

17 Tuesday
03 37 ☽ ☍ ☿
07 14 ☽ □ ♃
13 21 ☽ △ ♄
13 31 ☉ ⫲ ♄
13 41 ☽ ± ♆
22 59 ☽ ☍ ♅

18 Wednesday
00 37 ☽ ⊼ ♆
01 51 ☽ ⊼ ♇
03 58 ♀ St R
06 37 ☽ ☍ ♂
12 41 ☽ ☍ ♀
12 49 ☿ □ ♃
12 55 ☽ ± ♇
13 46 ☽ ⫲ ☿
16 23 ☽ Q ♃
16 40 ☽ ⊼ ☉
22 15 ☉ ∠ ♄

19 Thursday
04 48 ☽ ± ☉
05 45 ☽ ⚼ ♇
08 37 ☽ ⫲ ♅
15 06 ☽ ⚹ ♃
18 30 ☽ ⊼ ☿
21 32 ☽ □ ♄
23 20 ☽ ⚼ ☉

20 Friday
07 31 ☽ ⊼ ♅
07 37 ☽ ± ☿
09 06 ☽ ☍ ♆
10 28 ☽ △ ♇
18 45 ☽ ⊼ ♂
19 13 ☽ ± ♅
19 50 ☿ ⊥ ♆
20 14 ☽ ∠ ♃
20 53 ☿ ⚹ ♄
21 27 ☽ ⊼ ♀

21 Saturday
03 37 ☽ ⚼ ☿
07 00 ☽ △ ☉
07 13 ☽ ± ♂
08 46 ☽ Q ♄
09 07 ☽ ± ♀
12 58 ☽ ⚼ ♅
13 32 ☿ ∥ ♅
14 04 ☽ ⊥ ♃

22 Sunday
02 05 ☽ ⊻ ♃
02 09 ☽ ⚼ ♂
02 29 ♀ Q ♄
02 45 ☽ ⚼ ♀
08 48 ☽ ⚹ ♄
10 37 ♀ ☌ ♂
13 40 ☽ △ ☿
14 07 ♂ Q ♄
19 02 ☽ △ ♅
19 13 ♂ ⫲ ♃
20 31 ☽ ⊼ ♆
21 59 ☽ □ ♇

23 Monday
05 50 ☽ ⫲ ☿
08 25 ☽ △ ♀
08 38 ☽ ± ♆
09 29 ♂ ⊥ ♇
10 10 ☽ △ ♂
15 15 ☽ ∠ ♄
20 03 ☽ ⫲ ♅

24 Tuesday
00 34 ☽ □ ☉
02 54 ☽ ⚼ ♆
09 38 ☽ ⊥ ♄
09 56 ☿ ☌ ♅
10 31 ☽ Q ♇
15 06 ☽ ☌ ♃
21 07 ☿ ⊻ ♆
21 53 ☽ ⊻ ♄

25 Wednesday
05 03 ☉ ∠ ♆
08 00 ☽ □ ♅
09 19 ☽ △ ♆
09 36 ☿ ⊻ ♇
10 51 ☽ ⚹ ♇
11 02 ☽ □ ☿
14 37 ☽ ⫲ ☉
16 18 ☽ Q ☉
19 48 ☽ □ ♀

26 Thursday
02 37 ☽ □ ♂
06 33 ☿ ⫲ ♃
15 16 ☽ ∥ ♄
16 59 ☽ ∠ ♇
17 51 ☽ ∥ ♇
18 27 ☽ ⚹ ☉
20 13 ☽ Q ♅

27 Friday
03 42 ☽ ⊻ ♃
03 58 ☽ Q ☿
06 38 ☽ Q ♀
10 16 ☽ ☌ ♄
10 47 ☽ ⊥ ♇
15 33 ☽ ⊥ ♃
16 26 ☽ Q ♂
19 53 ☽ ⚹ ♅
21 00 ☽ □ ♆
22 19 ☿ ☌ ♀
22 31 ☽ ⊻ ♇

28 Saturday
02 26 ☽ ∠ ☉
05 18 ☽ ⚹ ♀
06 28 ☽ ⚹ ☿
09 05 ☽ ∠ ♃
13 46 ☿ ∥ ♂
16 57 ☽ ⚹ ♂
21 08 ☽ ⊥ ☉

29 Sunday
00 45 ☽ ∠ ♅
07 25 ☽ Q ♆
08 52 ☽ ∠ ♀
09 23 ☽ ⊻ ☉
12 11 ☿ Q ♄
13 37 ☽ ⚹ ♃
14 28 ☽ ∠ ☿
14 47 ☿ ⊥ ♇
17 39 ☽ ⊥ ♅
19 50 ☽ ⊻ ♄
22 41 ☽ ∠ ♂

30 Monday
01 00 ☽ ⊥ ♀
04 45 ☽ ⊻ ♅
05 40 ☽ ⚹ ♆
06 55 ☽ ⊥ ♄
07 08 ☽ ☌ ♇
09 05 ☽ ⊥ ☿
11 35 ☽ ⊻ ♀
11 48 ☽ Q ♃
16 03 ☽ ⊥ ♂
21 14 ☽ ⊻ ☿
23 19 ☽ ∠ ♄

31 Tuesday
03 26 ☽ ⊻ ♂
04 14 ☽ ⫲ ♇
05 55 ☽ ⫲ ♄
08 45 ☽ ∠ ♆
20 07 ☽ ☌ ☉
20 08 ☽ □ ♃
20 22 ☉ □ ♃

NOVEMBER 1978

LONGITUDES

Date	Sidereal time (h m s)	Sun ☉ (° ' ")	Moon ☽ (° ' ")	Moon ☽ 24.00 (° ' ")	Mercury ☿ (° ')	Venus ♀ (° ')	Mars ♂ (° ')	Jupiter ♃ (° ')	Saturn ♄ (° ')	Uranus ♅ (° ')	Neptune ♆ (° ')	Pluto ♇ (° ')
01	14 41 41	08 ♏ 42 40	17 ♏ 09 40	24 ♏ 07 18	27 ♏ 27	18 ♏ 50	29 ♏ 36	08 ♌ 07	11 ♍ 28	16 ♏ 13	16 ♐ 39	17 ♎ 28
02	14 45 38	09 42 44	01 ♐ 08 26	08 ♐ 12 29	28 ♏ 51	18 R 18	00 ♐ 19	08 11	11 33	16 17	16 41	17 30
03	14 49 34	10 42 50	15 ♐ 18 50	22 ♐ 26 51	00 ♐ 14	17 44	01 02	08 15	11 38	16 21	16 43	17 32
04	14 53 31	11 42 58	29 ♐ 35 52	06 ♑ 45 20	01 37	17 10	01 45	08 19	11 43	16 25	16 45	17 34
05	14 57 27	12 43 08	13 ♑ 54 40	21 ♑ 03 25	02 58	16 34	02 29	08 23	11 48	16 28	16 47	17 37
06	15 01 24	13 43 19	28 ♑ 11 10	05 ♒ 17 35	04 18	15 58	03 12	08 27	11 53	16 32	16 48	17 39
07	15 05 20	14 43 31	12 ♒ 22 26	19 ♒ 25 30	05 37	15 22	03 55	08 31	11 58	16 36	16 50	17 41
08	15 09 17	15 43 45	26 ♒ 26 41	03 ♓ 25 54	06 55	14 46	04 39	08 34	12 02	16 40	16 52	17 43
09	15 13 13	16 44 00	10 ♓ 23 03	17 ♓ 18 07	08 12	14 09	05 22	08 37	12 07	16 43	16 54	17 45
10	15 17 10	17 44 17	24 ♓ 11 03	01 ♈ 01 47	09 27	13 33	06 06	08 41	12 12	16 47	16 56	17 48
11	15 21 07	18 44 35	07 ♈ 50 15	14 ♈ 36 20	10 40	12 57	06 49	08 43	12 16	16 51	16 59	17 50
12	15 25 03	19 44 55	21 ♈ 19 56	28 ♈ 00 52	11 51	12 23	07 33	08 46	12 20	16 55	17 01	17 52
13	15 29 00	20 45 16	04 ♉ 39 00	11 ♉ 14 10	13 00	11 49	08 17	08 49	12 25	16 58	17 03	17 54
14	15 32 56	21 45 39	17 ♉ 46 11	24 ♉ 14 55	14 07	11 17	09 01	08 51	12 29	17 02	17 05	17 56
15	15 36 53	22 46 03	00 ♊ 40 13	07 ♊ 02 00	15 11	10 46	09 44	08 53	12 33	17 06	17 07	17 58
16	15 40 49	23 46 29	13 ♊ 20 15	19 ♊ 34 57	16 12	10 17	10 28	08 55	12 37	17 09	17 09	18 00
17	15 44 46	24 46 57	25 ♊ 46 13	01 ♋ 54 09	17 10	09 50	11 12	08 57	12 41	17 13	17 11	18 02
18	15 48 42	25 47 27	07 ♋ 59 00	14 ♋ 01 03	18 03	09 24	11 56	08 58	12 45	17 17	17 13	18 04
19	15 52 39	26 47 58	20 ♋ 00 39	25 ♋ 58 13	18 52	09 01	12 40	09 00	12 49	17 21	17 15	18 06
20	15 56 36	27 48 31	01 ♌ 54 13	07 ♌ 49 11	19 36	08 40	13 24	09 01	12 52	17 24	17 17	18 08
21	16 00 32	28 49 06	13 ♌ 43 42	19 ♌ 38 23	20 15	08 22	14 09	09 02	12 56	17 28	17 20	18 10
22	16 04 29	29 ♏ 49 42	25 ♌ 33 52	01 ♍ 30 51	20 47	08 05	14 53	09 02	12 59	17 32	17 22	18 12
23	16 08 25	00 ♐ 50 20	07 ♍ 30 00	13 ♍ 32 00	21 12	07 52	15 37	09 03	13 03	17 35	17 24	18 14
24	16 12 22	01 51 00	19 ♍ 37 32	25 ♍ 47 16	21 29	07 40	16 22	09 03	13 06	17 39	17 26	18 15
25	16 16 18	02 51 41	02 ♎ 01 47	08 ♎ 21 40	21 37	07 32	17 06	09 04	13 09	17 43	17 28	18 17
26	16 20 15	03 52 24	14 ♎ 47 24	21 ♎ 19 20	21 R 36	07 25	17 50	09 R 04	13 12	17 46	17 31	18 19
27	16 24 11	04 53 09	27 ♎ 57 45	04 ♏ 42 47	21 25	07 22	18 35	09 03	13 15	17 50	17 33	18 21
28	16 28 08	05 53 55	11 ♏ 34 22	18 ♏ 32 19	21 03	07 20	19 20	09 03	13 18	17 54	17 35	18 23
29	16 32 05	06 54 42	25 ♏ 36 15	02 ♐ 45 35	20 31	07 D 21	20 04	09 02	13 21	17 57	17 37	18 24
30	16 36 01	07 ♐ 55 31	09 ♐ 59 36	17 ♐ 17 26	19 ♐ 47	07 ♏ 25	20 ♐ 49	09 ♌ 01	13 ♍ 24	18 ♏ 01	17 ♐ 39	18 ♎ 26

Moon Nodes and Latitude

Date	Moon True ☊ (° ')	Moon Mean ☊ (° ')	Moon ☽ Latitude (° ')
01	26 ♍ 05	24 ♍ 26	04 N 00
02	25 R 58	24 23	04 40
03	25 51	24 19	05 03
04	25 44	24 16	05 07
05	25 39	24 13	04 53
06	25 36	24 10	04 20
07	25 34	24 07	03 31
08	25 D 35	24 04	02 30
09	25 36	24 00	01 21
10	25 37	23 57	00 N 08
11	25 R 36	23 54	01 S 05
12	25 33	23 51	02 14
13	25 28	23 48	03 13
14	25 20	23 45	04 02
15	25 10	23 41	04 37
16	24 58	23 38	04 58
17	24 47	23 35	05 04
18	24 37	23 32	04 56
19	24 29	23 29	04 35
20	24 23	23 25	04 02
21	24 19	23 22	03 19
22	24 18	23 19	02 27
23	24 D 18	23 16	01 28
24	24 18	23 13	00 S 25
25	24 R 18	23 10	00 N 41
26	24 16	23 06	01 47
27	24 12	23 03	02 49
28	24 05	23 00	03 43
29	23 55	22 57	04 26
30	23 ♍ 44	22 ♍ 54	04 N 53

DECLINATIONS

Date	Sun ☉	Moon ☽	Mercury ☿	Venus ♀	Mars ♂	Jupiter ♃	Saturn ♄	Uranus ♅	Neptune ♆	Pluto ♇
01	14 S 24	13 S 07	21 S 35	23 S 06	20 S 32	18 N 37	08 N 45	16 S 22	21 S 23	08 N 30
02	14 43	15 49	21 59	22 48	20 41	18 36	08 43	16 23	21 23	08 30
03	15 02	17 37	22 21	22 29	20 51	18 35	08 41	16 24	21 23	08 29
04	15 21	18 19	22 43	22 09	21 00	18 34	08 40	16 25	21 24	08 29
05	15 39	17 52	23 03	21 48	21 09	18 33	08 38	16 26	21 24	08 28
06	15 57	16 17	23 22	21 27	21 18	18 32	08 36	16 27	21 24	08 27
07	16 15	13 42	23 40	21 04	21 27	18 32	08 35	16 29	21 24	08 27
08	16 33	10 21	23 57	20 41	21 36	18 31	08 33	16 30	21 25	08 26
09	16 50	06 25	24 12	20 17	21 44	18 30	08 31	16 31	21 25	08 26
10	17 07	02 S 12	24 26	19 53	21 52	18 30	08 30	16 32	21 25	08 25
11	17 24	02 N 07	24 39	19 28	22 00	18 29	08 28	16 33	21 25	08 25
12	17 40	06 15	24 50	19 03	22 08	18 29	08 27	16 34	21 25	08 24
13	17 57	10 02	25 00	18 39	22 16	18 28	08 26	16 35	21 26	08 24
14	18 12	13 15	25 08	18 14	22 23	18 28	08 24	16 36	21 26	08 23
15	18 28	15 46	25 15	17 50	22 30	18 27	08 23	16 37	21 26	08 23
16	18 43	17 28	25 21	17 26	22 37	18 27	08 21	16 38	21 26	08 22
17	18 58	18 19	25 24	17 02	22 44	18 27	08 20	16 39	21 27	08 22
18	19 12	18 17	25 26	16 40	22 50	18 27	08 19	16 40	21 27	08 22
19	19 27	17 25	25 27	16 18	22 57	18 26	08 18	16 41	21 27	08 21
20	19 40	15 48	25 26	15 56	23 03	18 26	08 16	16 42	21 27	08 21
21	19 54	13 32	25 23	15 36	23 09	18 26	08 15	16 43	21 27	08 20
22	20 07	10 42	25 18	15 17	23 14	18 26	08 14	16 44	21 28	08 20
23	20 20	07 23	25 12	14 59	23 20	18 26	08 13	16 46	21 28	08 20
24	20 32	03 N 44	25 03	14 42	23 25	18 26	08 12	16 47	21 28	08 19
25	20 44	00 S 11	24 53	14 26	23 30	18 26	08 11	16 48	21 28	08 19
26	20 55	04 11	24 40	14 11	23 34	18 27	08 10	16 49	21 29	08 19
27	21 07	08 07	24 25	13 57	23 39	18 27	08 09	16 50	21 29	08 19
28	21 17	11 46	24 09	13 44	23 43	18 27	08 08	16 51	21 29	08 18
29	21 28	14 51	23 49	13 33	23 47	18 28	08 07	16 52	21 29	08 18
30	21 S 38	17 S 07	23 S 28	13 S 23	23 S 51	18 N 28	08 N 07	16 S 53	21 S 29	08 N 18

ZODIAC SIGN ENTRIES

Date	h m	Planets
02	01 20	♂ ♐
02	10 03	☽ ♐
03	07 48	☿ ♐
04	12 40	☽ ♑
06	15 04	☽ ♒
08	18 06	☽ ♓
10	22 11	☽ ♈
13	03 35	☽ ♉
15	10 45	☽ ♊
17	20 16	☽ ♋
20	08 09	☽ ♌
22	16 05	☉ ♐
22	20 57	☽ ♍
25	08 07	☽ ♎
27	15 39	☽ ♏
29	19 23	☽ ♐

LATITUDES

Date	Mercury ☿	Venus ♀	Mars ♂	Jupiter ♃	Saturn ♄	Uranus ♅	Neptune ♆	Pluto ♇
01	02 S 03	05 S 55	00 S 28	00 N 23	01 N 36	00 N 20	01 N 24	16 N 37
04	02 17	05 26	00 30	00 24	01 37	00 20	01 24	16 38
07	02 28	04 50	00 32	00 24	01 38	00 20	01 23	16 38
10	02 36	04 10	00 33	00 25	01 38	00 20	01 23	16 39
13	02 40	03 26	00 35	00 26	01 39	00 20	01 23	16 40
16	02 38	02 40	00 36	00 26	01 40	00 20	01 23	16 41
19	02 29	01 53	00 38	00 27	01 40	00 20	01 23	16 42
22	02 11	01 08	00 39	00 27	01 41	00 20	01 23	16 43
25	01 42	00 S 26	00 41	00 28	01 42	00 20	01 23	16 44
28	01 00	00 N 14	00 42	00 29	01 42	00 20	01 23	16 45
31	00 S 06	00 N 50	00 S 44	00 N 29	01 N 43	00 N 20	01 N 23	16 N 47

DATA

Julian Date	2443814
Delta T	+49 seconds
Ayanamsa	23° 33' 38"
Synetic vernal point	05° ♓ 33' 22"
True obliquity of ecliptic	23° 26' 22"

MOON'S PHASES, APSIDES AND POSITIONS ☽

Date	h m	Phase	Longitude (° ')	Eclipse Indicator
07	16 18	☽	14 ♒ 54	
14	20 00	○	22 ♉ 06	
22	21 24	☾	00 ♍ 13	
30	08 19	●	07 ♐ 46	

Day	h m		
05	11 27	Perigee	
20	21 54	Apogee	
04	14 37	Max dec	18° S 19'
11	00 12	ON	
17	23 02	Max dec	18° N 24'
25	10 55	OS	

LONGITUDES

Date	Chiron ⚷	Ceres ⚳	Pallas ⚴	Juno ⚵	Vesta ⚶	Black Moon Lilith ⚸
01	07 ♉ 19	19 ♑ 56	26 ♐ 18	00 ♒ 09	08 ♑ 41	02 ♌ 12
11	06 ♉ 51	22 ♑ 50	29 ♐ 50	03 ♒ 00	13 ♑ 18	03 ♌ 19
21	06 ♉ 23	25 ♑ 57	03 ♑ 27	06 ♒ 12	18 ♑ 01	04 ♌ 26
31	05 ♉ 58	29 ♑ 16	07 ♑ 08	09 ♒ 43	22 ♑ 51	05 ♌ 33

All ephemeris data is given at 12.00 UT and the Moon's longitude is additionally given for 24.00 UT

ASPECTARIAN

01 Wednesday
00 00 ☿ ∥ ♆
00 38 ☽ ⊥ ♆
02 02 ☽ ✶ ♄
10 22 ☽ ☌ ♅
11 07 ☽ ⊻ ♆
11 56 ♄ ⊥ ♇
12 31 ☽ ⊻ ♇
14 47 ☽ ☌ ♀
22 54 ☽ ⊥ ♇
22 56 ☽ Q ♄
23 59 ☽ ∥ ☉

02 Thursday
07 41 ☽ ☌ ☿
10 31 ☽ ☌ ♂
14 19 ☽ ∠ ♇
18 22 ☽ ∥ ♅

03 Friday
00 01 ☽ △ ♃
03 39 ☽ ⊻ ☉
05 45 ☽ □ ♄
11 56 ☉ ⊥ ♆
13 45 ☽ ⊻ ♅
14 22 ☽ ☌ ♆
14 32 ☽ ⊥ ☉
15 45 ☽ ✶ ♇
15 55 ☽ ⊻ ♀
16 34 ☿ ∥ ♀
19 53 ♀ ⊻ ♇
23 53 ☽ ⊥ ♅

04 Saturday
01 25 ☽ ꝑ ♃
01 37 ☽ ⊥ ♀
06 48 ☽ ∠ ☉
11 58 ☽ Q ♇
12 03 ☉ ✶ ♄
15 03 ☽ ∠ ♅
15 44 ☽ ⊻ ☿
15 49 ☽ ⊻ ♂
16 08 ☽ ∠ ♀
16 36 ☽ ± ♃
17 27 ☿ ☌ ♂

05 Sunday
02 24 ☽ ⊥ ♂
02 42 ☽ ⊼ ♃
02 51 ☽ ⊥ ☿
04 16 ♀ ⊻ ♆
05 30 ☿ ∠ ♇
08 26 ☽ △ ♄
09 51 ☽ ✶ ☉
15 44 ♀ ☌ ♅
16 17 ☽ ✶ ♀
16 19 ☽ ✶ ♅
16 40 ♂ ∠ ♇
16 49 ☽ ⊻ ♆
18 14 ☽ □ ♇
18 18 ☽ ∠ ♂
19 31 ☽ ∠ ☿

06 Monday
02 55 ☽ ⊥ ♆
07 32 ☽ Q ☉
09 47 ☽ ꝑ ♄
09 55 ☽ ∥ ♅
11 39 ☽ Q ♀
12 35 ☽ Q ♅
14 43 ♀ ∥ ♆
15 07 ☽ ∥ ☉
18 08 ☽ ∠ ♆
18 25 ♀ ∥ ♂
20 55 ☽ ✶ ♂
23 23 ☽ ✶ ☿

07 Tuesday
01 04 ☽ ± ♄
04 12 ♂ ∥ ♆
05 26 ☽ ☍ ♃
11 18 ☽ ⊼ ♄
16 18 ☽ □ ☉
16 53 ☽ □ ♀
18 22 ☽ Q ♂
19 13 ☽ □ ♅
19 37 ☽ ✶ ♆
21 03 ☽ △ ♇
21 34 ☉ ☌ ♀
21 50 ☽ Q ☿

08 Wednesday
07 13 ☉ ∥ ♅
16 11 ☽ Q ♆
22 48 ☽ ꝑ ♇
23 23 ☽ ⋕ ♄

09 Thursday
00 01 ☽ ⋕ ♇
02 52 ☽ □ ♂
07 50 ☽ □ ☿
08 57 ☽ ⊼ ♃
11 42 ☉ ☌ ♅
14 23 ☽ ± ♇
15 01 ☽ ☍ ♄
16 19 ☉ ⊻ ♆
18 15 ☽ △ ♀
19 23 ☽ ± ♃
20 32 ☿ △ ♃
23 02 ☽ △ ♅
23 21 ☽ □ ♆
23 53 ☽ △ ☉

10 Friday
00 49 ☽ ⊼ ♇
11 06 ☽ ꝑ ♃
13 20 ☉ ⊻ ♇
19 20 ☽ ꝑ ♀

11 Saturday
01 23 ☽ ꝑ ♅
04 12 ☽ ꝑ ☉
10 06 ☽ △ ♂
10 31 ☽ ± ♀
13 34 ☽ △ ♃
17 21 ☽ ± ♅
17 30 ☽ △ ☿
19 53 ☽ ⊼ ♄
20 42 ☽ ⊼ ♀
21 23 ☽ ± ☉

12 Sunday
04 04 ☽ ⊼ ♅
04 16 ☽ △ ♆
05 47 ☽ ☍ ♇
06 37 ☽ ± ♄
08 56 ☽ ⊼ ☉
13 29 ♀ ✶ ♄
14 19 ☽ ꝑ ♂
19 17 ☿ ⊻ ♀
22 41 ☿ □ ♄
22 51 ☽ ꝑ ♄
22 52 ☽ ꝑ ☿

13 Monday
01 15 ☽ ∥ ♇
01 29 ☽ ∥ ♄
07 16 ☽ ꝑ ♆
07 27 ☽ ± ♂
16 41 ☽ ± ☿
18 59 ☽ ⊼ ♂
19 35 ☽ □ ♃
22 19 ♀ ⋕ ♃
23 41 ☽ ± ♆

14 Tuesday
00 33 ☽ ☍ ♀
02 14 ☽ △ ♄
04 40 ☽ ⊼ ☿
06 25 ♂ △ ♃
10 38 ☽ ☍ ♅
10 43 ☽ ⊼ ♆
12 18 ☽ ⊼ ♇
12 58 ☉ ∥ ♀
20 00 ☽ ☍ ☉
20 44 ♀ ⊥ ♆
23 26 ☽ ± ♇

15 Wednesday
04 54 ☽ Q ♃
09 22 ♄ ∥ ♇
11 12 ☉ ⋕ ♃
16 20 ☽ ꝑ ♇
22 35 ☽ ⋕ ♅

16 Thursday
03 33 ☽ ✶ ♃
03 53 ♅ ⊻ ♆
06 12 ☽ ☍ ♂
06 23 ☽ ⊼ ♀
08 13 ♀ ⊻ ♂
10 37 ☽ □ ♄
11 22 ☽ ⋕ ♀
13 19 ☿ ⊥ ♀
17 26 ☽ ± ♀
17 32 ☉ ⊥ ♇
17 57 ☽ ☍ ☿
19 20 ☽ ☍ ♆
19 22 ☽ ⊼ ♅
20 58 ☽ △ ♇

17 Friday
07 01 ☽ ± ♅
08 26 ☽ ∠ ♃
09 29 ☉ Q ♄
09 54 ☽ ⊼ ☉
10 14 ☽ ꝑ ♀
12 39 ☿ ☌ ♆
13 41 ☿ ⊻ ♅
21 39 ☽ Q ♄
22 41 ☽ ± ☉

18 Saturday
00 41 ☽ ꝑ ♅
02 05 ☽ ⊥ ♃
11 21 ♀ ∥ ♅
12 26 ☿ ✶ ♇
13 58 ☽ ⊻ ♃
14 44 ☽ △ ♀
18 05 ☽ ꝑ ☉
20 22 ☽ ⊼ ♂
21 31 ☽ ✶ ♄

19 Sunday
06 27 ☽ ⊼ ♆
06 37 ☽ △ ♅
08 09 ☽ □ ♇
09 08 ☽ ± ♂
09 33 ☽ ⊼ ☿
13 39 ♀ □ ♃
16 56 ♂ □ ♄
18 33 ☽ ± ♆
22 27 ☽ ± ☿

20 Monday
00 02 ☽ ⋕ ♅
02 57 ☽ △ ☉
03 48 ☽ ∠ ♄
04 27 ☽ ꝑ ♂
09 56 ☽ ⋕ ♀
12 47 ☽ ꝑ ♆
17 48 ☽ ꝑ ☿
20 36 ☽ Q ♇
22 08 ☽ ⊥ ♄

21 Tuesday
01 22 ☽ □ ♀
02 27 ☽ ☌ ♃
10 22 ☽ ⊻ ♄
12 54 ☽ △ ♂
17 06 ♀ ⊥ ♂
19 20 ☽ △ ♆
19 38 ☽ □ ♅
21 02 ☽ ✶ ♇

22 Wednesday
01 53 ☽ △ ☿
13 03 ☽ Q ♀
21 24 ☽ □ ☉

23 Thursday
03 25 ☽ ∠ ♇
05 27 ☽ ∥ ♇
06 13 ☽ ∥ ♄
08 10 ☽ Q ♅
12 43 ☽ ✶ ♀
15 06 ☽ ⊻ ♃
21 26 ☽ ⊥ ♇
23 05 ☽ ☌ ♄

24 Friday
03 00 ☽ ⊥ ♃
05 10 ☽ □ ♂
07 41 ☽ □ ♆
08 06 ☽ ✶ ♅
09 19 ☽ ⊻ ♇
12 29 ☽ Q ☉
15 41 ☽ □ ☿
17 53 ☽ ∠ ♀
20 39 ☽ ∠ ♃

25 Saturday
11 03 ☽ ⊥ ♀
13 18 ☽ ∠ ♅
13 43 ☽ ✶ ☉
18 12 ☽ Q ♂
18 34 ☽ Q ♆
20 28 ♃ St R
21 41 ☿ St R
22 20 ☽ ⊻ ♀
22 27 ☉ ∠ ♇

26 Sunday
00 44 ♂ ☌ ♆
01 19 ☽ ✶ ♃
02 23 ☽ Q ☿
06 22 ☽ ⊥ ♅
09 03 ☽ ⊻ ♄
09 37 ♂ ⊻ ♅
17 02 ☽ ✶ ♆
17 32 ☽ ⊻ ♅
17 58 ☽ ✶ ♂
18 31 ☽ ☌ ♇
20 09 ☽ ∠ ☉
20 10 ☽ ⊥ ♄
23 31 ☽ Q ♃

27 Monday
00 22 ☽ ✶ ☿
04 08 ♂ ✶ ♇
12 11 ☽ ⋕ ♄
12 32 ☽ ∠ ♄
13 10 ☽ ⋕ ♇
13 47 ☽ ⊥ ☉
20 12 ☽ ∠ ♆
22 35 ☽ ∠ ♂

28 Tuesday
01 18 ☽ ⊻ ☉
02 39 ☽ ∠ ☿
04 37 ☽ ☌ ♀
07 36 ☽ □ ♃
12 01 ☽ ⊥ ♆
13 10 ♀ St D
15 00 ☽ ✶ ♄
15 12 ☽ ⊥ ♂
17 49 ☽ ⊥ ☿
22 24 ☽ ⊻ ♆
22 57 ☽ ☌ ♅
23 45 ☽ ⊻ ♇

29 Wednesday
01 48 ☽ ∥ ♀
02 05 ☽ ⊻ ♂
03 43 ☽ ⊻ ☿
09 59 ☽ ⊥ ♇
11 34 ☽ Q ♄
14 37 ☿ ∥ ♂
15 21 ☉ ∥ ♆
19 31 ☿ ☌ ♂
23 00 ☉ ⊻ ♀

30 Thursday
01 06 ☽ ∠ ♇
07 43 ☽ ⊻ ♀
08 19 ☽ ☌ ☉
08 46 ☽ ∥ ♅
10 24 ☽ △ ♃
17 37 ☽ □ ♄
17 40 ☽ ⊥ ♀

DECEMBER 1978

LONGITUDES

Date	Sidereal time h	m	s	Sun ☉ °	'	"	Moon ☽ °	'	"	Moon ☽ 24.00 °	'	"	Mercury ☿ °	'	Venus ♀ °	'	Mars ♂ °	'	Jupiter ♃ °	'	Saturn ♄ °	'	Uranus ♅ °	'	Neptune ♆ °	'	Pluto ♇ °	'
01	16	39	58	08 ♐	56	21	24 ♐	38	07	02 ♑	00	35	18 ♐	53	07 ♏	31	21 ♐	34	09 ♌	00	13 ♍	26	18 ♏	04	17 ♐	42	18 ♎	28
02	16	43	54	09	57	13	09 ♑	23	49	16 ♑	46	46	17 R	49	07	39	22	18	08 R	59	13	29	18	08	17	44	18	30
03	16	47	51	10	58	05	24 ♑	08	29	01 ♒	28	08	16	37	07	49	23	03	08	58	13	31	18	11	17	46	18	31
04	16	51	47	11	58	58	08 ♒	45	00	15 ♒	58	31	15	19	08	02	23	48	08	56	13	33	18	15	17	48	18	33
05	16	55	44	12	59	52	23 ♒	08	17	00 ♓	14	03	13	57	08	16	24	33	08	54	13	36	18	19	17	51	18	34
06	16	59	40	14	00	46	07 ♓	15	40	14 ♓	13	07	12	34	08	33	25	18	08	52	13	38	18	22	17	53	18	36
07	17	03	37	15	01	41	21 ♓	06	30	27 ♓	55	55	11	13	08	52	26	03	08	50	13	40	18	25	17	55	18	38
08	17	07	34	16	02	37	04 ♈	41	34	11 ♈	23	39	09	56	09	13	26	48	08	48	13	41	18	29	17	58	18	39
09	17	11	30	17	03	34	18 ♈	02	22	24 ♈	37	54	08	47	09	35	27	34	08	45	13	43	18	32	18	00	18	41
10	17	15	27	18	04	31	01 ♉	10	27	07 ♉	40	08	07	46	10	00	28	19	08	42	13	45	18	36	18	02	18	42
11	17	19	23	19	05	29	14 ♉	07	05	20 ♉	31	22	06	55	10	26	29	04	08	39	13	46	18	39	18	04	18	43
12	17	23	20	20	06	28	26 ♉	53	03	03 ♊	12	08	06	15	10	54	29 ♐	49	08	36	13	48	18	42	18	07	18	45
13	17	27	16	21	07	27	09 ♊	28	39	15 ♊	42	36	05	46	11	24	00 ♑	35	08	33	13	49	18	46	18	09	18	46
14	17	31	13	22	08	27	21 ♊	54	00	28 ♊	02	51	05	28	11	55	01	20	08	29	13	50	18	49	18	11	18	47
15	17	35	09	23	09	28	04 ♋	09	14	10 ♋	13	13	05	21	12	28	02	06	08	26	13	51	18	52	18	13	18	49
16	17	39	06	24	10	30	16 ♋	14	56	22 ♋	14	34	05 D	25	13	02	02	51	08	22	13	52	18	56	18	16	18	50
17	17	43	03	25	11	32	28 ♋	12	19	04 ♌	08	30	05	38	13	37	03	37	08	18	13	53	18	59	18	18	18	51
18	17	46	59	26	12	35	10 ♌	03	26	15 ♌	57	32	05	59	14	15	04	22	08	13	13	54	19	02	18	20	18	53
19	17	50	56	27	13	39	21 ♌	51	16	27 ♌	45	08	06	28	14	53	05	08	08	09	13	55	19	05	18	22	18	54
20	17	54	52	28	14	44	03 ♍	39	42	09 ♍	35	34	07	04	15	32	05	54	08	04	13	55	19	08	18	25	18	55
21	17	58	49	29 ♐	15	50	15 ♍	33	22	21 ♍	33	48	07	46	16	13	06	39	07	59	13	56	19	11	18	27	18	56
22	18	02	45	00 ♑	16	56	27 ♍	37	32	03 ♎	45	16	08	34	16	55	07	25	07	55	13	56	19	14	18	29	18	57
23	18	06	42	01	18	03	09 ♎	57	43	16 ♎	15	30	09	27	17	38	08	11	07	49	13	56	19	17	18	31	18	58
24	18	10	38	02	19	11	22 ♎	39	16	29 ♎	09	34	10	24	18	23	08	57	07	44	13	56	19	20	18	34	18	59
25	18	14	35	03	20	19	05 ♏	46	51	12 ♏	31	26	11	25	19	08	09	43	07	39	13 R	56	19	23	18	36	19	00
26	18	18	32	04	21	28	19 ♏	23	30	26 ♏	23	03	12	29	19	54	10	29	07	33	13	56	19	26	18	38	19	01
27	18	22	28	05	22	38	03 ♐	29	53	10 ♐	43	33	13	36	20	41	11	15	07	27	13	56	19	29	18	40	19	02
28	18	26	25	06	23	48	18 ♐	03	24	25 ♐	28	36	14	46	21	29	12	01	07	21	13	56	19	32	18	42	19	03
29	18	30	21	07	24	58	02 ♑	58	04	10 ♑	30	36	15	58	22	18	12	47	07	15	13	55	19	35	18	45	19	04
30	18	34	18	08	26	09	18 ♑	04	54	25 ♑	39	38	17	12	23	08	13	33	07	09	13	55	19	38	18	47	19	05
31	18	38	14	09 ♑	27	20	03 ♒	13	29	10 ♒	45	13	18 ♐	28	23 ♏	58	14 ♑	19	07 ♌	03	13 ♍	54	19 ♏	40	18 ♐	49	19 ♎	05

DECLINATIONS

Date	Moon True ☊ °	'	Moon Mean ☊ °	'	Moon ☽ Latitude °	'	Sun ☉ °	'	Moon ☽ °	'	Mercury ☿ °	'	Venus ♀ °	'	Mars ♂ °	'	Jupiter ♃ °	'	Saturn ♄ °	'	Uranus ♅ °	'	Neptune ♆ °	'	Pluto ♇ °	'
01	23 ♍	32	22 ♍	50	05 N	01	21 S	47	18 S	19	23 S	05	13 S	14	23 S	54	18 N	28	08 N	06	16 S	54	21 S	30	08 N	18
02	23 R	21	22	47	04	50	21	57	18	18	22	39	13	06	23	57	18	29	08	05	16	55	21	30	08	17
03	23	12	22	44	04	19	22	05	17	02	22	12	12	59	24	00	18	30	08	04	16	56	21	30	08	17
04	23	06	22	41	03	31	22	14	14	40	21	44	12	53	24	03	18	30	08	04	16	57	21	30	08	17
05	23	03	22	38	02	31	22	21	11	26	21	15	12	49	24	05	18	31	08	03	16	58	21	31	08	17
06	23	02	22	35	01	22	22	29	07	34	20	46	12	45	24	07	18	32	08	03	16	59	21	31	08	17
07	23 D	02	22	31	00 N	10	22	36	03 S	22	20	18	12	42	24	09	18	32	08	02	16	59	21	31	08	17
08	23 R	01	22	28	01 S	01	22	42	00 N	56	19	52	12	41	24	11	18	33	08	02	17	00	21	31	08	16
09	22	59	22	25	02	08	22	49	05	06	19	29	12	40	24	12	18	34	08	01	17	01	21	31	08	16
10	22	55	22	22	03	07	22	54	08	58	19	09	12	40	24	14	18	35	08	01	17	02	21	32	08	16
11	22	47	22	19	03	55	22	59	12	20	18	52	12	41	24	14	18	36	08	00	17	03	21	32	08	16
12	22	37	22	16	04	30	23	04	15	04	18	39	12	43	24	15	18	37	08	00	17	04	21	32	08	16
13	22	24	22	12	04	52	23	09	17	03	18	30	12	46	24	15	18	38	08	00	17	05	21	32	08	16
14	22	10	22	09	05	00	23	12	18	13	18	25	12	49	24	16	18	39	08	00	17	06	21	32	08	16
15	21	55	22	06	04	53	23	16	18	30	18	23	12	53	24	15	18	40	07	59	17	07	21	33	08	16
16	21	41	22	03	04	33	23	19	17	56	18	25	12	58	24	15	18	42	07	59	17	08	21	33	08	16
17	21	30	22	00	04	01	23	21	16	35	18	29	13	04	24	14	18	43	07	59	17	09	21	33	08	16
18	21	21	21	56	03	19	23	23	14	32	18	36	13	10	24	13	18	44	07	59	17	09	21	33	08	16
19	21	15	21	53	02	29	23	25	11	53	18	46	13	17	24	12	18	45	07	59	17	10	21	33	08	16
20	21	12	21	50	01	31	23	26	08	45	18	57	13	24	24	10	18	47	07	59	17	11	21	34	08	16
21	21	11	21	47	00 S	30	23	26	05	14	19	10	13	32	24	09	18	48	07	59	17	12	21	34	08	16
22	21 D	11	21	44	00 N	34	23	26	01 N	28	19	24	13	40	24	07	18	50	07	59	17	13	21	34	08	17
23	21 R	11	21	41	01	38	23	26	02 S	27	19	39	13	48	24	04	18	51	08	00	17	14	21	34	08	17
24	21	09	21	37	02	39	23	25	06	21	19	54	13	57	24	02	18	53	08	00	17	14	21	34	08	17
25	21	05	21	34	03	33	23	24	10	06	20	10	14	07	23	59	18	54	08	00	17	15	21	34	08	17
26	20	59	21	31	04	18	23	22	13	27	20	27	14	17	23	56	18	56	08	00	17	16	21	34	08	17
27	20	50	21	28	04	48	23	20	16	08	20	43	14	27	23	52	18	58	08	01	17	17	21	35	08	17
28	20	39	21	25	05	01	23	17	17	54	21	00	14	37	23	48	18	59	08	01	17	18	21	35	08	17
29	20	27	21	22	04	55	23	14	18	30	21	16	14	47	23	45	19	01	08	01	17	18	21	35	08	18
30	20	17	21	18	04	27	23	10	17	48	21	31	14	58	23	40	19	03	08	02	17	19	21	35	08	18
31	20 ♍	08	21 ♍	15	03 N	41	23 S	06	15 S	51	21 S	47	15 S	09	23 S	36	19 N	05	08 N	02	17 S	20	21 S	36	08 N	18

ZODIAC SIGN ENTRIES

Date	h	m	Planets
01	20	44	☽ ♑
03	21	35	☽ ♒
05	23	36	☽ ♓
08	03	40	☽ ♈
10	09	50	☽ ♉
12	17	39	♂ ♑
12	17	54	☽ ♊
15	03	50	☽ ♋
17	15	37	☽ ♌
20	04	34	☽ ♍
22	05	21	☉ ♑
22	16	40	☽ ♎
25	01	32	☽ ♏
27	06	07	☽ ♐
29	07	15	☽ ♑
31	06	53	☽ ♒

LATITUDES

Date	Mercury ☿ °	'	Venus ♀ °	'	Mars ♂ °	'	Jupiter ♃ °	'	Saturn ♄ °	'	Uranus ♅ °	'	Neptune ♆ °	'	Pluto ♇ °	'
01	00 S	06	00 N	50	00 S	44	00 N	29	01 N	43	00 N	20	01 N	23	16 N	47
04	00 N	54	01	22	00	45	00	30	01	44	00	20	01	23	16	48
07	01	50	01	51	00	47	00	30	01	45	00	20	01	23	16	49
10	02	29	02	15	00	48	00	31	01	45	00	20	01	23	16	51
13	02	49	02	37	00	49	00	32	01	46	00	20	01	23	16	52
16	02	50	02	55	00	50	00	32	01	47	00	20	01	23	16	54
19	02	40	03	10	00	52	00	33	01	48	00	20	01	23	16	55
22	02	22	03	22	00	53	00	33	01	48	00	20	01	23	16	57
25	02	00	03	31	00	54	00	34	01	49	00	20	01	23	16	59
28	01	35	03	39	00	55	00	35	01	50	00	20	01	23	17	00
31	01 N	10	03 N	44	00 S	56	00 N	35	01 N	51	00 N	20	01 N	23	17 N	02

DATA

Julian Date	2443844
Delta T	+49 seconds
Ayanamsa	23° 33' 42"
Synetic vernal point	05° ♓ 33' 18"
True obliquity of ecliptic	23° 26' 22"

MOON'S PHASES, APSIDES AND POSITIONS ☽

Date	h	m	Phase	Longitude °	'	Eclipse Indicator
07	00	34	☽	14 ♓	33	
14	12	31	○	22 ♊	10	
22	17	42	☾	00 ♎	31	
29	19	36	●	07 ♑	44	

Day	h	m		
02	16	01	Perigee	
18	16	19	Apogee	
30	21	32	Perigee	
01	23	32	Max dec	18° S 28'
08	06	48	ON	
15	07	57	Max dec	18° N 30'
22	21	03	OS	
29	11	14	Max dec	18° S 30'

LONGITUDES

Date	Chiron ⚷ °	'	Ceres ⚳ °	'	Pallas ⚴ °	'	Juno ⚵ °	'	Vesta ⚶ °	'	Black Moon Lilith ⚸ °	'
01	05 ♉	58	29 ♑	16	07 ♑	08	09 ♒	43	22 ♑	51	05 ♌	33
11	05 ♉	37	02 ♒	44	10 ♑	52	13 ♒	31	27 ♑	46	06 ♌	40
21	05 ♉	20	06 ♒	20	14 ♑	37	17 ♒	33	02 ♒	44	07 ♌	47
31	05 ♉	09	10 ♒	03	18 ♑	23	21 ♒	48	07 ♒	44	08 ♌	53

ASPECTARIAN

h m Aspects

01 Friday
00 38 ☽ ☌ ♆
01 14 ☽ ⊻ ♅
01 54 ☽ ✱ ♇
03 11 ☽ ☌ ☿
06 43 ☽ ☌ ♂
08 30 ☽ ∠ ♀
10 59 ☽ ⚼ ♃
11 05 ☽ ⊥ ♅
13 34 ☉ △ ♃
21 30 ☽ Q ♇
21 34 ☿ ✱ ♇

02 Saturday
01 36 ☽ ± ♃
01 47 ☽ ∠ ♅
05 36 ☿ ⊻ ♅
09 08 ☽ ✱ ♀
11 20 ☽ ⊼ ♃
12 58 ☽ ⊻ ☉
13 45 ☿ ☌ ♆
18 39 ☽ △ ♄
23 27 ☽ ⊥ ☉

03 Sunday
00 41 ☽ ⊻ ☿
01 35 ☽ ⊻ ♆
01 56 ♀ ∠ ♂
02 16 ☽ ✱ ♅
02 49 ☽ □ ♇
04 52 ☽ Q ♀
09 43 ☽ ⊥ ☿
10 08 ☽ ⊻ ♂
11 24 ☽ ⊥ ♆
13 22 ☽ ∥ ♅
15 12 ☽ ∠ ☉
16 25 ☉ ∥ ☿
19 11 ☽ ⚼ ♄
20 28 ☽ ⊥ ♂
21 57 ☽ Q ♅
23 16 ☽ ∠ ☿

04 Monday
02 11 ☽ ∠ ♆
10 01 ☽ ± ♄
10 47 ☽ □ ♀
12 06 ☽ ∠ ♂
12 18 ☽ ☍ ♃
16 04 ♂ ⚼ ♃
17 46 ☽ ✱ ☉
20 00 ☽ ⊼ ♄
21 58 ☽ ✱ ☿
23 03 ☿ ∥ ♆

05 Tuesday
02 12 ☽ ∥ ♀
03 06 ☽ ✱ ♆
03 30 ♂ ⊥ ♅
03 52 ☽ □ ♅
04 20 ☽ △ ♇
07 18 ☿ ⊥ ♀
14 31 ☽ ✱ ♂
15 22 ☽ Q ☉
16 20 ☽ Q ☿
18 05 ☿ □ ♄
21 33 ☉ ☌ ☿
23 22 ☽ Q ♆

06 Wednesday
02 35 ☉ □ ♄
05 43 ☽ ⚼ ♇
07 47 ☽ ⫲ ♇
09 11 ☽ ⫲ ♄
12 05 ☽ Q ♂
14 16 ☽ △ ♀
14 46 ☽ ⊼ ♃
20 20 ☽ □ ☿
21 13 ☽ ± ♇
23 00 ☽ ☍ ♄

07 Thursday
00 34 ☽ □ ☉
01 06 ☽ ± ♃
06 23 ☉ ⊥ ♀
06 25 ☽ □ ♆
07 17 ☽ △ ♅
07 39 ☽ ⊼ ♇
09 54 ♀ □ ♃
16 46 ☽ ⚼ ♃
16 58 ☽ ⚼ ♀
21 12 ☽ □ ♂

08 Friday
09 17 ☽ ± ♀
09 50 ☽ ⚼ ♅
19 19 ☽ △ ♃
20 19 ☽ ⊼ ♀
20 37 ☽ △ ☿
23 10 ☿ ⊻ ♀

09 Saturday
02 01 ☽ ± ♅
04 10 ☽ ⊼ ♄
10 05 ☽ △ ☉
11 55 ☽ △ ♆
12 37 ☿ △ ♃
12 55 ☽ ⊼ ♅
13 09 ☽ ☍ ♇
15 03 ☽ ± ♄
21 40 ☽ ⚼ ☿

10 Sunday
05 52 ☽ ∥ ♄
06 25 ☽ △ ♂
07 31 ☽ ∥ ♇
07 32 ☽ ⚼ ♄
11 00 ☉ ☌ ♆
13 01 ☽ ± ☿
15 26 ☽ ⚼ ♆
15 48 ☽ ⚼ ☉
23 23 ☽ ⊼ ☿

11 Monday
01 00 ☉ ⊻ ♅
01 52 ☽ □ ♃
03 06 ☉ ✱ ♇
04 53 ☽ ☍ ♀
08 10 ☽ ± ♆
09 55 ☽ ± ☉
11 21 ☽ △ ♄
11 54 ☽ ⚼ ♂
14 49 ☽ ⫲ ♀
19 25 ☽ ⊼ ♆
20 31 ☽ ☍ ♅
20 38 ☽ ⊼ ♇
22 07 ☽ ⊼ ☉

12 Tuesday
05 51 ☽ ± ♂
07 57 ☽ ± ♇
11 28 ☽ Q ♃
16 17 ☿ ⫲ ♃
17 55 ☽ ⊼ ♂

13 Wednesday
01 04 ☽ ⚼ ♇
05 07 ☽ ☍ ☿
10 13 ☽ ✱ ♃
12 25 ☽ ⫲ ♅
15 50 ☽ ⊼ ♀
17 12 ♅ ⊻ ♇
18 15 ♂ Q ♇
20 22 ☽ □ ♄

14 Thursday
03 55 ☽ ± ♀
04 46 ☽ ☍ ♆
05 57 ☽ △ ♇
05 59 ☽ ⊼ ♅
12 31 ☽ ☍ ☉
15 05 ☽ ∠ ♃
17 43 ☽ ± ♅
19 50 ☽ ⫲ ☿
22 14 ☽ ⚼ ♀

15 Friday
00 55 ♀ ⊥ ♆
07 28 ☽ Q ♄
07 40 ☽ ☍ ♂
08 37 ☽ ⊥ ♃
11 26 ☽ ⚼ ♅
14 22 ☽ ⊼ ☿
15 57 ☿ St D
17 57 ☉ ⚼ ♃
20 24 ☽ ⊻ ♃
20 37 ☽ ⫲ ☿
21 43 ♂ ± ♃

16 Saturday
02 17 ☽ ± ☿
05 16 ☽ △ ♀
07 15 ☽ ✱ ♄
16 02 ☽ ⊼ ♆
17 11 ☽ □ ♇
17 23 ☽ △ ♅
20 27 ☽ ⚼ ☿

17 Sunday
03 36 ☽ ⫲ ♅
04 06 ☽ ± ♆
05 22 ☽ ⊼ ☉
06 42 ☉ ⊥ ♅
13 23 ☽ ∠ ♄
18 36 ☽ ± ☉
22 19 ☽ ⚼ ♆
22 34 ♀ ✱ ♄
23 40 ☽ ⊼ ♂

18 Monday
00 33 ♂ ∠ ♅
03 26 ☽ △ ☿
05 32 ☽ Q ♇
07 37 ☽ ⊥ ♄
08 18 ☽ ☌ ♃
12 41 ☽ ± ♂
14 34 ☽ ⚼ ☉
19 49 ☽ ⊻ ♄
20 59 ☽ □ ♀

19 Tuesday
00 26 ☽ ⫲ ♀
04 53 ☽ △ ♆
05 58 ☽ ✱ ♇
06 20 ☽ □ ♅
08 15 ☽ ⚼ ♂
11 05 ☿ ⫲ ♃
23 58 ☽ △ ☉

20 Wednesday
11 44 ☽ Q ♀
12 31 ☽ ∠ ♇
15 22 ☽ ∥ ♇
16 50 ☽ △ ♂
17 22 ☽ ∥ ♄
19 04 ☽ Q ♅
19 19 ☽ □ ☿
20 52 ☽ ⊻ ♃

21 Thursday
06 43 ☽ ⊥ ♇
08 44 ☽ ☌ ♄
08 53 ☽ ⊥ ♃
13 25 ☽ ✱ ♀
17 48 ☽ □ ♆
18 11 ☿ △ ♃
18 46 ☽ ⊻ ♇
19 18 ☽ ✱ ♅

22 Friday
02 44 ☽ ∠ ♃
05 18 ♃ ⊥ ♄
09 46 ☽ Q ☿
17 42 ☽ □ ☉
20 57 ☽ ∠ ♀

23 Saturday
01 00 ☽ ∠ ♅
01 53 ♂ ⊼ ♃
04 02 ☉ Q ♇
05 21 ☽ Q ♆
07 55 ☽ ✱ ♃
08 21 ☽ □ ♂
10 56 ☽ ✱ ☿
15 25 ☽ ⊥ ♀
18 24 ☽ ⊥ ♅
19 36 ☽ ⊻ ♄
23 20 ☉ ± ♃

24 Sunday
03 30 ☽ ⊻ ♀
04 19 ☽ ✱ ♆
05 08 ☽ ☌ ♇
05 47 ☽ ⊻ ♅
06 35 ☽ Q ♃
06 56 ☽ ⊥ ♄
07 16 ☽ Q ☉
17 31 ☽ ∠ ☿
18 12 ♀ ⊻ ♆
20 26 ☽ Q ♂
21 13 ♄ St R
22 19 ☽ ⫲ ♄
23 36 ☽ ∠ ♄

25 Monday
00 08 ☽ ⫲ ♇
07 14 ☽ ✱ ☉
07 54 ♀ ⊻ ♇
08 03 ☽ ∠ ♆
11 17 ☽ ⊥ ☿
15 19 ☽ □ ♃
19 27 ☽ ✱ ♂
20 46 ♀ ☌ ♅
22 53 ☽ ⊻ ☿

26 Tuesday
00 10 ☽ ⊥ ♆
02 29 ☽ ✱ ♄
10 41 ☽ ⊻ ♆
11 21 ☽ ⊻ ♇
11 56 ☽ ∠ ☉
12 05 ☽ ☌ ♅
12 56 ☽ ☌ ♀
14 01 ☉ ∠ ♅
19 08 ☽ ∥ ♀
21 41 ☽ ⊥ ♇
23 03 ☽ ∠ ♂
23 14 ☽ Q ♄

27 Wednesday
04 32 ☽ ⊥ ☉
12 54 ☽ ∠ ♇
15 05 ☽ ⊥ ♂
15 23 ☽ ⊻ ☉
18 33 ☽ △ ♃
18 46 ☿ □ ♄

28 Thursday
01 34 ☽ ⊻ ♂
01 54 ☽ ∥ ♅
05 16 ☽ □ ♄
06 10 ☽ ☌ ☿
13 03 ☽ ☌ ♆
13 37 ☽ ✱ ♇
14 25 ☽ ⊻ ♅
17 53 ☽ ⊻ ♀
18 55 ☽ ⚼ ♃
22 20 ☉ ∠ ♀

29 Friday
00 08 ☽ ⊥ ♅
04 06 ☽ ⊥ ♀
08 35 ☉ ⊼ ♃
08 57 ☽ Q ♇
09 17 ☽ ± ♃
14 35 ☽ ∠ ♅
18 47 ☽ ⊼ ♃
19 18 ☽ ∠ ♀
19 36 ☽ ☌ ☉

30 Saturday
04 26 ☽ ☌ ♂
05 24 ☽ △ ♄
10 30 ☽ ⊻ ☿
13 06 ☽ ⊻ ♆
13 35 ☽ □ ♇
14 27 ☽ ✱ ♅
18 26 ☿ ∥ ♆
19 28 ☽ ∥ ♅
20 28 ☽ ✱ ♀
20 51 ☽ ⊥ ☿
22 38 ☽ ⊥ ♆
23 04 ♂ △ ♄

31 Sunday
03 55 ♃ Q ♇
05 08 ☽ ⚼ ♄
09 32 ☽ Q ♅
12 26 ☽ ∠ ☿
12 56 ☽ ∠ ♆
16 38 ☽ Q ♀
17 51 ☽ ∥ ♀
18 03 ☽ ☍ ♃
18 32 ☿ ☌ ♆
19 26 ☽ ± ♄
22 39 ☽ ⊻ ☉
23 35 ☿ ✱ ♇

All ephemeris data is given at 12.00 UT and the Moon's longitude is additionally given for 24.00 UT

JANUARY 1979

LONGITUDES

Date	Sidereal time h m s	Sun ☉ ° ' "	Moon ☽ ° ' "	Moon ☽ 24.00 ° ' "	Mercury ☿ ° '	Venus ♀ ° '	Mars ♂ ° '	Jupiter ♃ ° '	Saturn ♄ ° '	Uranus ♅ ° '	Neptune ♆ ° '	Pluto ♇ ° '
01	18 42 11	10 ♑ 28 30	18 ♒ 13 46	25 ♒ 38 11	19 ♐ 46	24 ♏ 50	15 ♑ 05	06 ♌ 56	13 ♍ 53	19 ♏ 43	18 ♐ 51	19 ♎ 06
02	18 46 07	11 29 41	02 ♓ 57 45	10 ♓ 11 58	21 05	25 42	15 52	06 R 50	13 R 52	19 46	18 53	19 07
03	18 50 04	12 30 51	17 ♓ 20 28	24 ♓ 23 08	22 25	26 35	16 38	06 43	13 51	19 49	18 55	19 07
04	18 54 01	13 32 01	01 ♈ 19 56	08 ♈ 11 00	23 47	27 28	17 24	06 36	13 50	19 51	18 57	19 08
05	18 57 57	14 33 11	14 ♈ 56 33	21 ♈ 36 53	25 09	28 22	18 11	06 29	13 49	19 54	18 59	19 09
06	19 01 54	15 34 20	28 ♈ 12 04	04 ♉ 43 17	26 33	29 ♏ 17	18 57	06 22	13 48	19 56	19 02	19 09
07	19 05 50	16 35 29	11 ♉ 10 05	17 ♉ 33 06	27 57	00 ♐ 12	19 43	06 15	13 46	19 59	19 04	19 10
08	19 09 47	17 36 37	23 ♉ 52 40	00 ♊ 09 07	29 ♐ 22	01 08	20 30	06 08	13 45	20 01	19 06	19 10
09	19 13 43	18 37 45	06 ♊ 22 46	12 ♊ 33 45	00 ♑ 48	02 05	21 16	06 01	13 43	20 03	19 08	19 11
10	19 17 40	19 38 53	18 ♊ 42 26	24 ♊ 48 56	02 15	03 02	22 03	05 53	13 41	20 06	19 10	19 11
11	19 21 36	20 40 00	00 ♋ 53 27	06 ♋ 56 08	03 42	04 00	22 50	05 46	13 39	20 08	19 12	19 12
12	19 25 33	21 41 07	12 ♋ 57 07	18 ♋ 56 34	05 10	04 58	23 36	05 38	13 37	20 10	19 14	19 12
13	19 29 30	22 42 14	24 ♋ 54 37	00 ♌ 51 25	06 39	05 56	24 23	05 31	13 35	20 13	19 16	19 12
14	19 33 26	23 43 20	06 ♌ 47 11	12 ♌ 42 07	08 08	06 56	25 09	05 23	13 33	20 15	19 18	19 13
15	19 37 23	24 44 26	18 ♌ 36 28	24 ♌ 30 30	09 38	07 55	25 56	05 15	13 31	20 17	19 20	19 13
16	19 41 19	25 45 31	00 ♍ 24 34	06 ♍ 19 02	11 08	08 55	26 43	05 07	13 28	20 19	19 22	19 13
17	19 45 16	26 46 36	12 ♍ 14 19	18 ♍ 10 52	12 39	09 56	27 30	04 59	13 26	20 21	19 23	19 13
18	19 49 12	27 47 41	24 ♍ 09 12	00 ♎ 09 52	14 10	10 57	28 16	04 52	13 23	20 23	19 25	19 13
19	19 53 09	28 48 45	06 ♎ 13 26	12 ♎ 20 30	15 42	11 58	29 03	04 44	13 21	20 25	19 27	19 13
20	19 57 05	29 ♑ 49 50	18 ♎ 31 41	24 ♎ 47 37	17 15	12 59	29 ♑ 50	04 36	13 18	20 27	19 29	19 13
21	20 01 02	00 ♒ 50 53	01 ♏ 08 53	07 ♏ 36 05	18 48	14 01	00 ♒ 37	04 28	13 15	20 29	19 31	19 13
22	20 04 59	01 51 57	14 ♏ 09 43	20 ♏ 50 14	20 22	15 04	01 24	04 20	13 12	20 30	19 33	19 R 13
23	20 08 55	02 52 59	27 ♏ 37 58	04 ♐ 33 07	21 56	16 07	02 11	04 12	13 09	20 32	19 34	19 13
24	20 12 52	03 54 02	11 ♐ 35 42	18 ♐ 45 33	23 31	17 10	02 58	04 04	13 06	20 34	19 36	19 13
25	20 16 48	04 55 04	26 ♐ 02 19	03 ♑ 25 23	25 06	18 13	03 45	03 55	13 03	20 36	19 38	19 13
26	20 20 45	05 56 06	10 ♑ 53 58	18 ♑ 27 01	26 43	19 17	04 32	03 47	13 00	20 37	19 40	19 13
27	20 24 41	06 57 06	26 ♑ 03 23	03 ♒ 41 44	28 19	20 21	05 19	03 39	12 56	20 39	19 41	19 13
28	20 28 38	07 58 06	11 ♒ 20 42	18 ♒ 58 55	29 ♑ 57	21 25	06 06	03 31	12 53	20 40	19 43	19 13
29	20 32 34	08 59 05	26 ♒ 35 03	04 ♓ 07 53	01 ♒ 35	22 29	06 53	03 23	12 50	20 42	19 45	19 12
30	20 36 31	10 00 03	11 ♓ 36 22	18 ♓ 59 37	03 13	23 34	07 40	03 15	12 46	20 43	19 46	19 12
31	20 40 28	11 ♒ 01 00	26 ♓ 16 59	03 ♈ 27 59	04 ♒ 53	24 ♐ 39	08 ♒ 27	03 ♌ 08	12 ♍ 42	20 ♏ 44	19 ♐ 48	19 ♎ 12

Date	Moon True ☊ ° '	Moon Mean ☊ ° '	Moon ☽ Latitude ° '
01	20 ♍ 01	21 ♍ 12	02 N 40
02	19 R 58	21 09	01 29
03	19 57	21 06	00 N 14
04	19 D 57	21 02	01 S 00
05	19 R 57	20 59	02 09
06	19 57	20 56	03 09
07	19 54	20 53	03 58
08	19 48	20 50	04 34
09	19 40	20 47	04 56
10	19 30	20 43	05 04
11	19 18	20 40	04 58
12	19 06	20 37	04 38
13	18 55	20 34	04 07
14	18 46	20 31	03 25
15	18 39	20 28	02 34
16	18 35	20 24	01 36
17	18 33	20 21	00 S 34
18	18 D 32	20 18	00 N 30
19	18 34	20 15	01 34
20	18 35	20 12	02 35
21	18 R 35	20 08	03 30
22	18 35	20 05	04 16
23	18 32	20 02	04 49
24	18 27	19 59	05 07
25	18 21	19 56	05 07
26	18 14	19 53	04 46
27	18 08	19 49	04 05
28	18 02	19 46	03 06
29	17 59	19 43	01 54
30	17 57	19 40	00 35
31	17 ♍ 57	19 ♍ 37	00 S 45

DECLINATIONS

Date	Sun ☉ ° '	Moon ☽ ° '	Mercury ☿ ° '	Venus ♀ ° '	Mars ♂ ° '	Jupiter ♃ ° '	Saturn ♄ ° '	Uranus ♅ ° '	Neptune ♆ ° '	Pluto ♇ ° '
01	23 S 02	12 S 50	22 S 01	15 S 20	23 S 31	19 N 06	08 N 03	17 S 21	21 S 36	08 N 18
02	22 57	09 02	22 15	15 32	23 26	19 08	08 04	17 21	21 36	08 19
03	22 51	04 47	22 28	15 43	23 21	19 10	08 04	17 22	21 36	08 19
04	22 45	00 S 24	22 41	15 54	23 15	19 12	08 05	17 23	21 36	08 19
05	22 39	03 N 54	22 52	16 06	23 09	19 14	08 06	17 23	21 36	08 20
06	22 32	07 53	23 03	16 18	23 03	19 16	08 06	17 24	21 37	08 20
07	22 25	11 24	23 13	16 29	22 57	19 18	08 07	17 25	21 37	08 20
08	22 17	14 19	23 21	16 41	22 50	19 20	08 08	17 25	21 37	08 21
09	22 09	16 31	23 29	16 52	22 43	19 22	08 09	17 26	21 37	08 21
10	22 00	17 55	23 36	17 04	22 36	19 24	08 10	17 27	21 37	08 21
11	21 51	18 28	23 41	17 15	22 29	19 26	08 11	17 27	21 37	08 22
12	21 42	18 11	23 45	17 26	22 21	19 28	08 12	17 28	21 37	08 22
13	21 32	17 06	23 48	17 38	22 13	19 30	08 13	17 28	21 38	08 23
14	21 21	15 16	23 50	17 49	22 05	19 32	08 14	17 29	21 38	08 23
15	21 11	12 49	23 51	17 59	21 57	19 34	08 15	17 30	21 38	08 24
16	21 00	09 50	23 50	18 10	21 48	19 36	08 16	17 30	21 38	08 24
17	20 48	06 27	23 48	18 21	21 39	19 38	08 17	17 31	21 38	08 25
18	20 36	02 N 47	23 45	18 31	21 30	19 40	08 18	17 31	21 38	08 25
19	20 24	01 S 02	23 40	18 41	21 21	19 42	08 19	17 32	21 38	08 26
20	20 11	04 52	23 34	18 51	21 11	19 44	08 21	17 32	21 38	08 26
21	19 58	08 36	23 27	19 00	21 01	19 46	08 22	17 33	21 39	08 27
22	19 45	12 01	23 18	19 10	20 51	19 48	08 23	17 33	21 39	08 27
23	19 31	14 56	23 08	19 18	20 41	19 50	08 25	17 34	21 39	08 28
24	19 17	17 06	22 57	19 27	20 30	19 52	08 26	17 34	21 39	08 28
25	19 02	18 16	22 44	19 35	20 20	19 54	08 28	17 34	21 39	08 29
26	18 47	18 15	22 30	19 43	20 09	19 56	08 29	17 35	21 39	08 29
27	18 32	16 56	22 15	19 51	19 57	19 58	08 30	17 35	21 39	08 30
28	18 17	14 24	21 57	19 58	19 46	20 00	08 32	17 36	21 39	08 31
29	18 01	10 52	21 39	20 05	19 34	20 02	08 34	17 36	21 39	08 31
30	17 44	06 41	21 19	20 11	19 22	20 04	08 35	17 36	21 40	08 32
31	17 S 28	02 S 10	20 S 58	20 S 17	19 S 10	20 N 06	08 N 37	17 S 37	21 S 40	08 N 32

ZODIAC SIGN ENTRIES

Date	h	m	Planets	
02	07	08	☽	♓
04	09	41	☽	♈
06	15	17	☽	♉
07	06	38	♀	♐
08	22	33	☿	♑
08	23	42	☽	♊
11	10	14	☽	♋
13	22	16	☽	♌
16	11	10	☽	♍
18	23	40	☽	♎
20	16	00	☉	♒
20	17	07	♂	♒
21	09	51	☽	♏
23	16	08	☽	♐
25	18	27	☽	♑
27	18	12	☽	♒
28	12	49	☿	♒
29	17	25	☽	♓
31	18	11	☽	♈

LATITUDES

Date	Mercury ☿ ° '	Venus ♀ ° '	Mars ♂ ° '	Jupiter ♃ ° '	Saturn ♄ ° '	Uranus ♅ ° '	Neptune ♆ ° '	Pluto ♇ ° '
01	01 N 02	03 N 45	00 S 56	00 N 35	01 N 51	00 N 20	01 N 23	17 N 03
04	00 37	03 47	00 57	00 36	01 52	00 20	01 23	17 04
07	00 N 13	03 48	00 58	00 36	01 53	00 20	01 23	17 06
10	00 S 10	03 46	00 59	00 37	01 53	00 20	01 23	17 08
13	00 32	03 44	01 00	00 37	01 54	00 20	01 23	17 10
16	00 52	03 40	01 01	00 38	01 55	00 20	01 23	17 11
19	01 10	03 34	01 01	00 38	01 56	00 20	01 23	17 13
22	01 26	03 28	01 02	00 39	01 56	00 20	01 23	17 15
25	01 39	03 20	01 02	00 39	01 57	00 20	01 23	17 17
28	01 50	03 12	01 03	00 39	01 58	00 20	01 23	17 18
31	01 S 58	03 N 03	01 S 04	00 N 40	01 N 58	00 N 20	01 N 23	17 N 20

DATA

Julian Date	2443875
Delta T	+50 seconds
Ayanamsa	23° 33' 47"
Synetic vernal point	05° ♓ 33' 12"
True obliquity of ecliptic	23° 26' 22"

MOON'S PHASES, APSIDES AND POSITIONS ☽

Date	h	m	Phase	Longitude ° '	Eclipse Indicator
05	11	15	☽	14 ♈ 31	
13	07	09	○	22 ♋ 30	
21	11	23	☾	00 ♏ 49	
28	06	20	●	07 ♒ 44	

Day	h	m		
15	02	39	Apogee	
28	09	43	Perigee	
04	14	09	ON	
11	15	51	Max dec	18° N 29'
19	05	34	OS	
25	23	32	Max dec	18° S 25'
31	23	29	ON	

LONGITUDES

Date	Chiron ⚷ ° '	Ceres ⚳ ° '	Pallas ⚴ ° '	Juno ⚵ ° '	Vesta ⚶ ° '	Black Moon Lilith ⚸ ° '
01	05 ♉ 08	10 ♒ 25	18 ♑ 46	22 ♒ 14	08 ♒ 14	09 ♌ 00
11	05 ♉ 03	14 ♒ 13	22 ♑ 31	26 ♒ 41	13 ♒ 17	10 ♌ 07
21	05 ♉ 04	18 ♒ 04	26 ♑ 15	01 ♓ 18	18 ♒ 20	11 ♌ 14
31	05 ♉ 10	21 ♒ 59	29 ♑ 56	06 ♓ 03	23 ♒ 24	12 ♌ 20

ASPECTARIAN

h m Aspects

01 Monday
05 01 ☽ ⚻ ♄
06 40 ☽ ⚺ ♂
08 58 ☽ ⊥ ☉
11 06 ☿ ⚺ ♅
13 00 ☽ ⚹ ♆
13 24 ☽ △ ♇
14 25 ☽ □ ♅
14 43 ☽ ⚹ ☿
16 52 ☽ ⊥ ♂
19 38 ♀ ⊥ ♇
23 21 ☽ □ ♀

02 Tuesday
00 37 ☽ ∠ ☉
08 21 ☽ ∠ ♂
08 34 ☽ Q ♆
12 13 ☽ Q ☿
13 54 ☽ bQ ♇
16 12 ☽ # ♇
16 38 ♀ Q ♄
17 39 ☽ # ♄
18 21 ☽ ⚻ ♃

03 Wednesday
00 25 ☿ bQ ♃
03 15 ☽ ⚹ ☉
04 17 ☽ ± ♃
04 53 ☽ ± ♇
06 08 ☽ ☍ ♄
10 44 ☽ ⚹ ♂
14 41 ☽ □ ♆
15 01 ☽ ⚻ ♇
16 12 ☽ △ ♅
19 23 ☽ bQ ♃
21 34 ☽ □ ☿

04 Thursday
01 11 ☽ Q ☉
04 51 ☽ △ ♀
08 27 ☽ Q ♂
17 35 ☉ ∥ ☿
18 10 ☽ bQ ♅
18 57 ☉ △ ♄
21 09 ☽ △ ♃

05 Friday
09 00 ☽ bQ ♀
09 59 ☽ ⚻ ♄
10 07 ☽ ± ♅
11 15 ☽ □ ☉
18 09 ☽ □ ♂
19 17 ☽ △ ♆
19 33 ☽ ☍ ♇
20 44 ☽ ± ♄
20 55 ☽ ⚻ ♅

06 Saturday
01 11 ☿ ⊥ ♅
02 21 ☽ ± ♀
08 36 ☽ △ ☿
12 15 ☿ ∥ ♂
13 04 ☽ bQ ♄
13 22 ☽ ∥ ♄
14 08 ☽ ⚻ ♀
14 26 ♂ ⚺ ♆
14 50 ☽ ∥ ♇
18 24 ♂ □ ♇
22 44 ☽ bQ ♆

07 Sunday
02 56 ☽ □ ♃
15 33 ☽ ± ♆
15 45 ☽ bQ ☿
16 52 ☽ △ ♄
20 18 ♂ ⚹ ♅
23 04 ☽ △ ☉

08 Monday
02 54 ☽ ⚻ ♆
03 03 ☽ ⚻ ♇
04 39 ☽ ☍ ♅
05 10 ☽ △ ♂
10 55 ☽ ± ☿
12 29 ☽ Q ♃
14 28 ☽ ± ♇
23 48 ☿ ± ♃
23 51 ☽ ⚻ ☿

09 Tuesday
03 02 ☽ ☍ ♀
06 13 ☽ bQ ☉
07 45 ☽ bQ ♇
11 18 ☽ ⚹ ♃
11 47 ☽ bQ ♂
17 38 ☽ # ♀
18 17 ☿ Q ♇

10 Wednesday
00 09 ☉ ⚺ ♆
01 03 ☉ □ ♇
01 13 ☽ ± ☉
02 11 ☽ # ♅
02 13 ☽ □ ♄
06 27 ☽ ± ♂
12 54 ☽ ☍ ♆
12 56 ☽ △ ♇
14 01 ☽ ⚻ ☉
14 44 ☽ ⚻ ♅
16 14 ☽ ∠ ♃
19 00 ☽ ⚻ ♂
23 00 ☉ ⚹ ♅

11 Thursday
02 34 ☽ ± ♅
09 42 ♆ ⚹ ♇
09 48 ☽ ⊥ ♃
13 31 ☽ Q ♄
16 57 ♀ ∠ ♇
18 21 ☽ ☍ ☿
18 41 ☽ ⚻ ♀
20 27 ☽ bQ ♅
21 34 ☽ ⚺ ♃

12 Friday
02 05 ☿ ⚺ ♀
07 41 ☽ ± ♀
12 04 ☿ ∠ ♅
13 20 ☽ ⚹ ♄
15 21 ♀ ∥ ♅
19 01 ☿ ⚻ ♃
22 05 ☉ ∥ ♆

13 Saturday
00 31 ☽ □ ♇
00 37 ☽ ⚻ ♆
02 31 ☽ △ ♅
02 40 ♀ △ ♃
03 18 ☽ bQ ♀
03 44 ☽ # ♀
05 28 ☽ # ♅
07 09 ☽ ☍ ☉
10 51 ☽ ☍ ♂
12 43 ☽ ± ♆
19 24 ☽ ∠ ♄

14 Sunday
06 56 ☽ bQ ♆
09 11 ☽ ☌ ♃
12 19 ☽ △ ♀
12 51 ☽ Q ♇
13 33 ☽ ⊥ ♄
15 07 ☽ ⚻ ☿
16 25 ♂ ⊥ ♆

15 Monday
01 41 ☽ ⚺ ♄
05 04 ☽ ± ☿
13 14 ☽ ⚹ ♇
13 28 ☽ △ ♆
15 25 ☽ □ ♅

16 Tuesday
01 39 ☽ ⚻ ☉
02 02 ☽ bQ ☿
02 16 ☉ ⊥ ♆
03 57 ☽ ⚻ ♂
15 00 ☽ ± ☉
17 01 ☽ ± ♂
19 44 ☽ ∠ ♇
21 28 ☽ ⚺ ♃
22 25 ☽ ∥ ♇
23 20 ☽ ∥ ♄

17 Wednesday
04 06 ☽ Q ♅
06 53 ☽ □ ♀
09 30 ☽ ⊥ ♃
10 59 ☽ bQ ☉
12 33 ☽ bQ ♂
12 57 ☽ △ ☿
13 59 ☽ ⊥ ♇
14 25 ☽ ☌ ♄
15 21 ♂ ∥ ♆

18 Thursday
00 03 ☿ △ ♄
02 06 ☽ ⚺ ♇
02 28 ☽ □ ♆
03 28 ☽ ∠ ♃
04 25 ☽ ⚹ ♅
15 28 ♂ bQ ♄
19 57 ☽ △ ☉
20 48 ☽ △ ♂
22 27 ☽ Q ♀

19 Friday
01 29 ☉ bQ ♄
09 05 ☽ ⚹ ♃
10 24 ☽ ∠ ♅
14 26 ☽ Q ♆

20 Saturday
00 17 ☽ ⚹ ♀
01 55 ☽ ⚺ ♄
04 05 ☽ ⊥ ♅
08 18 ☽ Q ♃
09 11 ☽ □ ☿
12 18 ☉ ☌ ♂
13 20 ☽ ☌ ♇
13 29 ☽ ⊥ ♄
13 51 ☽ ⚹ ♆
15 42 ☽ ⚺ ♅
18 56 ♀ □ ♄

21 Sunday
04 14 ♃ bQ ♆
06 35 ☽ ∠ ♄
07 39 ☽ ∠ ♀
10 29 ☽ # ♄
10 56 ☽ □ ♂
10 59 ☽ # ♇
11 23 ☽ □ ☉
18 07 ☽ □ ♃
18 18 ☽ ∠ ♆
18 31 ☿ □ ♇
20 46 ♇ St R
23 11 ☿ ⚺ ♆
23 57 ☽ Q ☿

22 Monday
01 53 ☽ ⊥ ♀
02 41 ♄ ⊥ ♇
06 37 ☉ # ♃
10 16 ☽ ⚹ ♄
10 53 ☽ ⊥ ♆
13 46 ☽ ⚺ ♀
14 15 ☿ ⚹ ♅
21 07 ☽ ⚺ ♇
21 43 ☽ ⚺ ♆
22 01 ☽ Q ♂
23 06 ☽ Q ☉
23 26 ☽ ☌ ♅

23 Tuesday
00 38 ☽ ⚹ ☿
03 36 ☉ Q ♅
07 40 ☽ Q ♄
07 46 ☽ ⊥ ♇
19 53 ☿ ⊥ ♀
20 23 ☽ ⚹ ♂
21 51 ☽ ⚹ ☉
23 16 ☽ △ ♃
23 25 ♂ Q ♅
23 26 ☽ ∠ ♇

24 Wednesday
01 04 ☉ ∥ ♀
06 07 ☽ ∠ ☿
14 32 ☽ □ ♄
15 18 ☉ ☍ ♃
19 13 ☽ ∥ ♅
22 05 ☽ ☌ ♀
23 10 ☽ ⊥ ☿
23 17 ☽ ∠ ♂

25 Thursday
00 23 ☽ bQ ♃
00 46 ☽ ⚹ ♇
01 10 ☽ ∠ ☉
01 26 ☽ ☌ ♆
03 01 ☽ ⚺ ♅
05 03 ☉ ∠ ♆
10 17 ☽ ⚺ ☿
12 54 ☽ ⊥ ♅
14 57 ☽ ⊥ ♂
15 03 ☽ ± ♃
16 44 ♂ ☍ ♃
17 03 ☽ ⊥ ☉
20 01 ☿ ⊥ ♆
20 26 ☽ Q ♇

26 Friday
00 42 ☽ ⚻ ♃
01 13 ☽ ⚺ ♂
02 16 ♀ bQ ♃
03 28 ☽ ⚺ ☉
03 31 ☽ ∠ ♅
10 42 ♀ ⚹ ♇
15 20 ☽ △ ♄
16 14 ♂ ∠ ♆
20 54 ♀ ☌ ♆
22 10 ♄ ∥ ♇

27 Saturday
01 13 ☽ □ ♇
01 56 ☽ ⚺ ♆
02 19 ☽ ⚺ ♀
02 49 ☽ ∥ ♅
03 27 ☽ ⚹ ♅
06 32 ☿ bQ ♄
09 21 ♀ ∠ ♂
10 30 ♂ # ♃
11 25 ☽ ⊥ ♆
11 46 ☉ ± ♄
12 29 ☽ ⊥ ♀
14 57 ☽ bQ ♄
15 59 ☽ ☌ ☿
18 57 ♀ ⚺ ♅
20 26 ♀ ∥ ♂
22 22 ☽ Q ♅
23 50 ☽ ☍ ♃

28 Sunday
01 35 ☽ ∠ ♆
03 19 ☽ ☌ ♂
03 41 ☽ ∠ ♀
05 02 ☽ ± ♄
06 20 ☽ ☌ ☉
14 24 ☽ ⚻ ♄
22 42 ♀ # ♃

29 Monday
00 22 ☽ △ ♇
01 11 ☽ ⚹ ♆
02 41 ☽ □ ♅
05 02 ☽ ⚹ ♀
10 27 ♂ ± ♄
11 30 ☿ ∥ ♆
20 12 ☽ Q ♆
20 54 ☽ ⚺ ☿
22 43 ☽ ⚻ ♃

30 Tuesday
00 07 ☽ bQ ♇
01 30 ☽ # ♄
01 33 ☽ Q ♀
01 46 ☽ # ♇
04 32 ☿ Q ♅
05 18 ☽ ⚺ ♂
07 41 ☽ ⊥ ☿
08 15 ☽ ± ♃
09 13 ☽ ⚺ ☉
12 28 ☿ ☍ ♃
13 52 ☽ ☍ ♄
14 35 ☽ ± ♇
15 31 ☽ ⊥ ♂
19 39 ☽ ⊥ ☉
22 42 ☽ bQ ♃
23 35 ☉ ∥ ♅

31 Wednesday
00 06 ☽ ∠ ☿
00 20 ☽ ⚻ ♇
01 18 ☽ □ ♆
02 51 ☽ △ ♅
07 03 ☽ ∠ ♂
09 05 ☽ □ ♀
10 48 ☿ ∠ ♆
11 31 ☽ ∠ ☉
23 19 ☽ △ ♃

All ephemeris data is given at 12.00 UT and the Moon's longitude is additionally given for 24.00 UT

FEBRUARY 1979

Date	Sidereal time h m s	LONGITUDES Sun ☉ ° ' "	Moon ☽ ° ' "	Moon ☽ 24.00 ° ' "	Mercury ☿ ° '	Venus ♀ ° '	Mars ♂ ° '	Jupiter ♃ ° '	Saturn ♄ ° '	Uranus ♅ ° '	Neptune ♆ ° '	Pluto ⯓ ° '
01	20 44 24	12 ♒ 01 55	10 ♈ 32 21	17 ♈ 29 58	06 ♒ 33	25 ♐ 44	09 ♒ 14	03 ♌ 00	12 ♍ 39	20 ♏ 46	19 ♐ 49	19 ♎ 11
02	20 48 21	13 02 49	24 ♈ 20 55	01 ♉ 05 22	08 14	26 50	10 01	02 R 52	12 R 35	20 47	19 51	19 R 11
03	20 52 17	14 03 42	07 ♉ 43 36	14 ♉ 15 59	09 55	27 55	10 48	02 44	12 31	20 48	19 52	19 11
04	20 56 14	15 04 33	20 ♉ 42 57	27 ♉ 04 55	11 37	29 ♐ 01	11 36	02 36	12 27	20 49	19 54	19 10
05	21 00 10	16 05 23	03 ♊ 22 23	09 ♊ 35 48	13 20	00 ♑ 08	12 23	02 29	12 23	20 50	19 55	19 10
06	21 04 07	17 06 11	15 ♊ 45 39	21 ♊ 52 22	15 04	01 14	13 10	02 21	12 19	20 51	19 57	19 09
07	21 08 03	18 06 58	27 ♊ 56 23	03 ♋ 58 07	16 48	02 20	13 57	02 14	12 15	20 52	19 58	19 09
08	21 12 00	19 07 44	09 ♋ 57 54	15 ♋ 56 07	18 34	03 27	14 44	02 06	12 11	20 53	20 00	19 08
09	21 15 57	20 08 28	21 ♋ 53 03	27 ♋ 49 02	20 20	04 34	15 32	01 59	12 06	20 54	20 01	19 07
10	21 19 53	21 09 11	03 ♌ 44 18	09 ♌ 39 07	22 06	05 41	16 19	01 52	12 02	20 55	20 02	19 07
11	21 23 50	22 09 52	15 ♌ 33 44	21 ♌ 28 23	23 54	06 49	17 06	01 45	11 58	20 55	20 04	19 06
12	21 27 46	23 10 32	27 ♌ 23 18	03 ♍ 18 44	25 42	07 56	17 53	01 38	11 53	20 56	20 05	19 05
13	21 31 43	24 11 11	09 ♍ 14 55	15 ♍ 12 08	27 31	09 04	18 41	01 31	11 49	20 57	20 06	19 05
14	21 35 39	25 11 48	21 ♍ 10 40	27 ♍ 10 49	29 ♒ 20	10 11	19 28	01 24	11 44	20 57	20 07	19 04
15	21 39 36	26 12 24	03 ♎ 12 55	09 ♎ 17 20	01 ♓ 11	11 19	20 15	01 17	11 40	20 58	20 09	19 03
16	21 43 32	27 12 58	15 ♎ 24 27	21 ♎ 34 39	03 01	12 28	21 03	01 10	11 35	20 58	20 10	19 02
17	21 47 29	28 13 31	27 ♎ 48 23	04 ♏ 06 05	04 52	13 36	21 50	01 04	11 31	20 59	20 11	19 01
18	21 51 26	29 ♒ 14 03	10 ♏ 28 11	16 ♏ 55 07	06 43	14 44	22 37	00 58	11 26	20 59	20 12	19 00
19	21 55 22	00 ♓ 14 34	23 ♏ 27 18	00 ♐ 05 06	08 35	15 53	23 25	00 51	11 21	20 59	20 13	18 59
20	21 59 19	01 15 04	06 ♐ 48 50	13 ♐ 38 44	10 26	17 01	24 12	00 45	11 17	21 00	20 14	18 58
21	22 03 15	02 15 32	20 ♐ 34 58	27 ♐ 37 30	12 18	18 10	24 59	00 39	11 12	21 00	20 15	18 57
22	22 07 12	03 15 59	04 ♑ 46 14	12 ♑ 00 53	14 08	19 19	25 47	00 34	11 07	21 00	20 16	18 56
23	22 11 08	04 16 25	19 ♑ 20 58	26 ♑ 45 52	15 58	20 28	26 34	00 28	11 03	21 00	20 17	18 55
24	22 15 05	05 16 49	04 ♒ 14 47	11 ♒ 46 45	17 47	21 37	27 21	00 23	10 58	21 R 00	20 18	18 54
25	22 19 01	06 17 12	19 ♒ 20 42	26 ♒ 55 28	19 35	22 47	28 09	00 17	10 53	21 00	20 19	18 53
26	22 22 58	07 17 33	04 ♓ 29 52	12 ♓ 02 41	21 21	23 56	28 56	00 12	10 48	21 00	20 20	18 52
27	22 26 55	08 17 52	19 ♓ 32 46	26 ♓ 59 06	23 04	25 06	29 ♒ 43	00 07	10 44	21 00	20 20	18 51
28	22 30 51	09 ♓ 18 09	04 ♈ 20 44	11 ♈ 36 55	24 ♓ 45	26 ♑ 15	00 ♓ 31	00 ♌ 02	10 ♍ 39	21 ♏ 00	20 ♐ 21	18 ♎ 50

Date	Moon True ☊ ° '	Moon Mean ☊ ° '	Moon ☽ Latitude ° '
01	17 ♍ 58	19 ♍ 34	02 S 00
02	18 D 00	19 30	03 06
03	18 01	19 27	03 59
04	18 R 01	19 24	04 38
05	17 59	19 21	05 03
06	17 56	19 18	05 13
07	17 52	19 14	05 08
08	17 47	19 11	04 50
09	17 42	19 08	04 19
10	17 37	19 05	03 38
11	17 33	19 02	02 47
12	17 30	18 59	01 48
13	17 29	18 55	00 S 45
14	17 D 29	18 52	00 N 20
15	17 29	18 49	01 26
16	17 31	18 46	02 28
17	17 32	18 43	03 25
18	17 34	18 39	04 13
19	17 35	18 36	04 49
20	17 R 35	18 33	05 11
21	17 34	18 30	05 16
22	17 33	18 27	05 03
23	17 31	18 24	04 29
24	17 29	18 20	03 37
25	17 28	18 17	02 30
26	17 27	18 14	01 N 11
27	17 D 27	18 11	00 S 12
28	17 ♍ 27	18 ♍ 08	01 S 33

DECLINATIONS

Date	Sun ☉ ° '	Moon ☽ ° '	Mercury ☿ ° '	Venus ♀ ° '	Mars ♂ ° '	Jupiter ♃ ° '	Saturn ♄ ° '	Uranus ♅ ° '	Neptune ♆ ° '	Pluto ⯓ ° '
01	17 S 11	02 N 20	20 S 35	20 S 23	18 S 58	20 N 08	08 N 38	17 S 37	21 S 40	08 N 33
02	16 54	06 34	20 11	20 28	18 46	20 10	08 40	17 37	21 40	08 34
03	16 37	10 19	19 45	20 33	18 33	20 12	08 42	17 38	21 40	08 34
04	16 19	13 27	19 17	20 37	18 20	20 14	08 43	17 38	21 40	08 35
05	16 01	15 52	18 49	20 41	18 07	20 16	08 45	17 38	21 40	08 36
06	15 43	17 30	18 19	20 44	17 54	20 17	08 47	17 38	21 40	08 36
07	15 24	18 17	17 47	20 47	17 40	20 19	08 48	17 39	21 40	08 37
08	15 05	18 15	17 14	20 49	17 26	20 21	08 50	17 39	21 40	08 38
09	14 46	17 24	16 39	20 51	17 13	20 23	08 52	17 39	21 40	08 39
10	14 27	15 47	16 03	20 52	16 58	20 24	08 54	17 39	21 40	08 39
11	14 07	13 31	15 25	20 53	16 44	20 26	08 56	17 40	21 40	08 40
12	13 48	10 41	14 46	20 53	16 30	20 28	08 57	17 40	21 40	08 41
13	13 28	07 24	14 06	20 52	16 15	20 29	08 59	17 40	21 41	08 41
14	13 07	03 49	13 24	20 52	16 01	20 31	09 01	17 40	21 41	08 42
15	12 47	00 N 02	12 41	20 50	15 46	20 32	09 03	17 40	21 41	08 43
16	12 26	03 S 47	11 57	20 48	15 31	20 34	09 05	17 40	21 41	08 44
17	12 05	07 30	11 11	20 46	15 15	20 35	09 07	17 40	21 41	08 44
18	11 44	10 57	10 24	20 43	15 00	20 37	09 08	17 40	21 41	08 45
19	11 23	13 58	09 36	20 39	14 44	20 38	09 10	17 40	21 41	08 46
20	11 02	16 20	08 47	20 35	14 29	20 39	09 12	17 41	21 41	08 47
21	10 40	17 51	07 57	20 30	14 13	20 41	09 14	17 41	21 41	08 48
22	10 18	18 19	07 07	20 25	13 57	20 42	09 16	17 41	21 41	08 48
23	09 57	17 36	06 16	20 19	13 41	20 43	09 18	17 41	21 41	08 49
24	09 35	15 40	05 24	20 13	13 25	20 45	09 20	17 41	21 41	08 50
25	09 12	12 39	04 32	20 06	13 08	20 46	09 22	17 41	21 41	08 51
26	08 50	08 45	03 40	19 58	12 52	20 47	09 24	17 41	21 41	08 52
27	08 28	04 19	02 48	19 50	12 35	20 48	09 26	17 41	21 41	08 52
28	08 S 05	00 N 19	01 S 57	19 S 42	12 S 18	20 N 49	09 N 27	17 S 40	21 S 41	08 N 53

ZODIAC SIGN ENTRIES

Date	h m	Planets
02	22 03	☽ ♉
05	05 33	☽ ♊
05	09 16	♀ ♑
07	16 06	☽ ♋
10	04 25	☽ ♌
12	17 18	☽ ♍
14	20 38	☿ ♓
15	05 37	☽ ♎
17	16 12	☽ ♏
19	06 13	☉ ♓
19	23 51	☽ ♐
22	04 00	☽ ♑
24	05 12	☽ ♒
26	04 52	☽ ♓
27	20 25	♂ ♓
28	04 54	☽ ♈
28	23 35	♃ ♋

LATITUDES

Date	Mercury ☿ ° '	Venus ♀ ° '	Mars ♂ ° '	Jupiter ♃ ° '	Saturn ♄ ° '	Uranus ♅ ° '	Neptune ♆ ° '	Pluto ⯓ ° '
01	02 S 01	03 N 00	01 S 04	00 N 40	01 N 58	00 N 20	01 N 24	17 N 21
04	02 04	02 49	01 04	00 40	01 59	00 20	01 24	17 22
07	02 05	02 39	01 04	00 40	01 59	00 20	01 24	17 24
10	02 01	02 27	01 05	00 41	02 00	00 20	01 24	17 25
13	01 53	02 16	01 05	00 41	02 00	00 20	01 24	17 27
16	01 40	02 04	01 05	00 41	02 01	00 20	01 24	17 28
19	01 21	01 51	01 05	00 41	02 01	00 20	01 24	17 30
22	01 00	01 39	01 05	00 41	02 02	00 21	01 24	17 31
25	00 S 27	01 26	01 05	00 41	02 02	00 21	01 24	17 33
28	00 N 09	01 14	01 05	00 41	02 02	00 21	01 24	17 34
31	00 N 49	01 N 01	01 S 05	00 N 41	02 N 02	00 N 21	01 N 25	17 N 35

DATA

Julian Date	2443906
Delta T	+50 seconds
Ayanamsa	23° 33' 52"
Synetic vernal point	05° ♓ 33' 08"
True obliquity of ecliptic	23° 26' 22"

MOON'S PHASES, APSIDES AND POSITIONS ☽

Date	h m	Phase	Longitude ° '	Eclipse Indicator
04	00 36	☽	14 ♉ 36	
12	02 39	○	22 ♌ 47	
20	01 17	☾	00 ♐ 48	
26	16 45	●	07 ♓ 29	Total

Day	h m		
11	02 33	Apogee	
25	22 10	Perigee	
07	22 43	Max dec	18° N 22'
15	12 13	0S	
22	09 46	Max dec	18° S 19'
28	10 23	0N	

LONGITUDES

Date	Chiron ⚷ ° '	Ceres ⚳ ° '	Pallas ⚴ ° '	Juno ⚵ ° '	Vesta ⚶ ° '	Black Moon Lilith ⚸ ° '
01	05 ♉ 11	22 ♒ 22	00 ♒ 18	06 ♓ 32	23 ♒ 54	12 ♌ 27
11	05 ♉ 25	26 ♒ 18	03 ♒ 54	11 ♓ 26	28 ♒ 57	13 ♌ 34
21	05 ♉ 43	00 ♓ 14	07 ♒ 25	16 ♓ 26	03 ♓ 59	14 ♌ 41
31	06 ♉ 06	04 ♓ 10	10 ♒ 50	21 ♓ 33	08 ♓ 59	15 ♌ 47

ASPECTARIAN

01 Thursday
03 52 ☽ ♇ ♅
04 18 ☽ ⚹ ☿
09 38 ☽ ⚹ ♂
13 19 ☿ ± ♄
14 45 ☽ ⚹ ☉
15 35 ☽ ⊼ ♄
19 16 ☽ ± ♅
21 56 ☿ ∥ ♀

02 Friday
01 35 ☉ ⊼ ♄
01 55 ☽ ± ♄
02 56 ☽ ☍ ⯓
03 44 ☽ Q ☿
04 05 ☽ △ ♆
05 43 ☽ ⊼ ♅
07 39 ☽ Q ♂
10 54 ♀ ⊥ ♅
12 31 ☿ # ♃
12 40 ♀ ± ♃
13 20 ☽ Q ☉
16 47 ☽ △ ♀
17 42 ☽ ♇ ♄

03 Saturday
00 23 ☽ ∥ ⯓
01 06 ☽ ∥ ♄
03 03 ☽ □ ♃
06 49 ☽ ♇ ♆
16 36 ☽ □ ☿
17 59 ☽ □ ♂
20 43 ☽ △ ♄
22 24 ☽ ♇ ♀
23 18 ☽ ± ♆

04 Sunday
00 36 ☽ □ ☉
09 07 ☽ ⊼ ⯓
10 28 ☽ ⊼ ♆
11 16 ☿ ☌ ♂
11 48 ☽ Q ♃
12 12 ☽ ☍ ♅
16 45 ☽ ± ♀
20 23 ☽ ± ⯓
23 10 ☿ ⊼ ♄

05 Monday
05 12 ☽ ⊼ ♀
10 18 ☽ ⚹ ♃
12 04 ♂ ⊼ ♄
13 28 ☽ # ☉
13 31 ☽ ♇ ⯓

06 Tuesday
05 19 ☽ □ ♄
06 35 ☽ △ ♂
10 19 ♀ Q ⯓
10 25 ☽ △ ☿
14 52 ☽ △ ☉
14 59 ☽ # ♅
15 05 ☽ ∠ ♃
18 39 ☽ △ ⯓
19 15 ☽ # ♂
20 14 ☽ ☍ ♆
22 00 ☽ ⊼ ♅

07 Wednesday
00 54 ☽ # ☿
08 38 ☽ ⊥ ♃
09 48 ♀ ⊼ ♃
09 52 ☽ ± ♅
14 09 ☽ ♇ ♂
14 23 ♂ ∥ ♅
16 33 ☽ Q ♄
17 56 ☿ ∥ ♅
20 26 ☽ ⊻ ♃
20 29 ☿ ∥ ♂
21 00 ☽ ♇ ☿
21 39 ☽ ☍ ♀
23 15 ☽ ♇ ☉

08 Thursday
03 49 ☽ ♇ ♅
07 59 ☿ ∠ ♀
09 22 ☽ ± ♂
12 07 ☉ △ ⯓
16 25 ☽ ⚹ ♄
18 07 ☽ ± ☿
18 57 ☽ ± ☉
19 45 ☿ △ ⯓
22 16 ☽ ⊼ ♂

09 Friday
06 06 ☉ ☌ ☿
06 26 ☽ □ ⯓
06 38 ☽ # ♅
07 44 ☿ ⚹ ♆
08 09 ☽ ⊼ ☉
08 13 ☽ ⊼ ♆
08 18 ☽ ⊼ ☿
08 59 ☉ ⚹ ♆
10 00 ☽ △ ♅
16 07 ☽ # ♂
19 46 ☿ □ ♅
20 22 ☽ ± ♆
22 30 ☽ ∠ ♄

10 Saturday
06 12 ☉ □ ♅
07 05 ☽ # ☿
08 14 ☽ ☌ ♃
14 39 ☽ ♇ ♆
16 22 ☽ ⊼ ♀
16 38 ☽ ⊥ ♄
16 51 ♀ ∠ ♅
18 50 ☽ Q ⯓

11 Sunday
04 44 ☽ ⊻ ♄
05 16 ☽ # ☉
05 49 ☽ ± ♀
15 21 ☽ ☍ ♂
19 11 ☽ ⚹ ⯓
21 09 ☽ △ ♆
22 54 ☽ □ ♅

12 Monday
02 01 ☽ ♇ ♀
02 39 ☽ ☍ ☉
07 58 ☽ ☍ ☿
20 30 ☽ ⊻ ♃

13 Tuesday
00 51 ☽ ∥ ♄
01 33 ☽ ∠ ⯓
02 56 ☽ ∥ ⯓
08 31 ☽ ⊥ ♃
11 23 ☽ Q ♅
11 35 ☽ △ ♀
17 09 ☽ ☌ ♄
19 43 ☽ ⊥ ⯓
23 56 ♂ △ ⯓

14 Wednesday
02 29 ☽ ∠ ♃
07 46 ☽ ⊻ ⯓
08 20 ☽ ⊼ ♂
09 53 ☽ □ ♆
11 33 ☽ ⚹ ♅
13 08 ☉ ∠ ♀
20 47 ☽ ⊼ ☉
21 11 ☽ ± ♂

15 Thursday
05 59 ☉ ∥ ☿
07 14 ☽ ⊼ ☿
08 12 ☽ ⚹ ♃
08 29 ♂ ⚹ ♆
09 49 ☽ ± ☉
13 20 ☿ ⊼ ♃
16 19 ☽ ♇ ♂
17 27 ☽ ∠ ♅
18 47 ♀ △ ♄
21 14 ☽ ± ☿
21 45 ☽ Q ♆

16 Friday
00 44 ☿ Q ♆
04 34 ☽ ⊻ ♄
05 11 ☽ ♇ ☉
05 38 ☽ □ ♀
07 40 ☽ Q ♃
09 49 ♂ □ ♅
11 09 ☽ ⊥ ♅
16 14 ☽ ⊥ ♄
17 59 ☽ ♇ ☿
19 04 ☽ ☌ ⯓
21 16 ☽ ⚹ ♆
22 50 ☽ ⊻ ♅
23 43 ☽ △ ♂

17 Saturday
01 06 ☿ ♇ ⯓
09 32 ☽ ∠ ♄
12 52 ☽ △ ☉
18 11 ☽ □ ♃
19 58 ☽ Q ♀
20 22 ☽ # ⯓
22 58 ☽ # ♄

18 Sunday
02 04 ☽ ∠ ♆
03 45 ☽ △ ☿
08 45 ☽ ∥ ☿
13 48 ☽ ⚹ ♄
14 54 ☿ ± ♃
17 17 ☽ ∥ ☉
18 58 ☽ ⊥ ♆
20 44 ☽ ⚹ ♀

19 Monday
03 50 ☽ ⊻ ⯓
06 04 ☽ ⊻ ♆
07 30 ☽ ☌ ♅
11 49 ☽ Q ♄
11 55 ☽ □ ♂
14 48 ☽ ⊥ ⯓
16 24 ♀ # ♃
18 25 ☽ ∥ ♂

20 Tuesday
00 17 ☿ # ♄
01 17 ☽ □ ☉
01 17 ☽ △ ♃
01 18 ☉ ⊼ ♃
02 41 ☽ ∠ ♀
06 58 ☽ ∠ ⯓
12 15 ☿ # ⯓
19 24 ☽ □ ☿
19 49 ☽ □ ♄
20 06 ☽ ⊥ ♀
22 03 ☽ Q ♂
22 27 ☿ ☍ ♄

21 Wednesday
03 34 ☽ ♇ ♃
07 29 ☽ ⊻ ♀
08 18 ☽ ∥ ♅
09 13 ☽ ⚹ ⯓
11 24 ☽ Q ☉
11 26 ☽ ☌ ♆
11 47 ☉ Q ♆
12 43 ☽ ⊻ ♅
18 55 ☽ ± ♃
19 58 ☽ ⚹ ♂
20 34 ☿ ± ⯓
22 56 ☽ ⊥ ♅

22 Thursday
04 13 ♀ □ ⯓
05 00 ☽ ⊼ ♃
05 36 ☽ Q ⯓
06 57 ☽ Q ☿
09 18 ☽ ⚹ ☉
14 03 ☽ ∠ ♅
22 28 ☽ △ ♄
22 32 ☽ ∠ ♂

23 Friday
03 49 ☉ ♇ ⯓
05 42 ☿ ♇ ♃
05 42 ☽ ⚹ ☿
08 00 ♀ ⊻ ♆
10 32 ☽ ∥ ♅
11 18 ☽ □ ⯓
11 52 ☽ ∠ ☉
13 31 ☽ ⊻ ♆
13 59 ☽ ☌ ♀
14 05 ☽ ⊥ ♂
14 41 ☽ ⚹ ♅
18 17 ♀ ⊥ ♂
22 47 ☽ ♇ ♄
23 02 ♀ ⚹ ♅
23 14 ☽ ⊥ ♆

24 Saturday
00 20 ☽ ⊻ ♂
03 28 ☽ ⊥ ☉
05 51 ☽ ☍ ♃
09 21 ☽ ∠ ☿
10 01 ☽ Q ♅
11 59 ♅ St R
13 08 ☽ ± ♄
13 41 ☽ ∠ ♆
13 46 ☽ ⊻ ☉
22 39 ☽ ⊼ ♄

25 Sunday
01 38 ☽ ⊥ ☿
02 35 ☉ # ♄
02 45 ☿ ⊼ ⯓
08 17 ☽ ∥ ♂
11 17 ☽ △ ⯓
12 02 ☉ ± ♃
12 26 ☽ ⊻ ☿
13 32 ☽ ⚹ ♆
14 37 ☽ □ ♅
17 53 ☽ ⊻ ♀
21 57 ☿ □ ♆

26 Monday
02 42 ☽ ☌ ♂
04 10 ☽ ⊥ ♀
05 14 ☽ ⊼ ♃
07 15 ☿ △ ♅
08 21 ☽ # ♄
08 33 ☽ Q ♆
10 24 ☉ # ⯓
11 00 ☽ ♇ ⯓
11 24 ☽ # ⯓
11 30 ☽ ∥ ☉
14 41 ☽ ± ♃
16 45 ☽ ☌ ☉
19 38 ☽ ∠ ♀
21 58 ☽ ☍ ♄

27 Tuesday
01 18 ☽ ± ⯓
04 57 ☽ ♇ ♃
10 53 ☽ ⊼ ⯓
13 17 ☽ □ ♆
14 20 ☽ △ ♅
18 24 ☽ ☌ ☿
21 40 ☽ ∥ ☿
21 42 ☽ ⚹ ♀
22 54 ♂ ⊼ ♃

28 Wednesday
00 13 ♀ ♇ ♄
05 00 ☽ △ ♃
05 23 ☽ ⊻ ♂
14 04 ♀ ⊥ ♆
14 42 ☽ ♇ ♅
15 46 ☽ ⊥ ♂
18 59 ☽ Q ♀
19 14 ☽ # ☿
20 46 ☽ ⊻ ☉
22 20 ☽ ⊼ ♄

All ephemeris data is given at 12.00 UT and the Moon's longitude is additionally given for 24.00 UT

MARCH 1979

LONGITUDES

Date	Sidereal time h m s	Sun ☉ ° ' "	Moon ☽ ° ' "	Moon ☽ 24.00 ° ' "	Mercury ☿ ° '	Venus ♀ ° '	Mars ♂ ° '	Jupiter ♃ ° '	Saturn ♄ ° '	Uranus ♅ ° '	Neptune ♆ ° '	Pluto ♇ ° '
01	22 34 48	10 ♓ 18 25	18 ♈ 47 04	25 ♈ 50 44	26 ♓ 23	27 ♑ 25	01 ♓ 18	29 ♋ 58	10 ♍ 34	20 ♏ 59	20 ♐ 22	18 ♎ 49
02	22 38 44	11 18 38	02 ♉ 47 41	09 ♉ 37 51	27 56	28 35	02 05	29 R 53	10 R 29	20 R 59	20 23	18 R 47
03	22 42 41	12 18 50	16 ♉ 21 16	22 ♉ 58 09	29 ♓ 26	29 ♑ 45	02 53	29 49	10 24	20 59	20 23	18 46
04	22 46 37	13 19 00	29 ♉ 28 47	05 ♊ 53 32	00 ♈ 50	00 ♒ 54	03 40	29 45	10 19	20 58	20 24	18 45
05	22 50 34	14 19 07	12 ♊ 12 53	18 ♊ 27 18	02 09	02 05	04 27	29 41	10 15	20 58	20 25	18 44
06	22 54 30	15 19 12	24 ♊ 37 19	00 ♋ 43 30	03 22	03 15	05 15	29 37	10 10	20 57	20 25	18 42
07	22 58 27	16 19 16	06 ♋ 46 23	12 ♋ 46 32	04 28	04 25	06 02	29 33	10 05	20 57	20 26	18 41
08	23 02 24	17 19 17	18 ♋ 44 28	24 ♋ 40 44	05 27	05 35	06 49	29 30	10 00	20 56	20 26	18 40
09	23 06 20	18 19 16	00 ♌ 35 48	06 ♌ 30 09	06 18	06 45	07 37	29 27	09 56	20 55	20 27	18 38
10	23 10 17	19 19 12	12 ♌ 24 14	18 ♌ 18 26	07 02	07 56	08 24	29 24	09 51	20 55	20 27	18 37
11	23 14 13	20 19 07	24 ♌ 13 08	00 ♍ 08 42	07 37	09 06	09 11	29 21	09 46	20 54	20 27	18 35
12	23 18 10	21 19 00	06 ♍ 05 25	12 ♍ 03 36	08 03	10 17	09 58	29 18	09 42	20 53	20 28	18 34
13	23 22 06	22 18 50	18 ♍ 03 30	24 ♍ 05 21	08 21	11 28	10 46	29 15	09 37	20 52	20 28	18 33
14	23 26 03	23 18 39	00 ♎ 09 23	06 ♎ 15 47	08 30	12 38	11 33	29 13	09 32	20 51	20 29	18 31
15	23 29 59	24 18 26	12 ♎ 24 46	18 ♎ 36 30	08 R 30	13 49	12 20	29 11	09 28	20 50	20 29	18 30
16	23 33 56	25 18 11	24 ♎ 51 10	01 ♏ 08 56	08 22	15 00	13 07	29 09	09 23	20 49	20 29	18 28
17	23 37 53	26 17 54	07 ♏ 30 00	13 ♏ 54 31	08 06	16 11	13 55	29 07	09 19	20 48	20 29	18 27
18	23 41 49	27 17 35	20 ♏ 22 40	26 ♏ 54 38	07 43	17 22	14 42	29 06	09 14	20 47	20 29	18 25
19	23 45 46	28 17 14	03 ♐ 30 34	10 ♐ 10 37	07 12	18 33	15 29	29 04	09 10	20 46	20 30	18 23
20	23 49 42	29 ♓ 16 52	16 ♐ 54 56	23 ♐ 43 38	06 35	19 44	16 16	29 03	09 05	20 45	20 30	18 22
21	23 53 39	00 ♈ 16 28	00 ♑ 36 45	07 ♑ 34 20	05 53	20 55	17 03	29 02	09 01	20 44	20 30	18 20
22	23 57 35	01 16 03	14 ♑ 36 18	21 ♑ 42 32	05 07	22 06	17 50	29 02	08 57	20 42	20 30	18 19
23	00 01 32	02 15 35	28 ♑ 52 49	06 ♒ 06 50	04 17	23 17	18 37	29 01	08 53	20 41	20 R 30	18 17
24	00 05 28	03 15 06	13 ♒ 24 09	20 ♒ 44 14	03 25	24 29	19 24	29 01	08 48	20 40	20 30	18 16
25	00 09 25	04 14 35	28 ♒ 06 27	05 ♓ 30 03	02 32	25 40	20 12	29 00	08 44	20 38	20 30	18 14
26	00 13 22	05 14 03	12 ♓ 54 13	20 ♓ 18 04	01 39	26 52	20 59	29 D 00	08 40	20 37	20 30	18 12
27	00 17 18	06 13 28	27 ♓ 40 42	05 ♈ 01 12	00 ♈ 47	28 03	21 46	29 01	08 36	20 35	20 30	18 11
28	00 21 15	07 12 51	12 ♈ 18 41	19 ♈ 32 20	29 ♓ 57	29 ♒ 14	22 33	29 01	08 32	20 33	20 29	18 09
29	00 25 11	08 12 12	26 ♈ 41 25	03 ♉ 45 19	29 10	00 ♓ 26	23 20	29 01	08 29	20 32	20 29	18 07
30	00 29 08	09 11 31	10 ♉ 43 34	17 ♉ 35 49	28 27	01 37	24 07	29 02	08 25	20 30	20 29	18 06
31	00 33 04	10 ♈ 10 48	24 ♉ 21 52	01 ♊ 01 38	27 ♓ 49	02 ♓ 49	24 ♓ 53	29 ♋ 03	08 ♍ 21	20 ♏ 29	20 ♐ 29	18 ♎ 04

Date	Moon True ☊ ° '	Moon Mean ☊ ° '	Moon ☽ Latitude ° '
01	17 ♍ 28	18 ♍ 05	02 S 46
02	17 D 29	18 01	03 46
03	17 29	17 58	04 32
04	17 30	17 55	05 02
05	17 30	17 52	05 17
06	17 R 30	17 49	05 15
07	17 30	17 45	05 00
08	17 30	17 42	04 32
09	17 D 30	17 39	03 52
10	17 30	17 36	03 03
11	17 30	17 33	02 06
12	17 30	17 30	01 S 03
13	17 R 30	17 26	00 N 03
14	17 30	17 23	01 10
15	17 30	17 20	02 14
16	17 29	17 17	03 13
17	17 28	17 14	04 03
18	17 26	17 11	04 43
19	17 25	17 07	05 08
20	17 24	17 04	05 17
21	17 D 24	17 01	05 09
22	17 24	16 58	04 42
23	17 25	16 55	03 58
24	17 26	16 51	02 58
25	17 28	16 48	01 45
26	17 28	16 45	00 N 25
27	17 R 28	16 42	00 S 56
28	17 27	16 39	02 13
29	17 25	16 36	03 20
30	17 22	16 32	04 13
31	17 ♍ 18	16 ♍ 29	04 S 50

DECLINATIONS

Date	Sun ☉ ° '	Moon ☽ ° '	Mercury ☿ ° '	Venus ♀ ° '	Mars ♂ ° '	Jupiter ♃ ° '	Saturn ♄ ° '	Uranus ♅ ° '	Neptune ♆ ° '	Pluto ♇ ° '
01	07 S 42	04 N 48	01 S 06	19 S 32	12 S 01	20 N 50	09 N 29	17 S 40	21 S 41	08 N 54
02	07 19	08 54	00 S 17	19 23	11 44	20 51	09 31	17 40	21 41	08 55
03	06 56	12 23	00 N 31	19 12	11 27	20 52	09 33	17 40	21 41	08 56
04	06 33	15 07	01 17	19 02	11 10	20 53	09 35	17 40	21 41	08 56
05	06 10	17 02	02 01	18 50	10 53	20 54	09 37	17 40	21 41	08 57
06	05 47	18 05	02 43	18 38	10 35	20 54	09 39	17 40	21 41	08 58
07	05 24	18 16	03 22	18 26	10 18	20 55	09 41	17 40	21 41	08 59
08	05 01	17 38	03 58	18 13	10 00	20 56	09 42	17 39	21 41	09 00
09	04 37	16 14	04 31	17 59	09 43	20 57	09 44	17 39	21 41	09 00
10	04 14	14 09	05 00	17 46	09 25	20 57	09 46	17 39	21 41	09 01
11	03 50	11 28	05 25	17 31	09 07	20 58	09 48	17 39	21 41	09 02
12	03 27	08 18	05 46	17 16	08 49	20 58	09 50	17 39	21 41	09 03
13	03 03	04 46	06 02	17 01	08 31	20 59	09 52	17 38	21 41	09 04
14	02 39	01 N 00	06 15	16 45	08 13	20 59	09 53	17 38	21 41	09 04
15	02 16	02 S 51	06 22	16 28	07 55	21 00	09 55	17 38	21 41	09 05
16	01 52	06 38	06 25	16 11	07 36	21 00	09 57	17 38	21 41	09 06
17	01 28	10 10	06 23	15 54	07 18	21 01	09 58	17 37	21 41	09 07
18	01 05	13 18	06 17	15 36	07 00	21 01	10 00	17 37	21 41	09 07
19	00 41	15 49	06 07	15 18	06 41	21 01	10 02	17 37	21 41	09 08
20	00 S 17	17 32	05 52	14 59	06 23	21 01	10 03	17 36	21 41	09 09
21	00 N 07	18 18	05 33	14 40	06 04	21 02	10 05	17 36	21 41	09 10
22	00 30	17 58	05 11	14 21	05 46	21 02	10 07	17 36	21 41	09 11
23	00 54	16 30	04 46	14 01	05 27	21 02	10 08	17 35	21 41	09 11
24	01 18	13 57	04 18	13 40	05 08	21 02	10 10	17 35	21 41	09 12
25	01 41	10 29	03 49	13 20	04 50	21 02	10 11	17 35	21 41	09 13
26	02 05	06 20	03 18	12 58	04 31	21 02	10 13	17 34	21 41	09 14
27	02 28	01 S 47	02 46	12 37	04 12	21 02	10 14	17 34	21 41	09 14
28	02 52	02 N 49	02 14	12 15	03 54	21 02	10 15	17 33	21 41	09 15
29	03 15	07 11	01 43	11 53	03 35	21 02	10 17	17 33	21 40	09 16
30	03 39	11 02	01 12	11 30	03 16	21 02	10 18	17 33	21 40	09 16
31	04 N 02	14 N 10	00 N 42	11 S 07	02 S 57	21 N 01	10 N 20	17 S 32	21 S 40	09 N 17

ZODIAC SIGN ENTRIES

Date	h	m	Planets
02	07	09	☽ ♉
03	17	18	♀ ♒
03	21	32	☿ ♈
04	12	58	☽ ♊
06	22	34	☽ ♋
09	10	47	☽ ♌
11	23	42	☽ ♍
14	11	42	☽ ♎
16	21	49	☽ ♏
19	05	38	☽ ♐
21	05	22	☉ ♈
21	10	56	☽ ♑
23	13	52	☽ ♒
25	15	04	☽ ♓
27	15	47	☽ ♈
28	10	39	☿ ♓
29	03	18	♀ ♓
29	17	36	☽ ♉
31	22	08	☽ ♊

LATITUDES

Date	Mercury ☿ ° '	Venus ♀ ° '	Mars ♂ ° '	Jupiter ♃ ° '	Saturn ♄ ° '	Uranus ♅ ° '	Neptune ♆ ° '	Pluto ♇ ° '
01	00 N 22	01 N 10	01 S 05	00 N 41	02 N 02	00 N 21	01 N 25	17 N 34
04	01 02	00 57	01 05	00 41	02 02	00 21	01 25	17 35
07	01 44	00 45	01 05	00 42	02 03	00 21	01 25	17 36
10	02 24	00 33	01 04	00 42	02 03	00 21	01 25	17 37
13	02 58	00 21	01 04	00 42	02 03	00 21	01 25	17 38
16	03 22	00 N 09	01 03	00 42	02 03	00 21	01 25	17 39
19	03 32	00 S 02	01 03	00 41	02 03	00 21	01 25	17 40
22	03 26	00 13	01 02	00 41	02 03	00 21	01 25	17 40
25	03 04	00 23	01 02	00 41	02 03	00 21	01 26	17 41
28	02 28	00 33	01 01	00 41	02 03	00 21	01 26	17 41
31	01 N 43	00 S 42	01 S 00	00 N 41	02 N 02	00 N 21	01 N 26	17 N 42

DATA

Julian Date	2443934
Delta T	+50 seconds
Ayanamsa	23° 33' 55"
Synetic vernal point	05° ♓ 33' 05"
True obliquity of ecliptic	23° 26' 23"

MOON'S PHASES, APSIDES AND POSITIONS ☽

Date	h	m	Phase	Longitude ° '	Eclipse Indicator
05	16	23	☽	14 ♊ 30	
13	21	14	○	22 ♍ 42	partial
21	11	22	☾	00 ♑ 15	
28	02	59	●	06 ♈ 51	

Day	h	m		
10	10	16	Apogee	
26	05	32	Perigee	
07	05	28	Max dec	18° N 18'
14	18	16	0S	
21	16	58	Max dec	18° S 19'
27	21	14	0N	

LONGITUDES

Date	Chiron ⚷ ° '	Ceres ⚳ ° '	Pallas ⚴ ° '	Juno ⚵ ° '	Vesta ⚶ ° '	Black Moon Lilith ⚸ ° '
01	06 ♉ 01	03 ♓ 23	10 ♒ 10	20 ♓ 31	07 ♓ 59	15 ♌ 34
11	06 ♉ 28	07 ♓ 17	13 ♒ 29	25 ♓ 42	12 ♓ 57	16 ♌ 41
21	06 ♉ 58	11 ♓ 10	16 ♒ 39	00 ♈ 57	17 ♓ 52	17 ♌ 47
31	07 ♉ 32	14 ♓ 59	19 ♒ 39	06 ♈ 18	22 ♓ 43	18 ♌ 54

All ephemeris data is given at 12.00 UT and the Moon's longitude is additionally given for 24.00 UT

ASPECTARIAN

h m	Aspects
01 Thursday	
05 38	☽ ± ♅
07 31	☽ ⊥ ☉
07 35	☽ ∠ ♂
08 17	☽ ± ♄
12 03	☽ ☍ ♇
14 40	☽ △ ♆
15 44	☽ ⊼ ♅
17 42	☉ ☍ ♄
23 27	☽ ⚼ ♄
23 56	☽ ∠ ☉
02 Friday	
02 33	☽ ⊻ ☿
03 14	☽ ⋕ ☉
04 02	☽ □ ♀
06 59	☽ □ ♃
10 42	☽ ⚹ ♂
12 06	☽ ∥ ♇
14 15	☽ ⊥ ☿
16 02	☽ ∥ ♄
16 31	☽ ⚼ ♆
20 49	♂ Q ♆
03 Saturday	
01 26	☽ △ ♄
04 11	☽ ⚹ ☉
05 42	☽ ⋕ ♂
08 08	☽ ∠ ☿
08 28	☽ ± ♆
09 11	☽ Q ♂
13 22	♀ ☍ ♃
14 37	☽ Q ♃
16 21	☽ ⊼ ♇
18 04	☿ △ ♃
19 18	☽ ⊼ ♆
20 22	☽ ☍ ♅
22 40	☉ ± ♇
04 Sunday	
03 16	☽ ± ♇
03 40	☽ Q ☉
12 29	☽ ⚹ ♃
14 21	♂ ⚼ ♇
14 49	☽ ⚹ ☿
14 55	☽ △ ♀
19 57	☽ ⚼ ♇
20 20	☽ □ ♂
21 33	☿ ⚹ ♀
05 Monday	
08 16	☽ □ ♄
16 07	☽ Q ☿
16 23	☽ □ ☉
16 42	☽ ∠ ♃
20 05	☉ ⚼ ♃
22 18	☽ ⚼ ♀
06 Tuesday	
00 08	☽ ⋕ ♅
00 30	☽ △ ♇
03 48	☽ ☍ ♆
04 51	☽ ⊼ ♅
06 08	♀ Q ♅
10 02	☽ ⊥ ♃
16 34	☽ ± ♅
17 41	☽ ± ♀
18 54	☽ Q ♄
21 46	☽ ⊻ ♃
22 25	♂ ± ♃
07 Wednesday	
05 41	♀ ± ♄
06 48	☽ ⊼ ♀
06 58	☽ □ ☿
10 21	☽ ⚼ ♅
10 26	☽ △ ♂
18 34	☽ ⚹ ♄
20 28	☿ ⚹ ♀
08 Thursday	
08 53	☽ △ ☉
08 55	♀ ∠ ♆
11 31	☽ ⋕ ♅
11 50	☽ □ ♇
15 25	☽ ⊼ ♆
16 26	☽ △ ♅
18 40	☽ ⚼ ♂
09 Friday	
00 35	☽ ∠ ♄
01 03	☿ ⚼ ♅
03 34	☽ ± ♆
09 40	☽ ☌ ♃
09 49	♂ ⋕ ♄
14 12	☽ ± ♂
14 48	☉ ⋕ ☿
18 03	☽ ⚼ ☉
18 43	☽ ⊥ ♄
19 25	☉ ⊼ ♇
21 51	☽ ⚼ ♆
10 Saturday	
00 15	☽ Q ♇
00 23	☽ △ ☿
01 54	☽ ☍ ♀
03 16	☽ ⊼ ♂
06 50	☽ ⊻ ♄
14 02	☽ ± ☉
22 53	♀ ∥ ♅
11 Sunday	
00 36	☽ ⚹ ♇
03 21	☽ ⊼ ☉
04 22	☽ △ ♆
05 16	☽ □ ♅
08 35	☽ ⚼ ☿
15 23	☉ □ ♆
16 59	♀ ⊻ ♂
18 21	♂ ⋕ ♇
22 20	☽ ⊻ ♃
12 Monday	
00 43	♀ ⊼ ♄
00 55	☽ ∥ ♄
01 48	☉ △ ♅
03 34	☽ ± ☿
04 11	♂ ☍ ♄
06 39	☽ ∥ ♇
06 55	☽ ∠ ♇
07 56	☽ ⋕ ♂
10 25	☽ ⊥ ♃
16 04	☽ ⊼ ☿
17 37	☽ Q ♅
19 12	☽ ☌ ♄
20 22	☽ ☍ ♂
21 21	☽ ⊼ ♀
13 Tuesday	
00 59	☽ ⊥ ♇
04 09	☽ ∥ ☿
04 26	☽ ∠ ♃
10 41	☽ ± ♀
12 58	☽ ⊻ ♇
16 49	☽ □ ♆
17 36	☽ ⚹ ♅
21 14	☽ ☍ ☉
14 Wednesday	
00 23	☽ ⋕ ☉
06 30	☽ ⚼ ♀
10 09	☽ ⚹ ♃
23 11	☽ ∠ ♅
15 Thursday	
01 16	☿ St R
04 20	☽ Q ♆
04 25	☽ ☍ ☿
06 17	☽ ⊻ ♄
08 41	☽ ∥ ☉
09 37	☽ Q ♃
11 50	☽ ⊼ ♂
15 01	☽ △ ♀
16 38	♂ ± ♇
16 42	☽ ⊥ ♅
17 53	☽ ⊥ ♄
23 45	☽ ☌ ♇
16 Friday	
00 15	☽ ± ♂
03 37	☽ ⚹ ♆
04 17	☽ ⊻ ♅
10 37	☽ ⋕ ☿
11 07	☽ ∠ ♄
12 56	☽ ⊼ ☉
17 55	☽ ∥ ♂
18 40	☽ ⚼ ♂
20 11	☽ □ ♃
17 Saturday	
01 21	☽ ± ☉
04 32	☽ ⋕ ♇
08 12	☽ ∠ ♆
10 34	☽ ⋕ ♄
13 07	☽ ⊼ ☿
15 23	☽ ⚹ ♄
18 18	♂ ⚼ ♃
19 43	☽ ⚼ ☉
18 Sunday	
00 02	☽ ± ☿
00 47	☽ △ ♂
01 05	☽ ⊥ ♆
05 52	☽ □ ♀
08 23	☽ ⊻ ♇
12 12	☽ ⊻ ♆
12 45	☽ ☌ ♅
13 35	☽ Q ♄
16 09	☽ ⚼ ☿
19 25	☽ ⊥ ♇
19 Monday	
01 44	☽ △ ☉
03 58	☽ △ ♃
07 06	☽ ∥ ♀
08 55	♀ △ ♇
11 47	☽ ∠ ♇
18 01	☽ Q ♀
18 23	☽ △ ☿
22 08	☽ □ ♄
20 Tuesday	
06 38	☉ △ ♃
06 56	☽ ⚼ ♃
10 47	☽ □ ♂
13 22	☽ ∥ ♅
14 34	☽ ⚹ ♇
17 27	☽ ⚹ ♀
18 19	☽ ☌ ♆
18 45	☽ ⊻ ♅
22 48	☽ ± ♃
21 Wednesday	
03 29	♀ ⚹ ♆
05 15	☽ ⊥ ♅
08 12	♀ □ ♅
09 16	☽ ⊼ ♃
11 22	☽ □ ☉
11 32	☽ Q ♇
11 37	☿ ∠ ♀
17 17	☿ ⚼ ♅
20 08	☽ Q ♂
20 38	☽ □ ☿
20 49	☽ ∠ ♅
22 00	☽ ∠ ♀
22 Thursday	
02 24	☽ △ ♄
14 46	☽ ⊥ ♀
17 48	☽ ⚹ ♂
18 16	☽ □ ♇
20 10	☽ ∥ ♅
20 29	☽ Q ☉
21 58	☽ ⊻ ♆
22 18	☽ ⚹ ♅
23 Friday	
01 35	☽ Q ☿
01 49	☽ ⊻ ♀
02 01	♂ ⊼ ♇
03 41	☽ ⚼ ♄
07 32	♆ St R
08 01	☽ ⊥ ♆
12 14	☽ ☍ ♃
18 02	☽ ⚹ ☉
18 19	☽ Q ♅
18 37	☽ ± ♄
20 20	☽ ∠ ♂
20 28	☽ ⚹ ☿
22 59	☽ ∠ ♆
24 Saturday	
04 29	☽ ⊼ ♄
12 01	☽ ⊥ ♂
14 05	☉ ☌ ☿
14 31	☽ ∥ ♀
19 45	☽ ∠ ☿
19 56	☽ △ ♇
20 31	☽ ∠ ☉
22 23	☽ ⊻ ♂
23 36	☽ ⚹ ♆
23 51	☽ □ ♅
25 Sunday	
07 41	☽ ☌ ♀
09 35	☽ ⊥ ☿
12 14	☽ ⊥ ☉
13 28	☽ ⊼ ♃
13 50	☽ ⋕ ♄
18 47	☽ ⊻ ☿
19 07	☽ Q ♆
19 40	☽ ⋕ ♇
20 18	☽ ⚼ ♇
21 16	♂ □ ♆
21 58	☿ ⊥ ♀
22 41	☽ ⊻ ☉
23 12	☽ ± ♃
26 Monday	
00 56	♃ St D
01 07	♂ △ ♅
05 10	☽ ☍ ♄
10 52	☽ ± ♇
13 47	☽ ⚼ ♃
20 35	☽ ⊼ ♇
20 53	☉ ⚼ ♅
22 25	☽ ∥ ♂
27 Tuesday	
00 19	☽ □ ♆
00 29	☽ △ ♅
01 50	☽ ☌ ♂
06 12	☽ ⋕ ☿
08 42	☽ ⋕ ☉
12 39	☽ ⊻ ♀
14 10	☽ △ ♃
16 48	☽ ☌ ☿
19 45	☉ ∥ ☿
23 20	☽ ⊥ ♀
28 Wednesday	
00 29	♄ Q ♅
00 54	☽ ⚼ ♅
02 59	☽ ☌ ☉
05 49	☽ ⊼ ♄
07 26	♀ ⊼ ♃
09 14	☽ ∥ ☿
12 14	☽ ∥ ☉
15 29	☽ ∠ ♀
15 40	☽ ± ♄
15 43	☽ ± ♅
17 20	☽ ⋕ ♂
20 37	☿ ⊻ ♀
21 40	☽ ☍ ♇
29 Thursday	
01 35	☽ △ ♆
01 41	☽ ⊼ ♅
06 01	☽ ⊻ ♂
06 37	☽ ⚼ ♄
15 57	☽ □ ♃
16 00	☽ ⊻ ☿
16 43	☽ ⊥ ♂
16 45	☿ △ ♃
18 12	☉ ⊼ ♄
18 55	☽ ⚹ ♀
23 04	☉ ⋕ ♂
30 Friday	
00 31	☽ ∥ ♇
01 43	☽ ⊥ ☿
02 58	☽ ⚼ ♆
07 08	☽ ∥ ♄
08 01	☽ △ ♄
09 02	☽ ∠ ♂
09 08	☽ ⊻ ☉
14 57	☽ ⋕ ♀
16 31	☽ ∠ ☿
17 31	☽ Q ♀
18 32	☽ ± ♆
20 23	☽ ⊥ ☉
23 02	☽ Q ♃
31 Saturday	
00 51	☽ ⊼ ♇
05 05	☽ ⊼ ♆
05 06	☽ ☍ ♅
05 11	♀ Q ♆
09 35	♅ ⊻ ♆
11 28	☽ ± ♇
13 00	☽ ⚹ ♂
13 35	☽ ∠ ☉
16 54	♀ ⚼ ♇
17 55	☽ ⚹ ☿
20 26	☽ ⚹ ♃

APRIL 1979

LONGITUDES

Date	Sidereal time h m s	Sun ☉ ° ' "	Moon ☽ ° ' "	Moon ☽ 24.00 ° ' "	Mercury ☿ ° '	Venus ♀ ° '	Mars ♂ ° '	Jupiter ♃ ° '	Saturn ♄ ° '	Uranus ♅ ° '	Neptune ♆ ° '	Pluto ♇ ° '
01	00 37 01	11 ♈ 10 03	07 ♊ 35 14	14 ♊ 02 49	27 ♓ 15	04 ♓ 01	25 ♓ 40	29 ♋ 04	08 ♍ 17	20 ♏ 27	20 ♐ 28	18 ♎ 02
02	00 40 57	12 09 15	20 ♊ 24 44	26 ♊ 41 20	26 R 46	05 12	26 27	29 06	08 R 14	20 R 25	20 R 28	18 R 01
03	00 44 54	13 08 26	02 ♋ 53 06	09 ♋ 00 34	26 22	06 24	27 14	29 07	08 10	20 23	20 28	17 59
04	00 48 51	14 07 33	15 ♋ 04 17	21 ♋ 04 52	26 04	07 36	28 01	29 09	08 07	20 21	20 27	17 57
05	00 52 47	15 06 39	27 ♋ 02 55	02 ♌ 59 04	25 52	08 48	28 48	29 11	08 03	20 20	20 27	17 56
06	00 56 44	16 05 42	08 ♌ 53 56	14 ♌ 48 07	25 45	09 59	29 ♓ 34	29 13	08 00	20 18	20 26	17 54
07	01 00 40	17 04 43	20 ♌ 42 13	26 ♌ 36 48	25 D 44	11 11	00 ♈ 21	29 15	07 57	20 16	20 26	17 52
08	01 04 37	18 03 41	02 ♍ 32 23	08 ♍ 29 28	25 48	12 23	01 08	29 18	07 54	20 14	20 25	17 51
09	01 08 33	19 02 38	14 ♍ 28 30	20 ♍ 29 54	25 58	13 35	01 54	29 20	07 51	20 12	20 25	17 49
10	01 12 30	20 01 32	26 ♍ 34 00	02 ♎ 41 08	26 12	14 47	02 41	29 23	07 48	20 10	20 24	17 47
11	01 16 26	21 00 24	08 ♎ 51 31	15 ♎ 05 21	26 31	15 59	03 28	29 26	07 45	20 08	20 24	17 46
12	01 20 23	21 59 14	21 ♎ 22 46	27 ♎ 43 50	26 55	17 11	04 14	29 29	07 42	20 05	20 23	17 44
13	01 24 20	22 58 02	04 ♏ 08 35	10 ♏ 36 59	27 24	18 23	05 01	29 33	07 40	20 03	20 22	17 42
14	01 28 16	23 56 48	17 ♏ 08 59	23 ♏ 44 28	27 57	19 35	05 47	29 36	07 37	20 01	20 22	17 41
15	01 32 13	24 55 32	00 ♐ 23 20	07 ♐ 05 24	28 33	20 47	06 34	29 40	07 35	19 59	20 21	17 39
16	01 36 09	25 54 15	13 ♐ 50 33	20 ♐ 38 36	29 14	21 59	07 20	29 44	07 32	19 57	20 20	17 37
17	01 40 06	26 52 55	27 ♐ 29 25	04 ♑ 22 49	29 ♓ 58	23 11	08 07	29 48	07 30	19 55	20 20	17 36
18	01 44 02	27 51 35	11 ♑ 18 42	18 ♑ 16 55	00 ♈ 46	24 23	08 53	29 52	07 28	19 52	20 19	17 34
19	01 47 59	28 50 12	25 ♑ 17 20	02 ♒ 19 49	01 38	25 35	09 39	29 ♋ 56	07 26	19 50	20 18	17 32
20	01 51 55	29 ♈ 48 48	09 ♒ 24 13	16 ♒ 30 23	02 32	26 47	10 26	00 ♌ 01	07 24	19 48	20 17	17 31
21	01 55 52	00 ♉ 47 22	23 ♒ 38 06	00 ♓ 47 08	03 29	28 00	11 12	00 05	07 22	19 45	20 16	17 29
22	01 59 49	01 45 54	07 ♓ 57 11	15 ♓ 07 53	04 30	29 ♓ 12	11 58	00 10	07 20	19 43	20 15	17 27
23	02 03 45	02 44 25	22 ♓ 18 50	29 ♓ 29 32	05 33	00 ♈ 24	12 44	00 15	07 18	19 41	20 14	17 26
24	02 07 42	03 42 54	06 ♈ 39 27	13 ♈ 48 00	06 39	01 36	13 30	00 20	07 16	19 38	20 13	17 24
25	02 11 38	04 41 22	20 ♈ 54 35	27 ♈ 58 34	07 47	02 49	14 16	00 26	07 15	19 36	20 12	17 22
26	02 15 35	05 39 47	04 ♉ 59 21	11 ♉ 56 22	08 58	04 01	15 03	00 31	07 14	19 33	20 11	17 21
27	02 19 31	06 38 11	18 ♉ 49 07	25 ♉ 37 10	10 11	05 13	15 49	00 37	07 12	19 31	20 10	17 19
28	02 23 28	07 36 33	02 ♊ 20 11	08 ♊ 57 58	11 26	06 25	16 35	00 43	07 11	19 29	20 09	17 18
29	02 27 24	08 34 53	15 ♊ 30 25	21 ♊ 57 31	12 44	07 38	17 20	00 49	07 10	19 26	20 08	17 16
30	02 31 21	09 ♉ 33 11	28 ♊ 19 25	04 ♋ 36 20	14 ♈ 04	08 ♈ 50	18 ♈ 06	00 ♌ 55	07 ♍ 09	19 ♏ 24	20 ♐ 07	17 ♎ 15

DECLINATIONS

Date	Moon True ☊ ° '	Moon Mean ☊ ° '	Moon ☽ Latitude ° '	Sun ☉ ° '	Moon ☽ ° '	Mercury ☿ ° '	Venus ♀ ° '	Mars ♂ ° '	Jupiter ♃ ° '	Saturn ♄ ° '	Uranus ♅ ° '	Neptune ♆ ° '	Pluto ♇ ° '
01	17 ♍ 15	16 ♍ 26	05 S 10	04 N 25	16 N 28	00 N 14	10 S 44	02 S 38	21 N 01	10 N 21	17 S 32	21 S 40	09 N 18
02	17 R 12	16 23	05 14	04 48	17 52	00 S 12	10 21	02 19	21 01	10 22	17 31	21 40	09 19
03	17 11	16 20	05 03	05 11	18 21	00 36	09 57	02 00	21 00	10 24	17 31	21 40	09 19
04	17 D 10	16 17	04 38	05 34	17 59	00 57	09 33	01 42	21 00	10 25	17 30	21 40	09 20
05	17 11	16 13	04 02	05 57	16 47	01 17	09 09	01 23	21 00	10 26	17 30	21 40	09 21
06	17 12	16 10	03 15	06 20	14 53	01 33	08 44	01 04	20 59	10 27	17 29	21 40	09 21
07	17 14	16 07	02 21	06 43	12 22	01 47	08 19	00 45	20 59	10 28	17 29	21 40	09 22
08	17 16	16 04	01 20	07 05	09 19	01 59	07 54	00 26	20 58	10 29	17 28	21 40	09 22
09	17 17	16 01	00 S 15	07 27	05 53	02 08	07 29	00 S 07	20 58	10 30	17 28	21 40	09 23
10	17 R 16	15 57	00 N 51	07 50	02 N 08	02 14	07 03	00 N 12	20 57	10 31	17 27	21 40	09 24
11	17 14	15 54	01 55	08 12	01 S 45	02 18	06 37	00 30	20 56	10 32	17 26	21 40	09 24
12	17 11	15 51	02 56	08 34	05 37	02 19	06 11	00 49	20 56	10 33	17 26	21 39	09 25
13	17 06	15 48	03 48	08 56	09 19	02 18	05 45	01 08	20 55	10 34	17 25	21 39	09 25
14	17 00	15 45	04 30	09 17	12 38	02 15	05 19	01 27	20 54	10 35	17 25	21 39	09 26
15	16 53	15 42	04 58	09 39	15 23	02 09	04 52	01 45	20 53	10 36	17 24	21 39	09 27
16	16 48	15 38	05 10	10 00	17 20	02 01	04 25	02 04	20 52	10 37	17 24	21 39	09 27
17	16 43	15 35	05 05	10 22	18 20	01 51	03 59	02 23	20 52	10 38	17 23	21 39	09 28
18	16 40	15 32	04 42	10 43	18 16	01 40	03 32	02 41	20 51	10 38	17 22	21 39	09 28
19	16 39	15 29	04 03	11 04	17 06	01 26	03 04	03 00	20 50	10 39	17 22	21 39	09 29
20	16 D 39	15 26	03 08	11 24	14 52	01 10	02 37	03 18	20 49	10 40	17 21	21 39	09 29
21	16 40	15 23	02 02	11 45	11 44	00 53	02 10	03 37	20 48	10 40	17 21	21 39	09 30
22	16 41	15 19	00 N 47	12 05	07 51	00 34	01 42	03 55	20 47	10 41	17 20	21 39	09 30
23	16 R 41	15 16	00 S 30	12 25	03 S 31	00 S 13	01 15	04 14	20 46	10 41	17 19	21 38	09 31
24	16 40	15 13	01 46	12 45	01 N 01	00 N 10	00 47	04 32	20 44	10 42	17 19	21 38	09 31
25	16 36	15 10	02 54	13 05	05 28	00 34	00 S 20	04 50	20 43	10 42	17 18	21 38	09 32
26	16 30	15 07	03 51	13 25	09 34	00 59	00 N 08	05 08	20 42	10 43	17 17	21 38	09 32
27	16 22	15 03	04 32	13 44	13 03	01 26	00 36	05 27	20 41	10 43	17 17	21 38	09 32
28	16 13	15 00	04 58	14 03	15 45	01 54	01 04	05 45	20 39	10 43	17 16	21 38	09 33
29	16 05	14 57	05 07	14 22	17 34	02 23	01 31	06 03	20 38	10 44	17 15	21 38	09 33
30	15 ♍ 57	14 ♍ 54	05 S 00	14 N 40	18 N 26	02 N 54	01 N 59	06 N 20	20 N 37	10 N 44	17 S 15	21 S 38	09 N 34

ZODIAC SIGN ENTRIES

Date	h	m	Planets
03	06	24	☽ ♋
05	17	58	☽ ♌
07	01	08	♂ ♈
08	06	52	☽ ♍
10	18	45	☽ ♎
13	04	16	☽ ♏
15	11	18	☽ ♐
17	12	48	☿ ♈
17	16	23	☽ ♑
19	20	02	☽ ♒
20	08	30	♃ ♌
20	16	35	☉ ♉
21	22	41	☽ ♓
23	04	02	♀ ♈
24	00	51	☽ ♈
26	03	27	☽ ♉
28	07	49	☽ ♊
30	15	11	☽ ♋

LATITUDES

Date	Mercury ☿ ° '	Venus ♀ ° '	Mars ♂ ° '	Jupiter ♃ ° '	Saturn ♄ ° '	Uranus ♅ ° '	Neptune ♆ ° '	Pluto ♇ ° '
01	01 N 27	00 S 45	01 S 00	00 N 41	02 N 02	00 N 21	01 N 26	17 N 42
04	00 N 40	00 54	00 59	00 41	02 02	00 21	01 26	17 42
07	00 S 06	01 01	00 58	00 41	02 02	00 21	01 26	17 42
10	00 47	01 09	00 57	00 41	02 02	00 21	01 26	17 42
13	01 23	01 15	00 56	00 41	02 01	00 21	01 26	17 42
16	01 52	01 21	00 55	00 41	02 01	00 21	01 26	17 42
19	02 16	01 26	00 54	00 41	02 01	00 21	01 26	17 42
22	02 33	01 31	00 53	00 41	02 00	00 21	01 27	17 41
25	02 45	01 34	00 51	00 41	02 00	00 21	01 27	17 41
28	02 51	01 37	00 50	00 41	02 00	00 21	01 27	17 40
31	02 S 52	01 S 40	00 S 49	00 N 40	01 N 59	00 N 21	01 N 27	17 N 40

DATA

Julian Date	2443965
Delta T	+50 seconds
Ayanamsa	23° 33' 57"
Synetic vernal point	05° ♓ 33' 02"
True obliquity of ecliptic	23° 26' 23"

LONGITUDES

Date	Chiron ⚷ ° '	Ceres ⚳ ° '	Pallas ⚴ ° '	Juno ⚵ ° '	Vesta ⚶ ° '	Black Moon Lilith ⚸ ° '
01	07 ♉ 35	15 ♓ 22	19 ♒ 56	06 ♈ 50	23 ♓ 12	19 ♌ 00
11	08 ♉ 12	19 ♓ 07	22 ♒ 43	12 ♈ 14	27 ♓ 59	20 ♌ 07
21	08 ♉ 49	22 ♓ 46	25 ♒ 15	17 ♈ 41	02 ♈ 41	21 ♌ 14
31	09 ♉ 27	26 ♓ 20	27 ♒ 30	23 ♈ 12	07 ♈ 17	22 ♌ 20

MOON'S PHASES, APSIDES AND POSITIONS ☽

Date	h	m	Phase	Longitude ° '	Eclipse Indicator
04	09	57	☽	14 ♋ 03	
12	13	15	○	22 ♎ 02	
19	18	30	☾	29 ♑ 06	
26	13	15	●	05 ♉ 43	

Day	h	m		
07	02	54	Apogee	
22	21	32	Perigee	
03	13	15	Max dec	18° N 21'
11	01	17	0S	
17	22	36	Max dec	18° S 26'
24	06	36	0N	
30	22	20	Max dec	18° N 31'

ASPECTARIAN

h m	Aspects
01 Sunday	
03 41	☽ bQ ♇
04 47	☽ □ ♀
12 10	☽ Q ♂
13 17	☽ □ ♄
14 56	☽ Q ☿
19 11	☽ ⚹ ☉
02 Monday	
00 04	☽ ∠ ♃
04 22	☽ # ♅
07 28	☽ △ ♇
09 43	♀ ± ♃
10 27	♀ # ♄
12 01	☽ ⊼ ♅
12 06	☽ ☍ ♆
17 07	☽ ⊥ ♃
18 06	☿ ☌ ♂
19 45	☽ Q ☉
23 04	☽ Q ♄
23 27	☽ ± ♅
23 45	☽ □ ☿
03 Tuesday	
00 19	☽ □ ♂
04 41	☽ ⊻ ♃
16 53	☽ bQ ♅
19 37	☽ △ ♀
22 18	☽ ⚹ ♄
04 Wednesday	
09 57	☽ □ ☉
11 44	☉ ± ♄
17 27	☉ ± ♅
17 44	☽ □ ♇
21 56	♀ ☍ ♄
22 31	☽ △ ♅
22 44	☽ ⊼ ♆
23 35	☽ # ♅
05 Thursday	
00 33	♀ # ♇
04 00	☽ ∠ ♄
04 43	☽ bQ ♀
09 39	☽ △ ☿
10 47	☽ ± ♆
15 46	☽ △ ♂
16 01	☿ ∥ ♂
16 19	☽ ☌ ♃
22 05	☽ ⊥ ♄
06 Friday	
00 26	♂ △ ♃
00 55	☽ ± ♀
04 59	☽ bQ ♆
05 56	☽ Q ♇
10 11	☽ ⊻ ♄
14 28	☽ ⊼ ♀
15 45	☽ bQ ☿
07 Saturday	
00 00	☉ ⊥ ♀
00 21	☽ bQ ♂
03 58	☽ △ ☉
05 22	☿ St D
06 15	☽ ⚹ ♇
10 01	☽ ± ☿
11 06	☽ □ ♅
11 27	☽ △ ♆
19 56	☽ ± ♂
22 15	☽ ⊼ ☿
08 Sunday	
01 27	♀ ± ♇
03 16	☽ ∥ ♄
05 25	☽ ⊻ ♃
06 50	☉ ☍ ♇
08 57	☽ ⊼ ♂
11 38	☽ ∥ ♇
12 37	☽ ∠ ♇
13 09	☽ bQ ☉
17 35	☽ ⊥ ♃
22 46	☽ ☌ ♄
23 26	☽ Q ♅
23 39	☽ # ♀
09 Monday	
02 19	☽ ∥ ☉
06 41	☽ ⊥ ♇
08 53	☽ ± ☉
10 01	☽ ☍ ♀
11 43	☽ ∠ ♃
12 33	☉ # ♀
18 39	☽ ⊻ ♇
21 55	☽ ⊼ ☉
23 22	☽ ⚹ ♅
23 50	☽ □ ♆
10 Tuesday	
03 46	♀ bQ ♃
11 16	☽ ☍ ☿
11 26	☽ # ☿
15 12	☉ ⊼ ♅
17 34	☽ ⚹ ♃
21 13	☉ △ ♆
23 11	☽ ∥ ♂
11 Wednesday	
00 49	☽ ☍ ♂
03 44	☽ # ♂
04 47	☽ ∠ ♅
09 52	☽ ⊻ ♄
11 06	☽ Q ♆
15 25	☽ ∥ ☿
17 00	☽ Q ♃
21 24	☽ ⊥ ♄
22 07	☽ ⊥ ♅
12 Thursday	
03 09	☽ ⊼ ♀
05 04	☽ ☌ ♇
09 34	☽ ⊻ ♅
10 07	☽ ⚹ ♆
13 15	☽ ☍ ☉
14 30	☽ ∠ ♄
15 11	☽ ∥ ♀
15 46	☽ ± ♀
22 50	♀ ⊼ ♇
22 52	☽ ⊼ ☿
13 Friday	
03 22	☽ □ ♃
04 50	☉ bQ ♄
09 04	☽ # ☉
10 26	☽ bQ ♀
10 33	☽ ± ☿
12 42	☽ # ♇
13 16	♂ bQ ♅
13 43	☽ ⊼ ♂
14 17	☽ ∠ ♆
18 31	☽ ⚹ ♄
20 42	☽ # ♄
14 Saturday	
01 32	☽ ± ♂
03 56	☽ bQ ☿
06 54	☽ ⊥ ♆
12 58	☽ ⊻ ♇
16 30	☽ Q ♄
16 53	☽ △ ♀
17 13	☽ ☌ ♅
17 51	☽ ⊻ ♆
19 03	☽ bQ ♂
20 35	♀ △ ♅
21 45	☉ ∥ ♇
23 51	☽ ⊥ ♇
15 Sunday	
01 22	☽ ⊼ ☉
03 34	♀ □ ♆
08 32	☽ △ ☿
10 41	☽ △ ♃
13 02	☽ ± ☉
16 03	☽ ∠ ♇
23 44	☽ △ ♂
16 Monday	
00 50	☽ □ ♄
06 23	☽ bQ ☉
09 31	☿ # ♂
12 59	☽ ∥ ♅
13 34	☽ bQ ♃
17 56	♂ ⊼ ♄
18 40	☽ ⚹ ♇
22 45	☽ ⊻ ♅
23 27	☽ ☌ ♆
17 Tuesday	
03 44	☽ □ ♀
05 30	☽ ± ♃
05 44	☿ △ ♃
09 15	☽ ⊥ ♅
10 51	☽ △ ☉
15 40	☽ Q ♇
16 02	☽ ⊼ ♃
16 36	☽ □ ☿
18 Wednesday	
00 53	☽ ∠ ♅
05 22	☽ △ ♄
06 44	☉ ∥ ♄
07 33	☽ □ ♂
14 01	☽ Q ♀
22 45	☽ □ ♇
19 Thursday	
01 40	☽ Q ☿
02 41	☽ ⚹ ♅
03 28	☽ ⊻ ♆
07 07	☽ bQ ♄
08 04	☽ ∥ ♅
12 33	☽ ⚹ ♀
13 43	☽ ⊥ ♆
14 22	♀ # ♂
16 16	☽ Q ♂
18 30	☽ □ ☉
19 58	☽ ☍ ♃
22 26	☽ ± ♄
23 08	☽ Q ♅
23 32	☽ ⚹ ☿
20 Friday	
05 02	☽ ∠ ♆
08 36	☽ ⊼ ♄
13 50	☽ ⚹ ♂
16 24	☽ ∠ ♀
17 17	☉ □ ♃
21 Saturday	
01 40	☽ △ ♇
02 42	☽ ∠ ☿
03 15	☽ Q ☉
05 30	☽ □ ♅
06 21	☽ ⚹ ♆
08 59	☽ ⊥ ♀
11 52	☽ # ☉
16 33	☽ ∠ ♂
18 55	☽ # ♄
18 57	☽ ⊥ ☿
19 59	☽ ⊻ ♀
22 54	☽ ⊼ ♃
22 Sunday	
00 53	☽ ⚹ ☉
02 15	☽ # ♇
02 28	☽ Q ♆
02 49	☽ bQ ♇
05 46	☽ ⊻ ☿
08 29	☽ ⊥ ♂
09 00	☽ ± ♃
10 58	☽ ☍ ♄
17 00	☿ bQ ♅
17 51	☽ ± ♇
19 06	☽ ⊻ ♂
23 Monday	
00 08	☽ bQ ♃
03 48	☽ ∠ ☉
03 51	☽ ⊼ ♇
07 36	☽ △ ♅
08 25	☽ # ♂
08 32	☽ □ ♆
08 52	♀ △ ♃
19 56	☽ ⊥ ☉
24 Tuesday	
01 21	☽ △ ♃
01 23	☽ ∥ ♀
02 45	☽ ☌ ♀
05 02	♂ ± ♄
06 14	☽ # ☿
06 43	☽ ⊻ ☉
07 02	☽ ∥ ☿
08 37	☽ bQ ♅
10 53	☽ # ♀
11 58	☽ ☌ ☿
13 02	☽ ⊼ ♄
15 55	♂ ± ♅
23 06	☽ ± ♄
23 42	☽ ± ♅
25 Wednesday	
00 10	☽ ☌ ♂
01 08	☿ ⊼ ♄
05 35	☿ # ♀
06 02	☽ ☍ ♇
08 14	☽ ∥ ♂
09 47	☽ ⊼ ♅
10 49	☽ △ ♆
14 16	☽ bQ ♄
26 Thursday	
00 31	☉ bQ ♆
04 17	☽ □ ♃
10 10	☽ ⊻ ♀
11 50	☽ ∥ ♇
12 21	☽ bQ ♆
13 15	☽ ☌ ☉
15 51	☽ △ ♄
19 26	☽ ∥ ♄
19 30	☽ ⊻ ☿
21 29	☽ ⊥ ♀
22 29	♀ bQ ♅
27 Friday	
03 53	☽ ± ♆
06 25	☽ ⊻ ♂
06 56	☽ ⊥ ☿
09 23	☽ ⊼ ♇
11 38	☽ Q ♃
13 13	☽ ☍ ♅
14 22	☽ ⊼ ♆
14 42	☽ ∠ ♀
17 34	☽ ⊥ ♂
18 03	☽ ∥ ☉
19 55	☽ ± ♇
28 Saturday	
00 22	☽ ∠ ☿
01 43	☉ △ ♄
09 04	☽ ⚹ ♃
10 33	☽ ∠ ♂
11 56	☽ bQ ♇
20 07	☽ ⚹ ♀
20 45	☽ □ ♄
22 17	☽ ⊻ ☉
29 Sunday	
02 53	♀ ⊼ ♄
06 20	☽ ⚹ ☿
06 50	☽ # ♅
09 52	♂ ☍ ♇
10 09	☽ ⊥ ☉
12 34	☽ ∠ ♃
15 15	☽ △ ♇
15 36	☽ ⚹ ♂
19 16	☽ ⊼ ♅
19 40	☿ ± ♄
20 26	☽ Q ♀
20 35	☽ ☍ ♆
30 Monday	
00 15	☿ ± ♅
04 17	☽ ∠ ☉
05 30	☽ ⊥ ♃
06 00	☽ Q ♄
06 29	☽ ± ♅
07 14	☽ Q ☿
15 36	☽ Q ♂
16 58	☽ ⊻ ♃
23 33	☽ bQ ♅

All ephemeris data is given at 12.00 UT and the Moon's longitude is additionally given for 24.00 UT

MAY 1979

LONGITUDES

Date	Sidereal time h m s	Sun ☉ ° ' "	Moon ☽ ° ' "	Moon ☽ 24.00 ° ' "	Mercury ☿ ° '	Venus ♀ ° '	Mars ♂ ° '	Jupiter ♃ ° '	Saturn ♄ ° '	Uranus ♅ ° '	Neptune ♆ ° '	Pluto ♇ ° '
01	02 35 18	10 ♉ 31 28	10 ♋ 48 35	16 ♋ 56 36	15 ♈ 26	10 ♈ 03	18 ♈ 52	01 ♌ 01	07 ♍ 08	19 ♏ 21	20 ♐ 06	17 ♎ 13
02	02 39 14	11 29 42	23 ♋ 00 52	29 ♋ 01 54	16 51	11 15	19 38	01 07	07 R 07	19 R 19	20 R 05	17 R 12
03	02 43 11	12 27 54	05 ♌ 00 19	10 ♌ 56 44	18 17	12 27	20 24	01 14	07 07	19 16	20 04	17 10
04	02 47 07	13 26 04	16 ♌ 51 48	22 ♌ 46 12	19 45	13 40	21 09	01 21	07 06	19 14	20 02	17 09
05	02 51 04	14 24 12	28 ♌ 40 34	04 ♍ 35 37	21 16	14 52	21 55	01 27	07 05	19 11	20 01	17 07
06	02 55 00	15 22 18	10 ♍ 31 57	16 ♍ 30 13	22 48	16 05	22 41	01 34	07 05	19 09	20 00	17 06
07	02 58 57	16 20 22	22 ♍ 31 00	28 ♍ 34 51	24 22	17 17	23 26	01 41	07 05	19 06	19 59	17 04
08	03 02 53	17 18 25	04 ♎ 42 16	10 ♎ 53 38	25 59	18 29	24 12	01 49	07 05	19 04	19 57	17 03
09	03 06 50	18 16 25	17 ♎ 09 21	23 ♎ 29 39	27 37	19 42	24 57	01 56	07 05	19 01	19 56	17 02
10	03 10 47	19 14 24	29 ♎ 54 42	06 ♏ 24 36	29 ♈ 17	20 54	25 43	02 03	07 D 05	18 59	19 55	17 00
11	03 14 43	20 12 21	12 ♏ 59 18	19 ♏ 38 41	01 ♉ 00	22 07	26 28	02 11	07 05	18 56	19 53	16 59
12	03 18 40	21 10 16	26 ♏ 22 32	03 ♐ 10 30	02 44	23 19	27 13	02 19	07 05	18 54	19 52	16 58
13	03 22 36	22 08 11	10 ♐ 02 14	16 ♐ 57 15	04 30	24 32	27 59	02 27	07 05	18 51	19 51	16 56
14	03 26 33	23 06 03	23 ♐ 55 06	00 ♑ 55 16	06 18	25 45	28 44	02 35	07 06	18 48	19 49	16 55
15	03 30 29	24 03 55	07 ♑ 57 14	15 ♑ 00 33	08 09	26 57	29 ♈ 29	02 43	07 06	18 46	19 48	16 54
16	03 34 26	25 01 45	22 ♑ 04 47	29 ♑ 09 31	10 01	28 10	00 ♉ 14	02 51	07 07	18 43	19 46	16 52
17	03 38 22	25 59 33	06 ♒ 14 27	13 ♒ 19 18	11 55	29 ♈ 22	00 59	02 59	07 08	18 41	19 45	16 51
18	03 42 19	26 57 21	20 ♒ 23 52	27 ♒ 27 58	13 51	00 ♉ 35	01 44	03 08	07 09	18 38	19 44	16 50
19	03 46 16	27 55 08	04 ♓ 31 29	11 ♓ 34 17	15 49	01 47	02 29	03 16	07 10	18 36	19 42	16 49
20	03 50 12	28 52 53	18 ♓ 36 17	25 ♓ 37 20	17 49	03 00	03 14	03 25	07 11	18 33	19 41	16 48
21	03 54 09	29 ♉ 50 37	02 ♈ 37 20	09 ♈ 36 06	19 51	04 13	03 59	03 34	07 12	18 31	19 39	16 47
22	03 58 05	00 ♊ 48 21	16 ♈ 33 26	23 ♈ 29 04	21 54	05 25	04 44	03 43	07 13	18 29	19 38	16 45
23	04 02 02	01 46 03	00 ♉ 22 45	07 ♉ 14 08	23 59	06 38	05 29	03 52	07 15	18 26	19 36	16 44
24	04 05 58	02 43 44	14 ♉ 02 53	20 ♉ 48 41	26 06	07 51	06 14	04 01	07 16	18 24	19 35	16 43
25	04 09 55	03 41 24	27 ♉ 31 10	04 ♊ 10 02	28 ♉ 14	09 04	06 58	04 10	07 18	18 21	19 33	16 42
26	04 13 51	04 39 03	10 ♊ 45 01	17 ♊ 15 55	00 ♊ 23	10 16	07 43	04 20	07 19	18 19	19 31	16 41
27	04 17 48	05 36 40	23 ♊ 42 35	00 ♋ 04 58	02 33	11 29	08 27	04 29	07 21	18 17	19 30	16 40
28	04 21 45	06 34 17	06 ♋ 23 04	12 ♋ 37 02	04 44	12 42	09 12	04 39	07 23	18 14	19 28	16 39
29	04 25 41	07 31 52	18 ♋ 47 02	24 ♋ 53 21	06 56	13 54	09 56	04 48	07 25	18 12	19 27	16 39
30	04 29 38	08 29 26	00 ♌ 56 23	06 ♌ 56 31	09 08	15 07	10 41	04 58	07 27	18 09	19 25	16 38
31	04 33 34	09 ♊ 26 58	12 ♌ 54 17	18 ♌ 50 13	11 ♊ 20	16 ♉ 20	11 ♉ 25	05 ♌ 08	07 ♍ 29	18 ♏ 07	19 ♐ 24	16 ♎ 37

Date	Moon True ☊ ° '	Moon Mean ☊ ° '	Moon ☽ Latitude ° '
01	15 ♍ 51	14 ♍ 51	04 S 38
02	15 R 47	14 48	04 05
03	15 45	14 44	03 21
04	15 D 45	14 41	02 29
05	15 45	14 38	01 30
06	15 46	14 35	00 S 28
07	15 R 46	14 32	00 N 36
08	15 44	14 28	01 40
09	15 40	14 25	02 40
10	15 33	14 22	03 33
11	15 25	14 19	04 16
12	15 14	14 16	04 47
13	15 04	14 13	05 02
14	14 53	14 09	04 59
15	14 45	14 06	04 39
16	14 39	14 03	04 01
17	14 35	14 00	03 09
18	14 34	13 57	02 05
19	14 D 34	13 54	00 N 53
20	14 R 34	13 50	00 S 21
21	14 33	13 47	01 34
22	14 30	13 44	02 41
23	14 24	13 41	03 37
24	14 15	13 38	04 21
25	14 04	13 34	04 49
26	13 52	13 31	05 00
27	13 39	13 28	04 56
28	13 28	13 25	04 38
29	13 18	13 22	04 06
30	13 11	13 19	03 23
31	13 ♍ 07	13 ♍ 15	02 S 33

DECLINATIONS

Date	Sun ☉ ° '	Moon ☽ ° '	Mercury ☿ ° '	Venus ♀ ° '	Mars ♂ ° '	Jupiter ♃ ° '	Saturn ♄ ° '	Uranus ♅ ° '	Neptune ♆ ° '	Pluto ♇ ° '
01	14 N 59	18 N 22	03 N 26	02 N 27	06 N 38	20 N 35	10 N 44	17 S 14	21 S 38	09 N 34
02	15 17	17 27	03 59	02 55	06 56	20 34	10 44	17 13	21 38	09 34
03	15 35	15 46	04 33	03 22	07 14	20 33	10 44	17 13	21 37	09 35
04	15 52	13 25	05 09	03 50	07 31	20 31	10 45	17 12	21 37	09 35
05	16 10	10 31	05 45	04 18	07 49	20 30	10 45	17 11	21 37	09 35
06	16 27	07 11	06 22	04 45	08 06	20 28	10 45	17 11	21 37	09 35
07	16 43	03 N 31	07 00	05 13	08 24	20 26	10 45	17 10	21 37	09 36
08	17 00	00 S 21	07 39	05 40	08 41	20 25	10 45	17 09	21 37	09 36
09	17 16	04 17	08 19	06 07	08 58	20 23	10 44	17 09	21 37	09 36
10	17 32	08 07	08 59	06 35	09 15	20 21	10 44	17 08	21 37	09 36
11	17 48	11 40	09 40	07 02	09 32	20 20	10 44	17 07	21 36	09 37
12	18 03	14 41	10 22	07 29	09 49	20 18	10 44	17 07	21 36	09 37
13	18 18	16 59	11 04	07 56	10 05	20 16	10 44	17 06	21 36	09 37
14	18 33	18 19	11 47	08 22	10 22	20 14	10 43	17 05	21 36	09 37
15	18 47	18 34	12 30	08 49	10 38	20 12	10 43	17 05	21 36	09 37
16	19 01	17 40	13 13	09 15	10 55	20 10	10 43	17 04	21 36	09 37
17	19 15	15 40	13 56	09 41	11 11	20 08	10 42	17 03	21 36	09 38
18	19 29	12 43	14 40	10 07	11 27	20 07	10 42	17 02	21 36	09 38
19	19 42	09 02	15 23	10 33	11 43	20 05	10 41	17 02	21 36	09 38
20	19 55	04 50	16 06	10 58	11 59	20 02	10 41	17 01	21 35	09 38
21	20 07	00 S 24	16 49	11 24	12 15	20 00	10 40	17 00	21 35	09 38
22	20 19	04 N 02	17 31	11 49	12 30	19 58	10 39	17 00	21 35	09 38
23	20 31	08 12	18 13	12 13	12 46	19 56	10 39	16 59	21 35	09 38
24	20 42	11 54	18 53	12 38	13 01	19 54	10 38	16 58	21 35	09 38
25	20 53	14 55	19 33	13 02	13 16	19 52	10 37	16 58	21 35	09 38
26	21 04	17 06	20 11	13 26	13 31	19 50	10 37	16 57	21 35	09 38
27	21 14	18 21	20 48	13 50	13 46	19 47	10 36	16 57	21 34	09 38
28	21 24	18 40	21 23	14 13	14 01	19 45	10 35	16 56	21 34	09 38
29	21 34	18 04	21 56	14 36	14 16	19 43	10 34	16 55	21 34	09 38
30	21 43	16 38	22 27	14 59	14 30	19 40	10 33	16 55	21 34	09 38
31	21 N 52	14 N 30	22 N 56	15 N 21	14 N 44	19 N 38	10 N 32	16 S 54	21 S 34	09 N 38

ZODIAC SIGN ENTRIES

Date	h	m	Planets
03	01	56	☽ ♌
05	14	41	☽ ♍
08	02	48	☽ ♎
10	12	10	☽ ♏
10	22	03	☿ ♉
12	18	25	☽ ♐
14	22	25	☽ ♑
16	04	25	♂ ♉
17	01	26	☽ ♒
18	00	29	♀ ♉
19	04	18	☽ ♓
21	07	30	☽ ♈
21	15	54	☉ ♊
23	11	20	☽ ♉
25	16	28	☽ ♊
26	07	44	☿ ♊
27	23	51	☽ ♋
30	10	08	☽ ♌

LATITUDES

Date	Mercury ☿ ° '	Venus ♀ ° '	Mars ♂ ° '	Jupiter ♃ ° '	Saturn ♄ ° '	Uranus ♅ ° '	Neptune ♆ ° '	Pluto ♇ ° '
01	02 S 52	01 S 40	00 S 49	00 N 40	01 N 59	00 N 21	01 N 27	17 N 40
04	02 47	01 41	00 47	00 40	01 59	00 21	01 27	17 39
07	02 38	01 42	00 46	00 40	01 58	00 21	01 27	17 38
10	02 24	01 42	00 44	00 40	01 58	00 21	01 27	17 37
13	02 05	01 42	00 43	00 40	01 58	00 21	01 27	17 36
16	01 41	01 41	00 41	00 40	01 57	00 21	01 27	17 35
19	01 15	01 39	00 40	00 40	01 57	00 21	01 27	17 34
22	00 45	01 37	00 38	00 40	01 56	00 21	01 27	17 33
25	00 S 13	01 34	00 36	00 40	01 56	00 21	01 27	17 32
28	00 N 18	01 30	00 35	00 40	01 55	00 20	01 27	17 31
31	00 N 48	01 S 26	00 S 33	00 N 40	01 N 55	00 N 20	01 N 27	17 N 29

DATA

Julian Date	2443995
Delta T	+50 seconds
Ayanamsa	23° 34' 00"
Synetic vernal point	05° ♓ 32' 59"
True obliquity of ecliptic	23° 26' 22"

MOON'S PHASES, APSIDES AND POSITIONS ☽

Date	h	m	Phase	Longitude ° '	Eclipse Indicator
04	04	25	☽	13 ♌ 08	
12	02	01	○	20 ♏ 46	
18	23	57	☾	27 ♒ 26	
26	00	00	●	04 ♊ 10	

Day	h	m		
04	22	07	Apogee	
18	09	13	Perigee	
08	09	53	0S	
15	05	10	Max dec	18° S 37'
21	14	09	0N	
28	07	54	Max dec	18° N 41'

LONGITUDES

Date	Chiron ⚷ ° '	Ceres ⚳ ° '	Pallas ⚴ ° '	Juno ⚵ ° '	Vesta ⚶ ° '	Black Moon Lilith ⚸ ° '
01	09 ♉ 27	26 ♓ 20	27 ♒ 30	23 ♈ 12	07 ♈ 17	22 ♌ 20
11	10 ♉ 05	29 ♓ 45	29 ♒ 27	28 ♈ 44	11 ♈ 47	23 ♌ 27
21	10 ♉ 43	03 ♈ 02	01 ♓ 01	04 ♉ 19	16 ♈ 10	24 ♌ 33
31	11 ♉ 19	06 ♈ 08	02 ♓ 10	09 ♉ 55	20 ♈ 25	25 ♌ 40

All ephemeris data is given at 12.00 UT and the Moon's longitude is additionally given for 24.00 UT

ASPECTARIAN

h m Aspects

01 Tuesday
04 53 ☽ ⚹ ♄
10 21 ☽ □ ♀
11 24 ☽ ⚹ ☉
22 13 ☽ □ ☿

02 Wednesday
00 31 ☽ □ ♇
02 24 ♂ ⚻ ♅
04 42 ☽ △ ♅
04 51 ☽ □ ♂
06 12 ☽ ⚻ ♆
10 14 ☽ ∠ ♄
11 41 ♃ ⊥ ♄
13 02 ☽ Q ☉
16 01 ☽ # ♅
17 47 ☿ ☍ ♇
18 05 ☽ ± ♆

03 Thursday
01 41 ♂ △ ♆
04 10 ☽ ⊥ ♄
04 20 ☽ ☌ ♃
12 07 ☽ ⚼ ♆
12 20 ☽ Q ♇
12 59 ☉ ⚺ ♀
13 54 ☽ ∥ ☉
16 14 ☽ ⚺ ♄

04 Friday
00 53 ♀ ± ♄
03 40 ♀ ± ♅
03 42 ☿ ⚻ ♅
04 25 ☽ □ ☉
04 46 ☽ △ ♀
12 34 ☽ ⚹ ♇
16 30 ☿ △ ♆
16 47 ☽ □ ♅
18 26 ☽ △ ♆
18 44 ☽ △ ☿
21 19 ☽ △ ♂

05 Saturday
02 41 ☉ ± ♆
10 17 ☽ ∥ ♄
14 42 ☽ ⚼ ♀
17 23 ♂ ⚼ ♄
17 42 ☽ ⚺ ♃
18 59 ☽ ∠ ♇
19 00 ☽ ∥ ♇

06 Sunday
00 55 ☿ ⚼ ♄
05 03 ☽ ☌ ♄
05 11 ☽ Q ♅
05 40 ☽ ⚼ ☿
05 51 ☽ ⚼ ♂
05 58 ☽ ⊥ ♃
06 09 ☽ ∥ ♂
08 17 ☿ ☌ ♂
10 59 ☽ ± ♀
13 08 ☽ ⊥ ♇
16 42 ☽ ∥ ☿
22 35 ☽ △ ☉

07 Monday
00 15 ☽ ∠ ♃
00 24 ☽ ⚻ ♀
01 10 ☽ ⚺ ♇
01 11 ☽ ± ♂
02 21 ☽ ∥ ♀
02 29 ☽ ± ☿
05 13 ☽ ⚹ ♅
06 57 ☽ □ ♆
07 54 ♀ ☍ ♇
13 57 ☽ ⚻ ♂
16 15 ☽ ⚻ ☿

08 Tuesday
01 55 ♄ Q ♅
05 46 ☉ ⚻ ♇
06 17 ☽ ⚹ ♃
06 55 ☽ ⚼ ☉
10 45 ☽ ∠ ♅
16 37 ☽ ⚺ ♄
18 19 ☽ Q ♆
22 56 ♀ ⚻ ♅

09 Wednesday
01 08 ☉ # ♅
01 52 ☽ ± ☉
04 07 ☽ ⊥ ♅
04 12 ☽ ⊥ ♄
05 47 ☽ Q ♃
11 45 ☽ ☌ ♇
14 18 ☽ ⚻ ☉
14 52 ♄ St D
15 32 ☽ ⚺ ♅
16 35 ♀ △ ♆
17 16 ☽ ⚹ ♆
17 21 ☽ ☍ ♀
21 20 ☽ ∠ ♄

10 Thursday
00 54 ☽ # ♀
03 40 ☽ ☍ ♂
05 43 ☉ ☍ ♅
10 40 ☽ ☍ ☿
16 01 ☽ □ ♃
18 56 ☽ # ☿
20 00 ☽ # ♂
21 14 ☽ ∠ ♆
21 47 ☽ # ♇

11 Friday
01 13 ☽ ⚹ ♄
03 50 ☿ ∥ ♂
04 20 ☉ ⚻ ♆
05 27 ☽ # ♄
10 01 ☿ ∥ ♇
11 16 ♀ ⚼ ♄
11 22 ☉ Q ♃
13 38 ☽ ⊥ ♆
19 05 ♂ ∥ ♇
19 12 ☽ ⚺ ♇
22 41 ☽ ☌ ♅
22 59 ☽ Q ♄

12 Saturday
00 25 ☽ ⚺ ♆
02 01 ☽ ☍ ☉
05 47 ☿ □ ♃
05 56 ☽ ⊥ ♇
06 02 ☽ ⚻ ♀
13 36 ☽ ⚻ ♂
17 44 ☽ ± ♀
21 51 ☽ ∠ ♇
22 35 ☽ △ ♃

13 Sunday
00 34 ☿ ∥ ♄
00 47 ☽ ± ♂
00 53 ☽ ⚻ ☿
06 51 ☽ □ ♄
11 02 ☽ ⚼ ♀
12 56 ☽ ± ☿
13 33 ☽ ∥ ♅
16 32 ☿ ⚼ ♆
17 24 ☽ ⚼ ♂
23 57 ☽ ⚹ ♇

14 Monday
00 58 ☽ ⚼ ♃
03 14 ☽ ⚺ ♅
04 58 ☽ ☌ ♆
06 50 ☽ ⚼ ☿
07 30 ☉ ± ♇
10 29 ☽ ⚻ ☉
13 31 ☽ ⊥ ♅
15 26 ☽ △ ♀
16 36 ☽ ± ♃
20 33 ☽ Q ♇
20 43 ☽ △ ♂
21 33 ☽ ± ☉
22 27 ☿ △ ♄

15 Tuesday
02 58 ☽ ⚻ ♃
04 53 ☽ ∠ ♅
10 33 ☽ △ ♄
12 22 ☽ △ ☿
14 02 ☽ ⚼ ☉
18 42 ♂ ∥ ♄

16 Wednesday
03 11 ☽ □ ♇
06 19 ☽ ⚹ ♅
08 06 ☽ ⚺ ♆
12 04 ☽ ⚼ ♄
17 22 ☽ △ ☉
18 15 ☽ ⊥ ♆
20 41 ☽ ∥ ♅
23 16 ☽ □ ♀

17 Thursday
02 36 ☽ □ ♂
02 37 ☽ Q ♅
03 20 ☽ ± ♄
06 26 ☽ ☍ ♃
08 42 ♀ ∥ ♇
09 29 ☽ ∠ ♆
13 30 ☽ ⚻ ♄
23 08 ☽ □ ☿

18 Friday
00 00 ☽ # ☿
05 58 ☽ △ ♇
08 38 ☽ Q ♀
09 02 ☽ □ ♅
10 31 ☿ ± ♆
10 49 ☽ Q ♂
10 52 ☽ ⚹ ♆
20 07 ☽ # ♂
23 57 ☽ □ ☉

19 Saturday
01 41 ☽ # ♄
03 33 ☽ # ♀
06 55 ☽ ⚹ ♀
07 12 ☽ Q ♆
07 24 ☽ ⚼ ♇
08 20 ☽ # ♇
08 21 ☽ ⚹ ♂
09 51 ☽ ⚻ ♃
10 36 ☽ Q ☿
16 30 ☽ ☍ ♄
19 41 ♀ ∥ ♄
20 10 ☽ ± ♃
22 42 ☽ ± ♇
23 56 ☿ ⚻ ♇

20 Sunday
08 50 ☽ Q ☉
08 55 ☽ ⚻ ♇
10 25 ☽ ⚹ ☿
10 52 ☽ ∠ ♀
11 20 ☽ ∠ ♂
11 41 ☽ ⚼ ♃
11 55 ☽ △ ♅
13 50 ☽ □ ♆
19 08 ♂ □ ♃
20 41 ☿ ☍ ♅
21 22 ♀ □ ♃

21 Monday
00 17 ♀ ☌ ♂
00 58 ☉ ∥ ♃
03 36 ☽ ⊥ ♂
03 43 ☽ ⊥ ♀
06 53 ☽ ⚹ ☉
09 48 ☿ ⚻ ♆
13 32 ☽ ⚼ ♅
13 38 ☽ △ ♃
14 29 ☽ ⚺ ♂
14 59 ☽ ⚺ ♀
16 28 ☽ ∠ ☿
18 25 ☿ # ♅
19 53 ☽ ⚻ ♄
20 32 ♀ ⚼ ♆

22 Tuesday
04 58 ☽ ± ♅
06 14 ☽ ± ♄
08 40 ♂ ⚼ ♆
09 40 ☿ Q ♃
10 36 ☽ ∠ ☉
10 40 ☽ ⊥ ☿
12 21 ☽ ☍ ♇
15 19 ☽ ⚻ ♅
17 18 ☽ △ ♆
21 49 ☽ ⚼ ♄
21 50 ☿ ± ♇
22 53 ☽ ⚺ ☿

23 Wednesday
03 22 ☽ ⊥ ☉
11 20 ☉ ⚼ ♇
14 36 ☽ ⚺ ☉
18 09 ☽ □ ♃
19 22 ☽ ⚼ ♆
20 49 ☽ ∥ ♇
21 26 ☽ ☌ ♂

24 Thursday
00 01 ☽ ☌ ♀
00 02 ☽ △ ♄
00 18 ♀ △ ♄
03 21 ☽ ∥ ♄
11 10 ☽ ± ♆
16 44 ☽ ⚻ ♇
18 03 ☽ ∥ ♀
19 41 ☽ ☍ ♅
20 54 ☽ ∥ ♂
21 47 ☽ ⚻ ♆

25 Friday
02 19 ☽ Q ♃
03 23 ☽ ± ♇
13 31 ☽ ☌ ☿
19 32 ☽ ⚼ ♇
22 51 ♂ △ ♄
23 03 ☿ ∥ ♃

26 Saturday
00 00 ☽ ☌ ☉
00 09 ☽ ⚹ ♃
02 19 ☉ ⚹ ♃
05 43 ☽ □ ♄
06 07 ☽ ⚺ ♂
10 01 ☽ # ♅
11 02 ☽ ⚺ ♀
17 46 ☽ ⊥ ♂
22 55 ☽ △ ♇
23 12 ☽ ⊥ ♀

27 Sunday
01 54 ☽ ⚻ ♅
02 04 ♀ ∥ ♂
02 20 ☿ ⚼ ♇
04 01 ☽ ∠ ♃
04 10 ☽ ☍ ♆
11 30 ☽ ∠ ♂
13 03 ☽ ± ♅
13 59 ♃ ⚼ ♆
15 05 ☽ Q ♄
17 45 ☽ ∠ ♀
21 05 ☽ ⊥ ♃

28 Monday
06 00 ☽ ⚼ ♅
08 12 ☽ ⚺ ☿
08 38 ☽ ⚺ ♃
10 50 ☿ ⚹ ♃
12 23 ☽ ⚺ ☉
13 15 ☉ ∥ ☿
13 55 ☽ ⚹ ♄
13 59 ♃ Q ♇
17 45 ☽ ⚹ ♂
19 56 ☿ # ♆
22 10 ☽ ⊥ ☿

29 Tuesday
00 55 ☽ ⊥ ☉
01 28 ☽ ⚹ ♀
03 05 ♀ ± ♆
07 49 ☽ □ ♇
09 05 ☉ □ ♄
10 51 ☽ △ ♅
12 37 ☉ # ♆
13 18 ☽ ⚻ ♆
17 21 ☿ □ ♄
18 35 ☽ Q ♂
19 09 ☽ ∠ ♄
19 32 ☽ ∠ ☿
19 59 ☽ ∠ ☉
23 31 ☉ ☌ ☿

30 Wednesday
01 04 ☽ ± ♆
03 34 ☽ Q ♀
08 13 ☽ # ♅
13 02 ☽ ⊥ ♄
18 56 ☽ ⚼ ♆
19 21 ☽ Q ♇
20 09 ☽ ☌ ♃

31 Thursday
01 04 ☽ ⚺ ♄
04 26 ☽ ⚹ ☉
04 31 ☽ ∥ ♀
08 07 ☽ ⚹ ☿
08 48 ☽ □ ♂
09 50 ☽ ∥ ♂
13 23 ☿ ⚺ ♂
17 30 ♀ ⚻ ♇
19 29 ☽ ⚹ ♇
19 43 ☽ □ ♀
22 31 ☽ □ ♅

JUNE 1979

LONGITUDES

Date	Sidereal time h m s	Sun ☉	Moon ☽	Moon ☽ 24.00	Mercury ☿	Venus ♀	Mars ♂	Jupiter ♃	Saturn ♄	Uranus ♅	Neptune ♆	Pluto ♇
01	04 37 31	10 ♊ 24 29	24 ♌ 44 56	00 ♍ 39 03	13 ♊ 32	17 ♉ 33	12 ♉ 10	05 ♌ 18	07 ♍ 32	18 ♏ 05	19 ♐ 22	16 ♎ 36
02	04 41 27	11 21 59	06 ♍ 33 14	12 ♍ 28 11	15 43	18 46	12 54	05 28	07 34	18 R 03	19 R 20	16 R 35
03	04 45 24	12 19 27	18 ♍ 24 34	24 ♍ 23 04	17 54	19 58	13 38	05 38	07 37	18 00	19 19	16 34
04	04 49 20	13 16 54	00 ♎ 24 23	06 ♎ 29 09	20 03	21 11	14 22	05 48	07 39	17 58	19 17	16 34
05	04 53 17	14 14 20	12 ♎ 37 59	18 ♎ 51 25	22 11	22 24	15 06	05 58	07 42	17 56	19 16	16 33
06	04 57 14	15 11 45	25 ♎ 09 57	01 ♏ 33 58	24 18	23 37	15 50	06 09	07 45	17 54	19 14	16 32
07	05 01 10	16 09 09	08 ♏ 03 46	14 ♏ 39 30	26 23	24 50	16 34	06 19	07 47	17 52	19 12	16 32
08	05 05 07	17 06 32	21 ♏ 21 14	28 ♏ 08 51	28 ♊ 26	26 02	17 18	06 30	07 50	17 50	19 11	16 31
09	05 09 03	18 03 53	05 ♐ 02 07	12 ♐ 00 37	00 ♋ 27	27 15	18 02	06 41	07 54	17 48	19 09	16 30
10	05 13 00	19 01 14	19 ♐ 03 50	26 ♐ 11 08	02 27	28 28	18 46	06 51	07 57	17 46	19 08	16 30
11	05 16 56	19 58 34	03 ♑ 21 46	10 ♑ 34 57	04 24	29 ♉ 41	19 29	07 02	08 00	17 44	19 06	16 29
12	05 20 53	20 55 54	17 ♑ 49 51	25 ♑ 05 40	06 18	00 ♊ 54	20 13	07 13	08 03	17 42	19 04	16 29
13	05 24 49	21 53 13	02 ♒ 21 36	09 ♒ 36 59	08 11	02 07	20 56	07 24	08 07	17 40	19 03	16 28
14	05 28 46	22 50 31	16 ♒ 51 10	24 ♒ 03 39	10 01	03 20	21 40	07 35	08 10	17 38	19 01	16 28
15	05 32 43	23 47 49	01 ♓ 14 01	08 ♓ 21 59	11 48	04 33	22 24	07 46	08 14	17 36	18 59	16 28
16	05 36 39	24 45 07	15 ♓ 27 19	22 ♓ 29 54	13 33	05 46	23 07	07 57	08 17	17 34	18 58	16 27
17	05 40 36	25 42 24	29 ♓ 29 40	06 ♈ 26 35	15 16	06 59	23 50	08 08	08 21	17 32	18 56	16 27
18	05 44 32	26 39 41	13 ♈ 20 40	20 ♈ 11 56	16 57	08 12	24 34	08 20	08 25	17 30	18 55	16 27
19	05 48 29	27 36 58	27 ♈ 00 24	03 ♉ 46 06	18 34	09 25	25 17	08 31	08 29	17 29	18 53	16 26
20	05 52 25	28 34 15	10 ♉ 28 59	17 ♉ 09 02	20 10	10 38	26 00	08 43	08 33	17 27	18 51	16 26
21	05 56 22	29 ♊ 31 31	23 ♉ 46 12	00 ♊ 20 26	21 43	11 51	26 43	08 54	08 37	17 25	18 50	16 26
22	06 00 18	00 ♋ 28 47	06 ♊ 51 36	13 ♊ 19 40	23 13	13 04	27 26	09 06	08 41	17 24	18 48	16 26
23	06 04 15	01 26 03	19 ♊ 44 31	26 ♊ 06 05	24 41	14 17	28 09	09 17	08 46	17 22	18 47	16 26
24	06 08 12	02 23 19	02 ♋ 24 21	08 ♋ 39 18	26 06	15 30	28 52	09 29	08 50	17 20	18 45	16 25
25	06 12 08	03 20 34	14 ♋ 50 59	20 ♋ 59 29	27 29	16 43	29 ♉ 35	09 41	08 54	17 19	18 44	16 25
26	06 16 05	04 17 49	27 ♋ 04 58	03 ♌ 07 37	28 ♋ 49	17 57	00 ♊ 18	09 53	08 59	17 18	18 42	16 25
27	06 20 01	05 15 03	09 ♌ 07 43	15 ♌ 05 35	00 ♌ 07	19 10	01 01	10 04	09 03	17 16	18 40	16 D 25
28	06 23 58	06 12 17	21 ♌ 01 38	26 ♌ 56 17	01 22	20 23	01 43	10 16	09 08	17 15	18 39	16 25
29	06 27 54	07 09 31	02 ♍ 50 03	08 ♍ 43 27	02 34	21 36	02 26	10 28	09 13	17 13	18 37	16 25
30	06 31 51	08 ♋ 06 44	14 ♍ 37 04	20 ♍ 31 33	03 ♌ 43	22 ♊ 49	03 ♊ 09	10 ♌ 40	09 ♍ 18	17 ♏ 12	18 ♐ 36	16 ♎ 26

MOON NODES, LATITUDE AND DECLINATIONS

Date	Moon True ☊	Moon Mean ☊	Moon ☽ Latitude	Sun ☉	Moon ☽	Mercury ☿	Venus ♀	Mars ♂	Jupiter ♃	Saturn ♄	Uranus ♅	Neptune ♆	Pluto ♇
01	13 ♍ 04	13 ♍ 12	01 S 36	22 N 01	11 N 46	23 N 23	15 N 43	14 N 59	19 N 35	10 N 31	16 S 53	21 S 34	09 N 37
02	13 D 04	13 09	00 S 34	22 09	08 34	23 47	16 04	15 13	19 33	10 30	16 53	21 34	09 37
03	13 R 04	13 06	00 N 28	22 16	05 01	24 08	16 26	15 26	19 31	10 29	16 52	21 34	09 37
04	13 03	13 03	01 31	22 24	01 N 13	24 26	16 46	15 40	19 28	10 28	16 52	21 33	09 37
05	13 01	13 00	02 30	22 31	02 S 41	24 42	17 07	15 54	19 25	10 27	16 51	21 33	09 37
06	12 57	12 56	03 23	22 37	06 35	24 55	17 27	16 07	19 23	10 26	16 50	21 33	09 37
07	12 50	12 53	04 08	22 43	10 17	25 06	17 46	16 20	19 20	10 25	16 50	21 33	09 36
08	12 40	12 50	04 41	22 49	13 35	25 13	18 05	16 33	19 18	10 23	16 49	21 33	09 36
09	12 29	12 47	04 58	22 54	16 15	25 18	18 24	16 46	19 15	10 22	16 49	21 33	09 36
10	12 18	12 44	04 59	22 59	18 01	25 20	18 42	16 59	19 12	10 21	16 48	21 33	09 36
11	12 07	12 40	04 41	23 04	18 43	25 20	19 00	17 11	19 09	10 19	16 48	21 33	09 35
12	11 57	12 37	04 05	23 08	18 13	25 17	19 17	17 24	19 07	10 18	16 47	21 32	09 35
13	11 51	12 34	03 12	23 11	16 30	25 12	19 33	17 36	19 04	10 17	16 47	21 32	09 35
14	11 46	12 31	02 08	23 15	13 45	25 05	19 50	17 48	19 01	10 15	16 46	21 32	09 34
15	11 45	12 28	00 N 55	23 18	10 10	24 56	20 05	18 00	18 58	10 14	16 46	21 32	09 34
16	11 D 45	12 25	00 S 20	23 20	06 02	24 45	20 20	18 11	18 55	10 12	16 45	21 32	09 34
17	11 R 45	12 21	01 33	23 22	01 S 37	24 32	20 35	18 23	18 52	10 11	16 45	21 32	09 33
18	11 44	12 18	02 39	23 24	02 N 49	24 18	20 49	18 34	18 50	10 09	16 44	21 32	09 33
19	11 41	12 15	03 36	23 25	07 03	24 02	21 02	18 45	18 47	10 08	16 44	21 32	09 33
20	11 35	12 12	04 19	23 26	10 51	23 44	21 15	18 56	18 44	10 06	16 43	21 32	09 32
21	11 27	12 09	04 48	23 26	14 04	23 26	21 28	19 07	18 41	10 04	16 43	21 31	09 32
22	11 17	12 06	05 01	23 26	16 30	23 06	21 39	19 17	18 38	10 03	16 42	21 31	09 31
23	11 05	12 02	04 59	23 26	18 05	22 45	21 50	19 28	18 34	10 01	16 42	21 31	09 31
24	10 54	11 59	04 42	23 25	18 44	22 23	22 01	19 38	18 31	09 59	16 41	21 31	09 30
25	10 43	11 56	04 11	23 24	18 27	22 00	22 11	19 48	18 28	09 57	16 41	21 31	09 30
26	10 33	11 53	03 29	23 22	17 19	21 36	22 20	19 57	18 25	09 56	16 41	21 31	09 29
27	10 27	11 50	02 39	23 20	15 25	21 12	22 29	20 07	18 22	09 54	16 40	21 31	09 29
28	10 23	11 46	01 41	23 18	12 53	20 48	22 37	20 16	18 19	09 52	16 40	21 31	09 28
29	10 21	11 43	00 40	23 15	09 50	20 22	22 44	20 26	18 15	09 50	16 40	21 30	09 28
30	10 ♍ 20	11 ♍ 40	00 N 23	23 N 11	06 N 25	19 N 57	22 N 51	20 N 35	18 N 12	09 N 48	16 S 39	21 S 30	09 N 27

ZODIAC SIGN ENTRIES

Date	h m	Planets
01	22 41	☽ ♍
04	11 12	☽ ♎
06	21 05	☽ ♏
09	03 15	☽ ♐
09	06 32	☿ ♋
11	06 23	☽ ♑
11	18 13	♀ ♊
13	08 06	☽ ♒
15	09 56	☽ ♓
17	12 52	☽ ♈
19	17 18	☽ ♉
21	23 23	☽ ♊
21	23 56	☉ ♋
24	07 24	☽ ♋
26	01 55	♂ ♊
26	17 47	☽ ♌
27	09 51	☿ ♌
29	06 14	☽ ♍

LATITUDES

Date	Mercury ☿	Venus ♀	Mars ♂	Jupiter ♃	Saturn ♄	Uranus ♅	Neptune ♆	Pluto ♇
01	00 N 58	01 S 24	00 S 32	00 N 40	01 N 55	00 N 20	01 N 27	17 N 29
04	01 23	01 20	00 30	00 40	01 54	00 20	01 27	17 27
07	01 42	01 14	00 28	00 40	01 54	00 20	01 27	17 26
10	01 55	01 09	00 26	00 40	01 54	00 20	01 27	17 24
13	02 01	01 03	00 24	00 40	01 53	00 20	01 27	17 23
16	02 01	00 57	00 22	00 40	01 53	00 20	01 27	17 21
19	01 54	00 50	00 20	00 40	01 52	00 20	01 27	17 20
22	01 41	00 43	00 18	00 40	01 52	00 20	01 27	17 18
25	01 22	00 36	00 16	00 40	01 52	00 20	01 27	17 16
28	00 58	00 29	00 14	00 40	01 51	00 20	01 27	17 15
31	00 N 28	00 S 21	00 S 12	00 N 40	01 N 51	00 N 20	01 N 27	17 N 13

DATA

Julian Date	2444026
Delta T	+50 seconds
Ayanamsa	23° 34' 05"
Synetic vernal point	05° ♓ 32' 55"
True obliquity of ecliptic	23° 26' 22"

MOON'S PHASES, APSIDES AND POSITIONS ☽

Date	h	m	Phase	Longitude	Eclipse Indicator
02	22	37	☽	11 ♍ 47	
10	11	55	○	19 ♐ 01	
17	05	01	☾	25 ♓ 26	
24	11	58	●	02 ♋ 23	

Day	h	m		
01	17	10	Apogee	
13	15	47	Perigee	
29	10	40	Apogee	
04	19	33	0S	
11	14	00	Max dec	18° S 43'
17	20	42	0N	
24	16	46	Max dec	18° N 45'

LONGITUDES

Date	Chiron ⚷	Ceres ⚳	Pallas ⚴	Juno ⚵	Vesta ⚶	Black Moon Lilith ⚸
01	11 ♉ 22	06 ♈ 26	02 ♓ 16	10 ♉ 29	20 ♈ 50	25 ♌ 46
11	11 ♉ 55	09 ♈ 19	02 ♓ 53	16 ♉ 05	24 ♈ 55	26 ♌ 53
21	12 ♉ 26	11 ♈ 56	02 ♓ 58	21 ♉ 42	28 ♈ 48	27 ♌ 59
31	12 ♉ 53	14 ♈ 15	02 ♓ 29	27 ♉ 18	02 ♉ 28	29 ♌ 06

ASPECTARIAN

h m	Aspects
01 Friday	
01 06	☽ △ ♆
06 50	☽ Q ☉
13 57	☽ Q ☿
21 49	☽ ‖ ♄
22 17	♀ ☍ ♅
02 Saturday	
01 55	☽ ∠ ♇
04 27	☽ ‖ ♇
09 45	☽ ⚺ ♃
10 58	☽ Q ♅
14 04	☽ ☌ ♄
20 10	☽ ⊥ ♇
21 29	☿ △ ♇
22 06	☽ ⊥ ♃
22 37	☽ □ ☉
23 16	♀ ⚻ ♆
03 Sunday	
01 43	☽ △ ♂
02 00	♂ ± ♆
08 18	☽ ⚺ ♇
10 44	☽ □ ☿
11 11	☽ ⚹ ♅
13 13	☿ ⚻ ♅
13 49	☽ □ ♆
15 30	☽ △ ♀
16 32	☽ ∠ ♃
04 Monday	
03 34	☿ ☍ ♆
09 48	☽ ⚼ ♂
17 04	☽ ∠ ♅
17 22	☿ ⊥ ♂
17 53	♀ # ♅
21 08	☿ ∠ ♃
22 49	☽ ⚹ ♃
05 Tuesday	
00 40	☽ ⚼ ♀
01 33	☽ Q ♆
02 20	☽ ⚺ ♄
04 41	☽ ± ♂
10 39	☽ ⊥ ♅
14 04	☽ ⊥ ♄
14 57	♀ ± ♇
15 22	☽ △ ☉
17 05	☽ ⚻ ♂
17 33	☿ ⚺ ♀
19 34	☽ ☌ ♇
20 04	☽ ± ♀
22 12	☽ ⚺ ♅
22 27	☽ Q ♃
06 Wednesday	
00 45	☽ ⚹ ♆
07 24	☽ ∠ ♄
07 29	☿ ± ♅
08 45	☽ ⚻ ♀
10 02	☽ △ ☿
22 12	☽ ⚼ ☉
07 Thursday	
00 21	♀ Q ♃
04 55	☽ ∠ ♆
04 59	☿ Q ♄
07 28	☽ # ♇
08 46	☽ □ ♃
10 42	♂ ⚻ ♇
11 30	☽ ⚹ ♄
12 52	☽ # ♄
16 07	☽ ± ☉
19 11	☽ ⚼ ☿
21 18	☉ △ ♇
21 21	☽ ⊥ ♆
08 Friday	
03 21	☽ ⚺ ♇
03 50	☽ ⚻ ☉
04 20	☽ ☍ ♂
05 43	☽ ☌ ♅
08 08	☽ ⚺ ♆
09 18	☽ Q ♄
14 04	☽ ⊥ ♇
14 16	☽ ± ☿
21 06	☽ ☍ ♀
09 Saturday	
02 41	☽ ⚻ ☿
04 34	♂ ☍ ♅
05 24	☉ ⚻ ♅
05 53	☽ ∠ ♇
08 13	☉ ⚺ ♂
14 52	☽ △ ♃
14 52	☿ ⊥ ♃
16 40	♂ # ♅
16 57	☽ □ ♄
18 21	☽ ‖ ♅
18 33	☽ # ♂
10 Sunday	
07 39	☽ ⚹ ♇
09 48	☽ ⚺ ♅
11 27	☽ ⚻ ♂
11 55	☽ ☍ ☉
12 06	☽ ☌ ♆
14 33	☉ ☍ ♆
15 47	☿ ⚼ ♅
16 46	☽ ⚼ ♃
19 54	☽ ⊥ ♅
22 07	☽ ± ♂
23 38	♂ ⚻ ♆
11 Monday	
03 52	☽ Q ♇
05 18	☽ ⚻ ♀
08 04	☽ ± ♃
10 56	☽ ∠ ♅
13 54	☿ ∠ ♂
13 58	☽ ⚼ ♂
13 59	☽ ☍ ☿
16 13	☽ ± ♀
18 11	☽ ⚻ ♃
19 44	☽ △ ♄
23 47	♀ ‖ ♃
12 Tuesday	
08 31	☽ ⚼ ♀
09 46	☽ □ ♇
11 46	☽ ⚹ ♅
14 03	☽ ⚺ ♆
16 09	☽ △ ♂
17 29	☽ ⚻ ☉
20 40	☽ ⚼ ♄
23 23	♀ ⚼ ♇
23 56	☽ ⊥ ♆
13 Wednesday	
00 04	☽ # ♂
00 53	☿ ⚺ ♃
04 06	☽ ± ☉
07 33	☽ Q ♅
09 04	☽ ‖ ♅
09 43	☿ ⊥ ♀
11 07	☿ ⚹ ♄
11 34	☽ △ ♀
11 35	☽ ± ♄
14 47	☽ ∠ ♆
20 01	☽ ⚼ ☉
20 26	☽ ☍ ♃
21 33	☽ ⚻ ♄
23 01	☽ ⚻ ☿
14 Thursday	
03 54	☉ ∠ ♃
10 24	☽ ± ☿
11 21	☽ △ ♇
13 17	☽ □ ♅
15 35	☽ ⚹ ♆
20 26	☽ □ ♂
22 41	☽ △ ☉
15 Friday	
03 32	☽ ⚼ ☿
07 07	☉ ± ♅
11 36	☽ Q ♆
11 39	☽ # ♄
12 23	☽ ⚼ ♇
14 12	♂ ± ♇
15 41	☽ # ♇
18 05	☽ □ ♀
23 08	☽ ⚻ ♃
23 49	☽ ☍ ♄
16 Saturday	
03 32	☽ ± ♇
04 15	☽ Q ♂
08 20	☽ △ ☿
09 25	☽ ± ♃
13 42	☽ ⚻ ♇
15 35	☽ △ ♅
17 57	☽ □ ♆
17 Sunday	
00 57	☽ ⚼ ♃
01 46	☽ ⚹ ♂
03 31	☽ Q ♀
05 01	☽ □ ☉
17 14	☽ ⚼ ♅
18 Monday	
02 11	☽ ⚹ ♀
03 09	☽ △ ♃
03 23	☽ ⚻ ♄
04 46	☿ □ ♇
05 03	☽ ∠ ♂
05 24	☉ Q ♄
08 48	☽ ± ♅
13 53	☽ ± ♄
14 28	☽ Q ☉
15 04	♀ ⚹ ♃
16 34	♀ □ ♄
17 25	☽ ☍ ♇
19 09	☽ □ ☿
19 16	☽ ⚻ ♅
20 04	☿ △ ♅
21 38	☽ ⊥ ♂
21 43	☽ △ ♆
19 Tuesday	
04 59	♃ ⚺ ♄
05 45	☽ ⚼ ♄
06 58	☽ ∠ ♀
08 47	☽ ⚺ ♂
13 10	☽ ⚹ ☉
14 28	♂ ‖ ♃
16 32	☿ ⚻ ♆
20 Wednesday	
00 11	☽ ⚼ ♆
00 30	☽ ⊥ ♀
03 18	☽ ‖ ♇
06 58	☽ ‖ ♄
07 18	☽ Q ☿
08 31	☽ △ ♄
08 47	☽ □ ♃
12 18	☽ ⚺ ♀
16 15	☽ ± ♆
17 58	☽ ∠ ☉
22 42	☽ ⚻ ♇
21 Thursday	
00 31	☽ ☍ ♅
03 03	☽ ⚻ ♆
07 46	☽ ⚹ ☿
09 34	☽ ± ♇
11 16	☉ ‖ ☿
11 31	☽ ⊥ ☉
17 41	☽ ☌ ♂
17 48	☽ Q ♃
19 29	♀ # ♆
20 15	♂ Q ♃
23 20	☽ ⚺ ☉
22 Friday	
02 00	☽ ⚼ ♇
14 26	☽ # ♅
14 50	☽ ∠ ☿
15 24	☽ □ ♄
16 12	☽ ⚹ ♃
19 57	☿ ∠ ♄
23 Saturday	
00 43	☽ ☌ ♀
05 47	☽ △ ♇
07 33	☽ ⚻ ♅
09 46	☽ ⊥ ☿
10 12	☽ ☍ ♆
13 30	☿ ± ♆
18 49	☽ ± ♅
20 42	☽ ∠ ♃
22 30	☽ ⚺ ☿
24 Sunday	
01 04	☽ ‖ ♃
01 19	☽ Q ♄
04 51	☽ ⚺ ♂
10 51	☉ ⚼ ♅
11 53	☽ ⚼ ♅
11 58	☽ ☌ ☉
14 06	☽ ⊥ ♃
17 00	☽ ⊥ ♂
25 Monday	
00 25	☽ ⚹ ♄
01 49	☽ ⚺ ♃
04 03	☿ ‖ ♀
06 03	♀ △ ♇
11 23	☽ ‖ ♃
11 27	☽ ∠ ♂
15 04	☽ □ ♇
16 03	☽ ⚺ ♀
16 48	☽ △ ♅
19 33	☽ ⚻ ♆
22 40	☉ ⊥ ♃
23 25	♀ ⚻ ♅
26 Tuesday	
05 07	☽ ⊥ ♀
05 50	☽ ∠ ♄
07 18	☽ ± ♆
15 52	☽ ☌ ☿
17 42	☿ # ♆
18 47	☽ ⚹ ♂
21 13	☽ # ♅
23 47	☽ ⊥ ♄
27 Wednesday	
00 57	☽ ∠ ♀
01 07	☽ ⚼ ♆
01 30	♇ St D
02 34	♀ ☍ ♆
02 35	☽ Q ♇
03 34	☽ ⚺ ☉
11 51	☽ ⚺ ♄
13 56	☽ ☌ ♃
16 38	☽ ⊥ ☉
20 18	☽ Q ♂
28 Thursday	
01 49	♂ ⚼ ♇
02 41	☽ ⚹ ♇
04 22	☽ □ ♅
07 12	☽ △ ♆
10 33	☽ ⚹ ♀
12 23	☽ ∠ ☉
29 Friday	
05 41	☿ ⚹ ♂
09 08	☽ ∠ ♇
09 49	☿ ‖ ♂
11 08	☽ □ ♂
11 23	☽ ⚺ ☿
12 03	☽ ‖ ♄
13 45	☽ Q ♀
14 46	☽ ‖ ♇
16 51	☽ Q ♅
21 35	☽ ⚹ ☉
30 Saturday	
00 56	☽ ⊥ ☿
01 05	☽ ☌ ♄
02 30	☿ ⊥ ♄
03 28	☽ ⊥ ♇
03 50	☽ ⚺ ♃
09 38	☿ ⚼ ♆
15 41	☽ ⚺ ♇
16 15	☽ ⊥ ♃
17 15	☽ ⚹ ♅
19 18	♀ ± ♅
20 04	☽ □ ♆
21 11	☽ ∠ ☿

All ephemeris data is given at 12.00 UT and the Moon's longitude is additionally given for 24.00 UT

JULY 1979

Date	Sidereal time h m s	Sun ☉ ° ' "	Moon ☽ ° ' "	Moon ☽ 24.00 ° ' "	Mercury ☿ ° '	Venus ♀ ° '	Mars ♂ ° '	Jupiter ♃ ° '	Saturn ♄ ° '	Uranus ♅ ° '	Neptune ♆ ° '	Pluto ♇ ° '
01	06 35 47	09 ♋ 03 56	26 ♍ 27 31	02 ♎ 25 38	04 ♌ 49	24 ♊ 03	03 ♊ 51	10 ♌ 53	09 ♍ 22	17 ♏ 11	18 ♐ 34	16 ♎ 26
02	06 39 44	10 01 09	08 ♎ 26 34	14 ♎ 31 00	05 52	25 16	04 34	11 05	09 27	17 R 10	18 R 33	16 26
03	06 43 41	10 58 21	20 ♎ 39 34	26 ♎ 52 55	06 52	26 29	05 16	11 17	09 32	17 08	18 31	16 26
04	06 47 37	11 55 33	03 ♏ 11 36	09 ♏ 36 08	07 49	27 42	05 58	11 29	09 37	17 07	18 30	16 26
05	06 51 34	12 52 44	16 ♏ 06 57	22 ♏ 44 21	08 43	28 ♊ 56	06 41	11 41	09 43	17 06	18 28	16 27
06	06 55 30	13 49 55	29 ♏ 28 31	06 ♐ 19 30	09 33	00 ♋ 09	07 23	11 54	09 48	17 05	18 27	16 27
07	06 59 27	14 47 06	13 ♐ 17 09	20 ♐ 21 10	10 20	01 22	08 05	12 06	09 53	17 04	18 26	16 27
08	07 03 23	15 44 18	27 ♐ 31 06	04 ♑ 46 16	11 03	02 36	08 47	12 18	09 59	17 03	18 24	16 28
09	07 07 20	16 41 29	12 ♑ 05 52	19 ♑ 29 00	11 43	03 49	09 29	12 31	10 04	17 03	18 23	16 28
10	07 11 16	17 38 40	26 ♑ 54 38	04 ♒ 21 43	12 18	05 03	10 11	12 43	10 09	17 02	18 21	16 28
11	07 15 13	18 35 51	11 ♒ 49 10	19 ♒ 16 00	12 50	06 16	10 53	12 56	10 15	17 01	18 20	16 29
12	07 19 10	19 33 03	26 ♒ 41 15	04 ♓ 04 06	13 17	07 29	11 34	13 08	10 21	17 00	18 19	16 29
13	07 23 06	20 30 15	11 ♓ 23 51	18 ♓ 39 55	13 40	08 43	12 16	13 21	10 26	17 00	18 17	16 30
14	07 27 03	21 27 28	25 ♓ 51 52	02 ♈ 59 26	13 59	09 56	12 58	13 34	10 32	16 59	18 16	16 30
15	07 30 59	22 24 41	10 ♈ 02 26	17 ♈ 00 47	14 13	11 10	13 39	13 46	10 38	16 58	18 15	16 31
16	07 34 56	23 21 55	23 ♈ 54 31	00 ♉ 43 41	14 22	12 23	14 21	13 59	10 44	16 58	18 14	16 32
17	07 38 52	24 19 09	07 ♉ 28 26	14 ♉ 08 55	14 27	13 37	15 02	14 12	10 50	16 57	18 12	16 32
18	07 42 49	25 16 24	20 ♉ 45 19	27 ♉ 17 49	14 R 27	14 50	15 44	14 25	10 56	16 57	18 11	16 33
19	07 46 45	26 13 40	03 ♊ 46 36	10 ♊ 11 51	14 21	16 04	16 25	14 37	11 02	16 56	18 10	16 34
20	07 50 42	27 10 57	16 ♊ 33 43	22 ♊ 52 23	14 11	17 18	17 06	14 50	11 08	16 56	18 09	16 35
21	07 54 39	28 08 14	29 ♊ 07 58	05 ♋ 20 39	13 57	18 31	17 48	15 03	11 14	16 56	18 08	16 35
22	07 58 35	29 ♋ 05 32	11 ♋ 30 32	17 ♋ 37 48	13 37	19 45	18 29	15 16	11 20	16 56	18 07	16 36
23	08 02 32	00 ♌ 02 51	23 ♋ 42 34	29 ♋ 45 01	13 13	20 59	19 10	15 29	11 26	16 55	18 05	16 37
24	08 06 28	01 00 10	05 ♌ 45 21	11 ♌ 43 45	12 45	22 12	19 51	15 42	11 33	16 55	18 04	16 38
25	08 10 25	01 57 30	17 ♌ 40 27	23 ♌ 35 46	12 13	23 26	20 32	15 55	11 39	16 55	18 03	16 39
26	08 14 21	02 54 50	29 ♌ 29 58	05 ♍ 23 24	11 37	24 40	21 13	16 08	11 45	16 D 55	18 02	16 40
27	08 18 18	03 52 11	11 ♍ 16 28	17 ♍ 09 34	10 58	25 54	21 54	16 21	11 52	16 55	18 01	16 41
28	08 22 14	04 49 32	23 ♍ 03 11	28 ♍ 57 49	10 17	27 07	22 34	16 34	11 58	16 55	18 00	16 42
29	08 26 11	05 46 54	04 ♎ 53 58	10 ♎ 52 13	09 34	28 21	23 15	16 47	12 05	16 55	17 59	16 43
30	08 30 08	06 44 16	16 ♎ 53 08	22 ♎ 57 19	08 49	29 ♋ 35	23 56	17 00	12 11	16 56	17 58	16 44
31	08 34 04	07 ♌ 41 39	29 ♎ 05 22	05 ♏ 17 53	08 ♌ 05	00 ♌ 49	24 ♊ 36	17 ♌ 13	12 ♍ 18	16 ♏ 56	17 ♐ 57	16 ♎ 45

DECLINATIONS

Date	Moon True ☊ ° '	Moon Mean ☊ ° '	Moon ☽ Latitude ° '	Sun ☉ ° '	Moon ☽ ° '	Mercury ☿ ° '	Venus ♀ ° '	Mars ♂ ° '	Jupiter ♃ ° '	Saturn ♄ ° '	Uranus ♅ ° '	Neptune ♆ ° '	Pluto ♇ ° '
01	10 ♍ 21	11 ♍ 37	01 N 25	23 N 08	02 N 43	19 N 31	22 N 57	20 N 43	18 N 09	09 N 46	16 S 39	21 S 30	09 N 27
02	10 D 22	11 34	02 25	23 04	01 S 08	19 05	23 02	20 52	18 06	09 44	16 39	21 30	09 26
03	10 R 21	11 31	03 19	22 59	05 00	18 40	23 07	21 00	18 02	09 42	16 38	21 30	09 26
04	10 19	11 27	04 05	22 54	08 44	18 14	23 11	21 08	17 59	09 40	16 38	21 30	09 25
05	10 15	11 24	04 40	22 49	12 11	17 48	23 15	21 16	17 56	09 38	16 38	21 30	09 24
06	10 08	11 21	05 01	22 43	15 08	17 23	23 17	21 24	17 52	09 36	16 37	21 30	09 24
07	10 01	11 18	05 06	22 37	17 20	16 58	23 19	21 32	17 49	09 34	16 37	21 30	09 23
08	09 52	11 15	04 52	22 31	18 33	16 34	23 21	21 39	17 45	09 32	16 37	21 30	09 22
09	09 44	11 12	04 19	22 24	18 35	16 10	23 22	21 46	17 42	09 30	16 37	21 30	09 22
10	09 37	11 08	03 29	22 17	17 22	15 47	23 22	21 53	17 38	09 27	16 36	21 29	09 21
11	09 32	11 05	02 23	22 09	14 57	15 24	23 21	22 00	17 35	09 25	16 36	21 29	09 20
12	09 30	11 02	01 N 09	22 01	11 33	15 03	23 20	22 06	17 31	09 23	16 36	21 29	09 20
13	09 D 29	10 59	00 S 10	21 52	07 27	14 43	23 17	22 13	17 28	09 21	16 36	21 29	09 19
14	09 30	10 56	01 27	21 44	02 S 59	14 23	23 15	22 19	17 24	09 19	16 36	21 29	09 18
15	09 31	10 52	02 38	21 35	01 N 34	14 06	23 11	22 24	17 20	09 16	16 36	21 29	09 18
16	09 R 32	10 49	03 37	21 25	05 55	13 49	23 07	22 30	17 17	09 14	16 36	21 29	09 17
17	09 31	10 46	04 23	21 15	09 52	13 34	23 02	22 36	17 13	09 12	16 36	21 29	09 16
18	09 28	10 43	04 53	21 05	13 14	13 21	22 57	22 41	17 09	09 09	16 35	21 29	09 15
19	09 24	10 40	05 08	20 54	15 52	13 09	22 51	22 46	17 06	09 07	16 35	21 29	09 14
20	09 18	10 37	05 07	20 43	17 40	13 00	22 44	22 51	17 02	09 05	16 35	21 29	09 14
21	09 11	10 33	04 51	20 32	18 35	12 52	22 36	22 55	16 58	09 02	16 35	21 29	09 13
22	09 03	10 30	04 22	20 20	18 36	12 46	22 28	23 00	16 55	09 00	16 35	21 29	09 12
23	08 57	10 27	03 41	20 08	17 44	12 43	22 19	23 04	16 51	08 57	16 35	21 29	09 11
24	08 51	10 24	02 50	19 56	16 05	12 41	22 10	23 08	16 47	08 55	16 35	21 28	09 10
25	08 47	10 21	01 53	19 43	13 45	12 42	22 00	23 12	16 43	08 52	16 35	21 28	09 10
26	08 45	10 17	00 S 50	19 30	10 52	12 45	21 49	23 15	16 39	08 50	16 35	21 28	09 09
27	08 D 45	10 14	00 N 14	19 17	07 33	12 49	21 38	23 19	16 35	08 47	16 35	21 28	09 08
28	08 46	10 11	01 18	19 04	03 57	12 56	21 26	23 22	16 32	08 45	16 35	21 28	09 07
29	08 47	10 08	02 18	18 50	00 N 10	13 05	21 13	23 25	16 28	08 42	16 35	21 28	09 06
30	08 49	10 05	03 14	18 35	03 S 39	13 16	21 00	23 28	16 24	08 40	16 35	21 28	09 05
31	08 ♍ 50	10 ♍ 02	04 N 02	18 N 21	07 S 23	13 N 28	20 N 46	23 N 30	16 N 20	08 N 37	16 S 35	21 S 28	09 N 04

ZODIAC SIGN ENTRIES

Date	h m	Planets
01	19 08	☽ ♎
04	05 57	☽ ♏
06	09 02	♀ ♋
06	12 56	☽ ♐
08	16 07	☽ ♑
10	16 59	☽ ♒
12	17 23	☽ ♓
14	18 57	☽ ♈
16	22 43	☽ ♉
19	05 00	☽ ♊
21	13 40	☽ ♋
23	10 49	☉ ♌
24	00 30	☽ ♌
26	13 01	☽ ♍
29	02 06	☽ ♎
30	20 07	♀ ♌
31	13 46	☽ ♏

LATITUDES

Date	Mercury ☿ ° '	Venus ♀ ° '	Mars ♂ ° '	Jupiter ♃ ° '	Saturn ♄ ° '	Uranus ♅ ° '	Neptune ♆ ° '	Pluto ♇ ° '
01	00 N 28	00 S 21	00 S 12	00 N 40	01 N 51	00 N 20	01 N 27	17 N 13
04	00 S 05	00 14	00 10	00 40	01 51	00 20	01 27	17 11
07	00 43	00 S 06	00 08	00 40	01 50	00 20	01 27	17 10
10	01 23	00 N 01	00 06	00 40	01 50	00 20	01 27	17 08
13	02 06	00 08	00 03	00 41	01 50	00 19	01 27	17 06
16	02 49	00 16	00 S 01	00 41	01 50	00 19	01 26	17 05
19	03 31	00 23	00 N 01	00 41	01 49	00 19	01 26	17 03
22	04 08	00 29	00 03	00 41	01 49	00 19	01 26	17 01
25	04 37	00 36	00 06	00 41	01 49	00 19	01 26	17 00
28	04 54	00 42	00 08	00 41	01 49	00 19	01 26	16 58
31	04 S 57	00 N 48	00 N 10	00 N 41	01 N 49	00 N 19	01 N 26	16 N 57

DATA

Julian Date	2444056
Delta T	+50 seconds
Ayanamsa	23° 34' 09"
Synetic vernal point	05° ♓ 32' 50"
True obliquity of ecliptic	23° 26' 22"

MOON'S PHASES, APSIDES AND POSITIONS ☽

Date	h m	Phase	Longitude ° '	Eclipse Indicator
02	15 24	☽	10 ♎ 09	
09	19 59	○	17 ♑ 56	
16	10 59	☾	23 ♈ 19	
24	01 41	●	00 ♌ 36	

Day	h m		
11	12 13	Perigee	
27	00 37	Apogee	
02	05 00	0S	
09	00 44	Max dec	18° S 43'
15	03 42	0N	
22	00 14	Max dec	18° N 42'
29	13 05	0S	

LONGITUDES

Date	Chiron ⚷ ° '	Ceres ⚳ ° '	Pallas ⚴ ° '	Juno ⚵ ° '	Vesta ⚶ ° '	Black Moon Lilith ° '
01	12 ♉ 53	14 ♈ 15	02 ♓ 29	27 ♉ 18	02 ♉ 28	29 ♌ 06
11	13 ♉ 15	16 ♈ 14	01 ♓ 24	01 ♊ 52	05 ♉ 54	00 ♍ 12
21	13 ♉ 33	17 ♈ 49	29 ♒ 46	08 ♊ 24	09 ♉ 03	01 ♍ 19
31	13 ♉ 46	18 ♈ 55	27 ♒ 39	13 ♊ 51	11 ♉ 51	02 ♍ 25

All ephemeris data is given at 12.00 UT and the Moon's longitude is additionally given for 24.00 UT

ASPECTARIAN

01 Sunday
00 08 ☽ Q ☉
03 25 ☿ Q ♆
06 34 ☽ □ ♀
10 48 ☽ ∠ ♃
20 29 ☉ ⚹ ♄
23 29 ☽ ∠ ♅

02 Monday
03 47 ☽ △ ♂
06 24 ☽ ⚹ ☿
08 14 ☽ Q ♆
14 01 ☽ ⊻ ♄
15 09 ☉ ∥ ♀
15 24 ☽ □ ☉
17 19 ☽ ⚹ ♃
17 23 ☽ ⊥ ♅

03 Tuesday
01 56 ☽ ⊥ ♄
03 45 ☽ ☌ ♇
05 09 ☽ ⊻ ♅
07 09 ♀ ∠ ♃
07 51 ☽ ⚹ ♆
08 13 ☽ Q ☿
11 11 ☽ ⚼ ♂
17 10 ☽ Q ♃
19 33 ☽ ∠ ♄
21 52 ☉ ⊻ ♃

04 Wednesday
00 27 ☽ △ ♀
05 32 ☽ ± ♂
10 16 ♀ Q ♄
12 34 ☽ ∠ ♆
16 22 ☉ ⊥ ♂
16 30 ☽ # ♇
17 32 ☽ ⊼ ♂
18 09 ☽ # ♄
21 21 ☽ □ ☿

05 Thursday
00 07 ☽ ⚹ ♄
03 44 ☽ □ ♃
04 03 ☿ ∥ ♃
05 20 ☽ ⊥ ♆
05 36 ☽ △ ☉
07 35 ☽ ⚼ ♀
12 36 ☽ ⊻ ♇
13 48 ☽ ☌ ♅
16 17 ☽ ⊻ ♆
22 13 ☽ Q ♄
23 28 ☽ ⊥ ♇

06 Friday
01 35 ☽ ± ♀
10 47 ☽ ⚼ ☉
13 19 ☽ ⊼ ♀
15 29 ☽ ∠ ♇
20 07 ☿ ⊻ ♄

07 Saturday
02 34 ☽ ☍ ♂
03 05 ☽ ∥ ♅
03 42 ☽ ± ☉
05 57 ♂ # ♆
06 08 ☽ □ ♄
06 39 ☽ △ ☿
08 02 ☽ # ☿
09 56 ☽ △ ♃
14 45 ☽ ⊼ ☉
17 24 ☽ ⚹ ♇
18 27 ☽ ⊻ ♅
18 53 ☽ # ♃
20 44 ☽ ☌ ♆

08 Sunday
01 33 ♀ ⚼ ♅
04 33 ☽ ⊥ ♅
08 35 ☿ # ♅
09 27 ☽ ⚼ ☿
11 39 ☽ ⚼ ♃
13 34 ☽ Q ♇
19 31 ☽ ∠ ♅
21 11 ☽ ☍ ♀

09 Monday
01 04 ☽ ± ☿
02 44 ☽ ± ♃
06 16 ☉ □ ♇
07 31 ☽ ⊼ ♂
08 40 ☽ △ ♄
11 21 ☽ ⊼ ☿
12 41 ☽ ⊼ ♃
17 47 ☽ ± ♂
19 07 ☽ □ ♇
19 59 ☽ ☍ ☉
20 02 ☽ ⚹ ♅
20 42 ☉ △ ♅
22 12 ☽ ⊻ ♆

10 Tuesday
07 53 ☽ ⊥ ♆
08 04 ☽ # ♃
09 04 ☽ ⚼ ♂
09 09 ☽ ⚼ ♄
11 10 ♂ □ ♄
15 25 ☽ Q ♅
20 47 ☽ ∥ ♅
22 22 ☽ ∠ ♆
23 45 ☽ ± ♄

11 Wednesday
02 16 ☽ ⊼ ♀
05 33 ☉ ⊼ ♆
07 36 ☽ # ☿
09 28 ☽ ⊼ ♄
10 25 ☽ △ ♂
12 47 ☽ ± ♀
13 41 ☽ ☍ ☿
13 49 ☽ ☍ ♃
19 31 ☽ △ ♆
20 22 ☽ □ ♅
20 50 ☿ ☌ ♃
22 29 ☽ ⚹ ♆
23 40 ☽ ⊼ ☉

12 Thursday
03 14 ☉ ∥ ♂
03 46 ♀ ⊥ ♃
04 36 ☽ ⚼ ♀
05 57 ☿ ⊥ ♀
10 02 ☽ ± ☉
17 53 ☽ Q ♆
19 48 ☽ ⚼ ♇

13 Friday
01 10 ☽ # ♄
01 24 ☽ # ♇
01 40 ☽ ⚼ ☉
07 12 ☽ △ ♀
10 25 ☽ ☍ ♄
10 31 ☽ ± ♇
13 30 ☽ □ ♂
15 16 ☽ ⊼ ♃
15 50 ☽ ⊼ ☿
20 25 ☽ ⊼ ♇
21 13 ☽ △ ♅
23 22 ☽ □ ♆

14 Saturday
01 20 ☽ ± ♃
01 59 ☽ ± ☿
04 07 ☽ △ ☉
04 22 ♄ ⊥ ♇
16 36 ☽ ⚼ ♃
17 13 ♄ ∥ ♇
17 20 ☽ ⚼ ☿
21 01 ☽ Q ♂
22 17 ☽ ⚼ ♅

15 Sunday
00 41 ♀ ⚹ ♄
13 01 ☽ ⊼ ♄
13 36 ☽ ± ♅
14 06 ☽ □ ♀
17 49 ♂ ⚹ ♃
18 30 ☽ △ ♃
18 32 ☽ ⚹ ♂
19 16 ☽ △ ☿
23 09 ☽ ☍ ♇
23 25 ☽ ± ♄
23 55 ☽ ⊼ ♅

16 Monday
02 07 ☽ △ ♆
02 08 ☉ # ♆
10 59 ☽ □ ☉
12 56 ☿ ⚹ ♂
15 13 ☽ ⚼ ♄
22 05 ☽ ∠ ♂

17 Tuesday
00 32 ☽ Q ♀
04 24 ☽ ⚼ ♆
07 45 ☽ ∥ ♄
08 11 ☽ ∥ ♇
09 14 ☉ ± ♆
14 57 ☽ ⊥ ♂
18 04 ☽ △ ♄
20 29 ☽ ± ♆
21 22 ☽ Q ☉
22 42 ☿ St R

18 Wednesday
00 09 ☽ ⚹ ♀
00 17 ☽ □ ♃
00 33 ☽ □ ☿
01 48 ♀ ⊻ ♃
02 22 ☽ ⊻ ♂
04 20 ☽ ⊼ ♇
04 25 ☿ ⊻ ♀
05 04 ☽ ☍ ♅
07 20 ☽ ⊼ ♆
12 56 ☽ ∥ ☿
15 01 ☿ ☌ ♃
15 17 ☽ ± ♇
20 56 ☽ ⚹ ☉

19 Thursday
06 22 ☉ ∠ ♄
06 27 ☽ ∠ ♀
07 53 ☽ ⚼ ♇
09 23 ☽ Q ☿
09 49 ☽ Q ♃
17 04 ♂ △ ♆
20 19 ☽ # ♅
21 47 ♀ □ ♆
22 07 ♀ ∥ ♂

20 Friday
01 00 ☽ ⊥ ♀
01 40 ☽ □ ♄
02 30 ☽ ∥ ♃
03 04 ☽ ∠ ☉
03 42 ♀ ⊻ ♂
05 00 ♀ △ ♅
06 01 ♂ ⊼ ♅
07 35 ☽ ⚹ ☿
08 41 ☽ ⚹ ♃
12 02 ☽ △ ♇
12 42 ☽ ⊼ ♅
13 06 ☽ ☌ ♂
13 32 ☽ ⊻ ♀
15 00 ☽ ☍ ♆
21 30 ☽ ⊥ ☉

21 Saturday
00 07 ☽ ± ♅
04 24 ♀ ⊼ ♆
09 56 ☽ ⊻ ☉
11 39 ☽ ∠ ☿
12 12 ☽ Q ♄
13 48 ☽ ∠ ♃
17 24 ☽ ⚼ ♅
23 19 ♂ ☍ ♆

22 Sunday
04 38 ☽ ⊥ ☿
07 33 ☽ ⊥ ♃
11 39 ☽ ⚹ ♄
16 00 ☽ ⊻ ☿
19 29 ☽ ⊻ ♃
22 00 ☽ □ ♇
22 37 ☽ △ ♅

23 Monday
00 55 ☽ ⊼ ♆
02 29 ☽ ⊻ ♂
05 59 ☽ ☌ ♀
12 45 ☽ ± ♆
15 03 ☽ ⊥ ♂
17 28 ☽ ∠ ♄

24 Tuesday
01 41 ☽ ☌ ☉
02 44 ☽ ∥ ♃
05 44 ☽ # ♅
06 38 ☽ ⚼ ♆
09 45 ☽ Q ♇
10 04 ☽ ∠ ♂
11 34 ☽ ⊥ ♄
23 44 ☽ ⊻ ♄

25 Wednesday
01 28 ☽ ☌ ☿
08 22 ☽ ☌ ♃
09 55 ☽ ⚹ ♇
10 29 ☽ □ ♅
12 46 ☽ △ ♆
18 08 ☽ ⚹ ♂
21 07 ☽ ∥ ☿
23 55 ♀ ± ♆

26 Thursday
01 02 ☽ ⊻ ♀
07 15 ☿ ⊻ ♄
10 58 ♅ St D
14 39 ☽ ⊥ ♀
15 02 ☉ ⚼ ♆
16 25 ☽ ∠ ♇
19 34 ☽ ⊻ ☉
20 02 ☽ Q ♂
23 03 ☽ Q ♅

27 Friday
00 48 ☽ ∥ ♇
03 12 ☽ ∥ ♄
08 53 ☽ ⊥ ☉
10 47 ☽ ⊥ ♇
11 08 ☽ ∠ ♀
11 24 ☽ ⊻ ☿
13 13 ☽ ☌ ♄
14 31 ♃ # ♅
22 32 ☽ ⊻ ♃
22 58 ☽ ⊥ ☿
23 02 ☽ ⊻ ♇
23 31 ☽ ⚹ ♅

28 Saturday
01 44 ☽ □ ♆
04 51 ☽ ∠ ☉
06 57 ♀ # ♆
08 43 ☉ Q ♇
08 46 ♀ ∠ ♄
10 58 ☽ □ ♂
10 59 ☽ ⊥ ♃
16 16 ☽ ∠ ☿
21 14 ☽ ⚹ ♀

29 Sunday
00 27 ♂ ± ♅
04 04 ♃ ⚹ ♇
05 35 ☽ ∠ ♃
05 59 ☽ ∠ ♅
13 56 ☽ ⚹ ☉
14 11 ☽ Q ♆
20 29 ☉ ⊥ ♄
20 50 ☽ ⚹ ☿

30 Monday
00 07 ☽ ⊥ ♅
00 13 ☽ Q ♀
02 33 ☽ ⊻ ♄
04 03 ♃ □ ♅
10 15 ☿ ∠ ♂
11 42 ☽ ☌ ♇
12 05 ☽ ⊻ ♅
12 14 ☽ ⚹ ♃
14 09 ☽ ⚹ ♆
14 37 ☽ ⊥ ♄
15 59 ☽ Q ☉
19 21 ☽ Q ☿

31 Tuesday
02 43 ☽ △ ♂
02 45 ♀ ⊥ ♂
08 29 ☽ ∠ ♄
12 15 ☽ Q ♃
15 43 ☽ □ ♀
17 27 ☉ ☌ ☿
19 29 ☽ ∠ ♆
20 11 ☽ # ♄
23 20 ☽ # ♇

AUGUST 1979

LONGITUDES

Date	Sidereal time h m s	Sun ☉ ° ' "	Moon ☽ ° ' "	Moon ☽ 24.00 ° ' "	Mercury ☿ ° '	Venus ♀ ° '	Mars ♂ ° '	Jupiter ♃ ° '	Saturn ♄ ° '	Uranus ♅ ° '	Neptune ♆ ° '	Pluto ♇ ° '
01	08 38 01	08 ♌ 39 03	11 ♏ 35 26	17 ♏ 58 34	07 ♌ 21	02 ♌ 03	25 ♊ 16	17 ♌ 26	12 ♍ 25	16 ♏ 56	17 ♐ 56	16 ♎ 46
02	08 41 57	09 36 27	24 ♏ 27 45	01 ♐ 03 25	06 R 38	03 17	25 57	17 39	12 31	16 56	17 R 56	16 47
03	08 45 54	10 33 52	07 ♐ 45 53	14 ♐ 35 20	05 57	04 31	26 37	17 52	12 38	16 57	17 55	16 49
04	08 49 50	11 31 17	21 ♐ 31 50	28 ♐ 35 17	05 19	05 44	27 17	18 05	12 45	16 57	17 54	16 50
05	08 53 47	12 28 44	05 ♑ 45 23	13 ♑ 01 41	04 45	06 58	27 58	18 18	12 52	16 58	17 53	16 51
06	08 57 43	13 26 11	20 ♑ 23 33	27 ♑ 50 08	04 15	08 12	28 38	18 32	12 59	16 58	17 52	16 52
07	09 01 40	14 23 38	05 ♒ 20 30	12 ♒ 53 33	03 51	09 26	29 18	18 45	13 06	16 59	17 52	16 54
08	09 05 37	15 21 07	20 ♒ 28 06	28 ♒ 02 59	03 31	10 40	29 ♊ 58	18 58	13 13	17 00	17 51	16 55
09	09 09 33	16 18 37	05 ♓ 37 01	13 ♓ 09 05	03 18	11 54	00 ♋ 37	19 11	13 19	17 00	17 50	16 56
10	09 13 30	17 16 08	20 ♓ 38 09	28 ♓ 03 20	03 11	13 08	01 17	19 24	13 26	17 01	17 50	16 58
11	09 17 26	18 13 40	05 ♈ 23 54	12 ♈ 39 15	03 D 10	14 22	01 57	19 37	13 34	17 02	17 49	16 59
12	09 21 23	19 11 14	19 ♈ 48 58	26 ♈ 52 47	03 17	15 36	02 37	19 50	13 41	17 03	17 49	17 01
13	09 25 19	20 08 49	03 ♉ 50 35	10 ♉ 42 21	03 30	16 51	03 16	20 04	13 48	17 04	17 48	17 02
14	09 29 16	21 06 26	17 ♉ 28 12	24 ♉ 08 19	03 51	18 05	03 56	20 17	13 55	17 05	17 47	17 04
15	09 33 12	22 04 04	00 ♊ 42 58	07 ♊ 12 27	04 19	19 19	04 35	20 30	14 02	17 06	17 47	17 05
16	09 37 09	23 01 44	13 ♊ 37 07	19 ♊ 57 21	04 54	20 33	05 15	20 43	14 09	17 07	17 47	17 07
17	09 41 06	23 59 25	26 ♊ 13 30	02 ♋ 25 57	05 36	21 47	05 54	20 56	14 16	17 08	17 46	17 08
18	09 45 02	24 57 08	08 ♋ 35 04	14 ♋ 41 13	06 25	23 01	06 33	21 09	14 24	17 09	17 46	17 10
19	09 48 59	25 54 52	20 ♋ 44 44	26 ♋ 45 57	07 20	24 16	07 12	21 23	14 31	17 10	17 45	17 12
20	09 52 55	26 52 38	02 ♌ 45 12	08 ♌ 42 45	08 23	25 30	07 51	21 36	14 38	17 11	17 45	17 13
21	09 56 52	27 50 25	14 ♌ 38 54	20 ♌ 33 57	09 31	26 44	08 30	21 49	14 45	17 13	17 45	17 15
22	10 00 48	28 48 14	26 ♌ 28 10	02 ♍ 21 49	10 46	27 58	09 09	22 02	14 53	17 14	17 44	17 17
23	10 04 45	29 ♌ 46 04	08 ♍ 15 12	14 ♍ 08 35	12 06	29 ♌ 13	09 48	22 15	15 00	17 16	17 44	17 19
24	10 08 41	00 ♍ 43 56	20 ♍ 02 17	25 ♍ 56 37	13 32	00 ♍ 27	10 27	22 28	15 08	17 17	17 44	17 20
25	10 12 38	01 41 49	01 ♎ 51 55	07 ♎ 48 31	15 03	01 41	11 05	22 41	15 15	17 19	17 44	17 22
26	10 16 35	02 39 43	13 ♎ 46 49	19 ♎ 47 12	16 38	02 56	11 44	22 54	15 22	17 20	17 44	17 24
27	10 20 31	03 37 38	25 ♎ 50 07	01 ♏ 55 58	18 18	04 10	12 23	23 07	15 30	17 22	17 44	17 26
28	10 24 28	04 35 35	08 ♏ 05 14	14 ♏ 18 22	20 01	05 24	13 01	23 20	15 37	17 23	17 43	17 28
29	10 28 24	05 33 34	20 ♏ 35 51	26 ♏ 58 07	21 47	06 39	13 39	23 33	15 45	17 25	17 43	17 29
30	10 32 21	06 31 33	03 ♐ 25 38	09 ♐ 58 47	23 36	07 53	14 18	23 46	15 52	17 27	17 D 43	17 31
31	10 36 17	07 ♍ 29 34	16 ♐ 37 56	23 ♐ 23 21	25 ♌ 27	09 ♍ 08	14 ♋ 56	23 ♌ 59	16 ♍ 00	17 ♏ 29	17 ♐ 43	17 ♎ 33

Moon True ☊, Mean ☊, Latitude / DECLINATIONS

Date	Moon True ☊ ° '	Moon Mean ☊ ° '	Moon ☽ Latitude ° '	Sun ☉ ° '	Moon ☽ ° '	Mercury ☿ ° '	Venus ♀ ° '	Mars ♂ ° '	Jupiter ♃ ° '	Saturn ♄ ° '	Uranus ♅ ° '	Neptune ♆ ° '	Pluto ♇ ° '
01	08 ♍ 51	09 ♍ 58	04 N 40	18 N 06	10 S 52	13 N 42	20 N 31	23 N 32	16 N 16	08 N 35	16 S 35	21 S 28	09 N 04
02	08 R 50	09 55	05 04	17 51	13 58	13 57	20 16	23 35	16 12	08 32	16 36	21 28	09 03
03	08 48	09 52	05 14	17 35	16 26	14 13	20 01	23 37	16 08	08 29	16 36	21 28	09 02
04	08 44	09 49	05 06	17 20	18 04	14 29	19 44	23 38	16 04	08 27	16 36	21 28	09 01
05	08 41	09 46	04 40	17 04	18 39	14 47	19 28	23 40	16 00	08 24	16 36	21 28	09 00
06	08 37	09 43	03 55	16 47	18 01	15 04	19 10	23 41	15 56	08 21	16 36	21 28	08 59
07	08 34	09 39	02 53	16 31	16 08	15 22	18 52	23 42	15 52	08 19	16 36	21 28	08 58
08	08 33	09 36	01 38	16 14	13 07	15 40	18 34	23 43	15 48	08 16	16 37	21 28	08 57
09	08 32	09 33	00 N 16	15 57	09 12	15 57	18 15	23 44	15 44	08 13	16 37	21 28	08 56
10	08 D 32	09 30	01 S 06	15 40	04 44	16 13	17 55	23 44	15 40	08 11	16 37	21 28	08 55
11	08 33	09 27	02 23	15 22	00 S 03	16 29	17 35	23 45	15 36	08 08	16 37	21 28	08 54
12	08 34	09 23	03 29	15 04	04 N 31	16 43	17 15	23 45	15 32	08 05	16 38	21 28	08 53
13	08 35	09 20	04 20	14 46	08 43	16 57	16 54	23 45	15 28	08 02	16 38	21 28	08 52
14	08 36	09 17	04 55	14 28	12 19	17 09	16 33	23 45	15 24	08 00	16 38	21 28	08 51
15	08 R 36	09 14	05 14	14 09	15 11	17 19	16 11	23 44	15 20	07 57	16 39	21 28	08 50
16	08 35	09 11	05 15	13 50	17 13	17 27	15 48	23 43	15 15	07 54	16 39	21 28	08 49
17	08 34	09 08	05 02	13 31	18 22	17 33	15 26	23 43	15 11	07 51	16 39	21 28	08 48
18	08 32	09 04	04 34	13 12	18 36	17 38	15 02	23 42	15 07	07 48	16 40	21 28	08 47
19	08 30	09 01	03 55	12 53	17 58	17 39	14 39	23 40	15 03	07 46	16 40	21 28	08 46
20	08 28	08 58	03 05	12 33	16 32	17 39	14 15	23 39	14 59	07 43	16 40	21 28	08 45
21	08 27	08 55	02 08	12 13	14 24	17 35	13 50	23 37	14 55	07 40	16 41	21 28	08 44
22	08 26	08 52	01 06	11 53	11 40	17 29	13 25	23 36	14 50	07 37	16 41	21 28	08 43
23	08 26	08 49	00 S 01	11 33	08 28	17 20	13 00	23 34	14 46	07 34	16 42	21 28	08 42
24	08 D 26	08 45	01 N 04	11 13	04 56	17 08	12 35	23 32	14 42	07 31	16 42	21 28	08 41
25	08 26	08 42	02 07	10 52	01 N 12	16 53	12 09	23 29	14 38	07 29	16 42	21 28	08 40
26	08 27	08 39	03 04	10 32	02 S 36	16 36	11 43	23 27	14 34	07 26	16 43	21 28	08 39
27	08 28	08 36	03 54	10 11	06 20	16 15	11 16	23 24	14 30	07 23	16 43	21 28	08 38
28	08 28	08 33	04 35	09 50	09 52	15 52	10 49	23 21	14 25	07 20	16 44	21 28	08 37
29	08 28	08 29	05 03	09 28	13 02	15 25	10 22	23 18	14 21	07 17	16 44	21 28	08 36
30	08 28	08 26	05 17	09 07	15 39	14 57	09 55	23 15	14 17	07 14	16 45	21 28	08 35
31	08 ♍ 28	08 ♍ 23	05 N 15	08 N 46	17 S 33	14 N 25	09 N 27	23 N 12	14 N 13	07 N 11	16 S 45	21 S 28	08 N 34

ZODIAC SIGN ENTRIES

Date	h	m	Planets
02	22	05	☽ ♐
05	02	23	☽ ♑
07	03	28	☽ ♒
08	13	28	♂ ♋
09	03	05	☽ ♓
11	03	10	☽ ♈
13	05	21	☽ ♉
15	10	41	☽ ♊
17	19	17	☽ ♋
20	06	28	☽ ♌
22	19	11	☽ ♍
23	17	47	☉ ♍
24	03	16	♀ ♍
25	08	13	☽ ♎
27	20	12	☽ ♏
30	05	39	☽ ♐

LATITUDES

Date	Mercury ☿ ° '	Venus ♀ ° '	Mars ♂ ° '	Jupiter ♃ ° '	Saturn ♄ ° '	Uranus ♅ ° '	Neptune ♆ ° '	Pluto ♇ ° '
01	04 S 54	00 N 50	00 N 11	00 N 41	01 N 49	00 N 19	01 N 26	16 N 56
04	04 35	00 56	00 14	00 42	01 48	00 19	01 26	16 55
07	04 02	01 01	00 16	00 42	01 48	00 19	01 26	16 53
10	03 19	01 06	00 18	00 42	01 48	00 19	01 25	16 52
13	02 30	01 10	00 21	00 42	01 48	00 19	01 25	16 50
16	01 38	01 14	00 23	00 42	01 48	00 19	01 25	16 49
19	00 49	01 17	00 26	00 43	01 48	00 19	01 25	16 48
22	00 S 03	01 20	00 28	00 43	01 48	00 18	01 25	16 46
25	00 N 36	01 22	00 31	00 43	01 48	00 18	01 25	16 45
28	01 06	01 23	00 33	00 43	01 48	00 18	01 24	16 44
31	01 N 28	01 N 24	00 N 36	00 N 44	01 N 48	00 N 18	01 N 24	16 N 43

DATA

Julian Date	2444087
Delta T	+50 seconds
Ayanamsa	23° 34' 14"
Synetic vernal point	05° ♓ 32' 46"
True obliquity of ecliptic	23° 26' 22"

MOON'S PHASES, APSIDES AND POSITIONS ☽

Date	h	m	Phase	Longitude ° '	Eclipse Indicator
01	05	57	☽	08 ♏ 25	
08	03	21	○	15 ♒ 00	
14	19	02	☾	21 ♉ 23	
22	17	10	●	29 ♌ 01	Annular
30	18	09	☽	06 ♐ 46	

Day	h	m		
08	18	54	Perigee	
23	07	21	Apogee	
05	11	49	Max dec	18° S 39'
11	12	14	0N	
18	06	28	Max dec	18° N 37'
25	19	33	0S	

LONGITUDES

Date	Chiron ⚷ ° '	Ceres ⚳ ° '	Pallas ⚴ ° '	Juno ⚵ ° '	Vesta ⚶ ° '	Black Moon Lilith ⚸ ° '
01	13 ♉ 47	19 ♈ 00	27 ♒ 25	14 ♊ 24	12 ♉ 06	02 ♍ 32
11	13 ♉ 54	19 ♈ 33	24 ♒ 57	19 ♊ 46	14 ♉ 28	03 ♍ 38
21	13 ♉ 55	19 ♈ 30	22 ♒ 23	25 ♊ 00	16 ♉ 21	04 ♍ 45
31	13 ♉ 51	18 ♈ 50	19 ♒ 56	00 ♋ 05	17 ♉ 41	05 ♍ 51

ASPECTARIAN

01 Wednesday
- 04 22 ☽ □ ☿
- 05 57 ☽ □ ☉
- 09 22 ☽ ⚼ ♂
- 12 40 ☽ ⊥ ♆
- 13 34 ☽ ⚹ ♄
- 21 46 ☽ ⊻ ♇
- 22 04 ☽ ☌ ♅
- 23 11 ☽ □ ♃
- 23 55 ☽ ⊻ ♆

02 Thursday
- 03 13 ☽ ± ♂
- 05 16 ♀ ⚼ ♆
- 08 55 ☽ ⊥ ♇
- 11 49 ☽ # ☿
- 12 07 ☽ Q ♄
- 14 52 ☽ ⊼ ♂
- 15 12 ☿ ⊥ ♄

03 Friday
- 01 20 ☽ ∠ ♇
- 05 37 ☽ △ ♀
- 08 45 ☽ # ♃
- 08 56 ☽ △ ☿
- 13 52 ☽ ∥ ♅
- 16 27 ♃ △ ♆
- 17 19 ☽ △ ☉
- 17 58 ♀ Q ♇
- 20 40 ☽ □ ♄

04 Saturday
- 01 06 ☽ # ☉
- 03 53 ☽ ⚹ ♇
- 04 07 ☽ ⊻ ♅
- 05 45 ☽ ☌ ♆
- 05 59 ☽ △ ♃
- 06 34 ☿ ☌ ♀
- 10 01 ☽ ⚼ ☿
- 10 31 ☽ ⚼ ♀
- 14 26 ☽ ⊥ ♅
- 21 07 ☽ ⚼ ☉
- 22 17 ☽ ☍ ♂

05 Sunday
- 00 26 ☽ Q ♇
- 00 43 ☽ ± ☿
- 03 16 ☽ ± ♀
- 05 40 ☽ ∠ ♅
- 07 51 ☿ Q ♇
- 07 51 ☽ ⚼ ♃
- 09 39 ♀ ⊥ ♄
- 10 24 ☽ ⊼ ☿
- 13 17 ☽ ± ☉
- 14 12 ☽ ⊼ ♀
- 22 57 ☉ ⊻ ♄
- 22 59 ☽ ± ♃
- 23 49 ☽ △ ♄
- 23 53 ☽ ⊼ ☉

06 Monday
- 04 12 ☿ ⊥ ♂
- 06 16 ☽ □ ♇
- 06 27 ☽ ⚹ ♅
- 07 55 ☽ ⊻ ♆
- 08 56 ☽ ⊼ ♃
- 17 37 ☽ ⊥ ♆

07 Tuesday
- 00 19 ☽ ⚼ ♄
- 01 50 ☽ Q ♅
- 01 53 ☽ ⊼ ♂
- 03 43 ☉ ∠ ♂
- 03 58 ☉ # ♅
- 07 15 ☽ ∥ ♅
- 07 42 ☽ # ☉
- 08 03 ☽ ∠ ♆
- 09 40 ☽ ☍ ☿
- 11 55 ☽ ± ♂
- 14 36 ☽ # ♃
- 14 49 ☽ ± ♄
- 18 20 ☽ # ☿
- 19 06 ☽ ☍ ♀

08 Wednesday
- 00 25 ☽ ⊼ ♄
- 02 53 ☽ ⚼ ♂
- 03 21 ☽ ☍ ☉
- 06 22 ☽ △ ♇
- 06 30 ☽ □ ♅
- 07 52 ☽ ⚹ ♆
- 09 35 ☽ ☍ ♃
- 21 38 ☿ ∥ ♃

09 Thursday
- 02 51 ☽ Q ♆
- 03 43 ☽ △ ♂
- 06 09 ☽ ⚼ ♇
- 08 21 ☽ ⊼ ☿
- 12 10 ☉ ∥ ☿
- 13 30 ☽ # ♇
- 17 31 ☽ # ♄
- 17 47 ☽ ± ☿
- 20 29 ☽ ± ♇
- 22 54 ☽ ⊼ ♀

10 Friday
- 00 22 ☽ ☍ ♄
- 04 11 ☉ ⚹ ♇
- 05 38 ☉ □ ♅
- 06 06 ☽ ⊼ ♇
- 06 11 ☽ △ ♅
- 06 13 ☽ ⊼ ☉
- 07 30 ☽ □ ♆
- 08 04 ☽ ⚼ ☿
- 09 23 ☽ ± ♀
- 09 59 ☽ ⊼ ♃
- 11 16 ☉ ∥ ♃
- 16 32 ☽ ± ☉
- 18 30 ♀ ⊻ ♄
- 18 48 ♂ Q ♄
- 19 49 ☽ ± ♃

11 Saturday
- 01 15 ☽ ⚼ ♀
- 01 32 ☿ St D
- 01 52 ☉ △ ♆
- 06 05 ☽ □ ♂
- 06 29 ☽ ⚼ ♅
- 08 11 ☽ ⚼ ☉
- 08 20 ☽ △ ☿
- 10 42 ☽ ⚼ ♃
- 15 03 ♂ ⚼ ♅
- 21 19 ☽ ± ♅

12 Sunday
- 01 37 ☽ ⊼ ♄
- 02 02 ☿ # ♅
- 04 16 ☽ △ ♀
- 07 16 ☽ ☍ ♇
- 07 20 ☽ ⊼ ♅
- 08 37 ☽ △ ♆
- 10 52 ☽ △ ☉
- 11 46 ☽ ± ♄
- 12 03 ☽ △ ♃
- 13 24 ☽ Q ♂

13 Monday
- 03 13 ☽ ⚼ ♄
- 07 58 ☽ ∥ ♄
- 09 12 ☉ ☌ ♃
- 10 03 ☿ ∥ ♀
- 10 12 ☽ ⚼ ♆
- 10 57 ☽ ⚹ ♂
- 11 24 ☽ □ ☿
- 12 58 ☽ ∥ ♇
- 15 51 ♀ ⚹ ♇
- 16 18 ♀ □ ♅

14 Tuesday
- 01 55 ☽ ± ♆
- 04 44 ☿ ⊻ ♂
- 05 37 ☽ △ ♄
- 05 50 ♀ # ♅
- 06 28 ♀ △ ♆
- 11 16 ☽ ⊼ ♇
- 11 18 ☽ ☍ ♅
- 12 34 ☽ ⊼ ♆
- 13 12 ☽ □ ♀
- 14 45 ☽ ∠ ♂
- 17 07 ☽ □ ♃
- 19 02 ☽ □ ☉
- 20 07 ☽ Q ☿
- 22 04 ☽ ± ♇

15 Wednesday
- 03 31 ☽ ∥ ☉
- 07 53 ☽ ⊥ ♂
- 13 20 ☽ ∥ ♃
- 14 32 ☽ ⚼ ♇
- 18 55 ☽ ⚹ ☿
- 19 31 ☽ ⊻ ♂
- 20 54 ☽ ∥ ♀
- 23 16 ♀ ∠ ♂

16 Thursday
- 01 30 ☽ Q ♀
- 02 39 ☽ Q ♃
- 04 06 ☽ # ♅
- 05 20 ♅ ⊻ ♇
- 06 44 ☽ Q ☉
- 13 01 ☽ □ ♄
- 15 58 ♀ ☌ ♃
- 16 02 ☽ ∥ ☿
- 18 36 ☽ ⊼ ♅
- 18 37 ☽ △ ♇
- 19 51 ☽ ☍ ♆
- 20 23 ☿ Q ♇

17 Friday
- 00 33 ☽ ∠ ☿
- 01 41 ☽ ⚹ ♃
- 02 34 ☽ ⚹ ♀
- 06 03 ☽ ± ♅
- 07 21 ☽ ⚹ ☉
- 14 12 ♂ ∠ ♃
- 18 56 ☽ ⊥ ☿
- 23 26 ☽ ⚼ ♅
- 23 48 ☽ Q ♄

18 Saturday
- 06 00 ♀ ∥ ♃
- 07 10 ☽ ∠ ♃
- 07 25 ☽ ⊻ ☿
- 07 48 ☽ ☌ ♂
- 10 47 ☽ ∠ ♀
- 14 55 ☽ ∠ ☉
- 23 32 ☽ ⚹ ♄

19 Sunday
- 01 10 ☽ ⊥ ♃
- 01 33 ☿ ⊻ ♂
- 04 54 ☽ △ ♅
- 04 57 ☽ □ ♇
- 06 04 ☽ ⊼ ♆
- 06 30 ☽ ⊥ ♀
- 10 12 ☽ ⊥ ☉
- 13 17 ☽ ⊻ ♃
- 17 59 ☽ ± ♆
- 18 32 ☽ ∥ ☿
- 19 48 ☽ ⊻ ♀
- 23 12 ☽ ⊻ ☉

20 Monday
- 05 41 ☽ ∠ ♄
- 10 11 ☽ # ♅
- 12 00 ☽ ⚼ ♆
- 16 59 ☽ Q ♇
- 18 19 ☿ ⊥ ♄
- 22 52 ☽ ⊻ ♂
- 23 58 ☽ ⊥ ♄

21 Tuesday
- 00 30 ☽ ☌ ☿
- 06 39 ☽ ∥ ♃
- 11 42 ☽ ⊥ ♂
- 12 13 ☽ ⊻ ♄
- 17 12 ☽ □ ♅
- 17 17 ☽ ⚹ ♇
- 18 16 ☽ △ ♆
- 18 17 ☽ ∥ ♀

22 Wednesday
- 02 48 ☽ ☌ ♃
- 07 01 ☽ ∠ ♂
- 09 53 ☽ ∥ ☉
- 15 25 ☽ ☌ ♀
- 17 10 ☽ ☌ ☉
- 23 51 ☽ ∠ ♇

23 Thursday
- 05 53 ☽ Q ♅
- 10 14 ☽ ∥ ♇
- 15 20 ☽ ⚹ ♂
- 18 14 ☽ ⊥ ♇
- 18 19 ☽ ∥ ♄
- 20 55 ☽ ⊻ ☿

24 Friday
- 01 54 ☽ ☌ ♄
- 06 23 ☽ ⚹ ♅
- 06 30 ☽ ⊻ ♇
- 07 19 ☽ □ ♆
- 10 50 ☽ ⊥ ☿
- 17 02 ☽ ⊻ ♃
- 17 11 ☽ Q ♂

25 Saturday
- 05 27 ☽ ⊥ ♃
- 07 46 ☽ ∠ ☿
- 11 36 ☽ ⊻ ♀
- 11 38 ☽ ⊻ ☉
- 12 38 ☉ ☌ ♀
- 12 54 ☽ ∠ ♅
- 15 19 ☿ ⊻ ♄
- 19 48 ☽ Q ♆
- 23 58 ☽ ∠ ♃

26 Sunday
- 00 49 ☽ ⊥ ☉
- 01 07 ☽ ⊥ ♀
- 01 28 ♀ ∠ ♇
- 02 58 ☿ # ♅
- 05 14 ☉ ∠ ♇
- 07 05 ☽ ⊥ ♅
- 07 40 ☽ □ ♂
- 15 13 ☽ ⊻ ♄
- 18 37 ☽ ⚹ ☿
- 19 08 ☽ ⊻ ♅
- 19 15 ☽ ☌ ♇
- 19 54 ☽ ⚹ ♆
- 20 26 ☽ ∠ ☉
- 21 15 ☽ ∠ ♀
- 22 23 ☿ □ ♅
- 23 21 ☿ ⚹ ♇

27 Monday
- 03 19 ☽ ⊥ ♄
- 03 52 ☿ △ ♆
- 06 32 ☽ ⚹ ♃
- 13 53 ☿ ⊥ ♂
- 18 47 ☽ # ♄
- 21 16 ☽ ∠ ♄
- 22 12 ☽ Q ☿

28 Tuesday
- 01 33 ☽ ∠ ♆
- 03 20 ☽ # ♇
- 04 37 ☽ ⚹ ☉
- 06 12 ☽ ⚹ ♀
- 06 34 ☽ Q ♃
- 11 38 ♀ Q ♅
- 11 44 ☽ # ☉
- 18 04 ☽ # ♀
- 19 02 ☽ ⊥ ♆
- 22 02 ☽ △ ♂

29 Wednesday
- 02 40 ☽ ⚹ ♄
- 05 45 ☽ Q ☉
- 05 56 ☽ ☌ ♅
- 06 05 ☽ ⊻ ♇
- 06 32 ☽ ⊻ ♆
- 07 54 ☽ Q ♀
- 08 22 ☉ Q ♅
- 14 36 ☽ □ ☿
- 17 29 ☽ ⊥ ♇
- 17 41 ☽ □ ♃
- 23 05 ☽ # ♃

30 Thursday
- 01 35 ☽ Q ♄
- 03 56 ☽ ⚼ ♂
- 06 05 ☽ # ☿
- 10 20 ☽ ∠ ♇
- 11 16 ♆ St D
- 14 39 ☿ ☌ ♃
- 18 09 ☽ □ ☉
- 21 02 ☽ □ ♀
- 21 23 ☽ ± ♂

31 Friday
- 00 37 ☽ ∥ ♅
- 08 48 ☽ ⊼ ♂
- 10 51 ☽ □ ♄
- 13 31 ☽ ⊻ ♅
- 13 39 ☽ ⚹ ♇
- 13 57 ☽ ☌ ♆
- 22 41 ☿ ∥ ♃

All ephemeris data is given at 12.00 UT and the Moon's longitude is additionally given for 24.00 UT

SEPTEMBER 1979

Date	Sidereal time h m s	Sun ☉ ° ' "	Moon ☽ ° ' "	Moon ☽ 24.00 ° ' "	Mercury ☿ ° '	Venus ♀ ° '	Mars ♂ ° '	Jupiter ♃ ° '	Saturn ♄ ° '	Uranus ♅ ° '	Neptune ♆ ° '	Pluto ♇ ° '
01	10 40 14	08 ♍ 27 37	00 ♑ 15 15	07 ♑ 13 42	27 ♌ 22	10 ♍ 22	15 ♋ 34	24 ♌ 12	16 ♍ 07	17 ♏ 30	17 ♐ 43	17 ♎ 35
02	10 44 10	09 25 40	14 ♑ 18 40	21 ♑ 29 56	29 ♌ 14	11 37	16 12	24 25	16 14	17 32	17 44	17 37
03	10 48 07	10 23 45	28 ♑ 47 08	06 ♒ 09 46	01 ♍ 09	12 51	16 50	24 38	16 22	17 34	17 44	17 39
04	10 52 04	11 21 52	13 ♒ 37 07	21 ♒ 08 19	03 05	14 05	17 28	24 51	16 29	17 36	17 44	17 41
05	10 56 00	12 20 00	28 ♒ 42 33	06 ♓ 18 13	05 01	15 20	18 05	25 04	16 37	17 38	17 44	17 43
06	10 59 57	13 18 10	13 ♓ 54 36	21 ♓ 30 21	06 58	16 34	18 43	25 17	16 45	17 40	17 44	17 45
07	11 03 53	14 16 21	29 ♓ 04 15	06 ♈ 35 11	08 54	17 49	19 21	25 29	16 52	17 42	17 44	17 47
08	11 07 50	15 14 34	14 ♈ 02 07	21 ♈ 24 08	10 50	19 03	19 58	25 42	17 00	17 45	17 45	17 49
09	11 11 46	16 12 49	28 ♈ 40 31	05 ♉ 50 42	12 46	20 18	20 36	25 55	17 07	17 47	17 45	17 51
10	11 15 43	17 11 06	12 ♉ 54 17	19 ♉ 51 03	14 41	21 32	21 13	26 08	17 15	17 49	17 45	17 54
11	11 19 39	18 09 25	26 ♉ 40 58	03 ♊ 24 06	16 35	22 47	21 50	26 20	17 22	17 51	17 46	17 56
12	11 23 36	19 07 47	10 ♊ 00 41	16 ♊ 31 02	18 28	24 01	22 27	26 33	17 30	17 54	17 46	17 58
13	11 27 32	20 06 10	22 ♊ 55 32	29 ♊ 14 38	20 21	25 16	23 04	26 45	17 37	17 56	17 47	18 00
14	11 31 29	21 04 36	05 ♋ 28 52	11 ♋ 38 43	22 13	26 31	23 41	26 58	17 45	17 58	17 47	18 02
15	11 35 26	22 03 04	17 ♋ 44 45	23 ♋ 47 29	24 03	27 45	24 18	27 11	17 52	18 01	17 48	18 04
16	11 39 22	23 01 34	29 ♋ 47 27	05 ♌ 45 11	25 53	29 ♍ 00	24 55	27 23	18 00	18 03	17 48	18 06
17	11 43 19	24 00 06	11 ♌ 41 09	17 ♌ 35 50	27 41	00 ♎ 14	25 32	27 35	18 07	18 06	17 49	18 09
18	11 47 15	24 58 39	23 ♌ 29 39	29 ♌ 23 01	29 ♍ 29	01 29	26 08	27 48	18 15	18 08	17 49	18 11
19	11 51 12	25 57 15	05 ♍ 16 18	11 ♍ 09 51	01 ♎ 15	02 44	26 45	28 00	18 22	18 11	17 50	18 13
20	11 55 08	26 55 53	17 ♍ 04 00	22 ♍ 59 02	03 01	03 58	27 21	28 12	18 30	18 14	17 51	18 15
21	11 59 05	27 54 33	28 ♍ 55 13	04 ♎ 52 49	04 45	05 13	27 58	28 25	18 37	18 16	17 51	18 18
22	12 03 01	28 53 15	10 ♎ 52 03	16 ♎ 53 11	06 29	06 28	28 34	28 37	18 45	18 19	17 52	18 20
23	12 06 58	29 ♍ 51 59	22 ♎ 56 24	29 ♎ 01 58	08 11	07 42	29 10	28 49	18 52	18 22	17 53	18 22
24	12 10 55	00 ♎ 50 45	05 ♏ 10 05	11 ♏ 21 00	09 53	08 57	29 ♋ 46	29 01	19 00	18 25	17 54	18 24
25	12 14 51	01 49 32	17 ♏ 34 56	23 ♏ 52 10	11 33	10 12	00 ♌ 22	29 13	19 07	18 27	17 54	18 27
26	12 18 48	02 48 22	00 ♐ 12 58	06 ♐ 37 34	13 13	11 26	00 58	29 25	19 14	18 30	17 55	18 29
27	12 22 44	03 47 13	13 ♐ 06 17	19 ♐ 39 23	14 52	12 41	01 33	29 37	19 22	18 33	17 56	18 31
28	12 26 41	04 46 06	26 ♐ 17 06	02 ♑ 59 42	16 29	13 56	02 09	29 ♌ 49	19 29	18 36	17 57	18 34
29	12 30 37	05 45 01	09 ♑ 47 23	16 ♑ 40 18	18 06	15 10	02 44	00 ♍ 01	19 36	18 39	17 58	18 36
30	12 34 34	06 ♎ 43 57	23 ♑ 38 33	00 ♒ 42 08	19 ♎ 42	16 ♎ 25	03 ♌ 20	00 ♍ 13	19 ♍ 44	18 ♏ 42	17 ♐ 59	18 ♎ 38

Date	Moon True ☊ ° '	Moon Mean ☊ ° '	Moon ☽ Latitude ° '
01	08 ♍ 28	08 ♍ 20	04 N 55
02	08 D 28	08 17	04 18
03	08 29	08 14	03 23
04	08 29	08 10	02 14
05	08 29	08 07	00 N 54
06	08 R 29	08 04	00 S 30
07	08 29	08 01	01 52
08	08 28	07 58	03 05
09	08 27	07 55	04 04
10	08 27	07 51	04 47
11	08 26	07 48	05 11
12	08 25	07 45	05 17
13	08 D 25	07 42	05 07
14	08 25	07 39	04 43
15	08 26	07 35	04 06
16	08 27	07 32	03 19
17	08 29	07 29	02 23
18	08 30	07 26	01 22
19	08 31	07 23	00 S 18
20	08 R 30	07 20	00 N 47
21	08 29	07 16	01 51
22	08 27	07 13	02 49
23	08 24	07 10	03 41
24	08 20	07 07	04 24
25	08 17	07 04	04 54
26	08 13	07 01	05 11
27	08 11	06 57	05 13
28	08 09	06 54	04 59
29	08 D 09	06 51	04 28
30	08 ♍ 10	06 ♍ 48	03 N 41

DECLINATIONS

Date	Sun ☉ ° '	Moon ☽ ° '	Mercury ☿ ° '	Venus ♀ ° '	Mars ♂ ° '	Jupiter ♃ ° '	Saturn ♄ ° '	Uranus ♅ ° '	Neptune ♆ ° '	Pluto ♇ ° '
01	08 N 24	18 S 31	13 N 52	08 N 59	23 N 08	14 N 09	07 N 08	16 S 46	21 S 28	08 N 33
02	08 02	18 24	13 16	08 31	23 05	14 04	07 06	16 46	21 29	08 32
03	07 40	17 05	12 39	08 03	23 01	14 00	07 03	16 47	21 29	08 31
04	07 18	14 36	12 00	07 34	22 57	13 56	07 00	16 47	21 29	08 30
05	06 56	11 05	11 19	07 05	22 53	13 52	06 57	16 48	21 29	08 29
06	06 34	06 48	10 37	06 36	22 49	13 47	06 54	16 49	21 29	08 28
07	06 11	02 S 05	09 53	06 07	22 44	13 43	06 51	16 49	21 29	08 27
08	05 49	02 N 41	09 09	05 37	22 40	13 39	06 48	16 50	21 29	08 27
09	05 26	07 12	08 24	05 08	22 35	13 35	06 45	16 51	21 29	08 25
10	05 04	11 09	07 38	04 38	22 30	13 31	06 42	16 51	21 29	08 24
11	04 41	14 22	06 52	04 08	22 25	13 26	06 40	16 52	21 29	08 23
12	04 18	16 43	06 05	03 38	22 20	13 22	06 37	16 53	21 29	08 22
13	03 55	18 08	05 18	03 08	22 15	13 18	06 34	16 53	21 29	08 21
14	03 32	18 37	04 30	02 38	22 09	13 14	06 31	16 54	21 30	08 20
15	03 09	18 12	03 43	02 08	22 04	13 10	06 28	16 55	21 30	08 19
16	02 46	16 57	02 55	01 37	21 58	13 05	06 25	16 55	21 30	08 18
17	02 23	14 59	02 08	01 07	21 52	13 01	06 22	16 56	21 30	08 17
18	02 00	12 24	01 20	00 36	21 46	12 57	06 19	16 57	21 30	08 16
19	01 36	09 18	00 N 33	00 N 06	21 40	12 53	06 16	16 58	21 30	08 15
20	01 13	05 50	00 S 14	00 S 25	21 34	12 49	06 13	16 58	21 30	08 14
21	00 50	02 N 07	01 01	00 55	21 28	12 45	06 11	16 59	21 30	08 13
22	00 27	01 S 42	01 48	01 26	21 21	12 41	06 08	17 00	21 30	08 12
23	00 N 03	05 30	02 34	01 56	21 15	12 36	06 05	17 01	21 30	08 11
24	00 S 20	09 06	03 20	02 27	21 08	12 32	06 02	17 01	21 31	08 10
25	00 44	12 22	04 05	02 57	21 01	12 28	05 59	17 02	21 31	08 09
26	01 07	15 07	04 50	03 28	20 55	12 24	05 57	17 03	21 31	08 08
27	01 30	17 11	05 35	03 58	20 48	12 20	05 54	17 04	21 31	08 07
28	01 54	18 24	06 19	04 29	20 41	12 16	05 51	17 05	21 31	08 06
29	02 17	18 37	07 03	04 59	20 33	12 12	05 48	17 06	21 31	08 05
30	02 S 40	17 S 45	07 S 46	05 S 29	20 N 26	12 N 08	05 N 45	17 S 06	21 S 31	08 N 04

ZODIAC SIGN ENTRIES

Date	h	m	Planets
01	11	34	☽ ♑
02	21	39	☿ ♍
03	13	59	☽ ♒
05	14	03	☽ ♓
07	13	29	☽ ♈
09	14	12	☽ ♉
11	17	54	☽ ♊
14	01	27	☽ ♋
16	12	25	☽ ♌
17	07	21	♀ ♎
18	18	59	☿ ♎
19	01	15	☽ ♍
21	14	11	☽ ♎
23	15	16	☉ ♎
24	01	54	☽ ♏
24	21	21	♂ ♌
26	11	36	☽ ♐
28	18	40	☽ ♑
29	10	24	♃ ♍
30	22	49	☽ ♒

LATITUDES

Date	Mercury ☿ ° '	Venus ♀ ° '	Mars ♂ ° '	Jupiter ♃ ° '	Saturn ♄ ° '	Uranus ♅ ° '	Neptune ♆ ° '	Pluto ♇ ° '
01	01 N 34	01 N 25	00 N 37	00 N 44	01 N 48	00 N 18	01 N 24	16 N 43
04	01 44	01 25	00 39	00 44	01 48	00 18	01 24	16 42
07	01 47	01 25	00 42	00 44	01 49	00 18	01 24	16 41
10	01 44	01 24	00 45	00 45	01 49	00 18	01 24	16 40
13	01 36	01 22	00 47	00 45	01 49	00 18	01 24	16 39
16	01 24	01 20	00 50	00 45	01 49	00 18	01 23	16 39
19	01 08	01 17	00 53	00 46	01 49	00 18	01 23	16 38
22	00 51	01 14	00 56	00 46	01 49	00 18	01 23	16 37
25	00 31	01 11	00 58	00 47	01 50	00 18	01 23	16 37
28	00 N 11	01 06	01 01	00 47	01 50	00 18	01 23	16 37
31	00 S 10	01 N 02	01 N 04	00 N 47	01 N 50	00 N 18	01 N 23	16 N 36

DATA

Julian Date	2444118
Delta T	+50 seconds
Ayanamsa	23° 34' 17"
Synetic vernal point	05° ♓ 32' 42"
True obliquity of ecliptic	23° 26' 23"

MOON'S PHASES, APSIDES AND POSITIONS ☽

Date	h	m	Phase	Longitude ° '	Eclipse Indicator
06	10	59	○	13 ♓ 16	total
13	06	15	☾	19 ♊ 52	
21	09	47	●	27 ♍ 49	
29	04	20	☽	05 ♑ 26	

Day	h	m		
06	05	08	Perigee	
19	10	08	Apogee	
01	21	32	Max dec	18° S 36'
07	22	24	0N	
14	12	38	Max dec	18° N 37'
22	01	20	0S	
29	04	56	Max dec	18° S 40'

LONGITUDES

Date	Chiron ⚷ ° '	Ceres ⚳ ° '	Pallas ⚴ ° '	Juno ⚵ ° '	Vesta ⚶ ° '	Black Moon Lilith ⚸ ° '
01	13 ♉ 50	18 ♈ 44	19 ♒ 42	00 ♋ 35	17 ♉ 47	05 ♍ 58
11	13 ♉ 40	17 ♈ 26	17 ♒ 38	05 ♋ 27	18 ♉ 25	07 ♍ 04
21	13 ♉ 24	15 ♈ 38	16 ♒ 26	10 ♋ 04	18 ♉ 20	08 ♍ 11
31	13 ♉ 03	13 ♈ 29	15 ♒ 00	14 ♋ 23	17 ♉ 30	09 ♍ 17

All ephemeris data is given at 12.00 UT and the Moon's longitude is additionally given for 24.00 UT

ASPECTARIAN

h m	Aspects
01 Saturday	
00 11	☽ ⊥ ♅
01 10	☉ ∥ ♇
01 16	☽ △ ♃
06 05	☽ △ ☿
10 50	☽ Q ♇
15 54	☽ ∠ ♅
02 Sunday	
03 08	☽ △ ☉
03 36	☽ bQ ♃
07 00	☽ △ ♀
10 58	♀ ∥ ♇
11 51	☽ bQ ☿
12 13	♀ ⊥ ♇
14 03	♂ ✶ ♄
15 16	☽ △ ♄
15 19	☽ ☍ ♂
17 25	☽ ✶ ♅
17 33	☽ □ ♇
17 43	☽ ⊻ ♆
18 59	☽ ± ♃
03 Monday	
03 41	☽ ⊥ ♆
05 05	☽ ⊼ ♃
05 08	☽ ± ☿
06 02	☽ bQ ☉
10 19	☽ bQ ♀
13 17	☽ Q ♅
15 34	☽ ∥ ♅
16 15	☽ bQ ♄
16 27	☽ ⊼ ☿
18 26	☽ ∠ ♆
21 47	☽ ± ☉
04 Tuesday	
00 32	☿ ∠ ♂
02 18	☽ ± ♀
06 56	☽ ± ♄
06 59	☿ ∠ ♇
08 08	☽ ⊼ ☉
12 49	☽ ⊼ ♀
16 38	☽ ⊼ ♄
17 11	☽ # ♃
17 45	♂ △ ♅
18 23	☽ □ ♅
18 25	☽ ⊼ ♂
18 31	☽ △ ♇
18 34	☽ ✶ ♆
20 15	☉ ⊥ ♇
21 03	♂ □ ♇
22 17	♂ ⊼ ♆
05 Wednesday	
04 22	☽ ± ♂
06 09	☽ ☍ ♃
10 15	☽ # ☿
10 56	☉ ∥ ♄
13 37	☽ Q ♆
18 21	☽ bQ ♇
19 14	☽ bQ ♂
19 35	♀ ∥ ♄
19 44	☿ Q ♅
22 25	♆ ✶ ♇
23 26	☽ ☍ ☿
06 Thursday	
02 58	☽ # ♇
08 35	☽ ± ♇
10 59	☽ ☍ ☉
11 26	☽ # ♄
13 08	☽ # ♀
13 19	☽ # ☉
15 39	♀ ☌ ♄
16 30	☽ ☍ ♄
16 35	☽ ☍ ♀
17 57	☽ △ ♅
18 03	☽ □ ♆
18 05	☽ ⊼ ♇
19 55	☽ △ ♂
20 07	☉ ∥ ♀
07 Friday	
06 14	☽ ⊼ ♃
09 53	♀ ✶ ♅
10 34	♀ □ ♆
11 29	♀ ⊻ ♇
15 54	☽ ± ♃
17 48	☽ bQ ♅
20 26	♂ ⊥ ♃
08 Saturday	
06 04	☽ ⊼ ☿
06 32	☽ bQ ♃
08 17	☽ ± ♅
13 23	♅ ⊻ ♆
14 06	☽ ⊼ ☉
16 50	☽ ⊼ ♄
17 14	☽ ± ☿
18 02	☽ △ ♆
18 02	☽ ⊼ ♅
18 10	☽ ☍ ♇
20 55	☽ ⊼ ♀
22 05	☽ □ ♂
09 Sunday	
00 29	☿ ⊥ ♇
00 34	☽ ± ☉
01 51	☽ ∥ ♀
02 44	☽ ± ♄
03 07	☽ ∥ ☉
07 21	☽ △ ♃
07 41	☽ ± ♀
09 35	☽ ∥ ♄
10 15	☽ bQ ☿
11 27	☿ ∥ ♇
16 32	☽ bQ ☉
17 47	☽ bQ ♄
17 50	☽ ∥ ☿
18 48	☽ bQ ♆
18 59	☽ ∥ ♇
23 28	♀ ✶ ♂
10 Monday	
00 08	☽ bQ ♀
05 25	☽ Q ♂
10 02	☽ ± ♆
13 39	☉ ☌ ♄
15 32	☽ △ ☿
19 32	☽ △ ♄
19 56	☽ △ ☉
20 22	☽ ⊼ ♆
20 29	☽ ☍ ♅
20 37	☽ ⊼ ♇
11 Tuesday	
02 11	☉ □ ♆
03 04	☽ ✶ ♂
04 14	☉ ✶ ♅
04 26	☽ △ ♀
04 34	☽ ∥ ♃
06 07	☉ ⊻ ♇
07 07	☽ ± ♇
11 23	☽ □ ♃
18 40	☿ ∥ ♄
22 40	☿ ☌ ♄
23 11	☽ bQ ♇
12 Wednesday	
03 00	☿ □ ♆
04 28	☿ ✶ ♅
05 23	☿ ⊻ ♇
07 07	☽ ∠ ♂
14 06	☽ # ♅
20 29	☽ Q ♃
13 Thursday	
00 29	☽ ⊥ ♂
01 57	☽ □ ♄
02 20	☽ ☍ ♆
02 36	☽ ⊼ ♅
02 44	☽ △ ♇
05 25	☉ ☌ ☿
06 15	☽ □ ☉
06 20	☽ □ ☿
12 18	☽ ⊻ ♂
13 55	☽ ± ♅
16 55	☽ □ ♀
19 23	☽ ✶ ♃
14 Friday	
07 09	☽ bQ ♅
12 31	☽ Q ♄
15 42	♂ ± ♆
19 35	☽ Q ☉
19 58	♄ □ ♆
22 35	♀ ⊻ ♃
22 49	☽ Q ☿
15 Saturday	
00 51	☽ ∠ ♃
07 37	☽ Q ♀
12 06	☽ ⊼ ♆
12 15	☽ ✶ ♄
12 32	☽ △ ♅
12 39	☽ □ ♇
16 59	☿ ✶ ♂
18 55	☽ ⊥ ♃
21 17	☽ ✶ ☉
16 Sunday	
00 01	☽ ± ♆
01 44	☽ ☌ ♂
02 46	☽ ✶ ☿
07 05	☽ ⊻ ♃
10 13	☽ ✶ ♀
12 29	☽ # ♅
18 03	☽ bQ ♆
18 31	☽ ∠ ♄
20 52	☉ ∥ ☿
17 Monday	
00 45	☽ Q ♇
05 12	♄ ✶ ♅
06 05	☽ ∠ ☉
10 31	☿ ⊻ ♃
12 53	☽ ⊥ ♄
14 24	☽ ∠ ☿
18 24	♄ ⊻ ♇
20 04	☽ ∠ ♀
18 Tuesday	
00 27	☽ △ ♆
01 04	☽ □ ♅
01 09	☽ ✶ ♇
01 12	☽ ⊻ ♄
01 59	☽ ⊥ ☉
07 06	☽ ∥ ♃
11 58	☽ ⊥ ☿
15 18	☽ ⊻ ☉
16 32	☽ ⊥ ♀
17 41	☽ ⊻ ♂
20 55	☽ ☌ ♃
19 Wednesday	
02 22	☽ ⊻ ☿
06 12	☽ ⊻ ♀
06 35	☽ ⊥ ♂
07 48	☽ ∠ ♇
13 52	☽ Q ♅
19 34	☽ ∥ ♇
21 06	♀ ∠ ♅
23 57	☿ # ♀
20 Thursday	
01 54	☽ ∠ ♂
02 11	☽ ⊥ ♇
09 20	☽ ∥ ♄
13 35	☽ □ ♆
14 22	☽ ✶ ♅
14 25	☽ ⊻ ♇
14 56	☽ ☌ ♄
14 59	☿ ∠ ♅
17 26	♀ ⊥ ♃
21 Friday	
02 49	♂ # ♆
02 55	☿ ∥ ♀
06 35	☿ ⊥ ♃
08 06	☉ # ☿
09 39	☉ # ♀
09 47	☽ ☌ ☉
09 58	☽ ✶ ♂
10 57	☽ ⊻ ♃
15 21	☉ ✶ ♂
17 45	☽ # ☿
18 41	☽ # ♀
20 48	☽ ∠ ♅
21 02	☽ ∥ ☉
23 15	☽ ⊥ ♃
22 Saturday	
00 25	♀ Q ♆
01 44	☽ ☌ ☿
01 58	☽ Q ♆
02 09	☽ ☌ ♀
03 21	☿ Q ♆
03 34	☉ ⊻ ♃
04 50	☽ # ☉
10 01	☽ ∥ ♀
10 56	☿ ☌ ♀
11 22	☽ Q ♂
12 45	☽ ∥ ☿
14 54	☽ ⊥ ♅
15 01	♂ ⊻ ♃
17 35	☽ ∠ ♃
23 Sunday	
01 58	☽ ✶ ♆
02 54	☽ ⊻ ♅
02 55	☽ ☌ ♇
03 51	☽ ⊻ ♄
15 47	☽ # ♄
15 51	☽ ⊥ ♄
23 46	☽ ✶ ♃
24 Monday	
00 54	☽ □ ♂
02 49	☽ ⊻ ☉
05 38	☽ # ♇
06 11	♅ ⊻ ♇
07 33	☽ ∠ ♆
09 41	☽ ∠ ♄
15 33	☽ ⊥ ☉
20 10	☽ ⊻ ♀
22 36	☽ ⊻ ☿
23 33	☽ Q ♃
25 Tuesday	
01 04	☽ ⊥ ♆
09 03	☽ ⊥ ♀
10 26	☽ ∠ ☉
11 57	☽ ⊥ ☿
12 37	☽ ⊻ ♆
12 50	☽ # ♃
13 40	☽ ⊻ ♇
13 41	☽ ☌ ♅
14 58	☽ ✶ ♄
26 Wednesday	
01 08	☽ ⊥ ♇
01 22	☿ ⊥ ♅
04 06	☽ ∠ ♀
06 12	☿ Q ♂
07 40	☽ ∠ ☿
10 29	☽ □ ♃
13 28	☽ △ ♂
13 57	☽ Q ♄
17 16	☽ ✶ ☉
18 09	☽ ∠ ♇
27 Thursday	
05 58	☉ ∠ ♅
07 57	☿ ∠ ♃
09 22	♀ ⊥ ♅
10 13	☽ ∥ ♅
11 09	☽ ✶ ♀
15 42	☽ ✶ ☿
17 20	☽ Q ☉
18 38	☽ bQ ♂
20 52	☽ ☌ ♆
21 35	☿ # ♄
21 58	☽ ✶ ♇
22 01	☽ ⊻ ♅
23 34	☽ □ ♄
28 Friday	
08 57	☽ ⊥ ♅
11 18	☽ Q ♀
11 45	☽ ± ♂
16 30	☽ Q ☿
18 26	☽ △ ♃
19 41	☽ Q ♇
20 07	♀ Q ♂
22 59	☽ ⊼ ♂
29 Saturday	
01 07	☽ ∠ ♅
04 20	☽ □ ☉
08 20	♀ ∠ ♃
09 55	☿ ✶ ♆
17 23	☉ Q ♆
19 36	☿ ☌ ♇
20 02	☉ ⊥ ♃
20 26	☿ ⊻ ♅
21 15	☽ bQ ♃
21 28	♂ bQ ♆
22 20	☽ □ ♀
30 Sunday	
02 15	☽ ⊻ ♆
03 23	☽ □ ♇
03 29	☽ ✶ ♅
04 22	☽ □ ☿
05 13	☽ △ ♄
12 26	☿ ⊻ ♄
12 35	☽ ⊥ ♆
12 59	☽ ± ♃
21 09	☽ ∥ ♅
22 16	☿ # ♇
23 19	☽ ⊼ ♃

OCTOBER 1979

LONGITUDES

Date	Sidereal time h m s	Sun ☉ ° ' "	Moon ☽ ° ' "	Moon ☽ 24.00 ° ' "	Mercury ☿ ° '	Venus ♀ ° '	Mars ♂ ° '	Jupiter ♃ ° '	Saturn ♄ ° '	Uranus ♅ ° '	Neptune ♆ ° '	Pluto ♇ ° '
01	12 38 30	07 ♎ 42 55	07 ♒ 50 56	15 ♒ 04 44	21 ♎ 17	17 ♎ 40	03 ♌ 55	00 ♍ 24	19 ♍ 51	18 ♏ 45	18 ♐ 00	18 ♎ 41
02	12 42 27	08 41 55	22 ♒ 23 11	29 ♒ 45 46	22 52	18 54	04 30	00 36	19 58	18 48	18 01	18 43
03	12 46 24	09 40 57	07 ♓ 11 51	14 ♓ 40 36	24 25	20 09	05 05	00 47	20 06	18 51	18 02	18 45
04	12 50 20	10 40 00	22 ♓ 11 08	29 ♓ 42 24	25 58	21 24	05 40	00 59	20 13	18 54	18 03	18 48
05	12 54 17	11 39 05	07 ♈ 13 18	14 ♈ 42 44	27 29	22 39	06 15	01 10	20 20	18 57	18 04	18 50
06	12 58 13	12 38 12	22 ♈ 09 32	29 ♈ 32 42	29 ♎ 00	23 53	06 50	01 22	20 27	19 01	18 06	18 53
07	13 02 10	13 37 22	06 ♉ 51 15	14 ♉ 04 22	00 ♏ 30	25 08	07 24	01 33	20 34	19 04	18 07	18 55
08	13 06 06	14 36 33	21 ♉ 11 24	28 ♉ 11 53	02 00	26 23	07 59	01 44	20 42	19 07	18 08	18 57
09	13 10 03	15 35 47	05 ♊ 05 30	11 ♊ 52 08	03 28	27 37	08 33	01 55	20 49	19 10	18 09	19 00
10	13 13 59	16 35 03	18 ♊ 31 49	25 ♊ 04 46	04 56	28 ♎ 52	09 07	02 06	20 56	19 14	18 10	19 02
11	13 17 56	17 34 22	01 ♋ 31 17	07 ♋ 51 47	06 23	00 ♏ 07	09 41	02 17	21 03	19 17	18 12	19 04
12	13 21 53	18 33 43	14 ♋ 06 47	20 ♋ 16 51	07 49	01 22	10 15	02 28	21 10	19 20	18 13	19 07
13	13 25 49	19 33 06	26 ♋ 22 35	02 ♌ 24 38	09 14	02 36	10 49	02 39	21 17	19 23	18 14	19 09
14	13 29 46	20 32 31	08 ♌ 23 39	14 ♌ 20 16	10 38	03 51	11 23	02 50	21 24	19 27	18 16	19 12
15	13 33 42	21 31 59	20 ♌ 15 07	26 ♌ 08 50	12 01	05 06	11 56	03 01	21 30	19 30	18 17	19 14
16	13 37 39	22 31 29	02 ♍ 02 01	07 ♍ 55 11	13 24	06 21	12 30	03 11	21 37	19 34	18 19	19 16
17	13 41 35	23 31 01	13 ♍ 48 52	19 ♍ 43 32	14 45	07 35	13 03	03 22	21 44	19 37	18 20	19 19
18	13 45 32	24 30 35	25 ♍ 39 36	01 ♎ 37 26	16 05	08 50	13 36	03 32	21 51	19 41	18 22	19 21
19	13 49 28	25 30 11	07 ♎ 37 23	13 ♎ 39 41	17 25	10 05	14 09	03 42	21 58	19 44	18 23	19 23
20	13 53 25	26 29 50	19 ♎ 44 35	25 ♎ 52 13	18 43	11 19	14 42	03 53	22 04	19 48	18 25	19 26
21	13 57 22	27 29 31	02 ♏ 02 44	08 ♏ 16 13	20 00	12 34	15 15	04 03	22 11	19 51	18 26	19 28
22	14 01 18	28 29 13	14 ♏ 32 43	20 ♏ 52 16	21 15	13 49	15 48	04 13	22 17	19 55	18 28	19 31
23	14 05 15	29 ♎ 28 58	27 ♏ 14 52	03 ♐ 40 31	22 29	15 04	16 20	04 23	22 24	19 58	18 29	19 33
24	14 09 11	00 ♏ 28 44	10 ♐ 09 14	16 ♐ 41 01	23 41	16 18	16 52	04 32	22 30	20 02	18 31	19 35
25	14 13 08	01 28 32	23 ♐ 15 53	29 ♐ 53 52	24 52	17 33	17 24	04 42	22 37	20 05	18 33	19 38
26	14 17 04	02 28 22	06 ♑ 35 01	13 ♑ 19 24	26 01	18 48	17 56	04 52	22 43	20 09	18 34	19 40
27	14 21 01	03 28 14	20 ♑ 07 06	26 ♑ 58 12	27 07	20 03	18 28	05 01	22 50	20 12	18 36	19 42
28	14 24 57	04 28 07	03 ♒ 52 45	10 ♒ 50 50	28 11	21 17	19 00	05 11	22 56	20 16	18 38	19 45
29	14 28 54	05 28 02	17 ♒ 52 27	24 ♒ 57 33	29 ♏ 13	22 32	19 31	05 20	23 02	20 20	18 40	19 47
30	14 32 51	06 27 59	02 ♓ 06 27	09 ♓ 17 40	00 ♐ 12	23 47	20 03	05 29	23 08	20 23	18 41	19 49
31	14 36 47	07 ♏ 27 57	16 ♓ 32 10	23 ♓ 49 05	01 ♐ 07	25 ♏ 02	20 ♌ 34	05 ♍ 38	23 ♍ 14	20 ♏ 27	18 ♐ 43	19 ♎ 52

DECLINATIONS

Date	Moon True ☊ ° '	Moon Mean ☊ ° '	Moon ☽ Latitude ° '	Sun ☉ ° '	Moon ☽ ° '	Mercury ☿ ° '	Venus ♀ ° '	Mars ♂ ° '	Jupiter ♃ ° '	Saturn ♄ ° '	Uranus ♅ ° '	Neptune ♆ ° '	Pluto ♇ ° '
01	08 ♍ 11	06 ♍ 45	02 N 39	03 S 04	15 S 45	08 S 28	05 S 59	20 N 19	12 N 04	05 N 43	17 S 07	21 S 32	08 N 03
02	08 D 13	06 41	01 26	03 27	12 42	09 10	06 29	20 11	12 00	05 40	17 08	21 32	08 02
03	08 13	06 38	00 N 06	03 50	08 47	09 51	06 59	20 04	11 56	05 37	17 09	21 32	08 01
04	08 R 13	06 35	01 S 16	04 13	04 S 16	10 31	07 28	19 56	11 52	05 34	17 10	21 32	08 00
05	08 11	06 32	02 32	04 36	00 N 32	11 11	07 58	19 49	11 48	05 32	17 11	21 32	07 59
06	08 07	06 29	03 37	05 00	05 16	11 50	08 27	19 41	11 44	05 29	17 12	21 32	07 59
07	08 02	06 26	04 27	05 23	09 36	12 29	08 56	19 33	11 41	05 26	17 12	21 32	07 58
08	07 56	06 22	04 58	05 46	13 15	13 07	09 25	19 25	11 37	05 23	17 13	21 33	07 57
09	07 50	06 19	05 11	06 08	16 03	13 44	09 54	19 17	11 33	05 21	17 14	21 33	07 56
10	07 46	06 16	05 06	06 31	17 52	14 20	10 22	19 09	11 29	05 18	17 15	21 33	07 55
11	07 43	06 13	04 45	06 54	18 41	14 56	10 51	19 01	11 25	05 15	17 16	21 33	07 54
12	07 42	06 10	04 11	07 16	18 32	15 30	11 19	18 53	11 21	05 13	17 17	21 33	07 53
13	07 D 42	06 06	03 26	07 39	17 30	16 04	11 46	18 44	11 18	05 10	17 18	21 33	07 52
14	07 43	06 03	02 33	08 01	15 42	16 37	12 14	18 36	11 14	05 08	17 19	21 34	07 52
15	07 44	06 00	01 34	08 24	13 15	17 09	12 41	18 28	11 10	05 05	17 20	21 34	07 51
16	07 46	05 57	00 S 31	08 46	10 16	17 40	13 08	18 19	11 07	05 02	17 21	21 34	07 50
17	07 R 45	05 54	00 N 33	09 08	06 52	18 10	13 34	18 11	11 03	05 00	17 22	21 34	07 49
18	07 44	05 51	01 35	09 30	03 N 11	18 40	14 01	18 02	10 59	04 57	17 23	21 34	07 48
19	07 40	05 47	02 34	09 52	00 S 40	19 08	14 27	17 54	10 56	04 55	17 24	21 34	07 47
20	07 33	05 44	03 27	10 13	04 32	19 35	14 52	17 45	10 52	04 52	17 25	21 35	07 47
21	07 25	05 41	04 10	10 35	08 16	20 01	15 17	17 37	10 49	04 50	17 25	21 35	07 46
22	07 16	05 38	04 43	10 56	11 42	20 26	15 42	17 28	10 45	04 47	17 26	21 35	07 45
23	07 06	05 35	05 02	11 17	14 39	20 50	16 07	17 19	10 42	04 45	17 27	21 35	07 44
24	06 57	05 32	05 06	11 38	16 56	21 13	16 31	17 11	10 38	04 43	17 28	21 35	07 44
25	06 49	05 28	04 54	11 59	18 23	21 34	16 54	17 02	10 35	04 40	17 29	21 35	07 43
26	06 44	05 25	04 26	12 20	18 51	21 54	17 17	16 53	10 32	04 38	17 30	21 36	07 42
27	06 41	05 22	03 43	12 40	18 15	22 13	17 40	16 44	10 28	04 35	17 31	21 36	07 41
28	06 40	05 19	02 46	13 01	16 35	22 30	18 02	16 35	10 25	04 33	17 32	21 36	07 41
29	06 D 40	05 16	01 39	13 21	13 54	22 46	18 24	16 27	10 22	04 31	17 33	21 36	07 40
30	06 41	05 12	00 25	13 41	10 21	23 01	18 45	16 18	10 19	04 29	17 34	21 36	07 39
31	06 ♍ 41	05 ♍ 09	00 S 52	14 S 00	06 S 07	23 S 13	19 S 06	16 N 09	10 N 15	04 N 26	17 S 35	21 S 37	07 N 39

ZODIAC SIGN ENTRIES

Date	h m	Planets
03	00 23	☽ ♓
05	00 28	☽ ♈
07	00 45	☽ ♉
07	03 55	☿ ♏
09	03 07	☽ ♊
11	09 09	☽ ♋
11	09 48	♀ ♏
13	19 12	☽ ♌
16	07 51	☽ ♍
18	20 44	☽ ♎
21	08 02	☽ ♏
23	17 09	☽ ♐
24	00 28	☉ ♏
26	00 11	☽ ♑
28	05 16	☽ ♒
30	07 06	☿ ♐
30	08 29	☽ ♓

LATITUDES

Date	Mercury ☿ ° '	Venus ♀ ° '	Mars ♂ ° '	Jupiter ♃ ° '	Saturn ♄ ° '	Uranus ♅ ° '	Neptune ♆ ° '	Pluto ♇ ° '
01	00 S 10	01 N 02	01 N 04	00 N 47	01 N 50	00 N 18	01 N 23	16 N 36
04	00 32	00 57	01 07	00 48	01 51	00 18	01 23	16 36
07	00 53	00 51	01 10	00 48	01 51	00 17	01 22	16 36
10	01 14	00 45	01 13	00 49	01 51	00 17	01 22	16 36
13	01 35	00 39	01 16	00 49	01 52	00 17	01 22	16 36
16	01 54	00 32	01 19	00 50	01 52	00 17	01 22	16 36
19	02 12	00 25	01 22	00 50	01 53	00 17	01 22	16 36
22	02 27	00 18	01 26	00 51	01 53	00 17	01 22	16 36
25	02 40	00 11	01 29	00 51	01 54	00 17	01 22	16 37
28	02 49	00 N 03	01 32	00 52	01 54	00 17	01 21	16 37
31	02 S 54	00 S 05	01 N 36	00 N 52	01 N 55	00 N 17	01 N 21	16 N 37

DATA

Julian Date	2444148
Delta T	+50 seconds
Ayanamsa	23° 34' 20"
Synetic vernal point	05° ♓ 32' 40"
True obliquity of ecliptic	23° 26' 23"

MOON'S PHASES, APSIDES AND POSITIONS ☽

Date	h m	Phase	Longitude ° '	Eclipse Indicator
05	19 35	○	11 ♈ 58	
12	21 24	☾	18 ♋ 57	
21	02 23	●	27 ♎ 06	
28	13 06	☽	04 ♒ 31	

Day	h m		
04	15 09	Perigee	
16	20 35	Apogee	
05	09 19	0N	
11	20 09	Max dec	18° N 44'
19	07 54	0S	
26	10 49	Max dec	18° S 51'

LONGITUDES

Date	Chiron ⚷ ° '	Ceres ⚳ ° '	Pallas ⚴ ° '	Juno ⚵ ° '	Vesta ⚶ ° '	Black Moon Lilith ⚸ ° '
01	13 ♉ 03	13 ♈ 29	15 ♒ 00	14 ♋ 23	17 ♉ 30	09 ♍ 17
11	12 ♉ 39	11 ♈ 14	14 ♒ 33	18 ♋ 19	15 ♉ 57	10 ♍ 24
21	12 ♉ 12	09 ♈ 05	14 ♒ 41	21 ♋ 48	13 ♉ 48	11 ♍ 30
31	11 ♉ 43	07 ♈ 18	15 ♒ 21	24 ♋ 43	11 ♉ 17	12 ♍ 37

ASPECTARIAN

h m Aspects

01 Monday
00 01 ♀ # ♄
00 02 ☽ Q ♅
03 52 ☽ ∠ ♆
05 08 ☽ ☍ ♂
06 57 ☽ bQ ♄
11 46 ☽ △ ☉
18 35 ♀ ✱ ♆
22 03 ☽ ± ♄

02 Tuesday
04 50 ☽ ✱ ♆
05 46 ☽ △ ♀
05 58 ☽ △ ♇
06 06 ☽ □ ♅
08 01 ☽ ⊼ ♄
08 13 ♀ ☌ ♇
09 51 ♀ ⊻ ♅
12 52 ☽ △ ☿
14 18 ☽ bQ ☉
16 43 ☽ # ♃

03 Wednesday
00 26 ☽ Q ♆
01 32 ☽ ☍ ♃
05 56 ☽ ± ☉
06 27 ☽ bQ ♇
06 45 ☽ ∥ ☿
08 25 ☽ bQ ♀
08 28 ☽ ⊼ ♂
10 44 ♀ ⊻ ♄
12 21 ♂ ∠ ♄
15 59 ☽ bQ ☿
16 14 ☽ # ♇
16 17 ☽ ⊼ ☉
18 30 ☽ ± ♂
20 53 ☽ ∥ ♀
20 57 ☽ ± ♇

04 Thursday
00 10 ☽ ± ♀
05 12 ☽ # ♄
05 24 ☽ □ ♆
06 34 ☽ ⊼ ♇
06 45 ☽ △ ♅
08 02 ☽ ± ☿
08 50 ☽ ☍ ♄
09 29 ☽ bQ ♂
10 38 ☽ ⊼ ♀
12 11 ☽ ∥ ☉
16 20 ☿ ⊥ ♄
18 42 ☽ ⊼ ☿

05 Friday
02 13 ☽ ⊼ ♃
06 46 ☽ bQ ♅
10 23 ☽ △ ♂
11 55 ☽ ± ♃
13 20 ♀ # ♇
19 35 ☽ ☍ ☉
21 13 ☽ ± ♅

06 Saturday
02 32 ☽ bQ ♃
05 26 ☽ △ ♆
06 41 ☽ ☍ ♇
06 54 ☽ ⊼ ♅
08 35 ☿ # ♃
09 13 ☽ ⊼ ♄
10 26 ☽ # ☉
13 07 ☽ ∥ ♄
14 09 ♂ Q ♇
15 03 ☽ ☍ ♀
19 01 ☽ ± ♄
21 35 ☉ ⊥ ♅

07 Sunday
00 23 ☽ ☍ ☿
02 36 ☽ ∥ ♇
03 10 ☽ △ ♃
05 50 ☽ bQ ♆
07 37 ☽ # ♀
09 52 ☽ bQ ♄
12 57 ☽ □ ♂
15 22 ☉ # ♄
20 44 ☽ ± ♆

08 Monday
00 05 ☽ ⊼ ☉
00 46 ☽ ∥ ♃
06 49 ☽ ⊼ ♆
07 15 ☿ ✱ ♃
08 12 ☽ ⊼ ♇
08 28 ☽ ☍ ♅
10 43 ☽ # ☿
10 56 ☽ ± ☉
11 09 ☽ △ ♄
18 26 ☽ ± ♇
18 40 ♀ ⊥ ♄
20 32 ☽ Q ♂
21 44 ☽ ⊼ ♀

09 Tuesday
03 33 ☽ bQ ☉
06 23 ☽ □ ♃
06 47 ☿ ∠ ♆
08 49 ☽ ⊼ ☿
09 10 ☽ ± ♀
10 04 ☽ bQ ♇
18 22 ☽ ✱ ♂
20 40 ☽ ± ☿

10 Wednesday
02 15 ☽ # ♅
02 43 ☽ bQ ♀
08 11 ☽ △ ☉
11 21 ☽ ☍ ♆
12 55 ☽ △ ♇
13 16 ☽ ⊼ ♅
14 52 ☽ bQ ☿
14 55 ☽ Q ♃
16 24 ☽ □ ♄
22 42 ☽ ∠ ♂

11 Thursday
00 19 ☽ ± ♅
03 34 ☉ ∠ ♃
06 00 ☿ ∠ ♄
09 05 ☽ △ ♀
13 28 ☽ ✱ ♃
16 16 ☽ ⊥ ♂
17 13 ☽ bQ ♅
22 21 ☽ △ ☿

12 Friday
02 23 ☽ Q ♄
03 28 ☉ ✱ ♆
04 13 ☽ ⊻ ♂
14 09 ♀ # ♃
18 37 ☽ ∠ ♃
19 59 ☽ ⊼ ♆
21 24 ☽ □ ☉
21 45 ☽ □ ♇
22 12 ☽ △ ♅

13 Saturday
01 51 ☽ ✱ ♄
01 56 ☉ ☌ ♇
07 46 ☽ ± ♆
07 53 ☉ ⊻ ♅
12 33 ☽ ⊥ ♃
13 03 ♀ ✱ ♃
15 18 ☽ # ♅

14 Sunday
00 29 ♀ ∠ ♆
00 40 ☽ ⊻ ♃
01 41 ☽ bQ ♆
01 45 ☉ # ♇
01 49 ☽ □ ♀
03 30 ☽ # ☿
07 56 ☽ ∠ ♄
09 35 ☽ Q ♇
12 19 ☽ Q ☉
17 07 ☽ □ ☿
18 19 ☽ ☌ ♂

15 Monday
02 16 ☽ ⊥ ♄
08 00 ☽ △ ♆
09 38 ☿ □ ♂
09 55 ☽ ✱ ♇
10 28 ☽ □ ♅
11 17 ☉ ⊻ ♄
14 35 ☽ ⊻ ♄
14 51 ☽ ✱ ☉
16 14 ☽ # ♀
16 42 ☿ ⊥ ♆
18 28 ☽ Q ♀
20 26 ☿ ∥ ♅

16 Tuesday
05 28 ☽ ∥ ♃
10 32 ☽ Q ☿
14 23 ☽ ☌ ♃
16 35 ☽ ∠ ♇
17 55 ♀ ∠ ♄
21 49 ☽ ✱ ♀
21 51 ☽ # ☉
23 19 ☽ Q ♅

17 Wednesday
00 13 ☽ ∠ ☉
05 32 ☽ ∥ ♇
10 23 ☽ ⊻ ♂
10 59 ☽ ⊥ ♇
12 18 ☿ # ♂
14 09 ☽ ✱ ☿
20 13 ☽ ⊥ ☉
21 12 ☽ □ ♆
23 10 ☽ ⊥ ♂
23 12 ☽ ⊻ ♇
23 50 ☽ ✱ ♅

18 Thursday
00 29 ☿ Q ♃
00 31 ☽ ∥ ♄
04 13 ☽ ☌ ♄
07 53 ☽ ∠ ♀
09 28 ☽ ⊻ ☉
18 13 ☽ ∠ ♂

19 Friday
00 17 ☽ ∠ ☿
04 03 ☽ ⊻ ♃
04 06 ☽ ⊥ ♀
06 12 ☽ ∠ ♅
09 32 ☽ Q ♆
16 12 ☽ ⊥ ♃
17 27 ☽ ⊻ ♀
20 27 ☽ ⊥ ☿

20 Saturday
00 12 ☽ ⊥ ♅
01 36 ☽ ✱ ♂
06 18 ☿ ⊻ ♆
09 22 ☽ ✱ ♆
09 44 ☽ ⊻ ☿
10 16 ☽ ∠ ♃
11 23 ☽ ☌ ♇
12 06 ☽ ⊻ ♅
14 08 ☽ # ♄
16 37 ☽ ⊻ ♄
23 07 ☉ Q ♂

21 Sunday
01 52 ☿ ⊻ ♇
02 16 ☽ Q ♂
02 23 ☽ ☌ ☉
04 26 ☽ ⊥ ♄
08 42 ☽ # ♇
09 12 ☿ ☌ ♅
09 23 ♀ ⊥ ♆
14 42 ☽ ∠ ♆
15 55 ☽ ✱ ♃
22 00 ☽ ∠ ♄

22 Monday
01 22 ☉ # ♃
05 14 ☽ # ♃
05 46 ☽ ∥ ☉
06 41 ☉ ⊥ ♄
08 01 ☽ ⊥ ♆
10 28 ☽ ☌ ♀
14 29 ☽ □ ♂
15 13 ☽ Q ♃
15 51 ♂ # ♅
19 28 ☽ ⊻ ♆
21 27 ☽ ⊻ ♇
22 14 ☽ ☌ ♅

23 Tuesday
02 05 ☽ ☌ ☿
02 48 ☽ ✱ ♄
08 48 ☽ ⊥ ♇
10 14 ☿ ✱ ♄
15 39 ♂ ⊥ ♄
16 32 ☽ ⊻ ☉

24 Wednesday
01 28 ☽ Q ♄
01 29 ☽ □ ♃
01 40 ☽ ∠ ♇
04 38 ☽ ⊥ ☉
05 40 ☽ ∥ ♀
14 57 ☽ # ♂
17 09 ♀ Q ♃
19 27 ☽ ∥ ♅
21 21 ♃ ∠ ♇
22 36 ☽ ∠ ☉

25 Thursday
00 30 ☽ ⊻ ♀
00 52 ☽ △ ♂
03 23 ☽ ☌ ♆
05 22 ☽ ✱ ♇
06 12 ☽ ⊻ ♅
07 04 ♀ □ ♂
10 49 ☽ □ ♄
12 35 ☽ ⊥ ♀
13 32 ☿ ∥ ♆
15 11 ☽ ⊻ ☿
17 08 ☽ ⊥ ♅
17 47 ♀ # ♂

26 Friday
03 02 ☽ ⊥ ☿
03 10 ☽ Q ♇
04 02 ☽ ✱ ☉
04 30 ☿ ⊥ ♇
05 12 ☽ bQ ♂
06 30 ☽ ∠ ♀
07 33 ♀ ⊻ ♆
08 53 ☽ △ ♃
09 25 ☽ ∠ ♅
20 36 ☽ ∠ ☿
21 56 ☽ ± ♂

27 Saturday
02 25 ♀ ∥ ♅
03 09 ☽ Q ☉
05 15 ♀ ⊻ ♇
08 59 ☽ ⊼ ♂
09 20 ☽ ⊻ ♆
11 16 ☽ □ ♇
11 50 ☽ bQ ♃
11 52 ☽ ✱ ♀
12 09 ☽ ✱ ♅
15 16 ☉ ∠ ♆
15 17 ♀ ☌ ♅
16 48 ☽ △ ♄
18 20 ♂ △ ♆
19 53 ☽ ⊥ ♆
20 22 ☽ ∥ ♀

28 Sunday
00 21 ☽ ∥ ♅
01 18 ☽ ✱ ☿
03 45 ☽ ± ♃
09 12 ☽ Q ♅
10 53 ☽ Q ♀
11 34 ☽ ∠ ♆
11 57 ☽ # ♂
13 06 ☽ □ ☉
14 16 ☽ ⊼ ♃
19 03 ☽ bQ ♄
23 45 ☽ Q ☿

29 Monday
08 13 ☉ ✱ ♃
10 34 ☽ ± ♄
13 20 ☽ ✱ ♆
14 55 ☽ ☍ ♂
15 15 ☽ △ ♇
15 46 ☽ ∥ ☉
16 11 ☽ □ ♅
20 40 ☽ □ ♀
20 49 ☽ ⊼ ♄
22 29 ♀ ✱ ♄

30 Tuesday
00 56 ♂ ✱ ♇
08 35 ☽ □ ☿
09 38 ☽ Q ♆
12 13 ☽ # ♃
16 34 ☽ bQ ♇
17 43 ☽ ☍ ♃
19 50 ☽ △ ☉

31 Wednesday
03 41 ☽ # ♇
06 03 ♂ □ ♅
07 34 ☽ ± ♇
15 37 ☽ □ ♆
17 30 ☽ ⊼ ♇
18 29 ☽ △ ♅
18 53 ☽ ⊼ ♂
20 57 ☽ # ♄
22 30 ☽ bQ ☉
23 08 ☽ ☍ ♄

All ephemeris data is given at 12.00 UT and the Moon's longitude is additionally given for 24.00 UT

NOVEMBER 1979

LONGITUDES

Date	Sidereal time h m s	Sun ☉ ° ' "	Moon ☽ ° ' "	Moon ☽ 24.00 ° ' "	Mercury ☿ ° '	Venus ♀ ° '	Mars ♂ ° '	Jupiter ♃ ° '	Saturn ♄ ° '	Uranus ♅ ° '	Neptune ♆ ° '	Pluto ♇ ° '
01	14 40 44	08 ♏ 27 56	01 ♈ 07 53	08 ♈ 27 52	01 ♐ 59	26 ♏ 16	21 ♌ 05	05 ♍ 47	23 ♍ 21	20 ♏ 31	18 ♐ 45	19 ♎ 54
02	14 44 40	09 27 57	15 ♈ 48 17	23 ♈ 08 16	02 47	27 31	21 36	05 56	23 27	20 34	18 47	19 56
03	14 48 37	10 28 01	00 ♉ 26 54	07 ♉ 43 16	03 31	28 ♏ 46	22 06	06 05	23 32	20 38	18 49	19 58
04	14 52 33	11 28 05	14 ♉ 56 27	22 ♉ 05 36	04 10	00 ♐ 01	22 37	06 13	23 38	20 42	18 51	20 01
05	14 56 30	12 28 12	29 ♉ 09 58	06 ♊ 08 56	04 43	01 15	23 07	06 22	23 44	20 45	18 52	20 03
06	15 00 26	13 28 21	13 ♊ 01 59	19 ♊ 48 50	05 10	02 30	23 37	06 30	23 50	20 49	18 54	20 05
07	15 04 23	14 28 32	26 ♊ 29 18	03 ♋ 03 23	05 31	03 45	24 07	06 39	23 56	20 53	18 56	20 07
08	15 08 20	15 28 44	09 ♋ 31 13	15 ♋ 53 06	05 44	04 59	24 37	06 47	24 01	20 57	18 58	20 10
09	15 12 16	16 28 59	22 ♋ 09 23	28 ♋ 20 35	05 49	06 14	25 06	06 55	24 07	21 00	19 00	20 12
10	15 16 13	17 29 16	04 ♌ 27 15	10 ♌ 30 00	05 R 45	07 29	25 35	07 03	24 12	21 04	19 02	20 14
11	15 20 09	18 29 34	16 ♌ 29 30	22 ♌ 26 25	05 32	08 43	26 04	07 10	24 18	21 08	19 04	20 16
12	15 24 06	19 29 55	28 ♌ 21 29	04 ♍ 15 23	05 09	09 58	26 33	07 18	24 23	21 12	19 06	20 18
13	15 28 02	20 30 17	10 ♍ 08 48	16 ♍ 02 25	04 36	11 13	27 02	07 25	24 28	21 15	19 08	20 21
14	15 31 59	21 30 41	21 ♍ 56 53	27 ♍ 52 48	03 53	12 28	27 30	07 33	24 34	21 19	19 10	20 23
15	15 35 55	22 31 08	03 ♎ 50 44	09 ♎ 51 10	02 59	13 42	27 59	07 40	24 39	21 23	19 12	20 25
16	15 39 52	23 31 36	15 ♎ 54 32	22 ♎ 01 14	01 57	14 57	28 27	07 47	24 44	21 26	19 14	20 27
17	15 43 49	24 32 05	28 ♎ 11 33	04 ♏ 25 41	00 ♐ 47	16 12	28 54	07 54	24 49	21 30	19 17	20 29
18	15 47 45	25 32 37	10 ♏ 43 47	17 ♏ 05 54	29 ♏ 31	17 26	29 22	08 01	24 54	21 34	19 19	20 31
19	15 51 42	26 33 10	23 ♏ 31 59	00 ♐ 01 57	28 11	18 41	29 ♌ 49	08 07	24 59	21 38	19 21	20 33
20	15 55 38	27 33 45	06 ♐ 35 38	13 ♐ 12 48	26 50	19 56	00 ♍ 16	08 14	25 03	21 41	19 23	20 35
21	15 59 35	28 34 21	19 ♐ 53 11	26 ♐ 36 32	25 30	21 10	00 43	08 20	25 08	21 45	19 25	20 37
22	16 03 31	29 ♏ 34 59	03 ♑ 22 33	10 ♑ 10 57	24 14	22 25	01 09	08 26	25 13	21 49	19 27	20 39
23	16 07 28	00 ♐ 35 38	17 ♑ 01 29	23 ♑ 53 56	23 04	23 40	01 36	08 32	25 17	21 53	19 29	20 41
24	16 11 24	01 36 18	00 ♒ 48 06	07 ♒ 43 52	22 03	24 54	02 02	08 38	25 22	21 56	19 32	20 43
25	16 15 21	02 36 59	14 ♒ 41 08	21 ♒ 39 49	21 11	26 09	02 27	08 44	25 26	22 00	19 34	20 45
26	16 19 18	03 37 41	28 ♒ 39 52	05 ♓ 41 15	20 31	27 24	02 53	08 50	25 30	22 04	19 36	20 47
27	16 23 14	04 38 24	12 ♓ 43 56	19 ♓ 47 49	20 02	28 38	03 18	08 55	25 34	22 07	19 38	20 49
28	16 27 11	05 39 09	26 ♓ 52 48	03 ♈ 58 42	19 45	29 ♐ 53	03 43	09 00	25 38	22 11	19 40	20 50
29	16 31 07	06 39 54	11 ♈ 05 16	18 ♈ 12 11	19 40	01 ♑ 07	04 07	09 05	25 42	22 15	19 43	20 52
30	16 35 04	07 ♐ 40 40	25 ♈ 19 02	02 ♉ 25 20	19 ♏ 45	02 ♑ 22	04 ♍ 32	09 ♍ 10	25 ♍ 46	22 ♏ 18	19 ♐ 45	20 ♎ 54

Date	Moon True ☊ ° '	Moon Mean ☊ ° '	Moon ☽ Latitude ° '
01	06 ♍ 39	05 ♍ 06	02 S 07
02	06 R 34	05 03	03 13
03	06 26	05 00	04 06
04	06 16	04 57	04 43
05	06 05	04 53	05 01
06	05 55	04 50	05 01
07	05 45	04 47	04 44
08	05 37	04 44	04 12
09	05 32	04 41	03 29
10	05 30	04 38	02 37
11	05 29	04 34	01 39
12	05 D 29	04 31	00 S 38
13	05 R 29	04 28	00 N 25
14	05 28	04 25	01 26
15	05 24	04 22	02 25
16	05 18	04 18	03 17
17	05 09	04 15	04 01
18	04 58	04 12	04 34
19	04 45	04 09	04 55
20	04 31	04 06	05 00
21	04 18	04 03	04 49
22	04 07	03 59	04 23
23	03 59	03 56	03 41
24	03 54	03 53	02 45
25	03 51	03 50	01 40
26	03 D 51	03 47	00 N 27
27	03 R 51	03 44	00 S 47
28	03 50	03 40	01 59
29	03 47	03 37	03 03
30	03 ♍ 41	03 ♍ 34	03 S 56

DECLINATIONS

Date	Sun ☉ ° '	Moon ☽ ° '	Mercury ☿ ° '	Venus ♀ ° '	Mars ♂ ° '	Jupiter ♃ ° '	Saturn ♄ ° '	Uranus ♅ ° '	Neptune ♆ ° '	Pluto ♇ ° '
01	14 S 20	01 S 29	23 S 24	19 S 26	16 N 00	10 N 12	04 N 24	17 S 36	21 S 37	07 N 38
02	14 39	03 N 15	23 34	19 46	15 51	10 09	04 22	17 37	21 37	07 37
03	14 58	07 47	23 41	20 05	15 42	10 06	04 20	17 38	21 37	07 37
04	15 16	11 49	23 46	20 24	15 34	10 03	04 17	17 39	21 37	07 36
05	15 35	15 04	23 49	20 42	15 25	10 00	04 15	17 40	21 38	07 35
06	15 53	17 23	23 50	20 59	15 16	09 57	04 13	17 41	21 38	07 35
07	16 11	18 40	23 49	21 16	15 07	09 54	04 11	17 42	21 38	07 34
08	16 29	18 54	23 44	21 33	14 58	09 52	04 09	17 43	21 38	07 34
09	16 46	18 11	23 37	21 48	14 50	09 49	04 07	17 44	21 38	07 33
10	17 03	16 36	23 27	22 03	14 41	09 46	04 05	17 45	21 38	07 32
11	17 20	14 19	23 14	22 18	14 32	09 43	04 03	17 46	21 39	07 32
12	17 36	11 27	22 58	22 32	14 23	09 41	04 01	17 47	21 39	07 31
13	17 53	08 09	22 38	22 45	14 15	09 38	03 59	17 48	21 39	07 31
14	18 08	04 31	22 14	22 57	14 06	09 36	03 57	17 49	21 39	07 30
15	18 24	00 N 41	21 47	23 09	13 58	09 33	03 56	17 50	21 39	07 30
16	18 39	03 S 14	21 16	23 20	13 49	09 31	03 54	17 51	21 40	07 29
17	18 54	07 05	20 43	23 31	13 41	09 28	03 52	17 52	21 40	07 29
18	19 09	10 42	20 07	23 41	13 32	09 26	03 50	17 53	21 40	07 28
19	19 23	13 54	19 29	23 50	13 24	09 24	03 49	17 54	21 40	07 28
20	19 37	16 29	18 51	23 58	13 16	09 22	03 47	17 55	21 40	07 28
21	19 51	18 15	18 13	24 06	13 07	09 20	03 45	17 56	21 41	07 27
22	20 04	19 01	17 37	24 13	12 59	09 17	03 44	17 57	21 41	07 27
23	20 17	18 43	17 03	24 19	12 51	09 15	03 42	17 58	21 41	07 26
24	20 29	17 17	16 33	24 24	12 43	09 13	03 40	17 59	21 41	07 26
25	20 41	14 50	16 07	24 29	12 35	09 12	03 39	18 00	21 41	07 26
26	20 53	11 31	15 46	24 33	12 27	09 10	03 38	18 01	21 42	07 25
27	21 04	07 30	15 30	24 36	12 19	09 08	03 36	18 02	21 42	07 25
28	21 15	03 S 03	15 19	24 39	12 12	09 06	03 35	18 03	21 42	07 25
29	21 25	01 N 35	15 13	24 40	12 04	09 04	03 33	18 04	21 42	07 24
30	21 S 35	06 N 07	15 S 11	24 S 41	11 N 56	09 N 03	03 N 32	18 S 04	21 S 42	07 N 24

ZODIAC SIGN ENTRIES

Date	h m	Planets
01	10 09	☽ ♈
03	11 16	☽ ♉
04	11 50	♀ ♐
05	13 25	☽ ♊
07	18 24	☽ ♋
10	03 14	☽ ♌
12	15 20	☽ ♍
15	04 16	☽ ♎
17	15 29	☽ ♏
18	03 09	☿ ♏
19	21 36	♂ ♍
19	23 56	☽ ♐
22	06 01	☽ ♑
22	21 54	☉ ♐
24	10 37	☽ ♒
26	14 17	☽ ♓
28	14 20	♀ ♑
28	17 17	☽ ♈
30	19 54	☽ ♉

LATITUDES

Date	Mercury ☿ ° '	Venus ♀ ° '	Mars ♂ ° '	Jupiter ♃ ° '	Saturn ♄ ° '	Uranus ♅ ° '	Neptune ♆ ° '	Pluto ♇ ° '
01	02 S 54	00 S 07	01 N 37	00 N 53	01 N 55	00 N 17	01 N 21	16 N 38
04	02 51	00 15	01 40	00 53	01 55	00 17	01 21	16 38
07	02 38	00 23	01 44	00 54	01 56	00 17	01 21	16 39
10	02 14	00 30	01 48	00 55	01 57	00 17	01 21	16 40
13	01 36	00 38	01 52	00 55	01 57	00 17	01 21	16 40
16	00 S 44	00 45	01 55	00 56	01 58	00 17	01 21	16 41
19	00 N 17	00 52	01 59	00 57	01 59	00 17	01 21	16 42
22	01 15	00 59	02 03	00 57	01 59	00 17	01 21	16 43
25	02 00	01 06	02 08	00 58	02 00	00 17	01 21	16 44
28	02 27	01 12	02 12	00 59	02 01	00 17	01 21	16 45
31	02 N 36	01 S 18	02 N 16	01 N 00	02 N 01	00 N 17	01 N 21	16 N 47

DATA

Julian Date	2444179
Delta T	+50 seconds
Ayanamsa	23° 34' 22"
Synetic vernal point	05° ♓ 32' 37"
True obliquity of ecliptic	23° 26' 23"

MOON'S PHASES, APSIDES AND POSITIONS ☽

Date	h m	Phase	Longitude ° '	Eclipse Indicator
04	05 47	○	11 ♉ 13	
11	16 24	☾	18 ♌ 41	
19	18 04	●	26 ♏ 48	
26	21 09	☽	04 ♓ 01	

Day	h m		
01	19 37	Perigee	
13	14 06	Apogee	
28	23 52	Perigee	
01	19 31	0N	
08	05 39	Max dec	18° N 56'
15	16 12	0S	
22	17 17	Max dec	19° S 03'
29	03 51	0N	

LONGITUDES

Date	Chiron ⚷ ° '	Ceres ⚳ ° '	Pallas ⚴ ° '	Juno ⚵ ° '	Vesta ⚶ ° '	Black Moon Lilith ⚸ ° '
01	11 ♉ 40	07 ♈ 09	15 ♒ 26	24 ♋ 58	11 ♉ 01	12 ♍ 43
11	11 ♉ 10	05 ♈ 57	16 ♒ 37	27 ♋ 10	08 ♉ 26	13 ♍ 50
21	10 ♉ 41	05 ♈ 22	18 ♒ 12	28 ♋ 34	06 ♉ 08	14 ♍ 56
31	10 ♉ 15	05 ♈ 26	20 ♒ 08	29 ♋ 04	04 ♉ 23	16 ♍ 03

ASPECTARIAN

h m	Aspects	h m	Aspects	h m	Aspects
01 Thursday		09 49	☽ ∥ ♂	21 26	☽ □ ♄
03 17	☽ △ ♀	15 40	☽ ⊥ ♄	22 53	☿ ∥ ♅
04 33	♀ ⊥ ♇	16 24	☽ □ ☉	23 47	♀ ⊻ ♅
05 07	☽ ± ♂	17 12	☽ △ ♆	**22 Thursday**	
08 42	☉ ∠ ♄	19 38	☽ ✶ ♇	02 06	☽ ⊥ ♅
13 29	☽ △ ☿	21 24	☽ □ ♅	04 44	☽ ⊻ ☉
14 21	☽ ± ☉	**12 Monday**		06 54	☽ ⊥ ☿
19 12	☽ Ⴒ ♅	02 16	☉ ⊻ ♆	07 57	☽ △ ♂
19 42	☽ ⊼ ♃	03 52	☿ ∠ ♇	10 43	☽ Q ♇
20 24	☽ Ⴒ ♂	03 53	☽ ⊻ ♄	16 13	☽ ⊥ ☉
02 Friday		06 35	☉ Q ♃	18 06	☽ ∠ ♅
00 53	☽ ⊼ ☉	08 11	☽ ☌ ♂	21 00	☽ △ ♃
05 37	☽ ± ♃	**13 Tuesday**		21 29	☽ ∠ ☿
06 08	☽ Ⴒ ♀	01 14	☽ □ ☿	**23 Friday**	
09 59	☽ ± ♅	01 23	☽ ∥ ♃	05 57	☿ ⊻ ♀
15 25	☽ Ⴒ ☿	02 11	☽ ∠ ♇	09 18	☽ ∠ ☉
16 53	☽ △ ♆	04 50	☉ ∥ ♅	11 13	☽ Ⴒ ♂
17 41	☽ ∥ ♄	06 24	☽ ☌ ♃	16 19	☽ ⊻ ♆
18 47	☽ ☍ ♇	07 00	☿ ∥ ♀	18 24	☽ □ ♇
19 50	☽ ⊼ ♅	07 59	☉ ⊻ ♇	20 31	☽ ✶ ♅
20 29	☽ Ⴒ ♃	08 21	☽ Q ☉	21 47	☽ ✶ ☿
21 49	☽ △ ♂	10 10	☽ Q ♅	23 27	☽ Ⴒ ♃
22 13	☽ ± ♀	14 26	☽ □ ♀	**24 Saturday**	
03 Saturday		16 20	☽ ∥ ♇	00 44	☽ ⊻ ♀
00 35	☽ ⊼ ♄	20 34	☽ ⊥ ♇	02 30	☽ △ ♄
06 57	☽ ± ☿	**14 Wednesday**		02 38	☽ ∥ ♅
08 58	☽ ⊼ ♀	06 21	☽ □ ♆	02 48	☽ ⊥ ♆
10 30	☽ ± ♄	07 02	☉ ☌ ♅	03 26	☽ ± ♂
11 02	☽ ∥ ♇	08 48	☽ ⊻ ♇	12 12	☽ ⊥ ♀
17 18	☽ ⊼ ☿	10 43	☽ ✶ ♅	13 30	☽ ✶ ☉
17 33	☽ Ⴒ ♆	11 02	☽ ✶ ☉	14 12	☽ ⊼ ♂
21 23	☽ △ ♃	11 52	☽ Q ☿	14 35	☿ ☌ ♅
04 Sunday		15 35	☽ ∥ ♄	15 12	☽ ± ♃
01 11	☽ ∥ ♃	17 20	☽ ☌ ♄	17 16	☽ Q ☿
01 27	☽ Ⴒ ♄	23 43	☽ ⊻ ♂	17 28	☽ Q ♅
05 47	☽ ☍ ☉	**15 Thursday**		18 28	☽ ∠ ♆
08 30	☽ ± ♆	07 12	☽ Q ♀	21 20	♀ □ ♄
18 33	☽ ⊼ ♆	10 25	☽ ✶ ☿	22 20	☽ ∥ ☿
20 31	☽ ⊼ ♇	12 16	☽ ⊥ ♂	**25 Sunday**	
21 41	☽ ☍ ♅	17 06	☽ ∠ ♅	01 40	☽ ⊼ ♃
05 Monday		17 59	☿ ∥ ♆	04 38	☽ Ⴒ ♄
01 21	☽ □ ♂	18 45	☽ Q ♆	05 18	☽ ∠ ♀
02 43	☽ △ ♄	19 43	☽ ⊻ ♃	05 26	☉ □ ♂
03 09	☉ ⋕ ♂	20 01	☽ ∠ ☉	11 52	☽ Q ☉
06 41	☽ ± ♇	**16 Friday**		20 13	☽ ± ♄
14 49	☽ ∥ ♂	06 56	☽ ∠ ♂	20 25	☽ ✶ ♆
15 55	☽ ☍ ♀	07 46	☽ ⊥ ♃	22 27	☽ △ ♇
17 04	☽ ⋕ ☉	09 53	☽ ✶ ♀	22 38	☽ □ ☿
21 52	☽ ☍ ☿	11 04	☽ ⊥ ♅	**26 Monday**	
22 01	☉ ⊥ ♆	13 53	☽ ∠ ☿	00 38	☽ □ ♅
22 08	☽ Ⴒ ♇	15 28	☽ ⊥ ☉	00 45	☿ ⊥ ♀
06 Tuesday		16 04	☽ ⋕ ♄	01 37	☿ Q ♃
00 30	☽ □ ♃	18 35	☽ ✶ ♆	02 20	☿ ⊻ ♇
06 43	☿ ∠ ♇	20 57	☽ ☌ ♇	05 31	☽ ⋕ ♂
09 25	☽ Q ♂	22 55	☽ ⊻ ♅	06 34	☽ ⊼ ♄
12 50	☽ ⊼ ☉	**17 Saturday**		09 37	☽ ✶ ♀
16 13	☽ ⋕ ♅	01 37	☽ ∠ ♃	17 02	☽ Q ♆
22 25	☽ ☍ ♆	04 16	☽ ⊻ ☉	19 26	☽ ☍ ♂
07 Wednesday		05 24	☽ ⊻ ♄	21 09	☽ □ ☉
00 19	☽ ± ☉	05 59	☽ ⊥ ☿	**27 Tuesday**	
00 31	☽ △ ♇	13 26	☽ ✶ ♂	00 11	☽ Ⴒ ♇
00 51	♂ ⊻ ♄	14 35	☽ ⋕ ♇	01 33	♀ ⊥ ♅
01 52	☽ ⊼ ♅	16 34	☽ ⊻ ☿	02 34	☽ ⋕ ♃
07 20	☽ □ ♄	17 06	☽ ⊥ ♄	05 28	☽ ☍ ♃
07 33	☽ ✶ ♂	18 26	☽ ∠ ♀	08 05	☽ Q ♀
08 38	☽ Q ♃	19 14	☉ ✶ ♄	12 29	☽ ⋕ ♇
12 43	☽ ± ♅	23 44	☽ ∠ ♆	15 32	☽ ± ♇
17 53	☽ Ⴒ ☉	**18 Sunday**		23 45	☽ □ ♆
08 Thursday		03 30	☽ ⋕ ♃	**28 Wednesday**	
02 40	☽ ⊼ ♀	06 48	☽ ✶ ♃	00 08	☽ △ ☿
04 50	☽ ⊼ ☿	10 25	☽ ∠ ♄	01 45	☽ ⊼ ♇
05 18	☽ Ⴒ ♅	13 15	☽ Q ♂	04 01	☽ △ ♅
06 50	☽ ✶ ♃	13 29	☽ ⊥ ♀	09 14	☽ ⋕ ♄
12 10	☽ ∠ ♂	14 09	☿ □ ♂	09 54	☽ ☍ ♄
15 03	☽ ± ♀	16 54	☽ ⊥ ♆	16 36	☉ ∠ ♇
15 25	♀ ∠ ♇	**19 Monday**		17 34	☽ □ ♀
16 11	☽ ± ☿	02 00	☽ ⊻ ♀	22 58	☿ ⊻ ♆
16 44	☽ Q ♄	04 11	☽ ⊻ ♆	23 54	☽ ⊼ ♂
20 29	♀ ∥ ♆	05 36	☽ Q ♃	**29 Thursday**	
09 Friday		06 26	☽ ⊻ ♇	01 11	☽ Ⴒ ☿
00 12	☽ △ ☉	08 09	☽ ⋕ ♂	03 58	☽ △ ☉
03 35	☿ ☌ ♀	08 27	☽ ☌ ♅	05 29	☽ Ⴒ ♅
05 54	☽ ⊥ ♂	11 57	☉ ⊥ ♇	08 36	☽ ⊼ ♃
05 56	☽ ⊼ ♆	14 41	☽ ✶ ♄	10 19	☽ ± ♂
08 13	☽ □ ♇	14 48	☉ ∥ ☿	12 42	☿ St D
09 24	☽ Ⴒ ☿	17 36	☽ ⊥ ♇	16 20	☽ ± ☿
09 30	♀ Q ♄	18 04	☽ ☌ ☉	18 47	☽ ± ♃
09 46	☽ △ ♅	19 48	☽ ☌ ☿	20 44	☽ ± ♅
10 02	☽ Ⴒ ♀	**20 Tuesday**		22 16	☽ ∥ ♄
11 32	☽ ∠ ♃	00 01	☽ □ ♂	**30 Friday**	
13 54	☿ St R	01 11	♀ ☌ ♆	01 57	☽ Ⴒ ♂
15 48	☽ ✶ ♄	04 36	☉ ☌ ☿	02 30	☽ ⊼ ☿
17 31	☽ ± ♆	10 09	☽ ∠ ♇	02 35	☽ △ ♆
17 56	☽ ⊻ ♂	12 51	☽ Q ♄	04 32	☽ ☍ ♇
19 59	☽ ⋕ ♅	15 00	☽ □ ♃	06 54	☽ ⊼ ♅
10 Saturday		16 19	☿ ⊥ ♇	07 12	☽ Ⴒ ☉
02 38	♀ □ ♃	**21 Wednesday**		10 03	☽ Ⴒ ♃
05 13	☽ ⊥ ♃	01 02	♀ ✶ ♇	12 43	☿ ⊻ ♆
07 01	☽ ⋕ ☉	06 37	☽ ∥ ♅	12 46	☽ ⊼ ♄
11 10	☽ Ⴒ ♆	11 09	☽ ☌ ♆	14 24	☉ Q ♄
14 31	☽ △ ☿	11 41	☽ ∥ ☿	19 03	☽ ∥ ♇
17 11	☽ ⊻ ♃	13 19	☽ ✶ ♇	22 35	♀ Q ♇
18 41	☽ △ ♀	14 32	☽ ☌ ♀	22 57	☽ ± ♄
19 31	☽ Q ♇	15 21	☽ ⊻ ♅	23 34	☽ ± ☉
21 30	☽ ∠ ♄	18 21	☿ ✶ ♄		
11 Sunday		21 09	☽ ⊻ ☿		

All ephemeris data is given at 12.00 UT and the Moon's longitude is additionally given for 24.00 UT

DECEMBER 1979

Date	Sidereal time h m s	Sun ☉ ° ' "	Moon ☽ ° ' "	Moon ☽ 24.00 ° ' "	Mercury ☿ ° '	Venus ♀ ° '	Mars ♂ ° '	Jupiter ♃ ° '	Saturn ♄ ° '	Uranus ♅ ° '	Neptune ♆ ° '	Pluto ⯓ ° '
01	16 39 00	08 ♐ 41 27	09 ♉ 30 33	16 ♉ 34 05	19 ♏ 59	03 ♑ 37	04 ♍ 56	09 ♍ 15	25 ♍ 50	22 ♏ 22	19 ♐ 47	20 ♎ 56
02	16 42 57	09 42 15	23 ♉ 35 17	00 ♊ 33 35	20 D 24	04 51	05 13	09 19	25 54	22 25	19 49	20 58
03	16 46 53	10 43 04	07 ♊ 28 21	14 ♊ 19 03	20 56	06 06	05 43	09 24	25 58	22 29	19 51	20 59
04	16 50 50	11 43 55	21 ♊ 05 15	27 ♊ 46 34	21 36	07 20	06 06	09 28	26 01	22 33	19 54	21 01
05	16 54 47	12 44 47	04 ♋ 22 46	10 ♋ 53 42	22 22	08 35	06 28	09 32	26 05	22 36	19 56	21 03
06	16 58 43	13 45 39	17 ♋ 19 23	23 ♋ 39 54	23 15	09 49	06 51	09 36	26 08	22 40	19 58	21 04
07	17 02 40	14 46 33	29 ♋ 55 29	06 ♌ 06 27	24 12	11 04	07 13	09 40	26 11	22 43	20 00	21 06
08	17 06 36	15 47 28	12 ♌ 13 12	18 ♌ 16 16	25 14	12 18	07 34	09 43	26 14	22 47	20 03	21 08
09	17 10 33	16 48 25	24 ♌ 16 10	00 ♍ 13 32	26 19	13 33	07 55	09 46	26 17	22 50	20 05	21 09
10	17 14 29	17 49 22	06 ♍ 08 59	12 ♍ 03 15	27 29	14 47	08 16	09 50	26 20	22 54	20 07	21 11
11	17 18 26	18 50 21	17 ♍ 56 59	23 ♍ 50 54	28 41	16 02	08 37	09 53	26 23	22 57	20 09	21 12
12	17 22 22	19 51 20	29 ♍ 45 42	05 ♎ 42 05	29 ♏ 55	17 16	08 57	09 55	26 26	23 01	20 12	21 14
13	17 26 19	20 52 21	11 ♎ 40 42	17 ♎ 42 10	01 ♐ 12	18 30	09 17	09 58	26 29	23 04	20 14	21 15
14	17 30 16	21 53 23	23 ♎ 47 05	29 ♎ 55 56	02 30	19 45	09 36	10 00	26 31	23 08	20 16	21 17
15	17 34 12	22 54 26	06 ♏ 09 11	12 ♏ 27 10	03 51	20 59	09 55	10 03	26 34	23 11	20 19	21 18
16	17 38 09	23 55 30	18 ♏ 50 10	25 ♏ 18 20	05 12	22 14	10 13	10 05	26 36	23 14	20 21	21 19
17	17 42 05	24 56 34	01 ♐ 51 43	08 ♐ 30 13	06 35	23 28	10 31	10 07	26 38	23 18	20 23	21 21
18	17 46 02	25 57 40	15 ♐ 13 39	22 ♐ 01 43	08 00	24 42	10 49	10 08	26 41	23 21	20 25	21 22
19	17 49 58	26 58 46	28 ♐ 54 03	05 ♑ 50 08	09 25	25 57	11 06	10 10	26 43	23 24	20 28	21 23
20	17 53 55	27 59 53	12 ♑ 49 26	19 ♑ 51 24	10 50	27 11	11 23	10 11	26 45	23 28	20 30	21 24
21	17 57 51	29 ♐ 01 00	26 ♑ 55 26	04 ♒ 00 57	12 17	28 25	11 39	10 12	26 46	23 31	20 32	21 26
22	18 01 48	00 ♑ 02 08	11 ♒ 07 27	18 ♒ 14 24	13 44	29 ♑ 40	11 55	10 13	26 48	23 34	20 34	21 27
23	18 05 45	01 03 16	25 ♒ 21 25	02 ♓ 28 07	15 12	00 ♒ 54	12 10	10 14	26 50	23 37	20 37	21 28
24	18 09 41	02 04 24	09 ♓ 34 13	16 ♓ 39 30	16 41	02 08	12 25	10 14	26 51	23 40	20 39	21 29
25	18 13 38	03 05 32	23 ♓ 43 47	00 ♈ 46 56	18 10	03 22	12 39	10 15	26 53	23 43	20 41	21 30
26	18 17 34	04 06 40	07 ♈ 48 50	14 ♈ 49 25	19 39	04 36	12 53	10 15	26 54	23 46	20 43	21 31
27	18 21 31	05 07 48	21 ♈ 48 34	28 ♈ 46 09	21 09	05 51	13 06	10 R 15	26 55	23 50	20 46	21 32
28	18 25 27	06 08 56	05 ♉ 42 05	12 ♉ 36 09	22 39	07 05	13 18	10 14	26 56	23 53	20 48	21 33
29	18 29 24	07 10 04	19 ♉ 28 12	26 ♉ 17 58	24 09	08 19	13 31	10 14	26 57	23 56	20 50	21 34
30	18 33 20	08 11 12	03 ♊ 05 15	09 ♊ 49 46	25 40	09 33	13 42	10 13	26 58	23 58	20 52	21 35
31	18 37 17	09 ♑ 12 20	16 ♊ 31 15	23 ♊ 09 28	27 ♐ 12	10 ♒ 47	13 ♍ 53	10 ♍ 12	26 ♍ 59	24 ♏ 01	20 ♐ 54	21 ♎ 36

DECLINATIONS

Date	Moon True ☊ ° '	Moon Mean ☊ ° '	Moon ☽ Latitude ° '	Sun ☉ ° '	Moon ☽ ° '	Mercury ☿ ° '	Venus ♀ ° '	Mars ♂ ° '	Jupiter ♃ ° '	Saturn ♄ ° '	Uranus ♅ ° '	Neptune ♆ ° '	Pluto ⯓ ° '
01	03 ♍ 32	03 ♍ 31	04 S 35	21 S 45	10 N 19	15 S 14	24 S 42	11 N 49	09 N 01	03 N 31	18 S 05	21 S 42	07 N 24
02	03 R 21	03 28	04 56	21 54	13 53	15 20	24 41	11 42	09 00	03 29	18 06	21 43	07 24
03	03 08	03 24	05 00	22 03	16 38	15 30	24 40	11 34	08 58	03 28	18 07	21 43	07 24
04	02 55	03 21	04 46	22 12	18 23	15 42	24 38	11 27	08 57	03 27	18 08	21 43	07 23
05	02 43	03 18	04 17	22 20	19 06	15 57	24 35	11 20	08 56	03 26	18 09	21 43	07 23
06	02 34	03 15	03 35	22 27	18 46	16 14	24 32	11 13	08 55	03 25	18 10	21 43	07 23
07	02 27	03 12	02 43	22 34	17 31	16 33	24 27	11 07	08 53	03 24	18 11	21 44	07 23
08	02 23	03 09	01 45	22 41	15 27	16 53	24 22	11 00	08 52	03 23	18 12	21 44	07 23
09	02 21	03 05	00 S 43	22 47	12 46	17 14	24 17	10 54	08 51	03 22	18 13	21 44	07 22
10	02 D 21	03 02	00 N 20	22 53	09 34	17 36	24 10	10 47	08 50	03 21	18 14	21 44	07 22
11	02 21	02 59	01 22	22 58	06 01	17 58	24 03	10 41	08 49	03 20	18 14	21 44	07 22
12	02 R 21	02 56	02 20	23 03	02 N 15	18 21	23 55	10 35	08 49	03 19	18 15	21 44	07 22
13	02 19	02 53	03 13	23 08	01 S 39	18 44	23 46	10 29	08 48	03 18	18 16	21 45	07 22
14	02 14	02 50	03 58	23 11	05 32	19 06	23 37	10 23	08 47	03 18	18 17	21 45	07 22
15	02 07	02 46	04 33	23 15	09 17	19 29	23 27	10 18	08 47	03 17	18 18	21 45	07 22
16	01 57	02 43	04 55	23 18	12 41	19 51	23 16	10 12	08 46	03 16	18 19	21 45	07 22
17	01 46	02 40	05 03	23 21	15 35	20 13	23 04	10 07	08 46	03 15	18 20	21 45	07 22
18	01 34	02 37	04 54	23 23	17 45	20 35	22 52	10 02	08 45	03 15	18 20	21 45	07 22
19	01 23	02 34	04 29	23 24	18 57	20 55	22 39	09 57	08 45	03 14	18 21	21 46	07 22
20	01 13	02 30	03 48	23 25	19 03	21 16	22 26	09 52	08 45	03 14	18 22	21 46	07 22
21	01 06	02 27	02 51	23 26	17 58	21 35	22 12	09 47	08 45	03 13	18 23	21 46	07 22
22	01 02	02 24	01 44	23 26	15 46	21 53	21 57	09 43	08 44	03 13	18 24	21 46	07 22
23	01 00	02 21	00 N 30	23 26	12 36	22 11	21 41	09 39	08 44	03 12	18 24	21 46	07 22
24	01 D 00	02 18	00 S 46	23 25	08 41	22 28	21 25	09 35	08 45	03 12	18 25	21 46	07 22
25	01 01	02 15	01 59	23 24	04 S 18	22 44	21 08	09 31	08 45	03 12	18 26	21 47	07 22
26	01 R 01	02 11	03 04	23 23	00 N 17	22 58	20 51	09 27	08 45	03 12	18 27	21 47	07 23
27	01 00	02 08	03 58	23 20	04 49	23 12	20 33	09 24	08 45	03 11	18 27	21 47	07 23
28	00 56	02 05	04 37	23 18	09 03	23 25	20 15	09 21	08 45	03 11	18 28	21 47	07 23
29	00 50	02 02	05 00	23 15	12 47	23 36	19 56	09 18	08 46	03 11	18 29	21 47	07 23
30	00 42	01 59	05 06	23 11	15 46	23 47	19 36	09 15	08 46	03 11	18 30	21 47	07 23
31	00 ♍ 33	01 ♍ 55	04 S 55	23 S 07	17 N 52	23 S 56	19 S 16	09 N 13	08 N 47	03 N 11	18 S 30	21 S 47	07 N 23

ZODIAC SIGN ENTRIES

Date	h	m	Planets
02	23	02	☽ ♊
05	04	01	☽ ♋
07	12	09	☽ ♌
09	23	33	☽ ♍
12	12	29	☽ ♎
12	13	34	☿ ♐
15	00	08	☽ ♏
17	08	36	☽ ♐
19	13	55	☽ ♑
21	17	13	☽ ♒
22	11	10	☉ ♑
22	18	35	♀ ♒
23	19	50	☽ ♓
25	22	40	☽ ♈
28	02	08	☽ ♉
30	06	32	☽ ♊

LATITUDES

Date	Mercury ☿ ° '	Venus ♀ ° '	Mars ♂ ° '	Jupiter ♃ ° '	Saturn ♄ ° '	Uranus ♅ ° '	Neptune ♆ ° '	Pluto ⯓ ° '
01	02 N 36	01 S 18	02 N 16	01 N 00	02 N 01	00 N 17	01 N 21	16 N 47
04	02 33	01 24	02 21	01 00	02 02	00 17	01 21	16 48
07	02 21	01 29	02 25	01 01	02 03	00 17	01 20	16 49
10	02 03	01 33	02 30	01 02	02 04	00 17	01 20	16 51
13	01 42	01 37	02 35	01 03	02 05	00 17	01 20	16 52
16	01 20	01 41	02 40	01 03	02 05	00 17	01 20	16 54
19	00 57	01 44	02 45	01 04	02 06	00 17	01 20	16 55
22	00 34	01 46	02 50	01 05	02 07	00 17	01 20	16 57
25	00 N 11	01 47	02 56	01 06	02 08	00 17	01 20	16 58
28	00 S 11	01 48	03 01	01 07	02 09	00 17	01 20	17 00
31	00 S 31	01 S 48	03 N 07	01 N 08	02 N 10	00 N 17	01 N 20	17 N 02

DATA

Julian Date	2444209
Delta T	+50 seconds
Ayanamsa	23° 34' 26"
Synetic vernal point	05° ♓ 32' 33"
True obliquity of ecliptic	23° 26' 22"

MOON'S PHASES, APSIDES AND POSITIONS ☽

Date	h	m	Phase	Longitude ° '	Eclipse Indicator
03	18	08	○	10 ♊ 59	
11	13	59	☾	18 ♍ 55	
19	08	23	●	26 ♐ 50	
26	05	11	☽	03 ♈ 49	

Day	h	m		
11	10	55	Apogee	
23	15	54	Perigee	
05	16	13	Max dec	19° N 07'
13	01	51	0S	
20	02	03	Max dec	19° S 09'
26	10	32	0N	

LONGITUDES

Date	Chiron ⚷ ° '	Ceres ⚳ ° '	Pallas ⚴ ° '	Juno ⚵ ° '	Vesta ⚶ ° '	Black Moon Lilith ⚸ ° '
01	10 ♉ 15	05 ♈ 26	20 ♒ 08	29 ♋ 04	04 ♉ 23	16 ♍ 03
11	09 ♉ 52	06 ♈ 06	22 ♒ 22	28 ♋ 39	03 ♉ 19	17 ♍ 09
21	09 ♉ 32	07 ♈ 20	24 ♒ 51	27 ♋ 20	03 ♉ 20	18 ♍ 16
31	09 ♉ 18	09 ♈ 04	27 ♒ 34	25 ♋ 16	03 ♉ 24	19 ♍ 22

ASPECTARIAN

h m Aspects

01 Saturday
01 03 ☽ △ ♀
03 58 ☽ ♇ ♆
04 01 ☽ △ ♂
04 21 ☽ ∥ ♃
05 11 ☉ ∥ ♆
10 30 ☽ ⊼ ☉
11 33 ☽ △ ♃
14 16 ☽ ♇ ♄
19 17 ☽ ± ♆
21 13 ☽ ∥ ♂

02 Sunday
02 15 ☉ □ ♃
04 59 ☽ ♇ ♀
05 32 ☽ ⊼ ♆
06 21 ☽ ☍ ☿
07 29 ☽ ⊼ ⯓
10 00 ☽ ☍ ♅
15 59 ☽ △ ♄
17 48 ☽ ± ⯓
21 56 ☽ ± ♀

03 Monday
00 19 ☽ # ☿
01 11 ♀ △ ♂
06 07 ☿ ∠ ♀
08 50 ☽ □ ♂
09 22 ☽ ⊼ ♀
09 25 ☽ ♇ ⯓
14 16 ☿ ⊻ ⯓
15 23 ☽ □ ♃
18 08 ☽ ☍ ☉

04 Tuesday
06 55 ♂ ∠ ⯓
07 04 ☿ Q ♃
07 26 ☽ # ♅
09 52 ☽ ☍ ♆
11 52 ☽ △ ⯓
12 58 ☽ ⊼ ☿
14 37 ☽ ⊼ ♅
16 14 ♀ ∠ ♅
17 32 ☽ Q ♂
20 52 ☽ □ ♄
23 30 ☽ Q ♃

05 Wednesday
00 23 ☽ ± ☿
01 27 ☽ ± ♅
15 57 ☽ ⚹ ♂
17 52 ☽ ♇ ☿
17 57 ☽ ♇ ♅
19 09 ☿ ☌ ♅
20 32 ☽ ☍ ♀
21 32 ☽ ⚹ ♃

06 Thursday
04 46 ☽ ⊼ ☉
06 00 ☽ Q ♄
07 29 ♀ △ ♃
17 00 ☽ ± ☉
17 00 ☽ ⊼ ♆
19 05 ☽ □ ⯓
20 47 ☽ ∠ ♂
22 09 ☽ △ ♅

07 Friday
00 06 ☽ △ ☿
01 23 ☽ # ♅
01 51 ☽ ∠ ♃
04 27 ☽ ± ♆
04 47 ☽ ⚹ ♄
11 41 ☽ ♇ ☉
14 33 ☽ ⊥ ♂
19 16 ☽ ⊥ ♃
21 53 ☽ ♇ ♆
22 36 ☽ # ☿

08 Saturday
02 35 ☽ ⊻ ♂
05 54 ☽ Q ⯓
07 03 ☽ ⊻ ♃
10 03 ☽ ∠ ♄
12 11 ☽ ⊼ ♀
19 43 ☽ △ ☉

09 Sunday
01 27 ☽ ± ♀
03 35 ☽ △ ♆
04 00 ☽ ⊥ ♄
05 44 ☽ ⚹ ⯓
09 07 ☽ □ ♅
11 14 ☿ ⚹ ♄
16 05 ☽ ⊻ ♄
16 34 ☽ □ ☿
21 37 ☽ ♇ ♀

10 Monday
02 54 ☽ ∥ ♂
05 46 ☿ ⊥ ⯓
12 03 ☽ ∠ ⯓
16 26 ☽ ☌ ♂
17 08 ☽ ∥ ♃
19 30 ☽ ☌ ♃
21 42 ☽ Q ♅

11 Tuesday
03 06 ☽ ∥ ⯓
06 24 ☽ ⊥ ⯓
07 38 ☽ △ ♀
09 06 ☽ Q ☿
13 59 ☽ □ ☉
16 31 ☽ □ ♆
18 38 ☽ ⊻ ⯓
22 14 ☽ ⚹ ♅

12 Wednesday
05 14 ☽ ☌ ♄
05 15 ☽ ∥ ♄
05 58 ☿ ∥ ♅
12 21 ☽ ⚹ ☿
20 21 ☉ ☌ ♆

13 Thursday
04 44 ☽ ∠ ♅
05 04 ☽ Q ♆
05 51 ☽ Q ☉
07 03 ☽ ⊻ ♂
08 34 ☽ ⊻ ♃
19 23 ☽ ⊥ ♂
20 35 ☽ ⊥ ♃
21 11 ☉ ⚹ ⯓
22 06 ☽ ∠ ☿
22 06 ☽ # ♄
22 48 ☽ ⊥ ♅

14 Friday
03 09 ☽ □ ♀
05 04 ☽ ⚹ ♆
07 03 ☽ ☌ ⯓
07 56 ☽ ⚹ ☉
10 42 ☽ ⊻ ♅
13 39 ☽ ∠ ♂
14 24 ☽ ∠ ♃
17 23 ☽ ⊻ ♄
17 59 ☽ ⊥ ☿
22 27 ♀ ⊻ ♆
23 33 ☽ # ⯓

15 Saturday
05 04 ☽ ⊥ ♄
07 02 ☽ ⊻ ☿
08 42 ☽ # ♃
10 23 ☽ ∠ ♆
15 39 ☽ ∠ ☉
18 01 ☽ Q ♀
18 08 ♀ □ ⯓
18 41 ☽ # ♂
18 54 ☉ ⊻ ♅
19 22 ☽ ⚹ ♂
19 27 ☽ ⚹ ♃
22 21 ☽ ∠ ♄
23 15 ♂ ☌ ♃

16 Sunday
03 33 ☽ ⊥ ♆
08 16 ☉ ∥ ♀
10 09 ☽ ⊥ ☉
14 50 ☽ ⊻ ♆
16 38 ☽ ⊻ ⯓
18 03 ☽ Q ♃
18 27 ☽ Q ♂
18 59 ☽ ⚹ ♀
20 13 ☽ ☌ ♅
22 15 ☽ ⊻ ☉

17 Monday
02 26 ☽ ⚹ ♄
03 44 ☽ ⊥ ⯓
07 41 ☿ ∠ ⯓
08 30 ♀ ⚹ ♅
20 08 ☽ ∠ ⯓
21 34 ☽ ☌ ☿

18 Tuesday
00 17 ☽ Q ♄
01 09 ☽ ∠ ♀
02 55 ☽ □ ♃
03 58 ☽ □ ♂
18 46 ☽ ⊥ ♀
20 31 ♀ ♇ ♃
21 12 ☽ ☌ ♆
21 31 ☽ ∥ ♅
22 51 ☽ ⚹ ⯓
23 54 ☿ Q ♄

19 Wednesday
02 22 ☽ ⊻ ♅
05 26 ☉ □ ♄
06 21 ☽ ⊻ ♀
08 11 ☽ □ ♄
08 23 ☽ ☌ ☉
12 53 ☽ ⊥ ♅
15 58 ♀ ♇ ♂
19 47 ☽ Q ⯓
22 16 ♀ ⊥ ♆

20 Thursday
00 51 ☿ □ ♃
03 12 ♀ △ ♄
04 29 ☽ ∠ ♅
07 28 ☽ △ ♃
08 13 ☽ ⊻ ☿
09 29 ☽ △ ♂
19 39 ☽ ⊥ ☿
20 29 ♂ Q ♅
23 07 ☿ □ ♂

21 Friday
01 08 ☽ ⊻ ♆
02 40 ☽ □ ⯓
05 29 ☽ ∥ ♅
06 12 ☽ ⚹ ♅
09 05 ☽ ♇ ♃
11 20 ☽ ⊥ ♆
11 32 ☽ ♇ ♂
11 45 ☽ △ ♄
12 41 ☽ ∠ ☿
14 47 ☽ ☌ ♀
15 49 ☽ ⊻ ☉

22 Saturday
00 20 ☽ ± ♃
00 21 ☉ ⊥ ♅
02 11 ☿ ∥ ♆
02 35 ☽ Q ♅
02 36 ☽ ∠ ♆
02 45 ☽ ⊥ ☉
03 03 ☽ ± ♂
10 28 ☽ ⊼ ♃
13 09 ☽ ♇ ♄
13 21 ☽ ⊼ ♂
14 20 ☿ ∥ ♀
16 55 ☽ ⚹ ☿
19 06 ☽ ∠ ☉

23 Sunday
03 58 ☽ ⚹ ♆
04 21 ☽ ± ♄
04 28 ♀ ∥ ♆
05 26 ☽ △ ⯓
09 04 ☽ □ ♅
14 29 ☽ ⊼ ♄
15 29 ☽ Q ☿
22 14 ☽ ⊻ ♀
22 21 ☽ ⚹ ☉

24 Monday
00 16 ☽ Q ♆
05 09 ☉ ⊻ ♀
06 44 ☽ # ♂
06 47 ☽ ♇ ⯓
09 21 ☽ ⊥ ♀
11 41 ☽ # ♃
13 08 ☽ ☍ ♃
16 54 ☽ ☍ ♂
19 25 ☽ # ⯓
20 13 ☽ Q ☉
22 02 ☽ ± ⯓

25 Tuesday
01 26 ☽ □ ☿
02 02 ☽ ∠ ♀
06 49 ☽ □ ♆
08 13 ☽ ⊼ ⯓
11 59 ☽ △ ♅
17 22 ☽ ☍ ♄
17 52 ☽ # ♄
21 52 ☉ Q ⯓

26 Wednesday
04 56 ♀ ± ♃
05 11 ☽ □ ☉
06 00 ☽ ⚹ ♀
08 11 ☿ ∠ ♀
13 39 ☽ ♇ ♅
14 59 ♃ St R
16 09 ☽ ⊼ ♃
20 48 ☽ ⊼ ♂

27 Thursday
02 26 ☽ ± ♃
03 17 ☽ ∥ ♄
04 31 ☽ Q ♀
05 08 ☽ ± ♅
05 39 ☿ ☌ ♆
07 16 ☽ ± ♂
10 11 ☽ △ ♆
10 17 ♀ ∠ ♆
10 43 ☽ △ ☿
11 32 ☽ ☍ ⯓
11 38 ♀ Q ♅
15 29 ☽ ⊼ ♅
17 55 ☽ ♇ ♃
18 21 ☿ ⚹ ⯓
20 49 ☽ ⊼ ♄
23 01 ☽ ♇ ♂

28 Friday
00 46 ☉ ∥ ☿
02 13 ☽ ∥ ⯓
07 12 ☽ ± ♄
10 13 ☽ ∥ ♃
12 10 ☽ ♇ ♆
12 50 ☽ △ ☉
13 43 ☽ ∥ ♂
14 38 ☽ □ ♀
15 48 ☽ ♇ ☿
17 21 ♀ ± ♂
19 53 ☽ △ ♃
22 51 ☽ ♇ ♄

29 Saturday
01 26 ☽ △ ♂
03 52 ☽ ± ♆
08 12 ☿ ⊻ ♅
09 25 ☽ ± ☿
14 24 ☽ ⊼ ♆
15 41 ☽ ⊼ ⯓
17 07 ☽ ♇ ☉
19 51 ☽ ☍ ♅
21 15 ☽ ⊼ ☿

30 Sunday
01 10 ☽ △ ♄
02 15 ☽ ± ⯓
10 16 ☽ ± ☉
18 13 ☽ ♇ ⯓
21 49 ☽ ⊼ ☉

31 Monday
00 40 ☽ △ ♀
00 41 ☽ □ ♃
00 57 ♀ ⊼ ♃
07 12 ☽ □ ♂
07 30 ☉ ∠ ♅
08 38 ☿ □ ♄
19 56 ☽ ☍ ♆
21 11 ☽ △ ⯓
23 07 ☽ # ♅

All ephemeris data is given at 12.00 UT and the Moon's longitude is additionally given for 24.00 UT

JANUARY 1980

Date	Sidereal time h m s	Sun ☉ ° ' "	Moon ☽ ° ' "	Moon ☽ 24.00 ° ' "	Mercury ☿ ° '	Venus ♀ ° '	Mars ♂ ° '	Jupiter ♃ ° '	Saturn ♄ ° '	Uranus ♅ ° '	Neptune ♆ ° '	Pluto ♇ ° '
01	18 41 14	10 ♑ 13 28	29 ♊ 44 10	06 ♋ 15 11	28 ♐ 43	12 ♒ 01	14 ♍ 03	10 ♍ 11	27 ♍ 00	24 ♏ 04	20 ♐ 56	21 ♎ 37
02	18 45 10	11 14 36	12 ♋ 42 20	19 ♋ 05 34	00 ♑ 15	13 15	14 13	10 R 10	27 00	24 07	20 59	21 38
03	18 49 07	12 15 44	25 ♋ 24 50	01 ♌ 40 13	01 48	14 29	14 22	10 09	27 01	24 10	21 01	21 38
04	18 53 03	13 16 52	07 ♌ 51 48	13 ♌ 59 48	03 20	15 43	14 31	10 07	27 01	24 13	21 03	21 39
05	18 57 00	14 18 01	20 ♌ 04 28	26 ♌ 06 09	04 53	16 56	14 39	10 05	27 01	24 15	21 05	21 40
06	19 00 56	15 19 09	02 ♍ 05 15	08 ♍ 02 12	06 27	18 10	14 46	10 03	27 01	24 18	21 07	21 41
07	19 04 53	16 20 18	13 ♍ 57 32	19 ♍ 51 48	08 00	19 24	14 53	10 01	27 R 01	24 21	21 09	21 41
08	19 08 49	17 21 26	25 ♍ 45 34	01 ♎ 39 28	09 35	20 38	14 59	09 59	27 01	24 23	21 11	21 42
09	19 12 46	18 22 35	07 ♎ 34 10	13 ♎ 30 17	11 09	21 51	15 04	09 56	27 01	24 26	21 13	21 42
10	19 16 43	19 23 43	19 ♎ 28 31	25 ♎ 29 29	12 44	23 05	15 08	09 53	27 00	24 29	21 15	21 43
11	19 20 39	20 24 52	01 ♏ 33 52	07 ♏ 42 15	14 20	24 18	15 12	09 50	27 00	24 31	21 18	21 43
12	19 24 36	21 26 01	13 ♏ 55 12	20 ♏ 13 14	15 55	25 32	15 15	09 47	26 59	24 34	21 20	21 44
13	19 28 32	22 27 09	26 ♏ 36 47	03 ♐ 06 11	17 32	26 45	15 18	09 44	26 59	24 36	21 22	21 44
14	19 32 29	23 28 18	09 ♐ 41 40	16 ♐ 23 21	19 09	27 59	15 20	09 40	26 58	24 38	21 24	21 45
15	19 36 25	24 29 26	23 ♐ 11 13	00 ♑ 05 06	20 46	29 ♒ 12	15 21	09 37	26 57	24 41	21 26	21 45
16	19 40 22	25 30 34	07 ♑ 04 40	14 ♑ 09 30	22 24	00 ♓ 26	15 R 21	09 33	26 56	24 43	21 28	21 45
17	19 44 18	26 31 42	21 ♑ 18 59	28 ♑ 32 26	24 02	01 39	15 20	09 29	26 55	24 45	21 29	21 46
18	19 48 15	27 32 49	05 ♒ 49 05	13 ♒ 08 04	25 41	02 52	15 19	09 25	26 54	24 47	21 31	21 46
19	19 52 12	28 33 56	20 ♒ 28 31	27 ♒ 49 34	27 20	04 05	15 17	09 20	26 53	24 50	21 33	21 46
20	19 56 08	29 ♑ 35 02	05 ♓ 10 24	12 ♓ 30 14	29 ♑ 00	05 18	15 14	09 16	26 51	24 52	21 35	21 46
21	20 00 05	00 ♒ 36 07	19 ♓ 48 23	27 ♓ 04 16	00 ♒ 41	06 32	15 10	09 11	26 50	24 54	21 37	21 46
22	20 04 01	01 37 11	04 ♈ 17 24	11 ♈ 27 22	02 22	07 45	15 06	09 06	26 48	24 56	21 39	21 46
23	20 07 58	02 38 14	18 ♈ 33 54	25 ♈ 36 46	04 03	08 58	15 00	09 01	26 46	24 58	21 41	21 47
24	20 11 54	03 39 16	02 ♉ 35 53	09 ♉ 31 09	05 45	10 10	14 54	08 56	26 44	25 00	21 43	21 47
25	20 15 51	04 40 17	16 ♉ 22 33	23 ♉ 10 09	07 28	11 23	14 48	08 50	26 43	25 01	21 44	21 R 47
26	20 19 47	05 41 17	29 ♉ 53 58	06 ♊ 34 05	09 12	12 36	14 40	08 45	26 40	25 03	21 46	21 46
27	20 23 44	06 42 16	13 ♊ 10 34	19 ♊ 43 32	10 55	13 49	14 31	08 39	26 38	25 05	21 48	21 46
28	20 27 41	07 43 13	26 ♊ 13 03	02 ♋ 39 11	12 40	15 01	14 22	08 33	26 36	25 07	21 50	21 46
29	20 31 37	08 44 10	09 ♋ 02 03	15 ♋ 21 44	14 25	16 14	14 12	08 28	26 34	25 08	21 51	21 46
30	20 35 34	09 45 06	21 ♋ 38 19	27 ♋ 51 54	16 10	17 26	14 01	08 22	26 31	25 10	21 53	21 46
31	20 39 30	10 ♒ 46 00	04 ♌ 02 36	10 ♌ 10 34	17 ♒ 55	18 ♓ 38	13 ♍ 50	08 ♍ 15	26 ♍ 29	25 ♏ 12	21 ♐ 55	21 ♎ 46

Date	Moon True ☊ ° '	Moon Mean ☊ ° '	Moon ☽ Latitude ° '
01	00 ♍ 23	01 ♍ 52	04 S 28
02	00 R 15	01 49	03 47
03	00 08	01 46	02 56
04	00 03	01 43	01 57
05	00 00	01 40	00 S 54
06	00 D 00	01 36	00 N 11
07	00 01	01 33	01 15
08	00 02	01 30	02 15
09	00 04	01 27	03 10
10	00 04	01 24	03 57
11	00 R 04	01 21	04 34
12	00 01	01 17	04 59
13	29 ♌ 57	01 14	05 11
14	29 52	01 11	05 07
15	29 45	01 08	04 46
16	29 40	01 05	04 08
17	29 35	01 01	03 14
18	29 31	00 58	02 06
19	29 29	00 55	00 N 49
20	29 D 29	00 52	00 S 31
21	29 30	00 49	01 49
22	29 32	00 46	03 00
23	29 33	00 42	03 58
24	29 34	00 39	04 41
25	29 R 33	00 36	05 07
26	29 31	00 33	05 15
27	29 29	00 30	05 06
28	29 25	00 27	04 41
29	29 22	00 23	04 03
30	29 19	00 20	03 13
31	29 ♌ 16	00 ♍ 17	02 S 15

DECLINATIONS

Date	Sun ☉ ° '	Moon ☽ ° '	Mercury ☿ ° '	Venus ♀ ° '	Mars ♂ ° '	Jupiter ♃ ° '	Saturn ♄ ° '	Uranus ♅ ° '	Neptune ♆ ° '	Pluto ♇ ° '
01	23 S 03	18 N 59	24 S 04	18 S 55	09 N 10	08 N 48	03 N 11	18 S 31	21 S 48	07 N 24
02	22 58	19 04	24 11	18 34	09 08	08 48	03 11	18 32	21 48	07 24
03	22 52	18 11	24 16	18 12	09 07	08 49	03 11	18 32	21 48	07 24
04	22 47	16 25	24 21	17 50	09 05	08 50	03 11	18 33	21 48	07 24
05	22 40	13 57	24 24	17 27	09 04	08 51	03 11	18 34	21 48	07 25
06	22 34	10 55	24 25	17 04	09 03	08 52	03 12	18 35	21 48	07 25
07	22 26	07 28	24 25	16 40	09 02	08 53	03 12	18 35	21 48	07 25
08	22 19	03 N 45	24 24	16 16	09 02	08 54	03 12	18 36	21 48	07 26
09	22 11	00 S 06	24 22	15 52	09 01	08 55	03 12	18 36	21 49	07 26
10	22 02	03 58	24 18	15 27	09 01	08 57	03 13	18 37	21 49	07 26
11	21 53	07 44	24 13	15 01	09 02	08 58	03 13	18 38	21 49	07 27
12	21 44	11 15	24 06	14 36	09 02	08 59	03 14	18 38	21 49	07 27
13	21 34	14 21	23 58	14 09	09 03	09 01	03 14	18 39	21 49	07 27
14	21 24	16 51	23 49	13 43	09 04	09 03	03 15	18 39	21 49	07 28
15	21 13	18 30	23 38	13 16	09 06	09 04	03 15	18 40	21 49	07 28
16	21 02	19 07	23 25	12 49	09 07	09 06	03 16	18 41	21 49	07 29
17	20 51	18 34	23 11	12 21	09 09	09 08	03 17	18 41	21 49	07 29
18	20 39	16 47	22 56	11 54	09 12	09 09	03 17	18 42	21 50	07 30
19	20 27	13 53	22 39	11 25	09 14	09 11	03 18	18 42	21 50	07 30
20	20 14	10 06	22 20	10 57	09 17	09 13	03 19	18 43	21 50	07 31
21	20 01	05 43	22 00	10 28	09 20	09 15	03 20	18 43	21 50	07 31
22	19 48	01 S 03	21 39	09 59	09 24	09 17	03 21	18 44	21 50	07 32
23	19 34	03 N 36	21 16	09 30	09 28	09 19	03 22	18 44	21 50	07 32
24	19 20	07 59	20 51	09 01	09 32	09 22	03 23	18 45	21 50	07 33
25	19 06	11 50	20 25	08 31	09 36	09 24	03 24	18 45	21 50	07 33
26	18 51	15 00	19 57	08 01	09 41	09 26	03 25	18 45	21 50	07 34
27	18 36	17 19	19 28	07 31	09 46	09 28	03 26	18 46	21 50	07 34
28	18 20	18 42	18 58	07 01	09 51	09 31	03 27	18 46	21 50	07 35
29	18 05	19 06	18 25	06 31	09 56	09 33	03 28	18 47	21 51	07 35
30	17 48	18 31	17 52	06 00	10 02	09 36	03 29	18 47	21 51	07 36
31	17 S 32	17 N 04	17 S 16	05 S 29	10 N 08	09 N 38	03 N 30	18 S 47	21 S 51	07 N 37

ZODIAC SIGN ENTRIES

Date	h	m	Planets
01	12	29	☽ ♋
02	08	02	☿ ♑
03	20	47	☽ ♌
06	07	48	☽ ♍
08	20	38	☽ ♎
11	08	55	☽ ♏
13	18	17	☽ ♐
15	23	51	☽ ♑
16	03	37	♀ ♓
18	02	25	☽ ♒
20	03	33	☽ ♓
20	21	49	☉ ♒
21	02	18	☿ ♒
22	04	52	☽ ♈
24	07	31	☽ ♉
26	12	11	☽ ♊
28	19	02	☽ ♋
31	04	08	☽ ♌

LATITUDES

Date	Mercury ☿ ° '	Venus ♀ ° '	Mars ♂ ° '	Jupiter ♃ ° '	Saturn ♄ ° '	Uranus ♅ ° '	Neptune ♆ ° '	Pluto ♇ ° '
01	00 S 38	01 S 48	03 N 09	01 N 08	02 N 10	00 N 17	01 N 20	17 N 02
04	00 57	01 47	03 15	01 09	02 11	00 17	01 20	17 04
07	01 14	01 45	03 20	01 09	02 12	00 17	01 21	17 06
10	01 29	01 43	03 26	01 10	02 12	00 17	01 21	17 07
13	01 42	01 40	03 32	01 11	02 13	00 17	01 21	17 09
16	01 52	01 36	03 38	01 12	02 14	00 17	01 21	17 11
19	02 00	01 31	03 44	01 12	02 15	00 17	01 21	17 13
22	02 04	01 26	03 50	01 13	02 16	00 17	01 21	17 14
25	02 05	01 19	03 55	01 14	02 16	00 17	01 21	17 16
28	02 02	01 12	04 00	01 15	02 17	00 17	01 21	17 18
31	01 S 54	01 S 05	04 N 05	01 N 15	02 N 18	00 N 17	01 N 21	17 N 20

DATA

Julian Date	2444240
Delta T	+51 seconds
Ayanamsa	23° 34' 32"
Synetic vernal point	05° ♓ 32' 28"
True obliquity of ecliptic	23° 26' 22"

MOON'S PHASES, APSIDES AND POSITIONS ☽

Date	h	m	Phase	Longitude ° '	Eclipse Indicator
02	09	02	○	11 ♋ 07	
10	11	50	☾	19 ♎ 23	
17	21	19	●	26 ♑ 55	
24	13	58	☽	03 ♉ 44	

Day	h	m		
08	08	00	Apogee	
20	01	47	Perigee	
02	02	00	Max dec	19° N 09'
09	11	25	0S	
16	12	53	Max dec	19° S 07'
22	17	20	0N	
29	09	36	Max dec	19° N 06'

LONGITUDES

Date	Chiron ⚷ ° '	Ceres ⚳ ° '	Pallas ⚴ ° '	Juno ⚵ ° '	Vesta ⚶ ° '	Black Moon Lilith ⚸ ° '
01	09 ♉ 17	09 ♈ 16	27 ♒ 51	25 ♋ 02	03 ♉ 29	19 ♍ 29
11	09 ♉ 10	11 ♈ 27	00 ♓ 44	22 ♋ 33	04 ♉ 37	20 ♍ 35
21	09 ♉ 08	14 ♈ 01	03 ♓ 47	20 ♋ 06	06 ♉ 19	21 ♍ 42
31	09 ♉ 12	16 ♈ 53	06 ♓ 57	18 ♋ 04	06 ♉ 31	22 ♍ 48

All ephemeris data is given at 12.00 UT and the Moon's longitude is additionally given for 24.00 UT

ASPECTARIAN

h m	Aspects
01 Tuesday	
01 37	☽ ⊼ ♅
06 30	☽ ⚼ ♀
06 59	☽ □ ♄
09 10	☽ Q ♃
09 54	☽ ☍ ☿
10 29	☽ # ♀
11 11	☉ △ ♃
11 35	♀ ⚼ ♄
12 37	☽ ± ♅
16 19	☽ Q ♂
02 Wednesday	
00 46	☽ ± ♀
05 17	☽ ⚼ ♅
07 17	☽ ⚹ ♃
09 02	☽ ☍ ☉
09 50	☿ ⊥ ♅
13 07	☽ ⊼ ♀
14 13	♀ ∥ ♅
14 52	☽ ⚹ ♂
16 18	☽ Q ♄
03 Thursday	
03 36	☽ ⊼ ♆
04 49	☽ □ ♇
04 51	☽ # ♅
09 37	☽ △ ♅
09 39	♀ ⊼ ♂
11 22	☽ # ♀
11 29	☽ ∠ ♃
15 03	☽ ⚹ ♄
15 04	☽ ± ♆
19 40	☽ ∠ ♂
04 Friday	
01 58	☽ ⊼ ☿
04 45	☽ ⊥ ♃
08 28	☽ ⚼ ♆
13 17	☽ ⊥ ♂
15 17	☽ ± ☿
15 30	☽ Q ♇
16 23	☽ ⊻ ♃
16 57	☿ Q ♇
20 07	☽ ∠ ♄
23 34	☽ ⊼ ☉
05 Saturday	
01 10	☽ ⊻ ♂
05 06	☽ ☍ ♀
11 34	☽ ⚼ ☿
12 29	☽ ± ☉
13 52	☽ ⊥ ♄
14 01	☽ △ ♆
15 10	☽ ⚹ ♇
20 21	☽ □ ♅
21 17	☉ △ ♂
06 Sunday	
01 50	☽ ⊻ ♄
08 07	☽ ⚼ ☉
21 16	☽ ∠ ♇
22 07	☽ △ ☿
22 41	♄ St R
07 Monday	
01 19	☽ ∥ ♂
02 27	☽ ∥ ♃
04 02	☽ ☌ ♃
08 43	☽ Q ♅
12 18	☽ ∥ ♇
13 53	☽ ☌ ♂
15 31	☽ ⊥ ♇
17 17	☽ △ ☉
08 Tuesday	
00 20	☽ ⊼ ♀
02 40	☽ □ ♆
03 43	☽ ⊻ ♇
09 05	☿ ∠ ♅
09 12	☽ ⚹ ♅
13 58	☽ ± ♀
14 34	☽ ☌ ♄
15 30	☽ ∥ ♄
17 57	☿ △ ♃
19 37	♀ ± ♄
23 19	♀ ⚹ ♆
09 Wednesday	
09 06	♀ △ ♇
10 23	☽ ⚼ ♀
15 22	☽ Q ♆
15 47	☽ ∠ ♅
16 46	☽ ⊻ ♃
20 22	☽ □ ☿
10 Thursday	
03 15	☽ ⊻ ♂
04 50	☽ ⊥ ♃
07 19	☽ # ♄
09 59	☽ ⊥ ♅
11 50	☽ □ ☉
15 21	☽ ⊥ ♂
15 35	☽ ⚹ ♆
16 29	☽ ☌ ♇
20 01	☽ △ ♀
22 01	☽ ⊻ ♅
22 46	☽ ∠ ♃
11 Friday	
03 00	☽ ⊻ ♄
09 19	☽ ∠ ♂
10 08	☽ # ♇
13 43	☽ Q ☿
14 49	☽ ⊥ ♄
16 16	♀ □ ♅
20 15	☽ # ♃
20 37	☽ # ♂
21 17	☽ ∠ ♆
23 27	☉ ∥ ♆
12 Saturday	
01 41	☿ △ ♂
02 35	☽ Q ☉
04 04	☽ ⚹ ♃
08 18	☽ ∠ ♄
09 23	☉ ⊻ ♆
14 34	☽ ⚹ ♂
14 42	☽ ⊥ ♆
16 24	☽ ⚹ ☿
19 03	☉ □ ♇
13 Sunday	
02 07	☽ ⊻ ♆
02 52	☽ ⊻ ♇
02 54	☽ Q ♃
03 32	☽ ⚹ ☉
08 14	☽ ☌ ♅
10 34	☽ ∥ ♀
12 18	☽ □ ♀
12 41	☽ ⚹ ♄
13 17	☽ Q ♂
14 06	☽ ⊥ ♇
16 19	♀ ⊼ ♄
14 Monday	
00 29	☽ ∠ ☿
06 39	☽ ∠ ♇
09 36	☽ ∠ ☉
10 41	☽ Q ♄
11 58	☽ □ ♃
19 04	☽ ⊥ ☿
22 07	☽ □ ♂
15 Tuesday	
00 24	☽ Q ♀
03 03	☽ ⊥ ☉
07 10	☽ ⊻ ☿
08 54	☽ ☌ ♆
09 29	☽ ⚹ ♇
14 28	☽ ⊻ ☉
14 37	☽ ⊻ ♅
14 41	☉ ⚼ ♃
15 37	☽ ∥ ♅
16 34	☉ ⚹ ♅
18 34	☽ □ ♄
21 56	☿ ⊻ ♆
23 29	☽ ⚹ ♀
16 Wednesday	
01 04	☽ ⊥ ♅
02 34	☿ □ ♇
06 18	♂ St R
06 19	☽ Q ♇
16 11	☽ △ ♃
16 30	☽ ∠ ♅
17 Thursday	
02 00	☽ △ ♂
03 28	☽ ∠ ♀
09 20	☽ ∥ ♅
12 18	☽ ⊻ ♆
12 44	☽ □ ♇
17 06	☽ ☌ ☿
17 15	☽ ⚼ ♃
17 44	☽ ⚹ ♅
18 15	☿ ⚼ ♃
19 52	☽ ⊥ ♀
21 00	☉ △ ♄
21 18	☽ △ ♄
21 19	☽ ☌ ☉
22 17	☽ ⊥ ♆
22 44	☿ ⚹ ♅
18 Friday	
02 57	☽ ⚼ ♂
06 42	☽ ⊻ ♀
08 03	☽ ± ♃
11 25	☉ ⊥ ♆
13 10	☽ ∠ ♆
13 36	☽ Q ♅
17 44	☽ ± ♂
17 52	☽ ⊼ ♃
21 58	☽ ⚼ ♄
19 Saturday	
01 12	♀ Q ♆
03 32	☽ ⊼ ♂
05 25	☿ △ ♄
12 39	☽ ± ♄
13 46	☽ ⚹ ♆
14 07	☽ △ ♇
15 13	☿ ⊥ ♆
19 07	☽ □ ♅
22 26	☽ ⊼ ♄
20 Sunday	
00 38	☽ ⊻ ☿
02 11	☽ ⊻ ☉
06 18	☽ ∥ ♀
09 24	☽ Q ♆
11 41	☽ ⊥ ☿
12 14	☽ ☌ ♀
12 43	☽ ⊥ ☉
14 37	☽ ⚼ ♇
16 35	☽ # ♂
16 59	☽ # ♃
18 39	☽ ☍ ♃
21 Monday	
02 25	☽ # ♇
02 25	☉ ⚼ ♂
04 06	☿ ⊥ ♀
04 20	☽ ∠ ☿
04 25	☽ ☍ ♂
04 34	☽ ∠ ☉
05 01	☿ ⚼ ♂
05 22	☽ ± ♇
09 15	☉ ☌ ☿
15 00	☽ □ ♆
15 15	☽ ⊼ ♇
16 52	♀ ⚼ ♇
20 25	☽ △ ♅
21 08	☉ ⊥ ♀
23 34	☽ ☍ ♄
23 58	☿ ∥ ♆
22 Tuesday	
00 17	☽ # ♄
07 13	☽ ⚹ ☉
08 22	☽ ⚹ ☿
18 18	☽ ⊻ ♀
20 00	☽ ⊼ ♃
21 27	☽ ⚼ ♅
22 00	☿ ± ♃
23 Wednesday	
04 51	☽ Q ☉
05 20	☽ ⊥ ♀
06 01	☽ ⊼ ♂
06 02	☽ ± ♃
07 11	☽ Q ☿
10 42	☽ ∥ ♄
12 40	☽ ± ♅
13 04	♀ ☍ ♃
13 46	♀ # ♂
16 07	☽ ± ♂
17 18	☽ △ ♆
17 27	☽ ☍ ♇
20 13	♀ # ♃
20 15	☉ ± ♃
21 13	☽ ⚼ ♃
22 02	☽ ∠ ♀
22 55	☽ ⊼ ♅
24 Thursday	
01 57	☽ ⊼ ♄
07 24	☽ ⚼ ♂
09 31	☽ ∥ ♇
12 15	☽ ± ♄
13 58	☽ □ ☉
15 52	♇ St R
17 26	☽ # ♀
18 14	☽ □ ☿
19 08	☽ ⚼ ♆
20 16	☽ ∥ ♃
21 24	☽ ∥ ♂
22 54	☽ △ ♃
25 Friday	
01 35	☿ ∠ ♆
02 25	☽ ⚹ ♀
03 50	☽ ⚼ ♄
05 39	☿ Q ♅
09 15	☽ △ ♂
10 53	☽ ± ♆
21 29	☽ ⊼ ♆
21 32	☽ ⊼ ♇
26 Saturday	
01 36	☽ Q ♀
03 20	☽ ☍ ♅
05 10	☿ ± ♂
06 09	☿ ⊼ ♃
06 15	☽ △ ♄
08 12	☽ ± ♇
16 49	♆ ⚹ ♇
20 41	☉ ∥ ♅
23 16	☽ △ ☉
27 Sunday	
00 22	☽ ⚼ ♇
03 50	☽ □ ♃
07 17	☽ △ ☿
09 33	♀ # ♇
13 16	☽ □ ♀
14 16	☉ ∠ ♆
14 26	☽ □ ♂
21 14	☉ Q ♅
21 42	☿ ⚼ ♄
28 Monday	
00 36	♀ ☍ ♂
03 46	☽ △ ♇
03 51	☽ ☍ ♆
04 58	☽ ⚼ ☉
05 17	☽ # ☉
09 57	☽ ⊼ ♅
12 38	☽ Q ♃
12 43	☽ □ ♄
14 01	☽ # ♅
15 06	☽ ⚼ ☿
16 43	☽ # ☿
20 27	☿ ∥ ♅
21 08	☽ ± ♅
23 08	☽ ± ☉
23 20	☽ Q ♂
29 Tuesday	
01 13	☉ ± ♂
02 55	♀ ± ♇
06 03	☉ ⊼ ♃
09 25	☿ ⊼ ♂
10 38	☽ ± ☿
10 55	☽ ⚹ ♃
11 23	☽ ⊼ ☉
14 06	☽ ⚼ ♅
21 40	☽ ⚹ ♂
22 27	☽ Q ♄
23 50	☽ ⊼ ☿
30 Wednesday	
03 06	☽ △ ♀
05 05	☽ # ♅
12 15	☽ □ ♇
12 28	☽ ⊼ ♆
15 17	☽ ∠ ♃
16 03	☉ ∥ ☿
18 48	☽ △ ♅
21 23	☽ ⚹ ♄
31 Thursday	
00 04	☽ ± ♆
02 02	☽ ∠ ♂
04 14	☽ # ☉
07 44	☽ # ☿
08 33	☽ ⊥ ♃
11 08	☽ ⚼ ♀
17 37	☽ ⚼ ♆
19 17	☽ ⊥ ♂
20 10	☽ ⊻ ♃
23 11	☽ Q ♇

FEBRUARY 1980

LONGITUDES

Date	Sidereal time h m s	Sun ☉ ° ' "	Moon ☽ ° ' "	Moon ☽ 24.00 ° ' "	Mercury ☿ ° '	Venus ♀ ° '	Mars ♂ ° '	Jupiter ♃ ° '	Saturn ♄ ° '	Uranus ♅ ° '	Neptune ♆ ° '	Pluto ♇ ° '
01	20 43 27	11 ♒ 46 54	16 ♌ 15 55	22 ♌ 18 52	19 ♒ 41	19 ♓ 51	13 ♍ 37	08 ♍ 09	26 ♍ 26	25 ♏ 13	21 ♐ 56	21 ♎ 45
02	20 47 23	12 47 46	28 ♌ 19 36	04 ♍ 18 23	21 28	21 03	13 R 24	08 R 03	26 R 24	25 15	21 58	21 R 45
03	20 51 20	13 48 37	10 ♍ 15 29	16 ♍ 11 12	23 14	22 15	13 10	07 56	26 21	25 16	21 59	21 45
04	20 55 16	14 49 28	22 ♍ 05 55	28 ♍ 00 00	25 00	23 27	12 56	07 49	26 18	25 17	22 01	21 44
05	20 59 13	15 50 17	03 ♎ 53 54	09 ♎ 48 03	26 46	24 39	12 41	07 43	26 15	25 19	22 03	21 44
06	21 03 10	16 51 05	15 ♎ 42 59	21 ♎ 39 11	28 ♒ 32	25 50	12 24	07 36	26 12	25 20	22 04	21 44
07	21 07 06	17 51 53	27 ♎ 37 14	03 ♏ 37 41	00 ♓ 17	27 02	12 08	07 29	26 09	25 21	22 05	21 43
08	21 11 03	18 52 39	09 ♏ 41 07	15 ♏ 48 08	02 01	28 14	11 50	07 22	26 06	25 22	22 07	21 43
09	21 14 59	19 53 25	21 ♏ 59 17	28 ♏ 15 10	03 44	29 ♓ 25	11 32	07 14	26 02	25 23	22 08	21 42
10	21 18 56	20 54 09	04 ♐ 36 16	11 ♐ 03 05	05 26	00 ♈ 37	11 14	07 07	25 59	25 25	22 10	21 42
11	21 22 52	21 54 53	17 ♐ 36 01	24 ♐ 15 24	07 05	01 48	10 54	07 00	25 55	25 26	22 11	21 41
12	21 26 49	22 55 35	01 ♑ 01 27	07 ♑ 54 15	08 42	02 59	10 34	06 52	25 52	25 26	22 12	21 40
13	21 30 45	23 56 17	14 ♑ 53 46	21 ♑ 59 47	10 16	04 10	10 14	06 45	25 48	25 27	22 14	21 40
14	21 34 42	24 56 57	29 ♑ 11 56	06 ♒ 29 40	11 47	05 21	09 53	06 37	25 45	25 28	22 15	21 39
15	21 38 39	25 57 36	13 ♒ 52 16	21 ♒ 18 55	13 14	06 32	09 32	06 30	25 41	25 29	22 16	21 38
16	21 42 35	26 58 13	28 ♒ 48 37	06 ♓ 20 18	14 35	07 43	09 10	06 22	25 37	25 30	22 18	21 38
17	21 46 32	27 58 49	13 ♓ 52 50	21 ♓ 25 07	15 52	08 54	08 47	06 14	25 33	25 30	22 19	21 37
18	21 50 28	28 59 23	28 ♓ 56 00	06 ♈ 24 30	17 02	10 04	08 25	06 06	25 29	25 31	22 20	21 36
19	21 54 25	29 ♒ 59 56	13 ♈ 49 39	21 ♈ 10 40	18 05	11 15	08 02	05 59	25 25	25 32	22 21	21 35
20	21 58 21	01 ♓ 00 27	28 ♈ 26 55	05 ♉ 37 53	19 01	12 25	07 39	05 51	25 21	25 32	22 22	21 34
21	22 02 18	02 00 55	12 ♉ 43 14	19 ♉ 42 47	19 49	13 35	07 15	05 43	25 17	25 33	22 23	21 33
22	22 06 14	03 01 23	26 ♉ 36 27	03 ♊ 24 19	20 28	14 45	06 52	05 35	25 13	25 33	22 24	21 32
23	22 10 11	04 01 48	10 ♊ 06 30	16 ♊ 43 16	20 58	15 55	06 28	05 27	25 09	25 33	22 25	21 31
24	22 14 08	05 02 11	23 ♊ 14 52	29 ♊ 41 39	21 18	17 05	06 04	05 19	25 04	25 34	22 26	21 30
25	22 18 04	06 02 32	06 ♋ 03 59	12 ♋ 22 13	21 29	18 14	05 40	05 11	25 00	25 34	22 27	21 29
26	22 22 01	07 02 52	18 ♋ 36 43	24 ♋ 47 52	21 R 29	19 23	05 17	05 04	24 56	25 34	22 28	21 28
27	22 25 57	08 03 09	00 ♌ 56 00	07 ♌ 01 29	21 20	20 33	04 53	04 56	24 51	25 34	22 29	21 27
28	22 29 54	09 03 25	13 ♌ 04 38	19 ♌ 05 44	21 02	21 42	04 29	04 48	24 47	25 34	22 30	21 26
29	22 33 50	10 ♓ 03 38	25 ♌ 05 06	01 ♍ 02 59	20 ♓ 34	22 ♈ 51	04 ♍ 06	04 ♍ 40	24 ♍ 42	25 ♏ 34	22 ♐ 31	21 ♎ 25

MOON NODES AND LATITUDE

Date	Moon True ☊ ° '	Moon Mean ☊ ° '	Moon ☽ Latitude ° '
01	29 ♌ 15	00 ♍ 14	01 S 11
02	29 R 14	00 11	00 S 05
03	29 D 15	00 07	01 N 01
04	29 16	00 04	02 03
05	29 17	00 ♍ 01	03 01
06	29 18	29 ♌ 58	03 50
07	29 19	29 55	04 30
08	29 20	29 52	04 59
09	29 R 20	29 48	05 15
10	29 20	29 45	05 16
11	29 19	29 42	05 02
12	29 19	29 39	04 31
13	29 18	29 36	03 43
14	29 17	29 33	02 40
15	29 17	29 29	01 25
16	29 17	29 26	00 N 03
17	29 D 17	29 23	01 S 20
18	29 17	29 20	02 38
19	29 R 17	29 17	03 43
20	29 17	29 13	04 33
21	29 17	29 10	05 05
22	29 16	29 07	05 18
23	29 D 17	29 04	05 12
24	29 17	29 01	04 51
25	29 17	28 58	04 15
26	29 18	28 54	03 28
27	29 19	28 51	02 31
28	29 20	28 48	01 29
29	29 ♌ 20	28 ♌ 45	00 S 24

DECLINATIONS

Date	Sun ☉ ° '	Moon ☽ ° '	Mercury ☿ ° '	Venus ♀ ° '	Mars ♂ ° '	Jupiter ♃ ° '	Saturn ♄ ° '	Uranus ♅ ° '	Neptune ♆ ° '	Pluto ♇ ° '
01	17 S 15	14 N 50	16 S 40	04 S 58	10 N 14	09 N 41	03 N 32	18 S 48	21 S 51	07 N 37
02	16 58	11 59	16 02	04 27	10 21	09 43	03 33	18 48	21 51	07 38
03	16 41	08 40	15 22	03 56	10 27	09 46	03 34	18 48	21 51	07 39
04	16 23	05 01	14 42	03 25	10 34	09 49	03 36	18 49	21 51	07 39
05	16 05	01 N 13	14 00	02 54	10 41	09 51	03 37	18 49	21 51	07 40
06	15 47	02 S 38	13 17	02 23	10 49	09 54	03 38	18 49	21 51	07 40
07	15 29	06 25	12 33	01 51	10 56	09 57	03 40	18 50	21 51	07 41
08	15 10	09 59	11 48	01 20	11 04	10 00	03 41	18 50	21 51	07 42
09	14 51	13 11	11 02	00 48	11 12	10 02	03 43	18 50	21 51	07 43
10	14 32	15 52	10 16	00 S 17	11 20	10 05	03 44	18 50	21 51	07 43
11	14 12	17 51	09 30	00 N 15	11 28	10 08	03 46	18 51	21 51	07 44
12	13 52	18 56	08 43	00 46	11 37	10 11	03 48	18 51	21 51	07 45
13	13 33	18 55	07 57	01 18	11 45	10 14	03 49	18 51	21 51	07 45
14	13 12	17 43	07 11	01 49	11 54	10 17	03 51	18 51	21 52	07 46
15	12 52	15 19	06 25	02 21	12 02	10 20	03 53	18 51	21 52	07 47
16	12 31	11 51	05 41	02 52	12 11	10 23	03 54	18 52	21 52	07 48
17	12 11	07 35	04 58	03 23	12 20	10 26	03 56	18 52	21 52	07 48
18	11 50	02 S 50	04 17	03 55	12 29	10 29	03 58	18 52	21 52	07 49
19	11 28	02 N 01	03 39	04 26	12 37	10 32	03 59	18 52	21 52	07 50
20	11 07	06 40	03 03	04 57	12 46	10 35	04 01	18 52	21 52	07 50
21	10 45	10 49	02 30	05 28	12 55	10 38	04 03	18 52	21 52	07 51
22	10 24	14 15	02 01	05 59	13 04	10 41	04 05	18 52	21 52	07 52
23	10 02	16 49	01 35	06 29	13 12	10 44	04 07	18 52	21 52	07 53
24	09 40	18 26	01 14	07 00	13 21	10 47	04 08	18 52	21 52	07 54
25	09 18	19 03	00 57	07 30	13 29	10 50	04 10	18 53	21 52	07 54
26	08 55	18 43	00 45	08 01	13 38	10 53	04 12	18 53	21 52	07 55
27	08 33	17 29	00 37	08 31	13 46	10 56	04 14	18 53	21 52	07 56
28	08 10	15 28	00 35	09 01	13 54	10 59	04 16	18 53	21 52	07 57
29	07 S 48	12 N 47	00 S 37	09 N 30	14 N 02	11 N 01	04 N 18	18 S 53	21 S 52	07 N 57

ZODIAC SIGN ENTRIES

Date	h m	Planets
02	15 21	☽ ♍
05	04 04	☽ ♎
07	08 07	☿ ♓
07	16 46	☽ ♏
09	23 39	♀ ♈
10	03 19	☽ ♐
12	10 12	☽ ♑
14	13 20	☽ ♒
16	13 54	☽ ♓
18	13 42	☽ ♈
19	12 02	☉ ♓
20	14 35	☽ ♉
22	17 58	☽ ♊
25	00 34	☽ ♋
27	10 10	☽ ♌
29	21 53	☽ ♍

LATITUDES

Date	Mercury ☿ ° '	Venus ♀ ° '	Mars ♂ ° '	Jupiter ♃ ° '	Saturn ♄ ° '	Uranus ♅ ° '	Neptune ♆ ° '	Pluto ♇ ° '
01	01 S 51	01 S 02	04 N 07	01 N 15	02 N 18	00 N 17	01 N 21	17 N 20
04	01 36	00 53	04 11	01 16	02 19	00 17	01 21	17 22
07	01 15	00 44	04 15	01 16	02 20	00 17	01 21	17 23
10	00 48	00 34	04 18	01 17	02 20	00 17	01 21	17 25
13	00 S 15	00 24	04 21	01 17	02 21	00 17	01 21	17 27
16	00 N 25	00 13	04 22	01 18	02 21	00 17	01 21	17 28
19	01 09	00 S 01	04 23	01 18	02 22	00 17	01 22	17 30
22	01 55	00 N 11	04 23	01 18	02 22	00 17	01 22	17 31
25	02 39	00 23	04 22	01 19	02 23	00 17	01 22	17 32
28	03 14	00 36	04 20	01 19	02 23	00 17	01 22	17 34
31	03 N 37	00 N 49	04 N 17	01 N 19	02 N 24	00 N 17	01 N 22	17 N 35

DATA

Julian Date	2444271
Delta T	+51 seconds
Ayanamsa	23° 34' 37"
Synetic vernal point	05° ♓ 32' 23"
True obliquity of ecliptic	23° 26' 23"

MOON'S PHASES, APSIDES AND POSITIONS ☽

Date	h m	Phase	Longitude ° '	Eclipse Indicator
01	02 21	○	11 ♌ 22	
09	07 35	☾	19 ♏ 42	
16	08 51	●	26 ♒ 50	Total
23	00 14	☽	03 ♊ 32	

Day	h m		
05	01 44	Apogee	
17	08 40	Perigee	
05	19 33	0S	
12	23 53	Max dec	19° S 04'
19	01 58	0N	
25	15 25	Max dec	19° N 04'

LONGITUDES

Date	Chiron ⚷ ° '	Ceres ⚳ ° '	Pallas ⚴ ° '	Juno ⚵ ° '	Vesta ⚶ ° '	Black Moon Lilith ⚸ ° '
01	09 ♉ 13	17 ♈ 11	07 ♓ 16	17 ♋ 54	08 ♉ 46	22 ♍ 55
11	09 ♉ 24	20 ♈ 20	10 ♓ 32	16 ♋ 37	11 ♉ 24	24 ♍ 02
21	09 ♉ 41	23 ♈ 41	13 ♓ 52	16 ♋ 09	14 ♉ 23	25 ♍ 08
31	10 ♉ 02	27 ♈ 13	17 ♓ 15	16 ♋ 29	17 ♉ 38	26 ♍ 15

All ephemeris data is given at 12.00 UT and the Moon's longitude is additionally given for 24.00 UT

ASPECTARIAN

h m	Aspects
01 Friday	
02 21	☽ ☍ ☉
02 31	☽ ∠ ♄
04 13	☉ ⚼ ♄
06 42	☽ ± ♀
06 52	☽ ⊻ ♂
18 29	☿ ⊻ ♀
19 52	☽ ⊼ ♀
19 57	☽ ☍ ☿
20 14	☽ ⊥ ♄
21 54	☿ ± ♄
22 53	☽ ⚹ ♇
23 17	☽ △ ♆
02 Saturday	
05 50	☽ □ ♅
08 09	☽ ⊻ ♄
15 57	☿ △ ♇
18 56	☿ ⚹ ♆
23 46	☽ ∥ ♂
23 46	☉ ⊼ ♂
03 Sunday	
02 04	♀ ⊼ ♇
03 19	♂ Q ♅
04 23	☽ ∥ ♃
04 55	☽ ∠ ♇
06 46	♀ □ ♆
07 21	☽ ☌ ♃
17 47	☽ ☌ ♂
18 06	☽ Q ♅
18 52	☽ ∥ ♇
19 51	☽ ⊼ ☉
23 06	☽ ⊥ ♇
04 Monday	
04 15	♀ # ♄
09 10	☽ ± ☉
11 16	☽ ⊻ ♇
11 50	☽ □ ♆
15 03	☽ ☍ ♀
15 58	☿ □ ♅
18 30	☽ ⚹ ♅
18 57	☽ ⊼ ☿
20 30	☽ ☌ ♄
21 02	☽ ∥ ♄
23 48	☽ # ♀
05 Tuesday	
05 07	☿ ⊼ ♄
05 11	☽ ⚼ ☉
09 18	☽ ± ☿
19 40	☽ ⊻ ♃
06 Wednesday	
00 31	☽ Q ♆
01 04	☽ ∠ ♅
01 38	♀ △ ♅
05 27	☽ ⊻ ♂
06 48	☽ ⚼ ☿
07 44	☽ ⊥ ♃
10 33	☽ ∥ ♀
14 31	☽ △ ☉
17 19	☽ ⊥ ♂
18 19	☽ # ♄
18 53	♀ ☍ ♄
19 20	☽ ⊥ ♅
07 Thursday	
00 09	☽ ☌ ♇
00 52	☽ ⚹ ♆
01 46	☽ ∠ ♃
07 27	☽ ⊻ ♅
09 03	☽ ⊻ ♄
10 42	☽ ⊼ ♀
11 02	☽ ∠ ♂
18 14	☽ △ ☿
20 22	☽ # ♇
21 01	☽ ⊥ ♄
08 Friday	
00 01	☽ ± ♀
06 55	☽ ∠ ♆
07 27	☽ ⚹ ♃
12 07	☽ # ♃
14 46	☽ ∠ ♄
16 08	☽ ⚹ ♂
19 43	☽ ⚼ ♀
20 08	☽ # ♂
22 42	☽ ∥ ☿
09 Saturday	
00 38	☽ ⊥ ♆
06 44	☽ Q ♃
07 35	☽ □ ☉
07 42	☿ # ♂
11 27	☽ ⊻ ♇
12 18	☽ ⊻ ♆
14 55	☽ Q ♂
15 20	☉ ± ♄
17 46	☿ Q ♆
18 33	☽ ☌ ♅
19 45	☽ ⚹ ♄
22 57	☽ ⊥ ♇
10 Sunday	
00 40	☽ ∥ ☉
03 42	☽ △ ♀
13 46	☽ □ ☿
15 54	☽ ∠ ♇
16 39	☽ □ ♃
17 18	☿ # ♃
18 17	☽ Q ♄
20 42	☽ Q ☉
11 Monday	
00 02	☽ □ ♂
06 10	☿ ⚼ ♇
06 34	☉ △ ♇
10 47	☿ ☍ ♃
18 34	☉ ⚹ ♆
19 23	☽ ⚹ ♇
20 18	☽ ☌ ♆
20 26	☽ ⚹ ☉
12 Tuesday	
02 06	☽ ⊻ ♅
02 55	☽ □ ♄
03 20	☽ Q ☿
08 56	☽ ∥ ♅
12 44	☽ ⊥ ♅
15 46	☽ □ ♀
16 38	☽ Q ♇
22 07	☽ △ ♃
13 Wednesday	
00 59	☽ ∠ ☉
03 06	☽ ⚹ ☿
04 12	☽ △ ♂
04 24	☽ ∠ ♅
11 29	☿ ☍ ♂
14 27	☽ ∥ ♅
17 34	☽ ⊥ ☉
17 45	☿ # ♇
23 26	☽ □ ♇
23 29	☽ ⚼ ♃
14 Thursday	
00 25	☽ ⊻ ♆
01 24	☽ Q ♀
04 24	☽ ⊻ ☉
05 00	☽ ⚼ ♂
05 48	☽ ⚹ ♅
05 52	♃ ∠ ♇
06 17	☽ △ ♄
07 32	☽ ∠ ☿
10 26	☽ ⊥ ♆
14 20	☽ ± ♃
19 32	☽ ± ♂
23 02	☽ ⚹ ♀
15 Friday	
00 02	☽ ⊥ ☿
00 06	☽ ⊼ ♃
00 32	☉ □ ♅
01 15	☽ ∠ ♆
01 36	☽ Q ♅
05 07	☽ ⊼ ♂
05 47	☉ ⊼ ♄
06 51	☽ ⚼ ♄
10 51	☽ ⊻ ☿
11 12	♀ ⊼ ♃
21 20	☽ ± ♄
16 Saturday	
00 30	☽ △ ♇
01 25	☽ ∠ ♀
01 33	☽ ⚹ ♆
06 42	☽ □ ♅
06 55	☽ ⊼ ♄
07 22	☽ ∥ ☉
08 51	☽ ☌ ☉
09 59	☽ # ♂
17 02	☽ ⊥ ♀
20 34	☽ # ♃
20 45	☽ Q ♆
23 57	☽ ☍ ♃
17 Sunday	
00 27	☽ ⚼ ♇
03 24	☽ ⊻ ♀
04 06	☽ ☍ ♂
04 28	☉ # ♂
07 13	☿ ± ♇
10 25	♀ ⊼ ♂
10 48	☽ # ♇
14 45	☽ ± ♇
15 26	☽ ☌ ☿
18 Monday	
00 18	☽ ⊼ ♇
01 27	☽ □ ♆
02 55	♄ ⚹ ♅
03 36	☽ ∥ ☿
06 27	☽ # ♄
06 31	☽ ☍ ♄
06 32	☽ △ ♅
07 12	☽ # ♀
12 06	☽ ⊻ ☉
14 31	♀ ∥ ♄
19 47	☿ # ♀
21 14	♀ ⚼ ♅
22 25	☽ ⊥ ☉
23 25	☽ ⊼ ♃
23 36	☿ # ♄
19 Tuesday	
02 51	☽ ⊼ ♂
06 39	☽ ⚼ ♅
07 27	☽ ☌ ♀
09 01	☽ ± ♃
12 20	☽ ± ♂
14 03	☽ ∠ ☉
19 15	☽ # ☿
19 26	☽ ⊻ ☿
21 18	☽ ± ♅
22 01	☽ ∥ ♄
23 34	☽ ⚼ ♃
20 Wednesday	
00 39	☽ ☍ ♇
01 33	♀ ± ♃
01 45	☽ ∥ ♀
01 57	☽ △ ♆
02 40	☽ ⚼ ♂
05 57	☽ ⊥ ☿
06 54	☽ ⊼ ♄
07 10	☽ ⊼ ♅
16 35	☽ ⚹ ☉
16 49	☽ ± ♄
18 30	☽ ∥ ♇
21 52	☽ ∠ ☿
21 Thursday	
00 15	☽ △ ♃
02 57	☽ ⚼ ♆
02 59	☽ △ ♂
06 59	♀ ± ♂
07 53	☽ ⚼ ♄
10 52	☽ ∥ ♃
11 42	☽ # ☉
13 36	☽ ⊻ ♀
14 23	☽ Q ☉
18 17	☽ ± ♆
19 29	☉ # ♃
22 Friday	
00 49	☽ ⚹ ☿
00 51	☽ ⊥ ♀
02 40	☽ ∥ ♂
03 10	☽ ⊼ ♇
04 39	☽ ⊼ ♆
09 35	☽ △ ♄
10 09	☽ ☍ ♅
13 38	☽ ± ♇
18 03	☽ ∠ ♀
22 46	☽ Q ☿
23 Saturday	
00 14	☽ □ ☉
03 44	☽ □ ♃
05 34	☽ ⚼ ♇
05 39	☽ □ ♂
08 35	♂ ∠ ♇
21 30	☉ Q ♆
23 33	☽ ⚹ ♀
24 Sunday	
08 20	☽ □ ☿
08 47	☽ △ ♇
10 30	☽ ☍ ♆
12 08	☽ Q ♃
13 29	☽ Q ♂
15 22	☽ □ ♄
16 17	☽ ⊼ ♅
18 01	☉ ☍ ♃
23 55	☽ Q ♀
25 Monday	
00 22	☽ # ♅
03 30	☽ ± ♅
05 43	☉ ☍ ♂
10 22	☽ ⚹ ♃
11 17	☽ ⚹ ♂
11 57	☽ △ ☉
14 14	☿ ⊼ ♇
20 33	☽ ⚼ ♅
22 28	☉ ⚼ ♇
26 Tuesday	
01 08	☽ Q ♄
01 33	☿ St R
06 36	☽ # ♅
07 25	♀ ∥ ♇
13 40	☽ □ ♀
14 46	☽ ∠ ♃
15 07	☽ ∠ ♂
15 39	♀ ± ♅
17 32	☽ □ ♇
17 32	☽ △ ☿
18 01	☿ ⊼ ♇
19 14	☽ ⚼ ☉
19 29	☽ ⊼ ♆
27 Wednesday	
00 11	☽ ⚹ ♄
00 28	♀ ⚼ ♃
01 30	☽ △ ♅
01 42	♀ ⚼ ♂
07 12	☽ ± ♆
07 44	♂ ☌ ♃
08 06	☽ ⊥ ♂
08 07	☽ ⊥ ♃
13 00	☉ # ♀
14 24	☽ ± ☉
19 31	☽ ⊻ ♂
19 47	☽ ⊻ ♃
22 25	☽ ⚼ ☿
28 Thursday	
00 55	☽ ⚼ ♆
01 18	☿ ⊻ ♀
03 18	☽ ⊼ ☉
04 47	☽ Q ♇
05 30	☽ ∠ ♄
06 37	♀ ☍ ♇
15 45	☽ ± ☿
23 18	☽ ⊥ ♄
29 Friday	
02 02	☽ ∥ ♂
02 07	☉ # ♇
03 18	☽ ⊼ ☿
04 39	☽ ⚹ ♇
04 57	♀ △ ♆
06 38	♅ St R
06 50	☽ △ ♆
07 02	☽ △ ♀
11 15	☽ ⊻ ♄
12 58	☽ □ ♅

MARCH 1980

LONGITUDES

Date	Sidereal time h m s	Sun ☉ ° ′ ″	Moon ☽ ° ′ ″	Moon ☽ 24.00 ° ′ ″	Mercury ☿ ° ′	Venus ♀ ° ′	Mars ♂ ° ′	Jupiter ♃ ° ′	Saturn ♄ ° ′	Uranus ♅ ° ′	Neptune ♆ ° ′	Pluto ♇ ° ′
01	22 37 47	11 ♓ 03 50	06 ♍ 59 41	12 ♍ 55 25	19 ♓ 58	23 ♈ 59	03 ♍ 42	04 ♍ 32	24 ♍ 38	25 ♏ 34	22 ♐ 32	21 ♎ 24
02	22 41 43	12 04 00	18 ♍ 50 27	24 ♍ 45 02	19 R 15	25 08	03 R 19	04 R 24	24 R 33	25 R 34	22 32	21 R 23
03	22 45 40	13 04 08	00 ♎ 39 26	06 ♎ 33 55	18 25	26 16	02 56	04 17	24 29	25 34	22 33	21 21
04	22 49 37	14 04 14	12 ♎ 28 47	18 ♎ 24 18	17 31	27 25	02 34	04 09	24 24	25 34	22 34	21 20
05	22 53 33	15 04 19	24 ♎ 20 49	00 ♏ 18 41	16 33	28 33	02 11	04 01	24 19	25 33	22 34	21 19
06	22 57 30	16 04 22	06 ♏ 18 16	12 ♏ 19 59	15 32	29 ♈ 41	01 49	03 54	24 15	25 33	22 35	21 18
07	23 01 26	17 04 23	18 ♏ 24 14	24 ♏ 31 29	14 31	00 ♉ 48	01 28	03 46	24 10	25 33	22 36	21 16
08	23 05 23	18 04 23	00 ♐ 42 12	06 ♐ 56 52	13 31	01 56	01 07	03 39	24 05	25 32	22 36	21 15
09	23 09 19	19 04 21	13 ♐ 15 57	19 ♐ 39 56	12 33	03 03	00 46	03 31	24 01	25 32	22 37	21 14
10	23 13 16	20 04 18	26 ♐ 09 17	02 ♑ 44 24	11 38	04 10	00 26	03 24	23 56	25 31	22 37	21 12
11	23 17 12	21 04 13	09 ♑ 25 40	16 ♑ 13 23	10 46	05 17	00 ♍ 07	03 17	23 51	25 31	22 38	21 11
12	23 21 09	22 04 06	23 ♑ 07 43	00 ♒ 08 45	10 00	06 23	29 ♌ 48	03 10	23 46	25 30	22 38	21 10
13	23 25 06	23 03 57	07 ♒ 16 26	14 ♒ 30 32	09 19	07 30	29 30	03 03	23 42	25 30	22 38	21 08
14	23 29 02	24 03 47	21 ♒ 50 37	29 ♒ 16 07	08 45	08 36	29 12	02 56	23 37	25 29	22 39	21 07
15	23 32 59	25 03 35	06 ♓ 46 14	14 ♓ 20 01	08 16	09 42	28 55	02 49	23 32	25 28	22 39	21 05
16	23 36 55	26 03 21	21 ♓ 56 21	29 ♓ 34 00	07 54	10 48	28 39	02 42	23 27	25 27	22 39	21 04
17	23 40 52	27 03 05	07 ♈ 11 40	14 ♈ 48 04	07 38	11 53	28 23	02 35	23 23	25 26	22 40	21 02
18	23 44 48	28 02 47	22 ♈ 21 55	29 ♈ 52 01	07 28	12 58	28 08	02 29	23 18	25 25	22 40	21 01
19	23 48 45	29 ♓ 02 27	07 ♉ 17 22	14 ♉ 37 04	07 24	14 03	27 54	02 22	23 13	25 24	22 40	20 59
20	23 52 41	00 ♈ 02 05	21 ♉ 50 28	28 ♉ 57 05	07 D 27	15 08	27 41	02 16	23 09	25 23	22 40	20 58
21	23 56 38	01 01 40	05 ♊ 56 39	12 ♊ 49 06	07 35	16 13	27 28	02 10	23 04	25 22	22 40	20 56
22	00 00 35	02 01 14	19 ♊ 34 29	26 ♊ 13 03	07 49	17 17	27 16	02 04	22 59	25 21	22 40	20 55
23	00 04 31	03 00 45	02 ♋ 45 08	09 ♋ 11 10	08 08	18 21	27 05	01 58	22 55	25 20	22 41	20 53
24	00 08 28	04 00 13	15 ♋ 31 38	21 ♋ 47 04	08 31	19 24	26 55	01 52	22 50	25 19	22 41	20 51
25	00 12 24	04 59 39	27 ♋ 58 03	04 ♌ 05 07	09 00	20 27	26 46	01 46	22 45	25 18	22 R 41	20 50
26	00 16 21	05 59 03	10 ♌ 08 51	16 ♌ 09 48	09 33	21 30	26 37	01 41	22 41	25 16	22 40	20 48
27	00 20 17	06 58 25	22 ♌ 08 29	28 ♌ 05 23	10 10	22 33	26 29	01 36	22 36	25 15	22 40	20 47
28	00 24 14	07 57 44	04 ♍ 00 57	09 ♍ 55 38	10 51	23 35	26 22	01 30	22 32	25 13	22 40	20 45
29	00 28 10	08 57 01	15 ♍ 49 47	21 ♍ 43 47	11 36	24 37	26 15	01 25	22 27	25 12	22 40	20 43
30	00 32 07	09 56 16	27 ♍ 37 54	03 ♎ 32 27	12 24	25 38	26 10	01 21	22 23	25 11	22 40	20 42
31	00 36 04	10 ♈ 55 29	09 ♎ 27 40	15 ♎ 23 47	13 ♓ 15	26 ♉ 40	26 ♌ 05	01 ♍ 16	22 ♍ 18	25 ♏ 09	22 ♐ 40	20 ♎ 40

Date	Moon True ☊ ° ′	Moon Mean ☊ ° ′	Moon ☽ Latitude ° ′
01	29 ♌ 20	28 ♌ 42	00 N 42
02	29 R 19	28 39	01 46
03	29 18	28 35	02 45
04	29 16	28 32	03 37
05	29 13	28 29	04 20
06	29 11	28 26	04 51
07	29 08	28 23	05 10
08	29 07	28 19	05 16
09	29 06	28 16	05 06
10	29 D 06	28 13	04 41
11	29 06	28 10	04 01
12	29 08	28 07	03 06
13	29 09	28 04	01 58
14	29 10	28 00	00 N 40
15	29 R 10	27 57	00 S 42
16	29 09	27 54	02 02
17	29 06	27 51	03 14
18	29 02	27 48	04 12
19	28 58	27 44	04 51
20	28 54	27 41	05 11
21	28 50	27 38	05 11
22	28 48	27 35	04 53
23	28 48	27 32	04 20
24	28 D 48	27 29	03 35
25	28 50	27 25	02 41
26	28 51	27 22	01 41
27	28 52	27 19	00 S 37
28	28 R 53	27 16	00 N 28
29	28 51	27 13	01 31
30	28 47	27 10	02 30
31	28 ♌ 42	27 ♌ 06	03 N 22

DECLINATIONS

Date	Sun ☉ ° ′	Moon ☽ ° ′	Mercury ☿ ° ′	Venus ♀ ° ′	Mars ♂ ° ′	Jupiter ♃ ° ′	Saturn ♄ ° ′	Uranus ♅ ° ′	Neptune ♆ ° ′	Pluto ♇ ° ′
01	07 S 25	09 N 36	00 S 44	10 00	14 N 09	11 N 04	04 N 20	18 S 53	21 S 52	07 N 58
02	07 02	06 03	00 56	10 29	14 17	11 07	04 22	18 53	21 52	07 59
03	06 39	02 N 16	01 12	10 58	14 24	11 10	04 24	18 53	21 52	08 00
04	06 16	01 S 36	01 32	11 27	14 31	11 13	04 25	18 52	21 52	08 01
05	05 53	05 25	01 55	11 56	14 38	11 16	04 27	18 52	21 52	08 01
06	05 30	09 02	02 21	12 24	14 44	11 19	04 29	18 52	21 52	08 02
07	05 06	12 20	02 49	12 52	14 51	11 21	04 31	18 52	21 52	08 03
08	04 43	15 09	03 18	13 20	14 57	11 24	04 33	18 52	21 52	08 04
09	04 19	17 19	03 48	13 47	15 03	11 27	04 35	18 52	21 52	08 05
10	03 56	18 42	04 19	14 14	15 08	11 29	04 37	18 52	21 52	08 05
11	03 32	19 06	04 49	14 41	15 13	11 32	04 39	18 52	21 52	08 06
12	03 09	18 24	05 18	15 08	15 18	11 35	04 41	18 52	21 52	08 07
13	02 45	16 33	05 45	15 34	15 23	11 37	04 43	18 51	21 52	08 08
14	02 21	13 35	06 11	16 00	15 27	11 40	04 45	18 51	21 52	08 09
15	01 58	09 40	06 35	16 25	15 31	11 42	04 47	18 51	21 52	08 09
16	01 34	05 04	06 57	16 50	15 35	11 44	04 49	18 51	21 52	08 10
17	01 10	00 S 07	07 17	17 15	15 38	11 47	04 51	18 51	21 52	08 11
18	00 47	04 N 49	07 35	17 40	15 42	11 49	04 53	18 50	21 52	08 12
19	00 S 23	09 21	07 50	18 04	15 44	11 51	04 54	18 50	21 52	08 12
20	00 N 01	13 13	08 02	18 27	15 47	11 54	04 56	18 50	21 52	08 13
21	00 25	16 12	08 12	18 50	15 49	11 56	04 58	18 50	21 52	08 14
22	00 48	18 10	08 20	19 13	15 51	11 58	05 00	18 49	21 52	08 15
23	01 12	19 05	08 26	19 35	15 53	12 00	05 02	18 49	21 52	08 15
24	01 35	18 59	08 29	19 57	15 54	12 02	05 04	18 49	21 51	08 16
25	01 59	17 56	08 30	20 19	15 55	12 04	05 06	18 48	21 51	08 17
26	02 23	16 05	08 28	20 40	15 55	12 06	05 07	18 48	21 51	08 18
27	02 46	13 33	08 25	21 00	15 56	12 08	05 09	18 48	21 51	08 18
28	03 10	10 28	08 20	21 20	15 56	12 09	05 11	18 48	21 51	08 19
29	03 33	06 59	08 12	21 40	15 56	12 11	05 13	18 47	21 51	08 20
30	03 56	03 14	08 03	21 59	15 55	12 13	05 14	18 47	21 51	08 21
31	04 N 19	00 S 39	07 S 52	22 N 18	15 N 55	12 N 14	05 N 16	18 S 46	21 S 51	08 N 21

ZODIAC SIGN ENTRIES

Date	h	m	Planets
03	10	40	☽ ♎
05	23	22	☽ ♏
06	18	54	♀ ♉
08	10	38	☽ ♐
10	19	02	☽ ♑
11	20	46	♂ ♌
12	23	45	☽ ♒
15	01	10	☽ ♓
17	00	41	☽ ♈
19	00	13	☽ ♉
20	11	10	☉ ♈
21	01	47	☽ ♊
23	06	55	☽ ♋
25	15	58	☽ ♌
28	03	52	☽ ♍
30	16	49	☽ ♎

LATITUDES

Date	Mercury ☿ ° ′	Venus ♀ ° ′	Mars ♂ ° ′	Jupiter ♃ ° ′	Saturn ♄ ° ′	Uranus ♅ ° ′	Neptune ♆ ° ′	Pluto ♇ ° ′
01	03 N 31	00 N 45	04 N 18	01 N 19	02 N 24	00 N 17	01 N 22	17 N 34
04	03 41	00 58	04 15	01 19	02 24	00 17	01 22	17 36
07	03 33	01 11	04 10	01 19	02 24	00 17	01 22	17 37
10	03 08	01 25	04 05	01 19	02 24	00 17	01 22	17 38
13	02 30	01 39	03 59	01 19	02 25	00 17	01 22	17 39
16	01 47	01 52	03 53	01 19	02 25	00 17	01 23	17 39
19	01 02	02 06	03 46	01 19	02 25	00 17	01 23	17 40
22	00 N 20	02 19	03 39	01 19	02 25	00 17	01 23	17 41
25	00 S 19	02 32	03 32	01 18	02 25	00 17	01 23	17 41
28	00 54	02 45	03 25	01 18	02 25	00 17	01 23	17 42
31	01 S 23	02 N 58	03 N 17	01 N 18	02 N 25	00 N 17	01 N 23	17 N 42

DATA

Julian Date	2444300
Delta T	+51 seconds
Ayanamsa	23° 34′ 40″
Synetic vernal point	05° ♓ 32′ 20″
True obliquity of ecliptic	23° 26′ 23″

MOON'S PHASES, APSIDES AND POSITIONS ☽

Date	h	m	Phase	Longitude ° ′	Eclipse Indicator
01	21	00	○	11 ♍ 26	
09	23	49	☾	19 ♐ 34	
16	18	56	●	26 ♓ 21	
23	12	31	☽	03 ♋ 02	
31	15	14	○	11 ♎ 03	

Day	h	m		
03	10	33	Apogee	
16	20	23	Perigee	
30	11	17	Apogee	
04	02	04	0S	
11	09	08	Max dec	19° S 06′
17	12	33	0N	
23	21	27	Max dec	19° N 09′
31	08	01	0S	

LONGITUDES

Date	Chiron ⚷ ° ′	Ceres ⚳ ° ′	Pallas ⚴ ° ′	Juno ⚵ ° ′	Vesta ⚶ ° ′	Black Moon Lilith ⚸ ° ′
01	10 ♉ 00	26 ♈ 51	16 ♓ 55	16 ♋ 25	17 ♉ 18	26 ♍ 08
11	10 ♉ 26	00 ♉ 31	20 ♓ 20	17 ♋ 24	20 ♉ 45	27 ♍ 15
21	10 ♉ 56	04 ♉ 17	23 ♓ 46	19 ♋ 01	24 ♉ 24	28 ♍ 21
31	11 ♉ 29	08 ♉ 10	27 ♓ 13	21 ♋ 10	28 ♉ 13	29 ♍ 28

All ephemeris data is given at 12.00 UT and the Moon's longitude is additionally given for 24.00 UT

ASPECTARIAN

h m Aspects

01 Saturday
01 29 ☽ ∥ ♃
05 34 ☽ ☌ ♂
07 05 ☽ ☌ ♃
09 31 ☽ ∥ ♀
10 48 ☽ ∠ ♇
16 28 ☽ ⚼ ♀
21 00 ☽ ☍ ☉
23 12 ☽ ∥ ♇

02 Sunday
00 36 ♀ ⊼ ♄
01 18 ☽ Q ♅
04 43 ☽ # ☉
04 59 ☽ ⊥ ♇
12 40 ☽ ± ♀
12 46 ☽ ☍ ☿
13 24 ☿ ⊥ ♀
17 08 ☽ ⊻ ♇
19 31 ☽ □ ♆
21 06 ♀ ⊼ ♅
22 43 ☽ ∥ ♄
23 32 ☽ ☌ ♄

03 Monday
01 39 ☽ ⚹ ♅
02 09 ☽ ⊼ ♀
16 29 ☽ ⊻ ♂
18 07 ☽ # ☿
19 17 ☽ ⊻ ♃
22 51 ♀ ∥ ♃

04 Tuesday
04 18 ☽ ⊥ ♂
07 19 ☽ ⊥ ♃
08 06 ☽ Q ♆
08 07 ☽ ∠ ♅
11 33 ☽ ∥ ☿
15 31 ☽ ⊼ ☉
21 26 ☽ ⊼ ☿
21 59 ☽ ∠ ♂

05 Wednesday
01 22 ☽ ∠ ♃
02 20 ☽ ⊥ ♅
04 47 ☽ ± ☉
05 51 ☽ # ♄
05 54 ☽ ☌ ♇
08 25 ☽ ⚹ ♆
08 39 ☽ ± ☿
11 57 ☽ ⊻ ♄
14 26 ☽ ⊻ ♅
14 44 ☽ ∥ ☉
17 43 ☉ ± ♇
21 20 ☽ ☍ ♀
23 57 ☽ ⊥ ♄

06 Thursday
00 34 ☽ ⚼ ☉
01 22 ☽ ⚼ ☿
03 18 ☽ ⚹ ♂
05 11 ☽ # ♇
05 39 ☉ ☌ ☿
07 14 ☽ ⚹ ♃
14 33 ☽ ∠ ♆
17 50 ☽ ∠ ♄
17 56 ☿ ± ♇
21 41 ☿ ∠ ♀
23 22 ♀ ± ♄

07 Friday
02 31 ☽ Q ♂
04 28 ☽ # ♃
04 56 ☽ △ ☿
06 51 ☽ Q ♃
08 26 ☽ ⊥ ♆
09 09 ☽ △ ☉
17 02 ☽ # ♀
17 38 ☽ ⊻ ♇
20 14 ☽ ⊻ ♆
22 46 ♀ △ ♂
23 14 ☽ ⚹ ♄

08 Saturday
01 59 ☽ ☌ ♅
05 19 ☽ ⊥ ♇
10 03 ☽ # ♂
12 47 ☽ □ ♂
14 36 ☽ ⊼ ♀
17 37 ☽ □ ♃
21 08 ☉ # ♄
22 18 ☽ Q ♄
22 39 ☽ ∠ ♇

09 Sunday
03 14 ☽ ± ♀
10 44 ☽ □ ☿
21 10 ♀ △ ♃
21 50 ☽ ⚼ ♀
23 49 ☽ □ ☉

10 Monday
01 50 ☉ ∥ ☿
02 53 ☽ ⚹ ♇
05 29 ☽ ☌ ♆
07 56 ☽ □ ♄
10 50 ☽ ⊻ ♅
17 02 ☽ ∥ ♅
17 57 ☽ Q ☿
19 38 ☽ △ ♂
21 47 ☽ ⊥ ♅

11 Tuesday
00 49 ☽ Q ♇
01 04 ☽ △ ♃
03 47 ☿ # ♄
03 54 ☽ △ ♀
11 19 ☽ Q ☉
13 56 ☽ ∠ ♅
14 15 ☽ ⚹ ☿
14 38 ☉ ⊼ ♇
21 50 ☽ ⚼ ♂

12 Wednesday
01 03 ☽ ∥ ♅
03 28 ☽ ⚼ ♃
08 36 ☽ □ ♇
10 02 ☽ ⚹ ☉
11 09 ☽ ⊻ ♆
13 06 ☽ △ ♄
13 08 ☽ ± ♂
15 04 ☽ ∠ ☿
16 05 ☽ ⚹ ♅
18 52 ☽ ± ♃
21 26 ☽ ⊥ ♆
23 10 ☽ ⊼ ♂
23 38 ♀ ∥ ♂

13 Thursday
01 42 ☉ □ ♆
04 57 ☽ ⊼ ♃
05 40 ☽ ⊥ ☿
12 22 ☽ Q ♅
12 24 ☽ □ ♀
12 37 ☽ ∠ ♆
13 25 ☽ ∠ ☉
14 22 ☽ ⚼ ♄
15 08 ♀ ⚼ ♆
15 17 ☽ ⊻ ☿
19 48 ☽ # ♀
22 17 ☽ # ♂

14 Friday
02 02 ☉ ☍ ♄
05 08 ☽ ± ♄
05 23 ☽ ⊥ ☉
10 49 ☽ △ ♇
12 19 ♀ ⚼ ♄
13 18 ☽ ⚹ ♆
14 06 ☿ ⚹ ♀
14 52 ☽ ⊼ ♄
15 52 ☽ ⊻ ☉
17 53 ☽ □ ♅
20 19 ☽ Q ♀
23 40 ☽ ☍ ♂

15 Saturday
00 20 ☽ # ♃
05 44 ☽ ☍ ♃
08 37 ☽ Q ♆
10 55 ☽ ⚼ ♇
14 19 ☽ ☌ ☿
17 01 ☽ ⚹ ♀
20 12 ☽ # ♇
21 40 ☉ △ ♅

16 Sunday
01 10 ☽ ± ♇
03 09 ☽ ∥ ☿
10 37 ☽ ⊼ ♇
13 08 ☽ □ ♆
13 15 ☽ # ♄
14 23 ☽ ☍ ♄
17 32 ☽ △ ♅
18 32 ☽ ∠ ♀
18 56 ☽ ☌ ☉
22 23 ☽ ⊼ ♂

17 Monday
04 48 ☽ ⊼ ♃
06 29 ☽ ∥ ☉
07 39 ☽ ± ♂
09 47 ☽ ⊥ ♀
12 40 ☽ ⊻ ☿
14 11 ☽ ± ♃
17 06 ☽ ⚼ ♅
17 44 ☽ # ☉
19 58 ☽ ⊻ ♀
21 36 ☽ ⚼ ♂
22 01 ☽ ⊥ ☿

18 Tuesday
04 18 ☽ ⚼ ♃
07 20 ☽ ± ♅
09 51 ☽ ☍ ♇
12 10 ☽ ∠ ☿
12 20 ☽ ∥ ♄
12 29 ☽ △ ♆
13 29 ☽ ⊼ ♄
13 49 ☉ ⊼ ♂
16 52 ☽ ⊼ ♅
21 05 ☽ △ ♂
21 43 ☽ ⊻ ☉
23 02 ☽ ± ♄

19 Wednesday
03 08 ☽ # ☿
04 05 ☽ △ ♃
05 38 ☽ ∥ ♇
07 34 ☉ ∠ ♀
08 05 ☽ ⊥ ☉
12 12 ☽ ⚹ ☿
12 37 ☽ ⚼ ♆
13 31 ☽ ⚼ ♄
13 59 ☿ St D
23 52 ☽ ∠ ☉
23 58 ☽ ☌ ♀

20 Thursday
03 07 ☽ ∥ ♃
03 23 ☽ ± ♆
07 59 ☽ Q ☿
10 32 ☽ ⊼ ♇
13 23 ☽ ⊼ ♆
14 10 ☽ △ ♄
17 57 ☽ ☍ ♅
20 36 ☽ ± ♇
21 42 ☽ □ ♂

21 Friday
02 54 ☽ ⚹ ☉
05 32 ☽ □ ♃
08 24 ☽ ∥ ♂
11 19 ♀ # ♅
11 59 ☽ ⚼ ♇
14 53 ☽ □ ☿
16 42 ☿ # ♇
22 25 ♀ ± ♆

22 Saturday
01 21 ☽ Q ☉
04 27 ☽ Q ♂
07 33 ☽ ⊻ ♀
12 52 ☽ Q ♃
12 56 ☉ ⊼ ♃
14 23 ☽ △ ♇
17 35 ☽ ☍ ♆
18 06 ☽ □ ♄
19 15 ☽ ⊥ ♀
22 25 ☽ ⊼ ♅

23 Sunday
01 44 ☽ ⚹ ♂
01 55 ☽ # ♅
09 23 ☽ ± ♅
10 33 ☽ ⚹ ♃
12 31 ☽ □ ☉
13 11 ☽ ∠ ♀
13 34 ☉ ± ♂
22 19 ☽ △ ☿

24 Monday
02 08 ☽ ⚼ ♅
03 09 ☽ Q ♄
05 14 ☽ ∠ ♂
14 32 ☽ ∠ ♃
17 39 ☽ # ♅
17 40 ♆ St R
20 06 ☽ ⚹ ♀
22 11 ☽ □ ♇
22 12 ☽ ⊥ ♂

25 Tuesday
01 43 ☽ ⊼ ♆
01 56 ☽ ⚹ ♄
03 57 ☽ ⚼ ☿
06 48 ☽ △ ♅
07 46 ☽ ⊥ ♃
09 40 ☽ ⊻ ♂
13 23 ☽ ± ♆
19 24 ☽ ⊻ ♃
20 21 ♀ ⊼ ♇
21 37 ☽ Q ♀
22 18 ☽ ± ☿

26 Wednesday
03 01 ☽ △ ☉
07 06 ☽ ⚼ ♆
07 08 ☽ ∠ ♄
09 20 ☽ Q ♇
10 45 ☽ ⊼ ☿
13 32 ♄ □ ♆
13 44 ☽ ∥ ♂
14 14 ☿ Q ♀

27 Thursday
00 57 ☽ ⊥ ♄
09 16 ☽ ⚹ ♇
11 38 ☽ ⚼ ☉
12 54 ☽ □ ♀
12 56 ☽ ⊻ ♄
13 04 ☽ △ ♆
13 14 ♀ △ ♄
14 55 ♀ ⊼ ♆
18 15 ☽ □ ♅
20 40 ☽ ☌ ♂
23 27 ☽ ∥ ♃

28 Friday
01 51 ☉ ± ♃
06 57 ☽ ☌ ♃
07 28 ☽ ± ☉
13 56 ☿ # ♇
15 31 ☽ ∠ ♇
20 44 ☽ ⊼ ☉

29 Saturday
02 47 ☽ ☍ ☿
03 02 ☽ ∥ ♇
03 32 ☽ # ☿
06 40 ☽ Q ♅
09 45 ☽ ⊥ ♇
21 56 ☽ ⊻ ♇
23 26 ☽ ∥ ♄

30 Sunday
01 21 ♀ ☍ ♅
01 23 ☽ ☌ ♄
01 54 ☽ □ ♆
02 13 ♀ # ♆
07 01 ☽ ⚹ ♅
07 34 ☽ △ ♀
08 00 ☽ ∥ ☉
09 02 ☽ ⊻ ♂
17 39 ☉ ⚼ ♅
19 29 ☽ ⊻ ♃
21 08 ☽ ⊥ ♂
23 22 ♀ □ ♂

31 Monday
07 35 ☽ ⊥ ♃
12 10 ♀ ± ♇
13 24 ☽ ∠ ♅
14 26 ☽ Q ♆
15 14 ☽ ☍ ☉
15 16 ☽ ∠ ♂
15 36 ☉ ⚼ ♂
16 52 ☽ ⚼ ♀
20 18 ☽ ⊼ ☿

APRIL 1980

LONGITUDES

Date	Sidereal time h m s	Sun ☉ ° ' "	Moon ☽ ° ' "	Moon ☽ 24.00 ° ' "	Mercury ☿ ° '	Venus ♀ ° '	Mars ♂ ° '	Jupiter ♃ ° '	Saturn ♄ ° '	Uranus ♅ ° '	Neptune ♆ ° '	Pluto ♇ ° '
01	00 40 00	11 ♈ 54 40	21 ♎ 21 01	27 ♎ 19 34	14 ♓ 10	27 ♉ 40	26 ♌ 01	01 ♍ 11	22 ♍ 14	25 ♏ 07	22 ♐ 40	20 ♎ 38
02	00 43 57	12 53 49	03 ♏ 19 37	09 ♏ 21 22	15 07	28 41	25 R 58	01 R 07	22 R 10	25 R 06	22 R 39	20 R 37
03	00 47 53	13 52 56	15 ♏ 25 02	21 ♏ 30 49	16 08	29 ♉ 41	25 55	01 03	22 06	25 04	22 39	20 35
04	00 51 50	14 52 01	27 ♏ 38 57	03 ♐ 49 43	17 11	00 ♊ 40	25 54	00 59	22 01	25 03	22 39	20 33
05	00 55 46	15 51 04	10 ♐ 03 22	16 ♐ 20 14	18 16	01 39	25 53	00 55	21 57	25 01	22 38	20 32
06	00 59 43	16 50 05	22 ♐ 40 38	29 ♐ 04 55	19 24	02 38	25 D 52	00 51	21 53	24 59	22 38	20 30
07	01 03 39	17 49 05	05 ♑ 33 27	12 ♑ 06 37	20 35	03 36	25 53	00 47	21 49	24 57	22 37	20 28
08	01 07 36	18 48 03	18 ♑ 44 44	25 ♑ 28 10	21 47	04 34	25 54	00 44	21 45	24 55	22 37	20 27
09	01 11 33	19 46 59	02 ♒ 17 12	09 ♒ 12 03	23 02	05 31	25 56	00 41	21 41	24 54	22 36	20 25
10	01 15 29	20 45 54	16 ♒ 12 51	23 ♒ 19 37	24 19	06 28	25 58	00 38	21 37	24 52	22 36	20 23
11	01 19 26	21 44 46	00 ♓ 32 13	07 ♓ 50 23	25 38	07 24	26 02	00 35	21 34	24 50	22 35	20 22
12	01 23 22	22 43 37	15 ♓ 13 38	22 ♓ 41 18	26 59	08 20	26 06	00 32	21 30	24 48	22 35	20 20
13	01 27 19	23 42 26	00 ♈ 12 33	07 ♈ 46 22	28 21	09 15	26 10	00 30	21 26	24 46	22 34	20 18
14	01 31 15	24 41 13	15 ♈ 21 35	22 ♈ 56 56	29 ♓ 46	10 10	26 16	00 28	21 23	24 44	22 34	20 17
15	01 35 12	25 39 59	00 ♉ 31 07	08 ♉ 02 50	01 ♈ 12	11 04	26 22	00 25	21 19	24 42	22 33	20 15
16	01 39 08	26 38 42	15 ♉ 30 50	22 ♉ 54 03	02 40	11 57	26 28	00 24	21 16	24 40	22 32	20 13
17	01 43 05	27 37 23	00 ♊ 11 32	07 ♊ 22 31	04 10	12 50	26 36	00 22	21 12	24 37	22 31	20 11
18	01 47 02	28 36 03	14 ♊ 26 31	21 ♊ 23 12	05 42	13 42	26 44	00 20	21 09	24 35	22 31	20 10
19	01 50 58	29 ♈ 34 40	28 ♊ 12 28	04 ♋ 54 24	07 15	14 33	26 52	00 19	21 06	24 33	22 30	20 08
20	01 54 55	00 ♉ 33 15	11 ♋ 29 14	17 ♋ 57 20	08 50	15 24	27 01	00 18	21 03	24 31	22 29	20 06
21	01 58 51	01 31 47	24 ♋ 19 13	00 ♌ 35 24	10 27	16 14	27 11	00 17	21 00	24 29	22 28	20 05
22	02 02 48	02 30 18	06 ♌ 46 32	12 ♌ 53 15	12 06	17 03	27 22	00 16	20 57	24 26	22 27	20 03
23	02 06 44	03 28 46	18 ♌ 56 12	24 ♌ 56 05	13 46	17 52	27 32	00 15	20 54	24 24	22 26	20 02
24	02 10 41	04 27 12	00 ♍ 53 31	06 ♍ 49 09	15 28	18 39	27 44	00 15	20 51	24 22	22 26	20 00
25	02 14 37	05 25 36	12 ♍ 43 35	18 ♍ 37 21	17 11	19 26	27 56	00 15	20 48	24 19	22 25	19 58
26	02 18 34	06 23 58	24 ♍ 31 00	00 ♎ 24 58	18 57	20 12	28 08	00 D 14	20 46	24 17	22 24	19 57
27	02 22 31	07 22 18	06 ♎ 19 42	12 ♎ 15 33	20 44	20 56	28 21	00 15	20 43	24 15	22 23	19 55
28	02 26 27	08 20 36	18 ♎ 12 50	24 ♎ 11 49	22 33	21 40	28 35	00 15	20 41	24 12	22 22	19 54
29	02 30 24	09 18 52	00 ♏ 12 43	06 ♏ 15 43	24 23	22 23	28 49	00 15	20 39	24 10	22 21	19 52
30	02 34 20	10 ♉ 17 06	12 ♏ 20 57	18 ♏ 28 32	26 ♈ 15	23 ♊ 05	29 ♌ 04	00 ♍ 16	20 ♍ 36	24 ♏ 08	22 ♐ 19	19 ♎ 50

Moon nodes and latitude

Date	Moon True ☊ ° '	Moon Mean ☊ ° '	Moon ☽ Latitude ° '
01	28 ♌ 35	27 ♌ 03	04 N 06
02	28 R 26	27 00	04 39
03	28 18	26 57	05 00
04	28 10	26 54	05 08
05	28 03	26 50	05 01
06	27 58	26 47	04 40
07	27 55	26 44	04 04
08	27 54	26 41	03 15
09	27 D 55	26 38	02 14
10	27 56	26 35	01 N 03
11	27 R 56	26 31	00 S 14
12	27 55	26 28	01 32
13	27 52	26 25	02 44
14	27 46	26 22	03 46
15	27 38	26 19	04 32
16	27 29	26 16	04 58
17	27 20	26 12	05 04
18	27 12	26 09	04 51
19	27 05	26 06	04 21
20	27 02	26 03	03 38
21	27 00	26 00	02 45
22	27 D 00	25 56	01 46
23	27 00	25 53	00 S 43
24	27 R 00	25 50	00 N 21
25	26 59	25 47	01 23
26	26 56	25 44	02 21
27	26 50	25 41	03 13
28	26 41	25 37	03 57
29	26 30	25 34	04 30
30	26 ♌ 17	25 ♌ 31	04 N 52

DECLINATIONS

Date	Sun ☉ ° '	Moon ☽ ° '	Mercury ☿ ° '	Venus ♀ ° '	Mars ♂ ° '	Jupiter ♃ ° '	Saturn ♄ ° '	Uranus ♅ ° '	Neptune ♆ ° '	Pluto ♇ ° '
01	04 N 43	04 S 31	07 S 39	22 N 36	15 N 54	12 N 16	05 N 18	18 S 46	21 S 51	08 N 22
02	05 06	08 15	07 24	22 53	15 52	12 17	05 19	18 46	21 51	08 23
03	05 29	11 40	07 08	23 10	15 51	12 19	05 21	18 45	21 51	08 23
04	05 51	14 38	06 50	23 27	15 49	12 20	05 23	18 45	21 51	08 24
05	06 14	16 59	06 30	23 43	15 47	12 21	05 24	18 44	21 51	08 25
06	06 37	18 35	06 09	23 59	15 45	12 23	05 26	18 44	21 51	08 25
07	06 59	19 15	05 46	24 14	15 42	12 24	05 27	18 44	21 51	08 26
08	07 22	18 55	05 22	24 28	15 40	12 25	05 29	18 43	21 51	08 27
09	07 44	17 29	04 56	24 42	15 37	12 26	05 30	18 43	21 51	08 27
10	08 06	14 59	04 29	24 56	15 33	12 27	05 32	18 42	21 51	08 28
11	08 29	11 30	04 01	25 08	15 30	12 28	05 33	18 42	21 51	08 28
12	08 50	07 14	03 31	25 21	15 26	12 28	05 35	18 41	21 50	08 29
13	09 12	02 S 26	03 01	25 33	15 23	12 29	05 36	18 41	21 50	08 30
14	09 34	02 N 34	02 29	25 44	15 18	12 30	05 37	18 40	21 50	08 30
15	09 55	07 24	01 55	25 55	15 14	12 31	05 39	18 40	21 50	08 31
16	10 17	11 43	01 21	26 05	15 10	12 31	05 40	18 39	21 50	08 31
17	10 38	15 14	00 45	26 15	15 05	12 32	05 41	18 39	21 50	08 32
18	10 59	17 43	00 S 08	26 24	15 00	12 32	05 42	18 38	21 50	08 32
19	11 19	19 04	00 N 29	26 33	14 55	12 32	05 43	18 38	21 50	08 33
20	11 40	19 19	01 08	26 41	14 50	12 33	05 44	18 37	21 50	08 33
21	12 00	18 32	01 48	26 48	14 44	12 33	05 46	18 37	21 50	08 34
22	12 21	16 52	02 29	26 55	14 39	12 33	05 47	18 36	21 50	08 34
23	12 41	14 28	03 10	27 02	14 33	12 33	05 48	18 36	21 50	08 35
24	13 00	11 29	03 53	27 08	14 27	12 33	05 49	18 35	21 50	08 35
25	13 20	08 03	04 36	27 13	14 20	12 33	05 50	18 34	21 49	08 36
26	13 39	04 20	05 21	27 18	14 14	12 33	05 50	18 34	21 49	08 36
27	13 58	00 N 26	06 06	27 23	14 08	12 33	05 51	18 33	21 49	08 37
28	14 17	03 S 29	06 51	27 27	14 01	12 32	05 52	18 33	21 49	08 37
29	14 36	07 19	07 38	27 30	13 54	12 32	05 53	18 32	21 49	08 37
30	14 N 54	10 S 54	08 N 25	27 N 33	13 N 47	12 N 32	05 N 54	18 S 32	21 S 49	08 N 38

ZODIAC SIGN ENTRIES

Date	h	m	Planets
02	05	21	☽ ♏
03	19	46	♀ ♊
04	16	35	☽ ♐
07	01	43	☽ ♑
09	08	00	☽ ♒
11	11	07	☽ ♓
13	11	40	☽ ♈
14	15	58	☿ ♈
15	11	11	☽ ♉
17	11	41	☽ ♊
19	15	11	☽ ♋
19	22	23	☉ ♉
21	22	52	☽ ♌
24	10	12	☽ ♍
26	23	09	☽ ♎
29	11	35	☽ ♏

LATITUDES

Date	Mercury ☿ ° '	Venus ♀ ° '	Mars ♂ ° '	Jupiter ♃ ° '	Saturn ♄ ° '	Uranus ♅ ° '	Neptune ♆ ° '	Pluto ♇ ° '
01	01 S 32	03 N 02	03 N 15	01 N 18	02 N 25	00 N 17	01 N 23	17 N 42
04	01 55	03 14	03 07	01 18	02 25	00 17	01 23	17 43
07	02 13	03 25	02 59	01 17	02 24	00 17	01 23	17 43
10	02 26	03 35	02 52	01 17	02 24	00 17	01 24	17 43
13	02 34	03 45	02 45	01 16	02 24	00 17	01 24	17 43
16	02 37	03 54	02 38	01 16	02 23	00 17	01 24	17 43
19	02 36	04 01	02 31	01 15	02 23	00 17	01 24	17 43
22	02 30	04 08	02 24	01 15	02 23	00 17	01 24	17 42
25	02 19	04 13	02 17	01 15	02 22	00 17	01 24	17 42
28	02 04	04 16	02 11	01 14	02 22	00 17	01 24	17 41
31	01 S 44	04 N 18	02 N 05	01 N 14	02 N 22	00 N 17	01 N 24	17 N 41

DATA

Julian Date	2444331
Delta T	+51 seconds
Ayanamsa	23° 34' 42"
Synetic vernal point	05° ♓ 32' 17"
True obliquity of ecliptic	23° 26' 24"

MOON'S PHASES, APSIDES AND POSITIONS ☽

Date	h	m	Phase	Longitude ° '	Eclipse Indicator
08	12	06	☾	18 ♑ 48	
15	03	46	●	25 ♈ 20	
22	02	59	☽	02 ♌ 08	
30	07	35	○	10 ♏ 06	

Day	h	m		
14	07	19	Perigee	
26	19	42	Apogee	
07	16	11	Max dec	19° S 16'
13	23	40	ON	
20	05	29	Max dec	19° N 21'
27	14	41	OS	

LONGITUDES

Date	Chiron ⚷ ° '	Ceres ⚳ ° '	Pallas ⚴ ° '	Juno ⚵ ° '	Vesta ⚶ ° '	Black Moon Lilith ⚸ ° '
01	11 ♉ 33	08 ♉ 33	27 ♓ 33	21 ♋ 25	28 ♉ 36	29 ♍ 35
11	12 ♉ 09	12 ♉ 31	00 ♈ 59	24 ♋ 02	02 ♊ 32	00 ♎ 41
21	12 ♉ 47	16 ♉ 32	04 ♈ 24	26 ♋ 59	06 ♊ 35	01 ♎ 48
31	13 ♉ 26	20 ♉ 35	07 ♈ 46	00 ♌ 12	10 ♊ 43	02 ♎ 55

ASPECTARIAN

h m	Aspects
01 Tuesday	
01 40	☽ ∠ ♃
07 32	☽ ⊥ ♅
09 25	☽ ± ☿
10 34	☽ ☌ ♇
12 42	☽ ± ♀
13 19	☽ # ☉
13 46	☽ ⊻ ♄
14 38	☽ ⚹ ♆
16 55	☽ # ♄
19 34	☽ ⊻ ♅
21 20	☽ ⚹ ♂
23 43	☿ ± ♇
02 Wednesday	
01 44	☽ ⊥ ♄
01 52	☽ ⊼ ♀
05 02	☽ bQ ☿
06 48	☽ ∥ ☿
07 37	☽ ⚹ ♃
12 54	☽ # ♇
19 36	☽ ∠ ♄
20 37	☽ ∠ ♆
21 12	☽ Q ♂
03 Thursday	
03 30	☉ ∥ ♄
07 20	☽ Q ♃
08 42	☽ ⊼ ☉
13 32	☽ △ ☿
14 26	☽ ⊥ ♆
16 56	☽ # ♃
21 34	☽ ± ☉
22 09	☽ ⊻ ♇
04 Friday	
01 04	☽ ⚹ ♄
02 13	☽ ⊻ ♆
06 55	☽ ☌ ♅
08 35	☽ □ ♂
09 52	☽ ⊥ ♇
16 41	☽ bQ ☉
18 23	☽ ☍ ♀
18 26	☽ □ ♃
18 59	♀ □ ♃
23 04	☽ # ♂
05 Saturday	
00 19	☽ Q ♄
03 18	☽ ∠ ♇
13 24	☉ bQ ♃
20 41	☉ # ☿
06 Sunday	
00 01	☽ △ ☉
05 13	☽ □ ☿
07 54	☽ ⚹ ♇
08 27	♂ St D
10 31	☽ □ ♄
11 55	☽ ☌ ♆
15 27	☽ ∥ ♅
16 20	☽ ⊻ ♅
18 00	☽ △ ♂
07 Monday	
03 13	☽ △ ♃
03 30	☽ ⊥ ♅
06 19	☽ Q ♇
08 06	☽ ⊼ ♀
09 52	☿ ⊼ ♇
18 07	☽ Q ☿
20 01	☽ ± ♀
20 03	☽ ∠ ♅
21 46	☽ bQ ♂
08 Tuesday	
05 40	☿ # ♄
06 36	☽ bQ ♃
11 18	☿ ☍ ♄
12 06	☽ □ ☉
13 35	☽ bQ ♀
14 04	☽ ± ♂
14 54	☉ ± ♅
15 02	☽ □ ♇
16 38	☽ ∥ ♅
17 22	☽ △ ♄
18 00	☽ ⚹ ☿
18 55	☽ ⊻ ♆
22 39	☽ ± ♃
23 00	☽ ⚹ ♅
09 Wednesday	
00 47	☽ ⊼ ♂
03 53	☿ □ ♆
05 33	☽ ⊥ ♆
09 12	☽ ⊼ ♃
09 26	♀ bQ ♇
18 03	☽ △ ♀
19 37	☽ bQ ♄
20 00	☽ Q ♅
21 15	☽ ∠ ♆
22 17	☽ Q ☉
23 00	☽ ∠ ☿
10 Thursday	
03 02	☉ ☍ ♇
07 08	☽ # ♂
11 00	☽ ± ♄
15 55	☽ ⊥ ☿
19 03	☽ △ ♇
20 16	☽ ⚹ ☉
21 06	☽ ⊼ ♄
21 46	☿ △ ♅
21 52	☿ Q ♀
22 46	☽ ⚹ ♆
11 Friday	
02 32	☽ □ ♅
03 01	☽ ⊻ ☿
04 29	☽ ☍ ♂
05 59	☽ # ♃
07 44	☉ ⊼ ♄
11 57	☉ ∥ ♇
12 05	☽ ☍ ♃
18 40	☽ Q ♆
19 31	☿ ⊼ ♂
19 56	☽ bQ ♇
22 57	☽ ∠ ☉
12 Saturday	
00 03	☽ □ ♀
04 02	☽ # ☉
05 20	☽ # ♇
08 27	☉ △ ♆
10 33	☽ ± ♇
14 36	☽ ⊥ ☉
20 12	☽ ⊼ ♇
20 27	☽ # ♄
22 03	☽ ☍ ♄
23 49	☽ □ ♆
13 Sunday	
00 54	☽ ⊻ ☉
03 20	☽ △ ♅
05 32	☽ ⊼ ♂
06 59	☽ Q ♀
08 45	☽ ☌ ☿
08 50	☽ ∥ ☿
12 27	☽ ⊼ ♃
15 08	☽ ± ♂
21 57	☽ ± ♃
14 Monday	
03 07	☽ bQ ♅
03 15	☽ ⚹ ♀
05 29	☽ bQ ♂
11 35	☽ # ☿
12 09	☽ bQ ♃
12 58	☉ ⊼ ♅
17 19	☽ ± ♅
19 45	☽ ☍ ♇
21 29	☽ ⊼ ♄
23 22	☿ ⊼ ♃
23 22	☽ △ ♆
15 Tuesday	
02 47	☽ ⊼ ♅
02 59	☽ ∥ ♄
03 46	☽ ☌ ☉
04 29	☽ ∠ ♀
05 22	☽ △ ♂
06 56	☽ ± ♄
11 51	☽ △ ♃
13 12	☽ ⊻ ☿
17 50	☽ ∥ ♇
19 41	☽ ⊥ ♀
21 12	☽ bQ ♄
23 12	☽ bQ ♆
23 48	☽ ⊥ ☿
16 Wednesday	
02 44	☽ ∥ ☉
05 54	☽ ⊻ ♀
07 15	☉ △ ♂
08 31	☿ ± ♂
13 39	☽ ± ♆
15 53	☽ ∠ ☿
16 56	☽ ∥ ♃
19 37	☽ ⊼ ♇
21 17	☽ △ ♄
23 24	☽ ⊼ ♆
17 Thursday	
02 19	☉ ± ♄
02 51	☽ ☍ ♅
05 24	☽ ± ♇
06 01	☽ □ ♂
07 27	☽ ⊻ ☉
10 53	☽ ∥ ♂
12 17	☽ □ ♃
18 08	☽ ⊥ ☉
19 24	☽ ⚹ ☿
20 19	☽ bQ ♇
18 Friday	
10 27	☽ ∠ ☉
10 39	☽ ☌ ♀
12 30	☽ Q ♂
18 18	☽ Q ☿
18 42	☽ Q ♃
21 45	☿ ± ♃
21 51	☽ △ ♇
23 33	☽ □ ♄
19 Saturday	
01 51	☽ # ♅
01 57	☽ ☍ ♆
05 33	☽ ⊼ ♅
07 45	☉ ∠ ♀
09 36	☽ ⚹ ♂
14 38	☽ ⚹ ☉
15 45	☽ ⚹ ♃
16 10	☽ ± ♅
22 51	♀ Q ♂
20 Sunday	
05 44	☉ △ ♃
06 29	☽ □ ☿
07 32	☽ Q ♄
08 23	☽ bQ ♅
13 00	☽ ∠ ♂
14 07	☽ Q ☉
19 02	☽ ∠ ♃
19 44	☽ ⊻ ♀
21 52	☿ bQ ♅
21 Monday	
04 00	☽ □ ♇
05 44	☽ ⚹ ♄
05 59	☽ ⊥ ♂
07 46	☽ ⊥ ♀
08 30	☽ ⊼ ♆
10 35	☽ # ♅
11 55	☽ ⊥ ♃
12 18	☽ △ ♅
17 32	☽ ⊻ ♂
19 55	☽ ± ♆
23 23	☽ ⊻ ♃
22 Tuesday	
02 11	☽ ∠ ♀
02 59	☽ □ ☉
10 23	☽ ∠ ♄
13 20	☽ bQ ♆
14 29	☽ Q ♇
16 17	☿ bQ ♂
23 Wednesday	
00 05	☽ △ ☿
02 55	☉ ∥ ♃
04 00	☽ ⊥ ♄
09 42	☽ ⚹ ♀
11 17	☽ ∥ ♂
14 10	☽ ⚹ ♇
15 54	☽ ⊻ ♄
18 59	☽ △ ♆
22 54	☽ □ ♅
23 46	♀ Q ♃
24 Thursday	
01 25	☽ ∥ ☉
03 51	☽ ∥ ♃
05 31	☽ ☌ ♂
08 59	☿ bQ ♃
10 42	☽ ☌ ♃
10 59	☽ bQ ☿
11 29	☽ Q ♀
19 51	☽ △ ☉
20 17	☽ ∠ ♇
25 Friday	
08 20	☽ ± ☿
08 23	☽ ∥ ♇
11 11	☽ Q ♅
14 32	☽ ⊥ ♇
20 58	☉ bQ ♄
22 40	☽ ⊼ ☿
26 Saturday	
02 31	☽ ∥ ♄
02 36	☽ □ ♀
02 43	☽ ⊻ ♇
03 13	☿ ± ♅
04 23	☽ ☌ ♄
04 25	♀ △ ♇
05 05	☽ bQ ☉
06 41	☽ ∥ ☿
07 41	☽ □ ♆
08 49	♃ St D
11 32	☽ ⚹ ♅
19 31	☽ ⊻ ♂
23 39	☽ ⊻ ♃
27 Sunday	
01 02	☽ ± ☉
01 17	☿ ☍ ♇
04 14	☿ ∥ ♄
05 17	♀ □ ♄
07 56	☽ ⊥ ♂
11 50	☽ ⊥ ♃
11 53	☿ ⊼ ♄
12 07	☉ bQ ♆
14 18	☽ ⊼ ☉
16 45	☿ ⚹ ♀
17 53	☽ ∠ ♅
20 11	☽ Q ♆
20 36	☉ ∥ ♂
28 Monday	
02 30	☽ ∠ ♂
06 02	☽ ∠ ♃
09 36	☿ △ ♆
11 59	☽ ⊥ ♅
15 22	☽ ☌ ♇
16 56	☽ ⊻ ♄
19 23	☽ △ ♀
20 19	☽ ⚹ ♆
22 16	☽ ☍ ☿
23 59	☽ ⊻ ♅
29 Tuesday	
02 49	☽ # ♄
04 54	☽ ⊥ ♄
09 10	☽ ⚹ ♂
09 14	☿ ⊼ ♅
10 30	♀ ☍ ♆
12 05	☽ ⚹ ♃
14 30	☽ # ☿
20 31	☽ # ♇
22 44	☽ ∠ ♄
30 Wednesday	
02 07	☽ ∠ ♆
03 06	☽ bQ ♀
07 35	☽ ☍ ☉
09 25	☽ Q ♂
11 50	☽ Q ♃
16 20	☿ ± ♄
18 48	☿ ∥ ♇
19 47	☽ ⊥ ♆
21 50	☽ ± ♀
23 52	☽ # ♃

All ephemeris data is given at 12.00 UT and the Moon's longitude is additionally given for 24.00 UT

MAY 1980

LONGITUDES

Date	Sidereal time h m s	Sun ☉ ° ' "	Moon ☽ ° ' "	Moon ☽ 24.00 ° ' "	Mercury ☿ ° '	Venus ♀ ° '	Mars ♂ ° '	Jupiter ♃ ° '	Saturn ♄ ° '	Uranus ♅ ° '	Neptune ♆ ° '	Pluto ♇ ° '
01	02 38 17	11 ♉ 15 19	24 ♏ 38 33	00 ♐ 51 03	28 ♈ 09	23 ♊ 46	29 ♌ 19	00 ♍ 17	20 ♍ 34	24 ♏ 05	22 ♐ 18	19 ♎ 49
02	02 42 13	12 13 30	07 ♐ 06 08	13 ♐ 23 50	00 ♉ 05	24 26	29 34	00 18	20 R 32	24 R 03	22 R 17	19 R 47
03	02 46 10	13 11 39	19 ♐ 44 16	26 ♐ 07 30	02 03	25 04	29 ♌ 50	00 19	20 30	24 00	22 16	19 46
04	02 50 06	14 09 47	02 ♑ 33 41	09 ♑ 02 57	04 02	25 41	00 ♍ 07	00 21	20 28	23 58	22 15	19 44
05	02 54 03	15 07 53	15 ♑ 35 27	22 ♑ 11 26	06 03	26 17	00 23	00 22	20 26	23 55	22 14	19 43
06	02 58 00	16 05 58	28 ♑ 51 04	05 ♒ 34 36	08 05	26 52	00 41	00 24	20 25	23 53	22 13	19 41
07	03 01 56	17 04 02	12 ♒ 22 15	19 ♒ 14 12	10 09	27 26	00 58	00 26	20 23	23 50	22 11	19 40
08	03 05 53	18 02 04	26 ♒ 10 37	03 ♓ 11 36	12 15	27 58	01 17	00 28	20 22	23 48	22 10	19 38
09	03 09 49	19 00 04	10 ♓ 17 08	17 ♓ 27 07	14 21	28 28	01 35	00 30	20 20	23 45	22 09	19 37
10	03 13 46	19 58 04	24 ♓ 41 20	01 ♈ 59 22	16 29	28 57	01 54	00 33	20 19	23 43	22 07	19 36
11	03 17 42	20 56 02	09 ♈ 20 42	16 ♈ 44 36	18 39	29 25	02 13	00 35	20 18	23 40	22 06	19 34
12	03 21 39	21 53 59	24 ♈ 10 15	01 ♉ 36 40	20 48	29 ♊ 51	02 33	00 38	20 17	23 38	22 05	19 33
13	03 25 35	22 51 54	09 ♉ 02 47	16 ♉ 27 30	22 59	00 ♋ 15	02 53	00 41	20 16	23 35	22 03	19 31
14	03 29 32	23 49 48	23 ♉ 49 41	01 ♊ 08 19	25 10	00 38	03 14	00 44	20 15	23 33	22 02	19 30
15	03 33 29	24 47 41	08 ♊ 22 25	15 ♊ 31 11	27 21	00 58	03 34	00 48	20 14	23 30	22 01	19 29
16	03 37 25	25 45 32	22 ♊ 33 58	29 ♊ 30 17	29 ♉ 32	01 17	03 56	00 51	20 13	23 28	21 59	19 28
17	03 41 22	26 43 21	06 ♋ 19 53	13 ♋ 02 38	01 ♊ 43	01 34	04 17	00 55	20 13	23 25	21 58	19 26
18	03 45 18	27 41 09	19 ♋ 38 39	26 ♋ 08 07	03 53	01 50	04 39	00 59	20 12	23 23	21 56	19 25
19	03 49 15	28 38 55	02 ♌ 31 25	08 ♌ 49 00	06 02	02 03	05 01	01 03	20 12	23 20	21 55	19 24
20	03 53 11	29 ♉ 36 40	15 ♌ 01 24	21 ♌ 09 14	08 10	02 13	05 24	01 07	20 12	23 18	21 54	19 23
21	03 57 08	00 ♊ 34 23	27 ♌ 13 08	03 ♍ 13 47	10 16	02 22	05 47	01 11	20 12	23 15	21 52	19 21
22	04 01 04	01 32 04	09 ♍ 11 52	15 ♍ 08 04	12 20	02 29	06 10	01 15	20 D 12	23 13	21 51	19 20
23	04 05 01	02 29 44	21 ♍ 03 03	26 ♍ 57 28	14 23	02 33	06 33	01 20	20 12	23 10	21 49	19 19
24	04 08 58	03 27 22	02 ♎ 51 54	08 ♎ 46 57	16 24	02 35	06 57	01 25	20 12	23 08	21 48	19 18
25	04 12 54	04 24 59	14 ♎ 43 09	20 ♎ 40 57	18 22	02 R 35	07 21	01 30	20 12	23 05	21 46	19 17
26	04 16 51	05 22 34	26 ♎ 40 47	02 ♏ 43 01	20 18	02 32	07 46	01 35	20 12	23 03	21 45	19 16
27	04 20 47	06 20 08	08 ♏ 47 55	14 ♏ 55 45	22 11	02 27	08 10	01 40	20 13	23 01	21 43	19 15
28	04 24 44	07 17 41	21 ♏ 06 39	27 ♏ 20 46	24 02	02 19	08 35	01 46	20 13	22 58	21 41	19 14
29	04 28 40	08 15 13	03 ♐ 38 07	09 ♐ 58 43	25 50	02 09	09 00	01 51	20 14	22 56	21 40	19 13
30	04 32 37	09 12 43	16 ♐ 22 32	22 ♐ 49 29	27 35	01 57	09 26	01 57	20 15	22 53	21 38	19 12
31	04 36 33	10 ♊ 10 13	29 ♐ 19 29	05 ♑ 52 27	29 ♊ 18	01 ♋ 42	09 ♍ 52	02 ♍ 03	20 ♍ 16	22 ♏ 51	21 ♐ 37	19 ♎ 11

Date	Moon True ☊ ° '	Moon Mean ☊ ° '	Moon ☽ Latitude ° '
01	26 ♌ 04	25 ♌ 28	05 N 01
02	25 R 52	25 25	04 55
03	25 41	25 22	04 35
04	25 33	25 18	04 01
05	25 28	25 15	03 14
06	25 25	25 12	02 16
07	25 D 24	25 09	01 N 08
08	25 R 24	25 06	00 S 04
09	25 23	25 02	01 18
10	25 21	24 59	02 28
11	25 17	24 56	03 30
12	25 09	24 53	04 18
13	25 00	24 50	04 49
14	24 48	24 47	05 00
15	24 37	24 43	04 52
16	24 26	24 40	04 26
17	24 18	24 37	03 44
18	24 12	24 34	02 52
19	24 09	24 31	01 52
20	24 08	24 27	00 S 48
21	24 D 08	24 24	00 N 16
22	24 R 07	24 21	01 19
23	24 06	24 18	02 18
24	24 03	24 15	03 10
25	23 57	24 12	03 54
26	23 49	24 08	04 28
27	23 39	24 05	04 51
28	23 27	24 02	05 01
29	23 14	23 59	04 56
30	23 02	23 56	04 37
31	22 ♌ 52	23 ♌ 53	04 N 03

DECLINATIONS

Date	Sun ☉ ° '	Moon ☽ ° '	Mercury ☿ ° '	Venus ♀ ° '	Mars ♂ ° '	Jupiter ♃ ° '	Saturn ♄ ° '	Uranus ♅ ° '	Neptune ♆ ° '	Pluto ♇ ° '
01	15 N 12	14 S 04	09 N 12	27 N 35	13 N 40	12 N 31	05 N 54	18 S 31	21 S 49	08 N 38
02	15 30	16 38	10 00	27 37	13 32	12 31	05 55	18 30	21 49	08 39
03	15 48	18 28	10 48	27 39	13 25	12 30	05 56	18 30	21 49	08 39
04	16 05	19 24	11 36	27 40	13 17	12 30	05 56	18 29	21 49	08 39
05	16 22	19 19	12 25	27 41	13 09	12 29	05 57	18 29	21 49	08 40
06	16 39	18 11	13 14	27 41	13 01	12 28	05 57	18 28	21 48	08 40
07	16 56	16 00	14 02	27 41	12 53	12 27	05 58	18 27	21 48	08 40
08	17 12	12 51	14 50	27 40	12 45	12 26	05 58	18 27	21 48	08 40
09	17 28	08 55	15 38	27 39	12 37	12 25	05 59	18 26	21 48	08 41
10	17 44	04 S 23	16 25	27 37	12 28	12 24	05 59	18 26	21 48	08 41
11	17 59	00 N 29	17 11	27 35	12 19	12 23	05 59	18 25	21 48	08 41
12	18 15	05 22	17 56	27 33	12 11	12 22	05 59	18 24	21 48	08 41
13	18 29	09 56	18 40	27 30	12 02	12 21	06 00	18 24	21 48	08 41
14	18 44	13 52	19 23	27 27	11 53	12 20	06 00	18 23	21 48	08 42
15	18 58	16 54	20 03	27 23	11 43	12 18	06 00	18 22	21 48	08 42
16	19 12	18 49	20 42	27 19	11 34	12 17	06 00	18 22	21 48	08 42
17	19 25	19 33	21 19	27 14	11 25	12 15	06 00	18 21	21 47	08 42
18	19 39	19 10	21 54	27 10	11 15	12 14	06 00	18 21	21 47	08 42
19	19 52	17 47	22 27	27 04	11 05	12 12	06 00	18 20	21 47	08 42
20	20 04	15 34	22 57	26 59	10 55	12 11	06 00	18 19	21 47	08 42
21	20 16	12 42	23 24	26 53	10 45	12 09	06 00	18 19	21 47	08 42
22	20 28	09 20	23 49	26 46	10 35	12 07	06 00	18 18	21 47	08 43
23	20 40	05 39	24 11	26 40	10 25	12 05	06 00	18 18	21 47	08 43
24	20 51	01 N 46	24 31	26 32	10 15	12 04	05 59	18 17	21 47	08 43
25	21 02	02 S 12	24 48	26 25	10 04	12 02	05 59	18 16	21 47	08 43
26	21 12	06 07	25 02	26 17	09 54	12 00	05 59	18 16	21 46	08 43
27	21 22	09 50	25 14	26 08	09 43	11 58	05 58	18 15	21 46	08 43
28	21 32	13 22	25 23	25 59	09 32	11 56	05 58	18 15	21 46	08 43
29	21 41	16 02	25 30	25 50	09 22	11 54	05 58	18 14	21 46	08 43
30	21 50	18 09	25 34	25 40	09 11	11 51	05 57	18 14	21 46	08 42
31	21 N 58	19 S 23	25 N 36	25 N 30	09 N 00	11 N 49	05 N 57	18 S 13	21 S 46	08 N 42

ZODIAC SIGN ENTRIES

Date	h	m	Planets
01	22	22	☽ ♐
02	10	56	☿ ♉
04	02	27	♂ ♍
04	07	14	☽ ♑
06	14	03	☽ ♒
08	18	33	☽ ♓
10	20	44	☽ ♈
12	20	53	♀ ♋
12	21	24	☽ ♉
14	22	07	☽ ♊
16	17	06	☿ ♊
17	00	52	☽ ♋
19	07	14	☽ ♌
20	21	42	☉ ♊
21	17	32	☽ ♍
24	06	11	☽ ♎
26	18	37	☽ ♏
29	05	05	☽ ♐
31	13	14	☽ ♑
31	22	05	☿ ♋

LATITUDES

Date	Mercury ☿ ° '	Venus ♀ ° '	Mars ♂ ° '	Jupiter ♃ ° '	Saturn ♄ ° '	Uranus ♅ ° '	Neptune ♆ ° '	Pluto ♇ ° '
01	01 S 44	04 N 18	02 N 05	01 N 14	02 N 22	00 N 17	01 N 24	17 N 41
04	01 20	04 18	01 59	01 13	02 21	00 17	01 24	17 40
07	00 53	04 16	01 53	01 13	02 21	00 17	01 24	17 39
10	00 S 22	04 11	01 47	01 12	02 20	00 17	01 24	17 39
13	00 N 09	04 03	01 42	01 12	02 19	00 17	01 24	17 38
16	00 40	03 53	01 37	01 11	02 19	00 17	01 24	17 37
19	01 09	03 39	01 31	01 11	02 18	00 17	01 25	17 35
22	01 34	03 22	01 27	01 10	02 18	00 17	01 25	17 34
25	01 52	03 00	01 22	01 10	02 17	00 17	01 25	17 33
28	02 05	02 34	01 17	01 09	02 17	00 17	01 25	17 32
31	02 N 10	02 N 04	01 N 13	01 N 09	02 N 16	00 N 17	01 N 25	17 N 31

DATA

Julian Date	2444361
Delta T	+51 seconds
Ayanamsa	23° 34' 45"
Synetic vernal point	05° ♓ 32' 14"
True obliquity of ecliptic	23° 26' 23"

MOON'S PHASES, APSIDES AND POSITIONS ☽

Date	h	m	Phase	Longitude ° '	Eclipse Indicator
07	20	51	☾	17 ♒ 25	
14	12	00	●	23 ♉ 50	
21	19	16	☽	00 ♍ 52	
29	21	28	○	08 ♐ 38	

Day	h	m		
12	12	49	Perigee	
24	11	24	Apogee	
04	22	13	Max dec	19° S 29'
11	09	38	0N	
17	15	30	Max dec	19° N 34'
24	22	41	0S	

LONGITUDES

Date	Chiron ⚷ ° '	Ceres ⚳ ° '	Pallas ⚴ ° '	Juno ⚵ ° '	Vesta ⚶ ° '	Black Moon Lilith ⚸ ° '
01	13 ♉ 26	20 ♉ 35	07 ♈ 46	00 ♌ 12	10 ♊ 43	02 ♎ 55
11	14 ♉ 05	24 ♉ 41	11 ♈ 05	03 ♌ 38	14 ♊ 55	04 ♎ 02
21	14 ♉ 44	28 ♉ 47	14 ♈ 19	07 ♌ 15	19 ♊ 10	05 ♎ 08
31	15 ♉ 21	02 ♊ 54	17 ♈ 27	10 ♌ 59	23 ♊ 27	06 ♎ 15

All ephemeris data is given at 12.00 UT and the Moon's longitude is additionally given for 24.00 UT

ASPECTARIAN

h m	Aspects
01 Thursday	
02 38	☽ ⊻ ♇
04 06	☽ ⚹ ♄
07 28	☽ ⊻ ♆
08 51	☽ # ♂
10 12	☽ ⊼ ♀
10 56	☽ ☌ ♅
14 16	☽ ⊥ ♇
20 03	☽ ⊼ ☿
21 13	☽ □ ♂
22 55	☽ □ ♃
22 56	♀ ⊼ ♅
23 11	☽ # ☉
02 Friday	
03 16	☽ Q ♄
04 37	☿ △ ♂
07 35	☽ ∠ ♇
09 42	☽ ± ☿
14 38	☿ △ ♃
22 35	☽ ⊼ ☉
03 Saturday	
05 59	☽ ⚼ ☿
10 53	☽ ± ☉
12 03	☽ ⚹ ♇
12 30	☽ ∥ ♅
13 26	☽ □ ♄
16 45	☽ ☌ ♆
20 00	☽ ⊻ ♅
22 32	☽ ☍ ♀
04 Sunday	
05 10	☽ ⚼ ☉
07 11	☽ ⊥ ♅
07 20	☽ △ ♂
07 52	☽ △ ♃
10 28	☽ Q ♇
15 14	☽ △ ☿
23 48	☽ ∠ ♅
05 Monday	
04 56	☿ ⚼ ♄
09 55	♂ ☌ ♃
11 06	☽ △ ☉
11 36	☽ ⚼ ♃
11 38	☽ ⚼ ♂
13 52	☿ ∥ ♃
19 30	☽ □ ♇
20 49	☽ △ ♄
06 Tuesday	
00 03	☽ ⊻ ♆
01 49	☿ ⚼ ♆
03 05	☽ ⚹ ♅
03 58	☽ ± ♃
04 20	☽ ± ♂
06 51	☿ ∥ ♂
07 34	☽ ∥ ♅
08 17	☽ ⊼ ♀
10 51	☽ ⊥ ♆
14 39	☉ ± ♆
14 47	☽ ⊼ ♃
15 21	☽ ⊼ ♂
19 30	☽ ± ♀
23 41	☽ ⚼ ♄
07 Wednesday	
00 30	☽ Q ♅
02 52	☽ ∠ ♆
03 57	☽ # ☉
07 24	☽ □ ☿
12 06	☽ ⚼ ♀
15 32	☽ ± ♄
20 51	☽ □ ☉
08 Thursday	
00 38	☽ # ☿
00 43	☽ △ ♇
01 58	☽ ⊼ ♄
05 06	☽ ⚹ ♆
07 55	☽ □ ♅
12 44	☽ # ♂
14 47	☽ # ♃
15 11	☽ △ ♀
19 22	☽ ☍ ♃
20 12	☽ Q ☿
20 56	☽ ☍ ♂
22 48	☿ ∠ ♀
09 Friday	
01 38	☽ Q ♆
02 26	☽ ⚼ ♇
06 03	☽ Q ☉
13 19	☽ # ♇
17 35	☽ ± ♇
20 01	☽ ⚹ ☿
10 Saturday	
02 55	☉ ⊼ ♇
03 35	☽ ⊼ ♇
03 37	☽ ⚹ ☉
03 48	☽ # ♄
04 46	☽ ☍ ♄
07 46	☽ □ ♆
07 56	☿ ± ♆
10 24	☽ △ ♅
19 15	☽ □ ♀
20 30	☉ △ ♄
21 40	☽ ⊼ ♃
11 Sunday	
00 07	☽ ⊼ ♂
00 10	♂ ∥ ♃
01 06	☽ ∠ ☿
06 03	☽ ∠ ☉
07 30	☽ ± ♃
10 08	☽ ± ♂
10 55	☽ ⚼ ♅
18 16	☽ ⊥ ☿
21 42	☽ ⊥ ☉
22 10	☽ ⚼ ♃
22 12	☿ ⊼ ♇
12 Monday	
00 56	♀ ± ♅
01 04	☽ ⚼ ♂
01 28	☽ ± ♅
01 29	☽ Q ♀
04 33	☽ ☍ ♇
05 38	☽ ⊻ ☿
05 43	☽ ⊼ ♄
06 12	☿ △ ♄
08 05	☽ ⊻ ☉
08 38	☽ △ ♆
11 08	☽ ⊼ ♅
15 08	☽ ∥ ♄
15 24	☽ ± ♄
16 23	☉ ⊼ ♆
21 25	☽ ⚹ ♀
22 28	☽ △ ♃
13 Tuesday	
01 50	☽ △ ♂
01 55	☿ ⊼ ♆
03 07	☉ ∥ ☿
03 08	☿ # ♅
03 11	☉ # ♅
05 10	☽ ∥ ♇
05 54	☽ ⚼ ♄
08 48	☽ ⚼ ♆
09 40	☉ ☌ ☿
18 33	☿ ☍ ♅
22 19	☽ ∠ ♀
23 20	☽ ± ♆
23 41	☽ ∥ ♂
14 Wednesday	
02 01	☽ ∥ ♃
04 54	☿ ⊥ ♀
04 57	☽ ⊼ ♇
05 16	☉ ☍ ♅
06 10	☽ △ ♄
09 05	☽ ⊼ ♆
11 33	☽ ☍ ♅
12 00	☽ ☌ ☉
13 20	☽ ⊥ ♀
14 34	☽ ☌ ☿
14 44	☽ ± ♇
15 39	☿ ± ♇
20 41	♀ ⚹ ♃
23 23	☽ □ ♃
23 26	☽ ⊻ ♀
15 Thursday	
03 50	☽ □ ♂
05 32	☽ ⚼ ♇
18 46	☉ ⊥ ♀
16 Friday	
04 42	☉ ± ♇
04 55	☽ # ♅
05 37	☽ Q ♃
06 42	☽ △ ♇
08 00	☽ □ ♄
10 53	☽ Q ♂
11 01	☽ ☍ ♆
13 32	☽ ⊼ ♅
17 54	☽ ⊻ ☉
23 11	☽ ∥ ☉
23 54	☽ ± ♅
17 Saturday	
02 20	☽ ⊻ ☿
02 25	☽ ⚹ ♃
02 54	☿ □ ♃
03 27	☽ ☌ ♀
05 09	☽ ⊥ ☉
08 17	☽ ⚹ ♂
10 15	☿ ⊻ ♀
14 55	☽ ⊥ ☿
15 21	☽ Q ♄
15 42	☽ ⚼ ♅
21 27	♂ ∠ ♇
22 22	☽ ∠ ☉
22 28	☽ ∥ ☉
18 Sunday	
05 16	☽ ∠ ♃
07 04	☿ # ♆
10 19	☽ ∠ ☿
11 35	☽ □ ♇
12 01	☽ ∠ ♂
13 02	☽ ⚹ ♄
16 13	☽ ⊼ ♆
17 56	☿ ⚼ ♇
18 52	☽ △ ♅
21 54	☽ ⊥ ♃
22 23	☿ □ ♂
19 Monday	
03 21	☽ ± ♆
04 06	☽ ⚹ ☉
04 09	☽ # ♅
05 12	☽ ⊥ ♂
09 11	☽ ⊻ ♃
11 04	☽ ⊻ ♀
16 53	☽ ⊻ ♂
17 05	☽ ∠ ♄
20 02	☽ ⚹ ☿
20 21	☽ ⚼ ♆
21 16	☽ Q ♇
22 41	☽ ⊥ ♀
20 Tuesday	
04 50	☽ Q ☉
10 24	☽ ⊥ ♄
16 21	☽ ∠ ♀
20 30	☽ ⚹ ♇
22 07	☽ ⊻ ♄
21 Wednesday	
00 09	☽ Q ☿
01 26	☽ △ ♆
04 10	☽ □ ♅
16 09	☽ ∥ ♃
19 16	☽ □ ☉
19 57	☽ ☌ ♃
22 23	☽ ⚹ ♀
22 Thursday	
02 14	☽ ∠ ♇
02 57	☽ ∥ ♂
04 31	☉ □ ♃
05 41	☽ ☌ ♂
11 49	♄ St D
16 03	☽ Q ♅
16 15	☽ ∥ ♇
19 40	☽ □ ☿
20 21	☽ ⊥ ♇
22 45	☽ Q ♀
23 Friday	
08 30	☽ ⊻ ♇
09 52	☽ ∥ ♄
10 16	☽ ☌ ♄
13 27	☉ ⊻ ♀
13 33	☽ □ ♆
16 18	☽ ⚹ ♅
24 Saturday	
09 02	☽ ⊻ ♃
11 26	☽ □ ♀
13 18	☽ △ ☉
20 10	♀ St R
20 35	☽ ⊻ ♂
21 18	☽ ⊥ ♃
22 39	☽ ∠ ♅
25 Sunday	
02 01	☽ Q ♆
08 45	☉ ⚼ ♇
09 09	☽ ⊥ ♂
15 37	☽ ∠ ♃
16 46	☽ ⊥ ♅
20 46	☽ △ ☿
21 11	☽ ☌ ♇
22 17	☽ ⚼ ☉
23 02	☽ ⊻ ♄
23 16	☿ △ ♇
26 Monday	
02 09	☽ ⚹ ♆
02 40	☿ Q ♃
03 54	☽ ∠ ♂
04 46	☽ ⊻ ♅
10 53	☿ □ ♄
11 03	☽ ⊥ ♄
11 10	☽ # ♄
17 50	☽ ± ☉
21 49	☽ ⚹ ♃
23 34	☽ △ ♀
27 Tuesday	
04 34	☽ # ♇
04 56	☽ ∠ ♄
06 05	☿ ☍ ♆
06 44	☽ ⊼ ☉
07 55	☽ ∠ ♆
08 15	☽ ⚼ ☿
10 44	☽ ⚹ ♂
11 18	☽ # ♂
21 37	☽ Q ♃
22 25	☿ ⊼ ♅
28 Wednesday	
01 30	☽ ⊥ ♆
02 40	☽ # ♃
04 44	☽ ⚼ ♀
05 00	☽ ± ☿
08 22	☽ ⊻ ♇
10 17	☽ ⚹ ♄
10 57	☽ Q ♂
13 07	☽ ⊻ ♆
15 35	☽ ☌ ♅
18 36	☽ ⊼ ☿
19 56	☽ ⊥ ♇
21 54	☽ ± ♀
29 Thursday	
08 35	☽ □ ♃
09 13	☽ ⊼ ♀
09 20	☽ Q ♄
13 06	☽ ∠ ♇
21 28	☽ ☍ ☉
22 31	☽ □ ♂
30 Friday	
01 40	☉ # ♆
09 10	☿ Q ♂
11 37	♀ ⚹ ♃
13 00	☽ ∥ ♅
17 16	☽ ⚹ ♇
19 14	☽ □ ♄
21 47	☽ ☌ ♆
21 59	☉ □ ♂
23 24	☿ ∥ ♀
31 Saturday	
00 05	☽ ⊻ ♅
05 49	☿ ± ♅
11 08	☽ ⊥ ♅
11 56	☽ ☍ ☿
15 25	☽ Q ♇
16 16	☽ ☍ ♀
17 02	☽ △ ♃

JUNE 1980

LONGITUDES

Date	Sidereal time h m s	Sun ☉ ° ' "	Moon ☽ ° ' "	Moon ☽ 24.00 ° ' "	Mercury ☿ ° '	Venus ♀ ° '	Mars ♂ ° '	Jupiter ♃ ° '	Saturn ♄ ° '	Uranus ♅ ° '	Neptune ♆ ° '	Pluto ♇ ° '
01	04 40 30	11 ♊ 07 41	12 ♑ 28 17	19 ♑ 06 52	00 ♋ 57	01 ♋ 24	10 ♍ 18	02 ♍ 09	20 ♍ 17	22 ♏ 49	21 ♐ 35	19 ♎ 10
02	04 44 27	12 05 09	25 ♑ 48 10	02 ♒ 32 08	02 34	01 R 05	10 44	02 15	20 18	22 R 46	21 R 34	19 R 09
03	04 48 23	13 02 36	09 ♒ 18 44	16 ♒ 07 57	04 08	00 43	11 10	02 21	20 19	22 44	21 32	19 09
04	04 52 20	14 00 02	22 ♒ 59 51	29 ♒ 54 25	05 39	00 ♋ 19	11 37	02 28	20 20	22 42	21 30	19 08
05	04 56 16	14 57 27	06 ♓ 51 41	13 ♓ 51 39	07 07	29 ♊ 53	12 04	02 34	20 22	22 39	21 29	19 07
06	05 00 13	15 54 52	20 ♓ 54 17	27 ♓ 59 29	08 33	29 25	12 31	02 41	20 23	22 37	21 27	19 06
07	05 04 09	16 52 16	05 ♈ 07 07	12 ♈ 16 54	09 55	28 55	12 59	02 48	20 25	22 35	21 26	19 06
08	05 08 06	17 49 40	19 ♈ 28 31	26 ♈ 41 31	11 14	28 24	13 26	02 54	20 26	22 33	21 24	19 05
09	05 12 02	18 47 03	03 ♉ 55 22	11 ♉ 09 26	12 30	27 51	13 54	03 02	20 28	22 31	21 22	19 04
10	05 15 59	19 44 26	18 ♉ 23 00	25 ♉ 35 19	13 43	27 17	14 22	03 09	20 30	22 28	21 21	19 04
11	05 19 56	20 41 48	02 ♊ 45 37	09 ♊ 53 08	14 52	26 41	14 51	03 16	20 32	22 26	21 19	19 03
12	05 23 52	21 39 09	16 ♊ 57 07	23 ♊ 56 57	15 59	26 05	15 19	03 24	20 34	22 24	21 17	19 03
13	05 27 49	22 36 30	00 ♋ 52 03	07 ♋ 42 00	17 02	25 28	15 48	03 31	20 36	22 22	21 16	19 02
14	05 31 45	23 33 50	14 ♋ 26 28	21 ♋ 05 18	18 01	24 51	16 17	03 39	20 39	22 20	21 14	19 02
15	05 35 42	24 31 10	27 ♋ 38 26	04 ♌ 05 58	18 58	24 13	16 46	03 47	20 41	22 18	21 13	19 01
16	05 39 38	25 28 28	10 ♌ 28 06	16 ♌ 45 09	19 50	23 36	17 15	03 55	20 43	22 16	21 11	19 01
17	05 43 35	26 25 46	22 ♌ 57 30	29 ♌ 05 38	20 39	22 58	17 45	04 03	20 46	22 14	21 09	19 00
18	05 47 31	27 23 03	05 ♍ 10 05	11 ♍ 11 26	21 25	22 22	18 15	04 11	20 49	22 12	21 08	19 00
19	05 51 28	28 20 19	17 ♍ 10 19	23 ♍ 07 20	22 06	21 46	18 44	04 19	20 51	22 10	21 06	19 00
20	05 55 25	29 ♊ 17 35	29 ♍ 03 10	04 ♎ 58 27	22 43	21 11	19 15	04 28	20 54	22 09	21 05	18 59
21	05 59 21	00 ♋ 14 50	10 ♎ 53 50	16 ♎ 49 56	23 17	20 37	19 45	04 36	20 57	22 07	21 03	18 59
22	06 03 18	01 12 04	22 ♎ 47 22	28 ♎ 46 41	23 46	20 04	20 15	04 45	21 00	22 05	21 01	18 59
23	06 07 14	02 09 17	04 ♏ 48 23	10 ♏ 52 58	24 10	19 33	20 46	04 54	21 03	22 03	21 00	18 59
24	06 11 11	03 06 30	17 ♏ 00 50	23 ♏ 12 19	24 31	19 04	21 17	05 03	21 06	22 02	20 58	18 58
25	06 15 07	04 03 42	29 ♏ 27 42	05 ♐ 47 11	24 47	18 37	21 48	05 11	21 10	22 00	20 57	18 58
26	06 19 04	05 00 54	12 ♐ 10 53	18 ♐ 38 49	24 58	18 11	22 19	05 21	21 13	21 58	20 55	18 58
27	06 23 00	05 58 06	25 ♐ 10 59	01 ♑ 47 15	25 05	17 48	22 50	05 30	21 17	21 57	20 54	18 58
28	06 26 57	06 55 18	08 ♑ 27 28	15 ♑ 11 23	25 R 07	17 27	23 22	05 39	21 20	21 55	20 52	18 58
29	06 30 54	07 52 29	21 ♑ 58 45	28 ♑ 49 15	25 05	17 08	23 53	05 48	21 24	21 54	20 50	18 D 58
30	06 34 50	08 ♋ 49 40	05 ♒ 42 34	12 ♒ 38 24	24 ♋ 58	16 ♊ 52	24 ♍ 25	05 ♍ 58	21 ♍ 27	21 ♏ 52	20 ♐ 49	18 ♎ 58

Date	Moon True ☊ ° '	Moon Mean ☊ ° '	Moon ☽ Latitude ° '
01	22 ♌ 44	23 ♌ 49	03 N 16
02	22 R 38	23 46	02 17
03	22 36	23 43	01 N 10
04	22 D 35	23 40	00 S 02
05	22 36	23 37	01 15
06	22 R 36	23 33	02 25
07	22 34	23 30	03 26
08	22 31	23 27	04 15
09	22 25	23 24	04 48
10	22 17	23 21	05 03
11	22 08	23 18	04 59
12	21 58	23 14	04 36
13	21 50	23 11	03 57
14	21 42	23 08	03 05
15	21 38	23 05	02 05
16	21 35	23 02	00 S 59
17	21 D 34	22 59	00 N 07
18	21 35	22 55	01 12
19	21 36	22 52	02 13
20	21 37	22 49	03 08
21	21 R 36	22 46	03 54
22	21 33	22 43	04 30
23	21 29	22 39	04 55
24	21 23	22 36	05 07
25	21 15	22 33	05 05
26	21 07	22 30	04 47
27	21 00	22 27	04 15
28	20 53	22 24	03 29
29	20 48	22 20	02 29
30	20 ♌ 45	22 ♌ 17	01 N 21

DECLINATIONS

Date	Sun ☉ ° '	Moon ☽ ° '	Mercury ☿ ° '	Venus ♀ ° '	Mars ♂ ° '	Jupiter ♃ ° '	Saturn ♄ ° '	Uranus ♅ ° '	Neptune ♆ ° '	Pluto ♇ ° '
01	22 N 07	19 S 36	25 N 36	25 N 19	08 N 48	11 N 47	05 N 56	18 S 12	21 S 46	08 N 42
02	22 14	18 44	25 35	25 08	08 37	11 44	05 55	18 12	21 46	08 42
03	22 22	16 48	25 31	24 57	08 26	11 42	05 55	18 11	21 46	08 42
04	22 29	13 53	25 25	24 44	08 14	11 40	05 54	18 10	21 46	08 42
05	22 35	10 09	25 18	24 32	08 03	11 37	05 53	18 10	21 45	08 42
06	22 42	05 49	25 09	24 19	07 51	11 35	05 53	18 09	21 45	08 41
07	22 48	01 S 07	24 59	24 05	07 39	11 32	05 52	18 09	21 45	08 41
08	22 53	03 N 41	24 47	23 52	07 27	11 29	05 51	18 08	21 45	08 41
09	22 58	08 18	24 34	23 37	07 15	11 27	05 50	18 08	21 45	08 41
10	23 03	12 26	24 21	23 23	07 03	11 24	05 49	18 07	21 45	08 41
11	23 07	15 50	24 06	23 08	06 51	11 21	05 48	18 07	21 45	08 40
12	23 11	18 13	23 50	22 53	06 39	11 18	05 47	18 06	21 45	08 40
13	23 14	19 29	23 33	22 37	06 26	11 15	05 46	18 06	21 45	08 40
14	23 17	19 35	23 16	22 22	06 14	11 13	05 45	18 05	21 45	08 39
15	23 20	18 36	22 58	22 06	06 01	11 10	05 44	18 05	21 44	08 39
16	23 22	16 40	22 40	21 51	05 49	11 07	05 43	18 04	21 44	08 39
17	23 23	13 59	22 21	21 35	05 36	11 04	05 42	18 04	21 44	08 38
18	23 25	10 44	22 02	21 20	05 23	11 00	05 40	18 03	21 44	08 38
19	23 26	07 07	21 43	21 04	05 10	10 57	05 39	18 03	21 44	08 38
20	23 26	03 N 15	21 24	20 49	04 57	10 54	05 38	18 02	21 44	08 37
21	23 26	00 S 44	21 05	20 35	04 44	10 51	05 37	18 02	21 44	08 37
22	23 26	04 41	20 46	20 20	04 31	10 48	05 35	18 01	21 44	08 36
23	23 25	08 29	20 27	20 06	04 18	10 44	05 34	18 01	21 44	08 36
24	23 24	12 01	20 09	19 53	04 05	10 41	05 32	18 01	21 44	08 36
25	23 23	15 05	19 51	19 40	03 52	10 38	05 31	18 00	21 44	08 35
26	23 21	17 30	19 33	19 28	03 38	10 34	05 29	18 00	21 43	08 35
27	23 18	19 06	19 17	19 17	03 25	10 31	05 28	17 59	21 43	08 34
28	23 16	19 42	19 01	19 06	03 11	10 27	05 26	17 59	21 43	08 34
29	23 12	19 11	18 45	18 56	02 58	10 24	05 25	17 59	21 43	08 33
30	23 N 09	17 S 32	18 N 31	18 N 46	02 N 44	10 N 20	05 N 23	17 S 58	21 S 43	08 N 33

ZODIAC SIGN ENTRIES

Date	h m	Planets
02	19 29	☽ ♒
05	00 10	☽ ♓
05	05 44	♀ ♊
07	03 23	☽ ♈
09	05 30	☽ ♉
11	07 22	☽ ♊
13	10 29	☽ ♋
15	16 22	☽ ♌
18	01 47	☽ ♍
20	13 55	☽ ♎
21	05 47	☉ ♋
23	02 26	☽ ♏
25	13 02	☽ ♐
27	20 46	☽ ♑
30	02 04	☽ ♒

LATITUDES

Date	Mercury ☿ ° '	Venus ♀ ° '	Mars ♂ ° '	Jupiter ♃ ° '	Saturn ♄ ° '	Uranus ♅ ° '	Neptune ♆ ° '	Pluto ♇ ° '
01	02 N 10	01 N 53	01 N 11	01 N 09	02 N 16	00 N 17	01 N 25	17 N 30
04	02 06	01 18	01 07	01 08	02 15	00 17	01 25	17 29
07	01 55	00 N 39	01 03	01 08	02 15	00 17	01 25	17 27
10	01 37	00 S 02	00 58	01 07	02 14	00 17	01 25	17 26
13	01 13	00 44	00 54	01 07	02 14	00 17	01 25	17 24
16	00 42	01 26	00 51	01 07	02 13	00 17	01 25	17 23
19	00 N 06	02 07	00 47	01 06	02 12	00 17	01 25	17 21
22	00 S 36	02 44	00 43	01 06	02 12	00 17	01 24	17 19
25	01 21	03 17	00 40	01 06	02 11	00 16	01 24	17 18
28	02 08	03 46	00 36	01 05	02 11	00 16	01 24	17 16
31	02 S 55	04 S 10	00 N 33	01 N 05	02 N 10	00 N 16	01 N 24	17 N 14

DATA

Julian Date	2444392
Delta T	+51 seconds
Ayanamsa	23° 34' 50"
Synetic vernal point	05° ♓ 32' 10"
True obliquity of ecliptic	23° 26' 23"

MOON'S PHASES, APSIDES AND POSITIONS ☽

Date	h m	Phase	Longitude ° '	Eclipse Indicator
06	02 53	☾	15 ♓ 33	
12	20 38	●	22 ♊ 00	
20	12 32	☽	29 ♍ 19	
28	09 02	○	06 ♑ 48	

Day	h m		
09	03 21	Perigee	
21	05 34	Apogee	
01	04 55	Max dec	19° S 39'
07	17 36	0N	
14	02 01	Max dec	19° N 41'
21	07 37	0S	
28	13 09	Max dec	19° S 42'

LONGITUDES

Date	Chiron ⚷ ° '	Ceres ⚳ ° '	Pallas ⚴ ° '	Juno ⚵ ° '	Vesta ⚶ ° '	Black Moon Lilith ⚸ ° '
01	15 ♉ 25	03 ♊ 19	17 ♈ 46	11 ♌ 21	23 ♊ 53	06 ♎ 22
11	16 ♉ 00	07 ♊ 25	20 ♈ 46	15 ♌ 12	28 ♊ 12	07 ♎ 29
21	16 ♉ 33	11 ♊ 31	23 ♈ 38	19 ♌ 07	02 ♋ 33	08 ♎ 36
31	17 ♉ 02	15 ♊ 35	26 ♈ 19	23 ♌ 06	06 ♋ 54	09 ♎ 42

All ephemeris data is given at 12.00 UT and the Moon's longitude is additionally given for 24.00 UT

ASPECTARIAN

h m	Aspects
01 Sunday	
03 33	☽ ∠ ♅
07 55	☽ △ ♂
09 22	☽ ⊼ ☉
17 33	☿ ☌ ♀
20 31	☽ ⅊ ♃
21 04	☽ ± ☉
02 Monday	
00 05	☽ □ ♇
01 10	♂ ∥ ♇
02 07	☽ △ ♄
04 25	☽ ⊻ ♆
06 35	☽ ⚹ ♅
06 47	☿ ⚹ ♃
11 52	☽ ⅊ ♂
12 48	☽ ± ♃
14 01	♂ Q ♅
14 28	☽ ⅊ ☉
15 08	☽ ⊥ ♆
20 14	☽ ∥ ♅
21 10	☽ ⊼ ♀
23 35	☽ ⊼ ♃
03 Tuesday	
01 39	☽ ⊼ ☿
03 55	☽ Q ♅
04 26	☽ ± ♂
04 55	☽ ⅊ ♄
07 06	☽ ∠ ♆
07 32	☽ ± ♀
13 39	☽ ± ☿
15 23	☽ ⊼ ♂
19 04	☽ △ ☉
20 50	☽ ± ♄
22 58	☽ ⅊ ♀
04 Wednesday	
05 15	☽ △ ♇
07 21	☽ ⊼ ♄
07 25	☽ ⅊ ☿
09 24	☽ ⚹ ♆
11 28	☽ □ ♅
05 Thursday	
00 20	☽ △ ♀
03 01	☽ ⫲ ♃
04 32	☽ ☍ ♃
06 11	☽ Q ♆
07 06	♀ Q ♂
07 17	☽ ⅊ ♇
12 30	☽ △ ☿
20 26	☽ ⫲ ♇
20 42	☿ ⅊ ♅
21 14	☽ ☍ ♂
22 43	☽ ± ♇
06 Friday	
00 37	☽ ⫲ ♂
02 53	☽ □ ☉
08 57	☽ ⊼ ♇
09 17	☿ Q ♄
11 07	☽ ☍ ♄
11 43	☽ ⫲ ♄
12 56	☽ □ ♆
14 54	☽ △ ♅
07 Saturday	
01 55	☽ □ ♀
08 04	☽ ⊼ ♃
11 33	☽ Q ☉
16 07	☽ ⅊ ♅
17 57	♂ ⊥ ♇
18 13	☽ ± ♃
20 51	☽ □ ☿
08 Sunday	
01 36	☽ ⊼ ♂
04 44	♀ ± ♅
07 03	☽ Q ♀
07 08	☽ ± ♅
09 04	☽ ⚹ ☉
09 22	☽ ⅊ ♃
11 21	☽ ☍ ♇
11 56	☽ ± ♂
13 37	☽ ⊼ ♄
15 12	☽ △ ♆
17 06	☽ ⊼ ♅
23 01	☽ ∥ ♄
23 36	☽ ± ♄
09 Monday	
02 17	☽ ⚹ ♀
03 24	☽ ⅊ ♂
05 47	☽ Q ☿
06 38	☽ ∥ ♂
10 30	☽ △ ♃
11 45	☽ ∠ ☉
14 04	☽ ∥ ♇
14 34	☽ ⅊ ♄
16 03	☽ ⅊ ♆
19 08	☉ △ ♇
10 Tuesday	
02 15	☽ ∠ ♀
03 32	☽ ⚹ ☿
03 44	☽ ⊥ ☉
05 07	☽ △ ♂
05 40	☽ ∥ ♃
06 58	☽ ± ♆
13 08	☽ ⊼ ♇
14 25	☽ ⊻ ☉
15 32	☽ △ ♄
16 37	☽ ⊥ ♀
16 55	☽ ⊼ ♆
18 47	☽ ☍ ♅
23 07	☽ ± ♇
11 Wednesday	
02 14	☽ ⊻ ♀
06 44	☽ ∠ ☿
07 46	☉ □ ♄
10 58	☿ ⚹ ♂
12 52	☽ □ ♃
13 24	☉ ∥ ♀
14 10	☽ ⅊ ♇
23 10	☽ ⊥ ☿
12 Thursday	
03 11	☉ ☍ ♆
04 30	☉ Q ♃
09 07	☽ □ ♂
10 12	☽ ⊻ ☿
10 27	☽ ⫲ ♅
15 34	☽ △ ♇
18 12	☽ □ ♄
19 25	☽ ☍ ♆
19 40	☽ Q ♃
20 38	☽ ☌ ☉
21 19	☽ ⊼ ♅
13 Friday	
03 01	☽ ☌ ♀
06 12	☉ ⊼ ♅
07 40	☽ ± ♅
16 41	☽ ⚹ ♃
17 19	☽ Q ♂
23 23	☽ ⅊ ♅
14 Saturday	
01 38	☽ Q ♄
10 56	☉ ∥ ☿
15 26	☽ ⚹ ♂
18 57	☽ ☌ ☿
19 39	☽ ∠ ♃
20 15	☽ □ ♇
23 13	☽ ⚹ ♄
15 Sunday	
00 15	☽ ⊼ ♆
02 14	☽ △ ♅
05 49	☽ ⊻ ☉
06 00	☽ ⊻ ♀
06 24	☿ ∠ ♃
07 27	☉ ☌ ♀
11 12	☽ ± ♆
12 15	☽ ⊥ ♃
13 28	☿ □ ♇
16 33	☽ ⊥ ♀
17 45	☽ ⊥ ☉
19 38	☽ ⫲ ♅
19 57	☽ ∠ ♂
23 31	☽ ⊻ ♃
16 Monday	
03 01	☽ ∠ ♄
03 56	☽ ⅊ ♆
05 28	☽ Q ♇
08 37	☽ ∠ ♀
12 01	☽ ∠ ☉
13 33	☽ ⊥ ♂
20 08	☽ ⊥ ♄
21 45	♀ ⫲ ♆
17 Tuesday	
00 23	♂ ∥ ♄
01 30	☽ ⊻ ♂
04 20	☽ ⚹ ♇
04 51	♃ ∠ ♇
07 14	☽ ⊻ ☿
07 44	☽ ⊻ ♄
08 31	☽ △ ♆
10 36	☽ □ ♅
12 02	☽ ⚹ ♀
15 35	☿ ⚹ ♄
19 21	☽ ⚹ ☉
19 42	☽ ⊥ ☿
18 Wednesday	
03 09	☿ ⊼ ♆
09 41	☽ ∠ ♇
10 01	☽ ☌ ♃
10 07	☽ ∥ ♃
10 29	☽ Q ♀
14 37	☽ ∠ ☿
17 47	♀ Q ♃
18 34	♀ ⊼ ♅
21 07	☽ Q ☉
22 00	☽ Q ♅
19 Thursday	
02 12	☽ ∥ ♇
03 37	☽ ⊥ ♇
05 35	☿ ⊻ ♀
07 58	☉ ± ♅
10 56	☿ ⫲ ♆
14 36	☿ △ ♅
15 18	☽ ☌ ♂
15 40	☽ ⊻ ♇
19 27	☽ ☌ ♄
19 54	☽ □ ♆
20 49	☽ □ ♀
21 14	☽ ∥ ♄
22 03	☽ ⚹ ♅
22 30	☽ ⚹ ☿
23 55	♂ ⊻ ♇
20 Friday	
00 53	☽ ∥ ♂
12 32	☽ □ ☉
16 29	♀ ☍ ♆
22 40	♀ □ ♄
23 06	☽ ⊻ ♃
21 Saturday	
00 04	☽ Q ☿
04 21	☽ ∠ ♅
08 16	☽ Q ♆
11 24	☽ ⊥ ♃
22 31	☽ ⊥ ♅
22 Sunday	
04 20	☽ ☌ ♇
05 48	☽ ∠ ♃
06 41	☽ ⊻ ♂
06 46	☽ △ ♀
07 46	♀ □ ♂
08 23	☽ ⊻ ♄
08 27	☽ ⚹ ♆
10 35	☽ ⊻ ♅
11 05	☽ ⫲ ♂
14 02	☽ □ ☿
17 33	☽ ⫲ ♄
18 21	♄ □ ♆
19 16	☽ ⊥ ♂
20 29	☽ ⊥ ♄
23 Monday	
06 17	☽ △ ☉
11 31	☽ ⅊ ♀
12 11	☽ ⚹ ♃
12 43	☽ ⫲ ♇
13 59	☽ ∠ ♂
14 21	☽ ∠ ♆
14 29	☽ ∠ ♄
22 12	♂ □ ♆
24 Tuesday	
02 48	☽ ⫲ ♃
02 58	♂ ☌ ♄
04 35	☽ ± ♀
08 01	☽ ⊥ ♆
12 03	☽ Q ♃
14 18	☽ ⅊ ☉
15 49	☽ ⊻ ♇
15 51	☽ ⊼ ♀
16 58	♀ △ ♇
19 40	☽ ⊻ ♆
19 59	☽ ⚹ ♄
20 39	☽ ⚹ ♂
21 42	☽ ☌ ♅
25 Wednesday	
02 51	☽ △ ☿
03 24	☽ ⊥ ♇
06 08	☿ ⊥ ♀
09 06	☽ ± ☉
19 04	☽ Q ♄
20 34	☽ ∠ ♇
20 36	☽ Q ♂
20 51	♂ ⚹ ♅
21 27	☽ ⊼ ☉
23 00	☽ □ ♃
26 Thursday	
07 49	☽ ⅊ ☿
18 04	☽ ∥ ♅
21 48	☉ ⚹ ♃
22 50	☽ ☍ ♀
27 Friday	
00 36	☽ ⚹ ♇
00 43	☽ ± ☿
04 09	☽ ☌ ♆
04 49	☽ □ ♄
06 05	☽ ⊻ ♅
07 32	☽ □ ♂
11 49	☽ ⊼ ☿
11 59	☿ ∥ ♀
15 09	☽ ⫲ ☿
15 23	☽ ⫲ ♀
17 01	☽ ⊥ ♅
22 31	☽ Q ♇
28 Saturday	
06 54	☽ △ ♃
09 02	☽ ☍ ☉
09 15	☽ ∠ ♅
11 13	☿ St R
11 57	☉ ⅊ ♅
20 09	♇ St D
29 Sunday	
03 38	☽ ⊼ ♀
06 41	☽ □ ♇
09 55	☽ ⅊ ♃
10 00	☽ ⊻ ♆
10 58	☽ △ ♄
11 51	☽ ⚹ ♅
14 00	☽ ± ♀
15 30	☽ △ ♂
17 25	☽ ☍ ☿
17 52	☽ ⫲ ♀
20 31	☽ ⊥ ♆
21 41	☽ ⫲ ☿
30 Monday	
01 53	☽ ± ♃
05 26	☽ ⅊ ♀
07 02	☽ ∥ ♅
08 49	☽ Q ♅
12 11	☽ ∠ ♆
12 27	☽ ⊼ ♃
13 18	☽ ⅊ ♄
17 48	☽ ⊼ ☉
18 42	☽ ⅊ ♂

JULY 1980

LONGITUDES

Date	Sidereal time h m s	Sun ☉ ° ' "	Moon ☽ ° ' "	Moon ☽ 24.00 ° ' "	Mercury ☿ ° '	Venus ♀ ° '	Mars ♂ ° '	Jupiter ♃ ° '	Saturn ♄ ° '	Uranus ♅ ° '	Neptune ♆ ° '	Pluto ♇ ° '
01	06 38 47	09 ♋ 46 51	19 ♒ 36 24	26 ♒ 36 17	24 ♋ 46	16 ♊ 38	24 ♍ 57	06 ♍ 07	21 ♍ 31	21 ♏ 51	20 ♐ 47	18 ♎ 58
02	06 42 43	10 44 02	03 ♓ 37 46	10 ♓ 40 34	24 R 30	16 R 26	25 29	06 17	21 35	21 R 49	20 R 46	18 58
03	06 46 40	11 41 14	17 ♓ 44 26	24 ♓ 49 10	24 10	16 17	26 02	06 27	21 39	21 48	20 44	18 58
04	06 50 36	12 38 25	01 ♈ 54 31	09 ♈ 00 16	23 47	16 10	26 34	06 37	21 43	21 47	20 43	18 59
05	06 54 33	13 35 37	16 ♈ 06 12	23 ♈ 12 05	23 19	16 06	27 07	06 47	21 47	21 46	20 41	18 59
06	06 58 29	14 32 50	00 ♉ 17 38	07 ♉ 22 35	22 49	16 04	27 39	06 57	21 52	21 44	20 40	18 59
07	07 02 26	15 30 02	14 ♉ 26 37	21 ♉ 29 22	22 15	16 D 04	28 12	07 07	21 56	21 43	20 39	18 59
08	07 06 23	16 27 15	28 ♉ 30 30	05 ♊ 29 35	21 40	16 06	28 45	07 17	22 00	21 42	20 37	19 00
09	07 10 19	17 24 29	12 ♊ 26 16	19 ♊ 20 08	21 03	16 11	29 18	07 27	22 05	21 41	20 36	19 00
10	07 14 16	18 21 43	26 ♊ 10 50	02 ♋ 57 59	20 25	16 18	29 ♍ 52	07 38	22 09	21 40	20 34	19 00
11	07 18 12	19 18 57	09 ♋ 41 20	16 ♋ 20 36	19 46	16 27	00 ♎ 25	07 48	22 14	21 39	20 33	19 01
12	07 22 09	20 16 11	22 ♋ 55 37	29 ♋ 26 17	19 08	16 39	00 59	07 59	22 18	21 38	20 32	19 01
13	07 26 05	21 13 26	05 ♌ 52 33	12 ♌ 14 28	18 30	16 52	01 32	08 09	22 23	21 37	20 30	19 02
14	07 30 02	22 10 41	18 ♌ 32 09	24 ♌ 45 48	17 54	17 07	02 06	08 20	22 28	21 36	20 29	19 02
15	07 33 58	23 07 56	00 ♍ 55 41	07 ♍ 02 07	17 21	17 24	02 40	08 31	22 33	21 36	20 28	19 03
16	07 37 55	24 05 11	13 ♍ 05 30	19 ♍ 06 17	16 50	17 43	03 14	08 41	22 38	21 35	20 26	19 03
17	07 41 52	25 02 27	25 ♍ 04 56	01 ♎ 01 59	16 22	18 04	03 49	08 52	22 43	21 34	20 25	19 04
18	07 45 48	25 59 42	06 ♎ 57 59	12 ♎ 53 32	15 59	18 27	04 23	09 03	22 48	21 34	20 24	19 05
19	07 49 45	26 56 58	18 ♎ 49 11	24 ♎ 45 34	15 40	18 51	04 58	09 14	22 53	21 33	20 23	19 05
20	07 53 41	27 54 14	00 ♏ 43 17	06 ♏ 42 55	15 25	19 17	05 32	09 26	22 58	21 32	20 21	19 06
21	07 57 38	28 51 30	12 ♏ 45 03	18 ♏ 50 14	15 16	19 44	06 07	09 37	23 04	21 32	20 20	19 07
22	08 01 34	29 ♋ 48 47	24 ♏ 58 59	01 ♐ 11 46	15 12	20 13	06 42	09 48	23 09	21 32	20 19	19 07
23	08 05 31	00 ♌ 46 04	07 ♐ 28 59	13 ♐ 51 00	15 D 14	20 43	07 17	09 59	23 15	21 31	20 18	19 08
24	08 09 27	01 43 21	20 ♐ 18 04	26 ♐ 50 23	15 22	21 15	07 52	10 11	23 20	21 31	20 17	19 09
25	08 13 24	02 40 39	03 ♑ 28 00	10 ♑ 10 56	15 35	21 48	08 27	10 22	23 26	21 31	20 16	19 10
26	08 17 21	03 37 58	16 ♑ 59 03	23 ♑ 52 08	15 55	22 23	09 03	10 34	23 31	21 30	20 15	19 11
27	08 21 17	04 35 17	00 ♒ 49 50	07 ♒ 51 46	16 20	22 58	09 38	10 45	23 37	21 30	20 14	19 12
28	08 25 14	05 32 36	14 ♒ 57 24	22 ♒ 06 13	16 52	23 35	10 14	10 57	23 43	21 30	20 13	19 13
29	08 29 10	06 29 57	29 ♒ 17 33	06 ♓ 30 48	17 29	24 13	10 50	11 09	23 48	21 30	20 11	19 14
30	08 33 07	07 27 18	13 ♓ 45 18	21 ♓ 00 24	18 13	24 53	11 25	11 20	23 54	21 D 30	20 11	19 15
31	08 37 03	08 ♌ 24 41	28 ♓ 15 29	05 ♈ 29 57	19 ♋ 02	25 ♊ 33	12 ♎ 01	11 ♍ 32	24 ♍ 00	21 ♏ 30	20 ♐ 10	19 ♎ 16

Date	Moon True ☊ ° '	Moon Mean ☊ ° '	Moon ☽ Latitude ° '
01	20 ♌ 44	22 ♌ 14	00 N 06
02	20 D 45	22 11	01 S 09
03	20 46	22 08	02 21
04	20 47	22 05	03 25
05	20 R 48	22 01	04 16
06	20 47	21 58	04 52
07	20 44	21 55	05 10
08	20 40	21 52	05 09
09	20 36	21 49	04 49
10	20 31	21 45	04 14
11	20 26	21 42	03 24
12	20 22	21 39	02 24
13	20 20	21 36	01 18
14	20 19	21 33	00 S 10
15	20 D 20	21 30	00 N 58
16	20 21	21 26	02 02
17	20 23	21 23	03 00
18	20 24	21 20	03 49
19	20 25	21 17	04 29
20	20 R 26	21 14	04 57
21	20 25	21 11	05 12
22	20 23	21 07	05 14
23	20 21	21 04	05 01
24	20 18	21 01	04 33
25	20 16	20 58	03 50
26	20 14	20 55	02 53
27	20 12	20 51	01 45
28	20 12	20 48	00 N 29
29	20 D 12	20 45	00 S 50
30	20 12	20 42	02 07
31	20 ♌ 13	20 ♌ 39	03 S 16

DECLINATIONS

Date	Sun ☉ ° '	Moon ☽ ° '	Mercury ☿ ° '	Venus ♀ ° '	Mars ♂ ° '	Jupiter ♃ ° '	Saturn ♄ ° '	Uranus ♅ ° '	Neptune ♆ ° '	Pluto ♇ ° '
01	23 N 05	14 S 50	18 N 18	18 N 38	02 N 30	10 N 16	05 N 21	17 S 58	21 S 43	08 N 32
02	23 00	11 15	18 05	18 30	02 17	10 13	05 20	17 58	21 43	08 31
03	22 56	07 01	17 54	18 23	02 03	10 09	05 18	17 57	21 43	08 31
04	22 50	02 S 23	17 45	18 16	01 49	10 05	05 16	17 57	21 43	08 30
05	22 45	02 N 23	17 36	18 10	01 35	10 01	05 14	17 57	21 43	08 30
06	22 39	07 01	17 29	18 06	01 21	09 58	05 13	17 56	21 43	08 29
07	22 32	11 14	17 23	18 01	01 07	09 54	05 11	17 56	21 43	08 28
08	22 26	14 49	17 19	17 58	00 53	09 50	05 09	17 56	21 43	08 28
09	22 18	17 30	17 16	17 55	00 39	09 46	05 07	17 56	21 42	08 27
10	22 11	19 10	17 14	17 53	00 24	09 42	05 05	17 55	21 42	08 26
11	22 03	19 42	17 14	17 51	00 N 10	09 38	05 03	17 55	21 42	08 26
12	21 55	19 07	17 15	17 50	00 S 04	09 34	05 01	17 55	21 42	08 25
13	21 46	17 32	17 18	17 50	00 18	09 30	04 59	17 55	21 42	08 24
14	21 37	15 07	17 22	17 50	00 33	09 26	04 57	17 54	21 42	08 24
15	21 27	12 03	17 27	17 50	00 47	09 22	04 55	17 54	21 42	08 23
16	21 18	08 31	17 33	17 51	01 02	09 18	04 53	17 54	21 42	08 22
17	21 07	04 42	17 41	17 53	01 16	09 14	04 51	17 54	21 42	08 22
18	20 57	00 N 44	17 49	17 55	01 31	09 09	04 49	17 54	21 42	08 21
19	20 46	03 S 14	17 58	17 57	01 45	09 05	04 47	17 54	21 42	08 20
20	20 35	07 05	18 08	17 59	02 00	09 01	04 44	17 54	21 42	08 19
21	20 23	10 42	18 19	18 02	02 15	08 57	04 42	17 53	21 42	08 18
22	20 11	13 56	18 30	18 05	02 29	08 52	04 40	17 53	21 42	08 18
23	19 59	16 36	18 41	18 09	02 44	08 48	04 38	17 53	21 42	08 17
24	19 47	18 33	18 53	18 12	02 59	08 44	04 35	17 53	21 42	08 16
25	19 34	19 34	19 04	18 16	03 14	08 39	04 33	17 53	21 42	08 15
26	19 20	19 30	19 16	18 20	03 28	08 35	04 31	17 53	21 42	08 14
27	19 07	18 16	19 27	18 24	03 43	08 30	04 28	17 53	21 42	08 14
28	18 53	15 53	19 38	18 28	03 58	08 26	04 26	17 53	21 41	08 13
29	18 39	12 30	19 49	18 32	04 13	08 21	04 23	17 53	21 41	08 12
30	18 24	08 20	19 58	18 36	04 28	08 17	04 21	17 53	21 41	08 11
31	18 N 10	03 S 41	20 N 07	18 N 40	04 S 43	08 N 12	04 N 19	17 S 53	21 S 41	08 N 10

ZODIAC SIGN ENTRIES

Date	h	m	Planets
02	05	48	☽ ♓
04	08	46	☽ ♈
06	11	30	☽ ♉
08	14	33	☽ ♊
10	17	59	♂ ♎
10	18	44	☽ ♋
13	01	03	☽ ♌
15	10	11	☽ ♍
17	21	55	☽ ♎
20	10	33	☽ ♏
22	16	42	☉ ♌
22	21	42	☽ ♐
25	05	45	☽ ♑
27	10	34	☽ ♒
29	13	11	☽ ♓
31	14	53	☽ ♈

LATITUDES

Date	Mercury ☿ ° '	Venus ♀ ° '	Mars ♂ ° '	Jupiter ♃ ° '	Saturn ♄ ° '	Uranus ♅ ° '	Neptune ♆ ° '	Pluto ♇ ° '
01	02 S 55	04 S 10	00 N 33	01 N 05	02 N 10	00 N 16	01 N 24	17 N 14
04	03 39	04 29	00 29	01 05	02 10	00 16	01 24	17 13
07	04 16	04 43	00 26	01 04	02 09	00 16	01 24	17 11
10	04 42	04 53	00 23	01 04	02 09	00 16	01 24	17 09
13	04 54	04 59	00 20	01 04	02 08	00 16	01 24	17 07
16	04 52	05 02	00 17	01 04	02 08	00 16	01 24	17 06
19	04 35	05 02	00 14	01 03	02 07	00 16	01 24	17 04
22	04 06	05 00	00 11	01 03	02 07	00 16	01 24	17 02
25	03 29	04 56	00 08	01 03	02 07	00 16	01 24	17 01
28	02 45	04 50	00 06	01 03	02 06	00 16	01 23	16 59
31	01 S 59	04 S 42	00 N 03	01 N 03	02 N 06	00 N 16	01 N 23	16 N 58

DATA

Julian Date	2444422
Delta T	+51 seconds
Ayanamsa	23° 34' 55"
Synetic vernal point	05° ♓ 32' 05"
True obliquity of ecliptic	23° 26' 23"

MOON'S PHASES, APSIDES AND POSITIONS ☽

Date	h	m	Phase	Longitude ° '	Eclipse Indicator
05	07	27	☾	13 ♈ 25	
12	06	46	●	20 ♋ 04	
20	05	51	☽	27 ♎ 40	
27	18	54	○	04 ♒ 52	

Day	h	m		
04	16	22	Perigee	
19	00	13	Apogee	
30	22	52	Perigee	
04	23	58	0N	
11	11	19	Max dec	19° N 42'
18	16	28	0S	
25	22	41	Max dec	19° S 40'

LONGITUDES

Date	Chiron ⚷ ° '	Ceres ⚳ ° '	Pallas ⚴ ° '	Juno ⚵ ° '	Vesta ⚶ ° '	Black Moon Lilith ⚸ ° '
01	17 ♉ 02	15 ♊ 35	26 ♈ 19	23 ♌ 06	06 ♋ 54	09 ♎ 42
11	17 ♉ 27	19 ♊ 36	28 ♈ 46	27 ♌ 06	11 ♋ 15	10 ♎ 49
21	17 ♉ 47	23 ♊ 34	00 ♉ 58	01 ♍ 09	15 ♋ 36	11 ♎ 56
31	18 ♉ 02	27 ♊ 28	02 ♉ 49	05 ♍ 12	19 ♋ 55	13 ♎ 03

All ephemeris data is given at 12.00 UT and the Moon's longitude is additionally given for 24.00 UT

ASPECTARIAN

h m	Aspects
01 Tuesday	
04 56	☽ ± ☉
04 56	☽ ± ♄
05 00	☉ Q ♄
06 07	☿ ⚹ ♂
06 58	☽ △ ♀
10 50	☽ ± ♂
10 54	☽ △ ♇
14 02	☽ ⚹ ♆
15 18	☽ ⊼ ♄
15 50	☽ □ ♅
20 42	☽ ⊼ ☿
21 32	☽ ⚼ ☉
21 32	☽ ⊼ ♂
02 Wednesday	
06 47	☽ ± ☿
10 32	☽ Q ♆
12 35	☽ ⚼ ♇
16 35	☽ ☍ ♃
18 19	☽ # ♃
21 48	☽ ⚼ ☿
03 Thursday	
00 58	☽ △ ☉
03 50	☽ # ♇
03 54	☽ ± ♇
05 26	☿ # ♅
09 33	☽ □ ♀
14 06	☽ ⊼ ♇
17 05	☽ □ ♆
18 40	☽ ☍ ♄
18 53	☽ △ ♅
21 06	☽ # ♄
22 37	☽ △ ☿
04 Friday	
02 36	☽ ☍ ♂
14 59	☽ # ♂
15 48	☽ Q ♀
20 03	☽ ⊼ ♃
20 14	☽ ⚼ ♅
05 Saturday	
04 16	♄ ⚹ ♅
06 19	☽ ± ♃
07 27	☽ □ ☉
08 07	☽ ∥ ♂
11 25	☽ ± ♅
11 59	☽ ⚹ ♀
16 52	☽ ☍ ♇
19 45	☽ △ ♆
21 33	☽ ⊼ ♅
21 40	☽ ⊼ ♄
21 42	☽ ⚼ ♃
23 47	☽ □ ☿
06 Sunday	
02 32	☽ ∥ ♄
07 21	☽ ⊼ ♂
07 51	☽ ± ♄
13 18	☽ ∠ ♀
16 05	☽ Q ☉
17 55	☽ ± ♂
20 00	☽ ∥ ♇
21 05	☽ ⚼ ♆
21 15	♀ St D
23 11	☽ ⚼ ♄
23 24	☽ △ ♃
07 Monday	
04 09	☽ ∥ ♃
04 33	☽ ⊥ ♀
05 10	☽ Q ☿
09 48	☽ ⚼ ♂
12 20	☽ ± ♆
13 56	☽ ⚹ ☉
14 46	☽ ⊻ ♀
16 38	☿ ∠ ♃
19 36	☿ ⊥ ♀
19 44	☽ ⊼ ♇
22 32	☽ ⊼ ♆
23 56	☿ ⚹ ♄
08 Tuesday	
00 23	☽ ☍ ♅
00 47	☽ ⚹ ☿
00 49	☽ △ ♄
02 42	☉ ⊻ ♀
05 59	☽ ± ♇
10 29	☿ △ ♅
12 26	☽ △ ♂
17 25	☽ ∠ ☉
21 25	☽ ⚼ ♇
09 Wednesday	
01 26	☽ ∠ ☿
03 16	☽ □ ♃
05 54	☉ Q ♂
06 05	♀ # ♅
09 28	☽ ∥ ☿
10 05	☽ ⊥ ☉
16 20	☽ ⊥ ☿
16 37	☽ ∥ ♀
16 49	☽ # ♅
18 34	☽ ☌ ♀
21 17	☽ ⊻ ☉
23 25	☽ △ ♇
10 Thursday	
02 11	☽ ☍ ♆
02 20	☽ ⊻ ☿
04 05	☽ ⊼ ♅
04 53	☽ □ ♄
05 39	☿ ⊼ ♆
11 01	☽ Q ♃
14 37	☽ ± ♅
18 46	☽ □ ♂
11 Friday	
04 19	☉ □ ♇
06 34	☽ ⚼ ♅
08 34	☽ ⚹ ♃
12 59	☽ Q ♄
18 45	☉ ☌ ☿
12 Saturday	
00 22	☽ ⊻ ♀
04 28	☽ Q ♂
04 52	☽ □ ♇
05 23	☽ ☌ ☿
06 46	☽ ☌ ☉
07 37	☽ ⊼ ♆
09 38	☽ △ ♅
10 51	☽ ⚹ ♄
11 28	☽ ⊥ ♀
12 05	☽ ∠ ♃
14 58	☿ Q ♂
15 58	☿ □ ♇
18 18	☉ ⊼ ♆
18 36	☽ ± ♆
13 Sunday	
03 32	☽ ⚹ ♂
04 22	☽ ∠ ♀
04 57	☽ ⊥ ♃
07 23	☽ # ♅
08 26	☽ ∥ ♀
11 18	☽ ⚼ ♆
14 10	☽ Q ♇
14 40	☽ ∥ ☿
14 51	☽ ∠ ♄
16 20	☽ ⊻ ♃
21 51	☉ △ ♅
22 00	☉ # ♆
14 Monday	
08 01	☽ ⊥ ♄
09 08	☽ ∠ ♂
09 14	☽ ⚹ ♀
10 51	☽ ⊻ ☿
12 58	☽ ⚹ ♇
15 44	☽ △ ♆
17 54	☽ □ ♅
19 35	☽ ⊻ ☉
19 37	☽ ⊻ ♄
19 56	☉ ⚹ ♄
21 53	☽ ⊥ ☿
15 Tuesday	
03 18	☽ ⊥ ♂
08 12	☽ ⊥ ☉
08 57	☽ Q ♀
10 14	☿ ⊻ ♀
14 39	☽ ∠ ☿
15 35	☽ ⊻ ♂
18 07	☽ ∠ ♇
22 10	☉ ⊥ ♀
23 43	☉ ∠ ♃
16 Wednesday	
03 08	☽ ☌ ♃
03 22	☽ ∠ ☉
05 02	☽ Q ♅
06 50	☽ ∥ ♃
11 56	☽ ⊥ ♇
12 58	☽ ∥ ♇
19 10	☽ ⚹ ☿
21 30	☽ □ ♀
23 55	☽ ⊻ ♇
17 Thursday	
02 39	☽ □ ♆
04 57	☽ ⚹ ♅
07 12	☽ ☌ ♄
11 06	☽ ∥ ♄
11 55	☽ ⚹ ☉
18 24	☽ Q ☿
18 Friday	
02 41	♀ # ♅
06 31	☽ ☌ ♂
07 37	☽ # ♂
11 11	☽ ∠ ♅
14 16	☽ Q ☉
14 53	☽ Q ♆
16 18	☽ ⊻ ♃
21 53	☉ ± ♆
19 Saturday	
00 49	☿ # ♅
02 29	☽ ∥ ♂
04 38	☽ ⊥ ♃
05 23	☽ ⊥ ♅
05 46	☽ □ ☿
07 39	☿ ∥ ♀
12 04	☽ △ ♀
12 33	☽ ☌ ♇
15 09	☽ ⚹ ♆
17 31	☽ ⊻ ♅
20 17	☽ ⊻ ♄
21 24	☽ # ♄
23 08	☽ ∠ ♃
20 Sunday	
01 51	♀ △ ♇
05 51	☽ □ ☉
08 28	☽ ⊥ ♄
19 25	☽ ⚼ ♀
19 57	☽ # ♇
21 16	☽ ∠ ♆
22 08	☽ ⊻ ♂
21 Monday	
00 18	☽ # ♃
02 19	♃ Q ♅
02 37	☽ ∠ ♄
05 40	☽ ⚹ ♃
10 41	☽ ⊥ ♂
14 02	☽ ± ♀
15 08	☽ ⊥ ♆
16 56	☽ △ ☿
22 Tuesday	
00 33	☽ ⊻ ♇
02 19	☽ ⊼ ♀
02 55	☽ ⊻ ♆
04 57	♂ ∠ ♅
05 16	☽ ☌ ♅
05 17	☽ ∠ ♂
05 42	☽ Q ♃
08 25	☽ ⚹ ♄
12 17	☽ ⊥ ♇
16 36	♀ ☍ ♆
16 36	☿ St D
22 06	☽ ⚼ ☿
22 07	☽ △ ☉
23 Wednesday	
05 38	☽ ∠ ♇
07 42	☽ Q ♄
11 36	☽ ⚹ ♂
15 20	☽ ± ☿
16 49	☽ □ ♃
24 Thursday	
02 39	☽ ∥ ♅
02 43	☽ ⊼ ☿
04 50	☽ ⚼ ☉
06 43	☽ # ♀
09 52	☽ ⚹ ♇
11 10	☽ Q ♂
11 58	☽ ☌ ♆
13 50	☽ ☍ ♀
14 14	☽ ⊻ ♅
17 38	☽ □ ♄
18 51	☽ # ☿
22 45	☽ ± ☉
23 23	♀ ⊼ ♅
25 Friday	
01 13	☽ ⊥ ♅
04 16	♂ Q ♆
07 51	☽ Q ♇
10 28	☽ ⊼ ☉
12 06	☽ # ☉
17 27	☽ ∠ ♅
21 20	☽ □ ♂
26 Saturday	
00 31	☽ △ ♃
10 04	☽ ☍ ☿
15 51	☽ □ ♇
16 13	☉ ∥ ☿
17 34	☽ # ☿
17 41	☽ ⊻ ♆
18 22	☽ # ☉
19 54	☽ ⚹ ♅
21 50	☽ ⊼ ♀
23 28	☽ △ ♄
27 Sunday	
03 08	☽ ⚼ ♃
04 05	☽ ⊥ ♆
08 40	☽ ± ♀
10 17	☽ # ♀
16 34	☽ Q ♅
16 42	☽ ∥ ♅
17 13	☉ ⊥ ♃
18 48	☽ ± ♃
18 54	☽ ☍ ☉
19 30	☽ ∠ ♆
28 Monday	
00 44	☽ ⚼ ♀
01 22	☽ ⚼ ♄
03 40	☽ △ ♂
03 44	☉ ⚼ ♆
05 08	☽ ⊼ ♃
15 20	☽ ⊼ ☿
16 40	☽ ± ♄
17 30	♀ □ ♄
19 10	☽ △ ♇
20 49	☽ ⚹ ♆
22 59	☽ □ ♅
29 Tuesday	
01 51	☽ ± ☿
02 47	☽ ⊼ ♄
03 09	☽ △ ♀
05 58	☽ ⚼ ♂
16 49	☽ Q ♆
17 35	☽ ⚼ ☿
20 13	☽ ⚼ ♇
21 04	☉ ∥ ♀
21 36	☽ ± ♂
30 Wednesday	
00 49	☽ ⊼ ☉
02 48	♂ # ♄
06 42	☉ Q ♇
06 55	♂ ⊻ ♃
07 57	☽ ☍ ♃
07 58	☽ ⊼ ♂
11 10	☽ ± ♇
11 28	☽ ± ☉
11 38	♅ St D
12 19	☽ # ♃
12 51	☽ # ♇
19 48	☽ △ ☿
21 06	☽ ⊼ ♇
22 37	☽ □ ♆
31 Thursday	
00 49	☽ △ ♅
03 25	☽ ⚼ ☉
04 55	☽ ☍ ♄
07 06	☽ ∥ ♂
07 18	☽ □ ♀
08 50	☽ # ♄
18 20	☿ □ ♇

AUGUST 1980

Date	Sidereal time h m s	Sun ☉ ° ' "	Moon ☽ ° ' "	Moon ☽ 24.00 ° ' "	Mercury ☿ ° '	Venus ♀ ° '	Mars ♂ ° '	Jupiter ♃ ° '	Saturn ♄ ° '	Uranus ♅ ° '	Neptune ♆ ° '	Pluto ⯓ ° '
01	08 41 00	09 ♌ 22 04	12 ♈ 43 18	19 ♈ 55 02	19 ♋ 58	26 ♊ 14	12 ♎ 37	11 ♍ 44	24 ♍ 06	21 ♏ 30	20 ♐ 09	19 ♎ 17
02	08 44 56	10 19 29	27 ♈ 04 45	04 ♉ 12 05	20 59	26 57	13 14	11 56	24 12	21 30	20 R 08	19 18
03	08 48 53	11 16 55	11 ♉ 16 44	18 ♉ 18 29	22 05	27 40	13 50	12 08	24 18	21 30	20 07	19 19
04	08 52 50	12 14 22	25 ♉ 17 08	02 ♊ 12 33	23 18	28 24	14 26	12 20	24 24	21 31	20 06	19 20
05	08 56 46	13 11 50	09 ♊ 04 36	15 ♊ 53 13	24 35	29 09	15 03	12 32	24 31	21 31	20 05	19 22
06	09 00 43	14 09 20	22 ♊ 38 21	29 ♊ 19 58	25 58	29 ♊ 55	15 39	12 44	24 37	21 31	20 04	19 23
07	09 04 39	15 06 51	05 ♋ 58 03	12 ♋ 32 35	27 26	00 ♋ 42	16 16	12 56	24 43	21 32	20 04	19 24
08	09 08 36	16 04 23	19 ♋ 03 36	25 ♋ 31 07	28 ♋ 58	01 30	16 53	13 08	24 49	21 32	20 03	19 25
09	09 12 32	17 01 57	01 ♌ 55 11	08 ♌ 15 51	00 ♌ 35	02 18	17 30	13 20	24 56	21 33	20 02	19 27
10	09 16 29	17 59 31	14 ♌ 33 13	20 ♌ 47 23	02 15	03 07	18 07	13 33	25 02	21 33	20 02	19 28
11	09 20 25	18 57 07	26 ♌ 58 29	03 ♍ 06 41	04 00	03 57	18 44	13 45	25 09	21 34	20 01	19 30
12	09 24 22	19 54 44	09 ♍ 12 12	15 ♍ 15 14	05 48	04 48	19 21	13 57	25 15	21 34	20 00	19 31
13	09 28 19	20 52 22	21 ♍ 16 05	27 ♍ 15 03	07 38	05 39	19 59	14 10	25 22	21 35	20 00	19 32
14	09 32 15	21 50 01	03 ♎ 12 27	09 ♎ 08 42	09 32	06 31	20 36	14 22	25 28	21 36	19 59	19 34
15	09 36 12	22 47 41	15 ♎ 04 12	20 ♎ 59 23	11 27	07 23	21 14	14 35	25 35	21 37	19 59	19 35
16	09 40 08	23 45 22	26 ♎ 54 45	02 ♏ 50 49	13 25	08 16	21 51	14 47	25 42	21 37	19 58	19 37
17	09 44 05	24 43 04	08 ♏ 48 05	14 ♏ 47 08	15 23	09 10	22 29	15 00	25 48	21 38	19 58	19 38
18	09 48 01	25 40 47	20 ♏ 48 30	26 ♏ 52 45	17 23	10 04	23 07	15 12	25 55	21 39	19 57	19 40
19	09 51 58	26 38 32	03 ♐ 00 28	09 ♐ 12 10	19 24	10 59	23 45	15 25	26 02	21 40	19 57	19 42
20	09 55 54	27 36 18	15 ♐ 28 25	21 ♐ 49 39	21 24	11 54	24 23	15 38	26 09	21 41	19 56	19 43
21	09 59 51	28 34 04	28 ♐ 16 21	04 ♑ 48 51	23 25	12 50	25 01	15 50	26 16	21 43	19 56	19 45
22	10 03 48	29 ♌ 31 52	11 ♑ 27 26	18 ♑ 12 18	25 26	13 46	25 40	16 03	26 23	21 44	19 56	19 47
23	10 07 44	00 ♍ 29 41	25 ♑ 03 29	02 ♒ 00 57	27 26	14 43	26 18	16 16	26 29	21 45	19 55	19 48
24	10 11 41	01 27 32	09 ♒ 04 28	16 ♒ 13 41	29 ♌ 26	15 40	26 56	16 28	26 36	21 46	19 55	19 50
25	10 15 37	02 25 23	23 ♒ 28 05	00 ♓ 47 00	01 ♍ 25	16 38	27 35	16 41	26 43	21 48	19 55	19 52
26	10 19 34	03 23 17	08 ♓ 09 40	15 ♓ 35 10	03 24	17 36	28 14	16 54	26 50	21 49	19 55	19 54
27	10 23 30	04 21 11	23 ♓ 02 31	00 ♈ 30 42	05 21	18 35	28 52	17 07	26 58	21 50	19 55	19 55
28	10 27 27	05 19 08	07 ♈ 58 40	15 ♈ 25 26	07 17	19 34	29 ♎ 31	17 19	27 05	21 52	19 54	19 57
29	10 31 23	06 17 06	22 ♈ 50 03	00 ♉ 11 40	09 12	20 33	00 ♏ 10	17 32	27 12	21 53	19 54	19 59
30	10 35 20	07 15 06	07 ♉ 29 34	14 ♉ 43 10	11 06	21 33	00 49	17 45	27 19	21 55	19 54	20 01
31	10 39 17	08 ♍ 13 07	21 ♉ 52 01	28 ♉ 55 48	12 ♍ 59	22 ♋ 33	01 ♏ 28	17 ♍ 58	27 ♍ 26	21 ♏ 56	19 ♐ 54	20 ♎ 03

Date	Moon True ☊ ° '	Moon Mean ☊ ° '	Moon ☽ Latitude ° '	DECLINATIONS Sun ☉ ° '	Moon ☽ ° '	Mercury ☿ ° '	Venus ♀ ° '	Mars ♂ ° '	Jupiter ♃ ° '	Saturn ♄ ° '	Uranus ♅ ° '	Neptune ♆ ° '	Pluto ⯓ ° '
01	20 ♌ 14	20 ♌ 36	04 S 12	17 N 55	01 N 10	20 N 15	18 N 44	04 S 57	08 N 08	04 N 16	17 S 53	21 S 41	08 N 09
02	20 D 15	20 32	04 51	17 39	05 54	20 21	18 49	05 12	08 03	04 14	17 53	21 41	08 08
03	20 15	20 29	05 13	17 24	10 15	20 26	18 53	05 27	07 59	04 11	17 53	21 41	08 07
04	20 R 15	20 26	05 16	17 08	13 58	20 29	18 57	05 42	07 54	04 09	17 53	21 41	08 07
05	20 14	20 23	05 00	16 51	16 52	20 30	19 00	05 57	07 49	04 06	17 54	21 41	08 06
06	20 13	20 20	04 28	16 35	18 47	20 29	19 04	06 12	07 45	04 04	17 54	21 41	08 05
07	20 13	20 16	03 41	16 18	19 37	20 26	19 08	06 27	07 40	04 01	17 54	21 41	08 04
08	20 12	20 13	02 44	16 01	19 22	20 21	19 11	06 42	07 35	03 58	17 54	21 41	08 03
09	20 12	20 10	01 40	15 44	18 07	20 13	19 14	06 57	07 30	03 56	17 54	21 41	08 02
10	20 12	20 07	00 S 31	15 26	15 58	20 02	19 18	07 12	07 26	03 53	17 54	21 41	08 01
11	20 D 12	20 04	00 N 38	15 09	13 07	19 49	19 20	07 27	07 21	03 50	17 54	21 41	08 00
12	20 12	20 01	01 44	14 51	09 43	19 34	19 23	07 41	07 16	03 48	17 55	21 41	07 59
13	20 R 12	19 57	02 44	14 32	05 58	19 15	19 25	07 56	07 11	03 45	17 55	21 41	07 58
14	20 12	19 54	03 37	14 14	02 N 02	18 54	19 27	08 11	07 07	03 42	17 55	21 41	07 57
15	20 11	19 51	04 20	13 55	01 S 57	18 30	19 29	08 26	07 02	03 40	17 55	21 41	07 56
16	20 11	19 48	04 51	13 36	05 50	18 04	19 30	08 41	06 57	03 37	17 55	21 41	07 55
17	20 11	19 45	05 11	13 17	09 31	17 35	19 31	08 56	06 52	03 34	17 56	21 41	07 54
18	20 11	19 42	05 17	12 58	12 51	17 04	19 32	09 11	06 47	03 32	17 56	21 41	07 53
19	20 D 11	19 38	05 09	12 38	15 42	16 31	19 33	09 25	06 42	03 29	17 56	21 41	07 52
20	20 11	19 35	04 47	12 18	17 54	15 56	19 33	09 40	06 37	03 26	17 57	21 41	07 51
21	20 12	19 32	04 10	11 58	19 16	15 19	19 32	09 55	06 32	03 23	17 57	21 41	07 50
22	20 12	19 29	03 19	11 38	19 39	14 41	19 32	10 09	06 27	03 20	17 57	21 41	07 49
23	20 13	19 26	02 15	11 18	18 54	14 01	19 31	10 24	06 22	03 18	17 58	21 41	07 48
24	20 14	19 22	01 N 01	10 57	17 00	13 19	19 29	10 39	06 18	03 15	17 58	21 41	07 47
25	20 R 14	19 19	00 S 18	10 37	13 59	12 37	19 27	10 53	06 13	03 12	17 58	21 42	07 46
26	20 13	19 16	01 38	10 16	10 01	11 54	19 25	11 08	06 08	03 09	17 59	21 42	07 45
27	20 12	19 13	02 52	09 55	05 23	11 09	19 22	11 22	06 03	03 06	17 59	21 42	07 44
28	20 11	19 10	03 54	09 34	00 S 25	10 23	19 19	11 37	05 58	03 04	17 59	21 42	07 43
29	20 09	19 07	04 41	09 12	04 N 32	09 39	19 15	11 51	05 53	03 01	18 00	21 42	07 42
30	20 07	19 03	05 08	08 51	09 09	08 53	19 11	12 06	05 48	02 58	18 00	21 42	07 41
31	20 ♌ 06	19 ♌ 00	05 S 16	08 N 29	13 N 09	08 N 06	19 N 07	12 S 20	05 N 43	02 N 55	18 S 01	21 S 42	07 N 40

ZODIAC SIGN ENTRIES

Date	h	m	Planets
02	16	55	☽ ♉
04	20	10	☽ ♊
06	14	25	♀ ♋
07	01	12	☽ ♋
09	03	31	☿ ♌
09	08	23	☽ ♌
11	17	54	☽ ♍
14	05	32	☽ ♎
16	18	15	☽ ♏
19	06	08	☽ ♐
21	15	11	☽ ♑
22	23	41	☉ ♍
23	20	32	☽ ♒
24	18	47	☿ ♍
25	22	43	☽ ♓
27	23	11	☽ ♈
29	05	50	♂ ♏
29	23	41	☽ ♉

LATITUDES

Date	Mercury ☿ ° '	Venus ♀ ° '	Mars ♂ ° '	Jupiter ♃ ° '	Saturn ♄ ° '	Uranus ♅ ° '	Neptune ♆ ° '	Pluto ⯓ ° '
01	01 S 44	04 S 39	00 N 02	01 N 03	02 N 06	00 N 16	01 N 23	16 N 57
04	00 58	04 29	00 S 01	01 03	02 06	00 16	01 23	16 56
07	00 S 15	04 18	00 03	01 02	02 05	00 15	01 23	16 54
10	00 N 24	04 07	00 06	01 02	02 05	00 15	01 23	16 53
13	00 55	03 54	00 08	01 02	02 05	00 15	01 23	16 51
16	01 20	03 41	00 11	01 02	02 05	00 15	01 23	16 50
19	01 36	03 27	00 13	01 02	02 05	00 15	01 22	16 48
22	01 44	03 13	00 15	01 02	02 04	00 15	01 22	16 47
25	01 46	02 58	00 18	01 02	02 04	00 15	01 22	16 46
28	01 42	02 44	00 20	01 02	02 04	00 15	01 22	16 45
31	01 N 32	02 S 28	00 S 22	01 N 02	02 N 04	00 N 15	01 N 22	16 N 44

DATA

Julian Date	2444453
Delta T	+51 seconds
Ayanamsa	23° 34' 59"
Synetic vernal point	05° ♓ 32' 00"
True obliquity of ecliptic	23° 26' 23"

LONGITUDES

Date	Chiron ⚷ ° '	Ceres ⚳ ° '	Pallas ⚴ ° '	Juno ⚵ ° '	Vesta ⚶ ° '	Black Moon Lilith ⚸ ° '
01	18 ♉ 04	27 ♊ 52	02 ♉ 59	05 ♍ 37	20 ♋ 21	13 ♎ 10
11	18 ♉ 13	01 ♋ 40	04 ♉ 23	09 ♍ 40	24 ♋ 40	14 ♎ 17
21	18 ♉ 17	05 ♋ 18	05 ♉ 18	13 ♍ 43	28 ♋ 56	15 ♎ 24
31	18 ♉ 14	08 ♋ 55	05 ♉ 37	17 ♍ 45	03 ♌ 09	16 ♎ 31

MOON'S PHASES, APSIDES AND POSITIONS ☽

Date	h	m	Phase	Longitude ° '	Eclipse Indicator
03	12	00	☾	11 ♉ 17	
10	19	09	●	18 ♌ 17	Annular
18	22	28	☽	26 ♏ 06	
26	03	42	○	03 ♓ 03	

Day	h	m		
15	18	09	Apogee	
27	19	15	Perigee	
01	06	15	0N	
07	18	28	Max dec	19° N 39'
15	00	16	0S	
22	08	29	Max dec	19° S 39'
28	14	02	0N	

ASPECTARIAN

01 Friday
01 40 ☽ ♇ ♅
02 03 ♃ ∥ ⯓
04 34 ☉ ∠ ♄
06 02 ☽ △ ☉
10 20 ☽ ⊼ ♃
11 50 ☽ ☍ ♂
14 10 ☉ # ♅
14 39 ☽ Q ♀
16 28 ☿ ⊼ ♆
16 38 ☽ ± ♅
20 28 ☽ ± ♃
22 57 ☽ ☍ ⯓

02 Saturday
00 22 ☽ △ ♆
00 58 ☽ □ ☿
02 39 ☽ ⊼ ♅
03 28 ☽ ∥ ♄
07 08 ☽ ⊼ ♄
08 12 ☽ # ♂
11 45 ☽ ♇ ♃
11 46 ☽ ✶ ♀
17 17 ☽ ± ♄
23 20 ☽ ∥ ♃
23 37 ☿ △ ♅
23 58 ☽ ∥ ⯓

03 Sunday
01 33 ☽ ♇ ♆
06 45 ♀ ± ♅
08 37 ☽ ♇ ♄
09 48 ☽ Q ☿
12 00 ☽ □ ☉
13 28 ☽ △ ♃
14 29 ☽ ∠ ♀
16 32 ☽ ⊼ ♂
16 50 ☽ ± ♆

04 Monday
01 45 ☽ ⊼ ⯓
03 05 ☽ ⊼ ♆
03 16 ☽ ± ♂
05 30 ☽ ☍ ♅
06 45 ☽ ⊥ ♀
08 14 ☽ ✶ ☿
10 28 ☽ △ ♄
12 06 ☽ ± ⯓
14 50 ☉ ⊻ ♃
17 42 ☽ ⊻ ♀
19 31 ☽ ♇ ♂
21 13 ☽ Q ☉

05 Tuesday
03 44 ☽ ♇ ⯓
10 27 ☿ ✶ ♄
11 54 ☽ ∥ ☉
13 00 ☽ ∠ ☿
18 10 ☽ □ ♃
19 48 ☽ ✶ ☉
23 00 ☽ △ ♂
23 14 ☽ # ♅

06 Wednesday
06 11 ☽ △ ⯓
06 41 ♂ ⊥ ♅
06 41 ☽ ⊥ ☿
07 26 ☽ ☍ ♆
10 00 ☽ ⊼ ♅
13 46 ☿ ± ♆
15 33 ☽ □ ♄
17 53 ☽ ∥ ♀
18 40 ☽ ⊻ ☿
20 45 ☽ ± ♅

07 Thursday
00 35 ☽ ∠ ☉
01 52 ☽ ☌ ♀
02 45 ☽ Q ♃
13 01 ☽ ♇ ♅
18 11 ☽ ⊥ ☉
21 14 ☿ ∠ ♃
21 26 ♀ Q ♃

08 Friday
00 25 ☽ Q ♄
00 55 ☽ ✶ ♃
06 03 ☽ ⊻ ☉
07 47 ☽ □ ♂
12 41 ☽ □ ⯓
13 50 ☽ ⊼ ♆
16 35 ☽ △ ♅
16 48 ☽ ∥ ♀
22 47 ☽ ✶ ♄

09 Saturday
00 59 ☽ ± ♆
05 10 ☽ ∠ ♃
09 06 ☽ ☌ ☿
12 46 ☽ ⊻ ♀
14 51 ☽ # ♅
17 53 ☽ ♇ ♆
19 06 ☽ Q ♂
22 25 ☽ ⊥ ♃
22 28 ☽ Q ⯓

10 Sunday
00 55 ☽ ⊥ ♀
02 04 ♃ ⊥ ⯓
03 18 ☽ ∠ ♄
10 02 ☽ ⊻ ♃
17 29 ☽ ∥ ☉
19 09 ☽ ☌ ☉
19 12 ☽ ✶ ♂
19 21 ☽ ∠ ♀
20 35 ☉ ✶ ♂
20 41 ☽ ⊥ ♄
21 28 ☽ ✶ ⯓
22 31 ☽ △ ♆

11 Monday
01 29 ☽ □ ♅
05 10 ♂ # ♃
08 24 ☽ ⊻ ♄
10 53 ☿ ⊻ ♀
12 28 ☉ ∠ ♀
17 25 ☉ ⊥ ♄

12 Tuesday
01 35 ☿ ♇ ♆
01 52 ☉ ✶ ⯓
01 56 ☽ ∠ ♂
02 40 ☽ ✶ ♀
02 45 ☽ ∠ ⯓
04 06 ☽ ⊻ ☿
12 44 ☽ Q ♅
14 16 ☉ △ ♆
18 03 ☽ ⊥ ☿
18 30 ♂ ☌ ⯓
20 34 ☽ ⊥ ⯓
20 40 ☽ ⊥ ♂
21 35 ☽ ☌ ♃
23 23 ☽ ∥ ⯓

13 Wednesday
00 24 ☽ # ♂
00 47 ☿ ∥ ♀
04 13 ☽ Q ♀
04 14 ☽ ∥ ♃
08 32 ☽ ⊻ ⯓
09 17 ☽ ⊻ ♂
09 27 ☽ □ ♆
10 42 ☿ Q ⯓
11 08 ☽ ⊻ ☉
12 37 ♂ ✶ ♆
12 38 ☽ ✶ ♅
14 44 ♂ # ⯓
15 15 ☽ ∠ ☿
18 27 ☿ Q ♂
19 31 ☿ ⊥ ♃
20 17 ☽ ☌ ♄
22 41 ♂ ⊥ ♃

14 Thursday
00 13 ☽ ⊥ ☉
01 47 ☽ ∥ ♄
06 01 ☉ □ ♅
14 18 ♀ ♇ ♅
18 51 ☽ ∠ ♅
19 13 ☽ □ ♀
19 58 ☽ ∠ ☉
21 39 ☽ Q ♆

15 Friday
00 31 ☿ ∠ ♄
03 15 ☽ ✶ ☿
10 59 ☽ ⊻ ♃
13 06 ☽ ⊥ ♅
21 11 ☽ ☌ ⯓
21 56 ☽ ✶ ♆
22 22 ☽ # ♄
23 22 ☽ ⊥ ♃

16 Saturday
01 11 ☽ ☌ ♂
01 16 ☽ ⊻ ♅
02 58 ♂ ⊻ ♅
05 02 ☽ ✶ ☉
08 21 ☽ Q ☿
09 31 ☽ ⊻ ♄
17 55 ☽ ∠ ♃
18 54 ☽ # ♃
19 16 ☿ # ♅
21 46 ☽ ⊥ ♄

17 Sunday
01 18 ☽ # ⯓
04 16 ☽ ∠ ♆
06 40 ☿ ⊻ ♃
07 07 ☿ ⊥ ♀
07 26 ☽ Q ☉
07 43 ☽ ∥ ♂
12 48 ☽ △ ♀
16 04 ☽ ∠ ♄
22 20 ☽ ⊥ ♆

18 Monday
00 38 ☽ ✶ ♃
03 50 ☽ □ ☿
09 44 ☽ ⊻ ⯓
10 18 ☽ ⊻ ♆
12 45 ☽ # ☉
13 41 ☽ ☌ ♅
16 50 ☽ ⊻ ♂
18 45 ☉ ⊻ ♄
21 07 ☽ ♇ ♀
21 38 ☽ ⊥ ⯓
22 12 ☽ ✶ ♄
22 28 ☽ □ ☉

19 Tuesday
00 52 ☽ Q ♃
05 17 ☽ ⊥ ♂
15 17 ☽ ∠ ⯓
15 39 ☿ ✶ ⯓
16 09 ☽ ± ♀
18 30 ☽ # ☿
18 34 ☿ △ ♆
20 05 ☿ ⊥ ♄
21 50 ☽ Q ♄
23 43 ☽ ∠ ♂

20 Wednesday
04 39 ☽ ⊼ ♀
12 18 ☽ □ ♃
12 40 ☽ ∥ ♅
15 25 ☿ □ ♅
20 04 ☽ ✶ ⯓
20 27 ☽ ☌ ♆
23 46 ☽ ⊻ ♅

21 Thursday
01 18 ☽ △ ☿
05 39 ☽ ✶ ♂
08 14 ☽ □ ♄
10 57 ☽ ⊥ ♅
12 35 ☽ △ ☉
18 25 ☽ Q ⯓
21 11 ☽ # ♀

22 Friday
03 28 ☽ ∠ ♅
04 49 ☽ Q ♂
09 50 ☽ ♇ ☿
15 56 ☿ ✶ ♂
16 27 ☽ ☍ ♀
17 55 ☽ ♇ ☉
20 15 ☽ # ♀
20 19 ☽ △ ♃
23 56 ☿ ⊻ ♄

23 Saturday
02 48 ☽ □ ⯓
03 02 ☽ ⊻ ♆
04 36 ☽ ± ☿
05 24 ♀ Q ♄
06 13 ☽ ✶ ♅
10 57 ☽ ± ☉
13 30 ☽ ⊥ ♆
14 15 ☽ □ ♂
14 30 ☽ △ ♄
16 49 ☽ ⊼ ☿
20 50 ♂ ⊻ ♄
22 05 ☽ ⊼ ☉
22 52 ☽ ♇ ♃

24 Sunday
01 40 ☽ ∥ ♅
02 59 ☽ Q ♅
04 57 ☽ ∠ ♆
14 23 ☽ ± ♃
16 18 ☽ ♇ ♄
23 52 ☽ ⊼ ♀

25 Monday
00 35 ☽ ⊼ ♃
00 43 ☉ # ♂
06 02 ☽ △ ⯓
06 08 ☽ ✶ ♆
07 26 ☽ ± ♄
09 14 ☽ □ ♅
10 32 ☽ ± ♀
13 32 ♀ ✶ ♃
17 05 ☿ ∠ ♀
17 24 ☽ ⊼ ♄
19 04 ☽ △ ♂
22 51 ☽ # ☿

26 Tuesday
01 51 ☽ Q ♆
02 20 ☽ ♇ ♀
03 04 ☽ ☍ ☿
03 42 ☽ ☍ ☉
06 04 ☽ ∥ ♂
06 41 ☽ ♇ ⯓
10 31 ☽ # ☉
11 52 ☉ ☌ ☿
20 34 ☽ ♇ ♂
21 17 ☽ ± ⯓

27 Wednesday
00 07 ☽ # ⯓
01 00 ♆ ✶ ⯓
02 19 ☽ ☍ ♃
04 19 ☽ △ ♀
06 41 ☿ ∠ ⯓
06 43 ☿ # ♂
06 58 ☽ □ ♆
06 59 ☽ ⊼ ⯓
08 42 ☽ # ♃
10 04 ☽ △ ♅
11 43 ☽ ± ♂
18 21 ☽ ☍ ♄
21 47 ☽ ⊼ ♂
23 15 ☽ # ♄

28 Thursday
02 38 ☉ ∠ ⯓
07 26 ☽ ⊼ ☉
10 12 ☽ ♇ ♅
10 43 ☽ ⊼ ☿
17 45 ☽ ± ☉
20 18 ♀ ⊼ ♆
21 46 ♀ □ ⯓
21 49 ☽ ± ☿

29 Friday
00 44 ☽ ± ♅
03 17 ☽ ⊼ ♃
04 36 ☽ ∥ ♄
07 15 ☽ △ ♆
07 22 ☽ ☍ ⯓
08 02 ☽ □ ♀
09 19 ☽ ♇ ☉
10 28 ☽ ⊼ ♅
13 10 ☽ ± ♃
14 34 ☽ ♇ ☿
18 37 ☽ ∥ ♃
19 09 ☽ ⊼ ♄
20 41 ☿ Q ♅

30 Saturday
00 31 ☽ ☍ ♂
04 05 ☽ ♇ ♃
04 08 ☽ ∥ ⯓
05 04 ☽ ± ♄
07 44 ☽ ♇ ♆
10 28 ☽ ∥ ☉
10 44 ☽ ∥ ☿
11 34 ☽ △ ☉
13 48 ☉ ∥ ☿
15 39 ☽ Q ♀
18 53 ☽ △ ☿
20 03 ☽ ♇ ♄
20 55 ♀ △ ♅
22 38 ☽ ± ♆

31 Sunday
05 20 ☽ △ ♃
06 20 ☽ # ♂
08 41 ☽ ⊼ ♆
08 56 ☽ ⊼ ⯓
12 07 ☽ ☍ ♅
13 15 ☽ ✶ ♀
19 06 ☽ ± ⯓
21 32 ☽ △ ♄
23 39 ♆ St D

All ephemeris data is given at 12.00 UT and the Moon's longitude is additionally given for 24.00 UT

SEPTEMBER 1980

LONGITUDES

Date	Sidereal time h m s	Sun ☉ ° ' "	Moon ☽ ° ' "	Moon ☽ 24.00 ° ' "	Mercury ☿ ° '	Venus ♀ ° '	Mars ♂ ° '	Jupiter ♃ ° '	Saturn ♄ ° '	Uranus ♅ ° '	Neptune ♆ ° '	Pluto ♇ ° '
01	10 43 13	09 ♍ 11 11	05 ♊ 54 22	12 ♊ 47 39	14 ♍ 51	23 ♋ 34	02 ♏ 07	18 ♍ 11	27 ♍ 33	21 ♏ 58	19 ♐ 54	20 ♎ 05
02	10 47 10	10 09 17	19 ♊ 35 42	26 ♊ 18 39	16 41	24 35	02 47	18 24	27 40	22 00	19 D 54	20 07
03	10 51 06	11 07 25	02 ♋ 56 42	09 ♋ 30 06	18 31	25 36	03 26	18 37	27 48	22 01	19 54	20 09
04	10 55 03	12 05 35	15 ♋ 59 06	22 ♋ 24 02	20 19	26 38	04 06	18 50	27 55	22 03	19 54	20 11
05	10 58 59	13 03 47	28 ♋ 45 11	05 ♌ 02 51	22 06	27 40	04 45	19 03	28 02	22 05	19 55	20 13
06	11 02 56	14 02 00	11 ♌ 17 21	17 ♌ 28 56	23 51	28 42	05 25	19 16	28 09	22 07	19 55	20 15
07	11 06 52	15 00 16	23 ♌ 37 52	29 ♌ 44 24	25 36	29 ♋ 44	06 05	19 29	28 17	22 09	19 55	20 17
08	11 10 49	15 58 33	05 ♍ 48 47	11 ♍ 51 13	27 19	00 ♌ 47	06 44	19 42	28 24	22 11	19 55	20 19
09	11 14 46	16 56 52	17 ♍ 51 56	23 ♍ 51 08	29 ♍ 01	01 50	07 24	19 55	28 31	22 13	19 55	20 21
10	11 18 42	17 55 13	29 ♍ 49 02	05 ♎ 45 52	00 ♎ 42	02 54	08 04	20 08	28 39	22 15	19 56	20 23
11	11 22 39	18 53 36	11 ♎ 41 52	17 ♎ 37 17	02 22	03 58	08 44	20 21	28 46	22 17	19 56	20 25
12	11 26 35	19 52 00	23 ♎ 32 24	29 ♎ 27 30	04 00	05 02	09 25	20 34	28 54	22 19	19 56	20 27
13	11 30 32	20 50 27	05 ♏ 22 57	11 ♏ 19 05	05 38	06 06	10 05	20 47	29 01	22 21	19 57	20 29
14	11 34 28	21 48 55	17 ♏ 16 20	23 ♏ 15 07	07 15	07 11	10 45	21 00	29 08	22 24	19 57	20 31
15	11 38 25	22 47 24	29 ♏ 15 55	05 ♐ 19 12	08 50	08 15	11 26	21 12	29 16	22 26	19 58	20 33
16	11 42 21	23 45 55	11 ♐ 25 30	17 ♐ 35 22	10 25	09 20	12 06	21 25	29 23	22 28	19 58	20 36
17	11 46 18	24 44 28	23 ♐ 49 20	00 ♑ 07 57	11 58	10 26	12 47	21 38	29 31	22 31	19 59	20 38
18	11 50 15	25 43 03	06 ♑ 31 45	13 ♑ 01 14	13 30	11 31	13 28	21 51	29 38	22 33	19 59	20 40
19	11 54 11	26 41 39	19 ♑ 36 52	26 ♑ 19 02	15 02	12 37	14 09	22 04	29 45	22 35	20 00	20 42
20	11 58 08	27 40 17	03 ♒ 08 00	10 ♒ 03 58	16 32	13 43	14 50	22 17	29 ♍ 53	22 38	20 00	20 45
21	12 02 04	28 38 57	17 ♒ 06 56	24 ♒ 16 47	18 01	14 49	15 31	22 30	00 ♎ 00	22 40	20 01	20 47
22	12 06 01	29 ♍ 37 38	01 ♓ 33 10	08 ♓ 55 35	19 29	15 55	16 12	22 43	00 08	22 43	20 02	20 49
23	12 09 57	00 ♎ 36 21	16 ♓ 23 15	23 ♓ 55 16	20 56	17 02	16 53	22 56	00 15	22 46	20 02	20 51
24	12 13 54	01 35 06	01 ♈ 30 31	09 ♈ 07 45	22 22	18 09	17 34	23 09	00 23	22 48	20 03	20 54
25	12 17 50	02 33 53	16 ♈ 45 39	24 ♈ 22 52	23 47	19 16	18 15	23 22	00 30	22 51	20 04	20 56
26	12 21 47	03 32 41	01 ♉ 58 02	09 ♉ 29 56	25 11	20 23	18 57	23 35	00 38	22 54	20 05	20 58
27	12 25 44	04 31 33	16 ♉ 57 28	24 ♉ 19 43	26 34	21 31	19 38	23 48	00 45	22 56	20 06	21 00
28	12 29 40	05 30 26	01 ♊ 35 58	08 ♊ 45 43	27 56	22 38	20 20	24 01	00 52	22 59	20 07	21 03
29	12 33 37	06 29 22	15 ♊ 48 39	22 ♊ 44 40	29 ♎ 16	23 46	21 01	24 14	01 00	23 02	20 07	21 05
30	12 37 33	07 ♎ 28 20	29 ♊ 33 52	06 ♋ 16 25	00 ♏ 35	24 ♌ 54	21 ♏ 43	24 ♍ 26	01 ♎ 07	23 ♏ 05	20 ♐ 08	21 ♎ 07

Date	Moon True ☊ ° '	Moon Mean ☊ ° '	Moon ☽ Latitude ° '
01	20 ♌ 05	18 ♌ 57	05 S 04
02	20 D 05	18 54	04 35
03	20 06	18 51	03 52
04	20 08	18 48	02 57
05	20 09	18 44	01 55
06	20 10	18 41	00 S 49
07	20 R 10	18 38	00 N 19
08	20 09	18 35	01 25
09	20 07	18 32	02 26
10	20 03	18 28	03 20
11	19 59	18 25	04 05
12	19 53	18 22	04 40
13	19 48	18 19	05 02
14	19 43	18 16	05 12
15	19 40	18 13	05 08
16	19 37	18 09	04 50
17	19 36	18 06	04 18
18	19 D 37	18 03	03 33
19	19 38	18 00	02 36
20	19 40	17 57	01 29
21	19 41	17 54	00 N 14
22	19 R 40	17 50	01 S 04
23	19 38	17 47	02 20
24	19 33	17 44	03 27
25	19 28	17 41	04 20
26	19 21	17 38	04 55
27	19 15	17 34	05 08
28	19 09	17 31	05 01
29	19 05	17 28	04 36
30	19 ♌ 03	17 ♌ 25	03 S 55

DECLINATIONS

Date	Sun ☉ ° '	Moon ☽ ° '	Mercury ☿ ° '	Venus ♀ ° '	Mars ♂ ° '	Jupiter ♃ ° '	Saturn ♄ ° '	Uranus ♅ ° '	Neptune ♆ ° '	Pluto ♇ ° '
01	08 N 08	16 N 18	07 N 20	19 N 02	12 S 34	05 N 38	02 N 52	18 S 01	21 S 42	07 N 39
02	07 46	18 28	06 33	18 56	12 48	05 32	02 49	18 02	21 42	07 38
03	07 24	19 33	05 46	18 50	13 02	05 27	02 46	18 02	21 42	07 37
04	07 02	19 33	04 59	18 44	13 16	05 22	02 43	18 03	21 42	07 36
05	06 39	18 32	04 12	18 37	13 30	05 17	02 40	18 03	21 42	07 35
06	06 17	16 37	03 25	18 30	13 44	05 12	02 38	18 04	21 42	07 34
07	05 54	13 57	02 38	18 22	13 58	05 07	02 35	18 04	21 42	07 33
08	05 32	10 42	01 51	18 13	14 12	05 02	02 32	18 05	21 42	07 32
09	05 09	07 02	01 05	18 05	14 26	04 57	02 29	18 05	21 42	07 31
10	04 47	03 N 08	00 N 18	17 55	14 39	04 52	02 26	18 06	21 42	07 30
11	04 24	00 S 52	00 S 28	17 45	14 53	04 47	02 23	18 06	21 42	07 29
12	04 01	04 48	01 13	17 35	15 06	04 42	02 20	18 07	21 43	07 28
13	03 38	08 34	01 58	17 24	15 20	04 37	02 17	18 07	21 43	07 27
14	03 15	12 00	02 43	17 13	15 33	04 32	02 14	18 08	21 43	07 26
15	02 52	14 59	03 27	17 01	15 46	04 26	02 11	18 09	21 43	07 25
16	02 29	17 22	04 11	16 49	15 59	04 21	02 08	18 09	21 43	07 24
17	02 05	19 00	04 55	16 36	16 13	04 16	02 05	18 10	21 43	07 23
18	01 42	19 44	05 38	16 23	16 25	04 11	02 02	18 10	21 43	07 22
19	01 19	19 26	06 20	16 10	16 38	04 06	01 59	18 11	21 43	07 21
20	00 56	18 01	07 02	15 55	16 51	04 01	01 56	18 12	21 43	07 20
21	00 32	15 29	07 43	15 41	17 04	03 56	01 54	18 12	21 43	07 19
22	00 N 09	11 55	08 24	15 26	17 16	03 51	01 51	18 13	21 43	07 18
23	00 S 14	07 31	09 04	15 10	17 28	03 46	01 48	18 14	21 44	07 17
24	00 38	02 S 34	09 43	14 54	17 41	03 41	01 45	18 14	21 44	07 16
25	01 01	02 N 35	10 21	14 38	17 53	03 36	01 42	18 15	21 44	07 15
26	01 25	07 33	10 59	14 21	18 05	03 31	01 39	18 16	21 44	07 14
27	01 48	11 58	11 36	14 04	18 17	03 26	01 36	18 17	21 44	07 13
28	02 11	15 34	12 13	13 46	18 28	03 21	01 33	18 17	21 44	07 12
29	02 35	18 07	12 48	13 28	18 40	03 16	01 30	18 18	21 44	07 11
30	02 S 58	19 N 31	13 S 23	13 N 09	18 S 52	03 N 10	01 N 27	18 S 19	21 S 44	07 N 10

ZODIAC SIGN ENTRIES

Date	h	m	Planets
01	01	50	☽ ♊
03	06	39	☽ ♋
05	14	22	☽ ♌
07	17	57	♀ ♌
08	00	31	☽ ♍
10	02	00	☿ ♎
10	12	22	☽ ♎
13	01	06	☽ ♏
15	13	28	☽ ♐
17	23	45	☽ ♑
20	06	31	☽ ♒
21	10	48	♄ ♎
22	09	27	☽ ♓
22	21	09	☉ ♎
24	09	37	☽ ♈
26	08	53	☽ ♉
28	09	21	☽ ♊
30	01	16	☿ ♏
30	12	46	☽ ♋

LATITUDES

Date	Mercury ☿ ° '	Venus ♀ ° '	Mars ♂ ° '	Jupiter ♃ ° '	Saturn ♄ ° '	Uranus ♅ ° '	Neptune ♆ ° '	Pluto ♇ ° '
01	01 N 28	02 S 23	00 S 23	01 N 02	02 N 04	00 N 15	01 N 22	16 N 43
04	01 14	02 08	00 25	01 02	02 04	00 15	01 22	16 42
07	00 57	01 53	00 27	01 02	02 04	00 15	01 21	16 41
10	00 38	01 38	00 29	01 02	02 04	00 15	01 21	16 40
13	00 N 17	01 23	00 31	01 02	02 04	00 15	01 21	16 39
16	00 S 04	01 08	00 33	01 02	02 04	00 15	01 21	16 39
19	00 27	00 54	00 35	01 03	02 04	00 15	01 21	16 38
22	00 50	00 40	00 37	01 03	02 04	00 14	01 21	16 37
25	01 13	00 26	00 38	01 03	02 04	00 14	01 21	16 37
28	01 35	00 13	00 40	01 03	02 04	00 14	01 20	16 36
31	01 S 56	00 00	00 S 42	01 N 03	02 N 04	00 N 14	01 N 20	16 N 36

DATA

Julian Date	2444484
Delta T	+51 seconds
Ayanamsa	23° 35' 02"
Synetic vernal point	05° ♓ 31' 57"
True obliquity of ecliptic	23° 26' 24"

MOON'S PHASES, APSIDES AND POSITIONS ☽

Date	h	m	Phase	Longitude ° '	Eclipse Indicator
01	18	08	☾	09 ♊ 26	
09	10	00	●	16 ♍ 52	
17	13	54	☽	24 ♐ 49	
24	12	08	○	01 ♈ 35	

Day	h	m		
12	08	52	Apogee	
25	02	41	Perigee	
03	23	58	Max dec	19° N 41'
11	06	51	0S	
18	17	21	Max dec	19° S 45'
24	23	58	0N	

LONGITUDES

Date	Chiron ⚷ ° '	Ceres ⚳ ° '	Pallas ⚴ ° '	Juno ⚵ ° '	Vesta ⚶ ° '	Black Moon Lilith ⚸ ° '
01	18 ♉ 14	09 ♋ 15	05 ♉ 37	18 ♍ 09	03 ♌ 34	16 ♎ 38
11	18 ♉ 05	12 ♋ 38	05 ♉ 11	22 ♍ 09	07 ♌ 43	17 ♎ 45
21	17 ♉ 51	15 ♋ 47	04 ♉ 01	26 ♍ 07	11 ♌ 47	18 ♎ 52
31	17 ♉ 32	18 ♋ 42	02 ♉ 05	00 ♎ 02	15 ♌ 44	19 ♎ 59

All ephemeris data is given at 12.00 UT and the Moon's longitude is additionally given for 24.00 UT

ASPECTARIAN

h m	Aspects
01 Monday	
01 45	☿ ∥ ♇
01 52	☿ ⊥ ♇
05 09	☽ ⊼ ♂
10 34	☽ ⚼ ♇
16 02	☽ ± ♂
16 59	☽ ∠ ♀
18 08	☽ □ ☉
02 Tuesday	
06 02	☽ # ♅
06 03	☽ □ ☿
07 56	☉ Q ♅
08 37	☽ ⚼ ♂
09 50	☽ □ ♃
10 03	☽ ⊥ ♀
12 33	☽ ☍ ♆
12 55	☽ △ ♇
16 17	☽ ⊼ ♅
19 17	☽ ∥ ♀
20 37	☉ ∥ ♇
21 37	☽ ⊻ ♀
03 Wednesday	
02 35	☽ □ ♄
03 04	☽ ± ♅
04 31	☽ Q ☉
10 24	☿ ∠ ♂
12 56	☽ △ ♂
13 32	☿ ☌ ♃
18 48	☽ Q ♃
19 09	♀ ± ♆
19 28	☽ ⚼ ♅
19 33	☽ Q ☿
21 43	♂ ∠ ♃
22 27	☿ ∥ ♃
04 Thursday	
04 05	♂ ⊥ ♄
04 12	☽ ⚹ ☉
06 34	☿ □ ♆
10 09	☿ ⊻ ♇
11 52	☽ Q ♄
17 24	☽ ⚹ ♃
19 19	☽ ⊼ ♆
19 51	☽ □ ♇
21 23	☽ ⚹ ☿
23 23	☽ △ ♅
05 Friday	
06 37	☽ ± ♆
09 45	☽ ☌ ♀
10 27	☽ ∥ ♀
10 35	☽ ∠ ☉
10 38	☽ ⚹ ♄
11 54	☿ ⚹ ♅
17 44	♂ ∠ ♆
19 02	☽ # ♅
21 52	♀ ⚹ ♄
22 15	☽ ∠ ♃
23 44	☽ ⚼ ♆
06 Saturday	
00 04	☽ □ ♂
05 12	☽ ⊥ ☉
06 07	☽ Q ♇
06 32	☽ ∠ ☿
15 39	☽ ∠ ♄
15 53	☽ ⊥ ♃
17 23	☉ ⊥ ♇
17 45	☽ ⊻ ☉
07 Sunday	
02 49	☽ ⊥ ☿
03 44	☽ ⊻ ♃
04 44	☽ △ ♆
05 26	☽ ⚹ ♇
09 06	☽ □ ♅
09 20	☽ ⊥ ♄
11 47	☽ # ♂
12 55	☽ Q ♂
13 43	☿ ∥ ♄
16 29	☽ ⊻ ☿
21 13	☽ ⊻ ♄
08 Monday	
01 08	☽ ⊻ ♀
11 00	☽ ∠ ♇
13 57	☽ ⚹ ♂
14 07	☽ ⊥ ♀
20 42	☽ Q ♅
09 Tuesday	
04 29	☿ ☌ ♄
04 57	☽ ⊥ ♇
08 58	☽ ∥ ♇
09 45	☽ ∠ ♀
10 00	☽ ☌ ☉
10 32	♀ # ♅
13 34	♃ □ ♆
16 07	☽ □ ♆
16 10	☽ ☌ ♃
16 59	☽ ⊻ ♇
20 44	☽ ⚹ ♅
21 38	☽ ∠ ♂
10 Wednesday	
00 59	☽ ∥ ☉
01 16	☽ ∥ ♃
04 40	☉ ∥ ♃
09 37	☽ ☌ ♄
14 04	☽ ☌ ☿
16 19	☽ ∥ ♄
16 49	☽ ⊥ ♂
17 46	☉ ∠ ♀
18 49	☽ ⚹ ♀
11 Thursday	
03 03	☽ ∠ ♅
04 23	☽ Q ♆
05 22	☽ # ☿
05 40	☽ ⊻ ♂
09 02	☽ ∥ ☿
21 04	☽ # ♄
21 16	☿ ⊥ ♂
21 19	☽ ⊥ ♅
21 29	☽ Q ♀
21 43	♃ ⊻ ♇
12 Friday	
03 53	☽ ⊻ ☉
04 42	☽ ⚹ ♆
05 43	☽ ☌ ♇
05 51	☽ ⊻ ♃
07 34	☽ # ☉
09 31	☽ ⊻ ♅
09 59	♀ ⚼ ♆
11 21	☽ # ♃
13 48	☉ □ ♆
17 08	☽ ⊥ ☉
18 14	☽ ⊥ ♃
22 58	☽ ⊻ ♄
13 Saturday	
02 54	♀ ∠ ♃
02 56	☉ ⊻ ♇
04 44	☽ # ♇
09 55	☉ ☌ ♃
11 07	☽ ∠ ♆
11 15	☽ ⊥ ♄
12 36	☽ ⊻ ☿
12 49	☽ ∠ ♃
13 01	☽ ∠ ☉
13 36	☽ □ ♀
21 25	☿ # ♄
22 05	☽ ☌ ♂
14 Sunday	
02 37	☽ ⊥ ☿
05 19	☽ ⊥ ♆
05 37	☽ ∠ ♄
08 50	☿ ⚹ ♀
14 19	☿ ∠ ♅
17 23	☽ ⊻ ♆
18 33	☽ ⊻ ♇
19 37	☽ ⚹ ♃
21 56	☽ ⚹ ☉
22 19	☽ ☌ ♅
22 42	☿ Q ♆
23 17	☉ # ☿
23 31	☽ ∠ ☿
15 Monday	
02 51	☉ ⚹ ♅
06 35	☽ ⊥ ♇
12 00	☽ ⚹ ♄
18 57	♀ Q ♇
19 55	☽ ∥ ♂
19 58	☽ Q ♃
23 55	☽ Q ☉
16 Tuesday	
00 30	☽ ∠ ♇
06 20	☽ # ♀
07 31	☽ △ ♀
09 43	☽ ⚹ ☿
11 55	☽ Q ♄
13 25	☽ ⊻ ♂
16 58	☿ # ♃
22 10	☽ ∥ ♅
17 Wednesday	
01 45	☽ ⊥ ♂
04 37	☽ ☌ ♆
05 51	☽ ⚹ ♇
07 45	☽ □ ♃
09 29	☽ ⊻ ♅
12 08	☉ ∥ ♄
12 19	☽ Q ☿
13 54	☽ □ ☉
15 21	☽ ⚼ ♀
19 59	☽ ∠ ♂
20 58	☽ ⊥ ♅
22 56	☽ □ ♄
18 Thursday	
04 45	☽ Q ♇
09 56	☽ ± ♀
09 59	♀ # ♂
10 52	☿ ⊻ ♂
13 54	☽ ∠ ♅
22 05	☽ ⊼ ♀
19 Friday	
01 31	☽ ⚹ ♂
02 35	☽ □ ☿
06 43	♂ ⊥ ♆
12 41	☽ ⊻ ♆
13 58	☽ □ ♇
16 30	☽ △ ♃
17 22	☽ ⚹ ♅
23 26	☽ ⊥ ♆
20 Saturday	
00 19	☽ Q ♂
01 39	☽ △ ☉
06 15	☽ △ ♄
09 47	☽ ∥ ♅
13 42	☿ ⊥ ♅
14 25	♂ ∠ ♄
14 37	☽ Q ♅
15 16	☽ ∠ ♆
19 20	☽ ⚼ ♃
22 14	☿ # ♇
23 25	☽ ∥ ♂
21 Sunday	
05 41	☽ ⚼ ☉
07 47	☽ ☍ ♀
08 24	☽ ⚼ ♄
09 09	☽ □ ♂
10 19	☽ # ♀
10 57	☽ ± ♃
13 42	☽ △ ☿
16 37	♀ ∠ ♄
16 53	☽ ⚹ ♆
18 11	☽ △ ♇
21 11	☽ ⊼ ♃
21 21	☽ □ ♅
21 58	☽ ± ☉
23 39	☽ ± ♄
22 Monday	
08 37	☽ ⊼ ☉
09 39	☽ ⊼ ♄
11 22	♃ ⚹ ♅
12 47	☽ Q ♆
17 19	☽ ⚼ ☿
18 59	☽ ⚼ ♇
21 01	☿ ⚹ ♆
23 Tuesday	
02 08	☉ ☌ ♄
03 13	♀ □ ♂
05 02	☽ ∥ ☿
09 25	♀ ⊥ ♃
09 26	☽ ± ☿
09 32	☽ ± ♇
10 34	☿ ☌ ♇
12 49	☽ △ ♂
13 07	☽ ⊼ ♀
13 12	☽ # ♇
17 50	☽ □ ♆
19 09	☽ ⊼ ♇
20 02	☽ ⊼ ☿
22 11	☽ △ ♅
22 35	☽ ☍ ♃
23 26	☽ ± ♀
24 Wednesday	
06 40	☽ # ♃
10 12	☽ ☍ ♄
12 08	☽ ☍ ☉
13 45	☽ ⚼ ♂
14 47	☽ ⚼ ♀
15 53	☽ # ♄
19 31	☿ ⊻ ♅
20 24	☽ ∥ ☉
21 57	☽ ⚼ ♅
25 Thursday	
03 33	☿ ⊻ ♃
04 07	☽ # ☉
04 35	☽ ± ♂
07 54	☽ ∥ ♄
12 08	☽ ± ♅
14 28	☽ ⊼ ♂
16 15	☽ △ ♀
16 42	☽ ∥ ♃
17 12	☽ △ ♆
18 35	☽ ☍ ♇
21 37	☽ ⊼ ♅
22 33	☽ ⊼ ♃
26 Friday	
00 11	☽ ☍ ☿
05 22	♀ △ ♆
08 10	☽ ± ♃
09 51	☽ ⊼ ♄
10 26	☽ ∥ ♇
14 41	☽ ⊼ ☉
16 57	☽ ⚼ ♆
19 28	☽ ± ♄
22 41	☽ ⚼ ♃
27 Saturday	
00 53	♀ ⚹ ♇
00 55	☽ ± ☉
01 04	☉ # ♄
07 22	☽ ± ♆
09 29	☽ # ☿
10 02	☽ ⚼ ♄
11 39	♂ ∥ ♅
16 28	☽ ⚼ ☉
16 33	☽ ☍ ♂
17 06	☽ ⊼ ♆
18 35	☽ ⊼ ♇
20 00	☽ □ ♀
21 05	☉ ∠ ♂
21 45	☽ ☍ ♅
23 18	☽ △ ♃
28 Sunday	
00 11	☽ ∥ ♀
04 16	♂ ⊻ ♆
04 27	☽ ± ♇
05 18	☽ ⊼ ☿
10 47	☽ △ ♄
16 17	☽ ± ☿
19 00	☽ △ ☉
19 27	☽ ⚼ ♇
19 43	♀ □ ♅
29 Monday	
04 30	☽ Q ♀
09 05	☽ ⚼ ☿
14 20	♂ ⊻ ♇
14 21	☽ # ♅
19 27	☽ ☍ ♆
20 22	☽ # ♂
21 08	☽ △ ♇
21 29	☽ ⊼ ♂
23 57	♀ ⊻ ♃
30 Tuesday	
00 33	☽ ⊼ ♅
02 49	☽ □ ♃
03 02	☽ ⚹ ♀
05 47	☿ # ♀
08 33	☽ ± ♂
08 48	☿ ⊥ ♃
11 08	☽ ± ♅
14 01	☽ △ ☿
14 48	☽ □ ♄
17 11	♀ ⊥ ♄
22 38	☉ # ♃
22 52	☿ ⊻ ♄

OCTOBER 1980

LONGITUDES

Date	Sidereal time h m s	Sun ☉ ° ' "	Moon ☽ ° ' "	Moon ☽ 24.00 ° ' "	Mercury ☿ ° '	Venus ♀ ° '	Mars ♂ ° '	Jupiter ♃ ° '	Saturn ♄ ° '	Uranus ♅ ° '	Neptune ♆ ° '	Pluto ♇ ° '
01	12 41 30	08 ♎ 27 20	12 ♋ 52 39	19 ♋ 22 59	01 ♏ 53	26 ♌ 02	22 ♏ 25	24 ♍ 39	01 ♎ 15	23 ♏ 08	20 ♐ 09	21 ♎ 10
02	12 45 26	09 26 23	25 ♋ 47 53	02 ♌ 07 51	03 10	27 11	23 07	24 52	01 22	23 11	20 10	21 12
03	12 49 23	10 25 28	08 ♌ 23 24	14 ♌ 35 04	04 25	28 19	23 49	25 05	01 30	23 14	20 11	21 15
04	12 53 19	11 24 35	20 ♌ 43 22	26 ♌ 48 45	05 38	29 ♌ 28	24 31	25 18	01 37	23 17	20 12	21 17
05	12 57 16	12 23 45	02 ♍ 51 42	08 ♍ 52 38	06 50	00 ♍ 37	25 13	25 30	01 44	23 20	20 14	21 19
06	13 01 13	13 22 56	14 ♍ 51 54	20 ♍ 49 53	08 01	01 46	25 55	25 43	01 52	23 23	20 15	21 22
07	13 05 09	14 22 10	26 ♍ 46 51	02 ♎ 43 06	09 09	02 55	26 37	25 56	01 59	23 26	20 16	21 24
08	13 09 06	15 21 26	08 ♎ 38 51	14 ♎ 34 20	10 15	04 05	27 20	26 08	02 06	23 29	20 17	21 26
09	13 13 02	16 20 44	20 ♎ 29 44	26 ♎ 25 15	11 20	05 14	28 02	26 21	02 14	23 32	20 18	21 29
10	13 16 59	17 20 04	02 ♏ 21 03	08 ♏ 17 20	12 22	06 24	28 45	26 34	02 21	23 35	20 19	21 31
11	13 20 55	18 19 26	14 ♏ 14 18	20 ♏ 12 11	13 21	07 34	29 ♏ 27	26 46	02 28	23 38	20 21	21 34
12	13 24 52	19 18 50	26 ♏ 11 14	02 ♐ 11 42	14 18	08 44	00 ♐ 10	26 59	02 35	23 42	20 22	21 36
13	13 28 48	20 18 16	08 ♐ 13 56	14 ♐ 18 16	15 11	09 54	00 53	27 11	02 43	23 45	20 23	21 38
14	13 32 45	21 17 44	20 ♐ 25 06	26 ♐ 34 53	16 01	11 04	01 35	27 23	02 50	23 48	20 25	21 41
15	13 36 42	22 17 13	02 ♑ 48 02	09 ♑ 05 05	16 48	12 14	02 18	27 36	02 57	23 51	20 26	21 43
16	13 40 38	23 16 45	15 ♑ 26 32	21 ♑ 52 54	17 30	13 25	03 01	27 48	03 04	23 55	20 27	21 46
17	13 44 35	24 16 18	28 ♑ 24 41	05 ♒ 02 22	18 08	14 35	03 44	28 00	03 11	23 58	20 29	21 48
18	13 48 31	25 15 53	11 ♒ 46 23	18 ♒ 37 05	18 41	15 46	04 28	28 13	03 19	24 01	20 30	21 50
19	13 52 28	26 15 29	25 ♒ 34 42	02 ♓ 39 21	19 09	16 57	05 11	28 25	03 26	24 05	20 32	21 53
20	13 56 24	27 15 07	09 ♓ 50 57	17 ♓ 09 14	19 31	18 08	05 54	28 37	03 33	24 08	20 33	21 55
21	14 00 21	28 14 47	24 ♓ 33 43	02 ♈ 03 41	19 46	19 19	06 37	28 49	03 40	24 12	20 35	21 57
22	14 04 17	29 ♎ 14 29	09 ♈ 38 11	17 ♈ 16 05	19 54	20 30	07 21	29 01	03 47	24 15	20 36	22 00
23	14 08 14	00 ♏ 14 12	24 ♈ 56 01	02 ♉ 36 35	19 R 55	21 42	08 04	29 13	03 54	24 18	20 38	22 02
24	14 12 11	01 13 58	10 ♉ 16 16	17 ♉ 53 37	19 47	22 53	08 48	29 25	04 01	24 22	20 40	22 05
25	14 16 07	02 13 46	25 ♉ 27 15	02 ♊ 55 57	19 31	24 05	09 31	29 37	04 07	24 25	20 41	22 07
26	14 20 04	03 13 35	10 ♊ 18 42	17 ♊ 34 43	19 06	25 16	10 15	29 ♍ 49	04 14	24 29	20 43	22 09
27	14 24 00	04 13 27	24 ♊ 43 28	01 ♋ 44 38	18 32	26 28	10 59	00 ♎ 01	04 21	24 33	20 45	22 12
28	14 27 57	05 13 21	08 ♋ 38 08	15 ♋ 24 06	17 48	27 40	11 43	00 13	04 28	24 36	20 46	22 14
29	14 31 53	06 13 18	22 ♋ 02 47	28 ♋ 34 38	16 56	28 ♍ 52	12 26	00 24	04 35	24 40	20 48	22 16
30	14 35 50	07 13 16	05 ♌ 00 07	11 ♌ 19 50	15 55	00 ♎ 04	13 10	00 36	04 41	24 43	20 50	22 19
31	14 39 46	08 ♏ 13 17	17 ♌ 34 23	23 ♌ 44 26	14 ♏ 48	01 ♎ 16	13 ♐ 54	00 ♎ 48	04 ♎ 48	24 ♏ 47	20 ♐ 51	22 ♎ 21

Date	Moon True ☊ ° '	Moon Mean ☊ ° '	Moon ☽ Latitude ° '
01	19 ♌ 03	17 ♌ 22	03 S 03
02	19 D 04	17 19	02 02
03	19 05	17 15	00 S 57
04	19 R 06	17 12	00 N 09
05	19 05	17 09	01 13
06	19 01	17 06	02 14
07	18 55	17 03	03 08
08	18 47	17 00	03 53
09	18 37	16 56	04 28
10	18 25	16 53	04 52
11	18 13	16 50	05 03
12	18 02	16 47	05 00
13	17 53	16 44	04 45
14	17 46	16 40	04 16
15	17 42	16 37	03 35
16	17 40	16 34	02 43
17	17 D 40	16 31	01 41
18	17 40	16 28	00 N 31
19	17 R 40	16 25	00 S 42
20	17 38	16 21	01 55
21	17 34	16 18	03 02
22	17 27	16 15	03 59
23	17 18	16 12	04 39
24	17 07	16 09	04 59
25	16 57	16 05	04 58
26	16 47	16 02	04 37
27	16 40	15 59	03 58
28	16 35	15 56	03 06
29	16 32	15 53	02 06
30	16 D 32	15 50	01 01
31	16 ♌ 32	15 ♌ 46	00 N 06

DECLINATIONS

Date	Sun ☉ ° '	Moon ☽ ° '	Mercury ☿ ° '	Venus ♀ ° '	Mars ♂ ° '	Jupiter ♃ ° '	Saturn ♄ ° '	Uranus ♅ ° '	Neptune ♆ ° '	Pluto ♇ ° '
01	03 S 21	19 N 47	13 S 57	12 N 50	19 S 03	03 N 05	01 N 24	18 S 19	21 S 45	07 N 09
02	03 44	18 59	14 30	12 31	19 14	03 00	01 21	18 20	21 45	07 08
03	04 08	17 15	15 02	12 11	19 25	02 55	01 19	18 21	21 45	07 07
04	04 31	14 43	15 33	11 51	19 36	02 50	01 16	18 22	21 45	07 06
05	04 54	11 36	16 03	11 30	19 47	02 45	01 13	18 23	21 45	07 05
06	05 17	08 01	16 32	11 09	19 57	02 40	01 10	18 23	21 45	07 04
07	05 40	04 09	16 59	10 48	20 08	02 36	01 07	18 24	21 45	07 04
08	06 03	00 N 09	17 26	10 27	20 18	02 31	01 04	18 25	21 45	07 03
09	06 26	03 S 52	17 51	10 05	20 28	02 26	01 01	18 26	21 46	07 02
10	06 48	07 43	18 15	09 42	20 38	02 21	00 59	18 26	21 46	07 01
11	07 11	11 17	18 38	09 20	20 48	02 16	00 56	18 27	21 46	07 00
12	07 34	14 26	18 59	08 57	20 58	02 11	00 53	18 28	21 46	06 59
13	07 56	17 00	19 18	08 33	21 07	02 06	00 50	18 29	21 46	06 58
14	08 18	18 50	19 36	08 10	21 16	02 01	00 47	18 30	21 46	06 57
15	08 41	19 50	19 52	07 46	21 26	01 56	00 45	18 31	21 46	06 57
16	09 03	19 51	20 06	07 22	21 34	01 52	00 42	18 31	21 47	06 56
17	09 25	18 50	20 18	06 58	21 43	01 47	00 39	18 32	21 47	06 55
18	09 46	16 45	20 28	06 33	21 52	01 42	00 36	18 33	21 47	06 54
19	10 08	13 39	20 35	06 08	22 00	01 37	00 34	18 34	21 47	06 53
20	10 30	09 39	20 40	05 43	22 08	01 32	00 31	18 35	21 47	06 52
21	10 51	04 S 57	20 41	05 18	22 16	01 28	00 28	18 36	21 47	06 52
22	11 12	00 N 09	20 40	04 52	22 24	01 23	00 26	18 37	21 47	06 51
23	11 33	05 19	20 35	04 26	22 31	01 18	00 23	18 37	21 48	06 50
24	11 54	10 10	20 27	04 00	22 39	01 14	00 20	18 38	21 48	06 49
25	12 15	14 18	20 15	03 34	22 46	01 09	00 18	18 39	21 48	06 49
26	12 35	17 26	19 59	03 08	22 53	01 04	00 15	18 40	21 48	06 48
27	12 56	19 22	19 39	02 41	22 59	01 00	00 13	18 41	21 48	06 47
28	13 16	20 04	19 14	02 15	23 06	00 55	00 10	18 42	21 48	06 46
29	13 36	19 34	18 45	01 48	23 12	00 51	00 08	18 43	21 48	06 46
30	13 55	18 02	18 12	01 21	23 18	00 46	00 05	18 43	21 49	06 45
31	14 S 15	15 N 39	17 S 35	00 N 54	23 S 24	00 N 42	00 03	18 S 44	21 S 49	06 N 44

ZODIAC SIGN ENTRIES

Date	h	m	Planets
02	19	57	☽ ♌
04	23	07	♀ ♍
05	06	19	☽ ♍
07	18	30	☽ ♎
10	07	15	☽ ♏
12	06	27	♂ ♐
12	19	37	☽ ♐
15	06	37	☽ ♑
17	14	54	☽ ♒
19	19	31	☽ ♓
21	20	43	☽ ♈
23	06	18	☉ ♏
23	19	55	☽ ♉
25	19	17	☽ ♊
27	10	10	♃ ♎
27	21	00	☽ ♋
30	02	38	☽ ♌
30	10	38	♀ ♎

LATITUDES

Date	Mercury ☿ ° '	Venus ♀ ° '	Mars ♂ ° '	Jupiter ♃ ° '	Saturn ♄ ° '	Uranus ♅ ° '	Neptune ♆ ° '	Pluto ♇ ° '
01	01 S 56	00 00	00 S 42	01 N 03	02 N 04	00 N 14	01 N 20	16 N 36
04	02 16	00 N 12	00 44	01 03	02 05	00 14	01 20	16 36
07	02 35	00 24	00 45	01 04	02 05	00 14	01 20	16 36
10	02 51	00 35	00 47	01 04	02 05	00 14	01 20	16 35
13	03 03	00 45	00 48	01 04	02 05	00 14	01 20	16 35
16	03 11	00 55	00 50	01 04	02 06	00 14	01 19	16 35
19	03 12	01 04	00 51	01 05	02 06	00 14	01 19	16 35
22	03 04	01 12	00 52	01 05	02 06	00 14	01 19	16 36
25	02 45	01 19	00 54	01 05	02 07	00 14	01 19	16 36
28	02 11	01 26	00 55	01 06	02 07	00 14	01 19	16 36
31	01 S 22	01 N 32	00 S 56	01 N 06	02 N 07	00 N 14	01 N 19	16 N 37

DATA

Julian Date	2444514
Delta T	+51 seconds
Ayanamsa	23° 35' 05"
Synetic vernal point	05° ♓ 31' 54"
True obliquity of ecliptic	23° 26' 24"

MOON'S PHASES, APSIDES AND POSITIONS ☽

Date	h	m	Phase	Longitude ° '	Eclipse Indicator
01	03	18	☾	08 ♋ 06	
09	02	50	●	15 ♎ 58	
17	03	47	☽	23 ♑ 56	
23	20	52	○	00 ♉ 36	
30	16	33	☾	07 ♌ 25	

Day	h	m		
09	15	18	Apogee	
23	14	06	Perigee	
01	05	41	Max dec	19° N 50'
08	12	51	0S	
16	00	40	Max dec	19° S 58'
22	11	16	0N	
28	13	36	Max dec	20° N 04'

LONGITUDES

Date	Chiron ⚷ ° '	Ceres ⚳ ° '	Pallas ⚴ ° '	Juno ⚵ ° '	Vesta ⚶ ° '	Black Moon Lilith ⚸ ° '
01	17 ♉ 32	18 ♋ 42	02 ♉ 05	00 ♎ 02	15 ♌ 44	19 ♎ 59
11	17 ♉ 08	21 ♋ 18	29 ♈ 30	03 ♎ 53	19 ♌ 34	21 ♎ 06
21	16 ♉ 41	23 ♋ 31	26 ♈ 29	07 ♎ 39	23 ♌ 14	22 ♎ 13
31	16 ♉ 12	25 ♋ 19	23 ♈ 21	11 ♎ 20	26 ♌ 43	23 ♎ 20

ASPECTARIAN

h m Aspects

01 Wednesday
01 30 ☽ bQ ♂
03 18 ☽ □ ☉
03 19 ☽ bQ ♅
03 37 ☉ ∠ ♅
04 34 ☉ Q ♆
08 20 ☽ ∠ ♀
11 35 ☽ Q ♃
23 51 ☽ Q ♄

02 Thursday
01 27 ☽ ⊼ ♆
02 30 ☽ ⊥ ♀
03 22 ☽ □ ♇
06 40 ☽ △ ♂
07 04 ☽ △ ♅
07 41 ☽ # ♂
10 13 ☽ ⚹ ♃
12 42 ☽ ± ♆
14 27 ♂ ☌ ♅
14 52 ☽ ⊻ ♀
15 21 ☽ Q ☉
22 19 ☽ # ♅
22 39 ☽ ⚹ ♄

03 Friday
03 31 ☽ □ ☿
05 50 ☽ bQ ♆
13 39 ☽ Q ♇
15 19 ☽ ∠ ♃
16 16 ☽ ⚹ ☉

04 Saturday
03 21 ☿ ∠ ♆
03 53 ☽ ∠ ♄
06 00 ☽ # ☿
09 09 ☽ ⊥ ♃
10 59 ☽ △ ♆
13 06 ☽ ⚹ ♇
17 03 ☽ □ ♅
18 22 ☽ Q ☿
19 55 ☽ □ ♂
21 09 ☽ ⊻ ♃
21 44 ☽ ⊥ ♄

05 Sunday
00 12 ☽ ∠ ☉
07 04 ☽ ☌ ♀
09 45 ☽ ⊻ ♄
12 42 ☽ ∥ ♀
18 55 ☽ ∠ ♇
19 40 ☽ ⊥ ☉
20 48 ☽ ⚹ ☿

06 Monday
02 16 ♂ ⚹ ♃
04 58 ☽ Q ♅
08 32 ☿ ⊥ ♄
08 45 ☽ ⊻ ☉
09 59 ☽ Q ♂
13 00 ☽ ⊥ ♇
14 10 ♀ ⊻ ♄
18 01 ☽ ∥ ♇
22 50 ☽ □ ♆

07 Tuesday
01 07 ☽ ⊻ ♇
03 35 ☽ # ☉
05 12 ☽ ⚹ ♅
06 08 ☽ ∠ ☿
10 15 ☽ ☌ ♃
11 39 ☽ ⚹ ♂
21 35 ☽ ∥ ♃
22 37 ☽ ☌ ♄

08 Wednesday
01 45 ☽ ⊻ ♀
02 13 ☽ ⊥ ☿
06 24 ☽ ∥ ♄
11 16 ☽ Q ♆
11 40 ☽ ∠ ♅
15 13 ☽ ⊥ ♀
15 35 ☽ ⊻ ☿
16 02 ♂ ⊥ ♇
19 09 ☽ # ♄
19 56 ☽ ∠ ♂

09 Thursday
02 50 ☽ ☌ ☉
03 32 ☽ # ♃
05 58 ☽ ⊥ ♅
11 25 ☽ ∠ ♀
11 37 ☽ ⚹ ♆
12 34 ☿ ∠ ♃
14 00 ☽ ☌ ♇
15 19 ☽ ⊥ ♂
18 11 ☽ ⊻ ♅

10 Friday
00 04 ☽ ⊻ ♃
04 14 ☽ ⊻ ♂
05 33 ☽ ∥ ☉
07 31 ☽ # ♇
12 00 ☽ ⊻ ♄
12 26 ☽ ⊥ ♃
14 35 ♀ ∠ ♇
18 01 ☽ ∠ ♆
18 26 ☉ ⊥ ♅
21 04 ☽ ⚹ ♀
23 48 ☽ # ♀

11 Saturday
00 15 ☽ ⊥ ♄
00 15 ☿ ∥ ♅
00 41 ☉ # ♇
06 56 ☽ ∠ ♃
10 03 ☽ ☌ ☿
12 13 ☽ ⊥ ♆
18 34 ☽ ∠ ♄
20 58 ☽ ⊻ ☉
23 52 ☽ Q ♀

12 Sunday
00 18 ☽ ⊻ ♆
02 46 ☽ ⊻ ♇
06 59 ☽ ☌ ♅
10 06 ☽ ⊥ ☉
13 36 ☽ ⚹ ♃
13 57 ☿ ⊥ ♆
14 50 ☽ ⊥ ♇
20 27 ☽ ☌ ♂

13 Monday
00 55 ☽ ⚹ ♄
05 40 ☽ ∠ ☉
08 50 ☽ ∠ ♇
13 55 ☽ Q ♃
14 05 ☉ ⚹ ♆
15 39 ☽ □ ♀

14 Tuesday
00 56 ☽ Q ♄
02 46 ☽ ⊻ ☿
06 37 ☽ ∥ ♅
07 33 ☉ # ♀
11 59 ☽ ☌ ♆
13 52 ☽ ⚹ ☉
14 28 ☽ ⚹ ♇
15 21 ☽ ⊥ ☿
18 38 ☽ ⊻ ♅
21 40 ☉ ☌ ♇

15 Wednesday
01 48 ☽ □ ♃
03 45 ♀ Q ♅
06 18 ☽ ⊥ ♅
09 57 ☽ ∠ ☿
11 00 ☽ ⊻ ♂
12 18 ☽ □ ♄
13 46 ☽ Q ♇
15 06 ☽ Q ☉
23 09 ☽ ⊥ ♂
23 37 ☽ ∠ ♅

16 Thursday
07 48 ☽ △ ♀
13 57 ♂ ⚹ ♄
16 04 ☽ ⚹ ☿
17 07 ☽ ∠ ♂
21 23 ☽ ⊻ ♆
23 49 ☽ □ ♇

17 Friday
03 47 ☽ □ ☉
03 49 ☽ ⚹ ♅
04 11 ☉ ⊻ ♅
08 28 ☽ ⊥ ♆
11 15 ☽ △ ♃
14 22 ☽ bQ ♀
14 43 ♀ ∥ ♇
14 45 ☿ ∠ ♄
15 17 ☽ Q ☿
16 20 ☽ ∥ ♅
20 45 ☽ △ ♄
21 53 ♂ ∥ ♆
22 13 ☽ ⚹ ♂

18 Saturday
00 49 ☽ ∠ ♆
01 43 ☽ Q ♅
08 07 ☽ ± ♀
13 26 ♀ ⊥ ♇
14 35 ☽ bQ ♃
19 42 ☽ ⊼ ♀
20 41 ☽ Q ♂
23 34 ☽ bQ ♄

19 Sunday
00 34 ☽ □ ☿
03 18 ☽ ⚹ ♆
05 38 ☽ △ ♇
06 30 ☽ ± ♃
09 25 ☽ □ ♅
13 15 ☽ △ ☉
15 11 ☽ ± ♄
16 54 ☽ ⊼ ♃
23 49 ☽ Q ♆

20 Monday
01 24 ☽ ⊼ ♄
05 05 ☽ □ ♂
07 07 ☽ bQ ♇
07 40 ☽ ∥ ☉
16 15 ☽ bQ ☉
22 01 ☽ ± ♇

21 Tuesday
02 34 ☽ # ♇
02 47 ☽ ☍ ♀
04 08 ☽ △ ☿
05 34 ☽ □ ♆
07 47 ☽ ⊼ ♇
08 00 ☽ ± ☉
10 11 ☽ # ♀
11 24 ☽ △ ♅
18 20 ☽ ⊼ ☉
18 55 ☽ ☍ ♃
22 31 ☿ ⚹ ♀
23 49 ♂ ∠ ♇

22 Wednesday
02 39 ☽ ☍ ♄
04 28 ☽ bQ ☿
04 45 ☽ # ♃
05 22 ☉ ⊻ ♃
08 12 ☽ △ ♂
09 16 ☽ # ♄
11 23 ☽ bQ ♅
13 15 ☽ ∥ ♄
14 04 ♀ □ ♆
17 35 ☽ ∥ ♃
18 44 ☽ ± ☿

23 Thursday
01 35 ☽ ± ♅
01 58 ☿ St R
04 10 ☽ ⊼ ☿
05 16 ☽ △ ♆
06 30 ☽ ⊼ ♀
07 27 ☽ ☍ ♇
08 09 ☽ ∥ ♀
08 56 ☽ bQ ♂
11 01 ☽ ⊼ ♅
16 41 ☽ ± ♀
18 48 ☽ ⊼ ♃
19 09 ♀ ⊻ ♇
19 13 ☽ ∥ ♇
20 52 ☽ ☍ ☉
23 43 ☽ ± ♂

24 Friday
02 07 ☽ ⊼ ♄
04 18 ☽ ± ♃
04 46 ☽ bQ ♆
07 57 ☽ bQ ♀
09 34 ☽ ⊼ ♂
11 35 ☽ ± ♄
18 37 ☽ bQ ♃
18 55 ☽ ± ♆
22 21 ☽ # ☉

25 Saturday
01 52 ☽ bQ ♄
02 46 ☽ ☍ ☿
04 24 ☽ ⊼ ♆
06 40 ☽ ⊼ ♇
09 37 ☽ △ ♀
10 21 ☽ ☍ ♅
16 16 ☽ ± ♇
18 45 ☽ △ ♃
19 19 ♀ ⚹ ♅
23 39 ☽ ⊼ ☉

26 Sunday
02 02 ☽ △ ♄
06 24 ☿ ∠ ♄
06 50 ☽ bQ ♇
10 06 ☽ ± ☉
11 54 ☽ ☍ ♂

27 Monday
01 28 ☽ # ♅
02 01 ☽ ⊼ ☿
02 03 ☽ bQ ☉
05 16 ☽ ☍ ♆
07 43 ☽ △ ♇
11 41 ☽ ± ☿
11 41 ☽ ⊼ ♅
15 14 ☽ □ ♀
15 26 ☉ ⊻ ♄
16 08 ☽ # ☿
21 09 ☽ □ ♃
21 58 ☽ ± ♅

28 Tuesday
02 22 ☽ bQ ☿
04 39 ☽ □ ♄
05 34 ☽ △ ☉
13 42 ☽ bQ ♅
17 44 ☽ ⊼ ♂

29 Wednesday
01 33 ☉ ∠ ♆
01 42 ☽ Q ♀
03 20 ☽ △ ☿
05 05 ☽ ± ♂
05 18 ☽ Q ♃
09 44 ☽ ⊼ ♆
10 25 ♂ Q ♃
12 25 ☽ □ ♇
12 58 ☽ Q ♄
13 44 ☿ ∥ ♅
16 48 ☽ △ ♅
17 29 ☉ ⊥ ♃
20 44 ☽ ± ♆
22 29 ☽ bQ ♂

30 Thursday
01 49 ☽ ⚹ ♀
03 00 ☽ # ♅
03 38 ☽ ⚹ ♃
07 44 ☉ ⊥ ♂
09 09 ☽ # ☿
11 24 ☽ ⚹ ♄
13 34 ☽ bQ ♆
16 33 ☽ □ ☉
18 04 ☿ ∠ ♃
20 56 ☿ ∠ ♀
22 05 ☽ Q ♇

31 Friday
00 37 ♀ ☌ ♃
04 30 ☽ △ ♂
07 06 ☽ □ ☿
08 30 ☽ ∠ ♃
09 13 ☽ ∠ ♀
10 46 ☿ ⊥ ♆
16 21 ☽ ∠ ♄
18 23 ☽ △ ♆
21 19 ☽ ⚹ ♇
22 33 ☽ # ☉
23 02 ☿ ⊻ ♂

All ephemeris data is given at 12.00 UT and the Moon's longitude is additionally given for 24.00 UT

NOVEMBER 1980

LONGITUDES

Date	Sidereal time h m s	Sun ☉ ° ' "	Moon ☽ ° ' "	Moon ☽ 24.00 ° ' "	Mercury ☿ ° '	Venus ♀ ° '	Mars ♂ ° '	Jupiter ♃ ° '	Saturn ♄ ° '	Uranus ♅ ° '	Neptune ♆ ° '	Pluto ♇ ° '
01	14 43 43	09 ♏ 13 19	29 ♌ 50 37	05 ♍ 53 34	13 ♏ 35	02 ♎ 29	14 ♐ 39	00 ♎ 59	04 ♎ 55	24 ♏ 50	20 ♐ 53	22 ♎ 23
02	14 47 40	10 13 24	11 ♍ 53 53	17 ♍ 52 09	12 R 18	03 41	15 23	01 10	05 01	24 54	20 55	22 26
03	14 51 36	11 13 31	23 ♍ 48 53	29 ♍ 44 35	10 59	04 53	16 07	01 22	05 08	24 58	20 57	22 28
04	14 55 33	12 13 40	05 ♎ 39 40	11 ♎ 34 32	09 42	06 06	16 51	01 33	05 14	25 01	20 59	22 30
05	14 59 29	13 13 51	17 ♎ 29 30	23 ♎ 24 53	08 28	07 19	17 36	01 44	05 21	25 05	21 01	22 33
06	15 03 26	14 14 03	29 ♎ 20 55	05 ♏ 17 48	07 20	08 31	18 20	01 56	05 27	25 09	21 02	22 35
07	15 07 22	15 14 18	11 ♏ 15 42	17 ♏ 14 46	06 20	09 44	19 05	02 07	05 33	25 12	21 04	22 37
08	15 11 19	16 14 34	23 ♏ 15 09	29 ♏ 16 57	05 29	10 57	19 49	02 18	05 39	25 16	21 06	22 39
09	15 15 15	17 14 53	05 ♐ 20 18	11 ♐ 25 20	04 49	12 10	20 34	02 29	05 46	25 20	21 08	22 42
10	15 19 12	18 15 12	17 ♐ 32 12	23 ♐ 41 06	04 21	13 23	21 18	02 39	05 52	25 23	21 10	22 44
11	15 23 09	19 15 34	29 ♐ 52 15	06 ♑ 05 52	04 04	14 36	22 03	02 50	05 58	25 27	21 12	22 46
12	15 27 05	20 15 57	12 ♑ 22 17	18 ♑ 41 47	03 D 59	15 49	22 48	03 01	06 04	25 31	21 14	22 48
13	15 31 02	21 16 21	25 ♑ 04 47	01 ♒ 31 37	04 05	17 03	23 33	03 12	06 10	25 35	21 16	22 50
14	15 34 58	22 16 47	08 ♒ 02 44	14 ♒ 38 31	04 21	18 16	24 18	03 22	06 16	25 38	21 18	22 53
15	15 38 55	23 17 14	21 ♒ 19 23	28 ♒ 05 41	04 48	19 29	25 03	03 32	06 22	25 42	21 20	22 55
16	15 42 51	24 17 43	04 ♓ 57 43	11 ♓ 55 42	05 23	20 43	25 48	03 43	06 28	25 46	21 22	22 57
17	15 46 48	25 18 12	18 ♓ 59 43	26 ♓ 09 44	06 05	21 56	26 33	03 53	06 33	25 50	21 24	22 59
18	15 50 44	26 18 43	03 ♈ 25 30	10 ♈ 46 36	06 55	23 10	27 18	04 03	06 39	25 53	21 27	23 01
19	15 54 41	27 19 15	18 ♈ 12 25	25 ♈ 42 06	07 52	24 23	28 03	04 13	06 45	25 57	21 29	23 03
20	15 58 38	28 19 48	03 ♉ 14 38	10 ♉ 48 51	08 53	25 37	28 49	04 23	06 50	26 01	21 31	23 05
21	16 02 34	29 ♏ 20 23	18 ♉ 23 26	25 ♉ 57 04	09 59	26 51	29 ♐ 34	04 33	06 56	26 04	21 33	23 07
22	16 06 31	00 ♐ 20 59	03 ♊ 28 24	10 ♊ 56 11	11 10	28 04	00 ♑ 19	04 43	07 01	26 08	21 35	23 09
23	16 10 27	01 21 37	18 ♊ 19 17	25 ♊ 36 43	12 23	29 ♎ 18	01 05	04 53	07 07	26 12	21 37	23 11
24	16 14 24	02 22 16	02 ♋ 47 46	09 ♋ 51 53	13 40	00 ♏ 32	01 50	05 02	07 12	26 16	21 39	23 13
25	16 18 20	03 22 57	16 ♋ 48 44	23 ♋ 38 15	15 00	01 46	02 36	05 12	07 17	26 19	21 41	23 15
26	16 22 17	04 23 39	00 ♌ 20 29	06 ♌ 55 40	16 21	03 00	03 22	05 21	07 22	26 23	21 44	23 17
27	16 26 13	05 24 23	13 ♌ 24 13	19 ♌ 46 34	17 44	04 14	04 07	05 30	07 27	26 27	21 46	23 19
28	16 30 10	06 25 08	26 ♌ 03 19	02 ♍ 15 04	19 09	05 28	04 53	05 39	07 32	26 30	21 48	23 21
29	16 34 07	07 25 55	08 ♍ 22 28	14 ♍ 26 12	20 36	06 42	05 39	05 48	07 37	26 34	21 50	23 23
30	16 38 03	08 ♐ 26 43	20 ♍ 26 55	26 ♍ 25 16	22 ♏ 03	07 ♏ 56	06 ♑ 25	05 ♎ 57	07 ♎ 42	26 ♏ 38	21 ♐ 52	23 ♎ 25

Date	Moon True ☊ ° '	Moon Mean ☊ ° '	Moon ☽ Latitude ° '
01	16 ♌ 31	15 ♌ 43	01 N 10
02	16 R 29	15 40	02 10
03	16 24	15 37	03 03
04	16 17	15 34	03 48
05	16 06	15 31	04 23
06	15 53	15 27	04 47
07	15 38	15 24	04 58
08	15 23	15 21	04 56
09	15 09	15 18	04 41
10	14 57	15 15	04 13
11	14 48	15 11	03 33
12	14 41	15 08	02 41
13	14 38	15 05	01 41
14	14 37	15 02	00 N 35
15	14 D 37	14 59	00 S 35
16	14 R 37	14 56	01 45
17	14 35	14 52	02 51
18	14 31	14 49	03 48
19	14 24	14 46	04 31
20	14 15	14 43	04 56
21	14 04	14 40	05 00
22	13 52	14 37	04 44
23	13 42	14 33	04 09
24	13 34	14 30	03 18
25	13 29	14 27	02 16
26	13 26	14 24	01 S 09
27	13 D 25	14 21	00 00
28	13 26	14 17	01 N 07
29	13 R 26	14 14	02 09
30	13 ♌ 25	14 ♌ 11	03 N 04

DECLINATIONS

Date	Sun ☉ ° '	Moon ☽ ° '	Mercury ☿ ° '	Venus ♀ ° '	Mars ♂ ° '	Jupiter ♃ ° '	Saturn ♄ ° '	Uranus ♅ ° '	Neptune ♆ ° '	Pluto ♇ ° '
01	14 S 34	12 N 37	16 S 55	00 N 27	23 S 29	00 N 37	00 00	18 S 45	21 S 49	06 N 44
02	14 53	09 06	16 12	00 00	23 35	00 33	00 S 02	18 46	21 49	06 43
03	15 12	05 16	15 28	00 S 28	23 40	00 29	00 05	18 47	21 49	06 42
04	15 30	01 N 15	14 44	00 55	23 45	00 24	00 07	18 48	21 49	06 42
05	15 49	02 S 48	14 02	01 22	23 49	00 20	00 10	18 49	21 50	06 41
06	16 07	06 46	13 21	01 50	23 54	00 16	00 12	18 50	21 50	06 40
07	16 24	10 29	12 44	02 17	23 58	00 11	00 14	18 51	21 50	06 40
08	16 42	13 48	12 12	02 45	24 02	00 07	00 17	18 52	21 50	06 39
09	16 59	16 35	11 44	03 12	24 05	00 N 03	00 19	18 52	21 50	06 39
10	17 16	18 39	11 22	03 40	24 09	00 S 01	00 21	18 53	21 50	06 38
11	17 32	19 54	11 06	04 07	24 12	00 06	00 24	18 54	21 51	06 37
12	17 49	20 11	10 56	04 35	24 15	00 10	00 26	18 55	21 51	06 37
13	18 05	19 27	10 51	05 02	24 18	00 14	00 28	18 56	21 51	06 36
14	18 20	17 42	10 51	05 29	24 20	00 18	00 30	18 57	21 51	06 36
15	18 36	14 57	10 56	05 57	24 22	00 22	00 32	18 58	21 51	06 35
16	18 51	11 20	11 05	06 24	24 24	00 26	00 34	18 59	21 51	06 35
17	19 05	06 59	11 18	06 51	24 26	00 30	00 36	19 00	21 51	06 34
18	19 20	02 S 07	11 34	07 18	24 27	00 33	00 39	19 00	21 52	06 34
19	19 34	02 N 58	11 53	07 45	24 28	00 37	00 41	19 01	21 52	06 33
20	19 47	07 58	12 15	08 11	24 29	00 41	00 43	19 02	21 52	06 33
21	20 01	12 29	12 38	08 38	24 29	00 45	00 45	19 03	21 52	06 33
22	20 13	16 12	13 03	09 04	24 30	00 49	00 47	19 04	21 52	06 32
23	20 26	18 48	13 30	09 31	24 30	00 52	00 48	19 05	21 52	06 32
24	20 38	20 07	13 57	09 57	24 29	00 56	00 50	19 06	21 53	06 31
25	20 50	20 08	14 26	10 23	24 29	00 59	00 52	19 07	21 53	06 31
26	21 01	18 57	14 54	10 48	24 28	01 03	00 54	19 07	21 53	06 31
27	21 12	16 48	15 23	11 14	24 27	01 06	00 56	19 08	21 53	06 30
28	21 23	13 53	15 53	11 39	24 26	01 10	00 58	19 09	21 53	06 30
29	21 33	10 25	16 22	12 04	24 24	01 13	00 59	19 10	21 53	06 30
30	21 S 43	06 N 36	16 S 51	12 S 29	24 S 22	01 S 17	01 S 01	19 S 11	21 S 53	06 N 29

ZODIAC SIGN ENTRIES

Date	h	m	Planets
01	12	19	☽ ♍
04	00	31	☽ ♎
06	13	19	☽ ♏
09	01	25	☽ ♐
11	12	15	☽ ♑
13	21	10	☽ ♒
16	03	21	☽ ♓
18	06	22	☽ ♈
20	06	51	☽ ♉
22	01	42	♂ ♑
22	03	41	☉ ♐
22	06	27	☽ ♊
24	01	35	♀ ♏
24	07	18	☽ ♋
26	11	23	☽ ♌
28	19	37	☽ ♍

LATITUDES

Date	Mercury ☿ ° '	Venus ♀ ° '	Mars ♂ ° '	Jupiter ♃ ° '	Saturn ♄ ° '	Uranus ♅ ° '	Neptune ♆ ° '	Pluto ♇ ° '
01	01 S 03	01 N 34	00 S 57	01 N 06	02 N 08	00 N 14	01 N 19	16 N 37
04	00 S 01	01 39	00 58	01 07	02 08	00 14	01 19	16 37
07	00 N 57	01 42	00 59	01 07	02 09	00 14	01 19	16 38
10	01 42	01 46	01 00	01 08	02 09	00 14	01 19	16 39
13	02 10	01 48	01 01	01 08	02 10	00 14	01 18	16 39
16	02 22	01 49	01 02	01 09	02 10	00 14	01 18	16 40
19	02 22	01 50	01 02	01 09	02 11	00 14	01 18	16 41
22	02 14	01 50	01 03	01 10	02 11	00 14	01 18	16 42
25	02 00	01 49	01 04	01 10	02 12	00 14	01 18	16 43
28	01 42	01 48	01 05	01 11	02 13	00 14	01 18	16 44
31	01 N 22	01 N 46	01 S 05	01 N 11	02 N 13	00 N 14	01 N 18	16 N 45

DATA

Julian Date	2444545
Delta T	+51 seconds
Ayanamsa	23° 35' 08"
Synetic vernal point	05° ♓ 31' 51"
True obliquity of ecliptic	23° 26' 24"

MOON'S PHASES, APSIDES AND POSITIONS ☽

Date	h	m	Phase	Longitude ° '	Eclipse Indicator
07	20	43	●	15 ♏ 36	
15	15	47	☽	23 ♒ 27	
22	06	39	○	00 ♊ 07	
29	09	59	☾	07 ♍ 21	

Day	h	m		
05	17	06	Apogee	
21	01	10	Perigee	
04	19	22	0S	
12	06	58	Max dec	20° S 12'
18	22	03	0N	
25	00	10	Max dec	20° N 17'

LONGITUDES

Date	Chiron ⚷ ° '	Ceres ⚳ ° '	Pallas ⚴ ° '	Juno ⚵ ° '	Vesta ⚶ ° '	Black Moon Lilith ⚸ ° '
01	16 ♉ 09	25 ♋ 28	23 ♈ 02	11 ♎ 42	27 ♌ 03	23 ♎ 27
11	15 ♉ 39	26 ♋ 40	20 ♈ 13	15 ♎ 15	00 ♍ 15	24 ♎ 34
21	15 ♉ 09	27 ♋ 16	18 ♈ 02	18 ♎ 41	03 ♍ 09	25 ♎ 41
31	14 ♉ 41	27 ♋ 12	16 ♈ 40	21 ♎ 57	05 ♍ 40	26 ♎ 48

All ephemeris data is given at 12.00 UT and the Moon's longitude is additionally given for 24.00 UT

ASPECTARIAN

h m	Aspects
01 Saturday	
00 58	♀ ∥ ♃
02 06	☽ □ ♅
02 17	☽ ⊥ ♃
04 38	☽ ⊥ ♀
06 22	☽ Q ☉
10 08	☽ ⊥ ♄
14 17	☽ ⊻ ♃
15 06	☽ Q ☿
17 47	☽ ⊻ ♀
20 29	♀ Q ♂
22 08	☽ ⊻ ♄
02 Sunday	
03 02	☽ ∠ ♇
08 20	☽ ✶ ☉
09 52	♀ # ♄
12 43	☽ ✶ ☿
14 01	☽ Q ♅
14 03	♀ ∥ ♄
19 27	☽ □ ♂
21 08	☽ ⊥ ♇
03 Monday	
03 08	☽ ∥ ♇
06 12	☽ □ ♆
09 16	☽ ⊻ ♇
09 21	☉ ⊥ ♄
09 32	☉ ☌ ☿
09 39	☿ ⊥ ♄
12 45	♀ # ♃
12 56	☿ ⊥ ♀
14 20	☽ ✶ ♅
15 58	☽ ∠ ☿
17 09	♀ ☌ ♄
17 20	☽ ∠ ☉
18 20	☉ ∥ ☿
22 17	☿ ⊥ ♂
04 Tuesday	
03 32	☽ ☌ ♃
08 24	☽ ⊥ ☿
10 15	☽ Q ♂
11 08	☽ ☌ ♄
12 59	☽ ☌ ♀
13 15	☽ ⊥ ☉
13 46	☽ # ♀
17 05	☽ ∥ ♃
18 35	☽ # ♄
18 45	☽ Q ♆
19 24	☽ ⊻ ☿
20 10	☽ ∥ ♄
20 54	☽ ∠ ♅
21 35	☽ # ♃
05 Wednesday	
02 24	☽ ∥ ♀
02 26	♂ Q ♄
02 33	☽ ⊻ ☉
02 48	☉ ⊥ ♀
12 13	☽ ✶ ♂
15 15	☽ ⊥ ♅
19 09	☽ ✶ ♆
22 16	☽ ☌ ♇
23 42	☿ ⊻ ♀
06 Thursday	
00 59	☿ ⊥ ♃
03 28	☽ ⊻ ♅
11 26	☽ # ♇
17 17	☽ ⊻ ♃
20 35	☽ ∠ ♂
22 33	♀ Q ♆
07 Friday	
00 25	☽ ⊻ ♄
00 46	♃ # ♄
01 32	☽ ∠ ♆
02 49	☽ ☌ ☿
05 34	☽ ⊥ ♃
07 56	☉ ⊥ ♆
08 35	☽ ⊻ ♀
12 35	☽ ⊥ ♄
15 53	☽ ⊥ ♂
18 37	☿ ∠ ♆
19 40	☽ ⊥ ♆
20 43	☽ ☌ ☉
21 48	♀ ∠ ♅
21 59	☽ ⊥ ♀
23 55	☽ ∠ ♃
08 Saturday	
01 37	☽ ∥ ☿
04 42	☽ ⊻ ♂
06 47	☽ ∠ ♄
07 22	☿ ⊻ ♄
07 42	☽ ⊻ ♆
10 48	☽ ⊻ ♇
16 02	☽ ☌ ♅
17 59	☽ ∠ ♀
22 47	☽ ⊥ ♇
23 01	☿ ∠ ♂
09 Sunday	
06 15	☽ ✶ ♃
11 02	☽ ⊻ ☿
12 51	☽ ✶ ♄
16 38	☽ ∥ ☉
16 40	☽ ∠ ♇
18 39	☉ ∠ ♃
22 22	☽ ⊥ ☿
10 Monday	
02 57	☽ ✶ ♀
06 16	☽ Q ♃
07 22	♂ ☌ ♆
12 39	☽ Q ♄
13 32	☽ ⊻ ☉
15 27	☽ ∠ ☿
15 28	☽ ∥ ♅
19 07	☽ ☌ ♆
19 51	☽ ☌ ♂
22 10	☽ ✶ ♇
11 Tuesday	
02 16	☽ ⊥ ☉
03 24	☽ ⊻ ♅
04 59	☽ Q ♀
15 04	☽ ⊥ ♅
17 49	☽ □ ♃
20 00	☽ ✶ ☿
21 13	☽ ∠ ☉
21 28	☽ Q ♇
23 51	☽ □ ♄
12 Wednesday	
08 27	☽ ∠ ♅
10 57	☿ St D
12 06	♂ ✶ ♇
18 53	☽ Q ☿
19 15	☽ □ ♀
13 Thursday	
04 15	☽ ✶ ☉
04 50	☽ ⊻ ♆
07 48	☽ □ ♇
08 57	☽ ⊻ ♂
09 13	☉ ∠ ♄
11 57	☉ ⊻ ♆
12 56	☽ ✶ ♅
16 06	☽ ⊥ ♆
20 41	☽ ∥ ♅
20 51	☽ ⊥ ♂
14 Friday	
03 17	☽ △ ♃
04 30	☽ Q ☉
05 03	☽ □ ☿
05 33	☽ ∥ ☉
08 43	☽ △ ♄
08 48	☽ ∠ ♆
11 15	☽ Q ♅
14 26	☽ ∠ ♂
15 Saturday	
02 45	☉ ⊻ ♇
06 57	☽ ⚼ ♃
08 23	☽ △ ♀
12 02	☽ ✶ ♆
12 04	☽ ⚼ ♄
14 50	☽ △ ♇
15 47	☽ □ ☉
16 27	♀ ⊥ ♅
19 01	☽ ✶ ♂
19 49	☽ □ ♅
23 10	☽ ± ♃
16 Sunday	
04 06	☽ ± ♄
09 14	☽ Q ♆
09 48	☽ ⊼ ♃
10 45	♂ ⊻ ♅
12 45	☽ △ ☿
13 24	☽ ∥ ☿
13 25	☽ ⚼ ♀
14 37	☽ ⊼ ♄
17 11	☽ ⚼ ♇
17 11	☽ Q ♂
21 36	♀ # ♇
17 Monday	
01 22	♀ ✶ ♆
01 44	☉ ∥ ♅
06 20	☽ ± ♀
08 36	☽ ± ♇
12 37	☽ ∥ ♀
14 05	☽ # ♇
15 44	☽ ⚼ ☿
16 04	☽ □ ♆
17 24	☽ ⊼ ♀
18 43	☽ ⊼ ♇
21 54	☿ ∠ ♆
23 22	☽ △ ☉
23 29	☽ △ ♅
18 Tuesday	
01 14	☉ ☌ ♅
01 21	☽ □ ♂
03 34	☿ ⊻ ♄
07 37	☽ ± ☿
09 08	♀ ☌ ♇
13 03	☽ ☍ ♃
17 19	☽ ☍ ♄
18 06	☽ ⊼ ☿
18 58	☽ ∥ ♄
19 20	☽ ∥ ♃
19 Wednesday	
00 14	☽ ⚼ ♅
00 50	☽ # ♃
01 10	☽ # ♄
01 49	☽ ⚼ ☉
14 49	☽ ± ♅
17 16	☽ △ ♆
17 21	☽ ± ☉
19 47	☽ ☍ ♇
22 47	☽ ☍ ♀
20 Thursday	
00 27	☽ ⊼ ♅
03 38	☽ ⊼ ☉
04 35	☽ △ ♂
05 05	☽ ∥ ♇
13 16	☽ # ♀
13 50	☽ ⊼ ♃
17 12	☽ ⚼ ♆
17 44	☽ ⊼ ♄
20 10	♀ ⊻ ♅
21 38	☽ ☍ ☿
23 27	☽ ± ♃
21 Friday	
03 18	☽ ± ♄
05 38	☽ ⚼ ♂
06 37	☉ ⊥ ♇
07 29	☽ ± ♆
08 59	♃ ∥ ♄
12 56	☽ # ☿
13 52	☽ ⚼ ♃
17 01	☽ ⊼ ♆
17 39	☽ ⚼ ♄
19 31	☽ ⊼ ♇
20 39	☽ ± ♂
22 Saturday	
00 15	☽ ☍ ♅
01 36	☿ ⊥ ♃
02 37	☽ ⊼ ♀
05 05	☽ ± ♇
06 39	☽ ☍ ☉
06 42	☽ ⊼ ♂
09 37	☉ ⊻ ♂
13 03	☽ ± ♀
14 01	☽ △ ♃
17 43	☽ △ ♄
19 32	☽ ⚼ ♇
23 Sunday	
01 28	☽ ⊼ ☿
04 52	☽ ⚼ ♀
12 07	☽ ± ☿
15 37	☽ # ♅
17 25	☽ ☍ ♆
20 01	☽ △ ♇
24 Monday	
01 02	☽ ⊼ ♅
02 36	☿ ⊥ ♄
04 24	☽ ⚼ ☿
07 51	☽ △ ♀
10 18	☽ ☍ ♂
11 06	☽ ± ♅
11 14	☽ ⊼ ☉
15 49	☽ □ ♃
19 30	☽ □ ♄
22 11	☽ ± ☉
25 Tuesday	
02 26	♂ ⊥ ♅
02 27	☽ ⚼ ♅
08 30	☽ △ ☿
14 58	☽ ⚼ ☉
20 35	☽ ⊼ ♆
23 21	☽ Q ♃
23 21	☽ □ ♇
26 Wednesday	
00 44	☿ ⊥ ♆
03 02	☽ Q ♄
04 51	☽ △ ♅
07 17	☽ ± ♆
09 34	☽ # ♅
17 19	☽ □ ♀
17 49	☽ ⊼ ♂
19 58	☽ △ ☉
21 13	☽ ✶ ♃
23 40	☽ ⚼ ♆
27 Thursday	
00 54	☽ ✶ ♄
05 31	☽ ± ♂
06 15	♀ ✶ ♂
08 06	☽ Q ♇
14 45	☉ ✶ ♃
21 10	☽ □ ☿
22 28	☽ # ☿
23 27	☽ ⚼ ♂
28 Friday	
01 33	☽ ∠ ♃
03 50	☽ △ ♆
05 13	☽ ∠ ♄
06 30	☽ Q ♀
06 48	☽ ✶ ♇
12 52	☽ □ ♅
16 11	♀ ⊻ ♃
19 03	☽ ⊥ ♃
22 41	☽ ⊥ ♄
29 Saturday	
02 08	☽ # ♀
03 20	♂ Q ♇
06 17	☽ △ ♂
06 54	☽ ⊻ ♃
08 21	☽ ✶ ♀
09 59	☽ □ ☉
10 30	☽ ⊻ ♄
12 01	☽ ∠ ♇
12 29	☽ Q ☿
13 52	☿ ∠ ♂
14 39	♀ ∠ ♆
15 59	☿ ∠ ♃
16 53	☉ ✶ ♄
18 21	♂ □ ♃
30 Sunday	
00 19	☽ Q ♅
05 55	☽ ⊥ ♇
07 03	♀ ⊻ ♄
09 04	☿ ⊻ ♆
11 11	☉ ∠ ♇
12 40	☽ ∥ ♇
14 52	☽ □ ♆
15 39	☽ ✶ ☿
17 35	☽ ∠ ♀
17 58	☽ ⊻ ♇
23 20	☿ ∠ ♄

DECEMBER 1980

Date	Sidereal time h m s	LONGITUDES Sun ☉ ° ′ ″	Moon ☽ ° ′ ″	Moon ☽ 24.00 ° ′ ″	Mercury ☿ ° ′	Venus ♀ ° ′	Mars ♂ ° ′	Jupiter ♃ ° ′	Saturn ♄ ° ′	Uranus ♅ ° ′	Neptune ♆ ° ′	Pluto ♇ ° ′
01	16 42 00	09 ♐ 27 32	02 ♎ 21 54	08 ♎ 17 25	23 ♏ 31	09 ♏ 11	07 ♑ 10	06 ♎ 06	07 ♎ 47	26 ♏ 41	21 ♐ 55	23 ♎ 27
02	16 45 56	10 28 23	14 ♎ 12 22	20 ♎ 07 16	25 00	10 25	07 56	06 15	07 52	26 45	21 57	23 28
03	16 49 53	11 29 16	26 ♎ 02 36	01 ♏ 58 47	26 30	11 39	08 42	06 24	07 56	26 49	21 59	23 30
04	16 53 49	12 30 10	07 ♏ 56 10	13 ♏ 55 04	28 00	12 54	09 28	06 32	08 01	26 52	22 01	23 32
05	16 57 46	13 31 04	19 ♏ 55 45	25 ♏ 58 24	29 ♏ 31	14 08	10 14	06 40	08 05	26 56	22 04	23 34
06	17 01 42	14 32 01	02 ♐ 03 12	08 ♐ 10 15	01 ♐ 02	15 22	11 01	06 49	08 10	27 00	22 06	23 35
07	17 05 39	15 32 58	14 ♐ 19 40	20 ♐ 31 29	02 33	16 37	11 47	06 57	08 14	27 03	22 08	23 37
08	17 09 36	16 33 56	26 ♐ 45 46	03 ♑ 02 34	04 05	17 51	12 33	07 05	08 18	27 07	22 10	23 39
09	17 13 32	17 34 55	09 ♑ 21 55	15 ♑ 43 51	05 37	19 06	13 19	07 12	08 22	27 10	22 13	23 40
10	17 17 29	18 35 55	22 ♑ 08 28	28 ♑ 35 52	07 09	20 20	14 06	07 20	08 26	27 14	22 15	23 42
11	17 21 25	19 36 56	05 ♒ 06 09	11 ♒ 39 28	08 41	21 35	14 52	07 28	08 30	27 17	22 17	23 43
12	17 25 22	20 37 57	18 ♒ 15 59	24 ♒ 55 54	10 13	22 50	15 38	07 35	08 34	27 21	22 19	23 45
13	17 29 18	21 38 59	01 ♓ 39 24	08 ♓ 26 39	11 46	24 04	16 25	07 42	08 38	27 24	22 22	23 47
14	17 33 15	22 40 01	15 ♓ 17 51	22 ♓ 13 05	13 19	25 19	17 11	07 50	08 41	27 28	22 24	23 48
15	17 37 11	23 41 03	29 ♓ 12 26	06 ♈ 15 52	14 51	26 33	17 58	07 57	08 45	27 31	22 26	23 49
16	17 41 08	24 42 06	13 ♈ 23 17	20 ♈ 34 26	16 24	27 48	18 44	08 03	08 49	27 35	22 28	23 51
17	17 45 05	25 43 09	27 ♈ 48 57	05 ♉ 06 20	17 58	29 ♏ 03	19 31	08 10	08 52	27 38	22 31	23 52
18	17 49 01	26 44 13	12 ♉ 25 56	19 ♉ 47 00	19 31	00 ♐ 18	20 17	08 17	08 55	27 41	22 33	23 54
19	17 52 58	27 45 17	27 ♉ 08 40	04 ♊ 30 01	21 04	01 32	21 04	08 23	08 58	27 45	22 35	23 55
20	17 56 54	28 46 21	11 ♊ 50 03	19 ♊ 07 50	22 38	02 47	21 51	08 29	09 02	27 48	22 38	23 56
21	18 00 51	29 ♐ 47 26	26 ♊ 22 25	03 ♋ 32 59	24 12	04 02	22 37	08 36	09 05	27 52	22 40	23 58
22	18 04 47	00 ♑ 48 32	10 ♋ 38 49	17 ♋ 39 18	25 45	05 17	23 24	08 42	09 07	27 55	22 42	23 59
23	18 08 44	01 49 38	24 ♋ 34 00	01 ♌ 22 38	27 20	06 32	24 11	08 47	09 10	27 58	22 44	24 00
24	18 12 40	02 50 44	08 ♌ 05 04	14 ♌ 41 21	28 ♐ 54	07 46	24 58	08 53	09 13	28 01	22 47	24 01
25	18 16 37	03 51 51	21 ♌ 11 35	27 ♌ 36 04	00 ♑ 28	09 01	25 45	08 59	09 16	28 05	22 49	24 02
26	18 20 34	04 52 58	03 ♍ 55 09	10 ♍ 09 18	02 03	10 16	26 32	09 04	09 18	28 08	22 51	24 03
27	18 24 30	05 54 06	16 ♍ 19 01	22 ♍ 24 52	03 38	11 31	27 18	09 09	09 21	28 11	22 53	24 05
28	18 28 27	06 55 14	28 ♍ 27 27	04 ♎ 27 23	05 14	12 46	28 05	09 14	09 23	28 14	22 55	24 06
29	18 32 23	07 56 23	10 ♎ 25 19	16 ♎ 21 51	06 49	14 01	28 52	09 19	09 25	28 17	22 58	24 07
30	18 36 20	08 57 32	22 ♎ 17 38	28 ♎ 13 15	08 25	15 16	29 ♑ 39	09 24	09 27	28 20	23 00	24 08
31	18 40 16	09 ♑ 58 42	04 ♏ 09 17	10 ♏ 06 16	10 ♑ 02	16 ♐ 31	00 ♒ 26	09 ♎ 28	09 ♎ 29	28 ♏ 23	23 ♐ 02	24 ♎ 09

Date	Moon True ☊ ° ′	Moon Mean ☊ ° ′	Moon ☽ Latitude ° ′
01	13 ♌ 22	14 ♌ 08	03 N 50
02	13 R 17	14 05	04 26
03	13 09	14 02	04 50
04	12 59	13 58	05 02
05	12 47	13 55	05 01
06	12 36	13 52	04 46
07	12 24	13 49	04 18
08	12 15	13 46	03 38
09	12 07	13 43	02 46
10	12 03	13 39	01 45
11	12 01	13 36	00 N 37
12	12 D 01	13 33	00 S 34
13	12 02	13 30	01 44
14	12 03	13 27	02 49
15	12 R 03	13 23	03 46
16	12 01	13 20	04 31
17	11 58	13 17	04 59
18	11 52	13 14	05 08
19	11 45	13 11	04 57
20	11 38	13 08	04 27
21	11 31	13 04	03 39
22	11 26	13 01	02 39
23	11 23	12 58	01 30
24	11 22	12 55	00 S 18
25	11 D 22	12 52	00 N 53
26	11 24	12 49	02 00
27	11 25	12 45	02 59
28	11 27	12 42	03 49
29	11 R 27	12 39	04 28
30	11 26	12 36	04 55
31	11 ♌ 23	12 ♌ 33	05 N 10

DECLINATIONS

Date	Sun ☉ ° ′	Moon ☽ ° ′	Mercury ☿ ° ′	Venus ♀ ° ′	Mars ♂ ° ′	Jupiter ♃ ° ′	Saturn ♄ ° ′	Uranus ♅ ° ′	Neptune ♆ ° ′	Pluto ♇ ° ′
01	21 S 52	02 N 35	17 S 20	12 S 53	24 S 20	01 S 20	01 S 03	19 S 12	21 S 54	06 N 29
02	22 01	01 S 31	17 49	13 17	24 17	01 23	01 04	19 13	21 54	06 29
03	22 10	05 33	18 17	13 41	24 15	01 26	01 06	19 13	21 54	06 29
04	22 18	09 23	18 44	14 05	24 12	01 30	01 08	19 14	21 54	06 28
05	22 25	12 53	19 11	14 28	24 08	01 33	01 09	19 15	21 54	06 28
06	22 33	15 53	19 37	14 51	24 05	01 36	01 11	19 16	21 54	06 28
07	22 39	18 14	20 02	15 13	24 01	01 39	01 12	19 17	21 55	06 28
08	22 46	19 46	20 27	15 35	23 57	01 42	01 13	19 18	21 55	06 28
09	22 52	20 21	20 51	15 57	23 53	01 45	01 15	19 18	21 55	06 27
10	22 57	19 54	21 14	16 18	23 48	01 47	01 16	19 19	21 55	06 27
11	23 02	18 24	21 35	16 39	23 43	01 50	01 18	19 20	21 55	06 27
12	23 07	15 53	21 56	17 00	23 38	01 53	01 19	19 21	21 55	06 27
13	23 11	12 30	22 16	17 20	23 32	01 56	01 20	19 22	21 55	06 27
14	23 14	08 24	22 35	17 39	23 27	01 58	01 21	19 22	21 56	06 27
15	23 17	03 S 47	22 53	17 59	23 21	02 01	01 22	19 23	21 56	06 27
16	23 20	01 N 07	23 09	18 17	23 14	02 03	01 24	19 24	21 56	06 27
17	23 22	06 02	23 25	18 35	23 08	02 06	01 25	19 25	21 56	06 27
18	23 24	10 40	23 39	18 53	23 01	02 08	01 26	19 26	21 56	06 27
19	23 25	14 42	23 52	19 10	22 54	02 10	01 27	19 26	21 56	06 27
20	23 26	17 48	24 04	19 27	22 47	02 13	01 28	19 27	21 56	06 27
21	23 26	19 44	24 15	19 43	22 39	02 15	01 29	19 28	21 56	06 27
22	23 26	20 22	24 24	19 59	22 31	02 17	01 30	19 29	21 57	06 27
23	23 26	19 44	24 32	20 14	22 23	02 19	01 30	19 29	21 57	06 27
24	23 25	17 58	24 39	20 28	22 15	02 21	01 31	19 30	21 57	06 27
25	23 23	15 17	24 45	20 42	22 06	02 23	01 32	19 31	21 57	06 27
26	23 21	11 56	24 49	20 55	21 58	02 25	01 33	19 31	21 57	06 27
27	23 18	08 09	24 52	21 08	21 48	02 27	01 33	19 32	21 57	06 27
28	23 16	04 N 07	24 53	21 20	21 39	02 28	01 34	19 33	21 57	06 28
29	23 12	00 S 01	24 53	21 32	21 29	02 30	01 35	19 34	21 57	06 28
30	23 08	04 07	24 52	21 43	21 20	02 32	01 35	19 34	21 57	06 28
31	23 S 04	08 S 02	24 S 49	21 S 53	21 S 10	02 S 33	01 S 36	19 S 35	21 S 58	06 N 28

ZODIAC SIGN ENTRIES

Date	h	m	Planets
01	07	13	☽ ♎
03	20	00	☽ ♏
05	19	45	☿ ♐
06	07	57	☽ ♐
08	18	12	☽ ♑
11	02	36	☽ ♒
13	09	03	☽ ♓
15	13	21	☽ ♈
17	15	36	☽ ♉
18	06	21	♀ ♐
19	16	39	☽ ♊
21	16	56	☉ ♑
21	18	03	☽ ♋
23	21	34	☽ ♌
25	04	46	☿ ♑
26	04	32	☽ ♍
28	15	05	☽ ♎
30	22	30	♂ ♒
31	03	36	☽ ♏

LATITUDES

Date	Mercury ☿ ° ′	Venus ♀ ° ′	Mars ♂ ° ′	Jupiter ♃ ° ′	Saturn ♄ ° ′	Uranus ♅ ° ′	Neptune ♆ ° ′	Pluto ♇ ° ′
01	01 N 22	01 N 46	01 S 05	01 N 11	02 N 13	00 N 14	01 N 18	16 N 45
04	01 00	01 43	01 06	01 12	02 14	00 14	01 18	16 47
07	00 39	01 39	01 06	01 13	02 15	00 14	01 18	16 48
10	00 N 17	01 35	01 07	01 13	02 15	00 14	01 18	16 49
13	00 S 04	01 30	01 07	01 14	02 16	00 14	01 18	16 51
16	00 25	01 25	01 07	01 15	02 17	00 14	01 18	16 52
19	00 44	01 19	01 08	01 15	02 18	00 14	01 18	16 54
22	01 02	01 13	01 08	01 16	02 19	00 13	01 18	16 55
25	01 18	01 07	01 08	01 17	02 19	00 13	01 18	16 57
28	01 33	01 00	01 08	01 17	02 20	00 13	01 18	16 58
31	01 S 45	00 N 53	01 S 08	01 N 18	02 N 21	00 N 13	01 N 18	17 N 00

DATA

Julian Date	2444575
Delta T	+51 seconds
Ayanamsa	23° 35' 12"
Synetic vernal point	05° ♓ 31' 47"
True obliquity of ecliptic	23° 26' 24"

MOON'S PHASES, APSIDES AND POSITIONS ☽

Date	h	m	Phase	Longitude ° ′	Eclipse Indicator
07	14	35	●	15 ♐ 40	
15	01	47	☽	23 ♓ 15	
21	18	08	○	00 ♋ 03	
29	06	32	☾	07 ♎ 42	

Day	h	m		
03	04	28	Apogee	
19	05	28	Perigee	
30	23	14	Apogee	
02	03	06	0S	
09	13	39	Max dec	20° S 21'
16	06	33	0N	
22	11	43	Max dec	20° N 22'
29	11	54	0S	

LONGITUDES

Date	Chiron ⚷ ° ′	Ceres ⚳ ° ′	Pallas ⚴ ° ′	Juno ⚵ ° ′	Vesta ⚶ ° ′	Black Moon Lilith ⚸ ° ′
01	14 ♉ 41	27 ♋ 12	16 ♈ 40	21 ♎ 57	05 ♍ 40	26 ♎ 48
11	14 ♉ 16	26 ♋ 26	16 ♈ 14	25 ♎ 01	07 ♍ 44	27 ♎ 55
21	13 ♉ 55	25 ♋ 00	16 ♈ 41	27 ♎ 53	09 ♍ 16	29 ♎ 02
31	13 ♉ 39	23 ♋ 01	17 ♈ 59	00 ♏ 28	10 ♍ 10	00 ♏ 10

All ephemeris data is given at 12.00 UT and the Moon's longitude is additionally given for 24.00 UT

ASPECTARIAN

h m	Aspects
01 Monday	
00 29	☽ ✶ ♅
01 10	☽ Q ☉
10 45	☿ ⊻ ♇
13 50	☽ ⊥ ♀
15 57	☉ ∥ ♆
19 12	☽ # ♃
19 40	☽ ☌ ♃
20 55	☽ # ♄
22 25	☽ □ ♂
23 03	☽ ☌ ♄
02 Tuesday	
02 15	☽ ∠ ☿
03 20	☽ Q ♆
03 25	☽ ⊻ ♀
03 43	☽ ✶ ☉
07 00	☽ ∠ ♅
09 15	♂ □ ♄
09 23	☽ ∥ ♄
11 14	☽ ∥ ♃
18 10	☉ ⊻ ♀
23 08	☽ ⊥ ☿
03 Wednesday	
01 21	☽ ⊥ ♅
03 45	☽ ✶ ♆
06 50	☽ ☌ ♇
12 59	☽ ∠ ☉
13 03	☽ ⊻ ☿
13 26	☽ Q ♂
13 34	☽ ⊻ ♅
17 15	☿ ☌ ♅
17 41	☽ # ♇
04 Thursday	
04 08	♀ ⊥ ♃
08 51	☽ ⊥ ☉
09 09	☽ ⊻ ♃
10 09	☽ ∠ ♆
12 09	☽ ⊻ ♄
15 18	☽ ✶ ♂
21 20	☽ ⊥ ♃
22 01	☽ ⊻ ☉
23 06	☽ ☌ ♀
05 Friday	
00 16	☽ ⊥ ♄
04 16	☽ ⊥ ♆
11 03	♀ ⊥ ♄
12 48	☿ ⊥ ♇
15 30	☽ ∠ ♃
16 01	☿ ∥ ♅
16 15	☽ ⊻ ♆
18 19	☽ ∠ ♄
19 14	☽ ⊻ ♇
23 16	☽ ∠ ♂
06 Saturday	
01 48	☽ ∥ ♀
01 58	☽ ☌ ♅
07 08	☽ ⊥ ♇
09 42	☽ ☌ ☿
18 12	☽ ⊥ ♂
21 26	☽ ✶ ♃
07 Sunday	
00 03	☽ ✶ ♄
00 51	☽ ∠ ♇
02 27	♀ ⊥ ♆
06 43	☽ ⊻ ♂
14 35	☽ ☌ ☉
16 56	☽ ⊻ ♀
21 03	☽ Q ♃
21 15	♂ ∠ ♅
23 30	☽ Q ♄
08 Monday	
02 45	☽ ∥ ♅
03 09	☽ ☌ ♆
05 48	☽ ⊥ ♀
06 00	☽ ✶ ♇
12 40	☽ ⊻ ♅
09 Tuesday	
00 11	☽ ⊥ ♅
00 55	☽ ∠ ♀
03 54	☽ ⊻ ☿
04 59	☽ Q ♇
07 52	☽ □ ♃
10 06	☽ □ ♄
16 49	☽ ⊥ ☿
17 19	☽ ∠ ♅
19 57	☽ ☌ ♂
10 Wednesday	
04 48	☽ ⊻ ☉
08 17	☽ ✶ ♀
12 00	☽ ∠ ☿
12 12	☽ ⊻ ♆
14 54	☽ □ ♇
15 15	☿ ✶ ♃
16 58	☽ ⊥ ☉
21 31	☽ ✶ ♅
23 10	☽ ∥ ♅
23 23	☽ ⊥ ♆
11 Thursday	
07 50	☉ Q ♃
08 55	☽ Q ♀
09 05	☿ ✶ ♄
11 02	☽ ∠ ☉
12 40	☿ ∠ ♇
16 01	☽ ∠ ♆
16 22	☽ △ ♃
17 43	☿ ⊥ ♂
18 16	☽ △ ♄
19 26	☽ ✶ ☿
19 43	☽ Q ♅
12 Friday	
02 00	♀ ⊻ ♆
03 34	☽ ∥ ♀
06 50	♀ ∠ ♃
06 57	☽ ⊻ ♂
10 21	☉ Q ♄
10 44	☿ ∥ ♆
16 37	☽ ✶ ☉
18 27	☽ ⊥ ♂
19 20	☽ ✶ ♆
19 51	☽ ᵽ ♃
20 04	☽ Q ☿
21 04	☽ □ ♀
21 36	☽ ᵽ ♄
21 54	☽ △ ♇
13 Saturday	
03 04	♀ ∠ ♄
04 24	☽ □ ♅
06 12	♀ ⊻ ♇
11 32	☽ ∠ ♂
12 05	☽ ± ♃
13 44	☽ ± ♄
15 49	☽ Q ☉
16 49	☽ Q ♆
22 48	☽ ⊼ ♃
14 Sunday	
00 23	☽ ⊼ ♄
00 36	☽ ᵽ ♇
05 26	☉ ☌ ♆
08 06	☽ □ ☿
15 29	☽ ✶ ♂
16 22	☽ ± ♇
22 23	☽ # ♇
15 Monday	
00 21	☽ □ ♆
01 47	☽ □ ☉
02 45	☽ ⊼ ♇
07 01	☽ △ ♀
09 06	☽ △ ♅
13 22	☽ Q ♂
15 23	☉ ✶ ♇
20 40	☽ ∥ ♃
20 50	☉ ∥ ♂
23 49	☽ ∥ ♄
16 Tuesday	
02 57	☽ ☍ ♃
04 16	☽ ☍ ♄
07 28	♀ ☌ ♅
10 38	☽ ᵽ ♅
10 55	☽ ᵽ ♀
13 18	☽ # ♄
16 32	☽ # ♃
17 25	☿ ∥ ♂
17 40	☽ △ ☿
21 27	☽ □ ♂
17 Wednesday	
01 43	☽ ± ♅
03 12	☽ △ ♆
03 22	☽ ± ♀
05 28	☽ ☍ ♇
07 10	☉ ∥ ☿
08 16	☽ △ ☉
11 42	☽ ⊼ ♅
14 02	☽ ∥ ♇
14 13	☽ ⊼ ♀
21 29	☽ ᵽ ☿
18 Thursday	
03 59	☽ ᵽ ♆
04 10	♀ ⊥ ♇
05 09	☽ ⊼ ♃
06 14	☽ ⊼ ♄
10 47	☽ ᵽ ☉
13 59	☽ ± ☿
15 02	☽ ± ♃
16 05	☽ ± ♄
18 45	☽ ± ♆
19 Friday	
00 40	☿ Q ♃
00 56	☽ ⊼ ☿
01 32	☽ △ ♂
02 34	☽ ± ☉
04 33	☽ ⊼ ♆
05 50	☽ ᵽ ♃
06 44	☽ ⊼ ♇
06 49	☽ ᵽ ♄
10 28	☿ Q ♄
11 49	☉ ⊻ ♅
11 55	☿ ⊻ ♂
12 59	☽ ☍ ♅
13 04	☽ ⊼ ☉
16 32	☽ ± ♇
19 50	☽ ☍ ♀
20 Saturday	
03 23	☽ ᵽ ♂
06 29	☽ △ ♃
07 15	☽ ᵽ ♇
07 23	☽ △ ♄
11 56	☿ ☌ ♆
12 08	♀ ∥ ♅
18 58	☽ ± ♂
21 Sunday	
05 25	☽ ⊼ ♂
05 49	☽ ☍ ♆
07 23	☽ # ♅
07 56	☽ ☍ ☿
07 59	☽ △ ♇
08 23	☿ ✶ ♇
11 37	☽ # ♀
13 15	♂ ⊻ ♆
14 29	☽ ⊼ ♅
18 08	☽ ☍ ☉
22 Monday	
00 34	☽ ± ♅
02 02	☽ ⊼ ♀
08 39	☽ □ ♃
09 24	☽ □ ♄
13 11	☽ ± ♀
15 53	☽ ᵽ ♅
23 Tuesday	
02 32	☽ # ♀
06 10	☽ ᵽ ♀
06 13	♂ □ ♇
08 48	☽ ⊼ ♆
11 01	☽ □ ♇
11 17	☽ ☍ ♂
15 55	☽ Q ♃
16 20	☽ # ♅
16 35	☽ Q ♄
17 28	☽ ⊼ ☿
18 00	☽ △ ♅
19 21	☽ ± ♆
22 09	☿ ⊻ ♅
24 Wednesday	
01 51	☽ ⊼ ☉
05 31	☽ ± ☿
11 23	☽ △ ♀
11 26	☽ ᵽ ♆
13 27	☽ ✶ ♃
13 29	☽ ± ☉
14 03	☽ ✶ ♄
19 08	☽ Q ♇
25 Thursday	
00 00	☽ ᵽ ☿
07 19	☽ ᵽ ☉
11 03	♀ ✶ ♃
12 21	♀ ∠ ♇
15 02	☽ △ ♆
16 45	♀ ✶ ♄
17 14	☽ ∠ ♃
17 15	☉ ⊥ ♅
17 19	☽ ✶ ♇
17 44	☽ ∠ ♄
21 04	☽ ⊼ ♂
26 Friday	
00 57	☽ □ ♅
07 56	☽ △ ☿
09 10	☽ ± ♂
10 21	☽ ⊥ ♃
10 49	☽ ⊥ ♄
13 17	♂ ∥ ♆
14 00	☽ △ ☉
21 54	☽ ∠ ♇
21 58	☽ ⊻ ♃
22 23	☽ ⊻ ♄
27 Saturday	
01 36	☽ □ ♀
03 39	☽ ᵽ ♂
11 44	☽ Q ♅
15 27	☽ ⊥ ♇
16 10	☉ Q ♇
20 28	☿ ⊥ ♅
22 10	☽ ∥ ♇
28 Sunday	
00 58	☽ □ ♆
03 19	☽ ⊻ ♇
11 13	☽ △ ♂
11 33	☽ ✶ ♅
16 41	♂ ✶ ♅
17 09	☽ Q ♀
21 29	☽ # ♃
29 Monday	
01 10	☿ Q ♇
02 46	☽ # ♄
03 38	☽ □ ☿
04 40	♀ ∠ ♂
06 32	☽ □ ☉
09 26	♀ ∥ ♂
09 45	☽ ☌ ♃
09 58	☽ ☌ ♄
13 05	☽ Q ♆
14 49	♂ ⊥ ♆
17 48	☽ ∠ ♅
20 06	☽ ✶ ♀
21 06	☽ ∥ ♄
30 Tuesday	
02 35	☽ ∥ ♃
12 05	☽ ⊥ ♅
13 26	☽ ✶ ♆
15 43	☽ ☌ ♇
21 40	☽ Q ☿
22 20	☽ Q ☉
23 05	☉ □ ♃
31 Wednesday	
00 02	☉ □ ♄
00 17	☽ ⊻ ♅
02 13	☽ # ♇
03 18	☿ □ ♃
03 46	☿ □ ♄
03 58	☽ □ ♂
06 02	☽ ∠ ♀
10 04	☉ ☌ ☿
19 51	☽ ∠ ♆
21 23	♃ ☌ ♄
22 47	☽ ⊻ ♄
22 47	☽ ⊻ ♃
23 53	♀ ∥ ♆

JANUARY 1981

LONGITUDES

Date	Sidereal time h m s	Sun ☉ ° ' "	Moon ☽ ° ' "	Moon ☽ 24.00 ° ' "	Mercury ☿ ° '	Venus ♀ ° '	Mars ♂ ° '	Jupiter ♃ ° '	Saturn ♄ ° '	Uranus ♅ ° '	Neptune ♆ ° '	Pluto ♇ ° '
01	18 44 13	10 ♑ 59 51	16 ♏ 04 43	22 ♏ 05 05	11 ♑ 38	17 ♐ 46	01 ♒ 14	09 ♎ 33	09 ♎ 31	28 ♏ 26	23 ♐ 04	24 ♎ 09
02	18 48 09	12 01 02	28 ♏ 07 47	04 ♐ 13 10	13 15	19 01	02 01	09 37	09 33	28 29	23 06	24 10
03	18 52 06	13 02 12	10 ♐ 21 32	16 ♐ 33 07	14 52	20 16	02 48	09 41	09 35	28 32	23 09	24 11
04	18 56 03	14 03 23	22 ♐ 48 06	29 ♐ 06 37	16 30	21 31	03 35	09 45	09 36	28 35	23 11	24 12
05	18 59 59	15 04 34	05 ♑ 28 42	11 ♑ 54 23	18 08	22 46	04 22	09 48	09 38	28 38	23 13	24 13
06	19 03 56	16 05 44	18 ♑ 23 38	24 ♑ 56 22	19 46	24 01	05 09	09 52	09 39	28 41	23 15	24 13
07	19 07 52	17 06 55	01 ♒ 32 28	08 ♒ 11 49	21 25	25 16	05 56	09 55	09 40	28 43	23 17	24 14
08	19 11 49	18 08 05	14 ♒ 54 16	21 ♒ 39 39	23 04	26 31	06 44	09 58	09 41	28 46	23 19	24 15
09	19 15 45	19 09 15	28 ♒ 27 48	05 ♓ 18 35	24 44	27 46	07 31	10 01	09 42	28 49	23 21	24 15
10	19 19 42	20 10 24	12 ♓ 11 49	19 ♓ 07 21	26 23	29 ♐ 01	08 18	10 04	09 43	28 52	23 23	24 16
11	19 23 38	21 11 33	26 ♓ 05 02	03 ♈ 04 44	28 04	00 ♑ 16	09 06	10 07	09 44	28 54	23 25	24 17
12	19 27 35	22 12 41	10 ♈ 06 16	17 ♈ 09 28	29 ♑ 44	01 31	09 53	10 09	09 45	28 57	23 28	24 17
13	19 31 32	23 13 49	24 ♈ 14 07	01 ♉ 20 00	01 ♒ 25	02 46	10 40	10 11	09 46	28 59	23 30	24 18
14	19 35 28	24 14 56	08 ♉ 26 50	15 ♉ 34 20	03 06	04 01	11 28	10 13	09 46	29 02	23 32	24 18
15	19 39 25	25 16 02	22 ♉ 42 07	29 ♉ 49 48	04 47	05 17	12 15	10 15	09 46	29 04	23 34	24 19
16	19 43 21	26 17 07	06 ♊ 56 56	14 ♊ 03 08	06 28	06 32	13 02	10 17	09 47	29 07	23 36	24 19
17	19 47 18	27 18 12	21 ♊ 07 41	28 ♊ 10 18	08 09	07 47	13 50	10 18	09 47	29 09	23 38	24 19
18	19 51 14	28 19 16	05 ♋ 10 25	12 ♋ 07 33	09 50	09 02	14 37	10 20	09 47	29 12	23 40	24 20
19	19 55 11	29 ♑ 20 19	19 ♋ 01 15	25 ♋ 51 09	11 31	10 17	15 24	10 21	09 R 47	29 14	23 41	24 20
20	19 59 07	00 ♒ 21 22	02 ♌ 36 55	09 ♌ 18 17	13 12	11 32	16 12	10 22	09 47	29 16	23 43	24 20
21	20 03 04	01 22 24	15 ♌ 55 05	22 ♌ 27 14	14 51	12 47	16 59	10 22	09 47	29 18	23 45	24 20
22	20 07 01	02 23 25	28 ♌ 54 42	05 ♍ 17 35	16 30	14 02	17 47	10 23	09 46	29 21	23 47	24 20
23	20 10 57	03 24 26	11 ♍ 36 03	17 ♍ 50 19	18 08	15 17	18 34	10 23	09 46	29 23	23 49	24 21
24	20 14 54	04 25 26	24 ♍ 00 42	00 ♎ 07 34	19 45	16 32	19 22	10 23	09 45	29 25	23 51	24 21
25	20 18 50	05 26 25	06 ♎ 11 21	12 ♎ 12 31	21 19	17 47	20 09	10 R 23	09 45	29 27	23 53	24 21
26	20 22 47	06 27 24	18 ♎ 11 36	24 ♎ 09 07	22 52	19 02	20 56	10 23	09 44	29 29	23 54	24 21
27	20 26 43	07 28 22	00 ♏ 05 39	06 ♏ 01 47	24 22	20 18	21 44	10 23	09 43	29 31	23 56	24 R 21
28	20 30 40	08 29 20	11 ♏ 58 07	17 ♏ 55 15	25 49	21 33	22 31	10 22	09 42	29 33	23 58	24 21
29	20 34 36	09 30 16	23 ♏ 53 45	29 ♏ 54 13	27 12	22 48	23 19	10 21	09 41	29 35	24 00	24 21
30	20 38 33	10 31 13	05 ♐ 57 11	12 ♐ 03 11	28 30	24 03	24 06	10 20	09 40	29 36	24 01	24 20
31	20 42 30	11 ♒ 32 08	18 ♐ 12 40	24 ♐ 26 06	29 ♒ 44	25 ♑ 18	24 ♒ 54	10 ♎ 19	09 ♎ 38	29 ♏ 38	24 ♐ 03	24 ♎ 20

Moon Nodes and Latitude

Date	Moon True ☊ ° '	Moon Mean ☊ ° '	Moon ☽ Latitude ° '
01	11 ♌ 19	12 ♌ 29	05 N 11
02	11 R 15	12 26	04 59
03	11 09	12 23	04 33
04	11 05	12 20	03 54
05	11 00	12 17	03 02
06	10 57	12 14	02 01
07	10 56	12 10	00 N 51
08	10 D 56	12 07	00 S 22
09	10 56	12 04	01 35
10	10 58	12 01	02 44
11	10 59	11 58	03 44
12	11 00	11 54	04 31
13	11 R 01	11 51	05 02
14	11 00	11 48	05 15
15	10 59	11 45	05 09
16	10 57	11 42	04 44
17	10 55	11 39	04 02
18	10 53	11 35	03 05
19	10 52	11 32	01 58
20	10 52	11 29	00 S 46
21	10 D 51	11 26	00 N 28
22	10 52	11 23	01 38
23	10 53	11 20	02 42
24	10 53	11 16	03 37
25	10 54	11 13	04 21
26	10 54	11 10	04 53
27	10 55	11 07	05 12
28	10 R 55	11 04	05 17
29	10 55	11 00	05 09
30	10 55	10 57	04 48
31	10 ♌ 55	10 ♌ 54	04 N 13

DECLINATIONS

Date	Sun ☉ ° '	Moon ☽ ° '	Mercury ☿ ° '	Venus ♀ ° '	Mars ♂ ° '	Jupiter ♃ ° '	Saturn ♄ ° '	Uranus ♅ ° '	Neptune ♆ ° '	Pluto ♇ ° '
01	22 S 59	11 S 41	24 S 44	22 S 02	20 S 59	02 S 35	01 S 36	19 S 36	21 S 58	06 N 28
02	22 54	14 53	24 38	22 11	20 49	02 36	01 37	19 36	21 58	06 28
03	22 48	17 30	24 31	22 20	20 38	02 37	01 37	19 37	21 58	06 29
04	22 42	19 21	24 22	22 27	20 27	02 39	01 37	19 38	21 58	06 29
05	22 35	20 17	24 12	22 34	20 16	02 40	01 38	19 38	21 58	06 29
06	22 28	20 11	24 00	22 40	20 04	02 41	01 38	19 39	21 58	06 29
07	22 21	18 59	23 47	22 46	19 53	02 42	01 38	19 39	21 58	06 30
08	22 13	16 43	23 32	22 51	19 41	02 43	01 38	19 40	21 58	06 30
09	22 04	13 30	23 15	22 55	19 29	02 44	01 38	19 41	21 58	06 30
10	21 55	09 31	22 57	22 59	19 16	02 45	01 39	19 41	21 59	06 31
11	21 46	04 59	22 37	23 01	19 04	02 46	01 39	19 42	21 59	06 31
12	21 37	00 S 09	22 16	23 04	18 51	02 46	01 39	19 42	21 59	06 31
13	21 26	04 N 42	21 53	23 05	18 38	02 47	01 39	19 43	21 59	06 32
14	21 16	09 20	21 29	23 06	18 25	02 47	01 39	19 43	21 59	06 32
15	21 05	13 28	21 03	23 06	18 12	02 48	01 39	19 44	21 59	06 32
16	20 54	16 48	20 36	23 05	17 58	02 48	01 38	19 44	21 59	06 33
17	20 42	19 08	20 07	23 03	17 44	02 49	01 38	19 45	21 59	06 33
18	20 30	20 16	19 37	23 01	17 30	02 49	01 38	19 46	21 59	06 34
19	20 17	20 09	19 05	22 58	17 16	02 49	01 38	19 46	21 59	06 34
20	20 05	18 50	18 32	22 55	17 02	02 49	01 37	19 47	21 59	06 35
21	19 51	16 31	17 57	22 51	16 47	02 49	01 37	19 47	21 59	06 35
22	19 38	13 23	17 22	22 46	16 33	02 49	01 37	19 47	22 00	06 36
23	19 24	09 43	16 45	22 40	16 18	02 49	01 36	19 48	22 00	06 36
24	19 09	05 42	16 08	22 34	16 03	02 49	01 36	19 48	22 00	06 37
25	18 55	01 N 32	15 29	22 27	15 48	02 49	01 35	19 49	22 00	06 37
26	18 40	02 S 37	14 50	22 19	15 32	02 48	01 35	19 49	22 00	06 38
27	18 24	06 38	14 11	22 11	15 17	02 48	01 34	19 50	22 00	06 38
28	18 08	10 23	13 31	22 01	15 01	02 47	01 33	19 50	22 00	06 39
29	17 52	13 45	12 52	21 52	14 46	02 47	01 33	19 50	22 00	06 39
30	17 36	16 35	12 13	21 41	14 30	02 46	01 32	19 51	22 00	06 40
31	17 S 19	18 S 43	11 S 35	21 S 30	14 S 14	02 S 45	01 S 31	19 S 51	22 S 00	06 N 40

ZODIAC SIGN ENTRIES

Date	h	m	Planets	
02	15	42	☽	♐
05	01	41	☽	♑
07	09	12	☽	♒
09	14	42	☽	♓
11	06	48	♀	♑
11	18	43	☽	♈
12	15	48	☿	♒
13	21	45	☽	♉
16	00	17	☽	♊
18	03	08	☽	♋
20	03	36	☉	♒
20	07	21	☽	♌
22	14	02	☽	♍
24	23	45	☽	♎
27	11	49	☽	♏
30	00	12	☽	♐
31	17	35	☿	♓

LATITUDES

Date	Mercury ☿ ° '	Venus ♀ ° '	Mars ♂ ° '	Jupiter ♃ ° '	Saturn ♄ ° '	Uranus ♅ ° '	Neptune ♆ ° '	Pluto ♇ ° '
01	01 S 49	00 N 50	01 S 08	01 N 19	02 N 21	00 N 13	01 N 18	17 N 00
04	01 58	00 43	01 08	01 19	02 22	00 13	01 18	17 02
07	02 04	00 35	01 07	01 20	02 23	00 13	01 18	17 04
10	02 07	00 28	01 07	01 21	02 24	00 13	01 18	17 06
13	02 06	00 20	01 07	01 22	02 25	00 13	01 18	17 07
16	02 00	00 12	01 07	01 23	02 26	00 13	01 18	17 09
19	01 50	00 N 04	01 06	01 23	02 27	00 13	01 18	17 11
22	01 33	00 S 04	01 06	01 24	02 28	00 13	01 18	17 13
25	01 09	00 11	01 05	01 25	02 28	00 13	01 18	17 14
28	00 39	00 19	01 04	01 26	02 29	00 13	01 18	17 16
31	00 S 01	00 S 26	01 S 04	01 N 27	02 N 30	00 N 14	01 N 18	17 N 18

DATA

Julian Date	2444606
Delta T	+51 seconds
Ayanamsa	23° 35' 18"
Synetic vernal point	05° ♓ 31' 42"
True obliquity of ecliptic	23° 26' 24"

MOON'S PHASES, APSIDES AND POSITIONS ☽

Date	h	m	Phase	Longitude ° '	Eclipse Indicator
06	07	24	●	15 ♑ 54	
13	10	10	☽	23 ♈ 09	
20	07	39	○	00 ♌ 10	
28	04	19	☾	08 ♏ 10	

Day	h	m		
15	03	42	Perigee	
27	20	28	Apogee	
05	21	42	Max dec	20° S 23'
12	12	46	0N	
18	21	42	Max dec	20° N 22'
25	20	50	0S	

LONGITUDES

Date	Chiron ⚷ ° '	Ceres ⚳ ° '	Pallas ⚴ ° '	Juno ⚵ ° '	Vesta ⚶ ° '	Black Moon Lilith ⚸ ° '
01	13 ♉ 38	22 ♋ 47	18 ♈ 09	00 ♏ 43	10 ♍ 14	00 ♏ 16
11	13 ♉ 28	20 ♋ 28	20 ♈ 15	02 ♏ 58	10 ♍ 20	01 ♏ 24
21	13 ♉ 24	18 ♋ 08	23 ♈ 00	04 ♏ 51	09 ♍ 40	02 ♏ 31
31	13 ♉ 27	16 ♋ 06	26 ♈ 17	06 ♏ 19	08 ♍ 13	03 ♏ 38

All ephemeris data is given at 12.00 UT and the Moon's longitude is additionally given for 24.00 UT

ASPECTARIAN

h m	Aspects
01 Thursday	
00 51	☽ ⚹ ☉
01 41	☽ ⚹ ☿
02 20	☽ ⊥ ♀
10 52	☽ ⊥ ♄
10 55	☽ ⊥ ♃
14 00	☽ ⊥ ♆
15 46	☽ ⊻ ♀
18 44	☽ Q ♂
02 Friday	
02 00	☽ ⊻ ♆
04 09	☽ ⊻ ♇
04 53	☽ ∠ ♄
04 59	☽ ∠ ♃
09 36	☽ ∠ ☉
12 17	☽ ∠ ☿
12 42	☽ ☌ ♅
15 36	☿ ∠ ♅
16 02	☽ ⊥ ♇
20 11	☽ ⚹ ♂
03 Saturday	
04 56	☽ ⊥ ☉
08 40	☽ ⊥ ☿
09 43	☽ ∠ ♇
10 28	☽ ⚹ ♄
10 40	☽ ⚹ ♃
17 40	☽ ⊻ ☉
22 04	☽ ⊻ ☿
04 Sunday	
00 18	☉ ∠ ♅
03 22	☽ ∠ ♂
09 16	☽ ☌ ♀
09 42	☽ Q ♄
09 58	☽ Q ♃
12 43	☽ ☌ ♆
13 43	♀ Q ♄
14 40	☽ ⚹ ♇
16 38	♀ Q ♃
16 59	☽ ‖ ♅
21 42	☽ ⊥ ♂
23 03	☽ ⊻ ♅
05 Monday	
09 47	☽ ⊻ ♂
10 24	☽ ⊥ ♅
11 01	☽ ‖ ♂
13 23	☽ Q ♇
13 57	☉ ‖ ♀
19 46	☽ □ ♄
20 08	☽ □ ♃
20 53	♀ ☌ ♆
06 Tuesday	
03 15	☽ ∠ ♅
07 24	☽ ☌ ☉
14 54	☽ ☌ ☿
16 03	♀ ⚹ ♇
16 44	☽ ‖ ♂
20 56	☽ ⊻ ♆
22 42	☽ □ ♇
23 24	☽ ⊻ ♀
07 Wednesday	
01 19	☽ ‖ ♅
06 52	☽ ⚹ ♅
07 54	☽ ⊥ ♆
11 27	☽ ⊥ ♀
20 27	☽ ☌ ♂
08 Thursday	
00 11	☽ ∠ ♆
02 40	☽ △ ♄
03 09	☽ △ ♃
04 35	☽ Q ♅
05 20	☽ ∠ ♀
13 22	♂ ‖ ♅
15 43	☿ ⊻ ♆
18 13	☽ ⊻ ☉
09 Friday	
02 58	☽ ⚹ ♆
04 30	☽ ⊻ ☿
04 35	☽ △ ♇
05 10	☿ □ ♇
05 23	☽ bQ ♄
05 42	☽ ⊥ ☉
05 55	☽ bQ ♃
10 39	☽ ⚹ ♀
12 37	☽ □ ♅
16 32	☽ ⊥ ☿
21 13	☽ ± ♄
21 47	☽ ± ♃
22 47	☽ ∠ ☉
10 Saturday	
00 07	☽ Q ♆
03 45	☉ ‖ ♆
04 49	☽ ⊻ ♂
06 54	☽ bQ ♇
07 42	☽ ⊼ ♄
08 17	☽ ⊼ ♃
08 50	♀ ⊻ ♅
09 45	☽ Q ♀
10 18	☿ ‖ ♀
10 25	☽ ∠ ☿
14 44	♂ ∠ ♆
15 53	☽ ⊥ ♂
22 32	☽ ± ♇
11 Sunday	
02 55	☽ ⚹ ☉
04 07	☽ # ♇
07 25	☽ □ ♆
08 22	☽ ∠ ♂
08 53	☽ ⊼ ♇
15 52	☽ ⚹ ☿
16 52	☽ △ ♅
19 54	☽ □ ♀
23 07	☽ ‖ ♃
12 Monday	
00 25	☿ ⚹ ♅
01 09	☽ Q ☉
04 39	☽ ‖ ♄
07 56	♂ △ ♄
07 58	☿ ⊥ ♆
11 24	☽ ☍ ♄
11 36	☽ ⚹ ♂
12 05	☽ ☍ ♃
15 09	☽ Q ☿
18 34	☽ bQ ♅
20 35	♂ △ ♃
20 51	☽ # ♄
13 Tuesday	
02 25	☽ # ♃
06 25	☿ ‖ ♆
09 12	☽ Q ♂
09 53	☽ ± ♅
10 10	☽ □ ☉
10 44	☽ △ ♆
12 06	☽ ☍ ♇
18 24	☉ ⊻ ♆
20 04	☽ ⊼ ♅
21 14	☽ ‖ ♇
22 17	♂ Q ♅
14 Wednesday	
01 46	☽ □ ☿
03 49	☽ △ ♀
12 08	☽ bQ ♆
13 16	☉ □ ♇
14 14	☽ ⊼ ♄
15 00	☽ ⊼ ♃
17 22	☽ □ ♂
15 Thursday	
00 20	☽ ± ♄
01 07	☽ ± ♃
03 20	☽ ± ♆
07 31	☽ bQ ♀
08 00	♀ ⊥ ♅
09 08	☉ ‖ ☿
13 27	☽ ⊼ ♆
14 42	☽ ⊼ ♇
15 29	☽ bQ ♄
16 18	☽ bQ ♃
16 39	☽ △ ☉
22 45	☽ ☍ ♅
16 Friday	
00 08	☽ ± ♀
00 49	☽ ± ♇
07 56	♀ Q ♇
11 05	☽ △ ☿
11 13	☽ ⊼ ♀
15 16	☿ ⊻ ♀
16 00	☽ bQ ♇
16 47	☽ △ ♄
17 38	☽ △ ♃
19 53	☽ bQ ☉
21 41	☽ # ♂
22 53	☽ △ ♂
17 Saturday	
12 19	☽ ± ☉
14 35	♀ ⊥ ♂
15 55	☽ bQ ☿
16 16	☽ ☍ ♆
17 26	☽ △ ♇
18 52	☿ ∠ ♆
22 05	☽ # ♅
23 20	☽ ⊼ ☉
18 Sunday	
00 15	☽ # ☿
01 43	☽ ⊼ ♅
01 54	☽ bQ ♂
05 10	☿ ‖ ♅
09 23	☽ ± ☿
11 14	☿ △ ♄
12 02	☽ ± ♅
16 58	♄ St R
18 17	☽ ± ♂
19 03	☿ △ ♃
19 18	☽ ☍ ♀
19 57	☽ □ ♄
20 54	☽ □ ♃
21 09	☽ ⊼ ☿
19 Monday	
02 29	♀ □ ♄
03 38	☽ bQ ♅
05 19	☽ ⊼ ♂
07 48	☿ Q ♅
09 23	☉ ⚹ ♅
13 15	♀ □ ♃
20 13	☽ ⊼ ♆
20 35	☉ ⊥ ♆
21 19	☽ # ♅
21 19	☽ □ ♇
20 Tuesday	
03 25	☽ Q ♄
04 26	☽ Q ♃
06 02	☽ △ ♅
06 50	☽ ± ♆
07 39	☽ ☍ ☉
17 17	☽ # ☿
22 59	☽ bQ ♆
21 Wednesday	
00 51	☽ ⚹ ♄
01 55	☽ ⚹ ♃
05 29	☽ Q ♇
05 42	☽ ⊼ ♀
09 18	☽ # ♂
09 47	☽ ☍ ☿
14 05	☽ ☍ ♂
17 48	☽ ± ♀
19 20	☉ ‖ ♅
22 Thursday	
02 26	☽ △ ♆
03 29	☽ ⚹ ♇
04 18	☽ ∠ ♄
05 25	☽ ∠ ♃
12 15	☽ bQ ♀
12 49	☽ □ ♅
18 05	♀ ∠ ♅
19 05	☽ ⊼ ☉
21 07	☽ ⊥ ♄
22 17	☽ ⊥ ♃
23 Friday	
07 27	☽ ± ☉
07 41	☽ ∠ ♇
08 30	☽ ⊻ ♄
09 41	☽ ⊻ ♃
19 52	☽ △ ♀
23 09	☽ Q ♅
24 Saturday	
00 33	☿ ☌ ♂
00 59	☽ ⊥ ♇
02 16	☽ bQ ☉
02 20	☽ ⊼ ♂
02 28	☽ ⊼ ☿
06 41	☽ ‖ ♇
11 41	☽ □ ♆
12 39	☽ ⊻ ♇
14 49	☽ ± ♂
15 54	☽ ± ☿
17 00	☿ ‖ ♂
19 28	♃ St R
22 38	☽ ⚹ ♅
25 Sunday	
04 41	☽ # ♃
09 47	☽ bQ ♂
10 23	☽ △ ☉
11 42	☽ # ♄
12 19	☽ bQ ☿
19 04	☽ ☌ ♄
20 22	☽ ☌ ♃
23 22	☽ Q ♆
26 Monday	
04 32	☽ ∠ ♅
05 57	☽ ‖ ♄
12 50	♇ St R
13 05	☽ ‖ ♃
13 54	☽ □ ♀
17 55	☽ △ ♂
22 41	☽ ⊥ ♅
22 47	☽ △ ☿
23 32	☽ ⚹ ♆
27 Tuesday	
00 23	☽ ☌ ♇
04 55	☿ ⚹ ♆
10 49	☽ ⊻ ♅
11 40	☿ △ ♇
12 00	☽ # ♇
17 38	☿ bQ ♄
28 Wednesday	
04 19	☽ □ ☉
04 34	☿ bQ ♃
05 55	☽ ∠ ♆
06 32	☽ Q ♀
07 25	☽ ⊻ ♄
08 46	☽ ⊻ ♃
15 59	♀ ‖ ♆
19 31	☽ ⊥ ♄
20 52	☽ ⊥ ♃
23 38	☉ ∠ ♆
29 Thursday	
00 07	☽ ⊥ ♆
06 28	☽ ‖ ☿
09 32	☽ ⚹ ♀
10 45	☽ □ ♂
12 12	☽ ⊻ ♆
12 54	☽ ⊻ ♇
13 34	☽ ∠ ♄
14 55	☽ ∠ ♃
16 03	☉ △ ♄
19 20	☽ ‖ ♂
19 26	☽ □ ☿
19 53	☽ Q ☉
23 23	☽ ☌ ♅
30 Friday	
00 52	☽ ⊥ ♇
07 45	☉ △ ♃
09 28	♂ ⚹ ♆
11 33	♀ ⊻ ♆
14 58	♀ ⊻ ♂
17 37	♀ □ ♇
18 41	☽ ∠ ♇
18 48	☽ ∠ ♀
19 09	♂ △ ♇
19 17	☽ ⚹ ♄
20 37	☽ ⚹ ♃
21 14	☽ ‖ ☉
21 49	☽ ⚹ ☉
31 Saturday	
00 56	☽ Q ♂
04 23	♂ bQ ♄
10 03	☿ □ ♅
10 58	☽ Q ☿
14 21	☽ ⊥ ♀
14 25	☉ Q ♅
18 37	☽ Q ♄
19 55	☽ Q ♃
23 18	☽ ☌ ♆
23 49	☽ ⚹ ♇

FEBRUARY 1981

LONGITUDES

Date	Sidereal time h m s	Sun ☉ ° ' "	Moon ☽ ° ' "	Moon ☽ 24.00 ° ' "	Mercury ☿ ° '	Venus ♀ ° '	Mars ♂ ° '	Jupiter ♃ ° '	Saturn ♄ ° '	Uranus ♅ ° '	Neptune ♆ ° '	Pluto ♇ ° '
01	20 46 26	12 ♒ 33 03	00 ♑ 43 49	07 ♑ 06 07	00 ♓ 51	26 ♑ 33	25 ♒ 41	10 ♎ 18	09 ♎ 37	29 ♏ 40	24 ♐ 05	24 ♎ 20
02	20 50 23	13 33 57	13 ♑ 33 15	20 ♑ 05 19	01 52	27 48	26 29	10 R 16	09 R 35	29 41	24 06	24 R 20
03	20 54 19	14 34 50	26 ♑ 42 23	03 ♒ 24 24	02 46	29 ♑ 03	27 16	10 14	09 34	29 43	24 08	24 20
04	20 58 16	15 35 42	10 ♒ 11 14	17 ♒ 02 37	03 31	00 ♒ 18	28 04	10 12	09 32	29 45	24 10	24 19
05	21 02 12	16 36 32	23 ♒ 58 14	00 ♓ 57 40	04 07	01 33	28 51	10 10	09 30	29 46	24 11	24 19
06	21 06 09	17 37 22	08 ♓ 00 27	15 ♓ 06 02	04 34	02 49	29 ♒ 39	10 08	09 28	29 48	24 13	24 19
07	21 10 05	18 38 10	22 ♓ 13 52	29 ♓ 23 21	04 50	04 04	00 ♓ 26	10 05	09 26	29 49	24 14	24 18
08	21 14 02	19 38 57	06 ♈ 33 54	13 ♈ 44 55	04 55	05 19	01 14	10 03	09 24	29 50	24 16	24 18
09	21 17 59	20 39 42	20 ♈ 55 53	28 ♈ 06 16	04 R 50	06 34	02 01	10 00	09 22	29 52	24 17	24 17
10	21 21 55	21 40 25	05 ♉ 15 39	12 ♉ 23 37	04 34	07 49	02 48	09 57	09 20	29 53	24 19	24 17
11	21 25 52	22 41 07	19 ♉ 29 51	26 ♉ 34 04	04 08	09 04	03 36	09 54	09 17	29 54	24 20	24 16
12	21 29 48	23 41 47	03 ♊ 36 02	10 ♊ 35 34	03 31	10 19	04 23	09 50	09 15	29 55	24 21	24 16
13	21 33 45	24 42 26	17 ♊ 32 32	24 ♊ 26 49	02 46	11 34	05 11	09 47	09 12	29 56	24 23	24 15
14	21 37 41	25 43 03	01 ♋ 18 19	08 ♋ 06 58	01 53	12 49	05 58	09 43	09 09	29 57	24 24	24 14
15	21 41 38	26 43 38	14 ♋ 52 40	21 ♋ 35 23	00 ♓ 53	14 04	06 45	09 39	09 07	29 58	24 25	24 14
16	21 45 34	27 44 12	28 ♋ 15 03	04 ♌ 51 35	29 ♒ 49	15 19	07 33	09 35	09 04	29 ♏ 59	24 27	24 13
17	21 49 31	28 44 44	11 ♌ 24 58	17 ♌ 55 07	28 42	16 34	08 20	09 31	09 01	00 ♐ 00	24 28	24 12
18	21 53 28	29 ♒ 45 14	24 ♌ 22 01	00 ♍ 45 40	27 34	17 49	09 08	09 26	08 58	00 01	24 29	24 12
19	21 57 24	00 ♓ 45 42	07 ♍ 06 02	13 ♍ 23 12	26 26	19 04	09 55	09 21	08 55	00 02	24 30	24 11
20	22 01 21	01 46 09	19 ♍ 37 12	25 ♍ 48 09	25 21	20 19	10 42	09 17	08 52	00 02	24 31	24 10
21	22 05 17	02 46 35	01 ♎ 56 13	08 ♎ 01 35	24 20	21 34	11 29	09 12	08 48	00 03	24 33	24 09
22	22 09 14	03 46 59	14 ♎ 04 31	20 ♎ 05 18	23 23	22 49	12 17	09 07	08 45	00 04	24 34	24 08
23	22 13 10	04 47 21	26 ♎ 04 16	02 ♏ 01 50	22 32	24 04	13 04	09 02	08 42	00 04	24 35	24 07
24	22 17 07	05 47 42	07 ♏ 58 25	13 ♏ 54 30	21 48	25 19	13 51	08 56	08 38	00 05	24 36	24 06
25	22 21 03	06 48 02	19 ♏ 50 35	25 ♏ 47 14	21 11	26 34	14 39	08 51	08 34	00 05	24 37	24 05
26	22 25 00	07 48 20	01 ♐ 45 00	07 ♐ 44 30	20 40	27 49	15 26	08 45	08 31	00 06	24 38	24 04
27	22 28 57	08 48 37	13 ♐ 46 19	19 ♐ 51 04	20 17	29 ♒ 04	16 13	08 39	08 27	00 06	24 39	24 03
28	22 32 53	09 ♓ 48 53	25 ♐ 59 21	02 ♑ 11 45	20 ♒ 02	00 ♓ 19	17 ♓ 00	08 ♎ 33	08 ♎ 23	00 ♐ 06	24 ♐ 40	24 ♎ 02

DECLINATIONS

Date	Moon True ☊ ° '	Moon Mean ☊ ° '	Moon ☽ Latitude ° '	Sun ☉ ° '	Moon ☽ ° '	Mercury ☿ ° '	Venus ♀ ° '	Mars ♂ ° '	Jupiter ♃ ° '	Saturn ♄ ° '	Uranus ♅ ° '	Neptune ♆ ° '	Pluto ♇ ° '
01	10 ♌ 55	10 ♌ 51	03 N 25	17 S 02	20 S 01	10 S 57	21 S 19	13 S 57	02 S 45	01 S 30	19 S 52	22 S 00	06 N 41
02	10 D 55	10 48	02 26	16 45	20 20	10 22	21 06	13 41	02 44	01 30	19 52	22 00	06 42
03	10 55	10 45	01 18	16 28	19 32	09 48	20 53	13 24	02 43	01 29	19 52	22 00	06 42
04	10 55	10 41	00 N 04	16 10	17 38	09 17	20 40	13 08	02 42	01 28	19 53	22 00	06 43
05	10 R 55	10 38	01 S 12	15 52	14 40	08 48	20 25	12 51	02 41	01 27	19 53	22 00	06 44
06	10 55	10 35	02 25	15 33	10 48	08 23	20 11	12 34	02 40	01 26	19 53	22 00	06 44
07	10 54	10 32	03 29	15 14	06 17	08 01	19 55	12 17	02 38	01 25	19 54	22 00	06 45
08	10 53	10 29	04 22	14 56	01 S 24	07 44	19 39	12 00	02 37	01 24	19 54	22 00	06 45
09	10 52	10 26	04 58	14 36	03 N 34	07 31	19 23	11 43	02 36	01 23	19 54	22 00	06 46
10	10 51	10 22	05 15	14 17	08 19	07 22	19 06	11 25	02 34	01 22	19 54	22 00	06 47
11	10 51	10 19	05 13	13 57	12 35	07 18	18 48	11 08	02 33	01 20	19 55	22 00	06 47
12	10 D 51	10 16	04 52	13 37	16 06	07 19	18 30	10 50	02 31	01 19	19 55	22 01	06 48
13	10 51	10 13	04 14	13 17	18 38	07 25	18 11	10 33	02 30	01 18	19 55	22 01	06 49
14	10 53	10 10	03 22	12 57	20 04	07 35	17 52	10 15	02 28	01 17	19 55	22 01	06 50
15	10 54	10 06	02 19	12 36	20 18	07 49	17 32	09 57	02 26	01 15	19 55	22 01	06 50
16	10 55	10 03	01 S 09	12 16	19 23	08 06	17 12	09 39	02 24	01 14	19 56	22 01	06 51
17	10 R 55	10 00	00 N 03	11 55	17 24	08 26	16 51	09 21	02 23	01 13	19 56	22 01	06 52
18	10 55	09 57	01 14	11 34	14 33	08 49	16 29	09 03	02 21	01 11	19 56	22 01	06 52
19	10 53	09 54	02 19	11 12	11 04	09 13	16 08	08 45	02 19	01 10	19 56	22 01	06 53
20	10 51	09 51	03 17	10 51	07 08	09 39	15 46	08 27	02 17	01 08	19 56	22 01	06 54
21	10 47	09 47	04 05	10 29	02 N 58	10 05	15 23	08 08	02 14	01 07	19 56	22 01	06 55
22	10 43	09 44	04 41	10 07	01 S 14	10 30	15 00	07 50	02 12	01 05	19 57	22 01	06 55
23	10 39	09 41	05 04	09 45	05 21	10 55	14 37	07 32	02 10	01 04	19 57	22 01	06 56
24	10 36	09 38	05 13	09 23	09 13	11 19	14 13	07 13	02 08	01 02	19 57	22 01	06 57
25	10 33	09 35	05 10	09 01	12 43	11 42	13 49	06 54	02 05	01 01	19 57	22 01	06 58
26	10 32	09 32	04 53	08 39	15 44	12 03	13 24	06 36	02 03	00 59	19 57	22 01	06 58
27	10 D 32	09 28	04 23	08 16	18 06	12 23	12 59	06 17	02 01	00 58	19 57	22 01	06 59
28	10 ♌ 33	09 ♌ 25	03 N 40	07 S 53	19 S 42	12 S 40	12 S 34	05 S 58	01 S 58	00 S 56	19 S 57	22 S 01	07 N 00

ZODIAC SIGN ENTRIES

Date	h	m	Planets
01	10	37	☽ ♑
03	17	55	☽ ♒
04	06	07	♀ ♒
05	22	21	☽ ♓
06	22	48	♂ ♓
08	01	01	☽ ♈
10	03	11	☽ ♉
12	05	51	☽ ♊
14	09	43	☽ ♋
16	08	02	☿ ♒
16	15	10	☽ ♌
17	09	02	♅ ♐
18	17	52	☉ ♓
18	22	34	☽ ♍
21	08	12	☽ ♎
23	19	54	☽ ♏
26	08	29	☽ ♐
28	06	01	♀ ♓
28	19	46	☽ ♑

LATITUDES

Date	Mercury ☿ ° '	Venus ♀ ° '	Mars ♂ ° '	Jupiter ♃ ° '	Saturn ♄ ° '	Uranus ♅ ° '	Neptune ♆ ° '	Pluto ♇ ° '
01	00 N 14	00 S 28	01 S 04	01 N 27	02 N 30	00 N 14	01 N 18	17 N 18
04	01 01	00 35	01 03	01 28	02 31	00 14	01 19	17 20
07	01 51	00 42	01 02	01 28	02 32	00 14	01 19	17 22
10	02 39	00 48	01 01	01 29	02 32	00 14	01 19	17 23
13	03 18	00 54	01 00	01 30	02 33	00 14	01 19	17 25
16	03 40	01 00	00 59	01 31	02 34	00 14	01 19	17 26
19	03 42	01 05	00 58	01 31	02 35	00 14	01 19	17 28
22	03 24	01 09	00 57	01 32	02 35	00 14	01 19	17 29
25	02 53	01 13	00 56	01 32	02 36	00 14	01 19	17 31
28	02 15	01 17	00 55	01 33	02 36	00 14	01 19	17 32
31	01 N 34	01 S 20	00 S 54	01 N 33	02 N 37	00 N 14	01 N 19	17 N 33

DATA

Julian Date	2444637
Delta T	+51 seconds
Ayanamsa	23° 35' 22"
Synetic vernal point	05° ♓ 31' 37"
True obliquity of ecliptic	23° 26' 24"

MOON'S PHASES, APSIDES AND POSITIONS ☽

Date	h	m	Phase	Longitude ° '	Eclipse Indicator
04	22	14	●	16 ♒ 02	Annular
11	17	49	☽	22 ♉ 56	
18	22	58	○	00 ♍ 13	
27	01	14	☾	08 ♐ 22	

Day	h	m		
08	22	28	Perigee	
24	16	42	Apogee	
02	06	56	Max dec	20° S 21'
08	18	44	0N	
15	04	45	Max dec	20° N 22'
22	04	56	0S	

LONGITUDES

Date	Chiron ⚷ ° '	Ceres ⚳ ° '	Pallas ⚴ ° '	Juno ⚵ ° '	Vesta ⚶ ° '	Black Moon Lilith ⚸ ° '
01	13 ♉ 27	15 ♋ 56	26 ♈ 38	06 ♏ 26	08 ♍ 02	03 ♏ 45
11	13 ♉ 36	14 ♋ 29	00 ♉ 26	07 ♏ 23	05 ♍ 53	04 ♏ 52
21	13 ♉ 52	13 ♋ 43	04 ♉ 38	07 ♏ 47	03 ♍ 21	05 ♏ 59
31	14 ♉ 12	13 ♋ 40	09 ♉ 10	07 ♏ 37	00 ♍ 44	07 ♏ 06

ASPECTARIAN

h m Aspects

01 Sunday
00 26 ♂ bQ ♃
01 45 ☽ ✶ ♂
03 10 ☽ ⊻ ♀
05 26 ☽ ∠ ☉
07 43 ☽ ∥ ♅
09 58 ☽ ⊻ ♅
12 16 ☽ ✶ ☿
21 19 ☽ ⊥ ♅
22 34 ☽ Q ♇
23 55 ☽ ⊥ ☉

02 Monday
04 39 ☽ □ ♄
05 55 ☽ □ ♃
07 54 ☽ ∠ ♂
12 01 ☽ ⊻ ☉
14 06 ☽ ∠ ♅
18 35 ☽ ∠ ☿

03 Tuesday
01 31 ☽ ⊥ ♂
05 21 ☽ ∥ ♅
07 21 ☽ ⊻ ♆
07 42 ☽ □ ♇
12 06 ☽ ⊥ ☿
13 05 ☽ ⊻ ♂
16 39 ☽ ☌ ♀
17 25 ☽ ✶ ♅
18 10 ☽ ⊥ ♆
23 32 ☽ ⊻ ☿

04 Wednesday
00 59 ♀ ✶ ♅
09 09 ♀ ⊥ ♆
10 11 ☽ ∠ ♆
10 51 ☽ △ ♄
12 02 ☽ △ ♃
12 39 ☿ ± ♄
14 44 ☽ Q ♅
22 14 ☽ ☌ ☉

05 Thursday
02 15 ☽ ∥ ☉
12 22 ☽ ✶ ♆
12 36 ☽ △ ♇
12 55 ☽ bQ ♄
14 04 ☽ bQ ♃
14 22 ☿ ± ♃
20 54 ☽ ☌ ♂
21 59 ☽ □ ♅

06 Friday
00 47 ☽ ∥ ♂
02 18 ☽ ⊻ ♀
04 18 ☽ ± ♄
05 26 ☽ ± ♃
05 59 ☽ ☌ ☿
08 57 ☽ Q ♆
13 30 ☽ ⊥ ♀
14 13 ☽ bQ ♇
14 29 ☽ ⊼ ♄
15 36 ☽ ⊼ ♃
16 40 ♂ □ ♅

07 Saturday
02 23 ☽ ∥ ☿
05 24 ☽ ± ♇
05 30 ☽ ⊻ ☉
06 10 ☽ ∠ ♀
09 41 ☽ # ♇
14 35 ♀ ∥ ♅
15 22 ☽ □ ♆
15 29 ☽ ⊼ ♇
16 21 ☽ ⊥ ☉

08 Sunday
00 44 ☽ △ ♅
02 33 ☽ ⊻ ♂
04 21 ☿ ⊻ ♀
06 05 ☽ ∥ ♃
08 33 ☽ ∠ ☉
09 15 ☽ ⊻ ☿
09 42 ☽ ✶ ♀
12 01 ☽ ∥ ♄
12 31 ☿ St R
13 10 ☽ ⊥ ♂
16 44 ☽ ☍ ♄
17 48 ☽ ☍ ♃
19 16 ☽ ⊥ ☿

09 Monday
01 25 ☽ # ♄
01 51 ☽ bQ ♅
05 05 ☽ ∠ ♂
07 18 ☽ # ♃
07 40 ☽ Q ♀
10 12 ☽ ∠ ☿
11 31 ☽ ✶ ☉
12 39 ♆ ✶ ♇
16 54 ☽ ± ♅
17 37 ☽ ☍ ♇
17 37 ☽ △ ♆

10 Tuesday
02 58 ☽ ⊼ ♅
03 59 ☽ ∥ ♇
07 09 ☽ # ☿
07 38 ☽ ✶ ♂
09 08 ☽ Q ☉
10 52 ☽ ✶ ☿
16 42 ☽ □ ♀
18 49 ☽ ⊼ ♄
18 49 ☽ bQ ♆
19 51 ☽ ⊼ ♃

11 Wednesday
02 59 ♂ ± ♄
03 59 ☽ # ♂
04 54 ☽ ± ♄
05 01 ☽ Q ♂
05 56 ☽ ± ♃
06 30 ☽ Q ☿
10 02 ☽ ± ♆
16 06 ♀ △ ♄
17 16 ♀ ∠ ♆
17 49 ☽ □ ☉
19 59 ☽ # ☉
20 05 ☽ ⊼ ♇
20 06 ☽ bQ ♄
20 13 ☽ ⊼ ♆
20 22 ♂ ± ♃
21 07 ☽ bQ ♃
21 26 ☿ ☌ ♂
23 08 ☿ ± ♃

12 Thursday
03 12 ♀ △ ♃
05 42 ☽ ☍ ♅
06 18 ☽ ± ♇
11 52 ☽ □ ☿
13 26 ☽ □ ♂
21 39 ☽ △ ♄
21 42 ☽ bQ ♇
21 55 ☿ ± ♄
22 39 ☽ △ ♃

13 Friday
00 29 ☉ bQ ♄
00 40 ☽ △ ♀
01 17 ☉ △ ♇
04 04 ☉ ✶ ♆
07 28 ☽ # ♀
13 33 ☉ bQ ♃
19 14 ♀ Q ♅
23 39 ☽ △ ♇
23 54 ☽ ☍ ♆

14 Saturday
01 26 ☽ △ ☉
05 16 ☽ bQ ♀
08 11 ☽ # ♅
09 38 ☽ ⊼ ♅
12 57 ☽ △ ☿
20 12 ☽ ± ♅
20 43 ☽ △ ♂
22 41 ☽ ± ♀

15 Sunday
01 33 ♂ Q ♆
01 48 ☽ □ ♄
02 45 ☽ □ ♃
05 57 ☽ bQ ☉
10 25 ☽ ⊼ ♀
12 10 ☽ bQ ♅
13 40 ☽ bQ ☿
23 18 ☽ ± ☉

16 Monday
01 04 ☽ bQ ♂
01 24 ☽ # ♅
04 37 ☽ ± ☿
04 44 ☽ □ ♇
05 07 ☽ ⊼ ♆
08 22 ☿ □ ♅
09 52 ☽ Q ♄
10 48 ☽ Q ♃
11 00 ☽ ⊼ ☉
14 37 ☽ ⊼ ☿
15 09 ☽ △ ♅
15 59 ☽ ± ♆
18 21 ☽ ± ♂

17 Tuesday
05 59 ☽ ⊼ ♂
07 37 ☽ ✶ ♄
08 25 ☽ bQ ♆
08 31 ☽ ✶ ♃
11 31 ☉ ☌ ☿
13 27 ☽ Q ♇
18 03 ☽ # ♀
22 31 ☽ ☍ ♀

18 Wednesday
07 21 ♂ ⊼ ♄
11 15 ☽ ∠ ♄
11 40 ☽ ✶ ♇
12 08 ☽ ∠ ♃
12 13 ☽ △ ♆
14 00 ♂ bQ ♇
17 30 ☽ ☍ ☿
18 19 ☉ □ ♅
20 01 ☿ ∥ ♂
20 34 ♂ ⊼ ♃
22 36 ☽ □ ♅
22 58 ☽ ☍ ☉

19 Thursday
04 06 ☽ ⊥ ♄
04 57 ☽ ⊥ ♃
10 59 ☽ # ☉
15 26 ☽ ⊻ ♄
15 57 ☽ ∠ ♇
16 16 ☽ ⊻ ♃
17 43 ☽ ☍ ♂
22 22 ☽ # ☿

20 Friday
03 33 ☽ # ♂
08 57 ☽ Q ♅
09 12 ☽ ⊥ ♇
13 23 ☽ ∥ ♇
13 30 ☽ ⊼ ♀
20 48 ☽ ⊻ ♇
21 32 ☽ □ ♆
22 15 ☽ ⊼ ☿

21 Saturday
02 28 ☽ ± ♀
06 55 ☿ ✶ ♆
08 18 ☽ ✶ ♅
09 05 ☽ ± ☿
12 38 ☉ ± ♄
13 48 ☽ ⊼ ☉
15 30 ☿ bQ ♃
16 13 ☽ # ♃
16 23 ☿ △ ♇
21 16 ☉ ± ♃
22 10 ☽ bQ ♀
22 38 ☽ # ♄

22 Sunday
00 19 ☉ ∥ ☿
01 29 ☽ ☌ ♄
01 30 ☽ bQ ☿
01 50 ☿ bQ ♄
02 13 ☽ ☌ ♃
02 42 ☽ ± ☉
08 11 ☽ ⊼ ♂
08 59 ☽ Q ♆
11 10 ☽ ∥ ♄
13 58 ☽ ∠ ♅
17 31 ☽ ∥ ♃
18 23 ☿ ☌ ♀
20 58 ☽ ± ♂
22 15 ☽ bQ ☉

23 Monday
05 08 ♀ bQ ♄
05 21 ☽ △ ☿
07 30 ☽ △ ♀
07 59 ☽ ⊥ ♅
08 05 ☽ ☌ ♇
09 00 ☽ ✶ ♆
11 17 ♀ bQ ♃
13 03 ♀ △ ♇
16 18 ☽ bQ ♂
20 03 ☽ ⊻ ♅
21 39 ☽ # ♇
22 00 ♀ ✶ ♆

24 Tuesday
00 17 ☽ ∥ ♂
07 11 ☽ △ ☉
12 59 ☽ ∥ ☉
13 20 ☽ ⊻ ♄
13 56 ☽ ⊻ ♃
15 17 ☽ ∠ ♆

25 Wednesday
00 44 ☽ △ ♂
01 24 ☽ ⊥ ♄
01 58 ☽ ⊥ ♃
03 48 ☽ ∥ ☿
07 26 ☉ Q ♆
08 04 ♂ # ♇
09 31 ☽ ⊥ ♆
14 34 ☽ □ ☿
15 45 ♅ ⊥ ♇
19 12 ☽ ∥ ♀
19 30 ☽ ∠ ♄
20 01 ☽ ∠ ♃
20 34 ☽ ⊻ ♇
21 39 ☽ ⊻ ♆

26 Thursday
03 09 ☽ □ ♀
08 38 ☽ ⊥ ♇
08 40 ☽ ☌ ♅

27 Friday
01 14 ☽ □ ☉
01 24 ☽ Q ☿
01 28 ☽ ✶ ♄
01 54 ☽ ✶ ♃
02 38 ☽ ∠ ♇
03 56 ☉ ⊼ ♄
08 36 ☉ ⊼ ♃
17 10 ☽ □ ♂
17 45 ☉ bQ ♇
19 15 ☽ Q ♀

28 Saturday
00 34 ☽ ✶ ☿
01 07 ☽ Q ♄
01 28 ☽ Q ♃
07 59 ♀ □ ♅
08 07 ☿ ∥ ♀
08 12 ☽ ✶ ♇
09 24 ☽ ☌ ♆
15 51 ☽ Q ☉
17 32 ☽ ∥ ♅
19 58 ☽ ⊻ ♅
21 18 ☽ ✶ ♀

All ephemeris data is given at 12.00 UT and the Moon's longitude is additionally given for 24.00 UT

MARCH 1981

Date	Sidereal time h m s	LONGITUDES Sun ☉ ° ' "	Moon ☽ ° ' "	Moon ☽ 24.00 ° ' "	Mercury ☿ ° '	Venus ♀ ° '	Mars ♂ ° '	Jupiter ♃ ° '	Saturn ♄ ° '	Uranus ♅ ° '	Neptune ♆ ° '	Pluto ♇ ° '
01	22 36 50	10 ♓ 49 06	08 ♑ 28 47	14 ♑ 51 00	19 ♒ 53	01 ♓ 34	17 ♓ 47	08 ♎ 27	08 ♎ 20	00 ♐ 06	24 ♐ 40	24 ♎ 01
02	22 40 46	11 49 19	21 ♑ 18 48	27 ♑ 52 34	19 D 51	02 48	18 34	08 R 21	08 R 16	00 07	24 41	24 R 00
03	22 44 43	12 49 30	04 ♒ 32 32	11 ♒ 18 50	19 56	04 03	19 21	08 15	08 12	00 07	24 42	23 59
04	22 48 39	13 49 39	18 ♒ 11 30	25 ♒ 10 21	20 06	05 18	20 09	08 08	08 08	00 07	24 43	23 58
05	22 52 36	14 49 46	02 ♓ 15 05	09 ♓ 25 14	20 23	06 33	20 56	08 02	08 04	00 R 07	24 44	23 57
06	22 56 32	15 49 52	16 ♓ 40 09	23 ♓ 59 05	20 44	07 48	21 43	07 55	07 59	00 07	24 44	23 55
07	23 00 29	16 49 55	01 ♈ 21 08	08 ♈ 45 17	21 11	09 03	22 30	07 48	07 55	00 07	24 45	23 54
08	23 04 26	17 49 57	16 ♈ 10 32	23 ♈ 35 48	21 43	10 18	23 17	07 42	07 51	00 06	24 46	23 53
09	23 08 22	18 49 57	01 ♉ 00 07	08 ♉ 22 31	22 19	11 33	24 04	07 35	07 47	00 06	24 46	23 52
10	23 12 19	19 49 54	15 ♉ 42 12	22 ♉ 58 27	22 59	12 47	24 51	07 28	07 42	00 06	24 47	23 50
11	23 16 15	20 49 50	00 ♊ 10 44	07 ♊ 18 38	23 43	14 02	25 37	07 21	07 38	00 06	24 47	23 49
12	23 20 12	21 49 43	14 ♊ 21 54	21 ♊ 20 25	24 31	15 17	26 24	07 13	07 34	00 05	24 48	23 48
13	23 24 08	22 49 34	28 ♊ 14 10	05 ♋ 03 13	25 22	16 32	27 11	07 06	07 29	00 05	24 48	23 46
14	23 28 05	23 49 23	11 ♋ 47 45	18 ♋ 27 57	26 17	17 47	27 58	06 59	07 25	00 04	24 49	23 45
15	23 32 01	24 49 09	25 ♋ 04 05	01 ♌ 36 23	27 13	19 01	28 45	06 51	07 20	00 04	24 49	23 43
16	23 35 58	25 48 53	08 ♌ 05 07	14 ♌ 30 33	28 13	20 16	29 ♓ 31	06 44	07 16	00 03	24 50	23 42
17	23 39 55	26 48 35	20 ♌ 52 54	27 ♌ 12 24	29 ♒ 15	21 31	00 ♈ 18	06 36	07 11	00 03	24 50	23 41
18	23 43 51	27 48 15	03 ♍ 29 13	09 ♍ 43 32	00 ♓ 20	22 45	01 05	06 29	07 07	00 02	24 50	23 39
19	23 47 48	28 47 53	15 ♍ 55 30	22 ♍ 05 13	01 28	24 00	01 51	06 21	07 02	00 01	24 50	23 38
20	23 51 44	29 ♓ 47 28	28 ♍ 12 50	04 ♎ 18 27	02 37	25 15	02 38	06 14	06 57	00 00	24 51	23 36
21	23 55 41	00 ♈ 47 02	10 ♎ 22 12	16 ♎ 24 11	03 49	26 29	03 25	06 06	06 53	00 ♐ 00	24 51	23 35
22	23 59 37	01 46 33	22 ♎ 24 35	28 ♎ 23 33	05 03	27 44	04 11	05 58	06 48	29 ♏ 59	24 51	23 33
23	00 03 34	02 46 03	04 ♏ 21 19	10 ♏ 18 07	06 18	28 ♓ 59	04 58	05 51	06 43	29 58	24 51	23 32
24	00 07 30	03 45 31	16 ♏ 14 15	22 ♏ 10 03	07 36	00 ♈ 13	05 44	05 43	06 39	29 57	24 51	23 30
25	00 11 27	04 44 57	28 ♏ 05 53	04 ♐ 02 12	08 55	01 28	06 31	05 35	06 34	29 56	24 51	23 28
26	00 15 24	05 44 21	09 ♐ 59 27	15 ♐ 58 09	10 16	02 43	07 17	05 27	06 29	29 55	24 51	23 27
27	00 19 20	06 43 43	21 ♐ 58 53	28 ♐ 02 11	11 39	03 57	08 03	05 20	06 25	29 54	24 R 51	23 25
28	00 23 17	07 43 04	04 ♑ 08 42	10 ♑ 19 03	13 03	05 12	08 50	05 12	06 20	29 52	24 51	23 24
29	00 27 13	08 42 23	16 ♑ 33 51	22 ♑ 53 43	14 29	06 26	09 36	05 04	06 15	29 51	24 51	23 22
30	00 31 10	09 41 40	29 ♑ 19 14	05 ♒ 50 56	15 56	07 41	10 22	04 56	06 11	29 50	24 51	23 20
31	00 35 06	10 ♈ 40 55	12 ♒ 29 16	19 ♒ 14 36	17 ♓ 25	08 ♈ 55	11 ♈ 08	04 ♎ 49	06 ♎ 06	29 ♏ 49	24 ♐ 51	23 ♎ 19

Date	Moon True ☊ ° '	Moon Mean ☊ ° '	Moon ☽ Latitude ° '
01	10 ♌ 35	09 ♌ 22	02 N 47
02	10 D 36	09 19	01 44
03	10 37	09 16	00 N 33
04	10 R 37	09 12	00 S 41
05	10 36	09 09	01 55
06	10 32	09 06	03 04
07	10 27	09 03	04 01
08	10 22	09 00	04 43
09	10 16	08 57	05 06
10	10 11	08 53	05 09
11	10 07	08 50	04 51
12	10 06	08 47	04 17
13	10 D 05	08 44	03 27
14	10 06	08 41	02 27
15	10 08	08 38	01 21
16	10 09	08 34	00 S 11
17	10 R 08	08 31	00 N 58
18	10 05	08 28	02 03
19	10 00	08 25	03 00
20	09 53	08 22	03 49
21	09 44	08 18	04 27
22	09 34	08 15	04 52
23	09 24	08 12	05 04
24	09 14	08 09	05 03
25	09 06	08 06	04 49
26	09 01	08 03	04 22
27	08 57	07 59	03 44
28	08 56	07 56	02 55
29	08 D 56	07 53	01 57
30	08 57	07 50	00 51
31	08 ♌ 57	07 ♌ 47	00 S 19

DECLINATIONS

Date	Sun ☉ ° '	Moon ☽ ° '	Mercury ☿ ° '	Venus ♀ ° '	Mars ♂ ° '	Jupiter ♃ ° '	Saturn ♄ ° '	Uranus ♅ ° '	Neptune ♆ ° '	Pluto ♇ ° '
01	07 S 31	20 S 24	12 S 56	12 S 08	05 S 40	01 S 56	00 S 54	19 S 57	22 S 01	07 N 01
02	07 08	20 03	13 10	11 42	05 21	01 53	00 53	19 57	22 01	07 01
03	06 45	18 35	13 21	11 16	05 02	01 50	00 51	19 57	22 01	07 02
04	06 22	16 02	13 31	10 49	04 43	01 48	00 49	19 57	22 01	07 03
05	05 59	12 28	13 38	10 22	04 24	01 45	00 47	19 57	22 01	07 04
06	05 35	08 05	13 44	09 55	04 05	01 42	00 46	19 57	22 01	07 05
07	05 12	03 S 09	13 48	09 28	03 46	01 39	00 44	19 57	22 01	07 05
08	04 49	02 N 00	13 50	09 00	03 27	01 37	00 42	19 57	22 01	07 06
09	04 25	07 02	13 49	08 32	03 08	01 34	00 40	19 57	22 01	07 07
10	04 02	11 37	13 48	08 04	02 49	01 31	00 38	19 57	22 01	07 08
11	03 38	15 26	13 44	07 35	02 30	01 28	00 37	19 57	22 01	07 09
12	03 14	18 17	13 38	07 07	02 11	01 25	00 35	19 57	22 01	07 09
13	02 51	19 58	13 31	06 38	01 52	01 22	00 33	19 57	22 01	07 10
14	02 27	20 28	13 23	06 09	01 33	01 19	00 31	19 57	22 01	07 11
15	02 04	19 48	13 12	05 40	01 14	01 16	00 29	19 57	22 01	07 12
16	01 40	18 04	13 00	05 11	00 55	01 13	00 27	19 56	22 01	07 12
17	01 16	15 27	12 47	04 41	00 36	01 10	00 25	19 56	22 01	07 13
18	00 52	12 08	12 32	04 12	00 S 17	01 07	00 24	19 56	22 00	07 14
19	00 29	08 19	12 15	03 42	00 N 02	01 04	00 22	19 56	22 00	07 15
20	00 S 05	04 N 13	11 57	03 12	00 21	01 01	00 20	19 56	22 00	07 15
21	00 N 19	00 S 01	11 37	02 43	00 40	00 58	00 18	19 56	22 00	07 16
22	00 42	04 12	11 17	02 13	00 59	00 55	00 16	19 55	22 00	07 17
23	01 06	08 12	10 54	01 43	01 18	00 52	00 14	19 55	22 00	07 18
24	01 30	11 51	10 31	01 13	01 37	00 48	00 12	19 55	22 00	07 18
25	01 53	15 02	10 06	00 43	01 55	00 45	00 10	19 55	22 00	07 19
26	02 17	17 37	09 39	00 S 13	02 14	00 42	00 08	19 55	22 00	07 20
27	02 40	19 28	09 12	00 N 18	02 33	00 39	00 07	19 54	22 00	07 21
28	03 04	20 28	08 43	00 48	02 52	00 36	00 05	19 54	22 00	07 21
29	03 27	20 29	08 13	01 18	03 10	00 33	00 03	19 54	22 00	07 22
30	03 50	19 27	07 41	01 48	03 29	00 30	00 01	19 54	22 00	07 23
31	04 N 14	17 S 22	07 S 09	02 N 18	03 N 48	00 S 27	00 N 01	19 S 53	22 S 00	07 N 24

ZODIAC SIGN ENTRIES

Date	h	m	Planets
03	03	51	☽ ♒
05	08	12	☽ ♓
07	09	48	☽ ♈
09	10	22	☽ ♉
11	11	42	☽ ♊
13	15	06	☽ ♋
15	21	02	☽ ♌
17	02	40	♂ ♈
18	04	33	☿ ♓
18	05	20	☽ ♍
20	15	31	☽ ♎
20	17	03	☉ ♈
20	23	15	♅ ♏
23	03	14	☽ ♏
24	07	43	♀ ♈
25	15	51	☽ ♐
28	03	52	☽ ♑
30	13	15	☽ ♒

LATITUDES

Date	Mercury ☿ ° '	Venus ♀ ° '	Mars ♂ ° '	Jupiter ♃ ° '	Saturn ♄ ° '	Uranus ♅ ° '	Neptune ♆ ° '	Pluto ♇ ° '
01	02 N 01	01 S 18	00 S 54	01 N 33	02 N 37	00 N 14	01 N 19	17 N 32
04	01 20	01 21	00 53	01 34	02 37	00 14	01 20	17 34
07	00 41	01 23	00 52	01 34	02 37	00 14	01 20	17 35
10	00 N 04	01 25	00 50	01 34	02 38	00 14	01 20	17 36
13	00 S 29	01 26	00 49	01 35	02 38	00 14	01 20	17 37
16	00 58	01 26	00 48	01 35	02 39	00 14	01 20	17 38
19	01 23	01 26	00 46	01 35	02 39	00 14	01 20	17 38
22	01 44	01 26	00 45	01 35	02 39	00 14	01 20	17 39
25	02 01	01 25	00 43	01 35	02 39	00 14	01 20	17 40
28	02 14	01 23	00 41	01 36	02 39	00 14	01 20	17 40
31	02 S 22	01 S 21	00 S 40	01 N 36	02 N 39	00 N 14	01 N 20	17 N 41

DATA

Julian Date	2444665
Delta T	+51 seconds
Ayanamsa	23° 35' 26"
Synetic vernal point	05° ♓ 31' 34"
True obliquity of ecliptic	23° 26' 25"

LONGITUDES

Date	Chiron ⚷ ° '	Ceres ⚳ ° '	Pallas ⚴ ° '	Juno ⚵ ° '	Vesta ⚶ ° '	Black Moon Lilith ⚸ ° '
01	14 ♉ 08	13 ♋ 37	08 ♉ 14	07 ♏ 42	01 ♍ 15	06 ♏ 53
11	14 ♉ 32	14 ♋ 08	13 ♉ 01	07 ♏ 03	28 ♌ 50	08 ♏ 00
21	15 ♉ 01	15 ♋ 18	18 ♉ 03	05 ♏ 50	26 ♌ 57	09 ♏ 07
31	15 ♉ 34	17 ♋ 02	23 ♉ 19	04 ♏ 07	25 ♌ 47	10 ♏ 15

MOON'S PHASES, APSIDES AND POSITIONS ☽

Date	h	m	Phase	Longitude ° '	Eclipse Indicator
06	10	31	●	15 ♓ 46	
13	01	51	☽	22 ♊ 24	
20	15	22	○	29 ♍ 56	
28	19	34	☾	08 ♑ 02	

Day	h	m		
08	12	03	Perigee	
24	08	39	Apogee	
01	16	13	Max dec	20° S 25'
08	02	42	0N	
14	09	57	Max dec	20° N 28'
21	11	54	0S	
29	00	33	Max dec	20° S 36'

All ephemeris data is given at 12.00 UT and the Moon's longitude is additionally given for 24.00 UT

ASPECTARIAN

h m	Aspects	h m	Aspects	h m	Aspects
01 Sunday		10 03	♂ □ ♆	10 12	☽ # ♄
05 12	☽ ∠ ☿	17 04	☽ ± ♆	10 18	☽ ∥ ☉
06 32	☽ Q ♂	19 18	☽ ✶ ☉	10 47	☽ ± ☿
07 20	☽ Q ♇	23 04	☽ bQ ♃	11 12	☉ # ♄
07 29	☽ ⊥ ♅	23 30	☽ bQ ♄	13 35	☽ ∥ ♄
11 43	☽ □ ♄	**11 Wednesday**		13 51	☽ # ☉
11 57	☽ □ ♃	00 39	☽ □ ☿	15 59	☽ # ♂
16 48	☽ ✶ ☉	00 48	☽ # ☿	16 56	☽ Q ♆
18 56	♂ ± ♇	01 25	☽ ⊼ ♇	17 18	☽ ∥ ♃
22 08	☽ ⊥ ☿	03 00	☽ ⊼ ♆	21 11	☽ ∠ ♅
02 Monday		03 58	☽ ✶ ♂	**22 Sunday**	
00 29	☽ ∠ ♅	04 26	☽ Q ♀	01 44	☽ ∥ ♀
02 00	♀ ± ♄	11 24	☽ ± ♇	06 43	☽ bQ ☿
03 54	♀ ± ♃	11 51	☽ ☍ ♅	07 22	♂ # ♃
04 49	☽ ∠ ♀	14 56	☿ △ ♇	14 17	☽ ☌ ♇
06 36	☽ ✶ ♂	16 47	☽ Q ☉	15 08	☽ ⊥ ♅
07 06	☿ St D	23 57	☽ △ ♃	16 54	☽ ✶ ♆
09 18	☽ ⊻ ☿	**12 Thursday**		22 58	☉ # ♃
14 18	☽ ∥ ♅	00 29	☽ △ ♄	23 55	☽ ⊼ ♀
16 56	☽ □ ♇	01 16	☽ Q ♂	**23 Monday**	
18 12	☽ ⊻ ♆	02 32	☽ bQ ♇	03 10	☽ ⊻ ♅
18 19	☉ # ♇	09 59	♀ # ♇	04 06	☿ ⊼ ♃
22 55	☽ ∠ ☉	13 44	☽ □ ♀	06 25	☽ # ♇
23 07	☽ ⊥ ♀	20 16	☿ ✶ ♆	08 31	☽ ⊼ ☉
03 Tuesday		**13 Friday**		13 18	☽ ⊼ ♂
04 03	☽ ✶ ♅	01 51	☽ □ ☉	13 24	☽ ± ♀
05 06	☽ ⊥ ♆	04 14	☽ △ ♇	14 58	☽ ⊻ ♃
11 03	☽ ⊻ ♀	06 01	☽ ☍ ♆	16 24	☽ △ ☿
11 39	☽ ∠ ♂	06 39	☽ △ ☿	16 45	☽ ⊻ ♄
16 24	☽ ⊥ ☉	10 03	☽ □ ♂	19 27	☿ ⊼ ♄
18 28	☽ △ ♄	11 25	☽ # ♅	21 43	☽ ± ☉
18 32	☽ △ ♃	15 14	☽ ⊼ ♅	22 20	☿ Q ♆
21 10	☽ ∠ ♆	**14 Saturday**		23 06	☽ ∠ ♆
04 Wednesday		01 49	☽ ± ♅	**24 Tuesday**	
01 24	☽ Q ♅	03 29	☽ □ ♃	00 16	♀ # ♂
03 49	☽ ⊻ ☉	04 14	☽ □ ♄	02 16	☽ ± ♂
04 32	☽ ⊥ ♂	10 14	☉ ⊼ ♇	02 57	☽ ⊥ ♃
10 16	☿ ⊻ ♂	10 59	☽ bQ ☿	03 47	☽ ∥ ☿
15 22	☽ ☌ ☿	11 28	♀ ± ♇	04 26	☉ # ♀
15 34	☽ ⊻ ♂	17 52	☽ bQ ♅	04 47	☽ ⊥ ♄
19 08	♃ ☌ ♄	23 52	☽ △ ♀	06 44	♀ △ ♅
20 28	☽ bQ ♃	**15 Sunday**		09 42	☽ bQ ♀
20 28	☽ bQ ♄	04 26	☽ ± ☿	11 25	♂ ☍ ♃
21 55	☽ △ ♇	08 58	☽ # ♅	17 18	☽ ⊥ ♆
23 14	☽ ✶ ♆	09 03	♂ ∥ ♃	17 34	☽ bQ ☉
05 Thursday		09 33	☽ □ ♇	20 57	☽ ∠ ♃
01 46	♅ St R	11 31	☽ △ ☉	21 44	☽ bQ ♂
04 57	☽ ∥ ☿	11 33	☽ ⊼ ♆	22 52	☽ ∠ ♄
08 23	☽ □ ♅	11 37	☽ Q ♃	**25 Wednesday**	
11 38	☽ ± ♃	12 01	☉ □ ♆	02 40	☽ ⊻ ♇
11 41	☽ ± ♄	12 29	☽ Q ♄	04 13	☿ bQ ♇
15 22	♀ Q ♆	16 15	☽ ⊼ ☿	05 27	☽ ⊻ ♆
19 31	☽ Q ♆	19 10	☽ △ ♂	09 36	♀ ∥ ♃
19 54	☽ ☌ ♀	21 09	☽ △ ♅	13 41	♂ ☍ ♄
21 37	☽ ⊼ ♃	22 33	☽ ± ♆	14 47	☽ ⊥ ♇
21 41	☽ ⊼ ♄	**16 Monday**		15 42	☽ ☌ ♅
23 11	☽ bQ ♇	06 13	☽ bQ ♀	19 36	☽ △ ♀
06 Friday		09 30	☽ ✶ ♃	23 24	☉ ∥ ♂
01 19	☽ ∥ ♀	10 29	☽ ✶ ♄	**26 Thursday**	
10 31	☽ ☌ ☉	15 15	☽ bQ ♆	02 39	☽ △ ☉
14 04	☽ ± ♇	17 31	☽ bQ ☉	02 58	☽ ✶ ♃
14 06	♀ ⊼ ♃	18 44	☽ Q ♇	05 00	☽ ✶ ♄
15 28	♀ ⊼ ♄	**17 Tuesday**		05 54	☉ ☍ ♃
17 03	☽ # ♇	00 47	☽ ± ♀	06 10	☽ △ ♂
18 53	☽ ⊻ ☿	00 48	☽ bQ ♂	08 54	☽ ∠ ♇
20 45	☽ ☌ ♂	04 06	♂ △ ♅	12 37	☽ □ ☿
23 53	☽ ⊼ ♇	11 51	☽ ± ☉	15 31	♀ ∥ ♄
07 Saturday		13 19	☽ ⊼ ♀	**27 Friday**	
01 14	☽ □ ♆	13 21	☽ ∠ ♃	02 49	☽ Q ♃
01 26	☽ ∥ ☉	14 27	☽ ∠ ♄	03 42	♀ # ♄
05 00	☽ ⊥ ☿	17 17	☽ ✶ ♇	04 53	☉ ☍ ♄
08 53	☽ ∥ ♂	18 54	☽ ± ♂	04 56	☽ Q ♄
09 14	♀ bQ ♇	19 07	☉ ∥ ♃	06 05	♆ St R
09 59	☽ △ ♅	19 29	☽ △ ♆	14 51	☽ ✶ ♇
19 04	☽ ∥ ♃	**18 Wednesday**		17 43	☽ ☌ ♆
20 07	☽ ∠ ☿	00 12	☽ ⊼ ☉	20 01	☽ ∥ ♅
22 23	☽ ☍ ♃	02 50	♂ ∥ ♄	**28 Saturday**	
22 36	☽ ☍ ♄	05 19	☿ □ ♅	03 36	♀ # ♃
23 22	☽ ∥ ♄	05 24	☽ □ ♅	03 38	☽ ⊻ ♅
08 Sunday		05 24	☽ ☍ ☿	05 08	☽ Q ☿
01 37	☽ ⊻ ♀	06 18	☽ ⊥ ♃	12 02	♀ ☍ ♃
05 59	☽ # ♄	07 05	☽ ⊼ ♂	14 02	☽ □ ♃
10 11	☽ # ♃	07 29	☽ ⊥ ♄	14 17	☽ □ ♀
10 16	☽ bQ ♅	09 08	☽ # ☿	14 26	☽ Q ♇
12 13	☽ ⊥ ♀	14 44	☿ ± ♃	15 22	☽ ⊥ ♅
13 09	☉ ± ♇	17 41	☽ ⊻ ♃	16 15	☽ □ ♄
14 52	☽ ⊻ ☉	18 55	☽ ⊻ ♄	19 34	☽ □ ☉
18 24	☽ # ♂	21 55	☽ ∠ ♇	21 43	☽ □ ♂
21 19	☽ ✶ ☿	**19 Thursday**		**29 Sunday**	
09 Monday		03 31	☿ ± ♄	07 30	☽ ✶ ☿
00 07	☽ ⊻ ♂	04 55	♀ ⊼ ♇	08 43	♀ ☍ ♄
00 15	☽ # ☉	15 18	☽ ⊥ ♇	08 44	☽ ∠ ♅
00 27	☽ ☍ ♇	16 04	☽ Q ♅	**30 Monday**	
00 49	☽ ± ♅	18 24	☽ ∥ ♇	00 52	☽ □ ♇
01 17	☽ ⊥ ☉	19 44	☉ ∥ ♄	03 41	☽ ⊻ ♆
01 54	☽ △ ♆	**20 Friday**		04 28	☽ ∥ ♅
04 06	☽ ∠ ♀	02 59	☽ ⊻ ♇	04 30	☽ Q ♀
06 01	♂ ⊼ ♇	03 00	☉ # ♂	08 44	☽ Q ☉
10 23	☽ ⊥ ♂	04 14	♀ □ ♆	10 07	☽ Q ♂
10 32	☽ ⊼ ♅	05 24	☽ □ ♆	12 57	☽ ✶ ♅
12 24	☽ ∥ ♇	05 31	☽ ☍ ♀	14 50	☽ ⊥ ♆
16 56	☽ ∠ ☉	10 36	♂ # ♄	15 23	☽ ∠ ☿
17 38	☽ Q ☿	12 48	☿ ⊻ ♂	22 14	☽ △ ♃
18 50	☽ # ♀	15 22	☽ ☍ ☉	**31 Tuesday**	
20 14	☿ bQ ♃	15 31	☽ ✶ ♅	00 32	☽ △ ♄
22 37	☽ ⊼ ♃	17 08	☉ △ ♅	01 28	☿ # ♇
22 58	☽ ⊼ ♄	18 30	☽ # ♀	04 55	☽ ✶ ♀
10 Tuesday		21 17	☽ ☍ ♂	07 16	☽ ∠ ♆
01 51	☽ ∠ ♂	21 36	☽ ⊼ ☿	08 30	☽ ✶ ☉
02 17	☽ bQ ♆	**21 Saturday**		09 26	☽ ✶ ♂
03 15	☿ bQ ♄	03 38	☽ ☌ ♃	09 51	☽ ⊥ ☿
06 46	☽ ✶ ♀	05 08	☽ ☌ ♄	10 16	☿ ± ♇
08 21	☽ ± ♃	06 24	☽ # ♃	10 47	☽ Q ♅
08 44	☽ ± ♄	08 24	☽ ∥ ♂	21 53	☽ ⊻ ☿

APRIL 1981

LONGITUDES

Date	Sidereal time h m s	Sun ☉ ° ' "	Moon ☽ ° ' "	Moon ☽ 24.00 ° ' "	Mercury ☿ ° '	Venus ♀ ° '	Mars ♂ ° '	Jupiter ♃ ° '	Saturn ♄ ° '	Uranus ♅ ° '	Neptune ♆ ° '	Pluto ♇ ° '
01	00 39 03	11 ♈ 40 09	26 ♒ 07 09	03 ♓ 06 59	18 ♓ 56	10 ♈ 10	11 ♈ 54	04 ♎ 41	06 ♎ 01	29 ♏ 47	24 ♐ 51	23 ♎ 17
02	00 42 59	12 39 20	10 ♓ 13 58	17 ♓ 27 48	20 28	11 24	12 41	04 R 33	05 R 57	29 R 46	24 R 51	23 R 16
03	00 46 56	13 38 30	24 ♓ 47 55	02 ♈ 13 34	22 02	12 39	13 27	04 26	05 52	29 44	24 51	23 14
04	00 50 53	14 37 38	09 ♈ 43 46	17 ♈ 17 22	23 37	13 53	14 13	04 18	05 47	29 43	24 50	23 12
05	00 54 49	15 36 43	24 ♈ 53 05	02 ♉ 29 32	25 13	15 08	14 59	04 11	05 43	29 41	24 50	23 11
06	00 58 46	16 35 47	10 ♉ 05 21	17 ♉ 39 12	26 51	16 22	15 45	04 03	05 38	29 40	24 50	23 09
07	01 02 42	17 34 49	25 ♉ 09 53	02 ♊ 36 19	28 ♓ 31	17 37	16 31	03 56	05 34	29 38	24 49	23 07
08	01 06 39	18 33 48	09 ♊ 57 38	17 ♊ 13 12	00 ♈ 12	18 51	17 16	03 48	05 29	29 37	24 49	23 06
09	01 10 35	19 32 45	24 ♊ 22 35	01 ♋ 25 32	01 54	20 05	18 02	03 41	05 25	29 35	24 49	23 04
10	01 14 32	20 31 40	08 ♋ 22 00	15 ♋ 12 06	03 38	21 20	18 48	03 34	05 20	29 33	24 48	23 02
11	01 18 28	21 30 33	21 ♋ 56 05	28 ♋ 34 15	05 24	22 34	19 34	03 27	05 16	29 31	24 48	23 00
12	01 22 25	22 29 23	05 ♌ 07 01	11 ♌ 34 49	07 11	23 48	20 19	03 20	05 11	29 30	24 47	22 59
13	01 26 22	23 28 11	17 ♌ 58 07	24 ♌ 17 24	09 00	25 03	21 05	03 13	05 07	29 28	24 47	22 57
14	01 30 18	24 26 57	00 ♍ 33 06	06 ♍ 45 39	10 50	26 17	21 51	03 06	05 03	29 26	24 46	22 55
15	01 34 15	25 25 40	12 ♍ 55 26	19 ♍ 02 50	12 41	27 31	22 36	02 59	04 59	29 24	24 45	22 54
16	01 38 11	26 24 21	25 ♍ 08 10	01 ♎ 11 42	14 34	28 ♈ 45	23 22	02 53	04 54	29 22	24 45	22 52
17	01 42 08	27 23 00	07 ♎ 13 41	13 ♎ 14 20	16 29	00 ♉ 00	24 07	02 46	04 50	29 20	24 44	22 50
18	01 46 04	28 21 37	19 ♎ 13 51	25 ♎ 12 23	18 25	01 14	24 52	02 40	04 46	29 18	24 43	22 49
19	01 50 01	29 ♈ 20 12	01 ♏ 10 06	07 ♏ 07 09	20 23	02 28	25 38	02 34	04 42	29 16	24 43	22 47
20	01 53 57	00 ♉ 18 46	13 ♏ 03 41	18 ♏ 59 54	22 23	03 42	26 23	02 27	04 38	29 14	24 42	22 45
21	01 57 54	01 17 17	24 ♏ 55 58	00 ♐ 52 07	24 23	04 56	27 08	02 21	04 34	29 12	24 41	22 44
22	02 01 50	02 15 47	06 ♐ 48 37	12 ♐ 45 46	26 25	06 10	27 54	02 15	04 30	29 10	24 40	22 42
23	02 05 47	03 14 15	18 ♐ 43 55	24 ♐ 43 26	28 ♈ 29	07 25	28 39	02 10	04 26	29 08	24 40	22 40
24	02 09 44	04 12 41	00 ♑ 44 46	06 ♑ 48 24	00 ♉ 34	08 39	29 ♈ 24	02 04	04 23	29 05	24 39	22 39
25	02 13 40	05 11 05	12 ♑ 54 50	19 ♑ 04 39	02 40	09 53	00 ♉ 09	01 58	04 19	29 03	24 38	22 37
26	02 17 37	06 09 28	25 ♑ 18 24	01 ♒ 36 41	04 47	11 07	00 54	01 53	04 15	29 01	24 37	22 35
27	02 21 33	07 07 50	08 ♒ 00 06	14 ♒ 29 13	06 55	12 21	01 39	01 48	04 12	28 59	24 36	22 34
28	02 25 30	08 06 10	21 ♒ 04 34	27 ♒ 46 36	09 03	13 35	02 24	01 43	04 08	28 57	24 35	22 32
29	02 29 26	09 04 28	04 ♓ 35 43	11 ♓ 32 07	11 12	14 49	03 09	01 38	04 05	28 54	24 34	22 30
30	02 33 23	10 ♉ 02 45	18 ♓ 35 55	25 ♓ 46 59	13 ♉ 22	16 ♉ 03	03 ♉ 53	01 ♎ 33	04 ♎ 01	28 ♏ 52	24 ♐ 33	22 ♎ 29

Date	Moon True ☊ ° '	Moon Mean ☊ ° '	Moon ☽ Latitude ° '
01	08 ♌ 56	07 ♌ 43	01 S 30
02	08 R 52	07 40	02 39
03	08 46	07 37	03 39
04	08 38	07 34	04 25
05	08 28	07 31	04 54
06	08 18	07 28	05 02
07	08 09	07 24	04 49
08	08 01	07 21	04 17
09	07 57	07 18	03 29
10	07 54	07 15	02 30
11	07 D 54	07 12	01 24
12	07 54	07 09	00 S 15
13	07 R 54	07 05	00 N 53
14	07 52	07 02	01 57
15	07 47	06 59	02 54
16	07 40	06 56	03 42
17	07 30	06 53	04 20
18	07 18	06 49	04 45
19	07 04	06 46	04 58
20	06 50	06 43	04 58
21	06 36	06 40	04 45
22	06 25	06 37	04 19
23	06 16	06 34	03 42
24	06 10	06 30	02 55
25	06 07	06 27	01 59
26	06 D 06	06 24	00 N 57
27	06 R 06	06 21	00 S 10
28	06 05	06 18	01 18
29	06 04	06 15	02 24
30	06 ♌ 00	06 ♌ 11	03 S 24

DECLINATIONS

Date	Sun ☉ ° '	Moon ☽ ° '	Mercury ☿ ° '	Venus ♀ ° '	Mars ♂ ° '	Jupiter ♃ ° '	Saturn ♄ ° '	Uranus ♅ ° '	Neptune ♆ ° '	Pluto ♇ ° '
01	04 N 37	14 S 14	06 S 35	02 N 48	04 N 06	00 S 24	00 N 03	19 S 53	22 S 00	07 N 24
02	05 00	10 11	06 00	03 18	04 25	00 21	00 05	19 53	22 00	07 25
03	05 23	05 25	05 24	03 48	04 43	00 18	00 06	19 52	22 00	07 26
04	05 46	00 S 13	04 47	04 18	05 02	00 15	00 08	19 52	22 00	07 26
05	06 09	05 N 04	04 08	04 48	05 20	00 12	00 10	19 52	22 00	07 27
06	06 31	10 03	03 29	05 17	05 38	00 09	00 12	19 51	22 00	07 28
07	06 54	14 22	02 48	05 47	05 56	00 06	00 14	19 51	22 00	07 28
08	07 17	17 42	02 07	06 16	06 14	00 03	00 15	19 51	22 00	07 29
09	07 39	19 50	01 24	06 46	06 32	00 S 01	00 17	19 50	22 00	07 30
10	08 01	20 41	00 S 40	07 15	06 50	00 N 02	00 19	19 50	22 00	07 30
11	08 23	20 17	00 N 04	07 44	07 08	00 05	00 21	19 50	22 00	07 31
12	08 45	18 45	00 50	08 13	07 26	00 08	00 22	19 49	21 59	07 31
13	09 07	16 17	01 36	08 41	07 43	00 10	00 24	19 49	21 59	07 32
14	09 29	13 06	02 24	09 10	08 01	00 13	00 26	19 48	21 59	07 33
15	09 50	09 23	03 12	09 38	08 19	00 15	00 27	19 48	21 59	07 33
16	10 11	05 20	04 01	10 06	08 36	00 18	00 29	19 48	21 59	07 34
17	10 33	01 N 06	04 51	10 34	08 53	00 20	00 30	19 47	21 59	07 34
18	10 54	03 S 08	05 41	11 01	09 10	00 23	00 32	19 47	21 59	07 35
19	11 14	07 13	06 32	11 28	09 28	00 25	00 33	19 46	21 59	07 35
20	11 35	11 01	07 24	11 55	09 45	00 28	00 35	19 46	21 59	07 36
21	11 55	14 24	08 15	12 22	10 01	00 30	00 36	19 45	21 59	07 36
22	12 16	17 11	09 08	12 49	10 18	00 32	00 38	19 45	21 59	07 37
23	12 36	19 16	10 00	13 15	10 35	00 34	00 39	19 45	21 59	07 37
24	12 55	20 31	10 53	13 40	10 51	00 36	00 41	19 44	21 59	07 38
25	13 15	20 50	11 45	14 06	11 08	00 38	00 42	19 44	21 59	07 38
26	13 34	20 09	12 38	14 31	11 24	00 40	00 43	19 43	21 59	07 39
27	13 54	18 26	13 29	14 56	11 40	00 42	00 45	19 43	21 59	07 39
28	14 13	15 43	14 21	15 20	11 56	00 44	00 46	19 42	21 58	07 40
29	14 31	12 04	15 11	15 44	12 12	00 46	00 47	19 42	21 58	07 40
30	14 N 50	07 S 39	16 N 01	16 N 08	12 N 28	00 N 48	00 N 48	19 S 41	21 S 58	07 N 41

ZODIAC SIGN ENTRIES

Date	h	m	Planets
01	18	41	☽ ♓
03	20	25	☽ ♈
05	20	04	☽ ♉
07	19	47	☽ ♊
08	09	11	☿ ♈
09	21	34	☽ ♋
12	02	36	☽ ♌
14	10	56	☽ ♍
16	21	38	☽ ♎
17	12	08	♀ ♉
19	09	39	☽ ♏
20	04	19	☉ ♉
21	22	15	☽ ♐
24	05	31	☿ ♉
24	10	31	☽ ♑
25	07	17	♂ ♉
26	20	57	☽ ♒
29	03	56	☽ ♓

LATITUDES

Date	Mercury ☿ ° '	Venus ♀ ° '	Mars ♂ ° '	Jupiter ♃ ° '	Saturn ♄ ° '	Uranus ♅ ° '	Neptune ♆ ° '	Pluto ♇ ° '
01	02 S 24	01 S 20	00 S 39	01 N 36	02 N 39	00 N 14	01 N 20	17 N 41
04	02 26	01 17	00 38	01 35	02 39	00 14	01 21	17 41
07	02 25	01 13	00 36	01 35	02 39	00 14	01 21	17 42
10	02 18	01 09	00 34	01 35	02 39	00 14	01 21	17 42
13	02 08	01 05	00 33	01 35	02 39	00 14	01 21	17 42
16	01 52	01 00	00 31	01 34	02 39	00 14	01 21	17 42
19	01 33	00 55	00 29	01 34	02 38	00 14	01 21	17 42
22	01 09	00 49	00 27	01 34	02 38	00 14	01 21	17 41
25	00 41	00 43	00 25	01 33	02 38	00 14	01 21	17 41
28	00 S 11	00 37	00 23	01 33	02 37	00 14	01 21	17 41
31	00 N 21	00 S 30	00 S 22	01 N 32	02 N 37	00 N 14	01 N 21	17 N 40

DATA

Julian Date	2444696
Delta T	+51 seconds
Ayanamsa	23° 35' 28"
Synetic vernal point	05° ♓ 31' 31"
True obliquity of ecliptic	23° 26' 25"

MOON'S PHASES, APSIDES AND POSITIONS ☽

Date	h	m	Phase	Longitude ° '	Eclipse Indicator
04	20	19	●	14 ♈ 58	
11	11	11	☽	21 ♋ 29	
19	07	59	○	29 ♎ 10	
27	10	14	☾	07 ♒ 04	

Day	h	m		
05	18	30	Perigee	
20	16	11	Apogee	
04	12	57	0N	
10	15	57	Max dec	20° N 42'
17	18	14	0S	
25	07	42	Max dec	20° S 51'

LONGITUDES

Date	Chiron ⚷ ° '	Ceres ⚳ ° '	Pallas ⚴ ° '	Juno ⚵ ° '	Vesta ⚶ ° '	Black Moon Lilith ⚸ ° '
01	15 ♉ 38	17 ♋ 14	23 ♉ 51	03 ♏ 55	25 ♌ 43	10 ♏ 21
11	16 ♉ 14	19 ♋ 30	29 ♉ 19	01 ♏ 48	25 ♌ 25	11 ♏ 29
21	16 ♉ 52	22 ♋ 11	04 ♊ 56	29 ♎ 30	25 ♌ 55	12 ♏ 36
31	17 ♉ 32	25 ♋ 13	10 ♊ 40	27 ♎ 16	27 ♌ 08	13 ♏ 43

All ephemeris data is given at 12.00 UT and the Moon's longitude is additionally given for 24.00 UT

ASPECTARIAN

h m Aspects

01 Wednesday
00 53 ☽ ꝑ ♃
03 10 ☽ ꝑ ♄
07 06 ☽ △ ♇
09 48 ☽ ⚹ ♆
10 11 ☽ ∠ ♀
13 01 ☽ ∠ ☉
13 27 ☽ ∠ ♂
16 23 ☽ ± ♃
18 18 ☽ □ ♅
18 41 ☽ ± ♄

02 Thursday
02 32 ☽ ⊼ ♃
03 06 ☽ ⊥ ♀
04 50 ☽ ⊼ ♄
05 33 ☽ ⊥ ☉
05 41 ☽ ⊥ ♂
06 19 ☽ Q ♆
08 42 ☽ ꝑ ♇
14 09 ☽ ⊻ ♀
14 13 ☉ ☌ ♂
16 18 ☽ ⊻ ♂
16 20 ☽ ⊻ ☉
23 38 ☽ ± ♇

03 Friday
02 13 ☽ ⋕ ♇
06 57 ☽ ☌ ☿
09 27 ☽ ⊼ ♇
12 04 ☽ ∥ ☿
12 04 ☽ □ ♆
12 07 ☽ ⋕ ☉
12 21 ☉ ⋕ ☿
15 05 ☽ ⋕ ♂
18 54 ☽ ⋕ ♀
19 59 ☽ △ ♅

04 Saturday
03 24 ☽ ☍ ♃
05 37 ☿ ⋕ ♂
05 45 ☽ ☍ ♄
05 56 ☿ ⊼ ♇
11 49 ☽ ∥ ♃
12 20 ☽ ⋕ ♄
13 35 ☽ ∥ ♄
14 04 ☽ ⋕ ♃
14 05 ☉ ꝑ ♅
19 12 ☽ ☌ ♀
19 30 ☽ ☌ ♂
19 55 ☽ ꝑ ♅
20 19 ☽ ☌ ☉
22 08 ☿ ⋕ ♀

05 Sunday
03 16 ♂ ꝑ ♅
03 42 ♀ ꝑ ♅
04 25 ♀ ☌ ♂
06 15 ☿ □ ♆
08 11 ☽ ⋕ ☿
09 18 ☽ ☍ ♇
10 07 ☽ ± ♅
10 36 ☽ ∥ ♀
11 55 ☽ △ ♆
12 36 ☽ ⊻ ☿
13 16 ☽ ∥ ♂
17 23 ☽ ∥ ☉
19 34 ☽ ⊼ ♅
22 09 ♃ ⋕ ♄
23 11 ☽ ∥ ♇
23 12 ☽ ⊥ ☿

06 Monday
02 32 ☽ ⊼ ♃
05 00 ☽ ⊼ ♄
11 35 ☽ ꝑ ♆
11 57 ☽ ± ♃
14 26 ☽ ± ♄
15 08 ☽ ∠ ☿
21 27 ☽ ⊻ ♂
22 51 ☽ ⊻ ♀
23 02 ☽ ⊻ ☉

07 Tuesday
01 52 ☽ ± ♆
02 07 ☽ ꝑ ♃
04 40 ☽ ꝑ ♄
07 31 ☽ ⊥ ♂
08 44 ☽ ⊼ ♇
09 17 ☽ ⊥ ♀
09 17 ☽ ⊥ ☉
09 22 ☉ ☌ ♀
11 27 ☽ ⊼ ♆
18 04 ☽ ⚹ ☿
18 21 ☽ ± ♇
19 11 ☽ ☍ ♅
22 47 ☽ ∠ ♂

08 Wednesday
00 49 ☽ ∠ ☉
01 06 ☽ ∠ ♀
02 02 ☽ △ ♃
03 46 ☿ △ ♅
04 43 ☽ △ ♄
07 29 ♀ ∥ ♂
08 57 ☽ ꝑ ♇
16 10 ☽ Q ☿

09 Thursday
00 46 ☽ ⚹ ♂
01 43 ☉ ∥ ♇
03 17 ☽ ⚹ ☉
04 06 ☽ ⚹ ♀
09 47 ☽ △ ♇
12 03 ☽ ⋕ ♅
12 44 ☽ ☍ ♆
20 49 ☽ ⊼ ♅
22 10 ☽ Q ♂

10 Friday
01 08 ☽ Q ☉
02 25 ☽ Q ♀
02 38 ☽ □ ☿
03 45 ☽ □ ♃
06 46 ☽ □ ♄
07 07 ☽ ± ♅
11 03 ☿ ☍ ♃
22 50 ☽ ꝑ ♅
23 09 ☿ ⋕ ♄

11 Saturday
00 58 ♀ ∥ ♇
07 21 ☿ ⋕ ♃
07 30 ☽ □ ♂
10 14 ☿ ☍ ♄
11 08 ☽ Q ♃
11 11 ☽ □ ☉
12 17 ☿ ∥ ♃
13 15 ☽ □ ♀
13 55 ☽ □ ♇
14 23 ☽ Q ♄
17 09 ☽ ⊼ ♆
20 22 ♀ ☍ ♇
20 55 ☿ ∥ ♄
21 01 ☽ ⋕ ♅

12 Sunday
01 42 ☽ △ ♅
04 03 ☽ ± ♆
06 06 ♀ ± ♅
08 44 ☽ ⚹ ♃
12 08 ☽ ⚹ ♄
16 26 ☽ △ ☿
19 54 ♂ ∥ ♇
20 39 ☽ ꝑ ♆
22 51 ☽ Q ♇
23 39 ☉ ☍ ♇

13 Monday
06 53 ♀ △ ♆
11 50 ☉ ± ♅
12 28 ☽ ∠ ♃
16 03 ☽ ∠ ♄
18 17 ☽ △ ♂
21 26 ☽ ⚹ ♇
23 19 ☽ △ ☉

14 Tuesday
00 55 ☽ △ ♆
01 23 ☽ ꝑ ☿
02 54 ☽ △ ♀
05 26 ☽ ⊥ ♃
09 07 ☽ ⊥ ♄
09 51 ☽ □ ♅
16 52 ☽ ⊻ ♃
19 44 ☉ △ ♆
20 38 ☽ ⊻ ♄
21 42 ☽ ± ☿

15 Wednesday
00 57 ☽ ꝑ ♂
02 13 ☽ ∠ ♇
06 43 ☽ ꝑ ☉
09 28 ☽ ∥ ☉
10 38 ☽ ∥ ♀
11 07 ☽ ꝑ ♀
11 27 ☽ ⊼ ☿
18 05 ☽ ∥ ♂
19 41 ☽ ± ♂
19 45 ☽ ⊥ ♇
20 45 ☽ Q ♅
20 56 ♂ ☍ ♇
23 00 ☽ ∥ ♇

16 Thursday
01 52 ☽ ± ☉
06 46 ☽ ± ♀
07 32 ☽ ⊻ ♇
08 16 ☽ ⊼ ♂
09 26 ☿ ꝑ ♅
11 14 ☽ □ ♆
12 13 ♂ ± ♅
14 44 ☽ ⊼ ☉
18 17 ☽ ∥ ☿
19 59 ☽ ⊼ ♀
20 21 ☽ ⚹ ♅
23 34 ♀ ⊼ ♅

17 Friday
03 13 ☽ ☌ ♃
07 16 ☽ ☌ ♄
07 44 ☉ ∥ ♀
15 22 ☽ ∥ ♄
16 17 ☽ ∥ ♃
20 14 ☽ ⋕ ♃
21 09 ☽ ⋕ ♄
22 59 ☽ Q ♆

18 Saturday
02 09 ☽ ∠ ♅
07 19 ♂ △ ♆
10 04 ☽ ☍ ☿
19 10 ☽ ☌ ♇
20 09 ☽ ⊥ ♅
23 01 ☽ ⚹ ♆

19 Sunday
00 06 ☽ ☍ ♂
06 46 ☽ ⋕ ☿
07 59 ☽ ☍ ☉
08 11 ☽ ⊻ ♅
10 22 ☉ ⊼ ♅
13 40 ♀ ⊼ ♃
14 15 ☽ ⋕ ♇
14 47 ☽ ⊻ ♃
14 55 ☽ ☍ ♀
19 05 ☽ ⊻ ♄

20 Monday
02 47 ☽ ⊥ ♃
02 59 ☽ ⋕ ♂
05 13 ☽ ∠ ♆
07 07 ☽ ⊥ ♄
16 09 ☽ ⋕ ☉
16 29 ☿ ☍ ♇
17 50 ☿ ∥ ♇
18 59 ☽ ⋕ ♀
20 48 ☽ ∠ ♃
22 06 ☿ ± ♅
23 23 ☽ ⊥ ♆

21 Tuesday
01 13 ☽ ∠ ♄
05 11 ♀ ⊼ ♄
07 33 ☽ ⊻ ♇
10 40 ☽ ⊼ ☿
11 30 ☽ ⊻ ♆
15 31 ☿ △ ♆
16 46 ☽ ⊼ ♂
19 39 ☽ ⊥ ♇
20 36 ☽ ☌ ♅

22 Wednesday
01 18 ☽ ± ☿
02 00 ☽ ⊼ ☉
02 53 ☽ ⚹ ♃
05 43 ☽ ± ♂
07 22 ☽ ⚹ ♄
10 34 ☽ ⊼ ♀
11 51 ☉ ⊼ ♃
13 47 ☽ ∠ ♇
15 12 ☽ ± ☉
23 14 ☽ ꝑ ☿

23 Thursday
00 04 ☽ ± ♀
01 05 ☽ ꝑ ♂
02 53 ☽ Q ♃
07 25 ☽ Q ♄
10 55 ☽ ꝑ ☉
14 56 ☿ ☌ ♂
19 13 ☽ ∥ ♅
19 20 ☿ ⊼ ♅
19 53 ☽ ⚹ ♇
20 13 ☽ ꝑ ♀
23 51 ☽ ☌ ♆

24 Friday
01 32 ♀ ± ♃
02 42 ♂ ⊼ ♅
08 43 ☽ ⊻ ♅
09 08 ☽ △ ♂
11 04 ☿ ∥ ♂
11 34 ☽ △ ☿
14 36 ☽ □ ♃
15 51 ☉ ⊼ ♄
19 10 ☽ □ ♄
19 28 ☽ △ ☉
19 43 ☽ Q ♇
20 35 ☽ ⊥ ♅

25 Saturday
04 27 ☿ ⊼ ♃
05 23 ☽ △ ♀
07 13 ♀ ꝑ ♆
14 13 ☽ ∠ ♅
20 05 ♀ ± ♄

26 Sunday
06 14 ☿ ⊼ ♄
06 48 ☽ □ ♇
10 41 ☽ ⊻ ♆
19 04 ☽ ⚹ ♅
19 34 ☽ ∥ ♅
22 06 ☽ ⊥ ♆
23 19 ☽ □ ♂

27 Monday
00 26 ☽ △ ♃
04 54 ☽ △ ♄
09 34 ☽ □ ☿
10 14 ☽ □ ☉
14 58 ☽ ∠ ♆
16 15 ♂ ⊼ ♃
16 30 ☉ ☌ ☿
17 31 ☽ Q ♅
20 55 ☽ □ ♀
21 31 ☿ ± ♃

28 Tuesday
03 04 ☉ ± ♃
04 07 ☽ ꝑ ♃
05 51 ☉ ∥ ☿
08 30 ☽ ꝑ ♄
10 42 ☽ Q ♂
14 28 ☽ ⋕ ♀
14 38 ☽ △ ♇
17 52 ☿ ꝑ ♆
18 18 ☽ ⚹ ♆
19 48 ☽ ⋕ ☿
20 16 ☽ ± ♃
21 43 ☽ Q ☉
21 44 ☽ ⋕ ☉
23 47 ☿ ± ♄

29 Wednesday
00 35 ☽ ± ♄
00 45 ☽ Q ☿
02 02 ☽ □ ♅
06 50 ☽ ⊼ ♃
08 35 ☽ Q ♀
09 19 ☽ ⚹ ♂
11 06 ☽ ⊼ ♄
11 14 ☽ ⋕ ♂
15 26 ☽ Q ♆
17 03 ☽ ꝑ ♇
20 21 ☽ ⚹ ☉
23 58 ☉ ꝑ ♆

30 Thursday
01 31 ☽ ⚹ ☿
07 17 ☽ ⚹ ♀
08 26 ☽ ± ♇
11 31 ☉ ± ♄
11 50 ☽ ⋕ ♇
12 31 ☽ ∠ ♂
16 01 ♂ ⊼ ♄
18 11 ☿ ∥ ♀
18 30 ☽ ⊼ ♇
21 05 ♀ ꝑ ♃
21 57 ☽ □ ♆
23 33 ☽ ∠ ☉

MAY 1981

LONGITUDES

Date	Sidereal time h m s	Sun ☉ ° ′ ″	Moon ☽ ° ′ ″	Moon ☽ 24.00 ° ′ ″	Mercury ☿ ° ′	Venus ♀ ° ′	Mars ♂ ° ′	Jupiter ♃ ° ′	Saturn ♄ ° ′	Uranus ♅ ° ′	Neptune ♆ ° ′	Pluto ♇ ° ′
01	02 37 19	11 ♉ 01 00	03 ♈ 04 59	10 ♈ 29 22	15 ♉ 31	17 ♉ 17	04 ♉ 38	01 ♎ 28	03 ♎ 58	28 ♏ 50	24 ♐ 32	22 ♎ 27
02	02 41 16	11 59 14	17 ♈ 59 19	25 ♈ 33 48	17 40	18 31	05 23	01 R 24	03 R 55	28 R 47	24 R 31	22 R 26
03	02 45 13	12 57 26	03 ♉ 11 35	10 ♉ 51 19	19 48	19 45	06 08	01 19	03 52	28 45	24 30	22 24
04	02 49 09	13 55 36	18 ♉ 31 30	26 ♉ 10 41	21 56	20 59	06 52	01 15	03 49	28 43	24 29	22 23
05	02 53 06	14 53 45	03 ♊ 47 24	11 ♊ 20 21	24 02	22 13	07 37	01 11	03 46	28 40	24 28	22 21
06	02 57 02	15 51 52	18 ♊ 48 25	26 ♊ 10 41	26 07	23 27	08 21	01 07	03 43	28 38	24 26	22 20
07	03 00 59	16 49 57	03 ♋ 26 26	10 ♋ 35 15	28 ♉ 10	24 41	09 06	01 04	03 40	28 35	24 25	22 18
08	03 04 55	17 48 01	17 ♋ 36 53	24 ♋ 31 18	00 ♊ 11	25 55	09 50	01 00	03 38	28 33	24 24	22 17
09	03 08 52	18 46 02	01 ♌ 18 39	07 ♌ 59 14	02 10	27 08	10 35	00 57	03 35	28 30	24 23	22 15
10	03 12 48	19 44 02	14 ♌ 33 25	21 ♌ 01 41	04 07	28 22	11 19	00 54	03 32	28 28	24 22	22 14
11	03 16 45	20 41 59	27 ♌ 24 33	03 ♍ 42 34	06 00	29 ♉ 36	12 03	00 51	03 30	28 25	24 20	22 12
12	03 20 42	21 39 55	09 ♍ 56 18	16 ♍ 06 18	07 51	00 ♊ 50	12 47	00 48	03 28	28 23	24 19	22 11
13	03 24 38	22 37 49	22 ♍ 13 06	28 ♍ 17 12	09 39	02 04	13 31	00 45	03 25	28 20	24 18	22 09
14	03 28 35	23 35 41	04 ♎ 19 05	10 ♎ 19 11	11 24	03 18	14 15	00 43	03 23	28 18	24 16	22 08
15	03 32 31	24 33 32	16 ♎ 17 52	22 ♎ 15 30	13 06	04 31	14 59	00 40	03 21	28 16	24 15	22 07
16	03 36 28	25 31 21	28 ♎ 12 24	04 ♏ 08 49	14 44	05 45	15 43	00 38	03 19	28 13	24 14	22 05
17	03 40 24	26 29 08	10 ♏ 04 59	16 ♏ 01 08	16 19	06 59	16 27	00 36	03 17	28 11	24 12	22 04
18	03 44 21	27 26 54	21 ♏ 57 28	27 ♏ 54 08	17 51	08 13	17 11	00 34	03 15	28 08	24 11	22 03
19	03 48 17	28 24 38	03 ♐ 51 20	09 ♐ 49 15	19 19	09 26	17 55	00 33	03 14	28 05	24 10	22 01
20	03 52 14	29 ♉ 22 21	15 ♐ 48 03	21 ♐ 47 58	20 44	10 40	18 39	00 31	03 12	28 03	24 08	22 00
21	03 56 11	00 ♊ 20 03	27 ♐ 49 13	03 ♑ 52 04	22 05	11 54	19 22	00 30	03 11	28 00	24 07	21 59
22	04 00 07	01 17 44	09 ♑ 56 50	16 ♑ 03 50	23 22	13 07	20 06	00 29	03 09	27 58	24 05	21 58
23	04 04 04	02 15 24	22 ♑ 13 26	28 ♑ 26 04	24 36	14 21	20 50	00 28	03 08	27 55	24 04	21 56
24	04 08 00	03 13 02	04 ♒ 42 08	11 ♒ 02 06	25 46	15 34	21 33	00 28	03 07	27 53	24 02	21 55
25	04 11 57	04 10 40	17 ♒ 26 28	23 ♒ 55 40	26 53	16 48	22 17	00 27	03 05	27 51	24 01	21 54
26	04 15 53	05 08 16	00 ♓ 30 10	07 ♓ 10 24	27 55	18 02	23 00	00 27	03 04	27 48	23 59	21 53
27	04 19 50	06 05 52	13 ♓ 56 41	20 ♓ 49 20	28 54	19 15	23 44	00 27	03 04	27 46	23 58	21 52
28	04 23 46	07 03 26	27 ♓ 48 28	04 ♈ 54 06	29 ♊ 49	20 29	24 27	00 D 27	03 03	27 43	23 56	21 51
29	04 27 43	08 01 00	12 ♈ 06 04	19 ♈ 24 02	00 ♋ 40	21 42	25 10	00 27	03 02	27 41	23 55	21 50
30	04 31 40	08 58 33	26 ♈ 47 26	04 ♉ 15 31	01 26	22 56	25 53	00 27	03 01	27 38	23 53	21 49
31	04 35 36	09 ♊ 56 05	11 ♉ 47 19	19 ♉ 21 43	02 ♋ 09	24 ♊ 09	26 ♉ 37	00 ♎ 28	03 ♎ 01	27 ♏ 36	23 ♐ 52	21 ♎ 48

Date	Moon True ☊ ° ′	Moon Mean ☊ ° ′	Moon ☽ Latitude ° ′
01	05 ♌ 54	06 ♌ 08	04 S 13
02	05 R 45	06 05	04 46
03	05 34	06 02	05 00
04	05 23	05 59	04 53
05	05 13	05 55	04 24
06	05 05	05 52	03 38
07	05 00	05 49	02 38
08	04 57	05 46	01 30
09	04 56	05 43	00 S 19
10	04 D 56	05 40	00 N 51
11	04 R 56	05 36	01 56
12	04 54	05 33	02 54
13	04 51	05 30	03 43
14	04 45	05 27	04 21
15	04 36	05 24	04 47
16	04 25	05 21	05 00
17	04 13	05 17	05 00
18	04 00	05 14	04 47
19	03 49	05 11	04 22
20	03 39	05 08	03 45
21	03 31	05 05	02 58
22	03 26	05 01	02 02
23	03 23	04 58	00 N 59
24	03 D 23	04 55	00 S 07
25	03 23	04 52	01 15
26	03 24	04 49	02 20
27	03 R 24	04 46	03 19
28	03 22	04 42	04 09
29	03 18	04 39	04 45
30	03 12	04 36	05 04
31	03 ♌ 05	04 ♌ 33	05 S 02

DECLINATIONS

Date	Sun ☉ ° ′	Moon ☽ ° ′	Mercury ☿ ° ′	Venus ♀ ° ′	Mars ♂ ° ′	Jupiter ♃ ° ′	Saturn ♄ ° ′	Uranus ♅ ° ′	Neptune ♆ ° ′	Pluto ♇ ° ′
01	15 N 08	02 S 39	16 N 50	16 N 31	12 N 44	00 N 49	00 N 49	19 S 41	21 S 58	07 N 41
02	15 26	02 N 38	17 37	16 54	12 59	00 51	00 51	19 40	21 58	07 41
03	15 44	07 52	18 22	17 16	13 15	00 53	00 52	19 40	21 58	07 42
04	16 01	12 39	19 06	17 38	13 30	00 54	00 53	19 39	21 58	07 42
05	16 18	16 35	19 48	17 59	13 45	00 55	00 54	19 39	21 58	07 42
06	16 35	19 21	20 27	18 20	14 00	00 57	00 55	19 38	21 58	07 43
07	16 52	20 45	21 05	18 41	14 15	00 58	00 56	19 38	21 58	07 43
08	17 08	20 47	21 40	19 01	14 29	00 59	00 57	19 37	21 58	07 43
09	17 24	19 33	22 12	19 20	14 44	01 00	00 57	19 37	21 58	07 44
10	17 40	17 17	22 42	19 39	14 58	01 01	00 58	19 36	21 58	07 44
11	17 56	14 11	23 10	19 58	15 13	01 02	00 59	19 35	21 58	07 44
12	18 11	10 32	23 35	20 15	15 27	01 03	01 00	19 35	21 57	07 44
13	18 26	06 30	23 57	20 33	15 41	01 04	01 01	19 34	21 57	07 45
14	18 40	02 N 16	24 17	20 50	15 54	01 05	01 01	19 34	21 57	07 45
15	18 55	02 S 00	24 35	21 06	16 08	01 06	01 02	19 33	21 57	07 45
16	19 09	06 10	24 50	21 21	16 21	01 06	01 03	19 33	21 57	07 45
17	19 22	10 05	25 03	21 36	16 35	01 07	01 03	19 32	21 57	07 45
18	19 35	13 38	25 13	21 51	16 48	01 07	01 04	19 32	21 57	07 45
19	19 48	16 38	25 21	22 05	17 01	01 08	01 04	19 31	21 57	07 46
20	20 01	18 57	25 27	22 18	17 13	01 08	01 05	19 31	21 57	07 46
21	20 13	20 35	25 32	22 31	17 26	01 08	01 05	19 30	21 57	07 46
22	20 25	21 03	25 34	22 42	17 38	01 09	01 05	19 29	21 57	07 46
23	20 37	20 38	25 35	22 54	17 51	01 09	01 06	19 29	21 57	07 46
24	20 48	19 12	25 33	23 04	18 03	01 09	01 06	19 28	21 57	07 46
25	20 59	16 47	25 31	23 14	18 14	01 09	01 06	19 28	21 57	07 46
26	21 09	13 29	25 26	23 24	18 26	01 09	01 06	19 27	21 56	07 46
27	21 20	09 23	25 21	23 32	18 38	01 08	01 07	19 27	21 56	07 46
28	21 29	04 S 41	25 14	23 40	18 49	01 08	01 07	19 26	21 56	07 46
29	21 39	00 N 24	25 05	23 48	19 00	01 08	01 07	19 26	21 56	07 46
30	21 48	05 36	24 56	23 54	19 11	01 08	01 07	19 25	21 56	07 46
31	21 N 56	10 N 34	24 N 46	24 N 00	19 N 22	01 N 07	01 N 07	19 S 25	21 S 56	07 N 46

ZODIAC SIGN ENTRIES

Date	h	m	Planets
01	06	57	☽ ♈
03	06	59	☽ ♉
05	06	01	☽ ♊
07	06	18	☽ ♋
08	09	42	☿ ♊
09	09	40	☽ ♌
11	16	55	☽ ♍
11	19	45	♀ ♊
14	03	24	☽ ♎
16	15	37	☽ ♏
19	04	14	☽ ♐
21	03	39	☉ ♊
21	16	20	☽ ♑
24	03	01	☽ ♒
26	11	05	☽ ♓
28	15	44	☽ ♈
28	17	04	☿ ♋
30	17	10	☽ ♉

LATITUDES

Date	Mercury ☿ ° ′	Venus ♀ ° ′	Mars ♂ ° ′	Jupiter ♃ ° ′	Saturn ♄ ° ′	Uranus ♅ ° ′	Neptune ♆ ° ′	Pluto ♇ ° ′
01	00 N 21	00 S 30	00 S 22	01 N 32	02 N 37	00 N 14	01 N 21	17 N 40
04	00 53	00 23	00 20	01 31	02 37	00 14	01 22	17 39
07	01 22	00 16	00 18	01 31	02 36	00 14	01 22	17 39
10	01 46	00 09	00 16	01 30	02 35	00 13	01 22	17 38
13	02 05	00 S 02	00 14	01 30	02 35	00 13	01 22	17 37
16	02 17	00 N 06	00 12	01 29	02 34	00 13	01 22	17 36
19	02 21	00 13	00 10	01 28	02 34	00 13	01 22	17 35
22	02 18	00 20	00 08	01 27	02 33	00 13	01 22	17 34
25	02 06	00 27	00 06	01 27	02 32	00 13	01 22	17 33
28	01 47	00 35	00 04	01 26	02 32	00 13	01 22	17 32
31	01 N 20	00 N 41	00 S 02	01 N 25	02 N 31	00 N 13	01 N 22	17 N 30

DATA

Julian Date	2444726
Delta T	+51 seconds
Ayanamsa	23° 35′ 31″
Synetic vernal point	05° ♓ 31′ 28″
True obliquity of ecliptic	23° 26′ 25″

MOON'S PHASES, APSIDES AND POSITIONS ☽

Date	h	m	Phase	Longitude ° ′	Eclipse Indicator
04	04	19	●	13 ♉ 37	
10	22	22	☽	20 ♌ 09	
19	00	04	○	27 ♏ 56	
26	21	00	☾	05 ♓ 30	

Day	h	m		
04	04	45	Perigee	
17	18	00	Apogee	
02	00	05	0N	
08	00	28	Max dec	20° N 56′
15	00	45	0S	
22	14	12	Max dec	21° S 03′
29	10	08	0N	

LONGITUDES

Date	Chiron ⚷ ° ′	Ceres ⚳ ° ′	Pallas ⚴ ° ′	Juno ⚵ ° ′	Vesta ⚶ ° ′	Black Moon Lilith ⚸ ° ′
01	17 ♉ 32	25 ♋ 13	10 ♊ 40	27 ♎ 16	27 ♌ 08	13 ♏ 43
11	18 ♉ 12	28 ♋ 32	16 ♊ 30	25 ♎ 16	29 ♌ 00	14 ♏ 50
21	18 ♉ 52	02 ♌ 06	22 ♊ 25	23 ♎ 41	01 ♍ 24	15 ♏ 58
31	19 ♉ 31	05 ♌ 51	28 ♊ 22	22 ♎ 36	04 ♍ 15	17 ♏ 05

ASPECTARIAN

01 Friday
04 19 ☽ ⊥ ♂
05 03 ☽ △ ♅
07 05 ☽ ∠ ☿
09 23 ☽ ☍ ♃
10 34 ☽ ∠ ♀
11 06 ♃ ∥ ♄
13 26 ☽ ☍ ♄
14 40 ☽ ⊻ ♂
15 22 ☽ ⊥ ☉
20 18 ☽ # ♃
20 19 ☽ # ♄
22 16 ☿ bQ ♃

02 Saturday
00 12 ☽ ⊥ ☿
01 44 ☽ ⊻ ☉
02 28 ☽ ⊥ ♀
03 52 ☽ ∥ ♄
03 53 ☽ ∥ ♃
05 19 ☽ bQ ♅
11 24 ☽ ⊻ ☿
11 58 ♀ ± ♆
12 55 ☽ ⊻ ♀
19 02 ☽ ☍ ♇
19 29 ♀ bQ ♄
19 36 ☽ ± ♅
21 26 ☿ ± ♆
22 20 ☽ △ ♆

03 Sunday
01 41 ☿ bQ ♄
05 02 ☽ ⊼ ♅
09 05 ☽ ⊼ ♃
10 29 ☿ ☌ ♀
11 10 ☽ ∥ ♇
13 03 ☽ ⊼ ♄
16 50 ☽ ☌ ♂
18 26 ☽ ± ♃
21 52 ☽ bQ ♆
22 25 ☽ ± ♄

04 Monday
04 19 ☽ ☌ ☉
08 27 ☽ bQ ♃
11 56 ☽ ± ♆
12 27 ☽ bQ ♄
16 11 ☽ ☌ ♀
16 59 ☿ ⊼ ♇
17 01 ☽ ∥ ♂
18 01 ☽ ⊼ ♇
18 12 ☽ ☌ ☿
21 19 ☽ ⊼ ♆
23 18 ♂ ± ♃

05 Tuesday
03 26 ☽ ± ♇
03 56 ☽ ☍ ♅
06 48 ☿ # ♅
07 54 ☽ △ ♃
09 56 ☽ ∥ ☉
11 58 ☽ △ ♄
14 36 ♀ ⊼ ♇
16 47 ☿ ⊼ ♆
17 38 ☽ bQ ♇
18 23 ☽ ⊻ ♂

06 Wednesday
00 23 ☽ ∥ ♀
04 27 ☽ ⊥ ♂
06 55 ☽ ⊻ ☉
15 26 ☽ # ♅
17 18 ☽ ⊥ ☉
17 42 ☽ △ ♇
17 57 ☉ bQ ♃
19 47 ☽ ∠ ♂
20 13 ☽ ⊻ ♀
21 09 ☽ ☍ ♆

07 Thursday
01 51 ☽ ⊻ ☿
03 59 ☽ ⊼ ♅
06 59 ☽ ⊥ ♀
07 03 ♀ ⊼ ♆
08 04 ☽ □ ♃
09 08 ☽ ∠ ☉
12 23 ☽ □ ♄
13 25 ☽ ⊥ ☿
13 30 ☿ ± ♇
13 54 ☽ ± ♅
16 48 ☿ ☍ ♅
22 00 ☽ ✶ ♂
22 15 ♂ bQ ♆
23 27 ☽ ∠ ♀

08 Friday
05 03 ☽ bQ ♅
05 34 ♂ ± ♄
07 09 ☽ ∠ ☿
12 21 ☽ ✶ ☉
14 23 ☽ Q ♃
18 55 ☽ Q ♄
19 43 ☽ Q ♂
20 04 ☽ □ ♇
21 28 ☿ △ ♃
23 46 ☽ ⊼ ♆

09 Saturday
01 01 ☿ # ♆
02 36 ☉ ± ♆
03 52 ☽ ✶ ♀
07 02 ☽ △ ♅
07 36 ☉ bQ ♄
10 21 ☽ ± ♆
10 57 ☽ Q ☉
11 18 ☽ # ♅
11 21 ☽ ✶ ♃
13 48 ☽ ✶ ☿
14 26 ☽ ∥ ♀
16 03 ☽ ✶ ♄

10 Sunday
02 30 ☽ bQ ♆
03 33 ☽ Q ♀
04 05 ☽ Q ♇
05 01 ☿ △ ♄
05 42 ☽ □ ♂
08 05 ♀ # ♅
08 46 ☽ ∥ ☉
09 14 ♀ ± ♇
13 45 ♀ ☍ ♅
14 27 ☽ ∠ ♃
15 22 ☽ Q ☿
19 21 ☽ ∠ ♄
22 22 ☽ □ ☉

11 Monday
02 13 ☽ ✶ ♇
05 07 ☽ ∥ ♂
06 13 ☽ △ ♆
07 10 ☽ ⊥ ♃
12 10 ☽ ⊥ ♄
13 55 ☽ □ ♅
16 37 ☽ □ ♀
18 30 ☽ ⊻ ♃
23 34 ☽ ⊻ ♄

12 Tuesday
03 15 ☿ bQ ♇
06 41 ☽ ∠ ♇
07 17 ☽ □ ☿
11 16 ♀ △ ♃
17 53 ☽ △ ♂

13 Wednesday
00 07 ☽ ⊥ ♇
00 29 ☉ ⊼ ♇
00 30 ☽ Q ♅
04 46 ☽ ∥ ♇
11 53 ☽ ⊻ ♇
12 53 ☽ △ ☉
16 05 ☽ □ ♆

14 Thursday
00 04 ☽ ✶ ♅
01 17 ☽ bQ ♂
04 50 ☽ ☌ ♃
09 43 ☽ △ ♀
10 09 ☽ ☌ ♄
13 46 ♀ △ ♄
18 39 ☽ ∥ ♃
19 01 ☽ ∥ ♄
20 23 ☽ ± ♂
21 18 ☽ bQ ☉

15 Friday
03 53 ☽ Q ♆
04 31 ☉ ⊼ ♆
04 31 ☽ △ ☿
05 55 ☽ ∠ ♅
06 33 ☽ # ♄
06 54 ☽ # ♃
09 12 ☽ ⊼ ♂
16 57 ☽ ± ☉
19 14 ☽ bQ ♀
23 41 ☽ ☌ ♇
23 57 ☽ ⊥ ♅

16 Saturday
03 59 ☽ ✶ ♆
06 06 ☽ ⊼ ☉
09 14 ♂ bQ ♃
12 01 ☽ ⊻ ♅
15 29 ☽ ± ♀
15 34 ☽ bQ ☿
16 53 ☽ ⊻ ♃
21 30 ☽ # ♇
22 18 ☽ ⊻ ♄

17 Sunday
04 59 ☽ ⊥ ♃
05 00 ☽ ⊼ ♀
10 14 ☽ ∠ ♆
10 24 ☽ ⊥ ♄
12 33 ☽ ± ☿
13 39 ♀ bQ ♇
15 59 ☿ ⊻ ♂
23 08 ☽ ∠ ♃

18 Monday
01 44 ☽ ☍ ♂
02 29 ☽ ⊼ ☿
04 23 ☽ ⊥ ♆
04 32 ☽ ∠ ♄
05 18 ☉ # ♅
12 10 ☽ ⊻ ♇
16 29 ☽ ⊻ ♆
22 25 ♀ # ♆

19 Tuesday
00 04 ☽ ☍ ☉
00 16 ☽ ⊥ ♇
00 25 ☽ ☌ ♅
02 32 ☉ ± ♇
04 22 ☉ ☍ ♅
05 21 ☽ ✶ ♃
10 44 ☽ ✶ ♄
15 45 ☽ # ♂
18 22 ☽ ∠ ♇
19 43 ♂ ± ♆
21 52 ♂ bQ ♄

20 Wednesday
00 31 ☽ ☍ ♀
05 26 ☽ Q ♃
10 48 ☽ Q ♄
18 04 ☽ ⊼ ♂
19 17 ☽ ∥ ♅
23 08 ☽ ☍ ☿

21 Thursday
00 23 ☽ ✶ ♇
04 38 ☽ ☌ ♆
06 02 ☽ # ☉
06 49 ☽ ± ♂
10 14 ☿ △ ♇
12 22 ☽ ⊻ ♅
16 09 ☉ △ ♃
17 19 ☽ □ ♃
17 26 ☽ ⊼ ☉
22 37 ☽ □ ♄

22 Friday
00 12 ☽ Q ♇
00 14 ☽ ⊥ ♅
01 50 ☽ bQ ♂
06 20 ☽ ± ☉
17 55 ☽ ∠ ♅
18 13 ♅ ⊥ ♇
18 56 ☽ ⊼ ♀

23 Saturday
01 31 ☽ bQ ☉
01 32 ☿ ☍ ♆
07 57 ☽ ± ♀
09 07 ☽ △ ♂
11 27 ☽ □ ♇
12 24 ☽ # ☉
15 33 ☽ ⊻ ♆
17 06 ☽ ⊼ ☿
22 59 ☽ ✶ ♅

24 Sunday
03 06 ☽ ⊥ ♆
03 15 ☽ bQ ♀
03 54 ☽ △ ♃
05 51 ☽ ± ☿
08 31 ☽ ∥ ♅
08 56 ☽ △ ☉
08 58 ☽ △ ♄
09 23 ☉ △ ♄
20 13 ☽ ∠ ♆
21 48 ☽ Q ♅
23 46 ☽ # ♂
23 50 ♂ ⊼ ♇

25 Monday
00 37 ☽ bQ ☿
08 18 ☽ bQ ♃
10 41 ☽ △ ♀
13 13 ☽ bQ ♄
20 16 ☽ △ ♇
21 30 ☽ □ ♂

26 Tuesday
00 08 ☽ ✶ ♆
00 57 ☽ ± ♃
05 46 ☽ ± ♄
06 55 ☽ △ ☿
07 06 ☽ □ ♅
09 12 ☿ ⊼ ♅
11 54 ☽ ⊼ ♃
16 39 ☽ ⊼ ♄
21 00 ☽ □ ☉
21 52 ☽ Q ♆
23 28 ☽ bQ ♇

27 Wednesday
07 52 ☽ Q ♂
15 22 ☽ ± ♇
18 27 ♃ St D
19 36 ♂ ⊼ ♆
20 33 ☽ # ♇
22 11 ☽ □ ♀

28 Thursday
01 47 ☽ ⊼ ♇
05 23 ☽ □ ♆
05 56 ☽ ✶ ♂
06 51 ☉ bQ ♇
06 57 ☽ Q ☉
11 51 ☽ △ ♅
15 38 ☽ □ ☿
16 29 ☽ ☍ ♃
20 52 ☽ ☍ ♄

29 Friday
04 44 ☽ ✶ ☉
04 52 ☽ # ♃
04 58 ☽ # ♄
05 45 ☿ □ ♃
07 39 ☽ Q ♀
08 38 ☽ ∠ ♂
12 57 ☽ bQ ♅
14 24 ♀ △ ♇
15 17 ☽ ∥ ♄
15 22 ☽ ∥ ♃
23 25 ☽ Q ☿

30 Saturday
00 13 ☽ ⊥ ♂
03 40 ☽ ± ♅
03 57 ☽ ☍ ♇
05 11 ☽ ✶ ♀
07 08 ☽ ∠ ☉
07 19 ☽ △ ♆
10 28 ☽ ⊻ ♂
13 22 ☽ ⊼ ♅
17 54 ☽ ⊼ ♃
19 52 ☽ ✶ ☿
22 01 ☽ ⊼ ♄
22 13 ☽ ∥ ♇
22 38 ☽ ⊥ ☉

31 Sunday
03 31 ☽ ± ♃
06 19 ♀ ☍ ♆
07 21 ☽ bQ ♆
07 27 ☽ ∠ ♀
07 35 ☽ ± ♄
08 51 ☽ ⊻ ☉
10 57 ☉ # ♆
17 50 ☽ bQ ♃
18 08 ♂ # ♅
20 53 ☽ ∠ ☿
21 37 ☽ ± ♆
21 52 ☽ bQ ♄
22 59 ☽ ⊥ ♀

All ephemeris data is given at 12.00 UT and the Moon's longitude is additionally given for 24.00 UT

JUNE 1981

LONGITUDES

Date	Sidereal time h m s	Sun ☉ ° ' "	Moon ☽ ° ' "	Moon ☽ 24.00 ° ' "	Mercury ☿ ° '	Venus ♀ ° '	Mars ♂ ° '	Jupiter ♃ ° '	Saturn ♄ ° '	Uranus ♅ ° '	Neptune ♆ ° '	Pluto ⯓ ° '
01	04 39 33	10 ♊ 53 37	26 ♉ 57 29	04 ♊ 33 18	02 ♋ 47	25 ♊ 23	27 ♉ 20	00 ♎ 29	03 ♎ 00	27 ♏ 33	23 ♐ 50	21 ♎ 47
02	04 43 29	11 51 07	12 ♊ 07 49	19 ♊ 39 46	03 21	26 36	28 03	00 30	03 R 00	27 R 31	23 R 49	21 R 46
03	04 47 26	12 48 37	27 ♊ 07 58	04 ♋ 31 23	03 51	27 50	28 46	00 31	03 00	27 29	23 47	21 45
04	04 51 22	13 46 05	11 ♋ 49 10	19 ♋ 00 41	04 16	29 ♊ 03	29 ♉ 29	00 32	03 00	27 26	23 45	21 44
05	04 55 19	14 43 32	26 ♋ 05 29	03 ♌ 03 20	04 36	00 ♋ 17	00 ♊ 12	00 34	03 D 00	27 24	23 44	21 43
06	04 59 15	15 40 58	09 ♌ 54 10	16 ♌ 38 05	04 52	01 30	00 55	00 35	03 00	27 21	23 42	21 42
07	05 03 12	16 38 23	23 ♌ 15 21	29 ♌ 46 18	05 04	02 44	01 37	00 37	03 00	27 19	23 41	21 42
08	05 07 09	17 35 47	06 ♍ 11 22	12 ♍ 31 04	05 10	03 57	02 20	00 39	03 00	27 17	23 39	21 41
09	05 11 05	18 33 10	18 ♍ 45 57	24 ♍ 56 34	05 R 13	05 11	03 03	00 41	03 01	27 15	23 37	21 40
10	05 15 02	19 30 32	01 ♎ 03 30	07 ♎ 07 19	05 10	06 24	03 45	00 44	03 01	27 12	23 36	21 39
11	05 18 58	20 27 52	13 ♎ 08 35	19 ♎ 07 50	05 04	07 37	04 28	00 46	03 02	27 10	23 34	21 39
12	05 22 55	21 25 12	25 ♎ 05 33	01 ♏ 02 15	04 53	08 51	05 10	00 49	03 03	27 08	23 33	21 38
13	05 26 51	22 22 30	06 ♏ 58 20	12 ♏ 54 13	04 37	10 04	05 53	00 52	03 03	27 06	23 31	21 38
14	05 30 48	23 19 48	18 ♏ 50 15	24 ♏ 46 46	04 18	11 17	06 35	00 55	03 04	27 03	23 29	21 37
15	05 34 44	24 17 05	00 ♐ 44 04	06 ♐ 42 23	03 56	12 30	07 18	00 58	03 05	27 01	23 28	21 36
16	05 38 41	25 14 22	12 ♐ 41 59	18 ♐ 43 02	03 30	13 44	08 00	01 01	03 06	26 59	23 26	21 36
17	05 42 38	26 11 38	24 ♐ 45 45	00 ♑ 50 19	03 02	14 57	08 42	01 05	03 08	26 57	23 24	21 35
18	05 46 34	27 08 53	06 ♑ 56 53	13 ♑ 05 40	02 31	16 10	09 24	01 09	03 09	26 55	23 23	21 35
19	05 50 31	28 06 08	19 ♑ 16 48	25 ♑ 30 31	01 59	17 23	10 07	01 12	03 10	26 53	23 21	21 35
20	05 54 27	29 ♊ 03 22	01 ♒ 46 59	08 ♒ 06 26	01 25	18 37	10 49	01 16	03 12	26 51	23 20	21 34
21	05 58 24	00 ♋ 00 36	14 ♒ 29 07	20 ♒ 55 15	00 50	19 50	11 31	01 21	03 13	26 49	23 18	21 34
22	06 02 20	00 57 50	27 ♒ 25 07	03 ♓ 58 57	00 ♋ 16	21 03	12 13	01 25	03 15	26 47	23 16	21 33
23	06 06 17	01 55 04	10 ♓ 37 00	17 ♓ 19 30	29 ♊ 41	22 16	12 55	01 29	03 17	26 45	23 15	21 33
24	06 10 13	02 52 18	24 ♓ 06 39	00 ♈ 58 34	29 08	23 29	13 36	01 34	03 19	26 43	23 13	21 33
25	06 14 10	03 49 31	07 ♈ 55 21	14 ♈ 56 59	28 37	24 42	14 18	01 39	03 21	26 42	23 12	21 33
26	06 18 07	04 46 45	22 ♈ 03 19	29 ♈ 14 09	28 08	25 55	15 00	01 44	03 23	26 40	23 10	21 32
27	06 22 03	05 43 58	06 ♉ 29 05	13 ♉ 47 37	27 41	27 09	15 42	01 49	03 25	26 38	23 08	21 32
28	06 26 00	06 41 12	21 ♉ 09 08	28 ♉ 32 51	27 17	28 22	16 23	01 54	03 27	26 36	23 07	21 32
29	06 29 56	07 38 26	05 ♊ 57 54	13 ♊ 23 21	26 58	29 ♋ 35	17 05	02 00	03 30	26 35	23 05	21 32
30	06 33 53	08 ♋ 35 39	20 ♊ 48 13	28 ♊ 11 30	26 ♊ 42	00 ♌ 48	17 ♊ 47	02 ♎ 05	03 ♎ 32	26 ♏ 33	23 ♐ 04	21 ♎ 32

Date	Moon True ☊ ° '	Moon Mean ☊ ° '	Moon ☽ Latitude ° '
01	02 ♌ 57	04 ♌ 30	04 S 40
02	02 R 50	04 27	03 57
03	02 44	04 23	02 59
04	02 40	04 20	01 50
05	02 38	04 17	00 S 35
06	02 D 38	04 14	00 N 39
07	02 39	04 11	01 49
08	02 40	04 07	02 51
09	02 R 41	04 04	03 43
10	02 40	04 01	04 24
11	02 38	03 58	04 52
12	02 33	03 55	05 07
13	02 28	03 52	05 08
14	02 21	03 48	04 57
15	02 14	03 45	04 32
16	02 07	03 42	03 56
17	02 02	03 39	03 08
18	01 58	03 36	02 12
19	01 56	03 32	01 08
20	01 55	03 29	00 N 01
21	01 D 55	03 26	01 S 08
22	01 57	03 23	02 15
23	01 58	03 20	03 16
24	01 59	03 17	04 08
25	02 R 00	03 13	04 46
26	01 59	03 10	05 09
27	01 57	03 07	05 13
28	01 54	03 04	04 56
29	01 50	03 01	04 20
30	01 ♌ 47	02 ♌ 58	03 S 26

DECLINATIONS

Date	Sun ☉ ° '	Moon ☽ ° '	Mercury ☿ ° '	Venus ♀ ° '	Mars ♂ ° '	Jupiter ♃ ° '	Saturn ♄ ° '	Uranus ♅ ° '	Neptune ♆ ° '	Pluto ⯓ ° '
01	22 N 05	14 N 56	24 N 34	24 N 05	19 N 32	01 N 07	01 N 07	19 S 24	21 S 56	07 N 46
02	22 13	18 19	24 22	24 10	19 43	01 06	01 07	19 23	21 56	07 46
03	22 20	20 26	24 09	24 13	19 53	01 05	01 07	19 23	21 56	07 46
04	22 27	21 06	23 55	24 17	20 03	01 04	01 06	19 22	21 56	07 46
05	22 34	20 21	23 41	24 19	20 13	01 04	01 06	19 22	21 56	07 45
06	22 40	18 24	23 26	24 20	20 22	01 03	01 06	19 21	21 56	07 45
07	22 46	15 29	23 10	24 21	20 31	01 02	01 06	19 21	21 56	07 45
08	22 52	11 54	22 54	24 22	20 41	01 01	01 05	19 20	21 55	07 45
09	22 57	07 52	22 38	24 21	20 50	01 00	01 05	19 20	21 55	07 45
10	23 02	03 N 37	22 22	24 20	20 58	00 59	01 05	19 19	21 55	07 44
11	23 06	00 S 42	22 05	24 18	21 07	00 57	01 04	19 19	21 55	07 44
12	23 10	04 57	21 49	24 15	21 15	00 56	01 04	19 18	21 55	07 44
13	23 13	08 59	21 32	24 12	21 23	00 55	01 03	19 18	21 55	07 44
14	23 16	12 40	21 16	24 08	21 31	00 53	01 02	19 17	21 55	07 43
15	23 19	15 52	21 00	24 03	21 39	00 52	01 02	19 17	21 55	07 43
16	23 21	18 25	20 44	23 57	21 47	00 50	01 01	19 16	21 55	07 43
17	23 23	20 12	20 29	23 51	21 54	00 49	01 01	19 16	21 55	07 43
18	23 25	21 04	20 15	23 44	22 01	00 47	01 00	19 16	21 55	07 42
19	23 26	20 55	20 01	23 36	22 08	00 45	00 59	19 15	21 55	07 42
20	23 26	19 45	19 48	23 28	22 15	00 43	00 58	19 15	21 55	07 42
21	23 26	17 34	19 36	23 19	22 21	00 41	00 57	19 14	21 55	07 41
22	23 26	14 29	19 25	23 09	22 27	00 39	00 57	19 14	21 54	07 41
23	23 26	10 37	19 15	22 59	22 34	00 37	00 56	19 13	21 54	07 40
24	23 25	06 08	19 06	22 48	22 39	00 35	00 55	19 13	21 54	07 40
25	23 23	01 S 15	18 59	22 36	22 45	00 33	00 54	19 13	21 54	07 39
26	23 21	03 N 49	18 53	22 24	22 50	00 31	00 53	19 12	21 54	07 39
27	23 19	08 45	18 48	22 11	22 56	00 29	00 52	19 12	21 54	07 39
28	23 16	13 17	18 45	21 58	23 01	00 27	00 50	19 11	21 54	07 38
29	23 13	17 02	18 44	21 43	23 05	00 24	00 49	19 11	21 54	07 38
30	23 N 10	19 N 41	18 N 44	21 N 29	23 N 10	00 N 22	00 N 48	19 S 11	21 S 54	07 N 37

ZODIAC SIGN ENTRIES

Date	h	m	Planets
01	16	48	☽ ♊
03	16	38	☽ ♋
05	05	26	♂ ♊
05	06	29	♀ ♋
05	18	43	☽ ♌
08	00	25	☽ ♍
10	09	55	☽ ♎
12	21	54	☽ ♏
15	10	31	☽ ♐
17	22	21	☽ ♑
20	08	36	☽ ♒
21	11	45	☉ ♋
22	16	44	☽ ♓
22	22	51	☿ ♊
24	22	18	☽ ♈
27	01	16	☽ ♉
29	02	21	☽ ♊
29	20	20	♀ ♌

LATITUDES

Date	Mercury ☿ ° '	Venus ♀ ° '	Mars ♂ ° '	Jupiter ♃ ° '	Saturn ♄ ° '	Uranus ♅ ° '	Neptune ♆ ° '	Pluto ⯓ ° '
01	01 N 10	00 N 44	00 S 02	01 N 25	02 N 31	00 N 13	01 N 22	17 N 30
04	00 N 33	00 50	00 00	01 24	02 30	00 13	01 22	17 29
07	00 S 10	00 57	00 N 02	01 24	02 30	00 13	01 22	17 27
10	00 59	01 03	00 04	01 23	02 29	00 13	01 22	17 26
13	01 49	01 08	00 06	01 22	02 28	00 13	01 22	17 24
16	02 39	01 14	00 08	01 22	02 27	00 13	01 22	17 23
19	03 25	01 19	00 10	01 21	02 27	00 13	01 22	17 21
22	04 02	01 23	00 12	01 20	02 26	00 13	01 22	17 19
25	04 27	01 27	00 14	01 19	02 25	00 13	01 22	17 18
28	04 40	01 30	00 16	01 19	02 25	00 13	01 22	17 16
31	04 S 38	01 N 33	00 N 18	01 N 18	02 N 24	00 N 13	01 N 22	17 N 14

DATA

Julian Date	2444757
Delta T	+51 seconds
Ayanamsa	23° 35' 36"
Synetic vernal point	05° ♓ 31' 24"
True obliquity of ecliptic	23° 26' 24"

MOON'S PHASES, APSIDES AND POSITIONS ☽

Date	h	m	Phase	Longitude ° '	Eclipse Indicator
02	11	32	●	11 ♊ 50	
09	11	33	☽	18 ♍ 32	
17	15	04	○	26 ♐ 19	
25	04	25	☾	03 ♈ 31	

Day	h	m		
01	14	13	Perigee	
14	02	32	Apogee	
29	18	49	Perigee	
04	11	05	Max dec	21° N 06'
11	08	03	0S	
18	20	49	Max dec	21° S 08'
25	17	56	0N	

LONGITUDES

Date	Chiron ⚷ ° '	Ceres ⚳ ° '	Pallas ⚴ ° '	Juno ⚵ ° '	Vesta ⚶ ° '	Black Moon Lilith ⚸ ° '
01	19 ♉ 35	06 ♌ 14	28 ♊ 58	22 ♎ 31	04 ♍ 34	17 ♏ 11
11	20 ♉ 13	10 ♌ 11	04 ♋ 56	22 ♎ 03	07 ♍ 51	18 ♏ 19
21	20 ♉ 48	14 ♌ 16	10 ♋ 55	22 ♎ 07	11 ♍ 27	19 ♏ 26
31	21 ♉ 19	18 ♌ 28	16 ♋ 52	22 ♎ 41	15 ♍ 21	20 ♏ 33

All ephemeris data is given at 12.00 UT and the Moon's longitude is additionally given for 24.00 UT

ASPECTARIAN

h m	Aspects
01 Monday	
00 32	♃ ∥ ♄
03 50	☽ ⊼ ⯓
07 05	☽ ⊼ ♆
09 18	☽ ⊻ ♀
11 43	☽ ⊥ ☿
12 37	☽ ☌ ♂
12 56	☽ ☍ ♅
13 18	☽ ± ⯓
17 34	☽ △ ♃
19 09	♂ ☍ ♅
20 55	☿ □ ♄
21 33	☽ △ ♄
21 35	☽ ⊻ ☿
02 Tuesday	
02 46	♂ ± ⯓
03 30	☽ ♇ ⯓
11 32	☽ ☌ ☉
19 01	☿ ± ♅
22 13	☽ # ♅
03 Wednesday	
03 12	☽ ∥ ♂
03 21	☽ △ ⯓
05 14	♀ ⊼ ♅
05 28	☿ ∥ ♀
06 37	☽ ☍ ♆
12 33	☽ ⊼ ♅
13 14	☽ ☌ ♀
14 46	☽ ⊻ ♂
17 29	☽ □ ♃
21 31	☽ □ ♄
22 16	☽ ± ♅
23 14	☽ ☌ ☿
04 Thursday	
01 02	☽ ⊥ ♂
13 01	☽ ♇ ♅
15 28	☽ ⊻ ☉
16 39	☽ ∠ ♂
23 13	☽ Q ♃
05 Friday	
02 12	♄ St D
02 14	☽ ⊥ ☉
03 21	☽ Q ♄
04 35	☽ □ ⯓
07 57	♀ ⊻ ♂
07 59	☽ ⊼ ♆
14 11	☽ ∥ ♂
14 14	☽ △ ♅
17 35	♀ □ ♃
18 14	☽ ± ♆
18 42	☽ ∠ ☉
19 26	☽ ⚹ ♂
19 41	☽ ⚹ ♃
19 53	☽ ⊻ ♀
23 54	☽ ⚹ ♄
06 Saturday	
00 43	♂ △ ♃
01 49	☽ # ♅
03 00	☽ ⊻ ☿
07 22	☽ ⊥ ♀
09 53	☽ ♇ ♆
11 39	☽ Q ⯓
13 45	☽ ⊥ ☿
17 38	☽ Q ♂
22 09	☽ ∠ ♃
23 05	☽ ⚹ ☉
07 Sunday	
00 57	☽ ∠ ♀
02 27	☽ ∠ ♄
06 07	☽ ∠ ☿
09 09	☽ ⚹ ⯓
12 46	☽ △ ♆
14 30	☽ ⊥ ♃
17 21	♀ □ ♄
18 53	☽ ⊥ ♄
19 26	☽ □ ♅
22 41	☽ Q ☉
23 13	♀ ± ♅
08 Monday	
01 36	☽ ⊻ ♃
04 21	☽ □ ♂
06 01	☽ ⊻ ♄
07 21	☽ ⚹ ♀
10 05	☽ ⚹ ☿
12 56	☽ ∠ ⯓
15 01	☉ ∥ ☿
09 Tuesday	
05 14	☽ Q ♅
06 02	☽ ⊥ ⯓
08 36	☽ Q ♀
09 00	☽ Q ☿
10 51	♂ △ ♄
11 33	☽ □ ☉
11 36	☿ St R
12 40	☿ ☌ ♀
12 42	☽ ∥ ⯓
17 37	☽ ⊻ ⯓
21 24	☽ □ ♆
10 Wednesday	
04 27	☽ ⚹ ♅
11 21	☽ ☌ ♃
15 53	☽ ☌ ♄
17 40	☽ △ ♂
20 04	☽ □ ☿
23 45	☽ □ ♀
11 Thursday	
02 06	☽ ∥ ♄
02 42	☽ ∥ ♃
08 52	☽ Q ♆
10 03	☽ ∠ ♅
13 22	☽ # ♃
14 00	☽ # ♄
12 Friday	
01 28	☽ ♇ ♂
02 49	☿ # ♆
03 58	☽ △ ☉
04 03	☽ ⊥ ♅
04 09	☿ ⊻ ♂
05 03	☽ ☌ ⯓
08 53	☽ ⚹ ♆
16 06	☽ ⊻ ♅
17 22	☉ △ ⯓
20 46	☽ ± ♂
23 36	☽ ⊻ ♃
13 Saturday	
04 04	☽ ⊻ ♄
04 22	☽ # ⯓
07 22	☽ △ ☿
09 39	☽ ⊼ ♂
11 47	☽ ⊥ ♃
12 53	☽ ♇ ☉
15 07	☽ ∠ ♆
16 13	☽ ⊥ ♄
18 59	☽ △ ♀
20 47	☿ ∥ ♂
14 Sunday	
06 04	☽ ∠ ♃
08 41	☽ ± ☉
09 17	☽ ⊥ ♆
10 27	☽ ∠ ♄
12 54	♂ ♇ ⯓
12 55	☽ ♇ ☿
15 51	☉ ☍ ♆
17 37	☽ ⊻ ⯓
21 22	☽ ⊻ ♆
21 52	☽ ⊼ ☉
15 Monday	
02 43	♀ ♇ ♅
04 33	☽ ☌ ♅
04 46	☽ ♇ ♀
05 42	☽ ⊥ ⯓
06 32	☽ ± ☿
12 28	☽ ⚹ ♃
16 44	☽ ⚹ ♄
18 13	☽ ⊼ ☿
23 47	☽ ∠ ⯓
16 Tuesday	
00 55	☽ ± ♀
02 00	☽ ☍ ♂
12 39	☽ Q ♃
14 17	☽ ⊼ ♀
16 49	☽ Q ♄
22 04	☽ ∥ ♅
17 Wednesday	
00 34	♀ ⊥ ♂
05 43	☽ ⚹ ⯓
07 36	☿ □ ♄
09 19	☽ ☌ ♆
14 38	♂ # ♆
15 04	☽ ☍ ☉
16 08	☿ ± ♅
16 19	☽ ⊻ ♅
16 36	☽ # ☿
18 Thursday	
00 32	☽ □ ♃
03 40	☽ ☍ ☿
04 07	☽ ⊥ ♅
04 32	☽ □ ♄
05 24	☽ Q ⯓
06 25	☉ ⊼ ♅
17 06	☽ ⊼ ♂
21 41	☽ ∠ ♅
19 Friday	
05 29	☽ ± ♂
07 57	☽ ☍ ♀
16 26	☽ □ ⯓
19 50	☽ ⊻ ♆
23 54	☽ ♇ ♂
20 Saturday	
02 36	☽ ⚹ ♅
06 22	☽ ⊼ ☉
07 19	☽ ⊥ ♆
11 02	☽ △ ♃
11 18	☽ # ☿
11 19	☽ ⊼ ☿
14 42	☽ △ ♄
17 01	☉ ∥ ♀
17 10	☿ □ ♃
18 43	☽ ∥ ♅
18 44	☽ ± ☉
22 13	☽ ± ☿
21 Sunday	
00 23	☽ ∠ ♆
01 22	☽ Q ♅
01 51	♀ Q ♃
06 06	☽ △ ♂
13 04	☽ ♇ ☉
14 25	☽ ♇ ☿
15 30	☽ ♇ ♃
19 00	☽ ♇ ♄
23 01	☽ ⊼ ♀
22 Monday	
00 57	☉ ☌ ☿
01 11	☽ △ ⯓
04 22	☽ ⚹ ♆
08 18	☽ ± ♃
10 50	☽ □ ♅
11 15	☽ ± ♀
11 41	☽ ± ♄
16 04	♀ Q ♄
16 59	☽ △ ☿
19 00	☽ △ ☉
19 22	☽ ⊼ ♃
21 58	♀ □ ⯓
22 42	☽ ⊼ ♄
23 Tuesday	
00 21	☉ □ ♃
02 19	☽ Q ♆
04 40	☽ ♇ ⯓
05 21	☽ ♇ ♀
15 33	☿ # ♅
16 21	☽ □ ♂
20 50	☽ ± ⯓
24 Wednesday	
04 03	☽ # ⯓
06 50	♀ ⊼ ♆
07 29	☽ ⊼ ⯓
08 23	☉ ± ♅
10 26	☽ □ ♆
10 48	☽ △ ♀
16 34	☽ △ ♅
20 28	☽ □ ☿
23 29	☉ □ ♄
25 Thursday	
00 12	♀ ∥ ♂
01 06	☽ ☍ ♃
01 47	☽ Q ♂
04 05	☽ ☍ ♄
04 25	☽ □ ☉
13 41	☽ # ♄
15 19	☽ # ♃
18 26	☽ ♇ ♅
20 31	☽ ∥ ♃
22 09	☽ ∥ ♄
23 28	☽ ⚹ ♂
26 Friday	
02 19	☽ Q ☿
09 40	☽ ± ♅
11 08	☽ ☍ ⯓
13 18	☽ Q ☉
13 52	☽ △ ♆
19 05	☽ □ ♀
19 42	☽ ⊼ ♅
21 50	☽ ⚹ ☿
27 Saturday	
01 56	☽ ∠ ♂
02 12	♀ △ ♅
04 14	☽ ⊼ ♃
06 29	☽ ∥ ⯓
06 55	☽ ⊼ ♄
10 40	☽ ⚹ ☉
14 12	☽ ± ♃
14 43	☽ ♇ ♆
16 50	☽ ± ♄
17 33	☽ ⊥ ♂
19 58	☿ ⊻ ♀
21 54	☽ ∠ ☿
28 Sunday	
03 29	☽ Q ♀
03 51	☽ ⊻ ♂
05 02	☽ ♇ ♃
05 26	☽ ± ♆
07 36	☽ ♇ ♄
12 13	☽ ⊥ ☿
12 38	☽ ⊼ ⯓
12 56	☽ ∠ ☉
15 11	☽ ⊼ ♆
18 08	♀ # ♆
20 50	☽ ☍ ♅
21 44	☽ ⊻ ☿
22 22	☽ ± ⯓
29 Monday	
00 45	☽ ⚹ ♀
02 33	♀ ± ♆
04 32	☽ ⊥ ☉
05 32	☽ △ ♃
08 00	☽ △ ♄
12 55	☽ ♇ ⯓
14 54	☽ ⊻ ☉
30 Tuesday	
01 55	☽ ∥ ☿
03 10	☽ ∠ ♀
06 20	☽ # ♅
06 51	☽ ☌ ♂
11 14	☉ ∥ ♂
13 11	☽ △ ⯓
15 39	☽ ☍ ♆
19 03	☽ ⊥ ♀
21 19	☽ ⊼ ♅
21 25	☽ ☌ ☿

JULY 1981

LONGITUDES

Date	Sidereal time h m s	Sun ☉ ° ' "	Moon ☽ ° ' "	Moon ☽ 24.00 ° ' "	Mercury ☿ ° '	Venus ♀ ° '	Mars ♂ ° '	Jupiter ♃ ° '	Saturn ♄ ° '	Uranus ♅ ° '	Neptune ♆ ° '	Pluto ♇ ° '
01	06 37 49	09 ♋ 32 53	05 ♋ 32 15	12 ♋ 49 34	26 ♊ 30	02 ♌ 01	18 ♊ 28	02 ♎ 11	03 ♎ 35	26 ♏ 31	23 ♐ 02	21 ♎ 32
02	06 41 46	10 30 07	20 ♋ 02 39	27 ♋ 10 50	26 R 23	03 14	19 09	02 17	03 37	26 R 30	23 R 01	21 D 32
03	06 45 42	11 27 21	04 ♌ 13 36	11 ♌ 10 32	26 20	04 27	19 51	02 23	03 40	26 28	22 59	21 32
04	06 49 39	12 24 34	18 ♌ 01 25	24 ♌ 46 09	26 D 22	05 40	20 32	02 29	03 43	26 27	22 58	21 32
05	06 53 36	13 21 47	01 ♍ 24 46	07 ♍ 57 25	26 29	06 53	21 14	02 35	03 46	26 25	22 56	21 32
06	06 57 32	14 19 00	14 ♍ 24 22	20 ♍ 45 58	26 42	08 06	21 55	02 42	03 49	26 24	22 55	21 32
07	07 01 29	15 16 13	27 ♍ 02 39	03 ♎ 14 52	26 59	09 18	22 36	02 48	03 52	26 22	22 53	21 33
08	07 05 25	16 13 25	09 ♎ 23 08	15 ♎ 28 00	27 22	10 31	23 17	02 55	03 55	26 21	22 52	21 33
09	07 09 22	17 10 38	21 ♎ 30 03	27 ♎ 29 49	27 49	11 44	23 58	03 02	03 58	26 20	22 50	21 33
10	07 13 18	18 07 50	03 ♏ 27 53	09 ♏ 24 47	28 22	12 57	24 39	03 09	04 02	26 19	22 49	21 33
11	07 17 15	19 05 02	15 ♏ 21 05	21 ♏ 17 18	29 00	14 10	25 20	03 16	04 05	26 17	22 47	21 34
12	07 21 11	20 02 15	27 ♏ 13 54	03 ♐ 11 21	29 ♊ 42	15 23	26 01	03 23	04 09	26 16	22 46	21 34
13	07 25 08	20 59 27	09 ♐ 10 04	15 ♐ 10 27	00 ♋ 30	16 35	26 42	03 30	04 12	26 15	22 45	21 34
14	07 29 05	21 56 40	21 ♐ 12 50	27 ♐ 17 32	01 23	17 48	27 23	03 38	04 16	26 14	22 43	21 35
15	07 33 01	22 53 52	03 ♑ 24 49	09 ♑ 34 54	02 21	19 01	28 03	03 45	04 20	26 13	22 42	21 35
16	07 36 58	23 51 05	15 ♑ 47 59	22 ♑ 04 12	03 23	20 13	28 44	03 53	04 24	26 12	22 41	21 36
17	07 40 54	24 48 19	28 ♑ 23 42	04 ♒ 46 32	04 30	21 26	29 ♊ 25	04 01	04 28	26 11	22 39	21 36
18	07 44 51	25 45 32	11 ♒ 12 47	17 ♒ 42 28	05 42	22 39	00 ♋ 05	04 09	04 32	26 10	22 38	21 37
19	07 48 47	26 42 47	24 ♒ 15 36	00 ♓ 52 11	06 59	23 51	00 46	04 17	04 36	26 10	22 37	21 37
20	07 52 44	27 40 02	07 ♓ 32 10	14 ♓ 15 31	08 20	25 04	01 26	04 25	04 40	26 09	22 35	21 38
21	07 56 40	28 37 17	21 ♓ 02 11	27 ♓ 52 06	09 46	26 16	02 07	04 33	04 44	26 08	22 34	21 39
22	08 00 37	29 ♋ 34 33	04 ♈ 45 10	11 ♈ 41 17	11 16	27 29	02 47	04 42	04 49	26 07	22 33	21 39
23	08 04 34	00 ♌ 31 50	18 ♈ 40 17	25 ♈ 42 02	12 50	28 41	03 27	04 50	04 53	26 07	22 32	21 40
24	08 08 30	01 29 08	02 ♉ 46 18	09 ♉ 52 51	14 28	29 ♌ 54	04 08	04 59	04 57	26 06	22 31	21 41
25	08 12 27	02 26 27	17 ♉ 01 23	24 ♉ 11 34	16 09	01 ♍ 06	04 48	05 08	05 02	26 06	22 29	21 42
26	08 16 23	03 23 47	01 ♊ 22 59	08 ♊ 35 12	17 55	02 19	05 28	05 16	05 07	26 05	22 28	21 42
27	08 20 20	04 21 08	15 ♊ 47 43	23 ♊ 00 01	19 43	03 31	06 08	05 25	05 11	26 05	22 27	21 43
28	08 24 16	05 18 30	00 ♋ 11 32	07 ♋ 21 40	21 35	04 43	06 48	05 34	05 16	26 04	22 26	21 44
29	08 28 13	06 15 53	14 ♋ 29 51	21 ♋ 35 30	23 30	05 56	07 28	05 44	05 21	26 04	22 25	21 45
30	08 32 09	07 13 17	28 ♋ 38 05	05 ♌ 37 05	25 27	07 08	08 08	05 53	05 26	26 04	22 24	21 46
31	08 36 06	08 ♌ 10 42	12 ♌ 32 04	19 ♌ 22 41	27 ♋ 26	08 ♍ 20	08 ♋ 48	06 ♎ 02	05 ♎ 31	26 ♏ 04	22 ♐ 23	21 ♎ 47

Date	Moon True ☊ ° '	Moon Mean ☊ ° '	Moon ☽ Latitude ° '
01	01 ♌ 45	02 ♌ 54	02 S 19
02	01 R 44	02 51	01 S 04
03	01 D 43	02 48	00 N 14
04	01 44	02 45	01 29
05	01 45	02 42	02 37
06	01 46	02 38	03 34
07	01 48	02 35	04 20
08	01 48	02 32	04 52
09	01 R 48	02 29	05 11
10	01 48	02 26	05 16
11	01 47	02 23	05 07
12	01 45	02 19	04 45
13	01 44	02 16	04 11
14	01 42	02 13	03 26
15	01 41	02 10	02 30
16	01 40	02 07	01 27
17	01 40	02 04	00 N 18
18	01 D 40	02 00	00 S 53
19	01 41	01 57	02 02
20	01 41	01 54	03 06
21	01 41	01 51	04 01
22	01 41	01 48	04 43
23	01 42	01 44	05 09
24	01 42	01 41	05 17
25	01 42	01 38	05 06
26	01 42	01 35	04 36
27	01 42	01 32	03 48
28	01 42	01 29	02 46
29	01 43	01 25	01 34
30	01 43	01 22	00 17
31	01 ♌ 43	01 ♌ 19	01 N 00

DECLINATIONS

Date	Sun ☉ ° '	Moon ☽ ° '	Mercury ☿ ° '	Venus ♀ ° '	Mars ♂ ° '	Jupiter ♃ ° '	Saturn ♄ ° '	Uranus ♅ ° '	Neptune ♆ ° '	Pluto ♇ ° '
01	23 N 06	21 N 00	18 N 45	21 N 13	23 N 14	00 N 19	00 N 47	19 S 10	21 S 54	07 N 37
02	23 01	20 53	18 48	20 57	23 18	00 17	00 46	19 10	21 54	07 36
03	22 57	19 26	18 52	20 41	23 22	00 14	00 44	19 10	21 54	07 36
04	22 52	16 50	18 58	20 23	23 26	00 12	00 43	19 09	21 54	07 35
05	22 46	13 25	19 05	20 06	23 30	00 09	00 42	19 09	21 54	07 34
06	22 40	09 26	19 13	19 47	23 33	00 06	00 40	19 09	21 54	07 34
07	22 34	05 09	19 22	19 29	23 36	00 N 03	00 39	19 08	21 54	07 33
08	22 27	00 N 46	19 32	19 09	23 39	00 00	00 37	19 08	21 54	07 33
09	22 20	03 S 34	19 43	18 49	23 41	00 S 02	00 36	19 08	21 53	07 32
10	22 13	07 43	19 54	18 29	23 44	00 05	00 34	19 07	21 53	07 31
11	22 05	11 32	20 06	18 08	23 46	00 08	00 33	19 07	21 53	07 31
12	21 57	14 54	20 18	17 47	23 48	00 11	00 31	19 07	21 53	07 30
13	21 48	17 41	20 31	17 25	23 50	00 15	00 30	19 07	21 53	07 29
14	21 39	19 41	20 44	17 03	23 51	00 18	00 28	19 06	21 53	07 29
15	21 30	20 54	20 56	16 40	23 53	00 21	00 26	19 06	21 53	07 28
16	21 20	21 04	21 08	16 17	23 54	00 24	00 25	19 06	21 53	07 27
17	21 10	20 11	21 20	15 53	23 55	00 27	00 23	19 06	21 53	07 27
18	21 00	18 15	21 31	15 29	23 55	00 31	00 21	19 06	21 53	07 26
19	20 49	15 21	21 42	15 05	23 56	00 34	00 19	19 05	21 53	07 25
20	20 38	11 37	21 51	14 40	23 56	00 38	00 17	19 05	21 53	07 24
21	20 26	07 14	21 59	14 15	23 56	00 41	00 15	19 05	21 53	07 24
22	20 14	02 S 26	22 06	13 49	23 56	00 45	00 14	19 05	21 53	07 23
23	20 02	02 N 33	22 11	13 24	23 56	00 48	00 12	19 05	21 53	07 22
24	19 50	07 28	22 14	12 57	23 55	00 52	00 10	19 05	21 53	07 21
25	19 37	12 02	22 16	12 31	23 55	00 55	00 08	19 05	21 53	07 21
26	19 24	15 56	22 15	12 04	23 54	00 59	00 06	19 05	21 53	07 20
27	19 10	18 54	22 12	11 37	23 53	01 03	00 04	19 05	21 53	07 19
28	18 57	20 40	22 06	11 09	23 51	01 06	00 N 02	19 04	21 53	07 18
29	18 42	21 06	21 58	10 42	23 50	01 10	00 00	19 04	21 53	07 17
30	18 28	20 09	21 47	10 14	23 48	01 14	00 S 03	19 04	21 53	07 16
31	18 N 13	18 N 00	21 N 34	09 N 45	23 N 46	01 S 18	00 S 05	19 S 04	21 S 53	07 N 16

ZODIAC SIGN ENTRIES

Date	h	m	Planets
01	02	57	☽ ♋
03	04	47	☽ ♌
05	09	26	☽ ♍
07	17	42	☽ ♎
10	05	02	☽ ♏
12	17	35	☽ ♐
12	21	08	☿ ♋
15	05	19	☽ ♑
17	15	02	☽ ♒
18	08	54	♂ ♋
19	22	26	☽ ♓
22	03	44	☽ ♈
22	22	40	☉ ♌
24	07	18	☽ ♉
24	14	04	♀ ♍
26	09	42	☽ ♊
28	11	41	☽ ♋
30	14	20	☽ ♌

LATITUDES

Date	Mercury ☿ ° '	Venus ♀ ° '	Mars ♂ ° '	Jupiter ♃ ° '	Saturn ♄ ° '	Uranus ♅ ° '	Neptune ♆ ° '	Pluto ♇ ° '
01	04 S 38	01 N 33	00 N 18	01 N 18	02 N 24	00 N 13	01 N 22	17 N 14
04	04 25	01 35	00 20	01 17	02 23	00 13	01 22	17 13
07	04 02	01 37	00 22	01 16	02 23	00 13	01 22	17 11
10	03 32	01 37	00 24	01 16	02 22	00 13	01 21	17 09
13	02 55	01 38	00 26	01 15	02 21	00 13	01 21	17 07
16	02 15	01 37	00 28	01 15	02 21	00 13	01 21	17 06
19	01 34	01 36	00 30	01 14	02 20	00 12	01 21	17 04
22	00 52	01 34	00 32	01 13	02 20	00 12	01 21	17 02
25	00 S 12	01 32	00 33	01 13	02 19	00 12	01 21	17 01
28	00 N 24	01 29	00 35	01 12	02 19	00 12	01 21	16 59
31	00 N 54	01 N 25	00 N 37	01 N 12	02 N 18	00 N 12	01 N 21	16 N 57

DATA

Julian Date	2444787
Delta T	+51 seconds
Ayanamsa	23° 35' 41"
Synetic vernal point	05° ♓ 31' 19"
True obliquity of ecliptic	23° 26' 25"

MOON'S PHASES, APSIDES AND POSITIONS ☽

Date	h	m	Phase	Longitude ° '	Eclipse Indicator
01	19	03	●	09 ♋ 50	
09	02	39	☽	16 ♎ 48	
17	04	39	○	24 ♑ 31	partial
24	09	40	☾	01 ♉ 24	
31	03	52	●	07 ♌ 51	Total

Day	h	m		
11	17	34	Apogee	
27	09	09	Perigee	
01	21	59	Max dec	21° N 08'
08	16	10	0S	
16	04	04	Max dec	21° S 08'
22	23	46	0N	
29	07	18	Max dec	21° N 07'

LONGITUDES

Date	Chiron ⚷ ° '	Ceres ⚳ ° '	Pallas ⚴ ° '	Juno ⚵ ° '	Vesta ⚶ ° '	Black Moon Lilith ⚸ ° '
01	21 ♉ 19	18 ♌ 28	16 ♋ 52	22 ♎ 41	15 ♍ 21	20 ♏ 33
11	21 ♉ 47	22 ♌ 46	22 ♋ 47	23 ♎ 43	19 ♍ 28	21 ♏ 40
21	22 ♉ 11	27 ♌ 09	28 ♋ 39	25 ♎ 09	23 ♍ 47	22 ♏ 47
31	22 ♉ 29	01 ♍ 36	04 ♌ 27	26 ♎ 56	28 ♍ 17	23 ♏ 54

All ephemeris data is given at 12.00 UT and the Moon's longitude is additionally given for 24.00 UT

ASPECTARIAN

h m	Aspects
01 Wednesday	
05 43	☽ ⊻ ♀
06 28	☽ □ ♃
07 04	☽ ± ♅
07 52	☿ ⊼ ♅
08 47	☽ □ ♄
15 39	♀ ✶ ♃
16 14	♇ St D
19 03	☽ ☌ ☉
20 22	☽ ∥ ♀
20 34	☿ ⊥ ♀
21 50	☽ ⚼ ♅
02 Thursday	
08 39	☽ ∥ ♀
10 27	☽ ⊻ ♂
12 24	☽ Q ♃
14 30	☽ □ ♇
14 39	☽ Q ♄
16 58	☽ ⊼ ♆
20 02	♀ ✶ ♄
21 01	☽ ⊥ ♂
22 36	☽ ⊻ ☿
22 49	☽ △ ♅
03 Friday	
03 04	☽ ± ♆
08 46	☽ ⊥ ☿
08 49	☽ ✶ ♃
11 02	☽ ✶ ♄
12 21	☉ ⚼ ♅
12 25	☽ ☌ ♀
12 57	☿ St D
13 07	☽ ∠ ♂
15 00	☽ # ♅
17 47	☽ ∥ ☿
18 27	☽ ⚼ ♆
21 09	☽ Q ♇
04 Saturday	
00 17	☽ ∠ ☿
01 25	☽ ⊻ ☉
06 21	♀ ∠ ♂
11 02	☽ ∠ ♃
12 44	☽ ⊥ ☉
13 13	☽ ∠ ♄
16 41	☽ ✶ ♂
18 14	☽ ✶ ♇
20 45	☽ △ ♆
05 Sunday	
02 01	☿ ⊼ ♅
02 59	☽ □ ♅
03 00	☽ ✶ ☿
03 12	☽ ⊥ ♃
05 22	☽ ⊥ ♄
06 02	☽ ∠ ☉
14 09	☽ ⊻ ♃
15 29	☽ Q ♂
16 18	☽ ⊻ ♄
21 23	☽ ∠ ♇
22 58	♂ △ ♇
23 02	☽ ⊻ ♀
23 54	☿ # ♅
06 Monday	
01 11	☽ Q ☿
08 29	♀ ⚼ ♆
11 21	☽ ⊥ ♀
11 49	☽ ✶ ☉
11 59	☽ Q ♅
14 08	☽ ⊥ ♇
22 38	☽ ∥ ♇
07 Tuesday	
01 28	☽ ⊻ ♇
03 00	☽ □ ♂
04 03	☽ □ ♆
06 11	☽ ∠ ♀
10 43	☽ ✶ ♅
11 53	☽ □ ☿
12 28	☽ Q ☉
16 41	♀ Q ♇
17 37	☿ ∥ ♀
21 44	♂ ☍ ♆
23 14	☽ ☌ ♃
08 Wednesday	
01 15	☽ ☌ ♄
12 45	☽ ∥ ♄
13 39	♀ # ♅
14 29	☽ ✶ ♀
14 54	☽ Q ♆
15 52	☽ ∠ ♅
16 09	☽ # ♃
16 10	☽ ∥ ♃
19 32	☽ # ♄
09 Thursday	
02 39	☽ □ ☉
09 40	☽ ⊥ ♅
12 06	☽ ☌ ♇
14 40	☽ ✶ ♆
16 58	☽ Q ♀
17 14	☽ △ ♂
21 39	☽ ⊻ ♅
10 Friday	
01 14	☽ △ ☿
10 51	☽ # ♇
11 21	☽ ⊻ ♃
13 08	☽ ⊻ ♄
20 45	☽ ∠ ♆
23 34	☽ ⊥ ♃
11 Saturday	
01 15	☽ ⚼ ♂
01 18	☽ ⊥ ♄
04 38	☿ ∠ ♀
09 05	☽ ⚼ ☿
09 19	☽ □ ♀
14 54	☽ ⊥ ♆
17 57	☽ ∠ ♃
19 35	☽ ∠ ♄
20 12	☽ △ ☉
20 32	☽ ± ♂
12 Sunday	
00 33	☽ ⊻ ♇
03 00	☽ ⊻ ♆
04 25	☽ ± ☿
09 24	☽ ⊼ ♂
10 04	☽ ☌ ♅
12 41	☽ ⊥ ♇
17 20	☽ ⊼ ☿
20 47	♂ ⊼ ♅
21 38	☉ # ♆
13 Monday	
00 31	☽ ✶ ♃
02 00	☽ ✶ ♄
05 05	☽ ⚼ ☉
06 48	☽ ∠ ♇
09 44	☽ # ♀
14 Tuesday	
00 38	☽ ± ☉
00 47	☽ Q ♃
02 08	☽ Q ♄
02 46	☉ □ ♇
02 50	☉ Q ♃
03 38	☽ ∥ ♅
04 28	☽ △ ♀
12 44	☽ ✶ ♇
13 34	☽ ⊼ ☉
14 59	☽ ☌ ♆
20 42	☉ Q ♄
21 54	☽ ⊻ ♅
15 Wednesday	
00 53	☽ ☍ ♂
06 18	♀ ∠ ♃
07 06	☉ ⊼ ♆
09 01	☿ ± ♅
09 40	☽ ⊥ ♅
09 44	☽ ☍ ☿
12 20	☽ Q ♇
12 40	☽ □ ♃
13 18	☽ ⚼ ♀
13 48	☽ □ ♄
14 20	☽ # ☿
18 40	♀ ∠ ♄
16 Thursday	
03 09	☽ ∠ ♅
08 18	☽ # ☿
08 38	☽ ± ♀
13 44	♃ # ♄
21 23	☽ ⊼ ♀
23 06	☽ □ ♇
17 Friday	
00 15	☿ □ ♃
00 47	☉ ∥ ☿
01 08	☽ ⊻ ♆
04 39	☽ ☍ ☉
07 50	☽ ✶ ♅
10 57	☿ □ ♄
12 29	☽ ⊥ ♆
14 02	☽ ⊼ ♂
15 24	♀ ✶ ♇
22 41	☽ △ ♃
23 28	☽ △ ♄
18 Saturday	
00 40	☽ ⊼ ☿
01 55	☽ ± ♂
03 03	☽ ∥ ♅
05 21	☽ ∠ ♆
06 21	☽ Q ♅
11 47	♀ △ ♆
13 01	☽ ± ☿
19 34	☽ ⚼ ♂
22 16	☉ △ ♅
19 Sunday	
02 48	☽ ⚼ ♃
03 25	☽ ⚼ ♄
07 11	☽ △ ♇
07 23	☽ ⚼ ☿
09 00	☽ ✶ ♆
11 11	☽ ☍ ♀
14 11	☽ # ♀
15 27	☽ □ ♅
16 49	☽ ⊼ ☉
19 23	☽ ± ♃
19 56	☽ ± ♄
20 Monday	
00 26	☽ △ ♂
04 31	☽ ± ☉
06 20	☽ ⊼ ♃
06 43	☽ Q ♆
06 49	☽ ⊼ ♄
10 23	☽ ⚼ ♇
13 36	☽ △ ☿
17 08	☿ # ♆
21 52	☽ ⚼ ☉
21 Tuesday	
02 27	☽ ± ♇
09 18	♀ □ ♅
10 44	☉ ± ♆
11 12	☽ # ♇
12 48	♂ ± ♅
13 04	☽ ⊼ ♇
14 42	☽ □ ♆
20 57	☽ △ ♅
22 06	☽ ⊼ ♀
22 Wednesday	
02 19	☽ △ ☉
08 24	☽ □ ♂
09 35	☽ ± ♀
09 53	☿ ⚼ ♅
11 54	☽ ☍ ♃
12 06	☽ ☍ ♄
20 06	☽ ∥ ♃
22 45	☽ # ♄
23 01	☽ ⚼ ♅
23 Thursday	
00 40	☽ □ ☿
00 46	☽ ∥ ♄
02 38	☽ ⚼ ♀
03 30	☽ # ♃
14 28	☽ ± ♅
15 20	♀ ⊥ ♃
16 07	♀ ⊥ ♄
17 00	☽ Q ♂
17 08	☽ ☍ ♇
18 35	☽ △ ♆
24 Friday	
00 42	☽ ⊼ ♅
04 15	♃ ☌ ♄
06 40	☽ △ ♀
09 40	☽ □ ☉
11 24	☽ Q ☿
11 27	☽ ∥ ♇
14 24	☽ ✶ ♂
15 43	☽ ⊼ ♄
15 46	☽ ⊼ ♃
20 00	☽ ⚼ ♆
25 Saturday	
01 53	☽ ± ♄
02 00	☽ ± ♃
09 36	☿ ∠ ♀
10 21	☽ ✶ ☿
11 06	☽ ± ♆
14 29	☽ ∥ ♀
16 53	☽ ∠ ♂
17 04	☽ ⚼ ♄
17 15	☽ ⚼ ♃
18 08	☽ Q ☉
19 50	☽ ⊼ ♇
21 09	☽ ⊼ ♆
21 33	♂ □ ♄
26 Sunday	
03 02	♂ □ ♃
03 10	☽ ☍ ♅
05 52	☽ ± ♇
08 39	☽ ⊥ ♂
13 41	☽ □ ♀
14 55	☽ ∠ ☿
15 36	☽ ✶ ☉
18 15	☽ △ ♄
18 33	☽ △ ♃
19 08	☽ ⊻ ♂
20 53	☽ ⚼ ♇
27 Monday	
08 03	☽ ⊥ ☿
13 44	☽ # ♅
14 30	☽ ∥ ☉
18 21	☽ ∠ ☉
19 30	☽ ⊻ ☿
21 53	☽ △ ♇
22 16	☉ # ♅
22 24	☽ Q ♀
23 04	☽ ☍ ♆
28 Tuesday	
05 08	☽ ⊼ ♅
10 25	☽ ⊥ ☉
10 57	☉ ✶ ♄
13 54	☿ □ ♇
15 08	☽ ± ♅
19 55	☉ ✶ ♃
20 16	☽ ✶ ♀
20 33	☽ □ ♄
21 06	☽ □ ♃
21 10	☽ ⊻ ☉
22 38	☿ ⊼ ♆
23 37	☽ ☌ ♂
23 41	♀ ⊻ ♄
29 Wednesday	
06 14	☽ ⚼ ♅
07 24	♀ ⊻ ♃
10 07	☿ Q ♄
15 06	☿ Q ♃
23 53	☽ ∠ ♀
30 Thursday	
00 14	☿ # ♆
00 17	☽ □ ♇
01 23	☽ ⊼ ♆
03 05	☽ Q ♄
03 48	☽ Q ♃
04 36	♀ ∠ ♇
05 41	☽ ☌ ☿
07 36	☽ △ ♅
11 36	☽ ± ♆
16 24	☉ ⚼ ♆
16 41	☽ ⊥ ♀
19 27	☿ △ ♅
20 34	☉ ⊻ ♀
23 45	☽ ✶ ♄
31 Friday	
00 36	☽ ✶ ♃
01 35	☽ # ♅
03 04	☽ ⚼ ♆
03 52	☽ ☌ ☉
04 01	☽ ⊻ ♀
05 11	☽ ⊻ ♂
07 12	☽ Q ♇
09 46	☽ ∥ ☉
16 10	☽ ⊥ ♂
23 15	☿ ± ♆

AUGUST 1981

LONGITUDES

Date	Sidereal time h m s	Sun ☉ ° ' "	Moon ☽ ° ' "	Moon ☽ 24.00 ° ' "	Mercury ☿ ° '	Venus ♀ ° '	Mars ♂ ° '	Jupiter ♃ ° '	Saturn ♄ ° '	Uranus ♅ ° '	Neptune ♆ ° '	Pluto ♇ ° '
01	08 40 03	09 ♌ 08 07	26 ♌ 08 36	02 ♍ 49 40	29 ♋ 27	09 ♍ 32	09 ♋ 28	06 ♎ 12	05 ♎ 36	26 ♏ 03	22 ♐ 22	21 ♎ 48
02	08 43 59	10 05 33	09 ♍ 25 44	15 ♍ 56 49	01 ♌ 29	10 45	10 08	06 21	05 41	26 R 03	22 R 21	21 49
03	08 47 56	11 03 00	22 ♍ 22 58	28 ♍ 44 22	03 33	11 57	10 48	06 31	05 46	26 03	22 20	21 50
04	08 51 52	12 00 28	05 ♎ 01 15	11 ♎ 13 55	05 37	13 09	11 27	06 41	05 52	26 D 03	22 19	21 51
05	08 55 49	12 57 56	17 ♎ 22 47	23 ♎ 28 15	07 42	14 21	12 07	06 50	05 57	26 03	22 18	21 52
06	08 59 45	13 55 26	29 ♎ 30 50	05 ♏ 31 02	09 46	15 33	12 47	07 00	06 02	26 03	22 17	21 53
07	09 03 42	14 52 55	11 ♏ 29 23	17 ♏ 26 29	11 51	16 45	13 26	07 10	06 08	26 03	22 17	21 55
08	09 07 38	15 50 26	23 ♏ 22 55	29 ♏ 19 14	13 55	17 57	14 06	07 21	06 13	26 04	22 16	21 56
09	09 11 35	16 47 58	05 ♐ 16 03	11 ♐ 13 55	15 59	19 09	14 45	07 31	06 19	26 04	22 15	21 57
10	09 15 32	17 45 30	17 ♐ 13 24	23 ♐ 15 01	18 02	20 21	15 24	07 41	06 25	26 04	22 14	21 58
11	09 19 28	18 43 04	29 ♐ 19 17	05 ♑ 26 38	20 04	21 33	16 04	07 51	06 30	26 04	22 14	22 00
12	09 23 25	19 40 38	11 ♑ 37 28	17 ♑ 52 10	22 05	22 45	16 43	08 02	06 36	26 05	22 13	22 01
13	09 27 21	20 38 14	24 ♑ 10 59	00 ♒ 34 10	24 04	23 57	17 22	08 12	06 42	26 05	22 12	22 02
14	09 31 18	21 35 50	07 ♒ 01 51	13 ♒ 34 06	26 03	25 08	18 02	08 23	06 48	26 06	22 12	22 04
15	09 35 14	22 33 28	20 ♒ 10 53	26 ♒ 52 08	28 00	26 20	18 41	08 34	06 54	26 06	22 11	22 05
16	09 39 11	23 31 06	03 ♓ 37 39	10 ♓ 27 12	29 ♌ 56	27 32	19 20	08 44	07 00	26 07	22 10	22 07
17	09 43 07	24 28 47	17 ♓ 20 26	24 ♓ 17 01	01 ♍ 51	28 43	19 59	08 55	07 06	26 08	22 10	22 08
18	09 47 04	25 26 28	01 ♈ 16 29	08 ♈ 18 24	03 44	29 ♍ 55	20 38	09 06	07 12	26 08	22 09	22 10
19	09 51 01	26 24 11	15 ♈ 22 19	22 ♈ 27 45	05 36	01 ♎ 06	21 17	09 17	07 18	26 09	22 09	22 11
20	09 54 57	27 21 56	29 ♈ 34 14	06 ♉ 41 22	07 26	02 18	21 55	09 28	07 24	26 10	22 08	22 13
21	09 58 54	28 19 42	13 ♉ 48 43	20 ♉ 55 56	09 15	03 29	22 34	09 39	07 30	26 11	22 08	22 14
22	10 02 50	29 ♌ 17 30	28 ♉ 02 41	05 ♊ 08 41	11 03	04 41	23 13	09 51	07 37	26 12	22 07	22 16
23	10 06 47	00 ♍ 15 20	12 ♊ 13 40	19 ♊ 17 26	12 49	05 52	23 52	10 02	07 43	26 13	22 07	22 18
24	10 10 43	01 13 12	26 ♊ 19 45	03 ♋ 20 26	14 34	07 03	24 30	10 13	07 49	26 14	22 07	22 19
25	10 14 40	02 11 06	10 ♋ 19 16	17 ♋ 16 05	16 18	08 15	25 09	10 25	07 56	26 15	22 06	22 21
26	10 18 36	03 09 01	24 ♋ 10 41	01 ♌ 02 51	18 00	09 26	25 48	10 36	08 02	26 16	22 06	22 23
27	10 22 33	04 06 58	07 ♌ 52 23	14 ♌ 39 04	19 41	10 37	26 26	10 48	08 09	26 17	22 06	22 24
28	10 26 30	05 04 56	21 ♌ 22 41	28 ♌ 03 02	21 20	11 48	27 05	10 59	08 15	26 18	22 06	22 26
29	10 30 26	06 02 57	04 ♍ 39 56	11 ♍ 13 13	22 59	12 59	27 43	11 11	08 22	26 19	22 06	22 28
30	10 34 23	07 00 58	17 ♍ 42 46	24 ♍ 08 30	24 36	14 10	28 21	11 22	08 29	26 21	22 05	22 30
31	10 38 19	07 ♍ 59 02	00 ♎ 30 24	06 ♎ 48 30	26 ♍ 11	15 ♎ 21	29 ♋ 00	11 ♎ 34	08 ♎ 35	26 ♏ 22	22 ♐ 05	22 ♎ 32

Date	Moon True ☊ ° '	Moon Mean ☊ ° '	Moon ☽ Latitude ° '
01	01 ♌ 42	01 ♌ 16	02 N 11
02	01 R 41	01 13	03 14
03	01 40	01 10	04 05
04	01 39	01 06	04 43
05	01 38	01 03	05 07
06	01 37	01 00	05 16
07	01 36	00 57	05 12
08	01 D 36	00 54	04 54
09	01 37	00 50	04 23
10	01 38	00 47	03 42
11	01 40	00 44	02 49
12	01 41	00 41	01 49
13	01 42	00 38	00 N 41
14	01 R 42	00 35	00 S 29
15	01 41	00 31	01 40
16	01 39	00 28	02 47
17	01 36	00 25	03 45
18	01 33	00 22	04 31
19	01 30	00 19	05 01
20	01 27	00 16	05 13
21	01 25	00 12	05 06
22	01 24	00 09	04 40
23	01 D 25	00 06	03 58
24	01 26	00 03	03 01
25	01 27	00 ♌ 00	01 53
26	01 28	29 ♋ 56	00 S 40
27	01 R 28	29 53	00 N 35
28	01 27	29 50	01 46
29	01 24	29 47	02 51
30	01 19	29 44	03 45
31	01 ♌ 13	29 ♋ 41	04 N 27

DECLINATIONS

Date	Sun ☉ ° '	Moon ☽ ° '	Mercury ☿ ° '	Venus ♀ ° '	Mars ♂ ° '	Jupiter ♃ ° '	Saturn ♄ ° '	Uranus ♅ ° '	Neptune ♆ ° '	Pluto ♇ ° '
01	17 N 58	14 N 52	21 N 18	09 N 17	23 N 44	01 S 22	00 S 07	19 S 04	21 S 53	07 N 15
02	17 43	11 02	20 59	08 48	23 42	01 26	00 09	19 04	21 53	07 14
03	17 27	06 46	20 38	08 19	23 39	01 30	00 11	19 04	21 53	07 13
04	17 12	02 N 20	20 14	07 50	23 36	01 34	00 13	19 04	21 53	07 12
05	16 55	02 S 06	19 48	07 20	23 33	01 38	00 16	19 04	21 53	07 11
06	16 39	06 22	19 19	06 51	23 30	01 42	00 18	19 04	21 53	07 10
07	16 22	10 20	18 49	06 21	23 27	01 46	00 20	19 04	21 53	07 09
08	16 05	13 53	18 16	05 51	23 24	01 50	00 23	19 04	21 53	07 08
09	15 48	16 52	17 42	05 21	23 20	01 54	00 25	19 05	21 53	07 08
10	15 31	19 09	17 06	04 51	23 16	01 58	00 27	19 05	21 53	07 07
11	15 13	20 37	16 28	04 21	23 12	02 03	00 30	19 05	21 53	07 06
12	14 55	21 08	15 49	03 50	23 08	02 07	00 32	19 05	21 53	07 05
13	14 37	20 36	15 09	03 20	23 04	02 11	00 35	19 05	21 53	07 04
14	14 18	18 59	14 28	02 49	22 59	02 15	00 37	19 05	21 53	07 03
15	14 00	16 20	13 46	02 18	22 54	02 20	00 39	19 05	21 53	07 02
16	13 41	12 46	13 03	01 47	22 50	02 24	00 42	19 05	21 53	07 01
17	13 22	08 27	12 20	01 17	22 45	02 28	00 44	19 06	21 53	07 00
18	13 02	03 S 38	11 35	00 46	22 39	02 33	00 47	19 06	21 53	06 59
19	12 43	01 N 25	10 51	00 N 15	22 34	02 37	00 49	19 06	21 53	06 58
20	12 23	06 26	10 06	00 S 16	22 28	02 42	00 52	19 06	21 53	06 57
21	12 03	11 07	09 21	00 47	22 23	02 46	00 55	19 06	21 53	06 56
22	11 43	15 10	08 35	01 18	22 17	02 51	00 57	19 07	21 53	06 55
23	11 23	18 20	07 50	01 49	22 11	02 55	01 00	19 07	21 53	06 54
24	11 02	20 23	07 04	02 20	22 05	03 00	01 02	19 07	21 53	06 53
25	10 42	21 10	06 18	02 51	21 58	03 04	01 05	19 07	21 53	06 52
26	10 21	20 37	05 33	03 22	21 52	03 09	01 08	19 08	21 53	06 51
27	10 00	18 52	04 47	03 53	21 45	03 14	01 10	19 08	21 53	06 50
28	09 39	16 03	04 02	04 24	21 38	03 18	01 13	19 08	21 53	06 49
29	09 18	12 27	03 17	04 54	21 31	03 23	01 16	19 09	21 53	06 48
30	08 56	08 19	02 32	05 25	21 24	03 28	01 18	19 09	21 53	06 47
31	08 N 35	03 N 53	01 N 47	05 S 56	21 N 17	03 S 32	01 S 21	19 S 09	21 S 53	06 N 46

ZODIAC SIGN ENTRIES

Date	h	m	Planets
01	18	30	☿ ♌
01	18	54	☽ ♍
04	02	24	☽ ♎
06	12	58	☽ ♏
09	01	22	☽ ♐
11	13	20	☽ ♑
13	22	56	☽ ♒
16	05	34	☽ ♓
16	12	47	☿ ♍
18	09	49	☽ ♈
18	13	44	♀ ♎
20	12	43	☽ ♉
22	15	18	☽ ♊
23	05	38	☉ ♍
24	18	17	☽ ♋
26	22	10	☽ ♌
29	03	32	☽ ♍
31	11	02	☽ ♎

LATITUDES

Date	Mercury ☿ ° '	Venus ♀ ° '	Mars ♂ ° '	Jupiter ♃ ° '	Saturn ♄ ° '	Uranus ♅ ° '	Neptune ♆ ° '	Pluto ♇ ° '
01	01 N 03	01 N 23	00 N 38	01 N 12	02 N 18	00 N 12	01 N 21	16 N 57
04	01 24	01 19	00 40	01 11	02 17	00 12	01 21	16 55
07	01 38	01 13	00 42	01 11	02 17	00 12	01 20	16 54
10	01 45	01 07	00 44	01 10	02 17	00 12	01 20	16 52
13	01 45	01 00	00 45	01 10	02 16	00 12	01 20	16 51
16	01 40	00 53	00 47	01 09	02 16	00 12	01 20	16 49
19	01 30	00 45	00 49	01 09	02 15	00 12	01 20	16 48
22	01 16	00 36	00 51	01 09	02 15	00 12	01 20	16 46
25	00 59	00 27	00 53	01 08	02 15	00 12	01 20	16 45
28	00 39	00 18	00 55	01 08	02 14	00 12	01 19	16 44
31	00 N 17	00 N 08	00 N 57	01 N 08	02 N 14	00 N 12	01 N 19	16 N 43

DATA

Julian Date	2444818
Delta T	+51 seconds
Ayanamsa	23° 35' 46"
Synetic vernal point	05° ♓ 31' 14"
True obliquity of ecliptic	23° 26' 25"

MOON'S PHASES, APSIDES AND POSITIONS ☽

Date	h	m	Phase	Longitude ° '	Eclipse Indicator
07	19	26	☽	15 ♏ 11	
15	16	37	○	22 ♒ 45	
22	14	16	☾	29 ♉ 23	
29	14	43	●	06 ♍ 10	

Day	h	m		
08	11	29	Apogee	
21	20	42	Perigee	
05	00	34	0S	
12	12	02	Max dec	21° S 08'
19	05	18	0N	
25	14	09	Max dec	21° N 10'

LONGITUDES

Date	Chiron ⚷ ° '	Ceres ⚳ ° '	Pallas ⚴ ° '	Juno ⚵ ° '	Vesta ⚶ ° '	Black Moon Lilith ⚸ ° '
01	22 ♉ 31	02 ♍ 03	05 ♌ 01	27 ♎ 07	28 ♍ 45	24 ♏ 01
11	22 ♉ 44	06 ♍ 33	10 ♌ 44	29 ♎ 14	03 ♎ 25	25 ♏ 08
21	22 ♉ 50	11 ♍ 05	16 ♌ 20	01 ♏ 36	08 ♎ 12	26 ♏ 15
31	22 ♉ 51	15 ♍ 39	21 ♌ 51	04 ♏ 12	13 ♎ 06	27 ♏ 22

All ephemeris data is given at 12.00 UT and the Moon's longitude is additionally given for 24.00 UT

ASPECTARIAN

h m	Aspects
01 Saturday	
02 06	☽ ∠ ♄
03 06	☽ ∠ ♃
04 16	☽ ⚹ ♇
05 18	☽ △ ♆
08 45	♀ ⚹ ♂
08 52	☽ ∠ ♂
11 51	☽ □ ♅
18 14	☽ ⊥ ♄
18 59	☽ ⊻ ☿
19 21	☽ ⊥ ♃
02 Sunday	
04 58	☉ Q ♇
05 08	☽ ⊻ ♄
06 19	☽ ⊻ ♃
07 14	☽ ∠ ♇
07 48	☽ ⊥ ☿
13 19	☽ ⊻ ☉
13 21	☽ ⚹ ♂
14 39	☽ ☌ ♀
15 12	☉ ⊻ ♂
20 30	☽ Q ♅
23 47	☽ ⊥ ♇
03 Monday	
01 15	☽ ⊥ ☉
02 24	☽ ∥ ♀
03 28	☽ ∠ ☿
09 34	☽ ∥ ♇
10 58	☽ ⊻ ♇
11 55	☽ □ ♆
12 49	☽ Q ♂
18 55	☽ ⚹ ♅
19 28	☽ ∠ ☉
21 20	♂ bQ ♅
04 Tuesday	
10 50	♅ St D
13 22	☽ ⚹ ☿
13 38	☽ ☌ ♄
14 58	☿ ⚹ ♄
15 14	☽ ☌ ♃
16 05	☽ # ♃
22 13	☽ Q ♆
23 15	☽ # ♄
23 39	☽ ∠ ♅
05 Wednesday	
01 08	☽ □ ♂
01 19	☿ ⚹ ♃
01 54	☽ ∥ ♄
02 39	☽ ⚹ ☉
05 26	☽ ⊻ ♀
06 01	♀ Q ♅
07 33	☿ bQ ♆
09 23	☽ ∥ ♃
17 15	☽ ⊥ ♅
17 29	☽ Q ☿
18 29	☽ ⊥ ♀
19 47	♀ ∥ ♇
20 51	☽ ☌ ♇
21 41	☽ ⚹ ♆
06 Thursday	
04 15	☽ Q ☉
05 07	☽ ⊻ ♅
06 01	☿ ⊥ ♀
13 23	☿ Q ♇
14 18	☽ ∠ ♀
14 29	☽ # ♀
16 40	☽ # ♇
18 53	♀ ⊥ ♇
23 47	☿ # ♅
07 Friday	
01 09	☽ ⊻ ♄
03 12	☽ ⊻ ♃
03 32	☽ ∠ ♆
12 53	☽ □ ☿
13 18	☽ ⊥ ♄
15 26	☽ ⊥ ♃
16 09	☽ △ ♂
19 26	☽ □ ☉
21 38	☽ ⊥ ♆
23 48	☽ ⚹ ♀
08 Saturday	
07 36	☽ ∠ ♄
09 04	☽ ⊻ ♇
09 44	☽ ⊻ ♆
09 52	☽ ∠ ♃
14 59	☿ ⊻ ♂
17 25	☽ ☌ ♅
21 13	☽ ⊥ ♇
09 Sunday	
00 13	☽ bQ ♂
02 46	☽ Q ♀
03 44	☽ # ☉
14 08	☽ ⚹ ♄
15 24	☽ ∠ ♇
16 35	☽ ⚹ ♃
18 20	☽ # ☿
19 25	☽ ± ♂
10 Monday	
06 01	☉ ☌ ☿
08 10	☽ ⚻ ♂
11 05	☽ ∥ ♅
13 10	☽ △ ☉
13 56	☽ △ ☿
14 23	☽ Q ♄
16 59	☽ Q ♃
18 56	☽ □ ♀
21 29	☽ ⚹ ♇
21 59	☽ ☌ ♆
11 Tuesday	
05 35	☽ ⊻ ♅
17 25	☽ ⊥ ♅
21 08	♀ ⊻ ♇
21 11	☽ Q ♇
21 22	☽ bQ ☉
12 Wednesday	
01 26	♀ □ ♆
01 28	☽ bQ ☿
02 11	☽ □ ♄
04 56	☽ □ ♃
06 02	☿ ∠ ♄
10 57	☽ ∠ ♅
11 17	☿ ⚹ ♇
13 38	☿ △ ♆
16 17	☽ ± ☉
22 12	☽ ± ☿
22 20	☽ ☍ ♂
23 25	☿ ⊥ ♂
13 Thursday	
00 32	☿ ∠ ♃
04 43	☽ ⚻ ☉
07 56	☽ □ ♇
08 01	☿ ⊻ ♀
08 15	☽ ⊻ ♆
11 30	☽ △ ♀
11 45	☽ ⚻ ☿
15 36	☽ ⚹ ♅
19 34	☽ ⊥ ♆
14 Friday	
10 55	☽ ∥ ♅
11 34	☽ △ ♄
12 18	☽ ∠ ♆
12 34	☿ □ ♅
13 58	☽ Q ♅
14 32	☽ △ ♃
17 33	☉ ∠ ♄
18 18	☽ bQ ♀
21 53	☉ ∥ ☿
23 58	☉ ⚹ ♇
15 Saturday	
02 43	☉ △ ♆
07 23	♀ ⚹ ♅
09 09	☽ ⚻ ♂
10 58	♀ # ♃
12 18	☽ ± ♀
15 07	☽ bQ ♄
15 26	☽ △ ♇
15 36	☽ ⚹ ♆
16 37	☽ ☍ ☉
18 10	☽ bQ ♃
20 30	☽ ± ♂
22 39	☽ □ ♅
16 Sunday	
00 07	☽ ⚻ ♀
04 23	☽ ☍ ☿
05 51	☽ # ☉
07 18	☽ ± ♄
09 53	☽ # ☿
10 25	☽ ± ♃
12 58	☽ Q ♆
13 18	☽ bQ ♂
17 59	☽ ⚻ ♄
18 09	☽ bQ ♇
18 50	☉ ∠ ♃
21 07	☽ ⚻ ♃
17 Monday	
02 00	☿ ⊥ ♄
09 54	☽ ± ♇
16 48	☽ △ ♂
19 29	☽ # ♇
20 19	☽ ⚻ ♇
20 20	☽ □ ♆
18 Tuesday	
01 15	☽ ⚻ ☉
03 05	☿ ⊥ ♃
03 11	☽ △ ♅
06 43	♆ ⚹ ♇
09 27	☽ ☍ ♀
11 06	♀ # ♄
12 18	☽ ± ☉
16 51	☽ ⚻ ☿
17 09	☽ ∥ ♃
22 11	☽ ☍ ♄
19 Wednesday	
01 30	☽ ∥ ♄
01 32	☽ ☍ ♃
03 16	☽ # ♀
04 38	☽ ± ☿
04 46	☽ bQ ☉
04 50	☽ bQ ♅
05 36	☉ □ ♅
06 59	☽ ∥ ♀
09 11	☽ # ♄
17 47	☽ # ♃
20 06	☽ ± ♅
22 10	☽ bQ ☿
22 28	☽ □ ♂
23 28	☽ △ ♆
23 33	☽ ☍ ♇
20 Thursday	
01 38	☿ ∠ ♂
06 15	☽ ⚻ ♅
08 01	☽ △ ☉
09 01	☿ ∠ ♇
11 30	☿ ⊻ ♄
14 33	☽ ∥ ♇
17 01	☽ ⚻ ♀
19 50	♂ ⚻ ♆
23 10	♂ □ ♇
21 Friday	
00 45	☽ bQ ♆
01 18	☽ ⚻ ♄
03 12	☽ △ ☿
03 58	☽ ∥ ☿
04 03	☽ ± ♀
04 54	☽ ⚻ ♃
06 17	☽ Q ♂
11 29	☽ ± ♄
15 09	☽ ± ♃
15 54	☽ ± ♆
16 51	☽ ∥ ☉
17 57	☿ ⊻ ♃
18 14	♀ ∥ ♄
20 36	☽ bQ ♀
22 Saturday	
02 01	☽ ⚻ ♆
02 14	☽ ⚻ ♇
02 46	☽ bQ ♄
03 28	☽ ⚹ ♂
06 26	☉ ⊥ ♂
06 31	☽ bQ ♃
08 52	☽ ☍ ♅
12 23	☽ ± ♇
14 16	☽ □ ☉
23 Sunday	
00 14	☽ △ ♀
03 37	☽ bQ ♇
04 18	☽ △ ♄
06 01	☽ ∠ ♂
08 13	☽ △ ♃
11 52	♀ Q ♂
13 09	☽ □ ☿
19 38	☽ # ♅
22 02	☽ ⊥ ♂
22 58	♃ Q ♆
22 59	☽ Q ☉
24 Monday	
04 49	☽ ☍ ♆
05 09	☽ △ ♇
07 13	☿ Q ♅
08 45	☽ ⊻ ♂
11 49	☽ ⚻ ♅
17 56	☿ ∥ ♇
20 59	☽ ⚹ ☉
22 06	☽ ± ♅
25 Tuesday	
00 11	☽ Q ☿
04 51	☉ ⊥ ♄
05 02	♀ ☌ ♄
05 44	☉ ⊥ ♀
07 51	☽ □ ♄
08 05	☽ □ ♀
12 09	☽ □ ♃
12 48	☿ ⊥ ♇
13 35	☽ bQ ♅
23 46	☽ ⚹ ☿
26 Wednesday	
00 10	♀ ∥ ♃
00 45	☽ ∠ ☉
07 36	♂ # ♆
08 24	☽ ⚻ ♆
08 52	☽ □ ♇
14 57	☽ ☌ ♂
15 16	☽ Q ♄
15 38	☽ △ ♅
17 34	☽ ⊥ ☉
18 12	☽ Q ♀
18 51	☽ ± ♆
19 50	☽ Q ♃
22 54	♂ Q ♄
27 Thursday	
01 35	♀ Q ♆
04 53	☽ ⊻ ☉
05 35	☽ ∠ ☿
06 00	♂ △ ♅
09 08	☽ # ♅
10 38	☽ bQ ♆
12 29	☽ ⚹ ♄
16 15	♀ ☌ ♃
16 29	☽ Q ♇
17 14	☽ ⚹ ♃
17 19	☽ ⚹ ♀
23 43	☽ ⊥ ☿
28 Friday	
01 41	♀ ∠ ♅
05 08	☿ # ♀
08 59	☉ ⊥ ♃
11 55	☽ ⊻ ☿
13 17	☽ △ ♆
13 54	☽ ⚹ ♇
15 24	☽ ∠ ♄
20 24	☽ ∠ ♃
20 51	☽ □ ♅
22 42	☽ ∠ ♀
22 46	☽ ⊻ ♂
23 02	☿ □ ♆
29 Saturday	
04 21	☿ ⊻ ♇
07 47	☽ ⊥ ♄
08 57	☿ # ♃
10 11	☽ ⊥ ♂
12 57	☽ ⊥ ♃
14 43	☽ ☌ ☉
16 39	☽ ⊥ ♀
17 08	☽ ∠ ♇
18 49	☽ ⊻ ♄
30 Sunday	
00 06	☽ ⊻ ♃
02 03	♂ ± ♆
03 31	☽ ∠ ♂
04 47	☽ ⊻ ♀
05 45	☽ Q ♅
07 47	♃ ∠ ♅
08 11	☽ ∥ ☉
09 44	☽ ⊥ ♇
20 09	☽ □ ♆
20 25	☽ ∥ ♇
20 56	☽ ⊻ ♇
31 Monday	
00 19	☉ ∠ ♇
02 09	☽ # ♀
02 41	☽ ☌ ☿
04 10	☽ ⚹ ♅
08 59	☽ ⚹ ♂
13 47	☽ # ♃
14 44	☿ ⚹ ♅

SEPTEMBER 1981

LONGITUDES

Date	Sidereal time h	m	s	Sun ☉ °	'	"	Moon ☽ °	'	"	Moon ☽ 24.00 °	'	"	Mercury ☿ °	'	Venus ♀ °	'	Mars ♂ °	'	Jupiter ♃ °	'	Saturn ♄ °	'	Uranus ♅ °	'	Neptune ♆ °	'	Pluto ♇ °	'
01	10	42	16	08 ♍	57	06	13 ♎	02	54	19 ♎	13	45	27 ♍	46	16 ♎	32	29 ♋	38	11 ♎	46	08 ♎	42	26 ♏	23	22 ♐	05	22 ♎	34
02	10	46	12	09	55	13	25 ♎	21	16	01 ♏	25	47	29 ♍	19	17	43	00 ♌	16	11	58	08	49	26	25	22 R	05	22	35
03	10	50	09	10	53	21	07 ♏	27	37	13 ♏	27	12	00 ♎	51	18	54	00	54	12	10	08	55	26	26	22 D	05	22	37
04	10	54	05	11	51	30	19 ♏	25	00	25 ♏	21	30	02	22	20	05	01	32	12	22	09	02	26	28	22	05	22	39
05	10	58	02	12	49	41	01 ♐	17	16	07 ♐	12	54	03	51	21	15	02	10	12	34	09	09	26	30	22	05	22	41
06	11	01	59	13	47	53	13 ♐	08	58	19 ♐	06	08	05	19	22	26	02	48	12	46	09	16	26	31	22	05	22	43
07	11	05	55	14	46	07	25 ♐	05	02	01 ♑	06	16	06	46	23	36	03	26	12	58	09	23	26	33	22	05	22	45
08	11	09	52	15	44	22	07 ♑	10	30	13 ♑	18	18	08	12	24	47	04	04	13	10	09	30	26	35	22	06	22	47
09	11	13	48	16	42	39	19 ♑	30	15	25 ♑	46	52	09	36	25	57	04	42	13	22	09	37	26	36	22	06	22	49
10	11	17	45	17	40	58	02 ♒	08	36	08 ♒	35	50	10	59	27	08	05	20	13	35	09	44	26	38	22	06	22	51
11	11	21	41	18	39	18	15 ♒	08	51	21 ♒	47	47	12	20	28	18	05	57	13	47	09	51	26	40	22	06	22	53
12	11	25	38	19	37	40	28 ♒	32	42	05 ♓	23	30	13	40	29 ♎	28	06	35	13	59	09	58	26	42	22	07	22	55
13	11	29	34	20	36	03	12 ♓	19	56	19 ♓	21	36	14	59	00 ♏	38	07	12	14	12	10	05	26	44	22	07	22	57
14	11	33	31	21	34	28	26 ♓	27	58	03 ♈	38	23	16	16	01	49	07	50	14	24	10	12	26	46	22	07	23	00
15	11	37	28	22	32	55	10 ♈	52	06	18 ♈	08	15	17	32	02	59	08	27	14	37	10	19	26	48	22	08	23	02
16	11	41	24	23	31	24	25 ♈	25	59	02 ♉	44	23	18	46	04	08	09	05	14	49	10	26	26	50	22	08	23	04
17	11	45	21	24	29	55	10 ♉	02	37	17 ♉	19	53	19	58	05	18	09	42	15	02	10	33	26	52	22	08	23	06
18	11	49	17	25	28	29	24 ♉	35	29	01 ♊	48	49	21	09	06	28	10	19	15	14	10	41	26	55	22	09	23	08
19	11	53	14	26	27	04	08 ♊	59	26	16 ♊	06	58	22	17	07	38	10	57	15	27	10	48	26	57	22	09	23	10
20	11	57	10	27	25	42	23 ♊	11	11	00 ♋	11	57	23	24	08	48	11	34	15	39	10	55	26	59	22	10	23	13
21	12	01	07	28	24	22	07 ♋	09	13	14 ♋	02	59	24	29	09	57	12	11	15	52	11	02	27	01	22	11	23	15
22	12	05	03	29 ♍	23	05	20 ♋	53	20	27 ♋	40	23	25	31	11	07	12	48	16	05	11	10	27	04	22	11	23	17
23	12	09	00	00 ♎	21	49	04 ♌	24	13	11 ♌	04	58	26	31	12	16	13	25	16	18	11	17	27	06	22	12	23	19
24	12	12	57	01	20	36	17 ♌	42	43	24 ♌	17	34	27	28	13	25	14	02	16	30	11	24	27	09	22	12	23	21
25	12	16	53	02	19	25	00 ♍	49	35	07 ♍	18	47	28	22	14	35	14	39	16	43	11	31	27	11	22	13	23	24
26	12	20	50	03	18	16	13 ♍	45	11	20 ♍	08	48	29 ♎	14	15	44	15	16	16	56	11	39	27	14	22	14	23	26
27	12	24	46	04	17	09	26 ♍	29	36	02 ♎	47	35	00 ♏	02	16	53	15	53	17	09	11	46	27	16	22	15	23	28
28	12	28	43	05	16	04	09 ♎	02	45	15 ♎	15	07	00	46	18	02	16	29	17	22	11	53	27	19	22	15	23	31
29	12	32	39	06	15	01	21 ♎	24	43	27 ♎	31	37	01	27	19	11	17	06	17	34	12	01	27	21	22	16	23	33
30	12	36	36	07 ♎	14	01	03 ♏	35	58	09 ♏	37	56	02 ♏	04	20 ♏	19	17 ♌	42	17 ♎	47	12 ♎	08	27 ♏	24	22 ♐	17	23 ♎	35

Date	Moon True ☊ °	'	Moon Mean ☊ °	'	Moon ☽ Latitude °	'
01	01 ♌	06	29 ♋	37	04 N	55
02	01 R	00	29	34	05	08
03	00	54	29	31	05	08
04	00	50	29	28	04	54
05	00	48	29	25	04	27
06	00	47	29	21	03	49
07	00 D	47	29	18	03	01
08	00	49	29	15	02	04
09	00	50	29	12	01 N	01
10	00 R	50	29	09	00 S	07
11	00	49	29	06	01	16
12	00	46	29	02	02	24
13	00	41	28	59	03	24
14	00	34	28	56	04	14
15	00	26	28	53	04	48
16	00	18	28	50	05	05
17	00	11	28	47	05	01
18	00	05	28	43	04	38
19	00	01	28	40	03	58
20	00	00	28	37	03	04
21	00 D	00	28	34	01	59
22	00	01	28	31	00 S	49
23	00 R	01	28	27	00 N	23
24	00	00	28	24	01	33
25	29 ♋	56	28	21	02	36
26	29	49	28	18	03	31
27	29	40	28	15	04	14
28	29	29	28	12	04	43
29	29	17	28	08	04	59
30	29 ♋	04	28 ♋	05	05 N	01

DECLINATIONS

Date	Sun ☉ °	'	Moon ☽ °	'	Mercury ☿ °	'	Venus ♀ °	'	Mars ♂ °	'	Jupiter ♃ °	'	Saturn ♄ °	'	Uranus ♅ °	'	Neptune ♆ °	'	Pluto ♇ °	'
01	08 N	13	00 S	38	01 N	02	06 S	26	21 N	10	03 S	37	01 S	24	19 S	10	21 S	53	06 N	45
02	07	51	05	01	00 N	18	06	56	21	02	03	42	01	27	19	10	21	53	06	44
03	07	29	09	09	00 S	25	07	26	20	54	03	46	01	29	19	10	21	53	06	43
04	07	07	12	52	01	09	07	57	20	47	03	51	01	32	19	11	21	53	06	42
05	06	45	16	03	01	52	08	26	20	39	03	56	01	35	19	11	21	53	06	41
06	06	22	18	35	02	34	08	56	20	31	04	01	01	38	19	12	21	53	06	40
07	06	00	20	20	03	16	09	26	20	22	04	06	01	40	19	12	21	54	06	39
08	05	37	21	10	03	57	09	55	20	14	04	10	01	43	19	12	21	54	06	38
09	05	15	21	01	04	38	10	24	20	06	04	15	01	46	19	13	21	54	06	37
10	04	52	19	48	05	18	10	53	19	57	04	20	01	49	19	13	21	54	06	36
11	04	29	17	31	05	58	11	22	19	48	04	25	01	51	19	14	21	54	06	35
12	04	06	14	13	06	36	11	50	19	40	04	30	01	54	19	14	21	54	06	34
13	03	43	10	05	07	15	12	19	19	31	04	35	01	57	19	15	21	54	06	33
14	03	20	05	17	07	52	12	47	19	22	04	39	02	00	19	15	21	54	06	32
15	02	57	00 S	07	08	29	13	14	19	12	04	44	02	03	19	16	21	54	06	31
16	02	34	05 N	06	09	04	13	42	19	03	04	49	02	06	19	16	21	54	06	30
17	02	11	10	04	09	39	14	09	18	54	04	54	02	08	19	17	21	54	06	29
18	01	48	14	25	10	13	14	36	18	44	04	59	02	11	19	17	21	54	06	28
19	01	25	17	52	10	46	15	03	18	35	05	04	02	14	19	18	21	54	06	27
20	01	01	20	12	11	18	15	29	18	25	05	09	02	17	19	18	21	54	06	26
21	00	38	21	16	11	49	15	55	18	15	05	14	02	20	19	19	21	55	06	25
22	00 N	15	21	01	12	18	16	20	18	05	05	19	02	23	19	19	21	55	06	24
23	00 S	09	19	32	12	47	16	45	17	55	05	24	02	26	19	20	21	55	06	23
24	00	32	17	00	13	14	17	10	17	45	05	29	02	28	19	20	21	55	06	22
25	00	55	13	37	13	39	17	35	17	35	05	34	02	31	19	21	21	55	06	21
26	01	19	09	38	14	03	17	59	17	25	05	38	02	34	19	22	21	55	06	20
27	01	42	05	16	14	26	18	22	17	15	05	43	02	37	19	22	21	55	06	19
28	02	06	00 N	45	14	47	18	46	17	04	05	48	02	40	19	23	21	55	06	18
29	02	29	03 S	43	15	05	19	09	16	54	05	53	02	43	19	23	21	55	06	17
30	02 S	52	08 S	00	15 S	22	19 S	31	16 N	43	05 S	58	02 S	46	19 S	24	21 S	55	06 N	16

ZODIAC SIGN ENTRIES

Date	h	m	Planets	
02	01	52	♂	♌
02	21	10	☽	♏
02	22	40	☿	♎
05	09	24	☽	♐
07	21	48	☽	♑
10	07	59	☽	♒
12	14	34	☽	♓
12	22	51	♀	♏
14	17	55	☽	♈
16	19	30	☽	♉
18	20	59	☽	♊
20	23	39	☽	♋
23	03	05	☉	♎
23	04	08	☽	♌
25	10	29	☽	♍
27	11	02	☿	♏
27	18	40	☽	♎
30	04	53	☽	♏

LATITUDES

Date	Mercury ☿ °	'	Venus ♀ °	'	Mars ♂ °	'	Jupiter ♃ °	'	Saturn ♄ °	'	Uranus ♅ °	'	Neptune ♆ °	'	Pluto ♇ °	'
01	00 N	10	00 N	04	00 N	57	01 N	08	02 N	14	00 N	12	01 N	19	16 N	42
04	00 S	13	00 S	06	00	59	01	07	02	14	00	12	01	19	16	41
07	00	38	00	17	01	01	01	07	02	14	00	11	01	19	16	40
10	01	02	00	28	01	03	01	07	02	13	00	11	01	19	16	39
13	01	27	00	39	01	05	01	06	02	13	00	11	01	19	16	38
16	01	52	00	51	01	06	01	06	02	13	00	11	01	18	16	37
19	02	15	01	03	01	08	01	06	02	13	00	11	01	18	16	37
22	02	37	01	14	01	10	01	06	02	13	00	11	01	18	16	36
25	02	57	01	26	01	12	01	06	02	13	00	11	01	18	16	35
28	03	14	01	37	01	14	01	06	02	13	00	11	01	18	16	35
31	03 S	27	01 S	49	01 N	16	01 N	05	02 N	13	00 N	11	01 N	18	16 N	34

DATA

Julian Date	2444849
Delta T	+51 seconds
Ayanamsa	23° 35' 49"
Synetic vernal point	05° ♓ 31' 10"
True obliquity of ecliptic	23° 26' 26"

MOON'S PHASES, APSIDES AND POSITIONS ☽

Date	h	m	Phase	Longitude °	'	Eclipse Indicator
06	13	26	☽	13 ♐	51	
14	03	09	○	21 ♓	13	
20	19	47	☾	27 ♊	45	
28	04	07	●	04 ♎	57	

Day	h	m		
05	06	35	Apogee	
17	04	05	Perigee	
01	08	39	0S	
08	20	24	Max dec	21° S 14'
15	12	33	0N	
21	19	21	Max dec	21° N 20'
28	16	00	0S	

LONGITUDES

Date	Chiron ⚷ °	'	Ceres ⚳ °	'	Pallas ⚴ °	'	Juno ⚵ °	'	Vesta ⚶ °	'	Black Moon Lilith ⚸ °	'
01	22 ♉	51	16 ♍	07	22 ♌	24	04 ♏	28	13 ♎	36	27 ♏	29
11	22 ♉	45	20 ♍	42	27 ♌	47	07 ♏	16	18 ♎	37	28 ♏	36
21	22 ♉	34	25 ♍	18	03 ♍	02	10 ♏	12	23 ♎	43	29 ♏	43
31	22 ♉	17	29 ♍	53	08 ♍	10	13 ♏	17	28 ♎	54	00 ♐	50

ASPECTARIAN

h m	Aspects
01 Tuesday	
01 08	☿ # ♄
01 19	☽ # ♄
01 21	☽ ∥ ☿
03 27	☽ ⚺ ☉
03 33	☽ ☌ ♄
04 53	☉ ⚺ ♄
06 18	☽ Q ♆
08 48	☽ ∠ ♅
09 07	☽ Q ♂
09 29	☽ ☌ ♃
13 55	☽ # ☿
16 00	☽ ⊥ ☉
16 11	☽ ∥ ♄
19 23	♂ Q ♃
19 28	☽ ☌ ♀
02 Wednesday	
02 18	☽ ⊥ ♅
02 34	♀ # ♇
04 31	☽ ∥ ♃
05 35	☽ ⚹ ♆
06 33	☽ ☌ ♇
11 04	☽ ∠ ☉
14 05	☽ ⚺ ♅
20 57	☽ ⚺ ☿
21 42	☽ # ♇
22 14	☽ □ ♂
03 Thursday	
00 26	☽ ∥ ♀
02 54	☽ # ☉
10 36	☽ ⊥ ☿
11 08	♆ St D
11 15	☽ ∠ ♆
13 10	☉ # ♀
13 30	☿ ⚹ ♂
14 57	☽ ⚺ ♄
19 28	☽ ⚹ ☉
21 34	☽ ⚺ ♃
04 Friday	
03 06	☽ ⊥ ♄
05 18	☽ ⊥ ♆
07 15	☽ ∠ ☿
09 50	☽ ⊥ ♃
13 29	☽ ⚺ ♀
17 23	☽ ⚺ ♆
18 33	☽ ⚺ ♇
20 07	♀ ⊥ ♅
21 25	☽ ∠ ♄
21 46	☽ Q ☉
05 Saturday	
01 55	☿ ∥ ♄
02 16	☽ ☌ ♅
02 56	☽ ⊥ ♀
03 46	☉ ⚺ ♃
04 20	☽ ∠ ♃
06 43	☽ ⊥ ♇
13 54	☽ △ ♂
16 13	☉ ∥ ♇
17 56	☽ ⚹ ☿
23 10	☽ ∠ ♀
06 Sunday	
00 59	☽ ∠ ♇
04 04	☽ ⚹ ♄
05 01	♀ ⚹ ♆
11 13	☽ ⚹ ♃
13 26	☽ □ ☉
18 03	♀ ☌ ♇
19 10	☽ ∥ ♅
21 35	☽ Q ☿
21 55	☽ ⚼ ♂
07 Monday	
04 30	☽ Q ♄
06 00	☽ ☌ ♆
06 23	☉ Q ♅
07 19	☽ ⚹ ♇
08 43	☽ ⚹ ♀
11 46	☽ Q ♃
12 43	☽ # ♂
14 56	☽ ⚺ ♅
16 58	☽ ± ♂
08 Tuesday	
02 54	☽ ⊥ ♅
05 32	☽ ⚻ ♂
07 17	☽ Q ♇
11 09	☽ Q ♀
14 16	☽ □ ☿
16 36	☽ □ ♄
20 39	☽ ∠ ♅
20 46	☿ ∥ ♃
23 56	☽ □ ♃
09 Wednesday	
06 09	☽ △ ☉
12 18	☿ ☌ ♄
14 46	☉ ⊥ ♇
16 59	☽ ⚺ ♆
18 22	☽ □ ♇
20 39	☿ Q ♆
10 Thursday	
01 36	☽ □ ♀
01 36	☽ ⚹ ♅
01 40	♀ ⚺ ♅
02 03	☉ # ☿
04 23	☽ ⊥ ♆
09 38	☽ # ♂
13 06	☽ ⚼ ☉
18 15	☽ ☍ ♂
19 12	☽ ∥ ♅
21 14	☽ ∠ ♆
23 51	☿ ∠ ♅
11 Friday	
00 06	☽ Q ♅
02 13	☽ △ ♄
06 17	☽ △ ☿
07 05	☽ ± ☉
09 29	☽ △ ♃
15 50	☉ # ♃
18 51	☽ ⚻ ☉
12 Saturday	
00 33	☽ ⚹ ♆
01 59	☽ △ ♇
05 36	☽ ⚼ ♄
08 44	☽ □ ♅
10 24	☿ # ♇
12 15	☽ ⚼ ☿
12 48	☽ ⚼ ♃
13 47	☽ △ ♀
18 48	☿ ☌ ♃
21 36	☽ ± ♄
21 46	☽ Q ♆
13 Sunday	
00 52	☽ ∥ ♀
02 44	☽ ⚻ ♂
04 26	☽ ⚼ ♇
04 46	☽ ± ♃
05 38	☽ ± ☿
08 06	☽ ⚻ ♄
08 24	♂ ⚼ ♆
13 34	☽ ± ♂
15 15	☽ ⚻ ♃
17 01	☽ ⚻ ☿
18 11	☽ ⚼ ♀
18 47	♄ Q ♆
19 56	☽ ± ♇
14 Monday	
00 50	☽ ∥ ☿
03 09	☽ ☍ ☉
04 40	☽ □ ♆
05 36	☽ ⚼ ♂
05 59	☽ # ♇
06 08	☽ ⚻ ♇
10 48	☽ ± ♀
12 30	☽ △ ♅
14 56	☽ ∥ ♃
21 44	☽ ⚻ ♀
21 54	☽ # ☉
15 Tuesday	
01 32	☉ □ ♆
03 15	☽ ∥ ♄
04 13	♂ # ♅
07 50	☽ △ ♂
11 05	☽ ☍ ♄
13 33	☽ ⚼ ♅
18 16	☽ ☍ ♃
21 59	☽ # ♄
16 Wednesday	
00 02	☽ ☍ ☿
00 14	☉ ⚺ ♇
01 07	☽ ∥ ☉
04 26	☽ ± ♅
06 34	☽ △ ♆
08 06	☽ ☍ ♇
08 38	☽ ⚻ ☉
10 39	☽ # ♃
14 19	☽ ⚻ ♅
18 31	☽ ∥ ♇
19 12	☽ ± ☉
17 Thursday	
03 32	☽ ☍ ♀
07 14	☽ ⚼ ♆
09 40	☽ # ☿
11 02	☽ ⚼ ☉
11 25	☽ □ ♂
12 51	☽ ⚻ ♄
14 28	☉ # ♄
20 19	☽ ⚻ ♃
22 03	☽ ± ♆
22 49	☽ ± ♄
18 Friday	
01 50	☉ ∠ ♂
05 48	☽ ⚻ ☿
06 22	☽ ± ♃
06 57	☿ ⊥ ♅
07 57	☽ ⚻ ♆
09 35	☽ ⚻ ♇
13 17	☽ # ♀
13 34	☽ △ ☉
13 49	☽ ⚼ ♄
15 51	☽ ☍ ♅
16 36	☽ ± ☿
18 28	☽ Q ♂
19 34	☽ ± ♇
21 31	☽ ⚼ ♃
19 Saturday	
02 08	♀ ∠ ♆
04 57	♂ ⚹ ♄
08 54	☽ ⚼ ☿
09 09	☿ ⚹ ♆
09 31	☽ ⚻ ♀
10 37	☽ ⚼ ♇
15 04	☽ △ ♄
15 26	☽ ⚹ ♂
17 44	☽ ∥ ♂
20 30	☽ ± ♀
21 20	♂ Q ♇
23 02	☽ △ ♃
20 Sunday	
00 39	☉ ⚹ ♅
01 06	☽ # ♅
07 38	☿ ☌ ♇
10 16	☽ ☍ ♆
12 02	☽ △ ♇
12 24	☽ △ ☿
13 07	☽ ⚼ ♀
18 02	☽ ∠ ♂
18 30	☽ ⚻ ♅
19 47	☽ □ ☉
20 22	☿ Q ♂
21 Monday	
04 50	☽ ± ♅
10 14	☽ ⊥ ♂
17 18	☽ △ ♀
18 49	☽ □ ♄
20 29	☽ ⚼ ♅
21 09	☽ ⚺ ♂
22 Tuesday	
03 25	☽ □ ♃
05 22	☽ Q ☉
13 11	♀ ⚺ ♄
14 17	☽ ⚻ ♆
16 14	☽ □ ♇
20 50	☽ □ ☿
22 57	☽ △ ♅
23 Wednesday	
00 55	☽ ± ♆
02 47	☽ Q ♄
04 13	☽ ⚹ ☉
11 48	☽ Q ♃
14 23	☽ # ♅
17 01	☽ ⚼ ♆
24 Thursday	
00 28	☽ Q ♇
00 28	☽ ⚹ ♄
03 22	☿ ⚺ ♅
03 29	☽ □ ♀
05 00	☽ ☌ ♂
05 19	☽ ∥ ♂
07 37	☽ Q ☿
09 19	☽ ∠ ☉
09 46	☽ ⚹ ♃
10 48	☽ # ♀
20 12	☽ △ ♆
22 19	☽ ⚹ ♇
25 Friday	
03 03	☽ ⊥ ☉
04 01	☽ ∠ ♄
05 17	☽ □ ♅
07 09	☽ ⚹ ☿
11 47	☽ # ☿
12 17	♀ # ♂
13 40	☽ ∠ ♃
14 59	☽ ⚺ ☉
15 14	♀ □ ♂
15 33	☽ Q ♀
20 46	☽ ⊥ ♄
26 Saturday	
02 03	☽ ∠ ♇
06 38	☽ ⊥ ♃
08 02	☽ ⚺ ♄
12 57	☽ ∠ ☿
14 46	☽ Q ♅
14 58	☽ ⚺ ♂
16 04	☽ ⚹ ♀
18 03	☽ ⚺ ♃
18 55	☽ ⊥ ♇
22 36	♀ ⊥ ♆
27 Sunday	
02 49	☽ ⊥ ♂
03 57	☽ □ ♆
06 15	☽ ⚺ ♇
06 21	☽ ∥ ♇
07 02	☽ ⊥ ☿
09 37	☽ # ♃
13 29	☽ ⚹ ♅
18 47	♀ ⚺ ♃
19 10	☽ ⚺ ☿
20 46	☽ ∠ ♂
23 17	☽ ∠ ♀
28 Monday	
01 59	☽ # ♄
04 07	☽ ☌ ☉
05 29	☽ # ☉
08 46	♀ ⊥ ♄
14 20	☽ Q ♆
17 33	☽ ☌ ♄
18 20	☽ ∠ ♅
18 21	☽ ⊥ ♀
29 Tuesday	
03 09	☽ ⚹ ♂
04 23	☽ ☌ ♃
04 38	☽ ∥ ☉
06 27	☽ ∥ ♄
07 11	☽ ⚺ ♀
11 53	☽ ⊥ ♅
13 41	☽ ⚹ ♆
16 12	☽ ☌ ♇
23 42	☽ ⚺ ♅
30 Wednesday	
00 12	☽ ∥ ♃
02 07	☽ # ♇
03 54	☽ Q ♂
04 12	☉ ∥ ♄
04 24	♀ ∥ ♅
08 48	☽ ☌ ☿
16 57	♂ ⚹ ♃
19 20	☽ ∠ ♆
19 52	☽ ⚺ ☉

All ephemeris data is given at 12.00 UT and the Moon's longitude is additionally given for 24.00 UT

OCTOBER 1981

Date	Sidereal time h m s	LONGITUDES Sun ☉ ° ' "	Moon ☽ ° ' "	Moon ☽ 24.00 ° ' "	Mercury ☿ ° '	Venus ♀ ° '	Mars ♂ ° '	Jupiter ♃ ° '	Saturn ♄ ° '	Uranus ♅ ° '	Neptune ♆ ° '	Pluto ♇ ° '
01	12 40 32	08 ♎ 13 02	15 ♏ 37 45	21 ♏ 35 41	02 ♏ 35	21 ♏ 28	18 ♌ 19	18 ♎ 00	12 ♎ 16	27 ♏ 27	22 ♐ 18	23 ♎ 38
02	12 44 29	09 12 05	27 ♏ 32 06	03 ♐ 27 23	03 02	22 37	18 55	18 13	12 23	27 29	22 19	23 40
03	12 48 26	10 11 09	09 ♐ 21 59	15 ♐ 16 26	03 24	23 45	19 32	18 26	12 30	27 32	22 20	23 42
04	12 52 22	11 10 16	21 ♐ 11 15	27 ♐ 07 04	03 40	24 54	20 08	18 39	12 38	27 35	22 21	23 45
05	12 56 19	12 09 25	03 ♑ 04 29	09 ♑ 04 11	03 49	26 02	20 44	18 52	12 45	27 38	22 22	23 47
06	13 00 15	13 08 35	15 ♑ 06 49	21 ♑ 13 05	03 R 52	27 10	21 20	19 05	12 52	27 41	22 23	23 49
07	13 04 12	14 07 47	27 ♑ 23 38	03 ♒ 39 09	03 47	28 18	21 57	19 18	13 00	27 44	22 24	23 52
08	13 08 08	15 07 01	10 ♒ 00 13	16 ♒ 27 23	03 35	29 ♏ 26	22 33	19 31	13 07	27 47	22 25	23 54
09	13 12 05	16 06 16	23 ♒ 01 07	29 ♒ 41 46	03 15	00 ♐ 34	23 09	19 44	13 14	27 50	22 26	23 56
10	13 16 01	17 05 34	06 ♓ 29 32	13 ♓ 24 29	02 47	01 41	23 44	19 57	13 22	27 53	22 27	23 59
11	13 19 58	18 04 53	20 ♓ 26 28	27 ♓ 35 09	02 11	02 49	24 20	20 10	13 29	27 56	22 29	24 01
12	13 23 55	19 04 14	04 ♈ 50 00	12 ♈ 10 15	01 26	03 56	24 56	20 23	13 37	27 59	22 30	24 04
13	13 27 51	20 03 37	19 ♈ 34 59	27 ♈ 03 06	00 ♏ 34	05 03	25 32	20 36	13 44	28 02	22 31	24 06
14	13 31 48	21 03 01	04 ♉ 33 25	12 ♉ 04 41	29 ♎ 35	06 10	26 07	20 49	13 51	28 05	22 32	24 08
15	13 35 44	22 02 29	19 ♉ 35 39	27 ♉ 05 09	28 30	07 17	26 43	21 02	13 59	28 08	22 34	24 11
16	13 39 41	23 01 58	04 ♊ 32 05	11 ♊ 55 32	27 20	08 24	27 18	21 15	14 06	28 11	22 35	24 13
17	13 43 37	24 01 30	19 ♊ 14 47	26 ♊ 29 14	26 07	09 31	27 54	21 28	14 13	28 15	22 36	24 16
18	13 47 34	25 01 04	03 ♋ 38 31	10 ♋ 42 27	24 53	10 37	28 29	21 41	14 20	28 18	22 38	24 18
19	13 51 30	26 00 40	17 ♋ 40 57	24 ♋ 34 08	23 40	11 43	29 04	21 54	14 28	28 21	22 39	24 20
20	13 55 27	27 00 19	01 ♌ 22 09	08 ♌ 05 15	22 30	12 49	29 ♌ 40	22 07	14 35	28 24	22 41	24 23
21	13 59 24	28 00 00	14 ♌ 43 45	21 ♌ 17 58	21 26	13 55	00 ♍ 15	22 20	14 42	28 28	22 42	24 25
22	14 03 20	28 59 43	27 ♌ 48 13	04 ♍ 14 51	20 29	15 01	00 50	22 33	14 50	28 31	22 44	24 28
23	14 07 17	29 ♎ 59 28	10 ♍ 38 09	16 ♍ 58 24	19 41	16 07	01 25	22 46	14 57	28 34	22 45	24 30
24	14 11 13	00 ♏ 59 15	23 ♍ 15 51	29 ♍ 30 41	19 03	17 12	02 00	22 59	15 04	28 38	22 47	24 32
25	14 15 10	01 59 05	05 ♎ 43 06	11 ♎ 53 14	18 36	18 17	02 34	23 12	15 11	28 41	22 48	24 35
26	14 19 06	02 58 57	18 ♎ 01 12	24 ♎ 07 07	18 20	19 22	03 09	23 25	15 18	28 45	22 50	24 37
27	14 23 03	03 58 50	00 ♏ 11 05	06 ♏ 13 12	18 D 15	20 27	03 44	23 38	15 25	28 48	22 51	24 40
28	14 26 59	04 58 46	12 ♏ 13 34	18 ♏ 12 20	18 22	21 31	04 18	23 51	15 33	28 51	22 53	24 42
29	14 30 56	05 58 44	24 ♏ 09 39	00 ♐ 05 44	18 40	22 36	04 53	24 04	15 40	28 55	22 55	24 44
30	14 34 53	06 58 43	06 ♐ 00 48	11 ♐ 55 08	19 07	23 40	05 27	24 17	15 47	28 58	22 57	24 47
31	14 38 49	07 ♏ 58 44	17 ♐ 49 04	23 ♐ 43 00	19 ♎ 44	24 ♐ 43	06 ♍ 01	24 ♎ 30	15 ♎ 54	29 ♏ 02	22 ♐ 58	24 ♎ 49

Date	Moon True ☊ ° '	Moon Mean ☊ ° '	Moon ☽ Latitude ° '
01	28 ♋ 53	28 ♋ 02	04 N 49
02	28 R 44	27 59	04 25
03	28 37	27 56	03 50
04	28 33	27 53	03 04
05	28 31	27 49	02 11
06	28 D 31	27 46	01 10
07	28 31	27 43	00 N 06
08	28 R 30	27 40	01 S 01
09	28 28	27 37	02 06
10	28 24	27 33	03 07
11	28 16	27 30	03 58
12	28 07	27 27	04 36
13	27 56	27 24	04 57
14	27 44	27 21	04 58
15	27 34	27 18	04 39
16	27 25	27 14	04 00
17	27 20	27 11	03 06
18	27 17	27 08	02 01
19	27 D 16	27 05	00 S 50
20	27 R 16	27 02	00 N 22
21	27 15	26 59	01 31
22	27 13	26 55	02 34
23	27 08	26 52	03 28
24	27 00	26 49	04 10
25	26 50	26 46	04 40
26	26 37	26 43	04 56
27	26 22	26 39	04 59
28	26 08	26 36	04 48
29	25 54	26 33	04 25
30	25 43	26 30	03 50
31	25 ♋ 35	26 ♋ 27	03 N 05

DECLINATIONS

Date	Sun ☉ ° '	Moon ☽ ° '	Mercury ☿ ° '	Venus ♀ ° '	Mars ♂ ° '	Jupiter ♃ ° '	Saturn ♄ ° '	Uranus ♅ ° '	Neptune ♆ ° '	Pluto ♇ ° '
01	03 S 16	11 S 54	15 S 36	19 S 53	16 N 32	06 S 03	02 S 48	19 S 25	21 S 55	06 N 15
02	03 39	15 18	15 48	20 14	16 22	06 08	02 51	19 25	21 56	06 14
03	04 02	18 04	15 58	20 35	16 11	06 13	02 54	19 26	21 56	06 13
04	04 25	20 05	16 04	20 56	16 00	06 18	02 57	19 27	21 56	06 12
05	04 48	21 14	16 08	21 16	15 49	06 23	03 00	19 27	21 56	06 11
06	05 11	21 25	16 08	21 36	15 38	06 28	03 03	19 28	21 56	06 10
07	05 34	20 35	16 04	21 55	15 27	06 33	03 06	19 29	21 56	06 09
08	05 57	18 43	15 57	22 13	15 16	06 38	03 08	19 29	21 56	06 08
09	06 20	15 50	15 45	22 31	15 04	06 43	03 11	19 30	21 56	06 08
10	06 43	12 01	15 30	22 49	14 53	06 48	03 14	19 31	21 56	06 07
11	07 06	07 26	15 09	23 05	14 42	06 53	03 17	19 31	21 57	06 06
12	07 28	02 S 18	14 44	23 22	14 30	06 57	03 20	19 32	21 57	06 05
13	07 51	03 N 04	14 15	23 38	14 19	07 02	03 22	19 33	21 57	06 04
14	08 13	08 21	13 41	23 53	14 07	07 07	03 25	19 34	21 57	06 03
15	08 35	13 09	13 03	24 07	13 56	07 12	03 28	19 34	21 57	06 02
16	08 57	17 07	12 22	24 21	13 44	07 17	03 31	19 35	21 57	06 01
17	09 19	19 55	11 38	24 35	13 33	07 22	03 34	19 36	21 57	06 01
18	09 41	21 22	10 53	24 48	13 21	07 27	03 36	19 36	21 57	06 00
19	10 03	21 26	10 07	25 00	13 09	07 32	03 39	19 37	21 57	05 59
20	10 24	20 13	09 22	25 12	12 57	07 36	03 42	19 38	21 58	05 58
21	10 46	17 52	08 39	25 23	12 46	07 41	03 45	19 39	21 58	05 57
22	11 07	14 38	07 59	25 33	12 34	07 46	03 47	19 39	21 58	05 56
23	11 28	10 47	07 23	25 43	12 22	07 51	03 50	19 40	21 58	05 56
24	11 49	06 30	06 52	25 52	12 10	07 56	03 53	19 41	21 58	05 55
25	12 10	02 N 01	06 26	26 00	11 58	08 01	03 55	19 42	21 58	05 54
26	12 30	02 S 30	06 06	26 08	11 46	08 05	03 58	19 42	21 58	05 53
27	12 51	06 52	05 52	26 15	11 34	08 10	04 01	19 43	21 58	05 53
28	13 11	10 56	05 44	26 22	11 22	08 15	04 03	19 44	21 59	05 52
29	13 31	14 32	05 41	26 28	11 10	08 20	04 06	19 45	21 59	05 51
30	13 51	17 32	05 44	26 33	10 58	08 24	04 09	19 45	21 59	05 50
31	14 S 10	19 S 49	05 S 52	26 S 38	10 N 46	08 S 29	04 S 11	19 S 46	21 S 59	05 N 50

ZODIAC SIGN ENTRIES

Date	h m	Planets
02	16 59	☽ ♐
05	05 49	☽ ♑
07	17 01	☽ ♒
09	00 04	♀ ♐
10	00 32	☽ ♓
12	04 01	☽ ♈
14	02 09	☿ ♎
14	04 43	☽ ♉
16	04 41	☽ ♊
18	05 52	☽ ♋
20	09 34	☽ ♌
21	01 56	♂ ♍
22	16 05	☽ ♍
23	12 13	☉ ♏
25	00 57	☽ ♎
27	11 38	☽ ♏
29	23 48	☽ ♐

LATITUDES

Date	Mercury ☿ ° '	Venus ♀ ° '	Mars ♂ ° '	Jupiter ♃ ° '	Saturn ♄ ° '	Uranus ♅ ° '	Neptune ♆ ° '	Pluto ♇ ° '
01	03 S 27	01 S 49	01 N 16	01 N 05	02 N 13	00 N 11	01 N 18	16 N 34
04	03 33	02 00	01 17	01 05	02 13	00 11	01 18	16 34
07	03 30	02 11	01 19	01 05	02 13	00 11	01 17	16 34
10	03 15	02 21	01 21	01 05	02 13	00 11	01 17	16 33
13	02 45	02 32	01 23	01 05	02 13	00 11	01 17	16 33
16	01 59	02 41	01 25	01 05	02 13	00 11	01 17	16 33
19	01 S 00	02 50	01 27	01 05	02 13	00 11	01 17	16 33
22	00 N 01	02 58	01 29	01 05	02 13	00 11	01 17	16 33
25	00 55	03 06	01 31	01 05	02 14	00 11	01 17	16 33
28	01 35	03 12	01 33	01 05	02 14	00 11	01 16	16 34
31	02 N 00	03 S 18	01 N 35	01 N 05	02 N 14	00 N 11	01 N 16	16 N 34

DATA

Julian Date	2444879
Delta T	+51 seconds
Ayanamsa	23° 35' 52"
Synetic vernal point	05° ♓ 31' 08"
True obliquity of ecliptic	23° 26' 26"

MOON'S PHASES, APSIDES AND POSITIONS ☽

Date	h m	Phase	Longitude ° '	Eclipse Indicator
06	07 45	☽	12 ♑ 58	
13	12 49	○	20 ♈ 06	
20	03 40	☾	26 ♋ 40	
27	20 13	●	04 ♏ 19	

Day	h m		
03	01 17	Apogee	
15	02 03	Perigee	
30	16 13	Apogee	
06	04 34	Max dec	21° S 28'
12	22 21	0N	
19	01 10	Max dec	21° N 35'
25	22 40	0S	

LONGITUDES

Date	Chiron ⚷ ° '	Ceres ⚳ ° '	Pallas ⚴ ° '	Juno ⚵ ° '	Vesta ⚶ ° '	Black Moon Lilith ⚸ ° '
01	22 ♉ 17	29 ♍ 53	08 ♍ 10	13 ♏ 17	28 ♎ 54	00 ♐ 50
11	21 ♉ 55	04 ♎ 27	13 ♍ 08	16 ♏ 28	04 ♏ 08	01 ♐ 57
21	21 ♉ 29	08 ♎ 59	17 ♍ 57	19 ♏ 44	09 ♏ 25	03 ♐ 04
31	21 ♉ 00	13 ♎ 28	22 ♍ 36	23 ♏ 04	14 ♏ 45	04 ♐ 11

All ephemeris data is given at 12.00 UT and the Moon's longitude is additionally given for 24.00 UT

ASPECTARIAN

01 Thursday
05 11 ☽ ⊻ ♄
08 55 ☽ ⊥ ☉
13 21 ☽ ⊥ ♆
16 51 ☽ ⊻ ♃
17 20 ☽ ⊥ ♄
17 41 ☽ □ ♂

02 Friday
01 00 ☽ ☌ ♀
01 26 ☽ ⊻ ♆
04 09 ☽ ⊻ ♇
04 39 ☽ ∠ ☉
05 10 ☽ ⊥ ♃
05 42 ♀ ⊻ ♆
11 41 ☽ ∠ ♄
11 55 ☽ ☌ ♅
16 12 ☽ ∥ ☿
16 20 ☽ ⊥ ♇
19 58 ☽ # ♂
23 32 ☽ ⊻ ☿
23 44 ☽ ∠ ♃

03 Saturday
10 39 ☽ ∠ ♇
10 55 ♀ ⊻ ♇
11 56 ♃ # ♇
12 04 ☽ ⊥ ☿
13 49 ☽ ✶ ☉
15 37 ☉ Q ♆
18 26 ☽ ✶ ♄
22 29 ♄ ∠ ♅

04 Sunday
03 12 ☽ ∥ ♅
05 20 ☿ # ♂
05 41 ♀ ⊥ ♃
06 45 ☽ ✶ ♃
06 47 ☽ ∠ ☿
09 45 ☽ △ ♂
14 21 ☽ ☌ ♆
16 23 ☽ Q ☉
17 12 ☽ ✶ ♇
19 02 ☽ Q ♄
20 18 ☽ ⊻ ♀

05 Monday
01 00 ☽ ⊻ ♅
07 29 ☽ Q ♃
09 41 ☽ ⊥ ♀
13 07 ☽ ⊥ ♅
13 30 ☽ ✶ ☿
15 09 ☽ ∥ ♀
17 27 ☽ Q ♇
17 37 ☽ ⚼ ♂

06 Tuesday
00 09 ☉ ∠ ♅
02 14 ☽ ∥ ♀
04 29 ☉ ☌ ♄
05 33 ☽ ∠ ♀
07 10 ☽ ∠ ♅
07 31 ☽ □ ♄
07 45 ☽ □ ☉
09 14 ☿ St R
12 28 ☽ ± ♂
13 29 ☽ Q ☿
19 57 ☽ □ ♃
23 21 ♀ ☌ ♅

07 Wednesday
00 52 ☽ ⊼ ♂
02 18 ☽ ⊻ ♆
04 46 ♀ ∠ ♄
05 08 ☽ □ ♇
06 53 ☿ Q ♂
12 39 ☽ ✶ ♅
13 55 ☽ ✶ ♀
13 57 ☽ ⊥ ♆
14 00 ♀ ∥ ♆

08 Thursday
00 06 ☽ □ ☿
03 37 ☽ ∥ ♅
06 54 ♂ △ ♆
07 08 ☽ ∠ ♆
11 34 ☽ Q ♅
14 56 ☽ Q ♀
17 52 ☽ △ ♄
22 19 ☽ △ ☉
22 20 ♀ ⊥ ♇
23 16 ☉ # ♇

09 Friday
05 55 ☽ △ ♃
10 57 ☽ ✶ ♆
12 14 ☽ ☍ ♂
12 32 ☽ ∥ ☿
13 41 ☽ △ ♇
17 31 ☽ # ♂
20 42 ☽ □ ♅
21 29 ☽ ⚼ ♄

10 Saturday
02 46 ☽ □ ♀
03 39 ☽ ⚼ ☉
05 43 ☽ △ ☿
08 25 ☽ Q ♆
09 15 ☽ ⚼ ♃
13 32 ☽ ± ♄
16 21 ☽ ⚼ ♇
18 27 ☉ ∥ ♃
20 37 ☽ ± ☉
22 20 ♂ ✶ ♇

11 Sunday
00 02 ☽ ⊼ ♄
01 08 ☽ ± ♃
03 25 ☿ ⊻ ♀
06 42 ☽ ⚼ ☿
07 42 ☽ ⊼ ☉
07 53 ☽ ± ♇
11 32 ☽ ⊼ ♃
13 33 ☽ ∥ ☉
14 40 ☽ ∥ ♃
15 27 ☽ □ ♆
18 03 ☽ ⊼ ♇
18 29 ☽ # ♇
18 51 ☽ ⊼ ♂
21 12 ☽ ± ☿

12 Monday
00 37 ☽ △ ♅
05 17 ☽ ± ♂
06 41 ☽ ⊼ ☿
07 24 ☽ ∥ ♄
10 24 ☽ △ ♀
20 43 ☽ ⚼ ♂

13 Tuesday
01 22 ☽ ⚼ ♅
02 27 ☽ ☍ ♄
07 05 ☿ # ♂
12 49 ☽ ⚼ ♀
12 49 ☽ ☍ ☉
12 50 ☉ ∠ ♀
13 21 ☽ # ♄
13 40 ☽ ☍ ♃
15 57 ☽ ± ♅
16 44 ☽ △ ♆
19 17 ☽ ☍ ♇
21 57 ☽ △ ♂

14 Wednesday
01 24 ☽ ∥ ♇
01 37 ☽ ⊼ ♅
02 30 ♀ ∠ ♃
04 26 ☽ ± ♀
04 33 ☽ ☍ ☿
04 46 ☉ ☌ ♃
05 22 ☿ ⊥ ♀
06 10 ☽ # ♃
11 18 ☽ # ☉
14 47 ☽ ⊼ ♀
16 46 ☽ ⚼ ♆

15 Thursday
02 57 ☽ ⊼ ♄
07 09 ☽ ± ♆
11 31 ☽ # ☿
12 37 ☽ ± ♄
14 20 ☽ ⊼ ♃
14 24 ☉ ⊥ ♅
16 05 ☽ ∥ ♂
16 11 ☽ ⊼ ☉
16 45 ☽ ⊼ ♆
19 12 ☿ ⊻ ♅
19 22 ☽ ⊼ ♇
23 52 ☽ □ ♂

16 Friday
00 06 ☽ ± ♃
00 54 ☉ ✶ ♆
01 15 ☽ ⊼ ☿
01 44 ☽ ☍ ♅
02 30 ☽ ± ☉
03 10 ☽ ⚼ ♄
05 01 ☽ ± ♇
10 12 ☽ ± ☿
12 18 ☿ ✶ ♂
14 49 ☽ ⚼ ♃
18 04 ☽ ⚼ ☉
18 46 ☽ ☍ ♀
19 37 ☽ ⚼ ♇
23 42 ☽ ⚼ ☿

17 Saturday
03 41 ☽ △ ♄
06 16 ☽ Q ♂
06 24 ♀ ∠ ♇
08 36 ☽ # ♅
15 43 ☽ △ ♃
17 34 ☽ ☍ ♆
17 57 ☉ ☌ ♇
20 19 ☽ △ ♇
20 29 ☽ △ ☉
22 29 ☽ △ ☿

18 Sunday
02 58 ☽ ✶ ♂
02 59 ☽ ⊼ ♅
03 29 ♂ □ ♅
04 26 ☿ ∠ ♀
10 31 ☉ ☌ ☿
13 07 ☽ ± ♅
22 59 ☿ ☌ ♇

19 Monday
00 52 ☽ ⊼ ♀
04 30 ☽ ⚼ ♅
05 30 ☽ ∠ ♂
06 24 ☽ □ ♄
12 04 ☽ ± ♀
13 36 ☉ ∥ ☿
19 27 ☽ □ ♃
20 40 ☽ ⊼ ♆
21 35 ☽ □ ☿
21 48 ☽ ⊥ ♂
23 38 ☽ □ ♇

20 Tuesday
03 40 ☽ □ ☉
05 10 ☽ ⚼ ♀
06 44 ☽ △ ♅
07 13 ☽ ± ♆
08 04 ♂ ∠ ♄
08 23 ☿ ✶ ♆
08 50 ☽ ⊻ ♂
14 01 ☿ ⊥ ♅
14 11 ☽ Q ♄
18 57 ☽ # ♅
18 57 ☿ ☌ ♃
23 17 ☽ ⚼ ♆

21 Wednesday
03 07 ☽ Q ☿
03 55 ☽ Q ♃
07 48 ☽ Q ♇
10 24 ☽ △ ♀
11 57 ☽ ✶ ♄
14 30 ☽ Q ☉
23 23 ☽ ✶ ☿
23 46 ☉ ⊻ ♅

22 Thursday
02 08 ☽ ✶ ♃
02 36 ☽ △ ♆
05 48 ☽ ✶ ♇
06 24 ♃ ⊥ ♅
07 16 ♀ ✶ ♄
13 20 ☽ □ ♅
14 24 ☽ ✶ ☉
15 47 ☽ ∠ ♄
17 53 ☽ ☌ ♂
19 18 ☿ ∥ ♃

23 Friday
01 26 ☽ ∠ ☿
02 01 ☽ ∥ ♂
06 31 ☽ ∠ ♃
08 15 ☽ # ☉
08 47 ☽ ⊥ ♄
09 49 ♃ ✶ ♆
09 51 ☽ ∠ ♇
17 27 ☽ ⊥ ☿
20 14 ☽ ⊻ ♄
20 56 ☽ ∠ ☉
23 17 ☽ Q ♅
23 20 ☽ □ ♀
23 49 ☽ ⊥ ♃

24 Saturday
02 57 ☽ ⊥ ♇
04 18 ☽ ⊻ ☿
04 19 ☽ # ♃
09 47 ☽ # ☿
11 04 ☽ □ ♆
11 28 ☽ ⊻ ♃
14 27 ☽ ⊻ ♇
15 11 ☽ ∥ ♇
15 35 ☽ ⊥ ☉
22 21 ☽ ✶ ♅

25 Sunday
01 57 ☽ # ♄
03 22 ☉ # ♂
04 09 ☽ ⊻ ☉
05 37 ☽ ⊻ ♂
13 12 ☽ Q ♀
17 12 ☿ ✶ ♀
17 49 ☽ ⊥ ♂
21 55 ☽ Q ♆

26 Monday
03 35 ☽ ∠ ♅
06 38 ☽ ☌ ♄
12 16 ☽ ∠ ♂
12 36 ☽ ☌ ☿
14 54 ☽ ✶ ♀
17 57 ☿ ∠ ♂
20 02 ☽ ∥ ♄
21 20 ☽ ⊥ ♅
21 29 ☽ ✶ ♆
21 39 ☉ ✶ ♂
22 49 ☽ ☌ ♃

27 Tuesday
01 02 ☽ ☌ ♇
06 27 ☽ # ♇
06 39 ☽ ∥ ☿
09 11 ☿ St D
09 14 ☽ ⊻ ♅
11 14 ☿ # ♇
19 24 ☽ ✶ ♂
19 36 ☽ ∥ ♃
20 13 ☽ ☌ ☉
23 29 ☽ ∠ ♀

28 Wednesday
03 18 ☽ ∠ ♆
14 38 ☽ # ♂
18 43 ☽ ⊻ ♄
19 15 ☽ ⊥ ♀
20 36 ☽ Q ♂
21 22 ☽ ⊥ ♆

29 Thursday
00 36 ☽ ⊻ ☿
04 06 ☽ ∥ ☉
06 54 ☽ ⊥ ♄
08 31 ☽ ⊻ ♀
09 29 ☽ ⊻ ♆
11 48 ☽ ⊻ ♃
13 03 ☽ ⊥ ☿
13 10 ☽ ⊻ ♇
19 24 ♀ ☌ ♆
21 39 ☽ ☌ ♅

30 Friday
00 10 ☽ ⊥ ♃
01 17 ☽ ∠ ♄
01 21 ☽ ⊥ ♇
07 59 ☽ ∠ ☿
10 48 ☽ □ ♂
14 08 ☽ ⊻ ☉
18 46 ☽ ∠ ♃
19 41 ☽ ∠ ♇

31 Saturday
03 28 ☽ ⊥ ☉
05 32 ♀ ✶ ♃
06 19 ☿ # ♇
08 03 ☽ ✶ ♄
11 30 ☽ ∥ ♅
11 47 ☉ ∠ ♆
14 15 ♀ ✶ ♇
16 09 ☽ ✶ ☿
22 30 ☽ ☌ ♆
23 28 ☽ ∠ ☉

NOVEMBER 1981

Date	Sidereal time h m s	Sun ☉ ° ' "	Moon ☽ ° ' "	Moon ☽ 24.00 ° ' "	Mercury ☿ ° '	Venus ♀ ° '	Mars ♂ ° '	Jupiter ♃ ° '	Saturn ♄ ° '	Uranus ♅ ° '	Neptune ♆ ° '	Pluto ♇ ° '
01	14 42 46	08 ♏ 58 47	29 ♐ 37 22	05 ♑ 32 38	20 ♎ 29	25 ♐ 47	06 ♍ 36	24 ♎ 43	16 ♎ 01	29 ♏ 05	23 ♐ 00	24 ♎ 51
02	14 46 42	09 58 52	11 ♑ 29 22	17 ♑ 28 07	21 22	26 50	07 10	24 55	16 08	29 09	23 02	24 54
03	14 50 39	10 58 58	23 ♑ 29 30	29 ♑ 34 11	22 22	27 53	07 44	25 08	16 15	29 13	23 04	24 56
04	14 54 35	11 59 06	05 ♒ 42 49	11 ♒ 56 03	23 27	28 56	08 18	25 21	16 22	29 16	23 05	24 59
05	14 58 32	12 59 15	18 ♒ 14 34	24 ♒ 38 59	24 37	29 ♐ 58	08 51	25 34	16 29	29 20	23 07	25 01
06	15 02 28	13 59 26	01 ♓ 09 53	07 ♓ 47 45	25 52	01 ♑ 00	09 25	25 46	16 35	29 23	23 09	25 03
07	15 06 25	14 59 39	14 ♓ 32 59	21 ♓ 25 49	27 10	02 02	09 59	25 59	16 42	29 27	23 11	25 05
08	15 10 22	15 59 52	28 ♓ 26 20	05 ♈ 34 25	28 31	03 04	10 32	26 12	16 49	29 31	23 13	25 08
09	15 14 18	17 00 08	12 ♈ 49 43	20 ♈ 11 40	29 ♎ 56	04 05	11 06	26 24	16 56	29 34	23 15	25 10
10	15 18 15	18 00 24	27 ♈ 39 27	05 ♉ 12 01	01 ♏ 22	05 05	11 39	26 37	17 02	29 38	23 17	25 12
11	15 22 11	19 00 43	12 ♉ 48 11	20 ♉ 26 34	02 50	06 06	12 12	26 50	17 09	29 42	23 18	25 15
12	15 26 08	20 01 03	28 ♉ 05 45	05 ♊ 44 18	04 20	07 06	12 45	27 02	17 15	29 45	23 20	25 17
13	15 30 04	21 01 25	13 ♊ 20 51	20 ♊ 54 08	05 51	08 05	13 18	27 15	17 22	29 49	23 22	25 19
14	15 34 01	22 01 49	28 ♊ 23 04	05 ♋ 46 46	07 23	09 04	13 51	27 27	17 29	29 53	23 24	25 21
15	15 37 57	23 02 14	13 ♋ 04 35	20 ♋ 16 04	08 56	10 03	14 24	27 39	17 35	29 ♏ 56	23 26	25 23
16	15 41 54	24 02 42	27 ♋ 20 57	04 ♌ 19 11	10 29	11 01	14 57	27 52	17 42	00 ♐ 00	23 28	25 26
17	15 45 51	25 03 11	11 ♌ 10 51	17 ♌ 56 12	12 03	11 59	15 29	28 04	17 48	00 04	23 30	25 28
18	15 49 47	26 03 42	24 ♌ 35 31	01 ♍ 09 12	13 37	12 56	16 02	28 16	17 54	00 07	23 33	25 30
19	15 53 44	27 04 15	07 ♍ 37 42	14 ♍ 01 26	15 12	13 53	16 34	28 29	18 01	00 11	23 35	25 32
20	15 57 40	28 04 50	20 ♍ 20 54	26 ♍ 36 31	16 46	14 49	17 06	28 41	18 07	00 15	23 37	25 34
21	16 01 37	29 ♏ 05 26	02 ♎ 48 44	08 ♎ 57 58	18 21	15 45	17 38	28 53	18 13	00 19	23 39	25 36
22	16 05 33	00 ♐ 06 04	15 ♎ 04 34	21 ♎ 08 54	19 56	16 40	18 10	29 05	18 19	00 22	23 41	25 38
23	16 09 30	01 06 44	27 ♎ 11 15	03 ♏ 11 54	21 31	17 35	18 42	29 17	18 25	00 26	23 43	25 40
24	16 13 26	02 07 25	09 ♏ 11 05	15 ♏ 09 02	23 06	18 29	19 14	29 29	18 31	00 30	23 45	25 42
25	16 17 23	03 08 07	21 ♏ 05 58	27 ♏ 02 02	24 41	19 22	19 45	29 41	18 37	00 33	23 47	25 44
26	16 21 20	04 08 52	02 ♐ 57 27	08 ♐ 52 23	26 16	20 15	20 17	29 ♎ 53	18 43	00 37	23 49	25 47
27	16 25 16	05 09 37	14 ♐ 47 03	20 ♐ 41 40	27 50	21 07	20 48	00 ♏ 05	18 49	00 41	23 52	25 48
28	16 29 13	06 10 24	26 ♐ 36 29	02 ♑ 31 44	29 ♏ 25	21 58	21 19	00 17	18 55	00 45	23 54	25 50
29	16 33 09	07 11 12	08 ♑ 27 46	14 ♑ 24 55	01 ♐ 00	22 49	21 50	00 28	19 01	00 48	23 56	25 52
30	16 37 06	08 ♐ 12 01	20 ♑ 23 33	26 ♑ 24 06	02 ♐ 34	23 ♑ 39	22 ♍ 21	00 ♏ 40	19 ♎ 06	00 ♐ 52	23 ♐ 58	25 ♎ 54

Date	Moon True ☊ ° '	Moon Mean ☊ ° '	Moon ☽ Latitude ° '
01	25 ♋ 29	26 ♋ 24	02 N 12
02	25 R 26	26 20	01 13
03	25 26	26 17	00 N 10
04	25 D 26	26 14	00 S 54
05	25 R 26	26 11	01 58
06	25 24	26 08	02 58
07	25 21	26 05	03 50
08	25 15	26 01	04 31
09	25 07	25 58	04 56
10	24 57	25 55	05 02
11	24 47	25 52	04 48
12	24 37	25 49	04 13
13	24 30	25 45	03 20
14	24 24	25 42	02 14
15	24 22	25 39	01 S 00
16	24 D 21	25 36	00 N 16
17	24 22	25 33	01 29
18	24 23	25 30	02 34
19	24 R 22	25 26	03 30
20	24 19	25 23	04 14
21	24 14	25 20	04 45
22	24 07	25 17	05 02
23	23 58	25 14	05 05
24	23 47	25 10	04 55
25	23 37	25 07	04 32
26	23 27	25 04	03 57
27	23 19	25 01	03 12
28	23 13	24 58	02 18
29	23 10	24 55	01 19
30	23 ♋ 08	24 ♋ 51	00 N 15

DECLINATIONS

Date	Sun ☉ ° '	Moon ☽ ° '	Mercury ☿ ° '	Venus ♀ ° '	Mars ♂ ° '	Jupiter ♃ ° '	Saturn ♄ ° '	Uranus ♅ ° '	Neptune ♆ ° '	Pluto ♇ ° '
01	14 S 29	21 S 14	06 S 04	26 S 42	10 N 34	08 S 34	04 S 14	19 S 47	21 S 59	05 N 49
02	14 49	21 44	06 21	26 45	10 22	08 38	04 17	19 48	21 59	05 48
03	15 07	21 14	06 40	26 48	10 10	08 43	04 19	19 49	21 59	05 48
04	15 26	19 43	07 03	26 50	09 58	08 48	04 22	19 49	21 59	05 47
05	15 44	17 14	07 29	26 51	09 46	08 52	04 24	19 50	22 00	05 46
06	16 02	13 50	07 57	26 52	09 34	08 57	04 27	19 51	22 00	05 46
07	16 20	09 37	08 26	26 52	09 22	09 01	04 29	19 52	22 00	05 45
08	16 38	04 S 46	08 58	26 52	09 09	09 06	04 32	19 53	22 00	05 44
09	16 55	00 N 31	09 30	26 51	08 57	09 10	04 34	19 53	22 00	05 44
10	17 12	05 56	10 03	26 49	08 45	09 15	04 37	19 54	22 00	05 43
11	17 28	11 06	10 38	26 47	08 33	09 19	04 39	19 55	22 00	05 43
12	17 45	15 37	11 12	26 44	08 21	09 24	04 41	19 56	22 01	05 42
13	18 01	19 05	11 47	26 40	08 09	09 28	04 44	19 57	22 01	05 41
14	18 17	21 12	12 23	26 36	07 57	09 33	04 46	19 57	22 01	05 41
15	18 32	21 48	12 58	26 31	07 45	09 37	04 48	19 58	22 01	05 40
16	18 47	20 57	13 33	26 26	07 33	09 41	04 51	19 59	22 01	05 40
17	19 02	18 51	14 08	26 20	07 21	09 46	04 53	20 00	22 01	05 39
18	19 16	15 45	14 42	26 14	07 09	09 50	04 55	20 00	22 01	05 39
19	19 30	11 57	15 16	26 07	06 57	09 54	04 58	20 01	22 01	05 38
20	19 44	07 43	15 50	25 59	06 45	09 59	05 00	20 02	22 02	05 38
21	19 57	03 N 14	16 23	25 51	06 33	10 03	05 02	20 03	22 02	05 37
22	20 10	01 S 18	16 55	25 43	06 21	10 07	05 04	20 04	22 02	05 37
23	20 23	05 43	17 27	25 34	06 10	10 11	05 06	20 04	22 02	05 37
24	20 35	09 54	17 58	25 24	05 58	10 15	05 08	20 05	22 02	05 36
25	20 47	13 40	18 28	25 14	05 46	10 20	05 10	20 06	22 02	05 36
26	20 59	16 53	18 57	25 04	05 34	10 24	05 13	20 07	22 02	05 35
27	21 10	19 24	19 26	24 53	05 23	10 28	05 15	20 08	22 02	05 35
28	21 20	21 05	19 53	24 42	05 11	10 32	05 17	20 08	22 03	05 35
29	21 31	21 52	20 20	24 30	04 59	10 36	05 19	20 09	22 03	05 34
30	21 S 41	21 S 39	20 S 45	24 S 18	04 N 48	10 S 40	05 S 21	20 S 10	22 S 03	05 N 34

ZODIAC SIGN ENTRIES

Date	h m	Planets
01	12 46	☽ ♑
04	00 51	☽ ♒
05	12 39	♀ ♑
06	09 52	☽ ♓
08	14 39	☽ ♈
09	13 14	☿ ♏
10	15 44	☽ ♉
12	14 59	☽ ♊
14	14 37	☽ ♋
16	12 05	♅ ♐
16	16 32	☽ ♌
18	21 53	☽ ♍
21	06 33	☽ ♎
22	09 36	☉ ♐
23	17 37	☽ ♏
26	06 00	☽ ♐
27	02 19	♃ ♏
28	18 53	☽ ♑
28	20 52	☿ ♐

LATITUDES

Date	Mercury ☿ ° '	Venus ♀ ° '	Mars ♂ ° '	Jupiter ♃ ° '	Saturn ♄ ° '	Uranus ♅ ° '	Neptune ♆ ° '	Pluto ♇ ° '
01	02 N 05	03 S 20	01 N 35	01 N 05	02 N 14	00 N 11	01 N 16	16 N 34
04	02 12	03 24	01 37	01 05	02 15	00 11	01 16	16 34
07	02 10	03 27	01 39	01 05	02 15	00 11	01 16	16 35
10	02 06	03 28	01 41	01 06	02 15	00 10	01 16	16 36
13	01 47	03 29	01 43	01 06	02 16	00 10	01 16	16 36
16	01 30	03 28	01 45	01 06	02 16	00 10	01 16	16 37
19	01 11	03 25	01 47	01 06	02 17	00 10	01 16	16 38
22	00 50	03 20	01 50	01 06	02 17	00 10	01 16	16 39
25	00 29	03 14	01 52	01 06	02 18	00 10	01 16	16 40
28	00 N 09	03 06	01 54	01 07	02 18	00 10	01 16	16 41
31	00 S 12	02 S 55	01 N 56	01 N 07	02 N 19	00 N 10	01 N 15	16 N 42

DATA

Julian Date	2444910
Delta T	+51 seconds
Ayanamsa	23° 35' 55"
Synetic vernal point	05° ♓ 31' 05"
True obliquity of ecliptic	23° 26' 26"

MOON'S PHASES, APSIDES AND POSITIONS ☽

Date	h m	Phase	Longitude ° '	Eclipse Indicator
05	01 09	☽	12 ♒ 32	
11	22 27	○	19 ♉ 27	
18	14 54	☾	26 ♌ 11	
26	14 38	●	04 ♐ 16	

Day	h m		
12	11 15	Perigee	
26	20 56	Apogee	
02	12 03	Max dec	21° S 44'
09	09 41	ON	
15	09 39	Max dec	21° N 49'
22	05 07	0S	
29	18 51	Max dec	21° S 54'

LONGITUDES

Date	Chiron ⚷ ° '	Ceres ⚳ ° '	Pallas ⚴ ° '	Juno ⚵ ° '	Vesta ⚶ ° '	Black Moon Lilith ⚸ ° '
01	20 ♉ 57	13 ♎ 55	23 ♍ 03	23 ♏ 24	15 ♏ 17	04 ♐ 18
11	20 ♉ 26	18 ♎ 20	27 ♍ 28	26 ♏ 48	20 ♏ 39	05 ♐ 25
21	19 ♉ 55	22 ♎ 40	01 ♎ 41	00 ♐ 13	26 ♏ 02	06 ♐ 32
31	19 ♉ 26	26 ♎ 54	05 ♎ 37	03 ♐ 39	01 ♐ 26	07 ♐ 39

ASPECTARIAN

h m Aspects

01 Sunday
01 50 ☽ ✶ ♃
02 17 ☽ ✶ ♇
03 26 ☽ ☌ ♀
08 42 ☽ Q ♄
10 55 ☽ ⊻ ♅
18 16 ☽ Q ☿
23 08 ☽ ⊥ ♅

02 Monday
02 37 ☽ Q ♃
02 42 ☽ Q ♇
02 50 ☽ △ ♂
08 27 ♃ ☌ ♇
08 41 ☽ ✶ ☉
17 23 ☽ ∠ ♅
21 24 ☽ □ ♄

03 Tuesday
09 32 ☽ □ ☿
10 24 ☽ bQ ♂
10 54 ☽ Q ☉
11 08 ☽ ⊻ ♆
14 52 ☽ □ ♇
15 19 ☽ □ ♃
21 12 ♀ Q ♄
21 30 ☽ ⊻ ♀
23 01 ☽ ⊥ ♆
23 21 ☽ ✶ ♅

04 Wednesday
04 07 ☿ ✶ ♆
05 01 ☽ ± ♂
05 21 ☿ ∠ ♂
08 02 ☿ ⊥ ♅
10 20 ☽ ⊥ ♀
10 47 ☽ ∥ ♅
16 37 ☽ ∠ ♆
17 14 ☽ ⊼ ♂
20 16 ♀ ⊻ ♅
22 47 ☽ Q ♅

05 Thursday
01 09 ☽ □ ☉
05 15 ☽ ∠ ♀
08 37 ☽ △ ♄
20 04 ☿ ☌ ♇
21 10 ☽ ✶ ♆
22 23 ☽ ∥ ☉

06 Friday
00 43 ☽ △ ♇
01 12 ☽ △ ☿
01 55 ☽ △ ♃
08 44 ☽ □ ♅
10 03 ☿ ☌ ♃
11 41 ☽ ✶ ♀
12 47 ☽ bQ ♄
19 15 ☽ Q ♆

07 Saturday
03 33 ☽ ☍ ♂
04 05 ☽ bQ ♇
05 08 ☽ ± ♄
05 36 ☽ bQ ♃
07 20 ☽ bQ ☿
11 01 ☽ Q ♀
12 51 ☽ △ ☉
13 25 ☽ # ♂
15 03 ☽ ∥ ♃
15 48 ☽ ⊼ ♄
17 14 ♂ ∠ ♇
17 30 ☽ ∥ ☿
19 58 ☽ ± ♇
21 39 ☽ ± ♃

08 Sunday
00 47 ☽ ± ☿
03 03 ☽ □ ♆
06 21 ☽ ⊼ ♇
07 21 ☽ # ♇
08 08 ☽ ⊼ ♃
12 10 ☽ ⊼ ☿
13 05 ☽ ∥ ♄
13 49 ☽ △ ♅
16 39 ☽ bQ ☉
17 13 ♂ # ♃
18 27 ☿ # ♂
19 12 ☿ ∥ ♃
20 24 ☽ □ ♀

09 Monday
03 00 ♂ ⊥ ♄
05 42 ☿ ⊻ ♅
08 46 ☽ ± ☉
09 02 ☽ ⊼ ♂
09 57 ☉ ⊻ ♄
14 52 ☽ bQ ♅
17 57 ☉ ⊥ ♆
18 45 ☽ ☍ ♄
19 14 ☽ ± ♂
19 19 ☽ ⊼ ☉

10 Tuesday
04 57 ☽ △ ♆
05 31 ☽ ± ♅
06 04 ☽ # ♄
08 04 ☽ ☍ ♇
09 53 ♂ ∠ ♃
10 19 ☽ ☍ ♃
10 19 ☽ bQ ♂
11 03 ☽ ∥ ♇
15 10 ☽ ⊼ ♅
18 33 ☽ ☍ ☿

11 Wednesday
00 22 ☽ ∥ ♂
00 40 ☽ △ ♀
01 47 ♀ ⊥ ♅
03 23 ☽ # ♃
04 54 ☽ bQ ♆
09 25 ☽ # ☿
11 01 ☽ △ ♂
18 53 ☽ ⊼ ♄
19 06 ☽ ± ♆
22 27 ☽ ☍ ☉

12 Thursday
01 56 ☽ bQ ♀
04 22 ☽ ± ♄
04 32 ☽ ⊼ ♆
07 34 ☽ ⊼ ♇
10 19 ☽ ⊼ ♃
14 37 ☽ ☍ ♅
16 40 ♀ Q ♇
17 00 ☽ ± ♇
17 02 ☽ ± ♀
18 35 ☽ bQ ♄
19 52 ☽ ± ♃
22 52 ☽ ⊼ ☿

13 Friday
02 51 ☽ # ☉
03 07 ☽ ⊼ ♀
07 12 ☽ bQ ♇
09 22 ☽ ± ☿
10 14 ☽ bQ ♃
11 56 ☽ □ ♂
18 25 ☽ △ ♄
19 55 ☽ # ♅

14 Saturday
01 04 ☽ ⊼ ☉
01 16 ☽ bQ ☿
03 59 ☽ ☍ ♆
07 07 ☽ △ ♇
10 28 ☽ △ ♃
11 23 ☽ ± ☉
14 25 ☽ ⊼ ♅
17 49 ☽ Q ♂
23 49 ♀ Q ♃

15 Sunday
00 13 ☽ ± ♅
03 05 ☽ bQ ☉
04 17 ☿ ∠ ♆
04 21 ☽ △ ☿
06 38 ☽ ☍ ♀
14 17 ☽ ✶ ♂
15 06 ☽ bQ ♅
19 33 ☽ □ ♄
21 55 ☉ ⊻ ♆

16 Monday
02 35 ☉ ⊥ ♄
05 24 ☽ ⊼ ♆
05 57 ☽ △ ☉
08 43 ☽ □ ♇
12 54 ☽ □ ♃
15 39 ☽ ± ♆
16 34 ☽ △ ♅
16 37 ☽ ∠ ♂

17 Tuesday
00 34 ☽ # ♅
02 30 ☽ Q ♄
07 17 ☽ bQ ♆
08 54 ☽ ⊥ ♂
09 20 ☿ ✶ ♀
10 26 ☽ # ☉
13 31 ☽ ⊼ ♀
13 44 ☽ □ ☿
16 03 ☽ Q ♇
19 57 ☽ ⊻ ♂
20 48 ☽ Q ♃
22 07 ☉ ⊻ ♇
23 51 ☽ ✶ ♄

18 Wednesday
01 01 ☽ ± ♀
10 05 ☽ △ ♆
13 39 ☽ ✶ ♇
14 54 ☽ □ ☉
18 05 ☽ # ☿
18 34 ☽ bQ ♀
18 49 ☽ ✶ ♃
22 09 ☽ □ ♅

19 Thursday
02 38 ☽ Q ☿
03 21 ☽ ∠ ♄
17 27 ☽ ∠ ♇
20 16 ☽ ⊥ ♄
23 09 ☽ ∠ ♃
23 39 ☽ # ♃

20 Friday
00 40 ☽ △ ♀
03 11 ☽ Q ☉
04 14 ☽ ✶ ☿
05 33 ☽ ☌ ♂
07 43 ☽ ⊻ ♄
07 59 ☽ Q ♅
10 31 ☽ ⊥ ♇
16 32 ☽ ⊥ ♃
17 28 ☽ ∥ ♂
18 15 ☽ □ ♆
19 33 ☿ ✶ ♂
22 02 ☽ ⊻ ♇
23 17 ☽ ∥ ♇
23 46 ♀ ∠ ♅

21 Saturday
01 00 ☿ ⊥ ♆
02 32 ☽ # ♄
04 09 ☽ ✶ ☉
04 16 ☽ ⊻ ♃
05 52 ☉ ⊻ ♃
07 07 ☽ ✶ ♅
09 45 ☿ ⊻ ♄
13 13 ☽ ∠ ☿
22 43 ☉ ∥ ♅

22 Sunday
05 18 ☽ Q ♆
09 25 ☽ ⊥ ☿
12 03 ☽ ∠ ☉
12 35 ☽ ∠ ♅
15 24 ☽ □ ♀
15 28 ☉ Q ♂
18 23 ☽ ⊻ ♂
18 27 ☽ ☌ ♄
18 49 ☉ ☌ ♅
20 16 ♂ ⊻ ♄
22 15 ♂ Q ♅
23 02 ☽ ⊻ ☿

23 Monday
05 04 ☽ ✶ ♆
06 29 ☽ ⊥ ♅
06 50 ☽ ⊥ ♂
07 30 ☽ ⊥ ☉
08 33 ☽ ∥ ♄
08 59 ☽ ☌ ♇
11 22 ☽ # ♇
14 20 ☽ # ♂
16 15 ☽ ☌ ♃
18 30 ☽ ⊻ ♅
20 33 ☽ ⊻ ☉

24 Tuesday
01 36 ☽ ∠ ♂
01 48 ☉ ⊥ ♇
06 08 ☽ Q ♀
11 08 ☽ ∠ ♆
13 15 ♀ □ ♄
14 14 ☽ ∥ ♃
22 08 ☿ ⊻ ♆

25 Wednesday
05 18 ☽ ⊥ ♆
06 57 ☽ ⊻ ♄
08 13 ☽ ✶ ♀
09 10 ☽ ✶ ♂
11 01 ☿ ⊥ ♄
11 17 ♂ ⊥ ♇
17 27 ☽ ⊻ ♆
19 11 ☽ ⊥ ♄
20 21 ☽ ☌ ☿
21 25 ☽ ⊻ ♇

26 Thursday
00 44 ☉ ∠ ♄
04 28 ☿ ⊻ ♇
05 40 ☽ ⊻ ♃
07 14 ☽ ☌ ♅
09 36 ☽ ⊥ ♇
09 47 ♂ ∥ ♇
10 34 ☽ Q ♂
13 33 ☽ ∠ ♄
14 19 ♀ △ ♂
14 38 ☽ ☌ ☉
17 01 ☽ ∠ ♀
18 02 ☽ ⊥ ♃

27 Friday
03 54 ☽ ∠ ♇
12 27 ☽ ∥ ☿
12 37 ☽ ∠ ♃
12 43 ☽ ⊥ ♀
20 15 ☽ ✶ ♄
20 57 ☽ ∥ ♅

28 Saturday
00 47 ☽ □ ♂
01 51 ☽ ⊻ ♀
02 08 ♂ # ♄
06 29 ☽ ☌ ♆
10 26 ☽ ✶ ♇
15 00 ☉ ⊥ ♃
18 15 ☽ ∥ ☉
18 34 ☽ ⊻ ☿
19 34 ☽ ✶ ♃
20 25 ☽ ⊻ ♅
20 48 ☽ Q ♄

29 Sunday
01 57 ☿ ∥ ♅
02 54 ☿ ⊻ ♃
08 35 ☽ ⊥ ☿
08 38 ☽ ⊥ ♅
08 59 ☿ ☌ ♅
09 11 ☽ ⊻ ☉
10 48 ☽ Q ♇
20 13 ☽ Q ♃
22 25 ☽ ⊥ ☉

30 Monday
01 40 ☿ ⊥ ♇
02 52 ☽ ∠ ♅
05 29 ☽ ∠ ☿
09 24 ☽ □ ♄
11 13 ☽ ∥ ☉
16 06 ☽ △ ♂
18 08 ☽ ∠ ☉
18 59 ☽ ☌ ♀
19 10 ☽ ⊻ ♆
21 59 ♀ ⊻ ♆
23 02 ☽ □ ♇

All ephemeris data is given at 12.00 UT and the Moon's longitude is additionally given for 24.00 UT

DECEMBER 1981

LONGITUDES

Date	Sidereal time h m s	Sun ☉ ° ' "	Moon ☽ ° ' "	Moon ☽ 24.00 ° ' "	Mercury ☿ ° '	Venus ♀ ° '	Mars ♂ ° '	Jupiter ♃ ° '	Saturn ♄ ° '	Uranus ♅ ° '	Neptune ♆ ° '	Pluto ♇ ° '
01	16 41 02	09 ♐ 12 51	02 ♒ 27 01	08 ♒ 32 48	04 ♐ 09	24 ♑ 28	22 ♍ 52	00 ♏ 51	19 ♎ 12	00 ♐ 56	24 ♐ 00	25 ♎ 56
02	16 44 59	10 13 42	14 ♒ 41 59	20 ♒ 55 07	05 43	25 16	23 23	01 03	19 17	00 59	24 03	25 58
03	16 48 55	11 14 33	27 ♒ 12 44	03 ♓ 35 24	07 17	26 03	23 53	01 14	19 23	01 03	24 05	26 00
04	16 52 52	12 15 26	10 ♓ 03 38	16 ♓ 37 57	08 52	26 49	24 23	01 26	19 28	01 07	24 07	26 02
05	16 56 49	13 16 19	23 ♓ 18 46	00 ♈ 06 26	10 26	27 35	24 53	01 37	19 34	01 10	24 09	26 04
06	17 00 45	14 17 13	07 ♈ 01 10	14 ♈ 03 02	12 00	28 19	25 23	01 48	19 39	01 14	24 11	26 05
07	17 04 42	15 18 08	21 ♈ 11 57	28 ♈ 27 37	13 34	29 02	25 53	01 59	19 44	01 18	24 14	26 07
08	17 08 38	16 19 04	05 ♉ 49 32	13 ♉ 17 00	15 08	29 ♑ 45	26 23	02 10	19 49	01 21	24 16	26 09
09	17 12 35	17 20 00	20 ♉ 49 05	28 ♉ 24 42	16 43	00 ♒ 26	26 52	02 21	19 54	01 25	24 18	26 11
10	17 16 31	18 20 57	06 ♊ 02 35	13 ♊ 41 26	18 17	01 06	27 22	02 32	19 59	01 28	24 20	26 12
11	17 20 28	19 21 55	21 ♊ 19 51	28 ♊ 56 31	19 51	01 44	27 51	02 43	20 04	01 32	24 23	26 14
12	17 24 24	20 22 54	06 ♋ 30 09	13 ♋ 59 38	21 26	02 22	28 20	02 54	20 09	01 36	24 25	26 16
13	17 28 21	21 23 54	21 ♋ 24 01	28 ♋ 42 31	23 00	02 58	28 49	03 04	20 14	01 39	24 27	26 17
14	17 32 18	22 24 54	05 ♌ 54 34	12 ♌ 59 50	24 35	03 33	29 17	03 15	20 18	01 43	24 30	26 19
15	17 36 14	23 25 56	19 ♌ 58 06	26 ♌ 49 23	26 09	04 06	29 ♍ 46	03 26	20 23	01 46	24 32	26 20
16	17 40 11	24 26 58	03 ♍ 33 50	10 ♍ 11 41	27 44	04 38	00 ♎ 14	03 36	20 28	01 50	24 34	26 22
17	17 44 07	25 28 01	16 ♍ 43 18	23 ♍ 09 08	29 ♐ 19	05 08	00 42	03 46	20 32	01 53	24 36	26 23
18	17 48 04	26 29 06	29 ♍ 29 38	05 ♎ 45 20	00 ♑ 54	05 37	01 10	03 56	20 36	01 57	24 39	26 25
19	17 52 00	27 30 11	11 ♎ 56 45	18 ♎ 04 25	02 29	06 04	01 37	04 07	20 41	02 00	24 41	26 26
20	17 55 57	28 31 17	24 ♎ 08 53	00 ♏ 10 37	04 05	06 29	02 05	04 17	20 45	02 04	24 43	26 28
21	17 59 53	29 ♐ 32 23	06 ♏ 10 08	12 ♏ 07 52	05 40	06 52	02 32	04 27	20 49	02 07	24 45	26 29
22	18 03 50	00 ♑ 33 31	18 ♏ 04 16	23 ♏ 59 42	07 16	07 14	02 59	04 36	20 53	02 10	24 48	26 30
23	18 07 47	01 34 39	29 ♏ 54 33	05 ♐ 49 09	08 52	07 33	03 26	04 46	20 57	02 14	24 50	26 32
24	18 11 43	02 35 47	11 ♐ 43 47	17 ♐ 38 45	10 28	07 51	03 52	04 56	21 01	02 17	24 52	26 33
25	18 15 40	03 36 56	23 ♐ 34 18	29 ♐ 30 40	12 04	08 07	04 19	05 05	21 05	02 20	24 54	26 34
26	18 19 36	04 38 06	05 ♑ 28 06	11 ♑ 26 49	13 41	08 20	04 45	05 15	21 08	02 24	24 57	26 35
27	18 23 33	05 39 16	17 ♑ 27 01	23 ♑ 28 58	15 17	08 32	05 10	05 24	21 12	02 27	24 59	26 36
28	18 27 29	06 40 26	29 ♑ 32 52	05 ♒ 39 00	16 54	08 41	05 36	05 33	21 15	02 30	25 01	26 38
29	18 31 26	07 41 36	11 ♒ 47 36	17 ♒ 58 57	18 30	08 48	06 01	05 42	21 19	02 33	25 03	26 39
30	18 35 22	08 42 46	24 ♒ 13 23	00 ♓ 31 11	20 07	08 52	06 26	05 51	21 22	02 37	25 06	26 40
31	18 39 19	09 ♑ 43 56	06 ♓ 52 41	13 ♓ 18 13	21 ♑ 43	08 ♒ 54	06 ♎ 51	06 ♏ 00	21 ♎ 25	02 ♐ 40	25 ♐ 08	26 ♎ 41

Date	Moon True ☊ ° '	Moon Mean ☊ ° '	Moon ☽ Latitude ° '
01	23 ♋ 09	24 ♋ 48	00 S 50
02	23 D 10	24 45	01 54
03	23 11	24 42	02 54
04	23 12	24 39	03 47
05	23 R 12	24 36	04 30
06	23 10	24 32	04 59
07	23 06	24 29	05 11
08	23 01	24 26	05 03
09	22 55	24 23	04 35
10	22 50	24 20	03 47
11	22 46	24 16	02 43
12	22 43	24 13	01 28
13	22 42	24 10	00 S 07
14	22 D 43	24 07	01 N 12
15	22 44	24 04	02 24
16	22 46	24 01	03 26
17	22 47	23 57	04 14
18	22 R 47	23 54	04 49
19	22 46	23 51	05 09
20	22 44	23 48	05 14
21	22 41	23 45	05 06
22	22 37	23 42	04 44
23	22 33	23 38	04 10
24	22 30	23 35	03 26
25	22 27	23 32	02 33
26	22 25	23 29	01 32
27	22 24	23 26	00 N 27
28	22 D 24	23 22	00 S 39
29	22 25	23 19	01 45
30	22 26	23 16	02 47
31	22 ♋ 28	23 ♋ 13	03 S 42

DECLINATIONS

Date	Sun ☉ ° '	Moon ☽ ° '	Mercury ☿ ° '	Venus ♀ ° '	Mars ♂ ° '	Jupiter ♃ ° '	Saturn ♄ ° '	Uranus ♅ ° '	Neptune ♆ ° '	Pluto ♇ ° '
01	21 S 50	20 S 26	21 S 10	24 S 06	04 N 36	10 S 44	05 S 23	20 S 11	22 S 03	05 N 34
02	21 59	18 15	21 34	23 53	04 25	10 48	05 24	20 11	22 03	05 33
03	22 08	15 10	21 56	23 40	04 14	10 51	05 26	20 12	22 03	05 33
04	22 16	11 18	22 18	23 27	04 02	10 55	05 28	20 13	22 03	05 33
05	22 24	06 47	22 38	23 13	03 51	10 59	05 30	20 14	22 03	05 33
06	22 31	01 S 48	22 58	22 59	03 40	11 03	05 32	20 14	22 03	05 32
07	22 38	03 N 28	23 16	22 45	03 29	11 07	05 34	20 15	22 04	05 32
08	22 44	08 41	23 33	22 30	03 18	11 10	05 35	20 16	22 04	05 32
09	22 50	13 32	23 49	22 15	03 07	11 14	05 37	20 17	22 04	05 32
10	22 56	17 35	24 03	22 01	02 56	11 18	05 39	20 17	22 04	05 32
11	23 01	20 27	24 16	21 45	02 45	11 21	05 40	20 18	22 04	05 31
12	23 06	21 49	24 28	21 30	02 34	11 25	05 42	20 19	22 04	05 31
13	23 10	21 37	24 39	21 15	02 23	11 28	05 44	20 20	22 04	05 31
14	23 13	19 57	24 49	20 59	02 13	11 32	05 45	20 20	22 04	05 31
15	23 17	17 06	24 57	20 44	02 02	11 35	05 47	20 21	22 04	05 31
16	23 19	13 24	25 03	20 28	01 51	11 39	05 48	20 22	22 04	05 31
17	23 22	09 09	25 09	20 12	01 41	11 42	05 50	20 22	22 05	05 31
18	23 24	04 37	25 13	19 57	01 31	11 45	05 51	20 23	22 05	05 31
19	23 25	00 N 01	25 15	19 41	01 20	11 49	05 52	20 24	22 05	05 31
20	23 26	04 S 29	25 16	19 25	01 10	11 52	05 54	20 25	22 05	05 31
21	23 26	08 46	25 16	19 09	01 00	11 55	05 55	20 25	22 05	05 31
22	23 26	12 40	25 14	18 54	00 50	11 58	05 56	20 26	22 05	05 31
23	23 26	16 03	25 11	18 38	00 40	12 01	05 58	20 27	22 05	05 31
24	23 25	18 47	25 06	18 23	00 31	12 04	05 59	20 27	22 05	05 31
25	23 23	20 45	25 00	18 08	00 21	12 07	06 00	20 28	22 05	05 31
26	23 22	21 47	24 52	17 53	00 11	12 10	06 01	20 29	22 05	05 31
27	23 19	21 51	24 42	17 38	00 N 02	12 13	06 02	20 29	22 05	05 31
28	23 16	20 53	24 31	17 23	00 S 07	12 16	06 03	20 30	22 06	05 31
29	23 13	18 56	24 19	17 09	00 17	12 19	06 04	20 30	22 06	05 31
30	23 09	16 05	24 05	16 55	00 26	12 22	06 05	20 31	22 06	05 31
31	23 S 05	12 S 26	23 S 49	16 S 41	00 S 35	12 S 25	06 S 06	20 S 32	22 S 06	05 N 31

ZODIAC SIGN ENTRIES

Date	h m	Planets
01	07 09	☽ ♒
03	17 16	☽ ♓
05	23 49	☽ ♈
08	02 31	☽ ♉
08	20 52	♀ ♒
10	02 30	☽ ♊
12	01 40	☽ ♋
14	02 08	☽ ♌
16	00 14	♂ ♎
16	05 38	☽ ♍
17	22 21	☿ ♑
18	12 58	☽ ♎
20	23 39	☽ ♏
21	22 51	☉ ♑
23	12 11	☽ ♐
26	00 59	☽ ♑
28	12 54	☽ ♒
30	23 01	☽ ♓

LATITUDES

Date	Mercury ☿ ° '	Venus ♀ ° '	Mars ♂ ° '	Jupiter ♃ ° '	Saturn ♄ ° '	Uranus ♅ ° '	Neptune ♆ ° '	Pluto ♇ ° '
01	00 S 12	02 S 55	01 N 56	01 N 07	02 N 19	00 N 10	01 N 15	16 N 42
04	00 32	02 42	01 58	01 07	02 19	00 10	01 15	16 43
07	00 50	02 27	02 00	01 07	02 20	00 10	01 15	16 44
10	01 08	02 09	02 03	01 08	02 21	00 10	01 15	16 45
13	01 24	01 48	02 05	01 08	02 21	00 10	01 15	16 47
16	01 38	01 24	02 07	01 08	02 22	00 10	01 15	16 48
19	01 50	00 57	02 10	01 09	02 23	00 10	01 15	16 50
22	02 00	00 S 27	02 12	01 09	02 23	00 10	01 15	16 51
25	02 07	00 N 07	02 15	01 10	02 24	00 10	01 15	16 53
28	02 10	00 44	02 17	01 10	02 25	00 10	01 15	16 54
31	02 S 10	01 N 24	02 N 20	01 N 10	02 N 25	00 N 10	01 N 15	16 N 56

DATA

Julian Date	2444940
Delta T	+51 seconds
Ayanamsa	23° 35' 59"
Synetic vernal point	05° ♓ 31' 00"
True obliquity of ecliptic	23° 26' 26"

MOON'S PHASES, APSIDES AND POSITIONS ☽

Date	h m	Phase	Longitude ° '	Eclipse Indicator
04	16 22	☽	12 ♓ 27	
11	08 41	○	19 ♊ 14	
18	05 47	☾	26 ♍ 13	
26	10 10	●	04 ♑ 33	

Day	h m		
11	00 07	Perigee	
23	23 00	Apogee	
06	20 16	ON	
12	20 51	Max dec	21° N 56'
19	12 05	0S	
27	01 25	Max dec	21° S 57'

LONGITUDES

Date	Chiron ⚷ ° '	Ceres ⚳ ° '	Pallas ⚴ ° '	Juno ⚵ ° '	Vesta ⚶ ° '	Black Moon Lilith ⚸ ° '
01	19 ♉ 26	26 ♎ 54	05 ♎ 37	03 ♐ 39	01 ♐ 26	07 ♐ 39
11	18 ♉ 58	01 ♏ 00	09 ♎ 16	07 ♐ 04	06 ♐ 50	08 ♐ 45
21	18 ♉ 35	04 ♏ 57	12 ♎ 35	10 ♐ 28	12 ♐ 13	09 ♐ 52
31	18 ♉ 16	08 ♏ 43	15 ♎ 29	13 ♐ 49	17 ♐ 34	10 ♐ 59

ASPECTARIAN

Day	h m	Aspects
01 Tuesday	02 49	☽ ∥ ☿
	07 09	☽ ⊥ ♆
	08 48	☽ □ ♃
	08 58	☽ ⚹ ♅
	12 51	☿ ∠ ♄
	15 24	☽ ∥ ♅
	15 50	☽ ⚹ ☿
	23 08	☽ bQ ♂
02 Wednesday	00 55	♃ ⊻ ♅
	00 56	☽ ∠ ♆
	02 30	☽ ⚹ ☉
	04 19	☿ Q ♂
	08 39	☽ Q ♅
	15 37	☉ ∠ ♀
	17 24	☽ ± ♂
	18 41	☽ Q ☿
	20 56	☽ △ ♄
	22 58	☉ ∥ ♆
03 Thursday	03 47	☽ Q ☉
	05 24	☽ ⊼ ♂
	06 02	☽ ⚹ ♆
	06 04	☉ ∠ ♇
	09 39	☽ ⊻ ♀
	09 42	☽ △ ♇
	10 25	♀ □ ♇
	11 08	☿ ⊥ ♃
	19 17	☽ □ ♅
	19 21	☿ ∥ ♆
	19 43	☽ △ ♃
	21 42	☽ ⊥ ♀
	22 00	♂ □ ♆
04 Friday	01 34	☽ bQ ♄
	04 41	☽ Q ♆
	08 09	☉ ∥ ☿
	09 29	☽ □ ☿
	13 47	☽ bQ ♇
	14 08	☽ ∥ ♃
	15 26	☽ ∠ ♀
	16 22	☽ □ ☉
	18 18	☽ ± ♄
	23 48	☽ bQ ♃
05 Saturday	05 14	☽ ⊼ ♄
	06 10	☽ ± ♇
	13 30	☽ □ ♆
	14 55	☽ ☍ ♂
	16 09	☽ ± ♃
	16 53	☽ ⊼ ♇
	18 10	☽ # ♇
	18 20	☽ ∥ ♄
	20 00	☽ ⚹ ♀
	21 49	☿ ∠ ♇
06 Sunday	01 55	☽ △ ♅
	02 50	☽ ⊼ ♃
	02 53	☽ # ♂
	12 56	☿ ∥ ♀
	17 58	☽ Q ♀
	18 49	♄ # ♇
	21 35	☽ △ ☿
07 Monday	01 21	☽ △ ☉
	03 45	☽ bQ ♅
	09 32	☽ ☍ ♄
	12 04	☽ ∥ ♂
	17 02	☽ △ ♆
	18 49	☽ ± ♅
	19 51	♂ ⊥ ♃
	19 51	☉ ∥ ♀
	20 02	☽ ⊼ ♂
	20 10	☽ ☍ ♇
	21 25	☽ ∥ ♇
	21 35	☽ # ♄
	23 57	♂ ⊻ ♇
08 Tuesday	01 05	☿ ∠ ♀
	01 37	☽ □ ♀
	01 39	☽ bQ ☿
	04 08	☽ bQ ☉
	04 42	☽ ⊼ ♅
	06 00	☽ ☍ ♃
	06 13	☽ ± ♂
	17 34	☽ bQ ♆
	17 59	☽ ± ☿
	19 46	☽ ± ☉
	21 15	☽ bQ ♂
09 Wednesday	00 06	☽ # ♃
	04 43	☽ ⊼ ☿
	06 04	☽ ⊼ ☉
	07 15	♀ ⊥ ♆
	08 00	☽ ± ♆
	10 32	☽ ⊼ ♄
	12 39	☉ ∠ ♃
	17 32	☽ ⊼ ♆
	20 05	☽ ± ♄
	20 29	☽ ⊼ ♇
	21 53	☽ △ ♂
	23 09	☿ ∠ ♃
10 Thursday	03 52	☽ △ ♀
	04 48	☽ ☍ ♅
	05 57	☽ ± ♇
	06 26	☽ ⊼ ♃
	06 44	♀ ∥ ♆
	10 20	☽ bQ ♄
	14 54	☉ ☌ ☿
	15 58	☽ ± ♃
	20 07	☽ bQ ♇
11 Friday	03 26	♀ ⚹ ♅
	04 28	☽ bQ ♀
	06 15	☽ bQ ♃
	08 41	☽ ☍ ☉
	09 25	☽ ☍ ☿
	10 00	☽ △ ♄
	10 27	☽ # ♅
	15 28	☿ ⚹ ♄
	16 49	☽ ☍ ♆
	19 15	☽ ± ♀
	19 44	☽ △ ♇
	22 36	☽ □ ♂
12 Saturday	04 10	☽ ⊼ ♅
	04 53	☽ # ♀
	05 08	☽ ⊼ ♀
	06 03	☉ ⚹ ♄
	06 12	☽ △ ♃
	13 45	☽ ± ♅
13 Sunday	04 16	☽ bQ ♅
	04 18	☽ Q ♂
	10 05	☽ □ ♄
	12 00	☽ ⊼ ☉
	14 56	☽ ⊼ ☿
	17 00	☽ ⊼ ♆
	18 24	♀ □ ♃
	20 01	☽ □ ♇
	21 06	☽ # ♀
	22 35	☽ ± ☉
14 Monday	00 35	☽ ⚹ ♂
	02 00	☽ ± ☿
	02 56	☽ ± ♆
	04 57	☽ △ ♅
	07 30	☽ □ ♃
	07 51	☽ # ♅
	07 52	☽ ☍ ♀
	10 40	☿ ☌ ♆
	14 44	☽ bQ ☉
	16 04	☽ Q ♄
	18 03	☽ bQ ♆
	18 58	☽ bQ ☿
15 Tuesday	02 16	☽ Q ♇
	02 42	☽ ∠ ♂
	12 44	☽ ⚹ ♄
	14 34	☽ Q ♃
	14 49	☿ ⚹ ♇
	18 31	☽ △ ☉
	18 51	☽ ⊥ ♂
	19 59	☽ △ ♆
	23 10	☽ ⚹ ♇
16 Wednesday	00 15	☽ △ ☿
	05 50	☽ ⊻ ♂
	08 53	☽ □ ♅
	12 04	☽ ⚹ ♃
	13 59	☽ ⊼ ♀
	14 53	☉ ☌ ♆
	15 26	☽ ∠ ♄
	21 09	♀ ∥ ♅
	22 02	☽ # ♃
17 Thursday	01 18	☽ ± ♀
	02 10	☽ ∠ ♇
	07 55	☿ ⊥ ♀
	07 56	☽ ⊥ ♄
	15 51	☽ ∠ ♃
	17 55	☽ Q ♅
	18 36	☽ bQ ♀
	18 50	☽ ⊥ ♇
	19 08	☽ ⊻ ♄
18 Friday	02 47	☽ □ ♆
	05 35	☽ # ♄
	05 47	☽ □ ☉
	06 08	☽ ⊻ ♇
	07 19	☽ ∥ ♇
	09 00	☽ ⊥ ♃
	10 14	☉ ⚹ ♇
	15 04	☽ □ ☿
	15 18	☽ ☌ ♂
	16 42	☽ ⚹ ♅
	17 32	☿ □ ♂
	20 37	☽ ⊻ ♃
19 Saturday	00 10	☽ △ ♀
	04 22	☿ ⊻ ♅
	04 49	☽ ∥ ♂
	13 26	☽ Q ♆
	14 58	☿ Q ♄
	18 53	☽ # ♂
	19 35	☽ Q ☉
	21 56	☽ ∠ ♅
20 Sunday	05 14	☽ ☌ ♄
	07 17	☽ Q ☿
	10 52	♂ ⚹ ♅
	13 08	☽ ⚹ ♆
	15 19	☿ ⚹ ♃
	15 49	☽ ⊥ ♅
	16 36	☽ ☌ ♇
	17 35	☽ # ♇
	19 45	☽ ∥ ♄
	21 30	☽ ⚹ ☉
21 Monday	03 50	☽ ⊻ ♅
	04 25	☽ ⊻ ♂
	08 29	☽ ☌ ♃
	10 51	☽ ⚹ ☿
	13 27	☽ □ ♀
	16 56	☽ ⊥ ♂
	19 14	☽ ∠ ♆
22 Tuesday	06 27	☽ ∠ ☉
	07 28	☽ ∥ ♃
	11 16	☿ ⊻ ♀
	11 49	☽ ∠ ♂
	13 28	☽ ⊥ ♆
	17 44	☽ ⊻ ♄
	21 50	☽ ∠ ☿
23 Wednesday	01 40	☽ ⊻ ♆
	02 06	☿ ⊥ ♅
	02 23	☽ ⊥ ☉
	02 56	☽ Q ♀
	05 07	☽ ⊻ ♇
	05 58	☽ ⊥ ♄
	06 49	☿ Q ♇
	11 20	☉ ⊥ ♀
	15 42	☽ ⊻ ☉
	16 44	☽ ☌ ♅
	17 19	☽ ⊥ ♇
	18 57	☽ ⊥ ☿
	19 25	☽ ⚹ ♂
	22 00	☽ ⊻ ♃
24 Thursday	00 20	☽ ∠ ♄
	03 56	☽ ⚹ ♀
	04 15	☉ ⊻ ♅
	08 20	☽ ∥ ♀
	09 02	☽ ⊻ ☿
	10 21	☽ ⊥ ♃
	11 38	☽ ∠ ♇
	20 43	☽ Q ♂
	22 28	☉ Q ♄
25 Friday	04 51	☽ ∠ ♃
	06 55	☽ ⚹ ♄
	07 48	☽ ∥ ♅
	11 03	☽ ∠ ♀
	14 43	☽ ☌ ♆
	18 04	☽ ⚹ ♇
26 Saturday	05 35	☽ ⊥ ♀
	05 47	☽ ⊻ ♅
	07 17	☽ Q ♄
	10 10	☽ ☌ ☉
	10 29	☽ □ ♂
	11 33	☽ ⚹ ♃
	16 26	☉ □ ♂
	17 52	☽ ⊻ ♀
	17 55	☽ ⊥ ♅
	18 17	☽ Q ♇
27 Sunday	04 55	☉ ⚹ ♃
	07 01	☽ ☌ ☿
	11 54	☽ Q ♃
	12 00	☽ ∠ ♅
	19 30	☽ □ ♄
28 Monday	03 01	☽ ⊻ ♆
	06 13	☽ □ ♇
	07 49	♂ ⊻ ♃
	14 55	☽ ⊥ ♆
	17 51	☽ ⚹ ♅
	17 54	☽ ∥ ♅
	21 25	☿ ∠ ♅
	22 50	☿ Q ♃
	23 57	☽ □ ♃
29 Tuesday	00 20	☽ △ ♂
	03 17	☽ ⊻ ☉
	06 06	☽ ☌ ♀
	08 36	☽ ∠ ♆
	16 02	☽ ⊥ ☉
	17 24	☽ Q ♅
30 Wednesday	02 56	☽ ⊻ ☿
	05 12	☽ ∥ ♀
	06 29	☽ bQ ♂
	06 30	☽ △ ♄
	09 29	☉ ⊥ ♅
	10 49	☉ Q ♇
	10 56	☽ ∠ ☉
	13 40	☽ ⚹ ♆
	15 47	☉ ⊻ ♀
	16 09	☽ ⊥ ☿
	16 40	☽ △ ♇
31 Thursday	00 15	☽ ± ♂
	04 02	☽ □ ♅
	07 22	☿ □ ♄
	10 20	☽ △ ♃
	11 08	☽ bQ ♄
	11 40	☽ ∠ ☿
	11 57	☽ ⊼ ♂
	12 05	☽ ∥ ♃
	12 28	☽ Q ♆
	15 47	☽ ⊻ ♀
	17 48	☽ ⚹ ☉
	19 45	♀ St R
	21 00	☽ bQ ♇
	23 11	♅ ⊥ ♇

All ephemeris data is given at 12.00 UT and the Moon's longitude is additionally given for 24.00 UT

JANUARY 1982

LONGITUDES

Date	Sidereal time h m s	Sun ☉ ° ' "	Moon ☽ ° ' "	Moon ☽ 24.00 ° ' "	Mercury ☿ ° '	Venus ♀ ° '	Mars ♂ ° '	Jupiter ♃ ° '	Saturn ♄ ° '	Uranus ♅ ° '	Neptune ♆ ° '	Pluto ♇ ° '
01	18 43 16	10 ♑ 45 06	19 ♓ 48 06	26 ♓ 22 41	23 ♑ 20	08 ♒ 54	07 ♎ 16	06 ♏ 09	21 ♎ 28	02 ♐ 43	25 ♐ 10	26 ♎ 42
02	18 47 12	11 46 15	03 ♈ 02 13	09 ♈ 46 58	24 56	08 R 51	07 40	06 17	21 31	02 46	25 12	26 43
03	18 51 09	12 47 24	16 ♈ 37 08	23 ♈ 32 48	26 31	08 45	08 04	06 26	21 34	02 49	25 14	26 44
04	18 55 05	13 48 33	00 ♉ 34 01	07 ♉ 40 39	28 07	08 37	08 28	06 34	21 37	02 52	25 17	26 45
05	18 59 02	14 49 42	14 ♉ 52 30	22 ♉ 09 10	29 ♑ 41	08 27	08 51	06 42	21 40	02 55	25 19	26 46
06	19 02 58	15 50 50	29 ♉ 30 08	06 ♊ 54 45	01 ♒ 15	08 14	09 14	06 51	21 43	02 58	25 21	26 46
07	19 06 55	16 51 58	14 ♊ 22 12	21 ♊ 51 33	02 47	07 58	09 37	06 59	21 45	03 01	25 23	26 47
08	19 10 51	17 53 06	29 ♊ 21 49	06 ♋ 51 54	04 19	07 40	09 59	07 06	21 47	03 04	25 25	26 48
09	19 14 48	18 54 14	14 ♋ 20 44	21 ♋ 47 16	05 48	07 20	10 21	07 14	21 50	03 07	25 27	26 49
10	19 18 45	19 55 21	29 ♋ 10 28	06 ♌ 29 27	07 16	06 57	10 43	07 22	21 52	03 10	25 29	26 49
11	19 22 41	20 56 28	13 ♌ 43 25	20 ♌ 51 46	08 41	06 32	11 05	07 29	21 54	03 13	25 32	26 50
12	19 26 38	21 57 34	27 ♌ 54 01	04 ♍ 49 51	10 03	06 05	11 26	07 37	21 56	03 15	25 34	26 51
13	19 30 34	22 58 41	11 ♍ 39 05	18 ♍ 21 45	11 22	05 36	11 47	07 44	21 58	03 18	25 36	26 51
14	19 34 31	23 59 47	24 ♍ 57 56	01 ♎ 27 54	12 37	05 06	12 07	07 51	22 00	03 21	25 38	26 52
15	19 38 27	25 00 53	07 ♎ 51 59	14 ♎ 10 37	13 47	04 34	12 27	07 58	22 02	03 23	25 40	26 52
16	19 42 24	26 01 59	20 ♎ 24 16	26 ♎ 33 28	14 51	04 00	12 47	08 04	22 03	03 26	25 42	26 53
17	19 46 20	27 03 04	02 ♏ 38 47	08 ♏ 40 49	15 49	03 25	13 07	08 11	22 05	03 29	25 44	26 53
18	19 50 17	28 04 10	14 ♏ 40 08	20 ♏ 37 19	16 40	02 50	13 26	08 18	22 06	03 31	25 46	26 54
19	19 54 14	29 ♑ 05 15	26 ♏ 32 59	02 ♐ 27 39	17 23	02 13	13 44	08 24	22 08	03 34	25 48	26 54
20	19 58 10	00 ♒ 06 19	08 ♐ 21 52	14 ♐ 16 08	17 57	01 37	14 02	08 30	22 09	03 36	25 50	26 54
21	20 02 07	01 07 24	20 ♐ 10 57	26 ♐ 06 43	18 22	01 00	14 20	08 36	22 10	03 39	25 52	26 55
22	20 06 03	02 08 28	02 ♑ 03 52	08 ♑ 02 44	18 35	00 ♒ 23	14 37	08 42	22 11	03 41	25 54	26 55
23	20 10 00	03 09 31	14 ♑ 03 39	20 ♑ 06 54	18 R 38	29 ♑ 46	14 54	08 48	22 12	03 43	25 56	26 55
24	20 13 56	04 10 34	26 ♑ 12 43	02 ♒ 21 17	18 30	29 10	15 11	08 53	22 12	03 46	25 57	26 55
25	20 17 53	05 11 35	08 ♒ 32 47	14 ♒ 47 21	18 10	28 35	15 27	08 59	22 13	03 48	25 59	26 56
26	20 21 49	06 12 36	21 ♒ 05 04	27 ♒ 26 01	17 38	28 01	15 42	09 04	22 14	03 50	26 01	26 56
27	20 25 46	07 13 36	03 ♓ 50 17	10 ♓ 17 52	16 56	27 28	15 57	09 09	22 14	03 52	26 03	26 56
28	20 29 43	08 14 35	16 ♓ 48 49	23 ♓ 23 09	16 04	26 57	16 12	09 14	22 14	03 54	26 05	26 56
29	20 33 39	09 15 33	00 ♈ 00 52	06 ♈ 42 00	15 04	26 27	16 26	09 19	22 15	03 56	26 07	26 R 56
30	20 37 36	10 16 30	13 ♈ 26 32	20 ♈ 14 27	13 57	25 59	16 40	09 23	22 15	03 58	26 09	26 56
31	20 41 32	11 ♒ 17 25	27 ♈ 05 46	04 ♉ 00 25	12 ♒ 46	25 ♑ 33	16 ♎ 53	09 ♏ 27	22 ♎ 15	04 ♐ 00	26 ♐ 10	26 ♎ 56

Date	Moon True ☊ ° '	Moon Mean ☊ ° '	Moon ☽ Latitude ° '
01	22 ♋ 29	23 ♋ 10	04 S 27
02	22 D 29	23 07	04 59
03	22 R 29	23 03	05 16
04	22 29	23 00	05 14
05	22 28	22 57	04 54
06	22 28	22 54	04 14
07	22 27	22 51	03 16
08	22 27	22 48	02 05
09	22 26	22 44	00 S 45
10	22 D 26	22 41	00 N 37
11	22 26	22 38	01 56
12	22 26	22 35	03 05
13	22 R 26	22 32	04 01
14	22 26	22 28	04 42
15	22 26	22 25	05 08
16	22 26	22 22	05 18
17	22 D 26	22 19	05 13
18	22 26	22 16	04 54
19	22 27	22 13	04 23
20	22 28	22 09	03 41
21	22 28	22 06	02 50
22	22 29	22 03	01 51
23	22 30	22 00	00 N 47
24	22 R 30	21 57	00 S 21
25	22 30	21 54	01 28
26	22 28	21 50	02 32
27	22 26	21 47	03 29
28	22 24	21 44	04 17
29	22 22	21 41	04 52
30	22 20	21 38	05 12
31	22 ♋ 18	21 ♋ 34	05 S 15

DECLINATIONS

Date	Sun ☉ ° '	Moon ☽ ° '	Mercury ☿ ° '	Venus ♀ ° '	Mars ♂ ° '	Jupiter ♃ ° '	Saturn ♄ ° '	Uranus ♅ ° '	Neptune ♆ ° '	Pluto ♇ ° '
01	23 S 00	08 S 08	23 S 32	16 S 27	00 S 44	12 S 28	06 S 07	20 S 32	22 S 06	05 N 32
02	22 55	03 S 22	23 13	16 14	00 52	12 30	06 08	20 33	22 06	05 32
03	22 50	01 N 40	22 53	16 01	01 01	12 33	06 09	20 34	22 06	05 32
04	22 43	06 46	22 32	15 49	01 09	12 35	06 10	20 35	22 06	05 32
05	22 37	11 37	22 09	15 37	01 18	12 38	06 10	20 35	22 06	05 33
06	22 30	15 55	21 44	15 26	01 26	12 41	06 11	20 35	22 06	05 33
07	22 23	19 17	21 18	15 15	01 34	12 43	06 12	20 36	22 06	05 33
08	22 15	21 22	20 51	15 05	01 42	12 45	06 13	20 36	22 06	05 33
09	22 06	21 56	20 23	14 55	01 50	12 48	06 13	20 37	22 06	05 33
10	21 58	20 56	19 54	14 45	01 58	12 50	06 14	20 38	22 07	05 34
11	21 49	18 33	19 24	14 36	02 05	12 52	06 15	20 38	22 07	05 34
12	21 39	15 05	18 54	14 28	02 13	12 55	06 15	20 39	22 07	05 34
13	21 29	10 54	18 23	14 21	02 20	12 57	06 15	20 39	22 07	05 35
14	21 19	06 19	17 51	14 13	02 27	12 59	06 16	20 40	22 07	05 35
15	21 08	01 N 35	17 20	14 07	02 34	13 01	06 16	20 40	22 07	05 35
16	20 57	03 S 04	16 49	14 01	02 41	13 03	06 16	20 41	22 07	05 36
17	20 45	07 29	16 19	13 56	02 48	13 05	06 17	20 41	22 07	05 36
18	20 33	11 33	15 49	13 51	02 54	13 07	06 17	20 42	22 07	05 37
19	20 21	15 07	15 21	13 47	03 00	13 09	06 17	20 42	22 07	05 37
20	20 08	18 03	14 55	13 43	03 06	13 10	06 17	20 43	22 07	05 37
21	19 55	20 15	14 32	13 40	03 12	13 12	06 18	20 43	22 07	05 38
22	19 41	21 34	14 10	13 37	03 18	13 14	06 18	20 44	22 07	05 38
23	19 27	21 55	13 52	13 36	03 24	13 15	06 18	20 44	22 07	05 39
24	19 13	21 15	13 38	13 35	03 29	13 17	06 18	20 44	22 07	05 39
25	18 58	19 32	13 27	13 34	03 35	13 19	06 18	20 45	22 07	05 40
26	18 43	16 52	13 20	13 34	03 40	13 20	06 18	20 45	22 07	05 40
27	18 28	13 21	13 17	13 34	03 45	13 22	06 18	20 46	22 07	05 41
28	18 12	09 09	13 18	13 35	03 49	13 23	06 17	20 46	22 07	05 41
29	17 56	04 S 28	13 23	13 36	03 54	13 24	06 17	20 46	22 07	05 42
30	17 40	00 N 31	13 31	13 37	03 58	13 25	06 17	20 47	22 07	05 42
31	17 S 23	05 N 33	13 S 42	13 S 39	04 S 02	13 S 27	06 S 17	20 S 47	22 S 07	05 N 43

ZODIAC SIGN ENTRIES

Date	h	m	Planets
02	06	33	☽ ♈
04	11	02	☽ ♉
05	16	49	☿ ♒
06	12	49	☽ ♊
08	13	01	☽ ♋
10	13	21	☽ ♌
12	15	37	☽ ♍
14	21	17	☽ ♎
17	06	46	☽ ♏
19	19	00	☽ ♐
20	09	31	☉ ♒
22	07	51	☽ ♑
23	02	56	♀ ♑
24	19	25	☽ ♒
27	04	49	☽ ♓
29	11	58	☽ ♈
31	17	03	☽ ♉

LATITUDES

Date	Mercury ☿ ° '	Venus ♀ ° '	Mars ♂ ° '	Jupiter ♃ ° '	Saturn ♄ ° '	Uranus ♅ ° '	Neptune ♆ ° '	Pluto ♇ ° '
01	02 S 08	01 N 38	02 N 21	01 N 11	02 N 26	00 N 10	01 N 15	16 N 56
04	02 02	02 22	02 23	01 11	02 27	00 10	01 15	16 58
07	01 49	03 08	02 26	01 12	02 27	00 10	01 15	17 00
10	01 30	03 54	02 29	01 12	02 28	00 10	01 15	17 01
13	01 03	04 40	02 32	01 13	02 29	00 10	01 15	17 03
16	00 S 27	05 23	02 34	01 13	02 30	00 10	01 15	17 05
19	00 N 17	06 02	02 37	01 14	02 31	00 10	01 15	17 06
22	01 08	06 35	02 40	01 14	02 32	00 10	01 16	17 08
25	02 02	07 01	02 43	01 15	02 32	00 10	01 16	17 10
28	02 51	07 19	02 46	01 15	02 33	00 10	01 16	17 12
31	03 N 25	07 N 30	02 N 49	01 N 16	02 N 34	00 N 10	01 N 16	17 N 13

DATA

Julian Date	2444971
Delta T	+52 seconds
Ayanamsa	23° 36' 05"
Synetic vernal point	05° ♓ 30' 55"
True obliquity of ecliptic	23° 26' 26"

MOON'S PHASES, APSIDES AND POSITIONS ☽

Date	h	m	Phase	Longitude ° '	Eclipse Indicator
03	04	45	☽	12 ♈ 29	
09	19	53	○	19 ♋ 14	total
16	23	58	☾	26 ♎ 32	
25	04	56	●	04 ♒ 54	Partial

Day	h	m		
08	11	38	Perigee	
20	12	36	Apogee	
03	04	08	ON	
09	08	36	Max dec	21° N 56'
15	20	07	OS	
23	08	18	Max dec	21° S 56'
30	09	33	ON	

LONGITUDES

Date	Chiron ⚷ ° '	Ceres ⚳ ° '	Pallas ⚴ ° '	Juno ⚵ ° '	Vesta ⚶ ° '	Black Moon Lilith ⚸ ° '
01	18 ♉ 14	09 ♏ 05	15 ♎ 45	14 ♐ 09	18 ♐ 06	11 ♐ 06
11	18 ♉ 02	12 ♏ 36	18 ♎ 08	17 ♐ 25	23 ♐ 25	12 ♐ 13
21	17 ♉ 55	15 ♏ 18	19 ♎ 58	20 ♐ 36	28 ♐ 41	13 ♐ 19
31	17 ♉ 54	18 ♏ 46	21 ♎ 08	23 ♐ 40	03 ♑ 53	14 ♐ 26

All ephemeris data is given at 12.00 UT and the Moon's longitude is additionally given for 24.00 UT

ASPECTARIAN

h m	Aspects
01 Friday	
02 58	☽ ⊥ ♀
03 59	☽ ± ♄
13 39	☽ ± ♇
14 30	☽ bQ ♃
15 05	☽ ⊼ ♄
17 51	☽ Q ☉
19 21	☽ ✱ ☿
19 28	☽ ∠ ♀
21 50	☽ □ ♆
22 21	☽ ∥ ♄
02 Saturday	
00 36	☽ ⊼ ♇
01 22	☽ # ♇
07 01	☽ ± ♃
11 31	☽ △ ♅
16 14	☿ ⊻ ♆
17 52	☽ ⊼ ♃
19 52	☽ Q ☿
20 30	☽ ☍ ♂
22 17	☽ ✱ ♀
23 40	☽ ∥ ♂
03 Sunday	
04 45	☽ □ ☉
08 50	☽ # ♂
14 06	☽ bQ ♅
15 09	☿ □ ♇
17 56	☉ ∥ ☿
19 07	☽ Q ♀
20 38	☽ ☍ ♄
04 Monday	
02 57	☽ △ ♆
05 29	☽ ☍ ♇
05 41	☽ ± ♅
06 11	☽ ∥ ♇
07 17	☽ □ ☿
09 09	☽ # ♄
15 55	☽ ⊼ ♅
18 59	♀ △ ♂
22 14	☽ ☍ ♃
05 Tuesday	
01 25	☽ □ ♀
01 41	☽ ⊼ ♂
04 23	☽ bQ ♆
11 55	☽ △ ☉
11 57	☽ ± ♂
14 26	☿ ∥ ♆
17 23	☽ # ♃
19 21	☽ ± ♆
23 14	☽ ⊼ ♄
06 Wednesday	
03 11	☽ bQ ♂
05 13	☽ ⊼ ♆
07 33	☽ ⊼ ♇
09 04	☽ ± ♄
09 12	☽ # ♀
13 36	☿ ⊥ ♆
14 21	☽ bQ ☉
15 10	☽ △ ☿
17 19	☽ ± ♇
17 39	☽ ☍ ♅
23 42	☽ bQ ♄
07 Thursday	
00 00	☽ ⊼ ♃
01 53	☽ △ ♀
04 09	☽ △ ♂
05 57	☽ ± ☉
07 51	☽ bQ ♇
09 45	☽ ± ♃
15 42	☿ ✱ ♅
16 18	☽ ⊼ ☉
18 07	☽ bQ ☿
23 52	☽ △ ♄
08 Friday	
00 18	☽ bQ ♃
01 15	☽ # ♅
01 31	☽ bQ ♀
05 41	☽ ☍ ♆
05 58	☽ # ☿
07 54	☽ △ ♇
10 08	☽ ± ☿
15 36	☽ ± ♀
16 31	☉ ∠ ♅
17 57	☽ ⊼ ♅
20 48	☽ ⊼ ☿
09 Saturday	
00 30	☽ △ ♃
00 42	☿ ∥ ♅
01 00	☽ ⊼ ♀
03 35	☽ ± ♅
05 26	☽ □ ♂
11 57	☉ ∥ ♆
16 32	♀ □ ♃
18 05	☽ bQ ♅
19 53	☽ ☍ ☉
20 57	☉ Q ♃
10 Sunday	
00 06	☽ □ ♄
05 59	☽ ⊼ ♆
07 54	☿ ☌ ♀
08 10	☽ □ ♇
11 15	☽ Q ♂
13 47	☿ □ ♃
15 48	☽ ± ♆
15 59	☽ # ♅
18 33	☽ △ ♅
11 Monday	
00 25	☽ ☍ ♀
01 33	☽ □ ♃
02 43	☽ ☍ ☿
02 57	☽ # ☿
05 38	☽ Q ♄
06 40	☽ bQ ♆
07 29	☽ ✱ ♂
13 51	☽ Q ♇
12 Tuesday	
01 04	☽ ⊼ ☉
01 48	☽ ✱ ♄
07 59	☽ △ ♆
08 02	☽ Q ♃
09 25	☽ ∠ ♂
10 11	☽ ✱ ♇
11 27	☉ □ ♄
12 07	☽ ± ☉
15 55	☽ # ♀
21 17	☽ □ ♅
21 22	☿ ∠ ♆
13 Wednesday	
00 47	☽ # ♃
01 24	☽ ⊥ ♂
01 43	☽ ⊼ ♀
03 43	☽ ∠ ♄
04 59	☽ bQ ☉
05 01	☽ ✱ ♃
11 27	☽ ⊼ ☿
11 55	☽ ± ♀
12 14	☽ ⊻ ♂
12 22	☽ ∠ ♇
19 43	☽ ⊥ ♄
22 47	☿ △ ♂
23 17	☽ ± ☿
14 Thursday	
03 28	☽ bQ ♀
04 31	☽ ⊥ ♇
05 23	☽ Q ♅
06 34	☽ ⊻ ♄
08 06	☽ ∠ ♃
10 05	☽ △ ☉
12 17	☽ # ♄
13 13	☽ □ ♆
15 29	☽ ⊻ ♇
15 44	☽ ∥ ♇
17 21	☽ bQ ☿
15 Friday	
00 50	☽ ⊥ ♃
03 34	☽ ✱ ♅
06 02	☽ △ ♀
07 09	☽ # ♂
12 11	☽ ⊻ ♃
20 57	☽ ☌ ♂
23 03	☽ Q ♆
16 Saturday	
00 19	☽ △ ☿
03 49	☉ ⊻ ♆
08 10	☽ ∠ ♅
09 56	☽ ∥ ♂
15 13	☽ ☌ ♄
22 21	☽ ✱ ♆
23 58	☽ □ ☉
17 Sunday	
00 39	☽ ☌ ♇
01 33	☽ # ♇
01 46	☽ ⊥ ♅
02 47	☿ Q ♅
05 15	☽ ∥ ♄
08 08	☉ □ ♇
09 52	♀ ✱ ♅
13 28	☽ □ ♀
13 39	☽ ⊻ ♅
19 21	☉ ∥ ♅
23 07	☽ ☌ ♃
18 Monday	
04 09	☽ ∠ ♆
09 26	☽ ⊻ ♂
15 05	☽ Q ☉
16 18	☽ □ ☿
21 51	☽ ⊥ ♂
22 10	☽ ∥ ♃
22 18	☽ ⊥ ♆
23 49	☽ Q ♀
19 Tuesday	
02 45	☽ ∥ ♀
03 01	☽ ⊻ ♄
10 28	☽ ⊻ ♆
12 43	☽ ⊻ ♇
13 34	☽ ∥ ☿
15 12	☽ ⊥ ♄
16 33	☽ ∠ ♂
17 20	♂ Q ♆
17 38	☽ ✱ ☉
22 57	☽ ✱ ♀
20 Wednesday	
00 54	☽ ⊥ ♇
02 17	☽ ☌ ♅
03 50	♀ ⊥ ♆
06 53	☽ Q ☿
09 31	☽ ∠ ♄
12 17	☽ ⊻ ♃
19 12	☽ ∠ ♇
23 50	☽ ✱ ♂
21 Thursday	
00 35	☽ ⊥ ♃
02 59	☽ ∠ ☉
03 55	☽ ∠ ♀
08 02	☽ ∥ ☉
08 12	☽ ✱ ☿
10 06	☉ ☌ ♀
16 01	☽ ✱ ♄
18 48	☽ ∥ ♅
18 59	☽ ∠ ♃
21 16	☽ ⊥ ♀
22 56	☽ ⊥ ☉
23 32	☽ ☌ ♆
22 Friday	
00 46	☽ Q ♂
01 37	☽ ✱ ♇
06 00	☉ ⊥ ♆
08 46	☽ ⊻ ♀
12 10	☽ ⊻ ☉
15 06	☽ ∠ ☿
15 16	☽ ⊻ ♅
16 15	☽ Q ♄
23 Saturday	
01 25	☽ ✱ ♃
01 45	☽ Q ♇
03 19	☽ ⊥ ♅
06 03	☿ St R
09 11	☽ ⊥ ☿
13 43	☽ □ ♂
21 01	☽ ⊻ ☿
21 16	☽ ∠ ♅
24 Sunday	
01 27	☽ Q ♃
01 48	☉ ✱ ♅
04 07	☽ □ ♄
11 30	☽ ⊻ ♆
13 24	☽ □ ♇
17 32	☽ ☌ ♀
18 32	☿ ∥ ♀
20 44	☽ ∥ ♅
23 15	☽ ⊥ ♆
25 Monday	
02 47	☽ ✱ ♅
04 56	☽ ☌ ☉
12 50	☽ □ ♃
16 43	☽ ∠ ♆
18 32	☽ ∥ ☉
26 Tuesday	
01 32	☽ △ ♂
01 58	☽ Q ♅
05 44	☽ ☌ ☿
12 12	☿ ∥ ♃
14 10	☽ △ ♄
21 22	☽ ✱ ♆
23 03	☽ △ ♇
27 Wednesday	
00 33	☽ ⊻ ♀
06 31	☽ bQ ♂
10 39	☽ ∥ ♀
11 21	☽ ⊥ ♀
11 57	☽ ∥ ♃
12 04	☽ □ ♅
12 24	☽ ∥ ☿
18 20	☽ bQ ♄
18 51	☽ ⊻ ☉
19 51	☽ Q ♆
21 56	☽ △ ♃
23 35	☽ ± ♂
28 Thursday	
03 01	☽ bQ ♇
03 24	☽ ∠ ♀
06 53	☽ ⊥ ☉
09 19	☿ △ ♂
10 44	☽ ⊻ ☿
10 51	☽ ⊼ ♂
10 57	☽ ± ♄
12 51	♀ □ ♇
16 02	☿ Q ♅
19 32	☽ ± ♇
20 57	☽ ⊥ ☿
21 55	☽ ⊼ ♄
29 Friday	
00 43	☽ ∠ ☉
01 37	☽ bQ ♃
02 53	☽ ∥ ♄
04 56	☽ □ ♆
05 48	☽ ✱ ♀
05 53	☽ # ♇
06 26	☽ ⊼ ♇
08 18	♇ St R
12 05	☽ ∠ ☿
13 16	☉ □ ♃
14 44	☽ ∥ ♂
17 58	☽ ± ♃
18 50	☿ ∥ ♃
19 04	☽ △ ♅
30 Saturday	
02 37	☽ Q ♀
04 34	♀ ⊻ ♆
04 45	☽ ⊼ ♃
05 55	☽ ✱ ☉
12 50	☽ ✱ ☿
17 47	☽ ☍ ♂
21 48	☽ bQ ♅
31 Sunday	
03 31	☽ ☍ ♄
03 46	♄ St R
04 41	☽ # ♂
04 49	☽ Q ☉
06 21	☿ ∥ ♀
08 16	☽ Q ☿
09 02	☉ ∠ ♆
09 24	☽ □ ♀
10 23	☽ △ ♆
11 43	☽ ☍ ♇
12 50	☽ ∥ ♇
13 35	☽ ± ♅
15 34	☽ # ♄

FEBRUARY 1982

Date	Sidereal time h m s	Sun ☉ ° ' "	Moon ☽ ° ' "	Moon ☽ 24.00 ° ' "	Mercury ☿ ° '	Venus ♀ ° '	Mars ♂ ° '	Jupiter ♃ ° '	Saturn ♄ ° '	Uranus ♅ ° '	Neptune ♆ ° '	Pluto ♇ ° '
01	20 45 29	12 ♒ 18 19	10 ♉ 58 22	17 ♉ 59 31	11 ♒ 32	25 ♑ 10	17 ♎ 05	09 ♏ 32	22 ♎ 15	04 ♐ 02	26 ♐ 12	26 ♎ 56
02	20 49 25	13 19 12	25 ♉ 03 43	02 ♊ 10 48	10 R 18	24 R 48	17 17	09 36	22 R 15	04 04	26 14	26 R 56
03	20 53 22	14 20 03	09 ♊ 20 30	16 ♊ 32 32	09 06	24 29	17 29	09 40	22 14	04 06	26 15	26 55
04	20 57 18	15 20 53	23 ♊ 46 28	01 ♋ 01 52	07 57	24 12	17 40	09 43	22 14	04 08	26 17	26 55
05	21 01 15	16 21 42	08 ♋ 18 10	15 ♋ 34 45	06 54	23 57	17 50	09 47	22 13	04 09	26 18	26 55
06	21 05 12	17 22 29	22 ♋ 50 57	00 ♌ 06 04	05 56	23 45	18 00	09 50	22 13	04 11	26 20	26 55
07	21 09 08	18 23 15	07 ♌ 19 21	14 ♌ 30 05	05 06	23 36	18 09	09 53	22 12	04 13	26 22	26 54
08	21 13 05	19 24 00	21 ♌ 37 33	28 ♌ 41 07	04 24	23 29	18 18	09 56	22 11	04 14	26 23	26 54
09	21 17 01	20 24 43	05 ♍ 40 12	12 ♍ 34 20	03 50	23 24	18 26	09 59	22 10	04 16	26 25	26 54
10	21 20 58	21 25 25	19 ♍ 23 09	26 ♍ 06 23	03 25	23 22	18 33	10 02	22 09	04 17	26 26	26 53
11	21 24 54	22 26 05	02 ♎ 43 55	09 ♎ 15 44	03 07	23 D 22	18 40	10 04	22 08	04 19	26 28	26 53
12	21 28 51	23 26 45	15 ♎ 41 59	22 ♎ 02 51	02 58	23 25	18 46	10 07	22 07	04 20	26 29	26 52
13	21 32 47	24 27 23	28 ♎ 18 40	04 ♏ 29 51	02 D 55	23 30	18 52	10 09	22 05	04 22	26 31	26 52
14	21 36 44	25 28 01	10 ♏ 36 51	16 ♏ 40 12	03 00	23 37	18 56	10 11	22 04	04 23	26 32	26 51
15	21 40 41	26 28 37	22 ♏ 40 31	28 ♏ 38 22	03 12	23 47	19 01	10 13	22 03	04 24	26 33	26 51
16	21 44 37	27 29 12	04 ♐ 34 24	10 ♐ 29 17	03 29	23 59	19 04	10 14	22 01	04 25	26 35	26 50
17	21 48 34	28 29 46	16 ♐ 23 37	22 ♐ 18 05	03 53	24 13	19 07	10 15	21 59	04 26	26 36	26 49
18	21 52 30	29 ♒ 30 18	28 ♐ 13 17	04 ♑ 09 49	04 22	24 29	19 09	10 17	21 57	04 27	26 37	26 49
19	21 56 27	00 ♓ 30 49	10 ♑ 08 16	16 ♑ 09 08	04 56	24 47	19 10	10 18	21 55	04 28	26 39	26 48
20	22 00 23	01 31 19	22 ♑ 12 54	28 ♑ 20 00	05 34	25 08	19 11	10 18	21 53	04 29	26 40	26 47
21	22 04 20	02 31 48	04 ♒ 30 46	10 ♒ 45 30	06 16	25 30	19 R 11	10 19	21 51	04 30	26 41	26 47
22	22 08 16	03 32 15	17 ♒ 04 25	23 ♒ 27 38	07 03	25 53	19 10	10 19	21 49	04 31	26 42	26 46
23	22 12 13	04 32 40	29 ♒ 55 12	06 ♓ 27 05	07 53	26 19	19 08	10 20	21 47	04 32	26 43	26 45
24	22 16 10	05 33 04	13 ♓ 03 11	19 ♓ 43 17	08 46	26 46	19 06	10 R 20	21 44	04 33	26 44	26 44
25	22 20 06	06 33 26	26 ♓ 27 09	03 ♈ 14 29	09 42	27 15	19 02	10 20	21 42	04 33	26 45	26 43
26	22 24 03	07 33 46	10 ♈ 04 56	16 ♈ 58 08	10 41	27 45	18 59	10 19	21 39	04 34	26 46	26 42
27	22 27 59	08 34 04	23 ♈ 53 44	00 ♉ 51 21	11 43	28 17	18 54	10 19	21 36	04 35	26 47	26 41
28	22 31 56	09 ♓ 34 20	07 ♉ 50 37	14 ♉ 51 16	12 ♒ 47	28 ♑ 50	18 ♎ 48	10 ♏ 18	21 ♎ 34	04 ♐ 35	26 ♐ 48	26 ♎ 40

Date	Moon True ☊ ° '	Moon Mean ☊ ° '	Moon ☽ Latitude ° '
01	22 ♋ 17	21 ♋ 31	04 S 59
02	22 D 18	21 28	04 26
03	22 19	21 25	03 35
04	22 20	21 22	02 31
05	22 22	21 19	01 S 17
06	22 R 22	21 15	00 N 03
07	22 22	21 12	01 21
08	22 19	21 09	02 34
09	22 16	21 06	03 36
10	22 11	21 03	04 23
11	22 06	20 59	04 55
12	22 02	20 56	05 11
13	21 58	20 53	05 11
14	21 55	20 50	04 56
15	21 54	20 47	04 29
16	21 D 54	20 44	03 50
17	21 55	20 40	03 02
18	21 57	20 37	02 06
19	21 58	20 34	01 N 04
20	21 R 59	20 31	00 S 01
21	21 58	20 28	01 08
22	21 55	20 25	02 12
23	21 50	20 21	03 11
24	21 44	20 18	04 01
25	21 36	20 15	04 39
26	21 28	20 12	05 02
27	21 21	20 09	05 08
28	21 ♋ 15	20 ♋ 05	04 S 55

DECLINATIONS

Date	Sun ☉ ° '	Moon ☽ ° '	Mercury ☿ ° '	Venus ♀ ° '	Mars ♂ ° '	Jupiter ♃ ° '	Saturn ♄ ° '	Uranus ♅ ° '	Neptune ♆ ° '	Pluto ♇ ° '
01	17 S 07	10 N 23	13 S 55	13 S 42	04 S 06	13 S 28	06 S 17	20 S 48	22 S 07	05 N 43
02	16 49	14 44	14 11	13 44	04 10	13 29	06 16	20 48	22 08	05 44
03	16 32	18 18	14 28	13 47	04 13	13 30	06 16	20 48	22 08	05 44
04	16 14	20 47	14 46	13 50	04 17	13 31	06 15	20 49	22 08	05 45
05	15 56	21 54	15 04	13 53	04 20	13 32	06 15	20 49	22 08	05 46
06	15 38	21 33	15 22	13 57	04 22	13 33	06 15	20 49	22 08	05 46
07	15 19	19 45	15 41	14 01	04 25	13 34	06 14	20 49	22 08	05 47
08	15 00	16 43	15 58	14 05	04 28	13 34	06 13	20 50	22 08	05 48
09	14 41	12 46	16 15	14 09	04 30	13 35	06 13	20 50	22 08	05 48
10	14 22	08 14	16 31	14 13	04 32	13 36	06 12	20 50	22 08	05 49
11	14 02	03 N 26	16 45	14 17	04 33	13 36	06 12	20 51	22 08	05 49
12	13 42	01 S 24	16 59	14 21	04 35	13 37	06 11	20 51	22 08	05 50
13	13 22	06 02	17 11	14 25	04 36	13 37	06 10	20 51	22 08	05 51
14	13 02	10 19	17 22	14 29	04 37	13 38	06 09	20 51	22 08	05 51
15	12 41	14 06	17 31	14 33	04 38	13 38	06 09	20 52	22 08	05 52
16	12 21	17 17	17 39	14 37	04 38	13 38	06 08	20 52	22 08	05 53
17	12 00	19 44	17 46	14 41	04 39	13 38	06 07	20 52	22 08	05 54
18	11 39	21 20	17 51	14 45	04 39	13 39	06 06	20 52	22 08	05 54
19	11 18	21 59	17 54	14 48	04 38	13 39	06 05	20 52	22 08	05 55
20	10 56	21 38	17 57	14 51	04 38	13 39	06 04	20 53	22 08	05 56
21	10 34	20 14	17 58	14 54	04 37	13 39	06 03	20 53	22 08	05 56
22	10 13	17 49	17 57	14 57	04 36	13 39	06 02	20 53	22 08	05 57
23	09 51	14 29	17 55	15 00	04 35	13 39	06 01	20 53	22 08	05 58
24	09 29	10 22	17 52	15 02	04 33	13 38	06 00	20 53	22 08	05 59
25	09 06	05 41	17 47	15 04	04 32	13 38	05 59	20 53	22 08	05 59
26	08 44	00 S 38	17 41	15 06	04 30	13 38	05 57	20 53	22 08	06 00
27	08 21	04 N 30	17 33	15 07	04 27	13 38	05 56	20 54	22 08	06 01
28	07 S 59	09 N 28	17 S 24	15 S 08	04 S 25	13 S 37	05 S 55	20 S 54	22 S 08	06 N 02

ZODIAC SIGN ENTRIES

Date	h	m	Planets
02	20	20	☽ ♊
04	22	18	☽ ♋
06	23	50	☽ ♌
09	02	15	☽ ♍
11	07	02	☽ ♎
13	15	16	☽ ♏
16	02	45	☽ ♐
18	15	36	☽ ♑
18	23	47	☉ ♓
21	03	15	☽ ♒
23	12	09	☽ ♓
25	18	17	☽ ♈
27	22	32	☽ ♉

LATITUDES

Date	Mercury ☿ ° '	Venus ♀ ° '	Mars ♂ ° '	Jupiter ♃ ° '	Saturn ♄ ° '	Uranus ♅ ° '	Neptune ♆ ° '	Pluto ♇ ° '
01	03 N 32	07 N 32	02 N 50	01 N 16	02 N 34	00 N 10	01 N 16	17 N 14
04	03 39	07 33	02 52	01 17	02 35	00 10	01 16	17 16
07	03 25	07 28	02 55	01 17	02 36	00 10	01 16	17 17
10	02 58	07 18	02 58	01 18	02 37	00 10	01 16	17 19
13	02 23	07 05	03 01	01 18	02 37	00 10	01 16	17 20
16	01 46	06 48	03 03	01 19	02 38	00 10	01 16	17 22
19	01 10	06 28	03 06	01 20	02 39	00 10	01 16	17 23
22	00 35	06 07	03 08	01 20	02 40	00 10	01 16	17 25
25	00 N 02	05 45	03 10	01 21	02 40	00 10	01 16	17 26
28	00 S 27	05 22	03 12	01 21	02 41	00 10	01 16	17 28
31	00 S 53	04 N 58	03 N 13	01 N 22	02 N 42	00 N 10	01 N 17	17 N 29

DATA

Julian Date	2445002
Delta T	+52 seconds
Ayanamsa	23° 36' 10"
Synetic vernal point	05° ♓ 30' 50"
True obliquity of ecliptic	23° 26' 27"

MOON'S PHASES, APSIDES AND POSITIONS ☽

Date	h	m	Phase	Longitude ° '	Eclipse Indicator
01	14	28	☽	12 ♉ 25	
08	07	57	○	19 ♌ 14	
15	20	21	☾	26 ♏ 50	
23	21	13	●	04 ♓ 56	

Day	h	m		
05	14	14	Perigee	
17	08	20	Apogee	
05	18	18	Max dec	21° N 57'
12	05	00	0S	
19	15	43	Max dec	22° S 00'
26	14	58	0N	

LONGITUDES

Date	Chiron ⚷ ° '	Ceres ⚳ ° '	Pallas ⚴ ° '	Juno ⚵ ° '	Vesta ⚶ ° '	Black Moon Lilith ⚸ ° '
01	17 ♉ 55	19 ♏ 02	21 ♎ 13	23 ♐ 58	04 ♑ 23	14 ♐ 33
11	18 ♉ 02	21 ♏ 32	21 ♎ 33	26 ♐ 52	09 ♑ 29	15 ♐ 40
21	18 ♉ 14	23 ♏ 35	21 ♎ 04	29 ♐ 35	14 ♑ 29	16 ♐ 46
31	18 ♉ 33	25 ♏ 07	19 ♎ 42	02 ♑ 06	19 ♑ 22	17 ♐ 53

All ephemeris data is given at 12.00 UT and the Moon's longitude is additionally given for 24.00 UT

ASPECTARIAN

h m	Aspects
01 Monday	
00 02	☽ ⊼ ♅
03 49	☉ ☌ ☿
09 30	☽ ☍ ♃
12 23	☽ ꝑ ♆
12 53	☽ □ ☿
14 28	☽ □ ☉
18 27	☿ ∠ ♆
22 37	☽ ⊼ ♂
02 Tuesday	
03 47	☽ ± ♆
04 42	☽ ⫲ ♃
06 07	☽ ⫲ ♀
07 14	☽ ⊼ ♄
08 30	☽ ⫲ ☿
08 57	☽ ± ♂
11 34	☽ △ ♀
13 58	☽ ⊼ ♆
15 09	☽ ⊼ ♇
17 22	☽ ± ♄
03 Wednesday	
00 16	☽ ⫲ ☉
00 21	☽ ꝑ ♂
01 14	☿ □ ♃
01 15	☽ ± ♇
03 12	☽ ☍ ♅
08 29	☽ ꝑ ♄
11 37	☽ △ ☿
12 13	☽ ꝑ ♀
12 32	☽ ⊼ ♃
16 19	☽ ꝑ ♇
20 57	☽ △ ☉
22 35	☽ ± ♃
04 Thursday	
01 44	☽ △ ♂
02 55	☽ ± ♀
09 27	☽ △ ♄
10 44	☽ ꝑ ☿
12 25	☽ ⫲ ♅
12 41	☽ ⊼ ♀
13 35	☽ ꝑ ♃
16 09	☽ ☍ ♆
17 12	☽ △ ♇
23 41	☽ ꝑ ☉
05 Friday	
00 35	☽ ± ☿
05 09	☽ ⊼ ♅
07 02	☉ Q ♅
09 49	☽ ⊼ ☿
14 27	☽ △ ♃
15 04	☽ ± ♅
15 39	☽ ± ☉
06 Saturday	
02 17	☽ ⊼ ☉
03 54	☽ □ ♂
05 56	☽ ꝑ ♅
10 57	☽ □ ♄
13 28	☽ ☍ ♀
17 46	☽ ⊼ ♆
18 43	☽ □ ♇
21 58	☉ ∥ ☿
23 59	☽ ⫲ ♅
07 Sunday	
03 44	☽ ± ♆
05 29	☉ △ ♂
06 49	☽ △ ♅
08 30	☽ ☍ ☿
10 02	☽ Q ♂
16 18	☽ □ ♃
16 48	☽ Q ♄
18 45	☽ ꝑ ♆
08 Monday	
00 41	☽ Q ♇
06 19	☽ ⚹ ♂
07 57	☽ ☍ ☉
12 57	☽ ⚹ ♄
15 07	☽ ⊼ ♀
16 38	☽ ⫲ ☿
18 14	☿ ⚹ ♅
20 06	☽ △ ♆
20 57	☽ ⚹ ♇
22 46	☽ Q ♃
23 59	☽ ⫲ ☉
09 Tuesday	
01 16	☽ ± ♀
04 17	☽ ⫲ ♀
07 26	☽ ⫲ ♃
08 06	☽ ∠ ♂
08 57	☽ ⊼ ☿
09 34	☽ □ ♅
14 36	☽ ∠ ♄
16 42	☽ ꝑ ♀
19 00	☽ ± ☿
19 31	☽ ⚹ ♃
22 49	☽ ∠ ♇
23 52	☽ ⊥ ♂
10 Wednesday	
06 18	☽ ⊥ ♄
10 19	☽ ꝑ ☿
10 31	☽ ⊻ ♂
14 40	☽ ⊥ ♇
15 55	☽ ⊼ ☉
16 55	☽ ⊻ ♄
17 10	☽ Q ♅
19 05	☽ △ ♀
20 38	♀ St D
21 05	☉ ∥ ♀
22 06	☽ ∠ ♃
22 15	☽ ⫲ ♄
11 Thursday	
00 09	☽ ∥ ♇
00 37	☽ □ ♆
01 24	☽ ⊻ ♇
03 33	☽ ± ☉
05 01	☉ △ ♄
06 27	☽ ⫲ ♂
12 42	☽ △ ☿
14 28	☽ ⊥ ♃
14 54	☽ ⚹ ♅
21 21	☽ ꝑ ☉
12 Friday	
01 33	☽ ⊻ ♃
09 43	☽ Q ♆
11 13	☉ ⊻ ♀
17 50	☽ ☌ ♂
18 32	☉ ∥ ♃
18 52	☽ ∠ ♅
13 Saturday	
00 06	☽ ☌ ♄
02 42	☽ □ ♀
03 57	☽ △ ☉
04 25	☽ ∥ ♂
07 18	☿ St D
08 32	☽ ⚹ ♆
09 13	☽ ☌ ♇
11 01	☽ ⫲ ♇
12 06	☽ ⊥ ♅
12 44	☽ ∥ ♄
20 58	☽ □ ☿
23 45	☽ ⊻ ♅
14 Sunday	
11 09	☽ ☌ ♃
13 49	☽ ∠ ♆
14 01	☽ Q ♀
15 Monday	
03 26	☽ ∥ ☉
04 38	☽ ⊻ ♂
07 45	☽ ⊥ ♆
08 49	☽ ∥ ♃
08 58	☽ Q ☿
10 44	☽ ⊻ ♄
13 55	☉ ⚹ ♆
14 16	☽ ⚹ ♀
15 11	☽ ∥ ♀
16 43	☽ ⊥ ♂
19 49	☽ ⊻ ♆
20 21	☽ □ ☉
20 23	☽ ⊻ ♇
20 40	☉ △ ♇
22 46	☽ ⊥ ♄
16 Tuesday	
08 29	☽ ⊥ ♇
09 45	☽ ⚹ ☿
10 58	☽ ∠ ♂
11 41	☽ ☌ ♅
15 19	☽ ∥ ☿
16 56	☽ ∠ ♄
21 07	☽ ∠ ♀
23 31	☽ ⊻ ♃
17 Wednesday	
02 43	☽ ∠ ♇
11 43	☽ ⊥ ♃
12 14	☽ Q ☉
15 48	☽ ⊥ ♀
17 15	☽ ∠ ☿
17 33	☽ ⚹ ♂
23 20	☽ ⚹ ♄
18 Thursday	
03 33	☽ ∥ ♅
04 15	☽ ⊻ ♀
06 02	☽ ∠ ♃
08 45	☽ ☌ ♆
09 09	☽ ⚹ ♇
12 18	☽ ⊥ ☿
14 50	☽ ⚹ ☉
16 23	☿ ⚹ ♅
17 56	☽ Q ♂
23 33	☽ Q ♄
19 Friday	
00 37	☽ ⊻ ♅
01 00	☽ ⊻ ☿
09 20	☽ Q ♇
12 19	☽ ⚹ ♃
12 40	☽ ⊥ ♅
21 46	☉ ⊥ ♀
23 43	☽ ∠ ☉
20 Saturday	
06 00	☽ □ ♂
06 37	☽ ∠ ♅
11 21	☽ □ ♄
12 11	☽ Q ♃
17 54	☽ ☌ ♀
19 05	☽ ⊥ ☉
19 13	♂ St R
20 45	☽ ⊻ ♆
20 58	☽ □ ♇
21 Sunday	
02 58	☽ ∥ ♅
07 49	☽ ⊻ ☉
08 27	☽ ⊥ ♆
11 59	☽ ⚹ ♅
15 37	☽ ☌ ☿
23 10	☽ □ ♃
22 Monday	
01 47	☽ ∠ ♆
10 51	☽ ∥ ☿
10 57	☽ Q ♅
15 56	☽ △ ♂
20 54	☽ △ ♄
23 Tuesday	
02 31	☉ ꝑ ♂
05 05	☽ ⊻ ♀
06 04	☽ ⚹ ♆
06 08	☽ △ ♇
08 40	☽ ∥ ♀
11 43	☉ □ ♅
16 34	☽ ⊥ ♀
17 13	☽ ∥ ♃
19 44	☽ ꝑ ♂
20 30	☽ □ ♅
21 13	☽ ☌ ☉
24 Wednesday	
00 33	☽ ꝑ ♄
03 39	☽ ⊻ ☿
04 10	☽ Q ♆
05 39	♃ St R
07 04	☽ △ ♃
09 04	♆ ⚹ ♇
09 35	☽ ∠ ♀
09 37	☽ ꝑ ♇
10 23	♀ □ ♇
10 29	♀ ⊻ ♆
12 04	☽ ± ♂
15 19	☽ ⊥ ☿
16 50	☽ ± ♄
17 11	☽ ∥ ☉
22 50	☽ ⊼ ♂
25 Thursday	
01 48	☽ ± ♇
03 02	♄ ⫲ ♇
03 33	☽ ⊼ ♄
08 39	☽ ∠ ☿
10 00	☽ ꝑ ♃
10 29	☽ ⫲ ♇
10 32	☽ ∥ ♄
12 28	☽ ⊼ ♇
12 32	☽ □ ♆
13 28	☽ ⚹ ♀
15 10	☉ ꝑ ♄
17 37	☽ ∥ ♂
26 Friday	
01 54	☽ ± ♃
02 20	☽ △ ♅
03 13	☿ □ ♃
07 14	☽ ⊻ ☉
11 24	☽ Q ♀
12 25	☽ ⊼ ♃
13 09	☽ ⚹ ☿
18 33	☽ ⊥ ☉
27 Saturday	
03 24	☽ ☍ ♂
04 31	☽ ꝑ ♅
08 03	☽ ☍ ♄
11 23	☽ ∠ ☉
11 40	☽ Q ☿
11 46	☽ ⫲ ♂
13 40	☿ ∠ ♆
16 49	☽ ☍ ♇
17 00	☽ △ ♆
17 25	☉ Q ♆
18 45	☽ ⫲ ♄
19 10	☽ ∥ ♇
19 53	☽ □ ♀
20 05	☽ ± ♅
28 Sunday	
05 10	☽ ⫲ ☉
06 24	☽ ⊼ ♅
15 11	☽ ⚹ ☉
16 12	☽ ☍ ♃
18 48	☽ ꝑ ♆
21 11	☽ □ ☿

MARCH 1982

LONGITUDES

Date	Sidereal time h m s	Sun ☉ ° ' "	Moon ☽ ° ' "	Moon ☽ 24.00 ° ' "	Mercury ☿ ° '	Venus ♀ ° '	Mars ♂ ° '	Jupiter ♃ ° '	Saturn ♄ ° '	Uranus ♅ ° '	Neptune ♆ ° '	Pluto ♇ ° '
01	22 35 52	10 ♓ 34 35	21 ♉ 52 58	28 ♉ 55 30	13 ♒ 54	29 ♑ 25	18 ♎ 42	10 ♏ 17	21 ♎ 31	04 ♐ 36	26 ♐ 49	26 ♎ 39
02	22 39 49	11 34 47	05 ♊ 58 40	13 ♊ 02 15	15 03	00 ♒ 01	18 R 35	10 R 16	21 R 28	04 36	26 50	26 R 38
03	22 43 45	12 34 57	20 ♊ 06 08	27 ♊ 10 09	16 13	00 38	18 27	10 15	21 25	04 36	26 51	26 37
04	22 47 42	13 35 06	04 ♋ 14 09	11 ♋ 17 59	17 26	01 17	18 19	10 13	21 22	04 37	26 52	26 36
05	22 51 39	14 35 12	18 ♋ 21 26	25 ♋ 24 19	18 41	01 56	18 09	10 12	21 19	04 37	26 53	26 35
06	22 55 35	15 35 16	02 ♌ 26 21	09 ♌ 27 12	19 57	02 37	17 59	10 10	21 15	04 37	26 54	26 34
07	22 59 32	16 35 17	16 ♌ 26 34	23 ♌ 24 01	21 15	03 19	17 48	10 08	21 12	04 37	26 54	26 33
08	23 03 28	17 35 17	00 ♍ 19 10	07 ♍ 11 34	22 34	04 02	17 37	10 06	21 09	04 37	26 55	26 32
09	23 07 25	18 35 15	14 ♍ 00 50	20 ♍ 46 32	23 55	04 46	17 24	10 03	21 05	04 38	26 56	26 30
10	23 11 21	19 35 10	27 ♍ 28 21	04 ♎ 05 58	25 18	05 31	17 11	10 01	21 02	04 R 38	26 56	26 29
11	23 15 18	20 35 04	10 ♎ 39 09	17 ♎ 07 47	26 42	06 16	16 57	09 58	20 58	04 37	26 57	26 28
12	23 19 14	21 34 56	23 ♎ 31 49	29 ♎ 51 17	28 07	07 03	16 43	09 55	20 54	04 37	26 58	26 27
13	23 23 11	22 34 47	06 ♏ 06 21	12 ♏ 17 14	29 ♒ 34	07 51	16 27	09 52	20 51	04 37	26 58	26 25
14	23 27 08	23 34 35	18 ♏ 24 15	24 ♏ 27 48	01 ♓ 02	08 39	16 11	09 49	20 47	04 37	26 59	26 24
15	23 31 04	24 34 22	00 ♐ 28 22	06 ♐ 26 29	02 31	09 28	15 55	09 45	20 43	04 37	26 59	26 23
16	23 35 01	25 34 07	12 ♐ 22 43	18 ♐ 17 41	04 02	10 18	15 37	09 42	20 39	04 36	27 00	26 21
17	23 38 57	26 33 51	24 ♐ 12 03	00 ♑ 06 30	05 34	11 09	15 20	09 38	20 35	04 36	27 00	26 20
18	23 42 54	27 33 33	06 ♑ 01 41	11 ♑ 58 20	07 07	12 00	15 01	09 34	20 31	04 36	27 01	26 18
19	23 46 50	28 33 13	17 ♑ 57 06	23 ♑ 58 38	08 42	12 52	14 42	09 30	20 27	04 35	27 01	26 17
20	23 50 47	29 ♓ 32 51	00 ♒ 03 35	06 ♒ 12 31	10 18	13 45	14 22	09 25	20 23	04 35	27 01	26 15
21	23 54 43	00 ♈ 32 27	12 ♒ 25 56	18 ♒ 44 17	11 55	14 38	14 02	09 21	20 19	04 34	27 02	26 14
22	23 58 40	01 32 02	25 ♒ 07 56	01 ♓ 37 06	13 33	15 32	13 42	09 16	20 14	04 33	27 02	26 13
23	00 02 37	02 31 35	08 ♓ 11 56	14 ♓ 52 26	15 13	16 27	13 21	09 12	20 10	04 33	27 02	26 11
24	00 06 33	03 31 06	21 ♓ 38 28	28 ♓ 29 44	16 54	17 22	12 59	09 07	20 06	04 32	27 02	26 10
25	00 10 30	04 30 34	05 ♈ 25 52	12 ♈ 26 20	18 36	18 17	12 37	09 02	20 01	04 31	27 02	26 08
26	00 14 26	05 30 01	19 ♈ 30 31	26 ♈ 37 44	20 20	19 14	12 15	08 56	19 57	04 30	27 02	26 06
27	00 18 23	06 29 26	03 ♉ 47 13	10 ♉ 58 15	22 05	20 10	11 52	08 51	19 53	04 29	27 03	26 05
28	00 22 19	07 28 49	18 ♉ 10 04	25 ♉ 22 01	23 51	21 07	11 30	08 46	19 48	04 28	27 03	26 03
29	00 26 16	08 28 09	02 ♊ 33 29	09 ♊ 43 56	25 39	22 05	11 07	08 40	19 44	04 27	27 03	26 02
30	00 30 12	09 27 27	16 ♊ 52 57	24 ♊ 00 26	27 28	23 03	10 44	08 34	19 40	04 26	27 R 03	26 00
31	00 34 09	10 ♈ 26 43	01 ♋ 05 28	08 ♋ 08 34	29 ♓ 18	24 ♒ 01	10 ♎ 21	08 ♏ 28	19 ♎ 35	04 ♐ 25	27 ♐ 03	25 ♎ 59

Date	Moon True ☊ ° '	Moon Mean ☊ ° '	Moon ☽ Latitude ° '
01	21 ♋ 11	20 ♋ 02	04 S 25
02	21 R 10	19 59	03 39
03	21 D 10	19 56	02 39
04	21 11	19 53	01 30
05	21 11	19 50	00 S 15
06	21 R 11	19 46	01 N 00
07	21 08	19 43	02 11
08	21 03	19 40	03 14
09	20 55	19 37	04 04
10	20 46	19 34	04 40
11	20 35	19 31	05 00
12	20 24	19 27	05 04
13	20 14	19 24	04 53
14	20 07	19 21	04 28
15	20 01	19 18	03 52
16	19 58	19 15	03 06
17	19 57	19 11	02 13
18	19 D 57	19 08	01 14
19	19 58	19 05	00 N 11
20	19 R 57	19 02	00 S 54
21	19 55	18 59	01 57
22	19 50	18 56	02 55
23	19 43	18 52	03 47
24	19 33	18 49	04 27
25	19 22	18 46	04 52
26	19 10	18 43	05 01
27	18 59	18 40	04 51
28	18 49	18 37	04 23
29	18 43	18 33	03 38
30	18 39	18 30	02 40
31	18 ♋ 38	18 ♋ 27	01 S 31

DECLINATIONS

Date	Sun ☉	Moon ☽	Mercury ☿	Venus ♀	Mars ♂	Jupiter ♃	Saturn ♄	Uranus ♅	Neptune ♆	Pluto ♇
01	07 S 36	13 N 58	17 S 14	15 S 09	04 S 22	13 S 37	05 S 54	20 S 54	22 S 08	06 N 02
02	07 13	17 43	17 02	15 09	04 19	13 36	05 52	20 54	22 08	06 03
03	06 50	20 25	16 49	15 09	04 15	13 36	05 51	20 54	22 08	06 04
04	06 27	21 52	16 35	15 09	04 12	13 35	05 50	20 54	22 08	06 05
05	06 04	21 56	16 19	15 08	04 08	13 34	05 48	20 54	22 08	06 05
06	05 41	20 36	16 02	15 07	04 04	13 34	05 47	20 54	22 08	06 06
07	05 18	18 00	15 44	15 05	04 00	13 33	05 46	20 54	22 08	06 07
08	04 54	14 23	15 24	15 03	03 55	13 32	05 44	20 54	22 08	06 08
09	04 31	10 02	15 03	15 01	03 50	13 31	05 43	20 54	22 08	06 08
10	04 07	05 17	14 41	14 58	03 45	13 30	05 41	20 54	22 08	06 09
11	03 44	00 N 23	14 17	14 54	03 40	13 29	05 40	20 54	22 07	06 10
12	03 20	04 S 26	13 52	14 51	03 34	13 28	05 38	20 54	22 07	06 11
13	02 57	08 57	13 26	14 46	03 29	13 27	05 37	20 54	22 07	06 11
14	02 33	13 00	12 59	14 41	03 23	13 26	05 35	20 54	22 07	06 12
15	02 09	16 28	12 30	14 36	03 17	13 24	05 33	20 54	22 07	06 13
16	01 46	19 12	12 00	14 30	03 10	13 23	05 32	20 54	22 07	06 14
17	01 22	21 06	11 29	14 24	03 04	13 22	05 30	20 54	22 07	06 14
18	00 58	22 05	10 57	14 18	02 57	13 20	05 29	20 54	22 07	06 15
19	00 35	22 04	10 23	14 10	02 50	13 19	05 27	20 54	22 07	06 16
20	00 S 11	21 01	09 48	14 03	02 43	13 17	05 25	20 54	22 07	06 17
21	00 N 13	18 56	09 12	13 55	02 36	13 16	05 24	20 53	22 07	06 18
22	00 37	15 54	08 35	13 46	02 29	13 14	05 22	20 53	22 07	06 18
23	01 00	12 00	07 57	13 37	02 22	13 13	05 20	20 53	22 07	06 19
24	01 24	07 24	07 17	13 27	02 14	13 11	05 19	20 53	22 07	06 20
25	01 48	02 S 19	06 37	13 17	02 07	13 09	05 17	20 53	22 07	06 20
26	02 11	02 N 59	05 55	13 07	01 59	13 07	05 15	20 53	22 07	06 21
27	02 35	08 13	05 12	12 56	01 51	13 06	05 13	20 53	22 07	06 22
28	02 58	13 02	04 28	12 44	01 44	13 04	05 12	20 52	22 07	06 23
29	03 22	17 06	03 43	12 32	01 36	13 02	05 10	20 52	22 07	06 23
30	03 45	20 09	02 56	12 20	01 28	13 00	05 08	20 52	22 07	06 24
31	04 N 08	21 N 55	02 S 09	12 S 07	01 S 21	12 S 58	05 S 06	20 S 52	22 S 07	06 N 25

ZODIAC SIGN ENTRIES

Date	h	m	Planets
02	01	50	☽ ♊
02	11	25	♀ ♒
04	04	48	☽ ♋
06	07	50	☽ ♌
08	11	27	☽ ♍
10	16	34	☽ ♎
13	00	17	☽ ♏
13	19	11	☿ ♓
15	11	03	☽ ♐
17	23	47	☽ ♑
20	11	53	☽ ♒
20	22	56	☉ ♈
22	21	01	☽ ♓
25	02	37	☽ ♈
27	05	39	☽ ♉
29	07	44	☽ ♊
31	10	09	☽ ♋
31	20	59	☿ ♈

LATITUDES

Date	Mercury ☿	Venus ♀	Mars ♂	Jupiter ♃	Saturn ♄	Uranus ♅	Neptune ♆	Pluto ♇
01	00 S 36	05 N 14	03 N 12	01 N 22	02 N 41	00 N 10	01 N 17	17 N 28
04	01 01	04 50	03 13	01 22	02 42	00 10	01 17	17 29
07	01 23	04 27	03 14	01 23	02 42	00 10	01 17	17 31
10	01 41	04 03	03 15	01 23	02 43	00 10	01 17	17 32
13	01 56	03 40	03 14	01 24	02 43	00 10	01 17	17 33
16	02 07	03 16	03 14	01 24	02 44	00 10	01 17	17 34
19	02 15	02 54	03 12	01 25	02 44	00 10	01 17	17 35
22	02 18	02 31	03 10	01 25	02 45	00 10	01 17	17 36
25	02 17	02 10	03 07	01 26	02 45	00 10	01 17	17 36
28	02 12	01 49	03 04	01 26	02 45	00 10	01 17	17 37
31	02 S 03	01 N 29	02 N 59	01 N 26	02 N 46	00 N 10	01 N 18	17 N 37

DATA

Julian Date	2445030
Delta T	+52 seconds
Ayanamsa	23° 36' 13"
Synetic vernal point	05° ♓ 30' 47"
True obliquity of ecliptic	23° 26' 27"

MOON'S PHASES, APSIDES AND POSITIONS ☽

Date	h	m	Phase	Longitude ° '	Eclipse Indicator
02	22	15	☽	12 ♊ 00	
09	20	45	○	18 ♍ 57	
17	17	15	☾	26 ♐ 47	
25	10	17	●	04 ♈ 26	

Day	h	m		
04	04	40	Perigee	
17	05	05	Apogee	
29	06	21	Perigee	
05	00	57	Max dec	22° N 05'
11	13	51	0S	
18	23	36	Max dec	22° S 12'
25	22	32	0N	

LONGITUDES

Date	Chiron ⚷ ° '	Ceres ⚳ ° '	Pallas ⚴ ° '	Juno ⚵ ° '	Vesta ⚶ ° '	Black Moon Lilith ⚸ ° '
01	18 ♉ 29	24 ♏ 52	20 ♎ 02	01 ♑ 37	18 ♑ 24	17 ♐ 40
11	18 ♉ 52	25 ♏ 56	18 ♎ 02	03 ♑ 55	23 ♑ 09	18 ♐ 46
21	19 ♉ 20	26 ♏ 20	15 ♎ 21	05 ♑ 55	27 ♑ 44	19 ♐ 53
31	19 ♉ 52	26 ♏ 08	12 ♎ 17	07 ♑ 35	02 ♒ 08	20 ♐ 59

All ephemeris data is given at 12.00 UT and the Moon's longitude is additionally given for 24.00 UT

ASPECTARIAN

h m	Aspects
01 Monday	
05 10	☉ △ ♃
06 37	☽ ⊼ ♂
09 59	☽ # ♃
10 11	☽ ± ♆
11 22	☽ ⊼ ♄
13 16	☽ Q ☉
16 46	☽ ± ♂
19 02	☽ # ♀
20 07	☽ ⊼ ♇
20 26	☽ ⊼ ♆
21 34	☽ ± ♄
02 Tuesday	
01 24	☽ △ ♀
06 20	☽ ± ♇
07 30	☽ # ☿
07 58	☽ bQ ♂
09 39	☽ ☍ ♅
12 50	☽ bQ ♄
13 23	☉ bQ ♇
19 17	☽ ⊼ ♃
21 37	☽ bQ ♇
22 15	☽ □ ☉
03 Wednesday	
04 04	☽ bQ ♀
04 48	☽ △ ☿
05 28	☽ ± ♃
09 14	☽ △ ♂
09 16	☉ ± ♂
14 13	☽ △ ♄
17 49	☽ # ♅
19 43	☿ Q ♅
20 04	☽ ± ♀
20 43	☽ bQ ♃
23 03	☽ △ ♇
23 28	☽ ☍ ♆
04 Thursday	
06 44	☽ ⊼ ♀
08 39	☽ bQ ☿
12 38	☽ ⊼ ♅
22 09	☽ △ ♃
22 50	☽ ± ♅
05 Friday	
01 24	☽ ± ☿
03 07	☿ △ ♂
05 06	☽ △ ☉
10 51	☉ # ♇
11 39	☽ □ ♂
12 36	☽ ⊼ ☿
14 09	☽ bQ ♅
17 00	☽ □ ♄
06 Saturday	
01 59	☽ □ ♇
02 32	☽ ⊼ ♆
04 27	☉ ± ♄
05 19	☉ ∥ ♄
08 07	☽ # ♅
08 36	☽ bQ ☉
12 19	☽ ☍ ♀
12 47	☽ ± ♆
15 44	☽ △ ♅
17 59	☽ Q ♂
21 49	♀ ⊥ ♆
23 37	☽ Q ♄
07 Sunday	
01 10	☽ ± ☉
01 11	☽ □ ♃
04 12	☽ bQ ♆
08 45	☽ Q ♇
11 14	☿ △ ♄
12 16	☽ ⊼ ☉
14 19	☽ ⚹ ♂
20 10	☽ ⚹ ♄
21 08	☽ ☍ ☿
08 Monday	
05 08	☽ # ☿
05 25	☽ ⚹ ♇
06 05	☽ △ ♆
07 52	☽ # ♀
08 09	☽ Q ♃
12 25	☉ ⊼ ♂
15 56	☽ ∠ ♂
16 59	☽ # ♃
18 50	☽ ⊼ ♀
19 30	☽ □ ♅
22 07	☽ ∠ ♄
09 Tuesday	
05 03	☽ ⚹ ♃
05 56	☽ ± ♀
07 28	☽ ⊥ ♂
07 35	☽ ∠ ♇
07 37	♀ ⚹ ♅
13 53	☽ ⊥ ♄
15 05	☿ ∥ ♀
17 54	☽ ⊻ ♂
19 42	♅ St R
20 45	☽ ☍ ☉
22 47	☽ bQ ♀
23 30	☽ ⊥ ♇
10 Wednesday	
00 30	☽ ⊻ ♄
03 18	☽ Q ♅
07 35	☽ ∠ ♃
07 38	☽ ⊼ ☿
07 43	☽ ∥ ♇
10 00	☽ # ♄
10 13	☽ ⊻ ♇
11 03	☽ □ ♆
18 11	☽ # ☉
19 38	☽ # ♂
19 43	☽ ± ☿
23 48	☽ ⊥ ♃
11 Thursday	
00 57	☽ ⚹ ♅
03 28	☽ △ ♀
08 08	☿ △ ♇
09 09	☉ ± ♇
10 44	☽ ⊻ ♃
14 09	☽ bQ ☿
16 24	☿ ⚹ ♆
17 16	☉ ∥ ♂
19 57	☽ Q ♆
20 40	☉ ⊼ ♄
23 27	☽ ☌ ♂
12 Friday	
04 39	☽ ∠ ♅
06 52	☽ ∥ ☉
07 05	☽ ☌ ♄
07 43	☽ ∥ ♂
08 01	☽ ⊼ ☉
17 30	☽ ☌ ♇
18 11	☽ ∥ ♄
18 30	☽ ⚹ ♆
20 20	☽ ± ☉
21 04	☽ # ♇
21 39	☽ ⊥ ♅
21 49	☽ △ ☿
13 Saturday	
09 08	☽ ⊻ ♅
11 36	☿ ∥ ♃
15 06	☽ bQ ☉
15 36	☽ □ ♀
19 15	☽ ☌ ♃
23 23	☽ ∠ ♆
14 Sunday	
07 44	☽ ⊻ ♂
11 52	☽ ∥ ☿
14 10	☿ bQ ♂
14 40	☽ ∥ ♃
16 40	☽ ⊻ ♄
17 05	☽ ⊥ ♆
19 19	☽ ⊥ ♂
21 46	☉ ∠ ♀
22 51	☽ ∥ ♀
23 09	☽ △ ☉
15 Monday	
03 49	☽ ⊻ ♇
04 32	☽ ⊥ ♄
05 01	☽ ⊻ ♆
05 33	☽ Q ♀
12 52	☽ ∠ ♂
15 49	☽ ⊥ ♇
16 06	☉ bQ ♃
16 42	☽ □ ☿
19 40	♀ □ ♃
20 19	☽ ☌ ♅
22 29	☽ ∠ ♄
16 Tuesday	
06 36	☽ ⊻ ♃
07 29	☽ ⚹ ♀
09 56	☽ ∠ ♇
18 25	☽ ⚹ ♂
18 41	☽ ⊥ ♃
20 57	☿ □ ♅
17 Wednesday	
04 41	☽ ⚹ ♄
06 29	☉ ⊼ ♇
08 46	☽ ∥ ♅
10 31	☽ Q ☿
12 14	☿ bQ ♄
12 52	☽ ∠ ♃
16 16	☽ ∠ ♀
16 19	☽ ⚹ ♇
17 15	☽ □ ☉
17 42	☽ ☌ ♆
18 11	☽ Q ♂
22 38	☉ □ ♆
18 Thursday	
04 56	☽ Q ♄
09 06	☽ ⊻ ♅
11 57	☽ ⊥ ♀
12 08	♀ ∠ ♆
12 42	♂ Q ♆
14 26	☽ ∥ ♆
14 33	☽ ⚹ ☿
16 36	☽ Q ♇
19 06	☽ ⚹ ♃
21 13	☽ ⊥ ♅
19 Friday	
01 00	☽ ⊻ ♀
05 39	☽ □ ♂
08 46	☽ ∥ ♆
08 57	☽ Q ☉
12 00	☿ ± ♂
15 16	☽ ∠ ♅
16 47	☿ Q ♆
16 57	☽ □ ♄
19 02	☽ Q ♃
23 29	☿ △ ♃
20 Saturday	
01 11	☽ ∠ ☿
04 32	☽ □ ♇
06 01	☽ ⊻ ♆
10 54	☽ ⚹ ☉
13 47	☽ ∥ ♅
17 48	☽ ⊥ ♆
20 49	☽ ⚹ ♅
21 32	☽ ⊥ ☿
21 Sunday	
00 15	♀ △ ♂
02 04	☿ bQ ♇
06 07	☽ □ ♃
10 52	☽ ⊻ ☿
11 13	☽ ∠ ♆
14 59	☽ △ ♂
16 33	☽ ☌ ♀
18 27	☽ ∠ ☉
19 53	☽ Q ♅
22 Monday	
02 53	☽ △ ♄
12 49	☽ ⊥ ☉
13 38	☿ ⊼ ♂
14 00	☽ △ ♇
15 32	☽ ⚹ ♆
18 26	☽ bQ ♂
21 29	☿ ± ♄
23 Tuesday	
00 49	☽ ⊻ ☉
02 14	☽ ∥ ♀
04 58	☽ ∥ ♃
05 22	☽ □ ♅
06 31	☽ bQ ♄
10 29	☽ ± ♂
13 31	☽ Q ♆
13 47	☽ △ ♃
14 30	♀ Q ♅
17 23	☽ bQ ♇
21 01	☽ ⊼ ♂
22 41	☽ ± ♄
24 Wednesday	
02 25	☽ ☌ ☿
02 55	☉ ± ♃
03 53	☽ ⊻ ♀
09 17	☽ ⊼ ♄
09 23	☽ ± ♇
12 38	☽ ∥ ☿
15 15	☽ ⊥ ♀
16 19	☽ bQ ♃
17 11	☽ # ♇
19 55	☽ ⊼ ♇
21 28	☽ □ ♆
22 08	☽ ∥ ♄
25 Thursday	
02 16	☿ ⊻ ♀
07 53	☽ ± ♃
08 03	☽ ∠ ♀
10 17	☽ ☌ ☉
10 26	☽ △ ♅
12 12	☉ △ ♅
12 58	☽ ∥ ♂
14 14	☽ # ☉
18 08	☽ ⊼ ♃
21 09	☿ # ♇
26 Friday	
00 00	☽ ☍ ♂
02 44	☉ # ♂
06 54	☿ ⊼ ♄
07 36	☽ # ♂
08 06	☽ ∥ ☉
08 54	☿ ± ♇
10 17	♀ ∥ ♃
11 29	☽ ⚹ ♀
12 00	☽ bQ ♅
12 45	☽ ☍ ♄
13 36	☽ ⊻ ☿
22 14	☽ # ♄
23 06	☽ ☍ ♇
23 43	☽ # ☿
27 Saturday	
00 42	☽ △ ♆
01 06	☽ ⊥ ☿
03 08	☽ ± ♅
03 23	☽ ∥ ♇
05 08	♀ △ ♄
09 06	☽ Q ♀
11 07	☿ ∥ ♄
13 10	☽ ⊼ ♅
16 51	☽ ⊻ ☉
18 17	☽ ∠ ☿
20 24	☽ ☍ ♃
28 Sunday	
01 10	☽ ⊼ ♂
01 47	☽ bQ ♆
03 37	☽ ⊥ ☉
10 32	☽ # ♀
10 45	☿ bQ ♃
10 55	☽ ± ♂
12 11	☽ # ♃
14 43	☽ ⊼ ♄
16 48	☽ ± ♆
17 16	☽ □ ♀
19 43	☽ ∠ ☉
22 49	☽ ⚹ ☿
29 Monday	
00 40	☽ ± ♄
01 07	☽ ⊼ ♇
01 31	☽ bQ ♂
02 48	☽ ⊼ ♆
11 07	☽ ± ♇
15 10	☽ ☍ ♅
15 36	☽ bQ ♄
16 19	☉ ⊼ ♃
16 35	♆ St R
16 58	☿ ⊼ ♇
19 18	☉ # ☿
21 45	☽ Q ☿
22 09	☽ ⊼ ♃
22 37	☽ ⚹ ☉
30 Tuesday	
01 56	☽ △ ♂
02 09	☽ bQ ♇
06 27	☿ □ ♆
08 08	☽ ± ♃
16 38	☽ △ ♄
19 45	☽ # ♅
20 17	☽ Q ☉
23 08	☽ △ ♀
23 11	☽ bQ ♃
31 Wednesday	
03 21	☽ △ ♇
05 08	☽ ☍ ♆
08 31	☽ □ ☿
10 13	☉ ☍ ♂
17 17	☽ # ♆
17 39	☽ ⊼ ♅

APRIL 1982

Date	Sidereal time h m s	Sun ☉ ° ′ ″	Moon ☽ ° ′ ″	Moon ☽ 24.00 ° ′ ″	Mercury ☿ ° ′	Venus ♀ ° ′	Mars ♂ ° ′	Jupiter ♃ ° ′	Saturn ♄ ° ′	Uranus ♅ ° ′	Neptune ♆ ° ′	Pluto ♇ ° ′
01	00 38 06	11 ♈ 25 56	15 ♋ 09 27	22 ♋ 08 03	01 ♈ 10	25 ♒ 00	09 ♎ 57	08 ♏ 22	19 ♎ 30	04 ♐ 24	27 ♐ 02	25 ♎ 57
02	00 42 02	12 25 08	29 ♋ 04 23	05 ♌ 58 27	03 03	25 59	09 R 34	08 R 16	19 R 26	04 R 23	27 R 02	25 R 55
03	00 45 59	13 24 16	12 ♌ 50 15	19 ♌ 39 48	04 58	26 58	09 11	08 10	19 21	04 22	27 02	25 54
04	00 49 55	14 23 23	26 ♌ 27 02	03 ♍ 11 55	06 53	27 58	08 48	08 03	19 16	04 21	27 02	25 52
05	00 53 52	15 22 27	09 ♍ 54 21	16 ♍ 34 12	08 51	28 59	08 25	07 57	19 12	04 19	27 02	25 50
06	00 57 48	16 21 29	23 ♍ 11 19	29 ♍ 45 33	10 49	29 ♒ 59	08 02	07 50	19 07	04 18	27 02	25 49
07	01 01 45	17 20 28	06 ♎ 16 45	12 ♎ 44 44	12 49	01 ♓ 00	07 39	07 44	19 02	04 16	27 01	25 47
08	01 05 41	18 19 26	19 ♎ 09 23	25 ♎ 30 36	14 50	02 01	07 17	07 37	18 58	04 15	27 01	25 45
09	01 09 38	19 18 21	01 ♏ 48 20	08 ♏ 02 34	16 52	03 03	06 55	07 30	18 53	04 14	27 01	25 44
10	01 13 35	20 17 15	14 ♏ 13 24	20 ♏ 20 56	18 55	04 05	06 33	07 23	18 49	04 12	27 01	25 42
11	01 17 31	21 16 07	26 ♏ 25 24	02 ♐ 27 02	21 00	05 07	06 11	07 16	18 44	04 10	27 00	25 40
12	01 21 28	22 14 57	08 ♐ 26 13	14 ♐ 23 19	23 05	06 09	05 50	07 09	18 39	04 09	26 59	25 39
13	01 25 24	23 13 45	20 ♐ 18 50	26 ♐ 13 16	25 11	07 12	05 29	07 02	18 35	04 07	26 59	25 37
14	01 29 21	24 12 32	02 ♑ 07 12	08 ♑ 01 15	27 17	08 15	05 09	06 54	18 30	04 06	26 59	25 35
15	01 33 17	25 11 17	13 ♑ 56 03	19 ♑ 52 16	29 ♈ 24	09 18	04 50	06 47	18 26	04 04	26 58	25 34
16	01 37 14	26 10 00	25 ♑ 50 37	01 ♒ 51 47	01 ♉ 30	10 22	04 30	06 40	18 21	04 02	26 57	25 32
17	01 41 10	27 08 41	07 ♒ 56 26	14 ♒ 05 16	03 37	11 26	04 12	06 32	18 16	04 00	26 57	25 30
18	01 45 07	28 07 21	20 ♒ 18 52	26 ♒ 37 51	05 43	12 30	03 54	06 25	18 12	03 58	26 56	25 29
19	01 49 04	29 ♈ 05 59	03 ♓ 02 42	09 ♓ 33 48	07 48	13 34	03 36	06 17	18 07	03 57	26 56	25 27
20	01 53 00	00 ♉ 04 35	16 ♓ 11 28	22 ♓ 55 49	09 52	14 38	03 20	06 10	18 03	03 55	26 55	25 25
21	01 56 57	01 03 09	29 ♓ 46 52	06 ♈ 44 24	11 54	15 43	03 04	06 02	17 58	03 53	26 54	25 23
22	02 00 53	02 01 42	13 ♈ 48 06	20 ♈ 57 24	13 55	16 48	02 48	05 54	17 54	03 51	26 53	25 22
23	02 04 50	03 00 13	28 ♈ 11 37	05 ♉ 29 53	15 53	17 53	02 34	05 47	17 50	03 49	26 53	25 20
24	02 08 46	03 58 42	12 ♉ 51 15	20 ♉ 14 42	17 50	18 58	02 20	05 39	17 45	03 47	26 52	25 18
25	02 12 43	04 57 09	27 ♉ 39 11	05 ♊ 03 41	19 43	20 03	02 07	05 31	17 41	03 45	26 51	25 17
26	02 16 39	05 55 35	12 ♊ 27 12	19 ♊ 48 55	21 34	21 09	01 54	05 24	17 37	03 43	26 50	25 15
27	02 20 36	06 53 58	27 ♊ 08 03	04 ♋ 24 02	23 21	22 14	01 42	05 16	17 32	03 40	26 49	25 13
28	02 24 33	07 52 19	11 ♋ 36 23	18 ♋ 44 48	25 05	23 20	01 32	05 08	17 28	03 38	26 49	25 12
29	02 28 29	08 50 38	25 ♋ 49 05	02 ♌ 49 10	26 45	24 26	01 22	05 01	17 24	03 36	26 48	25 10
30	02 32 26	09 ♉ 48 55	09 ♌ 45 02	16 ♌ 36 47	28 ♉ 22	25 ♓ 33	01 ♎ 12	04 ♏ 53	17 ♎ 20	03 ♐ 34	26 ♐ 47	25 ♎ 09

Date	Moon True ☊ ° ′	Moon Mean ☊ ° ′	Moon ☽ Latitude ° ′
01	18 ♋ 38	18 ♋ 24	00 S 18
02	18 R 37	18 21	00 N 55
03	18 36	18 17	02 04
04	18 32	18 14	03 06
05	18 25	18 11	03 56
06	18 16	18 08	04 32
07	18 04	18 05	04 54
08	17 50	18 02	05 00
09	17 37	17 58	04 51
10	17 25	17 55	04 28
11	17 14	17 52	03 54
12	17 07	17 49	03 09
13	17 02	17 46	02 16
14	16 59	17 43	01 18
15	16 59	17 39	00 N 16
16	16 D 59	17 36	00 S 47
17	16 R 58	17 33	01 49
18	16 56	17 30	02 47
19	16 52	17 27	03 39
20	16 46	17 23	04 20
21	16 37	17 20	04 49
22	16 26	17 17	05 01
23	16 15	17 14	04 55
24	16 05	17 11	04 30
25	15 56	17 08	03 46
26	15 50	17 04	02 47
27	15 46	17 01	01 37
28	15 45	16 58	00 S 22
29	15 D 45	16 55	00 N 53
30	15 ♋ 45	16 ♋ 52	02 N 04

DECLINATIONS

Date	Sun ☉ ° ′	Moon ☽ ° ′	Mercury ☿ ° ′	Venus ♀ ° ′	Mars ♂ ° ′	Jupiter ♃ ° ′	Saturn ♄ ° ′	Uranus ♅ ° ′	Neptune ♆ ° ′	Pluto ♇ ° ′
01	04 N 31	22 N 17	01 S 21	11 S 54	01 S 13	12 S 56	05 S 05	20 S 52	22 S 07	06 N 26
02	04 54	21 15	00 S 32	11 40	01 05	12 54	05 03	20 51	22 07	06 26
03	05 17	18 57	00 N 18	11 26	00 58	12 52	05 01	20 51	22 07	06 27
04	05 40	15 36	01 10	11 12	00 51	12 50	04 59	20 51	22 07	06 28
05	06 03	11 29	02 01	10 57	00 43	12 47	04 57	20 51	22 07	06 28
06	06 26	06 52	02 54	10 41	00 36	12 45	04 56	20 50	22 07	06 29
07	06 49	02 N 00	03 47	10 25	00 29	12 43	04 54	20 50	22 07	06 30
08	07 11	02 S 52	04 41	10 09	00 22	12 41	04 52	20 50	22 07	06 30
09	07 33	07 33	05 36	09 53	00 15	12 39	04 50	20 50	22 07	06 31
10	07 56	11 50	06 31	09 36	00 09	12 36	04 49	20 49	22 07	06 32
11	08 18	15 34	07 26	09 18	00 S 02	12 34	04 47	20 49	22 06	06 32
12	08 40	18 36	08 21	09 00	00 N 04	12 32	04 45	20 49	22 06	06 33
13	09 02	20 49	09 17	08 42	00 10	12 29	04 43	20 49	22 06	06 33
14	09 23	22 08	10 12	08 24	00 16	12 27	04 42	20 48	22 06	06 34
15	09 45	22 27	11 07	08 05	00 21	12 24	04 40	20 48	22 06	06 35
16	10 06	21 45	12 01	07 46	00 27	12 22	04 38	20 48	22 06	06 35
17	10 27	20 02	12 55	07 26	00 32	12 20	04 37	20 47	22 06	06 36
18	10 48	17 21	13 47	07 06	00 37	12 17	04 35	20 47	22 06	06 36
19	11 09	13 47	14 39	06 46	00 41	12 15	04 33	20 47	22 06	06 37
20	11 30	09 27	15 29	06 26	00 45	12 12	04 32	20 46	22 06	06 37
21	11 50	04 S 30	16 17	06 05	00 49	12 10	04 30	20 46	22 06	06 38
22	12 11	00 N 49	17 04	05 44	00 53	12 07	04 28	20 46	22 06	06 38
23	12 31	06 14	17 49	05 23	00 57	12 05	04 27	20 45	22 06	06 39
24	12 51	11 24	18 32	05 01	01 00	12 02	04 25	20 45	22 06	06 39
25	13 10	15 58	19 12	04 39	01 02	12 00	04 24	20 44	22 06	06 40
26	13 30	19 32	19 50	04 17	01 05	11 57	04 22	20 44	22 06	06 40
27	13 49	21 47	20 26	03 55	01 07	11 55	04 20	20 44	22 06	06 41
28	14 08	22 34	21 00	03 33	01 09	11 52	04 19	20 43	22 06	06 41
29	14 27	21 51	21 30	03 10	01 11	11 50	04 17	20 43	22 06	06 42
30	14 N 45	19 N 48	21 N 59	02 S 47	01 N 12	11 S 47	04 S 16	20 S 42	22 S 06	06 N 42

ZODIAC SIGN ENTRIES

Date	h	m	Planets
02	13	36	☽ ♌
04	18	18	☽ ♍
06	12	20	♀ ♓
07	00	26	☽ ♎
09	08	33	☽ ♏
11	19	07	☽ ♐
14	07	41	☽ ♑
15	18	54	☿ ♉
16	20	18	☽ ♒
19	06	20	☽ ♓
20	10	07	☉ ♉
21	12	23	☽ ♈
23	14	59	☽ ♉
25	15	29	☽ ♊
27	16	43	☽ ♋
29	19	09	☽ ♌

LATITUDES

Date	Mercury ☿ ° ′	Venus ♀ ° ′	Mars ♂ ° ′	Jupiter ♃ ° ′	Saturn ♄ ° ′	Uranus ♅ ° ′	Neptune ♆ ° ′	Pluto ♇ ° ′
01	01 S 59	01 N 22	02 N 58	01 N 26	02 N 46	00 N 10	01 N 18	17 N 38
04	01 43	01 03	02 53	01 27	02 46	00 10	01 18	17 38
07	01 23	00 45	02 47	01 27	02 46	00 10	01 18	17 38
10	00 58	00 27	02 40	01 27	02 46	00 10	01 18	17 38
13	00 S 30	00 N 10	02 34	01 27	02 46	00 10	01 18	17 39
16	00 N 01	00 S 05	02 26	01 27	02 46	00 10	01 18	17 39
19	00 34	00 20	02 19	01 27	02 46	00 10	01 18	17 39
22	01 06	00 34	02 11	01 27	02 45	00 10	01 18	17 38
25	01 36	00 47	02 03	01 27	02 45	00 10	01 18	17 38
28	02 01	00 59	01 55	01 27	02 45	00 10	01 18	17 38
31	02 N 20	01 S 10	01 N 47	01 N 27	02 N 45	00 N 10	01 N 19	17 N 37

DATA

Julian Date	2445061
Delta T	+52 seconds
Ayanamsa	23° 36′ 16″
Synetic vernal point	05° ♓ 30′ 43″
True obliquity of ecliptic	23° 26′ 27″

MOON'S PHASES, APSIDES AND POSITIONS ☽

Date	h	m	Phase	Longitude ° ′	Eclipse Indicator
01	05	08	☽	11 ♋ 09	
08	10	18	○	18 ♎ 15	
16	12	42	☾	26 ♑ 12	
23	20	29	●	03 ♉ 21	
30	12	07	☽	09 ♌ 49	

Day	h	m	
14	00	07	Apogee
25	21	11	Perigee

Day	h	m		
01	06	06	Max dec	22° N 19′
07	21	48	0S	
15	07	35	Max dec	22° S 28′
22	08	24	0N	
28	12	22	Max dec	22° N 34′

LONGITUDES

Date	Chiron ⚷ ° ′	Ceres ⚳ ° ′	Pallas ⚴ ° ′	Juno ⚵ ° ′	Vesta ⚶ ° ′	Black Moon Lilith ⚸ ° ′
01	19 ♉ 56	26 ♏ 04	11 ♎ 58	07 ♑ 43	02 ♒ 33	21 ♐ 06
11	20 ♉ 32	25 ♏ 06	08 ♎ 54	08 ♑ 56	06 ♒ 41	22 ♐ 12
21	21 ♉ 10	23 ♏ 32	06 ♎ 12	09 ♑ 42	10 ♒ 33	23 ♐ 19
31	21 ♉ 50	21 ♏ 31	04 ♎ 08	09 ♑ 56	14 ♒ 04	24 ♐ 26

ASPECTARIAN

h m	Aspects
01 Thursday	
00 28	☽ △ ♃
02 30	☽ bQ ♀
03 20	☽ □ ♂
03 52	☽ ± ♅
05 08	☽ □ ☉
07 21	☿ ⊥ ♀
11 19	♀ bQ ♂
16 34	☿ ∥ ♂
19 00	☽ ⋕ ♆
19 06	☽ ± ♀
19 17	☽ bQ ♅
19 25	☽ □ ♄
02 Friday	
02 33	☿ ± ♃
06 14	☽ ⊼ ♀
06 33	☽ □ ♇
08 29	☽ ⊼ ♆
09 28	☽ Q ♂
10 36	♀ △ ♇
17 04	☽ ⋕ ♅
18 53	☽ ± ♆
20 01	☽ △ ☿
20 02	☉ ⋕ ♄
21 13	☽ △ ♅
03 Saturday	
02 27	☽ Q ♄
03 53	☽ □ ♃
04 36	☿ △ ♅
05 46	☽ ⚹ ♂
05 59	♄ ∠ ♅
10 36	☽ bQ ♆
13 04	☽ △ ☉
13 32	♀ ⚹ ♆
13 51	☽ Q ♇
23 23	☽ ⚹ ♄
04 Sunday	
02 35	☽ bQ ☿
04 14	☿ ⋕ ♂
07 26	☽ ∠ ♂
10 58	☽ ⚹ ♇
11 18	☽ Q ♃
13 02	☽ △ ♆
14 55	☽ ☍ ♀
17 38	☽ bQ ☉
21 13	☽ ± ☿
22 58	☽ ⊥ ♂
05 Monday	
01 37	☿ ⊼ ♃
01 50	☽ ∠ ♄
02 01	☽ □ ♅
04 44	☽ ⋕ ♃
07 35	☿ ☍ ♂
08 31	☽ ⚹ ♃
09 24	☽ ⊻ ♂
09 46	☽ ⊼ ☿
10 58	☽ ± ☉
13 40	☽ ∠ ♇
15 07	☽ ⋕ ♀
17 53	☽ ⊥ ♄
22 38	☽ ⊼ ☉
06 Tuesday	
04 39	☽ ⊻ ♄
05 53	☽ ⊥ ♇
10 23	☽ Q ♅
11 22	☽ ∠ ♃
13 56	☽ ∥ ♇
14 02	☽ ∥ ☉
15 14	☉ ∥ ♇
16 46	☽ ⊻ ♇
19 00	☽ □ ♆
21 43	☽ ⋕ ♄
07 Wednesday	
01 28	☽ ⊼ ♀
03 41	☽ ⊥ ♃
04 36	☽ ∥ ☿
05 10	♂ ⊻ ♃
08 18	☽ ⚹ ♅
13 27	☽ ± ♀
14 28	☽ ☌ ♂
14 39	☽ ⊻ ♃
19 38	☽ ⋕ ♂
23 12	♀ ± ♂
23 53	☽ ∥ ♂
08 Thursday	
02 23	☽ ☍ ☿
04 15	☽ Q ♆
07 39	☽ bQ ♀
10 18	☽ ☍ ☉
11 38	☽ ☌ ♄
12 11	☽ ∠ ♅
16 38	☿ ⋕ ♄
22 01	☽ ∥ ♄
23 20	☽ ⋕ ☿
09 Friday	
00 26	☽ ☌ ♇
02 30	☉ ☍ ♄
02 51	☽ ⚹ ♆
05 11	☽ ⊥ ♅
06 33	☽ ⋕ ♇
10 05	☉ bQ ♅
12 04	☽ ⋕ ☉
14 36	☽ △ ♀
16 38	☽ ⊻ ♅
21 32	☽ ⊻ ♂
22 51	☽ ☌ ♃
23 56	☽ ∥ ♀
10 Saturday	
06 10	♀ bQ ♄
07 41	☽ ∠ ♆
08 49	☽ ⊥ ♂
10 43	☿ ☍ ♄
12 22	☿ ∥ ♇
14 44	♀ □ ♅
15 10	☿ bQ ♅
15 37	☿ ∠ ♀
16 36	☽ ∥ ♃
20 55	☽ ⊻ ♄
23 04	☽ ⊼ ☿
11 Sunday	
00 55	☽ ⊼ ☉
01 17	☽ ⊥ ♆
01 57	☽ ∠ ♂
08 40	☽ ⊥ ♄
10 31	☽ ⊻ ♇
13 08	☽ ⊻ ♆
13 22	☽ ± ☿
13 50	☽ ± ☉
17 58	☉ ☌ ☿
22 25	☽ ⊥ ♇
12 Monday	
02 28	☽ ∠ ♄
03 25	☽ ☌ ♅
06 27	♀ ⊼ ♂
06 55	☽ ⚹ ♂
06 59	☽ □ ♀
09 24	☽ bQ ☉
09 26	☽ ⊻ ♃
11 08	☽ bQ ☿
16 26	☽ ∠ ♇
21 24	☽ ⊥ ♃
13 Tuesday	
00 27	☉ ⋕ ♀
00 49	☿ ⋕ ♀
01 15	☉ ∥ ☿
06 26	☽ Q ♂
08 25	♀ △ ♃
08 30	☽ ⚹ ♄
11 49	☽ ∥ ♅
15 27	☽ ∠ ♃
16 56	☿ ☍ ♇
18 27	☽ △ ☉
22 45	☽ ⚹ ♇
22 54	☽ Q ♀
14 Wednesday	
00 01	☽ △ ☿
01 32	☽ ☌ ♆
08 31	☿ △ ♆
08 44	☽ Q ♄
11 26	☽ ∥ ♆
16 00	☽ ⊻ ♅
18 00	☽ □ ♂
21 05	☿ ± ♅
21 38	☽ ⚹ ♃
23 06	☽ Q ♇
15 Thursday	
01 41	☽ ⚹ ♀
04 10	☽ ⊥ ♅
04 21	♀ Q ♆
20 52	☉ ☍ ♇
21 02	☽ □ ♄
21 42	☽ Q ♃
22 21	☽ ∠ ♅
16 Friday	
03 40	☽ ∥ ♆
10 57	☽ ∠ ♀
11 23	☽ □ ♇
12 42	☽ □ ☉
14 14	☽ ⊻ ♆
15 41	♀ bQ ♇
20 57	☿ ⋕ ♃
17 Saturday	
01 40	☽ □ ☿
02 10	☽ ⊥ ♆
03 03	☽ ∥ ♅
04 16	☽ ⚹ ♅
04 48	☽ △ ♂
06 35	☽ ⊥ ♀
07 13	☉ △ ♆
09 16	☽ □ ♃
16 25	☿ ⊼ ♅
17 51	☿ ⊼ ♂
19 28	☽ ⊻ ♀
19 50	☽ ∠ ♆
18 Sunday	
03 15	☽ Q ☉
03 40	☽ Q ♅
05 08	♂ ⚹ ♅
05 48	♀ ± ♄
07 58	☽ △ ♄
08 27	☉ ± ♅
09 21	☽ bQ ♂
19 37	☿ ☍ ♃
19 45	☽ Q ☿
21 48	☽ △ ♇
19 Monday	
00 34	☽ ⚹ ♆
02 04	☽ ± ♂
04 02	☽ ⚹ ☉
07 39	☽ ⋕ ☿
12 09	☽ bQ ♄
13 01	☽ ⊼ ♂
13 40	☽ □ ♅
17 56	☽ △ ♃
21 03	☽ ∥ ♃
22 25	☽ ⚹ ☿
22 50	☽ Q ♆
22 53	♀ ⋕ ♇
20 Tuesday	
01 35	☽ bQ ♇
01 54	☽ ⋕ ☉
04 34	☽ ± ♄
06 31	☿ ± ♂
08 57	☽ ☌ ♀
09 50	☽ ∠ ☉
15 18	☽ ⊼ ♄
17 46	☽ ± ♇
20 47	☽ bQ ♃
21 Wednesday	
01 59	☽ ⋕ ♇
03 06	☽ ⊥ ☉
04 02	☽ ∥ ♀
04 21	☽ ⊼ ♇
06 07	☽ ∠ ☿
06 59	☽ □ ♆
12 01	☿ bQ ♆
12 02	☽ ∥ ♄
12 26	☽ ± ♃
14 22	☽ ⊻ ☉
17 34	☽ ☍ ♂
19 04	☽ △ ♅
22 42	☽ ⊼ ♃
22 Thursday	
00 21	☽ ⊥ ☿
04 32	☽ ⋕ ♂
08 19	☉ ⋕ ♃
12 13	☽ ⊻ ☿
12 19	☽ ∥ ♂
17 27	☽ ⊻ ♀
18 51	☽ ☍ ♄
20 27	☽ bQ ♅
23 Friday	
03 15	☉ ⊼ ♂
04 05	☽ ⋕ ♄
04 17	☽ ⊥ ♀
07 17	☽ ☍ ♇
08 26	☽ ⋕ ♀
09 50	☽ △ ♆
10 56	♀ ⊼ ♄
11 22	☽ ± ♅
13 53	☽ ∥ ♇
19 04	☽ ⊼ ♂
20 19	☽ ∠ ♀
20 29	☽ ☌ ☉
21 13	☽ ⊼ ♅
24 Saturday	
00 21	☽ ☍ ♃
04 44	☽ ± ♂
06 26	☿ bQ ♂
07 15	☉ ⊼ ♅
10 24	☽ bQ ♆
11 06	☿ ⊼ ♄
15 01	☉ ∠ ♀
15 04	☽ ⋕ ♃
19 10	☽ bQ ♂
19 22	♀ ± ♇
19 41	☽ ∥ ☉
19 55	☽ ⊼ ♄
21 16	☽ ☌ ☿
22 43	☽ ⚹ ♀
25 Sunday	
01 00	☽ ± ♆
05 36	☽ ± ♄
08 10	☽ ⊼ ♇
10 42	☽ ⊼ ♆
17 52	☽ ± ♇
19 07	☽ △ ♂
19 42	☽ Q ♀
20 06	☽ bQ ♄
21 17	♀ bQ ♃
21 51	☽ ☍ ♅
22 32	☿ ⚹ ♀
26 Monday	
00 28	☉ ☍ ♃
00 39	☽ ⊼ ♃
00 39	☽ ⊻ ☉
02 35	☿ ± ♆
06 36	♀ ∥ ♄
08 26	☽ bQ ♇
10 18	☽ ± ♃
11 05	☽ ⊥ ☉
15 21	☽ ∥ ☿
20 21	☽ △ ♄
22 54	☽ ⋕ ♅
27 Tuesday	
00 50	☽ bQ ♃
02 48	☽ ∠ ☉
03 19	☽ □ ♀
04 56	☽ ⊻ ☿
08 52	☽ △ ♇
11 29	☽ ☍ ♆
14 30	☿ ± ♄
16 09	☽ ⊥ ☿
17 24	☽ ⋕ ♆
19 27	☽ □ ♂
22 46	☽ ⊼ ♅
28 Wednesday	
00 08	☿ ⋕ ♅
01 19	☽ △ ♃
04 48	☉ ± ♂
05 19	☽ ⚹ ☉
08 43	☽ ± ♅
09 06	☽ ∠ ☿
13 37	☿ ⊼ ♇
21 48	☽ □ ♄
23 47	☽ bQ ♅
29 Thursday	
01 10	☽ Q ♂
02 55	☽ Q ☉
07 37	☽ ⋕ ♆
09 27	☽ △ ♀
10 54	☽ □ ♇
12 36	☿ ⊼ ♆
13 40	☽ ⊼ ♆
13 48	☽ ⚹ ☿
16 10	☽ ∥ ☿
21 23	☽ ⚹ ♂
23 56	☽ ± ♆
30 Friday	
01 19	☽ △ ♅
03 05	☽ ⋕ ♅
03 29	♀ ⊼ ♇
03 39	☽ □ ♃
04 22	☽ Q ♄
12 07	☽ □ ☉
13 12	☽ Q ☿
13 30	☽ bQ ♀
15 32	☽ bQ ♆
17 54	☽ Q ♇
18 11	☿ ⋕ ♆
23 10	☽ ∠ ♂

All ephemeris data is given at 12.00 UT and the Moon's longitude is additionally given for 24.00 UT

MAY 1982

LONGITUDES

Date	Sidereal time h m s	Sun ☉ ° ′ ″	Moon ☽ ° ′ ″	Moon ☽ 24.00 ° ′ ″	Mercury ☿ ° ′	Venus ♀ ° ′	Mars ♂ ° ′	Jupiter ♃ ° ′	Saturn ♄ ° ′	Uranus ♅ ° ′	Neptune ♆ ° ′	Pluto ♇ ° ′
01	02 36 22	10 ♉ 47 10	23 ♌ 24 31	00 ♍ 08 25	29 ♉ 54	26 ♓ 39	01 ♎ 04	04 ♏ 46	17 ♎ 16	03 ♐ 32	26 ♐ 46	25 ♎ 07
02	02 40 19	11 45 23	06 ♍ 48 37	13 ♍ 25 18	01 ♊ 23	27 45	00 R 56	04 R 38	17 R 12	03 R 30	26 R 45	25 R 05
03	02 44 15	12 43 34	19 ♍ 58 36	26 ♍ 28 41	02 48	28 52	00 49	04 30	17 08	03 27	26 44	25 04
04	02 48 12	13 41 43	02 ♎ 55 40	09 ♎ 19 37	04 08	29 ♓ 59	00 43	04 23	17 04	03 25	26 43	25 02
05	02 52 08	14 39 50	15 ♎ 40 39	21 ♎ 58 49	05 24	01 ♈ 06	00 38	04 16	17 00	03 23	26 42	25 01
06	02 56 05	15 37 55	28 ♎ 14 10	04 ♏ 26 47	06 36	02 13	00 34	04 08	16 56	03 20	26 40	24 59
07	03 00 02	16 35 58	10 ♏ 36 44	16 ♏ 44 05	07 43	03 20	00 30	04 01	16 53	03 18	26 39	24 57
08	03 03 58	17 34 00	22 ♏ 48 57	28 ♏ 51 29	08 46	04 27	00 27	03 53	16 49	03 16	26 38	24 56
09	03 07 55	18 32 00	04 ♐ 51 51	10 ♐ 50 16	09 44	05 35	00 25	03 46	16 45	03 13	26 37	24 54
10	03 11 51	19 29 59	16 ♐ 47 00	22 ♐ 42 21	10 38	06 42	00 23	03 39	16 42	03 11	26 36	24 53
11	03 15 48	20 27 56	28 ♐ 36 40	04 ♑ 30 23	11 27	07 50	00 23	03 32	16 38	03 08	26 35	24 51
12	03 19 44	21 25 52	10 ♑ 23 55	16 ♑ 17 48	12 11	08 58	00 D 23	03 25	16 35	03 06	26 33	24 50
13	03 23 41	22 23 46	22 ♑ 12 33	28 ♑ 08 45	12 51	10 05	00 24	03 18	16 31	03 04	26 32	24 48
14	03 27 37	23 21 40	04 ♒ 07 00	10 ♒ 07 56	13 25	11 13	00 26	03 11	16 28	03 01	26 31	24 47
15	03 31 34	24 19 31	16 ♒ 12 11	22 ♒ 20 24	13 55	12 22	00 28	03 04	16 25	02 59	26 30	24 46
16	03 35 31	25 17 22	28 ♒ 33 13	04 ♓ 51 14	14 20	13 30	00 31	02 58	16 22	02 56	26 28	24 44
17	03 39 27	26 15 12	11 ♓ 15 01	17 ♓ 45 04	14 40	14 38	00 35	02 51	16 19	02 54	26 27	24 43
18	03 43 24	27 13 00	24 ♓ 21 48	01 ♈ 05 30	14 55	15 46	00 39	02 45	16 16	02 51	26 26	24 41
19	03 47 20	28 10 47	07 ♈ 56 19	14 ♈ 54 16	15 05	16 55	00 45	02 38	16 13	02 49	26 24	24 40
20	03 51 17	29 ♉ 08 33	21 ♈ 59 09	29 ♈ 10 37	15 10	18 04	00 51	02 32	16 10	02 46	26 23	24 39
21	03 55 13	00 ♊ 06 18	06 ♉ 28 05	13 ♉ 50 48	15 R 10	19 12	00 57	02 26	16 08	02 44	26 21	24 37
22	03 59 10	01 04 02	21 ♉ 17 51	28 ♉ 48 10	15 06	20 21	01 05	02 20	16 05	02 41	26 20	24 36
23	04 03 06	02 01 44	06 ♊ 20 35	13 ♊ 53 55	14 57	21 30	01 13	02 14	16 03	02 39	26 19	24 35
24	04 07 03	02 59 25	21 ♊ 26 57	28 ♊ 58 33	14 44	22 39	01 21	02 08	16 00	02 36	26 17	24 34
25	04 11 00	03 57 05	06 ♋ 27 40	13 ♋ 52 17	14 27	23 48	01 31	02 03	15 58	02 34	26 16	24 32
26	04 14 56	04 54 44	21 ♋ 14 53	28 ♋ 31 38	14 07	24 57	01 40	01 57	15 56	02 31	26 14	24 31
27	04 18 53	05 52 21	05 ♌ 43 10	12 ♌ 49 12	13 43	26 06	01 51	01 52	15 54	02 29	26 13	24 30
28	04 22 49	06 49 57	19 ♌ 49 36	26 ♌ 44 20	13 16	27 15	02 01	01 47	15 51	02 26	26 11	24 29
29	04 26 46	07 47 31	03 ♍ 33 30	10 ♍ 17 27	12 47	28 25	02 12	01 42	15 49	02 24	26 10	24 28
30	04 30 42	08 45 03	16 ♍ 55 53	23 ♍ 29 36	12 15	29 ♈ 34	02 24	01 37	15 48	02 21	26 08	24 27
31	04 34 39	09 ♊ 42 34	29 ♍ 58 43	06 ♎ 23 34	11 ♊ 43	00 ♉ 43	02 ♎ 39	01 ♏ 32	15 ♎ 46	02 ♐ 19	26 ♐ 07	24 ♎ 26

Date	Moon True ☊ ° ′	Moon Mean ☊ ° ′	Moon ☽ Latitude ° ′
01	15 ♋ 45	16 ♋ 48	03 N 07
02	15 R 42	16 45	03 57
03	15 37	16 42	04 34
04	15 30	16 39	04 57
05	15 20	16 36	05 04
06	15 09	16 33	04 56
07	14 59	16 29	04 35
08	14 49	16 26	04 01
09	14 40	16 23	03 16
10	14 34	16 20	02 23
11	14 31	16 17	01 24
12	14 29	16 14	00 N 22
13	14 D 29	16 10	00 S 42
14	14 31	16 07	01 44
15	14 32	16 04	02 43
16	14 R 32	16 01	03 35
17	14 31	15 58	04 19
18	14 28	15 54	04 50
19	14 23	15 51	05 07
20	14 17	15 48	05 06
21	14 10	15 45	04 46
22	14 03	15 42	04 07
23	13 58	15 39	03 09
24	13 54	15 35	01 59
25	13 53	15 32	00 S 40
26	13 D 53	15 29	00 N 40
27	13 54	15 26	01 56
28	13 55	15 23	03 03
29	13 56	15 20	03 58
30	13 R 56	15 16	04 38
31	13 ♋ 54	15 ♋ 13	05 N 03

DECLINATIONS

Date	Sun ☉ ° ′	Moon ☽ ° ′	Mercury ☿ ° ′	Venus ♀ ° ′	Mars ♂ ° ′	Jupiter ♃ ° ′	Saturn ♄ ° ′	Uranus ♅ ° ′	Neptune ♆ ° ′	Pluto ♇ ° ′
01	15 N 04	16 N 39	22 N 24	02 S 24	01 N 13	11 S 45	04 S 15	20 S 42	22 S 06	06 N 43
02	15 22	12 41	22 48	02 00	01 13	11 42	04 13	20 42	22 05	06 43
03	15 40	08 10	23 09	01 37	01 14	11 40	04 12	20 41	22 05	06 43
04	15 57	03 N 22	23 27	01 13	01 14	11 38	04 10	20 41	22 05	06 44
05	16 14	01 S 30	23 43	00 50	01 13	11 35	04 09	20 40	22 05	06 44
06	16 31	06 14	23 57	00 26	01 13	11 33	04 08	20 40	22 05	06 44
07	16 48	10 39	24 08	00 S 02	01 12	11 30	04 06	20 39	22 05	06 45
08	17 04	14 36	24 18	00 N 23	01 11	11 28	04 05	20 39	22 05	06 45
09	17 21	17 54	24 25	00 47	01 09	11 26	04 04	20 39	22 05	06 45
10	17 36	20 25	24 30	01 11	01 07	11 23	04 03	20 38	22 05	06 46
11	17 52	22 02	24 33	01 36	01 05	11 21	04 02	20 38	22 05	06 46
12	18 07	22 40	24 34	02 00	01 03	11 19	04 00	20 37	22 05	06 46
13	18 22	22 18	24 33	02 25	01 00	11 17	03 59	20 37	22 05	06 46
14	18 37	20 55	24 31	02 50	00 57	11 15	03 58	20 36	22 05	06 47
15	18 51	18 34	24 26	03 14	00 54	11 12	03 57	20 36	22 05	06 47
16	19 05	15 20	24 20	03 39	00 51	11 10	03 56	20 35	22 05	06 47
17	19 19	11 20	24 13	04 04	00 47	11 08	03 55	20 35	22 05	06 47
18	19 32	06 41	24 03	04 29	00 43	11 06	03 54	20 34	22 05	06 48
19	19 45	01 S 33	23 52	04 53	00 39	11 04	03 53	20 34	22 05	06 48
20	19 58	03 N 50	23 40	05 18	00 35	11 02	03 52	20 33	22 05	06 48
21	20 10	09 10	23 26	05 43	00 30	11 00	03 52	20 33	22 05	06 48
22	20 22	14 07	23 11	06 08	00 25	10 58	03 51	20 32	22 04	06 48
23	20 34	18 15	22 55	06 32	00 20	10 57	03 50	20 32	22 04	06 48
24	20 45	21 11	22 37	06 57	00 15	10 55	03 49	20 32	22 04	06 48
25	20 56	22 37	22 19	07 22	00 09	10 53	03 49	20 31	22 04	06 48
26	21 07	22 25	21 59	07 46	00 N 03	10 51	03 48	20 31	22 04	06 48
27	21 17	20 43	21 39	08 10	00 S 03	10 50	03 47	20 30	22 04	06 48
28	21 27	17 46	21 19	08 35	00 10	10 48	03 47	20 30	22 04	06 48
29	21 37	13 54	20 58	08 59	00 16	10 47	03 46	20 29	22 04	06 48
30	21 46	09 26	20 36	09 23	00 23	10 45	03 46	20 29	22 04	06 48
31	21 N 54	04 N 38	20 N 15	09 N 47	00 S 30	10 S 44	03 S 45	20 S 28	22 S 04	06 N 48

ZODIAC SIGN ENTRIES

Date	h m	Planets
01	13 29	☿ ♊
01	23 45	☽ ♍
04	06 32	☽ ♎
04	12 27	♀ ♈
06	15 24	☽ ♏
09	02 17	☽ ♐
11	14 50	☽ ♑
14	03 44	☽ ♒
16	14 46	☽ ♓
18	22 04	☽ ♈
21	01 22	☽ ♉
21	09 23	☉ ♊
23	01 54	☽ ♊
25	01 38	☽ ♋
27	02 27	☽ ♌
29	05 43	☽ ♍
30	21 02	♀ ♉
31	12 02	☽ ♎

LATITUDES

Date	Mercury ☿ ° ′	Venus ♀ ° ′	Mars ♂ ° ′	Jupiter ♃ ° ′	Saturn ♄ ° ′	Uranus ♅ ° ′	Neptune ♆ ° ′	Pluto ♇ ° ′
01	02 N 20	01 S 10	01 N 47	01 N 27	02 N 45	00 N 10	01 N 19	17 N 37
04	02 31	01 19	01 39	01 26	02 44	00 10	01 19	17 37
07	02 35	01 28	01 35	01 26	02 44	00 10	01 19	17 36
10	02 29	01 36	01 24	01 26	02 43	00 10	01 19	17 35
13	02 14	01 43	01 16	01 25	02 43	00 10	01 19	17 35
16	01 50	01 49	01 09	01 25	02 42	00 10	01 19	17 34
19	01 16	01 54	01 01	01 24	02 42	00 10	01 19	17 33
22	00 N 35	01 58	00 55	01 23	02 41	00 10	01 19	17 32
25	00 S 14	02 01	00 49	01 23	02 41	00 10	01 19	17 31
28	01 06	02 03	00 42	01 22	02 40	00 10	01 19	17 29
31	01 S 57	02 S 05	00 N 36	01 N 22	02 N 39	00 N 10	01 N 19	17 N 28

DATA

Julian Date	2445091
Delta T	+52 seconds
Ayanamsa	23° 36′ 19″
Synetic vernal point	05° ♓ 30′ 40″
True obliquity of ecliptic	23° 26′ 27″

MOON'S PHASES, APSIDES AND POSITIONS ☽

Date	h m	Phase	Longitude ° ′	Eclipse Indicator
08	00 45	○	17 ♏ 07	
16	05 11	☾	25 ♒ 01	
23	04 40	●	01 ♊ 44	
29	20 07	☽	08 ♍ 07	

Day	h m		
11	15 03	Apogee	
24	02 40	Perigee	
05	04 37	0S	
12	15 12	Max dec	22° S 41′
19	18 59	0N	
25	21 06	Max dec	22° N 44′

LONGITUDES

Date	Chiron ⚷ ° ′	Ceres ⚳ ° ′	Pallas ⚴ ° ′	Juno ⚵ ° ′	Vesta ⚶ ° ′	Black Moon Lilith ⚸ ° ′
01	21 ♉ 50	21 ♏ 31	04 ♎ 08	09 ♑ 56	14 ♒ 04	24 ♐ 26
11	22 ♉ 32	19 ♏ 18	02 ♎ 51	09 ♑ 37	17 ♒ 13	25 ♐ 32
21	23 ♉ 13	17 ♏ 07	02 ♎ 23	08 ♑ 45	19 ♒ 54	26 ♐ 39
31	23 ♉ 55	15 ♏ 13	02 ♎ 41	07 ♑ 19	22 ♒ 42	27 ♐ 45

All ephemeris data is given at 12.00 UT and the Moon's longitude is additionally given for 24.00 UT

ASPECTARIAN

h m Aspects

01 Saturday
01 12 ☽ ✶ ♄
06 41 ☽ ± ♀
10 52 ☽ Q ♃
14 24 ♀ □ ♆
14 55 ☽ ⊥ ♂
15 01 ☽ ✶ ♇
17 57 ☽ △ ♆
18 17 ☽ ⊼ ♀
21 23 ☽ ∥ ☉

02 Sunday
01 02 ☽ □ ☿
01 31 ☽ ⊻ ♂
03 44 ☽ ∠ ♄
05 10 ☿ △ ♂
06 02 ☽ □ ♅
07 11 ☿ ± ♇
08 07 ☽ ✶ ♃
11 42 ☉ bQ ♆
17 27 ☽ # ♃
17 55 ☽ ∠ ♇
19 55 ☽ ⊥ ♄
21 41 ☽ △ ☉

03 Monday
01 03 ☿ bQ ♄
05 02 ♀ ± ♃
06 48 ☽ ⊻ ♄
10 19 ☽ ⊥ ♇
11 09 ☽ ∠ ♃
14 43 ☽ Q ♅
19 21 ☽ ∥ ♇
21 21 ☽ ⊻ ♇
23 20 ☿ ☍ ♅

04 Tuesday
00 27 ☽ □ ♆
03 28 ☽ bQ ☉
03 37 ☽ ⊥ ♃
05 59 ☽ ☍ ♀
07 55 ☽ ☌ ♂
08 02 ☽ # ♄
11 41 ♀ # ♂
12 55 ☽ ✶ ♅
14 31 ☽ △ ☿
14 42 ☽ ⊻ ♃
16 12 ☿ ⊼ ♃
21 40 ☽ ± ☉
22 34 ☽ ∥ ♂
23 32 ☽ # ♀

05 Wednesday
02 45 ♀ ☍ ♂
08 57 ☽ ∥ ♀
09 55 ☽ ⊼ ☉
10 08 ☽ Q ♆
10 40 ☽ # ♂
14 30 ☽ ☌ ♄
17 07 ☽ ∠ ♅
21 57 ☽ bQ ☿

06 Thursday
01 15 ☽ ∥ ♄
05 46 ☽ ☌ ♇
09 00 ☽ ✶ ♆
10 17 ☽ ⊥ ♅
10 17 ☉ bQ ♂
14 40 ☽ # ♇
16 27 ☽ ⊻ ♂
17 01 ☽ ± ☿
20 26 ☽ ⊼ ♀
21 49 ☽ ⊻ ♅
23 17 ☽ ☌ ♃

07 Friday
04 01 ☽ ⊥ ♂
05 49 ☽ ⊼ ☿
09 15 ☽ ± ♀
11 21 ♀ △ ♅
14 02 ☽ ∠ ♆
16 52 ☽ ∥ ♃
18 28 ☉ ⊼ ♄
21 32 ☽ ∠ ♂

08 Saturday
00 13 ☽ ⊻ ♄
00 45 ☽ ☍ ☉
01 10 ♀ ⊼ ♃
04 41 ☽ bQ ♀
07 42 ☽ ⊥ ♆
12 00 ☽ ⊥ ♄
16 11 ☽ ⊻ ♇
19 34 ☽ ⊻ ♆

09 Sunday
03 07 ☽ ✶ ♂
04 06 ☽ ⊥ ♇
05 49 ☽ ∠ ♄
07 10 ☽ # ☉
08 43 ☽ ☌ ♅
09 50 ☽ ⊻ ♃
12 43 ☿ ± ♃
13 34 ☽ △ ♀
16 13 ☿ bQ ♇
21 45 ☽ ⊥ ♃
22 06 ☽ ∠ ♇
22 36 ☽ ☍ ☿

10 Monday
03 08 ☽ Q ♂
08 28 ♀ ∥ ♂
11 50 ☽ ✶ ♄
14 38 ☽ ∥ ♅
15 45 ☽ ∠ ♃
17 59 ☽ ⊼ ☉

11 Tuesday
04 23 ☽ ✶ ♇
07 15 ☽ ± ☉
07 52 ☽ ☌ ♆
12 03 ☽ Q ♄
13 13 ☽ ∥ ♆
14 42 ☉ ± ♆
15 36 ☽ □ ♂
18 36 ♂ St D
21 11 ☽ ⊻ ♅
21 55 ☽ ✶ ♃

12 Wednesday
03 12 ☽ bQ ☉
04 45 ☽ Q ♇
08 45 ☽ □ ♀
09 22 ☽ ⊥ ♅
15 52 ☽ ⊼ ☿
22 07 ☽ Q ♃

13 Thursday
00 31 ☽ □ ♄
03 37 ☽ ∠ ♅
04 47 ☽ ± ☿
12 25 ☽ △ ☉
15 06 ☉ ± ♄
17 07 ☽ ∥ ♆
17 15 ☽ □ ♇
20 44 ☽ ⊻ ♆

14 Friday
00 00 ☽ bQ ☿
01 08 ☽ Q ♀
04 34 ☽ △ ♂
08 48 ☽ ⊥ ♆
09 49 ☽ ✶ ♅
10 09 ☽ □ ♃
15 48 ☽ ∥ ♅

15 Saturday
02 43 ☽ ∠ ♆
03 38 ☽ ✶ ♀
07 19 ☽ △ ☿
09 36 ☽ Q ♅
09 42 ☽ # ☉
10 33 ☽ bQ ♂
12 26 ☽ △ ♄
22 33 ☉ ⊼ ♇

16 Sunday
04 11 ☽ ± ♂
04 40 ☽ △ ♇
05 11 ☽ □ ☉
08 00 ☽ ✶ ♆
11 53 ☽ ∠ ♀
15 47 ☽ ⊼ ♂
17 22 ☽ bQ ♄
20 21 ☽ □ ♅
20 21 ☽ △ ♃
20 35 ♃ ⊻ ♅

17 Monday
03 55 ♀ # ♄
06 38 ☽ ⊥ ♀
06 47 ☽ Q ♆
09 08 ☽ bQ ♇
10 16 ☽ ± ♄
12 45 ☿ ✶ ♀
13 04 ☽ ∥ ♃
16 45 ☉ ⊼ ♆
18 01 ☽ Q ☉
18 28 ☽ □ ☿
18 52 ☽ ⊻ ♀
21 20 ☽ ⊼ ♄

18 Tuesday
00 05 ☽ bQ ♃
01 44 ☽ ± ♇
11 27 ☽ # ♇
12 35 ☽ ⊼ ♇
15 42 ☽ □ ♆
16 14 ☽ ± ♃
17 31 ☽ ✶ ☉
21 44 ☽ # ♀
22 01 ♀ ☍ ♄
23 18 ☽ ☍ ♂

19 Wednesday
01 15 ☽ ∥ ♄
02 48 ☽ ⊼ ♃
03 04 ☽ △ ♅
03 25 ☽ Q ☿
16 08 ☽ # ♂
21 43 ☽ ∠ ☉
21 46 ☽ ∥ ♂

20 Thursday
00 23 ☽ ✶ ☿
02 12 ☽ ☍ ♄
04 47 ☽ ☌ ♀
04 54 ☽ bQ ♅
06 12 ♀ bQ ♅
12 12 ☽ # ♄
14 05 ☽ ⊥ ☉
16 27 ☽ ☍ ♇
19 06 ☽ ∥ ♀
19 21 ☽ △ ♆
19 59 ☽ ± ♅

21 Friday
00 48 ☽ ⊻ ☉
01 13 ☽ ∥ ♇
01 39 ☽ ∠ ☿
02 06 ☿ St R
02 52 ☽ ⊼ ♂
05 25 ☽ ☍ ♃
05 53 ☽ ⊼ ♅
12 48 ☽ ± ♂
16 24 ☽ ⊥ ☿
19 57 ☽ bQ ♆
20 32 ☽ # ♃

22 Saturday
00 40 ☉ ± ♇
02 05 ☽ ⊻ ☿
03 32 ☽ bQ ♂
03 39 ☽ ⊼ ♄
10 21 ☽ ⊻ ♀
10 27 ☽ ± ♆
12 18 ☉ △ ♂
12 30 ☉ bQ ♄
13 16 ☽ ± ♄
17 17 ☽ ⊼ ♇
20 03 ☽ ⊼ ♆
20 45 ☽ ⊥ ♀

23 Sunday
02 51 ☽ ± ♇
03 36 ☽ bQ ♄
03 46 ☽ △ ♂
04 40 ☽ ☌ ☉
05 31 ☽ ⊼ ♃
06 08 ☽ ☍ ♅
07 43 ☉ # ♅
12 16 ☽ ∠ ♀
14 59 ☽ ± ♃
16 41 ☉ ⊼ ♃
17 08 ☽ bQ ♇

24 Monday
01 30 ☽ ☌ ☿
02 48 ☉ ☍ ♅
03 22 ☽ △ ♄
03 23 ♀ ∥ ♇
05 11 ☽ bQ ♃
05 27 ☽ # ♅
07 15 ☽ ∥ ☉
14 04 ☽ ✶ ♀
16 57 ☽ △ ♇
19 42 ☽ ☍ ♆
23 32 ☽ # ♆

25 Tuesday
03 58 ☽ □ ♂
04 57 ☽ △ ♃
05 40 ☽ ∥ ☿
05 46 ☽ ⊼ ♅
07 41 ☽ ⊻ ☉
10 50 ☽ Q ♀
15 23 ☽ ± ♅
18 01 ☽ ⊥ ☉

26 Wednesday
00 38 ☽ ⊻ ☿
03 16 ♀ ☍ ♇
03 20 ☽ □ ♄
05 54 ☿ # ♆
05 56 ☽ bQ ♅
09 24 ☽ Q ♂
09 40 ☽ ∠ ☉
10 11 ☽ ⊥ ☿
17 22 ☽ □ ♇
18 36 ☽ □ ♀
19 01 ☽ # ♆
20 12 ☽ ⊼ ♆
22 38 ☽ ∥ ☿

27 Thursday
00 38 ☽ ∠ ☿
05 27 ☽ ✶ ♂
05 35 ☽ □ ♃
06 08 ☽ ± ♆
06 19 ☽ ∥ ☉
06 36 ☽ △ ♅
08 57 ☽ Q ♄
12 17 ☽ ✶ ☉
13 20 ♂ ⊻ ♃
14 10 ☽ # ♅
14 17 ♀ △ ♆
19 40 ♀ ± ♅
21 15 ☽ bQ ♆
23 27 ☽ Q ♇

28 Friday
01 07 ☽ ✶ ☿
05 12 ☽ ✶ ♄
05 15 ☉ ∥ ☿
07 08 ☽ ∠ ♂
10 10 ☽ Q ☉
11 55 ☽ Q ♃
20 03 ☽ ✶ ♇
21 07 ☽ Q ☿
22 56 ☽ ⊥ ♂
23 01 ☽ △ ♆

29 Saturday
02 05 ☽ △ ♀
02 50 ☿ ∠ ♀
07 11 ☽ ∠ ♄
08 43 ☽ ✶ ♃
09 37 ☽ ⊻ ♂
09 45 ☉ ± ♃
09 57 ☽ □ ♅
20 07 ☽ □ ☉
22 30 ☽ ∠ ♇
23 09 ☽ ⊥ ♄

30 Sunday
03 52 ☽ □ ☿
04 18 ♂ ✶ ♅
05 06 ☽ # ♃
07 18 ☽ bQ ♀
09 56 ☽ ⊻ ♄
11 25 ☽ ∠ ♃
12 14 ☽ ∥ ♀
14 45 ☽ ⊥ ♇
18 14 ☽ Q ♅
20 54 ☿ # ♅

31 Monday
01 16 ☽ ∥ ♇
01 19 ☽ ± ♀
01 44 ☽ ⊻ ♇
03 48 ☽ ⊥ ♃
04 50 ☽ □ ♆
05 05 ☉ bQ ♇
13 31 ☽ ⊼ ♀
14 53 ☽ ⊻ ♃
16 21 ☽ ✶ ♅
16 23 ☽ # ♄
17 04 ☽ ☌ ♂

JUNE 1982

Longitudes

Date	Sidereal time h m s	Sun ☉ ° ' "	Moon ☽ ° ' "	Moon ☽ 24.00 ° ' "	Mercury ☿ ° '	Venus ♀ ° '	Mars ♂ ° '	Jupiter ♃ ° '	Saturn ♄ ° '	Uranus ♅ ° '	Neptune ♆ ° '	Pluto ♇ ° '
01	04 38 35	10 ♊ 40 04	12 ♎ 44 27	19 ♎ 01 42	11 ♊ 09	01 ♉ 53	02 ♎ 53	01 ♏ 27	15 ♎ 44	02 ♐ 17	26 ♐ 05	24 ♎ 25
02	04 42 32	11 37 33	25 ♎ 15 36	01 ♏ 26 26	10 R 36	03 02	03 07	01 R 23	15 R 42	02 R 14	26 R 04	24 R 24
03	04 46 29	12 35 00	07 ♏ 34 31	13 ♏ 40 04	10 02	04 12	03 21	01 19	15 41	02 12	26 02	24 23
04	04 50 25	13 32 27	19 ♏ 43 21	25 ♏ 44 37	09 30	05 22	03 36	01 14	15 40	02 09	26 00	24 22
05	04 54 22	14 29 52	01 ♐ 44 05	07 ♐ 41 58	08 59	06 31	03 52	01 10	15 38	02 07	25 59	24 21
06	04 58 18	15 27 16	13 ♐ 38 32	19 ♐ 33 59	08 30	07 41	04 08	01 07	15 37	02 04	25 57	24 20
07	05 02 15	16 24 40	25 ♐ 28 37	01 ♑ 22 40	08 04	08 51	04 25	01 03	15 36	02 02	25 56	24 19
08	05 06 11	17 22 02	07 ♑ 16 27	13 ♑ 10 17	07 40	10 01	04 42	00 59	15 35	02 00	25 54	24 18
09	05 10 08	18 19 24	19 ♑ 04 30	24 ♑ 59 29	07 20	11 11	04 59	00 56	15 34	01 57	25 53	24 17
10	05 14 04	19 16 45	00 ♒ 55 38	06 ♒ 53 23	07 03	12 21	05 17	00 53	15 33	01 55	25 51	24 17
11	05 18 01	20 14 06	12 ♒ 53 12	18 ♒ 55 35	06 50	13 31	05 36	00 50	15 32	01 53	25 49	24 16
12	05 21 58	21 11 26	25 ♒ 01 01	01 ♓ 10 03	06 42	14 42	05 55	00 47	15 32	01 50	25 48	24 15
13	05 25 54	22 08 45	07 ♓ 23 11	13 ♓ 40 58	06 38	15 52	06 14	00 45	15 31	01 48	25 46	24 14
14	05 29 51	23 06 05	20 ♓ 03 54	26 ♓ 32 27	06 D 38	17 02	06 34	00 42	15 31	01 46	25 44	24 14
15	05 33 47	24 03 23	03 ♈ 07 03	09 ♈ 48 03	06 42	18 13	06 54	00 40	15 30	01 43	25 43	24 13
16	05 37 44	25 00 42	16 ♈ 35 42	23 ♈ 30 09	06 52	19 23	07 15	00 38	15 30	01 41	25 41	24 13
17	05 41 40	25 58 00	00 ♉ 31 24	07 ♉ 39 18	07 05	20 34	07 36	00 36	15 30	01 39	25 40	24 12
18	05 45 37	26 55 17	14 ♉ 53 32	22 ♉ 13 36	07 24	21 44	07 58	00 34	15 D 30	01 37	25 38	24 11
19	05 49 33	27 52 35	29 ♉ 38 48	07 ♊ 08 18	07 47	22 55	08 20	00 32	15 30	01 35	25 36	24 11
20	05 53 30	28 49 52	14 ♊ 41 05	22 ♊ 16 04	08 14	24 05	08 42	00 31	15 30	01 32	25 35	24 10
21	05 57 27	29 ♊ 47 09	29 ♊ 52 02	07 ♋ 27 48	08 46	25 16	09 05	00 30	15 30	01 30	25 33	24 10
22	06 01 23	00 ♋ 44 26	15 ♋ 02 08	22 ♋ 33 56	09 23	26 27	09 28	00 29	15 31	01 28	25 32	24 09
23	06 05 20	01 41 42	00 ♌ 01 37	07 ♌ 25 53	10 04	27 38	09 51	00 28	15 31	01 26	25 30	24 09
24	06 09 16	02 38 57	14 ♌ 44 24	21 ♌ 57 08	10 49	28 48	10 15	00 27	15 32	01 24	25 28	24 09
25	06 13 13	03 36 12	29 ♌ 03 40	06 ♍ 03 47	11 38	29 ♉ 59	10 39	00 27	15 32	01 22	25 27	24 08
26	06 17 09	04 33 26	12 ♍ 57 23	19 ♍ 44 31	12 32	01 ♊ 10	11 04	00 26	15 33	01 20	25 25	24 08
27	06 21 06	05 30 40	26 ♍ 25 21	03 ♎ 00 08	13 29	02 21	11 28	00 26	15 34	01 18	25 24	24 08
28	06 25 02	06 27 53	09 ♎ 29 13	15 ♎ 52 58	14 31	03 32	11 54	00 D 26	15 35	01 16	25 22	24 08
29	06 28 59	07 25 05	22 ♎ 11 49	28 ♎ 26 13	15 37	04 43	12 19	00 26	15 36	01 15	25 20	24 07
30	06 32 56	08 ♋ 22 17	04 ♏ 36 39	10 ♏ 43 34	16 ♊ 46	05 ♊ 55	12 ♎ 45	00 ♏ 27	15 ♎ 37	01 ♐ 13	25 ♐ 19	24 ♎ 07

Moon

Date	Moon True ☊ ° '	Moon Mean ☊ ° '	Moon ☽ Latitude ° '
01	13 ♋ 51	15 ♋ 10	05 N 12
02	13 R 46	15 07	05 06
03	13 41	15 04	04 46
04	13 36	15 00	04 13
05	13 31	14 57	03 29
06	13 27	14 54	02 36
07	13 25	14 51	01 37
08	13 24	14 48	00 N 34
09	13 D 24	14 45	00 S 31
10	13 24	14 41	01 35
11	13 26	14 38	02 36
12	13 28	14 35	03 30
13	13 29	14 32	04 16
14	13 29	14 29	04 50
15	13 R 29	14 26	05 11
16	13 28	14 22	05 16
17	13 27	14 19	05 02
18	13 25	14 16	04 30
19	13 23	14 13	03 39
20	13 21	14 10	02 32
21	13 20	14 06	01 S 14
22	13 D 20	14 03	00 N 09
23	13 20	14 00	01 31
24	13 21	13 57	02 46
25	13 22	13 54	03 47
26	13 22	13 51	04 34
27	13 23	13 47	05 04
28	13 R 23	13 44	05 17
29	13 23	13 41	05 14
30	13 ♋ 22	13 ♋ 38	04 N 56

Declinations

Date	Sun ☉ ° '	Moon ☽ ° '	Mercury ☿ ° '	Venus ♀ ° '	Mars ♂ ° '	Jupiter ♃ ° '	Saturn ♄ ° '	Uranus ♅ ° '	Neptune ♆ ° '	Pluto ♇ ° '
01	22 N 03	00 S 14	19 N 54	10 N 11	00 S 37	10 S 42	03 S 45	20 S 28	22 S 04	06 N 48
02	22 11	05 01	19 34	10 34	00 44	10 41	03 44	20 27	22 04	06 48
03	22 18	09 32	19 14	10 58	00 52	10 40	03 44	20 27	22 04	06 48
04	22 26	13 37	18 55	11 21	01 00	10 39	03 44	20 26	22 04	06 48
05	22 32	17 06	18 38	11 44	01 08	10 37	03 44	20 26	22 04	06 48
06	22 39	19 51	18 21	12 07	01 16	10 36	03 43	20 25	22 04	06 48
07	22 45	21 45	18 06	12 29	01 24	10 35	03 43	20 25	22 04	06 48
08	22 50	22 41	17 53	12 52	01 32	10 34	03 43	20 24	22 04	06 48
09	22 56	22 36	17 42	13 14	01 41	10 33	03 42	20 24	22 04	06 47
10	23 00	21 30	17 32	13 36	01 50	10 32	03 42	20 23	22 04	06 47
11	23 05	19 26	17 24	13 57	01 59	10 32	03 42	20 23	22 03	06 47
12	23 09	16 29	17 18	14 19	02 08	10 31	03 42	20 23	22 03	06 47
13	23 12	12 45	17 14	14 40	02 17	10 30	03 42	20 22	22 03	06 47
14	23 16	08 23	17 12	15 01	02 26	10 30	03 42	20 22	22 03	06 46
15	23 18	03 S 31	17 13	15 21	02 36	10 29	03 43	20 21	22 03	06 46
16	23 21	01 N 40	17 14	15 41	02 45	10 29	03 43	20 21	22 03	06 46
17	23 23	06 56	17 18	16 01	02 55	10 28	03 43	20 21	22 03	06 46
18	23 24	12 00	17 24	16 21	03 05	10 28	03 43	20 20	22 03	06 46
19	23 25	16 31	17 31	16 40	03 15	10 27	03 43	20 20	22 03	06 45
20	23 26	20 02	17 40	16 59	03 25	10 27	03 44	20 19	22 03	06 45
21	23 26	22 12	17 50	17 17	03 36	10 27	03 44	20 19	22 03	06 44
22	23 26	22 45	18 02	17 35	03 46	10 27	03 44	20 18	22 03	06 44
23	23 26	21 38	18 15	17 53	03 57	10 27	03 45	20 18	22 03	06 44
24	23 25	19 03	18 29	18 10	04 07	10 27	03 45	20 18	22 03	06 43
25	23 24	15 21	18 44	18 27	04 18	10 27	03 46	20 17	22 03	06 43
26	23 22	10 55	19 00	18 43	04 29	10 27	03 46	20 17	22 03	06 42
27	23 20	06 04	19 16	18 59	04 40	10 27	03 47	20 16	22 03	06 42
28	23 17	01 N 06	19 34	19 14	04 51	10 28	03 47	20 16	22 03	06 41
29	23 14	03 S 47	19 51	19 29	05 02	10 28	03 48	20 16	22 03	06 41
30	23 N 11	08 S 24	20 N 09	19 N 44	05 S 14	10 S 28	03 S 49	20 S 15	22 S 03	06 N 41

Zodiac Sign Entries

Date	h	m	Planets
02	21	12	☽ ♏
05	08	31	☽ ♐
07	21	12	☽ ♑
10	10	08	☽ ♒
12	21	44	☽ ♓
15	06	20	☽ ♈
17	11	07	☽ ♉
19	12	34	☽ ♊
21	12	13	☽ ♋
21	17	23	☉ ♋
23	11	57	☽ ♌
25	12	13	♀ ♊
25	13	36	☽ ♍
27	18	30	☽ ♎
30	03	02	☽ ♏

Latitudes

Date	Mercury ☿ ° '	Venus ♀ ° '	Mars ♂ ° '	Jupiter ♃ ° '	Saturn ♄ ° '	Uranus ♅ ° '	Neptune ♆ ° '	Pluto ♇ ° '
01	02 S 14	02 S 05	00 N 34	01 N 22	02 N 39	00 N 10	01 N 19	17 N 28
04	02 59	02 05	00 29	01 21	02 38	00 10	01 19	17 26
07	03 36	02 04	00 23	01 20	02 38	00 09	01 19	17 25
10	04 01	02 03	00 18	01 19	02 37	00 09	01 19	17 24
13	04 14	02 00	00 13	01 19	02 36	00 09	01 19	17 22
16	04 17	01 58	00 08	01 18	02 35	00 09	01 19	17 21
19	04 09	01 54	00 N 03	01 17	02 35	00 09	01 19	17 19
22	03 52	01 50	00 S 01	01 16	02 34	00 09	01 19	17 18
25	03 29	01 45	00 06	01 16	02 33	00 09	01 19	17 16
28	03 00	01 39	00 10	01 15	02 33	00 09	01 19	17 14
31	02 S 27	01 S 33	00 S 14	01 N 14	02 N 32	00 N 09	01 N 19	17 N 12

Data

Julian Date	2445122
Delta T	+52 seconds
Ayanamsa	23° 36' 24"
Synetic vernal point	05° ♓ 30' 36"
True obliquity of ecliptic	23° 26' 27"

Moon's Phases, Apsides and Positions ☽

Date	h	m	Phase	Longitude ° '	Eclipse Indicator
06	15	59	○	15 ♐ 37	
14	18	06	☾	23 ♓ 21	
21	11	52	●	29 ♊ 47	Partial
28	05	56	☽	06 ♎ 13	

Day	h	m		
07	22	47	Apogee	
21	12	02	Perigee	
01	10	49	0S	
08	22	05	Max dec	22° S 46'
16	04	24	0N	
22	07	42	Max dec	22° N 47'
28	17	19	0S	

Longitudes

Date	Chiron ⚷ ° '	Ceres ⚳ ° '	Pallas ⚴ ° '	Juno ⚵ ° '	Vesta ⚶ ° '	Black Moon Lilith ⚸ ° '
01	23 ♉ 59	15 ♏ 03	02 ♎ 45	07 ♑ 09	22 ♒ 15	27 ♐ 52
11	24 ♉ 38	13 ♏ 42	03 ♎ 48	05 ♑ 14	23 ♒ 43	28 ♐ 59
21	25 ♉ 16	12 ♏ 57	05 ♎ 25	03 ♑ 01	24 ♒ 29	00 ♑ 05
31	25 ♉ 50	12 ♏ 50	07 ♎ 32	00 ♑ 43	24 ♒ 27	01 ♑ 12

Aspectarian

01 Tuesday
03 45 ♀ ☍ ♃
07 45 ☽ △ ☉
07 52 ☽ # ♂
09 07 ☽ △ ☿
13 55 ☽ ∥ ♂
14 33 ☽ Q ♆
15 16 ☉ # ♆
17 41 ☽ ☌ ♄
19 42 ☉ ☌ ☿
19 54 ♀ ⚻ ♅
20 37 ☽ ∠ ♅

02 Wednesday
05 28 ☽ ∥ ♄
10 20 ☽ ☌ ♇
12 37 ☽ ⚼ ☿
13 33 ☽ ⚹ ♆
13 52 ♀ ⚻ ♂
13 53 ☽ ⊥ ♅
14 52 ☽ ⚼ ☉
18 39 ♀ # ♃
21 16 ☽ # ♇
23 49 ☽ ☌ ♃

03 Thursday
01 30 ☽ ⚺ ♅
03 34 ☽ ⚺ ♂
04 42 ☽ ☍ ♀
05 22 ☽ ± ☿
09 45 ☿ ⊥ ♀
09 53 ☽ ± ☉
15 34 ☽ ⊥ ♂
16 38 ☽ ⚻ ☿
18 20 ☽ ∥ ♃
18 47 ☽ ∠ ♆
20 54 ☽ # ♀
22 42 ☽ ⚻ ☉

04 Friday
03 57 ☽ ⚺ ♄
09 44 ☽ ∠ ♂
12 34 ☽ ⊥ ♆
15 51 ☽ ⊥ ♄
18 30 ☿ ⚼ ♇
21 14 ☽ ⚺ ♇

05 Saturday
00 30 ☽ ⚺ ♆
09 13 ☽ ⊥ ♇
09 48 ☽ ∠ ♄
10 53 ☽ ⚺ ♃
12 45 ☽ ☌ ♅
16 23 ☽ ⚹ ♂
22 40 ☽ ⚻ ♀
22 53 ☽ ⊥ ♃
23 18 ☽ # ☿

06 Sunday
02 01 ☽ ☍ ☿
03 18 ☽ ∠ ♇
12 06 ☽ ± ♀
15 58 ☉ △ ♄
15 59 ☽ ⚹ ♄
15 59 ☽ ☍ ☉
16 58 ☽ ∠ ♃
17 10 ☽ Q ♂
18 04 ☽ ∥ ♅

07 Monday
00 04 ☿ ⚺ ♀
03 28 ☉ ⚼ ♃
08 20 ☽ ⚼ ♀
09 39 ☽ ⚹ ♇
12 55 ☽ ☌ ♆
16 18 ☽ Q ♄
17 43 ☽ ∥ ♆
23 17 ☽ ⚹ ♃

08 Tuesday
01 17 ☽ ⚺ ♅
06 37 ☽ □ ♂
10 01 ☽ Q ♇
12 47 ☽ ⚻ ☿
13 27 ☽ ⊥ ♅
18 12 ☽ △ ♀
23 35 ☽ Q ♃

09 Wednesday
00 38 ☽ ± ☿
04 53 ☽ □ ♄
05 45 ♀ ⚼ ♆
06 35 ♀ ± ♂
07 42 ☽ ∠ ♅
10 20 ☽ ⚻ ☉
18 27 ☽ ⚼ ☿
22 34 ☽ □ ♇
23 35 ☽ ± ☉

10 Thursday
01 46 ☽ ⚺ ♆
02 26 ☽ ∥ ♆
11 55 ☽ □ ♃
13 51 ☽ ⊥ ♆
13 59 ☽ ⚹ ♅
19 21 ☽ ⚼ ☉
21 01 ☽ △ ♂

11 Friday
00 06 ☽ △ ☿
02 13 ☽ ∥ ♅
07 53 ☽ ∠ ♆
13 23 ☿ ± ♃
13 24 ☽ □ ♀
13 58 ☽ Q ♅
17 16 ☽ △ ♄

12 Saturday
03 43 ☽ ⚼ ♂
03 50 ☽ △ ☉
05 45 ☽ # ☿
10 30 ☽ △ ♇
13 31 ☽ ⚹ ♆
21 49 ☽ ± ♂
22 45 ☽ ⚼ ♄
23 13 ☽ △ ♃

13 Sunday
01 15 ☽ # ♀
01 15 ☽ □ ♅
04 31 ☽ Q ♀
04 57 ♀ ⚻ ♄
09 44 ☽ ⚻ ♂
10 33 ☽ □ ☿
12 44 ☽ Q ♆
15 33 ☽ ⚼ ♇
16 05 ☽ ± ♄
23 20 ☿ St D

14 Monday
00 45 ☽ ∥ ♃
03 28 ☽ ⚻ ♄
03 50 ☽ ⚼ ♃
05 45 ☽ ⚹ ♀
08 34 ☽ ± ♇
16 53 ☿ △ ♂
18 06 ☽ □ ☉
19 07 ☉ ⊥ ♀
19 44 ☽ ⚻ ♇
20 10 ☽ # ♇
20 30 ☽ Q ☿
20 35 ☽ ± ♃
22 30 ☽ □ ♆

15 Tuesday
04 50 ☿ ± ♃
07 33 ☽ ⚻ ♃
09 29 ☽ △ ♅
11 05 ☽ ∥ ♄
12 11 ☽ ∠ ♀
15 59 ☉ △ ♇
16 12 ☽ ∥ ♂
18 32 ☽ ⚹ ☿
19 00 ☽ ☍ ♂

16 Wednesday
05 13 ☽ Q ☉
05 49 ☽ ⊥ ♀
10 05 ☽ ☍ ♄
12 10 ☽ ⚼ ♅
17 08 ☽ # ♂
17 19 ☽ ⚺ ♀
18 04 ♀ ± ♆
21 18 ☽ ∠ ☿
21 19 ☽ # ♄

17 Thursday
01 12 ☽ ☍ ♇
03 40 ☽ ⚹ ☉
03 43 ☽ ± ♅
03 43 ☽ △ ♆
04 31 ☉ ☍ ♆
11 12 ☽ ∥ ♇
12 07 ☽ ☍ ♃
12 59 ☽ ⊥ ☿
13 54 ☽ ⚻ ♅
23 17 ☽ ⚺ ☿

18 Friday
00 13 ☽ ⚻ ♂
04 31 ☽ # ♃
04 58 ☽ ⚼ ♆
06 45 ☽ ∠ ☉
07 10 ♀ ± ♄
10 26 ☽ ± ♂
11 05 ♄ St D
13 00 ☽ ⚻ ♄
19 46 ☽ ± ♆
22 34 ☽ ⊥ ☉
22 49 ☽ ± ♄

19 Saturday
00 10 ☽ ☌ ♀
01 32 ☽ ⚼ ♂
03 11 ☽ ⚻ ♇
05 30 ☽ ⚻ ♆
08 57 ☽ ⚺ ☉
12 51 ☽ ± ♇
13 00 ☽ ∥ ♀
13 22 ☽ ⚼ ♄
13 26 ☽ ⚻ ♃
15 06 ☽ ☍ ♅
18 20 ☽ ∥ ☿
23 02 ☽ ± ♃

20 Sunday
00 21 ♀ ⚼ ♂
01 25 ☽ ☌ ☿
02 15 ☽ △ ♂
03 15 ☽ ⚼ ♇
13 18 ☽ △ ♄
13 19 ☽ ⚼ ♃
13 40 ♀ ⚻ ♇
14 20 ☽ # ♅

21 Monday
03 00 ☽ △ ♇
04 08 ☽ ⚺ ♀
05 12 ☽ ☍ ♆
09 26 ☽ # ♆
11 52 ☽ ☌ ☉
12 59 ☽ △ ♃
14 24 ☽ ⊥ ♀
14 35 ☽ ⚻ ♅
17 40 ♀ ⚻ ♆

22 Tuesday
00 02 ☽ ± ♅
02 39 ☽ ⚺ ☿
02 56 ☽ □ ♂
03 33 ☿ ⚼ ♇
05 31 ☉ △ ♃
05 50 ☽ ∠ ♀
07 35 ♂ ∥ ♄
12 34 ☽ ⊥ ☿
12 45 ☽ □ ♄
14 17 ☽ ⚼ ♅
19 12 ☿ △ ♂

23 Wednesday
02 32 ☽ □ ♇
03 36 ☽ ∠ ☿
04 43 ☽ ⚻ ♆
05 45 ☉ ⚻ ♅
06 10 ☽ # ♆
07 47 ☽ ⚹ ♀
08 23 ☽ Q ♂
12 41 ☽ □ ♃
14 16 ☽ △ ♅
14 22 ☽ ± ♆
14 52 ☽ ⚺ ☉
17 38 ☽ Q ♄

24 Thursday
01 18 ☽ ⊥ ☉
01 55 ☽ # ♅
04 24 ☽ ⚹ ♂
04 57 ☽ Q ♀
04 59 ☽ ⚼ ♆
05 10 ☽ ⚹ ☿
07 43 ☽ Q ♇
13 18 ☽ ⚹ ♄
15 54 ☽ ∥ ☿
17 10 ☽ ∠ ☉
17 53 ☽ ∥ ♀
18 09 ☽ Q ♃

25 Friday
02 15 ☽ Q ☿
03 41 ☽ ⚹ ♇
05 53 ☽ △ ♆
06 03 ☽ ∠ ♂
13 44 ☽ □ ♀
14 21 ☽ ⚹ ♃
14 31 ☽ ∠ ♄
15 02 ♀ ± ♇
15 56 ☽ □ ♅
20 20 ☽ ⚹ ☉
21 11 ♀ ⚻ ♃
21 51 ☽ ⊥ ♂
23 17 ♀ ⚼ ♄

26 Saturday
05 20 ☽ ∠ ♇
06 03 ☽ ⊥ ♄
08 35 ☽ ⚺ ♂
11 12 ☽ □ ☿
14 20 ☽ # ♃
15 17 ♀ ☍ ♅
16 22 ☽ ∠ ♃
16 34 ☽ ⚺ ♄
18 49 ☽ Q ☉
21 08 ☽ ⊥ ♇
23 15 ☽ Q ♅

27 Sunday
07 52 ☽ ⚺ ♇
08 25 ☽ ⊥ ♃
08 56 ☽ ∥ ♇
10 08 ☽ □ ♆
18 16 ♃ St D
18 32 ☽ # ♂
19 18 ☽ ⚺ ♃
20 52 ☽ ⚹ ♅
23 02 ☽ # ♄
23 53 ☽ △ ♀

28 Monday
05 56 ☽ □ ☉
16 39 ☽ ☌ ♂
19 15 ☽ Q ♆
22 18 ☽ △ ☿
23 27 ☽ ☌ ♄

29 Tuesday
00 43 ☽ ∠ ♅
06 48 ☽ ⚼ ♀
07 45 ☉ ± ♅
08 17 ☿ ⚼ ♃
11 42 ☿ △ ♄
12 04 ☽ ∥ ♄
15 41 ☽ ☌ ♇
17 50 ☽ ⊥ ♅
18 01 ☽ ⚹ ♆
18 36 ☽ ∥ ♂

30 Wednesday
01 53 ☽ ± ♀
02 49 ☽ # ♇
03 53 ☽ ☌ ♃
05 24 ☽ ⚺ ♅
05 53 ☽ ⚼ ☿
14 49 ☽ ⚻ ♀
19 29 ☿ # ♅
19 59 ☽ △ ☉
23 00 ♀ ± ♃
23 10 ☽ ∠ ♆
23 28 ☽ ∥ ♃

All ephemeris data is given at 12.00 UT and the Moon's longitude is additionally given for 24.00 UT

JULY 1982

Date	Sidereal time h m s	Sun ☉ ° ' "	Moon ☽ ° ' "	Moon ☽ 24.00 ° ' "	Mercury ☿ ° '	Venus ♀ ° '	Mars ♂ ° '	Jupiter ♃ ° '	Saturn ♄ ° '	Uranus ♅ ° '	Neptune ♆ ° '	Pluto ♇ ° '
01	06 36 52	09 ♋ 19 29	16 ♏ 47 26	22 ♏ 48 42	18 ♊ 00	07 ♊ 06	13 ♎ 11	00 ♏ 27	15 ♎ 38	01 ♐ 11	25 ♐ 17	24 ♎ 07
02	06 40 49	10 16 40	28 ♏ 47 47	04 ♐ 45 06	19 17	08 17	13 38	00 28	15 40	01 R 09	25 R 16	24 R 07
03	06 44 45	11 13 52	10 ♐ 41 03	16 ♐ 35 59	20 39	09 28	14 04	00 29	15 41	01 08	25 14	24 07
04	06 48 42	12 11 03	22 ♐ 30 16	28 ♐ 24 12	22 04	10 39	14 32	00 30	15 43	01 06	25 13	24 07
05	06 52 38	13 08 14	04 ♑ 18 08	10 ♑ 12 19	23 32	11 51	14 59	00 32	15 44	01 04	25 11	24 D 07
06	06 56 35	14 05 25	16 ♑ 07 05	22 ♑ 02 41	25 04	13 02	15 26	00 33	15 46	01 03	25 10	24 07
07	07 00 31	15 02 36	27 ♑ 59 24	03 ♒ 57 32	26 40	14 14	15 54	00 35	15 48	01 01	25 08	24 07
08	07 04 28	15 59 47	09 ♒ 57 21	15 ♒ 59 08	28 ♊ 20	15 25	16 23	00 37	15 50	01 00	25 07	24 07
09	07 08 25	16 56 58	22 ♒ 03 12	28 ♒ 09 52	00 ♋ 02	16 37	16 51	00 39	15 52	00 58	25 05	24 07
10	07 12 21	17 54 10	04 ♓ 19 26	10 ♓ 32 15	01 48	17 48	17 20	00 41	15 54	00 57	25 04	24 08
11	07 16 18	18 51 22	16 ♓ 48 40	23 ♓ 09 00	03 38	19 00	17 49	00 43	15 56	00 55	25 02	24 08
12	07 20 14	19 48 34	29 ♓ 33 38	06 ♈ 02 53	05 30	20 12	18 18	00 46	15 59	00 54	25 01	24 08
13	07 24 11	20 45 47	12 ♈ 37 05	19 ♈ 16 29	07 25	21 23	18 48	00 48	16 01	00 52	24 59	24 08
14	07 28 07	21 43 00	26 ♈ 01 21	02 ♉ 51 52	09 22	22 35	19 17	00 51	16 03	00 51	24 58	24 09
15	07 32 04	22 40 14	09 ♉ 48 06	16 ♉ 50 04	11 22	23 47	19 47	00 54	16 06	00 50	24 56	24 09
16	07 36 00	23 37 29	23 ♉ 57 39	01 ♊ 10 37	13 24	24 59	20 18	00 57	16 09	00 49	24 55	24 09
17	07 39 57	24 34 45	08 ♊ 28 34	15 ♊ 50 59	15 28	26 11	20 48	01 01	16 11	00 48	24 54	24 10
18	07 43 54	25 32 01	23 ♊ 17 11	00 ♋ 46 22	17 33	27 23	21 19	01 04	16 14	00 47	24 52	24 10
19	07 47 50	26 29 17	08 ♋ 17 35	15 ♋ 49 49	19 39	28 35	21 50	01 08	16 17	00 45	24 51	24 11
20	07 51 47	27 26 35	23 ♋ 21 59	00 ♌ 52 57	21 46	29 ♊ 47	22 21	01 12	16 20	00 44	24 50	24 11
21	07 55 43	28 23 53	08 ♌ 21 38	15 ♌ 47 01	23 54	00 ♋ 59	22 52	01 16	16 23	00 44	24 49	24 12
22	07 59 40	29 ♋ 21 11	23 ♌ 08 08	00 ♍ 24 11	26 02	02 11	23 24	01 20	16 27	00 43	24 47	24 12
23	08 03 36	00 ♌ 18 29	07 ♍ 34 30	14 ♍ 38 35	28 ♋ 09	03 23	23 56	01 24	16 30	00 42	24 46	24 13
24	08 07 33	01 15 48	21 ♍ 36 06	28 ♍ 26 54	00 ♌ 17	04 36	24 28	01 29	16 33	00 41	24 45	24 14
25	08 11 29	02 13 08	05 ♎ 10 57	11 ♎ 48 23	02 24	05 48	25 00	01 33	16 37	00 40	24 44	24 14
26	08 15 26	03 10 27	18 ♎ 19 27	24 ♎ 44 29	04 30	07 00	25 33	01 38	16 40	00 39	24 42	24 15
27	08 19 23	04 07 47	01 ♏ 03 57	07 ♏ 18 18	06 35	08 13	26 06	01 43	16 44	00 39	24 41	24 16
28	08 23 19	05 05 08	13 ♏ 28 06	19 ♏ 33 54	08 39	09 25	26 39	01 48	16 48	00 38	24 40	24 17
29	08 27 16	06 02 29	25 ♏ 36 17	01 ♐ 35 50	10 42	10 37	27 12	01 54	16 52	00 38	24 39	24 17
30	08 31 12	06 59 51	07 ♐ 33 08	13 ♐ 28 44	12 44	11 50	27 45	01 59	16 55	00 37	24 38	24 18
31	08 35 09	07 ♌ 57 13	19 ♐ 23 12	25 ♐ 17 02	14 ♌ 44	13 ♋ 02	28 ♎ 19	02 ♏ 05	16 ♎ 59	00 ♐ 37	24 ♐ 37	24 ♎ 19

Date	Moon True ☊ ° '	Moon Mean ☊ ° '	Moon ☽ Latitude ° '
01	13 ♋ 22	13 ♋ 35	04 N 25
02	13 R 21	13 32	03 43
03	13 21	13 28	02 52
04	13 21	13 25	01 53
05	13 D 21	13 22	00 N 50
06	13 R 21	13 19	00 S 15
07	13 21	13 16	01 21
08	13 21	13 12	02 23
09	13 20	13 09	03 19
10	13 20	13 06	04 07
11	13 19	13 03	04 44
12	13 19	13 00	05 08
13	13 18	12 57	05 17
14	13 D 18	12 53	05 10
15	13 18	12 50	04 44
16	13 19	12 47	04 01
17	13 20	12 44	03 02
18	13 21	12 41	01 49
19	13 21	12 37	00 S 28
20	13 R 21	12 34	00 N 55
21	13 20	12 31	02 14
22	13 19	12 28	03 23
23	13 17	12 25	04 17
24	13 14	12 22	04 54
25	13 12	12 18	05 13
26	13 11	12 15	05 15
27	13 10	12 12	05 01
28	13 D 10	12 09	04 33
29	13 11	12 06	03 53
30	13 13	12 03	03 04
31	13 ♋ 15	11 ♋ 59	02 N 08

DECLINATIONS

Date	Sun ☉ ° '	Moon ☽ ° '	Mercury ☿ ° '	Venus ♀ ° '	Mars ♂ ° '	Jupiter ♃ ° '	Saturn ♄ ° '	Uranus ♅ ° '	Neptune ♆ ° '	Pluto ♇ ° '
01	23 N 07	12 S 37	20 N 28	19 N 58	05 S 25	10 S 29	03 S 49	20 S 15	22 S 03	06 N 40
02	23 03	16 16	20 46	20 11	05 37	10 29	03 50	20 15	22 03	06 40
03	22 58	19 13	21 04	20 24	05 48	10 30	03 51	20 14	22 03	06 39
04	22 53	21 21	21 22	20 37	06 00	10 31	03 52	20 14	22 03	06 39
05	22 48	22 32	21 39	20 49	06 11	10 31	03 53	20 14	22 02	06 38
06	22 42	22 43	21 56	21 00	06 23	10 32	03 54	20 13	22 02	06 37
07	22 36	21 53	22 11	21 11	06 35	10 33	03 54	20 13	22 02	06 37
08	22 29	20 03	22 26	21 21	06 47	10 34	03 55	20 13	22 02	06 36
09	22 22	17 18	22 39	21 31	06 59	10 35	03 56	20 12	22 02	06 36
10	22 15	13 45	22 51	21 40	07 11	10 36	03 57	20 12	22 02	06 35
11	22 07	09 34	23 01	21 49	07 23	10 37	03 59	20 12	22 02	06 34
12	21 59	04 S 53	23 09	21 57	07 36	10 38	04 00	20 12	22 02	06 34
13	21 50	00 N 07	23 16	22 04	07 48	10 39	04 01	20 11	22 02	06 33
14	21 41	05 14	23 20	22 11	08 00	10 40	04 02	20 11	22 02	06 33
15	21 32	10 15	23 21	22 17	08 12	10 42	04 03	20 11	22 02	06 32
16	21 22	14 52	23 20	22 23	08 25	10 43	04 04	20 11	22 02	06 31
17	21 12	18 44	23 17	22 28	08 37	10 44	04 06	20 10	22 02	06 31
18	21 02	21 27	23 11	22 33	08 50	10 46	04 07	20 10	22 02	06 30
19	20 51	22 43	23 02	22 36	09 02	10 47	04 08	20 10	22 02	06 29
20	20 40	22 20	22 50	22 39	09 15	10 49	04 10	20 10	22 02	06 28
21	20 29	20 20	22 36	22 42	09 28	10 51	04 11	20 10	22 02	06 28
22	20 17	17 00	22 19	22 44	09 40	10 52	04 13	20 09	22 02	06 27
23	20 05	12 42	21 59	22 45	09 53	10 54	04 14	20 09	22 02	06 26
24	19 53	07 50	21 37	22 46	10 06	10 56	04 16	20 09	22 02	06 25
25	19 40	02 N 44	21 13	22 46	10 18	10 58	04 17	20 09	22 02	06 25
26	19 27	02 S 20	20 47	22 45	10 31	11 00	04 19	20 09	22 02	06 24
27	19 14	07 09	20 18	22 43	10 44	11 02	04 20	20 09	22 02	06 23
28	19 00	11 32	19 48	22 41	10 57	11 04	04 22	20 09	22 02	06 22
29	18 46	15 23	19 16	22 39	11 10	11 06	04 24	20 08	22 02	06 21
30	18 31	18 32	18 42	22 35	11 22	11 08	04 25	20 08	22 02	06 21
31	18 N 17	20 S 54	18 N 07	22 N 31	11 S 35	11 S 10	04 S 27	20 S 08	22 S 02	06 N 20

ZODIAC SIGN ENTRIES

Date	h	m	Planets
02	14	25	☽ ♐
05	03	15	☽ ♑
07	16	03	☽ ♒
09	11	26	☿ ♋
10	03	35	☽ ♓
12	12	49	☽ ♈
14	19	00	☽ ♉
16	22	03	☽ ♊
18	22	46	☽ ♋
20	16	21	♀ ♋
20	22	35	☽ ♌
22	23	20	☽ ♍
23	04	15	☉ ♌
24	08	48	☿ ♌
25	02	45	☽ ♎
27	09	58	☽ ♏
29	20	48	☽ ♐

LATITUDES

Date	Mercury ☿ ° '	Venus ♀ ° '	Mars ♂ ° '	Jupiter ♃ ° '	Saturn ♄ ° '	Uranus ♅ ° '	Neptune ♆ ° '	Pluto ♇ ° '
01	02 S 27	01 S 33	00 S 14	01 N 14	02 N 32	00 N 09	01 N 19	17 N 12
04	01 51	01 27	00 18	01 13	02 31	00 09	01 19	17 11
07	01 13	01 20	00 21	01 12	02 30	00 09	01 19	17 09
10	00 S 35	01 13	00 25	01 11	02 29	00 09	01 19	17 07
13	00 N 02	01 05	00 28	01 11	02 29	00 09	01 19	17 05
16	00 35	00 58	00 31	01 10	02 28	00 09	01 18	17 04
19	01 02	00 50	00 34	01 09	02 27	00 09	01 18	17 02
22	01 24	00 42	00 37	01 08	02 26	00 09	01 18	17 00
25	01 38	00 33	00 40	01 07	02 26	00 09	01 18	16 59
28	01 46	00 25	00 43	01 07	02 25	00 09	01 18	16 57
31	01 N 47	00 S 17	00 S 46	01 N 06	02 N 24	00 N 09	01 N 18	16 N 55

DATA

Julian Date	2445152
Delta T	+52 seconds
Ayanamsa	23° 36' 29"
Synetic vernal point	05° ♓ 30' 31"
True obliquity of ecliptic	23° 26' 27"

MOON'S PHASES, APSIDES AND POSITIONS ☽

Date	h	m	Phase	Longitude ° '	Eclipse Indicator
06	07	32	○	13 ♑ 55	total
14	03	47	☾	21 ♈ 23	
20	18	57	●	27 ♋ 43	Partial
27	18	22	☽	04 ♏ 23	

Day	h	m		
05	01	15	Apogee	
19	21	14	Perigee	
06	04	21	Max dec	22° S 47'
13	11	29	0N	
19	18	24	Max dec	22° N 46'
26	00	51	0S	

LONGITUDES

Date	Chiron ⚷ ° '	Ceres ⚳ ° '	Pallas ⚴ ° '	Juno ⚵ ° '	Vesta ⚶ ° '	Black Moon Lilith ⚸ ° '
01	25 ♉ 50	12 ♏ 50	07 ♎ 32	00 ♑ 43	24 ♒ 27	01 ♑ 12
11	26 ♉ 22	13 ♏ 20	10 ♎ 03	28 ♐ 32	23 ♒ 38	02 ♑ 18
21	26 ♉ 48	14 ♏ 23	12 ♎ 55	26 ♐ 41	22 ♒ 05	03 ♑ 25
31	27 ♉ 11	15 ♏ 56	16 ♎ 04	25 ♐ 19	19 ♒ 56	04 ♑ 31

All ephemeris data is given at 12.00 UT and the Moon's longitude is additionally given for 24.00 UT

ASPECTARIAN

h m	Aspects
01 Thursday	
01 26	☽ ± ☿
04 35	☽ ⚺ ♂
09 43	☽ ⚺ ♄
14 41	☽ ⚻ ☿
16 57	☽ ⊥ ♂
16 57	☽ ⊥ ♆
17 11	♂ Q ♆
21 41	☽ ⊥ ♄
02 Friday	
02 37	☽ ⚺ ♇
04 20	☽ bQ ☉
04 55	☽ ⚺ ♆
11 39	☽ ∠ ♂
14 39	☽ ⊥ ♇
15 22	☽ ⚺ ♃
15 46	☽ ∠ ♄
16 44	☽ ☌ ♅
17 55	♀ # ♅
03 Saturday	
00 00	☽ ± ☉
03 29	☽ ⊥ ♃
04 52	♀ bQ ♇
08 49	☽ ∠ ♇
09 16	☽ ☍ ♀
13 12	☽ ⚻ ☉
19 09	☽ ⚹ ♂
21 45	☽ ∠ ♃
22 10	☽ ⚹ ♄
22 17	☽ ∥ ♅
04 Sunday	
01 30	☽ # ♀
10 58	☽ ☍ ☿
12 22	☽ # ☿
13 12	♇ St D
15 17	☽ ⚹ ♇
17 29	☽ ☌ ♆
20 30	☽ Q ♂
22 37	☽ Q ♄
23 35	☽ ∥ ♆
05 Monday	
04 18	☽ ⚹ ♃
05 26	☽ ⚺ ♅
15 41	☽ Q ♇
17 37	☽ ⊥ ♅
21 09	☿ △ ♇
22 56	☽ # ☉
06 Tuesday	
04 45	☽ Q ♃
05 03	☽ ⚻ ♀
07 32	☽ ☍ ☉
10 34	☽ □ ♂
11 17	☽ □ ♄
11 51	☽ ∠ ♅
13 16	☿ ☍ ♆
14 20	☽ # ☉
18 34	☽ ± ♀
22 12	☿ # ♆
07 Wednesday	
04 11	☽ □ ♇
06 04	♂ ☌ ♄
06 15	☽ ⚺ ♆
07 03	☽ # ☿
08 55	☽ ⚻ ☿
09 00	☽ ∥ ♆
14 46	☽ bQ ♀
15 22	♂ # ♇
17 13	☽ □ ♃
17 22	♂ ∠ ♅
18 05	☽ ⚹ ♅
18 19	☽ ⊥ ♆
21 42	☽ # ♀
22 55	☽ ± ☿
08 Thursday	
07 42	☉ □ ♄
10 13	☽ ∥ ♅
11 53	☉ bQ ♅
12 18	☽ ∠ ♆
15 29	☉ ∥ ☿
15 55	♀ bQ ♃
18 02	☽ Q ♅
19 49	☽ bQ ☿
20 32	♀ △ ♄
23 44	☽ △ ♄
09 Friday	
00 04	☽ △ ♀
01 03	☽ ⚻ ☉
01 18	☽ △ ♂
07 03	☉ □ ♂
13 55	☽ ± ☉
16 04	☽ △ ♇
17 57	☽ ⚹ ♆
20 01	♀ △ ♂
20 25	☿ △ ♃
10 Saturday	
00 30	☿ ⚻ ♅
04 53	☽ △ ♃
05 19	☽ bQ ♄
05 26	☽ □ ♅
06 17	☽ △ ☿
07 58	☽ bQ ♂
09 01	☽ bQ ☉
17 17	☽ Q ♆
21 17	☽ bQ ♇
21 42	☉ ⚺ ♀
22 48	☽ ± ♄
11 Sunday	
02 04	☽ ± ♂
05 01	♄ ∠ ♅
06 19	☽ ∥ ♃
09 55	☽ bQ ♃
10 20	☽ ⚻ ♄
13 59	☽ ⚻ ♂
14 30	☽ ± ♇
16 12	☽ △ ☉
16 35	☽ □ ♀
22 58	☽ ∥ ♂
12 Monday	
01 34	☉ # ♆
01 51	☽ ⚻ ♇
03 00	☽ ± ♃
03 31	☽ □ ♆
03 37	☽ # ♇
14 14	☽ ⚻ ♃
14 29	☽ △ ♅
14 40	☉ ∥ ♀
16 22	☽ ∥ ♄
13 Tuesday	
00 52	☽ □ ☿
04 59	♀ # ♆
05 24	☿ ± ♅
05 32	☽ Q ♀
17 53	☽ bQ ♅
18 10	☽ ☍ ♄
23 34	☽ ☍ ♂
14 Wednesday	
03 47	☽ □ ☉
05 19	☽ ⚹ ♀
06 20	☽ # ♄
08 40	☽ ☍ ♇
09 56	☽ ± ♅
10 08	☽ △ ♆
12 44	♃ ⚺ ♅
14 47	☽ Q ☿
18 08	☽ ∥ ♇
20 29	☽ ⚻ ♅
20 31	☽ ☍ ♃
15 Thursday	
01 38	☽ # ♂
10 05	☽ ∠ ♀
12 14	☽ bQ ♆
13 36	☽ Q ☉
14 11	☽ # ♃
15 08	☽ ⚹ ☿
19 22	♀ △ ♇
22 47	☽ ⚻ ♄
16 Friday	
02 52	☽ ⊥ ♀
03 32	☽ ± ♆
05 37	☽ ⚻ ♂
08 57	☽ ± ♄
10 45	♀ ☍ ♆
11 24	☽ ⚹ ☉
12 20	☽ ⚻ ♇
13 36	☽ ⚻ ♆
13 52	☽ ⚺ ♀
16 02	☽ ± ♂
20 37	☽ ∠ ☿
22 19	☽ ± ♇
23 23	☽ ☍ ♅
23 41	☽ ⚻ ♃
23 59	☽ bQ ♄
17 Saturday	
01 27	☉ □ ♇
07 28	☽ bQ ♂
09 36	☽ ± ♃
13 07	☽ bQ ♇
13 52	☽ ⊥ ☿
13 56	☽ ∠ ☉
15 49	☿ bQ ♅
19 47	☉ ⚻ ♆
20 37	☿ □ ♄
23 15	☽ # ♅
18 Sunday	
00 19	☽ bQ ♃
00 36	☽ △ ♄
01 14	☽ ⚺ ☿
05 32	☽ ⊥ ☉
07 42	☽ ∥ ☉
08 43	☽ △ ♂
13 25	☽ △ ♇
14 33	☽ ☍ ♆
15 51	☽ ⚺ ☉
19 09	☽ ☌ ♀
19 45	☽ # ♆
23 59	☽ ⚻ ♅
19 Monday	
00 32	☽ △ ♃
07 14	☽ ∥ ♀
09 33	☽ ± ♅
23 52	☽ bQ ♅
20 Tuesday	
00 46	☽ □ ♄
03 59	☽ ∥ ♀
09 02	☽ ☌ ☿
10 19	☽ □ ♂
13 19	☽ □ ♇
14 20	☽ ⚻ ♆
17 02	☽ # ♆
18 57	☽ ☌ ☉
20 41	☿ □ ♂
23 08	☽ ⚺ ♀
23 46	☽ △ ♅
23 54	☽ ± ♆
21 Wednesday	
00 33	☽ □ ♃
03 39	☿ ∥ ♀
05 36	☽ Q ♄
06 54	♀ ⚻ ♅
09 35	☽ ⊥ ♀
10 33	☽ ∥ ☉
13 31	☽ # ♅
14 20	☽ bQ ♆
15 24	☿ □ ♇
16 12	☽ Q ♂
17 54	♀ △ ♃
18 12	☽ Q ♇
22 10	☿ ⚻ ♆
22 Thursday	
01 02	☽ ⚹ ♄
01 25	☽ ∠ ♀
05 45	☽ Q ♃
12 27	☽ ⚹ ♂
13 46	☽ ⚹ ♇
14 43	☽ △ ♆
17 34	☽ ⚺ ☿
22 59	☽ ⚺ ☉
23 Friday	
00 30	☽ □ ♅
01 37	☽ ⚹ ♃
01 47	☽ ∠ ♄
03 53	☉ # ♅
04 20	☽ ⚹ ♀
05 16	☽ ⊥ ☿
08 54	☿ # ♆
09 43	☽ ⊥ ☉
14 23	☽ ∠ ♂
14 46	☽ ∠ ♇
16 58	☽ ⊥ ♄
21 02	☽ # ♃
21 36	☉ △ ♅
23 09	☽ ∠ ☿
23 16	☉ ± ♆
24 Saturday	
01 01	♂ ☌ ♇
01 31	☽ # ♂
02 06	☽ ∠ ☉
02 32	☽ Q ♀
03 06	☽ ∠ ♃
03 15	☽ ⚺ ♄
06 09	☽ ⊥ ♇
06 21	☽ ⊥ ♂
06 57	☽ Q ♅
16 30	☿ △ ♅
16 35	☽ ⚺ ♇
17 12	☿ ± ♆
17 12	☽ ⚺ ♂
17 29	☽ □ ♆
17 56	☉ □ ♃
18 40	☽ ∥ ♇
18 49	☽ ⊥ ♃
21 20	♂ ⊥ ♅
23 57	♂ ⚹ ♆
25 Sunday	
02 06	☿ □ ♃
03 57	☽ ⚹ ♅
04 44	☽ # ♄
05 29	☽ ⚺ ♃
06 05	☽ ⚹ ☿
06 17	☽ ⚹ ☉
08 17	☉ ☌ ☿
13 13	☽ □ ♀
26 Monday	
01 40	☽ Q ♆
05 10	♀ ± ♅
05 43	☽ Q ☉
07 04	☽ ∠ ♅
07 58	☽ Q ☿
08 56	☽ ☌ ♄
14 02	☿ Q ♄
21 44	☽ ∥ ♄
23 05	☽ ☌ ♇
23 50	☽ ⊥ ♅
23 55	☽ ⚹ ♆
27 Tuesday	
02 08	☽ ☌ ♂
08 07	☽ # ♇
11 12	☽ ⚺ ♅
13 16	☽ ☌ ♃
18 22	☽ □ ☉
19 39	☿ # ♅
28 Wednesday	
00 44	☽ □ ☿
03 14	☽ △ ♀
04 12	☉ Q ♄
04 36	☽ ∠ ♆
08 25	☽ ∥ ♂
09 13	☽ ∥ ♃
18 34	☽ ⚺ ♄
22 13	☽ ⊥ ♆
23 43	☿ bQ ♆
29 Thursday	
03 03	♂ ∥ ♃
06 30	☽ ⊥ ♄
09 23	☽ ⚺ ♇
09 41	☿ ⚺ ♀
10 06	☽ ⚺ ♆
12 03	☽ bQ ♀
15 20	☽ ⚺ ♂
21 23	☽ ⊥ ♇
22 03	☽ ☌ ♅
30 Friday	
00 36	☽ ∠ ♄
00 42	☽ ⚺ ♃
03 57	☽ ⊥ ♂
06 54	☿ Q ♇
08 08	☽ ± ♀
10 47	☽ △ ☉
11 53	☽ # ☉
12 53	☽ ⊥ ♃
13 09	☽ # ☿
15 33	☽ ∠ ♇
21 39	☽ ⚻ ♀
23 03	☽ ∠ ♂
31 Saturday	
00 19	☉ ∥ ☿
00 38	☽ △ ☿
03 18	☽ ∥ ♅
07 06	☽ ⚹ ♄
07 16	☽ ∠ ♃
19 54	☽ bQ ☉
22 03	☽ ⚹ ♇
22 37	☽ ☌ ♆

AUGUST 1982

LONGITUDES

Date	Sidereal time h m s	Sun ☉ ° ' "	Moon ☽ ° ' "	Moon ☽ 24.00 ° ' "	Mercury ☿ ° '	Venus ♀ ° '	Mars ♂ ° '	Jupiter ♃ ° '	Saturn ♄ ° '	Uranus ♅ ° '	Neptune ♆ ° '	Pluto ♇ ° '
01	08 39 05	08 ♌ 54 36	01 ♑ 10 43	07 ♑ 04 42	16 ♌ 43	14 ♋ 15	28 ♎ 52	02 ♏ 10	17 ♎ 03	00 ♐ 36	24 ♐ 36	24 ♎ 20
02	08 43 02	09 51 59	12 ♑ 59 25	18 ♑ 55 13	18 40	15 28	29 ♎ 26	02 16	17 08	00 R 36	24 R 35	24 21
03	08 46 58	10 49 24	24 ♑ 52 28	00 ♒ 51 28	20 36	16 40	00 ♏ 00	02 22	17 12	00 35	24 34	24 22
04	08 50 55	11 46 49	06 ♒ 52 28	12 ♒ 55 44	22 30	17 53	00 35	02 29	17 16	00 35	24 33	24 23
05	08 54 52	12 44 15	19 ♒ 01 27	25 ♒ 09 48	24 22	19 06	01 09	02 35	17 20	00 35	24 32	24 24
06	08 58 48	13 41 42	01 ♓ 20 56	07 ♓ 35 00	26 13	20 19	01 44	02 41	17 25	00 35	24 31	24 25
07	09 02 45	14 39 10	13 ♓ 52 07	20 ♓ 12 23	28 03	21 31	02 18	02 48	17 29	00 35	24 30	24 26
08	09 06 41	15 36 40	26 ♓ 35 55	03 ♈ 02 49	29 ♌ 51	22 44	02 53	02 55	17 34	00 35	24 29	24 27
09	09 10 38	16 34 10	09 ♈ 33 12	16 ♈ 07 11	01 ♍ 37	23 57	03 29	03 01	17 38	00 D 35	24 28	24 29
10	09 14 34	17 31 42	22 ♈ 44 51	29 ♈ 26 19	03 22	25 10	04 04	03 08	17 43	00 35	24 27	24 30
11	09 18 31	18 29 15	06 ♉ 11 42	13 ♉ 01 04	05 05	26 23	04 39	03 15	17 48	00 35	24 27	24 31
12	09 22 27	19 26 50	19 ♉ 54 30	26 ♉ 52 00	06 46	27 36	05 15	03 23	17 53	00 35	24 26	24 32
13	09 26 24	20 24 26	03 ♊ 53 34	10 ♊ 59 07	08 27	28 ♋ 49	05 51	03 30	17 58	00 35	24 25	24 34
14	09 30 21	21 22 04	18 ♊ 08 29	25 ♊ 21 25	10 05	00 ♌ 03	06 27	03 38	18 03	00 35	24 24	24 35
15	09 34 17	22 19 43	02 ♋ 37 33	09 ♋ 56 26	11 42	01 16	07 03	03 45	18 08	00 36	24 24	24 36
16	09 38 14	23 17 24	17 ♋ 17 28	24 ♋ 39 59	13 18	02 29	07 39	03 53	18 13	00 36	24 23	24 38
17	09 42 10	24 15 07	02 ♌ 03 10	09 ♌ 26 11	14 52	03 42	08 16	04 01	18 18	00 36	24 22	24 39
18	09 46 07	25 12 50	16 ♌ 48 05	24 ♌ 07 58	16 25	04 56	08 52	04 09	18 23	00 37	24 22	24 40
19	09 50 03	26 10 35	01 ♍ 24 54	08 ♍ 38 02	17 56	06 09	09 29	04 17	18 29	00 37	24 21	24 42
20	09 54 00	27 08 22	15 ♍ 46 35	22 ♍ 49 54	19 26	07 22	10 06	04 25	18 34	00 38	24 21	24 43
21	09 57 56	28 06 09	29 ♍ 47 27	06 ♎ 38 53	20 54	08 36	10 43	04 34	18 40	00 38	24 20	24 45
22	10 01 53	29 ♌ 03 58	13 ♎ 23 58	20 ♎ 02 39	22 21	09 49	11 20	04 42	18 45	00 39	24 20	24 46
23	10 05 50	00 ♍ 01 48	26 ♎ 35 01	03 ♏ 01 17	23 46	11 03	11 58	04 51	18 51	00 40	24 19	24 48
24	10 09 46	00 59 39	09 ♏ 21 47	15 ♏ 36 56	25 09	12 16	12 35	04 59	18 56	00 40	24 19	24 50
25	10 13 43	01 57 32	21 ♏ 47 15	27 ♏ 53 18	26 31	13 30	13 13	05 08	19 02	00 41	24 19	24 51
26	10 17 39	02 55 25	03 ♐ 55 41	09 ♐ 55 03	27 52	14 44	13 51	05 17	19 08	00 42	24 18	24 53
27	10 21 36	03 53 20	15 ♐ 52 03	21 ♐ 47 21	29 ♍ 11	15 57	14 29	05 26	19 14	00 43	24 18	24 55
28	10 25 32	04 51 17	27 ♐ 41 36	03 ♑ 35 26	00 ♎ 28	17 11	15 07	05 35	19 19	00 44	24 18	24 56
29	10 29 29	05 49 15	09 ♑ 29 28	15 ♑ 24 18	01 43	18 25	15 45	05 45	19 25	00 45	24 17	24 58
30	10 33 25	06 47 14	21 ♑ 20 28	27 ♑ 18 27	02 56	19 39	16 24	05 54	19 31	00 46	24 17	25 00
31	10 37 22	07 ♍ 45 14	03 ♒ 18 43	09 ♒ 21 39	04 ♎ 08	20 ♌ 52	17 ♏ 02	06 ♏ 04	19 ♎ 37	00 ♐ 47	24 ♐ 17	25 ♎ 02

Date	Moon True ☊ ° '	Moon Mean ☊ ° '	Moon ☽ Latitude ° '
01	13 ♋ 16	11 ♋ 56	01 N 06
02	13 D 17	11 53	00 N 02
03	13 R 16	11 50	01 S 03
04	13 14	11 47	02 06
05	13 11	11 43	03 04
06	13 06	11 40	03 53
07	13 01	11 37	04 33
08	12 56	11 34	04 59
09	12 51	11 31	05 11
10	12 48	11 28	05 07
11	12 45	11 24	04 46
12	12 D 45	11 21	04 09
13	12 45	11 18	03 16
14	12 47	11 15	02 10
15	12 48	11 12	00 S 55
16	12 R 48	11 09	00 N 24
17	12 47	11 05	01 43
18	12 43	11 02	02 54
19	12 38	10 59	03 53
20	12 31	10 56	04 36
21	12 24	10 53	05 01
22	12 17	10 49	05 08
23	12 11	10 46	04 59
24	12 07	10 43	04 35
25	12 05	10 40	03 58
26	12 D 04	10 37	03 11
27	12 05	10 34	02 17
28	12 06	10 30	01 17
29	12 07	10 27	00 N 14
30	12 R 06	10 24	00 S 49
31	12 ♋ 04	10 ♋ 21	01 S 51

DECLINATIONS

Date	Sun ☉ ° '	Moon ☽ ° '	Mercury ☿ ° '	Venus ♀ ° '	Mars ♂ ° '	Jupiter ♃ ° '	Saturn ♄ ° '	Uranus ♅ ° '	Neptune ♆ ° '	Pluto ♇ ° '
01	18 N 02	22 S 20	17 N 30	22 N 27	11 S 48	11 S 12	04 S 29	20 S 08	22 S 02	06 N 19
02	17 47	22 47	16 53	22 22	12 01	11 14	04 31	20 08	22 02	06 18
03	17 31	22 12	16 14	22 16	12 14	11 17	04 32	20 08	22 02	06 17
04	17 15	20 35	15 34	22 09	12 27	11 19	04 34	20 08	22 02	06 16
05	16 59	18 02	14 54	22 02	12 39	11 21	04 36	20 08	22 02	06 16
06	16 43	14 38	14 13	21 54	12 52	11 24	04 38	20 08	22 02	06 15
07	16 26	10 32	13 31	21 46	13 05	11 26	04 40	20 08	22 02	06 14
08	16 09	05 56	12 49	21 36	13 18	11 29	04 42	20 08	22 02	06 13
09	15 52	00 S 59	12 07	21 27	13 31	11 31	04 44	20 08	22 02	06 12
10	15 35	04 N 05	11 24	21 16	13 43	11 34	04 46	20 08	22 02	06 11
11	15 17	09 05	10 41	21 05	13 56	11 36	04 48	20 08	22 02	06 10
12	14 59	13 43	09 58	20 54	14 09	11 39	04 50	20 08	22 02	06 09
13	14 41	17 43	09 14	20 42	14 21	11 42	04 52	20 08	22 02	06 08
14	14 23	20 45	08 31	20 29	14 34	11 45	04 54	20 08	22 02	06 07
15	14 04	22 30	07 48	20 15	14 47	11 47	04 56	20 08	22 02	06 07
16	13 45	22 44	07 04	20 01	14 59	11 50	04 58	20 08	22 02	06 06
17	13 26	21 22	06 21	19 47	15 12	11 53	05 00	20 09	22 02	06 05
18	13 07	18 34	05 38	19 32	15 24	11 56	05 02	20 09	22 02	06 04
19	12 48	14 36	04 55	19 16	15 36	11 59	05 04	20 09	22 02	06 03
20	12 28	09 51	04 12	19 00	15 49	12 02	05 07	20 09	22 02	06 02
21	12 08	04 N 41	03 30	18 43	16 01	12 05	05 09	20 09	22 02	06 01
22	11 48	00 S 33	02 48	18 26	16 13	12 08	05 11	20 09	22 02	06 00
23	11 28	05 36	02 06	18 08	16 26	12 11	05 13	20 09	22 02	05 59
24	11 07	10 16	01 25	17 50	16 38	12 14	05 16	20 10	22 02	05 58
25	10 47	14 23	00 45	17 31	16 50	12 17	05 18	20 10	22 02	05 57
26	10 26	17 48	00 N 04	17 11	17 02	12 20	05 20	20 10	22 02	05 56
27	10 05	20 26	00 S 35	16 52	17 14	12 24	05 23	20 10	22 02	05 55
28	09 44	22 08	01 14	16 31	17 25	12 27	05 25	20 10	22 02	05 54
29	09 23	22 52	01 52	16 11	17 37	12 30	05 27	20 11	22 02	05 53
30	09 01	22 34	02 30	15 49	17 49	12 33	05 30	20 11	22 03	05 52
31	08 N 40	21 S 13	03 S 07	15 N 28	18 S 00	12 S 37	05 S 32	20 S 11	22 S 03	05 N 51

ZODIAC SIGN ENTRIES

Date	h m	Planets
01	09 36	☽ ♑
03	11 45	♂ ♏
03	22 17	☽ ♒
06	09 23	☽ ♓
08	14 06	☿ ♍
08	18 21	☽ ♈
11	01 00	☽ ♉
13	05 22	☽ ♊
14	11 09	♀ ♌
15	07 40	☽ ♋
17	08 40	☽ ♌
19	09 40	☽ ♍
21	12 22	☽ ♎
23	11 15	☉ ♍
23	18 21	☽ ♏
26	04 11	☽ ♐
28	03 22	☿ ♎
28	16 42	☽ ♑
31	05 23	☽ ♒

LATITUDES

Date	Mercury ☿ ° '	Venus ♀ ° '	Mars ♂ ° '	Jupiter ♃ ° '	Saturn ♄ ° '	Uranus ♅ ° '	Neptune ♆ ° '	Pluto ♇ ° '
01	01 N 46	00 S 14	00 S 47	01 N 06	02 N 24	00 N 09	01 N 18	16 N 55
04	01 39	00 S 06	00 49	01 05	02 24	00 09	01 18	16 53
07	01 28	00 N 02	00 52	01 04	02 23	00 09	01 18	16 51
10	01 13	00 10	00 54	01 04	02 22	00 08	01 18	16 50
13	00 54	00 18	00 56	01 03	02 22	00 08	01 17	16 48
16	00 33	00 26	00 58	01 02	02 21	00 08	01 17	16 47
19	00 N 10	00 33	01 00	01 02	02 21	00 08	01 17	16 45
22	00 S 15	00 40	01 02	01 01	02 20	00 08	01 17	16 44
25	00 42	00 46	01 04	01 00	02 20	00 08	01 17	16 42
28	01 09	00 53	01 06	01 00	02 19	00 08	01 17	16 41
31	01 S 36	00 N 58	01 S 08	00 N 59	02 N 19	00 N 08	01 N 17	16 N 40

DATA

Julian Date	2445183
Delta T	+52 seconds
Ayanamsa	23° 36' 34"
Synetic vernal point	05° ♓ 30' 26"
True obliquity of ecliptic	23° 26' 28"

MOON'S PHASES, APSIDES AND POSITIONS ☽

Date	h m	Phase	Longitude ° '	Eclipse Indicator
04	22 34	○	12 ♒ 12	
12	11 08	☾	19 ♉ 25	
19	02 45	●	25 ♌ 48	
26	09 49	☽	02 ♐ 50	

Day	h m		
01	09 25	Apogee	
17	02 21	Perigee	
29	00 19	Apogee	
02	10 27	Max dec	22° S 47'
09	16 41	0N	
16	03 30	Max dec	22° N 50'
22	09 27	0S	
29	17 00	Max dec	22° S 53'

LONGITUDES

Date	Chiron ⚷ ° '	Ceres ⚳ ° '	Pallas ⚴ ° '	Juno ⚵ ° '	Vesta ⚶ ° '	Black Moon Lilith ⚸ ° '
01	27 ♉ 12	16 ♏ 07	16 ♎ 23	25 ♐ 12	19 ♒ 42	04 ♑ 38
11	27 ♉ 29	18 ♏ 09	19 ♎ 48	24 ♐ 27	17 ♒ 15	05 ♑ 44
21	27 ♉ 39	20 ♏ 33	23 ♎ 25	24 ♐ 17	14 ♒ 52	06 ♑ 51
31	27 ♉ 44	23 ♏ 15	27 ♎ 11	24 ♐ 43	12 ♒ 53	07 ♑ 57

All ephemeris data is given at 12.00 UT and the Moon's longitude is additionally given for 24.00 UT

ASPECTARIAN

h m Aspects

01 Sunday
05 24 ☽ ∥ ♆
07 04 ☽ ⚹ ♂
07 39 ☽ Q ♄
10 50 ☽ ⊻ ♅
13 18 ☽ bQ ☿
14 03 ☽ ⚹ ♃
14 44 ☿ Q ♂
14 47 ☽ # ♀
15 50 ☽ ± ☉
16 21 ☿ ⚹ ♄
22 30 ☽ Q ♇
23 02 ☽ ⊥ ♅

02 Monday
04 53 ☉ bQ ♆
05 06 ☽ ⊼ ☉
08 42 ☽ Q ♂
11 13 ☽ ± ☿
14 37 ☽ Q ♃
14 40 ♀ bQ ♅
17 17 ☽ ∠ ♅
17 34 ☽ ☍ ♀
20 25 ☽ □ ♄

03 Tuesday
01 43 ☽ ⊼ ☿
09 03 ☿ Q ♃
10 21 ☽ # ♀
10 59 ☽ □ ♇
11 22 ☽ ⊻ ♆
15 18 ☽ ∥ ♆
22 49 ☽ □ ♂
23 01 ♀ □ ♄
23 23 ☽ ⊥ ♆
23 28 ☽ ⚹ ♅

04 Wednesday
03 09 ☽ □ ♃
12 23 ♂ ⊻ ♅
16 56 ☽ ∥ ♅
17 18 ☽ ∠ ♆
22 34 ☽ ☍ ☉
23 19 ☽ Q ♅

05 Thursday
03 26 ☉ Q ♇
08 40 ☽ △ ♄
11 38 ♀ # ♆
12 09 ☽ ⊼ ♀
12 23 ☿ ⚹ ♇
14 00 ☿ △ ♆
20 39 ☽ # ☉
22 32 ☽ △ ♇
22 45 ☽ ⚹ ♆

06 Friday
00 19 ☽ ☍ ☿
01 10 ☽ ± ♀
10 31 ☽ □ ♅
12 46 ☽ △ ♂
14 04 ☽ bQ ♄
14 36 ☽ △ ♃
15 09 ☽ # ☿
15 21 ☿ ⊥ ♀
20 27 ☽ bQ ♀
21 56 ☽ Q ♆
22 13 ☽ ∥ ♂

07 Saturday
03 32 ☽ bQ ♇
07 04 ☽ ∥ ♃
07 26 ☽ ± ♄
13 37 ☽ ⊼ ☉
18 50 ☽ bQ ♂
18 54 ☽ ⊼ ♄
19 31 ☽ bQ ♃
20 41 ☽ ± ♇
23 35 ☿ # ♂

08 Sunday
01 53 ☽ ± ☉
04 00 ☽ △ ♀
07 59 ☽ ⊼ ♇
08 03 ☽ □ ♆
10 34 ☽ # ♇
12 34 ☽ ± ♂
12 35 ☽ ± ♃
12 55 ♂ ☌ ♃
18 04 ☽ ∥ ♄
19 01 ☽ ⊼ ☿
19 25 ☽ △ ♅
20 05 ☽ bQ ☉
21 53 ☿ □ ♅
23 51 ☽ ⊼ ♃

09 Monday
00 16 ☽ ⊼ ♂
07 02 ♆ ⚹ ♇
07 53 ☽ ± ☿
10 22 ♅ St D
22 04 ♀ ⊼ ♆
22 30 ♀ □ ♇
23 01 ☽ bQ ♅

10 Tuesday
01 49 ☽ △ ☉
02 44 ☿ ∠ ♄
02 51 ☽ ☍ ♄
02 52 ☽ bQ ☿
06 51 ☿ # ♃
08 44 ☿ ⚹ ♃
15 04 ☽ △ ♆
15 09 ☽ ☍ ♇
15 12 ☽ # ♄
15 17 ☽ ± ♅
16 48 ☽ □ ♀
17 14 ☉ ⚹ ♄
21 56 ☽ ∥ ♇

11 Wednesday
02 02 ☽ ⊼ ♅
02 56 ☿ ⚹ ♂
06 45 ☽ ☍ ♃
09 09 ☽ ☍ ♂
09 45 ☽ △ ☿
17 43 ☽ bQ ♆
18 58 ☽ ∥ ☿

12 Thursday
00 53 ☽ # ♃
03 48 ☽ Q ♀
08 28 ☽ ⊼ ♄
09 26 ☽ ± ♆
11 08 ☽ □ ☉
14 28 ☽ # ♂
18 39 ☽ ∥ ☉
18 54 ☽ ± ♄
19 48 ☽ ⊼ ♆
20 01 ☽ ⊼ ♇

13 Friday
02 32 ☽ ⚹ ♀
06 19 ☽ ± ♇
06 22 ☽ ☍ ♅
10 25 ☽ bQ ♄
11 20 ☽ ⊼ ♃
15 28 ☽ ⊼ ♂
20 12 ☽ Q ☉
20 44 ☽ □ ☿
21 35 ☽ ± ♃
21 37 ☽ bQ ♇

14 Saturday
02 02 ☽ ± ♂
03 23 ☉ # ♂
04 29 ☿ ∠ ♇
06 19 ☽ # ♅
06 20 ☽ ∠ ♀
09 37 ☽ ∥ ♀
11 50 ☽ △ ♄
12 49 ☽ bQ ♃
17 45 ☽ bQ ♂
17 46 ☽ ⚹ ☉
18 57 ♅ ⊥ ♇
19 05 ♀ ± ♆
19 28 ☉ Q ♃
22 25 ☽ ☍ ♆
22 44 ☽ ⊥ ♀
22 44 ☽ △ ♇
22 45 ♀ △ ♅

15 Sunday
03 23 ☽ # ♆
06 36 ☽ Q ☿
08 39 ☽ ⊼ ♅
09 33 ☽ ⊻ ♀
13 52 ☽ △ ♃
18 31 ☽ ± ♅
18 43 ☿ ⊥ ♄
19 35 ☽ △ ♂
20 16 ☽ ∠ ☉

16 Monday
00 13 ♀ # ♅
04 42 ☽ ⚹ ☿
09 14 ☽ bQ ♅
12 00 ☽ ⊥ ☉
13 31 ☽ □ ♄
22 27 ☽ ⊻ ☉
23 32 ☽ ⊼ ♆
23 57 ☽ □ ♇

17 Tuesday
03 24 ☽ # ♆
08 02 ☽ ∠ ☿
09 16 ☽ ± ♆
09 39 ☽ △ ♅
14 56 ☽ ☌ ♀
15 00 ☉ △ ♆
15 13 ☽ □ ♃
18 49 ♀ □ ♃
18 57 ☽ Q ♄
21 16 ☿ ∥ ♇
22 12 ☉ ⚹ ♇
22 32 ☽ □ ♂
23 53 ☽ bQ ♆
23 55 ☽ # ♅

18 Wednesday
00 23 ☽ ⊥ ☿
04 15 ☽ ∥ ♀
05 16 ☽ Q ♇
11 18 ☽ ⊻ ☿
14 37 ☽ ⚹ ♄
20 50 ☽ Q ♃

19 Thursday
00 22 ☽ △ ♆
00 55 ☽ ⚹ ♇
02 45 ☽ ☌ ☉
05 14 ☽ Q ♂
06 39 ☽ # ♂
06 54 ☿ # ♄
06 55 ♂ ∠ ♆
10 41 ☽ □ ♅
15 26 ☽ ∠ ♄
16 48 ☽ ⚹ ♃
18 58 ♀ Q ♄
20 35 ☽ ⊻ ♀
21 15 ☿ ⊻ ♄
22 10 ☽ ∥ ☉
23 00 ☿ Q ♅

20 Friday
00 25 ☿ ⊥ ♇
01 26 ☽ # ♃
01 48 ☽ ∠ ♇
02 02 ☽ ⚹ ♂
06 34 ☽ ⊥ ♄
07 34 ☽ ⊥ ♀
11 51 ☿ ∠ ♃
16 46 ☽ ⊻ ♄
16 50 ☽ Q ♅
17 00 ☽ ⊥ ♇
18 15 ☽ ∠ ♃
18 55 ☽ ☌ ☿

21 Saturday
00 17 ☽ ∠ ♀
02 35 ☽ □ ♆
03 16 ☽ ⊻ ♇
04 38 ☽ ∠ ♂
05 54 ☽ ∥ ♇
08 51 ☽ ⊻ ☉
09 51 ☽ ⊥ ♃
09 54 ☽ # ♄
13 29 ☽ ⚹ ♅
15 19 ☉ # ♃
18 16 ☽ ∥ ☿
20 06 ☽ ⊥ ☉
20 25 ☽ ⊻ ♃
21 01 ☽ ⊥ ♂

22 Sunday
02 23 ♀ bQ ♆
04 59 ☽ ⚹ ♀
08 09 ☽ ⊻ ♂
10 05 ☽ Q ♆
13 17 ☽ ∠ ☉
16 03 ☽ ∠ ♅
21 15 ☽ # ☿
21 43 ☽ ☌ ♄

23 Monday
04 49 ☽ Q ♀
06 10 ☽ ⊻ ☿
07 20 ☉ Q ♂
07 50 ☽ ⚹ ♆
08 27 ☽ ⊥ ♅
08 42 ☽ ☌ ♇
10 07 ☽ ∥ ♄
13 51 ☽ # ♇
18 38 ☽ ⊥ ☿
18 55 ☽ ⚹ ☉
19 35 ☽ ⊻ ♅
21 30 ☿ □ ♆

24 Tuesday
03 37 ☽ ☌ ♃
03 56 ☉ □ ♅
06 10 ☿ ⊻ ♇
11 54 ☽ ∠ ♆
13 42 ☽ ∠ ☿
16 21 ☽ # ☉
18 10 ☽ □ ♀
18 30 ☽ ☌ ♂
19 32 ☽ Q ☉
23 04 ♀ Q ♇
23 09 ☽ ∥ ♃

25 Wednesday
00 37 ♀ □ ♂
05 13 ☽ ⊥ ♆
06 35 ☽ ⊻ ♄
16 56 ☽ ⊻ ♆
18 02 ☽ ⊻ ♇
18 25 ☽ ⊥ ♄
22 28 ☽ ⚹ ☿

26 Thursday
05 34 ☽ ☌ ♅
05 39 ☽ ∥ ♂
05 55 ☽ ⊥ ♇
07 42 ☽ # ♀
09 49 ☽ □ ☉
12 24 ☽ ∠ ♄
14 45 ☽ ⊻ ♃
19 27 ♀ # ♂
23 57 ☽ ∠ ♇

27 Friday
01 22 ☽ Q ☿
02 57 ☽ ⊥ ♃
09 02 ☽ ⊻ ♂
09 14 ☽ ∥ ♅
12 12 ☽ △ ♀
17 32 ☿ ⊥ ♃
18 51 ☽ ⚹ ♄
21 20 ☉ ∠ ♄
21 23 ☽ ∠ ♃
21 52 ☽ ⊥ ♂
23 10 ☿ ∠ ♂

28 Saturday
05 05 ☽ ☌ ♆
06 23 ☽ ⚹ ♇
10 09 ☽ ∥ ♆
17 13 ☽ ∠ ♂
17 17 ☿ ⚹ ♅
18 12 ☽ ⊻ ♅
18 18 ☽ □ ☿
19 27 ☽ Q ♄
22 12 ☽ bQ ♀

29 Sunday
03 52 ☽ △ ☉
04 17 ☽ ⚹ ♃
06 25 ☽ ⊥ ♅
06 51 ☽ Q ♇
09 46 ☉ ⚹ ♃
18 38 ☽ ± ♀

30 Monday
00 43 ☽ ∠ ♅
01 26 ☽ ⚹ ♂
04 58 ☽ Q ♃
08 11 ☽ ⊼ ♀
08 18 ☽ □ ♄
09 24 ♀ ⚹ ♄
12 59 ☽ bQ ☉
17 56 ☽ ⊻ ♆
19 23 ☽ □ ♇
23 36 ☽ ∥ ♆

31 Tuesday
02 59 ☽ Q ♂
05 57 ☽ ⊥ ♆
06 57 ☽ ⚹ ♅
08 37 ☽ ± ☉
13 48 ☽ △ ☿
17 32 ☽ □ ♃
21 35 ☽ ⊼ ☉
23 50 ☽ ∠ ♆
23 54 ☽ ∥ ♅

SEPTEMBER 1982

LONGITUDES

Date	Sidereal time h m s	Sun ☉ ° ' "	Moon ☽ ° ' "	Moon ☽ 24.00 ° ' "	Mercury ☿ ° '	Venus ♀ ° '	Mars ♂ ° '	Jupiter ♃ ° '	Saturn ♄ ° '	Uranus ♅ ° '	Neptune ♆ ° '	Pluto ♇ ° '
01	10 41 19	08 ♍ 43 16	15 ♒ 27 35	21 ♒ 36 46	05 ♎ 17	22 ♌ 06	17 ♏ 41	06 ♏ 13	19 ♎ 43	00 ♐ 48	24 ♐ 17	25 ♎ 03
02	10 45 15	09 41 19	27 ♒ 49 25	04 ♓ 05 40	06 25	23 20	18 19	06 23	19 50	00 50	24 R 17	25 05
03	10 49 12	10 39 24	10 ♓ 25 35	16 ♓ 49 10	07 30	24 34	18 58	06 33	19 56	00 51	24 17	25 07
04	10 53 08	11 37 31	23 ♓ 16 22	29 ♓ 47 05	08 33	25 48	19 37	06 42	20 02	00 52	24 16	25 09
05	10 57 05	12 35 39	06 ♈ 21 12	12 ♈ 58 33	09 33	27 02	20 16	06 52	20 08	00 53	24 16	25 11
06	11 01 01	13 33 50	19 ♈ 38 57	26 ♈ 22 14	10 31	28 16	20 56	07 02	20 14	00 55	24 D 16	25 13
07	11 04 58	14 32 02	03 ♉ 08 12	09 ♉ 56 43	11 26	29 ♌ 30	21 35	07 13	20 21	00 56	24 16	25 15
08	11 08 54	15 30 16	16 ♉ 47 39	23 ♉ 40 51	12 19	00 ♍ 44	22 15	07 23	20 27	00 58	24 17	25 17
09	11 12 51	16 28 32	00 ♊ 36 16	07 ♊ 33 48	13 08	01 59	22 54	07 33	20 34	00 59	24 17	25 19
10	11 16 48	17 26 51	14 ♊ 33 24	21 ♊ 35 00	13 54	03 13	23 34	07 43	20 40	01 01	24 17	25 21
11	11 20 44	18 25 11	28 ♊ 38 32	05 ♋ 43 53	14 36	04 27	24 14	07 54	20 47	01 03	24 17	25 23
12	11 24 41	19 23 34	12 ♋ 50 55	19 ♋ 59 25	15 15	05 41	24 54	08 05	20 53	01 04	24 17	25 25
13	11 28 37	20 21 59	27 ♋ 09 07	04 ♌ 19 38	15 49	06 56	25 34	08 15	21 00	01 06	24 17	25 27
14	11 32 34	21 20 26	11 ♌ 30 33	18 ♌ 41 20	16 19	08 10	26 14	08 26	21 06	01 08	24 18	25 29
15	11 36 30	22 18 55	25 ♌ 51 23	03 ♍ 00 03	16 44	09 24	26 55	08 37	21 13	01 10	24 18	25 31
16	11 40 27	23 17 26	10 ♍ 06 41	17 ♍ 10 34	17 04	10 39	27 35	08 48	21 20	01 12	24 18	25 33
17	11 44 23	24 15 59	24 ♍ 11 03	01 ♎ 07 32	17 19	11 53	28 16	08 59	21 27	01 14	24 19	25 35
18	11 48 20	25 14 33	07 ♎ 59 28	14 ♎ 46 25	17 28	13 08	28 56	09 10	21 33	01 15	24 19	25 37
19	11 52 17	26 13 10	21 ♎ 28 03	28 ♎ 04 13	17 R 31	14 22	29 ♏ 37	09 21	21 40	01 18	24 19	25 39
20	11 56 13	27 11 48	04 ♏ 34 49	10 ♏ 59 55	17 28	15 37	00 ♐ 18	09 32	21 47	01 20	24 20	25 41
21	12 00 10	28 10 29	17 ♏ 19 42	23 ♏ 34 28	17 17	16 51	00 59	09 43	21 54	01 22	24 20	25 44
22	12 04 06	29 ♍ 09 11	29 ♏ 44 36	05 ♐ 50 34	17 00	18 06	01 40	09 55	22 01	01 24	24 21	25 46
23	12 08 03	00 ♎ 07 54	11 ♐ 52 56	17 ♐ 52 17	16 36	19 20	02 22	10 06	22 08	01 26	24 21	25 48
24	12 11 59	01 06 40	23 ♐ 49 16	29 ♐ 44 32	16 04	20 35	03 03	10 18	22 15	01 28	24 22	25 50
25	12 15 56	02 05 27	05 ♑ 38 47	11 ♑ 32 44	15 25	21 49	03 45	10 29	22 22	01 31	24 23	25 53
26	12 19 52	03 04 16	17 ♑ 27 03	23 ♑ 22 26	14 39	23 04	04 26	10 41	22 29	01 33	24 23	25 55
27	12 23 49	04 03 07	29 ♑ 19 31	05 ♒ 18 55	13 47	24 19	05 08	10 53	22 36	01 35	24 24	25 57
28	12 27 46	05 01 59	11 ♒ 21 14	17 ♒ 26 58	12 49	25 34	05 50	11 04	22 43	01 38	24 25	25 59
29	12 31 42	06 00 53	23 ♒ 36 34	29 ♒ 50 24	11 46	26 48	06 31	11 16	22 50	01 40	24 26	26 02
30	12 35 39	06 ♎ 59 49	06 ♓ 08 46	12 ♓ 31 52	10 ♎ 40	28 ♍ 03	07 ♐ 13	11 ♏ 28	22 ♎ 57	01 ♐ 43	24 ♐ 26	26 ♎ 04

Date	Moon True ☊ ° '	Moon Mean ☊ ° '	Moon ☽ Latitude ° '
01	11 ♋ 59	10 ♋ 18	02 S 49
02	11 R 52	10 15	03 40
03	11 43	10 11	04 20
04	11 33	10 08	04 49
05	11 22	10 05	05 03
06	11 13	10 02	05 01
07	11 05	09 59	04 42
08	10 59	09 55	04 07
09	10 56	09 52	03 17
10	10 55	09 49	02 16
11	10 D 55	09 46	01 S 05
12	10 R 55	09 43	00 N 10
13	10 54	09 40	01 25
14	10 51	09 36	02 35
15	10 45	09 33	03 35
16	10 37	09 30	04 20
17	10 26	09 27	04 50
18	10 14	09 24	05 01
19	10 03	09 21	04 56
20	09 52	09 17	04 34
21	09 44	09 14	04 00
22	09 38	09 11	03 14
23	09 35	09 08	02 21
24	09 34	09 05	01 23
25	09 D 34	09 01	00 N 21
26	09 R 34	08 58	00 S 42
27	09 33	08 55	01 43
28	09 29	08 52	02 40
29	09 23	08 49	03 31
30	09 ♋ 15	08 ♋ 46	04 S 12

DECLINATIONS

Date	Sun ☉ ° '	Moon ☽ ° '	Mercury ☿ ° '	Venus ♀ ° '	Mars ♂ ° '	Jupiter ♃ ° '	Saturn ♄ ° '	Uranus ♅ ° '	Neptune ♆ ° '	Pluto ♇ ° '
01	08 N 18	18 S 53	03 S 43	15 N 05	18 S 12	12 S 40	05 S 35	20 S 11	22 S 03	05 N 50
02	07 56	15 40	04 28	14 43	18 23	12 43	05 37	20 12	22 03	05 49
03	07 34	11 40	04 52	14 20	18 34	12 47	05 39	20 12	22 03	05 48
04	07 12	07 05	05 25	13 57	18 46	12 50	05 42	20 12	22 03	05 47
05	06 50	02 S 07	05 57	13 33	18 57	12 53	05 44	20 12	22 03	05 46
06	06 28	03 N 03	06 28	13 09	19 08	12 57	05 47	20 13	22 03	05 45
07	06 05	08 08	06 58	12 44	19 19	13 00	05 49	20 13	22 03	05 44
08	05 43	12 54	07 26	12 19	19 29	13 04	05 52	20 13	22 03	05 43
09	05 20	17 04	07 53	11 54	19 40	13 07	05 54	20 14	22 03	05 42
10	04 58	20 18	08 18	11 29	19 50	13 11	05 57	20 14	22 03	05 41
11	04 35	22 21	08 42	11 03	20 01	13 15	05 59	20 14	22 03	05 40
12	04 12	22 59	09 03	10 37	20 11	13 18	06 02	20 15	22 03	05 39
13	03 49	22 07	09 23	10 10	20 21	13 22	06 05	20 15	22 03	05 38
14	03 26	19 49	09 40	09 43	20 31	13 25	06 07	20 16	22 03	05 37
15	03 03	16 16	09 56	09 16	20 41	13 29	06 10	20 16	22 03	05 36
16	02 40	11 48	10 08	08 49	20 51	13 32	06 12	20 16	22 03	05 35
17	02 17	06 44	10 18	08 22	21 00	13 36	06 15	20 17	22 04	05 34
18	01 53	01 N 27	10 25	07 54	21 10	13 40	06 18	20 17	22 04	05 33
19	01 30	03 S 48	10 29	07 26	21 19	13 43	06 20	20 18	22 04	05 32
20	01 07	08 44	10 29	06 57	21 28	13 47	06 23	20 18	22 04	05 31
21	00 44	13 10	10 26	06 29	21 37	13 51	06 25	20 18	22 04	05 30
22	00 N 20	16 56	10 18	06 00	21 46	13 55	06 28	20 19	22 04	05 29
23	00 S 03	19 53	10 07	05 32	21 54	13 58	06 31	20 19	22 04	05 28
24	00 27	21 55	09 51	05 03	22 03	14 02	06 33	20 20	22 04	05 27
25	00 50	22 58	09 31	04 33	22 11	14 06	06 36	20 20	22 04	05 26
26	01 13	23 00	09 07	04 04	22 19	14 10	06 39	20 21	22 04	05 25
27	01 37	21 58	08 38	03 35	22 27	14 13	06 41	20 21	22 04	05 24
28	02 00	19 56	08 05	03 05	22 35	14 17	06 44	20 22	22 04	05 23
29	02 23	16 58	07 29	02 36	22 43	14 21	06 47	20 22	22 04	05 22
30	02 S 47	13 S 10	06 S 49	02 N 06	22 S 50	14 S 25	06 S 49	20 S 23	22 S 04	05 N 21

ZODIAC SIGN ENTRIES

Date	h	m	Planets
02	16	11	☽ ♓
05	00	24	☽ ♈
07	06	27	☽ ♉
07	21	38	♀ ♍
09	10	57	☽ ♊
11	14	18	☽ ♋
13	16	46	☽ ♌
15	18	57	☽ ♍
17	22	03	☽ ♎
20	01	20	♂ ♐
20	03	32	☽ ♏
22	12	30	☽ ♐
23	08	46	☉ ♎
25	00	31	☽ ♑
27	13	21	☽ ♒
30	00	18	☽ ♓

LATITUDES

Date	Mercury ☿ ° '	Venus ♀ ° '	Mars ♂ ° '	Jupiter ♃ ° '	Saturn ♄ ° '	Uranus ♅ ° '	Neptune ♆ ° '	Pluto ♇ ° '
01	01 S 45	01 N 00	01 S 08	00 N 59	02 N 19	00 N 08	01 N 17	16 N 40
04	02 13	01 05	01 10	00 58	02 18	00 08	01 16	16 38
07	02 39	01 10	01 11	00 58	02 18	00 08	01 16	16 37
10	03 04	01 14	01 13	00 57	02 17	00 08	01 16	16 36
13	03 25	01 18	01 14	00 57	02 17	00 08	01 16	16 35
16	03 43	01 20	01 15	00 56	02 17	00 08	01 16	16 34
19	03 54	01 23	01 16	00 56	02 17	00 08	01 16	16 33
22	03 56	01 25	01 18	00 56	02 16	00 08	01 15	16 33
25	03 44	01 26	01 19	00 55	02 16	00 08	01 15	16 32
28	03 17	01 27	01 19	00 55	02 16	00 08	01 15	16 31
31	02 S 33	01 N 27	01 S 20	00 N 54	02 N 16	00 N 08	01 N 15	16 N 31

DATA

Julian Date	2445214
Delta T	+52 seconds
Ayanamsa	23° 36' 38"
Synetic vernal point	05° ♓ 30' 22"
True obliquity of ecliptic	23° 26' 28"

MOON'S PHASES, APSIDES AND POSITIONS ☽

Date	h	m	Phase	Longitude ° '	Eclipse Indicator
03	12	28	○	10 ♓ 41	
10	17	19	☾	17 ♊ 40	
17	12	09	●	24 ♍ 16	
25	04	07	☽	01 ♑ 46	

Day	h	m		
13	17	32	Perigee	
25	18	52	Apogee	
05	21	52	0N	
12	10	14	Max dec	23° N 00'
18	18	32	0S	
26	00	25	Max dec	23° S 07'

LONGITUDES

Date	Chiron ⚷ ° '	Ceres ⚳ ° '	Pallas ⚴ ° '	Juno ⚵ ° '	Vesta ⚶ ° '	Black Moon Lilith ⚸ ° '
01	27 ♉ 44	23 ♏ 32	27 ♎ 34	24 ♐ 47	12 ♒ 43	08 ♑ 04
11	27 ♉ 41	26 ♏ 32	01 ♏ 29	25 ♐ 49	11 ♒ 26	09 ♑ 10
21	27 ♉ 33	29 ♏ 45	05 ♏ 32	27 ♐ 19	10 ♒ 56	10 ♑ 17
31	27 ♉ 18	03 ♐ 09	08 ♏ 39	29 ♐ 15	11 ♒ 11	11 ♑ 23

All ephemeris data is given at 12.00 UT and the Moon's longitude is additionally given for 24.00 UT

ASPECTARIAN

h m	Aspects
01 Wednesday	
06 47	☽ Q ♅
16 35	☽ □ ♂
17 24	☽ ∥ ♂
20 24	☽ △ ♄
22 22	☽ bQ ☿
02 Thursday	
02 23	☽ ☍ ♀
05 10	☽ ⚹ ♆
06 43	☽ △ ♇
10 15	♂ ⊥ ♆
11 15	☿ ⊻ ♃
17 26	☽ ± ☿
17 46	☽ □ ♅
18 46	☽ # ♀
22 11	☉ ∠ ♇
03 Friday	
01 30	☽ bQ ♄
04 09	☽ Q ♆
04 33	☽ △ ♃
05 49	☽ ∥ ♃
05 57	☽ ⊼ ☿
06 16	♀ △ ♆
11 23	♀ Q ♃
11 25	☽ bQ ♇
12 28	☽ ☍ ☉
18 38	☽ ± ♄
22 56	♀ ⚹ ♇
04 Saturday	
04 19	☽ ± ♇
04 52	☽ △ ♂
05 56	☽ ⊼ ♄
09 04	☽ bQ ♃
11 23	☽ # ☉
13 51	☽ □ ♆
15 29	☽ ⊼ ♇
17 10	☽ ⊼ ♀
18 28	☽ # ♇
18 48	☽ ∥ ♄
19 24	☽ ∥ ☿
05 Sunday	
01 24	☿ ∥ ♄
01 52	☽ ± ♃
02 01	☽ △ ♅
03 38	☿ # ♇
05 19	☽ ± ♀
05 56	♂ ⊻ ♄
09 56	☽ bQ ♂
12 57	☽ ⊼ ♃
18 17	☽ ☍ ☿
22 32	♄ # ♇
23 21	☽ bQ ♀
23 38	♆ St D
06 Monday	
00 12	☽ ⊼ ☉
03 05	☽ ± ♂
05 17	☽ bQ ♅
11 50	☉ # ☿
11 50	☽ ± ☉
13 04	☽ ☍ ♄
14 25	☽ ⊼ ♂
20 16	☽ △ ♆
21 25	☽ ± ♅
21 58	☽ ☍ ♇
22 10	♀ # ♃
07 Tuesday	
00 35	☽ ∥ ♇
00 53	☽ # ♄
02 54	☽ ∥ ☉
04 55	☽ △ ♀
05 08	☽ bQ ☉
05 46	☽ # ☿
06 48	☉ ⊥ ♄
08 06	☽ ⊼ ♅
19 17	☽ ☍ ♃
22 49	☽ bQ ♆
08 Wednesday	
03 25	☉ # ♄
03 38	☽ ⊼ ☿
09 11	☽ ∥ ♀
09 34	☽ △ ☉
10 58	☿ Q ♆
12 02	☉ ∥ ♇
12 52	☽ # ♃
14 35	☽ ± ♆
14 49	☽ ± ☿
16 27	♀ □ ♅
18 26	☽ ⊼ ♄
21 59	☽ ☍ ♂
09 Thursday	
01 02	☽ ⊼ ♆
02 48	☽ ⊼ ♇
04 57	☽ ± ♄
07 28	☽ bQ ☿
12 40	☽ ☍ ♅
13 13	☽ ± ♇
14 36	☽ □ ♀
20 37	☽ bQ ♄
10 Friday	
00 08	☽ ⊼ ♃
04 45	☽ bQ ♇
07 46	☽ # ♂
10 33	☽ ± ♃
10 48	☽ △ ☿
11 24	☽ # ♅
17 19	☽ □ ☉
22 31	☽ △ ♄
11 Saturday	
00 28	☽ Q ♀
02 07	☽ bQ ♃
04 08	☽ ⊼ ♂
04 36	☽ ☍ ♆
06 26	☽ △ ♇
07 14	☽ # ♆
13 48	♂ ⊻ ♆
14 50	☽ ± ♂
16 05	☽ ⊼ ♅
22 46	☽ ⚹ ♀
12 Sunday	
02 08	☽ Q ☉
02 15	☽ ± ♅
03 51	☽ △ ♃
03 52	☉ Q ♅
06 47	☽ bQ ♂
12 25	☉ ⊥ ♇
16 12	☽ □ ☿
16 14	♀ ∠ ♄
17 26	☽ bQ ♅
21 12	♂ ∥ ♅
23 48	☽ ⚹ ☉
13 Monday	
01 36	☽ □ ♄
02 25	☽ ∠ ♀
07 12	☽ ⊼ ♆
07 19	♂ ⊻ ♇
09 08	☽ □ ♇
09 13	☽ △ ♂
13 01	☽ # ♆
17 15	☽ ± ♆
18 37	☽ △ ♅
18 55	☽ ⊥ ♀
23 34	☽ Q ☿
14 Tuesday	
02 05	☿ ∠ ♅
02 44	☽ ∠ ☉
05 31	☉ ⊻ ♄
05 53	☽ ⊻ ♀
06 20	☽ # ♂
06 48	☽ □ ♃
07 57	☽ Q ♄
08 17	☽ # ♅
08 18	☽ bQ ♆
13 40	☿ # ♀
15 18	☽ Q ♇
18 05	♀ ⚹ ♃
18 52	☽ ⊥ ☉
20 18	☽ ⚹ ☿
15 Wednesday	
04 10	☽ ⚹ ♄
05 38	☽ ⊻ ☉
09 23	☽ △ ♆
11 25	☽ ⚹ ♇
13 17	☽ Q ♃
13 51	☽ □ ♂
20 56	☽ □ ♅
22 08	☽ ∠ ☿
16 Thursday	
01 08	♂ ⊥ ♄
03 12	☽ # ♃
05 34	☽ ∠ ♄
09 45	☽ ⚹ ♃
10 05	♀ ∠ ♇
12 44	☽ ∠ ♇
12 59	☽ ☌ ♀
13 40	☽ ⊥ ☿
19 49	☽ # ☿
20 56	☽ ⊥ ♄
21 46	☽ Q ♂
22 41	☿ ⊥ ♀
17 Friday	
00 04	☽ ⊻ ☿
03 18	☉ ∠ ♃
03 29	☽ Q ♅
03 47	☽ ∥ ♀
04 05	☽ ⊥ ♇
07 15	☽ ⊻ ♄
11 39	☽ ∠ ♃
12 09	☽ ☌ ☉
12 13	☽ □ ♆
13 05	☉ □ ♆
14 14	☽ # ♄
14 25	☽ ⊻ ♇
17 24	☽ ∥ ♇
19 24	☽ ⚹ ♂
18 Saturday	
00 12	☽ ⚹ ♅
03 26	☽ ⊥ ♃
09 50	☽ ∥ ☉
14 06	☽ ⊻ ♃
19 38	☽ Q ♆
21 36	☉ ⊻ ♇
21 59	☽ ⊻ ♀
23 05	☽ ∠ ♂
19 Sunday	
02 07	☽ # ☉
02 41	☽ ∠ ♅
04 54	☽ ☌ ☿
08 35	♃ ∠ ♆
09 49	☽ ⊥ ♀
11 02	☿ St R
12 22	☽ ☌ ♄
16 06	☽ ⊥ ♂
17 10	☽ ⚹ ♆
18 57	☽ ⊥ ♅
19 37	☽ ☌ ♇
20 10	☽ # ♇
21 19	☽ ⊻ ☉
20 Monday	
00 11	☽ ∥ ♄
03 40	☽ ⊻ ♂
03 53	☽ ∠ ♀
03 54	☽ # ♀
05 58	☽ ⊻ ♅
09 14	☽ ⊥ ☉
15 43	♀ ⊥ ♄
20 52	☽ ∠ ♆
21 00	☽ ∥ ☿
21 23	☽ ☌ ♃
21 Tuesday	
03 27	☽ ∠ ☉
10 59	☽ ⚹ ♀
11 56	☽ ⊻ ☿
13 56	☽ ⊥ ♆
14 48	♀ # ♄
16 04	☽ ∥ ♃
19 04	☿ ⊻ ♀
20 51	☽ ⊻ ♄
23 13	☽ ⊥ ☿
22 Wednesday	
01 29	☽ ⊻ ♆
01 48	♂ ☌ ♅
04 13	☽ ⊻ ♇
08 35	☽ ⊥ ♄
10 45	☽ ⚹ ☉
12 46	☽ Q ♀
15 15	☽ ☌ ♅
15 20	♂ ⊥ ♇
15 58	☽ ⊥ ♇
16 01	☽ ☌ ♂
16 18	☽ ∠ ☿
19 27	☿ ∠ ♂
23 Thursday	
02 27	☽ ∠ ♄
08 24	☽ ⊻ ♃
09 50	☽ ∠ ♇
12 33	☽ Q ☉
13 55	♀ Q ♅
15 18	♀ ∥ ♇
16 22	☽ ∥ ♅
19 15	☿ ∠ ♅
20 35	☽ ⊥ ♃
21 03	☽ ⚹ ☿
21 13	♀ ⊥ ♇
24 Friday	
04 42	☽ □ ♀
08 47	☽ ⚹ ♄
13 06	☽ ☌ ♆
14 12	☽ ∥ ♂
14 17	☽ ∥ ♆
15 02	☽ ∠ ♃
15 10	♂ ∥ ♆
16 06	☽ ⚹ ♇
20 10	☽ Q ☿
21 10	☉ ⚹ ♅
25 Saturday	
03 34	☽ ⊻ ♅
04 07	☽ □ ☉
07 53	☽ ⊻ ♂
08 25	♀ Q ♂
09 22	☽ Q ♄
15 48	☽ ⊥ ♅
16 33	☽ Q ♇
20 51	☽ ⊥ ♂
22 01	☽ ⚹ ♃
23 25	♀ ⊻ ♄
26 Sunday	
06 40	☽ □ ☿
10 10	☽ ∠ ♅
16 17	☽ ∠ ♂
22 17	☽ □ ♄
22 47	☽ Q ♃
27 Monday	
00 43	☽ △ ♀
02 04	☽ ⊻ ♆
04 15	☽ ∥ ♂
05 11	☽ □ ♇
10 22	☽ ∥ ♆
13 41	♀ □ ♆
14 10	☽ ⊥ ♆
16 33	☽ ⚹ ♅
22 19	☽ △ ☉
28 Tuesday	
00 21	☽ ⚹ ♂
07 47	☽ ∥ ♅
08 09	☽ ∠ ♆
10 15	☽ bQ ♀
11 26	☽ □ ♃
13 15	☉ ⊥ ♃
14 39	☽ △ ☿
16 30	☽ Q ♅
20 32	♀ ⊻ ♇
21 11	☿ Q ♆
23 47	♀ ∠ ♃
29 Wednesday	
01 30	☽ Q ♂
05 56	☽ ± ♀
06 32	☽ bQ ☉
10 29	☽ △ ♄
13 35	☽ ⚹ ♆
16 41	☽ △ ♇
17 35	☉ # ♀
17 36	☽ bQ ☿
18 51	☽ ⊼ ♀
21 12	☿ ⊻ ♃
30 Thursday	
01 23	☽ ± ☉
03 33	☽ □ ♅
04 44	☽ ∥ ♃
09 25	☽ ± ☿
11 56	☿ ∥ ♄
12 33	☽ Q ♆
13 45	☽ ⊼ ☉
14 09	☽ □ ♂
15 26	☽ bQ ♄
19 49	☽ ⊼ ☿
21 17	☽ bQ ♇
22 10	☽ △ ♃

OCTOBER 1982

LONGITUDES

Date	Sidereal time h m s	Sun ☉ ° ' "	Moon ☽ ° ' "	Moon ☽ 24.00 ° ' "	Mercury ☿ ° '	Venus ♀ ° '	Mars ♂ ° '	Jupiter ♃ ° '	Saturn ♄ ° '	Uranus ♅ ° '	Neptune ♆ ° '	Pluto ♇ ° '
01	12 39 35	07 ♎ 58 47	18 ♓ 59 45	25 ♓ 32 27	09 ♎ 32	29 ♍ 18	07 ♐ 56	11 ♏ 40	23 ♎ 04	01 ♐ 45	24 ♐ 27	26 ♎ 06
02	12 43 32	08 57 47	02 ♈ 09 49	08 ♈ 51 38	08 R 23	00 ♎ 33	08 38	11 52	23 11	01 48	24 28	26 09
03	12 47 28	09 56 49	15 ♈ 37 37	22 ♈ 27 21	07 16	01 47	09 20	12 04	23 18	01 50	24 29	26 11
04	12 51 25	10 55 53	29 ♈ 20 26	06 ♉ 16 22	06 13	03 02	10 02	12 16	23 26	01 53	24 30	26 13
05	12 55 21	11 54 59	13 ♉ 14 40	20 ♉ 14 51	05 14	04 17	10 45	12 28	23 33	01 55	24 31	26 16
06	12 59 18	12 54 07	27 ♉ 16 28	04 ♊ 19 05	04 23	05 32	11 27	12 41	23 40	01 58	24 32	26 18
07	13 03 15	13 53 17	11 ♊ 22 11	18 ♊ 25 57	03 39	06 47	12 10	12 53	23 47	02 01	24 33	26 20
08	13 07 11	14 52 30	25 ♊ 29 40	02 ♋ 33 18	03 05	08 02	12 53	13 05	23 54	02 04	24 34	26 23
09	13 11 08	15 51 46	09 ♋ 36 42	16 ♋ 39 46	02 41	09 17	13 36	13 18	24 02	02 07	24 35	26 25
10	13 15 04	16 51 03	23 ♋ 42 24	00 ♌ 44 31	02 28	10 32	14 18	13 30	24 09	02 09	24 36	26 27
11	13 19 01	17 50 23	07 ♌ 46 01	14 ♌ 46 45	02 D 26	11 47	15 01	13 42	24 16	02 12	24 37	26 30
12	13 22 57	18 49 46	21 ♌ 46 34	28 ♌ 45 14	02 34	13 02	15 45	13 55	24 23	02 15	24 38	26 32
13	13 26 54	19 49 10	05 ♍ 42 31	12 ♍ 38 06	02 53	14 17	16 28	14 08	24 31	02 18	24 39	26 35
14	13 30 50	20 48 37	19 ♍ 31 37	26 ♍ 22 42	03 22	15 32	17 11	14 20	24 38	02 21	24 41	26 37
15	13 34 47	21 48 06	03 ♎ 10 59	09 ♎ 56 04	04 00	16 47	17 54	14 33	24 45	02 24	24 42	26 39
16	13 38 44	22 47 37	16 ♎ 37 37	23 ♎ 15 17	04 47	18 02	18 38	14 45	24 53	02 27	24 43	26 42
17	13 42 40	23 47 10	29 ♎ 48 51	06 ♏ 18 06	05 42	19 17	19 21	14 58	25 00	02 30	24 44	26 44
18	13 46 37	24 46 45	12 ♏ 42 57	19 ♏ 03 23	06 44	20 32	20 05	15 11	25 07	02 33	24 46	26 47
19	13 50 33	25 46 22	25 ♏ 19 28	01 ♐ 31 22	07 52	21 47	20 49	15 24	25 15	02 36	24 47	26 49
20	13 54 30	26 46 01	07 ♐ 39 22	13 ♐ 43 46	09 05	23 02	21 33	15 36	25 22	02 40	24 48	26 51
21	13 58 26	27 45 42	19 ♐ 44 59	25 ♐ 43 32	10 24	24 18	22 16	15 49	25 29	02 43	24 50	26 54
22	14 02 23	28 45 25	01 ♑ 39 55	07 ♑ 34 44	11 46	25 33	23 00	16 02	25 36	02 46	24 51	26 56
23	14 06 19	29 ♎ 45 09	13 ♑ 28 36	19 ♑ 22 12	13 12	26 48	23 44	16 15	25 44	02 49	24 53	26 59
24	14 10 16	00 ♏ 44 55	25 ♑ 16 12	01 ♒ 11 17	14 41	28 03	24 29	16 28	25 51	02 52	24 54	27 01
25	14 14 13	01 44 43	07 ♒ 08 09	13 ♒ 07 29	16 12	29 ♎ 18	25 13	16 41	25 58	02 56	24 56	27 03
26	14 18 09	02 44 33	19 ♒ 09 56	25 ♒ 16 09	17 45	00 ♏ 33	25 57	16 54	26 06	02 59	24 57	27 06
27	14 22 06	03 44 24	01 ♓ 26 40	07 ♓ 42 03	19 20	01 49	26 41	17 07	26 13	03 02	24 59	27 08
28	14 26 02	04 44 17	14 ♓ 02 42	20 ♓ 28 57	20 56	03 04	27 26	17 20	26 20	03 06	25 00	27 11
29	14 29 59	05 44 11	27 ♓ 01 04	03 ♈ 39 08	22 33	04 19	28 10	17 33	26 27	03 09	25 02	27 13
30	14 33 55	06 44 07	10 ♈ 23 07	17 ♈ 12 51	24 10	05 34	28 55	17 46	26 35	03 12	25 04	27 15
31	14 37 52	07 ♏ 44 05	24 ♈ 08 02	01 ♉ 08 14	25 ♎ 48	06 ♏ 50	29 ♐ 39	17 ♏ 59	26 ♎ 42	03 ♐ 16	25 ♐ 05	27 ♎ 18

Moon

Date	Moon True ☊ ° '	Moon Mean ☊ ° '	Moon ☽ Latitude ° '
01	09 ♋ 04	08 ♋ 42	04 S 42
02	08 R 52	08 39	04 58
03	08 39	08 36	04 58
04	08 27	08 33	04 41
05	08 17	08 30	04 07
06	08 10	08 26	03 18
07	08 05	08 23	02 16
08	08 03	08 20	01 S 06
09	08 D 03	08 17	00 N 08
10	08 R 03	08 14	01 22
11	08 02	08 11	02 30
12	07 59	08 07	03 29
13	07 52	08 04	04 16
14	07 43	08 01	04 46
15	07 32	07 58	05 00
16	07 20	07 55	04 57
17	07 08	07 52	04 38
18	06 56	07 48	04 05
19	06 47	07 45	03 21
20	06 41	07 42	02 28
21	06 37	07 39	01 29
22	06 36	07 36	00 N 26
23	06 D 36	07 32	00 S 37
24	06 36	07 29	01 38
25	06 R 36	07 26	02 36
26	06 35	07 23	03 27
27	06 31	07 20	04 10
28	06 25	07 17	04 42
29	06 17	07 13	05 01
30	06 07	07 10	05 04
31	05 ♋ 57	07 ♋ 07	04 S 50

DECLINATIONS

Date	Sun ☉ ° '	Moon ☽ ° '	Mercury ☿ ° '	Venus ♀ ° '	Mars ♂ ° '	Jupiter ♃ ° '	Saturn ♄ ° '	Uranus ♅ ° '	Neptune ♆ ° '	Pluto ♇ ° '
01	03 S 10	08 S 41	06 S 07	01 N 36	22 S 57	14 S 28	06 S 52	20 S 23	22 S 05	05 N 20
02	03 33	03 S 42	05 24	01 06	23 04	14 32	06 55	20 24	22 05	05 19
03	03 56	01 N 34	04 40	00 36	23 11	14 36	06 57	20 24	22 05	05 18
04	04 20	06 51	03 57	00 N 06	23 18	14 40	07 00	20 25	22 05	05 17
05	04 43	11 53	03 15	00 S 24	23 24	14 44	07 03	20 25	22 05	05 16
06	05 06	16 21	02 35	00 54	23 30	14 48	07 05	20 26	22 05	05 15
07	05 29	19 54	02 00	01 24	23 36	14 51	07 08	20 27	22 05	05 14
08	05 52	22 16	01 28	01 54	23 42	14 55	07 11	20 27	22 05	05 13
09	06 15	23 14	01 01	02 24	23 48	14 59	07 13	20 28	22 05	05 12
10	06 37	22 42	00 39	02 54	23 53	15 03	07 16	20 28	22 05	05 12
11	07 00	20 45	00 23	03 24	23 58	15 07	07 19	20 29	22 05	05 11
12	07 23	17 33	00 13	03 54	24 03	15 10	07 21	20 29	22 05	05 10
13	07 45	13 23	00 08	04 23	24 08	15 14	07 24	20 30	22 05	05 09
14	08 07	08 32	00 08	04 53	24 12	15 18	07 27	20 31	22 05	05 08
15	08 30	03 N 20	00 14	05 23	24 16	15 22	07 29	20 31	22 06	05 08
16	08 52	01 S 57	00 24	05 52	24 20	15 26	07 32	20 32	22 06	05 06
17	09 14	07 04	00 39	06 22	24 24	15 30	07 35	20 32	22 06	05 05
18	09 36	11 45	00 58	06 51	24 28	15 33	07 37	20 33	22 06	05 05
19	09 58	15 51	01 20	07 20	24 31	15 37	07 40	20 34	22 06	05 04
20	10 19	19 10	01 46	07 49	24 34	15 41	07 43	20 34	22 06	05 03
21	10 41	21 34	02 14	08 18	24 37	15 45	07 45	20 35	22 06	05 02
22	11 02	23 00	02 45	08 47	24 39	15 49	07 48	20 35	22 06	05 01
23	11 23	23 22	03 18	09 15	24 41	15 52	07 51	20 36	22 07	05 00
24	11 44	22 41	03 53	09 44	24 44	15 56	07 53	20 37	22 07	05 00
25	12 05	21 00	04 29	10 12	24 45	16 00	07 56	20 38	22 07	04 59
26	12 26	18 21	05 07	10 39	24 47	16 04	07 58	20 38	22 07	04 58
27	12 46	14 51	05 45	11 07	24 48	16 07	08 01	20 39	22 07	04 57
28	13 06	10 37	06 24	11 34	24 49	16 11	08 04	20 40	22 07	04 56
29	13 26	05 47	07 04	12 01	24 50	16 15	08 06	20 40	22 07	04 56
30	13 46	00 33	07 44	12 28	24 51	16 19	08 09	20 41	22 07	04 55
31	14 S 05	04 N 52	08 S 24	12 S 54	24 S 51	16 S 22	08 S 11	20 S 42	22 S 07	04 N 54

ZODIAC SIGN ENTRIES

Date	h	m	Planets
02	01	32	♀ ♎
02	08	06	☽ ♈
04	13	09	☽ ♉
06	16	39	☽ ♊
08	19	39	☽ ♋
10	22	44	☽ ♌
13	02	09	☽ ♍
15	06	23	☽ ♎
17	12	21	☽ ♏
19	21	02	☽ ♐
22	08	38	☽ ♑
23	17	58	☉ ♏
24	21	36	☽ ♒
26	01	19	♀ ♏
27	09	12	☽ ♓
29	17	25	☽ ♈
31	22	04	☽ ♉
31	23	05	♂ ♑

LATITUDES

Date	Mercury ☿ ° '	Venus ♀ ° '	Mars ♂ ° '	Jupiter ♃ ° '	Saturn ♄ ° '	Uranus ♅ ° '	Neptune ♆ ° '	Pluto ♇ ° '
01	02 S 33	01 N 27	01 S 20	00 N 54	02 N 16	00 N 08	01 N 15	16 N 31
04	01 36	01 26	01 21	00 54	02 16	00 08	01 15	16 30
07	00 S 35	01 25	01 22	00 54	02 15	00 08	01 15	16 30
10	00 N 21	01 23	01 22	00 53	02 15	00 08	01 15	16 29
13	01 06	01 21	01 23	00 53	02 15	00 07	01 14	16 29
16	01 38	01 18	01 23	00 53	02 15	00 07	01 14	16 29
19	01 56	01 15	01 24	00 52	02 15	00 07	01 14	16 29
22	02 04	01 11	01 24	00 52	02 15	00 07	01 14	16 29
25	02 02	01 06	01 24	00 52	02 15	00 07	01 14	16 29
28	01 54	01 01	01 24	00 51	02 15	00 07	01 14	16 29
31	01 N 42	00 N 56	01 S 24	00 N 51	02 N 16	00 N 07	01 N 14	16 N 29

DATA

Julian Date	2445244
Delta T	+52 seconds
Ayanamsa	23° 36' 40"
Synetic vernal point	05° ♓ 30' 19"
True obliquity of ecliptic	23° 26' 29"

MOON'S PHASES, APSIDES AND POSITIONS ☽

Date	h	m	Phase	Longitude ° '	Eclipse Indicator
03	01	09	○	09 ♈ 30	
09	23	26	☾	16 ♋ 20	
17	00	04	●	23 ♎ 18	
25	00	08	☽	01 ♒ 15	

Day	h	m	
09	00	40	Perigee
23	15	08	Apogee

Day	h	m		
03	04	56	0N	
09	15	33	Max dec	23° N 15'
16	03	05	0S	
23	08	31	Max dec	23° S 23'
30	14	27	0N	

LONGITUDES

Date	Chiron ⚷ ° '	Ceres ⚳ ° '	Pallas ⚴ ° '	Juno ⚵ ° '	Vesta ⚶ ° '	Black Moon Lilith ⚸ ° '
01	27 ♉ 18	03 ♐ 09	09 ♏ 39	29 ♐ 15	11 ♒ 11	11 ♑ 23
11	26 ♉ 58	06 ♐ 42	13 ♏ 51	01 ♑ 34	12 ♒ 11	12 ♑ 29
21	26 ♉ 34	10 ♐ 23	18 ♏ 07	04 ♑ 13	13 ♒ 49	13 ♑ 36
31	26 ♉ 05	14 ♐ 10	22 ♏ 25	07 ♑ 30	16 ♒ 01	14 ♑ 42

All ephemeris data is given at 12.00 UT and the Moon's longitude is additionally given for 24.00 UT

ASPECTARIAN

h m Aspects

01 Friday
07 22 ☉ ⚹ ♂
08 24 ☽ ± ♄
14 03 ☽ ± ♇
17 53 ♂ ∠ ♄
19 33 ☽ ⊼ ♄
20 54 ☽ ∥ ♄
22 01 ☽ □ ♆

02 Saturday
01 04 ☽ ⊼ ♇
02 16 ☽ bQ ♃
02 41 ☽ ∥ ☿
04 23 ☽ # ♇
05 28 ☉ ☌ ☿
08 46 ☽ ☍ ♀
08 50 ☿ ⚹ ♂
11 20 ☽ △ ♅
12 38 ☽ ∥ ☉
14 46 ☿ # ♇
18 45 ☽ ± ♃
22 18 ☽ ☍ ☿

03 Sunday
00 14 ☽ △ ♂
01 09 ☽ ☍ ☉
01 10 ☽ # ♀
05 37 ☽ ⊼ ♃
08 03 ☽ ∥ ♀
12 55 ♀ ⚹ ♅
14 08 ☽ bQ ♅
23 33 ☽ # ☉

04 Monday
00 19 ☽ # ☿
01 36 ☽ ☍ ♄
03 33 ☽ △ ♆
03 40 ☉ ∥ ☿
04 06 ☽ bQ ♂
04 50 ☽ ∥ ♇
05 58 ☽ ± ♅
06 33 ☽ ☍ ♇
10 52 ☿ ⊥ ♃
12 40 ☽ # ♄
16 25 ☽ ⊼ ♅
19 03 ☽ ⊼ ♀
20 35 ☽ ± ♂
23 06 ☽ ⊼ ☿

05 Tuesday
05 34 ☽ bQ ♆
06 25 ☽ ± ♀
07 29 ☽ ⊼ ♂
08 46 ☽ ± ☿
09 33 ☽ ⊼ ☉
10 39 ☽ ☍ ♃
20 37 ☽ ± ☉
21 02 ☽ ± ♆
22 41 ☿ ☌ ♀
23 16 ☽ bQ ☿
23 22 ☽ bQ ♀

06 Wednesday
02 44 ☉ Q ♆
03 02 ☽ # ♃
05 06 ☉ ⊻ ♃
05 47 ☽ ⊼ ♄
06 22 ♂ ∠ ♇
07 18 ☽ ⊼ ♆
10 20 ☽ ⊼ ♇
13 09 ☽ bQ ☉
16 07 ☽ ± ♄
20 02 ☽ ☍ ♅
20 35 ☽ ± ♇
21 29 ☉ # ♇
23 29 ☽ △ ☿

07 Thursday
03 26 ☽ △ ♀
07 34 ☽ bQ ♄
11 56 ☽ bQ ♇
13 25 ☽ ☍ ♂
14 17 ♀ ⊥ ♃
14 36 ☽ ⊼ ♃
16 32 ☽ # ♅
16 36 ☽ △ ☉

08 Friday
00 57 ☽ ± ♃
01 46 ☿ ∥ ♀
09 17 ☽ △ ♄
09 33 ☽ # ♆
10 25 ☽ ☍ ♆
13 30 ☽ △ ♇
16 28 ☽ bQ ♃
21 51 ♂ ⊻ ♃
23 12 ☽ ⊼ ♅

09 Saturday
00 31 ☽ □ ☿
09 26 ☽ ± ♅
11 23 ☽ □ ♀
18 21 ☽ △ ♃
19 08 ☽ ⊼ ♂
23 26 ☽ □ ☉

10 Sunday
00 48 ☽ bQ ♅
05 54 ☽ ± ♂
06 32 ☽ Q ☿
12 46 ☽ □ ♄
13 31 ☽ ⊼ ♆
16 36 ☿ Q ♂
16 42 ☽ □ ♇
19 47 ☉ ∠ ♅
21 02 ☽ Q ♀
21 34 ☽ # ♆
22 04 ☽ bQ ♂
23 46 ☽ ± ♆

11 Monday
02 28 ☽ △ ♅
02 52 ☽ ⚹ ☿
05 22 ☿ St D
08 27 ☽ Q ☉
14 24 ☽ # ♅
15 10 ☽ bQ ♆
19 32 ☽ ⚹ ♀
19 46 ☽ Q ♄
22 19 ☽ □ ♃
23 33 ☽ Q ♇

12 Tuesday
01 05 ☽ △ ♂
04 19 ♀ Q ♆
04 40 ☽ ∠ ☿
06 33 ☽ ⚹ ☉
10 37 ☉ ∥ ♄
16 32 ☽ ⚹ ♄
16 55 ☽ △ ♆
20 12 ☽ ⚹ ♇
20 23 ☽ ⊥ ☿
23 49 ☽ ∠ ♀

13 Wednesday
02 01 ☽ # ♃
05 43 ☽ Q ♃
06 06 ☽ □ ♅
06 59 ☽ ⊻ ☿
08 26 ♀ ⊻ ♃
10 21 ☽ ∠ ☉
16 53 ☽ ⊥ ♀
18 39 ☽ ∠ ♄
22 11 ☽ ∠ ♇

14 Thursday
02 49 ☽ ⚹ ♃
03 09 ☽ ⊥ ☉
04 20 ☽ ⊻ ♀
07 41 ☽ □ ♂
10 26 ☽ ⊥ ♄
13 27 ☽ Q ♅
13 48 ☽ # ☉
13 54 ☽ ⊥ ♇
14 25 ☽ ⊻ ☉
17 04 ☽ # ♄
21 01 ☽ ⊻ ♄
21 01 ☽ □ ♆
21 50 ♄ ⚹ ♆
23 37 ♀ # ♇

15 Friday
00 27 ☽ ⊻ ♇
03 27 ☽ # ♀
03 50 ☽ ∥ ♇
05 28 ☽ ∠ ♃
10 37 ☽ ⚹ ♅
13 32 ☽ ☌ ☿
17 06 ☽ Q ♂
21 40 ☽ ⊥ ♃

16 Saturday
00 22 ♀ ∠ ♅
01 38 ☽ # ☿
04 38 ☽ ∥ ☿
04 58 ☽ Q ♆
08 35 ☽ ⊻ ♃
13 30 ☽ ∠ ♅
14 48 ☽ ☌ ♀
15 50 ☽ ⚹ ♂

17 Sunday
00 04 ☽ ☌ ☉
02 35 ☽ # ♇
02 41 ☽ ⚹ ♆
03 06 ☽ ☌ ♄
05 54 ☽ ⊥ ♅
06 20 ☽ ☌ ♇
08 16 ☽ ∥ ♀
14 33 ☽ ∥ ♄
15 18 ♀ ⚹ ♂
16 59 ☽ ⊻ ♅
20 53 ☽ ∠ ♂
23 41 ☽ ∥ ☉
23 47 ☽ ⊻ ☿

18 Monday
06 27 ☽ ∠ ♆
11 33 ☉ ⚹ ♆
12 01 ☽ ⊥ ☿
14 44 ☽ ⊥ ♂
16 44 ☽ ☌ ♃
21 24 ☉ ☌ ♄
23 28 ☽ ⊥ ♆

19 Tuesday
02 49 ☽ ⊻ ♂
04 28 ☽ ⊻ ♀
06 47 ☽ ∠ ☿
10 32 ☽ ∥ ♃
10 57 ☽ ⊻ ♆
11 50 ☽ ⊻ ♄
12 56 ☽ ⊻ ☉
14 53 ☽ ⊻ ♇
17 17 ☽ ⊥ ♀
23 34 ☽ ⊥ ♄

20 Wednesday
01 35 ☽ ⊥ ☉
02 10 ☽ ☌ ♅
02 34 ☽ ⊥ ♇
05 57 ♀ ∥ ♄
09 15 ☉ ⊥ ♅
12 51 ☽ ∠ ♀
14 16 ☉ ☌ ♇
15 02 ♂ ⊥ ♃
15 09 ☽ ⚹ ☿
17 23 ☽ ∠ ♄
20 19 ☽ ∠ ♇
20 50 ☽ ∠ ☉
23 37 ☿ ⊥ ♃

21 Thursday
00 55 ☽ ∥ ♅
04 01 ☽ ⊻ ♃
07 12 ☿ Q ♂
16 14 ☽ ⊥ ♃
17 23 ☽ ☌ ♂
17 59 ☽ Q ☿
19 12 ☽ ∥ ♆
22 11 ☽ ⚹ ♀
22 13 ☽ ☌ ♆
22 29 ♀ ⚹ ♆
23 38 ☽ ⚹ ♄

22 Friday
02 25 ☽ ⚹ ♇
05 35 ☽ ⚹ ☉
10 42 ☽ ∠ ♃
13 18 ♀ ☌ ♄
14 14 ☽ ⊻ ♅

23 Saturday
00 11 ☽ Q ♄
01 21 ☽ Q ♀
02 29 ☽ ⊥ ♅
02 49 ☽ Q ♇
06 35 ☿ Q ♆
08 10 ☽ Q ☉
11 21 ☽ □ ☿
12 25 ♀ ⊥ ♅
15 33 ♀ ☌ ♇
17 45 ☽ ⚹ ♃
20 53 ☽ ∠ ♅

24 Sunday
10 17 ☽ ⊻ ♂
11 15 ☽ ⊻ ♆
13 11 ☽ □ ♄
15 34 ☽ □ ♇
18 19 ☽ □ ♀
18 37 ☽ Q ♃
21 55 ☽ ∥ ♆
23 15 ☽ ⊥ ♂
23 27 ☽ ⊥ ♆

25 Monday
00 08 ☽ □ ☉
02 25 ♂ ☌ ♆
03 29 ☽ ⚹ ♅
15 55 ☽ ∥ ♅
17 37 ☽ ∠ ♆
18 35 ☽ ∠ ♂
20 47 ☿ ⊻ ♃

26 Tuesday
03 40 ☽ Q ♅
06 28 ☿ # ♇
07 26 ☽ □ ♃
08 47 ☽ △ ☿
15 41 ☿ ∠ ♅
17 33 ♂ ⚹ ♄
18 09 ☉ ⊻ ♅
23 24 ☽ ⚹ ♆

27 Wednesday
01 45 ☽ △ ♄
02 11 ☽ ⚹ ♂
03 37 ☽ △ ♇
04 02 ☽ ∥ ♃
12 47 ☽ △ ♀
15 05 ☽ □ ♅
16 48 ☽ △ ☉
18 22 ☽ bQ ☿
22 39 ☽ Q ♆
23 23 ☽ ∥ ☉

28 Thursday
02 45 ☽ Q ♂
03 24 ♂ ⚹ ♇
06 51 ☽ bQ ♄
07 22 ☽ ∥ ♀
08 29 ☽ bQ ♇
12 36 ♀ ⊻ ♅
13 54 ☽ ± ☿
18 16 ☽ △ ♃
20 19 ☽ bQ ♀
23 30 ☽ bQ ☉
23 50 ☽ ± ♄

29 Friday
00 53 ☽ ∥ ♄
01 20 ☽ ± ♇
02 39 ☽ ⊼ ☿
06 36 ☽ ∥ ☿
08 22 ☽ □ ♆
10 58 ☽ ⊼ ♄
12 22 ☽ ⊼ ♇
14 13 ☽ □ ♂
14 37 ☽ ± ♀
16 02 ☽ # ♇
17 20 ☽ ± ☉
22 11 ☽ bQ ♃
23 09 ☽ △ ♅

30 Saturday
02 33 ☽ ⊼ ♀
05 00 ☽ ⊼ ☉
14 29 ☽ ± ♃

31 Sunday
01 11 ☽ ⊼ ♃
01 16 ☿ ⚹ ♆
01 47 ☽ bQ ♅
04 03 ☿ ∥ ♄
12 10 ☽ ∥ ♇
13 39 ☽ △ ♆
15 16 ☽ ☍ ☿
16 27 ☽ ☍ ♄
17 24 ☽ ± ♅
17 27 ☽ ☍ ♇
22 00 ☽ △ ♂

NOVEMBER 1982

Date	Sidereal time h m s	LONGITUDES Sun ☉ ° ' "	Moon ☽ ° ' "	Moon ☽ 24.00 ° '	Mercury ☿ ° '	Venus ♀ ° '	Mars ♂ ° '	Jupiter ♃ ° '	Saturn ♄ ° '	Uranus ♅ ° '	Neptune ♆ ° '	Pluto ♇ ° '
01	14 41 48	08 ♏ 44 05	08 ♉ 12 51	15 ♉ 21 15	27 ♎ 27	08 ♏ 05	00 ♑ 24	18 ♏ 13	26 ♎ 49	03 ♐ 19	25 ♐ 07	27 ♎ 20
02	14 45 45	09 44 07	22 ♉ 32 42	29 ♉ 46 24	29 ♎ 06	09 20	01 09	18 26	26 56	03 23	25 09	27 23
03	14 49 42	10 44 11	07 ♊ 01 33	14 ♊ 17 25	00 ♏ 45	10 35	01 54	18 39	27 03	03 26	25 10	27 25
04	14 53 38	11 44 17	21 ♊ 33 14	28 ♊ 48 22	02 23	11 51	02 39	18 52	27 11	03 30	25 12	27 27
05	14 57 35	12 44 24	06 ♋ 02 15	13 ♋ 14 25	04 02	13 06	03 23	19 05	27 18	03 33	25 14	27 30
06	15 01 31	13 44 34	20 ♋ 24 27	27 ♋ 32 06	05 41	14 21	04 08	19 18	27 25	03 37	25 16	27 32
07	15 05 28	14 44 46	04 ♌ 37 09	11 ♌ 39 27	07 19	15 36	04 54	19 32	27 32	03 40	25 18	27 34
08	15 09 24	15 45 00	18 ♌ 38 55	25 ♌ 35 30	08 58	16 52	05 39	19 45	27 39	03 44	25 19	27 37
09	15 13 21	16 45 16	02 ♍ 29 11	09 ♍ 19 58	10 36	18 07	06 24	19 58	27 46	03 47	25 21	27 39
10	15 17 17	17 45 35	16 ♍ 07 48	22 ♍ 52 42	12 13	19 22	07 09	20 11	27 53	03 51	25 23	27 41
11	15 21 14	18 45 55	29 ♍ 34 38	06 ♎ 13 33	13 51	20 38	07 54	20 25	28 00	03 55	25 25	27 44
12	15 25 11	19 46 16	12 ♎ 49 24	19 ♎ 22 07	15 28	21 53	08 40	20 38	28 07	03 58	25 27	27 46
13	15 29 07	20 46 40	25 ♎ 51 38	02 ♏ 17 54	17 05	23 08	09 25	20 51	28 14	04 02	25 29	27 48
14	15 33 04	21 47 06	08 ♏ 40 52	15 ♏ 00 30	18 42	24 24	10 11	21 04	28 21	04 05	25 31	27 50
15	15 37 00	22 47 33	21 ♏ 16 49	27 ♏ 29 51	20 18	25 39	10 56	21 18	28 28	04 09	25 33	27 53
16	15 40 57	23 48 02	03 ♐ 39 40	09 ♐ 46 25	21 54	26 54	11 42	21 31	28 35	04 13	25 35	27 55
17	15 44 53	24 48 33	15 ♐ 50 16	21 ♐ 51 29	23 30	28 10	12 28	21 44	28 41	04 16	25 37	27 57
18	15 48 50	25 49 05	27 ♐ 50 19	03 ♑ 47 07	25 05	29 ♏ 25	13 14	21 57	28 48	04 20	25 39	27 59
19	15 52 46	26 49 38	09 ♑ 42 18	15 ♑ 36 18	26 40	00 ♐ 40	13 59	22 11	28 55	04 24	25 41	28 02
20	15 56 43	27 50 13	21 ♑ 29 37	27 ♑ 22 45	28 15	01 56	14 45	22 24	29 02	04 27	25 43	28 04
21	16 00 40	28 50 49	03 ♒ 16 18	09 ♒ 10 52	29 ♏ 50	03 11	15 31	22 37	29 08	04 31	25 45	28 06
22	16 04 36	29 ♏ 51 26	15 ♒ 07 03	21 ♒ 05 30	01 ♐ 25	04 27	16 17	22 50	29 15	04 35	25 47	28 08
23	16 08 33	00 ♐ 52 05	27 ♒ 06 52	03 ♓ 11 47	02 59	05 42	17 03	23 04	29 22	04 39	25 49	28 10
24	16 12 29	01 52 44	09 ♓ 20 53	15 ♓ 34 46	04 34	06 57	17 49	23 17	29 28	04 42	25 51	28 12
25	16 16 26	02 53 25	21 ♓ 53 59	28 ♓ 19 01	06 08	08 13	18 35	23 30	29 35	04 46	25 53	28 14
26	16 20 22	03 54 06	04 ♈ 50 16	11 ♈ 28 03	07 42	09 28	19 21	23 43	29 41	04 50	25 55	28 16
27	16 24 19	04 54 49	18 ♈ 12 33	25 ♈ 03 48	09 15	10 43	20 07	23 56	29 48	04 53	25 58	28 18
28	16 28 15	05 55 33	02 ♉ 01 42	09 ♉ 05 58	10 49	11 59	20 54	24 09	29 ♎ 54	04 57	26 00	28 20
29	16 32 12	06 56 18	16 ♉ 16 10	23 ♉ 31 40	12 23	13 14	21 40	24 23	00 ♏ 00	05 01	26 02	28 23
30	16 36 09	07 ♐ 57 04	00 ♊ 51 45	08 ♊ 15 31	13 ♐ 56	14 ♐ 29	22 ♑ 26	24 ♏ 36	00 ♏ 07	05 ♐ 04	26 ♐ 04	28 ♎ 24

Date	Moon True ☊ ° '	Moon Mean ☊ ° '	Moon ☽ Latitude ° '
01	05 ♋ 48	07 ♋ 04	04 S 18
02	05 R 40	07 01	03 29
03	05 34	06 58	02 27
04	05 31	06 54	01 S 14
05	05 D 30	06 51	00 N 03
06	05 31	06 48	01 19
07	05 32	06 45	02 30
08	05 R 32	06 42	03 31
09	05 31	06 38	04 19
10	05 28	06 35	04 51
11	05 22	06 32	05 07
12	05 15	06 29	05 06
13	05 07	06 26	04 49
14	04 59	06 23	04 17
15	04 51	06 19	03 34
16	04 46	06 16	02 40
17	04 42	06 13	01 41
18	04 40	06 10	00 N 37
19	04 D 40	06 07	00 S 28
20	04 41	06 04	01 31
21	04 43	06 00	02 30
22	04 44	05 57	03 23
23	04 45	05 54	04 09
24	04 R 45	05 51	04 43
25	04 43	05 48	05 06
26	04 40	05 44	05 13
27	04 36	05 41	05 05
28	04 32	05 38	04 38
29	04 27	05 35	03 54
30	04 ♋ 24	05 ♋ 32	02 S 54

DECLINATIONS

Date	Sun ☉ ° '	Moon ☽ ° '	Mercury ☿ ° '	Venus ♀ ° '	Mars ♂ ° '	Jupiter ♃ ° '	Saturn ♄ ° '	Uranus ♅ ° '	Neptune ♆ ° '	Pluto ♇ ° '
01	14 S 25	10 N 10	09 S 04	13 S 21	24 S 51	16 S 26	08 S 14	20 S 42	22 S 07	04 N 54
02	14 44	15 02	09 44	13 46	24 50	16 30	08 16	20 43	22 08	04 53
03	15 03	19 04	10 24	14 12	24 50	16 33	08 19	20 44	22 08	04 52
04	15 21	21 56	11 03	14 37	24 49	16 37	08 21	20 44	22 08	04 51
05	15 40	23 21	11 42	15 02	24 48	16 41	08 24	20 45	22 08	04 51
06	15 58	23 12	12 21	15 26	24 47	16 44	08 26	20 46	22 08	04 50
07	16 16	21 32	13 00	15 50	24 45	16 48	08 29	20 46	22 08	04 49
08	16 33	18 35	13 37	16 14	24 43	16 52	08 31	20 47	22 08	04 49
09	16 51	14 37	14 16	16 37	24 41	16 55	08 34	20 48	22 08	04 48
10	17 08	09 56	14 51	16 59	24 39	16 59	08 36	20 48	22 08	04 48
11	17 24	04 N 51	15 27	17 22	24 36	17 03	08 39	20 49	22 08	04 47
12	17 41	00 S 22	16 02	17 43	24 33	17 06	08 41	20 50	22 09	04 46
13	17 57	05 31	16 36	18 05	24 30	17 10	08 43	20 50	22 09	04 46
14	18 13	10 20	17 09	18 26	24 26	17 13	08 46	20 51	22 09	04 45
15	18 28	14 39	17 42	18 46	24 23	17 17	08 48	20 52	22 09	04 45
16	18 43	18 16	18 14	19 06	24 18	17 20	08 50	20 52	22 09	04 44
17	18 58	21 01	18 45	19 25	24 14	17 24	08 53	20 53	22 09	04 44
18	19 13	22 48	19 14	19 44	24 10	17 27	08 55	20 54	22 09	04 43
19	19 27	23 33	19 43	20 02	24 05	17 31	08 57	20 54	22 09	04 43
20	19 41	23 13	20 12	20 20	24 00	17 34	09 00	20 55	22 09	04 42
21	19 54	21 52	20 39	20 37	23 54	17 37	09 02	20 56	22 09	04 42
22	20 07	19 32	21 05	20 53	23 49	17 41	09 04	20 57	22 09	04 41
23	20 20	16 22	21 30	21 09	23 43	17 44	09 06	20 57	22 10	04 41
24	20 32	12 26	21 53	21 24	23 37	17 47	09 09	20 58	22 10	04 40
25	20 44	07 53	22 16	21 39	23 30	17 51	09 11	20 59	22 10	04 40
26	20 56	02 S 52	22 38	21 53	23 23	17 54	09 13	20 59	22 10	04 39
27	21 07	02 N 27	22 59	22 07	23 16	17 57	09 15	21 00	22 10	04 39
28	21 18	07 49	23 18	22 19	23 09	18 01	09 17	21 01	22 10	04 39
29	21 28	12 58	23 36	22 31	23 02	18 04	09 19	21 01	22 10	04 38
30	21 S 38	17 N 30	23 S 53	22 S 43	22 S 54	18 S 07	09 S 21	21 S 02	22 S 10	04 N 38

ZODIAC SIGN ENTRIES

Date	h	m	Planets
03	00	23	☽ ♊
03	01	10	☿ ♏
05	01	59	☽ ♋
07	04	10	☽ ♌
09	07	40	☽ ♍
11	12	46	☽ ♎
13	19	42	☽ ♏
16	04	52	☽ ♐
18	16	21	♀ ♐
18	23	07	☽ ♑
21	05	20	☽ ♒
21	14	28	☿ ♐
22	15	23	☉ ♐
23	17	43	☽ ♓
26	03	07	☽ ♈
28	08	31	☽ ♉
29	10	29	♄ ♏
30	10	36	☽ ♊

LATITUDES

Date	Mercury ☿ ° '	Venus ♀ ° '	Mars ♂ ° '	Jupiter ♃ ° '	Saturn ♄ ° '	Uranus ♅ ° '	Neptune ♆ ° '	Pluto ♇ ° '
01	01 N 37	00 N 54	01 S 24	00 N 51	02 N 16	00 N 07	01 N 14	16 N 30
04	01 20	00 49	01 24	00 51	02 16	00 07	01 14	16 30
07	01 01	00 43	01 24	00 51	02 16	00 07	01 13	16 30
10	00 42	00 36	01 24	00 50	02 16	00 07	01 13	16 31
13	00 21	00 30	01 24	00 50	02 16	00 07	01 13	16 31
16	00 N 01	00 23	01 23	00 50	02 17	00 07	01 13	16 32
19	00 S 19	00 16	01 23	00 50	02 17	00 07	01 13	16 33
22	00 39	00 09	01 22	00 50	02 17	00 07	01 13	16 34
25	00 57	00 N 02	01 22	00 50	02 18	00 07	01 13	16 34
28	01 15	00 S 06	01 21	00 50	02 18	00 07	01 13	16 35
31	01 S 30	00 S 13	01 S 20	00 N 50	02 N 18	00 N 07	01 N 13	16 N 36

DATA

Julian Date	2445275
Delta T	+52 seconds
Ayanamsa	23° 36' 43"
Synetic vernal point	05° ♓ 30' 16"
True obliquity of ecliptic	23° 26' 29"

MOON'S PHASES, APSIDES AND POSITIONS ☽

Date	h	m	Phase	Longitude ° '	Eclipse Indicator
01	12	57	○	08 ♉ 46	
08	06	38	☾	15 ♌ 32	
15	15	10	●	22 ♏ 56	
23	20	06	☽	01 ♓ 13	

Day	h	m	
04	10	26	Perigee
20	10	54	Apogee

05	21	40	Max dec	23° N 29'
12	10	17	0S	
19	16	36	Max dec	23° S 34'
27	01	05	0N	

LONGITUDES

Date	Chiron ⚷ ° '	Ceres ⚳ ° '	Pallas ⚴ ° '	Juno ⚵ ° '	Vesta ⚶ ° '	Black Moon Lilith ⚸ ° '
01	26 ♉ 02	14 ♐ 34	22 ♏ 51	07 ♑ 27	16 ♒ 15	14 ♑ 49
11	25 ♉ 32	18 ♐ 27	27 ♏ 11	10 ♑ 40	18 ♒ 58	15 ♑ 55
21	25 ♉ 00	22 ♐ 24	01 ♐ 31	14 ♑ 05	22 ♒ 04	17 ♑ 02
31	24 ♉ 29	26 ♐ 24	05 ♐ 51	17 ♑ 42	25 ♒ 29	18 ♑ 08

All ephemeris data is given at 12.00 UT and the Moon's longitude is additionally given for 24.00 UT

ASPECTARIAN

h m	Aspects
01 Monday	
02 01	☿ ☌ ♄
03 00	☽ # ♄
03 41	☽ ⊼ ♅
06 07	☽ # ☿
10 04	☿ ⊥ ♅
10 19	☿ ☌ ♇
11 45	☽ ☍ ♀
12 57	☽ ☍ ☉
15 13	☽ bQ ♆
02 Tuesday	
00 44	☽ bQ ♂
04 48	☽ # ♀
05 02	☽ ☍ ♃
06 19	☽ ± ♆
08 56	♅ ⊥ ♇
10 18	☽ # ☉
16 20	☽ ⊼ ♆
16 34	☽ ± ♂
19 21	☽ ⊼ ♄
20 03	☽ ⊼ ♇
20 10	☽ # ♃
22 06	☉ ∠ ♆
03 Wednesday	
00 16	☽ ⊼ ☿
03 03	☽ ⊼ ♂
03 52	♀ ∠ ♆
05 23	☽ ± ♄
06 01	☽ ± ♇
06 03	☽ ☍ ♅
11 28	☽ ± ☿
18 27	☽ ⊼ ♀
18 35	☽ ⊼ ☉
20 23	☽ bQ ♄
20 56	☽ bQ ♇
04 Thursday	
00 30	☽ # ♅
02 02	☉ ☌ ♀
04 15	☽ bQ ☿
05 14	☽ ± ☉
05 17	☽ ± ♀
07 29	☽ ⊼ ♃
14 12	☽ # ♆
17 34	☽ ± ♃
18 03	☽ ☍ ♆
18 42	☿ ⚹ ♂
21 13	☽ bQ ☉
21 23	☽ △ ♄
21 35	☽ bQ ♀
21 47	☽ △ ♇
05 Friday	
04 42	☿ ⊻ ♅
07 22	☽ ☍ ♂
07 51	☽ ⊼ ♅
08 15	☽ △ ☿
08 42	☽ bQ ♃
17 40	♂ ⊻ ♅
17 53	☽ ± ♅
06 Saturday	
00 00	☽ △ ☉
00 53	☽ △ ♀
06 23	♃ ⊥ ♆
08 59	☽ bQ ♅
10 07	☽ △ ♃
19 29	♂ ∠ ♃
20 11	☽ ⊼ ♆
23 54	☽ □ ♄
07 Sunday	
00 02	☽ □ ♇
05 27	☽ # ♆
06 21	☽ ± ♆
10 23	☽ △ ♅
12 29	☽ ⊼ ♂
17 12	☽ □ ☿
19 14	☽ # ♅
21 41	☽ bQ ♆
23 18	☽ ± ♂
08 Monday	
00 44	♄ ☌ ♇
06 38	☽ □ ☉
06 46	☽ Q ♇
06 48	☽ Q ♄
08 37	☽ □ ♀
13 55	☽ □ ♃
15 38	☽ bQ ♂
22 52	☽ # ♃
23 34	☽ △ ♆
09 Tuesday	
00 01	☽ # ☉
01 33	☽ # ♀
03 33	☽ ⚹ ♇
03 42	☽ ⚹ ♄
04 18	☽ Q ☿
08 24	☿ ∠ ♆
13 47	☽ # ☿
14 17	☽ □ ♅
16 17	☽ Q ☉
19 00	☽ Q ♀
19 15	☽ △ ♂
20 19	☉ ∥ ♃
21 46	☽ Q ♃
22 00	♄ ⊥ ♅
10 Wednesday	
04 09	☽ ⚹ ☿
05 54	☽ ∠ ♇
06 13	☽ ∠ ♄
11 24	♀ ∥ ♃
12 15	♀ ⊥ ♆
15 07	☽ ⚹ ☉
18 21	☽ ⚹ ♀
18 23	☽ # ♄
19 20	☽ ⚹ ♃
21 54	☽ ⊥ ♇
22 13	☽ Q ♅
22 19	☽ ⊥ ♄
11 Thursday	
04 31	☽ □ ♆
06 56	♀ ☌ ♃
08 40	☽ ⊻ ♇
09 09	☽ ⊻ ♄
10 30	☽ ∠ ☿
12 21	☽ ∥ ♇
19 51	☽ ⚹ ♅
20 10	☽ ∠ ☉
22 42	☽ ∠ ♃
12 Friday	
00 04	☽ ∠ ♀
00 13	☉ ∥ ♀
03 58	☽ □ ♂
04 03	☉ ⊥ ♆
05 02	☽ ⊥ ☿
13 09	☽ Q ♆
13 53	☽ ⊥ ☉
15 22	☽ ⊥ ♃
17 31	☽ ⊻ ☿
18 12	☽ ⊥ ♀
23 19	☽ ∠ ♅
13 Saturday	
01 48	☽ ⊻ ☉
02 34	☽ ⊻ ♃
06 25	☽ ⊻ ♀
08 27	☽ # ♇
11 18	☽ ⚹ ♆
14 15	☉ ☌ ♃
15 05	☽ Q ♂
15 37	☽ ☌ ♇
16 03	☽ ⊥ ♅
16 27	☽ ☌ ♄
14 Sunday	
00 41	♂ Q ♇
03 19	☽ ⊻ ♅
03 53	☽ ∥ ♄
08 56	♂ ⊥ ♅
15 01	☽ ⚹ ♂
15 07	☿ ∥ ♃
15 28	☽ ∠ ♆
18 13	♂ Q ♄
15 Monday	
00 31	☿ ⊥ ♆
08 40	☽ ⊥ ♆
09 50	☽ ☌ ☿
09 57	♀ ⊻ ♆
12 02	☽ ☌ ♃
15 10	☽ ☌ ☉
20 15	☽ ⊻ ♆
21 22	☽ ☌ ♀
21 34	☽ ∠ ♂
16 Tuesday	
00 47	☽ ⊻ ♇
02 00	☽ ⊻ ♄
02 03	♀ ∠ ♂
05 14	☽ ∥ ♃
05 19	☿ ☌ ♃
11 43	☽ ∥ ☿
12 30	☽ ⊥ ♇
13 05	☽ ☌ ♅
13 49	☽ ⊥ ♄
15 52	☽ ∥ ☉
16 16	☽ ⊥ ♂
19 18	☽ ∥ ♀
17 Wednesday	
04 52	☽ ⊻ ♂
06 16	☽ ∠ ♇
07 42	☽ ∠ ♄
07 52	♀ ⊻ ♇
10 36	☽ ∥ ♅
23 06	♀ ⊻ ♄
23 59	☽ ⊻ ♃
18 Thursday	
01 31	☽ ∥ ♆
05 37	☽ ⊻ ☿
07 34	☽ ⊻ ☉
07 35	☽ ☌ ♆
07 46	☉ ⊻ ♆
09 21	☉ ∥ ☿
12 14	☽ ⊥ ♃
12 18	☽ ⚹ ♇
13 58	☽ ⚹ ♄
15 33	☽ ⊻ ♀
19 33	☽ ⊥ ☿
20 38	☿ ⊻ ♆
20 46	☽ ⊥ ☉
19 Friday	
01 11	☽ ⊻ ♅
05 07	☽ ⊥ ♀
06 46	☽ ∠ ♃
12 39	☽ Q ♇
13 25	☽ ⊥ ♅
14 29	☽ Q ♄
16 37	☽ ∠ ☿
16 43	☽ ∠ ☉
18 23	☉ ☌ ☿
21 19	☽ ☌ ♂
20 Saturday	
01 35	☽ ∠ ♀
07 50	☽ ∠ ♅
08 58	☿ ⊻ ♇
13 53	☽ ⚹ ♃
17 33	☉ ⊻ ♇
20 38	☽ ⊻ ♆
21 Sunday	
00 36	☿ ⊻ ♄
01 26	☽ □ ♇
02 09	☽ ⚹ ☉
03 31	☽ □ ♄
03 56	☽ ⚹ ☿
07 37	☿ ∥ ♀
07 58	☽ ∥ ♆
08 53	☽ ⊥ ♆
11 48	☽ ⚹ ♀
14 33	☽ ⚹ ♅
14 47	☽ Q ♃
19 50	☉ ⊻ ♄
22 43	☽ ∥ ♅
23 35	☽ ∥ ☿
22 Monday	
00 34	☽ ∥ ♀
03 13	☽ ∠ ♆
04 17	☿ ∥ ♅
04 48	☽ Q ☉
05 55	♀ ⊥ ♇
07 11	☽ ∥ ☉
08 02	☽ Q ☿
08 05	☿ ∠ ♂
14 31	☽ ⊻ ♂
14 47	♀ ☌ ♅
14 58	☽ Q ♅
14 59	☽ Q ♀
17 06	♀ ∥ ♅
23 Tuesday	
02 31	☽ ∥ ♃
03 22	☽ ⊥ ♂
03 47	☽ □ ♃
04 56	♀ ⊥ ♄
09 25	☽ ⚹ ♆
14 06	☽ △ ♇
16 29	☽ △ ♄
20 06	☽ □ ☉
22 24	☽ ∠ ♂
24 Wednesday	
01 18	☽ □ ☿
02 55	☽ □ ♅
06 27	☿ ⊥ ♇
06 49	☽ □ ♀
09 05	☽ Q ♆
14 18	☿ ☌ ♅
19 28	☽ bQ ♇
21 58	☽ bQ ♄
25 Thursday	
02 59	☿ ⊥ ♄
05 00	☿ ∥ ♆
05 19	☽ ⚹ ♂
05 32	☽ ∥ ♄
12 38	☽ ± ♇
15 04	☽ △ ♃
15 11	☽ ± ♄
19 30	☽ □ ♆
23 53	☽ ⊼ ♇
26 Friday	
02 28	☽ ⊼ ♄
03 40	☽ # ♇
05 13	☽ Q ♂
10 09	☽ △ ☉
11 59	☽ △ ♅
17 53	☽ △ ☿
19 10	☽ bQ ♃
19 37	☉ ∥ ♅
21 08	☉ ⊥ ♇
21 17	☽ △ ♀
27 Saturday	
04 04	♂ ∠ ♅
11 23	☉ ☌ ♅
11 31	☽ ± ♃
14 58	☽ bQ ♅
15 15	☽ bQ ☉
15 34	☽ □ ♂
18 19	♀ ∥ ♆
21 48	☽ ∥ ♇
22 12	☽ ⊼ ♃
23 57	☽ bQ ☿
28 Sunday	
01 35	☽ △ ♆
02 27	☽ bQ ♀
04 02	☿ ∥ ♂
05 39	☽ ☍ ♇
06 42	☽ ± ♅
08 07	☽ ± ☉
08 20	☽ ☍ ♄
08 49	☉ ∠ ♂
11 20	☉ ⊥ ♄
17 00	☽ ⊼ ♅
17 21	☽ ± ☿
18 42	☽ # ♄
19 09	☽ ⊼ ☉
19 23	☽ ± ♀
29 Monday	
03 14	☽ bQ ♆
04 43	☽ ⊼ ☿
06 27	☽ ⊼ ♀
14 45	♀ ∠ ♇
18 15	☽ ± ♆
21 26	☽ △ ♂
30 Tuesday	
01 36	☽ ☍ ♃
03 39	☿ ∠ ♇
04 09	☽ ⊼ ♆
07 59	☽ ⊼ ♇
10 46	☽ ⊼ ♄
15 47	☽ # ♃
17 47	☽ ± ♇
18 52	☽ ☍ ♅
20 35	☽ ± ♄
23 15	☽ bQ ♂

DECEMBER 1982

LONGITUDES

Date	Sidereal time h m s	Sun ☉ ° ' "	Moon ☽ ° ' "	Moon ☽ 24.00 ° ' "	Mercury ☿ ° '	Venus ♀ ° '	Mars ♂ ° '	Jupiter ♃ ° '	Saturn ♄ ° '	Uranus ♅ ° '	Neptune ♆ ° '	Pluto ♇ ° '
01	16 40 05	08 ♐ 57 51	15 ♊ 42 00	23 ♊ 10 10	15 ♐ 30	15 ♐ 45	23 ♑ 12	24 ♏ 49	00 ♏ 13	05 ♐ 08	26 ♐ 06	28 ♎ 26
02	16 44 02	09 58 40	00 ♋ 38 56	08 ♋ 07 19	17 03	17 00	23 59	25 02	00 19	05 12	26 08	28 28
03	16 47 58	10 59 30	15 ♋ 34 19	22 ♋ 59 02	18 37	18 16	24 45	25 15	00 25	05 16	26 11	28 30
04	16 51 55	12 00 21	00 ♌ 20 43	07 ♌ 38 42	20 10	19 31	25 32	25 28	00 31	05 19	26 13	28 32
05	16 55 51	13 01 14	14 ♌ 52 29	22 ♌ 01 41	21 43	20 46	26 18	25 41	00 37	05 23	26 15	28 34
06	16 59 48	14 02 07	29 ♌ 06 03	06 ♍ 05 25	23 16	22 02	27 05	25 54	00 43	05 27	26 17	28 36
07	17 03 44	15 03 03	12 ♍ 59 47	19 ♍ 49 09	24 50	23 17	27 51	26 07	00 49	05 30	26 19	28 38
08	17 07 41	16 03 59	26 ♍ 33 40	03 ♎ 13 29	26 23	24 32	28 38	26 20	00 55	05 34	26 22	28 40
09	17 11 37	17 04 56	09 ♎ 48 47	16 ♎ 19 49	27 56	25 48	29 ♑ 24	26 33	01 01	05 38	26 24	28 41
10	17 15 34	18 05 55	22 ♎ 46 48	29 ♎ 09 58	29 ♐ 29	27 03	00 ♒ 11	26 45	01 07	05 41	26 26	28 43
11	17 19 31	19 06 55	05 ♏ 29 35	11 ♏ 45 52	01 ♑ 02	28 18	00 58	26 58	01 12	05 45	26 28	28 45
12	17 23 27	20 07 56	17 ♏ 59 03	24 ♏ 09 20	02 34	29 ♐ 34	01 45	27 11	01 18	05 48	26 31	28 47
13	17 27 24	21 08 58	00 ♐ 16 57	06 ♐ 22 06	04 07	00 ♑ 49	02 31	27 24	01 24	05 52	26 33	28 48
14	17 31 20	22 10 01	12 ♐ 24 59	18 ♐ 25 48	05 39	02 05	03 18	27 36	01 29	05 56	26 35	28 50
15	17 35 17	23 11 04	24 ♐ 24 48	00 ♑ 22 10	07 12	03 20	04 05	27 49	01 35	05 59	26 38	28 52
16	17 39 13	24 12 09	06 ♑ 18 10	12 ♑ 13 02	08 43	04 35	04 52	28 02	01 40	06 03	26 40	28 53
17	17 43 10	25 13 14	18 ♑ 07 05	24 ♑ 00 37	10 15	05 51	05 39	28 14	01 45	06 06	26 42	28 55
18	17 47 06	26 14 19	29 ♑ 53 58	05 ♒ 47 31	11 46	07 06	06 25	28 27	01 51	06 10	26 44	28 56
19	17 51 03	27 15 25	11 ♒ 41 39	17 ♒ 36 49	13 16	08 21	07 12	28 39	01 56	06 14	26 47	28 58
20	17 55 00	28 16 32	23 ♒ 33 28	29 ♒ 32 06	14 45	09 37	07 59	28 52	02 01	06 17	26 49	28 59
21	17 58 56	29 ♐ 17 38	05 ♓ 33 15	11 ♓ 37 25	16 14	10 52	08 46	29 04	02 06	06 21	26 51	29 01
22	18 02 53	00 ♑ 18 45	17 ♓ 45 11	23 ♓ 57 05	17 41	12 07	09 33	29 16	02 11	06 24	26 53	29 02
23	18 06 49	01 19 52	00 ♈ 13 40	06 ♈ 35 27	19 07	13 23	10 20	29 29	02 16	06 27	26 56	29 04
24	18 10 46	02 20 59	13 ♈ 02 56	19 ♈ 36 33	20 32	14 38	11 07	29 41	02 21	06 31	26 58	29 05
25	18 14 42	03 22 06	26 ♈ 16 41	03 ♉ 03 34	21 54	15 53	11 54	29 ♏ 53	02 25	06 34	27 00	29 06
26	18 18 39	04 23 14	09 ♉ 57 22	16 ♉ 58 07	23 14	17 09	12 41	00 ♐ 05	02 30	06 38	27 02	29 08
27	18 22 35	05 24 21	24 ♉ 05 40	01 ♊ 19 42	24 31	18 24	13 28	00 17	02 34	06 41	27 05	29 09
28	18 26 32	06 25 29	08 ♊ 39 44	16 ♊ 05 06	25 45	19 39	14 15	00 29	02 39	06 44	27 07	29 10
29	18 30 29	07 26 36	23 ♊ 34 56	01 ♋ 08 16	26 55	20 55	15 02	00 41	02 43	06 48	27 09	29 11
30	18 34 25	08 27 44	08 ♋ 43 56	16 ♋ 20 46	28 01	22 10	15 49	00 53	02 48	06 51	27 11	29 12
31	18 38 22	09 ♑ 28 52	23 ♋ 57 31	01 ♌ 32 55	29 ♑ 02	23 ♑ 25	16 ♒ 37	01 ♐ 05	02 ♏ 52	06 ♐ 54	27 ♐ 14	29 ♎ 14

Moon nodes and latitude; DECLINATIONS

Date	Moon True ☊ ° '	Moon Mean ☊ ° '	Moon ☽ Latitude ° '	Sun ☉ ° '	Moon ☽ ° '	Mercury ☿ ° '	Venus ♀ ° '	Mars ♂ ° '	Jupiter ♃ ° '	Saturn ♄ ° '	Uranus ♅ ° '	Neptune ♆ ° '	Pluto ♇ ° '
01	04 ♋ 21	05 ♋ 29	01 S 41	21 S 48	21 N 00	24 S 09	22 S 53	22 S 46	18 S 10	09 S 23	21 S 03	22 S 10	04 N 38
02	04 R 20	05 25	00 S 20	21 57	23 06	24 24	23 04	22 38	18 13	09 25	21 03	22 10	04 37
03	04 D 21	05 22	01 N 02	22 06	23 33	24 37	23 13	22 29	18 17	09 27	21 04	22 10	04 37
04	04 22	05 19	02 19	22 14	22 20	24 49	23 22	22 20	18 20	09 29	21 05	22 10	04 37
05	04 23	05 16	03 26	22 22	19 39	25 00	23 30	22 11	18 23	09 31	21 05	22 11	04 36
06	04 24	05 13	04 18	22 29	15 49	25 09	23 37	22 02	18 26	09 33	21 06	22 11	04 36
07	04 25	05 10	04 54	22 36	11 13	25 17	23 43	21 53	18 29	09 35	21 07	22 11	04 36
08	04 R 25	05 06	05 13	22 43	06 09	25 23	23 49	21 43	18 32	09 37	21 07	22 11	04 36
09	04 24	05 03	05 15	22 49	00 N 56	25 29	23 54	21 33	18 35	09 39	21 08	22 11	04 35
10	04 22	05 00	05 00	22 54	04 S 13	25 32	23 58	21 23	18 38	09 41	21 08	22 11	04 35
11	04 19	04 57	04 31	23 00	09 06	25 35	24 02	21 12	18 41	09 43	21 09	22 11	04 35
12	04 17	04 54	03 49	23 04	13 31	25 36	24 05	21 02	18 44	09 44	21 10	22 11	04 35
13	04 15	04 50	02 57	23 09	17 19	25 35	24 07	20 51	18 47	09 46	21 11	22 11	04 35
14	04 14	04 47	01 58	23 13	20 20	25 33	24 08	20 40	18 50	09 48	21 11	22 11	04 34
15	04 13	04 44	00 N 54	23 16	22 25	25 30	24 09	20 28	18 52	09 50	21 12	22 11	04 34
16	04 D 13	04 41	00 S 12	23 19	23 29	25 25	24 09	20 17	18 55	09 51	21 12	22 11	04 34
17	04 13	04 38	01 17	23 21	23 29	25 18	24 08	20 05	18 58	09 53	21 13	22 11	04 34
18	04 14	04 35	02 18	23 23	22 25	25 10	24 06	19 53	19 01	09 55	21 14	22 11	04 34
19	04 14	04 31	03 14	23 25	20 23	25 00	24 04	19 41	19 04	09 56	21 14	22 11	04 34
20	04 15	04 28	04 01	23 26	17 28	24 49	24 01	19 28	19 06	09 58	21 15	22 12	04 34
21	04 16	04 25	04 39	23 26	13 48	24 37	23 57	19 15	19 09	09 59	21 15	22 12	04 34
22	04 16	04 22	05 05	23 26	09 31	24 23	23 52	19 03	19 12	10 01	21 16	22 12	04 34
23	04 16	04 19	05 17	23 26	04 S 46	24 08	23 47	18 50	19 14	10 02	21 16	22 12	04 34
24	04 R 16	04 15	05 14	23 25	00 N 19	23 51	23 41	18 36	19 17	10 04	21 17	22 12	04 34
25	04 16	04 12	04 55	23 24	05 33	23 33	23 34	18 23	19 19	10 05	21 18	22 12	04 34
26	04 D 16	04 09	04 19	23 22	10 42	23 14	23 27	18 09	19 22	10 06	21 18	22 12	04 34
27	04 16	04 06	03 26	23 20	15 28	22 54	23 18	17 55	19 25	10 08	21 19	22 12	04 34
28	04 16	04 03	02 18	23 17	19 29	22 33	23 09	17 41	19 27	10 09	21 20	22 12	04 34
29	04 16	04 00	00 S 59	23 14	22 18	22 11	23 00	17 27	19 29	10 11	21 20	22 12	04 34
30	04 R 16	03 56	00 N 25	23 10	23 34	21 49	22 49	17 13	19 32	10 12	21 21	22 12	04 34
31	04 ♋ 16	03 ♋ 53	01 N 47	23 S 06	23 N 05	21 S 26	22 S 38	16 S 58	19 S 34	10 S 13	21 S 21	22 S 12	04 N 34

ZODIAC SIGN ENTRIES

Date	h m	Planets
02	10 58	☽ ♋
04	11 26	☽ ♌
06	13 32	☽ ♍
08	18 11	☽ ♎
10	06 17	♂ ♒
10	20 04	☿ ♑
11	01 35	☽ ♏
12	20 20	♀ ♑
13	11 27	☽ ♐
15	23 15	☽ ♑
18	12 12	☽ ♒
21	00 56	☽ ♓
22	04 38	☉ ♑
23	11 34	☽ ♈
25	18 37	☽ ♉
26	01 57	♃ ♐
27	21 49	☽ ♊
29	22 12	☽ ♋
31	21 33	☽ ♌

LATITUDES

Date	Mercury ☿ ° '	Venus ♀ ° '	Mars ♂ ° '	Jupiter ♃ ° '	Saturn ♄ ° '	Uranus ♅ ° '	Neptune ♆ ° '	Pluto ♇ ° '
01	01 S 30	00 S 13	01 S 20	00 N 50	02 N 18	00 N 07	01 N 13	16 N 36
04	01 45	00 20	01 20	00 49	02 19	00 07	01 13	16 37
07	01 57	00 27	01 19	00 49	02 19	00 07	01 13	16 39
10	02 06	00 34	01 18	00 49	02 20	00 07	01 13	16 40
13	02 13	00 41	01 17	00 49	02 20	00 07	01 13	16 41
16	02 16	00 47	01 16	00 49	02 21	00 07	01 13	16 42
19	02 14	00 53	01 15	00 49	02 21	00 07	01 13	16 44
22	02 08	00 59	01 14	00 49	02 22	00 07	01 13	16 45
25	01 55	01 05	01 12	00 49	02 22	00 07	01 13	16 47
28	01 35	01 10	01 11	00 49	02 23	00 07	01 13	16 48
31	01 S 06	01 S 15	01 S 10	00 N 49	02 N 24	00 N 07	01 N 13	16 N 50

DATA

Julian Date	2445305
Delta T	+52 seconds
Ayanamsa	23° 36' 48"
Synetic vernal point	05° ♓ 30' 11"
True obliquity of ecliptic	23° 26' 28"

MOON'S PHASES, APSIDES AND POSITIONS ☽

Date	h m	Phase	Longitude ° '	Eclipse Indicator
01	00 21	○	08 ♊ 28	
07	15 53	☾	15 ♍ 13	
15	09 18	●	23 ♐ 04	Partial
23	14 17	☽	01 ♈ 26	
30	11 33	○	08 ♋ 27	total

Day	h m		
02	11 10	Perigee	
18	01 43	Apogee	
30	22 10	Perigee	
03	06 22	Max dec	23° N 36'
09	16 19	0S	
16	23 52	Max dec	23° S 37'
24	10 30	0N	
30	17 25	Max dec	23° N 37'

LONGITUDES

Date	Chiron ⚷ ° '	Ceres ⚳ ° '	Pallas ⚴ ° '	Juno ⚵ ° '	Vesta ⚶ ° '	Black Moon Lilith ⚸ ° '
01	24 ♉ 29	26 ♐ 24	05 ♐ 51	17 ♑ 42	25 ♒ 29	18 ♑ 08
11	23 ♉ 59	00 ♑ 26	10 ♐ 09	21 ♑ 28	29 ♒ 10	19 ♑ 15
21	23 ♉ 33	04 ♑ 29	14 ♐ 25	25 ♑ 23	03 ♓ 04	20 ♑ 21
31	23 ♉ 11	08 ♑ 32	18 ♐ 38	29 ♑ 25	07 ♓ 10	21 ♑ 28

ASPECTARIAN

h m Aspects

01 Wednesday
00 21 ☽ ☍ ☉
00 56 ♀ ∠ ♄
02 18 ♀ ∥ ♂
07 22 ☿ ∠ ♄
08 21 ☽ ⚼ ♇
11 13 ☽ ⚼ ♄
11 38 ☽ ☍ ☿
12 05 ☽ ☍ ♀
12 22 ☽ # ♅
14 33 ☽ ± ♂
19 43 ☽ # ☉
23 09 ☽ # ♆

02 Thursday
00 43 ☽ ⊼ ♂
02 51 ☽ ⊼ ♃
04 45 ☽ ☍ ♆
05 13 ☽ # ♂
07 58 ☿ ☌ ♀
08 30 ☽ △ ♇
11 09 ☽ # ♀
11 28 ☽ △ ♄
12 37 ☽ ± ♃
19 20 ☽ ⊼ ♅

03 Friday
03 18 ☽ ⚼ ♃
04 05 ☽ ⊼ ☉
05 01 ☽ ± ♅
14 28 ☽ ± ☉
16 28 ☿ ⊥ ♂
16 45 ☽ ⊼ ♀
17 29 ☽ ⊼ ☿
19 37 ☽ ⚼ ♅
21 18 ☽ # ♀

04 Saturday
01 54 ☉ ∥ ♆
03 23 ☽ ± ♀
03 42 ☽ ☍ ♂
03 55 ☽ △ ♃
04 22 ☽ ± ☿
05 14 ☽ ⊼ ♆
06 08 ☽ ⚼ ☉
09 02 ☽ □ ♇
09 14 ♂ ✶ ♃
11 56 ☽ # ♂
12 18 ☽ □ ♄
12 41 ♀ ⊥ ♂
13 09 ☽ # ☉
13 52 ☽ # ♆
15 04 ☽ ± ♆
19 29 ☽ ⚼ ♀
20 12 ☽ △ ♅
20 51 ☽ ⚼ ☿
21 19 ☉ ∥ ♂

05 Sunday
00 39 ☽ # ♅
05 57 ☽ ⚼ ♆
08 41 ☽ △ ☉
10 17 ♂ ⊻ ♆
14 19 ♂ ∥ ♆
14 50 ☽ Q ♇
18 19 ☽ Q ♄
20 31 ☽ # ♃
22 50 ☽ △ ♀

06 Monday
00 53 ☽ △ ☿
01 22 ☉ ∠ ♇
06 28 ☽ □ ♃
07 12 ☽ △ ♆
08 21 ☽ ⊼ ♂
11 09 ☽ ✶ ♇
14 48 ☽ ✶ ♄
19 13 ☽ ± ♂
22 56 ☽ □ ♅

07 Tuesday
11 44 ☽ ⚼ ♂
13 07 ☽ ∠ ♇
13 59 ☽ Q ♃
15 53 ☽ □ ☉
16 59 ☽ ∠ ♄
19 49 ☽ # ♄

08 Wednesday
05 01 ☽ ⊥ ♇
06 38 ☽ Q ♅
08 01 ☽ □ ♀
08 11 ☉ ∠ ♄
09 03 ☽ ⊥ ♄
11 06 ☿ ⊻ ♃
11 35 ☽ ✶ ♃
11 38 ☽ □ ☿
11 39 ☽ □ ♆
11 45 ☿ ☌ ♆
12 57 ♂ □ ♇
15 46 ☽ ⊻ ♇
15 56 ☽ △ ♂
16 39 ♃ ⊻ ♆
19 13 ☽ ∥ ♇
19 54 ☽ ⊻ ♄

09 Thursday
02 39 ☽ Q ☉
04 19 ☽ ✶ ♅
15 13 ☽ ∠ ♃
20 06 ☽ Q ♀
20 27 ☽ Q ♆
23 54 ♀ ☌ ♆

10 Friday
00 00 ☿ ✶ ♇
00 47 ☽ Q ☿
02 32 ☽ ✶ ☉
05 12 ♀ ⊻ ♃
08 04 ☽ ∠ ♅
08 09 ☽ ⊥ ♃
13 47 ☽ # ♇
18 53 ☽ ✶ ♆
19 35 ☽ ⊻ ♃
20 53 ☽ ✶ ♀
23 11 ☽ ☌ ♇

11 Saturday
01 03 ☽ ⊥ ♅
02 20 ☽ ✶ ☿
02 50 ☽ □ ♂
03 48 ☽ ☌ ♄
09 09 ☽ ∠ ☉
10 00 ☿ ⊻ ♂
12 29 ☽ ⊻ ♅
14 59 ☿ ✶ ♄
15 12 ☽ ∥ ♄
18 39 ♂ ∥ ♅
20 35 ♂ □ ♄
20 37 ♀ ✶ ♇
23 29 ☽ ∠ ♆

12 Sunday
03 54 ☽ ⊥ ☉
04 39 ☽ ∠ ♀
11 06 ☽ ∠ ☿
15 38 ☽ Q ♂
16 32 ☽ ⊻ ☉
16 55 ☽ ⊥ ♆
22 58 ☿ ⊥ ♃

13 Monday
00 05 ☽ ⊥ ♀
04 39 ☽ ⊻ ♆
06 14 ☽ ☌ ♃
07 08 ☽ ⊥ ☿
09 05 ☽ ⊻ ♇
12 55 ♂ ⊥ ♆
13 11 ☽ ⊻ ♀
14 12 ☽ ⊻ ♄
16 42 ☽ ✶ ♂
20 39 ☽ ⊻ ☿
20 56 ☽ ⊥ ♇
22 56 ☽ ∥ ♃
23 04 ☽ ☌ ♅
23 51 ♀ ✶ ♄

14 Tuesday
02 08 ☽ ⊥ ♄
14 50 ☽ ∠ ♇
14 53 ☽ ∥ ♂
16 24 ☿ ⊻ ♅
20 11 ☽ ∠ ♄
20 32 ☽ ∥ ♅

15 Wednesday
00 34 ☽ ∠ ♂
08 37 ☽ ∥ ♆
09 18 ☽ ☌ ☉
16 28 ☽ ☌ ♆
18 59 ☽ ⊻ ♃
19 54 ☽ ⊥ ♂
20 58 ☽ ✶ ♇
23 07 ♀ ⊥ ♃

16 Thursday
02 33 ☽ ✶ ♄
05 30 ☽ ∥ ☉
07 19 ☽ ⊥ ♃
08 07 ☽ ☌ ♀
08 53 ☽ ⊻ ♂
11 29 ☽ ⊻ ♅
17 38 ☽ ☌ ☿
21 19 ☽ Q ♇
23 43 ☽ ⊥ ♅

17 Friday
01 48 ♀ ⊻ ♂
01 54 ☽ ∠ ♃
03 04 ☽ Q ♄
16 26 ☽ ∥ ☉
17 16 ♀ ⊻ ♅
18 07 ☽ ∠ ♅
22 43 ☿ Q ♇

18 Saturday
03 26 ♂ ✶ ♅
03 50 ☽ ⊻ ☉
05 32 ☽ ⊻ ♆
08 59 ☽ ✶ ♃
10 02 ☽ □ ♇
15 22 ☽ ∥ ♆
15 59 ☽ □ ♄
17 13 ☽ ⊥ ☉
17 48 ☽ ⊥ ♆
18 42 ☿ ⊥ ♅

19 Sunday
00 14 ☉ ☌ ♆
00 50 ☽ ✶ ♅
02 14 ☽ ☌ ♂
03 14 ☽ ∥ ♅
04 25 ☽ ⊻ ♀
09 51 ☽ Q ♃
12 10 ☽ ∠ ♆
13 15 ☽ ∠ ☉
15 39 ☽ ⊻ ☿
18 03 ☽ ⊥ ♀
18 59 ☽ ∥ ♂
19 13 ☿ ∠ ♃
23 18 ☿ Q ♄
23 31 ☽ ∥ ♃

20 Monday
01 18 ☽ Q ♅
05 33 ☽ ⊥ ☿
14 22 ☽ ∠ ♀
18 34 ☽ ✶ ♆
22 21 ☽ ✶ ☉
22 50 ☽ □ ♃
22 56 ☽ △ ♇

21 Tuesday
02 11 ☽ ∠ ☿
04 50 ♃ ⊻ ♇
05 04 ☽ △ ♄
05 12 ☉ ✶ ♇
05 17 ☉ ⊻ ♃
13 34 ☽ □ ♅
14 49 ♀ Q ♇
18 33 ☽ Q ♆
18 49 ☽ ⊻ ♂
22 00 ♂ ∥ ♃
23 43 ☽ ✶ ♀

22 Wednesday
00 23 ☽ Q ☉
04 43 ☽ ⚼ ♇
07 25 ☽ ⊥ ♂
09 23 ☽ ∥ ♄
10 53 ☽ ⚼ ♄
11 51 ☽ ✶ ☿
17 32 ♀ ⊥ ♅
22 15 ☽ ± ♇

23 Thursday
01 43 ☽ Q ♀
02 02 ☽ ∠ ♂
04 23 ☽ ± ♄
05 41 ☽ □ ♆
09 46 ☽ ⊼ ♇
10 33 ☽ △ ♃
12 57 ☽ # ♇
13 54 ☽ Q ☿
14 17 ☽ □ ☉
15 53 ☽ ⊼ ♄
23 48 ☽ △ ♅

24 Friday
06 02 ♀ Q ♄
08 12 ☽ ✶ ♂
11 48 ☉ ✶ ♄
13 02 ♀ ∠ ♃
15 03 ☽ ⚼ ♃
15 14 ☽ □ ♀

25 Saturday
03 15 ☽ □ ☿
03 31 ☽ ⚼ ♅
05 59 ☿ ∠ ♅
07 27 ☽ ∥ ♇
07 29 ☽ Q ♂
07 39 ☽ ± ♃
10 18 ☿ ∥ ♀
11 09 ♂ Q ♃
13 18 ☽ △ ♆
15 13 ♂ ∠ ♆
17 02 ☽ ☍ ♇
18 30 ☽ ⊼ ♃
19 39 ☽ ± ♅
22 56 ☽ ☍ ♄

26 Sunday
01 18 ☉ ∥ ☿
01 33 ☽ △ ☉
06 13 ☽ ⊼ ♅
09 08 ☽ # ♄
15 36 ☽ ⚼ ♆
16 59 ☽ □ ♂

27 Monday
01 30 ☽ △ ♀
05 20 ☽ ⚼ ☉
06 10 ☉ ∥ ♀
06 56 ☽ ± ♆
12 47 ☽ △ ☿
16 59 ☽ ⊼ ♆
20 25 ☽ ⊼ ♇
21 30 ☽ ± ☉
22 25 ☽ ☍ ♃

28 Tuesday
01 10 ☽ # ♂
02 07 ☽ ⊼ ♄
04 51 ☽ ⚼ ♀
06 18 ☽ ± ♇
08 05 ☽ ⊼ ☉
08 51 ☽ ☍ ♅
11 48 ☽ # ♃
11 59 ☽ ± ♄
13 45 ☉ ⊥ ♃
15 42 ☽ ⚼ ☿
19 52 ☉ ⊻ ♅
20 50 ☽ ± ♀
20 56 ☽ ⚼ ♇
21 33 ☽ △ ♂

29 Wednesday
02 19 ☽ # ♅
02 35 ☽ ⚼ ♄
07 21 ☽ ⊼ ♀
07 24 ☽ ± ☿
10 53 ☽ # ♆
10 57 ☽ # ☿
11 17 ☿ ∥ ♆
17 03 ☿ ⊻ ♆
17 42 ☽ ☍ ♆
17 45 ☽ ⊼ ☿
20 25 ☽ # ♀
20 55 ☽ △ ♇
22 49 ☽ ⚼ ♂
23 26 ☽ ⊼ ♃

30 Thursday
01 03 ☽ # ☉
02 35 ☽ △ ♄
05 44 ♀ ∠ ♅
09 01 ☽ ⊼ ♅
09 02 ☽ ± ♃
11 33 ☽ ☍ ☉
13 49 ☽ ± ♂
18 31 ☽ ± ♅
23 25 ☽ ⚼ ♃
23 47 ☽ ⊼ ♂

31 Friday
08 45 ☽ ⚼ ♅
11 04 ☽ ☍ ♀
11 34 ☽ # ☉
17 00 ☿ □ ♇
17 02 ☿ ∥ ♅
17 11 ☽ ⊼ ♆
19 24 ☽ # ♀
20 20 ☽ □ ♇
20 33 ☽ ☍ ☿
23 24 ☽ △ ♃
23 50 ☽ # ♆

All ephemeris data is given at 12.00 UT and the Moon's longitude is additionally given for 24.00 UT

JANUARY 1983

LONGITUDES

Date	Sidereal time h m s	Sun ☉ ° ' "	Moon ☽ ° ' "	Moon ☽ 24.00 ° ' "	Mercury ☿ ° '	Venus ♀ ° '	Mars ♂ ° '	Jupiter ♃ ° '	Saturn ♄ ° '	Uranus ♅ ° '	Neptune ♆ ° '	Pluto ♇ ° '
01	18 42 18	10 ♑ 30 00	09 ♌ 05 48	16 ♌ 35 06	29 ♑ 57	24 ♑ 40	17 ♒ 24	01 ♐ 26	02 ♏ 56	06 ♐ 58	27 ♐ 16	29 ♎ 15
02	18 46 15	11 31 08	23 ♌ 59 53	01 ♍ 19 20	00 ♒ 45	25 56	18 11	01 28	03 00	07 01	27 18	29 16
03	18 50 11	12 32 17	08 ♍ 32 54	15 ♍ 40 07	01 25	27 11	18 58	01 40	03 04	07 04	27 20	29 17
04	18 54 08	13 33 26	22 ♍ 40 45	29 ♍ 34 44	01 57	28 26	19 45	01 51	03 08	07 07	27 23	29 18
05	18 58 04	14 34 34	06 ♎ 22 07	13 ♎ 03 05	02 20	29 ♑ 41	20 32	02 03	03 12	07 10	27 25	29 19
06	19 02 01	15 35 44	19 ♎ 37 55	26 ♎ 06 58	02 32	00 ♒ 57	21 19	02 14	03 16	07 13	27 27	29 20
07	19 05 58	16 36 53	02 ♏ 30 40	08 ♏ 49 28	02 R 34	02 12	22 06	02 25	03 19	07 17	27 29	29 21
08	19 09 54	17 38 03	15 ♏ 03 49	21 ♏ 14 14	02 24	03 27	22 53	02 36	03 23	07 20	27 31	29 22
09	19 13 51	18 39 12	27 ♏ 21 11	03 ♐ 25 06	02 04	04 42	23 41	02 48	03 26	07 23	27 33	29 22
10	19 17 47	19 40 22	09 ♐ 26 28	15 ♐ 25 42	01 28	05 57	24 28	02 59	03 30	07 26	27 36	29 23
11	19 21 44	20 41 31	21 ♐ 22 59	27 ♐ 19 16	00 ♒ 43	07 12	25 15	03 10	03 33	07 29	27 38	29 24
12	19 25 40	21 42 41	03 ♑ 14 19	09 ♑ 08 38	29 ♑ 47	08 28	26 02	03 21	03 36	07 31	27 40	29 25
13	19 29 37	22 43 50	15 ♑ 02 32	20 ♑ 56 15	28 43	09 43	26 49	03 31	03 39	07 34	27 42	29 25
14	19 33 33	23 44 59	26 ♑ 50 04	02 ♒ 44 14	27 32	10 58	27 36	03 42	03 43	07 37	27 44	29 26
15	19 37 30	24 46 07	08 ♒ 39 00	14 ♒ 34 35	26 16	12 13	28 23	03 53	03 45	07 40	27 46	29 27
16	19 41 27	25 47 15	20 ♒ 31 16	26 ♒ 29 17	24 57	13 28	29 11	04 03	03 48	07 43	27 48	29 27
17	19 45 23	26 48 22	02 ♓ 28 56	08 ♓ 30 30	23 39	14 43	29 ♒ 58	04 14	03 51	07 46	27 50	29 28
18	19 49 20	27 49 28	14 ♓ 34 18	20 ♓ 40 41	22 23	15 58	00 ♓ 45	04 24	03 54	07 48	27 52	29 28
19	19 53 16	28 50 34	26 ♓ 50 02	03 ♈ 02 42	21 11	17 13	01 32	04 34	03 56	07 51	27 54	29 29
20	19 57 13	29 ♑ 51 39	09 ♈ 19 36	15 ♈ 39 40	20 06	18 28	02 19	04 44	03 59	07 54	27 56	29 29
21	20 01 09	00 ♒ 52 43	22 ♈ 04 48	28 ♈ 34 55	19 08	19 43	03 06	04 54	04 01	07 56	27 58	29 30
22	20 05 06	01 53 46	05 ♉ 10 26	11 ♉ 51 40	18 19	20 58	03 53	05 04	04 03	07 59	28 00	29 30
23	20 09 02	02 54 48	18 ♉ 38 57	25 ♉ 32 31	17 39	22 13	04 40	05 14	04 05	08 01	28 02	29 30
24	20 12 59	03 55 49	02 ♊ 32 27	09 ♊ 38 48	17 08	23 28	05 27	05 24	04 07	08 04	28 04	29 31
25	20 16 56	04 56 49	16 ♊ 51 23	24 ♊ 09 55	16 46	24 43	06 14	05 34	04 09	08 06	28 06	29 31
26	20 20 52	05 57 48	01 ♋ 33 54	09 ♋ 02 39	16 33	25 58	07 01	05 43	04 11	08 09	28 08	29 31
27	20 24 49	06 58 46	16 ♋ 35 18	24 ♋ 10 51	16 29	27 13	07 48	05 53	04 13	08 11	28 10	29 32
28	20 28 45	07 59 43	01 ♌ 48 05	09 ♌ 25 46	16 D 32	28 28	08 35	06 02	04 14	08 13	28 12	29 32
29	20 32 42	09 00 39	17 ♌ 02 34	24 ♌ 37 10	16 43	29 ♒ 43	09 22	06 11	04 16	08 16	28 14	29 32
30	20 36 38	10 01 34	02 ♍ 08 20	09 ♍ 34 57	17 01	00 ♓ 58	10 09	06 20	04 17	08 18	28 15	29 32
31	20 40 35	11 ♒ 02 28	16 ♍ 56 00	24 ♍ 10 45	17 ♑ 25	02 ♓ 12	10 ♓ 56	06 ♐ 29	04 ♏ 19	08 ♐ 20	28 ♐ 17	29 ♎ 32

DECLINATIONS

Date	Moon True ☊ ° '	Moon Mean ☊ ° '	Moon ☽ Latitude ° '	Sun ☉ ° '	Moon ☽ ° '	Mercury ☿ ° '	Venus ♀ ° '	Mars ♂ ° '	Jupiter ♃ ° '	Saturn ♄ ° '	Uranus ♅ ° '	Neptune ♆ ° '	Pluto ♇ ° '
01	04 ♋ 16	03 ♋ 50	03 N 02	23 S 02	20 N 54	21 S 03	22 S 27	16 S 43	19 S 37	10 S 14	21 S 22	22 S 12	04 N 34
02	04 R 15	03 47	04 03	22 56	17 20	20 40	22 14	16 28	19 39	10 15	21 22	22 12	04 35
03	04 14	03 44	04 46	22 51	12 47	20 18	22 01	16 13	19 41	10 17	21 23	22 12	04 35
04	04 13	03 41	05 11	22 45	07 40	19 56	21 47	15 58	19 44	10 18	21 23	22 12	04 35
05	04 13	03 37	05 17	22 39	02 N 20	19 35	21 33	15 43	19 46	10 19	21 24	22 12	04 35
06	04 D 12	03 34	05 06	22 32	02 S 57	19 16	21 18	15 27	19 48	10 20	21 24	22 12	04 35
07	04 13	03 31	04 40	22 24	07 58	18 58	21 02	15 11	19 50	10 21	21 25	22 12	04 35
08	04 14	03 28	04 01	22 17	12 31	18 42	20 46	14 55	19 52	10 22	21 26	22 12	04 36
09	04 15	03 25	03 11	22 09	16 28	18 28	20 29	14 39	19 55	10 23	21 26	22 12	04 36
10	04 17	03 21	02 14	22 00	19 40	18 17	20 11	14 23	19 57	10 24	21 27	22 13	04 36
11	04 18	03 18	01 11	21 51	21 59	18 09	19 53	14 07	19 59	10 25	21 27	22 13	04 36
12	04 18	03 15	00 N 06	21 41	23 18	18 03	19 35	13 51	20 01	10 25	21 27	22 13	04 37
13	04 R 18	03 12	00 S 59	21 32	23 34	17 59	19 15	13 34	20 03	10 26	21 28	22 13	04 37
14	04 16	03 09	02 01	21 21	22 47	17 58	18 56	13 17	20 05	10 27	21 28	22 13	04 37
15	04 14	03 06	02 59	21 10	20 58	17 59	18 35	13 00	20 07	10 28	21 29	22 13	04 38
16	04 11	03 02	03 48	20 59	18 15	18 02	18 15	12 44	20 09	10 29	21 29	22 13	04 38
17	04 07	02 59	04 28	20 48	14 45	18 07	17 53	12 26	20 11	10 29	21 30	22 13	04 38
18	04 02	02 56	04 57	20 36	10 38	18 14	17 31	12 09	20 12	10 30	21 30	22 13	04 39
19	03 59	02 53	05 12	20 24	06 02	18 21	17 09	11 52	20 14	10 30	21 31	22 13	04 39
20	03 56	02 50	05 13	20 11	01 S 06	18 29	16 46	11 35	20 16	10 31	21 31	22 13	04 39
21	03 54	02 47	04 59	19 58	03 N 59	18 38	16 23	11 17	20 18	10 32	21 32	22 13	04 40
22	03 D 54	02 43	04 29	19 44	09 01	18 48	15 59	10 59	20 20	10 32	21 32	22 13	04 40
23	03 55	02 40	03 44	19 31	13 47	18 58	15 35	10 42	20 21	10 33	21 32	22 13	04 41
24	03 57	02 37	02 44	19 16	17 59	19 08	15 10	10 24	20 23	10 33	21 33	22 13	04 41
25	03 58	02 34	01 33	19 02	21 15	19 18	14 45	10 06	20 25	10 33	21 33	22 13	04 42
26	03 59	02 31	00 S 13	18 47	23 13	19 28	14 20	09 48	20 26	10 34	21 34	22 13	04 42
27	03 R 58	02 27	01 N 09	18 32	23 39	19 38	13 54	09 30	20 28	10 34	21 34	22 13	04 42
28	03 56	02 24	02 26	18 16	22 08	19 47	13 28	09 12	20 29	10 34	21 34	22 13	04 43
29	03 52	02 21	03 34	18 00	19 07	19 56	13 01	08 54	20 31	10 35	21 35	22 13	04 43
30	03 47	02 18	04 25	17 44	14 50	20 05	12 35	08 35	20 32	10 35	21 35	22 13	04 44
31	03 ♋ 41	02 ♋ 15	04 N 58	17 S 28	09 N 44	20 S 12	12 S 07	08 S 17	20 S 34	10 S 35	21 S 35	22 S 13	04 N 44

ZODIAC SIGN ENTRIES

Date	h	m	Planets
01	13	32	☿ ♒
02	21	49	☽ ♍
05	00	44	☽ ♎
05	17	58	♀ ♒
07	07	16	☽ ♏
09	17	14	☽ ♐
12	05	26	☽ ♑
12	06	55	☿ ♑
14	18	26	☽ ♒
17	07	02	☽ ♓
17	13	10	♂ ♓
19	18	08	☽ ♈
20	15	17	☉ ♒
22	02	36	☽ ♉
24	07	40	☽ ♊
26	09	28	☽ ♋
28	09	10	☽ ♌
29	17	31	♀ ♓
30	08	35	☽ ♍

LATITUDES

Date	Mercury ☿ ° '	Venus ♀ ° '	Mars ♂ ° '	Jupiter ♃ ° '	Saturn ♄ ° '	Uranus ♅ ° '	Neptune ♆ ° '	Pluto ♇ ° '
01	00 S 55	01 S 16	01 S 09	00 N 49	02 N 24	00 N 07	01 N 13	16 N 50
04	00 S 13	01 20	01 08	00 49	02 25	00 07	01 13	16 52
07	00 N 39	01 24	01 06	00 50	02 25	00 07	01 13	16 53
10	01 35	01 27	01 05	00 50	02 26	00 07	01 13	16 55
13	02 29	01 30	01 03	00 50	02 27	00 07	01 13	16 57
16	03 09	01 32	01 02	00 50	02 27	00 07	01 13	16 58
19	03 28	01 33	01 00	00 50	02 28	00 07	01 13	17 00
22	03 25	01 34	00 58	00 50	02 29	00 07	01 13	17 02
25	03 07	01 34	00 57	00 50	02 30	00 07	01 13	17 04
28	02 39	01 33	00 55	00 50	02 30	00 07	01 13	17 05
31	02 N 07	01 S 32	00 S 53	00 N 50	02 N 31	00 N 07	01 N 13	17 N 07

DATA

Julian Date	2445336
Delta T	+53 seconds
Ayanamsa	23° 36' 54"
Synetic vernal point	05° ♓ 30' 05"
True obliquity of ecliptic	23° 26' 28"

MOON'S PHASES, APSIDES AND POSITIONS ☽

Date	h	m	Phase	Longitude ° '	Eclipse Indicator
06	04	00	☾	15 ♎ 15	
14	05	08	●	23 ♑ 27	
22	05	33	☽	01 ♉ 37	
28	22	26	○	08 ♌ 26	

Day	h	m		
14	04	53	Apogee	
28	11	32	Perigee	
05	22	29	0S	
13	06	01	Max dec	23° S 36'
20	17	15	0N	
27	04	39	Max dec	23° N 38'

LONGITUDES

Date	Chiron ⚷ ° '	Ceres ⚳ ° '	Pallas ⚴ ° '	Juno ⚵ ° '	Vesta ⚶ ° '	Black Moon Lilith ⚸ ° '
01	23 ♉ 09	08 ♑ 56	19 ♐ 03	29 ♑ 49	07 ♓ 35	21 ♑ 35
11	22 ♉ 53	12 ♑ 59	23 ♐ 10	03 ♒ 58	11 ♓ 49	22 ♑ 41
21	22 ♉ 43	16 ♑ 59	27 ♐ 12	08 ♒ 12	16 ♓ 10	23 ♑ 48
31	22 ♉ 40	20 ♑ 56	01 ♑ 06	12 ♒ 30	20 ♓ 37	24 ♑ 54

ASPECTARIAN

h m	Aspects
01 Saturday	
02 09	☽ □ ♄
02 42	☽ ± ♆
08 06	☽ # ♅
08 35	☽ △ ♅
10 35	☽ # ☿
14 24	☽ ⊼ ☉
17 05	☽ bQ ♆
21 35	☽ # ♃
02 Sunday	
00 44	☽ ± ☉
01 05	☽ Q ♇
02 03	☽ ☍ ♂
05 52	☉ Q ♇
07 07	☽ Q ♄
15 24	♀ ∥ ♆
15 26	☽ ⊼ ♀
16 25	☽ bQ ☉
17 12	☽ # ♂
17 24	☽ △ ♆
20 37	☽ ⚹ ♇
23 37	☽ ⊼ ☿
03 Monday	
00 24	☽ □ ♃
02 14	☽ ± ♀
02 51	☽ ⚹ ♄
09 31	☽ □ ♅
10 02	☽ ± ☿
15 07	♀ ⊻ ♆
15 23	♂ Q ♅
18 41	☽ bQ ♀
19 13	☽ △ ☉
21 39	☽ ∠ ♇
23 57	☽ # ♄
04 Tuesday	
01 08	☉ ⊥ ♅
01 50	☽ bQ ☿
03 31	☿ ⚹ ♃
04 10	☽ ∠ ♄
06 40	☽ ⊼ ♂
07 04	☽ Q ♃
13 04	☽ ⊥ ♇
16 14	☽ Q ♅
17 38	☽ ± ♂
19 46	☽ ⊥ ♄
20 10	☽ □ ♆
23 00	☽ △ ♀
23 31	☽ ⊻ ♇
05 Wednesday	
00 39	☿ ∥ ♃
01 54	☽ ∥ ♇
04 13	☽ ⚹ ♃
04 40	☽ △ ☿
04 44	♀ □ ♇
06 21	☽ ⊻ ♄
07 40	☉ ⊥ ♂
10 25	☽ bQ ♂
13 26	☽ ⚹ ♅
06 Thursday	
01 49	♀ ∥ ♅
03 40	☉ Q ♄
04 00	☽ □ ☉
04 20	☽ Q ♆
07 32	☽ ∠ ♃
15 18	☽ △ ♂
16 48	☽ ∠ ♅
19 39	☽ # ♇
07 Friday	
00 24	☽ ⊥ ♃
02 32	☽ ⚹ ♆
02 59	☿ St R
06 02	☽ ☌ ♇
09 40	☽ ⊥ ♅
11 20	☽ □ ♀
11 50	☽ ⊻ ♃
12 06	☽ □ ☿
13 33	☽ ☌ ♄
16 20	☽ Q ☉
17 05	♀ ⚹ ♃
18 32	☿ ☌ ♀
21 05	☽ ⊻ ♅
23 11	☿ ⚹ ♃
08 Saturday	
00 15	☽ ∥ ♄
07 05	☽ ∠ ♆
10 42	♀ □ ♄
11 15	☉ ∠ ♃
13 26	♀ ⊥ ♆
17 26	☽ ⚹ ☉
22 06	☽ Q ☿
09 Sunday	
00 27	☉ ∥ ♆
00 35	☽ ⊥ ♆
01 12	☽ ∥ ♂
01 49	☽ Q ♀
04 17	☽ □ ♂
12 24	☽ ⊻ ♆
15 59	☽ ⊻ ♇
20 52	☽ ⚹ ☿
22 56	☽ ☌ ♃
10 Monday	
00 06	☽ ⊻ ♄
01 27	☽ ∥ ☿
01 37	☽ ∠ ☉
03 54	☽ ⊥ ♇
04 14	☽ ⚹ ♀
07 58	☽ ☌ ♅
12 07	☽ ⊥ ♄
14 28	☽ ∥ ♃
16 11	☽ ∥ ♀
18 28	☽ Q ♂
21 16	☽ ⊥ ☉
21 55	☽ ∠ ♇
11 Tuesday	
01 18	☽ ∠ ☿
05 30	☽ ∥ ♅
05 38	♀ ∥ ♃
06 16	☽ ∠ ♄
10 26	☽ ∥ ☉
10 28	☽ ⊻ ☉
13 51	☽ ∠ ♀
15 07	☽ ∥ ♆
17 20	♀ ⚹ ♅
18 16	☽ ⊥ ☿
20 21	☽ ⚹ ♂
12 Wednesday	
00 40	☽ ☌ ♆
04 14	☽ ⚹ ♇
05 32	☽ ⊻ ☿
10 14	☽ ⊥ ♀
12 13	☽ ⊻ ♃
12 45	☽ ⚹ ♄
20 45	☽ ⊻ ♅
20 45	☿ □ ♇
23 52	☽ ⊻ ♀
13 Thursday	
00 36	☽ ⊥ ♃
04 38	☽ Q ♇
04 58	☽ ∠ ♂
08 06	☉ ∠ ♅
09 00	☽ ⊥ ♅
13 16	☽ Q ♄
19 12	☽ ∠ ♃
20 01	☉ ∥ ♅
14 Friday	
00 36	☽ ⊥ ♂
03 23	☽ ∠ ♅
05 08	☽ ☌ ☉
08 08	☿ ⊻ ♆
11 07	☿ ⊻ ♂
13 17	☽ ☌ ☿
13 20	♃ ⊻ ♄
13 41	☽ ⊻ ♂
13 50	☽ ⊻ ♆
16 07	♂ ⚹ ♆
17 17	☽ □ ♇
21 01	☽ ∥ ♆
15 Saturday	
02 02	☽ □ ♄
02 04	☽ ⊥ ♆
02 10	☽ ⚹ ♃
06 22	☽ ∥ ♅
09 37	☽ ∥ ☉
10 00	☽ ⚹ ♅
20 05	☽ ☌ ♀
20 20	☽ ∥ ♃
20 22	☽ ∠ ♆
22 52	♀ ∠ ♆
16 Sunday	
02 51	☽ Q ♃
03 24	☉ ☌ ☿
10 22	☽ Q ♅
12 05	☽ ∥ ♀
13 34	☽ ∥ ☿
20 02	☽ ⊻ ☿
20 36	♂ △ ♇
23 15	☿ ∥ ♀
23 35	☽ ⊻ ☉
17 Monday	
02 41	☽ ⚹ ♆
05 58	☽ △ ♇
06 36	☽ ☌ ♂
06 53	☽ ⊥ ☿
08 20	☿ ⊥ ♂
12 42	☽ ⊥ ☉
14 44	☽ △ ♄
15 32	☽ □ ♃
22 33	☽ □ ♅
23 05	☽ ∠ ☿
18 Tuesday	
02 41	☽ Q ♆
02 53	☽ ∥ ♂
04 02	☿ ∠ ♅
08 14	☽ ∠ ☉
11 48	☽ bQ ♇
12 43	☽ ∥ ♄
13 08	☉ ⊻ ♆
15 05	☽ ⊻ ♀
20 32	☽ bQ ♄
21 32	♀ Q ♃
19 Wednesday	
01 57	☽ ⚹ ☿
04 11	☽ ⊥ ♀
05 28	☽ ± ♇
14 05	☽ □ ♆
14 09	☽ ± ♄
16 15	☽ ⚹ ☉
17 08	☽ ⊼ ♇
18 50	☽ # ♇
21 42	☽ ⊻ ♂
23 16	☽ Q ☿
23 35	☽ ∠ ♀
20 Thursday	
01 45	☽ ⊼ ♄
03 08	☽ △ ♃
03 08	☉ □ ♇
03 31	☉ ∥ ♃
09 17	☽ △ ♅
09 58	☽ ⊥ ♂
17 15	☽ Q ☉
19 17	☿ ∠ ♃
21 Friday	
04 06	☽ ∠ ♂
05 29	☿ ⊻ ♀
06 53	☽ □ ☿
07 08	☽ ⚹ ♀
07 54	☽ bQ ♃
13 36	☽ bQ ♅
15 14	☽ ∥ ♇
16 14	♀ Q ♅
22 54	☽ △ ♆
22 Saturday	
00 46	☽ ± ♃
01 41	☽ ☍ ♇
03 14	☿ ∠ ♂
05 33	☽ □ ☉
06 11	☽ ± ♅
07 36	☽ Q ♀
09 31	☽ ⚹ ♂
09 58	☽ ☍ ♄
11 49	☽ ⊼ ♃
17 04	☽ ⊼ ♅
17 17	♂ △ ♄
19 26	☽ # ♄
21 07	☽ # ♂
23 Sunday	
02 04	☽ bQ ♆
08 19	☽ Q ♂
10 19	☽ △ ☿
17 56	☽ ± ♆
18 52	☽ □ ♀
20 53	☽ # ♀
24 Monday	
00 03	♂ ∥ ♄
04 20	☽ ⊼ ♆
06 50	☽ ⊼ ♇
07 13	☿ ⊥ ♀
09 52	♂ □ ♃
11 20	☽ bQ ☿
14 33	☽ △ ☉
14 41	☽ ⊼ ♄
15 22	☉ ⊥ ♆
16 40	☉ □ ♄
16 54	☽ ☍ ♃
17 02	☽ ± ♇
17 14	☽ □ ♂
19 54	☽ # ☿
19 57	☽ # ☉
20 21	☉ ∥ ☿
21 22	☽ ☍ ♅
25 Tuesday	
00 49	☽ ± ♄
02 06	☽ ± ☿
04 50	☽ # ♃
04 58	♃ ⊥ ♇
08 07	☽ bQ ♇
11 52	☽ ⊼ ☿
14 44	☽ # ♅
15 48	☽ bQ ♄
17 28	☽ bQ ☉
21 35	☽ # ♆
26 Wednesday	
02 06	☽ △ ♀
05 12	☉ ⚹ ♃
06 26	☽ ☍ ♆
08 42	☽ △ ♇
09 13	☽ ± ☉
16 13	☽ △ ♄
18 45	☽ ⊼ ♃
19 35	☽ ⊼ ☉
21 15	☽ △ ♂
22 35	☽ ⊼ ♅
27 Thursday	
04 26	☽ bQ ♀
04 26	☽ ± ♃
08 10	☽ ± ♅
11 50	☽ ☍ ☿
13 26	☿ St D
18 51	☽ bQ ♃
19 59	☽ ± ♀
22 22	☽ bQ ♂
22 27	☽ bQ ♅
28 Friday	
00 04	♂ □ ♅
06 17	☽ ⊼ ♀
06 19	☽ ⊼ ♆
06 39	♀ ⚹ ♆
08 25	☽ □ ♇
11 09	☽ # ♆
13 19	☽ ± ♂
15 46	☽ ± ♆
15 51	☽ □ ♄
17 32	☉ ⚹ ♅
17 33	☽ # ♅
18 43	☽ △ ♃
22 07	☽ △ ♅
22 26	☽ ☍ ☉
23 16	☽ ⊼ ♂
29 Saturday	
02 22	☽ # ♃
05 58	☽ bQ ♆
06 44	☽ # ☿
08 25	♀ △ ♇
11 29	☽ ⊼ ☿
12 46	☽ Q ♇
19 21	☽ # ☉
20 17	☽ Q ♄
21 08	☽ ± ☿
30 Sunday	
05 47	☽ △ ♆
07 49	☽ ⚹ ♇
09 56	☽ ☍ ♀
11 47	☽ bQ ☿
15 09	♂ Q ♆
15 27	☽ ⚹ ♄
18 49	☽ □ ♃
21 56	☽ □ ♅
31 Monday	
00 03	☽ # ♀
01 29	☉ ⊻ ♂
01 39	☽ ☍ ♂
01 39	☽ ⊼ ☉
08 03	☽ ∠ ♇
08 10	☽ # ♄
12 11	☽ ± ☉
12 48	☽ △ ☿
15 55	☽ ∠ ♄
18 13	☿ ∠ ♀
18 48	☽ # ♂
22 55	☽ ⊥ ♇

All ephemeris data is given at 12.00 UT and the Moon's longitude is additionally given for 24.00 UT

FEBRUARY 1983

LONGITUDES

Date	Sidereal time h m s	Sun ☉ ° ' "	Moon ☽ ° ' "	Moon ☽ 24.00 ° ' "	Mercury ☿ ° '	Venus ♀ ° '	Mars ♂ ° '	Jupiter ♃ ° '	Saturn ♄ ° '	Uranus ♅ ° '	Neptune ♆ ° '	Pluto ♇ ° '
01	20 44 31	12 ♒ 03 21	01 ♎ 18 36	08 ♎ 19 11	17 ♑ 54	03 ♓ 27	11 ♓ 43	06 ♐ 38	04 ♏ 20	08 ♐ 22	28 ♐ 19	29 ♎ 32
02	20 48 28	13 04 13	15 ♎ 12 21	21 ♎ 58 08	18 29	04 42	12 30	06 47	04 21	08 24	28 21	29 R 32
03	20 52 25	14 05 05	28 ♎ 36 42	05 ♏ 08 23	19 09	05 57	13 17	06 55	04 22	08 26	28 22	29 32
04	20 56 21	15 05 56	11 ♏ 33 37	17 ♏ 52 55	19 53	07 11	14 04	07 04	04 23	08 28	28 24	29 32
05	21 00 18	16 06 46	24 ♏ 06 53	00 ♐ 16 05	20 41	08 26	14 51	07 12	04 24	08 30	28 26	29 31
06	21 04 14	17 07 35	06 ♐ 21 12	12 ♐ 22 50	21 33	09 41	15 38	07 20	04 24	08 32	28 27	29 31
07	21 08 11	18 08 23	18 ♐ 21 37	24 ♐ 18 10	22 28	10 55	16 24	07 29	04 25	08 34	28 29	29 31
08	21 12 07	19 09 10	00 ♑ 13 03	06 ♑ 06 49	23 27	12 10	17 11	07 37	04 25	08 36	28 31	29 31
09	21 16 04	20 09 56	11 ♑ 59 59	17 ♑ 52 59	24 28	13 24	17 58	07 44	04 26	08 37	28 32	29 31
10	21 20 00	21 10 41	23 ♑ 46 14	29 ♑ 40 08	25 31	14 39	18 45	07 52	04 26	08 39	28 34	29 30
11	21 23 57	22 11 25	05 ♒ 34 59	11 ♒ 31 04	26 37	15 53	19 31	08 00	04 26	08 41	28 35	29 30
12	21 27 54	23 12 07	17 ♒ 28 38	23 ♒ 27 53	27 45	17 08	20 18	08 07	04 R 26	08 42	28 37	29 30
13	21 31 50	24 12 48	29 ♒ 29 00	05 ♓ 32 08	28 ♑ 55	18 22	21 05	08 14	04 26	08 44	28 38	29 29
14	21 35 47	25 13 28	11 ♓ 37 25	17 ♓ 44 58	00 ♒ 07	19 37	21 51	08 22	04 26	08 46	28 40	29 29
15	21 39 43	26 14 06	23 ♓ 54 54	00 ♈ 07 23	01 21	20 51	22 38	08 29	04 26	08 47	28 41	29 28
16	21 43 40	27 14 43	06 ♈ 22 30	12 ♈ 40 28	02 36	22 05	23 25	08 35	04 25	08 48	28 43	29 28
17	21 47 36	28 15 17	19 ♈ 01 25	25 ♈ 25 34	03 53	23 20	24 11	08 42	04 25	08 50	28 44	29 27
18	21 51 33	29 ♒ 15 50	01 ♉ 53 09	08 ♉ 24 25	05 12	24 34	24 58	08 49	04 24	08 51	28 45	29 27
19	21 55 29	00 ♓ 16 22	14 ♉ 59 37	21 ♉ 39 03	06 32	25 48	25 44	08 55	04 24	08 52	28 47	29 26
20	21 59 26	01 16 51	28 ♉ 22 58	05 ♊ 11 37	07 53	27 02	26 31	09 01	04 23	08 54	28 48	29 25
21	22 03 23	02 17 19	12 ♊ 05 12	19 ♊ 03 52	09 16	28 16	27 17	09 08	04 22	08 55	28 49	29 25
22	22 07 19	03 17 45	26 ♊ 07 41	03 ♋ 16 36	10 39	29 ♓ 30	28 04	09 14	04 21	08 56	28 51	29 24
23	22 11 16	04 18 09	10 ♋ 30 26	17 ♋ 48 51	12 04	00 ♈ 44	28 50	09 19	04 20	08 57	28 52	29 23
24	22 15 12	05 18 31	25 ♋ 11 22	02 ♌ 37 19	13 30	01 58	29 ♓ 36	09 25	04 19	08 58	28 53	29 22
25	22 19 09	06 18 51	10 ♌ 05 51	17 ♌ 36 00	14 58	03 12	00 ♈ 23	09 31	04 17	08 59	28 54	29 22
26	22 23 05	07 19 09	25 ♌ 06 40	02 ♍ 36 39	16 26	04 26	01 09	09 36	04 16	09 00	28 55	29 21
27	22 27 02	08 19 25	10 ♍ 04 46	17 ♍ 29 50	17 56	05 40	01 55	09 41	04 14	09 00	28 56	29 20
28	22 30 58	09 ♓ 19 40	24 ♍ 50 44	02 ♎ 06 32	19 ♒ 26	06 ♈ 54	02 ♈ 41	09 ♐ 46	04 ♏ 13	09 ♐ 01	28 ♐ 57	29 ♎ 19

Date	Moon True ☊ ° '	Moon Mean ☊ ° '	Moon ☽ Latitude ° '
01	03 ♋ 35	02 ♋ 12	05 N 11
02	03 R 31	02 08	05 05
03	03 28	02 05	04 42
04	03 26	02 02	04 05
05	03 D 27	01 59	03 18
06	03 28	01 56	02 22
07	03 30	01 53	01 21
08	03 31	01 49	00 N 18
09	03 R 30	01 46	00 S 46
10	03 28	01 43	01 47
11	03 23	01 40	02 44
12	03 16	01 37	03 34
13	03 07	01 33	04 15
14	02 57	01 30	04 45
15	02 46	01 27	05 02
16	02 36	01 24	05 05
17	02 28	01 21	04 53
18	02 22	01 18	04 26
19	02 19	01 14	03 45
20	02 18	01 11	02 51
21	02 D 18	01 08	01 46
22	02 19	01 05	00 S 33
23	02 R 18	01 02	00 N 44
24	02 16	00 59	01 59
25	02 12	00 55	03 07
26	02 05	00 52	04 03
27	01 55	00 49	04 41
28	01 ♋ 45	00 ♋ 46	05 N 01

DECLINATIONS

Date	Sun ☉ ° '	Moon ☽ ° '	Mercury ☿ ° '	Venus ♀ ° '	Mars ♂ ° '	Jupiter ♃ ° '	Saturn ♄ ° '	Uranus ♅ ° '	Neptune ♆ ° '	Pluto ♇ ° '
01	17 S 11	04 N 14	20 S 20	11 S 40	07 S 59	20 S 35	10 S 35	21 S 36	22 S 13	04 N 45
02	16 54	01 S 18	20 26	11 12	07 40	20 37	10 35	21 36	22 13	04 45
03	16 36	06 35	20 32	10 44	07 22	20 38	10 36	21 36	22 13	04 46
04	16 19	11 25	20 36	10 15	07 03	20 39	10 36	21 37	22 13	04 47
05	16 01	15 37	20 40	09 47	06 44	20 41	10 36	21 37	22 13	04 47
06	15 42	19 02	20 43	09 18	06 26	20 42	10 36	21 37	22 13	04 48
07	15 24	21 35	20 44	08 49	06 07	20 43	10 36	21 38	22 13	04 48
08	15 05	23 09	20 45	08 19	05 48	20 45	10 35	21 38	22 13	04 49
09	14 46	23 40	20 44	07 50	05 29	20 46	10 35	21 38	22 13	04 50
10	14 26	23 07	20 43	07 20	05 10	20 47	10 35	21 38	22 13	04 50
11	14 07	21 32	20 40	06 50	04 51	20 48	10 35	21 39	22 13	04 51
12	13 47	19 00	20 36	06 20	04 33	20 49	10 35	21 39	22 13	04 51
13	13 27	15 38	20 31	05 49	04 14	20 50	10 35	21 39	22 13	04 52
14	13 07	11 36	20 24	05 19	03 55	20 51	10 34	21 39	22 13	04 53
15	12 46	07 02	20 17	04 48	03 36	20 52	10 34	21 40	22 13	04 53
16	12 26	02 S 08	20 08	04 17	03 17	20 53	10 34	21 40	22 13	04 54
17	12 05	02 N 56	19 58	03 47	02 58	20 54	10 33	21 40	22 13	04 55
18	11 44	07 58	19 46	03 16	02 39	20 55	10 33	21 40	22 13	04 55
19	11 23	12 45	19 34	02 45	02 19	20 56	10 32	21 41	22 13	04 56
20	11 01	17 01	19 20	02 13	02 00	20 57	10 32	21 41	22 13	04 57
21	10 40	20 29	19 05	01 42	01 41	20 58	10 31	21 41	22 13	04 57
22	10 18	22 50	18 48	01 11	01 22	20 59	10 31	21 41	22 13	04 58
23	09 56	23 45	18 31	00 40	01 03	21 00	10 30	21 41	22 13	04 59
24	09 34	23 03	18 12	00 S 08	00 44	21 00	10 29	21 41	22 13	04 59
25	09 12	20 43	17 52	00 N 23	00 25	21 01	10 29	21 42	22 13	05 00
26	08 49	16 58	17 30	00 54	00 S 06	21 02	10 28	21 42	22 13	05 01
27	08 27	12 08	17 07	01 26	00 N 13	21 03	10 27	21 42	22 13	05 02
28	08 S 04	06 N 39	16 S 43	01 N 57	00 N 32	21 S 03	10 S 27	21 S 42	22 S 13	05 N 02

ZODIAC SIGN ENTRIES

Date	h	m	Planets
01	09	47	☽ ♎
03	14	32	☽ ♏
05	23	28	☽ ♐
08	11	33	☽ ♑
11	00	40	☽ ♒
13	13	02	☽ ♓
14	09	36	☿ ♒
15	23	46	☽ ♈
18	08	30	☽ ♉
19	05	31	☉ ♓
20	14	52	☽ ♊
22	18	31	☽ ♋
22	21	35	♀ ♈
24	19	47	☽ ♌
25	00	19	♂ ♈
26	19	49	☽ ♍
28	20	30	☽ ♎

LATITUDES

Date	Mercury ☿ ° '	Venus ♀ ° '	Mars ♂ ° '	Jupiter ♃ ° '	Saturn ♄ ° '	Uranus ♅ ° '	Neptune ♆ ° '	Pluto ♇ ° '
01	01 N 56	01 S 32	00 S 53	00 N 51	02 N 31	00 N 07	01 N 13	17 N 08
04	01 23	01 30	00 51	00 51	02 32	00 07	01 13	17 09
07	00 51	01 27	00 49	00 51	02 33	00 07	01 13	17 11
10	00 N 20	01 24	00 47	00 51	02 34	00 07	01 13	17 13
13	00 S 08	01 20	00 45	00 51	02 34	00 06	01 13	17 14
16	00 34	01 15	00 43	00 51	02 35	00 06	01 13	17 16
19	00 57	01 10	00 41	00 52	02 36	00 06	01 13	17 17
22	01 17	01 05	00 39	00 52	02 37	00 06	01 13	17 19
25	01 35	00 58	00 37	00 52	02 37	00 06	01 14	17 20
28	01 49	00 52	00 35	00 52	02 38	00 06	01 14	17 22
31	02 S 00	00 S 44	00 S 33	00 N 53	02 N 39	00 N 06	01 N 14	17 N 23

DATA

Julian Date	2445367
Delta T	+53 seconds
Ayanamsa	23° 36' 59"
Synetic vernal point	05° ♓ 30' 00"
True obliquity of ecliptic	23° 26' 29"

MOON'S PHASES, APSIDES AND POSITIONS ☽

Date	h	m	Phase	Longitude ° '	Eclipse Indicator
04	19	17	☾	15 ♏ 24	
13	00	32	●	23 ♒ 44	
20	17	32	☽	01 ♊ 31	
27	08	58	○	08 ♍ 12	

Day	h	m		
10	08	12	Apogee	
25	21	59	Perigee	
02	06	18	0S	
09	11	38	Max dec	23° S 40'
16	22	10	0N	
23	13	49	Max dec	23° N 45'

LONGITUDES

Date	Chiron ⚷ ° '	Ceres ⚳ ° '	Pallas ⚴ ° '	Juno ⚵ ° '	Vesta ⚶ ° '	Black Moon Lilith ⚸ ° '
01	22 ♉ 40	21 ♑ 20	01 ♑ 29	12 ♒ 56	21 ♓ 04	25 ♑ 01
11	22 ♉ 44	25 ♑ 13	05 ♑ 13	17 ♒ 17	25 ♓ 35	26 ♑ 07
21	22 ♉ 54	29 ♑ 01	08 ♑ 46	21 ♒ 42	00 ♈ 09	27 ♑ 14
31	23 ♉ 11	02 ♒ 43	12 ♑ 06	26 ♒ 08	04 ♈ 45	28 ♑ 21

ASPECTARIAN

h m	Aspects
01 Tuesday	
00 39	☽ Q ♃
03 38	☽ Q ♅
04 16	☽ ⚼ ☉
05 55	♇ St R
06 56	☽ □ ♆
06 57	☽ ⊥ ♄
08 59	☽ ⚺ ♇
09 46	☽ ∥ ♇
16 00	☽ ⚻ ♀
17 09	☽ ⚺ ♄
21 11	☽ ⚹ ♃
02 Wednesday	
00 07	☽ ⚹ ♅
03 20	☽ ± ♀
05 11	♀ △ ♄
06 59	☽ ⚻ ♂
07 57	☽ △ ☉
14 01	☽ Q ♆
18 05	☽ □ ☿
18 11	☽ ± ♂
18 40	☉ ∠ ♆
20 45	☽ ⚼ ♀
23 47	☽ ∠ ♃
03 Thursday	
02 37	☽ ∠ ♅
03 32	☽ # ♇
11 22	☽ ⚼ ♂
11 34	☽ ⚹ ♆
13 40	☽ ☌ ♇
15 27	☽ ∥ ♂
16 16	☽ ⊥ ♃
19 01	☽ ⊥ ♅
19 03	♀ ∥ ♄
22 35	☽ ☌ ♄
04 Friday	
02 56	☽ △ ♀
03 29	☽ ⚺ ♃
04 41	☽ Q ☿
06 11	☽ ⚺ ♅
06 33	☽ ∥ ♀
07 43	☽ ∥ ♄
09 19	♀ □ ♃
15 29	☽ ∠ ♆
17 03	☽ △ ♂
19 17	☽ □ ☉
05 Saturday	
02 07	♂ ⚼ ♇
04 55	☽ ⚹ ☿
08 44	☽ ⊥ ♆
13 23	♀ □ ♅
14 20	☽ ∥ ☉
20 25	☽ ⚺ ♆
22 32	☽ ⚺ ♇
06 Sunday	
02 24	☿ ∥ ♃
08 09	☽ ⚺ ♄
09 21	☽ Q ☉
10 21	☽ ⊥ ♇
12 26	☽ ∠ ☿
13 59	☽ ☌ ♃
16 20	☽ ☌ ♅
19 22	☽ □ ♀
20 04	☽ ⊥ ♄
07 Monday	
02 46	☽ ∥ ♃
02 54	☽ ∥ ☿
03 25	♀ Q ♆
04 17	☽ ∠ ♇
07 48	☽ □ ♂
07 53	☽ ⊥ ☿
11 31	☽ ⚹ ☉
12 04	☿ ∠ ♃
12 32	☽ ∥ ♅
14 08	☽ ∠ ♄
20 01	☽ ∥ ♆
21 02	☽ ⚺ ☿
08 Tuesday	
08 32	☽ ☌ ♆
10 34	☽ ⚹ ♇
11 53	☽ Q ♀
15 45	☿ ∠ ♅
15 45	☿ ∥ ♃
20 34	☽ ⚹ ♄
20 46	☽ ∠ ☉
22 50	☽ Q ♂
09 Wednesday	
00 26	☉ Q ♃
03 13	☽ ⚺ ♃
05 06	☽ ⚺ ♅
11 00	☽ Q ♇
15 12	☽ ⚹ ♀
15 35	☽ ⊥ ♃
16 50	☽ ⊥ ☉
17 22	☽ ⊥ ♅
21 02	☽ Q ♄
23 11	☉ Q ♅
10 Thursday	
01 02	☽ ⚹ ♂
06 13	☽ ⚺ ☉
09 15	♀ ⚼ ♇
10 09	☽ ∠ ♃
11 46	☽ ∠ ♅
15 55	☽ ☌ ☿
21 47	☽ ⚺ ♆
23 40	☽ □ ♇
11 Friday	
01 22	☽ ∠ ♀
03 21	☽ ∥ ♆
09 16	♂ ⚼ ♄
09 41	☽ □ ♄
09 42	☽ ∠ ♂
09 59	☽ ⊥ ♆
10 40	☽ ∥ ♅
12 55	♂ # ♇
16 56	☽ ⚹ ♃
18 17	☽ ⚹ ♅
19 47	☽ ∥ ♃
21 27	☽ ∥ ☿
21 44	☽ ⊥ ♀
12 Saturday	
04 13	☽ ∠ ♆
05 10	☽ ⊥ ♂
11 13	☽ ⚺ ♀
11 18	♄ St R
17 21	☽ Q ♃
18 04	☽ ⚺ ♂
18 30	☽ Q ♅
13 Sunday	
00 32	☽ ☌ ☉
06 10	☿ ⚺ ♆
10 19	☽ ⚹ ♆
10 46	☽ ⚺ ☿
12 00	☽ △ ♇
21 49	☽ △ ♄
23 17	☿ □ ♇
23 58	☽ ⊥ ☿
14 Monday	
02 36	☽ ∥ ☉
05 31	☽ □ ♃
06 21	☽ □ ♅
08 35	♀ ⚼ ♄
10 07	☽ Q ♆
17 36	☽ ⚼ ♇
17 36	☽ ∥ ♄
19 37	☽ ∠ ☿
15 Tuesday	
03 17	☽ ⚼ ♄
05 23	☽ ☌ ♀
08 08	♀ # ♇
09 21	☽ ☌ ♂
11 08	☽ ± ♇
16 54	☽ ⚺ ☉
20 44	☽ ± ♄
21 15	☽ □ ♆
22 39	☽ # ♇
22 44	☽ ⚻ ♇
16 Wednesday	
00 23	☽ ∥ ♀
03 58	☽ ⚹ ☿
05 29	☽ ⊥ ☉
06 08	☽ ∥ ♂
08 16	☽ ⚻ ♄
13 31	♂ ± ♇
16 16	☽ △ ♃
16 39	☽ △ ♅
17 Thursday	
00 09	☽ ∠ ☉
05 25	☽ Q ☿
12 08	☽ # ♂
14 24	♀ ± ♇
15 38	☽ # ♀
20 51	☽ ⚼ ♃
20 56	☽ ⚺ ♀
21 02	☽ ⚼ ♅
21 24	☽ ∥ ♇
21 36	☿ □ ♄
22 19	☽ ⚺ ♂
23 39	☉ ⚹ ♆
18 Friday	
03 47	☿ ⊥ ♆
06 11	☽ △ ♆
06 44	☽ ⚹ ☉
07 29	☽ ☍ ♇
09 18	☽ ⊥ ♀
10 11	☽ ⊥ ♂
13 44	☽ ± ♃
13 47	☽ ± ♅
16 12	☉ △ ♇
16 39	☽ ☍ ♄
18 48	☽ □ ☿
22 41	♃ ☌ ♅
19 Saturday	
00 43	☽ # ♄
00 50	☽ ⚻ ♅
00 51	☽ ⚻ ♃
03 35	☽ ∠ ♀
03 47	☽ ∠ ♂
05 26	☽ # ☉
06 39	☽ Q ☉
08 42	♀ ☌ ♂
09 48	☽ ⚼ ♆
20 Sunday	
02 03	☽ ± ♆
08 29	☽ ⚹ ♂
09 22	☽ ⚹ ♀
12 44	☽ ⚻ ♆
13 50	☽ ⚻ ♇
17 32	☽ □ ☉
22 34	☽ ⚻ ♄
21 Monday	
00 23	☽ ± ♇
02 10	☽ # ☿
05 52	☿ ⚹ ♅
06 29	☽ ☍ ♅
06 33	☽ △ ☿
06 50	☽ ☍ ♃
06 51	☽ Q ♂
08 33	☽ Q ♀
09 01	☽ ± ♄
09 30	☿ ⚹ ♃
13 42	♀ ∥ ♂
13 47	♀ ± ♄
16 00	☽ # ♃
16 00	☽ ⚼ ♇
21 34	☉ ∥ ♄
22 35	☽ # ♅
22 50	♀ □ ♆
22 Tuesday	
00 30	☽ ⚼ ♄
04 10	☽ # ♆
09 53	♀ ⚻ ♇
11 07	☽ ⚼ ☿
15 27	☽ □ ♂
16 35	☽ ☍ ♆
17 30	☽ △ ♇
18 14	☽ □ ♀
20 48	♂ ± ♄
23 Wednesday	
00 56	☽ △ ☉
01 46	☽ △ ♄
03 51	☽ ± ☿
09 25	☽ ⚻ ♅
10 02	☽ ⚻ ♃
12 40	☉ △ ♄
12 58	♂ □ ♆
14 52	☽ ⚻ ☿
19 19	☽ ± ♅
19 58	☽ ± ♃
24 Thursday	
03 29	☽ ⚼ ☉
04 56	♂ ⚻ ♇
10 01	☽ ⚼ ♅
10 44	☽ ⚼ ♃
17 59	☽ ⚻ ♆
18 16	☿ ∠ ♆
18 45	☽ □ ♇
19 09	☽ ± ☉
19 32	☽ △ ♂
22 36	☽ # ♆
23 57	☽ △ ♀
25 Friday	
02 42	☽ □ ♄
03 39	☽ ± ♆
03 47	☽ # ♅
05 30	☽ ⚻ ☉
09 37	☽ # ♃
10 12	☽ △ ♅
11 03	☽ △ ♃
13 04	♀ # ♂
18 06	☽ ⚼ ♆
20 38	☽ ☍ ☿
20 54	☽ ⚼ ♂
23 36	☽ Q ♇
26 Saturday	
02 08	☽ ⚼ ♀
02 08	☿ ∠ ♂
07 28	☽ Q ♄
08 42	☽ # ☿
08 42	♀ ⚻ ♄
12 04	☽ ± ♂
17 48	☽ ± ♀
18 06	☽ △ ♆
18 46	☽ ⚹ ♇
22 11	☽ ⚻ ♂
27 Sunday	
02 38	☽ ⚹ ♄
04 16	☽ ⚻ ♀
08 58	☽ ☍ ☉
10 16	☽ □ ♅
11 22	☽ □ ♃
18 52	☽ ∠ ♇
19 33	☽ # ♄
28 Monday	
02 09	☽ ⚻ ☿
02 49	☽ ∠ ♄
04 35	☉ □ ♅
05 29	☽ # ☉
09 30	☽ ⊥ ♇
13 05	☽ ± ☿
15 35	☽ Q ♅
16 50	☽ Q ♃
17 32	☽ ⊥ ♄
18 45	☽ ∥ ♇
18 47	☽ □ ♆
19 22	☽ ⚺ ♇
23 26	☉ □ ♃

All ephemeris data is given at 12.00 UT and the Moon's longitude is additionally given for 24.00 UT

MARCH 1983

LONGITUDES

Date	Sidereal time h m s	Sun ☉ ° ' "	Moon ☽ ° ' "	Moon ☽ 24.00 ° ' "	Mercury ☿ ° '	Venus ♀ ° '	Mars ♂ ° '	Jupiter ♃ ° '	Saturn ♄ ° '	Uranus ♅ ° '	Neptune ♆ ° '	Pluto ♇ ° '
01	22 34 55	10 ♓ 19 53	09 ♎ 16 24	16 ♎ 19 43	20 ♒ 58	08 ♈ 08	03 ♈ 27	09 ♐ 51	04 ♏ 11	09 ♐ 02	28 ♐ 58	29 ♎ 18
02	22 38 52	11 20 04	23 ♎ 16 06	00 ♏ 05 17	22 31	09 21	04 13	09 56	04 R 09	09 03	28 59	29 R 17
03	22 42 48	12 20 14	06 ♏ 47 17	13 ♏ 22 15	24 05	10 35	04 59	10 00	04 08	09 03	29 00	29 16
04	22 46 45	13 20 22	19 ♏ 50 27	26 ♏ 12 19	25 40	11 49	05 45	10 04	04 06	09 04	29 01	29 15
05	22 50 41	14 20 28	02 ♐ 28 22	08 ♐ 39 12	27 16	13 02	06 31	10 09	04 03	09 04	29 02	29 14
06	22 54 38	15 20 34	14 ♐ 45 29	20 ♐ 47 51	28 ♒ 53	14 16	07 17	10 13	04 01	09 05	29 03	29 13
07	22 58 34	16 20 37	26 ♐ 47 02	02 ♑ 43 44	00 ♓ 31	15 29	08 03	10 16	03 59	09 05	29 04	29 12
08	23 02 31	17 20 39	08 ♑ 38 37	14 ♑ 32 22	02 11	16 43	08 49	10 20	03 57	09 06	29 05	29 11
09	23 06 27	18 20 39	20 ♑ 25 36	26 ♑ 18 56	03 51	17 56	09 35	10 24	03 54	09 06	29 05	29 10
10	23 10 24	19 20 38	02 ♒ 12 54	08 ♒ 08 02	05 33	19 09	10 21	10 27	03 52	09 06	29 06	29 08
11	23 14 21	20 20 35	14 ♒ 04 45	20 ♒ 03 27	07 15	20 23	11 07	10 30	03 49	09 06	29 07	29 07
12	23 18 17	21 20 30	26 ♒ 04 27	02 ♓ 08 01	08 59	21 36	11 52	10 33	03 46	09 07	29 07	29 06
13	23 22 14	22 20 23	08 ♓ 14 22	14 ♓ 23 37	10 44	22 49	12 38	10 36	03 44	09 07	29 08	29 05
14	23 26 10	23 20 14	20 ♓ 35 53	26 ♓ 51 10	12 30	24 02	13 24	10 38	03 41	09 07	29 09	29 04
15	23 30 07	24 20 04	03 ♈ 09 30	09 ♈ 30 51	14 18	25 15	14 09	10 41	03 38	09 R 07	29 09	29 02
16	23 34 03	25 19 51	15 ♈ 55 08	22 ♈ 22 19	16 06	26 28	14 55	10 43	03 35	09 07	29 10	29 01
17	23 38 00	26 19 36	28 ♈ 52 19	05 ♉ 25 07	17 56	27 41	15 40	10 45	03 32	09 06	29 10	29 00
18	23 41 56	27 19 19	12 ♉ 00 39	18 ♉ 38 56	19 47	28 ♈ 54	16 26	10 47	03 29	09 06	29 11	28 58
19	23 45 53	28 19 00	25 ♉ 20 00	02 ♊ 03 54	21 39	00 ♉ 07	17 11	10 49	03 25	09 06	29 11	28 57
20	23 49 50	29 ♓ 18 39	08 ♊ 50 43	15 ♊ 40 32	23 33	01 19	17 57	10 50	03 22	09 06	29 12	28 55
21	23 53 46	00 ♈ 18 16	22 ♊ 33 28	29 ♊ 29 36	25 27	02 32	18 42	10 51	03 19	09 05	29 12	28 54
22	23 57 43	01 17 50	06 ♋ 29 00	13 ♋ 31 41	27 23	03 44	19 27	10 53	03 15	09 05	29 12	28 53
23	00 01 39	02 17 22	20 ♋ 37 36	27 ♋ 46 35	29 ♓ 20	04 57	20 12	10 54	03 12	09 05	29 13	28 51
24	00 05 36	03 16 51	04 ♌ 58 24	12 ♌ 12 41	01 ♈ 18	06 09	20 58	10 54	03 08	09 04	29 13	28 50
25	00 09 32	04 16 19	19 ♌ 28 56	26 ♌ 46 32	03 17	07 22	21 43	10 55	03 04	09 03	29 13	28 48
26	00 13 29	05 15 43	04 ♍ 04 45	11 ♍ 22 46	05 17	08 34	22 28	10 55	03 01	09 03	29 13	28 47
27	00 17 25	06 15 06	18 ♍ 39 41	25 ♍ 54 37	07 18	09 46	23 13	10 55	02 57	09 02	29 14	28 45
28	00 21 22	07 14 27	03 ♎ 06 39	10 ♎ 14 57	09 20	10 58	23 58	10 R 55	02 53	09 02	29 14	28 44
29	00 25 19	08 13 45	17 ♎ 18 44	24 ♎ 17 22	11 22	12 10	24 43	10 55	02 49	09 01	29 14	28 42
30	00 29 15	09 13 01	01 ♏ 10 22	07 ♏ 57 22	13 25	13 22	25 28	10 55	02 45	09 00	29 14	28 41
31	00 33 12	10 ♈ 12 16	14 ♏ 38 12	21 ♏ 12 50	15 ♈ 28	14 ♉ 34	26 ♈ 13	10 ♐ 54	02 ♏ 41	08 ♐ 59	29 ♐ 14	28 ♎ 39

Date	Moon True ☊ ° '	Moon Mean ☊ ° '	Moon ☽ Latitude ° '
01	01 ♋ 34	00 ♋ 43	05 N 00
02	01 R 25	00 39	04 41
03	01 17	00 36	04 07
04	01 12	00 33	03 21
05	01 10	00 30	02 26
06	01 09	00 27	01 26
07	01 D 10	00 24	00 N 23
08	01 R 10	00 20	00 S 40
09	01 08	00 17	01 40
10	01 04	00 14	02 37
11	00 57	00 11	03 26
12	00 48	00 08	04 08
13	00 36	00 04	04 38
14	00 22	00 ♋ 01	04 56
15	00 08	29 ♊ 58	05 00
16	29 ♊ 55	29 55	04 49
17	29 44	29 52	04 23
18	29 35	29 49	03 42
19	29 30	29 45	02 50
20	29 27	29 42	01 47
21	29 D 26	29 39	00 S 36
22	29 R 26	29 36	00 N 37
23	29 25	29 33	01 50
24	29 23	29 30	02 56
25	29 18	29 26	03 52
26	29 10	29 23	04 33
27	28 59	29 20	04 56
28	28 48	29 17	05 00
29	28 36	29 14	04 45
30	28 25	29 10	04 13
31	28 ♊ 17	29 ♊ 07	03 N 29

DECLINATIONS

Date	Sun ☉ ° '	Moon ☽ ° '	Mercury ☿ ° '	Venus ♀ ° '	Mars ♂ ° '	Jupiter ♃ ° '	Saturn ♄ ° '	Uranus ♅ ° '	Neptune ♆ ° '	Pluto ♇ ° '
01	07 S 42	00 N 55	16 S 18	02 N 28	00 N 51	21 S 04	10 S 26	21 S 42	22 S 13	05 N 03
02	07 19	04 S 41	15 51	02 59	01 10	21 04	10 25	21 42	22 13	05 04
03	06 56	09 53	15 23	03 31	01 28	21 05	10 24	21 42	22 13	05 05
04	06 33	14 28	14 54	04 02	01 47	21 06	10 23	21 42	22 13	05 05
05	06 10	18 16	14 23	04 33	02 06	21 06	10 22	21 42	22 13	05 06
06	05 47	21 09	13 51	05 04	02 25	21 07	10 21	21 42	22 13	05 07
07	05 23	23 01	13 18	05 34	02 44	21 07	10 21	21 43	22 12	05 07
08	05 00	23 49	12 44	06 05	03 02	21 08	10 20	21 43	22 12	05 08
09	04 37	23 33	12 08	06 36	03 21	21 08	10 19	21 43	22 12	05 09
10	04 13	22 13	11 32	07 06	03 40	21 08	10 18	21 43	22 12	05 10
11	03 50	19 54	10 53	07 36	03 58	21 09	10 16	21 43	22 12	05 10
12	03 26	16 42	10 14	08 06	04 17	21 09	10 15	21 43	22 12	05 11
13	03 02	12 46	09 33	08 36	04 35	21 09	10 14	21 43	22 12	05 12
14	02 39	08 15	08 52	09 06	04 53	21 10	10 13	21 43	22 12	05 13
15	02 15	03 S 19	08 09	09 35	05 12	21 10	10 12	21 43	22 12	05 14
16	01 51	01 N 49	07 24	10 05	05 30	21 10	10 11	21 43	22 12	05 14
17	01 28	06 59	06 39	10 34	05 48	21 10	10 10	21 43	22 12	05 15
18	01 04	11 55	05 52	11 03	06 06	21 11	10 08	21 43	22 12	05 16
19	00 40	16 21	05 05	11 31	06 24	21 11	10 07	21 43	22 12	05 17
20	00 S 16	20 01	04 16	12 00	06 42	21 11	10 06	21 43	22 12	05 17
21	00 N 07	22 38	03 26	12 28	07 00	21 11	10 04	21 43	22 12	05 18
22	00 31	23 54	02 35	12 55	07 18	21 11	10 03	21 43	22 12	05 19
23	00 55	23 40	01 43	13 23	07 36	21 11	10 02	21 43	22 12	05 19
24	01 18	21 52	00 S 50	13 50	07 53	21 11	10 00	21 42	22 12	05 20
25	01 42	18 38	00 N 04	14 17	08 11	21 11	09 59	21 42	22 12	05 21
26	02 05	14 15	00 58	14 43	08 28	21 11	09 58	21 42	22 12	05 22
27	02 29	09 01	01 53	15 09	08 46	21 11	09 56	21 42	22 12	05 22
28	02 52	03 N 21	02 49	15 35	09 03	21 11	09 55	21 42	22 12	05 23
29	03 16	02 S 25	03 45	16 00	09 20	21 11	09 53	21 42	22 12	05 24
30	03 39	07 55	04 42	16 25	09 37	21 11	09 52	21 42	22 12	05 25
31	04 N 02	12 S 54	05 N 39	16 N 50	09 N 54	21 S 11	09 S 50	21 S 42	22 S 12	05 N 25

ZODIAC SIGN ENTRIES

Date	h	m	Planets
02	23	51	☽ ♏
05	07	15	☽ ♐
07	04	24	☿ ♓
07	18	29	☽ ♑
10	07	30	☽ ♒
12	19	47	☽ ♓
15	06	00	☽ ♈
17	14	04	☽ ♉
19	09	51	♀ ♉
19	20	20	☽ ♊
21	04	39	☉ ♈
22	00	52	☽ ♋
23	20	09	☿ ♈
24	03	43	☽ ♌
26	05	18	☽ ♍
28	06	48	☽ ♎
30	09	57	☽ ♏

LATITUDES

Date	Mercury ☿ ° '	Venus ♀ ° '	Mars ♂ ° '	Jupiter ♃ ° '	Saturn ♄ ° '	Uranus ♅ ° '	Neptune ♆ ° '	Pluto ♇ ° '
01	01 S 53	00 S 49	00 S 35	00 N 52	02 N 38	00 N 06	01 N 14	17 N 22
04	02 03	00 42	00 33	00 53	02 39	00 06	01 14	17 23
07	02 09	00 34	00 31	00 53	02 40	00 06	01 14	17 25
10	02 12	00 26	00 29	00 53	02 40	00 06	01 14	17 26
13	02 11	00 17	00 26	00 53	02 41	00 06	01 14	17 27
16	02 05	00 S 09	00 24	00 54	02 41	00 06	01 14	17 28
19	01 55	00 N 01	00 22	00 54	02 42	00 06	01 14	17 29
22	01 41	00 10	00 20	00 54	02 42	00 06	01 14	17 30
25	01 22	00 20	00 18	00 54	02 43	00 06	01 14	17 30
28	00 58	00 29	00 16	00 55	02 43	00 06	01 15	17 31
31	00 S 29	00 N 39	00 S 14	00 N 55	02 N 44	00 N 06	01 N 15	17 N 32

DATA

Julian Date	2445395
Delta T	+53 seconds
Ayanamsa	23° 37' 03"
Synetic vernal point	05° ♓ 29' 57"
True obliquity of ecliptic	23° 26' 30"

LONGITUDES

Date	Chiron ⚷ ° '	Ceres ⚳ ° '	Pallas ⚴ ° '	Juno ⚵ ° '	Vesta ⚶ ° '	Black Moon Lilith ⚸ ° '
01	23 ♉ 07	01 ♒ 59	11 ♑ 27	25 ♒ 15	03 ♈ 50	28 ♑ 07
11	23 ♉ 28	05 ♒ 35	14 ♑ 35	29 ♒ 43	08 ♈ 26	29 ♑ 14
21	23 ♉ 55	09 ♒ 03	17 ♑ 26	04 ♓ 11	13 ♈ 04	00 ♒ 21
31	24 ♉ 26	12 ♒ 20	19 ♑ 56	08 ♓ 40	17 ♈ 41	01 ♒ 27

MOON'S PHASES, APSIDES AND POSITIONS ☽

Date	h	m	Phase	Longitude ° '	Eclipse Indicator
06	13	16	☾	15 ♐ 24	
14	17	43	●	23 ♓ 35	
22	02	25	☽	00 ♋ 54	
28	19	27	○	07 ♎ 33	

Day	h	m		
09	22	45	Apogee	
25	21	53	Perigee	
01	15	53	0S	
08	17	49	Max dec	23° S 51'
16	03	33	0N	
22	20	20	Max dec	24° N 00'
29	01	54	0S	

All ephemeris data is given at 12.00 UT and the Moon's longitude is additionally given for 24.00 UT

ASPECTARIAN

h m	Aspects
01 Tuesday	
01 41	☽ ☍ ♂
03 29	☽ ⊻ ♄
05 47	☽ ⚼ ☿
06 04	☽ ∥ ♀
09 53	☽ ☍ ♀
11 36	☽ ✶ ♅
12 19	☽ ∥ ♂
12 59	☽ ✶ ♃
13 01	☿ Q ♅
13 55	☽ ⊼ ☉
19 54	☽ # ♂
02 Wednesday	
00 56	☽ ± ☉
01 07	☽ Q ♆
02 24	☿ Q ♃
03 35	☉ Q ♆
03 52	☽ # ♀
05 51	♀ △ ♅
10 01	♂ ⊼ ♄
10 31	☽ △ ☿
13 21	☽ ∠ ♅
13 41	☽ # ♇
14 55	☽ ∠ ♃
17 48	☽ ⚼ ☉
22 04	☽ ✶ ♆
22 34	☽ ☌ ♇
23 02	☽ ∥ ☉
23 52	♀ △ ♃
03 Thursday	
05 17	☽ ⊥ ♅
06 57	☽ ⊥ ♃
07 13	☽ ☌ ♄
08 34	☽ ⊼ ♂
14 31	☽ ∥ ♄
16 07	☽ ⊻ ♅
17 52	☽ ⊻ ♃
19 36	☽ ⊼ ♀
20 07	☽ ± ♂
22 56	☽ △ ☉
04 Friday	
01 11	☽ ∠ ♆
07 49	☽ ± ♀
13 50	☽ ⚼ ♂
14 11	☽ ∥ ☿
16 14	♃ ⊥ ♄
17 59	☽ ⊥ ♆
05 Saturday	
00 34	☽ □ ☿
02 34	☽ ⚼ ♀
05 23	☽ ⊻ ♆
05 47	☽ ⊻ ♇
09 28	☉ ⚼ ♇
15 03	☽ ⊻ ♄
17 20	☽ ⊥ ♇
20 22	☽ △ ♂
06 Sunday	
00 50	☽ ☌ ♅
02 42	☽ ⊥ ♄
03 00	☽ ☌ ♃
10 55	☽ △ ♀
10 56	☽ ∠ ♇
11 39	☽ ∥ ♃
13 16	☽ □ ☉
14 28	☿ ✶ ♆
14 34	♀ ∥ ♇
16 50	☿ △ ♇
16 52	☽ Q ☿
17 59	☽ ∥ ♅
20 26	☽ ∠ ♄
23 58	☽ ∥ ♆
07 Monday	
07 08	☉ # ♀
10 03	☿ ∠ ♀
16 36	☽ ☌ ♆
16 51	☽ ✶ ♇
20 45	☽ ✶ ☿
08 Tuesday	
02 30	☽ ✶ ♄
03 48	☉ # ♇
04 41	☽ Q ☉
12 23	☽ □ ♂
12 55	☽ ⊻ ♅
15 28	☽ ⊻ ♃
17 09	☽ Q ♇
20 39	♂ △ ♅
09 Wednesday	
01 08	☽ ⊥ ♅
02 49	☽ Q ♄
03 44	☽ ⊥ ♃
05 01	☿ ⊥ ♂
06 20	☽ □ ♀
07 22	☽ ✶ ☉
08 15	☽ ∠ ☿
12 45	☿ △ ♄
19 29	☽ ∠ ♅
22 10	☽ ∠ ♃
10 Thursday	
00 55	☉ ⚼ ♄
03 36	☽ Q ♂
05 39	☽ ⊥ ☿
05 40	☽ ⊻ ♆
05 46	☽ □ ♇
12 04	☽ ∥ ♆
15 20	☽ □ ♄
15 27	♂ △ ♃
16 43	☽ ∠ ☉
17 52	☽ ⊥ ♆
18 07	☽ ∥ ♅
19 53	☽ ⊻ ☿
23 10	☽ Q ♀
11 Friday	
00 18	☽ ∥ ♃
01 58	☽ ✶ ♅
04 45	☽ ✶ ♃
05 36	☽ ✶ ♂
07 07	☉ # ♂
08 28	☉ ⊻ ♀
12 04	☽ ∠ ♆
12 35	☽ ⊥ ☉
17 16	♆ ✶ ♇
12 Saturday	
01 43	☽ ⊻ ☉
02 04	☽ ✶ ♀
02 06	☽ Q ♅
04 57	☽ Q ♃
11 11	☿ ∥ ♄
13 41	☿ □ ♅
13 42	☽ ∠ ♂
18 00	☽ △ ♇
18 03	☽ ✶ ♆
13 Sunday	
03 11	☽ △ ♄
08 39	☽ ⊥ ♂
10 03	☿ □ ♃
11 05	☽ ∠ ♀
13 42	☽ □ ♅
16 38	☽ □ ♃
17 27	☿ Q ♆
17 40	☽ Q ♆
17 42	☽ ☌ ☿
21 09	☽ ⊻ ♂
23 22	☽ ⚼ ♇
14 Monday	
01 51	☽ ∥ ♄
05 26	☉ ± ♇
06 31	☽ ⊥ ♀
07 13	☿ # ♀
08 06	☽ # ♀
08 19	☽ ⚼ ♄
08 23	☽ ∥ ☿
13 03	♅ St R
13 32	♀ ⚼ ♅
16 43	☽ ± ♇
17 43	☽ ☌ ☉
19 19	☽ ⊻ ♀
15 Tuesday	
01 32	☽ ± ♄
02 57	☽ # ♇
03 35	☽ # ♂
04 11	☽ ⊼ ♇
04 23	☽ □ ♆
08 34	☿ ⚼ ♇
08 43	☿ ⊻ ♂
12 54	☽ ⊼ ♄
14 22	♂ ∥ ♇
17 29	☽ ∥ ☉
20 47	♀ ⚼ ♃
23 14	☽ △ ♅
16 Wednesday	
02 14	☽ △ ♃
10 00	☽ ☌ ♂
12 08	☽ # ☉
12 24	☽ ⊻ ☿
16 44	♀ # ♄
17 Thursday	
01 24	☽ ⊥ ☿
03 13	☽ ⚼ ♅
03 52	☽ ∥ ♇
06 08	☽ ∥ ♂
06 14	☽ ⚼ ♃
06 55	☽ ⊻ ☉
09 35	☽ ☌ ♀
10 39	☽ # ☿
12 13	☽ ☍ ♇
12 33	☽ △ ♆
18 52	☽ ⊥ ☉
19 31	☿ ⚼ ♄
19 46	☽ ± ♅
20 31	☽ ☍ ♄
20 41	☽ ∠ ☿
22 49	☽ ± ♃
18 Friday	
03 13	☽ # ♄
06 43	☽ ⊼ ♅
06 51	☿ # ♂
07 10	☽ ∥ ♀
09 46	☽ ⊼ ♃
12 37	☽ ∠ ☉
13 27	♀ ☍ ♇
15 31	☉ ± ♄
15 56	☽ ⚼ ♆
17 40	♀ △ ♆
20 29	☽ ⊻ ♂
19 Saturday	
04 20	☽ ✶ ☿
06 09	☿ # ♇
07 56	☽ ⊥ ♂
08 09	☽ ± ♆
17 45	☽ ✶ ☉
18 26	☽ ⊼ ♇
18 53	☽ ⊼ ♆
21 22	☽ ⊻ ♀
20 Sunday	
00 56	☽ ∠ ♂
02 21	☽ ⊼ ♄
02 52	☉ ⊼ ♇
04 14	☿ ± ♇
05 05	☽ ± ♇
05 13	☽ Q ☿
09 03	☽ ⊥ ♀
09 10	☉ □ ♆
12 26	☽ ☍ ♅
12 55	☽ ± ♄
15 31	☽ ☍ ♃
16 41	☽ Q ☉
20 55	☽ ⚼ ♇
21 22	☽ # ♃
21 Monday	
02 11	☽ # ♅
02 24	☽ ∠ ♀
04 38	☽ ⚼ ♄
04 54	☽ ✶ ♂
07 09	☽ # ♆
17 50	☽ □ ☿
22 58	☽ △ ♇
23 00	♀ ± ♅
23 30	☽ ☍ ♆
22 Tuesday	
02 25	☽ □ ☉
02 44	♀ ☍ ♄
02 53	☽ Q ♂
06 29	☽ △ ♄
06 52	☽ ✶ ♀
10 24	☿ ± ♄
16 26	☽ ⊼ ♅
19 30	☽ ⊼ ♃
23 37	☿ ⊥ ♀
23 Wednesday	
02 38	☽ ± ♅
05 13	☽ Q ♀
05 41	☽ ± ♃
06 10	☿ ⊼ ♇
10 30	☿ □ ♆
10 51	♀ ± ♃
11 15	☽ □ ♂
17 48	☽ ⚼ ♅
20 51	☽ ⚼ ♃
24 Thursday	
01 47	☽ □ ♇
02 24	☽ ⊼ ♆
03 13	☉ # ☿
04 55	☽ △ ☿
08 37	☉ ⊼ ♄
08 45	☽ # ♆
08 57	☽ □ ♄
08 59	☽ △ ☉
12 24	☽ ± ♆
13 30	☽ # ♅
14 09	☽ □ ♀
18 01	☽ # ♃
18 47	☽ △ ♅
21 51	☽ △ ♃
25 Friday	
03 19	☽ ⚼ ♆
07 35	☽ Q ♇
09 30	☿ ⊼ ♄
09 43	☽ ⚼ ☿
11 38	☽ ⚼ ☉
14 36	☽ Q ♄
15 52	☽ △ ♂
18 28	♄ ⊥ ♅
26 Saturday	
02 52	☽ ± ☿
03 18	☽ ✶ ♇
03 31	☽ ± ☉
04 01	☽ △ ♆
09 50	☽ ∥ ♀
10 15	☽ ✶ ♄
11 26	☉ ☌ ☿
14 05	☽ ⊼ ☉
14 18	☽ ⊼ ☿
17 52	☽ ⚼ ♂
20 02	☽ △ ♀
20 10	☽ □ ♅
21 29	♀ ⊼ ♅
23 15	☽ □ ♃
27 Sunday	
03 55	☽ ∠ ♇
07 57	☽ # ♄
09 29	☽ ± ♂
10 49	☽ ∠ ♄
13 04	☽ ∥ ♂
18 45	☽ ⊥ ♇
19 57	☽ ⊼ ♂
23 02	☽ ⚼ ♀
23 53	♃ St R
28 Monday	
01 40	☽ ⊥ ♄
01 52	☽ Q ♅
03 30	☽ ∥ ♇
04 42	☽ ⊻ ♇
05 01	☽ Q ♃
05 31	☽ □ ♆
08 25	☿ △ ♅
11 00	♀ ⊼ ♃
11 37	☽ ⊻ ♄
13 50	☽ ∥ ☉
13 53	♂ ⚼ ♅
13 54	☽ ∥ ☿
14 34	☉ ∥ ☿
15 24	☽ ± ♀
19 27	☽ ☍ ☉
21 56	☽ ✶ ♅
29 Tuesday	
00 12	☽ ☍ ☿
01 08	☽ ✶ ♃
02 27	☽ ⊼ ♀
06 43	☿ △ ♃
11 52	☽ Q ♆
15 54	☽ # ☉
18 52	☽ # ☿
23 31	☽ ∠ ♅
30 Wednesday	
00 50	☽ # ♇
01 28	☽ ☍ ♂
02 49	☽ ∠ ♃
06 47	☉ △ ♅
07 38	☽ ☌ ♇
08 36	☽ ✶ ♆
10 45	☿ ⊻ ♀
14 46	☽ ☌ ♄
15 13	☽ ⊥ ♅
18 36	☽ ⊥ ♃
20 17	☽ # ♂
20 58	☽ ∥ ♄
31 Thursday	
01 51	☽ ⊻ ♅
02 19	♂ ⚼ ♃
03 23	☽ ⊼ ☉
05 11	♀ ⚼ ♆
05 17	☽ ⊻ ♃
06 20	☿ ∥ ♇
07 37	♂ # ♄
11 16	☽ ∠ ♆
11 52	☽ ☍ ♀
13 47	☽ ⊼ ☿
15 04	☽ ± ☉

APRIL 1983

Date	Sidereal time h m s	LONGITUDES Sun ☉ ° ' "	Moon ☽ ° ' "	Moon ☽ 24.00 ° ' "	Mercury ☿ ° '	Venus ♀ ° '	Mars ♂ ° '	Jupiter ♃ ° '	Saturn ♄ ° '	Uranus ♅ ° '	Neptune ♆ ° '	Pluto ♇ ° '
01	00 37 08	11 ♈ 11 29	27 ♏ 41 22	04 ♐ 04 03	17 ♈ 31	15 ♉ 46	26 ♈ 57	10 ♐ 54	02 ♏ 37	08 ♐ 58	29 ♐ 14	28 ♎ 37
02	00 41 05	12 10 39	10 ♐ 21 15	16 ♐ 33 26	19 34	16 58	27 42	10 R 53	02 R 33	08 R 57	29 R 14	28 R 36
03	00 45 01	13 09 49	22 ♐ 41 06	28 ♐ 44 53	21 37	18 10	28 27	10 51	02 29	08 56	29 14	28 34
04	00 48 58	14 08 56	04 ♑ 45 24	10 ♑ 43 19	23 39	19 21	29 12	10 50	02 25	08 55	29 14	28 33
05	00 52 54	15 08 02	16 ♑ 39 20	22 ♑ 34 09	25 39	20 33	29 ♈ 56	10 49	02 20	08 54	29 14	28 31
06	00 56 51	16 07 05	28 ♑ 28 25	04 ♒ 22 49	27 38	21 44	00 ♉ 41	10 47	02 16	08 53	29 13	28 29
07	01 00 48	17 06 07	10 ♒ 18 00	16 ♒ 14 34	29 ♈ 36	22 55	01 25	10 45	02 12	08 52	29 13	28 28
08	01 04 44	18 05 08	22 ♒ 13 04	28 ♒ 14 00	01 ♉ 31	24 07	02 10	10 43	02 07	08 51	29 13	28 26
09	01 08 41	19 04 06	04 ♓ 17 51	10 ♓ 24 58	03 23	25 18	02 54	10 41	02 03	08 49	29 13	28 24
10	01 12 37	20 03 02	16 ♓ 35 40	22 ♓ 50 11	05 13	26 29	03 39	10 38	01 59	08 48	29 13	28 23
11	01 16 34	21 01 57	29 ♓ 08 40	05 ♈ 31 12	06 59	27 40	04 23	10 36	01 54	08 47	29 12	28 21
12	01 20 30	22 00 50	11 ♈ 57 45	18 ♈ 28 16	08 41	28 ♉ 51	05 07	10 33	01 50	08 45	29 12	28 19
13	01 24 27	22 59 40	25 ♈ 02 34	01 ♉ 40 29	10 20	00 ♊ 02	05 51	10 30	01 45	08 44	29 11	28 18
14	01 28 23	23 58 29	08 ♉ 21 46	15 ♉ 06 09	11 55	01 12	06 36	10 27	01 41	08 42	29 11	28 16
15	01 32 20	24 57 16	21 ♉ 53 21	28 ♉ 43 06	13 25	02 23	07 20	10 24	01 36	08 41	29 11	28 14
16	01 36 17	25 56 01	05 ♊ 35 08	12 ♊ 29 12	14 50	03 34	08 04	10 20	01 32	08 39	29 10	28 13
17	01 40 13	26 54 43	19 ♊ 25 06	26 ♊ 22 39	16 11	04 44	08 48	10 17	01 27	08 38	29 10	28 11
18	01 44 10	27 53 24	03 ♋ 21 43	10 ♋ 22 09	17 26	05 54	09 32	10 13	01 23	08 36	29 09	28 09
19	01 48 06	28 52 02	17 ♋ 23 51	24 ♋ 26 43	18 37	07 05	10 16	10 09	01 18	08 34	29 08	28 07
20	01 52 03	29 ♈ 50 38	01 ♌ 30 37	08 ♌ 35 25	19 42	08 15	11 00	10 05	01 13	08 33	29 08	28 06
21	01 55 59	00 ♉ 49 12	15 ♌ 40 56	22 ♌ 46 57	20 42	09 25	11 44	10 00	01 09	08 31	29 07	28 04
22	01 59 56	01 47 43	29 ♌ 53 10	06 ♍ 59 15	21 37	10 35	12 27	09 56	01 04	08 29	29 07	28 02
23	02 03 52	02 46 13	14 ♍ 04 47	21 ♍ 09 20	22 26	11 44	13 11	09 51	01 00	08 27	29 06	28 01
24	02 07 49	03 44 40	28 ♍ 12 23	05 ♎ 13 24	23 09	12 54	13 55	09 46	00 55	08 25	29 05	27 59
25	02 11 46	04 43 05	12 ♎ 11 53	19 ♎ 07 17	23 46	14 03	14 38	09 42	00 50	08 23	29 04	27 57
26	02 15 42	05 41 28	25 ♎ 59 07	02 ♏ 46 58	24 18	15 13	15 22	09 36	00 46	08 22	29 04	27 56
27	02 19 39	06 39 49	09 ♏ 30 29	16 ♏ 09 22	24 45	16 22	16 05	09 31	00 41	08 20	29 03	27 54
28	02 23 35	07 38 08	22 ♏ 43 28	29 ♏ 12 42	25 05	17 31	16 49	09 26	00 37	08 18	29 02	27 52
29	02 27 32	08 36 26	05 ♐ 37 04	11 ♐ 56 43	25 20	18 40	17 32	09 20	00 32	08 16	29 01	27 51
30	02 31 28	09 ♉ 34 42	18 ♐ 11 51	24 ♐ 22 46	25 ♉ 29	19 ♊ 49	18 ♉ 16	09 ♐ 15	00 ♏ 28	08 ♐ 13	29 ♐ 00	27 ♎ 49

Date	Moon True ☊ ° '	Moon Mean ☊ ° '	Moon ☽ Latitude ° '
01	28 ♊ 11	29 ♊ 04	02 N 34
02	28 R 07	29 01	01 33
03	28 06	28 58	00 N 29
04	28 D 06	28 55	00 S 35
05	28 R 06	28 51	01 37
06	28 05	28 48	02 34
07	28 03	28 45	03 24
08	27 58	28 42	04 06
09	27 50	28 39	04 38
10	27 41	28 36	04 57
11	27 29	28 32	05 02
12	27 18	28 29	04 53
13	27 07	28 26	04 28
14	26 57	28 23	03 48
15	26 50	28 20	02 55
16	26 46	28 16	01 50
17	26 44	28 13	00 S 39
18	26 D 44	28 10	00 N 35
19	26 44	28 07	01 48
20	26 R 45	28 04	02 55
21	26 44	28 01	03 51
22	26 41	27 57	04 33
23	26 35	27 54	04 59
24	26 28	27 51	05 06
25	26 20	27 48	04 55
26	26 11	27 45	04 26
27	26 04	27 42	03 43
28	25 58	27 38	02 49
29	25 53	27 35	01 47
30	25 ♊ 51	27 ♊ 32	00 N 41

DECLINATIONS

Date	Sun ☉ ° '	Moon ☽ ° '	Mercury ☿ ° '	Venus ♀ ° '	Mars ♂ ° '	Jupiter ♃ ° '	Saturn ♄ ° '	Uranus ♅ ° '	Neptune ♆ ° '	Pluto ♇ ° '
01	04 N 26	17 S 09	06 N 35	17 N 14	10 N 11	21 S 11	09 S 49	21 S 42	22 S 12	05 N 26
02	04 49	20 28	07 32	17 38	10 27	21 10	09 48	21 41	22 12	05 27
03	05 12	22 46	08 28	18 01	10 44	21 10	09 46	21 41	22 12	05 27
04	05 35	23 57	09 24	18 24	11 00	21 10	09 45	21 41	22 12	05 28
05	05 58	24 00	10 18	18 46	11 17	21 10	09 43	21 41	22 12	05 29
06	06 20	22 59	11 12	19 08	11 33	21 09	09 41	21 41	22 12	05 29
07	06 43	20 56	12 05	19 30	11 49	21 09	09 40	21 41	22 11	05 30
08	07 06	17 59	12 56	19 51	12 05	21 09	09 38	21 40	22 11	05 31
09	07 28	14 15	13 45	20 11	12 21	21 08	09 37	21 40	22 11	05 31
10	07 50	09 51	14 33	20 31	12 37	21 08	09 35	21 40	22 11	05 32
11	08 12	04 S 58	15 18	20 51	12 52	21 07	09 34	21 40	22 11	05 33
12	08 35	00 N 14	16 01	21 10	13 08	21 07	09 32	21 40	22 11	05 33
13	08 56	05 32	16 42	21 28	13 23	21 07	09 31	21 39	22 11	05 34
14	09 18	10 42	17 21	21 46	13 38	21 06	09 29	21 39	22 11	05 34
15	09 40	15 26	17 57	22 03	13 54	21 06	09 27	21 39	22 11	05 35
16	10 01	19 26	18 30	22 20	14 08	21 05	09 26	21 39	22 11	05 36
17	10 22	22 22	19 01	22 36	14 23	21 04	09 24	21 38	22 11	05 36
18	10 43	23 59	19 28	22 52	14 38	21 04	09 23	21 38	22 11	05 37
19	11 04	24 06	19 54	23 07	14 52	21 03	09 21	21 38	22 11	05 37
20	11 25	22 40	20 16	23 21	15 07	21 03	09 19	21 38	22 11	05 38
21	11 46	19 49	20 36	23 35	15 21	21 02	09 18	21 37	22 11	05 38
22	12 06	15 47	20 53	23 48	15 35	21 01	09 16	21 37	22 11	05 39
23	12 26	10 51	21 07	24 00	15 49	21 01	09 15	21 37	22 11	05 40
24	12 46	05 N 23	21 19	24 12	16 03	21 00	09 13	21 37	22 11	05 40
25	13 06	00 S 18	21 28	24 23	16 16	20 59	09 12	21 36	22 11	05 41
26	13 25	05 54	21 34	24 34	16 29	20 58	09 10	21 36	22 11	05 41
27	13 44	11 08	21 37	24 44	16 43	20 58	09 09	21 36	22 11	05 41
28	14 04	15 44	21 38	24 53	16 56	20 57	09 07	21 35	22 11	05 42
29	14 22	19 29	21 36	25 02	17 09	20 56	09 05	21 35	22 11	05 42
30	14 N 41	22 S 14	21 N 32	25 N 10	17 N 21	20 S 55	09 S 04	21 S 35	22 S 11	05 N 43

ZODIAC SIGN ENTRIES

Date	h	m	Planets
01	16	20	☽ ♐
04	02	30	☽ ♑
05	14	03	♂ ♉
06	15	06	☽ ♒
07	17	04	☿ ♉
09	03	30	☽ ♓
11	13	37	☽ ♈
13	11	26	♀ ♊
13	20	59	☽ ♉
16	02	15	☽ ♊
18	06	14	☽ ♋
20	09	26	☽ ♌
20	15	50	☉ ♉
22	12	12	☽ ♍
24	15	04	☽ ♎
26	19	04	☽ ♏
29	01	28	☽ ♐

LATITUDES

Date	Mercury ☿ ° '	Venus ♀ ° '	Mars ♂ ° '	Jupiter ♃ ° '	Saturn ♄ ° '	Uranus ♅ ° '	Neptune ♆ ° '	Pluto ♇ ° '
01	00 S 19	00 N 42	00 S 14	00 N 55	02 N 44	00 N 06	01 N 15	17 N 32
04	00 N 14	00 52	00 12	00 55	02 44	00 06	01 15	17 32
07	00 48	01 02	00 10	00 55	02 44	00 06	01 15	17 33
10	01 22	01 11	00 08	00 55	02 44	00 06	01 15	17 33
13	01 53	01 20	00 06	00 55	02 45	00 06	01 15	17 33
16	02 19	01 30	00 04	00 56	02 45	00 06	01 15	17 33
19	02 38	01 38	00 S 02	00 56	02 45	00 06	01 15	17 33
22	02 49	01 47	00 00	00 56	02 45	00 06	01 15	17 33
25	02 50	01 55	00 N 02	00 56	02 45	00 06	01 15	17 33
28	02 41	02 02	00 04	00 56	02 45	00 06	01 16	17 33
31	02 N 20	02 N 09	00 N 06	00 N 56	02 N 44	00 N 06	01 N 16	17 N 33

DATA

Julian Date	2445426
Delta T	+53 seconds
Ayanamsa	23° 37' 05"
Synetic vernal point	05° ♓ 29' 54"
True obliquity of ecliptic	23° 26' 30"

MOON'S PHASES, APSIDES AND POSITIONS ☽

Date	h	m	Phase	Longitude ° '	Eclipse Indicator
05	08	38	☾	15 ♑ 00	
13	07	58	●	22 ♈ 50	
20	08	58	☽	29 ♋ 43	
27	06	31	○	06 ♏ 26	

Day	h	m		
06	17	53	Apogee	
21	08	00	Perigee	
05	01	19	Max dec	24° S 07'
12	10	56	0N	
19	01	44	Max dec	24° N 15'
25	10	44	0S	

LONGITUDES

Date	Chiron ⚷ ° '	Ceres ⚳ ° '	Pallas ⚴ ° '	Juno ⚵ ° '	Vesta ⚶ ° '	Black Moon Lilith ⚸ ° '
01	24 ♉ 30	12 ♒ 39	20 ♑ 10	09 ♓ 07	18 ♈ 08	01 ♒ 34
11	25 ♉ 05	15 ♒ 44	22 ♑ 14	13 ♓ 35	22 ♈ 44	02 ♒ 41
21	25 ♉ 44	18 ♒ 33	23 ♑ 49	18 ♓ 02	27 ♈ 19	03 ♒ 47
31	26 ♉ 25	21 ♒ 07	24 ♑ 52	22 ♓ 28	01 ♉ 51	04 ♒ 54

All ephemeris data is given at 12.00 UT and the Moon's longitude is additionally given for 24.00 UT

ASPECTARIAN

h m	Aspects
01 Friday	
02 48	☽ ± ☿
03 43	☽ ⊥ ♆
04 24	♆ St R
04 49	☉ △ ♃
08 59	☽ ⚼ ☉
10 33	☽ ⊼ ♂
12 36	☽ ∦ ♀
13 44	☽ ⊻ ♇
14 53	☽ ⊻ ♆
21 12	☽ ⊻ ♄
22 31	☽ ± ♂
22 50	☽ ⚼ ☿
02 Saturday	
01 02	☽ ⊥ ♇
08 33	☽ ⊥ ♄
09 19	☽ ☌ ♅
13 00	☽ ☌ ♃
15 49	☽ △ ☉
16 49	☽ ⚼ ♂
18 14	☽ ∥ ♃
18 14	☽ ∠ ♇
23 21	☽ ∥ ♅
03 Sunday	
01 51	☽ ∠ ♄
02 10	☽ ⊼ ♀
04 51	☽ ∥ ♆
09 28	☽ △ ☿
15 13	☽ ± ♀
15 46	♂ ☍ ♇
23 37	☽ ⚹ ♇
04 Monday	
00 09	☽ △ ♂
00 58	☽ ☌ ♆
04 39	☉ ∥ ♇
07 20	☽ ⚹ ♄
11 06	☽ ⚼ ♀
13 11	♂ △ ♆
15 16	☿ ⚼ ♅
20 21	☽ ⊻ ♅
20 51	☿ ∦ ♄
23 37	☽ Q ♇
05 Tuesday	
00 12	☽ ⊻ ♃
07 20	☽ Q ♄
08 27	☽ ⊥ ♅
08 38	☽ □ ☉
12 19	☽ ⊥ ♃
13 52	☿ ⚼ ♃
20 46	☽ △ ♀
06 Wednesday	
02 41	☽ ∠ ♅
06 33	☽ ∠ ♃
09 58	☽ □ ☿
12 02	☽ □ ♇
13 32	☽ ⊻ ♆
16 47	☽ □ ♂
19 40	☽ □ ♄
22 15	☿ ☍ ♇
22 41	☽ ∥ ♆
07 Thursday	
00 31	☽ Q ☉
01 38	☿ ∥ ♂
01 43	☽ ⊥ ♆
04 32	☽ ∥ ♅
07 25	☿ △ ♆
09 06	☽ ⚹ ♅
09 58	☽ ∥ ♃
12 55	☽ ⚹ ♃
18 00	♀ ± ♆
19 55	☽ ∠ ♆
23 15	☽ ∦ ♀
08 Friday	
02 58	☽ ⚹ ☉
05 33	☽ Q ☿
07 37	☽ Q ♂
08 55	☉ ⊥ ♀
09 15	☽ Q ♅
10 50	♂ ☍ ♄
13 00	☽ Q ♃
16 12	☽ □ ♀
19 30	☿ ☍ ♄
09 Saturday	
00 22	☽ △ ♇
01 42	☿ ☌ ♂
01 57	☽ ⚹ ♆
04 50	☿ ± ♅
07 35	☽ △ ♄
09 04	☽ ⚹ ♂
09 28	♂ ± ♅
09 52	☽ ⚹ ☿
11 31	☽ ∠ ☉
14 23	☽ ∦ ☿
20 52	☽ □ ♅
22 07	☽ ∦ ♂
10 Sunday	
00 29	☽ □ ♃
01 33	☽ Q ♆
04 37	☿ ± ♃
05 47	☽ ⚼ ♇
06 39	☽ ⊥ ☉
07 29	☽ Q ♀
12 44	☽ ⚼ ♄
13 21	☽ ∥ ♄
16 12	☽ ∠ ♂
19 14	☽ ⊻ ☉
20 08	☽ ∠ ☿
21 24	☽ ∦ ☉
23 06	☽ ± ♇
11 Monday	
05 53	☽ ± ♄
08 55	☽ ⚹ ♀
09 15	☽ ∦ ♇
10 28	☽ ⊥ ♂
10 30	☽ ⊼ ♇
12 07	☽ □ ♆
16 01	☽ ⊥ ☿
17 11	☽ ⊼ ♄
18 36	♂ ± ♃
22 28	☽ ⊻ ♂
12 Tuesday	
01 34	♀ ⊼ ♇
05 00	☽ ⊻ ☿
06 03	☽ △ ♅
08 42	♀ ∦ ♃
09 23	☽ △ ♃
12 56	☿ ⊼ ♅
15 50	☽ ∠ ♀
19 04	♀ ⊼ ♆
13 Wednesday	
07 58	☽ ☌ ☉
09 37	☽ ⚼ ♅
09 58	☽ ⊥ ♀
12 08	☽ ∥ ♇
12 50	☽ ⚼ ♃
14 26	☿ ⊼ ♃
17 53	☽ ☍ ♇
19 31	☽ △ ♆
21 55	☽ ⊻ ♀
14 Thursday	
00 04	☽ ☍ ♄
01 53	☽ ± ♅
03 02	♀ ∦ ♅
04 53	☽ ∥ ☉
05 01	☽ ± ♃
05 35	☉ ⚼ ♅
06 16	☽ ∦ ♄
08 39	☽ ☌ ♂
12 37	☽ ⊼ ♅
15 43	☽ ⊼ ♃
19 08	☽ ☌ ☿
21 02	♀ ⊼ ♄
22 22	☽ ⚼ ♆
23 09	☉ ∦ ♄
15 Friday	
03 25	☽ ∥ ♂
14 16	☽ ± ♆
17 49	☽ ⊻ ☉
22 12	☉ ⚼ ♃
23 08	☽ ⊼ ♇
23 26	♀ ∦ ♆
16 Saturday	
00 40	☿ ⚼ ♆
00 48	☽ ⊼ ♆
04 54	☽ ∥ ☿
04 57	☽ ⊼ ♄
05 08	☽ ⊥ ☉
08 08	☽ ☌ ♀
09 36	☽ ± ♇
15 22	☽ ± ♄
16 34	☽ ⊻ ♂
17 20	☽ ☍ ♅
20 14	☽ ☍ ♃
22 01	☽ ∠ ☉
17 Sunday	
00 14	☽ ∦ ♃
00 58	♀ ± ♇
01 14	☽ ⚼ ♇
03 34	☽ ⊥ ♂
05 02	☽ ∦ ♅
05 49	☽ ⊻ ☿
06 37	♂ ⊼ ♅
06 54	☽ ⚼ ♄
10 07	☽ ∦ ♆
14 46	☽ ∥ ♀
17 15	☽ ⊥ ☿
19 59	☽ ∠ ♂
18 Monday	
01 54	☽ ⚹ ☉
03 04	☽ △ ♇
04 46	☽ ☍ ♆
08 37	☽ △ ♄
10 16	☽ ∠ ☿
16 45	☽ ⊻ ♀
18 16	☉ ☍ ♇
20 57	☽ ⊼ ♅
23 09	☽ ⚹ ♂
23 41	☽ ⊼ ♃
19 Tuesday	
00 01	☽ Q ☉
03 57	☽ ⊥ ♀
07 11	☽ ± ♅
08 29	♂ ⊼ ♃
09 53	☽ ± ♃
14 16	☽ ⚹ ☿
16 18	♀ ± ♄
18 40	☉ △ ♆
20 41	☽ ∠ ♀
20 45	☽ Q ♂
22 30	☽ ⚼ ♅
20 Wednesday	
01 08	☽ ⚼ ♃
04 33	☽ ∥ ♀
06 13	☽ □ ♇
07 58	☽ ⊼ ♆
08 58	☽ □ ☉
11 31	☽ □ ♄
12 21	☽ Q ☿
17 01	☽ ∦ ♆
18 00	♀ ☍ ♅
18 08	☽ ± ♆
22 08	☽ ∦ ♅
23 54	☽ △ ♅
21 Thursday	
00 26	☽ ⚹ ♀
02 27	☽ △ ♃
03 01	☽ ∦ ♃
04 57	☽ □ ♂
06 53	☽ ∥ ☿
09 22	☽ ⚼ ♆
12 39	☽ Q ♇
17 49	☽ Q ♄
19 27	☉ ☍ ♄
21 05	☽ □ ☿
22 33	☽ Q ♀
23 31	♀ ☍ ♃
22 Friday	
08 53	☽ ⚹ ♇
10 41	☽ △ ♆
12 58	☽ ∥ ♂
13 59	☽ ⚹ ♄
15 28	☽ △ ☉
23 Saturday	
00 53	☿ ∦ ♃
02 30	☽ □ ♅
04 27	☉ ± ♅
04 53	☽ □ ♃
05 07	☽ ∥ ☉
07 41	☽ □ ♀
10 12	☽ ∠ ♇
10 24	☽ △ ♂
15 14	☽ ∠ ♄
18 43	☽ ⚼ ☉
19 18	☽ ∦ ♄
24 Sunday	
01 26	☽ ⊥ ♇
02 56	☽ △ ☿
06 26	☽ ⊥ ♄
08 58	☽ Q ♅
09 53	☿ ± ♆
10 49	☽ ∥ ♇
11 09	☽ ± ☉
11 16	☽ Q ♃
11 37	☽ ⊻ ♇
12 41	☉ ± ♃
13 16	☽ ⚼ ♂
13 30	☽ □ ♆
13 43	♀ ⚼ ♇
16 36	☽ ⊻ ♄
17 41	♂ ⚼ ♆
22 10	☽ ⊼ ☉
25 Monday	
05 27	☽ ⚹ ♅
05 32	☽ ± ♂
05 50	☽ ⚼ ☿
07 42	☽ ⚹ ♃
15 30	☽ △ ♀
16 27	☽ ⊼ ♂
20 26	☽ Q ♆
22 04	☽ ± ☿
26 Tuesday	
07 24	☽ ∠ ♅
08 57	☽ ⊼ ☿
09 36	☽ ∠ ♃
11 02	☽ ∦ ♇
15 25	☽ ☌ ♇
17 25	☽ ⚹ ♆
20 09	☽ ⚼ ♀
20 23	☽ ☌ ♄
20 23	♀ ⊻ ♂
22 45	♀ ⚼ ♄
23 13	☽ ⊥ ♅
27 Wednesday	
00 13	☿ ∦ ♅
01 23	☽ ⊥ ♃
02 38	☽ ∥ ♄
06 31	☽ ☍ ☉
09 53	☽ ⊻ ♅
12 01	☽ ⊻ ♃
13 41	☽ ± ♀
20 10	☽ ∠ ♆
28 Thursday	
00 34	☽ ☍ ♂
01 34	☽ ⊼ ♀
02 08	☽ ∦ ☉
12 34	☽ ⊥ ♆
16 27	☽ ☍ ☿
19 27	☽ ∦ ♂
21 29	☽ ⊻ ♇
23 39	☽ ⊻ ♆
29 Friday	
02 32	☽ ⊻ ♄
03 41	☉ ⊼ ♅
08 40	☽ ⊥ ♇
13 44	☽ ⊥ ♄
16 59	☽ ☌ ♅
18 07	☽ ⊼ ☉
18 59	☽ ☌ ♃
20 32	☿ ∦ ♅
23 25	☽ ∥ ♃
30 Saturday	
01 41	☽ ∠ ♇
04 31	☉ ⊼ ♃
05 08	☽ ∦ ☿
05 22	☽ ∥ ♅
06 32	☽ ± ☉
06 46	☽ ∠ ♄
11 26	☽ ∥ ♆
12 08	☽ ⊼ ♂
15 27	☽ ☍ ♀

MAY 1983

Date	Sidereal time h m s	Sun ☉	Moon ☽	Moon ☽ 24.00	Mercury ☿	Venus ♀	Mars ♂	Jupiter ♃	Saturn ♄	Uranus ♅	Neptune ♆	Pluto ♇
01	02 35 25	10 ♉ 32 57	00 ♑ 29 52	06 ♑ 33 34	25 ♉ 33	20 ♊ 58	18 ♉ 59	09 ♐ 09	00 ♏ 23	08 ♐ 11	28 ♐ 59	27 ♎ 47
02	02 39 21	11 31 10	12 ♑ 34 23	18 ♑ 32 52	25 R 31	22 07	19 42	09 R 03	00 R 19	08 R 09	28 R 58	27 R 46
03	02 43 18	12 29 21	24 ♑ 29 35	00 ♒ 25 11	25 25	23 15	20 25	08 57	00 14	08 07	28 57	27 44
04	02 47 15	13 27 31	06 ♒ 20 16	12 ♒ 15 29	25 13	24 23	21 09	08 51	00 10	08 05	28 57	27 42
05	02 51 11	14 25 39	18 ♒ 11 28	24 ♒ 08 51	24 56	25 32	21 52	08 45	00 06	08 03	28 55	27 41
06	02 55 08	15 23 46	00 ♓ 08 16	06 ♓ 10 17	24 36	26 40	22 35	08 38	00 ♏ 01	08 01	28 54	27 39
07	02 59 04	16 21 52	12 ♓ 15 27	18 ♓ 24 16	24 12	27 48	23 18	08 32	29 ♎ 57	07 58	28 53	27 38
08	03 03 01	17 19 56	24 ♓ 37 11	00 ♈ 54 33	23 44	28 ♊ 55	24 01	08 25	29 53	07 56	28 52	27 36
09	03 06 57	18 17 59	07 ♈ 16 39	13 ♈ 43 43	23 13	00 ♋ 03	24 44	08 19	29 49	07 54	28 51	27 35
10	03 10 54	19 16 00	20 ♈ 15 50	26 ♈ 53 00	22 40	01 10	25 27	08 12	29 44	07 52	28 50	27 33
11	03 14 50	20 14 00	03 ♉ 35 09	10 ♉ 22 04	22 06	02 18	26 09	08 05	29 40	07 49	28 49	27 31
12	03 18 47	21 11 58	17 ♉ 13 28	24 ♉ 09 00	21 30	03 25	26 52	07 58	29 36	07 47	28 48	27 30
13	03 22 44	22 09 55	01 ♊ 08 12	08 ♊ 10 36	20 54	04 32	27 35	07 51	29 32	07 45	28 47	27 28
14	03 26 40	23 07 51	15 ♊ 15 39	22 ♊ 22 49	20 18	05 39	28 18	07 44	29 28	07 42	28 45	27 27
15	03 30 37	24 05 45	29 ♊ 31 33	06 ♋ 41 20	19 43	06 45	29 00	07 37	29 24	07 40	28 44	27 25
16	03 34 33	25 03 37	13 ♋ 51 38	21 ♋ 02 02	19 10	07 52	29 ♉ 43	07 29	29 20	07 37	28 43	27 24
17	03 38 30	26 01 27	28 ♋ 12 04	05 ♌ 21 22	18 38	08 58	00 ♊ 25	07 22	29 16	07 35	28 42	27 23
18	03 42 26	26 59 16	12 ♌ 29 38	19 ♌ 36 32	18 09	10 04	01 08	07 15	29 13	07 33	28 40	27 21
19	03 46 23	27 57 03	26 ♌ 41 50	03 ♍ 45 18	17 43	11 10	01 50	07 07	29 09	07 30	28 39	27 20
20	03 50 19	28 54 48	10 ♍ 46 42	17 ♍ 45 53	17 20	12 16	02 33	07 00	29 05	07 28	28 38	27 18
21	03 54 16	29 ♉ 52 32	24 ♍ 42 37	01 ♎ 36 45	17 01	13 21	03 15	06 52	29 02	07 25	28 36	27 17
22	03 58 13	00 ♊ 50 14	08 ♎ 28 07	15 ♎ 16 31	16 45	14 26	03 57	06 45	28 58	07 23	28 35	27 16
23	04 02 09	01 47 54	22 ♎ 01 49	28 ♎ 43 51	16 34	15 31	04 39	06 37	28 54	07 20	28 33	27 14
24	04 06 06	02 45 33	05 ♏ 22 29	11 ♏ 57 37	16 28	16 36	05 22	06 30	28 51	07 18	28 32	27 13
25	04 10 02	03 43 10	18 ♏ 29 08	24 ♏ 57 00	16 25	17 41	06 04	06 22	28 48	07 15	28 31	27 12
26	04 13 59	04 40 47	01 ♐ 21 11	07 ♐ 41 42	16 D 27	18 45	06 46	06 14	28 44	07 13	28 29	27 10
27	04 17 55	05 38 22	13 ♐ 58 37	20 ♐ 12 03	16 34	19 49	07 28	06 07	28 41	07 10	28 28	27 09
28	04 21 52	06 35 56	26 ♐ 22 10	02 ♑ 29 11	16 45	20 53	08 10	05 59	28 38	07 08	28 26	27 08
29	04 25 48	07 33 28	08 ♑ 33 22	14 ♑ 35 01	17 01	21 57	08 52	05 52	28 35	07 05	28 25	27 07
30	04 29 45	08 31 00	20 ♑ 34 30	26 ♑ 32 12	17 21	23 00	09 34	05 44	28 32	07 03	28 23	27 06
31	04 33 42	09 ♊ 28 31	02 ♒ 28 36	08 ♒ 24 09	17 ♉ 45	24 ♋ 03	10 ♊ 15	05 ♐ 36	28 ♎ 29	07 ♐ 00	28 ♐ 22	27 ♎ 05

Date	Moon True ☊	Moon Mean ☊	Moon ☽ Latitude
01	25 ♊ 51	27 ♊ 29	00 S 25
02	25 D 52	27 26	01 29
03	25 54	27 22	02 29
04	25 55	27 19	03 22
05	25 R 55	27 16	04 06
06	25 54	27 13	04 40
07	25 51	27 10	05 02
08	25 46	27 07	05 10
09	25 41	27 03	05 04
10	25 35	27 00	04 42
11	25 29	26 57	04 05
12	25 24	26 54	03 13
13	25 20	26 51	02 08
14	25 19	26 47	00 S 55
15	25 D 18	26 44	00 N 23
16	25 19	26 41	01 40
17	25 21	26 38	02 51
18	25 22	26 35	03 50
19	25 23	26 32	04 36
20	25 R 22	26 28	05 04
21	25 20	26 25	05 14
22	25 18	26 22	05 06
23	25 14	26 19	04 40
24	25 11	26 16	04 00
25	25 08	26 13	03 08
26	25 06	26 09	02 07
27	25 04	26 06	01 N 01
28	25 D 04	26 03	00 S 07
29	25 05	26 00	01 14
30	25 06	25 57	02 16
31	25 ♊ 07	25 ♊ 53	03 S 12

DECLINATIONS

Date	Sun ☉	Moon ☽	Mercury ☿	Venus ♀	Mars ♂	Jupiter ♃	Saturn ♄	Uranus ♅	Neptune ♆	Pluto ♇
01	14 N 59	23 S 52	21 N 25	25 N 17	17 N 34	20 S 54	09 S 02	21 S 34	22 S 11	05 N 43
02	15 17	24 20	21 15	25 24	17 46	20 53	09 01	21 34	22 11	05 44
03	15 35	23 40	21 04	25 29	17 58	20 52	09 00	21 34	22 11	05 44
04	15 53	21 57	20 49	25 35	18 11	20 52	08 58	21 33	22 11	05 44
05	16 10	19 17	20 33	25 39	18 22	20 51	08 57	21 33	22 11	05 45
06	16 27	15 47	20 15	25 43	18 34	20 50	08 55	21 33	22 11	05 45
07	16 44	11 36	19 54	25 47	18 46	20 49	08 54	21 32	22 11	05 46
08	17 00	06 53	19 32	25 49	18 57	20 48	08 52	21 32	22 10	05 46
09	17 17	01 S 46	19 09	25 51	19 08	20 47	08 51	21 32	22 10	05 46
10	17 33	03 N 33	18 44	25 52	19 19	20 46	08 50	21 31	22 10	05 47
11	17 48	08 52	18 19	25 53	19 30	20 45	08 48	21 31	22 10	05 47
12	18 04	13 54	17 53	25 53	19 40	20 43	08 47	21 31	22 10	05 47
13	18 19	18 18	17 26	25 52	19 51	20 42	08 46	21 30	22 10	05 47
14	18 33	21 43	17 00	25 50	20 01	20 41	08 44	21 30	22 10	05 48
15	18 48	23 50	16 34	25 48	20 11	20 40	08 43	21 29	22 10	05 48
16	19 02	24 23	16 09	25 46	20 21	20 39	08 42	21 29	22 10	05 48
17	19 16	23 18	15 45	25 42	20 30	20 38	08 41	21 29	22 10	05 48
18	19 29	20 44	15 22	25 38	20 40	20 37	08 39	21 28	22 10	05 49
19	19 42	16 56	15 00	25 33	20 49	20 36	08 38	21 28	22 10	05 49
20	19 55	12 12	14 41	25 28	20 58	20 35	08 37	21 28	22 10	05 49
21	20 08	06 54	14 23	25 22	21 07	20 33	08 36	21 27	22 10	05 49
22	20 20	01 N 19	14 07	25 16	21 15	20 32	08 35	21 27	22 10	05 49
23	20 31	04 S 15	13 54	25 09	21 24	20 31	08 34	21 26	22 10	05 49
24	20 43	09 32	13 43	25 01	21 32	20 30	08 33	21 26	22 10	05 49
25	20 54	14 19	13 34	24 52	21 40	20 29	08 32	21 26	22 10	05 50
26	21 05	18 22	13 27	24 43	21 48	20 27	08 31	21 25	22 10	05 50
27	21 15	21 28	13 23	24 34	21 55	20 26	08 30	21 25	22 10	05 50
28	21 25	23 31	13 21	24 24	22 02	20 25	08 29	21 24	22 10	05 50
29	21 34	24 24	13 21	24 13	22 10	20 24	08 28	21 24	22 10	05 50
30	21 44	24 07	13 24	24 02	22 17	20 23	08 27	21 24	22 10	05 50
31	21 N 52	22 S 44	13 N 28	23 N 50	22 N 23	20 S 21	08 S 26	21 S 23	22 S 10	05 N 50

ZODIAC SIGN ENTRIES

Date	h m	Planets
01	11 01	☽ ♑
03	23 09	☽ ♒
06	11 43	☽ ♓
06	19 29	♄ ♎
08	22 16	☽ ♈
09	10 56	♀ ♋
11	05 36	☽ ♉
13	10 03	☽ ♊
15	12 48	☽ ♋
16	21 43	♂ ♊
17	15 01	☽ ♌
19	17 37	☽ ♍
21	15 06	☉ ♊
21	21 11	☽ ♎
24	02 17	☽ ♏
26	09 27	☽ ♐
28	19 07	☽ ♑
31	07 00	☽ ♒

LATITUDES

Date	Mercury ☿	Venus ♀	Mars ♂	Jupiter ♃	Saturn ♄	Uranus ♅	Neptune ♆	Pluto ♇
01	02 N 20	02 N 09	00 N 06	00 N 56	02 N 44	00 N 06	01 N 16	17 N 33
04	01 48	02 15	00 08	00 56	02 44	00 06	01 16	17 32
07	01 07	02 21	00 10	00 56	02 44	00 06	01 16	17 32
10	00 N 18	02 26	00 12	00 56	02 44	00 06	01 16	17 31
13	00 S 34	02 30	00 14	00 56	02 43	00 06	01 16	17 30
16	01 25	02 33	00 15	00 55	02 43	00 06	01 16	17 29
19	02 12	02 36	00 17	00 55	02 42	00 06	01 16	17 28
22	02 50	02 37	00 19	00 55	02 42	00 06	01 16	17 27
25	03 19	02 37	00 21	00 55	02 41	00 06	01 16	17 26
28	03 39	02 37	00 23	00 54	02 41	00 06	01 16	17 25
31	03 S 48	02 N 35	00 N 24	00 N 54	02 N 40	00 N 06	01 N 16	17 N 24

DATA

Julian Date	2445456
Delta T	+53 seconds
Ayanamsa	23° 37' 09"
Synetic vernal point	05° ♓ 29' 51"
True obliquity of ecliptic	23° 26' 30"

MOON'S PHASES, APSIDES AND POSITIONS ☽

Date	h m	Phase	Longitude	Eclipse Indicator
05	03 43	☾	14 ♒ 06	
12	19 25	●	21 ♉ 30	
19	14 17	☽	28 ♌ 03	
26	18 48	○	04 ♐ 57	

Day	h m		
04	13 28	Apogee	
16	15 35	Perigee	
02	09 47	Max dec	24° S 20'
09	20 02	0N	
16	08 07	Max dec	24° N 24'
22	17 40	0S	
29	18 06	Max dec	24° S 26'

LONGITUDES

Date	Chiron ⚷	Ceres ⚳	Pallas ⚴	Juno ⚵	Vesta ⚶	Black Moon Lilith ⚸
01	26 ♉ 25	21 ♒ 07	24 ♑ 52	22 ♓ 28	01 ♉ 51	04 ♒ 54
11	27 ♉ 07	23 ♒ 21	25 ♑ 19	26 ♓ 51	06 ♉ 21	06 ♒ 01
21	27 ♉ 50	25 ♒ 13	25 ♑ 05	01 ♈ 10	10 ♉ 48	07 ♒ 08
31	28 ♉ 33	26 ♒ 39	24 ♑ 09	05 ♈ 25	15 ♉ 11	08 ♒ 14

ASPECTARIAN

h m	Aspects
01 Sunday	
00 30	☽ ± ♂
01 27	☽ Ꝑ ☉
02 15	☽ ⚻ ☿
06 41	☽ ⚹ ♇
09 02	☽ ☌ ♆
11 47	☽ ⚹ ♄
14 05	☽ ± ☿
16 37	☿ St R
19 19	☽ Ꝑ ♂
02 Monday	
03 12	☽ ⚺ ♅
05 01	☽ ⚺ ♃
06 24	☽ Q ♇
07 55	☽ Ꝑ ☿
09 42	☽ △ ☉
11 29	☽ Q ♄
15 10	☽ ⊥ ♅
16 56	☽ ⊥ ♃
03 Tuesday	
03 15	☽ △ ♂
09 13	☽ ⚻ ♀
09 14	☽ ∠ ♅
10 55	☽ ∠ ♃
13 50	☽ △ ☿
18 33	☽ □ ♇
21 02	☽ ⚺ ♆
22 39	☽ ± ♀
23 34	☽ □ ♄
04 Wednesday	
08 24	☿ # ♃
09 10	☽ ⊥ ♆
09 26	☽ ∥ ♆
15 32	☽ ⚹ ♅
16 03	☽ ∥ ♅
17 03	☽ ⚹ ♃
18 51	☽ Ꝑ ♀
22 49	☽ ∥ ♃
23 46	☉ Ꝑ ♆
05 Thursday	
00 18	☽ # ☿
02 11	☿ ⚺ ♀
03 23	☽ ∠ ♆
03 43	☽ □ ☉
15 44	☽ Q ♅
17 06	☽ Q ♃
18 22	☽ # ♂
19 53	☽ □ ♂
06 Friday	
01 14	☽ □ ☿
04 19	☽ △ ♀
07 03	☽ △ ♇
08 04	☽ # ☉
09 33	☽ ⚹ ♆
11 46	☽ △ ♄
19 04	☽ Q ☉
22 43	♂ ± ♆
07 Saturday	
03 35	☽ □ ♅
04 44	☽ □ ♃
08 34	♀ △ ♇
09 19	☽ Q ♆
10 00	☽ Q ♂
11 53	☽ Q ☿
12 43	☽ Ꝑ ♇
17 15	☽ Ꝑ ♄
20 43	☽ ⚹ ☉
08 Sunday	
02 08	☽ ∥ ♄
06 12	☽ ± ♇
06 19	☿ ☌ ♂
10 21	☽ ⚹ ☿
10 35	☽ ± ♄
10 46	☽ ⚹ ♂
10 56	♀ ☍ ♆
17 23	☽ # ♇
17 42	☽ ⚻ ♇
20 07	☽ □ ♆
21 02	☽ □ ♀
21 59	☽ ⚻ ♄
23 54	☿ ⊥ ♀
09 Monday	
03 54	☽ ∠ ☉
07 10	♀ △ ♄
12 36	☿ ∥ ♂
13 09	☽ △ ♅
13 41	☽ ∠ ☿
13 55	☽ △ ♃
16 51	☽ ∠ ♂
22 07	☽ ⊥ ☉
10 Tuesday	
04 44	☿ ± ♆
05 42	☽ ⊥ ☿
09 49	☽ Q ♀
10 02	☽ ⚺ ☉
10 25	☽ ⊥ ♂
16 12	☽ ⚺ ☿
16 43	☽ Ꝑ ♅
17 17	☽ Ꝑ ♃
21 56	☽ ⚺ ♂
21 58	☽ ∥ ♇
11 Wednesday	
01 10	☽ ☍ ♇
03 29	☽ △ ♆
03 46	♀ ⊥ ♂
05 03	☽ ☍ ♄
08 52	☽ ± ♅
09 21	☽ ± ♃
09 30	☽ ⚹ ♀
11 42	☽ # ♄
19 29	☽ ⚻ ♅
19 54	☽ ⚻ ♃
12 Thursday	
05 37	☉ ∥ ☿
06 01	☽ Ꝑ ♆
14 15	☽ ∠ ♀
16 37	☉ ☌ ☿
19 07	☽ ☌ ☿
19 25	☽ ☌ ☉
21 39	☽ ± ♆
13 Friday	
05 35	☽ ☌ ♂
05 44	☽ ⚻ ♇
07 09	☽ ⊥ ♀
07 25	☽ ∥ ☿
07 58	☽ ⚻ ♆
08 30	♂ ⚻ ♇
09 16	☽ ⚻ ♄
12 06	☽ ∥ ☉
15 59	☽ ± ♇
18 18	☽ ⚺ ♀
19 28	☽ ± ♄
22 22	☽ ∥ ♂
23 14	☽ ☍ ♅
23 21	☽ ☍ ♃
14 Saturday	
02 51	☉ ± ♆
03 53	☽ # ♃
07 11	☿ ∠ ♀
07 15	☽ Ꝑ ♇
10 08	☽ # ♅
10 40	☽ Ꝑ ♄
16 00	☽ # ♆
20 10	☽ ⚺ ☿
20 36	♃ ☌ ♅
15 Sunday	
02 13	☽ ⚺ ☉
03 11	♂ ⚻ ♆
05 52	☽ ⊥ ☿
08 29	☽ △ ♇
10 40	☽ ☍ ♆
11 05	☽ ⚺ ♂
11 47	☽ △ ♄
13 01	☽ ⊥ ☉
20 22	☽ ∠ ☿
21 39	☽ ⊥ ♂
16 Monday	
00 19	♂ ⚻ ♄
01 08	☽ ☌ ♀
01 26	☽ ⚻ ♃
01 36	☽ ⚻ ♅
04 43	♀ ⚻ ♃
05 11	☽ ∠ ☉
06 59	♀ ⚻ ♅
11 23	☽ ± ♃
11 36	☽ ± ♅
13 30	☽ ∠ ♂
20 33	☽ ⚹ ☿
17 Tuesday	
02 19	☽ Ꝑ ♃
02 37	☽ Ꝑ ♅
08 05	☽ ⚹ ☉
10 37	☽ □ ♇
12 49	☽ ⚻ ♆
13 47	☽ □ ♄
15 55	☽ ⚹ ♂
15 56	☽ Q ☿
22 52	☽ ± ♆
18 Wednesday	
00 14	☽ # ♆
03 15	☽ △ ♃
03 42	☽ △ ♅
05 40	☽ Q ☉
05 46	♂ # ♃
06 17	☽ # ♅
07 35	☽ ⚺ ♀
12 33	☽ ∥ ♂
12 55	☽ # ♃
13 08	☽ Q ♂
13 59	☽ Ꝑ ♆
16 48	☽ Q ♇
18 32	☽ ⊥ ♀
19 55	☽ Q ♄
20 11	☽ ∥ ☉
20 52	☉ ⚻ ♇
21 14	☽ □ ☿
19 Thursday	
11 01	☽ ∠ ♀
13 04	☽ ⚹ ♇
14 17	☽ □ ☉
15 18	☽ △ ♆
16 08	☽ ⚹ ♄
21 12	☽ □ ♂
23 01	☽ ∥ ☿
20 Friday	
05 00	☉ ⚻ ♆
05 35	☽ □ ♃
06 20	☽ □ ♅
14 37	☽ ∠ ♇
14 45	☽ ⚹ ♀
16 02	☉ ⚻ ♄
17 39	☽ ∠ ♄
22 59	☽ △ ☿
21 Saturday	
02 32	♀ ± ♃
04 29	☽ # ♄
06 05	☽ ⊥ ♇
09 06	☽ ⊥ ♄
12 17	☽ Q ♃
13 10	♂ ± ♇
13 12	☽ Q ♀
13 14	☽ Q ♅
13 27	♀ ± ♅
16 27	☽ ⚺ ♇
16 43	☽ ∥ ♇
18 45	☽ □ ♆
19 28	☽ ⚺ ♄
21 39	☽ △ ☉
22 Sunday	
00 27	☽ Ꝑ ☿
03 39	☽ △ ♂
09 00	☽ ⚹ ♃
10 06	☽ ⚹ ♅
15 58	☽ ± ☿
23 26	☽ □ ♀
23 Monday	
01 59	☽ Ꝑ ☉
02 17	☽ Q ♆
02 25	☽ ⚻ ☿
07 32	☽ Ꝑ ♂
11 12	☉ # ♃
11 17	☽ ∠ ♃
12 33	☽ ∠ ♅
18 59	☽ # ♇
19 15	☽ ± ☉
19 48	♂ # ♅
19 57	♂ ± ♄
21 18	☽ ☌ ♇
23 40	☽ ⚹ ♆
24 Tuesday	
00 16	☽ ☌ ♄
00 32	☽ ± ♂
03 16	☽ ⊥ ♃
04 39	☽ ⊥ ♅
06 54	☽ ⚻ ☉
07 23	☽ ∥ ♄
09 00	☿ ⚹ ♀
11 58	☽ ⚻ ♂
14 01	☽ ⚺ ♃
15 29	☽ ⚺ ♅
23 12	☉ ± ♇
25 Wednesday	
02 51	☽ ∠ ♆
08 06	☽ # ☿
08 11	☽ ☍ ☿
10 23	☽ △ ♀
12 48	☿ St D
19 27	☽ ⊥ ♆
20 55	♂ ☍ ♃
26 Thursday	
04 10	☽ ⚺ ♇
06 37	☽ ⚺ ♆
07 07	☽ ⚺ ♄
13 27	☉ ± ♄
15 26	☽ ⊥ ♇
16 56	☽ Ꝑ ♀
18 22	☽ ⊥ ♄
18 48	☽ ☍ ☉
21 09	☽ ☌ ♃
22 50	☽ ☍ ♂
23 03	☽ ☌ ♅
27 Friday	
02 39	♂ ☍ ♅
03 09	☽ ∥ ♃
08 31	☽ ∠ ♇
09 49	☽ # ☉
11 27	☽ ∥ ♅
11 27	☽ ∠ ♄
11 40	☽ ± ♀
16 31	☽ # ♂
17 02	☽ ⚻ ☿
18 48	☽ ∥ ♆
22 29	☉ ☍ ♃
28 Saturday	
00 19	☽ ⚻ ♀
04 49	☽ ± ☿
11 00	☉ # ♅
13 30	☽ ⚹ ♇
14 06	♀ Ꝑ ♃
16 02	☽ ☌ ♆
16 25	☽ ⚹ ♄
22 46	☽ Ꝑ ☿
29 Sunday	
00 47	☉ ☍ ♅
05 26	☽ # ♀
06 43	☽ ⚺ ♃
09 06	☽ ⚺ ♅
09 51	☽ ⚻ ☉
12 38	☽ ⚻ ♂
12 48	♂ # ♆
13 06	☽ Q ♇
15 14	♀ Ꝑ ♅
16 01	☽ Q ♄
18 30	☽ ⊥ ♃
20 59	☽ ⊥ ♅
22 49	☽ ± ☉
30 Monday	
01 20	☽ ± ♂
05 19	☽ △ ☿
12 19	☽ ∠ ♃
14 39	☽ # ♀
14 57	☽ ∠ ♅
17 21	☽ ☍ ♀
18 26	☽ Ꝑ ☉
20 31	☽ Ꝑ ♂
31 Tuesday	
01 06	☽ □ ♇
03 42	☽ ⚺ ♆
03 58	☽ □ ♄
15 49	☽ ⊥ ♆
15 55	☽ # ♂
18 16	☽ ⚹ ♃
18 40	☽ ∥ ♆
21 09	☽ ⚹ ♅
21 13	☽ # ☉

All ephemeris data is given at 12.00 UT and the Moon's longitude is additionally given for 24.00 UT

JUNE 1983

Date	Sidereal time h m s	Sun ☉ ° ' "	Moon ☽ ° ' "	Moon ☽ 24.00 ° ' "	Mercury ☿ ° '	Venus ♀ ° '	Mars ♂ ° '	Jupiter ♃ ° '	Saturn ♄ ° '	Uranus ♅ ° '	Neptune ♆ ° '	Pluto ♇ ° '
01	04 37 38	10 ♊ 26 01	14 ♒ 19 22	20 ♒ 14 47	18 ♉ 14	25 ♋ 06	10 ♊ 57	05 ♐ 29	28 ♎ 26	06 ♐ 58	28 ♐ 20	27 ♎ 03
02	04 41 35	11 23 31	26 ♒ 10 57	02 ♓ 08 28	18 46	26 08	11 39	05 R 21	28 R 24	06 R 55	28 R 19	27 R 02
03	04 45 31	12 20 59	08 ♓ 07 55	14 ♓ 09 51	19 23	27 10	12 21	05 14	28 21	06 53	28 17	27 01
04	04 49 28	13 18 27	20 ♓ 14 52	26 ♓ 23 31	20 04	28 12	13 02	05 06	28 18	06 51	28 16	27 00
05	04 53 24	14 15 54	02 ♈ 36 20	08 ♈ 53 48	20 49	29 ♋ 14	13 44	04 59	28 16	06 48	28 14	26 59
06	04 57 21	15 13 20	15 ♈ 16 20	21 ♈ 44 19	21 37	00 ♌ 15	14 25	04 51	28 14	06 46	28 13	26 58
07	05 01 17	16 10 46	28 ♈ 18 00	04 ♉ 57 39	22 30	01 16	15 07	04 44	28 11	06 43	28 11	26 57
08	05 05 14	17 08 11	11 ♉ 43 15	18 ♉ 34 48	23 26	02 17	15 48	04 36	28 09	06 41	28 09	26 56
09	05 09 11	18 05 35	25 ♉ 32 07	02 ♊ 34 55	24 25	03 17	16 30	04 29	28 07	06 38	28 08	26 55
10	05 13 07	19 02 59	09 ♊ 42 44	16 ♊ 55 03	25 28	04 17	17 11	04 22	28 05	06 36	28 06	26 55
11	05 17 04	20 00 22	24 ♊ 11 10	01 ♋ 30 21	26 34	05 17	17 52	04 15	28 03	06 34	28 05	26 54
12	05 21 00	20 57 44	08 ♋ 51 45	16 ♋ 14 32	27 44	06 16	18 34	04 08	28 01	06 31	28 03	26 53
13	05 24 57	21 55 06	23 ♋ 37 48	01 ♌ 00 42	28 ♉ 57	07 15	19 15	04 01	27 59	06 29	28 01	26 52
14	05 28 53	22 52 26	08 ♌ 22 24	15 ♌ 42 10	00 ♊ 13	08 13	19 56	03 54	27 57	06 26	28 00	26 51
15	05 32 50	23 49 46	22 ♌ 59 20	00 ♍ 13 19	01 32	09 11	20 37	03 47	27 56	06 24	27 58	26 51
16	05 36 46	24 47 05	07 ♍ 23 40	14 ♍ 30 02	02 54	10 08	21 18	03 40	27 54	06 22	27 57	26 50
17	05 40 43	25 44 23	21 ♍ 32 09	28 ♍ 29 51	04 20	11 06	21 59	03 34	27 53	06 19	27 55	26 49
18	05 44 40	26 41 39	05 ♎ 23 04	12 ♎ 11 48	05 48	12 02	22 40	03 27	27 52	06 17	27 53	26 49
19	05 48 36	27 38 55	18 ♎ 56 06	25 ♎ 36 04	07 20	12 58	23 21	03 21	27 50	06 15	27 52	26 48
20	05 52 33	28 36 11	02 ♏ 11 50	08 ♏ 43 34	08 55	13 54	24 02	03 14	27 49	06 13	27 50	26 47
21	05 56 29	29 ♊ 33 25	15 ♏ 11 26	21 ♏ 35 39	10 33	14 49	24 43	03 08	27 48	06 10	27 49	26 47
22	06 00 26	00 ♋ 30 39	27 ♏ 56 24	04 ♐ 13 53	12 13	15 44	25 24	03 02	27 47	06 08	27 47	26 46
23	06 04 22	01 27 53	10 ♐ 28 17	16 ♐ 39 49	13 57	16 38	26 05	02 56	27 46	06 06	27 45	26 46
24	06 08 19	02 25 06	22 ♐ 48 39	28 ♐ 55 01	15 43	17 31	26 46	02 50	27 46	06 04	27 44	26 45
25	06 12 15	03 22 18	04 ♑ 59 07	11 ♑ 01 09	17 33	18 24	27 26	02 44	27 45	06 02	27 42	26 45
26	06 16 12	04 19 31	17 ♑ 01 22	23 ♑ 00 00	19 25	19 17	28 07	02 39	27 45	06 00	27 41	26 45
27	06 20 09	05 16 43	28 ♑ 57 19	04 ♒ 53 36	21 20	20 08	28 48	02 33	27 44	05 58	27 39	26 44
28	06 24 05	06 13 55	10 ♒ 49 11	16 ♒ 44 22	23 17	20 59	29 ♊ 28	02 28	27 44	05 56	27 37	26 44
29	06 28 02	07 11 06	22 ♒ 39 34	28 ♒ 35 09	25 17	21 50	00 ♋ 09	02 23	27 44	05 54	27 36	26 44
30	06 31 58	08 ♋ 08 18	04 ♓ 31 32	10 ♓ 29 13	27 ♊ 18	22 ♌ 39	00 ♋ 49	02 ♐ 18	27 ♎ 43	05 ♐ 52	27 ♐ 34	26 ♎ 43

Date	Moon True ☊ ° '	Moon Mean ☊ ° '	Moon ☽ Latitude ° '	DECLINATIONS Sun ☉ ° '	Moon ☽ ° '	Mercury ☿ ° '	Venus ♀ ° '	Mars ♂ ° '	Jupiter ♃ ° '	Saturn ♄ ° '	Uranus ♅ ° '	Neptune ♆ ° '	Pluto ♇ ° '
01	25 ♊ 08	25 ♊ 50	04 S 00	22 N 01	20 S 21	13 N 35	23 N 38	22 N 30	20 S 20	08 S 26	21 S 23	22 S 10	05 N 50
02	25 D 09	25 47	04 37	22 09	17 08	13 44	23 25	22 36	20 19	08 25	21 22	22 10	05 50
03	25 10	25 44	05 03	22 17	13 12	13 54	23 12	22 42	20 18	08 24	21 22	22 10	05 50
04	25 R 10	25 41	05 15	22 24	08 42	14 06	22 58	22 48	20 16	08 23	21 22	22 10	05 50
05	25 09	25 38	05 14	22 31	03 S 46	14 20	22 44	22 54	20 15	08 23	21 21	22 10	05 50
06	25 08	25 34	04 57	22 37	01 N 26	14 36	22 29	22 59	20 14	08 22	21 21	22 10	05 50
07	25 07	25 31	04 25	22 43	06 44	14 53	22 14	23 05	20 13	08 21	21 20	22 10	05 49
08	25 06	25 28	03 38	22 49	11 54	15 11	21 59	23 10	20 12	08 21	21 20	22 10	05 49
09	25 05	25 25	02 37	22 54	16 37	15 31	21 43	23 15	20 11	08 20	21 20	22 10	05 49
10	25 05	25 22	01 24	22 59	20 31	15 52	21 26	23 19	20 09	08 20	21 19	22 10	05 49
11	25 05	25 19	00 S 05	23 04	23 14	16 13	21 09	23 24	20 08	08 19	21 19	22 10	05 49
12	25 D 05	25 15	01 N 16	23 08	24 24	16 36	20 52	23 28	20 07	08 19	21 18	22 10	05 49
13	25 05	25 12	02 32	23 12	23 52	17 00	20 34	23 32	20 06	08 18	21 18	22 10	05 49
14	25 05	25 09	03 38	23 15	21 41	17 24	20 17	23 36	20 05	08 18	21 18	22 10	05 48
15	25 05	25 06	04 29	23 18	18 05	17 48	19 58	23 39	20 04	08 18	21 17	22 09	05 48
16	25 05	25 03	05 02	23 20	13 28	18 14	19 40	23 42	20 03	08 17	21 17	22 09	05 48
17	25 R 05	24 59	05 17	23 22	08 12	18 39	19 21	23 46	20 02	08 17	21 16	22 09	05 48
18	25 D 05	24 56	05 12	23 24	02 N 38	19 05	19 01	23 48	20 00	08 17	21 16	22 09	05 47
19	25 06	24 53	04 50	23 25	02 S 56	19 31	18 42	23 51	19 59	08 17	21 16	22 09	05 47
20	25 06	24 50	04 13	23 26	08 16	19 56	18 22	23 54	19 58	08 17	21 15	22 09	05 47
21	25 06	24 47	03 24	23 26	13 08	20 21	18 02	23 56	19 57	08 16	21 15	22 09	05 46
22	25 07	24 44	02 25	23 26	17 21	20 46	17 42	23 58	19 56	08 16	21 15	22 09	05 46
23	25 07	24 40	01 20	23 26	20 42	21 11	17 21	24 00	19 55	08 16	21 14	22 09	05 46
24	25 07	24 37	00 N 13	23 25	23 02	21 34	17 00	24 01	19 55	08 16	21 14	22 09	05 45
25	25 R 07	24 34	00 S 54	23 24	24 15	21 57	16 39	24 03	19 54	08 16	21 13	22 09	05 45
26	25 07	24 31	01 58	23 22	24 19	22 18	16 18	24 04	19 53	08 16	21 13	22 09	05 45
27	25 06	24 28	02 57	23 20	23 15	22 38	15 57	24 05	19 52	08 16	21 13	22 09	05 44
28	25 04	24 25	03 47	23 18	21 09	22 56	15 35	24 06	19 51	08 17	21 12	22 09	05 44
29	25 03	24 21	04 27	23 15	18 10	23 13	15 14	24 06	19 50	08 17	21 12	22 09	05 43
30	25 ♊ 01	24 ♊ 18	04 S 56	23 N 11	14 S 27	23 N 28	14 N 52	24 N 07	19 S 49	08 S 17	21 S 12	22 S 09	05 N 43

ZODIAC SIGN ENTRIES

Date	h	m	Planets
02	19	42	☽ ♓
05	06	59	☽ ♈
06	06	04	♀ ♌
07	15	05	☽ ♉
09	19	37	☽ ♊
11	21	32	☽ ♋
13	22	21	☽ ♌
14	08	06	☿ ♊
15	23	38	☽ ♍
18	02	36	☽ ♎
20	07	59	☽ ♏
21	23	09	☉ ♋
22	15	55	☽ ♐
25	02	08	☽ ♑
27	14	07	☽ ♒
29	06	54	♂ ♋
30	02	52	☽ ♓

LATITUDES

Date	Mercury ☿ ° '	Venus ♀ ° '	Mars ♂ ° '	Jupiter ♃ ° '	Saturn ♄ ° '	Uranus ♅ ° '	Neptune ♆ ° '	Pluto ♇ ° '
01	03 S 49	02 N 34	00 N 25	00 N 54	02 N 40	00 N 06	01 N 16	17 N 24
04	03 47	02 30	00 27	00 53	02 39	00 06	01 16	17 22
07	03 38	02 25	00 28	00 53	02 39	00 06	01 16	17 21
10	03 22	02 19	00 30	00 53	02 38	00 06	01 16	17 20
13	03 00	02 11	00 31	00 52	02 37	00 06	01 16	17 18
16	02 34	02 02	00 33	00 52	02 37	00 06	01 16	17 17
19	02 03	01 51	00 35	00 51	02 36	00 06	01 16	17 15
22	01 30	01 39	00 36	00 51	02 35	00 06	01 16	17 13
25	00 55	01 25	00 38	00 50	02 34	00 06	01 16	17 12
28	00 S 20	01 09	00 39	00 49	02 34	00 06	01 16	17 10
31	00 N 14	00 N 51	00 N 41	00 N 49	02 N 33	00 N 05	01 N 16	17 N 08

DATA

Julian Date	2445487
Delta T	+53 seconds
Ayanamsa	23° 37' 14"
Synetic vernal point	05° ♓ 29' 46"
True obliquity of ecliptic	23° 26' 30"

LONGITUDES

Date	Chiron ⚷ ° '	Ceres ⚳ ° '	Pallas ⚴ ° '	Juno ⚵ ° '	Vesta ⚶ ° '	Black Moon Lilith ⚸ ° '
01	28 ♉ 37	26 ♒ 46	24 ♑ 01	05 ♈ 50	15 ♉ 37	08 ♒ 21
11	29 ♉ 19	27 ♒ 46	22 ♑ 19	09 ♈ 58	19 ♉ 55	09 ♒ 28
21	00 ♊ 00	28 ♒ 02	20 ♑ 04	13 ♈ 58	24 ♉ 07	10 ♒ 35
31	00 ♊ 37	27 ♒ 49	17 ♑ 24	17 ♈ 49	28 ♉ 14	11 ♒ 42

MOON'S PHASES, APSIDES AND POSITIONS ☽

Date	h	m	Phase	Longitude ° '	Eclipse Indicator
03	21	07	☾	12 ♓ 43	
11	04	38	●	19 ♊ 43	Total
17	19	46	☽	26 ♍ 03	
25	08	32	○	03 ♑ 14	partial

Day	h	m		
01	07	37	Apogee	
13	05	42	Perigee	
28	22	29	Apogee	
06	05	27	0N	
12	16	33	Max dec	24° N 26'
18	23	18	0S	
26	01	15	Max dec	24° S 26'

ASPECTARIAN

h m Aspects

01 Wednesday
02 44 ☽ ∥ ♅
03 25 ☽ △ ☉
04 44 ☽ △ ♂
10 01 ☽ ∠ ♆
12 11 ☽ ∥ ♃
18 20 ☽ Q ♃
20 17 ☽ □ ☿
21 23 ☽ Q ♅

02 Thursday
11 54 ☽ ⊼ ♀
13 44 ☽ △ ♇
14 34 ☉ # ♆
16 17 ☽ ⚹ ♆
16 27 ☽ △ ♄

03 Friday
01 08 ☽ ± ♀
01 10 ♂ ⚼ ♇
03 55 ☉ ⚼ ♇
06 15 ☽ □ ♃
08 09 ☽ # ☿
08 33 ♀ □ ♇
09 31 ☽ □ ♅
10 26 ☽ Q ☿
11 21 ☉ ☌ ♂
16 17 ☽ Q ♆
19 44 ☽ ⚼ ♇
20 48 ☽ ⚼ ♀
20 54 ☽ □ ♂
21 07 ☽ □ ☉
22 21 ☽ ⚼ ♄

04 Saturday
00 03 ♀ ∠ ♂
11 38 ☽ ⚹ ☿
12 00 ☉ ⚼ ♄
13 19 ♀ ⊼ ♆
13 29 ☽ ± ♇
13 33 ☽ ∥ ♄
14 18 ♀ □ ♄
16 01 ☽ ± ♄
20 50 ♂ ⚼ ♄

05 Sunday
00 06 ♀ ∥ ♂
01 10 ☽ ⊼ ♇
02 10 ☽ # ♇
03 36 ☽ □ ♆
03 40 ☽ ⊼ ♄
04 55 ☽ △ ♀
10 13 ☽ Q ♂
11 17 ☽ Q ☉
16 30 ☽ △ ♃
18 33 ☽ ∠ ☿
20 00 ☽ △ ♅

06 Monday
00 32 ☉ ∠ ♀
02 57 ☉ ∥ ♀
10 19 ☽ ⚹ ♂
11 54 ☽ ⚹ ☉
12 42 ☽ ⊥ ☿
20 26 ☽ ⚼ ♃

07 Tuesday
00 00 ☽ ⚼ ♅
00 37 ☽ ⚺ ☿
03 51 ☿ ± ♆
07 51 ☽ ∥ ♇
09 33 ☽ ☍ ♇
11 47 ☽ △ ♆
11 48 ☽ ☍ ♄
12 46 ☽ ± ♃
15 28 ☽ ∠ ♂
16 22 ☽ ± ♅
17 33 ♄ ⚹ ♆
17 37 ☽ ∠ ☉
17 49 ☽ □ ♀
19 02 ♀ # ♆
19 24 ☽ # ♄
23 29 ☽ ⊼ ♃

08 Wednesday
03 06 ☽ ⊼ ♅
08 26 ☽ ⊥ ♂
10 53 ☽ ⊥ ☉
14 31 ☽ ⚼ ♆
19 33 ☽ ⚺ ♂
22 12 ☽ ⚺ ☉

09 Thursday
04 07 ☽ Q ♀
05 39 ☽ ∥ ☿
06 10 ☽ ± ♆
09 56 ☽ ☌ ☿
14 22 ☽ ⊼ ♇
16 24 ☽ ⊼ ♄
16 26 ☽ ⊼ ♆

10 Friday
00 34 ☽ ± ♇
02 11 ☽ ⚹ ♀
02 33 ☽ ± ♄
03 05 ☽ ☍ ♃
06 47 ☽ ☍ ♅
07 50 ♀ ± ♆
09 28 ☽ # ♃
13 47 ♀ △ ♃
15 40 ☽ ⚼ ♇
17 37 ☽ ⚼ ♄
17 56 ☽ # ♅
18 18 ☽ ∥ ♀
22 18 ♀ # ♅

11 Saturday
00 59 ☽ # ♆
01 04 ☽ ☌ ♂
04 38 ☽ ☌ ☉
05 05 ☽ ∠ ♀
09 59 ☽ ∥ ☉
14 06 ☽ ∥ ♂
16 15 ☽ ⚺ ☿
16 27 ☽ △ ♇
18 19 ☽ △ ♄
18 23 ☽ ☍ ♆
18 48 ☿ ⊼ ♇
20 57 ☽ ⊥ ♀

12 Sunday
02 54 ☽ ⊥ ☿
04 21 ☽ ⊼ ♃
07 28 ☽ ⚺ ♀
08 12 ☽ ⊼ ♅
14 03 ☽ ± ♃
17 38 ☿ ⊼ ♄
17 56 ☽ ± ♅
18 02 ♀ △ ♅
18 22 ☿ ⊼ ♆
18 51 ☽ ∠ ☿

13 Monday
04 32 ☽ ⚺ ♂
04 33 ☽ ⚼ ♃
08 31 ☽ ⚼ ♅
09 02 ☽ ⚺ ☉
14 46 ☽ ⊥ ♂
16 59 ☽ ∥ ♂
17 15 ☽ □ ♇
19 04 ☽ □ ♄
19 07 ♃ ⊥ ♄
19 08 ☽ ⊼ ♆
19 27 ☽ ⊥ ☉
21 09 ☽ ∥ ☉
21 26 ☽ ⚹ ☿

14 Tuesday
04 45 ☽ △ ♃
04 53 ☽ ± ♆
06 07 ☽ ∠ ♂
07 54 ☽ # ♆
08 51 ☽ △ ♅
11 08 ☽ ∠ ☉
11 43 ☽ ☌ ♀
15 02 ☽ # ♅
18 53 ☽ Q ☿
19 33 ☽ ⚼ ♆
22 36 ☽ Q ♇
23 21 ☽ ∥ ♀
23 45 ☽ # ♃

15 Wednesday
00 24 ☽ Q ♄
04 28 ♀ # ♃
07 54 ☽ ⚹ ♂
13 27 ☽ ∥ ☿
13 29 ☽ ⚹ ☉
18 23 ☽ ⚹ ♇
20 10 ☽ ⚹ ♄
20 14 ☽ △ ♆

16 Thursday
03 40 ☽ □ ☿
04 48 ☽ Q ♂
05 48 ☽ □ ♃
10 16 ☽ □ ♅
10 46 ☿ ± ♇
10 54 ☽ Q ☉
16 57 ☽ ⚺ ♀
19 28 ☽ ∠ ♇
21 17 ☽ ∠ ♄

17 Friday
00 04 ☿ ☍ ♃
03 52 ☽ ⊥ ♀
04 40 ☿ ± ♄
10 47 ☽ ⊥ ♇
11 38 ☽ # ♄
12 02 ☽ Q ♃
12 36 ☽ ⊥ ♄
12 49 ☽ □ ♂
16 47 ☽ Q ♅
19 46 ☽ □ ☉
20 25 ☽ ∠ ♀
21 05 ☽ ⚺ ♇
22 29 ☽ ∥ ♇
22 55 ☽ ⚺ ♄
22 58 ☽ □ ♆

18 Saturday
08 39 ☽ ⚹ ♃
10 11 ☿ ∥ ♀
12 50 ☽ △ ☿
13 35 ☽ ⚹ ♅
14 53 ☉ △ ♇
19 25 ☿ ☍ ♅

19 Sunday
00 35 ☽ ⚹ ♀
06 31 ☽ Q ♆
09 15 ♀ ⚼ ♆
10 57 ☽ ∠ ♃
16 08 ☽ ∠ ♅
16 43 ☉ △ ♄
17 15 ☉ ☍ ♆
18 55 ☽ ⚼ ☿
20 23 ☽ △ ♂
23 41 ☽ Q ♀

20 Monday
00 36 ☽ # ♇
02 10 ☽ ☌ ♇
03 02 ☽ ⊥ ♃
04 02 ☽ ☌ ♄
04 04 ☽ ⚹ ♆
04 56 ☽ △ ☉
08 23 ☽ ⊥ ♅
12 02 ☽ ∥ ♄
13 30 ☽ ± ☿
13 53 ☽ ⚺ ♃
14 03 ☿ # ♃
19 20 ☽ ⚺ ♅

21 Tuesday
01 17 ☽ ⚼ ♂
02 06 ☽ ⊼ ☿
07 35 ☽ ∠ ♆
10 43 ☽ ⚼ ☉
11 00 ♀ Q ♇
11 15 ☽ □ ♀
18 58 ☽ ± ♂
22 59 ♄ ⚹ ♆

22 Wednesday
00 23 ☽ ⊥ ♆
04 58 ☽ ± ☉
05 41 ☿ ⚼ ♇
06 55 ☽ ⊼ ♂
09 47 ☽ ⚺ ♇
11 42 ☽ ⚺ ♆
11 43 ☽ ⚺ ♄
13 32 ♀ Q ♄
14 03 ☽ # ♀
17 18 ☽ ⊼ ☉
19 53 ☿ ⚼ ♄
21 12 ☽ ⊥ ♇
21 38 ☽ ☌ ♃
23 08 ☽ ⊥ ♄

23 Thursday
03 37 ☽ ☌ ♅
05 52 ☽ ∥ ♃
14 30 ☽ ∠ ♇
15 31 ☿ # ♅
16 27 ☽ ∠ ♄
16 42 ☽ ∥ ♅
16 54 ☽ # ☿
19 51 ☽ ☍ ☿

24 Friday
00 52 ☽ △ ♀
01 38 ☽ ∥ ♆
11 56 ♂ △ ♇
17 31 ☽ # ☉
19 45 ☽ ⚹ ♇
20 12 ☽ ☍ ♂
21 34 ☉ ⊼ ♃
21 38 ☽ ☌ ♆
21 43 ☽ ⚹ ♄

25 Saturday
05 35 ☽ # ♂
07 35 ☽ ⚺ ♃
08 32 ☽ ☍ ☉
08 37 ☽ ⚼ ♀
09 29 ♃ ⊥ ♇
14 04 ☽ ⚺ ♅
19 24 ☽ ⊥ ♃
19 29 ☽ Q ♇
21 00 ♂ ☍ ♆
21 28 ☽ Q ♄
21 55 ☉ ∠ ♀
22 59 ♂ △ ♄

26 Sunday
01 59 ☽ ⊥ ♅
02 01 ☿ # ♆
03 55 ☽ ± ♀
08 43 ☿ ⚹ ♀
13 15 ☽ ∠ ♃
16 52 ☽ ⊼ ♀
17 42 ☽ ⊼ ☿
19 57 ☽ ∠ ♅
20 15 ☽ # ♂

27 Monday
07 32 ☽ □ ♇
08 05 ☽ ± ☿
09 22 ☽ ⚺ ♆
09 32 ☽ □ ♄
10 39 ☽ # ☉
11 39 ☽ ⊼ ♂
19 09 ☽ # ☿
19 13 ☽ ⚹ ♃
21 28 ☽ ⊥ ♆

28 Tuesday
00 30 ☽ ± ♂
01 54 ☽ ∥ ♆
01 54 ☽ ⊼ ☉
02 07 ☽ ⚹ ♅
04 35 ☉ ⊼ ♅
05 50 ☽ ⚼ ☿
11 31 ☽ ∥ ♅
15 07 ☽ ± ☉
15 39 ☽ ∠ ♆
19 21 ☽ Q ♃
19 51 ☽ ⚼ ♂
23 21 ☽ ∥ ♃

29 Wednesday
02 22 ☽ Q ♅
10 11 ☽ ☍ ♀
10 57 ☽ ⚼ ☉
14 07 ☉ ∥ ☿
18 23 ☽ △ ☿
20 14 ☽ △ ♇
21 58 ☽ ⚹ ♆
22 15 ☽ △ ♄

30 Thursday
04 04 ☽ △ ♂
05 12 ☿ △ ♇
07 32 ☽ □ ♃
09 11 ☽ # ♀
14 41 ☽ □ ♅
15 03 ☿ ☍ ♆
15 44 ☉ ± ♃
16 54 ☿ △ ♄
19 55 ☽ △ ☉
22 08 ☽ Q ♆

All ephemeris data is given at 12.00 UT and the Moon's longitude is additionally given for 24.00 UT

JULY 1983

LONGITUDES

Date	Sidereal time h m s	Sun ☉ ° ' "	Moon ☽ ° ' "	Moon ☽ 24.00 ° ' "	Mercury ☿ ° '	Venus ♀ ° '	Mars ♂ ° '	Jupiter ♃ ° '	Saturn ♄ ° '	Uranus ♅ ° '	Neptune ♆ ° '	Pluto ♇ ° '
01	06 35 55	09 ♋ 05 30	16 ♓ 28 39	22 ♓ 30 20	29 ♊ 22	23 ♌ 28	01 ♋ 30	02 ♐ 13	27 ♎ 43	05 ♐ 50	27 ♐ 33	26 ♎ 43
02	06 39 51	10 02 42	28 ♓ 34 47	04 ♈ 42 33	01 ♋ 27	24 16	02 10	02 R 08	27 D 43	05 R 48	27 R 31	26 R 43
03	06 43 48	10 59 54	10 ♈ 54 09	17 ♈ 10 08	03 34	25 04	02 50	02 04	27 44	05 46	27 29	26 43
04	06 47 44	11 57 06	23 ♈ 30 59	29 ♈ 57 11	05 43	25 50	03 31	01 59	27 44	05 44	27 28	26 43
05	06 51 41	12 54 19	06 ♉ 29 10	13 ♉ 07 17	07 52	26 36	04 11	01 55	27 44	05 42	27 26	26 43
06	06 55 38	13 51 31	19 ♉ 51 50	26 ♉ 42 58	10 02	27 21	04 51	01 51	27 45	05 40	27 25	26 43
07	06 59 34	14 48 45	03 ♊ 40 44	10 ♊ 45 02	12 12	28 05	05 31	01 47	27 45	05 39	27 23	26 D 43
08	07 03 31	15 45 58	17 ♊ 55 38	25 ♊ 12 04	14 22	28 48	06 12	01 43	27 46	05 37	27 22	26 43
09	07 07 27	16 43 12	02 ♋ 33 44	09 ♋ 59 53	16 32	29 ♌ 30	06 52	01 40	27 47	05 35	27 20	26 43
10	07 11 24	17 40 26	17 ♋ 29 35	25 ♋ 01 46	18 41	00 ♍ 11	07 32	01 36	27 47	05 34	27 19	26 43
11	07 15 20	18 37 40	02 ♌ 35 17	10 ♌ 08 59	20 50	00 51	08 12	01 33	27 48	05 32	27 17	26 43
12	07 19 17	19 34 54	17 ♌ 41 38	25 ♌ 12 07	22 58	01 30	08 52	01 30	27 49	05 30	27 16	26 43
13	07 23 13	20 32 09	02 ♍ 39 23	10 ♍ 02 30	25 05	02 08	09 32	01 27	27 50	05 29	27 14	26 43
14	07 27 10	21 29 23	17 ♍ 20 42	24 ♍ 33 26	27 10	02 45	10 12	01 24	27 52	05 27	27 13	26 43
15	07 31 07	22 26 37	01 ♎ 40 14	08 ♎ 40 53	29 ♋ 14	03 20	10 52	01 21	27 53	05 26	27 11	26 44
16	07 35 03	23 23 51	15 ♎ 35 17	22 ♎ 23 30	01 ♌ 17	03 54	11 31	01 19	27 54	05 25	27 10	26 44
17	07 39 00	24 21 06	29 ♎ 05 41	05 ♏ 42 07	03 18	04 27	12 11	01 17	27 56	05 23	27 09	26 44
18	07 42 56	25 18 20	12 ♏ 13 07	18 ♏ 39 05	05 17	04 59	12 51	01 15	27 58	05 22	27 07	26 45
19	07 46 53	26 15 35	25 ♏ 00 25	01 ♐ 17 34	07 15	05 29	13 31	01 13	27 59	05 21	27 06	26 45
20	07 50 49	27 12 50	07 ♐ 30 57	13 ♐ 41 00	09 11	05 57	14 10	01 11	28 01	05 20	27 04	26 46
21	07 54 46	28 10 05	19 ♐ 48 08	25 ♐ 52 43	11 05	06 24	14 50	01 10	28 03	05 18	27 03	26 46
22	07 58 42	29 ♋ 07 20	01 ♑ 55 08	07 ♑ 55 43	12 58	06 50	15 30	01 08	28 05	05 17	27 02	26 46
23	08 02 39	00 ♌ 04 36	13 ♑ 54 47	19 ♑ 52 36	14 49	07 14	16 09	01 07	28 07	05 16	27 01	26 47
24	08 06 36	01 01 53	25 ♑ 49 27	01 ♒ 45 34	16 37	07 36	16 49	01 06	28 09	05 15	26 59	26 48
25	08 10 32	01 59 10	07 ♒ 41 13	13 ♒ 36 36	18 25	07 56	17 28	01 05	28 12	05 14	26 58	26 48
26	08 14 29	02 56 28	19 ♒ 31 57	25 ♒ 27 31	20 10	08 14	18 08	01 05	28 14	05 13	26 57	26 49
27	08 18 25	03 53 46	01 ♓ 23 33	07 ♓ 20 17	21 54	08 31	18 47	01 04	28 16	05 12	26 56	26 49
28	08 22 22	04 51 05	13 ♓ 18 01	19 ♓ 17 04	23 36	08 46	19 27	01 04	28 19	05 11	26 54	26 50
29	08 26 18	05 48 25	25 ♓ 17 46	01 ♈ 20 29	25 16	08 58	20 06	01 D 04	28 22	05 11	26 53	26 51
30	08 30 15	06 45 46	07 ♈ 25 37	13 ♈ 33 35	26 55	09 09	20 45	01 04	28 24	05 10	26 52	26 52
31	08 34 11	07 ♌ 43 08	19 ♈ 44 51	25 ♈ 59 53	28 ♌ 31	09 ♍ 17	21 ♋ 24	01 ♐ 05	28 ♎ 27	05 ♐ 09	26 ♐ 51	26 ♎ 52

Date	Moon True ☊ ° '	Moon Mean ☊ ° '	Moon ☽ Latitude ° '
01	24 ♊ 59	24 ♊ 15	05 S 13
02	24 R 58	24 12	05 15
03	24 58	24 09	05 04
04	24 D 58	24 05	04 38
05	24 59	24 02	03 57
06	25 01	23 59	03 02
07	25 02	23 56	01 55
08	25 03	23 53	00 S 39
09	25 R 03	23 50	00 N 41
10	25 02	23 46	02 01
11	24 59	23 43	03 12
12	24 56	23 40	04 10
13	24 53	23 37	04 51
14	24 50	23 34	05 11
15	24 47	23 31	05 11
16	24 46	23 27	04 53
17	24 D 46	23 24	04 19
18	24 47	23 21	03 32
19	24 48	23 18	02 36
20	24 50	23 15	01 33
21	24 51	23 11	00 N 28
22	24 R 51	23 08	00 S 39
23	24 49	23 05	01 42
24	24 46	23 02	02 41
25	24 41	22 59	03 32
26	24 34	22 56	04 14
27	24 27	22 52	04 45
28	24 20	22 49	05 03
29	24 14	22 46	05 09
30	24 08	22 43	05 01
31	24 ♊ 05	22 ♊ 40	04 S 39

DECLINATIONS

Date	Sun ☉ ° '	Moon ☽ ° '	Mercury ☿ ° '	Venus ♀ ° '	Mars ♂ ° '	Jupiter ♃ ° '	Saturn ♄ ° '	Uranus ♅ ° '	Neptune ♆ ° '	Pluto ♇ ° '
01	23 N 08	10 S 08	23 N 40	14 N 30	24 N 07	19 S 49	08 S 17	21 S 11	22 S 09	05 N 43
02	23 04	05 23	23 51	14 08	24 07	19 48	08 17	21 11	22 09	05 42
03	22 59	00 S 21	23 58	13 46	24 06	19 47	08 18	21 11	22 09	05 42
04	22 54	04 N 50	24 04	13 24	24 06	19 47	08 18	21 10	22 09	05 41
05	22 49	09 57	24 06	13 02	24 05	19 46	08 18	21 10	22 09	05 41
06	22 43	14 47	24 06	12 40	24 04	19 45	08 19	21 10	22 09	05 40
07	22 37	19 00	24 03	12 18	24 03	19 45	08 19	21 10	22 09	05 40
08	22 31	22 15	23 57	11 56	24 02	19 44	08 20	21 09	22 09	05 39
09	22 24	24 06	23 48	11 34	24 00	19 44	08 20	21 09	22 09	05 38
10	22 16	24 17	23 37	11 12	23 59	19 43	08 21	21 09	22 09	05 38
11	22 09	22 42	23 23	10 50	23 57	19 43	08 21	21 08	22 09	05 37
12	22 01	19 30	23 07	10 29	23 55	19 42	08 22	21 08	22 09	05 37
13	21 52	15 03	22 48	10 07	23 52	19 42	08 23	21 08	22 09	05 36
14	21 44	09 46	22 27	09 45	23 50	19 42	08 23	21 08	22 09	05 35
15	21 34	04 N 06	22 04	09 24	23 47	19 41	08 24	21 07	22 09	05 35
16	21 25	01 S 37	21 38	09 03	23 44	19 41	08 25	21 07	22 09	05 34
17	21 15	07 07	21 11	08 42	23 41	19 41	08 26	21 07	22 09	05 34
18	21 05	12 08	20 43	08 21	23 38	19 41	08 26	21 07	22 09	05 33
19	20 54	16 30	20 12	08 01	23 34	19 40	08 27	21 07	22 09	05 32
20	20 43	20 02	19 41	07 41	23 31	19 40	08 28	21 06	22 09	05 31
21	20 32	22 35	19 08	07 21	23 27	19 40	08 29	21 06	22 09	05 31
22	20 20	24 04	18 33	07 01	23 23	19 40	08 30	21 06	22 09	05 30
23	20 08	24 24	17 58	06 42	23 19	19 40	08 31	21 06	22 09	05 29
24	19 56	23 37	17 22	06 23	23 14	19 40	08 32	21 06	22 09	05 29
25	19 43	21 46	16 45	06 05	23 10	19 40	08 33	21 05	22 09	05 28
26	19 30	18 59	16 08	05 47	23 05	19 40	08 34	21 05	22 09	05 27
27	19 17	15 25	15 30	05 30	23 00	19 41	08 35	21 05	22 09	05 26
28	19 03	11 14	14 51	05 13	22 55	19 41	08 37	21 05	22 09	05 26
29	18 49	06 36	14 12	04 57	22 49	19 41	08 38	21 05	22 09	05 25
30	18 35	01 40	13 32	04 42	22 44	19 41	08 39	21 05	22 09	05 24
31	18 N 20	03 N 25	12 N 52	04 N 27	22 N 38	19 S 41	08 S 40	21 S 05	22 S 09	05 N 23

ZODIAC SIGN ENTRIES

Date	h m	Planets
01	19 18	☿ ♋
02	14 47	☽ ♈
05	00 05	☽ ♉
07	05 41	☽ ♊
09	07 50	☽ ♋
10	05 25	♀ ♍
11	07 54	☽ ♌
13	07 43	☽ ♍
15	09 10	☽ ♎
15	20 57	☿ ♌
17	13 38	☽ ♏
19	21 31	☽ ♐
22	08 11	☽ ♑
23	10 04	☉ ♌
24	20 26	☽ ♒
27	09 11	☽ ♓
29	21 21	☽ ♈

LATITUDES

Date	Mercury ☿ ° '	Venus ♀ ° '	Mars ♂ ° '	Jupiter ♃ ° '	Saturn ♄ ° '	Uranus ♅ ° '	Neptune ♆ ° '	Pluto ♇ ° '
01	00 N 14	00 N 51	00 N 41	00 N 49	02 N 33	00 N 05	01 N 16	17 N 08
04	00 45	00 32	00 42	00 48	02 32	00 05	01 16	17 07
07	01 10	00 N 10	00 44	00 47	02 31	00 05	01 16	17 05
10	01 30	00 S 13	00 45	00 47	02 30	00 05	01 16	17 03
13	01 43	00 39	00 47	00 46	02 30	00 05	01 16	17 02
16	01 49	01 06	00 48	00 45	02 29	00 05	01 16	17 00
19	01 48	01 36	00 49	00 45	02 28	00 05	01 15	16 58
22	01 42	02 08	00 51	00 44	02 27	00 05	01 15	16 56
25	01 31	02 42	00 52	00 43	02 26	00 05	01 15	16 55
28	01 16	03 18	00 53	00 43	02 26	00 05	01 15	16 53
31	00 N 57	03 S 56	00 N 55	00 N 42	02 N 25	00 N 05	01 N 15	16 N 51

DATA

Julian Date	2445517
Delta T	+53 seconds
Ayanamsa	23° 37' 19"
Synetic vernal point	05° ♓ 29' 41"
True obliquity of ecliptic	23° 26' 30"

MOON'S PHASES, APSIDES AND POSITIONS ☽

Date	h m	Phase	Longitude ° '	Eclipse Indicator
03	12 12	☾	11 ♈ 00	
10	12 18	●	17 ♋ 41	
17	02 50	☽	23 ♎ 59	
24	23 27	○	01 ♒ 29	

Day	h m		
11	10 17	Perigee	
26	06 56	Apogee	
03	13 37	0N	
10	02 31	Max dec	24° N 26'
16	05 08	0S	
23	07 04	Max dec	24° S 26'
30	19 53	0N	

LONGITUDES

Date	Chiron ⚷ ° '	Ceres ⚳ ° '	Pallas ⚴ ° '	Juno ⚵ ° '	Vesta ⚶ ° '	Black Moon Lilith ⚸ ° '
01	00 ♊ 37	27 ♒ 49	17 ♑ 24	17 ♈ 49	28 ♉ 14	11 ♒ 42
11	01 ♊ 12	27 ♒ 00	14 ♑ 37	21 ♈ 28	02 ♊ 14	12 ♒ 48
21	01 ♊ 43	25 ♒ 38	11 ♑ 58	24 ♈ 51	06 ♊ 05	13 ♒ 55
31	02 ♊ 08	23 ♒ 49	09 ♑ 42	27 ♈ 56	09 ♊ 47	15 ♒ 02

All ephemeris data is given at 12.00 UT and the Moon's longitude is additionally given for 24.00 UT

ASPECTARIAN

01 Friday
02 29 ☽ ⚼ ♇
04 29 ☽ ⚼ ♄
12 32 ♄ St D
20 27 ☽ ± ♇
21 33 ☽ ∥ ♄
22 27 ☽ ± ♄

02 Saturday
02 54 ☽ ⚻ ♀
08 20 ☽ ⚻ ♇
09 55 ☽ □ ♆
10 19 ☽ ⚻ ♄
10 27 ☽ # ♇
11 09 ♂ ⚻ ♃
15 33 ☽ ± ♀
18 49 ☽ □ ☿
18 56 ☽ △ ♃
19 27 ☽ □ ♂
19 28 ☿ ⚻ ♃
23 49 ☿ ☌ ♂

03 Sunday
02 05 ☽ △ ♅
10 16 ☽ ⚼ ♀
12 12 ☽ □ ☉
23 44 ☽ ⚼ ♃

04 Monday
06 41 ☉ ± ♅
06 46 ☽ ⚼ ♅
08 01 ☽ Q ♂
12 15 ☿ ⚻ ♅
12 26 ☽ Q ☿
15 58 ☽ ∥ ♇
16 36 ☽ ± ♃
16 37 ☽ △ ♀
17 59 ☽ ☍ ♇
19 22 ☽ △ ♆
19 53 ☽ ☍ ♄
23 34 ☽ ± ♅

05 Tuesday
00 57 ☽ Q ☉
03 18 ☿ ∥ ♂
03 40 ☽ ⚻ ♃
04 12 ☽ # ♄
07 34 ☽ ✶ ♂
10 35 ☽ ⚻ ♅
12 35 ☿ ± ♃
15 00 ☽ ✶ ☿
15 33 ♀ ✶ ♇
22 45 ☽ ⚼ ♆

06 Wednesday
00 30 ☽ ✶ ☉
02 01 ☽ ∥ ♀
11 59 ☽ ∠ ♂
14 01 ♀ △ ♆
14 43 ☽ ± ♆
22 45 ☽ ∠ ☿
23 59 ☽ ⚻ ♇

07 Thursday
01 01 ♀ ✶ ♄
01 11 ☽ ⚻ ♆
01 47 ☽ ⚻ ♄
01 50 ☽ □ ♀
04 30 ☽ ⊥ ♂
04 52 ☽ ∠ ☉
06 00 ☿ ± ♅
08 46 ☽ ☍ ♃
08 50 ☿ ∥ ♂
10 21 ☽ ± ♇
11 25 ♇ St D
12 08 ☽ ± ♄
15 18 ☽ ⊻ ♂
15 21 ☽ ☍ ♅
16 11 ♂ ⚻ ♅
16 48 ☽ # ♃
17 04 ☽ ⊥ ☿
21 21 ☽ ⊥ ☉

08 Friday
01 37 ☽ ⚼ ♇
02 43 ☿ ∠ ♀
02 54 ☽ # ♅
03 22 ☽ ⚼ ♄
05 00 ☽ ⊻ ☿
08 09 ☽ ⊻ ☉
10 02 ☽ Q ♀
11 11 ☽ # ♆
14 25 ☽ ∥ ☉

09 Saturday
02 28 ☽ △ ♇
03 31 ☽ ☍ ♆
04 12 ☽ △ ♄
06 47 ☽ ✶ ♀
06 54 ☽ ∥ ☿
09 59 ☽ ∥ ♂
10 32 ☽ ⚻ ♃
10 34 ☉ ⚼ ♃
13 25 ☿ ⚼ ♃
15 49 ☉ ☌ ☿
16 53 ☽ ⚻ ♅
19 17 ☽ ☌ ♂
20 12 ☽ ± ♃

10 Sunday
02 32 ☽ ± ♅
08 08 ☽ ∠ ♀
10 35 ☽ ⚼ ♃
12 18 ☽ ☌ ☉
14 13 ☽ ☌ ☿
14 25 ♂ ± ♃
16 53 ☽ ⚼ ♅
19 41 ☽ ∥ ♂
23 09 ☽ ⊥ ♀

11 Monday
02 41 ☽ □ ♇
03 16 ☽ ∥ ☿
03 36 ☽ ⚻ ♆
04 24 ☽ □ ♄
08 42 ☿ ⚼ ♅
09 07 ☽ ⊻ ♀
10 21 ☽ △ ♃
10 56 ☉ # ♆
13 06 ☽ ± ♆
16 40 ☽ △ ♅
17 02 ☽ # ♆
17 20 ☽ ∥ ☉
21 19 ☽ ⊻ ♂

12 Tuesday
01 03 ☽ # ♅
03 22 ☽ ⚼ ♆
07 16 ☽ Q ♇
07 17 ☽ ⊥ ♂
09 01 ☽ Q ♄
10 42 ☽ # ♃
11 42 ♀ □ ♃
15 13 ☽ ⊻ ☉
21 48 ☽ ⊻ ☿
22 19 ☽ ∠ ♂

13 Wednesday
01 28 ☽ ⊥ ☉
02 26 ☽ ✶ ♇
03 17 ☽ △ ♆
04 14 ☽ ✶ ♄
09 02 ☽ ⊥ ☿
10 03 ☽ □ ♃
10 42 ☉ ⚼ ♅
11 07 ☽ ☌ ♀
16 34 ☽ □ ♅
16 59 ☽ ∠ ☉
23 42 ☽ ✶ ♂

14 Thursday
02 04 ☽ ∠ ☿
02 45 ☽ ∠ ♇
04 36 ☽ ∠ ♄
05 11 ☿ ⊥ ♀
06 54 ☿ □ ♇
12 04 ☽ ∥ ♀
12 32 ☿ ⚻ ♆
15 24 ☽ Q ♃
17 36 ☽ ⊥ ♇
17 55 ☽ # ♄
19 22 ☽ ✶ ☉
19 30 ☽ ⊥ ♄
20 07 ☿ □ ♄
20 27 ☽ Q ♂
22 09 ☽ Q ♅

15 Friday
03 39 ☽ ⊻ ♇
04 26 ☽ □ ♆
05 35 ☽ ⊻ ♄
05 47 ☽ ∥ ♇
06 34 ☿ # ♆
07 10 ☽ ✶ ☿
11 28 ☽ ✶ ♃
14 58 ☽ ⊻ ♀
17 04 ☽ Q ☉
18 25 ☽ ✶ ♅

16 Saturday
01 42 ☽ ⊥ ♀
04 34 ☽ □ ♂
07 16 ☽ Q ☿
08 04 ♂ ± ♅
11 16 ☽ Q ♆
12 26 ☿ △ ♃
13 16 ☽ ∠ ♃
18 05 ☽ ∠ ♀
20 28 ☽ ∠ ♅

17 Sunday
02 50 ☽ □ ☉
05 04 ☽ # ♇
05 10 ☽ ⊥ ♃
07 12 ☉ ∥ ☿
07 46 ☽ ☌ ♇
08 30 ☽ ✶ ♆
09 54 ☽ ☌ ♄
10 11 ☿ ± ♆
12 32 ☽ ⊥ ♅
15 48 ☿ # ♅
15 56 ☽ ⊻ ♃
18 04 ☽ ∥ ♄
18 51 ☽ # ♀
20 58 ☽ □ ☿
22 08 ☽ ✶ ♀
23 24 ☽ ⊻ ♅

18 Monday
06 01 ♀ # ♄
06 57 ☿ ⊻ ♀
07 05 ☉ # ♅
11 49 ☽ ∠ ♆
12 57 ☿ △ ♅
13 14 ☽ △ ♂
21 14 ☽ Q ♀

19 Tuesday
04 37 ☽ ⊥ ♆
05 41 ♀ □ ♅
14 35 ☽ △ ☉
15 19 ☽ ⊻ ♇
15 58 ☽ ⊻ ♆
17 41 ☽ ⊻ ♄
19 03 ☽ ⚼ ♂
23 49 ☽ ☌ ♃

20 Wednesday
00 28 ☉ □ ♇
02 49 ☽ ⊥ ♇
05 14 ☽ ⊥ ♄
07 46 ☽ ☌ ♅
08 34 ☉ ⚻ ♆
08 52 ☽ □ ♀
09 16 ☽ ∥ ♃
09 41 ☽ # ☿
12 11 ☿ # ♃
13 21 ☽ ± ♂
15 50 ☽ △ ☿
17 15 ☽ # ☉
20 15 ☽ ∠ ♇
20 57 ☽ ∥ ♅
21 54 ☽ ⚼ ☉
22 44 ☽ ∠ ♄

21 Thursday
01 42 ☽ ⚻ ♂
07 05 ☽ ∥ ♆
08 57 ☉ □ ♄
17 04 ☽ ± ☉
23 09 ☽ # ♂

22 Friday
00 09 ☿ ⚼ ♆
01 46 ☽ ✶ ♇
02 18 ☽ ☌ ♆
02 42 ☽ ⚼ ☿
04 21 ☽ ✶ ♄
05 58 ☽ ⚻ ☉
10 27 ☽ ⊻ ♃
18 42 ☽ ⊻ ♅
22 09 ☽ △ ♀
22 24 ☽ ⊥ ♃
23 54 ☽ ± ☿

23 Saturday
01 42 ☽ Q ♇
04 22 ☽ Q ♄
06 42 ☽ ⊥ ♅
10 49 ♂ ⚼ ♃
11 40 ☿ Q ♇
14 07 ☽ ⚻ ☿
16 26 ☽ ∠ ♃
16 46 ☽ ☍ ♂

24 Sunday
00 46 ☽ ∠ ♅
05 17 ☽ ⚼ ♀
05 38 ☿ Q ♄
13 48 ☉ △ ♃
13 58 ☽ □ ♇
14 21 ☽ ⊻ ♆
15 57 ☿ ⊻ ♂
16 44 ☽ □ ♄
18 25 ☽ # ♂
22 40 ☽ ✶ ♃
23 27 ☽ ☍ ☉

25 Monday
00 01 ☽ ± ♀
02 28 ☽ ⊥ ♆
07 02 ☽ ✶ ♅
07 51 ☽ ∥ ♆
09 59 ☉ ⊥ ♀
12 31 ☽ ⚻ ♀
17 16 ☉ # ♃
18 31 ☽ ∥ ♅
20 39 ☽ ∠ ♆
22 56 ☽ Q ♃

26 Tuesday
06 37 ☽ ∥ ♃
07 19 ☽ Q ♅
07 39 ☽ # ☉
08 59 ☽ ⚻ ♂
12 07 ☉ ± ♆
13 31 ☽ ☍ ☿
21 51 ☽ ± ♂

27 Wednesday
02 45 ☽ △ ♇
02 59 ☽ ✶ ♆
05 40 ☽ △ ♄
11 21 ☽ □ ♃
11 24 ☽ # ☿
17 07 ☽ ⚼ ♂
17 30 ☽ ⚻ ☉
17 32 ♀ ∥ ♇
19 41 ☽ □ ♅

28 Thursday
02 41 ☽ ☍ ♀
03 10 ☽ Q ♆
06 39 ☽ ± ☉
09 03 ☽ ⚼ ♇
12 02 ☽ ⚼ ♄
20 21 ☉ △ ♅

29 Friday
01 02 ☽ △ ♂
01 45 ☽ ∥ ♄
02 16 ☽ ⚼ ☉
03 07 ☽ ± ♇
06 08 ☽ ± ♄
07 05 ♃ St D
11 56 ☽ ⚻ ☿
14 47 ♂ ⚼ ♅
15 05 ☽ ⚻ ♇
15 09 ☽ □ ♆
17 52 ☽ # ♇
18 07 ☽ ⚻ ♄
20 34 ☽ # ♀
23 28 ☽ △ ♃

30 Saturday
01 43 ☽ ± ☿
07 33 ☽ △ ♅
08 09 ☿ ⊥ ♂
10 35 ☽ △ ☉
11 16 ☿ ✶ ♇
11 22 ☿ △ ♆
15 25 ☽ ⚻ ♀
16 25 ♆ ✶ ♇
22 07 ☽ ⚼ ☿

31 Sunday
03 16 ☽ ± ♀
04 53 ☽ ⚼ ♃
10 56 ☿ ✶ ♄
12 47 ☽ ⚼ ♅
15 23 ☽ □ ♂
16 37 ☽ ∥ ♀
20 49 ☽ ⚼ ♀
21 16 ☽ ∥ ♇
22 15 ☽ ± ♃

AUGUST 1983

Date	Sidereal time h m s	Sun ☉ ° ' "	Moon ☽ ° ' "	Moon ☽ 24.00 ° ' "	Mercury ☿ ° '	Venus ♀ ° '	Mars ♂ ° '	Jupiter ♃ ° '	Saturn ♄ ° '	Uranus ♅ ° '	Neptune ♆ ° '	Pluto ♇ ° '
01	08 38 08	08 ♌ 40 31	02 ♉ 19 10	08 ♉ 43 12	00 ♍ 06	09 ♍ 24	22 ♋ 04	01 ♐ 05	28 ♎ 30	05 ♐ 08	26 ♐ 50	26 ♎ 53
02	08 42 05	09 37 56	15 ♉ 12 27	21 ♉ 47 23	01 40	09 28	22 43	01 06	28 33	05 R 08	26 R 49	26 54
03	08 46 01	10 35 21	28 ♉ 28 23	05 ♊ 15 49	03 11	09 30	23 22	01 07	28 36	05 07	26 48	26 55
04	08 49 58	11 32 48	12 ♊ 09 54	19 ♊ 10 47	04 41	09 R 29	24 01	01 08	28 39	05 07	26 47	26 56
05	08 53 54	12 30 16	26 ♊ 18 25	03 ♋ 32 37	06 09	09 27	24 40	01 09	28 43	05 06	26 46	26 57
06	08 57 51	13 27 45	10 ♋ 53 01	18 ♋ 18 59	07 36	09 21	25 19	01 10	28 46	05 06	26 45	26 58
07	09 01 47	14 25 16	25 ♋ 49 45	03 ♌ 24 18	09 00	09 14	25 59	01 12	28 49	05 05	26 44	26 59
08	09 05 44	15 22 47	11 ♌ 01 29	18 ♌ 39 58	10 23	09 04	26 38	01 14	28 53	05 05	26 43	27 00
09	09 09 40	16 20 20	26 ♌ 18 24	03 ♍ 55 24	11 44	08 52	27 17	01 16	28 57	05 05	26 42	27 01
10	09 13 37	17 17 53	11 ♍ 29 37	18 ♍ 59 49	13 03	08 38	27 55	01 18	29 00	05 05	26 41	27 02
11	09 17 34	18 15 28	26 ♍ 24 56	03 ♎ 44 05	14 20	08 21	28 34	01 20	29 04	05 04	26 40	27 03
12	09 21 30	19 13 03	10 ♎ 56 38	18 ♎ 02 08	15 35	08 02	29 13	01 23	29 08	05 04	26 39	27 04
13	09 25 27	20 10 39	25 ♎ 00 22	01 ♏ 51 18	16 48	07 40	29 ♋ 52	01 25	29 12	05 04	26 38	27 06
14	09 29 23	21 08 16	08 ♏ 35 06	15 ♏ 12 02	17 59	07 17	00 ♌ 31	01 28	29 16	05 D 04	26 38	27 07
15	09 33 20	22 05 55	21 ♏ 42 32	28 ♏ 07 03	19 07	06 51	01 10	01 31	29 20	05 04	26 37	27 08
16	09 37 16	23 03 34	04 ♐ 26 09	10 ♐ 40 25	20 14	06 23	01 48	01 34	29 24	05 04	26 36	27 09
17	09 41 13	24 01 14	16 ♐ 50 25	22 ♐ 56 46	21 18	05 54	02 27	01 38	29 28	05 04	26 35	27 11
18	09 45 09	24 58 55	29 ♐ 00 02	05 ♑ 00 47	22 19	05 23	03 06	01 41	29 33	05 05	26 35	27 12
19	09 49 06	25 56 38	10 ♑ 59 30	16 ♑ 56 42	23 18	04 50	03 44	01 45	29 37	05 05	26 34	27 14
20	09 53 03	26 54 21	22 ♑ 52 49	28 ♑ 48 14	24 14	04 16	04 23	01 49	29 41	05 05	26 33	27 15
21	09 56 59	27 52 06	04 ♒ 43 19	10 ♒ 38 22	25 07	03 41	05 02	01 53	29 46	05 06	26 33	27 16
22	10 00 56	28 49 52	16 ♒ 33 40	22 ♒ 29 28	25 57	03 05	05 40	01 57	29 51	05 06	26 32	27 18
23	10 04 52	29 ♌ 47 39	28 ♒ 25 57	04 ♓ 23 19	26 44	02 28	06 19	02 01	29 ♎ 55	05 06	26 32	27 19
24	10 08 49	00 ♍ 45 28	10 ♓ 21 45	16 ♓ 21 24	27 28	01 51	06 57	02 06	00 ♏ 00	05 07	26 31	27 21
25	10 12 45	01 43 18	22 ♓ 22 27	28 ♓ 25 03	28 07	01 14	07 35	02 10	00 05	05 07	26 31	27 22
26	10 16 42	02 41 10	04 ♈ 29 24	10 ♈ 35 43	28 43	00 37	08 14	02 15	00 10	05 08	26 30	27 24
27	10 20 38	03 39 04	16 ♈ 44 14	22 ♈ 55 13	29 15	00 ♍ 00	08 52	02 20	00 15	05 09	26 30	27 26
28	10 24 35	04 36 59	29 ♈ 08 59	05 ♉ 25 51	29 ♍ 42	29 ♌ 23	09 31	02 26	00 20	05 09	26 30	27 27
29	10 28 32	05 34 56	11 ♉ 46 11	18 ♉ 10 24	00 ♎ 05	28 47	10 09	02 31	00 25	05 10	26 29	27 29
30	10 32 28	06 32 55	24 ♉ 38 54	01 ♊ 12 06	00 23	28 13	10 47	02 36	00 30	05 11	26 29	27 31
31	10 36 25	07 ♍ 30 56	07 ♊ 50 24	14 ♊ 34 11	00 ♎ 36	27 ♌ 39	11 ♌ 25	02 ♐ 42	00 ♏ 35	05 ♐ 12	26 ♐ 29	27 ♎ 32

Date	Moon True ☊ ° '	Moon Mean ☊ ° '	Moon ☽ Latitude ° '
01	24 ♊ 03	22 ♊ 36	04 S 03
02	24 D 03	22 33	03 14
03	24 04	22 30	02 14
04	24 05	22 27	01 S 04
05	24 R 06	22 24	00 N 12
06	24 05	22 21	01 29
07	24 01	22 17	02 42
08	23 56	22 14	03 45
09	23 49	22 11	04 32
10	23 41	22 08	04 59
11	23 33	22 05	05 06
12	23 27	22 02	04 52
13	23 22	21 58	04 21
14	23 19	21 55	03 36
15	23 19	21 52	02 41
16	23 D 19	21 49	01 39
17	23 20	21 46	00 N 35
18	23 R 20	21 42	00 S 30
19	23 18	21 39	01 33
20	23 14	21 36	02 31
21	23 08	21 33	03 22
22	22 59	21 30	04 04
23	22 47	21 27	04 35
24	22 35	21 23	04 55
25	22 23	21 20	05 02
26	22 11	21 17	04 55
27	22 01	21 14	04 34
28	21 54	21 11	04 01
29	21 49	21 08	03 15
30	21 47	21 04	02 18
31	21 ♊ 47	21 ♊ 01	01 S 13

DECLINATIONS

Date	Sun ☉ ° '	Moon ☽ ° '	Mercury ☿ ° '	Venus ♀ ° '	Mars ♂ ° '	Jupiter ♃ ° '	Saturn ♄ ° '	Uranus ♅ ° '	Neptune ♆ ° '	Pluto ♇ ° '
01	18 N 06	08 N 28	12 N 12	04 N 12	22 N 32	19 S 42	08 S 41	21 S 05	22 S 09	05 N 22
02	17 50	13 18	11 32	03 59	22 26	19 42	08 43	21 04	22 09	05 22
03	17 35	17 39	10 52	03 46	22 20	19 43	08 44	21 04	22 09	05 21
04	17 19	21 12	10 12	03 34	22 14	19 43	08 45	21 04	22 09	05 20
05	17 03	23 35	09 32	03 23	22 07	19 43	08 47	21 04	22 09	05 19
06	16 47	24 29	08 52	03 13	22 01	19 44	08 48	21 04	22 09	05 18
07	16 30	23 38	08 12	03 03	21 54	19 45	08 50	21 04	22 09	05 17
08	16 14	21 04	07 32	02 55	21 47	19 45	08 51	21 04	22 09	05 16
09	15 56	17 00	06 53	02 48	21 40	19 46	08 53	21 04	22 09	05 16
10	15 39	11 52	06 14	02 41	21 32	19 46	08 54	21 04	22 09	05 15
11	15 22	06 06	05 36	02 36	21 25	19 47	08 56	21 04	22 09	05 14
12	15 04	00 N 09	04 58	02 31	21 17	19 48	08 57	21 04	22 09	05 13
13	14 46	05 S 38	04 20	02 28	21 09	19 49	08 59	21 04	22 09	05 12
14	14 27	10 57	03 44	02 26	21 02	19 49	09 00	21 04	22 09	05 11
15	14 09	15 36	03 07	02 25	20 53	19 50	09 02	21 04	22 09	05 10
16	13 50	19 24	02 32	02 25	20 45	19 51	09 04	21 04	22 09	05 09
17	13 31	22 13	01 57	02 26	20 37	19 52	09 05	21 04	22 09	05 08
18	13 12	23 57	01 24	02 28	20 28	19 53	09 07	21 04	22 10	05 07
19	12 52	24 32	00 51	02 32	20 20	19 54	09 09	21 04	22 10	05 07
20	12 33	23 59	00 N 19	02 36	20 11	19 55	09 11	21 04	22 10	05 06
21	12 13	22 21	00 S 11	02 42	20 02	19 56	09 12	21 04	22 10	05 05
22	11 53	19 45	00 41	02 48	19 53	19 57	09 14	21 04	22 10	05 04
23	11 33	16 19	01 08	02 56	19 44	19 58	09 16	21 05	22 10	05 03
24	11 12	12 14	01 35	03 04	19 34	19 59	09 18	21 05	22 10	05 02
25	10 52	07 38	02 00	03 13	19 25	20 00	09 20	21 05	22 10	05 01
26	10 31	02 S 43	02 23	03 23	19 15	20 01	09 22	21 05	22 10	05 00
27	10 10	02 N 21	02 44	03 34	19 06	20 02	09 24	21 05	22 10	04 59
28	09 49	07 25	03 03	03 45	18 56	20 04	09 26	21 05	22 10	04 58
29	09 28	12 16	03 20	03 56	18 46	20 05	09 27	21 05	22 10	04 57
30	09 07	16 42	03 35	04 09	18 36	20 06	09 29	21 05	22 10	04 56
31	08 N 45	20 N 25	03 S 47	04 N 21	18 N 26	20 S 07	09 S 31	21 S 06	22 S 10	04 N 55

ZODIAC SIGN ENTRIES

Date	h	m	Planets
01	07	37	☽ ♉
01	10	22	☿ ♍
03	14	43	☽ ♊
05	18	09	☽ ♋
07	18	37	☽ ♌
09	17	49	☽ ♍
11	17	51	☽ ♎
13	16	54	♂ ♌
13	20	44	☽ ♏
16	03	33	☽ ♐
18	13	59	☽ ♑
21	02	25	☽ ♒
23	15	10	☽ ♓
23	17	08	☉ ♍
24	11	54	♄ ♏
26	03	08	☽ ♈
27	11	43	♀ ♌
28	13	38	☽ ♉
29	06	07	☿ ♎
30	21	49	☽ ♊

LATITUDES

Date	Mercury ☿ ° '	Venus ♀ ° '	Mars ♂ ° '	Jupiter ♃ ° '	Saturn ♄ ° '	Uranus ♅ ° '	Neptune ♆ ° '	Pluto ♇ ° '
01	00 N 49	04 S 09	00 N 55	00 N 42	02 N 25	00 N 05	01 N 15	16 N 51
04	00 N 26	04 47	00 56	00 41	02 24	00 05	01 15	16 49
07	00 00	05 27	00 58	00 40	02 23	00 05	01 15	16 47
10	00 S 27	06 05	00 59	00 40	02 22	00 05	01 15	16 46
13	00 57	06 42	01 00	00 39	02 22	00 05	01 15	16 44
16	01 27	07 15	01 01	00 38	02 21	00 05	01 14	16 42
19	01 58	07 44	01 02	00 38	02 20	00 05	01 14	16 41
22	02 29	08 06	01 03	00 37	02 20	00 05	01 14	16 40
25	02 59	08 21	01 05	00 37	02 19	00 05	01 14	16 38
28	03 27	08 28	01 06	00 36	02 19	00 05	01 14	16 37
31	03 S 52	08 S 27	01 N 07	00 N 35	02 N 18	00 N 05	01 N 14	16 N 35

DATA

Julian Date	2445548
Delta T	+53 seconds
Ayanamsa	23° 37' 24"
Synetic vernal point	05° ♓ 29' 36"
True obliquity of ecliptic	23° 26' 30"

MOON'S PHASES, APSIDES AND POSITIONS ☽

Date	h	m	Phase	Longitude ° '	Eclipse Indicator
02	00	52	☾	09 ♉ 11	
08	19	18	●	15 ♌ 40	
15	12	47	☽	22 ♏ 08	
23	14	59	○	29 ♒ 55	
31	11	22	☾	07 ♊ 29	

Day	h	m		
08	19	23	Perigee	
22	09	13	Apogee	
06	12	37	Max dec	24° N 29'
12	12	36	0S	
19	12	16	Max dec	24° S 32'
27	00	55	0N	

LONGITUDES

Date	Chiron ⚷ ° '	Ceres ⚳ ° '	Pallas ⚴ ° '	Juno ⚵ ° '	Vesta ⚶ ° '	Black Moon Lilith ⚸ ° '
01	02 ♊ 11	23 ♒ 37	09 ♑ 30	28 ♈ 13	10 ♊ 09	15 ♒ 09
11	02 ♊ 31	21 ♒ 28	07 ♑ 52	00 ♉ 52	13 ♊ 38	16 ♒ 16
21	02 ♊ 46	19 ♒ 16	06 ♑ 53	03 ♉ 00	16 ♊ 53	17 ♒ 23
31	02 ♊ 54	17 ♒ 15	06 ♑ 33	04 ♉ 34	19 ♊ 53	18 ♒ 30

All ephemeris data is given at 12.00 UT and the Moon's longitude is additionally given for 24.00 UT

ASPECTARIAN

h m	Aspects
01 Monday	
01 36	☽ △ ♆
01 41	☽ ☍ ♇
04 45	☽ ☍ ♄
05 59	☽ ± ♅
07 13	☽ △ ☿
09 40	☽ ⊼ ♃
13 03	☽ # ♄
17 18	☽ ⊼ ♅
02 Tuesday	
00 52	☽ □ ☉
01 20	☽ △ ♀
03 09	☿ □ ♃
03 16	☽ Q ♂
04 08	☽ ∥ ☿
05 45	☽ bQ ♆
07 30	☉ ⊻ ♀
22 13	☽ ± ♆
03 Wednesday	
02 23	☽ ⚹ ♂
09 00	☽ ⊼ ♆
09 13	☽ ⊼ ♇
11 39	☽ ∥ ☉
12 13	☽ Q ☉
12 14	☽ ⊼ ♄
16 41	☽ ☍ ♃
19 43	♀ St R
19 53	☽ ± ♇
21 23	☽ □ ☿
22 53	☽ ± ♄
23 44	☽ ☍ ♅
04 Thursday	
01 10	☽ # ♃
06 17	☽ ∠ ♂
07 22	☽ □ ♀
10 51	☽ ⚹ ☉
11 01	☽ # ♅
11 36	☽ bQ ♇
14 35	☽ bQ ♄
17 39	☉ bQ ♆
18 51	☿ □ ♅
20 07	☽ # ♆
20 29	☽ ∥ ♂
22 31	☽ ⊥ ♂
05 Friday	
04 12	♀ ∠ ♂
05 10	♂ # ♆
07 59	☽ Q ☿
09 08	☽ ⊻ ♂
12 45	☽ ☍ ♆
13 04	☽ △ ♇
13 53	☽ Q ♀
14 08	☽ ∠ ☉
16 01	☽ △ ♄
20 04	☽ ⊼ ♃
06 Saturday	
02 33	☽ ⊼ ♅
05 56	☽ ± ♃
06 02	☽ ⊥ ☉
06 04	☽ ⚹ ☿
09 32	☽ ⚹ ♀
12 21	☽ ± ♅
14 09	☿ # ♄
16 28	☽ ⊻ ☉
20 34	☽ bQ ♃
07 Sunday	
02 51	☽ bQ ♅
08 48	☽ ∠ ☿
09 29	☽ ∠ ♀
12 15	☽ ☌ ♂
13 25	☽ ⊼ ♆
13 50	☽ □ ♇
15 36	☿ ☌ ♀
16 46	☽ □ ♄
20 32	☽ △ ♃
22 55	☽ ± ♆
23 37	☽ ⊥ ♀
08 Monday	
00 30	☽ ⊥ ☿
02 18	☉ Q ♇
02 39	☽ △ ♅
03 35	☽ # ♆
06 26	☽ ∥ ♂
08 58	☽ ⊻ ♀
10 53	☽ ⊻ ☿
12 00	☽ # ♅
13 05	☽ bQ ♆
15 05	♂ ⊼ ♆
18 15	☽ Q ♇
19 18	☽ ☌ ☉
20 39	☽ # ♃
21 14	☽ Q ♄
09 Tuesday	
02 12	♂ □ ♇
12 37	☽ △ ♆
13 07	☽ ⚹ ♇
13 35	☽ ⊻ ♂
16 10	☽ ⚹ ♄
17 16	☿ ∠ ♇
17 39	☽ ∥ ☉
19 49	☽ □ ♃
23 28	☽ ⊥ ♂
10 Wednesday	
01 49	☽ □ ♅
04 08	☉ Q ♄
07 31	☽ ☌ ♀
07 32	☿ ∠ ♂
12 52	☽ ∠ ♇
14 23	☽ ∠ ♂
14 42	☽ ☌ ☿
16 01	☽ ∠ ♄
21 54	☽ ⊻ ☉
11 Thursday	
00 28	☽ # ♄
00 31	☽ Q ♃
03 18	☽ ⊥ ♇
06 32	☽ ⊥ ♄
06 35	☽ Q ♅
06 46	☿ ∠ ♄
08 15	☽ ⊥ ☉
12 24	☽ □ ♆
13 03	☽ ⊻ ♇
14 16	☽ ∥ ☿
14 24	♄ ⊥ ♅
15 31	☽ ∥ ♇
15 41	☽ ⚹ ♂
16 21	☽ ⊻ ♄
20 04	☽ ⚹ ♃
12 Friday	
00 00	☽ ∠ ☉
02 11	☿ ∥ ♇
02 13	☽ ⚹ ♅
02 18	☽ ∥ ♀
07 14	☽ ⊻ ♀
08 18	♂ □ ♄
12 29	☽ Q ♂
17 04	☽ ⊥ ♀
18 15	☽ Q ♆
20 34	☽ ⊻ ☿
21 12	☽ ∠ ♃
22 49	☽ # ♀
13 Saturday	
03 03	☽ ⚹ ☉
03 29	☽ ∠ ♅
07 03	☽ # ☿
07 49	☽ ⊥ ☿
08 04	☽ ∠ ♀
10 10	☽ # ♇
12 43	☽ ⊥ ♃
14 50	☽ ⚹ ♆
15 39	☽ ☌ ♇
19 06	☽ ⊥ ♅
19 21	☽ ☌ ♄
20 55	☽ □ ♂
23 16	☽ ⊻ ♃
14 Sunday	
01 02	☽ ∠ ☿
01 32	☽ Q ☉
02 52	☽ ∥ ♄
04 38	♂ # ♅
05 42	☽ ⊻ ♅
07 12	♅ St D
09 43	☽ ⚹ ♀
17 29	☽ ∠ ♆
15 Monday	
04 32	☽ # ☉
05 06	♀ ⊥ ♂
06 45	☽ ⚹ ☿
06 53	☽ Q ♀
09 58	☽ ⊥ ♆
12 47	☽ □ ☉
20 53	☿ Q ♃
21 10	☽ ⊻ ♆
22 10	☽ ⊻ ♇
16 Tuesday	
02 22	☽ ⊻ ♄
02 26	♂ △ ♃
06 31	☽ ☌ ♃
06 43	☽ △ ♂
07 23	☽ Q ☿
09 33	☽ ⊥ ♇
13 13	☽ ☌ ♅
13 51	☽ ⊥ ♄
15 19	☽ ∥ ♃
15 36	☽ □ ♀
16 44	☿ ∥ ♀
21 57	☽ # ♂
17 Wednesday	
01 06	☽ ∥ ♅
02 54	☽ ∠ ♇
07 21	☽ ∠ ♄
09 19	☿ ⊥ ♇
11 24	☽ ∥ ♆
13 16	☽ bQ ♂
17 05	♂ ± ♆
21 33	☽ □ ☿
18 Thursday	
03 20	☽ △ ☉
07 12	☽ ☌ ♆
08 00	☽ ± ♂
08 25	☽ ⚹ ♇
13 05	☽ ⚹ ♄
17 23	☽ ⊻ ♃
20 38	☽ ⊼ ♂
19 Friday	
00 08	☽ ⊻ ♅
00 11	☽ △ ♀
01 20	♀ □ ♅
05 27	☽ ⊥ ♃
06 33	☿ Q ♅
08 27	☽ Q ♇
11 54	☽ bQ ☉
12 11	☽ ⊥ ♅
13 16	☽ Q ♄
20 41	☿ ⊥ ♄
23 40	☽ ∠ ♃
20 Saturday	
03 25	☉ △ ♆
05 02	☽ bQ ♀
06 21	☽ ∠ ♅
07 39	☽ ± ☉
09 46	♀ ⊻ ♂
14 58	☽ △ ☿
19 27	☽ ⊻ ♆
20 47	☉ ⚹ ♇
20 52	☽ □ ♇
20 52	☽ ⊼ ☉
22 24	☽ ± ♀
21 Sunday	
01 53	☽ □ ♄
06 12	☽ ⚹ ♃
07 36	☽ ⊥ ♆
10 00	☽ ⊼ ♀
12 39	☽ ☍ ♂
12 45	☽ ⚹ ♅
14 03	☽ ∥ ♆
14 32	♂ △ ♅
23 48	☽ bQ ☿
22 Monday	
00 47	☽ ∥ ♅
01 50	☽ ∠ ♆
02 47	♂ # ♃
06 41	☽ Q ♃
10 24	☽ ∥ ♃
10 51	☽ # ♂
13 05	☽ Q ♅
19 22	☽ ± ☿
23 Tuesday	
05 36	☿ □ ♆
08 10	☽ ⚹ ♆
08 21	☽ ⊼ ☿
09 45	☽ △ ♇
14 59	☽ ☍ ☉
15 01	☽ △ ♄
15 27	☉ ⚹ ♄
19 17	☽ □ ♃
19 45	☽ ☍ ♀
24 Wednesday	
01 27	☽ □ ♅
03 38	♀ □ ♃
04 46	☽ ⊼ ♂
08 04	☿ ⊻ ♇
08 19	☽ Q ♆
15 59	☽ bQ ♇
17 29	☽ ± ♂
18 00	☽ # ☉
21 21	☽ bQ ♄
25 Thursday	
03 28	☽ ∥ ♄
04 35	☉ ☌ ♀
10 00	☽ ± ♇
12 27	☽ bQ ♂
15 25	☽ ± ♄
20 13	☽ □ ♆
21 57	☽ ⊼ ♇
26 Friday	
00 01	☽ ☍ ☿
00 18	☉ □ ♃
01 01	☽ # ♇
03 24	☽ ⊼ ♄
04 43	☽ ⊼ ♀
07 34	☽ △ ♃
08 08	☽ ⊼ ☉
08 57	☽ # ♀
13 16	☽ △ ♅
13 32	☽ ∥ ☿
15 58	☽ ± ♀
19 46	☽ △ ♂
20 58	☽ ± ☉
27 Saturday	
03 20	♀ ⚹ ♄
08 46	☽ bQ ♀
13 11	☽ bQ ♃
13 55	☽ # ☿
16 02	☽ bQ ☉
17 53	☽ ∥ ♀
18 38	☽ bQ ♅
28 Sunday	
00 21	☽ ∥ ♇
04 34	☿ ⊻ ♀
06 44	☽ ± ♃
06 54	☽ △ ♆
08 44	☽ ☍ ♇
12 01	☽ ± ♅
12 26	☽ △ ♀
13 06	☽ ⊼ ☿
14 17	☽ ☍ ♄
18 19	☽ ⊼ ♃
21 49	☽ # ♄
22 54	☽ ∥ ☉
23 19	☽ △ ☉
23 29	☽ ⊼ ♅
29 Monday	
00 56	☽ ± ☿
01 36	☉ □ ♅
08 47	☽ □ ♂
11 28	☽ bQ ♆
12 29	☉ # ♄
18 24	☽ bQ ☿
30 Tuesday	
04 18	☽ ± ♆
15 23	☽ ⊼ ♆
17 16	☽ ⊼ ♇
18 16	☽ □ ♀
19 59	☽ Q ♂
22 42	☽ △ ☿
22 48	☽ ⊼ ♄
23 05	☽ ∥ ♂
31 Wednesday	
02 39	☽ ☍ ♃
04 14	☽ ± ♇
07 14	☽ ☍ ♅
09 02	☿ ⊻ ♄
09 44	☽ ± ♄
09 53	☽ # ♃
11 22	☽ □ ☉
14 07	♂ bQ ♆
16 46	♀ ⚹ ♇
17 09	☽ # ♅
18 44	☽ ⚹ ♂
20 25	☽ bQ ♇

SEPTEMBER 1983

LONGITUDES

Date	Sidereal time h m s	Sun ☉ ° ' "	Moon ☽ ° ' "	Moon ☽ 24.00 ° ' "	Mercury ☿ ° '	Venus ♀ ° '	Mars ♂ ° '	Jupiter ♃ ° '	Saturn ♄ ° '	Uranus ♅ ° '	Neptune ♆ ° '	Pluto ♇ ° '
01	10 40 21	08 ♍ 28 58	21 ♊ 23 46	28 ♊ 19 24	00 ♎ 43	27 ♌ 07	12 ♌ 04	02 ♐ 48	00 ♏ 40	05 ♐ 13	26 ♐ 29	27 ♎ 34
02	10 44 18	09 27 03	05 ♋ 21 14	12 ♋ 29 16	00 R 45	26 R 36	12 42	02 54	00 46	05 14	26 R 28	27 36
03	10 48 14	10 25 10	19 ♋ 43 20	27 ♋ 03 07	00 40	26 08	13 20	03 00	00 51	05 15	26 28	27 38
04	10 52 11	11 23 18	04 ♌ 28 03	11 ♌ 57 22	00 30	25 41	13 58	03 06	00 57	05 16	26 28	27 39
05	10 56 07	12 21 29	19 ♌ 30 08	27 ♌ 05 13	00 ♎ 13	25 16	14 36	03 12	01 02	05 17	26 28	27 41
06	11 00 04	13 19 41	04 ♍ 41 20	12 ♍ 17 08	29 ♍ 50	24 53	15 14	03 19	01 08	05 18	26 28	27 43
07	11 04 01	14 17 55	19 ♍ 51 15	27 ♍ 22 21	29 21	24 32	15 52	03 26	01 13	05 19	26 28	27 45
08	11 07 57	15 16 11	04 ♎ 49 14	12 ♎ 10 48	28 45	24 14	16 30	03 32	01 19	05 21	26 D 28	27 47
09	11 11 54	16 14 28	19 ♎ 26 13	26 ♎ 34 50	28 03	23 58	17 08	03 39	01 25	05 22	26 28	27 49
10	11 15 50	17 12 47	03 ♏ 36 12	10 ♏ 30 09	27 16	23 44	17 46	03 46	01 31	05 23	26 28	27 51
11	11 19 47	18 11 08	17 ♏ 16 41	23 ♏ 55 58	26 24	23 33	18 24	03 54	01 37	05 25	26 28	27 53
12	11 23 43	19 09 30	00 ♐ 28 19	06 ♐ 54 12	25 27	23 24	19 02	04 01	01 43	05 26	26 28	27 55
13	11 27 40	20 07 54	13 ♐ 14 08	19 ♐ 28 44	24 28	23 18	19 40	04 09	01 48	05 28	26 28	27 57
14	11 31 36	21 06 20	25 ♐ 38 39	01 ♑ 44 31	23 26	23 14	20 18	04 16	01 55	05 29	26 28	27 59
15	11 35 33	22 04 47	07 ♑ 47 00	13 ♑ 46 46	22 24	23 12	20 55	04 24	02 01	05 31	26 29	28 01
16	11 39 30	23 03 15	19 ♑ 44 27	25 ♑ 40 38	21 23	23 D 13	21 33	04 32	02 07	05 32	26 29	28 03
17	11 43 26	24 01 46	01 ♒ 35 54	07 ♒ 30 45	20 23	23 15	22 11	04 40	02 13	05 34	26 29	28 05
18	11 47 23	25 00 18	13 ♒ 25 40	19 ♒ 21 03	19 28	23 21	22 49	04 48	02 19	05 36	26 29	28 07
19	11 51 19	25 58 52	25 ♒ 17 16	01 ♓ 14 39	18 38	23 28	23 26	04 56	02 25	05 38	26 30	28 09
20	11 55 16	26 57 27	07 ♓ 13 27	13 ♓ 13 53	17 54	23 38	24 04	05 05	02 32	05 39	26 30	28 11
21	11 59 12	27 56 04	19 ♓ 16 07	25 ♓ 20 17	17 18	23 50	24 41	05 13	02 38	05 41	26 31	28 13
22	12 03 09	28 54 43	01 ♈ 26 30	07 ♈ 34 52	16 51	24 04	25 19	05 22	02 44	05 43	26 31	28 15
23	12 07 05	29 ♍ 53 25	13 ♈ 45 26	19 ♈ 58 17	16 32	24 20	25 57	05 31	02 51	05 45	26 32	28 18
24	12 11 02	00 ♎ 52 08	26 ♈ 13 29	02 ♉ 31 09	16 24	24 38	26 34	05 40	02 57	05 47	26 32	28 20
25	12 14 59	01 50 53	08 ♉ 51 23	15 ♉ 14 19	16 D 25	24 57	27 11	05 49	03 04	05 49	26 33	28 22
26	12 18 55	02 49 41	21 ♉ 40 09	28 ♉ 09 04	16 36	25 19	27 49	05 58	03 10	05 51	26 33	28 24
27	12 22 52	03 48 31	04 ♊ 41 17	11 ♊ 17 05	16 57	25 43	28 26	06 07	03 17	05 54	26 34	28 26
28	12 26 48	04 47 23	17 ♊ 56 44	24 ♊ 40 28	17 28	26 08	29 04	06 16	03 24	05 56	26 34	28 29
29	12 30 45	05 46 17	01 ♋ 28 33	08 ♋ 21 12	18 08	26 35	29 ♌ 41	06 26	03 30	05 58	26 35	28 31
30	12 34 41	06 ♎ 45 14	15 ♋ 18 33	22 ♋ 20 41	18 ♍ 57	27 ♌ 03	00 ♍ 18	06 ♐ 36	03 ♏ 37	06 ♐ 00	26 ♐ 36	28 ♎ 33

Date	Moon True ☊ ° '	Moon Mean ☊ ° '	Moon ☽ Latitude ° '
01	21 ♊ 47	20 ♊ 58	00 S 02
02	21 R 46	20 55	01 N 11
03	21 44	20 52	02 22
04	21 39	20 48	03 25
05	21 32	20 45	04 15
06	21 22	20 42	04 48
07	21 11	20 39	05 00
08	21 00	20 36	04 52
09	20 50	20 33	04 24
10	20 43	20 29	03 41
11	20 38	20 26	02 46
12	20 35	20 23	01 44
13	20 D 35	20 20	00 N 39
14	20 R 35	20 17	00 S 27
15	20 34	20 14	01 30
16	20 32	20 10	02 28
17	20 28	20 07	03 19
18	20 20	20 04	04 01
19	20 10	20 01	04 33
20	19 58	19 58	04 53
21	19 45	19 54	05 00
22	19 31	19 51	04 53
23	19 18	19 48	04 33
24	19 07	19 45	04 00
25	18 59	19 42	03 15
26	18 53	19 39	02 18
27	18 51	19 35	01 14
28	18 50	19 32	00 S 05
29	18 D 50	19 29	01 N 07
30	18 ♊ 50	19 ♊ 26	02 N 16

DECLINATIONS

Date	Sun ☉ ° '	Moon ☽ ° '	Mercury ☿ ° '	Venus ♀ ° '	Mars ♂ ° '	Jupiter ♃ ° '	Saturn ♄ ° '	Uranus ♅ ° '	Neptune ♆ ° '	Pluto ♇ ° '
01	08 N 23	23 N 08	03 S 56	04 N 34	18 N 15	20 S 09	09 S 33	21 S 06	22 S 10	04 N 54
02	08 02	24 31	04 02	04 47	18 05	20 10	09 35	21 06	22 10	04 53
03	07 40	24 20	04 05	05 00	17 54	20 12	09 38	21 06	22 10	04 52
04	07 18	22 28	04 05	05 13	17 44	20 13	09 40	21 06	22 10	04 51
05	06 55	19 01	04 01	05 26	17 33	20 14	09 42	21 07	22 10	04 50
06	06 33	14 15	03 54	05 39	17 22	20 16	09 44	21 07	22 10	04 49
07	06 11	08 37	03 42	05 52	17 11	20 17	09 46	21 07	22 10	04 48
08	05 48	02 N 33	03 27	06 04	17 00	20 19	09 48	21 07	22 10	04 47
09	05 26	03 S 32	03 08	06 17	16 49	20 20	09 50	21 08	22 10	04 46
10	05 03	09 15	02 45	06 29	16 37	20 22	09 52	21 08	22 10	04 45
11	04 40	14 20	02 18	06 40	16 26	20 23	09 54	21 08	22 10	04 44
12	04 17	18 33	01 48	06 51	16 15	20 25	09 57	21 08	22 11	04 43
13	03 55	21 45	01 15	07 02	16 03	20 26	09 59	21 09	22 11	04 42
14	03 32	23 49	00 40	07 12	15 51	20 28	10 01	21 09	22 11	04 41
15	03 09	24 42	00 S 02	07 22	15 40	20 30	10 03	21 09	22 11	04 40
16	02 45	24 25	00 N 36	07 31	15 28	20 31	10 05	21 09	22 11	04 39
17	02 22	23 02	01 15	07 40	15 16	20 33	10 08	21 10	22 11	04 38
18	01 59	20 38	01 54	07 48	15 04	20 35	10 10	21 10	22 11	04 37
19	01 36	17 22	02 31	07 55	14 52	20 36	10 12	21 10	22 11	04 36
20	01 13	13 23	03 06	08 02	14 40	20 38	10 15	21 11	22 11	04 35
21	00 49	08 50	03 39	08 08	14 27	20 40	10 17	21 11	22 11	04 34
22	00 26	03 S 55	04 08	08 14	14 15	20 41	10 19	21 11	22 11	04 33
23	00 N 03	01 N 14	04 33	08 19	14 03	20 43	10 21	21 12	22 11	04 32
24	00 S 21	06 24	04 54	08 23	13 50	20 45	10 24	21 12	22 11	04 31
25	00 44	11 23	05 10	08 26	13 38	20 46	10 26	21 13	22 11	04 30
26	01 07	15 57	05 22	08 29	13 25	20 48	10 28	21 13	22 11	04 29
27	01 31	19 52	05 29	08 31	13 12	20 50	10 31	21 13	22 11	04 28
28	01 54	22 49	05 31	08 33	13 00	20 52	10 33	21 14	22 11	04 27
29	02 18	24 33	05 28	08 34	12 47	20 54	10 35	21 14	22 12	04 26
30	02 S 41	24 N 48	05 N 20	08 N 34	12 N 34	20 S 55	10 S 38	21 S 15	22 S 12	04 N 25

ZODIAC SIGN ENTRIES

Date	h	m	Planets
02	02	53	☽ ♋
04	04	47	☽ ♌
06	02	30	☿ ♍
06	04	36	☽ ♍
08	04	13	☽ ♎
10	05	49	☽ ♏
12	11	08	☽ ♐
14	20	34	☽ ♑
17	08	46	☽ ♒
19	21	30	☽ ♓
22	09	10	☽ ♈
23	14	42	☉ ♎
24	19	12	☽ ♉
27	03	24	☽ ♊
29	09	24	☽ ♋
30	00	12	♂ ♍

LATITUDES

Date	Mercury ☿ ° '	Venus ♀ ° '	Mars ♂ ° '	Jupiter ♃ ° '	Saturn ♄ ° '	Uranus ♅ ° '	Neptune ♆ ° '	Pluto ♇ ° '
01	03 S 58	08 S 24	01 N 07	00 N 35	02 N 18	00 N 05	01 N 14	16 N 35
04	04 14	08 13	01 08	00 35	02 17	00 05	01 14	16 34
07	04 19	07 55	01 09	00 34	02 17	00 05	01 13	16 33
10	04 11	07 33	01 10	00 33	02 16	00 04	01 13	16 31
13	03 46	07 06	01 11	00 33	02 16	00 04	01 13	16 30
16	03 04	06 37	01 12	00 32	02 15	00 04	01 13	16 29
19	02 09	06 07	01 13	00 32	02 15	00 04	01 13	16 28
22	01 09	05 35	01 14	00 31	02 14	00 04	01 13	16 28
25	00 S 12	05 03	01 15	00 31	02 14	00 04	01 13	16 27
28	00 N 36	04 31	01 16	00 30	02 14	00 04	01 12	16 26
31	01 N 14	04 S 00	01 N 17	00 N 30	02 N 13	00 N 04	01 N 12	16 N 25

DATA

Julian Date	2445579
Delta T	+53 seconds
Ayanamsa	23° 37' 28"
Synetic vernal point	05° ♓ 29' 32"
True obliquity of ecliptic	23° 26' 31"

LONGITUDES

Date	Chiron ⚷ ° '	Ceres ⚳ ° '	Pallas ⚴ ° '	Juno ⚵ ° '	Vesta ⚶ ° '	Black Moon Lilith ⚸ ° '
01	02 ♊ 54	17 ♒ 04	06 ♑ 33	04 ♉ 41	20 ♊ 10	18 ♒ 37
11	02 ♊ 56	15 ♒ 27	06 ♑ 54	05 ♉ 27	22 ♊ 48	19 ♒ 44
21	02 ♊ 51	14 ♒ 22	07 ♑ 47	05 ♉ 25	25 ♊ 02	20 ♒ 51
31	02 ♊ 39	13 ♒ 52	09 ♑ 10	04 ♉ 34	26 ♊ 49	21 ♒ 58

MOON'S PHASES, APSIDES AND POSITIONS ☽

Date	h	m	Phase	Longitude ° '	Eclipse Indicator
07	02	35	●	13 ♍ 55	
14	02	24	☽	20 ♐ 43	
22	06	36	○	28 ♓ 42	
29	20	05	☾	06 ♋ 06	

Day	h	m		
06	05	03	Perigee	
18	16	37	Apogee	
02	21	22	Max dec	24° N 39'
08	21	59	OS	
15	18	07	Max dec	24° S 45'
23	06	19	ON	
30	04	08	Max dec	24° N 54'

ASPECTARIAN

h m	Aspects
01 Thursday	
01 23	☽ Q ♀
01 53	☽ bQ ♄
02 14	☽ # ♆
20 49	☽ ☍ ♆
21 29	☽ Q ☉
21 34	☽ ⚹ ♀
22 18	☽ ∠ ♂
22 43	☽ △ ♇
02 Friday	
04 08	☽ △ ♄
04 09	☽ □ ☿
06 41	☿ St R
07 48	☽ ⊼ ♃
07 53	☿ ⊻ ♄
11 47	☽ ⊼ ♅
14 23	☽ ⊥ ♂
18 01	☽ ± ♃
18 39	♀ △ ♆
19 25	☽ ⚹ ☉
21 54	☽ ± ♅
22 11	☽ ∠ ♀
22 33	♀ ∥ ♇
03 Saturday	
00 55	☽ ⊻ ♂
09 08	☽ bQ ♃
10 17	☽ Q ☿
12 39	☽ ⊥ ♀
12 52	☽ bQ ♅
22 00	☽ ∠ ☉
22 11	☽ ⊻ ♀
23 03	☽ ⊼ ♆
04 Sunday	
00 58	☽ □ ♇
05 41	☽ ⚹ ☿
06 17	☽ □ ♄
08 47	☽ ± ♆
09 47	☽ △ ♃
13 17	☽ △ ♅
13 35	☽ ⊥ ☉
14 33	☽ # ♆
22 42	☽ # ♅
23 13	☽ bQ ♆
23 52	☽ ⊻ ☉
05 Monday	
03 53	☽ ☌ ♂
04 36	☽ # ♃
05 20	☽ ∠ ☿
05 56	☽ Q ♇
11 15	☽ Q ♄
19 18	☽ ⊥ ☿
20 21	☽ ∥ ♂
20 24	☉ ∠ ♇
20 53	☽ ☌ ♀
23 01	☽ △ ♆
06 Tuesday	
00 58	☽ ⚹ ♇
02 48	☿ ∠ ♂
04 33	☽ ⊻ ☿
06 21	☽ ⚹ ♄
09 49	☽ □ ♃
12 58	☽ □ ♅
07 Wednesday	
00 43	☽ ∠ ♇
02 35	☽ ☌ ☉
05 24	☽ ⊻ ♂
06 12	☽ ∠ ♄
07 05	♂ Q ♇
07 21	☽ # ♄
14 31	☽ Q ♃
15 01	☽ ⊥ ♇
15 21	☽ ⊥ ♂
17 32	☽ Q ♅
19 19	☽ ⊻ ♀
20 37	☽ ⊥ ♄
22 23	☽ ∥ ☉
22 33	☽ □ ♆
22 38	☽ ∥ ♀
08 Thursday	
00 38	☽ ⊻ ♇
01 02	☉ ∥ ♀
02 36	☽ ☌ ☿
03 14	☽ ∥ ♇
04 44	☽ ⊥ ♀
06 18	☽ ⊻ ♄
06 24	☽ ∠ ♂
08 19	☽ # ☿
09 55	☽ ⚹ ♃
11 01	♆ St D
12 51	☽ ⚹ ♅
19 03	☽ ∠ ♀
09 Friday	
03 45	☽ Q ♆
06 19	☽ ⊻ ☉
08 00	☽ ⚹ ♂
10 30	☽ ∥ ☿
10 41	☽ ∠ ♃
13 33	☽ ∠ ♅
16 46	☉ ∠ ♄
17 01	☽ # ♇
17 02	☽ ⊥ ☉
19 15	☿ ⊻ ♇
19 17	☽ # ☉
19 28	☽ ⚹ ♀
23 40	☽ # ♀
23 48	☽ ⚹ ♆
10 Saturday	
01 45	☽ ⊻ ☿
01 56	☽ ⊥ ♃
02 07	☽ ☌ ♇
04 45	☽ ⊥ ♅
05 07	☽ Q ♂
08 23	☽ ☌ ♄
09 26	☽ ∠ ☉
11 27	☽ ⊥ ☿
12 18	☽ ⊻ ♃
14 46	☽ ∥ ♄
15 05	☽ ⊻ ♅
15 38	☽ Q ♀
11 Sunday	
01 41	☽ ∠ ♆
02 12	☽ ∠ ☿
07 02	♃ ⊥ ♇
08 01	☉ ∥ ♇
10 03	☿ □ ♆
13 45	☽ ⚹ ☉
14 07	☽ □ ♂
17 43	☽ ⊥ ♆
22 47	☽ # ♂
23 10	☽ □ ♀
12 Monday	
03 16	☉ ⊻ ♂
03 24	☽ ⚹ ☿
04 38	☽ ⊻ ♆
06 12	☿ ⊥ ♄
07 16	☽ ⊻ ♇
13 22	☽ Q ☉
14 19	☽ ⊻ ♄
18 15	☿ ⊥ ♂
18 24	☽ ⊥ ♇
18 39	☽ ☌ ♃
21 16	☽ ☌ ♅
13 Tuesday	
00 06	☽ Q ☿
01 04	☽ ∥ ♃
01 37	☽ ⊥ ♄
06 44	☽ ∥ ♅
11 26	☽ ∠ ♇
16 01	☽ ∥ ♆
18 30	♂ Q ♄
18 54	☽ ∠ ♄
14 Wednesday	
01 01	☽ △ ♂
02 24	☽ □ ☉
07 18	☽ △ ♀
08 01	☽ □ ☿
10 52	☿ Q ♅
13 37	☽ ☌ ♆
16 35	☽ ⚹ ♇
17 04	☿ ⊻ ♀
15 Thursday	
00 26	☽ ⚹ ♄
05 12	☽ ⊻ ♃
07 28	☽ ⊻ ♅
08 06	☽ bQ ♂
10 15	☉ ⊥ ♇
11 58	☿ Q ♃
12 50	☽ bQ ♀
15 49	☉ ☌ ☿
16 28	☽ Q ♇
17 17	☽ ⊥ ♃
17 22	♀ St D
19 28	☽ ⊥ ♅
20 45	☿ ⊥ ♇
21 09	☉ Q ♃
16 Friday	
00 34	☽ Q ♄
03 06	☽ ± ♂
06 53	☽ ± ♀
09 24	☿ ⊻ ♂
11 35	☽ ∠ ♃
13 37	☽ ∠ ♅
15 03	☽ △ ☿
15 52	☽ ⊼ ♂
15 56	☉ ⊻ ♀
19 02	☽ ⊼ ♀
19 18	☽ △ ☉
17 Saturday	
00 19	☉ Q ♅
01 38	☽ ⊻ ♆
04 51	☽ □ ♇
13 16	☽ □ ♄
13 48	☽ ⊥ ♆
18 18	☽ ⚹ ♃
19 07	☽ bQ ☿
20 04	☽ ⚹ ♅
21 39	☽ ∥ ♆
18 Sunday	
04 26	☽ bQ ☉
07 23	☽ ∥ ♅
08 04	☽ ∠ ♆
12 05	☽ ± ☿
12 29	☽ ∥ ♃
14 06	☉ ∥ ☿
18 55	☽ Q ♃
20 28	☽ Q ♅
23 25	☽ ⊼ ☿
19 Monday	
00 19	☽ ± ☉
08 03	☽ ☍ ♂
08 17	☽ ☍ ♀
13 31	☽ ⊼ ☉
13 39	♀ ☌ ♂
14 26	☽ ⚹ ♆
17 47	☽ △ ♇
20 Tuesday	
00 10	☉ ⊥ ♄
00 46	☉ □ ♆
02 30	☽ △ ♄
04 19	☽ # ♂
07 39	☽ □ ♃
08 51	☽ □ ♅
14 34	☽ Q ♆
23 57	☽ bQ ♇
21 Wednesday	
00 11	☿ ∠ ♄
04 43	☽ ∥ ♄
08 16	☽ ☍ ☿
08 44	☽ bQ ♄
15 27	☽ # ♀
17 52	☽ ± ♇
19 19	☉ ⊻ ♇
21 11	☽ ⊼ ♀
23 18	☽ ⊼ ♂
22 Thursday	
02 19	☽ □ ♆
02 41	☽ ± ♄
05 44	☽ ⊼ ♇
06 36	☽ ☍ ☉
08 58	☽ # ♇
09 14	☽ ± ♀
11 03	☽ # ☿
11 44	☽ ± ♂
14 34	☽ ⊼ ♄
19 46	☽ △ ♃
20 24	☽ △ ♅
23 Friday	
03 12	☽ bQ ♀
05 38	☽ # ☉
06 15	☽ bQ ♂
06 54	☽ ∥ ☉
10 53	☿ ∥ ♇
17 17	☽ ⊼ ☿
24 Saturday	
01 12	☽ bQ ♃
01 33	☽ bQ ♅
03 16	☽ ∥ ♇
03 18	☉ ⊥ ♀
04 34	☽ ∥ ☿
04 42	☽ ± ☿
08 52	☽ △ ♀
10 43	♂ △ ♆
12 35	☽ △ ♆
12 41	☽ △ ♂
16 02	☽ ☍ ♇
18 39	☽ ± ♃
18 49	☽ ± ♅
20 49	☿ St D
21 32	☽ ∥ ♀
21 36	☽ ⊼ ☉
21 50	☽ bQ ☿
25 Sunday	
00 56	☽ ☍ ♄
06 11	☽ ⊼ ♃
06 15	☽ ⊼ ♅
07 18	☽ # ♄
09 56	☽ ± ☉
13 54	♃ ☌ ♅
17 04	☽ bQ ♆
22 57	☽ ∥ ♂
26 Monday	
02 23	☽ △ ☿
04 15	☽ bQ ☉
09 55	☽ ± ♆
18 58	☽ □ ♀
21 03	☽ ⊼ ♆
21 31	☉ ⊻ ♄
23 57	☽ □ ♂
27 Tuesday	
00 30	☽ ⊼ ♇
09 24	☽ ⊼ ♄
10 16	☽ △ ☉
11 33	☽ ± ♇
12 06	♂ ⚹ ♇
14 12	☽ ☍ ♅
14 39	☽ ☍ ♃
19 03	☽ # ♃
20 26	☽ ± ♄
22 00	☽ # ♅
28 Wednesday	
03 57	☽ bQ ♇
04 55	☽ Q ♀
06 04	☽ # ♆
10 20	☽ Q ♂
11 07	☽ □ ☿
12 49	☽ bQ ♄
29 Thursday	
03 04	☽ ⚹ ♀
03 23	☽ ☍ ♆
06 10	☉ ⊥ ♂
06 47	☽ △ ♇
08 42	☽ ⚹ ♂
12 15	♀ △ ♆
15 35	☽ △ ♄
16 57	☉ ⚹ ♅
19 52	☽ ⊼ ♅
20 05	☽ □ ☉
20 38	☽ Q ☿
20 46	☽ ⊼ ♃
30 Friday	
01 14	☿ ∠ ♄
06 12	☽ ∠ ♀
06 18	☽ ± ♅
07 16	☉ ⚹ ♃
07 16	☽ ± ♃
12 00	☽ ∠ ♂
18 39	☽ ⚹ ☿
21 45	☽ bQ ♅
22 10	☽ ⊥ ♀
22 51	☽ bQ ♃

All ephemeris data is given at 12.00 UT and the Moon's longitude is additionally given for 24.00 UT

OCTOBER 1983

LONGITUDES

Date	Sidereal time h m s	Sun ☉ ° ' "	Moon ☽ ° ' "	Moon ☽ 24.00 ° ' "	Mercury ☿ ° '	Venus ♀ ° '	Mars ♂ ° '	Jupiter ♃ ° '	Saturn ♄ ° '	Uranus ♅ ° '	Neptune ♆ ° '	Pluto ♇ ° '
01	12 38 38	07 ♎ 44 13	29 ♋ 27 33	06 ♌ 38 59	19 ♍ 53	27 ♌ 33	00 ♍ 56	06 ♐ 45	03 ♏ 44	06 ♐ 03	26 ♐ 37	28 ♎ 36
02	12 42 34	08 43 15	13 ♌ 54 40	21 ♌ 14 06	20 57	28 05	01 33	06 55	03 50	06 05	26 37	28 38
03	12 46 31	09 42 18	28 ♌ 36 40	06 ♍ 01 32	22 07	28 38	02 10	07 05	03 57	06 07	26 38	28 40
04	12 50 28	10 41 24	13 ♍ 27 47	20 ♍ 54 22	23 24	29 13	02 47	07 15	04 04	06 10	26 39	28 42
05	12 54 24	11 40 32	28 ♍ 20 09	05 ♎ 44 02	24 45	29 ♌ 48	03 24	07 25	04 11	06 12	26 40	28 45
06	12 58 21	12 39 42	13 ♎ 04 55	20 ♎ 21 48	26 11	00 ♍ 26	04 02	07 35	04 18	06 15	26 41	28 47
07	13 02 17	13 38 54	27 ♎ 33 47	04 ♏ 40 09	27 41	01 04	04 39	07 45	04 25	06 17	26 42	28 49
08	13 06 14	14 38 09	11 ♏ 40 22	18 ♏ 34 03	29 ♍ 14	01 43	05 16	07 56	04 32	06 20	26 43	28 52
09	13 10 10	15 37 25	25 ♏ 21 02	02 ♐ 01 18	00 ♎ 49	02 24	05 53	08 06	04 39	06 23	26 44	28 54
10	13 14 07	16 36 43	08 ♐ 35 01	15 ♐ 02 26	02 27	03 06	06 30	08 17	04 46	06 25	26 45	28 57
11	13 18 03	17 36 03	21 ♐ 23 59	27 ♐ 40 09	04 07	03 49	07 07	08 28	04 53	06 28	26 46	28 59
12	13 22 00	18 35 25	03 ♑ 51 28	09 ♑ 58 34	05 47	04 32	07 44	08 38	05 00	06 31	26 47	29 01
13	13 25 57	19 34 48	16 ♑ 02 05	22 ♑ 02 40	07 29	05 17	08 20	08 49	05 07	06 33	26 48	29 04
14	13 29 53	20 34 13	28 ♑ 00 59	03 ♒ 57 41	09 12	06 03	08 57	09 00	05 14	06 36	26 49	29 06
15	13 33 50	21 33 40	09 ♒ 53 24	15 ♒ 48 45	10 55	06 50	09 34	09 11	05 21	06 39	26 50	29 09
16	13 37 46	22 33 09	21 ♒ 44 17	27 ♒ 40 34	12 39	07 38	10 11	09 22	05 28	06 42	26 51	29 11
17	13 41 43	23 32 40	03 ♓ 38 03	09 ♓ 37 11	14 22	08 26	10 48	09 33	05 35	06 45	26 53	29 13
18	13 45 39	24 32 12	15 ♓ 38 20	21 ♓ 41 51	16 06	09 15	11 24	09 45	05 42	06 48	26 54	29 16
19	13 49 36	25 31 46	27 ♓ 47 57	03 ♈ 56 53	17 50	10 05	12 01	09 56	05 49	06 51	26 55	29 18
20	13 53 32	26 31 22	10 ♈ 08 46	16 ♈ 23 42	19 33	10 56	12 38	10 08	05 57	06 54	26 56	29 21
21	13 57 29	27 31 00	22 ♈ 41 45	29 ♈ 02 54	21 16	11 48	13 14	10 19	06 04	06 57	26 58	29 23
22	14 01 26	28 30 40	05 ♉ 27 08	11 ♉ 54 23	22 58	12 40	13 51	10 31	06 11	07 00	26 59	29 25
23	14 05 22	29 ♎ 30 22	18 ♉ 24 38	24 ♉ 57 46	24 40	13 33	14 27	10 42	06 18	07 03	27 01	29 28
24	14 09 19	00 ♏ 30 07	01 ♊ 33 45	08 ♊ 12 31	26 22	14 27	15 04	10 54	06 25	07 06	27 02	29 30
25	14 13 15	01 29 53	14 ♊ 54 03	21 ♊ 38 18	28 03	15 22	15 40	11 06	06 33	07 09	27 03	29 33
26	14 17 12	02 29 42	28 ♊ 25 17	05 ♋ 14 59	29 ♎ 44	16 17	16 17	11 18	06 40	07 12	27 05	29 35
27	14 21 08	03 29 32	12 ♋ 07 27	19 ♋ 02 39	01 ♏ 24	17 12	16 53	11 30	06 47	07 16	27 06	29 37
28	14 25 05	04 29 26	26 ♋ 00 36	03 ♌ 01 14	03 04	18 09	17 29	11 42	06 54	07 19	27 08	29 40
29	14 29 01	05 29 21	10 ♌ 04 29	17 ♌ 10 11	04 43	19 05	18 06	11 54	07 01	07 22	27 09	29 42
30	14 32 58	06 29 18	24 ♌ 18 08	01 ♍ 28 00	06 22	20 03	18 42	12 06	07 09	07 25	27 11	29 45
31	14 36 55	07 ♏ 29 18	08 ♍ 39 25	15 ♍ 51 54	08 ♏ 00	21 ♍ 01	19 ♍ 18	12 ♐ 18	07 ♏ 16	07 ♐ 29	27 ♐ 13	29 ♎ 47

Date	Moon True ☊ ° '	Moon Mean ☊ ° '	Moon ☽ Latitude ° '
01	18 ♊ 48	19 ♊ 23	03 N 18
02	18 R 44	19 20	04 09
03	18 38	19 16	04 44
04	18 29	19 13	05 01
05	18 19	19 10	04 58
06	18 08	19 07	04 35
07	17 59	19 04	03 54
08	17 52	19 00	03 00
09	17 47	18 57	01 57
10	17 45	18 54	00 N 49
11	17 D 45	18 51	00 S 20
12	17 45	18 48	01 25
13	17 46	18 45	02 25
14	17 R 46	18 41	03 18
15	17 44	18 38	04 02
16	17 40	18 35	04 35
17	17 33	18 32	04 57
18	17 25	18 29	05 06
19	17 15	18 25	05 01
20	17 06	18 22	04 42
21	16 56	18 19	04 09
22	16 49	18 16	03 23
23	16 43	18 13	02 26
24	16 40	18 10	01 21
25	16 39	18 06	00 S 09
26	16 D 39	18 03	01 N 03
27	16 40	18 00	02 14
28	16 42	17 57	03 17
29	16 R 42	17 54	04 09
30	16 40	17 51	04 47
31	16 ♊ 37	17 ♊ 47	05 N 07

DECLINATIONS

Date	Sun ☉ ° '	Moon ☽ ° '	Mercury ☿ ° '	Venus ♀ ° '	Mars ♂ ° '	Jupiter ♃ ° '	Saturn ♄ ° '	Uranus ♅ ° '	Neptune ♆ ° '	Pluto ♇ ° '
01	03 S 04	23 N 29	05 N 08	08 N 33	12 N 21	20 S 57	10 S 40	21 S 15	22 S 12	04 N 24
02	03 27	20 37	04 52	08 32	12 08	20 59	10 43	21 15	22 12	04 23
03	03 51	16 24	04 32	08 30	11 55	21 01	10 45	21 16	22 12	04 22
04	04 14	11 08	04 08	08 28	11 42	21 03	10 47	21 16	22 12	04 21
05	04 37	05 N 13	03 41	08 25	11 29	21 04	10 50	21 17	22 12	04 20
06	05 00	00 S 57	03 11	08 21	11 15	21 06	10 52	21 17	22 12	04 19
07	05 23	06 58	02 39	08 17	11 02	21 08	10 54	21 18	22 12	04 18
08	05 46	12 29	02 04	08 12	10 49	21 10	10 57	21 18	22 12	04 18
09	06 09	17 13	01 27	08 06	10 35	21 12	10 59	21 18	22 12	04 17
10	06 32	20 56	00 49	08 00	10 22	21 13	11 02	21 19	22 12	04 16
11	06 55	23 29	00 N 09	07 53	10 08	21 15	11 04	21 19	22 12	04 15
12	07 17	24 48	00 S 31	07 46	09 55	21 17	11 06	21 20	22 12	04 14
13	07 40	24 53	01 13	07 37	09 41	21 19	11 09	21 20	22 13	04 13
14	08 02	23 48	01 56	07 29	09 28	21 21	11 11	21 21	22 13	04 12
15	08 24	21 39	02 39	07 20	09 14	21 23	11 14	21 21	22 13	04 11
16	08 47	18 36	03 22	07 10	09 01	21 24	11 16	21 22	22 13	04 10
17	09 09	14 47	04 06	07 00	08 47	21 26	11 19	21 22	22 13	04 09
18	09 31	10 21	04 50	06 49	08 33	21 28	11 21	21 23	22 13	04 09
19	09 52	05 28	05 34	06 37	08 19	21 30	11 23	21 23	22 13	04 08
20	10 14	00 S 18	06 17	06 25	08 06	21 32	11 26	21 24	22 13	04 07
21	10 35	04 N 59	07 01	06 13	07 52	21 33	11 28	21 24	22 13	04 06
22	10 57	10 09	07 44	06 00	07 38	21 35	11 31	21 25	22 13	04 05
23	11 18	14 58	08 27	05 47	07 24	21 37	11 33	21 25	22 13	04 04
24	11 39	19 10	09 10	05 33	07 10	21 39	11 35	21 26	22 13	04 04
25	12 00	22 26	09 52	05 18	06 57	21 41	11 38	21 27	22 13	04 03
26	12 20	24 29	10 33	05 03	06 43	21 42	11 40	21 27	22 13	04 02
27	12 41	25 06	11 14	04 48	06 29	21 44	11 42	21 28	22 14	04 01
28	13 01	24 10	11 54	04 32	06 15	21 46	11 45	21 28	22 14	04 00
29	13 21	21 43	12 34	04 16	06 01	21 48	11 47	21 29	22 14	04 00
30	13 41	17 56	13 13	04 00	05 47	21 49	11 50	21 29	22 14	03 59
31	14 S 01	13 N 04	13 S 52	03 N 42	05 N 33	21 S 51	11 S 52	21 S 30	22 S 14	03 N 58

ZODIAC SIGN ENTRIES

Date	h	m	Planets
01	12	54	☽ ♌
03	14	15	☽ ♍
05	14	42	☽ ♎
05	19	35	♀ ♍
07	16	06	☽ ♏
08	23	44	☿ ♎
09	20	21	☽ ♐
12	04	30	☽ ♑
14	16	00	☽ ♒
17	04	41	☽ ♓
19	16	18	☽ ♈
22	01	47	☽ ♉
23	23	54	☉ ♏
24	09	10	☽ ♊
26	14	47	☽ ♋
26	15	47	☿ ♏
28	18	50	☽ ♌
30	21	33	☽ ♍

LATITUDES

Date	Mercury ☿ ° '	Venus ♀ ° '	Mars ♂ ° '	Jupiter ♃ ° '	Saturn ♄ ° '	Uranus ♅ ° '	Neptune ♆ ° '	Pluto ♇ ° '
01	01 N 14	04 S 00	01 N 17	00 N 30	02 N 13	00 N 04	01 N 12	16 N 25
04	01 39	03 30	01 18	00 29	02 13	00 04	01 12	16 25
07	01 53	03 01	01 19	00 29	02 13	00 04	01 12	16 24
10	01 57	02 33	01 20	00 28	02 13	00 04	01 12	16 24
13	01 54	02 06	01 21	00 28	02 12	00 04	01 12	16 23
16	01 46	01 40	01 22	00 27	02 12	00 04	01 12	16 23
19	01 33	01 15	01 22	00 27	02 12	00 04	01 11	16 23
22	01 17	00 52	01 23	00 27	02 12	00 04	01 11	16 23
25	00 59	00 30	01 24	00 26	02 12	00 04	01 11	16 23
28	00 40	00 S 10	01 25	00 26	02 12	00 04	01 11	16 23
31	00 N 20	00 N 09	01 N 26	00 N 25	02 N 12	00 N 04	01 N 11	16 N 23

DATA

Julian Date	2445609
Delta T	+53 seconds
Ayanamsa	23° 37' 31"
Synetic vernal point	05° ♓ 29' 28"
True obliquity of ecliptic	23° 26' 31"

MOON'S PHASES, APSIDES AND POSITIONS ☽

Date	h	m	Phase	Longitude ° '	Eclipse Indicator
06	11	16	●	12 ♎ 38	
13	19	42	☽	19 ♑ 54	
21	21	53	○	27 ♈ 56	
29	03	37	☾	05 ♌ 08	

Day	h	m		
04	11	06	Perigee	
16	07	49	Apogee	
06	08	18	0S	
13	01	33	Max dec	25° S 00'
20	13	22	0N	
27	09	39	Max dec	25° N 07'

LONGITUDES

Date	Chiron ⚷ ° '	Ceres ⚳ ° '	Pallas ⚴ ° '	Juno ⚵ ° '	Vesta ⚶ ° '	Black Moon Lilith ⚸ ° '
01	02 ♊ 39	13 ♒ 52	09 ♑ 10	04 ♉ 34	26 ♊ 49	21 ♒ 58
11	02 ♊ 22	13 ♒ 59	10 ♑ 57	02 ♉ 57	28 ♊ 04	23 ♒ 05
21	01 ♊ 59	14 ♒ 40	13 ♑ 05	00 ♉ 49	28 ♊ 41	24 ♒ 12
31	01 ♊ 32	15 ♒ 53	15 ♑ 30	28 ♈ 31	28 ♊ 36	25 ♒ 19

ASPECTARIAN

h m	Aspects
01 Saturday	
04 01	☽ ⊥ ♂
05 16	☽ Q ☉
07 12	☽ ⊼ ♆
08 41	☽ ⊻ ♀
10 32	☽ □ ♇
14 34	☽ ⊻ ♂
17 17	☽ ± ♆
19 12	☽ □ ♄
21 46	☽ ∠ ☿
23 01	☽ △ ♅
02 Sunday	
00 19	☽ △ ♃
00 23	☽ # ♆
02 49	☽ ✶ ☉
07 37	☽ # ♅
08 14	☽ bQ ♆
09 34	☽ # ♃
13 51	☽ ⊥ ☿
16 29	☽ Q ♇
03 Monday	
00 31	☽ ⊻ ☿
01 05	☽ Q ♄
05 12	☽ ∠ ☉
08 47	☽ △ ♆
12 03	☽ ☌ ♀
12 06	☽ ✶ ♇
13 30	♀ ✶ ♇
18 01	☽ ☌ ♂
20 43	☽ ✶ ♄
20 50	☽ ⊥ ☉
22 25	☿ ∥ ♇
22 50	☿ ⊥ ♇
04 Tuesday	
00 11	☽ □ ♅
01 52	☽ □ ♃
07 13	☽ ⊻ ☉
09 05	☉ # ☿
09 30	☽ ∥ ♂
12 24	☽ ∠ ♇
13 27	☽ # ♄
19 22	☉ # ♇
21 06	☽ ∠ ♄
23 08	☽ ∥ ♀
05 Wednesday	
02 11	☿ Q ♅
02 57	☽ ⊥ ♇
05 18	☽ Q ♅
05 37	☽ ☌ ☿
07 14	☽ Q ♃
09 18	☽ □ ♆
11 45	☽ ⊥ ♄
12 40	☽ ⊻ ♇
14 12	☽ # ☉
14 29	☽ ⊻ ♀
15 26	☽ ∥ ♇
18 30	☽ ∥ ☿
20 35	☽ ⊻ ♂
21 33	☽ ⊻ ♄
06 Thursday	
00 39	☽ ⊥ ♀
00 48	☽ ✶ ♅
00 48	☿ Q ♃
02 55	☽ ✶ ♃
06 47	☽ ⊥ ♂
11 16	☽ ☌ ☉
14 37	☽ Q ♆
16 01	☽ ∠ ♀
20 05	☽ # ☿
20 09	☿ □ ♆
22 13	☽ ∠ ♂
07 Friday	
00 56	♂ ✶ ♄
01 16	☽ # ♇
01 30	☽ ∠ ♅
03 53	☽ ∠ ♃
05 07	☽ ∥ ☉
10 33	☽ ✶ ♆
12 13	☽ ⊻ ☿
14 07	☽ ☌ ♇
16 36	☽ ⊥ ♅
17 24	☽ # ♀
18 10	☽ ✶ ♀
19 09	☽ ⊥ ♃
23 32	♂ # ♄
23 34	☽ ⊥ ☿
23 40	☽ ☌ ♄
08 Saturday	
00 25	☿ ⊥ ♄
00 30	☽ ✶ ♂
02 48	☽ ⊻ ♅
04 42	☽ # ♂
04 58	☽ ∥ ♄
05 29	☽ ⊻ ♃
06 18	☿ ⊻ ♇
12 04	☽ ∠ ♆
13 50	☉ Q ♆
15 44	☽ Q ♀
16 59	☽ ∠ ☿
17 32	☽ ⊻ ☉
22 10	☽ Q ♂
09 Sunday	
03 47	☽ ⊥ ♆
04 52	☽ ⊥ ☉
14 28	☽ ⊻ ♆
18 23	☽ ⊻ ♇
22 14	☽ ∠ ☉
23 11	☽ ✶ ☿
10 Monday	
01 23	☽ □ ♀
04 55	☽ ⊻ ♄
05 18	☽ ⊥ ♇
07 58	☽ □ ♂
08 01	☽ ☌ ♅
08 51	♂ □ ♅
11 26	☽ ☌ ♃
14 16	☽ ∥ ♃
14 59	☽ ∥ ♅
16 04	☽ ⊥ ♄
21 59	☽ ∠ ♇
22 32	☽ ∥ ♆
11 Tuesday	
00 30	☽ Q ☿
04 12	☽ ✶ ☉
04 27	☿ ⊻ ♀
09 05	☽ ∠ ♄
22 16	☽ ☌ ♆
23 50	☿ ⊻ ♄
12 Wednesday	
02 35	☽ ✶ ♇
05 06	☽ Q ☉
13 25	☽ △ ♀
14 14	☽ ✶ ♄
16 23	☽ □ ☿
17 12	☽ ⊻ ♅
19 58	☽ △ ♂
21 30	☽ ⊻ ♃
22 29	☿ ✶ ♅
13 Thursday	
02 07	☽ Q ♇
05 04	☽ ⊥ ♅
05 20	♀ ✶ ♄
09 33	☽ ⊥ ♃
10 16	☉ # ♀
14 10	☽ Q ♄
19 42	☽ □ ☉
21 04	☽ bQ ♀
23 04	☽ ∠ ♅
14 Friday	
03 23	☽ bQ ♂
03 48	☽ ∠ ♃
06 38	☿ ⊻ ♂
08 53	☿ ✶ ♃
09 35	☽ ⊻ ♆
14 12	☽ □ ♇
14 26	♃ ∥ ♅
14 40	♂ □ ♃
16 23	☽ ± ♀
21 42	☽ ⊥ ♆
22 30	☽ ± ♂
15 Saturday	
02 43	☽ □ ♄
05 22	☽ ⊼ ♀
05 25	☽ ✶ ♅
06 08	♀ □ ♅
06 43	☽ ∥ ♆
10 33	☽ ✶ ♃
11 19	☽ ⊼ ♂
14 18	☉ ∠ ♅
14 27	☽ △ ☿
14 29	☽ ∥ ♃
14 41	☽ ∥ ♅
15 57	☽ ∠ ♆
16 Sunday	
05 49	☽ Q ♅
11 15	☽ Q ♃
13 48	☽ △ ☉
21 19	☉ ∠ ♀
21 25	☉ # ♂
22 22	☽ ✶ ♆
17 Monday	
01 58	☽ bQ ☿
03 05	☽ △ ♇
13 34	☿ ⊥ ♀
13 49	☿ # ♇
15 58	☽ △ ♄
18 17	☽ □ ♅
19 04	☿ Q ♆
22 20	☽ ☍ ♀
22 32	☽ Q ♆
22 44	☽ bQ ☉
23 06	☽ ± ☿
18 Tuesday	
00 04	☽ □ ♃
03 07	☽ ☍ ♂
06 53	☽ ∥ ♄
09 15	☽ bQ ♇
13 04	☽ ⊼ ☿
15 59	☽ ∥ ☉
18 15	☉ ∠ ♃
18 16	☽ ± ☉
21 31	☽ # ♂
22 08	☽ bQ ♄
19 Wednesday	
03 08	☽ ± ♇
06 16	♀ □ ♃
06 16	☽ # ♀
07 09	☽ ⊼ ☉
10 16	☽ □ ♆
11 37	☽ ∥ ☿
14 57	☽ ⊼ ♇
16 00	☽ ± ♄
16 06	☿ ⊥ ♂
18 20	☽ # ♇
20 Thursday	
03 48	☽ ⊼ ♄
05 42	☽ △ ♅
11 58	☽ △ ♃
13 38	☽ ⊼ ♀
15 26	☿ # ♀
17 01	☽ ⊼ ♂
22 19	☉ ✶ ♆
21 Friday	
02 00	☽ ± ♀
05 05	☽ ± ♂
08 02	☽ ∥ ♇
08 51	☽ ☍ ☿
10 34	☽ bQ ♅
17 02	☽ bQ ♃
17 28	☽ ∥ ♀
20 05	☽ △ ♆
20 20	☽ bQ ♀
21 53	☽ ☍ ☉
21 54	☿ ∠ ♆
22 52	☽ # ☿
23 00	☽ bQ ♂
22 Saturday	
00 40	☽ ☍ ♇
00 43	☽ ∥ ♂
03 38	☽ ± ♅
09 26	☿ # ♂
10 13	☽ ± ♃
13 23	☽ ☍ ♄
14 54	☽ ⊼ ♅
16 08	☽ # ☉
18 37	☽ # ♄
21 33	☽ ⊼ ♃
23 Sunday	
00 10	☽ bQ ♆
02 24	☽ △ ♀
04 21	☽ △ ♂
08 42	☉ ∠ ♂
10 56	☉ ☌ ♇
12 26	♂ ∠ ♇
16 47	☽ ± ♆
24 Monday	
01 10	☽ ⊼ ☿
03 45	☽ ⊼ ♆
04 29	☿ ∠ ♃
07 15	☉ ∥ ♄
08 15	☽ ⊼ ♇
09 55	☽ ⊼ ☉
13 24	♀ ∠ ♇
13 40	☽ ± ☿
19 09	☽ ± ♇
20 52	☽ ⊼ ♄
21 34	☿ ✶ ♆
21 39	☽ ± ☉
22 03	☽ ☍ ♅
25 Tuesday	
03 15	☉ ⊥ ♅
03 49	☽ # ♅
05 05	☽ ☍ ♃
05 37	☽ # ♃
07 45	☽ ± ♄
08 14	☽ bQ ☿
10 11	☽ # ♆
11 22	☽ bQ ♇
12 53	☽ □ ♀
13 26	☽ □ ♂
15 05	☽ bQ ☉
23 56	☽ bQ ♄
26 Wednesday	
09 38	☽ ☍ ♆
09 47	☿ ☌ ♇
11 47	♀ ☌ ♂
14 03	☽ △ ♇
14 38	☽ △ ☿
19 44	☽ △ ☉
22 46	☽ Q ♂
23 03	☽ Q ♀
27 Thursday	
02 36	☽ △ ♄
03 29	☽ ⊼ ♅
09 49	☿ ⊥ ♅
10 53	☽ ⊼ ♃
13 59	☽ ± ♅
20 38	☽ ✶ ♂
21 27	☽ ± ♃
21 27	☽ ✶ ♀
22 46	☿ ∠ ♂
28 Friday	
05 37	☽ bQ ♅
05 53	☿ ∥ ♄
13 12	☽ bQ ♃
13 56	☽ ⊼ ♆
14 30	☿ ∠ ♀
18 17	☽ □ ♇
23 35	☽ ∠ ♂
29 Saturday	
00 13	☽ ± ♆
01 05	☽ ∠ ♀
01 42	☽ □ ☿
03 37	☽ □ ☉
06 47	☽ □ ♄
07 23	☽ △ ♅
08 00	☽ # ♆
11 27	☽ # ♃
13 48	☽ # ♅
15 08	☽ △ ♃
15 32	☽ bQ ♆
15 34	☽ ⊥ ♂
17 29	☽ ⊥ ♀
30 Sunday	
00 15	☉ ⊥ ♃
00 56	☽ Q ♇
02 09	☽ ⊻ ♂
04 20	☽ ⊻ ♀
07 29	☿ ⊥ ♃
12 08	☽ Q ☿
12 20	☽ Q ☉
12 49	♀ ∥ ♇
13 26	☽ Q ♄
16 29	☉ ☌ ☿
16 51	☽ △ ♆
21 09	☽ ✶ ♇
31 Monday	
00 16	☿ ☌ ♄
03 58	☿ ⊻ ♅
05 56	☉ ☌ ♄
07 55	☽ # ☉
08 45	☽ # ☿
09 40	☽ ✶ ♄
09 54	☽ ✶ ☉
10 02	☽ □ ♅
10 47	☽ ✶ ☿
11 44	☉ ⊻ ♅
17 20	☽ # ♄
18 09	☽ □ ♃
22 14	☽ ∠ ♇
23 54	☉ ∥ ♃

All ephemeris data is given at 12.00 UT and the Moon's longitude is additionally given for 24.00 UT

NOVEMBER 1983

LONGITUDES

Date	Sidereal time h m s	Sun ☉ ° ' "	Moon ☽ ° ' "	Moon ☽ 24.00 ° ' "	Mercury ☿ ° '	Venus ♀ ° '	Mars ♂ ° '	Jupiter ♃ ° '	Saturn ♄ ° '	Uranus ♅ ° '	Neptune ♆ ° '	Pluto ♇ ° '
01	14 40 51	08 ♏ 29 20	23 ♍ 04 52	00 ♎ 17 41	09 ♏ 38	21 ♍ 59	19 ♍ 54	12 ♐ 30	07 ♏ 23	07 ♐ 32	27 ♐ 14	29 ♎ 50
02	14 44 48	09 29 24	07 ♎ 29 41	14 ♎ 40 08	11 15	22 58	20 30	12 43	07 30	07 35	27 16	29 52
03	14 48 44	10 29 29	21 ♎ 48 20	28 ♎ 53 33	12 52	23 57	21 06	12 55	07 38	07 39	27 18	29 54
04	14 52 41	11 29 37	05 ♏ 55 11	12 ♏ 52 37	14 29	24 57	21 42	13 08	07 45	07 42	27 19	29 57
05	14 56 37	12 29 47	19 ♏ 45 25	26 ♏ 33 11	16 05	25 58	22 18	13 20	07 52	07 46	27 21	29 ♎ 59
06	15 00 34	13 29 59	03 ♐ 15 41	09 ♐ 52 47	17 40	26 58	22 54	13 33	07 59	07 49	27 23	00 ♏ 01
07	15 04 30	14 30 12	16 ♐ 24 27	22 ♐ 50 50	19 15	27 59	23 30	13 45	08 07	07 52	27 25	00 04
08	15 08 27	15 30 27	29 ♐ 12 05	05 ♑ 28 33	20 50	29 ♍ 01	24 06	13 58	08 14	07 56	27 26	00 06
09	15 12 24	16 30 44	11 ♑ 40 34	17 ♑ 48 37	22 25	00 ♎ 03	24 42	14 11	08 21	07 59	27 28	00 09
10	15 16 20	17 31 02	23 ♑ 53 10	29 ♑ 54 48	23 59	01 05	25 18	14 23	08 28	08 03	27 30	00 11
11	15 20 17	18 31 22	05 ♒ 54 05	11 ♒ 51 37	25 32	02 08	25 54	14 36	08 35	08 06	27 32	00 13
12	15 24 13	19 31 43	17 ♒ 48 01	23 ♒ 43 54	27 06	03 11	26 29	14 49	08 43	08 10	27 34	00 16
13	15 28 10	20 32 05	29 ♒ 39 53	05 ♓ 36 35	28 ♏ 39	04 14	27 05	15 02	08 50	08 13	27 36	00 18
14	15 32 06	21 32 29	11 ♓ 34 33	17 ♓ 34 22	00 ♐ 12	05 18	27 40	15 15	08 57	08 17	27 37	00 20
15	15 36 03	22 32 54	23 ♓ 36 31	29 ♓ 41 29	01 44	06 22	28 16	15 28	09 04	08 21	27 39	00 22
16	15 39 59	23 33 20	05 ♈ 49 41	12 ♈ 01 29	03 17	07 26	28 51	15 41	09 11	08 24	27 41	00 25
17	15 43 56	24 33 48	18 ♈ 17 09	24 ♈ 36 56	04 49	08 31	29 ♍ 27	15 54	09 18	08 28	27 43	00 27
18	15 47 53	25 34 18	01 ♉ 00 59	07 ♉ 29 21	06 20	09 36	00 ♎ 02	16 07	09 25	08 31	27 45	00 29
19	15 51 49	26 34 48	14 ♉ 02 04	20 ♉ 39 02	07 52	10 41	00 38	16 20	09 32	08 35	27 47	00 31
20	15 55 46	27 35 21	27 ♉ 20 08	04 ♊ 05 08	09 23	11 47	01 13	16 33	09 39	08 39	27 49	00 34
21	15 59 42	28 35 54	10 ♊ 53 48	17 ♊ 45 50	10 54	12 53	01 48	16 47	09 46	08 42	27 51	00 36
22	16 03 39	29 ♏ 36 30	24 ♊ 40 52	01 ♋ 38 34	12 25	13 59	02 23	17 00	09 53	08 46	27 53	00 38
23	16 07 35	00 ♐ 37 07	08 ♋ 38 33	15 ♋ 40 27	13 55	15 05	02 58	17 13	10 00	08 50	27 55	00 40
24	16 11 32	01 37 46	22 ♋ 43 54	29 ♋ 48 32	15 25	16 12	03 34	17 26	10 07	08 53	27 57	00 42
25	16 15 28	02 38 26	06 ♌ 54 01	14 ♌ 00 02	16 55	17 18	04 09	17 40	10 14	08 57	28 00	00 45
26	16 19 25	03 39 08	21 ♌ 06 16	28 ♌ 12 25	18 25	18 25	04 44	17 53	10 21	09 01	28 02	00 47
27	16 23 22	04 39 51	05 ♍ 18 14	12 ♍ 23 25	19 54	19 33	05 19	18 07	10 28	09 04	28 04	00 49
28	16 27 18	05 40 36	19 ♍ 27 43	26 ♍ 30 53	21 23	20 40	05 53	18 20	10 34	09 08	28 06	00 51
29	16 31 15	06 41 23	03 ♎ 32 36	10 ♎ 32 39	22 51	21 48	06 28	18 33	10 41	09 12	28 08	00 53
30	16 35 11	07 ♐ 42 11	17 ♎ 30 43	24 ♎ 26 33	24 ♐ 18	22 ♎ 56	07 ♎ 03	18 ♐ 47	10 ♏ 48	09 ♐ 15	28 ♐ 10	00 ♏ 55

Date	Moon True ☊ ° '	Moon Mean ☊ ° '	Moon ☽ Latitude ° '
01	16 ♊ 33	17 ♊ 44	05 N 08
02	16 R 27	17 41	04 50
03	16 21	17 38	04 13
04	16 16	17 35	03 22
05	16 12	17 31	02 19
06	16 09	17 28	01 N 10
07	16 D 09	17 25	00 S 01
08	16 09	17 22	01 11
09	16 11	17 19	02 15
10	16 13	17 16	03 12
11	16 14	17 12	04 00
12	16 15	17 09	04 36
13	16 R 14	17 06	05 01
14	16 13	17 03	05 13
15	16 10	17 00	05 11
16	16 07	16 57	04 56
17	16 03	16 53	04 26
18	16 00	16 50	03 43
19	15 57	16 47	02 47
20	15 55	16 44	01 41
21	15 54	16 41	00 S 28
22	15 D 54	16 37	00 N 48
23	15 55	16 34	02 02
24	15 56	16 31	03 10
25	15 57	16 28	04 06
26	15 58	16 25	04 47
27	15 R 59	16 22	05 11
28	15 58	16 18	05 16
29	15 57	16 15	05 02
30	15 ♊ 57	16 ♊ 12	04 N 30

DECLINATIONS

Date	Sun ☉ ° '	Moon ☽ ° '	Mercury ☿ ° '	Venus ♀ ° '	Mars ♂ ° '	Jupiter ♃ ° '	Saturn ♄ ° '	Uranus ♅ ° '	Neptune ♆ ° '	Pluto ♇ ° '
01	14 S 20	07 N 28	14 S 29	03 N 25	05 N 19	21 S 53	11 S 54	21 S 30	22 S 14	03 N 57
02	14 39	01 N 28	15 06	03 07	05 05	21 54	11 57	21 31	22 14	03 57
03	14 58	04 S 35	15 42	02 49	04 51	21 56	11 59	21 31	22 14	03 56
04	15 17	10 19	16 17	02 31	04 37	21 58	12 01	21 32	22 14	03 55
05	15 35	15 27	16 52	02 12	04 23	22 00	12 04	21 33	22 14	03 55
06	15 54	19 40	17 25	01 52	04 09	22 01	12 06	21 33	22 14	03 54
07	16 11	22 46	17 58	01 33	03 55	22 03	12 08	21 34	22 14	03 53
08	16 29	24 37	18 30	01 13	03 41	22 04	12 10	21 34	22 14	03 53
09	16 47	25 11	19 01	00 53	03 27	22 06	12 13	21 35	22 14	03 52
10	17 04	24 29	19 31	00 33	03 13	22 08	12 15	21 35	22 15	03 51
11	17 20	22 40	20 00	00 N 12	02 59	22 09	12 17	21 36	22 15	03 51
12	17 37	19 53	20 28	00 S 09	02 45	22 11	12 20	21 37	22 15	03 50
13	17 53	16 17	20 55	00 30	02 31	22 12	12 22	21 37	22 15	03 49
14	18 09	12 03	21 21	00 52	02 17	22 14	12 24	21 38	22 15	03 49
15	18 25	07 18	21 46	01 13	02 03	22 15	12 26	21 38	22 15	03 48
16	18 40	02 S 13	22 10	01 35	01 49	22 17	12 28	21 39	22 15	03 48
17	18 55	03 N 04	22 32	01 57	01 36	22 18	12 31	21 39	22 15	03 47
18	19 09	08 21	22 54	02 19	01 22	22 20	12 33	21 40	22 15	03 47
19	19 24	13 24	23 14	02 42	01 08	22 21	12 35	21 41	22 15	03 46
20	19 37	17 56	23 34	03 04	00 54	22 23	12 37	21 41	22 15	03 46
21	19 51	21 37	23 52	03 27	00 40	22 24	12 39	21 42	22 15	03 45
22	20 04	24 08	24 09	03 50	00 27	22 26	12 41	21 42	22 15	03 45
23	20 17	25 12	24 25	04 13	00 N 13	22 27	12 44	21 43	22 15	03 44
24	20 29	24 39	24 39	04 36	00 S 01	22 28	12 46	21 43	22 15	03 44
25	20 41	22 31	24 52	04 59	00 15	22 30	12 48	21 44	22 15	03 43
26	20 53	19 00	25 04	05 22	00 28	22 31	12 50	21 45	22 15	03 43
27	21 04	14 23	25 14	05 45	00 42	22 32	12 52	21 45	22 16	03 42
28	21 15	09 01	25 24	06 08	00 56	22 34	12 54	21 46	22 16	03 42
29	21 26	03 12	25 31	06 32	01 09	22 35	12 56	21 46	22 16	03 42
30	21 S 36	02 S 43	25 S 38	06 S 55	01 S 23	22 S 36	12 S 58	21 S 47	22 S 16	03 N 41

ZODIAC SIGN ENTRIES

Date	h	m	Planets
01	23	31	☽ ♎
04	01	53	☽ ♏
05	21	07	♇ ♏
06	06	09	☽ ♐
08	13	31	☽ ♑
09	10	52	♀ ♎
11	00	10	☽ ♒
13	12	41	☽ ♓
14	08	56	☿ ♐
16	00	36	☽ ♈
18	10	06	☽ ♉
18	10	26	♂ ♎
20	16	45	☽ ♊
22	21	10	☽ ♋
22	21	18	☉ ♐
25	00	19	☽ ♌
27	03	02	☽ ♍
29	05	57	☽ ♎

LATITUDES

Date	Mercury ☿ ° '	Venus ♀ ° '	Mars ♂ ° '	Jupiter ♃ ° '	Saturn ♄ ° '	Uranus ♅ ° '	Neptune ♆ ° '	Pluto ♇ ° '
01	00 N 14	00 N 16	01 N 26	00 N 25	02 N 12	00 N 04	01 N 11	16 N 23
04	00 S 07	00 33	01 27	00 25	02 12	00 04	01 11	16 23
07	00 27	00 49	01 27	00 25	02 12	00 04	01 11	16 24
10	00 46	01 04	01 28	00 24	02 12	00 04	01 11	16 24
13	01 05	01 17	01 29	00 24	02 12	00 04	01 11	16 25
16	01 22	01 29	01 30	00 24	02 12	00 04	01 10	16 25
19	01 38	01 40	01 30	00 23	02 12	00 04	01 10	16 26
22	01 53	01 50	01 31	00 23	02 12	00 04	01 10	16 27
25	02 05	01 58	01 32	00 23	02 12	00 04	01 10	16 27
28	02 14	02 05	01 32	00 22	02 12	00 04	01 10	16 28
31	02 S 21	02 N 11	01 N 33	00 N 22	02 N 13	00 N 04	01 N 10	16 N 29

DATA

Julian Date	2445640
Delta T	+53 seconds
Ayanamsa	23° 37' 34"
Synetic vernal point	05° ♓ 29' 25"
True obliquity of ecliptic	23° 26' 31"

MOON'S PHASES, APSIDES AND POSITIONS ☽

Date	h	m	Phase	Longitude ° '	Eclipse Indicator
04	22	21	●	11 ♏ 56	
12	15	49	☽	19 ♒ 41	
20	12	29	○	27 ♉ 37	
27	10	50	☾	04 ♍ 37	

Day	h	m		
01	02	22	Perigee	
13	03	25	Apogee	
26	02	32	Perigee	
02	17	46	0S	
09	10	25	Max dec	25° S 11'
16	22	07	0N	
23	15	52	Max dec	25° N 13'
30	00	59	0S	

LONGITUDES

Date	Chiron ⚷ ° '	Ceres ⚳ ° '	Pallas ⚴ ° '	Juno ⚵ ° '	Vesta ⚶ ° '	Black Moon Lilith ⚸ ° '
01	01 ♊ 29	16 ♒ 02	15 ♑ 46	28 ♈ 18	28 ♊ 34	25 ♒ 26
11	00 ♊ 59	17 ♒ 46	18 ♑ 27	26 ♈ 17	27 ♊ 40	26 ♒ 33
21	00 ♊ 26	19 ♒ 54	21 ♑ 20	24 ♈ 56	26 ♊ 04	27 ♒ 40
31	29 ♉ 54	22 ♒ 23	24 ♑ 22	24 ♈ 27	23 ♊ 53	28 ♒ 47

ASPECTARIAN

h m	Aspects
01 Tuesday	
06 29	☽ ☌ ♂
10 03	☽ ☌ ♀
10 50	☽ ∠ ♄
12 44	☽ ∠ ☉
13 15	☽ ⊥ ♇
14 55	☽ ∠ ☿
16 06	☽ Q ♅
18 56	☽ □ ♆
21 02	☽ ∥ ♂
23 12	♀ ∠ ♄
23 15	☽ ⊻ ♇
02 Wednesday	
00 32	☽ Q ♃
01 56	☽ ⊥ ♄
02 09	☽ ∥ ♇
04 49	☽ ⊥ ☉
05 06	☽ ∥ ♀
07 48	☽ ⊥ ☿
12 01	☽ ⊻ ♄
12 09	☽ ⚹ ♅
15 35	☽ ⊻ ☉
19 05	☽ ⊻ ☿
20 51	☽ ⚹ ♃
03 Thursday	
01 02	☽ Q ♆
03 14	☿ ∠ ♆
05 16	☽ ⋕ ♀
09 24	☽ ⋕ ♇
10 41	♀ ⊥ ♇
10 46	☽ ⊻ ♂
12 48	☿ ⊻ ♃
13 03	☽ ⋕ ♂
13 25	☽ ∠ ♅
15 54	☽ ⊻ ♀
18 24	♄ ⊻ ♅
21 18	☽ ⚹ ♆
21 22	☽ ⊥ ♂
22 30	☽ ∠ ♃
04 Friday	
01 46	☽ ☌ ♇
02 52	☽ ⊥ ♀
04 45	☽ ⊥ ♅
13 25	☽ ∠ ♂
14 06	☽ ⊥ ♃
15 04	☽ ⊻ ♅
15 10	☽ ☌ ♄
19 29	☽ ∠ ♀
19 38	☽ ∥ ♄
22 21	☽ ☌ ☉
23 04	☽ ∠ ♆
05 Saturday	
00 37	☽ ⊻ ♃
04 43	☽ ☌ ☿
06 55	♀ Q ♅
08 24	☉ ∠ ♆
12 47	☽ ∥ ☉
14 48	☽ ⊥ ♆
16 42	☽ ⚹ ♂
20 33	☽ ∥ ☿
23 50	☽ ⚹ ♀
06 Sunday	
01 27	☽ ⊻ ♆
06 10	☽ ⊻ ♇
13 19	☉ ⊻ ♃
15 07	☽ Q ♂
16 07	♂ ∠ ♄
17 00	☽ ⊥ ♇
20 17	☽ ☌ ♅
20 38	☽ ⊻ ♄
21 54	♀ □ ♆
23 13	☽ Q ♀
07 Monday	
01 28	☽ ∥ ♅
05 24	☽ ∥ ♃
07 01	☽ ☌ ♃
07 07	☽ ∥ ♆
07 43	☽ ⊥ ♄
08 11	☽ ⊻ ☉
09 31	☽ ∠ ♇
15 10	♂ ∥ ♇
18 02	☽ ⊻ ☿
20 16	☽ ⊥ ☉
08 Tuesday	
00 37	☽ ∠ ♄
01 53	☽ □ ♂
06 53	☽ ⊥ ☿
08 39	☽ ☌ ♆
11 37	☽ □ ♀
11 58	♂ ⊥ ♇
13 43	☽ ⚹ ♇
14 42	☽ ∠ ☉
21 21	☿ ⊥ ♆
09 Wednesday	
02 32	☽ ∠ ☿
04 49	☽ ⊻ ♅
05 29	☽ ⚹ ♄
12 55	☽ Q ♇
14 16	♀ ⊻ ♇
16 32	☽ ⊥ ♅
16 58	☽ ⊻ ♃
22 18	☽ ⚹ ☉
10 Thursday	
04 57	☽ ⊥ ♃
05 10	☽ Q ♄
10 20	☽ ∠ ♅
12 13	☽ ⚹ ☿
14 57	☽ △ ♂
19 12	☽ ⊻ ♆
23 09	☽ ∠ ♃
11 Friday	
00 14	☽ Q ☉
00 35	☽ □ ♇
03 43	☽ △ ♀
07 14	☽ ⊥ ♆
08 30	♃ ⊥ ♄
15 48	☽ Q ☿
16 13	☽ ∥ ♆
16 27	☽ ⚹ ♅
17 01	☽ ∥ ♃
17 28	☽ □ ♄
20 46	☿ ⚹ ♂
21 34	♂ Q ♅
22 05	☽ ∥ ♅
22 35	☽ bQ ♂
23 49	♀ ⊥ ♄
12 Saturday	
01 23	☽ ∠ ♆
01 33	♀ Q ♃
05 52	☽ ⚹ ♃
08 06	☽ ∥ ☿
12 51	☽ bQ ♀
15 49	☽ □ ☉
16 49	☽ Q ♅
17 43	☽ ± ♂
19 18	☿ ⊻ ♆
13 Sunday	
02 35	☽ ∥ ☉
06 30	☽ ⊼ ♂
06 35	☽ Q ♃
07 48	☽ ⚹ ♆
08 50	☽ ± ♀
09 38	☽ □ ☿
13 17	☽ △ ♇
22 08	☽ ⊼ ♀
14 Monday	
05 21	☽ □ ♅
06 40	☽ △ ♄
08 04	☽ Q ♆
09 51	♂ □ ♆
10 06	☽ ∥ ♄
14 03	☉ ⊥ ♆
14 13	☿ ⊻ ♇
19 30	☽ □ ♃
19 33	☽ bQ ♇
23 49	♃ ∠ ♇
15 Tuesday	
02 27	♃ ∥ ♆
04 32	☿ ∥ ♅
09 42	☽ △ ☉
12 55	☽ bQ ♄
13 31	☽ ± ♇
20 01	☽ □ ♆
21 40	☽ ☍ ♂
16 Wednesday	
01 23	☽ ⊼ ♇
04 39	☽ ⋕ ♇
06 19	☽ △ ☿
06 48	☽ ± ♄
13 52	☽ ⋕ ♂
14 41	☽ ∥ ♀
15 26	☽ ☍ ♀
17 02	☽ △ ♅
17 30	☿ ∥ ♆
17 46	☽ bQ ☉
18 35	☽ ⊼ ♄
20 07	☿ ∥ ♃
21 36	♀ ⋕ ♂
17 Thursday	
05 37	☽ ∥ ♂
06 35	☽ ⋕ ♀
07 22	☽ △ ♃
10 45	♀ ⚹ ♅
12 34	☽ ± ☉
15 14	☽ ∥ ♇
15 18	☽ bQ ☿
21 52	☽ bQ ♅
18 Friday	
00 55	☽ ⊼ ☉
05 53	☽ △ ♆
07 32	♀ ⊻ ♄
10 05	☽ ⊼ ♂
10 34	☽ ± ☿
11 01	☽ ☍ ♇
12 12	☽ bQ ♃
14 22	☿ ⊥ ♇
14 49	☽ ± ♅
21 46	☽ ± ♂
23 12	☽ ⊼ ☿
19 Saturday	
01 58	☽ ⊼ ♅
03 42	☽ ☍ ♄
05 07	☽ ± ♃
05 19	☽ ⊼ ♀
07 31	♂ ⊻ ♇
07 58	☽ ⋕ ♄
09 43	☽ bQ ♆
15 02	☽ bQ ♂
16 16	☽ ⊼ ♃
17 16	☽ ± ♀
23 49	☿ ☌ ♅
20 Sunday	
02 05	☽ ± ♆
10 55	☽ bQ ♀
12 29	☽ ☍ ☉
12 52	☽ ⊼ ♆
16 35	☿ ⊻ ♄
17 43	☉ ⊻ ♆
17 46	☽ ⊼ ♇
19 14	☽ △ ♂
22 51	☽ ⋕ ☉
21 Monday	
04 25	☽ ± ♇
08 08	☽ ☍ ♅
10 00	☽ ⊼ ♄
12 01	☽ ☍ ☿
12 33	☽ ⋕ ♅
15 46	☽ △ ♀
16 59	☽ ⋕ ♆
18 17	☽ ⋕ ♃
20 15	☽ bQ ♇
20 36	☽ ± ♄
22 27	☽ ☍ ♃
22 Tuesday	
06 43	♀ ⋕ ♇
12 07	☽ ⋕ ☿
12 21	☽ bQ ♄
17 33	☽ ☍ ♆
21 10	☽ ⊼ ☉
22 18	☽ △ ♇
23 Wednesday	
01 52	☽ □ ♂
08 16	☽ ± ☉
12 19	☽ ⊼ ♅
13 18	☉ ⊻ ♇
14 21	☽ △ ♄
22 05	☽ ⊼ ☿
22 36	☽ ± ♅
23 56	☽ □ ♀
24 Thursday	
00 49	☽ bQ ☉
02 52	☽ ⊼ ♃
06 47	♀ Q ♆
09 31	☽ ± ☿
09 55	☽ Q ♂
11 56	☽ ⋕ ☿
13 13	☽ ± ♃
13 58	☽ bQ ♅
15 34	☿ Q ♂
16 39	☿ ∠ ♇
20 53	☽ ⊼ ♆
25 Friday	
00 02	☿ ⊥ ♄
01 33	☽ □ ♇
02 35	☽ bQ ☿
04 14	☽ △ ☉
04 43	☽ bQ ♃
07 04	☽ ± ♆
07 08	☽ ⚹ ♂
09 04	☽ Q ♀
12 10	☽ ⋕ ♃
14 06	☽ ⋕ ♆
15 29	☽ △ ♅
16 33	♂ ⊥ ♄
17 41	☽ □ ♄
18 08	☽ ⋕ ♅
21 35	♀ ⚹ ♃
22 19	☽ bQ ♆
26 Saturday	
00 47	☽ ⋕ ☉
02 02	☿ ☌ ♃
06 29	☽ △ ♃
06 55	☽ △ ☿
07 05	☽ ⚹ ♀
08 04	☽ Q ♇
09 34	☽ ∠ ♂
12 46	☿ ⚹ ♀
23 44	☽ △ ♆
27 Sunday	
00 20	☽ Q ♄
01 26	☽ ⊥ ♂
04 23	☽ ⚹ ♇
10 36	☽ ∠ ♀
10 50	☽ □ ☉
12 00	☽ ⊻ ♂
18 24	☽ □ ♅
19 02	☽ ⋕ ♄
20 48	☽ ⚹ ♄
28 Monday	
03 10	☽ ⊥ ♀
05 51	☽ ∠ ♇
10 03	☽ □ ♃
13 06	☉ ∠ ♀
14 14	☽ ⊻ ♀
15 38	☽ □ ☿
19 43	☽ Q ☉
21 11	☽ ⊥ ♇
22 29	☽ ∠ ♄
23 15	☽ ⋕ ♀
23 49	☉ ⚹ ♂
29 Tuesday	
01 07	☽ Q ♅
02 44	☽ □ ♆
07 26	☽ ⊻ ♇
10 02	☽ ∥ ♇
13 58	☽ ⊥ ♄
16 44	☉ ⊥ ♇
17 14	☽ ☌ ♂
17 15	☽ Q ♃
17 48	☽ ⚹ ☉
17 52	♂ Q ♃
20 01	☽ ⋕ ♂
21 44	☽ ⚹ ♅
30 Wednesday	
00 21	☽ ⊻ ♄
01 59	☽ Q ☿
06 20	☽ ∥ ♂
09 41	☽ Q ♆
14 14	☽ ⚹ ♃
15 58	☽ ⋕ ♇
21 42	☽ ∠ ☉
22 13	☽ ☌ ♀
23 44	☽ ∠ ♅

All ephemeris data is given at 12.00 UT and the Moon's longitude is additionally given for 24.00 UT

DECEMBER 1983

LONGITUDES

Date	Sidereal time	Sun ☉	Moon ☽	Moon ☽ 24.00	Mercury ☿	Venus ♀	Mars ♂	Jupiter ♃	Saturn ♄	Uranus ♅	Neptune ♆	Pluto ♇
	h m s	° ' "	° ' "	° ' "	° '	° '	° '	° '	° '	° '	° '	° '
01	16 39 08	08 ♐ 43 01	01 ♏ 19 51	08 ♏ 10 21	25 ♐ 46	24 ♎ 04	07 ♎ 38	19 ♐ 00	10 ♏ 55	09 ♐ 19	28 ♐ 12	00 ♏ 57
02	16 43 04	09 43 52	14 ♏ 57 47	21 ♏ 41 56	27 12	25 13	08 12	19 14	11 01	09 23	28 14	00 59
03	16 47 01	10 44 44	28 ♏ 22 35	04 ♐ 59 32	28 ♐ 38	26 21	08 47	19 27	11 08	09 26	28 17	01 01
04	16 50 57	11 45 38	11 ♐ 32 40	18 ♐ 01 53	00 ♑ 02	27 30	09 21	19 41	11 15	09 30	28 19	01 03
05	16 54 54	12 46 32	24 ♐ 27 10	00 ♑ 48 30	01 26	28 39	09 56	19 54	11 21	09 34	28 21	01 05
06	16 58 51	13 47 28	07 ♑ 06 00	13 ♑ 19 48	02 48	29 ♎ 48	10 30	20 08	11 28	09 38	28 23	01 07
07	17 02 47	14 48 25	19 ♑ 30 05	25 ♑ 37 08	04 09	00 ♏ 57	11 04	20 22	11 34	09 41	28 25	01 09
08	17 06 44	15 49 22	01 ♒ 41 15	07 ♒ 42 49	05 28	02 06	11 39	20 35	11 41	09 45	28 28	01 11
09	17 10 40	16 50 21	13 ♒ 42 14	19 ♒ 39 58	06 45	03 16	12 13	20 49	11 47	09 49	28 30	01 13
10	17 14 37	17 51 19	25 ♒ 36 32	01 ♓ 32 26	08 00	04 26	12 47	21 02	11 53	09 52	28 32	01 14
11	17 18 33	18 52 19	07 ♓ 28 15	13 ♓ 24 32	09 12	05 36	13 21	21 16	12 00	09 56	28 34	01 16
12	17 22 30	19 53 19	19 ♓ 21 54	25 ♓ 20 56	10 21	06 45	13 55	21 30	12 06	10 00	28 37	01 18
13	17 26 26	20 54 20	01 ♈ 22 13	07 ♈ 26 21	11 26	07 56	14 29	21 43	12 12	10 03	28 39	01 20
14	17 30 23	21 55 21	13 ♈ 33 53	19 ♈ 45 20	12 27	09 06	15 02	21 57	12 18	10 07	28 41	01 21
15	17 34 20	22 56 23	26 ♈ 01 12	02 ♉ 21 55	13 24	10 16	15 36	22 11	12 24	10 10	28 43	01 23
16	17 38 16	23 57 25	08 ♉ 47 49	15 ♉ 19 12	14 14	11 27	16 10	22 24	12 30	10 14	28 46	01 25
17	17 42 13	24 58 27	21 ♉ 56 15	28 ♉ 39 02	14 59	12 37	16 43	22 38	12 36	10 18	28 48	01 27
18	17 46 09	25 59 31	05 ♊ 27 31	12 ♊ 21 32	15 37	13 48	17 17	22 51	12 42	10 21	28 50	01 28
19	17 50 06	27 00 34	19 ♊ 20 49	26 ♊ 24 55	16 06	14 59	17 50	23 05	12 48	10 25	28 52	01 30
20	17 54 02	28 01 39	03 ♋ 33 21	10 ♋ 45 26	16 27	16 10	18 23	23 19	12 54	10 28	28 55	01 31
21	17 57 59	29 ♐ 02 44	18 ♋ 00 29	25 ♋ 17 42	16 38	17 21	18 56	23 32	13 00	10 32	28 57	01 33
22	18 01 55	00 ♑ 03 49	02 ♌ 36 16	09 ♌ 55 22	16 R 38	18 32	19 30	23 46	13 05	10 36	28 59	01 34
23	18 05 52	01 04 55	17 ♌ 14 10	24 ♌ 31 56	16 27	19 44	20 03	24 00	13 11	10 39	29 02	01 36
24	18 09 49	02 06 02	01 ♍ 47 59	09 ♍ 01 42	16 05	20 55	20 36	24 13	13 17	10 43	29 04	01 37
25	18 13 45	03 07 09	16 ♍ 12 36	23 ♍ 20 18	15 30	22 07	21 08	24 27	13 22	10 46	29 06	01 39
26	18 17 42	04 08 17	00 ♎ 24 28	07 ♎ 24 57	14 45	23 18	21 41	24 40	13 28	10 50	29 08	01 40
27	18 21 38	05 09 26	14 ♎ 21 36	21 ♎ 14 24	13 48	24 30	22 14	24 54	13 33	10 53	29 11	01 41
28	18 25 35	06 10 35	28 ♎ 03 22	04 ♏ 48 34	12 42	25 42	22 47	25 07	13 38	10 56	29 13	01 43
29	18 29 31	07 11 45	11 ♏ 30 05	18 ♏ 08 02	11 29	26 54	23 19	25 21	13 44	11 00	29 15	01 44
30	18 33 28	08 12 55	24 ♏ 42 33	01 ♐ 13 45	10 10	28 06	23 51	25 34	13 49	11 03	29 17	01 45
31	18 37 24	09 ♑ 14 06	07 ♐ 41 46	14 ♐ 06 42	08 ♑ 48	29 ♏ 18	24 ♎ 24	25 ♐ 48	13 ♏ 54	11 ♐ 07	29 ♐ 20	01 ♏ 47

MOON

Date	Moon True ☊	Moon Mean ☊	Moon ☽ Latitude
	° '	° '	° '
01	15 ♊ 56	16 ♊ 09	03 N 43
02	15 R 55	16 06	02 43
03	15 54	16 03	01 36
04	15 54	15 59	00 N 24
05	15 D 54	15 56	00 S 47
06	15 54	15 53	01 55
07	15 54	15 50	02 56
08	15 54	15 47	03 48
09	15 R 54	15 43	04 29
10	15 54	15 40	04 58
11	15 54	15 37	05 14
12	15 54	15 34	05 17
13	15 D 54	15 31	05 06
14	15 54	15 28	04 41
15	15 55	15 24	04 03
16	15 56	15 21	03 12
17	15 57	15 18	02 09
18	15 57	15 15	00 S 58
19	15 R 57	15 12	00 N 19
20	15 57	15 09	01 36
21	15 56	15 05	02 48
22	15 54	15 02	03 50
23	15 52	14 59	04 37
24	15 50	14 56	05 06
25	15 49	14 53	05 16
26	15 48	14 49	05 05
27	15 D 48	14 46	04 37
28	15 49	14 43	03 54
29	15 51	14 40	02 58
30	15 52	14 37	01 54
31	15 ♊ 53	14 ♊ 34	00 N 45

DECLINATIONS

Date	Sun ☉	Moon ☽	Mercury ☿	Venus ♀	Mars ♂	Jupiter ♃	Saturn ♄	Uranus ♅	Neptune ♆	Pluto ♇
	° '	° '	° '	° '	° '	° '	° '	° '	° '	° '
01	21 S 45	08 S 27	25 S 43	07 S 19	01 S 36	22 S 37	13 S 00	21 S 47	22 S 16	03 N 41
02	21 55	13 43	25 47	07 42	01 50	22 38	13 02	21 48	22 16	03 40
03	22 04	18 15	25 49	08 05	02 03	22 40	13 04	21 49	22 16	03 40
04	22 12	21 46	25 50	08 28	02 16	22 41	13 06	21 49	22 16	03 40
05	22 20	24 07	25 49	08 52	02 30	22 42	13 08	21 50	22 16	03 39
06	22 27	25 10	25 47	09 15	02 43	22 43	13 10	21 50	22 16	03 39
07	22 35	24 55	25 43	09 38	02 56	22 44	13 12	21 51	22 16	03 39
08	22 41	23 29	25 38	10 01	03 09	22 45	13 13	21 51	22 16	03 39
09	22 47	21 00	25 32	10 24	03 23	22 46	13 15	21 52	22 16	03 38
10	22 53	17 39	25 24	10 47	03 36	22 47	13 17	21 53	22 16	03 38
11	22 59	13 37	25 15	11 09	03 49	22 48	13 19	21 53	22 16	03 38
12	23 03	09 04	25 05	11 32	04 02	22 49	13 21	21 54	22 16	03 38
13	23 08	04 S 08	24 54	11 54	04 15	22 50	13 22	21 54	22 16	03 37
14	23 12	01 N 02	24 41	12 16	04 27	22 51	13 24	21 55	22 16	03 37
15	23 15	06 17	24 27	12 38	04 40	22 52	13 26	21 55	22 16	03 37
16	23 18	11 24	24 13	13 00	04 53	22 53	13 28	21 56	22 16	03 37
17	23 21	16 10	23 58	13 21	05 05	22 54	13 29	21 56	22 16	03 37
18	23 23	20 16	23 42	13 43	05 18	22 55	13 31	21 57	22 16	03 37
19	23 24	23 20	23 26	14 04	05 31	22 55	13 33	21 57	22 16	03 37
20	23 26	25 00	23 09	14 24	05 43	22 56	13 34	21 58	22 16	03 37
21	23 26	25 00	22 52	14 45	05 56	22 57	13 36	21 58	22 17	03 36
22	23 27	23 19	22 36	15 05	06 08	22 58	13 37	21 59	22 17	03 36
23	23 26	20 04	22 20	15 25	06 20	22 58	13 39	21 59	22 17	03 36
24	23 26	15 36	22 04	15 45	06 32	22 59	13 41	22 00	22 17	03 36
25	23 24	10 17	21 49	16 04	06 45	23 00	13 42	22 00	22 17	03 36
26	23 23	04 N 30	21 34	16 23	06 57	23 00	13 44	22 01	22 17	03 36
27	23 20	01 S 24	21 21	16 42	07 09	23 01	13 45	22 01	22 17	03 36
28	23 18	07 08	21 08	17 00	07 20	23 01	13 46	22 02	22 17	03 36
29	23 15	12 27	20 56	17 18	07 32	23 02	13 48	22 02	22 17	03 36
30	23 11	17 06	20 45	17 35	07 44	23 03	13 49	22 02	22 17	03 36
31	23 S 07	20 S 51	20 S 36	17 S 52	07 S 56	23 S 03	13 S 51	22 S 03	22 S 17	03 N 36

ZODIAC SIGN ENTRIES

Date	h	m	Planets
01	09	41	☽ ♏
03	14	56	☽ ♐
04	11	22	☿ ♑
05	22	28	☽ ♑
06	16	15	♀ ♏
08	08	39	☽ ♒
10	20	53	☽ ♓
13	09	17	☽ ♈
15	19	33	☽ ♉
18	02	24	☽ ♊
20	06	02	☽ ♋
22	07	44	☽ ♌
22	10	30	☉ ♑
24	09	01	☽ ♍
26	11	18	☽ ♎
28	15	27	☽ ♏
30	21	44	☽ ♐

LATITUDES

Date	Mercury ☿	Venus ♀	Mars ♂	Jupiter ♃	Saturn ♄	Uranus ♅	Neptune ♆	Pluto ♇
	° '	° '	° '	° '	° '	° '	° '	° '
01	02 S 21	02 N 11	01 N 33	00 N 22	02 N 13	00 N 04	01 N 10	16 N 29
04	02 23	02 16	01 34	00 22	02 13	00 04	01 10	16 30
07	02 21	02 19	01 34	00 22	02 13	00 04	01 10	16 31
10	02 13	02 21	01 35	00 21	02 13	00 03	01 10	16 32
13	01 57	02 23	01 35	00 21	02 14	00 03	01 10	16 34
16	01 33	02 23	01 36	00 21	02 14	00 03	01 10	16 35
19	00 58	02 23	01 37	00 20	02 15	00 03	01 10	16 36
22	00 S 12	02 21	01 37	00 20	02 15	00 03	01 10	16 37
25	00 N 44	02 18	01 38	00 20	02 15	00 03	01 10	16 39
28	01 43	02 15	01 38	00 20	02 16	00 03	01 10	16 40
31	02 N 33	02 N 11	01 N 38	00 N 19	02 N 16	00 N 03	01 N 10	16 N 42

DATA

Julian Date	2445670
Delta T	+53 seconds
Ayanamsa	23° 37' 39"
Synetic vernal point	05° ♓ 29' 21"
True obliquity of ecliptic	23° 26' 31"

MOON'S PHASES, APSIDES AND POSITIONS ☽

Date	h	m	Phase	Longitude ° '	Eclipse Indicator
04	12	26	●	11 ♐ 47	Annular
12	13	09	☽	19 ♓ 56	
20	02	00	○	27 ♊ 36	
26	18	52	☾	04 ♎ 26	

Day	h	m		
11	00	34	Apogee	
22	18	36	Perigee	
06	19	24	Max dec	25° S 14'
14	07	14	0N	
21	00	15	Max dec	25° N 13'
27	06	18	0S	

LONGITUDES

Date	Chiron ⚷	Ceres ⚳	Pallas ⚴	Juno ⚵	Vesta ⚶	Black Moon Lilith ⚸
	° '	° '	° '	° '	° '	° '
01	29 ♉ 54	22 ♒ 23	24 ♑ 36	24 ♈ 27	23 ♊ 53	28 ♒ 47
11	29 ♉ 12	25 ♒ 35	27 ♑ 32	24 ♈ 55	21 ♊ 20	29 ♒ 54
21	28 ♉ 53	28 ♒ 12	00 ♒ 48	26 ♈ 17	18 ♊ 44	01 ♓ 02
31	28 ♉ 28	01 ♓ 26	04 ♒ 09	28 ♈ 28	16 ♊ 23	02 ♓ 09

ASPECTARIAN

h m	Aspects
01 Thursday	
01 09	☽ ⚹ ☿
06 32	☽ ⚹ ♆
06 43	☽ ∥ ♀
11 20	☽ ☌ ♇
14 37	☽ ⊥ ☉
14 43	☿ ∠ ♄
15 30	☽ ⊥ ♅
16 46	☽ ∠ ♃
17 30	☉ ∥ ♅
17 33	♀ ∠ ♅
23 32	☽ ⊻ ♂
02 Friday	
02 00	☽ ⊻ ☉
02 05	☽ ⊻ ♅
03 09	☉ ☌ ♅
04 58	☽ ☌ ♄
06 32	☽ ∠ ☿
08 39	☽ ∥ ♄
08 53	☽ ⊥ ♃
08 56	☽ ∠ ♆
10 36	☽ ⊥ ♂
19 43	☽ ⊻ ♃
03 Saturday	
00 26	☽ ⊥ ☿
01 00	☽ ⊥ ♆
03 21	☽ ∠ ♂
05 58	☿ ☌ ♆
08 00	☽ ⊻ ♀
11 49	☽ ⊻ ♆
12 30	☽ ⊻ ☿
16 47	☽ ⊻ ♇
19 53	☽ ⊥ ♀
22 16	☉ ⊻ ♄
04 Sunday	
03 44	☽ ⊥ ♇
07 47	☽ ⚹ ♂
08 14	☽ ☌ ♅
11 26	☽ ⊻ ♄
12 24	☽ ∥ ♅
12 26	☽ ☌ ☉
13 55	☽ ∠ ♀
15 46	☽ ∥ ☉
16 09	☽ ∥ ♆
18 56	♂ ⚹ ♅
19 57	☽ ∥ ♃
20 21	☽ ∠ ♇
22 37	☽ ⊥ ♄
23 46	☉ ∥ ♆
05 Monday	
03 21	☽ ☌ ♃
05 40	♀ ⚹ ♆
05 52	☿ ⚹ ♇
07 03	☽ Q ♂
15 36	☽ ∠ ♄
19 22	☽ ☌ ♆
20 41	☽ ⚹ ♀
06 Tuesday	
00 33	☽ ⚹ ♇
02 47	☽ ☌ ☿
16 52	☽ ⊻ ♅
18 51	☽ □ ♂
20 28	☽ ⚹ ♄
21 58	☽ Q ♀
23 37	☽ Q ♇
07 Wednesday	
02 03	☽ ⊻ ☉
04 32	☽ ⊥ ♅
13 43	☽ ⊻ ♃
14 47	☽ ⊥ ☉
16 12	♀ ☌ ♇
20 02	☽ Q ♄
22 13	☽ ∠ ♅
08 Thursday	
01 43	☽ ⊥ ♃
05 35	☽ ⊻ ♆
10 08	☽ ∠ ☉
10 59	☽ □ ♇
12 55	☽ □ ♀
13 43	♂ ⊻ ♄
17 32	☽ ⊥ ♆
19 54	☽ ∠ ♃
20 01	☽ ∥ ♃
20 22	☽ ∥ ☉
20 25	☽ ⊻ ☿
20 40	☉ ∠ ♇
09 Friday	
00 52	☽ ∥ ♆
04 09	☽ ⚹ ♅
04 36	☽ ∥ ♅
06 01	☉ ∥ ♃
08 07	☽ □ ♄
08 51	☽ △ ♂
09 51	☽ ⊥ ☿
11 35	☽ ∠ ♆
18 53	☽ ⚹ ☉
23 52	♀ ⊥ ♅
10 Saturday	
02 36	☽ ⚹ ♃
04 25	☽ Q ♅
06 07	☽ ∠ ☿
12 51	☉ ⊥ ♄
16 37	☽ bQ ♂
16 40	♂ # ♇
17 56	☽ ⚹ ♆
21 24	☽ Q ☉
23 25	☽ △ ♇
11 Sunday	
03 20	☽ Q ♃
07 47	☽ △ ♀
11 44	☽ ± ♂
13 39	☽ ∥ ♄
15 52	☽ ⚹ ☿
17 00	☽ □ ♅
18 18	☽ Q ♆
21 13	☽ △ ♄
12 Monday	
00 15	☽ ∥ ♀
00 28	☽ ⊼ ♂
04 04	☿ ⊻ ♅
05 16	♀ ∠ ♃
05 49	☽ bQ ♇
13 09	☽ □ ☉
16 22	☽ □ ♃
17 20	☽ bQ ♀
18 36	☽ Q ☿
23 56	☽ ± ♇
13 Tuesday	
03 38	☽ bQ ♄
06 34	☽ □ ♆
11 29	☽ ∥ ♂
11 55	☽ ⊼ ♇
13 13	☽ ± ♀
14 23	☽ # ♇
21 38	☽ ± ♄
14 Wednesday	
02 20	☽ ⊼ ♀
05 13	☽ △ ♅
07 54	☿ ⚹ ♄
09 31	☽ ⊼ ♄
09 39	☽ □ ☿
12 48	☉ ☌ ♃
15 00	☽ ☍ ♂
23 50	☽ ∥ ♇
15 Thursday	
04 18	☽ # ♂
04 31	☽ △ ♃
05 36	☽ △ ☉
09 55	♀ ⊻ ♅
10 23	☽ bQ ♅
11 49	☿ Q ♇
17 09	☽ △ ♆
22 11	☽ ☍ ♇
16 Friday	
03 28	☽ ± ♅
09 22	☽ bQ ♃
12 19	☽ bQ ☉
14 40	☽ ⊼ ♅
17 23	☽ ☍ ♀
18 54	☽ ☍ ♄
20 22	☽ # ♀
21 11	☽ bQ ♆
22 06	☽ # ♄
22 39	☽ △ ☿
17 Saturday	
02 08	☽ ⊼ ♂
02 13	☽ ± ♃
06 12	☽ ± ☉
11 34	♀ ☌ ♄
13 16	☽ ⊼ ♃
13 28	☽ ± ♂
13 33	☽ ± ♆
15 43	♂ Q ♆
17 54	☽ ⊼ ☉
21 46	♀ ∥ ♄
18 Sunday	
00 18	☽ ⊼ ♆
03 05	☽ bQ ☿
04 58	☽ ⊼ ♇
06 10	☽ bQ ♂
12 43	♀ ∠ ♆
15 31	☽ ± ♇
19 32	☽ ± ☿
20 34	☽ ☍ ♅
19 Monday	
00 01	☽ # ♅
00 41	☽ ⊼ ♄
02 35	☽ # ♆
03 50	☽ ⊼ ♀
06 16	☽ ⊼ ☿
07 07	☽ bQ ♇
08 09	☽ # ♃
09 18	☽ △ ♂
11 04	☽ ± ♄
12 50	☽ # ☉
12 55	☽ # ☿
13 34	☉ ∥ ☿
15 03	☽ ± ♀
18 28	☽ ☍ ♃
20 Tuesday	
02 00	☽ ☍ ☉
02 26	☽ bQ ♄
04 11	☽ ☍ ♆
07 38	☽ bQ ♀
08 35	☽ △ ♇
08 38	☉ ∠ ♄
15 29	☿ ⊥ ♅
19 05	☿ ⚹ ♀
23 35	☽ ⊼ ♅
21 Wednesday	
03 39	☽ △ ♄
05 57	☿ ∥ ♃
09 33	☽ ± ♅
09 41	☉ ☌ ♆
09 42	☽ ☍ ☿
10 49	☽ △ ♀
13 36	☽ □ ♂
16 38	♀ ⊥ ♃
21 15	☽ ⊼ ♃
22 Thursday	
00 27	☽ bQ ♅
00 52	☿ St R
06 03	☽ ⊼ ♆
07 16	☽ ± ♃
07 31	☽ ⊼ ☉
10 18	☽ □ ♇
10 45	☽ # ☉
15 13	☽ # ♃
15 55	☽ ± ♆
17 59	☿ ⊥ ♅
18 06	☽ ± ☉
18 57	☽ # ☿
20 20	☽ Q ♂
20 54	☽ # ♆
22 16	☽ bQ ♃
23 07	☽ # ♅
23 Friday	
01 09	☽ △ ♅
05 18	☽ □ ♄
06 43	☽ bQ ♆
09 58	☽ bQ ☉
10 44	☽ ⊼ ☿
15 53	☽ Q ♇
16 28	☽ □ ♀
16 45	☿ ∥ ♆
16 48	☽ ⚹ ♂
20 24	☽ ± ☿
23 17	☽ △ ♃
23 48	♀ ⊻ ♂
24 Saturday	
00 27	☉ ⚹ ♇
07 28	☽ △ ♆
10 51	☽ bQ ☿
11 08	☽ Q ♄
11 19	☽ # ♀
11 42	☽ ⚹ ♇
12 32	☽ △ ☉
18 02	☿ ∥ ♅
18 32	☽ ∠ ♂
20 59	☽ # ♄
25 Sunday	
00 53	☽ Q ♀
02 52	☽ □ ♅
07 13	☽ ⚹ ♄
10 08	☽ ⊥ ♂
10 52	☽ △ ☿
12 44	☽ ∠ ♇
13 07	☉ Q ♂
20 37	☽ ⊻ ♂
22 50	☽ ⚹ ♀
26 Monday	
02 06	☽ □ ♃
02 21	☽ # ♂
03 56	☽ ⊥ ♇
08 33	♀ ⊥ ♆
08 40	☽ ∠ ♄
09 18	☽ Q ♅
09 50	☽ □ ♆
14 09	☽ ⊻ ♇
15 40	☽ ∥ ♇
18 52	☽ □ ☉
27 Tuesday	
00 09	☽ ⊥ ♄
02 48	☽ ∠ ♀
05 57	☽ ⚹ ♅
09 25	☽ Q ♃
10 35	☽ ⊻ ♄
11 06	☽ □ ☿
14 29	☿ Q ♇
16 55	☽ Q ♆
17 19	☿ ⚹ ♄
19 54	☽ ⊥ ♀
21 05	☽ # ♇
21 46	♀ ⊻ ♃
28 Wednesday	
02 18	☽ ☌ ♂
04 36	☽ Q ☉
06 44	☽ ⚹ ♃
07 26	☽ ⊻ ♀
08 15	☽ ∠ ♅
12 54	☽ ∥ ♂
14 03	☽ ⚹ ♆
16 19	☽ Q ☿
18 30	☽ ☌ ♇
29 Thursday	
00 17	☽ ⊥ ♅
03 38	☽ ⚹ ☉
07 54	☿ ∠ ♀
09 53	☽ ∠ ♃
11 05	☽ ⊻ ♅
11 58	☽ ⚹ ☿
16 02	☽ ☌ ♄
16 59	☽ ∠ ♆
18 33	☽ ∥ ♄
20 32	☿ ⊻ ♅
30 Friday	
02 27	☽ ⊥ ♃
09 02	☽ ∠ ☉
09 24	☽ ⊥ ♆
10 22	☽ ⊻ ♂
12 45	☽ ∠ ☿
13 37	☽ ⊻ ♃
14 59	☽ ∥ ♀
18 51	☽ ☌ ♀
20 27	☽ ⊻ ♆
21 52	☽ ⊥ ♂
31 Saturday	
01 00	☽ ⊻ ♇
03 00	☽ ⊥ ☉
03 47	☽ ⊥ ☿
07 41	☉ ☌ ☿
10 11	☽ ∥ ☿
12 09	☽ ⊥ ♇
12 36	♀ ⊻ ♆
13 52	☽ ⊻ ☿
15 07	☽ ⊻ ☉
15 18	☽ ∠ ♂
18 24	☽ ☌ ♅
21 35	☽ ∥ ♅
23 30	☽ ∥ ♆
23 41	☽ ⊻ ♄

All ephemeris data is given at 12.00 UT and the Moon's longitude is additionally given for 24.00 UT

JANUARY 1984

LONGITUDES

Date	Sidereal time h m s	Sun ☉ ° ' "	Moon ☽ ° ' "	Moon ☽ 24.00 ° ' "	Mercury ☿ ° '	Venus ♀ ° '	Mars ♂ ° '	Jupiter ♃ ° '	Saturn ♄ ° '	Uranus ♅ ° '	Neptune ♆ ° '	Pluto ♇ ° '
01	18 41 21	10 ♑ 15 17	20 ♐ 28 38	26 ♐ 47 42	07 ♑ 27	00 ♐ 30	24 ♎ 56	26 ♐ 01	13 ♏ 59	11 ♐ 10	29 ♐ 22	01 ♏ 48
02	18 45 18	11 16 28	03 ♑ 03 58	09 ♑ 17 32	06 R 08	01 42	25 28	26 15	14 04	11 13	29 24	01 49
03	18 49 14	12 17 39	15 ♑ 28 29	21 ♑ 36 57	04 54	02 55	26 00	26 28	14 09	11 17	29 26	01 50
04	18 53 11	13 18 50	27 ♑ 43 03	03 ♒ 46 56	03 47	04 07	26 32	26 42	14 14	11 20	29 29	01 51
05	18 57 07	14 20 01	09 ♒ 48 46	15 ♒ 48 45	02 49	05 19	27 03	26 55	14 18	11 23	29 31	01 52
06	19 01 04	15 21 12	21 ♒ 47 10	27 ♒ 44 18	02 00	06 32	27 35	27 08	14 23	11 26	29 33	01 53
07	19 05 00	16 22 23	03 ♓ 40 27	09 ♓ 36 01	01 21	07 44	28 06	27 22	14 28	11 30	29 35	01 54
08	19 08 57	17 23 33	15 ♓ 31 25	21 ♓ 27 07	00 52	08 57	28 38	27 35	14 32	11 33	29 37	01 55
09	19 12 53	18 24 43	27 ♓ 23 36	03 ♈ 21 24	00 34	10 10	29 09	27 48	14 37	11 36	29 40	01 56
10	19 16 50	19 25 52	09 ♈ 21 05	15 ♈ 23 15	00 24	11 22	29 ♎ 40	28 01	14 41	11 39	29 42	01 57
11	19 20 47	20 27 01	21 ♈ 28 30	27 ♈ 37 26	00 D 24	12 35	00 ♏ 11	28 14	14 45	11 42	29 44	01 58
12	19 24 43	21 28 09	03 ♉ 50 40	10 ♉ 08 46	00 32	13 48	00 42	28 28	14 50	11 45	29 46	01 59
13	19 28 40	22 29 17	16 ♉ 32 18	23 ♉ 01 47	00 48	15 01	01 13	28 41	14 54	11 48	29 48	02 00
14	19 32 36	23 30 24	29 ♉ 37 36	06 ♊ 20 07	01 11	16 14	01 43	28 54	14 58	11 51	29 50	02 00
15	19 36 33	24 31 30	13 ♊ 09 32	20 ♊ 05 54	01 41	17 27	02 14	29 07	15 02	11 54	29 52	02 01
16	19 40 29	25 32 36	27 ♊ 09 09	04 ♋ 18 58	02 16	18 40	02 44	29 20	15 06	11 57	29 55	02 02
17	19 44 26	26 33 41	11 ♋ 34 53	18 ♋ 56 14	02 56	19 53	03 14	29 33	15 09	12 00	29 57	02 03
18	19 48 22	27 34 46	26 ♋ 22 10	03 ♌ 51 39	03 42	21 06	03 44	29 46	15 13	12 03	29 ♐ 59	02 03
19	19 52 19	28 35 50	11 ♌ 23 32	18 ♌ 56 35	04 31	22 19	04 14	29 ♐ 58	15 17	12 06	00 ♑ 01	02 04
20	19 56 16	29 ♑ 36 54	26 ♌ 29 32	04 ♍ 01 10	05 25	23 32	04 44	00 ♑ 11	15 20	12 09	00 03	02 04
21	20 00 12	00 ♒ 37 56	11 ♍ 30 18	18 ♍ 55 56	06 22	24 46	05 14	00 24	15 24	12 11	00 05	02 05
22	20 04 09	01 38 59	26 ♍ 17 12	03 ♎ 33 23	07 22	25 59	05 43	00 37	15 27	12 14	00 07	02 05
23	20 08 05	02 40 01	10 ♎ 44 01	17 ♎ 48 48	08 24	27 12	06 12	00 49	15 30	12 17	00 09	02 06
24	20 12 02	03 41 02	24 ♎ 47 35	01 ♏ 40 24	09 30	28 26	06 41	01 02	15 34	12 19	00 11	02 06
25	20 15 58	04 42 03	08 ♏ 27 23	15 ♏ 08 49	10 37	29 ♐ 39	07 10	01 14	15 37	12 22	00 13	02 07
26	20 19 55	05 43 04	21 ♏ 44 59	28 ♏ 16 16	11 47	00 ♑ 52	07 39	01 27	15 40	12 25	00 15	02 07
27	20 23 51	06 44 04	04 ♐ 43 05	11 ♐ 05 50	12 59	02 06	08 08	01 39	15 43	12 27	00 17	02 07
28	20 27 48	07 45 04	17 ♐ 24 55	23 ♐ 40 44	14 13	03 19	08 36	01 52	15 45	12 30	00 19	02 07
29	20 31 45	08 46 02	29 ♐ 53 37	06 ♑ 03 55	15 28	04 33	09 04	02 04	15 48	12 32	00 20	02 08
30	20 35 41	09 47 00	12 ♑ 11 56	18 ♑ 17 54	16 44	05 47	09 32	02 16	15 51	12 34	00 22	02 08
31	20 39 38	10 ♒ 47 58	24 ♑ 22 04	00 ♒ 24 37	18 ♑ 03	07 ♑ 00	10 ♏ 00	02 ♑ 28	15 ♏ 53	12 ♐ 37	00 ♑ 24	02 ♏ 08

Date	Moon True ☊ ° '	Moon Mean ☊ ° '	Moon ☽ Latitude ° '
01	15 ♊ 54	14 ♊ 30	00 S 25
02	15 R 52	14 27	01 33
03	15 50	14 24	02 35
04	15 46	14 21	03 29
05	15 41	14 18	04 13
06	15 35	14 14	04 46
07	15 30	14 11	05 05
08	15 25	14 08	05 12
09	15 21	14 05	05 05
10	15 19	14 02	04 45
11	15 D 18	13 59	04 12
12	15 19	13 55	03 27
13	15 21	13 52	02 30
14	15 22	13 49	01 25
15	15 23	13 46	00 S 12
16	15 R 23	13 43	01 N 03
17	15 20	13 40	02 17
18	15 16	13 36	03 23
19	15 09	13 33	04 17
20	15 02	13 30	04 52
21	14 55	13 27	05 07
22	14 49	13 24	05 02
23	14 45	13 20	04 37
24	14 43	13 17	03 56
25	14 D 43	13 14	03 03
26	14 44	13 11	02 01
27	14 45	13 08	00 N 54
28	14 R 45	13 05	00 S 14
29	14 44	13 01	01 20
30	14 40	12 58	02 22
31	14 ♊ 34	12 ♊ 55	03 S 15

DECLINATIONS

Date	Sun ☉ ° '	Moon ☽ ° '	Mercury ☿ ° '	Venus ♀ ° '	Mars ♂ ° '	Jupiter ♃ ° '	Saturn ♄ ° '	Uranus ♅ ° '	Neptune ♆ ° '	Pluto ♇ ° '
01	23 S 03	23 S 31	20 S 28	18 S 09	08 S 07	23 S 04	13 S 52	22 S 04	22 S 17	03 N 37
02	22 58	24 57	20 21	18 25	08 19	23 04	13 53	22 04	22 17	03 37
03	22 52	25 07	20 16	18 41	08 30	23 05	13 54	22 05	22 17	03 37
04	22 47	24 02	20 12	18 56	08 42	23 05	13 56	22 05	22 17	03 37
05	22 40	21 51	20 10	19 11	08 53	23 05	13 57	22 06	22 17	03 37
06	22 33	18 45	20 10	19 26	09 04	23 06	13 58	22 06	22 17	03 37
07	22 26	14 54	20 11	19 39	09 15	23 06	13 59	22 07	22 17	03 37
08	22 19	10 30	20 14	19 53	09 26	23 06	14 01	22 07	22 17	03 38
09	22 11	05 42	20 19	20 06	09 37	23 07	14 02	22 07	22 17	03 38
10	22 02	00 S 40	20 24	20 18	09 48	23 07	14 03	22 08	22 17	03 38
11	21 53	04 N 29	20 31	20 30	09 59	23 07	14 04	22 08	22 17	03 38
12	21 44	09 33	20 38	20 41	10 09	23 07	14 05	22 09	22 17	03 38
13	21 34	14 23	20 46	20 52	10 20	23 08	14 06	22 09	22 17	03 39
14	21 24	18 42	20 55	21 02	10 30	23 08	14 07	22 10	22 17	03 39
15	21 13	22 11	21 04	21 11	10 41	23 08	14 08	22 10	22 17	03 39
16	21 02	24 28	21 13	21 20	10 51	23 08	14 09	22 10	22 17	03 40
17	20 51	25 13	21 22	21 29	11 01	23 08	14 10	22 11	22 17	03 40
18	20 39	24 12	21 30	21 36	11 11	23 08	14 11	22 11	22 17	03 40
19	20 27	21 28	21 39	21 44	11 21	23 08	14 11	22 12	22 17	03 41
20	20 14	17 16	21 47	21 50	11 31	23 09	14 12	22 12	22 17	03 41
21	20 01	11 59	21 54	21 56	11 41	23 09	14 13	22 12	22 17	03 41
22	19 48	06 N 06	22 01	22 02	11 51	23 09	14 14	22 13	22 17	03 42
23	19 34	00 00	22 07	22 06	12 00	23 09	14 15	22 13	22 17	03 42
24	19 20	05 S 56	22 12	22 10	12 10	23 09	14 15	22 13	22 17	03 42
25	19 05	11 26	22 16	22 14	12 19	23 09	14 16	22 14	22 17	03 43
26	18 51	16 16	22 20	22 16	12 28	23 09	14 17	22 14	22 17	03 43
27	18 35	20 12	22 22	22 19	12 38	23 08	14 17	22 14	22 17	03 44
28	18 20	23 05	22 24	22 20	12 47	23 08	14 18	22 15	22 17	03 44
29	18 04	24 47	22 24	22 21	12 56	23 08	14 18	22 15	22 17	03 45
30	17 48	25 14	22 24	22 21	13 04	23 08	14 19	22 15	22 17	03 45
31	17 S 32	24 S 27	22 S 22	22 S 21	13 S 13	23 S 08	14 S 19	22 S 16	22 S 17	03 N 46

ZODIAC SIGN ENTRIES

Date	h	m	Planets
01	02	00	♀ ♐
02	06	07	☽ ♑
04	16	30	☽ ♒
07	04	34	☽ ♓
09	17	15	☽ ♈
11	03	20	♂ ♏
12	04	36	☽ ♉
14	12	40	☽ ♊
16	16	47	☽ ♋
18	17	50	☽ ♌
19	02	55	♆ ♑
19	15	04	♃ ♑
20	17	35	☽ ♍
20	21	05	☉ ♒
22	18	07	☽ ♎
24	21	04	☽ ♏
25	18	51	♀ ♑
27	03	12	☽ ♐
29	12	12	☽ ♑
31	23	11	☽ ♒

LATITUDES

Date	Mercury ☿ ° '	Venus ♀ ° '	Mars ♂ ° '	Jupiter ♃ ° '	Saturn ♄ ° '	Uranus ♅ ° '	Neptune ♆ ° '	Pluto ♇ ° '
01	02 N 47	02 N 09	01 N 39	00 N 19	02 N 16	00 N 03	01 N 10	16 N 42
04	03 11	02 04	01 39	00 19	02 17	00 03	01 10	16 44
07	03 15	01 58	01 39	00 19	02 18	00 03	01 10	16 45
10	03 02	01 52	01 40	00 19	02 18	00 03	01 10	16 47
13	02 40	01 45	01 40	00 18	02 19	00 03	01 10	16 49
16	02 13	01 38	01 40	00 18	02 19	00 03	01 10	16 50
19	01 43	01 30	01 41	00 18	02 20	00 03	01 10	16 52
22	01 14	01 21	01 41	00 18	02 20	00 03	01 10	16 54
25	00 45	01 13	01 41	00 18	02 21	00 03	01 10	16 55
28	00 N 17	01 04	01 41	00 17	02 22	00 03	01 10	16 57
31	00 S 09	00 N 55	01 N 41	00 N 17	02 N 22	00 N 03	01 N 10	16 N 59

DATA

Julian Date	2445701
Delta T	+54 seconds
Ayanamsa	23° 37' 45"
Synetic vernal point	05° ♓ 29' 15"
True obliquity of ecliptic	23° 26' 31"

MOON'S PHASES, APSIDES AND POSITIONS ☽

Date	h	m	Phase	Longitude ° '	Eclipse Indicator
03	05	16	●	12 ♑ 00	
11	09	48	☽	20 ♈ 21	
18	14	05	○	27 ♋ 40	
25	04	48	☾	04 ♏ 24	

Day	h	m		
07	20	12	Apogee	
19	21	45	Perigee	
03	02	58	Max dec	25° S 12'
10	15	06	0N	
17	10	27	Max dec	25° N 13'
23	12	00	0S	
30	08	38	Max dec	25° S 15'

LONGITUDES

Date	Chiron ⚷ ° '	Ceres ⚳ ° '	Pallas ⚴ ° '	Juno ⚵ ° '	Vesta ⚶ ° '	Black Moon Lilith ⚸ ° '
01	28 ♉ 26	01 ♓ 46	04 ♒ 29	28 ♈ 43	16 ♊ 10	02 ♓ 16
11	28 ♉ 07	05 ♓ 12	07 ♒ 53	01 ♉ 39	14 ♊ 24	03 ♓ 23
21	27 ♉ 53	08 ♓ 47	11 ♒ 18	05 ♉ 09	13 ♊ 19	04 ♓ 30
31	27 ♉ 47	12 ♓ 28	14 ♒ 44	09 ♉ 07	13 ♊ 00	05 ♓ 37

All ephemeris data is given at 12.00 UT and the Moon's longitude is additionally given for 24.00 UT

ASPECTARIAN

h m	Aspects
01 Sunday	
05 02	☽ ∠ ♇
06 58	☽ ∥ ♃
07 00	☽ ∥ ☉
07 59	☉ ∥ ♃
11 03	☽ ⊥ ♄
18 38	☿ Q ♂
20 49	☽ ✶ ♂
20 57	☿ ⊥ ♀
22 43	☽ ☌ ♃
02 Monday	
03 24	♀ ⊥ ♂
04 17	☽ ∠ ♄
04 57	☽ ☌ ♆
09 07	☽ ⊻ ♀
09 36	☽ ✶ ♇
10 42	☉ ⊻ ♅
14 16	♀ ⊻ ♇
17 21	☽ ☌ ☿
20 51	☽ Q ♂
21 54	☽ ⊥ ♀
03 Tuesday	
03 48	☽ ⊻ ♅
05 16	☽ ☌ ☉
08 48	☽ Q ♇
09 24	☽ ✶ ♄
15 32	☽ ⊥ ♅
17 16	☽ ∠ ♀
04 Wednesday	
02 06	♂ ∠ ♅
08 33	☿ ⊻ ♀
09 03	☽ Q ♄
09 15	☽ ∠ ♅
09 33	☽ □ ♂
09 57	☽ ⊻ ♃
15 29	☽ ⊻ ♆
20 11	☽ □ ♇
22 01	☽ ⊥ ♃
23 05	☽ ⊻ ☿
23 53	☽ ∥ ♃
05 Thursday	
00 55	♂ ✶ ♃
00 57	☉ Q ♇
02 04	☽ ✶ ♀
03 25	☽ ⊥ ♆
03 53	☽ ∥ ☉
08 06	☽ ∥ ♆
09 50	☽ ∥ ♅
10 09	☽ ⊥ ☿
11 20	☉ ✶ ♄
15 09	☽ ✶ ♅
16 17	☽ ∠ ♃
21 03	☽ □ ♄
21 26	☽ ∠ ♆
21 53	☽ ⊻ ☉
06 Friday	
01 50	☽ ∥ ☿
02 58	☽ ∠ ☿
04 43	☽ Q ♀
07 34	☽ ∥ ♀
11 03	☽ ⊥ ☉
15 21	☽ Q ♅
15 40	☿ ✶ ♇
23 00	☽ ✶ ♃
07 Saturday	
00 14	☽ △ ♂
03 43	☽ ✶ ♆
06 54	☽ ∠ ☉
07 31	☽ ✶ ☿
08 25	☽ △ ♇
15 20	♀ ⊥ ♇
17 09	☽ ∥ ♄
21 10	☽ □ ♀
23 44	☽ Q ♃
08 Sunday	
03 54	☽ □ ♅
04 04	☽ Q ♆
06 49	☽ Q ☿
07 59	☽ Ꝑ ♂
09 59	☽ △ ♄
14 50	☽ Ꝑ ♇
15 49	☉ ⊥ ♅
16 08	☽ ✶ ☉
17 13	☽ ∥ ♂
17 43	☉ ∥ ♆
09 Monday	
03 03	☽ ± ♂
09 04	☽ ± ♇
12 50	☽ □ ♃
15 42	☽ ⊼ ♂
16 30	☽ Ꝑ ♄
16 35	☽ □ ♆
18 16	☽ □ ☿
18 39	☽ Q ☉
20 30	☉ ∥ ♅
21 10	☽ ⊼ ♇
21 57	☽ # ♇
10 Tuesday	
10 40	☽ ± ♄
13 19	♂ ✶ ♆
16 29	☽ △ ♀
16 36	☽ △ ♅
17 43	♀ ☌ ♅
19 00	♄ ∠ ♆
22 40	☽ ⊼ ♄
11 Wednesday	
00 38	☿ St D
08 05	☽ ∥ ♇
09 48	☽ □ ☉
18 39	☿ ∥ ♀
22 15	☽ Ꝑ ♅
12 Thursday	
00 27	☿ ✶ ♂
01 14	☽ Ꝑ ♀
01 26	☽ △ ♃
04 08	☽ △ ♆
05 32	☽ △ ☿
05 42	☽ ☍ ♂
08 25	☽ ☍ ♇
15 01	☽ # ♂
15 40	☽ ± ♅
20 21	☽ ± ♀
13 Friday	
03 06	☽ ⊼ ♅
06 34	☽ Ꝑ ♃
08 45	☽ Ꝑ ♆
08 52	☽ ⊼ ♀
08 55	☽ ☍ ♄
09 30	♀ ⊻ ♄
10 32	☽ # ♄
10 35	☽ Ꝑ ☿
23 33	☽ ± ♃
23 56	☽ △ ☉
14 Saturday	
01 27	☽ ± ♆
03 41	☽ ± ☿
10 39	☽ ⊼ ♃
12 23	☽ ⊼ ♆
14 54	☽ ⊼ ☿
15 55	☽ ⊼ ♂
16 17	☽ ⊼ ♇
15 Sunday	
01 49	♂ ☌ ♇
02 58	☽ ± ♇
03 01	☽ ± ♂
03 06	☽ # ☿
03 29	♀ ∠ ♇
04 02	☽ # ♀
04 38	♀ ∠ ♂
05 00	☽ # ☉
05 07	☽ Ꝑ ☉
09 48	☽ ☍ ♅
11 53	☽ # ♅
12 49	☽ # ♆
14 07	☉ ∥ ♀
15 16	☽ ⊼ ♄
18 42	☽ Ꝑ ♇
19 20	☽ Ꝑ ♂
20 09	☽ ☍ ♀
20 23	☽ # ♃
22 02	☽ ± ☉
23 17	☉ ∥ ☿
16 Monday	
01 40	☽ ± ♄
02 47	☿ ✶ ♇
09 04	☽ ⊼ ☉
15 43	☽ ☍ ♃
16 39	☽ ☍ ♆
16 58	☽ Ꝑ ♄
20 12	☽ △ ♇
20 59	☽ ☍ ☿
21 42	☽ △ ♂
17 Tuesday	
12 41	☽ ⊼ ♅
17 53	☽ △ ♄
22 31	☽ ± ♅
22 52	☉ ∠ ♅
18 Wednesday	
02 45	☽ ⊼ ♀
02 58	☉ Q ♄
13 06	☽ Ꝑ ♅
13 17	☽ ± ♀
14 05	☽ ☍ ☉
14 28	♀ ⊥ ♄
15 33	☿ ✶ ♂
17 31	☽ ⊼ ♃
17 48	☽ ⊼ ♆
21 07	☽ □ ♇
23 09	☽ # ♃
19 Thursday	
00 13	☽ □ ♂
00 24	☽ ⊼ ☿
03 15	☽ ± ♃
03 25	☽ ± ♆
04 57	☽ Ꝑ ♀
06 13	☽ # ♆
06 53	☽ # ♅
10 17	☽ # ♀
10 32	☽ ± ☿
10 52	☽ # ☿
13 07	☽ △ ♅
17 23	♃ ☌ ♆
17 46	☽ Ꝑ ♆
17 46	☽ Ꝑ ♃
18 12	☽ □ ♄
18 58	☽ # ☉
20 Friday	
01 43	☽ Ꝑ ☿
01 47	☽ Q ♇
05 49	☽ Q ♂
06 54	☽ △ ♀
17 20	☽ ⊼ ☉
17 40	☽ △ ♆
17 58	☽ △ ♃
20 48	☉ ⊥ ♀
20 54	☽ ✶ ♇
22 33	☉ ⊻ ♆
22 57	☽ Q ♄
21 Saturday	
01 35	☽ ✶ ♂
02 19	☽ # ♄
03 10	☽ △ ☿
03 37	☽ ± ☉
05 01	☉ ⊻ ♃
11 44	♃ ∠ ♄
13 06	☽ □ ♅
13 14	☽ # ♂
18 18	☽ ✶ ♄
19 09	☽ Ꝑ ☉
21 00	☽ ∠ ♇
22 Sunday	
02 35	☽ ∠ ♂
11 27	☽ □ ♀
11 40	☽ ⊥ ♇
17 50	☽ ⊥ ♂
18 19	☽ □ ♆
18 31	☽ Q ♅
18 53	☽ ∠ ♄
19 13	☽ □ ♃
21 27	☽ ∥ ♇
21 30	☽ △ ☉
21 34	☽ ⊻ ♇
22 25	☉ □ ♇
23 Monday	
02 37	☿ ∥ ♀
04 09	☽ ⊻ ♂
07 47	☽ □ ☿
09 56	☽ ⊥ ♄
14 37	☽ ✶ ♅
16 03	♂ ⊥ ♅
20 06	☽ ⊻ ♄
20 16	☽ Q ♀
24 Tuesday	
00 36	☽ Q ♆
01 56	☽ Q ♃
02 49	☽ # ♇
16 24	☽ ∠ ♅
17 06	☽ Q ☿
18 56	☽ ✶ ♀
19 34	☿ ∥ ♅
21 24	☽ ✶ ♆
23 02	☽ ✶ ♃
25 Wednesday	
00 46	☽ ☌ ♇
04 48	☽ □ ☉
08 16	☽ ⊥ ♅
09 38	☽ ☌ ♂
12 05	♀ ∥ ♅
13 38	☿ ∥ ♆
16 14	☽ ∥ ♂
16 14	☽ ✶ ☿
19 01	☽ ⊻ ♅
23 20	♀ ☌ ♆
26 Thursday	
00 09	☽ ∠ ♆
00 14	☽ ∠ ♀
00 53	☽ ☌ ♄
01 36	☽ ∥ ♄
02 12	☽ ∠ ♃
07 37	♀ ∠ ♄
13 39	♀ ∥ ♆
15 55	☽ Q ☉
16 35	☽ ⊥ ♆
18 20	☽ ⊥ ♀
18 54	☽ ⊥ ♃
22 11	☽ ∠ ☿
27 Friday	
00 52	☉ ⊥ ♆
01 01	☿ ⊻ ♅
01 29	♀ ☌ ♃
02 06	☽ ∥ ☉
03 42	☽ ⊻ ♆
06 11	☽ ⊻ ♃
06 36	☽ ⊻ ♀
07 09	☽ ⊻ ♇
12 23	♀ ✶ ♇
16 06	☽ ✶ ☉
16 42	☽ ⊥ ☿
18 23	☽ ⊥ ♇
18 39	☽ ⊻ ♂
28 Saturday	
02 37	☽ ☌ ♅
03 56	☽ ∥ ♅
04 14	☽ ∥ ♆
04 42	☽ ∥ ♀
05 14	☽ ⊻ ☿
05 17	☽ ∥ ☿
06 26	☽ ⊥ ♂
08 50	☽ ⊻ ♄
10 20	☿ Q ♇
11 27	☽ ∠ ♇
12 34	☽ ∥ ♃
15 12	☉ ⊥ ♃
20 20	☽ ⊥ ♄
23 07	☽ ∠ ☉
29 Sunday	
00 19	☽ ∠ ♂
12 52	☽ ☌ ♆
13 46	☽ ∠ ♄
16 17	☽ ☌ ♃
16 20	☽ ✶ ♇
18 05	☽ ⊥ ☉
18 38	☿ ✶ ♄
19 31	♃ ✶ ♇
22 03	☽ ☌ ♀
30 Monday	
01 20	☉ □ ♂
06 35	☽ ✶ ♂
06 50	☽ ⊻ ☉
12 44	☽ ⊻ ♅
15 48	☽ Q ♇
19 11	☽ ✶ ♄
22 00	☽ ☌ ☿
31 Tuesday	
00 35	☽ ⊥ ♅
07 08	☽ Q ♂
18 28	☽ ∠ ♅
19 00	☽ Q ♄
22 43	☿ ⊥ ♅

FEBRUARY 1984

LONGITUDES

Date	Sidereal time h m s	Sun ☉ ° ' "	Moon ☽ ° ' "	Moon ☽ 24.00 ° ' "	Mercury ☿ ° '	Venus ♀ ° '	Mars ♂ ° '	Jupiter ♃ ° '	Saturn ♄ ° '	Uranus ♅ ° '	Neptune ♆ ° '	Pluto ♇ ° '
01	20 43 34	11 ♒ 48 54	06 ♒ 25 44	12 ♒ 25 34	19 ♑ 22	08 ♑ 14	10 ♏ 28	02 ♑ 40	15 ♏ 56	12 ♐ 39	00 ♑ 26	02 ♏ 08
02	20 47 31	12 49 49	18 ♒ 24 16	24 ♒ 22 00	20 43	09 27	10 55	02 52	15 58	12 41	00 28	02 08
03	20 51 27	13 50 43	00 ♓ 18 54	06 ♓ 15 10	22 04	10 41	11 23	03 04	16 00	12 44	00 30	02 08
04	20 55 24	14 51 36	12 ♓ 10 58	18 ♓ 06 34	23 27	11 55	11 50	03 16	16 02	12 46	00 31	02 08
05	20 59 20	15 52 27	24 ♓ 02 12	29 ♓ 58 11	24 51	13 08	12 16	03 28	16 04	12 48	00 33	02 R 08
06	21 03 17	16 53 18	05 ♈ 54 52	11 ♈ 52 40	26 16	14 22	12 43	03 40	16 06	12 50	00 35	02 08
07	21 07 14	17 54 06	17 ♈ 52 00	23 ♈ 53 22	27 42	15 36	13 09	03 51	16 08	12 52	00 37	02 08
08	21 11 10	18 54 54	29 ♈ 57 18	06 ♉ 04 22	29 ♑ 09	16 50	13 35	04 03	16 10	12 54	00 38	02 08
09	21 15 07	19 55 39	12 ♉ 15 08	18 ♉ 30 14	00 ♒ 37	18 03	14 01	04 15	16 11	12 56	00 40	02 08
10	21 19 03	20 56 24	24 ♉ 50 17	01 ♊ 15 51	02 06	19 17	14 27	04 26	16 13	12 58	00 41	02 07
11	21 23 00	21 57 07	07 ♊ 47 32	14 ♊ 25 50	03 36	20 31	14 52	04 37	16 14	13 00	00 43	02 07
12	21 26 56	22 57 48	21 ♊ 11 10	28 ♊ 03 50	05 06	21 45	15 18	04 49	16 15	13 02	00 45	02 07
13	21 30 53	23 58 27	05 ♋ 04 01	12 ♋ 11 42	06 38	22 59	15 43	05 00	16 17	13 04	00 46	02 07
14	21 34 49	24 59 05	19 ♋ 26 38	26 ♋ 48 23	08 10	24 13	16 07	05 11	16 18	13 05	00 48	02 06
15	21 38 46	25 59 41	04 ♌ 16 14	11 ♌ 49 15	09 43	25 26	16 32	05 22	16 19	13 07	00 49	02 06
16	21 42 43	27 00 16	19 ♌ 26 17	27 ♌ 06 00	11 18	26 40	16 56	05 33	16 20	13 09	00 51	02 05
17	21 46 39	28 00 49	04 ♍ 46 56	12 ♍ 27 34	12 53	27 54	17 20	05 44	16 20	13 10	00 52	02 05
18	21 50 36	29 ♒ 01 20	20 ♍ 06 25	27 ♍ 42 05	14 28	29 ♑ 08	17 43	05 54	16 21	13 12	00 54	02 05
19	21 54 32	00 ♓ 01 50	05 ♎ 13 18	12 ♎ 39 03	16 05	00 ♒ 22	18 07	06 05	16 22	13 13	00 55	02 04
20	21 58 29	01 02 19	19 ♎ 58 30	27 ♎ 11 06	17 43	01 36	18 30	06 15	16 22	13 15	00 56	02 03
21	22 02 25	02 02 46	04 ♏ 16 30	11 ♏ 14 35	19 21	02 50	18 52	06 26	16 23	13 16	00 58	02 03
22	22 06 22	03 03 12	18 ♏ 05 27	24 ♏ 49 19	21 01	04 04	19 15	06 36	16 23	13 17	00 59	02 02
23	22 10 18	04 03 37	01 ♐ 26 33	07 ♐ 57 37	22 41	05 18	19 37	06 46	16 23	13 19	01 00	02 02
24	22 14 15	05 04 01	14 ♐ 23 01	20 ♐ 43 20	24 23	06 32	19 59	06 57	16 23	13 20	01 01	02 01
25	22 18 12	06 04 23	26 ♐ 59 06	03 ♑ 10 55	26 05	07 46	20 20	07 07	16 R 23	13 21	01 03	02 00
26	22 22 08	07 04 43	09 ♑ 19 19	15 ♑ 24 49	27 49	09 00	20 41	07 17	16 23	13 22	01 04	01 59
27	22 26 05	08 05 02	21 ♑ 27 54	27 ♑ 29 00	29 ♒ 33	10 14	21 02	07 26	16 23	13 23	01 05	01 59
28	22 30 01	09 05 20	03 ♒ 28 32	09 ♒ 26 50	01 ♓ 19	11 28	21 22	07 36	16 22	13 24	01 06	01 58
29	22 33 58	10 ♓ 05 36	15 ♒ 24 12	21 ♒ 20 54	03 ♓ 05	12 ♒ 42	21 ♏ 42	07 ♑ 46	16 ♏ 22	13 ♐ 25	01 ♑ 07	01 ♏ 57

Date	Moon True ☊ ° '	Moon Mean ☊ ° '	Moon ☽ Latitude ° '
01	14 ♊ 25	12 ♊ 52	04 S 00
02	14 R 14	12 49	04 33
03	14 01	12 46	04 54
04	13 49	12 42	05 03
05	13 38	12 39	04 58
06	13 28	12 36	04 40
07	13 21	12 33	04 10
08	13 17	12 30	03 29
09	13 15	12 26	02 37
10	13 D 15	12 23	01 36
11	13 15	12 20	00 S 29
12	13 R 15	12 17	00 N 42
13	13 13	12 14	01 53
14	13 09	12 11	02 59
15	13 02	12 07	03 55
16	12 53	12 04	04 36
17	12 42	12 01	04 58
18	12 31	11 58	04 58
19	12 21	11 55	04 37
20	12 13	11 52	03 59
21	12 08	11 48	03 06
22	12 06	11 45	02 03
23	12 D 05	11 42	00 N 56
24	12 R 05	11 39	00 S 12
25	12 05	11 36	01 18
26	12 02	11 32	02 18
27	11 57	11 29	03 12
28	11 49	11 26	03 56
29	11 ♊ 38	11 ♊ 23	04 S 29

DECLINATIONS

Date	Sun ☉	Moon ☽	Mercury ☿	Venus ♀	Mars ♂	Jupiter ♃	Saturn ♄	Uranus ♅	Neptune ♆	Pluto ♇
01	17 S 15	22 S 32	22 S 19	22 S 19	13 S 22	23 S 08	14 S 20	22 S 16	22 S 17	03 N 46
02	16 58	19 38	22 15	22 18	13 30	23 08	14 20	22 16	22 16	03 46
03	16 40	15 57	22 10	22 15	13 39	23 07	14 21	22 17	22 16	03 47
04	16 23	11 39	22 04	22 12	13 47	23 07	14 21	22 17	22 16	03 48
05	16 05	06 56	21 56	22 08	13 55	23 07	14 22	22 17	22 16	03 48
06	15 47	01 S 56	21 47	22 04	14 03	23 07	14 22	22 17	22 16	03 49
07	15 28	03 N 09	21 37	21 58	14 11	23 06	14 22	22 18	22 16	03 49
08	15 09	08 12	21 25	21 53	14 19	23 06	14 22	22 18	22 16	03 50
09	14 50	13 01	21 13	21 46	14 27	23 06	14 22	22 18	22 16	03 50
10	14 31	17 25	20 59	21 39	14 34	23 06	14 23	22 18	22 16	03 51
11	14 12	21 08	20 43	21 31	14 42	23 05	14 23	22 19	22 16	03 51
12	13 52	23 51	20 27	21 23	14 49	23 05	14 23	22 19	22 16	03 52
13	13 32	25 14	20 09	21 14	14 57	23 05	14 23	22 19	22 16	03 53
14	13 12	24 59	19 50	21 05	15 04	23 04	14 23	22 19	22 16	03 53
15	12 51	23 00	19 29	20 54	15 11	23 04	14 23	22 20	22 16	03 54
16	12 31	19 22	19 07	20 43	15 18	23 03	14 24	22 20	22 16	03 55
17	12 10	14 23	18 44	20 32	15 25	23 03	14 24	22 20	22 16	03 55
18	11 49	08 29	18 19	20 20	15 31	23 03	14 23	22 20	22 16	03 56
19	11 28	02 N 10	17 53	20 07	15 38	23 02	14 23	22 20	22 16	03 56
20	11 06	04 S 08	17 26	19 54	15 44	23 02	14 23	22 21	22 16	03 57
21	10 45	10 02	16 58	19 40	15 51	23 01	14 23	22 21	22 16	03 58
22	10 23	15 16	16 28	19 26	15 57	23 01	14 23	22 21	22 16	03 58
23	10 01	19 32	15 56	19 11	16 03	23 00	14 23	22 21	22 16	03 59
24	09 39	22 44	15 24	18 55	16 09	23 00	14 23	22 21	22 16	04 00
25	09 17	24 42	14 50	18 39	16 15	22 59	14 23	22 21	22 16	04 00
26	08 55	25 25	14 14	18 22	16 21	22 59	14 22	22 22	22 16	04 01
27	08 32	24 53	13 38	18 05	16 27	22 59	14 22	22 22	22 16	04 02
28	08 10	23 12	12 59	17 48	16 32	22 58	14 22	22 22	22 16	04 03
29	07 S 47	20 S 30	12 S 20	17 S 29	16 S 38	22 S 58	14 S 21	22 S 22	22 S 16	04 N 03

ZODIAC SIGN ENTRIES

Date	h	m	Planets
03	11	22	☽ ♓
06	00	04	☽ ♈
08	12	05	☽ ♉
09	01	50	☿ ♒
10	21	39	☽ ♊
13	03	20	☽ ♋
15	05	09	☽ ♌
17	04	32	☽ ♍
19	03	39	☽ ♎
19	04	53	♀ ♒
19	11	16	☉ ♓
21	04	44	☽ ♏
23	09	22	☽ ♐
25	17	49	☽ ♑
27	18	07	☿ ♓
28	05	02	☽ ♒

LATITUDES

Date	Mercury ☿	Venus ♀	Mars ♂	Jupiter ♃	Saturn ♄	Uranus ♅	Neptune ♆	Pluto ♇
01	00 S 17	00 N 52	01 N 41	00 N 17	02 N 23	00 N 03	01 N 10	16 N 59
04	00 40	00 43	01 41	00 17	02 24	00 03	01 10	17 01
07	01 01	00 34	01 41	00 17	02 24	00 03	01 10	17 03
10	01 19	00 24	01 41	00 17	02 25	00 03	01 10	17 04
13	01 35	00 15	01 40	00 16	02 25	00 03	01 10	17 06
16	01 48	00 N 06	01 40	00 16	02 26	00 03	01 10	17 07
19	01 58	00 S 03	01 40	00 16	02 27	00 03	01 10	17 09
22	02 05	00 11	01 39	00 16	02 27	00 03	01 10	17 11
25	02 08	00 20	01 38	00 16	02 28	00 03	01 11	17 12
28	02 07	00 28	01 37	00 15	02 29	00 03	01 11	17 13
31	02 S 03	00 S 36	01 N 36	00 N 15	02 N 29	00 N 03	01 N 11	17 N 15

DATA

Julian Date	2445732
Delta T	+54 seconds
Ayanamsa	23° 37' 50"
Synetic vernal point	05° ♓ 29' 09"
True obliquity of ecliptic	23° 26' 32"

LONGITUDES

Date	Chiron ⚷ ° '	Ceres ⚳ ° '	Pallas ⚴ ° '	Juno ⚵ ° '	Vesta ⚶ ° '	Black Moon Lilith ⚸ ° '
01	27 ♉ 46	12 ♓ 50	15 ♒ 05	09 ♉ 32	13 ♊ 00	05 ♓ 44
11	27 ♉ 47	16 ♓ 37	18 ♒ 30	13 ♉ 53	13 ♊ 28	06 ♓ 51
21	27 ♉ 54	20 ♓ 29	21 ♒ 54	18 ♉ 32	13 ♊ 36	07 ♓ 58
31	28 ♉ 08	24 ♓ 23	25 ♒ 14	23 ♉ 24	16 ♊ 17	09 ♓ 06

MOON'S PHASES, APSIDES AND POSITIONS ☽

Date	h	m	Phase	Longitude ° '	Eclipse Indicator
01	23	46	●	12 ♒ 19	
10	04	00	☽	20 ♉ 36	
17	00	41	○	27 ♌ 32	
23	17	12	☾	04 ♐ 17	

Day	h	m		
04	09	10	Apogee	
17	09	04	Perigee	
06	21	10	ON	
13	20	43	Max dec	25° N 20'
19	20	11	OS	
26	13	33	Max dec	25° S 25'

ASPECTARIAN

h m Aspects

01 Wednesday
00 01 ☽ ⊻ ♆
03 26 ☽ □ ♇
04 23 ☽ ⊻ ♃
05 49 ☽ ∥ ♃
10 16 ☿ ∥ ♀
12 01 ☽ ⊥ ♆
14 06 ☽ ∥ ♀
14 09 ☽ ∥ ☿
14 31 ☽ ∥ ♆
14 36 ☽ ∥ ♅
16 01 ☽ ⊻ ♀
16 34 ☽ ⊥ ♃
20 23 ☽ □ ♂
23 46 ☽ ☌ ☉

02 Thursday
00 30 ☽ ✱ ♅
04 57 ☿ ∥ ♆
05 24 ☽ ⊥ ♀
06 05 ☽ ∠ ♆
06 33 ☿ ∥ ♅
07 05 ☽ □ ♄
08 34 ☉ ✱ ♅
10 55 ☽ ∠ ♃
17 14 ☽ ⊻ ☿
22 58 ♀ ∥ ♆
23 33 ♀ ∥ ♅

03 Friday
00 42 ☽ Q ♅
01 35 ☽ ∠ ♀
03 58 ♅ ∥ ♆
06 53 ☽ ⊥ ☿
07 16 ☽ ∥ ☉
12 22 ☽ ✱ ♆
15 41 ☽ △ ♇
17 40 ☽ ✱ ♃
21 17 ☽ ∥ ♄

04 Saturday
00 49 ☽ ∥ ♂
02 04 ♇ St R
03 28 ☽ ∠ ☿
09 19 ♀ ✱ ♂
11 15 ☽ △ ♂
11 23 ☽ ✱ ♀
12 41 ☽ Q ♆
13 11 ☽ □ ♅
17 56 ☽ ⊻ ☉
18 21 ☽ Q ♃
19 50 ☽ △ ♄
21 21 ☿ Q ♂
22 02 ☽ bQ ♇

05 Sunday
04 08 ☉ ∠ ♆
05 09 ♀ ⊻ ♅
07 13 ☽ ⊥ ☉
13 53 ☽ ✱ ☿
14 30 ☽ Q ♀
16 15 ☽ ± ♇
16 47 ☉ □ ♄
18 48 ☽ bQ ♂

06 Monday
01 12 ☽ □ ♆
02 16 ☽ bQ ♄
03 07 ☽ ∠ ☉
03 07 ☽ # ♇
04 22 ☽ ⊼ ♇
07 23 ☽ □ ♃
07 25 ♀ Q ♇
13 41 ☽ ± ♂
17 24 ☽ Q ☿
19 06 ♂ ⊻ ♅
20 27 ☽ ± ♄

07 Tuesday
01 58 ☽ △ ♅
02 12 ☽ ⊼ ♂
06 57 ☽ □ ♀
08 31 ☽ ⊼ ♄
12 05 ☽ ✱ ☉
14 46 ☿ ∠ ♅
15 09 ☽ ∥ ♇
19 12 ☿ Q ♄
22 40 ♀ ✱ ♄

08 Wednesday
07 56 ☽ bQ ♅
10 13 ☽ □ ☿
13 21 ☽ △ ♆
14 04 ☽ Q ☉
15 57 ☉ ∠ ♃
16 17 ☽ ☍ ♇
20 10 ☽ △ ♃
22 46 ♂ ∥ ♄

09 Thursday
01 39 ☽ ± ♅
12 42 ☿ ⊻ ♆
13 19 ☽ ⊼ ♅
15 32 ☽ ☍ ♂
18 35 ☽ bQ ♆
19 05 ☽ # ♄
19 35 ☽ ☍ ♄
19 40 ☽ # ♂
20 55 ☽ # ☉

10 Friday
00 21 ☽ △ ♀
01 37 ☽ bQ ♃
04 00 ☽ □ ☉
05 37 ♀ ⊥ ♅
09 03 ☉ ∥ ♂
11 43 ☽ ± ♆
12 21 ☿ □ ♇
18 50 ☽ ± ♃
22 03 ☉ ∥ ♄
22 58 ☽ ⊼ ♆

11 Saturday
01 35 ☽ ⊼ ♇
03 19 ☽ △ ☿
06 07 ☽ ⊼ ♃
07 25 ☽ bQ ♀
09 15 ☽ # ☿
12 36 ☽ ± ♇
14 48 ☽ # ♀
20 52 ☽ # ♆
21 13 ☽ # ♅
21 27 ☽ ☍ ♅

12 Sunday
01 13 ☽ ⊼ ♂
01 23 ☽ ± ♀
03 15 ☽ ⊼ ♄
04 04 ☽ # ♃
04 48 ☽ bQ ♇
06 37 ☿ ⊻ ♃
09 52 ☽ bQ ☿
12 12 ☽ ± ♂
13 05 ☽ ⊼ ♀
13 53 ☽ ± ♄
15 22 ☽ △ ☉

13 Monday
03 30 ☽ ± ☿
04 20 ☽ bQ ♂
04 39 ☽ ☍ ♆
05 31 ☽ bQ ♄
06 58 ☽ △ ♇
11 53 ☽ ☍ ♃
14 12 ☿ ⊥ ♆
14 58 ☽ ⊼ ☿
15 44 ♂ ∠ ♆
19 06 ☽ bQ ☉

14 Tuesday
01 28 ☽ ⊼ ♅
06 22 ☽ △ ♂
06 48 ☽ △ ♄
11 11 ☽ ± ☉
11 25 ☽ ± ♅
14 33 ☉ Q ♅
20 30 ☽ ☍ ♀
21 43 ☽ ⊼ ☉
22 47 ♂ ☌ ♄

15 Wednesday
02 06 ☽ bQ ♅
06 28 ☽ ⊼ ♆
08 31 ☽ □ ♇
11 29 ☽ # ♃
13 46 ☽ ⊼ ♃
16 04 ☽ ± ♆
17 20 ☽ # ♅
17 47 ☽ # ♆
21 41 ☽ ☍ ☿
23 25 ☽ ± ♃

16 Thursday
02 04 ☽ △ ♅
03 44 ☽ # ♀
06 20 ☽ bQ ♆
07 06 ☽ □ ♄
07 57 ☽ □ ♂
13 01 ☽ Q ♇
13 27 ☽ # ☿
13 45 ☽ bQ ♃
16 20 ☿ ⊥ ♃

17 Friday
00 19 ☽ ⊼ ♀
00 41 ☽ ☍ ☉
05 53 ☽ △ ♆
07 31 ☽ # ♂
07 47 ☽ ✱ ♇
10 30 ☽ ± ♀
11 19 ☽ Q ♄
11 56 ☽ # ♄
12 52 ☽ Q ♂
13 29 ☽ △ ♃
16 32 ☿ ✱ ♅
17 21 ♀ ∠ ♅
20 38 ♀ Q ♄
21 56 ☽ # ☉

18 Saturday
00 00 ☉ ⊻ ♀
01 08 ☽ □ ♅
01 48 ☽ bQ ♀
02 07 ☽ ⊼ ☿
06 06 ☽ ✱ ♄
07 14 ☽ ∠ ♇
08 09 ☽ ✱ ♂
12 39 ☽ ± ☿
21 25 ☽ ⊥ ♇

19 Sunday
03 07 ☽ ⊼ ☉
03 33 ☽ △ ♀
04 35 ☽ bQ ☿
04 43 ♀ Q ♂
05 06 ☽ □ ♆
05 21 ☽ ∥ ♇
05 35 ☽ Q ♅
05 49 ☽ ∠ ♄
06 57 ☽ ⊻ ♇
08 32 ☽ ∠ ♂
09 26 ☿ ∠ ♆
13 24 ☽ ± ☉
13 24 ☽ □ ♃
16 06 ☿ □ ♄
20 17 ☽ ⊥ ♄
22 54 ♀ ⊻ ♆
23 25 ☽ ⊥ ♂

20 Monday
00 57 ☽ ✱ ♅
05 03 ☽ bQ ☉
06 04 ☽ ⊻ ♄
07 48 ☽ △ ☿
09 29 ☽ ⊻ ♂
09 33 ☉ ✱ ♆
10 17 ☽ Q ♆
11 19 ☽ # ♇
19 11 ☽ Q ♃
20 53 ♀ □ ♇

21 Tuesday
01 48 ☽ ∠ ♅
02 49 ☿ □ ♂
06 21 ☽ ✱ ♆
07 55 ☽ △ ☉
08 13 ☽ ☌ ♇
09 18 ☽ □ ♀
12 02 ☉ △ ♇
14 54 ☽ ∥ ☉
15 44 ☽ ✱ ♃
17 08 ☽ ⊥ ♅

22 Wednesday
03 33 ☽ ⊻ ♅
07 45 ☽ ∥ ♄
08 17 ☽ ∠ ♆
08 59 ☽ ☌ ♄
14 06 ☽ ☌ ♂
15 40 ☽ ∥ ♂
17 36 ☽ ∥ ☿
17 56 ☽ □ ☿
18 19 ☽ ∠ ♃
19 46 ☽ Q ♀
21 23 ☿ ∠ ♃

23 Thursday
00 18 ☽ ⊥ ♆
07 34 ☿ ∥ ♂
09 52 ☽ ∥ ♀
10 46 ☽ ⊥ ♃
11 12 ☽ ⊻ ♆
13 04 ☽ ⊻ ♇
17 12 ☽ □ ☉
19 49 ☽ ✱ ♀
21 56 ☽ ⊻ ♃

24 Friday
00 07 ☽ ⊥ ♇
07 40 ☽ Q ☿
07 52 ☽ ∥ ♆
08 39 ☽ ∥ ♅
10 01 ☽ ☌ ♅
14 33 ☽ ∥ ♃
14 36 ♄ St R
15 46 ☽ ⊻ ♄
16 58 ☽ ∠ ♇
21 20 ♀ ⊻ ♃
21 48 ♀ ⊥ ♆
22 53 ☽ ⊻ ♂

25 Saturday
01 33 ☿ Q ♅
03 01 ☽ ∠ ♀
03 10 ☽ ⊥ ♄
05 55 ☽ Q ☉
10 00 ☽ ✱ ☿
10 43 ☽ ⊥ ♂
19 52 ☽ ☌ ♆
20 30 ☽ ∠ ♄
21 42 ☽ ✱ ♇
22 16 ☽ ⊥ ♀

26 Sunday
04 40 ☽ ∠ ♂
06 34 ☿ ∥ ♄
07 13 ☽ ✱ ☉
07 56 ☽ ☌ ♃
11 17 ☽ ⊻ ♀
17 37 ☉ ✱ ♃
19 59 ☽ ⊻ ♅
20 00 ☽ ∠ ☿
21 11 ☽ Q ♇

27 Monday
01 55 ☽ ✱ ♄
07 52 ☽ ⊥ ♅
11 07 ☽ ✱ ♂
15 31 ☽ ∠ ☉
16 52 ☽ ⊥ ☿

28 Tuesday
01 47 ☽ Q ♄
01 50 ☽ ∠ ♅
06 55 ☽ ⊻ ☿
07 14 ☽ ⊻ ♆
08 58 ☽ □ ♇
09 08 ☿ ✱ ♆
11 09 ☽ ⊥ ☉
11 47 ☽ Q ♂
14 26 ☽ ∥ ♃
19 18 ☽ ⊥ ♆
20 20 ☽ ∥ ♅
20 24 ☽ ⊻ ♃
20 48 ☿ △ ♇
21 19 ☽ ∥ ♆

29 Wednesday
00 19 ☽ ⊻ ☉
05 55 ☽ ☌ ♀
08 00 ☽ ✱ ♅
08 39 ☽ ⊥ ♃
13 27 ☽ ∠ ♆
13 56 ☽ □ ♄

All ephemeris data is given at 12.00 UT and the Moon's longitude is additionally given for 24.00 UT

MARCH 1984

LONGITUDES

Date	Sidereal time h m s	Sun ☉ ° ′ ″	Moon ☽ ° ′ ″	Moon ☽ 24.00 ° ′ ″	Mercury ☿ ° ′	Venus ♀ ° ′	Mars ♂ ° ′	Jupiter ♃ ° ′	Saturn ♄ ° ′	Uranus ♅ ° ′	Neptune ♆ ° ′	Pluto ♇ ° ′
01	22 37 54	11 ♓ 05 50	27 ♒ 17 10	03 ♓ 13 11	04 ♓ 53	13 ♒ 56	22 ♏ 02	07 ♑ 55	16 ♏ 21	13 ♐ 26	01 ♑ 08	01 ♏ 56
02	22 41 51	12 06 03	09 ♓ 09 07	15 ♓ 05 09	06 41	15 10	22 21	08 05	16 R 21	13 27	01 09	01 R 55
03	22 45 47	13 06 13	21 ♓ 01 25	26 ♓ 58 05	08 31	16 24	22 40	08 14	16 20	13 28	01 10	01 54
04	22 49 44	14 06 22	02 ♈ 55 19	08 ♈ 53 18	10 21	17 38	22 59	08 23	16 19	13 29	01 11	01 53
05	22 53 41	15 06 29	14 ♈ 52 15	20 ♈ 52 25	12 13	18 52	23 17	08 32	16 18	13 29	01 12	01 52
06	22 57 37	16 06 34	26 ♈ 54 07	02 ♉ 57 39	14 06	20 06	23 34	08 41	16 17	13 30	01 13	01 51
07	23 01 34	17 06 37	09 ♉ 03 26	15 ♉ 11 51	15 59	21 20	23 52	08 50	16 16	13 31	01 14	01 50
08	23 05 30	18 06 37	21 ♉ 23 23	27 ♉ 38 32	17 54	22 34	24 08	08 58	16 14	13 31	01 15	01 49
09	23 09 27	19 06 36	03 ♊ 57 48	10 ♊ 21 45	19 49	23 48	24 25	09 07	16 13	13 32	01 16	01 48
10	23 13 23	20 06 33	16 ♊ 50 54	23 ♊ 25 46	21 46	25 02	24 41	09 15	16 12	13 32	01 17	01 47
11	23 17 20	21 06 27	00 ♋ 06 50	06 ♋ 54 28	23 43	26 16	24 56	09 23	16 10	13 33	01 17	01 46
12	23 21 16	22 06 19	13 ♋ 49 00	20 ♋ 50 33	25 41	27 30	25 11	09 32	16 08	13 33	01 18	01 45
13	23 25 13	23 06 09	27 ♋ 59 10	05 ♌ 14 36	27 39	28 44	25 25	09 40	16 07	13 33	01 19	01 43
14	23 29 10	24 05 57	12 ♌ 36 27	20 ♌ 04 04	29 ♓ 38	29 ♒ 58	25 39	09 47	16 05	13 33	01 19	01 42
15	23 33 06	25 05 42	27 ♌ 36 33	05 ♍ 12 49	01 ♈ 37	01 ♓ 12	25 53	09 55	16 03	13 34	01 20	01 41
16	23 37 03	26 05 25	12 ♍ 51 33	20 ♍ 31 23	03 36	02 26	26 06	10 03	16 01	13 34	01 21	01 40
17	23 40 59	27 05 06	28 ♍ 10 49	05 ♎ 48 24	05 35	03 40	26 18	10 10	15 59	13 34	01 21	01 38
18	23 44 56	28 04 45	13 ♎ 22 44	20 ♎ 52 34	07 33	04 54	26 30	10 17	15 57	13 R 34	01 22	01 37
19	23 48 52	29 ♓ 04 22	28 ♎ 16 53	05 ♏ 34 49	09 31	06 08	26 42	10 25	15 54	13 34	01 22	01 36
20	23 52 49	00 ♈ 03 57	12 ♏ 45 46	19 ♏ 49 23	11 27	07 22	26 53	10 32	15 52	13 34	01 23	01 34
21	23 56 45	01 03 31	26 ♏ 45 32	03 ♐ 34 15	13 22	08 36	27 03	10 39	15 49	13 34	01 23	01 33
22	00 00 42	02 03 03	10 ♐ 15 45	16 ♐ 50 24	15 15	09 50	27 13	10 45	15 47	13 33	01 23	01 32
23	00 04 39	03 02 33	23 ♐ 18 38	29 ♐ 41 00	17 05	11 04	27 22	10 52	15 44	13 33	01 24	01 30
24	00 08 35	04 02 01	05 ♑ 58 05	12 ♑ 10 28	18 53	12 19	27 30	10 58	15 41	13 33	01 24	01 29
25	00 12 32	05 01 28	18 ♑ 18 48	24 ♑ 23 40	20 37	13 33	27 38	11 05	15 39	13 32	01 24	01 27
26	00 16 28	06 00 53	00 ♒ 25 41	06 ♒ 25 24	22 18	14 47	27 45	11 11	15 36	13 32	01 25	01 26
27	00 20 25	07 00 16	12 ♒ 23 21	18 ♒ 20 02	23 55	16 01	27 52	11 17	15 33	13 32	01 25	01 24
28	00 24 21	07 59 37	24 ♒ 15 54	00 ♓ 11 19	25 28	17 15	27 58	11 22	15 30	13 31	01 25	01 23
29	00 28 18	08 58 56	06 ♓ 06 41	12 ♓ 02 17	26 55	18 29	28 03	11 28	15 27	13 31	01 25	01 21
30	00 32 14	09 58 13	17 ♓ 58 24	23 ♓ 55 16	28 18	19 43	28 08	11 34	15 23	13 30	01 25	01 20
31	00 36 11	10 ♈ 57 29	29 ♓ 53 06	05 ♈ 52 03	29 ♈ 34	20 ♓ 57	28 ♏ 12	11 ♑ 39	15 ♏ 20	13 ♐ 29	01 ♑ 25	01 ♏ 18

Date	Moon True ☊ ° ′	Moon Mean ☊ ° ′	Moon ☽ Latitude ° ′
01	11 ♊ 25	11 ♊ 20	04 S 50
02	11 R 10	11 17	04 59
03	10 55	11 13	04 54
04	10 41	11 10	04 37
05	10 29	11 07	04 08
06	10 19	11 04	03 27
07	10 13	11 01	02 36
08	10 10	10 58	01 37
09	10 09	10 54	00 S 33
10	10 D 09	10 51	00 N 35
11	10 R 08	10 48	01 44
12	10 07	10 45	02 48
13	10 03	10 42	03 44
14	09 57	10 38	04 28
15	09 48	10 35	04 54
16	09 38	10 32	05 01
17	09 27	10 29	04 45
18	09 17	10 26	04 10
19	09 09	10 23	03 18
20	09 04	10 19	02 15
21	09 02	10 16	01 N 05
22	09 D 01	10 13	00 S 07
23	09 02	10 10	01 15
24	09 R 02	10 07	02 18
25	09 01	10 03	03 13
26	08 58	10 00	03 58
27	08 52	09 57	04 32
28	08 44	09 54	04 54
29	08 34	09 51	05 03
30	08 23	09 48	04 59
31	08 ♊ 12	09 ♊ 44	04 S 42

DECLINATIONS

Date	Sun ☉ ° ′	Moon ☽ ° ′	Mercury ☿ ° ′	Venus ♀ ° ′	Mars ♂ ° ′	Jupiter ♃ ° ′	Saturn ♄ ° ′	Uranus ♅ ° ′	Neptune ♆ ° ′	Pluto ♇ ° ′
01	07 S 24	16 S 57	11 S 39	17 S 11	16 S 43	22 S 57	14 S 21	22 S 22	22 S 16	04 N 04
02	07 01	12 45	10 57	16 52	16 48	22 57	14 21	22 22	22 16	04 05
03	06 38	08 04	10 14	16 32	16 53	22 56	14 20	22 22	22 15	04 05
04	06 15	03 S 05	09 30	16 12	16 58	22 56	14 20	22 22	22 15	04 06
05	05 52	02 N 03	08 44	15 51	17 03	22 55	14 19	22 23	22 15	04 07
06	05 29	07 09	07 57	15 30	17 08	22 55	14 19	22 23	22 15	04 08
07	05 05	12 03	07 08	15 09	17 13	22 54	14 18	22 23	22 15	04 08
08	04 42	16 33	06 19	14 47	17 17	22 53	14 18	22 23	22 15	04 09
09	04 19	20 25	05 28	14 25	17 22	22 53	14 17	22 23	22 15	04 10
10	03 55	23 23	04 37	14 02	17 26	22 52	14 16	22 23	22 15	04 10
11	03 32	25 10	03 44	13 39	17 30	22 52	14 16	22 23	22 15	04 11
12	03 08	25 31	02 51	13 15	17 34	22 51	14 15	22 23	22 15	04 12
13	02 44	24 14	01 56	12 52	17 38	22 51	14 14	22 23	22 15	04 13
14	02 21	21 18	01 01	12 27	17 42	22 50	14 14	22 23	22 15	04 13
15	01 57	16 54	00 S 06	12 03	17 46	22 50	14 13	22 23	22 15	04 14
16	01 33	11 21	00 N 50	11 38	17 50	22 49	14 12	22 23	22 15	04 15
17	01 10	05 N 05	01 47	11 13	17 53	22 49	14 11	22 23	22 15	04 16
18	00 46	01 S 26	02 43	10 48	17 57	22 48	14 11	22 23	22 15	04 16
19	00 S 22	07 47	03 39	10 22	18 00	22 48	14 10	22 23	22 15	04 17
20	00 N 02	13 32	04 36	09 56	18 03	22 47	14 09	22 23	22 15	04 18
21	00 25	18 23	05 31	09 30	18 06	22 47	14 08	22 23	22 15	04 19
22	00 49	22 06	06 26	09 03	18 09	22 46	14 07	22 23	22 15	04 19
23	01 13	24 32	07 19	08 36	18 12	22 46	14 06	22 23	22 15	04 20
24	01 36	25 37	08 12	08 09	18 15	22 46	14 05	22 23	22 15	04 21
25	02 00	25 23	09 03	07 42	18 18	22 45	14 04	22 23	22 15	04 21
26	02 23	23 56	09 52	07 14	18 20	22 45	14 03	22 23	22 15	04 22
27	02 47	21 26	10 39	06 47	18 22	22 44	14 02	22 23	22 15	04 23
28	03 10	18 03	11 24	06 19	18 25	22 44	14 01	22 23	22 15	04 24
29	03 34	13 58	12 07	05 51	18 27	22 43	14 00	22 23	22 15	04 24
30	03 57	09 21	12 47	05 23	18 29	22 43	13 59	22 23	22 15	04 25
31	04 N 20	04 S 22	13 N 24	04 S 54	18 S 31	22 S 43	13 S 58	22 S 23	22 S 14	04 N 26

ZODIAC SIGN ENTRIES

Date	h	m	Planets
01	17	29	☽ ♓
04	06	07	☽ ♈
06	18	09	☽ ♉
09	04	30	☽ ♊
11	11	48	☽ ♋
13	15	21	☽ ♌
14	12	35	♀ ♓
14	16	27	☿ ♈
15	15	47	☽ ♍
17	14	51	☽ ♎
19	14	49	☽ ♏
20	10	24	☉ ♈
21	17	41	☽ ♐
24	00	36	☽ ♑
26	11	09	☽ ♒
28	23	37	☽ ♓
31	12	14	☽ ♈
31	20	25	☿ ♉

LATITUDES

Date	Mercury ☿ ° ′	Venus ♀ ° ′	Mars ♂ ° ′	Jupiter ♃ ° ′	Saturn ♄ ° ′	Uranus ♅ ° ′	Neptune ♆ ° ′	Pluto ♇ ° ′
01	02 S 05	00 S 33	01 N 37	00 N 15	02 N 29	00 N 03	01 N 11	17 N 14
04	01 57	00 41	01 36	00 15	02 30	00 03	01 11	17 16
07	01 45	00 48	01 35	00 15	02 30	00 03	01 11	17 17
10	01 28	00 54	01 33	00 15	02 31	00 03	01 11	17 18
13	01 06	01 00	01 32	00 15	02 32	00 03	01 11	17 19
16	00 39	01 06	01 30	00 14	02 32	00 03	01 11	17 20
19	00 S 07	01 11	01 28	00 14	02 33	00 03	01 11	17 21
22	00 N 28	01 16	01 25	00 14	02 33	00 03	01 11	17 22
25	01 04	01 20	01 23	00 14	02 34	00 03	01 11	17 23
28	01 40	01 23	01 20	00 14	02 34	00 03	01 12	17 24
31	02 N 13	01 S 26	01 N 17	00 N 13	02 N 35	00 N 03	01 N 12	17 N 24

DATA

Julian Date	2445761
Delta T	+54 seconds
Ayanamsa	23° 37′ 54″
Synetic vernal point	05° ♓ 29′ 05″
True obliquity of ecliptic	23° 26′ 32″

MOON'S PHASES, APSIDES AND POSITIONS ☽

Date	h	m	Phase	Longitude ° ′	Eclipse Indicator
02	18	31	●	12 ♓ 22	
10	18	27	☽	20 ♊ 23	
17	10	10	○	27 ♍ 01	
24	07	58	☾	03 ♑ 52	

Day	h	m		
02	11	10	Apogee	
16	21	17	Perigee	
29	16	19	Apogee	
05	02	26	0N	
12	05	17	Max dec	25° N 34′
18	06	42	0S	
24	19	37	Max dec	25° S 40′

LONGITUDES

Date	Chiron ⚷ ° ′	Ceres ⚳ ° ′	Pallas ⚴ ° ′	Juno ⚵ ° ′	Vesta ⚶ ° ′	Black Moon Lilith ⚸ ° ′
01	28 ♉ 07	23 ♓ 59	24 ♒ 54	22 ♉ 54	16 ♊ 06	08 ♓ 59
11	28 ♉ 27	27 ♓ 55	28 ♒ 11	27 ♉ 55	18 ♊ 14	10 ♓ 06
21	28 ♉ 52	01 ♈ 52	01 ♓ 23	03 ♊ 04	20 ♊ 48	11 ♓ 13
31	29 ♉ 23	05 ♈ 49	04 ♓ 29	08 ♊ 17	23 ♊ 43	12 ♓ 21

All ephemeris data is given at 12.00 UT and the Moon's longitude is additionally given for 24.00 UT

ASPECTARIAN

h m	Aspects
01 Thursday	
01 05	☽ □ ♂
02 20	♀ ✶ ♅
03 03	☽ ∠ ♃
08 16	☽ Q ♅
10 29	☽ ∥ ♀
11 47	♀ ⊥ ♃
13 24	☽ ∥ ♂
19 48	☽ ✶ ♆
21 23	☽ △ ♇
02 Friday	
03 15	☽ ∥ ♄
06 07	☽ ☌ ☿
09 48	☽ ✶ ♃
15 06	♀ ∥ ♂
18 31	☽ ☌ ☉
20 07	☽ Q ♆
20 43	☽ □ ♅
23 08	☽ ∥ ☿
03 Saturday	
01 34	☽ ⊻ ♀
02 32	☽ △ ♄
03 41	☽ ⚼ ♇
07 34	♀ ∠ ♆
07 58	☿ ✶ ♃
10 23	☽ Q ♃
10 43	♀ □ ♄
13 40	☉ Q ♆
15 06	☽ ⊥ ♀
15 25	☽ △ ♂
19 34	☽ ∥ ☉
20 48	☉ □ ♅
21 51	☽ ± ♇
04 Sunday	
07 10	☽ # ♇
08 30	☽ □ ♆
08 46	☽ ⚼ ♄
09 55	☽ ⊼ ♇
11 21	☽ ∠ ♀
22 26	☽ ⚼ ♂
23 07	☽ □ ♃
05 Monday	
02 51	☽ ± ♄
05 42	☽ ⊻ ☿
09 14	☽ △ ♅
12 31	☽ ⊻ ☉
14 51	☽ ⊼ ♄
16 56	☽ ± ♂
19 56	☽ ⊥ ☿
20 54	☽ ✶ ♀
21 40	☽ ∥ ♇
06 Tuesday	
00 46	☿ Q ♆
01 36	☽ ⊥ ☉
04 25	☿ □ ♅
04 38	☽ # ☉
05 13	☽ ⊼ ♂
15 11	☽ ⚼ ♅
15 16	☽ # ☿
16 03	☉ △ ♄
17 09	☽ ∠ ☿
20 34	☽ △ ♆
21 05	☽ ∠ ☉
21 48	☽ ☍ ♇
23 28	☽ Q ♀
07 Wednesday	
05 35	☉ ⚼ ♇
06 28	♂ ∠ ♃
08 58	☽ ± ♅
11 33	☽ △ ♃
15 25	☿ △ ♄
20 43	☽ ⊼ ♅
22 36	☿ ⚼ ♇
23 40	☽ # ♄
08 Thursday	
02 02	☽ ⚼ ♆
02 03	☽ ☍ ♄
03 00	☽ # ♀
04 01	☽ ✶ ☿
05 06	☽ ✶ ☉
07 11	♄ ∠ ♆
14 31	☽ □ ♀
16 22	☽ # ♂
17 02	☽ ⚼ ♃
17 25	☽ ☍ ♂
17 32	☉ ☌ ☿
19 26	☽ ± ♆
09 Friday	
06 08	☽ Q ☉
06 53	☽ ⊼ ♆
07 14	☽ Q ☿
07 55	☽ ⊼ ♇
10 22	☽ ± ♃
18 51	♀ ∠ ♃
19 12	☽ ± ♇
20 15	♀ ∥ ♄
20 47	☉ # ♇
21 47	☽ ⊼ ♃
10 Saturday	
01 53	☽ # ♆
02 56	☽ # ♅
03 08	♀ □ ♂
05 12	☿ Q ♃
05 53	☽ ☍ ♅
07 15	☽ # ♃
10 48	☽ ⊼ ♄
11 53	☽ ⚼ ♇
18 27	☽ □ ☉
21 45	☽ ± ♄
21 49	♀ Q ♅
22 32	☽ □ ☿
23 54	☿ # ♇
11 Sunday	
02 32	☽ ⊼ ♂
04 25	☽ △ ♀
13 29	☽ ± ♂
13 52	☽ ⚼ ♄
14 05	☽ ☍ ♆
14 56	☽ △ ♇
19 52	☉ Q ♃
22 15	☉ ∥ ☿
12 Monday	
04 30	☽ ☍ ♃
05 03	☿ △ ♂
05 36	☽ ⚼ ♂
09 31	☽ ⚼ ♀
11 32	☽ ⊼ ♅
12 46	☿ ± ♇
15 59	☽ △ ♄
21 49	☽ ± ♅
13 Tuesday	
00 09	♂ ⊥ ♆
02 22	☽ ± ♀
03 12	☽ △ ☉
07 39	☽ △ ♂
11 21	☽ △ ☿
12 57	☽ ⚼ ♅
13 22	☽ ⊼ ♀
17 32	☽ ⊼ ♆
18 12	☽ □ ♇
14 Wednesday	
01 00	☽ # ♃
03 24	☽ ± ♆
04 32	☽ # ♅
05 31	☽ # ♆
05 53	☽ ⚼ ☉
07 23	☽ ⊼ ♃
13 32	☽ △ ♅
15 46	☽ ⚼ ☿
17 11	☽ ± ♃
17 36	☽ □ ♄
18 00	☽ ⚼ ♆
21 29	☽ ± ☉
22 49	☿ ⊻ ♀
23 24	☽ Q ♇
15 Thursday	
05 15	☿ ⚼ ♄
07 42	☽ ⚼ ♃
07 44	☽ ⊼ ☉
07 52	☽ # ♂
08 22	☽ ± ☿
08 33	☿ □ ♆
09 13	☽ □ ♂
12 48	☿ ⊼ ♇
14 31	♀ ✶ ♆
17 53	☽ △ ♆
18 11	☽ ☍ ♀
18 26	☽ ✶ ♇
19 17	☽ ⊼ ☿
21 10	♀ △ ♇
22 09	☽ Q ♄
16 Friday	
00 10	☽ # ♄
01 53	☉ ± ♇
07 33	☽ △ ♃
10 48	☽ # ♀
12 15	☉ △ ♂
13 06	☽ □ ♅
13 58	☽ Q ♂
16 56	☽ ✶ ♄
17 57	☽ ∠ ♇
17 Saturday	
00 53	☉ # ☿
08 01	☽ ⊥ ♇
09 01	☽ ✶ ♂
10 10	☽ ☍ ☉
15 03	☽ ∥ ♇
16 23	☽ ∠ ♄
16 59	☽ □ ♆
17 19	☽ Q ♅
17 26	☽ ⊻ ♇
21 24	☽ ⊼ ♀
22 39	☽ ∥ ☿
18 Sunday	
01 23	☽ ☍ ☿
03 23	☽ # ☉
06 16	♅ St R
06 33	☽ ⊥ ♄
07 03	☽ □ ♃
07 43	☽ ± ♀
08 59	☽ ∠ ♂
09 39	☽ ∥ ☉
12 18	☽ ✶ ♅
16 05	☽ ⊻ ♄
17 33	☽ # ☿
21 34	☽ Q ♆
22 34	☽ # ♇
23 23	☽ ⚼ ♀
23 33	☽ ⊥ ♂
19 Monday	
09 23	☽ ⊻ ♂
12 13	☽ Q ♃
12 28	☽ ∠ ♅
13 23	☽ ⊼ ☉
16 46	☿ ± ♄
17 03	☽ ✶ ♆
17 25	☽ ☌ ♇
21 41	☽ ∥ ♀
23 51	☿ □ ♃
23 59	☽ ± ☉
20 Tuesday	
02 08	☽ △ ♀
03 18	☽ ⊥ ♅
04 18	☿ ∥ ♇
08 13	☽ ✶ ♃
09 27	☽ ⊼ ☿
13 21	☽ ⊻ ♅
14 47	☽ ∥ ♄
16 11	☽ ⚼ ☉
17 14	☽ ☌ ♄
17 53	☿ ⚼ ♂
18 07	☽ ∠ ♆
21 12	☽ ± ☿
21 Wednesday	
06 32	☉ ⚼ ♄
09 36	☽ ⊥ ♆
10 02	☽ ∠ ♃
10 26	☽ ∥ ♂
12 31	☽ ☌ ♂
14 30	☿ △ ♅
15 15	☽ ⚼ ☿
19 54	☉ □ ♆
20 08	☽ ⊻ ♆
20 09	☽ △ ☉
20 24	☽ ⊻ ♇
23 38	☉ ⊼ ♇
22 Thursday	
02 01	☽ ⊥ ♃
07 05	☽ ⊥ ♇
11 10	☽ □ ♀
12 54	☽ ⊻ ♃
13 09	☽ ∥ ♆
14 16	☽ ∥ ♅
17 30	☽ ∥ ♃
17 59	☽ ☌ ♅
18 47	☿ ⊼ ♄
22 01	☽ ⊻ ♄
22 34	☽ △ ☿
23 24	☽ ∠ ♇
23 Friday	
07 30	♀ ✶ ♃
09 04	☽ ⊥ ♄
11 36	☿ ⊥ ♀
19 42	☽ ⊻ ♂
24 Saturday	
00 01	☽ Q ♀
01 57	☽ ∠ ♄
03 16	☽ ☌ ♆
03 26	☽ ✶ ♇
07 13	☽ ⊥ ♂
07 58	☽ □ ☉
11 09	☿ # ♀
21 45	☽ ☌ ♃
25 Sunday	
00 47	☽ ∠ ♂
01 37	☽ ✶ ♀
02 31	☽ Q ♇
02 40	☽ ⊻ ♅
06 47	☽ ✶ ♄
09 19	♀ Q ♆
11 58	♀ □ ♅
14 25	☽ ⊥ ♅
17 17	☽ □ ☿
22 07	☽ Q ☉
26 Monday	
03 27	☿ ± ♂
06 23	☽ Q ♄
06 37	☽ ✶ ♂
08 14	☽ ∠ ♅
10 33	☽ ∠ ♀
13 58	☽ ⊻ ♆
14 00	☽ □ ♇
27 Tuesday	
00 11	☽ ✶ ☉
00 37	☽ ∥ ♃
01 59	☽ ⊥ ♆
03 19	♀ △ ♄
03 56	☽ ∥ ♅
05 10	☽ ∥ ♆
05 55	♆ ✶ ♇
06 39	☽ ⊥ ♀
06 52	☽ Q ♂
09 45	☽ ⊻ ♃
10 55	☽ Q ☿
14 18	☽ ✶ ♅
18 20	☽ □ ♄
19 33	♀ ⚼ ♇
20 07	☽ ∠ ♆
20 09	☽ ⊻ ♀
21 57	☽ ⊥ ♃
28 Wednesday	
09 12	☽ ∠ ☉
09 41	☽ ∥ ♂
14 32	☽ Q ♅
14 46	☽ ✶ ☿
16 18	☽ ∠ ♃
19 33	☽ □ ♂
29 Thursday	
02 23	☽ △ ♇
02 29	☽ ✶ ♆
05 05	☽ ⊥ ☉
11 45	☽ ∥ ♄
18 21	☽ ⊻ ☉
20 37	☽ # ☿
22 38	☉ ± ♄
22 56	☽ ✶ ♃
30 Friday	
01 19	☽ ∠ ☿
02 48	☽ Q ♆
02 58	☽ □ ♅
06 48	☽ △ ♄
08 41	☽ ⚼ ♇
08 56	☿ ⊼ ♂
15 44	☿ ⚼ ♅
15 55	☽ ☌ ♀
21 48	☽ ⊥ ☿
23 22	☽ Q ♃
31 Saturday	
02 48	☽ ± ♇
08 35	☽ △ ♂
09 11	☽ ∥ ♀
11 18	☽ ⊻ ☿
11 41	☽ # ♇
12 07	☽ # ☉
12 54	☽ ⚼ ♄
14 51	☽ ⊼ ♇
15 05	☽ □ ♆
17 50	☉ ∥ ♇

APRIL 1984

Date	Sidereal time h m s	Sun ☉ LONGITUDES	Moon ☽	Moon ☽ 24.00	Mercury ☿	Venus ♀	Mars ♂	Jupiter ♃	Saturn ♄	Uranus ♅	Neptune ♆	Pluto ♇
		° ' "	° ' "	° ' "	° '	° '	° '	° '	° '	° '	° '	° '
01	00 40 08	11 ♈ 56 42	11 ♈ 52 18	17 ♈ 54 00	00 ♉ 45	22 ♓ 11	28 ♏ 15	11 ♑ 44	15 ♏ 17	13 ♐ 29	01 ♑ 25	01 ♏ 17
02	00 44 04	12 55 53	23 ♈ 57 18	00 ♉ 02 21	01 51	23 25	28 18	11 49	15 R 13	13 R 28	01 25	01 R 15
03	00 48 01	13 55 02	06 ♉ 09 20	12 ♉ 18 25	02 49	24 39	28 19	11 54	15 10	13 27	01 R 25	01 14
04	00 51 57	14 54 09	18 ♉ 29 51	24 ♉ 43 50	03 42	25 53	28 20	11 59	15 06	13 26	01 25	01 12
05	00 55 54	15 53 14	01 ♊ 00 41	07 ♊ 20 40	04 28	27 07	28 21	12 03	15 03	13 25	01 25	01 10
06	00 59 50	16 52 17	13 ♊ 44 08	20 ♊ 11 26	05 07	28 21	28 R 20	12 07	14 59	13 24	01 25	01 09
07	01 03 47	17 51 18	26 ♊ 42 54	03 ♋ 18 54	05 39	29 ♓ 35	28 19	12 12	14 55	13 23	01 25	01 07
08	01 07 43	18 50 16	09 ♋ 59 46	16 ♋ 45 48	06 05	00 ♈ 49	28 18	12 16	14 52	13 22	01 25	01 05
09	01 11 40	19 49 12	23 ♋ 37 14	00 ♌ 34 14	06 24	02 03	28 15	12 19	14 48	13 21	01 25	01 04
10	01 15 37	20 48 05	07 ♌ 36 50	14 ♌ 44 56	06 36	03 17	28 12	12 23	14 44	13 20	01 24	01 02
11	01 19 33	21 46 57	21 ♌ 58 19	29 ♌ 16 34	06 41	04 30	28 08	12 27	14 40	13 19	01 24	01 00
12	01 23 30	22 45 45	06 ♍ 39 06	14 ♍ 05 10	06 R 40	05 44	28 03	12 30	14 36	13 18	01 24	00 59
13	01 27 26	23 44 32	21 ♍ 33 51	29 ♍ 04 07	06 33	06 58	27 57	12 33	14 32	13 16	01 24	00 57
14	01 31 23	24 43 16	06 ♎ 34 49	14 ♎ 04 47	06 20	08 12	27 51	12 36	14 28	13 15	01 23	00 55
15	01 35 19	25 41 59	21 ♎ 32 51	28 ♎ 57 52	06 02	09 26	27 44	12 39	14 24	13 14	01 23	00 54
16	01 39 16	26 40 39	06 ♏ 18 51	13 ♏ 34 55	05 38	10 40	27 36	12 41	14 19	13 12	01 22	00 52
17	01 43 12	27 39 18	20 ♏ 45 20	27 ♏ 49 34	05 10	11 54	27 27	12 44	14 15	13 11	01 22	00 50
18	01 47 09	28 37 54	04 ♐ 47 17	11 ♐ 38 17	04 37	13 08	27 18	12 46	14 11	13 09	01 21	00 49
19	01 51 06	29 ♈ 36 29	18 ♐ 22 35	25 ♐ 00 17	04 02	14 22	27 08	12 48	14 07	13 08	01 21	00 47
20	01 55 02	00 ♉ 35 03	01 ♑ 31 41	07 ♑ 57 07	03 23	15 36	26 57	12 50	14 02	13 06	01 20	00 45
21	01 58 59	01 33 34	14 ♑ 17 03	20 ♑ 31 59	02 43	16 49	26 45	12 51	13 58	13 05	01 20	00 44
22	02 02 55	02 32 04	26 ♑ 42 29	02 ♒ 49 06	02 02	18 03	26 33	12 53	13 54	13 03	01 19	00 42
23	02 06 52	03 30 33	08 ♒ 52 28	14 ♒ 53 10	01 21	19 17	26 20	12 54	13 49	13 01	01 18	00 40
24	02 10 48	04 29 00	20 ♒ 51 47	26 ♒ 48 53	00 40	20 31	26 06	12 55	13 45	13 00	01 18	00 39
25	02 14 45	05 27 25	02 ♓ 45 02	08 ♓ 40 44	00 ♉ 00	21 45	25 51	12 56	13 40	12 58	01 17	00 37
26	02 18 41	06 25 48	14 ♓ 36 29	20 ♓ 32 43	29 ♈ 22	22 59	25 36	12 57	13 36	12 56	01 16	00 35
27	02 22 38	07 24 10	26 ♓ 29 51	02 ♈ 28 15	28 46	24 13	25 20	12 57	13 31	12 54	01 16	00 33
28	02 26 35	08 22 30	08 ♈ 28 14	14 ♈ 30 05	28 13	25 27	25 04	12 58	13 27	12 52	01 15	00 32
29	02 30 31	09 20 48	20 ♈ 34 03	26 ♈ 40 21	27 44	26 40	24 47	12 58	13 22	12 50	01 14	00 30
30	02 34 28	10 ♉ 19 05	02 ♉ 49 07	09 ♉ 00 33	27 ♈ 19	27 ♈ 54	24 ♏ 29	12 ♑ 58	13 ♏ 18	12 ♐ 48	01 ♑ 13	00 ♏ 28

Date	Moon True ☊	Moon Mean ☊	Moon ☽ Latitude	DECLINATIONS Sun ☉	Moon ☽	Mercury ☿	Venus ♀	Mars ♂	Jupiter ♃	Saturn ♄	Uranus ♅	Neptune ♆	Pluto ♇
	° '	° '	° '	° '	° '	° '	° '	° '	° '	° '	° '	° '	° '
01	08 ♊ 01	09 ♊ 41	04 S 13	04 N 43	00 N 49	13 N 58	04 S 26	18 S 33	22 S 42	13 S 57	22 S 23	22 S 14	04 N 26
02	07 R 52	09 38	03 32	05 06	06 01	14 29	03 57	18 35	22 42	13 56	22 23	22 14	04 27
03	07 45	09 35	02 40	05 29	11 03	14 58	03 29	18 36	22 42	13 55	22 22	22 14	04 28
04	07 41	09 32	01 41	05 52	15 43	15 22	03 00	18 38	22 41	13 54	22 22	22 14	04 28
05	07 39	09 29	00 S 36	06 15	19 47	15 44	02 31	18 39	22 41	13 53	22 22	22 14	04 29
06	07 D 39	09 25	00 N 33	06 38	23 00	16 02	02 02	18 40	22 41	13 51	22 22	22 14	04 30
07	07 40	09 22	01 41	07 00	25 05	16 17	01 33	18 41	22 40	13 50	22 22	22 14	04 30
08	07 41	09 19	02 45	07 23	25 49	16 29	01 04	18 42	22 40	13 49	22 22	22 14	04 31
09	07 R 41	09 16	03 42	07 45	25 01	16 36	00 34	18 43	22 40	13 48	22 22	22 14	04 32
10	07 40	09 13	04 27	08 07	22 40	16 41	00 S 05	18 44	22 39	13 47	22 22	22 14	04 32
11	07 36	09 09	04 57	08 29	18 52	16 42	00 N 24	18 45	22 39	13 45	22 21	22 14	04 33
12	07 31	09 06	05 08	08 51	13 50	16 39	00 53	18 45	22 39	13 44	22 21	22 14	04 34
13	07 25	09 03	04 59	09 13	07 55	16 33	01 23	18 46	22 39	13 43	22 21	22 14	04 34
14	07 19	09 00	04 29	09 35	01 N 31	16 23	01 52	18 46	22 39	13 42	22 21	22 14	04 35
15	07 13	08 57	03 41	09 56	04 S 59	16 11	02 21	18 46	22 38	13 40	22 21	22 14	04 35
16	07 08	08 54	02 39	10 17	11 08	15 55	02 50	18 46	22 38	13 39	22 21	22 14	04 36
17	07 05	08 50	01 27	10 38	16 32	15 36	03 19	18 46	22 38	13 38	22 21	22 14	04 37
18	07 04	08 47	00 N 12	10 59	20 54	15 15	03 48	18 46	22 38	13 36	22 20	22 14	04 37
19	07 D 04	08 44	01 S 01	11 20	23 57	14 51	04 17	18 46	22 38	13 35	22 20	22 14	04 38
20	07 06	08 41	02 09	11 41	25 35	14 25	04 46	18 45	22 38	13 34	22 20	22 14	04 38
21	07 07	08 38	03 09	12 01	25 48	13 57	05 15	18 44	22 38	13 32	22 20	22 14	04 39
22	07 08	08 35	03 58	12 21	24 42	13 29	05 44	18 44	22 38	13 31	22 20	22 14	04 39
23	07 R 08	08 31	04 35	12 41	22 27	12 59	06 12	18 43	22 38	13 30	22 19	22 14	04 40
24	07 07	08 28	04 59	13 01	19 16	12 29	06 41	18 42	22 37	13 29	22 19	22 14	04 40
25	07 04	08 25	05 11	13 21	15 19	11 59	07 09	18 41	22 37	13 27	22 19	22 14	04 41
26	07 00	08 22	05 09	13 40	10 49	11 30	07 38	18 39	22 38	13 26	22 19	22 14	04 41
27	06 55	08 19	04 54	13 59	05 53	11 01	08 06	18 38	22 38	13 25	22 19	22 14	04 42
28	06 50	08 15	04 26	14 18	00 S 43	10 34	08 33	18 36	22 38	13 23	22 18	22 14	04 42
29	06 46	08 12	03 46	14 37	04 N 33	10 08	09 01	18 35	22 38	13 22	22 18	22 14	04 43
30	06 ♊ 42	08 ♊ 09	02 S 55	14 N 55	09 N 43	09 N 44	09 N 29	18 S 33	22 S 38	13 S 21	22 S 18	22 S 14	04 N 43

ZODIAC SIGN ENTRIES

Date	h	m	Planets
02	23	55	☽ ♉
05	10	04	☽ ♊
07	17	59	☽ ♋
07	20	13	♀ ♈
09	23	01	☽ ♌
12	01	11	☽ ♍
14	01	29	☽ ♎
16	01	41	☽ ♏
18	03	44	☽ ♐
19	21	38	☉ ♉
20	09	10	☽ ♑
22	18	27	☽ ♒
25	06	26	☽ ♓
25	11	49	☿ ♈
27	19	03	☽ ♈
30	06	30	☽ ♉

LATITUDES

Date	Mercury ☿	Venus ♀	Mars ♂	Jupiter ♃	Saturn ♄	Uranus ♅	Neptune ♆	Pluto ♇
	° '	° '	° '	° '	° '	° '	° '	° '
01	02 N 23	01 S 27	01 N 15	00 N 13	02 N 35	00 N 03	01 N 12	17 N 25
04	02 47	01 29	01 12	00 13	02 35	00 03	01 12	17 25
07	03 03	01 30	01 08	00 13	02 35	00 03	01 12	17 26
10	03 08	01 31	01 03	00 13	02 36	00 03	01 12	17 26
13	03 01	01 31	00 58	00 12	02 36	00 03	01 12	17 26
16	02 40	01 30	00 53	00 12	02 36	00 03	01 12	17 26
19	02 06	01 29	00 47	00 12	02 36	00 02	01 12	17 26
22	01 23	01 28	00 41	00 12	02 37	00 02	01 12	17 26
25	00 N 33	01 25	00 34	00 11	02 37	00 02	01 12	17 26
28	00 S 18	01 23	00 27	00 11	02 37	00 02	01 12	17 26
31	01 S 06	01 S 19	00 N 19	00 N 11	02 N 37	00 N 02	01 N 13	17 N 26

DATA

Julian Date	2445792
Delta T	+54 seconds
Ayanamsa	23° 37' 57"
Synetic vernal point	05° ♓ 29' 03"
True obliquity of ecliptic	23° 26' 33"

MOON'S PHASES, APSIDES AND POSITIONS ☽

Date	h	m	Phase	Longitude ° '	Eclipse Indicator
01	12	10	●	11 ♈ 57	
09	04	51	☽	19 ♋ 32	
15	19	11	○	26 ♎ 00	
23	00	26	☾	03 ♒ 02	

Day	h	m		
14	05	45	Perigee	
26	06	54	Apogee	
01	08	15	0N	
08	11	46	Max dec	25° N 49'
14	17	33	0S	
21	03	43	Max dec	25° S 53'
28	15	16	0N	

LONGITUDES

Date	Chiron ⚷	Ceres ⚳	Pallas ⚴	Juno ⚵	Vesta ⚶	Black Moon Lilith ⚸
	° '	° '	° '	° '	° '	° '
01	29 ♉ 26	06 ♈ 13	04 ♓ 47	08 ♊ 49	24 ♊ 01	12 ♓ 27
11	00 ♊ 01	10 ♈ 09	07 ♓ 44	14 ♊ 07	27 ♊ 16	13 ♓ 35
21	00 ♊ 40	14 ♈ 04	10 ♓ 32	19 ♊ 27	00 ♋ 44	14 ♓ 42
31	01 ♊ 22	17 ♈ 56	13 ♓ 09	24 ♊ 47	04 ♋ 25	15 ♓ 49

All ephemeris data is given at 12.00 UT and the Moon's longitude is additionally given for 24.00 UT

ASPECTARIAN

h m Aspects

01 Sunday
03 52 ☉ # ♀
06 25 ☉ □ ♃
06 51 ☽ ± ♄
11 19 ☿ # ♄
11 34 ♀ # ♇
11 44 ☽ □ ♃
12 10 ☽ ☌ ☉
14 46 ☽ bQ ♂
15 12 ☽ △ ♅
18 46 ☽ ⊼ ♄
22 58 ☿ ☍ ♇

02 Monday
02 27 ☿ △ ♆
03 16 ☽ # ♀
04 44 ☽ ∥ ♇
07 27 ☽ ∥ ☉
08 42 ☽ ± ♂
10 48 ☽ ⊻ ♀
14 05 ♆ St R
20 28 ♀ Q ♃
20 35 ☽ ⊼ ♂
20 53 ☽ bQ ♅
21 06 ☉ bQ ♂
23 59 ☽ ⊥ ♀

03 Tuesday
00 47 ☉ △ ♅
02 21 ☽ ☍ ♇
02 43 ☽ △ ♆
04 56 ☽ ☌ ☿
14 32 ☽ ± ♅
19 35 ☽ ∠ ♀
23 03 ♀ ± ♇
23 17 ☽ △ ♃

04 Wednesday
02 12 ☽ ⊼ ♅
02 22 ☽ # ♄
04 26 ☽ ⊻ ☉
05 28 ☽ ☍ ♄
07 59 ☽ bQ ♆
09 58 ☽ ∥ ☿
16 40 ☉ ⊼ ♄
17 02 ☽ ⊥ ☉

05 Thursday
01 20 ☽ ± ♆
03 45 ☽ ⚹ ♀
04 24 ☽ bQ ♃
04 49 ☽ # ♂
06 55 ☽ ☍ ♂
11 45 ☽ ∠ ☉
12 18 ☽ ⊼ ♇
12 22 ♂ St R
12 47 ☽ ⊼ ♆
18 56 ☽ ⊻ ☿
21 37 ☽ ± ♃
23 39 ☽ ± ♇

06 Friday
04 58 ☽ Q ♀
05 37 ☽ # ♆
06 40 ☽ # ♅
06 51 ☽ ⊥ ☿
08 58 ☽ ⊼ ♃
09 14 ☽ # ♃
11 23 ☽ ☍ ♅
11 56 ♀ △ ♂
14 19 ☽ ⊼ ♄
16 29 ☽ bQ ♇
18 20 ☽ ⚹ ☉

07 Saturday
00 24 ☽ ∠ ☿
01 24 ☽ ± ♄
14 28 ☿ ⊥ ♀
14 56 ☽ ⊼ ♂
17 46 ☽ □ ♀
17 49 ☽ bQ ♄
18 11 ☽ Q ☉
18 23 ♀ bQ ♄
20 00 ☽ △ ♇
20 34 ☽ ☍ ♆

08 Sunday
01 47 ☽ ± ♂
04 47 ☽ ⚹ ☿
16 03 ☽ ☍ ♃
17 20 ♀ ⊼ ♇
17 51 ☽ bQ ♂
18 00 ☽ ⊼ ♅
20 36 ☽ △ ♄
23 44 ♀ □ ♆

09 Monday
02 41 ☽ Q ☿
04 34 ☽ ± ♅
04 51 ☽ □ ☉
19 59 ☽ △ ♂
20 11 ☽ bQ ♅

10 Tuesday
00 49 ☽ □ ♇
01 26 ☽ ⊼ ♆
03 55 ☽ △ ♀
10 16 ☽ □ ☿
11 39 ☽ ± ♆
12 06 ☽ # ♃
14 20 ☽ # ♅
15 14 ☽ # ♆
20 04 ☽ ⊼ ♃
21 37 ☽ △ ♅
23 55 ☽ □ ♄

11 Wednesday
02 46 ☽ bQ ♆
06 08 ☽ ± ♃
07 06 ☽ Q ♇
07 33 ☽ bQ ♀
11 40 ☽ △ ☉
12 37 ☽ # ♂
19 47 ☉ ± ♂
20 24 ☿ St R
21 02 ☽ bQ ♃
22 04 ☽ □ ♂
23 04 ☽ ∥ ☿
23 44 ☽ ± ♀

12 Thursday
02 48 ☽ ⚹ ♇
03 28 ☽ △ ♆
05 27 ☽ Q ♄
10 23 ☽ ⊼ ♀
12 02 ☽ △ ☿
12 27 ☽ # ♄
13 56 ☽ bQ ☉
21 28 ☽ △ ♃
22 43 ☽ □ ♅

13 Friday
00 46 ☽ ⚹ ♄
03 01 ☽ ∠ ♇
03 03 ☽ Q ♂
04 47 ☿ ⊻ ♀
05 27 ☽ ± ☉
07 15 ☽ ∥ ☉
11 59 ☽ bQ ☿
15 44 ☽ ⊼ ☉
17 25 ☽ ⊥ ♇
22 09 ☽ ⚹ ♂

14 Saturday
00 40 ☽ ∥ ♇
00 41 ☽ ∠ ♄
02 11 ☽ ± ☿
02 59 ☽ ⊻ ♇
03 30 ☽ Q ♅
03 42 ☽ □ ♆
10 47 ☽ ∥ ♀
11 37 ☽ ⊼ ☿
14 50 ☽ ☍ ♀
15 00 ☽ ⊥ ♄
16 46 ♀ ± ♄
21 39 ☽ □ ♃
21 57 ☽ ∠ ♂
22 39 ☽ ⚹ ♅

15 Sunday
00 33 ☽ ⊻ ♄
01 25 ☽ # ♀
08 31 ☽ Q ♆
10 32 ☽ # ♇
12 17 ☽ ⊥ ♂
19 11 ☽ ☍ ☉
21 54 ☽ ⊻ ♂
22 47 ☽ ∠ ♅
23 40 ♂ ∠ ♃

16 Monday
02 46 ☽ Q ♃
03 07 ☽ ☌ ♇
03 55 ☽ ⚹ ♆
08 22 ☽ # ☉
10 55 ☽ ☍ ☿
13 27 ☽ ⊥ ♅
19 50 ☽ ⊼ ♀
22 32 ☽ ⚹ ♃
22 39 ☽ ∥ ♄
23 21 ☽ ⊻ ♅

17 Tuesday
01 10 ☽ ☌ ♄
04 38 ☽ ∠ ♆
06 45 ☽ ± ♀
07 39 ☉ ⊼ ♂
07 47 ☽ # ☿
19 48 ☽ ⊥ ♆
21 34 ♀ bQ ♂
23 14 ☽ ☌ ♂
23 25 ☽ bQ ♀
23 28 ☽ ∥ ♂
23 52 ☽ ∠ ♃

18 Wednesday
00 34 ☉ bQ ♅
00 35 ☽ ⊼ ☉
04 37 ♀ □ ♃
05 08 ☽ ⊻ ♇
06 04 ☽ ⊻ ♆
11 42 ☽ ± ☉
11 43 ☽ ⊼ ☿
12 28 ♀ △ ♅
15 27 ☽ ⊥ ♃
15 31 ☽ ⊥ ♇
21 14 ☽ ∥ ♆
21 47 ☽ ± ☿
22 02 ☽ ∥ ♅

19 Thursday
00 17 ☽ ∥ ♃
02 01 ☽ ⊻ ♃
02 39 ☽ ☌ ♅
04 06 ☽ △ ♀
04 25 ☽ ⊻ ♄
04 44 ☽ bQ ☉
07 02 ♃ Q ♇
07 22 ☽ ∠ ♇
07 23 ♀ ⊼ ♄
13 07 ☽ bQ ☿
15 06 ☽ ⊥ ♄

20 Friday
03 40 ☽ ⊻ ♂
05 11 ♀ ∥ ♇
07 25 ☽ ∠ ♄
10 07 ☽ △ ☉
10 34 ☽ ⚹ ♇
11 39 ☽ ☌ ♆
14 35 ☽ ⊥ ♂
15 18 ☽ △ ☿
16 05 ☉ ☍ ♇

21 Saturday
06 22 ☉ △ ♆
07 15 ☽ ∠ ♂
09 02 ☽ Q ♇
09 16 ☽ ☌ ♃
09 42 ☽ ⊻ ♅
11 24 ☽ ⚹ ♄
17 23 ☽ □ ♀
21 10 ☽ ⊥ ♅

22 Sunday
04 49 ☉ ☌ ☿
09 56 ☿ # ♄
10 25 ☽ Q ♄
11 41 ☽ ⚹ ♂
14 37 ☽ ∠ ♅
19 48 ☽ □ ♇
21 02 ☽ ⊻ ♆
21 54 ☽ □ ☿

23 Monday
00 26 ☽ □ ☉
08 29 ☽ Q ♀
08 53 ☽ ⊥ ♆
10 32 ☽ ∥ ♃
10 56 ☽ Q ♂
13 10 ☽ ∥ ♅
13 20 ☿ △ ♆
13 57 ☽ ∥ ♆
20 02 ☽ ⊻ ♃
20 15 ☽ ⚹ ♅
20 42 ☉ ∥ ☿
21 48 ☽ □ ♄

24 Tuesday
02 50 ☽ ∠ ♆
05 04 ♀ ± ♂
07 49 ☽ Q ☿
08 05 ☽ ⊥ ♃
11 14 ☽ ⚹ ♀
12 40 ☿ ☍ ♇
15 33 ☽ Q ☉
15 46 ☽ ∥ ♂
20 18 ☽ Q ♅
22 20 ☽ □ ♂

25 Wednesday
02 15 ☽ ∠ ♃
06 43 ☽ ⚹ ☿
07 41 ☽ △ ♇
09 02 ☽ ⚹ ♆
17 58 ☽ ⚹ ☉
19 41 ☉ # ♄
21 02 ☽ ∠ ♀
22 08 ☽ # ☉
22 19 ☽ ∥ ♄

26 Thursday
04 47 ♃ ⊻ ♅
08 05 ☽ # ☿
08 37 ☽ □ ♅
08 38 ☽ ⚹ ♃
09 18 ☽ Q ♆
09 58 ☽ △ ♄
11 31 ☽ ∠ ☿
13 58 ☽ bQ ♇
17 21 ☽ ⊥ ♀

27 Friday
00 02 ☉ ± ♅
02 21 ☽ # ♀
03 01 ☽ ∠ ☉
04 50 ☽ ⊥ ☿
06 52 ☽ ⊻ ♀
08 06 ☽ ± ♇
08 53 ☽ Q ♃
09 43 ☽ △ ♂
16 03 ☽ bQ ♄
16 21 ☽ ⊻ ☿
17 36 ☽ # ♇
19 20 ♂ ⊥ ♆
20 09 ☽ ⊼ ♇
21 34 ☽ □ ♆
22 44 ☽ ⊥ ☉

28 Saturday
05 59 ♀ ⊼ ♂
09 58 ☽ ± ♄
11 48 ☽ ⊻ ☉
15 06 ☽ bQ ♂
20 44 ☽ △ ♅
20 56 ☽ □ ♃
21 51 ☽ ⊼ ♄

29 Sunday
06 07 ☿ bQ ♅
08 33 ☽ ± ♂
12 46 ☽ ∥ ♇
18 41 ♃ St R
20 06 ☽ ⊼ ♂

30 Monday
01 20 ☽ ☌ ♀
01 36 ☽ ☌ ☿
02 15 ☽ bQ ♅
03 17 ☿ ☌ ♀
07 27 ☽ ☍ ♇
08 53 ☽ △ ♆
10 11 ♀ bQ ♅
10 46 ☽ ∥ ♀
12 04 ☽ ∥ ☿
19 13 ☿ ∥ ♀
19 43 ☽ ± ♅

LONGITUDES

Date	Sidereal time h m s	Sun ☉ ° ' "	Moon ☽ ° ' "	Moon ☽ 24.00 ° ' "	Mercury ☿ ° '	Venus ♀ ° '	Mars ♂ ° '	Jupiter ♃ ° '	Saturn ♄ ° '	Uranus ♅ ° '	Neptune ♆ ° '	Pluto ♇ ° '
01	02 38 24	11 ♉ 17 20	15 ♉ 14 43	21 ♉ 31 46	26 ♈ 57	29 ♈ 08	24 ♏ 11	12 ♑ 57	13 ♏ 13	12 ♐ 46	01 ♑ 12	00 ♏ 27
02	02 42 21	12 15 34	27 ♉ 51 46	04 ♊ 14 48	26 R 41	00 ♉ 22	23 R 52	12 R 57	13 R 09	12 R 44	01 R 11	00 R 25
03	02 46 17	13 13 45	10 ♊ 40 57	17 ♊ 10 18	26 28	01 36	23 33	12 56	13 04	12 42	01 10	00 23
04	02 50 14	14 11 55	23 ♊ 42 55	00 ♋ 18 53	26 21	02 49	23 14	12 56	13 00	12 40	01 09	00 22
05	02 54 10	15 10 03	06 ♋ 58 18	13 ♋ 41 13	26 18	04 03	22 54	12 55	12 55	12 38	01 08	00 20
06	02 58 07	16 08 09	20 ♋ 27 43	27 ♋ 17 50	26 D 20	05 17	22 33	12 53	12 51	12 36	01 07	00 19
07	03 02 04	17 06 13	04 ♌ 11 36	11 ♌ 08 58	26 27	06 31	22 13	12 52	12 46	12 34	01 06	00 17
08	03 06 00	18 04 15	18 ♌ 09 54	25 ♌ 14 14	26 38	07 45	21 52	12 51	12 42	12 32	01 05	00 15
09	03 09 57	19 02 15	02 ♍ 21 45	09 ♍ 32 09	26 54	08 58	21 31	12 49	12 37	12 30	01 04	00 14
10	03 13 53	20 00 14	16 ♍ 45 04	24 ♍ 00 00	27 14	10 12	21 09	12 47	12 33	12 28	01 03	00 12
11	03 17 50	20 58 10	01 ♎ 16 24	08 ♎ 33 38	27 39	11 26	20 48	12 45	12 28	12 25	01 02	00 11
12	03 21 46	21 56 04	15 ♎ 50 58	23 ♎ 07 41	28 08	12 40	20 26	12 43	12 24	12 23	01 01	00 09
13	03 25 43	22 53 57	00 ♏ 23 01	07 ♏ 36 12	28 41	13 53	20 05	12 40	12 19	12 21	01 00	00 08
14	03 29 39	23 51 48	14 ♏ 46 32	21 ♏ 53 19	29 18	15 07	19 43	12 37	12 15	12 19	00 59	00 06
15	03 33 36	24 49 38	28 ♏ 56 00	05 ♐ 54 03	29 ♈ 59	16 21	19 21	12 35	12 11	12 16	00 57	00 05
16	03 37 33	25 47 27	12 ♐ 47 05	19 ♐ 34 51	00 ♉ 44	17 35	19 00	12 32	12 06	12 14	00 56	00 03
17	03 41 29	26 45 14	26 ♐ 17 10	02 ♑ 53 59	01 32	18 48	18 38	12 28	12 02	12 12	00 55	00 02
18	03 45 26	27 42 59	09 ♑ 25 22	15 ♑ 51 28	02 23	20 02	18 17	12 25	11 58	12 09	00 54	00 ♏ 00
19	03 49 22	28 40 44	22 ♑ 12 03	28 ♑ 28 56	03 18	21 16	17 56	12 22	11 53	12 07	00 52	29 ♎ 59
20	03 53 19	29 ♉ 38 27	04 ♒ 41 01	10 ♒ 49 15	04 16	22 30	17 35	12 18	11 49	12 04	00 51	29 57
21	03 57 15	00 ♊ 36 10	16 ♒ 54 08	22 ♒ 56 11	05 18	23 43	17 14	12 14	11 45	12 02	00 50	29 56
22	04 01 12	01 33 51	28 ♒ 55 58	04 ♓ 54 03	06 22	24 57	16 54	12 10	11 41	12 00	00 48	29 55
23	04 05 08	02 31 31	10 ♓ 51 01	16 ♓ 47 25	07 30	26 11	16 34	12 06	11 37	11 57	00 47	29 53
24	04 09 05	03 29 10	22 ♓ 43 50	28 ♓ 40 49	08 40	27 24	16 15	12 01	11 33	11 55	00 46	29 52
25	04 13 02	04 26 48	04 ♈ 38 54	10 ♈ 38 34	09 53	28 38	15 55	11 57	11 29	11 52	00 44	29 50
26	04 16 58	05 24 25	16 ♈ 40 18	22 ♈ 44 31	11 09	29 ♉ 52	15 37	11 52	11 25	11 50	00 43	29 49
27	04 20 55	06 22 01	28 ♈ 51 36	05 ♉ 01 54	12 27	01 ♊ 06	15 19	11 47	11 21	11 47	00 41	29 48
28	04 24 51	07 19 36	11 ♉ 15 41	17 ♉ 33 11	13 49	02 19	15 01	11 42	11 17	11 45	00 40	29 47
29	04 28 48	08 17 10	23 ♉ 54 35	00 ♊ 19 58	15 12	03 33	14 44	11 37	11 13	11 42	00 39	29 45
30	04 32 44	09 14 43	06 ♊ 49 24	13 ♊ 22 52	16 39	04 47	14 28	11 32	11 09	11 40	00 37	29 44
31	04 36 41	10 ♊ 12 16	20 ♊ 00 19	26 ♊ 41 36	18 ♉ 08	06 ♊ 00	14 ♏ 13	11 ♑ 27	11 ♏ 06	11 ♐ 37	00 ♑ 36	29 ♎ 43

Date	Moon True ☊ ° '	Moon Mean ☊ ° '	Moon Latitude ° '
01	06 ♊ 39	08 ♊ 06	01 S 55
02	06 R 37	08 03	00 S 48
03	06 D 37	08 00	00 N 22
04	06 38	07 56	01 33
05	06 39	07 53	02 39
06	06 41	07 50	03 38
07	06 42	07 47	04 26
08	06 R 42	07 44	04 59
09	06 42	07 41	05 14
10	06 40	07 37	05 10
11	06 39	07 34	04 47
12	06 37	07 31	04 05
13	06 35	07 28	03 07
14	06 34	07 25	01 58
15	06 33	07 21	00 N 42
16	06 D 33	07 18	00 S 34
17	06 34	07 15	01 47
18	06 35	07 12	02 52
19	06 35	07 09	03 47
20	06 36	07 06	04 29
21	06 36	07 02	04 58
22	06 37	06 59	05 14
23	06 R 37	06 56	05 16
24	06 36	06 53	05 04
25	06 36	06 50	04 40
26	06 36	06 47	04 03
27	06 36	06 43	03 14
28	06 D 36	06 40	02 16
29	06 36	06 37	01 S 10
30	06 36	06 34	00 N 01
31	06 ♊ 36	06 ♊ 31	01 N 14

DECLINATIONS

Date	Sun ☉ ° '	Moon ☽ ° '	Mercury ☿ ° '	Venus ♀ ° '	Mars ♂ ° '	Jupiter ♃ ° '	Saturn ♄ ° '	Uranus ♅ ° '	Neptune ♆ ° '	Pluto ♇ ° '
01	15 N 13	14 N 35	09 N 21	09 N 56	18 S 31	22 S 38	13 S 19	22 S 18	22 S 14	04 N 44
02	15 31	18 54	09 01	10 23	18 29	22 38	13 18	22 17	22 14	04 44
03	15 49	22 25	08 44	10 50	18 26	22 38	13 17	22 17	22 14	04 44
04	16 06	24 50	08 29	11 16	18 24	22 38	13 15	22 17	22 14	04 45
05	16 23	25 55	08 16	11 42	18 22	22 39	13 14	22 17	22 14	04 45
06	16 40	25 29	08 06	12 08	18 19	22 39	13 13	22 16	22 14	04 46
07	16 57	23 31	07 58	12 34	18 17	22 39	13 11	22 16	22 14	04 46
08	17 13	20 07	07 53	13 00	18 14	22 39	13 10	22 16	22 14	04 46
09	17 29	15 31	07 50	13 25	18 11	22 39	13 09	22 16	22 14	04 47
10	17 45	10 00	07 50	13 50	18 08	22 40	13 07	22 15	22 14	04 47
11	18 00	03 N 53	07 52	14 14	18 05	22 40	13 06	22 15	22 14	04 47
12	18 15	02 S 28	07 57	14 38	18 02	22 40	13 05	22 15	22 14	04 48
13	18 30	08 41	08 03	15 02	17 59	22 41	13 04	22 14	22 14	04 48
14	18 44	14 24	08 12	15 25	17 56	22 41	13 02	22 14	22 14	04 48
15	18 59	19 14	08 23	15 48	17 53	22 41	13 01	22 14	22 14	04 48
16	19 12	22 54	08 36	16 11	17 50	22 42	13 00	22 14	22 14	04 49
17	19 26	25 10	08 50	16 33	17 47	22 42	12 59	22 13	22 14	04 49
18	19 39	25 58	09 07	16 55	17 44	22 43	12 57	22 13	22 14	04 49
19	19 52	25 20	09 25	17 16	17 40	22 43	12 56	22 13	22 14	04 49
20	20 05	23 27	09 45	17 37	17 37	22 44	12 55	22 12	22 13	04 49
21	20 17	20 30	10 06	17 58	17 34	22 44	12 54	22 12	22 13	04 50
22	20 29	16 45	10 29	18 18	17 31	22 44	12 53	22 12	22 13	04 50
23	20 40	12 22	10 53	18 37	17 28	22 45	12 52	22 11	22 13	04 50
24	20 51	07 33	11 18	18 56	17 25	22 45	12 51	22 11	22 13	04 50
25	21 02	02 S 26	11 45	19 15	17 22	22 46	12 49	22 11	22 13	04 50
26	21 12	02 N 49	12 12	19 33	17 19	22 46	12 48	22 10	22 13	04 50
27	21 22	08 03	12 41	19 50	17 17	22 47	12 47	22 10	22 13	04 50
28	21 32	13 03	13 10	20 07	17 14	22 48	12 46	22 10	22 13	04 50
29	21 41	17 38	13 41	20 24	17 11	22 48	12 45	22 09	22 13	04 50
30	21 50	21 28	14 12	20 40	17 09	22 49	12 44	22 09	22 13	04 50
31	21 N 59	24 N 17	14 N 43	20 N 55	17 S 07	22 S 49	12 S 43	22 S 09	22 S 13	04 N 50

ZODIAC SIGN ENTRIES

Date	h m	Planets
02	04 53	♀ ♉
02	16 02	☽ ♊
04	23 26	☽ ♋
07	04 43	☽ ♌
09	08 02	☽ ♍
11	09 54	☽ ♎
13	11 22	☽ ♏
15	12 33	☿ ♉
15	13 50	☽ ♐
17	18 43	☽ ♑
18	14 35	♇ ♎
20	02 55	☽ ♒
20	20 58	☉ ♊
22	14 09	☽ ♓
25	02 39	☽ ♈
26	14 40	♀ ♊
27	14 13	☽ ♉
29	23 23	☽ ♊

LATITUDES

Date	Mercury ☿ ° '	Venus ♀ ° '	Mars ♂ ° '	Jupiter ♃ ° '	Saturn ♄ ° '	Uranus ♅ ° '	Neptune ♆ ° '	Pluto ♇ ° '
01	01 S 06	01 S 19	00 N 19	00 N 11	02 N 37	00 N 02	01 N 13	17 N 26
04	01 49	01 16	00 11	00 11	02 37	00 02	01 13	17 25
07	02 24	01 11	00 N 03	00 10	02 36	00 02	01 13	17 25
10	02 51	01 07	00 S 05	00 10	02 36	00 02	01 13	17 24
13	03 10	01 02	00 14	00 10	02 36	00 02	01 13	17 23
16	03 20	00 56	00 22	00 09	02 36	00 02	01 13	17 23
19	03 24	00 50	00 31	00 09	02 35	00 02	01 13	17 22
22	03 21	00 44	00 39	00 09	02 35	00 02	01 13	17 21
25	03 12	00 38	00 48	00 08	02 35	00 02	01 13	17 20
28	02 57	00 31	00 56	00 08	02 34	00 02	01 13	17 19
31	02 S 37	00 S 24	01 S 03	00 N 08	02 N 34	00 N 02	01 N 13	17 N 17

DATA

Julian Date	2445822
Delta T	+54 seconds
Ayanamsa	23° 38' 00"
Synetic vernal point	05° ♓ 28' 59"
True obliquity of ecliptic	23° 26' 32"

MOON'S PHASES, APSIDES AND POSITIONS ☽

Date	h m	Phase	Longitude ° '	Eclipse Indicator
01	03 45	●	10 ♉ 57	
08	11 50	☽	18 ♌ 04	
15	04 29	○	24 ♏ 32	
22	17 45	☾	01 ♓ 48	
30	16 48	●	09 ♊ 26	Annular

Day	h m	
12	03 14	Perigee
24	01 02	Apogee

05	17 20	Max dec	25° N 57'
12	02 40	0S	
18	13 04	Max dec	25° S 58'
25	23 09	0N	

LONGITUDES

Date	Chiron ⚷ ° '	Ceres ⚳ ° '	Pallas ⚴ ° '	Juno ⚵ ° '	Vesta ⚶ ° '	Black Moon Lilith ⚸ ° '
01	01 ♊ 22	17 ♈ 56	13 ♓ 09	24 ♊ 47	04 ♋ 25	15 ♓ 49
11	02 ♊ 05	21 ♈ 45	15 ♓ 34	00 ♋ 07	08 ♋ 16	16 ♓ 56
21	02 ♊ 50	25 ♈ 30	17 ♓ 43	05 ♋ 26	12 ♋ 16	18 ♓ 03
31	03 ♊ 35	29 ♈ 09	19 ♓ 36	10 ♋ 43	16 ♋ 22	19 ♓ 11

ASPECTARIAN

h m	Aspects
01 Tuesday	
03 45	☽ ☌ ☉
05 36	☽ # ♄
07 16	☽ ⊼ ♅
07 37	☽ △ ♃
08 08	☽ ☍ ♄
13 50	☽ ⚼ ♆
15 33	☽ ∥ ☉
02 Wednesday	
04 38	☽ ☍ ♂
06 57	☽ ± ♆
09 27	☽ # ♂
09 48	☽ ⊻ ☿
12 10	☽ ⚼ ♃
13 02	♀ ☍ ♇
16 48	☽ ⊼ ♇
17 13	☽ ⊻ ♀
18 15	☽ ⊼ ♆
20 54	☽ ⊥ ☿
23 32	☉ ⊼ ♅
03 Thursday	
03 53	♀ △ ♆
04 02	☽ ± ♇
04 56	☉ △ ♃
05 02	☽ ± ♃
05 39	☽ ⊥ ♀
08 21	☉ ☍ ♄
10 29	☽ # ♆
10 56	☽ # ♅
13 27	☽ ∠ ☿
13 45	☽ # ♃
15 45	☽ ☍ ♅
16 11	☽ ⊼ ♃
16 24	☽ ⊼ ♄
17 06	☽ ⊻ ☉
20 42	☽ ⚼ ♇
04 Friday	
00 05	☽ ∠ ♀
03 24	☽ ± ♄
05 03	☽ ⊥ ☉
11 08	☽ ⊼ ♂
16 46	☽ ⚹ ☿
19 45	☽ ⚼ ♄
21 47	☽ ± ♂
22 46	☽ ∠ ☉
05 Saturday	
00 04	☽ △ ♇
01 31	☽ ☍ ♆
06 13	☽ ⚹ ♀
13 37	☽ ⚼ ♂
14 06	☿ St D
14 23	☽ Q ☿
15 38	♃ ⚹ ♄
22 06	☽ ⊼ ♅
22 34	☽ △ ♄
22 36	☽ ☍ ♃
06 Sunday	
03 46	☽ ⚹ ☉
05 50	☽ Q ♀
08 44	☽ ± ♅
11 44	☉ ⚼ ♆
15 36	☽ △ ♂
22 23	☽ □ ☿
07 Monday	
00 30	☽ ⚼ ♅
02 29	☽ Q ☉
05 13	☽ □ ♇
06 39	☽ ⊼ ♆
13 01	♀ ± ♅
16 24	☽ □ ♀
17 02	☽ ± ♆
19 08	☽ # ♃
22 00	☽ # ♅
22 17	☽ # ♆
08 Tuesday	
02 24	☽ △ ♅
02 42	☽ □ ♄
02 55	☽ ⊼ ♃
08 28	☽ ⚼ ♆
11 50	☽ □ ☉
12 09	☽ Q ♇
13 09	☽ ± ♃
18 08	☽ □ ♂
21 20	♀ # ♄
22 40	☽ # ♂
09 Wednesday	
02 37	☽ △ ☿
02 56	☽ ∥ ☉
04 22	☽ ⚼ ♃
08 25	☽ ⚹ ♇
09 05	☽ Q ♄
09 50	☽ △ ♆
20 53	☽ ∥ ♀
22 46	☽ # ♄
23 40	☽ Q ♂
10 Thursday	
00 06	☽ △ ♀
04 19	☽ ⚼ ☿
04 53	☽ □ ♅
05 03	☽ ⚹ ♄
05 25	☽ △ ♃
09 26	☽ ∠ ♇
17 47	☽ △ ☉
19 07	☽ ⚹ ♂
19 38	☽ ± ☿
20 39	☽ ∥ ☿
11 Friday	
00 12	☿ ⚼ ♅
00 19	☽ ⊥ ♇
03 17	☽ ⚼ ♀
05 46	☽ ∠ ♄
05 51	☽ ⊼ ☿
08 31	☽ ∥ ♇

h m	Aspects
08 52	☉ ☍ ♂
10 12	☽ ⊻ ♇
10 36	☽ Q ♅
11 36	☽ □ ♆
18 50	☉ # ♂
19 16	☽ ∠ ♂
19 29	☽ ± ♀
20 17	☽ ⚼ ☉
20 31	☽ ⊥ ♄
12 Saturday	
06 16	☽ ⊼ ♀
06 19	☽ ⚹ ♅
06 20	☽ ⊻ ♄
06 45	♀ ⊼ ♅
06 51	☽ □ ♃
07 06	♀ ☍ ♄
09 44	☽ ⊥ ♂
12 09	☽ ± ☉
12 54	♀ △ ♃
17 13	☽ Q ♆
19 12	♄ ⊻ ♅
19 23	☽ ⊻ ♂
20 49	☽ # ♇
22 45	☽ ⊼ ☉
13 Sunday	
06 59	☽ ∠ ♅
09 04	☽ ☍ ☿
09 25	☽ # ☿
11 34	☽ ☌ ♇
12 28	☽ Q ♃
13 01	☽ ⚹ ♆
21 53	☽ ⊥ ♅
14 Monday	
06 01	☽ ∥ ♄
07 47	☽ ☌ ♄
07 52	☽ ⊻ ♅
08 24	☽ ⚹ ♃
12 38	☽ ☍ ♀
14 01	☽ ∠ ♆
17 03	☽ # ♀
20 07	☽ ☌ ♂
15 Tuesday	
04 29	♀ ⚼ ♆
04 29	☽ ☍ ☉
04 48	☽ ∥ ♂
05 13	☽ ⊥ ♆
09 41	☽ ∠ ♃
10 28	☽ # ☉
13 54	☽ ⊼ ☿
13 57	☽ ⊻ ♇
15 00	☿ ☍ ♇
15 09	☉ ± ♆
15 28	☽ ⊻ ♆
16 Wednesday	
00 17	☽ ⊥ ♇
00 49	☽ ± ☿
01 08	☽ ⊥ ♃
06 50	☽ ∥ ♆
06 50	☽ ∥ ♅
10 22	☽ ∥ ♃
10 49	☽ ⊻ ♄
11 02	☽ ☌ ♅
11 33	☽ ⊻ ♃
11 56	♅ ∥ ♆
15 59	☽ ∠ ♇
17 29	☽ ⚼ ☿
18 18	☿ △ ♆
21 17	☽ ⊼ ♀
21 20	☽ ⊥ ♄
22 40	☽ ⊻ ♂
17 Thursday	
09 04	☽ ± ♀
09 07	☽ ⊥ ♂
09 25	♀ ☍ ♂
12 55	☽ ⊼ ☉
13 20	☽ ∠ ♄
18 45	☽ ⚹ ♇
20 22	☽ ☌ ♆
22 09	☽ △ ☿
18 Friday	
00 40	☽ ± ☉
01 00	☽ ∠ ♂
03 04	☽ ⚼ ♀
04 59	☉ ⚼ ♃
16 41	☽ ⚹ ♄
16 47	☽ Q ♇
17 03	☽ ⊻ ♅
17 32	☽ ☌ ♃
18 37	☽ ⚼ ☉
19 Saturday	
04 07	☽ ⚹ ♂
04 16	☽ ⊥ ♅
10 01	☽ △ ♀
15 11	☽ Q ♄
21 20	☽ ∠ ♅
20 Sunday	
01 25	☽ △ ☉
02 23	☽ Q ♂
02 51	☽ □ ♇
04 35	☽ ⊻ ♆
11 08	☽ □ ☿
12 00	♀ # ♂
16 13	☽ ⊥ ♆
18 38	☽ ∥ ♃
19 39	☉ ⊼ ♇
22 54	☽ ∥ ♆
23 05	☽ ∥ ♅
21 Monday	
01 53	☽ □ ♄
02 25	☽ ⚹ ♅
02 50	☽ ⊻ ♃
09 53	☽ ∠ ♆
12 39	☽ □ ♂

h m	Aspects
13 30	☽ # ☉
14 38	☽ ⊥ ♃
17 31	☉ ⊼ ♆
22 Tuesday	
01 57	☽ Q ☿
02 09	☽ Q ♅
03 07	☽ □ ♀
03 23	☽ # ♀
04 01	☿ ± ♅
07 21	☽ ∥ ♂
08 29	☽ ∠ ♃
09 16	♀ ± ♆
13 57	☽ △ ♇
15 45	☽ ⚹ ♆
17 45	☽ □ ☉
23 Wednesday	
04 30	☽ ⚹ ☿
09 24	☽ ∥ ♄
13 32	☽ △ ♄
14 13	☽ □ ♅
14 30	☽ ⚹ ♃
15 54	☽ Q ♆
18 58	☽ # ☿
19 30	☽ Q ♀
20 08	☽ ⚼ ♇
23 14	☽ △ ♂
24 Thursday	
04 57	♀ ⚼ ♃
09 16	☽ Q ☉
14 06	☽ ∠ ☿
14 17	☽ ± ♇
14 36	☽ Q ♃
19 39	☽ ⚼ ♄
22 31	☽ ⚹ ♀
25 Friday	
00 51	☽ # ♇
02 21	☽ ⊼ ♇
04 09	☽ □ ♆
04 43	☽ ⚼ ♂
10 17	☽ ⊥ ☿
11 34	☽ ⚹ ☉
13 39	☽ ± ♄
22 18	☽ ± ♂
23 42	☽ ⊻ ☿
26 Saturday	
01 36	☽ ⊼ ♄
02 24	☽ △ ♅
02 31	☽ □ ♃
03 35	♂ ∠ ♆
08 00	☽ ∠ ♀
09 57	☽ ⊼ ♂
11 09	♀ ⊼ ♇
16 44	☿ ☍ ♄
20 02	☽ ∠ ☉
21 13	☽ ∥ ♇
22 05	☉ ± ♇
22 43	☉ ± ♃
27 Sunday	
00 17	☿ ⊼ ♅
00 38	☿ △ ♃
03 48	☽ ⊥ ♀
04 18	♀ ⊼ ♆
07 58	☽ ⚼ ♅
08 54	♃ Q ♇
12 33	♃ ⊻ ♅
13 50	☽ ☍ ♇
15 11	☽ ⊥ ☉
15 34	☽ △ ♆
16 50	☽ ⊻ ♀
17 02	☿ # ♄
28 Monday	
01 25	☽ ± ♅
03 48	☽ ⊻ ☉
10 36	☽ # ♄
12 02	☽ ☍ ♄
12 39	☽ ∥ ☿
12 51	☽ △ ♃
12 56	☽ ⊼ ♅
17 28	☽ ☌ ☿
19 02	☽ ☍ ♂
20 24	☽ ⚼ ♆
29 Tuesday	
05 23	☿ ☍ ♂
09 35	☽ # ♂
13 22	☽ ± ♆
17 03	☽ ⚼ ♃
19 11	☿ ⚼ ♆
22 55	☽ ⊼ ♇
30 Wednesday	
00 33	☽ ⊼ ♆
06 01	☽ ∥ ♀
07 50	☽ ☌ ♀
09 39	☽ ± ♃
10 00	☽ ± ♇
14 46	☽ ∥ ☉
16 48	☽ ☌ ☉
16 59	☽ # ♅
17 32	☽ # ♆
19 54	☽ ⊼ ♄
20 34	☽ ⊼ ♃
20 51	☽ ☍ ♅
22 14	☽ # ♃
31 Thursday	
01 42	☽ ⊼ ♂
01 46	♀ ± ♃
02 27	☽ ⚼ ♇
06 26	♀ ± ♇
06 46	☽ ± ♄
08 12	☽ ⊻ ☿
12 22	☽ ± ♂
20 22	☽ ⊥ ☿
22 53	☽ ⚼ ♄

All ephemeris data is given at 12.00 UT and the Moon's longitude is additionally given for 24.00 UT

JUNE 1984

LONGITUDES

Date	Sidereal time h m s	Sun ☉ ° ' "	Moon ☽ ° ' "	Moon ☽ 24.00 ° ' "	Mercury ☿ ° '	Venus ♀ ° '	Mars ♂ ° '	Jupiter ♃ ° '	Saturn ♄ ° '	Uranus ♅ ° '	Neptune ♆ ° '	Pluto ♇ ° '
01	04 40 37	11 ♊ 09 47	03 ♋ 26 35	10 ♋ 15 03	19 ♉ 40	07 ♊ 14	13 ♏ 58	11 ♑ 21	11 ♏ 02	11 ♐ 35	00 ♑ 34	29 ♎ 42
02	04 44 34	12 07 17	17 ♋ 06 44	24 ♋ 01 24	21 14	08 28	13 R 43	11 R 15	10 R 59	11 R 33	00 R 33	29 R 41
03	04 48 31	13 04 45	00 ♌ 58 44	07 ♌ 58 26	22 51	09 42	13 30	11 10	10 55	11 30	00 31	29 40
04	04 52 27	14 02 13	15 ♌ 00 10	22 ♌ 03 39	24 30	10 55	13 17	11 04	10 52	11 28	00 30	29 39
05	04 56 24	14 59 39	29 ♌ 08 31	06 ♍ 14 29	26 12	12 09	13 05	10 57	10 48	11 25	00 28	29 37
06	05 00 20	15 57 04	13 ♍ 21 13	20 ♍ 28 24	27 56	13 23	12 54	10 51	10 45	11 23	00 26	29 36
07	05 04 17	16 54 28	27 ♍ 35 43	04 ♎ 42 51	29 ♉ 43	14 36	12 43	10 45	10 42	11 20	00 25	29 35
08	05 08 13	17 51 51	11 ♎ 49 29	18 ♎ 55 17	01 ♊ 32	15 50	12 34	10 39	10 39	11 18	00 23	29 34
09	05 12 10	18 49 12	25 ♎ 59 55	03 ♏ 03 02	03 24	17 04	12 25	10 32	10 36	11 15	00 22	29 33
10	05 16 06	19 46 33	10 ♏ 04 19	17 ♏ 03 25	05 18	18 17	12 17	10 25	10 33	11 13	00 20	29 33
11	05 20 03	20 43 53	24 ♏ 00 00	00 ♐ 53 43	07 15	19 31	12 10	10 19	10 30	11 10	00 19	29 32
12	05 24 00	21 41 11	07 ♐ 44 16	14 ♐ 31 23	09 13	20 45	12 03	10 12	10 27	11 08	00 17	29 31
13	05 27 56	22 38 30	21 ♐ 14 48	27 ♐ 54 19	11 14	21 58	11 58	10 05	10 24	11 06	00 15	29 30
14	05 31 53	23 35 47	04 ♑ 29 47	11 ♑ 01 05	13 17	23 12	11 53	09 58	10 21	11 03	00 14	29 29
15	05 35 49	24 33 04	17 ♑ 28 12	23 ♑ 51 08	15 22	24 26	11 49	09 51	10 19	11 01	00 12	29 28
16	05 39 46	25 30 20	00 ♒ 10 00	06 ♒ 24 58	17 28	25 40	11 46	09 44	10 16	10 58	00 11	29 28
17	05 43 42	26 27 36	12 ♒ 36 14	18 ♒ 44 08	19 36	26 53	11 44	09 36	10 14	10 56	00 09	29 27
18	05 47 39	27 24 52	24 ♒ 48 58	00 ♓ 51 10	21 45	28 07	11 43	09 29	10 11	10 54	00 07	29 26
19	05 51 35	28 22 07	06 ♓ 51 10	12 ♓ 49 28	23 55	29 ♊ 21	11 42	09 22	10 09	10 51	00 06	29 25
20	05 55 32	29 ♊ 19 22	18 ♓ 46 35	24 ♓ 43 06	26 06	00 ♋ 34	11 D 42	09 14	10 07	10 49	00 04	29 25
21	05 59 29	00 ♋ 16 36	00 ♈ 39 33	06 ♈ 36 33	28 ♊ 18	01 48	11 43	09 07	10 05	10 47	00 03	29 24
22	06 03 25	01 13 51	12 ♈ 34 41	18 ♈ 34 33	00 ♋ 29	03 02	11 45	08 59	10 03	10 44	00 ♑ 01	29 24
23	06 07 22	02 11 05	24 ♈ 36 44	00 ♉ 41 48	02 41	04 16	11 48	08 52	10 01	10 42	29 ♐ 59	29 23
24	06 11 18	03 08 20	06 ♉ 50 17	13 ♉ 02 39	04 52	05 29	11 51	08 44	09 59	10 40	29 58	29 23
25	06 15 15	04 05 34	19 ♉ 19 23	25 ♉ 40 50	07 02	06 43	11 55	08 37	09 57	10 38	29 56	29 22
26	06 19 11	05 02 48	02 ♊ 07 19	08 ♊ 39 03	09 12	07 57	12 00	08 29	09 56	10 36	29 54	29 22
27	06 23 08	06 00 02	15 ♊ 16 09	21 ♊ 58 37	11 21	09 11	12 06	08 21	09 54	10 33	29 53	29 21
28	06 27 04	06 57 17	28 ♊ 46 22	05 ♋ 39 09	13 28	10 24	12 12	08 14	09 53	10 31	29 51	29 21
29	06 31 01	07 54 31	12 ♋ 36 40	19 ♋ 38 27	15 34	11 38	12 19	08 06	09 51	10 29	29 50	29 21
30	06 34 58	08 ♋ 51 44	26 ♋ 43 58	03 ♌ 52 35	17 ♋ 39	12 ♋ 52	12 ♏ 27	07 ♑ 58	09 ♏ 50	10 ♐ 27	29 ♐ 48	29 ♎ 20

Date	Moon True ☊ ° '	Moon Mean ☊ ° '	Moon ☽ Latitude ° '
01	06 ♊ 36	06 ♊ 27	02 N 23
02	06 R 35	06 24	03 26
03	06 34	06 21	04 18
04	06 34	06 18	04 54
05	06 33	06 15	05 14
06	06 33	06 12	05 14
07	06 D 33	06 08	04 55
08	06 34	06 05	04 19
09	06 34	06 02	03 26
10	06 35	05 59	02 21
11	06 36	05 56	01 N 09
12	06 R 36	05 52	00 S 06
13	06 36	05 49	01 20
14	06 35	05 46	02 28
15	06 33	05 43	03 27
16	06 30	05 40	04 14
17	06 27	05 37	04 48
18	06 24	05 33	05 08
19	06 22	05 30	05 14
20	06 21	05 27	05 07
21	06 D 20	05 24	04 47
22	06 21	05 21	04 14
23	06 22	05 18	03 30
24	06 23	05 14	02 35
25	06 25	05 11	01 32
26	06 26	05 08	00 S 24
27	06 R 26	05 05	00 N 48
28	06 24	05 02	01 59
29	06 21	04 58	03 05
30	06 ♊ 17	04 ♊ 55	04 N 00

DECLINATIONS

Date	Sun ☉ ° '	Moon ☽ ° '	Mercury ☿ ° '	Venus ♀ ° '	Mars ♂ ° '	Jupiter ♃ ° '	Saturn ♄ ° '	Uranus ♅ ° '	Neptune ♆ ° '	Pluto ♇ ° '
01	22 N 07	25 N 47	15 N 16	21 N 10	17 S 05	22 S 50	12 S 42	22 S 08	22 S 13	04 N 50
02	22 15	25 45	15 48	21 24	17 03	22 51	12 41	22 08	22 13	04 50
03	22 22	24 08	16 21	21 38	17 01	22 51	12 41	22 08	22 13	04 50
04	22 29	21 01	16 54	21 50	16 59	22 52	12 40	22 08	22 13	04 50
05	22 36	16 40	17 28	22 03	16 58	22 52	12 39	22 07	22 13	04 50
06	22 42	11 22	18 01	22 15	16 56	22 53	12 38	22 07	22 13	04 50
07	22 48	05 N 28	18 34	22 26	16 55	22 54	12 37	22 07	22 13	04 50
08	22 53	00 S 43	19 06	22 36	16 54	22 54	12 36	22 06	22 13	04 50
09	22 58	06 50	19 39	22 46	16 54	22 55	12 36	22 06	22 13	04 50
10	23 03	12 36	20 10	22 55	16 53	22 56	12 35	22 06	22 13	04 50
11	23 07	17 39	20 41	23 04	16 53	22 56	12 34	22 05	22 13	04 50
12	23 11	21 42	21 11	23 11	16 53	22 57	12 33	22 05	22 13	04 50
13	23 14	24 29	21 39	23 19	16 53	22 58	12 33	22 05	22 14	04 49
14	23 17	25 50	22 06	23 25	16 53	22 58	12 32	22 04	22 14	04 49
15	23 20	25 43	22 32	23 31	16 54	22 59	12 32	22 04	22 14	04 49
16	23 22	24 15	22 55	23 36	16 55	23 00	12 31	22 04	22 14	04 49
17	23 24	21 37	23 17	23 40	16 56	23 00	12 30	22 03	22 14	04 48
18	23 25	18 05	23 36	23 44	16 57	23 01	12 30	22 03	22 14	04 48
19	23 26	13 51	23 54	23 47	16 58	23 01	12 29	22 03	22 14	04 48
20	23 26	09 09	24 08	23 49	17 00	23 02	12 29	22 02	22 14	04 48
21	23 27	04 S 07	24 20	23 51	17 01	23 03	12 29	22 02	22 14	04 47
22	23 26	01 N 04	24 29	23 52	17 03	23 03	12 28	22 02	22 14	04 47
23	23 25	06 17	24 36	23 52	17 06	23 04	12 28	22 01	22 14	04 47
24	23 24	11 21	24 39	23 52	17 08	23 05	12 27	22 01	22 14	04 46
25	23 23	16 05	24 40	23 51	17 11	23 05	12 27	22 01	22 14	04 46
26	23 21	20 12	24 38	23 49	17 13	23 06	12 27	22 00	22 14	04 46
27	23 18	23 26	24 34	23 46	17 16	23 07	12 27	22 00	22 14	04 45
28	23 16	25 25	24 26	23 43	17 20	23 07	12 26	22 00	22 14	04 45
29	23 12	25 55	24 16	23 39	17 23	23 08	12 26	22 00	22 14	04 45
30	23 N 09	24 N 44	24 N 04	23 N 34	17 S 27	23 S 08	12 S 26	21 S 59	22 S 14	04 N 44

ZODIAC SIGN ENTRIES

Date	h	m	Planets	
01	05	54	☽	♋
03	10	19	☽	♌
05	13	27	☽	♍
07	15	45	☿	♊
07	16	03	☽	♎
09	18	48	☽	♏
11	22	26	☽	♐
14	03	48	☽	♑
16	11	41	☽	♒
18	22	18	☽	♓
20	00	48	♀	♋
21	05	02	☉	♋
21	10	40	☽	♈
22	06	39	☿	♋
23	01	10	♆	♐
23	22	38	☽	♉
26	08	04	☽	♊
28	14	09	☽	♋
30	17	30	☽	♌

LATITUDES

Date	Mercury ☿ ° '	Venus ♀ ° '	Mars ♂ ° '	Jupiter ♃ ° '	Saturn ♄ ° '	Uranus ♅ ° '	Neptune ♆ ° '	Pluto ♇ ° '
01	02 S 29	00 S 22	01 S 06	00 N 08	02 N 34	00 N 02	01 N 13	17 N 17
04	02 03	00 15	01 13	00 07	02 33	00 02	01 13	17 16
07	01 34	00 S 08	01 20	00 07	02 33	00 02	01 13	17 14
10	01 02	00 00	01 26	00 06	02 32	00 02	01 13	17 12
13	00 S 29	00 N 07	01 32	00 06	02 31	00 02	01 13	17 10
16	00 N 04	00 14	01 37	00 06	02 31	00 02	01 13	17 10
19	00 35	00 21	01 42	00 05	02 30	00 02	01 13	17 09
22	01 03	00 28	01 47	00 05	02 29	00 02	01 13	17 07
25	01 25	00 34	01 51	00 04	02 29	00 02	01 13	17 05
28	01 41	00 41	01 55	00 04	02 28	00 02	01 13	17 04
31	01 N 50	00 N 47	01 S 59	00 N 04	02 N 27	00 N 02	01 N 13	17 N 02

DATA

Julian Date	2445853
Delta T	+54 seconds
Ayanamsa	23° 38' 05"
Synetic vernal point	05° ♓ 28' 54"
True obliquity of ecliptic	23° 26' 32"

MOON'S PHASES, APSIDES AND POSITIONS ☽

Date	h	m	Phase	Longitude ° '	Eclipse Indicator
06	16	42	☽	16 ♍ 08	
13	14	42	○	22 ♐ 45	
21	11	10	☾	00 ♈ 15	
29	03	18	●	07 ♋ 34	

Day	h	m		
07	11	26	Perigee	
20	19	42	Apogee	
01	23	32	Max dec	25° N 58'
08	09	16	0S	
14	22	02	Max dec	25° S 57'
22	07	04	0N	
29	07	13	Max dec	25° N 56'

LONGITUDES

Date	Chiron ⚷ ° '	Ceres ⚳ ° '	Pallas ⚴ ° '	Juno ⚵ ° '	Vesta ⚶ ° '	Black Moon Lilith ⚸ ° '
01	03 ♊ 39	29 ♈ 31	19 ♓ 46	11 ♋ 14	16 ♋ 47	19 ♓ 17
11	04 ♊ 24	03 ♉ 04	21 ♓ 15	16 ♋ 28	21 ♋ 01	20 ♓ 25
21	05 ♊ 07	06 ♉ 28	22 ♓ 21	21 ♋ 40	25 ♋ 19	21 ♓ 32
31	05 ♊ 48	09 ♉ 44	23 ♓ 00	26 ♋ 47	29 ♋ 43	22 ♓ 39

ASPECTARIAN

h m	Aspects
01 Friday	
04 11	☽ ⚼ ♂
05 22	☽ △ ♇
06 55	☽ ☍ ♆
09 00	☉ ⊼ ♄
14 26	☽ ∠ ☿
16 04	☉ # ♅
16 17	☉ ⊼ ♃
19 22	☽ ⊻ ♀
22 06	☉ ☍ ♅
02 Saturday	
01 19	☽ △ ♄
01 50	☽ ☍ ♃
02 18	☽ ⊼ ♅
02 37	☽ ⊻ ☉
06 11	☽ △ ♂
06 55	☽ ⊥ ♀
07 34	☉ # ♆
12 45	☽ ± ♅
13 53	☽ ⊥ ☉
20 06	☽ ✶ ☿
03 Sunday	
00 06	☽ ∠ ♀
04 18	☽ ⚼ ♅
06 38	☽ ∠ ☉
09 44	☽ □ ♇
11 12	☽ ⊼ ♆
19 31	☽ Q ☿
20 33	☉ ⊼ ♂
21 29	☽ ± ♆
23 09	☽ # ♃
04 Monday	
02 19	☽ ∥ ☉
03 54	☽ # ♆
04 22	☽ ✶ ♀
04 35	☽ # ♅
04 58	☽ □ ♄
05 19	☽ ⊼ ♃
05 59	☽ △ ♅
06 54	☽ ∥ ♀
09 07	☽ □ ♂
10 14	☽ ✶ ☉
10 54	♀ ⊼ ♄
11 52	☿ ± ♆
12 50	☽ ⚼ ♆
14 31	♀ ⊼ ♃
15 24	☿ # ♂
15 29	☽ ± ♃
16 29	☽ Q ♇
22 12	♀ ☍ ♅
05 Tuesday	
02 44	☽ Q ♀
02 53	☉ ⚼ ♇
06 20	☽ □ ☿
06 39	☽ ⚼ ♃
08 06	☽ Q ☉
08 27	☽ ∥ ☿
08 49	☿ ⚼ ♃
10 32	☽ # ♂
11 26	☽ Q ♄
12 49	☽ ✶ ♇
14 14	☽ △ ♆
15 15	☽ Q ♂
20 32	♀ # ♅
06 Wednesday	
03 51	♀ ⊼ ♂
06 34	☽ # ♄
07 38	☽ ✶ ♄
07 49	☽ △ ♃
08 41	☽ □ ♅
09 45	♀ # ♆
11 14	☽ ✶ ♂
12 03	☽ □ ♀
14 07	☽ ∠ ♇
16 42	☽ □ ☉
07 Thursday	
05 16	☽ ⊥ ♇
06 59	☉ ± ♄
08 49	☽ ∠ ♄
10 18	☿ ⊼ ♇
11 42	♀ ⚼ ♇
12 13	☽ ∠ ♂
14 30	☽ ∥ ♇
14 56	☽ Q ♅
15 21	☽ ⊻ ♇
16 06	☽ △ ☿
16 45	☽ □ ♆
21 06	☿ ⊼ ♆
23 56	☽ ⊥ ♄
08 Friday	
03 13	☽ ⊥ ♂
10 01	☽ ⊻ ♄
10 01	☽ □ ♃
11 04	♃ ✶ ♄
11 06	☽ ✶ ♅
13 14	☽ ⊻ ♂
17 11	☉ # ♃
19 25	☽ △ ♀
21 10	☽ ⚼ ☿
22 57	☽ △ ☉
23 05	☽ Q ♆
09 Saturday	
03 11	☉ ± ♂
03 13	♀ ± ♄
04 03	☽ # ♇
12 26	☽ ∠ ♅
14 45	☽ ± ☿
16 16	☽ Q ♃
18 03	☽ ☌ ♇
19 24	☽ ✶ ♆
23 18	☽ ⚼ ♀
10 Sunday	
01 33	☿ ± ♃
02 17	☽ ⚼ ☉
02 34	☽ ⊼ ☿
03 43	☽ ⊥ ♅
11 53	♀ ± ♂
11 55	☽ ∥ ♄
12 36	☽ ✶ ♃
12 48	☽ ☌ ♄
13 27	♀ # ♃
13 57	☽ ⊻ ♅
14 57	☿ ± ♇
15 45	☽ ☌ ♂
16 10	☽ ± ♀
18 49	☽ ± ☉
21 01	☽ ∠ ♆
11 Monday	
03 30	☽ ⊼ ♀
05 55	☽ ⊼ ☉
08 02	☽ ∥ ♂
12 32	☽ ⊥ ♆
14 15	☽ ∠ ♃
21 36	☽ ⊻ ♇
22 57	☽ ⊻ ♆
12 Tuesday	
05 50	☽ ⊥ ♃
07 52	☽ # ☿
08 05	☽ ⊥ ♇
08 46	☉ ∥ ♀
14 40	☽ ∥ ♅
15 04	☽ ☍ ☿
15 43	☽ ∥ ♆
16 18	☽ ⊻ ♃
16 46	☽ ⊻ ♄
17 58	☽ ☌ ♅
19 34	☽ ⊻ ♂
21 22	☽ ∥ ♃
23 02	☿ ⊼ ♃
23 31	☽ # ☉
23 52	☽ # ♀
23 58	☽ ∠ ♇
13 Wednesday	
02 19	☿ ⊼ ♄
03 22	☽ ⊥ ♄
06 10	☽ ⊥ ♂
10 20	☿ ☍ ♅
13 26	☽ ☍ ♀
14 42	☽ ☍ ☉
19 27	☽ ∠ ♄
20 15	☿ ⊼ ♂
22 14	☽ ∠ ♂
14 Thursday	
02 52	☽ ✶ ♇
04 14	☽ ☌ ♆
10 16	☿ # ♅
18 45	☿ # ♆
21 58	☽ ☌ ♃
22 44	☽ ✶ ♄
15 Friday	
00 02	☽ ⊻ ♅
00 51	☽ Q ♇
01 32	☽ ✶ ♂
01 49	☿ ⚼ ♇
07 18	☽ ⊼ ☿
11 09	☽ ⊥ ♅
20 45	☽ ± ☿
21 04	☽ Q ♄
22 32	☉ ☌ ♀
22 39	☿ ± ♄
23 54	☽ Q ♂
16 Saturday	
02 24	☽ ⊼ ☉
02 30	☽ ⊼ ♀
04 02	☽ ∠ ♅
04 40	♀ ⚼ ♄
06 21	☉ ⚼ ♄
10 39	☽ □ ♇
12 01	☽ ⊻ ♆
14 46	☽ ± ☉
15 10	☽ ± ♀
15 20	☿ ± ♂
16 36	☿ # ♃
17 19	☽ ⚼ ☿
18 41	☽ # ♀
21 04	☽ # ☉
23 31	☽ ⊥ ♆
23 39	☽ # ☿
17 Sunday	
00 36	☽ ∥ ♃
06 14	☽ ⊻ ♃
07 13	☽ ∥ ♆
07 24	☽ □ ♄
08 36	☽ ∥ ♅
08 46	☽ ✶ ♅
09 04	♀ ⚼ ♂
09 35	☽ ⚼ ☉
10 19	☽ □ ♂
10 27	☽ ⚼ ♀
16 58	☽ ∠ ♆
17 48	☽ ⊥ ♃
18 41	☉ ⚼ ♂
20 34	☉ ∥ ☿
18 Monday	
04 38	☽ △ ☿
08 13	☽ Q ♅
11 21	☽ ∠ ♃
17 36	☽ △ ☉
18 46	☽ ∥ ♂
19 18	☽ △ ♀
21 10	☽ △ ♇
22 31	☽ ✶ ♆
19 Tuesday	
00 41	☿ ∥ ♀
13 33	♀ △ ♇
16 59	☽ ✶ ♃
18 16	♂ St D
18 36	☽ △ ♄
19 11	☽ ∥ ♄
20 01	☽ □ ♅
21 44	☽ △ ♂
22 31	☽ Q ♆
20 Wednesday	
01 19	☿ ⚼ ♄
02 22	♀ ☍ ♆
03 13	☽ ⚼ ♇
14 16	☉ △ ♇
16 55	☽ Q ♃
18 35	☿ ⚼ ♂
21 21	☽ ± ♇
21 Thursday	
00 46	☽ ⚼ ♄
04 02	☽ ⚼ ♂
06 09	☽ □ ☿
06 15	☉ ☍ ♆
08 51	☽ # ♇
09 28	☽ ⊼ ♇
10 45	☽ □ ♆
11 10	☽ □ ☉
14 34	☽ □ ♀
18 53	☽ ± ♄
22 14	☽ ± ♂
22 Friday	
00 05	☿ △ ♇
04 52	☽ □ ♃
06 53	☿ ☍ ♆
06 56	☽ ⊼ ♄
08 19	☽ △ ♅
10 20	☽ ⊼ ♂
23 Saturday	
02 23	☉ ☌ ☿
02 27	☽ Q ☉
02 27	☽ Q ☿
05 02	☽ ∥ ♇
06 48	☽ Q ♀
14 09	☽ ⚼ ♅
21 25	☽ ☍ ♇
22 35	☽ △ ♆
24 Sunday	
04 11	☽ ✶ ☉
07 20	☽ ✶ ☿
07 47	☽ ± ♅
09 05	☽ ✶ ♀
15 39	☽ △ ♃
17 24	☽ # ♄
18 05	☽ ☍ ♄
19 24	☽ ⊼ ♅
21 45	☽ ☍ ♂
25 Monday	
03 39	☽ ⚼ ♆
03 45	☿ ☌ ♀
11 32	☽ ∠ ☉
17 01	☽ ∠ ♀
18 04	☽ # ♂
18 12	☽ ∠ ☿
20 02	☽ ⚼ ♃
20 42	☽ ± ♆
26 Tuesday	
04 26	☿ ☍ ♃
05 50	☽ ⊥ ☉
06 53	☽ ⊼ ♇
07 54	☽ ⊼ ♆
11 38	☽ ⊥ ♀
12 40	☽ ± ♃
14 24	☽ ⊥ ☿
17 49	☽ ⊻ ☉
17 58	☽ ± ♇
20 00	☿ △ ♄
21 29	♀ ☍ ♃
23 35	☽ ⊼ ♃
23 49	☽ ⊻ ♀
27 Wednesday	
00 26	☽ # ♅
02 06	☽ # ♆
02 18	☽ ⊼ ♄
03 17	☿ ⊼ ♅
03 30	☽ ☍ ♅
03 32	☽ ⊻ ☿
06 13	☽ ⊼ ♂
09 13	☽ # ♃
10 21	☽ ⚼ ♇
10 57	☽ ∥ ☉
13 08	☽ ± ♄
15 06	☽ ∥ ♀
17 07	☽ ± ♂
20 53	☿ △ ♂
22 57	☽ ∥ ☿
28 Thursday	
01 55	♀ △ ♄
05 09	☽ ⚼ ♄
09 13	☽ ⚼ ♂
13 00	☽ △ ♇
13 53	☽ ☍ ♆
14 13	♀ ⊼ ♅
29 Friday	
03 18	☽ ☌ ☉
04 18	☽ ☍ ♃
07 16	☽ △ ♄
08 21	☽ ⊼ ♅
10 10	☽ ☌ ♀
11 30	☽ △ ♂
16 12	☉ ☍ ♃
17 57	☽ ☌ ☿
18 37	☽ ± ♅
22 24	☿ ± ♅
30 Saturday	
03 05	♀ △ ♂
09 51	☽ ⚼ ♅
13 47	☉ # ♃
16 23	☽ □ ♇
17 09	☽ ⊼ ♆
20 02	☽ ∥ ☿

All ephemeris data is given at 12.00 UT and the Moon's longitude is additionally given for 24.00 UT

JULY 1984

LONGITUDES

Date	Sidereal time h m s	Sun ☉ ° ' "	Moon ☽ ° ' "	Moon ☽ 24.00 ° ' "	Mercury ☿ ° '	Venus ♀ ° '	Mars ♂ ° '	Jupiter ♃ ° '	Saturn ♄ ° '	Uranus ♅ ° '	Neptune ♆ ° '	Pluto ♇ ° '
01	06 38 54	09 ♋ 48 58	11 ♌ 03 39	18 ♌ 16 25	19 ♋ 41	14 ♋ 06	12 ♏ 36	07 ♑ 50	09 ♏ 49	10 ♐ 25	29 ♐ 46	29 ♎ 20
02	06 42 51	10 46 11	25 ♌ 30 12	02 ♍ 44 16	21 42	15 19	12 46	07 R 43	09 R 48	10 R 23	29 R 45	29 R 20
03	06 46 47	11 43 24	09 ♍ 57 58	17 ♍ 10 42	23 41	16 33	12 56	07 35	09 47	10 21	29 43	29 19
04	06 50 44	12 40 37	24 ♍ 21 58	01 ♎ 31 18	25 39	17 47	13 07	07 27	09 46	10 19	29 42	29 19
05	06 54 40	13 37 49	08 ♎ 38 22	15 ♎ 42 53	27 34	19 01	13 18	07 20	09 45	10 17	29 40	29 19
06	06 58 37	14 35 01	22 ♎ 44 40	29 ♎ 43 35	29 ♋ 28	20 14	13 30	07 12	09 44	10 15	29 39	29 19
07	07 02 33	15 32 12	06 ♏ 39 34	13 ♏ 32 34	01 ♌ 19	21 28	13 43	07 05	09 44	10 13	29 37	29 19
08	07 06 30	16 29 24	20 ♏ 22 35	27 ♏ 09 37	03 09	22 42	13 57	06 57	09 43	10 11	29 35	29 19
09	07 10 27	17 26 36	03 ♐ 53 41	10 ♐ 34 47	04 56	23 56	14 11	06 50	09 43	10 10	29 34	29 D 19
10	07 14 23	18 23 47	17 ♐ 12 55	23 ♐ 48 03	06 42	25 09	14 25	06 42	09 42	10 08	29 32	29 19
11	07 18 20	19 20 59	00 ♑ 20 12	06 ♑ 49 19	08 25	26 23	14 41	06 35	09 42	10 06	29 31	29 19
12	07 22 16	20 18 10	13 ♑ 15 22	19 ♑ 38 21	10 07	27 37	14 57	06 28	09 42	10 04	29 29	29 19
13	07 26 13	21 15 22	25 ♑ 58 15	02 ♒ 15 04	11 46	28 ♋ 51	15 13	06 20	09 D 42	10 03	29 28	29 19
14	07 30 09	22 12 34	08 ♒ 28 52	14 ♒ 39 43	13 24	00 ♌ 05	15 30	06 13	09 42	10 01	29 27	29 19
15	07 34 06	23 09 47	20 ♒ 47 44	26 ♒ 53 05	14 59	01 18	15 48	06 06	09 42	10 00	29 25	29 20
16	07 38 02	24 07 00	02 ♓ 55 59	08 ♓ 56 42	16 33	02 32	16 06	05 59	09 43	09 58	29 24	29 20
17	07 41 59	25 04 14	14 ♓ 55 35	20 ♓ 52 59	18 05	03 46	16 25	05 52	09 43	09 57	29 22	29 20
18	07 45 55	26 01 28	26 ♓ 49 20	02 ♈ 45 06	19 34	05 00	16 44	05 46	09 43	09 55	29 21	29 20
19	07 49 52	26 58 42	08 ♈ 40 49	14 ♈ 37 02	21 02	06 14	17 04	05 39	09 44	09 54	29 19	29 21
20	07 53 49	27 55 58	20 ♈ 34 21	26 ♈ 33 21	22 27	07 27	17 25	05 32	09 45	09 52	29 18	29 21
21	07 57 45	28 53 14	02 ♉ 34 42	08 ♉ 39 01	23 51	08 41	17 46	05 26	09 45	09 51	29 17	29 21
22	08 01 42	29 ♋ 50 31	14 ♉ 46 56	20 ♉ 59 05	25 12	09 55	18 07	05 19	09 46	09 50	29 15	29 22
23	08 05 38	00 ♌ 47 49	27 ♉ 16 03	03 ♊ 38 22	26 31	11 09	18 29	05 13	09 47	09 48	29 14	29 22
24	08 09 35	01 45 08	10 ♊ 06 31	16 ♊ 40 52	27 48	12 23	18 51	05 07	09 48	09 47	29 13	29 23
25	08 13 31	02 42 28	23 ♊ 21 42	00 ♋ 09 10	29 ♌ 03	13 36	19 14	05 01	09 49	09 46	29 11	29 23
26	08 17 28	03 39 49	07 ♋ 03 15	14 ♋ 03 46	00 ♍ 15	14 50	19 37	04 55	09 51	09 45	29 10	29 24
27	08 21 24	04 37 10	21 ♋ 10 23	28 ♋ 22 32	01 26	16 04	20 01	04 49	09 52	09 44	29 09	29 24
28	08 25 21	05 34 33	05 ♌ 39 32	13 ♌ 00 31	02 33	17 18	20 25	04 43	09 53	09 43	29 08	29 25
29	08 29 18	06 31 56	20 ♌ 24 29	27 ♌ 50 23	03 38	18 32	20 50	04 38	09 55	09 42	29 06	29 26
30	08 33 14	07 29 19	05 ♍ 17 05	12 ♍ 43 29	04 41	19 46	21 15	04 32	09 57	09 41	29 05	29 26
31	08 37 11	08 ♌ 26 44	20 ♍ 08 34	27 ♍ 31 21	05 ♍ 40	21 ♌ 00	21 ♏ 40	04 ♑ 27	09 ♏ 58	09 ♐ 40	29 ♐ 04	29 ♎ 27

Date	Moon True ☊ ° '	Moon Mean ☊ ° '	Moon ☽ Latitude ° '
01	06 ♊ 12	04 ♊ 52	04 N 42
02	06 R 07	04 49	05 06
03	06 03	04 46	05 10
04	06 00	04 43	04 55
05	05 59	04 39	04 22
06	05 D 59	04 36	03 33
07	06 00	04 33	02 33
08	06 01	04 30	01 24
09	06 02	04 27	00 N 12
10	06 R 02	04 24	01 S 00
11	05 59	04 20	02 08
12	05 55	04 17	03 07
13	05 49	04 14	03 56
14	05 41	04 11	04 33
15	05 32	04 08	04 57
16	05 24	04 04	05 06
17	05 16	04 01	05 02
18	05 11	03 58	04 45
19	05 07	03 55	04 16
20	05 05	03 52	03 36
21	05 D 04	03 49	02 45
22	05 05	03 45	01 47
23	05 06	03 42	00 S 42
24	05 R 06	03 39	00 N 27
25	05 05	03 36	01 36
26	05 01	03 33	02 42
27	04 55	03 30	03 40
28	04 47	03 26	04 25
29	04 38	03 23	04 54
30	04 29	03 20	05 03
31	04 ♊ 20	03 ♊ 17	04 N 52

DECLINATIONS

Date	Sun ☉ ° '	Moon ☽ ° '	Mercury ☿ ° '	Venus ♀ ° '	Mars ♂ ° '	Jupiter ♃ ° '	Saturn ♄ ° '	Uranus ♅ ° '	Neptune ♆ ° '	Pluto ♇ ° '
01	23 N 05	21 N 58	23 N 49	23 N 28	17 S 30	23 S 09	12 S 26	21 S 59	22 S 14	04 N 44
02	23 00	17 49	23 32	23 22	17 34	23 10	12 26	21 59	22 14	04 43
03	22 56	12 37	23 13	23 16	17 38	23 10	12 26	21 58	22 14	04 43
04	22 50	06 45	22 52	23 08	17 43	23 11	12 26	21 58	22 14	04 42
05	22 45	00 N 35	22 29	23 00	17 47	23 11	12 26	21 58	22 14	04 42
06	22 39	05 S 33	22 05	22 51	17 52	23 12	12 26	21 58	22 14	04 41
07	22 32	11 20	21 39	22 41	17 56	23 12	12 26	21 57	22 14	04 41
08	22 25	16 30	21 11	22 31	18 01	23 13	12 26	21 57	22 14	04 40
09	22 18	20 44	20 43	22 20	18 06	23 14	12 26	21 57	22 14	04 40
10	22 11	23 50	20 13	22 09	18 12	23 14	12 26	21 57	22 14	04 39
11	22 03	25 34	19 42	21 57	18 17	23 15	12 26	21 56	22 14	04 39
12	21 54	25 53	19 11	21 44	18 23	23 15	12 26	21 56	22 14	04 38
13	21 46	24 50	18 38	21 30	18 28	23 16	12 27	21 56	22 14	04 37
14	21 37	22 32	18 05	21 16	18 34	23 16	12 27	21 56	22 14	04 37
15	21 27	19 15	17 31	21 02	18 40	23 16	12 27	21 55	22 14	04 36
16	21 17	15 11	16 57	20 46	18 46	23 17	12 28	21 55	22 14	04 36
17	21 07	10 35	16 22	20 31	18 52	23 17	12 28	21 55	22 14	04 35
18	20 57	05 38	15 47	20 14	18 58	23 18	12 28	21 55	22 14	04 34
19	20 46	00 S 29	15 11	19 57	19 04	23 18	12 29	21 55	22 14	04 34
20	20 35	04 N 42	14 36	19 40	19 11	23 19	12 29	21 54	22 14	04 33
21	20 23	09 47	14 00	19 22	19 17	23 19	12 30	21 54	22 14	04 32
22	20 11	14 34	13 25	19 03	19 24	23 19	12 30	21 54	22 14	04 32
23	19 59	18 52	12 49	18 44	19 30	23 20	12 31	21 54	22 14	04 31
24	19 46	22 25	12 13	18 24	19 37	23 20	12 31	21 54	22 14	04 30
25	19 33	24 53	11 38	18 04	19 44	23 21	12 32	21 54	22 14	04 30
26	19 20	25 57	11 03	17 43	19 50	23 21	12 33	21 53	22 14	04 29
27	19 07	25 24	10 29	17 22	19 57	23 21	12 33	21 53	22 14	04 28
28	18 53	23 09	09 54	17 00	20 04	23 22	12 34	21 53	22 14	04 27
29	18 39	19 20	09 21	16 38	20 11	23 22	12 35	21 53	22 14	04 27
30	18 24	14 16	08 48	16 15	20 18	23 22	12 36	21 53	22 14	04 26
31	18 N 09	08 N 23	08 N 15	15 N 52	20 S 25	23 S 23	12 S 36	21 S 53	22 S 14	04 N 25

ZODIAC SIGN ENTRIES

Date	h	m	Planets
02	19	28	☽ ♍
04	21	27	☽ ♎
06	18	56	☿ ♌
07	00	28	☽ ♏
09	05	03	☽ ♐
11	11	23	☽ ♑
13	19	41	☽ ♒
14	10	30	♀ ♌
16	06	10	☽ ♓
18	18	26	☽ ♈
21	06	52	☽ ♉
22	15	58	☉ ♌
23	17	10	☽ ♊
25	23	44	☽ ♋
26	06	49	☿ ♍
28	02	41	☽ ♌
30	03	29	☽ ♍

LATITUDES

Date	Mercury ☿ ° '	Venus ♀ ° '	Mars ♂ ° '	Jupiter ♃ ° '	Saturn ♄ ° '	Uranus ♅ ° '	Neptune ♆ ° '	Pluto ♇ ° '
01	01 N 50	00 N 47	01 S 59	00 N 04	02 N 27	00 N 02	01 N 13	17 N 02
04	01 53	00 53	02 02	00 03	02 26	00 02	01 13	17 00
07	01 49	00 58	02 05	00 03	02 25	00 02	01 13	16 59
10	01 40	01 04	02 07	00 02	02 25	00 02	01 13	16 57
13	01 26	01 08	02 10	00 02	02 24	00 02	01 13	16 55
16	01 07	01 13	02 12	00 01	02 23	00 02	01 13	16 53
19	00 44	01 16	02 14	00 01	02 22	00 02	01 12	16 52
22	00 N 18	01 20	02 15	00 N 01	02 21	00 02	01 12	16 50
25	00 S 11	01 23	02 17	00 00	02 21	00 01	01 12	16 48
28	00 42	01 25	02 18	00 00	02 20	00 01	01 12	16 46
31	01 S 16	01 N 27	02 S 19	00 S 01	02 N 19	00 N 01	01 N 12	16 N 45

DATA

Julian Date	2445883
Delta T	+54 seconds
Ayanamsa	23° 38' 11"
Synetic vernal point	05° ♓ 28' 48"
True obliquity of ecliptic	23° 26' 32"

MOON'S PHASES, APSIDES AND POSITIONS ☽

Date	h	m	Phase	Longitude ° '	Eclipse Indicator
05	21	04	☽	13 ♎ 59	
13	02	20	○	20 ♑ 52	
21	04	01	☾	28 ♈ 34	
28	11	51	●	05 ♌ 34	

Day	h	m		
02	22	26	Perigee	
18	13	30	Apogee	
30	12	08	Perigee	
05	14	17	0S	
12	05	21	Max dec	25° S 56'
19	14	13	0N	
26	16	10	Max dec	25° N 59'

LONGITUDES

Date	Chiron ⚷ ° '	Ceres ⚳ ° '	Pallas ⚴ ° '	Juno ⚵ ° '	Vesta ⚶ ° '	Black Moon Lilith ⚸ ° '
01	05 ♊ 48	09 ♉ 44	23 ♓ 00	26 ♋ 47	29 ♋ 43	22 ♓ 39
11	06 ♊ 25	12 ♉ 49	23 ♓ 09	01 ♌ 52	04 ♌ 10	23 ♓ 46
21	06 ♊ 59	15 ♉ 40	22 ♓ 44	06 ♌ 52	08 ♌ 40	24 ♓ 53
31	07 ♊ 29	18 ♉ 16	21 ♓ 45	11 ♌ 48	13 ♌ 13	26 ♓ 00

ASPECTARIAN

h m	Aspects
01 Sunday	
00 12	☽ ∥ ♀
03 11	☽ ± ♆
03 18	☽ # ♃
03 39	☽ ∥ ☉
06 41	☽ ⊼ ♃
09 47	☽ ⊻ ☉
09 55	☽ □ ♄
10 11	☽ # ♆
10 56	☽ △ ♅
11 53	☽ # ♅
11 56	☉ △ ♄
14 36	☽ □ ♂
16 35	☽ ± ♃
17 31	☽ ⊻ ♀
18 10	☽ bQ ♆
20 29	☽ ⊥ ☉
22 26	☽ Q ♇
02 Monday	
02 36	☉ ⊼ ♅
04 25	☽ ⊥ ♀
04 41	☽ ⊻ ☿
07 25	☽ bQ ♃
12 28	☽ ∠ ☉
13 12	☽ # ♂
15 48	☽ Q ♄
16 14	☽ ⊥ ☿
18 21	☽ ⚹ ♇
19 02	☽ △ ♆
20 44	☽ ∠ ♀
20 49	☽ Q ♂
03 Tuesday	
06 58	☿ ∥ ♀
08 05	☽ △ ♃
08 10	♀ ± ♅
09 33	☽ ∠ ☿
11 41	☽ ⚹ ♄
12 38	☽ □ ♅
12 48	☽ # ♄
15 03	☿ # ♃
15 08	☽ ⚹ ☉
16 59	☽ ⚹ ♂
19 15	☽ ∠ ♇
23 59	☽ ⚹ ♀
04 Wednesday	
03 59	♀ # ♃
08 00	☿ bQ ♅
10 15	☽ ⊥ ♇
12 33	☽ Q ☉
12 40	☽ ∠ ♄
14 16	☉ ∥ ☿
14 29	☽ ⚹ ☿
18 21	☽ ∠ ♂
18 36	☽ Q ♅
20 03	☽ ∥ ♇
20 18	☽ ⊻ ♇
20 55	☽ □ ♆
21 56	☽ Q ♀
05 Thursday	
01 34	☉ △ ♂
03 45	☽ ⊥ ♄
09 42	☽ ⊥ ♂
09 49	☽ □ ♃
13 49	☽ Q ☿
13 53	☽ ⊻ ♄
14 47	☽ ⚹ ♅
20 01	☽ ⊻ ♂
21 04	☽ □ ☉
06 Friday	
03 15	☿ # ♆
03 18	☽ Q ♆
07 18	☽ □ ♀
08 36	☽ # ♇
10 11	☿ □ ♇
14 19	☿ ⊼ ♆
16 11	☽ Q ♃
16 18	☽ ∠ ♅
18 39	☿ # ♅
23 18	☽ ☌ ♇
23 50	☽ ⚹ ♆
07 Saturday	
01 20	☽ □ ☿
07 47	☽ ⊥ ♅
12 43	☽ ⚹ ♃
16 48	☽ ∥ ♄
17 20	☽ ☌ ♄
18 11	☽ ⊻ ♅
08 Sunday	
00 31	☽ ☌ ♂
01 51	☽ ∠ ♆
04 39	☽ △ ☉
04 42	☉ ± ♅
14 45	☽ ∠ ♃
16 30	☽ △ ♀
17 40	☽ ⊥ ♆
20 09	☽ ∥ ♂
09 Monday	
03 50	☽ ⊻ ♇
04 17	☽ ⊻ ♆
06 35	☽ ⊥ ♃
08 21	♇ St D
09 12	☽ bQ ☉
11 50	☽ # ☿
14 09	☽ △ ☿
14 32	☽ ⊥ ♇
17 12	☽ ⊻ ♃
20 17	☽ ∥ ♅
20 27	☿ ± ♆
21 57	☽ bQ ♀
22 23	☽ ∥ ♆
22 26	☽ ⊻ ♄
22 32	☽ # ☉
22 34	☽ # ♀
23 13	☽ ☌ ♅
10 Tuesday	
00 44	☉ ∥ ♀
01 48	♀ # ♆
02 21	☉ # ♆
02 36	☽ ± ☉
06 30	☽ ∥ ♃
06 45	☽ ∠ ♇
06 51	☽ ⊻ ♂
09 16	☽ ⊥ ♄
11 30	♀ bQ ♅
12 10	☿ ⊼ ♃
14 19	☽ ⊼ ☉
15 54	☽ ± ♀
17 57	☽ ⊥ ♂
21 23	☽ bQ ☿
22 01	♂ ∠ ♆
11 Wednesday	
01 39	☽ ∠ ♄
03 59	☽ ⊼ ♀
10 07	☽ ⚹ ♇
10 29	☽ ☌ ♆
10 46	☽ ∠ ♂
12 31	♀ # ♅
16 25	☽ ± ☿
23 27	☽ ☌ ♃
12 Thursday	
05 14	☽ ⊼ ☿
05 22	☽ ⚹ ♄
06 04	☽ ⊻ ♅
06 10	☿ □ ♄
06 58	☉ # ♅
08 22	☽ Q ♇
11 30	☿ △ ♅
15 14	☽ ⚹ ♂
17 16	☽ ⊥ ♅
13 Friday	
02 20	☽ ☍ ☉
03 54	☽ Q ♄
06 16	♄ St D
10 15	☽ ∠ ♅
14 26	☽ Q ♂
18 05	☽ ☍ ♀
18 07	☿ # ♂
18 23	☽ □ ♇
18 39	☽ ⊻ ♆
19 48	☿ ± ♃
21 15	♀ □ ♇
23 51	♀ ⊼ ♆
14 Saturday	
05 31	☽ ∥ ♃
06 09	☽ ⊥ ♆
07 41	☽ ⊻ ♃
14 22	☽ □ ♄
14 33	☽ ∥ ♆
14 58	☽ ⚹ ♅
17 00	☽ ∥ ♅
19 11	☽ ⊥ ♃
19 51	☽ # ☉
22 49	☽ # ♀
22 58	☽ ☍ ☿
23 33	☽ ∠ ♆
15 Sunday	
01 59	☽ □ ♂
03 27	☿ bQ ♆
12 36	☽ ∠ ♃
14 21	☽ Q ♅
15 37	☽ ∥ ♂
17 03	☽ ⊼ ☉
16 Monday	
00 27	☽ # ☿
03 27	☿ □ ♂
04 50	☽ △ ♇
04 59	☽ ⚹ ♆
05 55	☽ ± ☉
11 07	☽ ⊼ ♀
18 02	☽ ⚹ ♃
17 Tuesday	
00 13	☿ Q ♇
00 28	☽ ± ♀
01 25	☽ bQ ☉
01 32	☽ △ ♄
02 01	☽ □ ♅
02 28	☽ ∥ ♄
04 52	☽ Q ♆
08 51	☉ bQ ♅
10 48	☽ bQ ♇
15 05	☽ △ ♂
17 52	☽ Q ♃
19 15	☽ ⊼ ☿
20 37	☽ bQ ♀
18 Wednesday	
04 57	☽ ± ♇
07 45	☽ bQ ♄
09 06	☽ ± ☿
10 15	☽ △ ☉
16 58	☽ # ♇
17 06	☽ ⊼ ♇
17 06	☽ □ ♆
18 42	♀ ± ♆
18 49	♆ ⚹ ♇
22 14	☽ bQ ♂
19 Thursday	
01 38	♀ ⊼ ♃
01 59	☽ ± ♄
05 53	☽ bQ ☿
05 55	☽ □ ♃
06 06	☿ bQ ♃
06 28	☽ △ ♀
14 08	☽ ⊼ ♄
14 27	☽ △ ♅
16 59	☽ ± ♂
20 Friday	
05 27	☽ ⊼ ♂
11 17	☽ ∥ ♇
16 17	☽ △ ☿
20 37	☽ bQ ♅
21 Saturday	
04 01	☽ □ ☉
05 27	☽ △ ♆
05 35	☽ ☍ ♇
14 31	☽ ± ♅
16 19	♀ # ♂
17 36	☽ △ ♃
21 35	☉ ⊼ ♆
23 52	☉ □ ♇
22 Sunday	
01 24	☽ # ♄
01 25	☽ □ ♀
02 11	☽ ☍ ♄
02 20	☽ ⊼ ♅
06 40	☽ ∥ ☿
09 06	♀ □ ♄
10 16	♀ △ ♅
10 59	☽ bQ ♆
18 26	☽ Q ☉
18 40	☽ ☍ ♂
22 38	☽ bQ ♃
23 Monday	
04 20	☽ ± ♆
10 25	☽ □ ☿
11 12	☽ ∥ ♀
13 13	♀ ± ♃
15 39	☽ ± ♃
15 43	☽ ⊼ ♆
15 56	☽ Q ♀
15 59	☽ ⊼ ♇
16 00	☽ # ♂
18 33	☽ ∥ ☉
19 12	☽ ⚹ ☉
24 Tuesday	
00 01	☿ # ♄
00 26	♄ ⊻ ♅
02 49	☽ ⊼ ♃
03 14	☽ ± ♇
08 05	☽ # ♅
10 38	☽ # ♆
11 24	☽ ☍ ♅
11 26	☽ ⊼ ♄
11 59	☿ Q ♄
16 36	☽ ⚹ ♀
19 45	☽ # ♃
19 49	☽ bQ ♇
22 25	☽ ± ♄
23 30	☽ Q ☿
23 43	☉ # ♂
25 Wednesday	
01 04	☽ ∠ ☉
04 23	☽ ⊼ ♂
14 36	☽ bQ ♄
14 41	☿ △ ♆
15 25	☽ ± ♂
18 23	☽ ⊥ ☉
18 42	☿ ⚹ ♇
22 12	☽ ∠ ♀
22 18	☽ ☍ ♆
22 40	☽ △ ♇
23 03	☽ ⚹ ☿
23 09	♀ bQ ♆
26 Thursday	
05 41	☽ ⊻ ☉
07 40	☽ bQ ♂
08 19	☽ ☍ ♃
15 22	☽ ⊥ ♀
16 38	☽ ⊼ ♅
16 48	☽ △ ♄
27 Friday	
02 21	♂ ∠ ♃
02 35	☽ ⊻ ♀
02 50	☽ ± ♅
03 18	☽ ∠ ☿
10 00	☽ △ ♂
16 28	☉ ⊼ ♃
17 56	☽ bQ ♅
19 43	☽ ⊥ ☿
28 Saturday	
00 59	☉ ± ♆
01 16	☽ ⊼ ♆
01 43	☽ □ ♇
06 29	☽ ⊻ ☿
10 17	☽ # ♃
10 28	☽ ⊼ ♃
11 08	☽ ± ♆
11 51	☽ ☌ ☉
14 19	♀ Q ♇
18 37	☽ △ ♅
18 38	☽ # ♆
18 56	☽ □ ♄
20 13	☽ ± ♃
21 00	☽ # ♅
29 Sunday	
01 48	☽ bQ ♆
07 10	☽ Q ♇
07 27	☽ # ♂
08 41	☽ ☌ ♀
10 45	☽ bQ ♃
12 42	☽ □ ♂
15 47	☽ ∥ ☉
30 Monday	
00 09	☽ Q ♄
02 01	☽ △ ♆
02 28	☽ ∥ ♀
02 34	☽ ⚹ ♇
07 56	♀ bQ ♃
08 58	☿ △ ♃
10 48	☽ △ ♃
10 57	☽ ☌ ☿
15 48	☽ ⊻ ☉
18 34	☽ Q ♂
19 05	☽ □ ♅
19 06	☽ # ♄
19 31	☽ ⚹ ♄
31 Tuesday	
02 09	☽ ⊥ ☉
02 47	☽ ∠ ♇
12 32	☽ ∥ ☿
13 30	☽ ⊻ ♀
14 33	☽ ⚹ ♂
17 23	☽ ⊥ ♇
17 44	☽ ∠ ☉
19 51	☽ ∠ ♄

All ephemeris data is given at 12.00 UT and the Moon's longitude is additionally given for 24.00 UT

AUGUST 1984

LONGITUDES

Date	Sidereal time h m s	Sun ☉ ° ' "	Moon ☽ ° ' "	Moon ☽ 24.00 ° ' "	Mercury ☿ ° '	Venus ♀ ° '	Mars ♂ ° '	Jupiter ♃ ° '	Saturn ♄ ° '	Uranus ♅ ° '	Neptune ♆ ° '	Pluto ♇ ° '
01	08 41 07	09 ♌ 24 09	04 ♎ 51 04	12 ♎ 07 02	06 ♍ 37	22 ♌ 13	22 ♏ 06	04 ♑ 22	10 ♏ 00	09 ♐ 39	29 ♐ 03	29 ♎ 28
02	08 45 04	10 21 34	19 ♎ 18 45	26 ♎ 25 54	07 31	23 27	22 32	04 R 17	10 02	09 R 38	29 R 02	29 29
03	08 49 00	11 19 00	03 ♏ 28 16	10 ♏ 25 49	08 22	24 41	22 59	04 12	10 04	09 37	29 01	29 30
04	08 52 57	12 16 27	17 ♏ 18 36	24 ♏ 06 45	09 09	25 55	23 26	04 07	10 06	09 37	29 00	29 30
05	08 56 53	13 13 55	00 ♐ 50 28	07 ♐ 29 59	09 53	27 09	23 53	04 03	10 08	09 36	28 59	29 31
06	09 00 50	14 11 23	14 ♐ 05 34	20 ♐ 37 29	10 34	28 23	24 21	03 59	10 11	09 35	28 58	29 32
07	09 04 47	15 08 52	27 ♐ 05 59	03 ♑ 31 19	11 10	29 ♌ 36	24 49	03 55	10 13	09 35	28 57	29 33
08	09 08 43	16 06 22	09 ♑ 53 41	16 ♑ 13 17	11 43	00 ♍ 50	25 18	03 51	10 15	09 34	28 56	29 34
09	09 12 40	17 03 53	22 ♑ 30 15	28 ♑ 44 45	12 11	02 04	25 47	03 47	10 18	09 34	28 55	29 35
10	09 16 36	18 01 25	04 ♒ 56 51	11 ♒ 06 42	12 35	03 18	26 16	03 43	10 21	09 33	28 54	29 36
11	09 20 33	18 58 57	17 ♒ 14 20	23 ♒ 19 53	12 55	04 32	26 46	03 40	10 23	09 33	28 53	29 37
12	09 24 29	19 56 31	29 ♒ 23 26	05 ♓ 25 07	13 09	05 46	27 16	03 36	10 26	09 33	28 52	29 39
13	09 28 26	20 54 06	11 ♓ 25 04	17 ♓ 23 28	13 19	06 59	27 46	03 33	10 29	09 33	28 51	29 40
14	09 32 22	21 51 43	23 ♓ 20 33	29 ♓ 16 36	13 23	08 13	28 16	03 30	10 32	09 32	28 50	29 41
15	09 36 19	22 49 21	05 ♈ 11 54	11 ♈ 06 50	13 R 22	09 27	28 47	03 27	10 35	09 32	28 49	29 42
16	09 40 16	23 47 00	17 ♈ 01 50	22 ♈ 57 20	13 16	10 41	29 18	03 25	10 38	09 32	28 49	29 43
17	09 44 12	24 44 41	28 ♈ 53 52	04 ♉ 52 00	13 04	11 55	29 ♏ 50	03 22	10 42	09 32	28 48	29 45
18	09 48 09	25 42 23	10 ♉ 52 19	16 ♉ 55 26	12 46	13 08	00 ♐ 21	03 20	10 45	09 D 32	28 47	29 46
19	09 52 05	26 40 07	23 ♉ 02 00	29 ♉ 12 41	12 22	14 22	00 53	03 18	10 48	09 32	28 47	29 47
20	09 56 02	27 37 52	05 ♊ 28 08	11 ♊ 48 58	11 53	15 36	01 26	03 16	10 52	09 32	28 46	29 49
21	09 59 58	28 35 40	18 ♊ 15 48	24 ♊ 49 08	11 19	16 50	01 58	03 14	10 55	09 32	28 45	29 50
22	10 03 55	29 ♌ 33 29	01 ♋ 29 25	08 ♋ 16 57	10 40	18 04	02 31	03 13	10 59	09 32	28 45	29 51
23	10 07 51	00 ♍ 31 20	15 ♋ 11 55	22 ♋ 14 17	09 56	19 17	03 04	03 12	11 03	09 33	28 44	29 53
24	10 11 48	01 29 12	29 ♋ 23 51	06 ♌ 40 09	09 08	20 31	03 38	03 10	11 07	09 33	28 44	29 54
25	10 15 45	02 27 06	14 ♌ 02 34	21 ♌ 30 10	08 17	21 45	04 11	03 09	11 11	09 33	28 43	29 56
26	10 19 41	03 25 02	29 ♌ 01 54	06 ♍ 36 32	07 23	22 59	04 45	03 09	11 15	09 34	28 43	29 57
27	10 23 38	04 22 59	14 ♍ 12 42	21 ♍ 49 02	06 28	24 12	05 20	03 08	11 19	09 34	28 42	29 ♎ 59
28	10 27 34	05 20 58	29 ♍ 24 11	06 ♎ 56 52	05 33	25 26	05 54	03 08	11 23	09 35	28 42	00 ♏ 00
29	10 31 31	06 18 58	14 ♎ 25 57	21 ♎ 50 30	04 38	26 40	06 29	03 08	11 27	09 35	28 41	00 02
30	10 35 27	07 16 59	29 ♎ 09 46	06 ♏ 23 13	03 46	27 54	07 04	03 D 08	11 31	09 36	28 41	00 04
31	10 39 24	08 ♍ 15 02	13 ♏ 30 31	20 ♏ 31 33	02 ♍ 57	29 ♍ 07	07 ♐ 39	03 ♑ 08	11 ♏ 36	09 ♐ 37	28 ♐ 41	00 ♏ 05

Date	Moon True ☊ ° '	Moon Mean ☊ ° '	Moon ☽ Latitude ° '
01	04 ♊ 14	03 ♊ 14	04 N 22
02	04 R 10	03 10	03 35
03	04 09	03 07	02 35
04	04 D 09	03 04	01 28
05	04 09	03 01	00 N 18
06	04 R 09	02 58	00 S 53
07	04 06	02 55	01 59
08	04 02	02 51	02 57
09	03 55	02 48	03 46
10	03 45	02 45	04 24
11	03 33	02 42	04 48
12	03 20	02 39	05 00
13	03 07	02 36	04 57
14	02 55	02 32	04 42
15	02 45	02 29	04 14
16	02 37	02 26	03 36
17	02 33	02 23	02 48
18	02 30	02 20	01 52
19	02 D 30	02 16	00 S 50
20	02 R 30	02 13	00 N 16
21	02 29	02 10	01 22
22	02 27	02 07	02 27
23	02 23	02 04	03 25
24	02 15	02 01	04 13
25	02 06	01 57	04 46
26	01 55	01 54	05 00
27	01 44	01 51	04 53
28	01 34	01 48	04 26
29	01 26	01 45	03 40
30	01 21	01 41	02 41
31	01 ♊ 18	01 ♊ 38	01 N 33

DECLINATIONS

Date	Sun ☉ ° '	Moon ☽ ° '	Mercury ☿ ° '	Venus ♀ ° '	Mars ♂ ° '	Jupiter ♃ ° '	Saturn ♄ ° '	Uranus ♅ ° '	Neptune ♆ ° '	Pluto ♇ ° '
01	17 N 54	02 N 04	07 N 44	15 N 29	20 S 32	23 S 23	12 S 37	21 S 53	22 S 14	04 N 24
02	17 39	04 S 15	07 13	15 05	20 39	23 23	12 38	21 53	22 14	04 23
03	17 23	10 14	06 43	14 40	20 46	23 24	12 39	21 52	22 14	04 23
04	17 07	15 35	06 14	14 16	20 54	23 24	12 40	21 52	22 14	04 22
05	16 51	20 03	05 47	13 51	21 01	23 24	12 41	21 52	22 14	04 21
06	16 34	23 22	05 21	13 25	21 08	23 24	12 42	21 52	22 14	04 20
07	16 18	25 23	04 56	12 59	21 15	23 25	12 43	21 52	22 14	04 19
08	16 01	26 01	04 33	12 33	21 22	23 25	12 44	21 52	22 15	04 18
09	15 43	25 17	04 11	12 07	21 29	23 25	12 45	21 52	22 15	04 18
10	15 26	23 18	03 52	11 40	21 36	23 25	12 46	21 52	22 15	04 17
11	15 08	20 15	03 34	11 13	21 44	23 26	12 47	21 52	22 15	04 16
12	14 50	16 21	03 19	10 45	21 51	23 26	12 48	21 52	22 15	04 15
13	14 32	11 52	03 06	10 18	21 58	23 26	12 49	21 52	22 15	04 14
14	14 13	06 57	02 55	09 50	22 05	23 26	12 50	21 52	22 15	04 13
15	13 55	01 S 49	02 47	09 22	22 12	23 26	12 51	21 52	22 15	04 12
16	13 36	03 N 22	02 41	08 53	22 19	23 27	12 53	21 52	22 15	04 11
17	13 17	08 28	02 39	08 24	22 25	23 27	12 54	21 52	22 15	04 10
18	12 57	13 19	02 40	07 55	22 32	23 27	12 55	21 52	22 15	04 10
19	12 38	17 44	02 44	07 26	22 39	23 27	12 56	21 52	22 15	04 09
20	12 18	21 29	02 51	06 57	22 46	23 27	12 58	21 52	22 15	04 08
21	11 58	24 18	03 01	06 27	22 52	23 27	12 59	21 52	22 15	04 07
22	11 38	25 53	03 15	05 58	22 59	23 28	13 01	21 52	22 15	04 06
23	11 17	25 58	03 32	05 28	23 05	23 28	13 02	21 52	22 15	04 05
24	10 57	24 24	03 52	04 58	23 12	23 28	13 03	21 52	22 15	04 04
25	10 36	21 10	04 15	04 28	23 18	23 28	13 05	21 52	22 15	04 03
26	10 15	16 29	04 40	03 57	23 24	23 28	13 06	21 52	22 15	04 02
27	09 54	10 43	05 08	03 27	23 31	23 28	13 08	21 52	22 15	04 01
28	09 33	04 N 18	05 37	02 56	23 37	23 28	13 09	21 52	22 15	04 00
29	09 12	02 S 18	06 08	02 26	23 43	23 29	13 11	21 52	22 15	03 59
30	08 50	08 40	06 39	01 55	23 48	23 29	13 12	21 52	22 15	03 58
31	08 N 29	14 S 25	07 N 11	01 N 24	23 S 54	23 S 29	13 S 14	21 S 53	22 S 15	03 N 57

ZODIAC SIGN ENTRIES

Date	h m	Planets
01	04 03	☽ ♎
03	06 04	☽ ♏
05	10 30	☽ ♐
07	17 24	☽ ♑
07	19 40	♀ ♍
10	02 25	☽ ♒
12	13 13	☽ ♓
15	01 28	☽ ♈
17	14 13	☽ ♉
17	19 51	♂ ♐
20	01 31	☽ ♊
22	09 20	☽ ♋
22	23 00	☉ ♍
24	13 00	☽ ♌
26	13 32	☽ ♍
28	04 44	♇ ♏
28	12 57	☽ ♎
30	13 23	☽ ♏

LATITUDES

Date	Mercury ☿ ° '	Venus ♀ ° '	Mars ♂ ° '	Jupiter ♃ ° '	Saturn ♄ ° '	Uranus ♅ ° '	Neptune ♆ ° '	Pluto ♇ ° '
01	01 S 28	01 N 27	02 S 19	00 S 01	02 N 19	00 N 01	01 N 12	16 N 44
04	02 03	01 28	02 20	00 01	02 18	00 01	01 12	16 42
07	02 38	01 28	02 21	00 02	02 17	00 01	01 12	16 41
10	03 13	01 28	02 21	00 02	02 16	00 01	01 12	16 39
13	03 45	01 27	02 22	00 02	02 16	00 01	01 12	16 37
16	04 12	01 26	02 22	00 03	02 15	00 01	01 11	16 36
19	04 32	01 24	02 22	00 03	02 14	00 01	01 11	16 34
22	04 40	01 21	02 22	00 03	02 14	00 01	01 11	16 33
25	04 33	01 18	02 22	00 04	02 13	00 01	01 11	16 31
28	04 09	01 14	02 21	00 04	02 12	00 01	01 11	16 30
31	03 S 29	01 N 09	02 S 21	00 S 04	02 N 12	00 N 01	01 N 11	16 N 29

DATA

Julian Date	2445914
Delta T	+54 seconds
Ayanamsa	23° 38' 16"
Synetic vernal point	05° ♓ 28' 43"
True obliquity of ecliptic	23° 26' 33"

MOON'S PHASES, APSIDES AND POSITIONS ☽

Date	h m	Phase	Longitude ° '	Eclipse Indicator
04	02 33	☽	11 ♏ 54	
11	15 43	○	19 ♒ 08	
19	19 41	☾	26 ♉ 59	
26	19 26	●	03 ♍ 43	

Day	h m		
15	04 39	Apogee	
27	16 47	Perigee	
01	19 49	0S	
08	10 55	Max dec	26° S 01'
15	20 25	0N	
23	01 26	Max dec	26° N 08'
29	03 37	0S	

LONGITUDES

Date	Chiron ⚷ ° '	Ceres ⚳ ° '	Pallas ⚴ ° '	Juno ⚵ ° '	Vesta ⚶ ° '	Black Moon Lilith ⚸ ° '
01	07 ♊ 32	18 ♉ 30	21 ♓ 37	12 ♌ 17	13 ♌ 41	26 ♓ 07
11	07 ♊ 56	20 ♉ 45	20 ♓ 01	17 ♌ 08	18 ♌ 16	27 ♓ 14
21	08 ♊ 15	22 ♉ 38	17 ♓ 55	21 ♌ 53	22 ♌ 53	28 ♓ 21
31	08 ♊ 27	24 ♉ 04	15 ♓ 29	26 ♌ 34	27 ♌ 32	29 ♓ 28

All ephemeris data is given at 12.00 UT and the Moon's longitude is additionally given for 24.00 UT

ASPECTARIAN

h m	Aspects
01 Wednesday	
00 09	☽ ⊥ ♀
00 13	☽ Q ♅
02 30	☽ □ ♆
03 10	☽ ⊻ ♇
03 12	☽ ∥ ♇
08 16	♀ □ ♂
10 36	☽ ⊥ ♄
11 12	☽ □ ♃
15 06	☽ ⊻ ☿
15 49	☽ ∠ ♂
16 16	☽ ∠ ♀
18 07	☉ △ ♅
19 54	☽ ✶ ♅
20 02	☽ ✶ ☉
20 31	☽ ⊻ ♄
02 Thursday	
01 42	☽ ⊥ ☿
03 31	☉ □ ♄
07 13	☽ ⊥ ♂
08 11	☽ Q ♆
10 13	☉ ± ♃
12 33	☽ # ♇
16 58	☽ Q ♃
17 29	☽ Q ☉
17 36	☽ ⊻ ♂
17 44	☽ ∠ ☿
19 37	☽ ✶ ♀
20 57	☽ ∠ ♅
22 45	☽ # ☿
03 Friday	
04 24	☽ ✶ ♆
05 12	☽ ☌ ♇
12 16	☽ ⊥ ♅
13 15	☽ ✶ ♃
13 31	♂ ⊥ ♆
18 04	☽ Q ♀
20 57	☽ ✶ ☿
22 25	☽ ∥ ♄
22 36	☽ ⊻ ♅
23 24	☽ ☌ ♄
04 Saturday	
02 33	☽ □ ☉
06 10	☽ # ♀
06 12	☽ ∠ ♆
15 10	☽ ∠ ♃
19 10	☽ Q ☿
19 14	☽ # ☉
22 00	☽ ⊥ ♆
23 10	☽ ☌ ♂
05 Sunday	
02 31	☿ □ ♅
04 44	☽ □ ♀
07 02	☽ ⊥ ♃
08 40	☽ ⊻ ♆
09 38	☽ ⊻ ♇
17 44	☽ ⊻ ♃
18 23	☽ ∥ ♂
20 26	☽ ⊥ ♇
21 10	☿ ✶ ♄
06 Monday	
00 06	☽ ∥ ♅
02 50	☽ ∥ ♆
03 48	☽ ☌ ♅
04 50	☽ ⊻ ♄
05 13	☽ □ ☿
06 19	☉ Ᵽ ♆
07 58	♀ Q ♄
12 11	☽ △ ☉
12 22	☽ ∥ ♃
12 49	☽ ∠ ♇
15 50	☽ ⊥ ♄
23 12	♀ △ ♆
07 Tuesday	
07 37	☽ ⊻ ♂
08 29	☽ ∠ ♄
10 57	♀ ✶ ♇
15 26	☽ ☌ ♆
16 35	☽ ✶ ♇
17 10	☽ △ ♀
18 09	☽ Ᵽ ☉
19 13	☽ ⊥ ♂
08 Wednesday	
00 40	☽ ☌ ♃
02 48	♀ # ♄
11 24	☽ ⊻ ♅
12 26	☽ ± ☉
12 41	☽ ✶ ♄
12 48	☽ ∠ ♂
15 10	☽ Q ♇
15 35	☽ △ ☿
22 46	☽ ⊥ ♅
09 Thursday	
00 29	☽ Ᵽ ♀
00 45	☽ ⊼ ☉
04 41	☿ ∥ ♇
11 36	☽ Q ♄
15 57	☽ ∠ ♅
18 33	☽ ✶ ♂
19 35	☽ ± ♀
21 19	☽ Ᵽ ☿
10 Friday	
00 18	☽ ⊻ ♆
01 20	☉ Q ♇
01 39	☽ □ ♇
08 27	☽ ⊼ ♀
09 37	☽ ⊻ ♃
10 48	☽ ∥ ♃
11 54	☽ ⊥ ♆
15 17	☽ ± ☿
18 43	☽ Q ♂
19 48	♀ △ ♃
20 58	☽ ✶ ♅
21 10	☽ ∥ ♆
21 14	☽ ⊥ ♃
22 32	☽ □ ♄
11 Saturday	
00 11	☽ ∥ ♅
01 40	☽ ∥ ♂
03 19	☽ ⊼ ☿
04 21	☉ Ᵽ ♃
05 25	☽ ∠ ♆
14 47	☽ ∠ ♃
15 43	☽ ☍ ☉
20 29	☽ Q ♅
12 Sunday	
07 35	☽ □ ♂
10 58	☽ ✶ ♆
12 30	☽ △ ♇
16 07	♂ ∥ ♅
20 21	☽ ✶ ♃
21 04	☽ # ☉
13 Monday	
02 07	☽ ☍ ♀
02 53	♂ ⊥ ♃
07 07	☽ ∥ ♄
08 15	☽ □ ♅
10 07	☽ △ ♄
10 52	☽ Q ♆
15 51	☽ ☍ ☿
18 31	☽ Ᵽ ♇
20 16	☽ Q ♃
20 39	☽ # ♀
14 Tuesday	
08 45	☽ ⊼ ☉
12 41	☽ ± ♇
16 27	☽ Ᵽ ♄
19 33	☿ St R
21 56	☽ ± ☉
22 25	☽ △ ♂
23 06	☽ □ ♆
15 Wednesday	
00 50	☽ ⊼ ♇
00 55	☽ # ♇
07 28	☽ # ☿
08 29	☽ □ ♃
10 45	☽ ± ♄
13 42	♀ □ ♅
13 46	♂ ⊻ ♆
17 48	☽ Ᵽ ☉
20 48	☽ △ ♅
21 37	☽ ⊼ ♀
22 43	♂ ∥ ♆
22 59	☽ ⊼ ♄
16 Thursday	
04 27	☽ ⊼ ☿
06 13	☽ Ᵽ ♂
08 54	☽ ∥ ☿
11 11	♀ ✶ ♄
11 12	☽ ± ♀
15 48	☽ ∥ ♇
16 28	☽ ± ☿
17 Friday	
01 17	☽ ± ♂
02 53	☽ △ ☉
03 11	☽ Ᵽ ♅
07 32	☽ Ᵽ ♀
07 58	♂ ⊻ ♇
10 21	☽ Ᵽ ☿
11 43	☽ ∥ ♀
11 48	☽ △ ♆
13 42	☽ ☍ ♇
13 58	☽ ⊼ ♂
20 58	☽ △ ♃
21 19	☽ ± ♅
18 Saturday	
05 40	♅ St D
06 14	☿ ☌ ♀
09 20	☽ ⊼ ♅
09 57	☽ # ♄
10 15	☽ ∥ ☉
11 45	☽ ☍ ♄
14 16	☉ # ♄
15 39	☽ △ ☿
17 01	☽ △ ♀
17 47	☽ Ᵽ ♆
19 Sunday	
02 44	☽ Ᵽ ♃
11 30	☽ ± ♆
19 41	☽ □ ☉
20 17	☽ ± ♃
20 19	♀ ∠ ♇
23 09	☽ ⊼ ♆
20 Monday	
01 08	☽ ⊼ ♇
03 55	☽ ☍ ♂
07 48	☽ ⊼ ♃
12 39	☽ ± ♇
14 50	☽ # ♅
17 46	☽ # ♆
19 43	☽ ☍ ♅
22 15	☽ # ♂
22 16	☽ ⊼ ♄
23 38	☽ □ ☿
21 Tuesday	
03 50	☽ # ♃
05 38	☽ Ᵽ ♇
08 40	☽ Q ☉
09 04	☽ □ ♀
09 31	☽ ± ♄
15 57	☉ △ ♆
20 47	☉ Q ♄
22 Wednesday	
01 24	☿ ✶ ♄
02 04	☽ Ᵽ ♄
07 05	☽ ☍ ♆
07 11	☽ Q ☿
08 17	☽ ✶ ☉
09 05	☽ △ ♇
13 55	☽ ⊼ ♂
15 04	☽ ☍ ♃
19 37	☉ ✶ ♇
20 54	☽ Q ♀
23 Thursday	
00 56	☽ ± ♂
02 12	☽ ⊼ ♅
03 21	☽ ✶ ☿
04 48	☽ △ ♄
12 36	☽ ± ♅
12 36	☽ ∠ ☉
17 02	♂ ⊻ ♃
17 08	☽ Ᵽ ♂
19 40	☽ ✶ ♀
23 39	☿ □ ♅
24 Friday	
03 40	☽ ∠ ☿
03 54	☽ Ᵽ ♅
05 00	☽ ⊥ ☉
10 53	☽ ⊼ ♆
12 51	☽ □ ♇
15 43	☽ ⊻ ☉
17 51	☽ ⊥ ☿
18 15	☽ ⊼ ♃
19 17	☽ △ ♂
20 16	☽ # ♃
20 48	☽ ± ♆
22 02	☽ # ♂
23 02	☽ ∠ ♀
25 Saturday	
00 19	☿ ∥ ♇
03 09	☽ ⊻ ☿
04 05	☽ ± ♃
04 43	☽ △ ♅
05 09	☽ # ♆
07 20	☽ □ ♄
07 40	☽ # ♅
11 29	☽ Ᵽ ♆
15 00	☽ ⊥ ♀
17 35	☿ ∥ ♀
18 17	☽ Q ♇
18 38	☽ Ᵽ ♃
26 Sunday	
01 30	☽ ⊻ ♀
03 57	♀ Q ♂
05 19	☉ △ ♃
08 11	♀ ∥ ♇
11 29	☽ △ ♆
12 20	☽ Q ♄
13 28	☽ ✶ ♇
18 31	☽ △ ♃
19 26	☽ ☌ ☉
21 25	☽ □ ♂
27 Monday	
00 29	☽ ☌ ☿
02 27	☽ # ♄
02 52	♂ ∥ ♃
04 40	☽ □ ♅
07 24	☽ ✶ ♄
07 30	♀ ⊥ ♇
13 13	☽ ∠ ♇
15 20	☽ ∥ ☉
28 Tuesday	
02 57	☽ Q ♂
03 27	☽ ⊥ ♇
05 10	☽ ☌ ♀
06 22	☿ □ ♂
07 12	☽ ∠ ♄
07 31	☽ ∥ ☿
09 06	☽ Q ♅
10 53	☽ □ ♆
12 58	☽ ⊻ ♇
13 06	☽ ∥ ♇
14 32	☉ ☌ ☿
16 45	♂ ⊥ ♇
17 22	☽ ∥ ♀
17 55	☽ □ ♃
21 13	☽ ⊻ ☿
21 33	☽ ⊥ ♄
22 06	☽ ⊻ ☉
22 44	☽ ✶ ♂
29 Wednesday	
04 13	☽ ✶ ♅
06 15	☽ ⊥ ☿
07 11	☽ ⊻ ♄
07 35	♀ ∠ ♄
08 22	☽ ⊥ ☉
12 26	☽ # ♀
15 38	☽ Q ♆
18 11	☽ # ♇
19 57	☽ ∠ ☿
22 02	☉ □ ♂
22 50	☽ Q ♃
23 00	♃ St D
23 53	☽ ∠ ♂
23 56	☽ ∠ ☉
30 Thursday	
03 31	☽ # ☿
04 30	☽ ∠ ♅
06 09	♀ Q ♅
09 43	☽ ⊻ ♀
11 13	☽ ✶ ♆
12 38	☽ # ☉
13 29	☽ ☌ ♇
15 16	☽ ⊥ ♂
18 34	☽ ✶ ♃
19 12	☽ ✶ ☿
19 21	☽ ⊥ ♅
20 34	☽ ⊥ ♀
31 Friday	
01 41	☽ ⊻ ♂
02 29	☽ ✶ ☉
03 20	♀ □ ♆
05 24	☽ ⊻ ♅
06 27	☿ △ ♃
06 42	☽ ∥ ♄
08 45	☽ ☌ ♄
12 17	☽ ∠ ♆
13 09	☽ ∠ ♀
14 19	☽ Q ☿
19 53	☽ ∠ ♃

SEPTEMBER 1984

LONGITUDES

Date	Sidereal time h m s	Sun ☉ ° ′ ″	Moon ☽ ° ′ ″	Moon ☽ 24.00 ° ′ ″	Mercury ☿ ° ′	Venus ♀ ° ′	Mars ♂ ° ′	Jupiter ♃ ° ′	Saturn ♄ ° ′	Uranus ♅ ° ′	Neptune ♆ ° ′	Pluto ♇ ° ′
01	10 43 20	09 ♍ 13 06	27 ♏ 26 19	04 ♐ 14 59	02 ♍ 12	00 ♎ 21	08 ♐ 14	03 ♑ 08	11 ♏ 40	09 ♐ 37	28 ♐ 40	00 ♏ 07
02	10 47 17	10 11 12	10 ♐ 57 49	17 ♐ 35 10	01 R 32	01 35	08 50	03 09	11 45	09 38	28 R 40	00 09
03	10 51 14	11 09 19	24 ♐ 07 25	00 ♑ 35 00	00 59	02 49	09 26	03 10	11 49	09 39	28 40	00 11
04	10 55 10	12 07 27	06 ♑ 58 20	13 ♑ 17 50	00 32	04 02	10 02	03 11	11 54	09 40	28 40	00 12
05	10 59 07	13 05 37	19 ♑ 33 55	25 ♑ 46 57	00 14	05 16	10 38	03 12	11 59	09 41	28 40	00 14
06	11 03 03	14 03 49	01 ♒ 57 45	08 ♒ 05 13	00 04	06 30	11 15	03 13	12 04	09 42	28 39	00 16
07	11 07 00	15 02 02	14 ♒ 11 02	20 ♒ 14 57	00 D 02	07 43	11 51	03 15	12 09	09 43	28 39	00 18
08	11 10 56	16 00 16	26 ♒ 17 12	02 ♓ 17 58	00 10	08 57	12 28	03 16	12 13	09 44	28 39	00 20
09	11 14 53	16 58 32	08 ♓ 17 24	14 ♓ 15 41	00 26	10 11	13 05	03 18	12 18	09 45	28 39	00 21
10	11 18 49	17 56 50	20 ♓ 12 57	26 ♓ 09 24	00 51	11 24	13 43	03 20	12 24	09 46	28 D 39	00 23
11	11 22 46	18 55 10	02 ♈ 05 10	08 ♈ 00 30	01 26	12 38	14 20	03 22	12 29	09 47	28 39	00 25
12	11 26 43	19 53 32	13 ♈ 55 37	19 ♈ 50 46	02 08	13 51	14 58	03 25	12 34	09 48	28 39	00 27
13	11 30 39	20 51 55	25 ♈ 46 17	01 ♉ 42 30	02 59	15 05	15 36	03 28	12 39	09 50	28 39	00 29
14	11 34 36	21 50 21	07 ♉ 39 49	13 ♉ 38 40	03 57	16 19	16 14	03 30	12 44	09 51	28 40	00 31
15	11 38 32	22 48 49	19 ♉ 39 33	25 ♉ 42 59	05 03	17 32	16 52	03 33	12 50	09 52	28 40	00 33
16	11 42 29	23 47 19	01 ♊ 49 30	07 ♊ 59 43	06 15	18 46	17 30	03 37	12 55	09 54	28 40	00 35
17	11 46 25	24 45 51	14 ♊ 14 14	20 ♊ 33 37	07 33	19 59	18 09	03 40	13 01	09 55	28 40	00 37
18	11 50 22	25 44 25	26 ♊ 58 29	03 ♋ 29 24	08 57	21 13	18 48	03 43	13 06	09 57	28 40	00 39
19	11 54 18	26 43 02	10 ♋ 06 50	16 ♋ 51 13	10 25	22 26	19 26	03 47	13 12	09 59	28 41	00 41
20	11 58 15	27 41 40	23 ♋ 42 50	00 ♌ 41 50	11 58	23 40	20 05	03 51	13 18	10 00	28 41	00 43
21	12 02 12	28 40 21	07 ♌ 48 13	15 ♌ 01 44	13 34	24 54	20 45	03 55	13 23	10 02	28 41	00 46
22	12 06 08	29 ♍ 39 05	22 ♌ 21 56	29 ♌ 48 09	15 14	26 07	21 24	03 59	13 29	10 04	28 42	00 48
23	12 10 05	00 ♎ 37 50	07 ♍ 19 27	14 ♍ 54 46	16 55	27 21	22 04	04 04	13 35	10 06	28 42	00 50
24	12 14 01	01 36 37	22 ♍ 32 47	00 ♎ 12 09	18 39	28 34	22 43	04 08	13 41	10 07	28 43	00 52
25	12 17 58	02 35 27	07 ♎ 51 24	15 ♎ 29 10	20 25	29 ♎ 47	23 23	04 13	13 47	10 09	28 43	00 54
26	12 21 54	03 34 18	23 ♎ 04 04	00 ♏ 34 58	22 12	01 ♏ 01	24 03	04 18	13 53	10 11	28 44	00 56
27	12 25 51	04 33 11	08 ♏ 00 49	15 ♏ 20 50	23 59	02 14	24 44	04 23	13 59	10 13	28 44	00 59
28	12 29 47	05 32 06	22 ♏ 34 27	29 ♏ 41 17	25 48	03 28	25 24	04 28	14 05	10 15	28 45	01 01
29	12 33 44	06 31 03	06 ♐ 41 12	13 ♐ 34 10	27 36	04 41	26 04	04 34	14 11	10 17	28 46	01 03
30	12 37 41	07 ♎ 30 02	20 ♐ 20 22	27 ♐ 00 06	29 ♍ 25	05 ♏ 55	26 ♐ 45	04 ♑ 39	14 ♏ 18	10 ♐ 19	28 ♐ 46	01 ♏ 05

Date	Moon True ☊ ° ′	Moon Mean ☊ ° ′	Moon Latitude ° ′
01	01 ♊ 18	01 ♊ 35	00 N 20
02	01 R 18	01 32	00 S 51
03	01 17	01 29	01 58
04	01 15	01 26	02 57
05	01 10	01 22	03 46
06	01 03	01 19	04 23
07	00 53	01 16	04 48
08	00 41	01 13	05 00
09	00 28	01 10	04 58
10	00 15	01 07	04 43
11	00 03	01 03	04 16
12	29 ♉ 52	01 00	03 37
13	29 45	00 57	02 50
14	29 40	00 54	01 54
15	29 38	00 51	00 S 53
16	29 D 37	00 47	00 N 12
17	29 38	00 44	01 17
18	29 R 38	00 41	02 21
19	29 37	00 38	03 18
20	29 34	00 35	04 07
21	29 29	00 32	04 43
22	29 22	00 28	05 02
23	29 14	00 25	05 02
24	29 05	00 22	04 40
25	28 57	00 19	03 58
26	28 50	00 16	02 59
27	28 46	00 13	01 49
28	28 45	00 09	00 N 33
29	28 D 45	00 06	00 S 43
30	28 ♉ 46	00 ♊ 03	01 S 54

DECLINATIONS

Date	Sun ☉	Moon ☽	Mercury ☿	Venus ♀	Mars ♂	Jupiter ♃	Saturn ♄	Uranus ♅	Neptune ♆	Pluto ♇
01	08 N 07	19 S 16	07 N 42	00 N 54	24 S 00	23 S 29	13 S 15	21 S 53	22 S 15	03 N 56
02	07 45	22 56	08 12	00 N 23	24 05	23 29	13 17	21 53	22 15	03 55
03	07 23	25 16	08 41	00 S 08	24 11	23 29	13 18	21 53	22 15	03 54
04	07 01	26 12	09 08	00 39	24 16	23 29	13 20	21 53	22 16	03 53
05	06 39	25 44	09 32	01 10	24 21	23 29	13 21	21 53	22 16	03 52
06	06 16	24 00	09 54	01 41	24 26	23 29	13 23	21 53	22 16	03 51
07	05 54	21 10	10 12	02 12	24 31	23 29	13 25	21 54	22 16	03 50
08	05 31	17 27	10 27	02 43	24 35	23 29	13 27	21 54	22 16	03 49
09	05 09	13 04	10 38	03 14	24 40	23 30	13 28	21 54	22 16	03 48
10	04 46	08 13	10 46	03 44	24 44	23 30	13 30	21 54	22 16	03 47
11	04 23	03 S 05	10 50	04 15	24 49	23 30	13 32	21 54	22 16	03 46
12	04 00	02 N 09	10 49	04 46	24 53	23 30	13 34	21 55	22 16	03 45
13	03 37	07 20	10 45	05 16	24 57	23 30	13 35	21 55	22 16	03 44
14	03 14	12 16	10 36	05 47	25 00	23 30	13 37	21 55	22 16	03 43
15	02 51	16 48	10 24	06 17	25 04	23 30	13 39	21 55	22 16	03 42
16	02 28	20 43	10 06	06 47	25 07	23 30	13 41	21 55	22 16	03 41
17	02 05	23 47	09 49	07 17	25 11	23 30	13 43	21 56	22 16	03 40
18	01 42	25 45	09 26	07 47	25 14	23 30	13 44	21 56	22 16	03 39
19	01 18	26 21	09 00	08 17	25 16	23 30	13 46	21 56	22 16	03 38
20	00 55	25 25	08 31	08 46	25 19	23 30	13 48	21 56	22 16	03 37
21	00 32	22 53	07 59	09 16	25 22	23 30	13 50	21 57	22 16	03 36
22	00 N 08	18 49	07 25	09 45	25 24	23 30	13 52	21 57	22 16	03 36
23	00 S 15	13 29	06 48	10 14	25 26	23 30	13 54	21 57	22 16	03 35
24	00 38	07 14	06 10	10 43	25 28	23 30	13 56	21 57	22 17	03 34
25	01 02	00 N 31	05 30	11 11	25 30	23 30	13 58	21 58	22 17	03 33
26	01 25	06 S 12	04 49	11 40	25 32	23 29	14 00	21 58	22 17	03 32
27	01 49	12 38	04 06	12 08	25 33	23 29	14 02	21 58	22 17	03 31
28	02 12	17 53	03 23	12 36	25 33	23 29	14 03	21 59	22 17	03 30
29	02 35	22 08	02 38	13 03	25 34	23 29	14 05	21 59	22 17	03 29
30	02 S 59	24 S 59	01 N 53	13 S 30	25 S 35	23 S 29	14 S 07	21 S 59	22 S 17	03 N 28

ZODIAC SIGN ENTRIES

Date	h m	Planets
01	05 07	♀ ♎
01	16 30	☽ ♐
03	22 55	☽ ♑
06	08 11	☽ ♒
08	19 24	☽ ♓
11	07 47	☽ ♈
13	20 33	☽ ♉
16	08 26	☽ ♊
18	17 36	☽ ♋
20	22 49	☽ ♌
22	20 33	☉ ♎
23	00 19	☽ ♍
24	23 41	☽ ♎
25	16 05	♀ ♏
26	23 04	☽ ♏
29	00 32	☽ ♐
30	19 44	☿ ♎

LATITUDES

Date	Mercury ☿	Venus ♀	Mars ♂	Jupiter ♃	Saturn ♄	Uranus ♅	Neptune ♆	Pluto ♇
01	03 S 12	01 N 08	02 S 21	00 S 05	02 N 11	00 N 01	01 N 11	16 N 28
04	02 18	01 02	02 20	00 05	02 11	00 01	01 11	16 27
07	01 21	00 57	02 20	00 05	02 10	00 01	01 10	16 26
10	00 S 26	00 50	02 19	00 06	02 09	00 01	01 10	16 24
13	00 N 21	00 44	02 18	00 06	02 09	00 01	01 10	16 23
16	00 59	00 37	02 17	00 06	02 08	00 01	01 10	16 22
19	01 27	00 29	02 16	00 06	02 08	00 01	01 10	16 21
22	01 44	00 22	02 14	00 07	02 07	00 01	01 10	16 20
25	01 51	00 13	02 13	00 07	02 07	00 01	01 10	16 19
28	01 51	00 N 05	02 12	00 07	02 06	00 01	01 09	16 19
31	01 N 46	00 S 04	02 S 10	00 S 08	02 N 06	00 N 01	01 N 09	16 N 18

DATA

Julian Date	2445945
Delta T	+54 seconds
Ayanamsa	23° 38′ 20″
Synetic vernal point	05° ♓ 28′ 39″
True obliquity of ecliptic	23° 26′ 33″

MOON'S PHASES, APSIDES AND POSITIONS ☽

Date	h	m	Phase	Longitude ° ′	Eclipse Indicator
02	10	30	☽	10 ♐ 08	
10	07	01	○	17 ♓ 45	
18	09	31	☾	25 ♊ 38	
25	03	11	●	02 ♎ 14	

Day	h	m		
11	13	08	Apogee	
25	02	44	Perigee	
04	15	51	Max dec	26° S 13′
12	02	08	0N	
19	09	47	Max dec	26° N 22′
25	13	51	0S	

LONGITUDES

Date	Chiron ⚷ ° ′	Ceres ⚳ ° ′	Pallas ⚴ ° ′	Juno ⚵ ° ′	Vesta ⚶ ° ′	Black Moon Lilith ⚸ ° ′
01	08 ♊ 28	24 ♉ 11	15 ♓ 13	27 ♌ 31	28 ♌ 00	29 ♓ 35
11	08 ♊ 33	25 ♉ 04	12 ♓ 38	01 ♍ 35	02 ♍ 39	00 ♈ 42
21	08 ♊ 31	25 ♉ 23	10 ♓ 08	06 ♍ 02	07 ♍ 19	01 ♈ 49
31	08 ♊ 23	25 ♉ 04	07 ♓ 57	10 ♍ 21	11 ♍ 58	02 ♈ 56

All ephemeris data is given at 12.00 UT and the Moon's longitude is additionally given for 24.00 UT

ASPECTARIAN

h m Aspects

01 Saturday
00 23 ☽ Q ☉
03 43 ☽ ⊥ ♆
07 18 ♀ ⊻ ♇
11 28 ☽ ⊥ ♃
14 10 ☽ ⊻ ♆
16 43 ☽ ⊻ ♇
17 37 ☽ ⚹ ♀
19 57 ☽ □ ☿
22 02 ☽ ⊻ ♃
22 06 ☉ □ ♅
23 22 ☉ ∥ ☿

02 Sunday
03 21 ☽ ⊥ ♇
04 13 ☽ ∥ ♅
06 54 ☽ ∥ ♆
07 59 ☽ ☌ ♂
09 36 ☽ ☌ ♅
10 30 ☽ □ ☉
11 22 ☿ ⊻ ♀
13 25 ☽ ⊻ ♄
16 34 ☽ ∥ ♃
17 12 ☽ Q ♀
19 35 ☽ ∠ ♇
22 31 ☽ ∥ ♂

03 Monday
00 22 ☽ ⊥ ♄
17 02 ☽ ∠ ♄
18 54 ♀ □ ♃
20 25 ☽ ☌ ♆
20 58 ♂ ☌ ♅
23 16 ☽ ⚹ ♇

04 Tuesday
00 17 ☽ △ ☿
04 51 ☽ ☌ ♃
05 53 ☽ □ ♀
06 00 ☉ ⚹ ♄
17 06 ☽ ⊻ ♅
18 05 ☽ ⊻ ♂
21 24 ☽ ⚹ ♄
21 57 ☽ Q ♇
22 35 ☽ △ ☉

05 Wednesday
03 52 ☽ ⚼ ☿
04 32 ☽ ⊥ ♅
06 05 ☽ ⊥ ♂
11 43 ☿ ⚹ ♇
20 34 ☽ Q ♄
20 51 ☽ ± ☿
21 52 ☽ ∠ ♅

06 Thursday
00 19 ☽ ∠ ♂
02 55 ♀ ⊥ ♄
04 18 ☿ ⊥ ♀
05 35 ☽ ⊻ ♆
05 53 ☽ ⚼ ☉
07 29 ☽ ∥ ♂
08 21 ☽ ⊼ ☿
08 42 ☽ □ ♇
12 14 ☿ Q ♄
14 28 ☽ ⊻ ♃
17 03 ☽ ∥ ♃
17 17 ☽ ⊥ ♆
21 52 ☽ △ ♀

07 Friday
00 59 ☽ ± ☉
02 15 ☽ ⊥ ♃
03 10 ☽ ⚹ ♅
03 41 ☽ ∥ ♆
04 02 ☿ St D
06 35 ☽ ∥ ♅
07 10 ☽ ⚹ ♂
07 57 ☽ □ ♄
10 58 ☽ ∠ ♆
13 49 ☽ ⊼ ☉
18 41 ☉ ∠ ♇
20 02 ☽ ∠ ♃

08 Saturday
00 59 ♂ ⊻ ♄
02 55 ☽ Q ♅
06 49 ☽ ⚼ ♀
08 11 ☽ Q ♂
16 43 ☽ ⚹ ♆
19 52 ☽ ☍ ☿
20 05 ☽ △ ♇
22 15 ☿ Q ♄

09 Sunday
01 59 ☽ ⚹ ♃
02 49 ☽ ± ♀
03 26 ♀ ⚹ ♅
05 47 ☿ ⚹ ♇
09 53 ☽ ∥ ♄
14 56 ☽ □ ♅
16 13 ☽ ⊼ ♀
16 45 ☽ Q ♆
20 08 ☽ △ ♄
22 10 ☽ □ ♂
22 13 ♆ St D
23 52 ☽ # ☿

10 Monday
02 08 ☽ Q ♃
02 15 ☽ ⚼ ♇
07 01 ☽ ☍ ☉
14 21 ♀ # ♇
20 27 ☽ ± ♇

11 Tuesday
02 36 ☽ ⚼ ♄
05 03 ☽ □ ♆
05 30 ☽ # ☉
07 05 ☽ ∥ ♀
08 37 ☽ ⊼ ♇
08 47 ♀ ⊻ ♄
08 48 ☽ # ♇
10 35 ☽ ⊼ ☿
14 37 ☽ □ ♃
15 39 ☉ # ♀
20 58 ☽ ± ♄
23 28 ☽ ± ☿

12 Wednesday
03 38 ☽ △ ♅
09 13 ☽ ⊼ ♄
11 51 ☽ ☍ ♀
14 13 ☽ △ ♂
18 58 ☽ ⚼ ☿
19 21 ☽ ∥ ♇
19 56 ☽ ∥ ☉

13 Thursday
01 11 ☽ ⊼ ☉
01 19 ☽ # ♀
04 13 ☉ ∥ ♇
07 45 ♂ ∠ ♇
10 05 ☽ ⚼ ♅
14 25 ☽ ± ☉
17 50 ☽ △ ♆
21 34 ☽ ☍ ♇
22 18 ☽ ⚼ ♂

14 Friday
00 49 ☿ △ ♃
03 36 ☽ △ ♃
03 50 ☽ △ ☿
04 01 ☽ ∥ ☿
04 19 ☽ ± ♅
08 33 ♀ ⚹ ♂
10 12 ☽ ⚼ ☉
16 24 ☽ ⊼ ♅
17 26 ☽ ± ♂
18 51 ♀ Q ♆
18 55 ☽ # ♄
22 16 ☽ ☍ ♄

15 Saturday
00 02 ☽ ⚼ ♆
06 07 ☽ ⊼ ♂
07 18 ☽ ⊼ ♀
09 48 ☽ ⚼ ♃
17 58 ☽ ± ♆
18 49 ☽ △ ☉
20 34 ☽ ± ♀

16 Sunday
03 42 ☽ ± ♃
05 48 ☽ ⊼ ♆
09 34 ☽ ⊼ ♇
15 30 ☽ ⊼ ♃
16 12 ☽ ⚼ ♀
20 33 ☽ # ♅
21 18 ☽ ± ♇
21 36 ☽ □ ☿
23 10 ☽ # ♆

17 Monday
03 42 ☽ ☍ ♅
08 21 ☉ ⊥ ♇
09 19 ☽ # ♃
09 39 ☽ ⊼ ♄
14 39 ☽ ⚼ ♇
19 50 ☽ ☍ ♂
21 09 ☽ ± ♄

18 Tuesday
00 05 ☽ △ ♀
03 40 ☽ # ♂
09 31 ☽ □ ☉
11 56 ☽ Q ☿
14 07 ☽ ⚼ ♄
15 09 ☽ ☍ ♆
18 49 ☽ △ ♇
22 29 ♀ Q ♃

19 Wednesday
00 29 ☽ ☍ ♃
01 39 ♂ ⊥ ♄
04 46 ☿ □ ♅
11 45 ☽ ⊼ ♅
12 37 ☽ ⚹ ☿
17 34 ☽ △ ♄
20 52 ☽ Q ☉
22 28 ☽ ± ♅

20 Thursday
05 23 ☽ ⊼ ♂
05 40 ☿ # ♀
11 55 ☽ □ ♀
13 20 ☽ # ♂
14 14 ☽ ⚼ ♅
16 19 ☽ ± ♂
18 20 ☽ ∠ ☿
19 23 ☽ ⚹ ☉
19 50 ☉ Q ♅
20 34 ☽ ⊼ ♆

21 Friday
00 05 ☽ □ ♇
04 21 ☉ ∠ ♄
05 26 ☽ ⊼ ♃
06 46 ☽ ± ♆
07 21 ☽ # ♃
08 23 ☽ ⚼ ♂
09 12 ☿ ⚹ ♄
11 34 ☽ ⊥ ☿
12 28 ☉ □ ♆
14 50 ♀ ∠ ♅
15 33 ☽ ± ♃
15 44 ☽ △ ♅
16 12 ☽ # ♆
18 21 ☽ # ♅
21 15 ☽ Q ♀
21 22 ☽ □ ♄
21 48 ☽ ⚼ ♆
22 28 ☽ ∠ ☉
22 49 ☽ ⊻ ☿

22 Saturday
06 10 ☽ Q ♇
06 28 ☽ ⚼ ♃
10 22 ☽ △ ♂
14 14 ☽ ⊥ ☉
18 37 ☽ ⚹ ♀
20 17 ☿ ∠ ♇
22 14 ☽ △ ♆

23 Sunday
00 35 ☽ ⊻ ☉
01 37 ☽ ⚹ ♇
02 48 ☽ Q ♄
06 47 ☽ △ ♃
10 18 ☽ # ♄
16 24 ☽ □ ♅
17 07 ☉ ⊻ ♇
20 38 ☽ ∠ ♀
21 58 ☽ ⚹ ♄
23 54 ☽ # ♀

24 Monday
01 29 ☽ ∠ ♇
05 06 ☽ ☌ ☿
12 02 ☽ ⊥ ♀
12 18 ☽ □ ♂
14 53 ♀ ⚹ ♆
15 39 ☽ ⊥ ♇
16 19 ☽ ∥ ☿
19 42 ♄ ∠ ♆
20 46 ☽ Q ♅
21 40 ☽ □ ♆
21 41 ☽ ∠ ♄
22 15 ☽ ⊻ ♀

25 Tuesday
01 04 ☽ ⊻ ♇
01 17 ☽ ∥ ♇
03 11 ☽ ☌ ☉
06 16 ☽ □ ♃
10 19 ☽ # ☉
11 53 ☽ ⊥ ♄
15 37 ☽ ⚹ ♅
17 48 ☽ Q ♂
17 50 ☽ ∥ ☉
21 23 ☽ ⊻ ♄

26 Wednesday
01 57 ☽ Q ♆
02 23 ☽ # ♇
07 28 ☽ # ☿
10 26 ☽ ⊻ ☿
10 29 ♀ ☌ ♇
10 46 ☽ Q ♃
13 39 ☽ ⚹ ♂
15 23 ☽ ∠ ♅
21 02 ☽ ⚹ ♆
21 17 ☽ ⊥ ☿

27 Thursday
00 36 ☽ ☌ ♇
01 50 ☽ ☌ ♀
05 50 ☽ ⊥ ♅
06 00 ☽ ⊻ ☉
06 05 ☽ ⚹ ♃
07 30 ☉ □ ♃
10 32 ☽ ∥ ♀
13 49 ☽ ∠ ☿
14 55 ☽ ∠ ♂
15 36 ☽ ⊻ ♅
16 26 ☽ ⊥ ☉
18 31 ☽ ∥ ♄
21 22 ☽ ∠ ♆
21 49 ☽ ☌ ♄

28 Friday
01 25 ☿ ⊥ ♇
03 38 ☿ □ ♂
06 27 ☽ ⊥ ♂
06 48 ☽ ∠ ♃
08 04 ☿ ∥ ♇
08 21 ☽ ∠ ☉
12 18 ☽ ⊥ ♆
16 59 ☽ ⊻ ♂
18 12 ☽ ⚹ ☿
22 00 ☽ ⊥ ♃
22 25 ☽ ⊻ ♆

29 Saturday
02 18 ☽ ⊻ ♇
03 54 ♀ ⊥ ♅
08 13 ☽ ⊻ ♀
08 19 ☽ ⊻ ♃
09 21 ♀ ⚹ ♃
11 01 ☽ ∥ ♅
11 41 ☽ ⚹ ☉
12 38 ☽ ⊥ ♇
13 01 ☉ # ☿
13 01 ☽ ∥ ♆
17 50 ☽ Q ☿
18 16 ☽ ☌ ♅
19 38 ☽ ⊥ ♀
21 13 ☿ Q ♅
21 56 ☽ ∥ ♃

30 Sunday
01 11 ☽ ⊻ ♄
03 24 ☿ □ ♆
04 25 ☽ ∠ ♇
10 15 ☿ ∠ ♄
10 23 ☽ Q ☉
11 55 ☽ ⊥ ♄
13 08 ☽ ∠ ♀
19 49 ☽ ∥ ♂

OCTOBER 1984

LONGITUDES

Date	Sidereal time h m s	Sun ☉ ° ' "	Moon ☽ ° ' "	Moon ☽ 24.00 ° ' "	Mercury ☿ ° '	Venus ♀ ° '	Mars ♂ ° '	Jupiter ♃ ° '	Saturn ♄ ° '	Uranus ♅ ° '	Neptune ♆ ° '	Pluto ♇ ° '
01	12 41 37	08 ♎ 29 03	03 ♑ 33 42	10 ♑ 01 39	01 ♎ 14	07 ♏ 08	27 ♐ 26	04 ♑ 45	14 ♏ 24	10 ♐ 21	28 ♐ 47	01 ♏ 08
02	12 45 34	09 28 05	16 ♑ 24 24	22 ♑ 42 29	03 02	08 21	28 07	04 51	14 30	10 24	28 48	01 10
03	12 49 30	10 27 09	28 ♑ 56 24	05 ♒ 06 39	04 50	09 35	28 48	04 57	14 36	10 26	28 48	01 12
04	12 53 27	11 26 15	11 ♒ 13 45	17 ♒ 18 08	06 38	10 48	29 ♐ 29	05 03	14 42	10 28	28 49	01 15
05	12 57 23	12 25 22	23 ♒ 20 15	29 ♒ 20 29	08 25	12 01	00 ♑ 10	05 10	14 49	10 31	28 50	01 17
06	13 01 20	13 24 31	05 ♓ 19 14	11 ♓ 16 48	10 12	13 15	00 52	05 16	14 56	10 33	28 51	01 19
07	13 05 16	14 23 43	17 ♓ 13 30	23 ♓ 09 35	11 57	14 28	01 33	05 23	15 02	10 35	28 52	01 21
08	13 09 13	15 22 56	29 ♓ 05 20	05 ♈ 00 57	13 43	15 41	02 15	05 30	15 09	10 38	28 53	01 24
09	13 13 10	16 22 11	10 ♈ 56 39	16 ♈ 52 39	15 27	16 55	02 57	05 37	15 15	10 40	28 54	01 26
10	13 17 06	17 21 28	22 ♈ 49 09	28 ♈ 46 22	17 11	18 08	03 39	05 44	15 22	10 43	28 55	01 29
11	13 21 03	18 20 47	04 ♉ 44 32	10 ♉ 43 54	18 54	19 21	04 21	05 51	15 28	10 45	28 56	01 31
12	13 24 59	19 20 08	16 ♉ 44 43	22 ♉ 47 18	20 37	20 34	05 03	05 59	15 35	10 48	28 57	01 33
13	13 28 56	20 19 32	28 ♉ 51 58	04 ♊ 59 05	22 18	21 47	05 45	06 06	15 42	10 50	28 58	01 36
14	13 32 52	21 18 58	11 ♊ 09 01	17 ♊ 22 11	23 59	23 00	06 27	06 14	15 49	10 53	28 59	01 38
15	13 36 49	22 18 26	23 ♊ 39 01	29 ♊ 59 58	25 40	24 14	07 10	06 22	15 55	10 56	29 00	01 40
16	13 40 45	23 17 56	06 ♋ 25 29	12 ♋ 56 00	27 19	25 27	07 52	06 30	16 02	10 59	29 01	01 43
17	13 44 42	24 17 29	19 ♋ 31 56	26 ♋ 13 37	28 ♎ 58	26 40	08 35	06 38	16 09	11 01	29 02	01 45
18	13 48 39	25 17 04	03 ♌ 01 22	09 ♌ 55 23	00 ♏ 37	27 53	09 18	06 46	16 16	11 04	29 04	01 48
19	13 52 35	26 16 41	16 ♌ 55 44	24 ♌ 02 22	02 14	29 ♏ 06	10 00	06 55	16 23	11 07	29 05	01 50
20	13 56 32	27 16 21	01 ♍ 15 03	08 ♍ 33 23	03 51	00 ♐ 19	10 43	07 03	16 30	11 10	29 06	01 53
21	14 00 28	28 16 02	15 ♍ 56 47	23 ♍ 24 29	05 28	01 32	11 26	07 12	16 37	11 13	29 07	01 55
22	14 04 25	29 ♎ 15 46	00 ♎ 55 33	08 ♎ 28 53	07 04	02 45	12 10	07 21	16 43	11 16	29 09	01 57
23	14 08 21	00 ♏ 15 32	16 ♎ 03 19	23 ♎ 37 38	08 39	03 58	12 53	07 29	16 50	11 19	29 10	02 00
24	14 12 18	01 15 21	01 ♏ 10 34	08 ♏ 40 57	10 14	05 11	13 36	07 38	16 57	11 22	29 11	02 02
25	14 16 14	02 15 11	16 ♏ 07 40	23 ♏ 29 48	11 48	06 24	14 20	07 48	17 04	11 25	29 13	02 05
26	14 20 11	03 15 03	00 ♐ 46 30	07 ♐ 57 11	13 22	07 37	15 03	07 57	17 11	11 28	29 14	02 07
27	14 24 08	04 14 57	15 ♐ 01 25	21 ♐ 58 56	14 55	08 50	15 47	08 06	17 19	11 31	29 16	02 09
28	14 28 04	05 14 53	28 ♐ 49 38	05 ♑ 33 37	16 27	10 03	16 30	08 16	17 26	11 34	29 17	02 12
29	14 32 01	06 14 50	12 ♑ 11 04	18 ♑ 42 17	17 59	11 15	17 14	08 26	17 33	11 37	29 19	02 14
30	14 35 57	07 14 49	25 ♑ 07 39	01 ♒ 27 38	19 29	12 28	17 58	08 35	17 40	11 40	29 20	02 17
31	14 39 54	08 ♏ 14 50	07 ♒ 42 45	13 ♒ 53 31	21 ♏ 02	13 ♐ 41	18 ♑ 42	08 ♑ 45	17 ♏ 47	11 ♐ 44	29 ♐ 22	02 ♏ 19

DECLINATIONS

Date	Moon True ☊ ° '	Moon Mean ☊ ° '	Moon ☽ Latitude ° '	Sun ☉ ° '	Moon ☽ ° '	Mercury ☿ ° '	Venus ♀ ° '	Mars ♂ ° '	Jupiter ♃ ° '	Saturn ♄ ° '	Uranus ♅ ° '	Neptune ♆ ° '	Pluto ♇ ° '
01	28 ♉ 47	00 ♊ 00	02 S 56	03 S 22	26 S 20	01 N 08	13 S 57	25 S 35	23 S 29	14 S 09	22 S 00	22 S 17	03 N 27
02	28 R 46	29 ♉ 57	03 48	03 45	26 12	00 N 22	14 24	25 35	23 29	14 11	22 00	22 17	03 26
03	28 44	29 53	04 27	04 08	24 44	00 S 25	14 50	25 35	23 29	14 13	22 00	22 17	03 25
04	28 41	29 50	04 54	04 31	22 07	01 11	15 16	25 35	23 29	14 15	22 01	22 17	03 24
05	28 35	29 47	05 06	04 55	18 34	01 57	15 41	25 35	23 28	14 17	22 01	22 17	03 23
06	28 27	29 44	05 06	05 18	14 18	02 43	16 06	25 34	23 28	14 19	22 01	22 17	03 22
07	28 19	29 41	04 51	05 41	09 31	03 29	16 31	25 33	23 28	14 21	22 02	22 17	03 21
08	28 10	29 38	04 25	06 03	04 S 25	04 15	16 55	25 32	23 28	14 23	22 02	22 17	03 20
09	28 02	29 34	03 46	06 26	00 N 51	05 01	17 19	25 30	23 28	14 25	22 02	22 17	03 19
10	27 56	29 31	02 58	06 49	06 07	05 46	17 43	25 29	23 27	14 27	22 03	22 17	03 18
11	27 52	29 28	02 02	07 12	11 11	06 31	18 06	25 27	23 27	14 29	22 03	22 17	03 17
12	27 49	29 25	01 S 00	07 34	15 53	07 15	18 29	25 25	23 27	14 31	22 03	22 17	03 17
13	27 D 48	29 22	00 N 06	07 57	20 00	07 59	18 51	25 22	23 27	14 33	22 04	22 18	03 16
14	27 49	29 19	01 12	08 19	23 18	08 42	19 12	25 20	23 26	14 35	22 04	22 18	03 15
15	27 51	29 15	02 16	08 41	25 34	09 25	19 33	25 17	23 26	14 37	22 05	22 18	03 14
16	27 52	29 12	03 15	09 03	26 32	10 07	19 54	25 14	23 26	14 40	22 05	22 18	03 13
17	27 53	29 09	04 06	09 25	26 04	10 48	20 14	25 11	23 26	14 42	22 05	22 18	03 12
18	27 R 53	29 06	04 44	09 47	24 05	11 29	20 34	25 07	23 25	14 44	22 06	22 18	03 11
19	27 52	29 03	05 07	10 09	20 38	12 10	20 53	25 03	23 25	14 46	22 06	22 18	03 10
20	27 49	28 59	05 12	10 30	15 53	12 49	21 12	24 59	23 25	14 48	22 07	22 18	03 10
21	27 45	28 56	04 57	10 52	10 06	13 28	21 30	24 55	23 24	14 50	22 07	22 18	03 09
22	27 41	28 53	04 22	11 13	03 N 38	14 06	21 47	24 50	23 24	14 52	22 07	22 18	03 08
23	27 37	28 50	03 28	11 34	03 S 07	14 43	22 04	24 46	23 23	14 54	22 08	22 18	03 07
24	27 34	28 47	02 20	11 55	09 42	15 20	22 20	24 41	23 23	14 56	22 08	22 18	03 06
25	27 32	28 44	01 N 02	12 15	15 40	15 56	22 36	24 35	23 22	14 58	22 09	22 18	03 05
26	27 D 32	28 40	00 S 18	12 36	20 36	16 31	22 51	24 30	23 22	15 00	22 09	22 18	03 05
27	27 33	28 37	01 35	12 56	24 23	17 05	23 05	24 24	23 21	15 02	22 10	22 18	03 04
28	27 34	28 34	02 44	13 16	26 10	17 38	23 19	24 18	23 21	15 04	22 10	22 18	03 03
29	27 35	28 31	03 42	13 36	26 34	18 10	23 32	24 12	23 20	15 06	22 10	22 18	03 02
30	27 37	28 28	04 26	13 56	25 29	18 42	23 44	24 05	23 20	15 08	22 11	22 18	03 02
31	27 ♉ 37	28 ♉ 25	04 S 57	14 S 15	23 S 07	19 S 12	23 S 56	23 S 58	23 S 19	15 S 10	22 S 11	22 S 18	03 N 01

ZODIAC SIGN ENTRIES

Date	h	m	Planets
01	05	28	☽ ♑
03	14	03	☽ ♒
05	06	02	♂ ♑
06	01	19	☽ ♓
08	13	51	☽ ♈
11	02	28	☽ ♉
13	14	14	☽ ♊
16	00	00	☽ ♋
18	03	01	☿ ♏
18	06	41	☽ ♌
20	05	45	♀ ♐
20	09	56	☽ ♍
22	10	32	☽ ♎
23	05	46	☉ ♏
24	10	08	☽ ♏
26	10	43	☽ ♐
28	14	05	☽ ♑
30	21	13	☽ ♒

LATITUDES

Date	Mercury ☿ ° '	Venus ♀ ° '	Mars ♂ ° '	Jupiter ♃ ° '	Saturn ♄ ° '	Uranus ♅ ° '	Neptune ♆ ° '	Pluto ♇ ° '
01	01 N 46	00 S 04	02 S 10	00 S 08	02 N 06	00 N 01	01 N 09	16 N 18
04	01 35	00 12	02 09	00 08	02 05	00 01	01 09	16 17
07	01 21	00 21	02 07	00 08	02 05	00 01	01 09	16 17
10	01 04	00 30	02 05	00 08	02 05	00 01	01 09	16 16
13	00 46	00 39	02 04	00 09	02 04	00 01	01 09	16 16
16	00 26	00 48	02 02	00 09	02 04	00 01	01 09	16 16
19	00 N 06	00 57	02 00	00 09	02 04	00 01	01 09	16 15
22	00 S 14	01 05	01 58	00 09	02 03	00 01	01 09	16 15
25	00 35	01 14	01 56	00 10	02 03	00 N 01	01 08	16 15
28	00 55	01 22	01 53	00 10	02 03	00 00	01 08	16 15
31	01 S 14	01 S 30	01 S 51	00 S 10	02 N 03	00 00	01 N 08	16 N 15

DATA

Julian Date	2445975
Delta T	+54 seconds
Ayanamsa	23° 38' 23"
Synetic vernal point	05° ♓ 28' 36"
True obliquity of ecliptic	23° 26' 34"

MOON'S PHASES, APSIDES AND POSITIONS ☽

Date	h	m	Phase	Longitude ° '	Eclipse Indicator
01	21	52	☽	08 ♑ 53	
09	23	58	○	16 ♈ 52	
17	21	14	☾	24 ♋ 40	
24	12	08	●	01 ♏ 16	
31	13	07	☽	08 ♒ 18	

Day	h	m		
08	14	30	Apogee	
23	13	42	Perigee	
01	21	51	Max dec	26° S 27'
09	08	07	0N	
16	16	31	Max dec	26° N 34'
23	00	58	0S	
29	06	03	Max dec	26° S 37'

LONGITUDES

Date	Chiron ⚷ ° '	Ceres ⚳ ° '	Pallas ⚴ ° '	Juno ⚵ ° '	Vesta ⚶ ° '	Black Moon Lilith ⚸ ° '
01	08 ♊ 23	25 ♉ 04	07 ♓ 57	10 ♍ 21	11 ♍ 58	02 ♈ 56
11	08 ♊ 08	24 ♉ 07	06 ♓ 15	14 ♍ 33	16 ♍ 36	04 ♈ 07
21	07 ♊ 47	22 ♉ 34	05 ♓ 07	18 ♍ 35	21 ♍ 13	05 ♈ 10
31	07 ♊ 22	20 ♉ 33	04 ♓ 37	22 ♍ 27	25 ♍ 46	06 ♈ 17

ASPECTARIAN

h m	Aspects
01 Monday	
00 10	☽ ☌ ♂
03 14	☽ ☌ ♆
04 18	☽ ∠ ♄
07 01	☽ □ ☿
07 31	☽ ⚹ ♇
08 12	☿ ⊥ ♀
09 36	☉ ⊥ ♄
10 39	☿ ⊻ ♇
14 13	☽ ☌ ♃
16 50	☉ # ♇
19 18	☽ ⚹ ♀
21 52	☽ □ ☉
23 51	♀ ∥ ♄
02 Tuesday	
00 39	☽ ⊻ ♅
05 52	☽ Q ♇
08 22	☽ ⚹ ♄
11 59	☽ ⊥ ♅
20 19	☽ Q ♀
03 Wednesday	
00 34	☽ ∥ ♂
05 13	☽ ∠ ♅
07 27	☽ Q ♄
11 29	☉ ⚹ ♅
11 42	☽ ⊻ ♂
11 45	☽ ⊻ ♆
12 23	♂ ☌ ♆
13 38	☿ □ ♃
16 24	☽ □ ♇
23 25	☽ ⊥ ♆
23 47	☽ ⊻ ♃
04 Thursday	
00 04	☽ ⊥ ♂
00 41	☽ ∥ ♃
01 26	☽ △ ☿
05 15	♀ ⊻ ♅
10 30	☽ ⚹ ♅
10 40	☽ ∥ ♆
11 04	☽ □ ♀
11 39	☽ ⊥ ♃
12 27	☽ △ ☉
12 47	☽ ∥ ♅
17 07	☽ ∠ ♆
18 48	☽ ∠ ♂
18 56	☽ □ ♄
21 29	♂ ∠ ♄
05 Friday	
05 37	☽ ∠ ♃
10 21	☽ Q ♅
12 11	☽ bQ ☿
17 47	☿ ⊥ ♄
20 53	☽ bQ ☉
23 00	☽ ⚹ ♆
06 Saturday	
02 30	☽ ⚹ ♂
03 04	☽ ∥ ♀
03 56	☽ △ ♇
09 20	☽ ± ☿
11 52	☽ ∥ ♄
11 54	☽ ⚹ ♃
16 35	☽ ± ☉
16 56	☿ ⚹ ♅
22 34	☽ □ ♅
23 09	☽ Q ♆
23 31	☽ ⊼ ☿
23 58	♀ ∠ ♆
07 Sunday	
04 08	☽ Q ♂
04 35	☉ ⊻ ♀
04 49	♂ ⚹ ♇
05 46	☽ ⊼ ☉
05 48	☽ △ ♀
07 32	☽ △ ♄
07 49	☿ # ♇
10 15	☽ bQ ♇
12 19	☽ Q ♃
08 Monday	
00 14	♀ ☌ ♄
04 30	☽ ± ♇
04 54	☽ ∥ ☉
05 29	☉ ⊻ ♄
11 34	☽ □ ♆
12 38	☽ ∥ ☿
14 09	☽ bQ ♄
15 37	☽ bQ ♀
16 41	☽ ⊼ ♇
16 56	☽ # ♇
18 48	☽ □ ♂
09 Tuesday	
01 06	☽ □ ♃
08 33	☽ ± ♄
09 01	☿ ⊻ ♄
11 27	☽ △ ♅
11 55	☽ ± ♀
20 48	☽ ⊼ ♄
22 41	☽ ☍ ☿
23 10	☽ ∥ ♇
23 58	☽ ☍ ☉
10 Wednesday	
00 57	☉ Q ♆
01 27	☽ ⊼ ♀
08 08	☿ Q ♆
10 06	☽ # ☿
15 31	☽ # ☉
17 36	☉ ☌ ☿
17 51	☽ bQ ♅
11 Thursday	
00 18	☽ △ ♆
05 30	☽ ☍ ♇
11 09	☽ △ ♂
11 40	♀ ∠ ♂
12 01	☽ ± ♅
14 15	☽ △ ♃
12 Friday	
00 05	☽ ⊼ ♅
04 43	☽ # ♄
06 25	☽ bQ ♆
09 40	☽ ☍ ♄
09 51	☿ ⊻ ♀
17 37	☽ ⊼ ☉
18 58	☽ bQ ♂
20 27	☽ ☍ ♀
20 30	☽ bQ ♃
20 57	☽ ⊼ ☿
20 57	♀ ∠ ♃
13 Saturday	
00 20	☽ ± ♆
04 05	☽ # ♀
06 33	☽ ± ☉
09 40	☉ ∥ ☿
10 43	☽ ± ☿
12 11	☽ ⊼ ♆
13 50	☽ ± ♂
14 28	☽ ± ♃
17 23	☽ ⊼ ♇
14 Sunday	
01 47	☽ bQ ☉
02 10	☽ # ♅
02 19	☽ ⊼ ♂
02 20	☽ ⊼ ♃
02 47	♂ ☌ ♃
03 52	☽ # ♆
05 09	☽ ± ♇
07 09	☽ bQ ☿
11 27	♀ ⊥ ♆
11 29	☽ ☍ ♅
13 09	☽ # ♃
15 44	☿ Q ♃
21 05	☽ ⊼ ♄
22 37	☽ bQ ♇
23 27	☿ Q ♂
15 Monday	
08 18	☽ # ♂
08 41	☽ ± ♄
09 14	☽ △ ☉
13 13	☽ ⊼ ♀
15 58	☿ ∠ ♅
16 24	☽ △ ☿
22 08	☽ ☍ ♆
16 Tuesday	
01 44	☽ ± ♀
01 51	☽ bQ ♄
03 11	☽ △ ♇
12 08	☽ ☍ ♃
14 50	☽ ☍ ♂
20 12	☽ bQ ♀
20 26	☽ ⊼ ♅
17 Wednesday	
05 49	☽ △ ♄
07 26	☽ ± ♅
12 58	☿ ⚹ ♆
21 14	☽ □ ☉
21 32	☉ Q ♃
23 41	☽ bQ ♅
18 Thursday	
01 28	☽ # ♂
02 03	☽ △ ♀
05 01	☽ ⊼ ♆
07 11	☽ □ ☿
09 50	☽ □ ♇
15 34	☽ ± ♆
17 32	☽ # ♃
18 36	☽ ⊼ ♃
23 30	☽ ⊼ ♂
19 Friday	
01 39	☽ # ♆
02 01	☽ △ ♅
02 58	☽ # ♅
05 04	☽ ± ♃
05 52	☿ ☌ ♇
07 08	☽ bQ ♆
07 08	☽ Q ☉
07 54	☉ ∠ ♅
10 21	☽ ± ♂
10 40	☽ # ♀
11 03	☽ □ ♄
11 36	♀ ⊻ ♆
16 56	☽ Q ♇
18 20	☽ Q ☿
20 30	☽ bQ ♃
20 Saturday	
02 20	☽ bQ ♂
04 55	☽ ⚹ ☉
08 26	☽ △ ♆
10 19	☽ □ ♀
13 02	☽ ⚹ ♇
16 49	☽ # ♄
16 50	☽ ⚹ ☿
17 23	☽ Q ♄
21 38	☽ △ ♃
23 52	☽ # ☿
21 Sunday	
03 52	♂ ⊻ ♅
04 18	☽ □ ♅
04 19	☽ △ ♂
04 25	♀ ⊥ ♃
07 21	☽ ∠ ☉
08 07	☿ ⊥ ♅
09 13	☽ # ☉
13 05	☽ ⚹ ♄
13 34	☽ ∠ ♇
18 18	☽ Q ♀
19 47	♀ ⊻ ♇
20 09	☽ ∠ ☿
22 54	☽ ⊥ ☉
22 Monday	
04 03	☽ ⊥ ♇
09 05	☉ ⚹ ♆
09 10	☽ □ ♆
09 10	☽ ⊻ ☉
09 21	☽ Q ♅
12 15	☽ ⊥ ☿
13 17	☽ ∠ ♄
13 39	☽ ⊻ ♇
13 49	☽ ∥ ♇
15 09	☽ ⚹ ♀
16 39	☿ ⚹ ♃
22 18	☽ □ ♃
22 54	☽ ⊻ ☿
23 Tuesday	
03 41	☽ ⊥ ♄
04 28	☽ ⚹ ♅
06 43	☽ □ ♂
12 01	☽ # ♇
13 15	☽ ⊻ ♄
13 46	☽ Q ♆
17 01	☽ ∠ ♀
18 06	♀ ∥ ♅
19 17	☿ ∥ ♄
24 Wednesday	
03 06	☽ Q ♃
04 19	☽ ∠ ♅
08 33	☽ ⊥ ♀
08 50	☽ ⚹ ♆
09 11	♀ ∥ ♆
12 08	☽ ☌ ☉
12 43	☽ Q ♂
13 23	☽ ☌ ♇
18 42	☽ ⊥ ♅
18 58	☽ ⊻ ♀
21 02	☽ ∥ ☉
22 26	☽ ⚹ ♃
25 Thursday	
03 17	♂ Q ♇
04 11	☽ ☌ ☿
04 22	☽ ⊻ ♅
05 53	☿ ⊻ ♅
07 35	☉ ☌ ♇
08 54	☽ ∠ ♆
08 56	☽ ⚹ ♂
08 57	☽ ∥ ♄
13 15	☽ ∥ ☿
13 33	☽ ☌ ♄
18 22	☉ Q ♂
22 58	☽ ∠ ♃
23 33	☽ ⊥ ♆
26 Friday	
09 27	☽ ⊻ ♆
10 44	☽ ∠ ♂
13 58	☽ ⊥ ♃
14 14	☽ ⊻ ♇
16 26	☽ ⊻ ☉
19 35	♀ ⊻ ♃
21 14	☽ ∥ ♅
22 11	☽ ∥ ♆
22 17	♀ ⊥ ♇
27 Saturday	
00 08	☽ ⊻ ♃
00 19	☽ ⊥ ♇
00 30	☽ ☌ ♀
01 47	☿ ∠ ♆
02 36	☽ ⊥ ♂
02 52	☽ ∥ ♀
03 16	☽ ⊥ ☉
05 32	☽ ∥ ♃
06 00	☽ ☌ ♅
11 47	☽ ⊻ ☿
13 22	☽ ⊻ ♂
13 55	☽ ∥ ♂
15 40	☽ ∠ ♇
15 57	☽ ⊻ ♄
19 50	☽ ∠ ☉
23 25	☽ ⊥ ☿
28 Sunday	
02 26	☽ ⊥ ♄
12 49	☽ ☌ ♆
13 28	☿ ⚹ ♂
15 57	♀ ∥ ♃
17 16	☽ ∠ ☿
18 00	☽ ⚹ ♇
18 27	☽ ∠ ♄
20 05	☉ ⊥ ♅
29 Monday	
00 22	☽ ⚹ ☉
04 26	☿ ☌ ♄
05 05	☽ ☌ ♃
10 08	☽ ⊻ ♀
10 51	♀ ⊥ ♂
10 58	☽ ⊻ ♅
15 46	☽ Q ♇
19 28	♀ ☌ ♅
21 50	☽ ☌ ♂
21 57	☽ ⚹ ♄
22 02	☽ ⊥ ♅
22 17	☽ ⊥ ♀
30 Tuesday	
00 05	☽ Q ☉
00 07	☽ ⚹ ☿
00 10	♂ ⚹ ♄
01 37	♂ ⊥ ♅
14 55	☽ ∠ ♅
16 53	☽ ∠ ♀
19 59	☽ ⊻ ♆
20 40	☽ Q ♄
31 Wednesday	
01 36	☽ □ ♇
01 47	☽ Q ☿
04 11	☽ ∥ ♂
05 24	☽ ∥ ♀
07 28	☽ ⊥ ♆
10 18	☽ ∥ ♃
13 07	☽ □ ☉
14 02	☽ ⊻ ♃
15 16	♀ ∥ ♂
18 26	☽ ∥ ♆
19 19	☽ ∥ ♅
19 49	☽ ⚹ ♅

All ephemeris data is given at 12.00 UT and the Moon's longitude is additionally given for 24.00 UT

NOVEMBER 1984

LONGITUDES

Date	Sidereal time h m s	Sun ☉ ° ' "	Moon ☽ ° ' "	Moon ☽ 24.00 ° ' "	Mercury ☿ ° '	Venus ♀ ° '	Mars ♂ ° '	Jupiter ♃ ° '	Saturn ♄ ° '	Uranus ♅ ° '	Neptune ♆ ° '	Pluto ♇ ° '
01	14 43 50	09 ♏ 14 52	20 ♒ 00 30	26 ♒ 04 15	22 ♏ 33	14 ♐ 54	19 ♑ 26	08 ♑ 55	17 ♏ 54	11 ♐ 47	29 ♐ 23	02 ♏ 22
02	14 47 47	10 14 56	02 ♓ 05 19	08 ♓ 04 14	24 03	16 07	20 10	09 05	18 01	11 50	29 25	02 24
03	14 51 43	11 15 02	14 ♓ 01 32	19 ♓ 57 41	25 33	17 19	20 54	09 15	18 08	11 53	29 27	02 26
04	14 55 40	12 15 09	25 ♓ 53 09	01 ♈ 48 22	27 02	18 32	21 38	09 26	18 15	11 57	29 28	02 29
05	14 59 37	13 15 17	07 ♈ 43 44	13 ♈ 39 35	28 31	19 45	22 23	09 36	18 23	12 00	29 30	02 31
06	15 03 33	14 15 27	19 ♈ 36 17	25 ♈ 34 06	29 ♏ 59	20 57	23 07	09 47	18 30	12 03	29 32	02 34
07	15 07 30	15 15 39	01 ♉ 33 20	07 ♉ 34 11	01 ♐ 27	22 10	23 51	09 57	18 37	12 07	29 33	02 36
08	15 11 26	16 15 53	13 ♉ 36 54	19 ♉ 41 41	02 55	23 22	24 36	10 08	18 44	12 10	29 35	02 38
09	15 15 23	17 16 09	25 ♉ 48 43	01 ♊ 58 10	04 21	24 35	25 20	10 19	18 51	12 13	29 37	02 41
10	15 19 19	18 16 26	08 ♊ 10 14	14 ♊ 25 03	05 47	25 47	26 05	10 30	18 59	12 17	29 39	02 43
11	15 23 16	19 16 45	20 ♊ 42 50	27 ♊ 03 43	07 13	26 59	26 49	10 41	19 06	12 20	29 41	02 45
12	15 27 12	20 17 06	03 ♋ 27 54	09 ♋ 55 34	08 38	28 12	27 34	10 52	19 13	12 24	29 42	02 48
13	15 31 09	21 17 29	16 ♋ 26 55	23 ♋ 02 05	10 02	29 ♐ 24	28 19	11 03	19 20	12 27	29 44	02 50
14	15 35 06	22 17 54	29 ♋ 41 17	06 ♌ 24 39	11 25	00 ♑ 36	29 04	11 14	19 27	12 31	29 46	02 52
15	15 39 02	23 18 21	13 ♌ 12 17	20 ♌ 04 18	12 47	01 49	29 ♑ 49	11 26	19 34	12 34	29 48	02 55
16	15 42 59	24 18 49	27 ♌ 00 43	04 ♍ 01 29	14 08	03 01	00 ♒ 33	11 37	19 42	12 38	29 50	02 57
17	15 46 55	25 19 19	11 ♍ 06 29	18 ♍ 15 30	15 28	04 13	01 18	11 49	19 49	12 41	29 52	02 59
18	15 50 52	26 19 52	25 ♍ 28 14	02 ♎ 44 14	16 47	05 25	02 03	12 01	19 56	12 45	29 54	03 02
19	15 54 48	27 20 26	10 ♎ 02 58	17 ♎ 23 49	18 05	06 37	02 48	12 12	20 03	12 48	29 56	03 04
20	15 58 45	28 21 01	24 ♎ 46 03	02 ♏ 08 50	19 21	07 49	03 34	12 24	20 10	12 52	29 ♐ 58	03 06
21	16 02 41	29 ♏ 21 39	09 ♏ 31 20	16 ♏ 52 38	20 34	09 01	04 19	12 36	20 17	12 56	00 ♑ 00	03 08
22	16 06 38	00 ♐ 22 18	24 ♏ 11 53	01 ♐ 28 12	21 46	10 13	05 04	12 48	20 25	12 59	00 02	03 11
23	16 10 35	01 22 59	08 ♐ 40 49	15 ♐ 49 02	22 56	11 25	05 49	13 00	20 32	13 03	00 04	03 13
24	16 14 31	02 23 41	22 ♐ 52 16	29 ♐ 50 03	24 03	12 37	06 34	13 12	20 39	13 06	00 06	03 15
25	16 18 28	03 24 24	06 ♑ 42 03	13 ♑ 28 03	25 06	13 48	07 20	13 24	20 46	13 10	00 08	03 17
26	16 22 24	04 25 08	20 ♑ 08 02	26 ♑ 42 01	26 06	15 00	08 05	13 37	20 53	13 14	00 10	03 19
27	16 26 21	05 25 54	03 ♒ 10 12	09 ♒ 32 51	27 02	16 12	08 50	13 49	21 00	13 17	00 12	03 22
28	16 30 17	06 26 41	15 ♒ 50 20	22 ♒ 03 06	27 54	17 23	09 36	14 01	21 07	13 21	00 14	03 24
29	16 34 14	07 27 28	28 ♒ 11 37	04 ♓ 16 27	28 40	18 35	10 21	14 14	21 14	13 25	00 16	03 26
30	16 38 10	08 ♐ 28 16	10 ♓ 18 09	16 ♓ 17 20	29 ♐ 21	19 ♑ 46	11 ♒ 07	14 ♑ 26	21 ♏ 21	13 ♐ 28	00 ♑ 19	03 ♏ 28

Date	Moon True ☊ ° '	Moon Mean ☊ ° '	Moon ☽ Latitude ° '
01	27 ♉ 37	28 ♉ 21	05 S 13
02	27 R 35	28 18	05 15
03	27 33	28 15	05 03
04	27 31	28 12	04 38
05	27 29	28 09	04 01
06	27 27	28 05	03 14
07	27 25	28 02	02 18
08	27 24	27 59	01 16
09	27 24	27 56	00 S 09
10	27 D 24	27 53	00 N 59
11	27 24	27 50	02 06
12	27 25	27 46	03 07
13	27 26	27 43	04 00
14	27 26	27 40	04 41
15	27 27	27 37	05 08
16	27 R 27	27 34	05 17
17	27 27	27 30	05 08
18	27 26	27 27	04 40
19	27 26	27 24	03 54
20	27 D 26	27 21	02 52
21	27 27	27 18	01 38
22	27 27	27 15	00 N 18
23	27 R 27	27 11	01 S 02
24	27 26	27 08	02 17
25	27 26	27 05	03 21
26	27 25	27 02	04 13
27	27 24	26 59	04 49
28	27 23	26 56	05 11
29	27 22	26 52	05 17
30	27 ♉ 22	26 ♉ 49	05 S 09

DECLINATIONS

Date	Sun ☉ ° '	Moon ☽ ° '	Mercury ☿ ° '	Venus ♀ ° '	Mars ♂ ° '	Jupiter ♃ ° '	Saturn ♄ ° '	Uranus ♅ ° '	Neptune ♆ ° '	Pluto ♇ ° '
01	14 S 35	19 S 45	19 S 42	24 S 07	23 S 51	23 S 19	15 S 12	22 S 12	22 S 18	03 N 00
02	14 54	15 37	20 10	24 17	23 44	23 18	15 14	22 12	22 19	02 59
03	15 12	10 56	20 38	24 27	23 37	23 18	15 16	22 13	22 19	02 59
04	15 31	05 53	21 05	24 36	23 29	23 17	15 18	22 13	22 19	02 58
05	15 49	00 S 38	21 30	24 45	23 21	23 16	15 20	22 13	22 19	02 57
06	16 07	04 N 40	21 55	24 52	23 13	23 16	15 22	22 14	22 19	02 57
07	16 25	09 51	22 18	24 59	23 04	23 15	15 24	22 14	22 19	02 56
08	16 42	14 44	22 40	25 05	22 56	23 14	15 26	22 15	22 19	02 55
09	16 59	19 04	23 02	25 11	22 47	23 13	15 28	22 15	22 19	02 55
10	17 16	22 39	23 22	25 15	22 38	23 13	15 30	22 16	22 19	02 54
11	17 33	25 13	23 41	25 19	22 28	23 12	15 32	22 16	22 19	02 53
12	17 49	26 31	23 58	25 23	22 19	23 11	15 34	22 17	22 19	02 53
13	18 05	26 24	24 15	25 25	22 09	23 10	15 36	22 17	22 19	02 52
14	18 21	24 48	24 30	25 27	21 59	23 09	15 38	22 18	22 19	02 51
15	18 36	21 46	24 44	25 28	21 49	23 08	15 40	22 18	22 19	02 51
16	18 51	17 28	24 56	25 28	21 38	23 07	15 42	22 19	22 19	02 50
17	19 06	12 09	25 08	25 28	21 28	23 06	15 44	22 19	22 19	02 50
18	19 20	06 N 05	25 17	25 27	21 17	23 05	15 46	22 19	22 19	02 49
19	19 34	00 S 24	25 26	25 25	21 06	23 05	15 48	22 20	22 19	02 49
20	19 48	06 56	25 33	25 22	20 54	23 04	15 50	22 20	22 19	02 48
21	20 01	13 07	25 39	25 19	20 43	23 02	15 51	22 21	22 19	02 48
22	20 14	18 32	25 43	25 15	20 31	23 01	15 53	22 21	22 19	02 47
23	20 26	22 46	25 46	25 10	20 19	23 00	15 55	22 22	22 19	02 47
24	20 39	25 31	25 47	25 04	20 07	22 59	15 57	22 22	22 19	02 46
25	20 50	26 37	25 47	24 58	19 54	22 58	15 59	22 23	22 19	02 46
26	21 02	26 06	25 46	24 51	19 42	22 57	16 01	22 23	22 19	02 45
27	21 13	24 08	25 43	24 43	19 29	22 56	16 03	22 24	22 19	02 45
28	21 23	21 02	25 38	24 35	19 16	22 55	16 04	22 24	22 19	02 44
29	21 33	17 03	25 33	24 26	19 03	22 53	16 06	22 25	22 19	02 44
30	21 S 43	12 S 28	25 S 25	24 S 16	18 S 50	22 S 52	16 S 08	22 S 25	22 S 19	02 N 44

ZODIAC SIGN ENTRIES

Date	h	m	Planets
02	07	50	☽ ♓
04	20	20	☽ ♈
06	12	09	☿ ♐
07	08	53	☽ ♉
09	20	10	☽ ♊
12	05	31	☽ ♋
13	23	54	♀ ♑
14	12	34	☽ ♌
15	18	09	♂ ♒
16	17	08	☽ ♍
18	19	29	☽ ♎
20	20	31	☽ ♏
21	13	21	♆ ♑
22	03	11	☉ ♐
22	21	34	☽ ♐
25	00	17	☽ ♑
27	06	06	☽ ♒
29	15	33	☽ ♓

LATITUDES

Date	Mercury ☿ ° '	Venus ♀ ° '	Mars ♂ ° '	Jupiter ♃ ° '	Saturn ♄ ° '	Uranus ♅ ° '	Neptune ♆ ° '	Pluto ♇ ° '
01	01 S 20	01 S 32	01 S 51	00 S 10	02 N 03	00 00	01 N 08	16 N 15
04	01 37	01 40	01 48	00 10	02 03	00 00	01 08	16 15
07	01 53	01 47	01 46	00 11	02 02	00 00	01 08	16 16
10	02 07	01 53	01 43	00 11	02 02	00 00	01 08	16 16
13	02 19	01 59	01 41	00 11	02 02	00 00	01 08	16 16
16	02 27	02 04	01 39	00 11	02 02	00 00	01 08	16 17
19	02 32	02 08	01 36	00 12	02 02	00 00	01 07	16 17
22	02 32	02 12	01 33	00 12	02 02	00 00	01 07	16 18
25	02 26	02 15	01 31	00 12	02 02	00 00	01 07	16 19
28	02 13	02 17	01 28	00 12	02 02	00 00	01 07	16 20
31	01 S 50	02 S 19	01 S 26	00 S 13	02 N 02	00 00	01 N 07	16 N 21

DATA

Julian Date	2446006
Delta T	+54 seconds
Ayanamsa	23° 38' 27"
Synetic vernal point	05° ♓ 28' 32"
True obliquity of ecliptic	23° 26' 34"

MOON'S PHASES, APSIDES AND POSITIONS ☽

Date	h	m	Phase	Longitude ° '	Eclipse Indicator
08	17	43	○	16 ♉ 30	
16	06	59	☾	24 ♌ 06	
22	22	57	●	00 ♐ 50	Total
30	08	01	☽	08 ♓ 18	

Day	h	m		
04	22	32	Apogee	
20	20	42	Perigee	
05	14	51	ON	
12	22	08	Max dec	26° N 39'
19	10	34	0S	
25	15	56	Max dec	26° S 38'

LONGITUDES

Date	Chiron ⚷ ° '	Ceres ⚳ ° '	Pallas ⚴ ° '	Juno ⚵ ° '	Vesta ⚶ ° '	Black Moon Lilith ⚸ ° '
01	07 ♊ 19	20 ♉ 20	04 ♓ 36	22 ♍ 49	26 ♍ 13	06 ♈ 24
11	06 ♊ 49	18 ♉ 03	04 ♓ 45	26 ♍ 28	00 ♎ 42	07 ♈ 31
21	06 ♊ 16	15 ♉ 46	05 ♓ 27	29 ♍ 52	05 ♎ 06	08 ♈ 38
31	05 ♊ 42	13 ♉ 46	06 ♓ 41	03 ♎ 00	09 ♎ 23	09 ♈ 45

All ephemeris data is given at 12.00 UT and the Moon's longitude is additionally given for 24.00 UT

ASPECTARIAN

01 Thursday
00 52 ☽ ✶ ♀
00 57 ☽ ∠ ♆
01 52 ☽ ⊥ ♃
02 33 ☉ ✶ ♃
07 49 ☽ □ ♄
10 47 ☽ ⊻ ♂
12 20 ☽ ∥ ☿
17 44 ☽ □ ☿
19 29 ☽ Q ♅
19 50 ☽ ∠ ♃
23 25 ☽ ⊥ ♂

02 Friday
01 38 ☿ ⊥ ♆
03 10 ☽ Q ♀
06 39 ☽ ✶ ♆
12 36 ☿ ∠ ♃
12 38 ☽ △ ♇
14 03 ☽ ∥ ♄
15 38 ☽ ∥ ☉
18 34 ☽ ∠ ♂

03 Saturday
02 15 ☽ ✶ ♃
03 56 ☉ ⊥ ♀
05 53 ☽ △ ☉
06 47 ☽ Q ♆
07 40 ☽ □ ♅
14 26 ♀ ∠ ♇
17 28 ☉ ∥ ♄
18 55 ☽ ⚻ ♇
19 25 ☽ □ ♀
20 24 ☽ △ ♄

04 Sunday
02 49 ☽ ✶ ♂
02 50 ☽ Q ♃
04 09 ☉ ⊻ ♅
05 59 ♀ ⊻ ♄
13 12 ☽ ± ♇
14 40 ☽ △ ☿
15 02 ☽ ⚻ ☉
19 17 ☽ □ ♆

05 Monday
01 25 ☽ ⊼ ♇
01 26 ☽ # ♇
03 06 ☽ ⚻ ♄
04 45 ☽ Q ♂
10 57 ☽ ± ☉
15 51 ☽ □ ♃
20 41 ☽ △ ♅
21 30 ☽ ± ♄

06 Tuesday
00 13 ☽ ⊼ ☉
01 23 ☽ ⚻ ☿
03 43 ♂ ∥ ♃
04 10 ☽ ∥ ♇
04 19 ☿ ⊻ ♆
09 45 ☽ ⊼ ♄
15 01 ☽ △ ♀
18 42 ☉ ∠ ♆
19 32 ☽ □ ♂
22 04 ☽ ± ☿

07 Wednesday
03 03 ☽ ⚻ ♅
08 00 ☽ △ ♆
08 02 ☿ ∥ ♅
11 46 ☽ ⊼ ☿
12 37 ☿ ∥ ♆
14 06 ☽ ☍ ♇
21 08 ☽ ± ♅

08 Thursday
00 26 ☽ ⚻ ♀
05 00 ☽ △ ♃
07 25 ☿ ⊻ ♇
09 07 ☽ ⊼ ♅
13 56 ☽ ⚻ ♆
15 44 ☽ # ♄
17 43 ☽ ☍ ☉
20 14 ☽ ± ♀
22 13 ☽ ☍ ♄
23 14 ☽ # ☉

09 Friday
00 09 ☿ ∥ ♂
07 42 ☽ ± ♆
09 19 ☽ ⊼ ♀
11 01 ☽ △ ♂
11 01 ☽ ⚻ ♃
11 16 ☿ ⊥ ♃
18 08 ♀ ⊥ ♄
19 27 ☽ ⊼ ♆

10 Saturday
01 16 ☿ ∥ ♃
01 25 ☽ ⊼ ♇
04 48 ☽ ± ♃
06 48 ☽ ☍ ☿
09 04 ☽ # ♅
09 27 ☽ # ♆
11 51 ☽ # ♂
13 04 ☽ ± ♇
16 26 ☽ # ♃
16 33 ☽ ⊼ ♃
17 57 ☽ ⚻ ♂
18 27 ☽ # ☿
19 56 ☽ ☍ ♅

11 Sunday
03 19 ♀ ⊻ ♂
06 22 ☽ ⚻ ♇
07 01 ☉ ☌ ♄
08 54 ☽ ⊼ ♄
09 02 ☽ ⊼ ☉
12 13 ☽ ± ♂
13 28 ☽ # ♀
20 22 ☽ ± ♄
21 23 ☽ ± ☉

12 Monday
00 16 ☽ ⊼ ♂
01 06 ☽ ☍ ♀
04 57 ☽ ☍ ♆
05 57 ♂ ∠ ♅
10 45 ☽ △ ♇
11 52 ♂ ∥ ♆
13 25 ☽ ⚻ ♄
15 00 ☿ ⊥ ♇
15 41 ☽ ⚻ ☉
17 05 ♂ ∥ ♅
22 46 ☽ ⊼ ☿

13 Tuesday
01 56 ☽ ☍ ♃
04 38 ☽ ⊼ ♅
11 08 ☽ ± ☿
15 41 ☽ ± ♅
17 19 ☽ △ ♄
18 53 ♀ ☌ ♆
21 34 ☽ △ ☉

14 Wednesday
04 39 ☽ # ♀
05 25 ☽ ⚻ ☿
08 04 ☽ ⚻ ♅
08 32 ☿ ⊻ ♃
10 48 ☽ ☍ ♂
12 09 ☽ ⊼ ♆
13 49 ☽ ⊼ ♀
14 42 ☽ # ☿
17 43 ☽ □ ♇
22 53 ☽ ± ♆

15 Thursday
01 33 ☽ ± ♀
02 22 ☽ # ♃
08 04 ☿ ☌ ♅
08 20 ☽ # ♆
08 27 ☽ # ♅
08 50 ☽ ⊼ ♃
10 53 ☽ △ ♅
11 11 ☽ △ ☿
11 41 ☽ # ♂
11 47 ♂ ⊻ ♆
14 49 ☽ ⚻ ♆
18 55 ☽ ⚻ ♀
19 30 ☽ ± ♃
23 14 ☽ □ ♄

16 Friday
00 12 ☉ ⊥ ♆
01 30 ☽ Q ♇
05 19 ☽ # ☉
06 59 ☽ □ ☉
10 42 ♀ ✶ ♇
11 19 ☽ ⚻ ♃
16 51 ☽ △ ♆
18 25 ☽ ⊼ ♂
20 25 ☽ # ♄
22 12 ☽ ✶ ♇
23 14 ☽ △ ♀

17 Saturday
05 13 ☽ ± ♂
06 23 ☽ Q ♄
13 12 ☽ △ ♃
14 40 ☽ □ ♅
15 46 ♅ ∥ ♆
16 01 ☽ Q ☉
20 05 ☽ □ ☿
21 13 ☽ ⚻ ♂
23 35 ☽ ∠ ♇

18 Sunday
01 14 ♀ ∠ ♄
02 43 ☽ ✶ ♄
07 18 ♂ Q ♄
13 32 ☽ ✶ ☉
14 35 ☽ ⊥ ♇
19 20 ☽ □ ♆
20 45 ☽ Q ♅
23 28 ☽ △ ♂
23 46 ☿ ∠ ♂

19 Monday
00 14 ☽ ∥ ♇
00 31 ☽ ⊻ ♇
03 44 ☽ ∠ ♄
04 52 ☽ Q ☿
05 53 ☽ □ ♀
07 58 ☉ ∠ ♃
09 08 ☿ ∥ ♀
11 44 ☿ ∠ ♇
15 34 ☽ □ ♃
16 01 ☽ ∠ ☉
16 32 ☽ ✶ ♅
18 36 ☽ ⊥ ♄
20 40 ♂ □ ♇
20 49 ☽ # ♇

20 Tuesday
00 54 ☽ Q ♆
02 21 ☽ ✶ ☿
04 28 ☽ ⊻ ♄
07 47 ☽ ⊥ ☉
13 52 ☽ Q ♀
17 04 ☽ ∠ ♅
18 15 ☽ ⊻ ☉
20 28 ☽ ✶ ♆
21 17 ☽ Q ♃

21 Wednesday
01 35 ☽ ☌ ♇
03 04 ☽ □ ♂
05 00 ☽ ∠ ☿
05 48 ☿ ⊻ ♄
07 45 ☽ ⊥ ♅
11 06 ☽ ✶ ♀
17 05 ☽ ✶ ♃
17 34 ☽ ⊻ ♅
20 57 ☽ ∠ ♆
20 59 ☽ ⊥ ☿
23 40 ☽ ∥ ♄

22 Thursday
03 39 ☉ ⊻ ♆
05 44 ☽ ☌ ♄
07 40 ☽ ⊻ ☿
10 02 ☽ Q ♂
11 44 ☽ ⊥ ♆
13 50 ☽ ∠ ♀
18 01 ☽ ∠ ♃
21 10 ☽ ∥ ☉
21 39 ☽ ⊻ ♆
21 50 ☽ ∥ ♂
22 57 ☽ ☌ ☉

23 Friday
02 52 ☽ ⊻ ♇
04 37 ☉ ∥ ♂
06 03 ☽ ⊥ ♀
06 58 ☽ ✶ ♂
09 00 ☽ ∥ ♆
09 09 ☽ ⊥ ♃
09 16 ☽ ∥ ♅
12 54 ☽ ⊥ ♇
13 35 ☽ ∥ ♃
17 00 ☽ ⊻ ♀
19 21 ☽ ⊻ ♃
19 21 ☽ ☌ ♅
20 03 ♃ ⊻ ♅
20 15 ♂ ⊥ ♆

24 Saturday
04 06 ☽ ∠ ♇
07 05 ☽ ∥ ♀
08 10 ☽ ⊻ ♄
09 39 ☽ ∠ ♂
14 11 ☽ ☌ ☿
15 35 ☽ ∥ ☿
18 32 ☽ ⊥ ♄
22 28 ♀ ⊻ ♅

25 Sunday
00 30 ☽ ☌ ♆
02 03 ☽ ⊥ ♂
02 15 ♀ ☌ ♃
05 46 ☽ ⊻ ☉
06 00 ☽ ✶ ♇
09 03 ☉ ⊻ ♇
10 20 ☽ ∠ ♄
13 10 ☽ ⊻ ♂
17 10 ☽ ⊥ ☉
23 31 ☽ ⊻ ♅

26 Monday
00 04 ☽ ☌ ♃
01 50 ☽ ☌ ♀
03 18 ☽ Q ♇
10 21 ☽ ⊥ ♅
10 36 ☽ ∠ ☉
13 22 ☽ ✶ ♄
17 36 ☽ ∥ ☿
18 38 ♀ Q ♇
23 46 ☽ ⊻ ☿

27 Tuesday
02 53 ☽ ∠ ♅
05 54 ☽ ∥ ♀
06 28 ☽ ⊻ ♆
10 42 ☿ ⊥ ♄
11 41 ☽ Q ♄
11 45 ☽ ⊥ ☿
12 21 ☽ □ ♇
16 36 ☽ ✶ ☉
17 42 ☽ ⊥ ♆
22 21 ☽ ∥ ♃
23 20 ☽ ☌ ♂

28 Wednesday
02 20 ☽ ∥ ♅
02 54 ☽ ∥ ♆
05 59 ☽ ∠ ☿
07 13 ☽ ✶ ♅
08 28 ☽ ⊻ ♃
09 40 ☽ ∥ ☉
10 51 ☽ ∠ ♆
15 18 ☽ ⊻ ♀
17 28 ☽ Q ☉
20 12 ☽ ⊥ ♃
22 17 ☽ □ ♄
23 49 ☽ ∥ ♂

29 Thursday
04 10 ☽ ⊥ ♀
06 31 ☽ Q ♅
13 00 ☽ ✶ ☿
14 04 ☽ ∠ ♃
16 06 ☽ ✶ ♆
17 10 ☽ ∥ ♄
22 21 ☽ △ ♇
23 47 ☽ ∠ ♀

30 Friday
05 40 ♀ ⊥ ♅
08 01 ☽ □ ☉
11 03 ☉ ⊥ ♃
13 44 ☽ ⊻ ♂
14 12 ☽ Q ☿
16 02 ☽ Q ♆
18 23 ☽ □ ♅
20 26 ☽ ✶ ♃

DECEMBER 1984

LONGITUDES

Date	Sidereal time h m s	Sun ☉ ° ' "	Moon ☽ ° ' "	Moon ☽ 24.00 ° ' "	Mercury ☿ ° '	Venus ♀ ° '	Mars ♂ ° '	Jupiter ♃ ° '	Saturn ♄ ° '	Uranus ♅ ° '	Neptune ♆ ° '	Pluto ♇ ° '
01	16 42 07	09 ♐ 29 06	22 ♓ 34 34	28 ♓ 10 29	29 ♐ 54	20 ♑ 58	11 ♒ 52	14 ♑ 39	21 ♏ 28	13 ♐ 32	00 ♑ 21	03 ♏ 30
02	16 46 04	10 29 56	04 ♈ 05 39	10 ♈ 00 40	00 ♑ 21	22 09	12 38	14 52	21 35	13 36	00 23	03 32
03	16 50 00	11 30 47	15 ♈ 56 05	21 ♈ 52 24	00 39	23 20	13 23	15 04	21 42	13 39	00 25	03 34
04	16 53 57	12 31 39	27 ♈ 50 10	03 ♉ 49 47	00 48	24 31	14 09	15 17	21 49	13 43	00 27	03 36
05	16 57 53	13 32 32	09 ♉ 51 42	15 ♉ 55 15	00 R 47	25 42	14 55	15 30	21 56	13 47	00 29	03 38
06	17 01 50	14 33 26	22 ♉ 03 47	28 ♉ 14 32	00 35	26 53	15 40	15 43	22 02	13 50	00 32	03 40
07	17 05 46	15 34 21	04 ♊ 28 43	10 ♊ 46 29	00 ♑ 12	28 04	16 26	15 56	22 09	13 54	00 34	03 42
08	17 09 43	16 35 17	17 ♊ 07 55	23 ♊ 33 04	29 ♐ 38	29 ♑ 15	17 12	16 09	22 16	13 58	00 36	03 44
09	17 13 39	17 36 13	00 ♋ 01 56	06 ♋ 34 27	28 53	00 ♒ 25	17 57	16 22	22 23	14 01	00 38	03 46
10	17 17 36	18 37 11	13 ♋ 10 30	19 ♋ 50 00	27 57	01 36	18 43	16 35	22 29	14 05	00 40	03 48
11	17 21 33	19 38 10	26 ♋ 32 46	03 ♌ 18 37	26 51	02 46	19 29	16 48	22 36	14 09	00 43	03 49
12	17 25 29	20 39 10	10 ♌ 07 22	16 ♌ 58 51	25 38	03 56	20 15	17 02	22 43	14 12	00 45	03 51
13	17 29 26	21 40 10	23 ♌ 52 50	00 ♍ 49 10	24 18	05 07	21 00	17 15	22 49	14 16	00 47	03 53
14	17 33 22	22 41 12	07 ♍ 47 38	14 ♍ 48 04	22 56	06 17	21 46	17 28	22 56	14 20	00 49	03 55
15	17 37 19	23 42 15	21 ♍ 50 17	28 ♍ 54 06	21 33	07 27	22 32	17 42	23 02	14 23	00 52	03 57
16	17 41 15	24 43 18	05 ♎ 59 18	13 ♎ 05 41	20 13	08 37	23 18	17 55	23 09	14 27	00 54	03 58
17	17 45 12	25 44 23	20 ♎ 13 01	27 ♎ 21 00	18 57	09 46	24 03	18 08	23 15	14 31	00 56	04 00
18	17 49 08	26 45 29	04 ♏ 29 20	11 ♏ 37 39	17 49	10 56	24 49	18 22	23 22	14 34	00 58	04 02
19	17 53 05	27 46 35	18 ♏ 45 33	25 ♏ 52 34	16 50	12 05	25 35	18 35	23 28	14 38	01 01	04 03
20	17 57 02	28 47 42	02 ♐ 58 14	10 ♐ 02 02	16 00	13 15	26 21	18 49	23 35	14 41	01 03	04 05
21	18 00 58	29 ♐ 48 50	17 ♐ 03 28	24 ♐ 01 59	15 22	14 24	27 07	19 03	23 41	14 45	01 05	04 07
22	18 04 55	00 ♑ 49 59	00 ♑ 57 06	07 ♑ 48 23	14 54	15 33	27 53	19 16	23 47	14 49	01 08	04 08
23	18 08 51	01 51 08	14 ♑ 35 25	21 ♑ 17 54	14 37	16 42	28 39	19 30	23 53	14 52	01 10	04 10
24	18 12 48	02 52 17	27 ♑ 55 35	04 ♒ 28 20	14 30	17 51	29 ♒ 24	19 44	23 59	14 56	01 12	04 11
25	18 16 44	03 53 27	10 ♒ 56 06	17 ♒ 18 58	14 D 34	19 00	00 ♓ 10	19 57	24 06	14 59	01 14	04 13
26	18 20 41	04 54 37	23 ♒ 37 04	29 ♒ 50 39	14 46	20 08	00 56	20 11	24 12	15 03	01 17	04 14
27	18 24 37	05 55 46	06 ♓ 00 05	12 ♓ 05 45	15 06	21 16	01 42	20 25	24 18	15 06	01 19	04 16
28	18 28 34	06 56 56	18 ♓ 08 08	24 ♓ 07 46	15 33	22 24	02 28	20 39	24 23	15 10	01 21	04 17
29	18 32 31	07 58 06	00 ♈ 05 13	06 ♈ 01 07	16 08	23 32	03 14	20 53	24 29	15 13	01 23	04 18
30	18 36 27	08 59 15	11 ♈ 56 06	17 ♈ 50 48	16 48	24 40	04 00	21 07	24 35	15 17	01 26	04 20
31	18 40 24	10 ♑ 00 24	23 ♈ 45 55	29 ♈ 42 04	17 ♐ 34	25 ♒ 48	04 ♓ 45	21 ♑ 20	24 ♏ 41	15 ♐ 20	01 ♑ 28	04 ♏ 21

MOON

Date	Moon True ☊ ° '	Moon Mean ☊ ° '	Moon ☽ Latitude ° '
01	27 ♉ 23	26 ♉ 46	04 S 48
02	27 D 24	26 43	04 14
03	27 25	26 40	03 30
04	27 26	26 36	02 37
05	27 27	26 33	01 36
06	27 28	26 30	00 S 30
07	27 R 28	26 27	00 N 39
08	27 27	26 24	01 46
09	27 25	26 21	02 50
10	27 22	26 17	03 46
11	27 19	26 14	04 30
12	27 15	26 11	05 00
13	27 12	26 08	05 14
14	27 10	26 05	05 09
15	27 09	26 02	04 46
16	27 D 10	25 58	04 05
17	27 11	25 55	03 09
18	27 13	25 52	02 01
19	27 14	25 49	00 N 46
20	27 R 14	25 46	00 S 31
21	27 13	25 42	01 46
22	27 09	25 39	02 54
23	27 04	25 36	03 50
24	26 58	25 33	04 32
25	26 52	25 30	04 59
26	26 45	25 27	05 10
27	26 40	25 23	05 06
28	26 36	25 20	04 49
29	26 34	25 17	04 19
30	26 D 34	25 14	03 38
31	26 ♉ 35	25 ♉ 11	02 S 49

DECLINATIONS

Date	Sun ☉ ° '	Moon ☽ ° '	Mercury ☿ ° '	Venus ♀ ° '	Mars ♂ ° '	Jupiter ♃ ° '	Saturn ♄ ° '	Uranus ♅ ° '	Neptune ♆ ° '	Pluto ♇ ° '
01	21 S 53	07 S 29	25 S 17	24 S 05	18 S 36	22 S 51	16 S 10	22 S 25	22 S 19	02 N 43
02	22 01	02 S 16	25 06	23 54	18 22	22 49	16 11	22 26	22 19	02 43
03	22 10	03 N 02	24 55	23 42	18 08	22 48	16 13	22 26	22 19	02 42
04	22 18	08 16	24 42	23 30	17 54	22 47	16 15	22 27	22 19	02 42
05	22 26	13 15	24 28	23 17	17 40	22 45	16 17	22 27	22 19	02 42
06	22 33	17 48	24 12	23 03	17 26	22 44	16 18	22 28	22 19	02 41
07	22 40	21 40	23 55	22 49	17 11	22 42	16 20	22 28	22 19	02 41
08	22 46	24 35	23 37	22 34	16 57	22 41	16 22	22 29	22 19	02 41
09	22 52	26 17	23 17	22 18	16 42	22 39	16 23	22 29	22 19	02 41
10	22 57	26 32	22 57	22 02	16 27	22 38	16 25	22 30	22 19	02 40
11	23 02	25 16	22 35	21 45	16 11	22 36	16 27	22 30	22 19	02 40
12	23 07	22 31	22 13	21 28	15 56	22 35	16 28	22 30	22 19	02 40
13	23 11	18 29	21 51	21 10	15 41	22 33	16 30	22 31	22 19	02 40
14	23 14	13 25	21 28	20 51	15 25	22 32	16 31	22 31	22 19	02 40
15	23 18	07 36	21 07	20 32	15 09	22 30	16 33	22 32	22 19	02 39
16	23 20	01 N 22	20 46	20 13	14 53	22 28	16 35	22 32	22 19	02 39
17	23 22	04 S 59	20 28	19 53	14 37	22 27	16 36	22 33	22 19	02 39
18	23 24	11 07	20 12	19 32	14 21	22 25	16 38	22 33	22 19	02 39
19	23 25	16 40	19 58	19 11	14 05	22 23	16 39	22 33	22 19	02 39
20	23 26	21 16	19 48	18 50	13 48	22 21	16 41	22 34	22 19	02 39
21	23 27	24 35	19 41	18 28	13 32	22 19	16 42	22 34	22 19	02 39
22	23 26	26 20	19 36	18 05	13 15	22 18	16 44	22 35	22 19	02 38
23	23 26	26 27	19 35	17 43	12 58	22 16	16 45	22 35	22 19	02 38
24	23 25	25 01	19 36	17 19	12 41	22 14	16 46	22 35	22 19	02 38
25	23 23	22 16	19 40	16 56	12 24	22 12	16 48	22 36	22 19	02 38
26	23 21	18 31	19 45	16 32	12 07	22 10	16 49	22 36	22 19	02 38
27	23 19	14 03	19 53	16 07	11 50	22 08	16 51	22 37	22 19	02 38
28	23 16	09 07	20 02	15 42	11 33	22 06	16 52	22 37	22 19	02 38
29	23 12	03 S 56	20 13	15 17	11 15	22 04	16 53	22 37	22 19	02 38
30	23 08	01 N 22	20 24	14 52	10 58	22 02	16 55	22 38	22 19	02 38
31	23 S 04	06 N 37	20 S 36	14 S 26	10 S 40	22 S 00	16 S 56	22 S 38	22 S 19	02 N 38

ZODIAC SIGN ENTRIES

Date	h	m	Planets
01	16	30	☿ ♑
02	03	42	☽ ♈
04	16	20	☽ ♉
07	03	24	☽ ♊
07	21	46	☿ ♐
09	03	26	♀ ♒
09	11	56	☽ ♋
11	18	08	☽ ♌
13	22	35	☽ ♍
16	01	52	☽ ♎
18	04	27	☽ ♏
20	06	58	☽ ♐
21	16	23	☉ ♑
22	10	21	☽ ♑
24	15	47	☽ ♒
25	06	38	♂ ♓
27	00	18	☽ ♓
29	11	49	☽ ♈

LATITUDES

Date	Mercury ☿ ° '	Venus ♀ ° '	Mars ♂ ° '	Jupiter ♃ ° '	Saturn ♄ ° '	Uranus ♅ ° '	Neptune ♆ ° '	Pluto ♇ ° '
01	01 S 50	02 S 19	01 S 26	00 S 13	02 N 02	00 00	01 N 07	16 N 21
04	01 16	02 19	01 23	00 13	02 02	00 00	01 07	16 21
07	00 S 29	02 18	01 20	00 13	02 03	00 00	01 07	16 22
10	00 N 29	02 17	01 17	00 13	02 03	00 00	01 07	16 24
13	01 29	02 14	01 15	00 14	02 03	00 00	01 07	16 25
16	02 19	02 11	01 12	00 14	02 03	00 00	01 07	16 26
19	02 50	02 06	01 09	00 14	02 03	00 00	01 07	16 27
22	03 00	02 00	01 06	00 14	02 04	00 00	01 07	16 28
25	02 54	01 54	01 04	00 15	02 04	00 00	01 07	16 30
28	02 38	01 45	01 01	00 15	02 04	00 00	01 07	16 31
31	02 N 16	01 S 36	00 S 58	00 S 15	02 N 04	00 00	01 N 07	16 N 33

DATA

Julian Date	2446036
Delta T	+54 seconds
Ayanamsa	23° 38' 32"
Synetic vernal point	05° ♓ 28' 27"
True obliquity of ecliptic	23° 26' 33"

MOON'S PHASES, APSIDES AND POSITIONS ☽

Date	h	m	Phase	Longitude ° '	Eclipse Indicator
08	10	53	○	16 ♊ 32	
15	15	25	☾	23 ♍ 51	
22	11	47	●	00 ♑ 49	
30	05	27	☽	08 ♈ 43	

Day	h	m		
02	15	23	Apogee	
18	09	44	Perigee	
30	12	14	Apogee	
02	22	15	0N	
10	04	08	Max dec	26° N 37'
16	17	10	0S	
23	01	40	Max dec	26° S 36'
30	05	49	0N	

LONGITUDES

Date	Chiron ⚷ ° '	Ceres ⚳ ° '	Pallas ⚴ ° '	Juno ⚵ ° '	Vesta ⚶ ° '	Black Moon Lilith ⚸ ° '
01	05 ♊ 42	13 ♉ 46	06 ♓ 41	03 ♎ 00	09 ♎ 23	09 ♈ 45
11	05 ♊ 09	12 ♉ 16	08 ♓ 20	05 ♎ 50	13 ♎ 31	10 ♈ 51
21	04 ♊ 38	11 ♉ 22	10 ♓ 23	08 ♎ 17	17 ♎ 29	11 ♈ 58
31	04 ♊ 10	11 ♉ 09	12 ♓ 47	10 ♎ 19	21 ♎ 12	13 ♈ 05

ASPECTARIAN

h m Aspects

01 Saturday
02 35 ☽ ⊥ ♂
04 26 ☽ bQ ♇
09 07 ☽ ⚹ ♀
10 25 ☽ △ ♄
12 20 ☉ ⊥ ♇
21 04 ☽ Q ♃
22 00 ☽ ∠ ♂
22 40 ☽ ± ♇
23 20 ♀ ⚹ ♄

02 Sunday
04 08 ☽ □ ☿
04 27 ☽ □ ♆
09 57 ☽ # ♇
10 51 ☽ ⊼ ♇
12 07 ☽ Q ♀
14 34 ☿ ☌ ♆
17 06 ☽ bQ ♄

03 Monday
02 12 ☽ △ ☉
06 30 ☽ ⚹ ♂
07 22 ☽ △ ♅
10 14 ☽ □ ♃
10 31 ☽ ∥ ♇
11 31 ☽ ± ♄
21 05 ♂ ⚹ ♅
23 45 ☽ ⊼ ♄

04 Tuesday
04 36 ☽ □ ♀
08 23 ☽ Q ♂
11 19 ☽ bQ ☉
13 47 ☽ bQ ♅
16 11 ☉ ∥ ♆
17 16 ☽ △ ♆
17 58 ☽ △ ☿
21 45 ☿ St R
23 34 ☽ ☍ ♇

05 Wednesday
06 58 ☽ ± ☉
07 51 ☽ ± ♅
17 26 ☉ ∥ ♅
17 57 ☉ ☌ ♅
19 47 ☽ ⊼ ♅
19 57 ☽ ⊼ ☉
22 39 ☽ □ ♂
23 09 ☽ bQ ♆
23 20 ☽ △ ♃
23 33 ☽ bQ ☿

06 Thursday
03 42 ☽ # ♄
05 24 ♃ Q ♇
07 11 ♂ ∠ ♆
10 00 ☽ # ♂
10 01 ☿ ∠ ♂
11 57 ☽ ☍ ♄
13 56 ♂ ⊻ ♃
16 14 ☿ ☌ ♆
16 48 ☽ ± ☿
16 49 ☽ ± ♆
22 21 ☽ △ ♀

07 Friday
04 05 ☽ ⊼ ☿
04 27 ☽ ⊼ ♆
05 04 ☽ bQ ♃
10 30 ☽ ⊼ ♇
16 41 ☽ # ♆
17 47 ☽ # ♅
19 29 ☽ # ☉
19 31 ☽ # ♃
19 46 ☽ # ♀
20 21 ☉ ∥ ♃
21 59 ☽ ± ♇
22 12 ☉ ∥ ♀
22 35 ☽ ± ♃
22 49 ☉ ⊻ ♃
23 16 ♀ ∥ ♃

08 Saturday
03 51 ☽ # ☿
05 59 ♀ ∠ ♅
06 00 ☽ ☍ ♅
06 00 ☽ bQ ♀
10 07 ☽ ⊼ ♃
10 53 ☽ ☍ ☉
12 08 ☽ △ ♂
15 00 ☽ bQ ♇
17 07 ☿ ⊻ ♀
19 42 ♀ ∥ ♅
21 42 ☽ ⊼ ♄

09 Sunday
00 34 ☽ ± ♀
08 55 ☽ ± ♄
09 56 ♀ ∥ ♆
10 01 ☽ ☍ ☿
12 47 ☽ ⊼ ♀
13 07 ☽ ☍ ♆
16 35 ♀ ⊻ ♆
17 43 ☽ bQ ♂
18 52 ☽ △ ♇

10 Monday
00 05 ☿ ⊥ ♄
01 35 ☽ bQ ♄
11 07 ☽ ± ♂
11 35 ☉ ∥ ☿
13 39 ☽ ⊼ ♅
14 10 ♂ ∥ ♄
16 14 ☉ ∠ ♇
18 16 ☽ ☍ ♃
21 18 ☉ ⚹ ♂
22 36 ☽ ⊼ ♂
22 38 ☽ ⊼ ☉

11 Tuesday
00 31 ☽ ± ♅
04 54 ☽ △ ♄
10 15 ☽ ± ☉
10 45 ☿ ∥ ♃
12 30 ☽ ⊼ ☿
12 53 ☿ ⊥ ♀
16 39 ☽ bQ ♅
17 45 ☿ ∥ ♅
19 25 ☽ ⊼ ♆
22 18 ☽ ± ☿

12 Wednesday
00 05 ☽ ☍ ♀
00 56 ☽ □ ♇
03 30 ☽ bQ ☉
05 11 ☿ ∥ ♆
06 03 ☽ ± ♆
07 49 ☽ # ☉
10 12 ♀ □ ♇
11 36 ☽ # ♃
12 07 ☽ # ♅
12 49 ☽ bQ ☿
13 23 ☽ # ♆
14 21 ☽ # ☿
19 11 ☽ △ ♅
19 38 ☽ # ♀
21 53 ☽ bQ ♆

13 Thursday
00 17 ☽ ⊼ ♃
05 29 ♀ Q ♄
06 43 ☽ ☍ ♂
07 51 ☽ △ ☉
08 32 ☽ Q ♇
10 09 ☽ □ ♄
10 53 ☽ ± ♃
12 40 ☽ △ ☿
21 52 ☽ # ♄
23 59 ☽ △ ♆

14 Friday
02 33 ☽ # ♂
02 42 ☽ bQ ♃
05 19 ☽ ⚹ ♇
09 10 ☽ ⊼ ♀
12 01 ☿ ⊻ ♄
14 28 ☉ ☌ ☿
15 30 ♃ ∥ ♅
17 25 ☽ Q ♄
18 30 ☉ ⊻ ♄
20 23 ☽ ± ♀
23 15 ☽ □ ♅
23 35 ♀ ⊥ ♆

15 Saturday
01 01 ☿ ⚹ ♂
03 33 ☿ ∠ ♀
04 49 ☽ △ ♃
07 04 ☽ ∠ ♇
11 33 ☽ □ ☿
13 08 ☽ bQ ♀
13 15 ☽ ⊼ ♂
14 04 ☽ ⚹ ♄
15 25 ☽ □ ☉
22 24 ☽ ⊥ ♇

16 Sunday
00 01 ☽ ± ♂
03 22 ☽ □ ♆
05 59 ☽ Q ♅
06 43 ♂ □ ♄
07 08 ☽ ∥ ♇
08 35 ☽ ⊻ ♇
15 26 ☽ Q ☿
15 41 ☽ ∠ ♄
16 07 ☽ bQ ♂
16 50 ☽ △ ♀

17 Monday
00 15 ☽ Q ☉
02 21 ☽ ⚹ ♅
03 10 ☽ # ♇
06 59 ☽ ⊥ ♄
08 27 ☽ □ ♃
09 50 ☽ Q ♆
10 02 ☽ ⚹ ☿
11 03 ☿ ∠ ♇
15 41 ♂ ⊥ ♃
17 09 ☽ ⊻ ♄
18 50 ☽ △ ♂
22 00 ☽ ⚹ ☉

18 Tuesday
02 01 ☿ ⊻ ♃
03 42 ☽ ∠ ♅
06 05 ☽ ⚹ ♆
09 23 ☽ ∠ ☿
11 14 ☽ ☌ ♇
15 12 ☽ Q ♃
18 53 ☽ ⊥ ♅
23 29 ☽ ⊥ ☿
23 47 ☽ □ ♀

19 Wednesday
00 57 ☽ ∥ ♂
01 09 ☽ ∠ ☉
05 01 ☽ ⊻ ♅
07 22 ☽ ∠ ♆
08 57 ☽ ⊻ ☿
11 43 ☽ ⚹ ♃
11 57 ☽ ∥ ♄
17 29 ☽ ⊥ ☉
20 00 ☽ ☌ ♄
22 34 ☽ ⊥ ♆
23 32 ☽ ∥ ♀

20 Thursday
00 10 ☽ □ ♂
03 57 ☽ ∥ ☿
04 23 ☽ ⊻ ☉
08 44 ☽ ⊻ ♆
08 49 ☽ Q ♀
13 28 ☽ ∠ ♃
13 54 ☽ ⊻ ♇
18 38 ☽ ∥ ♆
18 46 ☽ ∥ ♃
20 15 ☽ ∥ ♅
23 40 ♂ Q ♅

21 Friday
00 07 ☽ ⊥ ♇
02 30 ☽ ∥ ☉
05 01 ☽ ⊥ ♃
07 03 ☽ ⚹ ♀
08 02 ☽ ☌ ♅
08 29 ☽ Q ♂
08 31 ☉ ⊥ ♄
09 12 ☽ ☌ ☿
11 27 ♃ ∥ ♆
15 28 ☽ ⊻ ♃
15 32 ☽ ∠ ♇
16 35 ☿ Q ♂
19 42 ♀ ⚹ ♅
23 29 ☽ ⊻ ♄

22 Saturday
01 58 ☿ ⚹ ♀
06 21 ☽ ⚹ ♂
09 57 ☽ ⊥ ♄
11 14 ☽ ∠ ♀
11 47 ☽ ☌ ☉
12 18 ☽ ☌ ♆
17 13 ☿ ☌ ♅
17 34 ☽ ⚹ ♇
19 10 ☉ ☌ ♆

23 Sunday
00 23 ♀ ∠ ♆
01 50 ☽ ∠ ♄
04 28 ☽ ⊥ ♀
10 13 ☽ ∠ ♂
12 03 ☽ ⊻ ☿
12 30 ☽ ⊻ ♅
14 48 ☽ Q ♇
16 07 ☽ ⊻ ♀
20 55 ☽ ☌ ♃
22 39 ☽ ⊥ ☿
23 17 ☽ ⊥ ♅

24 Monday
03 18 ☽ ⊥ ♂
04 48 ☽ ⚹ ♄
14 52 ☽ ⊻ ♂
14 53 ☽ ∠ ☿
15 40 ☽ ∠ ♅
16 12 ☿ St D
16 28 ☉ ∠ ♀
18 00 ☽ ⊻ ♆
21 49 ☽ ⊻ ☉
23 30 ☽ □ ♇

25 Tuesday
02 55 ☽ Q ♄
03 20 ☽ ∥ ☉
05 06 ☽ ⊥ ♆
09 36 ☽ ∥ ♅
09 53 ☽ ⊥ ☉
11 37 ☽ ∥ ♆
12 32 ☽ ∥ ♃
18 53 ☽ ⚹ ☿
19 30 ♀ ∥ ♄
19 38 ☽ ⚹ ♅
19 46 ☉ ⚹ ♇
22 00 ☽ ∠ ♆

26 Wednesday
04 18 ☽ ∠ ☉
04 41 ☽ ☌ ♀
04 51 ☽ ∥ ☿
05 20 ☽ ⊻ ♃
13 07 ☽ □ ♄
13 26 ♀ ⊻ ♃
17 02 ☽ ⊥ ♃
18 11 ☽ Q ☿
18 37 ☽ Q ♅
21 27 ☽ ∥ ♄
23 18 ♂ ⚹ ♆

27 Thursday
00 11 ☽ ∥ ♀
02 50 ☽ ⚹ ♆
03 03 ☽ ☌ ♂
08 35 ☽ △ ♇
10 50 ☽ ∠ ♃
11 51 ☽ ⚹ ☉
12 43 ☿ ☌ ♅
23 42 ☽ ∥ ♂

28 Friday
02 28 ☽ Q ♆
06 03 ☽ □ ♅
06 38 ☽ □ ☿
13 46 ☽ Q ☉
14 18 ☽ bQ ♇
17 07 ☽ ⚹ ♃
21 26 ☽ ⊻ ♀

29 Saturday
00 38 ☽ △ ♄
08 24 ☽ ± ♇
10 47 ☽ ⊥ ♀
14 39 ☽ □ ♆
17 45 ☽ Q ♃
17 52 ☽ # ♇
18 47 ☽ ⊻ ♂
20 33 ☽ ⊼ ♇

30 Sunday
05 27 ☽ □ ☉
06 55 ☽ ∠ ♀
07 12 ☽ bQ ♄
07 47 ☽ ⊥ ♂
10 07 ♀ □ ♄
17 47 ☽ ∥ ♇
18 49 ☽ △ ♅
22 32 ☽ △ ☿
22 50 ♂ △ ♇

31 Monday
01 37 ☽ ± ♄
03 19 ☽ ∠ ♂
03 35 ☉ ∠ ♄
06 59 ☽ □ ♃
11 24 ♃ ⊥ ♅
13 52 ☽ ⊼ ♄
16 32 ☽ ⚹ ♀

All ephemeris data is given at 12.00 UT and the Moon's longitude is additionally given for 24.00 UT

JANUARY 1985

LONGITUDES

Date	Sidereal time h m s	Sun ☉	Moon ☽	Moon ☽ 24.00	Mercury ☿	Venus ♀	Mars ♂	Jupiter ♃	Saturn ♄	Uranus ♅	Neptune ♆	Pluto ♇
01	18 44 20	11 ♑ 01 34	05 ♉ 39 54	11 ♉ 40 02	18 ♐ 24	26 ♒ 55	05 ♓ 31	21 ♑ 34	24 ♏ 47	15 ♐ 24	01 ♑ 30	04 ♏ 22
02	18 48 17	12 02 43	17 ♉ 43 04	23 ♉ 49 32	19 19	28 02	06 17	21 48	24 52	15 27	01 32	04 24
03	18 52 13	13 03 52	29 ♉ 59 56	06 ♊ 14 40	20 17	29 ♒ 09	07 03	22 02	24 58	15 30	01 35	04 25
04	18 56 10	14 05 01	12 ♊ 34 04	18 ♊ 58 25	21 19	00 ♓ 16	07 49	22 16	25 03	15 34	01 37	04 26
05	19 00 06	15 06 09	25 ♊ 27 51	02 ♋ 02 26	22 24	01 22	08 35	22 30	25 09	15 37	01 39	04 27
06	19 04 03	16 07 18	08 ♋ 42 05	15 ♋ 26 37	23 31	02 28	09 20	22 44	25 14	15 40	01 41	04 28
07	19 08 00	17 08 26	22 ♋ 15 47	29 ♋ 09 09	24 41	03 34	10 06	22 58	25 20	15 44	01 44	04 29
08	19 11 56	18 09 34	06 ♌ 06 17	13 ♌ 06 38	25 53	04 40	10 52	23 12	25 25	15 47	01 46	04 30
09	19 15 53	19 10 42	20 ♌ 09 36	27 ♌ 14 35	27 07	05 45	11 38	23 26	25 30	15 50	01 48	04 31
10	19 19 49	20 11 50	04 ♍ 20 59	11 ♍ 28 12	28 22	06 50	12 23	23 40	25 35	15 53	01 50	04 32
11	19 23 46	21 12 57	18 ♍ 35 44	25 ♍ 43 05	29 ♐ 39	07 55	13 09	23 54	25 40	15 57	01 52	04 33
12	19 27 42	22 14 05	02 ♎ 49 51	09 ♎ 55 44	00 ♑ 58	09 00	13 55	24 08	25 45	16 00	01 55	04 34
13	19 31 39	23 15 12	17 ♎ 00 29	24 ♎ 03 54	02 17	10 04	14 40	24 23	25 50	16 03	01 57	04 35
14	19 35 35	24 16 20	01 ♏ 05 52	08 ♏ 06 17	03 38	11 08	15 26	24 37	25 55	16 06	01 59	04 36
15	19 39 32	25 17 27	15 ♏ 05 07	22 ♏ 02 18	05 00	12 12	16 12	24 51	26 00	16 09	02 01	04 37
16	19 43 29	26 18 35	28 ♏ 57 46	05 ♐ 51 29	06 23	13 15	16 57	25 05	26 05	16 12	02 03	04 37
17	19 47 25	27 19 42	12 ♐ 43 19	19 ♐ 33 09	07 47	14 18	17 43	25 19	26 09	16 15	02 05	04 38
18	19 51 22	28 20 49	26 ♐ 20 49	03 ♑ 06 07	09 12	15 21	18 29	25 33	26 14	16 18	02 07	04 39
19	19 55 18	29 ♑ 21 55	09 ♑ 48 51	16 ♑ 28 47	10 38	16 23	19 14	25 47	26 18	16 21	02 09	04 39
20	19 59 15	00 ♒ 23 01	23 ♑ 05 38	29 ♑ 39 14	12 04	17 25	20 00	26 01	26 23	16 24	02 11	04 40
21	20 03 11	01 24 06	06 ♒ 09 20	12 ♒ 35 47	13 31	18 26	20 45	26 15	26 27	16 27	02 13	04 41
22	20 07 08	02 25 10	18 ♒ 58 28	25 ♒ 17 20	14 59	19 27	21 31	26 29	26 31	16 30	02 16	04 41
23	20 11 04	03 26 14	01 ♓ 32 25	07 ♓ 43 48	16 28	20 28	22 16	26 43	26 35	16 32	02 18	04 42
24	20 15 01	04 27 17	13 ♓ 51 39	19 ♓ 56 13	17 57	21 28	23 02	26 57	26 40	16 35	02 20	04 42
25	20 18 58	05 28 18	25 ♓ 57 50	01 ♈ 56 54	19 28	22 28	23 47	27 11	26 44	16 38	02 22	04 43
26	20 22 54	06 29 19	07 ♈ 53 52	13 ♈ 49 15	20 58	23 27	24 33	27 25	26 47	16 41	02 23	04 43
27	20 26 51	07 30 18	19 ♈ 43 39	25 ♈ 37 40	22 30	24 26	25 18	27 39	26 51	16 43	02 25	04 43
28	20 30 47	08 31 17	01 ♉ 31 58	07 ♉ 27 12	24 02	25 25	26 04	27 53	26 55	16 46	02 27	04 44
29	20 34 44	09 32 14	13 ♉ 24 06	19 ♉ 23 20	25 35	26 22	26 49	28 07	26 59	16 49	02 29	04 44
30	20 38 40	10 33 10	25 ♉ 25 37	01 ♊ 31 37	27 08	27 20	27 34	28 21	27 02	16 51	02 31	04 44
31	20 42 37	11 ♒ 34 05	07 ♊ 41 59	13 ♊ 57 19	28 ♑ 42	28 ♓ 16	28 ♓ 20	28 ♑ 35	27 ♏ 06	16 ♐ 54	02 ♑ 33	04 ♏ 44

Moon Nodes and Latitude

Date	Moon True ☊	Moon Mean ☊	Moon ☽ Latitude
01	26 ♉ 36	25 ♉ 08	01 S 51
02	26 D 37	25 04	00 S 48
03	26 R 38	25 01	00 N 18
04	26 36	24 58	01 25
05	26 33	24 55	02 29
06	26 27	24 52	03 27
07	26 19	24 48	04 14
08	26 10	24 45	04 47
09	26 01	24 42	05 04
10	25 53	24 39	05 02
11	25 46	24 36	04 42
12	25 42	24 33	04 04
13	25 40	24 29	03 12
14	25 D 40	24 26	02 08
15	25 41	24 23	00 N 56
16	25 R 41	24 20	00 S 17
17	25 40	24 17	01 30
18	25 36	24 14	02 36
19	25 30	24 10	03 32
20	25 20	24 07	04 16
21	25 09	24 04	04 46
22	24 56	24 01	05 00
23	24 43	23 58	05 00
24	24 32	23 54	04 45
25	24 23	23 51	04 18
26	24 17	23 48	03 40
27	24 13	23 45	02 52
28	24 11	23 42	01 57
29	24 D 11	23 39	00 S 57
30	24 R 11	23 35	00 N 07
31	24 ♉ 11	23 ♉ 32	01 N 11

DECLINATIONS

Date	Sun ☉	Moon ☽	Mercury ☿	Venus ♀	Mars ♂	Jupiter ♃	Saturn ♄	Uranus ♅	Neptune ♆	Pluto ♇
01	22 S 59	11 N 40	20 S 49	14 S 00	10 S 22	21 S 58	16 S 57	22 S 39	22 S 19	02 N 38
02	22 54	16 21	21 02	13 34	10 05	21 56	16 58	22 39	22 19	02 39
03	22 48	20 27	21 15	13 07	09 47	21 53	17 00	22 39	22 19	02 39
04	22 42	23 43	21 28	12 40	09 29	21 51	17 01	22 40	22 19	02 39
05	22 35	25 51	21 41	12 13	09 11	21 49	17 02	22 40	22 19	02 39
06	22 28	26 36	21 53	11 45	08 53	21 47	17 03	22 40	22 19	02 39
07	22 21	25 47	22 05	11 18	08 35	21 45	17 04	22 41	22 19	02 39
08	22 13	23 23	22 16	10 50	08 16	21 42	17 05	22 41	22 19	02 39
09	22 04	19 34	22 27	10 22	07 58	21 40	17 06	22 42	22 19	02 39
10	21 55	14 36	22 37	09 53	07 40	21 38	17 08	22 42	22 19	02 40
11	21 46	08 50	22 46	09 25	07 21	21 35	17 09	22 42	22 19	02 40
12	21 36	02 N 37	22 55	08 56	07 03	21 33	17 10	22 43	22 19	02 40
13	21 26	03 S 44	23 02	08 28	06 44	21 31	17 11	22 43	22 19	02 40
14	21 16	09 52	23 09	07 59	06 26	21 28	17 12	22 43	22 19	02 41
15	21 05	15 28	23 14	07 30	06 07	21 26	17 13	22 44	22 19	02 41
16	20 54	20 13	23 18	07 01	05 49	21 23	17 14	22 44	22 19	02 41
17	20 42	23 48	23 22	06 31	05 30	21 21	17 15	22 44	22 19	02 41
18	20 30	25 59	23 24	06 02	05 12	21 18	17 15	22 45	22 19	02 42
19	20 17	26 36	23 25	05 33	04 53	21 16	17 16	22 45	22 19	02 42
20	20 04	25 40	23 25	05 03	04 34	21 13	17 17	22 45	22 19	02 42
21	19 51	23 21	23 23	04 34	04 15	21 11	17 18	22 45	22 19	02 43
22	19 37	19 53	23 21	04 04	03 57	21 08	17 19	22 46	22 19	02 43
23	19 23	15 35	23 17	03 35	03 38	21 06	17 20	22 46	22 19	02 43
24	19 09	10 44	23 12	03 05	03 19	21 03	17 21	22 46	22 18	02 44
25	18 54	05 33	23 05	02 36	03 00	21 01	17 21	22 47	22 18	02 44
26	18 39	00 S 14	22 58	02 06	02 41	20 58	17 22	22 47	22 18	02 44
27	18 24	05 N 04	22 49	01 37	02 23	20 55	17 23	22 47	22 18	02 45
28	18 08	10 11	22 38	01 08	02 04	20 53	17 23	22 47	22 18	02 45
29	17 52	14 58	22 27	00 38	01 45	20 50	17 24	22 48	22 18	02 46
30	17 36	19 14	22 14	00 09	01 26	20 47	17 25	22 48	22 18	02 46
31	17 S 19	22 N 46	21 S 59	00 N 20	01 S 07	20 S 44	17 S 25	22 S 48	22 S 18	02 N 47

ZODIAC SIGN ENTRIES

Date	h	m	Planets
01	00	36	☽ ♉
03	12	00	☽ ♊
04	06	23	♀ ♓
05	20	18	☽ ♋
08	01	28	☽ ♌
10	04	40	☽ ♍
11	18	25	☿ ♑
12	07	13	☽ ♎
14	10	07	☽ ♏
16	13	48	☽ ♐
18	18	29	☽ ♑
20	02	58	☉ ♒
21	00	38	☽ ♒
23	09	02	☽ ♓
25	20	05	☽ ♈
28	08	53	☽ ♉
30	21	01	☽ ♊

LATITUDES

Date	Mercury ☿	Venus ♀	Mars ♂	Jupiter ♃	Saturn ♄	Uranus ♅	Neptune ♆	Pluto ♇
01	02 N 08	01 S 33	00 S 57	00 S 15	02 N 05	00 00	01 N 07	16 N 33
04	01 42	01 22	00 54	00 15	02 05	00 00	01 07	16 35
07	01 15	01 10	00 51	00 16	02 05	00 00	01 07	16 36
10	00 49	00 57	00 49	00 16	02 06	00 00	01 07	16 38
13	00 N 23	00 43	00 46	00 16	02 06	00 00	01 07	16 39
16	00 S 01	00 28	00 43	00 16	02 06	00 00	01 07	16 41
19	00 24	00 S 11	00 40	00 17	02 07	00 00	01 07	16 42
22	00 45	00 N 07	00 38	00 17	02 07	00 00	01 07	16 44
25	01 04	00 25	00 35	00 17	02 08	00 00	01 07	16 46
28	01 21	00 45	00 32	00 18	02 08	00 00	01 07	16 47
31	01 S 36	01 N 07	00 S 30	00 S 18	02 N 09	00 00	01 N 07	16 N 49

DATA

Julian Date	2446067
Delta T	+54 seconds
Ayanamsa	23° 38' 38"
Synetic vernal point	05° ♓ 28' 22"
True obliquity of ecliptic	23° 26' 33"

MOON'S PHASES, APSIDES AND POSITIONS ☽

Date	h	m	Phase	Longitude	Eclipse Indicator
07	02	16	○	16 ♋ 44	
13	23	27	☾	23 ♎ 44	
21	02	28	●	01 ♒ 00	
29	03	29	☽	09 ♉ 11	

Day	h	m		
12	03	36	Perigee	
27	09	33	Apogee	
06	11	43	Max dec	26° N 36'
12	21	51	0S	
19	09	23	Max dec	26° S 37'
26	13	02	0N	

LONGITUDES

Date	Chiron ⚷	Ceres ⚳	Pallas ⚴	Juno ⚵	Vesta ⚶	Black Moon Lilith ⚸
01	04 ♊ 08	11 ♉ 10	13 ♓ 02	10 ♎ 29	21 ♎ 34	13 ♈ 12
11	03 ♊ 45	11 ♉ 40	15 ♓ 44	11 ♎ 59	24 ♎ 58	14 ♈ 19
21	03 ♊ 29	12 ♉ 47	18 ♓ 41	12 ♎ 55	28 ♎ 02	15 ♈ 25
31	03 ♊ 19	14 ♉ 25	21 ♓ 51	13 ♎ 14	00 ♏ 40	16 ♈ 32

All ephemeris data is given at 12.00 UT and the Moon's longitude is additionally given for 24.00 UT

ASPECTARIAN

01 Tuesday
01 21 ☽ ⅊ ♅
03 36 ☽ △ ♆
06 05 ☽ # ♂
07 06 ☽ ⅊ ☿
09 24 ☽ ☍ ♇
11 42 ☽ ⚹ ♂
19 11 ☽ Q ♀
19 30 ☽ ± ♅
22 40 ☽ # ♀
22 49 ♀ Q ♅
23 43 ☽ △ ☉

02 Wednesday
02 33 ☽ ± ☿
05 49 ♀ ⊥ ♃
07 30 ☽ ⊼ ♅
09 40 ☽ ⅊ ♆
13 12 ☽ Q ♂
14 00 ☿ ∠ ♇
15 25 ☽ ⊼ ☿
15 26 ☽ # ♄
20 12 ☽ △ ♃

03 Thursday
02 09 ☽ ☍ ♄
03 24 ☽ ± ♆
07 55 ☽ ⅊ ☉
10 12 ☽ □ ♀
11 30 ♂ ∠ ♃
15 03 ☽ ⊼ ♆
17 36 ☽ # ☿
20 30 ☽ ⊼ ♇
21 38 ☽ # ♃

04 Friday
00 52 ☽ # ♆
01 46 ☽ ⅊ ♃
02 25 ☽ □ ♂
02 46 ☽ ± ☉
03 26 ☽ # ♅
03 59 ☽ # ☉
07 57 ☽ ± ♇
15 06 ☽ ⊼ ☉
17 39 ☽ ☍ ♅
19 05 ☽ ± ♃
19 26 ☉ ∥ ♅

05 Saturday
00 52 ☽ ⅊ ♇
05 50 ☽ ☍ ☿
06 27 ☽ ⊼ ♃
11 25 ☽ ⊼ ♄
14 56 ☿ ⊻ ♃
18 26 ♀ ⚹ ♆
22 27 ☽ ± ♄
23 20 ☽ ☍ ♆
23 46 ☽ △ ♀

06 Sunday
00 50 ☉ ⊻ ♅
02 00 ☿ ∥ ♃
04 23 ☽ △ ♇
13 13 ☽ △ ♂
14 46 ☽ ⅊ ♄
20 19 ☉ Q ♇

07 Monday
00 27 ☽ ⊼ ♅
02 16 ☽ ☍ ☉
04 57 ☽ ⅊ ♀
11 04 ☽ ± ♅
13 16 ☽ ☍ ♃
16 38 ☉ ∥ ♆
16 38 ☽ ⊼ ☿
17 15 ☽ ⅊ ♂
17 23 ☽ △ ♄
22 03 ☽ ± ♀

08 Tuesday
02 01 ☿ ⊻ ♄
02 47 ☽ ⅊ ♅
04 02 ☽ ± ☿
04 30 ☽ ⊼ ♆
07 28 ☉ ∥ ☿
08 26 ♀ △ ♇
09 15 ☽ □ ♇
09 19 ☽ ⊼ ♀
09 45 ☽ ± ♂
14 51 ☽ ± ♆
17 05 ☽ # ♅
18 05 ☿ ∥ ♅
19 32 ☽ # ☿
19 36 ☽ # ♆
20 38 ☽ ⊼ ♂
20 39 ☽ # ☉
20 58 ☽ ⅊ ☿
23 42 ☽ # ♃

09 Wednesday
04 37 ☽ △ ♅
06 16 ☽ ⅊ ♆
10 12 ☽ ⊼ ☉
16 01 ☽ Q ♇
17 39 ☽ ⊼ ♃
21 07 ☽ □ ♄
21 10 ☽ ± ☉

10 Thursday
00 26 ☽ # ♄
00 55 ☽ △ ☿
03 58 ☽ ± ♃
07 45 ☽ △ ♆
12 19 ☽ ⚹ ♇
13 32 ☽ ⅊ ☉
16 33 ☽ ☍ ♀
19 25 ☽ ⅊ ♃

11 Friday
00 43 ☿ ∥ ♅
02 19 ☽ ☍ ♂
03 40 ☽ Q ♄
07 31 ☽ □ ♅
09 29 ☽ # ♀
13 37 ☽ ∠ ♇
16 45 ☽ △ ☉
18 07 ☽ # ♂
21 06 ☽ △ ♃

12 Saturday
00 00 ☽ ⚹ ♄
04 48 ☽ ⊥ ♇
06 03 ☉ ⊥ ♅
08 31 ☽ □ ☿
10 26 ☽ □ ♆
11 47 ☽ ∥ ♇
11 51 ♂ Q ♆
13 58 ☽ Q ♅
14 56 ☽ ⊻ ♇
16 09 ♀ ∠ ♃
22 40 ☉ ∥ ♃
23 16 ☽ ⊼ ♀

13 Sunday
01 29 ☽ ∠ ♄
03 22 ☿ ⊥ ♄
05 38 ☿ ☌ ♆
07 49 ☽ ⊼ ♂
07 58 ☽ # ♇
10 16 ☽ ± ♀
10 22 ☽ ⚹ ♅
16 50 ☽ ⊥ ♄
17 00 ☽ Q ♆
18 10 ☽ Q ☿
18 35 ☽ ± ♂
23 03 ☽ ∥ ♂
23 27 ☽ □ ☉

14 Monday
00 44 ☽ □ ♃
02 50 ☽ ⅊ ♀
03 07 ☽ ⊻ ♄
03 50 ☿ Q ♂
05 01 ☽ ∥ ♀
10 48 ☽ ⅊ ♂
12 00 ☽ ∠ ♅
13 31 ☽ ⚹ ♆
16 49 ☽ ⚹ ☿
17 59 ☽ ☌ ♇
22 20 ☉ ☌ ♃

15 Tuesday
03 29 ☽ ⊥ ♅
05 01 ☿ ⚹ ♇
06 37 ☽ △ ♀
08 05 ☽ Q ♃
08 40 ☽ Q ☉
10 25 ♂ □ ♅
13 50 ☽ ⊻ ♅
14 02 ☽ △ ♂
15 20 ☽ ∠ ♆
20 17 ☽ ∥ ♄
21 25 ☽ ∠ ☿
21 29 ♄ ⊥ ♆

16 Wednesday
05 09 ☽ ⚹ ♃
06 04 ☉ ⚹ ♄
06 56 ☽ ⊥ ♆
06 58 ☽ ☌ ♄
07 02 ☽ ⚹ ☉
14 45 ☽ ⊥ ☿
15 46 ☽ ∥ ☉
17 23 ☽ ⊻ ♆
18 54 ☽ ∥ ♃
21 51 ☽ ⊻ ♇

17 Thursday
00 56 ☽ ∥ ♆
02 23 ☽ ⊻ ☿
03 51 ☽ ∥ ♅
06 58 ♀ Q ♆
07 43 ☽ ∠ ♃
08 20 ☽ ⊥ ♇
08 25 ☽ ∥ ☿
11 15 ☽ ∠ ☉
15 00 ☽ □ ♀
18 13 ☽ ☌ ♅
21 17 ☽ □ ♂
23 47 ☽ ⊥ ♃

18 Friday
00 09 ☽ ∠ ♇
04 21 ☽ ⊥ ☉
10 34 ☽ ⊻ ♃
11 48 ☽ ⊻ ♄
15 50 ☽ ⊻ ☉
22 17 ☽ ☌ ♆
22 30 ☽ ⊥ ♄

19 Saturday
01 28 ☽ Q ♀
02 46 ☽ ⚹ ♇
07 06 ☽ Q ♂
11 15 ♀ □ ♅
13 38 ☽ ☌ ☿
14 42 ☽ ∠ ♄
23 49 ☽ ⊻ ♅
23 55 ☿ ∠ ♄

20 Sunday
00 20 ☽ Q ♇
00 49 ☽ ⚹ ♀
01 24 ♂ ⅊ ♇
06 02 ☽ ⚹ ♂
10 44 ☽ ⊥ ♅
17 26 ☽ ☌ ♃
18 02 ☽ ⚹ ♄

21 Monday
02 28 ☽ ☌ ☉
03 16 ☽ ∠ ♅
04 43 ☽ ⊻ ♆
06 32 ☽ ∠ ♀
09 15 ☽ □ ♇
11 13 ☽ ∠ ♂
11 38 ☽ ∥ ☿
13 06 ☉ ∠ ♅
15 51 ☽ ⊥ ♆
16 17 ☽ Q ♄
16 38 ☽ ∥ ♅
19 58 ☽ ∥ ♆

22 Tuesday
00 43 ☽ ⊥ ♀
03 31 ☽ ⊻ ☿
03 58 ☽ ∥ ♃
05 04 ☽ ⊥ ♂
07 18 ☽ ⚹ ♅
08 05 ☉ ⊻ ♆
08 45 ☽ ∠ ♆
12 59 ☽ ⊻ ♀
13 42 ☽ ∥ ☉
16 19 ☽ ⊥ ☿
17 07 ☽ ⊻ ♂
17 26 ♃ ⚹ ♄
17 29 ♀ ⅊ ♇

23 Wednesday
02 26 ☽ □ ♄
02 33 ☽ ⊻ ♃
02 46 ☽ ∥ ♄
05 31 ♀ ∥ ♂
06 12 ☽ Q ♅
11 50 ☽ ∠ ☿
13 13 ☿ ⊻ ♅
13 27 ☽ ⚹ ♆
14 19 ☽ ⊥ ♃
15 40 ☿ Q ♇
16 00 ☽ ⊻ ☉
18 06 ☽ △ ♇

24 Thursday
04 43 ☽ ⊥ ☉
08 11 ☽ ∠ ♃
12 55 ☽ Q ♆
17 23 ☽ □ ♅
17 52 ☉ □ ♇
21 13 ☽ ⚹ ☿
23 32 ☽ ⅊ ♇

25 Friday
00 03 ☽ ∠ ☉
04 24 ☽ ☌ ♀
05 27 ♀ # ♇
07 22 ☽ ☌ ♂
13 32 ☽ △ ♄
14 30 ☽ ⚹ ♃
17 30 ☽ ± ♇

26 Saturday
00 14 ☽ ∥ ♂
00 37 ☽ Q ☿
00 43 ☽ # ♇
00 52 ☽ □ ♆
02 41 ☽ ∥ ♀
05 34 ☽ ⊼ ♇
08 07 ♂ # ♇
08 53 ☽ ⚹ ☉
15 08 ☽ Q ♃
19 56 ☽ ⅊ ♄
21 39 ☽ # ♀

27 Sunday
00 27 ☽ # ♂
01 26 ☽ ∥ ♇
05 52 ☽ △ ♅
11 30 ☽ Q ☉
14 18 ☽ ± ♄
15 40 ☿ ⊥ ♅
15 47 ☿ ∥ ♅
18 28 ☽ □ ☿
22 27 ☽ ⊻ ♀

28 Monday
00 07 ☽ ⊻ ♂
02 34 ☽ ⊼ ♄
04 27 ☽ □ ♃
10 25 ☉ ⊥ ♆
11 44 ☽ ⊥ ♀
12 28 ☽ ⅊ ♅
13 09 ☽ ⊥ ♂
13 53 ☽ △ ♆
18 29 ☽ ☍ ♇
21 58 ☉ Q ♄

29 Tuesday
03 29 ☽ □ ☉
06 45 ☽ ± ♅
07 34 ☽ ∠ ♀
08 35 ☽ ∠ ♂
17 39 ♂ △ ♄
18 52 ☽ ⊼ ♅
20 13 ☽ ⅊ ♆

30 Wednesday
01 20 ☽ # ♄
02 59 ☽ # ☉
03 52 ☿ ∥ ♆
04 11 ♀ △ ♄
10 29 ☿ ⚹ ♄
14 10 ☽ ± ♆
15 12 ☽ ☍ ♄
15 52 ☽ △ ☿
16 04 ☽ ⚹ ♀
16 31 ☽ ⚹ ♂
17 53 ☽ △ ♃
19 37 ☿ ⚹ ♀
21 45 ☽ # ♃

31 Thursday
00 55 ☿ ⚹ ♂
01 59 ☽ ⊼ ♆
03 11 ☉ ∥ ♄
06 16 ☽ ⊼ ♇
06 35 ☽ # ☿
08 29 ☽ # ♆
09 51 ☿ ☌ ♃
12 18 ☽ # ♅
17 22 ☽ Q ♀
17 23 ☽ Q ♂
17 51 ☽ ± ♇
18 36 ♀ ☌ ♂
20 06 ☽ △ ☉
22 28 ♀ ⚹ ♃
23 30 ☽ ⅊ ♃
23 49 ♂ ⚹ ♃
23 57 ♀ ± ♇

FEBRUARY 1985

LONGITUDES

Date	Sidereal time h m s	Sun ☉ ° ' "	Moon ☽ ° ' "	Moon ☽ 24.00 ° ' "	Mercury ☿ ° '	Venus ♀ ° '	Mars ♂ ° '	Jupiter ♃ ° '	Saturn ♄ ° '	Uranus ♅ ° '	Neptune ♆ ° '	Pluto ♇ ° '
01	20 46 33	12 ♒ 34 58	20 ♊ 18 08	26 ♊ 44 52	00 ♒ 17	29 ♓ 13	29 ♓ 05	28 ♑ 49	27 ♏ 09	16 ♐ 56	02 ♑ 35	04 ♏ 45
02	20 50 30	13 35 50	03 ♋ 17 51	09 ♋ 57 18	01 52	00 ♈ 08	29 ♓ 50	29 03	27 13	16 58	02 37	04 45
03	20 54 27	14 36 41	16 ♋ 43 15	23 ♋ 35 35	03 29	01 03	00 ♈ 35	29 17	27 16	17 01	02 39	04 45
04	20 58 23	15 37 31	00 ♌ 34 04	07 ♌ 38 13	05 06	01 57	01 20	29 30	27 19	17 03	02 40	04 45
05	21 02 20	16 38 19	14 ♌ 47 26	22 ♌ 00 56	06 43	02 51	02 05	29 44	27 22	17 05	02 42	04 45
06	21 06 16	17 39 06	29 ♌ 17 52	06 ♍ 37 16	08 22	03 44	02 51	29 ♑ 58	27 25	17 08	02 44	04 R 45
07	21 10 13	18 39 52	13 ♍ 58 08	21 ♍ 19 28	10 01	04 36	03 36	00 ♒ 12	27 28	17 10	02 46	04 45
08	21 14 09	19 40 37	28 ♍ 40 19	05 ♎ 59 50	11 41	05 27	04 21	00 25	27 30	17 12	02 47	04 45
09	21 18 06	20 41 21	13 ♎ 17 16	20 ♎ 32 01	13 22	06 18	05 06	00 39	27 33	17 14	02 49	04 45
10	21 22 02	21 42 03	27 ♎ 43 36	04 ♏ 51 42	15 03	07 08	05 51	00 53	27 36	17 16	02 51	04 45
11	21 25 59	22 42 45	11 ♏ 56 06	18 ♏ 56 43	16 46	07 56	06 35	01 06	27 38	17 18	02 52	04 44
12	21 29 56	23 43 25	25 ♏ 53 34	02 ♐ 46 44	18 29	08 45	07 20	01 20	27 41	17 20	02 54	04 44
13	21 33 52	24 44 05	09 ♐ 36 18	16 ♐ 22 27	20 13	09 32	08 05	01 33	27 43	17 22	02 55	04 44
14	21 37 49	25 44 43	23 ♐ 05 20	29 ♐ 45 07	21 58	10 18	08 50	01 47	27 45	17 24	02 57	04 44
15	21 41 45	26 45 21	06 ♑ 21 54	12 ♑ 55 50	23 44	11 03	09 35	02 00	27 47	17 26	02 59	04 43
16	21 45 42	27 45 57	19 ♑ 26 58	25 ♑ 55 21	25 31	11 48	10 20	02 14	27 49	17 28	03 00	04 43
17	21 49 38	28 46 31	02 ♒ 21 01	08 ♒ 43 57	27 18	12 31	11 04	02 27	27 51	17 30	03 02	04 43
18	21 53 35	29 ♒ 47 05	15 ♒ 04 09	21 ♒ 21 34	29 ♒ 07	13 13	11 49	02 41	27 53	17 31	03 03	04 42
19	21 57 31	00 ♓ 47 36	27 ♒ 36 12	03 ♓ 48 04	00 ♓ 56	13 54	12 34	02 54	27 54	17 33	03 04	04 42
20	22 01 28	01 48 06	09 ♓ 57 10	16 ♓ 03 34	02 46	14 34	13 18	03 07	27 56	17 35	03 06	04 41
21	22 05 25	02 48 35	22 ♓ 07 23	28 ♓ 08 47	04 37	15 12	14 03	03 20	27 57	17 36	03 07	04 41
22	22 09 21	03 49 01	04 ♈ 07 57	10 ♈ 05 11	06 29	15 50	14 47	03 33	27 59	17 38	03 09	04 40
23	22 13 18	04 49 26	16 ♈ 00 47	21 ♈ 55 11	08 21	16 26	15 32	03 47	28 00	17 39	03 10	04 39
24	22 17 14	05 49 49	27 ♈ 48 48	03 ♉ 42 08	10 14	17 00	16 16	04 00	28 01	17 41	03 11	04 39
25	22 21 11	06 50 11	09 ♉ 35 45	15 ♉ 30 14	12 08	17 33	17 01	04 13	28 02	17 42	03 12	04 38
26	22 25 07	07 50 30	21 ♉ 26 14	27 ♉ 24 23	14 02	18 05	17 45	04 25	28 03	17 43	03 14	04 38
27	22 29 04	08 50 47	03 ♊ 25 24	09 ♊ 29 58	15 56	18 35	18 29	04 38	28 04	17 45	03 15	04 37
28	22 33 00	09 ♓ 51 03	15 ♊ 38 46	21 ♊ 52 29	17 ♓ 50	19 ♈ 03	19 ♈ 14	04 ♒ 51	28 ♏ 05	17 ♐ 46	03 ♑ 16	04 ♏ 36

Date	Moon True ☊ ° '	Moon Mean ☊ ° '	Moon ☽ Latitude ° '
01	24 ♉ 08	23 ♉ 29	02 N 14
02	24 R 03	23 26	03 11
03	23 55	23 23	04 00
04	23 45	23 19	04 36
05	23 33	23 16	04 57
06	23 20	23 13	04 59
07	23 09	23 10	04 41
08	23 00	23 07	04 05
09	22 53	23 04	03 12
10	22 50	23 00	02 08
11	22 D 49	22 57	00 N 57
12	22 R 49	22 54	00 S 16
13	22 48	22 51	01 28
14	22 46	22 48	02 33
15	22 42	22 45	03 28
16	22 34	22 41	04 12
17	22 24	22 38	04 42
18	22 11	22 35	04 58
19	21 56	22 32	04 58
20	21 42	22 29	04 45
21	21 29	22 25	04 19
22	21 19	22 22	03 41
23	21 11	22 19	02 54
24	21 06	22 16	02 00
25	21 03	22 13	01 S 00
26	21 D 03	22 10	00 N 02
27	21 03	22 06	01 05
28	21 ♉ 03	22 ♉ 03	02 N 07

DECLINATIONS

Date	Sun ☉ ° '	Moon ☽ ° '	Mercury ☿ ° '	Venus ♀ ° '	Mars ♂ ° '	Jupiter ♃ ° '	Saturn ♄ ° '	Uranus ♅ ° '	Neptune ♆ ° '	Pluto ♇ ° '
01	17 S 02	25 N 19	21 S 43	00 N 49	00 S 48	20 S 42	17 S 26	22 S 48	22 S 18	02 N 47
02	16 45	26 35	21 26	01 18	00 30	20 39	17 27	22 49	22 18	02 48
03	16 27	26 22	21 08	01 47	00 S 11	20 36	17 27	22 49	22 18	02 48
04	16 09	24 32	20 48	02 15	00 N 08	20 33	17 28	22 49	22 18	02 49
05	15 51	21 07	20 26	02 44	00 27	20 31	17 28	22 49	22 18	02 49
06	15 33	16 23	20 03	03 12	00 45	20 28	17 29	22 50	22 18	02 50
07	15 14	10 37	19 39	03 40	01 04	20 25	17 29	22 50	22 18	02 50
08	14 55	04 N 16	19 14	04 08	01 23	20 22	17 30	22 50	22 18	02 51
09	14 36	02 S 17	18 47	04 35	01 41	20 19	17 30	22 50	22 18	02 51
10	14 16	08 40	18 18	05 02	02 00	20 16	17 30	22 51	22 18	02 52
11	13 57	14 31	17 48	05 30	02 18	20 14	17 31	22 51	22 18	02 52
12	13 37	19 30	17 17	05 56	02 37	20 11	17 31	22 51	22 17	02 53
13	13 17	23 20	16 44	06 23	02 55	20 08	17 31	22 51	22 17	02 53
14	12 56	25 48	16 10	06 49	03 14	20 05	17 32	22 51	22 17	02 54
15	12 36	26 45	15 34	07 15	03 32	20 02	17 32	22 51	22 17	02 55
16	12 15	26 11	14 57	07 40	03 51	19 59	17 32	22 52	22 17	02 55
17	11 54	24 13	14 19	08 06	04 09	19 56	17 33	22 52	22 17	02 56
18	11 33	21 03	13 39	08 30	04 27	19 53	17 33	22 52	22 17	02 56
19	11 12	16 58	12 57	08 55	04 45	19 50	17 33	22 52	22 17	02 57
20	10 50	12 14	12 15	09 19	05 03	19 47	17 33	22 52	22 17	02 58
21	10 28	07 05	11 31	09 42	05 21	19 45	17 33	22 52	22 17	02 58
22	10 07	01 S 45	10 45	10 05	05 39	19 42	17 33	22 53	22 17	02 59
23	09 45	03 N 37	09 59	10 28	05 57	19 39	17 33	22 53	22 17	03 00
24	09 22	08 50	09 11	10 50	06 15	19 36	17 34	22 53	22 17	03 00
25	09 00	13 44	08 22	11 11	06 33	19 33	17 34	22 53	22 17	03 01
26	08 38	18 09	07 32	11 32	06 51	19 30	17 34	22 53	22 17	03 02
27	08 15	21 55	06 41	11 53	07 08	19 27	17 34	22 53	22 17	03 02
28	07 S 53	24 N 46	05 S 49	12 N 12	07 N 26	19 S 24	17 S 34	22 S 53	22 S 17	03 N 03

ZODIAC SIGN ENTRIES

Date	h	m	Planets
01	07	43	☿ ♒
02	05	59	☽ ♋
02	08	29	♀ ♈
02	17	19	♂ ♈
04	11	02	☽ ♌
06	13	09	☽ ♍
06	15	35	♃ ♒
08	14	10	☽ ♎
10	15	49	☽ ♏
12	19	09	☽ ♐
15	00	27	☽ ♑
17	07	36	☽ ♒
18	17	07	☉ ♓
18	23	41	☿ ♓
19	16	38	☽ ♓
22	03	43	☽ ♈
24	16	27	☽ ♉
27	05	11	☽ ♊

LATITUDES

Date	Mercury ☿ ° '	Venus ♀ ° '	Mars ♂ ° '	Jupiter ♃ ° '	Saturn ♄ ° '	Uranus ♅ ° '	Neptune ♆ ° '	Pluto ♇ ° '
01	01 S 40	01 N 14	00 S 29	00 S 18	02 N 09	00 00	01 N 07	16 N 50
04	01 51	01 36	00 26	00 18	02 09	00 00	01 07	16 51
07	01 59	02 00	00 24	00 19	02 10	00 00	01 07	16 53
10	02 04	02 25	00 21	00 19	02 11	00 00	01 07	16 55
13	02 06	02 50	00 19	00 19	02 11	00 S 01	01 07	16 56
16	02 03	03 16	00 16	00 20	02 12	00 01	01 07	16 58
19	01 56	03 43	00 14	00 20	02 12	00 01	01 07	16 59
22	01 45	04 11	00 11	00 20	02 13	00 01	01 07	17 01
25	01 28	04 39	00 09	00 21	02 13	00 01	01 07	17 02
28	01 06	05 08	00 06	00 21	02 14	00 01	01 08	17 04
31	00 S 39	05 N 36	00 S 04	00 S 22	02 N 14	00 S 01	01 N 08	17 N 05

DATA

Julian Date	2446098
Delta T	+54 seconds
Ayanamsa	23° 38' 43"
Synetic vernal point	05° ♓ 28' 16"
True obliquity of ecliptic	23° 26' 34"

MOON'S PHASES, APSIDES AND POSITIONS ☽

Date	h	m	Phase	Longitude ° '	Eclipse Indicator
05	15	19	○	16 ♌ 47	
12	07	57	☾	23 ♏ 33	
19	18	43	●	01 ♓ 05	
27	23	41	☽	09 ♊ 20	

Day	h	m		
08	03	40	Perigee	
24	04	01	Apogee	
02	20	41	Max dec	26° N 41'
09	03	37	0S	
15	14	53	Max dec	26° S 46'
22	19	46	0N	

LONGITUDES

Date	Chiron ⚷ ° '	Ceres ⚳ ° '	Pallas ⚴ ° '	Juno ⚵ ° '	Vesta ⚶ ° '	Black Moon Lilith ⚸ ° '
01	03 ♊ 18	14 ♉ 37	22 ♓ 10	13 ♎ 14	00 ♏ 54	16 ♈ 39
11	03 ♊ 16	16 ♉ 46	25 ♓ 32	12 ♎ 50	02 ♏ 58	17 ♈ 45
21	03 ♊ 20	19 ♉ 19	29 ♓ 02	11 ♎ 46	04 ♏ 25	18 ♈ 52
31	03 ♊ 32	22 ♉ 12	02 ♈ 41	10 ♎ 07	05 ♏ 08	19 ♈ 59

ASPECTARIAN

h m Aspects

01 Friday
01 10 ☽ ⅊ ☿
01 14 ♂ ± ♇
05 38 ☽ ☍ ♅
10 57 ☽ ⅊ ♇
11 46 ♀ # ♂
16 47 ☽ ± ♃
20 28 ☽ ± ☿

02 Saturday
00 48 ☽ ⊼ ♄
02 41 ☽ ⅊ ☉
04 05 ☽ ⊼ ♃
05 17 ☽ □ ♂
05 48 ☽ □ ♀
09 03 ☽ ⊼ ☿
10 45 ☽ ☍ ♆
11 50 ☽ ± ♄
13 32 ☿ ∠ ♅
14 38 ☽ △ ♇
20 25 ☽ ± ☉
23 17 ☿ ⊻ ♆

03 Sunday
04 05 ☽ ⅊ ♄
07 59 ☽ ⊼ ☉
12 31 ☽ ⊼ ♅
23 02 ☽ ± ♅

04 Monday
06 24 ☽ △ ♄
06 52 ☿ □ ♇
10 09 ☽ ☍ ♃
13 23 ☽ △ ♂
14 32 ☽ △ ♀
14 32 ☽ ⅊ ♅
15 36 ☽ ⊼ ♆
19 07 ☽ □ ♇
20 42 ☽ ☍ ☿

05 Tuesday
01 19 ☽ # ♅
01 46 ☽ ± ♆
04 48 ☽ # ♆
06 32 ☿ ∥ ♃
07 58 ♀ □ ♆
15 19 ☽ ☍ ☉
15 31 ☽ # ♃
15 51 ☽ △ ♅
16 03 ☽ ⅊ ♂
16 15 ☽ # ☿
16 48 ♀ ∥ ♇
16 52 ☽ ⅊ ♆
17 25 ☽ ⅊ ♀
23 09 ☉ ✶ ♅
23 58 ♇ St R

06 Wednesday
01 13 ☽ Q ♇
06 56 ☽ # ♄
07 45 ☽ ± ♂
08 19 ♂ □ ♆
08 54 ☽ □ ♄
09 15 ☽ ± ♀
13 07 ☽ ⊼ ♃
13 57 ☉ ∠ ♆
15 55 ☽ # ☉
17 27 ☿ ⊥ ♆
17 39 ☽ △ ♆
18 08 ☽ ⊼ ♂
19 44 ☽ ⊼ ♀
20 56 ☽ ✶ ♇
23 06 ☽ ± ♃

07 Thursday
03 43 ☿ Q ♄
04 44 ☽ ⊼ ☿
05 27 ☉ ∠ ♂
14 02 ☽ ⅊ ♃
14 27 ☽ Q ♄
15 46 ☽ ± ☿
16 13 ♀ ⊼ ♇
17 14 ☽ □ ♅
20 14 ☽ ⊼ ☉
21 26 ☽ ∠ ♇

08 Friday
06 45 ☽ ± ☉
08 20 ☽ ⅊ ☿
10 05 ☽ ✶ ♄
12 07 ☽ ⊥ ♇
12 29 ☽ ∥ ♀
14 55 ☽ △ ♃
17 13 ☽ ∥ ♇
18 45 ☽ □ ♆
21 47 ☽ ☍ ♂
21 57 ☽ ⊻ ♇
22 06 ☽ ∥ ♂
22 34 ☽ ⅊ ☉
22 43 ☽ Q ♅
23 47 ☽ ☍ ♀

09 Saturday
00 53 ♂ ⊼ ♇
09 41 ☽ # ♂
10 47 ☽ ∠ ♄
12 08 ☽ △ ☿
14 04 ☽ # ♇
18 33 ☽ ✶ ♅
21 08 ☽ # ♀

10 Sunday
00 30 ☽ Q ♆
01 11 ☽ △ ☉
01 44 ☽ ⊥ ♄
11 47 ☽ ⊻ ♄
17 22 ☽ □ ♃
19 39 ☽ ∠ ♅
20 37 ☽ ✶ ♆
23 48 ☽ ☌ ♇

11 Monday
02 25 ☽ ⊼ ♂
04 48 ☽ ⊼ ♀
09 41 ☽ ∥ ☉
10 56 ☽ ⊥ ♅
13 11 ☽ ± ♂
15 38 ☽ ± ♀
19 44 ☿ ✶ ♅
21 12 ☽ ⊻ ♅
21 25 ☽ □ ☿
22 10 ☽ ∠ ♆

12 Tuesday
00 29 ☽ Q ♃
01 16 ☿ ∥ ♄
01 49 ☽ ∥ ☿
01 52 ☽ ∥ ♄
03 42 ☿ ∠ ♆
05 30 ☽ ⅊ ♂
07 57 ☽ □ ☉
08 03 ☽ ⅊ ♀
13 45 ☽ ⊥ ♆
14 04 ☉ ∠ ♀
15 06 ☽ ☌ ♄
15 43 ☽ ∥ ♃
21 38 ☽ ✶ ♃

13 Wednesday
00 14 ☽ ⊻ ♆
03 26 ☽ ⊻ ♇
04 37 ☽ ∥ ♆
08 27 ☽ ∥ ♅
09 10 ☽ △ ♂
09 12 ☽ Q ☿
09 22 ♂ ∥ ♇
11 51 ☽ △ ♀
13 59 ☽ ⊥ ♇
17 59 ☽ Q ☉

14 Thursday
00 32 ☽ ∠ ♃
01 49 ☽ ☌ ♅
05 59 ☽ ∠ ♇
09 42 ☽ ✶ ☿
16 55 ☽ ⊥ ♃
17 10 ☽ ✶ ☉
20 24 ☽ ⊻ ♄

15 Friday
03 57 ☽ ⊻ ♃
05 50 ☽ ☌ ♆
07 18 ☽ ⊥ ♄
09 01 ☽ ✶ ♇
17 00 ☽ ∠ ☿
18 13 ☽ □ ♂
21 05 ☽ □ ♀
22 40 ☽ ∠ ☉
23 46 ☽ ∠ ♄

16 Saturday
06 58 ☽ Q ♇
07 39 ☿ ∠ ♂
08 20 ☽ ⊻ ♅
12 08 ☽ ⊥ ☿
13 17 ☉ □ ♄
16 39 ☽ ⊥ ☉
19 27 ☽ ⊥ ♅

17 Sunday
01 03 ☽ ⊻ ☿
03 34 ☽ ✶ ♄
04 45 ☽ ⊻ ☉
05 29 ☽ Q ♂
08 21 ☽ Q ♀
12 12 ☽ ☌ ♃
12 16 ☽ ∠ ♅
13 16 ☽ ⊻ ♆
16 25 ☽ □ ♇
16 31 ☿ ∠ ♀
17 03 ♃ ∠ ♅
19 21 ☿ □ ♄
23 15 ☽ ∥ ♅
23 58 ♀ ⅊ ♄

18 Monday
00 35 ☽ ⊥ ♆
02 09 ☽ Q ♄
03 33 ☽ ∥ ♆
05 26 ☽ ✶ ♂
05 36 ☉ Q ♅
08 16 ☽ ✶ ♀
16 41 ☽ ✶ ♅
17 29 ☿ Q ♅
17 41 ☽ ∠ ♆
19 26 ☽ ∥ ♃

19 Tuesday
07 49 ☉ ☌ ☿
08 51 ☽ ∥ ♄
11 55 ☽ ∠ ♂
12 35 ☽ □ ♄
14 38 ☽ ∠ ♀
15 46 ☽ Q ♅
18 43 ☽ ☌ ☉
19 34 ☽ ☌ ☿
22 26 ☽ ⊻ ♃
22 37 ☽ ✶ ♆
23 40 ♂ ⅊ ♄

20 Wednesday
01 44 ☽ △ ♇
06 29 ☽ ⊥ ♂
09 07 ☽ ⊥ ♀
09 30 ♃ ⊻ ♆
10 20 ☽ ⊥ ♃
11 56 ☽ ∥ ☿
16 18 ☿ ✶ ♆
17 08 ☿ ⊻ ♃
19 00 ☽ ⊻ ♂
19 11 ☽ ∥ ☉
21 33 ☽ ⊻ ♀
22 07 ☽ Q ♆

21 Thursday
00 49 ☽ # ♀
03 02 ☽ □ ♅
04 22 ☽ ∠ ♃
07 09 ☽ ⅊ ♇
12 44 ☿ △ ♇
19 25 ☽ # ♂
19 35 ☉ ✶ ♆
19 52 ♀ Q ♃
23 39 ☽ △ ♄

22 Friday
01 03 ☽ ± ♇
04 05 ☉ ⊻ ♃
06 28 ☽ # ♇
10 00 ☽ □ ♆
10 49 ☽ ✶ ♃
11 18 ☽ ⊻ ☉
12 44 ☉ # ♀
13 05 ☽ ⊼ ♇
17 36 ☽ ⊻ ☿

23 Saturday
00 31 ☽ ⊥ ☉
02 00 ☿ # ♀
05 53 ☽ ⅊ ♄
08 00 ☽ ⊥ ☿
08 05 ☉ △ ♇
09 11 ☽ ∥ ♇
10 57 ☽ ☌ ♂
11 31 ☽ Q ♃
12 53 ☽ ☌ ♀
15 20 ☽ △ ♅
20 28 ☽ ∠ ☉
23 16 ☽ ∥ ♂
23 19 ♂ Q ♃

24 Sunday
00 11 ☽ ± ♄
01 33 ☉ ∥ ☿
05 45 ☽ ∠ ☿
08 29 ☿ ⊥ ♃
12 26 ☽ ⊼ ♄
12 40 ☿ ⊥ ♂
13 27 ☽ # ☿
14 24 ☽ # ☉
21 56 ☽ ⅊ ♅
22 19 ☽ ∥ ♀
22 58 ☽ △ ♆

25 Monday
00 50 ☽ □ ♃
01 46 ☿ ⊥ ♀
01 55 ☽ ☍ ♇
05 52 ☽ ✶ ☉
16 17 ☽ ± ♅
18 08 ☽ ✶ ☿
18 49 ♀ △ ♅

26 Tuesday
04 03 ☽ ⊻ ♂
04 29 ☽ ⊼ ♅
04 55 ☽ ⊻ ♀
05 30 ☽ ⅊ ♆
08 29 ☽ Q ☉
08 34 ☽ # ♄
11 05 ♂ △ ♅
16 58 ☽ ⊥ ♂
17 34 ☽ ⊥ ♀
19 55 ☽ # ♃
22 59 ☽ Q ☿
23 40 ☽ ± ♆

27 Wednesday
01 19 ☽ ☍ ♄
02 39 ☿ # ♂
03 17 ☿ Q ♆
09 16 ♃ □ ♇
11 39 ☽ ⊼ ♆
12 08 ☽ ∠ ♂
12 20 ☽ ∠ ♀
14 21 ☽ ⊼ ♇
14 27 ☽ △ ♃
14 40 ☽ # ♆
19 19 ☽ # ♅
20 55 ♀ ☌ ♂
23 41 ☽ □ ☉

28 Thursday
02 10 ☽ ± ♇
11 03 ☿ □ ♅
16 06 ☽ ☍ ♅
17 01 ☽ □ ☿
18 51 ☽ ✶ ♀
19 21 ☽ ✶ ♂
19 38 ☽ ⅊ ♇
20 16 ☽ ⅊ ♃

All ephemeris data is given at 12.00 UT and the Moon's longitude is additionally given for 24.00 UT

MARCH 1985

LONGITUDES

Date	Sidereal time h m s	Sun ☉ ° ' "	Moon ☽ ° ' "	Moon ☽ 24.00 ° ' "	Mercury ☿ ° '	Venus ♀ ° '	Mars ♂ ° '	Jupiter ♃ ° '	Saturn ♄ ° '	Uranus ♅ ° '	Neptune ♆ ° '	Pluto ♇ ° '
01	22 36 57	10 ♓ 51 16	28 ♊ 11 44	04 ♋ 37 07	19 ♓ 45	19 ♈ 30	19 ♈ 58	05 ♒ 04	28 ♏ 06	17 ♐ 47	03 ♑ 17	04 ♏ 35
02	22 40 54	11 51 27	11 ♋ 09 06	17 ♋ 48 05	21 39	19 55	20 42	05 17	28 06	17 48	03 18	04 R 34
03	22 44 50	12 51 36	24 ♋ 34 19	01 ♌ 27 53	23 32	20 18	21 26	05 29	28 07	17 49	03 19	04 34
04	22 48 47	13 51 43	08 ♌ 28 42	15 ♌ 36 29	25 24	20 40	22 10	05 42	28 07	17 50	03 20	04 33
05	22 52 43	14 51 49	22 ♌ 50 43	00 ♍ 10 42	27 16	20 59	22 54	05 54	28 07	17 51	03 21	04 32
06	22 56 40	15 51 52	07 ♍ 35 32	15 ♍ 04 09	29 ♓ 05	21 17	23 38	06 07	28 08	17 52	03 22	04 31
07	23 00 36	16 51 53	22 ♍ 35 24	00 ♎ 08 00	00 ♈ 53	21 32	24 22	06 19	28 08	17 53	03 23	04 30
08	23 04 33	17 51 52	07 ♎ 40 43	15 ♎ 12 01	02 38	21 45	25 06	06 31	28 R 08	17 54	03 24	04 29
09	23 08 29	18 51 49	22 ♎ 41 47	00 ♏ 08 01	04 20	21 56	25 50	06 44	28 07	17 54	03 25	04 28
10	23 12 26	19 51 45	07 ♏ 30 16	14 ♏ 47 54	05 59	22 05	26 34	06 56	28 07	17 55	03 26	04 27
11	23 16 23	20 51 39	22 ♏ 00 27	29 ♏ 07 39	07 34	22 12	27 18	07 08	28 07	17 56	03 27	04 26
12	23 20 19	21 51 31	06 ♐ 09 22	13 ♐ 05 35	09 04	22 16	28 02	07 20	28 06	17 56	03 28	04 25
13	23 24 16	22 51 22	19 ♐ 56 26	26 ♐ 42 05	10 29	22 18	28 45	07 32	28 06	17 57	03 28	04 24
14	23 28 12	23 51 11	03 ♑ 22 49	09 ♑ 58 54	11 49	22 R 17	29 ♈ 29	07 44	28 05	17 57	03 29	04 22
15	23 32 09	24 50 59	16 ♑ 30 39	22 ♑ 58 22	13 02	22 14	00 ♉ 13	07 55	28 04	17 58	03 30	04 21
16	23 36 05	25 50 45	29 ♑ 22 22	05 ♒ 42 55	14 10	22 09	00 56	08 07	28 04	17 58	03 31	04 20
17	23 40 02	26 50 29	12 ♒ 00 18	18 ♒ 14 45	15 10	22 01	01 40	08 19	28 03	17 58	03 31	04 19
18	23 43 58	27 50 11	24 ♒ 26 28	00 ♓ 35 41	16 03	21 50	02 23	08 30	28 01	17 59	03 32	04 18
19	23 47 55	28 49 51	06 ♓ 42 33	12 ♓ 47 13	16 49	21 37	03 07	08 42	28 00	17 59	03 32	04 16
20	23 51 52	29 ♓ 49 30	18 ♓ 49 52	24 ♓ 50 38	17 27	21 22	03 50	08 53	27 59	17 59	03 33	04 15
21	23 55 48	00 ♈ 49 06	00 ♈ 49 42	06 ♈ 47 13	17 57	21 04	04 33	09 04	27 58	17 59	03 33	04 14
22	23 59 45	01 48 40	12 ♈ 43 24	18 ♈ 38 28	18 19	20 44	05 17	09 15	27 56	17 59	03 34	04 12
23	00 03 41	02 48 13	24 ♈ 32 40	00 ♉ 26 18	18 33	20 21	06 00	09 26	27 55	17 R 59	03 34	04 11
24	00 07 38	03 47 43	06 ♉ 19 42	12 ♉ 13 15	18 39	19 57	06 43	09 37	27 53	17 59	03 35	04 10
25	00 11 34	04 47 11	18 ♉ 07 20	24 ♉ 02 26	18 R 38	19 30	07 26	09 48	27 51	17 59	03 35	04 08
26	00 15 31	05 46 36	29 ♉ 59 03	05 ♊ 57 42	18 28	19 01	08 10	09 59	27 50	17 59	03 35	04 07
27	00 19 27	06 46 00	11 ♊ 58 57	18 ♊ 03 24	18 12	18 31	08 53	10 10	27 48	17 59	03 36	04 05
28	00 23 24	07 45 21	24 ♊ 11 39	00 ♋ 24 19	17 49	17 58	09 36	10 20	27 46	17 58	03 36	04 04
29	00 27 21	08 44 40	06 ♋ 42 00	13 ♋ 05 16	17 20	17 25	10 19	10 31	27 44	17 58	03 36	04 02
30	00 31 17	09 43 57	19 ♋ 34 39	26 ♋ 10 37	16 46	16 50	11 02	10 41	27 41	17 58	03 36	04 01
31	00 35 14	10 ♈ 43 11	02 ♌ 53 32	09 ♌ 43 39	16 ♈ 06	16 ♈ 13	11 ♉ 45	10 ♒ 51	27 ♏ 39	17 ♐ 57	03 ♑ 37	03 ♏ 59

DECLINATIONS

Date	Moon True ☊ ° '	Moon Mean ☊ ° '	Moon ☽ Latitude ° '	Sun ☉ ° '	Moon ☽ ° '	Mercury ☿ ° '	Venus ♀ ° '	Mars ♂ ° '	Jupiter ♃ ° '	Saturn ♄ ° '	Uranus ♅ ° '	Neptune ♆ ° '	Pluto ♇ ° '
01	21 ♉ 01	22 ♉ 00	03 N 04	07 S 30	26 N 30	04 S 57	12 N 31	07 N 43	19 S 21	17 S 34	22 S 54	22 S 17	03 N 04
02	20 R 58	21 57	03 54	07 07	26 51	04 03	12 50	08 01	19 18	17 34	22 54	22 16	03 04
03	20 52	21 54	04 32	06 44	25 40	03 10	13 07	08 18	19 15	17 33	22 54	22 16	03 05
04	20 44	21 51	04 56	06 21	22 54	02 16	13 24	08 35	19 12	17 33	22 54	22 16	03 06
05	20 34	21 47	05 03	05 58	18 40	01 21	13 40	08 52	19 09	17 33	22 54	22 16	03 06
06	20 23	21 44	04 50	05 35	13 12	00 S 27	13 55	09 09	19 06	17 33	22 54	22 16	03 07
07	20 14	21 41	04 17	05 11	06 52	00 N 26	14 10	09 26	19 03	17 33	22 54	22 16	03 08
08	20 06	21 38	03 25	04 48	00 N 06	01 20	14 23	09 43	19 00	17 33	22 54	22 16	03 09
09	20 01	21 35	02 20	04 24	06 S 40	02 12	14 35	09 59	18 57	17 32	22 54	22 16	03 09
10	19 58	21 31	01 N 06	04 01	12 59	03 03	14 47	10 16	18 54	17 32	22 54	22 16	03 10
11	19 D 57	21 28	00 S 11	03 37	18 27	03 52	14 57	10 32	18 51	17 32	22 54	22 16	03 11
12	19 58	21 25	01 26	03 14	22 45	04 40	15 06	10 49	18 49	17 32	22 55	22 16	03 11
13	19 59	21 22	02 33	02 50	25 37	05 26	15 15	11 05	18 46	17 31	22 55	22 16	03 12
14	19 R 58	21 19	03 31	02 26	26 55	06 09	15 22	11 21	18 43	17 31	22 55	22 16	03 13
15	19 56	21 16	04 16	02 03	26 39	06 50	15 27	11 37	18 40	17 31	22 55	22 16	03 14
16	19 51	21 12	04 47	01 39	24 57	07 28	15 32	11 53	18 37	17 30	22 55	22 16	03 14
17	19 44	21 09	05 03	01 15	22 03	08 02	15 35	12 09	18 34	17 30	22 55	22 16	03 15
18	19 36	21 06	05 05	00 52	18 10	08 34	15 36	12 25	18 31	17 30	22 55	22 16	03 16
19	19 26	21 03	04 52	00 28	13 34	09 01	15 37	12 40	18 28	17 29	22 55	22 16	03 16
20	19 16	21 00	04 27	00 S 04	08 30	09 25	15 36	12 56	18 26	17 29	22 55	22 16	03 17
21	19 07	20 57	03 49	00 N 20	03 S 11	09 45	15 33	13 11	18 23	17 28	22 55	22 16	03 18
22	19 00	20 53	03 02	00 43	02 N 14	10 01	15 29	13 26	18 20	17 28	22 55	22 16	03 19
23	18 54	20 50	02 08	01 07	07 32	10 13	15 25	13 41	18 17	17 27	22 55	22 15	03 19
24	18 51	20 47	01 07	01 31	12 34	10 21	15 16	13 56	18 14	17 27	22 55	22 15	03 20
25	18 50	20 44	00 S 04	01 54	17 10	10 24	15 08	14 11	18 12	17 26	22 55	22 15	03 21
26	18 D 51	20 41	01 N 00	02 18	21 08	10 23	14 58	14 26	18 09	17 26	22 55	22 15	03 21
27	18 52	20 37	02 02	02 41	24 15	10 17	14 46	14 40	18 06	17 25	22 55	22 15	03 22
28	18 54	20 34	03 00	03 05	26 19	10 08	14 33	14 54	18 03	17 25	22 55	22 15	03 23
29	18 55	20 31	03 51	03 28	27 07	09 55	14 18	15 09	18 01	17 24	22 55	22 15	03 24
30	18 R 54	20 28	04 32	03 51	26 30	09 38	14 03	15 23	17 58	17 23	22 55	22 15	03 24
31	18 ♉ 52	20 ♉ 25	04 N 59	04 N 15	24 N 22	09 N 17	13 N 45	15 N 37	17 S 55	17 S 23	22 S 55	22 S 15	03 N 25

ZODIAC SIGN ENTRIES

Date	h	m	Planets
01	15	23	☽ ♋
03	21	28	☽ ♌
05	23	43	☽ ♍
07	00	07	☿ ♈
07	23	47	☽ ♎
09	23	47	☽ ♏
12	01	29	☽ ♐
14	05	36	☽ ♑
15	05	06	♂ ♉
16	13	11	☽ ♒
18	22	50	☽ ♓
20	16	14	☉ ♈
21	10	20	☽ ♈
23	20	06	☽ ♉
26	12	02	☽ ♊
28	23	13	☽ ♋
31	06	51	☽ ♌

LATITUDES

Date	Mercury ☿ ° '	Venus ♀ ° '	Mars ♂ ° '	Jupiter ♃ ° '	Saturn ♄ ° '	Uranus ♅ ° '	Neptune ♆ ° '	Pluto ♇ ° '
01	00 S 58	05 N 17	00 S 06	00 S 21	02 N 14	00 S 01	01 N 08	17 N 04
04	00 28	05 46	00 03	00 22	02 15	00 01	01 08	17 05
07	00 N 06	06 13	00 S 01	00 22	02 15	00 01	01 08	17 07
10	00 44	06 40	00 N 01	00 23	02 16	00 01	01 08	17 08
13	01 23	07 05	00 03	00 23	02 16	00 01	01 08	17 09
16	02 02	07 27	00 06	00 23	02 17	00 01	01 08	17 10
19	02 37	07 46	00 08	00 24	02 17	00 01	01 08	17 11
22	03 04	07 59	00 10	00 24	02 18	00 01	01 08	17 12
25	03 21	08 07	00 12	00 25	02 18	00 01	01 08	17 13
28	03 24	08 08	00 14	00 25	02 19	00 01	01 08	17 14
31	03 N 12	08 N 00	00 N 16	00 S 26	02 N 19	00 S 01	01 N 08	17 N 15

DATA

Julian Date	2446126
Delta T	+54 seconds
Ayanamsa	23° 38' 47"
Synetic vernal point	05° ♓ 28' 13"
True obliquity of ecliptic	23° 26' 34"

MOON'S PHASES, APSIDES AND POSITIONS ☽

Date	h	m	Phase	Longitude ° '	Eclipse Indicator
07	02	13	○	16 ♍ 27	
13	17	34	☾	23 ♐ 05	
21	11	59	●	00 ♈ 49	
29	16	11	☽	08 ♋ 55	

Day	h	m		
08	07	57	Perigee	
23	15	15	Apogee	
02	05	49	Max dec	26° N 54'
08	12	21	0S	
14	19	52	Max dec	27° S 00'
22	02	06	0N	
29	13	49	Max dec	27° N 07'

LONGITUDES

Date	Chiron ⚷ ° '	Ceres ⚳ ° '	Pallas ⚴ ° '	Juno ⚵ ° '	Vesta ⚶ ° '	Black Moon Lilith ⚸ ° '
01	03 ♊ 29	21 ♉ 36	01 ♈ 56	10 ♎ 29	05 ♏ 03	19 ♈ 46
11	03 ♊ 46	24 ♉ 42	05 ♈ 40	08 ♎ 27	05 ♏ 09	20 ♈ 52
21	04 ♊ 10	28 ♉ 03	09 ♈ 31	06 ♎ 06	04 ♏ 26	21 ♈ 59
31	04 ♊ 39	01 ♊ 36	13 ♈ 26	03 ♎ 41	02 ♏ 56	23 ♈ 05

All ephemeris data is given at 12.00 UT and the Moon's longitude is additionally given for 24.00 UT

ASPECTARIAN

h m	Aspects
01 Friday	
04 03	☉ # ♂
08 06	☿ ⊻ ♀
10 04	☿ P ♇
11 49	☽ ⊼ ♄
13 40	☽ ± ♃
16 33	☿ ⊻ ♂
16 35	☿ ∠ ♃
18 24	☉ ⊥ ♃
18 25	☽ Q ♀
19 30	☽ Q ♂
21 32	☽ ☍ ♆
23 02	☽ ± ♄
23 56	☽ △ ♇
02 Saturday	
01 02	☽ ⊼ ♃
13 23	☽ △ ☉
15 33	☽ P ♄
03 Sunday	
00 01	☽ ⊼ ♅
04 15	☽ □ ♀
06 09	☽ □ ♂
09 53	☽ △ ☿
10 40	☽ ± ♅
14 02	☿ # ♇
18 12	☽ △ ♄
18 12	☽ P ☉
04 Monday	
02 21	☽ P ♅
03 13	☽ ⊼ ♆
05 18	☽ □ ♇
07 12	☽ ☍ ♃
10 16	♂ ± ♄
10 52	☽ ± ☉
12 04	☽ # ♅
13 28	☽ ± ♆
15 45	☽ P ☿
16 10	☽ # ♆
21 46	☽ ⊼ ☉
05 Tuesday	
03 44	☽ △ ♅
04 34	☽ P ♆
08 52	☽ △ ♀
09 01	☽ ± ☿
09 34	☽ # ♃
11 29	☽ Q ♇
12 06	☽ △ ♂
16 16	☉ ⊥ ♀
17 18	☽ # ♄
20 17	☽ ⊼ ☿
20 39	☽ □ ♄
23 16	☿ △ ♄
06 Wednesday	
00 03	☉ Q ♆
04 28	☿ ± ♇
05 11	☽ △ ♆
07 02	☽ ⚹ ♇
09 11	☽ ∥ ♀
09 35	☽ ⊼ ♃
09 50	☽ P ♀
13 46	☽ P ♂
19 22	☽ ± ♃
07 Thursday	
00 33	☽ ± ♀
01 41	☽ Q ♄
02 13	☽ ☍ ☉
02 58	☽ ∥ ♂
04 29	☽ □ ♅
04 56	☽ ± ♂
07 05	☽ ∠ ♇
09 57	☽ P ♃
10 17	☽ ⊼ ♀
12 38	♄ St R
14 59	☽ ⊼ ♂
18 25	☽ # ☉
20 49	☽ ⚹ ♄
21 23	☽ ⊥ ♇
08 Friday	
01 19	☽ ∥ ♇
02 56	☽ ☍ ☿
05 12	☽ □ ♆
06 55	☽ ⊻ ♇
08 11	☽ ∥ ☿
09 09	☽ Q ♅
10 08	☽ △ ♃
12 42	☉ □ ♅
17 45	☽ # ☿
20 41	☽ ∠ ♄
22 51	☿ □ ♆
23 26	☽ # ♇
09 Saturday	
04 19	☽ ⚹ ♅
04 21	☽ ∥ ☉
05 25	☽ ⊼ ☉
09 57	☽ Q ♆
10 46	☽ ☍ ♀
11 05	☽ ⊥ ♄
13 47	☿ ⊼ ♇
15 44	☽ ± ☉
17 19	☽ ☍ ♂
20 45	☽ ⊻ ♄
10 Sunday	
00 55	☽ # ♂
02 12	☉ P ♇
04 31	☽ ∠ ♅
05 21	☽ ⚹ ♆
07 01	☽ ☌ ♇
07 22	☽ P ☉
09 12	☽ ⊼ ☿
11 03	☽ □ ♃
15 29	☿ ∥ ♇
18 15	♀ ± ♄
19 15	☽ ⊥ ♅
19 41	☽ # ♀
20 16	☽ ± ☿
11 Monday	
04 20	☿ ⚹ ♃
05 11	☽ ⊻ ♅
06 03	☽ ∠ ♆
07 02	☉ # ☿
07 39	☽ ∥ ♄
09 56	☽ △ ☉
12 19	☽ ⊼ ♀
13 03	☽ P ☿
13 59	☽ ∥ ♃
17 19	☽ Q ♃
21 10	☽ ⊥ ♆
21 23	☽ ⊼ ♂
22 17	☽ ☌ ♄
22 29	☽ ± ♀
12 Tuesday	
07 23	☽ ⊻ ♆
08 09	☽ ± ♂
08 54	☽ ∥ ♆
09 01	☽ ⊻ ♇
13 06	☽ ∥ ♅
13 55	☽ P ♀
14 03	☽ ⚹ ♃
14 21	☉ # ♇
14 37	♂ ⊼ ♄
17 36	☽ △ ☿
19 20	☽ ⊥ ♇
22 11	☉ ⊻ ♀
13 Wednesday	
00 33	☽ P ♂
02 11	☉ ∠ ♃
02 51	☉ ⊥ ♂
08 29	☽ ☌ ♅
11 02	☽ ∠ ♇
16 10	☽ △ ♀
16 39	☽ ∠ ♃
17 34	☽ □ ☉
18 17	♀ St R
14 Thursday	
02 29	☽ ⊻ ♄
04 34	☽ △ ♂
08 58	☽ ⊥ ♃
12 12	☽ ☌ ♆
13 16	☽ ⊥ ♄
13 48	☽ ⚹ ♇
20 00	☽ ⊻ ♃
15 Friday	
04 42	☽ Q ☉
04 58	☽ □ ☿
05 40	☽ ∠ ♄
11 43	☽ Q ♇
12 41	☿ P ♄
14 41	☽ ⊻ ♅
22 33	☽ □ ♀
16 Saturday	
01 51	☽ ⊥ ♅
04 49	☽ ⚹ ☉
09 32	☽ ⚹ ♄
15 08	☽ □ ♂
17 44	☽ Q ☿
18 48	☽ ∠ ♅
19 49	☽ ⊻ ♆
21 22	☽ □ ♇
17 Sunday	
04 50	☽ ☌ ♃
05 42	☽ ∥ ♅
06 12	♀ ± ♄
07 15	☽ ⊥ ♆
08 14	☽ Q ♀
08 15	☽ Q ♄
10 27	☽ ∥ ♆
11 40	☽ ∠ ☉
18 33	☽ ⚹ ☿
23 29	☽ ⚹ ♅
18 Monday	
00 32	☽ ∠ ♆
03 39	☽ Q ♂
06 30	☽ ⊥ ☉
07 01	☽ ⚹ ♀
09 57	☽ ∥ ♃
15 42	☽ ∥ ♄
16 28	☉ △ ♄
18 58	☽ □ ♄
19 12	☽ ⊻ ☉
22 47	☉ ± ♇
22 48	☽ Q ♅
19 Tuesday	
01 42	☽ # ♀
01 47	☽ ∠ ☿
04 29	☽ ⚹ ♂
05 46	☽ ⚹ ♆
07 13	☽ △ ♇
07 41	♂ P ♅
11 50	☽ ∠ ♀
15 58	☽ ⊻ ♃
16 11	☽ # ♂
20 34	☽ ⊥ ☿
20 Wednesday	
02 26	♂ △ ♆
04 02	☽ ⊥ ♃
05 16	☽ ⊥ ♀
05 28	☽ Q ♆
08 04	☽ # ☿
09 06	☽ ⊻ ☿
10 19	☽ □ ♅
12 00	☽ ∠ ♂
12 50	☽ P ♇
16 56	☽ ⊻ ♀
22 14	☽ ∠ ♃
21 Thursday	
01 23	♂ ☍ ♇
06 15	☽ △ ♄
06 47	☽ ± ♇
07 09	☽ ⊥ ♂
11 29	☽ # ♇
11 53	♀ Q ♃
11 59	☽ ☌ ☉
14 11	☿ △ ♅
17 30	☽ □ ♆
18 50	☽ ⊼ ♇
19 59	☽ ⊻ ♂
23 48	☽ # ☉
22 Friday	
04 47	☽ ∥ ☉
04 53	☽ ⚹ ♃
12 26	☽ P ♄
16 51	☽ ∥ ♇
22 04	♅ St R
22 40	☽ △ ♅
23 36	☽ ☌ ☿
23 Saturday	
03 45	☽ ☌ ♀
05 35	☽ Q ♃
06 40	☽ ± ♄
18 50	☽ ⊼ ♄
24 Sunday	
00 56	☽ ∥ ☿
05 11	☽ P ♅
05 58	♃ ⊥ ♆
06 22	☽ ⊻ ☉
06 24	☽ △ ♆
06 45	☉ □ ♆
07 35	☽ ☍ ♇
12 51	☽ ☌ ♂
18 49	☽ □ ♃
19 01	☿ St R
19 13	☽ ∥ ♂
19 43	☽ ⊥ ☉
20 37	☉ ⊼ ♇
23 31	☽ ± ♅
25 Monday	
01 21	☽ ∥ ♀
11 43	☽ ⊼ ♅
12 56	☽ P ♆
13 01	☽ ⊻ ☿
13 30	☽ # ♄
14 41	☽ ⊻ ♀
15 41	☽ ∠ ☉
17 45	☽ # ♃
17 51	♃ Q ♄
26 Tuesday	
01 03	☽ ⊥ ☿
02 23	☽ ⊥ ♀
07 10	☽ ± ♆
07 40	☽ ☍ ♄
18 53	☽ ∠ ☿
19 15	☽ ⊼ ♆
19 47	☽ ∠ ♀
19 52	☽ # ♆
20 17	☽ ⊼ ♇
27 Wednesday	
00 41	☽ ⚹ ☉
00 49	☽ # ♅
05 26	☽ ⊻ ♂
08 15	☽ ± ♇
08 20	☽ △ ♃
17 15	♀ ∥ ♂
18 06	☽ ⊥ ♂
23 50	☽ ☍ ♅
23 56	☽ ⚹ ☿
28 Thursday	
00 21	☽ ⚹ ♀
02 00	☽ P ♇
02 34	☽ Q ☉
03 10	☿ △ ♅
11 59	♀ △ ♅
12 50	☽ ∠ ♂
14 15	☽ P ♃
18 53	☽ ⊼ ♄
22 30	☽ Q ☿
22 41	☽ Q ♀
29 Friday	
06 07	☽ ☍ ♆
06 22	☽ ± ♄
06 58	☽ △ ♇
07 11	☉ ∥ ♇
07 48	☽ ± ♃
16 11	☽ □ ☉
19 13	☽ ⚹ ♂
19 17	☽ ⊼ ♃
20 50	♂ □ ♃
23 18	☽ P ♄
30 Saturday	
07 03	☽ □ ☿
07 09	☽ □ ♀
09 02	☽ ⊼ ♅
18 40	☽ Q ♂
19 59	☽ ± ♅
31 Sunday	
02 41	☽ △ ♄
12 07	☽ P ♅
13 16	☽ ⊼ ♆
13 56	☽ □ ♇
16 03	☉ ⚹ ♃
18 56	♂ ± ♅
22 54	☽ # ♅
23 48	☽ ± ♆

APRIL 1985

LONGITUDES

Date	Sidereal time h m s	Sun ☉ ° ' "	Moon ☽ ° ' "	Moon ☽ 24.00 ° ' "	Mercury ☿ ° '	Venus ♀ ° '	Mars ♂ ° '	Jupiter ♃ ° '	Saturn ♄ ° '	Uranus ♅ ° '	Neptune ♆ ° '	Pluto ♇ ° '
01	00 39 10	11 ♈ 42 23	16 ♌ 41 05	23 ♌ 45 46	15 ♈ 24	15 ♈ 37	12 ♉ 28	11 ♒ 41	27 ♏ 37	17 ♐ 57	03 ♑ 37	03 ♏ 58
02	00 43 07	12 41 33	00 ♍ 57 26	08 ♍ 15 38	14 R 38	14 R 59	13 10	11 12	27 R 34	17 R 56	03 37	03 R 56
03	00 47 03	13 40 40	15 ♍ 39 44	23 ♍ 08 49	13 50	14 21	13 53	11 22	27 32	17 56	03 37	03 55
04	00 51 00	14 39 45	00 ♎ 41 54	08 ♎ 17 47	13 01	13 43	14 36	11 31	27 29	17 55	03 37	03 53
05	00 54 56	15 38 48	15 ♎ 55 12	23 ♎ 32 51	12 13	13 05	15 19	11 41	27 26	17 54	03 R 37	03 52
06	00 58 53	16 37 49	01 ♏ 09 26	08 ♏ 43 43	11 25	12 28	16 01	11 51	27 24	17 54	03 37	03 50
07	01 02 50	17 36 48	16 ♏ 14 37	23 ♏ 41 09	10 39	11 51	16 44	12 00	27 21	17 53	03 37	03 48
08	01 06 46	18 35 45	01 ♐ 02 33	08 ♐ 18 12	09 56	11 16	17 26	12 10	27 18	17 52	03 37	03 47
09	01 10 43	19 34 41	15 ♐ 27 41	22 ♐ 30 47	09 16	10 41	18 09	12 19	27 15	17 51	03 37	03 45
10	01 14 39	20 33 35	29 ♐ 27 23	06 ♑ 17 33	08 40	10 08	18 51	12 28	27 12	17 50	03 37	03 44
11	01 18 36	21 32 27	13 ♑ 01 28	19 ♑ 39 23	08 09	09 36	19 34	12 37	27 09	17 49	03 36	03 42
12	01 22 32	22 31 17	26 ♑ 11 37	02 ♒ 38 33	07 42	09 07	20 16	12 46	27 05	17 48	03 36	03 40
13	01 26 29	23 30 06	09 ♒ 00 37	15 ♒ 18 14	07 20	08 39	20 58	12 55	27 02	17 47	03 36	03 39
14	01 30 25	24 28 52	21 ♒ 31 50	27 ♒ 41 50	07 03	08 13	21 41	13 03	26 59	17 46	03 36	03 37
15	01 34 22	25 27 37	03 ♓ 48 40	09 ♓ 52 44	06 51	07 49	22 23	13 12	26 55	17 45	03 35	03 35
16	01 38 19	26 26 21	15 ♓ 54 29	21 ♓ 54 02	06 45	07 27	23 05	13 20	26 52	17 44	03 35	03 34
17	01 42 15	27 25 02	27 ♓ 51 58	03 ♈ 48 31	06 D 44	07 08	23 47	13 28	26 48	17 43	03 35	03 32
18	01 46 12	28 23 42	09 ♈ 43 58	15 ♈ 38 36	06 48	06 51	24 29	13 37	26 45	17 41	03 34	03 30
19	01 50 08	29 ♈ 22 19	21 ♈ 32 42	27 ♈ 26 32	06 57	06 37	25 12	13 45	26 41	17 40	03 34	03 29
20	01 54 05	00 ♉ 20 55	03 ♉ 20 20	09 ♉ 14 24	07 11	06 24	25 54	13 52	26 37	17 39	03 33	03 27
21	01 58 01	01 19 29	15 ♉ 09 00	21 ♉ 04 25	07 29	06 15	26 36	14 00	26 33	17 37	03 33	03 25
22	02 01 58	02 18 01	27 ♉ 00 57	02 ♊ 58 56	07 52	06 08	27 17	14 08	26 30	17 36	03 32	03 23
23	02 05 54	03 16 31	08 ♊ 58 42	15 ♊ 00 37	08 20	06 03	27 59	14 15	26 26	17 34	03 32	03 22
24	02 09 51	04 14 59	21 ♊ 05 05	27 ♊ 12 30	08 52	06 00	28 41	14 22	26 22	17 33	03 31	03 20
25	02 13 48	05 13 25	03 ♋ 23 19	09 ♋ 37 57	09 28	06 D 00	29 ♉ 23	14 29	26 18	17 31	03 30	03 18
26	02 17 44	06 11 49	15 ♋ 56 51	22 ♋ 20 27	10 07	06 03	00 ♊ 05	14 36	26 14	17 30	03 30	03 17
27	02 21 41	07 10 11	28 ♋ 49 11	05 ♌ 23 27	10 51	06 07	00 47	14 43	26 10	17 28	03 29	03 15
28	02 25 37	08 08 31	12 ♌ 03 34	18 ♌ 49 48	11 38	06 14	01 28	14 50	26 05	17 26	03 28	03 13
29	02 29 34	09 06 49	25 ♌ 42 22	02 ♍ 41 18	12 28	06 23	02 10	14 56	26 01	17 24	03 28	03 12
30	02 33 30	10 ♉ 05 04	09 ♍ 46 35	16 ♍ 57 58	13 ♈ 22	06 ♈ 34	02 ♊ 51	15 ♒ 03	25 ♏ 57	17 ♐ 23	03 ♑ 27	03 ♏ 10

Date	Moon True ☊ ° '	Moon Mean ☊ ° '	Moon ☽ Latitude ° '
01	18 ♉ 48	20 ♉ 22	05 N 11
02	18 R 44	20 18	05 04
03	18 38	20 15	04 38
04	18 33	20 12	03 52
05	18 29	20 09	02 48
06	18 27	20 06	01 33
07	18 26	20 03	00 N 12
08	18 D 26	19 59	01 S 09
09	18 27	19 56	02 23
10	18 29	19 53	03 26
11	18 30	19 50	04 16
12	18 R 30	19 47	04 51
13	18 29	19 43	05 10
14	18 27	19 40	05 14
15	18 24	19 37	05 03
16	18 21	19 34	04 39
17	18 17	19 31	04 03
18	18 14	19 28	03 16
19	18 11	19 24	02 22
20	18 10	19 21	01 21
21	18 09	19 18	00 S 17
22	18 D 09	19 15	00 N 49
23	18 10	19 12	01 53
24	18 11	19 08	02 53
25	18 13	19 05	03 45
26	18 14	19 02	04 28
27	18 14	18 59	04 59
28	18 R 14	18 56	05 15
29	18 14	18 53	05 15
30	18 ♉ 13	18 ♉ 49	04 N 55

DECLINATIONS

Date	Sun ☉ ° '	Moon ☽ ° '	Mercury ☿ ° '	Venus ♀ ° '	Mars ♂ ° '	Jupiter ♃ ° '	Saturn ♄ ° '	Uranus ♅ ° '	Neptune ♆ ° '	Pluto ♇ ° '
01	04 N 38	20 N 47	08 N 54	13 N 27	15 N 50	17 S 53	17 S 22	22 S 55	22 S 15	03 N 26
02	05 01	15 52	08 27	13 08	16 04	17 50	17 21	22 55	22 15	03 26
03	05 24	09 55	07 59	12 47	16 17	17 48	17 21	22 55	22 15	03 27
04	05 47	03 N 16	07 29	12 26	16 31	17 45	17 20	22 55	22 15	03 28
05	06 10	03 S 40	06 58	12 04	16 44	17 42	17 19	22 55	22 15	03 28
06	06 32	10 25	06 26	11 42	16 57	17 40	17 18	22 55	22 15	03 29
07	06 55	16 30	05 54	11 19	17 10	17 38	17 18	22 54	22 15	03 30
08	07 17	21 29	05 23	10 55	17 22	17 35	17 17	22 54	22 15	03 30
09	07 40	25 01	04 53	10 32	17 35	17 33	17 16	22 54	22 15	03 31
10	08 02	26 53	04 23	10 09	17 47	17 30	17 15	22 54	22 15	03 31
11	08 24	27 03	03 56	09 45	17 59	17 28	17 14	22 54	22 15	03 32
12	08 46	25 40	03 30	09 22	18 11	17 26	17 14	22 54	22 15	03 33
13	09 08	22 59	03 06	08 59	18 23	17 23	17 13	22 54	22 15	03 33
14	09 29	19 17	02 45	08 37	18 35	17 21	17 12	22 54	22 15	03 34
15	09 51	14 49	02 26	08 15	18 46	17 19	17 11	22 54	22 15	03 35
16	10 12	09 51	02 10	07 54	18 58	17 16	17 10	22 54	22 15	03 35
17	10 33	04 S 34	01 57	07 34	19 09	17 14	17 09	22 54	22 15	03 36
18	10 54	00 N 51	01 46	07 14	19 20	17 12	17 08	22 54	22 15	03 36
19	11 15	06 13	01 37	06 56	19 30	17 10	17 07	22 53	22 15	03 37
20	11 36	11 22	01 31	06 38	19 41	17 08	17 06	22 53	22 15	03 37
21	11 56	16 07	01 28	06 21	19 51	17 06	17 06	22 53	22 15	03 38
22	12 16	20 17	01 27	06 06	20 02	17 04	17 05	22 53	22 15	03 39
23	12 36	23 39	01 29	05 51	20 12	17 02	17 04	22 53	22 15	03 39
24	12 56	26 01	01 33	05 38	20 22	17 00	17 03	22 53	22 15	03 40
25	13 16	27 09	01 39	05 25	20 31	16 58	17 02	22 53	22 15	03 40
26	13 35	26 56	01 48	05 14	20 41	16 56	17 01	22 53	22 15	03 41
27	13 54	25 17	01 58	05 03	20 50	16 54	17 00	22 52	22 15	03 41
28	14 13	22 13	02 11	04 54	20 59	16 53	16 59	22 52	22 15	03 42
29	14 32	17 53	02 25	04 46	21 08	16 51	16 58	22 52	22 15	03 42
30	14 N 51	12 N 27	02 N 42	04 N 39	21 N 17	16 S 49	16 S 57	22 S 52	22 S 15	03 N 42

ZODIAC SIGN ENTRIES

Date	h	m	Planets
02	10	25	☽ ♍
04	10	54	☽ ♎
06	10	10	☽ ♏
08	10	17	☽ ♐
10	12	57	☽ ♑
12	19	04	☽ ♒
15	04	30	☽ ♓
17	16	18	☽ ♈
20	03	26	☉ ♉
20	05	12	☽ ♉
22	18	01	☽ ♊
25	05	26	☽ ♋
26	09	13	♂ ♊
27	14	10	☽ ♌
29	19	24	☽ ♍

LATITUDES

Date	Mercury ☿ ° '	Venus ♀ ° '	Mars ♂ ° '	Jupiter ♃ ° '	Saturn ♄ ° '	Uranus ♅ ° '	Neptune ♆ ° '	Pluto ♇ ° '
01	03 N 04	07 N 56	00 N 16	00 S 26	02 N 19	00 S 01	01 N 09	17 N 15
04	02 33	07 38	00 18	00 26	02 20	00 01	01 09	17 15
07	01 50	07 12	00 20	00 27	02 20	00 01	01 09	17 16
10	01 02	06 40	00 22	00 27	02 21	00 01	01 09	17 16
13	00 N 13	06 03	00 24	00 28	02 21	00 01	01 09	17 17
16	00 S 33	05 23	00 26	00 28	02 21	00 01	01 09	17 17
19	01 14	04 41	00 27	00 29	02 21	00 01	01 09	17 17
22	01 49	03 59	00 29	00 30	02 22	00 01	01 09	17 17
25	02 17	03 18	00 31	00 30	02 22	00 01	01 09	17 17
28	02 38	02 38	00 32	00 31	02 22	00 01	01 09	17 17
31	02 S 52	02 N 01	00 N 34	00 S 31	02 N 22	00 S 01	01 N 09	17 N 17

DATA

Julian Date	2446157
Delta T	+54 seconds
Ayanamsa	23° 38' 50"
Synetic vernal point	05° ♓ 28' 09"
True obliquity of ecliptic	23° 26' 35"

MOON'S PHASES, APSIDES AND POSITIONS ☽

Date	h	m	Phase	Longitude ° '	Eclipse Indicator
05	11	32	○	15 ♎ 38	
12	04	41	☾	22 ♑ 13	
20	05	22	●	00 ♉ 05	
28	04	25	☽	07 ♌ 50	

Day	h	m	
05	18	34	Perigee
19	17	39	Apogee

Day	h	m		
04	23	19	0S	
11	02	28	Max dec	27° S 11'
18	08	15	0N	
25	20	16	Max dec	27° N 14'

LONGITUDES

Date	Chiron ⚷ ° '	Ceres ⚳ ° '	Pallas ⚴ ° '	Juno ⚵ ° '	Vesta ⚶ ° '	Black Moon Lilith ⚸ ° '
01	04 ♊ 42	01 ♊ 58	13 ♈ 50	03 ♎ 27	02 ♏ 44	23 ♈ 12
11	05 ♊ 08	05 ♊ 41	17 ♈ 51	01 ♎ 15	00 ♏ 35	24 ♈ 19
21	05 ♊ 55	09 ♊ 33	21 ♈ 55	29 ♍ 27	28 ♎ 06	25 ♈ 25
31	06 ♊ 37	13 ♊ 32	26 ♈ 04	28 ♍ 11	25 ♎ 41	26 ♈ 32

All ephemeris data is given at 12.00 UT and the Moon's longitude is additionally given for 24.00 UT

ASPECTARIAN

h m Aspects

01 Monday
02 08 ☽ ☍ ♃
02 47 ☽ △ ☉
03 14 ☽ ⋕ ♆
04 21 ☽ □ ♂
09 54 ☽ △ ☿
10 14 ☽ △ ♀
11 10 ♄ ⊥ ♆
14 09 ☽ △ ♅
15 17 ☽ ⚼ ♆
20 57 ☽ Q ♇
02 Tuesday
03 00 ☽ ⋕ ♃
05 19 ☽ ⋕ ♄
06 11 ☽ ⚼ ☉
06 24 ☽ □ ♄
09 10 ☉ ⚼ ♄
09 55 ☽ ⚼ ☿
10 27 ☽ ⚼ ♀
11 11 ☽ ∥ ♂
16 23 ☽ △ ♆
16 55 ☽ ✶ ♇
22 07 ☽ ± ☉
23 58 ☽ ± ☿
03 Wednesday
00 11 ☽ ∥ ♀
00 38 ☽ ± ♀
04 58 ☽ ⊼ ♃
08 34 ☽ ⊼ ☉
08 59 ☽ △ ♂
09 12 ☽ ⊼ ☿
09 58 ☽ ⊼ ♀
11 10 ☿ ⊻ ♂
11 47 ☽ Q ♄
14 05 ☉ ☌ ☿
14 46 ☽ ± ♃
15 39 ☽ □ ♅
17 13 ☽ ∠ ♇
19 45 ☽ ∥ ☿
20 19 ♀ ⊻ ♂
22 00 ☉ ☌ ♀
04 Thursday
03 36 ☽ ∥ ☉
05 18 ☽ ⚼ ♃
06 22 ☉ ⊻ ♂
06 55 ☽ ✶ ♄
07 33 ☽ ⊥ ♇
10 10 ☽ ⚼ ♂
11 19 ☽ ∥ ♇
16 37 ☽ □ ♆
17 02 ☽ ⊻ ♇
20 15 ☽ Q ♅
05 Friday
01 05 ☽ ± ♂
01 30 ♆ St R
04 49 ☿ ⚼ ♄
05 16 ☽ △ ♃
06 28 ☽ ☍ ☿
06 33 ☽ ∠ ♄
07 43 ☽ ☍ ♀
11 00 ☽ ⊼ ♂
11 18 ☽ ⋕ ♇
11 32 ☽ ☍ ☉
15 07 ☽ ✶ ♅
20 39 ☽ ⊥ ♄
20 58 ☽ Q ♆
21 11 ☽ ⋕ ☉
22 42 ☽ ⋕ ☿
06 Saturday
01 11 ☿ ✶ ♃
06 05 ☽ ⊻ ♄
09 22 ☉ ∥ ☿
14 45 ☽ ∠ ♅
15 07 ♀ ⚼ ♄
15 53 ☽ ✶ ♆
16 14 ☽ ☌ ♇
16 29 ☽ ⋕ ♀
07 Sunday
03 30 ☽ ⊼ ☿
05 02 ☽ ⊥ ♅
05 09 ☽ □ ♃
05 15 ☽ ⊼ ♀
07 23 ♀ ✶ ♃
10 46 ☿ ⊥ ♂
12 38 ☽ ± ☿
12 49 ☽ ☍ ♂
14 21 ☽ ⊼ ☉
14 29 ☽ ± ♀
14 38 ☽ ⊻ ♅
14 57 ☽ ⋕ ♂
15 24 ☽ ∥ ♄
15 49 ☽ ∠ ♆
16 51 ☽ ∥ ♃
18 28 ☉ △ ♅
08 Monday
00 44 ☽ ± ☉
02 12 ♂ ⋕ ♄
02 29 ☽ ⚼ ☿
04 29 ☽ ⚼ ♀
05 54 ☽ ☌ ♄
06 23 ☽ ⊥ ♆
08 44 ♀ ⊥ ♂
10 32 ☽ Q ♃
16 14 ☽ ⊻ ♆
16 22 ☽ ∥ ♆
16 30 ☽ ⊻ ♇
16 30 ☽ ⚼ ☉
20 25 ☽ ∥ ♅
09 Tuesday
02 04 ☽ △ ☿
02 13 ♂ ⊼ ♅
02 26 ☽ ⊥ ♇
04 17 ☽ △ ♀
06 39 ☽ ✶ ♃
08 25 ♂ ⋕ ♃
16 03 ☽ ☌ ♅
16 47 ☽ ⊼ ♂
17 34 ☽ ∠ ♇
19 30 ☽ △ ☉
10 Wednesday
03 36 ☽ ± ♂
03 40 ♂ ⚼ ♆
08 05 ☽ ⊻ ♄
08 30 ☽ ∠ ♃
18 31 ☽ ⊥ ♄
19 16 ☽ ☌ ♆
19 28 ☽ ✶ ♇
20 07 ☽ ⚼ ♂
11 Thursday
00 27 ☽ ⊥ ♃
02 47 ☉ ± ♄
03 36 ☽ □ ☿
06 07 ☽ □ ♀
10 26 ☽ ∠ ♄
11 16 ☽ ⊻ ♃
16 49 ☽ Q ♇
20 39 ☽ ⊻ ♅
12 Friday
00 30 ☽ △ ♂
04 41 ☽ □ ☉
07 36 ☽ ⊥ ♅
09 26 ☿ ∥ ♇
11 06 ☽ Q ☿
13 38 ☽ Q ♀
13 39 ☽ ✶ ♄
13 Saturday
00 17 ☽ ∠ ♅
01 48 ☽ ⊻ ♆
01 54 ☽ □ ♇
07 27 ☉ ∥ ♀
08 54 ☽ ✶ ☿
11 20 ☽ ✶ ♀
12 03 ☽ Q ♄
12 38 ☽ ∥ ♅
13 07 ☽ ⊥ ♆
17 08 ☽ Q ☉
17 17 ☽ ∥ ♆
19 31 ☽ ☌ ♃
14 Sunday
04 45 ☽ ✶ ♅
06 20 ☽ ∠ ♆
12 18 ☽ □ ♂
12 59 ☽ ∠ ☿
15 09 ☽ ∠ ♀
15 49 ☽ ⋕ ♂
18 13 ☽ ✶ ☉
21 33 ☿ ∠ ♂
22 33 ☽ □ ♄
22 53 ☽ ∥ ♃
23 38 ☽ ∥ ♄
15 Monday
04 02 ☽ Q ♅
04 27 ☉ Q ♃
06 15 ☽ ⊥ ☿
08 11 ☽ ⊥ ♀
11 34 ☽ ✶ ♆
11 34 ☽ △ ♇
12 59 ♆ ✶ ♇
17 56 ☽ ⊻ ☿
19 40 ☽ ⊻ ♀
21 37 ♀ ∠ ♂
16 Tuesday
01 48 ☽ Q ♂
02 19 ☽ ∠ ☉
06 49 ☽ ⊻ ♃
10 26 ☽ ⋕ ☉
11 21 ☽ Q ♆
15 38 ☽ □ ♅
17 18 ☽ ⚼ ♇
18 56 ☽ ⊥ ♃
21 33 ☽ ⋕ ♀
21 49 ☉ ⊼ ♄
21 53 ☽ ⊥ ☉
17 Wednesday
03 17 ☽ ✶ ♂
05 23 ☿ St D
09 52 ☽ △ ♄
11 01 ☽ ⊻ ☉
11 20 ☽ ± ♇
13 14 ☽ ∠ ♃
16 18 ☽ ⋕ ♇
23 25 ☽ ⊼ ♇
23 31 ☽ □ ♆
18 Thursday
00 05 ☽ ⋕ ☿
06 00 ☽ ☌ ☿
06 17 ☽ ☌ ♀
11 29 ☽ ∠ ♂
15 39 ☿ ☌ ♀
15 57 ☽ ∥ ☿
16 04 ☽ ⚼ ♄
19 57 ☽ ✶ ♃
19 Friday
00 18 ☽ ∥ ♇
04 08 ☽ △ ♅
06 55 ☽ ⊥ ♂
10 15 ☽ ± ♄
15 06 ☽ ∥ ♀
19 53 ☽ ⊻ ♂
20 38 ☽ Q ♃
22 24 ☽ ⊼ ♄
20 Saturday
05 22 ☽ ☌ ☉
10 35 ☽ ⚼ ♅
12 13 ☽ ☍ ♇
12 26 ☽ △ ♆
13 13 ☽ ∥ ☉
18 09 ☽ ⊻ ♀
20 00 ☽ ⊻ ☿
21 Sunday
04 51 ☽ ± ♅
06 11 ☽ ⊥ ♀
08 31 ☽ ⊥ ☿
09 39 ☽ □ ♃
10 52 ♂ ☍ ♄
17 00 ☽ ⊼ ♅
17 16 ☽ ⋕ ♄
17 16 ☽ ⋕ ♃
18 52 ☽ ⚼ ♆
19 33 ♃ ∥ ♄
22 Monday
00 13 ☽ ∠ ♀
03 20 ☽ ∠ ☿
10 18 ☽ ∥ ♂
10 57 ☽ ☍ ♄
12 35 ☽ ☌ ♂
13 03 ☽ ± ♆
19 05 ☉ ⚼ ♅
20 15 ♂ ± ♆
23 35 ☽ ⊻ ☉
23 Tuesday
00 47 ☽ ⊼ ♇
01 06 ☽ ⊼ ♆
01 08 ☽ ⋕ ♆
05 52 ☽ ⋕ ♅
06 10 ☽ ✶ ♀
10 39 ☽ ✶ ☿
12 39 ☽ ⊥ ☉
12 46 ☽ ± ♇
14 06 ☉ ☍ ♇
18 06 ☉ △ ♆
22 36 ☽ △ ♃
24 Wednesday
05 02 ☽ ☍ ♅
05 57 ☽ Q ♀
06 36 ☽ ⚼ ♇
08 04 ☽ ∠ ☉
11 33 ☽ Q ☿
22 17 ☽ ⊼ ♄
25 Thursday
00 09 ♀ St D
03 46 ☽ ⊻ ♂
04 22 ☽ ⚼ ♃
09 54 ☽ ± ♄
11 51 ☽ △ ♇
12 14 ☽ ☍ ♆
15 50 ☽ ✶ ☉
16 05 ☽ ⊥ ♂
17 03 ☽ □ ♀
21 54 ☽ ± ♃
26 Friday
00 18 ☽ □ ☿
03 05 ☽ ⚼ ♄
08 01 ☉ ⊻ ♀
09 26 ☽ ⊼ ♃
10 16 ☽ ∠ ♂
14 54 ☽ ⊼ ♅
16 35 ☽ Q ☉
27 Saturday
02 07 ☽ ± ♅
07 07 ☽ △ ♄
15 48 ☽ ✶ ♂
18 40 ☽ ⚼ ♅
20 06 ☽ □ ♇
20 32 ☽ ⊼ ♆
20 59 ☿ ⚼ ♄
28 Sunday
01 26 ☽ △ ♀
04 25 ☽ □ ☉
07 22 ☽ ± ♆
07 41 ☽ ⋕ ♅
11 11 ☽ △ ☿
11 52 ☽ ⋕ ♆
14 39 ☽ Q ♂
16 58 ☽ ☍ ♃
19 18 ☽ ∥ ♂
21 32 ☽ △ ♅
23 21 ☽ ⚼ ♆
29 Monday
04 10 ☽ Q ♇
04 24 ☽ ⚼ ♀
12 33 ☽ □ ♄
15 15 ☽ ⚼ ☿
16 25 ☽ ⋕ ♄
16 57 ☽ ⋕ ♃
20 10 ☽ ± ♀
23 41 ☽ □ ♂
30 Tuesday
00 50 ☽ ✶ ♇
01 18 ☽ △ ♆
02 29 ☽ ∥ ☉
06 31 ☽ ⊼ ♀
07 39 ☽ ± ☿
12 33 ☽ △ ☉
18 25 ☽ ⊼ ☿
18 57 ☽ Q ♄
20 52 ☽ ⊼ ♃
22 16 ♂ ⊼ ♇

MAY 1985

LONGITUDES

Date	Sidereal time h m s	Sun ☉ ° ' "	Moon ☽ ° ' "	Moon ☽ 24.00 ° ' "	Mercury ☿ ° '	Venus ♀ ° '	Mars ♂ ° '	Jupiter ♃ ° '	Saturn ♄ ° '	Uranus ♅ ° '	Neptune ♆ ° '	Pluto ♇ ° '
01	02 37 27	11 ♉ 03 17	24 ♍ 15 07	01 ♎ 37 30	14 ♈ 18	06 ♈ 48	03 ♊ 33	15 ♒ 09	25 ♏ 53	17 ♐ 21	03 ♑ 26	03 ♏ 08
02	02 41 23	12 01 29	09 ♎ 04 24	16 ♎ 34 58	15 18	07 03	04 15	15 15	25 R 49	17 R 19	03 R 25	03 R 07
03	02 45 20	12 59 38	24 ♎ 08 12	01 ♏ 43 01	16 21	07 20	04 56	15 21	25 44	17 17	03 24	03 05
04	02 49 17	13 57 46	09 ♏ 18 15	16 ♏ 52 42	17 26	07 39	05 37	15 27	25 40	17 15	03 23	03 03
05	02 53 13	14 55 52	24 ♏ 25 13	01 ♐ 54 41	18 34	08 00	06 19	15 32	25 36	17 13	03 22	03 02
06	02 57 10	15 53 56	09 ♐ 20 07	16 ♐ 40 38	19 45	08 23	07 00	15 38	25 31	17 11	03 22	03 00
07	03 01 06	16 51 59	23 ♐ 55 32	01 ♑ 04 17	20 58	08 47	07 41	15 43	25 27	17 09	03 21	02 58
08	03 05 03	17 50 01	08 ♑ 06 31	15 ♑ 02 01	22 14	09 13	08 23	15 48	25 22	17 07	03 20	02 57
09	03 08 59	18 48 01	21 ♑ 50 44	28 ♑ 32 46	23 32	09 41	09 04	15 53	25 18	17 05	03 19	02 55
10	03 12 56	19 46 00	05 ♒ 08 20	11 ♒ 37 44	24 52	10 10	09 45	15 58	25 13	17 03	03 18	02 53
11	03 16 52	20 43 57	18 ♒ 01 22	24 ♒ 19 41	26 15	10 41	10 26	16 02	25 09	17 01	03 16	02 52
12	03 20 49	21 41 53	00 ♓ 33 10	06 ♓ 42 21	27 40	11 13	11 07	16 06	25 04	16 59	03 15	02 50
13	03 24 46	22 39 48	12 ♓ 47 46	18 ♓ 49 58	29 ♈ 07	11 47	11 48	16 11	25 00	16 57	03 14	02 49
14	03 28 42	23 37 41	24 ♓ 49 29	00 ♈ 46 50	00 ♉ 37	12 22	12 29	16 15	24 56	16 55	03 13	02 47
15	03 32 39	24 35 34	06 ♈ 42 32	12 ♈ 37 04	02 09	12 58	13 10	16 19	24 51	16 53	03 12	02 46
16	03 36 35	25 33 25	18 ♈ 30 52	24 ♈ 24 22	03 42	13 35	13 51	16 22	24 47	16 50	03 11	02 44
17	03 40 32	26 31 14	00 ♉ 17 59	06 ♉ 12 04	05 18	14 14	14 32	16 26	24 42	16 48	03 10	02 43
18	03 44 28	27 29 03	12 ♉ 06 58	18 ♉ 03 00	06 57	14 53	15 13	16 29	24 38	16 46	03 08	02 41
19	03 48 25	28 26 50	24 ♉ 00 26	29 ♉ 59 33	08 37	15 34	15 54	16 32	24 33	16 43	03 07	02 40
20	03 52 21	29 ♉ 24 36	06 ♊ 00 36	12 ♊ 03 49	10 19	16 15	16 35	16 35	24 29	16 41	03 06	02 38
21	03 56 18	00 ♊ 22 20	18 ♊ 09 26	24 ♊ 17 38	12 04	16 58	17 15	16 38	24 24	16 39	03 05	02 37
22	04 00 15	01 20 03	00 ♋ 28 39	06 ♋ 42 41	13 51	17 42	17 56	16 41	24 20	16 37	03 03	02 35
23	04 04 11	02 17 45	12 ♋ 59 56	19 ♋ 20 38	15 40	18 26	18 37	16 43	24 15	16 34	03 02	02 34
24	04 08 08	03 15 25	25 ♋ 44 58	02 ♌ 13 10	17 31	19 11	19 17	16 45	24 11	16 32	03 01	02 32
25	04 12 04	04 13 04	08 ♌ 45 25	15 ♌ 21 54	19 24	19 58	19 58	16 48	24 06	16 29	02 59	02 31
26	04 16 01	05 10 41	22 ♌ 02 50	28 ♌ 48 21	21 19	20 45	20 38	16 49	24 02	16 27	02 58	02 30
27	04 19 57	06 08 17	05 ♍ 38 33	12 ♍ 33 31	23 16	21 32	21 19	16 51	23 58	16 25	02 57	02 28
28	04 23 54	07 05 51	19 ♍ 33 15	26 ♍ 37 40	25 16	22 21	21 59	16 53	23 53	16 22	02 55	02 27
29	04 27 50	08 03 24	03 ♎ 46 37	10 ♎ 59 49	27 17	23 10	22 40	16 54	23 49	16 20	02 54	02 26
30	04 31 47	09 00 55	18 ♎ 16 53	25 ♎ 37 20	29 ♉ 20	24 00	23 20	16 55	23 45	16 17	02 52	02 24
31	04 35 44	09 ♊ 58 25	03 ♏ 00 32	10 ♏ 25 47	01 ♊ 25	24 ♈ 51	24 ♊ 01	16 ♒ 56	23 ♏ 41	16 ♐ 15	02 ♑ 51	02 ♏ 23

Date	Moon True ☊ ° '	Moon Mean ☊ ° '	Moon ☽ Latitude ° '
01	18 ♉ 12	18 ♉ 46	04 N 17
02	18 R 12	18 43	03 21
03	18 11	18 40	02 10
04	18 11	18 37	00 N 49
05	18 D 11	18 34	00 S 35
06	18 11	18 30	01 55
07	18 11	18 27	03 06
08	18 11	18 24	04 03
09	18 R 11	18 21	04 45
10	18 11	18 18	05 10
11	18 11	18 14	05 18
12	18 D 11	18 11	05 10
13	18 11	18 08	04 49
14	18 12	18 05	04 15
15	18 12	18 02	03 31
16	18 13	17 59	02 38
17	18 14	17 55	01 38
18	18 14	17 52	00 S 34
19	18 R 14	17 49	00 N 32
20	18 14	17 46	01 37
21	18 13	17 43	02 38
22	18 11	17 40	03 33
23	18 09	17 36	04 19
24	18 07	17 33	04 52
25	18 05	17 30	05 12
26	18 04	17 27	05 16
27	18 04	17 24	05 01
28	18 D 04	17 20	04 30
29	18 05	17 17	03 41
30	18 06	17 14	02 37
31	18 ♉ 07	17 ♉ 11	01 N 22

DECLINATIONS

Date	Sun ☉	Moon ☽	Mercury ☿	Venus ♀	Mars ♂	Jupiter ♃	Saturn ♄	Uranus ♅	Neptune ♆	Pluto ♇
01	15 N 09	06 N 13	03 N 00	04 N 33	21 N 25	16 S 48	16 S 56	22 S 52	22 S 15	03 N 43
02	15 27	00 S 31	03 20	04 28	21 33	16 46	16 55	22 52	22 15	03 43
03	15 44	07 21	03 41	04 24	21 42	16 45	16 54	22 51	22 15	03 44
04	16 02	13 49	04 04	04 21	21 50	16 43	16 53	22 51	22 15	03 44
05	16 19	19 26	04 29	04 19	21 57	16 42	16 52	22 51	22 15	03 45
06	16 36	23 45	04 55	04 18	22 05	16 40	16 51	22 51	22 15	03 45
07	16 53	26 24	05 22	04 18	22 12	16 39	16 50	22 51	22 15	03 45
08	17 09	27 15	05 50	04 18	22 19	16 38	16 49	22 51	22 15	03 46
09	17 25	26 21	06 20	04 20	22 26	16 36	16 47	22 50	22 15	03 46
10	17 41	23 59	06 51	04 23	22 33	16 35	16 46	22 50	22 15	03 46
11	17 56	20 28	07 23	04 26	22 39	16 34	16 45	22 50	22 15	03 47
12	18 12	16 07	07 56	04 30	22 46	16 33	16 44	22 50	22 15	03 47
13	18 26	11 12	08 31	04 35	22 52	16 32	16 43	22 50	22 15	03 47
14	18 41	05 58	09 06	04 41	22 58	16 31	16 42	22 49	22 15	03 48
15	18 55	00 S 34	09 41	04 47	23 03	16 30	16 41	22 49	22 15	03 48
16	19 09	04 N 50	10 18	04 54	23 09	16 29	16 40	22 49	22 15	03 48
17	19 23	10 10	10 55	05 02	23 14	16 28	16 39	22 49	22 15	03 49
18	19 36	14 56	11 33	05 10	23 19	16 27	16 38	22 49	22 15	03 49
19	19 49	19 18	12 12	05 19	23 24	16 27	16 37	22 48	22 15	03 49
20	20 02	22 54	12 51	05 29	23 29	16 26	16 36	22 48	22 15	03 49
21	20 14	25 33	13 30	05 39	23 33	16 25	16 35	22 48	22 15	03 49
22	20 26	27 00	14 10	05 49	23 37	16 25	16 34	22 48	22 15	03 49
23	20 37	27 06	14 50	06 01	23 41	16 24	16 33	22 47	22 15	03 50
24	20 49	25 47	15 30	06 12	23 45	16 24	16 32	22 47	22 15	03 50
25	21 00	23 05	16 10	06 24	23 49	16 23	16 31	22 47	22 15	03 50
26	21 10	19 08	16 50	06 37	23 52	16 23	16 30	22 47	22 15	03 50
27	21 20	14 07	17 29	06 50	23 55	16 23	16 29	22 47	22 15	03 50
28	21 30	08 16	18 08	07 03	23 58	16 22	16 28	22 46	22 15	03 50
29	21 39	01 N 53	18 47	07 17	24 01	16 22	16 27	22 46	22 15	03 50
30	21 48	04 S 45	19 24	07 31	24 04	16 22	16 26	22 46	22 15	03 50
31	21 N 57	11 S 13	20 N 01	07 N 45	24 N 06	16 S 22	16 S 25	22 S 46	22 S 15	03 N 50

ZODIAC SIGN ENTRIES

Date	h	m	Planets
01	21	22	☽ ♎
03	21	17	☽ ♏
05	20	56	☽ ♐
07	22	11	☽ ♑
10	02	38	☽ ♒
12	10	56	☽ ♓
14	02	10	☿ ♉
14	22	25	☽ ♈
17	11	23	☽ ♉
20	00	01	☽ ♊
21	02	43	☉ ♊
22	11	05	☽ ♋
24	19	54	☽ ♌
27	02	06	☽ ♍
29	05	41	☽ ♎
30	19	44	☿ ♊
31	07	07	☽ ♏

LATITUDES

Date	Mercury ☿	Venus ♀	Mars ♂	Jupiter ♃	Saturn ♄	Uranus ♅	Neptune ♆	Pluto ♇
01	02 S 52	02 N 01	00 N 34	00 S 31	02 N 22	00 S 01	01 N 09	17 N 17
04	03 00	01 26	00 35	00 32	02 22	00 01	01 09	17 16
07	03 02	00 53	00 37	00 33	02 22	00 01	01 09	17 16
10	02 59	00 N 23	00 38	00 33	02 22	00 01	01 10	17 15
13	02 50	00 S 05	00 40	00 34	02 22	00 01	01 10	17 15
16	02 36	00 30	00 41	00 35	02 22	00 01	01 10	17 14
19	02 18	00 52	00 42	00 35	02 22	00 02	01 10	17 13
22	01 55	01 13	00 44	00 36	02 22	00 02	01 10	17 12
25	01 28	01 31	00 45	00 37	02 21	00 02	01 10	17 11
28	00 58	01 47	00 46	00 37	02 21	00 02	01 10	17 10
31	00 S 27	02 S 01	00 N 47	00 S 38	02 N 21	00 S 02	01 N 10	17 N 09

DATA

Julian Date	2446187
Delta T	+54 seconds
Ayanamsa	23° 38' 54"
Synetic vernal point	05° ♓ 28' 06"
True obliquity of ecliptic	23° 26' 35"

MOON'S PHASES, APSIDES AND POSITIONS ☽

Date	h	m	Phase	Longitude ° '	Eclipse Indicator
04	19	53	○	14 ♏ 17	total
11	17	34	☾	20 ♒ 57	
19	21	41	●	28 ♉ 50	Partial
27	12	56	☽	06 ♍ 11	

Day	h	m		
04	05	28	Perigee	
17	00	02	Apogee	
02	10	10	0S	
08	11	19	Max dec	27° S 15'
15	14	30	0N	
23	01	49	Max dec	27° N 14'
29	18	50	0S	

LONGITUDES

Date	Chiron ⚷ ° '	Ceres ⚳ ° '	Pallas ⚴ ° '	Juno ⚵ ° '	Vesta ⚶ ° '	Black Moon Lilith ⚸ ° '
01	06 ♊ 37	13 ♊ 32	26 ♈ 04	28 ♍ 11	25 ♎ 41	26 ♈ 32
11	07 ♊ 22	17 ♊ 37	00 ♉ 16	27 ♍ 31	23 ♎ 33	27 ♈ 38
21	08 ♊ 08	21 ♊ 46	04 ♉ 30	27 ♍ 26	22 ♎ 21	28 ♈ 45
31	08 ♊ 55	25 ♊ 59	08 ♉ 47	27 ♍ 54	21 ♎ 49	29 ♈ 51

ASPECTARIAN

h m	Aspects
01 Wednesday	
00 39	☽ □ ♅
01 58	☽ ∠ ♇
06 52	☽ ± ♃
07 59	♂ ⊼ ♆
14 39	☽ ⚹ ♄
15 09	☽ ⚼ ☉
16 42	☽ ⊥ ♇
18 08	☽ ∥ ♀
19 03	☉ ± ♅
21 01	☽ ∥ ♇
21 40	☽ ⚼ ♃
23 04	☽ ∥ ☿
02 Thursday	
02 25	☽ ⊻ ♇
02 54	☽ □ ♆
03 51	☽ △ ♂
05 59	☽ Q ♅
06 46	☽ ± ☉
08 41	☽ ☍ ♀
10 34	☿ ⚹ ♃
14 46	☽ ∠ ♄
17 03	☽ ⊼ ☉
21 56	☽ △ ♃
22 20	☽ # ☿
22 42	☽ ☍ ☿
23 12	☽ # ♇
03 Friday	
01 09	☽ ⚹ ♅
01 39	☽ # ♀
05 01	☽ ⚼ ♂
05 03	☽ ⊥ ♄
07 40	☽ Q ♆
14 31	☽ ⊻ ♄
14 53	☿ ∥ ♇
19 57	☽ ± ♂
04 Saturday	
00 24	☉ ⊥ ♀
00 52	☽ ∠ ♅
02 08	☽ ☌ ♇
02 39	☽ ⚹ ♆
05 54	☽ ⊼ ♂
08 11	☿ △ ♅
09 20	☽ ⊼ ♀
15 05	☽ ⊥ ♅
19 03	☽ ± ♀
19 53	☽ ☍ ☉
21 23	☽ # ☉
21 47	☽ □ ♃
23 45	☽ ∥ ♃
05 Sunday	
00 26	☽ ∥ ♄
00 34	☽ ⊻ ♅
01 55	☽ ⊼ ☿
02 23	☽ ∠ ♆
03 07	☿ ∥ ♀
09 41	☽ ⚼ ♀
12 16	☽ ± ☿
13 52	☽ ☌ ♄
16 43	☽ ⊥ ♆
06 Monday	
01 22	☽ # ♂
01 46	☽ ⊻ ♇
02 20	☽ ⊻ ♆
02 37	☽ ∥ ♆
02 42	☽ Q ♃
03 56	☽ ⚼ ☿
04 31	☉ □ ♃
06 13	☽ ∥ ♅
07 38	☿ ± ♄
08 02	☽ ☍ ♂
10 25	☽ △ ♀
11 27	☽ ⊥ ♇
17 34	☉ # ♃
22 20	☽ ⚹ ♃
23 29	☽ ⊼ ☉
07 Tuesday	
00 49	☽ ☌ ♅
02 09	☽ ∠ ♇
06 38	☽ △ ☿
07 46	☉ # ♄
10 07	☽ ± ☉
14 32	☽ ⊻ ♄
18 59	☉ ⊼ ♅
20 30	♂ # ♆
23 28	☽ ∠ ♃
08 Wednesday	
00 34	☽ ⊥ ♄
02 20	☽ ⚼ ☉
03 12	☽ ⚹ ♇
03 50	☽ ☌ ♆
12 29	☽ ⊼ ♂
13 59	☽ □ ♀
14 56	☽ ⊥ ♃
15 53	☽ ∠ ♄
23 25	☽ ± ♂
23 49	☽ Q ♇
09 Thursday	
00 01	☉ ⚼ ♆
01 25	☽ ⊻ ♃
03 37	☽ ⊻ ♅
06 12	☽ △ ☉
07 07	♂ ± ♇
14 12	☽ ⊥ ♅
15 20	☽ □ ☿
16 10	☽ ⚼ ♂
18 07	☽ ⚹ ♄
22 50	☽ Q ♀
10 Friday	
06 23	☽ ∠ ♅
07 33	☿ ∠ ♂
07 54	☽ □ ♇
08 37	☽ ⊻ ♆
14 05	♀ ⚼ ♄
15 49	☽ Q ♄
17 50	☿ ⊼ ♄
19 39	☽ ⊥ ♆
20 37	☽ ∥ ♅
20 59	☽ △ ♂
21 40	☽ ⚹ ♀
22 19	☽ # ♂
11 Saturday	
00 42	☽ ∥ ♆
04 02	☽ Q ☿
08 14	☽ ☌ ♃
10 07	☽ ⚹ ♅
12 28	☽ ∠ ♆
17 34	☽ □ ☉
12 Sunday	
01 30	☽ □ ♄
01 35	☽ # ☉
03 15	☽ ∠ ♀
05 42	☽ ⚹ ☿
08 44	☽ ∥ ♄
08 58	☽ Q ♅
09 45	☽ ∥ ♃
16 26	☽ △ ♇
17 15	☽ ⚹ ♆
19 40	☿ Q ♃
21 31	☽ ⊥ ♀
13 Monday	
03 56	♂ # ♅
07 12	♀ ⚹ ♂
07 25	☽ Q ☉
09 54	☽ ⊻ ♀
09 55	☽ □ ♂
15 00	☽ ∠ ☿
16 50	☽ Q ♆
18 45	☽ ⊻ ♃
20 13	☽ □ ♅
21 57	☽ ⚼ ♇
23 14	☽ # ☿
14 Tuesday	
06 48	☽ ⊥ ♃
09 23	☽ ⚹ ☉
11 31	☽ ⊥ ☿
12 12	☽ △ ♄
15 56	☽ ± ♇
17 37	☽ # ♀
21 40	☽ # ♇
15 Wednesday	
00 06	☽ Q ♂
01 01	☽ ∠ ♃
01 23	☽ ⊻ ☿
04 01	☽ ⊼ ♇
04 54	☽ □ ♆
07 56	☿ ⚼ ♅
17 57	☉ ☍ ♄
18 20	☽ ⚼ ♄
18 22	☽ ∠ ☉
21 23	☿ ☍ ♇
16 Thursday	
01 24	☽ ☌ ♀
01 56	☽ ⚹ ♂
04 05	☿ △ ♆
07 24	☽ ∥ ♇
07 37	☽ ⚹ ♃
08 36	☽ △ ♅
12 21	☽ ∥ ♀
12 32	☽ ± ♄
14 19	☽ ⊥ ☉
17 Friday	
00 40	☽ ⊼ ♄
03 37	☽ ⊻ ☉
08 11	☽ Q ♃
10 21	☽ ∠ ♂
15 03	☽ ⚼ ♅
16 45	☽ ∥ ☿
16 54	☽ ☍ ♇
17 49	☽ △ ♆
23 48	☽ ☌ ☿
18 Saturday	
03 36	☉ ± ♆
05 46	☽ ⊥ ♂
09 16	☽ ± ♅
17 56	☽ ⊻ ♀
18 39	☽ ⊻ ♂
19 58	☽ # ♃
20 53	☽ □ ♃
20 56	☽ # ♄
21 22	☽ ⊼ ♅
19 Sunday	
00 10	☽ ⚼ ♆
06 47	☽ ⊥ ♀
13 05	☽ ☍ ♄
15 21	☽ ∥ ☉
18 14	☽ ± ♆
21 41	☽ ☌ ☉
20 Monday	
01 56	☽ ∠ ♀
05 17	☽ ⊼ ♇
06 13	☽ ⊼ ♆
07 09	☽ # ♆
11 14	☽ # ♅
12 33	♂ △ ♃
15 43	♂ ☍ ♅
16 35	☽ ∥ ♂
16 57	☿ ± ♅
17 12	☽ ± ♇
17 48	☿ ⊥ ♂
21 59	☽ ⊻ ☿
21 Tuesday	
00 09	♀ ⚹ ♃
01 51	♀ △ ♅
09 00	☽ △ ♃
09 03	☽ ☍ ♅
09 31	☽ ⚹ ♀
10 07	☽ ☌ ♂
10 56	☽ ⚼ ♇
11 47	☽ ⊥ ☿
15 10	♃ ⚹ ♅
22 Wednesday	
00 08	☽ ⊼ ♄
00 10	♂ ⚼ ♇
08 18	☽ ∠ ☿
10 23	☽ Q ♀
11 43	☽ ± ♄
13 48	☽ ⊻ ☉
14 20	☽ ⚼ ♃
16 04	☽ △ ♇
16 58	☽ ☍ ♆
23 Thursday	
02 17	☽ ⊥ ☉
04 54	☽ ⚼ ♄
06 43	♀ ± ♄
07 39	☽ ± ♃
17 54	☽ ⚹ ☿
18 31	☉ ⊼ ♇
18 45	☽ ⊼ ♅
19 04	☽ ⊼ ♃
20 48	☽ ∠ ☉
22 56	☽ □ ♀
23 13	☽ ⊻ ♂
23 37	☿ ⊼ ♅
24 Friday	
02 07	☿ □ ♃
05 59	☉ ⊼ ♆
06 00	☽ ± ♅
09 05	☽ △ ♄
11 05	☽ ⊥ ♂
18 21	☿ ⚼ ♆
20 10	☽ Q ☿
22 42	☽ ⚼ ♅
25 Saturday	
00 34	☽ □ ♇
01 26	☽ ⊼ ♆
03 01	☽ ⚹ ☉
04 40	☽ ∠ ♂
06 40	☽ ∥ ♂
12 25	☽ ± ♆
13 01	♀ ⚹ ♂
14 07	☽ # ♅
17 44	☽ # ♆
19 55	☿ # ♃
23 03	☿ ⊻ ♂
23 58	☿ ⊻ ♀
26 Sunday	
00 26	☿ # ♄
00 59	☽ ∥ ☉
01 59	☽ △ ♅
02 35	☽ Q ☉
02 37	☽ ☍ ♃
04 42	☽ ⚼ ♆
09 14	☽ Q ♇
09 21	☽ ⚹ ♂
09 31	☽ △ ♀
10 28	☽ □ ☿
15 31	☽ □ ♄
22 12	☽ ∥ ☿
27 Monday	
01 09	☽ # ♄
01 42	☽ # ♃
06 28	☽ ⚹ ♇
07 17	☽ △ ♆
07 43	☽ Q ♂
12 56	☽ □ ☉
13 40	☽ ⚼ ♀
20 05	☿ ☍ ♄
22 55	☽ Q ♄
28 Tuesday	
06 11	☽ ± ♀
06 34	☽ □ ♅
07 25	☽ ⊼ ♃
08 25	☽ ∠ ♇
16 21	☽ □ ♂
16 33	☽ ∥ ♀
17 03	☽ ⊼ ♀
17 40	☽ ± ♃
19 20	☽ ⚹ ♄
23 17	☽ △ ☿
23 41	☽ ⊥ ♇
29 Wednesday	
04 48	☽ ∥ ♇
07 30	☿ ± ♆
08 51	☽ ⚼ ♃
09 45	☽ ⊻ ♇
10 32	☽ □ ♆
12 55	☽ Q ♅
19 38	☽ △ ☉
20 21	☽ ∠ ♄
21 06	☉ ± ♇
30 Thursday	
04 26	☽ ⚼ ☿
05 17	♀ ⊼ ♄
08 44	☽ # ♇
08 44	☽ ⚹ ♅
09 06	☉ ∠ ♀
09 45	☽ △ ♃
11 08	☽ ⊥ ♄
16 14	☽ Q ♆
20 40	☽ △ ♂
20 54	☽ ⊻ ♄
21 37	☽ ± ☿
21 55	☽ ☍ ♀
22 02	☽ ⚼ ☉
22 29	☽ # ♀
31 Friday	
01 00	☿ ⊥ ♀
01 13	♂ ⊼ ♄
08 59	☽ ⊼ ☿
09 09	☽ ∠ ♅
10 59	☽ ☌ ♇
11 44	☽ ⚹ ♆
13 40	☽ ± ☉
22 10	☽ ⚼ ♂
23 00	☿ ⊼ ♇
23 40	☽ ⊥ ♅

All ephemeris data is given at 12.00 UT and the Moon's longitude is additionally given for 24.00 UT

JUNE 1985

LONGITUDES

Date	Sidereal time h m s	Sun ☉ ° ' "	Moon ☽ ° ' "	Moon ☽ 24.00 ° ' "	Mercury ☿ ° '
01	04 39 40	10 ♊ 55 54	17 ♏ 52 14	25 ♏ 18 58	03 ♊ 31
02	04 43 37	11 53 21	02 ♐ 45 04	10 ♐ 09 31	05 39
03	04 47 33	12 50 48	17 ♐ 31 24	24 ♐ 49 47	07 48
04	04 51 30	13 48 14	02 ♑ 03 51	09 ♑ 12 52	09 59
05	04 55 26	14 45 39	16 ♑ 16 16	23 ♑ 13 38	12 10
06	04 59 23	15 43 03	00 ♒ 04 38	06 ♒ 49 11	14 22
07	05 03 19	16 40 27	13 ♒ 27 17	19 ♒ 59 03	16 34
08	05 07 16	17 37 49	26 ♒ 24 47	02 ♓ 44 50	18 46
09	05 11 13	18 35 12	08 ♓ 59 38	15 ♓ 09 43	20 58
10	05 15 09	19 32 33	21 ♓ 15 37	27 ♓ 17 57	23 09
11	05 19 06	20 29 54	03 ♈ 17 20	09 ♈ 14 22	25 20
12	05 23 02	21 27 15	15 ♈ 09 40	21 ♈ 03 53	27 29
13	05 26 59	22 24 35	26 ♈ 57 35	02 ♉ 51 20	29 ♊ 38
14	05 30 55	23 21 55	08 ♉ 45 40	14 ♉ 41 06	01 ♋ 45
15	05 34 52	24 19 14	20 ♉ 38 04	26 ♉ 37 01	03 50
16	05 38 48	25 16 33	02 ♊ 38 17	08 ♊ 42 11	05 54
17	05 42 45	26 13 52	14 ♊ 49 00	20 ♊ 58 56	07 56
18	05 46 42	27 11 10	27 ♊ 12 09	03 ♋ 28 46	09 56
19	05 50 38	28 08 28	09 ♋ 48 50	16 ♋ 12 23	11 53
20	05 54 35	29 ♊ 05 45	22 ♋ 39 25	29 ♋ 09 53	13 49
21	05 58 31	00 ♋ 03 01	05 ♌ 43 44	12 ♌ 20 53	15 43
22	06 02 28	01 00 17	19 ♌ 01 17	25 ♌ 44 49	17 34
23	06 06 24	01 57 32	02 ♍ 31 26	09 ♍ 21 03	19 23
24	06 10 21	02 54 46	16 ♍ 13 37	23 ♍ 09 02	21 10
25	06 14 17	03 52 00	00 ♎ 07 16	07 ♎ 08 14	22 55
26	06 18 14	04 49 13	14 ♎ 11 48	21 ♎ 17 52	24 37
27	06 22 11	05 46 26	28 ♎ 26 14	05 ♏ 36 40	26 17
28	06 26 07	06 43 38	12 ♏ 48 50	20 ♏ 02 23	27 55
29	06 30 04	07 40 50	27 ♏ 16 50	04 ♐ 31 38	29 ♋ 30
30	06 34 00	08 ♋ 38 01	11 ♐ 46 10	18 ♐ 59 47	01 ♌ 04

Date	Venus ♀ ° '	Mars ♂ ° '	Jupiter ♃ ° '	Saturn ♄ ° '	Uranus ♅ ° '	Neptune ♆ ° '	Pluto ♇ ° '
01	25 ♈ 42	24 ♊ 41	16 ♒ 57	23 ♏ 36	16 ♐ 12	02 ♑ 49	02 ♏ 22
02	26 34	25 21	16 57	23 R 32	16 R 10	02 R 48	02 R 21
03	27 26	26 01	16 58	23 28	16 07	02 46	02 20
04	28 19	26 42	16 58	23 24	16 05	02 45	02 18
05	29 ♈ 13	27 22	16 R 58	23 20	16 03	02 43	02 17
06	00 ♉ 07	28 02	16 58	23 16	16 00	02 42	02 16
07	01 02	28 42	16 57	23 12	15 58	02 40	02 15
08	01 57	29 ♊ 22	16 57	23 08	15 55	02 39	02 14
09	02 53	00 ♋ 02	16 56	23 04	15 53	02 37	02 13
10	03 49	00 42	16 55	23 01	15 50	02 36	02 12
11	04 45	01 22	16 54	22 57	15 48	02 34	02 11
12	05 42	02 02	16 52	22 53	15 45	02 33	02 10
13	06 40	02 42	16 51	22 50	15 43	02 31	02 09
14	07 38	03 22	16 49	22 46	15 40	02 29	02 08
15	08 36	04 02	16 47	22 42	15 38	02 28	02 07
16	09 35	04 42	16 45	22 39	15 36	02 26	02 06
17	10 34	05 21	16 43	22 36	15 33	02 25	02 06
18	11 33	06 01	16 40	22 32	15 31	02 23	02 05
19	12 33	06 41	16 38	22 29	15 28	02 21	02 04
20	13 33	07 21	16 35	22 26	15 26	02 20	02 03
21	14 33	08 00	16 32	22 23	15 24	02 18	02 03
22	15 34	08 40	16 28	22 20	15 21	02 17	02 02
23	16 35	09 20	16 25	22 17	15 19	02 15	02 01
24	17 36	09 59	16 22	22 14	15 17	02 13	02 01
25	18 37	10 39	16 18	22 11	15 14	02 12	02 00
26	19 39	11 18	16 14	22 09	15 12	02 10	02 00
27	20 41	11 58	16 10	22 06	15 10	02 08	01 59
28	21 44	12 37	16 06	22 03	15 07	02 07	01 59
29	22 46	13 17	16 01	22 01	15 05	02 05	01 58
30	23 ♉ 49	13 ♋ 56	15 ♒ 57	21 ♏ 59	15 ♐ 03	02 ♑ 04	01 ♏ 58

Date	Moon True ☊ ° '	Moon Mean ☊ ° '	Moon ☽ Latitude ° '
01	18 ♉ 08	17 ♉ 08	00 N 01
02	18 R 07	17 05	01 S 20
03	18 06	17 01	02 35
04	18 03	16 58	03 39
05	17 59	16 55	04 27
06	17 55	16 52	04 59
07	17 52	16 49	05 13
08	17 49	16 46	05 10
09	17 47	16 42	04 52
10	17 D 47	16 39	04 22
11	17 47	16 36	03 40
12	17 49	16 33	02 49
13	17 50	16 30	01 52
14	17 52	16 26	00 S 50
15	17 R 52	16 23	00 N 15
16	17 51	16 20	01 20
17	17 48	16 17	02 21
18	17 43	16 14	03 17
19	17 37	16 11	04 05
20	17 30	16 07	04 41
21	17 23	16 04	05 03
22	17 17	16 01	05 09
23	17 12	15 58	04 58
24	17 09	15 55	04 30
25	17 08	15 52	03 46
26	17 D 08	15 48	02 48
27	17 09	15 45	01 39
28	17 10	15 42	00 N 23
29	17 R 09	15 39	00 S 54
30	17 ♉ 07	15 ♉ 36	02 S 08

DECLINATIONS

Date	Sun ☉ ° '	Moon ☽ ° '	Mercury ☿ ° '	Venus ♀ ° '	Mars ♂ ° '	Jupiter ♃ ° '	Saturn ♄ ° '	Uranus ♅ ° '	Neptune ♆ ° '	Pluto ♇ ° '
01	22 N 05	17 S 08	20 N 36	08 N 00	24 N 08	16 S 22	16 S 25	22 S 45	22 S 15	03 N 50
02	22 13	22 01	21 10	08 15	24 10	16 22	16 24	22 45	22 15	03 50
03	22 21	25 26	21 42	08 30	24 12	16 22	16 23	22 45	22 15	03 50
04	22 28	27 04	22 13	08 45	24 13	16 22	16 22	22 45	22 15	03 50
05	22 34	26 52	22 41	09 01	24 14	16 23	16 21	22 44	22 15	03 50
06	22 41	25 00	23 08	09 17	24 15	16 23	16 20	22 44	22 15	03 50
07	22 47	21 46	23 32	09 33	24 16	16 23	16 19	22 44	22 15	03 50
08	22 52	17 34	23 53	09 49	24 17	16 24	16 18	22 44	22 15	03 50
09	22 57	12 43	24 12	10 06	24 17	16 24	16 18	22 43	22 15	03 50
10	23 02	07 28	24 28	10 22	24 18	16 25	16 17	22 43	22 15	03 50
11	23 06	02 S 04	24 42	10 39	24 18	16 25	16 16	22 43	22 15	03 50
12	23 10	03 N 22	24 52	10 56	24 18	16 26	16 15	22 43	22 15	03 50
13	23 14	08 39	25 00	11 13	24 17	16 27	16 15	22 42	22 15	03 50
14	23 17	13 38	25 05	11 30	24 17	16 27	16 14	22 42	22 15	03 50
15	23 19	18 09	25 07	11 47	24 16	16 28	16 13	22 42	22 15	03 49
16	23 22	22 00	25 07	12 04	24 15	16 29	16 12	22 42	22 15	03 49
17	23 23	24 55	25 04	12 21	24 14	16 30	16 12	22 41	22 15	03 49
18	23 25	26 42	24 58	12 38	24 12	16 31	16 11	22 41	22 15	03 49
19	23 26	27 09	24 51	12 55	24 11	16 32	16 11	22 41	22 15	03 49
20	23 26	26 09	24 41	13 12	24 09	16 33	16 10	22 41	22 16	03 48
21	23 27	23 44	24 28	13 29	24 07	16 34	16 09	22 40	22 16	03 48
22	23 26	20 00	24 14	13 46	24 05	16 35	16 09	22 40	22 16	03 48
23	23 26	15 12	23 58	14 03	24 02	16 37	16 08	22 40	22 16	03 47
24	23 25	09 35	23 41	14 20	24 00	16 38	16 08	22 40	22 16	03 47
25	23 23	03 N 25	23 21	14 36	23 57	16 39	16 07	22 39	22 16	03 47
26	23 21	03 S 01	23 00	14 53	23 54	16 41	16 07	22 39	22 16	03 46
27	23 19	09 22	22 38	15 09	23 51	16 42	16 06	22 39	22 16	03 46
28	23 16	15 19	22 15	15 26	23 48	16 43	16 06	22 39	22 16	03 46
29	23 13	20 26	21 51	15 42	23 44	16 45	16 05	22 38	22 16	03 45
30	23 N 10	24 S 19	21 N 25	15 N 58	23 N 40	16 S 47	16 S 05	22 S 38	22 S 16	03 N 45

ZODIAC SIGN ENTRIES

Date	h	m	Planets
02	07	33	☽ ♐
04	08	34	☽ ♑
06	08	53	♀ ♉
06	11	52	☽ ♒
08	18	46	☽ ♓
09	10	40	♂ ♋
11	05	24	☽ ♈
13	16	11	☿ ♋
13	18	11	☽ ♉
16	06	45	☽ ♊
18	17	22	☽ ♋
21	01	32	☽ ♌
21	10	44	☉ ♋
23	07	32	☽ ♍
25	11	48	☽ ♎
27	14	37	☽ ♏
29	16	30	☽ ♐
29	19	34	☿ ♌

LATITUDES

Date	Mercury ☿ ° '	Venus ♀ ° '	Mars ♂ ° '	Jupiter ♃ ° '	Saturn ♄ ° '	Uranus ♅ ° '	Neptune ♆ ° '	Pluto ♇ ° '
01	00 S 16	02 S 05	00 N 48	00 S 38	02 N 21	00 S 02	01 N 10	17 N 09
04	00 N 16	02 16	00 49	00 39	02 20	00 02	01 10	17 08
07	00 46	02 26	00 50	00 40	02 20	00 02	01 10	17 06
10	01 12	02 34	00 51	00 40	02 19	00 02	01 10	17 05
13	01 34	02 41	00 52	00 41	02 19	00 02	01 10	17 04
16	01 48	02 46	00 53	00 42	02 18	00 02	01 10	17 02
19	01 56	02 49	00 54	00 42	02 18	00 02	01 10	17 01
22	01 58	02 51	00 55	00 43	02 17	00 02	01 10	16 59
25	01 53	02 52	00 56	00 44	02 16	00 02	01 10	16 58
28	01 42	02 52	00 57	00 45	02 16	00 02	01 10	16 56
31	01 N 26	02 S 51	00 N 58	00 S 45	02 N 15	00 S 02	01 N 10	16 N 54

DATA

Julian Date	2446218
Delta T	+54 seconds
Ayanamsa	23° 38' 59"
Synetic vernal point	05° ♓ 28' 01"
True obliquity of ecliptic	23° 26' 34"

MOON'S PHASES, APSIDES AND POSITIONS ☽

Date	h	m	Phase	Longitude ° '	Eclipse Indicator
03	03	50	○	12 ♐ 31	
10	08	19	☾	19 ♓ 24	
18	11	58	●	27 ♊ 11	
25	18	53	☽	04 ♎ 08	

Day	h	m		
01	12	40	Perigee	
13	14	12	Apogee	
29	09	24	Perigee	
04	21	18	Max dec	27° S 12'
11	21	04	ON	
19	07	32	Max dec	27° N 10'
26	00	48	OS	

LONGITUDES

Date	Chiron ⚷ ° '	Ceres ⚳ ° '	Pallas ⚴ ° '	Juno ⚵ ° '	Vesta ⚶ ° '	Black Moon Lilith ⚸ ° '
01	08 ♊ 59	26 ♊ 24	09 ♉ 13	27 ♍ 58	21 ♎ 48	29 ♈ 58
11	09 ♊ 46	00 ♋ 40	13 ♉ 32	28 ♍ 59	22 ♎ 10	01 ♉ 05
21	10 ♊ 32	04 ♋ 59	17 ♉ 53	00 ♎ 25	23 ♎ 18	02 ♉ 11
31	11 ♊ 17	09 ♋ 19	22 ♉ 15	02 ♎ 13	25 ♎ 08	03 ♉ 18

All ephemeris data is given at 12.00 UT and the Moon's longitude is additionally given for 24.00 UT

ASPECTARIAN

01 Saturday
00 02 ☽ ⊼ ☉
04 10 ☿ ⊼ ♆
08 40 ☽ ∥ ♃
08 51 ☽ ∥ ♄
09 20 ☽ ⊻ ♅
10 30 ☽ □ ♃
11 55 ☽ ∠ ♆
13 22 ☽ ± ♂
21 12 ☽ ☌ ♄
23 30 ☽ ⊼ ♂

02 Sunday
01 23 ☽ ⊼ ♀
02 24 ☽ ⊥ ♆
06 39 ☽ # ☿
11 21 ☽ ⊻ ♇
11 41 ☽ ± ♀
12 05 ☽ ⊻ ♆
13 12 ☽ # ☉
13 21 ☽ ∥ ♆
15 34 ☽ Q ♃
16 20 ☽ ∥ ♅
17 29 ☽ ☍ ☿
18 09 ☉ # ♆
21 03 ☽ ⊥ ♇

03 Monday
01 49 ☽ # ♂
03 11 ☽ ꝑ ♀
03 50 ☽ ☍ ☉
09 43 ☽ ☌ ♅
11 05 ☽ ⚹ ♃
11 41 ☽ ∠ ♇
17 41 ☿ ± ♇
21 43 ☽ ⊻ ♄

04 Tuesday
00 15 ♃ ∥ ♄
02 39 ☽ ☍ ♂
05 22 ☽ △ ♀
07 35 ☽ ⊥ ♄
11 50 ☽ ∠ ♃
12 24 ☽ ⚹ ♇
13 08 ☽ ☌ ♆
13 48 ☿ # ♆
22 25 ♃ St R
22 35 ☽ ∠ ♄

05 Wednesday
02 58 ☽ ⊥ ♃
03 43 ☽ ⊼ ☿
04 02 ☉ ∥ ☿
05 16 ♀ Q ♃
08 37 ☽ Q ♇
09 14 ☽ ⊼ ☉
11 37 ☽ ⊻ ♅
13 11 ☽ ⊻ ♃
14 37 ☿ # ♅
15 51 ☽ ± ☿
20 18 ☽ ± ☉
21 55 ☽ ⊥ ♅

06 Thursday
00 08 ☽ ⚹ ♄
08 13 ☽ ⊼ ♂
10 30 ☽ ꝑ ☿
12 05 ☽ □ ♀
13 13 ☽ ꝑ ☉
13 38 ☽ ∠ ♅
15 52 ☽ □ ♇
16 38 ☽ ⊻ ♆
18 20 ☽ # ♂
18 50 ☉ ☍ ♅
19 23 ☽ ± ♂
21 10 ☽ Q ♄

07 Friday
01 17 ☽ # ☿
01 29 ☉ # ♅
02 05 ☿ ∠ ♀
03 21 ☽ ⊥ ♆
05 33 ☽ # ☉
05 34 ☿ ☍ ♅
05 40 ☽ ∥ ♅
08 54 ☽ ∥ ♆
10 16 ♀ ꝑ ♅
12 28 ☽ ꝑ ♂
14 10 ☉ ☌ ☿
16 15 ☿ △ ♃
16 34 ☽ ⚹ ♅
18 22 ☽ △ ☉
18 24 ☽ ☌ ♃
18 50 ☽ △ ☿
18 57 ☉ △ ♃
19 27 ☿ ꝑ ♇
19 43 ☽ ∠ ♆
23 01 ☽ Q ♀

08 Saturday
02 12 ☉ ꝑ ♇
04 22 ♂ ± ♄
05 54 ☽ □ ♄
14 50 ☽ Q ♅
17 53 ☽ △ ♂
18 02 ☽ ∥ ♃
18 30 ☽ ∥ ♄
19 13 ♀ ☍ ♇
23 00 ☽ △ ♇
23 19 ☽ ⚹ ♀
23 47 ☽ ⚹ ♆

09 Sunday
05 34 ♀ △ ♆
19 37 ☿ ∥ ♂
22 55 ☽ Q ♆
23 32 ☽ # ♀

10 Monday
01 21 ☽ □ ♅
03 27 ☽ ⊻ ♃
04 00 ☽ ꝑ ♇
06 46 ☽ ∠ ♀
08 19 ☽ □ ☉
10 29 ☿ ⊼ ♄
15 16 ☽ ⊥ ♃
15 27 ☽ △ ♄
16 34 ☽ □ ☿
21 47 ☽ ± ♇

11 Tuesday
02 08 ☽ ⊥ ♀
04 10 ☽ # ♇
07 55 ☽ □ ♂
09 12 ☽ ∠ ♃
09 47 ☽ ⊼ ♇
10 33 ☽ □ ♆
15 13 ☽ ⊻ ♀
21 20 ☽ ꝑ ♄
23 25 ☽ Q ☉

12 Wednesday
06 18 ♂ ꝑ ♃
12 49 ☽ Q ☿
13 12 ☽ △ ♅
14 04 ☽ ∥ ♇
15 28 ☽ ⚹ ♃
15 29 ☽ ± ♄
16 36 ♂ △ ♇
22 30 ☽ Q ♂

13 Thursday
01 55 ☽ ⚹ ☉
03 12 ☿ ± ♄
03 38 ☽ ⊼ ♄
05 35 ♂ ☍ ♆
15 50 ☽ Q ♃
18 38 ☽ ⚹ ☿
19 37 ☽ ꝑ ♅
21 50 ☉ ⊼ ♄
22 33 ☽ ☍ ♇
23 17 ☽ △ ♆

14 Friday
00 23 ☽ ⚹ ♂
00 49 ☽ ∥ ♀
09 30 ☽ ☌ ♀
11 08 ☽ ∠ ☉
12 48 ☿ ꝑ ♃
13 51 ☽ ± ♅
16 25 ☿ △ ♇
20 24 ☿ ☍ ♆

15 Saturday
01 22 ☽ # ♄
01 57 ☽ ⊼ ♅
02 38 ☽ # ♃
04 16 ☽ □ ♃
05 38 ☽ ꝑ ♆
06 56 ☽ ⊥ ☉
07 37 ☽ ∠ ☿
08 35 ☽ ∠ ♂
15 18 ☿ ☌ ♂
16 09 ☽ ☍ ♄
20 03 ☽ ⊻ ☉
23 40 ☽ ± ♆

16 Sunday
03 41 ☽ ⊥ ♂
05 26 ☽ ⊥ ☿
10 57 ☽ ⊼ ♇
11 36 ☽ ⊼ ♆
12 21 ♀ ± ♅
13 52 ☽ # ♆
16 19 ☽ ⊻ ♂
17 03 ☽ # ♅
19 46 ☽ ⊻ ☿
22 17 ☽ ∥ ☉
22 49 ☽ ± ♇

17 Monday
02 56 ☽ ⊻ ♀
05 32 ☽ ∥ ♂
08 08 ☿ ꝑ ♄
13 26 ☽ ∥ ☿
13 26 ☽ ☍ ♅
15 41 ☽ △ ♃
15 42 ☽ ⊥ ♀
16 26 ☽ ꝑ ♇

18 Tuesday
03 03 ☽ ⊼ ♄
10 38 ☽ ∠ ♀
11 58 ☽ ☌ ☉
14 33 ☽ ± ♄
20 31 ☽ ꝑ ♃
20 52 ☿ ± ♃
21 20 ☽ △ ♇
21 53 ☽ ☍ ♆

19 Wednesday
05 45 ☽ ☌ ♂
07 37 ☽ ꝑ ♄
13 32 ☽ ± ♃
16 36 ☽ ☌ ☿
17 35 ☽ ⚹ ♀
20 12 ☉ ± ♄
22 36 ☽ ⊼ ♅

20 Thursday
00 44 ☽ ⊼ ♃
04 53 ☿ ⚹ ♀
09 44 ☽ ± ♅
11 35 ☽ △ ♄
14 59 ♂ ꝑ ♄
17 47 ☽ Q ♀

21 Friday
00 49 ☽ ⊻ ☉
02 17 ☽ ꝑ ♅
05 17 ☽ □ ♇
05 22 ☽ ∥ ☿
05 46 ☽ ⊼ ♆
08 03 ☿ ⊼ ♅
08 52 ☽ ∥ ♂
12 38 ☽ ⊥ ☉
14 09 ☽ ∥ ☉
16 21 ☽ ⊻ ♂
16 40 ☽ ± ♆
19 39 ☽ # ♅
22 15 ☿ ⊼ ♃
22 25 ☽ # ♆

22 Saturday
03 46 ☽ ⊥ ♂
05 17 ☽ □ ♀
05 26 ☽ △ ♅
06 10 ☽ ∠ ☉
07 15 ♀ ⊼ ♅
07 27 ☽ ☍ ♃
08 53 ☽ ꝑ ♆
08 58 ☽ ⊻ ☿
13 49 ☽ Q ♇
17 54 ☽ □ ♄
20 44 ☽ ∠ ♂
21 24 ☽ ⊥ ☿
23 10 ☉ ꝑ ♃

23 Sunday
04 59 ☿ ∥ ♂
05 29 ☽ # ♃
07 41 ☽ # ♄
08 26 ♀ □ ♃
10 56 ☽ ⚹ ☉
11 07 ☽ ⚹ ♇
11 31 ☽ △ ♆
13 35 ☉ △ ♇
15 46 ☽ ∠ ☿
16 56 ☽ ∥ ♀
19 05 ☉ ☍ ♆

24 Monday
00 34 ☽ ⚹ ♂
01 35 ☽ Q ♄
03 23 ♀ ꝑ ♆
09 33 ☽ Q ☉
10 21 ☽ □ ♅
12 14 ☽ ⊼ ♃
13 22 ☽ ∠ ♇
13 29 ☿ ± ♅
14 34 ☽ △ ♀
21 48 ☽ ⚹ ☿
22 23 ☽ ⚹ ♄
22 29 ☽ Q ♂
22 35 ☽ ± ♃

25 Tuesday
00 24 ♂ ± ♃
02 17 ☿ △ ♄
04 56 ☽ ⊥ ♇
09 25 ☉ ∥ ☿
10 37 ☽ ∥ ♇
14 00 ☽ ꝑ ♃
15 13 ☽ ⊻ ♇
15 33 ☽ □ ♆
17 20 ☽ Q ♅
18 28 ☽ ꝑ ♀
18 53 ☽ □ ☉
21 20 ☽ Q ☿

26 Wednesday
00 03 ☽ ∠ ♄
06 51 ☽ □ ♂
11 00 ☽ ± ♀
13 42 ☽ ⚹ ♅
14 51 ☽ # ♇
15 17 ☽ ⊥ ♄
15 26 ☽ △ ♃
21 57 ☽ ⊼ ♀
22 04 ☽ Q ♆

27 Thursday
01 23 ☽ ⊻ ♄
07 55 ☽ □ ☿
11 21 ☿ # ♅
14 53 ☽ ∠ ♅
17 56 ☽ ☌ ♇
18 11 ☽ ⚹ ♆

28 Friday
01 09 ☽ △ ☉
05 53 ☽ ⊥ ♅
11 09 ☿ # ♆
11 40 ☽ △ ♂
12 31 ☉ ∠ ♀
12 31 ☽ # ♀
15 23 ☽ ∥ ♄
15 50 ☽ ⊻ ♅
17 25 ☽ □ ♃
18 12 ☽ ∥ ♃
19 08 ☽ ∠ ♆
19 20 ♀ ☍ ♄
19 56 ☉ ꝑ ♄

29 Saturday
03 18 ☽ ☌ ♄
03 50 ☽ ꝑ ☉
03 57 ☽ ☍ ♀
10 02 ☽ ⊥ ♆
13 44 ☽ ꝑ ♂
16 08 ☽ △ ☿
19 02 ☽ # ☿
19 46 ☽ ⊻ ♇
19 48 ☽ ± ☉
19 57 ☽ ⊻ ♆
20 43 ☿ ꝑ ♅
22 13 ☽ ∥ ♆
23 06 ☽ Q ♃

30 Sunday
00 30 ☽ ∥ ♅
03 57 ☽ # ☉
05 20 ☽ ± ♂
05 42 ☽ ⊥ ♇
06 26 ☽ ⊼ ☉
07 20 ☽ # ♂
15 46 ☽ ⊼ ♂
17 26 ☽ ☌ ♅
18 53 ☽ ⚹ ♃
19 58 ☽ ꝑ ☿
20 37 ☽ ∠ ♇
22 41 ♀ # ♄

JULY 1985

LONGITUDES

Date	Sidereal time h m s	Sun ☉ ° ' "	Moon ☽ ° ' "	Moon ☽ 24.00 ° ' "	Mercury ☿ ° '	Venus ♀ ° '	Mars ♂ ° '	Jupiter ♃ ° '	Saturn ♄ ° '	Uranus ♅ ° '	Neptune ♆ ° '	Pluto ♇ ° '
01	06 37 57	09 ♋ 35 13	26 ♐ 11 44	03 ♑ 21 19	02 ♌ 35	24 ♉ 52	14 ♋ 35	15 ♒ 52	21 ♏ 56	15 ♐ 01	02 ♑ 02	01 ♏ 57
02	06 41 53	10 32 24	10 ♑ 27 49	17 ♑ 30 33	04 03	25 55	15 15	15 R 47	21 R 54	14 R 59	02 R 00	01 R 57
03	06 45 50	11 29 35	24 ♑ 28 56	01 ♒ 22 26	05 29	26 59	15 54	15 42	21 52	14 57	01 59	01 57
04	06 49 46	12 26 46	08 ♒ 10 40	14 ♒ 53 20	06 53	28 03	16 33	15 37	21 50	14 55	01 57	01 57
05	06 53 43	13 23 57	21 ♒ 30 19	28 ♒ 01 34	08 15	29 ♉ 06	17 13	15 31	21 48	14 52	01 56	01 56
06	06 57 40	14 21 08	04 ♓ 27 11	10 ♓ 47 24	09 34	00 ♊ 11	17 52	15 26	21 46	14 50	01 54	01 56
07	07 01 36	15 18 19	17 ♓ 02 32	23 ♓ 13 00	10 50	01 15	18 31	15 20	21 44	14 48	01 52	01 56
08	07 05 33	16 15 31	29 ♓ 19 18	05 ♈ 21 58	12 04	02 20	19 10	15 14	21 43	14 46	01 51	01 56
09	07 09 29	17 12 43	11 ♈ 21 36	17 ♈ 18 51	13 16	03 24	19 49	15 08	21 41	14 44	01 49	01 56
10	07 13 26	18 09 56	23 ♈ 14 23	29 ♈ 08 50	14 24	04 29	20 29	15 02	21 39	14 42	01 48	01 56
11	07 17 22	19 07 09	05 ♉ 02 54	10 ♉ 57 15	15 30	05 35	21 08	14 56	21 38	14 41	01 46	01 55
12	07 21 19	20 04 22	16 ♉ 52 32	22 ♉ 49 20	16 34	06 40	21 47	14 50	21 37	14 39	01 45	01 D 55
13	07 25 15	21 01 36	28 ♉ 48 16	04 ♊ 49 52	17 34	07 46	22 26	14 44	21 35	14 37	01 43	01 55
14	07 29 12	21 58 51	10 ♊ 54 36	17 ♊ 02 54	18 31	08 51	23 05	14 37	21 34	14 35	01 42	01 56
15	07 33 09	22 56 06	23 ♊ 15 06	29 ♊ 31 29	19 25	09 57	23 44	14 30	21 33	14 33	01 40	01 56
16	07 37 05	23 53 22	05 ♋ 52 14	12 ♋ 17 26	20 16	11 03	24 23	14 24	21 32	14 32	01 39	01 56
17	07 41 02	24 50 38	18 ♋ 47 05	25 ♋ 21 07	21 03	12 09	25 02	14 17	21 31	14 30	01 37	01 56
18	07 44 58	25 47 54	01 ♌ 59 22	08 ♌ 41 35	21 47	13 16	25 41	14 10	21 31	14 28	01 36	01 56
19	07 48 55	26 45 11	15 ♌ 27 27	22 ♌ 16 38	22 27	14 22	26 20	14 03	21 30	14 27	01 34	01 56
20	07 52 51	27 42 28	29 ♌ 08 45	06 ♍ 03 24	23 04	15 29	26 59	13 56	21 29	14 25	01 33	01 57
21	07 56 48	28 39 46	13 ♍ 00 12	19 ♍ 58 47	23 36	16 36	27 38	13 48	21 29	14 24	01 32	01 57
22	08 00 44	29 ♋ 37 04	26 ♍ 58 48	03 ♎ 59 59	24 05	17 43	28 16	13 41	21 29	14 22	01 30	01 57
23	08 04 41	00 ♌ 34 22	11 ♎ 02 04	18 ♎ 04 51	24 29	18 50	28 55	13 34	21 28	14 21	01 29	01 58
24	08 08 38	01 31 40	25 ♎ 08 09	02 ♏ 11 51	24 48	19 57	29 ♋ 34	13 26	21 28	14 19	01 28	01 58
25	08 12 34	02 28 59	09 ♏ 15 49	16 ♏ 19 55	25 03	21 04	00 ♌ 13	13 19	21 28	14 18	01 26	01 58
26	08 16 31	03 26 18	23 ♏ 24 02	00 ♐ 28 01	25 13	22 12	00 52	13 11	21 D 28	14 16	01 25	01 59
27	08 20 27	04 23 38	07 ♐ 31 39	14 ♐ 34 42	25 18	23 20	01 30	13 04	21 28	14 15	01 24	01 59
28	08 24 24	05 20 58	21 ♐ 36 52	28 ♐ 37 49	25 R 18	24 27	02 09	12 56	21 28	14 14	01 22	02 00
29	08 28 20	06 18 19	05 ♑ 37 08	12 ♑ 34 24	25 14	25 35	02 48	12 48	21 29	14 13	01 21	02 01
30	08 32 17	07 15 40	19 ♑ 29 10	26 ♑ 20 59	25 03	26 43	03 26	12 41	21 29	14 12	01 20	02 01
31	08 36 13	08 ♌ 13 02	03 ♒ 09 24	09 ♒ 54 01	24 ♌ 48	27 ♊ 51	04 ♌ 05	12 ♒ 33	21 ♏ 30	14 ♐ 10	01 ♑ 19	02 ♏ 02

Date	Moon True ☊ ° '	Moon Mean ☊ ° '	Moon ☽ Latitude ° '
01	17 ♉ 03	15 ♉ 32	03 S 14
02	16 R 56	15 29	04 06
03	16 48	15 26	04 43
04	16 39	15 23	05 02
05	16 30	15 20	05 04
06	16 23	15 17	04 51
07	16 17	15 13	04 23
08	16 13	15 10	03 44
09	16 12	15 07	02 55
10	16 D 12	15 04	02 00
11	16 12	15 01	00 S 59
12	16 R 13	14 57	00 N 04
13	16 12	14 54	01 07
14	16 09	14 51	02 08
15	16 04	14 48	03 04
16	15 56	14 45	03 52
17	15 46	14 42	04 30
18	15 35	14 38	04 54
19	15 24	14 35	05 02
20	15 14	14 32	04 53
21	15 05	14 29	04 27
22	14 59	14 26	03 45
23	14 56	14 23	02 49
24	14 55	14 19	01 43
25	14 D 55	14 16	00 N 30
26	14 R 54	14 13	00 S 45
27	14 53	14 10	01 57
28	14 49	14 07	03 01
29	14 43	14 03	03 54
30	14 34	14 00	04 33
31	14 ♉ 22	13 ♉ 57	04 S 55

DECLINATIONS

Date	Sun ☉ ° '	Moon ☽ ° '	Mercury ☿ ° '	Venus ♀ ° '	Mars ♂ ° '	Jupiter ♃ ° '	Saturn ♄ ° '	Uranus ♅ ° '	Neptune ♆ ° '	Pluto ♇ ° '
01	23 N 06	26 S 37	20 N 59	16 N 13	23 N 36	16 S 48	16 S 05	22 S 38	22 S 16	03 N 45
02	23 01	27 07	20 32	16 29	23 32	16 50	16 04	22 38	22 16	03 44
03	22 57	25 52	20 04	16 44	23 28	16 52	16 04	22 38	22 16	03 44
04	22 52	23 05	19 36	16 59	23 23	16 53	16 04	22 37	22 16	03 43
05	22 46	19 08	19 07	17 14	23 19	16 55	16 03	22 37	22 16	03 43
06	22 40	14 23	18 38	17 29	23 14	16 57	16 03	22 37	22 16	03 42
07	22 34	09 09	18 08	17 43	23 09	16 59	16 03	22 37	22 16	03 42
08	22 27	03 S 41	17 38	17 57	23 04	17 01	16 03	22 36	22 16	03 41
09	22 20	01 N 48	17 09	18 11	22 58	17 03	16 02	22 36	22 16	03 41
10	22 13	07 11	16 39	18 25	22 53	17 05	16 02	22 36	22 16	03 40
11	22 05	12 16	16 09	18 38	22 47	17 07	16 02	22 36	22 16	03 40
12	21 57	16 56	15 40	18 50	22 41	17 09	16 02	22 36	22 16	03 39
13	21 48	20 59	15 10	19 03	22 35	17 11	16 02	22 35	22 16	03 39
14	21 39	24 11	14 42	19 15	22 29	17 13	16 02	22 35	22 16	03 38
15	21 30	26 20	14 13	19 27	22 22	17 15	16 02	22 35	22 16	03 38
16	21 20	27 11	13 45	19 38	22 16	17 17	16 02	22 35	22 17	03 37
17	21 10	26 34	13 18	19 49	22 09	17 19	16 02	22 35	22 17	03 36
18	20 59	24 29	12 52	20 00	22 02	17 22	16 02	22 34	22 17	03 36
19	20 49	21 00	12 27	20 10	21 55	17 24	16 02	22 34	22 17	03 35
20	20 37	16 21	12 02	20 19	21 48	17 26	16 02	22 34	22 17	03 35
21	20 26	10 47	11 39	20 29	21 40	17 28	16 02	22 34	22 17	03 34
22	20 14	04 N 39	11 17	20 37	21 33	17 31	16 02	22 34	22 17	03 33
23	20 02	01 S 46	10 56	20 46	21 25	17 33	16 03	22 34	22 17	03 33
24	19 49	08 08	10 37	20 54	21 17	17 35	16 03	22 33	22 17	03 32
25	19 37	14 07	10 20	21 01	21 09	17 37	16 03	22 33	22 17	03 31
26	19 23	19 21	10 04	21 08	21 01	17 40	16 03	22 33	22 17	03 31
27	19 10	23 29	09 50	21 14	20 53	17 42	16 04	22 33	22 17	03 30
28	18 56	26 11	09 38	21 20	20 44	17 44	16 04	22 33	22 17	03 29
29	18 42	27 13	09 29	21 26	20 35	17 47	16 04	22 33	22 17	03 28
30	18 28	26 31	09 21	21 30	20 27	17 49	16 05	22 33	22 17	03 28
31	18 N 13	24 S 14	09 N 17	21 N 35	20 N 18	17 S 51	16 S 05	22 S 32	22 S 17	03 N 27

ZODIAC SIGN ENTRIES

Date	h	m	Planets
01	18	22	☽ ♑
03	21	36	☽ ♒
06	03	40	☽ ♓
06	08	01	♀ ♊
08	13	20	☽ ♈
11	01	44	☽ ♉
13	14	23	☽ ♊
16	00	54	☽ ♋
18	08	25	☽ ♌
20	13	29	☽ ♍
22	17	10	☽ ♎
22	21	36	☉ ♌
24	20	16	☽ ♏
25	04	04	♂ ♌
26	23	12	☽ ♐
29	02	21	☽ ♑
31	06	25	☽ ♒

LATITUDES

Date	Mercury ☿ ° '	Venus ♀ ° '	Mars ♂ ° '	Jupiter ♃ ° '	Saturn ♄ ° '	Uranus ♅ ° '	Neptune ♆ ° '	Pluto ♇ ° '
01	01 N 26	02 S 51	00 N 58	00 S 45	02 N 15	00 S 02	01 N 10	16 N 54
04	01 04	02 48	00 59	00 46	02 15	00 02	01 10	16 53
07	00 39	02 45	01 00	00 47	02 14	00 02	01 10	16 51
10	00 N 09	02 41	01 01	00 47	02 13	00 02	01 10	16 49
13	00 S 25	02 35	01 01	00 48	02 12	00 02	01 09	16 47
16	01 02	02 30	01 02	00 48	02 12	00 02	01 09	16 46
19	01 41	02 23	01 03	00 49	02 11	00 02	01 09	16 44
22	02 21	02 16	01 03	00 49	02 10	00 02	01 09	16 42
25	03 01	02 08	01 04	00 50	02 09	00 02	01 09	16 40
28	03 39	02 00	01 05	00 50	02 09	00 02	01 09	16 39
31	04 S 13	01 S 51	01 N 05	00 S 51	02 N 08	00 S 02	01 N 09	16 N 37

DATA

Julian Date	2446248
Delta T	+54 seconds
Ayanamsa	23° 39' 04"
Synetic vernal point	05° ♓ 27' 55"
True obliquity of ecliptic	23° 26' 34"

LONGITUDES

Date	Chiron ⚷ ° '	Ceres ⚳ ° '	Pallas ⚴ ° '	Juno ⚵ ° '	Vesta ⚶ ° '	Black Moon Lilith ⚸ ° '
01	11 ♊ 17	09 ♋ 19	22 ♉ 15	02 ♎ 13	25 ♎ 08	03 ♉ 18
11	11 ♊ 59	13 ♋ 40	26 ♉ 37	04 ♎ 20	27 ♎ 32	04 ♉ 24
21	12 ♊ 37	18 ♋ 02	01 ♊ 00	06 ♎ 42	00 ♏ 27	05 ♉ 31
31	13 ♊ 11	22 ♋ 23	05 ♊ 21	09 ♎ 18	03 ♏ 47	06 ♉ 37

MOON'S PHASES, APSIDES AND POSITIONS ☽

Date	h	m	Phase	Longitude ° '	Eclipse Indicator
02	12	08	○	10 ♑ 33	
10	00	49	☾	17 ♈ 43	
17	23	56	●	25 ♋ 19	
24	23	39	☽	01 ♏ 59	
31	21	41	○	08 ♒ 36	

Day	h	m		
11	07	43	Apogee	
25	17	34	Perigee	
02	06	39	Max dec	27° S 10'
09	04	05	ON	
16	14	16	Max dec	27° N 11'
23	05	24	OS	
29	14	12	Max dec	27° S 14'

ASPECTARIAN

h m	Aspects
01 Monday	
02 11	☿ □ ♇
03 30	☿ ⊼ ♆
04 55	☽ ⊻ ♄
09 36	☽ ⊼ ♀
12 43	☽ ± ☿
14 54	☽ ⊥ ♄
18 25	☉ ± ♃
19 46	☽ ∠ ♃
20 26	☽ ± ♀
21 39	☽ ⚹ ♇
21 46	☽ ☌ ♆
23 56	☽ ⊼ ☿
02 Tuesday	
02 46	♂ ⊼ ♅
05 59	☽ ∠ ♄
10 51	☽ ⊥ ♃
12 08	☽ ☍ ☉
12 50	☽ ⚼ ♀
13 54	♀ ± ♆
17 55	☽ Q ♇
19 39	☽ ⊻ ♅
20 32	☽ ☍ ♂
21 00	☽ ⊻ ♃
03 Wednesday	
05 25	♂ ⊼ ♃
05 55	☽ ⊥ ♅
07 29	☽ ⚹ ♄
16 41	☽ △ ♀
21 28	☽ ∠ ♅
04 Thursday	
00 54	♀ # ♃
01 00	☽ □ ♇
01 02	☽ ⊻ ♆
04 20	☽ Q ♄
09 27	☽ ☍ ☿
09 46	☽ # ♂
11 36	☽ ⊥ ♆
13 36	☽ # ☉
15 10	☽ ∥ ♅
17 30	☽ ∥ ♆
20 12	☽ ⊼ ☉
05 Friday	
00 00	☽ ⚹ ♅
00 45	♆ ⚹ ♇
01 13	☽ ☌ ♃
03 42	☽ ∠ ♆
03 47	☽ ⊼ ♂
06 25	☿ ± ♆
07 52	☽ ± ☉
12 08	☽ # ☿
12 32	☽ □ ♄
15 17	☽ ± ♂
21 28	☽ # ♀
21 50	☽ Q ♅
23 31	☽ ∥ ♃
06 Saturday	
01 42	☽ ⚼ ☉
03 16	☽ □ ♀
03 55	☽ ∥ ♄
07 13	☽ ⚹ ♆
07 17	☽ △ ♇
08 51	☽ ⚼ ♂
22 46	☽ ⊼ ☿
23 52	☉ ⊼ ♅
07 Sunday	
01 17	☉ # ♅
05 55	☽ Q ♆
07 42	☽ □ ♅
08 22	☽ △ ☉
08 44	☽ ⊻ ♃
11 34	☽ ± ☿
11 47	☽ ⚼ ♇
12 43	☉ ⊼ ♃
15 01	☽ △ ♂
16 41	☽ Q ♀
20 16	☽ ⊥ ♃
21 05	☽ △ ♄
08 Monday	
01 34	☿ ∥ ♀
01 35	♀ ⊼ ♆
03 08	♀ ⊼ ♇
05 19	☽ ± ♇
07 04	☽ ⚼ ☿
12 00	☽ # ♇
13 48	☽ ∠ ♃
17 00	☽ □ ♆
17 10	☽ ⊼ ♇
18 32	☽ ⚹ ♀
09 Tuesday	
02 39	☽ ⚼ ♄
16 14	☽ △ ☿
16 27	☿ # ♃
18 47	☽ △ ♅
19 33	☽ ⚹ ♃
20 17	☽ ∥ ♇
20 41	☽ ± ♄
10 Wednesday	
00 23	☉ # ♆
00 49	☽ □ ☉
03 38	☽ ∠ ♀
06 04	☽ □ ♂
08 48	☽ ⊼ ♄
18 18	☿ △ ♅
19 39	☽ Q ♃
20 11	♂ ± ♅
23 45	☽ ⊥ ♀
11 Thursday	
00 32	☿ ☍ ♃
01 06	☽ ⚼ ♅
05 21	☽ △ ♆
05 39	☽ ☍ ♇
13 11	☽ ⊻ ♀
16 35	☽ Q ☉
17 41	☿ # ♄
19 21	☽ ± ♅
20 46	☽ Q ♂
12 Friday	
05 49	☽ ∥ ☿
06 01	♂ △ ♄
07 08	☽ # ♄
07 30	☽ ⊼ ♅
07 54	☽ □ ♃
08 40	♇ St D
11 18	☽ □ ☿
11 44	☽ ⚼ ♆
13 09	☽ # ♃
16 17	☿ ⚼ ♆
18 08	♀ ∠ ♂
19 01	☽ ⚹ ☉
21 33	☽ ☍ ♄
22 28	☽ ⚹ ♂
23 21	☽ ∥ ♀
13 Saturday	
01 58	☉ ± ♅
05 50	☽ ± ♆
10 32	♂ # ♅
15 39	♀ ± ♇
17 16	☽ ∥ ☉
17 48	☽ ⊼ ♆
18 14	☽ ⊼ ♇
20 51	☽ # ♆
22 45	☽ ∥ ♂
23 08	☽ # ♅
14 Sunday	
01 54	☉ △ ♄
02 37	☽ Q ☿
03 36	☽ ∠ ☉
06 07	☽ ∠ ♂
06 07	☽ ± ♇
07 33	☽ ☌ ♀
19 11	☽ ☍ ♅
19 12	☽ △ ♃
21 33	♃ ⚹ ♅
22 45	☽ ⊥ ☉
23 46	☽ ⚼ ♇
15 Monday	
00 44	☽ ⊥ ♂
04 02	☽ ⚹ ☿
08 44	☽ ⊼ ♄
11 20	☽ ⊻ ☉
12 58	☽ ⊻ ♂
20 14	☽ ± ♄
23 52	☽ ⚼ ♃
16 Tuesday	
02 17	☿ Q ♇
04 03	☽ ☍ ♆
04 34	☽ △ ♇
09 06	♂ # ♆
10 47	☽ ∠ ☿
13 15	☽ ⚼ ♄
16 42	☽ ± ♃
22 36	☽ ⊻ ♀
17 Wednesday	
03 46	☽ ⊼ ♃
04 07	☽ ⊼ ♅
04 41	☽ ⊥ ☿
10 44	☽ ⊥ ♀
15 08	☽ ± ♅
16 25	☽ ⊻ ☿
17 01	☽ △ ♄
23 56	☽ ☌ ☉
18 Thursday	
00 00	☽ ☌ ♂
02 41	☉ ☌ ♂
02 53	☿ □ ♄
04 40	☽ ∠ ♀
07 28	☽ ⚼ ♅
11 18	☽ ⊼ ♆
11 54	☽ □ ♇
22 02	☽ ± ♆
19 Friday	
02 16	☽ # ♅
04 12	☽ # ♆
05 39	♀ △ ♃
06 19	☽ ∥ ♂
09 32	☽ ☍ ♃
09 55	☽ ⚹ ♀
10 13	☽ △ ♅
13 11	☽ ∥ ☉
13 32	♀ ☍ ♅
13 58	☽ ⚼ ♆
16 37	☽ ∥ ♀
19 54	☽ Q ♇
22 38	☽ □ ♄
20 Saturday	
00 54	☽ ☌ ☿
06 53	☽ # ♃
08 02	☽ ⊻ ♂
08 51	☽ Q ♀
09 18	☽ ⊻ ☉
13 26	☽ # ♄
16 11	☽ △ ♆
16 52	☽ ⚹ ♇
18 59	☽ ⊥ ♂
20 31	☽ ⊥ ☉
21 Sunday	
05 56	☽ Q ♄
08 14	☽ ∥ ☿
08 53	☉ ∥ ♀
11 19	☽ ∠ ♂
13 13	☽ ∠ ☉
13 22	☽ ⊼ ♃
14 23	☽ □ ♅
18 43	☽ □ ♀
18 48	☽ ∠ ♇
19 37	♀ ⚼ ♇
23 36	☽ ± ♃
22 Monday	
02 34	☽ ⚹ ♄
05 52	☉ ⚼ ♅
06 52	☽ ⊻ ☿
10 15	☽ ⊥ ♇
14 19	☽ ⚹ ♂
14 54	☽ ⚼ ♃
16 07	☽ ∥ ♇
16 51	☽ ⚹ ☉
17 28	☽ ⊥ ☿
19 43	☽ □ ♆
20 31	☽ ⊻ ♇
21 12	☽ Q ♅
23 Tuesday	
04 13	☽ ∠ ♄
09 16	☽ ∠ ☿
11 48	☽ Q ♂
14 49	☽ Q ☉
16 16	☽ △ ♃
17 38	☽ ⚹ ♅
18 36	☽ # ♇
19 33	☽ ⊥ ♄
24 Wednesday	
02 22	☽ Q ♆
02 25	☽ △ ♀
03 07	♂ ⚼ ♅
05 46	☽ ⊻ ♄
10 18	☉ ⊼ ♆
11 25	☽ ⚹ ☿
19 06	☽ ∠ ♅
19 54	☽ □ ♂
21 15	☽ # ☿
22 44	☽ ⚹ ♆
23 06	☉ □ ♇
23 37	☽ ☌ ♇
23 39	☽ □ ☉
25 Thursday	
06 07	☽ ⚼ ♀
08 11	☽ Q ☿
10 22	☽ ⊥ ♅
18 49	☽ □ ♃
19 34	♄ St D
20 25	♀ ⊼ ♄
20 25	☽ ∥ ♄
20 32	☽ ⊻ ♅
22 43	☽ ± ♀
26 Friday	
00 10	☽ ∠ ♆
00 48	♀ ∥ ♂
03 43	☽ ∥ ♃
08 43	☽ ☌ ♄
09 47	☽ ⊼ ♀
12 11	☽ # ☉
15 07	☽ □ ☿
15 25	☽ ⊥ ♆
20 32	☽ # ♂
21 42	☽ # ♀
27 Saturday	
01 06	☽ Q ♃
01 16	☽ △ ♂
01 35	☽ ⊻ ♆
02 35	☽ ⊻ ♇
04 11	☽ ∥ ♆
05 50	☽ ∥ ♅
06 17	☽ △ ☉
07 57	♂ ⊼ ♆
12 47	☽ ⊥ ♇
21 20	☽ ⚹ ♃
23 26	☽ ☌ ♅
28 Sunday	
00 52	☿ St R
04 01	☽ ⚼ ♂
04 07	☽ ∠ ♇
06 15	♂ □ ♇
09 41	☽ ⚼ ☉
11 46	☽ ⊻ ♄
17 17	☽ ☍ ♀
18 17	☽ △ ☿
20 08	☽ ± ♂
22 01	☽ ⊥ ♄
22 43	☽ ∠ ♃
29 Monday	
02 12	☽ ± ☉
04 41	☽ ☌ ♆
05 01	☿ ⚹ ♀
05 47	☽ ⚹ ♇
06 55	☽ ⊼ ♂
13 16	☽ ⊼ ☉
13 29	☽ ∠ ♄
14 02	☽ ⊥ ♃
19 52	☽ ⚼ ☿
30 Tuesday	
00 17	☽ ⊻ ♃
02 30	☽ Q ♇
02 49	☽ ⊻ ♅
11 16	☽ ± ☿
13 14	☽ ⊥ ♅
13 42	☉ ± ♆
15 29	☽ ⚹ ♄
21 35	☽ ⊼ ☿
31 Wednesday	
01 48	☽ ⊼ ♀
04 17	♀ ± ♄
04 58	☽ ∠ ♅
06 10	♀ ⚼ ♃
08 44	☽ ⊻ ♆
10 00	☽ □ ♇
12 36	☽ Q ♄
13 21	☽ ± ♀
13 44	☽ ☍ ♂
19 22	☽ ⊥ ♆
21 41	☽ ☍ ☉
23 08	♀ ⊥ ♂

All ephemeris data is given at 12.00 UT and the Moon's longitude is additionally given for 24.00 UT

AUGUST 1985

Date	Sidereal time h m s	LONGITUDES Sun ☉ ° ' "	Moon ☽ ° ' "	Moon ☽ 24.00 ° ' "	Mercury ☿ ° '	Venus ♀ ° '	Mars ♂ ° '	Jupiter ♃ ° '	Saturn ♄ ° '	Uranus ♅ ° '	Neptune ♆ ° '	Pluto ⯓ ° '
01	08 40 10	09 ♌ 10 25	16 ♒ 34 29	23 ♒ 10 32	24 ♌ 28	29 ♊ 00	04 ♌ 44	12 ♒ 25	21 ♏ 30	14 ♐ 09	01 ♑ 17	02 ♏ 02
02	08 44 07	10 07 48	29 ♒ 41 59	06 ♓ 08 43	24 R 03	00 ♋ 08	05 22	12 R 17	21 31	14 R 08	01 R 16	02 03
03	08 48 03	11 05 13	12 ♓ 30 45	18 ♓ 48 11	23 33	01 17	06 01	12 10	21 32	14 07	01 15	02 04
04	08 52 00	12 02 38	25 ♓ 01 12	01 ♈ 10 06	22 59	02 25	06 40	12 02	21 33	14 06	01 14	02 05
05	08 55 56	13 00 05	07 ♈ 15 16	13 ♈ 17 09	22 21	03 34	07 18	11 54	21 34	14 05	01 13	02 05
06	08 59 53	13 57 33	19 ♈ 16 14	25 ♈ 13 07	21 39	04 43	07 57	11 46	21 35	14 05	01 12	02 06
07	09 03 49	14 55 02	01 ♉ 08 24	07 ♉ 02 43	20 55	05 52	08 35	11 38	21 36	14 04	01 11	02 07
08	09 07 46	15 52 33	12 ♉ 56 46	18 ♉ 51 13	20 09	07 01	09 14	11 31	21 37	14 03	01 10	02 08
09	09 11 42	16 50 05	24 ♉ 46 47	00 ♊ 44 07	19 21	08 10	09 52	11 23	21 39	14 02	01 09	02 09
10	09 15 39	17 47 38	06 ♊ 43 54	12 ♊ 46 46	18 32	09 20	10 31	11 15	21 40	14 02	01 08	02 10
11	09 19 36	18 45 13	18 ♊ 53 19	25 ♊ 04 05	17 44	10 29	11 09	11 07	21 42	14 01	01 07	02 11
12	09 23 32	19 42 49	01 ♋ 19 33	07 ♋ 40 06	16 58	11 39	11 48	11 00	21 43	14 01	01 06	02 12
13	09 27 29	20 40 26	14 ♋ 06 02	20 ♋ 37 30	16 13	12 48	12 26	10 52	21 45	14 00	01 05	02 13
14	09 31 25	21 38 05	27 ♋ 14 34	03 ♌ 57 10	15 32	13 58	13 05	10 45	21 47	14 00	01 04	02 14
15	09 35 22	22 35 46	10 ♌ 45 06	17 ♌ 38 02	14 55	15 08	13 43	10 37	21 49	13 59	01 03	02 15
16	09 39 18	23 33 27	24 ♌ 35 31	01 ♍ 37 00	14 22	16 18	14 21	10 30	21 51	13 59	01 02	02 16
17	09 43 15	24 31 10	08 ♍ 41 52	15 ♍ 49 25	13 55	17 28	15 00	10 22	21 53	13 59	01 01	02 18
18	09 47 11	25 28 54	22 ♍ 58 57	00 ♎ 09 47	13 35	18 38	15 38	10 15	21 55	13 58	01 01	02 19
19	09 51 08	26 26 39	07 ♎ 21 15	14 ♎ 32 44	13 21	19 48	16 17	10 08	21 57	13 58	01 00	02 20
20	09 55 05	27 24 25	21 ♎ 43 43	28 ♎ 53 44	13 14	20 59	16 55	10 01	22 00	13 58	00 59	02 21
21	09 59 01	28 22 12	06 ♏ 02 27	13 ♏ 09 34	13 D 14	22 09	17 33	09 54	22 02	13 58	00 58	02 23
22	10 02 58	29 ♌ 20 01	20 ♏ 14 52	27 ♏ 18 15	13 22	23 20	18 11	09 47	22 05	13 58	00 58	02 24
23	10 06 54	00 ♍ 17 51	04 ♐ 19 35	11 ♐ 18 48	13 38	24 30	18 50	09 40	22 08	13 58	00 57	02 25
24	10 10 51	01 15 42	18 ♐ 15 53	25 ♐ 10 45	14 01	25 41	19 28	09 33	22 10	13 D 58	00 57	02 27
25	10 14 47	02 13 34	02 ♑ 03 22	08 ♑ 53 38	14 33	26 52	20 06	09 26	22 13	13 58	00 56	02 28
26	10 18 44	03 11 27	15 ♑ 41 27	22 ♑ 26 42	15 12	28 03	20 45	09 20	22 16	13 58	00 55	02 30
27	10 22 40	04 09 22	29 ♑ 09 13	05 ♒ 48 52	15 59	29 ♋ 14	21 23	09 13	22 19	13 58	00 55	02 31
28	10 26 37	05 07 18	12 ♒ 25 27	18 ♒ 58 47	16 53	00 ♌ 25	22 01	09 07	22 22	13 59	00 54	02 33
29	10 30 34	06 05 15	25 ♒ 28 45	01 ♓ 55 12	17 54	01 36	22 39	09 01	22 26	13 59	00 54	02 34
30	10 34 30	07 03 14	08 ♓ 18 02	14 ♓ 37 13	19 01	02 47	23 17	08 55	22 29	13 59	00 54	02 36
31	10 38 27	08 ♍ 01 15	20 ♓ 52 45	27 ♓ 04 43	20 ♌ 15	03 ♌ 58	23 ♌ 55	08 ♒ 49	22 ♏ 32	14 ♐ 00	00 ♑ 53	02 ♏ 37

Date	Moon True ☊ ° '	Moon Mean ☊ ° '	Moon ☽ Latitude ° '
01	14 ♉ 10	13 ♉ 54	05 S 00
02	13 R 58	13 51	04 49
03	13 47	13 48	04 24
04	13 39	13 44	03 47
05	13 33	13 41	02 59
06	13 29	13 38	02 04
07	13 27	13 35	01 05
08	13 D 27	13 32	00 S 03
09	13 R 27	13 29	01 N 00
10	13 26	13 25	02 00
11	13 23	13 22	02 56
12	13 18	13 19	03 45
13	13 10	13 16	04 24
14	13 00	13 13	04 50
15	12 49	13 09	05 01
16	12 37	13 06	04 54
17	12 26	13 03	04 30
18	12 17	13 00	03 49
19	12 10	12 57	02 53
20	12 07	12 54	01 46
21	12 06	12 50	00 N 32
22	12 D 06	12 47	00 S 43
23	12 R 06	12 44	01 55
24	12 05	12 41	03 00
25	12 01	12 38	03 53
26	11 56	12 35	04 32
27	11 47	12 31	04 56
28	11 37	12 28	05 03
29	11 26	12 25	04 54
30	11 15	12 22	04 30
31	11 ♉ 05	12 ♉ 19	03 S 54

DECLINATIONS

Date	Sun ☉ ° '	Moon ☽ ° '	Mercury ☿ ° '	Venus ♀ ° '	Mars ♂ ° '	Jupiter ♃ ° '	Saturn ♄ ° '	Uranus ♅ ° '	Neptune ♆ ° '	Pluto ⯓ ° '
01	17 N 58	20 S 38	09 N 14	21 N 39	20 N 09	17 S 54	16 S 05	22 S 32	22 S 17	03 N 26
02	17 43	16 05	09 14	21 42	19 59	17 56	16 06	22 32	22 17	03 25
03	17 27	10 56	09 17	21 45	19 50	17 58	16 06	22 32	22 17	03 25
04	17 11	05 S 27	09 22	21 47	19 41	18 01	16 07	22 32	22 17	03 24
05	16 55	00 N 08	09 30	21 49	19 31	18 03	16 07	22 32	22 17	03 23
06	16 38	05 38	09 40	21 50	19 21	18 05	16 08	22 32	22 17	03 22
07	16 22	10 52	09 53	21 50	19 11	18 08	16 08	22 32	22 18	03 21
08	16 05	15 41	10 08	21 50	19 01	18 10	16 09	22 32	22 18	03 21
09	15 48	19 56	10 24	21 49	18 51	18 12	16 09	22 32	22 18	03 20
10	15 30	23 25	10 43	21 48	18 41	18 14	16 10	22 32	22 18	03 19
11	15 12	25 54	11 03	21 46	18 31	18 16	16 11	22 32	22 18	03 18
12	14 54	27 11	11 24	21 44	18 20	18 19	16 11	22 31	22 18	03 17
13	14 36	27 04	11 47	21 41	18 09	18 21	16 12	22 31	22 18	03 16
14	14 18	25 27	12 09	21 37	17 59	18 23	16 13	22 31	22 18	03 15
15	13 59	22 21	12 32	21 33	17 48	18 25	16 14	22 31	22 18	03 15
16	13 40	17 57	12 55	21 29	17 37	18 27	16 14	22 31	22 18	03 14
17	13 21	12 29	13 17	21 23	17 26	18 29	16 15	22 31	22 18	03 13
18	13 02	06 N 18	13 38	21 17	17 14	18 31	16 16	22 31	22 18	03 12
19	12 42	00 S 16	13 58	21 11	17 03	18 33	16 17	22 31	22 18	03 11
20	12 22	06 50	14 17	21 04	16 52	18 35	16 18	22 31	22 18	03 10
21	12 03	13 02	14 33	20 56	16 40	18 37	16 18	22 31	22 18	03 09
22	11 42	18 30	14 48	20 48	16 28	18 39	16 19	22 31	22 18	03 08
23	11 22	22 54	15 01	20 39	16 17	18 41	16 20	22 31	22 18	03 07
24	11 02	25 54	15 12	20 30	16 05	18 43	16 21	22 31	22 18	03 06
25	10 41	27 18	15 19	20 20	15 53	18 45	16 22	22 31	22 18	03 06
26	10 20	27 01	15 24	20 09	15 41	18 47	16 23	22 31	22 18	03 05
27	09 59	25 09	15 27	19 58	15 29	18 48	16 24	22 31	22 18	03 04
28	09 38	21 55	15 26	19 46	15 16	18 50	16 25	22 31	22 18	03 03
29	09 17	17 38	15 22	19 34	15 04	18 52	16 26	22 31	22 19	03 02
30	08 55	12 38	15 15	19 21	14 52	18 53	16 27	22 31	22 19	03 01
31	08 N 34	07 S 12	15 N 05	19 N 08	14 N 39	18 S 55	16 S 28	22 S 31	22 S 19	03 N 00

ZODIAC SIGN ENTRIES

Date	h	m	Planets
02	09	10	♀ ♋
02	12	33	☽ ♓
04	21	43	☽ ♈
07	09	41	☽ ♉
09	22	31	☽ ♊
12	09	28	☽ ♋
14	16	57	☽ ♌
16	21	15	☽ ♍
18	23	44	☽ ♎
21	01	51	☽ ♏
23	04	36	☉ ♍
23	04	36	☽ ♐
25	08	24	☽ ♑
27	13	31	☽ ♒
28	03	39	♀ ♌
29	20	25	☽ ♓

LATITUDES

Date	Mercury ☿ ° '	Venus ♀ ° '	Mars ♂ ° '	Jupiter ♃ ° '	Saturn ♄ ° '	Uranus ♅ ° '	Neptune ♆ ° '	Pluto ⯓ ° '
01	04 S 23	01 S 48	01 N 05	00 S 51	02 N 08	00 S 02	01 N 09	16 N 36
04	04 45	01 38	01 06	00 51	02 07	00 02	01 09	16 35
07	04 54	01 29	01 07	00 52	02 06	00 02	01 09	16 33
10	04 47	01 19	01 07	00 52	02 05	00 02	01 09	16 31
13	04 24	01 09	01 08	00 52	02 04	00 02	01 09	16 30
16	03 46	00 59	01 08	00 53	02 04	00 02	01 08	16 28
19	02 58	00 49	01 09	00 53	02 03	00 02	01 08	16 26
22	02 05	00 38	01 09	00 53	02 02	00 02	01 08	16 25
25	01 12	00 28	01 09	00 53	02 02	00 02	01 08	16 23
28	00 S 22	00 18	01 10	00 53	02 01	00 02	01 08	16 22
31	00 N 22	00 S 08	01 N 10	00 S 53	02 N 00	00 S 02	01 N 08	16 N 20

DATA

Julian Date	2446279
Delta T	+54 seconds
Ayanamsa	23° 39' 10"
Synetic vernal point	05° ♓ 27' 49"
True obliquity of ecliptic	23° 26' 35"

MOON'S PHASES, APSIDES AND POSITIONS ☽

Date	h	m	Phase	Longitude ° '	Eclipse Indicator
08	18	29	☾	16 ♉ 08	
16	10	06	●	23 ♌ 29	
23	04	36	☽	00 ♐ 00	
30	09	27	○	06 ♓ 57	

Day	h	m		
08	02	22	Apogee	
20	04	14	Perigee	
05	11	24	ON	
12	22	09	Max dec	27° N 19'
19	11	01	OS	
25	19	53	Max dec	27° S 24'

LONGITUDES

Date	Chiron ⚷ ° '	Ceres ⚳ ° '	Pallas ⚴ ° '	Juno ⚵ ° '	Vesta ⚶ ° '	Black Moon Lilith ⚸ ° '
01	13 ♊ 15	22 ♋ 49	05 ♊ 47	09 ♎ 35	04 ♏ 08	06 ♉ 44
11	13 ♊ 44	27 ♋ 10	10 ♊ 06	12 ♎ 23	07 ♏ 51	07 ♉ 50
21	14 ♊ 07	01 ♌ 29	14 ♊ 21	15 ♎ 21	11 ♏ 51	08 ♉ 57
31	14 ♊ 25	05 ♌ 46	18 ♊ 29	18 ♎ 26	16 ♏ 06	10 ♉ 03

ASPECTARIAN

h m Aspects

01 Thursday
00 14 ☽ ∥ ♅
01 54 ☽ ∥ ♆
04 35 ☽ ☌ ♃
06 03 ☽ # ♀
06 55 ☽ ♇ ♀
07 38 ☽ ⚹ ♅
11 29 ☽ ∠ ♆
14 56 ☽ # ♂
17 39 ☉ # ♃
20 57 ☽ □ ♄
02 Friday
01 56 ☽ ☍ ☿
02 49 ☽ ∥ ♃
03 26 ☽ # ☉
05 26 ☽ Q ♅
11 59 ☽ ∥ ♄
12 53 ☽ △ ♀
14 54 ☽ ⚹ ♆
16 22 ☽ △ ⯓
23 07 ☽ ⊼ ♂
03 Saturday
09 05 ☽ ⊼ ☉
11 01 ☽ ± ♂
11 20 ☽ ⊻ ♃
11 28 ♀ ☍ ♆
13 24 ☽ Q ♆
15 03 ☽ □ ♅
19 13 ☽ # ☿
20 41 ☽ ♇ ⯓
21 26 ☽ ± ☉
22 39 ☽ ⊥ ♃
04 Sunday
04 42 ♀ △ ⯓
05 09 ☽ ♇ ♂
05 16 ☽ △ ♄
08 14 ☽ ⊼ ☿
11 41 ☉ ☍ ♃
14 03 ☽ ± ⯓
15 52 ☽ ∠ ♃
16 16 ☽ ♇ ☉
19 21 ☽ ± ☿
20 50 ☽ # ⯓
05 Monday
00 06 ☽ □ ♆
01 48 ☽ ⊼ ⯓
03 58 ☽ □ ♀
08 46 ♂ ± ♆
10 37 ☽ ♇ ♄
12 06 ☽ △ ♂
12 11 ☽ ♇ ☿
21 08 ☽ ⚹ ♃
06 Tuesday
00 25 ☽ △ ☉
01 36 ☽ △ ♅
02 04 ☽ ∥ ⯓
04 34 ☽ ± ♄
14 34 ☿ □ ♄
14 55 ☉ △ ♅
16 32 ☽ △ ☿
16 39 ☽ ⊼ ♄
19 41 ☽ Q ♀
20 58 ☽ Q ♃
07 Wednesday
07 10 ☽ ∥ ☿
07 47 ♀ ± ♃
07 48 ☽ ♇ ♅
12 05 ☽ △ ♆
12 39 ☿ ∠ ♀
13 59 ☽ ☍ ⯓
22 38 ☽ ⚹ ♀
08 Thursday
02 03 ☽ ± ♅
03 33 ♀ ♇ ♄
04 00 ☽ □ ♂
06 32 ☉ # ♄
09 07 ☽ □ ♃
12 16 ☿ Q ⯓
13 58 ☽ ∥ ☉
14 15 ☽ ⊼ ♅
14 27 ☽ # ♄
18 29 ☽ □ ☉
18 31 ☽ ♇ ♆
19 00 ☉ ♇ ♆
09 Friday
01 40 ☽ # ♃
01 42 ☽ □ ☿
05 38 ☽ ☍ ♄
05 47 ☽ ∥ ♂
08 24 ☽ ∠ ♀
12 44 ☽ ± ♆
18 36 ☽ Q ♂
10 Saturday
00 14 ☽ ∥ ♀
00 48 ☽ ⊼ ♆
02 51 ☽ ⊼ ⯓
03 37 ☽ # ♆
04 28 ☽ ⊥ ♀
05 17 ☽ # ♅
09 58 ☽ Q ☉
11 38 ☽ Q ☿
14 52 ☽ ± ⯓
17 43 ☽ ⊻ ♀
19 56 ☽ ⚹ ♂
20 53 ☽ △ ♃
22 08 ☉ ☌ ☿
11 Sunday
02 27 ☽ ☍ ♅
08 39 ☽ ♇ ⯓
09 53 ☽ ⚹ ☿
11 04 ♂ ☍ ♃
11 43 ☽ ⚹ ☉
17 28 ☽ ⊼ ♄
23 56 ♀ ⊼ ♃
12 Monday
01 54 ☽ ♇ ♃
02 51 ☽ ∠ ♂
03 30 ☿ ⊥ ♀
05 05 ☽ ± ♄
11 34 ☽ ☍ ♆
13 08 ☽ ∠ ☿
13 40 ☽ △ ⯓
14 26 ♂ # ♃
18 54 ☽ ± ♃
18 57 ☽ ∠ ☉
18 59 ♀ ⊻ ♂
20 55 ☽ ⊥ ♂
22 14 ☽ ♇ ♄
13 Tuesday
00 22 ☉ Q ⯓
05 10 ☽ ⊥ ☿
06 03 ☽ ⊼ ♃
08 45 ☽ ⊻ ♂
09 22 ☽ ☌ ♀
11 49 ☽ ⊼ ♅
13 09 ☽ ⊥ ☉
15 43 ☽ ⊻ ☿
16 53 ☿ ♇ ♆
22 51 ☽ ± ♅
14 Wednesday
01 03 ☽ ⊻ ☉
02 05 ☽ △ ♄
12 31 ♀ ⊼ ♅
15 09 ☽ ♇ ♅
15 45 ☉ □ ♄
18 51 ☽ ⊼ ♆
20 57 ☽ □ ⯓
15 Thursday
05 30 ☽ ± ♆
09 00 ☿ ⊻ ♀
10 57 ☽ # ♅
11 46 ☽ ☍ ♃
12 23 ☽ # ♆
17 00 ☽ ∥ ♀
17 26 ☽ ☌ ♂
17 39 ☽ △ ♅
18 59 ☽ ☌ ☿
20 22 ☽ ⊻ ♀
21 15 ☽ ♇ ♆
22 01 ♂ △ ♅
16 Friday
04 33 ☽ Q ⯓
07 16 ☽ □ ♄
07 42 ☽ ⊥ ♀
09 33 ☽ # ♃
10 06 ☽ ☌ ☉
12 20 ☿ ☌ ♂
13 40 ☽ ∥ ♂
19 56 ☽ # ♄
23 00 ☽ △ ♆
17 Saturday
00 30 ☽ ∠ ♀
01 08 ☽ ⚹ ⯓
08 13 ☽ ∥ ☉
08 56 ☽ ∥ ☿
08 57 ☿ △ ♅
14 00 ☽ Q ♄
14 37 ☉ ∥ ☿
14 48 ☽ ⊼ ♃
20 35 ☽ ⊻ ☿
20 53 ☽ □ ♅
23 06 ☽ ⊻ ♂
18 Sunday
00 49 ☽ ± ♃
02 29 ☽ ∠ ⯓
04 04 ☽ ⚹ ♀
06 25 ☽ ⊥ ☿
09 38 ☽ ⊥ ♂
10 13 ☽ ⚹ ♄
15 46 ☽ ♇ ♃
16 29 ☽ ⊻ ☉
17 35 ☽ ⊥ ⯓
21 11 ☽ ∠ ☿
23 26 ☽ ∥ ⯓
19 Monday
01 23 ☽ ∠ ♂
01 24 ☽ □ ♆
01 46 ♂ ♇ ♆
01 55 ☽ Q ♀
03 01 ☽ Q ♅
03 13 ☽ ⊥ ☉
03 37 ☽ ⊻ ⯓
11 20 ☽ ∠ ♄
15 17 ♀ ± ♅
16 35 ☽ △ ♃
19 19 ☽ ∠ ☉
21 53 ☽ ⚹ ☿
22 33 ☽ # ⯓
23 02 ☽ ⚹ ♅
20 Tuesday
02 24 ☽ ⊥ ♄
03 35 ☽ ⚹ ♂
07 25 ☽ Q ♆
10 38 ☽ □ ♀
12 27 ☽ ⊻ ♄
17 50 ☽ Q ☿
22 11 ☽ ⚹ ☉
22 47 ☿ St D
21 Wednesday
00 07 ☽ ∠ ♅
00 36 ☽ Q ♂
03 30 ☽ ⚹ ♆
05 50 ☽ ☌ ⯓
08 13 ☽ # ☉
09 35 ♀ △ ♄
15 14 ☽ ⊥ ♅
18 26 ☽ □ ♃
18 37 ☽ # ☿
19 49 ☽ Q ☉
22 Thursday
00 13 ☽ □ ☿
01 21 ☽ ⊻ ♅
01 55 ☽ ∥ ♄
02 56 ☽ # ♂
04 45 ☽ ∠ ♆
08 21 ☽ □ ♂
12 35 ☉ ⊥ ♀
12 44 ☽ ∥ ♃
15 07 ☽ ☌ ♄
17 42 ☽ △ ♀
20 00 ☽ ⊥ ♆
23 18 ☽ # ♀
23 Friday
00 18 ♅ St D
00 42 ☽ Q ♃
04 36 ☽ □ ☉
05 09 ♂ # ♄
06 14 ☽ ⊻ ♆
08 18 ☽ ∥ ♆
08 44 ☽ ⊻ ⯓
09 37 ☽ ∥ ♅
19 02 ☽ ⊥ ⯓
21 05 ☽ ⚹ ♃
21 43 ☽ ♇ ♀
24 Saturday
04 08 ☉ △ ♆
04 26 ☽ △ ☿
04 34 ☽ ☌ ♅
08 47 ☿ △ ♅
10 35 ☽ ∠ ⯓
14 11 ☽ △ ♂
14 41 ☽ ± ♀
18 48 ☽ ⊻ ♄
22 49 ☽ ∠ ♃
25 Sunday
02 05 ☽ ⊼ ♀
05 17 ☽ ⊥ ♄
07 25 ☽ ♇ ☿
10 02 ☽ ☌ ♆
12 19 ☽ △ ☉
12 43 ☽ ⚹ ⯓
14 24 ☽ ⊥ ♃
17 36 ☽ ♇ ♂
18 12 ☉ ⚹ ⯓
21 05 ☽ ∠ ♄
23 56 ☽ ± ☿
26 Monday
00 51 ☽ ⊻ ♃
02 15 ♂ Q ⯓
08 57 ☽ ⊻ ♅
09 53 ☽ Q ⯓
10 14 ☽ ± ♂
11 05 ☽ ⊼ ☿
16 46 ☽ ♇ ☉
19 35 ☽ ⊥ ♅
21 25 ☽ ⊼ ♂
23 44 ☽ ⚹ ♄
27 Tuesday
06 46 ♀ ♇ ♅
10 04 ☽ ± ☉
10 14 ☿ ♇ ♆
11 40 ☽ ∠ ♅
12 09 ☽ ☍ ♀
15 10 ☽ ⊻ ♆
15 39 ☿ ∥ ♂
18 04 ☽ □ ⯓
21 20 ☽ Q ♄
21 43 ☽ ⊼ ☉
28 Wednesday
01 59 ☽ ⊥ ♆
06 02 ☽ ☌ ♃
08 05 ☽ ∥ ♅
09 29 ☽ ∥ ♆
14 50 ☽ ⚹ ♅
18 21 ☽ ∠ ♆
20 48 ☽ ☍ ☿
21 56 ♀ ⊼ ♆
29 Thursday
01 16 ☽ # ♀
02 40 ♂ □ ♄
05 37 ☽ ∥ ♃
06 20 ☽ □ ♄
06 30 ☽ ☍ ♂
12 56 ☽ Q ♅
18 00 ☽ ∥ ♄
22 05 ☽ ⚹ ♆
23 28 ☽ # ☿
30 Friday
00 34 ☽ ⊼ ♀
01 12 ☽ # ♂
01 14 ☽ △ ⯓
08 03 ♀ □ ⯓
09 27 ☽ ☍ ☉
13 01 ☽ ± ♀
13 09 ☽ ⊻ ♃
20 42 ☽ Q ♆
22 48 ☽ □ ♅
31 Saturday
00 28 ☽ ⊥ ♃
05 39 ☽ # ☉
05 44 ☽ ♇ ⯓
07 57 ☽ ♇ ♀
10 40 ☽ ⊼ ☿
15 13 ☽ △ ♄
17 37 ☽ ∠ ♃
18 12 ☽ ⊼ ♂
18 53 ☿ Q ⯓
23 08 ☽ ± ⯓
23 38 ☽ ± ☿

All ephemeris data is given at 12.00 UT and the Moon's longitude is additionally given for 24.00 UT

SEPTEMBER 1985

LONGITUDES

Date	Sidereal time h m s	Sun ☉ ° ' "	Moon ☽ ° ' "	Moon ☽ 24.00 ° ' "	Mercury ☿ ° '	Venus ♀ ° '	Mars ♂ ° '	Jupiter ♃ ° '	Saturn ♄ ° '	Uranus ♅ ° '	Neptune ♆ ° '	Pluto ♇ ° '
01	10 42 23	08 ♍ 59 17	03 ♈ 13 14	09 ♈ 18 31	21 ♌ 35	05 ♌ 10	24 ♌ 34	08 ♒ 43	22 ♏ 36	14 ♐ 00	00 ♑ 53	02 ♏ 39
02	10 46 20	09 57 21	15 ♈ 20 50	21 ♈ 20 32	23 01	06 21	25 12	08 R 37	22 39	14 01	00 R 52	02 41
03	10 50 16	10 55 26	27 ♈ 17 59	03 ♉ 13 39	24 31	07 33	25 50	08 32	22 43	14 01	00 52	02 42
04	10 54 13	11 53 34	09 ♉ 08 03	15 ♉ 01 43	26 06	08 45	26 28	08 27	22 47	14 02	00 52	02 44
05	10 58 09	12 51 44	20 ♉ 55 15	26 ♉ 49 17	27 45	09 56	27 06	08 21	22 50	14 02	00 52	02 46
06	11 02 06	13 49 55	02 ♊ 44 27	08 ♊ 41 26	29 ♌ 27	11 08	27 44	08 16	22 54	14 03	00 51	02 47
07	11 06 03	14 48 09	14 ♊ 40 54	20 ♊ 43 31	01 ♍ 12	12 20	28 22	08 11	22 58	14 04	00 51	02 49
08	11 09 59	15 46 24	26 ♊ 49 55	03 ♋ 00 45	02 59	13 32	29 00	08 07	23 02	14 05	00 51	02 51
09	11 13 56	16 44 42	09 ♋ 16 34	15 ♋ 37 53	04 49	14 44	29 ♌ 39	08 02	23 06	14 06	00 51	02 53
10	11 17 52	17 43 02	22 ♋ 05 09	28 ♋ 38 39	06 40	15 56	00 ♍ 17	07 58	23 10	14 06	00 51	02 55
11	11 21 49	18 41 24	05 ♌ 18 38	12 ♌ 05 08	08 32	17 09	00 55	07 54	23 15	14 07	00 51	02 56
12	11 25 45	19 39 48	18 ♌ 58 06	25 ♌ 57 15	10 25	18 21	01 33	07 49	23 19	14 08	00 D 51	02 58
13	11 29 42	20 38 13	03 ♍ 02 12	10 ♍ 12 23	12 19	19 33	02 11	07 46	23 23	14 10	00 51	03 00
14	11 33 38	21 36 41	17 ♍ 27 05	24 ♍ 45 28	14 12	20 46	02 49	07 42	23 28	14 11	00 51	03 02
15	11 37 35	22 35 11	02 ♎ 06 39	09 ♎ 29 40	16 06	21 58	03 27	07 38	23 32	14 12	00 51	03 04
16	11 41 32	23 33 42	16 ♎ 53 32	24 ♎ 17 19	18 00	23 11	04 05	07 35	23 37	14 13	00 51	03 06
17	11 45 28	24 32 15	01 ♏ 40 08	09 ♏ 01 12	19 53	24 23	04 43	07 32	23 42	14 14	00 51	03 08
18	11 49 25	25 30 50	16 ♏ 19 50	23 ♏ 35 28	21 45	25 36	05 21	07 29	23 46	14 16	00 51	03 10
19	11 53 21	26 29 27	00 ♐ 47 40	07 ♐ 56 07	23 37	26 49	05 59	07 26	23 51	14 17	00 52	03 12
20	11 57 18	27 28 05	15 ♐ 00 36	22 ♐ 01 00	25 29	28 02	06 37	07 23	23 56	14 18	00 52	03 14
21	12 01 14	28 26 45	28 ♐ 57 16	05 ♑ 49 25	27 19	29 ♌ 15	07 15	07 21	24 01	14 20	00 52	03 16
22	12 05 11	29 ♍ 25 27	12 ♑ 37 31	19 ♑ 21 39	29 ♍ 09	00 ♍ 28	07 53	07 19	24 06	14 21	00 53	03 18
23	12 09 07	00 ♎ 24 10	26 ♑ 01 57	02 ♒ 38 31	00 ♎ 58	01 41	08 31	07 17	24 11	14 23	00 53	03 20
24	12 13 04	01 22 55	09 ♒ 11 29	15 ♒ 40 57	02 46	02 54	09 08	07 15	24 16	14 25	00 53	03 23
25	12 17 01	02 21 42	22 ♒ 07 03	28 ♒ 29 52	04 33	04 07	09 46	07 13	24 22	14 26	00 54	03 25
26	12 20 57	03 20 30	04 ♓ 49 31	11 ♓ 06 05	06 19	05 20	10 24	07 12	24 27	14 28	00 54	03 27
27	12 24 54	04 19 20	17 ♓ 19 40	23 ♓ 30 23	08 04	06 34	11 02	07 11	24 32	14 30	00 55	03 29
28	12 28 50	05 18 12	29 ♓ 38 21	05 ♈ 43 43	09 49	07 47	11 40	07 10	24 38	14 32	00 55	03 31
29	12 32 47	06 17 06	11 ♈ 46 38	17 ♈ 47 18	11 32	09 00	12 18	07 09	24 43	14 33	00 56	03 33
30	12 36 43	07 ♎ 16 03	23 ♈ 45 56	29 ♈ 42 49	13 ♎ 14	10 ♍ 14	12 ♍ 56	07 ♒ 08	24 ♏ 49	14 ♐ 35	00 ♑ 56	03 ♏ 36

DECLINATIONS

Date	Moon True ☊ ° '	Moon Mean ☊ ° '	Moon ☽ Latitude ° '	Sun ☉ ° '	Moon ☽ ° '	Mercury ☿ ° '	Venus ♀ ° '	Mars ♂ ° '	Jupiter ♃ ° '	Saturn ♄ ° '	Uranus ♅ ° '	Neptune ♆ ° '	Pluto ♇ ° '
01	10 ♉ 57	12 ♉ 15	03 S 07	08 N 12	01 S 34	14 N 51	18 N 54	14 N 26	18 S 57	16 S 29	22 S 32	22 S 19	02 N 59
02	10 R 51	12 12	02 12	07 50	04 N 01	14 34	18 39	14 14	18 58	16 31	22 32	22 19	02 58
03	10 48	12 09	01 12	07 28	09 24	14 15	18 24	14 01	18 59	16 32	22 32	22 19	02 57
04	10 47	12 06	00 S 09	07 06	14 24	13 52	18 09	13 48	19 01	16 33	22 32	22 19	02 56
05	10 D 48	12 03	00 N 54	06 44	18 52	13 26	17 53	13 35	19 02	16 34	22 32	22 19	02 55
06	10 49	12 00	01 55	06 22	22 36	12 57	17 36	13 22	19 04	16 35	22 32	22 19	02 54
07	10 49	11 56	02 52	05 59	25 24	12 26	17 19	13 09	19 05	16 36	22 32	22 19	02 53
08	10 R 48	11 53	03 42	05 37	27 06	11 53	17 02	12 56	19 06	16 38	22 32	22 19	02 52
09	10 46	11 50	04 22	05 14	27 29	11 17	16 44	12 42	19 07	16 39	22 32	22 19	02 51
10	10 41	11 47	04 51	04 51	26 25	10 40	16 25	12 29	19 08	16 40	22 32	22 19	02 50
11	10 35	11 44	05 06	04 29	23 53	10 00	16 06	12 16	19 09	16 41	22 32	22 19	02 49
12	10 27	11 41	05 04	04 06	19 57	09 19	15 47	12 02	19 11	16 43	22 33	22 19	02 48
13	10 19	11 37	04 44	03 43	14 48	08 37	15 27	11 49	19 12	16 44	22 33	22 19	02 47
14	10 11	11 34	04 05	03 20	08 43	07 53	15 07	11 35	19 12	16 45	22 33	22 19	02 46
15	10 05	11 31	03 10	02 57	02 N 04	07 09	14 46	11 21	19 13	16 46	22 33	22 19	02 45
16	10 01	11 28	02 02	02 33	04 S 46	06 24	14 25	11 07	19 14	16 48	22 33	22 19	02 44
17	09 59	11 25	00 N 45	02 10	11 21	05 38	14 03	10 54	19 15	16 49	22 33	22 19	02 43
18	09 D 59	11 21	00 S 34	01 47	17 16	04 51	13 41	10 40	19 16	16 51	22 33	22 19	02 42
19	10 00	11 18	01 51	01 24	22 07	04 04	13 19	10 26	19 16	16 52	22 34	22 20	02 41
20	10 01	11 15	02 59	01 00	25 33	03 17	12 56	10 12	19 17	16 53	22 34	22 20	02 40
21	10 R 01	11 12	03 55	00 37	27 21	02 30	12 33	09 58	19 18	16 55	22 34	22 20	02 39
22	10 00	11 09	04 36	00 N 14	27 25	01 42	12 09	09 44	19 18	16 56	22 34	22 20	02 38
23	09 58	11 06	05 01	00 S 10	25 53	00 55	11 45	09 30	19 19	16 58	22 34	22 20	02 38
24	09 54	11 02	05 11	00 33	22 57	00 N 07	11 21	09 15	19 19	16 59	22 35	22 20	02 37
25	09 48	10 59	05 03	00 56	18 55	00 S 40	10 57	09 01	19 19	17 00	22 35	22 20	02 36
26	09 42	10 56	04 42	01 20	14 06	01 27	10 32	08 47	19 20	17 02	22 35	22 20	02 35
27	09 35	10 53	04 06	01 43	08 47	02 14	10 07	08 32	19 20	17 03	22 35	22 20	02 34
28	09 30	10 50	03 20	02 06	03 S 12	03 00	09 41	08 18	19 20	17 05	22 35	22 20	02 33
29	09 25	10 46	02 25	02 30	02 N 26	03 46	09 15	08 04	19 20	17 06	22 36	22 20	02 32
30	09 ♉ 22	10 ♉ 43	01 S 24	02 S 53	07 N 55	04 S 32	08 N 49	07 N 49	19 S 21	17 S 08	22 S 36	22 S 20	02 N 31

ZODIAC SIGN ENTRIES

Date	h m	Planets
01	05 42	☽ ♈
03	17 28	☽ ♉
06	06 27	☽ ♊
06	19 39	☿ ♍
08	18 10	☽ ♋
10	01 31	♂ ♍
11	02 27	☽ ♌
13	06 52	☽ ♍
15	08 34	☽ ♎
17	09 17	☽ ♏
19	10 40	☽ ♐
21	13 49	☽ ♑
22	02 53	♀ ♍
22	23 13	☿ ♎
23	02 07	☉ ♎
23	19 11	☽ ♒
26	02 50	☽ ♓
28	12 43	☽ ♈

LATITUDES

Date	Mercury ☿ ° '	Venus ♀ ° '	Mars ♂ ° '	Jupiter ♃ ° '	Saturn ♄ ° '	Uranus ♅ ° '	Neptune ♆ ° '	Pluto ♇ ° '
01	00 N 34	00 S 05	01 N 10	00 S 53	02 N 00	00 S 02	01 N 08	16 N 20
04	01 06	00 N 04	01 10	00 53	01 59	00 02	01 08	16 19
07	01 29	00 14	01 11	00 53	01 59	00 03	01 07	16 17
10	01 43	00 23	01 11	00 53	01 58	00 03	01 07	16 16
13	01 49	00 31	01 11	00 53	01 57	00 03	01 07	16 15
16	01 47	00 39	01 12	00 53	01 57	00 03	01 07	16 14
19	01 40	00 47	01 12	00 53	01 56	00 03	01 07	16 13
22	01 29	00 54	01 12	00 53	01 55	00 03	01 07	16 12
25	01 15	01 01	01 12	00 53	01 55	00 03	01 07	16 11
28	00 58	01 07	01 12	00 53	01 54	00 03	01 06	16 10
31	00 N 39	01 N 12	01 N 12	00 S 53	01 N 54	00 S 03	01 N 06	16 N 09

DATA

Julian Date	2446310
Delta T	+54 seconds
Ayanamsa	23° 39' 14"
Synetic vernal point	05° ♓ 27' 45"
True obliquity of ecliptic	23° 26' 35"

MOON'S PHASES, APSIDES AND POSITIONS ☽

Date	h m	Phase	Longitude ° '	Eclipse Indicator
07	12 16	☾	14 ♊ 49	
14	19 20	●	21 ♍ 55	
21	11 03	☽	28 ♐ 24	
29	00 08	○	05 ♈ 48	

Day	h m		
04	20 55	Apogee	
16	18 37	Perigee	
01	18 43	0N	
09	06 35	Max dec	27° N 31'
15	19 15	0S	
22	01 01	Max dec	27° S 36'
29	01 37	0N	

LONGITUDES

Date	Chiron ⚷ ° '	Ceres ⚳ ° '	Pallas ⚴ ° '	Juno ⚵ ° '	Vesta ⚶ ° '	Black Moon Lilith ⚸ ° '
01	14 ♊ 26	06 ♌ 11	18 ♊ 54	18 ♎ 45	16 ♏ 32	10 ♉ 10
11	14 ♊ 36	10 ♌ 24	22 ♊ 52	21 ♎ 57	21 ♏ 00	11 ♉ 16
21	14 ♊ 40	14 ♌ 33	26 ♊ 37	25 ♎ 14	25 ♏ 39	12 ♉ 23
31	14 ♊ 36	18 ♌ 36	00 ♋ 05	28 ♎ 34	00 ♐ 26	13 ♉ 29

All ephemeris data is given at 12.00 UT and the Moon's longitude is additionally given for 24.00 UT

ASPECTARIAN

01 Sunday
05 55 ☉ ⊼ ♃
06 00 ☽ # ♇
06 30 ☽ ± ♂
07 25 ☽ □ ♆
07 48 ♀ # ♃
10 52 ☽ ⊼ ♇
16 14 ☽ △ ♀
19 29 ☽ ⚻ ☿
20 39 ☽ ⚻ ♄
22 45 ☽ ⚹ ♃

02 Monday
00 21 ☽ ⊼ ☉
01 11 ☽ ⚻ ♂
05 50 ☿ □ ♄
07 27 ☽ ∥ ♇
09 20 ☽ △ ♅
13 19 ☽ ± ☉
14 37 ☽ ± ♄
22 22 ♀ ± ♆
22 29 ☽ Q ♃

03 Tuesday
02 43 ☽ ⊼ ♄
03 48 ☽ ∥ ☉
05 34 ☽ △ ☿
06 26 ☉ Q ♄
08 52 ☽ △ ♂
08 59 ☽ ⚻ ☉
15 29 ☽ ⚻ ♅
19 13 ☽ △ ♆
22 58 ☽ ☍ ♇

04 Wednesday
06 23 ♀ ☍ ♃
09 07 ☽ ∥ ♂
09 30 ☽ ∥ ☿
09 45 ☽ ± ♅
10 36 ☽ □ ♃
11 07 ☽ □ ♀
18 07 ☽ △ ☉
19 33 ☿ ∥ ♂
20 57 ☿ ☌ ♂
21 58 ☽ ⊼ ♅
23 10 ☽ # ♄

05 Thursday
01 42 ☽ ⚻ ♆
06 45 ☽ ∥ ♀
13 03 ☽ # ♃
15 55 ☽ ☍ ♄
20 01 ☽ ± ♆

06 Friday
01 17 ☽ □ ♂
03 53 ☽ Q ♀
04 12 ☽ □ ☿
08 11 ☽ ⊼ ♆
10 00 ☽ # ♆
11 33 ☽ # ♅
12 06 ☽ ⊼ ♇
17 30 ☉ □ ♅
22 02 ☉ ± ♃
23 05 ☽ △ ♃

07 Saturday
00 14 ☽ ± ♇
06 47 ☽ ⚹ ♀
07 20 ☿ △ ♆
10 46 ☽ ☍ ♅
12 16 ☽ □ ☉
15 33 ☽ Q ♂
18 15 ☽ ⚻ ♇
22 32 ☽ Q ☿

08 Sunday
04 31 ☽ ⊼ ♄
04 45 ☽ ⚻ ♃
10 05 ☿ ⚹ ♇
15 41 ☽ ∠ ♀
16 19 ☽ ± ♄
16 28 ☽ ⚹ ♂
19 49 ☽ ☍ ♆
22 12 ☽ ± ♃
22 59 ♀ △ ♅
23 43 ☽ △ ♇

09 Monday
02 00 ☽ ⚹ ☿
02 36 ☽ Q ☉
09 39 ☽ ⊼ ♃
09 45 ☽ ⚻ ♄
10 52 ☽ ⊥ ♀
18 07 ♀ # ♄
21 07 ☽ ⊼ ♅
22 40 ☽ ∠ ♂
23 24 ☽ ⊻ ♀

10 Tuesday
03 14 ☽ ⚹ ☉
08 20 ☽ ± ♅
10 11 ♀ ⚻ ♆
11 06 ☽ ∠ ☿
14 01 ☽ △ ♄
16 14 ☽ ⊥ ♂
16 53 ☉ ∠ ♇

11 Wednesday
00 51 ☽ ⚻ ♅
03 42 ☽ ⊻ ♂
03 59 ☽ ⊼ ♆
04 04 ☿ ⊼ ♃
06 13 ☽ ⊥ ☿
07 45 ☽ □ ♇
08 52 ☽ ∠ ☉
09 35 ♂ △ ♆
14 44 ☽ ± ♆
16 34 ☽ ☍ ♃
18 40 ☽ ⊻ ☿
21 08 ☽ # ♅
22 32 ☽ # ♆

12 Thursday
02 04 ☽ ⊥ ☉
03 36 ☽ △ ♅
06 35 ☽ ⚻ ♆
09 14 ♆ St D
10 49 ☽ ☌ ♀
13 18 ☽ ⊻ ☉
15 28 ☽ Q ♇
15 56 ☽ # ♃
19 32 ☽ □ ♄
23 51 ☿ Q ♄

13 Friday
03 35 ☽ # ♄
08 18 ☽ △ ♆
09 01 ☽ ∥ ♀
10 29 ☽ ☌ ♂
11 57 ☽ ⚹ ♇
19 53 ☽ ⊼ ♃

14 Saturday
00 38 ☽ ∥ ♂
02 02 ☽ Q ♄
05 46 ☿ ± ♃
05 49 ☽ ± ♃
05 50 ☽ ☌ ☿
06 35 ☽ □ ♅
11 39 ☿ □ ♅
12 58 ☽ ∠ ♇
15 28 ☽ ∥ ☿
17 35 ♀ Q ♇
17 56 ☽ ⊻ ♀
19 20 ☽ ☌ ☉
20 35 ☽ ⚻ ♃
20 52 ♂ ⚹ ♇
21 56 ☽ ⚹ ♄

15 Sunday
03 46 ☽ ⊥ ♇
04 39 ☽ ⊥ ♀
08 43 ☽ ∥ ☉
09 34 ☽ ∥ ♇
09 57 ☽ □ ♆
12 08 ☽ Q ♅
13 13 ☉ ⚻ ♃
13 34 ☽ ⊻ ♇
14 16 ☽ ⊻ ♂
20 36 ☽ ∠ ♀
20 57 ☽ △ ♃
22 30 ☽ ∠ ♄

16 Monday
00 11 ☉ ∥ ♇
00 27 ☽ ⊥ ♂
04 38 ☽ # ☉
04 52 ☽ # ♇
07 39 ☽ ⚹ ♅
13 11 ☽ ⊥ ♄
13 24 ☿ ∠ ♇
13 28 ☉ ⚹ ♄
14 03 ☽ ⊻ ☿
15 11 ☽ Q ♆
15 42 ☽ ∠ ♂
17 12 ☽ # ☿
21 14 ♀ □ ♄
22 58 ☽ ⊻ ♄
23 07 ☽ ⚹ ♀
23 35 ☽ ⊻ ☉

17 Tuesday
01 12 ☽ ⊥ ☿
08 02 ☽ ∠ ♅
10 02 ☽ ⊥ ☉
10 19 ☽ # ♂
10 40 ☽ ⚹ ♆
14 24 ☽ ☌ ♇
17 11 ☽ ⚹ ♂
18 00 ☽ ∠ ☿
20 24 ☽ Q ♀
21 32 ☽ □ ♃
21 54 ☽ # ♀
22 44 ☽ ⊥ ♅

18 Wednesday
01 46 ☽ ∠ ☉
02 53 ☉ ⊻ ♀
08 35 ☽ ⊻ ♅
10 06 ☽ ∥ ♄
11 13 ☽ ∠ ♆
13 45 ☽ Q ♂
21 04 ☿ ⚻ ♃
21 08 ☽ ∥ ♃
22 17 ☽ ⚹ ☿

19 Thursday
00 22 ☽ ☌ ♄
02 06 ☽ ⊥ ♆
03 05 ☽ Q ♃
04 18 ☽ ⚹ ☉
04 45 ☽ □ ♀
12 07 ☽ ⊻ ♆
13 11 ☽ ∥ ♆
14 34 ☽ ∥ ♅
15 07 ☿ ⚹ ♄
16 03 ☽ ⊻ ♇
21 06 ☽ □ ♂
21 19 ☽ Q ☿
23 07 ☽ ⚹ ♃

20 Friday
01 54 ☽ Q ☉
02 10 ☽ ⊥ ♇
06 05 ☉ ⊥ ♇
10 48 ☽ ☌ ♅
17 31 ☽ ∠ ♇

21 Saturday
00 36 ☽ ∠ ♃
03 24 ☽ ⊻ ♄
06 53 ☿ ∥ ♇
08 44 ☽ □ ☿
11 03 ☽ □ ☉
11 19 ☿ ⊥ ♇
12 33 ☽ △ ♀
13 52 ☽ ⊥ ♄
15 20 ☽ ☌ ♆
15 49 ♂ ⊼ ♃
16 10 ☽ ⊥ ♃
19 33 ☽ ⚹ ♇

22 Sunday
02 39 ☽ ⊻ ♃
03 12 ☽ △ ♂
05 44 ☽ ∠ ♄
15 05 ☽ ⊻ ♅
16 46 ☽ Q ♇
17 32 ☽ ⚻ ♀
19 49 ☉ ☌ ☿
20 11 ♀ △ ♆

23 Monday
01 49 ☽ ⊥ ♅
07 13 ☽ ⚻ ♂
08 39 ☽ ⚹ ♄
10 53 ☿ □ ♆
11 18 ☽ ± ♀
18 05 ☽ ∠ ♅
20 33 ☽ △ ☉
20 48 ☽ ⊻ ♆
22 22 ☽ △ ☿
23 17 ☽ ⊼ ♀
23 48 ☉ □ ♆

24 Tuesday
00 21 ☽ ± ♂
01 19 ☽ □ ♇
03 17 ☉ # ☿
06 36 ☽ Q ♄
07 12 ☿ Q ♅
07 46 ☽ ⊥ ♆
08 26 ☽ ☌ ♃
11 54 ☽ ⊼ ♂
14 28 ☽ ∥ ♅
16 05 ☽ ∥ ♆
17 39 ☿ ⊻ ♀
20 22 ☿ ⊻ ♇
21 40 ☽ ⚹ ♅
21 40 ♀ ⚹ ♇

25 Wednesday
00 23 ☽ ∠ ♆
02 23 ☽ ⚻ ☉
06 26 ☽ ⚻ ☿
09 49 ☽ ∥ ♃
13 56 ☉ Q ♅
16 14 ☽ □ ♄
20 08 ☽ Q ♅
20 38 ☽ ± ☉
21 52 ☽ ∥ ♄

26 Thursday
02 03 ☽ ± ☿
04 33 ☽ ⚹ ♆
04 41 ☉ ∥ ☿
08 56 ☽ ⊼ ☉
09 22 ☽ △ ♇
13 05 ☽ ☍ ♀
14 40 ☉ ⊻ ♇
15 18 ☽ ⊼ ☿
16 31 ☽ ⊻ ♃
23 14 ☽ ☍ ♂
23 53 ☿ △ ♃

27 Friday
03 28 ☽ Q ♆
04 00 ☽ ⊥ ♃
05 40 ☽ # ♀
06 31 ☽ □ ♅
13 07 ☽ # ♂
14 15 ☽ ⚻ ♇
21 24 ☽ ∠ ♃
22 05 ☿ # ♇
23 57 ♀ ⊼ ♃

28 Saturday
02 07 ☽ △ ♄
07 50 ☽ ± ♇
09 21 ☿ ∠ ♄
12 45 ☽ ∥ ☿
14 31 ☽ □ ♆
14 48 ☽ # ♇
16 21 ☽ ∥ ☉
19 40 ☽ ⊼ ♇

29 Sunday
00 08 ☽ ☍ ☉
02 49 ☽ ⚹ ♃
05 52 ☽ ⊼ ♀
07 53 ☽ ⚻ ♄
11 26 ☽ ☍ ☿
12 18 ☽ # ☉
12 25 ☽ ∥ ♇
13 06 ☽ ⊼ ♂
13 55 ☉ # ♇
17 33 ☽ △ ♅
18 42 ☽ # ☿
19 10 ☽ ± ♀

30 Monday
01 45 ☽ ± ♂
01 58 ☽ ± ♄
02 42 ☽ Q ♃
05 01 ☿ ⊻ ♂
06 47 ♂ Q ♄
08 47 ☉ △ ♃
11 34 ☽ ∥ ♂
14 07 ☽ ⊼ ♄
15 17 ☽ ⚻ ♀
15 46 ☽ ∥ ♀
19 44 ♂ ± ♃
20 52 ☽ ⚻ ♂
23 47 ☽ ⚻ ♅

OCTOBER 1985

LONGITUDES

Date	Sidereal time h m s	Sun ☉ ° ' "	Moon ☽ ° ' "	Moon ☽ 24.00 ° ' "	Mercury ☿ ° '	Venus ♀ ° '	Mars ♂ ° '	Jupiter ♃ ° '	Saturn ♄ ° '	Uranus ♅ ° '	Neptune ♆ ° '	Pluto ♇ ° '
01	12 40 40	08 ♎ 15 01	05 ♉ 38 14	11 ♉ 32 31	14 ♎ 56	11 ♍ 27	13 ♍ 34	07 ♒ 08	24 ♏ 54	14 ♐ 37	00 ♑ 57	03 ♏ 38
02	12 44 36	09 14 01	17 ♉ 26 04	23 ♉ 19 17	16 37	12 41	14 11	07 R 07	25 00	14 39	00 57	03 40
03	12 48 33	10 13 04	29 ♉ 12 38	05 ♊ 06 37	18 17	13 55	14 49	07 D 07	25 06	14 41	00 58	03 42
04	12 52 30	11 12 09	11 ♊ 01 46	16 ♊ 58 37	19 56	15 08	15 27	07 07	25 11	14 43	00 59	03 45
05	12 56 26	12 11 16	22 ♊ 57 46	28 ♊ 59 49	21 34	16 22	16 05	07 08	25 17	14 45	01 00	03 47
06	13 00 23	13 10 26	05 ♋ 05 22	11 ♋ 15 01	23 11	17 36	16 43	07 08	25 23	14 48	01 00	03 49
07	13 04 19	14 09 38	17 ♋ 29 22	23 ♋ 48 57	24 48	18 50	17 21	07 09	25 29	14 50	01 01	03 51
08	13 08 16	15 08 52	00 ♌ 14 18	06 ♌ 45 51	26 24	20 04	17 58	07 10	25 35	14 52	01 02	03 54
09	13 12 12	16 08 08	13 ♌ 23 59	20 ♌ 08 55	27 59	21 18	18 36	07 11	25 41	14 54	01 03	03 56
10	13 16 09	17 07 27	27 ♌ 00 49	03 ♍ 59 38	29 ♎ 33	22 32	19 14	07 12	25 47	14 57	01 04	03 58
11	13 20 05	18 06 48	11 ♍ 05 11	18 ♍ 17 05	01 ♏ 07	23 46	19 52	07 14	25 53	14 59	01 05	04 01
12	13 24 02	19 06 11	25 ♍ 34 49	02 ♎ 57 40	02 40	25 00	20 29	07 16	25 59	15 01	01 06	04 03
13	13 27 59	20 05 36	10 ♎ 24 45	17 ♎ 55 04	04 12	26 14	21 07	07 17	26 05	15 04	01 07	04 06
14	13 31 55	21 05 04	25 ♎ 27 31	03 ♏ 00 57	05 44	27 28	21 45	07 20	26 12	15 06	01 08	04 08
15	13 35 52	22 04 33	10 ♏ 34 12	18 ♏ 06 11	07 14	28 42	22 23	07 22	26 18	15 09	01 09	04 10
16	13 39 48	23 04 05	25 ♏ 35 48	03 ♐ 02 09	08 45	29 ♍ 57	23 00	07 24	26 24	15 11	01 10	04 13
17	13 43 45	24 03 38	10 ♐ 24 24	17 ♐ 41 55	10 14	01 ♎ 11	23 38	07 27	26 31	15 14	01 11	04 15
18	13 47 41	25 03 13	24 ♐ 54 12	02 ♑ 00 54	11 43	02 25	24 16	07 30	26 37	15 17	01 12	04 17
19	13 51 38	26 02 50	09 ♑ 01 49	15 ♑ 56 53	13 11	03 40	24 54	07 33	26 44	15 19	01 13	04 20
20	13 55 34	27 02 29	22 ♑ 46 08	29 ♑ 29 42	14 38	04 54	25 31	07 36	26 50	15 22	01 15	04 22
21	13 59 31	28 02 09	06 ♒ 07 49	12 ♒ 40 45	16 05	06 09	26 09	07 40	26 57	15 25	01 16	04 25
22	14 03 28	29 ♎ 01 51	19 ♒ 08 48	25 ♒ 32 19	17 31	07 23	26 47	07 43	27 03	15 27	01 17	04 27
23	14 07 24	00 ♏ 01 35	01 ♓ 51 38	08 ♓ 07 08	18 56	08 38	27 24	07 47	27 10	15 30	01 18	04 30
24	14 11 21	01 01 20	14 ♓ 19 08	20 ♓ 28 01	20 21	09 52	28 02	07 51	27 16	15 33	01 20	04 32
25	14 15 17	02 01 07	26 ♓ 34 05	02 ♈ 37 40	21 44	11 07	28 40	07 55	27 23	15 36	01 21	04 34
26	14 19 14	03 00 56	08 ♈ 39 01	14 ♈ 38 33	23 07	12 22	29 17	08 00	27 30	15 39	01 22	04 37
27	14 23 10	04 00 47	20 ♈ 36 25	26 ♈ 32 56	24 29	13 36	29 ♍ 55	08 04	27 36	15 42	01 24	04 39
28	14 27 07	05 00 39	02 ♉ 28 22	08 ♉ 22 58	25 49	14 51	00 ♎ 33	08 09	27 43	15 45	01 25	04 42
29	14 31 03	06 00 34	14 ♉ 17 01	20 ♉ 10 45	27 09	16 06	01 10	08 14	27 50	15 48	01 27	04 44
30	14 35 00	07 00 30	26 ♉ 04 30	01 ♊ 58 32	28 28	17 21	01 48	08 19	27 57	15 51	01 28	04 47
31	14 38 57	08 ♏ 00 29	07 ♊ 53 11	13 ♊ 48 47	29 ♏ 45	18 ♎ 35	02 ♎ 25	08 ♒ 24	28 ♏ 04	15 ♐ 54	01 ♑ 30	04 ♏ 49

Date	Moon True ☊ ° '	Moon Mean ☊ ° '	Moon ☽ Latitude ° '
01	09 ♉ 21	10 ♉ 40	00 S 20
02	09 D 21	10 37	00 N 44
03	09 23	10 34	01 47
04	09 24	10 31	02 45
05	09 26	10 27	03 37
06	09 27	10 24	04 20
07	09 R 28	10 21	04 52
08	09 27	10 18	05 11
09	09 25	10 15	05 14
10	09 22	10 12	05 00
11	09 19	10 08	04 28
12	09 16	10 05	03 38
13	09 14	10 02	02 33
14	09 13	09 59	01 N 15
15	09 D 12	09 56	00 S 08
16	09 13	09 52	01 30
17	09 14	09 49	02 44
18	09 15	09 46	03 47
19	09 16	09 43	04 34
20	09 16	09 40	05 04
21	09 R 16	09 37	05 17
22	09 15	09 33	05 12
23	09 14	09 30	04 53
24	09 13	09 27	04 20
25	09 12	09 24	03 35
26	09 12	09 21	02 42
27	09 11	09 18	01 41
28	09 11	09 14	00 S 37
29	09 D 11	09 11	00 N 28
30	09 11	09 08	01 33
31	09 ♉ 11	09 ♉ 05	02 N 33

DECLINATIONS

Date	Sun ☉ ° '	Moon ☽ ° '	Mercury ☿ ° '	Venus ♀ ° '	Mars ♂ ° '	Jupiter ♃ ° '	Saturn ♄ ° '	Uranus ♅ ° '	Neptune ♆ ° '	Pluto ♇ ° '
01	03 S 16	13 N 05	05 S 17	08 N 23	07 N 35	19 S 21	17 S 09	22 S 36	22 S 20	02 N 30
02	03 40	17 45	06 02	07 57	07 20	19 21	17 11	22 36	22 20	02 29
03	04 03	21 43	06 46	07 30	07 05	19 21	17 12	22 37	22 20	02 28
04	04 26	24 50	07 30	07 03	06 51	19 20	17 14	22 37	22 20	02 27
05	04 49	26 52	08 13	06 36	06 36	19 20	17 15	22 37	22 20	02 26
06	05 12	27 41	08 56	06 08	06 21	19 20	17 17	22 37	22 20	02 25
07	05 35	27 08	09 38	05 40	06 07	19 20	17 19	22 38	22 20	02 24
08	05 58	25 10	10 19	05 13	05 52	19 20	17 20	22 38	22 20	02 23
09	06 21	21 49	11 00	04 45	05 37	19 19	17 22	22 38	22 20	02 22
10	06 44	17 13	11 40	04 16	05 22	19 19	17 23	22 38	22 20	02 21
11	07 06	11 33	12 20	03 48	05 08	19 18	17 25	22 39	22 20	02 21
12	07 29	05 N 06	12 58	03 20	04 53	19 18	17 26	22 39	22 20	02 20
13	07 51	01 S 47	13 36	02 51	04 38	19 17	17 28	22 39	22 20	02 19
14	08 14	08 41	14 14	02 22	04 23	19 17	17 30	22 39	22 21	02 18
15	08 36	15 07	14 50	01 53	04 08	19 16	17 31	22 40	22 21	02 17
16	08 58	20 37	15 26	01 25	03 53	19 15	17 33	22 40	22 21	02 16
17	09 20	24 43	16 01	00 56	03 38	19 15	17 34	22 40	22 21	02 15
18	09 42	27 08	16 35	00 N 27	03 23	19 14	17 36	22 41	22 21	02 14
19	10 04	27 42	17 08	00 S 03	03 08	19 13	17 38	22 41	22 21	02 13
20	10 25	26 31	17 40	00 32	02 53	19 12	17 39	22 41	22 21	02 13
21	10 47	23 51	18 12	01 01	02 38	19 11	17 41	22 41	22 21	02 12
22	11 08	20 02	18 42	01 30	02 23	19 10	17 42	22 42	22 21	02 11
23	11 29	15 22	19 12	01 59	02 08	19 09	17 44	22 42	22 21	02 10
24	11 50	10 10	19 40	02 28	01 53	19 08	17 46	22 42	22 21	02 09
25	12 11	04 S 39	20 08	02 58	01 38	19 07	17 47	22 43	22 21	02 09
26	12 31	00 N 57	20 35	03 27	01 23	19 06	17 49	22 43	22 21	02 08
27	12 52	06 29	21 00	03 56	01 08	19 04	17 50	22 43	22 21	02 07
28	13 12	11 45	21 25	04 25	00 53	19 03	17 52	22 44	22 21	02 06
29	13 32	16 35	21 48	04 54	00 38	19 02	17 54	22 44	22 21	02 05
30	13 51	20 47	22 10	05 23	00 23	19 00	17 55	22 44	22 21	02 05
31	14 S 11	24 N 09	22 S 31	05 S 51	00 N 08	18 S 59	17 S 57	22 S 45	22 S 21	02 N 04

ZODIAC SIGN ENTRIES

Date	h	m	Planets
01	00	35	☽ ♉
03	13	36	☽ ♊
06	01	59	☽ ♋
08	11	33	☽ ♌
10	17	09	☽ ♍
10	18	50	☿ ♏
12	19	12	☽ ♎
14	19	13	☽ ♏
16	13	04	♀ ♎
16	19	05	☽ ♐
18	20	35	☽ ♑
21	00	54	☽ ♒
23	08	27	☽ ♓
23	11	22	☉ ♏
25	18	47	☽ ♈
27	15	16	♂ ♎
28	06	59	☽ ♉
30	19	59	☽ ♊
31	16	44	☿ ♐

LATITUDES

Date	Mercury ☿ ° '	Venus ♀ ° '	Mars ♂ ° '	Jupiter ♃ ° '	Saturn ♄ ° '	Uranus ♅ ° '	Neptune ♆ ° '	Pluto ♇ ° '
01	00 N 39	01 N 12	01 N 12	00 S 53	01 N 54	00 S 03	01 N 06	16 N 09
04	00 N 19	01 17	01 12	00 53	01 53	00 03	01 06	16 08
07	00 S 02	01 22	01 12	00 52	01 53	00 03	01 06	16 08
10	00 23	01 25	01 12	00 52	01 52	00 03	01 06	16 07
13	00 44	01 28	01 12	00 52	01 52	00 03	01 06	16 07
16	01 04	01 31	01 12	00 52	01 52	00 03	01 06	16 06
19	01 24	01 32	01 12	00 52	01 51	00 03	01 06	16 06
22	01 43	01 33	01 12	00 51	01 51	00 03	01 05	16 06
25	02 00	01 34	01 12	00 51	01 50	00 03	01 05	16 05
28	02 16	01 34	01 12	00 51	01 50	00 03	01 05	16 05
31	02 S 29	01 N 33	01 N 12	00 S 51	01 N 50	00 S 03	01 N 05	16 N 05

DATA

Julian Date	2446340
Delta T	+54 seconds
Ayanamsa	23° 39' 17"
Synetic vernal point	05° ♓ 27' 42"
True obliquity of ecliptic	23° 26' 36"

MOON'S PHASES, APSIDES AND POSITIONS ☽

Date	h	m	Phase	Longitude ° '	Eclipse Indicator
07	05	04	☾	13 ♋ 53	
14	04	33	●	20 ♎ 47	
20	20	13	☽	27 ♑ 23	
28	17	38	○	05 ♉ 15	total

Day	h	m		
02	13	08	Apogee	
15	00	44	Perigee	
29	21	33	Apogee	
06	14	31	Max dec	27° N 41'
13	05	51	0S	
19	07	29	Max dec	27° S 44'
26	07	55	0N	

LONGITUDES

Date	Chiron ⚷ ° '	Ceres ⚳ ° '	Pallas ⚴ ° '	Juno ⚵ ° '	Vesta ⚶ ° '	Black Moon Lilith ⚸ ° '
01	14 ♊ 36	18 ♌ 36	00 ♋ 05	28 ♎ 34	00 ♐ 26	13 ♉ 29
11	14 ♊ 25	22 ♌ 32	03 ♋ 07	01 ♏ 58	05 ♐ 21	14 ♉ 35
21	14 ♊ 08	26 ♌ 19	05 ♋ 38	05 ♏ 23	10 ♐ 22	15 ♉ 42
31	13 ♊ 45	29 ♌ 55	07 ♋ 26	08 ♏ 49	15 ♐ 28	16 ♉ 48

All ephemeris data is given at 12.00 UT and the Moon's longitude is additionally given for 24.00 UT

ASPECTARIAN

h m	Aspects
01 Tuesday	
02 29	☽ △ ♆
07 27	☿ ✶ ♅
07 55	☽ ☍ ♇
15 01	☽ □ ♃
17 47	☽ ⊼ ☉
18 04	☽ ± ♅
02 Wednesday	
00 29	♄ ⊥ ♆
01 12	☽ △ ♀
05 01	☽ △ ♂
06 19	☽ ⊼ ♅
07 06	☽ ± ☉
08 54	☽ ⋕ ♄
08 59	☽ ⚼ ♆
10 03	☽ ⊼ ☿
18 43	♀ Q ♄
20 37	♀ ± ♃
21 05	☽ ⋕ ♃
03 Thursday	
00 18	☽ ± ☿
03 07	☽ ⚼ ☉
03 21	☽ ± ♆
03 33	☽ ☍ ♄
06 37	♂ □ ♅
08 13	♃ St D
08 38	☉ ∠ ♄
15 35	☽ ⊼ ♆
16 12	☽ ⋕ ♆
18 08	☽ ⋕ ♅
19 09	♃ Q ♄
19 54	☿ ⋕ ♂
21 10	☽ ⊼ ♇
21 37	☽ ⚼ ☿
22 07	☿ Q ♆
04 Friday	
00 34	☿ ⊥ ♄
02 46	☿ ⋕ ♀
03 38	♀ □ ♅
04 05	☽ △ ♃
09 23	☽ ± ♇
12 23	☽ △ ☉
19 29	☽ ☍ ♅
21 15	☽ □ ♀
21 26	☽ □ ♂
05 Saturday	
00 34	♀ ☌ ♂
03 36	☽ ⚼ ♇
08 46	☽ △ ☿
10 20	☽ ⚼ ♃
10 46	♀ ∥ ♂
16 40	☽ ⊼ ♄
06 Sunday	
00 27	☿ ⊥ ♂
03 58	☽ ☍ ♆
04 13	☽ ± ♃
04 39	☽ ± ♄
09 30	☽ △ ♇
11 13	☽ Q ♂
13 06	☽ Q ♀
16 01	☽ ⊼ ♃
22 24	☽ ⚼ ♄
07 Monday	
05 04	☽ □ ☉
06 53	☽ ⊼ ♅
11 42	☽ ✶ ♂
12 36	♀ ∠ ♇
13 54	☿ ⊥ ♀
14 31	☉ ⋕ ♀
14 50	☽ ✶ ♀
18 22	☽ ± ♅
22 55	☿ ⊻ ♄
08 Tuesday	
03 15	☽ △ ♄
03 50	☽ □ ☿
04 53	☉ ✶ ♅
08 10	☉ ⋕ ♂
11 18	☽ ⚼ ♅
13 29	☽ ⊼ ♆
17 18	☽ ∠ ♂
17 49	☽ Q ☉
18 46	☽ □ ♇
21 48	☽ ∠ ♀
09 Wednesday	
00 30	☽ ± ♆
00 45	☽ ☍ ♃
06 56	☽ ⋕ ♅
08 48	☽ ⋕ ♆
10 30	☽ ⊥ ♂
14 42	☽ △ ♅
15 44	☽ ⊥ ♀
16 44	☽ ⚼ ♆
17 14	☽ Q ☿
17 17	☽ ✶ ☉
21 43	☽ ⊻ ♂
10 Thursday	
01 33	♂ ∠ ♇
01 46	☽ ⋕ ♃
03 11	☽ Q ♇
03 24	☽ ⊻ ♀
05 38	♀ ⚼ ♃
09 51	☽ □ ♄
11 11	☽ ⋕ ♄
16 57	☽ ✶ ☿
18 06	☿ ∠ ♅
19 00	☽ △ ♆
21 28	☽ ∠ ☉
11 Friday	
00 00	☽ ✶ ♇
05 29	☽ ⊼ ♃
09 12	☽ ⋕ ☿
11 26	☿ ✶ ♆
13 51	☽ ⊥ ☉
15 36	☽ ± ♃
16 43	☽ Q ♄
18 32	☽ □ ♅
21 25	☽ ∠ ☿
12 Saturday	
00 34	☽ ⊻ ☉
01 14	☽ ∠ ♇
03 16	☽ ☌ ♂
03 51	☽ ⋕ ☉
06 33	☽ ⚼ ♃
10 57	☽ ☌ ♀
11 48	☉ Q ♆
12 40	☽ ✶ ♄
12 48	☽ ∥ ♂
13 59	☽ ⊥ ☿
16 03	☽ ⊥ ♇
18 44	☽ ∥ ♀
20 59	☽ □ ♆
21 47	☽ ∥ ♇
13 Sunday	
00 08	☽ Q ♅
00 51	☽ ⊻ ☿
01 48	☽ ⊻ ♇
06 58	☽ △ ♃
08 59	♀ ✶ ♄
10 15	☿ ☌ ♇
11 54	☉ ⊥ ♄
13 06	☽ ∠ ♄
13 49	☽ ⋕ ♇
15 26	☽ ⋕ ♀
19 28	☽ ✶ ♅
21 29	☽ ⋕ ♂
14 Monday	
01 55	☽ Q ♆
03 34	☽ ⊥ ♄
04 33	☽ ☌ ☉
05 51	☽ ⊻ ♂
10 19	☽ ∥ ☉
13 11	☽ ⊻ ♄
15 29	☽ ⊻ ♀
15 48	☽ ⊥ ♂
15 48	♀ ∥ ♇
19 24	☽ ∠ ♅
21 01	☽ ✶ ♆
15 Tuesday	
01 16	♀ ⊥ ♆
01 48	☽ ☌ ♇
01 51	☽ ⊥ ♀
06 07	☽ ☌ ☿
06 42	☽ ∠ ♂
06 53	☽ □ ♃
09 44	☽ ⊥ ♅
10 46	☽ ∥ ☿
11 24	♂ ⚼ ♃
14 00	☿ □ ♃
15 45	☿ ∠ ♂
17 26	☽ ∠ ♀
19 18	☽ ⊻ ♅
20 53	☽ ∠ ♆
21 55	☽ ∥ ♄
16 Wednesday	
05 35	☽ ∥ ♃
07 39	☽ ⊻ ☉
07 40	☽ ✶ ♂
08 00	☉ ⊻ ♂
11 18	☽ ⊥ ♆
11 42	☽ Q ♃
13 18	☽ ☌ ♄
17 59	☽ ⊥ ☉
19 21	☿ ⊥ ♅
19 38	☽ ✶ ♀
20 59	☽ ⊻ ♆
21 02	☽ ∥ ♆
22 52	☽ ∥ ♅
17 Thursday	
01 57	☽ ⊻ ♇
03 52	☽ Q ♂
07 09	☽ ✶ ♃
09 38	☽ ∠ ☉
11 41	☽ ⊻ ☿
11 45	☽ ⊥ ♇
11 58	♀ □ ♆
16 59	☽ Q ♀
19 57	☽ ☌ ♅
22 40	☽ ⊥ ☿
18 Friday	
02 37	☽ ∠ ♇
07 58	☽ ∠ ♃
10 53	☽ □ ♂
12 16	☽ ✶ ☉
14 54	☽ ⊻ ♄
15 24	☽ ∠ ☿
22 38	☽ ☌ ♆
23 10	☽ ⊥ ♃
19 Saturday	
01 07	☽ ⊥ ♄
01 55	☽ □ ♀
03 55	☽ ✶ ♇
05 05	♀ Q ♅
09 27	☽ ⊻ ♃
10 11	☽ Q ☉
16 41	☽ ∠ ♄
20 02	☽ ✶ ☿
22 56	☽ ⊻ ♅
20 Sunday	
00 42	☽ Q ♇
01 20	♀ ⊻ ♇
06 22	☉ ⊻ ♄
09 30	☽ ⊥ ♅
11 04	☿ ∥ ♄
17 08	☽ △ ♂
19 17	☽ ✶ ♄
19 43	☽ Q ☿
20 13	☽ □ ☉
21 Monday	
00 25	☿ ⊻ ♅
01 37	☽ ∠ ♅
03 10	☽ ⊻ ♆
08 52	☽ □ ♇
12 02	☽ △ ♀
14 04	☽ ⊥ ♆
14 48	☽ ☌ ♃
15 01	☿ ∠ ♆
17 11	☽ Q ♄
19 58	☽ ∥ ♅
21 39	☽ ⚼ ♂
22 12	☽ ∥ ♆
22 Tuesday	
05 07	☽ ✶ ♅
06 40	☽ ∠ ♆
08 35	☽ □ ☿
15 12	☽ ± ♂
16 43	☽ ∥ ♃
18 28	☽ ∥ ☿
18 43	☽ ⚼ ♀
18 46	♀ △ ♃
23 Wednesday	
00 19	☽ ∥ ♄
00 42	♂ ✶ ♄
02 59	☽ □ ♄
03 05	☽ ⊼ ♂
03 41	☽ Q ♅
08 12	☽ △ ☉
08 50	♂ ∥ ♇
09 50	☿ ∥ ♃
10 56	☽ ✶ ♆
13 38	☽ ± ♀
16 53	♀ ⋕ ♂
17 03	☽ △ ♇
20 42	♀ ⋕ ♇
23 25	☽ ⊻ ♃
24 Thursday	
00 05	☉ ∠ ♅
02 26	☽ ⊼ ♀
04 58	☽ ∥ ☉
10 04	☽ Q ♆
11 05	☽ ⊥ ♃
14 24	☽ □ ♅
15 36	☽ ⚼ ☉
19 30	☉ ✶ ♆
22 12	☽ ⚼ ♇
25 Friday	
01 16	☽ △ ☿
04 47	☽ ∠ ♃
08 26	♂ ⊥ ♇
10 49	☽ ± ☉
13 37	☽ △ ♄
15 59	☽ ± ♇
16 22	☽ ☍ ♂
18 41	☽ ∥ ♀
21 29	☽ □ ♆
22 47	☽ ⋕ ♇
23 46	☽ ⊼ ☉
26 Saturday	
01 31	☽ ⋕ ♂
03 55	☽ ⊼ ♇
10 41	☽ ✶ ♃
10 47	☽ ⚼ ☿
13 47	☽ ∥ ♂
14 49	♀ ∠ ♄
17 02	☽ ∥ ♇
19 46	☽ ⚼ ♄
20 17	☽ ☍ ♀
23 45	☽ ⋕ ♀
27 Sunday	
02 05	☽ △ ♅
07 09	☽ ± ☿
10 55	☽ Q ♃
14 02	☽ ± ♄
20 49	☽ ⊼ ☿
28 Monday	
02 17	☽ ⊼ ♄
04 04	☉ ☌ ♇
04 40	☿ ⊥ ♆
07 52	☽ ⊼ ♂
08 29	☽ ⚼ ♅
09 52	☽ △ ♆
16 31	☽ ☍ ♇
17 38	☽ ☍ ☉
18 15	☿ Q ♃
19 24	☽ ⋕ ☉
20 43	☽ ± ♂
23 36	☽ □ ♃
29 Tuesday	
02 50	☽ ± ♅
05 56	♀ ✶ ♅
15 05	☽ ⊼ ♅
16 03	☽ ⚼ ♂
16 08	☽ ⊼ ♀
16 24	☽ ⚼ ♆
19 08	☽ ⋕ ♄
22 57	♂ □ ♆
30 Wednesday	
01 26	☽ ⋕ ♃
01 38	☿ ☌ ♄
05 47	☽ ± ♀
10 46	☽ ± ♆
15 51	☽ ☍ ♄
17 27	☽ ☍ ☿
22 12	☽ ⋕ ☿
22 25	☽ ⋕ ♆
23 00	☽ ⊼ ♆
31 Thursday	
00 17	☽ △ ♂
00 18	☿ ∥ ♆
01 12	☽ ⋕ ♅
02 15	☽ ⚼ ♀
05 45	☽ ⊼ ♇
12 16	☽ ⊼ ☉
13 03	☽ △ ♃
17 57	☽ ± ♇
22 24	☉ □ ♃

NOVEMBER 1985

LONGITUDES

Date	Sidereal time h m s	Sun ☉ ° ' "	Moon ☽ ° ' "	Moon ☽ 24.00 ° ' "	Mercury ☿ ° '	Venus ♀ ° '	Mars ♂ ° '	Jupiter ♃ ° '	Saturn ♄ ° '	Uranus ♅ ° '	Neptune ♆ ° '	Pluto ♇ ° '
01	14 42 53	09 ♏ 00 30	19 ♊ 45 42	25 ♊ 44 18	01 ♐ 01	19 ♎ 50	03 ♎ 03	08 ♒ 30	28 ♏ 10	15 ♐ 57	01 ♑ 31	04 ♏ 51
02	14 46 50	10 00 32	01 ♋ 45 02	07 ♋ 48 19	02 15	21 05	03 41	08 35	28 17	16 00	01 33	04 54
03	14 50 46	11 00 37	13 ♋ 54 36	20 ♋ 04 22	03 28	22 20	04 18	08 41	28 24	16 03	01 34	04 56
04	14 54 43	12 00 44	26 ♋ 18 05	02 ♌ 36 15	04 39	23 35	04 56	08 47	28 31	16 06	01 36	04 59
05	14 58 39	13 00 53	08 ♌ 59 20	15 ♌ 27 47	05 48	24 50	05 33	08 53	28 38	16 10	01 38	05 01
06	15 02 36	14 01 04	22 ♌ 02 00	28 ♌ 42 22	06 55	26 05	06 11	08 59	28 45	16 13	01 39	05 03
07	15 06 32	15 01 17	05 ♍ 29 08	12 ♍ 22 31	07 59	27 20	06 48	09 06	28 52	16 16	01 41	05 06
08	15 10 29	16 01 32	19 ♍ 22 34	26 ♍ 29 14	09 01	28 35	07 26	09 12	28 59	16 19	01 43	05 08
09	15 14 26	17 01 48	03 ♎ 42 15	11 ♎ 01 14	09 59	29 ♎ 50	08 03	09 19	29 06	16 23	01 44	05 11
10	15 18 22	18 02 07	18 ♎ 25 37	25 ♎ 54 38	10 54	01 ♏ 05	08 41	09 26	29 13	16 26	01 46	05 13
11	15 22 19	19 02 28	03 ♏ 27 22	11 ♏ 02 45	11 45	02 20	09 18	09 33	29 20	16 29	01 48	05 15
12	15 26 15	20 02 51	18 ♏ 39 37	26 ♏ 16 43	12 31	03 36	09 56	09 40	29 27	16 33	01 50	05 18
13	15 30 12	21 03 15	03 ♐ 52 46	11 ♐ 26 39	13 12	04 51	10 33	09 48	29 34	16 36	01 51	05 20
14	15 34 08	22 03 42	18 ♐ 57 07	26 ♐ 23 09	13 48	06 06	11 11	09 55	29 42	16 39	01 53	05 23
15	15 38 05	23 04 09	03 ♑ 43 54	10 ♑ 58 39	14 17	07 21	11 48	10 03	29 49	16 43	01 55	05 25
16	15 42 01	24 04 38	18 ♑ 06 55	25 ♑ 08 21	14 39	08 36	12 26	10 11	29 ♏ 56	16 46	01 57	05 27
17	15 45 58	25 05 09	02 ♒ 02 49	08 ♒ 50 20	14 54	09 52	13 03	10 19	00 ♐ 03	16 50	01 59	05 30
18	15 49 55	26 05 40	15 ♒ 31 04	22 ♒ 05 17	15 00	11 07	13 41	10 27	00 10	16 53	02 01	05 32
19	15 53 51	27 06 13	28 ♒ 33 22	04 ♓ 55 44	14 R 57	12 22	14 18	10 35	00 17	16 57	02 03	05 34
20	15 57 48	28 06 47	11 ♓ 12 55	17 ♓ 25 25	14 44	13 37	14 56	10 43	00 24	17 00	02 05	05 36
21	16 01 44	29 ♏ 07 22	23 ♓ 33 48	29 ♓ 38 36	14 21	14 53	15 33	10 52	00 31	17 04	02 07	05 39
22	16 05 41	00 ♐ 07 58	05 ♈ 40 21	11 ♈ 39 36	13 47	16 08	16 10	11 01	00 39	17 07	02 09	05 41
23	16 09 37	01 08 35	17 ♈ 36 50	23 ♈ 32 32	13 02	17 23	16 48	11 10	00 46	17 11	02 11	05 43
24	16 13 34	02 09 14	29 ♈ 27 08	05 ♉ 21 03	12 07	18 39	17 25	11 19	00 53	17 14	02 13	05 45
25	16 17 30	03 09 54	11 ♉ 14 39	17 ♉ 08 18	11 03	19 54	18 02	11 28	01 00	17 18	02 15	05 48
26	16 21 27	04 10 35	23 ♉ 02 17	28 ♉ 56 54	09 51	21 09	18 40	11 37	01 07	17 22	02 17	05 50
27	16 25 24	05 11 18	04 ♊ 52 25	10 ♊ 49 03	08 33	22 25	19 17	11 46	01 14	17 25	02 19	05 52
28	16 29 20	06 12 02	16 ♊ 47 33	22 ♊ 46 38	07 11	23 40	19 54	11 56	01 21	17 29	02 21	05 54
29	16 33 17	07 12 47	28 ♊ 47 59	04 ♋ 51 19	05 49	24 55	20 32	12 05	01 29	17 32	02 23	05 56
30	16 37 13	08 ♐ 13 34	10 ♋ 56 52	17 ♋ 04 51	04 ♐ 28	26 ♏ 11	21 ♎ 09	12 ♒ 15	01 ♐ 36	17 ♐ 36	02 ♑ 25	05 ♏ 59

Date	Moon True ☊ ° '	Moon Mean ☊ ° '	Moon ☽ Latitude ° '
01	09 ♉ 11	09 ♉ 02	03 N 27
02	09 R 11	08 58	04 13
03	09 11	08 55	04 48
04	09 10	08 52	05 10
05	09 10	08 49	05 18
06	09 D 10	08 46	05 10
07	09 11	08 43	04 45
08	09 11	08 39	04 03
09	09 12	08 36	03 05
10	09 13	08 33	01 53
11	09 13	08 30	00 N 32
12	09 R 13	08 27	00 S 52
13	09 12	08 24	02 13
14	09 11	08 20	03 23
15	09 09	08 17	04 18
16	09 07	08 14	04 56
17	09 06	08 11	05 14
18	09 05	08 08	05 15
19	09 D 04	08 04	04 59
20	09 05	08 01	04 28
21	09 06	07 58	03 46
22	09 08	07 55	02 55
23	09 09	07 52	01 56
24	09 11	07 49	00 S 53
25	09 R 11	07 45	00 N 11
26	09 10	07 42	01 16
27	09 08	07 39	02 17
28	09 04	07 36	03 12
29	09 00	07 33	03 59
30	08 ♉ 54	07 ♉ 30	04 N 36

DECLINATIONS

Date	Sun ☉	Moon ☽	Mercury ☿	Venus ♀	Mars ♂	Jupiter ♃	Saturn ♄	Uranus ♅	Neptune ♆	Pluto ♇
01	14 S 30	26 N 29	22 S 51	06 S 20	00 S 07	18 S 57	17 S 58	22 S 45	22 S 21	02 N 03
02	14 49	27 39	23 09	06 49	00 22	18 56	18 00	22 45	22 21	02 02
03	15 08	27 29	23 27	07 17	00 37	18 54	18 02	22 46	22 21	02 02
04	15 27	25 58	23 42	07 45	00 52	18 53	18 03	22 46	22 21	02 01
05	15 45	23 07	23 57	08 14	01 07	18 51	18 05	22 46	22 21	02 00
06	16 03	19 03	24 10	08 41	01 22	18 49	18 07	22 47	22 21	01 59
07	16 21	13 55	24 22	09 09	01 36	18 48	18 08	22 47	22 21	01 59
08	16 38	07 56	24 32	09 37	01 51	18 46	18 10	22 47	22 21	01 58
09	16 55	01 N 21	24 40	10 04	02 06	18 44	18 11	22 48	22 21	01 57
10	17 12	05 S 29	24 47	10 31	02 21	18 42	18 13	22 48	22 21	01 57
11	17 29	12 10	24 53	10 58	02 36	18 40	18 14	22 48	22 21	01 56
12	17 45	18 13	24 56	11 25	02 51	18 38	18 16	22 49	22 21	01 55
13	18 01	23 06	24 58	11 51	03 05	18 36	18 18	22 49	22 21	01 55
14	18 17	26 21	24 57	12 17	03 20	18 34	18 19	22 49	22 21	01 54
15	18 33	27 42	24 55	12 43	03 35	18 32	18 21	22 50	22 21	01 54
16	18 48	27 06	24 50	13 09	03 50	18 30	18 22	22 50	22 21	01 53
17	19 02	24 48	24 43	13 34	04 04	18 27	18 24	22 50	22 21	01 52
18	19 17	21 11	24 34	13 59	04 19	18 25	18 25	22 51	22 21	01 52
19	19 31	16 38	24 23	14 23	04 33	18 23	18 27	22 51	22 21	01 51
20	19 45	11 30	24 08	14 47	04 48	18 21	18 28	22 52	22 21	01 51
21	19 58	06 01	23 51	15 11	05 03	18 18	18 30	22 52	22 21	01 50
22	20 11	00 S 25	23 31	15 34	05 17	18 16	18 31	22 52	22 21	01 50
23	20 24	05 N 07	23 08	15 57	05 31	18 13	18 33	22 53	22 21	01 49
24	20 36	10 27	22 43	16 20	05 46	18 11	18 34	22 53	22 21	01 49
25	20 48	15 23	22 15	16 42	06 00	18 08	18 36	22 53	22 21	01 48
26	20 59	19 45	21 44	17 04	06 14	18 05	18 37	22 54	22 21	01 48
27	21 10	23 21	21 12	17 25	06 29	18 03	18 39	22 54	22 21	01 47
28	21 21	25 58	20 39	17 46	06 43	18 00	18 40	22 54	22 21	01 47
29	21 31	27 26	20 06	18 06	06 57	17 57	18 42	22 55	22 21	01 46
30	21 S 41	27 N 35	19 S 33	18 S 26	07 S 11	17 S 55	18 S 43	22 S 55	22 S 21	01 N 46

ZODIAC SIGN ENTRIES

Date	h	m	Planets
02	08	31	☽ ♋
04	19	04	☽ ♌
07	02	18	☽ ♍
09	05	52	☽ ♎
09	15	08	♀ ♏
11	06	31	☽ ♏
13	05	52	☽ ♐
15	05	53	☽ ♑
17	02	10	♄ ♐
17	08	25	☽ ♒
19	14	42	☽ ♓
22	00	42	☽ ♈
22	08	51	☉ ♐
24	13	07	☽ ♉
27	02	08	☽ ♊
29	14	23	☽ ♋

LATITUDES

Date	Mercury ☿	Venus ♀	Mars ♂	Jupiter ♃	Saturn ♄	Uranus ♅	Neptune ♆	Pluto ♇
01	02 S 32	01 N 32	01 N 12	00 S 51	01 N 50	00 S 03	01 N 05	16 N 05
04	02 41	01 31	01 12	00 50	01 49	00 03	01 05	16 05
07	02 45	01 28	01 12	00 50	01 49	00 03	01 05	16 06
10	02 44	01 25	01 11	00 50	01 49	00 03	01 05	16 06
13	02 36	01 22	01 11	00 50	01 48	00 03	01 05	16 06
16	02 18	01 18	01 11	00 50	01 48	00 03	01 05	16 07
19	01 48	01 13	01 10	00 50	01 48	00 03	01 04	16 07
22	01 04	01 08	01 10	00 49	01 48	00 03	01 04	16 08
25	00 S 08	01 03	01 10	00 49	01 48	00 03	01 04	16 08
28	00 N 53	00 57	01 09	00 49	01 48	00 03	01 04	16 09
31	01 N 47	00 N 51	01 N 09	00 S 49	01 N 48	00 S 03	01 N 04	16 N 10

DATA

Julian Date	2446371
Delta T	+54 seconds
Ayanamsa	23° 39' 21"
Synetic vernal point	05° ♓ 27' 39"
True obliquity of ecliptic	23° 26' 35"

MOON'S PHASES, APSIDES AND POSITIONS ☽

Date	h	m	Phase	Longitude ° '	Eclipse Indicator
05	20	07	☾	13 ♌ 21	
12	14	20	●	20 ♏ 09	Total
19	09	04	☽	26 ♒ 59	
27	12	42	○	05 ♊ 13	

Day	h	m		
12	12	24	Perigee	
25	21	48	Apogee	
02	21	11	Max dec	27° N 44'
09	16	47	0S	
15	16	22	Max dec	27° S 44'
22	13	48	0N	
30	02	47	Max dec	27° N 41'

LONGITUDES

Date	Chiron ⚷	Ceres ⚳	Pallas ⚴	Juno ⚵	Vesta ⚶	Black Moon Lilith ⚸
01	13 ♊ 42	00 ♍ 16	07 ♋ 34	09 ♏ 10	15 ♐ 58	16 ♉ 55
11	13 ♊ 14	03 ♍ 36	08 ♋ 24	12 ♏ 35	21 ♐ 09	18 ♉ 01
21	12 ♊ 41	06 ♍ 40	08 ♋ 09	16 ♏ 00	26 ♐ 22	19 ♉ 08
31	12 ♊ 07	09 ♍ 23	06 ♋ 45	19 ♏ 22	01 ♑ 39	20 ♉ 14

ASPECTARIAN

h m	Aspects
01 Friday	
01 32	☽ ± ☉
04 17	☽ ☍ ♅
04 39	☿ ∥ ♅
05 42	♀ Q ♆
12 10	☽ △ ♀
12 11	☽ ⚼ ♇
14 36	☉ ⊥ ♂
19 33	☽ ⚼ ♃
21 19	☽ ⚼ ☉
21 53	☿ ⊻ ♆
02 Saturday	
05 02	☽ ⊼ ♄
11 35	☽ ☍ ♆
11 47	☉ ⊥ ♅
13 07	☽ ⊼ ☿
13 40	☽ ± ♃
16 02	☽ □ ♂
17 05	☽ ± ♄
18 16	☽ △ ♇
03 Sunday	
01 36	♂ Q ♅
01 39	☽ ⊼ ♃
02 20	☽ ± ☿
05 48	☽ △ ☉
11 00	☽ ⚼ ♄
13 27	♀ ⊥ ♄
16 12	☽ ⊼ ♅
21 51	☽ ⚼ ☿
04 Monday	
03 54	☽ ± ♅
05 10	☽ Q ♂
06 12	☽ □ ♀
14 02	♂ ⊻ ♇
16 17	☽ △ ♄
18 53	☿ ⊻ ♇
21 12	☽ ⚼ ♅
22 07	☽ ⊼ ♆
05 Tuesday	
00 17	☿ ✶ ♂
04 32	☽ □ ♇
05 14	☽ ✶ ♂
05 27	☽ △ ☿
06 24	☽ # ☿
09 27	☽ ± ♆
11 48	☽ ☍ ♃
14 20	☽ # ♅
17 05	☽ # ♆
19 55	☽ Q ♀
20 07	☽ □ ☉
06 Wednesday	
01 20	☽ △ ♅
02 10	☽ ⚼ ♆
10 22	☽ ∠ ♂
13 10	☽ # ♃
13 52	☽ Q ♇
16 44	☽ # ♄
20 04	☽ ✶ ♀
07 Thursday	
00 11	☽ □ ♄
01 47	☽ # ☉
03 20	☽ ⊥ ♂
05 16	☽ △ ♆
07 19	☽ Q ☉
11 19	☽ ✶ ♇
14 25	☽ ⊻ ♂
16 45	☽ □ ☿
18 22	☽ ⊼ ♃
08 Friday	
01 07	☽ ∠ ♀
04 49	☽ ± ♃
05 50	☽ ✶ ☉
05 59	☽ # ♀
06 46	☽ □ ♅
07 53	☽ Q ♄
13 18	☽ ∠ ♇
17 11	☿ ✶ ♃
17 58	☽ ⊥ ♀
19 29	☉ ⊻ ♅
20 14	☽ ⚼ ♃
20 28	♀ ⊻ ♄
22 27	♂ # ♇
09 Saturday	
01 50	☽ Q ☿
04 18	☽ ✶ ♄
04 28	☽ ⊥ ♇
04 49	☉ ∠ ♆
04 59	☽ ⊻ ♀
08 44	☽ □ ♆
09 02	☽ ∠ ☉
09 26	☽ # ♂
09 51	☽ ∥ ♇
13 07	☽ Q ♅
14 26	☽ ⊻ ♇
19 29	☽ ☌ ♂
21 17	☽ △ ♃
23 00	☽ ✶ ☿
23 38	☽ # ♇
10 Sunday	
00 37	☽ ∥ ♂
00 54	☽ ⊥ ☉
05 09	☽ ∠ ♄
08 46	☽ ✶ ♅
11 19	☽ ⊻ ☉
14 10	☽ Q ♆
18 53	♀ ∠ ♅
19 46	☽ ⊥ ♄
21 17	☿ ⊥ ♇
11 Monday	
00 42	☽ ∠ ☿
01 19	♀ ✶ ♆
05 25	☽ ⊻ ♄
07 15	☽ ∥ ♀
08 52	☽ ∠ ♅
09 22	☽ ✶ ♆
10 04	☽ ☌ ♀
14 52	☽ ☌ ♇
15 50	☽ ⊥ ☿
21 39	☽ ⊻ ♂
21 43	☽ □ ♃
23 10	☽ ⊥ ♅
23 31	♂ △ ♃
12 Tuesday	
01 49	☽ ⊻ ☿
07 31	☽ ⊥ ♂
08 39	☽ ⊻ ♅
09 06	☽ ∠ ♆
09 56	☽ ∥ ☉
12 13	☽ ∥ ♄
13 49	☽ ∥ ♃
14 20	☽ ☌ ☉
22 18	☽ ∠ ♂
23 19	☽ ⊥ ♆
13 Wednesday	
02 19	☽ Q ♃
05 09	☽ ☌ ♄
07 48	☽ ∥ ♆
08 48	☽ ⊻ ♆
10 23	☽ ∥ ♅
13 40	☽ ⊻ ♀
14 19	☽ ⊻ ♇
21 27	☽ ✶ ♃
21 42	♀ ☌ ♇
23 03	☽ ✶ ♂
23 52	☽ ⊥ ♇
14 Thursday	
00 02	☽ ∥ ☿
00 03	☽ ⊥ ♀
03 26	☽ ☌ ☿
08 18	☽ ☌ ♅
14 18	☽ ∠ ♇
15 26	☉ ∥ ♄
15 46	☽ ∠ ♀
17 22	☽ ⊻ ☉
19 07	☽ Q ♂
21 42	☽ ∠ ♃
15 Friday	
03 48	☽ ⊥ ☉
05 31	☽ ⊻ ♄
09 01	☽ ☌ ♆
10 57	☉ ∥ ♃
12 31	☽ ⊥ ♃
14 47	☽ ✶ ♇
15 27	☽ ⊥ ♄
18 32	☽ ✶ ♀
19 42	☽ ∠ ☉
22 33	☽ ⊻ ♃
16 Saturday	
02 00	☽ □ ♂
06 02	☽ ⊻ ☿
06 35	☽ ∠ ♄
09 43	☽ ⊻ ♅
10 53	☽ Q ♇
13 18	☿ ⊥ ♀
16 25	☽ ⊥ ☿
16 39	☽ Q ♀
19 58	☽ ⊥ ♅
22 58	☽ ✶ ☉
17 Sunday	
08 12	☽ ∠ ☿
08 29	☽ ✶ ♄
11 37	☽ ∠ ♅
11 53	☽ ⊻ ♆
12 40	☽ ∥ ☿
18 05	☽ □ ♇
21 36	☽ Q ☉
21 41	♀ □ ♃
22 30	☽ ⊥ ♆
18 Monday	
01 54	☽ ∥ ♅
02 47	☽ ☌ ♃
03 15	☽ □ ♀
05 01	☽ ∥ ♆
05 54	☽ Q ♄
07 27	♀ ⊥ ♅
08 31	☽ △ ♂
09 59	☉ ⊥ ♆
10 38	♃ ∥ ♄
11 04	☽ ✶ ☿
14 30	☽ ✶ ♅
14 43	☽ ∠ ♆
16 08	☿ St R
22 02	☽ ∥ ☉
19 Tuesday	
02 58	☽ ∥ ♄
03 11	☽ ∥ ♃
09 02	☽ Q ☿
09 04	☽ □ ☉
12 44	☽ Q ♅
13 28	☽ ⚼ ♂
15 16	☽ □ ♄
18 34	☽ ✶ ♆
21 59	☽ ∥ ♀
20 Wednesday	
01 15	☽ △ ♇
06 55	☿ ✶ ♂
07 23	☽ ± ♂
11 03	☽ ⊻ ♃
17 09	☽ △ ♀
17 32	☽ Q ♆
18 37	☽ □ ☿
19 32	☽ ⊼ ♂
22 46	☽ ⊥ ♃
23 14	☽ □ ♅
21 Thursday	
04 30	☿ ⊻ ♀
04 56	☉ Q ♃
06 16	☽ ⚼ ♇
10 52	♂ ∠ ♄
16 01	☽ ∥ ♂
16 35	☽ ∠ ♃
23 58	☽ △ ☉
22 Friday	
00 03	☽ ± ♇
01 53	☽ △ ♄
01 54	☽ ⚼ ♀
04 57	☽ □ ♆
05 57	☽ # ♇
12 01	☽ ⊼ ♇
13 30	♀ ⊻ ♂
21 39	☽ ∥ ♇
21 59	☽ ± ♀
22 50	☽ ✶ ♃
23 Saturday	
01 45	☉ ☌ ♄
03 21	☽ △ ☿
07 51	♀ ⊻ ♅
07 52	♀ ∠ ♆
08 13	☽ ⚼ ♄
08 45	☽ ⚼ ☉
10 15	☽ ☍ ♂
11 07	☽ △ ♅
11 29	☽ ⊼ ♀
13 51	☽ # ♂
23 22	☽ Q ♃
24 Sunday	
02 37	☽ ± ♄
02 54	☿ ∥ ♅
04 28	♂ ✶ ♅
04 40	☽ ± ☉
07 37	☽ ⚼ ☿
13 22	☉ ⊻ ♆
14 56	☽ ⊼ ♄
17 37	☽ △ ♆
17 42	☽ ⚼ ♅
18 01	☽ ⊼ ☉
20 13	☿ ⊥ ♇
25 Monday	
00 28	☽ ± ☿
00 52	☽ ☍ ♇
04 11	☉ ∠ ♂
04 16	☿ ✶ ♃
06 35	☿ ∥ ♆
11 38	☽ ⊼ ☿
12 07	☽ ± ♅
12 27	☽ □ ♃
19 28	☽ # ♀
26 Tuesday	
00 15	☽ ⚼ ♆
00 23	☽ ⊼ ♅
02 31	☽ # ♃
02 36	☽ ⊼ ♂
05 23	☽ # ♄
07 43	☽ ☍ ♀
15 29	☽ ± ♂
18 36	☽ ± ♆
19 59	☽ # ☉
22 54	☽ # ☿
27 Wednesday	
04 34	☽ ☍ ♄
04 42	☽ # ♆
06 48	☽ ⊼ ♆
08 37	☽ # ♅
10 44	☽ ⚼ ♂
12 42	☽ ☍ ☉
13 03	☉ ∥ ☿
14 01	☽ ⊼ ♇
18 40	☽ ☍ ☿
28 Thursday	
02 06	☽ △ ♃
02 10	☽ ± ♇
04 44	☉ ⊻ ♇
13 24	☽ ☍ ♅
18 36	☽ △ ♂
20 17	☽ ⚼ ♇
21 56	☉ ☌ ☿
29 Friday	
02 46	♀ ∥ ♃
03 23	☽ ⊼ ♀
06 06	♂ Q ♆
08 33	☽ ⚼ ♃
09 51	☿ ⊻ ♇
15 30	☿ ∠ ♂
16 42	☽ ± ♀
17 22	☽ ⊼ ♄
19 08	☽ ☍ ♆
30 Saturday	
00 30	☽ ⊼ ☿
02 11	☽ △ ♇
02 38	☽ ± ♃
05 20	☽ ± ♄
06 10	☽ ⊼ ☉
11 09	☽ ± ☿
12 30	☽ ⚼ ♀
14 35	☽ ⊼ ♃
16 42	♀ ⊥ ♆
19 00	☽ ± ☉
23 10	☽ ⚼ ♄

All ephemeris data is given at 12.00 UT and the Moon's longitude is additionally given for 24.00 UT

DECEMBER 1985

Date	Sidereal time h m s	Sun ☉ ° ′ ″	Moon ☽ ° ′ ″	Moon ☽ 24.00 ° ′ ″	Mercury ☿ ° ′	Venus ♀ ° ′	Mars ♂ ° ′	Jupiter ♃ ° ′	Saturn ♄ ° ′	Uranus ♅ ° ′	Neptune ♆ ° ′	Pluto ♇ ° ′
01	16 41 10	09 ♐ 14 22	23 ♋ 15 30	29 ♋ 29 05	03 ♐ 12	27 ♏ 26	21 ♎ 46	12 ♒ 25	01 ♐ 43	17 ♐ 40	02 ♑ 27	06 ♏ 01
02	16 45 06	10 15 11	05 ♌ 45 51	12 ♌ 06 07	02 R 03	28 41	22 23	12 35	01 50	17 43	02 29	06 03
03	16 49 03	11 16 02	18 ♌ 30 10	24 ♌ 58 20	01 03	29 ♏ 57	23 01	12 45	01 57	17 47	02 31	06 05
04	16 52 59	12 16 53	01 ♍ 30 55	08 ♍ 08 14	00 ♐ 13	01 ♐ 12	23 38	12 55	02 04	17 51	02 34	06 07
05	16 56 56	13 17 47	14 ♍ 50 35	21 ♍ 38 12	29 ♏ 34	02 28	24 15	13 06	02 11	17 54	02 36	06 09
06	17 00 53	14 18 41	28 ♍ 31 18	05 ♎ 30 00	29 07	03 43	24 52	13 16	02 18	17 58	02 38	06 11
07	17 04 49	15 19 37	12 ♎ 34 20	19 ♎ 44 12	28 51	04 59	25 29	13 26	02 25	18 01	02 40	06 13
08	17 08 46	16 20 34	26 ♎ 59 24	04 ♏ 19 31	28 D 46	06 14	26 07	13 37	02 32	18 05	02 42	06 15
09	17 12 42	17 21 33	11 ♏ 44 02	19 ♏ 12 13	28 51	07 30	26 44	13 48	02 39	18 09	02 44	06 17
10	17 16 39	18 22 33	26 ♏ 43 12	04 ♐ 15 58	29 06	08 45	27 21	13 59	02 46	18 12	02 47	06 19
11	17 20 35	19 23 33	11 ♐ 49 21	19 ♐ 22 11	29 ♏ 30	10 01	27 58	14 10	02 53	18 16	02 49	06 21
12	17 24 32	20 24 35	26 ♐ 53 14	04 ♑ 21 18	00 ♐ 01	11 16	28 35	14 21	03 00	18 20	02 51	06 23
13	17 28 28	21 25 37	11 ♑ 45 17	19 ♑ 04 11	00 40	12 32	29 12	14 32	03 07	18 23	02 53	06 25
14	17 32 25	22 26 41	26 ♑ 17 11	03 ♒ 23 39	01 25	13 47	29 ♎ 49	14 43	03 14	18 27	02 56	06 27
15	17 36 22	23 27 44	10 ♒ 23 10	17 ♒ 15 28	02 15	15 03	00 ♏ 26	14 55	03 21	18 31	02 58	06 28
16	17 40 18	24 28 48	24 ♒ 00 30	00 ♓ 38 26	03 10	16 18	01 03	15 06	03 28	18 34	03 00	06 30
17	17 44 15	25 29 53	07 ♓ 09 29	13 ♓ 34 04	04 10	17 34	01 40	15 18	03 34	18 38	03 02	06 32
18	17 48 11	26 30 57	19 ♓ 52 41	26 ♓ 05 54	05 13	18 49	02 17	15 30	03 41	18 42	03 05	06 34
19	17 52 08	27 32 02	02 ♈ 14 18	08 ♈ 18 35	06 20	20 05	02 54	15 41	03 48	18 45	03 07	06 35
20	17 56 04	28 33 07	14 ♈ 19 22	20 ♈ 17 22	07 30	21 20	03 31	15 53	03 55	18 49	03 09	06 37
21	18 00 01	29 ♐ 34 13	26 ♈ 13 12	02 ♉ 07 31	08 42	22 36	04 08	16 05	04 01	18 53	03 11	06 39
22	18 03 57	00 ♑ 35 19	08 ♉ 00 56	13 ♉ 54 01	09 56	23 51	04 45	16 17	04 08	18 56	03 14	06 40
23	18 07 54	01 36 25	19 ♉ 47 17	25 ♉ 41 14	11 12	25 07	05 22	16 29	04 15	19 00	03 16	06 42
24	18 11 51	02 37 31	01 ♊ 36 17	07 ♊ 32 50	12 30	26 22	05 59	16 41	04 21	19 03	03 18	06 44
25	18 15 47	03 38 38	13 ♊ 31 12	19 ♊ 31 39	13 49	27 38	06 35	16 54	04 28	19 07	03 20	06 45
26	18 19 44	04 39 45	25 ♊ 34 24	01 ♋ 39 39	15 10	28 ♐ 53	07 12	17 06	04 34	19 11	03 23	06 47
27	18 23 40	05 40 52	07 ♋ 47 31	13 ♋ 58 06	16 32	00 ♑ 09	07 49	17 18	04 41	19 14	03 25	06 48
28	18 27 37	06 42 00	20 ♋ 11 26	26 ♋ 27 34	17 55	01 24	08 26	17 31	04 47	19 18	03 27	06 50
29	18 31 33	07 43 07	02 ♌ 46 32	09 ♌ 08 19	19 19	02 40	09 03	17 44	04 54	19 21	03 30	06 51
30	18 35 30	08 44 15	15 ♌ 32 58	22 ♌ 00 29	20 44	03 55	09 39	17 56	05 00	19 25	03 32	06 53
31	18 39 26	09 ♑ 45 24	28 ♌ 30 57	05 ♍ 04 23	22 ♐ 10	05 ♑ 11	10 ♏ 16	18 ♒ 09	05 ♐ 06	19 ♐ 28	03 ♑ 34	06 ♏ 54

Date	Moon True ☊ ° ′	Moon Mean ☊ ° ′	Moon ☽ Latitude ° ′
01	08 ♉ 49	07 ♉ 26	05 N 01
02	08 R 44	07 23	05 12
03	08 41	07 20	05 08
04	08 39	07 17	04 48
05	08 D 38	07 14	04 12
06	08 39	07 10	03 22
07	08 40	07 07	02 18
08	08 42	07 04	01 N 03
09	08 R 42	07 01	00 S 17
10	08 41	06 58	01 36
11	08 37	06 55	02 50
12	08 32	06 51	03 52
13	08 26	06 48	04 36
14	08 18	06 45	05 02
15	08 11	06 42	05 09
16	08 06	06 39	04 57
17	08 02	06 35	04 30
18	08 00	06 32	03 51
19	08 D 00	06 29	03 01
20	08 01	06 26	02 04
21	08 02	06 23	01 S 03
22	08 03	06 20	00 00
23	08 R 02	06 16	01 N 03
24	07 59	06 13	02 03
25	07 53	06 10	02 58
26	07 45	06 07	03 46
27	07 35	06 04	04 24
28	07 23	06 01	04 50
29	07 12	05 57	05 03
30	07 01	05 54	05 00
31	06 ♉ 52	05 ♉ 51	04 N 43

DECLINATIONS

Date	Sun ☉ ° ′	Moon ☽ ° ′	Mercury ☿ ° ′	Venus ♀ ° ′	Mars ♂ ° ′	Jupiter ♃ ° ′	Saturn ♄ ° ′	Uranus ♅ ° ′	Neptune ♆ ° ′	Pluto ♇ ° ′
01	21 S 50	26 N 23	19 S 03	18 S 45	07 S 25	17 S 52	18 S 45	22 S 55	22 S 21	01 N 46
02	21 59	23 52	18 35	19 04	07 39	17 49	18 46	22 56	22 21	01 45
03	22 08	20 09	18 11	19 22	07 53	17 46	18 48	22 56	22 21	01 45
04	22 16	15 25	17 51	19 40	08 07	17 43	18 49	22 56	22 21	01 44
05	22 24	09 51	17 35	19 57	08 21	17 40	18 50	22 57	22 21	01 44
06	22 31	03 N 40	17 23	20 14	08 35	17 37	18 52	22 57	22 21	01 44
07	22 38	02 S 51	17 16	20 30	08 48	17 34	18 53	22 57	22 21	01 43
08	22 45	09 25	17 13	20 46	09 02	17 31	18 55	22 58	22 21	01 43
09	22 51	15 37	17 13	21 00	09 16	17 28	18 56	22 58	22 21	01 43
10	22 56	20 59	17 18	21 15	09 29	17 25	18 57	22 58	22 21	01 43
11	23 01	25 01	17 25	21 28	09 43	17 21	18 59	22 59	22 21	01 42
12	23 06	27 16	17 35	21 41	09 56	17 18	19 00	22 59	22 21	01 42
13	23 10	27 31	17 47	21 54	10 09	17 15	19 01	22 59	22 21	01 42
14	23 14	25 50	18 01	22 06	10 23	17 12	19 03	23 00	22 21	01 42
15	23 17	22 35	18 16	22 17	10 36	17 08	19 04	23 00	22 21	01 41
16	23 20	18 11	18 33	22 27	10 49	17 05	19 05	23 00	22 21	01 41
17	23 22	13 04	18 51	22 37	11 02	17 01	19 06	23 01	22 21	01 41
18	23 24	07 32	19 09	22 46	11 15	16 58	19 08	23 01	22 21	01 41
19	23 25	01 S 53	19 28	22 54	11 28	16 54	19 09	23 01	22 21	01 41
20	23 26	03 N 44	19 47	23 02	11 40	16 51	19 10	23 02	22 21	01 40
21	23 27	09 08	20 06	23 09	11 53	16 47	19 11	23 02	22 20	01 40
22	23 27	14 11	20 25	23 15	12 06	16 44	19 12	23 02	22 20	01 40
23	23 26	18 42	20 43	23 21	12 18	16 40	19 14	23 03	22 20	01 40
24	23 25	22 30	21 02	23 26	12 31	16 36	19 15	23 03	22 20	01 40
25	23 24	25 23	21 19	23 30	12 43	16 32	19 16	23 03	22 20	01 40
26	23 22	27 08	21 37	23 33	12 55	16 29	19 17	23 03	22 20	01 40
27	23 19	27 37	21 53	23 36	13 07	16 25	19 18	23 04	22 20	01 40
28	23 16	26 42	22 09	23 38	13 20	16 21	19 19	23 04	22 20	01 40
29	23 13	24 27	22 24	23 39	13 32	16 17	19 20	23 04	22 20	01 40
30	23 09	20 57	22 39	23 39	13 43	16 13	19 22	23 05	22 20	01 40
31	23 S 05	16 N 24	22 S 52	23 S 39	13 S 55	16 S 09	19 S 23	23 S 05	22 S 20	01 N 40

ZODIAC SIGN ENTRIES

Date	h	m	Planets
02	00	59	☽ ♌
03	13	00	♀ ♐
04	09	14	☽ ♍
04	19	23	☿ ♏
06	14	33	☽ ♎
08	16	56	☽ ♏
10	17	13	☽ ♐
12	11	05	☿ ♐
12	16	59	☽ ♑
14	18	15	☽ ♒
14	18	59	♂ ♏
16	22	50	☽ ♓
19	07	37	☽ ♈
21	19	41	☽ ♉
21	22	08	☉ ♑
24	08	45	☽ ♊
26	20	44	☽ ♋
27	09	17	♀ ♑
29	06	44	☽ ♌
31	14	43	☽ ♍

LATITUDES

Date	Mercury ☿ ° ′	Venus ♀ ° ′	Mars ♂ ° ′	Jupiter ♃ ° ′	Saturn ♄ ° ′	Uranus ♅ ° ′	Neptune ♆ ° ′	Pluto ♇ ° ′
01	01 N 47	00 N 51	01 N 09	00 S 49	01 N 48	00 S 03	01 N 04	16 N 10
04	02 24	00 45	01 08	00 49	01 48	00 03	01 04	16 11
07	02 42	00 38	01 08	00 49	01 48	00 03	01 04	16 11
10	02 44	00 31	01 07	00 49	01 48	00 03	01 04	16 12
13	02 34	00 24	01 07	00 49	01 48	00 03	01 04	16 14
16	02 17	00 17	01 06	00 48	01 48	00 03	01 04	16 15
19	01 56	00 10	01 05	00 48	01 48	00 03	01 04	16 16
22	01 33	00 N 03	01 04	00 48	01 48	00 03	01 04	16 17
25	01 09	00 S 04	01 04	00 48	01 48	00 03	01 04	16 18
28	00 45	00 12	01 03	00 48	01 48	00 03	01 04	16 20
31	00 N 21	00 S 19	01 N 02	00 S 48	01 N 48	00 S 03	01 N 04	16 N 21

DATA

Julian Date	2446401
Delta T	+54 seconds
Ayanamsa	23° 39′ 26″
Synetic vernal point	05° ♓ 27′ 33″
True obliquity of ecliptic	23° 26′ 35″

MOON'S PHASES, APSIDES AND POSITIONS ☽

Date	h	m	Phase	Longitude ° ′	Eclipse Indicator
05	09	01	☾	13 ♍ 10	
12	00	54	●	19 ♐ 56	
19	01	58	☽	27 ♓ 06	
27	07	30	○	05 ♋ 29	

Day	h	m		
11	00	35	Perigee	
23	07	09	Apogee	
07	01	36	0S	
13	02	54	Max dec	27° S 39′
19	19	58	0N	
27	08	22	Max dec	27° N 38′

LONGITUDES

Date	Chiron ⚷ ° ′	Ceres ⚳ ° ′	Pallas ⚴ ° ′	Juno ⚵ ° ′	Vesta ⚶ ° ′	Black Moon Lilith ⚸ ° ′
01	12 ♊ 07	09 ♍ 23	06 ♋ 45	19 ♏ 22	01 ♑ 39	20 ♉ 14
11	11 ♊ 32	11 ♍ 41	04 ♋ 17	22 ♏ 40	06 ♑ 56	21 ♉ 21
21	10 ♊ 58	13 ♍ 30	01 ♋ 06	25 ♏ 53	12 ♑ 15	22 ♉ 27
31	10 ♊ 27	14 ♍ 44	27 ♊ 49	29 ♏ 00	17 ♑ 35	23 ♉ 34

All ephemeris data is given at 12.00 UT and the Moon's longitude is additionally given for 24.00 UT

ASPECTARIAN

01 Sunday
01 04 ☽ ⊼ ♅
03 05 ☽ bQ ☿
08 58 ☽ □ ♂
11 08 ♀ ∥ ♄
12 47 ☽ ± ♅
14 04 ☽ bQ ☉
20 53 ☿ ∥ ♀
20 58 ☽ △ ♀

02 Monday
00 36 ♀ ⊥ ♂
02 45 ☿ ∥ ♄
02 54 ☿ ⊻ ♆
04 26 ☽ △ ♄
05 30 ☽ △ ☿
05 44 ☽ ⊼ ♆
06 10 ☽ bQ ♅
12 32 ☽ □ ♇
16 29 ☿ ☌ ♄
17 11 ☽ ± ♆
18 49 ☽ # ♅
21 13 ☽ Q ♂
21 15 ☽ △ ☉
22 43 ☽ # ♆

03 Tuesday
00 34 ☽ # ☉
01 05 ☽ ☍ ♃
10 10 ☽ bQ ♆
10 39 ☽ △ ♅
16 03 ☽ # ♀
18 47 ☿ Q ♃
19 23 ☽ # ♄
20 48 ☽ ⚹ ♂
22 23 ☽ Q ♇
23 27 ☽ # ☿

04 Wednesday
00 25 ☿ ☌ ♀
00 49 ☽ # ♃
05 42 ♀ Q ♃
07 58 ☉ ⊥ ♇
09 46 ☽ □ ☿
11 22 ☽ □ ♀
13 01 ☽ □ ♄
13 54 ☽ △ ♆
20 23 ☽ ⚹ ♇
22 43 ☿ ⊥ ♂

05 Thursday
01 03 ☿ ∥ ♃
01 31 ☽ ∠ ♂
02 49 ☉ ∥ ♆
06 10 ☉ ⚹ ♃
06 10 ♀ ☌ ♄
08 50 ☽ ⊼ ♃
09 01 ☽ □ ☉
14 36 ♀ ⊻ ♆
16 39 ☽ Q ☿
17 27 ☽ □ ♅
17 48 ☽ # ♂
18 20 ☽ ⊥ ♂
19 37 ☽ ± ♃
21 32 ☽ Q ♄
22 57 ☽ Q ♀
23 10 ☽ ∠ ♇

06 Friday
05 21 ☽ ⊻ ♂
11 33 ☽ bQ ♃
13 00 ☽ ⚹ ☿
14 53 ☽ ⊥ ♇
18 35 ☽ ⚹ ♄
19 03 ☽ Q ☉
19 06 ☽ □ ♆
19 14 ☽ ∥ ♇
21 50 ☽ ⚹ ♀

07 Saturday
00 51 ☽ Q ♅
01 12 ☽ ⊻ ♇
07 53 ☽ # ♇
13 29 ☽ △ ♃
14 08 ☽ ∠ ☿
16 59 ☽ ⚹ ☉
20 12 ☽ ∠ ♄
21 11 ☽ ⚹ ♅

08 Sunday
01 35 ☽ Q ♆
01 35 ☽ ∠ ♀
05 02 ☽ ⊥ ☿
10 29 ☽ ☌ ♂
10 32 ☽ ∥ ♂
11 15 ☽ ⊥ ♄
11 24 ☿ St D
12 19 ♀ ⊻ ♇
14 55 ☽ ⊻ ☿
17 49 ☽ ⊥ ♀
19 40 ☽ ∠ ☉
21 10 ☽ ⊻ ♄
21 23 ☽ ⚹ ♆
22 01 ☽ ∠ ♅

09 Monday
03 10 ☽ ☌ ♇
04 30 ☽ ⊻ ♀
08 27 ♂ ⊥ ♄
11 21 ☽ ⊥ ☉
12 40 ☽ ⊥ ♅
15 22 ☽ □ ♃
18 46 ☽ ∥ ☿
19 39 ☽ ∥ ♃
21 41 ☽ ∠ ♆
21 42 ☽ ⊻ ☉
22 21 ☽ ⊻ ♅

10 Tuesday
02 16 ☽ ∥ ♄
07 48 ☉ ☌ ♅
12 05 ☽ ⊥ ♆
13 03 ☽ ⊻ ♂
13 22 ☽ ∥ ♀
14 06 ♄ ⊻ ♆
15 53 ☽ ☌ ☿
19 07 ☽ ∥ ♆
20 28 ☽ Q ♃
21 40 ☽ ⊻ ♆
21 42 ☽ ☌ ♄
22 39 ☽ ∥ ☉
22 40 ☽ ∥ ♅
22 59 ☽ ⊥ ♂
23 35 ☉ ∥ ♅

11 Wednesday
03 17 ☽ ⊻ ♇
04 55 ☿ ∥ ♃
08 52 ☽ ☌ ♀
12 50 ☽ ⊥ ♇
13 54 ☽ ∠ ♂
15 46 ☽ ⚹ ♃
22 17 ☽ ☌ ♅

12 Thursday
00 54 ☽ ☌ ☉
03 11 ☽ ∠ ♇
14 50 ☽ ⚹ ♂
16 00 ☽ ∠ ♃
17 14 ☽ ⊻ ☿
21 36 ☽ ☌ ♆
21 54 ☽ ⊻ ♄

13 Friday
03 18 ☽ ⚹ ♇
03 20 ☽ ⊥ ☿
06 42 ☽ ⊥ ♃
07 41 ☽ ⊥ ♄
09 47 ♀ ⊥ ♇
11 04 ☽ Q ♂
11 38 ☉ ∠ ♇
13 23 ☽ ⊻ ♀
16 36 ☽ ⊻ ♃
18 43 ☽ ∠ ☿
22 31 ☽ ∠ ♄
22 56 ☽ ⊻ ♅
22 56 ☽ Q ♇

14 Saturday
00 09 ☽ ⊥ ♀
05 06 ☽ ⊻ ☉
08 55 ☽ ⊥ ♅
15 54 ☽ ⊥ ☉
16 36 ☽ ∠ ♀
18 13 ☽ □ ♂
21 10 ☽ ⚹ ☿
23 14 ☽ ⊻ ♆
23 49 ☽ ⚹ ♄

15 Sunday
00 09 ☽ ∠ ♅
05 15 ☽ □ ♇
07 40 ☽ ∥ ☉
08 25 ☽ ∠ ☉
09 05 ♀ ⚹ ♃
09 25 ☽ ∥ ♅
09 32 ☽ ⊥ ♆
13 29 ☽ ∥ ♆
13 48 ☽ ∥ ♀
19 12 ☽ Q ☿
20 00 ☽ ☌ ♃
20 43 ☽ Q ♄
20 56 ☽ ⚹ ♀
21 06 ♀ ∥ ♆

16 Monday
01 17 ☽ ∠ ♆
02 17 ☽ ⚹ ♅
02 47 ♀ ∠ ♂
07 26 ☿ ⊻ ♆
07 30 ☽ ∥ ♄
09 47 ☿ Q ♃
10 18 ☽ ∥ ☿
12 55 ☽ ⚹ ☉
17 30 ☽ ∥ ♃
20 02 ☿ ☌ ♄
20 33 ☽ Q ♀
23 56 ☽ Q ♅

17 Tuesday
01 23 ☽ △ ♂
04 22 ☽ ⚹ ♆
05 19 ☽ □ ♄
06 00 ☽ □ ☿
10 50 ☽ △ ♇
12 41 ☽ Q ☉
20 38 ☽ ∥ ♂

18 Wednesday
02 49 ☽ Q ♆
03 31 ☽ ⊻ ♃
06 48 ☽ bQ ♂
09 34 ♀ ☌ ♅
09 44 ☽ □ ♅
09 45 ☽ □ ♀
10 14 ☿ ∥ ♄
15 09 ☽ ⊥ ♃
15 14 ☽ bQ ♇

19 Thursday
01 01 ☽ ± ♂
01 58 ☽ □ ☉
08 46 ☽ ± ♇
08 54 ☽ ∠ ♃
12 51 ☽ # ♇
13 23 ☽ ⊼ ♂
13 44 ☽ □ ♆
15 06 ☽ △ ♄
17 33 ☿ ⊻ ♇
20 37 ☽ ⊼ ♇
20 45 ♂ ⚹ ♆
20 55 ☽ △ ☿

20 Friday
03 08 ☽ ∥ ♇
11 01 ♀ ∥ ♅
15 11 ☽ ⚹ ♃
17 34 ♀ ∠ ♇
19 38 ♃ Q ♄
21 05 ☽ △ ♅
21 19 ☽ bQ ♄

21 Saturday
00 55 ♂ ∠ ♅
03 47 ☽ △ ♀
06 18 ☽ bQ ☿
06 48 ♂ ⊻ ♄
15 42 ☽ ± ♄
15 51 ☽ Q ♃
19 27 ☽ △ ☉

22 Sunday
01 22 ☽ # ♂
02 13 ☽ △ ♆
02 41 ☽ ± ☿
03 39 ☽ bQ ♅
04 01 ☽ ⊼ ♄
04 58 ☽ ☍ ♂
09 16 ☽ ☍ ♇
13 54 ☽ bQ ♀
16 22 ☽ ⊼ ☿
22 05 ☽ ± ♅

23 Monday
00 57 ☽ # ♃
04 54 ☽ bQ ☉
05 09 ☽ □ ♃
08 28 ☉ ∠ ♃
08 53 ☽ bQ ♆
10 23 ☽ ⊼ ♅
10 27 ☽ ± ♀
15 06 ☽ # ♄
17 44 ☿ ⊥ ♂

24 Tuesday
00 07 ☽ ⊼ ♀
00 57 ☽ ± ☉
01 13 ☽ # ☿
03 15 ☽ ± ♆
09 28 ☉ ∥ ♀
10 54 ☽ # ♆
14 16 ☽ ⊼ ☉
15 27 ☽ ⊼ ♆
16 02 ☽ # ♅
16 15 ☿ ⊥ ♇
17 37 ☽ ☍ ♄
18 47 ☽ # ☉
19 04 ☽ # ♀
21 19 ☽ ⊼ ♂
22 22 ☽ ⊼ ♇

25 Wednesday
04 35 ☉ ☌ ♆
10 02 ☽ ± ♂
10 28 ☽ ± ♇
12 41 ☽ ☍ ☿
18 42 ♂ ☌ ♇
18 52 ☽ △ ♃
23 14 ☽ ☍ ♅

26 Thursday
04 28 ☽ bQ ♇
04 58 ☽ bQ ♂
09 40 ☉ ⊻ ♄
19 17 ☽ ☍ ♀

27 Friday
01 05 ☽ bQ ♃
03 25 ☽ ☍ ♆
05 52 ☽ ⊼ ♄
07 30 ☽ ☍ ☉
10 04 ☽ △ ♇
12 03 ☽ △ ♂
17 41 ☽ ± ♄
18 57 ☽ ± ♃

28 Saturday
03 46 ☿ ⚹ ♃
06 46 ☽ ⊼ ♃
07 05 ☽ ⊼ ☿
10 16 ☽ ⊼ ♅
11 13 ☽ bQ ♄
15 08 ☉ ⚹ ♇
20 03 ☽ ± ☿
21 49 ☽ ± ♅

29 Sunday
05 11 ☿ ∥ ♆
11 45 ☽ ⊼ ♀
12 36 ☿ ☌ ♅
13 22 ☽ ⊼ ♆
13 31 ♀ ∠ ♃
15 00 ☽ bQ ♅
15 17 ☽ bQ ☿
16 03 ☽ △ ♄
18 14 ☽ # ♀
19 43 ☽ □ ♇
21 34 ☽ # ☉
22 08 ☽ ⊼ ☉
22 24 ☽ # ♅

30 Monday
00 19 ☽ ± ♀
00 25 ☽ □ ♂
00 42 ☽ ± ♆
02 01 ☽ # ☿
03 25 ☽ # ♆
04 23 ♀ ☌ ♆
10 21 ☽ ± ☉
16 31 ☽ ☍ ♃
17 34 ☽ bQ ♆
18 57 ☽ bQ ♀
19 13 ☽ △ ♅
20 57 ☽ # ♄
22 50 ☽ △ ☿

31 Tuesday
04 29 ☽ bQ ☉
05 20 ☽ Q ♇
07 33 ☿ ∠ ♇
10 35 ♀ ⊻ ♄
11 31 ☽ Q ♂
12 24 ☉ ∥ ♅
13 12 ☽ # ♃
21 17 ☽ △ ♆
23 04 ☽ # ♂

JANUARY 1986

Date	Sidereal time h m s	Sun ☉ ° ' "	Moon ☽ ° ' "	Moon ☽ 24.00 ° ' "	Mercury ☿ ° '	Venus ♀ ° '	Mars ♂ ° '	Jupiter ♃ ° '	Saturn ♄ ° '	Uranus ♅ ° '	Neptune ♆ ° '	Pluto ♇ ° '
01	18 43 23	10 ♑ 46 32	11 ♍ 40 55	18 ♍ 20 39	23 ♐ 36	06 ♑ 26	10 ♏ 53	18 ♒ 22	05 ♐ 13	19 ♐ 32	03 ♑ 36	06 ♏ 55
02	18 47 20	11 47 41	25 ♍ 03 44	01 ♎ 50 17	25 03	07 42	11 29	18 35	05 19	19 35	03 39	06 57
03	18 51 16	12 48 50	08 ♎ 40 29	15 ♎ 34 27	26 31	08 57	12 06	18 47	05 25	19 39	03 41	06 58
04	18 55 13	13 50 00	22 ♎ 32 18	29 ♎ 34 04	27 59	10 13	12 42	19 00	05 31	19 42	03 43	06 59
05	18 59 09	14 51 10	06 ♏ 39 44	13 ♏ 49 12	29 ♐ 28	11 28	13 19	19 14	05 37	19 46	03 45	07 01
06	19 03 06	15 52 20	21 ♏ 02 15	28 ♏ 18 29	00 ♑ 57	12 44	13 56	19 27	05 43	19 49	03 48	07 02
07	19 07 02	16 53 30	05 ♐ 37 27	12 ♐ 58 30	02 27	13 59	14 32	19 40	05 49	19 52	03 50	07 03
08	19 10 59	17 54 41	20 ♐ 20 52	27 ♐ 43 40	03 57	15 15	15 08	19 53	05 55	19 56	03 52	07 04
09	19 14 55	18 55 51	05 ♑ 05 56	12 ♑ 26 39	05 28	16 30	15 45	20 06	06 01	19 59	03 54	07 05
10	19 18 52	19 57 01	19 ♑ 44 48	26 ♑ 59 24	06 59	17 46	16 21	20 20	06 07	20 02	03 56	07 06
11	19 22 49	20 58 11	04 ♒ 09 34	11 ♒ 14 32	08 31	19 01	16 58	20 33	06 13	20 06	03 59	07 07
12	19 26 45	21 59 21	18 ♒ 13 40	25 ♒ 06 31	10 03	20 17	17 34	20 47	06 19	20 09	04 01	07 08
13	19 30 42	23 00 30	01 ♓ 52 49	08 ♓ 32 27	11 36	21 32	18 10	21 00	06 24	20 12	04 03	07 09
14	19 34 38	24 01 38	15 ♓ 05 30	21 ♓ 32 11	13 09	22 47	18 47	21 14	06 30	20 15	04 05	07 10
15	19 38 35	25 02 46	27 ♓ 53 49	04 ♈ 07 52	14 42	24 03	19 23	21 27	06 35	20 18	04 07	07 11
16	19 42 31	26 03 53	10 ♈ 17 53	16 ♈ 23 28	16 16	25 18	19 59	21 41	06 41	20 22	04 09	07 12
17	19 46 28	27 04 59	22 ♈ 25 17	28 ♈ 24 00	17 51	26 34	20 35	21 55	06 46	20 25	04 12	07 13
18	19 50 24	28 06 05	04 ♉ 20 21	10 ♉ 15 02	19 26	27 49	21 11	22 08	06 52	20 28	04 14	07 13
19	19 54 21	29 ♑ 07 09	16 ♉ 08 46	22 ♉ 02 12	21 02	29 ♑ 05	21 47	22 22	06 57	20 31	04 16	07 14
20	19 58 18	00 ♒ 08 13	27 ♉ 56 01	03 ♊ 50 49	22 38	00 ♒ 20	22 23	22 36	07 02	20 34	04 18	07 15
21	20 02 14	01 09 16	09 ♊ 47 12	15 ♊ 45 39	24 15	01 36	22 59	22 50	07 07	20 37	04 20	07 16
22	20 06 11	02 10 18	21 ♊ 46 39	27 ♊ 50 35	25 52	02 51	23 35	23 04	07 12	20 40	04 22	07 16
23	20 10 07	03 11 19	03 ♋ 58 16	10 ♋ 08 28	27 30	04 06	24 11	23 18	07 17	20 43	04 24	07 17
24	20 14 04	04 12 20	16 ♋ 22 49	22 ♋ 40 56	29 ♑ 08	05 22	24 47	23 32	07 22	20 46	04 26	07 17
25	20 18 00	05 13 19	29 ♋ 02 49	05 ♌ 28 25	00 ♒ 47	06 37	25 23	23 46	07 27	20 49	04 28	07 18
26	20 21 57	06 14 18	11 ♌ 57 37	18 ♌ 30 15	02 27	07 53	25 59	24 00	07 32	20 52	04 30	07 19
27	20 25 53	07 15 16	25 ♌ 06 07	01 ♍ 44 59	04 07	09 08	26 35	24 14	07 37	20 54	04 32	07 19
28	20 29 50	08 16 13	08 ♍ 26 38	15 ♍ 10 49	05 48	10 23	27 10	24 28	07 41	20 57	04 34	07 19
29	20 33 47	09 17 09	21 ♍ 57 21	28 ♍ 46 02	07 30	11 39	27 46	24 42	07 46	21 00	04 36	07 20
30	20 37 44	10 18 04	05 ♎ 36 45	12 ♎ 29 22	09 12	12 54	28 22	24 56	07 50	21 03	04 38	07 20
31	20 41 40	11 ♒ 18 58	19 ♎ 23 50	26 ♎ 20 06	10 ♒ 56	14 ♒ 09	28 ♏ 57	25 ♒ 10	07 ♐ 55	21 ♐ 05	04 ♑ 40	07 ♏ 20

DECLINATIONS

Date	Moon True ☊ ° '	Moon Mean ☊ ° '	Moon ☽ Latitude ° '	Sun ☉ ° '	Moon ☽ ° '	Mercury ☿ ° '	Venus ♀ ° '	Mars ♂ ° '	Jupiter ♃ ° '	Saturn ♄ ° '	Uranus ♅ ° '	Neptune ♆ ° '	Pluto ♇ ° '
01	06 ♉ 46	05 ♉ 48	04 N 10	23 S 00	11 N 02	23 S 04	23 S 38	14 S 07	16 S 05	19 S 24	23 S 05	22 S 20	01 N 40
02	06 R 42	05 45	03 23	22 55	05 N 04	23 16	23 37	14 19	16 01	19 25	23 06	22 20	01 40
03	06 41	05 41	02 23	22 50	01 S 15	23 26	23 34	14 30	15 57	19 26	23 06	22 20	01 40
04	06 D 41	05 38	01 N 15	22 44	07 37	23 35	23 31	14 41	15 53	19 27	23 06	22 20	01 40
05	06 41	05 35	00 00	22 37	13 44	23 43	23 27	14 53	15 49	19 28	23 06	22 20	01 40
06	06 R 41	05 32	01 S 16	22 30	19 14	23 50	23 22	15 04	15 45	19 29	23 07	22 20	01 40
07	06 38	05 29	02 27	22 23	23 40	23 56	23 17	15 15	15 41	19 30	23 07	22 20	01 40
08	06 32	05 26	03 30	22 15	26 35	24 00	23 11	15 26	15 37	19 31	23 07	22 19	01 40
09	06 23	05 22	04 18	22 06	27 39	24 04	23 04	15 37	15 33	19 32	23 07	22 19	01 41
10	06 12	05 19	04 49	21 58	26 45	24 06	22 57	15 48	15 28	19 33	23 08	22 19	01 41
11	06 00	05 16	05 01	21 48	24 05	24 07	22 48	15 59	15 24	19 33	23 08	22 19	01 41
12	05 48	05 13	04 54	21 39	20 02	24 06	22 39	16 09	15 20	19 34	23 08	22 19	01 41
13	05 38	05 10	04 31	21 29	15 01	24 04	22 30	16 20	15 16	19 35	23 08	22 19	01 41
14	05 30	05 07	03 53	21 18	09 27	24 01	22 20	16 30	15 11	19 36	23 09	22 19	01 42
15	05 25	05 03	03 05	21 08	03 S 40	23 56	22 09	16 40	15 07	19 37	23 09	22 19	01 42
16	05 22	05 00	02 09	20 56	02 N 06	23 51	21 57	16 50	15 02	19 38	23 09	22 19	01 42
17	05 21	04 57	01 08	20 45	07 41	23 43	21 45	17 00	14 58	19 39	23 09	22 19	01 42
18	05 D 21	04 54	00 S 05	20 33	12 53	23 35	21 32	17 10	14 54	19 39	23 10	22 19	01 42
19	05 R 21	04 51	00 N 57	20 20	17 35	23 25	21 18	17 20	14 49	19 40	23 10	22 19	01 43
20	05 19	04 47	01 57	20 07	21 36	23 13	21 04	17 30	14 45	19 41	23 10	22 19	01 43
21	05 15	04 44	02 51	19 54	24 45	23 00	20 49	17 39	14 40	19 42	23 10	22 19	01 43
22	05 08	04 41	03 39	19 41	26 50	22 46	20 33	17 49	14 36	19 42	23 11	22 19	01 44
23	04 58	04 38	04 18	19 27	27 41	22 30	20 17	17 58	14 31	19 43	23 11	22 18	01 44
24	04 46	04 35	04 45	19 13	27 09	22 13	20 01	18 07	14 26	19 44	23 11	22 18	01 44
25	04 32	04 32	04 58	18 58	25 13	21 54	19 43	18 16	14 22	19 45	23 11	22 18	01 45
26	04 18	04 28	04 57	18 43	21 57	21 34	19 26	18 25	14 17	19 45	23 11	22 18	01 45
27	04 05	04 25	04 40	18 28	17 33	21 12	19 07	18 34	14 13	19 46	23 12	22 18	01 45
28	03 54	04 22	04 08	18 12	12 14	20 49	18 48	18 43	14 08	19 47	23 12	22 18	01 46
29	03 46	04 19	03 21	17 56	06 N 16	20 24	18 29	18 51	14 03	19 47	23 12	22 18	01 46
30	03 41	04 16	02 22	17 40	00 S 03	19 58	18 09	19 00	13 58	19 48	23 12	22 18	01 47
31	03 ♉ 39	04 ♉ 13	01 N 15	17 S 23	06 S 27	19 S 30	17 S 48	19 S 08	13 S 54	19 S 48	23 S 12	22 S 18	01 N 47

ZODIAC SIGN ENTRIES

Date	h	m	Planets
02	20	45	☽ ♎
05	00	44	☽ ♏
05	20	42	☿ ♑
07	02	47	☽ ♐
09	03	42	☽ ♑
11	05	01	☽ ♒
13	08	39	☽ ♓
15	16	03	☽ ♈
18	03	14	☽ ♉
20	05	36	♀ ♒
20	08	46	☉ ♒
20	16	12	☽ ♊
23	04	15	☽ ♋
25	00	33	☿ ♒
25	13	47	☽ ♌
27	20	51	☽ ♍
30	02	10	☽ ♎

LATITUDES

Date	Mercury ☿ ° '	Venus ♀ ° '	Mars ♂ ° '	Jupiter ♃ ° '	Saturn ♄ ° '	Uranus ♅ ° '	Neptune ♆ ° '	Pluto ♇ ° '
01	00 N 13	00 S 21	01 N 02	00 S 48	01 N 48	00 S 03	01 N 04	16 N 21
04	00 S 09	00 28	01 01	00 48	01 49	00 03	01 04	16 23
07	00 31	00 35	01 00	00 48	01 49	00 03	01 04	16 24
10	00 50	00 41	00 58	00 48	01 49	00 04	01 04	16 26
13	01 08	00 47	00 57	00 48	01 49	00 04	01 04	16 27
16	01 24	00 53	00 56	00 48	01 49	00 04	01 04	16 29
19	01 38	00 59	00 55	00 48	01 50	00 04	01 04	16 31
22	01 49	01 04	00 54	00 48	01 50	00 04	01 04	16 32
25	01 58	01 08	00 52	00 48	01 50	00 04	01 04	16 34
28	02 03	01 12	00 51	00 48	01 51	00 04	01 04	16 35
31	02 S 05	01 S 16	00 N 49	00 S 49	01 N 51	00 S 04	01 N 04	16 N 37

DATA

Julian Date	2446432
Delta T	+55 seconds
Ayanamsa	23° 39' 32"
Synetic vernal point	05° ♓ 27' 27"
True obliquity of ecliptic	23° 26' 35"

MOON'S PHASES, APSIDES AND POSITIONS ☽

Date	h	m	Phase	Longitude ° '	Eclipse Indicator
03	19	47	☾	13 ♎ 09	
10	12	22	●	19 ♑ 58	
17	22	13	☽	27 ♈ 31	
26	00	31	○	05 ♌ 45	

Day	h	m	
08	07	18	Perigee
20	01	19	Apogee

Day	h	m		
03	07	19	OS	
09	13	01	Max dec	27° S 39'
16	03	11	ON	
23	14	57	Max dec	27° N 41'
30	11	48	OS	

LONGITUDES

Date	Chiron ⚷ ° '	Ceres ⚳ ° '	Pallas ⚴ ° '	Juno ⚵ ° '	Vesta ⚶ ° '	Black Moon Lilith ⚸ ° '
01	10 ♊ 24	14 ♍ 49	27 ♊ 30	29 ♏ 18	18 ♑ 06	23 ♉ 41
11	09 ♊ 58	15 ♍ 20	24 ♊ 49	02 ♐ 16	23 ♑ 25	24 ♉ 47
21	09 ♊ 37	15 ♍ 07	23 ♊ 40	05 ♐ 04	28 ♑ 43	25 ♉ 54
31	09 ♊ 23	14 ♍ 11	22 ♊ 41	07 ♐ 40	03 ♒ 59	27 ♉ 00

ASPECTARIAN

h m Aspects

01 Wednesday
00 10 ☽ □ ♄
01 28 ☽ △ ♀
03 21 ☽ ✶ ♇
06 30 ☉ ∥ ☿
10 13 ☽ △ ☉
10 28 ☽ ✶ ♂
14 11 ☿ ∥ ♅
17 55 ☉ ✶ ♂
21 31 ♀ ✶ ♇
23 27 ☉ ⊥ ♄

02 Thursday
00 14 ☽ ⊼ ♃
02 11 ☽ □ ♅
06 26 ☽ ∠ ♇
08 52 ☽ Q ♄
11 07 ☽ ± ♃
11 59 ☽ □ ☿
14 39 ☽ ∠ ♂
21 06 ♃ ∠ ♆
22 27 ☽ ⊥ ♇

03 Friday
01 01 ☽ ∥ ♇
03 13 ☽ □ ♆
03 18 ☽ ꝑ ♃
06 15 ☽ ✶ ♄
07 17 ☽ ⊥ ♂
09 01 ☽ ⊻ ♇
10 11 ☽ Q ♅
11 19 ☉ ⊥ ♃
12 32 ☽ □ ♀
13 35 ☽ ⋕ ♇
18 15 ☽ ⊻ ♂
19 47 ☽ □ ☉
23 22 ☽ Q ☿

04 Saturday
03 52 ☿ ∥ ♀
04 21 ☿ ∠ ♂
05 50 ☽ △ ♃
07 06 ☽ ✶ ♅
08 31 ☽ ∠ ♄
10 35 ☽ Q ♆
22 24 ☽ ✶ ☿
22 38 ☽ Q ♀

05 Sunday
00 00 ☽ ⊥ ♄
05 04 ☽ Q ☉
07 05 ☽ ✶ ♆
08 47 ☽ ∠ ♅
10 14 ☽ ⊻ ♄
12 35 ☽ ☌ ♇
15 14 ♀ ⊥ ♄
16 52 ☽ ∥ ♂
20 35 ☽ ∥ ♃
20 51 ☽ ✶ ♀
23 39 ☽ ☌ ♂
23 57 ☽ ⊥ ♅

06 Monday
02 35 ☽ ∠ ☿
02 46 ☽ ✶ ☉
07 13 ♂ ⊥ ♅
08 16 ☽ ∠ ♆
09 19 ☽ □ ♃
09 58 ☽ ⊻ ♅
13 11 ☽ ∥ ♄
19 12 ☽ ⊥ ☿
23 11 ☽ ⊥ ♆

07 Tuesday
00 05 ☽ ∠ ♀
03 58 ☽ ∥ ♆
04 28 ☽ ∥ ☉
04 36 ♀ ⊥ ♃
05 26 ☽ ∠ ☉
06 12 ☽ ⊻ ☿
08 35 ☽ ∥ ♅
09 03 ☽ ⊻ ♆
09 41 ☽ ∥ ♀
12 20 ☽ ☌ ♄
13 48 ☽ ∥ ☿
14 20 ☽ ⊻ ♇
15 23 ☽ Q ♃
16 13 ☽ ⊥ ♀
21 15 ☽ ⊥ ☉
21 19 ☉ ∥ ♆

08 Wednesday
00 08 ☽ ⊥ ♇
02 55 ☽ ⊻ ♀
03 10 ☽ ⊻ ♂
07 45 ☽ ⊻ ☉
08 17 ♀ ✶ ♂
10 37 ☿ ☌ ♆
11 14 ☽ ✶ ♃
11 19 ☽ ☌ ♅
13 21 ☽ ⊥ ♂
14 48 ☽ ∠ ♇
18 17 ♃ ✶ ♅

09 Thursday
00 58 ♀ ∥ ♅
04 37 ☽ ∠ ♂
04 59 ♂ ∥ ♃
05 22 ☿ ∠ ♃
10 03 ☽ ☌ ♆
12 01 ☽ ∠ ♃
12 40 ☽ ☌ ☿
13 31 ☽ ⊻ ♄
15 15 ☽ ✶ ♇
15 42 ☉ Q ♇
21 24 ☿ ⊻ ♄
23 23 ☽ ⊥ ♄

10 Friday
02 57 ☽ ⊥ ♃
06 11 ☽ ✶ ♂
08 25 ☽ ☌ ♀
10 56 ☽ Q ♇
12 22 ☽ ☌ ☉
12 29 ☽ ⊻ ♅
12 59 ☽ ⊻ ♃
13 54 ☿ ✶ ♇
14 11 ☉ ⊻ ♅
14 17 ☽ ∠ ♄
22 27 ☽ ⊥ ♅
23 24 ☉ ⊻ ♃

11 Saturday
02 54 ☽ Q ♂
11 42 ☽ ⊻ ♆
11 53 ☽ ∥ ☿
13 35 ☽ ∠ ♅
14 00 ♀ Q ♇
15 29 ☽ ✶ ♄
17 00 ☽ □ ♇
18 20 ☉ ∠ ♄
18 21 ☽ ∥ ♅
20 15 ☽ ⊻ ☿
20 43 ☽ ∥ ♀
21 52 ☽ ⊥ ♆
23 20 ☽ ∥ ♆

12 Sunday
02 53 ☽ ∥ ☉
07 46 ☽ ⊥ ☿
09 27 ♀ ⊻ ♅
10 48 ☽ □ ♂
12 08 ☽ Q ♄
13 22 ☽ ∠ ♆
14 21 ☽ ∥ ♄
15 21 ☽ ✶ ♅
15 54 ☽ ⊻ ♀
16 30 ☽ ☌ ♃
19 04 ☽ ⊻ ☉
23 39 ♀ ⊻ ♃

13 Monday
01 25 ☽ ∠ ☿
03 29 ☽ ⊥ ♀
06 14 ☽ ∥ ♂
06 28 ☽ ⊥ ☉
09 19 ♀ ∠ ♄
10 54 ☽ ∥ ♃
12 35 ☽ Q ♅
15 54 ☽ ✶ ♆
20 11 ☽ □ ♄
21 14 ☽ ∠ ♀
21 30 ☽ △ ♇
23 57 ☽ ∠ ☉

14 Tuesday
01 21 ☿ ⊥ ♄
07 56 ☽ ✶ ☿
13 03 ♀ ∥ ♆
13 51 ☽ Q ♆
19 11 ☽ △ ♂
21 38 ☽ □ ♅
23 38 ☽ ⊻ ♃

15 Wednesday
01 04 ♂ ∠ ♆
01 12 ☽ ꝑ ♇
03 56 ☽ ✶ ♀
06 09 ☽ ✶ ☉
09 27 ☽ Q ☿
11 10 ☽ ⊥ ♃
18 20 ☽ ± ♇
20 08 ☽ ⋕ ♇

16 Thursday
00 01 ☽ □ ♆
01 07 ☽ ꝑ ♂
01 27 ☿ ⊥ ♃
04 49 ☽ ∠ ♃
04 53 ☽ △ ♄
05 30 ☽ Q ♀
05 56 ☽ ⊼ ♇
07 15 ☽ Q ☉
10 17 ☽ ∥ ♇
19 21 ☉ ⊥ ♅
19 37 ☽ ± ♂

17 Friday
01 31 ☽ □ ☿
04 25 ♂ ⊻ ♅
07 58 ☽ △ ♅
08 09 ☽ ⊼ ♂
08 59 ♀ ⊥ ♅
10 41 ☽ ꝑ ♄
10 58 ☽ ✶ ♃
21 17 ☽ □ ♀
22 13 ☽ □ ☉

18 Saturday
04 54 ☽ ± ♄
08 48 ☿ Q ♇
11 35 ☽ Q ♃
11 47 ☽ △ ♆
14 17 ☽ ꝑ ♅
17 09 ☽ ⊼ ♄
17 51 ☽ ☍ ♇
21 45 ☽ ⋕ ♃

19 Sunday
04 03 ☿ ⊻ ♅
08 40 ☽ ± ♅
10 36 ☽ ⋕ ♂
16 05 ☉ ☌ ♀
18 22 ☽ ꝑ ♆
20 56 ☽ ⊼ ♅
23 31 ☽ △ ☿
23 57 ☽ ⋕ ♄

20 Monday
00 07 ☽ ☍ ♂
00 56 ☽ □ ♃
02 34 ☿ ∠ ♄
03 09 ☽ ⋕ ☉
06 16 ☿ ✶ ♂
08 44 ☽ ⋕ ♀
11 28 ☿ ⊻ ♃
12 45 ☽ ± ♆
16 52 ☽ ⋕ ♆
16 54 ☽ △ ☉
17 28 ☽ △ ♀
17 35 ☿ ∥ ♅
22 44 ☽ ⋕ ☿
23 05 ☽ ⋕ ♅

21 Tuesday
00 57 ☽ ⊼ ♆
01 39 ♂ □ ♃
06 35 ☽ ☍ ♄
06 54 ☽ ⊼ ♇
10 44 ☽ ꝑ ☿
19 00 ☽ ± ♇

22 Wednesday
01 58 ☽ ꝑ ☉
03 16 ☽ ꝑ ♀
07 36 ☽ ± ☿
09 09 ☉ ∥ ♄
09 47 ☽ ☍ ♅
12 59 ☽ ꝑ ♆
14 36 ☽ △ ♃
15 47 ☽ ⊼ ♂
21 21 ☽ ⊼ ☿
21 30 ☽ ± ☉
23 12 ☽ ± ♀

23 Thursday
00 11 ☿ ⊥ ♅
04 14 ☽ ± ♂
09 58 ♄ ⊻ ♇
10 21 ☽ ⊼ ☉
12 19 ☽ ⊼ ♀
12 52 ☽ ☍ ♆
17 50 ♀ ⊻ ♆
18 28 ☽ △ ♇
18 31 ☽ ⊼ ♄
20 35 ☽ ꝑ ♃
22 40 ☽ ꝑ ♂

24 Friday
04 11 ☿ ∥ ♆
06 11 ☽ ± ♄
14 14 ☽ ± ♃
17 39 ☉ ⊻ ♆
19 59 ♀ ∠ ♅
20 24 ☽ ⊼ ♅
23 29 ☽ ꝑ ♄

25 Saturday
01 51 ☽ ⊼ ♃
04 46 ☽ △ ♂
07 47 ☽ ± ♅
10 23 ♀ ∥ ♄
15 45 ☽ ☍ ☿
22 10 ☽ ⊼ ♆

26 Sunday
00 31 ☽ ☍ ☉
00 41 ☽ ꝑ ♅
01 06 ♀ □ ♇
02 37 ☉ ∠ ♅
03 24 ☽ □ ♇
03 39 ☽ ☍ ♀
03 46 ☽ △ ♄
03 57 ☽ ⋕ ♅
05 00 ♀ ✶ ♄
09 19 ☽ ± ♆
09 49 ☽ ⋕ ♆
14 40 ☽ ⋕ ☿
15 50 ♀ Q ♂

27 Monday
00 40 ☽ ⋕ ♄
01 52 ☽ ꝑ ♆
03 30 ☽ ⋕ ♀
04 21 ☽ △ ♅
05 50 ☉ ∥ ♂
07 05 ☽ ⋕ ♂
07 12 ☽ ⋕ ☉
10 24 ☽ ☍ ♃
12 23 ☽ Q ♇
13 29 ☉ □ ♇
14 48 ☽ □ ♂
18 02 ☿ ⊻ ♆
21 07 ☉ ✶ ♄

28 Tuesday
03 42 ☽ ⋕ ♃
05 03 ☽ △ ♆
06 36 ☽ ⊼ ☿
10 00 ☽ ✶ ♇
10 38 ☽ □ ♄
11 40 ☽ ⊼ ☉
14 08 ☿ ∠ ♅
15 36 ♀ ⊥ ♆
15 50 ☽ ⊼ ♀
16 46 ♀ ∥ ♂
18 52 ☽ ± ☿
23 14 ☽ ± ☉

29 Wednesday
00 32 ☽ Q ♂
03 35 ☽ ± ♀
09 34 ☿ □ ♇
10 18 ☽ □ ♅
12 40 ☽ ∠ ♇
13 06 ☽ ꝑ ☿
15 52 ☿ ✶ ♄
16 27 ☽ ꝑ ☉
16 56 ☽ ⊼ ♃
18 45 ☽ Q ♄
21 06 ☽ ꝑ ♀
22 42 ☽ ✶ ♂

30 Thursday
03 40 ☽ ± ♃
04 30 ☽ ⊥ ♇
05 08 ☽ ∥ ♇
10 17 ☽ □ ♆
15 01 ☽ ⊻ ♇
15 21 ☉ Q ♂
15 55 ☽ ✶ ♄
18 01 ☽ Q ♅
18 28 ☽ ⋕ ♇
19 10 ☽ △ ☿
19 41 ☽ ꝑ ♃
20 09 ☉ ⊥ ♆
20 22 ☿ ∥ ♄
20 50 ☽ △ ☉
23 47 ♂ ⊥ ♆

31 Friday
01 59 ☽ △ ♀
02 07 ☽ ∠ ♂
08 20 ☿ ⊥ ♆
12 35 ☿ Q ♂
14 56 ☽ ✶ ♅
17 41 ☽ Q ♆
18 07 ☽ ∠ ♄
18 26 ☽ ⊥ ♂
22 10 ☽ △ ♃

All ephemeris data is given at 12.00 UT and the Moon's longitude is additionally given for 24.00 UT

FEBRUARY 1986

Date	Sidereal time h m s	Sun ☉ ° ′ ″	Moon ☽ ° ′ ″	Moon ☽ 24.00 ° ′ ″	Mercury ☿ ° ′	Venus ♀ ° ′	Mars ♂ ° ′	Jupiter ♃ ° ′	Saturn ♄ ° ′	Uranus ♅ ° ′	Neptune ♆ ° ′	Pluto ♇ ° ′
01	20 45 36	12 ♒ 19 52	03 ♏ 18 10	10 ♏ 17 59	12 ♒ 39	15 ♒ 25	29 ♏ 33	25 ♒ 25	07 ♐ 59	21 ♐ 08	04 ♑ 42	07 ♏ 21
02	20 49 33	13 20 46	17 ♏ 19 34	24 ♏ 22 52	14 24	16 40	00 ♐ 08	25 39	08 03	21 11	04 44	07 21
03	20 53 29	14 21 38	01 ♐ 27 45	08 ♐ 34 06	16 09	17 55	00 44	25 53	08 08	21 13	04 46	07 21
04	20 57 26	15 22 30	15 ♐ 41 40	22 ♐ 50 08	17 55	19 10	01 19	26 07	08 12	21 16	04 48	07 21
05	21 01 22	16 23 21	29 ♐ 59 05	07 ♑ 08 02	19 41	20 26	01 54	26 22	08 16	21 18	04 49	07 22
06	21 05 19	17 24 11	14 ♑ 16 25	21 ♑ 23 35	21 28	21 41	02 30	26 36	08 20	21 21	04 51	07 22
07	21 09 16	18 25 00	28 ♑ 28 52	05 ♒ 31 36	23 16	22 56	03 05	26 50	08 24	21 23	04 53	07 22
08	21 13 12	19 25 47	12 ♒ 31 06	19 ♒ 26 45	25 04	24 12	03 40	27 05	08 27	21 25	04 55	07 22
09	21 17 09	20 26 34	26 ♒ 18 02	03 ♓ 04 28	26 52	25 27	04 15	27 19	08 31	21 28	04 56	07 R 22
10	21 21 05	21 27 19	09 ♓ 45 46	16 ♓ 21 41	28 ♒ 41	26 42	04 50	27 33	08 35	21 30	04 58	07 22
11	21 25 02	22 28 02	22 ♓ 52 09	29 ♓ 17 14	00 ♓ 30	27 57	05 25	27 48	08 38	21 32	05 00	07 22
12	21 28 58	23 28 44	05 ♈ 37 05	11 ♈ 51 58	02 20	29 ♒ 12	06 00	28 02	08 42	21 34	05 02	07 22
13	21 32 55	24 29 25	18 ♈ 02 16	24 ♈ 08 26	04 09	00 ♓ 28	06 35	28 17	08 45	21 36	05 03	07 21
14	21 36 51	25 30 04	00 ♉ 11 00	06 ♉ 10 34	05 58	01 43	07 10	28 31	08 48	21 39	05 05	07 21
15	21 40 48	26 30 41	12 ♉ 07 45	18 ♉ 03 12	07 46	02 58	07 44	28 45	08 51	21 41	05 06	07 21
16	21 44 45	27 31 16	23 ♉ 57 37	29 ♉ 51 41	09 34	04 13	08 19	29 00	08 55	21 43	05 08	07 21
17	21 48 41	28 31 50	05 ♊ 46 06	11 ♊ 41 31	11 21	05 28	08 54	29 14	08 57	21 45	05 09	07 20
18	21 52 38	29 ♒ 32 22	17 ♊ 38 38	23 ♊ 38 02	13 07	06 43	09 28	29 29	09 00	21 47	05 11	07 20
19	21 56 34	00 ♓ 32 52	29 ♊ 40 19	05 ♋ 46 01	14 51	07 58	10 03	29 43	09 03	21 48	05 13	07 20
20	22 00 31	01 33 21	11 ♋ 55 35	18 ♋ 09 26	16 32	09 13	10 37	29 ♒ 58	09 06	21 50	05 14	07 19
21	22 04 27	02 33 48	24 ♋ 27 50	00 ♌ 51 01	18 11	10 28	11 11	00 ♓ 12	09 09	21 52	05 15	07 19
22	22 08 24	03 34 12	07 ♌ 19 06	13 ♌ 52 05	19 47	11 43	11 45	00 26	09 11	21 54	05 17	07 18
23	22 12 20	04 34 35	20 ♌ 29 52	27 ♌ 12 16	21 19	12 58	12 20	00 41	09 13	21 55	05 18	07 18
24	22 16 17	05 34 57	03 ♍ 59 00	10 ♍ 49 41	22 47	14 13	12 54	00 55	09 16	21 57	05 20	07 17
25	22 20 13	06 35 16	17 ♍ 43 55	24 ♍ 41 12	24 09	15 28	13 28	01 10	09 18	21 59	05 21	07 17
26	22 24 10	07 35 34	01 ♎ 41 04	08 ♎ 42 59	25 27	16 43	14 02	01 24	09 20	22 00	05 22	07 16
27	22 28 07	08 35 50	15 ♎ 46 30	22 ♎ 51 10	26 37	17 58	14 35	01 38	09 22	22 02	05 23	07 16
28	22 32 03	09 ♓ 36 05	29 ♎ 56 32	07 ♏ 02 17	27 ♓ 41	19 ♓ 13	15 ♐ 09	01 ♓ 53	09 ♐ 24	22 ♐ 03	05 ♑ 25	07 ♏ 15

Date	Moon True ☊ ° ′	Moon Mean ☊ ° ′	Moon ☽ Latitude ° ′
01	03 ♉ 38	04 ♉ 09	00 N 02
02	03 R 38	04 06	01 S 12
03	03 37	04 03	02 22
04	03 34	04 00	03 23
05	03 29	03 57	04 12
06	03 20	03 53	04 45
07	03 09	03 50	05 00
08	02 57	03 47	04 57
09	02 45	03 44	04 36
10	02 34	03 41	04 01
11	02 26	03 38	03 13
12	02 20	03 34	02 17
13	02 16	03 31	01 15
14	02 15	03 28	00 S 11
15	02 D 16	03 25	00 N 52
16	02 16	03 22	01 53
17	02 R 16	03 19	02 49
18	02 14	03 15	03 38
19	02 10	03 12	04 17
20	02 03	03 09	04 46
21	01 55	03 06	05 02
22	01 45	03 03	05 03
23	01 34	02 59	04 48
24	01 24	02 56	04 17
25	01 16	02 53	03 31
26	01 10	02 50	02 31
27	01 07	02 47	01 21
28	01 ♉ 06	02 ♉ 44	00 N 06

DECLINATIONS

Date	Sun ☉ ° ′	Moon ☽ ° ′	Mercury ☿ ° ′	Venus ♀ ° ′	Mars ♂ ° ′	Jupiter ♃ ° ′	Saturn ♄ ° ′	Uranus ♅ ° ′	Neptune ♆ ° ′	Pluto ♇ ° ′
01	17 S 06	12 S 35	19 S 00	17 S 27	19 S 16	13 S 49	19 S 49	23 S 13	22 S 18	01 N 47
02	16 49	18 09	18 30	17 05	19 24	13 44	19 50	23 13	22 18	01 48
03	16 32	22 46	17 57	16 43	19 32	13 39	19 50	23 13	22 18	01 48
04	16 14	26 02	17 24	16 21	19 40	13 35	19 51	23 13	22 17	01 49
05	15 56	27 38	16 48	15 58	19 48	13 30	19 51	23 13	22 17	01 49
06	15 37	27 24	16 12	15 34	19 55	13 25	19 52	23 13	22 17	01 50
07	15 19	25 22	15 34	15 11	20 03	13 20	19 52	23 14	22 17	01 50
08	15 00	21 48	14 54	14 46	20 10	13 15	19 53	23 14	22 17	01 51
09	14 41	17 05	14 13	14 22	20 17	13 10	19 53	23 14	22 17	01 51
10	14 21	11 37	13 31	13 57	20 24	13 05	19 53	23 14	22 17	01 52
11	14 02	05 S 47	12 47	13 31	20 31	13 00	19 54	23 14	22 17	01 52
12	13 42	00 N 08	12 02	13 05	20 38	12 56	19 54	23 14	22 17	01 53
13	13 22	05 55	11 17	12 39	20 44	12 51	19 55	23 15	22 17	01 53
14	13 01	11 22	10 30	12 13	20 51	12 46	19 55	23 15	22 17	01 54
15	12 41	16 19	09 42	11 46	20 57	12 41	19 55	23 15	22 17	01 54
16	12 20	20 36	08 53	11 19	21 03	12 36	19 56	23 15	22 17	01 55
17	11 59	24 03	08 04	10 52	21 09	12 31	19 56	23 15	22 16	01 56
18	11 38	26 29	07 14	10 24	21 15	12 26	19 56	23 15	22 16	01 56
19	11 17	27 44	06 24	09 56	21 21	12 22	19 57	23 15	22 16	01 57
20	10 55	27 39	05 34	09 28	21 27	12 16	19 57	23 16	22 16	01 57
21	10 34	26 11	04 45	08 59	21 33	12 11	19 57	23 16	22 16	01 58
22	10 12	23 19	03 56	08 30	21 38	12 06	19 57	23 16	22 16	01 59
23	09 50	19 13	03 07	08 01	21 43	12 01	19 58	23 16	22 16	01 59
24	09 28	14 02	02 20	07 32	21 48	11 56	19 58	23 16	22 16	02 00
25	09 06	08 05	01 35	07 03	21 54	11 51	19 58	23 16	22 16	02 00
26	08 43	01 N 38	00 51	06 33	21 58	11 46	19 58	23 16	22 16	02 01
27	08 21	04 S 57	00 10	06 04	22 03	11 40	19 58	23 16	22 16	02 02
28	07 S 58	11 S 21	00 N 29	05 S 34	22 S 08	11 S 35	19 S 59	23 S 16	22 S 16	02 N 02

ZODIAC SIGN ENTRIES

Date	h m	Planets
01	06 19	☽ ♏
02	06 27	♂ ♐
03	09 32	☽ ♐
05	12 02	☽ ♑
07	14 35	☽ ♒
09	18 32	☽ ♓
11	05 21	☿ ♓
12	01 21	☽ ♈
13	03 11	♀ ♓
14	11 38	☽ ♉
17	00 17	☽ ♊
18	22 58	☉ ♓
19	12 39	☽ ♋
20	16 05	♃ ♓
21	22 25	☽ ♌
24	04 58	☽ ♍
26	09 07	☽ ♎
28	12 06	☽ ♏

LATITUDES

Date	Mercury ☿ ° ′	Venus ♀ ° ′	Mars ♂ ° ′	Jupiter ♃ ° ′	Saturn ♄ ° ′	Uranus ♅ ° ′	Neptune ♆ ° ′	Pluto ♇ ° ′
01	02 S 05	01 S 17	00 N 48	00 S 49	01 N 51	00 S 04	01 N 04	16 N 38
04	02 02	01 20	00 47	00 49	01 51	00 04	01 04	16 39
07	01 54	01 23	00 45	00 49	01 52	00 04	01 04	16 41
10	01 41	01 25	00 43	00 49	01 52	00 04	01 04	16 42
13	01 23	01 26	00 41	00 49	01 53	00 04	01 04	16 44
16	00 59	01 27	00 39	00 49	01 53	00 04	01 04	16 46
19	00 S 28	01 27	00 37	00 50	01 53	00 04	01 04	16 47
22	00 N 08	01 27	00 34	00 50	01 54	00 04	01 04	16 49
25	00 49	01 26	00 32	00 50	01 54	00 04	01 04	16 50
28	01 32	01 25	00 29	00 50	01 55	00 04	01 04	16 51
31	02 N 15	01 S 22	00 N 27	00 S 50	01 N 55	00 S 04	01 N 04	16 N 53

DATA

Julian Date	2446463
Delta T	+55 seconds
Ayanamsa	23° 39′ 37″
Synetic vernal point	05° ♓ 27′ 22″
True obliquity of ecliptic	23° 26′ 36″

MOON'S PHASES, APSIDES AND POSITIONS ☽

Date	h m	Phase	Longitude ° ′	Eclipse Indicator
02	04 41	☾	13 ♏ 02	
09	00 55	●	19 ♒ 59	
16	19 55	☽	27 ♉ 51	
24	15 02	○	05 ♍ 43	

Day	h m		
04	16 01	Perigee	
16	22 22	Apogee	
05	20 53	Max dec	27° S 46′
12	11 26	0N	
19	22 44	Max dec	27° N 52′
26	17 58	0S	

LONGITUDES

Date	Chiron ⚷ ° ′	Ceres ⚳ ° ′	Pallas ⚴ ° ′	Juno ⚵ ° ′	Vesta ⚶ ° ′	Black Moon Lilith ⚸ ° ′
01	09 ♊ 22	14 ♍ 03	22 ♊ 43	07 ♐ 55	04 ♒ 31	27 ♉ 07
11	09 ♊ 15	12 ♍ 24	23 ♊ 30	10 ♐ 15	09 ♒ 44	28 ♉ 13
21	09 ♊ 16	12 ♍ 17	25 ♊ 16	12 ♐ 18	14 ♒ 54	29 ♉ 20
31	09 ♊ 24	07 ♍ 58	27 ♊ 49	14 ♐ 00	20 ♒ 01	00 ♊ 27

ASPECTARIAN

01 Saturday
01 11 ☉ ☌ ☿
02 06 ☿ ∥ ♂
05 15 ☽ ⊻ ♂
09 43 ☽ ⊥ ♄
14 24 ☽ ✶ ♆
16 52 ☽ ∠ ♅
16 58 ☽ ∥ ♃
18 57 ☽ ☌ ♇
20 05 ☽ ⊻ ♄

02 Sunday
04 41 ☽ □ ☉
06 13 ☽ ∥ ☉
06 17 ☽ □ ☿
07 26 ☽ ∥ ♀
08 19 ☽ ⊥ ♅
10 46 ☽ □ ♀
13 27 ☽ ∥ ☿
16 06 ☽ ∠ ♆
18 07 ☽ ∥ ♂
18 34 ☽ ⊻ ♅
20 03 ☽ ∥ ♄

03 Monday
02 24 ☽ □ ♃
07 25 ☽ ⊥ ♆
09 13 ☽ ∥ ♆
10 42 ☽ ☌ ♂
13 38 ☽ Q ☉
14 45 ☽ ∥ ♅
17 11 ☽ Q ☿
17 35 ☽ ⊻ ♆
20 16 ☽ Q ♀
21 57 ☽ ⊻ ♇
23 19 ☽ ☌ ♄

04 Tuesday
08 04 ☽ ⊥ ♇
09 19 ☽ Q ♃
11 25 ☽ ✶ ☉
16 15 ☽ ✶ ☿
18 25 ☽ ✶ ♀
21 23 ☽ ☌ ♅
23 12 ☽ ∠ ♇

05 Wednesday
00 07 ♀ ∠ ♆
05 49 ☽ ✶ ♃
08 38 ♀ Q ♄
13 56 ☿ ∠ ♆
14 32 ☽ ∠ ☉
15 22 ☽ ⊻ ♂
20 08 ☿ Q ♄
20 08 ☽ ☌ ♆
21 00 ☽ ∠ ☿
22 01 ☽ ∠ ♀
22 50 ☉ ∥ ♀

06 Thursday
00 00 ♂ ∥ ♄
00 23 ☽ ✶ ♇
01 52 ☽ ⊥ ♂
01 58 ☽ ⊻ ♄
05 15 ♀ ✶ ♅
06 48 ☽ ⊥ ☉
07 26 ☽ ∠ ♃
10 19 ☿ ✶ ♅
12 06 ☽ ⊥ ♄
14 18 ☽ ⊥ ☿
14 36 ☽ ⊥ ♀
17 39 ☽ ∠ ♂
17 40 ☽ ⊻ ☉
20 34 ☽ Q ♇
21 46 ☿ ☌ ♀
22 51 ☽ ⊥ ♃
23 57 ☽ ⊻ ♅

07 Friday
01 42 ☽ ⊻ ♀
01 52 ☽ ⊻ ☿
03 20 ☽ ∠ ♄
09 10 ☽ ⊻ ♃
10 08 ☽ ⊥ ♅
20 10 ☽ ✶ ♂
22 55 ☽ ⊻ ♆

08 Saturday
01 30 ☽ ∠ ♅
03 08 ☽ □ ♇
03 23 ☽ ∥ ♅
04 59 ☽ ✶ ♄
05 27 ☉ ∥ ☿
09 09 ☽ ∥ ♆
09 14 ☽ ⊥ ♆
17 41 ☽ Q ♂
20 18 ♇ St R
20 39 ☽ ∥ ♂
22 21 ☽ ∥ ♄
23 22 ☿ ∥ ♀
23 46 ☉ ∠ ♆

09 Sunday
00 50 ☽ ∠ ♆
00 55 ☽ ☌ ☉
01 49 ☽ Q ♄
03 29 ☽ ✶ ♅
10 21 ☽ ☌ ♀
13 09 ☽ ☌ ☿
13 50 ☽ ☌ ♃
13 54 ☉ Q ♄
18 51 ☿ ☌ ♃
23 34 ☽ ∥ ☉

10 Monday
00 43 ☽ Q ♅
01 16 ☽ ∥ ♀
02 45 ☽ □ ♂
02 46 ☽ ∥ ☿
03 22 ☽ ✶ ♆
05 41 ☽ ∥ ♃
07 40 ☽ △ ♇
09 51 ☽ □ ♄
13 04 ☉ ✶ ♅
17 46 ♂ ⊻ ♆

11 Tuesday
01 09 ☽ Q ♆
03 52 ☿ ∥ ♃
08 17 ♀ ☌ ♃
09 31 ☽ □ ♅
11 03 ☽ ⚼ ♇
11 11 ☽ ⊻ ☉
21 23 ☽ ⊻ ♃
22 32 ☽ ⊻ ♀
23 21 ☽ ⊥ ☉

12 Wednesday
03 48 ☽ # ♇
03 55 ☽ ± ♇
04 41 ☽ ⊻ ☿
08 56 ☽ ⊥ ♃
10 52 ☽ □ ♆
11 08 ☽ ⊥ ♀
12 46 ☽ △ ♂
15 20 ☽ ⊼ ♇
17 55 ☽ △ ♄
17 57 ☽ ∠ ☉
18 04 ☽ ⊥ ☿
19 08 ☽ ∥ ♇
23 17 ♀ ∥ ♃

13 Thursday
02 33 ☽ ∠ ♃
04 46 ☿ Q ♅
06 24 ☽ ∠ ♀
14 33 ☽ ∠ ☿
19 02 ☽ △ ♅
19 18 ☽ ⚼ ♂
23 17 ☽ ⚼ ♄

14 Friday
00 09 ☿ ✶ ♆
01 51 ☽ ✶ ☉
08 33 ☽ # ☿
08 37 ☽ ✶ ♃
14 03 ☽ ± ♂
15 25 ☽ ✶ ♀
15 37 ☽ # ♀
17 16 ☽ ± ♄
18 24 ☽ # ♃
19 15 ☽ # ☉
19 54 ♂ ⊻ ♇
21 49 ☽ △ ♆

15 Saturday
00 59 ☽ ⚼ ♅
01 39 ☽ ✶ ☿
02 22 ☽ ☍ ♇
02 42 ☽ ⊼ ♂
04 01 ☽ Q ☉
05 22 ☽ ⊼ ♄
06 23 ☿ △ ♇
09 10 ☽ Q ♃
11 21 ☿ □ ♂
12 13 ☉ ∥ ♃
18 25 ☽ Q ♀
19 12 ☽ ± ♅

16 Sunday
02 01 ♀ Q ♅
02 52 ☿ □ ♄
04 12 ☽ ⚼ ♆
06 17 ☽ Q ☿
07 25 ☽ ⊼ ♅
07 59 ☽ # ♄
14 59 ☽ # ♂
19 55 ☽ □ ☉
22 27 ☽ □ ♃
22 32 ☽ ± ♆
22 55 ☽ # ♆

17 Monday
05 53 ☽ # ♅
05 54 ♀ ✶ ♆
10 46 ☽ ⊼ ♆
11 19 ☽ □ ♀
14 57 ♂ ☌ ♄
15 11 ☽ ⊼ ♇
18 30 ☽ ☍ ♄
18 40 ☽ ☍ ♂

18 Tuesday
01 17 ☽ □ ☿
03 19 ☽ ± ♇
10 06 ☉ ☌ ♃
20 19 ☽ ☍ ♅
21 24 ☽ ⚼ ♇
23 42 ♀ △ ♇

19 Wednesday
12 06 ☽ △ ♃
13 53 ☽ △ ☉
22 56 ☽ ☍ ♆

20 Thursday
03 03 ☽ △ ♇
06 09 ☽ △ ♀
06 29 ☽ ⊼ ♄
09 20 ☽ ⊼ ♂
09 31 ♀ □ ♄
17 58 ☽ ⚼ ♃
18 09 ☽ ± ♄
21 29 ☽ ± ♂
21 43 ☽ ⚼ ☉
22 11 ☿ Q ♆
22 16 ☽ △ ☿

21 Friday
07 04 ☽ ⊼ ♅
11 23 ☽ ⚼ ♄
11 29 ☽ ± ♃
14 07 ☽ ⚼ ♀
15 24 ☽ ⚼ ♂
16 18 ☽ ± ☉
18 26 ☽ ± ♅
22 59 ☽ ⊼ ♃

22 Saturday
04 29 ☽ ⊼ ☉
06 40 ☽ ⚼ ☿
08 14 ☽ ⊼ ♆
08 45 ☽ ± ♀
11 13 ☽ ⚼ ♅
11 59 ☽ □ ♇
12 25 ☽ # ♅
13 10 ♀ □ ♂
15 27 ☽ △ ♄
18 52 ☽ # ♆
19 17 ☽ ± ♆
19 59 ☉ Q ♅
20 31 ☽ △ ♂
20 56 ☽ ⊼ ♀
22 29 ☽ # ♂

23 Sunday
01 25 ☽ ± ☿
08 05 ☽ # ♄
11 39 ☽ ⚼ ♆
13 40 ☽ ⊼ ☿
14 34 ☽ △ ♅
20 36 ☽ Q ♇
22 00 ☿ □ ♅

24 Monday
03 53 ☿ ⚼ ♇
05 45 ☉ ✶ ♆
06 30 ☽ ☍ ♃
14 22 ☽ △ ♆
15 02 ☽ ☍ ☉
17 48 ☽ ✶ ♇
20 57 ☽ # ♃
21 18 ☽ □ ♄
22 29 ☿ # ♇

25 Tuesday
04 16 ☽ □ ♂
04 29 ♂ ⊥ ♇
07 42 ☽ ☍ ♀
07 50 ☽ # ☉
16 17 ☽ # ♀
19 21 ☽ □ ♅
19 51 ☽ ∠ ♇

26 Wednesday
00 14 ☽ ☍ ☿
04 23 ☉ △ ♇
04 32 ☽ Q ♄
10 37 ☽ ∥ ♇
11 18 ☽ ⊥ ♇
11 30 ☽ ⊼ ♃
12 37 ☽ Q ♂
15 14 ☽ # ☿
18 18 ☽ □ ♆
20 11 ☽ ∥ ☿
21 32 ☽ ⊻ ♇
21 55 ☽ ± ♃
22 52 ☽ ⊼ ☉

27 Thursday
00 40 ♀ Q ♆
01 05 ☽ ✶ ♄
01 19 ☽ # ♇
02 13 ☽ Q ♅
09 51 ☽ ± ☉
09 54 ☽ ✶ ♂
13 29 ☽ ⚼ ♃
15 46 ☽ ∥ ♀
16 05 ☽ ⊼ ♀
22 37 ☽ ✶ ♅
23 51 ☽ ∥ ☉

28 Friday
00 56 ☽ Q ♆
02 16 ☽ ⚼ ☉
02 36 ☽ ∠ ♄
03 14 ☽ ± ♀
07 06 ☉ □ ♄
07 54 ☽ ⊼ ☿
12 22 ☽ ∠ ♂
12 54 ☽ ∥ ♃
15 20 ☽ △ ♃
17 52 ☽ ⊥ ♄
18 49 ☽ ± ☿
19 56 ☽ ⚼ ♀
21 16 ☽ ✶ ♆

All ephemeris data is given at 12.00 UT and the Moon's longitude is additionally given for 24.00 UT

MARCH 1986

LONGITUDES

Date	Sidereal time h m s	Sun ☉ ° ' "	Moon ☽ ° ' "	Moon ☽ 24.00 ° ' "	Mercury ☿ ° '	Venus ♀ ° '	Mars ♂ ° '	Jupiter ♃ ° '	Saturn ♄ ° '	Uranus ♅ ° '	Neptune ♆ ° '	Pluto ♇ ° '
01	22 36 00	10 ♓ 36 18	14 ♏ 08 05	21 ♏ 13 41	28 ♓ 38	20 ♓ 28	15 ♐ 43	02 ♓ 07	09 ♐ 26	22 ♐ 05	05 ♑ 26	07 ♏ 14
02	22 39 56	11 36 30	28 ♏ 18 51	05 ♐ 23 26	29 ♓ 27	21 43	16 16	02 21	09 28	22 06	05 27	07 R 14
03	22 43 53	12 36 41	12 ♐ 27 15	19 ♐ 30 09	00 ♈ 07	22 58	16 50	02 36	09 29	22 07	05 28	07 13
04	22 47 49	13 36 50	26 ♐ 32 00	03 ♑ 32 37	00 39	24 13	17 23	02 50	09 31	22 08	05 29	07 12
05	22 51 46	14 36 57	10 ♑ 31 50	17 ♑ 29 26	01 01	25 27	17 57	03 04	09 32	22 10	05 30	07 11
06	22 55 42	15 37 03	24 ♑ 25 11	01 ♒ 18 50	01 15	26 42	18 30	03 19	09 34	22 11	05 32	07 10
07	22 59 39	16 37 07	08 ♒ 10 05	14 ♒ 58 39	01 R 19	27 57	19 03	03 33	09 35	22 12	05 33	07 09
08	23 03 36	17 37 10	21 ♒ 44 15	28 ♒ 26 35	01 14	29 ♓ 12	19 36	03 47	09 36	22 13	05 34	07 09
09	23 07 32	18 37 10	05 ♓ 05 25	11 ♓ 40 30	01 00	00 ♈ 26	20 09	04 01	09 37	22 14	05 35	07 08
10	23 11 29	19 37 09	18 ♓ 11 41	24 ♓ 38 50	00 38	01 41	20 42	04 15	09 38	22 15	05 36	07 07
11	23 15 25	20 37 06	01 ♈ 01 54	07 ♈ 20 54	00 ♈ 08	02 56	21 14	04 30	09 39	22 16	05 36	07 06
12	23 19 22	21 37 01	13 ♈ 35 56	19 ♈ 47 10	29 ♓ 31	04 10	21 47	04 44	09 40	22 16	05 37	07 05
13	23 23 18	22 36 54	25 ♈ 54 49	01 ♉ 59 12	28 48	05 25	22 19	04 58	09 40	22 17	05 38	07 04
14	23 27 15	23 36 44	08 ♉ 00 42	13 ♉ 59 43	28 00	06 40	22 52	05 12	09 41	22 18	05 39	07 02
15	23 31 11	24 36 33	19 ♉ 56 45	25 ♉ 52 18	27 08	07 54	23 24	05 26	09 41	22 19	05 40	07 01
16	23 35 08	25 36 20	01 ♊ 46 57	07 ♊ 41 16	26 13	09 09	23 56	05 40	09 42	22 19	05 40	07 00
17	23 39 05	26 36 04	13 ♊ 35 54	19 ♊ 31 27	25 17	10 23	24 28	05 54	09 42	22 20	05 41	06 59
18	23 43 01	27 35 46	25 ♊ 28 34	01 ♋ 27 54	24 20	11 38	25 00	06 08	09 42	22 20	05 42	06 58
19	23 46 58	28 35 26	07 ♋ 30 02	13 ♋ 35 37	23 25	12 52	25 31	06 22	09 R 42	22 21	05 43	06 57
20	23 50 54	29 ♓ 35 03	19 ♋ 45 10	25 ♋ 59 14	22 31	14 07	26 03	06 36	09 42	22 21	05 43	06 55
21	23 54 51	00 ♈ 34 39	02 ♌ 18 16	08 ♌ 42 39	21 41	15 21	26 34	06 49	09 42	22 21	05 44	06 54
22	23 58 47	01 34 12	15 ♌ 12 39	21 ♌ 48 29	20 55	16 35	27 06	07 03	09 42	22 22	05 44	06 53
23	00 02 44	02 33 42	28 ♌ 30 13	05 ♍ 17 47	20 13	17 50	27 37	07 17	09 41	22 22	05 45	06 52
24	00 06 40	03 33 11	12 ♍ 11 02	19 ♍ 09 39	19 36	19 04	28 08	07 31	09 41	22 22	05 46	06 50
25	00 10 37	04 32 37	26 ♍ 13 12	03 ♎ 21 09	19 05	20 18	28 39	07 44	09 40	22 22	05 46	06 49
26	00 14 34	05 32 01	10 ♎ 32 51	17 ♎ 47 35	18 39	21 33	29 09	07 58	09 40	22 22	05 46	06 48
27	00 18 30	06 31 23	25 ♎ 04 36	02 ♏ 23 06	18 19	22 47	29 ♐ 40	08 11	09 39	22 22	05 47	06 46
28	00 22 27	07 30 43	09 ♏ 42 19	17 ♏ 01 28	18 06	24 01	00 ♑ 10	08 25	09 38	22 R 22	05 47	06 45
29	00 26 23	08 30 02	24 ♏ 19 51	01 ♐ 36 49	17 58	25 15	00 41	08 38	09 37	22 22	05 47	06 43
30	00 30 20	09 29 19	08 ♐ 51 49	16 ♐ 04 21	17 D 56	26 29	01 11	08 52	09 36	22 22	05 48	06 42
31	00 34 16	10 ♈ 28 34	23 ♐ 14 03	00 ♑ 20 34	17 ♓ 59	27 ♈ 43	01 ♑ 41	09 ♓ 05	09 ♐ 35	22 ♐ 22	05 ♑ 48	06 ♏ 41

Date	Moon True ☊ ° '	Moon Mean ☊ ° '	Moon ☽ Latitude ° '
01	01 ♉ 06	02 ♉ 40	01 S 10
02	01 D 07	02 37	02 21
03	01 08	02 34	03 24
04	01 R 07	02 31	04 14
05	01 04	02 28	04 49
06	01 00	02 24	05 06
07	00 53	02 21	05 06
08	00 46	02 18	04 48
09	00 38	02 15	04 15
10	00 31	02 12	03 29
11	00 26	02 09	02 32
12	00 23	02 05	01 30
13	00 21	02 02	00 S 24
14	00 D 21	01 59	00 N 42
15	00 23	01 56	01 45
16	00 24	01 53	02 43
17	00 26	01 50	03 34
18	00 27	01 46	04 17
19	00 R 26	01 43	04 48
20	00 25	01 40	05 08
21	00 22	01 37	05 13
22	00 18	01 34	05 03
23	00 13	01 30	04 36
24	00 09	01 27	03 53
25	00 05	01 24	02 55
26	00 03	01 21	01 45
27	00 02	01 18	00 N 27
28	00 D 02	01 15	00 S 53
29	00 03	01 11	02 10
30	00 05	01 08	03 18
31	00 ♉ 06	01 ♉ 05	04 S 13

DECLINATIONS

Date	Sun ☉ ° '	Moon ☽ ° '	Mercury ☿ ° '	Venus ♀ ° '	Mars ♂ ° '	Jupiter ♃ ° '	Saturn ♄ ° '	Uranus ♅ ° '	Neptune ♆ ° '	Pluto ♇ ° '
01	07 S 36	17 S 12	01 N 05	05 S 04	22 S 12	11 S 30	19 S 59	23 S 17	22 S 16	02 N 03
02	07 13	22 05	01 38	04 34	22 17	11 25	19 59	23 17	22 15	02 04
03	06 50	25 40	02 06	04 03	22 21	11 20	19 59	23 17	22 15	02 04
04	06 27	27 38	02 31	03 33	22 25	11 15	19 59	23 17	22 15	02 05
05	06 04	27 49	02 52	03 02	22 29	11 10	19 59	23 17	22 15	02 06
06	05 40	26 16	03 08	02 32	22 33	11 05	19 59	23 17	22 15	02 06
07	05 17	23 09	03 20	02 01	22 37	11 00	19 59	23 17	22 15	02 07
08	04 54	18 48	03 27	01 31	22 41	10 55	19 59	23 17	22 15	02 08
09	04 30	13 36	03 29	01 00	22 44	10 50	19 59	23 17	22 15	02 08
10	04 07	07 52	03 26	00 S 29	22 47	10 45	19 59	23 17	22 15	02 09
11	03 43	01 S 55	03 19	00 N 02	22 51	10 40	19 59	23 17	22 15	02 10
12	03 20	03 N 59	03 07	00 33	22 54	10 34	19 59	23 17	22 15	02 10
13	02 56	09 38	02 50	01 04	22 57	10 29	19 59	23 17	22 15	02 11
14	02 32	14 50	02 31	01 34	23 00	10 24	19 59	23 17	22 15	02 12
15	02 09	19 25	02 07	02 05	23 03	10 19	19 59	23 18	22 15	02 13
16	01 45	23 11	01 41	02 36	23 05	10 14	19 59	23 18	22 15	02 13
17	01 21	25 59	01 13	03 07	23 08	10 09	19 59	23 18	22 14	02 14
18	00 57	27 39	00 43	03 37	23 10	10 04	19 59	23 18	22 14	02 15
19	00 34	28 02	00 N 11	04 08	23 13	09 59	19 59	23 18	22 14	02 15
20	00 S 10	27 04	00 S 20	04 38	23 15	09 54	19 58	23 18	22 14	02 16
21	00 N 14	24 43	00 52	05 09	23 17	09 49	19 58	23 18	22 14	02 17
22	00 37	21 05	01 22	05 39	23 19	09 44	19 58	23 18	22 14	02 17
23	01 01	16 18	01 52	06 09	23 21	09 39	19 58	23 18	22 14	02 18
24	01 25	10 35	02 20	06 39	23 23	09 34	19 58	23 18	22 14	02 19
25	01 48	04 N 11	02 46	07 09	23 25	09 29	19 57	23 18	22 14	02 20
26	02 12	02 S 34	03 11	07 39	23 26	09 24	19 57	23 18	22 14	02 20
27	02 35	09 17	03 33	08 08	23 28	09 19	19 57	23 18	22 14	02 21
28	02 59	15 34	03 52	08 38	23 29	09 14	19 57	23 18	22 14	02 22
29	03 22	20 57	04 09	09 07	23 31	09 09	19 56	23 18	22 14	02 22
30	03 46	25 03	04 24	09 36	23 32	09 04	19 56	23 18	22 14	02 23
31	04 N 09	27 S 29	04 S 36	10 N 04	23 S 33	08 S 59	19 S 56	23 S 18	22 S 14	02 N 24

ZODIAC SIGN ENTRIES

Date	h	m	Planets
02	14	51	☽ ♐
03	07	22	☿ ♈
04	17	56	☽ ♑
06	21	42	☽ ♒
09	02	48	☽ ♓
09	03	32	♀ ♈
11	10	03	☽ ♈
11	17	36	☿ ♓
13	20	04	☽ ♉
16	08	23	☽ ♊
18	21	04	☽ ♋
20	22	03	☉ ♈
21	07	38	☽ ♌
23	14	39	☽ ♍
25	18	22	☽ ♎
27	20	05	☽ ♏
28	03	47	♂ ♑
29	21	20	☽ ♐
31	23	25	☽ ♑

LATITUDES

Date	Mercury ☿ ° '	Venus ♀ ° '	Mars ♂ ° '	Jupiter ♃ ° '	Saturn ♄ ° '	Uranus ♅ ° '	Neptune ♆ ° '	Pluto ♇ ° '
01	01 N 46	01 S 24	00 N 28	00 S 50	01 N 55	00 S 04	01 N 04	16 N 52
04	02 28	01 22	00 26	00 51	01 55	00 04	01 05	16 53
07	03 04	01 19	00 23	00 51	01 56	00 04	01 05	16 55
10	03 28	01 15	00 20	00 51	01 56	00 04	01 05	16 56
13	03 37	01 11	00 16	00 51	01 56	00 04	01 05	16 57
16	03 28	01 07	00 13	00 52	01 57	00 04	01 05	16 58
19	03 03	01 02	00 09	00 52	01 57	00 04	01 05	16 59
22	02 26	00 57	00 05	00 52	01 58	00 04	01 05	17 00
25	01 41	00 51	00 N 01	00 53	01 58	00 05	01 05	17 01
28	00 55	00 45	00 S 03	00 53	01 59	00 05	01 05	17 02
31	00 N 10	00 S 38	00 S 07	00 S 54	01 N 59	00 S 05	01 N 05	17 N 03

DATA

Julian Date	2446491
Delta T	+55 seconds
Ayanamsa	23° 39' 41"
Synetic vernal point	05° ♓ 27' 18"
True obliquity of ecliptic	23° 26' 36"

MOON'S PHASES, APSIDES AND POSITIONS ☽

Date	h	m	Phase	Longitude ° '	Eclipse Indicator
03	12	17	☾	12 ♐ 37	
10	14	52	●	19 ♓ 44	
18	16	39	☽	27 ♊ 47	
26	03	02	○	05 ♎ 10	

Day	h	m		
01	09	42	Perigee	
16	18	51	Apogee	
28	14	02	Perigee	
05	02	32	Max dec	27° S 58'
11	19	45	ON	
19	06	59	Max dec	28° N 04'
26	02	57	OS	

LONGITUDES

Date	Chiron ⚷ ° '	Ceres ⚳ ° '	Pallas ⚴ ° '	Juno ⚵ ° '	Vesta ⚶ ° '	Black Moon Lilith ⚸ ° '
01	09 ♊ 22	08 ♍ 25	27 ♊ 15	13 ♐ 42	19 ♒ 00	00 ♊ 13
11	09 ♊ 36	06 ♍ 10	00 ♋ 19	15 ♐ 06	24 ♒ 00	01 ♊ 20
21	09 ♊ 56	04 ♍ 15	03 ♋ 52	16 ♐ 04	29 ♒ 01	02 ♊ 27
31	10 ♊ 24	02 ♍ 54	07 ♋ 47	16 ♐ 34	03 ♓ 53	03 ♊ 33

ASPECTARIAN

h m	Aspects
01 Saturday	
00 03	☽ ∠ ♅
00 21	☽ ☌ ♇
04 02	☽ ⚺ ♄
04 13	☽ ⊥ ♂
05 35	☽ △ ☉
11 06	☽ bQ ☿
14 47	☽ ⚺ ♂
15 17	☽ ⊥ ♅
22 40	☽ ∠ ♆
23 45	☽ △ ♀
02 Sunday	
00 59	☽ ∥ ♄
01 27	☽ ⚺ ♅
04 38	♂ ∥ ♆
12 59	☽ ∥ ♆
13 08	☽ ∥ ♂
13 56	☽ ⊥ ♆
14 01	☽ △ ☿
18 58	☽ □ ♃
19 00	☽ ∥ ♅
19 30	♀ □ ♅
21 44	♀ bQ ♇
03 Monday	
00 07	☽ ⚺ ♆
03 06	☽ ⚺ ♇
06 57	☽ ☌ ♄
10 09	☿ ∥ ♇
12 17	☽ □ ☉
13 17	☽ ⊥ ♇
19 45	☽ ☌ ♂
04 Tuesday	
02 06	☽ Q ♃
04 29	☽ ☌ ♅
04 36	☽ ∠ ♇
07 39	☽ □ ♀
19 15	☽ □ ☿
21 22	☽ Q ☉
22 58	☽ ⚹ ♃
05 Wednesday	
03 22	☽ ☌ ♆
06 15	☽ ⚹ ♇
10 18	☽ ⚺ ♄
17 10	☿ # ♀
17 32	☽ Q ♀
19 35	☽ ⚹ ☉
20 39	☽ ⊥ ♄
06 Thursday	
01 14	☽ ∠ ♃
01 19	☽ ⚺ ♂
02 51	☿ ± ♇
02 55	☽ Q ♇
02 55	☽ Q ☿
07 23	♅ ∠ ♇
08 06	☽ ⚺ ♅
12 08	☽ ⊥ ♂
12 15	☽ ∠ ♄
16 22	☽ ⚹ ♀
17 07	☽ ⊥ ♃
18 33	☽ ⊥ ♅
23 38	☽ ∠ ☉
23 58	☽ ⚹ ☿
07 Friday	
03 46	☽ ⚺ ♃
04 29	☽ ∠ ♂
07 23	☽ ⚺ ♆
07 31	♀ # ♇
10 14	☽ □ ♇
10 18	☽ ∠ ♅
10 57	☿ St R
11 07	☽ ∥ ♅
14 30	☽ ⚹ ♄
15 14	☽ ∥ ♂
16 39	☽ ⊥ ☉
17 28	☽ ∥ ♆
17 57	☽ ⊥ ♆
21 16	☽ ∠ ♀
08 Saturday	
02 19	☽ ∠ ☿
04 06	☽ ⚺ ☉
05 59	☽ ∥ ♄
08 02	☽ ⚹ ♂
09 54	☽ ∠ ♆
10 33	☉ Q ♆
11 46	☽ Q ♄
12 51	☽ ⚹ ♅
14 52	☽ ⊥ ♀
18 10	☽ ⊥ ☿
09 Sunday	
00 06	☿ ± ♇
02 44	☽ ⚺ ♀
04 46	☽ ⚺ ☿
06 27	☽ Q ♂
10 02	☽ ☌ ♃
10 26	☽ Q ♅
12 53	☽ ⚹ ♆
15 42	☽ △ ♇
20 15	☽ □ ♄
20 36	☿ ☌ ♀
23 58	☽ ∥ ♃
10 Monday	
01 04	♀ ± ♇
10 46	♃ Q ♅
10 53	☽ Q ♆
14 52	☽ ☌ ☉
16 50	☽ □ ♂
19 15	☽ bQ ♇
19 32	☽ □ ♅
11 Tuesday	
04 16	☽ ∥ ☉
06 17	☽ # ☿
10 23	☽ ☌ ☿
11 02	☽ # ♇
12 07	☽ ± ♇
15 59	☽ ☌ ♀
18 41	☽ ⚺ ♃
19 01	☽ # ♀
20 37	☽ ∥ ♀
20 41	☽ □ ♆
23 30	☽ ⚻ ♇
12 Wednesday	
04 25	☽ △ ♄
04 33	☽ ∥ ♇
06 22	☽ ⊥ ♃
08 32	☽ ∥ ☿
09 27	☽ # ☉
20 37	☉ □ ♂
22 52	☉ bQ ♇
13 Thursday	
00 07	☽ ∠ ♃
00 42	♂ ∠ ♇
01 12	♀ ⚺ ♃
04 00	☉ □ ♅
04 38	☽ △ ♂
04 53	☽ △ ♅
04 57	☽ ⚺ ☉
09 34	☽ bQ ♄
10 24	♂ ☌ ♅
15 43	☽ # ♃
16 16	♀ □ ♆
17 21	☽ ⚺ ☿
17 48	☽ ⊥ ☉
14 Friday	
03 22	☽ ± ♄
04 30	☽ ⊥ ☿
06 16	☽ ⚹ ♃
07 17	☽ △ ♆
08 59	☽ ⚺ ♀
10 04	☽ ☍ ♇
10 34	☽ bQ ♅
11 41	☽ bQ ♂
13 19	☽ ∠ ☉
15 21	☽ ⚻ ♄
19 15	♀ ⚻ ♇
21 19	☽ ∠ ☿
22 24	☽ ⊥ ♀
15 Saturday	
04 39	☽ ± ♅
06 37	☽ ± ♂
06 49	☽ Q ♃
06 50	☿ ∥ ♇
07 59	☉ # ♇
12 52	☿ ∥ ♀
13 27	☽ bQ ♆
13 29	☉ # ♀
14 46	♂ Q ♃
15 18	☽ # ♄
16 47	☽ ⚻ ♅
17 57	♀ ∥ ♇
18 41	☽ ∠ ♀
19 19	☽ ⚻ ♂
22 18	☽ ⚹ ☉
16 Sunday	
01 31	☽ ⚹ ☿
05 26	☽ # ♆
07 43	☽ ± ♆
11 18	☽ # ♂
12 47	☽ # ♅
13 02	♃ ⚹ ♆
19 34	☉ ☌ ☿
19 55	☽ ⚻ ♆
20 03	☽ □ ♃
22 36	☽ ⚻ ♇
22 42	♀ △ ♄
17 Monday	
00 07	☽ Q ☿
00 55	☽ Q ☉
04 05	☽ ☍ ♄
04 43	☽ ⚹ ♀
10 45	☽ ± ♇
18 Tuesday	
01 13	☿ □ ♂
04 56	☽ bQ ♇
05 41	☽ ☍ ♅
07 51	☽ Q ♀
09 52	☽ □ ☿
10 59	☽ ☍ ♂
16 39	☽ □ ☉
23 56	♀ ⊥ ♃
19 Wednesday	
08 27	☽ ☍ ♆
09 27	♄ St R
09 42	☽ △ ♃
10 54	☽ △ ♇
16 21	☽ ⚻ ♄
23 46	☽ □ ♀
20 Thursday	
04 08	☽ ± ♄
07 34	☉ ∥ ☿
15 38	☽ bQ ♃
16 42	☿ □ ♅
17 00	☽ △ ☿
17 01	☽ ⚻ ♅
21 32	☽ bQ ♄
21 Friday	
00 39	☽ ⚻ ♂
04 31	☽ ± ♅
05 22	☿ bQ ♇
08 27	☽ △ ☉
09 09	☽ ± ♃
12 32	☽ ± ♂
18 26	☽ ⚻ ♆
19 00	♂ ∥ ♅
19 38	♃ △ ♇
19 43	☉ ± ♇
19 44	☽ bQ ☿
20 37	☽ □ ♇
20 38	☽ ⚻ ♃
21 29	☽ bQ ♅
22 24	☽ # ♂
22 26	☽ # ♅
22 Saturday	
01 50	☽ △ ♄
05 15	☽ # ♆
05 36	☽ ± ♆
06 02	☽ bQ ♂
11 29	☽ ± ☿
14 41	☽ bQ ☉
14 47	☽ △ ♀
18 09	☽ # ♄
21 50	☽ ⚻ ☿
22 04	☽ bQ ♆
23 Sunday	
01 00	☽ △ ♅
05 30	☽ Q ♇
08 16	☽ ± ☉
10 21	☽ △ ♂
19 45	☽ ⚻ ☉
20 26	☽ bQ ♀
24 Monday	
00 48	☽ △ ♆
02 43	☽ ⚹ ♇
03 44	☽ ☍ ♃
07 40	☽ □ ♄
10 55	☿ # ♇
13 41	☽ ± ♀
16 00	☽ # ♃
19 08	☿ ⚺ ♀
25 Tuesday	
00 16	☽ ☍ ☿
00 59	☽ ⚻ ♀
01 53	☽ ∥ ♀
04 32	☽ ∠ ♇
05 28	☽ □ ♅
14 27	☽ Q ♄
16 15	☽ □ ♂
16 47	☽ # ☿
18 40	☽ ∥ ♇
19 44	☽ ⊥ ♇
20 04	☽ ∥ ☉
26 Wednesday	
03 02	☽ ☍ ☉
04 02	☽ □ ♆
05 46	☽ ⚺ ♇
07 38	☽ ⚻ ♃
10 32	☽ ⚹ ♄
10 38	☽ # ☉
11 12	☽ # ♇
11 43	☽ Q ♅
14 18	☽ ∥ ☿
17 46	☽ ± ♃
17 46	☉ □ ♆
20 46	☉ ∥ ♇
23 21	☽ Q ♂
27 Thursday	
01 06	☽ ⚻ ☿
04 09	♀ △ ♅
07 30	☽ # ♀
07 33	☽ ⚹ ♅
07 52	☽ ☍ ♀
08 51	☽ bQ ♃
09 52	☽ Q ♆
10 47	☽ ± ☿
11 18	☽ ∠ ♄
12 08	☽ ∥ ♃
14 15	♅ St R
17 52	☉ ⚻ ♇
19 49	☽ ⚹ ♂
21 48	♀ ∠ ♃
28 Friday	
01 19	☽ bQ ☿
02 04	☽ ⊥ ♄
05 34	☽ ⚹ ♆
07 10	☽ ☌ ♇
08 09	☽ ⚻ ☉
08 11	☽ ∠ ♅
09 51	☽ △ ♃
11 53	☽ ⚺ ♄
13 21	☿ ⊥ ♀
18 42	☽ ± ☉
21 17	☽ ∠ ♂
22 56	☽ ⊥ ♅
23 53	♀ bQ ♄
29 Saturday	
01 37	☽ △ ☿
06 10	☽ ∠ ♆
07 04	☽ ∥ ♄
08 47	☽ ⚺ ♅
10 32	☽ bQ ☉
12 35	☽ ⊥ ♂
13 39	☽ ⚻ ♀
13 49	♀ # ♃
16 25	☉ ⚺ ♃
18 37	☽ ∥ ♆
20 59	☽ ⊥ ♆
22 50	☽ ⚺ ♂
30 Sunday	
00 28	☽ ± ♀
00 40	☽ ∥ ♅
02 01	☽ ∥ ♂
06 54	☽ ⚺ ♆
08 25	☽ ⚺ ♇
08 42	☿ St D
12 00	☽ □ ♃
13 07	☽ △ ☉
13 13	☽ ☌ ♄
14 41	☉ △ ♄
16 46	☽ bQ ♀
18 22	☽ ⊥ ♇
31 Monday	
03 09	☽ □ ☿
09 23	☽ ∠ ♇
10 33	☽ ☌ ♅
18 36	☽ Q ♃
20 17	☽ △ ♀

All ephemeris data is given at 12.00 UT and the Moon's longitude is additionally given for 24.00 UT

APRIL 1986

LONGITUDES

Date	Sidereal time h m s	Sun ☉ ° ' "	Moon ☽ ° ' "	Moon ☽ 24.00 ° ' "	Mercury ☿ ° '	Venus ♀ ° '	Mars ♂ ° '	Jupiter ♃ ° '	Saturn ♄ ° '	Uranus ♅ ° '	Neptune ♆ ° '	Pluto ⯓ ° '
01	00 38 13	11 ♈ 27 47	07 ♑ 23 41	14 ♑ 23 15	18 ♓ 08	28 ♈ 57	02 ♑ 10	09 ♓ 19	09 ♐ 34	22 ♐ 22	05 ♑ 48	06 ♏ 39
02	00 42 09	12 26 58	21 ♑ 19 08	28 ♑ 11 16	18 23	00 ♉ 11	02 40	09 32	09 R 32	22 R 22	05 48	06 R 38
03	00 46 06	13 26 08	04 ♒ 59 40	11 ♒ 44 20	18 42	01 25	03 09	09 45	09 31	22 21	05 48	06 36
04	00 50 03	14 25 16	18 ♒ 25 16	25 ♒ 02 33	19 06	02 39	03 39	09 58	09 29	22 21	05 48	06 35
05	00 53 59	15 24 22	01 ♓ 36 14	08 ♓ 06 22	19 34	03 53	04 08	10 11	09 28	22 20	05 49	06 33
06	00 57 56	16 23 26	14 ♓ 33 03	20 ♓ 56 20	20 07	05 07	04 36	10 24	09 26	22 20	05 49	06 31
07	01 01 52	17 22 29	27 ♓ 16 20	03 ♈ 33 07	20 44	06 21	05 05	10 37	09 24	22 19	05 49	06 30
08	01 05 49	18 21 29	09 ♈ 46 50	15 ♈ 57 35	21 25	07 35	05 33	10 50	09 22	22 19	05 R 49	06 28
09	01 09 45	19 20 27	22 ♈ 05 32	28 ♈ 10 51	22 10	08 49	06 02	11 03	09 20	22 18	05 49	06 27
10	01 13 42	20 19 24	04 ♉ 13 44	10 ♉ 14 26	22 58	10 03	06 30	11 16	09 18	22 17	05 48	06 25
11	01 17 38	21 18 18	16 ♉ 13 12	22 ♉ 10 21	23 49	11 16	06 57	11 29	09 16	22 17	05 48	06 23
12	01 21 35	22 17 10	28 ♉ 06 12	04 ♊ 01 08	24 44	12 30	07 25	11 41	09 14	22 16	05 48	06 22
13	01 25 32	23 16 00	09 ♊ 55 34	15 ♊ 49 56	25 41	13 44	07 52	11 54	09 11	22 15	05 48	06 20
14	01 29 28	24 14 48	21 ♊ 44 41	27 ♊ 40 22	26 42	14 57	08 19	12 06	09 09	22 14	05 48	06 19
15	01 33 25	25 13 34	03 ♋ 37 28	09 ♋ 36 33	27 45	16 11	08 46	12 19	09 07	22 13	05 48	06 17
16	01 37 21	26 12 18	15 ♋ 38 10	21 ♋ 42 54	28 50	17 25	09 12	12 31	09 04	22 12	05 47	06 15
17	01 41 18	27 10 59	27 ♋ 51 19	04 ♌ 03 57	29 ♓ 58	18 38	09 39	12 43	09 01	22 11	05 47	06 14
18	01 45 14	28 09 38	10 ♌ 21 21	16 ♌ 44 00	01 ♈ 09	19 52	10 05	12 56	08 58	22 10	05 47	06 12
19	01 49 11	29 ♈ 08 15	23 ♌ 12 21	29 ♌ 46 45	02 22	21 05	10 30	13 08	08 56	22 09	05 46	06 10
20	01 53 07	00 ♉ 06 50	06 ♍ 27 29	13 ♍ 14 45	03 37	22 18	10 56	13 20	08 53	22 08	05 46	06 09
21	01 57 04	01 05 22	20 ♍ 08 35	27 ♍ 08 53	04 54	23 32	11 21	13 32	08 50	22 07	05 45	06 07
22	02 01 01	02 03 52	04 ♎ 15 26	11 ♎ 27 50	06 14	24 45	11 46	13 44	08 47	22 06	05 45	06 05
23	02 04 57	03 02 20	18 ♎ 45 32	26 ♎ 07 48	07 35	25 58	12 11	13 55	08 44	22 04	05 44	06 04
24	02 08 54	04 00 47	03 ♏ 33 50	11 ♏ 02 38	08 59	27 12	12 35	14 07	08 40	22 03	05 44	06 02
25	02 12 50	04 59 11	18 ♏ 33 12	26 ♏ 04 25	10 24	28 25	12 59	14 19	08 37	22 02	05 43	06 00
26	02 16 47	05 57 34	03 ♐ 35 12	11 ♐ 04 31	11 52	29 ♉ 38	13 23	14 30	08 34	22 00	05 43	05 58
27	02 20 43	06 55 55	18 ♐ 31 22	25 ♐ 54 52	13 21	00 ♊ 51	13 46	14 42	08 30	21 59	05 42	05 57
28	02 24 40	07 54 15	03 ♑ 14 16	10 ♑ 28 56	14 52	02 04	14 09	14 53	08 27	21 57	05 42	05 55
29	02 28 36	08 52 33	17 ♑ 38 26	24 ♑ 42 27	16 25	03 17	14 32	15 04	08 23	21 56	05 41	05 53
30	02 32 33	09 ♉ 50 49	01 ♒ 40 47	08 ♒ 33 25	18 ♈ 00	04 ♊ 31	14 ♑ 54	15 ♓ 16	08 ♐ 20	21 ♐ 54	05 ♑ 40	05 ♏ 52

Date	Moon True ☊ ° '	Moon Mean ☊ ° '	Moon ☽ Latitude ° '
01	00 ♉ 06	01 ♉ 02	04 S 51
02	00 R 06	00 59	05 12
03	00 05	00 56	05 15
04	00 04	00 52	05 00
05	00 02	00 49	04 29
06	00 00	00 46	03 46
07	29 ♈ 58	00 43	02 51
08	29 57	00 40	01 49
09	29 56	00 36	00 S 43
10	29 D 56	00 33	00 N 24
11	29 56	00 30	01 29
12	29 57	00 27	02 30
13	29 58	00 24	03 24
14	29 58	00 21	04 10
15	29 59	00 17	04 45
16	29 59	00 14	05 08
17	29 R 59	00 11	05 17
18	29 59	00 08	05 12
19	29 59	00 05	04 52
20	29 D 59	00 ♉ 02	04 16
21	29 59	29 ♈ 58	03 24
22	29 59	29 55	02 18
23	29 59	29 52	01 N 02
24	29 R 59	29 49	00 S 20
25	29 59	29 46	01 41
26	29 59	29 42	02 56
27	29 58	29 39	03 58
28	29 57	29 36	04 43
29	29 57	29 33	05 10
30	29 ♈ 56	29 ♈ 30	05 S 17

DECLINATIONS

Date	Sun ☉ ° '	Moon ☽ ° '	Mercury ☿ ° '	Venus ♀ ° '	Mars ♂ ° '	Jupiter ♃ ° '	Saturn ♄ ° '	Uranus ♅ ° '	Neptune ♆ ° '	Pluto ⯓ ° '
01	04 N 32	28 S 05	04 S 45	10 N 33	23 S 34	08 S 55	19 S 56	23 S 18	22 S 14	02 N 24
02	04 55	26 53	04 52	11 01	23 35	08 50	19 55	23 18	22 14	02 25
03	05 18	24 06	04 57	11 29	23 36	08 45	19 55	23 18	22 14	02 26
04	05 41	20 03	04 59	11 57	23 37	08 40	19 55	23 18	22 14	02 26
05	06 04	15 06	04 59	12 24	23 38	08 35	19 54	23 18	22 14	02 27
06	06 27	09 33	04 56	12 51	23 39	08 30	19 54	23 18	22 14	02 28
07	06 49	03 S 42	04 51	13 18	23 39	08 26	19 53	23 18	22 14	02 28
08	07 12	02 N 12	04 44	13 45	23 40	08 21	19 53	23 18	22 14	02 29
09	07 34	07 56	04 35	14 11	23 40	08 16	19 53	23 18	22 14	02 30
10	07 57	13 18	04 24	14 37	23 41	08 11	19 52	23 18	22 14	02 30
11	08 19	18 07	04 12	15 02	23 41	08 07	19 52	23 18	22 13	02 31
12	08 41	22 11	03 57	15 27	23 42	08 02	19 51	23 18	22 13	02 31
13	09 03	25 18	03 41	15 52	23 42	07 57	19 51	23 18	22 13	02 32
14	09 24	27 20	03 22	16 16	23 42	07 53	19 50	23 18	22 13	02 33
15	09 46	28 08	03 03	16 40	23 43	07 48	19 50	23 18	22 13	02 33
16	10 07	27 37	02 41	17 04	23 43	07 44	19 49	23 18	22 13	02 34
17	10 28	25 46	02 18	17 27	23 43	07 39	19 49	23 18	22 13	02 34
18	10 49	22 39	01 54	17 49	23 43	07 34	19 48	23 18	22 13	02 35
19	11 10	18 22	01 28	18 11	23 43	07 30	19 48	23 17	22 13	02 36
20	11 31	13 06	01 01	18 33	23 43	07 25	19 47	23 17	22 13	02 36
21	11 51	07 02	00 32	18 54	23 43	07 21	19 47	23 17	22 13	02 37
22	12 11	00 N 25	00 S 02	19 15	23 43	07 17	19 46	23 17	22 13	02 37
23	12 32	06 S 24	00 N 29	19 35	23 43	07 12	19 46	23 17	22 13	02 38
24	12 52	13 01	01 01	19 55	23 43	07 08	19 45	23 17	22 13	02 38
25	13 11	18 58	01 35	20 14	23 43	07 04	19 45	23 17	22 13	02 39
26	13 31	23 45	02 10	20 33	23 43	06 59	19 44	23 17	22 13	02 39
27	13 50	26 54	02 45	20 51	23 43	06 55	19 43	23 17	22 13	02 40
28	14 09	28 08	03 22	21 08	23 43	06 51	19 43	23 17	22 13	02 40
29	14 28	27 24	04 00	21 25	23 43	06 47	19 42	23 17	22 13	02 41
30	14 N 46	24 S 56	04 N 39	21 N 41	23 S 43	06 S 43	19 S 42	23 S 17	22 S 13	02 N 41

ZODIAC SIGN ENTRIES

Date	h	m	Planets
02	08	19	♀ ♉
03	03	11	☽ ♒
05	09	03	☽ ♓
07	17	12	☽ ♈
10	03	36	☽ ♉
12	15	51	☽ ♊
15	04	42	☽ ♋
17	12	33	☿ ♈
17	16	10	☽ ♌
20	00	24	☽ ♍
20	09	12	☉ ♉
22	04	50	☽ ♎
24	06	15	☽ ♏
26	06	16	☽ ♐
26	19	10	♀ ♊
28	06	41	☽ ♑
30	09	06	☽ ♒

LATITUDES

Date	Mercury ☿ ° '	Venus ♀ ° '	Mars ♂ ° '	Jupiter ♃ ° '	Saturn ♄ ° '	Uranus ♅ ° '	Neptune ♆ ° '	Pluto ⯓ ° '
01	00 S 04	00 S 36	00 S 09	00 S 54	01 N 59	00 S 05	01 N 05	17 N 03
04	00 43	00 29	00 14	00 54	01 59	00 05	01 05	17 04
07	01 17	00 21	00 19	00 55	02 00	00 05	01 05	17 04
10	01 46	00 14	00 24	00 55	02 00	00 05	01 06	17 05
13	02 08	00 S 06	00 29	00 55	02 00	00 05	01 06	17 05
16	02 26	00 N 02	00 35	00 56	02 01	00 05	01 06	17 05
19	02 37	00 10	00 42	00 56	02 01	00 05	01 06	17 06
22	02 44	00 18	00 48	00 57	02 01	00 05	01 06	17 06
25	02 45	00 26	00 55	00 58	02 01	00 05	01 06	17 06
28	02 42	00 34	01 02	00 58	02 02	00 05	01 06	17 06
31	02 S 33	00 N 42	01 S 10	00 S 59	02 N 02	00 S 05	01 N 06	17 N 06

DATA

Julian Date	2446522
Delta T	+55 seconds
Ayanamsa	23° 39' 45"
Synetic vernal point	05° ♓ 27' 14"
True obliquity of ecliptic	23° 26' 36"

MOON'S PHASES, APSIDES AND POSITIONS ☽

Date	h	m	Phase	Longitude ° '	Eclipse Indicator
01	19	30	☾	11 ♑ 46	
09	06	08	●	19 ♈ 06	Partial
17	10	35	☽	27 ♋ 08	
24	12	46	○	04 ♏ 03	total

Day	h	m		
13	12	05	Apogee	
25	17	47	Perigee	
01	07	51	Max dec	28° S 07'
08	03	01	0N	
15	14	41	Max dec	28° N 09'
22	13	29	0S	
28	14	51	Max dec	28° S 08'

LONGITUDES

Date	Chiron ⚷ ° '	Ceres ⚳ ° '	Pallas ⚴ ° '	Juno ⚵ ° '	Vesta ⚶ ° '	Black Moon Lilith ⚸ ° '
01	10 ♊ 27	02 ♍ 48	08 ♋ 12	16 ♐ 35	04 ♓ 22	03 ♊ 40
11	11 ♊ 00	02 ♍ 12	12 ♋ 25	16 ♐ 30	09 ♓ 07	04 ♊ 47
21	11 ♊ 38	02 ♍ 18	16 ♋ 50	15 ♐ 51	13 ♓ 44	05 ♊ 53
31	12 ♊ 20	03 ♍ 05	21 ♋ 24	14 ♐ 40	18 ♓ 12	07 ♊ 00

ASPECTARIAN

h m	Aspects
01 Tuesday	
02 47	☽ ☌ ♂
09 17	☽ ☌ ♆
09 50	☽ Q ☿
10 44	☽ ⚹ ⯓
15 20	☽ ⚹ ♃
15 42	☽ ⊻ ♄
19 30	☽ □ ☉
02 Wednesday	
02 00	☽ ⊥ ♄
06 47	☽ ⚹ ☿
07 20	☽ Q ⯓
07 57	☉ # ☿
12 49	♃ □ ♄
13 49	☽ ⊻ ♅
17 36	☽ ∠ ♄
17 41	☽ ∠ ♃
03 Thursday	
00 18	☽ ⊥ ♅
05 04	☽ □ ♀
05 13	☽ Q ☉
08 38	☽ ⊻ ♂
09 39	☽ ∠ ☿
09 46	☽ ⊥ ♃
13 26	☽ ⊻ ♆
14 51	☽ □ ⯓
15 21	☽ ‖ ♂
16 11	☽ ∠ ♅
17 21	☽ ‖ ♅
19 40	☽ ⊥ ♂
20 01	☽ ⚹ ♄
20 35	☽ ⊻ ♃
23 56	☽ ‖ ♆
04 Friday	
00 07	☽ ⊥ ♆
02 07	☽ ⊥ ☿
04 14	☽ ⚹ ☉
12 25	☽ ∠ ♂
12 46	☽ ‖ ♄
13 16	☽ ⊻ ☿
16 19	☽ ∠ ♆
16 27	☽ Q ♀
17 32	☽ Q ♄
19 06	☽ ⚹ ♅
05 Saturday	
03 53	♀ ± ♄
09 37	☽ ∠ ☉
16 39	☽ ⚹ ♀
16 49	☽ ⚹ ♂
17 02	☽ Q ♅
19 39	♀ △ ♂
19 45	☽ ⚹ ♆
21 06	☽ △ ⯓
22 59	☽ # ♀
06 Sunday	
02 29	☽ □ ♄
03 36	☽ ⊥ ☉
04 08	☽ ☌ ♃
12 06	☿ ∠ ♀
12 28	☉ ⊥ ♃
15 44	☽ ⊻ ☉
16 00	☽ Q ♂
16 24	☽ ‖ ♃
18 07	☽ Q ♆
22 59	☽ ☌ ☿
23 35	☽ ∠ ♀
07 Monday	
00 03	☽ # ☉
01 05	☽ ♇ ⯓
01 27	♀ △ ♆
02 37	☽ □ ♅
07 14	☽ ‖ ☿
12 51	♆ St R
14 47	♀ ☍ ⯓
16 59	☽ # ⯓
18 08	☽ ± ⯓
18 31	☽ ⊥ ♀
08 Tuesday	
03 32	☽ □ ♂
04 20	☽ □ ♆
05 38	☽ ⊼ ⯓
06 47	♀ ♇ ♅
07 17	☽ ⊻ ♀
11 13	☽ △ ♄
13 10	☽ ‖ ⯓
13 39	☿ ♇ ⯓
14 05	☽ ⊻ ♃
22 15	☽ # ☿
09 Wednesday	
00 53	♂ ☌ ♆
01 57	☽ ⊥ ♃
06 08	☽ ☌ ☉
10 20	☽ ‖ ☉
12 09	☽ ⊻ ☿
12 25	☽ △ ♅
13 25	☽ # ♃
16 15	☿ □ ♅
16 24	☽ ♇ ♄
19 56	☽ ∠ ♃
21 59	♀ ⊼ ♄
10 Thursday	
00 48	☽ ⊥ ☿
08 21	♂ ⚹ ⯓
10 10	☽ ± ♄
15 09	☽ △ ♆
16 21	☽ ☍ ⯓
16 42	☽ △ ♂
18 06	☽ ♇ ♅
18 48	☽ ‖ ♀
20 00	☽ ∠ ☿
22 06	☽ ⊼ ♄
11 Friday	
00 56	☽ ☌ ♀
01 15	☉ # ♃
02 18	☽ ⚹ ♃
12 07	☽ ± ♅
16 49	♀ ⚹ ♃
21 14	☽ ♇ ♆
21 43	☽ # ♄
23 10	☽ ⊻ ☉
12 Saturday	
00 02	☽ ♇ ♂
00 12	☽ ⊼ ♅
02 54	☽ Q ♃
04 35	☽ ⚹ ☿
11 30	☉ △ ♅
12 19	☽ # ♆
12 24	☽ ⊥ ☉
15 27	☽ ± ♆
18 59	☽ ± ♂
19 44	☽ # ♅
22 41	☽ # ♂
13 Sunday	
03 37	☽ ⊼ ♆
04 43	☽ ⊼ ⯓
07 02	☽ Q ☿
07 39	☽ ⊼ ♂
08 19	☽ ∠ ☉
10 31	☽ ☍ ♄
16 05	☽ □ ♃
16 53	☽ ± ⯓
20 37	☽ ⊻ ♀
14 Monday	
09 44	☉ ♇ ♄
10 13	☽ ⊥ ♀
11 07	☽ ♇ ⯓
13 00	☽ ☍ ♅
17 32	☽ ⚹ ☉
22 59	☽ □ ☿
15 Tuesday	
06 31	☽ ∠ ♀
12 45	♀ ± ♅
16 21	☽ ☍ ♆
17 20	☽ △ ⯓
19 52	☽ Q ☉
22 42	☽ ☍ ♂
22 58	☽ ⊼ ♄
16 Wednesday	
05 00	♂ ⊻ ♄
05 42	☽ △ ♃
10 52	☽ ± ♄
15 54	☽ ⚹ ♀
19 40	☿ # ⯓
17 Thursday	
00 57	☽ ⊼ ♅
04 33	☽ ♇ ♄
10 35	☽ □ ☉
11 44	☽ ♇ ♃
12 39	☽ ± ♅
16 32	☽ △ ☿
17 06	☿ ± ⯓
17 59	☽ Q ♀
18 Friday	
03 17	☽ ⊼ ♆
04 06	☽ □ ⯓
04 45	☉ ∠ ♃
04 49	☽ # ♂
05 22	☽ ± ♃
05 57	☽ ♇ ♅
07 46	☽ # ♅
09 23	☽ △ ♄
11 27	☽ ⊼ ♂
14 41	☽ ± ♆
14 44	☽ # ♆
16 56	☽ ⊼ ♃
23 09	☽ ± ♂
19 Saturday	
00 02	☽ ♇ ☿
04 38	☽ # ♄
05 54	♀ ♇ ♆
07 31	☽ ♇ ♆
07 41	☽ □ ♀
10 04	☽ △ ♅
12 51	☽ ‖ ♀
13 46	☽ Q ⯓
16 22	☽ ♇ ♂
18 24	☽ ± ☿
23 42	☽ △ ☉
20 Sunday	
06 23	☽ ⊼ ☿
08 37	♀ ⊼ ♅
10 46	☽ △ ♆
11 26	☽ ⚹ ⯓
16 17	☽ □ ♄
18 10	☽ ‖ ☉
20 11	☽ △ ♂
21 Monday	
00 20	☽ ☍ ♃
04 26	☽ ♇ ☉
10 46	☽ # ♃
13 40	☽ ∠ ⯓
15 23	☽ □ ♅
18 23	☽ △ ♀
21 08	☽ ± ☉
23 25	☽ Q ♄
22 Tuesday	
03 25	☿ □ ♆
04 10	☽ ‖ ⯓
05 00	☽ ⊥ ⯓
08 03	☽ ⊼ ☉
09 29	☿ ⊼ ⯓
13 29	☽ # ☿
13 30	☽ ‖ ☿
14 30	☽ □ ♆
15 03	☽ ⊻ ⯓
15 38	☽ ☍ ☿
19 31	☽ ⚹ ♄
21 43	☽ Q ♅
22 01	☽ ♇ ♀
22 44	☽ # ⯓
23 Wednesday	
00 52	☽ □ ♂
03 57	☽ ⊼ ♃
04 42	☉ ± ♄
10 48	♀ Q ♃
13 56	☽ ± ♃
14 10	☽ ± ♀
14 50	☽ ‖ ♃
17 24	☽ ⚹ ♅
20 04	☽ ∠ ♄
20 07	☽ Q ♆
24 Thursday	
00 29	♀ # ♄
00 48	☽ ⊼ ♀
04 44	☽ ♇ ♃
06 57	☿ △ ♄
07 04	☽ Q ♂
10 34	☽ ⊥ ♄
11 22	☽ # ☉
12 46	☽ ☍ ☉
15 29	☽ ⚹ ♆
15 57	☽ ☌ ⯓
17 35	☽ ∠ ♅
20 10	☽ ⊻ ♄
21 36	☽ ⊼ ☿
23 21	♀ ♇ ♂
25 Friday	
02 52	☽ ⚹ ♂
05 08	☽ △ ♃
07 58	☽ ⊥ ♅
08 12	☽ ± ☿
15 27	☽ ‖ ♄
15 28	☽ ∠ ♆
17 32	☽ ⊻ ♅
18 05	☽ # ♀
26 Saturday	
00 06	☽ ♇ ☿
03 27	☽ ∠ ♂
03 29	☽ ‖ ♆
05 08	☽ ☍ ♀
05 49	☽ ⊥ ♆
06 00	☉ △ ♆
09 17	☽ ‖ ♅
11 50	☽ ‖ ♂
12 21	☉ ☍ ⯓
13 31	♀ ± ♆
15 24	☽ ⊻ ♆
15 49	☽ ⊻ ⯓
16 04	☽ ⊼ ☉
18 14	☽ ⊥ ♂
19 56	☽ ☌ ♄
27 Sunday	
01 25	☽ ⊥ ⯓
02 21	☽ ± ☉
02 44	☽ △ ☿
04 08	☽ ⊻ ♂
05 45	☽ □ ♃
08 15	☿ ‖ ⯓
13 06	☉ ♇ ♅
15 55	☽ ∠ ⯓
17 35	☽ ☌ ♅
17 55	☽ ♇ ☉
20 55	☿ □ ♂
28 Monday	
09 55	☽ ⊼ ♀
11 25	☽ Q ♃
12 15	☿ ⊻ ♃
16 03	☽ ☌ ♆
16 25	☽ ⚹ ⯓
20 16	☽ △ ☉
20 35	☽ ⊻ ♄
20 44	☽ ± ♀
29 Tuesday	
00 41	☉ ⊼ ♄
04 10	♂ ⊥ ♄
06 33	☽ ⊥ ♄
06 38	☽ ☌ ♂
07 37	☽ ⚹ ♃
09 42	☽ □ ☿
12 25	☽ Q ⯓
13 12	☽ ♇ ♀
19 15	☽ ⊻ ♅
21 43	☽ ∠ ♄
30 Wednesday	
05 29	☽ ⊥ ♅
09 31	☽ ∠ ♃
17 24	☽ △ ♀
18 56	☽ ⊻ ♆
19 16	☽ □ ⯓
20 25	☽ ‖ ♂
20 32	☽ Q ☿
21 05	☽ ∠ ♅
23 17	☽ ‖ ♅
23 33	☽ ⚹ ♄

All ephemeris data is given at 12.00 UT and the Moon's longitude is additionally given for 24.00 UT

MAY 1986

Date	Sidereal time h m s	Sun ☉ ° ' "	Moon ☽ ° ' "	Moon ☽ 24.00 ° ' "	Mercury ☿ ° '	Venus ♀ ° '	Mars ♂ ° '	Jupiter ♃ ° '	Saturn ♄ ° '	Uranus ♅ ° '	Neptune ♆ ° '	Pluto ♇ ° '
01	02 36 30	10 ♉ 49 04	15 ♒ 20 23	22 ♒ 01 51	19 ♈ 37	05 ♊ 44	15 ♑ 16	15 ♓ 27	08 ♐ 16	21 ♐ 52	05 ♑ 39	05 ♏ 50
02	02 40 26	11 47 17	28 ♒ 38 03	05 ♓ 09 15	21 16	06 56	15 38	15 38	08 R 12	21 R 51	05 R 39	05 R 48
03	02 44 23	12 45 29	11 ♓ 35 45	17 ♓ 57 56	22 57	08 09	15 59	15 49	08 09	21 49	05 38	05 47
04	02 48 19	13 43 40	24 ♓ 16 07	00 ♈ 30 41	24 39	09 22	16 20	15 59	08 05	21 47	05 37	05 45
05	02 52 16	14 41 49	06 ♈ 41 57	12 ♈ 50 17	26 23	10 35	16 41	16 10	08 01	21 46	05 36	05 43
06	02 56 12	15 39 56	18 ♈ 55 59	24 ♈ 59 22	28 09	11 48	17 01	16 21	07 57	21 44	05 35	05 42
07	03 00 09	16 38 02	01 ♉ 00 43	07 ♉ 00 20	29 ♈ 57	13 01	17 20	16 31	07 53	21 42	05 34	05 40
08	03 04 05	17 36 06	12 ♉ 58 26	18 ♉ 55 20	01 ♉ 47	14 13	17 40	16 42	07 49	21 40	05 34	05 38
09	03 08 02	18 34 09	24 ♉ 51 14	00 ♊ 46 25	03 39	15 26	17 58	16 52	07 45	21 38	05 33	05 37
10	03 11 59	19 32 10	06 ♊ 41 07	12 ♊ 35 38	05 33	16 39	18 17	17 02	07 41	21 36	05 32	05 35
11	03 15 55	20 30 09	18 ♊ 30 13	24 ♊ 25 11	07 28	17 51	18 35	17 12	07 37	21 34	05 31	05 33
12	03 19 52	21 28 07	00 ♋ 20 52	06 ♋ 17 36	09 26	19 04	18 52	17 22	07 33	21 32	05 30	05 32
13	03 23 48	22 26 03	12 ♋ 15 45	18 ♋ 15 43	11 25	20 16	19 09	17 32	07 28	21 30	05 29	05 30
14	03 27 45	23 23 58	24 ♋ 17 57	00 ♌ 22 53	13 26	21 29	19 25	17 41	07 24	21 28	05 27	05 29
15	03 31 41	24 21 50	06 ♌ 30 59	12 ♌ 42 45	15 28	22 41	19 41	17 51	07 20	21 26	05 26	05 27
16	03 35 38	25 19 41	18 ♌ 58 40	25 ♌ 19 14	17 32	23 54	19 57	18 00	07 16	21 24	05 25	05 25
17	03 39 34	26 17 30	01 ♍ 44 56	08 ♍ 16 13	19 38	25 06	20 12	18 10	07 11	21 22	05 24	05 24
18	03 43 31	27 15 18	14 ♍ 53 29	21 ♍ 37 06	21 45	26 18	20 26	18 19	07 07	21 20	05 23	05 22
19	03 47 28	28 13 03	28 ♍ 27 18	05 ♎ 24 15	23 54	27 30	20 40	18 28	07 03	21 18	05 22	05 21
20	03 51 24	29 ♉ 10 47	12 ♎ 27 58	19 ♎ 38 18	26 03	28 43	20 53	18 37	06 58	21 15	05 21	05 19
21	03 55 21	00 ♊ 08 30	26 ♎ 54 56	04 ♏ 17 24	28 ♉ 14	29 ♊ 55	21 06	18 46	06 54	21 13	05 19	05 18
22	03 59 17	01 06 11	11 ♏ 44 59	19 ♏ 16 49	00 ♊ 25	01 ♋ 07	21 18	18 55	06 49	21 11	05 18	05 16
23	04 03 14	02 03 50	26 ♏ 51 51	04 ♐ 28 56	02 37	02 19	21 30	19 03	06 45	21 09	05 17	05 15
24	04 07 10	03 01 29	12 ♐ 06 46	19 ♐ 44 03	04 48	03 31	21 41	19 12	06 40	21 06	05 16	05 13
25	04 11 07	03 59 06	27 ♐ 19 29	04 ♑ 51 49	07 00	04 43	21 51	19 20	06 36	21 04	05 14	05 12
26	04 15 03	04 56 42	12 ♑ 19 57	19 ♑ 42 55	09 11	05 54	22 01	19 28	06 32	21 02	05 13	05 10
27	04 19 00	05 54 17	26 ♑ 59 57	04 ♒ 10 30	11 22	07 06	22 10	19 36	06 27	20 59	05 12	05 09
28	04 22 57	06 51 51	11 ♒ 14 11	18 ♒ 10 50	13 32	08 18	22 19	19 44	06 23	20 57	05 10	05 08
29	04 26 53	07 49 25	25 ♒ 00 28	01 ♓ 43 13	15 40	09 30	22 26	19 52	06 18	20 55	05 09	05 06
30	04 30 50	08 46 57	08 ♓ 19 23	14 ♓ 49 21	17 47	10 41	22 34	20 00	06 14	20 52	05 07	05 05
31	04 34 46	09 ♊ 44 28	21 ♓ 13 34	27 ♓ 32 32	19 ♊ 53	11 ♋ 53	22 ♑ 40	20 ♓ 07	06 ♐ 09	20 ♐ 50	05 ♑ 06	05 ♏ 04

Date	Moon True ☊ ° '	Moon Mean ☊ ° '	Moon ☽ Latitude ° '
01	29 ♈ 56	29 ♈ 27	05 S 06
02	29 D 56	29 23	04 38
03	29 57	29 20	03 57
04	29 59	29 17	03 05
05	00 ♉ 00	29 14	02 06
06	00 01	29 11	01 S 01
07	00 R 01	29 07	00 N 06
08	00 00	29 04	01 11
09	29 ♈ 59	29 01	02 13
10	29 57	28 58	03 09
11	29 54	28 55	03 56
12	29 50	28 52	04 34
13	29 46	28 48	05 00
14	29 43	28 45	05 13
15	29 41	28 42	05 12
16	29 40	28 39	04 57
17	29 D 40	28 36	04 26
18	29 41	28 33	03 42
19	29 42	28 29	02 43
20	29 44	28 26	01 33
21	29 44	28 23	00 N 15
22	29 R 44	28 20	01 S 05
23	29 42	28 17	02 23
24	29 39	28 13	03 31
25	29 34	28 10	04 24
26	29 29	28 07	04 58
27	29 25	28 04	05 11
28	29 21	28 01	05 05
29	29 18	27 58	04 41
30	29 18	27 54	04 02
31	29 ♈ 18	27 ♈ 51	03 S 12

DECLINATIONS

Date	Sun ☉ ° '	Moon ☽ ° '	Mercury ☿ ° '	Venus ♀ ° '	Mars ♂ ° '	Jupiter ♃ ° '	Saturn ♄ ° '	Uranus ♅ ° '	Neptune ♆ ° '	Pluto ♇ ° '
01	15 N 04	21 S 06	05 N 18	21 N 57	23 S 44	06 S 38	19 S 41	23 S 17	22 S 13	02 N 42
02	15 22	16 18	05 59	22 12	23 44	06 34	19 40	23 17	22 13	02 42
03	15 40	10 52	06 40	22 27	23 44	06 30	19 40	23 16	22 13	02 43
04	15 58	05 S 07	07 22	22 41	23 44	06 26	19 39	23 16	22 13	02 43
05	16 15	00 N 44	08 05	22 54	23 44	06 22	19 39	23 16	22 13	02 43
06	16 32	06 29	08 49	23 07	23 44	06 18	19 38	23 16	22 13	02 44
07	16 49	11 55	09 33	23 19	23 45	06 14	19 37	23 16	22 13	02 44
08	17 05	16 52	10 17	23 30	23 45	06 11	19 37	23 16	22 13	02 45
09	17 21	21 08	11 02	23 41	23 46	06 07	19 36	23 16	22 13	02 45
10	17 37	24 32	11 48	23 51	23 46	06 03	19 35	23 16	22 13	02 45
11	17 53	26 52	12 33	24 00	23 47	05 59	19 35	23 16	22 14	02 46
12	18 08	28 00	13 19	24 09	23 47	05 56	19 34	23 15	22 14	02 46
13	18 23	27 51	14 05	24 17	23 48	05 52	19 33	23 15	22 14	02 46
14	18 38	26 23	14 50	24 24	23 49	05 48	19 32	23 15	22 14	02 47
15	18 52	23 41	15 36	24 31	23 50	05 45	19 32	23 15	22 14	02 47
16	19 06	19 50	16 20	24 37	23 51	05 41	19 31	23 15	22 14	02 47
17	19 20	15 00	17 05	24 42	23 52	05 38	19 30	23 15	22 14	02 47
18	19 33	09 21	17 48	24 46	23 53	05 35	19 30	23 15	22 14	02 48
19	19 46	03 N 07	18 31	24 50	23 54	05 31	19 29	23 15	22 14	02 48
20	19 59	03 S 30	19 13	24 53	23 56	05 28	19 28	23 14	22 14	02 48
21	20 11	10 08	19 53	24 56	23 57	05 25	19 28	23 14	22 14	02 48
22	20 23	16 24	20 31	24 57	23 59	05 22	19 27	23 14	22 14	02 49
23	20 35	21 47	21 08	24 58	24 01	05 18	19 26	23 14	22 14	02 49
24	20 46	25 44	21 43	24 58	24 03	05 15	19 26	23 14	22 14	02 49
25	20 57	27 49	22 16	24 58	24 05	05 12	19 25	23 14	22 14	02 49
26	21 07	27 49	22 46	24 57	24 07	05 09	19 24	23 14	22 14	02 49
27	21 18	25 51	23 14	24 55	24 09	05 06	19 23	23 13	22 14	02 49
28	21 28	22 17	23 40	24 52	24 11	05 03	19 23	23 13	22 14	02 49
29	21 37	17 35	24 03	24 49	24 14	05 01	19 22	23 13	22 14	02 50
30	21 46	12 11	24 23	24 45	24 17	04 58	19 21	23 13	22 14	02 50
31	21 N 55	06 S 25	24 N 40	24 N 40	24 S 20	04 S 55	19 S 21	23 S 13	22 S 14	02 N 50

ZODIAC SIGN ENTRIES

Date	h	m	Planets
02	14	30	☽ ♓
04	23	01	☽ ♈
07	09	59	☽ ♉
07	12	33	☿ ♉
09	22	26	☽ ♊
12	11	18	☽ ♋
14	23	15	☽ ♌
17	08	45	☽ ♍
19	14	41	☽ ♎
21	08	28	☉ ♊
21	13	46	♀ ♋
21	17	02	☽ ♏
22	07	26	☿ ♊
23	16	57	☽ ♐
25	16	15	☽ ♑
27	17	00	☽ ♒
29	20	54	☽ ♓

LATITUDES

Date	Mercury ☿ ° '	Venus ♀ ° '	Mars ♂ ° '	Jupiter ♃ ° '	Saturn ♄ ° '	Uranus ♅ ° '	Neptune ♆ ° '	Pluto ♇ ° '
01	02 S 33	00 N 42	01 S 10	00 S 59	02 N 02	00 S 05	01 N 06	17 N 06
04	02 20	00 50	01 18	00 59	02 02	00 05	01 06	17 05
07	02 03	00 57	01 26	01 00	02 02	00 05	01 06	17 05
10	01 40	01 05	01 35	01 01	02 02	00 05	01 06	17 04
13	01 15	01 12	01 45	01 01	02 02	00 05	01 06	17 04
16	00 45	01 19	01 54	01 02	02 02	00 05	01 06	17 03
19	00 S 14	01 25	02 04	01 03	02 02	00 05	01 06	17 03
22	00 N 17	01 31	02 15	01 03	02 02	00 05	01 06	17 02
25	00 48	01 36	02 26	01 04	02 02	00 05	01 06	17 01
28	01 15	01 41	02 38	01 05	02 02	00 05	01 06	17 00
31	01 N 37	01 N 46	02 S 50	01 S 06	02 N 01	00 S 05	01 N 07	16 N 59

DATA

Julian Date	2446552
Delta T	+55 seconds
Ayanamsa	23° 39' 49"
Synetic vernal point	05° ♓ 27' 10"
True obliquity of ecliptic	23° 26' 36"

MOON'S PHASES, APSIDES AND POSITIONS ☽

Date	h	m	Phase	Longitude ° '	Eclipse Indicator
01	03	22	☾	10 ♒ 28	
08	22	10	●	18 ♉ 01	
17	01	00	☽	25 ♌ 51	
23	20	45	○	02 ♐ 25	
30	12	55	☾	08 ♓ 49	

Day	h	m		
10	22	51	Apogee	
24	03	07	Perigee	
05	08	58	ON	
12	21	10	Max dec	28° N 06'
19	23	24	OS	
25	23	56	Max dec	28° S 04'

LONGITUDES

Date	Chiron ⚷ ° '	Ceres ⚳ ° '	Pallas ⚴ ° '	Juno ⚵ ° '	Vesta ⚶ ° '	Black Moon Lilith ⚸ ° '
01	12 ♊ 20	03 ♍ 05	21 ♋ 24	14 ♐ 40	18 ♓ 12	07 ♊ 00
11	13 ♊ 05	04 ♍ 28	26 ♋ 03	13 ♐ 01	22 ♓ 31	08 ♊ 07
21	13 ♊ 53	06 ♍ 23	00 ♌ 47	10 ♐ 59	26 ♓ 37	09 ♊ 14
31	14 ♊ 41	08 ♍ 45	05 ♌ 33	08 ♐ 45	00 ♈ 30	10 ♊ 20

All ephemeris data is given at 12.00 UT and the Moon's longitude is additionally given for 24.00 UT

ASPECTARIAN

h m	Aspects
01 Thursday	
01 25	☽ ⊥ ♃
03 22	☽ □ ☉
05 28	☽ ⊥ ♆
05 40	☽ ∥ ♆
07 30	☽ # ♀
10 41	♀ ⊼ ♆
11 53	☽ ⊻ ♂
12 11	☽ ⊻ ♃
14 05	♀ ⊼ ♇
19 32	☽ ∥ ♄
20 44	☽ ⚹ ☿
20 47	☽ Q ♄
21 31	☽ ∠ ♆
22 56	☽ ⊥ ♂
23 42	☽ ⚹ ♅
02 Friday	
11 22	♂ ⚹ ♃
13 38	♀ # ♆
14 17	☽ Q ☉
15 46	☽ ∠ ♂
16 00	☽ # ☉
17 51	☿ ⊥ ♃
20 13	☿ △ ♅
21 34	☽ Q ♅
03 Saturday	
00 54	☽ ⚹ ♆
01 11	☽ △ ♇
04 09	☽ ∠ ☿
04 54	☽ □ ♀
05 35	☽ □ ♄
06 41	☿ # ♃
11 47	♀ ☍ ♄
14 22	☽ ⚹ ☉
14 45	☿ ⚼ ♄
20 02	☽ ☌ ♃
20 30	☽ ⚹ ♂
22 36	☿ ∠ ♀
23 21	☽ Q ♆
23 37	☽ ⊥ ☿
04 Sunday	
03 40	☽ # ☿
05 18	☽ ⚼ ♇
06 28	☽ ∥ ♃
07 17	☽ □ ♅
12 51	☽ ⊻ ☿
18 35	☽ Q ♀
20 02	☽ Q ♂
21 17	☽ ∠ ☉
21 48	☽ # ♇
22 30	☽ ± ♇
05 Monday	
09 52	☽ □ ♆
10 06	☽ ⊼ ♇
14 30	♀ ± ♂
14 33	☽ △ ♄
16 14	☽ ⊥ ☉
20 13	☽ ∥ ♇
20 25	☽ ⚹ ♀
06 Tuesday	
05 00	☽ ⊻ ☉
06 49	☽ ⊻ ♃
08 06	☽ □ ♂
09 57	♀ ± ♇
11 17	☽ # ♃
13 32	☉ ± ♅
17 31	☽ △ ♅
18 51	☽ ⊥ ♃
19 55	☽ ⚼ ♄
23 40	☽ ∥ ☿
07 Wednesday	
05 21	☽ ∠ ♀
06 25	♀ # ♅
08 32	☉ ⚹ ♃
09 31	☽ ☌ ☿
13 02	☽ ∠ ♃
13 44	☽ ± ♄
21 07	☽ △ ♆
21 17	☽ ☍ ♇
23 21	☽ ⚼ ♅
08 Thursday	
01 22	☽ ⊥ ♀
01 41	☽ ⊼ ♄
10 31	♂ Q ♇
10 36	☿ ∠ ♃
12 22	☿ ± ♄
13 13	☽ ∥ ☉
14 06	☉ △ ♂
14 48	☽ ⊻ ♀
17 25	☽ ± ♅
19 36	☽ ⚹ ♃
21 42	☽ △ ♂
22 10	☽ ☌ ☉
09 Friday	
02 53	☽ # ♄
03 17	☽ ⚼ ♆
05 30	☽ ⊼ ♅
19 01	☽ # ♆
20 15	☽ Q ♃
21 30	☽ ± ♆
23 36	♀ # ♂
10 Saturday	
02 13	☽ # ♅
04 54	☽ ⚼ ♂
05 57	☽ # ♂
06 17	☽ ∥ ♀
09 14	☽ ⊻ ☿
09 39	☽ ⊼ ♆
09 46	☽ ⊼ ♇
11 46	☿ △ ♆
12 27	☿ ☍ ♇
14 01	☽ ☍ ♄
20 55	♀ □ ♃
21 55	☽ ± ♇
23 39	☽ ± ♂
23 47	☽ ⊥ ☿
11 Sunday	
01 01	☿ ⚼ ♅
09 19	☽ □ ♃
10 32	☽ ☌ ♀
12 09	☽ ⊼ ♂
12 12	☉ ⚼ ♆
13 42	☿ ⊼ ♄
16 09	☽ ⚼ ♇
16 25	☽ ⊻ ☉
18 12	☽ ☍ ♅
21 38	☽ ∠ ☿
12 Monday	
05 40	☽ ⊥ ☉
06 49	♀ ⊼ ♂
13 40	☉ ⊼ ♅
22 22	☽ ☍ ♆
22 26	☽ △ ♇
13 Tuesday	
01 27	☽ ∠ ☉
02 26	☽ ⊼ ♄
04 58	♃ Q ♆
09 57	☽ ⚹ ☿
14 25	☽ ± ♄
16 27	♀ ⚼ ♇
22 41	☽ △ ♃
14 Wednesday	
02 05	☽ ☍ ♂
05 47	☽ ⊻ ♀
06 24	☽ ⊼ ♅
08 16	☽ ⚼ ♄
10 04	☽ ⚹ ☉
11 48	♀ ☍ ♅
14 41	☽ Q ☿
18 15	☽ ± ♅
18 59	☽ ⊥ ♀
15 Thursday	
04 45	☽ ⚼ ♃
05 50	☽ ∥ ♀
09 54	☽ ⊼ ♆
09 55	☽ □ ♇
10 53	☽ # ♂
11 36	☿ ± ♅
11 41	☽ Q ☉
11 51	☽ ⚼ ♅
13 35	☽ △ ♄
14 31	☽ ∠ ♀
15 01	☽ # ♅
21 32	☽ ± ♆
21 51	☽ # ♆
22 28	☽ ± ♃
16 Friday	
08 43	☽ □ ☿
10 08	☽ ⊼ ♃
13 42	☽ # ♄
13 53	☽ ⊼ ♂
14 44	☽ ⚼ ♆
15 45	☽ ∥ ☉
16 35	☽ △ ♅
17 10	♆ ⚹ ♇
17 49	☿ ⚹ ♃
20 25	☽ Q ♇
21 35	☿ ⊥ ♀
22 17	☽ ⚹ ♀
17 Saturday	
01 00	☽ □ ☉
01 26	☽ ± ♂
03 28	☽ ∥ ☿
18 29	☽ ⚼ ♂
18 43	☽ ⚹ ♇
18 44	☽ △ ♆
19 08	☿ △ ♂
20 37	☿ ⚼ ♆
21 58	☽ □ ♄
22 51	☽ Q ♀
18 Sunday	
06 24	☉ # ♄
07 16	☿ ⊼ ♅
18 12	☽ ☍ ♃
21 46	☽ ∠ ♇
22 04	☽ △ ♂
23 27	☽ □ ♅
19 Monday	
02 32	☽ △ ☿
02 53	☽ # ♃
06 04	☽ Q ♄
10 11	☽ □ ♀
11 33	☽ △ ☉
13 09	☽ ∥ ♇
13 33	☽ ⊥ ♇
23 53	☽ ⊻ ♇
23 55	☽ □ ♆
20 Tuesday	
02 43	☽ ⚹ ♄
06 35	☽ Q ♅
09 11	☽ ⚼ ☿
09 31	☽ # ♇
15 06	☽ ⚼ ☉
15 59	☉ ± ♆
19 01	☽ ∥ ♃
21 09	☿ # ♄
22 25	☽ ⊼ ♃
21 Wednesday	
02 17	☽ □ ♂
02 39	☽ ⚹ ♅
02 56	☽ ± ☿
03 47	☽ ∠ ♄
06 06	☽ Q ♆
07 07	☽ ± ☉
08 26	☽ ± ♃
14 32	☽ ⊼ ☿
17 20	☽ △ ♀
17 38	☽ ⊼ ☉
18 28	☽ ⊥ ♄
23 16	☽ ⚼ ♃
23 53	☿ ± ♆
23 59	♂ ⊻ ♅
22 Thursday	
01 36	☽ ☌ ♇
01 39	☽ ⚹ ♆
03 05	☽ ∠ ♅
04 08	☽ ⊻ ♄
04 22	☉ ∥ ☿
06 21	☉ Q ♃
08 01	☽ Q ♂
11 02	☉ ⊻ ♀
17 28	☽ ⊥ ♅
17 48	☿ Q ♃
19 34	☽ ⚼ ♀
23 31	☽ △ ♃
23 Friday	
00 57	☽ ∥ ♄
01 24	☉ ☌ ☿
01 36	☽ ∠ ♆
02 59	☽ ⊻ ♅
03 24	☽ ⚹ ♂
04 52	☿ ⊻ ♀
05 56	☽ # ☉
08 24	☽ # ☿
11 03	☽ ± ♀
14 20	☽ ∥ ♆
15 48	☽ ⊥ ♆
19 45	☽ ∥ ♅
20 45	☽ ☍ ☉
21 19	☽ ⊼ ♀
22 34	☽ ☍ ☿
24 Saturday	
00 23	☽ ∥ ♂
01 11	☽ ⊻ ♇
01 14	☽ ⊻ ♆
03 21	☽ ∠ ♂
03 29	☽ ☌ ♄
06 29	☽ # ♀
10 36	☽ ⊥ ♇
11 39	♂ ∠ ♄
16 31	☿ ⊼ ♇
16 55	☿ ⊼ ♆
17 41	☽ ⊥ ♂
23 15	☽ □ ♃
25 Sunday	
00 45	☽ ∠ ♇
02 08	☽ ☌ ♅
03 15	☽ ⊻ ♂
07 47	☿ ☍ ♄
10 16	☿ ⚼ ♂
10 35	☿ # ♆
21 34	♀ △ ♇
22 22	♀ ☍ ♆
23 19	☽ ⊼ ☉
26 Monday	
00 31	☽ ⚹ ♇
00 35	☽ ☌ ♆
00 46	☽ ☍ ♀
02 42	☽ ⊻ ♄
04 06	☽ Q ♃
06 04	☽ ⊼ ☿
09 36	☽ ± ☉
12 19	☽ ⊥ ♄
17 25	☽ ± ☿
17 36	☉ ⊼ ♇
18 36	☉ ⊼ ♆
19 50	☽ Q ♇
23 40	♀ ⊼ ♄
23 42	☽ ⚹ ♃
27 Tuesday	
01 15	☽ ⚼ ☉
02 07	☽ ⊻ ♅
02 54	☽ ∠ ♄
03 56	☽ ☌ ♂
09 41	☿ ± ♇
10 46	☽ ⚼ ☿
11 17	☿ # ♅
11 59	☽ ⊥ ♅
19 22	☽ # ♀
28 Wednesday	
00 25	☽ ∥ ♂
00 42	☉ ☍ ♄
00 51	☽ ∠ ♃
01 37	☽ □ ♇
01 42	☽ ⊻ ♆
03 01	☽ ∠ ♅
03 46	☽ ⚹ ♄
04 00	☽ △ ☉
04 32	☽ # ☿
06 28	☽ ∥ ♅
06 31	☽ ⊼ ♀
11 53	☽ ⊥ ♆
12 17	☽ ∥ ♆
16 20	☽ ⊥ ♃
16 26	☽ # ☉
16 39	☽ △ ☿
17 46	☽ ± ♀
29 Thursday	
00 17	☽ Q ♄
00 58	☉ ⚼ ♂
02 51	☽ ⊻ ♃
03 25	☽ ∥ ♄
03 27	☽ ∠ ♆
04 48	☽ ⚹ ♅
07 25	☽ ⊻ ♂
11 00	☽ ⚼ ♀
18 10	☽ ⊥ ♂
21 17	☿ ± ♂
30 Friday	
02 07	☽ Q ♅
03 24	☿ # ♂
06 06	☽ △ ♇
06 10	☽ ⚹ ♆
08 12	☽ □ ♄
10 36	☽ ∠ ♂
12 55	☽ □ ☉
16 47	☽ △ ♀
31 Saturday	
02 38	♃ ⚼ ♇
04 16	☽ Q ♆
08 58	☽ □ ☿
09 48	☽ ⚼ ♇
09 53	☽ ☌ ♃
11 16	☽ □ ♅
11 46	☿ ∥ ♀
14 06	☿ ⚼ ♇
14 45	☽ ⚹ ♂
14 57	☿ □ ♃
17 09	♀ ± ♄
18 12	☽ ∥ ♃
22 51	☿ ☍ ♅

JUNE 1986

Date	Sidereal time h m s	Sun ☉ ° ' "	Moon ☽ ° ' "	Moon ☽ 24.00 ° ' "	Mercury ☿ ° '	Venus ♀ ° '	Mars ♂ ° '	Jupiter ♃ ° '	Saturn ♄ ° '	Uranus ♅ ° '	Neptune ♆ ° '	Pluto ♇ ° '
01	04 38 43	10 ♊ 41 59	03 ♈ 46 48	09 ♈ 56 55	21 ♊ 56	13 ♋ 04	22 ♑ 46	20 ♓ 14	06 ♐ 05	20 ♐ 48	05 ♑ 05	05 ♏ 02
02	04 42 39	11 39 29	16 ♈ 03 24	22 ♈ 06 48	23 58	14 16	22 51	20 22	06 R 00	20 R 45	05 R 03	05 R 01
03	04 46 36	12 36 58	28 ♈ 07 39	04 ♉ 06 23	25 58	15 27	22 56	20 29	05 56	20 43	05 02	05 00
04	04 50 32	13 34 26	10 ♉ 03 29	15 ♉ 59 22	27 55	16 39	22 59	20 36	05 52	20 40	05 00	04 59
05	04 54 29	14 31 54	21 ♉ 54 23	27 ♉ 48 53	29 ♊ 50	17 50	23 02	20 42	05 47	20 38	04 59	04 57
06	04 58 26	15 29 21	03 ♊ 43 10	09 ♊ 37 32	01 ♋ 43	19 01	23 04	20 49	05 43	20 35	04 57	04 56
07	05 02 22	16 26 47	15 ♊ 32 13	21 ♊ 27 27	03 33	20 13	23 06	20 55	05 38	20 33	04 56	04 55
08	05 06 19	17 24 12	27 ♊ 23 28	03 ♋ 20 27	05 20	21 24	23 07	21 02	05 34	20 30	04 54	04 54
09	05 10 15	18 21 36	09 ♋ 18 37	15 ♋ 18 11	07 06	22 35	23 R 07	21 08	05 30	20 28	04 53	04 53
10	05 14 12	19 18 59	21 ♋ 19 22	27 ♋ 22 26	08 48	23 46	23 06	21 14	05 25	20 26	04 51	04 52
11	05 18 08	20 16 22	03 ♌ 27 36	09 ♌ 35 12	10 28	24 57	23 04	21 19	05 21	20 23	04 50	04 51
12	05 22 05	21 13 43	15 ♌ 45 32	21 ♌ 58 57	12 05	26 08	23 02	21 25	05 17	20 21	04 48	04 50
13	05 26 01	22 11 04	28 ♌ 15 48	04 ♍ 36 30	13 40	27 18	22 59	21 31	05 13	20 18	04 47	04 49
14	05 29 58	23 08 23	11 ♍ 01 28	17 ♍ 31 05	15 12	28 29	22 55	21 36	05 09	20 16	04 45	04 48
15	05 33 55	24 05 42	24 ♍ 05 48	00 ♎ 45 58	16 41	29 ♋ 40	22 51	21 41	05 05	20 13	04 43	04 47
16	05 37 51	25 03 00	07 ♎ 31 56	14 ♎ 23 59	18 08	00 ♌ 50	22 45	21 46	05 01	20 11	04 42	04 46
17	05 41 48	26 00 16	21 ♎ 22 19	28 ♎ 27 00	19 32	02 01	22 39	21 51	04 57	20 08	04 40	04 45
18	05 45 44	26 57 32	05 ♏ 37 57	12 ♏ 54 57	20 53	03 11	22 33	21 55	04 53	20 06	04 39	04 44
19	05 49 41	27 54 47	20 ♏ 17 32	27 ♏ 45 06	22 11	04 22	22 25	22 00	04 49	20 03	04 37	04 43
20	05 53 37	28 52 02	05 ♐ 16 49	12 ♐ 51 39	23 26	05 32	22 17	22 04	04 45	20 01	04 35	04 42
21	05 57 34	29 ♊ 49 16	20 ♐ 28 26	28 ♐ 05 53	24 39	06 42	22 08	22 08	04 41	19 59	04 34	04 42
22	06 01 30	00 ♋ 46 30	05 ♑ 42 36	13 ♑ 17 16	25 48	07 52	21 59	22 12	04 37	19 56	04 32	04 41
23	06 05 27	01 43 43	20 ♑ 48 34	28 ♑ 15 20	26 55	09 02	21 49	22 16	04 33	19 54	04 31	04 40
24	06 09 24	02 40 56	05 ♒ 36 32	12 ♒ 51 22	27 58	10 12	21 38	22 19	04 30	19 51	04 29	04 39
25	06 13 20	03 38 09	19 ♒ 59 15	26 ♒ 59 49	28 58	11 22	21 26	22 23	04 26	19 49	04 27	04 39
26	06 17 17	04 35 21	03 ♓ 52 53	10 ♓ 38 31	29 ♋ 55	12 32	21 14	22 26	04 23	19 47	04 26	04 38
27	06 21 13	05 32 34	17 ♓ 16 55	23 ♓ 48 26	00 ♌ 48	13 42	21 02	22 29	04 19	19 44	04 24	04 37
28	06 25 10	06 29 46	00 ♈ 13 31	06 ♈ 32 43	01 38	14 51	20 48	22 32	04 16	19 42	04 22	04 37
29	06 29 06	07 26 59	12 ♈ 46 37	18 ♈ 55 52	02 25	16 01	20 34	22 34	04 12	19 40	04 21	04 36
30	06 33 03	08 ♋ 24 11	25 ♈ 01 07	01 ♉ 03 00	03 ♌ 07	17 ♌ 10	20 ♑ 20	22 ♓ 37	04 ♐ 09	19 ♐ 37	04 ♑ 19	04 ♏ 36

Date	Moon True ☊ ° '	Moon Mean ☊ ° '	Moon ☽ Latitude ° '	DECLINATIONS Sun ☉ ° '	Moon ☽ ° '	Mercury ☿ ° '	Venus ♀ ° '	Mars ♂ ° '	Jupiter ♃ ° '	Saturn ♄ ° '	Uranus ♅ ° '	Neptune ♆ ° '	Pluto ♇ ° '
01	29 ♈ 20	27 ♈ 48	02 S 14	22 N 03	00 S 33	24 N 55	24 N 34	24 S 23	04 S 53	19 S 20	23 S 13	22 S 14	02 N 50
02	29 D 21	27 45	01 12	22 11	05 N 13	25 07	24 28	24 26	04 50	19 19	23 13	22 14	02 50
03	29 22	27 42	00 S 07	22 19	10 42	25 16	24 21	24 29	04 47	19 19	23 12	22 14	02 50
04	29 R 21	27 39	00 N 58	22 26	15 45	25 23	24 14	24 33	04 45	19 18	23 12	22 14	02 50
05	29 19	27 35	01 59	22 33	20 10	25 27	24 06	24 37	04 42	19 17	23 12	22 14	02 50
06	29 14	27 32	02 55	22 39	23 45	25 29	23 57	24 40	04 40	19 17	23 12	22 15	02 50
07	29 07	27 29	03 43	22 45	26 21	25 28	23 47	24 44	04 38	19 16	23 12	22 15	02 50
08	28 59	27 26	04 22	22 51	27 47	25 25	23 37	24 48	04 36	19 15	23 12	22 15	02 50
09	28 50	27 23	04 49	22 56	27 55	25 20	23 26	24 53	04 33	19 15	23 11	22 15	02 50
10	28 40	27 19	05 04	23 01	26 45	25 13	23 15	24 57	04 31	19 14	23 11	22 15	02 50
11	28 32	27 16	05 05	23 05	24 19	25 05	23 03	25 02	04 29	19 13	23 11	22 15	02 50
12	28 25	27 13	04 52	23 09	20 46	24 54	22 50	25 07	04 27	19 13	23 11	22 15	02 50
13	28 20	27 10	04 26	23 13	16 14	24 42	22 37	25 12	04 25	19 12	23 11	22 15	02 49
14	28 17	27 07	03 46	23 16	10 54	24 28	22 23	25 17	04 24	19 12	23 11	22 15	02 49
15	28 D 17	27 04	02 53	23 19	04 N 59	24 13	22 08	25 22	04 22	19 11	23 10	22 15	02 49
16	28 17	27 00	01 49	23 21	01 S 19	23 57	21 53	25 27	04 20	19 10	23 10	22 15	02 49
17	28 18	26 57	00 N 37	23 23	07 46	23 40	21 37	25 32	04 19	19 10	23 10	22 15	02 49
18	28 R 18	26 54	00 S 39	23 24	14 01	23 21	21 21	25 38	04 17	19 09	23 10	22 15	02 49
19	28 16	26 51	01 55	23 26	19 40	23 02	21 04	25 44	04 16	19 09	23 10	22 15	02 48
20	28 12	26 48	03 04	23 26	24 12	22 42	20 47	25 49	04 14	19 08	23 10	22 15	02 48
21	28 06	26 45	04 01	23 27	27 07	22 21	20 29	25 55	04 13	19 07	23 09	22 15	02 48
22	27 57	26 41	04 41	23 26	28 00	21 59	20 10	26 01	04 11	19 07	23 09	22 15	02 48
23	27 48	26 38	05 01	23 26	26 48	21 37	19 52	26 07	04 10	19 06	23 09	22 15	02 47
24	27 39	26 35	05 01	23 25	23 43	21 15	19 32	26 13	04 09	19 06	23 09	22 16	02 47
25	27 31	26 32	04 41	23 24	19 15	20 52	19 12	26 19	04 08	19 05	23 09	22 16	02 47
26	27 26	26 29	04 05	23 22	13 53	20 30	18 52	26 25	04 07	19 05	23 09	22 16	02 47
27	27 22	26 25	03 16	23 20	08 02	20 07	18 31	26 31	04 06	19 04	23 08	22 16	02 46
28	27 21	26 22	02 19	23 17	02 S 02	19 44	18 10	26 37	04 05	19 04	23 08	22 16	02 46
29	27 D 21	26 19	01 17	23 14	03 N 52	19 21	17 48	26 43	04 05	19 04	23 08	22 16	02 46
30	27 ♈ 21	26 ♈ 16	00 S 12	23 N 11	09 N 30	18 N 59	17 N 26	26 S 49	04 S 04	19 S 03	23 S 08	22 S 16	02 N 45

ZODIAC SIGN ENTRIES

Date	h	m	Planets
01	04	43	☽ ♈
03	15	45	☽ ♉
05	14	06	☿ ♋
06	04	26	☽ ♊
08	17	16	☽ ♋
11	05	11	☽ ♌
13	15	18	☽ ♍
15	18	52	♀ ♌
15	22	38	☽ ♎
18	02	36	☽ ♏
20	03	36	☽ ♐
21	16	30	☉ ♋
22	03	00	☽ ♑
24	02	50	☽ ♒
26	05	12	☽ ♓
26	14	15	☿ ♌
28	11	35	☽ ♈
30	21	54	☽ ♉

LATITUDES

Date	Mercury ☿ ° '	Venus ♀ ° '	Mars ♂ ° '	Jupiter ♃ ° '	Saturn ♄ ° '	Uranus ♅ ° '	Neptune ♆ ° '	Pluto ♇ ° '
01	01 N 43	01 N 47	02 S 54	01 S 06	02 N 01	00 S 05	01 N 07	16 N 58
04	01 57	01 50	03 06	01 07	02 01	00 05	01 07	16 57
07	02 04	01 53	03 19	01 07	02 01	00 05	01 07	16 56
10	02 05	01 55	03 32	01 08	02 01	00 05	01 07	16 55
13	01 58	01 57	03 46	01 09	02 00	00 05	01 07	16 53
16	01 45	01 57	03 59	01 10	02 00	00 06	01 07	16 52
19	01 26	01 57	04 12	01 11	01 59	00 06	01 07	16 51
22	01 01	01 56	04 26	01 12	01 59	00 06	01 07	16 49
25	00 N 31	01 54	04 38	01 13	01 58	00 06	01 07	16 48
28	00 S 04	01 52	04 51	01 13	01 58	00 06	01 06	16 46
31	00 S 44	01 N 48	05 S 02	01 S 14	01 N 57	00 S 06	01 N 06	16 N 44

DATA

Julian Date	2446583
Delta T	+55 seconds
Ayanamsa	23° 39' 54"
Synetic vernal point	05° ♓ 27' 06"
True obliquity of ecliptic	23° 26' 36"

MOON'S PHASES, APSIDES AND POSITIONS ☽

Date	h	m	Phase	Longitude ° '	Eclipse Indicator
07	14	00	●	16 ♊ 32	
15	12	00	☽	24 ♍ 06	
22	03	42	○	00 ♑ 27	
29	00	53	☾	07 ♈ 00	

Day	h	m		
07	02	11	Apogee	
21	12	54	Perigee	
01	14	17	0N	
09	02	38	Max dec	28° N 01'
16	07	03	0S	
22	10	04	Max dec	28° S 01'
28	20	11	0N	

LONGITUDES

Date	Chiron ⚷ ° '	Ceres ⚳ ° '	Pallas ⚴ ° '	Juno ⚵ ° '	Vesta ⚶ ° '	Black Moon Lilith ⚸ ° '
01	14 ♊ 46	09 ♍ 01	06 ♌ 02	08 ♐ 32	00 ♈ 53	10 ♊ 27
11	15 ♊ 36	11 ♍ 48	10 ♌ 49	06 ♐ 28	04 ♈ 28	11 ♊ 34
21	16 ♊ 25	14 ♍ 53	15 ♌ 37	04 ♐ 19	07 ♈ 46	12 ♊ 41
31	17 ♊ 13	18 ♍ 14	20 ♌ 26	02 ♐ 42	10 ♈ 41	13 ♊ 48

ASPECTARIAN

01 Sunday
01 24 ☽ Q ☉
02 41 ☽ # ♇
02 53 ☽ ± ♇
13 56 ☽ Q ♂
14 26 ☽ ⚻ ♇
14 30 ☽ □ ♆
16 26 ☽ △ ♄
20 19 ☉ ± ♇
22 11 ☿ ⚻ ♂

02 Monday
01 59 ☽ ∥ ♇
02 22 ☽ Q ☿
02 37 ☽ ⚹ ☉
08 05 ☽ □ ♀
10 23 ☽ # ♃
17 41 ♀ # ♂
20 36 ☽ ⊻ ♃
21 16 ☽ △ ♅
21 44 ☽ ⚼ ♄
21 46 ☉ # ♆

03 Tuesday
01 33 ☽ □ ♂
06 49 ☽ ⚹ ☿
08 40 ☽ ⊥ ♃
10 53 ☽ ∠ ☉
15 36 ☽ ± ♄
23 53 ☽ Q ♀

04 Wednesday
01 46 ☽ ☍ ♇
01 50 ☽ △ ♆
02 54 ☽ ∠ ♃
03 11 ☽ ⚼ ♅
03 35 ☽ ⚻ ♄
06 33 ☽ ⊥ ☉
18 54 ☽ ∠ ☿
19 44 ☽ ⊻ ☉
21 18 ☽ ± ♅

05 Thursday
00 13 ♃ □ ♅
02 49 ☽ ⚹ ♀
06 56 ☽ # ♄
08 06 ☽ ⚼ ♆
09 25 ☽ ⚻ ♅
09 32 ☽ ⚹ ♃
14 18 ☽ △ ♂
16 40 ☽ ⊥ ☿

06 Friday
01 08 ☽ # ♆
02 20 ☽ ± ♆
03 41 ☽ ∥ ☉
07 09 ☽ ⊻ ☿
07 50 ☽ # ♅
10 09 ☽ Q ♃
12 41 ☽ ∠ ♀
13 24 ☽ ∥ ♀
14 28 ☽ ⚻ ♇
14 30 ☽ ⚻ ♆
16 02 ☽ ☍ ♄
19 34 ☽ # ♂
20 52 ☽ ⚼ ♂

07 Saturday
02 38 ☽ ± ♇
02 40 ☽ ∥ ☿
09 00 ☽ ⊥ ♀
14 00 ☽ ☌ ☉
15 10 ☽ ± ♂
18 38 ♀ ⚻ ♅
20 14 ♀ ⚼ ♄
20 52 ☽ ⚼ ♇
22 07 ☽ ☍ ♅
22 31 ☽ ⊻ ♀
23 01 ☽ □ ♃

08 Sunday
03 20 ☽ ⚻ ♂
03 50 ♀ △ ♃
04 36 ☉ ± ♂
06 05 ☿ △ ♇
06 11 ☿ ☍ ♆
14 57 ☿ ⚻ ♄
23 25 ♂ St R

09 Monday
03 07 ☽ △ ♇
03 07 ☽ ☍ ♆
04 23 ☽ ⚻ ♄
06 47 ☽ ☌ ☿
08 02 ♆ ⚹ ♇
16 21 ☽ ± ♄
22 41 ♀ ☍ ♂

10 Tuesday
07 40 ☽ ⊻ ☉
10 13 ☽ ⚻ ♅
10 13 ☽ ⚼ ♄
11 49 ☽ △ ♃
15 31 ☽ ☍ ♂
17 22 ☽ ☌ ♀
19 17 ♀ # ♅
20 36 ☽ ⊥ ☉
22 05 ☽ ± ♅

11 Wednesday
01 26 ☉ ⚼ ♇
05 19 ☽ ∥ ☿
06 16 ☽ # ♂
08 25 ☉ ∥ ♀
14 41 ☽ ⚻ ♆
14 41 ☉ ☍ ♅
14 43 ☽ □ ♇
15 42 ☽ △ ♄
15 46 ☽ ⚼ ♅
15 52 ☽ ∠ ☉
16 27 ☿ # ♂
17 40 ☽ ⚼ ♃
20 31 ☽ # ♅
21 01 ☽ ∥ ☉
22 05 ☽ ∥ ♀

12 Thursday
00 31 ☿ ± ♄
02 23 ☽ ± ♆
02 50 ☽ # ♆
03 48 ☽ ⊻ ☿
11 20 ☽ ± ♃
16 15 ♀ ± ♅
17 10 ☽ ⊥ ☿
17 16 ☉ □ ♃
19 47 ☽ ⚼ ♆
20 47 ☽ # ♄
20 49 ☽ △ ♅
22 56 ☉ # ♅
23 00 ☽ ⚻ ♃
23 26 ☽ ⚹ ☉

13 Friday
01 36 ☽ Q ♇
01 58 ☽ ⚻ ♂
10 00 ☽ ⊻ ♀
12 52 ☽ ∠ ☿
13 22 ☽ ± ♂
22 31 ☽ ⊥ ♀

14 Saturday
00 07 ☽ Q ☉
00 17 ☽ △ ♆
00 22 ☽ ⚹ ♇
01 04 ☽ □ ♄
06 15 ☽ ⚼ ♂
06 52 ☉ ⚻ ♂
17 01 ☽ ∠ ♀
20 44 ☽ ⚹ ☿

15 Sunday
00 50 ♀ # ♆
04 09 ☽ ∠ ♇
04 58 ☽ □ ♅
07 35 ☽ ☍ ♃
09 45 ☽ △ ♂
10 10 ☽ Q ♄
12 00 ☽ □ ☉
14 26 ☽ # ♃
20 22 ☽ ∥ ♇
20 26 ☽ ⊥ ♇
21 17 ☽ Q ☿
23 00 ☽ ⚹ ♀

16 Monday
07 00 ☽ □ ♆
07 07 ☽ ⊻ ♇
07 34 ☽ ⚹ ♄
13 08 ☽ Q ♅
17 34 ☽ # ♇
22 09 ☽ Q ♀
23 11 ☽ ∥ ♃

17 Tuesday
08 30 ☽ □ ☿
09 34 ☽ ∠ ♄
09 54 ☽ ⚹ ♅
10 49 ☉ ⊥ ♀
12 49 ☽ ⚻ ♃
14 11 ☽ □ ♂
14 13 ☽ Q ♆
18 59 ☿ ⚼ ♄
20 27 ☽ △ ☉
22 29 ☿ ⚻ ♅
23 02 ☽ ± ♃

18 Wednesday
00 46 ☽ ⊥ ♄
07 35 ☽ □ ♀
08 14 ☉ ∥ ☿
10 22 ☽ ⚹ ♆
10 30 ☽ ☌ ♇
10 45 ☽ ⊻ ♄
11 07 ☽ ∠ ♅
14 09 ☽ ⚼ ♃
20 03 ☽ Q ♂
23 10 ☽ ⚼ ☉

19 Thursday
01 54 ☽ ⊥ ♅
02 24 ☿ # ♅
08 22 ☿ △ ♃
09 37 ☽ ∥ ♄
10 54 ☽ ∠ ♆
11 37 ☽ ⊻ ♅
14 46 ☽ △ ♃
14 48 ☽ ± ☉
15 20 ☽ △ ☿
15 25 ☽ ⚹ ♂
16 08 ☿ ☍ ♂
17 05 ♀ ⚻ ♆
18 21 ☽ # ♀
19 13 ♀ □ ♇
20 42 ♀ △ ♄

20 Friday
00 47 ☽ ∥ ♆
01 05 ☽ ⚻ ☉
01 22 ☽ ⊥ ♆
01 46 ♀ ⚼ ♅
03 49 ☽ # ☿
05 46 ☽ ∥ ♅
07 22 ☽ # ☉
10 54 ☽ ⊻ ♆
11 05 ☽ ⊻ ♇
11 09 ☽ ☌ ♄
12 26 ☽ △ ♀
15 09 ☽ ∠ ♂
17 27 ☽ ⚼ ☿
20 35 ☽ ⊥ ♇
23 50 ☽ ∥ ♂

21 Saturday
05 14 ☽ ⊥ ♂
07 05 ♄ ⊻ ♇
08 52 ☽ ± ☿
10 46 ☽ ∠ ♇
11 13 ☽ ☌ ♅
12 26 ♂ ⚹ ♃
14 06 ☽ ⚼ ♀
14 36 ☽ ⊻ ♂
14 38 ☽ □ ♃
18 16 ☿ # ♆
19 07 ☽ ⚻ ☿
21 24 ♀ ⚼ ♃

22 Sunday
03 42 ☽ ☍ ☉
05 27 ☽ ± ♀
10 09 ☽ ☌ ♆
10 17 ☽ ⊻ ♄
10 22 ☽ ⚹ ♇
14 46 ☿ ± ♅
15 42 ☽ ⚻ ♀
19 08 ☽ Q ♃
19 44 ☽ ⊥ ♄

23 Monday
05 23 ☽ Q ♇
10 00 ☽ ∠ ♄
10 33 ☽ ⊻ ♅
13 35 ☽ ☌ ♂
14 21 ☽ ⚹ ♃
18 23 ☽ ∥ ♂
20 10 ☽ ⊥ ♅
22 35 ☽ ☍ ☿

24 Tuesday
06 52 ☽ ⚻ ☉
10 09 ☽ ⊻ ♆
10 11 ☽ ⚹ ♄
10 26 ☽ □ ♇
10 46 ☽ ∠ ♅
13 54 ☽ # ☉
14 50 ☽ ∠ ♃
15 30 ☽ ∥ ♅
17 25 ☽ ± ☉
17 34 ♀ ± ♆
20 02 ☽ ⊥ ♆
20 15 ☽ ☍ ♀
20 35 ☽ ∥ ♆
21 49 ♄ ⊻ ♆

25 Wednesday
03 16 ☽ # ☿
05 53 ☽ ⊥ ♃
06 01 ☽ Q ♄
09 33 ☽ ⚼ ☉
11 06 ☽ ∠ ♆
11 43 ☽ ⚹ ♅
12 16 ☽ # ♀
12 47 ☽ ∥ ♄
14 26 ☽ ⊻ ♂
16 05 ☽ ⊻ ♃
20 12 ♀ # ♄

26 Thursday
00 35 ☽ ⊥ ♂
04 33 ☽ ⚻ ☿
06 57 ☉ ⚻ ♄
08 03 ☉ ☍ ♆
08 19 ☽ Q ♅
12 52 ☽ □ ♄
12 58 ☽ ⚹ ♆
13 08 ☉ △ ♇
13 20 ☽ △ ♇
13 20 ☽ △ ☉
15 51 ☽ ± ☿
16 06 ☽ ∠ ♂

27 Friday
04 52 ☽ ⚻ ♀
09 07 ☽ ⚼ ☿
10 24 ☽ Q ♆
16 17 ☽ ⚼ ♇
16 29 ☽ □ ♅
16 51 ☽ ± ♀
18 45 ☽ ⚹ ♂
21 35 ☽ ☌ ♃

28 Saturday
03 46 ☽ ∥ ♃
08 58 ☽ ± ♇
09 04 ☽ # ♇
11 14 ☽ ⚼ ♀
14 51 ☽ △ ☿
16 47 ☽ Q ♂
19 36 ☽ △ ♄
19 50 ☽ □ ♆
20 19 ☽ ⚻ ♇

29 Sunday
00 53 ☽ □ ☉
07 26 ☽ ∥ ♇
12 51 ☽ # ♃
18 57 ☽ △ ♀

30 Monday
00 04 ♀ ± ♃
00 29 ☽ ⚼ ♄
01 24 ☽ △ ♅
02 56 ☽ □ ♂
06 57 ☿ # ♄
07 14 ☽ ⊻ ♃
14 59 ☽ Q ☉
18 11 ☽ ± ♄
19 10 ☽ ⊥ ♃

All ephemeris data is given at 12.00 UT and the Moon's longitude is additionally given for 24.00 UT

JULY 1986

LONGITUDES

Date	Sidereal time h m s	Sun ☉ ° ' "	Moon ☽ ° ' "	Moon ☽ 24.00 ° ' "	Mercury ☿ ° '	Venus ♀ ° '	Mars ♂ ° '	Jupiter ♃ ° '	Saturn ♄ ° '	Uranus ♅ ° '	Neptune ♆ ° '	Pluto ♇ ° '
01	06 36 59	09 ♋ 21 24	07 ♉ 02 10	12 ♉ 59 13	03 ♌ 46	18 ♌ 20	20 ♑ 05	22 ♓ 39	04 ♐ 06	19 ♐ 35	04 ♑ 18	04 ♏ 35
02	06 40 56	10 18 37	18 ♉ 54 45	24 ♉ 49 17	04 21	19 29	19 R 50	22 41	04 R 03	19 R 33	04 R 16	04 R 35
03	06 44 53	11 15 50	00 ♊ 43 19	06 ♊ 37 20	04 52	20 38	19 34	22 43	03 59	19 31	04 14	04 35
04	06 48 49	12 13 04	12 ♊ 31 41	18 ♊ 26 46	05 19	21 47	19 18	22 45	03 56	19 28	04 13	04 34
05	06 52 46	13 10 17	24 ♊ 22 52	00 ♋ 20 14	05 41	22 56	19 02	22 46	03 53	19 26	04 11	04 34
06	06 56 42	14 07 31	06 ♋ 19 07	12 ♋ 19 40	05 59	24 05	18 45	22 48	03 51	19 24	04 10	04 33
07	07 00 39	15 04 44	18 ♋ 22 04	24 ♋ 26 26	06 12	25 14	18 28	22 49	03 48	19 22	04 08	04 33
08	07 04 35	16 01 58	00 ♌ 32 52	06 ♌ 41 30	06 21	26 22	18 11	22 50	03 45	19 20	04 06	04 33
09	07 08 32	16 59 12	12 ♌ 52 26	19 ♌ 05 48	06 25	27 31	17 54	22 50	03 42	19 18	04 05	04 33
10	07 12 28	17 56 25	25 ♌ 21 43	01 ♍ 40 23	06 R 24	28 39	17 36	22 51	03 40	19 16	04 03	04 33
11	07 16 25	18 53 39	08 ♍ 01 58	14 ♍ 26 41	06 19	29 ♌ 48	17 19	22 51	03 37	19 14	04 02	04 32
12	07 20 22	19 50 53	20 ♍ 54 48	27 ♍ 26 35	06 09	00 ♍ 56	17 01	22 51	03 35	19 12	04 00	04 32
13	07 24 18	20 48 06	04 ♎ 02 19	10 ♎ 42 18	05 54	02 04	16 44	22 R 51	03 33	19 10	03 59	04 32
14	07 28 15	21 45 20	17 ♎ 26 49	24 ♎ 16 08	05 35	03 12	16 26	22 51	03 30	19 08	03 57	04 32
15	07 32 11	22 42 33	01 ♏ 10 27	08 ♏ 09 56	05 11	04 20	16 09	22 51	03 28	19 06	03 56	04 D 32
16	07 36 08	23 39 47	15 ♏ 14 37	22 ♏ 24 24	04 44	05 27	15 52	22 50	03 26	19 04	03 54	04 32
17	07 40 04	24 37 01	29 ♏ 39 06	06 ♐ 58 18	04 13	06 35	15 35	22 49	03 24	19 02	03 53	04 32
18	07 44 01	25 34 15	14 ♐ 21 27	21 ♐ 47 49	03 38	07 42	15 18	22 48	03 22	19 00	03 51	04 32
19	07 47 57	26 31 29	29 ♐ 16 29	06 ♑ 46 26	03 01	08 50	15 02	22 47	03 21	18 58	03 50	04 33
20	07 51 54	27 28 44	14 ♑ 16 29	21 ♑ 45 28	02 22	09 57	14 46	22 46	03 19	18 57	03 48	04 33
21	07 55 51	28 25 59	29 ♑ 12 10	06 ♒ 35 26	01 41	11 04	14 30	22 44	03 17	18 55	03 47	04 33
22	07 59 47	29 ♋ 23 14	13 ♒ 54 14	21 ♒ 07 39	00 59	12 10	14 15	22 42	03 16	18 53	03 45	04 33
23	08 03 44	00 ♌ 20 30	28 ♒ 15 00	05 ♓ 15 45	00 ♌ 17	13 17	14 01	22 40	03 14	18 52	03 44	04 33
24	08 07 40	01 17 47	12 ♓ 09 34	18 ♓ 56 20	29 ♋ 36	14 24	13 46	22 38	03 13	18 50	03 43	04 34
25	08 11 37	02 15 05	25 ♓ 36 05	02 ♈ 09 04	28 56	15 30	13 33	22 36	03 12	18 48	03 41	04 34
26	08 15 33	03 12 23	08 ♈ 35 35	14 ♈ 56 07	28 18	16 36	13 19	22 33	03 10	18 47	03 40	04 34
27	08 19 30	04 09 42	21 ♈ 11 11	27 ♈ 21 24	27 43	17 42	13 07	22 30	03 09	18 45	03 38	04 35
28	08 23 26	05 07 03	03 ♉ 27 24	09 ♉ 29 52	27 11	18 48	12 55	22 27	03 08	18 44	03 37	04 35
29	08 27 23	06 04 24	15 ♉ 29 27	21 ♉ 26 49	26 44	19 54	12 44	22 24	03 07	18 43	03 36	04 36
30	08 31 20	07 01 46	27 ♉ 22 39	03 ♊ 17 33	26 23	20 59	12 33	22 21	03 07	18 41	03 34	04 36
31	08 35 16	07 ♌ 59 10	09 ♊ 12 07	15 ♊ 06 54	26 ♋ 03	22 ♍ 05	12 ♑ 23	22 ♓ 17	03 ♐ 06	18 ♐ 40	03 ♑ 33	04 ♏ 37

MOON

Date	Moon True ☊ ° '	Moon Mean ☊ ° '	Moon ☽ Latitude ° '
01	27 ♈ 21	26 ♈ 13	00 N 51
02	27 R 19	26 10	01 52
03	27 14	26 06	02 47
04	27 07	26 03	03 35
05	26 57	26 00	04 14
06	26 46	25 57	04 42
07	26 32	25 54	04 57
08	26 18	25 51	04 59
09	26 07	25 47	04 48
10	25 57	25 44	04 22
11	25 49	25 41	03 43
12	25 44	25 38	02 53
13	25 42	25 35	01 52
14	25 D 41	25 31	00 N 43
15	25 R 41	25 28	00 S 29
16	25 40	25 25	01 41
17	25 38	25 22	02 49
18	25 33	25 19	03 47
19	25 25	25 16	04 30
20	25 15	25 12	04 55
21	25 04	25 09	05 00
22	24 53	25 06	04 44
23	24 44	25 03	04 11
24	24 36	25 00	03 24
25	24 32	24 56	02 26
26	24 29	24 53	01 23
27	24 28	24 50	00 S 17
28	24 D 28	24 47	00 N 47
29	24 R 28	24 44	01 49
30	24 26	24 41	02 45
31	24 ♈ 23	24 ♈ 37	03 N 33

DECLINATIONS

Date	Sun ☉ ° '	Moon ☽ ° '	Mercury ☿ ° '	Venus ♀ ° '	Mars ♂ ° '	Jupiter ♃ ° '	Saturn ♄ ° '	Uranus ♅ ° '	Neptune ♆ ° '	Pluto ♇ ° '
01	23 N 07	14 N 40	18 N 36	17 N 03	26 S 55	04 S 03	19 S 03	23 S 08	22 S 16	02 N 45
02	23 03	19 14	18 15	16 40	27 01	04 03	19 02	23 08	22 16	02 45
03	22 58	23 02	17 53	16 17	27 07	04 02	19 02	23 07	22 16	02 44
04	22 53	25 51	17 33	15 53	27 13	04 02	19 02	23 07	22 16	02 44
05	22 47	27 33	17 13	15 29	27 19	04 01	19 01	23 07	22 16	02 43
06	22 42	27 59	16 54	15 04	27 24	04 01	19 01	23 07	22 16	02 43
07	22 35	27 05	16 36	14 39	27 30	04 01	19 01	23 07	22 16	02 42
08	22 29	24 54	16 19	14 14	27 35	04 01	19 00	23 07	22 16	02 42
09	22 22	21 33	16 04	13 49	27 40	04 01	19 00	23 06	22 17	02 42
10	22 14	17 11	15 50	13 23	27 45	04 01	19 00	23 06	22 17	02 41
11	22 07	12 01	15 37	12 57	27 50	04 01	18 59	23 06	22 17	02 41
12	21 59	06 15	15 25	12 30	27 55	04 01	18 59	23 06	22 17	02 40
13	21 50	00 N 06	15 16	12 03	28 00	04 02	18 59	23 06	22 17	02 40
14	21 41	06 S 11	15 08	11 37	28 04	04 02	18 59	23 06	22 17	02 39
15	21 32	12 20	15 01	11 09	28 08	04 03	18 59	23 05	22 17	02 38
16	21 22	18 02	14 57	10 42	28 12	04 03	18 58	23 05	22 17	02 38
17	21 12	22 50	14 54	10 14	28 16	04 04	18 58	23 05	22 17	02 37
18	21 02	26 17	14 53	09 46	28 19	04 04	18 58	23 05	22 17	02 37
19	20 51	27 56	14 54	09 18	28 22	04 05	18 58	23 05	22 17	02 36
20	20 40	27 34	14 56	08 50	28 25	04 06	18 58	23 05	22 17	02 36
21	20 29	25 12	15 01	08 21	28 28	04 07	18 58	23 05	22 17	02 35
22	20 17	21 11	15 07	07 53	28 30	04 08	18 58	23 04	22 17	02 34
23	20 05	16 00	15 14	07 24	28 33	04 09	18 58	23 04	22 17	02 34
24	19 52	10 08	15 23	06 55	28 35	04 10	18 58	23 04	22 18	02 33
25	19 40	03 S 59	15 33	06 26	28 37	04 11	18 58	23 04	22 18	02 32
26	19 27	02 N 08	15 45	05 57	28 38	04 13	18 58	23 04	22 18	02 32
27	19 13	08 00	15 57	05 27	28 40	04 14	18 58	23 04	22 18	02 31
28	18 59	13 25	16 10	04 58	28 41	04 15	18 58	23 04	22 18	02 30
29	18 45	18 13	16 24	04 28	28 42	04 17	18 58	23 04	22 18	02 30
30	18 31	22 15	16 38	03 59	28 42	04 18	18 58	23 03	22 18	02 29
31	18 N 16	25 N 21	16 N 52	03 N 29	28 S 43	04 S 20	18 S 58	23 S 03	22 S 18	02 N 28

ZODIAC SIGN ENTRIES

Date	h	m	Planets
03	10	32	☽ ♊
05	23	19	☽ ♋
08	10	56	☽ ♌
10	20	50	☽ ♍
11	16	23	♀ ♍
13	04	40	☽ ♎
15	09	58	☽ ♏
17	12	34	☽ ♐
19	13	10	☽ ♑
21	13	17	☽ ♒
23	03	24	☉ ♌
23	14	59	☽ ♓
23	21	51	☿ ♋
25	20	02	☽ ♈
28	05	11	☽ ♉
30	17	19	☽ ♊

LATITUDES

Date	Mercury ☿ ° '	Venus ♀ ° '	Mars ♂ ° '	Jupiter ♃ ° '	Saturn ♄ ° '	Uranus ♅ ° '	Neptune ♆ ° '	Pluto ♇ ° '
01	00 S 44	01 N 48	05 S 02	01 S 14	01 N 57	00 S 06	01 N 06	16 N 44
04	01 26	01 44	05 13	01 15	01 57	00 06	01 06	16 43
07	02 11	01 38	05 23	01 16	01 56	00 06	01 06	16 41
10	02 56	01 32	05 31	01 17	01 56	00 06	01 06	16 39
13	03 39	01 25	05 39	01 18	01 55	00 06	01 06	16 38
16	04 16	01 16	05 44	01 19	01 54	00 06	01 06	16 36
19	04 43	01 07	05 49	01 20	01 54	00 06	01 06	16 34
22	04 57	00 57	05 52	01 21	01 53	00 06	01 06	16 32
25	04 55	00 47	05 53	01 22	01 53	00 06	01 06	16 31
28	04 39	00 35	05 53	01 22	01 52	00 06	01 06	16 29
31	04 S 09	00 N 22	05 S 52	01 S 23	01 N 51	00 S 06	01 N 06	16 N 27

DATA

Julian Date	2446613
Delta T	+55 seconds
Ayanamsa	23° 39' 59"
Synetic vernal point	05° ♓ 27' 00"
True obliquity of ecliptic	23° 26' 35"

MOON'S PHASES, APSIDES AND POSITIONS ☽

Date	h	m	Phase	Longitude ° '	Eclipse Indicator
07	04	55	●	14 ♋ 48	
14	20	10	☽	22 ♎ 05	
21	10	40	○	28 ♑ 23	
28	15	34	☾	05 ♉ 16	

Day	h	m		
04	08	14	Apogee	
19	19	37	Perigee	
31	21	27	Apogee	
06	07	52	Max dec	28° N 00'
13	12	24	0S	
19	19	38	Max dec	28° S 03'
26	03	33	0N	

LONGITUDES

Date	Chiron ⚷ ° '	Ceres ⚳ ° '	Pallas ⚴ ° '	Juno ⚵ ° '	Vesta ⚶ ° '	Black Moon Lilith ⚸ ° '
01	17 ♊ 13	18 ♍ 14	20 ♌ 26	02 ♐ 42	10 ♈ 41	13 ♊ 48
11	17 ♊ 59	21 ♍ 48	25 ♌ 42	01 ♐ 34	13 ♈ 12	15 ♊ 55
21	18 ♊ 43	25 ♍ 33	00 ♍ 01	01 ♐ 59	15 ♈ 12	16 ♊ 11
31	19 ♊ 22	29 ♍ 26	04 ♍ 48	00 ♐ 58	16 ♈ 38	17 ♊ 08

All ephemeris data is given at 12.00 UT and the Moon's longitude is additionally given for 24.00 UT

ASPECTARIAN

(h m — Aspects)

01 Tuesday
05 05 ☽ □ ☿
06 07 ☽ ⚻ ♄
06 18 ☉ # ♅
06 30 ☽ △ ♆
07 05 ☽ ☍ ♇
07 06 ☽ ⚼ ♅
08 54 ♅ ∠ ♇
13 15 ☽ ∠ ♃
17 05 ☽ ⚹ ☉
23 07 ☽ ∥ ♀
23 54 ☿ △ ♄

02 Wednesday
01 10 ☽ ± ♅
05 37 ☉ ± ♄
06 51 ☽ ∥ ☿
07 39 ♀ ⚼ ♆
08 22 ☿ ⚻ ♆
10 51 ☽ # ♄
12 43 ☽ ⚼ ♆
13 17 ☽ □ ♀
13 17 ☽ ⚻ ♅
13 22 ♀ △ ♅
13 50 ☽ △ ♂
18 00 ♀ ⚻ ♂
19 19 ☽ Q ☿
19 41 ☽ ⚹ ♃
20 06 ☿ ⚼ ♅
22 07 ☿ □ ♇

03 Thursday
02 08 ☽ ∠ ☉
06 42 ☽ # ♆
06 58 ☽ ± ♆
11 33 ☽ ∥ ☉
12 40 ☽ # ♅
18 20 ♂ ⚺ ♅
18 37 ☽ ☍ ♄
19 08 ☽ ⚻ ♆
19 39 ☽ ⚼ ♂
19 50 ☽ ⚻ ♇
20 09 ☽ Q ♃
20 47 ☽ ⚹ ☿
22 03 ☽ ⊥ ☉

04 Friday
05 49 ☽ Q ♀
08 01 ☽ ± ♇
11 19 ☽ ⚺ ☉
13 32 ☽ ± ♂

05 Saturday
01 26 ☽ ⚻ ♂
02 02 ☽ ☍ ♅
02 16 ☽ ⚼ ♇
04 18 ☽ ∠ ☿
04 19 ♀ Q ♇
06 52 ☽ # ♂
08 34 ♀ ⚻ ♃
08 45 ☽ □ ♃
08 46 ☽ ⚹ ♀
22 59 ☽ ⊥ ☿

06 Sunday
02 25 ♂ ∠ ♄
07 04 ☽ ⚻ ♄
07 41 ☽ ☍ ♆
08 29 ☽ △ ♇
11 19 ☽ ⚺ ☿
18 06 ☽ ∠ ♀
19 01 ☽ ± ♄
23 15 ♀ ± ♂

07 Monday
04 55 ☽ ☌ ☉
05 18 ☽ # ♂
12 12 ☽ ☍ ♂
12 51 ☽ ⚼ ♄
13 53 ☽ ⊥ ♀
13 58 ☽ ⚻ ♅
20 48 ☽ △ ♃

08 Tuesday
01 47 ☽ ± ♅
02 57 ☽ ⚺ ♀
18 15 ☽ △ ♄
18 57 ☽ ⚻ ♆
19 23 ☽ ⚼ ♅
19 49 ☽ □ ♇
23 25 ☽ ☌ ☿

09 Wednesday
01 48 ☽ # ♅
02 14 ☽ ⚼ ♃
06 36 ☽ ± ♆
06 38 ☽ ∥ ☉
07 23 ☽ # ♆
19 40 ☽ ± ♃
20 27 ☿ St R
20 36 ☽ ⚺ ☉
21 28 ☽ ⚻ ♂
23 57 ☽ ⚼ ♆

10 Thursday
00 21 ☽ △ ♅
02 36 ☽ # ♄
05 15 ☉ # ♆
05 28 ☉ ☍ ♂
06 37 ☽ Q ♇
07 12 ☽ ⚻ ♃
08 43 ☽ ± ♂
09 04 ☽ ⊥ ☉
18 54 ☽ ☌ ♀
18 55 ☽ ∥ ☿

11 Friday
01 27 ☽ ⚼ ♂
03 34 ☽ ∠ ☉
03 43 ☽ □ ♄
04 28 ☽ △ ♆
05 25 ☽ ⚹ ♇
05 27 ☉ ⚼ ♄
07 31 ☽ ∥ ♀
08 48 ☽ ⚺ ☿
19 56 ☽ ⊥ ☿
20 06 ☉ ⚻ ♅

12 Saturday
04 57 ☽ △ ♂
08 50 ☽ □ ♅
09 28 ☽ ∠ ♇
09 52 ☽ ⚹ ☉
12 25 ☽ ∠ ☿
13 14 ☽ Q ♄
15 35 ☽ ☍ ♃
17 01 ♃ St R
20 48 ☽ # ♃

13 Sunday
02 00 ☽ ⊥ ♇
02 07 ☽ ∥ ♇
06 18 ♀ ⚼ ♂
08 05 ☽ ⚺ ♀
09 35 ☽ Q ☉
11 06 ☽ ⚹ ♄
11 53 ☽ □ ♆
12 54 ☽ ⚺ ♇
15 18 ☽ ⚹ ☿
17 37 ☽ Q ♅
19 56 ☽ ⊥ ♀
22 33 ☽ # ♇

14 Monday
03 35 ♂ Q ♇
03 48 ☽ ∥ ♃
10 15 ☽ □ ♂
12 14 ☽ Q ☿
13 27 ☽ ∠ ♀
13 52 ☽ ∠ ♄
14 58 ☽ ⚹ ♅
18 23 ♀ □ ♄
19 56 ☽ Q ♆
20 10 ☽ □ ☉
21 31 ☽ ⚻ ♃

15 Tuesday
03 41 ♀ △ ♆
05 36 ☽ ⊥ ♄
06 34 ♇ St D
07 37 ☽ # ♀
07 58 ☽ ± ♃
15 22 ☉ △ ♃
15 57 ☽ ⚺ ♄
16 28 ♀ ⚹ ♇
16 44 ☽ ⚹ ♆
17 01 ☽ Q ♂
17 01 ☽ ∠ ♅
17 47 ☽ ☌ ♇
17 54 ☽ ⚹ ♀
18 42 ☽ □ ☿
22 49 ☽ # ☿
23 27 ☽ ⚼ ♃

16 Wednesday
01 08 ☿ ⚺ ♀
08 20 ☽ ⊥ ♅
13 01 ☽ ⚹ ♂
16 02 ☽ Q ♀
16 20 ☽ ∥ ♄
18 08 ☽ ∠ ♆
18 24 ☽ ⚺ ♅
21 09 ☿ □ ♇

17 Thursday
00 42 ☽ △ ♃
03 05 ☽ △ ☉
03 31 ☽ # ☉
08 57 ☽ ∥ ♆
09 05 ☽ ⊥ ♆
13 28 ☽ ∥ ♅
13 30 ☽ ∠ ♂
18 09 ☽ ☌ ♄
18 56 ☽ ⚺ ♆
19 13 ☽ △ ☿
19 59 ☿ ⚼ ♅
20 01 ☽ ⚺ ♇
22 10 ☉ ± ♅

18 Friday
00 18 ☽ □ ♀
02 47 ☿ ⚻ ♆
03 57 ☽ ⊥ ♂
05 26 ☽ ⚼ ☉
05 49 ☽ ⊥ ♇
13 30 ☽ ⚺ ♂
18 38 ☽ ⚼ ☿
19 29 ☽ ☌ ♅
20 22 ☽ ∠ ♇
21 00 ☽ ± ☉
22 56 ☿ △ ♄

19 Saturday
01 36 ☽ □ ♃
07 18 ☽ ⚻ ☉
08 32 ☽ ± ☿
14 35 ☿ ⊥ ♀
17 45 ☽ ⚻ ☿
18 30 ☽ ⚺ ♄
19 17 ☽ ☌ ♆
20 26 ☽ ⚹ ♇

20 Sunday
04 05 ☽ ⊥ ♄
04 31 ☽ △ ♀
06 23 ☽ Q ♃
12 46 ☽ ☌ ♂
15 38 ☽ Q ♇
18 28 ☽ ∠ ♄
19 28 ☽ ⚺ ♅

21 Monday
01 35 ☽ ⚹ ♃
05 06 ☽ ⊥ ♅
06 31 ☽ ⚼ ♀
10 40 ☽ ☍ ☉
15 50 ☽ ☍ ☿
18 36 ☽ ⚹ ♄
19 25 ☽ ⚺ ♆
19 38 ☽ ∠ ♅
20 40 ☽ □ ♇
22 17 ☽ ± ♀

22 Tuesday
01 44 ☽ ∥ ♅
01 50 ☽ ∠ ♃
05 11 ☽ ⊥ ♆
06 10 ☽ ∥ ♆
08 55 ☽ ⚻ ♀
12 34 ☽ ⚺ ♂
14 14 ☽ Q ♄
16 37 ☽ ⊥ ♃
16 44 ☽ # ☉
20 02 ☽ ∠ ♆
20 15 ☽ ⚹ ♅
22 22 ☽ ⊥ ♂
22 51 ☽ ∥ ♄

23 Wednesday
02 36 ☽ ⚺ ♃
11 10 ☉ ☌ ☿
13 16 ☽ ∠ ♂
15 12 ☽ # ☿
15 18 ☽ ⚻ ☿
15 49 ☽ ⚻ ☉
16 26 ☽ Q ♅
20 30 ☽ □ ♄
21 21 ☽ ⚹ ♆
22 47 ☽ △ ♇

24 Thursday
00 53 ♀ △ ♂
01 07 ☽ ± ☿
02 53 ☽ ± ☉
14 42 ☿ ∠ ♀
14 47 ☽ ⚹ ♂
16 05 ☽ ⚼ ☿
16 17 ☽ ☍ ♀
18 15 ☽ Q ♆
19 51 ☽ ⚼ ☉
23 47 ☽ □ ♅

25 Friday
01 07 ☽ ⚼ ♇
01 42 ☽ # ♀
06 34 ☽ ☌ ♃
11 13 ☽ ∥ ♃
11 54 ☽ Q ♂
17 25 ☽ ± ♇
17 38 ☽ # ♇
17 47 ☽ △ ☿

26 Saturday
01 09 ☽ △ ☉
01 54 ☽ △ ♄
02 49 ☽ □ ♆
04 29 ☽ ⚻ ♇
11 11 ☉ △ ♄
13 34 ☽ ∥ ♇
20 21 ☽ # ♃
20 47 ☽ □ ♂
23 12 ☉ ⚻ ♆

27 Sunday
02 06 ☉ ⚼ ♅
02 14 ☽ ∥ ♀
04 39 ☽ ⚻ ♀
06 10 ☽ ⚼ ♄
07 20 ☽ △ ♅
14 32 ☽ ⚺ ♃
17 21 ☽ ± ♀
22 34 ☉ □ ♇
23 35 ☽ ± ♄

28 Monday
00 10 ☽ □ ☿
02 11 ☽ ⊥ ♃
10 31 ♀ □ ♅
11 22 ☽ ⚻ ♄
12 19 ☽ △ ♆
12 33 ☽ ⚼ ♅
12 45 ☽ ⚼ ♀
14 14 ☽ ☍ ♇
15 16 ☉ # ♄
15 34 ☽ □ ☉
19 54 ☽ ∠ ♃

29 Tuesday
01 59 ☽ ∥ ☿
05 18 ♀ ∠ ♇
06 26 ☽ ± ♅
06 33 ☽ △ ♂
10 31 ☽ Q ☿
14 47 ☽ ∥ ☉
16 04 ☽ # ♄
18 14 ☽ ⚼ ♆
18 28 ☽ ⚻ ♅
20 52 ♀ # ♃
21 46 ☽ △ ♀

30 Wednesday
01 52 ☽ ⚹ ♃
06 50 ☽ Q ☉
09 58 ☽ ⚹ ☿
12 19 ☽ # ♆
12 21 ☽ ⚼ ♂
12 24 ☽ ± ♆
14 39 ♀ Q ♄
17 33 ☽ # ♅
19 32 ☉ ⚼ ♃
23 37 ☽ ☍ ♄

31 Thursday
00 33 ☽ ⚻ ♆
02 04 ☽ Q ♃
02 40 ☽ ⚻ ♇
06 21 ☽ ± ♂
09 19 ☽ ⚹ ☉
14 52 ☽ ± ♇
15 41 ☽ ∠ ☿
16 25 ♀ ☍ ♃
18 23 ☽ ⚻ ♂

AUGUST 1986

LONGITUDES

Date	Sidereal time h m s	Sun ☉ ° ' "	Moon ☽ ° ' "	Moon ☽ 24.00 ° ' "	Mercury ☿ ° '	Venus ♀ ° '	Mars ♂ ° '	Jupiter ♃ ° '	Saturn ♄ ° '	Uranus ♅ ° '	Neptune ♆ ° '	Pluto ♇ ° '
01	08 39 13	08 ♌ 56 35	21 ♊ 02 26	26 ♊ 59 08	25 ♋ 51	23 ♍ 10	12 ♑ 14	22 ♓ 14	03 ♐ 05	18 ♐ 39	03 ♑ 32	04 ♏ 37
02	08 43 09	09 54 00	02 ♋ 57 26	08 ♋ 57 39	25 R 45	24 15	12 R 05	22 R 10	03 R 05	18 R 37	03 R 31	04 38
03	08 47 06	10 51 27	15 ♋ 00 06	21 ♋ 05 00	25 D 44	25 20	11 58	22 06	03 04	18 36	03 30	04 39
04	08 51 02	11 48 55	27 ♋ 12 31	03 ♌ 22 48	25 51	26 24	11 51	22 02	03 04	18 35	03 28	04 39
05	08 54 59	12 46 24	09 ♌ 35 54	15 ♌ 51 52	26 03	27 29	11 45	21 57	03 04	18 34	03 27	04 40
06	08 58 55	13 43 53	22 ♌ 10 43	28 ♌ 32 27	26 23	28 33	11 39	21 53	03 04	18 33	03 26	04 41
07	09 02 52	14 41 24	04 ♍ 57 02	11 ♍ 24 27	26 48	29 ♍ 37	11 35	21 48	03 D 04	18 32	03 25	04 41
08	09 06 49	15 38 56	17 ♍ 54 41	24 ♍ 27 45	27 21	00 ♎ 40	11 31	21 43	03 04	18 31	03 24	04 42
09	09 10 45	16 36 28	01 ♎ 03 40	07 ♎ 42 30	28 00	01 44	11 28	21 38	03 04	18 30	03 23	04 43
10	09 14 42	17 34 01	14 ♎ 24 17	21 ♎ 09 08	28 46	02 47	11 26	21 33	03 04	18 29	03 22	04 44
11	09 18 38	18 31 36	27 ♎ 57 09	04 ♏ 48 26	29 ♋ 38	03 50	11 25	21 27	03 05	18 28	03 21	04 45
12	09 22 35	19 29 11	11 ♏ 43 04	18 ♏ 41 06	00 ♌ 37	04 53	11 D 25	21 22	03 05	18 27	03 20	04 46
13	09 26 31	20 26 47	25 ♏ 42 32	02 ♐ 47 20	01 42	05 55	11 26	21 16	03 06	18 27	03 19	04 47
14	09 30 28	21 24 24	09 ♐ 55 21	17 ♐ 06 19	02 53	06 58	11 27	21 10	03 06	18 26	03 18	04 48
15	09 34 24	22 22 02	24 ♐ 19 53	01 ♑ 35 34	04 09	08 00	11 29	21 04	03 07	18 25	03 17	04 49
16	09 38 21	23 19 41	08 ♑ 52 46	16 ♑ 10 48	05 31	09 01	11 32	20 58	03 08	18 25	03 16	04 50
17	09 42 18	24 17 21	23 ♑ 28 51	00 ♒ 46 04	06 58	10 03	11 36	20 52	03 09	18 24	03 15	04 51
18	09 46 14	25 15 02	08 ♒ 01 36	15 ♒ 14 32	08 30	11 04	11 41	20 46	03 10	18 24	03 14	04 52
19	09 50 11	26 12 44	22 ♒ 24 05	29 ♒ 29 28	10 06	12 04	11 47	20 39	03 11	18 23	03 13	04 53
20	09 54 07	27 10 28	06 ♓ 30 03	13 ♓ 25 19	11 47	13 05	11 53	20 33	03 12	18 23	03 13	04 54
21	09 58 04	28 08 12	20 ♓ 14 55	26 ♓ 58 36	13 31	14 05	12 00	20 26	03 14	18 22	03 12	04 56
22	10 02 00	29 ♌ 05 59	03 ♈ 36 17	10 ♈ 08 03	15 18	15 05	12 08	20 19	03 15	18 22	03 11	04 57
23	10 05 57	00 ♍ 03 47	16 ♈ 34 04	22 ♈ 54 38	17 08	16 04	12 17	20 12	03 17	18 22	03 10	04 58
24	10 09 53	01 01 37	29 ♈ 10 09	05 ♉ 21 04	19 00	17 03	12 26	20 05	03 18	18 22	03 10	04 59
25	10 13 50	01 59 28	11 ♉ 27 56	17 ♉ 31 20	20 55	18 02	12 36	19 58	03 20	18 22	03 09	05 01
26	10 17 47	02 57 21	23 ♉ 31 51	29 ♉ 30 09	22 51	19 00	12 47	19 51	03 22	18 22	03 08	05 02
27	10 21 43	03 55 16	05 ♊ 26 52	11 ♊ 22 38	24 48	19 58	12 59	19 43	03 24	18 21	03 08	05 04
28	10 25 40	04 53 13	17 ♊ 18 06	23 ♊ 13 52	26 45	20 55	13 11	19 36	03 26	18 D 21	03 07	05 05
29	10 29 36	05 51 12	29 ♊ 10 32	05 ♋ 08 38	28 ♌ 44	21 52	13 24	19 29	03 28	18 22	03 07	05 06
30	10 33 33	06 49 12	11 ♋ 08 41	17 ♋ 11 09	00 ♍ 42	22 49	13 38	19 21	03 30	18 22	03 06	05 08
31	10 37 29	07 ♍ 47 15	23 ♋ 16 25	29 ♋ 24 51	02 ♍ 40	23 ♎ 45	13 ♑ 53	19 ♓ 13	03 ♐ 32	18 ♐ 22	03 ♑ 06	05 ♏ 09

Date	Moon True ☊ ° '	Moon Mean ☊ ° '	Moon ☽ Latitude ° '
01	24 ♈ 16	24 ♈ 34	04 N 13
02	24 R 07	24 31	04 41
03	23 56	24 28	04 58
04	23 44	24 25	05 01
05	23 32	24 22	04 50
06	23 20	24 18	04 24
07	23 10	24 15	03 46
08	23 03	24 12	02 55
09	22 59	24 09	01 54
10	22 57	24 06	00 N 45
11	22 D 57	24 02	00 S 27
12	22 57	23 59	01 38
13	22 R 57	23 56	02 45
14	22 56	23 53	03 43
15	22 52	23 50	04 28
16	22 46	23 47	04 56
17	22 39	23 43	05 04
18	22 30	23 40	04 53
19	22 21	23 37	04 24
20	22 13	23 34	03 39
21	22 07	23 31	02 42
22	22 04	23 28	01 37
23	22 02	23 24	00 S 29
24	22 D 02	23 21	00 N 38
25	22 03	23 18	01 43
26	22 04	23 15	02 41
27	22 R 05	23 12	03 32
28	22 04	23 08	04 14
29	22 01	23 05	04 44
30	21 56	23 02	05 03
31	21 ♈ 50	22 ♈ 59	05 N 08

DECLINATIONS

Date	Sun ☉ ° '	Moon ☽ ° '	Mercury ☿ ° '	Venus ♀ ° '	Mars ♂ ° '	Jupiter ♃ ° '	Saturn ♄ ° '	Uranus ♅ ° '	Neptune ♆ ° '	Pluto ♇ ° '
01	18 N 02	27 N 21	17 N 07	02 N 59	28 S 43	04 S 22	18 S 58	23 S 03	22 S 18	02 N 27
02	17 46	28 06	17 21	02 29	28 43	04 23	18 58	23 03	22 18	02 27
03	17 31	27 32	17 35	01 59	28 43	04 25	18 58	23 03	22 18	02 26
04	17 15	25 38	17 49	01 30	28 42	04 27	18 58	23 03	22 18	02 25
05	16 59	22 30	18 02	01 00	28 42	04 29	18 59	23 03	22 18	02 24
06	16 42	18 17	18 14	00 N 30	28 41	04 31	18 59	23 03	22 18	02 24
07	16 26	13 12	18 25	00 00	28 40	04 33	18 59	23 03	22 18	02 23
08	16 09	07 28	18 35	00 S 30	28 39	04 36	18 59	23 03	22 18	02 22
09	15 52	01 N 19	18 43	01 00	28 38	04 38	19 00	23 03	22 19	02 21
10	15 34	04 S 59	18 50	01 30	28 36	04 40	19 00	23 03	22 19	02 20
11	15 17	11 10	18 55	02 00	28 35	04 42	19 00	23 02	22 19	02 20
12	14 59	16 55	18 59	02 30	28 33	04 45	19 00	23 02	22 19	02 19
13	14 41	21 52	19 00	03 00	28 31	04 47	19 01	23 02	22 19	02 18
14	14 22	25 37	18 59	03 29	28 29	04 50	19 01	23 02	22 19	02 17
15	14 04	27 47	18 55	03 59	28 27	04 52	19 02	23 02	22 19	02 16
16	13 45	28 04	18 49	04 28	28 24	04 55	19 02	23 02	22 19	02 15
17	13 26	26 24	18 41	04 58	28 22	04 58	19 02	23 02	22 19	02 15
18	13 06	22 59	18 29	05 27	28 19	05 00	19 03	23 02	22 19	02 14
19	12 47	18 12	18 15	05 56	28 16	05 03	19 03	23 02	22 19	02 13
20	12 27	12 31	17 58	06 25	28 13	05 06	19 04	23 02	22 19	02 12
21	12 07	06 21	17 39	06 54	28 10	05 09	19 04	23 02	22 19	02 11
22	11 47	00 S 03	17 16	07 23	28 07	05 11	19 05	23 02	22 19	02 10
23	11 27	06 N 04	16 51	07 52	28 04	05 14	19 05	23 02	22 19	02 09
24	11 07	11 47	16 23	08 20	28 01	05 17	19 06	23 02	22 19	02 08
25	10 46	16 54	15 53	08 48	27 57	05 20	19 06	23 02	22 19	02 08
26	10 25	21 16	15 21	09 16	27 54	05 23	19 07	23 02	22 20	02 07
27	10 04	24 42	14 46	09 44	27 50	05 26	19 07	23 02	22 20	02 06
28	09 43	27 03	14 10	10 12	27 46	05 29	19 08	23 02	22 20	02 05
29	09 22	28 11	13 32	10 39	27 42	05 32	19 09	23 02	22 20	02 04
30	09 01	28 00	12 52	11 07	27 38	05 35	19 09	23 02	22 20	02 03
31	08 N 39	26 N 30	12 N 10	11 S 34	27 S 34	05 S 39	19 S 10	23 S 02	22 S 20	02 N 02

ZODIAC SIGN ENTRIES

Date	h m	Planets
02	06 04	☽ ♋
04	17 26	☽ ♌
07	02 44	☽ ♍
07	20 46	♀ ♎
09	10 05	☽ ♎
11	15 36	☽ ♏
11	21 09	☿ ♌
13	19 17	☽ ♐
15	21 22	☽ ♑
17	22 44	☽ ♒
20	00 52	☽ ♓
22	05 27	☽ ♈
23	10 26	☉ ♍
24	13 36	☽ ♉
27	01 00	☽ ♊
29	13 40	☽ ♋
30	03 28	☿ ♍

LATITUDES

Date	Mercury ☿ ° '	Venus ♀ ° '	Mars ♂ ° '	Jupiter ♃ ° '	Saturn ♄ ° '	Uranus ♅ ° '	Neptune ♆ ° '	Pluto ♇ ° '
01	03 S 56	00 N 18	05 S 52	01 S 23	01 N 51	00 S 06	01 N 06	16 N 26
04	03 13	00 N 04	05 49	01 24	01 50	00 06	01 06	16 25
07	02 26	00 S 10	05 45	01 25	01 49	00 06	01 06	16 23
10	01 36	00 26	05 41	01 26	01 49	00 06	01 05	16 21
13	00 49	00 42	05 35	01 27	01 48	00 06	01 05	16 20
16	00 S 05	00 59	05 29	01 27	01 47	00 06	01 05	16 18
19	00 N 34	01 16	05 22	01 28	01 47	00 06	01 05	16 16
22	01 04	01 34	05 15	01 28	01 46	00 06	01 05	16 15
25	01 26	01 53	05 08	01 29	01 45	00 06	01 05	16 13
28	01 40	02 12	05 00	01 29	01 45	00 06	01 05	16 12
31	01 N 46	02 S 31	04 S 52	01 S 30	01 N 44	00 S 06	01 N 05	16 N 10

DATA

Julian Date	2446644
Delta T	+55 seconds
Ayanamsa	23° 40' 05"
Synetic vernal point	05° ♓ 26' 55"
True obliquity of ecliptic	23° 26' 36"

MOON'S PHASES, APSIDES AND POSITIONS ☽

Date	h m	Phase	Longitude ° '	Eclipse Indicator
05	18 36	●	13 ♌ 02	
13	02 21	☽	20 ♏ 04	
19	18 54	○	26 ♒ 29	
27	08 39	☾	03 ♊ 47	

Day	h m		
16	16 44	Perigee	
28	14 44	Apogee	
02	13 47	Max dec	28° N 06'
09	17 02	0S	
16	03 27	Max dec	28° S 11'
22	12 13	0N	
29	20 52	Max dec	28° N 16'

LONGITUDES

Date	Chiron ⚷ ° '	Ceres ⚳ ° '	Pallas ⚴ ° '	Juno ⚵ ° '	Vesta ⚶ ° '	Black Moon Lilith ⚸ ° '
01	19 ♊ 26	29 ♍ 50	05 ♍ 17	00 ♐ 59	16 ♈ 45	17 ♊ 15
11	20 ♊ 00	03 ♎ 52	10 ♍ 02	01 ♐ 32	17 ♈ 27	18 ♊ 22
21	20 ♊ 30	08 ♎ 00	14 ♍ 46	02 ♐ 34	17 ♈ 25	19 ♊ 29
31	20 ♊ 53	12 ♎ 13	19 ♍ 29	04 ♐ 02	16 ♈ 37	20 ♊ 36

ASPECTARIAN

h m	Aspects
01 Friday	
07 09	☽ ☍ ♅
09 08	☽ ᕈ ♇
09 37	☽ ⊥ ☿
14 23	☽ □ ♃
16 43	☽ □ ♀
18 23	☽ ∠ ☉
21 36	☽ ⊻ ☿
02 Saturday	
02 29	☉ ± ♆
12 15	☽ ⊼ ♄
13 07	☽ ☍ ♆
14 03	☽ ⊥ ☉
14 10	♀ ∥ ♇
15 21	☽ △ ♇
03 Sunday	
00 14	☽ ± ♄
00 47	☿ St D
03 04	☽ ⊻ ☉
06 02	☽ ☍ ♂
08 21	☽ Q ♀
08 29	☉ ∥ ☿
18 04	☽ ᕈ ♄
19 06	☽ ⊼ ♅
21 51	☿ ⚹ ♀
04 Monday	
01 55	☽ △ ♃
06 53	☽ ± ♅
09 18	☽ ☌ ☿
10 17	☽ ⚹ ♀
12 43	☉ ⊼ ♂
23 24	☽ △ ♄
05 Tuesday	
00 10	☽ ⊼ ♆
00 23	☽ ᕈ ♅
02 29	☽ □ ♇
06 56	☽ ᕈ ♃
08 21	☽ # ♅
11 43	☽ ± ♆
13 15	☽ # ♆
16 05	☽ ⊼ ♂
18 02	☽ ∠ ♀
18 36	☽ ☌ ☉
06 Wednesday	
00 06	☽ ± ♃
03 28	☽ ± ♂
04 54	☽ ᕈ ♆
05 07	☽ △ ♅
08 24	☽ # ♄
11 26	☽ ⊼ ♃
12 17	☽ ∥ ☿
12 46	☽ ⊥ ♀
12 57	☽ Q ♇
14 59	♀ ⊥ ♇
20 11	☽ ⊻ ☿
20 18	☽ ∥ ☉
20 24	☽ ᕈ ♂
07 Thursday	
01 06	☽ ⊻ ♀
04 49	♄ St D
07 51	☽ ⊥ ☿
08 29	☽ □ ♄
09 08	☽ △ ♆
11 31	☽ ⚹ ♇
08 Friday	
00 16	☽ △ ♂
01 16	☽ ∠ ☿
06 03	☉ ∠ ♀
07 30	☽ ⊻ ☉
13 06	☽ □ ♅
13 34	☉ ± ♃
15 18	☽ ∠ ♇
17 47	☽ Q ♄
18 56	☽ ☍ ♃
19 24	☽ ⊥ ☉
23 16	☽ # ♃
09 Saturday	
06 09	☽ ⚹ ☿
07 45	☽ ⊥ ♇
08 00	☽ ∥ ♇
13 04	☽ ∠ ☉
13 06	☽ # ♀
13 19	☽ ☌ ♀
15 38	☽ ⚹ ♄
16 11	☽ □ ♆
18 37	☽ ⊻ ♇
21 37	☽ ∥ ♀
21 49	☽ Q ♅
10 Sunday	
01 58	☽ # ♇
05 05	☽ Q ☿
06 43	☽ □ ♂
08 55	☉ ± ♂
10 47	☽ ∥ ♃
18 04	☽ ⚹ ☉
18 32	☽ ∠ ♄
18 33	♀ ⚹ ♄
19 15	☽ ⚹ ♅
11 Monday	
00 21	☽ Q ♆
00 37	☽ ⊼ ♃
00 54	♀ □ ♆
07 31	☉ ᕈ ♆
10 28	☽ ⊥ ♄
10 36	☉ △ ♅
11 08	☽ ± ♃
14 35	☽ Q ♂
15 11	☽ □ ☿
16 51	☽ Q ☉
20 59	☽ ⊻ ♄
21 26	☽ ⚹ ♆
21 39	☽ ∠ ♅
23 09	☽ ⊻ ♀
23 55	☽ ☌ ♇
12 Tuesday	
02 46	☽ ᕈ ♃
03 23	♀ # ♇
04 03	☽ # ☉
07 46	♂ St D
09 12	♀ ⊻ ♇
10 26	☽ ⊥ ♀
11 29	☽ ⚹ ♂
13 17	☽ ⊥ ♅
21 30	☽ # ☿
21 36	☽ ∥ ♄
23 22	☽ ∠ ♆
23 36	☽ ⊻ ♅
13 Wednesday	
02 21	☽ □ ☉
03 11	☽ ∠ ♀
04 29	☽ △ ♃
13 13	☽ ∠ ♂
14 28	☽ ∥ ♆
14 43	☽ ⊥ ♆
18 37	☽ ∥ ♅
23 03	☽ △ ☿
23 53	♀ Q ♅
14 Thursday	
00 32	☽ ☌ ♄
00 52	☽ ⊻ ♆
03 22	☽ ⊻ ♇
04 28	☽ ⊥ ♂
06 38	☽ ⚹ ♀
06 42	☉ ⊼ ♃
13 28	☽ ⊥ ♇
14 34	☽ ⊻ ♂
16 32	☿ △ ♄
20 00	☿ ⊼ ♆
22 38	☿ ᕈ ♅
15 Friday	
02 12	☽ ☌ ♅
02 33	☽ ᕈ ☿
04 15	☽ Q ♀
04 30	☽ ∠ ♇
06 38	☽ □ ♃
08 31	☽ △ ☉
18 57	☽ ± ☿
23 20	☉ Q ♇
16 Saturday	
00 01	☿ □ ♇
02 32	☽ ⊻ ♄
02 46	☽ ☌ ♆
05 20	☽ ⚹ ♇
05 53	☽ ⊼ ☿
11 02	☽ ᕈ ☉
12 09	☽ Q ♃
12 15	☽ □ ♀
12 25	☽ ⊥ ♄
16 24	☽ ☌ ♂
19 12	☿ ᕈ ♃
17 Sunday	
01 05	☽ Q ♇
02 52	☽ ± ☉
03 14	☽ ∠ ♄
03 39	☽ ⊻ ♅
07 44	☽ ⚹ ♃
11 41	♀ ∥ ♃
13 25	☽ ⊼ ☉
13 31	☽ ⊥ ♅
18 Monday	
03 57	☽ ⚹ ♄
04 05	☽ ⊻ ♆
04 20	☽ ∠ ♅
06 46	☽ □ ♇
08 17	☽ ∠ ♃
11 42	☽ ∥ ♅
12 53	☽ ☍ ☿
14 00	☽ ⊥ ♆
15 44	☽ ∥ ♆
17 25	☽ △ ♀
18 07	☽ ⊻ ♂
23 04	☿ ± ♆
23 07	☽ ⊥ ♃
23 53	☽ Q ♄
19 Tuesday	
04 10	♀ □ ♂
04 11	☽ ⊥ ♂
04 59	☽ ∠ ♆
05 16	☽ ⚹ ♅
08 07	☽ ∥ ♄
09 05	☽ ⊻ ♃
11 46	☽ # ☿
18 54	☽ ☍ ☉
19 27	☽ ∠ ♂
20 30	☽ ᕈ ♀
20 Wednesday	
01 31	☽ Q ♅
03 46	☉ ᕈ ♂
06 20	☽ □ ♄
06 21	☽ ⚹ ♆
09 15	☽ △ ♇
12 15	☽ # ☉
13 05	☽ ± ♀
13 14	♄ ⊻ ♆
13 35	☿ ⊼ ♂
21 24	☽ ⚹ ♂
22 26	☽ ⊼ ☿
21 Thursday	
00 18	☽ ⊼ ♀
03 07	☽ Q ♆
08 41	☽ □ ♅
10 01	☽ ∥ ♀
10 30	☽ ± ☿
11 26	☽ ᕈ ♇
12 19	☽ ☌ ♃
16 33	☽ ∥ ♃
18 44	☽ Q ♂
23 43	☿ ± ♃
22 Friday	
03 11	☽ ⊼ ☉
03 32	☽ ± ♇
03 53	☽ # ♇
05 03	☽ ᕈ ☿
05 28	☿ ⚹ ♀
11 14	☽ □ ♆
11 21	☽ △ ♄
14 27	☽ ⊼ ♇
14 57	☽ ± ☉
20 35	☽ ∥ ♇
23 Saturday	
03 53	☽ □ ♂
08 40	☽ # ♃
08 57	☽ ᕈ ☉
10 59	☽ ☍ ♀
13 15	☽ △ ☿
15 14	☽ ᕈ ♄
15 23	☽ △ ♅
18 48	☽ ⊻ ♃
19 59	☽ # ♀
24 Sunday	
01 17	☿ ᕈ ♆
03 49	☿ △ ♅
04 02	☿ ± ♂
06 07	☽ ⊥ ♃
08 24	☽ ± ♄
09 15	☽ ∥ ☉
15 54	☽ △ ☉
19 43	☽ △ ♆
20 02	☽ ⊼ ♄
20 07	☽ ᕈ ♅
23 19	☽ ☍ ♇
23 22	☽ ∠ ♃
25 Monday	
00 50	☿ ⊼ ♃
04 13	♃ ᕈ ♇
07 27	☽ ∥ ☿
13 46	☽ ± ♅
14 17	☽ △ ♂
19 48	♀ ∠ ♄
20 10	♀ ⚹ ♅
23 34	☽ # ♄
26 Tuesday	
01 14	☽ ᕈ ♆
01 40	☽ ⊼ ♅
02 09	☽ ⊼ ♀
04 42	☽ ⚹ ♃
10 21	☽ □ ☿
14 24	☿ Q ♇
15 12	☽ ± ♀
16 31	☉ △ ♆
18 44	☽ # ♆
19 14	☽ ± ♆
20 41	☽ ᕈ ♂
22 30	☉ □ ♄
23 30	☽ # ♅
27 Wednesday	
04 33	☽ Q ♃
06 43	♀ ⊼ ♃
07 19	☽ ⊼ ♆
07 51	☽ ☍ ♄
08 39	☽ □ ☉
10 56	☽ ᕈ ♀
11 13	☽ ⊼ ♇
15 09	☽ ± ♂
21 15	♅ St D
21 53	☉ # ♀
23 23	☽ ± ♇
28 Thursday	
03 31	☽ ⊼ ♂
05 49	☽ Q ☿
14 08	☽ ☍ ♅
16 36	☽ □ ♃
16 58	♀ Q ♆
17 00	☉ ⚹ ♇
17 39	☽ ᕈ ♇
19 58	☽ △ ♀
23 14	☽ # ♂
29 Friday	
00 18	☽ Q ☉
07 35	☿ ᕈ ♂
10 55	☽ ⚹ ☿
19 55	☽ ☍ ♆
20 39	☽ ⊼ ♄
23 57	☽ △ ♇
30 Saturday	
02 36	☽ ⚹ ☉
08 43	☽ ± ♄
17 04	☽ ☍ ♂
20 47	☽ # ♂
22 49	☽ ∠ ☿
31 Sunday	
02 20	☽ ⊼ ♅
02 39	☽ ᕈ ♄
04 07	☽ △ ♃
10 58	☽ ∠ ☉
13 01	☽ □ ♀
14 08	☽ ± ♅
17 06	☿ △ ♆
19 56	☽ ⊥ ☿
22 45	☿ □ ♄

All ephemeris data is given at 12.00 UT and the Moon's longitude is additionally given for 24.00 UT

SEPTEMBER 1986

LONGITUDES

Date	Sidereal time h m s	Sun ☉ ° ' "	Moon ☽ ° ' "	Moon ☽ 24.00 ° ' "	Mercury ☿ ° '	Venus ♀ ° '	Mars ♂ ° '	Jupiter ♃ ° '	Saturn ♄ ° '	Uranus ♅ ° '	Neptune ♆ ° '	Pluto ♇ ° '
01	10 41 26	08 ♍ 45 19	05 ♌ 36 43	11 ♌ 52 14	04 ♍ 39	24 ♎ 41	14 ♑ 08	19 ♓ 06	03 ♐ 35	18 ♐ 22	03 ♑ 05	05 ♏ 11
02	10 45 22	09 43 25	18 ♌ 11 32	24 ♌ 34 42	06 36	25 36	14 24	18 R 58	03 37	18 22	03 R 05	05 13
03	10 49 19	10 41 32	01 ♍ 01 44	07 ♍ 32 35	08 33	26 30	14 40	18 50	03 40	18 23	03 04	05 14
04	10 53 16	11 39 42	14 ♍ 07 09	20 ♍ 45 16	10 29	27 25	14 57	18 42	03 42	18 23	03 04	05 16
05	10 57 12	12 37 53	27 ♍ 26 45	04 ♎ 11 25	12 25	28 18	15 15	18 34	03 45	18 23	03 04	05 17
06	11 01 09	13 36 05	10 ♎ 59 01	17 ♎ 49 20	14 19	29 ♎ 11	15 34	18 27	03 48	18 24	03 03	05 19
07	11 05 05	14 34 20	24 ♎ 42 08	01 ♏ 37 12	16 13	00 ♏ 04	15 53	18 19	03 51	18 24	03 03	05 21
08	11 09 02	15 32 35	08 ♏ 34 21	15 ♏ 33 22	18 05	00 56	16 12	18 11	03 54	18 25	03 03	05 22
09	11 12 58	16 30 53	22 ♏ 34 06	29 ♏ 36 20	19 57	01 47	16 33	18 03	03 57	18 26	03 03	05 24
10	11 16 55	17 29 12	06 ♐ 39 56	13 ♐ 44 40	21 47	02 37	16 53	17 55	04 00	18 26	03 03	05 26
11	11 20 51	18 27 32	20 ♐ 50 21	27 ♐ 56 44	23 36	03 27	17 15	17 47	04 03	18 27	03 03	05 28
12	11 24 48	19 25 54	05 ♑ 03 34	12 ♑ 10 30	25 24	04 17	17 37	17 39	04 07	18 28	03 02	05 30
13	11 28 45	20 24 18	19 ♑ 17 12	26 ♑ 23 17	27 11	05 05	17 59	17 31	04 10	18 29	03 02	05 31
14	11 32 41	21 22 43	03 ♒ 28 17	10 ♒ 31 46	28 ♍ 57	05 53	18 22	17 23	04 14	18 29	03 02	05 33
15	11 36 38	22 21 10	17 ♒ 33 14	24 ♒ 32 14	00 ♎ 41	06 39	18 46	17 15	04 17	18 30	03 D 02	05 35
16	11 40 34	23 19 38	01 ♓ 28 18	08 ♓ 20 59	02 25	07 25	19 10	17 07	04 21	18 31	03 02	05 37
17	11 44 31	24 18 08	15 ♓ 09 55	21 ♓ 54 46	04 08	08 11	19 34	16 59	04 25	18 32	03 02	05 39
18	11 48 27	25 16 40	28 ♓ 35 17	05 ♈ 11 16	05 49	08 55	19 59	16 51	04 28	18 33	03 03	05 41
19	11 52 24	26 15 14	11 ♈ 42 39	18 ♈ 09 24	07 30	09 38	20 24	16 44	04 32	18 35	03 03	05 43
20	11 56 20	27 13 50	24 ♈ 31 37	00 ♉ 49 26	09 09	10 21	20 50	16 36	04 36	18 36	03 03	05 45
21	12 00 17	28 12 28	07 ♉ 03 06	13 ♉ 12 55	10 48	11 02	21 17	16 28	04 40	18 37	03 03	05 47
22	12 04 14	29 ♍ 11 08	19 ♉ 19 15	25 ♉ 22 32	12 25	11 42	21 43	16 21	04 45	18 38	03 03	05 49
23	12 08 10	00 ♎ 09 50	01 ♊ 23 14	07 ♊ 21 53	14 01	12 21	22 10	16 13	04 49	18 40	03 04	05 51
24	12 12 07	01 08 34	13 ♊ 19 00	19 ♊ 15 12	15 37	12 59	22 38	16 06	04 53	18 41	03 04	05 53
25	12 16 03	02 07 21	25 ♊ 11 01	01 ♋ 07 06	17 12	13 36	23 06	15 58	04 57	18 42	03 04	05 55
26	12 20 00	03 06 10	07 ♋ 04 01	13 ♋ 02 23	18 45	14 12	23 34	15 51	05 02	18 44	03 05	05 57
27	12 23 56	04 05 02	19 ♋ 02 46	25 ♋ 05 44	20 18	14 46	24 03	15 44	05 06	18 45	03 05	05 59
28	12 27 53	05 03 55	01 ♌ 11 47	07 ♌ 21 26	21 50	15 19	24 32	15 37	05 11	18 47	03 05	06 02
29	12 31 49	06 02 51	13 ♌ 35 05	19 ♌ 53 08	23 21	15 51	25 02	15 30	05 16	18 49	03 06	06 04
30	12 35 46	07 ♎ 01 49	26 ♌ 15 51	02 ♍ 43 28	24 ♎ 51	16 ♏ 21	25 ♑ 31	15 ♓ 23	05 ♐ 20	18 ♐ 50	03 ♑ 06	06 ♏ 06

Moon Nodes, Latitude and DECLINATIONS

Date	Moon True ☊ ° '	Moon Mean ☊ ° '	Moon ☽ Latitude ° '	Sun ☉ ° '	Moon ☽ ° '	Mercury ☿ ° '	Venus ♀ ° '	Mars ♂ ° '	Jupiter ♃ ° '	Saturn ♄ ° '	Uranus ♅ ° '	Neptune ♆ ° '	Pluto ♇ ° '
01	21 ♈ 43	22 ♈ 56	04 N 59	08 N 17	23 N 42	11 N 28	12 S 00	27 S 30	05 S 42	19 S 11	23 S 02	22 S 20	02 N 01
02	21 R 35	22 53	04 36	07 55	19 45	10 44	12 27	27 25	05 45	19 11	23 02	22 20	02 00
03	21 28	22 49	03 59	07 34	14 49	10 00	12 53	27 21	05 48	19 12	23 02	22 20	01 59
04	21 22	22 46	03 08	07 11	09 08	09 14	13 19	27 16	05 51	19 12	23 02	22 20	01 58
05	21 18	22 43	02 06	06 49	02 N 56	08 28	13 45	27 11	05 54	19 13	23 02	22 20	01 57
06	21 16	22 40	00 N 56	06 27	03 S 30	07 42	14 10	27 06	05 57	19 14	23 02	22 20	01 56
07	21 D 16	22 37	00 S 19	06 05	09 52	06 55	14 35	27 01	06 01	19 15	23 02	22 20	01 55
08	21 16	22 34	01 33	05 42	15 50	06 08	15 00	26 56	06 04	19 16	23 02	22 20	01 54
09	21 18	22 30	02 42	05 19	21 02	05 21	15 24	26 51	06 07	19 16	23 02	22 20	01 53
10	21 19	22 27	03 42	04 57	25 05	04 33	15 48	26 45	06 10	19 17	23 02	22 20	01 53
11	21 R 19	22 24	04 29	04 34	27 36	03 46	16 12	26 40	06 13	19 18	23 02	22 20	01 52
12	21 19	22 21	05 00	04 11	28 20	02 58	16 35	26 34	06 17	19 19	23 03	22 20	01 51
13	21 16	22 18	05 12	03 48	27 12	02 11	16 58	26 28	06 20	19 20	23 03	22 20	01 50
14	21 13	22 14	05 05	03 25	24 19	01 24	17 21	26 23	06 23	19 21	23 03	22 20	01 49
15	21 08	22 11	04 40	03 02	20 01	00 N 37	17 43	26 17	06 26	19 21	23 03	22 21	01 48
16	21 04	22 08	03 58	02 39	14 40	00 S 10	18 05	26 10	06 29	19 22	23 03	22 21	01 47
17	21 01	22 05	03 04	02 16	08 40	00 57	18 26	26 04	06 32	19 23	23 03	22 21	01 46
18	20 58	22 02	02 00	01 53	02 S 23	01 43	18 47	25 58	06 35	19 24	23 03	22 21	01 45
19	20 56	21 59	00 S 50	01 29	03 N 51	02 29	19 07	25 51	06 38	19 25	23 03	22 21	01 44
20	20 D 56	21 55	00 N 20	01 06	09 49	03 14	19 27	25 44	06 41	19 26	23 03	22 21	01 43
21	20 57	21 52	01 28	00 43	15 15	03 59	19 47	25 37	06 44	19 27	23 03	22 21	01 42
22	20 58	21 49	02 30	00 N 19	19 58	04 44	20 06	25 31	06 47	19 28	23 04	22 21	01 41
23	21 00	21 46	03 28	00 S 04	23 47	05 28	20 24	25 23	06 50	19 29	23 04	22 21	01 40
24	21 01	21 43	04 10	00 27	26 32	06 11	20 42	25 16	06 53	19 30	23 04	22 21	01 39
25	21 02	21 40	04 44	00 51	28 05	06 54	21 00	25 09	06 56	19 31	23 04	22 21	01 38
26	21 R 02	21 36	05 06	01 14	28 21	07 37	21 17	25 01	06 58	19 32	23 04	22 21	01 37
27	21 02	21 33	05 15	01 37	27 18	08 19	21 33	24 53	07 01	19 33	23 04	22 21	01 36
28	21 00	21 30	05 11	02 01	24 57	09 00	21 49	24 46	07 04	19 34	23 04	22 21	01 35
29	20 58	21 27	04 51	02 24	21 24	09 41	22 04	24 38	07 06	19 35	23 04	22 21	01 34
30	20 ♈ 56	21 ♈ 24	04 N 18	02 S 47	16 N 48	10 S 21	22 S 18	24 S 29	07 S 09	19 S 36	23 S 05	22 S 21	01 N 33

ZODIAC SIGN ENTRIES

Date	h	m	Planets
01	01	08	☽ ♌
03	10	06	☽ ♍
05	16	33	☽ ♎
07	10	15	♀ ♏
07	21	12	☽ ♏
10	00	40	☽ ♐
12	03	28	☽ ♑
14	06	07	☽ ♒
15	02	28	☿ ♎
16	09	27	☽ ♓
18	14	33	☽ ♈
20	22	25	☽ ♉
23	07	59	☉ ♎
23	09	13	☽ ♊
25	21	44	☽ ♋
28	09	39	☽ ♌
30	18	57	☽ ♍

LATITUDES

Date	Mercury ☿ ° '	Venus ♀ ° '	Mars ♂ ° '	Jupiter ♃ ° '	Saturn ♄ ° '	Uranus ♅ ° '	Neptune ♆ ° '	Pluto ♇ ° '
01	01 N 47	02 S 38	04 S 50	01 S 30	01 N 44	00 S 06	01 N 05	16 N 10
04	01 44	02 58	04 41	01 30	01 43	00 06	01 04	16 08
07	01 36	03 18	04 33	01 30	01 42	00 06	01 04	16 07
10	01 24	03 38	04 25	01 31	01 42	00 06	01 04	16 06
13	01 09	03 59	04 17	01 31	01 41	00 06	01 04	16 04
16	00 52	04 19	04 08	01 31	01 40	00 06	01 04	16 03
19	00 32	04 40	04 00	01 31	01 40	00 06	01 04	16 02
22	00 N 12	05 00	03 52	01 31	01 39	00 06	01 04	16 01
25	00 S 10	05 19	03 44	01 31	01 39	00 06	01 04	16 00
28	00 32	05 38	03 36	01 31	01 38	00 06	01 04	15 59
31	00 S 54	05 S 55	03 S 28	01 S 30	01 N 37	00 S 06	01 N 03	15 N 58

DATA

Julian Date	2446675
Delta T	+55 seconds
Ayanamsa	23° 40' 09"
Synetic vernal point	05° ♓ 26' 50"
True obliquity of ecliptic	23° 26' 36"

MOON'S PHASES, APSIDES AND POSITIONS ☽

Date	h	m	Phase	Longitude ° '	Eclipse Indicator
04	07	10	●	11 ♍ 28	
11	07	41	☽	18 ♐ 17	
18	05	34	○	25 ♓ 01	
26	03	17	☾	02 ♋ 45	

Day	h	m	
12	00	07	Perigee
25	10	00	Apogee

Day	h	m		
05	23	01	0S	
12	09	24	Max dec	28° S 21'
18	21	07	0N	
26	04	48	Max dec	28° N 25'

LONGITUDES

Date	Chiron ⚷ ° '	Ceres ⚳ ° '	Pallas ⚴ ° '	Juno ⚵ ° '	Vesta ⚶ ° '	Black Moon Lilith ⚸ ° '
01	20 ♊ 55	12 ♎ 38	19 ♍ 58	04 ♐ 12	16 ♈ 30	20 ♊ 43
11	21 ♊ 11	16 ♎ 55	24 ♍ 39	06 ♐ 05	14 ♈ 54	21 ♊ 50
21	21 ♊ 21	21 ♎ 16	29 ♍ 19	08 ♐ 18	12 ♈ 43	22 ♊ 57
31	21 ♊ 22	25 ♎ 38	03 ♎ 56	10 ♐ 48	10 ♈ 12	24 ♊ 04

All ephemeris data is given at 12.00 UT and the Moon's longitude is additionally given for 24.00 UT

ASPECTARIAN

h m	Aspects
01 Monday	
00 47	☿ # ♀
06 01	☽ ⊥ ☉
07 08	☽ ⊼ ♆
07 40	☽ ⚼ ♅
08 04	☽ △ ♄
09 06	☽ ⚼ ♃
09 47	☽ ⊻ ☿
11 10	☽ □ ♇
16 35	☽ # ♅
18 33	☽ ⊻ ☉
18 40	☽ ± ♆
18 41	☿ ⚹ ♇
21 07	☽ # ♆
21 33	♀ ± ♃
02 Tuesday	
02 11	☽ ± ♃
02 36	☽ Q ♀
04 39	☽ ⊼ ♂
11 47	☽ ⚼ ♆
12 20	☽ △ ♅
13 27	☽ ⊼ ♃
14 59	☽ # ♄
16 15	☽ ± ♂
21 27	☽ Q ♇
03 Wednesday	
02 58	☽ ⚹ ♀
09 26	☽ ⚼ ♂
15 47	☽ △ ♆
16 53	☽ □ ♄
19 47	☽ ⚹ ♇
19 53	☽ # ♀
04 Thursday	
04 15	☽ ☌ ☿
07 10	☽ ☌ ☉
08 40	☽ ∠ ♀
11 31	☽ ∥ ☿
13 33	☽ △ ♂
19 44	☽ □ ♅
20 12	☽ ∥ ☉
20 14	☽ ☍ ♃
20 19	♀ ⊥ ♄
23 08	☽ ∠ ♇
05 Friday	
00 47	☽ # ♃
01 46	☽ Q ♄
02 08	☽ ⊥ ♀
13 38	☽ ⊻ ♀
15 18	☽ ⊥ ♇
15 43	☽ ∥ ♇
17 32	☉ ☌ ☿
22 00	☽ □ ♆
23 16	☽ ⚹ ♄
06 Saturday	
01 59	☽ ⊻ ♇
03 54	☽ Q ♅
06 14	☽ # ♇
08 52	☿ ∠ ♀
16 57	☽ ⊻ ☉
18 49	☽ ⊻ ☿
19 52	♃ □ ♅
20 14	☽ □ ♂
21 16	☽ ∥ ♃
22 26	☽ # ☉
07 Sunday	
00 58	☽ ⊼ ♃
01 01	☽ ⚹ ♅
01 45	☽ ∠ ♄
02 02	☽ # ☿
04 15	☽ ⊥ ☉
05 39	☽ Q ♆
06 51	☿ △ ♂
06 59	☽ ⊥ ☿
11 20	☽ ± ♃
15 37	☉ # ♃
17 29	☽ ⊥ ♄
21 05	☽ ∠ ☉
21 56	☽ ☌ ♀
08 Monday	
01 04	☽ ∠ ☿
02 29	☽ ⚹ ♆
02 47	☽ ⚼ ♃
03 06	☽ ∠ ♅
03 54	☽ ⊻ ♄
04 17	☽ Q ♂
06 29	☽ ☌ ♇
08 13	☽ ∥ ♀
13 06	☿ ☍ ♃
14 01	☿ # ♃
16 15	☿ □ ♅
18 37	☽ ⊥ ♅
09 Tuesday	
00 52	☽ ⚹ ☉
01 26	☽ ⚹ ♂
03 21	☽ ∥ ♄
04 16	☽ ∠ ♆
04 20	☽ △ ♃
04 55	☽ ⊻ ♅
06 50	☽ ⚹ ☿
13 03	☉ △ ♂
13 17	☉ ∥ ☿
18 06	☿ ∠ ♇
18 56	☽ ∥ ♆
19 38	☽ ⊥ ♆
22 54	☽ Q ☉
22 55	☽ ∥ ♅
10 Wednesday	
03 41	☽ ∠ ♂
04 42	☽ ⊻ ♀
05 51	☽ ⊻ ♆
06 22	☽ Q ☿
07 28	☽ ☌ ♄
09 54	☽ ⊻ ♇
14 59	☿ Q ♄
15 32	☽ ⊥ ♀
19 09	♀ ⚼ ♃
19 21	☽ ⊥ ♂
20 06	☽ ⊥ ♇
21 15	☉ ☍ ♃
11 Thursday	
00 03	♀ ⚹ ♆
01 28	☽ ∥ ♂
05 46	☽ ⊻ ♂
06 53	☽ □ ♃
07 41	☽ □ ☉
07 43	☽ ∠ ♀
07 57	☽ ☌ ♅
11 22	☽ ∠ ♇
11 30	☉ ∠ ♀
11 45	☉ □ ♅
11 48	♀ ∠ ♅
17 21	☽ □ ☿
12 Friday	
03 40	♂ Q ♇
06 48	♀ ⊻ ♄
08 36	☽ ☌ ♆
10 24	☽ ⊻ ♄
10 36	☽ ⚹ ♀
12 44	☽ ⚹ ♇
12 59	☽ Q ♃
13 44	♂ ⚹ ♃
20 33	☽ ⊥ ♄
13 Saturday	
08 03	☽ Q ♀
09 01	☽ Q ♇
09 02	☽ ⚹ ♃
09 45	☽ ☌ ♂
10 38	☽ ⊻ ♅
11 48	☽ ∠ ♄
14 02	☽ △ ☉
15 02	☉ ∠ ♇
19 50	☽ ∥ ♂
20 47	☽ ⊥ ♅
23 00	☿ ∥ ♇
14 Sunday	
01 50	♀ ☌ ♇
03 15	☽ △ ☿
10 10	☽ ∠ ♃
11 16	☽ ⊻ ♆
12 02	☽ ∠ ♅
13 17	☽ ⚹ ♄
15 33	☽ □ ♇
16 20	☽ □ ♀
17 18	☽ ⚼ ☉
19 38	♆ St D
19 50	♂ ⊻ ♅
19 54	☽ ∥ ♅
20 30	☿ ⊥ ♇
21 28	☽ ⊥ ♆
23 54	☽ ∥ ♆
15 Monday	
01 20	☽ ⊥ ♃
08 22	☽ ⚼ ☿
09 47	☽ ± ☉
09 49	☽ Q ♄
10 16	☉ Q ♄
11 10	☿ ⊥ ♀
11 29	☽ ⊻ ♃
12 50	☽ ∠ ♆
13 38	☽ ⚹ ♅
14 08	☽ ⊻ ♂
15 10	☽ ∥ ♄
18 34	♀ Q ♂
20 51	☽ ⊼ ☉
22 06	☽ ∥ ♀
16 Tuesday	
00 45	☽ ⊥ ♂
02 00	☽ ± ☿
10 21	☽ Q ♅
13 53	☽ ⊼ ☿
14 44	☽ ⚹ ♆
16 49	☽ ∠ ♂
17 02	☽ □ ♄
19 14	☽ △ ♇
20 42	☿ □ ♆
22 59	☽ △ ♀
17 Wednesday	
01 06	♂ ∠ ♄
11 47	☽ Q ♆
15 12	☽ ☌ ♃
16 09	☿ ⚹ ♄
18 00	☽ □ ♅
20 04	☽ ⚹ ♂
20 09	☽ ∥ ♃
21 46	☽ ⚼ ♇
18 Thursday	
03 06	☽ ⚼ ♀
05 34	☽ ☍ ☉
10 02	☿ ⊻ ♇
13 00	☿ # ♇
13 59	☽ ± ♇
14 05	☽ # ☉
14 18	☽ ∥ ☿
14 27	☽ # ♇
15 23	☉ # ☿
18 22	☽ Q ♂
20 05	☽ □ ♆
20 19	☽ ± ♀
20 22	☉ ∥ ♇
22 40	☿ Q ♅
22 45	☽ △ ♄
19 Friday	
00 56	☽ ⊼ ♇
03 05	☽ ☍ ☿
03 22	☽ ∥ ☉
03 47	☽ ∥ ♇
05 53	☽ # ☿
07 57	☽ ⊼ ♀
21 14	☽ ⊻ ♃
23 05	☽ # ♃
20 Saturday	
00 48	☽ △ ♅
02 40	☽ ⚼ ♄
04 47	☽ □ ♂
08 24	☽ ⊥ ♃
10 29	♀ ∥ ♄
17 34	☽ ⊼ ☉
19 48	☽ ± ♄
21 Sunday	
01 21	☽ ∠ ♃
04 16	☽ △ ♆
05 21	☽ ⚼ ♅
06 02	☽ ± ☉
07 23	☽ ⊼ ♄
09 32	☽ ☍ ♇
18 01	☿ ⊻ ♀
20 11	☽ ☍ ♀
20 23	☽ ⊼ ☿
22 51	☽ ± ♅
22 Monday	
01 01	☽ ⚼ ☉
06 12	☽ ⚹ ♃
09 13	☽ # ♄
09 30	☽ ⚼ ♆
09 57	☽ ± ☿
10 39	☽ ⊼ ♅
12 47	☽ # ♀
16 56	☽ △ ♂
23 Tuesday	
02 14	☽ # ♆
03 21	☽ ± ♆
04 03	☉ ⊥ ♇
05 44	☽ Q ♃
06 33	☽ ⚼ ☿
06 57	☽ # ♅
09 20	☽ △ ☉
15 21	☽ ⊼ ♆
18 55	☽ ☍ ♄
20 59	☽ ⊼ ♇
23 52	♀ ⊥ ♅
24 Wednesday	
00 05	☽ ⚼ ♂
00 15	☽ # ♂
09 06	☽ ± ♇
11 18	☽ ⊼ ♀
17 22	☽ △ ☿
17 33	☽ □ ♃
18 45	☿ ⊼ ♃
18 58	☽ ± ♂
22 52	☽ ☍ ♅
25 Thursday	
00 06	☽ ± ♀
03 21	☽ ⚼ ♇
07 36	☽ ⊼ ♂
12 46	☿ ∥ ♃
19 17	☽ ⚼ ♀
26 Friday	
03 17	☽ □ ☉
03 57	☽ ☍ ♆
07 52	☽ ⊼ ♄
09 45	☽ △ ♇
11 21	☉ □ ♆
11 38	☿ ⚹ ♅
20 01	☽ ± ♄
27 Saturday	
03 03	☽ △ ♀
05 27	☽ △ ♃
08 49	☿ ∠ ♄
10 50	☉ # ♇
11 25	☽ ⊼ ♅
14 07	☽ ⚼ ♄
14 52	☽ □ ☿
18 34	☽ Q ☉
22 21	☽ ☍ ♂
23 21	☽ ± ♅
28 Sunday	
00 17	☿ Q ♆
08 49	☿ ± ♃
10 52	☽ ⚼ ♃
13 32	☽ # ♂
15 08	☉ ⚹ ♄
15 42	☽ ⊼ ♆
17 04	☽ ⚼ ♅
19 50	☽ △ ♄
20 12	☽ ⚹ ☉
21 27	☽ □ ♇
22 42	♀ △ ♃
29 Monday	
00 51	♂ ⊥ ♅
01 36	☽ # ♅
03 22	☽ ± ♆
04 13	☽ ± ♃
06 14	☽ # ♆
07 07	☽ Q ☿
08 15	☽ # ♀
12 23	☉ ⊻ ♇
15 37	☽ ⊼ ♃
16 31	☽ □ ♀
20 37	☽ ⚼ ♆
21 59	☽ △ ♅
22 05	☽ # ♄
30 Tuesday	
03 23	☽ ∠ ☉
07 09	☉ Q ♅
07 56	☽ Q ♇
09 00	☽ ⚹ ☿
10 33	☽ ⊼ ♂
16 26	♀ ∥ ♆
21 36	☽ ⊥ ☉
22 10	☽ ± ♂

OCTOBER 1986

Date	Sidereal time h m s	Sun ☉ ° ' "	Moon ☽ ° ' "	Moon ☽ 24.00 ° ' "	Mercury ☿ ° '	Venus ♀ ° '	Mars ♂ ° '	Jupiter ♃ ° '	Saturn ♄ ° '	Uranus ♅ ° '	Neptune ♆ ° '	Pluto ♇ ° '
01	12 39 43	08 ♎ 00 49	09 ♍ 16 08	15 ♍ 53 52	26 ♎ 20	16 ♏ 50	26 ♑ 02	15 ♓ 16	05 ♐ 25	18 ♐ 52	03 ♑ 07	06 ♏ 08
02	12 43 39	08 59 51	22 ♍ 36 37	29 ♍ 24 15	27 48	17 17	26 32	15 R 09	05 30	18 54	03 07	06 10
03	12 47 36	09 58 55	06 ♎ 13 26	13 ♎ 13 02	29 ♎ 16	17 43	27 03	15 03	05 35	18 56	03 08	06 13
04	12 51 32	10 58 02	20 ♎ 13 26	27 ♎ 17 15	00 ♏ 42	18 07	27 34	14 57	05 40	18 57	03 09	06 15
05	12 55 29	11 57 10	04 ♏ 23 55	11 ♏ 32 52	02 07	18 29	28 06	14 50	05 45	18 59	03 09	06 17
06	12 59 25	12 56 21	18 ♏ 43 31	25 ♏ 55 18	03 32	18 50	28 38	14 44	05 50	19 01	03 10	06 19
07	13 03 22	13 55 33	03 ♐ 07 36	10 ♐ 19 53	04 55	19 09	29 10	14 38	05 55	19 03	03 11	06 22
08	13 07 18	14 54 47	17 ♐ 31 39	24 ♐ 42 25	06 18	19 25	29 ♑ 42	14 32	06 01	19 05	03 12	06 24
09	13 11 15	15 54 03	01 ♑ 51 47	08 ♑ 59 23	07 39	19 40	00 ♒ 15	14 26	06 06	19 07	03 12	06 26
10	13 15 12	16 53 21	16 ♑ 04 55	23 ♑ 08 07	09 00	19 53	00 48	14 21	06 11	19 09	03 13	06 28
11	13 19 08	17 52 40	00 ♒ 08 45	07 ♒ 06 39	10 19	20 04	01 21	14 16	06 17	19 12	03 14	06 31
12	13 23 05	18 52 01	14 ♒ 01 41	20 ♒ 53 42	11 37	20 12	01 55	14 10	06 22	19 14	03 15	06 33
13	13 27 01	19 51 24	27 ♒ 42 36	04 ♓ 28 20	12 53	20 18	02 29	14 05	06 28	19 16	03 16	06 35
14	13 30 58	20 50 48	11 ♓ 10 47	17 ♓ 49 56	14 08	20 22	03 03	14 00	06 34	19 18	03 17	06 38
15	13 34 54	21 50 15	24 ♓ 25 42	00 ♈ 58 04	15 22	20 24	03 37	13 56	06 39	19 21	03 18	06 40
16	13 38 51	22 49 43	07 ♈ 27 02	13 ♈ 52 34	16 34	20 R 23	04 12	13 51	06 45	19 23	03 19	06 43
17	13 42 47	23 49 13	20 ♈ 14 43	26 ♈ 33 31	17 45	20 20	04 47	13 47	06 51	19 25	03 20	06 45
18	13 46 44	24 48 45	02 ♉ 49 04	09 ♉ 01 27	18 53	20 14	05 22	13 42	06 57	19 28	03 21	06 47
19	13 50 41	25 48 19	15 ♉ 10 51	21 ♉ 17 25	20 00	20 06	05 57	13 38	07 02	19 30	03 22	06 50
20	13 54 37	26 47 56	27 ♉ 21 25	03 ♊ 23 05	21 04	19 56	06 32	13 34	07 08	19 33	03 23	06 52
21	13 58 34	27 47 34	09 ♊ 22 45	15 ♊ 20 46	22 06	19 43	07 08	13 31	07 14	19 35	03 24	06 55
22	14 02 30	28 47 15	21 ♊ 17 32	27 ♊ 13 29	23 05	19 28	07 44	13 27	07 20	19 38	03 25	06 57
23	14 06 27	29 ♎ 46 58	03 ♋ 09 04	09 ♋ 04 48	24 01	19 11	08 20	13 24	07 26	19 41	03 27	06 59
24	14 10 23	00 ♏ 46 43	15 ♋ 01 13	20 ♋ 58 51	24 54	18 51	08 56	13 21	07 33	19 43	03 28	07 02
25	14 14 20	01 46 30	26 ♋ 58 18	03 ♌ 00 07	25 43	18 29	09 33	13 18	07 39	19 46	03 29	07 04
26	14 18 16	02 46 20	09 ♌ 04 54	15 ♌ 13 13	26 28	18 05	10 09	13 15	07 45	19 49	03 31	07 07
27	14 22 13	03 46 12	21 ♌ 25 37	27 ♌ 42 38	27 09	17 38	10 46	13 13	07 51	19 51	03 32	07 09
28	14 26 10	04 46 06	04 ♍ 04 46	10 ♍ 32 25	27 45	17 10	11 23	13 10	07 58	19 54	03 33	07 11
29	14 30 06	05 46 02	17 ♍ 05 56	23 ♍ 45 35	28 15	16 41	12 00	13 08	08 04	19 57	03 35	07 14
30	14 34 03	06 46 00	00 ♎ 31 30	07 ♎ 23 43	28 39	16 09	12 38	13 06	08 10	20 00	03 36	07 16
31	14 37 59	07 ♏ 46 00	14 ♎ 22 07	21 ♎ 26 25	28 ♏ 57	15 ♏ 37	13 ♒ 15	13 ♓ 05	08 ♐ 17	20 ♐ 03	03 ♑ 37	07 ♏ 19

DECLINATIONS

Date	Moon True ☊ ° '	Moon Mean ☊ ° '	Moon ☽ Latitude ° '	Sun ☉ ° '	Moon ☽ ° '	Mercury ☿ ° '	Venus ♀ ° '	Mars ♂ ° '	Jupiter ♃ ° '	Saturn ♄ ° '	Uranus ♅ ° '	Neptune ♆ ° '	Pluto ♇ ° '
01	20 ♈ 54	21 ♈ 20	03 N 30	03 S 11	11 N 20	11 S 00	22 S 32	24 S 21	07 S 12	19 S 37	23 S 05	22 S 21	01 N 32
02	20 R 53	21 17	02 30	03 34	05 N 14	11 39	22 46	24 13	07 14	19 38	23 05	22 21	01 31
03	20 52	21 14	01 20	03 57	01 S 16	12 17	22 58	24 04	07 17	19 39	23 05	22 21	01 31
04	20 52	21 11	00 N 04	04 20	07 51	12 54	23 10	23 55	07 19	19 40	23 05	22 21	01 30
05	20 D 52	21 08	01 S 14	04 44	14 09	13 30	23 22	23 46	07 21	19 41	23 05	22 21	01 29
06	20 52	21 05	02 28	05 07	19 46	14 06	23 32	23 37	07 24	19 42	23 05	22 21	01 28
07	20 53	21 01	03 33	05 30	24 16	14 41	23 42	23 28	07 26	19 43	23 06	22 21	01 27
08	20 54	20 58	04 25	05 53	27 15	15 15	23 51	23 19	07 28	19 44	23 06	22 21	01 26
09	20 54	20 55	05 00	06 15	28 26	15 48	23 59	23 09	07 30	19 45	23 06	22 21	01 25
10	20 54	20 52	05 16	06 38	27 42	16 20	24 06	22 59	07 32	19 46	23 06	22 21	01 24
11	20 R 54	20 49	05 13	07 01	25 13	16 52	24 12	22 50	07 34	19 47	23 06	22 21	01 23
12	20 54	20 45	04 52	07 24	21 16	17 22	24 17	22 40	07 36	19 48	23 07	22 21	01 22
13	20 54	20 42	04 14	07 46	16 14	17 51	24 22	22 29	07 38	19 50	23 07	22 21	01 21
14	20 54	20 39	03 23	08 08	10 30	18 20	24 25	22 19	07 40	19 51	23 07	22 21	01 21
15	20 D 54	20 36	02 22	08 31	04 S 23	18 47	24 27	22 09	07 41	19 52	23 07	22 21	01 20
16	20 54	20 33	01 14	08 53	01 N 50	19 13	24 28	21 58	07 43	19 53	23 07	22 22	01 19
17	20 54	20 30	00 S 04	09 15	07 51	19 38	24 28	21 47	07 44	19 54	23 08	22 22	01 18
18	20 R 54	20 26	01 N 06	09 37	13 29	20 01	24 27	21 36	07 46	19 55	23 08	22 22	01 17
19	20 54	20 23	02 11	09 58	18 28	20 24	24 25	21 25	07 47	19 56	23 08	22 22	01 16
20	20 53	20 20	03 09	10 20	22 38	20 45	24 21	21 14	07 49	19 57	23 08	22 22	01 15
21	20 52	20 17	03 57	10 41	25 46	21 04	24 16	21 03	07 50	19 59	23 08	22 22	01 14
22	20 51	20 14	04 35	11 03	27 44	21 23	24 10	20 51	07 51	20 00	23 09	22 22	01 14
23	20 50	20 11	05 02	11 24	28 26	21 39	24 02	20 39	07 52	20 01	23 09	22 22	01 13
24	20 49	20 07	05 15	11 45	27 49	21 54	23 53	20 28	07 53	20 02	23 09	22 22	01 12
25	20 49	20 04	05 15	12 06	25 54	22 07	23 43	20 16	07 54	20 03	23 09	22 22	01 11
26	20 D 49	20 01	05 01	12 26	22 49	22 19	23 31	20 03	07 55	20 04	23 09	22 22	01 10
27	20 49	19 58	04 32	12 47	18 40	22 28	23 18	19 51	07 56	20 05	23 10	22 22	01 10
28	20 50	19 55	03 51	13 07	13 36	22 35	23 04	19 39	07 56	20 06	23 10	22 22	01 09
29	20 51	19 51	02 56	13 27	07 48	22 40	22 48	19 26	07 57	20 08	23 10	22 22	01 08
30	20 53	19 48	01 50	13 47	01 29	22 42	22 32	19 13	07 58	20 09	23 10	22 22	01 07
31	20 ♈ 53	19 ♈ 45	00 N 36	14 S 06	05 S 07	22 S 42	22 S 13	19 S 01	07 S 58	20 S 10	23 S 10	22 S 22	01 N 06

ZODIAC SIGN ENTRIES

Date	h	m	Planets	
03	01	03	☽	♎
04	00	19	☿	♏
05	04	35	☽	♏
07	06	48	☽	♐
09	01	01	♂	♒
09	08	52	☽	♑
11	11	45	☽	♒
13	16	03	☽	♓
15	22	13	☽	♈
18	06	35	☽	♉
20	17	15	☽	♊
23	05	37	☽	♋
23	17	14	☉	♏
25	18	02	☽	♌
28	04	20	☽	♍
30	11	05	☽	♎

LATITUDES

Date	Mercury ☿ ° '	Venus ♀ ° '	Mars ♂ ° '	Jupiter ♃ ° '	Saturn ♄ ° '	Uranus ♅ ° '	Neptune ♆ ° '	Pluto ♇ ° '
01	00 S 54	05 S 55	03 S 28	01 S 30	01 N 37	00 S 06	01 N 03	15 N 58
04	01 16	06 12	03 20	01 30	01 37	00 06	01 03	15 57
07	01 37	06 26	03 13	01 30	01 36	00 06	01 03	15 57
10	01 57	06 38	03 05	01 29	01 36	00 06	01 03	15 56
13	02 15	06 47	02 58	01 29	01 35	00 06	01 03	15 56
16	02 32	06 53	02 50	01 28	01 35	00 06	01 03	15 55
19	02 45	06 54	02 43	01 28	01 34	00 06	01 03	15 55
22	02 56	06 50	02 36	01 27	01 34	00 06	01 02	15 54
25	03 01	06 39	02 29	01 27	01 34	00 06	01 02	15 54
28	03 00	06 22	02 22	01 26	01 33	00 06	01 02	15 54
31	02 S 51	05 S 58	02 S 16	01 S 26	01 N 33	00 S 06	01 N 02	15 N 54

DATA

Julian Date	2446705
Delta T	+55 seconds
Ayanamsa	23° 40' 13"
Synetic vernal point	05° ♓ 26' 47"
True obliquity of ecliptic	23° 26' 37"

MOON'S PHASES, APSIDES AND POSITIONS ☽

Date	h	m	Phase	Longitude ° '	Eclipse Indicator
03	18	55	●	10 ♎ 16	Ann-Total
10	13	28	☽	16 ♑ 57	
17	19	22	○	24 ♈ 07	total
25	22	26	☾	02 ♌ 12	

Day	h	m		
07	09	45	Perigee	
23	05	30	Apogee	
03	07	22	0S	
09	14	45	Max dec	28° S 26'
16	04	55	0N	
23	12	39	Max dec	28° N 26'
30	17	25	0S	

LONGITUDES

Date	Chiron ⚷ ° '	Ceres ⚳ ° '	Pallas ⚴ ° '	Juno ⚵ ° '	Vesta ⚶ ° '	Black Moon Lilith ⚸ ° '
01	21 ♊ 22	25 ♎ 38	03 ♎ 56	10 ♐ 48	10 ♈ 12	24 ♊ 04
11	21 ♊ 17	00 ♏ 02	08 ♎ 31	13 ♐ 32	07 ♈ 40	25 ♊ 11
21	21 ♊ 04	04 ♏ 26	13 ♎ 04	16 ♐ 30	05 ♈ 26	26 ♊ 18
31	20 ♊ 44	08 ♏ 50	17 ♎ 33	19 ♐ 39	03 ♈ 46	27 ♊ 25

All ephemeris data is given at 12.00 UT and the Moon's longitude is additionally given for 24.00 UT

ASPECTARIAN

h m Aspects

01 Wednesday
00 43 ☽ △ ♆
03 35 ☽ Q ♀
04 27 ☿ □ ♂
04 55 ☽ □ ♄
06 15 ☽ ⚹ ♇
09 32 ☽ ⊻ ☉
13 15 ☽ # ☿
15 19 ☽ ₽ ♂
16 14 ☽ ∠ ☿
22 47 ☽ ☍ ♃

02 Thursday
02 10 ☽ ⚹ ♀
04 23 ☽ # ♃
05 22 ☽ □ ♅
09 26 ☽ ∠ ♇
10 24 ☽ ⊥ ☿
13 35 ☽ Q ♄
17 53 ☽ # ☉
18 41 ♃ Q ♆
19 13 ☽ △ ♂
22 17 ☽ ⊻ ☿

03 Friday
01 23 ☽ ⊥ ♇
01 49 ☽ ∥ ♇
05 37 ☽ ∠ ♀
06 32 ☽ □ ♆
10 47 ☽ ⚹ ♄
11 53 ☽ ⊻ ♇
12 52 ☽ # ♇
13 08 ☽ Q ♅
17 43 ☿ ⊥ ♄
18 55 ☽ ☌ ☉
21 42 ☽ ⊥ ♀
22 22 ☽ ∥ ☉

04 Saturday
00 13 ☿ ₽ ♃
01 19 ♀ ∥ ♅
03 02 ☽ ⊼ ♃
08 18 ☽ ⊻ ♀
09 50 ☽ ⚹ ♅
10 01 ☽ ∥ ♃
12 46 ☽ ∠ ♄
13 13 ☽ ± ♃
13 34 ☽ Q ♆
13 39 ♀ ∠ ♆

05 Sunday
00 57 ☽ □ ♂
04 07 ☽ ⊥ ♄
04 22 ☽ ₽ ♃
07 45 ☽ ☌ ☿
09 09 ☽ ∥ ☿
09 54 ☽ ⚹ ♆
11 18 ☽ ∠ ♅
14 17 ☽ ⊻ ♄
15 11 ☽ ☌ ♇

06 Monday
01 37 ☽ ⊻ ☉
02 27 ☽ ⊥ ♅
05 23 ☽ △ ♃
05 43 ☿ ⚹ ♆
08 08 ☉ ⊥ ♀
08 22 ☽ Q ♂
11 04 ☽ ∠ ♆
11 39 ☽ ∥ ♄
12 11 ☽ ☌ ♀
12 23 ☽ ⊥ ☉
12 30 ☽ ⊻ ♅
18 43 ♀ ∥ ♂
20 36 ☿ ∠ ♅

07 Tuesday
00 53 ☽ ∥ ♆
02 05 ☽ ⊥ ♆
03 48 ♀ ⊻ ♅
04 29 ☽ ∠ ☉
04 57 ☽ ∥ ♅
05 09 ☽ ⚹ ♂
07 17 ☽ ∥ ♂
08 18 ☽ ∥ ♀
12 05 ☽ ⊻ ♆
15 18 ☽ ⊻ ☿
16 41 ☽ ☌ ♄
17 24 ☽ ⊻ ♇

08 Wednesday
02 22 ☽ ⊥ ☿
03 25 ☽ ⊥ ♇
03 40 ☉ ⊼ ♃
05 43 ♂ ∠ ♃
06 38 ☿ ⊻ ♄
07 03 ☽ □ ♃
07 06 ☽ ∠ ♂
07 19 ☽ ⚹ ☉
13 49 ☿ ☌ ♇
14 37 ☽ ☌ ♅
15 13 ☽ ⊻ ♀
18 29 ☽ ∠ ♇
18 57 ☽ ∠ ☿
22 44 ☽ ⊥ ♂

09 Thursday
01 26 ☽ ⊥ ♀
04 51 ☽ Q ☉
09 11 ☽ ⊻ ♂
12 58 ☽ Q ♃
14 16 ☽ ☌ ♆
16 48 ☽ ∠ ♀
19 10 ☽ ⊻ ♄
19 43 ☽ ⚹ ♇
19 46 ♂ ∥ ♅
22 46 ☽ ⚹ ☿

10 Friday
05 22 ☽ ⊥ ♄
09 05 ☽ ⚹ ♃
13 28 ☽ □ ☉
16 04 ☽ Q ♇
17 14 ☽ ⊻ ♅
18 33 ☽ ⚹ ♀
20 44 ☽ ∠ ♄
21 13 ☽ Q ☿

11 Saturday
03 30 ☽ ⊥ ♅
10 29 ☽ ∠ ♃
14 10 ☽ ☌ ♂
15 20 ☽ Q ♀
17 19 ☽ ⊻ ♆
18 49 ☽ ∥ ♀
18 59 ☽ ∠ ♅
22 38 ☽ ⚹ ♄
23 00 ☽ □ ♇

12 Sunday
01 45 ☽ ∥ ♅
01 54 ☽ ⊥ ♃
03 41 ☽ ⊥ ♆
04 03 ☽ ∥ ♂
06 05 ☽ ∥ ♆
07 22 ☽ □ ☿
12 15 ☽ ⊻ ♃
19 22 ☽ ∠ ♆
19 25 ☽ ∥ ♄
19 38 ☽ Q ♄
21 06 ☽ △ ☉
21 06 ☽ ⚹ ♅
21 08 ☉ ⚹ ♅
22 52 ☽ □ ♀

13 Monday
02 26 ☉ ∥ ♃
03 01 ♀ Q ♂
05 21 ☽ ∥ ☿
17 09 ☉ ± ♃
18 19 ☽ Q ♅
19 27 ☿ ⊥ ♅
20 50 ☽ ⊻ ♂
21 52 ☽ ⚹ ♆
23 44 ☉ ⊻ ♀

14 Tuesday
01 42 ☽ ₽ ☉
03 40 ☽ □ ♄
03 50 ☽ △ ♇
06 37 ♂ ∥ ♆
08 00 ☽ ⊥ ♂
09 32 ☿ △ ♃
17 03 ☽ ☌ ♃
17 53 ☽ △ ☿
19 08 ☽ ± ☉
19 24 ☽ Q ♆
20 50 ☽ ∥ ☉
21 58 ♂ ⊻ ♆
22 40 ☉ Q ♆
23 12 ☽ ∥ ♃

15 Wednesday
00 57 ☽ ∠ ♂
02 43 ☽ □ ♅
04 39 ☽ △ ♀
06 53 ☽ ⊼ ☉
06 57 ☽ ₽ ♇
07 05 ☉ ∠ ♄
16 33 ♀ St R
19 04 ♄ ⊻ ♇
23 29 ☽ ± ♇
23 49 ☽ # ♇

16 Thursday
00 01 ☽ ₽ ☿
04 19 ☽ □ ♆
05 41 ☽ ⚹ ♂
08 10 ☽ ₽ ♀
10 01 ☽ ∥ ♇
10 37 ☽ ⊼ ♇
10 41 ☽ △ ♄
18 24 ☽ ± ☿
20 14 ♂ ∠ ♅
23 53 ☽ ⊻ ♃

17 Friday
00 55 ☽ ± ♀
05 08 ☽ Q ♂
06 48 ☽ ⊼ ☿
10 26 ☽ △ ♅
11 07 ☽ ⊥ ♃
11 31 ☽ # ♃
12 10 ☽ ⊼ ♀
15 03 ☽ ₽ ♄
18 07 ☽ # ☉
19 22 ☽ ☍ ☉

18 Saturday
00 24 ☿ ∠ ♆
04 09 ☽ ∠ ♃
05 13 ☿ ∥ ♄
08 22 ☽ ± ♄
13 01 ☽ △ ♆
15 11 ☽ ₽ ♅
17 09 ☽ □ ♂
19 41 ☽ ☍ ♇
20 02 ☽ ⊼ ♄

19 Sunday
00 51 ☿ ⊻ ♅
08 43 ☽ ± ♅
09 00 ☽ ⚹ ♃
14 08 ☿ ☌ ♀
18 15 ☽ ₽ ♆
19 54 ☽ # ♄
20 31 ☽ ⊼ ♅
21 33 ☽ ☍ ♀
22 22 ☽ ☍ ☿
23 25 ☽ # ☿

20 Monday
03 42 ☽ # ♂
08 29 ☽ Q ♃
10 15 ☽ # ♆
10 48 ☽ ⊼ ☉
12 03 ☽ ± ♆
15 21 ☽ # ♅
23 47 ☽ # ♀
23 48 ☽ ± ☉

21 Tuesday
00 01 ☽ ⊼ ♆
02 13 ♂ □ ♇
07 02 ☽ ⊼ ♇
07 16 ☽ △ ♂
07 40 ☽ ☍ ♄
10 32 ☿ ∥ ♂
17 03 ♂ ⚹ ♄
19 07 ☽ ± ♇
19 29 ☽ ₽ ☉
20 16 ☽ □ ♃
23 00 ♀ ⊻ ♅

22 Wednesday
01 48 ♂ ⊥ ♃
04 24 ☉ ₽ ♃
08 24 ☽ ⊼ ♀
08 38 ☽ ☍ ♅
13 20 ☽ ₽ ♇
15 04 ☽ ₽ ♂
15 56 ☽ ⊼ ☿
20 15 ☽ ± ♀

23 Thursday
04 33 ☽ △ ☉
05 08 ☽ ± ☿
10 15 ☽ ± ♂
12 36 ☽ ☍ ♆
14 01 ☽ ₽ ♀
19 48 ☽ △ ♇
20 46 ☽ ⊼ ♄
23 03 ☽ ⊼ ♂

24 Friday
00 51 ☽ ₽ ☿
08 38 ☽ △ ♃
09 00 ☽ ± ♄
19 29 ☽ △ ♀
21 30 ☽ ⊼ ♅

25 Saturday
03 16 ☽ ₽ ♄
08 33 ☉ ⊥ ♄
09 19 ☽ △ ☿
09 34 ♂ ⊥ ♆
09 35 ☽ ± ♅
11 27 ♀ ∠ ♆
14 38 ☽ ₽ ♃
22 26 ☽ □ ☉

26 Sunday
00 59 ☽ ⊼ ♆
03 33 ☽ ₽ ♅
06 55 ☽ # ♀
08 06 ☽ □ ♇
08 25 ☽ ± ♃
09 22 ☽ △ ♄
09 43 ☽ # ♅
10 36 ♂ ∥ ♄
12 50 ☽ ± ♆
14 13 ☽ ☍ ♂
14 56 ☽ # ♆
15 07 ☽ # ☿
19 16 ☿ ∥ ♆
20 08 ☽ ⊼ ♃

27 Monday
04 23 ☽ # ♄
04 57 ☽ □ ♀
05 21 ☽ # ♂
06 07 ☉ ⚹ ♆
06 25 ☽ ₽ ♆
08 58 ☽ △ ♅
12 43 ☽ Q ☉
19 09 ☽ Q ♇
23 30 ☽ □ ☿

28 Tuesday
02 40 ♀ ∥ ♅
03 25 ☿ ⊥ ♆
11 01 ☽ △ ♆
13 24 ☽ ⚹ ☉
13 58 ☽ Q ♀
13 59 ☽ # ☉
15 27 ☉ ∠ ♅
17 49 ☽ ⚹ ♇
19 17 ☽ □ ♄

29 Wednesday
02 14 ☽ ⊼ ♂
04 47 ☽ ☍ ♃
10 24 ☽ Q ☿
11 16 ☽ ⚹ ♀
11 24 ☽ # ♃
13 44 ☽ ± ♂
17 11 ☽ □ ♅
19 10 ☽ ∠ ☉
21 17 ☽ ∠ ♇
22 22 ☿ ∥ ♀

30 Thursday
04 14 ☽ Q ♄
06 38 ☽ ₽ ♂
08 37 ☽ ⚹ ☿
12 28 ☽ ⊥ ☉
13 04 ☽ ∠ ♀
13 19 ☽ ∥ ♇
13 19 ☽ ⊥ ♇
17 24 ☽ □ ♆
21 30 ☽ # ♇
23 46 ☽ ⊻ ☉
23 49 ☽ ⊻ ♇

31 Friday
00 38 ☉ ☌ ♇
01 06 ☽ Q ♅
01 19 ♀ ∥ ♆
01 27 ☽ ⚹ ♄
04 08 ☽ ⊥ ♀
05 25 ♂ ⊻ ♃
09 48 ☽ ⊼ ♃
10 00 ☽ △ ♂
11 16 ☽ ∠ ☿
14 02 ☽ ⊻ ♀
19 59 ☽ ± ♃
21 41 ☽ ⚹ ♅
22 24 ☽ ∥ ♃

NOVEMBER 1986

Date	Sidereal time h m s	Sun ☉ ° ' "	Moon ☽ ° ' "	Moon ☽ 24.00 ° ' "	Mercury ☿ ° '	Venus ♀ ° '	Mars ♂ ° '	Jupiter ♃ ° '	Saturn ♄ ° '	Uranus ♅ ° '	Neptune ♆ ° '	Pluto ♇ ° '
01	14 41 56	08 ♏ 46 03	28 ♎ 36 12	05 ♏ 50 54	29 ♏ 07	15 ♏ 03	13 ♒ 53	13 ♓ 03	08 ♐ 23	20 ♐ 06	03 ♑ 39	07 ♏ 21
02	14 45 52	09 46 07	13 ♏ 09 48	20 ♏ 32 04	29 R 10	14 R 28	14 31	13 R 02	08 30	20 09	03 40	07 24
03	14 49 49	10 46 13	27 ♏ 56 44	05 ♐ 22 50	29 03	13 52	15 09	13 01	08 36	20 12	03 42	07 26
04	14 53 45	11 46 21	12 ♐ 49 20	20 ♐ 15 12	28 48	13 16	15 47	13 00	08 43	20 15	03 43	07 28
05	14 57 42	12 46 31	27 ♐ 39 30	05 ♑ 01 20	28 23	12 40	16 25	12 59	08 50	20 18	03 45	07 31
06	15 01 39	13 46 42	12 ♑ 19 58	19 ♑ 34 44	27 49	12 03	17 04	12 58	08 56	20 21	03 47	07 33
07	15 05 35	14 46 55	26 ♑ 45 10	03 ♒ 50 55	27 04	11 27	17 42	12 58	09 03	20 24	03 48	07 36
08	15 09 32	15 47 09	10 ♒ 51 45	17 ♒ 47 36	26 11	10 51	18 21	12 D 58	09 10	20 27	03 50	07 38
09	15 13 28	16 47 25	24 ♒ 38 28	01 ♓ 24 27	25 08	10 16	19 00	12 58	09 16	20 30	03 52	07 41
10	15 17 25	17 47 42	08 ♓ 05 44	14 ♓ 42 31	23 59	09 42	19 39	12 59	09 23	20 34	03 53	07 43
11	15 21 21	18 48 00	21 ♓ 15 04	27 ♓ 43 40	22 43	09 09	20 18	12 59	09 30	20 37	03 55	07 45
12	15 25 18	19 48 20	04 ♈ 08 34	10 ♈ 30 03	21 24	08 37	20 57	13 00	09 37	20 40	03 57	07 48
13	15 29 14	20 48 41	16 ♈ 48 23	23 ♈ 03 48	20 04	08 07	21 36	13 01	09 44	20 43	03 59	07 50
14	15 33 11	21 49 04	29 ♈ 16 32	05 ♉ 26 48	18 45	07 39	22 16	13 02	09 50	20 47	04 00	07 53
15	15 37 08	22 49 28	11 ♉ 34 46	17 ♉ 40 38	17 30	07 12	22 55	13 03	09 57	20 50	04 02	07 55
16	15 41 04	23 49 54	23 ♉ 44 34	29 ♉ 46 44	16 21	06 48	23 35	13 05	10 04	20 53	04 04	07 57
17	15 45 01	24 50 22	05 ♊ 47 17	11 ♊ 46 26	15 21	06 26	24 15	13 07	10 11	20 57	04 06	08 00
18	15 48 57	25 50 51	17 ♊ 44 21	23 ♊ 41 16	14 31	06 06	24 54	13 09	10 18	21 00	04 08	08 02
19	15 52 54	26 51 22	29 ♊ 37 25	05 ♋ 33 05	13 52	05 48	25 34	13 11	10 25	21 03	04 10	08 04
20	15 56 50	27 51 55	11 ♋ 28 35	17 ♋ 24 15	13 25	05 33	26 14	13 13	10 32	21 07	04 11	08 07
21	16 00 47	28 52 29	23 ♋ 20 29	29 ♋ 17 42	13 09	05 20	26 54	13 16	10 39	21 10	04 13	08 09
22	16 04 43	29 ♏ 53 05	05 ♌ 16 24	11 ♌ 17 03	13 D 05	05 10	27 34	13 18	10 46	21 14	04 15	08 11
23	16 08 40	00 ♐ 53 42	17 ♌ 20 12	23 ♌ 26 25	13 12	05 02	28 15	13 21	10 53	21 17	04 17	08 14
24	16 12 37	01 54 22	29 ♌ 36 16	05 ♍ 50 22	13 29	04 57	28 55	13 25	11 00	21 21	04 19	08 16
25	16 16 33	02 55 02	12 ♍ 09 16	18 ♍ 33 33	13 55	04 54	29 ♒ 35	13 28	11 07	21 24	04 21	08 18
26	16 20 30	03 55 45	25 ♍ 03 45	01 ♎ 40 19	14 29	04 D 54	00 ♓ 16	13 31	11 14	21 28	04 23	08 20
27	16 24 26	04 56 29	08 ♎ 23 40	15 ♎ 14 03	15 12	04 56	00 56	13 35	11 21	21 31	04 25	08 23
28	16 28 23	05 57 15	22 ♎ 11 38	29 ♎ 16 22	16 00	05 01	01 37	13 39	11 28	21 35	04 27	08 25
29	16 32 19	06 58 02	06 ♏ 28 04	13 ♏ 46 18	16 55	05 07	02 18	13 43	11 36	21 38	04 29	08 27
30	16 36 16	07 ♐ 58 50	21 ♏ 10 27	28 ♏ 39 39	17 ♏ 55	05 ♏ 17	02 ♓ 58	13 ♓ 47	11 ♐ 43	21 ♐ 42	04 ♑ 31	08 ♏ 29

Date	Moon True ☊ ° '	Moon Mean ☊ ° '	Moon ☽ Latitude ° '
01	20 ♈ 53	19 ♈ 42	00 S 42
02	20 R 52	19 39	02 00
03	20 50	19 36	03 10
04	0 48	19 32	04 08
05	20 45	19 29	04 49
06	20 42	19 26	05 11
07	20 40	19 23	05 12
08	20 39	19 20	04 55
09	20 D 39	19 17	04 21
10	20 40	19 13	03 33
11	20 42	19 10	02 35
12	20 43	19 07	01 30
13	20 44	19 04	00 S 22
14	20 R 44	19 01	00 N 47
15	20 42	18 57	01 51
16	20 38	18 54	02 50
17	20 33	18 51	03 41
18	20 27	18 48	04 21
19	20 20	18 45	04 50
20	20 13	18 42	05 06
21	20 07	18 38	05 09
22	20 02	18 35	04 59
23	19 59	18 32	04 35
24	19 58	18 29	03 59
25	19 D 58	18 26	03 10
26	19 59	18 23	02 11
27	20 01	18 19	01 N 02
28	20 R 01	18 16	00 S 12
29	20 00	18 13	01 28
30	19 ♈ 57	18 ♈ 10	02 S 40

DECLINATIONS

Date	Sun ☉ ° '	Moon ☽ ° '	Mercury ☿ ° '	Venus ♀ ° '	Mars ♂ ° '	Jupiter ♃ ° '	Saturn ♄ ° '	Uranus ♅ ° '	Neptune ♆ ° '	Pluto ♇ ° '
01	14 S 26	11 S 38	22 S 39	21 S 54	18 S 47	07 S 58	20 S 11	23 S 11	22 S 22	01 N 06
02	14 45	17 42	22 33	21 34	18 34	07 59	20 12	23 11	22 22	01 05
03	15 04	22 48	22 23	21 13	18 21	07 59	20 13	23 11	22 22	01 04
04	15 22	26 26	22 10	20 50	18 08	07 59	20 14	23 11	22 22	01 03
05	15 41	28 15	21 53	20 28	17 54	07 59	20 16	23 12	22 22	01 03
06	15 59	28 02	21 33	20 04	17 40	07 59	20 17	23 12	22 22	01 02
07	16 16	25 55	21 08	19 40	17 26	07 59	20 18	23 12	22 22	01 01
08	16 34	22 14	20 39	19 15	17 13	07 59	20 19	23 12	22 22	01 01
09	16 51	17 24	20 07	18 51	16 58	07 59	20 20	23 13	22 22	01 00
10	17 08	11 49	19 31	18 26	16 44	07 58	20 21	23 13	22 22	00 59
11	17 25	05 S 50	18 53	18 01	16 30	07 58	20 22	23 13	22 21	00 59
12	17 42	00 N 16	18 13	17 36	16 15	07 57	20 23	23 13	22 21	00 58
13	17 58	06 16	17 32	17 12	16 01	07 57	20 24	23 13	22 21	00 57
14	18 13	11 57	16 51	16 48	15 46	07 56	20 26	23 14	22 21	00 57
15	18 29	17 04	16 11	16 25	15 31	07 55	20 27	23 14	22 21	00 56
16	18 44	21 28	15 35	16 02	15 16	07 55	20 28	23 14	22 21	00 55
17	18 59	24 54	15 02	15 40	15 01	07 54	20 29	23 14	22 21	00 55
18	19 13	27 13	14 33	15 19	14 46	07 53	20 30	23 15	22 21	00 54
19	19 27	28 17	14 10	14 59	14 31	07 52	20 31	23 15	22 21	00 54
20	19 41	28 02	13 52	14 40	14 15	07 51	20 32	23 15	22 21	00 53
21	19 55	26 30	13 39	14 22	14 00	07 49	20 33	23 15	22 21	00 53
22	20 08	23 47	13 31	14 05	13 44	07 48	20 34	23 16	22 21	00 52
23	20 20	20 01	13 29	13 50	13 29	07 47	20 35	23 16	22 21	00 52
24	20 33	15 20	13 31	13 35	13 13	07 45	20 36	23 16	22 21	00 51
25	20 45	09 56	13 37	13 22	12 57	07 44	20 37	23 16	22 21	00 51
26	20 56	03 N 58	13 47	13 09	12 41	07 42	20 38	23 17	22 21	00 50
27	21 07	02 S 22	14 00	12 58	12 25	07 41	20 39	23 17	22 21	00 50
28	21 18	08 49	14 16	12 48	12 09	07 39	20 40	23 17	22 21	00 49
29	21 29	15 03	14 35	12 40	11 52	07 37	20 41	23 17	22 21	00 49
30	21 S 39	20 S 37	14 S 55	12 S 32	11 S 36	07 S 35	20 S 42	23 S 17	22 S 21	00 N 48

ZODIAC SIGN ENTRIES

Date	h	m	Planets
01	14	19	☽ ♏
03	15	19	☽ ♐
05	15	48	☽ ♑
07	17	28	☽ ♒
09	21	30	☽ ♓
12	04	14	☽ ♈
14	13	24	☽ ♉
17	00	26	☽ ♊
19	12	46	☽ ♋
22	01	25	☽ ♌
22	14	44	☉ ♐
24	12	46	☽ ♍
26	02	35	♂ ♓
26	20	59	☽ ♎
29	01	13	☽ ♏

LATITUDES

Date	Mercury ☿ ° '	Venus ♀ ° '	Mars ♂ ° '	Jupiter ♃ ° '	Saturn ♄ ° '	Uranus ♅ ° '	Neptune ♆ ° '	Pluto ♇ ° '
01	02 S 45	05 S 49	02 S 13	01 S 25	01 N 33	00 S 06	01 N 02	15 N 54
04	02 20	05 16	02 07	01 25	01 32	00 06	01 02	15 54
07	01 40	04 38	02 01	01 24	01 32	00 06	01 02	15 54
10	00 S 47	03 55	01 54	01 23	01 32	00 06	01 02	15 54
13	00 N 15	03 09	01 48	01 23	01 31	00 06	01 02	15 54
16	01 12	02 23	01 42	01 22	01 31	00 06	01 01	15 54
19	01 56	01 38	01 36	01 21	01 31	00 06	01 01	15 55
22	02 21	00 54	01 31	01 21	01 31	00 06	01 01	15 55
25	02 31	00 S 13	01 25	01 20	01 30	00 06	01 01	15 56
28	02 28	00 N 25	01 20	01 19	01 30	00 06	01 01	15 56
31	02 N 17	00 N 59	01 S 15	01 S 18	01 N 30	00 S 06	01 N 01	15 N 57

DATA

Julian Date	2446736
Delta T	+55 seconds
Ayanamsa	23° 40' 16"
Synetic vernal point	05° ♓ 26' 43"
True obliquity of ecliptic	23° 26' 36"

MOON'S PHASES, APSIDES AND POSITIONS ☽

Date	h	m	Phase	Longitude ° '	Eclipse Indicator
02	06	02	●	09 ♏ 31	
08	21	11	☽	16 ♒ 10	
16	12	12	○	23 ♉ 50	
24	16	50	☾	02 ♍ 07	

Day	h	m		
04	02	23	Perigee	
19	22	01	Apogee	
05	21	25	Max dec	28° S 24'
12	10	55	ON	
19	19	30	Max dec	28° N 20'
27	03	07	OS	

LONGITUDES

Date	Chiron ⚷ ° '	Ceres ⚳ ° '	Pallas ⚴ ° '	Juno ⚵ ° '	Vesta ⚶ ° '	Black Moon Lilith ⚸ ° '
01	20 ♊ 42	09 ♏ 17	17 ♎ 59	19 ♐ 58	03 ♈ 38	27 ♊ 32
11	20 ♊ 16	13 ♏ 40	22 ♎ 24	23 ♐ 17	02 ♈ 45	28 ♊ 39
21	19 ♊ 45	18 ♏ 02	26 ♎ 44	26 ♐ 44	02 ♈ 38	29 ♊ 46
31	19 ♊ 11	22 ♏ 21	00 ♏ 57	00 ♑ 17	03 ♈ 14	00 ♋ 53

All ephemeris data is given at 12.00 UT and the Moon's longitude is additionally given for 24.00 UT

ASPECTARIAN

h m Aspects

01 Saturday
00 20 ☽ Q ♆
01 47 ☉ ⊻ ♄
02 45 ☽ ⊥ ☿
03 13 ☽ ∠ ♄
11 05 ☽ ⚼ ♃
12 52 ☽ ⊻ ☿
18 20 ☽ ⊥ ♄
20 23 ☽ ⚹ ♆
22 48 ☽ ∠ ♅
23 17 ☽ ∥ ☉

02 Sunday
02 31 ☽ ☌ ♇
04 18 ☽ ⊻ ♄
06 02 ☽ ☌ ☉
06 46 ☿ St R
11 00 ♀ □ ♂
11 47 ☽ △ ♃
13 37 ☽ ⊥ ♅
14 02 ☽ ☌ ♀
14 18 ☽ □ ♂
15 38 ☽ ∥ ♂
21 00 ☽ ∠ ♆
23 08 ☽ ∥ ♄
23 25 ☽ ⊻ ♅
23 49 ♀ ⊥ ♅

03 Monday
04 27 ☽ ∥ ♀
09 42 ☽ ∥ ♆
09 55 ☽ ∥ ☿
11 36 ☽ ⊥ ♆
13 46 ☽ ☌ ☿
14 09 ☽ ∥ ♅
15 21 ☿ ∥ ♆
20 46 ☽ Q ♂
21 18 ☽ ⊻ ♆

04 Tuesday
03 21 ☽ ⊻ ♇
05 20 ☽ ☌ ♄
10 11 ☽ ⊻ ☉
12 17 ☽ □ ♃
12 41 ☽ ⊻ ♀
13 03 ☽ ⊥ ♇
16 59 ☽ ⚹ ♂
20 34 ☽ ⊥ ☉
22 00 ☽ ⊥ ♀
22 59 ♀ △ ♃

05 Wednesday
00 02 ☽ ☌ ♅
03 38 ☽ ∠ ♇
10 16 ☉ ☌ ♀
12 00 ☽ ∠ ♀
12 12 ☽ ∠ ☉
13 09 ☽ ⊻ ☿
16 55 ☉ △ ♃
17 24 ☽ Q ♃
18 24 ☽ ∠ ♂
21 56 ☽ ☌ ♆
22 35 ☽ ⊥ ☿
23 44 ♀ ∥ ♄

06 Thursday
04 08 ☽ ⚹ ♇
06 22 ☽ ⊻ ♄
09 49 ☽ ⊥ ♂
11 33 ☽ ⚹ ♀
12 45 ☽ ∠ ☿
13 03 ☽ ⚹ ♃
13 11 ☿ ⊥ ♆
14 34 ☽ ⚹ ☉
16 20 ☽ ⊥ ♄
20 11 ☽ ⊻ ♂

07 Friday
00 00 ☽ Q ♇
01 20 ☽ ⊻ ♅
02 26 ☉ ⊥ ♅
06 41 ☽ Q ♀
07 25 ☽ ∠ ♄
11 25 ☽ ⊥ ♅
12 03 ☽ Q ☉
12 31 ☽ ⚹ ☿
14 03 ☽ ∠ ♃
23 57 ☽ ⊻ ♆

08 Saturday
02 42 ☽ ∠ ♅
05 19 ☽ ⊥ ♃
06 27 ☽ □ ♇
06 27 ☽ ∥ ♅
07 41 ☽ Q ☿
09 03 ☽ ⚹ ♄
09 22 ♃ St D
10 14 ☽ ⊥ ♆
11 18 ☽ ∥ ♆
11 59 ☽ □ ♀
15 38 ☽ ⊻ ♃
21 11 ☽ □ ☉
21 25 ☽ ∥ ☿
22 04 ☽ ∥ ♄

09 Sunday
01 37 ☽ ☌ ♂
01 50 ☽ ∠ ♆
03 05 ☿ ∥ ♄
04 42 ☽ ∥ ♀
04 43 ☽ ⚹ ♅
06 02 ☽ Q ♄
06 42 ♂ ∠ ♆
12 49 ☽ □ ☿
14 02 ☽ ∥ ♂
14 21 ☽ ∥ ☉
17 24 ☉ ∥ ♂

10 Monday
02 01 ☽ Q ♅
04 25 ☽ ⚹ ♆
11 19 ☽ △ ♇
14 21 ☽ □ ♄
14 46 ☽ △ ♀
20 51 ☽ ☌ ♃
23 10 ♀ ⊻ ♄

11 Tuesday
02 11 ☽ Q ♆
03 34 ☽ ∥ ♃
07 07 ☽ △ ☉
10 09 ☽ ⊻ ♂
10 49 ☽ □ ♅
14 28 ☽ △ ☿
14 47 ☽ ⚼ ♇
14 53 ☉ ∠ ♆
17 08 ☽ ⚼ ♀
21 50 ☽ ⊥ ♂

12 Wednesday
00 41 ♂ ⚹ ♅
07 07 ☽ # ♇
07 35 ☽ ± ♇
09 00 ☉ ∥ ♀
09 15 ☽ ± ♀
11 38 ☽ □ ♆
13 21 ☽ ⚼ ☉
14 43 ☽ ∥ ♇
15 35 ☽ ∠ ♂
15 51 ☽ ⚼ ☿
17 24 ☿ □ ♂
18 54 ☽ ⊼ ♇
20 07 ☽ ⊼ ♀
22 24 ☽ △ ♄

13 Thursday
00 35 ☿ ⊻ ♅
01 07 ☉ ∥ ☿
04 19 ☉ ☌ ☿
04 46 ☽ ⊻ ♃
07 16 ☽ ± ☿
07 52 ☽ ± ☉
09 46 ☉ ⊻ ♅
16 14 ☽ ⊥ ♃
17 16 ♂ Q ♄
17 38 ☽ ⊼ ☿
18 53 ☽ # ♃
19 32 ☽ △ ♅
20 21 ☽ ⊼ ☉
21 42 ☽ ⚹ ♂

14 Friday
00 58 ♀ ☌ ♇
03 21 ☽ ⚼ ♄
07 18 ☿ ∠ ♆
09 35 ☽ ∠ ♃
15 39 ☿ ∥ ♀
20 57 ☽ ± ♄
21 13 ☽ △ ♆
22 14 ☽ Q ♂

15 Saturday
00 42 ☽ ⚼ ♅
03 43 ☽ ☍ ♀
04 45 ☽ # ♂
04 48 ☽ ☍ ♇
08 08 ☽ # ☿
08 47 ☽ ⊼ ♄
08 58 ☽ # ♀
14 54 ☽ ⚹ ♃
18 25 ☽ ± ♅
18 42 ☉ □ ♂
19 39 ☽ # ☉
22 37 ☽ ☍ ☿

16 Sunday
02 43 ☽ ⚼ ♆
06 05 ☽ # ♄
06 19 ☽ ⊼ ♅
11 40 ☽ □ ♂
12 12 ☽ ☍ ☉
14 40 ☽ Q ♃
17 37 ☽ # ♆
20 37 ☽ ± ♆
23 30 ☽ # ♅

17 Monday
08 36 ☽ ⊼ ♆
12 58 ☿ ∥ ♂
13 14 ☽ ⊼ ♀
16 26 ☽ ⊼ ♇
20 54 ☽ ☍ ♄
22 30 ☿ ⊥ ♅

18 Tuesday
00 56 ☽ ± ♀
02 43 ☽ □ ♃
04 31 ☽ ± ♇
05 55 ☽ ⊼ ☿
17 17 ☽ ± ☿
18 05 ♀ ∠ ♅
18 36 ☽ ⚼ ♀
18 36 ☽ ☍ ♅
22 43 ☽ ⚼ ♇

19 Wednesday
03 19 ☽ △ ♂
05 53 ☽ ⊼ ☉
10 33 ☽ ⚼ ☿
19 09 ☽ ± ☉
21 12 ☽ ☍ ♆

20 Thursday
00 14 ☽ △ ♀
05 10 ☽ △ ♇
10 04 ☽ ⊼ ♄
11 29 ☽ ⚼ ♂
15 04 ☽ ⚼ ☉
15 32 ☽ △ ♃
15 49 ☽ △ ☿
19 59 ☉ ⊥ ♆
22 20 ☽ ± ♄

21 Friday
01 55 ☿ △ ♃
06 47 ☽ ± ♂
07 36 ☽ ⊼ ♅
16 42 ☽ ⚼ ♄
19 37 ☽ ⊼ ♂
19 46 ☽ ± ♅
21 57 ☽ ⚼ ♃

22 Saturday
00 11 ☽ △ ☉
09 03 ☿ St D
09 57 ☽ ⊼ ♆
11 47 ☽ □ ♀
13 55 ☽ ⚼ ♅
15 44 ☽ # ♅
16 05 ☽ ± ♃
17 51 ☽ □ ♇
21 50 ☽ # ♆
21 59 ☽ ± ♆
23 05 ☽ △ ♄

23 Sunday
03 41 ☽ □ ☿
04 05 ☽ ⊼ ♃
08 42 ☽ # ♄
10 12 ☽ # ☉
11 23 ☿ ∥ ♂
15 51 ☽ ⚼ ♆
19 49 ☽ △ ♅
23 07 ☽ Q ♀

24 Monday
05 05 ♀ ⊥ ♄
05 30 ☽ Q ♇
06 10 ☿ △ ♃
10 36 ☽ ☍ ♂
15 44 ☽ Q ☿
16 50 ☽ □ ☉
17 07 ☿ ∥ ♀
19 42 ☉ ∥ ♄
20 18 ☽ # ☿
20 29 ☽ # ♀
21 07 ☽ △ ♆
22 15 ☽ ⚹ ♀
22 18 ☽ # ♂

25 Tuesday
04 41 ☽ ⚹ ♇
10 02 ☽ □ ♄
14 29 ☽ ☍ ♃
15 27 ☽ ⚹ ☿
21 07 ☽ # ♃

26 Wednesday
02 29 ☽ ∠ ♀
02 47 ♀ St D
05 21 ☽ □ ♅
05 45 ☽ Q ☉
08 50 ☽ ∠ ♇
19 00 ☽ ⊥ ♀
19 40 ☽ Q ♄
20 29 ☽ ∠ ☿
21 58 ☽ ⊼ ♂
23 14 ☉ ⊻ ♆
23 59 ☽ ∥ ♇

27 Thursday
01 14 ☽ ⊥ ♇
04 55 ☽ □ ♆
05 22 ☽ ⚹ ☉
05 49 ☽ ⊻ ♀
06 14 ☽ # ♇
09 17 ☽ ± ♂
11 47 ☉ ⊻ ♀
11 58 ☽ ⊻ ♇
13 30 ☽ ⊥ ☿
14 00 ☽ Q ♅
17 16 ☽ ⚹ ♄
21 10 ☽ ⊼ ♃
22 53 ☿ ⊥ ♅

28 Friday
00 39 ☽ ⊻ ☿
01 54 ☽ ⚼ ♂
07 37 ☽ ± ♃
07 37 ☽ ∥ ♃
09 43 ☽ ∠ ☉
10 57 ☽ ⚹ ♅
12 27 ☽ Q ♆
19 20 ☽ ∠ ♄
23 00 ☽ ⚼ ♃

29 Saturday
00 03 ☽ ∥ ♂
02 09 ☽ ⊥ ☉
02 47 ☽ ∥ ♀
04 43 ☽ △ ♂
08 43 ☽ ⚹ ♆
09 45 ☽ ☌ ♀
09 59 ☽ ∥ ☿
10 32 ☽ ⊥ ♄
12 17 ☽ ∠ ♅
12 53 ☽ ⊻ ☉
15 17 ☽ ☌ ♇
20 30 ☽ ⊻ ♄
23 58 ☽ △ ♃

30 Sunday
03 06 ☽ ⊥ ♅
06 20 ☽ ☌ ☿
09 20 ☽ ∠ ♆
12 24 ☽ ∥ ♄
12 51 ☽ ⊻ ♅
17 08 ☽ ∥ ☉
20 33 ☽ ∥ ♆
23 48 ☽ ⊥ ♆

DECEMBER 1986

LONGITUDES

Date	Sidereal time h m s	Sun ☉ ° ' "	Moon ☽ ° ' "	Moon ☽ 24.00 ° ' "	Mercury ☿ ° '	Venus ♀ ° '	Mars ♂ ° '	Jupiter ♃ ° '	Saturn ♄ ° '	Uranus ♅ ° '	Neptune ♆ ° '	Pluto ♇ ° '
01	16 40 12	08 ♐ 59 40	06 ♐ 12 53	13 ♐ 48 55	18 ♏ 59	05 ♏ 28	03 ♓ 39	13 ♓ 52	11 ♐ 50	21 ♐ 45	04 ♑ 33	08 ♏ 31
02	16 44 09	10 00 32	21 ♐ 26 26	29 ♐ 04 02	20 07	05 42	04 20	13 57	11 57	21 49	04 36	08 34
03	16 48 06	11 01 24	06 ♑ 40 22	14 ♑ 14 05	21 19	05 57	05 01	14 02	12 04	21 53	04 38	08 36
04	16 52 02	12 02 18	21 ♑ 44 00	29 ♑ 09 07	22 33	06 15	05 42	14 07	12 11	21 56	04 40	08 38
05	16 55 59	13 03 12	06 ♒ 28 35	13 ♒ 41 49	23 50	06 35	06 23	14 12	12 18	22 00	04 42	08 40
06	16 59 55	14 04 07	20 ♒ 48 26	27 ♒ 48 15	25 09	06 56	07 04	14 17	12 25	22 04	04 44	08 42
07	17 03 52	15 05 02	04 ♓ 41 17	11 ♓ 27 41	26 30	07 20	07 46	14 23	12 32	22 07	04 46	08 44
08	17 07 48	16 05 58	18 ♓ 07 44	24 ♓ 41 49	27 52	07 45	08 27	14 29	12 40	22 11	04 48	08 46
09	17 11 45	17 06 55	01 ♈ 10 23	07 ♈ 33 55	29 ♏ 16	08 12	09 08	14 35	12 47	22 14	04 51	08 48
10	17 15 41	18 07 53	13 ♈ 52 56	20 ♈ 07 56	00 ♐ 41	08 41	09 49	14 41	12 54	22 18	04 53	08 50
11	17 19 38	19 08 51	26 ♈ 19 26	02 ♉ 27 54	02 06	09 11	10 31	14 47	13 01	22 22	04 55	08 52
12	17 23 35	20 09 50	08 ♉ 33 46	14 ♉ 37 27	03 33	09 43	11 12	14 53	13 08	22 25	04 57	08 54
13	17 27 31	21 10 49	20 ♉ 39 18	26 ♉ 39 39	05 01	10 16	11 54	15 00	13 15	22 29	04 59	08 56
14	17 31 28	22 11 50	02 ♊ 38 47	08 ♊ 36 57	06 29	10 51	12 35	15 07	13 22	22 33	05 02	08 58
15	17 35 24	23 12 51	14 ♊ 34 21	20 ♊ 31 11	07 58	11 27	13 17	15 14	13 29	22 36	05 04	09 00
16	17 39 21	24 13 52	26 ♊ 27 38	02 ♋ 23 50	09 27	12 05	13 58	15 21	13 36	22 40	05 06	09 02
17	17 43 17	25 14 54	08 ♋ 19 58	14 ♋ 16 11	10 56	12 44	14 40	15 28	13 43	22 44	05 08	09 04
18	17 47 14	26 15 58	20 ♋ 12 41	26 ♋ 09 39	12 26	13 24	15 21	15 36	13 50	22 47	05 11	09 05
19	17 51 10	27 17 01	02 ♌ 07 21	08 ♌ 06 01	13 57	14 05	16 03	15 43	13 57	22 51	05 13	09 07
20	17 55 07	28 18 06	14 ♌ 05 59	20 ♌ 07 36	15 28	14 48	16 44	15 51	14 04	22 55	05 15	09 09
21	17 59 04	29 ♐ 19 11	26 ♌ 11 16	02 ♍ 17 26	16 59	15 31	17 26	15 59	14 11	22 58	05 17	09 11
22	18 03 00	00 ♑ 20 17	08 ♍ 26 34	14 ♍ 39 11	18 30	16 16	18 08	16 07	14 18	23 02	05 20	09 13
23	18 06 57	01 21 23	20 ♍ 55 50	27 ♍ 17 06	20 01	17 01	18 50	16 15	14 25	23 05	05 22	09 14
24	18 10 53	02 22 30	03 ♎ 43 31	10 ♎ 15 38	21 33	17 48	19 31	16 23	14 32	23 09	05 24	09 16
25	18 14 50	03 23 38	16 ♎ 53 59	23 ♎ 38 58	23 05	18 36	20 13	16 32	14 39	23 13	05 26	09 18
26	18 18 46	04 24 47	00 ♏ 30 59	07 ♏ 30 13	24 38	19 24	20 55	16 41	14 46	23 16	05 29	09 19
27	18 22 43	05 25 56	14 ♏ 36 44	21 ♏ 50 24	26 10	20 13	21 37	16 49	14 53	23 20	05 31	09 21
28	18 26 39	06 27 06	29 ♏ 10 53	06 ♐ 37 35	27 43	21 04	22 18	16 58	14 59	23 23	05 33	09 22
29	18 30 36	07 28 16	14 ♐ 09 39	21 ♐ 46 02	29 ♐ 17	21 54	23 00	17 07	15 06	23 27	05 36	09 24
30	18 34 33	08 29 27	29 ♐ 25 27	07 ♑ 06 30	00 ♑ 50	22 46	23 42	17 17	15 13	23 31	05 38	09 25
31	18 38 29	09 ♑ 30 38	14 ♑ 47 38	22 ♑ 27 21	02 ♑ 24	23 ♏ 39	24 ♓ 24	17 ♓ 26	15 ♐ 20	23 ♐ 34	05 ♑ 40	09 ♏ 27

Date	Moon True ☊ ° '	Moon Mean ☊ ° '	Moon ☽ Latitude ° '
01	19 ♈ 51	18 ♈ 07	03 S 42
02	19 R 44	18 03	04 30
03	19 35	18 00	04 58
04	19 27	17 57	05 06
05	19 21	17 54	04 53
06	19 16	17 51	04 21
07	19 14	17 48	03 35
08	19 D 13	17 44	02 39
09	19 14	17 41	01 35
10	19 15	17 38	00 S 29
11	19 R 14	17 35	00 N 38
12	19 12	17 32	01 41
13	19 08	17 29	02 39
14	19 00	17 25	03 29
15	18 50	17 22	04 10
16	18 38	17 19	04 39
17	18 25	17 16	04 57
18	18 11	17 13	05 01
19	17 59	17 09	04 52
20	17 49	17 06	04 31
21	17 41	17 03	03 57
22	17 36	17 00	03 11
23	17 34	16 57	02 16
24	17 D 34	16 54	01 13
25	17 34	16 50	00 N 04
26	17 R 33	16 47	01 S 08
27	17 31	16 44	02 18
28	17 26	16 41	03 21
29	17 18	16 38	04 12
30	17 08	16 34	04 46
31	16 ♈ 56	16 ♈ 31	05 S 00

DECLINATIONS

Date	Sun ☉ ° '	Moon ☽ ° '	Mercury ☿ ° '	Venus ♀ ° '	Mars ♂ ° '	Jupiter ♃ ° '	Saturn ♄ ° '	Uranus ♅ ° '	Neptune ♆ ° '	Pluto ♇ ° '
01	21 S 48	25 S 00	15 S 17	12 S 25	11 S 20	07 S 33	20 S 43	23 S 18	22 S 21	00 N 48
02	21 57	27 39	15 40	12 20	11 03	07 31	20 44	23 18	22 21	00 47
03	22 06	28 14	16 04	12 16	10 46	07 29	20 45	23 18	22 21	00 47
04	22 14	26 43	16 29	12 12	10 30	07 27	20 46	23 18	22 21	00 47
05	22 22	23 22	16 54	12 10	10 13	07 25	20 47	23 19	22 21	00 46
06	22 30	18 41	17 19	12 09	09 56	07 22	20 48	23 19	22 21	00 46
07	22 37	13 08	17 45	12 08	09 39	07 20	20 49	23 19	22 21	00 45
08	22 43	07 08	18 10	12 09	09 22	07 18	20 50	23 19	22 20	00 45
09	22 49	00 S 59	18 36	12 10	09 05	07 15	20 51	23 19	22 20	00 45
10	22 55	05 N 02	19 01	12 12	08 48	07 13	20 52	23 20	22 20	00 45
11	23 00	10 45	19 25	12 15	08 31	07 10	20 53	23 20	22 20	00 44
12	23 05	15 57	19 49	12 19	08 14	07 07	20 54	23 20	22 20	00 44
13	23 09	20 29	20 13	12 23	07 57	07 05	20 55	23 20	22 20	00 44
14	23 13	24 07	20 36	12 29	07 39	07 02	20 56	23 21	22 20	00 43
15	23 16	26 41	20 58	12 34	07 22	06 59	20 57	23 21	22 20	00 43
16	23 19	28 03	21 19	12 41	07 05	06 56	20 58	23 21	22 20	00 43
17	23 21	28 07	21 39	12 48	06 47	06 53	20 58	23 21	22 20	00 43
18	23 23	26 53	21 59	12 55	06 30	06 50	20 59	23 21	22 20	00 43
19	23 25	24 26	22 17	13 03	06 12	06 47	21 00	23 22	22 20	00 42
20	23 26	20 55	22 35	13 12	05 55	06 43	21 01	23 22	22 20	00 42
21	23 26	16 30	22 51	13 21	05 37	06 40	21 02	23 22	22 20	00 42
22	23 27	11 22	23 07	13 30	05 19	06 37	21 03	23 22	22 19	00 42
23	23 26	05 N 41	23 21	13 40	05 02	06 34	21 03	23 22	22 19	00 42
24	23 25	00 S 22	23 34	13 50	04 44	06 30	21 04	23 23	22 19	00 42
25	23 24	06 35	23 46	14 01	04 26	06 27	21 05	23 23	22 19	00 42
26	23 22	12 43	23 57	14 11	04 09	06 23	21 06	23 23	22 19	00 41
27	23 20	18 25	24 06	14 22	03 51	06 20	21 07	23 23	22 19	00 41
28	23 17	23 15	24 15	14 33	03 33	06 16	21 07	23 23	22 19	00 41
29	23 14	26 40	24 22	14 45	03 15	06 12	21 08	23 24	22 19	00 41
30	23 10	28 13	24 28	14 56	02 58	06 08	21 09	23 24	22 19	00 41
31	23 S 06	27 S 35	24 S 32	15 S 08	02 S 40	06 S 05	21 S 10	23 S 24	22 S 19	00 N 41

ZODIAC SIGN ENTRIES

Date	h	m	Planets
01	02	08	☽ ♐
03	01	28	☽ ♑
05	01	23	☽ ♒
07	03	48	☽ ♓
09	09	49	☽ ♈
10	00	34	☿ ♐
11	19	10	☽ ♉
14	06	41	☽ ♊
16	19	09	☽ ♋
19	07	44	☽ ♌
21	19	30	☽ ♍
22	04	02	☉ ♑
24	05	05	☽ ♎
26	11	06	☽ ♏
28	13	20	☽ ♐
29	23	09	☿ ♑
30	12	54	☽ ♑

LATITUDES

Date	Mercury ☿ ° '	Venus ♀ ° '	Mars ♂ ° '	Jupiter ♃ ° '	Saturn ♄ ° '	Uranus ♅ ° '	Neptune ♆ ° '	Pluto ♇ ° '
01	02 N 17	00 N 59	01 S 15	01 S 18	01 N 30	00 S 06	01 N 01	15 N 57
04	02 00	01 29	01 09	01 18	01 30	00 07	01 01	15 58
07	01 40	01 56	01 04	01 17	01 30	00 07	01 01	15 59
10	01 19	02 19	00 59	01 17	01 29	00 07	01 01	16 00
13	00 56	02 39	00 55	01 16	01 29	00 07	01 01	16 01
16	00 34	02 55	00 50	01 15	01 29	00 07	01 01	16 02
19	00 N 12	03 09	00 46	01 15	01 29	00 07	01 01	16 03
22	00 S 10	03 20	00 41	01 14	01 29	00 07	01 01	16 04
25	00 30	03 29	00 37	01 13	01 29	00 07	01 01	16 05
28	00 49	03 36	00 33	01 13	01 29	00 07	01 01	16 06
31	01 S 07	03 N 40	00 S 29	01 S 12	01 N 29	00 S 07	01 N 01	16 N 08

DATA

Julian Date	2446766
Delta T	+55 seconds
Ayanamsa	23° 40' 21"
Synetic vernal point	05° ♓ 26' 38"
True obliquity of ecliptic	23° 26' 36"

MOON'S PHASES, APSIDES AND POSITIONS ☽

Date	h	m	Phase	Longitude ° '	Eclipse Indicator
01	16	43	●	09 ♐ 12	
08	08	01	☽	15 ♓ 56	
16	07	04	○	24 ♊ 01	
24	09	17	☾	02 ♎ 16	
31	03	10	●	09 ♑ 08	

Day	h	m		
02	10	25	Perigee	
17	04	47	Apogee	
30	23	08	Perigee	
03	06	30	Max dec	28° S 18'
09	15	53	0N	
17	01	10	Max dec	28° N 15'
24	10	34	0S	
30	17	11	Max dec	28° S 16'

LONGITUDES

Date	Chiron ⚷ ° '	Ceres ⚳ ° '	Pallas ⚴ ° '	Juno ⚵ ° '	Vesta ⚶ ° '	Black Moon Lilith ⚸ ° '
01	19 ♊ 11	22 ♏ 21	00 ♏ 57	00 ♑ 17	03 ♈ 14	00 ♋ 53
11	18 ♊ 35	26 ♏ 37	05 ♏ 04	03 ♑ 56	04 ♈ 30	02 ♋ 01
21	17 ♊ 59	00 ♐ 48	09 ♏ 02	07 ♑ 40	06 ♈ 20	03 ♋ 08
31	17 ♊ 24	04 ♐ 54	12 ♏ 50	11 ♑ 26	08 ♈ 40	04 ♋ 15

ASPECTARIAN

h m	Aspects
01 Monday	
00 25	☉ ⊻ ♇
01 39	☽ ∥ ♅
07 45	☽ □ ♂
09 22	☽ ⊻ ♆
10 48	☽ ⊻ ♀
15 40	☽ ⊻ ♇
15 59	♂ Q ♅
16 43	☽ ☌ ☉
20 25	☽ ⊥ ♀
20 56	☽ ☌ ♄
02 Tuesday	
00 09	☽ □ ♃
00 42	☿ ∠ ♆
01 09	☽ ⊥ ♇
09 45	☽ ⊻ ☿
10 48	☽ ∠ ♀
12 36	☽ ☌ ♅
13 29	☽ Q ♂
15 20	☽ ∠ ♇
19 58	☽ ⊥ ☿
21 31	♂ ⚹ ♆
03 Wednesday	
04 37	☽ Q ♃
08 46	☽ ☌ ♆
09 16	☽ ⚹ ♂
10 51	☽ ⚹ ♀
11 23	☽ ∠ ☿
15 03	☽ ⚹ ♇
19 23	☽ ⊻ ☉
20 37	☽ ⊻ ♄
23 40	☿ ⊻ ♅
23 44	☽ ⚹ ♃
04 Thursday	
03 47	♀ ⊥ ♄
05 39	☽ ⊥ ☉
06 16	☽ ⊥ ♄
06 18	☽ Q ♀
10 13	☽ Q ♇
10 16	☽ ∠ ♂
12 20	☽ ⊻ ♅
13 26	☽ ⚹ ☿
15 56	☉ ☌ ♄
19 16	☉ ⊥ ♀
20 53	☽ ∠ ♄
21 12	☽ ∠ ☉
22 04	☽ ⊥ ♅
05 Friday	
00 00	☽ ∠ ♃
01 32	☽ ⊥ ♂
07 20	☉ ∥ ♆
09 04	☽ ⊻ ♆
10 50	☽ Q ☿
11 51	☽ ⊻ ♂
12 10	☽ □ ♀
12 23	☽ ∥ ♅
12 52	☽ ∠ ♅
14 52	☽ ⊥ ♃
15 38	☽ □ ♇
17 31	☽ ∥ ☉
17 49	☽ ∥ ♆
19 01	☽ ⊥ ♆
21 45	☽ ⚹ ♄
23 45	☽ ⚹ ☉
06 Saturday	
00 55	☽ ⊻ ♃
01 49	♀ △ ♂
01 51	☽ ∥ ♄
10 10	☽ ∠ ♆
14 08	☽ ⚹ ♅
17 42	☉ □ ♃
17 43	☽ ∥ ☿
18 14	☽ Q ♄
20 12	☽ □ ☿
20 59	♀ ∠ ♅
21 43	☽ Q ☉
07 Sunday	
03 28	☉ ⊥ ♇
11 00	☽ Q ♅
12 09	☽ ⚹ ♆
16 03	☽ ∥ ♀
16 48	☽ △ ♀
17 42	☽ ☌ ♂
19 10	☽ △ ♇
08 Monday	
02 03	☽ □ ♄
02 42	☽ ∥ ♂
05 21	☽ ☌ ♃
08 01	☽ □ ☉
09 36	☽ Q ♆
11 20	☽ ∥ ♃
19 25	☽ □ ♅
20 43	☽ ⚼ ♀
22 19	☽ ⚼ ♇
23 50	♂ △ ♇
09 Tuesday	
04 39	☿ ⊥ ♆
08 00	☽ △ ☿
12 57	☽ # ♇
14 00	☽ ± ♀
15 03	☽ ± ♇
18 50	☽ ∥ ♇
18 54	☽ □ ♆
10 Wednesday	
01 43	☽ ⊼ ♀
02 23	☽ ⊼ ♇
03 50	☽ ⊻ ♂
10 06	☽ △ ♄
13 32	☽ ⊻ ♃
15 52	☽ ⚼ ☿
15 56	☽ ⊥ ♂
20 07	♀ ☌ ♇
20 52	☽ △ ☉
20 52	☽ # ♃
11 Thursday	
01 10	☽ ⊥ ♃
02 53	☽ # ♂
04 16	☽ △ ♅
10 20	☽ ∠ ♂
11 31	☽ ± ☿
15 20	☽ ⚼ ♄
18 45	☽ # ♀
18 48	☽ ∠ ♃
12 Friday	
00 49	☽ ⊼ ☿
04 42	☽ ⚼ ☉
04 52	☽ △ ♆
09 09	☽ ± ♄
09 44	☽ ⚼ ♅
12 40	☽ ☍ ♇
14 23	☽ ☍ ♀
17 32	☽ ⚹ ♂
21 08	☽ ⊼ ♄
13 Saturday	
00 06	☽ ± ☉
00 39	☽ ⚹ ♃
03 39	☽ ± ♅
10 20	☽ # ☿
10 40	☽ ⚼ ♆
11 39	☿ ⊻ ♆
13 09	☽ ⊼ ☉
14 37	☽ # ♄
15 40	☽ ⊼ ♅
18 51	☽ Q ♂
23 28	☽ # ♆
14 Sunday	
00 48	☽ Q ♃
04 43	☽ ± ♆
05 17	☽ # ☉
06 19	☽ # ♅
16 48	☽ ⊼ ♆
20 43	☉ ☌ ♅
20 47	☽ ☍ ☿
15 Monday	
00 44	☽ ⊼ ♇
05 23	☽ ⊼ ♀
09 13	☽ □ ♂
09 47	☽ ☍ ♄
11 00	☿ ∥ ♄
12 52	☽ ± ♇
13 20	☽ □ ♃
18 08	☽ ± ♀
20 47	♂ □ ♄
16 Tuesday	
04 18	☽ ☍ ♅
05 09	☿ ⊻ ♇
07 04	☽ ☍ ☉
07 05	☽ ⚼ ♇
07 05	☉ ∠ ♇
13 20	☽ ⚼ ♀
17 Wednesday	
02 36	♂ ∥ ♃
05 31	☽ ☍ ♆
09 04	☉ ∥ ♅
13 28	☽ △ ♇
18 02	☽ ⊼ ☿
21 25	☽ △ ♀
23 00	☽ ⊼ ♄
18 Thursday	
01 35	☽ △ ♂
02 35	☽ △ ♃
07 54	☽ ± ☿
11 14	☽ ± ♄
17 14	☽ ⊼ ♅
22 10	♂ ☌ ♃
19 Friday	
01 21	☽ ⊼ ☉
04 41	☽ ⚼ ☿
05 23	☽ ± ♅
05 34	☽ ⚼ ♄
06 32	♀ ⊻ ♄
09 09	☽ ⚼ ♃
09 42	☽ ⚼ ♂
12 08	☿ ☌ ♄
14 33	☽ ± ☉
15 25	☿ ∥ ♆
16 08	☿ ⊻ ♀
18 14	☽ ⊼ ♆
19 38	☽ # ☉
20 04	☽ # ♅
23 33	☽ ⚼ ♅
20 Saturday	
02 05	☽ □ ♇
02 15	☽ # ☿
03 08	☽ # ♆
03 25	☽ ± ♃
04 53	☽ ± ♂
06 18	☽ ± ♆
07 01	☿ ⊥ ♇
10 16	☽ ⚼ ☉
11 23	☽ # ♄
11 56	☽ △ ♄
13 28	☽ □ ♀
15 06	☽ △ ☿
15 32	☽ ⊼ ♃
17 35	☽ ⊼ ♂
18 47	☿ □ ♃
21 Sunday	
00 17	☽ ⚼ ♆
05 36	☽ △ ♅
06 40	♂ Q ♆
13 58	☽ Q ♇
18 44	☽ △ ☉
22 Monday	
01 22	☿ □ ♂
02 37	☽ # ♀
03 20	☽ Q ♀
05 55	☽ △ ♆
06 15	♀ △ ♃
13 29	☽ ⚹ ♇
23 26	☽ □ ♄
23 Tuesday	
02 58	☽ ☍ ♃
04 04	☽ ⚹ ♀
07 45	☽ ☍ ♂
08 22	☽ # ♃
10 02	☽ □ ☿
14 17	♀ ⊥ ♅
14 46	☽ # ♂
14 50	☿ ∥ ♅
16 07	☽ □ ♅
18 17	☽ ∠ ♇
20 58	☉ ∥ ☿
24 Wednesday	
07 51	☽ ∥ ♇
09 17	☽ □ ☉
09 46	☽ Q ♄
10 11	☽ ∠ ♀
11 09	☽ ⊥ ♇
13 15	☽ # ♇
15 06	☽ □ ♆
22 12	☽ ⊻ ♇
25 Thursday	
00 08	☽ Q ☿
01 41	☽ Q ♅
03 44	☽ ⊥ ♀
04 06	☽ ∥ ♂
07 55	☽ ⚹ ♄
11 20	☽ ⊼ ♃
11 27	☽ ∥ ♃
13 58	☿ ☌ ♅
15 13	☽ ⊻ ♀
18 15	☽ ⊼ ♂
20 40	☽ Q ☉
22 08	☽ ± ♃
23 17	☽ ⚹ ♅
23 40	☽ Q ♆
26 Friday	
00 25	☽ ⚹ ☿
02 26	☉ ∥ ♅
05 24	☽ ± ♂
07 04	☿ ∠ ♇
10 34	☉ ∠ ♀
10 41	☽ ∠ ♄
14 02	☽ ⚼ ♃
18 09	☽ ∥ ♀
19 14	☉ Q ♃
19 15	☽ ⚹ ☉
20 34	☽ ⚹ ♆
21 46	☽ ⚼ ♂
27 Saturday	
01 22	☽ ∠ ♅
02 15	☽ ⊥ ♄
03 07	☽ ☌ ♇
05 31	☽ ∠ ☿
12 27	☽ ⊻ ♄
13 38	☿ ⊥ ♀
14 04	☉ ☌ ♆
15 44	☽ △ ♃
16 33	☽ ⊥ ♅
20 52	♀ ∠ ♆
21 51	☽ ∠ ♆
21 54	☽ ☌ ♀
22 21	☽ ⊥ ☿
22 24	☽ ∠ ☉
28 Sunday	
00 12	☽ △ ♂
00 43	☽ ∥ ♄
02 31	☽ ⊻ ♅
06 54	☽ ∥ ♆
09 21	☽ ⊻ ☿
12 14	☽ ∥ ☉
12 36	☽ ⊥ ♆
12 50	☽ ∥ ♅
14 13	☽ ⊥ ☉
18 10	☽ ∥ ☿
22 18	☽ ⊻ ♆
29 Monday	
00 35	☽ ⊻ ☉
04 25	☽ ⊻ ♇
13 30	☽ ☌ ♄
13 58	☽ ⊥ ♇
16 44	☽ □ ♃
30 Tuesday	
00 57	☽ ⊻ ♀
02 37	☽ □ ♂
02 42	☽ ☌ ♅
04 09	☽ ∠ ♇
04 47	♂ □ ♅
10 55	☽ ⊥ ♀
14 27	☽ ☌ ☿
21 14	☽ Q ♃
21 43	☽ ☌ ♆
31 Wednesday	
01 49	☽ ∠ ♀
03 10	☽ ☌ ☉
03 38	☽ ⚹ ♇
08 05	☽ Q ♂
09 49	♀ ⊻ ♅
10 28	☉ ⚹ ♇
12 50	☽ ⊻ ♄
13 40	♂ ⚼ ♇
16 10	☽ ⚹ ♃
22 18	☽ ⊥ ♄
22 26	☽ Q ♇

All ephemeris data is given at 12.00 UT and the Moon's longitude is additionally given for 24.00 UT

JANUARY 1987

LONGITUDES

Date	Sidereal time h m s	Sun ☉ ° ' "	Moon ☽ ° ' "	Moon ☽ 24.00 ° ' "	Mercury ☿ ° '	Venus ♀ ° '	Mars ♂ ° '	Jupiter ♃ ° '	Saturn ♄ ° '	Uranus ♅ ° '	Neptune ♆ ° '	Pluto ♇ ° '
01	18 42 26	10 ♑ 31 49	00 ♒ 04 10	07 ♒ 36 44	03 ♑ 58	24 ♏ 32	25 ♓ 06	17 ♓ 35	15 ♐ 26	23 ♐ 38	05 ♑ 42	09 ♏ 28
02	18 46 22	11 33 00	15 ♒ 03 55	22 ♒ 24 47	05 32	25 26	25 48	17 45	15 33	23 41	05 45	09 30
03	18 50 19	12 34 10	29 ♒ 38 39	06 ♓ 45 05	07 07	26 20	26 30	17 55	15 40	23 45	05 47	09 31
04	18 54 15	13 35 21	13 ♓ 43 55	20 ♓ 35 08	08 42	27 15	27 12	18 04	15 46	23 48	05 49	09 32
05	18 58 12	14 36 31	27 ♓ 18 56	03 ♈ 55 42	10 17	28 11	27 54	18 14	15 53	23 52	05 51	09 34
06	19 02 08	15 37 40	10 ♈ 25 51	16 ♈ 49 58	11 53	29 ♏ 07	28 36	18 25	15 59	23 55	05 54	09 35
07	19 06 05	16 38 50	23 ♈ 08 36	29 ♈ 22 24	13 29	00 ♐ 04	29 17	18 35	16 06	23 59	05 56	09 36
08	19 10 02	17 39 59	05 ♉ 32 01	11 ♉ 38 02	15 06	01 01	29 ♓ 59	18 45	16 12	24 02	05 58	09 37
09	19 13 58	18 41 07	17 ♉ 41 04	23 ♉ 41 42	16 43	01 59	00 ♈ 41	18 56	16 19	24 05	06 00	09 39
10	19 17 55	19 42 15	29 ♉ 40 27	05 ♊ 37 49	18 20	02 58	01 23	19 06	16 25	24 09	06 03	09 40
11	19 21 51	20 43 23	11 ♊ 34 13	17 ♊ 30 02	19 58	03 56	02 05	19 17	16 31	24 12	06 05	09 41
12	19 25 48	21 44 30	23 ♊ 25 38	29 ♊ 21 16	21 37	04 56	02 47	19 28	16 37	24 16	06 07	09 42
13	19 29 44	22 45 37	05 ♋ 17 12	11 ♋ 13 39	23 15	05 56	03 29	19 38	16 44	24 19	06 09	09 43
14	19 33 41	23 46 43	17 ♋ 10 47	23 ♋ 08 45	24 55	06 56	04 11	19 49	16 50	24 22	06 11	09 44
15	19 37 37	24 47 49	29 ♋ 07 42	05 ♌ 07 45	26 34	07 56	04 53	20 01	16 56	24 26	06 14	09 45
16	19 41 34	25 48 55	11 ♌ 09 02	17 ♌ 11 41	28 15	08 58	05 35	20 12	17 02	24 29	06 16	09 46
17	19 45 31	26 50 00	23 ♌ 15 54	29 ♌ 21 51	29 ♑ 55	09 59	06 17	20 23	17 08	24 32	06 18	09 47
18	19 49 27	27 51 04	05 ♍ 29 45	11 ♍ 39 53	01 ♒ 36	11 01	06 59	20 34	17 14	24 35	06 20	09 48
19	19 53 24	28 52 08	17 ♍ 52 33	24 ♍ 08 05	03 18	12 03	07 41	20 46	17 20	24 38	06 22	09 49
20	19 57 20	29 ♑ 53 12	00 ♎ 26 52	06 ♎ 49 20	05 00	13 05	08 23	20 58	17 26	24 42	06 24	09 50
21	20 01 17	00 ♒ 54 15	13 ♎ 15 53	19 ♎ 46 58	06 43	14 08	09 04	21 09	17 32	24 45	06 26	09 50
22	20 05 13	01 55 18	26 ♎ 23 03	03 ♏ 04 32	08 26	15 12	09 46	21 21	17 38	24 48	06 29	09 51
23	20 09 10	02 56 21	09 ♏ 51 46	16 ♏ 45 04	10 09	16 15	10 28	21 33	17 43	24 51	06 31	09 52
24	20 13 06	03 57 23	23 ♏ 44 35	00 ♐ 50 23	11 53	17 19	11 10	21 45	17 49	24 54	06 33	09 53
25	20 17 03	04 58 25	08 ♐ 02 21	15 ♐ 20 10	13 37	18 23	11 52	21 57	17 54	24 57	06 35	09 53
26	20 21 00	05 59 26	22 ♐ 43 19	00 ♑ 11 05	15 21	19 28	12 34	22 09	18 00	25 00	06 37	09 54
27	20 24 56	07 00 27	07 ♑ 42 33	15 ♑ 16 34	17 05	20 32	13 16	22 21	18 06	25 03	06 39	09 54
28	20 28 53	08 01 27	22 ♑ 51 55	00 ♒ 27 14	18 49	21 37	13 58	22 34	18 11	25 06	06 41	09 55
29	20 32 49	09 02 27	08 ♒ 01 10	15 ♒ 32 23	20 33	22 42	14 39	22 46	18 16	25 09	06 43	09 55
30	20 36 46	10 03 25	22 ♒ 59 39	00 ♓ 21 53	22 16	23 48	15 21	22 59	18 22	25 12	06 45	09 56
31	20 40 42	11 ♒ 04 22	07 ♓ 38 13	14 ♓ 47 58	23 ♒ 59	24 ♐ 53	16 ♈ 03	23 ♓ 11	18 ♐ 27	25 ♐ 14	06 ♑ 47	09 ♏ 56

Date	Moon True ☊ ° '	Moon Mean ☊ ° '	Moon ☽ Latitude ° '
01	16 ♈ 45	16 ♈ 28	04 S 52
02	16 R 34	16 25	04 24
03	16 27	16 22	03 40
04	16 22	16 19	02 43
05	16 19	16 15	01 39
06	16 D 19	16 12	00 S 31
07	16 R 19	16 09	00 N 36
08	16 18	16 06	01 40
09	16 15	16 03	02 37
10	16 10	16 00	03 27
11	16 02	15 56	04 08
12	15 50	15 53	04 37
13	15 37	15 50	04 54
14	15 22	15 47	04 59
15	15 07	15 44	04 50
16	14 53	15 40	04 29
17	14 42	15 37	03 55
18	14 33	15 34	03 10
19	14 27	15 31	02 15
20	14 24	15 28	01 13
21	14 24	15 25	00 N 06
22	14 D 24	15 21	01 S 03
23	14 R 24	15 18	02 11
24	14 22	15 15	03 13
25	14 18	15 12	04 04
26	14 12	15 09	04 42
27	14 03	15 06	05 00
28	13 53	15 02	04 59
29	13 42	14 59	04 36
30	13 33	14 56	03 54
31	13 ♈ 26	14 ♈ 53	02 S 58

DECLINATIONS

Date	Sun ☉ ° '	Moon ☽ ° '	Mercury ☿ ° '	Venus ♀ ° '	Mars ♂ ° '	Jupiter ♃ ° '	Saturn ♄ ° '	Uranus ♅ ° '	Neptune ♆ ° '	Pluto ♇ ° '
01	23 S 02	24 S 54	24 S 36	15 S 20	02 S 22	06 S 01	21 S 10	23 S 24	22 S 19	00 N 41
02	22 56	20 31	24 37	15 32	02 04	05 57	21 11	23 24	22 19	00 41
03	22 51	15 01	24 38	15 44	01 46	05 53	21 12	23 25	22 18	00 41
04	22 45	08 54	24 37	15 56	01 28	05 49	21 12	23 25	22 18	00 41
05	22 39	02 S 34	24 35	16 08	01 11	05 45	21 13	23 25	22 18	00 41
06	22 32	03 N 39	24 31	16 20	00 53	05 41	21 14	23 25	22 18	00 41
07	22 24	09 33	24 26	16 32	00 35	05 37	21 14	23 25	22 18	00 41
08	22 17	14 56	24 19	16 44	00 S 17	05 32	21 15	23 25	22 18	00 42
09	22 08	19 38	24 11	16 56	00 N 01	05 28	21 16	23 26	22 18	00 42
10	22 00	23 27	24 02	17 08	00 18	05 24	21 16	23 26	22 18	00 42
11	21 51	26 16	23 51	17 20	00 36	05 20	21 17	23 26	22 18	00 42
12	21 41	27 54	23 38	17 31	00 54	05 15	21 17	23 26	22 18	00 42
13	21 31	28 14	23 24	17 43	01 12	05 11	21 18	23 26	22 17	00 42
14	21 21	27 17	23 09	17 54	01 29	05 06	21 19	23 26	22 17	00 42
15	21 10	25 04	22 51	18 05	01 47	05 02	21 19	23 26	22 17	00 43
16	20 59	21 44	22 33	18 16	02 05	04 57	21 20	23 27	22 17	00 43
17	20 48	17 28	22 13	18 27	02 22	04 53	21 20	23 27	22 17	00 43
18	20 36	12 27	21 51	18 37	02 40	04 48	21 21	23 27	22 17	00 43
19	20 23	06 52	21 28	18 47	02 58	04 43	21 21	23 27	22 17	00 43
20	20 11	00 N 56	21 03	18 57	03 15	04 39	21 22	23 27	22 17	00 44
21	19 58	05 S 09	20 36	19 07	03 33	04 34	21 22	23 27	22 17	00 44
22	19 44	11 10	20 08	19 16	03 50	04 29	21 23	23 27	22 17	00 44
23	19 30	16 50	19 39	19 25	04 07	04 24	21 23	23 28	22 16	00 45
24	19 16	21 49	19 08	19 34	04 25	04 19	21 24	23 28	22 16	00 45
25	19 02	25 40	18 36	19 42	04 42	04 15	21 24	23 28	22 16	00 45
26	18 47	27 56	18 02	19 50	04 59	04 10	21 25	23 28	22 16	00 45
27	18 31	28 13	17 26	19 58	05 17	04 05	21 25	23 28	22 16	00 46
28	18 16	26 25	16 50	20 05	05 34	04 00	21 26	23 28	22 16	00 46
29	18 00	22 42	16 12	20 11	05 51	03 55	21 26	23 28	22 16	00 46
30	17 44	17 32	15 32	20 18	06 08	03 50	21 26	23 28	22 16	00 47
31	17 S 27	11 S 27	14 S 52	20 S 24	06 N 25	03 S 45	21 S 27	23 S 29	22 S 16	00 N 47

ZODIAC SIGN ENTRIES

Date	h	m	Planets
01	11	53	☽ ♒
03	12	36	☽ ♓
05	16	51	☽ ♈
07	10	20	♀ ♐
08	01	13	☽ ♉
08	12	20	♂ ♈
10	12	39	☽ ♊
13	01	18	☽ ♋
15	13	45	☽ ♌
17	13	08	☿ ♒
18	01	15	☽ ♍
20	11	09	☽ ♎
20	14	40	☉ ♒
22	18	30	☽ ♏
24	22	35	☽ ♐
26	23	42	☽ ♑
28	23	17	☽ ♒
30	23	24	☽ ♓

LATITUDES

Date	Mercury ☿ ° '	Venus ♀ ° '	Mars ♂ ° '	Jupiter ♃ ° '	Saturn ♄ ° '	Uranus ♅ ° '	Neptune ♆ ° '	Pluto ♇ ° '
01	01 S 13	03 N 41	00 S 27	01 S 12	01 N 29	00 S 07	01 N 01	16 N 08
04	01 28	03 43	00 23	01 12	01 29	00 07	01 01	16 09
07	01 41	03 43	00 20	01 11	01 29	00 07	01 01	16 11
10	01 52	03 42	00 16	01 11	01 29	00 07	01 01	16 12
13	01 59	03 39	00 12	01 10	01 29	00 07	01 01	16 14
16	02 04	03 34	00 09	01 10	01 29	00 07	01 01	16 15
19	02 06	03 29	00 06	01 09	01 29	00 07	01 01	16 17
22	02 03	03 22	00 S 02	01 09	01 30	00 07	01 01	16 18
25	01 56	03 15	00 N 01	01 09	01 30	00 07	01 01	16 20
28	01 44	03 06	00 04	01 08	01 30	00 07	01 01	16 21
31	01 S 25	02 N 57	00 N 07	01 S 08	01 N 30	00 S 07	01 N 01	16 N 23

DATA

Julian Date	2446797
Delta T	+55 seconds
Ayanamsa	23° 40' 28"
Synetic vernal point	05° ♓ 26' 32"
True obliquity of ecliptic	23° 26' 36"

MOON'S PHASES, APSIDES AND POSITIONS ☽

Date	h	m	Phase	Longitude ° '	Eclipse Indicator
06	22	34	☽	16 ♈ 05	
15	02	30	○	24 ♋ 24	
22	22	45	☾	02 ♏ 23	
29	13	45	●	09 ♒ 07	

Day	h	m		
13	04	46	Apogee	
28	11	04	Perigee	
05	21	49	0N	
13	06	23	Max dec	28° N 17'
20	15	44	0S	
27	03	21	Max dec	28° S 21'

LONGITUDES

Date	Chiron ⚷ ° '	Ceres ⚳ ° '	Pallas ⚴ ° '	Juno ⚵ ° '	Vesta ⚶ ° '	Black Moon Lilith ⚸ ° '
01	17 ♊ 21	05 ♐ 18	13 ♏ 13	11 ♑ 49	08 ♈ 55	04 ♋ 22
11	16 ♊ 51	09 ♐ 16	16 ♏ 47	15 ♑ 38	11 ♈ 41	05 ♋ 29
21	16 ♊ 25	13 ♐ 06	20 ♏ 06	19 ♑ 29	14 ♈ 47	06 ♋ 36
31	16 ♊ 06	16 ♐ 46	23 ♏ 08	23 ♑ 20	18 ♈ 09	07 ♋ 43

All ephemeris data is given at 12.00 UT and the Moon's longitude is additionally given for 24.00 UT

ASPECTARIAN

h m Aspects

01 Thursday
01 48 ☽ ⊻ ♅
02 43 ☽ ✱ ♀
03 47 ☽ ✱ ♂
11 18 ☽ ⊥ ♅
12 35 ☽ ∠ ♄
13 56 ☽ ∥ ☿
16 02 ☽ ∠ ♃
18 54 ☽ ⊻ ☿
20 59 ☽ ⊻ ♆
21 02 ♄ ⊥ ♇
21 06 ☽ ∥ ♅
22 55 ☽ Q ♀
23 26 ☽ ∥ ☉

02 Friday
01 41 ☽ ∠ ♅
03 00 ☽ □ ♇
03 05 ☽ ∥ ♆
04 46 ☽ ∠ ♂
05 38 ☽ ⊥ ☿
05 54 ☽ ⊻ ☉
06 35 ☽ ⊥ ♃
06 37 ☽ ⊥ ♆
08 49 ☽ ∥ ♄
11 03 ♃ Q ♆
12 47 ☽ ✱ ♄
15 12 ☿ ☌ ♆
15 34 ☿ Q ♃
16 20 ☽ ⊥ ☉
16 25 ☽ ⊻ ♃
20 05 ☽ ⊥ ♂
21 17 ☽ ∠ ♆
22 00 ☽ ∠ ☿

03 Saturday
02 09 ☽ ✱ ♅
06 07 ☽ □ ♀
06 29 ☽ ⊻ ♂
08 16 ☽ ∠ ☉
08 40 ☽ Q ♄
09 12 ☽ ∥ ♀
22 20 ☽ Q ♅
22 23 ☽ ✱ ♆

04 Sunday
02 13 ☽ ✱ ☿
04 45 ☽ △ ♇
05 38 ♀ △ ♂
11 44 ☽ ✱ ☉
15 34 ☽ □ ♄
19 09 ☽ Q ♆
19 40 ☽ ☌ ♃
23 51 ☽ ∥ ♃

05 Monday
00 51 ☿ ✱ ♇
01 49 ☽ Q ☿
05 47 ☽ □ ♅
07 03 ☽ bQ ♇
10 37 ☽ Q ☉
13 06 ☽ ☌ ♂
13 41 ☽ △ ♀
17 35 ☽ ∥ ♂
19 11 ☽ # ♇
23 21 ☽ ± ♇

06 Tuesday
00 28 ☽ ∥ ♇
01 42 ☽ # ♂
03 35 ☽ □ ♆
10 25 ☽ ⊼ ♇
15 06 ☽ □ ☿
19 26 ☽ bQ ♀
19 57 ☽ # ♃
21 27 ☉ ⊻ ♄
22 30 ☽ △ ♄
22 34 ☽ □ ☉

07 Wednesday
03 11 ☽ ⊻ ♃
03 19 ♂ # ♇
08 28 ♀ ⊥ ♆
13 36 ☽ △ ♅
13 55 ☽ ± ♀
14 47 ☽ ⊥ ♃

08 Thursday
00 33 ☽ ⊻ ♂
02 28 ☽ ⊼ ♀
03 29 ☽ bQ ♄
07 47 ☉ ∥ ♆
08 28 ☽ ∠ ♃
12 51 ☽ △ ♆
12 57 ☽ ⊥ ♂
18 54 ☽ bQ ♅
20 03 ☽ ☍ ♇
21 08 ☽ # ♀
21 15 ☽ ± ♄

09 Friday
05 32 ☿ ⊻ ♄
07 47 ☽ ∠ ♂
09 14 ☽ ⊼ ♄
09 47 ☽ △ ☿
12 18 ☉ Q ♂
12 49 ☽ ± ♅
14 11 ☽ △ ☉
14 30 ☽ ✱ ♃
18 39 ☽ bQ ♆
18 50 ☉ ✱ ♃
21 34 ☽ # ♄
21 59 ☿ ∠ ♀

10 Saturday
00 51 ☽ ⊼ ♅
02 30 ☽ # ☉
04 04 ☽ # ♆
11 47 ☽ # ♅
12 45 ☽ ± ♆
14 55 ☽ Q ♃
15 40 ☽ ✱ ♂
15 59 ☽ # ☿
19 12 ☽ ☍ ♀
20 33 ☽ bQ ☿
23 05 ☽ bQ ☉

11 Sunday
00 36 ☿ ✱ ♃
00 52 ☽ ⊼ ♆
08 11 ☽ ⊼ ♇
15 00 ☿ Q ♂
17 25 ☽ Q ♂
17 38 ☽ ± ☿
18 59 ☽ ± ☉
19 44 ♂ ∥ ♇
20 20 ☽ ± ♇
22 06 ☽ ☍ ♄

12 Monday
03 50 ☽ □ ♃
07 44 ☽ ⊼ ☿
08 16 ☽ ⊼ ☉
11 01 ☉ Q ♇
13 21 ☿ Q ♇
13 42 ☽ ☍ ♅
14 35 ☽ bQ ♇
17 06 ☉ ☌ ☿

13 Tuesday
03 49 ☿ ⊥ ♄
08 08 ☽ □ ♂
08 43 ☿ ∥ ♅
11 09 ☉ ⊥ ♄
13 25 ☽ ⊼ ♀
13 45 ☽ ☍ ♆
17 40 ♀ ⊻ ♆
20 12 ♂ ± ♇
20 58 ☽ △ ♇

14 Wednesday
02 39 ☽ ± ♀
03 56 ☿ ⊻ ♅
11 17 ☽ ⊼ ♄
17 03 ☉ ∥ ♄
17 24 ☽ △ ♃
22 26 ☽ bQ ♀
23 28 ☽ ± ♄

15 Thursday
02 30 ☽ ☍ ☉
02 32 ☽ ⊼ ♅
02 45 ☉ ⊻ ♅
06 03 ☽ ☍ ☿
14 37 ☽ ± ♅
17 40 ☽ bQ ♄
23 57 ☽ bQ ♃

16 Friday
00 13 ☽ △ ♂
00 39 ☽ # ♅
02 14 ☽ ⊼ ♆
06 17 ☽ # ☿
07 14 ☽ △ ♀
08 30 ☽ # ♆
08 40 ☽ bQ ♅
09 15 ☽ □ ♇
14 13 ☽ ± ♆
14 31 ☽ # ♄
16 50 ☽ # ☉
18 09 ☽ ± ♃
23 47 ☽ △ ♄

17 Saturday
06 13 ☽ ⊼ ♃
06 52 ☿ ∥ ♆
07 05 ☽ # ♀
07 17 ♀ ⊻ ♇
07 51 ☽ bQ ♂
08 06 ☽ bQ ♆
12 39 ♂ □ ♆
14 31 ☽ △ ♅
19 40 ☽ ⊼ ☉
20 54 ☽ Q ♇
21 02 ☿ ⊥ ♅

18 Sunday
02 38 ☽ ± ♂
03 11 ☽ ⊼ ☿
08 30 ☽ ± ☉
13 38 ☽ △ ♆
15 04 ☽ ⊼ ♂
16 46 ☽ ± ☿
20 23 ☽ ✱ ♇
21 28 ☿ ∠ ♄
23 43 ☽ □ ♀

19 Monday
03 35 ☽ bQ ☉
10 57 ☽ □ ♄
12 57 ☽ bQ ☿
17 38 ☽ ☍ ♃
18 01 ☿ ∥ ♄
20 56 ☽ # ♃

20 Tuesday
01 01 ☽ □ ♅
01 19 ☽ ∠ ♇
03 12 ☽ ∥ ♂
10 51 ☽ △ ☉
12 51 ☽ ∥ ♇
13 20 ☽ Q ♀
18 23 ☽ ⊥ ♇
18 37 ☽ # ♇
21 28 ☽ Q ♄
21 55 ☽ △ ☿
23 15 ☽ □ ♆

21 Wednesday
03 10 ☿ ∠ ♃
03 46 ☽ ☍ ♂
05 23 ☽ # ♂
05 38 ☽ ⊻ ♇
08 03 ☉ ⊥ ♅
08 07 ☿ ⊻ ♆
09 45 ☽ ∥ ♃
11 02 ☽ Q ♅
13 46 ☽ ✱ ♀
19 56 ☽ ✱ ♄

22 Thursday
02 44 ☽ ⊼ ♃
08 32 ☽ Q ♆
09 07 ☽ ✱ ♅
13 46 ☽ ± ♃
14 47 ♂ ⊼ ♇
19 26 ☽ ∠ ♀
22 45 ☽ □ ☉
23 17 ☽ ∠ ♄

23 Friday
03 07 ♀ ⊥ ♇
06 05 ☽ bQ ♃
06 05 ☽ ✱ ♆
06 19 ☉ ∠ ♄
07 39 ☿ ∠ ♅
07 58 ☿ □ ♇
11 58 ☽ ∠ ♅
12 00 ☽ ☌ ♇
12 35 ☽ □ ☿
12 44 ☽ ⊥ ♀
13 07 ☽ ⊼ ♂
15 17 ☽ ⊥ ♄
17 29 ☉ ∥ ♀
19 27 ☿ ✱ ♂
20 29 ☿ ∥ ♀
21 30 ♅ ∠ ♇
23 45 ☽ ∥ ☉
23 48 ☽ ∥ ☿

24 Saturday
00 04 ☽ ⊻ ♀
00 07 ☽ ± ♂
00 15 ☽ ∥ ♀
00 34 ☉ ∥ ☿
01 46 ☽ ⊻ ♄
03 41 ☽ ⊥ ♅
06 14 ♂ # ♃
08 14 ☽ ∠ ♆
08 33 ☽ △ ♃
08 43 ☽ Q ☉
09 46 ☽ ∥ ♄
13 58 ☽ ⊻ ♅
14 27 ☽ ∥ ♆
16 20 ☽ bQ ♂
21 15 ☽ ∥ ♅
21 26 ☿ ⊥ ♆
23 32 ☽ ⊥ ♆
23 49 ☽ Q ☿

25 Sunday
00 17 ♀ ☌ ♄
06 31 ☽ ✱ ☉
09 35 ☽ ⊻ ♆
15 03 ☽ ⊻ ♇
18 38 ☽ △ ♂
22 24 ☽ ✱ ☿

26 Monday
00 54 ☽ ⊥ ♇
04 18 ☽ ☌ ♄
06 18 ☽ ☌ ♀
09 00 ☽ ∠ ☉
11 04 ☽ □ ♃
15 31 ☽ ∠ ♇
15 41 ☽ ☌ ♅

27 Tuesday
00 32 ☽ ⊥ ☉
00 39 ☿ ⊥ ♃
01 52 ☽ ∠ ☿
03 14 ☉ ⊻ ♆
10 19 ☽ ☌ ♆
10 48 ☽ ⊻ ☉
15 29 ☽ ✱ ♇
16 16 ☽ Q ♃
18 03 ☽ ⊥ ☿
21 14 ☽ □ ♂
22 19 ☉ ∠ ♃

28 Wednesday
02 46 ☿ ✱ ♄
04 33 ☽ ⊻ ♄
04 46 ☽ ⊻ ☿
09 53 ☽ ⊻ ♀
10 30 ☽ Q ♇
11 31 ☽ ✱ ♃
14 06 ☽ ⊥ ♄
15 32 ☽ ⊻ ♅
20 06 ☽ ⊥ ♀

29 Thursday
01 04 ☽ ⊥ ♅
03 05 ☽ Q ♂
04 25 ☽ ∠ ♄
07 48 ☽ ∥ ♅
09 55 ☽ ⊻ ♆
11 28 ☽ ∠ ♀
11 36 ☽ ∠ ♃
13 43 ♀ □ ♃
13 45 ☽ ☌ ☉
14 18 ☽ ∥ ♆
15 02 ☽ □ ♇
15 24 ☽ ∠ ♅
18 27 ☽ ∥ ♄
19 30 ☽ ⊥ ♆
23 06 ☽ ✱ ♂

30 Friday
00 07 ☽ ∥ ♀
02 10 ☽ ⊥ ♃
04 29 ☽ ✱ ♄
04 36 ☿ ∠ ♆
09 00 ☉ □ ♇
09 59 ☽ ∠ ♆
10 40 ☽ ☌ ☿
11 10 ☽ ∥ ☉
11 58 ☽ ⊻ ♃
13 24 ☽ ✱ ♀
15 23 ☉ ∠ ♅
15 35 ☽ ✱ ♅
21 13 ☽ ∥ ☿
23 16 ☿ ⊻ ♃

31 Saturday
00 04 ☽ Q ♄
00 35 ☽ ∠ ♂
10 35 ☽ ✱ ♆
10 39 ☽ Q ♀
11 20 ☽ Q ♅
13 02 ♀ ∠ ♇
15 50 ☽ △ ♇
16 14 ☽ ⊥ ♂
18 10 ☽ ⊻ ☉
20 00 ♀ ☌ ♅

FEBRUARY 1987

LONGITUDES

Date	Sidereal time h m s	Sun ☉ ° ' "	Moon ☽ ° ' "	Moon ☽ 24.00 ° ' "	Mercury ☿ ° '	Venus ♀ ° '	Mars ♂ ° '	Jupiter ♃ ° '	Saturn ♄ ° '	Uranus ♅ ° '	Neptune ♆ ° '	Pluto ♇ ° '
01	20 44 39	12 ♒ 05 18	21 ♓ 50 41	28 ♓ 46 08	25 ♒ 41	25 ♐ 59	16 ♈ 45	23 ♓ 24	18 ♐ 32	25 ♐ 17	06 ♑ 49	09 ♏ 57
02	20 48 35	13 06 12	05 ♈ 34 16	12 ♈ 15 15	27 22	27 05	17 27	23 36	18 37	25 20	06 51	09 57
03	20 52 32	14 07 06	18 ♈ 49 21	25 ♈ 16 58	29 ♒ 01	28 12	18 08	23 49	18 42	25 23	06 53	09 57
04	20 56 29	15 07 58	01 ♉ 38 38	07 ♉ 54 53	00 ♓ 38	29 ♐ 18	18 50	24 02	18 47	25 25	06 55	09 58
05	21 00 25	16 08 48	14 ♉ 06 21	20 ♉ 13 40	02 13	00 ♑ 25	19 32	24 15	18 52	25 28	06 56	09 58
06	21 04 22	17 09 37	26 ♉ 17 29	02 ♊ 18 25	03 45	01 32	20 14	24 28	18 57	25 31	06 58	09 58
07	21 08 18	18 10 25	08 ♊ 17 07	14 ♊ 14 11	05 14	02 39	20 55	24 41	19 02	25 33	07 00	09 58
08	21 12 15	19 11 11	20 ♊ 10 09	26 ♊ 05 34	06 39	03 46	21 37	24 54	19 06	25 36	07 02	09 58
09	21 16 11	20 11 56	02 ♋ 00 54	07 ♋ 56 34	07 58	04 54	22 19	25 07	19 11	25 38	07 04	09 58
10	21 20 08	21 12 39	13 ♋ 52 59	19 ♋ 50 27	09 13	06 01	23 00	25 21	19 15	25 41	07 05	09 58
11	21 24 04	22 13 21	25 ♋ 49 17	01 ♌ 49 41	10 21	07 09	23 42	25 34	19 20	25 43	07 07	09 59
12	21 28 01	23 14 02	07 ♌ 51 54	13 ♌ 56 03	11 23	08 17	24 23	25 47	19 24	25 45	07 09	09 R 59
13	21 31 58	24 14 40	20 ♌ 02 17	26 ♌ 10 43	12 17	09 25	25 05	26 01	19 28	25 48	07 11	09 58
14	21 35 54	25 15 18	02 ♍ 21 25	08 ♍ 34 30	13 02	10 33	25 47	26 14	19 33	25 50	07 12	09 58
15	21 39 51	26 15 54	14 ♍ 50 03	21 ♍ 08 08	13 39	11 41	26 28	26 28	19 37	25 52	07 14	09 58
16	21 43 47	27 16 28	27 ♍ 28 53	03 ♎ 52 25	14 06	12 50	27 10	26 41	19 41	25 55	07 16	09 58
17	21 47 44	28 17 02	10 ♎ 18 52	16 ♎ 48 25	14 23	13 59	27 51	26 55	19 45	25 57	07 17	09 58
18	21 51 40	29 ♒ 17 34	23 ♎ 21 16	29 ♎ 57 35	14 30	15 07	28 32	27 09	19 49	25 59	07 19	09 58
19	21 55 37	00 ♓ 18 04	06 ♏ 37 36	13 ♏ 21 31	14 R 27	16 16	29 14	27 22	19 53	26 01	07 20	09 57
20	21 59 33	01 18 34	20 ♏ 09 30	27 ♏ 01 43	14 13	17 25	29 ♈ 55	27 36	19 56	26 03	07 22	09 57
21	22 03 30	02 19 02	03 ♐ 58 16	10 ♐ 59 08	13 50	18 35	00 ♉ 37	27 50	20 00	26 05	07 23	09 57
22	22 07 27	03 19 29	18 ♐ 04 17	25 ♐ 13 31	13 17	19 44	01 18	28 04	20 04	26 07	07 25	09 56
23	22 11 23	04 19 55	02 ♑ 26 31	09 ♑ 42 51	12 36	20 53	01 59	28 18	20 07	26 09	07 26	09 56
24	22 15 20	05 20 19	17 ♑ 01 57	24 ♑ 23 06	11 47	22 03	02 41	28 32	20 10	26 11	07 28	09 56
25	22 19 16	06 20 42	01 ♒ 45 29	09 ♒ 08 12	10 52	23 12	03 22	28 46	20 14	26 12	07 29	09 55
26	22 23 13	07 21 03	16 ♒ 30 18	23 ♒ 50 47	09 53	24 22	04 03	29 00	20 17	26 14	07 31	09 55
27	22 27 09	08 21 23	01 ♓ 08 44	08 ♓ 23 15	08 51	25 32	04 45	29 14	20 20	26 16	07 32	09 54
28	22 31 06	09 ♓ 21 41	15 ♓ 33 30	22 ♓ 38 51	07 ♓ 47	26 ♑ 42	05 ♉ 26	29 ♓ 28	20 ♐ 23	26 ♐ 17	07 ♑ 33	09 ♏ 54

Moon Nodes and Latitude

Date	Moon True ☊ ° '	Moon Mean ☊ ° '	Moon ☽ Latitude ° '
01	13 ♈ 21	14 ♈ 50	01 S 52
02	13 R 19	14 46	00 S 41
03	13 D 19	14 43	00 N 29
04	13 20	14 40	01 36
05	13 21	14 37	02 37
06	13 R 20	14 34	03 29
07	13 18	14 31	04 11
08	13 13	14 27	04 41
09	13 06	14 24	05 00
10	12 57	14 21	05 05
11	12 47	14 18	04 58
12	12 37	14 15	04 37
13	12 28	14 12	04 03
14	12 20	14 08	03 18
15	12 14	14 05	02 22
16	12 11	14 02	01 19
17	12 10	13 59	00 N 10
18	12 D 10	13 56	01 S 00
19	12 12	13 52	02 09
20	12 13	13 49	03 11
21	12 R 13	13 46	04 04
22	12 12	13 43	04 43
23	12 10	13 40	05 06
24	12 06	13 37	05 09
25	12 00	13 33	04 52
26	11 55	13 30	04 16
27	11 50	13 27	03 23
28	11 ♈ 46	13 ♈ 24	02 S 18

DECLINATIONS

Date	Sun ☉ ° '	Moon ☽ ° '	Mercury ☿ ° '	Venus ♀ ° '	Mars ♂ ° '	Jupiter ♃ ° '	Saturn ♄ ° '	Uranus ♅ ° '	Neptune ♆ ° '	Pluto ♇ ° '
01	17 S 10	04 S 57	14 S 11	20 S 29	06 N 42	03 S 40	21 S 27	23 S 29	22 S 15	00 N 48
02	16 53	01 N 35	13 29	20 34	06 59	03 34	21 28	23 29	22 15	00 48
03	16 36	07 50	12 46	20 39	07 16	03 29	21 28	23 29	22 15	00 48
04	16 18	13 33	12 02	20 43	07 32	03 24	21 28	23 29	22 15	00 49
05	16 00	18 34	11 18	20 46	07 49	03 19	21 29	23 29	22 15	00 49
06	15 42	22 43	10 35	20 49	08 06	03 14	21 29	23 29	22 15	00 50
07	15 23	25 49	09 51	20 52	08 22	03 08	21 29	23 29	22 15	00 50
08	15 04	27 45	09 08	20 54	08 39	03 03	21 30	23 29	22 15	00 51
09	14 45	28 26	08 26	20 56	08 55	02 58	21 30	23 30	22 15	00 51
10	14 26	27 47	07 44	20 57	09 11	02 52	21 30	23 30	22 14	00 52
11	14 06	25 51	07 05	20 57	09 27	02 47	21 30	23 30	22 14	00 52
12	13 47	22 46	06 28	20 57	09 44	02 42	21 31	23 30	22 14	00 53
13	13 27	18 39	05 53	20 57	10 00	02 36	21 31	23 30	22 14	00 53
14	13 06	13 42	05 21	20 55	10 16	02 31	21 31	23 30	22 14	00 54
15	12 46	08 09	04 52	20 54	10 31	02 25	21 32	23 30	22 14	00 54
16	12 25	02 N 12	04 27	20 52	10 47	02 20	21 32	23 30	22 14	00 55
17	12 04	03 S 56	04 06	20 49	11 03	02 14	21 32	23 30	22 14	00 55
18	11 43	10 00	03 49	20 45	11 18	02 09	21 32	23 30	22 14	00 56
19	11 22	15 45	03 37	20 42	11 34	02 03	21 32	23 30	22 14	00 56
20	11 01	20 51	03 30	20 37	11 49	01 58	21 33	23 30	22 13	00 57
21	10 39	24 56	03 28	20 32	12 04	01 52	21 33	23 31	22 13	00 57
22	10 17	27 37	03 30	20 27	12 19	01 47	21 33	23 31	22 13	00 58
23	09 55	28 31	03 38	20 20	12 35	01 41	21 33	23 31	22 13	00 59
24	09 33	27 28	03 50	20 14	12 49	01 36	21 33	23 31	22 13	00 59
25	09 11	24 31	04 06	20 06	13 04	01 30	21 33	23 31	22 13	01 00
26	08 49	19 57	04 26	19 59	13 19	01 24	21 34	23 31	22 13	01 00
27	08 26	14 04	04 48	19 50	13 34	01 19	21 34	23 31	22 13	01 01
28	08 S 04	07 S 49	05 S 14	19 S 41	13 N 48	01 S 13	21 S 34	23 S 31	22 S 13	01 N 02

ZODIAC SIGN ENTRIES

Date	h	m	Planets
02	02	09	☽ ♈
04	02	31	☿ ♓
04	08	53	☽ ♉
05	03	03	♀ ♑
06	19	23	☽ ♊
09	07	55	☽ ♋
11	20	21	☽ ♌
14	07	26	☽ ♍
16	16	44	☽ ♎
19	00	04	☽ ♏
19	04	50	☉ ♓
20	14	44	♂ ♉
21	05	09	☽ ♐
23	07	57	☽ ♑
25	09	08	☽ ♒
27	10	07	☽ ♓

LATITUDES

Date	Mercury ☿ ° '	Venus ♀ ° '	Mars ♂ ° '	Jupiter ♃ ° '	Saturn ♄ ° '	Uranus ♅ ° '	Neptune ♆ ° '	Pluto ♇ ° '
01	01 S 18	02 N 54	00 N 08	01 S 08	01 N 30	00 S 07	01 N 01	16 N 23
04	00 51	02 44	00 10	01 07	01 30	00 07	01 01	16 25
07	00 S 16	02 33	00 13	01 07	01 30	00 07	01 01	16 27
10	00 N 24	02 22	00 16	01 07	01 31	00 07	01 01	16 28
13	01 10	02 10	00 18	01 07	01 31	00 07	01 01	16 30
16	01 58	01 58	00 21	01 06	01 31	00 07	01 01	16 31
19	02 43	01 46	00 23	01 06	01 31	00 07	01 01	16 33
22	03 19	01 34	00 25	01 06	01 32	00 07	01 01	16 34
25	03 40	01 22	00 28	01 06	01 32	00 07	01 01	16 36
28	03 41	01 09	00 30	01 06	01 32	00 08	01 01	16 37
31	03 N 24	00 N 57	00 N 32	01 S 05	01 N 32	00 S 08	01 N 01	16 N 39

DATA

Julian Date	2446828
Delta T	+55 seconds
Ayanamsa	23° 40' 33"
Synetic vernal point	05° ♓ 26' 26"
True obliquity of ecliptic	23° 26' 36"

MOON'S PHASES, APSIDES AND POSITIONS ☽

Date	h	m	Phase	Longitude ° '	Eclipse Indicator
05	16	21	☽	16 ♉ 20	
13	20	58	○	24 ♌ 37	
21	08	56	☾	02 ♐ 11	
28	00	51	●	08 ♓ 54	

Day	h	m		
09	15	56	Apogee	
25	15	44	Perigee	
02	06	08	ON	
09	12	15	Max dec	28° N 26'
16	20	39	0S	
23	11	17	Max dec	28° S 31'

LONGITUDES

Date	Chiron ⚷ ° '	Ceres ⚳ ° '	Pallas ⚴ ° '	Juno ⚵ ° '	Vesta ⚶ ° '	Black Moon Lilith ⚸ ° '
01	16 ♊ 04	17 ♐ 07	23 ♏ 25	23 ♑ 43	18 ♈ 30	07 ♋ 50
11	15 ♊ 53	20 ♐ 34	26 ♏ 02	27 ♑ 33	22 ♈ 07	08 ♋ 57
21	15 ♊ 49	23 ♐ 47	28 ♏ 13	01 ♒ 21	25 ♈ 54	10 ♋ 05
31	15 ♊ 53	26 ♐ 43	29 ♏ 53	05 ♒ 07	29 ♈ 50	11 ♋ 12

All ephemeris data is given at 12.00 UT and the Moon's longitude is additionally given for 24.00 UT

ASPECTARIAN

h m	Aspects
01 Sunday	
02 51	☽ ⊻ ♂
05 05	☽ ⊥ ☉
05 52	☽ # ♂
06 15	☿ ⚹ ♅
06 18	☽ □ ♄
06 48	☽ Q ♆
14 43	☽ ☌ ♃
16 46	☽ ∥ ♃
17 21	☽ ⚼ ♇
17 58	☽ □ ♅
19 33	☽ ⊻ ☿
19 46	☽ □ ♀
21 47	☽ ∠ ☉
02 Monday	
00 38	☿ ⚹ ♀
03 11	☽ # ♇
05 43	☉ ⊥ ♆
07 32	☽ ⊥ ☿
09 05	☽ ∥ ♇
09 07	☽ ± ♇
14 17	☽ □ ♆
19 23	☽ # ♃
19 51	☽ ⊼ ♇
03 Tuesday	
01 57	☽ ∠ ☿
02 40	☽ ⚹ ☉
09 38	☽ ∥ ♂
10 40	☽ ☌ ♂
11 47	☽ △ ♄
21 26	☽ ⊻ ♃
04 Wednesday	
00 13	☽ △ ♅
02 44	☽ Q ☉
06 08	☽ # ☿
07 08	☽ △ ♀
08 54	☽ ⊥ ♃
09 49	☽ ⚹ ☿
10 04	♂ △ ♄
14 21	☿ Q ♄
16 06	☽ ⚼ ♄
22 05	☽ △ ♆
23 57	☽ # ☉
05 Thursday	
02 25	☽ ∠ ♃
03 57	☽ ☍ ♇
04 54	☽ ⚼ ♅
09 34	☽ ± ♄
12 15	☽ Q ☿
14 49	☽ ⚼ ♀
16 21	☽ □ ☉
21 23	☽ ⊼ ♄
22 32	☽ ± ♅
23 16	☽ ⊻ ♂
06 Friday	
00 13	☽ # ♀
03 25	☽ ⚼ ♆
04 16	☽ # ♄
08 19	☽ ⚹ ♃
09 01	☽ # ♆
10 20	☽ ± ♀
10 27	☽ ⊼ ♅
11 52	☽ ⊥ ♂
17 17	☽ # ♅
21 21	☽ ± ♆
23 31	☽ ⊼ ♀
07 Saturday	
05 01	☽ □ ☿
06 57	☽ ∠ ♂
08 43	☽ Q ♃
09 25	☽ ⊼ ♆
15 24	☽ ⊼ ♇
08 Sunday	
03 27	☉ ⊥ ♃
03 30	☽ ± ♇
09 49	☽ △ ☉
09 50	☽ ☍ ♄
09 52	☉ ⚹ ♄
11 03	☿ ∠ ♂
15 07	☽ ⚹ ♂
19 02	☿ ⚹ ♆
19 34	♃ ⚼ ♇
21 44	☽ ⚼ ♇
21 46	☽ □ ♃
23 02	☽ ☍ ♅
23 54	☿ # ♂
09 Monday	
05 35	☿ Q ♅
16 56	☽ Q ♂
18 26	☽ ☍ ♀
19 03	☽ ⚼ ☉
22 15	☽ ☍ ♆
10 Tuesday	
01 30	☽ △ ☿
04 06	☽ △ ♇
14 56	☽ ± ☉
22 53	☽ ⊼ ♄
11 Wednesday	
03 45	☿ △ ♇
04 07	☽ ⊼ ☉
07 29	☽ □ ♂
09 31	☉ ∠ ♆
10 58	☽ ⚼ ☿
11 01	☽ ± ♄
11 24	♀ ☌ ♆
11 29	☽ △ ♃
11 48	☽ ⊼ ♅
16 58	♇ St R
23 49	☽ ± ♅
12 Thursday	
02 55	☉ ∠ ♀
05 05	☽ ⚼ ♄
06 38	☽ ± ☿
06 59	☽ # ♅
07 56	♃ □ ♅
10 35	☽ ⊼ ♆
12 54	☽ ⊼ ♀
15 23	☽ # ♆
16 11	☽ □ ♇
17 45	☽ ⚼ ♅
17 54	☽ ⚼ ♃
19 33	☽ ⊼ ☿
19 52	☽ # ♄
22 29	☽ ± ♆
23 14	☽ # ♀
13 Friday	
01 59	☽ ± ♀
10 53	☽ △ ♄
11 57	☽ ± ♃
16 12	☽ ⚼ ♆
20 58	☽ ☍ ☉
21 26	☽ ⚼ ♀
22 27	☽ △ ♂
23 17	☽ △ ♅
23 49	♀ ⚹ ♇
23 54	☽ ⊼ ♃
14 Saturday	
03 30	☽ Q ♇
14 10	♂ △ ♅
14 55	☽ # ☉
21 23	☽ △ ♆
15 Sunday	
02 18	☉ ⚹ ♅
02 29	☽ ∥ ♂
02 41	☽ ⚹ ♇
05 11	☽ ⚼ ♂
05 23	☽ △ ♀
09 38	☽ ☍ ☿
11 44	♂ ⊻ ♃
18 02	☉ ⊻ ♃
21 10	☽ □ ♄
23 21	☽ ± ♂
16 Monday	
02 27	☽ # ☿
03 15	☉ ⚹ ♂
07 16	☽ ∠ ♇
09 01	☽ □ ♅
10 29	☽ ☍ ♃
11 21	☽ ⊼ ♂
11 29	☽ # ♃
11 35	☽ ⊼ ☉
17 04	☽ ∥ ♇
23 49	☽ ± ☉
17 Tuesday	
00 10	☽ ⊥ ♇
00 14	☽ # ♇
05 30	☽ ∥ ♃
06 22	☽ □ ♆
07 12	☽ Q ♄
11 21	☽ ⊻ ♇
12 37	☽ ∥ ☿
17 58	☽ ⚼ ☉
18 44	☽ Q ♅
19 26	☽ □ ♀
19 38	☽ ⊼ ☿
21 58	☿ ⚹ ♀
18 Wednesday	
05 29	☽ ⚹ ♄
06 46	☽ ± ☿
12 31	♀ Q ♃
15 35	☽ Q ♆
16 07	☿ St R
16 48	☽ ⚹ ♅
17 30	☽ # ♂
18 34	☽ ∥ ☉
19 01	☽ ⊼ ♃
21 57	☽ ☍ ♂
23 09	☽ ⚼ ☿
23 41	☽ △ ☉
19 Thursday	
04 22	☉ # ♂
06 03	☽ ± ♃
07 23	☽ Q ♀
08 51	☽ ∠ ♄
13 17	☽ ⚹ ♆
17 57	☽ ☌ ♇
17 57	☿ ∠ ♂
19 51	☽ ∠ ♅
22 26	☽ ⚼ ♃
20 Friday	
00 59	☽ ⊥ ♄
01 44	☽ △ ☿
06 45	☽ ⚹ ♀
10 49	☽ ∥ ♀
11 37	☽ ⊻ ♄
11 48	☽ ⊥ ♅
15 37	☽ ∥ ♄
15 52	☽ ∠ ♆
19 18	☽ ∥ ♆
22 19	☽ ⊻ ♅
21 Saturday	
01 13	☽ △ ♃
02 44	☽ ∥ ♅
03 57	☉ Q ♄
05 54	☽ ⊼ ♂
07 33	☽ ⊥ ♆
08 56	☽ □ ☉
11 15	☽ ∠ ♀
16 46	☽ ± ♂
17 52	☽ ⊻ ♆
22 14	☽ ⊻ ♇
22 Sunday	
04 01	☽ ⊥ ♀
04 13	☽ □ ☿
08 24	☽ ⊥ ♇
08 52	☽ ⚼ ♂
15 02	☽ ⊻ ♀
15 22	☽ ☌ ♄
17 53	☽ Q ☉
19 11	♀ ⊻ ♄
23 31	☽ ∠ ♇
23 Monday	
01 31	☽ ☌ ♅
05 00	☽ □ ♃
09 05	☽ Q ☿
11 13	☽ △ ♂
15 22	☽ ⚹ ☉
20 16	☽ ☌ ♆
24 Tuesday	
00 21	☽ ⚹ ♇
03 52	☽ ⚹ ☿
09 34	♀ Q ♇
11 10	☽ Q ♃
17 09	☽ ⊻ ♄
17 48	☽ ∠ ☉
19 59	☽ Q ♇
20 53	☽ ☌ ♀
25 Wednesday	
02 57	☽ ⊻ ♅
02 58	☽ ⊥ ♄
03 00	☽ ∠ ☿
07 03	☽ ⚹ ♃
09 32	☽ ⊥ ☉
12 44	☽ ⊥ ♅
14 45	☽ □ ♂
16 45	☽ ⊥ ☿
17 40	☽ ∠ ♄
17 59	☽ ∥ ♅
20 00	☽ ⊻ ☉
21 20	☽ ⊻ ♆
26 Thursday	
01 02	☽ ∥ ♆
01 16	☽ □ ♇
01 54	☽ ⊻ ☿
03 24	☽ ∠ ♅
04 21	☽ ∥ ♄
07 07	☽ ⊥ ♆
07 50	☽ ∠ ♃
11 22	☿ △ ♇
11 55	☽ ∥ ♀
15 51	☉ ⚹ ♆
17 40	☿ ∠ ♀
18 11	☽ ⚹ ♄
21 31	☽ Q ♂
21 49	☽ ∠ ♆
22 46	☽ ⊥ ♃
27 Friday	
01 58	☽ ⊻ ♀
03 57	☽ ⚹ ♅
08 47	☽ ⊻ ♃
09 43	☉ Q ♅
12 41	☽ ⊥ ♀
13 58	☽ Q ♄
14 31	☽ # ♂
17 40	☉ ☌ ☿
18 15	☽ ⚹ ♂
22 35	☽ ⚹ ♆
23 49	☽ Q ♅
23 53	☽ ☌ ☿
28 Saturday	
00 47	☿ Q ♅
00 51	☽ ☌ ☉
02 31	☽ △ ♇
03 28	♀ ⊻ ♅
04 57	☽ ∠ ♀
05 19	♀ ⊥ ♄
10 18	♂ ⚼ ♄
11 03	☽ ∥ ☉
13 44	♂ ⊥ ♃
16 59	☿ ⚹ ♆
18 45	☽ Q ♆
20 11	☽ □ ♄
20 39	☽ ∠ ♂
20 45	☽ ∥ ☿

MARCH 1987

LONGITUDES

Date	Sidereal time h m s	Sun ☉ ° ' "	Moon ☽ ° ' "	Moon ☽ 24.00 ° ' "	Mercury ☿ ° '	Venus ♀ ° '	Mars ♂ ° '	Jupiter ♃ ° '	Saturn ♄ ° '	Uranus ♅ ° '	Neptune ♆ ° '	Pluto ♇ ° '
01	22 35 02	10 ♓ 21 57	29 ♓ 38 44	06 ♈ 32 47	06 ♓ 43	27 ♑ 52	06 ♉ 07	29 ♓ 42	20 ♐ 26	26 ♐ 19	07 ♑ 34	09 ♏ 53
02	22 38 59	11 22 11	13 ♈ 20 45	20 ♈ 02 34	05 R 41	29 ♑ 02	06 48	29 ♓ 56	20 29	26 21	07 36	09 R 52
03	22 42 56	12 22 23	26 ♈ 38 16	03 ♉ 08 01	04 41	00 ♒ 12	07 29	00 ♈ 10	20 32	26 22	07 37	09 52
04	22 46 52	13 22 34	09 ♉ 32 08	15 ♉ 50 59	03 46	01 22	08 11	00 24	20 34	26 24	07 38	09 51
05	22 50 49	14 22 42	22 ♉ 05 01	28 ♉ 14 44	02 55	02 32	08 52	00 39	20 37	26 25	07 39	09 50
06	22 54 45	15 22 48	04 ♊ 20 42	10 ♊ 23 30	02 10	03 43	09 33	00 53	20 39	26 26	07 40	09 49
07	22 58 42	16 22 52	16 ♊ 23 44	22 ♊ 22 00	01 32	04 53	10 14	01 07	20 42	26 28	07 42	09 49
08	23 02 38	17 22 54	28 ♊ 18 55	04 ♋ 15 03	00 59	06 04	10 55	01 22	20 44	26 29	07 43	09 48
09	23 06 35	18 22 54	10 ♋ 11 00	16 ♋ 07 18	00 34	07 14	11 36	01 36	20 46	26 30	07 44	09 47
10	23 10 31	19 22 52	22 ♋ 04 28	28 ♋ 02 58	00 15	08 25	12 17	01 50	20 48	26 31	07 45	09 46
11	23 14 28	20 22 47	04 ♌ 03 16	10 ♌ 05 44	00 ♓ 03	09 36	12 58	02 05	20 50	26 32	07 46	09 45
12	23 18 25	21 22 41	16 ♌ 10 43	22 ♌ 18 30	29 ♒ 57	10 47	13 39	02 19	20 52	26 33	07 47	09 44
13	23 22 21	22 22 32	28 ♌ 29 20	04 ♍ 43 24	29 ♒ 58	11 57	14 20	02 34	20 54	26 34	07 48	09 43
14	23 26 18	23 22 21	11 ♍ 00 50	17 ♍ 21 45	00 ♓ 05	13 08	15 00	02 48	20 56	26 35	07 48	09 42
15	23 30 14	24 22 08	23 ♍ 46 09	00 ♎ 14 05	00 D 17	14 19	15 41	03 02	20 57	26 36	07 49	09 41
16	23 34 11	25 21 53	06 ♎ 45 29	13 ♎ 20 18	00 35	15 30	16 22	03 17	20 59	26 37	07 50	09 40
17	23 38 07	26 21 37	19 ♎ 58 28	26 ♎ 39 51	00 58	16 41	17 03	03 31	21 00	26 38	07 51	09 39
18	23 42 04	27 21 18	03 ♏ 24 22	10 ♏ 11 52	01 26	17 53	17 44	03 46	21 02	26 39	07 52	09 38
19	23 46 00	28 20 58	17 ♏ 02 14	23 ♏ 55 19	01 58	19 04	18 24	04 00	21 03	26 39	07 52	09 37
20	23 49 57	29 ♓ 20 36	00 ♐ 50 58	07 ♐ 49 03	02 35	20 15	19 05	04 15	21 04	26 40	07 53	09 36
21	23 53 54	00 ♈ 20 12	14 ♐ 49 22	21 ♐ 51 45	03 16	21 26	19 46	04 29	21 05	26 41	07 54	09 35
22	23 57 50	01 19 46	28 ♐ 56 00	06 ♑ 01 52	04 00	22 38	20 26	04 44	21 06	26 41	07 54	09 33
23	00 01 47	02 19 19	13 ♑ 09 03	20 ♑ 17 17	04 48	23 49	21 07	04 58	21 07	26 42	07 55	09 32
24	00 05 43	03 18 50	27 ♑ 26 11	04 ♒ 35 22	05 39	25 01	21 47	05 13	21 08	26 42	07 56	09 31
25	00 09 40	04 18 20	11 ♒ 44 22	18 ♒ 52 45	06 34	26 12	22 28	05 27	21 08	26 43	07 56	09 30
26	00 13 36	05 17 47	26 ♒ 00 01	03 ♓ 05 38	07 31	27 24	23 08	05 42	21 09	26 43	07 57	09 28
27	00 17 33	06 17 13	10 ♓ 09 07	17 ♓ 09 56	08 31	28 35	23 49	05 56	21 09	26 43	07 57	09 27
28	00 21 29	07 16 36	24 ♓ 07 39	01 ♈ 01 48	09 34	29 ♒ 47	24 29	06 11	21 09	26 43	07 58	09 26
29	00 25 26	08 15 58	07 ♈ 52 02	14 ♈ 38 01	10 39	00 ♓ 59	25 10	06 26	21 10	26 43	07 58	09 24
30	00 29 23	09 15 18	21 ♈ 19 31	27 ♈ 56 24	11 47	02 10	25 50	06 40	21 10	26 44	07 58	09 23
31	00 33 19	10 ♈ 14 35	04 ♉ 28 33	10 ♉ 56 01	12 ♓ 57	03 ♓ 22	26 ♉ 31	06 ♈ 55	21 ♐ 10	26 ♐ 44	07 ♑ 59	09 ♏ 21

Date	Moon True ☊ ° '	Moon Mean ☊ ° '	Moon ☽ Latitude ° '
01	11 ♈ 44	13 ♈ 21	01 S 06
02	11 D 43	13 18	00 N 09
03	11 44	13 14	01 21
04	11 46	13 11	02 26
05	11 47	13 08	03 23
06	11 49	13 05	04 09
07	11 R 49	13 02	04 44
08	11 49	12 58	05 05
09	11 47	12 55	05 14
10	11 44	12 52	05 09
11	11 41	12 49	04 51
12	11 38	12 46	04 19
13	11 34	12 43	03 35
14	11 32	12 39	02 40
15	11 30	12 36	01 36
16	11 29	12 33	00 N 26
17	11 D 29	12 30	00 S 47
18	11 30	12 27	01 58
19	11 31	12 23	03 04
20	11 32	12 20	04 00
21	11 33	12 17	04 43
22	11 33	12 14	05 09
23	11 R 33	12 11	05 16
24	11 33	12 08	05 04
25	11 32	12 04	04 34
26	11 31	12 01	03 46
27	11 31	11 58	02 45
28	11 30	11 55	01 35
29	11 30	11 52	00 S 20
30	11 D 30	11 49	00 N 54
31	11 ♈ 30	11 ♈ 45	02 N 04

DECLINATIONS

Date	Sun ☉ ° '	Moon ☽ ° '	Mercury ☿ ° '	Venus ♀ ° '	Mars ♂ ° '	Jupiter ♃ ° '	Saturn ♄ ° '	Uranus ♅ ° '	Neptune ♆ ° '	Pluto ♇ ° '
01	07 S 41	01 S 09	05 S 41	19 S 32	14 N 02	01 S 07	21 S 34	23 S 31	22 S 13	01 N 02
02	07 18	05 N 24	06 09	19 22	14 17	01 02	21 34	23 31	22 12	01 03
03	06 55	11 32	06 38	19 11	14 31	00 56	21 34	23 31	22 12	01 04
04	06 32	16 59	07 06	19 00	14 45	00 50	21 34	23 31	22 12	01 04
05	06 09	21 34	07 34	18 48	14 59	00 45	21 34	23 31	22 12	01 05
06	05 46	25 06	08 01	18 36	15 12	00 39	21 34	23 31	22 12	01 06
07	05 23	27 27	08 27	18 23	15 26	00 33	21 35	23 31	22 12	01 06
08	04 59	28 31	08 51	18 10	15 39	00 27	21 35	23 31	22 12	01 07
09	04 36	28 16	09 13	17 56	15 53	00 22	21 35	23 32	22 12	01 07
10	04 12	26 43	09 33	17 42	16 06	00 16	21 35	23 32	22 12	01 08
11	03 49	23 57	09 50	17 27	16 19	00 10	21 35	23 32	22 12	01 09
12	03 25	20 06	10 06	17 12	16 32	00 S 04	21 35	23 32	22 12	01 09
13	03 02	15 22	10 19	16 56	16 45	00 N 01	21 35	23 32	22 12	01 10
14	02 38	09 54	10 30	16 40	16 57	00 07	21 35	23 32	22 11	01 11
15	02 14	03 N 57	10 39	16 23	17 10	00 13	21 35	23 32	22 11	01 11
16	01 51	02 S 17	10 45	16 06	17 22	00 19	21 35	23 32	22 11	01 12
17	01 27	08 32	10 49	15 48	17 34	00 24	21 35	23 32	22 11	01 13
18	01 03	14 30	10 51	15 30	17 46	00 30	21 35	23 32	22 11	01 13
19	00 39	19 52	10 51	15 12	17 58	00 36	21 35	23 32	22 11	01 14
20	00 S 16	24 15	10 49	14 53	18 10	00 42	21 35	23 32	22 11	01 15
21	00 N 08	27 16	10 45	14 33	18 22	00 47	21 35	23 32	22 11	01 16
22	00 32	28 35	10 39	14 13	18 33	00 53	21 35	23 32	22 11	01 16
23	00 55	28 02	10 32	13 53	18 44	00 59	21 35	23 32	22 11	01 17
24	01 19	25 39	10 22	13 32	18 56	01 05	21 35	23 32	22 11	01 18
25	01 43	21 39	10 11	13 11	19 07	01 10	21 35	23 32	22 11	01 18
26	02 06	16 24	09 58	12 50	19 17	01 16	21 35	23 32	22 11	01 19
27	02 30	10 19	09 43	12 28	19 28	01 22	21 35	23 32	22 11	01 20
28	02 53	03 S 47	09 27	12 06	19 38	01 28	21 34	23 32	22 11	01 20
29	03 17	02 N 49	09 09	11 44	19 49	01 33	21 34	23 32	22 10	01 21
30	03 40	09 09	08 49	11 21	19 59	01 39	21 34	23 32	22 10	01 22
31	04 N 03	14 N 58	08 S 28	10 S 58	20 N 09	01 N 45	21 S 34	23 S 32	22 S 10	01 N 22

ZODIAC SIGN ENTRIES

Date	h	m	Planets
01	12	37	☽ ♈
02	18	41	♃ ♈
03	07	55	♀ ♒
03	18	11	☽ ♉
06	03	26	☽ ♊
08	15	24	☽ ♋
11	03	54	☽ ♌
11	21	55	☿ ♒
13	14	55	☽ ♍
13	21	09	☿ ♓
15	23	34	☽ ♎
18	05	57	☽ ♏
20	10	32	☽ ♐
21	03	52	☉ ♈
22	13	48	☽ ♑
24	16	18	☽ ♒
26	18	46	☽ ♓
28	16	20	♀ ♓
28	22	12	☽ ♈
31	03	46	☽ ♉

LATITUDES

Date	Mercury ☿ ° '	Venus ♀ ° '	Mars ♂ ° '	Jupiter ♃ ° '	Saturn ♄ ° '	Uranus ♅ ° '	Neptune ♆ ° '	Pluto ♇ ° '
01	03 N 37	01 N 05	00 N 30	01 S 06	01 N 32	00 S 08	01 N 01	16 N 38
04	03 15	00 53	00 32	01 05	01 32	00 08	01 01	16 39
07	02 39	00 41	00 34	01 05	01 33	00 08	01 01	16 40
10	01 58	00 29	00 36	01 05	01 33	00 08	01 01	16 42
13	01 15	00 17	00 38	01 05	01 33	00 08	01 01	16 43
16	00 N 33	00 N 06	00 40	01 05	01 34	00 08	01 02	16 44
19	00 S 05	00 S 05	00 41	01 05	01 34	00 08	01 02	16 45
22	00 40	00 16	00 43	01 05	01 34	00 08	01 02	16 46
25	01 09	00 26	00 45	01 05	01 34	00 08	01 02	16 47
28	01 35	00 35	00 46	01 05	01 35	00 08	01 02	16 48
31	01 S 55	00 S 44	00 N 48	01 S 05	01 N 35	00 S 08	01 N 02	16 N 49

DATA

Julian Date	2446856
Delta T	+55 seconds
Ayanamsa	23° 40' 37"
Synetic vernal point	05° ♓ 26' 22"
True obliquity of ecliptic	23° 26' 37"

MOON'S PHASES, APSIDES AND POSITIONS ☽

Date	h	m	Phase	Longitude ° '	Eclipse Indicator
07	11	58	☽	16 ♊ 23	
15	13	13	○	24 ♍ 25	
22	16	22	☾	01 ♑ 31	
29	12	46	●	08 ♈ 18	Ann-Total

Day	h	m		
09	10	26	Apogee	
24	18	52	Perigee	
01	16	07	0N	
08	19	23	Max dec	28° N 35'
16	03	16	0S	
22	17	05	Max dec	28° S 38'
29	01	43	0N	

LONGITUDES

Date	Chiron ⚷ ° '	Ceres ⚳ ° '	Pallas ⚴ ° '	Juno ⚵ ° '	Vesta ⚶ ° '	Black Moon Lilith ⚸ ° '
01	15 ♊ 51	26 ♐ 09	29 ♏ 36	04 ♒ 22	29 ♈ 02	10 ♋ 58
11	16 ♊ 01	28 ♐ 50	00 ♐ 46	08 ♒ 05	03 ♉ 04	12 ♋ 06
21	16 ♊ 19	01 ♑ 09	01 ♐ 16	11 ♒ 43	07 ♉ 12	13 ♋ 13
31	16 ♊ 43	03 ♑ 04	01 ♐ 00	15 ♒ 15	11 ♉ 24	14 ♋ 20

All ephemeris data is given at 12.00 UT and the Moon's longitude is additionally given for 24.00 UT

ASPECTARIAN

01 Sunday
00 34 ☉ △ ♇
03 49 ☽ bQ ♇
06 16 ☽ □ ♅
08 39 ☽ ⚹ ♀
12 05 ☽ ∥ ♃
12 06 ☽ ☌ ♃
12 23 ☽ # ♇
12 51 ☽ ⊥ ♂
19 20 ☽ ± ♇
19 54 ☽ ∥ ♇
20 05 ☽ # ♃
20 16 ☿ ⚹ ♂
23 25 ☽ ⊻ ☿
23 50 ☽ ⊻ ♂

02 Monday
01 49 ☽ □ ♆
05 51 ☽ ⚻ ♇
07 07 ♃ # ♇
07 30 ☽ Q ♀
08 13 ☽ ⊻ ☉
09 15 ☽ ⊥ ☿
15 02 ☽ # ☿
18 48 ☽ # ☉
19 07 ☿ ⊥ ♀
19 46 ☽ ⊥ ☉

03 Tuesday
00 13 ☽ ∠ ☿
00 50 ☽ △ ♄
11 16 ♀ ⚹ ♃
11 30 ☽ △ ♅
13 28 ☽ ∠ ☉
16 33 ♂ △ ♆
18 38 ☽ ⊻ ♃
19 12 ☽ □ ♀
20 08 ☉ ∥ ☿

04 Wednesday
01 16 ☽ ∥ ♂
01 53 ☽ ⚹ ☿
04 31 ☽ bQ ♄
06 00 ☽ ⊥ ♃
08 25 ☽ △ ♆
09 17 ☽ ☌ ♂
12 35 ☽ ☍ ♇
15 31 ☽ bQ ♅
19 55 ☽ ⚹ ☉
21 33 ☽ # ♀
21 35 ☽ ± ♄
23 04 ☽ Q ☿
23 22 ☽ ∠ ♃

05 Thursday
08 46 ☽ ± ♅
09 09 ☽ ⚻ ♄
09 24 ♀ ⊥ ♅
12 03 ☽ # ♄
13 07 ☽ bQ ♆
15 50 ☽ # ♆
16 38 ☿ ⊻ ♀
20 26 ☽ ⚻ ♅
20 54 ☿ Q ♄
21 05 ☽ Q ☉

06 Friday
00 21 ☽ # ♅
05 02 ☽ ⚹ ♃
06 43 ☽ ± ♆
07 57 ☽ □ ☿
10 37 ☽ △ ♀
18 36 ☽ ⚻ ♆
21 35 ♂ ☍ ♇
22 51 ☽ ⚻ ♇
22 56 ☽ ⊻ ♂

07 Saturday
05 19 ☽ Q ♃
10 50 ☽ ± ♇
11 39 ☽ ⊥ ♂
11 58 ☽ □ ☉
19 47 ☽ bQ ♀
20 40 ☽ ☍ ♄

08 Sunday
00 08 ☿ ⊻ ♃
04 54 ☽ bQ ♇
05 03 ♀ ∠ ♄
06 51 ☽ ∠ ♂
08 18 ☽ ☍ ♅
15 55 ☽ ± ♀
17 12 ☽ △ ☿
18 17 ☽ □ ♃

09 Monday
05 24 ☽ ⚻ ♀
07 02 ☽ ☍ ♆
08 35 ♂ bQ ♅
11 11 ☽ △ ♇
15 02 ☽ ⚹ ♂
22 05 ♀ ⊻ ♆
22 34 ☽ bQ ☿

10 Tuesday
06 05 ☽ △ ☉
09 26 ☽ ⚻ ♄
11 20 ☿ Q ♂
16 18 ☽ ± ☿
16 42 ☽ Q ♂
20 54 ☉ Q ♆
20 57 ☽ ⚻ ♅
21 32 ☽ ± ♄

11 Wednesday
04 07 ☽ ⚻ ☿
07 59 ☽ △ ♃
08 59 ☽ ± ♅
14 53 ☽ bQ ☉
14 58 ☽ # ♅
15 08 ♀ □ ♇
15 34 ☽ bQ ♄
19 23 ☽ ⚻ ♆
23 18 ☽ □ ♇
23 25 ☉ □ ♄
23 44 ☽ # ♆

12 Thursday
00 12 ☽ ☍ ♀
02 53 ☽ bQ ♅
03 29 ☽ # ♄
06 43 ☽ □ ♂
07 16 ☽ ± ♆
10 17 ☽ ± ☉
14 17 ☽ bQ ♃
21 13 ☽ △ ♄
21 23 ☿ St D
23 05 ☽ ⚻ ☉

13 Friday
00 56 ☽ bQ ♆
04 03 ☽ # ♀
04 07 ♀ ∠ ♅
05 40 ☽ ∥ ♂
08 11 ☽ ± ♃
08 17 ☽ △ ♅
10 31 ☽ Q ♇
14 52 ☽ ☍ ☿
20 00 ☽ ⚻ ♃
21 27 ♀ # ♂

14 Saturday
05 53 ☽ △ ♆
09 11 ♂ ± ♄
09 31 ☽ ⚹ ♇
09 34 ☽ # ☿
16 27 ☽ ⚻ ♀
19 59 ☽ △ ♂

15 Sunday
01 43 ♀ ⊥ ♆
04 54 ☽ ± ♀
06 44 ☽ □ ♄
13 13 ☽ ☍ ☉
13 42 ☽ ∠ ♇
17 17 ☽ □ ♅
19 06 ☽ # ☉
19 32 ☉ bQ ♇
22 40 ☽ ∥ ♇
23 21 ☽ bQ ♀

16 Monday
00 22 ☽ ⚻ ☿
01 33 ☽ bQ ♂
02 14 ☽ ∥ ♃
04 20 ☽ # ♃
05 30 ☽ ☍ ♃
06 20 ☽ ⊥ ♇
07 52 ☽ # ♇
10 25 ☽ ∥ ☉
11 40 ☽ ± ☿
13 58 ☽ □ ♆
16 04 ☽ Q ♄
17 19 ☽ ⊻ ♇
18 57 ☽ ± ♂

17 Tuesday
02 20 ☽ Q ♅
04 32 ☽ bQ ☿
05 30 ☽ △ ♀
06 26 ☽ ⚻ ♂
13 52 ☽ ⚹ ♄
18 40 ☉ □ ♅
21 04 ☽ ∥ ☿
22 33 ☽ Q ♆
23 57 ☽ ⚹ ♅
23 58 ♃ ± ♇

18 Wednesday
00 22 ☽ ⚻ ☉
01 48 ☉ # ♇
04 48 ♀ □ ♂
08 22 ☽ △ ☿
11 54 ☽ ± ☉
12 39 ☽ ⚻ ♃
16 00 ☽ ∥ ♀
16 39 ☽ ∠ ♄
19 53 ☽ ⚹ ♆
22 59 ☽ ☌ ♇
23 26 ☽ ± ♃

19 Thursday
02 34 ☽ ∠ ♅
02 46 ☽ # ♂
05 02 ☽ bQ ☉
08 31 ☽ ⊥ ♄
10 29 ♀ ∠ ♃
14 31 ☽ ☍ ♂
14 53 ☉ # ♃
15 30 ☽ bQ ♃
15 53 ☽ □ ♀
18 19 ☽ ⊥ ♅
19 01 ☽ ⊻ ♄
20 40 ☽ ∥ ♄
22 11 ☽ ∠ ♆
23 54 ☽ ∥ ♆

20 Friday
04 46 ☽ ⊻ ♅
07 36 ☽ ∥ ♅
09 12 ☽ △ ☉
13 47 ☽ ⊥ ♆
15 08 ☽ □ ☿
17 58 ☽ △ ♃
21 05 ♂ ∠ ♃

21 Saturday
00 08 ☽ ⊻ ♆
01 56 ☽ Q ♀
03 02 ☽ ⊻ ♇
04 42 ♀ ⚹ ♄
05 49 ☿ Q ♄
13 17 ☽ ⊥ ♇
20 51 ☽ ⚻ ♂
22 41 ☽ ☌ ♄
23 34 ☽ Q ☿

22 Sunday
00 19 ☽ ⚹ ♀
04 35 ☽ ∠ ♇
07 33 ☽ ± ♂
08 11 ☽ ☌ ♅
16 22 ☽ □ ☉
17 38 ♀ ∠ ♆
20 59 ♂ ± ♅
21 04 ☽ ⚹ ☿
21 58 ☽ □ ♃
23 33 ☽ bQ ♂

23 Monday
03 11 ☽ ☌ ♆
04 02 ☽ ∠ ♀
05 55 ☽ ⚹ ♇
12 00 ♂ ⚻ ♄
16 43 ☉ ∥ ♃
18 57 ☿ ⊻ ♃
20 34 ☽ ⊥ ♀
23 53 ☽ ∠ ☿

24 Tuesday
00 57 ☽ Q ☉
01 24 ☽ ⊻ ♄
02 03 ☽ △ ♂
02 04 ☽ Q ♇
04 48 ☽ Q ♃
07 34 ☽ ⊻ ♀
10 25 ☉ ∥ ♇
10 46 ☽ ⊻ ♅
11 29 ☽ ⊥ ♄
15 58 ☽ ⊥ ☿
16 44 ☉ ± ♇
20 50 ☽ ⊥ ♅
22 36 ☽ ⚹ ☉

25 Wednesday
01 16 ☽ ⚹ ♃
01 44 ☽ ∥ ♅
02 35 ☽ ∠ ♄
02 43 ☽ ⊻ ☿
05 37 ☽ ⊻ ♆
08 14 ☽ □ ♇
09 14 ☽ ∥ ♆
11 57 ☽ ∠ ♅
12 21 ☽ ∥ ♄
15 41 ☽ ⊥ ♆
22 11 ♀ ⚹ ♅
23 48 ☽ # ♂

26 Thursday
01 40 ☽ ∠ ☉
02 55 ☽ ∠ ♃
03 49 ☽ ⚹ ♄
04 55 ♂ bQ ♆
06 51 ☽ ∠ ♆
06 56 ☽ □ ♂
13 12 ☽ ⚹ ♅
14 35 ☽ ☌ ♀
17 59 ☽ ⊥ ☉
18 22 ☽ ⊥ ♃
22 23 ☿ ⚹ ♆

27 Friday
00 06 ☽ Q ♄
00 47 ♃ ∥ ♇
00 55 ☉ ☌ ♃
03 14 ☽ ∥ ♀
04 42 ☽ ⊻ ♃
04 55 ☽ ⊻ ☉
08 15 ☽ ⚹ ♆
09 00 ☽ ☌ ☿
09 33 ☽ Q ♅
10 48 ☽ △ ♇
14 19 ☽ ∥ ☿
14 59 ☽ Q ♂
16 35 ☿ Q ♅

28 Saturday
04 48 ☽ Q ♆
06 52 ☽ □ ♄
08 50 ☿ △ ♇
12 31 ☽ bQ ♇
12 40 ☽ ⚹ ♂
15 04 ☽ # ☉
16 30 ☽ □ ♅
20 18 ☽ # ♃
20 51 ☽ # ♇
22 03 ♀ ⊥ ♃
22 46 ☽ ⊻ ♀

29 Sunday
04 10 ☽ ± ♇
04 40 ☉ □ ♆
06 37 ☽ ∥ ♇
07 19 ☽ ∥ ♃
09 25 ☽ ☌ ♃
10 17 ☽ ⊥ ♀
12 10 ☽ □ ♆
12 46 ☽ ☌ ☉
13 50 ☽ ∥ ☉
14 43 ☽ ⚻ ♇
16 16 ☽ ∠ ♂
17 23 ☽ ⊻ ☿

30 Monday
03 49 ☽ ∠ ♀
05 03 ☽ ⊥ ☿
09 11 ☽ ⊥ ♂
10 46 ☽ # ☿
11 42 ☽ △ ♄
14 58 ☉ ⚻ ♇
20 13 ☽ # ♀
20 37 ☽ ⊻ ♂
21 47 ☽ △ ♅
22 51 ☽ ∠ ☿

31 Tuesday
04 42 ♄ St R
07 51 ♀ Q ♄
09 45 ☽ ⚹ ♀
15 07 ☽ bQ ♄
16 35 ☽ ⊻ ♃
18 29 ☽ △ ♆
19 43 ♂ ⚻ ♅
21 02 ☽ ☍ ♇
23 36 ☽ ⊻ ☉

APRIL 1987

Date	Sidereal time h m s	Sun ☉ ° ' "	Moon ☽ ° ' "	Moon ☽ 24.00 ° ' "	Mercury ☿ ° '	Venus ♀ ° '	Mars ♂ ° '	Jupiter ♃ ° '	Saturn ♄ ° '	Uranus ♅ ° '	Neptune ♆ ° '	Pluto ♇ ° '
01	00 37 16	11 ♈ 13 50	17 ♉ 18 53	23 ♉ 37 19	14 ♓ 09	04 ♓ 34	27 ♉ 11	07 ♈ 09	21 ♐ 10	26 ♐ 44	07 ♑ 59	09 ♏ 20
02	00 41 12	12 13 04	29 ♉ 51 34	06 ♊ 01 57	15 24	05 46	27 51	07 24	21 R 10	26 R 44	07 59	09 R 19
03	00 45 09	13 12 15	12 ♊ 08 52	18 ♊ 12 44	16 40	06 58	28 32	07 38	21 09	26 44	07 59	09 17
04	00 49 05	14 11 23	24 ♊ 14 01	00 ♋ 13 15	17 58	08 10	29 12	07 53	21 09	26 43	08 00	09 16
05	00 53 02	15 10 30	06 ♋ 10 58	12 ♋ 07 43	19 18	09 21	29 ♉ 52	08 07	21 08	26 43	08 00	09 14
06	00 56 58	16 09 34	18 ♋ 04 07	24 ♋ 00 44	20 40	10 33	00 ♊ 32	08 22	21 08	26 43	08 00	09 13
07	01 00 55	17 08 36	29 ♋ 58 08	05 ♌ 56 55	22 04	11 45	01 13	08 36	21 07	26 43	08 00	09 11
08	01 04 52	18 07 36	11 ♌ 57 38	18 ♌ 00 48	23 29	12 57	01 53	08 50	21 06	26 42	08 00	09 10
09	01 08 48	19 06 33	24 ♌ 06 57	00 ♍ 16 32	24 56	14 09	02 33	09 05	21 06	26 42	08 00	09 08
10	01 12 45	20 05 28	06 ♍ 29 57	12 ♍ 47 33	26 25	15 21	03 13	09 19	21 05	26 41	08 R 00	09 06
11	01 16 41	21 04 20	19 ♍ 09 39	25 ♍ 36 26	27 55	16 33	03 53	09 34	21 03	26 41	08 00	09 05
12	01 20 38	22 03 11	02 ♎ 08 02	08 ♎ 44 31	29 ♓ 27	17 45	04 33	09 48	21 02	26 40	08 00	09 03
13	01 24 34	23 01 59	15 ♎ 25 50	22 ♎ 11 50	01 ♈ 01	18 58	05 13	10 02	21 01	26 40	08 00	09 02
14	01 28 31	24 00 46	29 ♎ 02 18	05 ♏ 56 55	02 36	20 10	05 53	10 17	21 00	26 39	08 00	09 00
15	01 32 27	24 59 30	12 ♏ 55 19	19 ♏ 57 01	04 13	21 22	06 33	10 31	20 58	26 38	08 00	08 58
16	01 36 24	25 58 13	27 ♏ 01 32	04 ♐ 08 18	05 52	22 34	07 13	10 45	20 57	26 38	07 59	08 57
17	01 40 21	26 56 54	11 ♐ 16 47	18 ♐ 26 24	07 32	23 46	07 53	10 59	20 55	26 37	07 59	08 55
18	01 44 17	27 55 33	25 ♐ 36 36	02 ♑ 46 52	09 14	24 58	08 33	11 14	20 53	26 36	07 59	08 53
19	01 48 14	28 54 11	09 ♑ 56 43	17 ♑ 05 44	10 57	26 11	09 13	11 28	20 52	26 35	07 59	08 52
20	01 52 10	29 ♈ 52 47	24 ♑ 13 31	01 ♒ 19 45	12 42	27 23	09 53	11 42	20 50	26 34	07 58	08 50
21	01 56 07	00 ♉ 51 21	08 ♒ 24 09	15 ♒ 26 29	14 29	28 35	10 32	11 56	20 48	26 33	07 58	08 49
22	02 00 03	01 49 53	22 ♒ 26 35	29 ♒ 24 17	16 17	29 ♓ 48	11 12	12 10	20 46	26 32	07 58	08 47
23	02 04 00	02 48 24	06 ♓ 19 26	13 ♓ 11 56	18 07	01 ♈ 00	11 52	12 24	20 43	26 31	07 57	08 45
24	02 07 56	03 46 54	20 ♓ 01 41	26 ♓ 48 34	19 59	02 12	12 32	12 38	20 41	26 30	07 57	08 43
25	02 11 53	04 45 21	03 ♈ 32 29	10 ♈ 13 22	21 52	03 25	13 11	12 52	20 39	26 29	07 56	08 42
26	02 15 50	05 43 47	16 ♈ 51 05	23 ♈ 25 35	23 47	04 37	13 51	13 06	20 36	26 28	07 56	08 40
27	02 19 46	06 42 11	29 ♈ 56 47	06 ♉ 24 37	25 44	05 49	14 31	13 20	20 34	26 26	07 55	08 38
28	02 23 43	07 40 33	12 ♉ 49 03	19 ♉ 10 06	27 42	07 02	15 10	13 34	20 31	26 25	07 55	08 37
29	02 27 39	08 38 54	25 ♉ 27 46	01 ♊ 42 09	29 ♈ 42	08 14	15 50	13 48	20 29	26 24	07 54	08 35
30	02 31 36	09 ♉ 37 13	07 ♊ 53 21	14 ♊ 01 32	01 ♉ 43	09 ♈ 27	16 ♊ 29	14 ♈ 02	20 ♐ 26	26 ♐ 22	07 ♑ 53	08 ♏ 33

Date	Moon True ☊ ° '	Moon Mean ☊ ° '	Moon ☽ Latitude ° '
01	11 ♈ 30	11 ♈ 42	03 N 06
02	11 R 30	11 39	03 57
03	11 30	11 36	04 36
04	11 30	11 33	05 03
05	11 30	11 29	05 15
06	11 D 30	11 26	05 15
07	11 30	11 23	05 01
08	11 30	11 20	04 33
09	11 31	11 17	03 53
10	11 32	11 14	03 02
11	11 33	11 10	02 01
12	11 33	11 07	00 N 52
13	11 R 33	11 04	00 S 21
14	11 33	11 01	01 35
15	11 32	10 58	02 45
16	11 30	10 55	03 45
17	11 28	10 51	04 33
18	11 26	10 48	05 03
19	11 25	10 45	05 15
20	11 24	10 42	05 07
21	11 D 24	10 39	04 41
22	11 25	10 35	03 58
23	11 26	10 32	03 02
24	11 27	10 29	01 55
25	11 28	10 26	00 S 44
26	11 R 28	10 23	00 N 30
27	11 27	10 20	01 40
28	11 25	10 16	02 43
29	11 21	10 13	03 38
30	11 ♈ 17	10 ♈ 10	04 N 21

DECLINATIONS

Date	Sun ☉ ° '	Moon ☽ ° '	Mercury ☿ ° '	Venus ♀ ° '	Mars ♂ ° '	Jupiter ♃ ° '	Saturn ♄ ° '	Uranus ♅ ° '	Neptune ♆ ° '	Pluto ♇ ° '
01	04 N 27	19 N 58	08 S 06	10 S 34	20 N 19	01 N 51	21 S 34	23 S 32	22 S 10	01 N 23
02	04 50	23 59	07 42	10 11	20 28	01 56	21 34	23 32	22 10	01 24
03	05 13	26 49	07 17	09 47	20 38	02 02	21 34	23 32	22 10	01 24
04	05 36	28 21	06 51	09 22	20 47	02 08	21 34	23 32	22 10	01 25
05	05 59	28 33	06 23	08 58	20 56	02 13	21 34	23 32	22 10	01 26
06	06 21	27 25	05 53	08 33	21 05	02 19	21 34	23 32	22 10	01 26
07	06 44	25 03	05 23	08 08	21 14	02 25	21 33	23 32	22 10	01 27
08	07 07	21 35	04 51	07 42	21 23	02 30	21 33	23 32	22 10	01 27
09	07 29	17 09	04 18	07 17	21 31	02 36	21 33	23 32	22 10	01 28
10	07 51	11 57	03 44	06 51	21 39	02 42	21 33	23 32	22 10	01 29
11	08 13	06 N 09	03 09	06 25	21 47	02 47	21 33	23 32	22 10	01 29
12	08 35	00 S 03	02 33	05 59	21 55	02 53	21 33	23 32	22 10	01 30
13	08 57	06 24	01 55	05 33	22 03	02 58	21 32	23 32	22 10	01 31
14	09 19	12 37	01 17	05 06	22 10	03 04	21 32	23 32	22 10	01 31
15	09 41	18 20	00 S 37	04 39	22 17	03 10	21 32	23 32	22 10	01 32
16	10 02	23 09	00 N 03	04 13	22 25	03 15	21 32	23 32	22 10	01 32
17	10 23	26 38	00 45	03 46	22 31	03 21	21 32	23 32	22 10	01 33
18	10 44	28 25	01 28	03 18	22 38	03 26	21 32	23 32	22 10	01 33
19	11 05	28 18	02 11	02 51	22 45	03 32	21 31	23 32	22 10	01 34
20	11 26	26 19	02 55	02 24	22 51	03 37	21 31	23 32	22 10	01 35
21	11 46	22 41	03 41	01 56	22 57	03 43	21 31	23 32	22 10	01 35
22	12 07	17 47	04 27	01 29	23 03	03 48	21 31	23 32	22 10	01 36
23	12 27	12 00	05 14	01 01	23 09	03 53	21 31	23 32	22 10	01 36
24	12 47	05 S 43	06 01	00 33	23 15	03 59	21 30	23 32	22 10	01 37
25	13 07	00 N 45	06 49	00 S 06	23 20	04 04	21 30	23 32	22 10	01 37
26	13 26	07 05	07 38	00 N 22	23 25	04 10	21 30	23 32	22 10	01 38
27	13 45	13 01	08 27	00 50	23 30	04 15	21 30	23 32	22 10	01 38
28	14 04	18 17	09 17	01 18	23 35	04 20	21 29	23 32	22 10	01 39
29	14 23	22 39	10 07	01 46	23 40	04 26	21 29	23 32	22 10	01 39
30	14 N 42	25 N 55	10 N 57	02 N 13	23 N 44	04 N 31	21 S 29	23 S 32	22 S 10	01 N 40

ZODIAC SIGN ENTRIES

Date	h	m	Planets
02	12	16	☽ ♊
04	23	33	☽ ♋
05	16	37	♂ ♊
07	12	04	☽ ♌
09	23	28	☽ ♍
12	08	06	☽ ♎
12	20	23	☿ ♈
14	13	41	☽ ♏
16	17	02	☽ ♐
18	19	21	☽ ♑
20	14	58	☉ ♉
20	21	45	☽ ♒
22	16	07	♀ ♈
23	01	02	☽ ♓
25	05	41	☽ ♈
27	12	06	☽ ♉
29	15	39	☿ ♉
29	20	43	☽ ♊

LATITUDES

Date	Mercury ☿ ° '	Venus ♀ ° '	Mars ♂ ° '	Jupiter ♃ ° '	Saturn ♄ ° '	Uranus ♅ ° '	Neptune ♆ ° '	Pluto ♇ ° '
01	02 S 01	00 S 47	00 N 48	01 S 05	01 N 35	00 S 08	01 N 02	16 N 49
04	02 16	00 56	00 49	01 05	01 35	00 08	01 02	16 50
07	02 26	01 03	00 51	01 05	01 36	00 08	01 02	16 51
10	02 31	01 10	00 52	01 05	01 36	00 08	01 02	16 51
13	02 32	01 17	00 53	01 05	01 36	00 08	01 02	16 52
16	02 29	01 22	00 54	01 06	01 36	00 08	01 02	16 52
19	02 20	01 27	00 55	01 06	01 37	00 08	01 02	16 52
22	02 07	01 31	00 56	01 06	01 37	00 08	01 03	16 52
25	01 50	01 35	00 57	01 06	01 37	00 08	01 03	16 52
28	01 28	01 38	00 58	01 06	01 37	00 08	01 03	16 52
31	01 S 02	01 S 40	00 N 59	01 S 06	01 N 37	00 S 09	01 N 03	16 N 52

DATA

Julian Date	2446887
Delta T	+55 seconds
Ayanamsa	23° 40' 40"
Synetic vernal point	05° ♓ 26' 19"
True obliquity of ecliptic	23° 26' 37"

MOON'S PHASES, APSIDES AND POSITIONS ☽

Date	h	m	Phase	Longitude ° '	Eclipse Indicator
06	07	48	☽	15 ♋ 59	
14	02	31	○	23 ♎ 38	
20	22	15	☾	00 ♒ 18	
28	01	34	●	07 ♉ 15	

Day	h	m		
06	06	37	Apogee	
18	16	45	Perigee	
05	03	27	Max dec	28° N 38'
12	11	47	0S	
18	22	33	Max dec	28° S 37'
25	09	14	0N	

LONGITUDES

Date	Chiron ⚷ ° '	Ceres ⚳ ° '	Pallas ⚴ ° '	Juno ⚵ ° '	Vesta ⚶ ° '	Black Moon Lilith ⚸ ° '
01	16 ♊ 46	03 ♑ 14	00 ♐ 56	15 ♒ 35	11 ♉ 49	14 ♋ 27
11	17 ♊ 17	04 ♑ 36	29 ♏ 47	18 ♒ 59	16 ♉ 05	15 ♋ 34
21	17 ♊ 54	05 ♑ 26	27 ♏ 51	22 ♒ 13	20 ♉ 24	16 ♋ 41
31	18 ♊ 36	05 ♑ 41	25 ♏ 17	25 ♒ 15	24 ♉ 44	17 ♋ 48

All ephemeris data is given at 12.00 UT and the Moon's longitude is additionally given for 24.00 UT

ASPECTARIAN

h m	Aspects
01 Wednesday	
01 29	☽ ⚼ ♅
04 00	☽ ⊥ ♃
04 36	♅ St R
05 25	☽ ✶ ☿
07 56	☽ ± ♄
10 26	☽ Q ♀
11 50	☽ ⊥ ☉
13 53	☽ ∥ ♂
18 29	☽ ± ♅
19 18	☽ ⊼ ♄
20 48	☽ # ♄
21 22	☽ ∠ ♃
22 47	☽ ⚼ ♆
02 Thursday	
00 21	☽ # ♆
05 58	☽ ⊼ ♅
06 28	☽ ∠ ☉
06 42	☽ Q ☿
07 55	☽ ☌ ♂
08 58	☽ # ♅
16 07	☽ ± ♆
03 Friday	
00 43	☽ □ ♀
02 57	☽ ✶ ♃
03 50	☽ ⊼ ♆
06 23	☽ ⊼ ♇
06 42	☿ Q ♂
14 16	☽ ✶ ☉
18 11	☽ ± ♇
22 00	☽ □ ☿
04 Saturday	
03 08	☽ Q ♃
04 55	♀ ⊻ ♃
05 51	☽ ☍ ♄
08 42	♀ ✶ ♆
12 03	☽ ⚼ ♇
12 47	☉ ∠ ♂
15 33	☉ ⊥ ♀
16 16	☽ Q ☉
16 59	☽ ☍ ♅
22 33	☽ ⊻ ♂
23 17	♀ Q ♅
23 53	♃ □ ♆
05 Sunday	
09 37	♀ △ ♇
11 20	☽ ⊥ ♂
15 40	☽ ☍ ♆
15 59	☽ □ ♃
18 09	☽ △ ♇
19 07	☽ △ ♀
23 09	☉ # ☿
06 Monday	
00 19	☿ Q ♆
06 35	☽ ∠ ♂
07 48	☽ □ ☉
17 56	☽ △ ☿
18 11	☽ ⊼ ♄
19 59	☿ □ ♄
07 Tuesday	
04 48	☽ ⚼ ♀
05 27	☽ ⊼ ♅
06 16	☽ ± ♄
14 39	☽ ✶ ♂
17 30	☽ ± ♅
23 20	☽ # ♅
08 Wednesday	
00 20	☽ ⚼ ♄
00 54	☽ ± ♀
04 06	☽ ⊼ ♆
04 08	☽ ⚼ ☿
05 39	☽ △ ♃
06 26	☽ □ ♇
08 22	☽ # ♆
11 30	☽ ⚼ ♅
12 08	☽ # ♄
13 10	☽ ∥ ♂
14 12	☽ ⊼ ♀
16 02	☽ Q ♂
16 03	☽ ± ♆
16 22	♂ ± ♆
23 01	☿ ⚼ ♇
09 Thursday	
00 26	☽ ± ☿
01 18	☽ △ ☉
05 56	☉ # ♀
06 04	☽ △ ♄
09 49	☽ ⚼ ♆
11 56	☽ ⚼ ♃
13 49	☽ ⊼ ☿
16 49	♃ ⊼ ♇
17 03	☽ △ ♅
17 53	☽ Q ♇
18 15	♂ # ♄
10 Friday	
00 13	♆ St R
05 20	☽ □ ♂
05 46	☽ ± ♃
09 04	☽ ⚼ ☉
14 53	☽ △ ♆
16 23	☿ □ ♅
16 59	☽ ✶ ♇
17 30	☽ ⊼ ♃
11 Saturday	
03 40	☽ ± ☉
04 07	☽ ∥ ☉
06 36	☽ ☍ ♀
10 49	☽ # ♀
11 39	☉ △ ♄
15 33	☽ □ ♄
15 52	☽ ⊼ ☉
21 09	☽ ∠ ♇
12 Sunday	
00 39	☿ # ♃
00 56	☽ ∥ ♃
00 58	☽ # ☿
01 59	☽ □ ♅
06 05	☽ ∥ ♇
06 26	☽ ☍ ☿
13 41	☽ ⊥ ♇
16 39	☽ △ ♂
17 29	☽ # ♇
20 36	☽ ∥ ☿
22 40	☽ □ ♆
22 51	☽ # ♃
13 Monday	
00 31	☽ Q ♄
00 32	☽ ⊻ ♇
02 10	☽ ☍ ♃
08 57	☽ ∥ ♀
10 38	☽ Q ♅
18 53	☽ ⊼ ♀
20 57	☽ ⚼ ♂
21 54	☽ ✶ ♄
22 19	☽ # ☉
14 Tuesday	
02 31	☽ ☍ ☉
03 20	☿ # ♇
06 29	☽ ± ♀
06 41	☽ Q ♆
07 50	☽ ✶ ♅
08 45	♀ Q ♆
11 19	♂ # ♆
13 33	☽ ± ♂
17 46	☿ ± ♇
19 02	☽ ⊼ ☿
23 39	☽ ⚼ ♀
15 Wednesday	
00 04	☽ ∠ ♄
00 29	☽ ⊼ ♂
03 32	☽ ✶ ♆
04 20	♀ □ ♄
05 14	☽ ☌ ♇
06 45	☽ ± ☿
07 48	☽ ⊼ ♃
09 48	☽ ∠ ♅
15 30	☽ ⊥ ♄
18 15	☽ ± ♃
16 Thursday	
00 11	☽ ⚼ ☿
01 10	☽ ⊥ ♅
01 43	☽ ⊻ ♄
03 17	☽ ∥ ♄
03 45	☽ △ ♀
05 10	☽ ∠ ♆
06 35	☽ ∥ ♆
07 47	☽ # ♂
09 48	☽ ⚼ ♃
10 05	☽ ⊼ ☉
11 20	☽ ⊻ ♅
14 15	☽ ∥ ♅
20 23	☽ ⊥ ♆
20 58	☽ ± ☉
17 Friday	
03 55	☉ △ ♅
04 53	☽ △ ☿
06 01	☽ ☍ ♂
06 28	☽ ⊻ ♆
08 03	☽ ⊻ ♇
11 30	☽ △ ♃
13 12	☽ ⚼ ☉
14 55	♀ ⚼ ♇
15 45	♂ ⊼ ♆
18 05	☽ ⊥ ♇
18 25	☿ □ ♆
20 09	☿ ✶ ♂
18 Saturday	
04 07	☽ ☌ ♄
06 19	♀ # ♃
07 18	☿ ⊼ ♇
09 08	☽ ∠ ♇
10 50	☽ □ ♀
13 39	☽ ☌ ♅
15 20	☿ ∥ ♇
16 09	☽ △ ☉
23 56	♂ ⊼ ♇
19 Sunday	
08 42	☽ ☌ ♆
10 11	☽ ✶ ♇
10 43	☽ ⊼ ♂
13 56	☽ □ ☿
14 35	☽ □ ♃
19 45	☽ Q ♀
20 01	♀ □ ♅
20 05	☿ ☌ ♃
21 16	☽ ± ♂
20 Monday	
01 27	☿ # ♀
06 17	☽ ⊻ ♄
06 18	☽ Q ♇
13 09	☽ ⚼ ♂
15 57	☽ ⊻ ♅
16 23	☽ ⊥ ♄
17 49	☽ ✶ ♀
21 24	☽ Q ♃
22 15	☽ □ ☉
21 Tuesday	
00 30	☽ Q ☿
02 05	☽ ⊥ ♅
07 07	☽ ∥ ♅
07 35	☽ ∠ ♄
09 50	♀ Q ♂
10 31	☽ # ♂
11 16	☽ ⊻ ♆
12 41	☽ □ ♇
13 07	☿ ∥ ♃
14 50	☽ ∥ ♆
15 49	☽ △ ♂
17 22	☽ ∠ ♅
18 07	☽ ✶ ♃
18 14	☽ ∥ ♄
21 29	☽ ⊥ ♆
21 40	☽ ∠ ♀
23 53	☽ ✶ ☿
22 Wednesday	
06 07	♀ # ♇
07 11	☽ Q ☉
09 07	☽ ✶ ♄
12 53	☽ ∠ ♆
14 32	☽ ⊥ ♀
19 02	☽ ✶ ♅
20 17	☽ ∠ ♃
23 Thursday	
01 53	☽ ⊻ ♀
05 26	☽ ✶ ☉
05 35	☽ ∠ ☿
05 46	☽ Q ♄
10 20	☽ # ☉
12 08	☽ ⊥ ♃
14 50	☽ ✶ ♆
15 49	☽ Q ♅
16 13	☽ △ ♇
22 09	☽ □ ♂
22 47	☽ ⊻ ♃
23 42	☽ ⊥ ☿
24 Friday	
09 38	☽ ∠ ☉
11 00	☽ # ☿
11 51	☽ Q ♆
11 54	☽ ⊻ ☿
13 10	☽ □ ♄
18 12	♂ ✶ ♃
18 23	☽ # ♃
18 31	☽ ⚼ ♇
20 50	☿ △ ♄
22 06	♀ ± ♇
23 26	☽ □ ♅
25 Saturday	
02 48	☽ ⊥ ☉
03 14	☽ # ♇
07 35	☽ Q ♂
08 39	☽ ∥ ♀
09 45	☽ # ♀
10 30	☽ ± ♇
11 45	☽ ☌ ♀
14 21	☽ ⊻ ☉
15 17	☽ ∥ ♇
19 53	☽ □ ♆
21 14	☽ ⊼ ♇
26 Sunday	
00 40	☽ ∥ ♃
05 05	☽ ☌ ♃
06 16	☽ ✶ ♂
09 07	☉ ⚼ ♄
14 30	☽ ∥ ☿
18 49	☽ △ ♄
27 Monday	
02 51	☽ ☌ ☿
05 33	☽ △ ♅
11 09	☽ ∠ ♂
15 24	☽ ∥ ☉
16 32	♂ ± ♇
20 37	☿ △ ♅
21 53	♂ # ♅
22 24	☽ ⚼ ♄
28 Tuesday	
00 02	☽ ⊻ ♀
01 34	☽ ☌ ☉
02 49	☽ △ ♆
04 08	☽ ☍ ♇
04 47	☽ ⊥ ♂
09 22	☽ ⚼ ♅
12 27	☽ ⊥ ♀
13 26	☽ ⊻ ♃
15 12	☽ ± ♄
16 41	☽ ⊻ ♂
17 45	☉ △ ♆
29 Wednesday	
01 00	☽ ⊥ ♃
02 21	☽ ± ♅
02 31	☽ ⊼ ♄
05 01	☽ # ♄
05 21	♀ □ ♆
06 25	♀ ∥ ♇
07 06	☽ ⚼ ♆
07 18	☽ ∠ ♀
09 00	☽ # ♆
10 26	☉ ☍ ♇
13 47	☽ ⊼ ♅
17 41	☽ # ♅
18 31	☽ ∠ ♃
18 40	☽ ∥ ♂
18 41	♀ ⊼ ♇
21 42	☽ ⊻ ☿
30 Thursday	
00 22	☽ ± ♆
08 02	☿ ∠ ♂
11 36	☽ ⊥ ☿
12 00	☽ ⊼ ♆
13 18	☽ ⊼ ♇
15 22	☽ ✶ ♀
15 40	☽ ⊻ ☉

MAY 1987

Date	Sidereal time h m s	Sun ☉ ° ' "	Moon ☽ ° ' "	Moon ☽ 24.00 ° ' "	Mercury ☿ ° '	Venus ♀ ° '	Mars ♂ ° '	Jupiter ♃ ° '	Saturn ♄ ° '	Uranus ♅ ° '	Neptune ♆ ° '	Pluto ♇ ° '
01	02 35 32	10 ♉ 35 30	20 ♊ 06 56	26 ♊ 09 49	03 ♉ 46	10 ♈ 39	17 ♊ 09	14 ♈ 15	20 ♐ 23	26 ♐ 21	07 ♑ 53	08 ♏ 32
02	02 39 29	11 33 45	02 ♋ 10 30	08 ♋ 09 21	05 50	11 52	17 48	14 29	20 R 20	26 R 19	07 R 52	08 R 30
03	02 43 25	12 31 57	14 ♋ 06 49	20 ♋ 03 20	07 56	13 04	18 28	14 43	20 17	26 18	07 51	08 28
04	02 47 22	13 30 08	25 ♋ 59 25	01 ♌ 55 36	10 03	14 17	19 07	14 56	20 14	26 16	07 51	08 27
05	02 51 19	14 28 17	07 ♌ 52 28	13 ♌ 50 35	12 11	15 29	19 47	15 10	20 11	26 15	07 50	08 25
06	02 55 15	15 26 24	19 ♌ 50 34	25 ♌ 53 02	14 20	16 42	20 26	15 23	20 08	26 13	07 49	08 23
07	02 59 12	16 24 29	01 ♍ 58 35	08 ♍ 07 49	16 29	17 54	21 06	15 37	20 05	26 11	07 48	08 21
08	03 03 08	17 22 32	14 ♍ 21 17	20 ♍ 39 31	18 39	19 07	21 45	15 50	20 01	26 10	07 47	08 20
09	03 07 05	18 20 34	27 ♍ 02 59	03 ♎ 32 05	20 50	20 19	22 24	16 04	19 58	26 08	07 46	08 18
10	03 11 01	19 18 33	10 ♎ 07 08	16 ♎ 48 20	23 00	21 32	23 04	16 17	19 55	26 06	07 46	08 17
11	03 14 58	20 16 30	23 ♎ 35 46	00 ♏ 29 21	25 10	22 44	23 43	16 30	19 51	26 04	07 45	08 15
12	03 18 54	21 14 26	07 ♏ 28 53	14 ♏ 34 00	27 20	23 57	24 22	16 43	19 47	26 03	07 44	08 13
13	03 22 51	22 12 20	21 ♏ 44 10	28 ♏ 58 44	29 ♉ 29	25 10	25 02	16 57	19 44	26 01	07 43	08 12
14	03 26 48	23 10 13	06 ♐ 16 53	13 ♐ 37 44	01 ♊ 37	26 22	25 41	17 10	19 40	25 59	07 42	08 10
15	03 30 44	24 08 05	21 ♐ 00 18	28 ♐ 23 37	03 43	27 35	26 20	17 23	19 36	25 57	07 41	08 08
16	03 34 41	25 05 55	05 ♑ 46 40	13 ♑ 08 34	05 47	28 ♈ 48	26 59	17 36	19 33	25 55	07 40	08 07
17	03 38 37	26 03 44	20 ♑ 28 27	27 ♑ 45 36	07 50	00 ♉ 00	27 38	17 49	19 29	25 53	07 38	08 05
18	03 42 34	27 01 31	04 ♒ 59 25	12 ♒ 09 26	09 51	01 13	28 17	18 01	19 25	25 51	07 37	08 04
19	03 46 30	27 59 18	19 ♒ 15 21	26 ♒ 16 57	11 49	02 26	28 56	18 14	19 21	25 49	07 36	08 02
20	03 50 27	28 57 03	03 ♓ 14 10	10 ♓ 07 01	13 45	03 38	29 ♊ 36	18 27	19 17	25 47	07 35	08 00
21	03 54 23	29 ♉ 54 47	16 ♓ 55 35	23 ♓ 40 01	15 38	04 51	00 ♋ 15	18 39	19 13	25 45	07 34	07 59
22	03 58 20	00 ♊ 52 30	00 ♈ 20 30	06 ♈ 57 15	17 29	06 04	00 54	18 52	19 09	25 43	07 33	07 57
23	04 02 17	01 50 12	13 ♈ 30 27	20 ♈ 00 19	19 17	07 16	01 33	19 04	19 05	25 40	07 31	07 56
24	04 06 13	02 47 53	26 ♈ 27 03	02 ♉ 50 48	21 02	08 29	02 12	19 17	19 01	25 38	07 30	07 54
25	04 10 10	03 45 33	09 ♉ 11 43	15 ♉ 29 56	22 44	09 42	02 51	19 29	18 56	25 36	07 29	07 53
26	04 14 06	04 43 12	21 ♉ 45 34	27 ♉ 58 41	24 22	10 55	03 30	19 42	18 52	25 34	07 28	07 51
27	04 18 03	05 40 50	04 ♊ 09 25	10 ♊ 17 49	25 58	12 08	04 09	19 54	18 48	25 32	07 26	07 50
28	04 21 59	06 38 26	16 ♊ 24 00	22 ♊ 28 04	27 31	13 20	04 48	20 06	18 44	25 29	07 25	07 48
29	04 25 56	07 36 02	28 ♊ 30 11	04 ♋ 30 30	29 ♊ 01	14 33	05 26	20 18	18 39	25 27	07 24	07 47
30	04 29 52	08 33 36	10 ♋ 29 14	16 ♋ 26 38	00 ♋ 27	15 46	06 05	20 30	18 35	25 25	07 22	07 46
31	04 33 49	09 ♊ 31 09	22 ♋ 22 59	28 ♋ 18 38	01 ♋ 50	16 ♉ 59	06 ♋ 44	20 ♈ 42	18 ♐ 31	25 ♐ 22	07 ♑ 21	07 ♏ 44

Date	Moon True ☊ ° '	Moon Mean ☊ ° '	Moon ☽ Latitude ° '
01	11 ♈ 12	10 ♈ 07	04 N 51
02	11 R 07	10 04	05 08
03	11 03	10 01	05 11
04	11 00	09 57	05 01
05	10 58	09 54	04 38
06	10 D 58	09 51	04 03
07	10 59	09 48	03 17
08	11 01	09 45	02 20
09	11 02	09 41	01 15
10	11 03	09 38	00 N 05
11	11 R 02	09 35	01 S 08
12	11 00	09 32	02 18
13	10 56	09 29	03 22
14	10 50	09 26	04 14
15	10 43	09 22	04 50
16	10 37	09 19	05 07
17	10 32	09 16	05 04
18	10 28	09 13	04 41
19	10 26	09 10	04 01
20	10 D 25	09 07	03 07
21	10 26	09 03	02 03
22	10 27	09 00	00 S 54
23	10 R 28	08 57	00 N 16
24	10 26	08 54	01 25
25	10 23	08 51	02 28
26	10 17	08 47	03 22
27	10 09	08 44	04 07
28	09 59	08 41	04 39
29	09 48	08 38	04 58
30	09 38	08 35	05 04
31	09 ♈ 28	08 ♈ 32	04 N 56

DECLINATIONS

Date	Sun ☉ ° '	Moon ☽ ° '	Mercury ☿ ° '	Venus ♀ ° '	Mars ♂ ° '	Jupiter ♃ ° '	Saturn ♄ ° '	Uranus ♅ ° '	Neptune ♆ ° '	Pluto ♇ ° '
01	15 N 00	27 N 55	11 N 48	02 N 41	23 N 48	04 N 36	21 S 29	23 S 32	22 S 10	01 N 40
02	15 18	28 34	12 38	03 09	23 52	04 41	21 28	23 32	22 10	01 41
03	15 36	27 51	13 28	03 37	23 56	04 47	21 28	23 32	22 10	01 41
04	15 54	25 53	14 18	04 05	24 00	04 52	21 28	23 32	22 10	01 41
05	16 11	22 47	15 07	04 32	24 03	04 57	21 28	23 32	22 10	01 42
06	16 28	18 43	15 56	05 00	24 06	05 02	21 27	23 32	22 10	01 42
07	16 45	13 50	16 44	05 27	24 09	05 07	21 27	23 32	22 10	01 43
08	17 01	08 19	17 30	05 55	24 12	05 12	21 27	23 32	22 10	01 43
09	17 18	02 N 20	18 16	06 22	24 15	05 17	21 27	23 32	22 10	01 43
10	17 33	03 S 56	19 00	06 49	24 17	05 22	21 26	23 32	22 10	01 44
11	17 49	10 13	19 42	07 16	24 19	05 27	21 26	23 32	22 10	01 44
12	18 04	16 11	20 22	07 43	24 21	05 32	21 26	23 32	22 10	01 45
13	18 19	21 27	21 01	08 10	24 23	05 37	21 25	23 32	22 10	01 45
14	18 34	25 32	21 37	08 37	24 25	05 42	21 25	23 32	22 10	01 45
15	18 49	27 58	22 10	09 03	24 26	05 47	21 25	23 32	22 10	01 46
16	19 03	28 26	22 42	09 29	24 27	05 52	21 25	23 32	22 10	01 46
17	19 16	26 53	23 10	09 56	24 29	05 56	21 24	23 32	22 10	01 46
18	19 30	23 33	23 36	10 21	24 29	06 01	21 24	23 31	22 11	01 46
19	19 43	18 51	24 00	10 47	24 30	06 06	21 24	23 31	22 11	01 47
20	19 56	13 13	24 21	11 12	24 30	06 11	21 23	23 31	22 11	01 47
21	20 08	07 03	24 39	11 38	24 31	06 15	21 23	23 31	22 11	01 47
22	20 20	00 S 42	24 55	12 03	24 31	06 20	21 23	23 31	22 11	01 47
23	20 32	05 N 35	25 08	12 27	24 30	06 25	21 22	23 31	22 11	01 48
24	20 43	11 31	25 19	12 52	24 30	06 29	21 22	23 31	22 11	01 48
25	20 54	16 54	25 27	13 16	24 30	06 34	21 22	23 31	22 11	01 48
26	21 05	21 28	25 33	13 40	24 29	06 38	21 21	23 31	22 11	01 48
27	21 15	25 01	25 37	14 03	24 28	06 43	21 21	23 31	22 11	01 48
28	21 25	27 22	25 39	14 26	24 27	06 47	21 21	23 31	22 11	01 48
29	21 35	28 24	25 39	14 49	24 25	06 52	21 20	23 31	22 11	01 48
30	21 44	28 04	25 38	15 12	24 24	06 56	21 20	23 31	22 11	01 49
31	21 N 53	26 N 27	25 N 34	15 N 34	24 N 22	07 N 00	21 S 20	23 S 31	22 S 11	01 N 49

ZODIAC SIGN ENTRIES

Date	h	m	Planets
02	07	39	☽ ♋
04	20	06	☽ ♌
07	08	07	☽ ♍
09	17	29	☽ ♎
11	23	09	☽ ♏
13	17	50	☿ ♊
14	01	41	☽ ♐
16	02	37	☽ ♑
17	11	56	♀ ♉
18	03	42	☽ ♒
20	06	24	☽ ♓
21	03	01	♂ ♋
21	14	10	☉ ♊
22	11	23	☽ ♈
24	18	39	☽ ♉
27	03	55	☽ ♊
29	14	59	☽ ♋
30	04	21	☿ ♋

LATITUDES

Date	Mercury ☿ ° '	Venus ♀ ° '	Mars ♂ ° '	Jupiter ♃ ° '	Saturn ♄ ° '	Uranus ♅ ° '	Neptune ♆ ° '	Pluto ♇ ° '
01	01 S 02	01 S 40	00 N 59	01 S 06	01 N 37	00 S 09	01 N 03	16 N 52
04	00 34	01 41	01 00	01 07	01 37	00 09	01 03	16 52
07	00 S 03	01 42	01 01	01 07	01 37	00 09	01 03	16 52
10	00 N 29	01 42	01 02	01 07	01 38	00 09	01 03	16 51
13	00 59	01 41	01 02	01 08	01 38	00 09	01 03	16 51
16	01 26	01 40	01 03	01 08	01 38	00 09	01 03	16 50
19	01 48	01 38	01 04	01 08	01 38	00 09	01 03	16 50
22	02 04	01 35	01 04	01 09	01 38	00 09	01 03	16 49
25	02 13	01 32	01 05	01 09	01 38	00 09	01 03	16 48
28	02 14	01 28	01 05	01 09	01 37	00 09	01 03	16 47
31	02 N 08	01 S 24	01 N 06	01 S 10	01 N 37	00 S 09	01 N 03	16 N 46

DATA

Julian Date	2446917
Delta T	+55 seconds
Ayanamsa	23° 40' 44"
Synetic vernal point	05° ♓ 26' 15"
True obliquity of ecliptic	23° 26' 36"

MOON'S PHASES, APSIDES AND POSITIONS ☽

Date	h	m	Phase	Longitude ° '	Eclipse Indicator
06	02	26	☽	15 ♌ 03	
13	12	50	○	22 ♏ 14	
20	04	02	☾	28 ♒ 38	
27	15	13	●	05 ♊ 49	

Day	h	m		
04	01	44	Apogee	
15	22	43	Perigee	
31	17	54	Apogee	
02	11	21	Max dec	28° N 34'
09	21	00	0S	
16	05	27	Max dec	28° S 30'
22	14	37	0N	
29	18	10	Max dec	28° N 27'

LONGITUDES

Date	Chiron ⚷ ° '	Ceres ⚳ ° '	Pallas ⚴ ° '	Juno ⚵ ° '	Vesta ⚶ ° '	Black Moon Lilith ⚸ ° '
01	18 ♊ 36	05 ♑ 41	25 ♏ 17	25 ♒ 15	24 ♉ 44	17 ♋ 48
11	19 ♊ 21	05 ♑ 18	22 ♏ 20	28 ♒ 04	29 ♉ 04	18 ♋ 56
21	20 ♊ 10	04 ♑ 18	19 ♏ 20	00 ♓ 36	03 ♊ 26	20 ♋ 03
31	21 ♊ 01	02 ♑ 45	16 ♏ 38	02 ♓ 48	07 ♊ 47	21 ♋ 10

All ephemeris data is given at 12.00 UT and the Moon's longitude is additionally given for 24.00 UT

ASPECTARIAN

01 Friday
00 14 ☽ ⚹ ♃
01 01 ☽ ± ♇
04 27 ☽ ⊥ ☉
05 39 ☉ ⊻ ♀
05 49 ☽ ☌ ♂
08 47 ☽ ∠ ☿
12 32 ☽ ☍ ♄
17 35 ☽ Q ♀
18 45 ☽ ⚼ ♇
23 49 ☽ ∠ ☉

02 Saturday
00 21 ☽ ☍ ♅
00 25 ☽ Q ♃
06 15 ☉ ⚼ ♅
06 23 ☿ ⚼ ♄
20 54 ☽ ⚹ ☿
23 25 ☽ ☍ ♆

03 Sunday
00 40 ☽ △ ♇
06 53 ☉ ⊥ ♂
08 32 ☽ ⚹ ☉
09 39 ☽ □ ♀
11 09 ☿ △ ♆
13 14 ☽ □ ♃
18 04 ☿ ☍ ♇
21 18 ☽ ⊻ ♂

04 Monday
00 25 ☽ ⊼ ♄
02 17 ☽ Q ☿
10 09 ☽ ⊥ ♂
10 56 ☽ Q ☉
12 30 ☽ ± ♄
12 34 ☽ ⊼ ♅

05 Tuesday
00 40 ☽ ± ♅
01 40 ☿ ⚼ ♅
03 16 ☽ ∥ ♂
04 08 ♀ ☌ ♃
05 17 ☉ ± ♄
05 24 ☽ ∠ ♂
06 36 ☽ ⚼ ♄
06 52 ☽ # ♅
11 55 ☽ ⊼ ♆
13 05 ☽ □ ♇
15 59 ☽ # ♆
18 46 ☽ ⚼ ♅
20 23 ☽ # ♄
22 33 ☽ □ ☿
23 58 ☽ ± ♆

06 Wednesday
01 41 ♂ ☍ ♄
02 26 ☽ □ ☉
02 56 ☽ △ ♃
05 01 ☽ △ ♀
09 54 ☿ ± ♄
10 24 ☉ ⊻ ♃
12 35 ☽ △ ♄
13 15 ☽ ⚹ ♂
13 47 ☿ ⊥ ♂
14 18 ♀ ∥ ♃
17 55 ☽ ⚼ ♆
22 51 ☽ ∥ ☉

07 Thursday
00 07 ☽ ∥ ☿
00 38 ☽ △ ♅
00 58 ☽ Q ♇
01 12 ☿ ⊻ ♃
09 17 ☽ ⚼ ♃
10 25 ☉ ☌ ☿
12 49 ☉ ∥ ☿
14 01 ☽ ⚼ ♀
14 19 ☽ Q ♂
23 21 ☽ △ ♆

08 Friday
00 25 ☽ ⚹ ♇
03 09 ☽ ± ♃
09 22 ☽ ± ♀
14 53 ☽ ⊼ ♃
18 15 ☽ △ ☉
21 07 ☽ ∥ ♀
21 55 ☽ △ ☿
22 02 ☽ ⊼ ♀
22 45 ☽ □ ♄
23 23 ☿ ⊻ ♀

09 Saturday
00 29 ☽ ∥ ♃
02 43 ☿ ⊼ ♄
02 50 ☽ □ ♂
04 25 ☿ ± ♅
05 00 ☽ ∠ ♇
05 15 ♀ △ ♄
10 18 ☽ □ ♅
14 20 ☽ ∥ ♇
21 43 ☽ ⊥ ♇

10 Sunday
00 35 ☽ ⚼ ☉
03 08 ☿ ⊥ ♃
03 37 ☽ # ♇
07 25 ☽ ⚼ ☿
07 44 ☽ □ ♆
08 01 ☽ Q ♄
08 40 ☽ ⊻ ♇
09 19 ☿ ⚼ ♆
12 55 ☿ ⊻ ♂
17 32 ☽ # ♃
18 12 ☽ ± ☉
19 10 ☽ Q ♅
19 29 ♂ ⚼ ♇
23 15 ☽ ☍ ♃
23 50 ☽ # ♀

11 Monday
02 04 ☉ ⊼ ♄
02 43 ☽ ± ☿
05 26 ☽ ⚹ ♄
05 42 ☽ ⊼ ☉
07 10 ☉ ± ♅
10 21 ☽ ☍ ♀
12 13 ☽ △ ♂
15 17 ☽ ⊼ ☿
15 45 ☽ Q ♆
16 20 ☽ ⚹ ♅
21 51 ☿ ⊼ ♅

12 Tuesday
07 25 ☽ ∠ ♄
12 25 ☽ ⚹ ♆
13 15 ☽ ☌ ♇
15 22 ☽ ⚼ ♂
18 02 ☽ ∠ ♅
20 31 ☽ # ☉
22 39 ☽ ⊥ ♄

13 Wednesday
03 52 ☽ ⊼ ♃
06 08 ♀ ⚹ ♂
07 16 ☽ ± ♂
08 40 ☽ ⊻ ♄
09 08 ☽ ⊥ ♅
09 27 ☽ # ☿
11 50 ☽ ∥ ♄
12 50 ☽ ☍ ☉
13 37 ☽ ∠ ♆
14 02 ☽ ± ♃
15 41 ☽ ∥ ♆
17 43 ☽ ⊼ ♂
18 12 ☽ ⊼ ♀
19 05 ☽ ⊻ ♅
23 09 ☽ ∥ ♅

14 Thursday
00 21 ☉ ⚼ ♆
03 02 ☽ ☍ ☿
04 17 ☿ # ♄
04 26 ☽ # ♂
04 28 ♀ △ ♅
04 29 ☽ ⊥ ♆
05 00 ☽ ± ♀
05 08 ☽ ⚼ ♃
11 41 ☉ ⊥ ♃
12 57 ☿ ± ♆
14 19 ☽ ⊻ ♆
15 05 ☽ ⊻ ♇
18 59 ☿ ∠ ♃
21 04 ☽ ⚼ ♀
22 35 ♂ ☍ ♅

15 Friday
00 51 ☽ ⊥ ♇
06 01 ☽ △ ♃
08 25 ☿ ⊥ ♀
09 44 ☽ ☌ ♄
12 04 ☿ # ♆
15 28 ☽ ∠ ♇
17 26 ☽ ⊼ ☉
20 01 ☽ ☌ ♅
21 03 ☽ ☍ ♂
23 38 ☽ △ ♀

16 Saturday
03 52 ☽ ± ☉
12 02 ☽ ⊼ ☿
15 03 ☽ ☌ ♆
15 48 ☽ ⚹ ♇
19 32 ☽ ⚼ ☉
23 23 ☽ ± ☿

17 Sunday
07 34 ☽ □ ♃
07 42 ☉ ⊼ ♅
09 42 ☿ ⊼ ♆
10 23 ☽ ⊻ ♄
11 22 ☽ Q ♇
14 54 ☿ ⊼ ♇
16 31 ☽ ⚼ ☿
20 12 ☽ ⊥ ♄
20 53 ☽ ⊻ ♅
21 51 ☽ △ ☉

18 Monday
00 21 ☽ ⊼ ♂
05 09 ☽ □ ♀
06 19 ☽ # ♂
06 47 ☽ ⊥ ♅
07 20 ☿ # ♅
10 47 ☽ ± ♂
11 03 ☽ ∠ ♄
11 44 ☽ # ☿
12 11 ☽ ∥ ♅
13 45 ☽ Q ♃
16 23 ☽ ⊻ ♆
17 07 ☽ □ ♇
19 42 ☽ ∥ ♆
21 26 ☽ △ ☿
21 47 ☽ ∠ ♅
23 45 ☽ ∥ ♄

19 Tuesday
02 27 ☽ ⊥ ♆
02 34 ☽ ⚼ ♂
08 10 ☽ # ☉
10 14 ☽ ⚹ ♃
12 10 ☽ ⚹ ♄
14 11 ☽ Q ♀
17 42 ☽ ∠ ♆
23 10 ☽ ⚹ ♅

20 Wednesday
04 02 ☽ □ ☉
05 23 ☽ △ ♂
08 38 ☽ Q ♄
12 22 ☽ ∠ ♃
12 46 ☽ ⚹ ♀
15 10 ☿ ± ♇
19 29 ☽ # ♀
19 34 ☽ ⚹ ♆
19 53 ☽ Q ♅
20 17 ☽ △ ♇

21 Thursday
00 07 ♀ ⚼ ♄
00 28 ☿ ∥ ♂
04 21 ☽ ⊥ ♃
09 22 ☽ □ ☿
13 53 ☽ Q ☉
14 59 ☽ # ♃
15 07 ☽ ⊻ ♃
16 03 ☽ □ ♄
16 41 ☽ Q ♆
17 42 ☽ ∠ ♀
22 45 ☽ ⚼ ♇

22 Friday
03 41 ☽ □ ♅
07 53 ☽ # ♇
10 30 ♂ Q ♃
11 27 ☽ ⊥ ♀
13 02 ☽ ⚹ ☉
13 03 ☽ □ ♂
13 29 ☉ ⊻ ♂
14 55 ☽ ± ♇
21 24 ☽ ∥ ♇
22 48 ☽ Q ☿
23 26 ☽ ⊻ ♀

23 Saturday
01 03 ☽ □ ♆
01 48 ☽ ⊼ ♇
04 23 ☉ ± ♆
08 52 ☿ ⚹ ♃
09 24 ☿ ☍ ♄
12 26 ♃ △ ♄
15 17 ☽ ∥ ♃
16 53 ♀ △ ♆
18 38 ☽ ∠ ☉
22 14 ☽ △ ♄
22 27 ☽ ☌ ♃
23 44 ☽ Q ♂

24 Sunday
00 20 ☽ ⚹ ☿
00 44 ♀ ☍ ♇
10 29 ☽ △ ♅
12 42 ☽ ⊥ ☉
18 09 ☽ ∥ ♀
23 21 ☽ ⚹ ♂

25 Monday
00 53 ☽ ⊻ ☉
02 07 ☽ ⚼ ♄
08 46 ☽ △ ♆
08 47 ☽ ∠ ☿
09 31 ☽ ☍ ♇
13 03 ☽ ☌ ♀
14 11 ☿ ⚼ ♇
14 40 ☽ ⚼ ♅
19 05 ☽ ± ♄

26 Tuesday
04 32 ☽ ⊥ ☿
05 19 ♀ ⚼ ♅
05 23 ☽ ∠ ♂
06 29 ☽ ⊼ ♄
07 48 ☽ ± ♅
07 58 ☽ ⊻ ♃
09 42 ☽ ∥ ☉
11 06 ☉ ∠ ♃
11 22 ☽ # ♄
13 21 ☽ ⚼ ♆
16 18 ☽ # ♆
17 47 ☽ ⊻ ☿
19 19 ☽ ⊼ ♅
19 42 ☽ ⊥ ♃
23 40 ☽ ⊥ ♂

27 Wednesday
00 55 ☽ # ♅
05 23 ☿ ☍ ♅
06 43 ☽ ± ♆
07 43 ☽ ∥ ♂
11 58 ☽ ⊻ ♂
13 28 ☽ ∠ ♃
15 13 ☽ ☌ ☉
17 11 ☽ ∥ ☿
18 24 ☽ ⊼ ♆
19 10 ☽ ⊼ ♇

28 Thursday
00 36 ♀ ± ♄
01 18 ☉ # ♄
05 18 ☽ ⊻ ♀
06 54 ☽ ± ♇
16 34 ☽ ☍ ♄
18 27 ☽ ⊥ ♀
19 26 ☽ ⚹ ♃

29 Friday
00 39 ☽ ⚼ ♇
05 57 ☽ ☍ ♅
07 00 ☉ ⊼ ♆
13 09 ☽ ☌ ☿
14 20 ☽ ∠ ♀
16 30 ☉ ⊼ ♇
19 42 ☽ Q ♃

30 Saturday
02 40 ☽ ☌ ♂
05 45 ☽ ☍ ♆
06 32 ☽ △ ♇
07 47 ☽ ⊻ ☉
20 55 ☽ ⊥ ☉
23 50 ☽ ⚹ ♀

31 Sunday
04 13 ☽ ⊼ ♄
08 32 ☽ □ ♃
16 17 ☽ ± ♄
16 42 ☽ ∠ ☉
18 02 ☽ ⊼ ♅
20 58 ☽ ∥ ☿

JUNE 1987

LONGITUDES

Date	Sidereal time h m s	Sun ☉ ° ' "	Moon ☽ ° ' "	Moon ☽ 24.00 ° ' "	Mercury ☿ ° '	Venus ♀ ° '	Mars ♂ ° '	Jupiter ♃ ° '	Saturn ♄ ° '	Uranus ♅ ° '	Neptune ♆ ° '	Pluto ♇ ° '
01	04 37 46	10 ♊ 28 40	04 ♌ 14 00	10 ♌ 09 29	03 ♋ 10	18 ♉ 12	07 ♋ 23	20 ♈ 53	18 ♐ 26	25 ♐ 20	07 ♑ 20	07 ♏ 43
02	04 41 42	11 26 11	16 ♌ 05 35	22 ♌ 02 50	04 27	19 24	08 02	21 05	18 R 22	25 R 18	07 R 18	07 R 42
03	04 45 39	12 23 40	28 ♌ 01 48	04 ♍ 03 05	05 41	20 37	08 41	21 17	18 18	25 15	07 17	07 40
04	04 49 35	13 21 08	10 ♍ 07 17	16 ♍ 15 03	06 51	21 50	09 19	21 28	18 13	25 13	07 15	07 39
05	04 53 32	14 18 34	22 ♍ 27 02	28 ♍ 43 51	07 57	23 03	09 58	21 40	18 09	25 11	07 14	07 38
06	04 57 28	15 15 59	05 ♎ 06 06	11 ♎ 34 19	09 00	24 16	10 37	21 51	18 04	25 08	07 12	07 36
07	05 01 25	16 13 23	18 ♎ 09 01	24 ♎ 50 33	10 00	25 29	11 16	22 02	18 00	25 06	07 11	07 35
08	05 05 21	17 10 46	01 ♏ 39 11	08 ♏ 35 02	10 55	26 42	11 54	22 13	17 55	25 03	07 10	07 34
09	05 09 18	18 08 08	15 ♏ 38 03	22 ♏ 47 57	11 47	27 55	12 33	22 24	17 51	25 01	07 08	07 33
10	05 13 15	19 05 29	00 ♐ 04 15	07 ♐ 26 18	12 36	29 ♉ 08	13 12	22 35	17 47	24 59	07 07	07 32
11	05 17 11	20 02 50	14 ♐ 53 11	22 ♐ 23 51	13 20	00 ♊ 21	13 50	22 46	17 42	24 56	07 05	07 31
12	05 21 08	21 00 09	29 ♐ 57 05	07 ♑ 31 34	14 00	01 33	14 29	22 57	17 38	24 54	07 03	07 29
13	05 25 04	21 57 28	15 ♑ 06 00	22 ♑ 39 05	14 36	02 46	15 08	23 08	17 33	24 51	07 02	07 28
14	05 29 01	22 54 46	00 ♒ 09 37	07 ♒ 36 32	15 08	03 59	15 46	23 18	17 29	24 49	07 00	07 27
15	05 32 57	23 52 04	14 ♒ 58 57	22 ♒ 16 11	15 36	05 12	16 25	23 29	17 24	24 46	06 59	07 26
16	05 36 54	24 49 21	29 ♒ 27 44	06 ♓ 33 20	15 59	06 25	17 03	23 39	17 20	24 44	06 57	07 25
17	05 40 50	25 46 38	13 ♓ 32 51	20 ♓ 26 20	16 18	07 39	17 42	23 49	17 16	24 41	06 56	07 24
18	05 44 47	26 43 54	27 ♓ 13 58	03 ♈ 55 59	16 33	08 52	18 21	23 59	17 11	24 39	06 54	07 23
19	05 48 44	27 41 11	10 ♈ 32 45	17 ♈ 04 37	16 42	10 05	18 59	24 09	17 07	24 36	06 52	07 22
20	05 52 40	28 38 27	23 ♈ 32 01	29 ♈ 55 21	16 48	11 18	19 38	24 19	17 03	24 34	06 51	07 22
21	05 56 37	29 ♊ 35 43	06 ♉ 15 00	12 ♉ 31 23	16 R 48	12 31	20 16	24 29	16 58	24 32	06 49	07 21
22	06 00 33	00 ♋ 32 58	18 ♉ 44 49	24 ♉ 55 37	16 45	13 44	20 55	24 39	16 54	24 29	06 48	07 20
23	06 04 30	01 30 14	01 ♊ 04 04	07 ♊ 10 26	16 36	14 57	21 33	24 48	16 50	24 27	06 46	07 19
24	06 08 26	02 27 29	13 ♊ 14 55	19 ♊ 17 42	16 24	16 10	22 12	24 58	16 46	24 24	06 44	07 18
25	06 12 23	03 24 44	25 ♊ 18 57	01 ♋ 18 49	16 07	17 23	22 50	25 07	16 41	24 22	06 43	07 17
26	06 16 19	04 21 59	07 ♋ 17 29	13 ♋ 15 03	15 47	18 37	23 29	25 16	16 37	24 19	06 41	07 17
27	06 20 16	05 19 14	19 ♋ 11 44	25 ♋ 07 41	15 22	19 50	24 07	25 25	16 33	24 17	06 40	07 16
28	06 24 13	06 16 28	01 ♌ 03 08	06 ♌ 58 18	14 55	21 03	24 45	25 34	16 29	24 15	06 38	07 15
29	06 28 09	07 13 42	12 ♌ 53 31	18 ♌ 49 04	14 25	22 16	25 24	25 43	16 25	24 12	06 36	07 15
30	06 32 06	08 ♋ 10 55	24 ♌ 45 21	00 ♍ 42 47	13 ♋ 52	23 ♊ 30	26 ♋ 02	25 ♈ 52	16 ♐ 21	24 ♐ 10	06 ♑ 35	07 ♏ 14

Date	Moon True ☊ ° '	Moon Mean ☊ ° '	Moon ☽ Latitude ° '
01	09 ♈ 20	08 ♈ 28	04 N 36
02	09 R 15	08 25	04 04
03	09 12	08 22	03 21
04	09 11	08 19	02 28
05	09 D 11	08 16	01 28
06	09 11	08 12	00 N 22
07	09 R 11	08 09	00 S 48
08	09 09	08 06	01 56
09	09 05	08 03	03 01
10	08 58	08 00	03 55
11	08 49	07 57	04 36
12	08 39	07 53	04 58
13	08 29	07 50	04 59
14	08 20	07 47	04 40
15	08 14	07 44	04 02
16	08 09	07 41	03 09
17	08 07	07 38	02 06
18	08 D 07	07 34	00 S 57
19	08 R 07	07 31	00 N 13
20	08 06	07 28	01 21
21	08 04	07 25	02 23
22	07 59	07 22	03 17
23	07 51	07 18	04 01
24	07 40	07 15	04 33
25	07 27	07 12	04 53
26	07 14	07 09	04 59
27	07 00	07 06	04 53
28	06 47	07 03	04 33
29	06 37	06 59	04 02
30	06 ♈ 29	06 ♈ 56	03 N 21

DECLINATIONS

Date	Sun ☉ ° '	Moon ☽ ° '	Mercury ☿ ° '	Venus ♀ ° '	Mars ♂ ° '	Jupiter ♃ ° '	Saturn ♄ ° '	Uranus ♅ ° '	Neptune ♆ ° '	Pluto ♇ ° '
01	22 N 01	23 N 40	25 N 29	15 N 56	24 N 20	07 N 05	21 S 19	23 S 31	22 S 11	01 N 49
02	22 09	19 53	25 22	16 17	24 18	07 09	21 19	23 31	22 11	01 49
03	22 17	15 18	25 14	16 38	24 16	07 13	21 19	23 30	22 11	01 49
04	22 24	10 04	25 05	16 58	24 13	07 17	21 18	23 30	22 12	01 49
05	22 31	04 N 21	24 54	17 19	24 11	07 21	21 18	23 30	22 12	01 49
06	22 38	01 S 42	24 43	17 38	24 08	07 25	21 18	23 30	22 12	01 49
07	22 44	07 51	24 30	17 58	24 05	07 29	21 17	23 30	22 12	01 49
08	22 50	13 52	24 17	18 16	24 01	07 33	21 17	23 30	22 12	01 49
09	22 55	19 24	24 02	18 35	23 58	07 37	21 17	23 30	22 12	01 49
10	23 00	24 00	23 47	18 53	23 54	07 41	21 17	23 30	22 12	01 49
11	23 04	27 09	23 32	19 10	23 50	07 45	21 16	23 30	22 12	01 49
12	23 08	28 25	23 16	19 27	23 46	07 49	21 16	23 30	22 12	01 49
13	23 12	27 33	22 59	19 43	23 42	07 53	21 16	23 30	22 12	01 49
14	23 15	24 41	22 42	19 59	23 38	07 57	21 15	23 30	22 12	01 49
15	23 18	20 12	22 25	20 15	23 33	08 00	21 15	23 30	22 12	01 48
16	23 21	14 37	22 08	20 30	23 29	08 04	21 15	23 29	22 12	01 48
17	23 23	08 25	21 51	20 44	23 24	08 08	21 14	23 29	22 13	01 48
18	23 24	01 S 59	21 34	20 58	23 19	08 11	21 14	23 29	22 13	01 48
19	23 25	04 N 30	21 17	21 11	23 13	08 15	21 14	23 29	22 13	01 48
20	23 26	10 23	21 01	21 23	23 08	08 18	21 13	23 29	22 13	01 48
21	23 27	15 51	20 44	21 35	23 02	08 21	21 13	23 29	22 13	01 48
22	23 27	20 33	20 29	21 47	22 57	08 25	21 13	23 29	22 13	01 47
23	23 26	24 18	20 13	21 57	22 51	08 28	21 12	23 29	22 13	01 47
24	23 25	26 55	19 59	22 08	22 45	08 31	21 12	23 29	22 13	01 47
25	23 24	28 14	19 45	22 17	22 38	08 35	21 12	23 29	22 13	01 47
26	23 22	28 13	19 31	22 26	22 32	08 38	21 12	23 29	22 13	01 46
27	23 20	26 54	19 19	22 34	22 25	08 41	21 11	23 28	22 13	01 46
28	23 18	24 22	19 08	22 42	22 19	08 44	21 11	23 28	22 14	01 46
29	23 15	20 49	18 57	22 49	22 12	08 47	21 11	23 28	22 14	01 45
30	23 N 11	16 N 25	18 N 48	22 N 56	22 N 04	08 N 50	21 S 10	23 S 28	22 S 14	01 N 45

ZODIAC SIGN ENTRIES

Date	h	m	Planets
01	03	25	☽ ♌
03	15	56	☽ ♍
06	02	24	☽ ♎
08	09	06	☽ ♏
10	11	53	☽ ♐
11	05	15	♀ ♊
12	12	05	☽ ♑
14	11	45	☽ ♒
16	12	54	☽ ♓
18	16	56	☽ ♈
21	00	09	☽ ♉
21	22	11	☉ ♋
23	09	54	☽ ♊
25	21	22	☽ ♋
28	09	52	☽ ♌
30	22	34	☽ ♍

LATITUDES

Date	Mercury ☿ ° '	Venus ♀ ° '	Mars ♂ ° '	Jupiter ♃ ° '	Saturn ♄ ° '	Uranus ♅ ° '	Neptune ♆ ° '	Pluto ♇ ° '
01	02 N 05	01 S 23	01 N 06	01 S 10	01 N 37	00 S 09	01 N 03	16 N 46
04	01 49	01 18	01 06	01 10	01 37	00 09	01 03	16 45
07	01 26	01 13	01 07	01 11	01 37	00 09	01 03	16 44
10	00 57	01 07	01 07	01 11	01 37	00 09	01 03	16 43
13	00 N 21	01 01	01 08	01 12	01 37	00 09	01 03	16 41
16	00 S 21	00 54	01 08	01 12	01 36	00 09	01 03	16 40
19	01 07	00 48	01 08	01 13	01 36	00 09	01 03	16 39
22	01 56	00 41	01 08	01 13	01 36	00 09	01 03	16 37
25	02 45	00 34	01 09	01 14	01 35	00 09	01 03	16 36
28	03 30	00 26	01 09	01 14	01 35	00 09	01 03	16 35
31	04 S 08	00 S 19	01 N 09	01 S 15	01 N 35	00 S 09	01 N 03	16 N 34

DATA

Julian Date	2446948
Delta T	+55 seconds
Ayanamsa	23° 40' 50"
Synetic vernal point	05° ♓ 26' 10"
True obliquity of ecliptic	23° 26' 36"

MOON'S PHASES, APSIDES AND POSITIONS ☽

Date	h	m	Phase	Longitude ° '	Eclipse Indicator
04	18	53	☽	13 ♍ 38	
11	20	49	○	20 ♐ 24	
18	11	03	☾	26 ♓ 42	
26	05	37	●	04 ♋ 07	

Day	h	m		
13	01	21	Perigee	
28	04	26	Apogee	
06	05	21	OS	
12	14	17	Max dec	28° S 25'
18	19	23	ON	
25	23	45	Max dec	28° N 24'

LONGITUDES

Date	Chiron ⚷ ° '	Ceres ⚳ ° '	Pallas ⚴ ° '	Juno ⚵ ° '	Vesta ⚶ ° '	Black Moon Lilith ⚸ ° '
01	21 ♊ 06	02 ♑ 35	16 ♏ 23	03 ♓ 00	08 ♊ 13	21 ♋ 17
11	21 ♊ 58	00 ♑ 35	14 ♏ 20	04 ♓ 45	12 ♊ 33	22 ♋ 24
21	22 ♊ 51	28 ♐ 23	13 ♏ 00	06 ♓ 02	16 ♊ 52	23 ♋ 31
31	23 ♊ 43	26 ♐ 14	12 ♏ 27	06 ♓ 46	21 ♊ 09	24 ♋ 38

ASPECTARIAN

h m	Aspects
01 Monday	
02 53	☽ Q ♀
05 54	☿ Q ♃
06 09	☽ ± ♅
06 57	☽ ∥ ♂
09 35	☽ ⊻ ☿
09 59	♂ ☍ ♆
10 24	☽ bQ ♄
13 08	☽ # ♅
16 35	♀ ⊼ ♄
17 16	☿ ∠ ♀
18 15	☽ ⊼ ♆
18 45	☽ ⊻ ♂
19 02	☽ □ ♇
22 05	☽ # ♆
22 46	☽ ∥ ☉
23 14	☽ ⊥ ☿
23 53	♂ △ ♇
02 Tuesday	
00 19	☽ bQ ♅
01 46	☽ ✶ ☉
03 33	☽ # ♄
06 23	☽ ± ♆
07 36	☽ ⊥ ♂
09 52	♀ ± ♅
16 34	☽ △ ♄
18 19	☉ # ♆
19 27	☽ □ ♀
19 34	☽ ∠ ☿
22 14	☽ △ ♃
03 Wednesday	
00 30	☽ bQ ♆
02 47	☽ ∠ ♂
04 05	☽ Q ☉
05 51	☽ ∥ ♀
06 28	☽ △ ♅
07 17	☽ Q ♇
04 Thursday	
03 26	♀ ⊻ ♃
04 41	☽ bQ ♃
04 51	☽ ✶ ☿
06 22	☽ △ ♆
07 08	☽ ✶ ♇
10 21	☽ ✶ ♂
14 03	☿ ∠ ♀
18 53	☽ □ ☉
19 18	☉ ± ♇
20 09	♀ bQ ♆
20 37	☿ ☍ ♆
22 39	☽ ± ♃
23 43	☽ ∥ ♃
05 Friday	
03 44	☽ □ ♄
05 00	☿ △ ♇
06 43	☽ Q ☿
10 27	☽ ⊼ ♃
11 02	☽ Q ♂
12 20	☽ ∠ ♇
13 17	☽ △ ♀
17 13	☽ □ ♅
22 09	☽ ∥ ♇
06 Saturday	
05 27	☽ ⊥ ♇
12 29	☽ # ♇
13 48	☽ Q ♄
15 55	☽ □ ♆
16 40	☽ ⊻ ♇
19 52	☽ □ ☿
20 33	☽ bQ ♀
22 46	☽ □ ♂
07 Sunday	
02 50	☽ Q ♅
04 40	♀ ⊼ ♅
08 14	☽ △ ☉
10 35	☽ # ♃
11 44	☽ ✶ ♄
14 39	☽ ± ♀
19 06	☽ ☍ ♃
08 Monday	
00 25	☽ ✶ ♅
00 35	☽ Q ♆
02 26	☽ ⊼ ♀
12 59	☽ bQ ☉
14 12	☽ ∠ ♄
20 55	♀ ∠ ♂
21 32	☽ ✶ ♆
22 14	☽ ☌ ♇
09 Tuesday	
02 29	☽ ∠ ♅
05 04	☽ △ ☿
05 20	☉ ☍ ♄
05 37	☽ ⊥ ♄
05 38	☽ ± ☉
06 32	☽ △ ♂
07 59	☽ # ♀
15 43	☽ ⊻ ♄
16 31	☽ ⊼ ☉
17 40	☽ ⊥ ♅
21 05	☽ ∥ ♄
21 26	☿ ∥ ♂
22 52	☽ ∠ ♆
23 30	☽ ⊼ ♃
23 34	♀ ⊥ ♃
10 Wednesday	
01 50	☽ ∥ ♆
03 38	☽ ⊻ ♅
06 03	☽ # ☉
07 42	☽ bQ ☿
08 47	☽ bQ ♂
09 02	☽ ∥ ♅
09 32	☽ ± ♃
10 19	☽ ☍ ♀
10 48	☽ # ☿
11 25	☽ # ♂
13 42	☽ ⊥ ♆
23 12	☽ ± ☿
23 27	☽ ⊻ ♆
11 Thursday	
00 08	☽ ⊻ ♇
00 08	☽ ± ♂
00 24	☽ bQ ♃
09 23	☽ ⊼ ☿
09 48	☽ ⊥ ♇
10 15	☽ ⊼ ♂
14 49	☿ # ♅
16 29	☽ ☌ ♄
20 49	☽ ☍ ☉
12 Friday	
00 10	☽ ∠ ♇
00 45	☽ △ ♃
02 20	♀ ± ♆
04 00	☽ ☌ ♅
14 46	☽ ⊼ ♀
20 47	☉ ∥ ☿
23 14	☽ ☌ ♆
23 56	☽ ✶ ♇
13 Saturday	
01 06	☽ ± ♀
11 11	☽ ☍ ☿
12 03	☽ ☍ ♂
15 53	☽ ⊻ ♄
16 37	☽ bQ ♀
18 56	☽ Q ♇
23 38	☽ ⊼ ☉
14 Sunday	
00 42	☉ bQ ♇
00 55	☽ □ ♃
01 22	☽ ⊥ ♄
03 28	☽ ⊻ ♅
09 52	☽ ± ☉
13 03	☽ ⊥ ♅
15 42	☽ ∠ ♄
18 27	☽ # ♂
18 42	☽ △ ♀
19 08	☽ ∥ ♅
20 23	☽ # ☉
23 00	☽ ⊻ ♆
23 44	☽ □ ♇
15 Monday	
00 03	☉ ✶ ♃
00 15	☽ # ☿
01 21	☽ bQ ☉
02 07	☽ ∥ ♆
03 32	☽ ∠ ♅
06 13	☽ Q ♃
06 57	☽ ∥ ♄
08 44	☽ ⊥ ♆
11 47	☽ # ♀
13 03	☽ ⊼ ☿
14 27	☽ ⊼ ♂
15 57	☽ ✶ ♄
23 12	☽ ± ☿
23 30	☽ ∠ ♆
16 Tuesday	
00 49	☽ ± ♂
02 11	☽ ✶ ♃
03 41	☽ △ ☉
04 07	☽ ✶ ♅
06 08	☿ # ♆
08 04	♂ # ♅
09 48	☉ ☍ ♅
11 47	☽ Q ♄
14 38	☽ bQ ☿
16 35	☽ bQ ♂
21 15	♂ ⊼ ♄
22 13	♀ ⊼ ♆
17 Wednesday	
00 16	☽ Q ♅
00 39	☽ ✶ ♆
00 53	☽ □ ♀
01 28	☽ △ ♇
03 46	☽ ∠ ♃
07 23	♀ ⊼ ♇
13 04	☽ # ♃
16 19	☉ ∥ ♂
16 52	☽ △ ☿
18 25	☽ □ ♄
19 31	☽ ⊥ ♃
19 34	☽ △ ♂
21 20	☽ Q ♆
18 Thursday	
03 26	☽ bQ ♇
06 11	☽ ⊻ ♃
07 26	☽ □ ♅
11 03	☽ □ ☉
11 16	☽ Q ♀
12 39	☽ # ♇
15 00	♀ ∠ ♃
19 25	☽ ± ♇
19 Friday	
02 10	☽ ∥ ♇
05 20	☽ □ ♆
06 14	☽ ⊼ ♇
11 03	☽ ✶ ♀
17 19	♀ # ♄
17 20	☿ ∥ ♀
17 21	☿ # ♄
22 11	☽ Q ☉
23 25	☽ □ ☿
20 Saturday	
00 00	☽ △ ♄
01 34	☿ ⊥ ♀
03 23	☽ ∥ ♃
04 21	☽ □ ♂
13 30	☽ ☌ ♃
13 56	☽ △ ♅
17 43	☽ ∠ ♀
22 22	☽ ✶ ☉
21 Sunday	
03 44	☿ St R
03 55	☽ bQ ♄
09 15	☽ Q ☿
12 33	☽ ⊥ ♀
13 05	☽ △ ♆
14 05	☽ ☍ ♇
16 04	☽ Q ♂
16 52	♃ △ ♅
18 14	☽ bQ ♅
20 58	☽ ± ♄
22 Monday	
01 17	☽ ⊻ ♀
04 10	♀ ± ♇
05 19	☽ ∠ ☉
08 09	☽ ✶ ☿
08 27	☽ ⊼ ♄
11 30	☽ ± ♅
11 35	☽ ∥ ☿
15 46	☽ # ♄
16 25	☽ ✶ ♂
17 54	☽ bQ ♆
19 29	☽ ∥ ♀
21 50	☽ # ♆
23 06	☽ ⊼ ♅
23 36	☽ ⊻ ♃
23 Tuesday	
00 13	☽ ⊥ ☉
02 07	☽ ∥ ♂
05 48	☽ ∥ ☉
06 06	☽ # ♅
11 25	☽ ± ♆
11 29	☽ ⊥ ♃
12 56	☽ ⊻ ☉
13 02	☽ ∠ ☿
23 11	☽ ⊼ ♆
23 22	☽ ∠ ♂
24 Wednesday	
00 16	☽ ⊼ ♇
05 25	☽ ∠ ♃
06 28	☽ ⊥ ☿
12 07	☽ ± ♇
12 56	♀ ⊥ ♂
15 41	☿ ⊻ ♀
18 07	☽ ⊻ ☿
18 10	☽ ⊥ ♂
18 27	☽ ☌ ♀
18 55	☽ ☍ ♄
22 58	♀ ☍ ♄
25 Thursday	
01 49	♀ # ♆
05 58	☽ bQ ♇
06 46	☽ ⊻ ♂
07 09	♂ ± ♄
10 06	☽ ☍ ♅
11 36	☽ ✶ ♃
26 Friday	
05 37	☽ ☌ ☉
10 47	☽ ☍ ♆
11 58	☽ Q ♃
11 59	☽ △ ♇
21 05	♀ ∥ ♂
27 Saturday	
04 33	☽ ☌ ☿
06 42	☽ ⊼ ♄
13 26	☽ ⊻ ♀
17 56	♂ ⊼ ♅
18 45	☽ ± ♄
22 15	☽ ⊼ ♅
22 31	☽ ☌ ♂
28 Sunday	
00 45	☽ □ ♃
02 58	☽ ⊥ ♀
10 22	☽ ± ♅
12 53	☽ bQ ♄
18 43	☽ # ♅
20 05	☽ ∥ ☉
20 47	☉ ☍ ♆
23 17	☽ ⊼ ♆
23 18	☽ ∠ ♀
23 31	☽ ⊻ ☉
23 40	☽ ∥ ♀
29 Monday	
00 34	☽ □ ♇
03 09	☽ ∥ ♂
03 12	☽ # ♆
04 33	☽ bQ ♅
04 55	♂ # ♆
09 48	☽ # ♄
11 25	☽ ± ♆
11 29	♀ bQ ♇
12 27	☉ △ ♇
12 44	☽ ⊥ ☉
14 57	☽ ⊻ ☿
19 06	☽ △ ♄
23 06	☽ ∥ ☿
30 Tuesday	
02 33	☉ Q ♃
02 34	☽ ⊥ ☿
03 35	♂ □ ♃
05 36	☽ bQ ♆
08 33	☽ ∠ ☉
09 10	☽ ✶ ♀
10 49	☽ △ ♅
12 58	☽ Q ♇
14 16	☽ △ ♃
14 44	☽ ⊻ ♂
19 55	☽ ∠ ☿

All ephemeris data is given at 12.00 UT and the Moon's longitude is additionally given for 24.00 UT

LONGITUDES

Date	Sidereal time h m s	Sun ☉ ° ' "	Moon ☽ ° ' "	Moon ☽ 24.00 ° ' "	Mercury ☿ ° '	Venus ♀ ° '	Mars ♂ ° '	Jupiter ♃ ° '	Saturn ♄ ° '	Uranus ♅ ° '	Neptune ♆ ° '	Pluto ♇ ° '
01	06 36 02	09 ♋ 08 08	06 ♍ 41 50	12 ♍ 43 00	13 ♋ 18	24 ♊ 43	26 ♋ 41	26 ♈ 00	16 ♐ 17	24 ♐ 08	06 ♑ 33	07 ♏ 14
02	06 39 59	10 05 21	18 ♍ 46 51	24 ♍ 53 57	12 R 42	25 56	27 19	26 09	16 R 13	24 R 05	06 R 32	07 R 13
03	06 43 55	11 02 33	01 ♎ 04 55	07 ♎ 20 22	12 05	27 09	27 57	26 17	16 10	24 03	06 30	07 13
04	06 47 52	11 59 45	13 ♎ 40 55	20 ♎ 07 10	11 29	28 23	28 36	26 25	16 06	24 01	06 28	07 12
05	06 51 48	12 56 56	26 ♎ 39 39	03 ♏ 18 52	10 53	29 ♊ 36	29 14	26 33	16 02	23 58	06 27	07 12
06	06 55 45	13 54 08	10 ♏ 05 14	16 ♏ 58 59	10 18	00 ♋ 49	29 ♋ 52	26 41	15 58	23 56	06 25	07 11
07	06 59 42	14 51 19	24 ♏ 00 13	01 ♐ 08 53	09 45	02 03	00 ♌ 31	26 48	15 55	23 54	06 23	07 11
08	07 03 38	15 48 30	08 ♐ 24 41	15 ♐ 47 04	09 14	03 16	01 09	26 56	15 51	23 52	06 22	07 11
09	07 07 35	16 45 41	23 ♐ 15 18	00 ♑ 48 22	08 46	04 30	01 47	27 04	15 48	23 49	06 20	07 10
10	07 11 31	17 42 52	08 ♑ 25 05	16 ♑ 04 07	08 21	05 43	02 26	27 11	15 44	23 47	06 19	07 10
11	07 15 28	18 40 03	23 ♑ 44 02	01 ♒ 23 22	08 01	06 56	03 04	27 18	15 41	23 45	06 17	07 10
12	07 19 24	19 37 15	09 ♒ 00 41	16 ♒ 34 42	07 45	08 10	03 42	27 25	15 38	23 43	06 16	07 09
13	07 23 21	20 34 26	24 ♒ 04 15	01 ♓ 28 23	07 33	09 23	04 20	27 32	15 34	23 41	06 14	07 09
14	07 27 17	21 31 38	08 ♓ 46 25	15 ♓ 57 49	07 26	10 37	04 59	27 39	15 31	23 39	06 12	07 09
15	07 31 14	22 28 51	23 ♓ 02 19	29 ♓ 59 50	07 D 24	11 50	05 37	27 45	15 28	23 37	06 11	07 09
16	07 35 11	23 26 04	06 ♈ 50 28	13 ♈ 34 27	07 28	13 04	06 15	27 52	15 25	23 35	06 09	07 09
17	07 39 07	24 23 18	20 ♈ 12 07	26 ♈ 43 52	07 37	14 18	06 53	27 58	15 22	23 33	06 08	07 09
18	07 43 04	25 20 32	03 ♉ 10 13	09 ♉ 31 37	07 51	15 31	07 32	28 04	15 19	23 31	06 06	07 D 09
19	07 47 00	26 17 47	15 ♉ 48 38	22 ♉ 01 44	08 12	16 45	08 10	28 10	15 16	23 29	06 05	07 09
20	07 50 57	27 15 03	28 ♉ 11 26	04 ♊ 18 12	08 37	17 58	08 48	28 15	15 14	23 27	06 03	07 09
21	07 54 53	28 12 20	10 ♊ 22 27	16 ♊ 24 36	09 09	19 12	09 26	28 21	15 11	23 25	06 02	07 09
22	07 58 50	29 ♋ 09 37	22 ♊ 25 00	28 ♊ 23 59	09 46	20 26	10 04	28 26	15 08	23 23	06 00	07 09
23	08 02 46	00 ♌ 06 55	04 ♋ 21 50	10 ♋ 18 47	10 28	21 39	10 43	28 32	15 06	23 21	05 59	07 09
24	08 06 43	01 04 14	16 ♋ 15 06	22 ♋ 10 58	11 16	22 53	11 21	28 37	15 04	23 20	05 58	07 10
25	08 10 40	02 01 34	28 ♋ 06 35	04 ♌ 02 10	12 10	24 07	11 59	28 42	15 01	23 18	05 56	07 10
26	08 14 36	02 58 54	09 ♌ 57 52	15 ♌ 53 55	13 09	25 21	12 37	28 46	14 59	23 16	05 55	07 10
27	08 18 33	03 56 14	21 ♌ 50 31	27 ♌ 47 55	14 13	26 34	13 15	28 51	14 57	23 14	05 53	07 10
28	08 22 29	04 53 36	03 ♍ 46 24	09 ♍ 46 14	15 23	27 48	13 54	28 55	14 55	23 13	05 52	07 11
29	08 26 26	05 50 57	15 ♍ 47 48	21 ♍ 51 27	16 37	29 ♋ 02	14 32	28 59	14 53	23 11	05 51	07 11
30	08 30 22	06 48 20	27 ♍ 57 37	04 ♎ 06 44	17 57	00 ♌ 16	15 10	29 03	14 51	23 10	05 49	07 12
31	08 34 19	07 ♌ 45 43	10 ♎ 19 17	16 ♎ 35 46	19 ♋ 21	01 ♌ 30	15 ♌ 48	29 ♈ 07	14 ♐ 49	23 ♐ 08	05 ♑ 48	07 ♏ 12

Date	Moon True ☊ ° '	Moon Mean ☊ ° '	Moon ☽ Latitude ° '
01	06 ♈ 24	06 ♈ 53	02 N 30
02	06 R 21	06 50	01 31
03	06 D 21	06 47	00 N 28
04	06 R 21	06 44	00 S 39
05	06 20	06 40	01 45
06	06 18	06 37	02 48
07	06 14	06 34	03 44
08	06 07	06 31	04 27
09	05 59	06 28	04 54
10	05 48	06 24	05 01
11	05 38	06 21	04 46
12	05 29	06 18	04 11
13	05 22	06 15	03 20
14	05 17	06 12	02 15
15	05 15	06 09	01 S 04
16	05 D 14	06 05	00 N 09
17	05 15	06 02	01 19
18	05 R 14	05 59	02 22
19	05 13	05 56	03 18
20	05 09	05 53	04 02
21	05 02	05 50	04 35
22	04 54	05 46	04 55
23	04 43	05 43	05 02
24	04 31	05 40	04 56
25	04 19	05 37	04 37
26	04 08	05 34	04 06
27	04 00	05 30	03 24
28	03 53	05 27	02 33
29	03 49	05 24	01 35
30	03 47	05 21	00 31
31	03 ♈ 47	05 ♈ 18	00 S 35

DECLINATIONS

Date	Sun ☉ ° '	Moon ☽ ° '	Mercury ☿ ° '	Venus ♀ ° '	Mars ♂ ° '	Jupiter ♃ ° '	Saturn ♄ ° '	Uranus ♅ ° '	Neptune ♆ ° '	Pluto ♇ ° '
01	23 N 08	11 N 22	18 N 40	23 N 01	21 N 57	08 N 53	21 S 10	23 S 28	22 S 14	01 N 45
02	23 04	05 N 50	18 33	23 06	21 50	08 56	21 10	23 28	22 14	01 44
03	22 59	00 00	18 27	23 11	21 42	08 58	21 10	23 28	22 14	01 44
04	22 54	06 S 00	18 22	23 14	21 34	09 01	21 09	23 28	22 14	01 44
05	22 49	11 55	18 19	23 17	21 27	09 04	21 09	23 28	22 14	01 43
06	22 43	17 30	18 17	23 20	21 19	09 06	21 09	23 28	22 14	01 43
07	22 37	22 23	18 17	23 21	21 10	09 09	21 09	23 27	22 14	01 43
08	22 30	26 06	18 18	23 22	21 02	09 12	21 09	23 27	22 14	01 42
09	22 23	28 09	18 20	23 23	20 54	09 14	21 08	23 27	22 15	01 42
10	22 16	28 10	18 23	23 22	20 45	09 16	21 08	23 27	22 15	01 41
11	22 09	26 03	18 28	23 21	20 36	09 19	21 08	23 27	22 15	01 41
12	22 00	22 03	18 33	23 20	20 27	09 21	21 08	23 27	22 15	01 40
13	21 52	16 38	18 40	23 17	20 18	09 23	21 08	23 27	22 15	01 40
14	21 43	10 22	18 48	23 14	20 09	09 26	21 07	23 27	22 15	01 39
15	21 34	03 S 45	18 56	23 10	20 00	09 28	21 07	23 27	22 15	01 39
16	21 25	02 N 51	19 06	23 06	19 50	09 30	21 07	23 27	22 15	01 38
17	21 15	09 07	19 16	23 01	19 40	09 32	21 07	23 27	22 15	01 38
18	21 04	14 48	19 26	22 55	19 31	09 34	21 07	23 26	22 15	01 37
19	20 54	19 44	19 37	22 48	19 21	09 36	21 07	23 26	22 15	01 37
20	20 43	23 42	19 48	22 41	19 11	09 37	21 07	23 26	22 16	01 36
21	20 31	26 33	20 00	22 33	19 01	09 39	21 06	23 26	22 16	01 35
22	20 20	28 08	20 11	22 24	18 50	09 41	21 06	23 26	22 16	01 35
23	20 08	28 24	20 22	22 15	18 40	09 42	21 06	23 26	22 16	01 34
24	19 55	27 21	20 32	22 05	18 29	09 44	21 06	23 26	22 16	01 34
25	19 43	25 04	20 42	21 55	18 19	09 45	21 06	23 26	22 16	01 33
26	19 30	21 42	20 52	21 44	18 08	09 47	21 06	23 26	22 16	01 32
27	19 16	17 27	21 00	21 32	17 57	09 48	21 06	23 26	22 16	01 32
28	19 03	12 30	21 07	21 20	17 46	09 50	21 06	23 26	22 16	01 31
29	18 49	07 03	21 13	21 07	17 35	09 51	21 06	23 25	22 16	01 30
30	18 34	01 17	21 17	20 53	17 24	09 52	21 06	23 25	22 16	01 30
31	18 N 20	04 S 37	21 N 20	20 N 39	17 N 12	09 N 53	21 S 06	23 S 25	22 S 17	01 N 29

ZODIAC SIGN ENTRIES

Date	h	m	Planets
03	09	55	☽ ♎
05	18	03	☽ ♏
05	19	50	♀ ♋
06	16	46	♂ ♌
07	22	05	☽ ♐
09	22	43	☽ ♑
11	21	49	☽ ♒
13	21	36	☽ ♓
16	00	00	☽ ♈
18	06	04	☽ ♉
20	15	33	☽ ♊
23	03	13	☽ ♋
23	09	06	☉ ♌
25	15	50	☽ ♌
28	04	26	☽ ♍
30	06	49	♀ ♌
30	15	59	☽ ♎

LATITUDES

Date	Mercury ☿ ° '	Venus ♀ ° '	Mars ♂ ° '	Jupiter ♃ ° '	Saturn ♄ ° '	Uranus ♅ ° '	Neptune ♆ ° '	Pluto ♇ ° '
01	04 S 08	00 S 19	01 N 09	01 S 15	01 N 35	00 S 09	01 N 03	16 N 32
04	04 35	00 12	01 09	01 16	01 34	00 09	01 03	16 31
07	04 49	00 S 04	01 09	01 16	01 34	00 09	01 03	16 29
10	04 48	00 N 03	01 09	01 17	01 33	00 09	01 03	16 28
13	04 34	00 10	01 09	01 18	01 33	00 09	01 03	16 26
16	04 09	00 18	01 10	01 18	01 32	00 09	01 03	16 24
19	03 35	00 25	01 10	01 19	01 32	00 09	01 03	16 22
22	02 55	00 31	01 09	01 20	01 31	00 09	01 03	16 21
25	02 11	00 38	01 09	01 20	01 31	00 09	01 03	16 19
28	01 27	00 44	01 09	01 21	01 30	00 09	01 03	16 17
31	00 S 43	00 N 50	01 N 09	01 S 22	01 N 29	00 S 09	01 N 03	16 N 15

DATA

Julian Date	2446978
Delta T	+55 seconds
Ayanamsa	23° 40' 55"
Synetic vernal point	05° ♓ 26' 04"
True obliquity of ecliptic	23° 26' 36"

MOON'S PHASES, APSIDES AND POSITIONS ☽

Date	h	m	Phase	Longitude ° '	Eclipse Indicator
04	08	34	☽	11 ♎ 52	
11	03	33	○	18 ♑ 20	
17	20	17	☾	24 ♈ 43	
25	20	38	●	02 ♌ 22	

Day	h	m		
11	09	46	Perigee	
25	08	03	Apogee	
03	11	59	0S	
10	00	13	Max dec	28° S 26'
16	01	33	0N	
23	04	47	Max dec	28° N 28'
30	17	15	0S	

LONGITUDES

Date	Chiron ⚷ ° '	Ceres ⚳ ° '	Pallas ⚴ ° '	Juno ⚵ ° '	Vesta ⚶ ° '	Black Moon Lilith ⚸ ° '
01	23 ♊ 43	26 ♐ 14	12 ♏ 27	06 ♓ 46	21 ♊ 09	24 ♋ 38
11	24 ♊ 33	24 ♐ 21	12 ♏ 40	06 ♓ 52	25 ♊ 23	25 ♋ 46
21	25 ♊ 21	22 ♐ 54	13 ♏ 32	06 ♓ 18	29 ♊ 35	26 ♋ 53
31	26 ♊ 06	22 ♐ 01	14 ♏ 59	05 ♓ 03	03 ♋ 43	28 ♋ 00

All ephemeris data is given at 12.00 UT and the Moon's longitude is additionally given for 24.00 UT

ASPECTARIAN

h m	Aspects
01 Wednesday	
00 49	♀ ☍ ♅
03 29	☽ ⊥ ♂
11 43	☽ △ ♆
12 02	☽ Q ♀
13 03	☽ ✶ ♇
17 17	☽ ✶ ☉
20 42	☽ bQ ♃
22 29	☽ ∠ ♂
22 56	☽ ∥ ♃
02 Thursday	
00 32	☽ ✶ ☿
04 48	☉ ∥ ♀
06 59	☽ □ ♄
14 43	☽ ± ♃
16 39	♀ ✶ ♃
18 45	☽ ∠ ♇
19 03	☽ Q ☉
22 23	☽ □ ♅
23 04	☽ Q ☿
03 Friday	
02 36	☽ ⊼ ♃
03 34	☽ □ ♀
04 57	☽ ∥ ♇
05 37	☽ ✶ ♂
12 15	☽ ⊥ ♇
17 53	☽ Q ♄
18 57	☽ # ♇
22 22	☽ □ ♆
23 45	☽ ⊻ ♇
04 Saturday	
04 03	☉ ☌ ☿
05 52	☽ Q ♂
08 02	☽ □ ☿
08 34	☽ □ ☉
08 52	☽ Q ♅
16 30	☽ ✶ ♄
20 53	♀ ⊻ ♂
05 Sunday	
00 15	☽ # ♃
07 06	☽ ✶ ♅
07 58	☽ Q ♆
11 48	☽ ☍ ♃
16 54	☽ □ ♂
17 52	☽ △ ♀
19 52	☽ ∠ ♄
06 Monday	
05 32	☽ ✶ ♆
06 53	☽ ☌ ♇
09 59	☽ ∠ ♅
11 48	☽ ⊥ ♄
12 21	☽ △ ☿
14 12	♂ ± ♅
15 35	☽ # ☿
19 10	☽ △ ☉
22 13	☽ ⊻ ♄
22 58	☽ bQ ♀
07 Tuesday	
01 36	☽ ⊥ ♅
05 27	☽ ∥ ♄
05 46	☽ # ♂
07 34	☽ ∠ ♆
11 13	☽ ∥ ♆
11 49	☽ ⊻ ♅
13 12	☽ bQ ☿
13 15	☽ # ☉
15 47	☽ ± ♀
16 47	☽ ⊼ ♃
16 50	♂ # ♄
17 32	☽ # ♀
18 05	☽ ∥ ♅
22 33	☽ bQ ☉
22 43	☽ ⊥ ♆
23 27	☽ △ ♂
08 Wednesday	
01 48	♂ bQ ♄
02 45	☽ ⊼ ♀
02 54	☽ ± ♃
03 45	☽ ± ☿
08 39	☽ ⊻ ♆
09 58	☽ ⊻ ♇
13 04	☉ ⊼ ♄
13 18	☽ ⊼ ☿
14 27	☽ ± ☉
17 49	☽ bQ ♃
19 46	☽ ⊥ ♇
23 35	☿ Q ♃
09 Thursday	
00 04	☽ ☌ ♄
00 52	☽ ⊼ ☉
01 09	☽ bQ ♂
10 16	☽ ∠ ♇
12 54	☽ ☌ ♅
16 13	☽ ± ♂
18 07	☽ △ ♃
10 Friday	
02 09	☽ ⊼ ♂
07 23	☽ ☍ ♀
08 42	☽ ☌ ♆
10 02	☽ ✶ ♇
11 54	☽ ☍ ☿
16 54	☉ # ♆
23 25	♀ ☍ ♆
23 26	☽ ⊻ ♄
11 Saturday	
03 33	☽ ☍ ☉
04 51	☽ Q ♇
08 48	☽ ⊥ ♄
12 02	☽ ⊻ ♅
16 19	♀ △ ♇
17 38	☽ □ ♃
21 24	☽ ⊥ ♅
22 51	☽ ∠ ♄
12 Sunday	
03 16	☽ ☍ ♂
04 38	☽ ∥ ♅
05 05	☿ ☌ ♀
05 16	☽ # ♀
07 40	☽ ⊻ ♆
09 05	☽ □ ♇
10 02	☽ ⊼ ☿
10 33	☽ ⊼ ♀
10 59	☽ ∥ ♆
11 32	☽ ∠ ♅
12 11	☽ # ☉
16 28	☽ ∥ ♄
17 08	☽ ⊥ ♆
19 23	☽ ± ☿
19 51	☽ # ♂
20 53	☽ ± ♀
22 14	☽ Q ♃
22 27	☽ ✶ ♄
13 Monday	
03 43	☽ # ☿
06 00	☽ ⊼ ☉
07 27	☽ ∠ ♆
09 34	☽ bQ ☿
11 22	☽ ✶ ♅
12 34	☽ bQ ♀
15 05	♀ Q ♃
16 19	☽ ± ☉
17 38	☽ ✶ ♃
17 38	☽ Q ♄
14 Tuesday	
05 27	☽ ⊼ ♂
06 51	☽ Q ♅
07 46	☽ ✶ ♆
08 02	☽ bQ ☉
09 19	☽ △ ♇
09 48	☽ △ ☿
11 50	☉ ± ♄
15 20	☽ △ ♀
15 27	☽ # ♃
15 50	☽ ± ♂
18 29	☽ ∠ ♃
23 13	☽ □ ♄
15 Wednesday	
03 45	☽ Q ♆
07 40	☽ bQ ♂
07 51	☿ St D
09 47	☽ ⊥ ♃
10 29	☽ bQ ♇
10 59	☽ △ ☉
12 59	☽ □ ♅
19 35	☽ # ♇
20 10	☽ ⊻ ♃
16 Thursday	
02 00	☽ ± ♇
07 33	☽ ∥ ♇
08 32	♂ ⊼ ♆
10 48	☽ □ ♆
10 54	☽ △ ♂
12 33	☽ ⊼ ♇
13 07	☽ □ ☿
15 30	☉ ⊼ ♅
17 Friday	
00 12	☽ □ ♀
03 16	☽ △ ♄
13 41	☽ ∥ ♃
18 06	☽ △ ♅
20 17	☽ □ ☉
21 47	♂ □ ♇
22 06	☽ Q ☿
18 Saturday	
02 24	☽ ☌ ♃
06 10	♇ St D
06 23	☉ # ♄
06 41	☽ bQ ♄
08 16	♀ ⊼ ♄
12 43	☽ Q ♀
17 14	☿ ∥ ♂
17 31	☽ △ ♆
19 30	☽ ☍ ♇
20 38	☽ □ ♂
21 03	☽ ✶ ☿
22 03	☽ bQ ♅
23 34	☽ ± ♄
19 Sunday	
08 51	☽ Q ☉
10 03	☽ ∥ ♂
10 58	☽ ⊼ ♄
11 24	☽ ∥ ☿
14 00	☽ ✶ ♀
14 52	☿ ⊻ ♂
15 12	☽ ± ♅
18 12	☽ ∥ ☉
19 41	☽ # ♄
22 09	☽ bQ ♆
23 22	♂ bQ ♅
20 Monday	
02 33	☽ # ♆
02 45	☽ ∠ ☿
02 47	☽ ⊼ ♅
05 26	☽ ∥ ♀
09 08	☽ Q ♂
10 01	☽ ✶ ☉
10 13	☽ # ♅
12 08	☽ ⊻ ♃
15 39	☽ ± ♆
21 04	☽ ⊥ ☿
22 26	☽ ∠ ♀
21 Tuesday	
00 00	☽ ⊥ ♃
03 25	☽ ⊼ ♆
05 37	☽ ⊼ ♇
09 27	☽ ⊻ ☿
10 02	☽ ✶ ♂
16 02	☉ □ ♃
17 31	☽ ± ♇
17 57	☽ ∠ ♃
18 06	☽ ∠ ☉
18 15	☽ ⊥ ♀
21 31	☽ ☍ ♄
22 Wednesday	
07 34	☽ ⊻ ♀
11 28	☽ bQ ♇
13 37	☽ ⊥ ☉
13 56	☽ ☍ ♅
17 30	☉ ± ♅
17 37	☽ ∠ ♂
21 21	☉ ∥ ☿
23 Thursday	
00 10	☽ ✶ ♃
01 27	♀ ± ♄
02 42	☽ ⊻ ☉
10 39	♀ # ♆
11 37	☉ bQ ♄
12 44	☽ ⊥ ♂
13 57	☿ Q ♃
15 15	☽ ☍ ♆
17 38	☽ △ ♇
24 Friday	
00 31	☽ Q ♃
01 11	☽ ☌ ☿
01 32	☽ ⊻ ♂
09 36	☽ ⊼ ♄
19 46	☿ ⊻ ♂
20 24	♀ ⊼ ♅
21 42	☽ ± ♄
25 Saturday	
02 17	☽ ⊼ ♅
02 59	☽ ☌ ♀
10 13	♂ ± ♆
13 11	☽ □ ♃
14 24	☽ ± ♅
15 51	☽ bQ ♄
20 38	☽ ☌ ☉
26 Sunday	
00 33	☽ # ♅
03 49	☽ ⊼ ♆
06 20	☽ □ ♇
08 24	☽ # ♆
08 35	☽ bQ ♅
11 49	☽ ∥ ♀
15 42	☽ # ♄
15 56	☽ ± ♆
16 57	☽ ∥ ☿
17 41	☽ ☌ ♂
19 04	☽ ⊻ ☿
22 07	☽ △ ♄
27 Monday	
01 43	☽ ∥ ☉
08 24	☽ ⊥ ☿
09 16	☽ ∥ ♂
10 05	☽ bQ ♆
14 49	☽ △ ♅
18 43	☽ Q ♇
22 38	☽ ⊻ ♀
28 Tuesday	
02 12	☽ △ ♃
02 46	☿ ⊼ ♄
04 26	☽ ∠ ☿
07 34	☿ # ♄
12 04	☽ ⊥ ♀
14 26	☽ ⊻ ☉
16 11	☽ △ ♆
18 49	☽ ✶ ♇
23 59	☽ ∥ ♃
29 Wednesday	
03 28	☽ ⊥ ☉
03 42	☿ ∥ ♀
08 06	☽ ∠ ♀
08 23	☽ bQ ♃
09 21	☽ ⊻ ♂
10 11	☽ □ ♄
11 04	♀ □ ♃
11 51	☉ ⊼ ♆
13 15	♀ # ♄
13 50	☽ ✶ ☿
14 55	♀ ± ♅
21 54	☽ ⊥ ♂
22 52	☽ ∠ ☉
30 Thursday	
00 31	♂ △ ♄
00 39	☽ ∠ ♇
02 19	☽ ± ♃
02 35	☽ □ ♅
04 02	♀ bQ ♄
11 09	☽ ∥ ♇
14 09	☽ ⊼ ♃
16 23	☽ Q ☿
16 33	☽ ∠ ♂
17 01	☽ ✶ ♀
18 19	☽ ⊥ ♇
21 31	☽ Q ♄
21 46	☉ □ ♇
23 18	☽ # ♇
31 Friday	
03 17	☽ □ ♆
05 59	☽ ⊻ ♇
06 39	☽ ✶ ☉
13 34	☽ Q ♅
18 45	☽ Q ♀
20 35	☽ ✶ ♄
21 09	☉ bQ ♅
23 03	☽ ✶ ♂

AUGUST 1987

LONGITUDES

Date	Sidereal time h m s	Sun ☉ ° ' "	Moon ☽ ° ' "	Moon ☽ 24.00 ° ' "	Mercury ☿ ° '	Venus ♀ ° '	Mars ♂ ° '	Jupiter ♃ ° '	Saturn ♄ ° '	Uranus ♅ ° '	Neptune ♆ ° '	Pluto ♇ ° '
01	08 38 15	08 ♌ 43 06	22 ♎ 56 42	29 ♎ 22 36	20 ♋ 50	02 ♌ 44	16 ♌ 26	29 ♈ 11	14 ♐ 47	23 ♐ 07	05 ♑ 47	07 ♏ 12
02	08 42 12	09 40 31	05 ♏ 53 58	12 ♏ 31 15	22 24	03 58	17 04	29 14	14 R 46	23 R 05	05 R 45	07 13
03	08 46 09	10 37 56	19 ♏ 14 50	26 ♏ 05 03	24 02	05 11	17 43	29 18	14 44	23 04	05 44	07 13
04	08 50 05	11 35 21	03 ♐ 02 04	10 ♐ 05 56	25 43	06 25	18 21	29 21	14 43	23 02	05 43	07 14
05	08 54 02	12 32 47	17 ♐ 16 33	24 ♐ 33 34	27 28	07 39	18 59	29 23	14 41	23 01	05 42	07 15
06	08 57 58	13 30 14	01 ♑ 56 27	09 ♑ 24 29	29 ♋ 17	08 53	19 37	29 26	14 40	23 00	05 40	07 15
07	09 01 55	14 27 42	16 ♑ 56 44	24 ♑ 32 03	01 ♌ 09	10 07	20 15	29 29	14 39	22 59	05 39	07 16
08	09 05 51	15 25 11	02 ♒ 09 12	09 ♒ 46 50	03 03	11 21	20 53	29 31	14 38	22 58	05 38	07 17
09	09 09 48	16 22 40	17 ♒ 23 37	24 ♒ 58 12	04 59	12 35	21 32	29 33	14 37	22 56	05 37	07 17
10	09 13 44	17 20 11	02 ♓ 29 22	09 ♓ 56 03	06 58	13 49	22 10	29 35	14 36	22 55	05 36	07 18
11	09 17 41	18 17 42	17 ♓ 17 19	24 ♓ 32 29	08 58	15 03	22 48	29 37	14 35	22 54	05 35	07 19
12	09 21 38	19 15 15	01 ♈ 41 02	08 ♈ 42 41	10 59	16 18	23 26	29 38	14 34	22 53	05 34	07 20
13	09 25 34	20 12 50	15 ♈ 37 19	22 ♈ 24 59	13 01	17 32	24 04	29 40	14 34	22 52	05 33	07 21
14	09 29 31	21 10 25	29 ♈ 05 54	05 ♉ 40 22	15 03	18 46	24 42	29 41	14 33	22 51	05 32	07 22
15	09 33 27	22 08 03	12 ♉ 08 47	18 ♉ 31 37	17 06	20 00	25 20	29 42	14 33	22 50	05 31	07 23
16	09 37 24	23 05 42	24 ♉ 49 23	01 ♊ 02 39	19 08	21 14	25 59	29 43	14 32	22 50	05 30	07 24
17	09 41 20	24 03 22	07 ♊ 11 55	13 ♊ 17 47	21 10	22 28	26 37	29 43	14 32	22 49	05 29	07 25
18	09 45 17	25 01 04	19 ♊ 20 45	25 ♊ 21 21	23 12	23 42	27 15	29 44	14 32	22 48	05 28	07 26
19	09 49 13	25 58 48	01 ♋ 20 04	07 ♋ 17 22	25 13	24 57	27 53	29 44	14 D 32	22 47	05 27	07 27
20	09 53 10	26 56 33	13 ♋ 13 40	19 ♋ 09 22	27 13	26 11	28 31	29 R 44	14 32	22 47	05 26	07 28
21	09 57 07	27 54 20	25 ♋ 04 49	01 ♌ 00 21	29 ♌ 13	27 25	29 09	29 44	14 32	22 46	05 25	07 29
22	10 01 03	28 52 08	06 ♌ 56 14	12 ♌ 52 46	01 ♍ 11	28 39	29 ♌ 48	29 43	14 32	22 46	05 24	07 30
23	10 05 00	29 ♌ 49 58	18 ♌ 50 09	24 ♌ 48 39	03 08	29 ♌ 54	00 ♍ 26	29 43	14 33	22 45	05 24	07 31
24	10 08 56	00 ♍ 47 49	00 ♍ 48 27	06 ♍ 49 46	05 04	01 ♍ 08	01 04	29 42	14 33	22 45	05 23	07 32
25	10 12 53	01 45 42	12 ♍ 52 49	18 ♍ 57 47	06 58	02 22	01 42	29 41	14 34	22 44	05 22	07 34
26	10 16 49	02 43 36	25 ♍ 04 54	01 ♎ 14 24	08 52	03 37	02 20	29 39	14 34	22 44	05 21	07 35
27	10 20 46	03 41 31	07 ♎ 26 31	13 ♎ 41 32	10 44	04 51	02 58	29 38	14 35	22 44	05 21	07 36
28	10 24 42	04 39 28	19 ♎ 59 44	26 ♎ 21 24	12 35	06 06	03 37	29 36	14 36	22 44	05 20	07 38
29	10 28 39	05 37 26	02 ♏ 46 52	09 ♏ 16 25	14 24	07 20	04 15	29 35	14 37	22 43	05 19	07 39
30	10 32 36	06 35 26	15 ♏ 50 22	22 ♏ 29 01	16 12	08 34	04 53	29 33	14 38	22 43	05 19	07 40
31	10 36 32	07 ♍ 33 27	29 ♏ 12 36	06 ♐ 01 19	17 ♍ 59	09 ♍ 49	05 ♍ 31	29 ♈ 30	14 ♐ 39	22 ♐ 43	05 ♑ 18	07 ♏ 42

Date	Moon True ☊ ° '	Moon Mean ☊ ° '	Moon ☽ Latitude ° '
01	03 ♈ 48	05 ♈ 15	01 S 41
02	03 D 49	05 11	02 44
03	03 R 49	05 08	03 39
04	03 47	05 05	04 24
05	03 43	05 02	04 55
06	03 37	04 59	05 07
07	03 30	04 56	04 59
08	03 23	04 52	04 30
09	03 16	04 49	03 41
10	03 12	04 46	02 38
11	03 09	04 43	01 25
12	03 08	04 40	00 S 08
13	03 D 08	04 36	01 N 07
14	03 09	04 33	02 16
15	03 11	04 30	03 16
16	03 R 11	04 27	04 04
17	03 10	04 24	04 39
18	03 07	04 21	05 01
19	03 03	04 17	05 10
20	02 58	04 14	05 05
21	02 51	04 11	04 48
22	02 45	04 08	04 17
23	02 39	04 05	03 36
24	02 35	04 01	02 44
25	02 32	03 58	01 45
26	02 30	03 55	00 N 41
27	02 D 30	03 52	00 S 27
28	02 31	03 49	01 35
29	02 32	03 46	02 39
30	02 34	03 42	03 36
31	02 ♈ 35	03 ♈ 39	04 S 23

DECLINATIONS

Date	Sun ☉ ° '	Moon ☽ ° '	Mercury ☿ ° '	Venus ♀ ° '	Mars ♂ ° '	Jupiter ♃ ° '	Saturn ♄ ° '	Uranus ♅ ° '	Neptune ♆ ° '	Pluto ♇ ° '
01	18 N 05	10 S 29	21 N 21	20 N 24	17 N 01	09 N 54	21 S 06	23 S 25	22 S 17	01 N 28
02	17 50	16 04	21 19	20 08	16 49	09 55	21 06	23 25	22 17	01 28
03	17 34	21 03	21 16	19 52	16 37	09 56	21 06	23 25	22 17	01 27
04	17 19	25 05	21 10	19 36	16 26	09 57	21 06	23 25	22 17	01 26
05	17 03	27 43	21 02	19 19	16 14	09 58	21 06	23 25	22 17	01 25
06	16 46	28 33	20 51	19 01	16 02	09 58	21 06	23 25	22 17	01 25
07	16 30	27 18	20 37	18 43	15 50	09 59	21 06	23 25	22 17	01 24
08	16 13	24 03	20 21	18 24	15 37	10 00	21 06	23 25	22 17	01 23
09	15 56	19 08	20 02	18 05	15 25	10 00	21 06	23 25	22 17	01 22
10	15 38	13 02	19 40	17 45	15 13	10 01	21 06	23 25	22 17	01 22
11	15 21	06 S 19	19 16	17 25	15 00	10 01	21 06	23 25	22 17	01 21
12	15 03	00 N 33	18 50	17 04	14 47	10 01	21 07	23 25	22 18	01 20
13	14 45	07 11	18 21	16 43	14 35	10 02	21 07	23 25	22 18	01 19
14	14 27	13 16	17 50	16 21	14 22	10 02	21 07	23 24	22 18	01 18
15	14 08	18 35	17 16	15 59	14 09	10 02	21 07	23 24	22 18	01 18
16	13 49	22 55	16 41	15 36	13 56	10 02	21 07	23 24	22 18	01 17
17	13 30	26 06	16 05	15 13	13 43	10 02	21 07	23 24	22 18	01 16
18	13 11	28 01	15 26	14 50	13 30	10 02	21 08	23 24	22 18	01 15
19	12 52	28 36	14 47	14 26	13 17	10 02	21 08	23 24	22 18	01 14
20	12 32	27 51	14 06	14 02	13 03	10 01	21 08	23 24	22 18	01 13
21	12 12	25 50	13 23	13 37	12 50	10 01	21 08	23 24	22 18	01 13
22	11 52	22 41	12 40	13 12	12 36	10 01	21 08	23 24	22 18	01 12
23	11 32	18 36	11 57	12 47	12 23	10 00	21 09	23 24	22 18	01 11
24	11 12	13 45	11 12	12 21	12 09	10 00	21 09	23 24	22 18	01 10
25	10 51	08 21	10 27	11 55	11 55	09 59	21 09	23 24	22 18	01 09
26	10 30	02 N 34	09 41	11 28	11 42	09 58	21 10	23 24	22 19	01 08
27	10 09	03 S 22	08 55	11 02	11 28	09 58	21 10	23 24	22 19	01 07
28	09 48	09 17	08 09	10 35	11 14	09 57	21 10	23 24	22 19	01 06
29	09 27	14 55	07 23	10 07	11 00	09 56	21 10	23 24	22 19	01 06
30	09 06	20 01	06 36	09 40	10 46	09 55	21 11	23 24	22 19	01 05
31	08 N 44	24 S 16	05 N 50	09 N 12	10 N 32	09 N 54	21 S 11	23 S 24	22 S 19	01 N 04

ZODIAC SIGN ENTRIES

Date	h	m	Planets
02	01	09	☽ ♏
04	06	47	☽ ♐
06	08	52	☽ ♑
06	21	20	☿ ♌
08	08	37	☽ ♒
10	08	01	☽ ♓
12	09	09	☽ ♈
14	13	38	☽ ♉
16	21	59	☽ ♊
19	09	19	☽ ♋
21	21	36	☿ ♍
21	21	58	☽ ♌
22	19	51	♂ ♍
23	14	00	♀ ♍
23	16	10	☉ ♍
24	10	23	☽ ♍
26	21	35	☽ ♎
29	06	49	☽ ♏
31	13	24	☽ ♐

LATITUDES

Date	Mercury ☿ ° '	Venus ♀ ° '	Mars ♂ ° '	Jupiter ♃ ° '	Saturn ♄ ° '	Uranus ♅ ° '	Neptune ♆ ° '	Pluto ♇ ° '
01	00 S 29	00 N 52	01 N 09	01 S 22	01 N 29	00 S 09	01 N 02	16 N 15
04	00 N 10	00 57	01 09	01 23	01 29	00 09	01 02	16 13
07	00 44	01 02	01 09	01 24	01 28	00 09	01 02	16 11
10	01 10	01 07	01 09	01 24	01 27	00 09	01 02	16 10
13	01 30	01 11	01 09	01 25	01 27	00 09	01 02	16 08
16	01 41	01 15	01 08	01 26	01 26	00 09	01 02	16 06
19	01 46	01 18	01 08	01 27	01 26	00 09	01 02	16 04
22	01 44	01 20	01 08	01 27	01 25	00 09	01 02	16 03
25	01 37	01 22	01 08	01 28	01 24	00 09	01 02	16 01
28	01 25	01 24	01 07	01 29	01 24	00 09	01 02	16 00
31	01 N 10	01 N 25	01 N 07	01 S 30	01 N 23	00 S 10	01 N 01	15 N 58

DATA

Julian Date	2447009
Delta T	+55 seconds
Ayanamsa	23° 41' 00"
Synetic vernal point	05° ♓ 25' 59"
True obliquity of ecliptic	23° 26' 36"

MOON'S PHASES, APSIDES AND POSITIONS ☽

Date	h	m	Phase	Longitude ° '	Eclipse Indicator
02	19	24	☽	09 ♏ 58	
09	10	17	○	16 ♒ 19	
16	08	25	☾	22 ♉ 57	
24	11	59	●	00 ♍ 48	

Day	h	m		
08	19	25	Perigee	
21	13	49	Apogee	
06	09	48	Max dec	28° S 33'
12	10	04	0N	
19	10	21	Max dec	28° N 37'
26	22	26	0S	

LONGITUDES

Date	Chiron ⚷ ° '	Ceres ⚳ ° '	Pallas ⚴ ° '	Juno ⚵ ° '	Vesta ⚶ ° '	Black Moon Lilith ⚸ ° '
01	26 ♊ 11	21 ♐ 57	15 ♏ 10	04 ♓ 53	04 ♋ 07	28 ♋ 07
11	26 ♊ 51	21 ♐ 44	17 ♏ 09	02 ♓ 59	08 ♋ 10	29 ♋ 14
21	27 ♊ 27	22 ♐ 05	19 ♏ 33	00 ♓ 39	12 ♋ 06	00 ♌ 21
31	27 ♊ 58	23 ♐ 00	22 ♏ 18	28 ♒ 11	15 ♋ 56	01 ♌ 28

All ephemeris data is given at 12.00 UT and the Moon's longitude is additionally given for 24.00 UT

ASPECTARIAN

h m	Aspects
01 Saturday	
07 28	☽ Q ☉
07 30	☽ □ ☿
09 35	☽ # ♃
11 11	☿ ± ♄
12 19	☽ ⚹ ♅
13 34	☽ Q ♆
22 47	☽ Q ♂
23 41	☽ ☍ ♃
02 Sunday	
00 44	☽ ∠ ♄
08 05	☽ □ ♀
11 44	☽ ⚹ ♆
14 24	☽ ☌ ♇
15 19	☽ # ♂
15 59	☽ ∠ ♅
17 12	☽ ⊥ ♄
19 24	☽ □ ☉
19 44	☽ # ☉
22 09	☿ ⊼ ♅
03 Monday	
03 59	☽ ⊻ ♄
06 18	☽ # ♀
08 08	☽ ⊥ ♅
09 08	☽ □ ♂
12 14	☽ ∥ ♄
13 05	☽ # ☿
14 37	☽ ∠ ♆
18 37	☽ ∥ ♆
18 43	☽ ⊻ ♅
21 35	☽ △ ☿
22 24	♀ ⊼ ♆
04 Tuesday	
01 12	☽ ∥ ♅
05 38	☽ ⊼ ♃
06 18	☽ ⊥ ♆
15 04	☉ ± ♆
15 57	☽ ± ♃
16 34	☽ ⊻ ♆
18 20	☽ △ ♀
19 10	☽ ⊻ ♇
05 Wednesday	
00 19	☿ # ♄
02 51	☽ Ꝑ ☿
03 33	☽ △ ☉
03 55	♀ □ ♇
05 17	☽ ⊥ ♇
07 11	☽ Ꝑ ♃
07 42	☽ ☌ ♄
14 57	☽ △ ♂
18 57	♀ Ꝑ ♅
19 54	☽ ± ☿
20 13	☽ ∠ ♇
21 28	☽ ☌ ♅
21 42	☽ Ꝑ ♀
06 Thursday	
06 03	☽ Ꝑ ☉
07 05	☽ ⊼ ☿
07 56	☽ △ ♃
08 19	☿ ± ♅
13 40	☽ ± ♀
14 04	☿ □ ♃
16 31	☽ Ꝑ ♂
16 57	☿ Ꝑ ♄
18 00	☽ ☌ ♆
20 34	☽ ⚹ ♇
21 34	☽ ± ☉
07 Friday	
00 10	☽ ⊼ ♀
07 32	☽ ± ♂
07 47	☽ ⊼ ☉
08 21	☽ ⊻ ♄
15 41	☽ Q ♇
16 33	☉ △ ♄
17 28	☽ ⊼ ♂
17 51	☽ ⊥ ♄
21 32	☽ ⊻ ♅
08 Saturday	
02 39	♂ Ꝑ ♆
06 59	☽ ⊥ ♅
07 51	☽ □ ♃
08 02	☽ ∠ ♄
13 37	☽ ☍ ☿
15 36	☽ ∥ ♅
17 21	♀ ± ♆
17 28	☽ ⊻ ♆
20 04	☽ □ ♇
21 07	☽ ∠ ♅
21 28	☽ ∥ ♆
09 Sunday	
02 54	☽ ⊥ ♆
03 12	☽ ∥ ♄
03 45	☽ ☍ ♀
07 37	☽ ⚹ ♄
07 49	☽ # ☿
10 17	☽ ☍ ☉
12 15	☽ Q ♃
16 42	☽ # ♀
17 05	☽ ∠ ♆
18 49	☽ ☍ ♂
19 36	☿ ⊼ ♆
20 46	☽ ⚹ ♅
10 Monday	
01 41	☽ # ☉
02 36	☽ Q ♄
03 33	☽ # ♂
07 21	☽ ⚹ ♃
15 54	☽ Q ♅
16 09	☿ □ ♇
16 59	☽ ⚹ ♆
18 19	♃ Ꝑ ♄
19 45	☽ △ ♇
20 18	☽ ⊼ ☿
22 59	☽ # ♃
23 27	☿ Ꝑ ♅
11 Tuesday	
02 53	♀ △ ♄
07 34	☽ □ ♄
07 35	☽ ± ☿
07 37	☽ ∠ ♃
08 00	☽ ⊼ ♀
12 29	☽ Q ♆
13 46	☽ ⊼ ☉
15 55	♂ △ ♅
18 48	☽ ± ♀
20 18	☽ Ꝑ ♇
21 16	☽ □ ♅
21 31	☽ ⊼ ♂
22 29	☽ ⊥ ♃
12 Wednesday	
00 25	☽ ± ☉
00 50	☽ Ꝑ ☿
05 22	☽ # ♇
08 01	☽ ± ♂
08 32	☽ ⊻ ♃
11 16	☽ Ꝑ ♀
11 24	☽ ± ♇
14 46	☽ ∥ ♇
16 41	☽ Ꝑ ☉
18 35	☽ □ ♆
18 50	☿ ± ♆
21 38	☽ ⊼ ♇
13 Thursday	
00 04	☽ Ꝑ ♂
06 39	☽ △ ☿
10 09	☽ △ ♄
15 41	☽ △ ♀
20 06	☉ Ꝑ ♆
20 42	☽ △ ☉
22 53	☽ ∥ ♃
14 Friday	
00 48	☽ △ ♅
03 41	☽ △ ♂
06 10	☿ △ ♄
12 49	☽ Ꝑ ♄
13 04	☽ ☌ ♃
16 27	☽ ∥ ♂
16 42	☽ ∥ ☉
23 43	☽ △ ♆
15 Saturday	
00 32	☽ ∥ ♀
03 08	☽ ☍ ♇
04 00	☽ Ꝑ ♅
05 18	☽ ± ♄
06 19	☽ ∥ ☿
08 00	☉ ∥ ♂
13 24	♂ Q ♇
16 29	☽ ⊼ ♄
20 48	☽ ± ♅
21 48	♀ Ꝑ ♆
23 04	☽ □ ☿
16 Sunday	
01 19	☽ # ♄
03 45	☽ Ꝑ ♆
04 24	☽ □ ♀
05 24	☉ △ ♅
08 10	☽ # ♆
08 11	☽ ⊼ ♅
08 25	☽ □ ☉
14 20	☽ □ ♂
15 11	☽ # ♅
20 59	☽ ± ♆
21 26	☽ ⊻ ♃
17 Monday	
03 51	☿ Ꝑ ♆
08 38	☽ ⊼ ♆
09 06	☽ ⊥ ♃
12 25	☽ ⊼ ♇
16 39	☽ Q ☿
18 35	♀ △ ♅
19 09	☽ Q ♀
22 22	☽ Q ☉
18 Tuesday	
00 14	☽ ± ♇
02 27	☽ ☍ ♄
02 50	☽ ∠ ♃
03 25	☽ Q ♂
07 16	☿ △ ♅
18 09	☽ Ꝑ ♇
18 53	☽ ☍ ♅
21 15	☽ ⚹ ☿
21 42	☽ ⚹ ♀
22 22	☉ Q ♇
19 Wednesday	
00 19	☽ ⚹ ☉
03 29	☿ ☌ ♀
04 41	☽ ⚹ ♂
08 47	☽ ⚹ ♃
08 54	♄ St D
14 42	☿ Q ♇
20 17	☽ ☍ ♆
21 08	♃ St R
21 49	♀ Q ♇
20 Thursday	
00 20	☽ △ ♇
05 29	☉ ☌ ☿
07 23	☽ ∠ ♀
08 58	☽ Q ♃
09 10	☽ ∠ ☉
09 33	☽ ∠ ☿
12 37	☽ ∠ ♂
14 39	☽ ⊼ ♄
17 29	☿ ∥ ♀
21 Friday	
02 48	☽ ± ♄
03 43	☽ ⊥ ♀
05 00	☽ ⊥ ☉
07 19	☽ ⊼ ♅
07 27	☽ ⊥ ☿
07 53	☽ ⊥ ♂
11 01	☿ ☌ ♂
17 18	☽ ⊻ ♀
18 14	☽ ⊻ ☉
18 15	☿ △ ♃
19 28	☽ ± ♅
20 43	☽ ⊻ ♂
21 02	☽ Ꝑ ♄
21 24	☽ □ ♃
22 02	☽ ⊻ ☿
22 Saturday	
07 10	☽ # ♅
08 54	☽ ⊼ ♆
09 18	♂ △ ♃
13 08	☽ □ ♇
13 40	☽ Ꝑ ♅
14 30	☽ # ♆
15 18	☿ ∥ ♂
21 01	☽ ± ♆
21 40	☽ # ♄
23 Sunday	
03 22	☽ △ ♄
06 25	☉ ☌ ♀
08 24	♀ △ ♃
08 57	☉ △ ♃
15 08	☽ Ꝑ ♆
19 52	☽ △ ♅
24 Monday	
01 27	☽ Q ♇
09 10	♀ ☌ ♂
09 47	☽ △ ♃
11 59	☽ ☌ ☉
12 29	☉ ∥ ☿
12 32	☽ ☌ ♂
12 44	☽ ☌ ♀
15 58	☿ △ ♆
19 01	☽ ∥ ♀
19 39	☽ ∥ ♂
21 07	☽ △ ♆
22 05	☽ ☌ ☿
25 Tuesday	
00 24	☽ ∥ ☉
01 26	☽ ⚹ ♇
01 26	☽ ∥ ☿
04 55	☽ ∥ ♃
07 32	☉ ☌ ♂
10 46	♀ ∥ ♂
15 20	☽ □ ♄
15 33	☽ Ꝑ ♃
19 33	☿ ⚹ ♇
26 Wednesday	
02 54	☿ ∥ ♃
07 06	☽ ∠ ♇
07 24	☽ □ ♅
09 13	☽ ± ♃
17 51	☽ ∥ ♇
20 54	☽ ⊼ ♃
27 Thursday	
00 41	☽ ⊥ ♇
02 36	☽ Q ♄
02 54	☽ ⊻ ♂
02 59	☽ # ♇
04 09	☽ ⊻ ☉
06 27	☽ ⊻ ♀
07 57	☽ □ ♆
12 19	☽ ⊻ ♇
15 06	☽ ⊥ ♂
16 42	☽ ⊥ ☉
18 19	☽ Q ♅
19 17	☽ ⊥ ♀
19 25	☽ ⊻ ☿
21 28	♀ △ ♆
28 Friday	
01 44	☽ ⚹ ♄
01 44	☉ ∥ ♃
07 55	☽ # ☿
08 51	☽ ⊥ ☿
09 14	☽ ∠ ♂
11 18	☽ ∠ ☉
14 03	☽ # ☉
14 18	☽ ∠ ♀
14 43	♃ Ꝑ ♄
14 46	☽ # ♃
17 00	☽ # ♀
17 10	☽ ⚹ ♅
18 19	☽ Q ♆
19 48	☽ # ♂
29 Saturday	
04 39	☉ △ ♆
04 40	☽ ∠ ☿
06 03	☽ ☍ ♃
06 06	☽ ∠ ♄
14 17	☿ Ꝑ ♃
14 52	☽ ⚹ ♂
14 54	☿ □ ♄
16 43	☽ ⚹ ♆
17 42	☽ ⚹ ☉
18 16	♀ ⚹ ♇
21 02	☽ ☌ ♇
21 09	☽ ∠ ♅
21 19	☽ ⚹ ♀
22 07	♀ ∥ ♃
22 48	☽ ⊥ ♄
30 Sunday	
09 48	☽ ⊻ ♄
12 46	☽ ⚹ ☿
13 36	☽ ⊥ ♅
13 59	☽ Q ♂
17 23	☽ Q ☉
18 01	☽ ∥ ♄
20 06	☽ ∠ ♆
21 26	☽ Q ♀
31 Monday	
00 14	☽ ∥ ♆
00 25	☽ ⊻ ♅
04 04	♂ △ ♆
06 36	☽ ∥ ♅
12 10	☽ ⊥ ♆
12 31	☽ ⊼ ♃
13 35	☽ Q ☿
15 33	☉ ⚹ ♇
22 44	☽ ⊻ ♆
23 04	☽ ± ♃
23 40	☽ □ ♂

SEPTEMBER 1987

LONGITUDES

Date	Sidereal time h m s	Sun ☉ ° ' "	Moon ☽ ° ' "	Moon ☽ 24.00 ° ' "	Mercury ☿ ° '	Venus ♀ ° '	Mars ♂ ° '	Jupiter ♃ ° '	Saturn ♄ ° '	Uranus ♅ ° '	Neptune ♆ ° '	Pluto ♇ ° '
01	10 40 29	08 ♍ 31 29	12 ♐ 55 20	19 ♐ 54 40	19 ♍ 45	11 ♍ 03	06 ♍ 09	29 ♈ 28	14 ♐ 40	22 ♐ 43	05 ♑ 18	07 ♏ 43
02	10 44 25	09 29 33	26 ♐ 59 15	04 ♑ 08 55	21 29	12 18	06 48	29 R 25	14 42	22 D 43	05 R 17	07 45
03	10 48 22	10 27 38	11 ♑ 23 19	18 ♑ 42 00	23 12	13 32	07 26	29 22	14 43	22 43	05 17	07 46
04	10 52 18	11 25 45	26 ♑ 04 20	03 ♒ 29 34	24 54	14 46	08 04	29 19	14 45	22 43	05 16	07 48
05	10 56 15	12 23 53	10 ♒ 56 50	18 ♒ 25 08	26 35	16 01	08 42	29 16	14 46	22 43	05 16	07 49
06	11 00 11	13 22 02	25 ♒ 53 27	03 ♓ 20 43	28 14	17 15	09 20	29 13	14 48	22 44	05 16	07 51
07	11 04 08	14 20 13	10 ♓ 45 54	18 ♓ 08 00	29 ♍ 52	18 30	09 59	29 09	14 50	22 44	05 15	07 53
08	11 08 05	15 18 26	25 ♓ 26 09	02 ♈ 39 34	01 ♎ 30	19 44	10 37	29 06	14 52	22 44	05 15	07 54
09	11 12 01	16 16 40	09 ♈ 47 37	16 ♈ 49 50	03 05	20 59	11 15	29 02	14 54	22 45	05 15	07 56
10	11 15 58	17 14 57	23 ♈ 45 52	00 ♉ 35 34	04 40	22 13	11 53	28 58	14 56	22 45	05 15	07 58
11	11 19 54	18 13 15	07 ♉ 18 54	13 ♉ 55 58	06 14	23 28	12 32	28 53	14 58	22 46	05 14	07 59
12	11 23 51	19 11 36	20 ♉ 27 00	26 ♉ 52 17	07 46	24 42	13 10	28 49	15 00	22 46	05 14	08 01
13	11 27 47	20 09 59	03 ♊ 12 15	09 ♊ 27 19	09 18	25 57	13 48	28 44	15 03	22 47	05 14	08 03
14	11 31 44	21 08 24	15 ♊ 38 00	21 ♊ 44 51	10 48	27 12	14 26	28 39	15 05	22 47	05 14	08 05
15	11 35 40	22 06 51	27 ♊ 48 26	03 ♋ 49 17	12 17	28 26	15 05	28 34	15 08	22 48	05 14	08 06
16	11 39 37	23 05 20	09 ♋ 47 59	15 ♋ 45 07	13 45	29 ♍ 41	15 43	28 29	15 10	22 49	05 14	08 08
17	11 43 34	24 03 51	21 ♋ 41 11	27 ♋ 36 45	15 12	00 ♎ 55	16 21	28 24	15 13	22 50	05 D 14	08 10
18	11 47 30	25 02 24	03 ♌ 32 17	09 ♌ 28 15	16 38	02 10	17 00	28 18	15 16	22 50	05 14	08 12
19	11 51 27	26 01 00	15 ♌ 25 06	21 ♌ 23 14	18 02	03 25	17 38	28 13	15 19	22 51	05 14	08 14
20	11 55 23	26 59 38	27 ♌ 22 59	03 ♍ 24 42	19 26	04 39	18 16	28 07	15 22	22 52	05 14	08 16
21	11 59 20	27 58 17	09 ♍ 28 40	15 ♍ 35 06	20 48	05 54	18 55	28 01	15 25	22 53	05 14	08 18
22	12 03 16	28 56 59	21 ♍ 44 15	27 ♍ 56 16	22 08	07 08	19 33	27 55	15 28	22 54	05 14	08 20
23	12 07 13	29 ♍ 55 42	04 ♎ 11 19	10 ♎ 29 30	23 28	08 23	20 11	27 49	15 31	22 55	05 14	08 22
24	12 11 09	00 ♎ 54 28	16 ♎ 50 55	23 ♎ 15 38	24 46	09 38	20 50	27 42	15 34	22 56	05 15	08 24
25	12 15 06	01 53 15	29 ♎ 43 43	06 ♏ 15 12	26 03	10 52	21 28	27 36	15 38	22 58	05 15	08 26
26	12 19 03	02 52 05	12 ♏ 50 07	19 ♏ 28 29	27 18	12 07	22 07	27 29	15 41	22 59	05 15	08 28
27	12 22 59	03 50 56	26 ♏ 10 17	02 ♐ 55 32	28 32	13 22	22 45	27 23	15 45	23 00	05 15	08 30
28	12 26 56	04 49 49	09 ♐ 44 13	16 ♐ 36 17	29 ♎ 44	14 36	23 23	27 16	15 49	23 01	05 16	08 32
29	12 30 52	05 48 44	23 ♐ 31 39	00 ♑ 30 15	00 ♏ 55	15 51	24 02	27 09	15 52	23 03	05 16	08 34
30	12 34 49	06 ♎ 47 41	07 ♑ 31 56	14 ♑ 36 31	02 ♏ 03	17 ♎ 06	24 ♍ 40	27 ♈ 02	15 ♐ 56	23 ♐ 04	05 ♑ 17	08 ♏ 36

Moon

Date	Moon True ☊ ° '	Moon Mean ☊ ° '	Moon ☽ Latitude ° '
01	02 ♈ 35	03 ♈ 36	04 S 56
02	02 R 34	03 33	05 13
03	02 32	03 30	05 11
04	02 30	03 27	04 48
05	02 27	03 23	04 07
06	02 25	03 20	03 08
07	02 23	03 17	01 56
08	02 22	03 14	00 S 38
09	02 D 22	03 11	00 N 41
10	02 23	03 07	01 56
11	02 24	03 04	03 01
12	02 25	03 01	03 56
13	02 26	02 58	04 36
14	02 27	02 55	05 03
15	02 R 27	02 52	05 15
16	02 26	02 48	05 14
17	02 25	02 45	04 59
18	02 24	02 42	04 31
19	02 23	02 39	03 52
20	02 23	02 36	03 02
21	02 22	02 33	02 04
22	02 22	02 29	00 N 59
23	02 D 22	02 26	00 S 10
24	02 22	02 23	01 19
25	02 22	02 20	02 26
26	02 R 22	02 17	03 26
27	02 22	02 13	04 16
28	02 22	02 10	04 53
29	02 21	02 07	05 13
30	02 ♈ 21	02 ♈ 04	05 S 16

DECLINATIONS

Date	Sun ☉ ° '	Moon ☽ ° '	Mercury ☿ ° '	Venus ♀ ° '	Mars ♂ ° '	Jupiter ♃ ° '	Saturn ♄ ° '	Uranus ♅ ° '	Neptune ♆ ° '	Pluto ♇ ° '
01	08 N 22	27 S 15	05 N 03	08 N 44	10 N 17	09 N 53	21 S 11	23 S 24	22 S 19	01 N 03
02	08 01	28 38	04 17	08 15	10 03	09 52	21 12	23 24	22 19	01 02
03	07 39	28 07	03 30	07 47	09 49	09 51	21 12	23 24	22 19	01 01
04	07 17	25 39	02 44	07 18	09 34	09 50	21 12	23 24	22 19	01 00
05	06 55	21 26	01 58	06 49	09 20	09 48	21 13	23 24	22 19	00 59
06	06 32	15 50	01 12	06 20	09 05	09 47	21 13	23 24	22 19	00 58
07	06 10	09 20	00 N 27	05 51	08 51	09 45	21 14	23 24	22 19	00 57
08	05 47	02 S 24	00 S 19	05 21	08 36	09 44	21 14	23 24	22 19	00 56
09	05 25	04 N 30	01 03	04 51	08 22	09 42	21 15	23 24	22 19	00 55
10	05 02	11 01	01 48	04 22	08 07	09 41	21 15	23 24	22 19	00 54
11	04 40	16 49	02 32	03 52	07 52	09 39	21 15	23 24	22 19	00 54
12	04 17	21 39	03 15	03 22	07 37	09 37	21 16	23 24	22 20	00 53
13	03 54	25 19	03 59	02 52	07 22	09 35	21 16	23 24	22 20	00 52
14	03 31	27 41	04 41	02 21	07 07	09 34	21 17	23 24	22 20	00 51
15	03 08	28 41	05 23	01 51	06 52	09 32	21 17	23 24	22 20	00 50
16	02 45	28 18	06 05	01 21	06 37	09 30	21 18	23 24	22 20	00 49
17	02 21	26 37	06 46	00 50	06 22	09 28	21 18	23 24	22 20	00 48
18	01 58	23 07	07 26	00 N 20	06 07	09 25	21 19	23 24	22 20	00 47
19	01 35	19 54	08 06	00 S 11	05 52	09 23	21 19	23 25	22 20	00 46
20	01 12	15 14	08 45	00 42	05 37	09 21	21 20	23 25	22 20	00 45
21	00 48	09 55	09 23	01 12	05 22	09 19	21 20	23 25	22 20	00 44
22	00 25	04 N 11	10 01	01 43	05 07	09 17	21 21	23 25	22 20	00 43
23	00 N 02	01 S 49	10 37	02 13	04 51	09 14	21 21	23 25	22 20	00 42
24	00 S 22	07 51	11 13	02 44	04 36	09 12	21 22	23 25	22 20	00 41
25	00 45	13 39	11 48	03 14	04 21	09 09	21 22	23 25	22 20	00 40
26	01 08	18 58	12 22	03 45	04 06	09 07	21 23	23 25	22 20	00 39
27	01 32	23 27	12 56	04 15	03 50	09 04	21 23	23 25	22 20	00 38
28	01 55	26 44	13 28	04 46	03 35	09 02	21 24	23 25	22 20	00 38
29	02 19	28 30	13 59	05 16	03 19	08 59	21 25	23 25	22 20	00 37
30	02 S 42	28 S 29	14 S 30	05 S 46	03 N 04	08 N 57	21 S 25	23 S 25	22 S 20	00 N 36

ZODIAC SIGN ENTRIES

Date	h	m	Planets
02	17	04	☽ ♑
04	18	22	☽ ♒
06	18	37	☽ ♓
07	13	52	☿ ♎
08	19	34	☽ ♈
10	22	57	☽ ♉
13	05	54	☽ ♊
15	16	22	☽ ♋
16	18	12	♀ ♎
18	04	50	☽ ♌
20	17	13	☽ ♍
23	03	58	☽ ♎
23	13	45	☉ ♎
25	12	30	☽ ♏
27	18	49	☽ ♐
28	17	21	☿ ♏
29	23	08	☽ ♑

LATITUDES

Date	Mercury ☿ ° '	Venus ♀ ° '	Mars ♂ ° '	Jupiter ♃ ° '	Saturn ♄ ° '	Uranus ♅ ° '	Neptune ♆ ° '	Pluto ♇ ° '
01	01 N 05	01 N 25	01 N 07	01 S 30	01 N 23	00 S 10	01 N 01	15 N 58
04	00 46	01 25	01 06	01 30	01 22	00 10	01 01	15 56
07	00 26	01 24	01 06	01 31	01 22	00 10	01 01	15 55
10	00 N 04	01 23	01 06	01 31	01 21	00 10	01 01	15 54
13	00 S 19	01 22	01 05	01 32	01 20	00 10	01 01	15 52
16	00 42	01 20	01 05	01 33	01 20	00 10	01 01	15 51
19	01 06	01 17	01 04	01 33	01 19	00 10	01 01	15 50
22	01 29	01 13	01 04	01 34	01 19	00 10	01 00	15 49
25	01 52	01 09	01 03	01 34	01 18	00 10	01 00	15 48
28	02 14	01 05	01 03	01 34	01 18	00 10	01 00	15 47
31	02 S 34	01 N 00	01 N 02	01 S 35	01 N 17	00 S 10	01 N 00	15 N 46

DATA

Julian Date	2447040
Delta T	+55 seconds
Ayanamsa	23° 41' 05"
Synetic vernal point	05° ♓ 25' 55"
True obliquity of ecliptic	23° 26' 37"

MOON'S PHASES, APSIDES AND POSITIONS ☽

Date	h	m	Phase	Longitude ° '	Eclipse Indicator
01	03	48	☽	08 ♐ 12	
07	18	13	○	14 ♓ 35	
14	23	44	☾	21 ♊ 37	
23	03	08	●	29 ♍ 34	Annular
30	10	39	☽	06 ♑ 44	

Day	h	m		
06	02	48	Perigee	
18	03	11	Apogee	
02	17	46	Max dec	28° S 41'
08	20	15	0N	
15	17	12	Max dec	28° N 43'
23	04	46	0S	
29	23	48	Max dec	28° S 43'

LONGITUDES

Date	Chiron ⚷ ° '	Ceres ⚳ ° '	Pallas ⚴ ° '	Juno ⚵ ° '	Vesta ⚶ ° '	Black Moon Lilith ⚸ ° '
01	28 ♊ 00	23 ♐ 07	22 ♏ 35	27 ♒ 56	16 ♋ 19	01 ♌ 35
11	28 ♊ 23	24 ♐ 34	25 ♏ 38	25 ♒ 40	19 ♋ 59	02 ♌ 42
21	28 ♊ 40	26 ♐ 27	28 ♏ 55	23 ♒ 52	23 ♋ 29	03 ♌ 49
31	28 ♊ 48	28 ♐ 42	02 ♐ 57	22 ♒ 44	26 ♋ 46	04 ♌ 56

ASPECTARIAN

h m	Aspects
01 Tuesday	
02 57	☽ ⊻ ♇
03 48	☽ □ ☉
08 27	☽ □ ♀
08 32	♅ ∠ ♇
13 23	☽ ⊥ ♇
14 23	♅ St D
14 39	☽ ⚼ ♃
15 02	☽ ☌ ♄
02 Wednesday	
01 22	☽ □ ☿
04 47	☽ ☌ ♅
04 49	☽ ∠ ♇
16 05	☽ △ ♃
03 Thursday	
01 54	☽ ☌ ♆
05 09	☽ △ ♂
05 12	☿ □ ♅
05 51	☿ ∠ ♇
06 01	☽ ⚹ ♇
08 10	♂ ∥ ♃
10 22	☽ △ ☉
14 21	☿ ± ♃
15 52	☽ △ ♀
17 29	☽ ⊻ ♄
04 Friday	
01 25	♂ ⚹ ♇
01 46	☽ Q ♇
03 19	☽ ⊥ ♄
03 39	♀ ⚼ ♃
06 33	☽ ⊻ ♅
06 54	☽ ⚼ ♂
09 51	☽ △ ☿
11 25	♀ □ ♄
12 37	☽ ⚼ ☉
16 18	☽ ⊥ ♅
16 54	☉ ∥ ♀
17 15	☽ □ ♃
17 57	☽ ∠ ♄
18 33	☽ ⚼ ♀
22 08	☽ ± ♂
05 Saturday	
01 56	☽ ∥ ♅
02 52	☽ ⊻ ♆
04 10	☽ ± ☉
06 49	☽ ∠ ♅
06 58	☽ □ ♇
07 39	☽ ∥ ♆
08 14	☽ ⊼ ♂
10 22	☽ ± ♀
12 31	☽ ⊥ ♆
13 04	☽ ∥ ♄
13 09	☽ ⚼ ☿
14 30	☽ ⊼ ☉
18 09	☽ ⚹ ♄
20 53	☽ ⊼ ♀
22 07	☽ Q ♃
06 Sunday	
02 58	☽ ∠ ♆
05 24	☽ ± ☿
06 55	☽ ⚹ ♅
13 28	☽ Q ♄
16 14	☽ ⊼ ☿
17 20	☽ ⚹ ♃
19 27	☿ ∥ ♇
07 Monday	
01 49	☿ ⊼ ♃
02 14	☽ Q ♅
03 05	☽ ⚹ ♆
07 19	☽ △ ♇
07 47	☉ ⚼ ♃
10 28	☽ # ♃
10 40	☽ ☍ ♂
13 45	☽ # ♂
17 29	☽ ∠ ♃
18 13	☽ ☍ ☉
18 37	☽ □ ♄
22 34	☽ Q ♆
23 40	☽ # ☉
08 Tuesday	
00 36	☉ □ ♄
01 04	☽ # ♀
01 46	☽ ☍ ♀
07 33	☽ □ ♅
07 49	☽ ⚼ ♇
08 09	☽ ⊥ ♃
17 02	☽ # ♇
18 02	☽ ⊻ ♃
18 16	☿ ⊥ ♇
18 29	☽ ∥ ☿
22 27	☽ # ☿
22 46	☽ ± ♇
23 19	☽ ☍ ☿
23 29	☽ ∥ ♇
09 Wednesday	
04 20	☽ □ ♆
07 48	☿ # ♇
08 51	☽ ⊼ ♇
08 59	☿ Q ♄
13 10	☽ ∥ ♀
14 35	☽ ⊼ ♂
15 04	☽ ∥ ☉
20 42	☽ △ ♄
23 53	☽ ⊼ ☉
10 Thursday	
01 20	☽ ± ♂
01 26	☽ ∥ ♂
06 54	☽ ∥ ♃
09 03	☽ ⊼ ♀
10 14	☽ △ ♅
11 02	☽ ± ☉
17 44	☽ ⚼ ♂
20 36	☽ ± ♀
20 45	☿ □ ♆
21 04	☽ ☌ ♃
22 16	♀ □ ♅
22 51	☽ ⚼ ♄
11 Friday	
01 28	♀ ± ♃
02 35	♀ ∠ ♇
04 06	☽ ⚼ ☉
08 17	☽ △ ♆
09 48	☽ ⊼ ☿
12 48	☽ ⚼ ♅
13 13	☽ ☍ ♇
14 17	☽ ⚼ ♀
14 59	☽ ± ♄
21 55	☽ △ ♂
22 05	☽ ± ☿
12 Saturday	
01 56	☽ ⊼ ♄
05 12	☽ ± ♅
09 29	☽ △ ☉
09 53	☽ # ♄
11 36	☽ ⚼ ♆
14 03	☿ # ♀
15 54	☽ # ♆
15 55	☿ ⊻ ♇
16 19	☽ ⊼ ♅
16 54	☽ ⚼ ☿
20 47	☽ △ ♀
22 30	☽ # ♅
13 Sunday	
03 34	☽ ⊻ ♃
04 28	☽ ± ♆
09 48	♂ ⚼ ♃
10 16	☉ # ☿
14 55	☽ ⊥ ♃
15 53	☽ ⊼ ♆
21 18	☽ ⊼ ♇
14 Monday	
01 19	☽ △ ☿
08 10	☽ ∠ ♃
08 58	☽ ± ♇
09 33	☽ □ ♂
10 55	☽ ☍ ♄
11 49	☿ Q ♅
23 44	☽ □ ☉
15 Tuesday	
02 04	☽ ☍ ♅
02 40	☽ ⚼ ♇
13 24	☽ □ ♀
13 31	☽ ⚹ ♃
13 54	♂ □ ♄
14 29	♀ ⊼ ♃
22 24	☉ ± ♃
23 06	☽ Q ♂
16 Wednesday	
02 49	☽ ☍ ♆
05 06	☉ □ ♅
08 39	☽ △ ♇
13 16	☉ ∠ ♇
13 22	☽ Q ♃
14 50	☽ Q ☉
21 04	☽ □ ☿
22 52	☽ ⊼ ♄
17 Thursday	
00 36	☽ ⚹ ♂
02 03	☿ # ♂
05 45	☽ Q ♀
08 23	♆ St D
11 03	☽ ± ♄
12 16	☿ ⚹ ♄
13 44	♀ ∥ ♇
14 18	☽ ⊼ ♅
17 15	☽ ⚹ ☉
18 Friday	
01 29	☽ □ ♃
02 28	☽ ± ♅
05 20	☽ ⚼ ♄
08 42	☽ ∠ ♂
08 54	☽ ⚹ ♀
12 43	♀ ⊥ ♇
14 27	☽ # ♅
14 30	☽ Q ☿
15 26	☽ ⊼ ♆
20 43	☽ ⚼ ♅
21 28	☽ □ ♇
21 35	☽ # ♆
23 20	☿ ⊻ ♂
19 Saturday	
02 20	☽ ∠ ☉
03 33	☽ ± ♆
03 50	☽ # ♄
03 56	☽ ⊥ ♂
10 00	♀ Q ♄
11 47	☽ △ ♄
16 43	☽ ⊻ ♂
17 58	☽ ⚹ ☿
18 43	☽ ∠ ♀
21 41	☽ ⚼ ♆
22 04	☽ ⊥ ☉
20 Sunday	
02 58	☽ △ ♅
09 46	☽ Q ♇
11 09	☽ ⊻ ☉
13 27	☽ △ ♃
14 38	♀ # ♇
14 50	☽ ⊥ ♀
23 12	♀ □ ♆
21 Monday	
01 25	☉ # ♀
03 37	☽ △ ♆
03 48	☽ ∠ ☿
04 07	☽ ⊻ ♀
09 40	☽ ⚹ ♇
09 41	☿ # ♃
13 03	☉ ⊼ ♃
14 06	☽ # ☿
14 38	☽ ∥ ♃
16 36	☉ ∥ ♇
18 55	☽ ⚼ ♃
23 43	☽ □ ♄
23 45	☽ ⊥ ☿
22 Tuesday	
07 31	☽ ☌ ♂
08 00	☽ ∥ ♂
12 21	☽ ± ♃
12 53	☽ ⊻ ☿
14 16	☽ □ ♅
15 06	☽ ∠ ♇
21 09	☽ # ♀
23 52	☽ ⊼ ♃
23 Wednesday	
01 56	☿ ⚹ ♅
01 56	☽ ∥ ♇
03 08	☽ ☌ ☉
04 09	☽ ∥ ☉
05 19	☽ # ☉
07 35	☽ # ♇
07 51	☿ Q ♆
08 30	☽ ⊥ ♇
10 43	☽ Q ♄
11 35	♀ ⊻ ♇
13 45	☽ ∥ ♀
14 01	☽ □ ♆
19 59	☽ ⊻ ♇
20 53	☽ ☌ ♀
23 34	☽ # ♂
24 Thursday	
00 50	☽ Q ♅
09 35	☽ ⚹ ♄
17 26	☽ # ♃
19 51	☽ ⊻ ♂
23 25	☽ ⚹ ♅
23 58	☽ Q ♆
25 Friday	
03 18	☽ ∥ ☿
04 27	☽ ☌ ☿
07 22	☉ # ♇
07 36	☽ ⊥ ♂
08 06	☽ ☍ ♃
13 40	☽ ∠ ♄
13 42	♀ Q ♅
16 08	♂ ± ♃
16 18	☽ ⊻ ☉
22 10	☽ ⚹ ♆
26 Saturday	
01 02	☽ ∠ ♂
01 48	☉ ⊥ ♇
03 09	☽ ∠ ♅
04 02	☽ ☌ ♇
04 12	☽ ⊥ ☉
06 15	☽ ⊥ ♄
10 34	☽ ⊻ ♀
15 15	☿ ☍ ♃
17 12	☽ ⊻ ♄
19 31	☽ ⊥ ♅
21 50	☽ ∠ ☉
22 32	☽ ⊥ ♀
22 53	♀ # ♂
27 Sunday	
00 18	☽ ∥ ♄
01 24	☽ ∠ ♆
05 31	☽ ∥ ♆
05 34	☽ ⚹ ♂
06 19	☽ ⊻ ♅
09 23	☉ Q ♄
11 50	☽ ∥ ♅
14 08	☽ ⊼ ♃
16 18	☽ ∠ ♀
16 38	☽ ⊻ ☿
17 30	☽ ⊥ ♆
21 04	☿ ⊥ ♂
21 49	♂ □ ♅
28 Monday	
00 41	☽ ± ♃
02 42	☽ ⚹ ☉
03 59	☽ Q ♂
04 08	☽ ⊻ ♆
04 17	☽ ⊥ ☿
09 53	☽ ⊻ ♇
16 23	☽ ⚼ ♃
17 47	♂ ∠ ♇
20 25	☽ ⊥ ♇
21 22	☽ ⚹ ♀
21 34	☽ ∠ ☿
22 40	☽ ☌ ♄
22 40	☉ □ ♆
29 Tuesday	
01 21	☽ Q ☉
11 10	☽ ☌ ♅
11 11	☿ ∠ ♄
12 05	☽ ∠ ♇
12 25	♀ ⚹ ♄
12 55	☽ □ ♂
18 11	☽ △ ♃
20 11	☽ Q ♀
30 Wednesday	
01 50	☽ ⚹ ☿
08 09	☽ ☌ ♆
10 39	☽ □ ☉
13 50	☽ ⚹ ♇

All ephemeris data is given at 12.00 UT and the Moon's longitude is additionally given for 24.00 UT

OCTOBER 1987

LONGITUDES

Date	Sidereal time h m s	Sun ☉ ° ' "	Moon ☽ ° ' "	Moon ☽ 24.00 ° ' "	Mercury ☿ ° '	Venus ♀ ° '	Mars ♂ ° '	Jupiter ♃ ° '	Saturn ♄ ° '	Uranus ♅ ° '	Neptune ♆ ° '	Pluto ♇ ° '
01	12 38 45	07 ♎ 46 39	21 ♑ 43 46	28 ♑ 53 24	03 ♏ 10	18 ♎ 20	25 ♍ 19	26 ♈ 54	16 ♐ 00	23 ♐ 06	05 ♑ 17	08 ♏ 39
02	12 42 42	08 45 39	06 ♒ 05 03	13 ♒ 18 18	04 14	19 35	25 57	26 R 47	16 04	23 07	05 18	08 41
03	12 46 38	09 44 41	20 ♒ 32 40	27 ♒ 47 36	05 16	20 50	26 36	26 40	16 08	23 09	05 18	08 43
04	12 50 35	10 43 44	05 ♓ 02 30	12 ♓ 16 43	06 16	22 04	27 14	26 32	16 12	23 11	05 19	08 45
05	12 54 31	11 42 49	19 ♓ 29 37	26 ♓ 40 32	07 13	23 19	27 53	26 25	16 16	23 12	05 19	08 47
06	12 58 28	12 41 56	03 ♈ 48 49	10 ♈ 53 50	08 07	24 34	28 31	26 17	16 21	23 14	05 20	08 49
07	13 02 25	13 41 05	17 ♈ 55 03	24 ♈ 51 57	08 57	25 49	29 10	26 09	16 25	23 16	05 21	08 52
08	13 06 21	14 40 17	01 ♉ 44 08	08 ♉ 31 19	09 45	27 03	29 ♍ 48	26 02	16 30	23 17	05 21	08 54
09	13 10 18	15 39 30	15 ♉ 13 15	21 ♉ 49 52	10 28	28 18	00 ♎ 27	25 54	16 34	23 19	05 22	08 56
10	13 14 14	16 38 46	28 ♉ 21 09	04 ♊ 47 13	11 08	29 ♎ 33	01 05	25 46	16 39	23 21	05 23	08 59
11	13 18 11	17 38 04	11 ♊ 08 15	17 ♊ 24 33	11 43	00 ♏ 47	01 44	25 38	16 43	23 23	05 23	09 01
12	13 22 07	18 37 24	23 ♊ 36 28	29 ♊ 44 26	12 12	02 02	02 22	25 30	16 48	23 25	05 24	09 03
13	13 26 04	19 36 46	05 ♋ 48 55	11 ♋ 50 28	12 37	03 17	03 01	25 22	16 53	23 27	05 25	09 05
14	13 30 00	20 36 11	17 ♋ 49 38	23 ♋ 47 00	12 55	04 31	03 39	25 14	16 58	23 29	05 26	09 08
15	13 33 57	21 35 38	29 ♋ 43 10	05 ♌ 38 44	13 07	05 46	04 18	25 06	17 03	23 31	05 27	09 10
16	13 37 54	22 35 08	11 ♌ 34 19	17 ♌ 30 31	13 13	07 01	04 57	24 58	17 08	23 34	05 28	09 12
17	13 41 50	23 34 39	23 ♌ 27 54	29 ♌ 27 01	13 R 10	08 16	05 35	24 49	17 13	23 36	05 29	09 15
18	13 45 47	24 34 13	05 ♍ 28 23	11 ♍ 32 30	13 00	09 30	06 14	24 41	17 18	23 38	05 30	09 17
19	13 49 43	25 33 49	17 ♍ 39 47	23 ♍ 50 36	12 41	10 45	06 53	24 33	17 23	23 40	05 31	09 20
20	13 53 40	26 33 27	00 ♎ 05 18	06 ♎ 24 06	12 14	12 00	07 31	24 25	17 28	23 43	05 32	09 22
21	13 57 36	27 33 07	12 ♎ 47 11	19 ♎ 14 39	11 38	13 14	08 10	24 17	17 34	23 45	05 33	09 24
22	14 01 33	28 32 50	25 ♎ 46 32	02 ♏ 22 45	10 53	14 29	08 49	24 09	17 39	23 47	05 34	09 27
23	14 05 29	29 ♎ 32 34	09 ♏ 03 10	15 ♏ 47 36	10 00	15 44	09 27	24 01	17 44	23 50	05 35	09 29
24	14 09 26	00 ♏ 32 21	22 ♏ 35 45	29 ♏ 27 18	09 00	16 59	10 06	23 53	17 50	23 52	05 36	09 31
25	14 13 23	01 32 09	06 ♐ 21 54	13 ♐ 19 08	07 52	18 13	10 45	23 45	17 56	23 55	05 38	09 34
26	14 17 19	02 31 59	20 ♐ 18 37	27 ♐ 19 57	06 40	19 28	11 24	23 37	18 01	23 57	05 39	09 36
27	14 21 16	03 31 51	04 ♑ 22 43	11 ♑ 26 34	05 25	20 43	12 02	23 29	18 07	24 00	05 40	09 39
28	14 25 12	04 31 45	18 ♑ 31 10	25 ♑ 36 11	04 08	21 57	12 41	23 21	18 13	24 03	05 41	09 41
29	14 29 09	05 31 40	02 ♒ 41 22	09 ♒ 46 28	02 53	23 12	13 20	23 13	18 18	24 05	05 43	09 44
30	14 33 05	06 31 37	16 ♒ 51 15	23 ♒ 55 31	01 41	24 27	13 59	23 06	18 24	24 08	05 44	09 46
31	14 37 02	07 ♏ 31 35	00 ♓ 59 05	08 ♓ 01 45	00 ♏ 35	25 ♏ 42	14 ♎ 38	22 ♈ 58	18 ♐ 30	24 ♐ 11	05 ♑ 45	09 ♏ 48

Date	Moon True ☊ ° '	Moon Mean ☊ ° '	Moon ☽ Latitude ° '
01	02 ♈ 21	02 ♈ 01	04 S 59
02	02 D 22	01 58	04 24
03	02 22	01 55	03 32
04	02 23	01 51	02 26
05	02 24	01 48	01 S 11
06	02 R 24	01 45	00 N 08
07	02 24	01 42	01 25
08	02 23	01 39	02 35
09	02 21	01 35	03 35
10	02 19	01 32	04 22
11	02 17	01 29	04 54
12	02 15	01 26	05 12
13	02 14	01 23	05 15
14	02 13	01 19	05 04
15	02 D 13	01 16	04 40
16	02 14	01 13	04 04
17	02 16	01 10	03 18
18	02 17	01 07	02 22
19	02 19	01 04	01 20
20	02 20	01 00	00 N 12
21	02 R 19	00 57	00 S 57
22	02 18	00 54	02 05
23	02 14	00 51	03 08
24	02 10	00 48	04 01
25	02 06	00 45	04 41
26	02 01	00 41	05 06
27	01 58	00 38	05 12
28	01 55	00 35	04 59
29	01 54	00 32	04 28
30	01 D 54	00 29	03 41
31	01 ♈ 56	00 ♈ 25	02 S 41

DECLINATIONS

Date	Sun ☉	Moon ☽	Mercury ☿	Venus ♀	Mars ♂	Jupiter ♃	Saturn ♄	Uranus ♅	Neptune ♆	Pluto ♇
01	03 S 05	26 S 37	14 S 59	06 S 16	02 N 49	08 N 54	21 S 26	23 S 25	22 S 20	00 N 35
02	03 28	23 01	15 27	06 46	02 33	08 51	21 26	23 25	22 20	00 34
03	03 52	17 59	15 53	07 15	02 18	08 49	21 27	23 25	22 20	00 33
04	04 15	11 56	16 19	07 45	02 02	08 46	21 28	23 26	22 20	00 32
05	04 38	05 S 15	16 43	08 14	01 47	08 43	21 28	23 26	22 20	00 31
06	05 01	01 N 38	17 05	08 44	01 31	08 40	21 29	23 26	22 20	00 30
07	05 24	08 20	17 26	09 13	01 16	08 37	21 29	23 26	22 20	00 29
08	05 47	14 30	17 45	09 41	01 00	08 35	21 30	23 26	22 20	00 28
09	06 10	19 50	18 02	10 10	00 45	08 32	21 31	23 26	22 20	00 27
10	06 33	24 03	18 17	10 38	00 29	08 29	21 31	23 26	22 20	00 26
11	06 55	26 58	18 30	11 07	00 N 13	08 26	21 32	23 26	22 20	00 26
12	07 18	28 29	18 41	11 34	00 S 02	08 23	21 32	23 26	22 20	00 25
13	07 40	28 34	18 49	12 02	00 18	08 20	21 33	23 26	22 20	00 24
14	08 03	27 17	18 54	12 29	00 33	08 17	21 34	23 27	22 20	00 23
15	08 25	24 46	18 56	12 56	00 49	08 14	21 34	23 27	22 20	00 22
16	08 47	21 14	18 55	13 23	01 04	08 11	21 35	23 27	22 20	00 21
17	09 09	16 49	18 51	13 50	01 20	08 08	21 36	23 27	22 20	00 20
18	09 31	11 43	18 43	14 16	01 35	08 06	21 36	23 27	22 20	00 19
19	09 53	06 06	18 31	14 41	01 51	08 03	21 37	23 27	22 20	00 19
20	10 15	00 N 09	18 14	15 07	02 06	08 00	21 38	23 27	22 20	00 18
21	10 36	05 S 56	17 53	15 32	02 21	07 57	21 38	23 27	22 20	00 17
22	10 58	11 54	17 28	15 56	02 37	07 54	21 39	23 27	22 20	00 16
23	11 19	17 29	16 58	16 20	02 53	07 51	21 40	23 28	22 20	00 15
24	11 40	22 18	16 24	16 44	03 08	07 48	21 40	23 28	22 20	00 14
25	12 01	26 00	15 46	17 08	03 24	07 45	21 41	23 28	22 20	00 14
26	12 21	28 10	15 05	17 30	03 39	07 42	21 41	23 28	22 20	00 13
27	12 42	28 34	14 21	17 53	03 55	07 40	21 42	23 28	22 20	00 12
28	13 02	27 06	13 36	18 15	04 10	07 37	21 43	23 28	22 20	00 11
29	13 22	23 55	12 51	18 36	04 25	07 34	21 43	23 28	22 20	00 10
30	13 42	19 18	12 07	18 57	04 41	07 31	21 44	23 28	22 20	00 10
31	14 S 01	13 S 38	11 S 25	19 S 18	04 S 56	07 N 29	21 S 45	23 S 29	22 S 20	00 N 09

ZODIAC SIGN ENTRIES

Date	h	m	Planets
02	01	51	☽ ♒
04	03	39	☽ ♓
06	05	35	☽ ♈
08	08	57	☽ ♉
08	19	27	♂ ♎
10	15	03	☽ ♊
10	20	49	♀ ♏
13	00	31	☽ ♋
15	12	34	☽ ♌
18	01	06	☽ ♍
20	11	50	☽ ♎
22	19	41	☽ ♏
23	23	01	☉ ♏
25	00	57	☽ ♐
27	04	33	☽ ♑
29	07	27	☽ ♒
31	10	19	☽ ♓

LATITUDES

Date	Mercury ☿	Venus ♀	Mars ♂	Jupiter ♃	Saturn ♄	Uranus ♅	Neptune ♆	Pluto ♇
01	02 S 34	01 N 00	01 N 02	01 S 35	01 N 17	00 S 10	01 N 00	15 N 46
04	02 52	00 55	01 01	01 35	01 17	00 10	01 00	15 45
07	03 07	00 49	01 01	01 35	01 16	00 10	01 00	15 44
10	03 17	00 43	01 00	01 35	01 16	00 10	01 00	15 43
13	03 21	00 37	00 59	01 35	01 15	00 10	01 00	15 43
16	03 16	00 30	00 58	01 35	01 15	00 10	00 59	15 42
19	03 00	00 23	00 58	01 35	01 14	00 10	00 59	15 42
22	02 30	00 16	00 57	01 35	01 14	00 10	00 59	15 41
25	01 43	00 08	00 56	01 34	01 13	00 10	00 59	15 41
28	00 S 45	00 N 01	00 55	01 34	01 13	00 10	00 59	15 40
31	00 N 17	00 S 07	00 N 54	01 S 34	01 N 12	00 S 10	00 N 59	15 N 40

DATA

Julian Date	2447070
Delta T	+55 seconds
Ayanamsa	23° 41' 08"
Synetic vernal point	05° ♓ 25' 51"
True obliquity of ecliptic	23° 26' 37"

MOON'S PHASES, APSIDES AND POSITIONS ☽

Date	h	m	Phase	Longitude ° '	Eclipse Indicator
07	04	12	○	13 ♈ 22	
14	18	06	☾	20 ♋ 51	
22	17	28	●	28 ♎ 46	
29	17	10	☽	05 ♒ 45	

Day	h	m		
04	00	41	Perigee	
15	21	24	Apogee	
30	02	38	Perigee	
06	06	17	ON	
13	01	16	Max dec	28° N 42'
20	12	36	OS	
27	05	06	Max dec	28° S 38'

LONGITUDES

Date	Chiron ⚷ ° '	Ceres ⚳ ° '	Pallas ⚴ ° '	Juno ⚵ ° '	Vesta ⚶ ° '	Black Moon Lilith ⚸ ° '
01	28 ♊ 48	28 ♐ 42	02 ♐ 23	22 ♒ 44	26 ♋ 46	04 ♌ 56
11	28 ♊ 49	01 ♑ 17	06 ♐ 00	22 ♒ 33	29 ♋ 40	06 ♌ 03
21	28 ♊ 42	04 ♑ 08	09 ♐ 44	22 ♒ 49	02 ♌ 31	07 ♌ 10
31	28 ♊ 28	07 ♑ 13	13 ♐ 34	24 ♒ 00	04 ♌ 52	08 ♌ 17

All ephemeris data is given at 12.00 UT and the Moon's longitude is additionally given for 24.00 UT

ASPECTARIAN

h m Aspects

01 Thursday
00 00 ☽ Q ☿
01 42 ☉ # ♂
02 18 ☽ ⊻ ♄
05 45 ☽ □ ♀
10 10 ☽ Q ♇
12 28 ☽ ⊥ ♄
14 18 ☽ ⊻ ♅
18 18 ☽ △ ♂
20 37 ☽ □ ♃

02 Friday
00 22 ☽ ⊥ ♅
03 36 ☽ ∠ ♄
08 40 ☽ □ ☿
09 42 ☽ ∥ ♅
09 55 ☉ ⊻ ♇
10 41 ☽ ⊻ ♆
15 24 ☽ ∠ ♅
15 37 ☽ ∥ ♆
16 20 ☽ □ ♇
16 47 ☽ △ ☉
20 10 ☽ ∥ ♄
20 28 ☽ ꝗ ♂
20 40 ☽ ⊥ ♆

03 Saturday
02 20 ☽ Q ♃
04 40 ☽ ⚹ ♄
11 36 ☽ ∠ ♆
12 05 ☽ ± ♂
12 31 ☽ △ ♀
12 48 ☿ ⚹ ♆
14 09 ♂ ⊼ ♃
16 19 ☽ ⚹ ♅
19 28 ☽ ꝗ ☉
20 06 ☽ ∥ ☿
22 02 ☽ ⚹ ♃
22 29 ☽ ⊼ ♂

04 Sunday
00 38 ☽ Q ♄
11 27 ☽ ± ☉
12 13 ☽ Q ♅
12 27 ☽ ⚹ ♆
14 10 ☽ △ ☿
15 41 ☽ ꝗ ♀
18 10 ☽ △ ♇
22 07 ☽ ⊼ ☉
22 41 ☽ ∠ ♃
23 12 ☉ Q ♅
23 37 ☽ # ♃

05 Monday
02 08 ☽ ∥ ♀
06 37 ☽ □ ♄
08 02 ☽ ± ♀
08 23 ☽ Q ♆
09 42 ♀ ⚹ ♅
12 02 ♀ Q ♆
13 31 ☽ ⊥ ♃
14 02 ☽ ∥ ☉
16 51 ☽ ꝗ ☿
18 12 ☽ □ ♅
18 59 ☽ ⊼ ♀
19 11 ☽ ꝗ ♇
23 27 ☽ ⊻ ♃

06 Tuesday
00 34 ☽ # ♂
02 40 ☽ ☍ ♂
04 31 ☽ # ♇
08 03 ☽ ∥ ♇
08 56 ☽ ± ☿
09 31 ♀ # ♃
10 20 ☽ ± ♇
11 36 ☽ ∥ ♂
14 34 ☽ □ ♆
15 27 ☿ ∠ ♅
19 44 ☽ ⊼ ☿
20 30 ☽ ⊼ ♇

07 Wednesday
00 40 ☽ # ☉
04 12 ☽ ☍ ☉
09 05 ☿ ☌ ♇
09 25 ☽ △ ♄
13 03 ☽ ∥ ♃
15 30 ☽ # ♀
18 03 ♀ ☍ ♃
21 14 ☽ △ ♅

08 Thursday
02 07 ☽ ☌ ♃
03 00 ☽ ☍ ♀
08 26 ☽ ⊼ ♂
11 34 ☽ ꝗ ♄
18 23 ☽ △ ♆
19 32 ☽ ± ♂
23 37 ☽ ꝗ ♅

09 Friday
00 42 ☽ ☍ ♇
02 54 ☽ # ☿
03 00 ☽ ☍ ☿
03 36 ☽ ± ♄
12 25 ☽ ꝗ ♂
12 51 ☽ ⊼ ☉
14 27 ☽ ⊼ ♄
15 45 ☿ ⊥ ♄
15 48 ☽ ± ♅
20 46 ☽ # ♄
21 20 ☽ ꝗ ♆

10 Saturday
00 38 ☽ ± ☉
01 24 ☽ # ♆
02 46 ☽ ⊼ ♅
07 16 ☽ ⊻ ♃
08 00 ☽ # ♅
11 58 ☉ ⚹ ♄
13 54 ☽ ± ♆
14 27 ☽ ⊼ ♀
16 10 ♂ ∥ ♇
17 21 ☽ △ ♂
18 17 ☽ ⊥ ♃
18 38 ☽ ꝗ ☉

11 Sunday
01 07 ☽ ⊼ ♆
02 52 ☽ ± ♀
07 57 ☽ ⊼ ♇
11 03 ☽ ∠ ♃
13 08 ☽ ⊼ ☿
19 25 ☽ ± ♇
21 52 ☽ ꝗ ♀
22 45 ☽ ☍ ♄

12 Monday
01 07 ☽ ± ☿
01 30 ☽ △ ☉
07 14 ♀ ∠ ♄
11 38 ☽ ☍ ♅
12 52 ☽ ꝗ ♇
15 39 ☽ ⚹ ♃
19 17 ☽ ꝗ ☿

13 Tuesday
01 28 ♀ ⊻ ♂
06 09 ☽ □ ♂
06 24 ☽ △ ♀
11 13 ☽ ☍ ♆
15 02 ☽ Q ♃
15 03 ♂ ⊥ ♇
18 32 ☽ △ ♇
21 04 ♂ # ♇

14 Wednesday
01 56 ☽ △ ☿
10 15 ☽ ⊼ ♄
18 06 ☽ □ ☉
20 09 ☽ Q ♂
22 25 ☽ ± ♄
23 26 ☽ ⊼ ♅

15 Thursday
01 41 ☉ # ♃
02 45 ☽ □ ♃
05 45 ♀ ⚹ ♆
11 36 ☽ ± ♅
16 44 ☽ ꝗ ♄
21 46 ☽ # ♅
21 49 ☽ ⚹ ♂
23 37 ☽ ⊼ ♆

16 Friday
01 41 ☽ □ ♀
05 07 ☽ # ♆
05 53 ☽ ꝗ ♅
07 12 ☽ □ ♇
09 49 ☽ Q ☉
09 50 ☽ # ♄
11 47 ☽ ± ♆
15 19 ☽ □ ☿
16 45 ☿ St R
19 52 ♂ Q ♄
23 19 ☽ △ ♄

17 Saturday
01 13 ☽ # ☿
05 53 ☽ ∠ ♂
05 59 ☽ ꝗ ♆
07 50 ♂ □ ♆
09 35 ☉ Q ♆
12 15 ☽ ⚹ ☉
12 16 ☽ △ ♅
12 28 ☉ ⚹ ♅
14 42 ☽ △ ♃
18 16 ☽ Q ♀
18 43 ♀ ∠ ♅
19 37 ☽ Q ♇

18 Sunday
00 58 ☽ ⊥ ♂
01 17 ☽ # ♀
03 15 ☽ Q ☿
07 40 ♀ ☌ ♇
12 03 ☽ △ ♆
13 36 ☽ ⊻ ♂
14 32 ☉ ☍ ♃
19 35 ☽ ⚹ ♇
20 15 ☽ ꝗ ♃
20 50 ☽ ∠ ☉
20 54 ☽ ⚹ ♀
21 00 ☽ # ☉

19 Monday
02 32 ☽ ⚹ ☿
03 49 ☽ ∥ ♃
07 35 ☿ ⊥ ♂
11 27 ☽ □ ♄
13 43 ☽ ± ♃
16 02 ☽ ⊥ ☉
23 42 ☽ □ ♅

20 Tuesday
00 58 ☽ ∠ ♇
01 11 ♀ ⊥ ♄
01 14 ☽ ⊼ ♃
04 33 ☽ # ♂
04 39 ☽ ⊻ ☉
05 25 ☽ ∠ ♀
06 45 ☽ ∠ ☿
11 26 ☽ ∥ ♇
13 47 ☽ # ♇
15 16 ☿ ☌ ♀
18 16 ☽ ⊥ ♇
21 20 ☽ ∥ ♂
22 19 ☽ Q ♄
22 22 ☽ □ ♆
23 12 ☽ ⊥ ☿

21 Wednesday
00 27 ☽ ⊥ ♀
02 52 ☽ ☌ ♂
05 38 ☽ ⊻ ♇
09 57 ☽ ⊻ ☿
10 03 ☽ Q ♅
12 56 ☽ ⊻ ♀
14 21 ☿ ⊥ ♄
19 56 ☽ # ♃
20 57 ☽ ⚹ ♄

22 Thursday
07 53 ☽ ∥ ☉
07 57 ☽ Q ♆
08 21 ☽ ⚹ ♅
09 03 ☽ ☍ ♃
17 28 ☽ ☌ ☉

23 Friday
00 35 ☽ ∠ ♄
01 05 ♀ ⊥ ♂
05 46 ☽ ⚹ ♆
06 28 ☽ ∥ ♀
09 53 ☽ ∥ ☿
11 36 ☽ ∠ ♅
12 46 ☽ ⊻ ♂
12 46 ☽ ☌ ♇
13 05 ♂ ⊻ ♇
13 36 ☽ ☌ ☿
16 50 ☽ ⊥ ♄
20 09 ☿ ⊻ ♂
23 59 ☽ ⊥ ♂

24 Saturday
00 16 ☿ ☌ ♇
01 05 ☽ ☌ ♀
03 33 ☽ ⊻ ♄
03 39 ☽ ⊥ ♅
03 45 ☿ ∥ ♀
08 30 ☽ ∠ ♆
08 31 ☽ ∥ ♄
12 11 ☽ ∥ ♆
13 04 ♃ △ ♅
14 14 ☽ ⊼ ♃
14 15 ☽ ⊻ ♅
14 38 ☿ ∠ ♅
16 37 ☽ ∠ ♂
18 40 ☽ ∥ ♅

25 Sunday
00 17 ☽ ⊥ ♆
00 37 ☽ ± ♃
02 58 ☽ ⊻ ☉
05 52 ♀ ⊻ ♄
05 54 ♀ ⊥ ♅
10 43 ☽ ⊻ ♆
14 11 ☽ ⊥ ☉
14 25 ☽ ⊻ ☿
16 05 ☽ ꝗ ♃
17 33 ☽ ⊻ ♇
19 56 ☽ ⚹ ♂
23 56 ☽ ⊥ ☿

26 Monday
03 54 ☽ ⊥ ♇
06 52 ☽ ∠ ☉
08 03 ☽ ☌ ♄
10 25 ☽ ⊻ ♀
14 08 ☽ ∠ ☿
17 32 ☽ Q ♂
17 36 ☽ △ ♃
18 15 ☽ ☌ ♅
19 22 ☽ ∠ ♇
21 40 ☽ ⊥ ♀

27 Tuesday
00 56 ☉ ∠ ♄
07 14 ☿ ⚹ ♆
10 24 ♂ Q ♅
10 27 ☽ ⚹ ☉
11 09 ♀ ∠ ♆
13 37 ☽ ⚹ ☿
14 12 ☽ ☌ ♆
14 29 ☽ ∠ ♀
20 58 ☽ ⚹ ♇

28 Wednesday
01 38 ☽ □ ♂
07 50 ☉ ☌ ☿
08 17 ☽ Q ☿
08 22 ☽ Q ☉
11 28 ☽ ⊻ ♄
17 23 ☽ Q ♇
18 23 ☽ ⚹ ♀
20 07 ☽ □ ♃
21 23 ☽ ⊻ ♅
21 42 ☽ ⊥ ♄

29 Thursday
00 34 ☉ ∥ ☿
04 19 ☿ ∠ ♄
07 35 ☽ ⊥ ♅
12 18 ☽ □ ☿
12 23 ♀ ⊼ ♃
13 03 ☽ ∠ ♄
14 38 ☽ ∥ ♅
16 31 ☉ ⚹ ♆
16 40 ☽ Q ♀
17 08 ☽ ⊻ ♆
17 10 ☽ □ ☉
20 58 ☽ ∥ ♆
22 52 ☽ ∠ ♅
23 57 ☽ □ ♇

30 Friday
00 10 ☽ ∥ ♄
02 20 ☽ Q ♃
03 18 ☽ ⊥ ♆
05 45 ♀ ⊻ ♅
06 54 ☽ △ ♂
13 28 ☽ ∥ ♀
14 39 ☽ ⚹ ♄
18 36 ☽ ∠ ♆
22 30 ☽ ⚹ ♃

31 Saturday
00 24 ☽ ⚹ ♅
02 08 ☽ □ ♀
09 35 ☽ ꝗ ♂
10 31 ☽ ∥ ☉
11 10 ☽ Q ♄
11 22 ☽ △ ☿
20 08 ☽ ⚹ ♆
20 53 ☽ Q ♅
21 35 ☽ ∥ ☿
23 47 ☽ ∠ ♃

NOVEMBER 1987

LONGITUDES

Date	Sidereal time h m s	Sun ☉ ° ' "	Moon ☽ ° ' "	Moon ☽ 24.00 ° ' "	Mercury ☿ ° '	Venus ♀ ° '	Mars ♂ ° '	Jupiter ♃ ° '	Saturn ♄ ° '	Uranus ♅ ° '	Neptune ♆ ° '	Pluto ♇ ° '
01	14 40 58	08 ♏ 31 35	15 ♓ 03 18	22 ♓ 03 31	29 ♎ 37	26 ♏ 56	15 ♎ 17	22 ♈ 51	18 ♐ 36	24 ♐ 14	05 ♑ 47	09 ♏ 51
02	14 44 55	09 31 37	29 ♓ 02 09	05 ♈ 58 56	28 R 48	28 11	15 55	22 R 43	18 42	24 16	05 48	09 53
03	14 48 52	10 31 40	12 ♈ 53 35	19 ♈ 45 48	28 10	29 ♏ 26	16 34	22 36	18 48	24 19	05 50	09 56
04	14 52 48	11 31 45	26 ♈ 35 17	03 ♉ 21 42	27 43	00 ♐ 40	17 13	22 28	18 54	24 22	05 51	09 58
05	14 56 45	12 31 51	10 ♉ 04 46	16 ♉ 44 12	27 28	01 55	17 52	22 21	19 00	24 25	05 53	10 01
06	15 00 41	13 32 00	23 ♉ 19 48	29 ♉ 51 22	27 D 24	03 10	18 31	22 14	19 06	24 28	05 54	10 03
07	15 04 38	14 32 10	06 ♊ 18 48	12 ♊ 42 02	27 32	04 24	19 10	22 07	19 13	24 31	05 56	10 05
08	15 08 34	15 32 23	19 ♊ 01 06	25 ♊ 16 07	27 50	05 39	19 49	22 01	19 19	24 34	05 57	10 08
09	15 12 31	16 32 37	01 ♋ 27 16	07 ♋ 34 49	28 18	06 54	20 28	21 54	19 25	24 37	05 59	10 10
10	15 16 27	17 32 53	13 ♋ 39 06	19 ♋ 40 30	28 54	08 08	21 07	21 47	19 32	24 40	06 01	10 13
11	15 20 24	18 33 11	25 ♋ 39 36	01 ♌ 36 37	29 ♎ 39	09 23	21 46	21 41	19 38	24 43	06 02	10 15
12	15 24 21	19 33 31	07 ♌ 32 25	13 ♌ 27 29	00 ♏ 31	10 38	22 25	21 35	19 44	24 46	06 04	10 18
13	15 28 17	20 33 53	19 ♌ 22 28	25 ♌ 18 01	01 30	11 52	23 04	21 28	19 51	24 49	06 06	10 20
14	15 32 14	21 34 17	01 ♍ 14 47	07 ♍ 13 26	02 33	13 07	23 43	21 22	19 57	24 53	06 07	10 22
15	15 36 10	22 34 43	13 ♍ 14 38	19 ♍ 19 00	03 42	14 21	24 22	21 17	20 04	24 56	06 09	10 25
16	15 40 07	23 35 10	25 ♍ 27 09	01 ♎ 39 38	04 55	15 36	25 01	21 11	20 11	24 59	06 11	10 27
17	15 44 03	24 35 40	07 ♎ 56 56	14 ♎ 19 29	06 11	16 51	25 40	21 05	20 17	25 02	06 13	10 29
18	15 48 00	25 36 11	20 ♎ 47 36	27 ♎ 21 31	07 30	18 05	26 19	21 00	20 24	25 05	06 15	10 32
19	15 51 56	26 36 44	04 ♏ 01 17	10 ♏ 46 54	08 52	19 20	26 58	20 55	20 30	25 09	06 16	10 34
20	15 55 53	27 37 19	17 ♏ 38 10	24 ♏ 34 45	10 15	20 35	27 37	20 50	20 37	25 12	06 18	10 37
21	15 59 50	28 37 55	01 ♐ 36 11	08 ♐ 41 52	11 41	21 49	28 17	20 45	20 44	25 15	06 20	10 39
22	16 03 46	29 ♏ 38 33	15 ♐ 51 07	23 ♐ 03 09	13 08	23 04	28 56	20 40	20 51	25 19	06 22	10 41
23	16 07 43	00 ♐ 39 12	00 ♑ 17 07	07 ♑ 32 13	14 37	24 19	29 ♎ 35	20 36	20 57	25 22	06 24	10 44
24	16 11 39	01 39 52	14 ♑ 47 37	22 ♑ 02 33	16 06	25 33	00 ♏ 14	20 31	21 04	25 26	06 26	10 46
25	16 15 36	02 40 34	29 ♑ 16 22	06 ♒ 28 28	17 36	26 48	00 53	20 27	21 11	25 29	06 28	10 48
26	16 19 32	03 41 17	13 ♒ 38 25	20 ♒ 45 51	19 07	28 02	01 33	20 23	21 18	25 32	06 30	10 51
27	16 23 29	04 42 00	27 ♒ 50 32	04 ♓ 52 20	20 38	29 ♐ 17	02 12	20 19	21 25	25 36	06 32	10 53
28	16 27 25	05 42 45	11 ♓ 51 10	18 ♓ 47 03	22 10	00 ♑ 32	02 51	20 16	21 32	25 39	06 34	10 55
29	16 31 22	06 43 30	25 ♓ 40 02	02 ♈ 30 10	23 43	01 46	03 30	20 12	21 39	25 43	06 36	10 57
30	16 35 19	07 ♐ 44 17	09 ♈ 17 33	16 ♈ 02 15	25 ♏ 15	03 ♑ 01	04 ♏ 10	20 ♈ 09	21 ♐ 46	25 ♐ 46	06 ♑ 38	11 ♏ 00

Moon nodes and latitude

Date	Moon True ☊ ° '	Moon Mean ☊ ° '	Moon ☽ Latitude ° '
01	01 ♈ 57	00 ♈ 22	01 S 31
02	01 D 58	00 19	00 S 16
03	01 R 57	00 16	00 N 59
04	01 55	00 13	02 10
05	01 51	00 10	03 12
06	01 44	00 06	04 02
07	01 37	00 03	04 39
08	01 29	00 ♈ 00	05 01
09	01 21	29 ♓ 57	05 08
10	01 14	29 54	05 01
11	01 10	29 50	04 41
12	01 07	29 47	04 08
13	01 06	29 44	03 26
14	01 D 06	29 41	02 34
15	01 07	29 38	01 35
16	01 08	29 35	00 N 31
17	01 R 08	29 31	00 S 37
18	01 06	29 28	01 43
19	01 02	29 25	02 47
20	00 55	29 22	03 42
21	00 46	29 19	04 26
22	00 36	29 16	04 54
23	00 26	29 12	05 04
24	00 18	29 09	04 54
25	00 11	29 06	04 26
26	00 07	29 03	03 41
27	00 05	29 00	02 43
28	00 D 05	28 56	01 36
29	00 06	28 53	00 24
30	00 ♈ 05	28 ♓ 50	00 N 49

DECLINATIONS

Date	Sun ☉ ° '	Moon ☽ ° '	Mercury ☿ ° '	Venus ♀ ° '	Mars ♂ ° '	Jupiter ♃ ° '	Saturn ♄ ° '	Uranus ♅ ° '	Neptune ♆ ° '	Pluto ♇ ° '
01	14 S 21	07 S 17	10 S 47	19 S 38	05 S 11	07 N 26	21 S 45	23 S 29	22 S 20	00 N 08
02	14 40	00 S 38	10 13	19 57	05 26	07 24	21 46	23 29	22 20	00 07
03	14 59	06 N 00	09 44	20 16	05 42	07 21	21 47	23 29	22 20	00 07
04	15 18	12 16	09 21	20 35	05 57	07 18	21 47	23 29	22 20	00 06
05	15 36	17 52	09 04	20 52	06 12	07 16	21 48	23 29	22 20	00 05
06	15 54	22 31	08 52	21 10	06 27	07 13	21 48	23 29	22 20	00 04
07	16 12	25 57	08 46	21 26	06 42	07 11	21 49	23 29	22 20	00 04
08	16 30	27 59	08 46	21 42	06 57	07 09	21 50	23 30	22 20	00 03
09	16 47	28 34	08 50	21 58	07 12	07 06	21 50	23 30	22 20	00 02
10	17 04	27 44	09 00	22 12	07 27	07 04	21 51	23 30	22 20	00 02
11	17 21	25 36	09 13	22 26	07 42	07 02	21 52	23 30	22 20	00 N 01
12	17 37	22 23	09 30	22 40	07 57	07 00	21 52	23 30	22 20	00 00
13	17 54	18 16	09 50	22 53	08 11	06 58	21 53	23 30	22 20	00 00
14	18 10	13 26	10 12	23 05	08 26	06 56	21 53	23 30	22 20	00 S 01
15	18 25	08 03	10 37	23 16	08 41	06 54	21 54	23 30	22 20	00 01
16	18 40	02 N 17	11 04	23 27	08 55	06 52	21 55	23 31	22 20	00 02
17	18 55	03 S 43	11 32	23 37	09 10	06 50	21 55	23 31	22 20	00 03
18	19 10	09 43	12 01	23 47	09 24	06 48	21 56	23 31	22 20	00 03
19	19 24	15 28	12 32	23 55	09 39	06 46	21 56	23 31	22 19	00 04
20	19 38	20 39	13 03	24 03	09 53	06 45	21 57	23 31	22 19	00 05
21	19 51	24 49	13 34	24 11	10 08	06 43	21 57	23 31	22 19	00 05
22	20 05	27 34	14 06	24 17	10 22	06 42	21 58	23 31	22 19	00 06
23	20 17	28 30	14 38	24 23	10 36	06 40	21 59	23 31	22 19	00 06
24	20 30	27 30	15 10	24 28	10 50	06 39	21 59	23 32	22 19	00 07
25	20 42	24 38	15 41	24 33	11 04	06 37	22 00	23 32	22 19	00 07
26	20 54	20 15	16 13	24 36	11 18	06 36	22 00	23 32	22 19	00 08
27	21 05	14 46	16 44	24 39	11 32	06 35	22 01	23 32	22 19	00 08
28	21 16	08 35	17 14	24 41	11 46	06 34	22 01	23 32	22 19	00 09
29	21 26	02 05	17 44	24 42	12 00	06 33	22 02	23 32	22 19	00 09
30	21 S 36	04 N 26	18 S 13	24 S 43	12 S 13	06 N 32	22 S 02	23 S 32	22 S 19	00 S 10

ZODIAC SIGN ENTRIES

Date	h	m	Planets	
01	01	57	☿	♎
02	13	40	☽	♈
03	23	04	♀	♐
04	18	02	☽	♉
07	00	16	☽	♊
09	09	10	☽	♋
11	20	45	☽	♌
11	21	57	☿	♏
14	09	29	☽	♍
16	20	48	☽	♎
19	04	47	☽	♏
21	09	16	☽	♐
22	20	29	☉	♐
23	11	32	☽	♑
24	03	19	♂	♏
25	13	13	☽	♒
27	15	40	☽	♓
28	01	51	♀	♑
29	19	36	☽	♈

LATITUDES

Date	Mercury ☿ ° '	Venus ♀ ° '	Mars ♂ ° '	Jupiter ♃ ° '	Saturn ♄ ° '	Uranus ♅ ° '	Neptune ♆ ° '	Pluto ♇ ° '
01	00 N 36	00 S 10	00 N 54	01 S 34	01 N 12	00 S 10	00 N 59	15 N 40
04	01 24	00 17	00 53	01 33	01 12	00 10	00 59	15 40
07	01 57	00 25	00 52	01 33	01 11	00 10	00 59	15 40
10	02 14	00 33	00 51	01 32	01 11	00 10	00 59	15 40
13	02 19	00 40	00 50	01 31	01 11	00 10	00 58	15 40
16	02 14	00 48	00 49	01 31	01 10	00 10	00 58	15 40
19	02 02	00 55	00 48	01 30	01 10	00 10	00 58	15 41
22	01 46	01 02	00 47	01 29	01 10	00 10	00 58	15 41
25	01 27	01 08	00 46	01 28	01 09	00 10	00 58	15 42
28	01 07	01 15	00 45	01 28	01 09	00 10	00 58	15 42
31	00 N 46	01 S 20	00 N 43	01 S 27	01 N 09	00 S 10	00 N 58	15 N 43

DATA

Julian Date	2447101
Delta T	+55 seconds
Ayanamsa	23° 41' 12"
Synetic vernal point	05° ♓ 25' 47"
True obliquity of ecliptic	23° 26' 36"

MOON'S PHASES, APSIDES AND POSITIONS ☽

Date	h	m	Phase	Longitude ° '	Eclipse Indicator
05	16	46	○	12 ♉ 44	
13	14	38	☾	20 ♌ 41	
21	06	33	●	28 ♏ 24	
28	00	37	☽	05 ♓ 14	

Day	h	m		
12	17	51	Apogee	
24	14	42	Perigee	
02	14	15	0N	
09	09	37	Max dec	28° N 35'
16	21	11	0S	
23	11	39	Max dec	28° S 30'
29	19	37	0N	

LONGITUDES

Date	Chiron ⚷ ° '	Ceres ⚳ ° '	Pallas ⚴ ° '	Juno ⚵ ° '	Vesta ⚶ ° '	Black Moon Lilith ⚸ ° '
01	28 ♊ 26	07 ♑ 32	13 ♐ 58	24 ♒ 10	05 ♌ 05	08 ♌ 23
11	28 ♊ 03	10 ♑ 50	17 ♐ 53	26 ♒ 06	06 ♌ 56	09 ♌ 30
21	27 ♊ 35	14 ♑ 17	21 ♐ 51	28 ♒ 38	08 ♌ 14	10 ♌ 37
31	27 ♊ 02	17 ♑ 52	25 ♐ 51	01 ♓ 41	08 ♌ 55	11 ♌ 44

All ephemeris data is given at 12.00 UT and the Moon's longitude is additionally given for 24.00 UT

ASPECTARIAN

h m Aspects

01 Sunday
00 00 ☽ △ ☉
01 39 ☽ ± ♂
03 05 ☽ △ ♇
11 17 ☽ ⚼ ☿
11 27 ☽ # ♃
12 24 ☽ ⊼ ♂
15 02 ☽ ⊥ ♃
16 40 ☽ Q ♆
18 07 ☽ □ ♄
19 21 ☽ ∥ ♂

02 Monday
01 14 ☽ ⊻ ♃
01 50 ☽ ± ☿
03 38 ☽ ⚼ ☉
03 47 ☽ □ ♅
04 50 ☽ ⚼ ♇
05 38 ☉ ∠ ♅
10 23 ☽ △ ♀
11 37 ☽ ⊼ ☿
13 49 ☽ # ♇
14 41 ☽ ∥ ♇
19 38 ☿ ⊻ ♀
20 21 ☽ ± ☉
20 24 ☽ ± ♇
21 02 ☉ ☌ ♇
21 25 ♀ ± ♃
23 43 ☽ □ ♆

03 Tuesday
06 50 ☽ ⊼ ♇
07 34 ☽ ⊼ ☉
10 49 ☽ # ♂
14 56 ☽ ⚼ ♀
16 59 ☽ ∥ ♃
18 44 ☽ ☍ ♂
19 55 ♀ ⊥ ♆
22 24 ☽ △ ♄

04 Wednesday
01 11 ☽ # ☿
04 49 ☽ ☌ ♃
08 04 ☽ △ ♅
08 17 ☽ ± ♀
13 57 ☽ ☍ ☿
19 57 ☽ ⊼ ♀

05 Thursday
01 04 ☽ ⚼ ♄
01 15 ☽ # ☉
04 28 ☽ △ ♆
10 49 ☽ ⚼ ♅
11 52 ☽ ☍ ♇
16 46 ☽ ☍ ☉
17 18 ☽ ± ♄

06 Friday
00 39 ☉ ⊥ ♄
02 46 ☽ ⊼ ♂
03 06 ☽ ± ♅
03 52 ☽ # ♀
04 14 ☽ ⊼ ♄
07 34 ☽ ⚼ ♆
07 40 ☿ St D
07 56 ☽ # ♄
10 01 ☽ ⊻ ♃
10 57 ☽ # ♆
14 05 ☽ ⊼ ♅
14 17 ☽ ± ♂
16 54 ☿ ⊥ ♀
17 59 ☽ # ♅
19 31 ☽ ⊼ ☿
20 56 ☽ ⊥ ♃

07 Saturday
00 07 ☽ ± ♆
02 20 ♀ ∠ ♂
06 44 ☽ ± ☿
07 47 ☽ ⚼ ♂
08 03 ☽ ☍ ♀
11 17 ☽ ⊼ ♆
13 30 ☽ ∠ ♃
14 06 ♂ ✶ ♄
19 06 ☽ ⊼ ♇
23 55 ☽ ⚼ ☿

08 Sunday
04 48 ☽ ⊼ ☉
06 29 ☽ ± ♇
12 34 ☽ ☍ ♄
13 36 ☽ △ ♂
17 15 ☽ ± ☉
17 40 ☽ ✶ ♃
18 04 ♀ ⊻ ♆
22 41 ☽ ☍ ♅
23 46 ☽ ⚼ ♇

09 Monday
00 05 ♀ ∥ ♄
04 17 ♂ # ♃
05 35 ☽ △ ☿
12 05 ♀ ⚼ ♃
12 11 ☽ ⚼ ☉
16 44 ☽ Q ♃
20 53 ☽ ☍ ♆
23 51 ☽ ⊼ ♀

10 Tuesday
05 10 ☽ △ ♇
13 05 ☽ ± ♀
20 28 ☽ △ ☉
23 49 ☽ ⊼ ♄

11 Wednesday
00 42 ♀ ∥ ♆
03 44 ☽ □ ♂
04 05 ☽ □ ♃
09 08 ☽ ⚼ ♀
09 25 ♂ ☍ ♃
10 06 ☽ ⊼ ♅
11 57 ☽ ± ♄
16 10 ☉ ⊥ ♅
20 39 ☽ □ ☿
22 15 ☽ ± ♅

12 Thursday
04 30 ☽ # ♅
05 21 ♀ ⊻ ♇
06 17 ☽ ⚼ ♄
09 00 ☽ ⊼ ♆
10 17 ☽ # ♀
12 22 ☽ # ♆
15 17 ☽ # ♄
16 32 ☽ ⚼ ♅
16 52 ☉ ⊻ ♄
17 36 ☽ □ ♇
18 10 ☽ Q ♂
18 59 ☽ △ ♀
21 12 ☽ ± ♆

13 Friday
12 16 ☽ Q ☿
12 58 ☽ △ ♄
13 51 ☽ # ☉
14 38 ☽ □ ☉
15 30 ☽ ⚼ ♆
16 13 ☽ △ ♃
19 54 ☽ ✶ ♂
23 05 ☽ △ ♅

14 Saturday
01 01 ☉ ∠ ♆
06 11 ☽ Q ♇
07 42 ☉ ⊼ ♃
14 54 ☽ ✶ ☿
21 49 ☽ △ ♆
22 13 ☽ ⚼ ♃

15 Sunday
01 36 ☽ # ☿
03 49 ♂ Q ♆
03 50 ☽ ∠ ♂
06 13 ☽ Q ☉
06 21 ☽ ✶ ♇
09 23 ☽ # ♂
14 28 ☽ □ ♀
16 00 ☽ ± ♃
16 55 ☽ ∥ ♃
22 42 ☽ ⊥ ♂
23 57 ☽ ∠ ☿

16 Monday
01 36 ☽ □ ♄
03 43 ☽ ⊼ ♃
08 02 ☽ ✶ ☉
10 40 ♂ ✶ ♅
11 05 ☽ □ ♅
11 06 ☽ ⊻ ♂
12 00 ☽ ∠ ♇
17 37 ☿ ∠ ♄
19 27 ☽ ⊥ ☿
19 43 ♀ ∥ ♅
21 01 ☽ # ♇
21 21 ☽ ∥ ♇

17 Tuesday
04 56 ♀ ⊥ ♇
05 24 ☽ ⊥ ♇
05 27 ☽ Q ♀
08 15 ☽ ⊻ ☿
08 42 ☽ □ ♆
12 38 ☿ ✶ ♆
12 39 ☽ Q ♄
15 23 ☽ ∠ ☉
16 49 ☽ ⊻ ♇
21 38 ☽ Q ♅
23 08 ☉ ⊻ ♅

18 Wednesday
00 21 ☽ # ♃
06 29 ☽ ✶ ♀
09 37 ☽ ⊥ ☉
10 42 ☽ ∥ ♂
11 16 ☽ ✶ ♄
12 22 ☽ ☍ ♃
18 20 ☽ Q ♆
19 55 ☽ ✶ ♅
21 32 ☽ ⊻ ☉
22 19 ☽ ∥ ☿
22 38 ☽ ☌ ♂

19 Thursday
12 37 ☽ ∠ ♀
14 41 ☽ ∠ ♄
16 02 ☽ ✶ ♆
18 33 ☉ ± ♃
21 35 ☽ ☌ ☿
22 55 ☽ ∠ ♅
23 40 ☽ ☌ ♇

20 Friday
06 08 ☽ ⊥ ♀
06 42 ☽ ⊥ ♄
06 44 ☽ ∥ ☉
11 01 ☿ ∠ ♅
12 09 ☉ ⊻ ♂
12 53 ♀ ☌ ♄
14 44 ☽ ⊥ ♅
16 30 ♀ △ ♃
17 13 ☽ ⊻ ♄
17 30 ☽ ⊼ ♃
17 36 ☽ ⊻ ♀
18 08 ☿ ☌ ♇
18 22 ☽ ∠ ♆
18 48 ☽ ∥ ♄
20 50 ☽ ∥ ♆

21 Saturday
01 07 ☽ ⊻ ♅
03 42 ☽ ∥ ♅
03 46 ☽ ± ♃
06 03 ☽ ⊻ ♂
06 33 ☽ ☌ ☉
07 39 ☽ ∥ ♀
09 50 ☽ ⊥ ♆
13 39 ♃ △ ♄
16 45 ☽ ⊥ ♂
18 59 ☽ ⚼ ♃
20 02 ☽ ⊻ ♆

22 Sunday
03 19 ☽ ⊻ ♇
06 56 ☽ ⊻ ☿
08 38 ☽ ∠ ♂
13 24 ☽ ⊥ ♇
18 06 ☽ ⊥ ☿
19 59 ☽ △ ♃
20 24 ☽ ☌ ♄

23 Monday
01 09 ☽ ☌ ♀
03 49 ☽ ☌ ♅
04 25 ☽ ∠ ♇
05 46 ☉ ⊥ ♆
10 45 ☽ ∠ ☿
10 47 ☽ ✶ ♂
12 39 ☽ ⊻ ☉
18 07 ☿ ⊥ ♄
22 08 ☽ ☌ ♆
23 20 ☽ ⊥ ☉

24 Tuesday
05 19 ☽ ✶ ♇
07 34 ☽ Q ♂
09 26 ♀ ☌ ♅
14 24 ☽ ✶ ☿
15 20 ☽ ∠ ☉
16 14 ♀ ∠ ♇
21 26 ☽ □ ♃
22 28 ☽ ⊻ ♄

25 Wednesday
01 14 ☽ Q ♇
05 41 ☽ ⊻ ♅
07 30 ☽ ⊻ ♀
08 30 ☽ ⊥ ♄
12 37 ☽ ∥ ♀
12 37 ☽ Q ☿
14 49 ☽ □ ♂
15 41 ☽ ⊥ ♅
18 06 ☽ ✶ ☉
18 25 ☽ ⊥ ♀
18 50 ☽ ∥ ♅
23 37 ☽ ∠ ♄

26 Thursday
00 01 ☽ ⊻ ♆
01 36 ☽ ∥ ♆
03 14 ☽ Q ♃
03 17 ☽ ∥ ♄
06 47 ☽ ∠ ♅
07 18 ☽ □ ♇
09 02 ☽ ∥ ☉
10 05 ☽ ⊥ ♆
10 54 ☽ ∠ ♀
15 42 ☽ Q ☉
18 56 ☿ ⊥ ♅
22 19 ☽ □ ☿
23 18 ☽ ✶ ♃

27 Friday
01 01 ☽ ✶ ♄
01 16 ☽ ∠ ♆
04 33 ☽ ∥ ☿
07 10 ☿ ⊼ ♃
08 10 ☽ ✶ ♅
14 41 ☽ ✶ ♀
19 47 ☽ △ ♂
21 35 ☽ Q ♄

28 Saturday
00 21 ☽ ∥ ♂
00 37 ☽ □ ☉
00 43 ☽ ∠ ♃
01 07 ☿ ⊻ ♄
01 55 ☉ ⚼ ♃
02 15 ☿ ∠ ♆
02 53 ☽ ✶ ♆
04 45 ☽ Q ♅
10 23 ☽ △ ♇
13 16 ☽ Q ♀
16 09 ☽ ⊥ ♃
19 33 ☽ # ♃
22 54 ☽ ⚼ ♂
23 39 ☽ Q ♆

29 Sunday
02 31 ☽ ⊻ ♃
04 55 ☽ □ ♄
08 09 ☽ △ ☿
08 51 ☉ ⊻ ♆
12 05 ☽ □ ♅
12 30 ☽ ⚼ ♇
15 23 ☽ ± ♂
19 03 ☽ ∥ ♇
20 12 ☽ # ♇
23 47 ☽ □ ♀

30 Monday
02 28 ☽ ⊼ ♂
04 22 ☽ ± ♇
07 16 ☽ □ ♆
09 01 ☽ △ ☉
13 55 ☽ ⚼ ☿
15 01 ☽ ⊼ ♇
19 55 ☽ ∥ ♃
20 22 ☿ ⊻ ♅

DECEMBER 1987

LONGITUDES

Date	Sidereal time h m s	Sun ☉ ° ' "	Moon ☽ ° ' "	Moon ☽ 24.00 ° ' "	Mercury ☿ ° '	Venus ♀ ° '	Mars ♂ ° '	Jupiter ♃ ° '	Saturn ♄ ° '	Uranus ♅ ° '	Neptune ♆ ° '	Pluto ♇ ° '
01	16 39 15	08 ♐ 45 04	22 ♈ 44 19	29 ♈ 23 47	26 ♏ 48	04 ♑ 15	04 ♏ 49	20 ♈ 06	21 ♐ 53	25 ♐ 50	06 ♑ 40	11 ♏ 02
02	16 43 12	09 45 53	06 ♉ 00 38	12 ♉ 34 52	28 21	05 30	05 28	20 R 03	22 00	25 53	06 42	11 04
03	16 47 08	10 46 42	19 ♉ 06 25	25 ♉ 35 11	29 ♏ 54	06 44	06 08	20 01	22 07	25 57	06 44	11 06
04	16 51 05	11 47 33	02 ♊ 01 07	08 ♊ 24 07	01 ♐ 27	07 59	06 47	19 58	22 14	26 00	06 46	11 08
05	16 55 01	12 48 24	14 ♊ 44 06	21 ♊ 01 01	03 00	09 13	07 26	19 56	22 21	26 04	06 48	11 10
06	16 58 58	13 49 17	27 ♊ 14 51	03 ♋ 25 38	04 34	10 28	08 06	19 54	22 28	26 08	06 50	11 13
07	17 02 54	14 50 11	09 ♋ 33 26	15 ♋ 38 22	06 07	11 42	08 45	19 53	22 35	26 11	06 52	11 15
08	17 06 51	15 51 06	21 ♋ 40 39	27 ♋ 40 33	07 41	12 57	09 25	19 51	22 42	26 15	06 55	11 17
09	17 10 48	16 52 02	03 ♌ 38 22	09 ♌ 34 31	09 14	14 11	10 04	19 50	22 49	26 18	06 57	11 19
10	17 14 44	17 52 59	15 ♌ 29 25	21 ♌ 23 36	10 48	15 25	10 43	19 48	22 56	26 22	06 59	11 21
11	17 18 41	18 53 57	27 ♌ 17 37	03 ♍ 12 04	12 21	16 40	11 23	19 48	23 03	26 26	07 01	11 23
12	17 22 37	19 54 56	09 ♍ 07 35	15 ♍ 04 51	13 55	17 54	12 02	19 47	23 10	26 29	07 03	11 25
13	17 26 34	20 55 56	21 ♍ 04 33	27 ♍ 07 24	15 29	19 08	12 42	19 46	23 17	26 33	07 06	11 27
14	17 30 30	21 56 58	03 ♎ 14 04	09 ♎ 25 16	17 03	20 23	13 21	19 46	23 24	26 36	07 08	11 29
15	17 34 27	22 58 00	15 ♎ 41 38	22 ♎ 03 46	18 37	21 37	14 01	19 46	23 31	26 40	07 10	11 31
16	17 38 23	23 59 03	28 ♎ 32 09	05 ♏ 07 14	20 11	22 52	14 41	19 D 46	23 39	26 44	07 12	11 33
17	17 42 20	25 00 08	11 ♏ 49 17	18 ♏ 38 26	21 45	24 06	15 20	19 46	23 46	26 47	07 14	11 35
18	17 46 17	26 01 13	25 ♏ 34 40	02 ♐ 37 43	23 19	25 20	16 00	19 47	23 53	26 51	07 17	11 37
19	17 50 13	27 02 19	09 ♐ 47 11	17 ♐ 02 24	24 53	26 34	16 39	19 48	24 00	26 55	07 19	11 39
20	17 54 10	28 03 26	24 ♐ 22 34	01 ♑ 46 41	26 28	27 49	17 19	19 48	24 07	26 58	07 21	11 41
21	17 58 06	29 ♐ 04 34	09 ♑ 13 39	16 ♑ 42 16	28 03	29 ♑ 03	17 59	19 50	24 14	27 02	07 23	11 42
22	18 02 03	00 ♑ 05 42	24 ♑ 11 19	01 ♒ 39 40	29 ♐ 38	00 ♒ 17	18 38	19 51	24 21	27 05	07 26	11 44
23	18 05 59	01 06 50	09 ♒ 06 12	16 ♒ 29 59	01 ♑ 13	01 31	19 18	19 53	24 28	27 09	07 28	11 46
24	18 09 56	02 07 58	23 ♒ 50 14	01 ♓ 06 18	02 48	02 45	19 58	19 54	24 35	27 13	07 30	11 48
25	18 13 52	03 09 07	08 ♓ 17 45	15 ♓ 24 20	04 24	04 00	20 37	19 56	24 42	27 16	07 32	11 49
26	18 17 49	04 10 15	22 ♓ 25 54	29 ♓ 22 30	05 59	05 14	21 17	19 58	24 49	27 20	07 35	11 51
27	18 21 46	05 11 23	06 ♈ 14 12	13 ♈ 01 14	07 35	06 28	21 57	20 01	24 56	27 24	07 37	11 53
28	18 25 42	06 12 32	19 ♈ 43 49	26 ♈ 22 14	09 12	07 42	22 36	20 03	25 03	27 27	07 39	11 54
29	18 29 39	07 13 40	02 ♉ 56 47	09 ♉ 27 44	10 48	08 56	23 16	20 06	25 10	27 31	07 41	11 56
30	18 33 35	08 14 48	15 ♉ 55 22	22 ♉ 19 54	12 25	10 10	23 56	20 09	25 17	27 34	07 44	11 58
31	18 37 32	09 ♑ 15 57	28 ♉ 41 32	05 ♊ 00 29	14 ♑ 02	11 ♒ 24	24 ♏ 36	20 ♈ 12	25 ♐ 24	27 ♐ 38	07 ♑ 46	11 ♏ 59

Moon

Date	Moon True ☊ ° '	Moon Mean ☊ ° '	Moon ☽ Latitude ° '
01	00 ♈ 03	28 ♓ 47	01 N 57
02	29 ♓ 58	28 44	02 58
03	29 51	28 41	03 49
04	29 40	28 37	04 27
05	29 27	28 34	04 51
06	29 14	28 31	05 00
07	29 00	28 28	04 56
08	28 48	28 25	04 37
09	28 38	28 22	04 07
10	28 31	28 18	03 26
11	28 27	28 15	02 37
12	28 25	28 12	01 40
13	28 D 25	28 09	00 N 39
14	28 R 25	28 06	00 S 25
15	28 24	28 02	01 30
16	28 21	27 59	02 32
17	28 16	27 56	03 28
18	28 08	27 53	04 14
19	27 57	27 50	04 45
20	27 45	27 47	04 59
21	27 33	27 43	04 54
22	27 22	27 40	04 28
23	27 13	27 37	03 44
24	27 07	27 34	02 46
25	27 04	27 31	01 37
26	27 D 03	27 28	00 S 24
27	27 R 03	27 24	00 N 48
28	27 03	27 21	01 57
29	27 00	27 18	02 57
30	26 55	27 15	03 48
31	26 ♓ 48	27 ♓ 12	04 N 26

DECLINATIONS

Date	Sun ☉ ° '	Moon ☽ ° '	Mercury ☿ ° '	Venus ♀ ° '	Mars ♂ ° '	Jupiter ♃ ° '	Saturn ♄ ° '	Uranus ♅ ° '	Neptune ♆ ° '	Pluto ♇ ° '
01	21 S 46	10 N 40	18 S 42	24 S 43	12 S 27	06 N 31	22 S 03	23 S 32	22 S 19	00 S 10
02	21 55	16 20	19 10	24 42	12 40	06 30	22 03	23 32	22 19	00 11
03	22 04	21 10	19 37	24 40	12 54	06 30	22 04	23 33	22 18	00 11
04	22 12	24 55	20 03	24 38	13 07	06 29	22 04	23 33	22 18	00 11
05	22 20	27 23	20 29	24 35	13 20	06 29	22 05	23 33	22 18	00 12
06	22 28	28 25	20 53	24 31	13 33	06 28	22 05	23 33	22 18	00 12
07	22 35	28 01	21 17	24 26	13 46	06 28	22 06	23 33	22 18	00 13
08	22 41	26 16	21 39	24 21	13 59	06 27	22 06	23 33	22 18	00 13
09	22 48	23 21	22 01	24 15	14 12	06 27	22 07	23 33	22 18	00 13
10	22 53	19 29	22 21	24 08	14 25	06 27	22 07	23 33	22 18	00 14
11	22 59	14 52	22 40	24 00	14 38	06 27	22 08	23 33	22 18	00 14
12	23 04	09 42	22 59	23 52	14 50	06 27	22 08	23 34	22 18	00 14
13	23 08	04 N 08	23 16	23 43	15 03	06 27	22 08	23 34	22 18	00 15
14	23 12	01 S 41	23 32	23 33	15 15	06 27	22 09	23 34	22 17	00 15
15	23 15	07 34	23 46	23 22	15 27	06 28	22 09	23 34	22 17	00 15
16	23 18	13 19	24 00	23 11	15 39	06 28	22 10	23 34	22 17	00 15
17	23 21	18 41	24 12	22 59	15 51	06 28	22 10	23 34	22 17	00 16
18	23 23	23 16	24 23	22 47	16 03	06 29	22 10	23 34	22 17	00 16
19	23 25	26 37	24 33	22 34	16 15	06 29	22 11	23 34	22 17	00 16
20	23 26	28 19	24 41	22 20	16 27	06 30	22 11	23 34	22 17	00 16
21	23 26	28 00	24 48	22 05	16 38	06 31	22 12	23 34	22 17	00 16
22	23 27	25 41	24 54	21 50	16 50	06 31	22 12	23 35	22 17	00 17
23	23 26	21 35	24 59	21 34	17 01	06 32	22 12	23 35	22 17	00 17
24	23 26	16 11	25 02	21 18	17 12	06 33	22 13	23 35	22 16	00 17
25	23 24	09 58	25 03	21 01	17 23	06 34	22 13	23 35	22 16	00 17
26	23 23	03 S 23	25 03	20 43	17 34	06 35	22 13	23 35	22 16	00 17
27	23 20	03 N 13	25 02	20 25	17 45	06 37	22 14	23 35	22 16	00 17
28	23 18	09 31	24 59	20 06	17 56	06 38	22 14	23 35	22 16	00 17
29	23 15	15 16	24 55	19 47	18 06	06 39	22 14	23 35	22 16	00 17
30	23 11	20 14	24 49	19 27	18 17	06 41	22 14	23 35	22 16	00 17
31	23 S 07	24 N 11	24 S 42	19 S 06	18 S 27	06 N 42	22 S 15	23 S 35	22 S 16	00 S 18

ZODIAC SIGN ENTRIES

Date	h	m	Planets
02	01	06	☽ ♉
03	13	33	☿ ♐
04	08	13	☽ ♊
06	17	20	☽ ♋
09	04	40	☽ ♌
11	17	30	☽ ♍
14	05	40	☽ ♎
16	14	41	☽ ♏
18	19	33	☽ ♐
20	21	08	☽ ♑
22	06	29	♀ ♒
22	09	46	☉ ♑
22	17	40	☿ ♑
22	21	20	☽ ♒
24	22	10	☽ ♓
27	01	05	☽ ♈
29	06	37	☽ ♉
31	14	29	☽ ♊

LATITUDES

Date	Mercury ☿ ° '	Venus ♀ ° '	Mars ♂ ° '	Jupiter ♃ ° '	Saturn ♄ ° '	Uranus ♅ ° '	Neptune ♆ ° '	Pluto ♇ ° '
01	00 N 46	01 S 20	00 N 43	01 S 27	01 N 09	00 S 10	00 N 58	15 N 43
04	00 24	01 26	00 42	01 26	01 09	00 10	00 58	15 43
07	00 N 03	01 31	00 41	01 25	01 08	00 10	00 58	15 44
10	00 S 17	01 35	00 40	01 24	01 08	00 10	00 58	15 45
13	00 37	01 39	00 38	01 23	01 08	00 10	00 58	15 46
16	00 55	01 42	00 37	01 22	01 08	00 10	00 58	15 47
19	01 12	01 45	00 35	01 21	01 08	00 10	00 58	15 48
22	01 28	01 47	00 34	01 20	01 08	00 10	00 58	15 49
25	01 41	01 48	00 32	01 19	01 07	00 10	00 58	15 50
28	01 52	01 49	00 31	01 18	01 07	00 10	00 57	15 51
31	02 S 01	01 S 49	00 N 29	01 S 17	01 N 07	00 S 10	00 N 57	15 N 52

DATA

Julian Date	2447131
Delta T	+55 seconds
Ayanamsa	23° 41' 17"
Synetic vernal point	05° ♓ 25' 42"
True obliquity of ecliptic	23° 26' 36"

MOON'S PHASES, APSIDES AND POSITIONS ☽

Date	h	m	Phase	Longitude ° '	Eclipse Indicator
05	08	01	○	12 ♊ 38	
13	11	41	☾	20 ♍ 55	
20	18	25	●	28 ♐ 20	
27	10	01	☽	05 ♈ 06	

Day	h	m		
10	14	01	Apogee	
22	11	05	Perigee	
06	17	05	Max dec	28° N 27'
14	05	08	0S	
20	20	30	Max dec	28° S 26'
27	00	13	0N	

LONGITUDES

Date	Chiron ⚷ ° '	Ceres ⚳ ° '	Pallas ⚴ ° '	Juno ⚵ ° '	Vesta ⚶ ° '	Black Moon Lilith ⚸ ° '
01	27 ♊ 02	17 ♑ 52	25 ♐ 51	01 ♓ 41	08 ♌ 55	11 ♌ 44
11	26 ♊ 26	21 ♑ 34	29 ♐ 52	05 ♓ 11	08 ♌ 53	12 ♌ 51
21	25 ♊ 48	25 ♑ 21	03 ♑ 53	09 ♓ 04	08 ♌ 06	13 ♌ 58
31	25 ♊ 11	29 ♑ 13	07 ♑ 52	13 ♓ 18	06 ♌ 34	15 ♌ 05

All ephemeris data is given at 12.00 UT and the Moon's longitude is additionally given for 24.00 UT

ASPECTARIAN

h m Aspects

01 Tuesday
01 32 ☿ ± ♃
07 17 ☽ ☌ ♃
08 04 ☽ ± ☿
10 26 ☽ △ ♄
13 58 ☽ ᵽ ☉
17 35 ☽ △ ♅
19 34 ☽ # ♂
20 16 ☽ ⊼ ☿

02 Wednesday
07 35 ☽ ± ☉
10 58 ☽ △ ♀
10 58 ☽ ☍ ♂
11 03 ♀ ⚹ ♂
13 15 ☽ △ ♆
13 48 ☽ ᵽ ♄
19 25 ☽ ⊼ ☉
20 56 ☽ ᵽ ♅
21 15 ☽ ☍ ♇

03 Thursday
02 54 ☽ # ☿
06 26 ☽ ± ♄
11 51 ☉ ∥ ♄
11 57 ♀ ☌ ♆
13 34 ☽ ± ♅
13 40 ☽ ⊻ ♃
16 52 ☽ ᵽ ♆
17 06 ☽ # ♄
17 17 ☽ # ☉
17 22 ☽ ᵽ ♀
17 36 ☽ ⊼ ♄
18 32 ☽ # ♆
19 58 ☉ ⊻ ♇

04 Friday
00 44 ☽ ⊼ ♅
00 45 ☽ ⊥ ♃
01 11 ☿ ⊥ ♆
02 15 ☽ # ♅
09 39 ☽ ± ♆
09 50 ☽ # ♀
10 48 ☽ ☍ ☿
11 27 ♂ ⚹ ♆
11 55 ☽ ± ♀
17 32 ☽ ∠ ♃
20 57 ☽ ⊼ ♆
21 26 ☽ ⊼ ♂

05 Saturday
00 25 ☽ ⊼ ♀
05 13 ☽ ⊼ ♇
06 11 ☉ ∥ ♆
07 48 ♂ ∠ ♄
08 01 ☽ ☍ ☉
09 24 ☽ ± ♂
16 40 ☽ ± ♇
21 54 ☽ ⚹ ♃

06 Sunday
02 41 ☽ ☍ ♄
03 33 ☽ ᵽ ♂
04 08 ☿ ⊥ ♀
09 49 ☽ ☍ ♅
09 59 ☽ ᵽ ♇
17 09 ☿ ᵽ ♃
21 01 ☽ Q ♃

07 Monday
02 56 ♀ ⚹ ♇
04 17 ☽ ⊼ ☿
06 22 ☉ ⊥ ♂
06 43 ☽ ☍ ♆
10 20 ☽ △ ♂
15 20 ☽ △ ♇
16 42 ☽ ☍ ♀
17 47 ☽ ± ☿
23 22 ☽ ⊼ ☉
23 54 ☿ ⊻ ♆

08 Tuesday
08 22 ☽ □ ♃
12 23 ☽ ± ☉
14 03 ☽ ⊼ ♄
14 18 ☽ ᵽ ☿
21 11 ☽ ⊼ ♅

09 Wednesday
02 11 ☽ ± ♄
05 13 ☽ # ♀
08 06 ☽ ᵽ ☉
09 18 ☽ ± ♅
10 30 ☽ # ♅
15 40 ☽ # ☉
18 42 ☽ ⊼ ♆
19 02 ☽ # ♆
19 08 ☿ ∥ ♄
20 08 ☽ # ☿
20 14 ☽ # ♄
20 31 ☽ ᵽ ♄
22 20 ♅ ∠ ♇
22 58 ☉ ⊥ ♇

10 Thursday
01 02 ☽ △ ☿
01 46 ☽ □ ♂
03 34 ☽ □ ♇
03 35 ☽ ᵽ ♅
06 54 ☽ ± ♆
08 11 ☿ ∥ ♆
10 04 ☿ ⊻ ♂
11 51 ☽ ⊼ ♀
17 19 ☽ △ ☉
20 43 ☿ ⊻ ♇
20 46 ☽ △ ♃

11 Friday
01 14 ☽ ᵽ ♆
01 29 ☽ ± ♀
03 17 ☽ △ ♄
10 13 ☽ △ ♅
12 06 ♂ ☌ ♇
13 07 ☽ # ♂
13 44 ♂ ∠ ♅
16 16 ☽ Q ♇
16 30 ☽ Q ♂
21 55 ☽ ᵽ ♀

12 Saturday
03 13 ☽ ᵽ ♃
07 48 ☽ △ ♆
08 50 ☉ △ ♃
16 38 ☽ ⚹ ♇
18 13 ☽ ⚹ ♂
21 13 ☉ ∥ ☿
21 23 ☽ ± ♃
23 07 ☽ □ ☿

13 Sunday
02 10 ☽ ∥ ♃
07 42 ☽ △ ♀
09 24 ☽ ⊼ ♃
11 41 ☽ □ ☉
16 27 ☽ □ ♄
22 42 ☽ ∠ ♇
22 55 ☽ □ ♅

14 Monday
00 08 ♀ □ ♃
01 53 ☽ ∠ ♂
04 08 ☽ # ♇
06 09 ☽ ∥ ♇
09 48 ♀ ∥ ♅
13 10 ☿ ∥ ♀
15 26 ☿ ∥ ♅
16 02 ☽ Q ☿
16 24 ☽ ⊥ ♇
18 54 ☿ ⊥ ♇
19 36 ☽ □ ♆
20 28 ☽ ⊥ ♂

15 Tuesday
02 10 ☽ Q ☉
03 58 ☽ Q ♄
04 00 ☽ ⊻ ♇
07 29 ☽ # ♃
08 38 ☽ ⊻ ♂
10 02 ☽ Q ♅
12 19 ♃ St D
18 17 ☽ ⚹ ☿
19 41 ☽ ☍ ♃
23 56 ☉ ∥ ♀

16 Wednesday
00 22 ☽ □ ♀
02 52 ☽ ⚹ ☉
02 52 ☽ ⚹ ♄
02 53 ☉ ☌ ♄
05 42 ☿ △ ♃
05 50 ☽ Q ♆
08 39 ☽ ⚹ ♅
22 33 ☽ ∥ ♂

17 Thursday
01 11 ☿ ⊥ ♂
01 43 ☽ ∠ ☿
01 45 ♀ Q ♇
03 48 ☽ ⚹ ♆
04 48 ♀ ⊻ ♄
06 30 ☽ ∠ ♄
08 30 ☽ ∠ ☉
11 34 ☽ ☌ ♇
11 56 ☽ ∠ ♅
12 32 ☽ Q ♀
18 31 ☽ ☌ ♂
19 50 ☽ ⊥ ☿
22 33 ☽ ⊥ ♄

18 Friday
01 38 ☽ ⊥ ☉
01 59 ☽ ⊼ ♃
03 48 ☽ ⊥ ♅
05 48 ☽ ∥ ♄
06 18 ☽ ∠ ♆
06 25 ☽ ∥ ♆
07 37 ☽ ⊻ ☿
09 03 ☽ ⊻ ♄
09 21 ☽ ∥ ♀
11 33 ☽ ⚹ ♀
12 21 ☽ ± ♃
12 45 ☽ ∥ ☉
12 49 ☽ ⊻ ☉
13 52 ☽ ∥ ♅
14 11 ☽ ⊻ ♅
19 20 ☽ ∥ ☿
21 18 ☿ ☌ ♄
21 44 ☽ ⊥ ♆

19 Saturday
02 25 ☉ ∠ ♇
03 38 ☽ ᵽ ♃
07 52 ☽ ⊻ ♆
08 45 ☉ ☌ ♅
15 06 ☽ ⊻ ♇
15 15 ☽ ∠ ♀
18 51 ♀ ⊻ ♅
23 54 ☽ ⊻ ♂

20 Sunday
01 01 ☽ ⊥ ♇
04 32 ☽ △ ♃
07 26 ☽ ⊥ ♀
10 11 ☽ ⊥ ♂
11 34 ☽ ☌ ♄
15 15 ☿ ∠ ♇
15 45 ☽ ∠ ♇
15 48 ☽ ☌ ☿
16 14 ☽ ☌ ♅
16 54 ♀ ∥ ♆
18 05 ☽ ⊻ ♀
18 25 ☽ ☌ ☉
19 58 ☿ ☌ ♅

21 Monday
01 28 ☽ ∠ ♂
01 54 ♀ ∥ ♄
09 02 ☽ ☌ ♆
15 08 ☉ ⊻ ♀
15 59 ☽ ⚹ ♇
23 26 ♂ ⊥ ♄

22 Tuesday
02 41 ☽ ⚹ ♂
05 02 ☽ □ ♃
11 16 ☽ Q ♇
12 16 ☽ ⊻ ♄
13 27 ♀ ⊥ ♄
16 41 ☽ ⊻ ♅
17 14 ☽ ∥ ☿
21 46 ☽ ⊻ ☿
21 58 ☽ ⊥ ♄
22 11 ☽ ⊻ ☉
22 40 ☽ ☌ ♀
22 50 ☽ Q ♂

23 Wednesday
01 26 ☽ ∥ ♅
02 12 ☽ ∥ ☉
02 22 ☽ ⊥ ♅
02 41 ♀ Q ♂
07 52 ☉ ☌ ☿
08 30 ☽ ∥ ♆
08 33 ☽ ⊥ ☉
08 35 ☽ ⊥ ☿
08 52 ☽ ∥ ♄
09 21 ☽ ⊻ ♆
10 01 ☽ Q ♃
12 04 ☽ ∥ ♀
12 36 ☽ ∠ ♄
16 19 ☽ □ ♇
16 57 ☽ ∠ ♅
19 05 ☽ ⊥ ♆

24 Thursday
00 13 ☽ ∠ ☉
00 55 ☽ ∠ ☿
05 21 ☽ □ ♂
05 33 ☽ ⚹ ♃
07 54 ☽ ∥ ♂
09 02 ☿ ⊻ ♀
09 48 ☽ ∠ ♆
09 57 ♂ ⊼ ♃
13 15 ☽ ⚹ ♄
17 35 ☽ ⚹ ♅
21 18 ♀ ⊥ ♅

25 Friday
02 45 ☽ ⚹ ☉
04 08 ☽ ⊻ ♀
04 39 ☽ ⚹ ☿
06 22 ☽ ∠ ♃
09 19 ☽ Q ♄
10 44 ☽ ⚹ ♆
13 39 ☽ Q ♅
15 07 ☽ ⊥ ♀
17 57 ☽ △ ♇
21 32 ☽ ⊥ ♃

26 Saturday
00 24 ☽ # ♃
00 28 ☽ Q ☉
03 25 ☽ Q ☿
07 06 ☽ Q ♆
07 47 ☽ ⊻ ♃
07 51 ☽ ∠ ♀
09 56 ☽ △ ♂
14 02 ♂ ⊥ ♅
16 09 ☽ □ ♄
19 29 ☿ ∠ ♂
19 38 ☽ ᵽ ♇
20 30 ☽ □ ♅
23 11 ☽ ∥ ♇

27 Sunday
01 16 ☽ # ♇
10 01 ☽ □ ☉
11 22 ☽ ± ♇
12 23 ☿ ☌ ♆
12 26 ☽ ⚹ ♀
13 18 ☽ ᵽ ♂
14 26 ☽ □ ♆
14 42 ☽ □ ☿
22 00 ☽ ⊼ ♇

28 Monday
00 46 ☽ ∥ ♃
06 06 ☽ ± ♂
11 09 ♀ ⊻ ♆
11 56 ☽ Q ♀
12 35 ☽ ☌ ♃
13 50 ♂ ∠ ♆
17 27 ☽ ⊼ ♂
19 18 ♀ Q ♃
21 42 ☽ △ ♄

29 Tuesday
02 02 ☽ △ ♅
20 33 ☽ △ ☉
20 45 ☽ △ ♆
23 20 ☉ ☌ ♆

30 Wednesday
00 10 ☽ □ ♀
01 26 ☽ ᵽ ♄
01 38 ☽ # ♂
04 33 ☽ △ ☿
04 37 ☽ ☍ ♇
05 03 ☿ ⚹ ♇
05 44 ☽ ᵽ ♅
08 08 ☽ # ♀
14 46 ♀ ∠ ♄
18 21 ☽ ± ♄
19 57 ☽ ⊻ ♃
22 37 ☽ ± ♅
23 20 ☽ # ♄
23 27 ☽ # ♆

31 Thursday
00 47 ☽ ᵽ ♆
02 55 ☽ ᵽ ☉
03 50 ☽ ☍ ♂
04 55 ☽ # ☉
05 44 ☽ ⊼ ♄
07 17 ☽ ⊥ ♃
07 52 ☽ # ♅
09 59 ☽ ⊼ ♅
12 45 ☽ ᵽ ☿
15 38 ☽ # ☿
17 51 ☽ ± ♆
21 27 ☽ ± ☉
23 46 ♀ □ ♇

JANUARY 1988

LONGITUDES

Date	Sidereal time h m s	Sun ☉ ° ' "	Moon ☽ ° ' "	Moon ☽ 24.00 ° ' "	Mercury ☿ ° '	Venus ♀ ° '	Mars ♂ ° '	Jupiter ♃ ° '	Saturn ♄ ° '	Uranus ♅ ° '	Neptune ♆ ° '	Pluto ♇ ° '
01	18 41 28	10 ♑ 17 05	11 ♊ 16 51	17 ♊ 30 45	15 ♑ 40	12 ♒ 38	25 ♏ 15	20 ♈ 16	25 ♐ 31	27 ♐ 42	07 ♑ 48	12 ♏ 01
02	18 45 25	11 18 14	23 ♊ 42 17	29 ♊ 51 33	17 17	13 51	25 55	20 19	25 38	27 45	07 51	12 02
03	18 49 21	12 19 22	05 ♋ 58 35	12 ♋ 03 30	18 55	15 05	26 35	20 23	25 45	27 49	07 53	12 04
04	18 53 18	13 20 30	18 ♋ 06 21	24 ♋ 07 16	20 34	16 19	27 15	20 27	25 52	27 52	07 55	12 05
05	18 57 15	14 21 38	00 ♌ 06 22	06 ♌ 03 52	22 12	17 33	27 55	20 31	25 59	27 56	07 57	12 06
06	19 01 11	15 22 47	11 ♌ 59 58	17 ♌ 54 55	23 51	18 46	28 35	20 36	26 06	27 59	08 00	12 08
07	19 05 08	16 23 55	23 ♌ 49 05	29 ♌ 42 47	25 30	20 00	29 14	20 40	26 13	28 03	08 02	12 09
08	19 09 04	17 25 03	05 ♍ 36 29	11 ♍ 30 39	27 09	21 14	29 ♏ 54	20 45	26 19	28 06	08 04	12 11
09	19 13 01	18 26 11	17 ♍ 25 48	23 ♍ 22 30	28 ♑ 48	22 27	00 ♐ 34	20 50	26 26	28 10	08 06	12 12
10	19 16 57	19 27 19	29 ♍ 21 22	05 ♎ 23 02	00 ♒ 27	23 41	01 14	20 55	26 33	28 13	08 09	12 13
11	19 20 54	20 28 28	11 ♎ 28 10	17 ♎ 37 27	02 06	24 54	01 54	21 00	26 40	28 17	08 11	12 14
12	19 24 50	21 29 36	23 ♎ 51 32	00 ♏ 11 04	03 45	26 08	02 34	21 06	26 46	28 20	08 13	12 15
13	19 28 47	22 30 44	06 ♏ 36 41	13 ♏ 08 54	05 24	27 21	03 14	21 11	26 53	28 23	08 15	12 17
14	19 32 44	23 31 52	19 ♏ 48 11	26 ♏ 34 51	07 02	28 34	03 54	21 17	27 00	28 27	08 18	12 18
15	19 36 40	24 33 00	03 ♐ 29 06	10 ♐ 30 55	08 39	29 ♒ 48	04 34	21 23	27 06	28 30	08 20	12 19
16	19 40 37	25 34 08	17 ♐ 40 07	24 ♐ 56 16	10 16	01 ♓ 01	05 14	21 29	27 13	28 34	08 22	12 20
17	19 44 33	26 35 16	02 ♑ 18 42	09 ♑ 46 35	11 51	02 14	05 54	21 35	27 19	28 37	08 24	12 21
18	19 48 30	27 36 23	17 ♑ 18 49	24 ♑ 54 10	13 26	03 27	06 34	21 42	27 26	28 40	08 26	12 22
19	19 52 26	28 37 30	02 ♒ 31 19	10 ♒ 08 53	14 58	04 40	07 14	21 49	27 32	28 44	08 28	12 23
20	19 56 23	29 ♑ 38 36	17 ♒ 45 29	25 ♒ 19 51	16 28	05 53	07 54	21 55	27 39	28 47	08 31	12 24
21	20 00 19	00 ♒ 39 42	02 ♓ 50 49	10 ♓ 17 25	17 56	07 06	08 34	22 02	27 45	28 50	08 33	12 25
22	20 04 16	01 40 46	17 ♓ 38 52	24 ♓ 54 34	19 20	08 19	09 14	22 09	27 51	28 53	08 35	12 25
23	20 08 13	02 41 50	02 ♈ 03 46	09 ♈ 07 26	20 41	09 32	09 55	22 17	27 58	28 56	08 37	12 26
24	20 12 09	03 42 52	16 ♈ 04 22	22 ♈ 55 04	21 58	10 45	10 35	22 24	28 04	29 00	08 39	12 27
25	20 16 06	04 43 53	29 ♈ 39 45	06 ♉ 18 43	23 09	11 57	11 15	22 32	28 10	29 03	08 41	12 28
26	20 20 02	05 44 54	12 ♉ 52 20	19 ♉ 20 59	24 15	13 10	11 55	22 40	28 16	29 06	08 43	12 29
27	20 23 59	06 45 53	25 ♉ 45 06	02 ♊ 05 04	25 14	14 22	12 35	22 48	28 22	29 09	08 45	12 29
28	20 27 55	07 46 51	08 ♊ 21 20	14 ♊ 34 15	26 05	15 35	13 15	22 56	28 28	29 12	08 47	12 30
29	20 31 52	08 47 49	20 ♊ 44 12	26 ♊ 51 31	26 49	16 47	13 55	23 04	28 34	29 15	08 49	12 30
30	20 35 48	09 48 44	02 ♋ 56 31	08 ♋ 59 27	27 22	17 59	14 35	23 12	28 40	29 18	08 51	12 31
31	20 39 45	10 ♒ 49 39	15 ♋ 00 35	21 ♋ 00 10	27 ♒ 47	19 ♓ 12	15 ♐ 16	23 ♈ 21	28 ♐ 46	29 ♐ 21	08 ♑ 54	12 ♏ 32

DECLINATIONS

Date	Moon True ☊ ° '	Moon Mean ☊ ° '	Moon ☽ Latitude ° '	Sun ☉ ° '	Moon ☽ ° '	Mercury ☿ ° '	Venus ♀ ° '	Mars ♂ ° '	Jupiter ♃ ° '	Saturn ♄ ° '	Uranus ♅ ° '	Neptune ♆ ° '	Pluto ♇ ° '
01	26 ♓ 37	27 ♓ 08	04 N 50	23 S 03	26 N 55	24 S 33	18 S 45	18 S 37	06 N 44	22 S 15	23 S 35	22 S 15	00 S 18
02	26 R 24	27 05	05 00	22 58	28 17	24 23	18 24	18 47	06 45	22 15	23 35	22 15	00 18
03	26 10	27 02	04 56	22 52	28 14	24 11	18 02	18 57	06 47	22 16	23 36	22 15	00 18
04	25 56	26 59	04 39	22 46	26 49	23 58	17 39	19 07	06 49	22 16	23 36	22 15	00 18
05	25 43	26 56	04 09	22 40	24 11	23 43	17 16	19 16	06 51	22 16	23 36	22 15	00 18
06	25 33	26 53	03 28	22 33	20 32	23 27	16 53	19 26	06 53	22 16	23 36	22 15	00 18
07	25 26	26 49	02 39	22 26	16 05	23 09	16 29	19 35	06 55	22 16	23 36	22 15	00 17
08	25 21	26 46	01 43	22 19	11 03	22 49	16 05	19 44	06 57	22 17	23 36	22 15	00 17
09	25 19	26 43	00 N 42	22 10	05 N 37	22 28	15 40	19 53	06 59	22 17	23 36	22 14	00 17
10	25 D 19	26 40	00 S 22	22 02	00 S 04	22 05	15 15	20 02	07 01	22 17	23 36	22 14	00 17
11	25 19	26 37	01 25	21 53	05 51	21 41	14 49	20 11	07 03	22 17	23 36	22 14	00 17
12	25 R 20	26 34	02 26	21 44	11 31	21 15	14 23	20 19	07 05	22 17	23 36	22 14	00 17
13	25 19	26 30	03 22	21 34	16 54	20 48	13 57	20 28	07 08	22 18	23 36	22 14	00 17
14	25 15	26 27	04 09	21 24	21 41	20 19	13 30	20 36	07 10	22 18	23 36	22 14	00 17
15	25 10	26 24	04 43	21 13	25 29	19 49	13 03	20 44	07 13	22 18	23 36	22 14	00 17
16	25 02	26 21	05 02	21 02	27 53	19 18	12 36	20 52	07 15	22 18	23 36	22 13	00 16
17	24 53	26 18	05 02	20 50	28 27	18 46	12 08	20 59	07 18	22 18	23 36	22 13	00 16
18	24 43	26 14	04 42	20 39	26 59	18 13	11 40	21 07	07 21	22 18	23 36	22 13	00 16
19	24 35	26 11	04 01	20 26	23 31	17 38	11 12	21 14	07 23	22 19	23 36	22 13	00 16
20	24 28	26 08	03 03	20 14	18 25	17 03	10 43	21 21	07 26	22 19	23 36	22 13	00 16
21	24 24	26 05	01 50	20 01	12 15	16 27	10 15	21 28	07 29	22 19	23 37	22 13	00 16
22	24 22	26 02	00 S 36	19 47	05 S 26	15 51	09 45	21 35	07 32	22 19	23 37	22 13	00 15
23	24 D 22	25 59	00 N 41	19 34	01 N 27	15 15	09 16	21 42	07 35	22 19	23 37	22 12	00 15
24	24 23	25 55	01 54	19 20	08 05	14 39	08 47	21 48	07 38	22 19	23 37	22 12	00 15
25	24 24	25 52	02 58	19 05	14 08	14 03	08 17	21 55	07 41	22 19	23 37	22 12	00 14
26	24 R 23	25 49	03 51	18 50	19 22	13 28	07 47	22 01	07 44	22 19	23 37	22 12	00 14
27	24 21	25 46	04 31	18 35	23 35	12 55	07 17	22 07	07 47	22 19	23 37	22 12	00 14
28	24 17	25 43	04 56	18 20	26 35	12 23	06 46	22 13	07 50	22 19	23 37	22 12	00 14
29	24 11	25 39	05 07	18 04	28 14	11 53	06 16	22 18	07 54	22 20	23 37	22 12	00 13
30	24 03	25 36	05 04	17 48	28 29	11 26	05 45	22 24	07 57	22 20	23 37	22 12	00 13
31	23 ♓ 55	25 ♓ 33	04 N 48	17 S 31	27 N 22	11 S 01	05 S 14	22 S 29	08 N 00	22 S 20	23 S 37	22 S 11	00 S 13

ZODIAC SIGN ENTRIES

Date	h	m	Planets
03	00	17	☽ ♋
05	11	47	☽ ♌
08	00	35	☽ ♍
08	15	24	♂ ♐
10	05	28	☿ ♒
10	13	17	☽ ♎
12	23	39	☽ ♏
15	05	58	☽ ♐
15	16	04	♀ ♓
17	08	15	☽ ♑
19	08	02	☽ ♒
20	20	24	☉ ♒
21	07	27	☽ ♓
23	08	31	☽ ♈
25	12	36	☽ ♉
27	20	02	☽ ♊
30	06	11	☽ ♋

LATITUDES

Date	Mercury ☿ ° '	Venus ♀ ° '	Mars ♂ ° '	Jupiter ♃ ° '	Saturn ♄ ° '	Uranus ♅ ° '	Neptune ♆ ° '	Pluto ♇ ° '
01	02 S 03	01 S 48	00 N 28	01 S 17	01 N 07	00 S 10	00 N 57	15 N 53
04	02 07	01 47	00 27	01 16	01 07	00 10	00 57	15 54
07	02 08	01 45	00 25	01 15	01 07	00 10	00 57	15 55
10	02 05	01 43	00 23	01 14	01 07	00 10	00 57	15 57
13	01 56	01 39	00 21	01 13	01 07	00 10	00 57	15 58
16	01 42	01 35	00 19	01 13	01 07	00 10	00 57	15 59
19	01 21	01 30	00 17	01 12	01 07	00 10	00 57	16 01
22	00 52	01 24	00 15	01 11	01 07	00 10	00 57	16 02
25	00 S 16	01 18	00 13	01 11	01 07	00 10	00 57	16 04
28	00 N 28	01 11	00 11	01 10	01 07	00 10	00 57	16 05
31	01 N 18	01 S 03	00 N 09	01 S 09	01 N 07	00 S 10	00 N 57	16 N 07

DATA

Julian Date	2447162
Delta T	+56 seconds
Ayanamsa	23° 41' 23"
Synetic vernal point	05° ♓ 25' 36"
True obliquity of ecliptic	23° 26' 36"

MOON'S PHASES, APSIDES AND POSITIONS ☽

Date	h	m	Phase	Longitude ° '	Eclipse Indicator
04	01	40	○	12 ♋ 54	
12	07	04	☾	21 ♎ 17	
19	05	26	●	28 ♑ 21	
25	21	54	☽	05 ♉ 09	

Day	h	m		
07	05	40	Apogee	
19	20	45	Perigee	
02	23	07	Max dec	28° N 26'
10	11	42	0S	
17	06	53	Max dec	28° S 30'
23	06	53	0N	
30	04	17	Max dec	28° N 33'

LONGITUDES

Date	Chiron ⚷ ° '	Ceres ⚳ ° '	Pallas ⚴ ° '	Juno ⚵ ° '	Vesta ⚶ ° '	Black Moon Lilith ☾ ° '
01	25 ♊ 07	29 ♑ 56	08 ♑ 16	13 ♓ 44	06 ♌ 23	15 ♌ 11
11	24 ♊ 33	03 ♒ 30	12 ♑ 13	18 ♓ 17	04 ♌ 21	16 ♌ 18
21	24 ♊ 02	07 ♒ 26	16 ♑ 07	23 ♓ 08	01 ♌ 37	17 ♌ 25
31	23 ♊ 37	11 ♒ 23	19 ♑ 55	28 ♓ 04	29 ♋ 00	18 ♌ 32

All ephemeris data is given at 12.00 UT and the Moon's longitude is additionally given for 24.00 UT

ASPECTARIAN

h m Aspects

01 Friday
00 26 ☽ ∠ ♃
05 19 ☽ ⊼ ♆
08 26 ☽ ± ☿
09 55 ☽ ⊼ ☉
10 06 ☉ ∠ ♂
13 23 ♀ ∠ ♅
13 24 ☽ ⊼ ♇
14 52 ☽ △ ♀
18 11 ♀ ∥ ♂
21 42 ☽ ⊼ ☿
23 40 ♂ ⊻ ♄

02 Saturday
00 59 ☽ ± ♇
05 24 ☽ ✶ ♃
11 44 ♀ ⊥ ♆
14 13 ♄ ∥ ♆
15 48 ☽ ☍ ♄
16 33 ☽ ⊼ ♂
18 30 ☽ bQ ♇
19 55 ☽ ☍ ♅
23 09 ☽ bQ ♀

03 Sunday
04 05 ♂ ± ♃
04 55 ☽ Q ♃
04 57 ☽ ± ♂
05 40 ☉ ✶ ♇
15 46 ☽ ☍ ♆
18 49 ☽ ± ♀
23 42 ☽ bQ ♂

04 Monday
00 02 ☽ △ ♇
01 40 ☽ ☍ ☉
08 02 ☽ ⊼ ♀
10 20 ☿ □ ♃
16 42 ☽ □ ♃
17 40 ☽ ☍ ☿

05 Tuesday
03 39 ☽ ⊼ ♄
07 20 ☽ △ ♂
07 36 ☽ ⊼ ♅
12 42 ♂ ⊻ ♅
15 44 ☽ # ☿
15 49 ☽ ± ♄
16 19 ☽ # ♅
19 44 ☽ ± ♅
23 02 ☽ # ☉
23 03 ☿ ∥ ♅

06 Wednesday
01 19 ☽ # ♄
01 27 ☽ # ♆
03 52 ☽ ⊼ ♆
10 09 ☽ bQ ♄
12 16 ☽ □ ♇
14 01 ☽ bQ ♅
16 03 ☽ ± ♆
16 12 ☿ Q ♇
18 05 ☽ # ♂
19 30 ☽ ⊼ ☉

07 Thursday
03 20 ☽ ☍ ♀
05 33 ☽ △ ♃
08 50 ☽ ± ☉
09 48 ☽ # ♀
10 24 ☽ bQ ♆
15 58 ☽ ⊼ ☿
16 55 ☽ △ ♄
20 39 ☽ △ ♅
23 11 ☿ ⊻ ♄
23 42 ☽ □ ♂

08 Friday
00 55 ☽ Q ♇
01 58 ♀ ✶ ♃
04 53 ☽ bQ ☉
06 10 ☽ ± ☿
12 17 ☽ bQ ♃
17 01 ☽ △ ♆
17 23 ☉ ∥ ♄

09 Saturday
00 11 ☉ ∥ ♆
01 22 ☽ ✶ ♇
02 26 ☿ ⊻ ♅
03 27 ☽ bQ ☿
06 08 ☽ ∥ ♃
06 42 ☽ ± ♃
14 14 ☽ △ ☉
14 27 ☽ Q ♂
18 55 ☽ ⊼ ♃
23 18 ☽ ⊼ ♀
23 34 ☿ ∥ ♄

10 Sunday
01 11 ♀ ∠ ♆
02 23 ☿ ∥ ♆
06 20 ☽ □ ♄
07 43 ☽ ∠ ♇
09 43 ☽ □ ♅
10 29 ☽ # ♇
12 43 ☽ ± ♀
12 54 ☽ ∥ ♇
14 32 ☽ △ ☿
15 58 ☽ ✶ ♂
17 04 ☉ ∥ ☿

11 Monday
01 40 ☽ ⊥ ♇
05 31 ☽ □ ♆
07 10 ☿ ✶ ♂
08 35 ☽ bQ ♀
13 30 ☽ ⊻ ♇
17 05 ☽ # ♃
18 18 ☽ Q ♄
20 45 ☿ ⊥ ♄
21 26 ☽ Q ♅
22 39 ♂ ⊥ ♆
23 12 ☽ ∠ ♂

12 Tuesday
01 37 ☉ □ ♃
06 39 ☽ ☍ ♃
07 04 ☽ □ ☉
16 31 ☽ Q ♆
16 47 ☽ △ ♀
17 27 ☽ ⊥ ♂
17 36 ☽ ✶ ♄
20 33 ☽ ✶ ♅
20 51 ☿ ⊥ ♅
23 36 ☽ ∥ ♀

13 Wednesday
01 57 ♀ ✶ ♄
05 23 ☽ ⊻ ♂
09 25 ☽ □ ☿
15 03 ☽ ✶ ♆
19 48 ☽ Q ☉
21 47 ☽ ∠ ♄
22 26 ☽ ☌ ♇

14 Thursday
00 30 ☽ ∠ ♅
01 27 ☿ ∥ ♂
05 27 ☽ ∥ ☿
06 01 ☽ ∥ ♂
09 27 ♀ ✶ ♅
10 29 ☽ ∥ ☉
14 09 ☽ ⊥ ♄
14 39 ☽ ⊼ ♃
15 04 ☽ ∥ ♆
15 27 ☽ ∥ ♄
16 44 ☽ ⊥ ♅
18 13 ☽ ∠ ♆
19 10 ☽ ✶ ☉
22 32 ☽ Q ☿
23 12 ☽ ∥ ♅

15 Friday
00 50 ☽ ⊻ ♄
01 20 ☽ ± ♃
03 20 ☽ ⊻ ♅
04 59 ☽ □ ♀
06 20 ☉ Q ♇
07 06 ☿ ⊻ ♆
10 00 ☽ ⊥ ♆
13 57 ☽ ☌ ♂
17 00 ☽ bQ ♃
20 19 ☽ ⊻ ♆
21 59 ☽ ✶ ☿
23 10 ☽ ∠ ☉
23 34 ☿ Q ♃

16 Saturday
03 03 ☽ ⊻ ♇
13 06 ☽ ⊥ ♇
14 26 ☽ Q ♀
15 24 ☽ ⊥ ☉
18 22 ☽ △ ♃

17 Sunday
00 55 ☉ ∥ ♂
02 00 ☽ ⊻ ☉
02 04 ☽ ∠ ☿
03 50 ☽ ☌ ♄
03 56 ☽ ∠ ♇
05 59 ☽ ☌ ♅
11 52 ☽ ✶ ♀
18 04 ☽ ⊻ ♂
18 23 ☽ ⊥ ☿
18 49 ♄ ∠ ♇
19 34 ☿ □ ♇
19 37 ☿ ∠ ♄
21 49 ☽ ☌ ♆

18 Monday
04 07 ☽ ⊥ ♂
04 07 ☽ ✶ ♇
05 06 ☽ ⊻ ☿
07 21 ☉ ⊻ ♄
13 58 ☽ ∠ ♀
15 56 ☿ ∠ ♅
17 32 ♂ bQ ♃
18 59 ☽ □ ♃
19 03 ☽ ∠ ♂
23 10 ☽ Q ♇

19 Tuesday
04 06 ☽ ⊻ ♄
04 07 ☿ ⊥ ♆
05 25 ☽ ⊥ ♀
05 26 ☽ ☌ ☉
06 00 ☽ ⊻ ♅
06 35 ☉ ⊥ ♀
11 28 ☽ ∥ ♅
13 37 ☽ ⊥ ♄
14 30 ☉ ⊻ ♅
15 29 ☽ ⊥ ♅
15 40 ☽ ⊻ ♀
18 18 ☽ ∥ ♄
18 46 ☽ ∥ ♆
19 46 ☽ ✶ ♂
21 23 ☽ ⊻ ♆
23 16 ☽ ∥ ♂
23 33 ☽ Q ♃

20 Wednesday
03 32 ☽ □ ♇
03 50 ☽ ∥ ☉
03 53 ☽ ∠ ♄
05 42 ☽ ∠ ♅
06 52 ☽ ⊥ ♆
09 45 ☽ ☌ ☿
15 33 ☽ Q ♂
18 13 ☽ ∥ ☿
18 38 ☽ ✶ ♃
21 08 ☽ ∠ ♆

21 Thursday
03 48 ☽ ✶ ♄
05 33 ☽ ✶ ♅
08 15 ☽ ⊻ ☉
10 37 ♀ ∠ ♃
11 00 ♂ ⊻ ♆
18 35 ☽ ⊥ ☉
18 48 ☽ ∠ ♃
19 27 ☽ ☌ ♀
19 40 ☽ ∥ ♀
21 12 ☽ ✶ ♆
21 39 ☽ □ ♂
23 12 ☽ Q ♄

22 Friday
00 56 ☽ Q ♅
03 27 ☽ △ ♇
04 45 ☽ # ♃
09 32 ☽ ⊥ ♃
10 18 ☽ ∠ ☉
15 04 ☽ ⊻ ☿
16 50 ☽ Q ♆
17 24 ♀ ✶ ♆
19 30 ☽ ⊻ ♃

23 Saturday
02 03 ☽ ⊥ ☿
04 13 ☽ bQ ♇
05 02 ☽ □ ♄
06 01 ☽ ∥ ♇
06 43 ☽ □ ♅
07 46 ☽ # ♇
13 09 ☽ ✶ ☉
18 45 ☽ ∠ ☿
19 25 ☽ ± ♇
21 18 ♀ Q ♄
23 10 ☽ □ ♆

24 Sunday
01 54 ☽ ⊻ ♀
02 01 ☽ △ ♂
04 44 ♀ □ ♂
05 43 ☽ ⊼ ♇
10 18 ☽ ∥ ♃
11 20 ☽ Q ☉
13 17 ☽ ⊥ ♀
14 27 ☽ # ♀
17 11 ♀ Q ♅
21 11 ☉ ⊥ ♄
21 43 ☿ ✶ ♃
23 12 ☽ ☌ ♃
23 19 ☽ ✶ ☿

25 Monday
05 35 ☽ bQ ♂
06 41 ☽ ∠ ♀
09 18 ☽ △ ♄
10 53 ☽ △ ♅
11 41 ☽ # ☿
16 50 ☿ Q ♂
19 48 ☉ ⊥ ♅
21 54 ☽ □ ☉
22 13 ♀ △ ♇
22 36 ☽ ± ♂
22 49 ☽ Q ☿
23 50 ☿ ∠ ♆

26 Tuesday
04 22 ☽ △ ♆
09 28 ☽ # ☉
10 09 ☽ ⊼ ♂
11 16 ☽ ☍ ♇
12 35 ☽ ✶ ♀
12 44 ☽ bQ ♄
14 06 ♀ # ♃
14 16 ☽ bQ ♅

27 Wednesday
02 39 ☽ # ♂
03 23 ☽ # ♆
04 07 ☽ # ♄
05 36 ☽ ± ♄
06 23 ☽ ⊻ ♃
07 05 ☽ ± ♅
08 14 ☽ bQ ♆
08 25 ♂ ⊻ ♇
10 56 ☽ □ ☿
12 13 ☽ # ♅
13 18 ☽ Q ♀
16 59 ☽ ⊼ ♄
17 48 ☽ ⊥ ♃
18 27 ☽ ⊼ ♅

28 Thursday
01 19 ☽ ± ♆
08 17 ♂ ∥ ♆
10 48 ☽ △ ☉
11 10 ☽ ∠ ♃
12 50 ☽ ⊼ ♆
19 59 ☽ ⊼ ♇
21 59 ☽ ☍ ♂

29 Friday
03 28 ☽ □ ♀
07 39 ☽ ± ♇
12 41 ☉ ⊻ ♆
16 36 ☽ ✶ ♃
17 21 ♂ ∥ ♄
18 17 ♀ ⊥ ♃
18 32 ☽ bQ ☉

30 Saturday
00 31 ☽ △ ☿
01 17 ☽ bQ ♇
03 30 ☽ ☍ ♄
04 47 ☽ ☍ ♅
13 53 ☽ ± ☉
16 32 ☽ Q ♃
23 46 ☽ ☍ ♆

31 Sunday
02 53 ☽ ⊼ ☉
07 02 ☽ △ ♇
07 25 ☽ bQ ☿
12 32 ☽ ⊼ ♂
21 18 ☽ △ ♀

FEBRUARY 1988

Date	Sidereal time h m s	Sun ☉ ° ' "	Moon ☽ ° ' "	Moon ☽ 24.00 ° ' "	Mercury ☿ ° '	Venus ♀ ° '	Mars ♂ ° '	Jupiter ♃ ° '	Saturn ♄ ° '	Uranus ♅ ° '	Neptune ♆ ° '	Pluto ⯓ ° '
01	20 43 42	11 ♒ 50 33	26 ♋ 58 22	02 ♌ 55 25	28 ♒ 01	20 ♓ 24	15 ♐ 56	23 ♈ 29	28 ♐ 52	29 ♐ 24	08 ♑ 56	12 ♏ 32
02	20 47 38	12 51 26	08 ♌ 51 31	14 ♌ 46 51	28 R 04	21 36	16 36	23 38	28 58	29 27	08 57	12 32
03	20 51 35	13 52 17	20 ♌ 41 39	26 ♌ 36 06	27 55	22 48	17 16	23 47	29 03	29 30	08 59	12 33
04	20 55 31	14 53 07	02 ♍ 30 29	08 ♍ 25 02	27 36	23 59	17 57	23 56	29 09	29 32	09 01	12 33
05	20 59 28	15 53 57	14 ♍ 20 06	20 ♍ 15 58	27 06	25 11	18 37	24 05	29 15	29 35	09 03	12 34
06	21 03 24	16 54 45	26 ♍ 13 03	02 ♎ 11 44	26 25	26 23	19 17	24 15	29 20	29 38	09 05	12 34
07	21 07 21	17 55 32	08 ♎ 12 28	14 ♎ 15 44	25 36	27 34	19 57	24 24	29 26	29 41	09 07	12 34
08	21 11 17	18 56 18	20 ♎ 22 02	26 ♎ 31 55	24 39	28 46	20 38	24 34	29 31	29 43	09 09	12 34
09	21 15 14	19 57 03	02 ♏ 45 54	09 ♏ 04 32	23 35	29 ♓ 57	21 18	24 43	29 37	29 46	09 11	12 35
10	21 19 11	20 57 47	15 ♏ 28 22	21 ♏ 57 54	22 27	01 ♈ 08	21 58	24 53	29 42	29 49	09 13	12 35
11	21 23 07	21 58 30	28 ♏ 33 36	05 ♐ 15 51	21 17	02 19	22 39	25 03	29 47	29 51	09 14	12 35
12	21 27 04	22 59 13	12 ♐ 04 57	19 ♐ 01 03	20 06	03 30	23 19	25 13	29 52	29 54	09 16	12 35
13	21 31 00	23 59 54	26 ♐ 04 10	03 ♑ 14 10	18 57	04 41	23 59	25 23	29 ♐ 57	29 56	09 18	12 35
14	21 34 57	25 00 34	10 ♑ 30 42	17 ♑ 53 13	17 50	05 52	24 40	25 34	00 ♑ 03	29 ♐ 59	09 20	12 35
15	21 38 53	26 01 13	25 ♑ 20 57	02 ♒ 53 00	16 48	07 02	25 20	25 44	00 08	00 ♑ 01	09 21	12 R 35
16	21 42 50	27 01 50	10 ♒ 28 15	18 ♒ 05 28	15 52	08 13	26 01	25 55	00 12	00 04	09 23	12 35
17	21 46 46	28 02 27	25 ♒ 43 32	03 ♓ 20 40	15 03	09 23	26 41	26 05	00 17	00 06	09 25	12 35
18	21 50 43	29 ♒ 03 01	10 ♓ 56 02	18 ♓ 28 18	14 20	10 34	27 21	26 16	00 22	00 08	09 26	12 35
19	21 54 40	00 ♓ 03 34	25 ♓ 56 23	03 ♈ 19 24	13 45	11 44	28 02	26 27	00 27	00 10	09 28	12 35
20	21 58 36	01 04 05	10 ♈ 36 38	17 ♈ 47 33	13 18	12 54	28 42	26 38	00 31	00 13	09 30	12 34
21	22 02 33	02 04 35	24 ♈ 51 49	01 ♉ 49 15	12 58	14 04	29 ♐ 23	26 49	00 36	00 15	09 31	12 34
22	22 06 29	03 05 02	08 ♉ 39 53	15 ♉ 23 49	12 46	15 13	00 ♑ 03	27 00	00 40	00 17	09 33	12 34
23	22 10 26	04 05 28	22 ♉ 01 19	28 ♉ 32 44	12 41	16 23	00 43	27 11	00 45	00 19	09 34	12 34
24	22 14 22	05 05 52	04 ♊ 58 27	11 ♊ 18 57	12 D 43	17 32	01 24	27 23	00 49	00 21	09 36	12 33
25	22 18 19	06 06 14	17 ♊ 34 42	23 ♊ 46 12	12 51	18 42	02 04	27 34	00 53	00 23	09 37	12 33
26	22 22 15	07 06 34	29 ♊ 54 00	05 ♋ 58 34	13 06	19 51	02 45	27 46	00 58	00 25	09 39	12 33
27	22 26 12	08 06 52	12 ♋ 00 23	17 ♋ 59 57	13 26	21 00	03 25	27 57	01 02	00 27	09 40	12 32
28	22 30 09	09 07 08	23 ♋ 57 42	29 ♋ 54 02	13 51	22 08	04 06	28 09	01 06	00 29	09 41	12 32
29	22 34 05	10 ♓ 07 22	05 ♌ 49 21	11 ♌ 44 01	14 ♒ 22	23 ♈ 17	04 ♑ 46	28 ♈ 21	01 ♑ 10	00 ♑ 31	09 ♑ 43	12 ♏ 31

Date	Moon True ☊ ° '	Moon Mean ☊ ° '	Moon ☽ Latitude ° '
01	23 ♓ 46	25 ♓ 30	04 N 18
02	23 R 39	25 27	03 38
03	23 32	25 24	02 48
04	23 28	25 20	01 51
05	23 26	25 17	00 N 49
06	23 D 25	25 14	00 S 15
07	23 26	25 11	01 20
08	23 28	25 08	02 22
09	23 30	25 05	03 19
10	23 31	25 01	04 07
11	23 R 31	24 58	04 44
12	23 29	24 55	05 07
13	23 27	24 52	05 13
14	23 23	24 49	05 00
15	23 19	24 45	04 26
16	23 15	24 42	03 33
17	23 12	24 39	02 25
18	23 10	24 36	01 S 07
19	23 D 10	24 33	00 N 15
20	23 11	24 30	01 35
21	23 12	24 26	02 46
22	23 13	24 23	03 45
23	23 15	24 20	04 30
24	23 15	24 17	05 00
25	23 R 15	24 14	05 14
26	23 13	24 11	05 14
27	23 11	24 07	04 59
28	23 09	24 04	04 32
29	23 ♓ 07	24 ♓ 01	03 N 53

DECLINATIONS

Date	Sun ☉ ° '	Moon ☽ ° '	Mercury ☿ ° '	Venus ♀ ° '	Mars ♂ ° '	Jupiter ♃ ° '	Saturn ♄ ° '	Uranus ♅ ° '	Neptune ♆ ° '	Pluto ⯓ ° '
01	17 S 14	24 N 59	10 S 40	04 S 44	22 S 34	08 N 04	22 S 20	23 S 37	22 S 11	00 S 12
02	16 57	21 33	10 23	04 13	22 39	08 07	22 20	23 37	22 11	00 12
03	16 40	17 15	10 10	03 41	22 44	08 11	22 20	23 37	22 11	00 11
04	16 22	12 19	10 01	03 10	22 48	08 14	22 20	23 37	22 11	00 11
05	16 04	06 56	09 57	02 39	22 52	08 18	22 20	23 37	22 11	00 11
06	15 46	01 N 16	09 57	02 08	22 57	08 22	22 20	23 37	22 11	00 10
07	15 28	04 S 29	10 01	01 36	23 01	08 25	22 20	23 37	22 10	00 10
08	15 09	10 09	10 10	01 05	23 04	08 29	22 20	23 37	22 10	00 09
09	14 50	15 32	10 22	00 33	23 08	08 33	22 20	23 37	22 10	00 09
10	14 31	20 25	10 37	00 S 02	23 11	08 36	22 20	23 37	22 10	00 09
11	14 11	24 28	10 56	00 N 30	23 14	08 40	22 20	23 37	22 10	00 08
12	13 51	27 19	11 16	01 01	23 17	08 44	22 20	23 37	22 10	00 08
13	13 31	28 36	11 38	01 33	23 20	08 48	22 20	23 37	22 10	00 07
14	13 11	28 00	12 01	02 04	23 23	08 52	22 20	23 37	22 09	00 07
15	12 51	25 26	12 24	02 36	23 25	08 56	22 20	23 37	22 09	00 06
16	12 30	21 02	12 47	03 07	23 27	09 00	22 20	23 37	22 09	00 06
17	12 09	15 14	13 09	03 38	23 29	09 04	22 20	23 37	22 09	00 05
18	11 48	08 30	13 30	04 09	23 31	09 08	22 20	23 37	22 09	00 05
19	11 27	01 S 23	13 50	04 41	23 33	09 12	22 20	23 37	22 09	00 04
20	11 06	05 N 39	14 09	05 12	23 34	09 16	22 20	23 37	22 09	00 04
21	10 44	12 12	14 26	05 43	23 35	09 21	22 20	23 37	22 09	00 03
22	10 23	17 57	14 42	06 13	23 36	09 25	22 19	23 37	22 08	00 03
23	10 01	22 38	14 56	06 44	23 37	09 29	22 19	23 37	22 08	00 02
24	09 39	26 03	15 08	07 15	23 38	09 33	22 19	23 37	22 08	00 01
25	09 16	28 05	15 18	07 45	23 38	09 38	22 19	23 37	22 08	00 S 01
26	08 54	28 40	15 27	08 15	23 38	09 42	22 19	23 37	22 08	00 00
27	08 32	27 52	15 33	08 45	23 38	09 46	22 19	23 37	22 08	00 00
28	08 09	25 47	15 38	09 15	23 38	09 51	22 19	23 37	22 08	00 N 01
29	07 S 46	22 N 35	15 S 42	09 N 45	23 S 38	09 N 55	22 S 19	23 S 37	22 S 08	00 N 01

ZODIAC SIGN ENTRIES

Date	h	m	Planets
01	18	06	☽ ♌
04	06	54	☽ ♍
06	19	36	☽ ♎
09	06	42	☽ ♏
09	13	04	♀ ♈
11	14	36	☽ ♐
13	18	36	☽ ♑
13	23	51	♄ ♑
15	00	11	♅ ♑
15	19	25	☽ ♒
17	18	44	☽ ♓
19	10	35	☉ ♓
19	18	35	☽ ♈
21	20	50	☽ ♉
22	10	15	♂ ♑
24	02	42	☽ ♊
26	12	12	☽ ♋
29	00	12	☽ ♌

LATITUDES

Date	Mercury ☿ ° '	Venus ♀ ° '	Mars ♂ ° '	Jupiter ♃ ° '	Saturn ♄ ° '	Uranus ♅ ° '	Neptune ♆ ° '	Pluto ⯓ ° '
01	01 N 36	01 S 00	00 N 08	01 S 08	01 N 07	00 S 10	00 N 57	16 N 08
04	02 26	00 51	00 06	01 08	01 07	00 10	00 57	16 09
07	03 09	00 42	00 03	01 07	01 07	00 10	00 58	16 11
10	03 36	00 31	00 N 01	01 06	01 07	00 11	00 58	16 12
13	03 42	00 21	00 S 02	01 06	01 07	00 11	00 58	16 14
16	03 28	00 S 09	00 04	01 05	01 07	00 11	00 58	16 15
19	02 59	00 N 02	00 07	01 04	01 07	00 11	00 58	16 17
22	02 23	00 15	00 10	01 04	01 07	00 11	00 58	16 18
25	01 44	00 27	00 13	01 03	01 07	00 11	00 58	16 20
28	01 05	00 40	00 15	01 03	01 07	00 11	00 58	16 21
31	00 N 28	00 N 54	00 S 18	01 S 02	01 N 07	00 S 11	00 N 58	16 N 23

DATA

Julian Date	2447193
Delta T	+56 seconds
Ayanamsa	23° 41' 29"
Synetic vernal point	05° ♓ 25' 30"
True obliquity of ecliptic	23° 26' 36"

MOON'S PHASES, APSIDES AND POSITIONS ☽

Date	h	m	Phase	Longitude ° '	Eclipse Indicator
02	20	51	○	13 ♌ 14	
10	23	01	☾	21 ♏ 26	
17	15	54	●	28 ♒ 12	
24	12	15	☽	05 ♊ 07	

Day	h	m		
03	10	10	Apogee	
17	09	36	Perigee	
06	17	19	0S	
13	16	45	Max dec	28° S 38'
19	16	38	0N	
26	09	58	Max dec	28° N 41'

LONGITUDES

Date	Chiron ⚷ ° '	Ceres ⚳ ° '	Pallas ⚴ ° '	Juno ⚵ ° '	Vesta ⚶ ° '	Black Moon Lilith ⚸ ° '
01	23 ♊ 35	11 ♒ 47	20 ♑ 18	28 ♓ 35	28 ♋ 44	18 ♌ 38
11	23 ♊ 18	15 ♒ 43	24 ♑ 01	03 ♈ 47	26 ♋ 26	19 ♌ 45
21	23 ♊ 09	19 ♒ 38	27 ♑ 36	09 ♈ 07	24 ♋ 42	20 ♌ 52
31	23 ♊ 07	23 ♒ 31	01 ♒ 03	14 ♈ 39	23 ♋ 41	21 ♌ 58

All ephemeris data is given at 12.00 UT and the Moon's longitude is additionally given for 24.00 UT

ASPECTARIAN

h m	Aspects
01 Monday	
01 16	☽ ± ♂
01 53	☽ ± ☿
02 15	☉ Q ♃
04 54	☽ □ ♃
14 07	☽ ⊼ ☿
15 31	☿ Q ♂
15 51	☽ ⊼ ♄
16 54	☽ ⊼ ♅
20 27	☽ ♇ ♂
22 26	☽ ♯ ♅
22 56	♀ Q ♆
02 Tuesday	
04 03	☽ ± ♄
04 28	☉ □ ⯓
05 04	☽ ± ♅
05 11	☽ ♯ ♂
06 17	☿ St R
06 54	☽ ♇ ♀
07 07	☽ ♯ ♄
08 02	☽ ♯ ♆
12 12	☽ ⊼ ♆
19 28	☽ □ ⯓
20 51	☽ ☍ ☉
22 25	☽ ♇ ♄
23 22	☽ ♇ ♅
03 Wednesday	
00 24	☽ ± ♆
03 11	☽ ± ♀
04 38	☽ △ ♂
15 13	☽ ♯ ☉
16 44	☽ ⊼ ♀
16 52	☉ ∠ ♄
18 21	☽ △ ♃
18 43	☽ ♇ ♆
04 Thursday	
02 20	☽ ☍ ☿
03 26	☉ ∠ ♅
05 08	☽ △ ♄
05 57	☽ △ ♅
08 02	☽ Q ⯓
10 46	♀ ⊻ ♃
15 22	☉ ⊥ ♆
22 37	☽ ♯ ☿
05 Friday	
01 13	☽ ♇ ♃
01 16	☽ △ ♆
06 06	☽ ∥ ♃
08 24	☽ ⚹ ⯓
10 01	♂ ⊥ ⯓
15 28	☽ ⊼ ☉
19 42	☽ ± ♃
21 11	☽ □ ♂
06 Saturday	
04 43	☽ ± ☉
07 58	☽ ⊼ ♃
08 03	☽ ♯ ♀
12 21	☽ ☍ ♀
12 23	☽ ⊼ ☿
12 34	☿ ⊻ ♀
14 43	☽ ∠ ⯓
16 36	☽ ♯ ⯓
18 01	☽ ∥ ⯓
18 19	☽ □ ♄
18 53	☽ □ ♅
23 41	☽ ± ☿
07 Sunday	
00 29	☽ ♇ ☉
01 00	☽ ∥ ♀
08 44	☽ ⊥ ⯓
11 28	☽ Q ♂
12 01	♀ ♇ ⯓
13 49	☽ □ ♆
16 25	☽ ♇ ☿
20 39	☽ ⊻ ⯓
08 Monday	
04 47	☽ ♯ ♃
06 23	☽ Q ♄
06 48	☽ Q ♅
08 57	☽ △ ☉
12 03	☽ ∥ ☿
12 33	☽ ⚹ ♂
13 41	☿ ⚹ ♃
19 42	☽ △ ☿
20 17	☽ ☍ ♃
23 06	☿ ∠ ♆
09 Tuesday	
01 14	☽ Q ♆
04 38	♀ □ ♄
05 54	☽ ⚹ ♄
06 01	☽ ⊼ ♀
06 14	☽ ⚹ ♅
08 10	☿ ⊥ ♀
08 13	♀ □ ♅
08 55	☽ ∥ ☉
18 42	☽ ± ♀
19 07	☽ ∠ ♂
10 Wednesday	
00 14	☽ ⚹ ♆
06 36	☽ ☌ ⯓
06 45	♀ ∥ ⯓
10 33	☽ ∠ ♄
10 46	☽ ∠ ♅
12 59	☽ ⊥ ♂
13 21	☽ ♇ ♀
18 15	☿ ⚹ ♂
19 46	♀ ♯ ♆
21 41	☽ ∥ ♆
22 39	☽ ∥ ♄
23 01	☽ □ ☉
23 50	☽ □ ☿
11 Thursday	
00 40	☽ ⊻ ♂
03 17	☽ ⊥ ♄
03 26	☽ ⊥ ♅
03 59	☽ ∥ ♂
04 09	☽ ∠ ♆
04 23	☉ ☌ ☿
05 33	☽ ⊼ ♃
06 28	☽ ∥ ♅
14 13	☽ ⊻ ♄
14 20	☽ ⊻ ♅
16 32	☽ ± ♃
19 24	☽ △ ♀
20 25	☽ ⊥ ♆
12 Friday	
05 34	☽ Q ☿
07 04	☽ ⊻ ♆
08 42	☽ ♇ ♃
09 56	☽ Q ☉
12 52	☽ ⊻ ⯓
23 15	☽ ⊥ ⯓
13 Saturday	
00 47	☽ ⚹ ☿
00 58	♄ ☌ ♅
04 21	☿ ∠ ♀
08 14	☽ ⚹ ☉
08 18	☽ ☌ ♂
10 50	☽ △ ♃
11 27	☉ ⚹ ♂
14 33	☽ ∠ ⯓
18 31	☽ ☌ ♅
18 34	☽ ☌ ♄
19 21	☉ ∠ ♆
14 Sunday	
00 13	☽ ∠ ☿
03 41	☽ □ ♀
10 03	☽ ☌ ♆
11 07	☽ ∠ ☉
14 01	☽ ⊥ ☿
14 52	⯓ St R
15 23	☽ ⚹ ⯓
23 08	☽ ⊻ ☿
15 Monday	
02 43	♀ ± ⯓
02 49	☽ ⊥ ☉
03 46	☉ ⚹ ♃
10 47	☽ Q ⯓
11 28	☽ Q ♀
11 59	☽ ⊻ ♂
12 37	☽ □ ♃
13 09	☽ ⊻ ☉
19 28	☽ ⊻ ♅
19 40	☽ ⊻ ♄
21 59	☽ ⊥ ♂
23 01	☽ ∥ ♅
16 Tuesday	
00 01	☽ ∥ ♂
02 59	☉ ∥ ☿
05 01	☽ ⊥ ♅
05 14	☽ ⊥ ♄
05 48	☽ ∥ ♄
06 40	☽ ∥ ♆
07 09	♂ △ ♃
08 09	☽ ⚹ ♀
10 17	☽ ⊻ ♆
12 53	☽ ∠ ♂
15 20	☽ □ ⯓
17 29	☽ Q ♃
19 15	☽ ∠ ♅
19 30	☽ ∠ ♄
19 45	☽ ⊥ ♆
20 03	☽ ☌ ☿
17 Wednesday	
01 14	☿ ⊥ ♆
05 13	☿ ∠ ♄
09 44	☽ ∠ ♀
09 56	☽ ∠ ♆
10 23	☿ ∠ ♅
12 30	♀ □ ♆
12 35	☽ ⚹ ♃
13 35	☽ ⚹ ♂
15 54	☽ ☌ ☉
18 54	☽ ⚹ ♅
19 13	☽ ⚹ ♄
19 17	☽ ∥ ☿
23 51	☽ ∥ ☉
18 Thursday	
01 05	☽ ⊥ ♀
09 23	☽ Q ♂
09 38	☽ ⚹ ♆
09 50	☽ ♯ ♃
11 21	☽ ⊻ ♀
12 32	☽ ∠ ♃
13 55	☽ Q ♅
14 06	☿ Q ♃
14 17	☽ Q ♄
14 37	☽ △ ⯓
17 11	☽ ⊻ ☿
19 59	♂ ∠ ⯓
19 Friday	
01 41	☽ ♯ ♀
02 24	☽ ⊥ ☿
03 03	☽ ⊥ ♃
04 47	☽ Q ♆
12 50	☽ ⊻ ♆
14 39	☽ ♇ ⯓
14 51	☉ ⚹ ♅
15 32	☽ □ ♂
16 24	☽ ∠ ☿
16 25	☽ ∥ ⯓
16 52	☽ ♯ ⯓
18 53	☽ □ ♅
19 10	☽ ⊻ ☉
19 21	☽ □ ♄
21 58	☉ ⚹ ♄
20 Saturday	
03 07	☿ ∠ ♂
05 20	☽ ± ⯓
05 24	♀ ⊼ ⯓
05 43	☽ ⊥ ☉
10 09	☽ □ ♆
10 16	☽ ∥ ♀
15 16	☽ ⊼ ⯓
16 08	☽ ☌ ♀
16 21	☽ ⚹ ☿
18 17	☿ ⚹ ♀
21 48	☽ ∠ ☉
21 Sunday	
01 06	☽ ∥ ♃
06 42	☽ ♯ ☉
12 11	☽ Q ☿
15 23	☽ ☌ ♃
20 09	☽ △ ♂
21 18	☽ △ ♅
21 18	☽ ♯ ☿
21 56	☽ △ ♄
22 Monday	
01 26	☽ ⚹ ☉
13 34	☽ △ ♆
18 56	☽ ☍ ⯓
19 13	☽ □ ☿
20 48	♂ ☌ ♅
23 50	☽ ♇ ♅
23 59	☽ ♇ ♂
23 Tuesday	
00 22	☽ Q ☉
00 34	☽ ♇ ♄
00 48	☽ ⊻ ♀
09 11	☽ ♯ ♆
10 14	☽ ♯ ♄
12 43	☽ ⊥ ♀
12 59	♂ ☌ ♄
16 13	☽ ± ♅
16 40	☽ ♇ ♆
17 01	☽ ± ♄
17 13	☽ ± ♂
17 30	☿ St D
18 06	☽ ♯ ♂
18 06	☽ ♯ ♅
18 43	♂ ∥ ♅
21 37	☽ ⊻ ♃
24 Wednesday	
03 20	☽ ⊼ ♅
04 11	☽ ⊼ ♄
04 56	☽ ⊼ ♂
06 59	☽ ∠ ♀
08 57	☽ ⊥ ♃
09 24	☽ ± ♆
12 15	☽ □ ☉
16 52	☉ ♯ ♃
20 45	☽ ⊼ ♆
25 Thursday	
02 14	☽ ∠ ♃
02 21	☽ ⊼ ⯓
02 48	☽ △ ☿
13 52	☽ ± ⯓
14 22	☽ ⚹ ♀
26 Friday	
07 22	☽ ♇ ⯓
07 44	☽ ⚹ ♃
08 22	☽ ♇ ☿
13 01	☽ ☍ ♅
14 06	☽ ☍ ♄
16 14	☽ Q ♀
17 56	☽ ☍ ♂
27 Saturday	
02 36	☽ ± ☿
03 33	☽ △ ☉
05 52	☉ ♯ ♀
07 20	☽ ☍ ♆
07 50	☽ Q ♃
13 03	☽ △ ⯓
14 56	☽ ⊼ ☿
28 Sunday	
07 56	☽ □ ♀
12 21	☽ ♇ ☉
20 36	☽ □ ♃
29 Monday	
01 13	☽ ⊼ ♅
01 58	☉ ⚹ ♆
02 30	☽ ⊼ ♄
04 51	☽ ♯ ♂
04 55	☽ ♯ ♅
08 14	☽ ± ☉
09 44	☽ ⊼ ♂
13 24	☽ ± ♅
13 40	☽ ♯ ♄
14 44	☽ ± ♄
14 53	☽ ♯ ♆
19 55	☽ ⊼ ♆
21 32	☽ ⊼ ☉
21 57	♀ ± ♅
22 39	☽ ⊥ ♂

MARCH 1988

LONGITUDES

Date	Sidereal time h m s	Sun ☉ ° ' "	Moon ☽ ° ' "	Moon ☽ 24.00 ° ' "	Mercury ☿ ° '	Venus ♀ ° '	Mars ♂ ° '	Jupiter ♃ ° '	Saturn ♄ ° '	Uranus ♅ ° '	Neptune ♆ ° '	Pluto ♇ ° '
01	22 38 02	11 ♓ 07 34	17 ♌ 38 22	23 ♌ 32 43	14 ♒ 57	24 ♈ 25	05 ♑ 26	28 ♈ 33	01 ♑ 14	00 ♑ 32	09 ♑ 44	12 ♏ 31
02	22 41 58	12 07 45	29 ♌ 27 20	05 ♍ 22 31	15 36	25 34	06 07	28 45	01 17	00 34	09 45	12 R 30
03	22 45 55	13 07 53	11 ♍ 18 31	17 ♍ 15 35	16 19	26 42	06 47	28 57	01 21	00 36	09 47	12 29
04	22 49 51	14 07 59	23 ♍ 13 58	29 ♍ 13 54	17 06	27 50	07 28	29 09	01 25	00 37	09 48	12 29
05	22 53 48	15 08 04	05 ♎ 15 39	11 ♎ 19 27	17 57	28 ♈ 57	08 08	29 21	01 28	00 39	09 49	12 28
06	22 57 44	16 08 07	17 ♎ 25 36	23 ♎ 34 21	18 51	00 ♉ 05	08 49	29 33	01 32	00 41	09 50	12 27
07	23 01 41	17 08 08	29 ♎ 46 00	06 ♏ 00 52	19 47	01 12	09 29	29 46	01 35	00 42	09 51	12 27
08	23 05 38	18 08 08	12 ♏ 19 16	18 ♏ 41 31	20 47	02 19	10 10	29 ♈ 58	01 38	00 43	09 53	12 26
09	23 09 34	19 08 06	25 ♏ 07 56	01 ♐ 38 52	21 49	03 26	10 50	00 ♉ 11	01 41	00 45	09 54	12 25
10	23 13 31	20 08 02	08 ♐ 14 35	14 ♐ 55 22	22 54	04 32	11 31	00 23	01 44	00 46	09 55	12 24
11	23 17 27	21 07 57	21 ♐ 41 26	28 ♐ 32 57	24 00	05 39	12 11	00 36	01 47	00 47	09 56	12 23
12	23 21 24	22 07 50	05 ♑ 29 59	12 ♑ 32 32	25 10	06 45	12 52	00 49	01 50	00 49	09 57	12 23
13	23 25 20	23 07 42	19 ♑ 40 26	26 ♑ 53 28	26 21	07 51	13 32	01 01	01 53	00 50	09 58	12 22
14	23 29 17	24 07 32	04 ♒ 11 12	11 ♒ 33 05	27 34	08 57	14 13	01 14	01 56	00 51	09 59	12 21
15	23 33 13	25 07 20	18 ♒ 58 28	26 ♒ 26 31	28 ♒ 49	10 02	14 54	01 27	01 58	00 52	10 00	12 20
16	23 37 10	26 07 06	03 ♓ 56 18	11 ♓ 26 48	00 ♓ 06	11 07	15 34	01 40	02 01	00 53	10 00	12 19
17	23 41 07	27 06 50	18 ♓ 56 59	26 ♓ 25 44	01 25	12 12	16 15	01 53	02 03	00 54	10 01	12 18
18	23 45 03	28 06 33	03 ♈ 52 02	11 ♈ 14 54	02 45	13 17	16 55	02 06	02 06	00 55	10 02	12 17
19	23 49 00	29 ♓ 06 13	18 ♈ 33 25	25 ♈ 46 51	04 07	14 21	17 36	02 19	02 08	00 56	10 03	12 16
20	23 52 56	00 ♈ 05 51	02 ♉ 54 35	09 ♉ 56 09	05 30	15 25	18 16	02 33	02 10	00 57	10 04	12 14
21	23 56 53	01 05 27	16 ♉ 51 15	23 ♉ 39 45	06 55	16 29	18 57	02 46	02 12	00 58	10 04	12 13
22	00 00 49	02 05 01	00 ♊ 21 39	06 ♊ 57 06	08 22	17 33	19 37	02 59	02 14	00 58	10 05	12 12
23	00 04 46	03 04 33	13 ♊ 26 21	19 ♊ 49 44	09 50	18 36	20 18	03 13	02 16	00 59	10 06	12 11
24	00 08 42	04 04 02	26 ♊ 07 43	02 ♋ 20 46	11 19	19 39	20 58	03 26	02 18	01 00	10 06	12 10
25	00 12 39	05 03 29	08 ♋ 29 25	14 ♋ 34 14	12 50	20 41	21 39	03 40	02 20	01 00	10 07	12 08
26	00 16 36	06 02 54	20 ♋ 35 48	26 ♋ 34 43	14 22	21 44	22 19	03 53	02 21	01 01	10 07	12 07
27	00 20 32	07 02 17	02 ♌ 31 34	08 ♌ 26 54	15 56	22 45	22 59	04 07	02 23	01 01	10 08	12 06
28	00 24 29	08 01 37	14 ♌ 21 18	20 ♌ 15 16	17 31	23 47	23 40	04 20	02 24	01 01	10 08	12 05
29	00 28 25	09 00 55	26 ♌ 09 18	02 ♍ 03 54	19 08	24 48	24 20	04 34	02 25	01 02	10 09	12 03
30	00 32 22	10 00 10	07 ♍ 59 28	13 ♍ 56 24	20 45	25 49	25 01	04 48	02 27	01 02	10 09	12 02
31	00 36 18	10 ♈ 59 24	19 ♍ 55 03	25 ♍ 55 45	22 ♓ 25	26 ♉ 49	25 ♑ 41	05 ♉ 01	02 ♑ 28	01 ♑ 02	10 ♑ 10	12 ♏ 01

Date	Moon True ☊ ° '	Moon Mean ☊ ° '	Moon ☽ Latitude ° '
01	23 ♓ 05	23 ♓ 58	03 N 04
02	23 R 04	23 55	02 07
03	23 03	23 51	01 N 05
04	23 D 03	23 48	00 S 01
05	23 03	23 45	01 07
06	23 04	23 42	02 11
07	23 04	23 39	03 10
08	23 05	23 36	04 01
09	23 05	23 32	04 41
10	23 05	23 29	05 07
11	23 R 06	23 26	05 18
12	23 06	23 23	05 10
13	23 D 05	23 20	04 44
14	23 06	23 17	04 00
15	23 06	23 13	02 59
16	23 06	23 10	01 45
17	23 06	23 07	00 S 23
18	23 R 06	23 04	00 N 59
19	23 06	23 01	02 17
20	23 05	22 57	03 24
21	23 04	22 54	04 16
22	23 03	22 51	04 53
23	23 02	22 48	05 13
24	23 01	22 45	05 17
25	23 D 01	22 42	05 06
26	23 02	22 38	04 41
27	23 03	22 35	04 05
28	23 04	22 32	03 19
29	23 06	22 29	02 24
30	23 07	22 26	01 23
31	23 ♓ 07	22 ♓ 23	00 N 18

DECLINATIONS

Date	Sun ☉ ° '	Moon ☽ ° '	Mercury ☿ ° '	Venus ♀ ° '	Mars ♂ ° '	Jupiter ♃ ° '	Saturn ♄ ° '	Uranus ♅ ° '	Neptune ♆ ° '	Pluto ♇ ° '
01	07 S 24	18 N 28	15 S 43	10 N 14	23 S 37	09 N 59	22 S 19	23 S 37	22 S 07	00 N 02
02	07 01	13 39	15 43	10 43	23 37	10 04	22 19	23 37	22 07	00 03
03	06 38	08 19	15 41	11 12	23 36	10 08	22 19	23 37	22 07	00 03
04	06 15	02 N 40	15 38	11 41	23 34	10 13	22 19	23 37	22 07	00 04
05	05 51	03 S 07	15 33	12 09	23 33	10 17	22 19	23 37	22 07	00 05
06	05 28	08 52	15 26	12 37	23 32	10 22	22 19	23 37	22 07	00 05
07	05 05	14 21	15 18	13 05	23 30	10 26	22 19	23 37	22 07	00 06
08	04 42	19 21	15 09	13 33	23 28	10 31	22 18	23 37	22 07	00 06
09	04 18	23 35	14 58	14 00	23 26	10 35	22 18	23 37	22 07	00 07
10	03 55	26 44	14 45	14 27	23 23	10 40	22 18	23 38	22 06	00 08
11	03 31	28 28	14 31	14 54	23 21	10 44	22 18	23 38	22 06	00 08
12	03 07	28 30	14 15	15 20	23 18	10 49	22 18	23 38	22 06	00 09
13	02 44	26 41	13 58	15 46	23 15	10 53	22 18	23 38	22 06	00 10
14	02 20	23 06	13 40	16 12	23 12	10 58	22 18	23 38	22 06	00 10
15	01 56	17 58	13 20	16 37	23 09	11 03	22 18	23 38	22 06	00 11
16	01 33	11 41	12 59	17 02	23 05	11 07	22 18	23 38	22 06	00 12
17	01 09	04 S 44	12 37	17 27	23 02	11 12	22 18	23 38	22 06	00 12
18	00 45	02 N 27	12 13	17 51	22 58	11 17	22 18	23 38	22 06	00 13
19	00 S 21	09 23	11 48	18 15	22 54	11 21	22 18	23 38	22 06	00 13
20	00 N 02	15 40	11 22	18 38	22 50	11 26	22 18	23 38	22 06	00 14
21	00 26	20 58	10 54	19 01	22 45	11 31	22 17	23 38	22 06	00 15
22	00 50	25 00	10 25	19 24	22 41	11 35	22 17	23 38	22 05	00 15
23	01 13	27 35	09 55	19 46	22 36	11 40	22 17	23 38	22 05	00 16
24	01 37	28 40	09 24	20 07	22 31	11 45	22 17	23 38	22 05	00 17
25	02 01	28 15	08 51	20 29	22 26	11 49	22 17	23 38	22 05	00 17
26	02 24	26 30	08 18	20 50	22 20	11 54	22 17	23 38	22 05	00 18
27	02 48	23 34	07 43	21 10	22 15	11 59	22 17	23 38	22 05	00 19
28	03 11	19 41	07 07	21 30	22 09	12 03	22 17	23 38	22 05	00 19
29	03 34	15 03	06 29	21 49	22 04	12 08	22 17	23 38	22 05	00 20
30	03 58	09 51	05 51	22 08	21 58	12 13	22 17	23 38	22 05	00 21
31	04 N 21	04 N 16	05 S 11	22 N 26	21 S 51	12 N 17	22 S 17	23 S 38	22 S 05	00 N 21

ZODIAC SIGN ENTRIES

Date	h	m	Planets
02	13	06	☽ ♍
05	01	32	☽ ♎
06	10	21	♀ ♉
07	12	27	☽ ♏
08	15	44	♃ ♉
09	20	59	☽ ♐
12	02	31	☽ ♑
14	05	08	☽ ♒
16	05	42	☽ ♓
16	10	09	☿ ♓
18	05	45	☽ ♈
20	07	05	☽ ♉
20	09	39	☉ ♈
22	11	21	☽ ♊
24	19	27	☽ ♋
27	06	54	☽ ♌
29	19	49	☽ ♍

LATITUDES

Date	Mercury ☿ ° '	Venus ♀ ° '	Mars ♂ ° '	Jupiter ♃ ° '	Saturn ♄ ° '	Uranus ♅ ° '	Neptune ♆ ° '	Pluto ♇ ° '
01	00 N 40	00 N 49	00 S 17	01 S 02	01 N 07	00 S 11	00 N 58	16 N 22
04	00 N 05	01 03	00 20	01 02	01 07	00 11	00 58	16 23
07	00 S 27	01 16	00 24	01 01	01 07	00 11	00 58	16 25
10	00 55	01 30	00 27	01 01	01 08	00 11	00 58	16 26
13	01 19	01 44	00 30	01 00	01 08	00 11	00 58	16 27
16	01 39	01 58	00 33	01 00	01 08	00 11	00 58	16 28
19	01 56	02 12	00 37	00 59	01 08	00 11	00 58	16 30
22	02 09	02 25	00 41	00 59	01 08	00 11	00 58	16 31
25	02 17	02 39	00 44	00 59	01 08	00 11	00 58	16 32
28	02 22	02 52	00 48	00 58	01 08	00 11	00 58	16 32
31	02 S 22	03 N 05	00 S 52	00 S 58	01 N 09	00 S 11	00 N 59	16 N 33

DATA

Julian Date	2447222
Delta T	+56 seconds
Ayanamsa	23° 41' 33"
Synetic vernal point	05° ♓ 25' 26"
True obliquity of ecliptic	23° 26' 37"

MOON'S PHASES, APSIDES AND POSITIONS ☽

Date	h	m	Phase	Longitude ° '	Eclipse Indicator
03	16	01	○	13 ♍ 18	partial
11	10	56	☾	21 ♐ 05	
18	02	02	●	27 ♓ 42	Total
25	04	42	☽	04 ♋ 45	

Day	h	m		
01	11	40	Apogee	
16	20	19	Perigee	
29	00	22	Apogee	
04	23	06	OS	
12	00	28	Max dec	28° S 42'
18	03	48	ON	
24	17	12	Max dec	28° N 42'

LONGITUDES

Date	Chiron ⚷ ° '	Ceres ⚳ ° '	Pallas ⚴ ° '	Juno ⚵ ° '	Vesta ⚶ ° '	Black Moon Lilith ⚸ ° '
01	23 ♊ 07	23 ♒ 08	00 ♒ 43	14 ♈ 05	23 ♋ 45	21 ♌ 52
11	23 ♊ 13	26 ♒ 58	04 ♒ 01	19 ♈ 41	23 ♋ 26	22 ♌ 58
21	23 ♊ 27	00 ♓ 45	07 ♒ 07	25 ♈ 24	23 ♋ 52	24 ♌ 05
31	23 ♊ 48	04 ♓ 25	10 ♒ 00	01 ♉ 11	25 ♋ 00	25 ♌ 12

All ephemeris data is given at 12.00 UT and the Moon's longitude is additionally given for 24.00 UT

ASPECTARIAN

h m Aspects

01 Tuesday
01 35 ☽ □ ♇
06 13 ☽ ☍ ☿
07 43 ☽ ⚼ ♅
08 07 ☽ ± ♆
08 37 ♂ ∥ ♅
09 07 ☽ ⚼ ♄
18 02 ☽ ⚼ ♂

02 Wednesday
02 02 ☽ # ☿
02 27 ☽ ⚼ ♆
03 15 ☽ △ ♀
10 32 ☽ △ ♃
10 52 ☿ ∠ ♅
14 07 ☽ Q ♇
14 16 ☽ △ ♅
15 44 ☽ △ ♄
17 32 ☿ ⊥ ♆
20 48 ☉ △ ♇
22 51 ☉ Q ♅

03 Thursday
00 19 ☽ ∥ ♀
02 19 ☽ △ ♂
04 08 ☽ ∥ ♃
08 54 ☽ △ ♆
12 52 ☽ ⚼ ♀
12 56 ☿ ∠ ♄
14 23 ☽ ⚹ ♇
16 01 ☽ ☍ ☉
17 25 ☽ ⚼ ♃
17 34 ☉ Q ♄
19 50 ☽ # ☉
22 49 ☽ ⚻ ☿

04 Friday
08 53 ☽ ± ♀
11 44 ☽ ± ☿
11 49 ☽ ± ♃
12 22 ☉ ∠ ♃
13 31 ☿ Q ♃
20 30 ☽ ∠ ♇
22 09 ☽ ⚻ ♀
22 48 ☽ ∥ ♇
23 23 ☽ # ♇

05 Saturday
00 02 ☽ ⚻ ♃
02 49 ☽ □ ♅
04 25 ☽ □ ♄
07 03 ☽ ⚼ ☿
14 24 ☽ ⊥ ♇
18 03 ☽ □ ♂
21 02 ☽ □ ♆
22 19 ♀ ☌ ♃
22 39 ☽ ∥ ☉

06 Sunday
02 15 ☽ ⚺ ♇
09 15 ☽ ⚻ ☉
14 27 ☽ Q ♅
15 00 ☽ △ ☿
16 08 ☽ Q ♄
18 28 ☽ # ♃
22 01 ☽ ± ☉

07 Monday
01 06 ♀ △ ♅
05 46 ☽ # ♀
07 21 ☽ Q ♂
08 18 ☽ Q ♆
11 59 ☽ ☍ ♃
13 48 ☽ ⚹ ♅
15 02 ☽ ☍ ♀
15 31 ☽ ⚹ ♄
16 14 ☽ ∥ ☿
16 58 ☽ ⚼ ☉
20 39 ♀ △ ♄

08 Tuesday
01 27 ♂ ☌ ♆
07 21 ☽ ⚹ ♆
07 41 ☽ ⚹ ♂
12 13 ☽ ☌ ♇
18 26 ☽ ∠ ♅
20 10 ☽ ∠ ♄
23 53 ☽ △ ☉

09 Wednesday
03 00 ☽ ∥ ♆
04 09 ☽ ∥ ♄
05 17 ☽ □ ☿
10 59 ☽ ∥ ♂
11 17 ☽ ⊥ ♅
11 33 ☽ ∠ ♆
12 14 ☽ ∥ ♅
13 02 ☽ ⊥ ♄
13 23 ☽ ∠ ♂
21 27 ☽ ⚻ ♃
22 22 ☽ ⚺ ♅

10 Thursday
00 07 ☽ ⚺ ♄
04 08 ☽ ⊥ ♆
04 40 ☽ ⚻ ♀
06 47 ☽ ⊥ ♂
08 35 ☽ ± ♃
15 01 ☽ ⚺ ♆
16 31 ☽ ± ♀
17 12 ☽ Q ☿
18 13 ☽ ⚺ ♂
19 29 ☽ ⚺ ♇
22 33 ☿ # ♀

11 Friday
01 02 ☽ ⚼ ♃
06 10 ☽ ⊥ ♇
10 00 ☽ ⚼ ♀
10 56 ☽ □ ☉
16 26 ☽ ⚹ ☿
18 57 ♂ ⚹ ♇
21 58 ☽ ∠ ♇

12 Saturday
03 48 ☽ △ ♃
03 55 ☽ ☌ ♅
05 41 ☽ ☌ ♄
07 28 ☉ Q ♆
07 31 ☿ ∠ ♆
12 19 ♃ △ ♅
14 19 ☽ △ ♀
19 36 ☽ ☌ ♆
20 31 ☽ Q ☉
20 41 ☽ ∠ ☿
23 42 ☽ ⚹ ♇

13 Sunday
01 10 ☽ ☌ ♂
13 14 ☽ ⊥ ☿
18 11 ☽ ⚹ ☉
19 48 ☽ Q ♇

14 Monday
00 07 ☽ ⚺ ☿
06 31 ☽ ⚺ ♅
07 05 ☽ □ ♃
08 17 ☽ ⚺ ♄
09 02 ☽ ∥ ♅
11 24 ☽ ∥ ♂
16 11 ☽ ∥ ♄
16 21 ☽ ⊥ ♅
17 12 ☽ ∥ ♆
18 08 ☽ ⊥ ♄
20 23 ☽ □ ♀
20 39 ☽ ∠ ☉
21 27 ☽ ⚺ ♆

15 Tuesday
01 16 ☽ □ ♇
05 06 ☽ ⚺ ♂
06 59 ☽ ∠ ♅
07 11 ☽ ⊥ ♆
08 46 ☽ ∠ ♄
11 05 ♀ △ ♆
12 15 ☽ ⊥ ☉
12 47 ☽ Q ♃
15 14 ☽ ⊥ ♂
17 07 ☽ # ♀
21 41 ☽ ∠ ♆
22 10 ☉ ⊥ ♃
22 35 ☽ ⚺ ☉

16 Wednesday
03 41 ☽ Q ♀
05 17 ☽ ☌ ☿
06 21 ☽ ∠ ♂
07 01 ☽ ∥ ☿
07 07 ☽ ⚹ ♅
08 19 ☽ ⚹ ♃
08 55 ☽ ⚹ ♄
11 35 ☉ ∠ ♀
14 00 ☽ # ♃
21 42 ☽ ⚹ ♆

17 Thursday
00 22 ☽ ⚹ ♀
01 22 ☽ △ ♇
02 19 ☽ Q ♅
02 39 ☿ ⚹ ♅
04 09 ☽ Q ♄
05 47 ☿ ∠ ♂
07 28 ☽ ⚹ ♂
08 39 ☽ ∠ ♃
14 01 ♀ ☍ ♇
16 17 ☉ ⚼ ♇
16 56 ☽ Q ♆
22 17 ☿ ⚹ ♃
23 18 ☽ ⊥ ♃

18 Friday
00 02 ☿ ⚹ ♄
00 40 ☽ ∥ ☉
01 23 ☽ ⚼ ♇
02 02 ☽ ☌ ☉
02 17 ☽ ∠ ♀
03 06 ☽ # ♇
03 38 ☽ Q ♂
04 30 ☽ ∥ ♇
06 36 ☽ # ☉
07 14 ☽ □ ♅
09 06 ☽ ⚺ ♃
09 08 ☽ □ ♄
10 00 ☽ ⚺ ☿
10 57 ♃ △ ♄
15 54 ☽ ± ♇
17 58 ☽ ⊥ ♀
20 44 ☽ ⊥ ☿
22 02 ☽ □ ♆

19 Saturday
01 40 ☽ ⚻ ♇
04 32 ☽ ⚺ ♀
10 20 ☽ □ ♂
13 01 ☽ ∠ ☿
19 17 ☽ ∥ ♃
19 47 ☉ # ♇
20 19 ☽ # ☿

20 Sunday
06 53 ☽ ⚺ ☉
08 40 ☽ △ ♅
08 56 ☿ # ♃
10 45 ☽ △ ♄
11 22 ☽ ☌ ♃
16 54 ☽ ⚹ ☿
17 50 ☽ ⊥ ☉
23 57 ♀ ⚼ ♅

21 Monday
00 13 ☽ △ ♆
00 16 ☉ ∥ ♇
00 50 ☉ Q ♂
01 49 ☽ ∥ ♀
03 57 ☽ ☍ ♇
08 46 ☉ □ ♅
10 26 ☽ ⚼ ♅
10 34 ☽ ∠ ☉
11 18 ☽ ☌ ♀
12 37 ☽ ⚼ ♄
15 51 ☽ △ ♂
16 03 ☽ Q ☿
17 58 ☽ # ♆
19 03 ☽ # ♄
21 30 ☽ # ♂

22 Tuesday
02 19 ☽ ± ♅
02 31 ☽ ⚼ ♆
02 53 ☽ # ♅
04 34 ☽ ± ♄
04 49 ♀ ⚼ ♄
13 06 ☽ ⚻ ♅
15 22 ☽ ⚹ ☉
15 24 ☽ ⚻ ♄
15 51 ☉ □ ♄
16 51 ☽ ⚺ ♃
18 46 ☽ ± ♆
20 09 ☽ ⚼ ♂

23 Wednesday
04 01 ☽ ⊥ ♃
04 27 ☽ □ ☿
05 47 ☽ ⚻ ♆
09 40 ☽ ⚻ ♇
13 41 ☽ ± ♂
15 19 ☽ Q ☉
16 14 ☉ ⚺ ♃
16 18 ☿ ⚹ ♆
20 53 ☽ ± ♇
21 06 ☽ ∠ ♃
22 33 ☽ ⚺ ♀

24 Thursday
01 36 ☽ ⚻ ♂
10 59 ☽ ⊥ ♀
13 59 ☽ ⚼ ♇
21 23 ☽ ☍ ♅
23 56 ☽ ☍ ♄

25 Friday
01 12 ☿ △ ♇
02 23 ☽ ⚹ ♃
04 42 ☽ □ ☉
06 00 ☽ ∠ ♀
14 38 ☿ Q ♅
15 12 ☽ ☍ ♆
19 11 ☽ △ ♇
21 48 ☽ △ ☿

26 Saturday
02 26 ☽ Q ♃
11 42 ☿ Q ♄
13 43 ☉ ± ♇
14 28 ☽ ⚹ ♀
15 39 ☽ ☍ ♂

27 Sunday
03 20 ♂ ∥ ♄
08 17 ☽ ⚼ ☿
08 57 ☽ ⚻ ♅
11 37 ☽ # ♅
11 42 ☽ ⚻ ♄
15 16 ☽ □ ♃
16 57 ☽ Q ♀
20 35 ☽ # ♄
21 01 ☽ # ♂
21 06 ☽ ± ♅
21 51 ☽ # ♆
21 58 ☽ △ ☉
23 53 ☽ ± ♄

28 Monday
02 18 ☽ ∥ ♀
03 26 ☽ ⚻ ♆
03 58 ♀ △ ♂
05 20 ☽ ± ☿
07 23 ☽ □ ♇
15 24 ☽ ⚼ ♅
15 38 ☽ ± ♆
18 12 ☽ ⚼ ♄
19 26 ☽ ⚻ ☿

29 Tuesday
02 14 ♂ Q ♇
06 09 ♂ ∥ ♆
07 15 ☽ ⚼ ☉
08 05 ☽ ⚻ ♂
08 59 ☽ □ ♀
09 57 ☽ ⚼ ♆
17 29 ♀ ± ♅
19 35 ☿ ∠ ♃
19 54 ☽ Q ♇
20 17 ♀ ⚼ ♆
21 01 ☽ ± ♂
21 54 ☽ △ ♅

30 Wednesday
00 45 ☽ △ ♄
01 33 ☽ ∥ ♃
01 58 ♀ # ♂
03 12 ☽ ± ☉
05 24 ☽ △ ♃
08 09 ♀ # ♆
15 42 ☉ □ ♆
16 20 ☽ ⚼ ♂
16 22 ☽ △ ♆
16 26 ☽ ⚻ ☉
20 09 ☽ ⚹ ♇
23 25 ♀ # ♄

31 Thursday
03 20 ♀ ± ♄
07 36 ☽ # ☿
08 24 ☿ Q ♆
11 40 ☽ ∥ ☉
12 13 ☽ ⚼ ♃
17 47 ☽ ☍ ☿

APRIL 1988

LONGITUDES

Date	Sidereal time h m s	Sun ☉ ° ' "	Moon ☽ ° ' "	Moon ☽ 24.00 ° ' "	Mercury ☿ ° '	Venus ♀ ° '	Mars ♂ ° '	Jupiter ♃ ° '	Saturn ♄ ° '	Uranus ♅ ° '	Neptune ♆ ° '	Pluto ♇ ° '
01	00 40 15	11 ♈ 58 35	01 ♎ 58 45	08 ♎ 04 19	24 ♓ 05	27 ♉ 49	26 ♑ 22	05 ♉ 15	02 ♑ 29	01 ♑ 03	10 ♑ 10	11 ♏ 59
02	00 44 11	12 57 44	14 ♎ 12 39	20 ♎ 23 54	25 47	28 48	27 02	05 29	02 30	01 03	10 11	11 R 58
03	00 48 08	13 56 51	26 ♎ 38 12	02 ♏ 55 41	27 31	29 ♉ 47	27 42	05 43	02 30	01 03	10 11	11 56
04	00 52 05	14 55 56	09 ♏ 16 26	15 ♏ 40 30	29 ♓ 16	00 ♊ 46	28 23	05 57	02 31	01 03	10 11	11 55
05	00 56 01	15 54 59	22 ♏ 07 57	28 ♏ 38 49	01 ♈ 02	01 44	29 03	06 11	02 32	01 03	10 11	11 53
06	00 59 58	16 54 01	05 ♐ 13 08	11 ♐ 50 57	02 50	02 42	29 ♑ 44	06 24	02 32	01 R 03	10 11	11 52
07	01 03 54	17 53 01	18 ♐ 32 16	25 ♐ 17 07	04 39	03 39	00 ♒ 24	06 38	02 33	01 03	10 11	11 51
08	01 07 51	18 51 59	02 ♑ 05 30	08 ♑ 57 24	06 30	04 36	01 04	06 52	02 33	01 02	10 11	11 49
09	01 11 47	19 50 55	15 ♑ 52 49	22 ♑ 51 42	08 22	05 32	01 45	07 07	02 33	01 02	10 12	11 48
10	01 15 44	20 49 49	29 ♑ 53 56	06 ♒ 59 24	10 15	06 27	02 25	07 21	02 33	01 02	10 12	11 46
11	01 19 40	21 48 42	14 ♒ 07 54	21 ♒ 19 10	12 11	07 22	03 05	07 35	02 R 33	01 02	10 12	11 44
12	01 23 37	22 47 33	28 ♒ 32 52	05 ♓ 48 33	14 07	08 17	03 46	07 49	02 33	01 01	10 12	11 43
13	01 27 34	23 46 22	13 ♓ 05 44	20 ♓ 23 49	16 05	09 11	04 26	08 03	02 33	01 01	10 R 12	11 41
14	01 31 30	24 45 10	27 ♓ 42 08	04 ♈ 59 56	18 05	10 04	05 06	08 17	02 33	01 00	10 12	11 40
15	01 35 27	25 43 55	12 ♈ 16 29	19 ♈ 30 58	20 06	10 56	05 46	08 31	02 32	01 00	10 12	11 38
16	01 39 23	26 42 39	26 ♈ 42 39	03 ♉ 50 47	22 08	11 48	06 27	08 45	02 32	00 59	10 11	11 36
17	01 43 20	27 41 21	10 ♉ 54 42	17 ♉ 53 49	24 11	12 40	07 07	09 00	02 31	00 59	10 11	11 35
18	01 47 16	28 40 01	24 ♉ 47 40	01 ♊ 35 56	26 16	13 30	07 47	09 14	02 30	00 58	10 11	11 33
19	01 51 13	29 ♈ 38 38	08 ♊ 18 22	14 ♊ 54 54	28 ♈ 21	14 20	08 27	09 28	02 30	00 57	10 11	11 32
20	01 55 09	00 ♉ 37 14	21 ♊ 25 35	27 ♊ 50 35	00 ♉ 28	15 09	09 07	09 42	02 29	00 57	10 11	11 30
21	01 59 06	01 35 48	04 ♋ 10 10	10 ♋ 24 42	02 35	15 57	09 47	09 57	02 28	00 56	10 11	11 28
22	02 03 03	02 34 19	16 ♋ 34 39	22 ♋ 40 30	04 43	16 44	10 27	10 11	02 27	00 55	10 10	11 27
23	02 06 59	03 32 48	28 ♋ 42 51	04 ♌ 42 17	06 51	17 31	11 07	10 25	02 26	00 54	10 09	11 25
24	02 10 56	04 31 16	10 ♌ 39 26	16 ♌ 34 56	09 00	18 17	11 47	10 40	02 24	00 53	10 09	11 23
25	02 14 52	05 29 40	22 ♌ 29 27	28 ♌ 23 38	11 08	19 01	12 27	10 54	02 23	00 52	10 08	11 22
26	02 18 49	06 28 03	04 ♍ 18 05	10 ♍ 13 26	13 16	19 45	13 07	11 08	02 22	00 51	10 08	11 20
27	02 22 45	07 26 24	16 ♍ 10 15	22 ♍ 09 03	15 22	20 28	13 47	11 22	02 20	00 50	10 08	11 18
28	02 26 42	08 24 43	28 ♍ 10 22	04 ♎ 14 36	17 28	21 09	14 27	11 37	02 18	00 49	10 07	11 17
29	02 30 38	09 22 59	10 ♎ 22 09	16 ♎ 33 20	19 33	21 50	15 06	11 51	02 17	00 48	10 06	11 15
30	02 34 35	10 ♉ 21 14	22 ♎ 48 23	29 ♎ 07 30	21 ♉ 36	22 ♊ 29	15 ♒ 46	12 ♉ 05	02 ♑ 15	00 ♑ 46	10 ♑ 06	11 ♏ 13

DECLINATIONS

Date	Moon True ☊ ° '	Moon Mean ☊ ° '	Moon ☽ Latitude ° '	Sun ☉	Moon ☽	Mercury ☿	Venus ♀	Mars ♂	Jupiter ♃	Saturn ♄	Uranus ♅	Neptune ♆	Pluto ♇
01	23 ♓ 07	22 ♓ 19	00 S 49	04 N 44	01 S 32	04 S 31	22 N 44	21 S 45	12 N 22	22 S 17	23 S 38	22 S 05	00 N 22
02	23 R 06	22 16	01 54	05 07	07 21	03 49	23 02	21 39	12 27	22 17	23 38	22 05	00 23
03	23 04	22 13	02 54	05 30	12 59	03 06	23 19	21 32	12 31	22 17	23 38	22 05	00 23
04	23 01	22 10	03 48	05 53	18 11	02 22	23 35	21 25	12 36	22 16	23 38	22 05	00 24
05	22 57	22 07	04 30	06 16	22 39	01 38	23 51	21 18	12 41	22 16	23 38	22 05	00 24
06	22 54	22 03	04 59	06 38	26 05	00 52	24 06	21 11	12 46	22 16	23 38	22 05	00 25
07	22 50	22 00	05 13	07 01	28 09	00 S 05	24 21	21 04	12 50	22 16	23 38	22 05	00 26
08	22 48	21 57	05 10	07 23	28 36	00 N 43	24 35	20 56	12 55	22 16	23 38	22 05	00 26
09	22 47	21 54	04 49	07 46	27 17	01 32	24 49	20 49	13 00	22 16	23 38	22 05	00 27
10	22 D 48	21 51	04 11	08 08	24 15	02 21	25 02	20 41	13 04	22 16	23 38	22 04	00 28
11	22 49	21 48	03 17	08 30	19 44	03 11	25 15	20 33	13 09	22 16	23 38	22 04	00 28
12	22 50	21 44	02 09	08 52	14 00	04 03	25 27	20 25	13 14	22 16	23 38	22 04	00 29
13	22 51	21 41	00 S 53	09 14	07 28	04 54	25 39	20 17	13 18	22 16	23 38	22 04	00 29
14	22 R 51	21 38	00 N 27	09 35	00 S 31	05 47	25 50	20 08	13 23	22 16	23 38	22 04	00 30
15	22 50	21 35	01 44	09 57	06 N 27	06 40	26 01	20 00	13 27	22 16	23 38	22 04	00 30
16	22 47	21 32	02 54	10 18	13 01	07 33	26 11	19 51	13 32	22 16	23 38	22 04	00 31
17	22 43	21 28	03 52	10 39	18 47	08 27	26 20	19 43	13 37	22 16	23 38	22 04	00 32
18	22 37	21 25	04 35	11 00	23 25	09 20	26 29	19 34	13 41	22 16	23 38	22 04	00 32
19	22 31	21 22	05 01	11 21	26 39	10 14	26 38	19 25	13 46	22 16	23 38	22 04	00 33
20	22 26	21 19	05 11	11 41	28 20	11 08	26 46	19 16	13 50	22 16	23 38	22 04	00 33
21	22 21	21 16	05 04	12 02	28 27	12 02	26 53	19 06	13 55	22 16	23 38	22 04	00 34
22	22 18	21 13	04 44	12 22	27 06	12 55	27 00	18 57	14 00	22 16	23 38	22 04	00 34
23	22 16	21 09	04 10	12 42	24 30	13 47	27 07	18 47	14 04	22 16	23 38	22 04	00 35
24	22 D 16	21 06	03 27	13 02	20 53	14 39	27 12	18 38	14 09	22 16	23 38	22 04	00 35
25	22 18	21 03	02 35	13 21	16 27	15 29	27 18	18 28	14 13	22 16	23 38	22 04	00 36
26	22 19	21 00	01 36	13 41	11 26	16 19	27 23	18 18	14 18	22 16	23 38	22 04	00 36
27	22 20	20 57	00 N 33	14 00	05 58	17 07	27 27	18 08	14 22	22 16	23 38	22 04	00 37
28	22 R 20	20 54	00 S 32	14 19	00 N 15	17 53	27 31	17 58	14 27	22 16	23 38	22 04	00 37
29	22 19	20 50	01 36	14 37	05 S 35	18 37	27 34	17 48	14 31	22 16	23 38	22 04	00 38
30	22 ♓ 15	20 ♓ 47	02 S 37	14 N 56	11 S 18	19 N 20	27 N 37	17 S 38	14 N 35	22 S 16	23 S 38	22 S 04	00 N 38

ZODIAC SIGN ENTRIES

Date	h	m	Planets	
01	08	05	☽	♎
03	17	07	♀	♊
03	18	26	☽	♏
04	22	04	☿	♈
06	02	29	☽	♐
06	21	44	♂	♒
08	08	19	☽	♑
10	12	10	☽	♒
12	14	24	☽	♓
14	15	47	☽	♈
16	17	31	☽	♉
18	21	10	☽	♊
19	20	45	☉	♉
20	06	42	☿	♉
21	04	04	☽	♋
23	14	34	☽	♌
26	03	16	☽	♍
28	15	37	☽	♎

LATITUDES

Date	Mercury ☿ ° '	Venus ♀ ° '	Mars ♂ ° '	Jupiter ♃ ° '	Saturn ♄ ° '	Uranus ♅ ° '	Neptune ♆ ° '	Pluto ♇ ° '
01	02 S 21	03 N 09	00 S 53	00 S 58	01 N 09	00 S 11	00 N 59	16 N 34
04	02 16	03 21	00 57	00 57	01 09	00 12	00 59	16 34
07	02 06	03 32	01 01	00 57	01 09	00 12	00 59	16 35
10	01 52	03 42	01 05	00 57	01 09	00 12	00 59	16 36
13	01 33	03 52	01 09	00 57	01 09	00 12	00 59	16 36
16	01 09	04 01	01 14	00 56	01 09	00 12	00 59	16 36
19	00 42	04 08	01 18	00 56	01 09	00 12	00 59	16 37
22	00 S 12	04 15	01 23	00 56	01 09	00 12	00 59	16 37
25	00 N 20	04 19	01 27	00 56	01 09	00 12	00 59	16 37
28	00 52	04 23	01 32	00 55	01 10	00 12	00 59	16 37
31	01 N 22	04 N 24	01 S 37	00 S 55	01 N 10	00 S 12	00 N 59	16 N 37

DATA

Julian Date	2447253
Delta T	+56 seconds
Ayanamsa	23° 41' 36"
Synetic vernal point	05° ♓ 25' 23"
True obliquity of ecliptic	23° 26' 37"

MOON'S PHASES, APSIDES AND POSITIONS ☽

Date	h	m	Phase	Longitude ° '	Eclipse Indicator
02	09	21	○	12 ♎ 51	
09	19	21	☾	20 ♑ 09	
16	12	00	●	26 ♈ 43	
23	22	32	☽	03 ♌ 58	

Day	h	m		
13	22	39	Perigee	
25	18	31	Apogee	
01	05	42	0S	
08	06	12	Max dec	28° S 39'
14	13	44	0N	
21	01	42	Max dec	28° N 35'
28	13	01	0S	

LONGITUDES

Date	Chiron ⚷ ° '	Ceres ⚳ ° '	Pallas ⚴ ° '	Juno ⚵ ° '	Vesta ⚶ ° '	Black Moon Lilith ⚸ ° '
01	23 ♊ 51	04 ♓ 47	10 ♒ 16	01 ♉ 46	25 ♋ 09	25 ♌ 18
11	24 ♊ 20	08 ♓ 22	12 ♒ 52	07 ♉ 37	26 ♋ 56	26 ♌ 25
21	24 ♊ 56	11 ♓ 48	15 ♒ 09	13 ♉ 31	29 ♋ 14	27 ♌ 31
31	25 ♊ 37	15 ♓ 05	17 ♒ 04	19 ♉ 27	01 ♌ 59	28 ♌ 38

ASPECTARIAN

h m Aspects

01 Friday
00 12 ☽ △ ♂
02 07 ☽ ∠ ♇
03 01 ☽ △ ♀
04 12 ☽ ∥ ♇
06 30 ☽ ± ♃
07 00 ☉ # ☿
07 12 ☽ # ♇
10 09 ☽ □ ♅
12 15 ☉ ⊼ ♇
12 59 ☽ □ ♄
18 35 ☽ ⊼ ♃
19 53 ☽ ⊥ ♇
22 57 ☽ ∥ ☿

02 Saturday
02 06 ☽ # ☉
04 07 ☽ □ ♆
07 38 ☽ ⊻ ♇
09 21 ☽ ☍ ☉
11 09 ☽ bQ ♀
21 23 ☽ Q ♅

03 Sunday
00 12 ☽ Q ♄
04 11 ☿ bQ ♇
06 04 ☽ ± ♀
09 57 ☽ # ♃
13 57 ☽ ⊼ ☿
14 10 ☽ □ ♂
14 57 ☽ Q ♆
16 24 ☿ ✶ ♂
18 32 ☽ ⊼ ♀
20 25 ☽ ✶ ♅
23 13 ☽ ✶ ♄

04 Monday
03 12 ☽ ± ☿
05 36 ☽ ☍ ♃
13 42 ☽ ✶ ♆
16 21 ♀ # ♅
16 57 ☽ ☌ ♇
18 55 ♀ ⊼ ♅
19 25 ♅ St R
22 42 ☿ ⊥ ♃
22 51 ☽ bQ ☿
23 30 ☽ ⊼ ☉

05 Tuesday
00 42 ☽ ∠ ♅
02 03 ☽ Q ♂
03 27 ☽ ∠ ♄
04 27 ☽ ∥ ♂
08 38 ☽ ∥ ♆
09 46 ☽ ∥ ♄
11 34 ☽ ± ☉
12 12 ☿ □ ♅
17 23 ☽ ⊥ ♅
17 38 ☽ ∠ ♆
18 05 ☽ ∥ ♅
20 03 ☽ # ♀
20 07 ☽ ⊥ ♄

06 Wednesday
01 26 ☽ ✶ ♂
04 24 ☽ ⊻ ♅
05 28 ☽ bQ ☉
06 57 ☽ △ ☿
07 03 ☽ ☍ ♀
07 07 ☽ ⊻ ♄
07 56 ♀ ⊼ ♄
08 06 ☿ □ ♄
08 17 ☿ ✶ ♀
10 07 ☽ ⊥ ♆
14 12 ☽ ⊼ ♃
21 00 ☽ ⊻ ♆

07 Thursday
00 01 ☽ ⊻ ♇
01 14 ☽ ± ♃
01 30 ☿ # ♇
06 05 ☽ ∠ ♂
10 44 ☽ △ ☉
10 46 ☽ ⊥ ♇
17 38 ☽ bQ ♃
22 59 ☽ ⊥ ♂

08 Friday
01 44 ♀ ± ♆
02 44 ☽ ∠ ♇
03 21 ☿ ± ♇
03 37 ☿ ∥ ♇
10 07 ☽ ⊻ ♂
10 09 ☽ ☌ ♅
10 54 ♂ ⊻ ♅
12 48 ☽ ☌ ♄
16 42 ☽ ⊼ ♀
17 37 ☿ ⊻ ♃
20 31 ☽ △ ♃
20 55 ☽ □ ☿

09 Saturday
02 09 ☽ ☌ ♆
03 56 ☽ ± ♀
04 56 ☽ ✶ ♇
19 21 ☽ □ ☉
20 34 ☽ bQ ♀

10 Sunday
01 34 ☽ Q ♇
07 10 ☽ # ♀
08 46 ☽ Q ☿
11 12 ☿ □ ♆
13 56 ☽ ⊻ ♅
15 48 ☽ ∥ ♅
16 29 ☽ ☌ ♂
16 30 ☽ ⊻ ♄
16 52 ♂ ⊻ ♄
23 26 ☽ ∥ ♄
23 52 ☽ △ ♀

11 Monday
00 04 ☽ ⊥ ♅
00 28 ☽ ∥ ♆
00 48 ☽ □ ♃
02 09 ♄ St R
02 38 ☽ ⊥ ♄
04 13 ☽ Q ☉
05 24 ☽ ⊻ ♆
06 38 ☿ ⊼ ♇
08 00 ☽ □ ♇
08 00 ☽ ∥ ♂
08 13 ☽ ✶ ☿
13 19 ♆ St R
15 10 ☽ ∠ ♅
15 27 ☽ ⊥ ♆
17 43 ☽ ∠ ♄
19 18 ♀ ⊻ ♃

12 Tuesday
01 45 ☽ ✶ ☉
06 26 ☽ ∠ ♆
07 24 ☽ Q ♃
13 06 ☽ ∠ ☿
14 57 ☽ # ♃
16 06 ☽ ✶ ♅
18 37 ☽ ✶ ♄
21 02 ☽ ⊻ ♂

13 Wednesday
03 33 ☽ ✶ ♃
04 22 ☽ ∠ ☉
05 08 ☽ □ ♀
06 03 ☽ # ☉
06 16 ☽ ⊥ ☿
07 13 ☽ ✶ ♆
07 24 ☽ ⊥ ♂
09 41 ☽ △ ♇
11 52 ☽ Q ♅
14 23 ☽ Q ♄
17 42 ☽ ⊻ ☿
18 18 ☿ Q ♂
19 56 ☽ # ☿
20 15 ☽ ⊥ ☉
22 55 ☽ ∠ ♂

14 Thursday
02 57 ☽ Q ♆
04 37 ☽ ∠ ♃
06 48 ☽ ⊻ ☉
10 18 ☽ bQ ♇
12 02 ☽ # ♇
12 38 ☽ Q ♀
15 27 ☽ ∥ ♇
15 28 ♀ ⊼ ♆
17 26 ☽ □ ♅
19 39 ☽ ⊥ ♃
19 57 ☽ □ ♄

15 Friday
00 45 ☽ ✶ ♂
01 04 ☽ ± ♇
05 42 ☽ ⊻ ♃
08 33 ☽ □ ♆
09 39 ☽ ✶ ♀
10 57 ☽ ⊼ ♇
12 50 ☽ ∥ ☿
21 33 ☽ Q ♂

16 Saturday
01 14 ☽ ∥ ☉
03 05 ☽ ☌ ☿
06 39 ♀ ⊼ ♇
12 00 ☽ ☌ ☉
12 10 ☽ ∠ ♀
14 03 ☽ ∥ ♃
19 11 ☽ △ ♅
21 46 ☽ △ ♄

17 Sunday
04 18 ☽ ⊥ ♀
05 13 ☽ □ ♂
06 32 ☉ ∠ ♀
07 17 ♂ ⊥ ♅
08 40 ☽ ☌ ♃
10 46 ☽ △ ♆
13 08 ☽ ☍ ♇
15 11 ☽ ⊻ ♀
16 13 ☽ # ♂
20 41 ☽ bQ ♅
23 20 ☽ bQ ♄

18 Monday
04 18 ☽ # ♆
05 22 ☽ # ♄
12 18 ☽ ± ♅
12 41 ☽ bQ ♆
13 20 ☽ # ♅
15 00 ☽ ± ♄
15 02 ☽ ⊻ ☿
19 20 ☽ ⊻ ☉
22 52 ☽ ⊼ ♅

19 Tuesday
01 36 ☽ ⊼ ♄
03 36 ☽ ⊥ ☿
04 36 ☽ ± ♆
06 50 ☽ ⊥ ☉
11 47 ☽ ∥ ♀
12 16 ☽ △ ♂
13 35 ♂ ⊥ ♄
14 08 ☽ ⊻ ♃
15 23 ☽ ⊼ ♆
17 49 ☽ ⊼ ♇
22 54 ☽ ∠ ☿
23 40 ☽ ☌ ♀

20 Wednesday
00 26 ☽ ∠ ☉
01 15 ☽ ⊥ ♃
04 45 ☽ ± ♇
06 07 ☿ ∠ ♀
15 14 ☉ ☌ ☿
17 17 ☽ bQ ♂
17 21 ☿ △ ♅
18 14 ☽ ∠ ♃
19 49 ☉ △ ♅
21 27 ☽ bQ ♇

21 Thursday
05 50 ☽ ☍ ♅
06 41 ☽ ✶ ☉
08 23 ☽ ✶ ☿
08 45 ☽ ☍ ♄
10 36 ☿ △ ♄
11 14 ☽ ± ♂
11 40 ♀ ⊥ ♃
12 11 ☉ ∥ ☿
20 51 ♂ □ ♃
23 19 ☽ ✶ ♃
23 24 ☽ ⊼ ♂
23 31 ☽ ☍ ♆

22 Friday
01 36 ♂ ⊻ ♆
02 01 ☽ △ ♇
07 45 ☽ Q ☉
08 58 ☉ △ ♄
09 59 ♃ △ ♆
12 20 ☽ Q ☿
12 21 ☽ ⊻ ♀
13 07 ☽ ∥ ♀
23 15 ☽ Q ♃

23 Saturday
00 58 ☽ ⊥ ♀
08 58 ♀ ± ♇
16 22 ☽ ⊼ ♅
18 20 ☽ # ♅
19 25 ☽ ⊼ ♄
20 08 ☽ ∠ ♀
20 32 ☿ ∥ ♃
22 15 ♂ □ ♇
22 32 ☽ □ ☉

24 Sunday
03 31 ☽ # ♄
04 24 ☽ ± ♅
04 44 ☽ # ♆
07 28 ☽ ± ♄
07 55 ☽ □ ☿
10 58 ☽ ⊼ ♆
12 00 ☽ □ ♃
13 28 ☽ □ ♇
14 25 ☽ ☍ ♂
22 34 ☽ bQ ♅
23 07 ☽ ± ♆

25 Monday
00 54 ☿ △ ♆
01 10 ☽ # ♂
01 39 ☽ bQ ♄
04 29 ☽ ✶ ♀
09 02 ☿ ☌ ♃
14 32 ☿ ☍ ♇
16 07 ☽ ∥ ☿
17 23 ☽ bQ ♆
22 50 ☽ ∥ ♃

26 Tuesday
01 56 ☽ Q ♇
02 11 ☽ ∥ ☉
05 00 ☽ △ ♅
06 29 ☽ Q ♀
08 04 ☽ △ ♄
09 38 ☿ □ ♂
16 47 ☽ △ ☉
20 22 ☿ ⊥ ♀
23 49 ☽ △ ♆

27 Wednesday
02 08 ☽ △ ♃
02 12 ☽ ✶ ♇
05 39 ♃ ☍ ♇
06 54 ☽ ⊼ ♂
10 03 ☽ △ ☿
17 10 ☿ bQ ♅
19 41 ☽ ± ♂
21 09 ☽ □ ♀

28 Thursday
01 41 ☽ bQ ☉
08 14 ☽ ∠ ♇
08 50 ☽ bQ ♃
10 08 ☿ bQ ♄
10 26 ☽ ∥ ♇
14 14 ☿ # ♂
14 40 ☽ bQ ♂
15 36 ☽ # ♇
17 13 ☽ □ ♅
20 10 ☽ □ ♄
21 07 ☽ ± ☉
22 16 ☽ bQ ☿

29 Friday
01 19 ☉ ∥ ♃
02 00 ☽ ⊥ ♇
02 59 ☽ ± ♃
09 55 ☽ ⊼ ☉
11 29 ☽ □ ♆
13 42 ☽ ⊻ ♇
14 57 ☽ ⊼ ♃
19 25 ☽ ± ☿
21 43 ☽ △ ♂

30 Saturday
04 17 ☽ Q ♅
05 43 ☉ △ ♆
07 07 ☽ Q ♄
09 14 ☽ ⊼ ☿
11 21 ☽ △ ♀
12 09 ♂ ∠ ♅
22 03 ☽ Q ♆
23 46 ♂ ⊥ ♆

All ephemeris data is given at 12.00 UT and the Moon's longitude is additionally given for 24.00 UT

MAY 1988

LONGITUDES

Date	Sidereal time h m s	Sun ☉ ° ' "	Moon ☽ ° ' "	Moon ☽ 24.00 ° ' "	Mercury ☿ ° '	Venus ♀ ° '
01	02 38 32	11 ♉ 19 27	05 ♏ 30 45	11 ♏ 58 10	23 ♉ 36	23 ♊ 07
02	02 42 28	12 17 38	18 ♏ 29 41	25 ♏ 05 12	25 35	23 44
03	02 46 25	13 15 48	01 ♐ 44 31	08 ♐ 27 24	27 31	24 20
04	02 50 21	14 13 56	15 ♐ 13 35	22 ♐ 02 46	29 ♉ 24	24 54
05	02 54 18	15 12 02	28 ♐ 54 39	05 ♑ 48 55	01 ♊ 15	25 27
06	02 58 14	16 10 07	12 ♑ 45 16	19 ♑ 43 26	03 02	25 59
07	03 02 11	17 08 11	26 ♑ 43 11	03 ♒ 44 17	04 46	26 29
08	03 06 07	18 06 13	10 ♒ 46 33	17 ♒ 49 50	06 27	26 58
09	03 10 04	19 04 14	24 ♒ 53 58	01 ♓ 58 48	08 04	27 25
10	03 14 01	20 02 13	09 ♓ 04 11	16 ♓ 09 57	09 38	27 50
11	03 17 57	21 00 11	23 ♓ 15 52	00 ♈ 21 42	11 08	28 14
12	03 21 54	21 58 08	07 ♈ 27 07	14 ♈ 31 46	12 34	28 36
13	03 25 50	22 56 03	21 ♈ 35 16	28 ♈ 37 07	13 57	28 56
14	03 29 47	23 53 58	05 ♉ 36 52	12 ♉ 34 01	15 16	29 15
15	03 33 43	24 51 51	19 ♉ 28 03	26 ♉ 18 32	16 30	29 31
16	03 37 40	25 49 42	03 ♊ 05 00	09 ♊ 47 08	17 41	29 46
17	03 41 36	26 47 32	16 ♊ 24 37	22 ♊ 57 18	18 48	29 ♊ 58
18	03 45 33	27 45 21	29 ♊ 25 05	05 ♋ 47 58	19 51	00 ♋ 08
19	03 49 30	28 43 08	12 ♋ 06 07	18 ♋ 19 43	20 49	00 16
20	03 53 26	29 ♉ 40 53	24 ♋ 29 06	00 ♌ 34 39	21 44	00 22
21	03 57 23	00 ♊ 38 37	06 ♌ 36 51	12 ♌ 36 14	22 34	00 26
22	04 01 19	01 36 19	18 ♌ 33 22	24 ♌ 28 53	23 20	00 27
23	04 05 16	02 34 00	00 ♍ 23 26	06 ♍ 17 42	24 01	00 R 26
24	04 09 12	03 31 39	12 ♍ 12 21	18 ♍ 08 05	24 38	00 23
25	04 13 09	04 29 16	24 ♍ 05 33	00 ♎ 05 25	25 10	00 17
26	04 17 05	05 26 52	06 ♎ 08 18	12 ♎ 14 45	25 38	00 ♋ 09
27	04 21 02	06 24 27	18 ♎ 25 18	24 ♎ 40 24	26 01	29 ♊ 58
28	04 24 59	07 22 00	01 ♏ 00 23	07 ♏ 25 32	26 19	29 45
29	04 28 55	08 19 32	13 ♏ 55 59	20 ♏ 31 49	26 33	29 29
30	04 32 52	09 17 03	27 ♏ 12 55	03 ♐ 59 06	26 42	29 11
31	04 36 48	10 ♊ 14 33	10 ♐ 50 02	17 ♐ 45 19	26 ♊ 47	28 ♊ 51

Date	Mars ♂ ° '	Jupiter ♃ ° '	Saturn ♄ ° '	Uranus ♅ ° '	Neptune ♆ ° '	Pluto ♇ ° '
01	16 ♒ 26	12 ♉ 20	02 ♑ 13	00 ♑ 45	10 ♑ 05	11 ♏ 11
02	17 05	12 34	02 R 11	00 R 44	10 R 05	11 R 10
03	17 45	12 48	02 09	00 42	10 04	11 08
04	18 25	13 03	02 07	00 41	10 03	11 06
05	19 04	13 17	02 05	00 40	10 02	11 05
06	19 44	13 31	02 02	00 38	10 02	11 03
07	20 23	13 46	02 00	00 37	10 01	11 01
08	21 03	14 00	01 57	00 35	10 00	11 00
09	21 42	14 14	01 55	00 33	09 59	10 58
10	22 21	14 29	01 52	00 32	09 58	10 56
11	23 00	14 43	01 49	00 30	09 57	10 55
12	23 39	14 57	01 47	00 28	09 57	10 53
13	24 19	15 11	01 44	00 27	09 56	10 51
14	24 58	15 26	01 41	00 25	09 55	10 50
15	25 37	15 40	01 38	00 23	09 54	10 48
16	26 15	15 54	01 35	00 21	09 53	10 46
17	26 54	16 08	01 31	00 19	09 52	10 45
18	27 33	16 22	01 28	00 17	09 50	10 43
19	28 12	16 36	01 25	00 16	09 49	10 42
20	28 50	16 51	01 21	00 14	09 48	10 40
21	29 ♒ 29	17 05	01 18	00 12	09 47	10 38
22	00 ♓ 07	17 19	01 15	00 10	09 46	10 37
23	00 45	17 33	01 11	00 08	09 45	10 35
24	01 23	17 47	01 07	00 05	09 44	10 34
25	02 01	18 01	01 04	00 03	09 42	10 32
26	02 39	18 15	01 00	00 ♑ 01	09 41	10 31
27	03 17	18 29	00 56	29 ♐ 59	09 40	10 29
28	03 55	18 43	00 52	29 57	09 39	10 28
29	04 33	18 57	00 48	29 55	09 37	10 26
30	05 10	19 10	00 45	29 52	09 36	10 25
31	05 ♓ 48	19 ♉ 24	00 ♑ 41	29 ♐ 50	09 ♑ 35	10 ♏ 24

Date	Moon True ☊ ° '	Moon Mean ☊ ° '	Moon ☽ Latitude ° '
01	22 ♓ 09	20 ♓ 44	03 S 31
02	22 R 01	20 41	04 16
03	21 53	20 38	04 48
04	21 44	20 34	05 04
05	21 36	20 31	05 04
06	21 30	20 28	04 46
07	21 26	20 25	04 10
08	21 24	20 22	03 20
09	21 D 24	20 19	02 17
10	21 25	20 15	01 S 06
11	21 R 25	20 12	00 N 10
12	21 24	20 09	01 25
13	21 21	20 06	02 34
14	21 15	20 03	03 33
15	21 06	20 00	04 19
16	20 56	19 56	04 49
17	20 45	19 53	05 02
18	20 34	19 50	04 59
19	20 25	19 47	04 42
20	20 18	19 44	04 11
21	20 13	19 40	03 30
22	20 10	19 37	02 40
23	20 09	19 34	01 43
24	20 D 10	19 31	00 N 42
25	20 R 10	19 28	00 S 21
26	20 09	19 25	01 24
27	20 06	19 21	02 24
28	20 00	19 18	03 19
29	19 52	19 15	04 04
30	19 42	19 12	04 38
31	19 ♓ 30	19 ♓ 09	04 S 57

DECLINATIONS

Date	Sun ☉ ° '	Moon ☽ ° '	Mercury ☿ ° '	Venus ♀ ° '	Mars ♂ ° '	Jupiter ♃ ° '	Saturn ♄ ° '	Uranus ♅ ° '	Neptune ♆ ° '	Pluto ♇ ° '
01	15 N 14	16 S 41	20 00	27 N 39	17 S 27	14 N 40	22 S 16	23 S 38	22 S 05	00 N 39
02	15 32	21 26	20 38	27 41	17 17	14 44	22 16	23 39	22 05	00 39
03	15 49	25 12	21 14	27 43	17 06	14 49	22 16	23 39	22 05	00 40
04	16 07	27 40	21 47	27 44	16 56	14 53	22 16	23 39	22 05	00 40
05	16 24	28 30	22 17	27 44	16 45	14 57	22 16	23 39	22 05	00 41
06	16 41	27 34	22 46	27 44	16 34	15 02	22 16	23 39	22 05	00 41
07	16 57	24 55	23 11	27 44	16 23	15 06	22 16	23 39	22 05	00 41
08	17 14	20 44	23 34	27 43	16 12	15 10	22 16	23 39	22 05	00 42
09	17 30	15 23	23 55	27 42	16 01	15 15	22 16	23 39	22 05	00 42
10	17 45	09 11	24 13	27 40	15 50	15 19	22 16	23 39	22 05	00 42
11	18 01	02 S 31	24 29	27 38	15 38	15 23	22 16	23 39	22 05	00 43
12	18 16	04 N 15	24 43	27 35	15 27	15 27	22 16	23 39	22 05	00 43
13	18 31	10 48	24 54	27 33	15 16	15 31	22 16	23 39	22 05	00 44
14	18 45	16 44	25 03	27 29	15 04	15 36	22 16	23 39	22 05	00 44
15	18 59	21 45	25 10	27 25	14 53	15 40	22 16	23 39	22 05	00 44
16	19 13	25 30	25 15	27 21	14 41	15 44	22 16	23 39	22 05	00 44
17	19 27	27 45	25 19	27 16	14 29	15 48	22 16	23 39	22 05	00 45
18	19 40	28 26	25 20	27 11	14 18	15 52	22 16	23 39	22 05	00 45
19	19 53	27 34	25 20	27 06	14 06	15 56	22 16	23 39	22 05	00 45
20	20 05	25 21	25 18	27 00	13 54	16 00	22 16	23 39	22 05	00 46
21	20 17	22 00	25 15	26 53	13 42	16 04	22 17	23 39	22 06	00 46
22	20 29	17 48	25 10	26 46	13 31	16 08	22 17	23 39	22 06	00 46
23	20 41	12 57	25 03	26 39	13 19	16 12	22 17	23 39	22 06	00 46
24	20 52	07 38	24 56	26 31	13 07	16 16	22 17	23 39	22 06	00 46
25	21 03	02 N 02	24 47	26 23	12 55	16 20	22 17	23 39	22 06	00 47
26	21 13	03 S 43	24 37	26 14	12 43	16 24	22 17	23 39	22 06	00 47
27	21 23	09 26	24 25	26 05	12 31	16 28	22 17	23 39	22 06	00 47
28	21 33	14 55	24 13	25 55	12 19	16 32	22 17	23 39	22 06	00 47
29	21 42	19 54	24 00	25 45	12 07	16 35	22 17	23 39	22 06	00 47
30	21 51	24 03	23 46	25 34	11 55	16 39	22 17	23 39	22 06	00 47
31	21 N 59	26 S 59	23 N 31	25 N 23	11 S 43	16 N 43	22 S 17	23 S 39	22 S 06	00 N 47

ZODIAC SIGN ENTRIES

Date	h	m	Planets
01	01	39	☽ ♏
03	08	52	☽ ♐
04	19	40	☿ ♊
05	13	54	☽ ♑
07	17	37	☽ ♒
09	20	39	☽ ♓
11	23	23	☽ ♈
14	02	22	☽ ♉
16	06	31	☽ ♊
17	16	26	♀ ♋
18	13	05	☽ ♋
20	19	57	☉ ♊
20	22	51	☽ ♌
22	07	42	♂ ♓
23	11	12	☽ ♍
25	23	49	☽ ♎
27	01	17	♅ ♐
27	07	36	♀ ♊
28	10	06	☽ ♏
30	16	57	☽ ♐

LATITUDES

Date	Mercury ☿ ° '	Venus ♀ ° '	Mars ♂ ° '	Jupiter ♃ ° '	Saturn ♄ ° '	Uranus ♅ ° '	Neptune ♆ ° '	Pluto ♇ ° '
01	01 N 22	04 N 24	01 S 37	00 S 55	01 N 10	00 S 12	00 N 59	16 N 37
04	01 48	04 23	01 42	00 55	01 10	00 12	00 59	16 37
07	02 08	04 20	01 47	00 55	01 10	00 12	00 59	16 37
10	02 21	04 15	01 52	00 55	01 10	00 12	00 59	16 36
13	02 26	04 06	01 57	00 55	01 10	00 12	00 59	16 36
16	02 24	03 54	02 02	00 54	01 10	00 12	01 00	16 35
19	02 13	03 39	02 08	00 54	01 10	00 12	01 00	16 35
22	01 53	03 20	02 14	00 54	01 10	00 12	01 00	16 34
25	01 25	02 56	02 19	00 54	01 10	00 12	01 00	16 33
28	00 50	02 28	02 24	00 54	01 10	00 12	01 00	16 32
31	00 N 06	01 N 56	02 S 30	00 S 54	01 N 09	00 S 12	01 N 00	16 N 31

DATA

Julian Date	2447283
Delta T	+56 seconds
Ayanamsa	23° 41' 40"
Synetic vernal point	05° ♓ 25' 20"
True obliquity of ecliptic	23° 26' 36"

MOON'S PHASES, APSIDES AND POSITIONS ☽

Date	h	m	Phase	Longitude ° '	Eclipse Indicator
01	23	41	○	11 ♏ 48	
09	01	23	☾	18 ♒ 39	
15	22	11	●	25 ♉ 16	
23	16	49	☽	02 ♍ 46	
31	10	53	○	10 ♐ 12	

Day	h	m		
10	22	18	Perigee	
23	13	44	Apogee	
05	11	29	Max dec	28° S 30'
11	20	55	0N	
18	10	15	Max dec	28° N 26'
25	20	30	0S	

LONGITUDES

Date	Chiron ⚷ ° '	Ceres ⚳ ° '	Pallas ⚴ ° '	Juno ⚵ ° '	Vesta ⚶ ° '	Black Moon Lilith ⚸ ° '
01	25 ♊ 37	15 ♓ 05	17 ♒ 04	19 ♉ 27	01 ♌ 59	28 ♌ 38
11	26 ♊ 23	18 ♓ 11	18 ♒ 35	25 ♉ 24	05 ♌ 29	29 ♌ 45
21	27 ♊ 12	21 ♓ 05	19 ♒ 38	01 ♊ 22	08 ♌ 25	00 ♍ 51
31	28 ♊ 05	23 ♓ 44	20 ♒ 09	07 ♊ 19	12 ♌ 10	01 ♍ 58

All ephemeris data is given at 12.00 UT and the Moon's longitude is additionally given for 24.00 UT

ASPECTARIAN

h m	Aspects
01 Sunday	
02 38	☽ # ♃
03 05	☽ ✶ ♅
03 31	☿ ⊻ ♀
04 55	☽ # ☉
05 50	☽ ✶ ♄
08 48	☉ ☍ ♇
15 33	☽ ∥ ♂
17 07	☽ bQ ♀
20 30	☽ ✶ ♆
22 32	☽ ☌ ♇
23 41	☽ ☍ ☉
02 Monday	
00 54	☽ ☍ ♃
01 42	☿ ± ♅
05 50	☿ bQ ♆
06 56	☽ ∠ ♅
06 59	☽ # ☿
09 18	☽ □ ♂
09 36	☽ ∠ ♄
10 33	☽ ± ♀
15 12	♂ ∠ ♄
15 39	☽ ∥ ♆
16 45	☽ ∥ ♄
19 16	☿ ± ♄
20 59	☉ ☌ ♃
22 01	☽ ⊼ ♀
23 20	☽ ⊥ ♅
23 58	☽ ∠ ♆
03 Tuesday	
01 09	☽ ∥ ♅
01 57	☽ ⊥ ♄
03 07	☽ ☍ ☿
10 09	☽ ⊻ ♅
12 44	☽ ⊻ ♄
16 10	☽ ⊥ ♆
19 33	☽ Q ♂
04 Wednesday	
02 51	☽ ⊻ ♆
04 43	☽ ⊻ ♇
08 04	☽ ⊼ ♃
10 07	☽ ⊼ ☉
13 00	☽ # ♀
15 19	☽ ⊥ ♇
17 54	☽ ✶ ♂
18 51	☽ ± ♃
21 29	☽ ± ☉
05 Thursday	
01 44	☿ # ♆
04 22	☿ ⊼ ♅
05 43	☽ ☍ ♀
07 04	☽ ∠ ♇
10 49	☿ # ♄
10 53	☽ bQ ♃
14 25	☽ bQ ☉
15 02	☽ ☌ ♅
16 41	☽ ⊼ ☿
17 30	☽ ☌ ♄
21 25	☽ ∠ ♂
22 48	☿ ⊼ ♄
23 06	☉ bQ ♅
06 Friday	
04 38	☽ ± ☿
06 06	☉ # ♂
07 18	☽ ☌ ♆
09 04	☽ ✶ ♇
09 37	☽ # ♀
13 21	☽ △ ♃
13 46	☽ ⊥ ♂
14 57	♀ bQ ♇
18 19	☽ △ ☉
22 24	☽ bQ ☿
07 Saturday	
00 36	☽ ⊻ ♂
01 32	☿ ± ♆
05 41	☽ Q ♇
08 41	☉ bQ ♄
11 35	☽ ⊼ ♀
18 39	☽ ⊻ ♅
20 10	☽ ∥ ♅
21 00	☽ ⊻ ♄
21 55	☽ # ☿
22 13	☽ ± ♀
08 Sunday	
03 39	☽ △ ☿
04 03	☽ ∥ ♄
04 52	☽ ⊥ ♅
05 04	☽ ∥ ♆
07 13	☽ ⊥ ♄
10 41	☽ ⊻ ♆
12 22	☽ □ ♇
14 05	☽ bQ ♀
16 52	☿ # ♅
17 35	☽ □ ♃
20 10	☽ ∠ ♅
20 53	☽ ⊥ ♆
22 29	☽ ∠ ♄
09 Monday	
01 23	☽ □ ☉
03 28	☽ # ☉
06 18	☽ ☌ ♂
09 16	☽ ∥ ♂
12 09	☽ ∠ ♆
12 32	☽ # ♃
16 24	☽ △ ♀
21 34	☽ ✶ ♅
23 51	☽ ✶ ♄
10 Tuesday	
00 39	☽ Q ♃
10 07	☽ Q ☉
13 04	☽ □ ☿
13 32	☽ ✶ ♆
15 09	☽ △ ♇
17 18	☿ ⊼ ♆
17 51	☽ Q ♅
20 05	☽ Q ♄
21 18	☽ ✶ ♃
11 Wednesday	
07 54	☽ ✶ ☉
08 26	☿ ⊼ ♇
09 48	☽ Q ♆
11 32	☽ ⊻ ♂
16 28	☽ bQ ♇
18 23	☽ # ♇
20 38	☽ □ ♀
22 10	☽ ⊥ ♂
23 03	☽ Q ☿
23 05	☽ ∠ ♃
23 27	☽ ∥ ♇
12 Thursday	
00 13	☽ □ ♅
02 25	☽ □ ♄
07 40	☽ ± ♇
11 07	☽ ∠ ☉
11 37	♂ # ♃
14 08	☽ ∠ ♂
14 35	☽ ⊥ ♃
16 13	☽ □ ♆
17 48	☽ ⊼ ♇
21 38	☽ ✶ ☿
13 Friday	
00 56	☽ ⊻ ♃
03 30	☽ ⊥ ☉
03 54	☽ Q ♀
12 09	☉ ⊥ ♀
14 28	☽ ⊻ ☉
16 52	☽ ✶ ♂
14 Saturday	
00 50	☽ ✶ ♀
01 52	☽ ∠ ☿
03 06	☽ △ ♅
05 09	☽ # ♂
05 16	☽ △ ♄
07 03	☽ ∥ ♃
10 14	♂ ∠ ♆
11 05	♃ bQ ♅
14 26	☽ Q ♂
15 52	☿ ⊻ ♃
18 55	☽ ⊥ ☿
19 24	☽ △ ♆
20 58	☽ ☍ ♇
21 28	☽ ∥ ☉
15 Sunday	
00 27	☉ ± ♅
03 13	☽ ∠ ♀
04 54	☽ bQ ♅
05 15	☽ ☌ ♃
06 20	☽ ⊻ ☿
07 04	☽ bQ ♄
12 44	☉ bQ ♆
13 52	☽ # ♆
14 54	☽ # ♄
17 45	☿ ± ♇
19 13	☽ ⊥ ♀
20 36	☽ ± ♅
21 30	☽ bQ ♆
22 11	☽ ☌ ☉
22 46	☽ ± ♄
23 05	☽ # ♅
23 18	☽ □ ♂
16 Monday	
06 00	☽ ⊻ ♀
06 02	☉ ± ♄
07 10	☽ ⊼ ♅
09 20	☽ ⊼ ♄
10 08	☽ ∥ ☿
13 25	☽ ± ♆
17 Tuesday	
00 09	☽ ⊼ ♆
01 45	☽ ⊼ ♇
05 26	☽ ∥ ♀
11 29	☽ ⊻ ♃
12 37	☽ ± ♇
16 45	☽ ☌ ☿
20 21	☉ □ ♂
22 41	☽ ⊥ ♃
18 Wednesday	
05 08	☽ bQ ♇
08 20	☽ △ ♂
08 39	☽ ⊻ ☉
13 22	☽ ☌ ♀
13 38	☽ ☍ ♅
15 44	☽ ∠ ♃
15 49	☽ ☍ ♄
19 59	♃ bQ ♄
20 48	☽ ⊥ ☉
19 Thursday	
07 39	☽ ☍ ♆
09 19	☽ △ ♇
09 52	♀ ☍ ♅
14 12	☽ bQ ♂
15 22	☽ ∠ ☉
18 49	☽ ∥ ♀
20 50	☽ ✶ ♃
20 Friday	
06 12	☽ ⊻ ☿
08 35	☽ ± ♂
12 23	☽ ∥ ☿
18 52	☽ ⊥ ☿
20 45	☽ Q ♃
21 02	☽ ⊼ ♂
23 06	☽ ✶ ☉
23 16	☽ ⊼ ♅
23 40	☽ ⊻ ♀
21 Saturday	
01 05	☽ # ♅
01 09	☉ ⊼ ♅
01 29	☽ ⊼ ♄
06 27	☉ ⊻ ♀
10 18	☽ # ♄
11 10	☽ ± ♅
11 28	☽ # ♆
11 38	☽ ⊥ ♀
13 22	☽ ± ♄
14 02	☽ ∠ ☿
18 20	☽ ⊼ ♆
20 02	☽ □ ♇
21 51	☽ ∥ ☉
22 Sunday	
01 08	☽ Q ☉
03 27	☉ ⊼ ♄
05 10	☽ bQ ♅
05 44	☽ ∠ ♀
06 23	☽ ± ♆
07 21	☽ bQ ♄
09 26	☽ □ ♃
11 18	☿ ⊥ ♃
13 26	♀ St R
13 36	♂ ✶ ♅
20 25	☽ ∥ ♃
22 17	☽ ✶ ☿
23 Monday	
00 33	☽ bQ ♆
00 36	♀ △ ♂
08 21	☽ Q ♇
10 12	☽ # ♂
11 28	☽ △ ♅
12 06	☽ ✶ ♀
12 47	☽ ☍ ♂
13 36	☽ △ ♄
16 49	☽ □ ☉
24 Tuesday	
00 05	☽ Q ☿
02 49	♂ ✶ ♄
06 59	☽ △ ♆
08 41	☽ ✶ ♇
12 21	☽ Q ♀
16 51	☉ ± ♆
23 31	☽ △ ♃
25 Wednesday	
14 15	☽ □ ☿
14 54	☽ ∠ ♇
17 15	☽ ∥ ♇
23 45	☽ # ♇
23 54	☽ □ ♅
26 Thursday	
00 15	☽ □ ♀
01 52	☽ □ ♄
04 44	☽ ⊼ ♂
05 45	☿ bQ ♇
06 10	☽ bQ ♃
08 48	☽ ⊥ ♇
10 31	☽ △ ☉
17 14	☽ ± ♂
18 58	☽ □ ♆
20 36	☽ ⊻ ♇
27 Friday	
00 14	☽ ± ♃
09 02	♀ ☍ ♅
11 09	☽ Q ♅
11 44	☽ bQ ♂
12 07	☽ ⊼ ♃
12 59	☽ Q ♄
18 14	☽ bQ ☉
28 Saturday	
00 47	☽ ∥ ♂
02 57	☽ △ ☿
05 39	☽ Q ♆
09 40	☽ △ ♀
10 01	☽ ✶ ♅
11 45	☽ ✶ ♄
12 44	☽ ± ☉
17 45	☽ △ ♂
19 29	☽ # ♃
29 Sunday	
00 51	☽ ⊼ ☉
04 05	☽ ✶ ♆
05 36	☽ ☌ ♇
07 34	☽ bQ ☿
12 59	☽ bQ ♀
13 47	☽ ∠ ♅
15 25	☽ ∠ ♄
21 17	☽ ☍ ♃
22 01	☽ # ☉
23 59	☽ ∥ ♆
30 Monday	
00 12	☽ ± ☿
01 01	☽ ∥ ♄
04 57	☽ ± ♀
06 02	☽ ⊥ ♅
07 20	☽ ∠ ♆
07 36	☽ ⊥ ♄
09 23	☽ ∥ ♅
10 13	☽ # ☿
11 05	☽ ⊼ ☿
15 26	☽ ⊼ ♀
16 43	☽ ⊻ ♅
18 14	☽ ⊻ ♄
19 41	☉ ⊼ ♆
22 21	☽ # ♀
22 40	☿ # ♅
23 18	☽ ⊥ ♆
31 Tuesday	
02 46	☽ □ ♂
09 49	☽ ⊻ ♆
10 53	☽ ☍ ☉
11 14	☽ ⊻ ♇
15 41	☉ ⊼ ♇
21 38	☽ ⊥ ♇
22 44	☿ St R

JUNE 1988

LONGITUDES

Date	Sidereal time h m s	Sun ☉ ° ' "	Moon ☽ ° ' "	Moon ☽ 24.00 ° ' "	Mercury ☿ ° '	Venus ♀ ° '	Mars ♂ ° '	Jupiter ♃ ° '	Saturn ♄ ° '	Uranus ♅ ° '	Neptune ♆ ° '	Pluto ♇ ° '
01	04 40 45	11 ♊ 12 02	24 ♐ 44 25	01 ♑ 46 44	26 ♊ 46	28 ♊ 28	06 ♓ 25	19 ♉ 38	00 ♑ 37	29 ♐ 48	09 ♑ 33	10 ♏ 22
02	04 44 41	12 09 29	08 ♑ 51 40	15 ♑ 58 33	26 R 42	28 R 04	07 03	19 52	00 R 32	29 R 46	09 R 32	10 R 21
03	04 48 38	13 06 56	23 ♑ 06 46	00 ♒ 15 41	26 33	27 37	07 40	20 05	00 28	29 43	09 30	10 19
04	04 52 34	14 04 23	07 ♒ 24 47	14 ♒ 33 34	26 19	27 08	08 17	20 19	00 24	29 41	09 29	10 18
05	04 56 31	15 01 48	21 ♒ 41 39	28 ♒ 48 43	26 02	26 38	08 54	20 33	00 20	29 39	09 28	10 17
06	05 00 28	15 59 13	05 ♓ 54 31	12 ♓ 58 53	25 41	26 06	09 30	20 46	00 16	29 36	09 26	10 16
07	05 04 24	16 56 37	20 ♓ 01 40	27 ♓ 02 49	25 17	25 32	10 07	21 00	00 12	29 34	09 25	10 14
08	05 08 21	17 54 00	04 ♈ 02 15	10 ♈ 59 55	24 51	24 58	10 44	21 13	00 07	29 32	09 23	10 13
09	05 12 17	18 51 23	17 ♈ 55 44	24 ♈ 49 39	24 21	24 22	11 20	21 27	00 ♑ 03	29 29	09 22	10 12
10	05 16 14	19 48 46	01 ♉ 41 31	08 ♉ 31 12	23 50	23 45	11 56	21 40	29 ♐ 59	29 27	09 20	10 11
11	05 20 10	20 46 08	15 ♉ 18 32	22 ♉ 03 18	23 17	23 08	12 32	21 54	29 54	29 24	09 19	10 09
12	05 24 07	21 43 29	28 ♉ 45 17	05 ♊ 24 16	22 44	22 31	13 08	22 07	29 50	29 22	09 17	10 08
13	05 28 03	22 40 50	12 ♊ 00 00	18 ♊ 32 17	22 10	21 53	13 44	22 20	29 46	29 20	09 16	10 07
14	05 32 00	23 38 10	25 ♊ 00 56	01 ♋ 25 50	21 37	21 16	14 19	22 34	29 41	29 17	09 14	10 06
15	05 35 57	24 35 30	07 ♋ 46 54	14 ♋ 04 07	21 04	20 39	14 55	22 47	29 37	29 15	09 13	10 05
16	05 39 53	25 32 49	20 ♋ 17 32	26 ♋ 27 19	20 33	20 02	15 30	23 00	29 33	29 12	09 11	10 04
17	05 43 50	26 30 08	02 ♌ 33 39	08 ♌ 36 50	20 04	19 27	16 05	23 13	29 28	29 10	09 09	10 03
18	05 47 46	27 27 25	14 ♌ 37 13	20 ♌ 35 12	19 37	18 52	16 40	23 26	29 24	29 07	09 08	10 02
19	05 51 43	28 24 42	26 ♌ 31 18	02 ♍ 26 02	19 13	18 19	17 14	23 39	29 19	29 05	09 06	10 01
20	05 55 39	29 ♊ 21 58	08 ♍ 19 59	14 ♍ 13 46	18 52	17 47	17 48	23 52	29 15	29 03	09 05	10 00
21	05 59 36	00 ♋ 19 13	20 ♍ 08 03	26 ♍ 03 30	18 36	17 16	18 23	24 05	29 10	29 00	09 03	09 59
22	06 03 32	01 16 28	02 ♎ 00 48	08 ♎ 00 38	18 23	16 48	18 57	24 17	29 06	28 58	09 02	09 58
23	06 07 29	02 13 42	14 ♎ 03 40	20 ♎ 10 34	18 14	16 21	19 30	24 30	29 02	28 55	09 00	09 57
24	06 11 26	03 10 55	26 ♎ 21 56	02 ♏ 38 21	18 10	15 56	20 04	24 43	28 57	28 53	08 58	09 56
25	06 15 22	04 08 08	09 ♏ 00 17	15 ♏ 28 09	18 D 10	15 34	20 37	24 55	28 53	28 50	08 57	09 56
26	06 19 19	05 05 21	22 ♏ 02 12	28 ♏ 42 38	18 15	15 14	21 10	25 08	28 48	28 48	08 55	09 55
27	06 23 15	06 02 32	05 ♐ 29 27	12 ♐ 22 31	18 24	14 55	21 43	25 20	28 44	28 45	08 53	09 54
28	06 27 12	06 59 44	19 ♐ 21 31	26 ♐ 26 01	18 39	14 40	22 16	25 33	28 40	28 43	08 52	09 53
29	06 31 08	07 56 56	03 ♑ 35 24	10 ♑ 48 55	18 58	14 26	22 48	25 45	28 35	28 41	08 50	09 53
30	06 35 05	08 ♋ 54 07	18 ♑ 05 45	25 ♑ 24 59	19 ♊ 23	14 ♊ 15	23 ♓ 20	25 ♉ 57	28 ♐ 31	28 ♐ 38	08 ♑ 49	09 ♏ 52

Date	Moon True ☊ ° '	Moon Mean ☊ ° '	Moon ☽ Latitude ° '
01	19 ♓ 19	19 ♓ 06	04 S 59
02	19 R 08	19 02	04 43
03	18 59	18 59	04 10
04	18 53	18 56	03 20
05	18 50	18 53	02 18
06	18 D 49	18 50	01 S 08
07	18 R 49	18 46	00 N 06
08	18 48	18 43	01 20
09	18 47	18 40	02 27
10	18 42	18 37	03 26
11	18 35	18 34	04 12
12	18 26	18 31	04 43
13	18 14	18 27	04 59
14	18 01	18 24	04 58
15	17 49	18 21	04 43
16	17 37	18 18	04 14
17	17 28	18 15	03 33
18	17 22	18 12	02 44
19	17 18	18 08	01 48
20	17 17	18 05	00 N 47
21	17 D 16	18 02	00 S 15
22	17 R 17	17 59	01 18
23	17 16	17 56	02 17
24	17 14	17 52	03 12
25	17 09	17 49	03 59
26	17 03	17 46	04 35
27	16 54	17 43	04 57
28	16 44	17 40	05 02
29	16 33	17 37	04 49
30	16 ♓ 24	17 ♓ 33	04 S 18

DECLINATIONS

Date	Sun ☉ ° '	Moon ☽ ° '	Mercury ☿ ° '	Venus ♀ ° '	Mars ♂ ° '	Jupiter ♃ ° '	Saturn ♄ ° '	Uranus ♅ ° '	Neptune ♆ ° '	Pluto ♇ ° '
01	22 N 07	28 S 19	23 N 15	25 N 11	11 S 31	16 N 47	22 S 17	23 S 39	22 S 06	00 N 47
02	22 15	27 51	22 59	24 58	11 19	16 50	22 17	23 39	22 06	00 48
03	22 23	25 34	22 42	24 45	11 06	16 54	22 17	23 39	22 07	00 48
04	22 30	21 39	22 24	24 32	10 54	16 58	22 17	23 39	22 07	00 48
05	22 36	16 27	22 07	24 18	10 42	17 01	22 17	23 39	22 07	00 48
06	22 42	10 24	21 49	24 04	10 30	17 05	22 17	23 39	22 07	00 48
07	22 48	03 S 51	21 31	23 49	10 18	17 08	22 17	23 39	22 07	00 48
08	22 54	02 N 49	21 13	23 34	10 06	17 12	22 17	23 39	22 07	00 48
09	22 59	09 18	20 55	23 19	09 54	17 15	22 18	23 39	22 07	00 48
10	23 03	15 17	20 37	23 03	09 42	17 19	22 18	23 39	22 07	00 48
11	23 07	20 26	20 22	22 47	09 30	17 22	22 18	23 39	22 07	00 48
12	23 11	24 29	20 04	22 31	09 18	17 26	22 18	23 39	22 07	00 48
13	23 14	27 10	19 48	22 14	09 06	17 29	22 18	23 39	22 08	00 47
14	23 17	28 19	19 34	21 58	08 54	17 33	22 18	23 39	22 08	00 47
15	23 20	27 55	19 20	21 42	08 43	17 36	22 18	23 39	22 08	00 47
16	23 22	26 05	19 08	21 26	08 31	17 39	22 18	23 39	22 08	00 47
17	23 24	23 03	18 57	21 10	08 19	17 42	22 18	23 39	22 08	00 47
18	23 25	19 04	18 48	20 54	08 07	17 46	22 18	23 39	22 08	00 47
19	23 26	14 22	18 40	20 39	07 56	17 49	22 18	23 39	22 08	00 47
20	23 26	09 11	18 34	20 24	07 44	17 52	22 18	23 39	22 08	00 47
21	23 27	03 N 41	18 29	20 09	07 33	17 55	22 18	23 39	22 08	00 46
22	23 26	01 S 59	18 26	19 55	07 21	17 58	22 18	23 39	22 09	00 46
23	23 25	07 39	18 25	19 42	07 10	18 01	22 18	23 39	22 09	00 46
24	23 24	13 09	18 25	19 29	06 59	18 04	22 18	23 39	22 09	00 46
25	23 23	18 16	18 27	19 17	06 47	18 07	22 18	23 39	22 09	00 46
26	23 21	22 42	18 31	19 06	06 36	18 10	22 18	23 39	22 09	00 45
27	23 18	26 05	18 36	18 55	06 25	18 13	22 18	23 39	22 09	00 45
28	23 15	28 02	18 43	18 45	06 14	18 16	22 19	23 39	22 09	00 45
29	23 12	28 13	18 51	18 36	06 03	18 19	22 19	23 39	22 09	00 45
30	23 N 09	26 S 29	19 N 00	18 N 28	05 S 53	18 N 22	22 S 19	23 S 39	22 S 09	00 N 44

ZODIAC SIGN ENTRIES

Date	h	m	Planets
01	20	58	☽ ♑
03	23	34	☽ ♒
06	02	00	☽ ♓
08	05	04	☽ ♈
10	05	22	♄ ♐
10	09	02	☽ ♉
12	14	14	☽ ♊
14	21	19	☽ ♋
17	06	57	☽ ♌
19	19	03	☽ ♍
21	03	57	☉ ♋
22	07	57	☽ ♎
24	18	58	☽ ♏
27	02	18	☽ ♐
29	06	00	☽ ♑

LATITUDES

Date	Mercury ☿ ° '	Venus ♀ ° '	Mars ♂ ° '	Jupiter ♃ ° '	Saturn ♄ ° '	Uranus ♅ ° '	Neptune ♆ ° '	Pluto ♇ ° '
01	00 S 09	01 N 45	02 S 32	00 S 54	01 N 09	00 S 13	01 N 00	16 N 31
04	00 59	01 07	02 38	00 54	01 09	00 13	01 00	16 30
07	01 51	00 N 27	02 44	00 54	01 09	00 13	01 00	16 29
10	02 41	00 S 15	02 50	00 54	01 09	00 13	01 00	16 28
13	03 25	00 57	02 56	00 54	01 09	00 13	01 00	16 27
16	03 59	01 39	03 02	00 54	01 09	00 13	01 00	16 25
19	04 21	02 18	03 08	00 54	01 08	00 13	01 00	16 24
22	04 31	02 53	03 14	00 54	01 08	00 13	01 00	16 23
25	04 28	03 24	03 21	00 54	01 08	00 13	01 00	16 21
28	04 15	03 50	03 27	00 54	01 08	00 13	01 00	16 20
31	03 S 53	04 S 11	03 S 33	00 S 54	01 N 07	00 S 13	01 N 00	16 N 18

DATA

Julian Date	2447314
Delta T	+56 seconds
Ayanamsa	23° 41' 45"
Synetic vernal point	05° ♓ 25' 14"
True obliquity of ecliptic	23° 26' 35"

MOON'S PHASES, APSIDES AND POSITIONS ☽

Date	h	m	Phase	Longitude ° '	Eclipse Indicator
07	06	22	☾	16 ♓ 43	
14	09	14	●	23 ♊ 32	
22	10	23	☽	01 ♎ 13	
29	19	46	○	08 ♑ 15	

Day	h	m		
04	23	55	Perigee	
20	08	13	Apogee	
01	18	01	Max dec	28° S 23'
08	01	50	0N	
14	17	42	Max dec	28° N 21'
22	03	37	0S	
29	02	20	Max dec	28° S 22'

LONGITUDES

Date	Chiron ⚷ ° '	Ceres ⚳ ° '	Pallas ⚴ ° '	Juno ⚵ ° '	Vesta ⚶ ° '	Black Moon Lilith ⚸ ° '
01	28 ♊ 10	23 ♓ 59	20 ♒ 11	07 ♊ 55	12 ♌ 33	02 ♍ 04
11	29 ♊ 05	26 ♓ 18	20 ♒ 04	13 ♊ 51	16 ♌ 28	03 ♍ 11
21	00 ♋ 01	28 ♓ 17	19 ♒ 19	18 ♊ 45	20 ♌ 34	04 ♍ 17
31	00 ♋ 56	29 ♓ 52	17 ♒ 58	25 ♊ 35	24 ♌ 50	05 ♍ 24

All ephemeris data is given at 12.00 UT and the Moon's longitude is additionally given for 24.00 UT

ASPECTARIAN

h m Aspects

01 Wednesday
03 05 ☽ ⊼ ♃
08 51 ☉ # ♆
11 26 ☽ Q ♂
13 05 ☽ ∠ ♇
13 33 ☽ ± ♃
15 28 ☽ ☍ ☿
18 12 ☽ ☍ ♀
20 37 ☽ ☌ ♅
21 58 ☽ ☌ ♄

02 Thursday
05 07 ☽ ꝑ ♃
08 47 ☽ ⚹ ♂
13 08 ☽ ☌ ♆
14 30 ☽ ⚹ ♇
17 58 ☽ ⊼ ☉
18 10 ☉ # ♄

03 Friday
04 48 ☽ ± ☉
06 50 ☽ △ ♃
10 41 ☽ Q ♇
11 13 ☽ ∠ ♂
17 41 ☽ ⊼ ☿
18 09 ☽ # ♀
19 19 ☽ ⊼ ♀
21 00 ☽ ꝑ ☉
23 04 ☽ ⊻ ♅

04 Saturday
00 18 ☽ ⊻ ♄
00 49 ☽ ∥ ♅
03 00 ☽ ⊥ ♂
03 36 ☽ ± ☿
05 04 ☽ ± ♀
06 53 ☉ ∥ ☿
07 35 ☽ # ☉
07 39 ☽ # ☿
08 36 ☽ ∥ ♄
09 06 ☽ ⊥ ♅
09 32 ☽ ∥ ♆
10 19 ☽ ⊥ ♄
12 07 ☿ ⊥ ♃
13 31 ☽ ⊻ ♂
14 27 ♂ Q ♃
15 28 ☽ ⊻ ♆
16 51 ☽ □ ♇
18 27 ☽ ꝑ ☿
19 40 ☽ ꝑ ♀
21 41 ☿ # ♄
23 59 ☽ △ ☉

05 Sunday
00 11 ☽ ∠ ♅
01 21 ☽ ∠ ♄
01 32 ☽ ⊥ ♆
09 37 ☽ # ♃
10 02 ☽ □ ♃
11 51 ☿ # ♆
14 43 ♀ ⊥ ♃
16 39 ☽ ∠ ♆
19 09 ☽ △ ☿
20 01 ☽ △ ♀

06 Monday
01 22 ☽ ⚹ ♅
02 30 ☽ ⚹ ♄
09 17 ♂ ⚹ ♆
11 35 ☽ ∥ ♂
16 56 ☽ Q ♃
17 58 ☽ ⚹ ♆
18 23 ☽ ☌ ♂
18 40 ☉ ± ♇
19 22 ☽ △ ♇
21 38 ☽ Q ♅
22 44 ☽ Q ♄

07 Tuesday
06 22 ☽ □ ☉
13 41 ☽ ⚹ ♃
14 21 ☽ Q ♆
15 07 ☿ ꝑ ♇
16 33 ♂ △ ♇
20 43 ☽ □ ☿
20 53 ☽ ꝑ ♇
21 03 ☽ □ ♀
22 58 ☽ # ♇

08 Wednesday
01 08 ♀ ꝑ ♇
04 08 ♀ # ♅
04 16 ☽ □ ♅
04 41 ☽ ∥ ♇
05 19 ☽ □ ♄
12 18 ☽ ± ♇
15 27 ☽ Q ☉
15 50 ☽ ∠ ♃
21 12 ☽ □ ♆
22 38 ☽ ⊼ ♇

09 Thursday
00 03 ☽ ⊻ ♂
02 41 ☽ Q ☿
02 46 ☽ Q ♀
07 38 ☽ ⊥ ♃
10 55 ☽ ⊥ ♂
13 44 ☽ ⚹ ☉
14 13 ☽ # ♂
15 10 ☿ ☌ ♀
17 48 ♂ Q ♅
18 13 ☽ ⊻ ♃
22 44 ☽ ⚹ ♀
22 46 ☽ ⚹ ☿

10 Friday
03 18 ☽ ∠ ♂
08 05 ☽ △ ♅
09 01 ☽ △ ♄
11 46 ☉ ∥ ♀
13 36 ♂ Q ♄
17 53 ☽ ∠ ☉
21 03 ☽ ∥ ♃
23 53 ☽ ∠ ♀

11 Saturday
00 04 ☽ ∠ ☿
01 25 ☽ △ ♆
02 54 ☽ ☍ ♇
06 52 ☽ ⚹ ♂
10 24 ☽ ꝑ ♅
10 58 ☽ ⊥ ☉
11 18 ☽ ꝑ ♄
11 31 ☽ ∥ ☿
15 06 ☽ ⊥ ♀
15 23 ☽ ⊥ ☿
21 08 ☽ # ♆
22 06 ☽ # ♄
22 27 ☽ ⊻ ☉
23 55 ☽ ☌ ♃

12 Sunday
00 10 ☽ ∥ ♀
01 19 ☽ ⊻ ♀
01 38 ☽ ⊻ ☿
02 22 ☽ ± ♅
03 14 ☽ ± ♄
03 18 ☽ ∥ ☉
04 00 ☽ ꝑ ♆
05 12 ☽ Q ♂
06 23 ☽ # ♅
13 06 ☽ ⊼ ♅
13 56 ☽ ⊼ ♄
20 09 ☽ ± ♆
23 18 ♀ ⊻ ♃

13 Monday
00 00 ☉ ☌ ♀
00 49 ☉ ⊻ ♃
03 55 ☉ ☌ ☿
06 49 ☿ ⊻ ♃
07 01 ☽ ⊼ ♆
07 16 ♀ # ♄
08 34 ☽ ⊼ ♇
15 19 ☽ □ ♂
19 32 ☽ ± ♇
22 07 ♀ # ♆

14 Tuesday
05 21 ☽ ☌ ♀
05 56 ☽ ☌ ☿
07 21 ☽ ⊻ ♃
09 14 ☽ ☌ ☉
12 09 ☽ ꝑ ♇
18 44 ☽ ⊥ ♃
19 57 ☽ ☍ ♅
20 41 ☽ ☍ ♄

15 Wednesday
12 00 ☽ ∠ ♃
14 43 ☽ ☍ ♆
16 22 ☽ △ ♇

16 Thursday
00 05 ☉ ꝑ ♇
02 17 ☽ △ ♂
11 32 ☽ ⊻ ♀
12 29 ☽ ⊻ ☿
17 21 ☽ ⚹ ♃
22 40 ☽ ⊥ ♀
23 05 ☽ ⊻ ☉
23 42 ☽ ⊥ ☿

17 Friday
05 20 ☽ ⊼ ♅
05 57 ☽ ⊼ ♄
07 22 ♃ ± ♅
07 51 ☽ # ♅
08 56 ☽ ꝑ ♂
09 39 ☽ ∥ ☉
11 52 ☽ ⊥ ☉
15 33 ☽ ∠ ♀
16 46 ☽ ∠ ☿
16 56 ☽ # ♄
17 08 ☽ ± ♅
17 21 ☽ Q ♃
17 43 ☽ ± ♄
17 59 ☽ # ♆

18 Saturday
00 49 ☽ ∥ ♀
01 03 ☽ ⊼ ♆
02 50 ☽ □ ♇
03 41 ☽ ± ♂
07 18 ☽ ∠ ☉
09 00 ♃ ± ♄
11 01 ☽ ꝑ ♅
11 33 ☽ ꝑ ♄
13 01 ☽ ± ♆
13 30 ☽ ∥ ☿
16 18 ☽ ⊼ ♂
18 53 ☽ ∥ ♃
20 09 ☽ ⚹ ♀
21 42 ☽ ⚹ ☿

19 Sunday
06 04 ☽ □ ♃
07 07 ☽ ꝑ ♆
15 01 ☽ Q ♇
16 10 ☽ ⚹ ☉
17 11 ☽ △ ♅
17 39 ☽ △ ♄
19 21 ☽ Q ♀
21 15 ☽ Q ☿

20 Monday
04 11 ☉ ☍ ♅
09 13 ☉ ☍ ♄
11 22 ♀ □ ♂
13 31 ☽ △ ♆
15 23 ☽ ⚹ ♇
18 38 ☽ # ♂
18 43 ☽ Q ☉

21 Tuesday
04 04 ☉ ⊥ ♃
06 25 ☽ □ ♀
08 15 ☽ ☍ ♂
08 56 ☽ □ ☿
09 41 ♃ ꝑ ♆
18 22 ☿ □ ♂
20 08 ☽ △ ♃
21 49 ☽ ∠ ♇

22 Wednesday
00 21 ☽ ∥ ♇
05 53 ☽ □ ♅
06 10 ☽ □ ♄
06 53 ☽ # ♇
10 23 ☽ □ ☉
15 55 ☽ ⊥ ♇

23 Thursday
01 59 ☽ □ ♆
02 48 ☽ ꝑ ♃
03 53 ☽ ⊻ ♇
09 59 ☽ ∥ ♂
16 22 ☽ △ ♀
17 36 ☽ Q ♅
17 48 ☽ Q ♄
20 08 ☽ △ ☿
20 53 ☽ ± ♃
23 12 ☽ ⊼ ♂

24 Friday
08 45 ☽ ⊼ ♃
11 24 ☽ ± ♂
12 03 ♀ ± ♇
13 10 ☽ Q ♆
16 49 ☽ ⚹ ♅
16 56 ☽ ⚹ ♄
20 30 ☽ ꝑ ♀
22 40 ☿ St D

25 Saturday
00 58 ☽ ꝑ ☿
02 05 ☽ △ ☉
05 21 ☽ ꝑ ♂
11 17 ☽ # ♃
11 53 ☽ ⚹ ♆
12 58 ☽ # ☿
13 01 ☽ ± ♀
13 43 ☽ ☌ ♇
16 57 ☽ # ♀
17 55 ☽ ± ☿
20 58 ☽ ∠ ♅
21 01 ☽ ∠ ♄
23 51 ☽ ⊼ ♀

26 Sunday
01 26 ♂ Q ♆
05 02 ☽ ⊼ ☿
08 11 ☽ ꝑ ☉
08 45 ☽ ∥ ♆
09 41 ☽ ∥ ♄
10 22 ☽ △ ♂
13 22 ☽ ⊥ ♅
13 23 ☽ ⊥ ♄
15 24 ☽ ∠ ♆
15 58 ☽ # ☉
17 06 ♄ ☌ ♅
17 40 ☽ ☍ ♃
17 58 ☽ ∥ ♅

27 Monday
00 06 ☽ ⊻ ♄
00 07 ☽ ⊻ ♅
01 38 ☽ ± ☉
07 26 ☽ ⊥ ♆
13 02 ☽ ⊼ ☉
17 56 ☽ ⊻ ♆
19 42 ☽ ⊻ ♇

28 Tuesday
04 05 ☽ ☍ ♀
06 04 ☽ ⊥ ♇
10 46 ☽ ☍ ☿
15 08 ☿ ∥ ♀
17 08 ☽ □ ♂
21 23 ☽ ∠ ♇
22 39 ☽ ⊼ ♃

29 Wednesday
03 40 ☽ ☌ ♄
03 48 ☽ ☌ ♅
08 53 ☽ ± ♃
19 46 ☽ ☍ ☉
20 43 ☽ ☌ ♆
22 26 ☽ ⚹ ♇

30 Thursday
00 03 ☽ ꝑ ♃
00 26 ☽ Q ♂
05 45 ☽ ⊼ ♀
09 46 ☉ ☍ ♆
14 10 ☽ ⊼ ☿
15 31 ☽ ± ♀
18 11 ☽ Q ♇
20 55 ☽ ⚹ ♂

JULY 1988

LONGITUDES

Date	Sidereal time h m s	Sun ☉ ° ' "	Moon ☽ ° ' "	Moon ☽ 24.00 ° ' "	Mercury ☿ ° '	Venus ♀ ° '	Mars ♂ ° '	Jupiter ♃ ° '	Saturn ♄ ° '	Uranus ♅ ° '	Neptune ♆ ° '	Pluto ♇ ° '
01	06 39 01	09 ♋ 51 18	02 ♒ 45 40	10 ♒ 06 54	19 ♊ 52	14 ♊ 07	23 ♓ 52	26 ♉ 09	28 ♐ 27	28 ♐ 36	08 ♑ 47	09 ♏ 51
02	06 42 58	10 48 29	17 ♒ 27 47	24 ♒ 47 33	20 26	14 R 01	24 24	26 21	28 R 22	28 R 33	08 R 45	09 R 51
03	06 46 55	11 45 40	02 ♓ 05 28	09 ♓ 20 58	21 04	13 57	24 55	26 33	28 18	28 31	08 44	09 50
04	06 50 51	12 42 51	16 ♓ 33 36	23 ♓ 43 00	21 48	13 56	25 26	26 45	28 14	28 29	08 42	09 50
05	06 54 48	13 40 03	00 ♈ 48 58	07 ♈ 51 19	22 36	13 D 57	25 57	26 57	28 10	28 26	08 41	09 49
06	06 58 44	14 37 14	14 ♈ 50 02	21 ♈ 45 06	23 28	14 00	26 27	27 09	28 05	28 24	08 39	09 49
07	07 02 41	15 34 27	28 ♈ 36 33	05 ♉ 24 28	24 26	14 05	26 57	27 21	28 01	28 22	08 37	09 48
08	07 06 37	16 31 39	12 ♉ 08 56	18 ♉ 50 02	25 27	14 13	27 27	27 32	27 57	28 19	08 36	09 48
09	07 10 34	17 28 53	25 ♉ 27 49	02 ♊ 02 22	26 33	14 23	27 57	27 44	27 53	28 17	08 34	09 47
10	07 14 30	18 26 06	08 ♊ 33 45	15 ♊ 01 57	27 44	14 35	28 26	27 55	27 49	28 15	08 33	09 47
11	07 18 27	19 23 20	21 ♊ 27 02	27 ♊ 49 01	28 ♊ 59	14 49	28 54	28 06	27 45	28 13	08 31	09 47
12	07 22 24	20 20 34	04 ♋ 07 54	10 ♋ 23 44	00 ♋ 18	15 05	29 23	28 18	27 41	28 11	08 29	09 47
13	07 26 20	21 17 49	16 ♋ 36 32	22 ♋ 46 24	01 41	15 22	29 ♓ 51	28 29	27 37	28 08	08 28	09 46
14	07 30 17	22 15 04	28 ♋ 53 26	04 ♌ 57 45	03 08	15 42	00 ♈ 18	28 40	27 34	28 06	08 26	09 46
15	07 34 13	23 12 19	10 ♌ 59 33	16 ♌ 59 03	04 40	16 03	00 46	28 51	27 30	28 04	08 25	09 46
16	07 38 10	24 09 34	22 ♌ 56 32	28 ♌ 52 20	06 15	16 27	01 12	29 02	27 26	28 02	08 23	09 46
17	07 42 06	25 06 50	04 ♍ 46 48	10 ♍ 40 23	07 54	16 51	01 39	29 13	27 22	28 00	08 22	09 46
18	07 46 03	26 04 06	16 ♍ 33 33	22 ♍ 26 48	09 36	17 18	02 05	29 23	27 19	27 58	08 20	09 46
19	07 49 59	27 01 22	28 ♍ 20 41	04 ♎ 15 48	11 23	17 46	02 30	29 34	27 15	27 56	08 19	09 46
20	07 53 56	27 58 38	10 ♎ 12 46	16 ♎ 12 13	13 12	18 15	02 56	29 44	27 12	27 54	08 17	09 D 46
21	07 57 53	28 55 55	22 ♎ 14 46	28 ♎ 21 06	15 04	18 46	03 20	29 ♉ 55	27 08	27 52	08 16	09 46
22	08 01 49	29 ♋ 53 12	04 ♏ 31 49	10 ♏ 47 32	16 59	19 18	03 44	00 ♊ 05	27 05	27 50	08 14	09 46
23	08 05 46	00 ♌ 50 29	17 ♏ 08 49	23 ♏ 36 10	18 57	19 52	04 08	00 15	27 02	27 48	08 13	09 46
24	08 09 42	01 47 46	00 ♐ 09 58	06 ♐ 50 33	20 57	20 26	04 31	00 25	26 58	27 46	08 11	09 46
25	08 13 39	02 45 04	13 ♐ 38 05	20 ♐ 32 36	22 58	21 02	04 54	00 35	26 55	27 44	08 10	09 46
26	08 17 35	03 42 23	27 ♐ 33 56	04 ♑ 41 48	25 01	21 40	05 17	00 45	26 52	27 42	08 08	09 46
27	08 21 32	04 39 42	11 ♑ 55 41	19 ♑ 14 54	27 06	22 18	05 38	00 55	26 49	27 40	08 07	09 46
28	08 25 28	05 37 02	26 ♑ 38 38	04 ♒ 05 53	29 ♋ 11	22 58	06 00	01 05	26 46	27 39	08 05	09 47
29	08 29 25	06 34 22	11 ♒ 35 37	19 ♒ 06 41	01 ♌ 17	23 38	06 20	01 14	26 43	27 37	08 04	09 47
30	08 33 22	07 31 43	26 ♒ 37 58	04 ♓ 08 22	03 23	24 20	06 41	01 23	26 40	27 35	08 03	09 47
31	08 37 18	08 ♌ 29 05	11 ♓ 36 53	19 ♓ 02 36	05 ♌ 29	25 ♊ 02	07 ♈ 00	01 ♊ 33	26 ♐ 38	27 ♐ 34	08 ♑ 01	09 ♏ 48

Date	Moon True ☊ ° '	Moon Mean ☊ ° '	Moon ☽ Latitude ° '
01	16 ♓ 16	17 ♓ 30	03 S 29
02	16 R 11	17 27	02 26
03	16 08	17 24	01 S 14
04	16 D 07	17 21	00 N 02
05	16 08	17 17	01 18
06	16 R 08	17 14	02 27
07	16 07	17 11	03 27
08	16 05	17 08	04 14
09	16 00	17 05	04 46
10	15 52	17 02	05 03
11	15 43	16 58	05 04
12	15 33	16 55	04 49
13	15 24	16 52	04 22
14	15 15	16 49	03 42
15	15 08	16 46	02 53
16	15 03	16 43	01 56
17	15 01	16 39	00 N 55
18	15 D 00	16 36	00 S 08
19	15 01	16 33	01 12
20	15 02	16 30	02 12
21	15 03	16 27	03 08
22	15 R 04	16 23	03 56
23	15 02	16 20	04 34
24	14 59	16 17	05 00
25	14 54	16 14	05 10
26	14 49	16 11	05 02
27	14 42	16 08	04 36
28	14 37	16 04	03 51
29	14 32	16 01	02 49
30	14 30	15 58	01 36
31	14 ♓ 28	15 ♓ 55	00 S 16

DECLINATIONS

Date	Sun ☉ ° '	Moon ☽ ° '	Mercury ☿ ° '	Venus ♀ ° '	Mars ♂ ° '	Jupiter ♃ ° '	Saturn ♄ ° '	Uranus ♅ ° '	Neptune ♆ ° '	Pluto ♇ ° '
01	23 N 05	22 S 56	19 N 11	18 N 20	05 S 42	18 N 25	22 S 19	23 S 39	22 S 10	00 N 44
02	23 00	17 56	19 22	18 14	05 31	18 27	22 19	23 39	22 10	00 44
03	22 55	11 53	19 35	18 08	05 21	18 30	22 19	23 39	22 10	00 43
04	22 50	05 S 16	19 48	18 03	05 10	18 33	22 19	23 39	22 10	00 43
05	22 44	01 N 31	20 01	17 58	05 00	18 36	22 19	23 39	22 10	00 43
06	22 38	08 06	20 15	17 55	04 50	18 38	22 19	23 39	22 10	00 42
07	22 32	14 12	20 30	17 51	04 40	18 41	22 19	23 39	22 10	00 42
08	22 25	19 30	20 45	17 49	04 30	18 43	22 19	23 39	22 10	00 41
09	22 18	23 45	20 59	17 47	04 20	18 46	22 19	23 39	22 11	00 41
10	22 10	26 43	21 14	17 46	04 11	18 49	22 19	23 39	22 11	00 41
11	22 02	28 13	21 28	17 46	04 01	18 51	22 19	23 39	22 11	00 40
12	21 54	28 12	21 41	17 46	03 52	18 53	22 19	23 39	22 11	00 40
13	21 45	26 44	21 54	17 46	03 43	18 56	22 19	23 39	22 11	00 39
14	21 36	24 00	22 06	17 47	03 34	18 58	22 19	23 39	22 11	00 39
15	21 27	20 14	22 16	17 49	03 25	19 01	22 19	23 39	22 11	00 38
16	21 17	15 42	22 26	17 51	03 16	19 03	22 20	23 39	22 11	00 38
17	21 07	10 37	22 34	17 53	03 08	19 05	22 20	23 39	22 11	00 37
18	20 56	05 N 11	22 40	17 56	02 59	19 07	22 20	23 39	22 12	00 37
19	20 45	00 S 26	22 44	17 58	02 51	19 10	22 20	23 39	22 12	00 36
20	20 34	06 04	22 46	18 02	02 43	19 12	22 20	23 39	22 12	00 36
21	20 23	11 34	22 46	18 05	02 35	19 14	22 20	23 38	22 12	00 35
22	20 11	16 44	22 44	18 09	02 28	19 16	22 20	23 38	22 12	00 34
23	19 58	21 20	22 39	18 13	02 20	19 18	22 20	23 38	22 12	00 34
24	19 46	25 04	22 31	18 17	02 13	19 20	22 20	23 38	22 12	00 33
25	19 33	27 34	22 21	18 21	02 06	19 22	22 20	23 38	22 12	00 33
26	19 20	28 28	22 08	18 25	01 59	19 24	22 20	23 38	22 12	00 32
27	19 06	27 29	21 53	18 30	01 52	19 26	22 20	23 38	22 13	00 31
28	18 52	24 36	21 34	18 34	01 46	19 28	22 20	23 38	22 13	00 31
29	18 38	20 01	21 14	18 39	01 39	19 30	22 20	23 38	22 13	00 30
30	18 23	14 08	20 50	18 43	01 33	19 32	22 21	23 38	22 13	00 30
31	18 N 09	07 S 27	20 N 25	18 N 48	01 S 27	19 N 33	22 S 21	23 S 38	22 S 13	00 N 29

ZODIAC SIGN ENTRIES

Date	h m	Planets
01	07 30	☽ ♒
03	08 33	☽ ♓
05	10 37	☽ ♈
07	14 27	☽ ♉
09	20 16	☽ ♊
12	04 08	☽ ♋
12	06 42	☿ ♋
13	20 00	♂ ♈
14	14 11	☽ ♌
17	02 17	☽ ♍
19	15 22	☽ ♎
22	00 00	♃ ♊
22	03 13	☽ ♏
22	14 51	☉ ♌
24	11 42	☽ ♐
26	16 07	☽ ♑
28	17 25	☽ ♒
28	21 19	☿ ♌
30	17 23	☽ ♓

LATITUDES

Date	Mercury ☿ ° '	Venus ♀ ° '	Mars ♂ ° '	Jupiter ♃ ° '	Saturn ♄ ° '	Uranus ♅ ° '	Neptune ♆ ° '	Pluto ♇ ° '
01	03 S 53	04 S 11	03 S 33	00 S 54	01 N 07	00 S 13	01 N 00	16 N 18
04	03 24	04 28	03 40	00 55	01 07	00 13	01 00	16 16
07	02 50	04 40	03 46	00 55	01 07	00 13	01 00	16 15
10	02 12	04 49	03 53	00 55	01 06	00 13	01 00	16 13
13	01 32	04 54	03 59	00 55	01 06	00 13	00 59	16 11
16	00 52	04 56	04 05	00 55	01 06	00 13	00 59	16 10
19	00 S 13	04 55	04 12	00 55	01 05	00 13	00 59	16 08
22	00 N 22	04 53	04 18	00 56	01 05	00 13	00 59	16 06
25	00 53	04 48	04 24	00 56	01 04	00 13	00 59	16 05
28	01 17	04 42	04 31	00 56	01 04	00 13	00 59	16 03
31	01 N 33	04 S 34	04 S 37	00 S 56	01 N 03	00 S 13	00 N 59	16 N 01

DATA

Julian Date	2447344
Delta T	+56 seconds
Ayanamsa	23° 41' 51"
Synetic vernal point	05° ♓ 25' 08"
True obliquity of ecliptic	23° 26' 35"

MOON'S PHASES, APSIDES AND POSITIONS ☽

Date	h m	Phase	Longitude ° '	Eclipse Indicator
06	11 36	☾	14 ♈ 36	
13	21 53	●	21 ♋ 41	
22	02 14	☽	29 ♎ 30	
29	03 25	○	06 ♒ 14	

Day	h m		
02	05 49	Perigee	
18	00 35	Apogee	
30	08 04	Perigee	
05	06 38	0N	
11	23 43	Max dec	28° N 24'
19	10 09	0S	
26	11 51	Max dec	28° S 28'

LONGITUDES

Date	Chiron ⚷ ° '	Ceres ⚳ ° '	Pallas ⚴ ° '	Juno ⚵ ° '	Vesta ⚶ ° '	Black Moon Lilith ⚸ ° '
01	00 ♋ 56	29 ♓ 52	17 ♒ 58	25 ♊ 35	24 ♌ 50	05 ♍ 24
11	01 ♋ 51	00 ♈ 59	16 ♒ 03	01 ♋ 22	29 ♌ 15	06 ♍ 30
21	02 ♋ 45	01 ♈ 35	13 ♒ 43	07 ♋ 05	03 ♍ 47	07 ♍ 37
31	03 ♋ 36	01 ♈ 38	11 ♒ 07	12 ♋ 42	08 ♍ 25	08 ♍ 43

All ephemeris data is given at 12.00 UT and the Moon's longitude is additionally given for 24.00 UT

ASPECTARIAN

h m	Aspects
01 Friday	
00 20	☽ ± ☿
01 03	☽ △ ♃
02 00	♀ ∥ ♃
04 59	☽ ⚺ ♄
05 13	☽ ⚺ ♅
06 06	☽ bQ ♀
07 58	☽ ∥ ♅
11 13	☽ # ☉
12 02	☉ △ ♇
14 44	☽ ⊥ ♄
14 59	☽ ⊥ ♅
15 23	☽ ∥ ♄
15 34	☽ bQ ☿
16 11	☽ ∥ ♆
21 49	☽ ⚺ ♆
22 20	☽ ∠ ♂
23 34	☽ □ ♇
02 Saturday	
00 23	☽ ⚻ ☉
05 21	☽ ∠ ♄
05 38	☽ ∠ ♅
05 53	☽ # ☿
06 24	☽ △ ♀
07 35	☽ ⊥ ♆
09 43	☽ # ♃
10 40	☽ # ♀
10 51	☽ ± ☉
13 35	☽ ⊥ ♂
17 04	☽ △ ☿
22 17	☽ ∠ ♆
23 46	☽ ⚺ ♂
03 Sunday	
02 37	☽ bQ ☉
02 46	☽ □ ♃
05 26	☉ ∠ ♃
05 47	☽ ✶ ♄
06 08	☽ ✶ ♅
08 21	♂ bQ ♇
22 57	☽ ✶ ♆
04 Monday	
00 48	☽ △ ♇
01 31	☽ Q ♄
01 54	☽ Q ♅
05 08	☽ △ ☉
07 36	☽ □ ♀
08 57	☽ Q ♃
12 21	☽ ∥ ♂
14 09	♀ St D
18 55	☽ Q ♆
21 16	☽ □ ☿
05 Tuesday	
01 52	☽ bQ ♇
03 27	☽ ☌ ♂
04 07	☽ # ♇
05 22	☽ ✶ ♃
07 31	☽ □ ♄
07 59	☽ □ ♅
09 09	☽ ∥ ♇
13 55	☽ Q ♀
17 06	☽ ± ♇
19 13	☉ ⚺ ♀
06 Wednesday	
00 14	☽ # ♂
01 23	☽ □ ♆
03 21	☽ ⚻ ♇
05 48	☽ Q ☿
07 18	☽ ∠ ♃
10 33	☽ ✶ ♀
11 36	☽ □ ☉
23 06	☽ ⊥ ♃
07 Thursday	
04 06	☽ ✶ ☿
08 59	☽ ⚺ ♂
09 45	☽ ⚺ ♃
10 58	☽ △ ♄
11 34	☽ △ ♅
12 51	☽ ∠ ♀
19 57	☽ ⊥ ♂
20 56	☿ bQ ♇
21 25	☽ Q ☉
08 Friday	
03 57	☽ ∥ ♀
04 55	☽ ⊥ ♀
05 40	☽ △ ♆
07 48	☽ ☍ ♇
08 09	☽ ∥ ♃
08 43	☽ ∠ ☿
12 34	☽ ∠ ♂
13 26	☽ bQ ♄
14 06	☽ bQ ♅
15 44	☽ ⚺ ♀
18 43	☽ ∥ ☿
18 47	♂ ✶ ♃
20 27	☽ ✶ ☉
09 Saturday	
02 16	☽ # ♆
02 17	☽ ⊥ ☿
03 07	☽ # ♄
03 16	☽ ∥ ☉
05 33	☽ ± ♄
06 15	☽ ± ♅
08 28	☉ # ♄
08 34	☽ bQ ♆
09 34	♂ □ ♄
11 18	☽ # ♅
14 11	☽ ⚺ ☿
16 11	☽ ☌ ♃
16 23	☽ ⚻ ♄
16 41	☽ ✶ ♂
17 08	☽ ⚻ ♅
10 Sunday	
00 57	☽ ± ♆
01 49	☽ ∠ ☉
02 38	♃ ⚻ ♄
03 49	♂ □ ♅
11 24	☉ # ♆
11 58	☽ ⚻ ♆
13 37	☿ ☍ ♄
14 16	☽ ⚻ ♇
15 35	☽ Q ♂
16 21	☿ ⚺ ♃
19 45	☽ ⊥ ☉
21 49	☿ ☍ ♅
23 21	☽ ☌ ♀
11 Monday	
01 24	☽ ± ♇
07 49	☽ ⚺ ☉
09 48	☿ □ ♂
18 16	☽ bQ ♇
23 04	♃ ⚻ ♅
23 49	☽ ☍ ♄
12 Tuesday	
00 43	☽ ☍ ♅
00 44	☽ ⚺ ♃
02 37	☽ □ ♂
03 50	☽ ☌ ☿
12 19	☽ ⊥ ♃
20 19	☽ ☍ ♆
22 48	☽ △ ♇
13 Wednesday	
02 31	☉ ∥ ☿
05 51	☽ ∠ ♃
09 33	☽ ⚺ ♀
14 51	☉ ⊥ ♀
21 31	☽ ⊥ ♀
21 53	☽ ☌ ☉
14 Thursday	
09 24	☽ ⚻ ♄
10 27	☽ ⚻ ♅
11 33	☽ ✶ ♃
14 34	☽ # ♅
14 54	☽ △ ♂
15 40	☽ ∠ ♀
16 42	♀ ± ♇
21 10	☽ ± ♄
21 35	☽ ⚺ ☿
22 16	☽ ± ♅
23 25	☽ # ♄
23 53	☿ # ♆
15 Friday	
00 17	☽ ∥ ☿
00 18	☽ # ♆
04 35	☽ ∥ ☉
06 52	☽ ⚻ ♆
09 33	☽ □ ♇
11 15	☽ ⊥ ☿
11 43	☽ Q ♃
14 59	☽ bQ ♄
15 17	☿ ⊥ ♃
16 08	☽ bQ ♅
18 48	☽ ∥ ♃
18 49	☽ ± ♆
19 26	☿ # ♄
21 55	☽ bQ ♂
22 28	☽ ✶ ♀
16 Saturday	
01 10	☽ ∥ ♀
08 03	☽ ∠ ☿
12 54	☽ bQ ♆
14 40	☽ ⚺ ☉
16 45	☽ ± ♂
21 02	☽ △ ♄
21 45	☽ Q ♇
22 16	☽ △ ♅
23 31	☽ Q ♀
17 Sunday	
00 31	☽ □ ♃
03 54	☽ ⊥ ☉
05 23	☽ ⚻ ♂
18 27	☿ ☍ ♆
19 16	☽ △ ♆
19 25	☽ ✶ ☿
22 08	☽ ✶ ♇
23 49	☽ ∠ ☉
18 Monday	
13 34	☽ □ ♀
14 05	☿ △ ♇
21 38	☽ # ♂
19 Tuesday	
00 06	☽ Q ☿
04 43	☽ ∠ ♇
07 35	☽ ∥ ♇
09 05	☽ ✶ ☉
09 48	☽ □ ♄
11 09	☽ □ ♅
12 42	☽ # ♇
14 31	☽ △ ♃
17 27	☉ ⚻ ♄
20 45	☽ ☍ ♂
22 02	☽ ∥ ♂
22 59	☽ ⊥ ♇
20 Wednesday	
04 19	♇ St D
08 07	☽ □ ♆
09 59	☉ ⚻ ♅
11 05	☽ ⚺ ♇
11 29	☽ Q ☉
19 05	☽ □ ☿
21 13	☽ bQ ♃
21 56	☽ Q ♄
23 21	☽ Q ♅
21 Thursday	
04 47	☽ △ ♀
09 50	☿ ∠ ♃
15 20	☽ ± ♃
19 53	☽ Q ♆
21 35	☽ ✶ ♄
23 01	☽ ✶ ♅
22 Friday	
02 14	☽ □ ☉
03 16	☽ ⚻ ♃
10 25	☽ ⚻ ♂
11 32	☽ bQ ♀
13 13	♂ ± ♇
18 05	☉ ✶ ♃
19 06	☽ # ♀
19 06	☽ ✶ ♆
22 02	☽ ☌ ♇
22 19	☽ ± ♂
23 Saturday	
00 51	☽ # ♃
02 23	☽ ∠ ♄
03 50	☽ ∠ ♅
04 49	☽ # ☉
05 32	☽ ± ♀
15 50	☽ bQ ♂
15 59	☽ △ ☿
17 02	☽ ∥ ♆
17 18	☽ ⚻ ♀
17 49	☽ ∥ ♄
19 12	☽ ⊥ ♄
19 30	☽ # ☿
20 38	☽ ⊥ ♅
23 15	☽ ∠ ♆
24 Sunday	
01 59	☽ ∥ ♅
03 29	☿ ⚺ ♀
06 13	☽ ⚺ ♄
07 39	☽ ⚺ ♅
12 28	☽ ☍ ♃
15 11	☽ △ ☉
15 39	☽ ⊥ ♆
20 05	☽ △ ♂
25 Monday	
00 14	☽ bQ ☿
02 22	☽ ⚺ ♆
05 11	☽ ⚺ ♇
13 49	☿ # ♄
15 44	☽ ⊥ ♇
16 00	☉ ± ♄
18 50	☽ ± ☿
19 42	☽ bQ ☉
26 Tuesday	
01 27	☽ ☍ ♀
04 30	☿ # ♆
04 58	☉ ∥ ♃
06 56	☽ ⚻ ☿
07 15	☽ ∠ ♇
10 49	☽ ☌ ♄
11 56	☉ ± ♅
12 14	☽ ☌ ♅
12 15	☽ ± ☉
17 27	☽ ⚻ ♃
23 05	☽ ⚻ ☉
27 Wednesday	
01 18	☽ □ ♂
03 36	☽ ± ♃
05 42	☽ ☌ ♆
08 27	☽ ✶ ♇
08 50	☿ ⚻ ♄
18 32	☿ ⚻ ♅
18 38	☽ bQ ♃
28 Thursday	
04 07	☽ Q ♇
05 45	☽ ⚻ ♀
07 36	☽ Q ♂
08 11	☿ ⊥ ♀
12 12	☽ ⚺ ♄
13 37	☽ ⚺ ♅
15 55	☽ ± ♀
16 46	☽ ☍ ☿
17 49	☽ ∥ ♅
19 13	☽ △ ♃
21 50	☽ ⊥ ♄
23 15	☽ ⊥ ♅
29 Friday	
00 51	☽ ∥ ♄
01 31	☽ ∥ ♆
02 51	☉ △ ♂
03 24	☽ ✶ ♂
03 25	☽ ☍ ☉
05 57	☽ # ☿
06 22	☽ ⚺ ♆
07 03	☽ bQ ♀
09 06	☽ □ ♇
11 03	☉ ∥ ♀
11 23	☿ ✶ ♃
12 12	☽ ∠ ♄
13 38	☽ ∠ ♅
14 20	☽ # ♃
15 57	☽ ⊥ ♆
17 58	☽ # ♀
18 22	☽ # ☉
30 Saturday	
03 55	☽ ∠ ♂
04 02	☿ ± ♄
06 17	☽ ∠ ♆
08 09	☽ △ ♀
12 04	☽ ✶ ♄
13 31	☽ ✶ ♅
14 16	☿ ± ♅
18 36	☽ ⊥ ♂
19 41	☽ □ ♃
31 Sunday	
00 33	☽ ⚻ ☿
00 39	☉ ⚻ ♆
03 46	♀ bQ ♇
04 25	☽ ⚺ ♂
06 14	☽ ✶ ♆
06 37	☽ ⚻ ☉
07 13	☽ Q ♄
08 42	☽ Q ♅
09 04	☽ △ ♇
11 46	☽ ± ☿
16 57	☽ ± ☉

AUGUST 1988

Date	Sidereal time h m s	Sun ☉ ° ' "	Moon ☽ ° ' "	Moon ☽ 24.00 ° ' "	Mercury ☿ ° '	Venus ♀ ° '	Mars ♂ ° '	Jupiter ♃ ° '	Saturn ♄ ° '	Uranus ♅ ° '	Neptune ♆ ° '	Pluto ♇ ° '
01	08 41 15	09 ♌ 26 28	26 ♓ 24 44	03 ♈ 42 40	07 ♌ 35	25 ♊ 46	07 ♈ 19	01 ♊ 42	26 ♐ 35	27 ♐ 32	08 ♑ 00	09 ♏ 48
02	08 45 11	10 23 51	10 ♈ 55 54	18 ♈ 04 06	09 40	26 30	07 38	01 51	26 R 32	27 R 30	07 R 59	09 49
03	08 49 08	11 21 17	25 ♈ 07 05	02 ♉ 04 45	11 45	27 16	07 55	02 00	26 30	27 29	07 57	09 49
04	08 53 04	12 18 43	08 ♉ 57 06	15 ♉ 44 16	13 49	28 02	08 13	02 09	26 27	27 27	07 56	09 50
05	08 57 01	13 16 11	22 ♉ 26 24	29 ♉ 03 42	15 51	28 49	08 29	02 17	26 25	27 26	07 55	09 50
06	09 00 57	14 13 40	05 ♊ 36 24	12 ♊ 04 47	17 53	29 ♊ 37	08 45	02 26	26 23	27 24	07 54	09 51
07	09 04 54	15 11 10	18 ♊ 29 07	24 ♊ 49 39	19 53	00 ♋ 26	09 00	02 34	26 21	27 23	07 52	09 51
08	09 08 51	16 08 42	01 ♋ 06 40	07 ♋ 20 25	21 52	01 15	09 15	02 42	26 19	27 22	07 51	09 52
09	09 12 47	17 06 15	13 ♋ 31 09	19 ♋ 39 05	23 50	02 05	09 29	02 50	26 17	27 20	07 50	09 53
10	09 16 44	18 03 49	25 ♋ 44 28	01 ♌ 47 31	25 46	02 56	09 42	02 58	26 15	27 19	07 49	09 53
11	09 20 40	19 01 25	07 ♌ 48 28	13 ♌ 47 31	27 41	03 47	09 55	03 06	26 13	27 18	07 48	09 54
12	09 24 37	19 59 01	19 ♌ 44 55	25 ♌ 40 54	29 ♌ 34	04 39	10 06	03 14	26 11	27 17	07 47	09 55
13	09 28 33	20 56 39	01 ♍ 35 43	07 ♍ 29 40	01 ♍ 26	05 32	10 17	03 22	26 09	27 16	07 46	09 56
14	09 32 30	21 54 18	13 ♍ 23 02	19 ♍ 16 09	03 17	06 25	10 27	03 29	26 08	27 15	07 45	09 57
15	09 36 26	22 51 58	25 ♍ 09 22	01 ♎ 03 06	05 05	07 19	10 37	03 36	26 06	27 13	07 44	09 57
16	09 40 23	23 49 39	06 ♎ 57 44	12 ♎ 53 44	06 53	08 13	10 46	03 43	26 05	27 12	07 42	09 58
17	09 44 20	24 47 21	18 ♎ 51 34	24 ♎ 51 45	08 39	09 08	10 53	03 50	26 04	27 12	07 42	09 59
18	09 48 16	25 45 04	00 ♏ 54 49	07 ♏ 01 16	10 23	10 04	11 01	03 57	26 03	27 11	07 41	10 00
19	09 52 13	26 42 49	13 ♏ 11 41	19 ♏ 26 36	12 07	11 00	11 07	04 04	26 01	27 10	07 40	10 01
20	09 56 09	27 40 34	25 ♏ 46 33	02 ♐ 12 02	13 48	11 56	11 12	04 10	26 00	27 09	07 39	10 02
21	10 00 06	28 38 21	08 ♐ 43 29	15 ♐ 21 19	15 29	12 53	11 17	04 17	26 00	27 08	07 38	10 03
22	10 04 02	29 ♌ 36 09	22 ♐ 05 48	28 ♐ 57 09	17 08	13 50	11 21	04 23	25 59	27 07	07 37	10 04
23	10 07 59	00 ♍ 33 58	05 ♑ 55 25	13 ♑ 00 30	18 45	14 48	11 24	04 29	25 58	27 07	07 36	10 06
24	10 11 55	01 31 48	20 ♑ 12 08	27 ♑ 29 55	20 21	15 46	11 26	04 35	25 57	27 06	07 35	10 07
25	10 15 52	02 29 40	04 ♒ 53 12	12 ♒ 21 12	21 56	16 45	11 27	04 40	25 57	27 06	07 35	10 08
26	10 19 49	03 27 33	19 ♒ 53 00	27 ♒ 27 29	23 29	17 44	11 28	04 46	25 56	27 05	07 34	10 09
27	10 23 45	04 25 27	05 ♓ 03 32	12 ♓ 39 55	25 01	18 43	11 R 27	04 51	25 56	27 05	07 33	10 10
28	10 27 42	05 23 23	20 ♓ 15 25	27 ♓ 48 53	26 32	19 43	11 26	04 57	25 56	27 04	07 32	10 12
29	10 31 38	06 21 20	05 ♈ 19 14	12 ♈ 45 30	28 02	20 43	11 24	05 02	25 56	27 04	07 32	10 13
30	10 35 35	07 19 19	20 ♈ 06 53	27 ♈ 22 44	29 ♍ 29	21 44	11 21	05 06	25 D 56	27 04	07 31	10 14
31	10 39 31	08 ♍ 17 21	04 ♉ 32 33	11 ♉ 36 03	00 ♎ 56	22 ♋ 45	11 ♈ 18	05 ♊ 11	25 ♐ 56	27 ♐ 03	07 ♑ 31	10 ♏ 16

Date	Moon True ☊ ° '	Moon Mean ☊ ° '	Moon ☽ Latitude ° '
01	14 ♓ 29	15 ♓ 52	01 N 05
02	14 D 30	15 49	02 20
03	14 31	15 45	03 24
04	14 32	15 42	04 15
05	14 R 32	15 39	04 50
06	14 30	15 36	05 09
07	14 27	15 33	05 12
08	14 23	15 29	05 00
09	14 19	15 26	04 34
10	14 14	15 23	03 55
11	14 11	15 20	03 07
12	14 08	15 17	02 10
13	14 06	15 14	01 08
14	14 05	15 10	00 N 04
15	14 D 06	15 07	01 S 01
16	14 07	15 04	02 03
17	14 08	15 01	03 01
18	14 10	14 58	03 51
19	14 11	14 55	04 32
20	14 11	14 51	05 01
21	14 R 11	14 48	05 15
22	14 10	14 45	05 14
23	14 09	14 42	04 54
24	14 07	14 39	04 16
25	14 06	14 35	03 21
26	14 05	14 32	02 10
27	14 04	14 29	00 S 50
28	14 D 04	14 26	00 N 34
29	14 04	14 23	01 55
30	14 05	14 20	03 07
31	14 ♓ 06	14 ♓ 16	04 N 05

DECLINATIONS

Date	Sun ☉ ° '	Moon ☽ ° '	Mercury ☿ ° '	Venus ♀ ° '	Mars ♂ ° '	Jupiter ♃ ° '	Saturn ♄ ° '	Uranus ♅ ° '	Neptune ♆ ° '	Pluto ♇ ° '
01	17 N 54	00 S 26	19 N 57	18 N 52	01 S 22	19 N 35	22 S 21	23 S 38	22 S 13	00 N 28
02	17 38	06 N 28	19 27	18 56	01 16	19 37	22 21	23 38	22 13	00 27
03	17 23	12 53	18 55	19 01	01 11	19 39	22 21	23 38	22 13	00 27
04	17 07	18 30	18 21	19 05	01 06	19 40	22 21	23 38	22 13	00 26
05	16 50	23 03	17 46	19 09	01 01	19 42	22 21	23 38	22 14	00 25
06	16 34	26 19	17 09	19 13	00 57	19 43	22 21	23 38	22 14	00 25
07	16 17	28 08	16 31	19 16	00 52	19 45	22 21	23 38	22 14	00 24
08	16 00	28 26	15 52	19 20	00 48	19 47	22 21	23 38	22 14	00 23
09	15 43	27 18	15 12	19 23	00 44	19 48	22 21	23 38	22 14	00 22
10	15 25	24 51	14 31	19 26	00 41	19 49	22 22	23 38	22 14	00 22
11	15 07	21 19	13 49	19 29	00 38	19 51	22 22	23 38	22 14	00 21
12	14 49	16 57	13 06	19 31	00 35	19 52	22 22	23 38	22 14	00 20
13	14 31	11 58	12 23	19 34	00 32	19 54	22 22	23 38	22 14	00 19
14	14 13	06 36	11 39	19 36	00 29	19 55	22 22	23 38	22 14	00 19
15	13 54	01 N 00	10 55	19 37	00 27	19 56	22 22	23 38	22 15	00 18
16	13 35	04 S 39	10 11	19 39	00 25	19 57	22 22	23 38	22 15	00 17
17	13 16	10 10	09 26	19 39	00 23	19 59	22 22	23 38	22 15	00 16
18	12 56	15 24	08 42	19 40	00 22	20 00	22 23	23 38	22 15	00 15
19	12 37	20 07	07 57	19 40	00 20	20 01	22 23	23 38	22 15	00 15
20	12 17	24 04	07 12	19 40	00 19	20 02	22 23	23 38	22 15	00 14
21	11 57	26 57	06 27	19 40	00 19	20 03	22 23	23 38	22 15	00 13
22	11 37	28 26	05 42	19 39	00 18	20 04	22 23	23 38	22 15	00 12
23	11 16	28 13	04 58	19 38	00 18	20 05	22 23	23 38	22 15	00 11
24	10 56	26 09	04 13	19 36	00 18	20 06	22 23	23 38	22 15	00 10
25	10 35	22 17	03 29	19 34	00 18	20 07	22 24	23 38	22 15	00 10
26	10 14	16 55	02 45	19 31	00 19	20 08	22 24	23 38	22 15	00 09
27	09 53	10 26	02 01	19 28	00 19	20 09	22 24	23 38	22 16	00 08
28	09 32	03 S 20	01 18	19 25	00 20	20 10	22 24	23 38	22 16	00 07
29	09 11	03 N 53	00 N 35	19 21	00 22	20 11	22 24	23 38	22 16	00 06
30	08 49	10 45	00 S 07	19 16	00 23	20 11	22 24	23 38	22 16	00 05
31	08 N 28	16 N 53	00 S 49	19 N 12	00 S 25	20 N 12	22 S 25	23 S 38	22 S 16	00 N 04

ZODIAC SIGN ENTRIES

Date	h	m	Planets
01	17	53	☽ ♈
03	20	24	☽ ♉
06	01	43	☽ ♊
06	23	24	♀ ♋
08	09	52	☽ ♋
10	20	26	☽ ♌
12	17	29	☿ ♍
13	08	46	☽ ♍
15	21	52	☽ ♎
18	10	12	☽ ♏
20	19	55	☽ ♐
22	21	54	☉ ♍
23	01	49	☽ ♑
25	04	05	☽ ♒
27	04	01	☽ ♓
29	03	29	☽ ♈
30	20	25	☿ ♎
31	04	22	☽ ♉

LATITUDES

Date	Mercury ☿ ° '	Venus ♀ ° '	Mars ♂ ° '	Jupiter ♃ ° '	Saturn ♄ ° '	Uranus ♅ ° '	Neptune ♆ ° '	Pluto ♇ ° '
01	01 N 37	04 S 31	04 S 39	00 S 56	01 N 03	00 S 13	00 N 59	16 N 00
04	01 45	04 21	04 44	00 56	01 03	00 13	00 59	15 59
07	01 46	04 10	04 50	00 57	01 02	00 13	00 59	15 57
10	01 41	03 59	04 55	00 57	01 02	00 13	00 59	15 55
13	01 31	03 46	05 00	00 57	01 01	00 13	00 59	15 54
16	01 17	03 33	05 05	00 57	01 01	00 13	00 59	15 52
19	01 00	03 20	05 09	00 58	01 00	00 13	00 59	15 50
22	00 40	03 06	05 12	00 58	01 00	00 13	00 58	15 49
25	00 N 19	02 51	05 15	00 58	00 59	00 13	00 58	15 47
28	00 S 05	02 37	05 17	00 59	00 59	00 13	00 58	15 45
31	00 S 30	02 S 22	05 S 18	00 S 59	00 N 58	00 S 13	00 N 58	15 N 44

DATA

Julian Date	2447375
Delta T	+56 seconds
Ayanamsa	23° 41' 57"
Synetic vernal point	05° ♓ 25' 03"
True obliquity of ecliptic	23° 26' 36"

MOON'S PHASES, APSIDES AND POSITIONS ☽

Date	h	m	Phase	Longitude ° '	Eclipse Indicator
04	18	22	☾	12 ♉ 34	
12	12	31	●	20 ♌ 00	
20	15	51	☽	27 ♏ 50	
27	10	56	○	04 ♓ 23	partial

Day	h	m		
14	11	55	Apogee	
27	16	55	Perigee	
01	13	29	0N	
08	04	56	Max dec	28° N 30'
15	16	14	0S	
22	21	11	Max dec	28° S 33'
28	23	03	0N	

LONGITUDES

Date	Chiron ⚷ ° '	Ceres ⚳ ° '	Pallas ⚴ ° '	Juno ⚵ ° '	Vesta ⚶ ° '	Black Moon Lilith ⚸ ° '
01	03 ♋ 41	01 ♈ 36	10 ♒ 51	13 ♋ 15	08 ♍ 54	08 ♍ 50
11	04 ♋ 28	00 ♈ 59	08 ♒ 16	18 ♋ 45	13 ♍ 39	09 ♍ 56
21	05 ♋ 10	29 ♓ 46	05 ♒ 55	24 ♋ 09	18 ♍ 29	11 ♍ 03
31	05 ♋ 48	28 ♓ 04	03 ♒ 59	29 ♋ 23	23 ♍ 23	12 ♍ 09

ASPECTARIAN

01 Monday
00 57 ☽ Q ♃
01 34 ☽ Q ♆
04 43 ☽ ⚼ ☿
08 25 ☿ △ ♂
08 33 ☽ ⚼ ☉
08 48 ☽ ∥ ♂
09 22 ☽ ⚼ ♇
10 53 ☽ □ ♀
11 53 ☽ # ♇
12 17 ☽ □ ♄
13 50 ☽ □ ♅
15 05 ☽ ∥ ♇
16 43 ☿ ⊼ ♆
18 04 ☽ # ♂
20 46 ☽ ⚹ ♃
21 09 ☉ □ ♇

02 Tuesday
00 09 ☽ ± ♇
04 35 ☿ ∥ ♃
06 22 ☽ ☌ ♂
07 05 ☽ □ ♆
09 33 ☽ △ ☿
10 08 ☽ ⊼ ♇
11 03 ☽ △ ☉
12 58 ♀ ☍ ♄
13 35 ☿ □ ♇
18 20 ☽ Q ♀
22 03 ☽ ∠ ♃

03 Wednesday
03 29 ☉ ☌ ☿
06 22 ♃ ± ♆
08 19 ☿ ∥ ♀
09 08 ☿ ⚼ ♄
13 31 ☽ ⊥ ♃
14 22 ☽ △ ♄
14 27 ♂ □ ♆
15 26 ☉ ⚼ ♄
15 54 ☽ ⚹ ♀
16 03 ☽ △ ♅
18 35 ♀ ☍ ♅
20 23 ☿ ⚼ ♅
21 29 ☿ ∠ ♀
23 59 ☽ ⊻ ♃

04 Thursday
05 56 ☽ ∥ ☉
10 13 ☽ △ ♆
10 40 ☽ ⊻ ♂
11 23 ☽ ∥ ☿
13 25 ☿ ± ♆
13 32 ☽ ☍ ♇
14 46 ☽ ∥ ♀
15 30 ☉ ⚼ ♅
16 09 ☿ Q ♃
16 24 ☽ ⚼ ♄
17 38 ☽ ∥ ♃
18 10 ☽ ⚼ ♅
18 22 ☽ □ ☉
19 38 ☽ ∠ ♀
21 29 ☽ ⊥ ♂
22 07 ☽ □ ☿

05 Friday
07 06 ☽ # ♆
07 49 ☽ # ♄
08 22 ☽ ± ♄
10 11 ☽ ± ♅
12 44 ☽ ⊥ ♀
12 51 ☽ ⚼ ♆
13 56 ☽ ∠ ♂
15 37 ☽ # ♅
19 10 ☽ ⊼ ♄
21 01 ☽ ⊼ ♅

06 Saturday
00 18 ☽ ⊻ ♀
03 49 ☉ ± ♆
05 11 ☽ ± ♆
05 18 ☽ Q ☉
06 06 ☽ ☌ ♃
12 36 ☽ Q ☿
16 13 ☽ ⊼ ♆
17 54 ☉ Q ♃
17 56 ☽ ⚹ ♂
19 51 ☽ ⊼ ♇

07 Sunday
05 18 ☽ ⚹ ☉
07 03 ☽ ± ♇
15 08 ☽ ⚹ ☿
16 51 ☽ Q ♂

08 Monday
00 04 ☽ ⚼ ♇
02 51 ☽ ☍ ♄
03 38 ☉ ∥ ☿
04 50 ☽ ☍ ♅
12 04 ☽ ∠ ☉
12 17 ☽ ☌ ♀
15 06 ☽ ⊻ ♃
23 51 ☿ ⚼ ♆

09 Tuesday
00 58 ☽ ☍ ♆
01 11 ☽ ∠ ☿
02 48 ☽ ⊥ ♃
04 00 ☽ □ ♂
04 55 ☽ △ ♇
06 54 ☽ ⊥ ☉
08 05 ☉ ∠ ♀
19 36 ☽ ⊻ ☉
20 33 ☽ ∠ ♃
21 03 ☿ ⚼ ♂
22 02 ☽ ⊥ ☿

10 Wednesday
12 04 ☽ ⊻ ☿
13 00 ☽ ⊼ ♄
13 27 ♀ ⊻ ♃
15 07 ☽ ⊼ ♅
17 49 ☿ △ ♄
21 04 ☽ # ♅

11 Thursday
00 52 ☽ ± ♄
02 31 ☽ ⚹ ♃
03 01 ☽ ± ♅
03 21 ☽ ⊻ ♀
05 34 ☽ # ♄
06 22 ☽ # ♆
07 12 ☿ △ ♅
11 01 ♂ ⊼ ♇
11 59 ☽ ⊼ ♆
14 46 ☿ Q ♇
16 12 ☽ □ ♇
16 16 ☽ ⊥ ♀
16 17 ☽ △ ♂
18 48 ☽ ⚼ ♄
20 31 ☽ ∥ ♃
20 59 ☽ ⚼ ♅
22 30 ☽ ∥ ♀
23 59 ☽ ± ♆

12 Friday
02 48 ☽ Q ♃
11 48 ☽ ∠ ♀
12 31 ☽ ☌ ☉
18 07 ☽ ⚼ ♆
23 00 ☽ ⚼ ♂
23 16 ☽ ∥ ☉

13 Saturday
00 59 ☽ △ ♄
03 13 ☽ △ ♅
04 33 ☽ Q ♇
09 48 ☽ ∥ ☿
11 37 ☽ ☌ ☿
15 37 ☽ □ ♃
17 33 ☽ ± ♂
20 39 ☽ ⚹ ♀

14 Sunday
00 31 ☽ △ ♆
04 59 ☽ ⚹ ♇
05 57 ☽ ⊼ ♂
14 54 ☿ □ ♃
23 07 ☽ Q ♀

15 Monday
05 05 ☿ ± ♂
06 55 ☽ ⊻ ☉
08 33 ☉ ⚼ ♆
11 36 ☽ ∠ ♇
13 56 ☽ □ ♄
14 20 ☽ # ♂
14 58 ☽ ∥ ♇
16 12 ☽ □ ♅
17 28 ☽ # ♇
18 06 ☽ ∥ ♂
20 13 ☽ ⊥ ☉
22 38 ♀ ☍ ♆

16 Tuesday
05 22 ☽ △ ♃
05 56 ☽ ⊥ ♇
11 49 ☽ ⊻ ☿
13 31 ☽ □ ♆
14 46 ☽ □ ♀
16 07 ☽ ∠ ☉
18 06 ☽ ⊻ ♇
19 46 ☽ ☍ ♂
23 05 ☿ △ ♆

17 Wednesday
02 04 ☽ ⊥ ☿
02 22 ☽ Q ♄
04 38 ☽ Q ♅
09 09 ☽ # ☿
11 57 ☽ ⚼ ♃
23 13 ☽ ∠ ☿

18 Thursday
00 53 ☽ ⚹ ☉
01 09 ☽ # ☉
01 38 ☽ Q ♆
02 14 ☿ ⚹ ♀
02 22 ☽ ⚹ ♄
04 37 ☽ ⚹ ♅
06 05 ☽ ± ♃
06 35 ☿ ⚹ ♇
08 45 ♀ ⊥ ♃
10 27 ♀ △ ♇
18 02 ☽ ⊼ ♃
19 08 ☉ △ ♄
19 14 ☉ ⚼ ♂
21 09 ☿ ⊼ ♂

19 Friday
01 16 ☽ ⚹ ♆
02 34 ☽ Q ☉
05 50 ☽ ☌ ♇
07 23 ☽ △ ♀
07 48 ☽ ∠ ♄
07 56 ☽ ⊼ ♂
09 34 ☽ ⚹ ☿
09 36 ☽ # ♀
10 00 ☽ ∠ ♅
11 28 ☽ # ♃
15 24 ♀ □ ♂
19 36 ☽ ± ♂
23 03 ☉ △ ♅

20 Saturday
00 16 ☽ ∥ ♆
01 04 ☽ ∥ ♄
01 05 ☽ ⊥ ♄
03 16 ☽ ⊥ ♅
06 06 ☽ ∠ ♆
09 01 ☽ ∥ ♅
12 04 ☽ Q ☿
12 26 ☽ ⊻ ♄
12 49 ☽ ⚼ ♂
14 21 ☽ ⚼ ♀
14 35 ☽ ⊻ ♅
15 51 ☽ □ ☉
21 11 ☉ Q ♇
22 57 ☽ ⊥ ♆

21 Sunday
03 46 ☽ ☍ ♃
08 22 ☽ ± ♀
10 00 ☽ ⊻ ♆
14 26 ☽ ⊻ ♇
16 41 ☽ △ ♂
20 08 ☽ ⊼ ♀

22 Monday
01 16 ☽ ⊥ ♇
01 56 ☽ □ ☿
17 15 ☽ ∠ ♇
18 49 ☽ ☌ ♄
20 49 ☽ ☌ ♅

23 Tuesday
02 06 ☽ △ ☉
09 31 ☽ ⊼ ♃
14 52 ☽ ☌ ♆
19 06 ☽ ⚹ ♇
19 48 ☽ ± ♃
21 19 ☽ □ ♂

24 Wednesday
04 06 ☽ ☍ ♀
05 27 ☽ ⚼ ☉
10 58 ☽ ⚼ ♃
12 17 ☽ △ ☿
15 10 ☽ Q ♇
21 24 ☽ ± ☉
21 28 ☽ ⊻ ♄
23 21 ☽ ⊻ ♅

25 Thursday
03 11 ☽ Q ♂
04 47 ☽ ∥ ♅
07 15 ☽ ⊥ ♄
07 52 ☽ ⊼ ☉
09 06 ☽ ⊥ ♅
11 29 ☽ ∥ ♄
11 39 ☽ △ ♃
12 11 ☽ ∥ ♆
15 41 ☽ ⚼ ☿
16 20 ☽ ⊻ ♆
20 27 ☽ □ ♇
21 45 ☽ ∠ ♄
22 24 ☽ # ♃
22 34 ☽ ⚹ ♂
23 35 ☽ ∠ ♅

26 Friday
01 01 ☽ # ♀
01 57 ☽ ⊥ ♆
07 46 ☽ ± ☿
08 21 ☽ ⊼ ♀
14 40 ♂ St R
16 15 ☽ ∠ ♆
18 22 ☽ ⊼ ☿
18 32 ☽ ± ♀
21 36 ☽ ⚹ ♄
22 26 ☽ ∠ ♂
23 24 ☽ ⚹ ♅

27 Saturday
09 45 ☽ ⚼ ♀
10 56 ☽ ☍ ☉
11 41 ☽ □ ♃
12 38 ☽ ⊥ ♂
13 59 ☽ # ☉
14 23 ☿ ∠ ♇
15 56 ☽ ⚹ ♆
16 32 ☽ Q ♄
18 20 ☽ Q ♅
20 05 ☽ △ ♇
22 05 ☽ ⊻ ♂
23 47 ☉ □ ♃

28 Sunday
02 23 ☿ □ ♄
10 52 ☽ Q ♆
11 05 ☽ △ ♀
13 11 ☉ ± ♂
16 17 ☽ Q ♃
17 47 ♀ ∠ ♃
19 28 ☽ # ☿
19 51 ☽ ⚼ ♇
20 32 ☿ □ ♅
21 00 ☽ □ ♄
21 53 ☽ ∥ ♂
22 41 ☽ # ♇
22 49 ☽ □ ♅
23 04 ☽ ☍ ☿
23 24 ☽ ∥ ♇

29 Monday
00 13 ☽ # ♂
01 59 ☽ ∥ ☿
10 14 ☽ ± ♇
11 31 ☽ ⚹ ♃
13 47 ☽ ⊼ ☉
15 33 ☽ □ ♆
19 21 ☿ # ♂
19 54 ☽ ⊼ ♇
21 47 ☽ ☌ ♂

30 Tuesday
00 08 ☽ ± ☉
04 46 ☿ ∥ ♇
05 24 ☽ ∥ ☉
10 06 ♄ St D
10 47 ☿ # ♇
11 59 ☽ ∠ ♃
14 52 ☽ □ ♀
15 53 ☽ ⚼ ☉
16 52 ☉ △ ♆
21 21 ☿ ∥ ♂
21 36 ☽ △ ♄
23 28 ☽ △ ♅

31 Wednesday
02 58 ☽ ⊥ ♃
05 16 ☽ ⊼ ☿
13 05 ☽ ⊻ ♃
16 30 ☽ ± ☿
17 01 ☽ △ ♆
18 49 ☽ △ ☉
21 44 ☽ ☍ ♇
22 06 ☽ ∥ ♀
22 51 ☽ ⚼ ♄
23 22 ☽ Q ♀
23 25 ☽ ⊻ ♂

All ephemeris data is given at 12.00 UT and the Moon's longitude is additionally given for 24.00 UT

SEPTEMBER 1988

LONGITUDES

Date	Sidereal time h m s	Sun ☉ ° ' "	Moon ☽ ° ' "	Moon ☽ 24.00 ° ' "	Mercury ☿ ° '	Venus ♀ ° '	Mars ♂ ° '	Jupiter ♃ ° '	Saturn ♄ ° '	Uranus ♅ ° '	Neptune ♆ ° '	Pluto ♇ ° '
01	10 43 28	09 ♍ 15 24	18 ♉ 33 05	25 ♉ 23 37	02 ♎ 21	23 ♋ 46	11 ♈ 13	05 ♊ 16	25 ♐ 56	27 ♐ 03	07 ♑ 30	10 ♏ 17
02	10 47 24	10 13 29	02 ♊ 07 46	08 ♊ 45 45	03 45	24 48	11 R 08	05 20	25 56	27 R 03	07 R 29	10 19
03	10 51 21	11 11 36	15 ♊ 17 53	21 ♊ 44 29	05 07	25 50	11 02	05 24	25 57	27 03	07 29	10 20
04	10 55 18	12 09 45	28 ♊ 06 00	04 ♋ 22 51	06 28	26 52	10 54	05 28	25 57	27 03	07 29	10 21
05	10 59 14	13 07 56	10 ♋ 35 30	16 ♋ 44 23	07 48	27 55	10 47	05 32	25 58	27 D 03	07 28	10 23
06	11 03 11	14 06 08	22 ♋ 49 58	28 ♋ 52 41	09 05	28 ♋ 58	10 38	05 36	25 58	27 03	07 28	10 25
07	11 07 07	15 04 23	04 ♌ 52 58	10 ♌ 51 14	10 22	00 ♌ 01	10 28	05 39	25 59	27 03	07 27	10 26
08	11 11 04	16 02 40	16 ♌ 47 51	22 ♌ 43 10	11 36	01 04	10 18	05 42	26 00	27 03	07 27	10 28
09	11 15 00	17 00 59	28 ♌ 37 34	04 ♍ 31 20	12 49	02 08	10 07	05 45	26 01	27 03	07 27	10 29
10	11 18 57	17 59 19	10 ♍ 24 47	16 ♍ 18 14	14 00	03 12	09 56	05 48	26 02	27 03	07 27	10 31
11	11 22 53	18 57 41	22 ♍ 11 56	28 ♍ 06 11	15 09	04 17	09 43	05 51	26 03	27 04	07 26	10 33
12	11 26 50	19 56 06	04 ♎ 01 15	09 ♎ 57 26	16 16	05 21	09 30	05 53	26 04	27 04	07 26	10 34
13	11 30 47	20 54 31	15 ♎ 54 59	21 ♎ 54 17	17 21	06 26	09 17	05 56	26 05	27 04	07 26	10 36
14	11 34 43	21 52 59	27 ♎ 55 29	03 ♏ 59 03	18 23	07 31	09 03	05 58	26 07	27 05	07 25	10 38
15	11 38 40	22 51 29	10 ♏ 05 17	16 ♏ 14 32	19 24	08 36	08 48	06 00	26 08	27 05	07 25	10 40
16	11 42 36	23 50 00	22 ♏ 26 49	28 ♏ 43 37	20 21	09 42	08 33	06 01	26 10	27 06	07 25	10 41
17	11 46 33	24 48 33	05 ♐ 04 12	11 ♐ 29 20	21 16	10 48	08 17	06 03	26 12	27 06	07 25	10 43
18	11 50 29	25 47 07	17 ♐ 59 23	24 ♐ 34 42	22 08	11 54	08 01	06 04	26 14	27 07	07 25	10 45
19	11 54 26	26 45 43	01 ♑ 15 35	08 ♑ 02 19	22 57	13 00	07 44	06 05	26 15	27 08	07 D 25	10 47
20	11 58 22	27 44 21	14 ♑ 55 03	21 ♑ 53 52	23 42	14 06	07 28	06 06	26 17	27 08	07 25	10 49
21	12 02 19	28 43 01	28 ♑ 58 46	06 ♒ 09 35	24 24	15 13	07 11	06 07	26 20	27 09	07 25	10 51
22	12 06 16	29 ♍ 41 42	13 ♒ 26 00	20 ♒ 47 34	25 01	16 20	06 53	06 07	26 22	27 10	07 25	10 53
23	12 10 12	00 ♎ 40 25	28 ♒ 13 40	05 ♓ 43 30	25 35	17 27	06 36	06 08	26 24	27 11	07 26	10 55
24	12 14 09	01 39 09	13 ♓ 16 09	20 ♓ 50 35	26 03	18 34	06 18	06 08	26 26	27 12	07 26	10 57
25	12 18 05	02 37 56	28 ♓ 25 38	06 ♈ 00 07	26 27	19 41	06 01	06 R 08	26 29	27 13	07 26	10 59
26	12 22 02	03 36 44	13 ♈ 32 51	21 ♈ 02 41	26 45	20 49	05 43	06 07	26 31	27 14	07 26	11 01
27	12 25 58	04 35 35	28 ♈ 28 33	05 ♉ 49 31	26 57	21 57	05 25	06 07	26 34	27 15	07 26	11 03
28	12 29 55	05 34 27	13 ♉ 04 48	20 ♉ 13 47	27 03	23 05	05 07	06 06	26 37	27 16	07 27	11 05
29	12 33 51	06 33 22	27 ♉ 16 03	04 ♊ 11 21	27 R 02	24 13	04 50	06 05	26 40	27 18	07 27	11 07
30	12 37 48	07 ♎ 32 20	10 ♊ 59 36	17 ♊ 40 54	26 ♎ 55	25 ♌ 21	04 ♈ 32	06 ♊ 04	26 ♐ 42	27 ♐ 19	07 ♑ 27	11 ♏ 09

DECLINATIONS

Date	Moon True ☊ ° '	Moon Mean ☊ ° '	Moon ☽ Latitude ° '	Sun ☉ ° '	Moon ☽ ° '	Mercury ☿ ° '	Venus ♀ ° '	Mars ♂ ° '	Jupiter ♃ ° '	Saturn ♄ ° '	Uranus ♅ ° '	Neptune ♆ ° '	Pluto ♇ ° '
01	14 ♓ 06	14 ♓ 13	04 N 47	08 N 06	21 N 56	01 S 31	19 N 06	00 S 27	20 N 13	22 S 25	23 S 38	22 S 16	00 N 03
02	14 D 06	14 10	05 11	07 44	25 40	02 12	19 01	00 29	20 13	22 25	23 38	22 16	00 02
03	14 R 06	14 07	05 18	07 22	27 54	02 53	18 54	00 31	20 14	22 25	23 38	22 16	00 02
04	14 06	14 04	05 08	07 00	28 34	03 33	18 48	00 34	20 15	22 25	23 38	22 16	00 N 01
05	14 06	14 01	04 45	06 38	27 45	04 12	18 40	00 37	20 15	22 25	23 38	22 16	00 00
06	14 D 06	13 57	04 08	06 15	25 35	04 51	18 33	00 40	20 16	22 26	23 38	22 16	00 S 01
07	14 06	13 54	03 21	05 53	22 18	05 28	18 24	00 43	20 16	22 26	23 38	22 16	00 02
08	14 06	13 51	02 26	05 30	18 07	06 06	18 16	00 46	20 17	22 26	23 38	22 16	00 03
09	14 06	13 48	01 25	05 08	13 17	06 42	18 07	00 50	20 17	22 26	23 38	22 16	00 04
10	14 07	13 45	00 N 21	04 45	07 59	07 17	17 57	00 54	20 18	22 26	23 38	22 16	00 05
11	14 R 07	13 41	00 S 45	04 22	02 N 25	07 52	17 47	00 57	20 18	22 27	23 38	22 17	00 06
12	14 06	13 38	01 48	03 59	03 S 15	08 25	17 36	01 01	20 18	22 27	23 38	22 17	00 07
13	14 06	13 35	02 48	03 36	08 50	08 58	17 25	01 05	20 19	22 27	23 38	22 17	00 08
14	14 05	13 32	03 40	03 13	14 10	09 29	17 13	01 09	20 19	22 27	23 38	22 17	00 08
15	14 03	13 29	04 23	02 50	19 00	10 00	17 01	01 14	20 19	22 27	23 38	22 17	00 09
16	14 02	13 26	04 55	02 27	23 08	10 29	16 49	01 18	20 19	22 28	23 38	22 17	00 10
17	14 01	13 22	05 13	02 04	26 17	10 56	16 36	01 22	20 20	22 28	23 38	22 17	00 11
18	14 01	13 19	05 16	01 41	28 09	11 22	16 22	01 27	20 20	22 28	23 38	22 17	00 12
19	14 D 01	13 16	05 03	01 17	28 29	11 47	16 08	01 31	20 20	22 28	23 38	22 17	00 13
20	14 01	13 13	04 32	00 54	27 07	12 10	15 54	01 35	20 20	22 29	23 38	22 17	00 14
21	14 02	13 10	03 45	00 31	24 02	12 31	15 39	01 40	20 20	22 29	23 38	22 17	00 15
22	14 03	13 06	02 42	00 N 07	19 22	12 50	15 23	01 44	20 20	22 29	23 38	22 17	00 16
23	14 04	13 03	01 27	00 S 16	13 27	13 07	15 07	01 48	20 20	22 29	23 38	22 17	00 17
24	14 05	13 00	00 S 04	00 39	06 S 39	13 22	14 51	01 52	20 20	22 29	23 38	22 17	00 18
25	14 R 04	12 57	01 N 19	01 03	00 N 35	13 34	14 34	01 56	20 20	22 30	23 38	22 17	00 19
26	14 03	12 54	02 36	01 26	07 44	13 43	14 17	02 00	20 19	22 30	23 38	22 17	00 20
27	14 00	12 51	03 41	01 50	14 14	13 50	13 59	02 04	20 19	22 30	23 38	22 17	00 20
28	13 58	12 47	04 31	02 13	20 04	13 53	13 41	02 08	20 19	22 31	23 38	22 17	00 21
29	13 55	12 44	05 01	02 36	24 27	13 53	13 23	02 11	20 19	22 31	23 38	22 17	00 22
30	13 ♓ 52	12 ♓ 41	05 N 15	03 S 00	27 N 17	13 S 49	13 N 04	02 S 15	20 N 19	22 S 31	23 S 38	22 S 17	00 S 23

ZODIAC SIGN ENTRIES

Date	h	m	Planets
02	08	11	☽ ♊
04	15	37	☽ ♋
07	02	14	☽ ♌
07	11	37	♀ ♌
09	14	48	☽ ♍
12	03	51	☽ ♎
14	16	07	☽ ♏
17	02	25	☽ ♐
19	09	45	☽ ♑
21	13	43	☽ ♒
22	19	29	☉ ♎
23	14	51	☽ ♓
25	14	29	☽ ♈
27	14	29	☽ ♉
29	16	43	☽ ♊

LATITUDES

Date	Mercury ☿ ° '	Venus ♀ ° '	Mars ♂ ° '	Jupiter ♃ ° '	Saturn ♄ ° '	Uranus ♅ ° '	Neptune ♆ ° '	Pluto ♇ ° '
01	00 S 38	02 S 17	05 S 18	00 S 59	00 N 58	00 S 13	00 N 58	15 N 43
04	01 04	02 02	05 18	00 59	00 58	00 13	00 58	15 42
07	01 29	01 47	05 17	01 00	00 57	00 13	00 58	15 40
10	01 55	01 32	05 15	01 00	00 57	00 13	00 58	15 39
13	02 20	01 18	05 11	01 00	00 56	00 13	00 58	15 38
16	02 43	01 03	05 06	01 00	00 56	00 13	00 57	15 36
19	03 05	00 49	04 59	01 01	00 55	00 13	00 57	15 35
22	03 23	00 35	04 52	01 01	00 55	00 13	00 57	15 34
25	03 36	00 22	04 43	01 01	00 54	00 13	00 57	15 33
28	03 43	00 S 09	04 32	01 02	00 54	00 13	00 57	15 32
31	03 S 39	00 N 04	04 S 21	01 S 02	00 N 53	00 S 13	00 N 57	15 N 31

DATA

Julian Date	2447406
Delta T	+56 seconds
Ayanamsa	23° 42' 01"
Synetic vernal point	05° ♓ 24' 59"
True obliquity of ecliptic	23° 26' 36"

MOON'S PHASES, APSIDES AND POSITIONS ☽

Date	h	m	Phase	Longitude ° '	Eclipse Indicator
03	03	50	☾	10 ♊ 52	
11	04	49	●	18 ♍ 40	Annular
19	03	18	☽	26 ♐ 24	
25	19	07	○	02 ♈ 55	

Day	h	m	
10	15	23	Apogee
25	03	42	Perigee

Day	h	m		
04	10	33	Max dec	28° N 34'
11	22	13	0S	
19	04	55	Max dec	28° S 33'
25	10	06	0N	

LONGITUDES

Date	Chiron ⚷ ° '	Ceres ⚳ ° '	Pallas ⚴ ° '	Juno ⚵ ° '	Vesta ⚶ ° '	Black Moon Lilith ⚸ ° '
01	05 ♋ 51	27 ♓ 52	03 ♒ 49	29 ♋ 54	23 ♍ 53	12 ♍ 16
11	06 ♋ 22	25 ♓ 46	02 ♒ 30	04 ♌ 59	28 ♍ 51	13 ♍ 22
21	06 ♋ 45	23 ♓ 31	01 ♒ 46	09 ♌ 53	03 ♎ 54	14 ♍ 29
31	07 ♋ 01	21 ♓ 23	01 ♒ 38	14 ♌ 35	08 ♎ 59	15 ♍ 35

All ephemeris data is given at 12.00 UT and the Moon's longitude is additionally given for 24.00 UT

ASPECTARIAN

h m	Aspects
01 Thursday	
00 46	☽ ⚼ ♅
03 06	☽ ‖ ♃
09 41	☽ ⚼ ☿
09 42	☽ ⊥ ♂
13 47	☽ # ♆
14 24	☽ ± ♄
14 37	☽ # ♄
16 22	☽ ± ♅
18 54	☽ ⚼ ♆
21 45	☽ # ♅
21 53	☽ ✶ ♀
02 Friday	
00 57	☽ ⊼ ♄
01 23	☽ ∠ ♂
02 56	☽ ⊼ ♅
10 51	☽ ± ♆
14 09	☉ ✶ ♇
15 15	☽ △ ☿
17 48	☽ ☌ ♃
21 41	☽ ⊼ ♆
21 53	☿ ⊥ ♇
03 Saturday	
02 51	☽ ⊼ ♇
03 05	☽ ∠ ♀
03 50	☽ □ ☉
04 12	☽ ✶ ♂
08 16	☉ ⊼ ♂
13 55	☽ ± ♇
14 34	♀ ⊼ ♄
17 11	☿ △ ♃
21 10	☽ ⊥ ♀
04 Sunday	
02 17	☽ Q ♂
06 48	☽ ⚼ ♇
07 55	☽ ☍ ♄
09 28	☽ ⊻ ♀
10 00	☽ ☍ ♅
16 00	♀ ⊼ ♅
16 15	☽ Q ☉
05 Monday	
02 10	☽ ⊻ ♃
05 56	☽ □ ☿
05 57	☽ ☍ ♆
06 04	☿ □ ♆
09 42	♅ St D
11 36	☽ △ ♇
12 21	☽ □ ♂
13 50	☽ ⊥ ♃
17 22	☽ ✶ ☉
06 Tuesday	
07 33	☽ ∠ ♃
18 13	☽ ⊼ ♄
20 21	☽ ⊼ ♅
21 27	☽ Q ☿
07 Wednesday	
01 20	☽ ☌ ♀
01 32	☽ ∠ ☉
03 10	☽ # ♅
06 11	☽ ± ♄
08 19	☽ ± ♅
11 10	☽ # ♄
12 11	☽ # ♆
13 27	☿ ⊻ ♇
13 33	☽ ✶ ♃
13 55	☿ ☍ ♂
16 52	♂ ⊼ ♇
17 09	☽ ⊼ ♆
21 10	☽ ⊥ ☉
21 49	☉ # ☿
23 05	☽ △ ♂
23 11	☽ □ ♇
08 Thursday	
00 12	☽ ‖ ♃
00 16	☽ ⚼ ♄
00 18	☽ ✶ ☿
02 24	☽ ⚼ ♅
03 44	☉ ∠ ♀
05 14	☽ ± ♆
10 20	☽ ⊻ ☉
11 15	☽ ‖ ♀
13 50	☽ Q ♃
23 27	☽ ⚼ ♆
09 Friday	
04 59	☽ ⚼ ♂
06 41	☽ △ ♄
08 48	☽ △ ♅
09 07	♀ ± ♄
10 10	☽ ∠ ☿
11 43	☽ Q ♇
19 51	☽ ⊻ ♀
23 00	☽ ± ♂
10 Saturday	
02 34	☽ □ ♃
05 57	☽ △ ♆
06 32	☽ ⊥ ☿
08 37	♀ ± ♅
09 18	☽ ⊥ ♀
11 02	☽ ⊼ ♂
12 13	☽ ✶ ♇
12 39	☿ Q ♄
14 44	☽ # ☿
20 06	☽ ⊻ ☿
11 Sunday	
03 02	☽ ‖ ☉
04 49	☽ ☌ ☉
05 27	☽ ∠ ♀
10 06	☿ Q ♅
18 06	☽ # ♂
18 49	☽ ∠ ♇
19 50	☽ □ ♄
21 47	☽ # ♇
21 53	☽ □ ♅
22 38	☽ ‖ ♇
12 Monday	
02 25	☽ ‖ ♂
13 07	☽ ⊥ ♇
14 54	☽ # ☉
14 58	☽ ✶ ♀
15 47	☽ △ ♃
18 54	☽ □ ♆
22 53	☽ ☍ ♂
13 Tuesday	
00 21	♀ ✶ ♃
01 16	☽ ⊻ ♇
08 19	☽ Q ♄
10 18	☽ Q ♅
12 36	☽ ‖ ☿
15 09	☽ ☌ ☿
17 33	☽ Q ♀
22 04	☽ ⚼ ♃
22 54	☽ ⊻ ☉
14 Wednesday	
07 02	☽ Q ♆
08 24	☽ ✶ ♄
09 57	♀ ⊼ ♆
10 19	☽ ✶ ♅
11 55	☽ ⊥ ☉
16 03	☽ ± ♃
15 Thursday	
02 14	☽ # ♀
03 56	☽ ⊼ ♃
06 46	☽ ✶ ♆
07 15	☽ ∠ ☉
08 49	☽ □ ♀
09 31	☽ ⊼ ♂
13 07	☽ ☌ ♇
14 04	☽ ∠ ♄
15 25	♀ △ ♂
15 55	☽ ∠ ♅
19 09	☽ # ♃
21 00	☽ ± ♂
16 Friday	
06 36	☽ ‖ ♆
07 35	☽ ⊥ ♄
07 38	☽ ⊻ ☿
07 43	☽ ‖ ♄
09 23	☽ ⊥ ♅
11 56	☽ ∠ ♆
14 03	☽ ⚼ ♂
14 52	☽ ✶ ☉
15 16	☽ ‖ ♅
19 08	☽ ⊻ ♄
20 05	☽ ⊥ ☿
20 54	☽ ⊻ ♅
17 Saturday	
05 07	☽ ⊥ ♆
05 50	☿ ⚼ ♃
10 21	♀ □ ♇
13 50	☽ ☍ ♃
14 25	☽ ∠ ☿
15 32	☽ Q ☉
16 25	☽ ⊻ ♆
17 54	☽ △ ♂
21 01	♀ ⚼ ♄
22 36	☽ ⊻ ♇
23 42	☽ △ ♀
18 Sunday	
09 43	☽ ⊥ ♇
11 08	☉ ∠ ♇
16 56	♀ ⚼ ♅
18 16	♆ St D
20 05	☽ ✶ ☿
23 10	☉ □ ♄
19 Monday	
00 10	☉ # ♂
02 09	☽ ∠ ♇
03 01	☽ ☌ ♄
03 18	☽ □ ☉
04 36	☽ ☌ ♅
05 38	☽ ⚼ ♀
18 57	☽ Q ☿
20 34	☽ ⊼ ♃
21 07	☉ □ ♅
21 12	♀ ± ♆
22 55	☽ ☌ ♆
23 04	☽ ± ♀
23 15	☽ □ ♂
20 Tuesday	
04 51	☽ ✶ ♇
07 06	☽ ± ♃
10 28	☽ ⊼ ♀
15 21	♂ □ ♆
22 39	☽ ⚼ ♃
21 Wednesday	
01 36	☽ Q ♇
03 52	☽ □ ☿
05 42	☽ Q ♂
07 31	☽ ⊻ ♄
08 55	☽ ⊻ ♅
11 32	☽ △ ☉
14 25	☽ ‖ ♅
17 38	☽ ⊥ ♄
19 00	☽ ⊥ ♅
20 51	☽ ‖ ♄
21 54	☽ ‖ ♆
23 56	☽ △ ♃
22 Thursday	
01 25	☽ ✶ ♂
02 06	☽ ⊻ ♆
03 32	☉ # ♇
07 37	☽ # ♃
07 48	☽ □ ♇
08 35	☽ ∠ ♄
09 55	☽ ∠ ♅
11 59	☽ ⊥ ♆
14 13	☽ ⚼ ☉
17 07	☽ ☍ ♀
23 Friday	
01 31	☽ ∠ ♂
02 39	☽ ∠ ♆
05 05	☿ Q ♆
05 21	☽ # ♀
05 52	☽ ± ☉
07 35	☽ △ ☿
09 03	☽ ✶ ♄
10 19	☽ ✶ ♅
12 43	☉ ‖ ♇
13 11	☽ ‖ ☿
15 44	☽ ⊥ ♂
16 12	☽ ⊼ ☉
24 Saturday	
00 39	☽ □ ♃
01 08	☽ ⊻ ♂
02 41	♀ Q ♃
02 43	☽ ✶ ♆
04 18	☽ Q ♄
05 32	☽ Q ♅
08 18	☽ △ ♇
08 23	☽ ⚼ ☿
14 00	♃ St R
21 04	☽ ⊼ ♀
21 46	☽ Q ♆
23 03	☽ ± ☿
25 Sunday	
02 11	♂ ✶ ♃
03 47	☽ ‖ ♂
05 12	☽ Q ♃
06 55	☽ ‖ ☉
07 19	☽ ± ♀
08 07	☽ ⚼ ♇
08 47	☽ ⊼ ☿
08 55	☽ □ ♄
09 05	☽ ‖ ♇
10 05	☽ □ ♅
11 07	☽ # ♇
13 39	☽ # ☉
14 46	☿ ✶ ♄
16 33	☽ # ♂
19 07	☽ ☍ ☉
22 24	☽ ± ♇
22 43	☽ ⚼ ♀
23 47	☽ ☌ ♂
26 Monday	
00 12	☽ ✶ ♃
02 16	☽ □ ♆
07 57	☽ ⊼ ♇
10 17	♀ ⚼ ♂
27 Tuesday	
00 07	☽ ∠ ♃
00 35	☽ △ ♀
08 54	☽ △ ♄
09 30	☽ ☍ ☿
09 52	☽ # ☿
10 01	☽ △ ♅
10 35	☽ ‖ ♀
14 40	☽ ⊥ ♃
22 32	♀ ⚼ ♆
22 34	☿ # ♀
22 42	☽ ⊼ ☉
23 06	☽ ⊻ ♂
23 27	☉ ⊥ ♇
28 Wednesday	
00 28	☽ ⊻ ♃
02 40	☽ △ ♆
03 31	☉ ☍ ♂
05 52	☉ ‖ ♂
08 40	☽ ☍ ♇
08 49	☽ ⊥ ♂
09 19	☽ ± ☉
09 33	☽ ⚼ ♄
10 39	☽ ⚼ ♅
13 12	☽ ‖ ♃
15 13	♂ ± ♇
21 36	☿ St R
23 11	☽ # ♆
23 35	☽ ∠ ♂
29 Thursday	
00 24	☽ # ♄
00 41	☽ ± ♄
00 47	☉ △ ♃
01 31	☽ ⚼ ☉
01 47	☽ ± ♅
03 46	☽ ⚼ ♆
06 19	☽ □ ♀
06 51	☽ # ♅
10 57	☽ ⊼ ♄
11 36	☽ ⊼ ☿
12 02	☽ ⊼ ♅
19 14	☽ ± ♆
21 55	☽ ± ☿
30 Friday	
00 51	☽ ✶ ♂
03 19	☽ ☌ ♃
05 24	☽ △ ☉
05 44	☽ ⊼ ♆
10 00	☉ □ ♆
12 17	☽ ⊼ ♇
13 36	☽ ⚼ ☿
16 36	☽ Q ♀
21 43	☽ Q ♂
23 04	☽ ± ♇

OCTOBER 1988

LONGITUDES

Date	Sidereal time h m s	Sun ☉ ° ' "	Moon ☽ ° ' "	Moon ☽ 24.00 ° ' "	Mercury ☿ ° '	Venus ♀ ° '	Mars ♂ ° '	Jupiter ♃ ° '	Saturn ♄ ° '	Uranus ♅ ° '	Neptune ♆ ° '	Pluto ♇ ° '
01	12 41 45	08 ♎ 31 19	24 ♊ 15 27	00 ♋ 43 37	26 ♎ 39	26 ♌ 30	04 ♈ 15	06 ♊ 03	26 ♐ 45	27 ♐ 20	07 ♑ 28	11 ♏ 11
02	12 45 41	09 30 21	07 ♋ 05 48	13 ♋ 22 30	26 R 17	27 39	03 R 58	06 R 02	26 49	27 21	07 28	11 13
03	12 49 38	10 29 25	19 ♋ 34 16	25 ♋ 41 40	25 46	28 47	03 41	06 00	26 52	27 23	07 29	11 15
04	12 53 34	11 28 32	01 ♌ 45 19	07 ♌ 45 49	25 08	29 ♌ 56	03 25	05 58	26 55	27 24	07 29	11 17
05	12 57 31	12 27 40	13 ♌ 43 44	19 ♌ 39 40	24 22	01 ♍ 06	03 09	05 56	26 58	27 26	07 30	11 20
06	13 01 27	13 26 51	25 ♌ 34 10	01 ♍ 27 44	23 29	02 15	02 53	05 54	27 02	27 27	07 30	11 22
07	13 05 24	14 26 05	07 ♍ 20 53	13 ♍ 14 04	22 30	03 24	02 38	05 51	27 05	27 29	07 31	11 24
08	13 09 20	15 25 20	19 ♍ 07 41	25 ♍ 02 06	21 25	04 34	02 23	05 48	27 09	27 30	07 32	11 26
09	13 13 17	16 24 37	00 ♎ 57 41	06 ♎ 54 43	20 16	05 44	02 09	05 45	27 12	27 32	07 32	11 29
10	13 17 14	17 23 57	12 ♎ 53 28	18 ♎ 54 09	19 05	06 54	01 55	05 42	27 16	27 34	07 33	11 31
11	13 21 10	18 23 19	24 ♎ 56 59	01 ♏ 02 08	17 54	08 04	01 42	05 39	27 20	27 36	07 34	11 33
12	13 25 07	19 22 42	07 ♏ 09 46	13 ♏ 20 01	16 43	09 14	01 30	05 36	27 24	27 37	07 34	11 35
13	13 29 03	20 22 08	19 ♏ 33 01	25 ♏ 48 54	15 37	10 25	01 18	05 32	27 28	27 39	07 35	11 38
14	13 33 00	21 21 36	02 ♐ 07 50	08 ♐ 29 55	14 35	11 35	01 07	05 28	27 32	27 41	07 36	11 40
15	13 36 56	22 21 05	14 ♐ 55 20	21 ♐ 24 15	13 41	12 46	00 57	05 24	27 36	27 43	07 37	11 42
16	13 40 53	23 20 37	27 ♐ 56 50	04 ♑ 33 16	12 56	13 56	00 47	05 20	27 41	27 45	07 38	11 45
17	13 44 49	24 20 10	11 ♑ 13 44	17 ♑ 58 25	12 20	15 07	00 38	05 15	27 45	27 47	07 39	11 47
18	13 48 46	25 19 45	24 ♑ 47 28	01 ♒ 41 02	11 55	16 18	00 30	05 11	27 49	27 49	07 40	11 49
19	13 52 43	26 19 22	08 ♒ 39 12	15 ♒ 41 57	11 41	17 29	00 23	05 06	27 54	27 51	07 41	11 52
20	13 56 39	27 19 00	22 ♒ 49 15	00 ♓ 00 55	11 D 39	18 40	00 16	05 01	27 58	27 54	07 42	11 54
21	14 00 36	28 18 40	07 ♓ 16 40	14 ♓ 36 05	11 48	19 52	00 11	04 56	28 03	27 56	07 43	11 56
22	14 04 32	29 ♎ 18 21	21 ♓ 58 36	29 ♓ 23 32	12 07	21 03	00 06	04 51	28 08	27 58	07 44	11 59
23	14 08 29	00 ♏ 18 05	06 ♈ 50 02	14 ♈ 17 11	12 36	22 15	00 ♈ 01	04 45	28 12	28 00	07 45	12 01
24	14 12 25	01 17 50	21 ♈ 43 58	29 ♈ 09 20	13 14	23 26	29 ♓ 58	04 40	28 17	28 03	07 46	12 03
25	14 16 22	02 17 37	06 ♉ 32 12	13 ♉ 51 34	14 01	24 38	29 56	04 34	28 22	28 05	07 47	12 06
26	14 20 18	03 17 27	21 ♉ 06 31	28 ♉ 16 13	14 56	25 50	29 54	04 28	28 27	28 08	07 48	12 08
27	14 24 15	04 17 18	05 ♊ 20 01	12 ♊ 17 26	15 57	27 02	29 53	04 22	28 32	28 10	07 50	12 11
28	14 28 12	05 17 11	19 ♊ 08 10	25 ♊ 52 03	17 04	28 14	29 D 53	04 16	28 37	28 13	07 51	12 13
29	14 32 08	06 17 07	02 ♋ 29 09	08 ♋ 59 38	18 16	29 ♍ 26	29 53	04 10	28 42	28 15	07 52	12 16
30	14 36 05	07 17 05	15 ♋ 23 49	21 ♋ 42 07	19 33	00 ♎ 38	29 55	04 03	28 47	28 18	07 53	12 18
31	14 40 01	08 ♏ 17 05	27 ♋ 55 03	04 ♌ 03 13	20 ♎ 54	01 ♎ 51	29 ♓ 57	03 ♊ 57	28 ♐ 53	28 ♐ 20	07 ♑ 55	12 ♏ 20

Moon and DECLINATIONS

Date	Moon True ☊ ° '	Moon Mean ☊ ° '	Moon ☽ Latitude ° '	Sun ☉ ° '	Moon ☽ ° '	Mercury ☿ ° '	Venus ♀ ° '	Mars ♂ ° '	Jupiter ♃ ° '	Saturn ♄ ° '	Uranus ♅ ° '	Neptune ♆ ° '	Pluto ♇ ° '
01	13 ♓ 50	12 ♓ 38	05 N 10	03 S 23	28 N 29	13 S 41	12 N 45	02 S 18	20 N 18	22 S 31	23 S 38	22 S 17	00 S 24
02	13 R 49	12 35	04 50	03 46	28 04	13 29	12 25	02 21	20 18	22 31	23 38	22 17	00 25
03	13 D 50	12 32	04 16	04 09	26 14	13 12	12 05	02 24	20 17	22 32	23 38	22 17	00 26
04	13 51	12 28	03 31	04 32	23 12	12 52	11 44	02 27	20 17	22 32	23 38	22 17	00 27
05	13 53	12 25	02 38	04 55	19 14	12 27	11 22	02 29	20 17	22 32	23 38	22 17	00 28
06	13 54	12 22	01 39	05 19	14 33	11 57	11 02	02 32	20 16	22 32	23 38	22 17	00 29
07	13 55	12 19	00 N 36	05 41	09 22	11 23	10 41	02 34	20 16	22 32	23 38	22 17	00 30
08	13 R 55	12 16	00 S 28	06 04	03 N 52	10 45	10 19	02 36	20 15	22 33	23 38	22 17	00 30
09	13 54	12 12	01 32	06 27	01 S 47	10 04	09 56	02 37	20 15	22 33	23 38	22 17	00 31
10	13 51	12 09	02 32	06 50	07 25	09 21	09 34	02 38	20 14	22 33	23 38	22 17	00 32
11	13 46	12 06	03 25	07 13	12 51	08 36	09 11	02 39	20 13	22 33	23 38	22 17	00 33
12	13 40	12 03	04 10	07 35	17 50	07 50	08 48	02 40	20 13	22 34	23 38	22 17	00 34
13	13 34	12 00	04 44	07 58	22 10	07 06	08 24	02 40	20 12	22 34	23 38	22 17	00 35
14	13 27	11 57	05 04	08 20	25 33	06 23	08 00	02 41	20 11	22 34	23 38	22 17	00 36
15	13 21	11 53	05 10	08 42	27 44	05 43	07 36	02 41	20 11	22 34	23 38	22 17	00 37
16	13 17	11 50	05 00	09 04	28 26	05 07	07 12	02 40	20 10	22 34	23 38	22 17	00 37
17	13 14	11 47	04 34	09 26	27 31	04 36	06 47	02 39	20 09	22 35	23 38	22 17	00 38
18	13 13	11 44	03 53	09 48	24 59	04 10	06 22	02 38	20 08	22 35	23 38	22 17	00 39
19	13 D 14	11 41	02 56	10 10	20 56	03 50	05 57	02 37	20 07	22 35	23 38	22 17	00 40
20	13 15	11 38	01 49	10 31	15 37	03 35	05 32	02 35	20 06	22 35	23 38	22 17	00 41
21	13 16	11 34	00 S 32	10 53	09 20	03 27	05 06	02 33	20 05	22 36	23 38	22 17	00 42
22	13 R 16	11 31	00 N 47	11 14	02 S 28	03 24	04 41	02 31	20 04	22 36	23 39	22 17	00 42
23	13 14	11 28	02 04	11 35	04 N 37	03 26	04 15	02 28	20 03	22 36	23 39	22 17	00 43
24	13 09	11 25	03 12	11 56	11 27	03 34	03 49	02 25	20 02	22 36	23 39	22 17	00 44
25	13 03	11 22	04 07	12 16	17 35	03 46	03 23	02 22	20 01	22 36	23 39	22 17	00 45
26	12 55	11 18	04 45	12 37	22 37	04 02	02 56	02 19	20 00	22 37	23 39	22 17	00 46
27	12 47	11 15	05 04	12 57	26 11	04 22	02 29	02 15	19 59	22 37	23 39	22 17	00 47
28	12 39	11 12	05 05	13 17	28 04	04 45	02 02	02 11	19 58	22 37	23 39	22 17	00 47
29	12 32	11 09	04 49	13 37	28 14	05 11	01 35	02 07	19 57	22 37	23 39	22 17	00 48
30	12 27	11 06	04 18	13 57	26 49	05 40	01 08	02 02	19 56	22 37	23 39	22 17	00 49
31	12 ♓ 25	11 ♓ 03	03 N 36	14 S 16	24 N 06	06 S 11	00 N 41	01 S 57	19 N 54	22 S 37	23 S 39	22 S 17	00 S 50

ZODIAC SIGN ENTRIES

Date	h	m	Planets
01	22	39	☽ ♋
04	08	31	☽ ♌
04	13	15	♀ ♍
06	21	01	☽ ♍
09	10	03	☽ ♎
11	21	58	☽ ♏
14	07	58	☽ ♐
16	15	44	☽ ♑
18	21	05	☽ ♒
20	23	58	☽ ♓
23	00	59	☽ ♈
23	04	44	☉ ♏
23	22	01	♂ ♓
25	01	22	☽ ♉
27	02	55	☽ ♊
29	07	28	☽ ♋
29	23	20	♀ ♎
31	16	03	☽ ♌

LATITUDES

Date	Mercury ☿ ° '	Venus ♀ ° '	Mars ♂ ° '	Jupiter ♃ ° '	Saturn ♄ ° '	Uranus ♅ ° '	Neptune ♆ ° '	Pluto ♇ ° '
01	03 S 39	00 N 04	04 S 21	01 S 02	00 N 53	00 S 13	00 N 57	15 N 31
04	03 23	00 16	04 09	01 02	00 53	00 13	00 57	15 30
07	02 50	00 27	03 56	01 02	00 52	00 13	00 57	15 29
10	02 01	00 38	03 43	01 03	00 52	00 13	00 56	15 28
13	01 02	00 48	03 29	01 03	00 51	00 13	00 56	15 28
16	00 S 01	00 57	03 15	01 03	00 51	00 13	00 56	15 27
19	00 N 52	01 06	03 01	01 03	00 50	00 13	00 56	15 26
22	01 31	01 14	02 47	01 03	00 50	00 13	00 56	15 26
25	01 55	01 21	02 33	01 03	00 50	00 13	00 56	15 26
28	02 07	01 27	02 20	01 03	00 49	00 13	00 56	15 25
31	02 N 08	01 N 33	02 S 07	01 S 03	00 N 49	00 S 13	00 N 56	15 N 25

DATA

Julian Date	2447436
Delta T	+56 seconds
Ayanamsa	23° 42' 04"
Synetic vernal point	05° ♓ 24' 55"
True obliquity of ecliptic	23° 26' 36"

MOON'S PHASES, APSIDES AND POSITIONS ☽

Date	h	m	Phase	Longitude ° '	Eclipse Indicator
02	16	58	☾	09 ♋ 43	
10	21	49	●	17 ♎ 48	
18	13	01	☽	25 ♑ 22	
25	04	36	○	01 ♉ 59	

Day	h	m		
07	20	42	Apogee	
23	12	25	Perigee	
01	17	39	Max dec	28° N 31'
09	04	00	0S	
16	10	42	Max dec	28° S 26'
22	20	22	0N	
29	02	21	Max dec	28° N 22'

LONGITUDES

Date	Chiron ⚷ ° '	Ceres ⚳ ° '	Pallas ⚴ ° '	Juno ⚵ ° '	Vesta ⚶ ° '	Black Moon Lilith ⚸ ° '
01	07 ♋ 01	21 ♓ 23	01 ♒ 38	14 ♌ 35	08 ♎ 59	15 ♍ 35
11	07 ♋ 09	19 ♓ 35	02 ♒ 04	19 ♌ 03	14 ♎ 06	16 ♍ 42
21	07 ♋ 09	18 ♓ 16	02 ♒ 59	23 ♌ 14	19 ♎ 15	17 ♍ 48
31	07 ♋ 01	17 ♓ 33	04 ♒ 21	27 ♌ 07	24 ♎ 26	18 ♍ 54

All ephemeris data is given at 12.00 UT and the Moon's longitude is additionally given for 24.00 UT

ASPECTARIAN

h m	Aspects
01 Saturday	
05 05	☿ ⚹ ♄
14 38	☿ ⚹ ♀
15 34	☽ ⚼ ♇
16 19	☽ △ ☿
16 32	☽ ⚹ ♀
16 38	☽ ☍ ♄
17 42	☽ ☍ ♅
17 43	♀ △ ♄
02 Sunday	
05 54	♀ △ ♅
06 13	☽ □ ♂
09 58	☽ ⊻ ♃
12 43	☽ ☍ ♆
16 58	☽ □ ☉
17 28	♀ ± ♂
19 53	☽ △ ♇
21 23	☽ ⊥ ♃
23 40	☽ ∠ ♀
03 Monday	
14 47	☽ ∠ ♃
18 57	☽ ⊥ ♀
22 02	♀ Q ♇
23 23	☿ Q ♆
23 34	☽ □ ☿
04 Tuesday	
02 22	☽ ⊼ ♄
03 21	☽ ⊼ ♅
07 04	☽ Q ☉
07 20	☉ ⊻ ♇
08 01	☽ ⊻ ♀
09 04	☽ # ♅
14 19	☽ ± ♄
15 14	☽ △ ♂
15 17	☽ ± ♅
16 28	☽ # ♄
18 01	☽ # ♆
20 23	☽ ⚹ ♃
23 27	☽ ⊼ ♆
05 Wednesday	
06 08	☽ ∥ ♃
07 09	☽ □ ♇
08 26	☽ ⚼ ♄
09 13	☽ ⚹ ☉
09 22	☽ ⚼ ♅
09 26	☽ Q ☿
11 32	☽ ± ♆
20 28	☽ Q ♃
20 44	☽ ⚼ ♂
06 Thursday	
05 46	☽ ⚼ ♆
08 04	☽ ⚹ ☿
14 37	☽ ± ♂
14 59	☽ △ ♄
15 51	☽ △ ♅
18 24	☽ ∠ ☉
19 45	☽ Q ♇
22 47	♀ ⊼ ♂
07 Friday	
01 45	☽ # ☿
02 35	☽ ⊼ ♂
03 05	☽ ☌ ♀
05 41	☽ ∥ ♀
08 57	☽ □ ♃
12 17	☽ ∠ ☿
12 21	☽ △ ♆
14 25	☽ ⊥ ☉
20 17	☽ ⚹ ♇
08 Saturday	
03 07	☽ # ☉
03 46	☽ ⊻ ☉
04 53	☉ Q ♄
05 06	☽ ⊥ ☿
14 09	☉ Q ♅
16 15	☽ ⊻ ☿
17 25	☽ # ♂
09 Sunday	
01 32	☿ ⚼ ♃
02 15	☽ # ♇
02 53	☽ ∠ ♇
04 22	☽ □ ♄
05 03	☽ □ ♅
06 38	☽ ∥ ♇
07 16	☿ ∠ ♀
12 30	♀ □ ♃
14 21	☽ ☍ ♂
15 31	☽ ∥ ♂
21 08	☽ ⊥ ♇
21 09	☿ # ♀
21 38	☽ △ ♃
22 40	☽ ⊻ ♀
10 Monday	
01 16	☽ □ ♆
09 14	☽ ⊻ ♇
09 17	☽ ∥ ☉
12 01	☽ ⊥ ♀
16 47	☽ Q ♄
17 22	☽ Q ♅
19 22	☽ ∥ ☿
20 43	☽ # ♀
21 49	☽ ☌ ☉
23 15	☽ ☌ ☿
11 Tuesday	
01 31	♀ △ ♆
03 31	☽ ⚼ ♃
06 33	☉ ☌ ☿
07 52	☽ ∠ ♀
13 13	☽ Q ♆
16 44	☽ ⚹ ♄
17 14	☽ ⚹ ♅
21 14	☽ ± ♃
12 Wednesday	
01 05	☽ ⊼ ♂
08 57	☽ ⊼ ♃
12 39	☽ ± ♂
12 48	☽ ⚹ ♆
16 28	☽ ⚹ ♀
17 26	☉ ∥ ☿
20 39	☽ ☌ ♇
22 15	☽ ∠ ♄
22 39	☽ ∠ ♅
13 Thursday	
00 33	☽ # ♃
03 32	☿ ⊥ ♀
05 01	☽ ⊻ ☿
05 50	☽ ⚼ ♂
11 02	☿ Q ♅
12 44	☽ ∥ ♆
13 43	☽ ⊻ ☉
14 27	☽ ∥ ♄
15 03	☿ Q ♄
15 39	☽ ⊥ ☿
15 42	☽ ⊥ ♄
15 44	☉ ⚼ ♃
16 03	☽ ⊥ ♅
17 50	☽ ∠ ♆
18 03	☽ Q ♀
21 29	☽ ∥ ♅
14 Friday	
01 51	☉ # ♀
02 10	☽ ⊥ ☉
03 14	☽ ⊻ ♄
03 33	☽ ⊻ ♅
07 32	☽ ∠ ☿
10 07	☽ △ ♂
11 00	☽ ⊥ ♆
13 44	♀ ⚹ ♇
18 16	☽ ☍ ♃
20 39	☽ ∠ ☉
22 20	☽ ⊻ ♆
15 Saturday	
05 59	☽ ⊻ ♇
07 34	☽ □ ♀
09 51	☽ ⚹ ☿
17 11	☽ ⊥ ♇
23 16	☿ ⊻ ♀
16 Sunday	
02 53	☽ ⚹ ☉
06 46	☽ Q ☿
09 48	☽ ∠ ♇
11 30	☽ ☌ ♄
11 39	☽ ☌ ♅
17 07	☽ □ ♂
17 Monday	
01 20	☽ ⊼ ♃
02 30	☽ Q ☉
05 34	☽ ☌ ♆
12 03	☽ ± ♃
12 59	☽ ⚹ ♇
13 55	☽ □ ☿
19 36	☽ △ ♀
18 Tuesday	
01 03	☽ Q ♂
03 57	☽ ⚼ ♃
10 18	☽ Q ♇
13 01	☽ □ ☉
13 26	♄ ☌ ♅
17 18	☽ ⊻ ♅
17 19	☽ ⊻ ♄
19 32	☿ ⊻ ♇
20 10	☉ Q ♆
20 57	☽ ∥ ♅
21 52	☽ ⚹ ♂
19 Wednesday	
00 24	☽ ⚼ ♀
03 12	☽ ∥ ♄
03 44	☽ ⊥ ♅
03 47	☽ ⊥ ♄
04 51	☽ ∥ ♆
05 56	☽ △ ♃
10 20	☽ ⊻ ♆
16 05	☽ # ♃
17 09	☽ △ ☿
17 17	☽ ± ♀
17 30	☽ □ ♇
19 12	☽ ∠ ♅
19 17	☽ ∠ ♄
20 35	☽ ⊥ ♆
23 22	☽ ∠ ♂
20 Thursday	
04 23	☽ ⊼ ♀
05 22	☿ St D
11 47	☽ ∠ ♆
14 25	☽ ⊥ ♂
18 27	☽ ⚼ ☿
20 04	☽ △ ☉
20 30	☽ ⚹ ♅
20 39	☽ ⚹ ♄
21 Friday	
00 21	☽ ⊻ ♂
02 29	☉ ⚹ ♅
05 06	☉ ⚹ ♄
06 41	☽ ∥ ☉
08 10	☽ □ ♃
09 31	☽ ± ☿
12 43	☽ ⚹ ♆
16 22	☽ Q ♅
16 34	☽ Q ♄
19 33	☽ ⊼ ☿
19 40	☽ △ ♇
22 37	☽ ⚼ ☉
22 Saturday	
01 51	☉ ± ♃
02 00	☿ ⊻ ♇
03 55	☽ # ♀
08 21	☽ Q ♆
08 49	☽ ∥ ☿
10 22	☽ ☍ ♀
11 50	☽ ∥ ♂
13 24	☽ Q ♃
14 19	☽ ± ☉
17 57	☽ ∥ ♇
20 07	☽ ⚼ ♇
21 44	☽ □ ♅
22 00	☽ □ ♄
22 47	☽ # ♇
23 Sunday	
00 43	☽ ⊼ ☉
01 04	☽ ☌ ♂
04 46	☽ # ♂
05 44	☉ ⊼ ♂
07 58	☽ # ☿
08 41	☽ ⚹ ♃
10 41	☽ ± ♇
10 50	☽ ∥ ♀
13 28	☽ □ ♆
20 22	☽ ⊼ ♇
21 40	☽ ☍ ☿
24 Monday	
08 41	☽ ∠ ♃
13 53	☽ # ☉
14 59	☽ ⊼ ♀
21 46	☿ # ♀
22 14	☽ △ ♅
22 39	☽ △ ♄
23 08	☽ ⊥ ♃
25 Tuesday	
01 17	☽ ⊻ ♂
01 33	☽ ± ♀
04 36	☽ ☍ ☉
08 49	☽ ⊻ ♃
11 00	☽ ⊥ ♂
14 03	☽ △ ♆
17 31	☽ ⚼ ♀
21 08	☽ ☍ ♇
22 45	☽ ⚼ ♅
22 48	☽ ∥ ♃
23 15	☽ ⚼ ♄
26 Wednesday	
01 03	☽ ⊼ ☿
01 44	☽ ∠ ♂
10 11	☽ # ♆
11 41	☽ ± ☿
11 57	☽ # ♄
13 42	☽ ± ♅
14 15	☽ ± ♄
14 50	☽ ⚼ ♆
17 55	☽ # ♅
20 37	☽ △ ♀
23 47	☽ ⊼ ♅
27 Thursday	
00 22	☽ ⊼ ♄
02 44	☽ ⚹ ♂
03 56	☽ ⚼ ☿
06 01	☽ ± ♆
10 05	☽ ⊼ ☉
10 22	☽ ☌ ♃
13 47	☉ ⊼ ♃
15 05	♀ ∠ ♇
16 17	☽ ⊼ ♆
17 01	☿ Q ♅
21 11	☽ ± ☉
23 17	☽ Q ♂
23 50	☽ ⊼ ♇
28 Friday	
01 47	☿ Q ♄
02 53	♀ # ♂
05 07	♂ St D
08 01	☽ △ ☿
10 22	☽ ± ♇
11 34	♀ □ ♅
14 12	☽ ⚼ ☉
20 20	♀ □ ♄
29 Saturday	
02 20	☉ ± ♂
02 29	☽ ⚼ ♇
04 16	☽ ☍ ♅
05 04	☽ ☍ ♄
05 53	☽ □ ♀
07 16	☽ □ ♂
15 03	☽ ⊻ ♃
19 34	☽ △ ☉
21 16	♀ ☍ ♂
21 56	☽ ☍ ♆
30 Sunday	
02 03	☽ ⊥ ♃
03 30	☿ ⚼ ♃
06 09	☽ △ ♇
18 48	☽ Q ♀
18 53	☽ ∠ ♃
20 49	☽ □ ☿
31 Monday	
02 53	☉ ⚹ ♆
04 48	♀ # ♇
12 49	☽ ⊼ ♅
13 53	☽ ⊼ ♄
15 16	☽ # ♅
15 58	☽ △ ♂
20 30	☽ ⚹ ♀
22 09	☽ # ♄
23 41	☽ ⚹ ♃

NOVEMBER 1988

	Sidereal time	LONGITUDES Sun ☉	Moon ☽	Moon ☽ 24.00	Mercury ☿	Venus ♀	Mars ♂	Jupiter ♃	Saturn ♄	Uranus ♅	Neptune ♆	Pluto ♇
Date	h m s	° ' "	° ' "	° ' "	° '	° '	° '	° '	° '	° '	° '	° '
01	14 43 58	09 ♏ 17 07	10 ♌ 07 13	16 ♌ 07 44	22 ♎ 18	03 ♎ 03	00 ♈ 00	03 ♊ 50	28 ♐ 58	28 ♐ 23	07 ♑ 56	12 ♏ 23
02	14 47 54	10 17 11	22 ♌ 05 26	28 ♌ 01 01	23 44	04 15	00 04	03 R 43	29 03	28 26	07 58	12 25
03	14 51 51	11 17 17	03 ♍ 55 09	09 ♍ 48 30	25 13	05 28	00 08	03 36	29 09	28 28	07 59	12 28
04	14 55 47	12 17 25	15 ♍ 41 41	21 ♍ 35 18	26 43	06 41	00 13	03 29	29 14	28 31	08 00	12 30
05	14 59 44	13 17 36	27 ♍ 29 55	03 ♎ 26 02	28 15	07 54	00 19	03 22	29 20	28 34	08 02	12 33
06	15 03 41	14 17 48	09 ♎ 24 06	15 ♎ 24 31	29 ♎ 48	09 06	00 26	03 14	29 26	28 37	08 03	12 35
07	15 07 37	15 18 02	21 ♎ 27 36	27 ♎ 33 38	01 ♏ 23	10 19	00 33	03 07	29 31	28 39	08 05	12 37
08	15 11 34	16 18 18	03 ♏ 42 49	09 ♏ 55 17	02 58	11 32	00 42	03 00	29 37	28 42	08 06	12 40
09	15 15 30	17 18 36	16 ♏ 11 07	22 ♏ 30 21	04 33	12 45	00 50	02 52	29 43	28 45	08 08	12 42
10	15 19 27	18 18 56	28 ♏ 52 55	05 ♐ 18 48	06 09	13 59	01 00	02 44	29 49	28 48	08 10	12 45
11	15 23 23	19 19 18	11 ♐ 47 51	18 ♐ 19 59	07 45	15 12	01 10	02 36	29 ♐ 55	28 51	08 11	12 47
12	15 27 20	20 19 41	24 ♐ 55 04	01 ♑ 32 58	09 21	16 25	01 21	02 29	00 ♑ 01	28 54	08 13	12 49
13	15 31 16	21 20 05	08 ♑ 13 35	14 ♑ 56 50	10 58	17 38	01 33	02 21	00 07	28 57	08 15	12 52
14	15 35 13	22 20 31	21 ♑ 42 39	28 ♑ 31 00	12 34	18 52	01 45	02 13	00 13	29 00	08 16	12 54
15	15 39 10	23 20 59	05 ♒ 21 54	12 ♒ 15 20	14 11	20 05	01 58	02 05	00 19	29 03	08 18	12 57
16	15 43 06	24 21 27	19 ♒ 11 21	26 ♒ 09 58	15 47	21 19	02 12	01 57	00 25	29 06	08 20	12 59
17	15 47 03	25 21 57	03 ♓ 11 11	10 ♓ 14 57	17 23	22 32	02 26	01 49	00 31	29 10	08 22	13 02
18	15 50 59	26 22 28	17 ♓ 21 12	24 ♓ 29 47	18 59	23 46	02 41	01 41	00 37	29 13	08 23	13 04
19	15 54 56	27 23 01	01 ♈ 40 25	08 ♈ 52 46	20 35	25 00	02 56	01 32	00 44	29 16	08 25	13 06
20	15 58 52	28 23 34	16 ♈ 06 23	23 ♈ 20 43	22 11	26 14	03 12	01 24	00 50	29 19	08 27	13 09
21	16 02 49	29 ♏ 24 09	00 ♉ 35 05	07 ♉ 48 46	23 47	27 27	03 28	01 16	00 56	29 22	08 29	13 11
22	16 06 45	00 ♐ 24 45	15 ♉ 00 59	22 ♉ 10 56	25 22	28 41	03 45	01 08	01 03	29 26	08 31	13 13
23	16 10 42	01 25 22	29 ♉ 17 47	06 ♊ 20 49	26 58	29 ♎ 55	04 03	01 00	01 09	29 29	08 33	13 16
24	16 14 39	02 26 01	13 ♊ 19 21	20 ♊ 12 50	28 ♏ 33	01 ♏ 09	04 21	00 52	01 16	29 32	08 34	13 18
25	16 18 35	03 26 42	27 ♊ 00 49	03 ♋ 43 00	00 ♐ 08	02 23	04 39	00 43	01 22	29 36	08 36	13 20
26	16 22 32	04 27 24	10 ♋ 19 15	16 ♋ 49 33	01 42	03 37	04 58	00 35	01 29	29 39	08 38	13 23
27	16 26 28	05 28 07	23 ♋ 14 03	29 ♋ 32 59	03 17	04 51	05 18	00 27	01 35	29 42	08 40	13 25
28	16 30 25	06 28 52	05 ♌ 46 44	11 ♌ 55 46	04 52	06 06	05 38	00 19	01 42	29 46	08 42	13 27
29	16 34 21	07 29 38	18 ♌ 00 37	24 ♌ 01 54	06 26	07 20	05 58	00 11	01 49	29 49	08 44	13 30
30	16 38 18	08 ♐ 30 26	00 ♍ 00 15	05 ♍ 56 21	08 ♐ 01	08 ♏ 34	06 ♈ 19	00 ♊ 03	01 ♑ 55	29 ♐ 53	08 ♑ 46	13 ♏ 32

Date	Moon True ☊ ° '	Moon Mean ☊ ° '	Moon ☽ Latitude ° '
01	12 ♓ 24	10 ♓ 59	02 N 44
02	12 D 25	10 56	01 47
03	12 25	10 53	00 N 46
04	12 R 26	10 50	00 S 18
05	12 24	10 47	01 20
06	12 21	10 44	02 19
07	12 14	10 40	03 13
08	12 05	10 37	03 58
09	11 54	10 34	04 33
10	11 42	10 31	04 55
11	11 30	10 28	05 02
12	11 19	10 24	04 54
13	11 10	10 21	04 30
14	11 04	10 18	03 50
15	11 00	10 15	02 57
16	10 59	10 12	01 53
17	10 D 59	10 09	00 S 41
18	10 R 59	10 05	00 N 34
19	10 57	10 02	01 47
20	10 54	09 59	02 54
21	10 47	09 56	03 50
22	10 37	09 53	04 31
23	10 26	09 50	04 55
24	10 13	09 46	05 00
25	10 01	09 43	04 48
26	09 50	09 40	04 20
27	09 42	09 37	03 39
28	09 36	09 34	02 48
29	09 33	09 30	01 51
30	09 ♓ 32	09 ♓ 27	00 N 50

DECLINATIONS

Date	Sun ☉	Moon ☽	Mercury ☿	Venus ♀	Mars ♂	Jupiter ♃	Saturn ♄	Uranus ♅	Neptune ♆	Pluto ♇
01	14 S 35	20 N 21	06 S 43	00 N 14	01 S 52	19 N 53	22 S 38	23 S 39	22 S 17	00 S 50
02	14 54	15 50	07 17	00 S 13	01 47	19 52	22 38	23 39	22 17	00 51
03	15 13	10 47	07 52	00 41	01 42	19 51	22 38	23 39	22 17	00 52
04	15 32	05 N 22	08 28	01 08	01 36	19 49	22 38	23 39	22 17	00 53
05	15 50	00 S 14	09 04	01 36	01 30	19 48	22 38	23 39	22 17	00 53
06	16 08	05 51	09 41	02 03	01 23	19 47	22 38	23 39	22 17	00 54
07	16 26	11 21	10 19	02 31	01 17	19 45	22 39	23 39	22 17	00 55
08	16 43	16 29	10 56	02 58	01 10	19 44	22 39	23 39	22 16	00 55
09	17 00	21 02	11 33	03 26	01 03	19 42	22 39	23 39	22 16	00 56
10	17 17	24 42	12 11	03 53	00 56	19 41	22 39	23 39	22 16	00 57
11	17 34	27 12	12 48	04 21	00 49	19 39	22 39	23 39	22 16	00 57
12	17 50	28 14	13 24	04 49	00 41	19 38	22 39	23 39	22 16	00 58
13	18 06	27 40	14 00	05 16	00 33	19 37	22 39	23 39	22 16	00 59
14	18 21	25 29	14 36	05 43	00 25	19 35	22 39	23 39	22 16	00 59
15	18 37	21 47	15 11	06 11	00 17	19 34	22 39	23 39	22 16	01 00
16	18 52	16 51	15 46	06 38	00 S 08	19 32	22 40	23 39	22 16	01 01
17	19 06	10 59	16 20	07 05	00 00	19 31	22 40	23 39	22 16	01 01
18	19 21	04 S 29	16 53	07 32	00 N 09	19 29	22 40	23 39	22 16	01 02
19	19 35	02 N 18	17 25	07 59	00 18	19 27	22 40	23 39	22 16	01 02
20	19 48	09 01	17 57	08 25	00 27	19 26	22 40	23 39	22 16	01 03
21	20 02	15 16	18 28	08 52	00 37	19 24	22 40	23 39	22 15	01 04
22	20 14	20 40	18 58	09 18	00 46	19 23	22 40	23 39	22 15	01 04
23	20 27	24 48	19 26	09 45	00 56	19 21	22 40	23 39	22 15	01 05
24	20 39	27 22	19 55	10 11	01 05	19 20	22 40	23 39	22 15	01 05
25	20 51	28 12	20 22	10 36	01 15	19 18	22 40	23 39	22 15	01 06
26	21 02	27 21	20 48	11 02	01 25	19 17	22 40	23 39	22 15	01 06
27	21 13	25 02	21 13	11 27	01 36	19 15	22 40	23 39	22 15	01 07
28	21 24	21 33	21 37	11 53	01 46	19 14	22 40	23 39	22 15	01 07
29	21 34	17 12	22 00	12 17	01 56	19 12	22 40	23 39	22 15	01 08
30	21 S 44	12 N 15	22 S 22	12 S 42	02 N 07	19 N 10	22 S 40	23 S 39	22 S 15	01 S 08

ZODIAC SIGN ENTRIES

Date	h m	Planets
01	12 57	♂ ♈
03	04 02	☽ ♍
05	17 04	☽ ♎
06	14 57	☿ ♏
08	04 46	☽ ♏
10	14 06	☽ ♐
12	09 26	♄ ♑
12	21 12	☽ ♑
15	02 36	☽ ♒
17	06 34	☽ ♓
19	09 12	☽ ♈
21	11 02	☽ ♉
22	02 12	☉ ♐
23	13 12	☽ ♊
23	13 34	♀ ♏
25	10 04	☿ ♐
25	17 19	☽ ♋
28	00 52	☽ ♌
30	12 00	☽ ♍
30	20 53	♃ ♉

LATITUDES

Date	Mercury ☿	Venus ♀	Mars ♂	Jupiter ♃	Saturn ♄	Uranus ♅	Neptune ♆	Pluto ♇
01	02 N 07	01 N 35	02 S 02	01 S 03	00 N 49	00 S 13	00 N 56	15 N 25
04	01 58	01 39	01 50	01 03	00 48	00 13	00 55	15 25
07	01 45	01 43	01 38	01 03	00 48	00 13	00 55	15 25
10	01 29	01 46	01 27	01 03	00 48	00 13	00 55	15 25
13	01 10	01 48	01 16	01 03	00 47	00 13	00 55	15 25
16	00 50	01 49	01 06	01 02	00 47	00 13	00 55	15 25
19	00 30	01 50	00 56	01 02	00 47	00 13	00 55	15 25
22	00 N 09	01 49	00 47	01 02	00 46	00 13	00 55	15 25
25	00 S 11	01 48	00 39	01 01	00 46	00 13	00 55	15 26
28	00 31	01 47	00 31	01 01	00 46	00 13	00 55	15 26
31	00 S 50	01 N 44	00 S 23	01 S 01	00 N 45	00 S 13	00 N 55	15 N 27

DATA

Julian Date	2447467
Delta T	+56 seconds
Ayanamsa	23° 42' 08"
Synetic vernal point	05° ♓ 24' 51"
True obliquity of ecliptic	23° 26' 35"

MOON'S PHASES, APSIDES AND POSITIONS ☽

Date	h m	Phase	Longitude ° '	Eclipse Indicator
01	10 11	☾	09 ♌ 13	
09	14 20	●	17 ♏ 24	
16	21 35	☽	24 ♒ 46	
23	15 53	○	01 ♊ 35	

Day	h m		
04	10 57	Apogee	
20	10 35	Perigee	
05	11 02	0S	
12	15 44	Max dec	28° S 15'
19	03 53	0N	
25	11 38	Max dec	28° N 12'

LONGITUDES

Date	Chiron ⚷	Ceres ⚳	Pallas ⚴	Juno ⚵	Vesta ⚶	Black Moon Lilith ⚸
01	06 ♋ 59	17 ♓ 31	04 ♒ 31	27 ♌ 30	24 ♎ 57	19 ♍ 01
11	06 ♋ 42	17 ♓ 29	06 ♒ 18	00 ♍ 59	00 ♏ 08	20 ♍ 08
21	06 ♋ 17	18 ♓ 04	08 ♒ 24	04 ♍ 03	05 ♏ 19	21 ♍ 14
31	05 ♋ 47	19 ♓ 12	10 ♒ 46	06 ♍ 37	10 ♏ 29	22 ♍ 21

All ephemeris data is given at 12.00 UT and the Moon's longitude is additionally given for 24.00 UT

ASPECTARIAN

h m Aspects

01 Tuesday
00 21 ☽ # ♆
00 36 ☽ ± ♅
01 43 ☽ ± ♄
07 39 ☽ ⊼ ♆
10 11 ☽ □ ☉
12 23 ☽ Q ☿
14 40 ☽ ∥ ♃
16 31 ☽ □ ♇
18 31 ☽ bQ ♅
19 37 ☽ ± ♆
19 44 ☽ bQ ♄
21 47 ☽ bQ ♂
23 18 ☽ Q ♃

02 Wednesday
02 12 ♀ △ ♃
05 39 ☽ ∠ ♀
13 46 ☽ bQ ♆
15 19 ☉ ⊥ ♀
15 47 ☽ ✶ ☿
16 00 ☽ ± ♂
16 18 ☽ # ☉

03 Thursday
00 53 ☽ △ ♅
01 43 ☽ Q ☉
01 55 ☽ ⊥ ♀
02 13 ☽ △ ♄
04 15 ☽ ⊼ ♂
04 56 ☽ Q ♇
11 21 ☽ □ ♃
15 31 ☽ ⊻ ♀
20 18 ☽ △ ♆
22 00 ♀ ∥ ♇
23 47 ☽ # ☿

04 Friday
00 32 ☿ Q ♆
02 42 ☽ ∠ ☿
04 25 ☽ ✶ ☉
05 28 ☽ ✶ ♇
08 21 ♀ ⊥ ♇
17 16 ☉ ☌ ♇
23 06 ☿ ± ♃
23 45 ☽ ⊥ ☿

05 Saturday
04 32 ☽ # ♂
04 50 ☽ # ♀
07 16 ☽ # ♇
07 44 ♀ ∥ ♂
12 05 ☽ ∠ ♇
13 46 ☽ ∠ ☉
13 46 ☽ ⊻ ☿
14 10 ☽ □ ♅
14 49 ♀ □ ♆
14 50 ☽ ∥ ♇
15 45 ☽ □ ♄
16 56 ☿ ✶ ♅
17 18 ☽ ∥ ♂
17 46 ☽ ☍ ♂
18 20 ☽ ∥ ♀
18 46 ☉ ∠ ♅
23 44 ☽ △ ♃

06 Sunday
05 46 ☿ ✶ ♄
06 19 ☽ ⊥ ♇
09 18 ☽ □ ♆
09 35 ☽ ⊥ ☉
11 21 ☽ ☌ ♀
15 27 ☉ ∠ ♄
18 23 ☽ ⊻ ♇
22 23 ☿ ⊼ ♂
22 40 ☽ ⊻ ☉

07 Monday
02 27 ☽ Q ♅
04 08 ☽ Q ♄
05 27 ☽ bQ ♃
06 48 ☽ ∥ ☿
19 04 ☉ bQ ♂
21 07 ☽ Q ♆
23 01 ☽ ± ♃

08 Tuesday
02 12 ☽ ✶ ♅
03 58 ☽ ✶ ♄
06 03 ☽ ⊼ ♂
10 19 ☽ ☌ ☿
10 37 ☽ ⊼ ♃
12 27 ☿ ⊼ ♃
13 13 ☽ ∥ ☉
17 50 ☽ ± ♂
20 31 ☽ ✶ ♆

09 Wednesday
04 37 ☽ # ♃
04 44 ☽ ⊻ ♀
05 19 ☽ ☌ ♇
07 20 ☽ ∠ ♅
09 10 ☽ ∠ ♄
10 55 ♀ ⊻ ♇
11 20 ☽ bQ ♂
14 20 ☽ ☌ ☉
17 25 ☽ ⊥ ♀
19 25 ☽ ∥ ♆
21 46 ☽ ∥ ♄

10 Thursday
00 31 ☽ ⊥ ♅
01 13 ☽ ∠ ♆
02 24 ☽ ⊥ ♄
04 23 ☽ ∥ ♅
09 32 ♂ ∥ ♇
11 51 ☽ ⊻ ♅
12 12 ☽ ∠ ♀
13 45 ☽ ⊻ ♄
16 01 ☽ △ ♂
18 09 ☽ ⊥ ♆
19 08 ☽ ☍ ♃

11 Friday
02 17 ☿ ± ♂
03 28 ☽ ⊻ ☿
05 19 ☽ ⊻ ♆
13 49 ☽ ⊻ ♇
16 06 ☽ ⊥ ☿
18 39 ☿ ✶ ♆
18 54 ☽ ✶ ♀

12 Saturday
00 52 ☽ ⊥ ♇
02 57 ☽ ⊻ ☉
10 50 ☽ ∠ ☿
14 46 ☽ ⊥ ☉
17 17 ☽ ∠ ♇
18 59 ☽ Q ♀
19 15 ☽ ☌ ♅
21 18 ☽ ☌ ♄
21 55 ♀ Q ♅
23 49 ☽ □ ♂

13 Sunday
01 32 ☽ ⊼ ♃
06 47 ♀ bQ ♃
08 20 ☽ ∠ ☉
12 02 ☽ ☌ ♆
12 13 ☽ ± ♃
17 34 ☽ ✶ ☿
20 19 ☽ ✶ ♇
22 03 ♀ Q ♄

14 Monday
03 22 ♃ ± ♆
04 06 ☽ bQ ♃
06 28 ☽ □ ♀
08 29 ☽ Q ♂
13 12 ☽ ✶ ☉
17 06 ☿ ☌ ♇
17 40 ☽ Q ♇
17 44 ☽ Q ☿

15 Tuesday
00 54 ☽ ⊻ ♅
00 59 ☽ ∥ ♅
03 05 ☽ ⊻ ♄
04 38 ☉ ⊥ ♅
05 58 ☽ ✶ ♂
06 19 ☽ △ ♃
07 04 ☽ ∥ ♄
09 20 ☽ ∥ ♆
10 06 ☿ ∠ ♅
10 48 ☉ ∠ ♆
11 27 ☽ ⊥ ♅
11 58 ☽ Q ☉
13 40 ☽ ⊥ ♄
17 08 ☽ ⊻ ♆
19 35 ♂ ✶ ♃
23 35 ☽ # ♃

16 Wednesday
01 14 ☽ □ ♇
03 11 ☽ ∠ ♅
03 17 ☽ ∥ ☉
03 35 ☽ ⊥ ♆
05 21 ☽ □ ☿
05 26 ☽ ∠ ♄
06 07 ☿ ∠ ♄
08 30 ☽ ∠ ♂
13 34 ☉ ⊥ ♄
16 01 ☽ △ ♀
16 18 ☽ ∥ ☿
19 09 ☽ ∠ ♆
21 35 ☽ □ ☉

17 Thursday
00 15 ☽ ⊥ ♂
05 06 ☽ ✶ ♅
07 25 ☽ ✶ ♄
09 41 ☽ □ ♃
10 41 ☽ ⊻ ♂
12 44 ☿ bQ ♂
20 07 ☽ bQ ♀
20 49 ☽ ✶ ♆

18 Friday
01 36 ☽ Q ♅
01 39 ☽ ∥ ♀
03 58 ☽ Q ♄
04 45 ☽ △ ♇
12 46 ☽ ± ♀
15 06 ☽ △ ☿
15 52 ☽ Q ♃
17 07 ☽ Q ♆
23 48 ☽ ⊼ ♀

19 Saturday
00 14 ☽ ∥ ♇
00 42 ☉ # ♃
03 01 ☽ # ♂
04 17 ☽ △ ☉
04 47 ☽ ∥ ♂
06 01 ☽ bQ ♇
07 32 ☽ # ♇
07 58 ☽ □ ♅
10 25 ☽ □ ♄
11 47 ☽ ✶ ♃
14 08 ☽ ☌ ♂
19 20 ☽ bQ ☿
21 04 ☽ ± ♇
21 34 ♀ ± ♃
23 15 ☽ □ ♆

20 Sunday
07 04 ☽ ⊼ ♇
07 10 ☽ bQ ☉
09 40 ☽ # ♀
12 09 ☽ ± ☿
12 29 ☽ ∠ ♃
16 28 ♀ Q ♆
23 12 ☽ ± ☉
23 20 ☽ ⊼ ☿

21 Monday
03 16 ☽ ⊥ ♃
05 40 ☿ ⊥ ♅
06 20 ☽ ☍ ♀
07 24 ☿ ∠ ♆
09 54 ☽ ⊼ ☉
09 59 ☽ △ ♅
11 17 ☉ ⊻ ♅
12 36 ☽ △ ♄
13 07 ☽ ⊻ ♃
16 53 ☽ ⊻ ♂

22 Tuesday
01 08 ☽ △ ♆
03 01 ☽ # ☿
03 03 ☽ ⊥ ♂
05 51 ☽ ∥ ♃
06 46 ☿ ⊥ ♄
09 00 ☽ ☍ ♇
09 49 ☽ # ☉
11 01 ☽ bQ ♅
13 44 ☽ bQ ♄
18 23 ☽ ∠ ♂
20 19 ☽ # ♆
20 25 ♃ ⊼ ♄
22 37 ☽ # ♄

23 Wednesday
02 09 ☽ ± ♅
02 16 ☽ bQ ♆
03 04 ☉ ☍ ♃
03 06 ♀ ✶ ♅
04 28 ☽ # ♅
04 52 ☉ ⊻ ♄
04 57 ☽ ± ♄
07 33 ☽ ☍ ☿
07 51 ☿ # ♃
12 19 ☽ ⊼ ♅
13 09 ☽ ⊼ ♀
14 51 ☽ ☌ ♃
15 11 ☽ ⊼ ♄
15 53 ☽ ☍ ☉
17 31 ☽ ± ♆
20 15 ☽ ✶ ♂

24 Thursday
00 21 ☽ ± ♀
03 48 ☽ ⊼ ♆
06 52 ♀ ⊼ ♃
11 16 ♂ # ♇
11 58 ☽ ⊼ ♇
14 21 ♀ ✶ ♄
15 27 ☉ ⊥ ♆
17 22 ☽ Q ♂
17 23 ☽ bQ ♀
22 26 ☽ ± ♇

25 Friday
03 37 ☿ ⊻ ♅
14 22 ☽ bQ ♇
16 37 ☽ ☍ ♅
18 18 ☽ ⊼ ☿
18 33 ☽ ⊻ ♃
19 51 ☽ ☍ ♄
20 20 ☿ ☍ ♃
22 35 ☽ △ ♀

26 Saturday
00 27 ☽ ⊼ ☉
02 01 ☽ □ ♂
05 16 ☽ ⊥ ♃
06 35 ☽ ± ☿
08 18 ☿ ⊻ ♄
08 55 ☽ ☍ ♆
12 16 ☽ ± ☉
17 38 ☽ △ ♇
21 36 ☽ ∠ ♃

27 Sunday
01 25 ☽ bQ ☿
02 28 ☿ ⊥ ♆
05 56 ☉ △ ♂
06 21 ☽ bQ ☉
12 27 ☉ ∥ ☿
22 21 ☽ # ♅
23 39 ♀ ⊼ ♂

28 Monday
00 21 ☽ ⊼ ♅
01 35 ☽ ✶ ♃
04 03 ☽ ⊼ ♄
05 00 ☽ # ♄
07 42 ☽ # ♆
09 58 ☽ △ ☿
11 39 ☽ # ☿
11 42 ☽ △ ♂
11 58 ☽ ± ♅
12 41 ☽ □ ♀
12 54 ☽ # ☉
13 29 ☽ △ ☉
15 46 ☽ ± ♄
17 42 ☽ ⊼ ♆

29 Tuesday
00 37 ☽ Q ♃
01 24 ☽ ∥ ♃
02 52 ☿ △ ♂
03 03 ☽ □ ♇
05 31 ☽ ± ♆
05 39 ☽ bQ ♅
09 36 ☽ bQ ♄
18 03 ☽ bQ ♂
23 27 ☽ bQ ♆

30 Wednesday
04 01 ☿ ∥ ♆
04 17 ☽ Q ♀
05 39 ☉ ⊻ ♀
10 05 ☽ # ♀
11 44 ☽ △ ♅
12 05 ☽ □ ♃
12 39 ☽ ± ♂
15 05 ☽ Q ♇
15 55 ☽ △ ♄
16 05 ♀ ✶ ♆
18 29 ☉ ⊻ ♆
23 53 ☿ ⊻ ♆

DECEMBER 1988

Date	Sidereal time h m s	Sun ☉ ° ' "	Moon ☽ ° ' "	Moon ☽ 24.00 ° ' "	Mercury ☿ ° '	Venus ♀ ° '	Mars ♂ ° '	Jupiter ♃ ° '	Saturn ♄ ° '	Uranus ♅ ° '	Neptune ♆ ° '	Pluto ♇ ° '
01	16 42 14	09 ♐ 31 15	11 ♍ 50 55	17 ♍ 44 39	09 ♐ 35	09 ♏ 48	06 ♈ 40	29 ♉ 55	02 ♑ 02	29 ♐ 56	08 ♑ 48	13 ♏ 34
02	16 46 11	10 32 06	23 ♍ 38 16	29 ♍ 32 27	11 09	11 03	07 02	29 R 47	02 09	29 ♐ 59	08 50	13 36
03	16 50 08	11 32 57	05 ♎ 27 52	11 ♎ 25 09	12 43	12 17	07 24	29 39	02 16	00 ♑ 03	08 52	13 38
04	16 54 04	12 33 51	17 ♎ 24 53	23 ♎ 27 37	14 18	13 31	07 47	29 31	02 22	00 06	08 55	13 41
05	16 58 01	13 34 45	29 ♎ 33 48	05 ♏ 43 51	15 52	14 46	08 09	29 24	02 29	00 10	08 57	13 43
06	17 01 57	14 35 41	11 ♏ 58 03	18 ♏ 16 39	17 26	16 00	08 33	29 16	02 36	00 13	08 59	13 45
07	17 05 54	15 36 38	24 ♏ 39 45	01 ♐ 07 23	19 00	17 15	08 56	29 08	02 43	00 17	09 01	13 47
08	17 09 50	16 37 36	07 ♐ 39 30	14 ♐ 15 56	20 34	18 29	09 21	29 01	02 50	00 21	09 03	13 49
09	17 13 47	17 38 36	20 ♐ 56 25	27 ♐ 40 40	22 08	19 44	09 45	28 54	02 57	00 24	09 05	13 51
10	17 17 43	18 39 36	04 ♑ 28 17	11 ♑ 18 54	23 43	20 59	10 10	28 46	03 04	00 28	09 07	13 54
11	17 21 40	19 40 37	18 ♑ 12 04	25 ♑ 07 24	25 17	22 13	10 35	28 39	03 11	00 31	09 09	13 56
12	17 25 37	20 41 38	02 ♒ 04 29	09 ♒ 02 59	26 51	23 28	11 00	28 32	03 18	00 35	09 12	13 58
13	17 29 33	21 42 40	16 ♒ 02 36	23 ♒ 03 03	28 ♐ 26	24 43	11 26	28 25	03 25	00 38	09 14	14 00
14	17 33 30	22 43 43	00 ♓ 04 09	07 ♓ 05 46	00 ♑ 00	25 57	11 52	28 18	03 32	00 42	09 16	14 02
15	17 37 26	23 44 46	14 ♓ 07 45	21 ♓ 10 02	01 35	27 12	12 19	28 12	03 39	00 46	09 18	14 04
16	17 41 23	24 45 49	28 ♓ 12 33	05 ♈ 15 11	03 10	28 27	12 45	28 05	03 46	00 49	09 20	14 06
17	17 45 19	25 46 53	12 ♈ 17 52	19 ♈ 20 26	04 45	29 ♏ 42	13 12	27 59	03 53	00 53	09 23	14 08
18	17 49 16	26 47 57	26 ♈ 22 42	03 ♉ 24 26	06 19	00 ♐ 56	13 40	27 52	04 00	00 56	09 25	14 10
19	17 53 12	27 49 01	10 ♉ 25 20	17 ♉ 25 02	07 54	02 11	14 07	27 46	04 07	01 00	09 27	14 12
20	17 57 09	28 50 06	24 ♉ 23 08	01 ♊ 19 11	09 29	03 26	14 35	27 40	04 14	01 04	09 29	14 13
21	18 01 06	29 ♐ 51 11	08 ♊ 12 44	15 ♊ 03 19	11 04	04 41	15 03	27 35	04 21	01 07	09 32	14 15
22	18 05 02	00 ♑ 52 16	21 ♊ 50 30	28 ♊ 33 52	12 39	05 56	15 32	27 29	04 28	01 11	09 34	14 17
23	18 08 59	01 53 22	05 ♋ 13 06	11 ♋ 47 54	14 13	07 11	16 00	27 23	04 35	01 15	09 36	14 19
24	18 12 55	02 54 28	18 ♋ 18 07	24 ♋ 43 40	15 48	08 25	16 29	27 18	04 42	01 18	09 38	14 21
25	18 16 52	03 55 35	01 ♌ 04 33	07 ♌ 20 54	17 22	09 40	16 58	27 13	04 49	01 22	09 41	14 23
26	18 20 48	04 56 42	13 ♌ 32 55	19 ♌ 40 54	18 56	10 55	17 27	27 08	04 57	01 25	09 43	14 24
27	18 24 45	05 57 50	25 ♌ 45 14	01 ♍ 46 22	20 30	12 10	17 57	27 03	05 04	01 29	09 45	14 26
28	18 28 41	06 58 58	07 ♍ 44 50	13 ♍ 41 11	22 03	13 25	18 27	26 58	05 11	01 33	09 47	14 28
29	18 32 38	08 00 06	19 ♍ 36 02	25 ♍ 30 01	23 35	14 40	18 57	26 54	05 18	01 36	09 50	14 29
30	18 36 35	09 01 15	01 ♎ 23 50	07 ♎ 18 09	25 06	15 55	19 27	26 50	05 25	01 40	09 52	14 31
31	18 40 31	10 ♑ 02 24	13 ♎ 13 40	19 ♎ 11 02	26 ♑ 37	17 ♐ 10	19 ♈ 57	26 ♉ 46	05 ♑ 32	01 ♑ 43	09 ♑ 54	14 ♏ 33

Date	Moon True ☊ ° '	Moon Mean ☊ ° '	Moon ☽ Latitude ° '
01	09 ♓ 32	09 ♓ 24	00 S 12
02	09 R 32	09 21	01 14
03	09 30	09 18	02 13
04	09 26	09 15	03 06
05	09 19	09 11	03 52
06	09 10	09 08	04 28
07	08 58	09 05	04 51
08	08 44	09 02	05 00
09	08 31	08 59	04 53
10	08 18	08 55	04 30
11	08 08	08 52	03 51
12	08 01	08 49	02 58
13	07 57	08 46	01 53
14	07 55	08 43	00 S 42
15	07 D 55	08 40	00 N 33
16	07 R 55	08 36	01 45
17	07 54	08 33	02 52
18	07 51	08 30	03 47
19	07 45	08 27	04 29
20	07 36	08 24	04 54
21	07 26	08 21	05 02
22	07 14	08 17	04 53
23	07 02	08 14	04 27
24	06 52	08 11	03 48
25	06 44	08 08	02 58
26	06 39	08 05	02 00
27	06 36	08 01	00 N 58
28	06 D 35	07 58	00 S 06
29	06 36	07 55	01 09
30	06 37	07 52	02 09
31	06 ♓ 37	07 ♓ 49	03 S 04

DECLINATIONS

Date	Sun ☉ ° '	Moon ☽ ° '	Mercury ☿ ° '	Venus ♀ ° '	Mars ♂ ° '	Jupiter ♃ ° '	Saturn ♄ ° '	Uranus ♅ ° '	Neptune ♆ ° '	Pluto ♇ ° '
01	21 S 53	06 N 56	22 S 43	13 S 06	02 N 18	19 N 09	22 S 40	23 S 39	22 S 14	01 S 09
02	22 02	01 N 24	23 02	13 30	02 29	19 07	22 40	23 40	22 14	01 09
03	22 10	04 S 12	23 21	13 54	02 40	19 06	22 40	23 40	22 14	01 09
04	22 18	09 42	23 38	14 18	02 51	19 04	22 40	23 40	22 14	01 10
05	22 26	14 56	23 54	14 41	03 02	19 03	22 40	23 40	22 14	01 10
06	22 33	19 40	24 09	15 03	03 13	19 01	22 40	23 40	22 14	01 11
07	22 40	23 38	24 23	15 26	03 24	19 00	22 40	23 40	22 14	01 11
08	22 46	26 31	24 35	15 48	03 36	18 59	22 40	23 40	22 14	01 11
09	22 52	28 01	24 46	16 09	03 47	18 57	22 40	23 40	22 14	01 12
10	22 58	27 52	24 56	16 30	03 59	18 56	22 40	23 39	22 13	01 12
11	23 02	26 01	25 04	16 51	04 11	18 54	22 40	23 39	22 13	01 12
12	23 07	22 35	25 11	17 11	04 23	18 53	22 40	23 39	22 13	01 13
13	23 11	17 50	25 17	17 31	04 34	18 52	22 40	23 39	22 13	01 13
14	23 15	12 06	25 21	17 51	04 46	18 50	22 39	23 39	22 13	01 13
15	23 18	05 S 44	25 24	18 10	04 58	18 49	22 39	23 39	22 13	01 14
16	23 20	00 N 54	25 25	18 28	05 11	18 48	22 39	23 39	22 13	01 14
17	23 23	07 30	25 25	18 46	05 23	18 47	22 39	23 39	22 12	01 14
18	23 24	13 43	25 23	19 03	05 35	18 46	22 39	23 39	22 12	01 14
19	23 25	19 12	25 20	19 20	05 47	18 44	22 39	23 39	22 12	01 15
20	23 26	23 37	25 16	19 37	06 00	18 43	22 39	23 39	22 12	01 15
21	23 27	26 39	25 09	19 53	06 12	18 42	22 39	23 39	22 12	01 15
22	23 26	28 04	25 02	20 08	06 24	18 41	22 38	23 39	22 12	01 15
23	23 26	27 47	24 53	20 23	06 37	18 40	22 38	23 39	22 12	01 15
24	23 25	25 57	24 42	20 37	06 49	18 39	22 38	23 39	22 11	01 16
25	23 23	22 49	24 30	20 50	07 02	18 38	22 38	23 39	22 11	01 16
26	23 21	18 40	24 16	21 03	07 14	18 37	22 38	23 39	22 11	01 16
27	23 19	13 51	24 01	21 16	07 27	18 36	22 37	23 39	22 11	01 16
28	23 16	08 34	23 44	21 28	07 40	18 36	22 37	23 39	22 11	01 16
29	23 12	03 N 03	23 26	21 39	07 52	18 35	22 37	23 39	22 11	01 16
30	23 08	02 S 32	23 06	21 49	08 05	18 34	22 37	23 39	22 11	01 16
31	23 S 04	08 S 02	22 S 45	21 S 59	08 N 18	18 N 33	22 S 37	23 S 39	22 S 10	01 S 16

ZODIAC SIGN ENTRIES

Date	h m	Planets
02	15 35	♅ ♑
03	00 56	☽ ♎
05	12 51	☽ ♏
07	21 55	☽ ♐
10	04 07	☽ ♑
12	08 25	☽ ♒
14	11 53	☽ ♓
14	11 53	☿ ♑
16	15 03	☽ ♈
17	17 56	♀ ♐
18	18 11	☽ ♉
20	21 43	☽ ♊
21	15 28	☉ ♑
23	02 35	☽ ♋
25	09 57	☽ ♌
27	20 27	☽ ♍
30	09 09	☽ ♎

LATITUDES

Date	Mercury ☿ ° '	Venus ♀ ° '	Mars ♂ ° '	Jupiter ♃ ° '	Saturn ♄ ° '	Uranus ♅ ° '	Neptune ♆ ° '	Pluto ♇ ° '
01	00 S 50	01 N 44	00 S 23	01 S 01	00 N 45	00 S 13	00 N 55	15 N 27
04	01 07	01 41	00 16	01 00	00 45	00 13	00 55	15 27
07	01 24	01 37	00 09	01 00	00 45	00 13	00 55	15 28
10	01 38	01 33	00 S 03	00 59	00 45	00 13	00 54	15 29
13	01 51	01 28	00 N 03	00 58	00 44	00 13	00 54	15 29
16	02 01	01 23	00 09	00 58	00 44	00 13	00 54	15 30
19	02 08	01 17	00 14	00 57	00 44	00 13	00 54	15 31
22	02 12	01 11	00 19	00 56	00 44	00 13	00 54	15 32
25	02 12	01 05	00 24	00 56	00 43	00 13	00 54	15 33
28	02 07	00 58	00 28	00 55	00 43	00 13	00 54	15 34
31	01 S 57	00 N 50	00 N 32	00 S 54	00 N 43	00 S 13	00 N 54	15 N 36

DATA

Julian Date	2447497
Delta T	+56 seconds
Ayanamsa	23° 42' 13"
Synetic vernal point	05° ♓ 24' 46"
True obliquity of ecliptic	23° 26' 35"

MOON'S PHASES, APSIDES AND POSITIONS ☽

Date	h m	Phase	Longitude ° '	Eclipse Indicator
01	06 49	☾	09 ♍ 18	
09	05 36	●	17 ♐ 22	
16	05 40	☽	24 ♓ 30	
23	05 29	○	01 ♋ 37	
31	04 57	☾	09 ♎ 44	

Day	h m		
02	06 31	Apogee	
16	03 47	Perigee	
30	03 53	Apogee	
02	17 59	0S	
09	21 59	Max dec	28° S 09'
16	08 46	0N	
22	20 01	Max dec	28° N 09'
30	01 08	0S	

LONGITUDES

Date	Chiron ⚷ ° '	Ceres ⚳ ° '	Pallas ⚴ ° '	Juno ⚵ ° '	Vesta ⚶ ° '	Black Moon Lilith ⚸ ° '
01	05 ♋ 47	19 ♓ 12	10 ♒ 46	06 ♍ 37	10 ♏ 29	22 ♍ 21
11	05 ♋ 11	20 ♓ 50	13 ♒ 22	08 ♍ 37	15 ♏ 37	23 ♍ 27
21	04 ♋ 33	22 ♓ 54	16 ♒ 09	09 ♍ 57	20 ♏ 43	24 ♍ 34
31	03 ♋ 54	25 ♓ 21	19 ♒ 06	10 ♍ 33	25 ♏ 46	25 ♍ 40

All ephemeris data is given at 12.00 UT and the Moon's longitude is additionally given for 24.00 UT

ASPECTARIAN

01 Thursday
01 09 ☽ ⚻ ♂
05 48 ☽ △ ♆
06 41 ☽ □ ☿
06 49 ☽ □ ☉
07 22 ☽ ⚹ ♀
09 14 ☿ ∥ ♄
09 23 ☉ ☌ ☿
09 48 ♃ ⚻ ♅
15 30 ☽ ⚹ ♇

02 Friday
04 08 ☿ ⊻ ♀
07 30 ☽ ∥ ♂
13 03 ☽ # ♇
17 28 ☽ ∠ ♀
22 08 ☽ ∠ ♇
22 54 ☽ Q ☉
22 55 ☽ ∥ ♇

03 Saturday
00 21 ☽ △ ♃
00 56 ☽ Q ☿
00 59 ☽ □ ♅
05 09 ☽ # ♂
05 27 ☽ □ ♄
13 51 ☽ ⊥ ♀
16 02 ☽ ☍ ♂
16 25 ☽ ⊥ ♇
18 54 ☽ □ ♆
23 22 ☉ ∥ ♆

04 Sunday
01 24 ☽ ⚹ ☉
02 23 ☿ ⊻ ♇
03 19 ☽ ⊻ ♀
04 31 ☽ ⊻ ♇
04 49 ☽ ⚹ ☿
05 08 ♂ ± ♇
06 17 ☽ ⚼ ♃
13 23 ☽ Q ♅
14 13 ☿ ∥ ♅
15 05 ♀ ☌ ♇
17 57 ☽ Q ♄
19 02 ♀ ± ♂

05 Monday
00 00 ☽ ± ♃
06 51 ☽ Q ♆
09 54 ☽ ∠ ☉
10 41 ☽ ∥ ♀
11 40 ☽ ⚻ ♃
13 11 ☽ ⚹ ♅
14 54 ☽ ∠ ☿
15 19 ☉ ⊻ ♇
17 46 ☽ ⚹ ♄
20 09 ♀ ∠ ♅

06 Tuesday
04 57 ☽ ⊥ ☉
05 13 ☽ ⚻ ♂
06 15 ☽ ⚹ ♆
08 33 ☽ # ♃
10 49 ☽ ⊥ ☿
15 25 ☽ ☌ ♇
17 05 ☽ ± ♂
17 27 ☽ ⊻ ☉
18 14 ☽ ∠ ♅
20 32 ☽ ☌ ♀
22 49 ☽ ∠ ♄
23 52 ☽ ⊻ ☿

07 Wednesday
02 49 ☽ ∥ ♆
05 22 ☽ ∥ ☉
05 34 ☽ ∥ ♄
10 37 ☽ ⚼ ♂
10 47 ☽ ∠ ♆
11 17 ☽ ⊥ ♅
12 08 ☽ ∥ ♅
12 18 ☉ ∥ ♄
15 52 ☽ ⊥ ♄
16 46 ♂ □ ♆
17 38 ☽ ∥ ☿
20 15 ☽ ☍ ♃
21 59 ♀ ∠ ♄
22 30 ☽ ⊻ ♅

08 Thursday
00 19 ☿ ⊥ ♇
03 04 ☽ ⊻ ♄
03 32 ☽ ⊥ ♆
14 33 ☽ ⊻ ♆
15 10 ☽ △ ♂
23 14 ☽ ⊻ ♇

09 Friday
05 36 ☽ ☌ ☉
09 37 ☽ ⊻ ♀
10 03 ☽ ⊥ ♇
14 26 ☽ ☌ ☿
21 25 ☽ ⊥ ♀

10 Saturday
02 02 ☽ ⚻ ♃
02 08 ☽ ∠ ♇
04 54 ☽ ☌ ♅
09 30 ☽ ☌ ♄
12 31 ☽ ± ♃
14 55 ☽ ∠ ♀
20 11 ☽ ☌ ♆
22 18 ☽ □ ♂

11 Sunday
04 09 ☽ ⚼ ♃
04 33 ☽ ⚹ ♇
14 46 ☽ ⊻ ☉
18 07 ☉ ⊥ ♇
19 29 ☽ ∥ ☿
19 40 ☽ ⚹ ♀

12 Monday
01 25 ☽ Q ♇
01 51 ☽ ⊻ ☿
01 59 ☽ ⊥ ☉
05 29 ☽ ∥ ♅
05 57 ☽ △ ♃
06 32 ☽ Q ♂
08 54 ☽ ∥ ☉
09 25 ☽ ⊻ ♅
11 33 ☽ ∥ ♄
13 31 ☽ ⊥ ☿
14 05 ☽ ∥ ♆
14 07 ☽ ⊻ ♄
18 25 ☽ Q ♀
18 43 ☽ ∠ ☉
19 47 ☽ ⊥ ♅

13 Tuesday
00 17 ☽ ⊻ ♆
00 32 ☽ ⊥ ♄
02 27 ♀ ∠ ♆
03 51 ☽ ⚹ ♂
06 57 ☽ ∠ ☿
07 13 ☽ # ♃
08 29 ☽ □ ♇
10 35 ♀ ⊥ ♅
10 36 ☽ ⊥ ♆
11 18 ☽ ∠ ♅
11 50 ☿ ⚻ ♃
13 19 ☽ ∥ ♀
16 06 ☽ ∠ ♄
20 47 ☿ ∠ ♇
22 28 ☽ ⚹ ☉

14 Wednesday
02 03 ☽ ∠ ♆
04 17 ☽ □ ♀
06 22 ☽ ∠ ♂
09 01 ☽ □ ♃
11 53 ☽ ⚹ ☿
13 05 ☽ ⚹ ♅
17 58 ☽ ⚹ ♄
20 35 ☽ Q ☉
22 14 ☽ ⊥ ♂
22 57 ☿ ☌ ♅

15 Thursday
03 45 ☽ ⚹ ♆
08 48 ☽ ⊻ ♂
09 39 ☽ Q ♅
10 57 ☽ Q ☿
11 53 ☽ △ ♇
14 37 ☽ Q ♄
14 43 ☽ # ♂
15 20 ♀ ⚼ ♂
15 30 ☽ Q ♃
21 28 ♀ ⊥ ♄

16 Friday
00 16 ☽ Q ♆
04 20 ☽ ∥ ♇
05 37 ♀ ☍ ♃
05 40 ☽ □ ☉
11 47 ☽ ⚹ ♃
12 27 ☽ △ ♀
13 12 ☽ # ♇
13 31 ☽ ⚼ ♇
16 28 ☽ □ ♅
21 30 ☽ □ ☿
21 32 ☽ □ ♄
21 52 ☿ ☌ ♄

17 Saturday
01 08 ☿ ± ♃
02 09 ♂ ∠ ♃
03 58 ☽ ∥ ♂
04 53 ☽ ± ♇
07 01 ☽ □ ♆
12 59 ♀ # ♃
13 09 ☽ ∠ ♃
13 36 ☽ ☌ ♂
15 08 ☽ ⚻ ♇
16 29 ☽ ⚼ ♀
22 34 ♃ ± ♄

18 Sunday
04 23 ☽ ⊥ ♃
09 18 ☽ ± ♀
12 02 ♀ ⊻ ♅
12 46 ☽ △ ☉
14 32 ☽ ⊻ ♃
19 49 ☽ △ ♅
20 32 ☽ ⚻ ♀

19 Monday
01 07 ☽ △ ♄
07 08 ☽ △ ☿
09 49 ☽ ∥ ♃
10 20 ☽ △ ♆
11 02 ☉ ⚻ ♃
12 42 ☽ # ♀
15 54 ♂ ⚻ ♇
16 26 ☽ ⚼ ☉
18 29 ☽ ☍ ♇
18 34 ☽ ⊻ ♂
21 36 ☽ ⚼ ♅

20 Tuesday
03 03 ☽ ⚼ ♄
03 30 ☽ # ♆
05 13 ☽ ⊥ ♂
06 03 ☽ # ♄
08 16 ☿ ⊥ ♀
09 07 ☽ ± ☉
10 50 ☽ # ☉
12 03 ☿ ☌ ♆
12 11 ☽ ⚼ ♆
12 12 ☽ ⚼ ☿
12 13 ☽ # ♅
13 06 ♀ ⊥ ♆
13 10 ☽ ± ♅
17 38 ☽ ☌ ♃
18 43 ☽ ± ♄
20 18 ☽ ⚻ ☉
21 18 ☽ ∠ ♂
21 28 ☉ ∠ ♇
23 07 ☽ # ☿
23 36 ☽ ⚻ ♅

21 Wednesday
03 49 ☽ ± ♆
05 02 ♀ ⊻ ♄
05 13 ☽ ⚻ ♄
05 14 ☽ ☍ ♀
05 48 ☽ ± ☿
14 18 ☽ ⚻ ♆
17 39 ☽ ⚻ ☿
22 37 ☽ ⚻ ♇

22 Thursday
00 26 ☽ ⚹ ♂
09 14 ☽ ± ♇
09 39 ☿ ⚼ ♃
19 47 ☉ ☌ ♅
21 59 ☽ ⊻ ♃
22 31 ☽ Q ♂

23 Friday
01 20 ☽ ⚼ ♇
04 47 ☽ ☍ ♅
05 29 ☽ ☍ ☉
08 43 ☽ ⊥ ♃
10 51 ☽ ☍ ♄
13 26 ☿ ⚹ ♇
15 56 ☽ ⚻ ♀
20 00 ☽ ☍ ♆

24 Saturday
01 00 ☽ ∠ ♃
04 04 ☽ ± ♀
04 40 ☽ △ ♇
06 44 ☽ ☍ ☿
08 30 ☽ □ ♂
20 32 ☉ ± ♃
22 35 ☽ ⚼ ♀
23 23 ☽ # ☿

25 Sunday
03 06 ☿ □ ♂
04 44 ☽ ⚹ ♃
06 20 ☽ # ♅
08 09 ☽ # ☉
12 04 ♀ ⊻ ♆
12 33 ☽ ⚻ ♅
13 09 ☽ # ♄
15 57 ☽ # ♆
17 55 ☽ ⚻ ☉
19 13 ☽ ⚻ ♄
23 25 ☽ # ♀

26 Monday
00 05 ☽ ± ♅
03 30 ☽ Q ♃
04 32 ☽ ⚻ ♆
06 20 ☽ △ ♀
06 30 ☽ ± ☉
06 54 ☽ ± ♄
11 57 ☉ ☌ ♄
12 16 ☽ ∥ ♃
13 40 ☽ □ ♇
16 14 ☽ ± ♆
17 38 ☽ ⚼ ♅
19 57 ☽ △ ♂

27 Tuesday
00 05 ☽ ⚻ ☿
00 38 ☽ ⚼ ♄
01 40 ☽ ⚼ ☉
10 00 ☽ ⚼ ♆
13 41 ☽ ± ☿
14 34 ☽ □ ♃
23 29 ☽ △ ♅

28 Wednesday
01 21 ☽ Q ♇
02 58 ☽ ⚼ ♂
06 47 ☽ △ ♄
10 19 ☽ △ ☉
10 23 ☽ ⚼ ☿
15 51 ☽ ∥ ♂
16 08 ☽ △ ♆
18 31 ☿ ∥ ♅
21 54 ☽ ± ♂

29 Thursday
00 49 ☽ □ ♀
01 36 ☽ ⚹ ♇
08 28 ♀ ⊻ ♇
10 36 ☽ ⚻ ♂
19 41 ☽ # ♇
21 18 ☽ △ ☿

30 Friday
02 45 ☽ △ ♃
06 35 ☽ ∥ ♇
08 10 ☽ ∠ ♇
08 59 ☉ ∥ ☿
12 33 ☽ □ ♅
17 44 ☽ Q ♀
20 15 ☽ □ ♄

31 Saturday
02 30 ☽ ⊥ ♇
04 57 ☽ □ ☉
05 15 ☽ □ ♆
08 39 ☉ ☌ ♆
09 03 ☽ ⚼ ♃
10 56 ☿ Q ♇
13 11 ☽ # ♂
14 20 ☿ △ ♃
14 40 ☽ ⊻ ♇
20 53 ☽ ⚹ ♀
21 30 ☿ ∥ ♄

JANUARY 1989

LONGITUDES

Date	Sidereal time h m s	Sun ☉ ° ' "	Moon ☽ ° ' "	Moon ☽ 24.00 ° ' "	Mercury ☿ ° '	Venus ♀ ° '	Mars ♂ ° '	Jupiter ♃ ° '	Saturn ♄ ° '	Uranus ♅ ° '	Neptune ♆ ° '	Pluto ♇ ° '
01	18 44 28	11 ♑ 03 33	25 ♎ 10 57	01 ♏ 14 02	28 ♑ 05	18 ♐ 25	20 ♈ 27	26 ♉ 42	05 ♑ 39	01 ♑ 47	09 ♑ 56	14 ♏ 34
02	18 48 24	12 04 43	07 ♏ 20 52	13 ♏ 32 01	29 ♑ 33	19 40	20 58	26 R 38	05 46	01 51	09 59	14 36
03	18 52 21	13 05 54	19 ♏ 47 56	26 ♏ 08 59	00 ♒ 58	20 55	21 29	26 34	05 53	01 54	10 01	14 37
04	18 56 17	14 07 04	02 ♐ 35 30	09 ♐ 07 38	02 20	22 10	22 00	26 31	06 00	01 58	10 03	14 39
05	19 00 14	15 08 15	15 ♐ 45 27	22 ♐ 28 52	03 40	23 25	22 31	26 28	06 07	02 01	10 06	14 40
06	19 04 10	16 09 25	29 ♐ 17 43	06 ♑ 11 38	04 57	24 40	23 03	26 25	06 14	02 05	10 08	14 42
07	19 08 07	17 10 36	13 ♑ 10 13	20 ♑ 12 55	06 09	25 56	23 34	26 22	06 21	02 08	10 10	14 43
08	19 12 04	18 11 47	27 ♑ 19 05	04 ♒ 28 04	07 17	27 11	24 06	26 20	06 28	02 12	10 12	14 44
09	19 16 00	19 12 57	11 ♒ 39 09	18 ♒ 51 37	08 19	28 26	24 38	26 17	06 35	02 15	10 15	14 46
10	19 19 57	20 14 07	26 ♒ 04 47	03 ♓ 18 02	09 15	29 ♐ 41	25 10	26 15	06 42	02 19	10 17	14 47
11	19 23 53	21 15 16	10 ♓ 30 48	17 ♓ 42 33	10 04	00 ♑ 56	25 42	26 13	06 49	02 22	10 19	14 48
12	19 27 50	22 16 25	24 ♓ 52 55	02 ♈ 01 31	10 45	02 11	26 15	26 12	06 56	02 26	10 21	14 50
13	19 31 46	23 17 33	09 ♈ 08 07	16 ♈ 12 31	11 17	03 26	26 47	26 10	07 03	02 29	10 24	14 51
14	19 35 43	24 18 40	23 ♈ 14 33	00 ♉ 14 07	11 40	04 41	27 20	26 09	07 10	02 33	10 26	14 52
15	19 39 39	25 19 46	07 ♉ 11 08	14 ♉ 05 32	11 52	05 56	27 53	26 08	07 17	02 36	10 28	14 53
16	19 43 36	26 20 52	20 ♉ 57 15	27 ♉ 46 14	11 R 53	07 11	28 26	26 07	07 24	02 40	10 30	14 54
17	19 47 33	27 21 57	04 ♊ 32 24	11 ♊ 15 40	11 42	08 27	28 59	26 06	07 31	02 43	10 32	14 55
18	19 51 29	28 23 02	17 ♊ 55 57	24 ♊ 33 08	11 19	09 42	29 ♈ 32	26 06	07 38	02 46	10 35	14 56
19	19 55 26	29 ♑ 24 05	01 ♋ 07 09	07 ♋ 37 53	10 45	10 57	00 ♉ 05	26 06	07 44	02 50	10 37	14 57
20	19 59 22	00 ♒ 25 08	14 ♋ 05 16	20 ♋ 29 15	10 01	12 12	00 39	26 D 05	07 51	02 53	10 39	14 58
21	20 03 19	01 26 10	26 ♋ 49 48	03 ♌ 06 55	09 06	13 27	01 12	26 06	07 58	02 56	10 41	14 59
22	20 07 15	02 27 12	09 ♌ 20 40	15 ♌ 31 08	08 03	14 42	01 46	26 06	08 05	03 00	10 43	15 00
23	20 11 12	03 28 12	21 ♌ 38 28	27 ♌ 42 53	06 53	15 57	02 20	26 07	08 11	03 03	10 45	15 01
24	20 15 08	04 29 12	03 ♍ 44 37	09 ♍ 43 59	05 39	17 12	02 53	26 07	08 18	03 06	10 48	15 02
25	20 19 05	05 30 11	15 ♍ 41 21	21 ♍ 37 07	04 22	18 27	03 27	26 08	08 24	03 09	10 50	15 03
26	20 23 02	06 31 10	27 ♍ 31 44	03 ♎ 25 43	03 06	19 43	04 01	26 09	08 31	03 13	10 52	15 03
27	20 26 58	07 32 08	09 ♎ 19 35	15 ♎ 13 55	01 53	20 58	04 35	26 11	08 38	03 16	10 54	15 04
28	20 30 55	08 33 05	21 ♎ 09 17	27 ♎ 06 19	00 ♒ 43	22 13	05 10	26 12	08 44	03 19	10 56	15 05
29	20 34 51	09 34 01	03 ♏ 05 39	09 ♏ 07 53	29 ♑ 40	23 28	05 44	26 14	08 50	03 22	10 58	15 06
30	20 38 48	10 34 57	15 ♏ 13 39	21 ♏ 23 33	28 43	24 43	06 18	26 16	08 57	03 25	11 00	15 06
31	20 42 44	11 ♒ 35 53	27 ♏ 38 10	03 ♐ 57 59	27 ♑ 55	25 ♑ 58	06 ♉ 53	26 ♉ 18	09 ♑ 03	03 ♑ 28	11 ♑ 02	15 ♏ 07

Moon True ☊ / Moon Mean ☊ / Moon ☽ Latitude

Date	Moon True ☊ ° '	Moon Mean ☊ ° '	Moon ☽ Latitude ° '
01	06 ♓ 36	07 ♓ 46	03 S 51
02	06 R 32	07 42	04 28
03	06 27	07 39	04 54
04	06 19	07 36	05 06
05	06 10	07 33	05 02
06	06 01	07 30	04 42
07	05 53	07 27	04 05
08	05 46	07 23	03 12
09	05 41	07 20	02 06
10	05 39	07 17	00 S 52
11	05 D 38	07 14	00 N 26
12	05 39	07 11	01 42
13	05 41	07 07	02 52
14	05 41	07 04	03 50
15	05 R 41	07 01	04 33
16	05 38	06 58	05 00
17	05 34	06 55	05 10
18	05 28	06 52	05 03
19	05 22	06 48	04 40
20	05 15	06 45	04 03
21	05 10	06 42	03 14
22	05 06	06 39	02 16
23	05 03	06 36	01 13
24	05 02	06 33	00 N 07
25	05 D 03	06 29	00 S 58
26	05 04	06 26	02 00
27	05 06	06 23	02 57
28	05 08	06 20	03 47
29	05 09	06 17	04 27
30	05 R 09	06 13	04 56
31	05 ♓ 08	06 ♓ 10	05 S 12

DECLINATIONS

Date	Sun ☉ ° '	Moon ☽ ° '	Mercury ☿ ° '	Venus ♀ ° '	Mars ♂ ° '	Jupiter ♃ ° '	Saturn ♄ ° '	Uranus ♅ ° '	Neptune ♆ ° '	Pluto ♇ ° '
01	22 S 59	13 S 19	22 S 23	22 S 09	08 N 30	18 N 33	22 S 36	23 S 39	22 S 10	01 S 16
02	22 54	18 11	21 59	22 17	08 43	18 32	22 36	23 39	22 10	01 16
03	22 48	22 24	21 35	22 25	08 56	18 32	22 36	23 39	22 10	01 16
04	22 42	25 41	21 09	22 32	09 09	18 31	22 36	23 39	22 10	01 16
05	22 35	27 41	20 43	22 39	09 21	18 31	22 35	23 39	22 10	01 16
06	22 28	28 08	20 16	22 45	09 34	18 30	22 35	23 39	22 09	01 16
07	22 20	26 51	19 48	22 50	09 47	18 30	22 35	23 39	22 09	01 16
08	22 12	23 50	19 20	22 54	10 00	18 29	22 35	23 39	22 09	01 16
09	22 04	19 19	18 53	22 58	10 13	18 29	22 34	23 39	22 09	01 16
10	21 55	13 38	18 25	23 01	10 25	18 29	22 34	23 39	22 09	01 16
11	21 46	07 13	17 58	23 04	10 38	18 29	22 34	23 39	22 09	01 16
12	21 36	00 S 28	17 33	23 06	10 51	18 29	22 33	23 39	22 08	01 16
13	21 26	06 N 15	17 08	23 07	11 04	18 29	22 33	23 38	22 08	01 16
14	21 15	12 35	16 46	23 07	11 16	18 29	22 33	23 38	22 08	01 16
15	21 04	18 13	16 26	23 07	11 29	18 29	22 32	23 38	22 08	01 16
16	20 53	22 49	16 08	23 05	11 42	18 29	22 32	23 38	22 08	01 16
17	20 41	26 08	15 53	23 04	11 54	18 29	22 32	23 38	22 08	01 15
18	20 29	27 56	15 41	23 01	12 07	18 29	22 31	23 38	22 07	01 15
19	20 17	28 06	15 33	22 58	12 19	18 29	22 31	23 38	22 07	01 15
20	20 04	26 43	15 28	22 54	12 32	18 29	22 31	23 38	22 07	01 15
21	19 51	23 58	15 27	22 49	12 44	18 30	22 30	23 38	22 07	01 15
22	19 37	20 06	15 29	22 44	12 57	18 30	22 30	23 38	22 07	01 14
23	19 23	15 26	15 34	22 38	13 09	18 30	22 30	23 38	22 06	01 14
24	19 09	10 15	15 41	22 31	13 22	18 31	22 29	23 38	22 06	01 14
25	18 54	04 N 45	15 51	22 24	13 34	18 31	22 29	23 38	22 06	01 14
26	18 39	00 S 52	16 03	22 16	13 46	18 32	22 28	23 38	22 06	01 13
27	18 23	06 25	16 16	22 07	13 59	18 32	22 28	23 38	22 06	01 13
28	18 08	11 46	16 30	21 57	14 11	18 33	22 28	23 38	22 06	01 13
29	17 52	16 44	16 44	21 47	14 23	18 34	22 27	23 38	22 05	01 13
30	17 35	21 07	16 59	21 36	14 35	18 34	22 27	23 37	22 05	01 12
31	17 S 18	24 S 42	17 S 13	21 S 25	14 N 47	18 N 35	22 S 27	23 S 37	22 S 05	01 S 12

ZODIAC SIGN ENTRIES

Date	h	m	Planets
01	21	34	☽ ♏
02	19	41	☿ ♒
04	07	12	☽ ♐
06	13	14	☽ ♑
08	16	31	☽ ♒
10	18	08	♀ ♑
10	18	31	☽ ♓
12	20	36	☽ ♈
14	23	36	☽ ♉
17	03	57	☽ ♊
19	08	11	♂ ♉
19	09	57	☽ ♋
20	02	07	☉ ♒
21	18	02	☽ ♌
24	04	32	☽ ♍
26	17	01	☽ ♎
29	04	06	☿ ♑
29	05	49	☽ ♏
31	16	30	☽ ♐

LATITUDES

Date	Mercury ☿ ° '	Venus ♀ ° '	Mars ♂ ° '	Jupiter ♃ ° '	Saturn ♄ ° '	Uranus ♅ ° '	Neptune ♆ ° '	Pluto ♇ ° '
01	01 S 52	00 N 48	00 N 33	00 S 54	00 N 43	00 S 13	00 N 54	15 N 36
04	01 33	00 40	00 37	00 53	00 43	00 13	00 54	15 37
07	01 06	00 33	00 41	00 52	00 43	00 13	00 54	15 38
10	00 S 30	00 25	00 44	00 52	00 43	00 13	00 54	15 40
13	00 N 16	00 17	00 47	00 51	00 42	00 13	00 54	15 41
16	01 09	00 09	00 50	00 50	00 42	00 13	00 54	15 42
19	02 04	00 N 02	00 52	00 49	00 42	00 13	00 54	15 44
22	02 52	00 S 06	00 55	00 48	00 42	00 13	00 54	15 45
25	03 25	00 14	00 57	00 48	00 42	00 13	00 54	15 47
28	03 35	00 21	01 00	00 47	00 42	00 13	00 54	15 48
31	03 N 26	00 S 28	01 N 02	00 S 46	00 N 42	00 S 14	00 N 54	15 N 50

DATA

Julian Date	2447528
Delta T	+56 seconds
Ayanamsa	23° 42' 19"
Synetic vernal point	05° ♓ 24' 40"
True obliquity of ecliptic	23° 26' 35"

MOON'S PHASES, APSIDES AND POSITIONS ☽

Date	h	m	Phase	Longitude ° '	Eclipse Indicator
07	19	22	●	17 ♑ 29	
14	13	58	☽	24 ♈ 24	
21	21	34	○	01 ♌ 50	
30	02	02	☾	10 ♏ 10	

Day	h	m		
10	22	40	Perigee	
26	23	55	Apogee	
06	06	24	Max dec	28° S 11'
12	13	39	0N	
19	02	36	Max dec	28° N 14'
26	08	20	0S	

LONGITUDES

Date	Chiron ⚷ ° '	Ceres ⚳ ° '	Pallas ⚴ ° '	Juno ⚵ ° '	Vesta ⚶ ° '	Black Moon Lilith ⚸ ° '
01	03 ♋ 50	25 ♓ 37	19 ♒ 24	10 ♍ 34	26 ♏ 16	25 ♍ 47
11	03 ♋ 12	28 ♓ 24	22 ♒ 28	10 ♍ 18	01 ♐ 13	26 ♍ 53
21	02 ♋ 37	01 ♈ 27	25 ♒ 38	09 ♍ 14	06 ♐ 03	28 ♍ 00
31	02 ♋ 06	04 ♈ 43	28 ♒ 52	07 ♍ 27	10 ♐ 46	29 ♍ 06

All ephemeris data is given at 12.00 UT and the Moon's longitude is additionally given for 24.00 UT

ASPECTARIAN

h m	Aspects
01 Sunday	
01 09	☽ Q ♅
02 08	☽ ☍ ♂
03 05	☽ ± ♃
08 55	☽ Q ♄
15 00	☽ ⊼ ♃
16 30	♀ ∥ ♆
17 30	☽ Q ♆
18 34	☽ □ ☿
20 24	☽ Q ☉
21 55	♂ ⊥ ♃
22 49	☿ ∥ ♀
02 Monday	
01 09	☽ ✶ ♅
01 13	☿ ∥ ♆
02 06	☉ bQ ♃
06 10	☽ ∠ ♀
08 53	☽ ✶ ♄
13 51	☽ # ♃
17 08	☽ ✶ ♆
22 01	☽ ✶ ☉
03 Tuesday	
01 38	☽ ⊥ ♀
02 04	☽ ☌ ♇
06 07	♀ ⊥ ♇
06 27	☽ ∠ ♅
07 25	☽ ∥ ☿
10 12	☽ Q ☿
10 31	☽ ∥ ♆
12 06	☽ ∥ ♀
13 15	☽ ∥ ♄
14 05	☽ ∠ ♄
14 22	☽ ⊻ ♀
14 28	☽ ∥ ☉
15 20	☽ ⊼ ♂
20 14	☽ ∥ ♅
21 54	☽ ∠ ♆
23 35	☽ ⊥ ♅
04 Wednesday	
00 44	☽ ☍ ♃
03 07	☽ ± ♂
05 00	☽ ∠ ☉
05 04	☿ ⊻ ♅
06 26	♀ △ ♂
07 10	☽ ⊥ ♄
10 50	☽ ⊻ ♅
11 29	☽ ✶ ☿
14 43	☽ ⊥ ♆
18 21	☽ ⊻ ♄
20 27	☽ bQ ♂
23 01	☽ ⊥ ☉
23 22	♀ ∥ ♄
05 Thursday	
00 43	☉ ✶ ♇
01 44	☽ ⊻ ♆
05 00	☉ ∥ ♀
10 02	☽ ⊻ ♇
10 48	☽ ⊻ ☉
10 50	☉ ∥ ♄
17 47	☽ ∠ ☿
20 48	☽ ⊥ ♇
06 Friday	
00 34	☽ △ ♂
03 03	☽ ☌ ♀
06 58	☽ ⊼ ♃
11 19	☽ ⊥ ☿
12 42	☽ ∠ ♇
16 53	☽ ☌ ♅
17 26	☽ ± ♃
22 47	☽ ⊻ ☿
07 Saturday	
00 11	☽ ☌ ♄
06 50	☽ ☌ ♆
08 56	☽ bQ ♃
14 39	☽ ✶ ♇
16 46	☿ ⊻ ♄
19 22	☽ ☌ ☉
20 16	♀ ⊼ ♃
08 Sunday	
06 22	☽ □ ♂
10 20	☽ △ ♃
11 01	☽ Q ♇
11 44	☽ ⊻ ♀
13 10	☽ ∥ ♅
17 28	☽ ∥ ♀
19 27	☽ ∥ ♄
20 14	☽ ⊻ ♅
21 48	☽ ∥ ☉
21 48	☽ ∥ ♆
21 48	☉ ∥ ♆
22 47	☽ ⊥ ♀
09 Monday	
03 29	☽ ⊻ ♄
06 01	☽ ☌ ☿
06 19	☽ ⊥ ♅
09 39	☽ ⊻ ♆
10 28	☿ ⊥ ♅
13 35	☽ ⊥ ♄
13 42	☽ Q ♂
14 11	☽ ∥ ☿
15 15	☽ ∠ ♀
15 46	☽ # ♃
17 11	☽ □ ♇
19 40	☽ ⊥ ♆
20 25	♃ ± ♅
21 22	☽ ∠ ♅
10 Tuesday	
01 33	☽ ⊻ ☉
04 41	☽ ∠ ♄
08 36	☿ # ♃
10 26	☽ ✶ ♂
10 40	☽ ∠ ♆
12 17	☽ ⊥ ☉
12 17	☽ □ ♃
14 01	♀ ∠ ♇
18 33	☽ ✶ ♀
22 24	☽ ✶ ♅
23 53	☽ # ♂
11 Wednesday	
04 22	☽ ∠ ☉
05 49	☽ ✶ ♄
11 13	☽ ⊻ ☿
11 40	☽ ✶ ♆
12 20	☽ ∠ ♂
16 25	☽ Q ♀
18 10	☽ Q ♃
18 28	☽ Q ♅
19 10	☽ △ ♇
20 37	☿ ⊻ ♆
21 45	☽ ⊥ ☿
12 Thursday	
01 58	☽ Q ♄
03 56	☽ ⊥ ♂
07 18	☽ ✶ ☉
07 45	☽ Q ♆
09 10	☽ ∥ ♇
09 48	♂ ⊻ ♃
12 13	♀ ± ♃
13 32	☽ ∠ ☿
14 12	☽ ✶ ♃
14 23	☽ ⊻ ♂
16 59	♀ ☌ ♅
18 08	☽ # ♇
20 19	☽ bQ ♇
13 Friday	
00 44	☽ □ ♅
01 27	☽ □ ♀
05 00	☽ Q ☉
08 27	☽ □ ♄
11 31	☽ ± ♇
14 08	☽ □ ♆
15 26	☽ ∠ ♃
15 46	☽ ✶ ☿
21 42	☽ ⊼ ♇
14 Saturday	
06 40	☽ ∥ ♂
06 43	☽ ⊥ ♃
12 44	☽ Q ☿
13 58	☽ □ ☉
16 58	☽ ⊻ ♃
19 18	☽ ☌ ♂
15 Sunday	
04 03	☽ △ ♅
04 25	☽ # ☿
09 38	☽ △ ♀
10 26	☿ ⊥ ♀
12 10	☽ △ ♄
13 14	☽ ∥ ♃
17 42	☽ △ ♆
20 10	☽ □ ☿
16 Monday	
01 24	☽ ☍ ♇
01 36	☽ # ☉
01 45	☿ St R
06 12	☽ bQ ♅
06 35	☉ △ ♃
07 57	☽ # ♆
10 18	☽ # ♄
13 37	☽ # ♀
14 23	☽ bQ ♀
14 33	☽ bQ ♄
16 23	♀ ☌ ♄
17 04	☽ # ♅
20 01	☽ bQ ♆
21 04	☽ ☌ ♃
22 05	☽ ± ♅
22 15	☽ △ ☉
17 Tuesday	
01 20	☉ Q ♇
01 43	☽ ⊻ ♂
06 35	☽ ± ♄
07 54	☽ ± ♀
08 45	☽ ⊼ ♅
12 00	☽ ± ♆
12 49	☽ ⊥ ♂
17 21	☽ ⊼ ♄
19 40	☽ ⊼ ♀
22 44	☽ ⊼ ♆
18 Wednesday	
00 29	☽ △ ☿
03 08	☽ bQ ☉
05 37	☽ ∠ ♂
06 36	☽ ⊼ ♇
17 27	☽ ± ♇
20 44	☽ ± ☉
19 Thursday	
02 38	☽ bQ ☿
02 48	☽ ⊻ ♃
05 25	♀ ☌ ♆
08 35	☽ ⊼ ☉
09 37	☿ ⊻ ♀
09 52	☽ bQ ♇
10 02	☽ ✶ ♂
13 47	☽ ⊥ ♃
14 47	♀ bQ ♃
15 09	☽ ☍ ♅
16 50	☿ ⊻ ♆
18 21	☽ ± ☿
20 Friday	
00 18	☽ ☍ ♄
04 51	☽ ⊼ ☿
05 35	☽ ☍ ♆
06 10	♃ St D
06 25	☽ ∠ ♃
08 06	☽ ☍ ♀
09 11	☽ Q ♂
13 39	☽ △ ♇
23 49	☉ □ ♂
21 Saturday	
10 36	☽ ✶ ♃
14 18	☽ # ♅
15 33	☿ ⊥ ♅
19 55	☽ # ♀
20 44	☽ □ ♂
21 34	☽ ☍ ☉
21 47	☽ # ♄
23 43	☽ ⊼ ♅
22 Sunday	
00 14	☽ # ♆
09 31	☽ ⊼ ♄
09 36	☽ Q ♃
09 41	☽ ☍ ☿
11 19	☽ ± ♅
11 20	☿ ⊻ ♄
14 40	☽ ⊼ ♆
14 50	☽ # ☉
17 51	♀ ✶ ♇
20 39	☽ ∥ ♃
21 16	☽ ± ♄
23 00	☽ □ ♇
23 35	☽ ⊼ ♀
23 Monday	
01 29	☉ ⊻ ♅
02 23	☽ ± ♆
04 55	☽ bQ ♅
11 26	☽ # ☿
12 41	☽ ± ♀
15 04	☽ bQ ♄
20 09	☽ bQ ♆
20 50	☽ □ ♃
22 23	☽ ∥ ♂
24 Tuesday	
08 34	☽ bQ ♀
10 13	☽ △ ♂
10 35	☽ Q ♇
10 43	☽ △ ♅
13 37	☽ ⊼ ☉
15 26	☽ ⊼ ☿
18 43	♀ ∥ ♄
21 12	☽ △ ♄
21 58	♂ △ ♅
25 Wednesday	
00 10	☉ ☌ ☿
02 10	☽ △ ♆
02 19	☽ ± ☿
02 46	☽ ± ☉
10 42	☽ ✶ ♇
17 52	☽ bQ ♂
18 15	☽ △ ♀
18 44	☽ bQ ☿
22 39	☽ bQ ☉
26 Thursday	
00 00	☿ □ ♂
03 05	☽ # ♇
09 12	☽ △ ♃
10 08	☿ ⊻ ♅
13 03	☽ ± ♂
13 34	☽ ∥ ♇
17 09	☽ ∠ ♇
22 16	☽ △ ☿
22 38	☉ # ♃
23 36	☽ □ ♅
27 Friday	
01 53	☽ ⊼ ♂
08 01	☽ △ ☉
10 34	☽ □ ♄
11 29	☽ ⊥ ♇
14 46	♀ ∥ ♆
15 12	☽ □ ♆
15 47	☽ bQ ♃
23 41	☽ ⊻ ♇
28 Saturday	
10 05	☽ ± ♃
12 19	☽ Q ♅
14 24	☽ □ ♀
16 49	☉ ⊻ ♄
22 13	☽ ⊼ ♃
23 21	☽ Q ♄
23 53	☽ # ♂
29 Sunday	
03 43	☽ Q ♆
05 40	☽ □ ☿
06 58	☉ ⊥ ♅
12 02	☽ ∥ ☿
12 33	☽ ✶ ♅
17 31	☽ ∥ ☉
17 31	☽ ☍ ♂
21 35	☽ # ♃
23 32	☽ ✶ ♄
30 Monday	
02 02	☽ □ ☉
03 40	☽ ✶ ♆
06 31	☽ Q ♀
11 45	☽ ☌ ♇
14 44	☽ Q ☿
14 48	☽ ∥ ♀
17 53	☽ ∥ ♆
18 15	☽ ∠ ♅
20 10	☽ ∥ ♄
22 13	☉ ⊻ ♆
31 Tuesday	
04 03	☽ ∥ ♅
05 05	☽ ∠ ♄
08 28	☽ ✶ ♀
08 56	☽ ∠ ♆
09 27	☽ ☍ ♃
11 41	☽ ⊥ ♅
12 30	☽ ✶ ☿
15 51	☉ ∥ ☿
16 04	☽ Q ☉
18 38	♀ △ ♃
22 22	☽ ⊥ ♄
23 06	☽ ⊻ ♅

FEBRUARY 1989

LONGITUDES

Date	Sidereal time h m s	Sun ☉ ° ' "	Moon ☽ ° ' "	Moon ☽ 24.00 ° ' "	Mercury ☿ ° '	Venus ♀ ° '	Mars ♂ ° '	Jupiter ♃ ° '	Saturn ♄ ° '	Uranus ♅ ° '	Neptune ♆ ° '	Pluto ♇ ° '
01	20 46 41	12 ♒ 36 47	10 ♐ 23 28	16 ♐ 55 01	27 ♑ 15	27 ♑ 13	07 ♉ 27	26 ♉ 21	09 ♑ 10	03 ♑ 31	11 ♑ 04	15 ♏ 07
02	20 50 37	13 37 41	23 ♐ 32 52	00 ♑ 17 11	26 R 43	28 28	08 02	26 23	09 16	03 34	11 06	15 08
03	20 54 34	14 38 34	07 ♑ 08 00	14 ♑ 05 12	26 20	29 ♑ 44	08 37	26 26	09 22	03 37	11 08	15 08
04	20 58 31	15 39 26	21 ♑ 08 29	28 ♑ 17 26	26 06	00 ♒ 59	09 12	26 29	09 28	03 40	11 10	15 09
05	21 02 27	16 40 16	05 ♒ 31 28	12 ♒ 49 52	25 59	02 14	09 46	26 32	09 35	03 43	11 12	15 09
06	21 06 24	17 41 06	20 ♒ 11 48	27 ♒ 36 21	26 D 01	03 29	10 21	26 36	09 41	03 46	11 14	15 10
07	21 10 20	18 41 54	05 ♓ 02 32	12 ♓ 29 22	26 09	04 44	10 56	26 39	09 47	03 49	11 16	15 10
08	21 14 17	19 42 41	19 ♓ 55 52	27 ♓ 21 07	26 24	05 59	11 32	26 43	09 53	03 51	11 18	15 10
09	21 18 13	20 43 27	04 ♈ 44 17	12 ♈ 04 35	26 45	07 14	12 07	26 47	09 59	03 54	11 20	15 11
10	21 22 10	21 44 11	19 ♈ 21 25	26 ♈ 34 16	27 12	08 29	12 42	26 51	10 05	03 57	11 21	15 11
11	21 26 06	22 44 53	03 ♉ 42 45	10 ♉ 46 36	27 44	09 44	13 17	26 55	10 10	04 00	11 23	15 11
12	21 30 03	23 45 34	17 ♉ 45 41	24 ♉ 39 54	28 21	10 59	13 53	27 00	10 16	04 02	11 25	15 11
13	21 34 00	24 46 13	01 ♊ 29 19	08 ♊ 13 59	29 02	12 15	14 28	27 04	10 22	04 05	11 27	15 11
14	21 37 56	25 46 51	14 ♊ 54 04	21 ♊ 29 44	29 ♑ 48	13 30	15 04	27 09	10 28	04 07	11 29	15 11
15	21 41 53	26 47 26	28 ♊ 01 11	04 ♋ 28 39	00 ♒ 37	14 45	15 39	27 14	10 33	04 10	11 30	15 11
16	21 45 49	27 48 01	10 ♋ 52 21	17 ♋ 12 30	01 29	16 00	16 15	27 19	10 39	04 12	11 32	15 R 11
17	21 49 46	28 48 33	23 ♋ 29 21	29 ♋ 43 06	02 25	17 15	16 50	27 25	10 44	04 15	11 34	15 11
18	21 53 42	29 ♒ 49 03	05 ♌ 53 58	12 ♌ 02 11	03 24	18 30	17 26	27 30	10 50	04 17	11 35	15 11
19	21 57 39	00 ♓ 49 32	18 ♌ 07 56	24 ♌ 11 25	04 25	19 45	18 02	27 36	10 55	04 20	11 37	15 11
20	22 01 35	01 49 59	00 ♍ 12 53	06 ♍ 12 31	05 29	21 00	18 38	27 42	11 01	04 22	11 39	15 11
21	22 05 32	02 50 25	12 ♍ 10 34	18 ♍ 07 15	06 35	22 15	19 13	27 48	11 06	04 24	11 40	15 11
22	22 09 29	03 50 49	24 ♍ 02 52	29 ♍ 57 42	07 43	23 30	19 49	27 54	11 11	04 27	11 42	15 11
23	22 13 25	04 51 11	05 ♎ 52 02	11 ♎ 46 13	08 54	24 45	20 25	28 01	11 16	04 29	11 43	15 10
24	22 17 22	05 51 32	17 ♎ 40 39	23 ♎ 35 42	10 06	26 00	21 01	28 07	11 21	04 31	11 45	15 10
25	22 21 18	06 51 52	29 ♎ 31 48	05 ♏ 29 25	11 20	27 15	21 37	28 14	11 26	04 33	11 46	15 10
26	22 25 15	07 52 09	11 ♏ 29 02	17 ♏ 31 09	12 36	28 30	22 13	28 21	11 31	04 35	11 48	15 10
27	22 29 11	08 52 26	23 ♏ 36 18	29 ♏ 45 01	13 53	29 ♒ 44	22 49	28 28	11 36	04 37	11 49	15 09
28	22 33 08	09 ♓ 52 41	05 ♐ 57 51	12 ♐ 15 18	15 ♒ 12	00 ♓ 59	23 ♉ 25	28 ♉ 35	11 ♑ 41	04 ♑ 39	11 ♑ 51	15 ♏ 09

Moon Nodes and Latitude; DECLINATIONS

Date	Moon True ☊ ° '	Moon Mean ☊ ° '	Moon ☽ Latitude ° '	Sun ☉ ° '	Moon ☽ ° '	Mercury ☿ ° '	Venus ♀ ° '	Mars ♂ ° '	Jupiter ♃ ° '	Saturn ♄ ° '	Uranus ♅ ° '	Neptune ♆ ° '	Pluto ♇ ° '
01	05 ♓ 06	06 ♓ 07	05 S 14	17 S 01	27 S 11	17 S 28	21 S 13	14 N 59	18 N 36	22 S 26	23 S 37	22 S 05	01 S 12
02	05 R 03	06 04	04 59	16 44	28 16	17 42	21 00	15 11	18 37	22 26	23 37	22 05	01 11
03	05 00	06 01	04 27	16 27	27 42	17 55	20 47	15 23	18 38	22 25	23 37	22 05	01 11
04	04 57	05 58	03 39	16 09	25 23	18 08	20 33	15 34	18 39	22 25	23 37	22 04	01 11
05	04 55	05 54	02 35	15 51	21 24	18 19	20 18	15 46	18 40	22 25	23 37	22 04	01 10
06	04 53	05 51	01 S 20	15 32	16 02	18 30	20 03	15 58	18 41	22 24	23 37	22 04	01 10
07	04 D 53	05 48	00 N 01	15 13	09 39	18 40	19 48	16 09	18 42	22 24	23 37	22 04	01 09
08	04 53	05 45	01 22	14 54	02 S 43	18 49	19 31	16 21	18 43	22 23	23 37	22 04	01 09
09	04 54	05 42	02 38	14 35	04 N 18	18 56	19 14	16 32	18 44	22 23	23 37	22 04	01 09
10	04 55	05 39	03 42	14 16	11 00	19 03	18 57	16 43	18 45	22 22	23 37	22 03	01 08
11	04 56	05 35	04 31	13 56	17 00	19 08	18 39	16 55	18 46	22 22	23 37	22 03	01 08
12	04 57	05 32	05 03	13 36	21 58	19 12	18 20	17 06	18 48	22 22	23 37	22 03	01 07
13	04 R 57	05 29	05 17	13 16	25 37	19 15	18 01	17 17	18 49	22 21	23 37	22 03	01 07
14	04 56	05 26	05 12	12 56	27 46	19 17	17 42	17 28	18 50	22 21	23 36	22 03	01 06
15	04 55	05 23	04 52	12 35	28 05	19 17	17 22	17 39	18 52	22 20	23 36	22 03	01 06
16	04 54	05 19	04 17	12 14	27 16	19 16	17 01	17 49	18 53	22 20	23 36	22 03	01 06
17	04 53	05 16	03 30	11 53	24 51	19 14	16 40	18 00	18 54	22 20	23 36	22 03	01 05
18	04 53	05 13	02 34	11 32	21 17	19 11	16 18	18 11	18 56	22 19	23 36	22 03	01 05
19	04 53	05 10	01 32	11 11	16 51	19 06	15 56	18 21	18 57	22 19	23 36	22 02	01 04
20	04 52	05 07	00 N 26	10 49	11 48	19 01	15 34	18 31	18 59	22 18	23 36	22 02	01 04
21	04 D 52	05 04	00 S 41	10 28	06 22	18 53	15 11	18 42	19 01	22 18	23 36	22 02	01 03
22	04 52	05 00	01 45	10 06	00 N 46	18 45	14 48	18 52	19 02	22 17	23 36	22 02	01 03
23	04 52	04 57	02 44	09 44	04 S 50	18 35	14 24	19 02	19 04	22 17	23 36	22 01	01 02
24	04 R 52	04 54	03 36	09 22	10 16	18 24	14 00	19 12	19 06	22 17	23 36	22 01	01 02
25	04 52	04 51	04 19	09 00	15 21	18 11	13 36	19 22	19 07	22 16	23 36	22 01	01 01
26	04 52	04 48	04 52	08 37	19 54	17 58	13 11	19 31	19 09	22 16	23 36	22 01	01 00
27	04 51	04 44	05 12	08 15	23 42	17 43	12 46	19 41	19 11	22 15	23 36	22 01	01 00
28	04 ♓ 51	04 ♓ 41	05 S 18	07 S 52	26 S 31	17 S 26	12 S 20	19 N 50	19 N 13	22 S 15	23 S 36	22 S 01	00 S 59

ZODIAC SIGN ENTRIES

Date	h	m	Planets
02	23	30	☽ ♑
03	17	15	♀ ♒
05	02	51	☽ ♒
07	03	52	☽ ♓
09	04	18	☽ ♈
11	05	45	☽ ♉
13	09	22	☽ ♊
14	18	11	☿ ♒
15	15	40	☽ ♋
18	00	33	☽ ♌
18	16	21	☉ ♓
20	11	34	☽ ♍
23	00	05	☽ ♎
25	12	57	☽ ♏
27	16	59	♀ ♓
28	00	29	☽ ♐

LATITUDES

Date	Mercury ☿ ° '	Venus ♀ ° '	Mars ♂ ° '	Jupiter ♃ ° '	Saturn ♄ ° '	Uranus ♅ ° '	Neptune ♆ ° '	Pluto ♇ ° '
01	03 N 19	00 S 30	01 N 02	00 S 46	00 N 42	00 S 14	00 N 54	15 N 50
04	02 51	00 37	01 04	00 45	00 41	00 14	00 54	15 52
07	02 18	00 44	01 06	00 44	00 41	00 14	00 54	15 53
10	01 43	00 50	01 07	00 43	00 41	00 14	00 54	15 55
13	01 08	00 56	01 09	00 43	00 41	00 14	00 54	15 56
16	00 34	01 01	01 10	00 42	00 41	00 14	00 54	15 58
19	00 N 03	01 06	01 12	00 41	00 41	00 14	00 54	15 59
22	00 S 25	01 10	01 13	00 41	00 41	00 14	00 54	16 01
25	00 50	01 14	01 14	00 40	00 41	00 14	00 54	16 02
28	01 13	01 18	01 15	00 39	00 41	00 14	00 54	16 03
31	01 S 32	01 S 21	01 N 16	00 S 39	00 N 41	00 S 14	00 N 55	16 N 05

DATA

Julian Date	2447559
Delta T	+56 seconds
Ayanamsa	23° 42' 25"
Synetic vernal point	05° ♓ 24' 35"
True obliquity of ecliptic	23° 26' 35"

MOON'S PHASES, APSIDES AND POSITIONS ☽

Date	h	m	Phase	Longitude ° '	Eclipse Indicator
06	07	37	●	17 ♒ 30	
12	23	15	☽	24 ♉ 14	
20	15	32	○	01 ♍ 59	total
28	20	08	☾	10 ♐ 13	

Day	h	m	
07	21	52	Perigee
23	14	16	Apogee

Day	h	m		
02	16	08	Max dec	28° S 17'
08	21	16	0N	
15	07	58	Max dec	28° N 19'
22	15	15	0S	

LONGITUDES

Date	Chiron ⚷ ° '	Ceres ⚳ ° '	Pallas ⚴ ° '	Juno ⚵ ° '	Vesta ⚶ ° '	Black Moon Lilith ⚸ ° '
01	02 ♋ 04	05 ♈ 03	29 ♒ 12	07 ♍ 15	11 ♐ 14	29 ♍ 13
11	01 ♋ 41	08 ♈ 31	02 ♓ 30	04 ♍ 53	15 ♐ 46	00 ♎ 20
21	01 ♋ 25	12 ♈ 08	05 ♓ 49	02 ♍ 19	20 ♐ 07	01 ♎ 26
31	01 ♋ 18	15 ♈ 52	09 ♓ 09	29 ♌ 51	24 ♐ 13	02 ♎ 33

ASPECTARIAN

h m	Aspects
01 Wednesday	
02 03	☽ ⊥ ♆
06 17	☽ ⊼ ♂
09 42	☽ ⊻ ♄
10 05	♀ Q ♇
12 17	☿ ☌ ♀
13 15	☽ ⊻ ♆
15 17	☽ ∠ ☿
15 45	☽ ∠ ♀
16 27	☽ ✱ ☉
16 57	☿ Q ♇
17 55	☽ ± ♂
20 44	☽ ⊻ ♇
02 Thursday	
07 05	☽ ⊥ ☿
07 39	☽ ⊥ ♇
09 52	☽ ⊥ ♀
11 02	☽ bQ ♂
17 06	☽ ⊼ ♃
17 30	☽ ⊻ ☿
21 41	☽ ⊻ ♀
21 48	☽ ∠ ☉
23 44	☽ ∠ ♇
03 Friday	
03 45	☽ ± ♃
05 50	☽ ☌ ♅
05 50	☿ △ ♃
14 41	☽ △ ♂
14 50	☽ ⊥ ☉
15 54	☽ ☌ ♄
18 57	☽ ☌ ♆
19 28	☽ bQ ♃
23 51	☉ □ ♇
04 Saturday	
01 49	☽ ✱ ♇
01 58	☽ ⊻ ☉
07 08	☉ ⊥ ♄
20 15	☽ ☌ ☿
21 01	☽ △ ♃
22 06	☽ Q ♇
23 48	☽ ∥ ♅
05 Sunday	
02 04	♂ △ ♄
06 02	☽ ☌ ♀
06 42	☽ ∥ ♄
08 32	☽ ∥ ♆
09 00	☽ ⊻ ♅
15 36	☉ ♯ ♂
17 39	☽ ∥ ♀
18 43	☽ ⊻ ♄
18 55	☽ ⊥ ♅
19 17	☽ □ ♂
20 08	☿ St D
21 21	☽ ⊻ ♆
06 Monday	
00 52	☽ ♯ ♃
00 55	☉ ⊥ ♆
01 56	☽ ∥ ☿
03 48	☽ □ ♇
04 36	☽ ⊥ ♄
07 10	☽ ⊥ ♆
07 37	☽ ☌ ☉
09 40	☽ ∠ ♅
12 15	☽ ♯ ♂
14 04	☽ ∥ ☉
17 35	♀ ⊻ ♅
19 19	☽ ∠ ♄
21 29	☽ ⊻ ☿
21 48	☽ ∠ ♆
22 24	☽ □ ♃
07 Tuesday	
01 45	☽ Q ♂
07 16	☽ ⊥ ☿
10 00	☽ ✱ ♅
11 27	☽ ⊻ ♀
14 45	☉ ∠ ♅
17 30	☿ ♯ ♃
19 41	☽ ✱ ♄
21 54	☽ ✱ ♂
21 59	☽ ∠ ☿
22 01	☽ ⊥ ♀
22 03	☽ ✱ ♆
08 Wednesday	
02 01	♂ △ ♆
03 33	☽ Q ♃
04 19	☽ △ ♇
05 24	☽ Q ♅
11 37	☽ ⊻ ☉
13 52	☽ ∠ ♀
15 10	☽ Q ♄
17 21	☽ ∥ ♇
17 27	☽ Q ♆
22 02	☽ ⊥ ☉
22 41	☽ ✱ ☿
23 01	☽ ✱ ♃
23 06	☽ ∠ ♂
09 Thursday	
01 11	☽ ♯ ♇
04 35	☽ bQ ♇
10 38	☽ □ ♅
13 44	☽ ∠ ☉
14 05	☿ △ ♃
14 20	☽ ⊥ ♂
16 27	☽ ✱ ♀
18 44	☽ Q ☿
19 15	☽ ± ♇
20 37	☽ □ ♄
22 47	☽ □ ♆
23 34	☽ ∠ ♃
10 Friday	
00 34	☽ ⊻ ♂
05 06	☽ ⊼ ♇
06 06	☿ ∥ ♀
11 03	☿ Q ♇
14 03	☽ Q ♀
14 29	☽ ⊥ ♃
16 15	☽ ✱ ☉
23 57	☽ ♯ ☉
11 Saturday	
00 32	☽ ⊻ ♃
01 32	☽ □ ☿
02 44	♀ ♯ ♃
11 34	☽ ∥ ♂
12 28	☽ △ ♅
13 53	☽ Q ☉
17 02	♀ ⊥ ♅
18 55	☽ ♯ ♀
19 56	☽ ∥ ♃
21 02	♀ ⊻ ♄
21 44	☽ ♯ ☿
23 03	☽ △ ♄
23 14	☽ □ ♀
12 Sunday	
01 04	☽ △ ♆
05 01	☽ ☌ ♂
07 33	☽ ☍ ♇
12 26	☽ ♯ ♆
14 10	☽ ♯ ♄
14 13	☽ bQ ♅
20 23	♀ ⊻ ♆
21 38	☽ ♯ ♅
23 15	☽ □ ☉
13 Monday	
01 09	☽ bQ ♄
03 06	☽ bQ ♆
04 10	☽ ☌ ♃
05 58	☽ ± ♅
07 26	☽ △ ☿
16 37	☽ ⊼ ♅
17 08	☽ ± ♄
19 03	☽ ± ♆
14 Tuesday	
03 38	☉ ∠ ♄
03 56	☽ ⊼ ♄
05 48	☽ ⊼ ♆
09 11	☽ △ ♀
11 48	☽ bQ ☿
12 18	☽ ⊻ ♂
12 31	☽ ⊼ ♇
17 15	♂ ☍ ♇
22 50	♀ ♯ ♂
23 26	☽ ± ♇
23 44	☽ ⊥ ♂
15 Wednesday	
05 01	☉ ∠ ♆
05 17	☽ ± ☿
09 32	☽ △ ☉
10 33	☽ ⊻ ♃
15 32	☽ bQ ♀
16 01	☽ bQ ♇
17 07	☽ ∠ ♂
17 09	☽ ⊼ ☿
20 34	♀ □ ♇
21 45	☽ ⊥ ♃
23 27	☽ ☍ ♅
23 36	☉ □ ♃
16 Thursday	
09 43	♇ St R
10 10	☽ ± ♀
11 35	☽ ☍ ♄
13 15	☽ ☍ ♆
14 46	☽ ∠ ♃
15 57	☽ bQ ☉
20 10	☽ △ ♇
21 10	♀ □ ♂
22 40	☽ ✱ ♂
22 45	☽ ⊼ ♀
17 Friday	
01 34	♀ ⊥ ♄
10 35	☽ ± ☉
18 15	♀ ⊥ ♆
19 36	☽ ✱ ♃
21 12	☽ ♯ ♅
22 49	☽ Q ♂
23 09	☽ ⊼ ☉
18 Saturday	
05 41	☽ ♯ ♄
06 42	☽ ☍ ☿
07 28	☽ ♯ ♆
08 51	☽ ⊼ ♅
19 06	☽ Q ♃
20 36	☽ ± ♅
21 43	☽ ⊼ ♄
23 09	☽ ⊼ ♆
19 Sunday	
00 05	☽ ♯ ☿
01 10	☽ ∥ ♃
03 46	♀ ∠ ♅
04 36	☽ ∥ ♂
06 12	☽ □ ♇
09 36	☽ ± ♄
09 55	☿ ⊻ ♅
10 59	☽ ± ♆
11 47	☽ □ ♂
14 22	☽ bQ ♅
15 33	☽ ☍ ♀
16 53	☽ ♯ ♀
20 Monday	
03 33	☽ bQ ♄
04 52	☽ bQ ♆
06 56	☽ □ ♃
15 32	☽ ☍ ☉
16 20	☿ ♯ ♃
16 44	☽ ♯ ☉
17 56	☽ Q ♇
20 20	☽ △ ♅
23 36	☽ ⊼ ☿
21 Tuesday	
09 49	☽ △ ♄
10 59	☽ △ ♆
12 54	☽ ± ☿
18 04	☽ ✱ ♇
19 56	♂ bQ ♅
22 Wednesday	
02 59	☽ △ ♂
03 09	☿ ♯ ♂
09 01	☽ bQ ☿
10 45	☽ ⊼ ♀
10 49	☽ ♯ ♇
19 41	☽ ∥ ♇
19 53	☽ △ ♃
23 Thursday	
00 21	☽ ± ♀
00 26	☽ ∠ ♇
02 49	☉ ✱ ♅
09 11	☽ □ ♅
09 45	☽ ⊼ ☉
11 02	☽ bQ ♂
17 51	♂ ∥ ♃
18 43	☽ ⊥ ♇
18 51	☽ △ ☿
20 49	☽ bQ ♀
23 04	☽ □ ♄
23 05	☽ ± ☉
23 56	☽ □ ♆
24 Friday	
02 39	☽ bQ ♃
06 19	☽ ± ♂
06 55	☽ ⊻ ♇
08 11	☽ ∥ ☉
19 03	☽ bQ ☉
19 08	☽ ⊼ ♂
19 29	♀ ∠ ♄
20 30	☿ ⊥ ♅
21 06	☽ ± ♃
21 51	☽ Q ♅
25 Saturday	
02 49	♀ ∠ ♆
04 07	☽ ∥ ♀
06 50	☽ △ ♀
09 21	☽ ⊼ ♃
11 49	☽ Q ♄
12 30	☽ Q ♆
14 13	☿ ⊻ ♄
20 38	☿ ⊻ ♆
22 09	☽ ✱ ♅
26 Sunday	
01 54	☽ ∥ ☿
04 07	☽ △ ☉
07 47	☽ ♯ ♃
08 53	♀ □ ♃
09 48	☽ ♯ ♂
12 05	☽ ✱ ♄
12 38	☽ ✱ ♆
14 29	☽ □ ☿
19 19	☽ ☌ ♇
27 Monday	
00 41	☽ ∥ ♆
02 15	☽ ∥ ♄
04 08	☽ ∠ ♅
10 23	☽ ☍ ♂
11 15	☽ ∥ ♅
17 55	☽ ∠ ♄
18 19	☽ ∠ ♆
21 35	☽ ☍ ♃
21 50	☽ ⊥ ♅
28 Tuesday	
01 20	☽ □ ♀
06 03	☽ Q ☿
09 29	☽ ⊻ ♅
11 00	☿ □ ♇
11 28	☽ ⊥ ♄
11 46	☽ ⊥ ♆
20 08	☽ □ ☉
22 59	☽ ⊻ ♄
23 15	☽ ⊻ ♆

All ephemeris data is given at 12.00 UT and the Moon's longitude is additionally given for 24.00 UT

MARCH 1989

LONGITUDES

Date	Sidereal time h m s	Sun ☉ ° ' "	Moon ☽ ° ' "	Moon ☽ 24.00 ° ' "	Mercury ☿ ° '	Venus ♀ ° '	Mars ♂ ° '	Jupiter ♃ ° '	Saturn ♄ ° '	Uranus ♅ ° '	Neptune ♆ ° '	Pluto ♇ ° '
01	22 37 04	10 ♓ 52 55	18 ♐ 37 53	25 ♐ 06 06	16 ♒ 33	02 ♓ 14	24 ♉ 01	28 ♉ 42	11 ♑ 46	04 ♑ 41	11 ♑ 52	15 ♏ 08
02	22 41 01	11 53 07	01 ♑ 40 21	08 ♑ 20 59	17 54	03 29	24 38	28 50	11 50	04 43	11 54	15 R 08
03	22 44 58	12 53 17	15 ♑ 08 15	22 ♑ 02 18	19 18	04 44	25 14	28 57	11 55	04 45	11 55	15 07
04	22 48 54	13 53 27	29 ♑ 03 08	06 ♒ 10 37	20 42	05 59	25 50	29 05	12 00	04 47	11 56	15 07
05	22 52 51	14 53 34	13 ♒ 24 26	20 ♒ 44 05	22 08	07 14	26 26	29 13	12 04	04 49	11 57	15 06
06	22 56 47	15 53 40	28 ♒ 08 54	05 ♓ 38 04	23 35	08 29	27 03	29 21	12 08	04 51	11 59	15 06
07	23 00 44	16 53 43	13 ♓ 10 34	20 ♓ 45 18	25 04	09 44	27 39	29 29	12 13	04 52	12 00	15 05
08	23 04 40	17 53 45	28 ♓ 21 04	05 ♈ 56 38	26 33	10 59	28 15	29 37	12 17	04 54	12 01	15 04
09	23 08 37	18 53 45	13 ♈ 30 48	21 ♈ 02 24	28 04	12 13	28 52	29 46	12 21	04 55	12 02	15 04
10	23 12 33	19 53 43	28 ♈ 30 21	05 ♉ 53 46	29 ♒ 36	13 28	29 ♉ 28	29 ♉ 54	12 25	04 57	12 03	15 03
11	23 16 30	20 53 39	13 ♉ 11 53	20 ♉ 24 07	01 ♓ 10	14 43	00 ♊ 05	00 ♊ 03	12 29	04 59	12 05	15 02
12	23 20 27	21 53 33	27 ♉ 30 05	04 ♊ 29 32	02 44	15 58	00 41	00 12	12 33	05 00	12 06	15 01
13	23 24 23	22 53 25	11 ♊ 22 26	18 ♊ 08 51	04 20	17 13	01 18	00 21	12 37	05 01	12 07	15 00
14	23 28 20	23 53 14	24 ♊ 48 59	01 ♋ 23 07	05 56	18 27	01 54	00 30	12 41	05 03	12 08	15 00
15	23 32 16	24 53 01	07 ♋ 51 38	14 ♋ 14 56	07 34	19 42	02 31	00 39	12 44	05 04	12 09	14 59
16	23 36 13	25 52 46	20 ♋ 33 29	26 ♋ 47 46	09 14	20 57	03 08	00 49	12 48	05 05	12 10	14 58
17	23 40 09	26 52 29	02 ♌ 58 13	09 ♌ 05 20	10 54	22 12	03 44	00 58	12 51	05 07	12 11	14 57
18	23 44 06	27 52 09	15 ♌ 09 33	21 ♌ 11 19	12 36	23 26	04 21	01 08	12 55	05 08	12 11	14 56
19	23 48 02	28 51 47	27 ♌ 11 01	03 ♍ 09 03	14 19	24 41	04 57	01 17	12 58	05 09	12 12	14 55
20	23 51 59	29 ♓ 51 23	09 ♍ 05 46	15 ♍ 01 29	16 03	25 56	05 34	01 27	13 01	05 10	12 13	14 54
21	23 55 56	00 ♈ 50 57	20 ♍ 56 30	26 ♍ 51 05	17 48	27 10	06 11	01 37	13 04	05 11	12 14	14 53
22	23 59 52	01 50 29	02 ♎ 45 31	08 ♎ 40 03	19 35	28 25	06 48	01 47	13 07	05 12	12 15	14 52
23	00 03 49	02 49 58	14 ♎ 34 54	20 ♎ 30 20	21 23	29 ♓ 40	07 24	01 57	13 10	05 13	12 15	14 51
24	00 07 45	03 49 26	26 ♎ 26 35	02 ♏ 23 53	23 12	00 ♈ 54	08 01	02 08	13 13	05 14	12 16	14 49
25	00 11 42	04 48 52	08 ♏ 22 31	14 ♏ 22 46	25 02	02 09	08 38	02 18	13 16	05 14	12 17	14 48
26	00 15 38	05 48 16	20 ♏ 24 56	26 ♏ 29 21	26 54	03 23	09 15	02 28	13 19	05 15	12 17	14 47
27	00 19 35	06 47 38	02 ♐ 36 22	08 ♐ 46 21	28 ♓ 47	04 38	09 51	02 39	13 21	05 16	12 18	14 46
28	00 23 31	07 46 59	14 ♐ 59 42	21 ♐ 16 51	00 ♈ 42	05 53	10 28	02 50	13 24	05 16	12 19	14 45
29	00 27 28	08 46 17	27 ♐ 38 13	04 ♑ 04 15	02 38	07 07	11 05	03 00	13 26	05 17	12 19	14 43
30	00 31 25	09 45 34	10 ♑ 35 22	17 ♑ 11 57	04 35	08 22	11 42	03 11	13 29	05 18	12 20	14 42
31	00 35 21	10 ♈ 44 49	23 ♑ 54 23	00 ♒ 42 57	06 ♈ 33	09 ♈ 36	12 ♊ 19	03 ♊ 22	13 ♑ 31	05 ♑ 18	12 ♑ 20	14 ♏ 41

Date	Moon True ☊ ° '	Moon Mean ☊ ° '	Moon ☽ Latitude ° '
01	04 ♓ 51	04 ♓ 38	05 S 09
02	04 D 52	04 35	04 44
03	04 52	04 32	04 03
04	04 53	04 29	03 06
05	04 54	04 25	01 57
06	04 54	04 22	00 S 37
07	04 R 54	04 19	00 N 46
08	04 54	04 16	02 06
09	04 52	04 13	03 18
10	04 50	04 10	04 15
11	04 48	04 06	04 54
12	04 47	04 03	05 11
13	04 46	04 00	05 14
14	04 D 45	03 57	04 57
15	04 46	03 54	04 25
16	04 47	03 50	03 41
17	04 48	03 47	02 47
18	04 50	03 44	01 47
19	04 51	03 41	00 N 42
20	04 R 51	03 38	00 S 23
21	04 50	03 35	01 27
22	04 48	03 31	02 27
23	04 44	03 28	03 21
24	04 39	03 25	04 06
25	04 34	03 22	04 40
26	04 28	03 19	05 03
27	04 23	03 16	05 12
28	04 20	03 12	05 07
29	04 18	03 09	04 47
30	04 D 17	03 06	04 12
31	04 ♓ 18	03 ♓ 03	03 S 23

DECLINATIONS

Date	Sun ☉ ° '	Moon ☽ ° '	Mercury ☿ ° '	Venus ♀ ° '	Mars ♂ ° '	Jupiter ♃ ° '	Saturn ♄ ° '	Uranus ♅ ° '	Neptune ♆ ° '	Pluto ♇ ° '
01	07 S 29	28 S 05	17 S 09	11 S 54	20 00	19 N 14	22 S 15	23 S 36	22 S 01	00 S 59
02	07 06	28 10	16 50	11 28	20 09	19 16	22 14	23 36	22 00	00 58
03	06 43	26 36	16 30	11 01	20 18	19 18	22 14	23 36	22 00	00 58
04	06 20	23 23	16 08	10 35	20 27	19 20	22 13	23 36	22 00	00 57
05	05 57	18 40	15 45	10 08	20 36	19 22	22 13	23 35	22 00	00 56
06	05 34	12 42	15 21	09 40	20 45	19 24	22 13	23 35	22 00	00 56
07	05 11	05 S 54	14 56	09 13	20 53	19 26	22 12	23 35	22 00	00 55
08	04 47	01 N 17	14 29	08 45	21 02	19 28	22 12	23 35	22 00	00 55
09	04 24	08 22	14 01	08 17	21 10	19 30	22 12	23 35	21 59	00 54
10	04 00	14 55	13 32	07 48	21 18	19 32	22 11	23 35	21 59	00 53
11	03 37	20 28	13 02	07 20	21 26	19 34	22 11	23 35	21 59	00 53
12	03 13	24 42	12 30	06 51	21 34	19 36	22 11	23 35	21 59	00 52
13	02 49	27 20	11 57	06 22	21 42	19 38	22 10	23 35	21 59	00 52
14	02 26	28 17	11 23	05 53	21 50	19 40	22 10	23 35	21 59	00 51
15	02 02	27 37	10 48	05 24	21 57	19 42	22 09	23 35	21 59	00 50
16	01 38	25 30	10 12	04 55	22 05	19 44	22 09	23 35	21 59	00 50
17	01 15	22 12	09 34	04 25	22 12	19 46	22 09	23 35	21 59	00 49
18	00 51	17 59	08 55	03 56	22 19	19 49	22 09	23 35	21 58	00 48
19	00 27	13 07	08 15	03 26	22 26	19 51	22 08	23 35	21 58	00 48
20	00 S 03	07 48	07 34	02 56	22 33	19 53	22 08	23 35	21 58	00 47
21	00 N 20	02 N 15	06 51	02 26	22 39	19 55	22 08	23 35	21 58	00 47
22	00 44	03 S 21	06 07	01 56	22 46	19 57	22 07	23 35	21 58	00 46
23	01 08	08 50	05 23	01 26	22 52	19 59	22 07	23 35	21 58	00 45
24	01 31	14 01	04 37	00 56	22 58	20 02	22 07	23 35	21 58	00 45
25	01 55	18 43	03 50	00 S 26	23 04	20 04	22 07	23 35	21 58	00 44
26	02 18	22 43	03 02	00 N 04	23 10	20 06	22 06	23 35	21 58	00 43
27	02 42	25 47	02 13	00 34	23 16	20 08	22 06	23 35	21 58	00 43
28	03 05	27 41	01 23	01 05	23 21	20 11	22 06	23 35	21 58	00 42
29	03 29	28 12	00 S 32	01 35	23 27	20 13	22 06	23 35	21 58	00 42
30	03 52	27 12	00 N 20	02 05	23 32	20 15	22 05	23 35	21 57	00 41
31	04 N 15	24 S 39	01 N 13	02 N 35	23 N 37	20 N 17	22 S 05	23 S 35	21 S 57	00 S 40

ZODIAC SIGN ENTRIES

Date	h	m	Planets
02	08	58	☽ ♑
04	13	36	☽ ♒
06	14	59	☽ ♓
08	14	36	☽ ♈
10	14	25	☽ ♉
10	18	07	☿ ♓
11	03	26	♃ ♊
11	08	51	♂ ♊
12	16	16	☽ ♊
14	21	27	☽ ♋
17	06	13	☽ ♌
19	17	39	☽ ♍
20	15	28	☉ ♈
22	06	24	☽ ♎
23	18	32	♀ ♈
24	19	10	☽ ♏
27	06	54	☽ ♐
28	03	16	☿ ♈
29	16	25	☽ ♑
31	22	45	☽ ♒

LATITUDES

Date	Mercury ☿ ° '	Venus ♀ ° '	Mars ♂ ° '	Jupiter ♃ ° '	Saturn ♄ ° '	Uranus ♅ ° '	Neptune ♆ ° '	Pluto ♇ ° '
01	01 S 20	01 S 19	01 N 15	00 S 39	00 N 41	00 S 14	00 N 54	16 N 04
04	01 38	01 21	01 16	00 38	00 41	00 14	00 55	16 05
07	01 52	01 23	01 17	00 38	00 41	00 14	00 55	16 07
10	02 03	01 25	01 18	00 37	00 41	00 14	00 55	16 08
13	02 11	01 26	01 18	00 36	00 41	00 14	00 55	16 09
16	02 15	01 26	01 19	00 36	00 41	00 14	00 55	16 10
19	02 15	01 26	01 20	00 35	00 41	00 14	00 55	16 11
22	02 10	01 25	01 20	00 35	00 41	00 15	00 55	16 13
25	02 02	01 24	01 20	00 34	00 41	00 15	00 55	16 14
28	01 48	01 22	01 21	00 34	00 41	00 15	00 55	16 15
31	01 S 30	01 S 20	01 N 21	00 S 33	00 N 41	00 S 15	00 N 55	16 N 15

DATA

Julian Date	2447587
Delta T	+56 seconds
Ayanamsa	23° 42' 28"
Synetic vernal point	05° ♓ 24' 31"
True obliquity of ecliptic	23° 26' 35"

MOON'S PHASES, APSIDES AND POSITIONS ☽

Date	h	m	Phase	Longitude ° '	Eclipse Indicator
07	18	19	●	17 ♓ 10	Partial
14	10	11	☽	23 ♊ 49	
22	09	58	○	01 ♎ 45	
30	10	21	☾	09 ♑ 42	

Day	h	m		
08	07	45	Perigee	
22	18	03	Apogee	
02	01	17	Max dec	28° S 19'
08	07	46	0N	
14	13	50	Max dec	28° N 18'
21	21	38	0S	
29	08	30	Max dec	28° S 13'

LONGITUDES

Date	Chiron ⚷ ° '	Ceres ⚳ ° '	Pallas ⚴ ° '	Juno ⚵ ° '	Vesta ⚶ ° '	Black Moon Lilith ⚸ ° '
01	01 ♋ 19	15 ♈ 07	08 ♓ 29	00 ♍ 19	23 ♐ 25	02 ♎ 20
11	01 ♋ 19	18 ♈ 56	11 ♓ 48	28 ♌ 09	27 ♐ 17	03 ♎ 26
21	01 ♋ 27	22 ♈ 49	15 ♓ 07	26 ♌ 31	00 ♑ 50	04 ♎ 33
31	01 ♋ 44	26 ♈ 46	18 ♓ 23	25 ♌ 34	03 ♑ 59	05 ♎ 39

All ephemeris data is given at 12.00 UT and the Moon's longitude is additionally given for 24.00 UT

ASPECTARIAN

h m	Aspects
01 Wednesday	
05 27	☽ ⊻ ♇
07 37	☽ ✱ ☿
08 51	♃ ± ♅
15 19	☽ Q ♀
16 40	☽ ⊥ ♇
22 30	☽ ⊼ ♂
02 Thursday	
06 47	☽ ⊼ ♃
08 29	☽ Q ☉
09 12	☽ ∠ ♇
10 01	☽ ± ♂
10 47	☿ ⊥ ♄
10 51	☉ ✱ ♄
11 45	☿ ⊥ ♆
12 11	☉ ✱ ♆
14 29	☽ ∠ ☿
15 37	☽ ✱ ♀
17 31	☽ ☌ ♅
17 45	☽ ± ♃
03 Friday	
02 56	☽ ⚼ ♂
06 18	☽ ☌ ♄
06 19	☽ ☌ ♆
07 44	☽ ✱ ☉
08 23	☽ ⊥ ☿
09 54	☽ ⚼ ♃
10 44	♄ ☌ ♆
11 58	☽ ✱ ♇
12 21	♀ ✱ ♅
20 02	☿ ∠ ♅
20 04	☽ ⊻ ☿
20 49	☽ ∠ ♀
04 Saturday	
06 16	☽ △ ♂
08 36	☉ Q ♂
08 42	☽ Q ♇
10 46	☽ ∥ ♅
11 42	☽ ∠ ☉
12 03	☽ △ ♃
13 44	☽ ⊥ ♀
18 38	☽ ∥ ♄
19 50	☽ ∥ ♆
21 41	☽ ⊻ ♅
05 Sunday	
00 47	☽ ⊻ ♀
03 14	☽ ⋕ ♂
03 58	☽ ⊥ ☉
07 42	☽ ⊥ ♅
08 50	☽ ⋕ ♃
09 36	☽ ⊻ ♆
09 47	☽ ⊻ ♄
14 38	☽ ⊻ ☉
14 47	☽ □ ♇
17 00	☉ △ ♇
19 29	☽ ⊥ ♆
19 41	☽ ⊥ ♄
22 31	☽ ∠ ♅
06 Monday	
01 06	☽ ∥ ☿
03 50	☽ ☌ ☿
09 19	♂ ⚼ ♆
10 07	☽ ∠ ♆
10 09	☽ □ ♂
10 22	☽ ∠ ♄
13 57	☽ □ ♃
16 19	♂ ⚼ ♄
22 45	☽ ✱ ♅
23 47	☽ ∥ ♀
07 Tuesday	
06 02	☽ ☌ ♀
10 08	☽ ✱ ♆
10 28	☽ ✱ ♄
11 24	☉ Q ♅
14 38	☽ ∥ ☉
15 01	☽ △ ♇
16 05	☽ Q ♂
17 52	☽ Q ♅
18 19	☽ ☌ ☉
18 53	☽ Q ♃
08 Wednesday	
04 23	☉ Q ♃
04 44	☽ ∥ ♇
05 09	☽ Q ♆
05 33	☽ Q ♄
08 51	☽ ⊻ ☿
10 47	☽ ⋕ ♇
11 51	☽ ✱ ♂
14 02	☽ ✱ ♃
14 43	☽ ⚼ ♇
19 23	☽ ⊥ ☿
19 28	☿ ∠ ♆
22 22	☽ □ ♅
23 07	☽ ⋕ ☉
09 Thursday	
00 06	☿ ∠ ♄
04 56	☽ ± ♇
08 23	♀ ✱ ♆
09 39	☽ □ ♆
09 46	☽ ⊻ ♀
10 09	☽ □ ♄
11 13	☽ ∠ ☿
11 41	☽ ⋕ ♀
12 35	☽ ∠ ♂
14 00	☽ ∠ ♃
14 27	☽ ⊼ ♇
14 31	♂ ± ♅
14 36	♀ ✱ ♄
20 11	☽ ⊥ ♀
21 11	☽ ⊻ ☉
10 Friday	
03 33	☽ ⊥ ♂
04 31	☽ ⊥ ♃
07 05	☽ ⋕ ☿
07 29	☽ ⊥ ☉
08 34	☿ □ ♂
11 56	☽ ∠ ♀
13 38	☽ ⊻ ♂
13 59	☽ ✱ ☿
14 17	☽ ⊻ ♃
17 10	☿ □ ♃
22 28	☽ △ ♅
23 07	☽ ∠ ☉
11 Saturday	
07 40	☽ ∥ ♃
10 09	☽ △ ♆
10 34	♂ ☌ ♃
10 49	☽ △ ♄
11 56	☽ Q ☿
14 45	☽ ✱ ♀
15 02	☽ ☍ ♇
16 59	☽ ∥ ♂
18 02	♀ △ ♇
19 42	☽ ⋕ ♆
20 44	☽ ⋕ ♄
23 18	☽ ⚼ ♅
12 Sunday	
01 47	☽ ✱ ☉
04 49	☽ ⋕ ♅
11 18	☽ ⚼ ♆
12 05	☽ ⚼ ♄
12 52	☽ Q ♀
14 34	☽ ± ♅
16 40	☽ ☌ ♃
17 42	☽ ☌ ♂
22 07	☽ □ ☿
23 49	☽ Q ☉
13 Monday	
00 54	☽ ⊼ ♅
02 48	☽ ± ♆
03 39	☽ ± ♄
08 19	♀ Q ♅
13 18	☽ ⊼ ♆
14 12	☽ ⊼ ♄
18 24	☽ ⊼ ♇
22 32	☿ ✱ ♅
23 23	☽ □ ♀
14 Tuesday	
05 06	☽ ± ♇
10 11	☽ □ ☉
12 57	♀ Q ♃
17 54	☉ Q ♆
21 26	☽ ⚼ ♇
22 30	☽ ⊻ ♃
15 Wednesday	
01 36	☽ ⊻ ♂
06 48	☽ ☍ ♅
08 15	☉ Q ♄
09 44	☽ ⊥ ♃
11 23	☽ △ ☿
13 17	☽ ⊥ ♂
16 41	♂ ⋕ ♆
20 02	☽ ☍ ♆
21 11	☽ ☍ ♄
16 Thursday	
01 22	☽ △ ♇
02 51	☽ ∠ ♃
07 07	☽ ∠ ♂
12 50	☽ △ ♀
18 41	♀ Q ♂
20 08	☽ ⚼ ☿
23 07	☽ △ ☉
17 Friday	
02 13	♂ ⋕ ♄
02 52	☽ ⋕ ♅
08 03	☽ ✱ ♃
12 03	☽ ∥ ♂
12 22	☽ ⋕ ♄
13 26	☽ ⋕ ♆
13 35	☽ ✱ ♂
16 11	☽ ⊼ ♅
16 23	☽ ± ☿
21 13	☽ ⚼ ♀
18 Saturday	
02 15	☽ ∥ ♃
04 00	☽ ± ♅
06 06	☽ ⊼ ☿
06 07	☽ ⊼ ♆
06 14	☿ ✱ ♆
07 03	☽ ⚼ ☉
07 32	☽ ⊼ ♄
07 55	☽ Q ♃
11 33	☽ □ ♇
14 28	☉ ∥ ♇
14 29	☽ Q ♂
16 36	☿ ✱ ♄
17 03	☽ ± ♀
18 02	☽ ± ♆
19 30	☽ ± ♄
21 54	☽ ⚼ ♅
19 Sunday	
02 34	☽ ± ☉
02 39	♀ Q ♆
06 25	☽ ⊼ ♀
12 03	☽ ⚼ ♆
13 35	☽ ⚼ ♄
15 41	☽ ⊼ ☉
17 42	♀ Q ♄
19 36	♂ ⊼ ♅
20 18	☿ △ ♇
20 22	☽ □ ♃
23 30	☽ Q ♇
20 Monday	
04 03	☽ △ ♅
04 29	☽ □ ♂
12 58	☉ ⚼ ♇
13 13	☽ ⋕ ☿
18 19	☽ △ ♆
19 59	☽ △ ♄
23 43	☽ ✱ ♇
21 Tuesday	
03 27	☿ Q ♅
04 31	☽ ☍ ☿
11 08	☽ ⋕ ♀
14 03	♂ ± ♆
18 20	☽ ⋕ ♇
19 39	☽ ∥ ☉
23 55	☽ ⋕ ☉
22 Wednesday	
00 56	☽ ∥ ♇
02 09	☽ ☍ ♀
06 07	☽ ∠ ♇
06 25	☽ ∥ ♀
09 58	☽ ☍ ☉
10 00	☽ △ ♃
10 24	☉ ✱ ♃
14 00	☉ ⋕ ♇
15 04	☿ Q ♃
16 57	☽ □ ♅
20 38	☽ △ ♂
22 36	☽ ∥ ☿
23 Thursday	
00 22	☽ ⊥ ♇
02 10	♂ ± ♄
07 17	☽ □ ♆
09 08	☽ □ ♄
12 32	☽ ⊻ ♇
15 27	♀ ⚼ ♇
16 53	☽ ⚼ ♃
20 17	☉ ⋕ ♀
24 Friday	
04 15	☽ ⊼ ☿
04 42	☽ ⚼ ♂
05 30	☽ Q ♅
11 21	☽ ± ♃
18 34	☽ ± ☿
19 43	☽ Q ♆
21 14	♀ ∥ ♇
21 40	☽ Q ♄
22 03	☽ ⊼ ♀
23 37	☽ ⊼ ♃
23 50	☽ ± ♂
25 Saturday	
02 03	☿ Q ♆
04 13	☽ ⊼ ☉
05 42	☽ ✱ ♅
11 30	☽ ± ♀
12 32	☽ ⊼ ♂
15 01	☿ Q ♄
15 23	♀ ✱ ♃
15 57	☽ ⚼ ☿
17 19	☽ ± ☉
19 38	☽ ⋕ ♃
19 49	☽ ✱ ♆
21 49	☽ ✱ ♄
22 25	☉ □ ♅
26 Sunday	
00 50	☽ ☌ ♇
07 05	☽ ∥ ♆
07 31	☽ ⚼ ♀
07 59	☽ ∥ ♄
11 40	☽ ∠ ♅
12 50	☽ ⚼ ☉
15 11	☽ ⋕ ♂
18 02	☽ ∥ ♅
18 25	☿ Q ♂
27 Monday	
01 35	☽ ∠ ♆
02 26	☉ ⋕ ☿
03 09	☽ △ ☿
03 39	☽ ∠ ♄
05 27	☽ ⊥ ♅
12 05	☽ ☍ ♃
16 24	☽ △ ♀
17 11	☽ ⊻ ♅
18 34	♀ ⋕ ♇
19 12	☽ ⊥ ♆
20 52	☽ △ ☉
21 17	☽ ⊥ ♄
28 Tuesday	
00 10	☿ ⚼ ♇
00 14	♀ □ ♅
02 50	☽ ☍ ♂
06 50	☽ ⊻ ♆
08 55	☽ ⊻ ♄
11 31	☽ ⊻ ♇
17 24	☿ ⋕ ♀
22 58	☽ ⊥ ♇
29 Wednesday	
07 21	☿ ∥ ♇
10 51	☉ ± ♇
15 54	☽ ∠ ♇
17 11	☿ ✱ ♃
22 10	☽ ⊼ ♃
22 58	☽ □ ☿
30 Thursday	
02 15	☽ ☌ ♅
07 29	☽ □ ♀
09 24	☽ ± ♃
10 21	☽ □ ☉
14 07	☽ ⊼ ♂
15 10	☽ ☌ ♆
17 17	☽ ☌ ♄
18 27	♀ ± ♇
19 28	☽ ✱ ♇
20 47	☿ □ ♅
21 22	☿ ⋕ ♇
31 Friday	
01 31	☽ ± ♂
01 58	☽ ⚼ ♃
02 21	♂ ⋕ ♅
12 58	♂ ⊼ ♆
13 20	☽ Q ☿
16 54	☽ Q ♇
18 18	☽ ⚼ ♂
18 54	☽ ⋕ ♂
19 12	☽ Q ♀
19 16	☽ ∥ ♅
21 13	☽ Q ☉

APRIL 1989

Date	Sidereal time h m s	Sun ☉ ° ' "	Moon ☽ ° ' "	Moon ☽ 24.00 ° ' "	Mercury ☿ ° '	Venus ♀ ° '	Mars ♂ ° '	Jupiter ♃ ° '	Saturn ♄ ° '	Uranus ♅ ° '	Neptune ♆ ° '	Pluto ♇ ° '
01	00 39 18	11 ♈ 44 03	07 ♒ 37 52	14 ♒ 39 15	08 ♈ 32	10 ♈ 51	12 ♊ 55	03 ♊ 33	13 ♑ 33	05 ♑ 18	12 ♑ 20	14 ♏ 39
02	00 43 14	12 43 14	21 ♒ 47 05	29 ♒ 01 09	10 33	12 05	13 32	03 44	13 35	05 19	12 21	14 R 38
03	00 47 11	13 42 24	06 ♓ 21 07	13 ♓ 46 26	12 35	13 20	14 09	03 56	13 37	05 19	12 21	14 37
04	00 51 07	14 41 32	21 ♓ 16 21	28 ♓ 49 57	14 37	14 34	14 46	04 07	13 39	05 19	12 22	14 35
05	00 55 04	15 40 38	06 ♈ 26 08	14 ♈ 03 40	16 41	15 49	15 23	04 18	13 41	05 20	12 22	14 34
06	00 59 00	16 39 42	21 ♈ 41 14	29 ♈ 17 31	18 45	17 03	16 00	04 30	13 42	05 20	12 22	14 33
07	01 02 57	17 38 44	06 ♉ 51 11	14 ♉ 21 01	20 50	18 17	16 37	04 41	13 44	05 20	12 22	14 31
08	01 06 54	18 37 44	21 ♉ 45 56	29 ♉ 05 02	22 55	19 32	17 14	04 53	13 45	05 20	12 23	14 30
09	01 10 50	19 36 42	06 ♊ 17 35	13 ♊ 23 07	25 00	20 46	17 51	05 05	13 47	05 R 20	12 23	14 28
10	01 14 47	20 35 37	20 ♊ 21 21	27 ♊ 12 12	27 06	22 01	18 28	05 17	13 48	05 20	12 23	14 27
11	01 18 43	21 34 30	03 ♋ 55 45	10 ♋ 32 17	29 ♈ 10	23 15	19 05	05 28	13 49	05 20	12 23	14 25
12	01 22 40	22 33 21	17 ♋ 02 09	23 ♋ 25 49	01 ♉ 14	24 29	19 42	05 40	13 50	05 20	12 23	14 24
13	01 26 36	23 32 10	29 ♋ 43 51	05 ♌ 56 48	03 17	25 44	20 19	05 52	13 51	05 20	12 23	14 22
14	01 30 33	24 30 57	12 ♌ 05 19	18 ♌ 10 00	05 19	26 58	20 56	06 05	13 52	05 19	12 R 23	14 21
15	01 34 29	25 29 41	24 ♌ 11 29	00 ♍ 10 22	07 19	28 13	21 33	06 17	13 53	05 19	12 23	14 19
16	01 38 26	26 28 22	06 ♍ 07 12	12 ♍ 02 34	09 17	29 ♈ 26	22 10	06 29	13 54	05 19	12 23	14 17
17	01 42 23	27 27 02	17 ♍ 56 57	23 ♍ 50 48	11 12	00 ♉ 41	22 47	06 41	13 54	05 18	12 23	14 16
18	01 46 19	28 25 40	29 ♍ 44 34	05 ♎ 38 36	13 05	01 55	23 24	06 54	13 55	05 18	12 23	14 14
19	01 50 16	29 ♈ 24 15	11 ♎ 33 15	17 ♎ 28 48	14 55	03 09	24 01	07 06	13 55	05 17	12 23	14 13
20	01 54 12	00 ♉ 22 49	23 ♎ 25 30	29 ♎ 23 34	16 41	04 23	24 38	07 18	13 55	05 17	12 22	14 11
21	01 58 09	01 21 20	05 ♏ 23 11	11 ♏ 24 32	18 24	05 37	25 15	07 31	13 56	05 16	12 22	14 09
22	02 02 05	02 19 50	17 ♏ 27 45	23 ♏ 33 00	20 04	06 51	25 52	07 44	13 56	05 16	12 22	14 08
23	02 06 02	03 18 18	29 ♏ 40 25	05 ♐ 50 08	21 39	08 06	26 29	07 56	13 R 56	05 15	12 22	14 06
24	02 09 58	04 16 44	12 ♐ 02 20	18 ♐ 17 11	23 10	09 20	27 06	08 09	13 56	05 14	12 21	14 04
25	02 13 55	05 15 09	24 ♐ 34 54	00 ♑ 55 42	24 37	10 34	27 43	08 22	13 55	05 13	12 21	14 03
26	02 17 52	06 13 32	07 ♑ 19 52	13 ♑ 47 40	25 59	11 48	28 20	08 35	13 55	05 13	12 21	14 01
27	02 21 48	07 11 53	20 ♑ 19 23	26 ♑ 55 22	27 17	13 02	28 57	08 48	13 55	05 12	12 20	13 59
28	02 25 45	08 10 13	03 ♒ 35 53	10 ♒ 21 15	28 30	14 16	29 ♊ 34	09 00	13 54	05 11	12 20	13 58
29	02 29 41	09 08 32	17 ♒ 11 43	24 ♒ 07 30	29 ♉ 39	15 30	00 ♋ 11	09 13	13 54	05 10	12 19	13 56
30	02 33 38	10 ♉ 06 48	01 ♓ 08 41	08 ♓ 15 19	00 ♊ 42	16 ♉ 44	00 ♋ 48	09 ♊ 26	13 ♑ 53	05 ♑ 09	12 ♑ 19	13 ♏ 54

Date	Moon True ☊ ° '	Moon Mean ☊ ° '	Moon ☽ Latitude ° '	DECLINATIONS Sun ☉ ° '	Moon ☽ ° '	Mercury ☿ ° '	Venus ♀ ° '	Mars ♂ ° '	Jupiter ♃ ° '	Saturn ♄ ° '	Uranus ♅ ° '	Neptune ♆ ° '	Pluto ♇ ° '
01	04 ♓ 19	03 ♓ 00	02 S 20	04 N 38	20 S 38	02 N 07	03 N 05	23 N 42	20 N 20	22 S 05	23 S 35	21 S 57	00 S 40
02	04 D 20	02 56	01 S 08	05 02	15 19	03 01	03 35	23 47	20 22	22 05	23 35	21 57	00 39
03	04 R 21	02 53	00 N 11	05 25	09 01	03 56	04 05	23 51	20 24	22 04	23 35	21 57	00 38
04	04 20	02 50	01 31	05 47	02 S 04	04 51	04 35	23 55	20 26	22 04	23 35	21 57	00 38
05	04 17	02 47	02 45	06 10	05 N 05	05 47	05 05	24 00	20 29	22 04	23 35	21 57	00 37
06	04 12	02 44	03 48	06 33	11 59	06 43	05 34	24 04	20 31	22 04	23 35	21 57	00 37
07	04 06	02 41	04 35	06 56	18 08	07 39	06 04	24 08	20 33	22 04	23 35	21 57	00 36
08	03 59	02 37	05 02	07 18	23 04	08 35	06 33	24 11	20 35	22 04	23 35	21 57	00 35
09	03 52	02 34	05 09	07 40	26 25	09 31	07 02	24 15	20 38	22 04	23 35	21 57	00 35
10	03 47	02 31	04 56	08 03	28 01	10 26	07 31	24 18	20 40	22 03	23 35	21 57	00 34
11	03 43	02 28	04 27	08 25	27 50	11 21	08 00	24 21	20 42	22 03	23 35	21 57	00 34
12	03 42	02 25	03 45	08 47	26 05	12 15	08 29	24 24	20 45	22 03	23 35	21 57	00 33
13	03 D 42	02 22	02 53	09 08	23 02	13 08	08 58	24 27	20 47	22 03	23 35	21 57	00 32
14	03 43	02 18	01 54	09 30	19 00	14 00	09 26	24 30	20 49	22 03	23 35	21 57	00 32
15	03 44	02 15	00 N 51	09 52	14 16	14 50	09 54	24 32	20 51	22 03	23 35	21 57	00 31
16	03 R 44	02 12	00 S 13	10 13	09 04	15 38	10 22	24 35	20 53	22 03	23 35	21 57	00 31
17	03 43	02 09	01 16	10 34	03 N 36	16 25	10 50	24 37	20 56	22 03	23 35	21 57	00 30
18	03 40	02 06	02 15	10 55	01 S 58	17 09	11 17	24 39	20 58	22 03	23 35	21 57	00 30
19	03 34	02 02	03 08	11 16	07 28	17 52	11 44	24 41	21 00	22 03	23 35	21 57	00 29
20	03 24	01 59	03 54	11 36	12 43	18 32	12 11	24 42	21 02	22 03	23 35	21 57	00 29
21	03 16	01 56	04 29	11 57	17 33	19 09	12 38	24 44	21 05	22 03	23 35	21 57	00 28
22	03 05	01 53	04 53	12 17	21 43	19 45	13 04	24 45	21 07	22 03	23 35	21 57	00 27
23	02 54	01 50	05 03	12 37	25 01	20 17	13 30	24 46	21 09	22 03	23 36	21 57	00 27
24	02 43	01 47	05 00	12 57	27 11	20 47	13 55	24 47	21 11	22 03	23 36	21 57	00 26
25	02 35	01 43	04 42	13 16	28 02	21 15	14 21	24 48	21 13	22 03	23 36	21 57	00 26
26	02 29	01 40	04 10	13 36	27 24	21 40	14 45	24 48	21 15	22 03	23 36	21 57	00 25
27	02 26	01 37	03 24	13 55	25 16	22 03	15 10	24 48	21 18	22 03	23 36	21 57	00 25
28	02 24	01 34	02 27	14 14	21 44	22 23	15 34	24 49	21 20	22 03	23 36	21 57	00 24
29	02 D 24	01 31	01 20	14 33	16 57	22 40	15 58	24 49	21 22	22 03	23 36	21 57	00 24
30	02 ♓ 25	01 ♓ 27	00 S 07	14 N 51	11 S 10	22 N 55	16 N 21	24 N 49	21 N 24	22 S 03	23 S 36	21 S 57	00 S 24

ZODIAC SIGN ENTRIES

Date	h	m	Planets
03	01	37	☽ ♓
05	01	51	☽ ♈
07	01	07	☽ ♉
09	01	31	☽ ♊
11	04	58	☽ ♋
11	21	36	☿ ♉
13	12	31	☽ ♌
15	23	39	☽ ♍
16	22	52	♀ ♉
18	12	31	☽ ♎
20	02	39	☉ ♉
21	01	13	☽ ♏
23	12	38	☽ ♐
25	22	15	☽ ♑
28	05	33	☽ ♒
29	04	37	♂ ♋
29	19	53	☿ ♊
30	10	03	☽ ♓

LATITUDES

Date	Mercury ☿ ° '	Venus ♀ ° '	Mars ♂ ° '	Jupiter ♃ ° '	Saturn ♄ ° '	Uranus ♅ ° '	Neptune ♆ ° '	Pluto ♇ ° '
01	01 S 23	01 S 19	01 N 21	00 S 33	00 N 41	00 S 15	00 N 55	16 N 16
04	00 59	01 16	01 22	00 33	00 41	00 15	00 55	16 16
07	00 S 31	01 12	01 22	00 32	00 40	00 15	00 55	16 17
10	00 00	01 08	01 22	00 32	00 40	00 15	00 55	16 18
13	00 N 33	01 03	01 22	00 31	00 40	00 15	00 55	16 18
16	01 06	00 58	01 22	00 31	00 40	00 15	00 55	16 19
19	01 37	00 53	01 22	00 30	00 40	00 15	00 55	16 19
22	02 04	00 47	01 22	00 30	00 40	00 15	00 56	16 19
25	02 24	00 41	01 22	00 29	00 40	00 15	00 56	16 20
28	02 37	00 35	01 22	00 29	00 40	00 15	00 56	16 20
31	02 N 41	00 S 28	01 N 22	00 S 29	00 N 40	00 S 15	00 N 56	16 N 20

DATA

Julian Date	2447618
Delta T	+56 seconds
Ayanamsa	23° 42' 32"
Synetic vernal point	05° ♓ 24' 27"
True obliquity of ecliptic	23° 26' 35"

MOON'S PHASES, APSIDES AND POSITIONS ☽

Date	h	m	Phase	Longitude ° '	Eclipse Indicator
06	03	33	●	16 ♈ 19	
12	23	13	☽	23 ♋ 01	
21	03	13	○	01 ♏ 00	
28	20	46	☾	08 ♒ 32	

Day	h	m		
05	19	27	Perigee	
18	21	00	Apogee	
04	18	57	ON	
10	21	26	Max dec	28° N 09'
18	03	33	OS	
25	13	56	Max dec	28° S 02'

LONGITUDES

Date	Chiron ⚷ ° '	Ceres ⚳ ° '	Pallas ⚴ ° '	Juno ⚵ ° '	Vesta ⚶ ° '	Black Moon Lilith ☾ ° '
01	01 ♋ 46	27 ♈ 10	18 ♓ 42	25 ♌ 30	04 ♑ 16	05 ♎ 46
11	02 ♋ 11	01 ♉ 16	21 ♓ 55	25 ♌ 16	06 ♑ 53	06 ♎ 53
21	02 ♋ 44	05 ♉ 12	25 ♓ 03	25 ♌ 45	08 ♑ 56	07 ♎ 59
31	03 ♋ 23	09 ♉ 14	28 ♓ 05	26 ♌ 47	10 ♑ 19	09 ♎ 06

ASPECTARIAN

01 Saturday
04 13 ☽ ∥ ♄
04 51 ☽ △ ♃
04 56 ☽ ∥ ♆
07 59 ☽ ⊻ ♅
13 25 ☿ ± ♇
13 30 ☽ # ♃
13 49 ☽ ✱ ☿
18 03 ☽ ✱ ♀
18 18 ☽ ⊥ ♅
19 34 ☽ ✱ ☉
20 04 ☽ ⊻ ♆
21 28 ☽ △ ♂
22 09 ☽ ⊻ ♄
23 59 ☽ □ ♇

02 Sunday
02 52 ☉ □ ♆
06 14 ☽ ⊥ ♆
08 19 ☽ ⊥ ♄
09 32 ☽ ∠ ♅
13 55 ♂ ⊼ ♄
17 04 ♀ □ ♆
19 17 ☽ ∠ ☿
21 15 ☽ ∠ ♆
21 38 ☽ ∠ ♀
22 35 ☽ ∠ ☉
23 19 ☽ ∠ ♄

03 Monday
08 00 ☽ □ ♃
09 22 ☿ □ ♆
09 45 ☉ □ ♄
10 19 ☽ ✱ ♅
12 25 ☽ ⊥ ☿
13 44 ☽ ⊥ ♀
14 21 ☽ ⊥ ☉
17 44 ♀ □ ♄
20 45 ☿ ∥ ♀
21 43 ☽ ✱ ♆
23 40 ☽ ⊻ ☿
23 46 ☽ ✱ ♄
23 59 ☽ # ☉

04 Tuesday
00 18 ☽ ⊻ ♀
00 26 ☿ □ ♄
00 44 ☽ ⊻ ☉
01 09 ☽ □ ♂
01 20 ☽ △ ♇
03 40 ☽ # ☿
04 03 ☽ # ♀
05 17 ♂ ⊼ ♇
05 42 ☽ Q ♅
09 33 ☉ ⊼ ♇
10 29 ☿ ☌ ♀
11 38 ☿ ⊼ ♇
12 23 ♀ ⊼ ♇
13 22 ☽ Q ♃
13 36 ☉ ☌ ☿
14 27 ☿ ✱ ♂
16 51 ☽ ∥ ♇
16 55 ☽ Q ♆
16 56 ☉ ✱ ♂
18 58 ☽ Q ♄
19 37 ♀ ✱ ♂
21 03 ☽ # ♇
23 29 ☉ ☌ ♀

05 Wednesday
01 11 ☽ ⊑ ♇
06 59 ☽ Q ♂
08 36 ☽ ✱ ♃
10 15 ☽ □ ♅
11 58 ☽ ∥ ♀
14 43 ☽ ∥ ☿
15 21 ☽ ± ♇
15 53 ☽ ∥ ☉
21 20 ☽ □ ♆
23 25 ☽ □ ♄

06 Thursday
00 46 ☽ ⊼ ♇
02 40 ☽ ✱ ♂
03 33 ☽ ☌ ☉
04 03 ☽ ☌ ♀
04 54 ☉ ∥ ☿
06 39 ☽ ☌ ☿
08 30 ☽ ∠ ♃
21 29 ☿ ∠ ♃
22 53 ☽ ⊥ ♃

07 Friday
03 19 ☽ ∠ ♂
08 31 ☽ ⊻ ♃
09 35 ☽ △ ♅
18 16 ☽ ⊥ ♂
20 50 ☽ △ ♆
23 01 ☽ △ ♄
23 02 ☽ ∥ ♃

08 Saturday
00 15 ☽ ☍ ♇
04 20 ☽ ⊻ ♂
05 55 ☽ # ♆
06 30 ☽ # ♄
06 33 ☽ ⊻ ☉
08 02 ☽ ⊻ ♀
09 40 ☽ ⊑ ♅
14 11 ☽ ⊻ ☿
15 03 ☽ # ♅
17 01 ☽ ⊥ ☉
18 43 ☽ ⊥ ♀
18 55 ☽ ∥ ♂
20 05 ♀ ∠ ♃
21 11 ☽ ⊑ ♆
23 29 ☽ ⊑ ♄

09 Sunday
00 25 ☽ ± ♅
01 42 ☽ ⊥ ☿
08 52 ♅ St R
08 59 ☽ ∠ ☉
09 56 ☽ ☌ ♃
10 24 ☽ ⊼ ♅
11 03 ☽ ∠ ♀
12 09 ☽ ± ♆
14 30 ☽ ± ♄
19 20 ☽ ∠ ☿
22 17 ☽ ⊼ ♆

10 Monday
00 42 ☽ ⊼ ♄
01 50 ☽ ⊼ ♇
02 16 ☉ ∠ ♃
08 34 ☽ ☌ ♂
12 09 ☽ ± ♇
12 27 ☽ ✱ ☉
15 10 ☽ ✱ ♀
18 59 ♃ ⊼ ♅

11 Tuesday
01 57 ☽ ✱ ☿
03 57 ☽ ⊑ ♇
11 19 ☽ Q ☉
14 32 ☽ ☍ ♅
14 38 ☽ Q ♀
14 50 ☽ ⊻ ♃
15 52 ☿ ⊥ ♃

12 Wednesday
01 56 ☽ ⊥ ♃
03 23 ☽ ☍ ♆
03 38 ☽ Q ☿
06 04 ☽ ☍ ♄
07 06 ☽ △ ♇
17 13 ☽ ⊻ ♂
18 55 ☽ ∠ ♃
23 13 ☽ □ ☉

13 Thursday
02 01 ☽ ∥ ♂
03 31 ☽ □ ♀
05 08 ☽ ⊥ ♂
08 12 ☽ # ♅
14 06 ♂ ± ♇
18 18 ☽ # ♄
18 58 ☽ # ♆
20 12 ☽ □ ☿
22 48 ☽ ⊼ ♅
23 20 ☽ ∠ ♂
23 38 ♆ St R

14 Friday
00 03 ☽ ✱ ♃
01 52 ☽ ∥ ♃
10 30 ☽ ± ♅
12 04 ☿ △ ♅
12 35 ☽ ⊼ ♆
15 30 ☽ ⊼ ♄
16 26 ☽ □ ♇
22 04 ☿ ⊻ ♃
22 33 ☿ ∠ ♂

15 Saturday
00 01 ☽ Q ♃
00 26 ☽ ± ♆
03 05 ☉ ∥ ♀
03 24 ☽ ± ♄
04 17 ☽ ⊑ ♅
06 26 ☽ ✱ ♂
09 40 ☽ ∥ ☿
14 50 ☽ △ ☉
18 24 ☽ ⊑ ♆
20 58 ☽ △ ♀
21 25 ☽ ⊑ ♄

16 Sunday
00 25 ♃ ± ♆
04 17 ☽ Q ♇
06 38 ☽ ∥ ♀
07 10 ☽ ∥ ☉
07 50 ☽ Q ♂
10 22 ☽ △ ♅
12 45 ☽ □ ♃
19 39 ☽ △ ☿
23 49 ☽ ⊑ ☉

17 Monday
00 41 ☽ △ ♆
03 46 ☽ △ ♄
04 32 ☽ ✱ ♇
06 50 ☽ ⊑ ♀
12 14 ♀ ⊥ ♃
19 46 ☽ ± ☉
22 22 ☽ □ ♂

18 Tuesday
01 24 ☽ # ♇
02 55 ☿ △ ♆
03 18 ☽ ± ♀
05 41 ☽ ∥ ♇
08 00 ☽ ⊑ ☿
09 05 ☽ ⊼ ☉
10 58 ☽ ∠ ♇
16 56 ☽ ⊼ ♀
22 47 ☿ △ ♄
23 17 ☽ □ ♅

19 Wednesday
02 48 ☽ △ ♃
02 48 ☿ ☍ ♇
05 14 ☽ ⊥ ♇
05 41 ☽ ± ☿
13 40 ☽ □ ♆
16 48 ☽ □ ♄
17 22 ☽ ⊻ ♇
20 01 ☽ ⊼ ☿

20 Thursday
06 26 ☽ # ☉
09 15 ☽ # ♀
09 43 ☽ ⊑ ♃
11 43 ☽ Q ♅
14 34 ☽ △ ♂

21 Friday
01 58 ☽ Q ♆
03 13 ☽ ☍ ☉
04 07 ☽ ± ♃
05 05 ☽ Q ♄
05 15 ♀ △ ♅
11 46 ☽ ✱ ♅
12 31 ☽ ☍ ♀
16 20 ☽ ⊼ ♃
17 04 ☉ ⊥ ♃
22 12 ☽ # ☿
22 13 ☽ ⊑ ♂

22 Saturday
01 54 ☽ ✱ ♆
05 00 ☽ ✱ ♄
05 25 ☽ ☌ ♇
07 25 ☿ ⊥ ♂
08 10 ☽ # ♃
13 26 ☽ ∥ ♆
14 05 ☽ ∥ ♄
14 59 ☿ ⊑ ♅
17 00 ☽ ± ♂
17 31 ☽ ∠ ♅
17 54 ☽ ☍ ☿
23 37 ♄ St R

23 Sunday
00 42 ☽ ∥ ♅
05 26 ☽ ⊼ ♂
07 29 ☽ ∠ ♆
08 24 ♀ ⊻ ♃
09 51 ☽ # ♂
10 33 ☽ ∠ ♄
10 43 ♃ ± ♄
11 10 ☽ ⊥ ♅
19 41 ☽ ⊼ ☉
22 51 ☽ ⊻ ♅

24 Monday
01 01 ☽ ⊥ ♆
04 03 ☽ ⊥ ♄
04 21 ☽ ☍ ♃
06 11 ☽ ⊼ ♀
08 19 ☽ ± ☉
12 36 ☽ ⊻ ♆
15 38 ☽ ⊻ ♄
15 55 ☽ ⊻ ♇
19 01 ☽ ± ♀

25 Tuesday
03 04 ☽ ⊑ ☉
03 23 ☽ ⊥ ♇
10 12 ☿ ∥ ♃
11 19 ☉ △ ♅
12 04 ☽ ⊼ ☿
14 04 ☽ ⊑ ♀
18 15 ☽ ☍ ♂
20 26 ☽ ∠ ♇

26 Wednesday
00 48 ☽ ± ☿
08 02 ☽ ☌ ♅
09 46 ☽ △ ☉
14 22 ☽ ⊼ ♃
19 34 ☽ ⊑ ☿
21 11 ☽ △ ♀
21 19 ☽ ☌ ♆
22 33 ♀ △ ♆

27 Thursday
00 13 ☽ ☌ ♄
00 23 ☽ ✱ ♇
01 40 ☽ ± ♃
05 33 ☿ # ♆
12 22 ☿ # ♄
13 02 ☿ ⊑ ♆
13 22 ♂ ⊑ ♇
15 46 ☽ # ♂
18 26 ☽ ⊑ ♃
22 18 ☽ Q ♇

28 Friday
00 31 ☽ ∥ ♅
01 57 ☽ △ ☿
04 26 ☽ ⊼ ♂
05 00 ♀ △ ♄
06 15 ♀ ☍ ♇
08 31 ☽ # ☿
10 11 ☽ ∥ ♄
10 47 ☽ ∥ ♆
14 17 ☽ # ♃
14 49 ☽ ⊻ ♅
15 41 ☽ ± ♂
20 12 ☿ ⊑ ♄
20 46 ☽ □ ☉
21 47 ☽ △ ♃
23 58 ♀ ∠ ♂

29 Saturday
01 26 ☽ ⊥ ♅
01 53 ☿ ± ♅
03 28 ☽ ⊻ ♆
06 14 ☽ ⊻ ♄
06 19 ☽ □ ♇
08 20 ☽ ⊑ ♂
08 45 ☽ □ ♀
13 57 ☽ ⊥ ♆
14 37 ☉ ⊻ ♃
16 06 ☽ # ♀
16 41 ☽ ⊥ ♄
17 09 ☽ ∠ ♅
21 55 ☽ # ☉

30 Sunday
05 28 ☽ ∠ ♆
06 27 ☽ Q ☉
08 09 ☽ ∠ ♄
11 12 ☽ □ ☿
11 24 ☽ △ ♂
18 23 ☿ ⊻ ♂
18 39 ☽ Q ♀
18 46 ☽ ✱ ♅

All ephemeris data is given at 12.00 UT and the Moon's longitude is additionally given for 24.00 UT

MAY 1989

LONGITUDES

Date	Sidereal time h m s	Sun ☉ ° ' "	Moon ☽ ° ' "	Moon ☽ 24.00 ° ' "	Mercury ☿ ° '	Venus ♀ ° '	Mars ♂ ° '	Jupiter ♃ ° '	Saturn ♄ ° '	Uranus ♅ ° '	Neptune ♆ ° '	Pluto ♇ ° '
01	02 37 34	11 ♉ 05 04	15 ♓ 27 16	22 ♓ 44 15	01 ♊ 41	17 ♉ 58	01 ♋ 26	09 ♊ 40	13 ♑ 52	05 ♑ 08	12 ♑ 18	13 ♏ 53
02	02 41 31	12 03 17	00 ♈ 05 51	07 ♈ 31 25	02 35	19 12	02 03	09 53	13 R 51	05 R 07	12 R 18	13 R 51
03	02 45 27	13 01 29	15 ♈ 00 10	22 ♈ 31 08	03 24	20 26	02 40	10 06	13 50	05 06	12 17	13 49
04	02 49 24	13 59 40	00 ♉ 03 12	07 ♉ 35 10	04 07	21 40	03 17	10 19	13 49	05 04	12 16	13 48
05	02 53 21	14 57 49	15 ♉ 05 47	22 ♉ 33 50	04 46	22 54	03 54	10 32	13 48	05 03	12 16	13 46
06	02 57 17	15 55 57	29 ♉ 58 08	07 ♊ 17 38	05 20	24 08	04 31	10 46	13 47	05 02	12 15	13 44
07	03 01 14	16 54 03	14 ♊ 31 27	21 ♊ 38 53	05 48	25 22	05 08	10 59	13 45	05 00	12 14	13 43
08	03 05 10	17 52 07	28 ♊ 39 27	05 ♋ 32 51	06 11	26 35	05 45	11 12	13 44	04 59	12 13	13 41
09	03 09 07	18 50 09	12 ♋ 19 01	18 ♋ 58 01	06 29	27 49	06 22	11 26	13 42	04 58	12 13	13 39
10	03 13 03	19 48 10	25 ♋ 30 08	01 ♌ 55 43	06 41	29 ♉ 03	07 00	11 39	13 41	04 56	12 12	13 38
11	03 17 00	20 46 08	08 ♌ 15 17	14 ♌ 29 23	06 49	00 ♊ 17	07 37	11 53	13 39	04 55	12 11	13 36
12	03 20 56	21 44 05	20 ♌ 38 38	26 ♌ 43 43	06 R 51	01 31	08 14	12 06	13 37	04 53	12 10	13 34
13	03 24 53	22 41 59	02 ♍ 45 18	08 ♍ 44 04	06 49	02 45	08 51	12 20	13 35	04 52	12 09	13 33
14	03 28 50	23 39 52	14 ♍ 40 41	20 ♍ 35 48	06 42	03 58	09 28	12 33	13 33	04 50	12 08	13 31
15	03 32 46	24 37 44	26 ♍ 30 02	02 ♎ 23 58	06 30	05 12	10 05	12 47	13 31	04 48	12 07	13 29
16	03 36 43	25 35 33	08 ♎ 18 09	14 ♎ 13 04	06 14	06 26	10 43	13 00	13 29	04 47	12 07	13 28
17	03 40 39	26 33 21	20 ♎ 09 08	26 ♎ 06 46	05 54	07 40	11 20	13 14	13 27	04 45	12 06	13 26
18	03 44 36	27 31 07	02 ♏ 06 16	08 ♏ 07 56	05 30	08 53	11 57	13 28	13 25	04 43	12 05	13 24
19	03 48 32	28 28 52	14 ♏ 11 57	20 ♏ 18 29	05 04	10 07	12 34	13 41	13 22	04 42	12 03	13 23
20	03 52 29	29 ♉ 26 36	26 ♏ 27 40	02 ♐ 39 35	04 35	11 21	13 11	13 55	13 20	04 40	12 02	13 21
21	03 56 25	00 ♊ 24 18	08 ♐ 54 15	15 ♐ 11 42	04 03	12 34	13 48	14 09	13 17	04 38	12 01	13 19
22	04 00 22	01 21 59	21 ♐ 31 56	27 ♐ 54 57	03 30	13 48	14 26	14 23	13 14	04 36	12 00	13 18
23	04 04 18	02 19 39	04 ♑ 20 46	10 ♑ 49 24	02 56	15 02	15 03	14 36	13 12	04 34	11 59	13 16
24	04 08 15	03 17 17	17 ♑ 20 52	23 ♑ 55 16	02 22	16 15	15 40	14 50	13 09	04 32	11 58	13 15
25	04 12 12	04 14 55	00 ♒ 32 39	07 ♒ 13 10	01 48	17 29	16 17	15 04	13 06	04 30	11 57	13 13
26	04 16 08	05 12 31	13 ♒ 56 56	20 ♒ 44 05	01 15	18 42	16 54	15 18	13 03	04 28	11 56	13 12
27	04 20 05	06 10 07	27 ♒ 34 47	04 ♓ 29 09	00 43	19 56	17 32	15 32	13 00	04 26	11 54	13 10
28	04 24 01	07 07 42	11 ♓ 27 17	18 ♓ 29 12	00 ♊ 13	21 09	18 09	15 46	12 57	04 24	11 53	13 09
29	04 27 58	08 05 15	25 ♓ 34 53	02 ♈ 44 10	29 ♉ 45	22 23	18 46	15 59	12 54	04 22	11 52	13 07
30	04 31 54	09 02 48	09 ♈ 56 50	17 ♈ 12 29	29 20	23 36	19 23	16 13	12 51	04 20	11 51	13 06
31	04 35 51	10 ♊ 00 21	24 ♈ 30 36	01 ♉ 50 31	28 ♉ 59	24 ♊ 50	20 ♋ 00	16 ♊ 27	12 ♑ 47	04 ♑ 18	11 ♑ 49	13 ♏ 04

DECLINATIONS

Date	Moon True ☊ ° '	Moon Mean ☊ ° '	Moon ☽ Latitude ° '	Sun ☉ ° '	Moon ☽ ° '	Mercury ☿ ° '	Venus ♀ ° '	Mars ♂ ° '	Jupiter ♃ ° '	Saturn ♄ ° '	Uranus ♅ ° '	Neptune ♆ ° '	Pluto ♇ ° '
01	02 ♓ 24	01 ♓ 24	01 N 09	15 N 09	04 S 41	23 N 08	16 N 44	24 N 48	21 N 26	22 S 03	23 S 36	21 S 57	00 S 23
02	02 R 22	01 21	02 21	15 27	02 N 12	23 18	17 07	24 48	21 28	22 03	23 36	21 57	00 23
03	02 16	01 18	03 26	15 45	09 04	23 26	17 29	24 47	21 30	22 03	23 36	21 57	00 22
04	02 09	01 15	04 16	16 02	15 29	23 32	17 51	24 46	21 32	22 04	23 36	21 57	00 22
05	01 59	01 12	04 49	16 20	20 58	23 35	18 12	24 45	21 34	22 04	23 36	21 57	00 21
06	01 48	01 08	05 01	16 37	25 03	23 36	18 33	24 44	21 36	22 04	23 36	21 57	00 21
07	01 38	01 05	04 54	16 53	27 25	23 35	18 53	24 42	21 38	22 04	23 36	21 57	00 21
08	01 29	01 02	04 29	17 10	27 55	23 32	19 12	24 41	21 40	22 04	23 36	21 57	00 20
09	01 22	00 59	03 49	17 26	26 40	23 27	19 32	24 39	21 42	22 04	23 36	21 57	00 20
10	01 18	00 56	02 57	17 41	23 57	23 20	19 50	24 37	21 44	22 05	23 37	21 57	00 19
11	01 16	00 53	01 59	17 57	20 07	23 11	20 09	24 35	21 46	22 05	23 37	21 57	00 19
12	01 D 15	00 49	00 N 56	18 12	15 30	23 00	20 26	24 32	21 48	22 05	23 37	21 57	00 19
13	01 R 15	00 46	00 S 08	18 27	10 22	22 47	20 43	24 30	21 50	22 05	23 37	21 58	00 18
14	01 15	00 43	01 11	18 42	04 N 57	22 32	21 00	24 27	21 52	22 05	23 37	21 58	00 18
15	01 12	00 40	02 09	18 56	00 S 35	22 16	21 16	24 24	21 53	22 06	23 37	21 58	00 18
16	01 08	00 37	03 03	19 10	06 05	21 58	21 31	24 21	21 55	22 06	23 37	21 58	00 17
17	01 01	00 33	03 48	19 23	11 24	21 39	21 46	24 18	21 57	22 06	23 37	21 58	00 17
18	00 51	00 30	04 24	19 37	16 19	21 18	22 00	24 15	21 59	22 06	23 37	21 58	00 17
19	00 39	00 27	04 48	19 49	20 41	20 57	22 13	24 11	22 01	22 07	23 37	21 58	00 17
20	00 26	00 24	04 59	20 02	24 12	20 34	22 26	24 08	22 02	22 07	23 37	21 58	00 16
21	00 12	00 21	04 56	20 14	26 40	20 11	22 38	24 04	22 04	22 07	23 37	21 58	00 16
22	00 00	00 18	04 39	20 26	27 49	19 48	22 50	24 00	22 06	22 08	23 38	21 58	00 16
23	29 ♒ 49	00 14	04 08	20 38	27 30	19 24	23 01	23 56	22 07	22 08	23 38	21 58	00 16
24	29 42	00 11	03 23	20 49	25 40	19 01	23 11	23 51	22 09	22 08	23 38	21 58	00 15
25	29 37	00 08	02 27	21 00	22 26	18 38	23 21	23 47	22 11	22 08	23 38	21 58	00 15
26	29 35	00 05	01 22	21 10	17 57	18 16	23 30	23 42	22 12	22 09	23 38	21 59	00 15
27	29 D 34	00 ♓ 02	00 S 11	21 20	12 29	17 54	23 38	23 37	22 14	22 09	23 38	21 59	00 15
28	29 R 34	29 ♒ 59	01 N 02	21 30	06 S 18	17 34	23 46	23 32	22 16	22 09	23 38	21 59	00 15
29	29 34	29 55	02 13	21 40	00 N 17	17 15	23 52	23 27	22 17	22 10	23 38	21 59	00 14
30	29 31	29 52	03 16	21 49	06 57	16 58	23 59	23 21	22 19	22 10	23 38	21 59	00 14
31	29 ♒ 26	29 ♒ 49	04 N 07	21 N 57	13 N 20	16 N 42	24 N 04	23 N 16	22 N 20	22 S 11	23 S 38	21 S 59	00 S 14

ZODIAC SIGN ENTRIES

Date	h	m	Planets
02	11	51	☽ ♈
04	11	55	☽ ♉
06	12	03	☽ ♊
08	14	19	☽ ♋
10	20	23	☽ ♌
11	06	28	♀ ♊
13	06	30	☽ ♍
15	19	07	☽ ♎
18	07	48	☽ ♏
20	18	52	☽ ♐
21	01	54	☉ ♊
23	03	54	☽ ♑
25	11	01	☽ ♒
27	16	13	☽ ♓
28	22	53	☿ ♉
29	19	25	☽ ♈
31	20	59	☽ ♉

LATITUDES

Date	Mercury ☿ ° '	Venus ♀ ° '	Mars ♂ ° '	Jupiter ♃ ° '	Saturn ♄ ° '	Uranus ♅ ° '	Neptune ♆ ° '	Pluto ♇ ° '
01	02 N 41	00 S 28	01 N 22	00 S 29	00 N 40	00 S 15	00 N 56	16 N 20
04	02 36	00 21	01 22	00 28	00 40	00 15	00 56	16 20
07	02 21	00 14	01 22	00 28	00 40	00 15	00 56	16 19
10	01 56	00 S 07	01 22	00 27	00 40	00 16	00 56	16 19
13	01 21	00 00	01 21	00 27	00 40	00 16	00 56	16 18
16	00 N 38	00 N 08	01 21	00 27	00 40	00 16	00 56	16 18
19	00 S 12	00 15	01 21	00 26	00 40	00 16	00 56	16 17
22	01 05	00 22	01 21	00 26	00 40	00 16	00 56	16 17
25	01 56	00 30	01 20	00 26	00 40	00 16	00 56	16 16
28	02 41	00 37	01 20	00 25	00 39	00 16	00 56	16 16
31	03 S 18	00 N 44	01 N 19	00 S 25	00 N 39	00 S 16	00 N 56	16 N 15

DATA

Julian Date	2447648
Delta T	+56 seconds
Ayanamsa	23° 42' 36"
Synetic vernal point	05° ♓ 24' 24"
True obliquity of ecliptic	23° 26' 35"

MOON'S PHASES, APSIDES AND POSITIONS ☽

Date	h	m	Phase	Longitude ° '	Eclipse Indicator
05	11	46	●	14 ♉ 57	
12	14	20	☽	21 ♌ 50	
20	18	16	○	29 ♏ 42	
28	04	01	☾	06 ♓ 49	

Day	h	m	
04	04	25	Perigee
16	09	19	Apogee

Day	h	m		
02	04	23	ON	
08	06	32	Max dec	27° N 58'
15	09	27	OS	
22	18	59	Max dec	27° S 53'
29	11	00	ON	

LONGITUDES

Date	Chiron ⚷ ° '	Ceres ⚳ ° '	Pallas ⚴ ° '	Juno ⚵ ° '	Vesta ⚶ ° '	Black Moon Lilith ⚸ ° '
01	03 ♋ 23	09 ♉ 14	28 ♓ 05	26 ♌ 47	10 ♑ 19	09 ♎ 06
11	04 ♋ 35	13 ♉ 17	01 ♈ 01	28 ♌ 18	10 ♑ 57	10 ♎ 13
21	04 ♋ 58	17 ♉ 19	03 ♈ 48	00 ♍ 14	10 ♑ 47	11 ♎ 20
31	05 ♋ 51	21 ♉ 20	06 ♈ 25	02 ♍ 31	09 ♑ 48	12 ♎ 27

All ephemeris data is given at 12.00 UT and the Moon's longitude is additionally given for 24.00 UT

ASPECTARIAN

h m Aspects

01 Monday
02 12 ☽ □ ♃
04 12 ☽ ⚹ ☉
06 46 ☽ ⚹ ♆
09 22 ☽ ⚹ ♄
09 23 ☽ △ ♇
14 46 ☽ Q ♅
16 32 ☽ ⚹ ♀
19 28 ☽ Q ☿

02 Tuesday
02 33 ☽ Q ♆
03 04 ☽ ∥ ♇
05 06 ☽ Q ♄
05 26 ♄ ⚹ ♇
05 42 ☽ # ♇
06 42 ☽ ∠ ☉
08 20 ☽ Q ♃
09 59 ☽ bQ ♇
15 18 ☽ □ ♂
16 16 ☽ ⚹ ☿
17 49 ☉ △ ♆
19 14 ☽ ∠ ♀
20 06 ☽ □ ♅
22 18 ☽ ⊥ ☉

03 Wednesday
00 30 ☽ ± ♇
04 01 ☽ ⚹ ♃
05 30 ♀ bQ ♅
07 39 ☽ □ ♆
08 37 ☽ ⊻ ☉
10 07 ☽ ⊼ ♇
10 08 ☽ □ ♄
11 00 ☽ ⊥ ♀
17 43 ☽ ∠ ☿
21 26 ☽ Q ♂
21 27 ☽ ⊻ ♀

04 Thursday
04 21 ☽ ∠ ♃
07 11 ☉ ☍ ♇
07 45 ☉ △ ♄
08 47 ☽ ⊥ ☿
14 21 ☽ ∥ ☉
17 22 ☽ ⚹ ♂
18 47 ☽ ⊻ ☿
18 54 ☽ ⊥ ♃
19 59 ☽ △ ♅
22 23 ☽ ∥ ♀

05 Friday
04 36 ☽ ⊻ ♃
07 28 ☽ △ ♆
09 52 ☽ ☍ ♇
09 56 ☽ △ ♄
11 46 ☽ ☌ ☉
15 04 ☽ ∥ ♃
17 02 ☽ # ♆
17 38 ☽ # ♄
18 22 ☽ ∠ ♂
19 56 ☽ bQ ♅
23 17 ☿ ⊼ ♅

06 Saturday
01 40 ☽ ☌ ♀
02 22 ☽ ∥ ☿
02 22 ☽ # ♅
04 34 ☿ # ♅
07 35 ☽ bQ ♆
09 32 ☽ ⊥ ♂
09 42 ☽ ∥ ♂
10 04 ☽ bQ ♄
10 28 ☽ ± ♅
19 46 ☽ ⊻ ♂
20 16 ☽ ⊼ ♅
21 00 ☿ # ♅
21 04 ☽ ☌ ☿
22 16 ☽ ± ♆

07 Sunday
00 47 ☽ ± ♄
06 01 ☽ ☌ ♃
07 08 ♂ ☍ ♅
08 11 ☽ ⊼ ♆
10 39 ☽ ⊼ ♇
10 43 ☽ ⊼ ♄
16 16 ☽ ⊻ ☉
20 42 ☽ ± ♇

08 Monday
03 10 ☽ ⊥ ☉
08 06 ☽ ⊻ ♀
12 02 ☽ bQ ♇
15 00 ☿ ± ♆
19 30 ☽ ⊥ ♀
19 52 ☽ ∠ ☉
23 00 ☽ ☍ ♅

09 Tuesday
00 14 ♀ bQ ♆
00 57 ☽ ☌ ♂
01 26 ☽ ⊻ ☿
10 23 ☽ ⊻ ♃
11 49 ☽ ☍ ♆
12 18 ☽ ⊥ ☿
13 00 ☽ ∠ ♀
14 23 ☽ △ ♇
14 29 ☽ ☍ ♄
18 31 ☿ ⊻ ♂
21 22 ☽ ⊥ ♃

10 Wednesday
00 42 ☽ ⚹ ☉
04 53 ☽ ∠ ☿
04 53 ♀ bQ ♄
06 58 ☽ ∥ ♂
09 48 ♀ ± ♅
14 10 ☽ ∠ ♃
14 24 ☽ # ♅
15 16 ☉ bQ ♅
16 28 ☽ ∥ ☿
19 18 ☽ ⚹ ♀

11 Thursday
00 26 ☽ # ♄
00 44 ☽ Q ☉
01 12 ☽ # ♆
02 27 ☽ ∥ ♃
05 39 ☽ ⊼ ♅
09 14 ☽ ⚹ ☿
10 43 ☽ ⊻ ♂
11 51 ☽ ∥ ♀
17 05 ☽ ± ♅
19 05 ☽ ⚹ ♃
19 32 ☽ ⊼ ♆
20 35 ☽ Q ♀
22 15 ☽ □ ♇
22 21 ☽ ⊼ ♄
22 51 ☽ ⊥ ♂
23 04 ☽ ∥ ☉

12 Friday
07 10 ☽ ± ♆
08 30 ☽ Q ☿
10 00 ☽ ± ♄
10 31 ☽ bQ ♅
11 53 ☿ St R
14 20 ☽ □ ☉
17 22 ☽ ∠ ♂
18 45 ♃ ⊼ ♆
18 56 ☽ Q ♃

13 Saturday
00 52 ☽ bQ ♆
03 43 ☽ bQ ♄
09 35 ☽ Q ♇
11 58 ☽ □ ♀
16 13 ☽ △ ♅
16 18 ♀ ⊥ ♂
20 04 ☽ □ ☿

14 Sunday
00 55 ☽ ⚹ ♂
06 53 ☽ △ ♆
07 37 ☽ □ ♃
09 39 ☽ ⚹ ♇
09 44 ☽ △ ♄

15 Monday
02 32 ☽ Q ♂
04 28 ♀ ⊼ ♅
07 51 ☽ △ ☉
08 11 ☽ # ♇
10 44 ☽ ∥ ♇
16 02 ☽ ∠ ♇

16 Tuesday
01 57 ☿ # ♄
04 51 ☽ □ ♅
05 47 ♀ ± ♆
07 45 ☽ △ ♀
07 54 ☽ △ ☿
08 52 ☿ ☌ ♀
10 17 ☽ ⊥ ♇
12 34 ☿ # ♆
15 26 ☿ ∥ ♃
17 04 ☽ bQ ☉
17 10 ☽ □ ♂
19 43 ☽ □ ♆
20 15 ☉ ∠ ♂
21 44 ☽ △ ♃
21 56 ☿ ± ♆
22 26 ☽ ⊻ ♇
22 29 ☽ □ ♄

17 Wednesday
07 19 ☿ ∥ ♀
08 00 ♀ ± ♄
12 53 ☽ ± ☉
13 28 ☽ bQ ☿
17 14 ☽ Q ♅
17 38 ☽ bQ ♀
22 45 ♃ # ♆

18 Thursday
01 08 ☉ bQ ♆
01 41 ☿ ⊥ ♂
02 01 ☽ ⊼ ☉
04 34 ☽ bQ ♃
06 30 ♃ ⊼ ♇
06 59 ☽ ± ☿
07 13 ♃ ⊼ ♄
07 57 ☽ Q ♆
08 51 ♀ # ♆
10 17 ♀ ∥ ♃
10 37 ☽ Q ♄
13 45 ☽ ± ♀
16 45 ♂ ☍ ♆
17 12 ☽ ⚹ ♅
18 33 ☽ ⊼ ☿
21 51 ♄ ⚹ ♇
22 53 ☽ ± ♃

19 Friday
00 07 ♀ # ♄
03 01 ☽ ⊼ ♀
06 42 ☽ # ☉
07 47 ☽ ⚹ ♆
08 37 ☽ △ ♂
09 20 ☉ bQ ♄
10 22 ☽ ⚹ ♄
10 23 ☽ ☌ ♇
10 59 ☽ ⊼ ♃
13 30 ☽ # ☿
17 06 ☉ ± ♅
20 04 ☽ ∥ ♆
20 25 ☽ # ♃
21 01 ☽ ∥ ♄
22 21 ☽ # ♀
22 46 ☽ ∠ ♅

20 Saturday
07 31 ☽ ∥ ♅
07 41 ☿ ⊼ ♅
11 23 ☽ # ♂
13 07 ☽ ∠ ♆
15 32 ☽ bQ ♂
15 37 ☽ ∠ ♄
16 16 ☽ ⊥ ♅
17 05 ♂ ☍ ♄
18 02 ♂ △ ♇
18 16 ☽ ☍ ☉

21 Sunday
01 25 ♀ ⊼ ♆
03 04 ☽ ☍ ☿
03 49 ☽ ⊻ ♅
06 29 ☽ ⊥ ♆
08 55 ☽ ⊥ ♄
09 47 ☽ ± ♂
09 59 ☉ ∥ ☿
17 57 ☽ ⊻ ♆
19 46 ☽ ☍ ♀
20 20 ☽ ⊻ ♄
20 25 ☽ ⊻ ♇
21 51 ☽ ⊼ ♂
22 12 ☽ ☍ ♃

22 Monday
01 29 ♀ ⊼ ♄
02 25 ♀ ⊼ ♇
07 47 ☽ ⊥ ♇
08 57 ♂ ⊻ ♃

23 Tuesday
00 41 ☽ ∠ ♇
01 57 ♀ ☌ ♃
07 56 ☽ ⊼ ☉
09 30 ☽ ⊼ ☿
12 25 ☽ ☌ ♅
12 52 ♀ ⊻ ♂
19 11 ♃ # ♄
19 59 ☽ ± ☉
20 10 ☽ ± ☿
21 35 ☉ ☌ ☿

24 Wednesday
02 07 ☽ ☌ ♆
04 19 ☽ ☌ ♄
04 29 ☽ ⚹ ♇
07 19 ☽ ⊼ ♃
08 46 ☽ ☍ ♂
09 47 ☽ ⊼ ♀
12 02 ☽ bQ ☿
13 51 ☽ bQ ☉
18 30 ☽ ± ♃
21 53 ☽ ± ♀

25 Thursday
02 23 ☽ Q ♇
02 59 ☽ # ♂
04 12 ☽ ∥ ♅
06 22 ☽ # ♀
11 07 ☽ bQ ♃
13 30 ☽ # ♃
13 44 ☽ ∥ ♄
14 10 ☽ △ ☿
14 44 ☽ ∥ ♆
15 50 ☽ bQ ♀
18 09 ☉ ⊼ ♅
19 07 ☽ ⊻ ♅
19 11 ☽ △ ☉
20 01 ☽ # ☉
22 27 ☿ ∠ ♂

26 Friday
05 49 ☽ ⊥ ♅
08 25 ☽ ⊻ ♆
10 22 ☽ # ☿
10 25 ☽ ⊻ ♄
10 40 ☽ □ ♇
14 26 ☽ △ ♃
17 30 ☽ ⊼ ♂
19 02 ☽ ⊥ ♆
21 00 ☽ ⊥ ♄
21 15 ☽ △ ♀
21 23 ♀ ± ♇
21 45 ☽ ∠ ♅

27 Saturday
04 34 ☽ ± ♂
05 34 ☉ ± ♆
07 12 ♂ # ♅
10 09 ♀ ∥ ♂
10 50 ☽ ∠ ♆
12 02 ♀ # ♅
12 44 ☽ ∠ ♄
17 16 ☽ □ ☿
21 01 ☽ bQ ♂
23 53 ☽ ⚹ ♅

28 Sunday
04 01 ☽ □ ☉
07 47 ☉ ± ♄
12 44 ☽ ⚹ ♆
14 33 ☽ ⚹ ♄
14 53 ☽ △ ♇
19 29 ☽ □ ♃
20 26 ☽ Q ♅
23 10 ☽ Q ☿
23 57 ☽ △ ♂

29 Monday
06 05 ☽ □ ♀
09 07 ☽ Q ♆
10 08 ☽ ∥ ♇
10 51 ☽ Q ♄
11 52 ☽ # ♇
12 55 ☽ Q ☉
16 16 ☽ bQ ♇
18 48 ☽ ⚹ ☿

30 Tuesday
02 19 ☽ Q ♃
02 41 ☽ □ ♅
07 16 ☽ ± ♇
10 24 ☽ ⚹ ☉
15 00 ☽ Q ♀
15 08 ☽ □ ♆
16 47 ☽ □ ♄
17 12 ☽ ⊼ ♇
19 05 ☽ ∠ ☿
22 32 ☽ ⚹ ♃

31 Wednesday
04 17 ☽ □ ♂
09 33 ☽ ⊥ ☿
12 35 ☽ ⚹ ♀
12 52 ☽ ∠ ☉
17 25 ☉ # ♆
19 10 ☽ ⊻ ☿
23 33 ☽ ∠ ♃

JUNE 1989

LONGITUDES

Date	Sidereal time h m s	Sun ☉ ° ' "	Moon ☽ ° ' "	Moon ☽ 24.00 ° ' "	Mercury ☿ ° '	Venus ♀ ° '	Mars ♂ ° '	Jupiter ♃ ° '	Saturn ♄ ° '	Uranus ♅ ° '	Neptune ♆ ° '	Pluto ♇ ° '
01	04 39 47	10 ♊ 57 52	09 ♉ 11 29	16 ♉ 32 37	28 ♉ 41	26 ♊ 03	20 ♋ 38	16 ♊ 41	12 ♑ 44	04 ♑ 16	11 ♑ 48	13 ♏ 03
02	04 43 44	11 55 23	23 ♉ 52 59	01 ♊ 11 35	28 R 27	27 17	21 15	16 55	12 R 41	04 R 13	11 R 47	13 R 01
03	04 47 41	12 52 52	08 ♊ 27 28	15 ♊ 39 42	28 17	28 30	21 52	17 09	12 37	04 11	11 45	13 00
04	04 51 37	13 50 21	22 ♊ 47 30	29 ♊ 50 08	28 11	29 ♊ 44	22 29	17 23	12 33	04 09	11 44	12 58
05	04 55 34	14 47 49	06 ♋ 47 05	13 ♋ 37 56	28 D 09	00 ♋ 57	23 07	17 37	12 30	04 07	11 43	12 57
06	04 59 30	15 45 16	20 ♋ 22 29	27 ♋ 00 40	28 12	02 11	23 44	17 51	12 26	04 05	11 41	12 56
07	05 03 27	16 42 43	03 ♌ 32 35	09 ♌ 58 28	28 20	03 24	24 21	18 05	12 22	04 02	11 40	12 54
08	05 07 23	17 40 07	16 ♌ 18 38	22 ♌ 33 34	28 32	04 37	24 58	18 18	12 19	04 00	11 38	12 53
09	05 11 20	18 37 30	28 ♌ 43 46	04 ♍ 49 49	28 48	05 51	25 36	18 32	12 15	03 58	11 37	12 52
10	05 15 16	19 34 52	10 ♍ 52 22	16 ♍ 52 02	29 09	07 04	26 13	18 46	12 11	03 55	11 35	12 51
11	05 19 13	20 32 14	22 ♍ 49 31	28 ♍ 45 28	29 ♉ 34	08 17	26 50	19 00	12 07	03 53	11 34	12 49
12	05 23 10	21 29 34	04 ♎ 40 34	10 ♎ 35 27	00 ♊ 04	09 31	27 27	19 14	12 03	03 51	11 32	12 48
13	05 27 06	22 26 54	16 ♎ 30 43	22 ♎ 26 58	00 38	10 44	28 05	19 28	11 59	03 48	11 31	12 47
14	05 31 03	23 24 12	28 ♎ 24 45	04 ♏ 24 31	01 16	11 57	28 42	19 42	11 55	03 46	11 29	12 46
15	05 34 59	24 21 30	10 ♏ 26 44	16 ♏ 31 45	01 58	13 11	29 19	19 56	11 51	03 43	11 28	12 45
16	05 38 56	25 18 47	22 ♏ 39 53	28 ♏ 51 22	02 45	14 24	29 ♋ 57	20 10	11 47	03 41	11 26	12 43
17	05 42 52	26 16 03	05 ♐ 06 22	11 ♐ 25 00	03 35	15 37	00 ♌ 34	20 23	11 43	03 39	11 25	12 42
18	05 46 49	27 13 19	17 ♐ 47 17	24 ♐ 13 13	04 29	16 50	01 11	20 37	11 38	03 36	11 23	12 41
19	05 50 45	28 10 34	00 ♑ 42 42	07 ♑ 15 37	05 28	18 03	01 49	20 51	11 34	03 34	11 22	12 40
20	05 54 42	29 ♊ 07 49	13 ♑ 51 50	20 ♑ 31 10	06 29	19 17	02 26	21 05	11 30	03 31	11 20	12 39
21	05 58 39	00 ♋ 05 03	27 ♑ 13 26	03 ♒ 58 28	07 35	20 30	03 03	21 19	11 26	03 29	11 19	12 38
22	06 02 35	01 02 17	10 ♒ 46 04	17 ♒ 36 07	08 44	21 43	03 41	21 33	11 21	03 26	11 17	12 37
23	06 06 32	01 59 30	24 ♒ 28 28	01 ♓ 23 00	09 57	22 56	04 18	21 46	11 17	03 24	11 15	12 36
24	06 10 28	02 56 44	08 ♓ 19 38	15 ♓ 18 16	11 14	24 09	04 55	22 00	11 13	03 22	11 14	12 35
25	06 14 25	03 53 57	22 ♓ 18 50	29 ♓ 21 13	12 34	25 22	05 33	22 14	11 08	03 19	11 12	12 34
26	06 18 21	04 51 11	06 ♈ 25 19	13 ♈ 30 58	13 57	26 35	06 10	22 28	11 04	03 17	11 11	12 34
27	06 22 18	05 48 24	20 ♈ 37 58	27 ♈ 46 02	15 24	27 48	06 47	22 41	11 00	03 14	11 09	12 33
28	06 26 14	06 45 38	04 ♉ 54 51	12 ♉ 04 01	16 55	29 ♋ 01	07 25	22 55	10 55	03 12	11 07	12 32
29	06 30 11	07 42 51	19 ♉ 13 04	26 ♉ 21 26	18 29	00 ♌ 14	08 02	23 09	10 51	03 09	11 06	12 31
30	06 34 08	08 ♋ 40 05	03 ♊ 28 36	10 ♊ 33 55	20 ♊ 06	01 ♌ 27	08 ♌ 39	23 ♊ 22	10 ♑ 46	03 ♑ 07	11 ♑ 04	12 ♏ 30

Date	Moon True ☊ ° '	Moon Mean ☊ ° '	Moon ☽ Latitude ° '
01	29 ♒ 18	29 ♒ 46	04 N 43
02	29 R 08	29 43	05 00
03	28 57	29 39	04 57
04	28 46	29 36	04 35
05	28 37	29 33	03 58
06	28 30	29 30	03 07
07	28 25	29 27	02 08
08	28 22	29 24	01 N 04
09	28 D 22	29 20	00 S 02
10	28 22	29 17	01 06
11	28 R 22	29 14	02 06
12	28 21	29 11	03 01
13	28 19	29 08	03 47
14	28 13	29 05	04 24
15	28 06	29 01	04 49
16	27 57	28 58	05 02
17	27 46	28 55	05 01
18	27 36	28 52	04 45
19	27 26	28 49	04 14
20	27 18	28 45	03 30
21	27 12	28 42	02 33
22	27 08	28 39	01 26
23	27 07	28 36	00 S 14
24	27 D 08	28 33	01 N 00
25	27 08	28 30	02 11
26	27 R 09	28 26	03 15
27	27 08	28 23	04 07
28	27 05	28 20	04 45
29	27 00	28 17	05 04
30	26 ♒ 54	28 ♒ 14	05 N 05

DECLINATIONS

Date	Sun ☉ ° '	Moon ☽ ° '	Mercury ☿ ° '	Venus ♀ ° '	Mars ♂ ° '	Jupiter ♃ ° '	Saturn ♄ ° '	Uranus ♅ ° '	Neptune ♆ ° '	Pluto ♇ ° '
01	22 N 05	19 N 01	16 N 29	24 N 09	23 N 10	22 N 22	22 S 11	23 S 38	21 S 59	00 S 14
02	22 13	23 35	16 17	24 13	23 04	22 23	22 11	23 38	21 59	00 14
03	22 21	26 36	16 07	24 16	22 58	22 25	22 12	23 39	21 59	00 14
04	22 28	27 50	16 00	24 19	22 52	22 26	22 12	23 39	22 00	00 14
05	22 35	27 13	15 55	24 21	22 45	22 27	22 12	23 39	22 00	00 14
06	22 41	24 59	15 51	24 22	22 39	22 29	22 13	23 39	22 00	00 14
07	22 47	21 26	15 50	24 23	22 32	22 30	22 13	23 39	22 00	00 14
08	22 52	16 58	15 51	24 22	22 25	22 31	22 14	23 39	22 00	00 14
09	22 57	11 53	15 54	24 22	22 18	22 33	22 14	23 39	22 00	00 14
10	23 02	06 28	15 59	24 20	22 11	22 34	22 14	23 39	22 00	00 14
11	23 06	00 N 55	16 06	24 17	22 04	22 35	22 15	23 39	22 00	00 14
12	23 10	04 S 37	16 14	24 14	21 56	22 36	22 15	23 39	22 00	00 14
13	23 14	09 59	16 25	24 11	21 49	22 38	22 16	23 39	22 01	00 14
14	23 17	15 01	16 36	24 06	21 41	22 39	22 16	23 39	22 01	00 14
15	23 19	19 32	16 50	24 01	21 33	22 40	22 17	23 40	22 01	00 14
16	23 22	23 18	17 04	23 55	21 25	22 41	22 17	23 40	22 01	00 14
17	23 23	26 05	17 20	23 48	21 17	22 42	22 17	23 40	22 01	00 14
18	23 25	27 37	17 37	23 41	21 09	22 43	22 18	23 40	22 01	00 14
19	23 26	27 41	17 55	23 33	21 00	22 44	22 18	23 40	22 01	00 14
20	23 26	26 12	18 14	23 24	20 52	22 45	22 19	23 40	22 02	00 14
21	23 27	23 13	18 34	23 15	20 43	22 46	22 19	23 40	22 02	00 15
22	23 26	18 55	18 54	23 05	20 34	22 47	22 20	23 40	22 02	00 15
23	23 26	13 35	19 14	22 54	20 25	22 48	22 20	23 40	22 02	00 15
24	23 25	07 31	19 36	22 43	20 16	22 49	22 20	23 40	22 02	00 15
25	23 23	01 S 03	19 57	22 31	20 06	22 50	22 21	23 40	22 02	00 15
26	23 21	05 N 32	20 18	22 18	19 57	22 51	22 21	23 40	22 02	00 16
27	23 19	11 52	20 39	22 05	19 47	22 52	22 22	23 40	22 02	00 16
28	23 16	17 38	21 00	21 51	19 38	22 52	22 22	23 40	22 03	00 16
29	23 13	22 24	21 21	21 36	19 28	22 53	22 23	23 41	22 03	00 16
30	23 N 10	25 N 50	21 N 40	21 N 21	19 N 18	22 N 54	22 S 23	23 S 41	22 S 03	00 S 17

ZODIAC SIGN ENTRIES

Date	h m	Planets
02	22 02	☽ ♊
04	17 17	♀ ♋
05	00 17	☽ ♋
07	05 28	☽ ♌
09	14 29	☽ ♍
12	02 31	☽ ♎
12	08 56	☿ ♊
14	15 11	☽ ♏
16	14 10	♂ ♌
17	02 12	☽ ♐
19	10 41	☽ ♑
21	09 53	☉ ♋
21	16 57	☽ ♒
23	21 36	☽ ♓
26	01 06	☽ ♈
28	03 45	☽ ♉
29	07 21	♀ ♌
30	06 08	☽ ♊

LATITUDES

Date	Mercury ☿ ° '	Venus ♀ ° '	Mars ♂ ° '	Jupiter ♃ ° '	Saturn ♄ ° '	Uranus ♅ ° '	Neptune ♆ ° '	Pluto ♇ ° '
01	03 S 28	00 N 46	01 N 19	00 S 25	00 N 39	00 S 16	00 N 56	16 N 15
04	03 51	00 52	01 19	00 25	00 39	00 16	00 56	16 14
07	04 03	00 59	01 18	00 24	00 39	00 16	00 56	16 13
10	04 05	01 05	01 18	00 24	00 39	00 16	00 56	16 11
13	03 58	01 10	01 18	00 24	00 38	00 16	00 56	16 10
16	03 43	01 15	01 17	00 24	00 38	00 16	00 56	16 09
19	03 21	01 20	01 16	00 23	00 38	00 16	00 56	16 08
22	02 54	01 24	01 16	00 23	00 38	00 16	00 56	16 06
25	02 23	01 28	01 15	00 23	00 38	00 16	00 56	16 05
28	01 48	01 31	01 15	00 23	00 38	00 16	00 56	16 04
31	01 S 12	01 N 34	01 N 14	00 S 22	00 N 37	00 S 16	00 N 56	16 N 02

DATA

Julian Date	2447679
Delta T	+56 seconds
Ayanamsa	23° 42' 41"
Synetic vernal point	05° ♓ 24' 19"
True obliquity of ecliptic	23° 26' 34"

MOON'S PHASES, APSIDES AND POSITIONS ☽

Date	h m	Phase	Longitude ° '	Eclipse Indicator
03	19 53	●	13 ♊ 12	
11	06 59	☽	20 ♍ 20	
19	06 57	○	27 ♐ 59	
26	09 09	☾	04 ♈ 44	

Day	h m		
01	05 08	Perigee	
13	02 21	Apogee	
28	04 14	Perigee	
04	15 51	Max dec	27° N 51'
11	15 56	0S	
19	01 07	Max dec	27° S 50'
25	15 48	0N	

LONGITUDES

Date	Chiron ⚷ ° '	Ceres ⚳ ° '	Pallas ⚴ ° '	Juno ⚵ ° '	Vesta ⚶ ° '	Black Moon Lilith ⚸ ° '
01	05 ♋ 57	21 ♉ 44	06 ♈ 40	02 ♍ 46	09 ♑ 40	12 ♎ 33
11	06 ♋ 54	25 ♉ 43	09 ♈ 03	05 ♍ 21	07 ♑ 53	13 ♎ 40
21	07 ♋ 52	29 ♉ 39	11 ♈ 11	08 ♍ 11	05 ♑ 39	14 ♎ 47
31	08 ♋ 52	03 ♊ 31	13 ♈ 02	11 ♍ 12	03 ♑ 15	15 ♎ 54

All ephemeris data is given at 12.00 UT and the Moon's longitude is additionally given for 24.00 UT

ASPECTARIAN

h m	Aspects
01 Thursday	
01 12	☽ ∥ ☿
03 58	☽ △ ♅
04 37	☽ ⊥ ☉
11 02	☽ Q ♂
14 28	☽ ⊥ ♃
15 06	☽ ⊻ ☉
15 19	☽ ∠ ♀
16 15	☽ △ ♆
17 45	☽ △ ♄
18 17	☽ ☍ ♇
02 Friday	
00 26	☽ ⊻ ♃
02 45	☽ # ♆
03 48	☽ ∥ ☉
03 50	☽ # ♄
04 24	☽ bQ ♅
04 54	☽ ∥ ♃
05 19	☉ # ♄
07 21	☽ ⊥ ♀
07 30	☽ ✶ ♂
08 26	☉ ⊼ ♆
08 57	☽ ∥ ♂
12 22	☽ # ♅
16 10	☽ ∥ ♀
16 44	☽ bQ ♆
18 05	☽ ⊻ ♀
18 12	☽ bQ ♄
19 06	☽ ± ♅
19 23	☽ ☌ ☿
03 Saturday	
02 14	♀ bQ ♇
04 57	☽ ⊼ ♅
05 46	☉ ⊼ ♄
07 32	☽ ± ♆
08 01	☿ ⊻ ♀
08 58	☽ ± ♄
09 15	☽ ∠ ♂
14 51	☉ ⊼ ♇
17 28	☽ ⊼ ♆
18 53	☽ ⊼ ♄
19 32	☽ ⊼ ♇
19 53	☽ ☌ ☉
04 Sunday	
00 54	☽ ⊥ ♂
02 44	☽ ☌ ♃
03 44	☉ ∥ ♃
05 34	☽ ± ♇
11 28	☽ ⊻ ♂
20 48	☽ bQ ♇
21 08	☽ ⊻ ☿
05 Monday	
00 57	☽ ☌ ♀
07 23	☽ ☍ ♅
07 26	☽ ⊥ ☿
08 07	☿ St D
20 36	☽ ☍ ♆
21 57	☽ ☍ ♄
22 47	☽ △ ♇
23 11	☽ ∠ ☿
06 Tuesday	
03 08	☽ ⊻ ☉
07 24	☽ ⊻ ♃
08 19	☉ ∥ ♂
14 40	☽ ⊥ ☉
16 44	☽ ∥ ♀
18 20	☽ ☌ ♂
18 22	☽ ⊥ ♃
18 57	♂ ⊥ ♃
21 56	☽ # ♅
07 Wednesday	
02 18	☽ ✶ ☿
03 56	☽ ∥ ☉
05 04	☽ ∥ ♂
05 32	☽ ∥ ♃
07 16	☽ # ♄
08 21	☽ ∠ ☉
08 37	☽ # ♆
11 07	☽ ∠ ♃
11 43	☽ ⊻ ♀
12 55	☽ ⊼ ♅
18 21	♂ ∥ ♃
08 Thursday	
00 05	☽ ⊥ ♀
00 05	☽ ± ♅
00 06	♀ ☍ ♅
00 51	☽ Q ☿
03 09	☽ ⊼ ♆
04 27	☽ ⊼ ♄
05 30	☽ □ ♇
09 42	☿ ⊥ ♀
14 32	☽ ± ♆
14 48	☽ ✶ ☉
15 48	☽ ± ♄
15 53	☽ ✶ ♃
17 08	☽ bQ ♅
17 25	☽ ∥ ☿
19 02	☽ ∠ ♀
09 Friday	
05 34	☽ ⊻ ♂
07 53	☽ bQ ♆
09 07	☽ bQ ♄
09 10	☉ ☌ ♃
12 09	☽ □ ☿
15 37	☽ Q ♃
16 02	☽ Q ☉
16 11	☽ Q ♇
17 52	☉ ± ♇
17 55	☽ ⊥ ♂
22 15	☽ △ ♅
10 Saturday	
01 55	♂ # ♄
03 35	☽ ✶ ♀
12 43	☽ ∠ ♂
13 26	☽ △ ♆
14 36	☽ △ ♄
15 56	☽ ✶ ♇
18 49	♃ ± ♇
11 Sunday	
04 08	☽ □ ♃
06 18	☽ Q ♀
06 59	☽ □ ☉
14 57	☽ # ♇
16 55	☽ ∥ ♇
20 33	☽ ✶ ♂
22 05	☽ ∠ ♇
23 18	♂ # ♆
12 Monday	
02 14	☽ △ ☿
09 29	☉ ⊥ ♂
10 19	☽ □ ♅
16 18	☽ ⊥ ♇
22 14	☽ Q ♂
22 57	☽ □ ♀
13 Tuesday	
01 54	☽ □ ♆
02 53	☽ □ ♄
04 27	☽ ⊻ ♇
10 08	☽ bQ ☿
18 06	☽ △ ♃
22 40	☽ Q ♅
14 Wednesday	
01 03	☽ △ ☉
03 01	♀ ☍ ♆
05 19	☽ ± ☿
11 17	♀ ☍ ♄
12 37	☽ □ ♂
14 09	☽ Q ♆
15 00	☽ Q ♄
18 04	☽ ⊼ ☿
20 29	☽ # ☿
22 41	☽ ✶ ♅
15 Thursday	
00 49	☽ bQ ♃
03 36	♀ △ ♇
09 40	☽ bQ ☉
14 01	☽ ✶ ♆
14 46	☽ ✶ ♄
16 32	☽ ☌ ♇
18 00	☽ △ ♀
19 01	☽ ± ♃
23 46	☽ # ♂
16 Friday	
03 09	☽ ∥ ♆
04 15	☽ ∠ ♅
04 54	☽ ∥ ♄
04 54	☽ ± ☉
07 01	☽ ⊼ ♃
07 36	☽ # ♃
12 24	☽ # ☉
14 36	☽ ∥ ♅
16 23	☽ # ♀
17 35	☽ ⊼ ☉
19 19	☽ ∠ ♆
19 57	☽ ∠ ♄
21 42	☽ ⊥ ♅
17 Saturday	
02 28	☽ bQ ♀
02 50	☽ △ ♂
08 53	☽ ☍ ☿
09 13	☽ ⊻ ♅
12 35	☽ ⊥ ♆
13 09	☽ ⊥ ♄
13 31	☿ ⊼ ♅
21 30	☽ ± ♀
23 58	☽ ⊻ ♆
18 Sunday	
00 29	☽ ⊻ ♄
02 25	☽ ⊻ ♇
08 51	☽ bQ ♂
10 02	☽ ⊼ ♀
13 41	☽ ⊥ ♇
15 37	♀ # ♅
17 24	☽ ☍ ♃
23 30	☉ bQ ♇
19 Monday	
02 29	☽ ± ♂
06 24	☽ ∠ ♇
06 57	☽ ☍ ☉
09 42	☿ ± ♆
14 07	☽ ⊼ ♂
14 30	☿ ± ♄
17 13	☽ ☌ ♅
21 26	☽ ⊼ ☿
20 Tuesday	
06 30	☉ ∥ ♀
07 25	☽ ☌ ♆
07 44	☽ ☌ ♄
09 18	☽ ± ☿
09 49	☽ ✶ ♇
22 45	☽ ☍ ♀
21 Wednesday	
01 14	☽ ⊼ ♃
02 57	☽ bQ ☿
07 23	☽ Q ♇
09 00	☽ ∥ ♅
10 30	☽ # ☉
11 47	☽ # ♀
12 10	☽ ± ♃
14 50	☽ # ♃
17 29	☽ ⊼ ☉
17 37	☽ ∥ ♄
19 21	☽ ∥ ♆
22 52	☽ ☍ ♂
23 06	☽ ⊻ ♅
22 Thursday	
03 17	☽ # ♂
03 32	♂ ⊼ ♅
04 25	☽ bQ ♃
04 55	☽ ± ☉
07 52	♀ ⊻ ♃
08 05	☽ △ ☿
09 40	☽ ⊥ ♅
12 07	☽ # ☿
12 54	☽ ⊻ ♆
13 02	☽ ⊻ ♄
15 15	☽ □ ♇
21 57	☽ bQ ☉
23 25	☽ ⊥ ♆
23 31	☽ ⊥ ♄
23 Friday	
01 26	☽ ∠ ♅
07 13	☽ △ ♃
09 03	☽ ⊼ ♀
15 06	☽ ∠ ♆
15 08	☽ ∠ ♄
20 30	☽ ± ♀
23 41	♀ ∥ ♃
24 Saturday	
02 01	☽ △ ☉
03 11	♄ ☌ ♆
03 27	☽ ✶ ♅
05 51	☽ ⊼ ♂
11 41	☿ ⊼ ♄
11 59	☿ ⊼ ♆
13 33	☽ bQ ♀
16 40	☽ ± ♂
16 57	☽ ✶ ♄
16 59	☽ ✶ ♆
17 31	☽ □ ☿
19 20	☽ △ ♇
22 01	☉ ☍ ♅
25 Sunday	
00 04	☽ Q ♅
08 50	☽ bQ ♂
11 51	☽ □ ♃
12 11	☿ ⊼ ♇
13 24	☽ Q ♄
13 31	☽ Q ♆
14 52	☽ ∥ ♇
16 44	☽ # ♇
17 42	☽ △ ♀
19 27	☿ ∥ ♂
20 58	☽ bQ ♇
26 Monday	
03 35	☽ Q ☿
06 11	♀ # ♄
06 41	☽ □ ♅
09 09	☽ □ ☉
11 33	☽ △ ♂
12 14	☽ ± ♇
18 57	☽ Q ♃
19 49	☽ □ ♄
20 02	☽ □ ♆
22 22	☽ ⊼ ♇
27 Tuesday	
02 11	☽ ✶ ☿
15 31	☽ ✶ ♃
15 56	♀ # ♆
17 43	☽ Q ☉
28 Wednesday	
01 11	☽ □ ♀
06 22	☽ ∠ ☿
09 08	☽ △ ♅
09 31	♀ ⊥ ♃
15 19	☽ ✶ ☉
16 23	☽ □ ♂
17 07	☽ ∠ ♃
21 03	☽ ∥ ♂
22 01	☽ △ ♄
22 24	☽ △ ♆
23 17	☽ ⊥ ☿
29 Thursday	
00 46	☽ ☍ ♇
05 38	☽ ∥ ☿
07 45	☽ ∥ ♀
08 28	☽ ⊥ ♃
09 57	☽ # ♆
10 12	☽ Q ♀
10 13	☽ bQ ♅
10 36	☽ ⊻ ☿
11 50	☽ # ♄
12 40	☿ ± ♇
14 51	☽ ∥ ♃
16 46	☽ ∥ ☉
18 18	☽ ∠ ☉
18 39	♂ ∠ ♃
18 42	☽ ⊻ ♃
19 44	☽ # ♅
22 48	☿ ∥ ♀
23 05	☽ bQ ♄
23 32	☽ bQ ♆
23 59	☽ Q ♂
30 Friday	
01 18	☽ ± ♅
08 16	☽ ✶ ♀
10 32	☽ ⊥ ☉
11 15	☉ ⊻ ♂
11 24	☽ ⊼ ♅
14 11	☽ ± ♄
14 41	☽ ± ♆
21 10	☽ ✶ ♂
21 25	☽ ⊻ ☉

JULY 1989

Date	Sidereal time h m s	LONGITUDES Sun ☉	Moon ☽	Moon ☽ 24.00	Mercury ☿	Venus ♀	Mars ♂	Jupiter ♃	Saturn ♄	Uranus ♅	Neptune ♆	Pluto ♇
01	06 38 04	09 ♋ 37 19	17 ♊ 36 48	24 ♊ 36 39	21 ♊ 46	02 ♌ 40	09 ♌ 17	23 ♊ 36	10 ♑ 42	03 ♑ 05	11 ♑ 02	12 ♏ 30
02	06 42 01	10 34 33	01 ♋ 32 56	08 ♋ 25 08	23 30	03 53	09 54	23 49	10 R 37	03 R 02	11 R 01	12 R 29
03	06 45 57	11 31 47	15 ♋ 12 53	21 ♋ 55 50	25 16	05 06	10 32	24 03	10 33	03 00	10 59	12 28
04	06 49 54	12 29 01	28 ♋ 33 47	05 ♌ 06 39	27 06	06 19	11 09	24 17	10 29	02 57	10 58	12 28
05	06 53 50	13 26 14	11 ♌ 34 25	17 ♌ 57 12	28 ♊ 58	07 32	11 47	24 30	10 24	02 55	10 56	12 27
06	06 57 47	14 23 28	24 ♌ 15 13	00 ♍ 28 45	00 ♋ 54	08 45	12 24	24 44	10 20	02 53	10 54	12 27
07	07 01 43	15 20 41	06 ♍ 38 11	12 ♍ 43 58	02 51	09 57	13 02	24 57	10 15	02 50	10 53	12 26
08	07 05 40	16 17 54	18 ♍ 46 35	24 ♍ 46 35	04 51	11 10	13 39	25 10	10 11	02 48	10 51	12 26
09	07 09 37	17 15 07	00 ♎ 44 32	06 ♎ 41 04	06 54	12 23	14 17	25 24	10 07	02 45	10 50	12 25
10	07 13 33	18 12 20	12 ♎ 36 48	18 ♎ 32 20	08 58	13 36	14 54	25 37	10 02	02 43	10 48	12 25
11	07 17 30	19 09 33	24 ♎ 28 18	00 ♏ 25 20	11 03	14 49	15 32	25 50	09 58	02 41	10 46	12 24
12	07 21 26	20 06 46	06 ♏ 24 01	12 ♏ 24 55	13 10	16 01	16 09	26 04	09 53	02 39	10 45	12 24
13	07 25 23	21 03 59	18 ♏ 28 34	24 ♏ 35 26	15 18	17 14	16 47	26 17	09 49	02 36	10 43	12 24
14	07 29 19	22 01 12	00 ♐ 45 58	07 ♐ 00 31	17 26	18 27	17 24	26 30	09 45	02 34	10 42	12 23
15	07 33 16	22 58 25	13 ♐ 19 23	19 ♐ 42 48	19 35	19 39	18 02	26 43	09 41	02 32	10 40	12 23
16	07 37 12	23 55 38	26 ♐ 10 52	02 ♑ 43 38	21 44	20 52	18 39	26 56	09 36	02 30	10 38	12 23
17	07 41 09	24 52 52	09 ♑ 21 06	16 ♑ 03 06	23 53	22 04	19 17	27 09	09 32	02 27	10 37	12 23
18	07 45 06	25 50 06	22 ♑ 49 26	29 ♑ 39 51	26 01	23 17	19 54	27 22	09 28	02 25	10 35	12 22
19	07 49 02	26 47 20	06 ♒ 33 58	13 ♒ 31 26	28 ♋ 09	24 29	20 32	27 35	09 24	02 23	10 34	12 22
20	07 52 59	27 44 35	20 ♒ 31 48	27 ♒ 34 38	00 ♌ 15	25 42	21 10	27 48	09 20	02 21	10 32	12 22
21	07 56 55	28 41 50	04 ♓ 39 29	11 ♓ 45 53	02 21	26 54	21 47	28 01	09 16	02 19	10 31	12 22
22	08 00 52	29 ♋ 39 06	18 ♓ 53 24	26 ♓ 01 37	04 26	28 07	22 25	28 14	09 12	02 17	10 29	12 22
23	08 04 48	00 ♌ 36 23	03 ♈ 10 10	10 ♈ 18 39	06 29	29 ♌ 19	23 02	28 26	09 08	02 15	10 28	12 D 22
24	08 08 45	01 33 40	17 ♈ 26 46	24 ♈ 34 12	08 30	00 ♍ 32	23 40	28 39	09 04	02 13	10 26	12 22
25	08 12 41	02 30 59	01 ♉ 40 39	08 ♉ 45 51	10 31	01 44	24 18	28 52	09 00	02 11	10 25	12 22
26	08 16 38	03 28 19	15 ♉ 49 33	22 ♉ 51 30	12 29	02 56	24 55	29 04	08 56	02 09	10 23	12 22
27	08 20 35	04 25 39	29 ♉ 51 26	06 ♊ 49 08	14 27	04 09	25 33	29 17	08 52	02 07	10 22	12 22
28	08 24 31	05 23 01	13 ♊ 44 20	20 ♊ 36 48	16 22	05 21	26 11	29 29	08 49	02 05	10 20	12 22
29	08 28 28	06 20 24	27 ♊ 26 20	04 ♋ 12 41	18 16	06 33	26 49	29 42	08 45	02 03	10 19	12 23
30	08 32 24	07 17 47	10 ♋ 55 40	17 ♋ 35 06	20 08	07 45	27 26	29 ♊ 54	08 41	02 01	10 17	12 23
31	08 36 21	08 ♌ 15 12	24 ♋ 10 51	00 ♌ 42 49	21 ♌ 59	08 ♍ 57	28 ♌ 04	00 ♋ 06	08 ♑ 38	01 ♑ 59	10 ♑ 16	12 ♏ 23

Date	Moon True ☊	Moon Mean ☊	Moon Latitude
01	26 ♒ 46	28 ♒ 11	04 N 47
02	26 R 39	28 07	04 13
03	26 32	28 04	03 24
04	26 27	28 01	02 25
05	26 24	27 58	01 19
06	26 23	27 55	00 N 12
07	26 D 24	27 51	00 S 55
08	26 25	27 48	01 59
09	26 27	27 45	02 56
10	26 28	27 42	03 45
11	26 R 28	27 39	04 24
12	26 26	27 36	04 53
13	26 23	27 32	05 08
14	26 19	27 29	05 10
15	26 14	27 26	04 58
16	26 09	27 23	04 30
17	26 04	27 20	03 47
18	26 00	27 16	02 51
19	25 57	27 13	01 44
20	25 56	27 10	00 S 30
21	25 D 56	27 07	00 N 48
22	25 57	27 04	02 03
23	25 59	27 01	03 10
24	26 00	26 57	04 06
25	26 01	26 54	04 47
26	26 R 00	26 51	05 10
27	25 59	26 48	05 14
28	25 56	26 45	05 00
29	25 54	26 42	04 28
30	25 51	26 38	03 43
31	25 ♒ 49	26 ♒ 35	02 N 46

Date	DECLINATIONS Sun ☉	Moon ☽	Mercury ☿	Venus ♀	Mars ♂	Jupiter ♃	Saturn ♄	Uranus ♅	Neptune ♆	Pluto ♇
01	23 N 06	27 N 38	21 N 59	21 N 05	19 N 08	22 N 55	22 S 23	23 S 41	22 S 03	00 S 17
02	23 01	27 39	22 17	20 49	18 57	22 56	22 24	23 41	22 03	00 17
03	22 57	25 57	22 34	20 32	18 47	22 56	22 24	23 41	22 03	00 17
04	22 51	22 49	22 49	20 15	18 37	22 57	22 25	23 41	22 04	00 18
05	22 46	18 35	23 03	19 57	18 26	22 58	22 25	23 41	22 04	00 18
06	22 40	13 37	23 15	19 38	18 15	22 58	22 26	23 41	22 04	00 18
07	22 34	08 13	23 25	19 19	18 04	22 59	22 26	23 41	22 04	00 19
08	22 27	02 N 37	23 33	18 59	17 53	22 59	22 27	23 41	22 04	00 19
09	22 20	02 S 59	23 38	18 39	17 42	23 00	22 27	23 41	22 04	00 20
10	22 12	08 26	23 41	18 19	17 31	23 00	22 27	23 41	22 04	00 20
11	22 04	13 35	23 42	17 57	17 20	23 01	22 28	23 41	22 05	00 20
12	21 56	18 15	23 39	17 36	17 08	23 01	22 28	23 41	22 05	00 21
13	21 48	22 16	23 34	17 14	16 57	23 02	22 29	23 41	22 05	00 21
14	21 39	25 22	23 27	16 51	16 45	23 02	22 29	23 41	22 05	00 22
15	21 29	27 20	23 16	16 28	16 33	23 03	22 29	23 42	22 05	00 22
16	21 19	27 53	23 03	16 05	16 22	23 03	22 30	23 42	22 05	00 23
17	21 09	26 54	22 47	15 41	16 10	23 03	22 30	23 42	22 05	00 23
18	20 59	24 20	22 29	15 17	15 57	23 04	22 31	23 42	22 05	00 24
19	20 48	20 19	22 08	14 52	15 45	23 04	22 31	23 42	22 06	00 24
20	20 37	15 07	21 45	14 27	15 33	23 04	22 31	23 42	22 06	00 25
21	20 25	09 04	21 20	14 02	15 21	23 05	22 32	23 42	22 06	00 25
22	20 14	02 S 31	20 52	13 36	15 08	23 05	22 32	23 42	22 06	00 26
23	20 01	04 N 10	20 23	13 10	14 56	23 05	22 33	23 42	22 06	00 26
24	19 49	10 38	19 52	12 44	14 43	23 05	22 33	23 42	22 06	00 27
25	19 36	16 32	19 20	12 17	14 30	23 06	22 33	23 42	22 06	00 27
26	19 23	21 31	18 46	11 50	14 17	23 06	22 34	23 42	22 07	00 28
27	19 09	25 14	18 11	11 23	14 04	23 06	22 34	23 42	22 07	00 28
28	18 56	27 24	17 34	10 55	13 51	23 06	22 34	23 42	22 07	00 29
29	18 41	27 53	16 57	10 27	13 38	23 06	22 35	23 42	22 07	00 30
30	18 27	26 41	16 19	09 59	13 25	23 06	22 35	23 42	22 07	00 30
31	18 N 12	24 N 00	15 N 40	09 N 31	13 N 12	23 N 06	22 S 35	23 S 42	22 S 07	00 S 31

ZODIAC SIGN ENTRIES

Date	h	m	Planets
02	09	19	☽ ♋
04	14	37	☽ ♌
06	00	55	☿ ♋
06	23	04	☽ ♍
09	10	30	☽ ♎
11	23	09	☽ ♏
14	10	31	☽ ♐
16	19	01	☽ ♑
19	00	35	☽ ♒
20	09	04	☿ ♌
21	04	07	☽ ♓
22	20	45	☉ ♌
23	06	41	☽ ♈
24	01	31	♀ ♍
25	09	10	☽ ♉
27	12	15	☽ ♊
29	16	32	☽ ♋
30	23	50	♃ ♋
31	22	41	☽ ♌

LATITUDES

Date	Mercury ☿	Venus ♀	Mars ♂	Jupiter ♃	Saturn ♄	Uranus ♅	Neptune ♆	Pluto ♇
01	01 S 12	01 N 34	01 N 14	00 S 22	00 N 37	00 S 16	00 N 56	16 N 02
04	00 S 35	01 36	01 14	00 22	00 37	00 16	00 56	16 00
07	00 00	01 37	01 13	00 22	00 37	00 16	00 56	15 59
10	00 N 33	01 38	01 12	00 22	00 37	00 16	00 56	15 57
13	01 01	01 38	01 12	00 22	00 36	00 16	00 56	15 56
16	01 23	01 37	01 11	00 21	00 36	00 16	00 56	15 54
19	01 38	01 36	01 10	00 21	00 36	00 16	00 56	15 52
22	01 46	01 34	01 09	00 21	00 35	00 16	00 56	15 51
25	01 48	01 31	01 09	00 21	00 35	00 16	00 56	15 49
28	01 43	01 28	01 08	00 21	00 35	00 16	00 56	15 47
31	01 N 34	01 N 24	01 N 07	00 S 20	00 N 34	00 S 16	00 N 56	15 N 45

DATA

Julian Date	2447709
Delta T	+56 seconds
Ayanamsa	23° 42' 47"
Synetic vernal point	05° ♓ 24' 13"
True obliquity of ecliptic	23° 26' 33"

MOON'S PHASES, APSIDES AND POSITIONS ☽

Date	h	m	Phase	Longitude	Eclipse Indicator
03	04	59	●	11 ♋ 15	
11	00	19	☽	18 ♎ 42	
18	17	42	○	26 ♑ 04	
25	13	31	☾	02 ♉ 35	

Day	h	m		
10	20	50	Apogee	
23	07	09	Perigee	
02	00	06	Max dec	27° N 52'
08	23	11	0S	
16	08	55	Max dec	27° S 54'
22	21	01	0N	
29	06	43	Max dec	27° N 56'

LONGITUDES

Date	Chiron ⚷	Ceres ⚳	Pallas ⚴	Juno ⚵	Vesta ⚶	Black Moon Lilith ⚸
01	08 ♋ 52	03 ♊ 31	13 ♈ 02	11 ♍ 12	03 ♑ 15	15 ♎ 54
11	09 ♋ 52	07 ♊ 18	14 ♈ 32	14 ♍ 22	01 ♑ 03	17 ♎ 01
21	10 ♋ 51	11 ♊ 00	15 ♈ 38	17 ♍ 40	29 ♐ 22	18 ♎ 08
31	11 ♋ 48	14 ♊ 34	16 ♈ 14	21 ♍ 04	28 ♐ 24	19 ♎ 15

All ephemeris data is given at 12.00 UT and the Moon's longitude is additionally given for 24.00 UT

ASPECTARIAN

01 Saturday
00 17 ☽ ⊼ ♄
00 50 ☽ ⊼ ♆
03 17 ☽ ⊼ ♇
04 35 ♂ ± ♅
12 06 ☽ ∠ ♀
13 30 ☽ ± ♇
16 52 ☿ # ♆
19 48 ♀ ⊼ ♅
20 06 ☽ ☌ ☿
22 26 ☽ ☌ ♃
23 58 ☽ ∠ ♂

02 Sunday
04 58 ☽ bQ ♇
05 02 ☽ ⊥ ♀
13 08 ☉ ☍ ♄
14 35 ☽ ☍ ♅
16 18 ☽ ⊥ ♂
16 28 ☽ ⊻ ♀
17 13 ☿ ☌ ♃
21 26 ☿ # ♄
22 43 ☉ ☍ ♆

03 Monday
03 19 ☽ ⊻ ♂
03 48 ☽ ☍ ♄
04 32 ☽ ☍ ♆
04 59 ☽ ☌ ☉
07 09 ☽ △ ♇
12 44 ♂ ⊼ ♄
13 07 ☉ ∥ ♃
17 16 ☿ ∠ ♂

04 Tuesday
04 06 ☽ ⊻ ♃
04 52 ♂ ⊼ ♆
06 14 ☽ # ♅
08 54 ☽ ⊻ ☿
11 08 ☽ ∥ ♃
11 28 ☉ △ ♇
11 43 ☽ ∥ ☉
11 56 ☽ ∥ ☿
14 18 ☉ ∥ ☿
14 31 ☽ # ♄
15 11 ☽ ⊥ ♃
16 42 ☿ bQ ♇
16 42 ☽ # ♆
20 01 ☽ ⊼ ♅
21 40 ☽ ⊥ ☿

05 Wednesday
01 17 ☿ ∥ ♃
03 42 ☽ ☌ ♀
04 17 ☽ ∥ ♀
07 04 ☽ ± ♅
08 04 ☽ ∠ ♃
09 50 ☽ ⊼ ♄
10 48 ☽ ⊼ ♆
12 24 ☽ ☌ ♂
12 50 ☽ ∥ ♂
13 39 ☽ □ ♇
15 46 ☽ ⊻ ☉
17 17 ☽ ∠ ☿
21 01 ☽ ± ♄
22 03 ☽ ± ♆
23 54 ☽ bQ ♅

06 Thursday
04 01 ☽ ⊥ ☉
12 55 ☽ ⚹ ♃
13 34 ♂ □ ♇
14 03 ☽ bQ ♄
14 32 ♀ ± ♅
15 10 ☽ bQ ♆
22 43 ☽ ∠ ☉
23 55 ☽ Q ♇

07 Friday
03 12 ☽ ⚹ ☿
04 36 ☽ △ ♅
11 47 ☿ ☍ ♅
11 48 ♀ ∠ ♃
12 37 ☽ Q ♃
17 33 ♀ ⊼ ♄
19 04 ☽ △ ♄
19 15 ☽ ⊻ ♀
20 19 ☽ △ ♆
23 24 ☽ ⚹ ♇

08 Saturday
01 16 ☽ ⊻ ♂
05 49 ♀ ⊼ ♆
06 39 ☽ ⚹ ☉
07 25 ☽ Q ☿
08 27 ☽ ⊥ ♀
13 08 ☉ # ♄
13 50 ☽ ⊥ ♂
21 17 ☿ ⊥ ♀
21 49 ☽ # ♇

09 Sunday
00 34 ☽ ∥ ♇
01 02 ☽ □ ♃
04 28 ☽ ∠ ♀
05 19 ☽ ∠ ♇
08 44 ☽ Q ☉
08 53 ☽ ∠ ♂
12 40 ♀ □ ♇
16 03 ☽ □ ♅
23 27 ☽ ⊥ ♇

10 Monday
03 02 ☽ □ ☿
06 49 ☽ □ ♄
08 20 ☽ □ ♆
09 49 ☿ # ♅
11 01 ☿ ⊥ ♂
11 35 ☽ ⊻ ♇
14 13 ☽ ⚹ ♀
16 53 ☽ ⚹ ♂
23 57 ☿ ☍ ♄

11 Tuesday
00 19 ☽ □ ☉
04 22 ☽ Q ♅
08 49 ☿ ☍ ♆
11 41 ☉ # ♆
14 49 ☽ △ ♃
17 15 ☽ Q ♀
18 30 ☽ Q ♂
19 00 ☽ Q ♄
20 16 ☿ # ♅
20 39 ☽ Q ♆

12 Wednesday
03 03 ♂ ± ♄
03 19 ☿ △ ♇
04 29 ☽ ⚹ ♅
06 14 ☽ # ♂
08 42 ☽ # ♀
09 34 ♀ ± ♄
17 20 ♀ ☌ ♂
18 56 ☽ ⚹ ♄
20 39 ☽ ⚹ ♆
21 29 ☽ bQ ♃
23 58 ☽ ☌ ♇

13 Thursday
02 02 ♀ ± ♆
04 23 ☽ △ ☿
08 28 ☽ □ ♂
09 02 ☽ # ☉
09 16 ☽ □ ♀
09 51 ♂ ± ♆
10 17 ☽ ∠ ♅
10 47 ☽ ∥ ♆
13 27 ☽ ∥ ♄
15 37 ☽ ± ♃
17 17 ☽ # ♃
17 32 ☽ △ ☉
19 09 ♀ bQ ♅
20 53 ☽ # ☿
22 04 ☽ ∥ ♅

14 Friday
00 22 ☽ ∠ ♄
02 10 ☽ ∠ ♆
03 34 ☽ ⊼ ♃
03 53 ☽ ⊥ ♅
11 25 ☿ ⊻ ♂
15 28 ☽ ⊻ ♅
15 54 ☽ bQ ☿
17 43 ☽ ⊥ ♄
17 57 ♂ bQ ♅
19 33 ☽ ⊥ ♆

15 Saturday
01 01 ☽ bQ ☉
01 11 ♀ ∥ ♂
05 08 ☽ ⊻ ♄
06 59 ☽ ⊻ ♆
10 13 ☽ ⊻ ♇
12 36 ☽ ± ☿
13 43 ☿ ⊻ ♀
19 26 ☽ ± ☉
21 18 ☽ △ ♂
21 31 ☽ ⊥ ♇

16 Sunday
01 07 ☽ △ ♀
02 07 ☽ ⊼ ☿
07 30 ☽ ⊼ ☉
11 47 ☿ ∥ ♃
13 25 ☽ ☍ ♃
14 12 ☽ ∠ ♇
23 32 ☽ ☌ ♅

17 Monday
02 22 ☽ bQ ♂
07 29 ☽ bQ ♀
12 20 ☽ ☌ ♄
14 16 ☽ ☌ ♆
17 26 ☽ ⚹ ♇
19 24 ☽ ± ♂

18 Tuesday
01 14 ☽ ± ♀
06 36 ☽ ⊼ ♂
08 16 ☉ ☌ ☿
09 41 ☿ # ♄
12 24 ♃ bQ ♇
12 53 ☽ ⊼ ♀
14 44 ☽ Q ♇
16 23 ☽ ∥ ♅
17 42 ☽ ☍ ☉
18 39 ☽ ☍ ☿
20 07 ☽ ⊼ ♃
20 28 ☽ # ♃
23 50 ☽ ∥ ♄

19 Wednesday
01 08 ☽ # ☿
02 19 ☽ ∥ ♆
04 46 ☽ ⊻ ♅
04 57 ☿ ⊻ ♃
06 45 ☽ ± ♃
09 21 ☽ # ☉
10 15 ♀ bQ ♄
14 28 ☿ # ♆
15 08 ☽ ⊥ ♅
16 52 ☽ ⊻ ♄
18 53 ☽ ⊻ ♆
22 01 ☽ □ ♇
22 33 ☽ bQ ♃

20 Thursday
03 09 ☽ ⊥ ♄
05 11 ☽ ⊥ ♆
06 34 ☽ ∠ ♅
08 50 ♀ bQ ♆
10 05 ☽ # ♂
13 07 ☽ ☍ ♂
13 50 ☉ ⊻ ♃
14 59 ☽ # ♀
18 27 ☽ ∠ ♄
20 31 ☽ ∠ ♆
21 38 ☽ ☍ ♀

21 Friday
00 34 ☽ △ ♃
01 10 ☽ ⊼ ☉
07 25 ☽ ⊼ ☿
08 02 ☽ ⚹ ♅
11 34 ☿ ⊼ ♅
12 04 ☽ ± ☉
19 19 ☽ ± ☿
19 45 ☽ ⚹ ♄
21 52 ☽ ⚹ ♆

22 Saturday
01 01 ☽ △ ♇
04 15 ☽ Q ♅
04 21 ☽ bQ ☉
09 25 ☿ ⊥ ♃
13 03 ☽ bQ ☿
14 43 ♀ ⚹ ♃
15 52 ☽ Q ♄
18 02 ☽ Q ♆
18 12 ☽ ⊼ ♂
19 29 ☽ ∥ ♇
22 33 ☽ # ♇

23 Sunday
02 15 ☽ bQ ♇
03 56 ☽ □ ♃
03 58 ♇ St D
04 45 ☽ ± ♂
04 56 ☽ ⊼ ♀
07 23 ☽ △ ☉
10 27 ☽ □ ♅
15 57 ☽ ± ♀
17 22 ☽ ± ♇
18 29 ☽ △ ☿
20 34 ☽ bQ ♂
21 58 ☽ □ ♄

24 Monday
00 14 ☽ □ ♆
03 27 ☽ ⊼ ♇
08 28 ☽ bQ ♀
08 31 ☿ ± ♅
08 50 ♀ Q ♇
10 38 ☽ Q ♃
16 08 ☉ ∥ ☿
18 25 ☿ ⊼ ♄
19 37 ☽ ∥ ♀
22 58 ☽ △ ♂

25 Tuesday
01 41 ♂ bQ ♄
03 40 ☽ ∥ ♂
03 45 ☉ ⊼ ♅
07 10 ☽ ⚹ ♃
10 48 ☿ ⊼ ♆
12 06 ☽ △ ♀
12 51 ☽ △ ♅
13 31 ☽ □ ☉
20 37 ♀ △ ♅
23 31 ☽ ∥ ☿

26 Wednesday
00 20 ☽ △ ♄
01 32 ☽ ∥ ☉
02 46 ☽ △ ♆
05 24 ☽ □ ☿
06 07 ☽ ☍ ♇
08 58 ☽ ∠ ♃
10 31 ☿ □ ♇
14 14 ☽ bQ ♅
15 20 ☽ # ♆
17 57 ☽ # ♄
21 10 ☽ ∥ ♃
22 20 ☽ Q ☉

27 Thursday
00 33 ☽ ⊥ ♃
00 59 ☽ # ♅
01 47 ☽ bQ ♄
04 16 ☽ □ ♂
04 18 ☽ bQ ♆
05 01 ♂ bQ ♆
05 35 ☽ ± ♅
09 44 ☿ ∠ ♃
10 59 ☽ ⊻ ♃
15 52 ☽ ⊼ ♅
17 09 ☿ ± ♄
17 10 ☽ ± ♄
17 10 ☽ Q ☿
19 45 ☽ ± ♆
20 05 ☽ □ ♀
20 27 ☽ ⚹ ☉

28 Friday
03 29 ☽ ⊼ ♄
06 06 ☽ ⊼ ♆
09 38 ☽ ⊼ ♇
11 38 ☿ ± ♆
12 48 ☽ Q ♂
15 18 ☉ ⊥ ♃
15 36 ☉ ⊻ ♀
17 19 ☽ ⚹ ☿
20 05 ☽ ± ♇
20 48 ☿ bQ ♅

29 Saturday
00 28 ☽ ∠ ☉
06 25 ☽ Q ♀
10 50 ☽ ⚹ ♂
11 54 ☽ bQ ♇
16 03 ☽ ☌ ♃
17 31 ☽ ⊥ ☉
20 08 ☽ ☍ ♅
23 59 ☽ ∠ ☿

30 Sunday
05 00 ☽ ⊻ ☉
05 45 ☽ ⚹ ♀
08 00 ☽ ☍ ♄
10 52 ☽ ☍ ♆
14 37 ☽ △ ♇
14 51 ☽ ∠ ♂
18 42 ☽ ⊥ ☿

31 Monday
05 32 ☉ ± ♅
05 45 ♀ △ ♄
07 20 ☽ ⊻ ☿
07 57 ☽ ⊥ ♂
11 33 ☽ ∠ ♀
14 05 ☽ # ♅
18 07 ☽ ∥ ♃
19 29 ☽ ⊻ ♂
20 51 ☉ ⊼ ♄
21 25 ☽ # ♄
23 03 ☽ ⊻ ♃

AUGUST 1989

LONGITUDES

Date	Sidereal time h m s	Sun ☉ ° ' "	Moon ☽ ° ' "	Moon ☽ 24.00 ° ' "	Mercury ☿ ° '	Venus ♀ ° '	Mars ♂ ° '	Jupiter ♃ ° '	Saturn ♄ ° '	Uranus ♅ ° '	Neptune ♆ ° '	Pluto ♇ ° '
01	08 40 17	09 ♌ 12 37	07 ♌ 10 55	13 ♌ 35 10	23 ♌ 48	10 ♍ 10	28 ♌ 42	00 ♋ 18	08 ♑ 34	01 ♑ 57	10 ♑ 15	12 ♏ 24
02	08 44 14	10 10 04	19 ♌ 55 34	26 ♌ 12 13	25 35	11 22	29 20	00 31	08 R 31	01 R 56	10 R 13	12 24
03	08 48 10	11 07 31	02 ♍ 25 17	08 ♍ 34 56	27 21	12 34	29 ♌ 57	00 43	08 27	01 54	10 12	12 24
04	08 52 07	12 04 59	14 ♍ 41 27	20 ♍ 45 07	29 ♌ 05	13 46	00 ♍ 35	00 55	08 24	01 52	10 11	12 25
05	08 56 04	13 02 27	26 ♍ 46 17	02 ♎ 45 22	00 ♍ 47	14 58	01 13	01 07	08 21	01 51	10 09	12 25
06	09 00 00	13 59 57	08 ♎ 42 48	14 ♎ 39 03	02 28	16 10	01 51	01 18	08 18	01 49	10 08	12 26
07	09 03 57	14 57 27	20 ♎ 34 38	26 ♎ 30 06	04 07	17 22	02 29	01 30	08 14	01 47	10 07	12 26
08	09 07 53	15 54 58	02 ♏ 25 59	08 ♏ 22 53	05 45	18 33	03 07	01 42	08 11	01 46	10 06	12 27
09	09 11 50	16 52 30	14 ♏ 21 22	20 ♏ 22 01	07 21	19 45	03 45	01 54	08 08	01 44	10 04	12 27
10	09 15 46	17 50 03	26 ♏ 25 24	02 ♐ 32 05	08 55	20 57	04 23	02 05	08 06	01 43	10 03	12 28
11	09 19 43	18 47 36	08 ♐ 42 36	14 ♐ 57 26	10 28	22 09	05 01	02 17	08 03	01 42	10 02	12 28
12	09 23 39	19 45 11	21 ♐ 17 02	27 ♐ 41 46	11 59	23 21	05 38	02 28	08 00	01 40	10 01	12 29
13	09 27 36	20 42 47	04 ♑ 11 57	10 ♑ 47 48	13 29	24 32	06 16	02 39	07 57	01 39	10 00	12 30
14	09 31 33	21 40 23	17 ♑ 29 25	24 ♑ 16 50	14 57	25 44	06 54	02 51	07 55	01 38	09 59	12 31
15	09 35 29	22 38 01	01 ♒ 09 55	08 ♒ 08 27	16 23	26 55	07 32	03 02	07 52	01 36	09 57	12 31
16	09 39 26	23 35 40	15 ♒ 12 04	22 ♒ 20 17	17 48	28 07	08 10	03 13	07 50	01 35	09 56	12 32
17	09 43 22	24 33 19	29 ♒ 32 32	06 ♓ 48 08	19 11	29 ♍ 18	08 48	03 24	07 47	01 34	09 55	12 33
18	09 47 19	25 31 01	14 ♓ 06 19	21 ♓ 26 18	20 32	00 ♎ 30	09 27	03 35	07 45	01 33	09 54	12 34
19	09 51 15	26 28 43	28 ♓ 47 15	06 ♈ 08 21	21 52	01 41	10 05	03 45	07 43	01 32	09 53	12 35
20	09 55 12	27 26 27	13 ♈ 28 48	20 ♈ 47 52	23 10	02 53	10 43	03 56	07 41	01 31	09 52	12 36
21	09 59 08	28 24 13	28 ♈ 04 52	05 ♉ 19 14	24 26	04 04	11 21	04 07	07 39	01 30	09 51	12 37
22	10 03 05	29 ♌ 22 01	12 ♉ 30 28	19 ♉ 38 10	25 40	05 15	11 59	04 17	07 37	01 29	09 51	12 38
23	10 07 02	00 ♍ 19 50	26 ♉ 42 02	03 ♊ 41 54	26 52	06 26	12 37	04 28	07 35	01 28	09 50	12 39
24	10 10 58	01 17 41	10 ♊ 37 36	17 ♊ 29 07	28 02	07 37	13 15	04 38	07 33	01 27	09 49	12 40
25	10 14 55	02 15 34	24 ♊ 16 26	00 ♋ 59 38	29 ♍ 10	08 49	13 53	04 48	07 32	01 26	09 48	12 41
26	10 18 51	03 13 29	07 ♋ 38 47	14 ♋ 14 01	00 ♎ 16	10 00	14 32	04 58	07 30	01 26	09 47	12 42
27	10 22 48	04 11 25	20 ♋ 45 28	27 ♋ 13 16	01 19	11 11	15 10	05 08	07 29	01 25	09 46	12 43
28	10 26 44	05 09 23	03 ♌ 37 36	09 ♌ 58 37	02 20	12 22	15 48	05 18	07 27	01 24	09 46	12 44
29	10 30 41	06 07 23	16 ♌ 16 27	22 ♌ 31 16	03 18	13 32	16 27	05 28	07 26	01 24	09 45	12 46
30	10 34 37	07 05 24	28 ♌ 43 14	04 ♍ 52 31	04 14	14 43	17 05	05 38	07 25	01 23	09 44	12 47
31	10 38 34	08 ♍ 03 27	10 ♍ 59 17	17 ♍ 03 43	05 ♎ 07	15 ♎ 54	17 ♍ 43	05 ♋ 47	07 ♑ 24	01 ♑ 23	09 ♑ 43	12 ♏ 48

Date	Moon True ☊ ° '	Moon Mean ☊ ° '	Moon ☽ Latitude ° '
01	25 ♒ 47	26 ♒ 32	01 N 41
02	25 R 46	26 29	00 N 32
03	25 D 46	26 26	00 S 37
04	25 47	26 22	01 43
05	25 48	26 19	02 43
06	25 49	26 16	03 36
07	25 50	26 13	04 19
08	25 51	26 10	04 51
09	25 51	26 07	05 10
10	25 R 51	26 03	05 17
11	25 51	26 00	05 09
12	25 50	25 57	04 46
13	25 50	25 54	04 09
14	25 49	25 51	03 17
15	25 49	25 48	02 12
16	25 49	25 44	00 S 59
17	25 D 49	25 41	00 N 21
18	25 49	25 38	01 40
19	25 R 49	25 35	02 53
20	25 48	25 32	03 55
21	25 48	25 28	04 41
22	25 48	25 25	05 09
23	25 48	25 22	05 17
24	25 D 48	25 19	05 07
25	25 48	25 16	04 39
26	25 49	25 13	03 56
27	25 50	25 09	03 02
28	25 50	25 06	02 00
29	25 51	25 03	00 N 53
30	25 R 51	25 00	00 S 16
31	25 ♒ 51	24 ♒ 57	01 S 23

DECLINATIONS

Date	Sun ☉ ° '	Moon ☽ ° '	Mercury ☿ ° '	Venus ♀ ° '	Mars ♂ ° '	Jupiter ♃ ° '	Saturn ♄ ° '	Uranus ♅ ° '	Neptune ♆ ° '	Pluto ♇ ° '
01	17 N 57	20 N 06	15 N 00	09 N 02	12 N 58	23 N 06	22 S 36	23 S 42	22 S 07	00 S 32
02	17 42	15 21	14 20	08 33	12 45	23 06	22 36	23 42	22 08	00 32
03	17 26	10 03	13 39	08 04	12 31	23 06	22 36	23 42	22 08	00 33
04	17 10	04 N 27	12 58	07 35	12 18	23 06	22 37	23 42	22 08	00 34
05	16 54	01 S 13	12 16	07 05	12 04	23 06	22 37	23 42	22 08	00 34
06	16 38	06 46	11 34	06 36	11 50	23 06	22 37	23 42	22 08	00 35
07	16 21	12 02	10 52	06 06	11 36	23 06	22 38	23 42	22 08	00 36
08	16 04	16 52	10 10	05 36	11 22	23 06	22 38	23 42	22 08	00 36
09	15 47	21 05	09 28	05 06	11 08	23 06	22 38	23 42	22 08	00 37
10	15 29	24 29	08 46	04 35	10 54	23 06	22 39	23 42	22 09	00 38
11	15 12	26 50	08 04	04 05	10 40	23 06	22 39	23 42	22 09	00 38
12	14 54	27 55	07 21	03 34	10 26	23 06	22 39	23 42	22 09	00 39
13	14 35	27 31	06 40	03 04	10 11	23 05	22 40	23 42	22 09	00 40
14	14 17	25 33	05 58	02 33	09 57	23 05	22 40	23 42	22 09	00 41
15	13 58	22 03	05 17	02 02	09 43	23 05	22 40	23 42	22 09	00 41
16	13 39	17 13	04 36	01 32	09 28	23 05	22 40	23 42	22 09	00 42
17	13 20	11 19	03 55	01 01	09 14	23 05	22 41	23 42	22 09	00 43
18	13 01	04 S 43	03 15	00 N 30	08 59	23 04	22 41	23 42	22 09	00 44
19	12 41	02 N 10	02 35	00 S 01	08 44	23 04	22 41	23 42	22 10	00 45
20	12 22	08 55	01 55	00 32	08 30	23 04	22 41	23 42	22 10	00 45
21	12 02	15 10	01 17	01 03	08 15	23 04	22 42	23 42	22 10	00 46
22	11 42	20 29	00 39	01 34	08 00	23 03	22 42	23 42	22 10	00 47
23	11 21	24 33	00 N 01	02 05	07 45	23 03	22 42	23 42	22 10	00 48
24	11 01	27 06	00 S 35	02 36	07 30	23 03	22 42	23 42	22 10	00 49
25	10 40	27 58	01 11	03 07	07 15	23 03	22 43	23 42	22 10	00 49
26	10 19	27 09	01 46	03 38	07 00	23 02	22 43	23 42	22 10	00 50
27	09 58	24 51	02 20	04 09	06 45	23 02	22 43	23 42	22 10	00 51
28	09 37	21 17	02 53	04 39	06 30	23 02	22 43	23 42	22 10	00 52
29	09 16	16 48	03 25	05 10	06 15	23 01	22 43	23 42	22 11	00 53
30	08 55	11 40	03 56	05 41	06 00	23 01	22 44	23 42	22 11	00 54
31	08 N 33	06 N 10	04 S 26	06 S 11	05 N 44	23 N 01	22 S 44	23 S 42	22 S 11	00 S 54

ZODIAC SIGN ENTRIES

Date	h m	Planets
03	07 19	☽ ♍
03	13 35	♂ ♍
05	00 54	☿ ♍
05	18 28	☽ ♎
08	07 05	☽ ♏
10	19 02	☽ ♐
13	04 16	☽ ♑
15	09 59	☽ ♒
17	12 46	☽ ♓
18	01 58	♀ ♎
19	13 59	☽ ♈
21	15 10	☽ ♉
23	03 46	☉ ♍
23	17 39	☽ ♊
25	22 13	☽ ♋
26	06 14	☿ ♎
28	05 12	☽ ♌
30	14 29	☽ ♍

LATITUDES

Date	Mercury ☿ ° '	Venus ♀ ° '	Mars ♂ ° '	Jupiter ♃ ° '	Saturn ♄ ° '	Uranus ♅ ° '	Neptune ♆ ° '	Pluto ♇ ° '
01	01 N 30	01 N 22	01 N 07	00 S 20	00 N 34	00 S 16	00 N 56	15 N 45
04	01 15	01 17	01 06	00 20	00 34	00 16	00 56	15 43
07	00 56	01 12	01 05	00 20	00 33	00 16	00 55	15 41
10	00 35	01 05	01 04	00 20	00 33	00 16	00 55	15 40
13	00 N 11	00 58	01 03	00 20	00 33	00 16	00 55	15 38
16	00 S 15	00 51	01 02	00 19	00 32	00 16	00 55	15 36
19	00 42	00 43	01 01	00 19	00 32	00 16	00 55	15 34
22	01 11	00 34	01 01	00 19	00 32	00 16	00 55	15 33
25	01 39	00 25	01 00	00 19	00 31	00 16	00 55	15 31
28	02 08	00 15	00 59	00 19	00 31	00 16	00 55	15 30
31	02 S 37	00 N 05	00 N 58	00 S 18	00 N 30	00 S 16	00 N 55	15 N 28

DATA

Julian Date	2447740
Delta T	+56 seconds
Ayanamsa	23° 42' 52"
Synetic vernal point	05° ♓ 24' 07"
True obliquity of ecliptic	23° 26' 34"

MOON'S PHASES, APSIDES AND POSITIONS ☽

Date	h m	Phase	Longitude ° '	Eclipse Indicator
01	16 06	●	09 ♌ 22	
09	17 29	☽	17 ♏ 06	
17	03 07	○	24 ♒ 12	total
23	18 40	☾	00 ♊ 36	
31	05 45	●	07 ♍ 48	Partial

Day	h m		
07	15 25	Apogee	
19	12 39	Perigee	
05	06 51	0S	
12	17 49	Max dec	27° S 58'
19	04 29	0N	
25	12 13	Max dec	27° N 58'

LONGITUDES

Date	Chiron ⚷ ° '	Ceres ⚳ ° '	Pallas ⚴ ° '	Juno ⚵ ° '	Vesta ⚶ ° '	Black Moon Lilith ⚸ ° '
01	11 ♋ 54	14 ♊ 55	16 ♈ 16	21 ♍ 24	28 ♐ 21	19 ♎ 21
11	12 ♋ 48	18 ♊ 20	16 ♈ 16	24 ♍ 54	28 ♐ 15	20 ♎ 28
21	13 ♋ 38	21 ♊ 35	15 ♈ 40	28 ♍ 26	28 ♐ 56	21 ♎ 35
31	14 ♋ 24	24 ♊ 36	14 ♈ 24	02 ♎ 02	00 ♑ 19	22 ♎ 42

All ephemeris data is given at 12.00 UT and the Moon's longitude is additionally given for 24.00 UT

ASPECTARIAN

01 Tuesday
00 21 ☽ # ♆
02 19 ☽ ⚻ ♅
05 48 ☽ ⊥ ♀
09 05 ☿ ⚼ ♄
10 21 ☽ ⊥ ♃
13 27 ☽ ± ♅
13 42 ♀ △ ♆
14 35 ☽ ⚻ ♄
16 06 ☽ ☌ ☉
17 43 ☽ ⚻ ♆
18 08 ☽ ⊻ ♀
21 45 ☽ □ ♇
23 56 ☽ ∥ ☉

02 Wednesday
01 47 ☽ ± ♄
03 30 ☽ ∠ ♃
04 59 ☽ ± ♆
06 19 ☽ ⚼ ♅
07 11 ☿ ⚼ ♆
13 21 ☉ ⚻ ♆
17 29 ☽ ∥ ☿
18 49 ☽ ⚼ ♄
22 06 ☽ ⚼ ♆

03 Thursday
00 32 ☽ ∥ ♂
00 35 ☽ ☌ ☿
06 59 ☽ ☌ ♂
08 05 ☽ Q ♇
08 38 ☽ ⚹ ♃
08 50 ♀ ⚹ ♇
10 59 ☽ △ ♅
15 35 ♀ Q ♃
21 23 ☽ ∥ ♀
23 42 ☽ △ ♄

04 Friday
03 08 ☽ △ ♆
05 08 ♂ Q ♇
06 26 ☽ ⊻ ☉
07 30 ☽ ⚹ ♇
08 26 ☽ Q ♃
09 58 ☽ ☌ ♀
19 17 ☽ ⊥ ☉
20 16 ☉ □ ♇

05 Saturday
04 26 ☽ # ♇
05 55 ♂ ⚹ ♃
06 46 ☿ Q ♇
09 16 ☽ ∥ ♇
13 18 ☽ ∠ ♇
14 46 ☽ ∠ ☉
17 13 ☿ ⚹ ♃
20 50 ☽ □ ♃
21 22 ☽ ⊻ ☿
21 25 ☽ ⊻ ♂
21 51 ☿ ☌ ♂
22 09 ☽ □ ♅
22 29 ☿ ∥ ♂

06 Sunday
02 51 ☿ △ ♅
07 23 ☽ ⊥ ♇
10 10 ☽ ⊥ ♂
10 48 ♂ △ ♅
11 09 ☽ □ ♄
11 19 ☽ # ♀
11 25 ☽ ⊥ ☿
14 52 ☽ □ ♆
18 59 ☉ ± ♄
19 30 ☽ ⊻ ♇
23 37 ☽ ⚹ ☉

07 Monday
04 45 ☽ ⊻ ♀
05 23 ☽ ∠ ♂
07 13 ☽ # ☿
08 34 ☽ ∠ ☿
10 04 ☽ # ♂
10 25 ☽ Q ♅
18 16 ☽ ⊥ ♀
23 25 ☽ Q ♄

08 Tuesday
02 04 ☽ Q ☉
03 14 ☽ Q ♆
08 05 ☽ # ☉
10 29 ☽ △ ♃
10 39 ☽ ⚹ ♅
13 27 ☽ ⚹ ♂
14 32 ☽ ∠ ♀
16 20 ☉ ± ♆
19 18 ♃ ☍ ♅
19 44 ☽ ⚹ ☿
23 34 ☽ ⚹ ♄

09 Wednesday
03 25 ☽ ⚹ ♆
08 11 ☽ ☌ ♇
08 44 ☉ ⚼ ♅
12 34 ☉ ∠ ♃
14 56 ☽ Q ♂
16 46 ☽ ∠ ♅
17 09 ☽ ⚼ ♃
17 29 ☽ □ ☉
18 48 ☽ ∥ ♆
22 12 ☽ ∥ ♄
23 28 ☽ Q ☿
23 44 ☿ △ ♄
23 58 ☽ ⚹ ♀

10 Thursday
01 24 ☽ # ♃
05 26 ☽ ∠ ♄
05 51 ☽ ∥ ♅
09 18 ☽ ∠ ♆
10 36 ☽ ⊥ ♅
11 19 ☽ ± ♃
22 23 ☽ ⊻ ♅
23 06 ☽ ⊥ ♄
23 18 ☽ ⚻ ♃

11 Friday
02 12 ☽ Q ♀
02 56 ☽ ⊥ ♆
04 26 ☽ □ ♂
05 21 ☿ △ ♆
10 43 ☽ ⊻ ♄
14 33 ☽ ⊻ ♆
15 52 ☽ □ ☿
19 15 ☽ ⊻ ♇

12 Saturday
06 42 ☽ ⊥ ♇
08 52 ☽ △ ☉
16 16 ☽ □ ♀
20 03 ☿ ⚹ ♇
23 37 ☽ ∠ ♇

13 Sunday
07 19 ☽ ☌ ♅
09 08 ☽ ☍ ♃
14 59 ☽ ⚼ ☉
15 59 ☽ △ ♂
18 50 ☽ ☌ ♄
22 32 ☽ ☌ ♆

14 Monday
03 05 ☽ ⚹ ♇
06 54 ☽ △ ☿
08 31 ☽ ± ☉
10 04 ☿ Q ♃
19 58 ☽ ⚻ ☉
20 12 ☽ ⚼ ♂

15 Tuesday
00 25 ☽ Q ♇
01 54 ☽ ∥ ♅
03 56 ☽ △ ♀
05 52 ☽ # ♃
08 26 ☽ ∥ ♄
11 27 ☽ ∥ ♆
12 25 ☽ ⚼ ☿
12 41 ☽ ± ♂
12 46 ☽ ⊻ ♅
15 16 ☽ ⚻ ♃
17 41 ☉ ⚼ ♄
23 04 ☽ ⊥ ♅
23 30 ☽ ⚻ ♂
23 30 ☽ ⊻ ♄
23 45 ♂ △ ♄

16 Wednesday
00 11 ♀ ∠ ♇
01 42 ☽ ± ♃
03 05 ☽ ⊻ ♆
05 36 ☽ ± ☿
07 29 ☽ □ ♇
08 09 ☽ ⚼ ♀
09 41 ☽ ⊥ ♄
13 15 ☽ ⊥ ♆
14 20 ☽ ∠ ♅
16 51 ☽ ⚻ ☿
17 09 ☽ ⚼ ♃

17 Thursday
00 41 ☽ ± ♀
00 47 ☽ ∠ ♄
03 07 ☽ ☍ ☉
03 42 ☽ # ☉
04 20 ☽ ∠ ♆
11 35 ☽ ⚻ ♀
15 21 ☽ ⚹ ♅
18 28 ☽ △ ♃
20 06 ☽ # ♂
21 01 ☉ ⚼ ♆

18 Friday
01 24 ♀ # ♇
01 36 ☽ ⚹ ♄
04 00 ☽ ☍ ♂
05 07 ☽ ⚹ ♆
09 28 ☽ △ ♇
11 05 ☽ Q ♅
17 45 ☽ # ☿
21 13 ☽ Q ♄
23 35 ☽ ☍ ☿

19 Saturday
00 45 ☽ Q ♆
01 55 ☽ ∥ ♇
03 57 ☽ # ♀
04 56 ☽ ∥ ♀
05 05 ♂ △ ♆
07 03 ☽ # ♇
07 58 ☽ ⚻ ☉
08 53 ♀ □ ♅
10 02 ☽ ⚼ ♇
13 20 ☽ ∥ ☿
16 28 ☽ □ ♅
17 09 ☽ ☍ ♀
18 27 ☽ ± ☉
20 13 ☽ □ ♃

20 Sunday
00 44 ☽ ± ♇
02 33 ☽ □ ♄
06 06 ☽ □ ♆
07 16 ☽ ⚻ ♂
10 11 ☽ ⚼ ☉
10 30 ☽ ∥ ♂
10 33 ☽ ⚻ ♇
17 32 ☽ ± ♂
22 30 ♀ ∥ ♇

21 Monday
00 14 ☽ ∥ ☉
02 02 ☽ Q ♃
05 24 ☽ ⚻ ☿
09 00 ☽ ⚼ ♂
12 34 ☽ △ ☉
13 09 ♀ □ ♃
16 15 ☽ ± ☿
16 43 ☿ # ♀
17 39 ☽ △ ♅
22 07 ☽ ⚹ ♃
22 48 ☽ ⚻ ♀

22 Tuesday
03 50 ☽ △ ♄
06 52 ☿ # ♇
07 33 ☽ △ ♆
08 37 ☽ ⚼ ☿
09 42 ☽ ± ♀
11 05 ☽ △ ♂
12 12 ☽ ☍ ♇
18 40 ☽ ⚼ ♅
20 56 ☽ # ♆
23 33 ☽ ∠ ♃

23 Wednesday
00 01 ☽ # ♄
00 22 ☉ ⊥ ♀
02 08 ☽ ∥ ♃
02 14 ☽ ⚼ ♀
05 01 ☽ ⚼ ♄
06 13 ☽ # ♅
08 49 ☽ ⚼ ♆
09 54 ☽ ± ♅
12 18 ☽ △ ☿
13 04 ♂ ⚹ ♇
15 03 ☽ ⊥ ♃
16 15 ♀ ⊥ ♇
18 40 ☽ □ ☉
20 00 ☉ Q ♇
20 09 ☽ ⚻ ♅
20 21 ☽ ± ♄

24 Thursday
00 13 ☽ ± ♆
01 29 ☽ ⊻ ♃
04 16 ☿ ∠ ♇
06 18 ☽ △ ♀
06 41 ☽ ⚻ ♄
10 35 ☽ ⚻ ♆
10 40 ♀ □ ♄
15 33 ☽ ⚻ ♇
15 52 ☉ △ ♅
16 48 ☽ □ ♂
20 59 ☿ ∥ ♇

25 Friday
02 06 ☽ ± ♇
04 21 ☽ Q ☉
18 05 ☽ ⚼ ♇
21 30 ☽ □ ☿

26 Saturday
00 47 ☽ ☍ ♅
02 18 ☽ Q ♂
03 23 ☽ ⚹ ☉
07 06 ☽ ☌ ♃
07 49 ♀ □ ♆
11 45 ☽ ☍ ♄
15 53 ☽ ☍ ♆
16 41 ☽ □ ♀
21 13 ☽ △ ♇

27 Sunday
01 11 ☽ ⚹ ♂
08 52 ☽ ∠ ☉
09 07 ☽ Q ☿
14 14 ☿ □ ♅
20 32 ☽ # ♅

28 Monday
01 10 ☽ ∥ ♃
02 56 ☽ ⊥ ☉
03 12 ☽ # ♄
05 15 ☽ Q ♀
06 25 ☽ ∠ ♂
06 39 ☽ # ♆
07 50 ☽ ⚻ ♅
09 22 ☽ ⚹ ☿
15 07 ☽ ⊻ ☉
15 12 ☽ ⊻ ♃
16 26 ☉ ⚹ ♃
19 07 ☽ ± ♅
19 13 ☽ ⚻ ♄
19 52 ♀ ⊻ ♇
23 35 ☽ ⚻ ♆

29 Tuesday
00 17 ☽ ⊥ ♂
02 43 ☽ ⊥ ♃
05 17 ☽ □ ♇
06 14 ☽ ⚹ ♀
06 35 ☽ ± ♄
11 00 ☽ ± ♆
12 14 ☽ ⚼ ♅
12 20 ☽ ⊻ ♂
16 13 ☽ ∠ ☿
20 09 ☽ ∠ ♃
23 49 ☽ ⚼ ♄

30 Wednesday
04 17 ☽ ⚼ ♆
10 59 ☽ ⊥ ☿
14 09 ☽ ∠ ♀
16 01 ☽ Q ♇
17 11 ☽ △ ♅
19 54 ☉ △ ♄
21 55 ♀ # ♂
23 36 ☽ ⊻ ☿

31 Thursday
01 02 ☽ ∥ ☉
01 39 ☽ ⚹ ♃
04 57 ☽ △ ♄
05 45 ☽ ☌ ☉
09 31 ☽ △ ♆
09 38 ☽ ⊥ ♀
11 56 ☽ # ♀
13 56 ☽ ∥ ♂
15 30 ♂ Q ♃
15 35 ☽ ⚹ ♇
18 51 ☽ # ☿
22 45 ☽ ⊻ ♀

SEPTEMBER 1989

LONGITUDES

Date	Sidereal time h m s	Sun ☉ ° ' "	Moon ☽ ° ' "	Moon ☽ 24.00 ° ' "	Mercury ☿ ° '	Venus ♀ ° '	Mars ♂ ° '	Jupiter ♃ ° '	Saturn ♄ ° '	Uranus ♅ ° '	Neptune ♆ ° '	Pluto ♇ ° '
01	10 42 31	09 ♍ 01 32	23 ♍ 06 02	29 ♍ 06 26	05 ♎ 57	17 ♎ 05	18 ♍ 22	05 ♋ 57	07 ♑ 23	01 ♑ 22	09 ♑ 43	12 ♏ 49
02	10 46 27	09 59 38	05 ♎ 05 11	11 ♎ 02 33	06 43	18 16	19 00	06 06	07 R 22	01 R 22	09 R 42	12 51
03	10 50 24	10 57 46	16 ♎ 58 49	22 ♎ 54 20	07 26	19 26	19 38	06 15	07 21	01 21	09 42	12 52
04	10 54 20	11 55 55	28 ♎ 49 28	04 ♏ 44 36	08 06	20 37	20 17	06 25	07 20	01 21	09 41	12 54
05	10 58 17	12 54 06	10 ♏ 40 12	16 ♏ 36 41	08 41	21 47	20 55	06 34	07 20	01 21	09 41	12 55
06	11 02 13	13 52 18	22 ♏ 34 35	28 ♏ 34 24	09 12	22 58	21 34	06 42	07 19	01 21	09 40	12 57
07	11 06 10	14 50 32	04 ♐ 36 40	10 ♐ 41 57	09 39	24 08	22 12	06 51	07 19	01 20	09 40	12 58
08	11 10 06	15 48 48	16 ♐ 50 48	23 ♐ 03 46	10 00	25 18	22 51	07 00	07 18	01 20	09 39	13 00
09	11 14 03	16 47 05	29 ♐ 21 24	05 ♑ 44 13	10 17	26 29	23 29	07 08	07 18	01 20	09 39	13 01
10	11 18 00	17 45 23	12 ♑ 12 41	18 ♑ 47 12	10 28	27 39	24 08	07 17	07 18	01 D 20	09 38	13 03
11	11 21 56	18 43 43	25 ♑ 28 07	02 ♒ 15 39	10 33	28 49	24 46	07 25	07 D 18	01 20	09 38	13 04
12	11 25 53	19 42 05	09 ♒ 09 53	16 ♒ 10 49	10 R 32	29 ♎ 59	25 25	07 33	07 18	01 20	09 38	13 06
13	11 29 49	20 40 28	23 ♒ 18 15	00 ♓ 31 47	10 25	01 ♏ 09	26 04	07 41	07 18	01 21	09 38	13 08
14	11 33 46	21 38 53	07 ♓ 50 55	15 ♓ 14 53	10 11	02 19	26 42	07 49	07 19	01 21	09 37	13 09
15	11 37 42	22 37 19	22 ♓ 42 50	00 ♈ 13 44	09 51	03 28	27 21	07 56	07 19	01 21	09 37	13 11
16	11 41 39	23 35 48	07 ♈ 46 26	15 ♈ 19 47	09 23	04 38	28 00	08 04	07 19	01 21	09 37	13 13
17	11 45 35	24 34 18	22 ♈ 52 33	00 ♉ 23 34	08 49	05 48	28 38	08 11	07 20	01 22	09 37	13 14
18	11 49 32	25 32 51	07 ♉ 51 46	15 ♉ 16 11	08 08	06 57	29 17	08 19	07 21	01 22	09 37	13 16
19	11 53 29	26 31 26	22 ♉ 36 00	29 ♉ 50 34	07 21	08 07	29 ♍ 56	08 26	07 21	01 23	09 37	13 18
20	11 57 25	27 30 03	06 ♊ 59 28	14 ♊ 02 22	06 28	09 16	00 ♎ 35	08 33	07 22	01 23	09 37	13 20
21	12 01 22	28 28 42	20 ♊ 59 11	27 ♊ 49 54	05 30	10 26	01 13	08 40	07 23	01 24	09 D 37	13 22
22	12 05 18	29 ♍ 27 24	04 ♋ 34 42	11 ♋ 13 48	04 28	11 35	01 52	08 46	07 24	01 24	09 37	13 23
23	12 09 15	00 ♎ 26 07	17 ♋ 47 31	24 ♋ 16 12	03 23	12 44	02 31	08 53	07 25	01 25	09 37	13 25
24	12 13 11	01 24 53	00 ♌ 40 17	07 ♌ 00 10	02 18	13 53	03 10	08 59	07 27	01 26	09 37	13 27
25	12 17 08	02 23 42	13 ♌ 16 14	19 ♌ 28 56	01 12	15 02	03 49	09 06	07 28	01 26	09 37	13 29
26	12 21 04	03 22 32	25 ♌ 38 37	01 ♍ 45 40	00 ♎ 09	16 11	04 28	09 12	07 29	01 27	09 37	13 31
27	12 25 01	04 21 24	07 ♍ 50 24	13 ♍ 53 08	29 ♍ 09	17 19	05 07	09 17	07 31	01 28	09 37	13 33
28	12 28 58	05 20 19	19 ♍ 54 09	25 ♍ 53 42	28 15	18 28	05 46	09 23	07 33	01 29	09 37	13 35
29	12 32 54	06 19 15	01 ♎ 52 00	07 ♎ 49 18	27 27	19 37	06 25	09 29	07 34	01 30	09 38	13 37
30	12 36 51	07 ♎ 18 14	13 ♎ 45 46	19 ♎ 41 39	26 ♍ 47	20 ♏ 45	07 ♎ 04	09 ♋ 34	07 ♑ 36	01 ♑ 31	09 ♑ 38	13 ♏ 39

Moon / DECLINATIONS

Date	Moon True ☊ ° '	Moon Mean ☊ ° '	Moon ☽ Latitude ° '	Sun ☉ ° '	Moon ☽ ° '	Mercury ☿ ° '	Venus ♀ ° '	Mars ♂ ° '	Jupiter ♃ ° '	Saturn ♄ ° '	Uranus ♅ ° '	Neptune ♆ ° '	Pluto ♇ ° '
01	25 ♒ 49	24 ♒ 54	02 S 25	08 N 11	00 N 31	04 S 54	06 S 42	05 N 29	23 N 00	22 S 44	23 S 42	22 S 11	00 S 55
02	25 R 48	24 50	03 20	07 49	05 S 05	05 20	07 12	05 14	23 00	22 44	23 42	22 11	00 56
03	25 46	24 47	04 06	07 27	10 28	05 46	07 42	04 58	22 59	22 44	23 42	22 11	00 57
04	25 43	24 44	04 41	07 05	15 26	06 09	08 12	04 43	22 59	22 44	23 42	22 11	00 58
05	25 41	24 41	05 05	06 43	19 50	06 30	08 42	04 27	22 59	22 45	23 42	22 11	00 59
06	25 39	24 38	05 15	06 21	23 29	06 50	09 11	04 12	22 58	22 45	23 42	22 11	01 00
07	25 38	24 34	05 11	05 58	26 10	07 07	09 41	03 56	22 58	22 45	23 42	22 11	01 01
08	25 D 37	24 31	04 54	05 36	27 40	07 22	10 10	03 41	22 58	22 45	23 42	22 11	01 01
09	25 38	24 28	04 22	05 13	27 49	07 34	10 39	03 25	22 57	22 45	23 42	22 11	01 02
10	25 39	24 25	03 37	04 50	26 29	07 43	11 08	03 10	22 57	22 45	23 42	22 11	01 03
11	25 40	24 22	02 39	04 28	23 39	07 50	11 37	02 54	22 56	22 45	23 42	22 11	01 04
12	25 42	24 19	01 30	04 05	19 24	07 53	12 05	02 38	22 56	22 46	23 42	22 12	01 05
13	25 42	24 15	00 S 13	03 42	13 58	07 52	12 33	02 23	22 56	22 46	23 42	22 12	01 06
14	25 R 42	24 12	01 N 06	03 19	07 36	07 48	13 01	02 07	22 55	22 46	23 42	22 12	01 07
15	25 40	24 09	02 23	02 56	00 S 42	07 40	13 29	01 51	22 55	22 46	23 42	22 12	01 08
16	25 37	24 06	03 30	02 33	06 N 18	07 28	13 56	01 36	22 54	22 46	23 42	22 12	01 09
17	25 33	24 03	04 23	02 09	12 58	07 12	14 23	01 20	22 54	22 46	23 42	22 12	01 10
18	25 29	24 00	04 58	01 46	18 49	06 52	14 50	01 04	22 54	22 46	23 42	22 12	01 10
19	25 26	23 56	05 12	01 23	23 27	06 28	15 16	00 48	22 53	22 46	23 42	22 12	01 11
20	25 23	23 53	05 06	01 00	26 30	05 59	15 42	00 33	22 53	22 46	23 42	22 12	01 12
21	25 21	23 50	04 42	00 36	27 49	05 27	16 08	00 17	22 52	22 46	23 42	22 12	01 13
22	25 D 21	23 47	04 02	00 N 13	27 24	04 51	16 33	00 N 01	22 52	22 47	23 42	22 12	01 14
23	25 22	23 44	03 11	00 S 10	25 25	04 13	16 58	00 S 15	22 52	22 47	23 42	22 12	01 15
24	25 24	23 40	02 11	00 34	22 08	03 32	17 23	00 31	22 51	22 47	23 42	22 12	01 16
25	25 25	23 37	01 N 06	00 57	17 53	02 51	17 47	00 46	22 51	22 47	23 42	22 12	01 17
26	25 R 26	23 34	00 S 01	01 21	12 57	02 09	18 11	01 02	22 51	22 47	23 42	22 12	01 18
27	25 25	23 31	01 07	01 44	07 36	01 27	18 35	01 18	22 50	22 47	23 42	22 12	01 19
28	25 22	23 28	02 09	02 07	02 N 01	00 47	18 58	01 34	22 50	22 47	23 42	22 12	01 19
29	25 17	23 25	03 04	02 31	03 S 34	00 10	19 20	01 50	22 50	22 47	23 42	22 12	01 20
30	25 ♒ 11	23 ♒ 21	03 S 52	02 S 54	08 S 59	00 N 25	19 S 42	02 S 06	22 N 49	22 S 47	23 S 42	22 S 12	01 S 21

ZODIAC SIGN ENTRIES

Date	h	m	Planets
02	01	47	☽ ♎
04	14	23	☽ ♏
07	02	51	☽ ♐
09	13	13	☽ ♑
11	20	02	☽ ♒
12	12	22	♀ ♏
13	23	08	☽ ♓
15	23	38	☽ ♈
17	23	22	☽ ♉
19	14	38	♂ ♎
20	00	16	☽ ♊
22	03	50	☽ ♋
23	01	20	☉ ♎
24	10	44	☽ ♌
26	15	28	☿ ♍
26	20	32	☽ ♍
29	08	15	☽ ♎

LATITUDES

Date	Mercury ☿ ° '	Venus ♀ ° '	Mars ♂ ° '	Jupiter ♃ ° '	Saturn ♄ ° '	Uranus ♅ ° '	Neptune ♆ ° '	Pluto ♇ ° '
01	02 S 46	00 N 01	00 N 57	00 S 18	00 N 30	00 S 16	00 N 55	15 N 28
04	03 12	00 S 09	00 56	00 18	00 30	00 16	00 55	15 26
07	03 35	00 20	00 55	00 18	00 30	00 16	00 54	15 25
10	03 53	00 32	00 54	00 18	00 29	00 16	00 54	15 23
13	04 04	00 43	00 53	00 18	00 29	00 16	00 54	15 22
16	04 05	00 55	00 52	00 17	00 28	00 16	00 54	15 20
19	03 52	01 07	00 51	00 17	00 28	00 16	00 54	15 19
22	03 21	01 19	00 50	00 17	00 28	00 16	00 54	15 18
25	02 35	01 30	00 49	00 17	00 27	00 16	00 54	15 17
28	01 37	01 42	00 47	00 17	00 27	00 16	00 54	15 16
31	00 S 36	01 S 54	00 N 46	00 S 17	00 N 27	00 S 16	00 N 53	15 N 15

DATA

Julian Date	2447771
Delta T	+56 seconds
Ayanamsa	23° 42' 56"
Synetic vernal point	05° ♓ 24' 03"
True obliquity of ecliptic	23° 26' 34"

MOON'S PHASES, APSIDES AND POSITIONS ☽

Date	h	m	Phase	Longitude ° '	Eclipse Indicator
08	09	49	☽	15 ♐ 44	
15	11	51	○	22 ♓ 37	
22	02	10	☾	29 ♊ 03	
29	21	47	●	06 ♎ 43	

Day	h	m		
04	08	31	Apogee	
16	15	26	Perigee	
01	14	12	0S	
09	02	31	Max dec	27° S 56'
15	14	25	0N	
21	17	57	Max dec	27° N 53'
28	20	39	0S	

LONGITUDES

Date	Chiron ⚷ ° '	Ceres ⚳ ° '	Pallas ⚴ ° '	Juno ⚵ ° '	Vesta ⚶ ° '	Black Moon Lilith ⚸ ° '
01	14 ♋ 29	24 ♊ 54	14 ♈ 15	02 ♎ 23	00 ♑ 30	22 ♎ 49
11	15 ♋ 08	27 ♊ 38	12 ♈ 18	06 ♎ 01	02 ♑ 35	23 ♎ 56
21	15 ♋ 41	00 ♋ 09	09 ♈ 51	09 ♎ 39	05 ♑ 10	25 ♎ 03
31	16 ♋ 07	02 ♋ 06	07 ♈ 05	13 ♎ 17	08 ♑ 13	26 ♎ 10

All ephemeris data is given at 12.00 UT and the Moon's longitude is additionally given for 24.00 UT

ASPECTARIAN

h m	Aspects
01 Friday	
01 37	☽ Q ♃
02 03	☽ ☌ ♂
10 17	☽ # ♇
12 01	☿ □ ♃
18 08	☽ ∥ ♇
21 27	☽ ∠ ♇
02 Saturday	
04 31	☽ □ ♅
04 53	☉ △ ♆
08 03	☿ # ♂
12 36	☽ # ♂
13 13	☽ ∥ ☿
14 04	☽ □ ♃
15 30	☽ ☌ ☿
15 33	☽ ⊥ ♇
16 12	☿ ⊥ ♇
16 35	☽ □ ♄
21 17	☽ □ ♆
22 13	☽ ∥ ♀
22 46	☽ ⊻ ☉
23 17	☽ # ☉
03 Sunday	
03 40	☽ ⊻ ♇
05 19	☉ # ♀
09 02	☿ □ ♄
10 22	♀ Q ♅
11 58	☽ ⊥ ☉
16 48	☽ Q ♅
17 31	☽ ☌ ♀
17 41	☽ ⊻ ♂
21 02	♀ ⊻ ♂
04 Monday	
04 56	☽ Q ♄
06 33	☽ ⊥ ♂
07 49	☽ ∠ ☉
09 41	☽ Q ♆
17 07	☽ ✶ ♅
05 Tuesday	
01 50	☽ ∠ ♂
03 34	☽ △ ♃
05 15	☽ ✶ ♄
07 47	☽ ⊻ ☿
09 59	☽ ✶ ♆
12 24	☉ ✶ ♇
16 33	☽ ☌ ♇
16 55	☽ ✶ ☉
19 13	☉ # ☿
20 30	☽ ⊥ ☿
23 28	☽ ∠ ♅
06 Wednesday	
02 46	☽ ∥ ♆
06 38	☽ ∥ ♄
08 15	☽ # ♃
09 51	☽ ✶ ♂
10 14	☽ ⚼ ♃
11 29	☽ ∠ ♄
12 51	☽ ⊻ ♀
13 42	☽ ∥ ♅
15 23	☽ ∠ ☿
16 11	☽ ∠ ♆
17 33	☽ ⊥ ♅
19 11	☽ Q ☉
07 Thursday	
02 09	☽ ⊥ ♀
04 27	☽ ± ♃
05 28	☽ ⊥ ♄
05 31	☽ ⊻ ♅
10 07	☽ ⊥ ♆
11 09	☽ Q ♂
13 03	☿ □ ♆
16 29	☽ ⊼ ♃
17 20	☽ ⊻ ♄
21 52	☽ ∠ ♀
21 57	☽ ⊻ ♆
22 15	☽ ✶ ☿
08 Friday	
04 28	☽ ⊻ ♇
09 49	☽ □ ☉
12 02	♀ Q ♄
16 10	☽ ⊥ ♇
22 13	☽ Q ☿
09 Saturday	
00 12	☽ □ ♂
05 58	☽ ✶ ♀
09 27	☽ ∠ ♇
15 45	☽ ☌ ♅
10 Sunday	
01 12	♅ St D
02 46	☽ ☍ ♃
02 55	☽ ☌ ♄
06 48	☽ Q ♀
07 16	☽ ☌ ♆
08 45	☽ □ ☿
11 53	♀ Q ♆
13 32	☽ ✶ ♇
16 19	♃ ☍ ♄
22 56	☽ △ ☉
11 Monday	
07 11	♄ St D
10 42	☽ △ ♂
11 17	☽ Q ♇
11 37	☽ ∥ ♅
16 37	☽ # ♃
17 44	☽ ∥ ♄
18 30	☽ □ ♀
20 57	☿ St R
21 08	☽ ∥ ♆
22 23	☽ ⊻ ♅
12 Tuesday	
03 41	☽ ⚼ ☉
07 39	☉ Q ♃
08 47	☽ ⊻ ♄
08 51	☽ ⊥ ♅
09 11	☽ ⊼ ♃
12 48	☽ ⊻ ♆
14 15	☽ ⚼ ♂
14 21	☽ △ ☿
18 46	☽ □ ♇
19 06	☽ ⊥ ♄
19 36	☽ ± ♃
20 22	☽ ± ☉
23 04	☽ ⊥ ♆
13 Wednesday	
00 16	☽ ∠ ♅
06 18	☽ ± ♂
07 16	☽ ⊼ ☉
10 20	☽ ∠ ♄
10 57	☽ ⚼ ♃
14 12	☽ ∠ ♆
15 29	☽ ⚼ ☿
16 02	♀ ✶ ♅
16 48	☽ ⊼ ♂
17 11	☽ ∥ ♀
14 Thursday	
01 21	☽ ✶ ♅
02 09	☽ △ ♀
06 08	☽ ± ☿
11 07	☽ ✶ ♄
11 16	☽ ∥ ☿
11 56	☽ △ ♃
14 53	☽ ✶ ♆
15 44	☽ ⊼ ☿
20 38	☽ △ ♇
20 56	☽ Q ♅
15 Friday	
03 56	☽ # ☉
04 37	☽ ⚼ ♀
06 04	♀ ⊥ ♂
06 33	☽ Q ♄
07 55	☽ # ♂
10 15	☽ Q ♆
10 34	☽ ∥ ♇
11 51	☽ ☍ ☉
18 16	☽ # ♇
19 44	☽ ☍ ♂
20 15	☽ ± ♀
20 25	☽ ∥ ♂
20 45	☽ ⚼ ♇
23 45	☽ ∥ ☉
16 Saturday	
00 50	☿ □ ♆
01 47	☽ □ ♅
06 36	☽ ⊼ ♀
11 06	☽ ± ♇
11 17	☽ □ ♄
12 28	☽ □ ♃
14 29	☽ ☍ ☿
14 55	☽ □ ♆
15 56	☽ # ☿
20 30	♂ ∠ ♇
20 39	☽ ⊼ ♇
17 Sunday	
14 53	☽ ⊼ ☉
17 20	☽ Q ♃
17 52	☽ # ♀
21 36	☽ ⊼ ♂
18 Monday	
01 09	☽ ± ☉
01 33	☽ △ ♅
02 58	♂ # ♇
06 58	☿ □ ♃
07 39	☽ ± ♂
10 25	☽ ☍ ♀
11 10	☽ △ ♄
12 25	☽ ⊼ ☿
12 44	☽ ✶ ♃
14 49	☽ △ ♆
16 39	☽ ⚼ ☉
20 06	♀ ✶ ♄
20 46	☽ ☍ ♇
21 39	☽ ± ☿
22 52	☽ ⚼ ♂
19 Tuesday	
01 48	☽ ⚼ ♅
02 41	☿ ⊻ ♀
04 43	☽ # ♆
07 58	☽ # ♄
08 38	☽ ∥ ♃
11 36	☽ ⚼ ♄
11 36	☽ ⚼ ☿
11 44	☿ □ ♄
13 19	☿ ⊥ ♇
13 23	☽ ∠ ♃
13 37	☽ # ♅
15 19	☽ ⚼ ♆
16 35	☽ ± ♅
18 57	☽ △ ☉
19 19	♀ △ ♃
23 32	☉ # ♇
20 Wednesday	
00 43	☽ △ ♂
02 32	☽ ± ♄
02 34	☽ ⊼ ♅
04 28	☽ ⊥ ♃
06 18	☽ ± ♆
11 10	☽ △ ☿
12 39	☽ ⊼ ♄
14 39	☽ ⊻ ♃
16 12	☽ ⊼ ♀
16 26	☽ ⊼ ♆
19 02	♀ ✶ ♆
22 48	☽ ⊼ ♇
21 Thursday	
03 23	☽ ± ♀
06 57	♆ St D
09 00	☉ ∠ ♇
09 10	☽ ± ♇
18 23	♂ □ ♅
20 29	☽ ⚼ ♀
23 52	☿ ⊥ ♀
22 Friday	
00 58	☽ ⚼ ♇
02 10	☽ □ ☉
06 20	☽ ☍ ♅
06 55	☽ □ ♂
11 49	☽ □ ☿
17 05	☽ ☍ ♄
19 37	☽ ☌ ♃
20 34	☉ # ♂
21 04	☽ ☍ ♆
23 Saturday	
01 51	☽ △ ♀
03 58	☽ △ ♇
13 17	☽ Q ☉
17 18	☽ Q ♂
18 08	☽ Q ☿
24 Sunday	
00 01	☿ ☌ ♂
01 32	☽ # ♅
02 01	☉ ∥ ♂
02 50	♀ ☌ ♇
07 21	☽ ∥ ♃
07 53	☽ # ♄
11 36	☽ # ♆
12 17	☉ □ ♅
13 25	☽ ⊼ ♅
13 31	☽ ✶ ☉
14 49	☽ ✶ ☿
16 58	☽ ✶ ♂
22 11	☉ ☌ ☿
25 Monday	
00 49	☽ ± ♅
00 52	☽ ⊼ ♄
03 55	☽ ⊻ ♃
04 59	☽ ⊼ ♆
06 53	☿ □ ♅
12 23	☽ ± ♄
12 25	☽ □ ♇
12 29	☽ # ♀
15 32	☽ ⊥ ♃
15 44	☽ □ ♀
16 31	☽ ± ♆
17 12	☽ ∠ ☿
18 07	☽ ⚼ ♅
20 38	☽ ∠ ☉
23 18	☽ ∠ ♂
26 Tuesday	
00 43	☿ ∠ ♀
05 50	☽ ⚼ ♄
08 55	☉ ∥ ♇
09 09	☽ ∠ ♃
09 18	☽ ⊥ ☿
10 00	☽ ⚼ ♆
15 41	☽ ⊥ ☉
17 48	♀ ∠ ♅
17 50	☽ ⊥ ♂
20 09	☽ ⊻ ☿
23 24	☽ △ ♅
23 33	☽ Q ♇
27 Wednesday	
04 31	☽ ⊻ ☉
05 45	☉ ∥ ☿
06 19	☽ ⊻ ♂
06 30	☽ Q ♀
11 21	☽ △ ♄
12 45	♂ ∥ ♇
14 54	☽ ✶ ♃
15 32	☽ △ ♆
15 50	☿ ∥ ♂
16 56	☿ ∥ ♇
23 22	☽ ✶ ♇
28 Thursday	
03 00	☿ ∠ ♇
08 50	☽ ✶ ♀
11 36	☽ # ☉
13 52	☽ # ♂
14 59	☽ # ♇
15 00	☽ Q ♃
17 59	☽ # ☿
22 47	☽ ∥ ☿
29 Friday	
02 23	☽ ∥ ♇
03 39	☽ ☌ ☿
04 08	☽ ∥ ♂
05 27	☽ ∠ ♇
07 06	☽ ∥ ☉
11 15	☽ □ ♅
18 07	☽ ∠ ♀
19 00	☉ ☌ ♂
21 42	☽ ☌ ♂
21 47	☽ ☌ ☉
23 31	☽ □ ♄
23 38	☽ ⊥ ♇
30 Saturday	
03 28	☽ □ ♃
03 39	☽ □ ♆
11 47	☽ ⊻ ♇
14 13	☽ ⊥ ♀
19 30	☉ □ ♄
20 50	☉ ⊥ ♇
23 39	☽ Q ♅

OCTOBER 1989

LONGITUDES

Date	Sidereal time h m s	Sun ☉ ° ′ ″	Moon ☽ ° ′ ″	Moon ☽ 24.00 ° ′ ″	Mercury ☿ ° ′	Venus ♀ ° ′	Mars ♂ ° ′	Jupiter ♃ ° ′	Saturn ♄ ° ′	Uranus ♅ ° ′	Neptune ♆ ° ′	Pluto ♇ ° ′
01	12 40 47	08 ♎ 17 15	25 ♎ 37 07	01 ♏ 32 24	26 ♍ 17	21 ♏ 54	07 ♎ 43	09 ♋ 39	07 ♑ 38	01 ♑ 32	09 ♑ 38	13 ♏ 41
02	12 44 44	09 16 17	07 ♏ 27 44	13 ♏ 23 23	25 R 56	23 02	08 22	09 45	07 40	01 33	09 39	13 43
03	12 48 40	10 15 22	19 ♏ 19 38	25 ♏ 16 46	25 46	24 10	09 02	09 49	07 42	01 34	09 39	13 45
04	12 52 37	11 14 29	01 ♐ 15 11	07 ♐ 15 14	25 D 46	25 18	09 41	09 54	07 44	01 35	09 39	13 48
05	12 56 33	12 13 37	13 ♐ 17 21	19 ♐ 22 00	25 56	26 26	10 20	09 59	07 47	01 37	09 40	13 50
06	13 00 30	13 12 47	25 ♐ 29 40	01 ♑ 40 52	26 17	27 34	10 59	10 03	07 49	01 38	09 40	13 52
07	13 04 27	14 11 59	07 ♑ 56 08	14 ♑ 16 02	26 48	28 41	11 39	10 07	07 52	01 39	09 41	13 54
08	13 08 23	15 11 13	20 ♑ 41 05	27 ♑ 11 48	27 28	29 ♏ 49	12 18	10 11	07 54	01 41	09 41	13 56
09	13 12 20	16 10 29	03 ♒ 48 42	10 ♒ 32 09	28 16	00 ♐ 56	12 57	10 15	07 57	01 42	09 42	13 58
10	13 16 16	17 09 46	17 ♒ 22 31	24 ♒ 19 59	29 ♍ 13	02 03	13 37	10 19	08 00	01 44	09 43	14 01
11	13 20 13	18 09 05	01 ♓ 24 36	08 ♓ 36 17	00 ♎ 16	03 10	14 16	10 22	08 02	01 45	09 43	14 03
12	13 24 09	19 08 25	15 ♓ 54 42	23 ♓ 19 18	01 26	04 17	14 55	10 26	08 05	01 47	09 44	14 05
13	13 28 06	20 07 48	00 ♈ 49 20	08 ♈ 23 50	02 41	05 24	15 35	10 29	08 08	01 49	09 45	14 07
14	13 32 02	21 07 12	16 ♈ 01 37	23 ♈ 41 21	04 01	06 31	16 14	10 32	08 11	01 50	09 45	14 10
15	13 35 59	22 06 39	01 ♉ 21 37	09 ♉ 00 58	05 26	07 37	16 54	10 34	08 15	01 52	09 46	14 12
16	13 39 56	23 06 07	16 ♉ 37 58	24 ♉ 11 18	06 54	08 43	17 33	10 37	08 18	01 54	09 47	14 14
17	13 43 52	24 05 38	01 ♊ 39 47	09 ♊ 02 28	08 25	09 50	18 13	10 39	08 21	01 56	09 48	14 16
18	13 47 49	25 05 11	16 ♊ 18 37	23 ♊ 27 43	09 58	10 55	18 53	10 42	08 25	01 57	09 49	14 19
19	13 51 45	26 04 47	00 ♋ 29 31	07 ♋ 23 54	11 33	12 01	19 32	10 44	08 28	01 59	09 50	14 21
20	13 55 42	27 04 25	14 ♋ 11 02	20 ♋ 51 09	13 10	13 07	20 12	10 45	08 32	02 01	09 51	14 23
21	13 59 38	28 04 05	27 ♋ 24 40	03 ♌ 52 02	14 49	14 12	20 52	10 47	08 36	02 03	09 52	14 26
22	14 03 35	29 ♎ 03 47	10 ♌ 13 48	16 ♌ 30 34	16 28	15 17	21 31	10 48	08 39	02 05	09 53	14 28
23	14 07 31	00 ♏ 03 31	22 ♌ 42 53	28 ♌ 51 22	18 08	16 22	22 11	10 50	08 43	02 08	09 54	14 30
24	14 11 28	01 03 18	04 ♍ 56 35	10 ♍ 59 05	19 48	17 27	22 51	10 51	08 47	02 10	09 55	14 33
25	14 15 25	02 03 07	16 ♍ 59 21	22 ♍ 57 54	21 29	18 32	23 31	10 51	08 51	02 12	09 56	14 35
26	14 19 21	03 02 58	28 ♍ 55 07	04 ♎ 51 23	23 09	19 36	24 11	10 52	08 55	02 14	09 57	14 38
27	14 23 18	04 02 51	10 ♎ 47 03	16 ♎ 42 25	24 50	20 40	24 50	10 52	09 00	02 16	09 58	14 40
28	14 27 14	05 02 46	22 ♎ 37 42	28 ♎ 33 08	26 31	21 44	25 30	10 53	09 04	02 19	09 59	14 42
29	14 31 11	06 02 43	04 ♏ 28 56	10 ♏ 25 14	28 12	22 47	26 10	10 R 53	09 08	02 21	10 00	14 45
30	14 35 07	07 02 42	16 ♏ 22 13	22 ♏ 20 02	29 ♎ 52	23 51	26 50	10 52	09 13	02 23	10 02	14 47
31	14 39 04	08 ♏ 02 43	28 ♏ 18 50	04 ♐ 18 48	01 ♏ 32	24 ♐ 54	27 ♎ 30	10 ♋ 52	09 ♑ 17	02 ♑ 26	10 ♑ 03	14 ♏ 50

Date	Moon True ☊ ° ′	Moon Mean ☊ ° ′	Moon ☽ Latitude ° ′
01	25 ♒ 03	23 ♒ 18	04 S 28
02	24 R 54	23 15	04 54
03	24 46	23 12	05 06
04	24 38	23 09	05 05
05	24 32	23 05	04 51
06	24 28	23 02	04 24
07	24 27	22 59	03 44
08	24 D 27	22 56	02 51
09	24 28	22 53	01 49
10	24 28	22 50	00 S 38
11	24 R 28	22 46	00 N 37
12	24 27	22 43	01 52
13	24 22	22 40	03 02
14	24 16	22 37	04 00
15	24 08	22 34	04 41
16	23 59	22 31	05 02
17	23 50	22 27	05 01
18	23 43	22 24	04 41
19	23 38	22 21	04 04
20	23 35	22 18	03 14
21	23 34	22 15	02 15
22	23 D 34	22 11	01 11
23	23 35	22 08	00 N 05
24	23 R 34	22 05	01 S 00
25	23 32	22 02	02 01
26	23 26	21 59	02 56
27	23 18	21 56	03 43
28	23 07	21 52	04 20
29	22 54	21 49	04 45
30	22 40	21 46	04 58
31	22 ♒ 26	21 ♒ 43	04 S 59

DECLINATIONS

Date	Sun ☉ ° ′	Moon ☽ ° ′	Mercury ☿ ° ′	Venus ♀ ° ′	Mars ♂ ° ′	Jupiter ♃ ° ′	Saturn ♄ ° ′	Uranus ♅ ° ′	Neptune ♆ ° ′	Pluto ♇ ° ′
01	03 S 17	14 S 04	00 N 55	20 S 04	02 S 21	22 N 49	22 S 47	23 S 42	22 S 12	01 S 22
02	03 41	18 37	01 21	20 25	02 37	22 49	22 47	23 42	22 12	01 23
03	04 04	22 28	01 42	20 46	02 53	22 48	22 47	23 42	22 12	01 24
04	04 27	25 24	01 58	21 07	03 09	22 48	22 47	23 42	22 12	01 25
05	04 50	27 13	02 09	21 26	03 25	22 48	22 47	23 42	22 12	01 26
06	05 13	27 46	02 14	21 46	03 40	22 48	22 47	23 42	22 12	01 27
07	05 36	26 55	02 14	22 05	03 56	22 47	22 47	23 42	22 12	01 27
08	05 59	24 40	02 09	22 23	04 12	22 47	22 47	23 42	22 12	01 28
09	06 22	21 04	01 59	22 41	04 27	22 47	22 47	23 42	22 12	01 29
10	06 45	16 14	01 45	22 58	04 43	22 47	22 47	23 42	22 12	01 30
11	07 07	10 24	01 27	23 15	04 59	22 46	22 47	23 42	22 12	01 31
12	07 30	03 S 50	01 05	23 31	05 14	22 46	22 47	23 42	22 12	01 32
13	07 52	03 N 06	00 39	23 46	05 30	22 46	22 46	23 42	22 12	01 33
14	08 14	09 59	00 N 11	24 01	05 46	22 46	22 46	23 42	22 12	01 34
15	08 37	16 20	00 S 20	24 16	06 01	22 46	22 46	23 42	22 12	01 34
16	08 59	21 37	00 54	24 30	06 17	22 46	22 46	23 42	22 12	01 35
17	09 21	25 25	01 29	24 43	06 32	22 46	22 46	23 42	22 12	01 36
18	09 43	27 24	02 06	24 55	06 48	22 45	22 46	23 42	22 12	01 37
19	10 04	27 30	02 44	25 07	07 03	22 45	22 46	23 42	22 12	01 38
20	10 26	25 54	03 24	25 19	07 18	22 45	22 46	23 42	22 12	01 39
21	10 47	22 53	04 04	25 30	07 34	22 45	22 46	23 42	22 12	01 39
22	11 09	18 49	04 45	25 40	07 49	22 45	22 46	23 42	22 12	01 40
23	11 30	14 01	05 26	25 49	08 04	22 45	22 45	23 42	22 12	01 41
24	11 51	08 46	06 08	25 58	08 19	22 45	22 45	23 42	22 12	01 42
25	12 11	03 N 17	06 49	26 07	08 35	22 45	22 45	23 42	22 12	01 43
26	12 32	02 S 16	07 31	26 14	08 50	22 45	22 45	23 41	22 12	01 43
27	12 52	07 41	08 13	26 21	09 05	22 45	22 45	23 41	22 12	01 44
28	13 12	12 49	08 54	26 28	09 20	22 45	22 45	23 41	22 12	01 45
29	13 32	17 29	09 36	26 33	09 35	22 45	22 44	23 41	22 12	01 46
30	13 52	21 30	10 17	26 39	09 50	22 45	22 44	23 41	22 12	01 47
31	14 S 12	24 S 38	10 S 57	26 S 43	10 S 04	22 N 46	22 S 44	23 S 41	22 S 12	01 S 47

ZODIAC SIGN ENTRIES

Date	h	m	Planets
01	20	53	☽ ♏
04	09	29	☽ ♐
06	20	45	☽ ♑
08	16	00	♀ ♐
09	05	07	☽ ♒
11	06	11	☿ ♎
11	09	37	☽ ♓
13	10	41	☽ ♈
15	09	52	☽ ♉
17	09	19	☽ ♊
19	11	09	☽ ♋
21	16	47	☽ ♌
23	10	35	☉ ♏
24	02	15	☽ ♍
26	14	11	☽ ♎
29	02	56	☽ ♏
30	13	53	☿ ♏
31	15	23	☽ ♐

LATITUDES

Date	Mercury ☿ ° ′	Venus ♀ ° ′	Mars ♂ ° ′	Jupiter ♃ ° ′	Saturn ♄ ° ′	Uranus ♅ ° ′	Neptune ♆ ° ′	Pluto ♇ ° ′
01	00 S 36	01 S 54	00 N 46	00 S 17	00 N 27	00 S 16	00 N 53	15 N 15
04	00 N 19	02 05	00 45	00 16	00 26	00 16	00 53	15 14
07	01 03	02 16	00 44	00 16	00 26	00 16	00 53	15 13
10	01 34	02 26	00 43	00 16	00 25	00 16	00 53	15 12
13	01 52	02 36	00 41	00 16	00 25	00 16	00 53	15 11
16	02 00	02 46	00 40	00 15	00 25	00 16	00 53	15 10
19	02 00	02 55	00 39	00 15	00 24	00 16	00 53	15 10
22	01 52	03 03	00 37	00 15	00 24	00 16	00 53	15 09
25	01 40	03 11	00 36	00 15	00 24	00 16	00 53	15 09
28	01 25	03 17	00 34	00 15	00 23	00 16	00 52	15 08
31	01 N 08	03 S 23	00 N 33	00 S 14	00 N 23	00 S 16	00 N 52	15 N 08

DATA

Julian Date	2447801
Delta T	+56 seconds
Ayanamsa	23° 42′ 59″
Synetic vernal point	05° ♓ 24′ 00″
True obliquity of ecliptic	23° 26′ 34″

LONGITUDES

Date	Chiron ⚷ ° ′	Ceres ⚳ ° ′	Pallas ⚴ ° ′	Juno ⚵ ° ′	Vesta ⚶ ° ′	Black Moon Lilith ⚸ ° ′
01	16 ♋ 07	02 ♋ 06	07 ♈ 05	13 ♎ 17	08 ♑ 13	26 ♎ 10
11	16 ♋ 25	03 ♋ 41	04 ♈ 14	16 ♎ 55	11 ♑ 38	27 ♎ 17
21	16 ♋ 35	04 ♋ 46	01 ♈ 38	20 ♎ 31	15 ♑ 21	28 ♎ 24
31	16 ♋ 35	05 ♋ 14	29 ♓ 29	24 ♎ 04	19 ♑ 19	29 ♎ 31

MOON'S PHASES, APSIDES AND POSITIONS ☽

Date	h	m	Phase	Longitude ° ′	Eclipse Indicator
08	00	52	☽	14 ♑ 44	
14	20	32	○	21 ♈ 28	
21	13	19	☾	28 ♋ 07	
29	15	27	●	06 ♏ 11	

Day	h	m		
01	20	19	Apogee	
15	01	25	Perigee	
28	22	38	Apogee	
06	09	42	Max dec	27° S 46′
13	01	19	ON	
19	01	21	Max dec	27° N 41′
26	02	12	OS	

ASPECTARIAN

h m	Aspects
01 Sunday	
03 39	☽ ⊻ ♀
05 52	♃ ☍ ♆
08 36	♂ □ ♄
10 43	♂ ⊥ ♇
12 02	☽ Q ♄
13 18	☽ ⊻ ☿
16 06	☽ Q ♆
02 Monday	
00 00	☽ ✶ ♅
01 05	☽ ⊥ ☿
04 06	♀ ∠ ♄
12 25	☽ ✶ ♄
13 57	☽ ⊻ ♂
13 59	☿ # ♇
16 00	☽ ⊻ ☉
16 25	☽ ✶ ♆
16 39	☽ △ ♃
18 54	☽ ∠ ☿
21 07	☉ □ ♆
23 41	☽ ∥ ♀
03 Tuesday	
00 32	☉ □ ♃
00 43	☽ ☌ ♇
02 48	☽ ⊥ ♂
05 04	♀ ∠ ♂
05 14	☽ ⊥ ☉
06 25	☽ ∠ ♅
10 12	☽ ∥ ♆
14 15	☽ ∥ ♄
14 26	☽ # ♃
18 50	☽ ∠ ♄
21 10	☽ ∥ ♅
22 02	☽ ∠ ♂
22 18	♀ ∠ ♆
22 44	☽ ∠ ♆
22 47	☽ ☌ ♀
23 09	☽ ⚼ ♃
23 48	☿ St D
04 Wednesday	
00 36	☽ ⊥ ♅
00 56	☽ ✶ ☿
01 02	☽ ∠ ☉
03 00	♀ ⚼ ♃
11 08	♂ □ ♆
12 41	☽ ⊻ ♅
12 59	☽ ⊥ ♄
16 49	☽ ⊥ ♆
17 20	☽ ± ♃
18 18	♀ ⊥ ♅
21 18	♂ □ ♃
23 07	☿ ✶ ♀
05 Thursday	
01 00	☽ ⊻ ♄
01 10	☽ Q ☿
04 48	☽ ⊻ ♆
05 23	☽ ⊼ ♃
05 48	☽ ✶ ♂
09 42	☽ ✶ ☉
13 04	☽ ⊻ ♇
06 Friday	
00 57	☽ ⊥ ♇
06 50	☽ Q ♂
11 24	☽ Q ☉
13 36	☽ □ ☿
16 26	☽ ⊻ ♀
18 34	☽ ∠ ♇
23 56	☽ ☌ ♅
07 Saturday	
04 26	☉ ⊻ ♇
05 10	☽ ⊥ ♀
11 51	☽ ☌ ♄
15 20	☽ ☌ ♆
16 11	☽ ☍ ♃
19 26	☽ □ ♂
21 59	♀ ∥ ♆
23 20	☽ ✶ ♇
23 58	☽ ∠ ♀
08 Sunday	
00 52	☽ □ ☉
19 24	☽ ∥ ♅
21 43	☽ Q ♇
09 Monday	
01 16	☽ △ ☿
01 38	☽ # ♃
01 39	☽ ∥ ♄
03 00	☽ ∥ ♀
05 17	☽ ∥ ♆
06 20	☽ ✶ ♀
08 11	☽ ⊻ ♅
18 59	☽ ⊥ ♅
19 26	☽ ⊻ ♄
20 14	♀ ∥ ♄
20 21	♀ # ♃
22 32	☽ ⊻ ♆
23 33	☽ ⊼ ♃
10 Tuesday	
04 50	♀ ⊻ ♅
05 05	☽ △ ♂
05 41	☽ Q ♀
06 02	☽ ⚼ ☿
06 04	☽ ⊥ ♄
06 06	☽ □ ♇
06 58	☿ ∠ ♇
07 55	♃ # ♄
09 06	☽ ⊥ ♆
10 09	☽ ± ♃
10 33	♀ ⊥ ♄
10 52	☽ ∠ ♅
11 36	☽ △ ☉
21 44	☽ ∠ ♄
22 52	☽ ± ☿
11 Wednesday	
00 39	☽ ∠ ♆
01 44	☽ ⚼ ♃
03 31	♂ ⊻ ♇
06 54	☿ # ♇
07 41	☉ ∠ ♀
08 13	☽ ⚼ ♂
09 55	☽ ⊼ ☿
12 35	☽ ✶ ♅
15 08	☽ ⚼ ☉
15 13	☽ □ ♀
23 06	☽ ✶ ♄
23 35	☽ ∥ ☉
23 53	♀ ⊥ ♆
23 59	☽ ± ♂
12 Thursday	
01 51	☽ ✶ ♆
02 59	☽ △ ♃
07 08	☽ ± ☉
07 12	☽ ∥ ♂
08 31	☽ Q ♅
09 00	☽ △ ♇
10 19	☽ ⊼ ♂
15 08	♀ ± ♃
17 37	☽ ⊼ ☉
18 48	☽ Q ♄
19 03	☿ □ ♅
20 01	☽ ∥ ♇
21 27	☽ Q ♆
22 13	☽ # ☿
13 Friday	
04 00	☉ Q ♅
04 04	☽ ∥ ☿
05 06	♀ ∥ ♅
06 38	☽ # ♇
09 17	☽ ⚼ ♇
13 34	☽ □ ♅
15 15	☽ ☍ ☿
19 51	☽ △ ♀
20 34	☽ # ♂
23 36	☽ ± ♇
23 38	☽ □ ♄
14 Saturday	
02 08	☽ □ ♆
03 20	☽ □ ♃
05 26	☽ # ☉
09 04	☽ ⊼ ♇
12 21	☽ ☍ ♂
20 32	☽ ☍ ☉
21 16	☽ ⚼ ♀
15 Sunday	
07 38	☽ Q ♃
12 26	☽ ± ♀
12 48	☽ △ ♅
19 02	☽ ⊼ ☿
22 34	☽ ⊼ ♀
22 50	☽ △ ♄
16 Monday	
01 12	☽ △ ♆
02 15	♀ ⊻ ♄
02 29	☽ ✶ ♃
05 28	☽ ± ☿
08 12	☽ ☍ ♇
12 25	☽ ⚼ ♅
13 32	☽ ⊼ ♂
15 05	☽ # ♆
18 11	☽ ∥ ♃
18 14	☽ # ♄
21 16	☽ ⚼ ☿
22 37	☽ ⚼ ♄
23 00	☽ ⊼ ☉
23 30	☽ ± ♂
23 47	☽ # ♅
17 Tuesday	
00 58	☽ ⚼ ♆
02 19	☽ ∠ ♃
02 46	☽ ± ♅
06 15	☽ # ♀
09 17	☽ ± ☉
09 49	☿ ⊥ ♇
11 06	☿ □ ♄
11 22	♀ ⊻ ♆
12 26	☽ ⊼ ♅
13 07	☽ ± ♄
14 38	☽ ⚼ ♂
15 27	☽ ± ♆
16 49	☿ ∥ ♇
16 52	☽ ⊥ ♃
22 55	☽ ⊼ ♄
18 Wednesday	
00 16	☽ △ ☿
00 58	☽ ⚼ ☉
01 15	☽ ⊼ ♆
02 22	☽ ☍ ♀
02 41	☽ ⊻ ♃
06 46	♀ ⊼ ♃
08 40	☽ ⊼ ♇
09 37	☿ □ ♆
16 29	☽ △ ♂
18 43	☽ ± ♇
23 16	☿ □ ♃
19 Thursday	
03 52	☽ △ ☉
10 02	☽ ⚼ ♇
14 36	☽ ☍ ♅
22 04	☉ Q ♄
20 Friday	
01 57	☽ ☍ ♄
04 17	☽ ☍ ♆
05 21	♂ Q ♅
05 54	☽ ☌ ♃
09 21	☿ ✶ ♀
09 55	☽ ⊼ ♀
09 57	☽ □ ☿
12 22	☽ △ ♇
17 11	☽ # ♀
21 38	☽ ± ♀
23 23	☽ □ ♂
21 Saturday	
06 19	☿ ⊻ ♇
06 24	☽ # ♅
06 53	☉ Q ♆
12 48	☽ # ♄
12 51	☽ ∥ ♃
13 19	☽ □ ☉
15 37	☽ ⚼ ♀
16 26	☽ # ♆
17 12	♀ ⊻ ♇
20 38	☽ ⊼ ♅
23 30	☽ Q ☿
22 Sunday	
07 56	☽ ± ♅
09 00	☽ ⊼ ♄
10 35	☽ Q ♂
11 20	☽ ⊼ ♆
13 06	☽ ⊻ ♃
20 07	☽ □ ♇
20 29	☽ ± ♄
22 34	☽ △ ♀
22 48	☽ ± ♆
23 Monday	
00 36	☽ ⊥ ♃
01 09	☽ ⚼ ♅
01 44	☽ ✶ ☿
02 12	☽ Q ☉
10 55	☽ ✶ ♂
13 58	☽ ⚼ ♄
16 15	☽ ⚼ ♆
18 04	☽ ∠ ♃
23 00	☽ # ☉
24 Tuesday	
03 38	☽ ✶ ☉
06 29	☽ △ ♅
07 15	☽ Q ♇
11 40	☽ ∠ ☿
13 53	☽ # ♂
17 18	☿ Q ♅
18 06	☽ ∠ ♂
19 40	☽ △ ♄
21 53	☽ △ ♆
22 19	☽ # ☿
23 44	☽ ✶ ♃
25 Wednesday	
02 30	♃ # ♄
07 11	☽ ✶ ♇
08 29	☽ ⊥ ☿
12 08	☽ ∠ ☉
13 07	☽ ⊥ ♂
15 23	☽ □ ♀
15 39	☉ ✶ ♅
18 46	☽ # ♇
22 29	☽ ⊻ ☿
23 48	☽ Q ♃
26 Thursday	
01 52	☽ ⊻ ♂
07 53	☽ ⊥ ☉
09 40	☽ ∥ ♇
13 26	☽ ∠ ♇
18 43	☽ □ ♅
21 07	☽ ⊻ ☉
27 Friday	
07 17	☽ Q ♀
07 42	☽ ⊥ ♇
08 21	☽ □ ♄
10 20	☽ □ ♆
12 02	♀ ⊥ ♇
12 05	☿ ☌ ♂
12 11	☽ □ ♃
14 48	☽ ∥ ☿
18 42	☽ ∥ ♂
19 54	☽ ⊻ ♇
28 Saturday	
07 17	☽ Q ♅
10 00	☽ ✶ ♀
14 03	☽ ∥ ☉
18 11	☽ ☌ ♂
20 10	☿ Q ♄
21 02	☽ Q ♄
21 11	☽ ☌ ☿
22 52	☽ Q ♆
23 59	♃ St R
29 Sunday	
07 40	☽ ✶ ♅
09 16	☿ Q ♆
11 00	☿ ∥ ♂
15 27	☽ ☌ ☉
19 20	☽ ∠ ♀
21 28	☽ ✶ ♄
23 11	☽ ✶ ♆
30 Monday	
00 55	☽ △ ♃
08 48	☽ ☌ ♇
14 04	☽ ∠ ♅
15 15	☽ ⊥ ♀
16 46	☽ ∥ ♆
20 38	☽ ∥ ♄
20 49	☽ # ♃
31 Tuesday	
03 00	♂ Q ♄
03 52	☽ ∠ ♄
03 55	☽ ∥ ♅
04 29	☽ ⊻ ♀
05 26	☽ ∠ ♆
07 06	☽ ⚼ ♃
08 13	☽ ⊥ ♅
10 17	☽ ⊻ ♂
19 30	☽ ⊻ ☿
20 16	☽ ⊻ ♅
22 01	☽ ⊥ ♄
23 00	☽ ⊥ ♂
23 29	☽ ⊥ ♆

All ephemeris data is given at 12.00 UT and the Moon's longitude is additionally given for 24.00 UT

NOVEMBER 1989

LONGITUDES

Date	Sidereal time h m s	Sun ☉ ° ' "	Moon ☽ ° ' "	Moon ☽ 24.00 ° ' "	Mercury ☿ ° '	Venus ♀ ° '	Mars ♂ ° '	Jupiter ♃ ° '	Saturn ♄ ° '	Uranus ♅ ° '	Neptune ♆ ° '	Pluto ♇ ° '
01	14 43 00	09 ♏ 02 46	10 ♐ 20 09	16 ♐ 23 05	03 ♏ 12	25 ♐ 56	28 ♎ 11	10 ♋ 51	09 ♑ 22	02 ♑ 28	10 ♑ 04	14 ♏ 52
02	14 46 57	10 02 50	22 ♐ 27 52	28 ♐ 34 49	04 52	26 59	28 51	10 R 51	09 26	02 31	10 06	14 54
03	14 50 54	11 02 56	04 ♑ 44 15	10 ♑ 56 34	06 31	28 01	29 ♎ 31	10 50	09 31	02 33	10 07	14 57
04	14 54 50	12 03 04	17 ♑ 12 10	23 ♑ 31 30	08 09	29 ♐ 03	00 ♏ 11	10 48	09 36	02 36	10 08	14 59
05	14 58 47	13 03 14	29 ♑ 55 04	06 ♒ 23 19	09 48	00 ♑ 05	00 51	10 47	09 41	02 39	10 10	15 02
06	15 02 43	14 03 25	12 ♒ 56 45	19 ♒ 35 50	11 26	01 06	01 31	10 45	09 46	02 41	10 11	15 04
07	15 06 40	15 03 37	26 ♒ 20 59	03 ♓ 12 31	13 03	02 07	02 12	10 44	09 51	02 44	10 13	15 07
08	15 10 36	16 03 51	10 ♓ 10 41	17 ♓ 15 36	14 41	03 07	02 52	10 42	09 56	02 47	10 14	15 09
09	15 14 33	17 04 06	24 ♓ 27 12	01 ♈ 45 12	16 18	04 07	03 32	10 39	10 01	02 49	10 16	15 11
10	15 18 29	18 04 23	09 ♈ 09 09	16 ♈ 38 21	17 54	05 07	04 13	10 37	10 06	02 52	10 17	15 14
11	15 22 26	19 04 41	24 ♈ 11 52	01 ♉ 48 33	19 30	06 06	04 53	10 34	10 11	02 55	10 19	15 16
12	15 26 23	20 05 00	09 ♉ 27 08	17 ♉ 06 10	21 06	07 05	05 33	10 32	10 17	02 58	10 20	15 19
13	15 30 19	21 05 22	24 ♉ 44 11	02 ♊ 19 47	22 41	08 03	06 14	10 29	10 22	03 01	10 22	15 21
14	15 34 16	22 05 45	09 ♊ 51 34	17 ♊ 18 22	24 17	09 01	06 54	10 25	10 27	03 04	10 24	15 24
15	15 38 12	23 06 10	24 ♊ 39 11	01 ♋ 53 14	25 51	09 59	07 35	10 22	10 33	03 07	10 25	15 26
16	15 42 09	24 06 37	09 ♋ 00 00	15 ♋ 59 11	27 26	10 56	08 15	10 18	10 38	03 10	10 27	15 28
17	15 46 05	25 07 06	22 ♋ 50 43	29 ♋ 34 43	29 ♏ 00	11 52	08 56	10 15	10 44	03 13	10 29	15 31
18	15 50 02	26 07 36	06 ♌ 11 29	12 ♌ 41 24	00 ♐ 35	12 48	09 37	10 11	10 50	03 16	10 30	15 33
19	15 53 58	27 08 08	19 ♌ 05 01	25 ♌ 22 55	02 09	13 44	10 17	10 07	10 55	03 19	10 32	15 36
20	15 57 55	28 08 42	01 ♍ 35 43	07 ♍ 44 06	03 42	14 38	10 58	10 02	11 01	03 22	10 34	15 38
21	16 01 52	29 ♏ 09 17	13 ♍ 48 43	19 ♍ 50 14	05 16	15 33	11 39	09 58	11 07	03 25	10 36	15 40
22	16 05 48	00 ♐ 09 55	25 ♍ 49 17	01 ♎ 46 27	06 49	16 26	12 20	09 53	11 13	03 28	10 37	15 43
23	16 09 45	01 10 34	07 ♎ 42 20	13 ♎ 37 25	08 22	17 19	13 00	09 48	11 19	03 31	10 39	15 45
24	16 13 41	02 11 14	19 ♎ 32 12	25 ♎ 27 05	09 55	18 12	13 41	09 43	11 25	03 34	10 41	15 47
25	16 17 38	03 11 56	01 ♏ 22 26	07 ♏ 18 34	11 28	19 03	14 22	09 38	11 31	03 38	10 43	15 50
26	16 21 34	04 12 40	13 ♏ 15 45	19 ♏ 14 11	13 01	19 54	15 03	09 33	11 37	03 41	10 45	15 52
27	16 25 31	05 13 25	25 ♏ 14 05	01 ♐ 15 33	14 33	20 44	15 44	09 27	11 43	03 44	10 47	15 54
28	16 29 27	06 14 12	07 ♐ 18 44	13 ♐ 23 43	16 06	21 34	16 25	09 22	11 49	03 47	10 49	15 57
29	16 33 24	07 15 00	19 ♐ 30 35	25 ♐ 39 28	17 38	22 22	17 06	09 16	11 56	03 51	10 51	15 59
30	16 37 21	08 ♐ 15 49	01 ♑ 50 25	08 ♑ 03 36	19 ♐ 10	23 ♑ 10	17 ♏ 47	09 ♋ 10	12 ♑ 02	03 ♑ 54	10 ♑ 53	16 ♏ 01

Date	Moon True ☊ ° '	Moon Mean ☊ ° '	Moon ☽ Latitude ° '
01	22 ♒ 13	21 ♒ 40	04 S 46
02	22 R 03	21 37	04 20
03	21 55	21 33	03 41
04	21 50	21 30	02 52
05	21 48	21 27	01 53
06	21 D 48	21 24	00 S 47
07	21 R 48	21 21	00 N 24
08	21 47	21 17	01 35
09	21 44	21 14	02 43
10	21 39	21 11	03 42
11	21 30	21 08	04 27
12	21 20	21 05	04 54
13	21 08	21 02	04 59
14	20 57	20 58	04 44
15	20 47	20 55	04 10
16	20 40	20 52	03 20
17	20 35	20 49	02 21
18	20 33	20 46	01 15
19	20 D 32	20 43	00 N 08
20	20 R 32	20 39	00 S 58
21	20 32	20 36	02 00
22	20 29	20 33	02 55
23	20 24	20 30	03 42
24	20 16	20 27	04 19
25	20 05	20 24	04 45
26	19 52	20 20	04 58
27	19 38	20 17	04 59
28	19 24	20 14	04 46
29	19 11	20 11	04 20
30	19 ♒ 00	20 ♒ 08	03 S 41

DECLINATIONS

Date	Sun ☉ ° '	Moon ☽ ° '	Mercury ☿ ° '	Venus ♀ ° '	Mars ♂ ° '	Jupiter ♃ ° '	Saturn ♄ ° '	Uranus ♅ ° '	Neptune ♆ ° '	Pluto ♇ ° '
01	14 S 31	26 S 43	11 S 37	26 S 47	10 S 19	22 N 46	22 S 44	23 S 41	22 S 12	01 S 48
02	14 50	27 33	12 17	26 50	10 34	22 46	22 44	23 41	22 11	01 49
03	15 09	27 03	12 56	26 53	10 49	22 46	22 43	23 41	22 11	01 50
04	15 27	25 11	13 34	26 55	11 03	22 46	22 43	23 41	22 11	01 50
05	15 46	22 01	14 12	26 56	11 18	22 46	22 43	23 41	22 11	01 51
06	16 04	17 41	14 49	26 56	11 32	22 47	22 43	23 41	22 11	01 52
07	16 21	12 21	15 25	26 56	11 46	22 47	22 42	23 41	22 11	01 52
08	16 39	06 S 17	16 00	26 56	12 01	22 47	22 42	23 41	22 11	01 53
09	16 56	00 N 18	16 35	26 55	12 15	22 47	22 42	23 41	22 11	01 54
10	17 13	07 02	17 09	26 53	12 29	22 48	22 41	23 41	22 11	01 55
11	17 30	13 31	17 42	26 50	12 43	22 48	22 41	23 41	22 11	01 55
12	17 46	19 17	18 14	26 47	12 57	22 48	22 41	23 41	22 11	01 56
13	18 02	23 47	18 45	26 44	13 11	22 49	22 41	23 40	22 11	01 57
14	18 18	26 37	19 15	26 40	13 25	22 49	22 40	23 40	22 10	01 57
15	18 33	27 30	19 44	26 35	13 38	22 49	22 40	23 40	22 10	01 58
16	18 48	26 28	20 12	26 30	13 52	22 50	22 40	23 40	22 10	01 58
17	19 03	23 49	20 40	26 24	14 05	22 50	22 39	23 40	22 10	01 59
18	19 17	19 56	21 06	26 18	14 19	22 51	22 39	23 40	22 10	02 00
19	19 31	15 14	21 31	26 11	14 32	22 51	22 38	23 40	22 10	02 00
20	19 45	10 00	21 55	26 03	14 45	22 51	22 38	23 40	22 10	02 01
21	19 58	04 N 31	22 18	25 55	14 58	22 52	22 38	23 40	22 10	02 01
22	20 11	01 S 01	22 40	25 47	15 11	22 52	22 37	23 40	22 10	02 02
23	20 24	06 27	23 01	25 38	15 24	22 53	22 37	23 40	22 09	02 03
24	20 36	11 38	23 21	25 29	15 37	22 53	22 36	23 40	22 09	02 03
25	20 48	16 24	23 39	25 19	15 50	22 54	22 36	23 40	22 09	02 04
26	20 59	20 33	23 57	25 09	16 02	22 54	22 36	23 39	22 09	02 04
27	21 10	23 54	24 13	24 58	16 15	22 55	22 35	23 39	22 09	02 05
28	21 21	26 14	24 28	24 47	16 27	22 55	22 35	23 39	22 09	02 05
29	21 31	27 21	24 41	24 36	16 39	22 56	22 34	23 39	22 09	02 06
30	21 S 41	27 S 07	24 S 53	24 S 24	16 S 51	22 N 56	22 S 34	23 S 39	22 S 09	02 S 06

ZODIAC SIGN ENTRIES

Date	h	m	Planets
03	02	46	☽ ♑
04	05	29	♂ ♏
05	10	13	♀ ♑
05	12	09	☽ ♒
07	18	25	☽ ♓
09	21	08	☽ ♈
11	21	09	☽ ♉
13	20	19	☽ ♊
15	20	51	☽ ♋
18	00	45	☽ ♌
18	03	10	☿ ♐
20	08	54	☽ ♍
22	08	05	☉ ♐
22	20	25	☽ ♎
25	09	13	☽ ♏
27	21	30	☽ ♐
30	08	26	☽ ♑

LATITUDES

Date	Mercury ☿ ° '	Venus ♀ ° '	Mars ♂ ° '	Jupiter ♃ ° '	Saturn ♄ ° '	Uranus ♅ ° '	Neptune ♆ ° '	Pluto ♇ ° '
01	01 N 01	03 S 24	00 N 33	00 S 14	00 N 23	00 S 16	00 N 52	15 N 08
04	00 42	03 28	00 31	00 14	00 23	00 16	00 52	15 08
07	00 22	03 31	00 30	00 14	00 22	00 16	00 52	15 07
10	00 N 02	03 32	00 28	00 13	00 22	00 16	00 52	15 07
13	00 S 18	03 32	00 27	00 13	00 22	00 16	00 52	15 07
16	00 38	03 31	00 25	00 13	00 21	00 16	00 52	15 07
19	00 57	03 28	00 24	00 13	00 21	00 16	00 52	15 08
22	01 14	03 23	00 22	00 12	00 21	00 16	00 52	15 08
25	01 31	03 16	00 20	00 12	00 21	00 16	00 52	15 08
28	01 45	03 07	00 19	00 12	00 20	00 16	00 51	15 08
31	01 S 58	02 S 55	00 N 17	00 S 11	00 N 20	00 S 16	00 N 51	15 N 09

DATA

Julian Date	2447832
Delta T	+56 seconds
Ayanamsa	23° 43' 03"
Synetic vernal point	05° ♓ 23' 56"
True obliquity of ecliptic	23° 26' 33"

MOON'S PHASES, APSIDES AND POSITIONS ☽

Date	h	m	Phase	Longitude ° '	Eclipse Indicator
06	14	11	☽	14 ♒ 09	
13	05	51	○	20 ♉ 50	
20	04	44	☾	27 ♌ 50	
28	09	41	●	06 ♐ 08	

Day	h	m		
12	13	25	Perigee	
25	03	59	Apogee	
02	15	09	Max dec	27° S 34'
09	10	57	0N	
15	10	47	Max dec	27° N 30'
22	07	35	0S	
29	20	04	Max dec	27° S 25'

LONGITUDES

Date	Chiron ⚷ ° '	Ceres ⚳ ° '	Pallas ⚴ ° '	Juno ⚵ ° '	Vesta ⚶ ° '	Black Moon Lilith ⚸ ° '
01	16 ♋ 35	05 ♋ 14	29 ♓ 19	24 ♎ 25	19 ♑ 43	29 ♎ 38
11	16 ♋ 25	04 ♋ 59	27 ♓ 53	27 ♎ 55	23 ♑ 56	00 ♏ 45
21	16 ♋ 08	04 ♋ 03	27 ♓ 12	01 ♏ 20	28 ♑ 18	01 ♏ 52
31	15 ♋ 42	02 ♋ 29	27 ♓ 15	04 ♏ 38	02 ♒ 49	03 ♏ 00

ASPECTARIAN

h m	Aspects
01 Wednesday	
01 06	☽ ± ♃
01 11	☿ ⚹ ♅
08 06	♂ Q ♆
09 12	☽ ⊻ ☉
09 23	☽ ⊥ ☿
10 03	☽ ⊻ ♄
11 28	☽ ⊻ ♆
13 02	☽ ⊼ ♃
13 15	☽ ‖ ♀
17 58	☽ ∠ ♂
20 12	☉ ⚹ ♄
21 01	☽ ⊻ ♇
22 11	☽ ⊥ ☉
02 Thursday	
06 03	☽ ∠ ☿
08 55	☽ ⊥ ♇
13 06	☉ ⚹ ♆
17 32	☽ ∠ ☉
21 42	☽ ☌ ♀
03 Friday	
01 14	☽ ⚹ ♂
02 39	☽ ∠ ♇
06 46	☉ △ ♃
07 45	☽ ☌ ♅
15 02	☽ ‖ ♀
15 58	☽ ⚹ ☿
21 19	☽ ☌ ♄
22 25	☽ ☌ ♆
23 45	☽ ☍ ♃
04 Saturday	
01 16	☽ ⚹ ☉
01 50	☽ Q ♂
07 45	☽ ⚹ ♇
18 28	☽ Q ☿
05 Sunday	
00 29	☽ ‖ ♅
02 06	☽ Q ☉
06 35	☽ Q ♇
06 59	☽ # ♃
07 23	☽ ‖ ♄
10 10	☿ ⚹ ♄
10 51	♀ ∠ ♇
10 51	☽ ‖ ♆
12 19	☽ ⊻ ♀
13 50	☽ □ ♂
17 05	☽ ⊻ ♅
17 25	☿ ⚹ ♆
06 Monday	
00 24	☽ ⊥ ♀
02 14	☿ △ ♃
04 12	☽ ⊥ ♅
06 09	☽ ⊻ ♄
06 58	☽ ⊻ ♆
08 01	☽ ⊼ ♃
08 51	☽ □ ☿
14 11	☽ □ ☉
15 52	☽ □ ♇
17 08	☽ ⊥ ♄
17 53	☽ ⊥ ♆
18 11	☽ ∠ ♀
18 53	☽ ± ♃
19 19	☽ ‖ ☉
20 36	☽ ∠ ♅
07 Tuesday	
00 03	☽ ‖ ☿
09 19	☽ ∠ ♄
09 59	☽ ∠ ♆
10 54	☽ Ꝑ ♃
13 14	☉ ☌ ♇
14 20	☽ ‖ ♂
17 51	♀ ⚹ ♂
22 46	☽ △ ♂
22 53	☽ ⚹ ♀
23 12	☽ ⚹ ♅
08 Wednesday	
03 27	♀ ☌ ♅
08 37	♂ ⚹ ♅
11 34	☽ ⚹ ♄
12 06	☽ ⚹ ♆
12 53	☽ △ ♃
19 12	☿ ☌ ♇
19 50	☽ Q ♅
20 28	☽ △ ♇
20 38	☽ △ ☿
21 02	☽ Q ♀
22 45	☽ △ ☉
09 Thursday	
01 39	☽ Ꝑ ♂
04 09	☽ ‖ ♇
07 56	☽ Q ♄
08 21	☽ Q ♆
17 20	☽ ± ♂
17 44	☽ # ♇
21 28	☽ Ꝑ ♇
10 Friday	
00 38	☽ Ꝑ ☿
01 26	☽ Ꝑ ☉
01 47	☽ □ ♅
03 37	☽ ⊼ ♂
05 00	☽ □ ♀
06 53	☉ ∠ ♅
11 31	☿ ∠ ♅
12 08	☽ ± ♇
13 32	☽ □ ♄
13 50	☽ □ ♆
14 21	☽ □ ♃
16 57	☽ ± ☿
17 02	☽ ± ☉
18 25	☉ ‖ ☿
18 56	☉ ☌ ☿
21 47	☽ ⊼ ♇
11 Saturday	
03 18	☽ ⊼ ☉
03 40	☽ ⊼ ☿
08 48	☽ # ♂
18 53	☽ Q ♃
12 Sunday	
01 47	☽ △ ♅
04 56	☽ # ☉
05 36	☽ ☍ ♂
06 46	☽ # ☿
08 02	☽ △ ♀
13 18	☽ △ ♄
13 24	☽ △ ♆
13 41	☽ ⚹ ♃
21 13	☽ ☍ ♇
13 Monday	
01 24	☽ Ꝑ ♅
02 28	☽ # ♆
05 16	☽ # ♄
05 51	☽ ☍ ☉
06 02	☽ ‖ ♃
08 24	☽ ☍ ☿
09 11	☽ Ꝑ ♀
11 16	☽ # ♅
11 39	♄ ☌ ♆
13 00	☽ Ꝑ ♆
13 00	☽ Ꝑ ♄
13 10	☽ ∠ ♃
15 36	☽ ± ♅
14 Tuesday	
00 22	☽ ± ♀
01 07	☽ ⊼ ♅
02 13	☿ ∠ ♀
03 16	☽ ± ♆
03 20	☽ ± ♄
03 21	☽ ⊥ ♃
06 27	♃ ☍ ♄
07 03	☽ ⊼ ♂
10 34	☽ ⊼ ♀
12 39	☽ # ♀
12 51	☽ ⊼ ♆
12 54	☽ ⊻ ♃
12 58	☽ ⊼ ♄
17 07	☽ ± ♂
20 54	♃ ☍ ♆
20 56	☽ ⊼ ♇
15 Wednesday	
04 48	☿ Ꝑ ♃
05 14	☿ ∠ ♆
06 42	☽ ± ♇
07 00	☿ ∠ ♄
08 26	☽ Ꝑ ♂
09 16	☽ ⊼ ☉
14 14	☽ ⊼ ☿
19 55	☽ ± ☉
21 10	♀ ☍ ♃
21 36	☽ Ꝑ ♇
23 24	♀ ☌ ♆
16 Thursday	
01 26	☽ ± ☿
02 06	☽ ☍ ♅
03 51	♀ ☌ ♄
07 39	☿ ⊥ ♅
10 40	☽ △ ♂
11 38	☽ # ♀
12 12	☽ Ꝑ ☉
14 13	☽ ☌ ♃
14 28	☽ ☍ ♆
14 49	☽ ☍ ♄
15 32	☽ ☍ ♀
18 37	☽ Ꝑ ☿
23 09	☽ △ ♇
17 Friday	
13 04	☽ # ♅
14 51	☉ Ꝑ ♃
16 21	☽ △ ☉
18 44	☽ ‖ ♃
19 56	☽ # ♄
20 47	☉ ∠ ♆
23 00	☽ # ♆
18 Saturday	
00 26	☽ △ ☿
04 12	☉ ∠ ♄
06 03	☽ # ☿
06 38	☽ ⊼ ♅
15 22	☽ # ☉
17 40	☽ ± ♅
18 38	☽ □ ♂
19 18	☽ ⊻ ♃
19 58	☽ ⊼ ♆
20 36	☽ ⊼ ♄
19 Sunday	
01 10	☽ ⊼ ♀
05 24	☽ □ ♇
06 19	♂ △ ♃
06 26	☽ ⊥ ♃
07 11	☽ ± ♆
07 54	☽ ± ♄
10 32	☽ Ꝑ ♅
13 19	☽ ± ♀
15 09	☽ # ♂
16 23	☉ ⊥ ♅
21 07	♂ ⚹ ♆
23 25	☽ ∠ ♃
20 Monday	
00 19	☽ Ꝑ ♆
01 09	☽ Ꝑ ♄
04 44	☽ □ ☉
06 34	☿ ⊻ ♅
06 37	☽ Q ♂
07 55	☽ Ꝑ ♀
14 14	♂ ⚹ ♄
15 27	☽ △ ♅
15 59	☽ Q ♇
16 42	☽ □ ☿
16 57	☿ ± ♃
21 Tuesday	
01 31	☿ ⊥ ♆
02 48	☿ ‖ ♆
04 26	☽ ⚹ ♃
05 37	☽ △ ♆
06 37	☽ △ ♄
07 27	☽ ⚹ ♂
09 38	☿ ⊥ ♄
15 34	♀ ⚹ ♇
15 42	☽ ⚹ ♇
15 43	☽ △ ♀
19 15	☽ Q ☉
22 49	☽ # ♇
22 Wednesday	
04 09	☽ Q ♃
08 32	☿ ‖ ♄
09 41	☽ Q ☿
15 13	☽ ∠ ♂
16 27	☽ ‖ ♇
21 34	☽ ⚹ ☉
21 53	☽ ∠ ♇
23 Thursday	
01 47	☿ # ♃
03 29	☽ □ ♅
10 30	☽ ⊥ ♂
13 33	☽ ⚹ ☿
16 10	☽ ⊥ ♇
16 14	☽ □ ♃
18 00	☽ □ ♆
19 23	☽ □ ♄
23 24	☽ ⊻ ♂
24 Friday	
04 22	☽ ⊻ ♇
06 47	☽ ∠ ☉
09 03	☽ □ ♀
09 07	☿ ⊼ ♃
16 09	☽ Q ♅
25 Saturday	
00 07	☿ ⊻ ♆
00 34	☽ ∠ ☿
02 45	☽ ⊥ ☉
06 36	☽ Q ♆
08 12	☽ Q ♄
08 51	☽ ‖ ♂
12 10	☿ ‖ ♅
12 49	☿ ⊻ ♄
16 02	☽ ⊻ ☉
16 35	☽ ⚹ ♅
21 31	☽ ⊥ ☿
21 35	☉ ± ♃
22 42	☉ ⊻ ♅
26 Sunday	
00 22	☽ Q ♀
04 35	☽ △ ♃
06 56	☽ ⚹ ♆
08 40	☽ ⚹ ♄
11 25	☽ ⊻ ☿
14 57	☽ ‖ ☉
15 49	☽ ☌ ♂
17 16	☽ ☌ ♇
22 38	☽ ‖ ♆
22 56	☽ ∠ ♅
27 Monday	
01 11	☉ ⊥ ♆
01 46	☽ ‖ ♄
02 20	☽ ⚹ ♀
04 08	☽ # ♃
09 58	☽ ‖ ♅
10 28	☽ Ꝑ ♃
13 06	☽ ∠ ♆
14 55	☽ ‖ ☿
14 59	☽ ∠ ♄
17 01	☽ ⊥ ♅
18 04	☿ ⊥ ♀
18 29	♂ ☌ ♇
20 57	☽ ‖ ♀
28 Tuesday	
01 07	☉ ⊥ ♄
04 14	☽ ± ♃
05 00	☽ ⊻ ♅
07 02	☽ ⊥ ♆
09 02	☽ ⊥ ♄
09 35	☿ ⊻ ♇
09 41	☽ ☌ ☉
10 24	☽ ∠ ♀
16 01	☽ ⊼ ♃
18 56	☽ ⊻ ♆
20 58	☿ ⊻ ♂
20 59	☽ ⊻ ♄
29 Wednesday	
05 04	☽ ⊻ ♇
05 25	☽ ⊥ ♀
06 39	☿ ‖ ♀
07 00	☽ ⊻ ♂
07 48	☽ ☌ ☿
16 51	☽ ⊥ ♇
17 59	☽ ⊻ ♀
19 26	☽ ⊥ ♂
30 Thursday	
01 16	☉ ∠ ♀
10 25	☽ ∠ ♇
13 56	☽ ∠ ♂
16 00	☽ ☌ ♅

All ephemeris data is given at 12.00 UT and the Moon's longitude is additionally given for 24.00 UT

DECEMBER 1989

LONGITUDES

Date	Sidereal time h m s	Sun ☉ ° ' "	Moon ☽ ° ' "	Moon ☽ 24.00 ° ' "	Mercury ☿ ° '	Venus ♀ ° '	Mars ♂ ° '	Jupiter ♃ ° '	Saturn ♄ ° '	Uranus ♅ ° '	Neptune ♆ ° '	Pluto ♇ ° '
01	16 41 17	09 ♐ 16 39	14 ♑ 19 09	20 ♑ 37 15	20 ♐ 42	23 ♑ 57	18 ♏ 28	09 ♋ 04	12 ♑ 08	03 ♑ 57	10 ♑ 55	16 ♏ 04
02	16 45 14	10 17 30	26 ♑ 58 06	03 ♒ 21 58	22 14	24 43	19 09	08 R 58	12 15	04 01	10 57	16 06
03	16 49 10	11 18 22	09 ♒ 49 07	16 ♒ 19 52	23 46	25 27	19 51	08 51	12 21	04 04	10 59	16 08
04	16 53 07	12 19 15	22 ♒ 54 32	29 ♒ 33 26	25 18	26 11	20 32	08 45	12 28	04 08	11 01	16 10
05	16 57 03	13 20 08	06 ♓ 16 55	13 ♓ 05 15	26 49	26 54	21 13	08 38	12 34	04 11	11 03	16 13
06	17 01 00	14 21 02	19 ♓ 58 39	26 ♓ 57 18	28 20	27 36	21 54	08 31	12 41	04 14	11 05	16 15
07	17 04 56	15 21 57	04 ♈ 01 13	11 ♈ 10 19	29 ♐ 51	28 16	22 36	08 24	12 47	04 18	11 07	16 17
08	17 08 53	16 22 53	18 ♈ 24 23	25 ♈ 43 00	01 ♑ 21	28 56	23 17	08 17	12 54	04 21	11 09	16 19
09	17 12 50	17 23 49	03 ♉ 05 04	10 ♉ 31 20	02 51	29 ♑ 34	23 58	08 10	13 00	04 25	11 11	16 21
10	17 16 46	18 24 47	17 ♉ 59 21	25 ♉ 28 36	04 20	00 ♒ 11	24 40	08 03	13 07	04 28	11 13	16 24
11	17 20 43	19 25 45	02 ♊ 57 53	10 ♊ 26 03	05 49	00 46	25 21	07 56	13 14	04 32	11 16	16 26
12	17 24 39	20 26 43	17 ♊ 51 52	25 ♊ 14 15	07 17	01 20	26 03	07 48	13 20	04 35	11 18	16 28
13	17 28 36	21 27 43	02 ♋ 32 11	09 ♋ 44 47	08 44	01 53	26 44	07 41	13 27	04 39	11 20	16 30
14	17 32 32	22 28 43	16 ♋ 51 24	23 ♋ 51 31	10 11	02 24	27 26	07 33	13 34	04 42	11 22	16 32
15	17 36 29	23 29 44	00 ♌ 44 50	07 ♌ 31 16	11 36	02 53	28 07	07 26	13 41	04 46	11 24	16 34
16	17 40 25	24 30 46	14 ♌ 10 51	20 ♌ 43 47	12 59	03 21	28 49	07 18	13 48	04 50	11 26	16 36
17	17 44 22	25 31 49	27 ♌ 10 26	03 ♍ 31 14	14 21	03 47	29 ♏ 31	07 10	13 54	04 53	11 29	16 38
18	17 48 19	26 32 53	09 ♍ 46 41	15 ♍ 57 23	15 41	04 12	00 ♐ 12	07 02	14 01	04 57	11 31	16 40
19	17 52 15	27 33 57	22 ♍ 03 57	28 ♍ 07 01	16 58	04 34	00 54	06 54	14 08	05 00	11 33	16 42
20	17 56 12	28 35 03	04 ♎ 07 15	10 ♎ 05 18	18 13	04 55	01 36	06 46	14 15	05 04	11 35	16 44
21	18 00 08	29 ♐ 36 09	16 ♎ 01 47	21 ♎ 57 21	19 25	05 14	02 18	06 38	14 22	05 07	11 37	16 46
22	18 04 05	00 ♑ 37 16	27 ♎ 52 33	03 ♏ 47 56	20 32	05 30	02 59	06 30	14 29	05 11	11 40	16 48
23	18 08 01	01 38 23	09 ♏ 44 00	15 ♏ 41 13	21 36	05 45	03 41	06 22	14 36	05 15	11 42	16 50
24	18 11 58	02 39 32	21 ♏ 39 59	27 ♏ 40 38	22 34	05 57	04 23	06 14	14 43	05 18	11 44	16 52
25	18 15 54	03 40 41	03 ♐ 43 29	09 ♐ 48 46	23 26	06 08	05 05	06 06	14 50	05 22	11 46	16 53
26	18 19 51	04 41 50	15 ♐ 56 39	22 ♐ 07 18	24 12	06 16	05 47	05 58	14 57	05 25	11 49	16 55
27	18 23 48	05 43 00	28 ♐ 20 48	04 ♑ 37 12	24 51	06 21	06 29	05 50	15 04	05 29	11 51	16 57
28	18 27 44	06 44 10	10 ♑ 56 32	17 ♑ 18 48	25 21	06 25	07 11	05 41	15 11	05 33	11 53	16 59
29	18 31 41	07 45 20	23 ♑ 44 00	00 ♒ 12 06	25 42	06 R 25	07 53	05 33	15 18	05 36	11 55	17 01
30	18 35 37	08 46 31	06 ♒ 43 06	13 ♒ 16 59	25 52	06 24	08 36	05 25	15 25	05 40	11 58	17 02
31	18 39 34	09 ♑ 47 41	19 ♒ 53 45	26 ♒ 33 26	25 ♑ 52	06 ♒ 20	09 ♐ 18	05 ♋ 17	15 ♑ 32	05 ♑ 44	12 ♑ 00	17 ♏ 04

Moon nodes and latitude

Date	Moon True ☊ ° '	Moon Mean ☊ ° '	Moon ☽ Latitude ° '
01	18 ♒ 52	20 ♒ 04	02 S 52
02	18 R 47	20 01	01 53
03	18 45	19 58	00 S 47
04	18 D 45	19 55	00 N 22
05	18 45	19 52	01 32
06	18 R 45	19 49	02 38
07	18 44	19 45	03 37
08	18 40	19 42	04 23
09	18 34	19 39	04 53
10	18 25	19 36	05 04
11	18 16	19 33	04 54
12	18 06	19 29	04 24
13	17 58	19 26	03 37
14	17 52	19 23	02 38
15	17 48	19 20	01 30
16	17 47	19 17	00 N 19
17	17 D 47	19 14	00 S 50
18	17 48	19 10	01 55
19	17 49	19 07	02 54
20	17 R 49	19 04	03 43
21	17 47	19 01	04 20
22	17 44	18 58	04 50
23	17 38	18 54	05 05
24	17 30	18 51	05 07
25	17 22	18 48	04 55
26	17 13	18 45	04 30
27	17 05	18 42	03 52
28	16 59	18 39	03 02
29	16 54	18 35	02 02
30	16 52	18 32	00 55
31	16 ♒ 51	18 ♒ 29	00 N 17

DECLINATIONS

Date	Sun ☉	Moon ☽	Mercury ☿	Venus ♀	Mars ♂	Jupiter ♃	Saturn ♄	Uranus ♅	Neptune ♆	Pluto ♇
01	21 S 51	25 S 31	25 S 04	24 S 12	17 S 03	22 N 57	22 S 33	23 S 39	22 S 08	02 S 07
02	22 00	22 37	25 14	23 59	17 15	22 57	22 33	23 39	22 08	02 07
03	22 08	18 33	25 22	23 46	17 27	22 58	22 32	23 39	22 08	02 07
04	22 16	13 32	25 29	23 33	17 38	22 58	22 32	23 39	22 08	02 08
05	22 24	07 47	25 35	23 20	17 50	22 59	22 31	23 39	22 08	02 08
06	22 31	01 S 33	25 39	23 06	18 01	23 00	22 31	23 38	22 08	02 09
07	22 38	04 N 55	25 41	22 53	18 12	23 00	22 30	23 38	22 08	02 09
08	22 45	11 16	25 42	22 38	18 23	23 01	22 30	23 38	22 08	02 09
09	22 51	17 08	25 42	22 24	18 34	23 01	22 29	23 38	22 07	02 10
10	22 56	22 03	25 40	22 10	18 45	23 02	22 29	23 38	22 07	02 10
11	23 01	25 34	25 37	21 55	18 55	23 02	22 28	23 38	22 07	02 11
12	23 06	27 16	25 33	21 40	19 06	23 03	22 27	23 38	22 07	02 11
13	23 10	27 02	25 26	21 26	19 16	23 04	22 27	23 38	22 07	02 11
14	23 14	24 59	25 19	21 11	19 26	23 04	22 26	23 38	22 06	02 12
15	23 17	21 27	25 10	20 56	19 36	23 05	22 26	23 37	22 06	02 12
16	23 20	16 53	24 59	20 41	19 46	23 05	22 25	23 37	22 06	02 12
17	23 22	11 40	24 48	20 25	19 56	23 06	22 24	23 37	22 06	02 12
18	23 24	06 07	24 35	20 10	20 05	23 06	22 24	23 37	22 06	02 13
19	23 25	00 N 29	24 20	19 55	20 14	23 07	22 23	23 37	22 06	02 13
20	23 26	05 S 03	24 05	19 40	20 23	23 07	22 22	23 37	22 05	02 13
21	23 27	10 20	23 48	19 25	20 32	23 08	22 22	23 37	22 05	02 14
22	23 26	15 13	23 30	19 11	20 41	23 09	22 21	23 37	22 05	02 14
23	23 26	19 33	23 12	18 56	20 50	23 09	22 20	23 36	22 05	02 14
24	23 25	23 07	22 53	18 41	20 58	23 10	22 20	23 36	22 05	02 14
25	23 23	25 44	22 33	18 27	21 06	23 10	22 19	23 36	22 05	02 14
26	23 22	27 11	22 14	18 13	21 15	23 11	22 18	23 36	22 04	02 14
27	23 19	27 18	21 54	17 59	21 22	23 11	22 18	23 36	22 04	02 14
28	23 16	26 01	21 34	17 45	21 30	23 12	22 17	23 36	22 04	02 15
29	23 13	23 22	21 15	17 31	21 38	23 12	22 16	23 36	22 04	02 15
30	23 09	19 29	20 57	17 18	21 45	23 13	22 15	23 35	22 04	02 15
31	23 S 05	14 S 35	20 S 39	17 S 05	21 S 52	23 N 13	22 S 15	23 S 35	22 S 03	02 S 15

ZODIAC SIGN ENTRIES

Date	h	m	Planets
02	17	42	☽ ♒
05	00	48	☽ ♓
07	05	11	☽ ♈
07	14	30	☿ ♑
09	06	59	☽ ♉
10	04	54	♀ ♒
11	07	15	☽ ♊
13	07	49	☽ ♋
15	10	41	☽ ♌
17	17	19	☽ ♍
18	04	57	♂ ♐
20	03	45	☽ ♎
21	21	22	☉ ♑
22	16	18	☽ ♏
25	04	37	☽ ♐
27	15	10	☽ ♑
29	23	38	☽ ♒

LATITUDES

Date	Mercury ☿	Venus ♀	Mars ♂	Jupiter ♃	Saturn ♄	Uranus ♅	Neptune ♆	Pluto ♇
01	01 S 58	02 S 55	00 N 17	00 S 11	00 N 20	00 S 16	00 N 51	15 N 09
04	02 08	02 41	00 15	00 11	00 20	00 16	00 51	15 09
07	02 15	02 25	00 14	00 10	00 20	00 16	00 51	15 10
10	02 18	02 06	00 12	00 10	00 19	00 16	00 51	15 10
13	02 18	01 44	00 10	00 10	00 19	00 16	00 51	15 11
16	02 12	01 18	00 08	00 09	00 19	00 16	00 51	15 12
19	01 59	00 49	00 06	00 09	00 19	00 16	00 51	15 13
22	01 39	00 S 17	00 04	00 08	00 18	00 16	00 51	15 14
25	01 10	00 N 18	00 03	00 08	00 18	00 16	00 51	15 15
28	00 S 30	00 57	00 N 01	00 08	00 18	00 16	00 51	15 16
31	00 N 19	01 N 39	00 S 01	00 S 07	00 N 18	00 S 16	00 N 51	15 N 17

DATA

Julian Date	2447862
Delta T	+56 seconds
Ayanamsa	23° 43' 08"
Synetic vernal point	05° ♓ 23' 51"
True obliquity of ecliptic	23° 26' 33"

MOON'S PHASES, APSIDES AND POSITIONS ☽

Date	h	m	Phase	Longitude	Eclipse Indicator
06	01	26	☽	13 ♓ 54	
12	16	30	○	20 ♊ 38	
19	23	55	☾	28 ♍ 04	
28	03	20	●	06 ♑ 22	

Day	h	m		
10	22	41	Perigee	
22	19	34	Apogee	
06	17	47	ON	
12	21	04	Max dec	27° N 25'
19	14	05	0S	
27	02	09	Max dec	27° S 25'

LONGITUDES

Date	Chiron ⚷	Ceres ⚳	Pallas ⚴	Juno ⚵	Vesta ⚶	Black Moon Lilith ⚸
01	15 ♋ 42	02 ♋ 29	27 ♓ 15	04 ♏ 38	02 ♒ 49	03 ♏ 00
11	15 ♋ 10	00 ♋ 25	28 ♓ 00	07 ♏ 50	07 ♒ 27	04 ♏ 07
21	14 ♋ 33	28 ♊ 05	29 ♓ 23	10 ♏ 52	12 ♒ 10	05 ♏ 14
31	13 ♋ 53	25 ♊ 46	01 ♈ 20	13 ♏ 43	16 ♒ 58	06 ♏ 21

All ephemeris data is given at 12.00 UT and the Moon's longitude is additionally given for 24.00 UT

ASPECTARIAN

h m Aspects

01 Friday
01 29 ☽ ⚺ ☉
02 01 ☽ ☍ ♃
05 28 ☽ ☌ ♆
07 26 ☉ ⚻ ♃
07 48 ☽ ☌ ♄
13 59 ☽ ⊥ ☉
15 20 ☽ ⚹ ♇
16 11 ☽ ∥ ☿
20 22 ☽ ⚹ ♂

02 Saturday
01 08 ☽ ∥ ♀
01 50 ☽ ⚺ ☿
04 32 ☽ ∥ ♅
06 31 ♂ ∠ ♅
07 28 ☽ ☌ ♀
08 34 ☽ ∠ ☉
09 38 ☽ ⋕ ♃
09 46 ☿ ⊥ ♇
12 27 ☽ ∥ ♄
14 08 ☽ Q ♇
14 43 ☽ ⊥ ☿
15 11 ☽ ∥ ♆
15 57 ☽ ∥ ☉
20 19 ☽ Q ♂

03 Sunday
01 16 ☽ ⚺ ♅
04 01 ☉ ⚺ ♆
09 47 ☽ ∠ ☿
10 14 ☽ ⚻ ♃
11 37 ☉ ∥ ♆
12 28 ☽ ⊥ ♅
14 09 ☽ ⚺ ♆
14 59 ☽ ⚹ ☉
16 43 ☽ ⚺ ♄
17 26 ☽ ∥ ♂
21 13 ☽ ± ♃
23 41 ☽ □ ♇

04 Monday
01 13 ☽ ⊥ ♆
02 16 ♀ ∥ ♅
03 50 ☽ ⊥ ♄
05 05 ☽ ∠ ♅
07 26 ☽ □ ♂
13 30 ☽ ⚼ ♃
14 46 ☽ Q ☉
15 39 ☉ ⚺ ♄
16 53 ☽ ⚹ ☿
17 38 ☽ ∠ ♆
18 17 ☽ ⚺ ♀
20 18 ☽ ∠ ♄

05 Tuesday
05 39 ☽ ⊥ ♀
08 15 ☽ ⚹ ♅
14 36 ☿ ⚺ ♀
16 08 ☽ △ ♃
17 03 ☽ Q ☿
20 26 ☽ ⚹ ♆
22 27 ☽ ∠ ♀
23 11 ☽ ⚹ ♄
23 37 ☿ ⊥ ♂

06 Wednesday
01 26 ☽ □ ☉
05 29 ☽ Q ♅
05 30 ☽ △ ♇
09 38 ☉ ∥ ♄
09 45 ☽ ∥ ♇
15 30 ☽ △ ♂
17 22 ☽ Q ♆
20 10 ☽ Q ♄
23 26 ♀ ⋕ ♃

07 Thursday
01 45 ☽ ⋕ ♇
01 46 ☽ ⚹ ♀
04 05 ☽ □ ☿
07 22 ☽ ⚼ ♇
12 25 ♀ Q ♇
12 28 ☽ □ ♅
18 19 ☽ ⚼ ♂
19 19 ☽ □ ♃
22 33 ☽ ± ♇
23 00 ☽ Q ♀
23 56 ☽ □ ♆

08 Friday
02 48 ☽ □ ♄
04 37 ☉ ∥ ♀
08 24 ☽ △ ☉
08 33 ☽ ⚻ ♇
10 03 ☽ ± ♂
10 31 ☉ ⚺ ♇
11 32 ☿ ∠ ♇
12 16 ♂ ⚼ ♃
20 25 ☽ ⚻ ♂

09 Saturday
00 50 ☽ Q ♃
03 28 ♀ ∥ ♄
06 01 ☽ □ ♀
10 47 ☽ ⚼ ☉
11 34 ☽ △ ☿
14 09 ☽ △ ♅
18 42 ☽ ⋕ ♂
20 09 ☽ ⚹ ♃

10 Sunday
01 06 ☽ △ ♆
02 23 ☽ ± ☉
04 07 ☽ △ ♄
09 26 ☽ ☍ ♇
12 23 ☽ ⋕ ♆
12 36 ☽ ⋕ ♀
12 44 ☽ ⚻ ☉
14 14 ☿ ☌ ♅
14 23 ☽ ⚼ ♅
14 24 ☽ ⚼ ☿
14 25 ☽ ⋕ ♄
16 24 ♀ ∥ ♆
17 16 ☽ ⋕ ☉
17 44 ☽ ∥ ♃
20 03 ☽ ∠ ♃
21 30 ☽ ⋕ ♅
23 13 ☽ ☍ ♂

11 Monday
01 14 ☽ ⚼ ♆
04 21 ☽ ⚼ ♄
04 52 ☽ ± ♅
06 24 ☽ ± ☿
08 20 ☽ △ ♀
10 21 ☽ ⊥ ♃
12 33 ☽ ⋕ ☿
14 31 ☽ ⚻ ♅
15 41 ☽ ± ♆
17 05 ☽ ⚻ ☿
18 21 ☉ ⋕ ♃
18 54 ☽ ± ♄
19 54 ☽ ⚺ ♃

12 Tuesday
01 22 ☽ ⚻ ♆
04 38 ☽ ⚻ ♄
09 26 ☽ ⚼ ♀
09 44 ☽ ⚻ ♇
16 30 ☽ ☍ ☉
19 29 ☽ ± ♇
19 52 ☿ ☍ ♃
21 11 ♂ ∠ ♆

13 Wednesday
00 38 ☽ ± ♀
01 59 ☽ ⚻ ♂
10 17 ☽ ⚼ ♇
10 53 ☽ ⚻ ♀
12 21 ☽ ± ♂
15 31 ☽ ☍ ♅
20 28 ☽ ☌ ♃
23 28 ☽ ☍ ☿

14 Thursday
02 42 ☽ ☍ ♆
04 07 ☽ ⚼ ♂
06 23 ☽ ☍ ♄
09 01 ☽ ⋕ ☿
11 27 ☽ △ ♇
13 20 ☉ ⊥ ♇
22 19 ☽ ⋕ ♅
22 23 ☽ ⚻ ☉

15 Friday
00 49 ☽ ⋕ ☉
02 02 ☽ ∥ ♃
06 10 ☽ ⋕ ♄
07 09 ☽ △ ♂
08 09 ☽ ⋕ ♆
08 39 ☿ ☌ ♆
09 38 ☽ ± ☉
15 14 ☽ ⋕ ♀
15 55 ☽ ☍ ♀
19 08 ☽ ⚻ ♅
21 55 ☽ ⋕ ♂
23 43 ☽ ⚺ ♃

16 Saturday
02 52 ☽ ⚼ ☉
05 54 ☽ ± ♅
07 01 ☽ ⚻ ♆
09 35 ☽ ⚻ ☿
10 25 ☽ ⊥ ♃
11 03 ♂ ∠ ♄
11 17 ☽ ⚻ ♄
12 24 ♂ ⊥ ♅
16 25 ☽ □ ♇
17 58 ☽ ± ♆
21 49 ☽ ± ☿
22 22 ☽ ± ♄
22 23 ☽ ⚼ ♅

17 Sunday
02 45 ☽ ∠ ♃
03 25 ☿ ☌ ♄
08 39 ☽ △ ☉
10 41 ☽ ⚼ ♆
15 17 ☽ ⚼ ♄
16 35 ☽ ⚼ ☿
16 39 ☽ □ ♂
17 53 ☿ ∠ ♂

18 Monday
00 56 ☽ ⚻ ♀
02 10 ☽ Q ♇
02 40 ☽ △ ♅
06 47 ☽ ⚹ ♃
12 50 ☽ ± ♀
15 22 ☽ △ ♆
17 20 ♀ ∥ ♂
20 18 ☽ △ ♄

19 Tuesday
00 49 ☽ △ ☿
01 26 ☽ ⚹ ♇
04 38 ☽ ⋕ ♇
05 23 ☽ Q ♂
05 50 ☽ Q ♃
06 46 ☿ ⚹ ♇
06 56 ☽ ⚼ ♀
12 07 ♂ ± ♃
23 38 ☽ ∥ ♇
23 55 ☽ □ ☉

20 Wednesday
06 38 ☽ ⚹ ♂
07 12 ☽ ∠ ♇
13 38 ☽ △ ♀
13 54 ☽ □ ♅
17 16 ☽ □ ♃
23 22 ☉ ⊥ ♀

21 Thursday
01 20 ☽ ⊥
01 52 ♀
03 04
0[…]
14[…]
15[…]
19 3[…]

22 Frid[ay]
02 26 ☽ Q ♅
03 38 ☿ ∥ ♅
10 06 ☽ ⊥ ♂
15 38 ☽ Q ♆
17 17 ☉ ∥ ☿
18 05 ☽ ⚹ ☉
21 26 ☽ Q ♄
23 01 ☽ ⚺ ♂

23 Saturday
02 53 ☽ ⚹ ♅
03 47 ☽ □ ♀
05 17 ☽ △ ♃
08 34 ☽ ∥ ♀
11 41 ☽ Q ☿
15 38 ☿ ⋕ ♃
15 59 ☽ ⚹ ♆
16 36 ☉ ∠ ♇
20 20 ☽ ∥ ♂
21 54 ☽ ⚹ ♄

24 Sunday
02 20 ☽ ☌ ♇
03 13 ☽ ∠ ☉
04 24 ☽ ∥ ♆
06 11 ☽ ∥ ♄
09 15 ☽ ∠ ♅
10 23 ☽ ∥ ☿
11 09 ☽ ⚼ ♃
12 20 ☽ ⋕ ♃
13 57 ☽ ⚹ ☿
14 19 ☽ ∥ ☉
15 50 ☽ ∥ ♅
16 39 ☽ Q ♀
22 09 ☽ ∠ ♆
22 54 ☽ ⊥ ☉

25 Monday
03 19 ☽ ⊥ ♅
04 13 ☽ ∠ ♄
04 53 ☽ ± ♃
09 36 ♀ ⚻ ♃
11 54 ☽ ⚺ ☉
14 51 ☽ ☌ ♂
15 15 ☽ ⚺ ♅
16 04 ☽ ⊥ ♆
16 38 ☽ ⚻ ♃
16 48 ☽ ⚹ ♀
21 57 ☽ ∠ ☿
22 10 ☽ ⊥ ♄
22 23 ♂ ⚺ ♅

26 Tuesday
03 54 ☽ ⚺ ♆
06 04 ☿ ∥ ♄
10 03 ☽ ⚺ ♄
12 51 ♂ ⊥ ♆
13 54 ☽ ⚺ ♇
16 40 ☽ ⊥ ☿
17 00 ♂ ⚻ ♃
22 25 ☽ ∠ ♀
23 17 ☿ ∥ ♆

27 Wednesday
01 35 ☽ ⊥ ♇
04 56 ☽ ⚺ ☿
06 12 ☉ ☌ ♅
06 50 ♀ ⚹ ♂
14 16 ☉ ☍ ♃
15 52 ☽ ⊥ ♀
18 55 ☽ ∠ ♇

28 Thursday
01 43 ☽ ☌ ♅
02 09 ☽ ☍ ♃
03 20 ☽ ☌ ☉
03 23 ☽ ⚺ ♀
03 59 ☉ ⚺ ♀
04 28 ☽ ⚺ ♂
13 47 ☽ ☌ ♆
15 34 ☿ ∥ ♂
16 29 ☽ ⊥ ♂
20 05 ☽ ☌ ♄
23 24 ☽ ⚹ ♇

29 Friday
05 46 ♃ ☍ ♅
08 50 ♀ St R
10 18 ☽ ∥ ♅
10 20 ☽ ∠ ♂
13 05 ☽ ∥ ☉
13 10 ☽ ⋕ ♃
15 43 ☽ ☌ ☿
16 50 ☉ ⋕ ♃
19 33 ☽ ∥ ♄
20 52 ☽ ∥ ♆
21 49 ☽ Q ♇
22 15 ☉ ⚺ ♂
23 15 ☽ ∥ ♂

30 Saturday
03 04 ☽ ∥ ☿
09 38 ☽ ⚻ ♃
10 03 ☽ ⚺ ♅
11 25 ☽ ☌ ♀
15 38 ☽ ⚹ ♂
16 05 ☽ ⚺ ☉
20 31 ☽ ± ♃
21 06 ☽ ⊥ ♅
21 37 ☽ ⚺ ♆
23 28 ☿ St R
23 45 ☽ ∥ ♀

31 Sunday
03 57 ☽
04 02
0[…]

JANUARY 1990

Date	Sidereal time h m s	Sun ☉ ° ' "	Moon ☽ ° ' "	Moon ☽ 24.00 ° ' "	Mercury ☿ ° '	Venus ♀ ° '	Mars ♂ ° '	Jupiter ♃ ° '	Saturn ♄ ° '	Uranus ♅ ° '	Neptune ♆ ° '	Pluto ♇ ° '
01	18 43 30	10 ♑ 48 51	03 ♓ 16 04	10 ♓ 01 40	25 ♑ 40	06 ♒ 13	10 ♐ 00	05 ♋ 09	15 ♑ 39	05 ♑ 47	12 ♑ 02	17 ♏ 06
02	18 47 27	11 50 01	16 ♓ 50 18	23 ♓ 42 00	25 R 17	06 R 04	10 42	05 R 01	15 47	05 51	12 05	17 07
03	18 51 23	12 51 11	00 ♈ 36 49	07 ♈ 34 44	24 41	05 53	11 25	04 53	15 54	05 54	12 07	17 09
04	18 55 20	13 52 20	14 ♈ 35 41	21 ♈ 39 36	23 54	05 39	12 07	04 45	16 01	05 58	12 09	17 10
05	18 59 17	14 53 30	28 ♈ 46 17	05 ♉ 55 28	22 57	05 22	12 49	04 37	16 08	06 01	12 11	17 12
06	19 03 13	15 54 38	13 ♉ 06 50	20 ♉ 19 54	21 51	05 03	13 32	04 29	16 15	06 05	12 14	17 13
07	19 07 10	16 55 47	27 ♉ 34 09	04 ♊ 48 57	20 38	04 42	14 14	04 21	16 22	06 09	12 16	17 15
08	19 11 06	17 56 55	12 ♊ 03 37	19 ♊ 17 26	19 20	04 18	14 56	04 14	16 29	06 12	12 18	17 16
09	19 15 03	18 58 03	26 ♊ 29 37	03 ♋ 39 26	17 59	03 52	15 39	04 06	16 36	06 16	12 20	17 18
10	19 18 59	19 59 10	10 ♋ 46 10	17 ♋ 49 10	16 40	03 24	16 21	03 58	16 43	06 19	12 23	17 19
11	19 22 56	21 00 17	24 ♋ 47 51	01 ♌ 41 45	15 23	02 55	17 04	03 51	16 51	06 23	12 25	17 21
12	19 26 52	22 01 24	08 ♌ 30 30	15 ♌ 13 53	14 11	02 23	17 47	03 44	16 58	06 26	12 27	17 22
13	19 30 49	23 02 30	21 ♌ 51 47	28 ♌ 24 11	13 06	01 50	18 29	03 36	17 05	06 30	12 29	17 23
14	19 34 46	24 03 36	04 ♍ 51 12	11 ♍ 13 04	12 09	01 16	19 12	03 29	17 12	06 33	12 32	17 24
15	19 38 42	25 04 42	17 ♍ 30 04	23 ♍ 42 36	11 21	00 41	19 55	03 22	17 19	06 37	12 34	17 26
16	19 42 39	26 05 47	29 ♍ 51 06	05 ♎ 56 05	10 42	00 ♒ 05	20 37	03 15	17 26	06 40	12 36	17 27
17	19 46 35	27 06 52	11 ♎ 58 06	17 ♎ 57 43	10 13	29 ♑ 29	21 20	03 08	17 33	06 44	12 38	17 28
18	19 50 32	28 07 58	23 ♎ 55 31	29 ♎ 52 07	09 54	28 52	22 03	03 01	17 40	06 47	12 41	17 29
19	19 54 28	29 ♑ 09 02	05 ♏ 48 08	11 ♏ 44 10	09 44	28 15	22 46	02 55	17 47	06 51	12 43	17 30
20	19 58 25	00 ♒ 10 07	17 ♏ 40 47	23 ♏ 38 35	09 D 42	27 38	23 29	02 48	17 54	06 54	12 45	17 31
21	20 02 21	01 11 11	29 ♏ 38 05	05 ♐ 39 47	09 49	27 02	24 12	02 42	18 01	06 57	12 47	17 32
22	20 06 18	02 12 15	11 ♐ 44 10	17 ♐ 51 37	10 03	26 26	24 54	02 36	18 08	07 01	12 49	17 33
23	20 10 15	03 13 18	24 ♐ 02 29	00 ♑ 17 05	10 24	25 51	25 37	02 30	18 15	07 04	12 52	17 34
24	20 14 11	04 14 20	06 ♑ 35 38	12 ♑ 58 18	10 51	25 18	26 20	02 24	18 22	07 07	12 54	17 35
25	20 18 08	05 15 22	19 ♑ 25 09	25 ♑ 56 14	11 23	24 46	27 04	02 18	18 29	07 11	12 56	17 36
26	20 22 04	06 16 24	02 ♒ 31 28	09 ♒ 10 47	12 02	24 15	27 47	02 12	18 36	07 14	12 58	17 37
27	20 26 01	07 17 24	15 ♒ 53 58	22 ♒ 40 48	12 44	23 47	28 30	02 07	18 43	07 17	13 00	17 38
28	20 29 57	08 18 23	29 ♒ 31 02	06 ♓ 24 21	13 31	23 20	29 13	02 02	18 50	07 20	13 02	17 39
29	20 33 54	09 19 22	13 ♓ 20 26	20 ♓ 18 56	14 22	22 55	29 ♐ 56	01 56	18 56	07 24	13 04	17 40
30	20 37 50	10 20 19	27 ♓ 19 30	04 ♈ 21 47	15 17	22 32	00 ♑ 39	01 52	19 03	07 27	13 06	17 40
31	20 41 47	11 ♒ 21 15	11 ♈ 25 27	18 ♈ 30 10	16 ♑ 14	22 ♑ 12	01 ♑ 23	01 ♋ 47	19 ♑ 10	07 ♑ 30	13 ♑ 08	17 ♏ 41

Date	Moon True ☊ ° '	Moon Mean ☊ ° '	Moon ☽ Latitude ° '	DECLINATIONS Sun ☉ ° '	Moon ☽ ° '	Mercury ☿ ° '	Venus ♀ ° '	Mars ♂ ° '	Jupiter ♃ ° '	Saturn ♄ ° '	Uranus ♅ ° '	Neptune ♆ ° '	Pluto ♇ ° '
01	16 ♒ 52	18 ♒ 26	01 N 28	23 S 00	08 S 56	20 S 24	16 S 53	21 S 59	23 N 14	22 S 14	23 S 35	22 S 03	02 S 15
02	16 D 54	18 23	02 36	22 55	02 S 48	20 09	16 40	22 06	23 14	22 13	23 35	22 03	02 15
03	16 55	18 20	03 36	22 49	03 N 33	19 56	16 29	22 12	23 14	22 12	23 35	22 03	02 15
04	16 R 56	18 16	04 24	22 43	09 49	19 45	16 17	22 19	23 15	22 12	23 35	22 03	02 15
05	16 55	18 13	04 57	22 37	15 39	19 36	16 06	22 25	23 15	22 11	23 35	22 02	02 15
06	16 53	18 10	05 12	22 30	20 43	19 29	15 56	22 31	23 16	22 10	23 34	22 02	02 15
07	16 49	18 07	05 07	22 22	24 36	19 24	15 46	22 37	23 16	22 09	23 34	22 02	02 15
08	16 45	18 04	04 43	22 14	26 54	19 21	15 36	22 42	23 16	22 09	23 34	22 02	02 15
09	16 40	18 00	04 00	22 06	27 24	19 19	15 27	22 47	23 17	22 08	23 34	22 02	02 15
10	16 37	17 57	03 04	21 57	26 03	19 19	15 18	22 53	23 17	22 07	23 34	22 01	02 15
11	16 34	17 54	01 57	21 48	23 05	19 20	15 10	22 58	23 18	22 06	23 34	22 01	02 15
12	16 32	17 51	00 N 44	21 38	18 51	19 22	15 03	23 02	23 18	22 05	23 33	22 01	02 15
13	16 D 32	17 48	00 S 29	21 28	13 46	19 26	14 56	23 07	23 18	22 04	23 33	22 01	02 15
14	16 33	17 45	01 39	21 18	08 12	19 30	14 49	23 11	23 19	22 04	23 33	22 01	02 15
15	16 35	17 41	02 42	21 07	02 N 27	19 36	14 43	23 15	23 19	22 03	23 33	22 00	02 15
16	16 36	17 38	03 36	20 56	03 S 15	19 43	14 38	23 19	23 19	22 02	23 33	22 00	02 15
17	16 38	17 35	04 20	20 44	08 43	19 50	14 33	23 22	23 20	22 01	23 33	22 00	02 14
18	16 39	17 32	04 51	20 32	13 47	19 58	14 29	23 26	23 20	22 00	23 32	22 00	02 14
19	16 R 38	17 29	05 10	20 20	18 19	20 06	14 25	23 29	23 20	21 59	23 32	22 00	02 14
20	16 38	17 26	05 15	20 07	22 09	20 14	14 22	23 32	23 21	21 59	23 32	21 59	02 14
21	16 36	17 22	05 07	19 54	25 05	20 23	14 19	23 35	23 21	21 58	23 32	21 59	02 14
22	16 34	17 19	04 46	19 40	26 55	20 31	14 17	23 37	23 21	21 57	23 32	21 59	02 14
23	16 32	17 16	04 11	19 26	27 29	20 39	14 15	23 39	23 21	21 56	23 32	21 59	02 13
24	16 30	17 13	03 23	19 12	26 39	20 48	14 14	23 41	23 22	21 55	23 32	21 59	02 13
25	16 28	17 10	02 24	18 57	24 25	20 55	14 14	23 43	23 22	21 54	23 31	21 58	02 13
26	16 27	17 06	01 16	18 42	20 50	21 02	14 13	23 45	23 22	21 53	23 31	21 58	02 13
27	16 27	17 03	00 S 03	18 27	16 07	21 09	14 13	23 46	23 22	21 52	23 31	21 58	02 13
28	16 D 27	17 00	01 N 12	18 11	10 31	21 15	14 14	23 47	23 23	21 52	23 31	21 58	02 12
29	16 28	16 57	02 24	17 55	04 S 20	21 20	14 15	23 48	23 23	21 51	23 31	21 58	02 12
30	16 28	16 54	03 28	17 39	02 N 07	21 25	14 16	23 49	23 23	21 50	23 31	21 57	02 12
31	16 ♒ 29	16 ♒ 51	04 N 20	17 S 22	08 N 30	21 S 28	14 S 18	23 S 49	23 N 23	21 S 49	23 S 30	21 S 57	02 S 12

ZODIAC SIGN ENTRIES

Date	h	m	Planets
01	06	10	☽ ♓
03	10	56	☽ ♈
05	14	04	☽ ♉
07	16	52	☽ ♊
09	17	52	☽ ♋
11	21	02	☽ ♌
14	02	57	☽ ♍
16	12	17	☽ ♎
16	15	23	♀ ♑
19	00	16	☽ ♏
20	08	02	☉ ♒
21	12	44	☽ ♐
23	23	27	☽ ♑
26	07	25	☽ ♒
28	12	51	☽ ♓
29	14	10	♂ ♑
30	16	34	☽ ♈

LATITUDES

Date	Mercury ☿ ° '	Venus ♀ ° '	Mars ♂ ° '	Jupiter ♃ ° '	Saturn ♄ ° '	Uranus ♅ ° '	Neptune ♆ ° '	Pluto ♇ ° '
01	00 N 38	01 N 54	00 S 02	00 S 07	00 N 18	00 S 16	00 N 51	15 N 17
04	01 36	02 39	00 06	00 07	00 17	00 16	00 51	15 18
07	02 29	03 26	00 06	00 06	00 17	00 16	00 51	15 19
10	03 07	04 12	00 08	00 06	00 17	00 16	00 51	15 21
13	03 23	04 57	00 10	00 05	00 17	00 16	00 51	15 22
16	03 19	05 37	00 12	00 05	00 17	00 16	00 51	15 23
19	03 00	06 13	00 14	00 04	00 16	00 16	00 51	15 24
22	02 33	06 42	00 17	00 04	00 16	00 16	00 51	15 26
25	02 02	07 04	00 19	00 03	00 16	00 16	00 51	15 27
28	01 31	07 18	00 21	00 03	00 16	00 17	00 51	15 29
31	01 N 00	07 N 24	00 S 23	00 S 03	00 N 16	00 S 17	00 N 51	15 N 30

DATA

Julian Date	2447893
Delta T	+57 seconds
Ayanamsa	23° 43' 14"
Synetic vernal point	05° ♓ 23' 45"
True obliquity of ecliptic	23° 26' 33"

MOON'S PHASES, APSIDES AND POSITIONS ☽

Date	h	m	Phase	Longitude ° '	Eclipse Indicator
04	10	40	☽	13 ♈ 49	
11	04	57	○	20 ♋ 42	
18	21	17	☾	28 ♎ 32	
26	19	20	●	06 ♒ 35	Annular

Day	h	m	
07	18	59	Perigee
19	16	00	Apogee

Day	h	m		
02	22	37	0N	
09	06	20	Max dec	27° N 27'
15	22	14	0S	
23	10	00	Max dec	27° S 29'
30	04	11	0N	

LONGITUDES

Date	Chiron ⚷ ° '	Ceres ⚳ ° '	Pallas ⚴ ° '	Juno ⚵ ° '	Vesta ⚶ ° '	Black Moon Lilith ⚸ ° '
01	13 ♋ 49	25 ♊ 33	01 ♈ 34	14 ♏ 00	17 ♒ 27	06 ♏ 28
11	13 ♋ 08	23 ♊ 35	04 ♈ 03	16 ♏ 36	22 ♒ 18	07 ♏ 35
21	12 ♋ 28	22 ♊ 10	06 ♈ 57	18 ♏ 57	27 ♒ 11	08 ♏ 42
31	11 ♋ 52	21 ♊ 24	10 ♈ 14	21 ♏ 00	02 ♓ 05	09 ♏ 50

is data is given at 12.00 UT and the Moon's longitude is additionally given for 24.00 UT

ASPECTARIAN

h m	Aspects
01 Monday	
00 50	☽ ∠ ♆
07 18	☽ ∠ ♄
09 13	☽ ⊥ ☿
15 19	☽ △ ♃
16 30	☽ ⚹ ♅
17 12	☽ ⊻ ♀
02 Tuesday	
00 36	☽ □ ♂
00 49	☽ ∠ ☿
02 05	♂ ∥ ♆
02 29	☽ ⚹ ☉
03 13	☿ ∠ ♂
03 36	☽ ⚹ ♆
03 43	☽ ⊥ ♀
10 07	☽ ⚹ ♄
12 30	☽ △ ♇
13 46	☽ Q ♅
14 06	☽ ∥ ♇
17 55	☉ ☌ ♆
19 19	☽ ∠ ♀
03 Wednesday	
00 41	☽ Q ♆
01 13	☽ Q ☉
02 11	☽ ⚹ ☿
07 06	☽ # ♇
07 15	☽ Q ♄
09 32	♀ ⊻ ♅
12 05	♂ ∥ ♄
14 39	☽ bQ ♇
19 18	☽ □ ♃
20 56	☽ ⚹ ♀
21 10	☽ □ ♅
21 57	☽ Q ☿
04 Thursday	
06 09	☽ ± ♇
07 33	☽ △ ♂
07 49	☽ □ ♆
10 40	☽ □ ☉
13 22	♂ ⊻ ♆
14 26	☽ □ ♄
16 24	☽ ⊼ ♇
17 06	☽ Q ♀
05 Friday	
01 43	☽ Q ♃
02 50	☽ □ ☿
10 19	☽ bQ ♂
13 53	☽ # ♀
21 43	☽ ⚹ ♃
22 50	☽ □ ♀
06 Saturday	
00 13	☽ △ ♅
02 12	☽ ± ♂
05 50	☽ # ☿
09 45	☉ ∥ ♂
10 31	☽ △ ♆
12 43	☽ ⊼ ♂
17 01	☽ △ ☉
17 16	☽ △ ♄
18 51	☽ ☍ ♇
19 15	☽ # ♆
19 59	☽ # ♄
21 02	☉ ☌ ♄
21 39	☽ # ☉
22 19	☽ # ♂
22 30	☽ ∠ ♃
07 Sunday	
01 18	☽ bQ ♅
01 24	☽ △ ☿
02 44	☽ ∥ ♃
04 43	☽ # ♅
11 30	☽ bQ ♆
13 17	☽ ⊥ ♃
16 17	☽ ± ♅
16 47	☿ ⊥ ♂
18 21	☽ bQ ♄
19 42	☉ ⚹ ♇
19 46	☽ bQ ☉
23 08	☽ ⊻ ♃
23 30	☽ △ ♀
08 Monday	
00 15	☽ bQ ☿
02 15	☽ ⊼ ♅
02 26	☽ ± ♆
09 22	☽ ± ♄
11 48	☽ ± ☉
12 24	☽ ⊼ ♆
13 55	☽ ± ☿
17 01	☽ ☍ ♂
18 00	♀ ⊼ ♃
19 24	☽ ⊼ ♄
20 40	☽ ⊼ ♇
22 31	☽ ⊼ ☉
23 02	☽ ⊼ ☿
23 40	☽ bQ ♀
09 Tuesday	
02 04	☉ ☌ ☿
06 38	☉ ∥ ♄
06 39	☽ ± ♇
14 14	☽ ± ♀
21 44	☽ bQ ♇
23 58	☽ ⊼ ♀
10 Wednesday	
00 14	☿ ⚹ ♇
00 29	☉ ∥ ♆
00 38	☽ ☌ ♃
04 27	☽ ☍ ♅
10 55	☿ ☌ ♄
14 44	☽ ☍ ♆
15 36	☿ ⊻ ♂
21 10	☽ ☍ ☿
22 00	☽ ⊼ ♂
22 13	☽ ☍ ♄
23 10	☽ △ ♇
11 Thursday	
02 54	♂ ⊻ ♄
04 57	☽ ☍ ☉
08 48	☽ # ♅
08 51	☽ ± ♂
10 36	☽ ∥ ♃
12 47	☽ # ♂
18 09	☽ # ♄
18 38	☽ # ♆
20 15	☽ # ☉
21 39	♂ ⊻ ♇
12 Friday	
01 20	☽ bQ ♂
01 37	☽ ☍ ♀
03 38	☽ ⊻ ♃
04 31	♀ ∠ ♂
08 19	☽ ⊼ ♅
09 21	☽ # ☿
14 08	☽ ⊥ ♃
19 02	☽ ± ♅
19 03	☽ ⊼ ♆
21 20	☽ ⊼ ☿
13 Saturday	
03 15	☽ ⊼ ♄
03 52	☽ □ ♇
05 31	☽ △ ♂
05 52	☽ ± ♆
06 08	☽ ∠ ♃
06 37	☽ # ♀
07 20	☽ ± ☿
11 20	☽ bQ ♅
14 14	☽ ± ♄
14 20	☽ ⊼ ☉
22 21	☽ bQ ♆
22 37	☽ bQ ☿
14 Sunday	
02 12	☿ ☌ ♆
02 18	☽ ± ☉
05 36	☽ ⊼ ♀
07 00	☽ bQ ♄
09 28	☽ ⚹ ♃
13 02	☽ Q ♇
15 12	☽ △ ♅
16 21	☽ ± ♀
20 36	☽ bQ ☉
15 Monday	
00 54	☽ △ ☿
02 32	☽ △ ♆
07 57	☽ Q ♃
08 41	☽ bQ ♀
11 38	☽ △ ♄
11 52	☽ ⚹ ♇
12 51	☽ # ♇
16 55	☽ □ ♂
16 Tuesday	
03 59	☽ △ ☉
07 43	☽ ∥ ♇
12 26	☽ △ ♀
15 28	♂ # ♃
15 56	♄ ⚹ ♇
17 07	☽ ∠ ♇
18 38	☽ □ ♃
17 Wednesday	
01 31	☽ □ ♅
06 25	☽ Q ♂
08 38	☽ □ ☿
11 00	☽ ⊥ ♇
12 18	♀ Q ♇
13 21	☽ □ ♆
23 02	☽ ⊻ ♇
23 17	☽ □ ♄
18 Thursday	
05 18	☉ ⊥ ♂
07 59	☽ ⚹ ♂
13 45	☽ Q ♅
15 25	☽ ∥ ♀
19 53	☽ Q ☿
21 17	☽ □ ☉
21 28	☽ □ ♀
22 42	☉ ☌ ♀
19 Friday	
01 41	☽ Q ♆
02 40	♀ ⊥ ♂
06 12	☽ △ ♃
06 26	♄ ∥ ♆
11 58	☽ Q ♄
14 07	☽ ⚹ ♅
16 13	☽ ∠ ♂
19 54	☽ ⚹ ☿
20 31	☉ Q ♇
22 54	☽ ∥ ☿
23 21	☽ ∥ ☉
20 Saturday	
02 01	☽ ⚹ ♆
04 08	☉ ∥ ☿
04 33	☿ St D
08 04	☽ Q ♀
10 48	☽ ∥ ♄
10 54	☽ ∥ ♆
11 34	☽ ⊥ ♂
11 41	☽ ☌ ♇
12 15	☽ bQ ♃
12 27	☽ ⚹ ♄
13 05	☽ Q ☉
13 36	♂ ⊥ ♇
13 36	♂ ∥ ♅
20 32	☽ ∠ ♅
20 52	☽ # ♃
22 23	☽ ∥ ♅
22 31	☽ ∥ ♂
21 Sunday	
00 24	☽ ⊻ ♂
02 14	☽ ∠ ☿
06 11	☽ ± ♃
07 02	☽ ⚹ ♀
08 18	☽ ∠ ♆
14 39	☽ ⊥ ♅
15 23	☽ ⚹ ☉
18 03	☽ ⊼ ♃
18 49	☽ ∠ ♄
20 28	☽ ⊥ ☿
22 Monday	
02 16	☽ ⊥ ♆
02 38	☽ ⊻ ♅
08 35	☽ ⊻ ☿
11 26	☽ ∠ ♀
12 47	☽ ⊥ ♄
14 09	☽ ⊻ ♆
19 42	♃ bQ ♇
20 23	☉ ⊼ ♃
23 25	☽ ⊻ ♇
23 41	☽ ∠ ☉
23 Tuesday	
00 39	☽ ⊻ ♄
04 15	☽ ⊥ ♀
11 06	☽ ⊥ ♇
15 14	☽ ☌ ♂
15 21	☽ ⊻ ♀
16 22	♀ ⊻ ♂
18 40	☽ ⊥ ☉
24 Wednesday	
04 05	☽ ☍ ♃
04 23	☽ ∠ ♇
07 09	☽ ⊻ ☉
13 00	☽ ☌ ♅
20 21	☽ ☌ ☿
23 53	☽ ☌ ♆
25 Thursday	
08 38	☽ ⚹ ♇
10 15	☽ ☌ ♄
17 20	☽ ∥ ♂
18 51	☽ ∥ ♅
19 58	☽ # ♃
21 29	☽ ☌ ♀
26 Friday	
02 52	☽ ⊻ ♂
05 15	☽ ∥ ♆
05 43	☽ ∥ ♄
06 43	☽ Q ♇
10 52	☽ ∥ ☿
11 26	☽ ⊼ ♃
14 24	☽ ⊥ ♂
19 20	☽ ☌ ☉
20 32	☽ ⊻ ♅
22 11	☽ ± ♃
27 Saturday	
00 09	☽ ∥ ☉
06 02	☽ ⊻ ☿
06 50	☽ ⊻ ♆
07 20	☽ ⊥ ♅
07 29	☽ ∠ ♂
11 55	☉ ⊻ ♅
14 09	☽ bQ ♃
15 05	☽ □ ♇
17 02	☽ ⊻ ♄
17 20	☽ ⊥ ☿
17 31	☽ ⊥ ♆
20 29	☽ ∥ ♀
20 44	☿ ☌ ♆
23 21	☽ ∠ ♅
28 Sunday	
01 29	☽ ⊻ ♀
03 42	☽ ⊥ ♄
05 55	☉ ± ♃
09 24	☽ ∠ ♆
10 09	☽ ∠ ☿
11 27	☽ ⚹ ♂
11 41	☽ ⊥ ♀
16 21	☽ △ ♃
19 35	☽ ∠ ♄
29 Monday	
01 40	☽ ⚹ ♅
02 53	☽ ∠ ♀
04 30	☽ ⊻ ☉
09 26	☽ Q ♂
11 32	☽ ⚹ ♆
13 54	☽ ⚹ ☿
15 41	☽ ⊥ ☉
19 27	☽ △ ♇
20 00	☽ ∥ ♇
21 43	☽ ⚹ ♄
22 27	☽ Q ♅
30 Tuesday	
04 01	☽ ⚹ ♀
08 12	☽ Q ♆
08 20	☽ ∠ ☉
11 55	☽ Q ☿
12 19	☽ # ♇
17 59	☽ □ ♂
18 25	☽ Q ♄
19 41	☽ □ ♃
21 08	☽ bQ ♇
31 Wednesday	
00 00	☽ Q ♀
05 19	☽ □ ♅
11 52	☽ ⚹ ☉
12 26	☽ ± ♇
14 55	☽ □ ♆
20 46	☽ □ ☿
22 37	☽ ⊼ ♇

FEBRUARY 1990

LONGITUDES

Date	Sidereal time h m s	Sun ☉ ° ′ ″	Moon ☽ ° ′ ″	Moon ☽ 24.00 ° ′ ″	Mercury ☿ ° ′	Venus ♀ ° ′	Mars ♂ ° ′	Jupiter ♃ ° ′	Saturn ♄ ° ′	Uranus ♅ ° ′	Neptune ♆ ° ′	Pluto ♇ ° ′
01	20 45 44	12 ♒ 22 09	25 ♈ 35 37	02 ♉ 41 29	17 ♑ 15	21 ♑ 54	02 ♑ 06	01 ♋ 42	19 ♑ 17	07 ♑ 33	13 ♑ 11	17 ♏ 42
02	20 49 40	13 23 03	09 ♉ 47 30	16 ♉ 53 21	18 18	21 R 38	02 49	01 R 38	19 23	07 36	13 13	17 42
03	20 53 37	14 23 55	23 ♉ 58 46	01 ♊ 03 29	19 23	21 25	03 32	01 33	19 30	07 39	13 15	17 43
04	20 57 33	15 24 45	08 ♊ 07 13	15 ♊ 09 41	20 31	21 14	04 16	01 29	19 37	07 42	13 17	17 43
05	21 01 30	16 25 35	22 ♊ 10 35	29 ♊ 09 39	21 41	21 05	04 59	01 25	19 43	07 45	13 19	17 44
06	21 05 26	17 26 22	06 ♋ 06 35	13 ♋ 01 05	22 53	21 00	05 43	01 22	19 50	07 48	13 21	17 44
07	21 09 23	18 27 09	19 ♋ 52 52	26 ♋ 41 40	24 06	20 56	06 26	01 18	19 57	07 51	13 22	17 45
08	21 13 19	19 27 54	03 ♌ 27 13	10 ♌ 09 18	25 21	20 D 55	07 10	01 15	20 03	07 54	13 24	17 45
09	21 17 16	20 28 37	16 ♌ 47 43	23 ♌ 22 19	26 38	20 57	07 53	01 12	20 10	07 57	13 26	17 46
10	21 21 13	21 29 19	29 ♌ 53 00	06 ♍ 19 42	27 56	21 01	08 37	01 09	20 16	08 00	13 28	17 46
11	21 25 09	22 30 00	12 ♍ 42 26	19 ♍ 01 16	29 ♑ 16	21 07	09 20	01 06	20 22	08 03	13 30	17 46
12	21 29 06	23 30 40	25 ♍ 16 20	01 ♎ 27 49	00 ♒ 37	21 15	10 04	01 04	20 29	08 06	13 32	17 46
13	21 33 02	24 31 18	07 ♎ 35 58	13 ♎ 41 06	01 59	21 26	10 48	01 01	20 35	08 08	13 34	17 47
14	21 36 59	25 31 55	19 ♎ 43 35	25 ♎ 43 49	03 22	21 39	11 31	00 59	20 41	08 11	13 36	17 47
15	21 40 55	26 32 31	01 ♏ 42 16	07 ♏ 39 25	04 46	21 54	12 15	00 57	20 48	08 14	13 37	17 47
16	21 44 52	27 33 06	13 ♏ 35 48	19 ♏ 31 59	06 12	22 11	12 59	00 55	20 54	08 17	13 39	17 47
17	21 48 48	28 33 39	25 ♏ 28 33	01 ♐ 26 03	07 39	22 30	13 43	00 54	21 00	08 19	13 41	17 47
18	21 52 45	29 ♒ 34 11	07 ♐ 25 08	13 ♐ 26 21	09 06	22 51	14 26	00 52	21 06	08 22	13 43	17 47
19	21 56 42	00 ♓ 34 43	19 ♐ 30 18	25 ♐ 37 32	10 35	23 14	15 10	00 51	21 12	08 24	13 44	17 R 47
20	22 00 38	01 35 12	01 ♑ 48 36	08 ♑ 03 58	12 05	23 39	15 54	00 50	21 18	08 27	13 46	17 47
21	22 04 35	02 35 41	14 ♑ 24 05	20 ♑ 49 17	13 35	24 05	16 38	00 50	21 24	08 29	13 48	17 47
22	22 08 31	03 36 08	27 ♑ 19 53	03 ♒ 56 02	15 07	24 33	17 22	00 49	21 30	08 32	13 49	17 47
23	22 12 28	04 36 33	10 ♒ 37 50	17 ♒ 25 15	16 40	25 03	18 06	00 49	21 36	08 34	13 51	17 47
24	22 16 24	05 36 57	24 ♒ 18 07	01 ♓ 16 10	18 14	25 34	18 50	00 48	21 42	08 37	13 53	17 47
25	22 20 21	06 37 20	08 ♓ 18 58	15 ♓ 26 00	19 48	26 07	19 34	00 D 49	21 47	08 39	13 54	17 47
26	22 24 17	07 37 40	22 ♓ 36 39	29 ♓ 50 11	21 24	26 41	20 18	00 49	21 53	08 41	13 56	17 46
27	22 28 14	08 37 59	07 ♈ 05 52	14 ♈ 22 52	23 01	27 16	21 02	00 49	21 59	08 43	13 57	17 46
28	22 32 11	09 ♓ 38 16	21 ♈ 40 23	28 ♈ 57 39	24 ♒ 39	27 ♑ 53	21 ♑ 46	00 ♋ 50	22 ♑ 04	08 ♑ 46	13 ♑ 59	17 ♏ 46

Date	Moon True ☊ ° ′	Moon Mean ☊ ° ′	Moon ☽ Latitude ° ′
01	16 ♒ 29	16 ♒ 47	04 N 56
02	16 D 30	16 44	05 15
03	16 R 29	16 41	05 14
04	16 29	16 38	04 55
05	16 29	16 35	04 18
06	16 D 29	16 32	03 26
07	16 30	16 28	02 22
08	16 30	16 25	01 N 12
09	16 R 30	16 22	00 S 02
10	16 30	16 19	01 14
11	16 29	16 16	02 20
12	16 28	16 12	03 19
13	16 27	16 09	04 07
14	16 26	16 06	04 43
15	16 25	16 03	05 06
16	16 24	16 00	05 16
17	16 24	15 57	05 12
18	16 D 24	15 53	04 55
19	16 24	15 50	04 25
20	16 26	15 47	03 42
21	16 27	15 44	02 48
22	16 28	15 41	01 44
23	16 29	15 38	00 S 32
24	16 R 29	15 34	00 N 43
25	16 28	15 31	01 57
26	16 26	15 28	03 06
27	16 23	15 25	04 03
28	16 ♒ 20	15 ♒ 22	04 N 45

DECLINATIONS

Date	Sun ☉ ° ′	Moon ☽ ° ′	Mercury ☿ ° ′	Venus ♀ ° ′	Mars ♂ ° ′	Jupiter ♃ ° ′	Saturn ♄ ° ′	Uranus ♅ ° ′	Neptune ♆ ° ′	Pluto ♇ ° ′
01	17 S 06	14 N 29	21 S 31	14 S 20	23 S 49	23 N 23	21 S 48	23 S 30	21 S 57	02 S 11
02	16 48	19 43	21 32	14 23	23 49	23 24	21 47	23 30	21 57	02 11
03	16 31	23 50	21 33	14 25	23 49	23 24	21 46	23 30	21 56	02 11
04	16 13	26 31	21 33	14 28	23 49	23 24	21 45	23 30	21 56	02 10
05	15 55	27 30	21 31	14 31	23 48	23 24	21 44	23 30	21 56	02 10
06	15 37	26 44	21 28	14 34	23 47	23 24	21 44	23 29	21 56	02 10
07	15 18	24 19	21 25	14 38	23 46	23 25	21 43	23 29	21 56	02 09
08	14 59	20 33	21 20	14 41	23 44	23 25	21 42	23 29	21 55	02 09
09	14 40	15 47	21 13	14 45	23 42	23 25	21 41	23 29	21 55	02 09
10	14 21	10 22	21 06	14 48	23 40	23 25	21 40	23 29	21 55	02 08
11	14 01	04 N 38	20 57	14 52	23 38	23 25	21 39	23 29	21 55	02 08
12	13 41	01 S 10	20 48	14 56	23 36	23 25	21 38	23 28	21 55	02 07
13	13 21	06 47	20 37	14 59	23 33	23 25	21 37	23 28	21 55	02 07
14	13 01	12 05	20 24	15 03	23 30	23 26	21 36	23 28	21 54	02 07
15	12 40	16 51	20 11	15 06	23 27	23 26	21 36	23 28	21 54	02 06
16	12 19	20 56	19 56	15 10	23 24	23 26	21 35	23 28	21 54	02 06
17	11 59	24 11	19 40	15 13	23 20	23 26	21 34	23 28	21 54	02 05
18	11 37	26 24	19 22	15 16	23 16	23 26	21 33	23 28	21 54	02 05
19	11 16	27 26	19 03	15 19	23 12	23 26	21 32	23 27	21 53	02 04
20	10 55	27 08	18 43	15 22	23 08	23 26	21 31	23 27	21 53	02 04
21	10 33	25 27	18 22	15 24	23 04	23 27	21 30	23 27	21 53	02 03
22	10 11	22 24	17 59	15 26	22 59	23 27	21 29	23 27	21 53	02 03
23	09 49	18 05	17 36	15 28	22 54	23 27	21 29	23 27	21 53	02 02
24	09 27	12 45	17 10	15 30	22 49	23 27	21 28	23 27	21 52	02 02
25	09 05	06 38	16 44	15 31	22 43	23 27	21 27	23 27	21 52	02 01
26	08 43	00 S 05	16 16	15 33	22 37	23 27	21 26	23 27	21 52	02 01
27	08 20	06 N 33	15 47	15 33	22 32	23 27	21 25	23 26	21 52	02 00
28	07 S 57	12 N 51	15 S 16	15 S 34	22 S 25	23 N 27	21 S 24	23 S 26	21 S 52	02 S 00

ZODIAC SIGN ENTRIES

Date	h m	Planets
01	19 27	☽ ♉
03	22 12	☽ ♊
06	01 27	☽ ♋
08	05 51	☽ ♌
10	12 13	☽ ♍
12	01 11	☿ ♒
12	21 09	☽ ♎
15	08 34	☽ ♏
17	21 07	☽ ♐
18	22 14	☉ ♓
20	08 30	☽ ♑
22	16 52	☽ ♒
24	21 49	☽ ♓
27	00 16	☽ ♈

LATITUDES

Date	Mercury ☿ ° ′	Venus ♀ ° ′	Mars ♂ ° ′	Jupiter ♃ ° ′	Saturn ♄ ° ′	Uranus ♅ ° ′	Neptune ♆ ° ′	Pluto ♇ ° ′
01	00 N 49	07 N 25	00 S 24	00 S 02	00 N 16	00 S 17	00 N 51	15 N 31
04	00 N 20	07 23	00 26	00 02	00 15	00 17	00 51	15 32
07	00 S 07	07 16	00 28	00 02	00 15	00 17	00 51	15 33
10	00 32	07 05	00 31	00 01	00 15	00 17	00 51	15 35
13	00 55	06 50	00 33	00 S 01	00 15	00 17	00 51	15 36
16	01 15	06 32	00 35	00 00	00 15	00 17	00 51	15 38
19	01 32	06 13	00 38	00 00	00 14	00 17	00 51	15 39
22	01 46	05 52	00 40	00 00	00 14	00 17	00 51	15 41
25	01 57	05 30	00 43	00 N 01	00 14	00 17	00 51	15 42
28	02 05	05 08	00 45	00 01	00 14	00 17	00 51	15 44
31	02 S 09	04 N 45	00 S 47	00 N 01	00 N 14	00 S 17	00 N 51	15 N 45

DATA

Julian Date	2447924
Delta T	+57 seconds
Ayanamsa	23° 43′ 20″
Synetic vernal point	05° ♓ 23′ 40″
True obliquity of ecliptic	23° 26′ 33″

MOON'S PHASES, APSIDES AND POSITIONS ☽

Date	h m	Phase	Longitude ° ′	Eclipse Indicator
02	18 32	☽	13 ♉ 40	
09	19 16	○	20 ♌ 47	total
17	18 48	☾	28 ♏ 51	
25	08 54	●	06 ♓ 30	

Day	h m		
02	02 40	Perigee	
16	13 05	Apogee	
28	07 59	Perigee	
05	13 24	Max dec	27° N 30′
12	07 10	0S	
19	18 50	Max dec	27° S 29′
26	12 19	0N	

LONGITUDES

Date	Chiron ⚷ ° ′	Ceres ⚳ ° ′	Pallas ⚴ ° ′	Juno ⚵ ° ′	Vesta ⚶ ° ′	Black Moon Lilith ⚸ ° ′
01	11 ♋ 49	21 ♊ 22	10 ♈ 35	21 ♏ 11	02 ♓ 35	09 ♏ 56
11	11 ♋ 18	21 ♊ 23	14 ♈ 13	22 ♏ 51	07 ♓ 30	11 ♏ 03
21	10 ♋ 55	22 ♊ 06	18 ♈ 08	24 ♏ 05	12 ♓ 25	12 ♏ 11
31	10 ♋ 40	23 ♊ 25	22 ♈ 18	24 ♏ 51	17 ♓ 18	13 ♏ 18

All ephemeris data is given at 12.00 UT and the Moon's longitude is additionally given for 24.00 UT

ASPECTARIAN

h m	Aspects
01 Thursday	
00 07	♂ ☍ ♃
01 14	☽ □ ♄
02 05	☽ Q ♃
05 52	☽ □ ♀
09 46	☽ Q ☉
11 21	☽ # ♀
22 16	☽ ⚹ ♃
22 29	☿ ⚹ ♇
22 49	☽ # ☉
23 35	☽ △ ♂
02 Friday	
07 43	☉ ⊻ ♆
08 07	♂ ∠ ♇
08 17	☽ △ ♅
17 28	☉ ⊥ ♅
17 47	☽ △ ♆
18 32	☽ □ ☉
21 44	☽ # ☿
23 03	☽ # ♄
23 30	☽ ∠ ♃
03 Saturday	
00 00	☽ # ♆
01 23	☽ ☍ ♇
02 18	☽ ⚼ ♂
03 34	☽ △ ☿
04 22	☽ △ ♄
07 43	☽ △ ♀
08 59	☽ ‖ ♃
09 40	☽ # ♅
09 45	☽ ⚼ ♅
11 51	☽ # ♂
14 40	☽ ⊥ ♃
14 44	☿ ☌ ♄
18 22	☽ ± ♂
19 14	☽ ⚼ ♆
04 Sunday	
00 47	☽ ⊻ ♃
01 04	☽ ± ♅
05 05	☽ ⊼ ♂
05 59	☽ ⚼ ♄
07 11	☽ ⚼ ☿
08 49	☽ ⚼ ♀
10 34	☽ ± ♆
11 18	☽ ⊼ ♅
20 48	☽ ⊼ ♆
21 26	☽ ± ♄
23 53	☽ ± ☿
23 59	☽ ± ♀
05 Monday	
01 06	☿ ☌ ♀
01 24	☽ △ ☉
04 23	☽ ⊼ ♇
07 46	☽ ⊼ ♄
10 09	☽ ⊼ ♀
11 05	☽ ⊼ ☿
11 58	☉ ⚼ ♃
14 40	☽ ± ♇
06 Tuesday	
03 50	☽ ☌ ♃
05 09	☽ ⚼ ☉
06 10	☽ ⚼ ♇
11 16	☽ ☍ ♂
14 57	☽ ☍ ♅
19 09	☉ □ ♇
21 59	☽ ± ☉
07 Wednesday	
00 36	☽ ☍ ♆
08 15	☽ △ ♇
09 18	☽ ⊼ ☉
12 07	☽ ☍ ♄
13 51	☽ ☍ ♀
16 07	☽ # ♂
18 00	☽ # ♅
18 32	☽ ‖ ♃
20 10	☽ ☍ ☿
08 Thursday	
04 03	☽ # ♆
05 24	☽ # ♄
07 30	☽ # ☿
08 06	☽ ⊻ ♃
09 17	♀ St D
10 35	☉ ⊥ ♆
18 45	☽ ⊥ ♃
19 00	☽ ⊼ ♂
19 59	☽ ⊼ ♅
09 Friday	
03 34	☉ ⊻ ♄
05 54	☽ ⊼ ♆
06 25	☽ ± ♂
06 50	☽ ± ♅
07 00	☉ ‖ ♀
10 55	☽ ∠ ♃
13 45	☽ □ ♇
14 22	♂ ☌ ♅
16 42	☽ # ♀
16 49	☽ ± ♆
17 27	☽ # ☉
18 10	☽ ⊼ ♄
19 16	☽ ☍ ☉
19 35	☽ ⊼ ♀
23 16	☽ ⚼ ♅
23 45	☉ ⊻ ♀
23 46	☽ ⚼ ♂
10 Saturday	
05 16	☽ ± ♄
06 39	☽ ± ♀
08 00	☽ ⊼ ☿
09 23	☽ ⚼ ♆
14 20	☽ ⚹ ♃
20 24	☽ ± ☿
22 06	☽ ⚼ ♄
22 57	☽ Q ♇
23 29	☽ ⚼ ♀
11 Sunday	
03 11	☽ △ ♅
05 16	☽ △ ♂
12 45	☽ Q ♃
13 30	☽ △ ♆
15 18	☽ ⚼ ☿
21 07	☿ Q ♇
21 37	☽ ⚹ ♇
22 21	☽ # ♇
12 Monday	
01 38	☉ ∠ ♅
02 43	☽ △ ♄
04 11	☽ △ ♀
08 18	☽ ⊼ ☉
16 01	☽ ‖ ♇
19 43	☿ ⊼ ♃
20 56	☽ ± ☉
23 11	☽ □ ♃
23 37	☽ △ ☿
13 Tuesday	
02 34	☽ ∠ ♇
13 04	☽ □ ♅
16 07	☽ ⚼ ☉
18 41	☽ □ ♂
20 14	☽ ⊥ ♇
23 47	☽ □ ♆
14 Wednesday	
08 08	☽ ⊻ ♇
13 56	☽ □ ♄
15 55	☽ □ ♀
16 12	☽ ‖ ☉
15 Thursday	
00 40	☽ △ ☉
00 58	☽ Q ♅
02 46	☽ ‖ ♀
05 40	♂ ‖ ♅
08 53	☽ Q ♂
10 29	☽ △ ♃
11 50	☽ Q ♆
18 40	☉ ⊥ ♄
19 01	☽ □ ☿
22 06	♂ # ♃
16 Friday	
01 13	☽ ⚹ ♅
02 25	☽ Q ♄
04 56	☽ Q ♀
06 02	☽ ‖ ☿
10 40	☽ ⚹ ♂
12 07	☽ ⚹ ♆
16 13	☽ ‖ ♄
16 42	☽ ⚼ ♃
18 26	☽ ‖ ♆
20 28	☽ ☌ ♇
23 52	☿ ± ♃
17 Saturday	
02 53	☽ ⚹ ♄
05 09	☽ ‖ ♂
05 47	☽ # ♃
05 50	☽ ⚹ ♀
06 01	☽ ‖ ♅
07 38	☽ ∠ ♅
10 02	☉ ⊥ ♀
10 50	☽ ± ♃
11 05	♂ ☌ ♆
12 23	☽ Q ☿
14 58	☉ ∠ ♆
18 29	☽ ∠ ♆
18 48	☽ □ ☉
18 57	☽ ∠ ♂
22 54	☽ ⊼ ♃
23 30	☿ ⊻ ♅
18 Sunday	
00 46	☉ ∠ ♂
01 50	☽ ⊥ ♅
09 20	☽ ∠ ♄
12 35	☽ ⊥ ♆
12 54	☽ ∠ ♀
13 54	☽ ⊻ ♅
14 10	☽ ⊥ ♂
15 50	☽ ⚹ ☿
19 Monday	
00 34	☽ ⊻ ♆
02 53	☽ ⊻ ♂
03 26	☽ ⊥ ♄
06 29	♇ St R
07 22	☽ ⊥ ♀
08 37	☽ ⊻ ♇
10 01	☽ Q ☉
15 22	☽ ⊻ ♄
18 28	☉ △ ♃
19 35	☽ ⊻ ♀
20 25	☽ ⊥ ♇
20 Tuesday	
01 33	☽ ∠ ☿
10 08	☽ ☍ ♃
11 32	☽ ⚹ ☉
13 53	☽ ∠ ♇
21 19	☽ ⊥ ☿
21 Wednesday	
00 46	☽ ☌ ♅
10 16	☽ ⊻ ☿
10 51	☽ ☌ ♆
15 18	☿ ⊻ ♆
16 27	☽ ☌ ♂
18 21	☽ ⚹ ♇
18 30	☽ ∠ ☉
22 Thursday	
01 11	☽ ☌ ♄
02 31	☿ ⊥ ♅
04 46	☽ ‖ ♅
04 49	☽ # ♃
06 42	☽ ☌ ♀
08 00	☽ ‖ ♂
12 32	☽ ⊥ ☉
15 15	☽ ‖ ♆
16 29	☽ Q ♇
17 39	☽ ‖ ♄
18 21	☽ ⊼ ♃
22 49	☿ ⚼ ♃
23 Friday	
00 20	☽ ⊻ ☉
01 40	♂ ⚹ ♇
05 11	☽ ± ♃
08 19	☽ ⊻ ♅
14 39	☽ ‖ ☿
14 57	♃ # ♅
17 43	☽ ⊻ ♆
19 00	☽ ⊥ ♅
21 10	☽ ⚼ ♃
24 Saturday	
00 03	☽ ☌ ☿
00 12	☽ ‖ ♀
00 38	☽ □ ♇
01 56	☽ ⊻ ♂
04 17	☽ ⊥ ♆
05 10	☿ □ ♇
07 26	☽ ⊻ ♄
10 48	☽ ∠ ♅
12 58	☽ ⊥ ♂
14 17	☽ ⊻ ♀
17 54	☽ ⊥ ♄
19 11	♃ St D
19 55	☽ ∠ ♆
23 13	☽ △ ♃
25 Sunday	
01 01	☽ ⊥ ♀
02 05	☽ ‖ ☉
05 14	☿ ⊻ ♂
05 17	☽ ∠ ♂
08 54	☽ ☌ ☉
09 24	☽ ∠ ♄
12 34	☽ ⚹ ♅
13 29	☿ ⊥ ♆
16 25	☉ ∠ ♄
16 55	☽ ∠ ♀
21 27	☽ ⚹ ♆
26 Monday	
03 55	☽ △ ♇
05 01	☽ ‖ ♇
07 56	☽ ⚹ ♂
08 47	☽ Q ♅
09 44	☽ ⊻ ☿
10 47	☽ ⚹ ♄
17 32	☽ Q ♆
19 03	☽ ⚹ ♀
19 34	☽ # ♇
19 40	☿ ⊻ ♄
20 57	☽ ⊥ ☿
27 Tuesday	
01 37	☽ □ ♃
04 51	☽ ⚼ ♇
04 56	☽ Q ♂
06 49	☽ Q ♄
13 42	☽ ∠ ☿
14 13	☉ ⚹ ♅
14 41	☽ □ ♅
14 43	☽ ⊻ ☉
15 44	☽ Q ♀
18 16	☽ # ☉
19 42	☽ ± ♇
22 35	☿ ‖ ♀
22 44	☿ ∠ ♅
23 19	☽ □ ♆
28 Wednesday	
01 20	☽ ⊥ ☉
05 34	☽ ⊼ ♇
07 19	☽ Q ♃
12 10	☽ □ ♂
12 40	☽ □ ♄
17 14	☽ ∠ ☉
17 30	☽ ⚹ ☿
21 04	☽ # ☿
22 41	☽ □ ♀
23 07	☽ # ♀
23 15	♂ ☌ ♄

MARCH 1990

LONGITUDES

Date	Sidereal time h m s	Sun ☉ ° ' "	Moon ☽ ° ' "	Moon ☽ 24.00 ° ' "	Mercury ☿ ° '	Venus ♀ ° '	Mars ♂ ° '	Jupiter ♃ ° '	Saturn ♄ ° '	Uranus ♅ ° '	Neptune ♆ ° '	Pluto ♇ ° '
01	22 36 07	10 ♓ 38 31	06 ♉ 13 56	13 ♉ 28 35	26 ♒ 17	28 ♑ 31	22 ♑ 30	00 ♋ 51	22 ♑ 10	08 ♑ 48	14 ♑ 00	17 ♏ 45
02	22 40 04	11 38 43	20 ♉ 41 02	27 ♉ 50 48	27 57	29 10	23 14	00 52	22 15	08 50	14 02	17 R 45
03	22 44 00	12 38 54	04 ♊ 57 33	12 ♊ 01 00	29 ♒ 38	29 ♑ 50	23 59	00 53	22 21	08 52	14 03	17 45
04	22 47 57	13 39 03	19 ♊ 01 00	25 ♊ 57 25	01 ♓ 20	00 ♒ 31	24 43	00 54	22 26	08 54	14 04	17 44
05	22 51 53	14 39 10	02 ♋ 50 14	09 ♋ 39 29	03 03	01 14	25 27	00 56	22 31	08 56	14 06	17 44
06	22 55 50	15 39 14	16 ♋ 25 13	23 ♋ 07 32	04 47	01 57	26 11	00 58	22 36	08 58	14 07	17 43
07	22 59 46	16 39 17	29 ♋ 46 30	06 ♌ 22 14	06 32	02 42	26 56	01 00	22 42	09 00	14 08	17 43
08	23 03 43	17 39 17	12 ♌ 54 49	19 ♌ 24 20	08 18	03 27	27 40	01 02	22 47	09 02	14 10	17 42
09	23 07 40	18 39 15	25 ♌ 50 53	02 ♍ 14 30	10 05	04 13	28 24	01 04	22 52	09 03	14 11	17 41
10	23 11 36	19 39 11	08 ♍ 35 17	14 ♍ 53 15	11 54	05 00	29 08	01 07	22 57	09 05	14 12	17 41
11	23 15 33	20 39 05	21 ♍ 08 29	27 ♍ 21 02	13 43	05 48	29 ♑ 53	01 10	23 01	09 07	14 13	17 40
12	23 19 29	21 38 58	03 ♎ 31 01	09 ♎ 38 30	15 34	06 37	00 ♒ 37	01 13	23 06	09 08	14 14	17 39
13	23 23 26	22 38 48	15 ♎ 43 39	21 ♎ 46 38	17 26	07 27	01 22	01 16	23 11	09 10	14 15	17 39
14	23 27 22	23 38 36	27 ♎ 47 38	03 ♏ 46 56	19 19	08 17	02 06	01 19	23 16	09 12	14 17	17 38
15	23 31 19	24 38 23	09 ♏ 44 49	15 ♏ 41 37	21 13	09 08	02 50	01 23	23 20	09 13	14 18	17 37
16	23 35 15	25 38 08	21 ♏ 37 45	27 ♏ 33 38	23 08	10 00	03 35	01 26	23 25	09 15	14 19	17 36
17	23 39 12	26 37 51	03 ♐ 29 45	09 ♐ 26 36	25 05	10 52	04 19	01 30	23 29	09 16	14 20	17 35
18	23 43 09	27 37 33	15 ♐ 24 46	21 ♐ 24 48	27 02	11 45	05 04	01 34	23 33	09 17	14 21	17 34
19	23 47 05	28 37 13	27 ♐ 27 20	03 ♑ 32 58	29 ♓ 00	12 39	05 49	01 39	23 38	09 19	14 22	17 34
20	23 51 02	29 ♓ 36 51	09 ♑ 42 19	15 ♑ 56 00	00 ♈ 59	13 33	06 33	01 43	23 42	09 20	14 22	17 33
21	23 54 58	00 ♈ 36 27	22 ♑ 14 36	28 ♑ 38 40	02 59	14 28	07 18	01 48	23 46	09 21	14 23	17 32
22	23 58 55	01 36 02	05 ♒ 08 42	11 ♒ 45 05	05 00	15 24	08 02	01 52	23 50	09 22	14 24	17 31
23	00 02 51	02 35 34	18 ♒ 28 07	25 ♒ 17 59	07 01	16 20	08 47	01 57	23 54	09 24	14 25	17 30
24	00 06 48	03 35 05	02 ♓ 14 42	09 ♓ 18 08	09 02	17 16	09 32	02 02	23 58	09 25	14 26	17 29
25	00 10 44	04 34 34	16 ♓ 27 56	23 ♓ 43 34	11 04	18 13	10 16	02 07	24 02	09 26	14 26	17 27
26	00 14 41	05 34 01	01 ♈ 04 19	08 ♈ 29 19	13 05	19 10	11 01	02 13	24 06	09 27	14 27	17 26
27	00 18 38	06 33 26	15 ♈ 57 30	23 ♈ 27 43	15 06	20 08	11 46	02 18	24 09	09 28	14 28	17 25
28	00 22 34	07 32 49	00 ♉ 58 45	08 ♉ 29 24	17 07	21 06	12 30	02 24	24 13	09 28	14 28	17 24
29	00 26 31	08 32 10	15 ♉ 58 30	23 ♉ 24 57	19 06	22 05	13 15	02 30	24 16	09 29	14 29	17 23
30	00 30 27	09 31 29	00 ♊ 47 51	08 ♊ 06 25	21 04	23 04	14 00	02 36	24 20	09 30	14 30	17 22
31	00 34 24	10 ♈ 30 45	15 ♊ 20 03	22 ♊ 28 22	23 ♈ 00	24 ♒ 03	14 ♒ 44	02 ♋ 42	24 ♑ 23	09 ♑ 31	14 ♑ 30	17 ♏ 20

Date	Moon True ☊ ° '	Moon Mean ☊ ° '	Moon ☽ Latitude ° '
01	16 ♒ 16	15 ♒ 18	05 N 09
02	16 R 14	15 15	05 13
03	16 13	15 12	04 57
04	16 D 13	15 09	04 24
05	16 13	15 06	03 36
06	16 15	15 03	02 36
07	16 17	14 59	01 29
08	16 17	14 56	00 N 18
09	16 R 17	14 53	00 S 52
10	16 15	14 50	01 59
11	16 11	14 47	02 58
12	16 06	14 43	03 49
13	15 59	14 40	04 28
14	15 52	14 37	04 54
15	15 45	14 34	05 08
16	15 39	14 31	05 08
17	15 35	14 28	04 55
18	15 32	14 24	04 29
19	15 31	14 21	03 51
20	15 D 32	14 18	03 02
21	15 33	14 15	02 03
22	15 34	14 12	00 S 56
23	15 R 35	14 09	00 N 16
24	15 33	14 05	01 29
25	15 30	14 02	02 38
26	15 24	13 59	03 40
27	15 17	13 56	04 27
28	15 09	13 53	04 57
29	15 01	13 49	05 06
30	14 54	13 46	04 54
31	14 ♒ 49	13 ♒ 43	04 N 24

DECLINATIONS

Date	Sun ☉ ° '	Moon ☽ ° '	Mercury ☿ ° '	Venus ♀ ° '	Mars ♂ ° '	Jupiter ♃ ° '	Saturn ♄ ° '	Uranus ♅ ° '	Neptune ♆ ° '	Pluto ♇ ° '
01	07 S 35	18 N 27	14 S 45	15 S 34	22 S 19	23 N 28	21 S 23	23 S 26	21 S 52	01 S 59
02	07 12	22 57	14 12	15 34	22 12	23 28	21 23	23 26	21 51	01 59
03	06 49	26 00	13 37	15 33	22 06	23 28	21 22	23 26	21 51	01 58
04	06 26	27 22	13 02	15 32	21 59	23 28	21 21	23 26	21 51	01 58
05	06 03	27 01	12 25	15 30	21 51	23 28	21 20	23 26	21 51	01 57
06	05 39	25 01	11 46	15 29	21 44	23 28	21 19	23 26	21 51	01 57
07	05 16	21 39	11 07	15 26	21 36	23 28	21 19	23 25	21 51	01 56
08	04 53	17 14	10 26	15 24	21 28	23 28	21 18	23 25	21 51	01 55
09	04 29	12 05	09 44	15 20	21 20	23 28	21 17	23 25	21 50	01 55
10	04 06	06 31	09 01	15 17	21 12	23 29	21 16	23 25	21 50	01 54
11	03 42	00 N 47	08 17	15 13	21 04	23 29	21 16	23 25	21 50	01 54
12	03 19	04 S 54	07 31	15 08	20 55	23 29	21 15	23 25	21 50	01 53
13	02 55	10 18	06 44	15 03	20 46	23 29	21 14	23 25	21 50	01 53
14	02 31	15 16	05 56	14 58	20 37	23 29	21 13	23 25	21 50	01 52
15	02 08	19 36	05 07	14 52	20 27	23 29	21 13	23 25	21 50	01 51
16	01 44	23 07	04 16	14 46	20 18	23 29	21 12	23 25	21 49	01 51
17	01 20	25 40	03 25	14 39	20 08	23 29	21 11	23 25	21 49	01 50
18	00 57	27 06	02 33	14 31	19 58	23 29	21 11	23 24	21 49	01 50
19	00 33	27 16	01 39	14 24	19 48	23 29	21 10	23 24	21 49	01 49
20	00 S 09	26 06	00 S 45	14 15	19 38	23 29	21 09	23 24	21 49	01 48
21	00 N 14	23 37	00 N 10	14 07	19 27	23 29	21 09	23 24	21 49	01 48
22	00 38	19 53	01 05	13 57	19 17	23 29	21 08	23 24	21 49	01 47
23	01 02	15 03	02 01	13 48	19 06	23 29	21 07	23 24	21 49	01 47
24	01 26	09 18	02 58	13 38	18 55	23 29	21 07	23 24	21 49	01 46
25	01 49	02 S 55	03 55	13 27	18 43	23 29	21 06	23 24	21 48	01 45
26	02 13	03 N 47	04 51	13 16	18 32	23 29	21 05	23 24	21 48	01 45
27	02 36	10 23	05 48	13 04	18 20	23 29	21 05	23 24	21 48	01 44
28	03 00	16 27	06 45	12 52	18 09	23 29	21 04	23 24	21 48	01 44
29	03 23	21 30	07 41	12 40	17 57	23 29	21 04	23 24	21 48	01 43
30	03 46	25 07	08 36	12 27	17 44	23 29	21 03	23 24	21 48	01 42
31	04 N 10	27 N 00	09 N 30	12 S 14	17 S 32	23 N 29	21 S 03	23 S 24	21 S 48	01 S 42

ZODIAC SIGN ENTRIES

Date	h	m	Planets
01	01	43	☽ ♉
03	03	37	☽ ♊
03	17	14	☿ ♓
03	17	52	♀ ♒
05	07	02	☽ ♋
07	12	24	☽ ♌
09	19	47	☽ ♍
11	15	54	♂ ♒
12	05	09	☽ ♎
14	16	25	☽ ♏
17	04	56	☽ ♐
19	17	01	☽ ♑
20	00	04	☿ ♈
20	21	19	☉ ♈
22	02	31	☽ ♒
24	08	09	☽ ♓
26	10	15	☽ ♈
28	10	26	☽ ♉
30	10	42	☽ ♊

LATITUDES

Date	Mercury ☿ ° '	Venus ♀ ° '	Mars ♂ ° '	Jupiter ♃ ° '	Saturn ♄ ° '	Uranus ♅ ° '	Neptune ♆ ° '	Pluto ♇ ° '
01	02 S 07	05 N 00	00 S 46	00 N 01	00 N 14	00 S 17	00 N 51	15 N 44
04	02 10	04 37	00 48	00 02	00 14	00 17	00 51	15 45
07	02 09	04 14	00 51	00 02	00 13	00 17	00 51	15 47
10	02 04	03 51	00 53	00 02	00 13	00 17	00 51	15 48
13	01 55	03 28	00 56	00 03	00 13	00 17	00 51	15 49
16	01 41	03 06	00 58	00 03	00 13	00 17	00 51	15 50
19	01 22	02 44	01 01	00 03	00 13	00 18	00 51	15 52
22	00 59	02 22	01 03	00 04	00 13	00 18	00 51	15 53
25	00 S 31	02 01	01 06	00 04	00 12	00 18	00 51	15 54
28	00 N 01	01 41	01 08	00 04	00 12	00 18	00 51	15 55
31	00 N 36	01 N 21	01 S 10	00 N 05	00 N 12	00 S 18	00 N 52	15 N 56

DATA

Julian Date	2447952
Delta T	+57 seconds
Ayanamsa	23° 43' 23"
Synetic vernal point	05° ♓ 23' 36"
True obliquity of ecliptic	23° 26' 33"

MOON'S PHASES, APSIDES AND POSITIONS ☽

Date	h	m	Phase	Longitude ° '	Eclipse Indicator
04	02	05	☽	13 ♊ 14	
11	10	59	○	20 ♍ 37	
19	14	30	☾	28 ♐ 43	
26	19	48	●	05 ♈ 53	

Day	h	m		
16	07	41	Apogee	
28	08	01	Perigee	
04	18	54	Max dec	27° N 27'
11	15	15	0S	
19	03	07	Max dec	27° S 21'
25	22	29	0N	

LONGITUDES

Date	Chiron ⚷ ° '	Ceres ⚳ ° '	Pallas ⚴ ° '	Juno ⚵ ° '	Vesta ⚶ ° '	Black Moon Lilith ⚸ ° '
01	10 ♋ 43	23 ♊ 06	21 ♈ 27	24 ♏ 44	16 ♓ 20	13 ♏ 04
11	10 ♋ 35	24 ♊ 52	25 ♈ 48	25 ♏ 06	21 ♓ 12	14 ♏ 12
21	10 ♋ 36	27 ♊ 06	00 ♉ 22	24 ♏ 54	26 ♓ 03	15 ♏ 19
31	10 ♋ 46	29 ♊ 45	05 ♉ 07	24 ♏ 09	00 ♈ 51	16 ♏ 26

All ephemeris data is given at 12.00 UT and the Moon's longitude is additionally given for 24.00 UT

ASPECTARIAN

h m Aspects

01 Thursday
03 06 ☽ ⚹ ♃
15 50 ☽ Q ☿
16 15 ☽ △ ♅
19 50 ☽ ⚹ ☉
02 Friday
00 54 ☽ △ ♆
02 49 ☽ # ♄
03 57 ☽ ∠ ♃
05 29 ☽ # ♆
07 07 ☽ ☍ ♇
07 38 ☽ # ♂
14 38 ☽ △ ♄
15 08 ☽ # ♅
15 19 ☽ ∥ ♃
16 30 ☽ △ ♂
16 35 ☿ ⊥ ♄
17 16 ☽ ⚼ ♅
17 19 ☽ Q ☉
19 00 ☽ ⊥ ♃
03 Saturday
01 48 ☽ □ ☿
02 00 ☽ ⚼ ♆
02 55 ☽ △ ♀
03 36 ☿ ∠ ♆
05 06 ☽ ⊻ ♃
08 27 ☽ ± ♅
08 50 ♀ Q ♇
16 04 ☽ ⚼ ♄
16 49 ☿ ⊻ ♀
17 15 ☽ ± ♆
18 39 ☽ ⊼ ♅
19 12 ☽ ⚼ ♂
20 41 ☿ ⊥ ♂
04 Sunday
02 05 ☽ □ ☉
03 30 ☽ ⊼ ♆
05 41 ☽ ⚼ ♀
05 58 ☿ △ ♃
07 32 ☽ ± ♄
09 48 ☽ ⊼ ♇
11 27 ☽ ± ♂
17 56 ☽ ⊼ ♄
20 09 ☽ ± ♇
22 01 ☽ ± ♀
22 20 ☉ ⚹ ♆
22 24 ☽ ⊼ ♂
05 Monday
01 38 ♀ ⊼ ♃
08 40 ☽ ☌ ♃
09 02 ☽ ⊼ ♀
11 48 ☽ ⚼ ♇
12 25 ☽ △ ☿
13 28 ♂ ∥ ♆
22 45 ☽ ☍ ♅
06 Tuesday
07 54 ☽ ☍ ♆
10 32 ☽ △ ☉
14 19 ☽ △ ♇
18 54 ☽ ⚼ ☿
23 08 ☽ ☍ ♄
07 Wednesday
00 08 ☽ ∥ ♃
00 27 ☽ # ♅
06 33 ☽ ☍ ♂
10 50 ☽ # ♆
12 18 ☽ # ♂
13 35 ☽ ± ☿
14 04 ☽ # ♄
14 13 ☽ ⊻ ♃
15 41 ☽ ⚼ ☉
17 37 ☽ ☍ ♀
08 Thursday
01 11 ☽ ⊥ ♃
02 12 ☽ ⊼ ☿
04 35 ☿ ∠ ♄
04 51 ☽ ⊼ ♅
09 30 ☽ ± ☉
13 05 ☉ △ ♇
14 18 ☽ ⊼ ♆
15 54 ☽ ± ♅
17 46 ☽ ∠ ♃
20 50 ☽ □ ♇
21 00 ☽ # ♀
21 29 ☽ ⊼ ☉
21 57 ☿ ⚹ ♅
09 Friday
01 25 ☽ ± ♆
06 23 ☽ ⊼ ♄
08 39 ☽ ⚼ ♅
15 08 ☿ ⊥ ♀
17 05 ☽ ⊼ ♂
17 41 ☽ ± ♄
18 15 ☽ ⚼ ♆
21 50 ☽ ⚹ ♃
22 22 ♂ ∥ ♄
23 48 ☽ # ☿
10 Saturday
04 46 ☽ ⊼ ♀
05 04 ☽ ± ♂
06 30 ☽ Q ♇
10 46 ☽ ⚼ ♄
12 57 ☽ △ ♅
16 54 ☽ ± ♀
19 21 ☽ ☍ ☿
20 39 ☽ Q ♃
22 42 ☽ △ ♆
22 53 ☽ # ☉
23 14 ☽ ⚼ ♂
11 Sunday
05 13 ♂ Q ♇
05 20 ☽ ⚹ ♇
07 20 ☽ # ♇
10 59 ☽ ☍ ☉
11 19 ☽ ⚼ ♀
15 39 ☽ △ ♄
18 34 ☿ ⚹ ♆
23 12 ☽ ∥ ♇
23 26 ☉ Q ♅
12 Monday
05 40 ☽ ∥ ☉
05 59 ☽ △ ♂
07 29 ☽ □ ♃
08 09 ☉ ∠ ♀
10 19 ☽ ∠ ♇
13 07 ☿ ∠ ♂
18 30 ☽ △ ♀
21 59 ☽ ∥ ☿
23 02 ☽ □ ♅
13 Tuesday
03 57 ☽ ⊥ ♇
06 24 ♀ ± ♃
08 42 ♂ ⊼ ♃
09 05 ☽ □ ♆
14 41 ☿ △ ♇
15 47 ☽ ⊻ ♇
16 00 ☽ ⊼ ☿
14 Wednesday
01 58 ☉ ⚹ ♄
02 54 ☽ □ ♄
02 58 ☽ ⊼ ☉
06 08 ☽ ± ☿
10 29 ☽ ∥ ♀
10 48 ☽ Q ♅
16 02 ☽ ± ☉
19 06 ☽ △ ♃
20 59 ☽ Q ♆
21 12 ☽ □ ♂
15 Thursday
03 33 ☽ ⚼ ☿
10 41 ☽ □ ♀
10 56 ☽ ⚹ ♅
11 46 ☽ ⚼ ☉
12 01 ☿ Q ♅
14 23 ♀ ⊻ ♅
15 13 ☽ Q ♄
17 07 ☽ ∥ ♂
21 11 ☽ ⚹ ♆
22 13 ☽ ∥ ♄
16 Friday
01 27 ☽ ⚼ ♃
02 26 ☽ ∥ ♆
03 52 ☽ ☌ ♇
05 03 ☉ ∥ ♇
11 54 ☽ Q ♂
14 17 ☽ ∥ ♅
14 53 ☽ # ♃
15 32 ☿ ⚹ ♄
15 38 ☽ ⚹ ♄
15 39 ☽ △ ☿
17 18 ☽ ∠ ♅
19 45 ☽ ± ♃
20 51 ☽ △ ☉
17 Saturday
01 55 ☽ Q ♀
03 34 ☽ ∠ ♆
04 33 ☉ Q ♆
07 57 ☽ ⊼ ♃
11 32 ☽ ⊥ ♅
13 47 ☽ ⚹ ♂
21 46 ☽ ⊥ ♆
22 08 ☽ ∠ ♄
23 40 ☽ ⊻ ♅
18 Sunday
03 30 ☿ Q ♆
04 04 ☽ ⚹ ♀
05 53 ☿ ∠ ♀
09 51 ☽ ⊻ ♆
16 19 ☽ ⊥ ♄
16 20 ☽ ⊻ ♇
21 56 ☽ ∠ ♂
19 Monday
02 39 ☉ ☌ ☿
04 17 ☽ ⊥ ♇
04 22 ☽ ⊻ ♄
07 39 ☿ ∥ ♇
12 25 ☽ ∠ ♀
14 30 ☽ □ ☉
15 39 ☽ □ ☿
16 57 ☽ ⊥ ♂
20 18 ☽ ☍ ♃
22 03 ☽ ∠ ♇
20 Tuesday
05 29 ☽ ⊻ ♂
07 30 ☽ ⊥ ♀
11 17 ☽ ☌ ♅
20 01 ☽ ⊻ ♀
21 01 ☽ ☌ ♆
21 06 ☿ □ ♃
21 Wednesday
03 03 ☽ ⚹ ♇
04 31 ☽ Q ☉
04 39 ☉ # ☿
06 32 ☿ ⚼ ♇
09 11 ☽ Q ☿
09 48 ♀ ⊻ ♆
13 02 ☽ # ♃
13 40 ☽ ∥ ♅
14 53 ☽ ☌ ♄
15 42 ☉ ∥ ☿
22 Thursday
00 36 ☽ ∥ ♆
01 37 ☽ Q ♇
04 48 ☽ ∥ ♄
04 56 ☽ ⚹ ☉
05 57 ♂ ± ♃
05 57 ☽ ⊼ ♃
11 28 ♀ ⊥ ♅
11 41 ☽ ⚹ ☿
15 30 ☽ ∥ ♂
17 00 ☽ ± ♃
17 36 ☽ ☌ ♂
19 07 ☉ □ ♃
19 43 ☽ ⊻ ♅
22 20 ☿ Q ♄
23 Friday
04 47 ☽ ⊻ ♆
05 48 ☿ # ♇
06 31 ☽ ⊥ ♅
07 55 ☽ ☌ ♀
09 18 ☽ ⚼ ♃
09 38 ☉ ⚼ ♇
10 16 ☽ □ ♇
10 19 ☽ ∠ ☉
15 27 ☽ ⊥ ♆
17 41 ☽ ∥ ♀
19 20 ☽ ∠ ☿
21 36 ☽ ⊻ ♄
22 26 ☽ ∠ ♅
24 Saturday
03 21 ☽ ⊥ ☉
05 36 ♀ ⚼ ♃
07 09 ☽ ∠ ♆
08 04 ☽ ⊥ ♄
08 11 ♂ ⊻ ♅
11 39 ☽ △ ♃
13 35 ☽ ⊥ ☿
14 28 ☽ ⊻ ☉
16 27 ☿ □ ♅
17 13 ♀ □ ♇
21 09 ☿ ⚹ ♂
23 29 ☽ ∠ ♄
25 Sunday
00 12 ☽ ⚹ ♅
01 04 ☽ ⊻ ♂
01 28 ☽ ⊻ ☿
08 17 ☉ # ♇
08 37 ☽ ⚹ ♆
08 49 ☽ # ☿
11 40 ☽ ⊥ ♂
13 39 ☽ △ ♇
15 07 ☽ ⊻ ♀
15 44 ☽ # ☉
16 11 ☽ ∥ ♇
16 37 ☿ ± ♇
20 14 ☽ Q ♅
26 Monday
00 33 ☽ ⚹ ♄
01 42 ☽ ⊥ ♀
03 19 ☽ ∠ ♂
04 28 ☽ Q ♆
04 44 ☽ # ♇
06 02 ☽ ∥ ☉
13 52 ☽ □ ♃
14 13 ☽ ⚼ ♇
16 28 ☽ ∥ ☿
17 22 ☽ ∠ ♀
19 48 ☽ ☌ ☉
20 10 ☽ Q ♄
27 Tuesday
01 33 ☽ □ ♅
01 34 ☉ Q ♄
04 20 ☿ □ ♆
04 44 ☽ ± ♇
04 55 ☽ ⚹ ♂
09 36 ☽ □ ♆
10 25 ☽ ☌ ☿
14 20 ☽ ⊼ ♇
19 00 ☽ Q ♃
19 09 ☽ ⚹ ♀
20 18 ♀ ⊥ ♆
21 57 ☽ # ♀
28 Wednesday
01 08 ☽ Q ♂
01 09 ☽ □ ♄
11 22 ♃ ⚼ ♇
14 17 ☽ ⚹ ♃
15 27 ☿ ⊼ ♇
15 38 ☽ Q ♀
19 11 ☽ # ♂
23 14 ☽ ⊻ ☉
29 Thursday
01 35 ☽ △ ♅
07 24 ☽ □ ♂
09 32 ☽ ⊥ ☉
09 36 ☽ △ ♆
09 41 ☽ # ♄
13 42 ☽ # ♆
14 15 ☽ ☍ ♇
14 28 ☽ ∠ ♃
17 48 ☽ ⊻ ☿
22 32 ☽ □ ♀
23 18 ☽ # ♅
23 54 ☽ ∥ ♃
30 Friday
01 04 ☽ ∠ ☉
01 26 ☽ △ ♄
01 45 ☽ ⚼ ♅
04 59 ☽ ⊥ ☿
05 07 ☽ ⊥ ♃
06 02 ☿ Q ♃
09 52 ☽ ⚼ ♆
11 24 ☉ □ ♅
14 58 ☽ ⊻ ♃
16 26 ☽ ± ♅
21 58 ☽ ∠ ☿
31 Saturday
00 39 ☽ ± ♆
02 04 ☽ ⚼ ♄
02 19 ☽ ⊼ ♅
03 24 ☽ ⚹ ☉
04 16 ♂ ⊻ ♆
10 37 ☽ ⊼ ♆
10 57 ☽ △ ♂
15 21 ☽ ⊼ ♇
17 07 ☽ ± ♄
20 20 ♀ ⊻ ♄
23 09 ♀ ∠ ♅

APRIL 1990

LONGITUDES

Date	Sidereal time h m s	Sun ☉ ° ' "	Moon ☽ ° ' "	Moon ☽ 24.00 ° ' "	Mercury ☿ ° '	Venus ♀ ° '	Mars ♂ ° '	Jupiter ♃ ° '	Saturn ♄ ° '	Uranus ♅ ° '	Neptune ♆ ° '	Pluto ♇ ° '
01	00 38 20	11 ♈ 30 00	29 ♊ 31 07	06 ♋ 28 15	24 ♈ 54	25 ♒ 03	15 ♒ 29	02 ♋ 49	24 ♑ 26	09 ♑ 31	14 ♑ 31	17 ♏ 19
02	00 42 17	12 29 11	13 ♋ 19 48	20 ♋ 05 57	26 45	26 03	16 14	02 55	24 29	09 32	14 31	17 R 18
03	00 46 13	13 28 21	26 ♋ 46 58	03 ♌ 23 08	28 ♈ 34	27 04	16 59	03 02	24 32	09 33	14 32	17 17
04	00 50 10	14 27 28	09 ♌ 54 49	16 ♌ 22 23	00 ♉ 19	28 05	17 44	03 09	24 35	09 33	14 32	17 15
05	00 54 07	15 26 33	22 ♌ 46 11	29 ♌ 06 35	02 01	29 ♒ 06	18 28	03 16	24 38	09 34	14 32	17 14
06	00 58 03	16 25 35	05 ♍ 23 55	11 ♍ 38 28	03 38	00 ♓ 07	19 13	03 23	24 41	09 34	14 33	17 13
07	01 02 00	17 24 35	17 ♍ 50 30	24 ♍ 00 16	05 11	01 09	19 58	03 30	24 44	09 34	14 33	17 11
08	01 05 56	18 23 33	00 ♎ 07 58	06 ♎ 13 47	06 40	02 11	20 43	03 37	24 46	09 35	14 33	17 10
09	01 09 53	19 22 29	12 ♎ 17 53	18 ♎ 20 23	08 03	03 13	21 28	03 45	24 49	09 35	14 34	17 08
10	01 13 49	20 21 23	24 ♎ 21 26	00 ♏ 21 10	09 21	04 16	22 12	03 53	24 51	09 35	14 34	17 07
11	01 17 46	21 20 15	06 ♏ 19 45	12 ♏ 17 19	10 34	05 19	22 57	04 00	24 53	09 35	14 34	17 05
12	01 21 42	22 19 05	18 ♏ 14 05	24 ♏ 10 15	11 42	06 22	23 42	04 08	24 56	09 35	14 34	17 04
13	01 25 39	23 17 53	00 ♐ 06 06	06 ♐ 01 56	12 44	07 25	24 27	04 16	24 58	09 35	14 34	17 02
14	01 29 36	24 16 39	11 ♐ 58 05	17 ♐ 54 57	13 40	08 29	25 12	04 24	25 00	09 R 35	14 34	17 01
15	01 33 32	25 15 24	23 ♐ 53 00	29 ♐ 52 41	14 30	09 33	25 57	04 33	25 02	09 35	14 34	16 59
16	01 37 29	26 14 07	05 ♑ 54 32	11 ♑ 59 09	15 14	10 37	26 41	04 41	25 04	09 35	14 34	16 58
17	01 41 25	27 12 48	18 ♑ 07 06	24 ♑ 18 59	15 52	11 41	27 26	04 50	25 05	09 35	14 R 34	16 56
18	01 45 22	28 11 27	00 ♒ 35 28	06 ♒ 57 08	16 24	12 46	28 11	04 58	25 07	09 35	14 34	16 55
19	01 49 18	29 ♈ 10 05	13 ♒ 24 35	19 ♒ 58 21	16 49	13 51	28 56	05 07	25 09	09 35	14 34	16 53
20	01 53 15	00 ♉ 08 41	26 ♒ 38 55	03 ♓ 26 37	17 09	14 56	29 ♒ 41	05 16	25 10	09 34	14 34	16 52
21	01 57 11	01 07 15	10 ♓ 21 42	17 ♓ 24 14	17 22	16 01	00 ♓ 26	05 25	25 11	09 34	14 34	16 50
22	02 01 08	02 05 48	24 ♓ 34 04	01 ♈ 50 53	17 30	17 06	01 11	05 34	25 13	09 33	14 34	16 48
23	02 05 05	03 04 19	09 ♈ 14 04	16 ♈ 42 50	17 R 31	18 12	01 56	05 44	25 14	09 33	14 34	16 47
24	02 09 01	04 02 48	24 ♈ 16 07	01 ♉ 52 43	17 27	19 17	02 41	05 53	25 15	09 33	14 33	16 45
25	02 12 58	05 01 16	09 ♉ 31 15	17 ♉ 10 17	17 18	20 23	03 25	06 02	25 16	09 32	14 33	16 44
26	02 16 54	05 59 42	24 ♉ 48 21	02 ♊ 24 05	17 03	21 29	04 10	06 12	25 17	09 31	14 33	16 42
27	02 20 51	06 58 06	09 ♊ 56 11	17 ♊ 23 34	16 44	22 36	04 55	06 22	25 18	09 31	14 32	16 40
28	02 24 47	07 56 27	24 ♊ 45 22	02 ♋ 00 55	16 20	23 42	05 40	06 32	25 18	09 30	14 32	16 39
29	02 28 44	08 54 47	09 ♋ 09 48	16 ♋ 11 49	15 53	24 48	06 25	06 41	25 19	09 29	14 32	16 37
30	02 32 40	09 ♉ 53 05	23 ♋ 06 56	29 ♋ 55 18	15 ♉ 22	25 ♓ 55	07 ♓ 10	06 ♋ 51	25 ♑ 19	09 ♑ 28	14 ♑ 31	16 ♏ 35

Date	Moon True ☊ ° '	Moon Mean ☊ ° '	Moon ☽ Latitude ° '
01	14 ♒ 47	13 ♒ 40	03 N 38
02	14 D 46	13 37	02 40
03	14 47	13 34	01 35
04	14 47	13 30	00 N 26
05	14 R 47	13 27	00 S 43
06	14 45	13 24	01 48
07	14 40	13 21	02 47
08	14 33	13 18	03 37
09	14 23	13 15	04 16
10	14 12	13 11	04 44
11	13 59	13 08	04 59
12	13 46	13 05	05 01
13	13 35	13 02	04 49
14	13 26	12 59	04 25
15	13 19	12 55	03 50
16	13 15	12 52	03 04
17	13 13	12 49	02 09
18	13 D 13	12 46	01 S 06
19	13 R 13	12 43	00 N 01
20	13 12	12 40	01 11
21	13 10	12 36	02 18
22	13 05	12 33	03 20
23	12 58	12 30	04 10
24	12 48	12 27	04 45
25	12 37	12 24	05 00
26	12 26	12 21	04 53
27	12 17	12 17	04 26
28	12 09	12 14	03 42
29	12 05	12 11	02 44
30	12 ♒ 03	12 ♒ 08	01 N 38

DECLINATIONS

Date	Sun ☉ ° '	Moon ☽ ° '	Mercury ☿ ° '	Venus ♀ ° '	Mars ♂ ° '	Jupiter ♃ ° '	Saturn ♄ ° '	Uranus ♅ ° '	Neptune ♆ ° '	Pluto ♇ ° '
01	04 N 33	27 N 04	10 N 23	12 S 00	17 S 20	23 N 29	21 S 02	23 S 24	21 S 48	01 S 41
02	04 56	25 26	11 15	11 45	17 07	23 29	21 02	23 24	21 48	01 41
03	05 19	22 21	12 04	11 31	16 54	23 29	21 01	23 24	21 48	01 40
04	05 42	18 11	12 53	11 16	16 41	23 29	21 01	23 24	21 48	01 39
05	06 05	13 15	13 38	11 00	16 28	23 29	21 00	23 24	21 48	01 39
06	06 28	07 52	14 22	10 44	16 15	23 29	21 00	23 24	21 48	01 38
07	06 50	02 N 15	15 04	10 28	16 02	23 29	20 59	23 24	21 47	01 38
08	07 13	03 S 22	15 42	10 11	15 48	23 29	20 59	23 24	21 47	01 37
09	07 35	08 48	16 19	09 54	15 35	23 29	20 59	23 24	21 47	01 36
10	07 57	13 50	16 52	09 37	15 21	23 29	20 58	23 24	21 47	01 36
11	08 19	18 20	17 23	09 19	15 07	23 28	20 58	23 24	21 47	01 35
12	08 41	22 04	17 50	09 01	14 53	23 28	20 58	23 24	21 47	01 35
13	09 03	24 53	18 15	08 42	14 39	23 28	20 57	23 24	21 47	01 34
14	09 25	26 37	18 37	08 23	14 24	23 28	20 57	23 24	21 47	01 34
15	09 46	27 08	18 56	08 04	14 10	23 28	20 57	23 24	21 47	01 33
16	10 08	26 22	19 12	07 44	13 55	23 28	20 56	23 24	21 47	01 32
17	10 29	24 21	19 25	07 24	13 41	23 27	20 56	23 24	21 47	01 32
18	10 50	21 06	19 35	07 04	13 26	23 27	20 56	23 24	21 47	01 31
19	11 11	16 47	19 42	06 44	13 11	23 27	20 56	23 24	21 47	01 31
20	11 31	11 32	19 46	06 23	12 56	23 27	20 55	23 24	21 47	01 30
21	11 52	05 S 33	19 46	06 02	12 41	23 26	20 55	23 24	21 47	01 30
22	12 12	00 N 54	19 44	05 40	12 25	23 26	20 55	23 24	21 47	01 29
23	12 32	07 30	19 39	05 18	12 10	23 26	20 55	23 24	21 47	01 29
24	12 52	13 49	19 31	04 57	11 55	23 25	20 55	23 24	21 47	01 28
25	13 12	19 24	19 20	04 34	11 39	23 25	20 55	23 24	21 47	01 28
26	13 31	23 43	19 07	04 12	11 23	23 25	20 55	23 24	21 47	01 27
27	13 51	26 20	18 51	03 49	11 08	23 24	20 55	23 25	21 47	01 27
28	14 10	27 02	18 32	03 26	10 52	23 24	20 54	23 25	21 47	01 26
29	14 28	25 51	18 11	03 03	10 36	23 24	20 54	23 25	21 47	01 26
30	14 N 47	23 N 04	17 N 48	02 S 40	10 S 20	23 N 23	20 S 54	23 S 25	21 S 47	01 S 25

ZODIAC SIGN ENTRIES

Date	h m	Planets
01	12 50	☽ ♋
03	17 50	☽ ♌
04	07 35	☿ ♉
06	01 42	☽ ♍
06	09 13	♀ ♓
08	11 44	☽ ♎
10	23 18	☽ ♏
13	11 48	☽ ♐
16	00 15	☽ ♑
18	10 53	☽ ♒
20	08 27	☉ ♉
20	17 57	☽ ♓
20	22 09	♂ ♓
22	20 58	☽ ♈
24	21 03	☽ ♉
26	20 12	☽ ♊
28	20 39	☽ ♋

LATITUDES

Date	Mercury ☿ ° '	Venus ♀ ° '	Mars ♂ ° '	Jupiter ♃ ° '	Saturn ♄ ° '	Uranus ♅ ° '	Neptune ♆ ° '	Pluto ♇ ° '
01	00 N 48	01 N 15	01 S 11	00 N 05	00 N 12	00 S 18	00 N 52	15 N 56
04	01 23	00 56	01 14	00 05	00 12	00 18	00 52	15 57
07	01 55	00 39	01 16	00 05	00 12	00 18	00 52	15 57
10	02 23	00 22	01 18	00 05	00 12	00 18	00 52	15 58
13	02 44	00 N 05	01 21	00 06	00 11	00 18	00 52	15 59
16	02 56	00 S 10	01 23	00 06	00 11	00 18	00 52	15 59
19	02 57	00 24	01 26	00 06	00 11	00 18	00 52	15 59
22	02 48	00 37	01 28	00 07	00 11	00 18	00 52	16 00
25	02 26	00 50	01 30	00 07	00 11	00 18	00 52	16 00
28	01 53	01 01	01 32	00 07	00 10	00 19	00 52	16 00
31	01 N 10	01 S 12	01 S 34	00 N 07	00 N 10	00 S 19	00 N 52	16 N 00

DATA

Julian Date	2447983
Delta T	+57 seconds
Ayanamsa	23° 43' 27"
Synetic vernal point	05° ♓ 23' 32"
True obliquity of ecliptic	23° 26' 33"

MOON'S PHASES, APSIDES AND POSITIONS ☽

Date	h m	Phase	Longitude ° '	Eclipse Indicator
02	10 24	☽	12 ♋ 25	
10	03 18	○	20 ♎ 00	
18	07 03	☾	27 ♑ 59	
25	04 27	●	04 ♉ 43	

Day	h m		
12	20 16	Apogee	
25	16 44	Perigee	
01	00 49	Max dec	27° N 16'
07	21 34	0S	
15	09 51	Max dec	27° S 08'
22	08 42	0N	
28	08 38	Max dec	27° N 03'

LONGITUDES

Date	Chiron ⚷ ° '	Ceres ⚳ ° '	Pallas ⚴ ° '	Juno ⚵ ° '	Vesta ⚶ ° '	Black Moon Lilith ⚸ ° '
01	10 ♋ 48	00 ♋ 02	05 ♉ 36	24 ♏ 03	01 ♈ 20	16 ♏ 33
11	11 ♋ 07	03 ♋ 03	10 ♉ 32	22 ♏ 42	06 ♈ 04	17 ♏ 40
21	11 ♋ 36	06 ♋ 22	15 ♉ 38	20 ♏ 54	10 ♈ 45	18 ♏ 47
31	12 ♋ 11	09 ♋ 54	20 ♉ 53	18 ♏ 47	15 ♈ 21	19 ♏ 55

All ephemeris data is given at 12.00 UT and the Moon's longitude is additionally given for 24.00 UT

ASPECTARIAN

h m Aspects

01 Sunday
00 58 ☽ Q ☉
01 27 ☽ ± ♇
02 54 ☽ ⚹ ☿
03 18 ☽ ⊼ ♄
03 48 ☽ △ ♀
05 56 ☿ □ ♄
07 41 ☉ ± ♇
13 11 ♂ ⊥ ♅
13 45 ☽ bQ ♂
16 10 ☿ ⚹ ♀
16 48 ☽ bQ ♇
17 43 ☽ ☌ ♃

02 Monday
02 44 ☽ Q ☿
05 20 ☽ ☍ ♅
06 15 ☽ ± ♂
07 41 ☽ bQ ♀
10 24 ☽ □ ☉
14 06 ☽ ☍ ♆
17 26 ☽ ⊼ ♂
19 00 ☽ △ ♇
23 26 ☿ # ♀

03 Tuesday
00 53 ☽ ± ♀
04 13 ☽ ∥ ♃
04 52 ☽ # ♅
07 56 ☽ ☍ ♄
12 33 ☽ ⊼ ♀
15 34 ☽ # ♆
15 44 ☽ □ ☿
20 18 ☽ # ♄
21 16 ♂ □ ♇
21 42 ☿ Q ♂
23 27 ☽ ⊻ ♃

04 Wednesday
10 34 ☽ ⊥ ♃
11 20 ☽ ⊼ ♅
13 53 ☉ □ ♆
19 58 ☽ # ♂
20 35 ☽ ⊼ ♆
21 08 ☽ △ ☉
22 28 ☽ ± ♅

05 Thursday
01 37 ☽ □ ♇
03 25 ☽ ☍ ♂
03 27 ☽ ∠ ♃
03 59 ♂ bQ ♃
07 48 ☽ ± ♆
10 27 ☽ ∥ ☿
15 23 ☽ bQ ♅
15 32 ☽ ⊼ ♄
22 32 ♀ ∠ ♆
22 44 ☽ # ♀

06 Friday
00 50 ☽ bQ ♆
01 02 ☽ ☍ ♀
02 57 ☽ ± ♄
03 46 ☽ bQ ☉
07 52 ☿ ⚹ ♃
08 06 ☽ ⚹ ♃
08 08 ☽ △ ☿
11 38 ☽ Q ♇
17 40 ☽ ∥ ☉
20 00 ☽ △ ♅
20 15 ☽ bQ ♄
22 29 ☽ ± ☉

07 Saturday
01 44 ♀ ⊥ ♄
05 37 ☽ △ ♆
06 39 ☉ ⊼ ♇
07 25 ☽ Q ♃
10 44 ☽ ⚹ ♇
11 05 ☽ ⊼ ☉
14 39 ☽ # ♇
16 24 ☽ ⊼ ♂
17 11 ☽ bQ ☿

08 Sunday
01 27 ☽ △ ♄
04 29 ☽ ∥ ♇
04 51 ☽ ± ♂
07 00 ♂ ⊥ ♆
13 10 ☽ ± ☿
14 48 ☿ # ♂
15 59 ☽ ∠ ♇
16 24 ☽ ⊼ ♀
18 56 ☽ □ ♃
23 42 ☽ bQ ♂

09 Monday
02 32 ☽ ⊼ ☿
05 20 ☽ ± ♀
06 08 ☽ # ☉
06 37 ☽ □ ♅
09 42 ☽ ⊥ ♇
16 29 ☽ □ ♆
16 50 ☽ ∥ ♀
21 35 ☽ ⊻ ♇

10 Tuesday
00 53 ☽ bQ ♀
01 48 ♀ △ ♃
03 18 ☽ ☍ ☉
07 25 ☽ △ ♂
12 59 ☽ □ ♄
16 22 ☿ △ ♅
18 27 ☽ Q ♅
19 19 ☽ ∥ ♂

11 Wednesday
04 26 ☽ Q ♆
05 54 ☽ # ☿
07 17 ☽ △ ♃
09 46 ☽ △ ♀
18 33 ☽ ⚹ ♅
21 28 ☽ ☍ ☿

12 Thursday
01 15 ☽ Q ♄
04 18 ☽ ∥ ♄
04 36 ☽ ⚹ ♆
06 53 ☉ Q ♃
09 39 ☽ ☌ ♇
09 59 ☽ ∥ ♆
13 51 ☽ bQ ♃
21 00 ☽ ⊼ ☉
22 16 ☽ ∥ ♅
22 53 ☽ # ♃
23 30 ☉ # ♀
23 47 ☽ □ ♂

13 Friday
00 51 ☽ ∠ ♅
01 34 ☽ ⚹ ♄
08 15 ☽ ± ♃
10 14 ☽ ± ☉
10 56 ☽ ∠ ♆
16 32 ♂ ∠ ♅
19 03 ☽ ⊥ ♅
20 32 ☽ ⊼ ♃
22 21 ♅ St R

14 Saturday
04 16 ☽ □ ♀
05 08 ☽ ⊥ ♆
05 19 ♂ ⊻ ♄
06 04 ☽ bQ ☉
07 12 ☽ ⊻ ♅
08 00 ☽ ∠ ♄
14 39 ☽ Q ♂
15 41 ☽ ⊼ ☿
17 16 ☽ ⊻ ♆
22 10 ☽ ⊻ ♇

15 Sunday
02 13 ☽ ⊥ ♄
04 42 ☽ ± ☿
06 14 ☉ □ ♄
10 13 ☽ ⊥ ♇
12 53 ♀ ⚹ ♅
14 18 ☽ ⊻ ♄
14 26 ☿ △ ♆
15 00 ☽ △ ☉
16 24 ☽ ⚹ ♂
20 04 ☽ Q ♀
23 07 ♀ ∠ ♄

16 Monday
00 00 ☽ bQ ☿
04 11 ☽ ∠ ♇
09 33 ☽ ☍ ♃
12 56 ♆ St R
19 16 ☽ ☌ ♅
22 12 ☽ ⚹ ♀

17 Tuesday
00 10 ☽ ∠ ♂
05 05 ☽ ☌ ♆
07 23 ☽ △ ☿
09 42 ☽ ⚹ ♇
18 52 ☽ ⊥ ♂
19 31 ☽ # ♃
19 57 ☽ ∥ ♅

18 Wednesday
01 31 ☽ ☌ ♄
06 07 ☽ ∠ ♀
07 03 ☽ □ ☉
07 08 ☽ ⊻ ♂
07 36 ☽ ∥ ♆
08 49 ☽ Q ♇
11 40 ☉ ⚹ ♂
13 06 ☽ ∥ ♄
20 23 ☽ ⊼ ♃
20 51 ☽ # ☿

19 Thursday
00 43 ☽ ⊥ ♀
04 54 ☽ ⊻ ♅
07 43 ☽ ± ♃
12 53 ☽ ⊻ ♀
14 08 ☽ ⊻ ♆
15 59 ☽ ⊥ ♅
16 02 ☿ ☍ ♇
18 22 ☽ □ ♇
18 26 ☽ □ ☿
19 27 ☽ Q ☉

20 Friday
00 24 ☽ bQ ♃
01 05 ☽ ⊥ ♆
04 05 ♀ ⚹ ♆
05 39 ☽ ∥ ♂
08 17 ☽ ∠ ♅
08 22 ♂ ∠ ♆
09 21 ☽ ⊻ ♄
12 01 ☽ # ☉
17 11 ☽ ∠ ♆
17 42 ☽ ☌ ♂
18 41 ☽ ⚹ ☉
20 01 ☽ ⊥ ♄

21 Saturday
03 15 ☽ Q ☿
03 21 ☽ △ ♃
10 03 ☽ ∥ ♀
10 38 ☽ ⚹ ♅
11 42 ☽ ∠ ♄
19 11 ☽ ⚹ ♆
22 27 ☽ ☌ ♀
22 33 ☽ ∠ ☉
23 01 ☽ △ ♇

22 Sunday
00 04 ☽ ⚹ ☿
03 14 ☽ ∥ ♇
05 37 ♀ △ ♇
06 59 ☽ Q ♅
10 32 ☉ ∠ ♀
13 03 ♂ ⊥ ♄
13 04 ☽ ⚹ ♄
14 08 ☽ # ♇
14 43 ☽ ⊥ ☉
15 19 ☽ Q ♆
20 56 ☉ # ♂
21 07 ☿ ⚹ ♀
23 30 ☽ ⊻ ♂
23 55 ☽ bQ ♇

23 Monday
01 06 ☽ ∠ ☿
01 17 ☽ ⊻ ☉
04 25 ☽ # ♀
06 16 ☽ □ ♃
06 55 ☿ St R
08 45 ☽ Q ♄
09 47 ☽ ⊥ ♂
12 31 ☽ □ ♅
14 29 ☽ ± ♇
15 41 ☽ ⊥ ☿
20 33 ☽ □ ♆

24 Tuesday
00 05 ☽ ⊼ ♇
00 59 ☽ ∠ ♂
01 15 ☽ ⊻ ☿
03 29 ☽ ⊻ ♀
04 50 ☽ # ♂
08 02 ☽ ∥ ☉
11 23 ☽ Q ♃
13 33 ☽ □ ♄
13 45 ☽ ⊥ ♀

25 Wednesday
01 56 ☽ ⚹ ♂
04 27 ☽ ☌ ☉
05 01 ☽ ∠ ♀
06 29 ☽ ⚹ ♃
11 44 ☽ ∥ ☿
12 01 ☽ △ ♅
19 32 ☽ # ♄
19 53 ☽ △ ♆
21 44 ☽ Q ♂
23 17 ☽ ☍ ♇

26 Thursday
00 02 ☽ ☌ ☿
00 14 ☽ # ♆
06 16 ☽ ∠ ♃
06 23 ☽ ⚹ ♀
09 58 ☽ # ♅
10 00 ☽ ∥ ♃
11 33 ☽ bQ ♅
12 43 ♀ Q ♅
12 45 ☽ △ ♄
18 05 ☉ ⚹ ♃
19 29 ☽ bQ ♆
20 36 ☽ ⊥ ♃

27 Friday
01 46 ☽ ± ♅
02 48 ☽ Q ♀
03 35 ☽ □ ♂
04 59 ♃ # ♅
06 14 ☽ ⊻ ♃
06 56 ☽ ⊻ ☉
07 59 ☿ Q ♂
09 46 ☽ ± ♆
11 19 ☽ ⊼ ♅
12 34 ☽ bQ ♄
16 32 ☿ ☍ ♇
17 12 ☽ ⊥ ☉
19 23 ☽ ⊼ ♆
22 40 ☽ ⊻ ☿
22 49 ☽ ⊼ ♇

28 Saturday
03 06 ☽ ± ♄
08 09 ☽ ⊥ ☿
08 33 ☽ ± ♇
08 49 ☽ ∠ ☉
10 07 ☽ □ ♀
12 54 ☽ ⊼ ♄
22 33 ☽ ∠ ☿
23 21 ☽ bQ ♇

29 Sunday
07 06 ☽ △ ♂
07 47 ☽ ☌ ♃
11 33 ☽ ⚹ ☉
12 33 ☽ ☍ ♅
21 08 ☽ ☍ ♆
23 03 ☽ ⚹ ☿
23 05 ♀ ⚹ ♄
23 31 ♂ △ ♃

30 Monday
00 42 ☽ △ ♇
01 59 ☉ △ ♅
09 37 ☽ # ♅
09 41 ☽ Q ☉
09 48 ☽ ∥ ♃
10 14 ☽ bQ ♂
15 52 ☽ ☍ ♄
17 21 ☽ △ ♀
19 11 ☽ Q ☿
20 27 ☽ # ♆

MAY 1990

LONGITUDES

Date	Sidereal time h m s	Sun ☉ ° ' "	Moon ☽ ° ' "	Moon ☽ 24.00 ° ' "	Mercury ☿ ° '	Venus ♀ ° '	Mars ♂ ° '	Jupiter ♃ ° '	Saturn ♄ ° '	Uranus ♅ ° '	Neptune ♆ ° '	Pluto ♇ ° '
01	02 36 37	10 ♉ 51 21	06 ♌ 37 11	13 ♌ 12 59	14 ♉ 48	27 ♓ 02	07 ♓ 54	07 ♋ 02	25 ♑ 20	09 ♑ 28	14 ♑ 31	16 ♏ 34
02	02 40 34	11 49 34	19 ♌ 43 09	26 ♌ 08 08	14 R 12	28 09	08 39	07 12	25 20	09 R 27	14 R 30	16 R 32
03	02 44 30	12 47 46	02 ♍ 28 30	08 ♍ 44 44	13 35	29 ♓ 16	09 24	07 22	25 20	09 26	14 30	16 30
04	02 48 27	13 45 55	14 ♍ 57 20	21 ♍ 06 48	12 57	00 ♈ 23	10 09	07 33	25 20	09 25	14 29	16 29
05	02 52 23	14 44 03	27 ♍ 13 33	03 ♎ 18 02	12 19	01 30	10 53	07 43	25 R 20	09 24	14 29	16 27
06	02 56 20	15 42 08	09 ♎ 20 35	15 ♎ 21 32	11 42	02 37	11 38	07 54	25 20	09 23	14 28	16 25
07	03 00 16	16 40 12	21 ♎ 21 10	27 ♎ 19 46	11 05	03 45	12 23	08 04	25 20	09 22	14 27	16 23
08	03 04 13	17 38 14	03 ♏ 17 31	09 ♏ 14 37	10 31	04 53	13 08	08 15	25 20	09 20	14 27	16 22
09	03 08 09	18 36 14	15 ♏ 11 16	21 ♏ 07 37	09 59	06 00	13 52	08 26	25 19	09 19	14 26	16 20
10	03 12 06	19 34 13	27 ♏ 03 51	03 ♐ 00 07	09 30	07 08	14 37	08 37	25 19	09 18	14 25	16 18
11	03 16 03	20 32 11	08 ♐ 56 37	14 ♐ 53 33	09 04	08 16	15 22	08 48	25 18	09 17	14 24	16 17
12	03 19 59	21 30 07	20 ♐ 51 11	26 ♐ 49 47	08 42	09 24	16 06	08 59	25 18	09 15	14 24	16 15
13	03 23 56	22 28 01	02 ♑ 49 39	08 ♑ 51 10	08 24	10 32	16 51	09 10	25 17	09 14	14 23	16 13
14	03 27 52	23 25 54	14 ♑ 54 43	21 ♑ 00 44	08 10	11 41	17 35	09 21	25 16	09 12	14 22	16 12
15	03 31 49	24 23 46	27 ♑ 09 43	03 ♒ 22 10	08 01	12 49	18 20	09 33	25 15	09 11	14 21	16 10
16	03 35 45	25 21 36	09 ♒ 38 38	15 ♒ 59 39	07 56	13 58	19 05	09 44	25 14	09 10	14 20	16 08
17	03 39 42	26 19 26	22 ♒ 25 47	28 ♒ 57 33	07 D 55	15 06	19 49	09 55	25 13	09 08	14 19	16 07
18	03 43 38	27 17 14	05 ♓ 35 27	12 ♓ 19 53	08 00	16 15	20 34	10 07	25 11	09 06	14 18	16 05
19	03 47 35	28 15 01	19 ♓ 11 12	26 ♓ 09 34	08 08	17 24	21 18	10 18	25 10	09 05	14 17	16 03
20	03 51 32	29 ♉ 12 47	03 ♈ 15 03	10 ♈ 27 29	08 22	18 32	22 03	10 30	25 09	09 03	14 16	16 02
21	03 55 28	00 ♊ 10 32	17 ♈ 46 31	25 ♈ 11 32	08 39	19 41	22 47	10 42	25 07	09 01	14 15	16 00
22	03 59 25	01 08 15	02 ♉ 41 45	10 ♉ 16 07	09 02	20 50	23 31	10 54	25 06	09 00	14 14	15 59
23	04 03 21	02 05 58	17 ♉ 53 25	25 ♉ 32 18	09 28	21 59	24 16	11 06	25 04	08 58	14 13	15 57
24	04 07 18	03 03 40	03 ♊ 11 21	10 ♊ 49 07	09 59	23 09	25 00	11 18	25 02	08 56	14 12	15 55
25	04 11 14	04 01 20	18 ♊ 24 13	25 ♊ 55 25	10 33	24 18	25 44	11 30	25 00	08 54	14 11	15 54
26	04 15 11	04 58 59	03 ♋ 21 36	10 ♋ 41 55	11 12	25 27	26 28	11 42	24 58	08 53	14 10	15 52
27	04 19 07	05 56 37	17 ♋ 55 43	25 ♋ 02 33	11 55	26 37	27 13	11 54	24 56	08 51	14 09	15 51
28	04 23 04	06 54 13	02 ♌ 02 13	08 ♌ 54 42	12 41	27 46	27 57	12 06	24 54	08 49	14 08	15 49
29	04 27 01	07 51 48	15 ♌ 40 08	22 ♌ 18 48	13 31	28 ♈ 56	28 41	12 18	24 52	08 47	14 06	15 48
30	04 30 57	08 49 21	28 ♌ 51 06	05 ♍ 17 30	14 24	00 ♉ 05	29 ♓ 25	12 30	24 49	08 45	14 05	15 46
31	04 34 54	09 ♊ 46 53	11 ♍ 38 30	17 ♍ 54 42	15 ♉ 21	01 ♉ 15	00 ♈ 09	12 ♋ 43	24 ♑ 47	08 ♑ 43	14 ♑ 04	15 ♏ 45

DECLINATIONS

Date	Moon True ☊ ° '	Moon Mean ☊ ° '	Moon ☽ Latitude ° '	Sun ☉ ° '	Moon ☽ ° '	Mercury ☿ ° '	Venus ♀ ° '	Mars ♂ ° '	Jupiter ♃ ° '	Saturn ♄ ° '	Uranus ♅ ° '	Neptune ♆ ° '	Pluto ♇ ° '
01	12 ♒ 02	12 ♒ 05	00 N 29	15 N 05	19 N 05	17 N 24	02 S 17	10 S 04	23 N 23	20 S 54	23 S 25	21 S 47	01 S 25
02	12 R 02	12 01	00 S 41	15 23	14 16	16 58	01 53	09 48	23 22	20 54	23 25	21 47	01 24
03	12 01	11 58	01 46	15 41	08 57	16 31	01 29	09 32	23 22	20 54	23 25	21 47	01 24
04	11 59	11 55	02 44	15 58	03 N 24	16 03	01 06	09 15	23 21	20 54	23 25	21 47	01 24
05	11 53	11 52	03 34	16 16	02 S 10	15 35	00 42	08 59	23 21	20 55	23 25	21 47	01 23
06	11 45	11 49	04 14	16 33	07 35	15 07	00 S 17	08 43	23 20	20 55	23 25	21 47	01 23
07	11 34	11 46	04 41	16 49	12 40	14 39	00 N 07	08 26	23 20	20 55	23 25	21 48	01 22
08	11 22	11 42	04 57	17 06	17 15	14 12	00 31	08 10	23 19	20 55	23 25	21 48	01 22
09	11 08	11 39	04 59	17 22	21 09	13 46	00 56	07 53	23 18	20 55	23 26	21 48	01 22
10	10 54	11 36	04 48	17 38	24 10	13 22	01 20	07 36	23 18	20 55	23 26	21 48	01 21
11	10 41	11 33	04 25	17 53	26 09	12 59	01 45	07 20	23 17	20 55	23 26	21 48	01 21
12	10 31	11 30	03 49	18 08	26 57	12 38	02 10	07 03	23 17	20 56	23 26	21 48	01 20
13	10 23	11 27	03 04	18 23	26 29	12 19	02 35	06 46	23 16	20 56	23 26	21 48	01 20
14	10 18	11 23	02 10	18 38	24 45	12 02	03 00	06 29	23 15	20 56	23 26	21 48	01 20
15	10 15	11 20	01 09	18 52	21 51	11 48	03 24	06 12	23 14	20 56	23 26	21 48	01 19
16	10 15	11 17	00 S 03	19 06	17 53	11 36	03 49	05 55	23 14	20 56	23 26	21 48	01 19
17	10 D 15	11 14	01 N 04	19 20	13 02	11 26	04 14	05 39	23 13	20 57	23 27	21 48	01 19
18	10 R 15	11 11	02 10	19 33	07 27	11 19	04 39	05 22	23 12	20 57	23 27	21 48	01 19
19	10 13	11 07	03 11	19 46	01 S 21	11 14	05 04	05 05	23 11	20 57	23 27	21 48	01 18
20	10 09	11 04	04 02	19 59	05 N 00	11 12	05 29	04 48	23 11	20 58	23 27	21 49	01 18
21	10 03	11 01	04 40	20 11	11 17	11 12	05 54	04 31	23 10	20 58	23 27	21 49	01 18
22	09 55	10 58	05 00	20 23	17 06	11 14	06 19	04 14	23 09	20 58	23 27	21 49	01 17
23	09 45	10 55	04 59	20 35	21 57	11 18	06 44	03 56	23 08	20 59	23 27	21 49	01 17
24	09 36	10 52	04 37	20 46	25 20	11 25	07 08	03 39	23 07	20 59	23 27	21 49	01 17
25	09 27	10 48	03 56	20 57	26 51	11 34	07 33	03 22	23 06	21 00	23 28	21 49	01 17
26	09 21	10 45	02 59	21 08	26 23	11 44	07 57	03 05	23 05	21 00	23 28	21 49	01 17
27	09 17	10 42	01 51	21 18	24 04	11 57	08 22	02 48	23 04	21 00	23 28	21 49	01 16
28	09 15	10 39	00 N 38	21 28	20 20	12 11	08 46	02 31	23 03	21 01	23 28	21 49	01 16
29	09 D 15	10 36	00 S 34	21 37	15 36	12 27	09 10	02 14	23 02	21 01	23 28	21 49	01 16
30	09 15	10 32	01 43	21 46	10 16	12 44	09 35	01 57	23 01	21 02	23 28	21 50	01 16
31	09 ♒ 16	10 ♒ 29	02 S 44	21 N 55	04 N 40	13 N 03	09 N 58	01 S 40	23 N 00	21 S 02	23 S 28	21 S 50	01 S 16

ZODIAC SIGN ENTRIES

Date	h	m	Planets
01	00	08	☽ ♌
03	07	18	☽ ♍
04	03	52	♀ ♈
05	17	28	☽ ♎
08	05	22	☽ ♏
10	17	56	☽ ♐
13	06	21	☽ ♑
15	17	30	☽ ♒
18	01	54	☽ ♓
20	06	31	☽ ♈
21	07	37	☉ ♊
22	07	42	☽ ♉
24	07	00	☽ ♊
26	06	34	☽ ♋
28	08	29	☽ ♌
30	10	13	♀ ♉
30	14	08	☽ ♍
31	07	11	♂ ♈

LATITUDES

Date	Mercury ☿ ° '	Venus ♀ ° '	Mars ♂ ° '	Jupiter ♃ ° '	Saturn ♄ ° '	Uranus ♅ ° '	Neptune ♆ ° '	Pluto ♇ ° '
01	01 N 10	01 S 12	01 S 34	00 N 07	00 N 10	00 S 19	00 N 52	16 N 00
04	00 N 20	01 21	01 37	00 08	00 10	00 19	00 52	16 00
07	00 S 32	01 30	01 39	00 08	00 10	00 19	00 52	16 00
10	01 22	01 37	01 41	00 08	00 10	00 19	00 52	16 00
13	02 06	01 44	01 43	00 08	00 09	00 19	00 52	16 00
16	02 42	01 50	01 44	00 09	00 09	00 19	00 52	15 59
19	03 09	01 54	01 46	00 09	00 09	00 19	00 52	15 59
22	03 27	01 58	01 48	00 09	00 09	00 19	00 52	15 58
25	03 37	02 01	01 50	00 09	00 09	00 19	00 52	15 58
28	03 38	02 03	01 51	00 10	00 08	00 19	00 52	15 57
31	03 S 32	02 S 04	01 S 53	00 N 10	00 N 08	00 S 19	00 N 53	15 N 56

DATA

Julian Date	2448013
Delta T	+57 seconds
Ayanamsa	23° 43' 31"
Synetic vernal point	05° ♓ 23' 28"
True obliquity of ecliptic	23° 26' 32"

MOON'S PHASES, APSIDES AND POSITIONS ☽

Date	h	m	Phase	Longitude ° '	Eclipse Indicator
01	20	18	☽	11 ♌ 11	
09	19	31	○	18 ♏ 54	
17	19	45	☾	26 ♒ 38	
24	11	47	●	03 ♊ 03	
31	08	11	☽	09 ♍ 38	

Day	h	m		
10	00	00	Apogee	
24	02	55	Perigee	
05	02	36	0S	
12	15	15	Max dec	26° S 57'
19	17	09	0N	
25	18	11	Max dec	26° N 55'

LONGITUDES

Date	Chiron ⚷ ° '	Ceres ⚳ ° '	Pallas ⚴ ° '	Juno ⚵ ° '	Vesta ⚶ ° '	Black Moon Lilith ⚸ ° '
01	12 ♋ 11	09 ♋ 54	20 ♉ 53	18 ♏ 47	15 ♈ 21	19 ♏ 55
11	12 ♋ 54	13 ♋ 39	26 ♉ 16	16 ♏ 32	19 ♈ 52	21 ♏ 02
21	13 ♋ 42	17 ♋ 33	01 ♊ 46	14 ♏ 21	24 ♈ 17	22 ♏ 09
31	14 ♋ 36	21 ♋ 36	07 ♊ 23	12 ♏ 25	28 ♈ 35	23 ♏ 16

All ephemeris data is given at 12.00 UT and the Moon's longitude is additionally given for 24.00 UT

ASPECTARIAN

h m	Aspects
01 Tuesday	
00 58	♀ Q ♆
01 46	☽ ⋕ ♄
03 02	☽ ± ♂
12 45	☽ ⊻ ♃
14 28	☽ ⊼ ♂
17 08	☽ ⊼ ♅
20 18	☽ □ ☉
21 39	☽ ∥ ☿
22 44	☽ Ꝑ ♀
23 48	☽ ⊥ ♃
23 55	☿ △ ♆
02 Wednesday	
02 16	☽ □ ☿
02 22	☽ ⊼ ♆
04 06	☽ ± ♅
06 07	☽ □ ♇
06 59	☽ ∥ ☉
13 28	☽ ± ♆
16 41	☽ ∠ ♃
16 57	☽ ± ♀
20 49	☽ Ꝑ ♅
22 30	☽ ⊼ ♄
03 Thursday	
02 42	☿ ∠ ♀
05 18	☽ ⊼ ♀
06 21	☽ Ꝑ ♆
09 20	☽ ⋕ ♂
09 50	☽ ± ♄
12 57	♂ ∗ ♅
15 52	☽ Q ♇
17 35	♀ ∥ ♇
21 29	☽ ∗ ♃
23 49	☉ ☌ ☿
04 Friday	
01 18	☽ △ ♅
02 06	☽ ☍ ♂
03 04	☽ Ꝑ ♄
08 18	☽ △ ☿
09 30	☽ △ ☉
11 05	☽ △ ♆
14 26	☉ ∥ ☿
14 57	☽ ∗ ♇
18 14	♂ ∠ ♄
20 37	☽ ⋕ ♇
21 03	☽ Q ♃
22 40	☽ ⋕ ♀
22 43	♄ St R
05 Saturday	
05 42	☉ △ ♆
06 01	☽ ∥ ♀
08 17	☽ △ ♄
08 35	☽ ∥ ♇
10 53	♀ Ꝑ ♇
12 10	☽ Ꝑ ☿
17 22	☽ Ꝑ ☉
20 19	☽ ∠ ♇
21 18	☽ ☍ ♀
06 Sunday	
05 06	☽ ± ☿
09 04	☽ □ ♃
12 04	☽ □ ♅
12 47	☽ ± ☉
12 59	☿ ∗ ♂
14 08	☽ ⊥ ♇
16 27	☽ ⊼ ☿
16 52	☽ ⊼ ♂
16 52	☽ ∥ ♂
22 13	☽ □ ♆
07 Monday	
01 48	☽ ⊼ ☉
02 05	☽ ⊻ ♇
05 17	☉ ☍ ♇
05 39	☽ ± ♂
19 59	☽ □ ♄
21 06	☽ ⋕ ☿
08 Tuesday	
00 02	☽ Q ♅
00 55	☽ Ꝑ ♂
06 52	☿ ⊥ ♀
10 18	☽ Q ♆
11 03	☽ ⋕ ☉
15 32	☽ ⊼ ♀
22 09	☽ △ ♃
09 Wednesday	
00 10	☽ ∗ ♅
01 56	☽ ☍ ☿
04 54	☽ ± ♀
08 14	☽ Q ♄
09 10	☽ △ ♂
10 25	☽ ∥ ♄
10 29	☽ ∗ ♆
14 19	☽ ☌ ♇
16 33	☽ ∥ ♆
19 31	☽ ☍ ☉
10 Thursday	
00 59	☽ Ꝑ ♀
04 15	☽ ⋕ ♃
04 55	☽ Ꝑ ♃
05 19	☽ ∥ ♅
05 49	♂ ∗ ♆
06 25	☽ ∠ ♅
08 28	☽ ∗ ♄
12 40	♀ ⋕ ♇
15 44	♀ Q ♄
16 46	☽ ∠ ♆
23 20	☿ △ ♅
23 23	☽ ± ♃
11 Friday	
00 35	☽ ⊥ ♅
10 30	☽ △ ♀
10 55	☽ ⊥ ♆
11 42	☽ ⊼ ♃
12 15	☽ ⊼ ☿
12 40	☽ ⊻ ♅
14 45	☽ ∠ ♄
23 01	☽ ⊻ ♆
23 34	☿ ∗ ♃
23 58	☽ ± ☿
12 Saturday	
00 39	☿ ⊻ ♀
01 17	♀ □ ♃
01 48	☽ □ ♂
02 46	☽ ⊻ ♇
08 52	☽ ⊥ ♄
08 54	♀ □ ♅
13 25	☽ ⊼ ☉
14 48	☽ ⊥ ♇
16 34	♂ △ ♇
17 34	☽ Ꝑ ☿
20 55	☽ ⊻ ♄
13 Sunday	
02 31	☽ ± ☉
05 28	♀ ± ♇
08 48	☽ ∠ ♇
16 18	☽ Q ♂
19 31	♃ ☍ ♅
22 03	☽ Ꝑ ☉
22 53	☽ △ ☿
14 Monday	
00 44	☽ ☌ ♅
00 49	☽ ☍ ♃
04 57	☽ □ ♀
10 56	☽ ☌ ♆
14 32	☽ ∗ ♇
17 37	☽ ∗ ♂
15 Tuesday	
00 05	☽ ∥ ♅
01 38	☽ ⋕ ♃
06 09	☽ △ ☉
06 51	☉ Ꝑ ♅
08 17	☽ ☌ ♄
12 22	☽ ∥ ♆
13 57	☽ Q ♇
16 31	☉ ∠ ♃
18 07	☽ ∥ ♄
19 48	☽ Q ♀
16 Wednesday	
00 41	☽ ∠ ♂
05 35	☽ ⋕ ☉
08 45	☽ □ ☿
08 52	☉ △ ♄
11 05	☽ ⊻ ♅
12 10	☽ ⊼ ♃
18 55	☽ ⊥ ♂
19 50	♀ □ ♆
20 52	☽ ⊻ ♆
20 59	☽ ∗ ♀
22 25	☽ ⊥ ♅
23 41	☽ ± ♃
17 Thursday	
00 15	☽ □ ♇
02 02	☿ St D
06 51	☽ ⊻ ♂
08 06	☽ ⊥ ♆
15 09	☽ ∠ ♅
16 40	☽ Ꝑ ♃
17 08	☽ ⊻ ♄
18 28	☽ Q ☿
19 19	☽ ⋕ ☿
19 45	☽ □ ☉
18 Friday	
00 39	☽ ∠ ♆
03 25	☽ ∠ ♀
04 04	☽ ⊥ ♄
08 40	♀ ⊼ ♇
16 20	☽ ∗ ☿
18 16	☽ ∗ ♅
20 12	☽ △ ♃
20 12	☽ ∠ ♄
20 48	☽ ∥ ♂
21 05	☽ ⊥ ♀
22 28	☽ ⋕ ♀
19 Saturday	
03 28	☽ ∗ ♆
05 06	♂ Q ♅
06 30	☽ Q ☉
06 33	☽ △ ♇
08 36	☽ ⊻ ♀
12 11	☽ ∥ ♇
12 14	♀ ⋕ ♂
15 16	☽ Q ♅
15 52	☽ ☌ ♂
18 55	☽ ∠ ☿
22 05	☽ ⋕ ♇
22 17	☽ ∗ ♄
20 Sunday	
00 13	☽ Q ♆
04 42	☽ ∗ ☉
08 16	☽ Ꝑ ♇
10 29	☽ ⊥ ☿
11 16	☽ ⋕ ♂
13 29	☉ Ꝑ ♆
13 57	☽ ∥ ♀
18 30	☽ Q ♄
20 42	☽ ⊻ ☿
21 39	☽ □ ♅
23 16	☽ ± ♇
21 Monday	
00 14	☽ □ ♃
06 15	☽ □ ♆
07 27	☽ ∠ ☉
09 07	☽ ⊼ ♇
11 38	☽ ∥ ☿
15 22	☽ ☌ ♀
20 33	☽ ⊻ ♂
23 05	☽ ⊥ ☉
23 52	☽ □ ♄
22 Tuesday	
05 51	☽ Q ♃
06 41	☽ ⊥ ♂
09 21	☽ ⊻ ☉
10 16	☿ △ ♅
21 43	☽ ∠ ♂
21 58	☽ △ ♅
22 20	☽ ☌ ☿
23 Wednesday	
01 10	☽ ∗ ♃
04 16	☽ ∥ ☉
06 14	☽ △ ♆
06 39	☽ ⋕ ♄
08 57	☽ ☍ ♇
11 14	☽ ⋕ ♆
18 58	☽ ⊻ ♀
19 07	☽ ∥ ♃
21 17	☽ ⋕ ♅
21 31	☽ Ꝑ ♅
22 30	☽ ∗ ♂
23 14	☽ △ ♄
24 Thursday	
01 02	☽ ∠ ♃
05 08	☽ ⊥ ♀
05 45	☽ Ꝑ ♆
09 00	☉ ± ♅
09 30	☿ ∠ ♂
11 36	☽ ± ♅
11 47	☽ ☌ ☉
13 06	♂ ∗ ♄
15 21	☽ ⊥ ♃
18 17	☽ Q ♂
19 52	☽ ± ♆
20 26	☽ ∠ ♀
21 01	☽ ⊼ ♅
22 44	☽ Ꝑ ♄
23 05	☽ ⊻ ☿
25 Friday	
00 55	☽ ⊻ ♃
05 19	☽ ⊼ ♆
08 02	☽ ⊼ ♇
08 57	☽ ⊥ ☿
12 57	☽ ± ♄
17 17	☉ ⋕ ♄
17 33	☽ ± ♇
22 11	☽ ∗ ♀
22 30	☽ ⊼ ♄
23 55	☽ ∠ ☿
26 Saturday	
00 18	☽ □ ♂
02 13	♂ Q ♆
02 14	♀ □ ♄
07 58	☽ Ꝑ ♇
14 49	☽ ⊻ ☉
19 15	☽ Q ♀
20 59	☽ ☍ ♅
27 Sunday	
01 21	☽ ⊥ ☉
01 28	☽ ∗ ☿
01 50	☽ ☌ ♃
05 42	☽ ☍ ♆
08 32	☽ △ ♇
10 29	☉ ⊥ ♃
11 25	☿ ∗ ♃
16 32	☽ ⋕ ♅
17 26	☽ ∠ ☉
19 20	☽ ∥ ♃
22 38	☽ Q ☿
23 47	☽ ☍ ♄
28 Monday	
03 20	☽ ⋕ ♆
03 59	☽ □ ♀
04 34	☽ △ ♂
05 44	☽ ∥ ☉
08 08	☽ ⋕ ♄
21 07	☽ ∗ ☉
22 04	♀ ⊻ ♂
23 48	☽ ⊼ ♅
29 Tuesday	
05 54	☽ ⊻ ♃
07 53	☽ □ ☿
08 15	☽ Ꝑ ♂
09 13	☽ ⊼ ♆
10 25	☽ ± ♅
12 13	☽ □ ♇
16 48	☽ ⊥ ♃
17 56	☉ ± ♆
19 59	☽ ± ♆
20 08	☽ Q ☉
30 Wednesday	
01 25	☽ ± ♂
01 40	☽ ∥ ☿
02 38	☽ Ꝑ ♅
03 55	☿ △ ♆
04 36	☽ ⊼ ♄
09 29	☽ ∠ ♃
10 10	☉ ⊼ ♅
12 26	☽ Ꝑ ♆
13 06	☽ ⊼ ♂
14 01	☿ ∠ ♂
14 31	☽ △ ♀
14 49	☽ ∥ ♀
15 39	☽ ± ♄
20 39	☉ ⋕ ♆
21 08	☽ Q ♇
22 36	♀ Q ♃
31 Thursday	
06 28	☽ △ ♅
08 11	☽ □ ☉
08 29	☽ Ꝑ ♄
12 03	☉ Ꝑ ♄
14 04	☽ ∗ ♃
16 37	☽ △ ♆
19 40	☽ △ ☿
19 49	☽ ∗ ♇
21 27	☿ ☍ ♇
21 42	☽ Ꝑ ♀

JUNE 1990

LONGITUDES

Date	Sidereal time h m s	Sun ☉ ° ' "	Moon ☽ ° ' "	Moon ☽ 24.00 ° ' "	Mercury ☿ ° '	Venus ♀ ° '	Mars ♂ ° '	Jupiter ♃ ° '	Saturn ♄ ° '	Uranus ♅ ° '	Neptune ♆ ° '	Pluto ♇ ° '
01	04 38 50	10 ♊ 44 24	24 ♍ 06 38	00 ♎ 14 52	16 ♉ 21	02 ♉ 24	00 ♈ 53	12 ♋ 55	24 ♑ 45	08 ♑ 41	14 ♑ 03	15 ♏ 43
02	04 42 47	11 41 53	06 ♎ 19 59	12 ♎ 22 28	17 24	03 34	01 37	13 08	24 R 42	08 R 39	14 R 01	15 R 42
03	04 46 43	12 39 21	18 ♎ 22 51	24 ♎ 21 35	18 31	04 44	02 21	13 20	24 39	08 37	14 00	15 40
04	04 50 40	13 36 48	00 ♏ 19 05	06 ♏ 15 45	19 40	05 54	03 04	13 33	24 37	08 35	13 59	15 39
05	04 54 36	14 34 14	12 ♏ 11 54	18 ♏ 07 52	20 53	07 04	03 48	13 45	24 34	08 32	13 57	15 37
06	04 58 33	15 31 38	24 ♏ 03 55	00 ♐ 00 17	22 09	08 14	04 32	13 58	24 31	08 30	13 56	15 36
07	05 02 30	16 29 02	05 ♐ 57 12	11 ♐ 54 51	23 28	09 24	05 15	14 11	24 28	08 28	13 55	15 34
08	05 06 26	17 26 25	17 ♐ 53 26	23 ♐ 53 08	24 49	10 34	05 59	14 23	24 25	08 26	13 53	15 33
09	05 10 23	18 23 47	29 ♐ 54 09	05 ♑ 56 40	26 14	11 44	06 43	14 36	24 22	08 24	13 52	15 32
10	05 14 19	19 21 09	12 ♑ 00 54	18 ♑ 07 04	27 42	12 55	07 26	14 49	24 19	08 21	13 50	15 30
11	05 18 16	20 18 29	24 ♑ 15 28	00 ♒ 26 21	29 ♉ 12	14 05	08 09	15 02	24 15	08 19	13 49	15 29
12	05 22 12	21 15 49	06 ♒ 40 03	12 ♒ 56 53	00 ♊ 45	15 15	08 53	15 15	24 12	08 17	13 47	15 28
13	05 26 09	22 13 09	19 ♒ 17 15	25 ♒ 41 30	02 21	16 26	09 36	15 27	24 09	08 15	13 46	15 27
14	05 30 05	23 10 28	02 ♓ 10 03	08 ♓ 43 15	04 00	17 36	10 19	15 40	24 05	08 12	13 45	15 25
15	05 34 02	24 07 46	15 ♓ 21 29	22 ♓ 05 03	05 41	18 47	11 02	15 53	24 02	08 10	13 43	15 24
16	05 37 59	25 05 05	28 ♓ 54 14	05 ♈ 49 13	07 26	19 57	11 46	16 06	23 58	08 08	13 42	15 23
17	05 41 55	26 02 22	12 ♈ 50 02	19 ♈ 56 39	09 13	21 08	12 29	16 19	23 55	08 05	13 40	15 22
18	05 45 52	26 59 40	27 ♈ 08 51	04 ♉ 26 15	11 03	22 19	13 12	16 33	23 51	08 03	13 38	15 21
19	05 49 48	27 56 58	11 ♉ 48 17	19 ♉ 14 14	12 55	23 29	13 54	16 46	23 47	08 00	13 37	15 19
20	05 53 45	28 54 15	26 ♉ 43 12	04 ♊ 14 08	14 50	24 40	14 37	16 59	23 44	07 58	13 35	15 18
21	05 57 41	29 ♊ 51 32	11 ♊ 45 56	19 ♊ 17 24	16 47	25 51	15 20	17 12	23 40	07 56	13 34	15 17
22	06 01 38	00 ♋ 48 49	26 ♊ 47 21	04 ♋ 14 38	18 46	27 02	16 03	17 25	23 36	07 53	13 32	15 16
23	06 05 34	01 46 05	11 ♋ 38 13	18 ♋ 57 09	20 48	28 13	16 45	17 38	23 32	07 51	13 31	15 15
24	06 09 31	02 43 21	26 ♋ 10 42	03 ♌ 18 16	22 52	29 ♉ 24	17 28	17 52	23 28	07 48	13 29	15 14
25	06 13 28	03 40 37	10 ♌ 19 27	17 ♌ 14 02	24 57	00 ♊ 35	18 10	18 05	23 24	07 46	13 28	15 13
26	06 17 24	04 37 51	24 ♌ 01 56	00 ♍ 43 15	27 04	01 46	18 52	18 18	23 20	07 44	13 26	15 12
27	06 21 21	05 35 06	07 ♍ 18 12	13 ♍ 47 06	29 ♊ 13	02 57	19 35	18 31	23 16	07 41	13 24	15 11
28	06 25 17	06 32 20	20 ♍ 10 21	26 ♍ 28 25	01 ♋ 22	04 08	20 17	18 45	23 12	07 39	13 23	15 10
29	06 29 14	07 29 33	02 ♎ 41 50	08 ♎ 51 09	03 32	05 19	20 59	18 58	23 08	07 36	13 21	15 10
30	06 33 10	08 ♋ 26 46	14 ♎ 56 56	20 ♎ 59 44	05 ♋ 43	06 ♊ 31	21 ♈ 41	19 ♋ 11	23 ♑ 03	07 ♑ 34	13 ♑ 19	15 ♏ 09

Date	Moon True ☊ ° '	Moon Mean ☊ ° '	Moon ☽ Latitude ° '
01	09 ♒ 15	10 ♒ 26	03 S 36
02	09 R 12	10 23	04 17
03	09 06	10 20	04 45
04	08 59	10 17	05 01
05	08 50	10 13	05 04
06	08 40	10 10	04 54
07	08 30	10 07	04 31
08	08 21	10 04	03 56
09	08 14	10 01	03 10
10	08 09	09 58	02 16
11	08 05	09 54	01 14
12	08 04	09 51	00 S 08
13	08 D 05	09 48	01 N 00
14	08 06	09 45	02 07
15	08 07	09 42	03 08
16	08 R 08	09 38	04 00
17	08 06	09 35	04 40
18	08 03	09 32	05 04
19	07 59	09 29	05 09
20	07 54	09 26	04 53
21	07 48	09 23	04 17
22	07 43	09 19	03 24
23	07 39	09 16	02 17
24	07 37	09 13	01 N 02
25	07 D 37	09 10	00 S 15
26	07 38	09 07	01 28
27	07 39	09 04	02 35
28	07 41	09 00	03 32
29	07 42	08 57	04 17
30	07 ♒ 42	08 ♒ 54	04 S 49

DECLINATIONS

Date	Sun ☉ ° '	Moon ☽ ° '	Mercury ☿ ° '	Venus ♀ ° '	Mars ♂ ° '	Jupiter ♃ ° '	Saturn ♄ ° '	Uranus ♅ ° '	Neptune ♆ ° '	Pluto ♇ ° '
01	22 N 04	00 S 58	13 N 24	10 N 22	01 S 23	22 N 59	21 S 03	23 S 29	21 S 50	01 S 16
02	22 11	06 27	13 46	10 46	01 06	22 58	21 03	23 29	21 50	01 16
03	22 19	11 36	14 08	11 09	00 49	22 57	21 04	23 29	21 50	01 15
04	22 26	16 17	14 32	11 33	00 32	22 55	21 04	23 29	21 50	01 15
05	22 33	20 19	14 57	11 56	00 S 15	22 54	21 05	23 29	21 50	01 15
06	22 39	23 32	15 23	12 18	00 N 02	22 53	21 06	23 29	21 51	01 15
07	22 45	25 45	15 50	12 41	00 19	22 52	21 06	23 30	21 51	01 15
08	22 51	26 49	16 17	13 03	00 36	22 50	21 07	23 30	21 51	01 15
09	22 56	26 37	16 45	13 25	00 52	22 49	21 07	23 30	21 51	01 15
10	23 01	25 09	17 13	13 47	01 09	22 48	21 08	23 30	21 51	01 15
11	23 05	22 29	17 42	14 09	01 26	22 46	21 09	23 30	21 51	01 15
12	23 09	18 44	18 11	14 30	01 43	22 45	21 09	23 30	21 51	01 15
13	23 13	14 05	18 40	14 51	01 59	22 44	21 10	23 30	21 52	01 15
14	23 16	08 44	19 09	15 12	02 16	22 42	21 11	23 31	21 52	01 15
15	23 19	02 S 53	19 37	15 32	02 33	22 41	21 11	23 31	21 52	01 15
16	23 21	03 N 14	20 05	15 52	02 49	22 39	21 12	23 31	21 52	01 15
17	23 23	09 22	20 33	16 12	03 06	22 38	21 13	23 31	21 52	01 15
18	23 24	15 11	21 00	16 32	03 22	22 36	21 14	23 31	21 52	01 15
19	23 26	20 16	21 26	16 51	03 38	22 35	21 14	23 31	21 52	01 15
20	23 26	24 11	21 51	17 09	03 54	22 33	21 15	23 32	21 53	01 16
21	23 27	26 27	22 15	17 27	04 11	22 32	21 16	23 32	21 53	01 16
22	23 26	26 48	22 37	17 45	04 27	22 30	21 16	23 32	21 53	01 16
23	23 26	25 12	22 58	18 03	04 43	22 28	21 17	23 32	21 53	01 16
24	23 25	21 56	23 17	18 20	04 59	22 27	21 18	23 32	21 53	01 16
25	23 23	17 25	23 33	18 36	05 15	22 25	21 19	23 32	21 53	01 16
26	23 22	12 07	23 48	18 52	05 31	22 23	21 20	23 33	21 54	01 17
27	23 19	06 26	24 00	19 08	05 46	22 22	21 20	23 33	21 54	01 17
28	23 17	00 N 39	24 09	19 23	06 02	22 20	21 21	23 33	21 54	01 17
29	23 14	05 S 00	24 16	19 38	06 17	22 18	21 22	23 33	21 54	01 17
30	23 N 10	10 S 20	24 N 20	19 N 52	06 N 33	22 N 16	21 S 23	23 S 33	21 S 54	01 S 17

ZODIAC SIGN ENTRIES

Date	h	m	Planets
01	23	31	☽ ♎
04	11	22	☽ ♏
06	23	59	☽ ♐
09	12	12	☽ ♑
11	23	09	☽ ♒
12	00	29	☿ ♊
14	08	00	☽ ♓
16	13	55	☽ ♈
18	16	43	☽ ♉
20	17	15	☽ ♊
21	15	33	☉ ♋
22	17	09	☽ ♋
24	18	25	☽ ♌
25	00	14	♀ ♊
26	22	42	☽ ♍
27	20	46	☿ ♋
29	06	47	☽ ♎

LATITUDES

Date	Mercury ☿ ° '	Venus ♀ ° '	Mars ♂ ° '	Jupiter ♃ ° '	Saturn ♄ ° '	Uranus ♅ ° '	Neptune ♆ ° '	Pluto ♇ ° '
01	03 S 28	02 S 04	01 S 53	00 N 10	00 N 08	00 S 19	00 N 53	15 N 56
04	03 14	02 04	01 55	00 10	00 08	00 19	00 53	15 55
07	02 54	02 03	01 56	00 10	00 08	00 19	00 53	15 54
10	02 29	02 01	01 57	00 11	00 07	00 19	00 53	15 53
13	02 01	01 59	01 58	00 11	00 07	00 19	00 53	15 52
16	01 29	01 56	01 59	00 11	00 07	00 19	00 53	15 51
19	00 55	01 52	02 00	00 11	00 07	00 19	00 53	15 50
22	00 S 21	01 47	02 01	00 12	00 06	00 20	00 53	15 48
25	00 N 13	01 42	02 02	00 12	00 06	00 20	00 53	15 47
28	00 43	01 37	02 03	00 12	00 06	00 20	00 53	15 46
31	01 N 09	01 S 31	02 S 03	00 N 13	00 N 06	00 S 20	00 N 53	15 N 44

DATA

Julian Date	2448044
Delta T	+57 seconds
Ayanamsa	23° 43' 36"
Synetic vernal point	05° ♓ 23' 24"
True obliquity of ecliptic	23° 26' 32"

MOON'S PHASES, APSIDES AND POSITIONS ☽

Date	h	m	Phase	Longitude ° '	Eclipse Indicator
08	11	01	○	17 ♐ 24	
16	04	48	☾	24 ♓ 48	
22	18	55	●	01 ♋ 05	
29	22	07	☽	07 ♎ 54	

Day	h	m		
06	03	52	Apogee	
21	10	47	Perigee	
01	07	52	OS	
08	20	22	Max dec	26° S 53'
15	23	23	ON	
22	04	11	Max dec	26° N 54'
28	14	43	OS	

LONGITUDES

Date	Chiron ⚷ ° '	Ceres ⚳ ° '	Pallas ⚴ ° '	Juno ⚵ ° '	Vesta ⚶ ° '	Black Moon Lilith ⚸ ° '
01	14 ♋ 42	22 ♋ 00	07 ♊ 57	12 ♏ 15	29 ♈ 01	23 ♏ 23
11	15 ♋ 40	26 ♋ 11	13 ♊ 40	10 ♏ 47	03 ♉ 11	24 ♏ 30
21	16 ♋ 41	00 ♌ 27	19 ♊ 28	09 ♏ 49	07 ♉ 12	25 ♏ 37
31	17 ♋ 44	04 ♌ 48	25 ♊ 20	09 ♏ 23	11 ♉ 03	26 ♏ 45

ASPECTARIAN

h m	Aspects
01 Friday	
01 26	☽ # ♂
02 29	☽ # ♇
06 51	♂ ⚼ ♇
13 14	☽ △ ♄
13 17	☽ ∥ ♇
13 36	☽ Q ♃
13 43	☽ ∥ ♂
16 57	☽ ± ♀
22 24	♂ ∥ ♇
02 Saturday	
00 54	☽ ∠ ♇
02 05	☽ ☍ ♂
03 29	☽ ⚼ ☿
05 58	☽ ⊼ ♀
16 34	☽ □ ♅
18 39	☽ ⊥ ♇
23 04	☽ ± ☿
23 34	☽ △ ☉
03 Sunday	
01 44	☽ □ ♃
03 15	☽ □ ♆
06 35	☽ ⊻ ♇
09 39	☽ # ♀
12 17	☽ ⊼ ☿
04 Monday	
00 33	☽ □ ♄
01 50	☽ # ☿
04 29	☽ Q ♅
08 16	☽ ⚼ ☉
09 47	☉ ⊻ ♃
15 21	☽ Q ♆
17 55	☽ ⊼ ♂
20 56	☉ ⊼ ♆
05 Tuesday	
00 30	☽ ☍ ♀
04 01	☽ ± ☉
04 37	☽ ✶ ♅
06 50	☽ ± ♂
12 44	☽ Q ♄
15 12	☽ △ ♃
15 33	☽ ✶ ♆
17 08	☽ ∥ ♄
17 13	☽ ⊼ ☉
18 54	☽ ☌ ♇
22 30	☽ ∥ ♆
06 Wednesday	
02 14	☽ ⚼ ♂
04 26	☽ # ☉
06 32	☽ # ♃
07 39	☽ ☍ ☿
08 45	♃ ☍ ♆
10 52	☽ ∠ ♅
11 35	☽ ∥ ♅
12 54	☽ ✶ ♄
13 43	☉ ⊼ ♇
17 25	♀ △ ♅
21 49	☽ ∠ ♆
22 05	☽ ⚼ ♃
07 Thursday	
05 00	☽ ⊥ ♅
10 30	☽ △ ♂
12 05	☿ ⚼ ♅
15 56	☽ ⊥ ♆
16 33	☽ ± ♃
17 03	☽ ⊻ ♅
19 03	☽ ∠ ♄
19 42	☽ ⊼ ♀
08 Friday	
03 59	☽ ⊻ ♆
04 51	☽ ⊼ ♃
05 09	☿ △ ♄
07 19	☽ ⊻ ♇
09 04	☽ ± ♀
10 06	☉ ∥ ♃
11 01	☽ ☍ ☉
13 03	☽ ⊥ ♄
19 19	☽ ⊥ ♇
09 Saturday	
01 00	☽ ⊻ ♄
01 22	♂ Q ♄
03 42	☽ ⊼ ☿
05 01	☽ ⚼ ♀
11 14	☉ ± ♄
13 15	☽ ∠ ♇
17 16	☽ ± ☿
10 Sunday	
02 22	☽ □ ♂
04 48	☽ ☌ ♅
13 31	☽ ⚼ ☿
13 57	☽ △ ♀
15 35	☽ ☌ ♆
17 37	☽ ☍ ♃
18 52	☽ ✶ ♇
20 17	♂ # ♇
20 22	☉ Q ♂
11 Monday	
03 38	☽ ⊼ ☉
03 57	☽ ∥ ♅
06 04	☿ ⚼ ♆
06 40	♀ △ ♆
07 26	☽ # ☉
09 45	☽ # ♃
12 00	☽ ☌ ♄
15 56	☽ Q ♂
16 03	♀ ⊥ ♂
16 19	☽ ± ☉
16 29	☽ ∥ ♆
17 06	♂ □ ♅
18 16	☽ Q ♇
21 16	☽ ∥ ♄
22 58	☽ △ ☿
12 Tuesday	
02 58	☿ ∠ ♃
11 10	☽ ⚼ ☉
11 42	♀ ✶ ♃
13 06	☉ ⊥ ♀
14 47	☽ # ☿
15 05	☽ ⊻ ♅
16 13	♀ ☍ ♇
16 30	☽ ✶ ♂
16 55	☉ ± ♇
13 Wednesday	
01 35	☽ ⊻ ♆
02 29	☽ ⊥ ♅
04 38	☽ ⊼ ♃
04 45	☽ □ ♇
06 03	☽ □ ♀
06 52	♂ ± ♇
08 32	☽ # ♀
10 26	☿ ± ♅
10 30	♃ △ ♇
12 54	☽ ⊥ ♆
16 09	☽ ± ♃
17 57	☽ △ ☉
19 24	☽ ∠ ♅
21 05	☽ ⊻ ♄
22 33	☽ ∠ ♂
14 Thursday	
05 41	☽ ∠ ♆
08 11	☽ ⊥ ♄
09 12	☽ ⚼ ♃
15 52	☽ □ ☿
16 12	☽ ⊥ ♂
18 56	☽ Q ♀
23 02	☽ ✶ ♅
15 Friday	
00 37	☽ ∠ ♄
03 46	☽ ⊻ ♂
09 03	☽ ✶ ♆
09 41	☉ ⊼ ♄
12 05	☽ △ ♇
12 58	☽ △ ♃
13 18	☽ # ♂
18 28	☽ ∥ ♇
18 42	☽ ✶ ♀
20 34	☽ Q ♅
16 Saturday	
03 23	☽ ✶ ♄
04 16	☽ # ♇
04 48	☽ □ ☉
05 00	☽ Q ☿
06 23	☽ Q ♆
10 17	☽ ∥ ♂
14 34	☽ ⚼ ♇
15 31	☿ ± ♆
21 14	☿ ⊼ ♅
23 29	☽ ∠ ♀
17 Sunday	
00 13	☽ Q ♄
03 55	☽ □ ♅
04 55	☽ ✶ ☿
06 05	☽ ± ♇
08 08	☿ ⚼ ♄
11 22	☽ ☌ ♂
13 25	☽ □ ♆
14 12	☽ Q ☉
16 15	☽ ⊥ ♀
16 17	☽ ⊼ ♇
18 00	☽ □ ♃
18 Monday	
03 14	☽ ⊻ ♀
04 37	☿ ⊥ ♃
06 33	☽ □ ♄
09 54	☽ ∠ ☿
11 44	☽ ✶ ☉
18 19	☽ ∥ ♀
19 Tuesday	
00 21	☽ Q ♃
00 30	☿ # ♄
02 31	♂ □ ♆
02 32	♀ ⚼ ♅
02 53	☽ ⊥ ☿
05 51	☽ △ ♅
13 59	☽ ∠ ☉
14 04	☽ ⊻ ☿
14 56	☽ △ ♆
15 35	☽ ⊻ ♂
17 14	☽ # ♄
17 41	☽ ☍ ♇
17 49	♀ △ ♄
19 03	☽ ∥ ☿
20 08	☽ ✶ ♃
20 45	☿ ⊼ ♆
20 51	☽ # ♆
20 Wednesday	
01 01	☽ ∥ ♃
01 44	☽ ⊥ ♂
05 28	☽ ⊥ ☉
06 00	☽ ⚼ ♅
06 40	☽ ∥ ☉
07 14	☽ △ ♄
07 17	☽ # ♅
07 56	☿ ✶ ♂
08 26	☽ ☌ ♀
13 13	☿ # ♆
14 59	☽ ⚼ ♆
15 44	☽ ⊻ ☉
16 52	☽ ∠ ♂
17 52	☿ ⊼ ♇
20 22	☽ ± ♅
20 32	☽ ∠ ♃
21 Thursday	
05 19	☽ ± ♆
05 54	☽ ⊼ ♅
07 05	☽ ⚼ ♄
10 30	♂ ⊼ ♇
11 05	☽ ⊥ ♃
14 52	☽ ⊼ ♆
17 36	☽ ⊼ ♇
17 41	☿ ⊻ ♃
17 58	☽ ✶ ♂
20 47	☽ ⊻ ♃
21 13	☽ ☌ ☿
21 22	☽ ± ♄
22 20	☿ ± ♄
22 35	☉ ⚼ ♇
22 Friday	
03 10	☽ ± ♇
04 27	☿ ∥ ♃
06 55	☽ ⊼ ♄
12 25	☽ ⊻ ♀
14 07	☽ Q ♂
17 35	☽ ⚼ ♇
18 55	☽ ☌ ☉
22 55	☽ ⊥ ♀
23 Saturday	
05 51	☽ ☍ ♅
14 48	☽ ∠ ♀
15 03	☽ ☍ ♆
17 13	☿ ± ♇
17 54	☽ △ ♇
17 55	♀ ⚼ ♆
20 48	☽ □ ♂
21 59	☽ ☌ ♃
24 Sunday	
01 33	☽ # ♅
02 21	☽ ∥ ☉
04 01	☽ ∥ ☿
05 33	☽ ⊻ ☿
07 30	☽ ☍ ♄
08 48	☽ ∥ ♃
12 16	☽ # ♆
15 43	☽ # ♄
17 16	☽ ⊥ ☿
17 53	☽ ✶ ♀
18 43	☿ ⊼ ♄
22 26	☉ ∥ ☿
23 49	☽ ⊻ ☉
25 Monday	
06 29	☽ ∥ ♀
07 37	☽ ⊼ ♅
07 43	♂ □ ♃
10 20	☿ # ♅
10 48	☽ ⊥ ☉
11 15	☽ ∠ ☿
16 16	☽ Q ♀
17 25	☽ ⊼ ♆
17 56	☽ ± ♅
20 28	☽ □ ♇
26 Tuesday	
01 42	☽ ⊻ ♃
02 23	☽ △ ♂
03 37	☽ ∠ ☉
03 52	☽ ± ♆
09 41	☽ ⚼ ♅
10 46	☽ ⊼ ♄
11 15	♀ ± ♅
12 29	☽ ⊥ ♃
18 28	☽ ✶ ☿
19 51	☽ ⚼ ♆
21 27	☽ ± ♄
27 Wednesday	
03 15	☽ □ ♀
04 29	☽ Q ♇
04 58	☽ ∠ ♃
06 43	☽ ⚼ ♂
08 36	☽ ✶ ☉
12 42	☽ △ ♅
13 46	☽ ⚼ ♄
14 37	☽ ∥ ♂
20 40	☽ Q ☿
22 48	☿ ⚼ ♇
23 16	☽ △ ♆
28 Thursday	
00 17	☽ ± ♂
02 18	♀ ∠ ♃
02 36	☽ ✶ ♇
08 40	☽ Q ☉
09 15	☽ ✶ ♃
09 21	☽ # ♇
12 13	☽ ⊼ ♂
17 42	☽ △ ♄
20 06	☽ ∥ ♇
29 Friday	
02 49	☿ Q ♂
07 06	☽ ∠ ♇
08 36	☽ Q ♃
13 59	☽ □ ☿
14 43	☉ ☍ ♅
17 39	☽ △ ♀
17 57	☽ # ♂
21 32	☽ □ ♅
22 07	☽ □ ☉
30 Saturday	
00 35	☽ ⊥ ♇
08 48	☽ □ ♆
12 23	☽ ⊻ ♇
20 06	♀ ∠ ♂
20 34	☽ □ ♃

All ephemeris data is given at 12.00 UT and the Moon's longitude is additionally given for 24.00 UT

JULY 1990

LONGITUDES

Date	Sidereal time h m s	Sun ☉ ° ' "	Moon ☽ ° ' "	Moon ☽ 24.00 ° ' "	Mercury ☿ ° '	Venus ♀ ° '	Mars ♂ ° '	Jupiter ♃ ° '	Saturn ♄ ° '	Uranus ♅ ° '	Neptune ♆ ° '	Pluto ♇ ° '
01	06 37 07	09 ♋ 23 58	27 ♎ 00 07	02 ♏ 58 39	07 ♋ 54	07 ♊ 42	22 ♈ 22	19 ♋ 25	22 ♑ 59	07 ♑ 31	13 ♑ 18	15 ♏ 08
02	06 41 03	10 21 10	08 ♏ 55 51	14 ♏ 52 12	10 05	08 53	23 04	19 38	22 R 55	07 R 29	13 R 16	15 R 07
03	06 45 00	11 18 22	20 ♏ 48 12	26 ♏ 44 16	12 15	10 05	23 46	19 52	22 51	07 27	13 15	15 06
04	06 48 57	12 15 34	02 ♐ 40 49	08 ♐ 38 11	14 25	11 16	24 27	20 05	22 46	07 24	13 13	15 05
05	06 52 53	13 12 45	14 ♐ 36 44	20 ♐ 36 45	16 34	12 27	25 09	20 18	22 42	07 22	13 11	15 05
06	06 56 50	14 09 56	26 ♐ 38 28	02 ♑ 42 10	18 42	13 39	25 50	20 32	22 38	07 19	13 10	15 04
07	07 00 46	15 07 08	08 ♑ 48 00	14 ♑ 56 11	20 49	14 51	26 31	20 45	22 33	07 17	13 08	15 04
08	07 04 43	16 04 19	21 ♑ 06 53	27 ♑ 20 13	22 54	16 02	27 12	20 59	22 29	07 15	13 07	15 03
09	07 08 39	17 01 31	03 ♒ 36 22	09 ♒ 55 26	24 58	17 14	27 53	21 12	22 24	07 12	13 05	15 03
10	07 12 36	17 58 42	16 ♒ 17 35	22 ♒ 42 56	27 01	18 26	28 34	21 26	22 20	07 10	13 03	15 02
11	07 16 32	18 55 54	29 ♒ 11 37	05 ♓ 43 46	29 ♋ 01	19 37	29 15	21 39	22 16	07 07	13 02	15 02
12	07 20 29	19 53 06	12 ♓ 19 32	18 ♓ 59 01	01 ♌ 00	20 49	29 ♈ 55	21 53	22 11	07 05	13 00	15 01
13	07 24 26	20 50 19	25 ♓ 42 21	02 ♈ 29 38	02 57	22 01	00 ♉ 36	22 06	22 07	07 03	12 58	15 01
14	07 28 22	21 47 32	09 ♈ 20 54	16 ♈ 16 10	04 53	23 13	01 16	22 20	22 02	07 00	12 57	15 00
15	07 32 19	22 44 46	23 ♈ 15 25	00 ♉ 18 32	06 46	24 25	01 57	22 33	21 58	06 58	12 55	15 00
16	07 36 15	23 42 01	07 ♉ 25 18	14 ♉ 35 28	08 38	25 37	02 37	22 46	21 53	06 56	12 54	15 00
17	07 40 12	24 39 16	21 ♉ 48 40	29 ♉ 04 24	10 28	26 49	03 17	23 00	21 49	06 53	12 52	14 59
18	07 44 08	25 36 32	06 ♊ 22 06	13 ♊ 41 09	12 16	28 01	03 56	23 13	21 45	06 51	12 51	14 59
19	07 48 05	26 33 49	21 ♊ 00 47	28 ♊ 20 14	14 02	29 ♊ 13	04 36	23 27	21 40	06 49	12 49	14 59
20	07 52 01	27 31 06	05 ♋ 38 42	12 ♋ 55 23	15 47	00 ♋ 25	05 16	23 40	21 36	06 47	12 47	14 59
21	07 55 58	28 28 24	20 ♋ 09 29	27 ♋ 20 17	17 29	01 37	05 55	23 54	21 31	06 44	12 46	14 58
22	07 59 55	29 ♋ 25 43	04 ♌ 27 06	11 ♌ 29 23	19 10	02 49	06 34	24 07	21 27	06 42	12 44	14 58
23	08 03 51	00 ♌ 23 02	18 ♌ 26 41	25 ♌ 18 39	20 49	04 02	07 14	24 21	21 23	06 40	12 43	14 58
24	08 07 48	01 20 21	02 ♍ 05 04	08 ♍ 45 50	22 26	05 14	07 52	24 34	21 18	06 38	12 41	14 58
25	08 11 44	02 17 41	15 ♍ 20 58	21 ♍ 50 36	24 01	06 26	08 31	24 47	21 14	06 36	12 40	14 58
26	08 15 41	03 15 01	28 ♍ 14 56	04 ♎ 34 17	25 34	07 39	09 10	25 01	21 10	06 34	12 38	14 D 58
27	08 19 37	04 12 22	10 ♎ 49 01	16 ♎ 59 35	27 06	08 51	09 48	25 14	21 05	06 32	12 37	14 58
28	08 23 34	05 09 43	23 ♎ 06 28	29 ♎ 10 10	28 ♌ 35	10 04	10 27	25 28	21 01	06 30	12 35	14 58
29	08 27 30	06 07 04	05 ♏ 11 16	11 ♏ 10 18	00 ♍ 03	11 16	11 05	25 41	20 57	06 28	12 34	14 58
30	08 31 27	07 04 27	17 ♏ 07 50	23 ♏ 04 28	01 29	12 29	11 43	25 54	20 53	06 26	12 32	14 58
31	08 35 24	08 ♌ 01 49	29 ♏ 00 44	04 ♐ 57 12	02 ♍ 53	13 ♋ 41	12 ♉ 20	26 ♋ 07	20 ♑ 48	06 ♑ 24	12 ♑ 31	14 ♏ 59

DECLINATIONS

Date	Moon True ☊ ° '	Moon Mean ☊ ° '	Moon ☽ Latitude ° '	Sun ☉ ° '	Moon ☽ ° '	Mercury ☿ ° '	Venus ♀ ° '	Mars ♂ ° '	Jupiter ♃ ° '	Saturn ♄ ° '	Uranus ♅ ° '	Neptune ♆ ° '	Pluto ♇ ° '
01	07 ♒ 40	08 ♒ 51	05 S 08	23 N 07	15 S 11	24 N 22	20 N 06	06 N 48	22 N 15	21 S 23	23 S 33	21 S 54	01 S 18
02	07 R 38	08 48	05 13	23 02	19 25	24 20	20 19	07 04	22 13	21 24	23 33	21 55	01 18
03	07 34	08 44	05 05	22 58	22 51	24 16	20 32	07 19	22 11	21 25	23 34	21 55	01 18
04	07 30	08 41	04 44	22 53	25 20	24 09	20 44	07 34	22 09	21 26	23 34	21 55	01 18
05	07 26	08 38	04 10	22 47	26 41	23 59	20 56	07 49	22 07	21 27	23 34	21 55	01 19
06	07 22	08 35	03 25	22 41	26 49	23 46	21 07	08 04	22 05	21 27	23 34	21 55	01 19
07	07 19	08 32	02 30	22 35	25 39	23 31	21 18	08 18	22 03	21 28	23 34	21 55	01 19
08	07 18	08 29	01 28	22 28	23 14	23 14	21 28	08 33	22 01	21 29	23 34	21 56	01 20
09	07 17	08 25	00 S 20	22 21	19 41	22 55	21 37	08 48	21 59	21 30	23 34	21 56	01 20
10	07 D 17	08 22	00 N 50	22 14	15 10	22 33	21 46	09 02	21 57	21 31	23 35	21 56	01 20
11	07 18	08 19	01 58	22 06	09 55	22 09	21 55	09 16	21 55	21 31	23 35	21 56	01 21
12	07 19	08 16	03 01	21 58	04 S 08	21 44	22 02	09 31	21 53	21 32	23 35	21 56	01 21
13	07 20	08 13	03 56	21 50	01 N 55	21 17	22 09	09 45	21 51	21 33	23 35	21 56	01 22
14	07 21	08 10	04 39	21 41	07 59	20 48	22 16	09 59	21 49	21 34	23 35	21 57	01 22
15	07 R 22	08 06	05 06	21 31	13 47	20 18	22 22	10 13	21 47	21 35	23 35	21 57	01 22
16	07 21	08 03	05 16	21 22	18 58	19 47	22 27	10 26	21 45	21 35	23 35	21 57	01 23
17	07 20	08 00	05 06	21 12	23 08	19 15	22 32	10 40	21 43	21 36	23 35	21 57	01 23
18	07 19	07 57	04 36	21 01	25 55	18 41	22 36	10 54	21 40	21 37	23 36	21 57	01 24
19	07 18	07 54	03 49	20 51	26 56	18 07	22 39	11 07	21 38	21 38	23 36	21 57	01 24
20	07 17	07 50	02 46	20 40	26 05	17 32	22 42	11 20	21 36	21 39	23 36	21 58	01 25
21	07 16	07 47	01 33	20 28	23 28	16 56	22 44	11 33	21 34	21 39	23 36	21 58	01 25
22	07 16	07 44	00 N 16	20 16	19 24	16 19	22 46	11 46	21 32	21 40	23 36	21 58	01 26
23	07 D 16	07 41	01 S 02	20 04	14 19	15 42	22 47	11 59	21 29	21 41	23 36	21 58	01 26
24	07 16	07 38	02 13	19 52	08 40	15 05	22 47	12 12	21 27	21 42	23 36	21 58	01 27
25	07 17	07 35	03 16	19 39	02 N 46	14 27	22 46	12 24	21 25	21 42	23 36	21 58	01 27
26	07 17	07 31	04 06	19 26	03 S 04	13 49	22 45	12 37	21 22	21 43	23 36	21 59	01 28
27	07 17	07 28	04 44	19 13	08 38	13 10	22 44	12 49	21 20	21 44	23 37	21 59	01 28
28	07 18	07 25	05 07	18 59	13 44	12 32	22 41	13 01	21 18	21 45	23 37	21 59	01 29
29	07 18	07 22	05 17	18 45	18 13	11 53	22 38	13 13	21 15	21 45	23 37	21 59	01 30
30	07 R 18	07 19	05 12	18 30	21 56	11 15	22 35	13 25	21 13	21 46	23 37	21 59	01 30
31	07 ♒ 18	07 ♒ 16	04 S 55	18 N 16	24 S 44	10 N 37	22 N 30	13 N 37	21 N 11	21 S 47	23 S 37	21 S 59	01 S 31

ZODIAC SIGN ENTRIES

Date	h	m	Planets
01	18	01	☽ ♏
04	06	35	☽ ♐
06	18	39	☽ ♑
09	05	07	☽ ♒
11	13	29	☽ ♓
11	23	48	☿ ♌
12	14	44	♂ ♉
13	19	36	☽ ♈
15	23	29	☽ ♉
18	01	32	☽ ♊
20	02	44	☽ ♋
20	03	41	♀ ♋
22	04	29	☽ ♌
23	02	22	☉ ♌
24	08	17	☽ ♍
26	15	18	☽ ♎
29	01	39	☽ ♏
29	11	10	☿ ♍
31	14	00	☽ ♐

LATITUDES

Date	Mercury ☿ ° '	Venus ♀ ° '	Mars ♂ ° '	Jupiter ♃ ° '	Saturn ♄ ° '	Uranus ♅ ° '	Neptune ♆ ° '	Pluto ♇ ° '
01	01 N 09	01 S 31	02 S 03	00 N 13	00 N 06	00 S 20	00 N 53	15 N 44
04	01 29	01 24	02 04	00 13	00 05	00 20	00 52	15 43
07	01 43	01 18	02 04	00 13	00 05	00 20	00 52	15 41
10	01 50	01 10	02 04	00 13	00 05	00 20	00 52	15 39
13	01 50	01 03	02 04	00 14	00 05	00 20	00 52	15 38
16	01 44	00 55	02 04	00 14	00 04	00 20	00 52	15 36
19	01 34	00 47	02 04	00 14	00 04	00 20	00 52	15 35
22	01 18	00 39	02 03	00 14	00 04	00 20	00 52	15 33
25	00 59	00 31	02 03	00 15	00 04	00 20	00 52	15 31
28	00 36	00 22	02 02	00 15	00 03	00 20	00 52	15 30
31	00 N 10	00 S 14	02 S 01	00 N 15	00 N 03	00 S 20	00 N 52	15 N 28

DATA

Julian Date	2448074
Delta T	+57 seconds
Ayanamsa	23° 43' 41"
Synetic vernal point	05° ♓ 23' 18"
True obliquity of ecliptic	23° 26' 31"

MOON'S PHASES, APSIDES AND POSITIONS ☽

Date	h	m	Phase	Longitude ° '	Eclipse Indicator
08	01	23	○	15 ♑ 39	
15	11	04	☾	22 ♈ 43	
22	02	54	●	29 ♋ 04	Total
29	14	01	☽	06 ♏ 12	

Day	h	m		
03	15	43	Apogee	
19	11	07	Perigee	
31	08	17	Apogee	
06	02	20	Max dec	26° S 55'
13	04	29	0N	
19	13	09	Max dec	26° N 57'
25	23	17	0S	

LONGITUDES

Date	Chiron ⚷ ° '	Ceres ⚳ ° '	Pallas ⚴ ° '	Juno ⚵ ° '	Vesta ⚶ ° '	Black Moon Lilith ⚸ ° '
01	17 ♋ 44	04 ♌ 48	25 ♊ 20	09 ♏ 23	11 ♉ 03	26 ♏ 45
11	18 ♋ 36	09 ♌ 13	01 ♋ 15	09 ♏ 29	14 ♉ 42	27 ♏ 52
21	19 ♋ 52	13 ♌ 41	07 ♋ 58	10 ♏ 05	18 ♉ 07	28 ♏ 59
31	20 ♋ 56	18 ♌ 12	13 ♋ 10	11 ♏ 09	21 ♉ 16	00 ♐ 06

All ephemeris data is given at 12.00 UT and the Moon's longitude is additionally given for 24.00 UT

ASPECTARIAN

h m	Aspects
01 Sunday	
02 11	☽ ☍ ♂
02 27	☽ bQ ♀
04 01	☽ □ ♄
04 06	♀ ± ♆
07 08	☿ ⊻ ♀
07 56	☿ ☍ ♅
08 35	♀ ⊼ ♅
09 03	☽ Q ♅
17 28	♀ bQ ♄
20 36	☽ Q ♆
22 28	☽ ± ♀
02 Monday	
07 11	♂ □ ♄
09 05	☽ ⚹ ♅
11 54	☽ ⊼ ♀
14 50	☽ △ ☿
15 07	☽ △ ☉
15 59	☽ Q ♄
17 21	☉ ☌ ☿
18 11	☽ # ♀
20 45	☽ ⚹ ♆
03 Tuesday	
00 30	☽ ☌ ♇
01 12	☽ ∥ ♄
04 48	☽ ∥ ♆
06 52	☽ # ♃
10 03	☽ △ ♃
12 52	☽ # ☉
15 18	☽ ∠ ♅
16 06	☽ ⚹ ♄
17 57	☽ ∥ ♅
18 21	☽ ⊼ ♂
22 50	☿ ☍ ♆
23 53	☽ # ☿
04 Wednesday	
00 06	☽ bQ ☉
03 01	☽ ∠ ♆
03 57	☽ bQ ☿
07 14	☽ ± ♂
09 26	☽ ⊥ ♅
16 56	☽ bQ ♃
19 31	☿ △ ♇
19 51	☽ ± ☉
21 07	☽ ⊥ ♆
21 29	☽ ⊻ ♅
22 12	☽ ∠ ♄
05 Thursday	
02 06	☽ ± ☿
02 29	☽ bQ ♂
07 12	☽ ☍ ♀
08 57	☽ ⊼ ☉
09 09	☽ ⊻ ♆
11 23	☽ ± ♃
11 27	☉ ☍ ♆
12 57	☽ ⊻ ♇
16 09	☽ ⊥ ♄
16 46	☽ ⊼ ☿
23 36	☽ ⊼ ♃
06 Friday	
00 56	☽ ⊥ ♇
02 25	♀ ⊼ ♆
04 04	☽ ⊻ ♄
10 18	☽ △ ♂
18 48	☽ ∠ ♇
07 Saturday	
08 00	☿ # ♅
09 02	☽ ☌ ♅
09 50	♀ ⊥ ♃
10 37	☉ △ ♇
11 15	☿ ☌ ♃
12 46	☿ ⊥ ♀
16 24	♀ ⊼ ♇
20 28	☽ ☌ ♆
08 Sunday	
00 14	☽ ⚹ ♇
01 05	☽ ⊼ ♀
01 23	☽ ☍ ☉
07 17	☿ ☍ ♄
09 14	☽ ∥ ♅
11 44	☽ ☍ ♃
11 57	☽ # ☿
13 58	☽ ± ♀
14 38	☽ ☌ ♄
15 08	♀ # ♄
15 35	☉ ⊻ ♀
16 09	☽ ☍ ☿
17 58	☽ # ☉
20 25	♀ ± ♄
21 06	☽ # ♃
21 40	☽ ∥ ♆
23 27	☽ Q ♇
09 Monday	
00 16	☽ # ♀
00 25	☽ □ ♂
00 38	☽ ∥ ♄
09 06	☽ bQ ♀
18 49	☽ ⊻ ♅
10 Tuesday	
05 55	☽ ⊻ ♆
06 08	☽ ⊥ ♅
09 38	☽ □ ♇
12 33	☽ Q ♂
15 25	☽ ⊼ ☉
16 24	☽ △ ♀
17 10	☽ ⊥ ♆
21 46	☽ ⊼ ♃
22 56	☽ ∠ ♅
23 13	☽ ⊻ ♄
11 Wednesday	
02 59	♃ # ♆
03 29	☽ ± ☉
09 06	☽ ± ♃
09 51	☽ ∠ ♆
10 17	☽ ⊥ ♄
11 37	☽ ⊼ ☿
12 06	☽ ⚹ ♂
13 37	♀ ∥ ♃
14 38	☽ # ♂
16 06	☿ □ ♂
16 33	☉ ∥ ☿
16 37	♀ # ♆
21 24	☽ bQ ☉
22 53	☿ ∥ ♀
12 Thursday	
00 37	☽ ± ☿
00 45	☿ # ♆
01 55	☽ bQ ♃
02 30	☽ ⚹ ♅
02 43	☽ ∠ ♄
02 48	☿ ∥ ♃
05 40	☉ ∥ ♀
13 13	☽ ⚹ ♆
16 02	♀ ± ♇
16 52	☽ △ ♇
16 57	☽ ∠ ♂
17 17	☉ # ♆
19 47	☽ bQ ☿
22 20	☿ # ♄
23 07	☽ ∥ ♇
13 Friday	
00 09	☽ Q ♅
02 40	☽ △ ☉
04 47	☽ □ ♀
05 28	☽ △ ♃
05 38	☽ ⚹ ♄
06 26	☉ ∥ ♃
09 50	☽ # ♇
09 56	☽ ⊥ ♂
10 42	☽ Q ♆
12 53	♃ ☍ ♄
13 50	♀ ⊼ ♄
14 08	♀ ⊻ ♃
19 37	☽ bQ ♇
21 07	☽ ⊻ ♂
14 Saturday	
02 46	☽ Q ♄
02 55	☽ △ ☿
07 55	☽ □ ♅
11 24	☽ ± ♇
15 33	☽ Q ♀
17 44	☉ ☍ ♄
18 15	☽ □ ♆
20 25	☽ ∥ ♂
21 49	☽ ⊼ ♇
15 Sunday	
04 24	☉ # ♄
05 32	☉ ☌ ♃
09 48	☽ □ ♄
10 46	☽ □ ♃
11 04	☽ □ ☉
14 09	☽ ⚹ ♀
14 26	☿ ⊼ ♅
16 Monday	
03 30	☽ ☌ ♂
11 10	☽ △ ♅
14 20	☽ □ ☿
15 48	☽ ∥ ☿
17 43	☽ Q ♃
17 50	☽ ∠ ♀
19 41	☽ Q ☉
21 09	☽ △ ♆
17 Tuesday	
00 26	☽ ∥ ☉
00 40	☽ ☍ ♇
02 18	☽ # ♄
03 02	☽ ∥ ♃
04 24	☽ # ♆
07 55	☽ ∥ ♀
10 12	☽ ⊥ ♀
12 00	☽ △ ♄
12 08	☽ bQ ♅
14 00	☽ ⚹ ♃
15 08	☽ # ♅
17 02	☽ ⚹ ☉
21 01	☽ ⊻ ♀
22 00	☽ bQ ♆
18 Wednesday	
00 34	☽ Q ☿
02 57	☽ ± ♅
07 49	☽ ⊻ ♂
12 37	☽ bQ ♄
12 47	☽ ± ♆
12 48	☽ ⊼ ♅
15 05	☽ ∠ ♃
18 08	☽ ⊥ ♂
19 27	☽ ∠ ☉
19 38	☿ ⊼ ♆
19 43	☿ ± ♅
22 36	☽ ⊼ ♆
23 01	☽ ⚹ ☿
19 Thursday	
02 07	☽ ⊼ ♇
03 17	☽ ± ♄
06 04	☽ ⊥ ♃
09 35	☽ ∠ ♂
11 13	☽ ⊥ ☉
11 57	☽ ± ♇
13 04	☽ ⊼ ♄
15 51	♃ # ♄
16 03	☽ ⊻ ♃
19 46	☿ ∠ ♀
21 44	☽ ⊻ ☉
20 Friday	
00 56	☿ □ ♇
02 39	☽ ☌ ♀
02 41	☽ bQ ♇
02 56	☽ ∠ ☿
03 15	♀ bQ ♇
11 20	☽ ⚹ ♂
13 52	☽ ☍ ♅
19 43	☽ ⊥ ☿
23 46	☽ ☍ ♆
21 Saturday	
03 24	☽ △ ♇
06 58	☽ ⊻ ☿
08 06	☽ Q ♂
11 04	☽ # ♅
14 16	☽ ☍ ♄
16 52	☽ ∥ ♀
18 20	☽ ☌ ♃
21 46	☽ # ♆
23 34	☽ # ♄
22 Sunday	
00 15	☽ ∥ ♃
02 54	☽ ☌ ☉
05 59	☿ ± ♆
07 11	☽ ∥ ☉
08 59	☽ ⊻ ♀
15 47	☽ □ ♂
15 49	☽ ⊼ ♅
16 33	♂ △ ♅
20 08	☽ ⊥ ♀
23 Monday	
02 03	☽ ± ♅
02 07	☽ ⊼ ♆
05 01	☽ ∥ ☿
05 59	☽ □ ♇
12 28	☽ ± ♆
13 07	☽ ∠ ♀
16 40	☽ ☌ ☿
17 05	☽ ⊼ ♄
17 36	☽ bQ ♅
19 59	☿ ⊼ ♄
21 45	☽ ∥ ♂
22 28	☽ ⊻ ♃
24 Tuesday	
00 24	☿ bQ ♅
03 33	☽ ± ♄
04 12	☽ bQ ♆
09 15	☽ ⊥ ♃
10 34	☽ ⊻ ☉
13 35	☽ Q ♇
18 12	☽ ⚹ ♀
19 31	☽ bQ ♄
20 08	☽ △ ♅
22 09	☽ ⊥ ☉
22 55	☽ △ ♂
25 Wednesday	
01 41	☽ ∠ ♃
07 06	☽ △ ♆
11 18	☽ ⚹ ♇
15 02	♀ ☍ ♅
15 52	☽ ∠ ☉
17 19	☽ # ♇
18 16	☽ Q ♀
22 48	☽ △ ♄
26 Thursday	
01 25	♇ St D
01 55	☿ ⊻ ♃
03 55	☽ bQ ♂
05 18	☽ ∥ ♇
05 49	☽ ⚹ ♃
06 16	☽ ⊻ ☿
15 15	☽ ∠ ♇
19 09	☽ ⊥ ☿
21 49	☽ ± ♂
22 16	☽ ⚹ ☉
27 Friday	
03 46	☽ □ ♅
04 59	☽ Q ♃
07 49	☽ □ ♀
08 26	☽ ⊥ ♇
09 56	☽ ⊼ ♂
11 54	☿ ± ♄
14 49	☽ ∠ ☿
15 28	☽ □ ♆
20 03	☽ ⊻ ♇
20 09	☿ bQ ♆
22 06	☿ ∥ ♂
23 21	☽ Q ☉
28 Saturday	
06 47	☽ # ☿
07 54	☽ □ ♄
08 20	☽ # ♂
14 44	☽ Q ♅
16 44	☽ □ ♃
29 Sunday	
00 21	☽ ⚹ ☿
02 47	☽ Q ♆
03 54	♀ ⚹ ♂
14 01	☽ □ ☉
14 32	☽ ⚹ ♅
14 55	☽ # ☉
19 29	☽ Q ♄
20 18	☉ ⊼ ♅
30 Monday	
00 28	☽ ☍ ♂
01 35	☽ △ ♀
02 46	☽ ⚹ ♆
03 39	☽ Q ☿
06 56	☽ # ♃
07 39	☽ ☌ ♇
10 46	☽ ∥ ♄
12 22	☽ ∥ ♆
13 08	♀ ☍ ♆
16 42	☽ # ♀
19 31	☽ ⚹ ♄
20 31	☿ ⊥ ♃
20 39	☽ ∠ ♅
31 Tuesday	
01 22	☽ ∥ ♅
06 03	☽ △ ♃
08 59	☽ ∠ ♆
11 17	☽ bQ ♀
13 41	☿ Q ♇
14 47	☽ ⊥ ♅
18 31	♂ △ ♆
20 50	☽ □ ☿

AUGUST 1990

Date	Sidereal time h m s	Sun ☉ ° ' "	Moon ☽ ° ' "	Moon ☽ 24.00 ° ' "	Mercury ☿ ° '	Venus ♀ ° '	Mars ♂ ° '	Jupiter ♃ ° '	Saturn ♄ ° '	Uranus ♅ ° '	Neptune ♆ ° '	Pluto ♇ ° '
01	08 39 20	08 ♌ 59 13	10 ♐ 54 23	16 ♐ 52 47	04 ♍ 15	14 ♋ 54	12 ♉ 58	26 ♋ 21	20 ♑ 44	06 ♑ 22	12 ♑ 30	14 ♏ 59
02	08 43 17	09 56 37	22 ♐ 52 53	28 ♐ 55 06	05 35	16 07	13 35	26 34	20 R 40	06 R 20	12 R 28	14 59
03	08 47 13	10 54 02	04 ♑ 59 49	11 ♑ 07 24	06 53	17 20	14 12	26 47	20 36	06 18	12 27	14 59
04	08 51 10	11 51 27	17 ♑ 18 08	23 ♑ 32 17	08 09	18 32	14 50	27 00	20 32	06 16	12 25	15 00
05	08 55 06	12 48 53	29 ♑ 50 02	06 ♒ 11 31	09 23	19 45	15 26	27 14	20 28	06 14	12 24	15 00
06	08 59 03	13 46 21	12 ♒ 36 51	19 ♒ 06 02	10 34	20 58	16 03	27 27	20 24	06 13	12 23	15 00
07	09 02 59	14 43 49	25 ♒ 39 04	02 ♓ 15 54	11 43	22 11	16 39	27 40	20 21	06 11	12 21	15 01
08	09 06 56	15 41 18	08 ♓ 56 24	15 ♓ 40 27	12 50	23 24	17 16	27 53	20 17	06 09	12 20	15 01
09	09 10 53	16 38 48	22 ♓ 27 51	29 ♓ 18 24	13 55	24 37	17 52	28 06	20 13	06 07	12 19	15 02
10	09 14 49	17 36 20	06 ♈ 11 52	13 ♈ 08 01	14 57	25 50	18 27	28 19	20 09	06 06	12 18	15 02
11	09 18 46	18 33 53	20 ♈ 06 34	27 ♈ 07 18	15 56	27 03	19 03	28 32	20 06	06 04	12 16	15 03
12	09 22 42	19 31 27	04 ♉ 09 54	11 ♉ 14 07	16 53	28 16	19 38	28 45	20 02	06 03	12 15	15 03
13	09 26 39	20 29 03	18 ♉ 19 39	25 ♉ 26 14	17 46	29 ♋ 29	20 14	28 58	19 59	06 01	12 14	15 04
14	09 30 35	21 26 40	02 ♊ 33 33	09 ♊ 41 19	18 37	00 ♌ 42	20 48	29 11	19 55	06 00	12 13	15 05
15	09 34 32	22 24 19	16 ♊ 49 11	23 ♊ 56 50	19 24	01 56	21 23	29 24	19 52	05 58	12 12	15 05
16	09 38 28	23 22 00	01 ♋ 03 53	08 ♋ 10 00	20 08	03 09	21 58	29 37	19 48	05 57	12 10	15 06
17	09 42 25	24 19 42	15 ♋ 14 45	22 ♋ 17 45	20 48	04 22	22 32	29 ♋ 50	19 45	05 56	12 09	15 07
18	09 46 22	25 17 26	29 ♋ 18 35	06 ♌ 16 52	21 25	05 36	23 06	00 ♌ 02	19 42	05 54	12 08	15 08
19	09 50 18	26 15 11	13 ♌ 12 11	20 ♌ 04 10	21 57	06 49	23 39	00 15	19 39	05 53	12 07	15 08
20	09 54 15	27 12 57	26 ♌ 52 30	03 ♍ 36 53	22 25	08 03	24 13	00 28	19 36	05 52	12 06	15 09
21	09 58 11	28 10 44	10 ♍ 17 04	16 ♍ 52 54	22 49	09 16	24 46	00 40	19 33	05 51	12 05	15 10
22	10 02 08	29 ♌ 08 33	23 ♍ 24 16	29 ♍ 51 09	23 08	10 30	25 19	00 53	19 30	05 49	12 04	15 11
23	10 06 04	00 ♍ 06 24	06 ♎ 13 37	12 ♎ 31 47	23 22	11 43	25 51	01 06	19 27	05 48	12 03	15 12
24	10 10 01	01 04 15	18 ♎ 45 52	24 ♎ 56 07	23 31	12 57	26 23	01 18	19 24	05 47	12 02	15 13
25	10 13 57	02 02 08	01 ♏ 02 55	07 ♏ 06 38	23 34	14 11	26 55	01 31	19 22	05 46	12 01	15 14
26	10 17 54	03 00 02	13 ♏ 07 45	19 ♏ 06 46	23 R 32	15 24	27 27	01 43	19 19	05 45	12 01	15 15
27	10 21 51	03 57 57	25 ♏ 04 12	01 ♐ 00 38	23 23	16 38	27 58	01 55	19 16	05 44	12 00	15 16
28	10 25 47	04 55 54	06 ♐ 56 40	12 ♐ 52 53	23 09	17 52	28 29	02 08	19 14	05 43	11 59	15 17
29	10 29 44	05 53 52	18 ♐ 49 54	24 ♐ 48 20	22 49	19 06	29 00	02 20	19 12	05 43	11 58	15 18
30	10 33 40	06 51 52	00 ♑ 48 46	06 ♑ 51 46	22 23	20 19	29 ♉ 30	02 32	19 09	05 43	11 58	15 19
31	10 37 37	07 ♍ 49 52	12 ♑ 57 54	19 ♑ 07 40	21 ♍ 50	21 ♌ 33	00 ♊ 00	02 ♌ 44	19 ♑ 07	05 ♑ 41	11 ♑ 57	15 ♏ 21

Date	Moon True ☊ ° '	Moon Mean ☊ ° '	Moon ☽ Latitude ° '
01	07 ♒ 18	07 ♒ 12	04 S 24
02	07 D 18	07 09	03 42
03	07 18	07 06	02 50
04	07 19	07 03	01 49
05	07 19	07 00	00 S 41
06	07 R 19	06 56	00 N 29
07	07 19	06 53	01 40
08	07 18	06 50	02 46
09	07 17	06 47	03 45
10	07 16	06 44	04 31
11	07 15	06 41	05 02
12	07 14	06 37	05 16
13	07 13	06 34	05 10
14	07 D 14	06 31	04 46
15	07 14	06 28	04 04
16	07 15	06 25	03 07
17	07 17	06 21	01 59
18	07 17	06 18	00 N 44
19	07 R 17	06 15	00 S 33
20	07 16	06 12	01 46
21	07 15	06 09	02 52
22	07 12	06 06	03 47
23	07 08	06 02	04 29
24	07 05	05 59	04 58
25	07 01	05 56	05 12
26	06 59	05 53	05 12
27	06 57	05 50	04 58
28	06 D 57	05 47	04 32
29	06 57	05 43	03 54
30	06 59	05 40	03 05
31	07 ♒ 01	05 ♒ 37	02 S 08

DECLINATIONS

Date	Sun ☉ ° '	Moon ☽ ° '	Mercury ☿ ° '	Venus ♀ ° '	Mars ♂ ° '	Jupiter ♃ ° '	Saturn ♄ ° '	Uranus ♅ ° '	Neptune ♆ ° '	Pluto ♇ ° '
01	18 N 01	26 S 26	09 N 58	22 N 25	13 N 48	21 N 08	21 S 48	23 S 37	22 S 00	01 S 31
02	17 45	26 57	09 20	22 20	14 00	21 06	21 48	23 37	22 00	01 32
03	17 30	26 10	08 43	22 13	14 11	21 03	21 49	23 37	22 00	01 33
04	17 14	24 07	08 05	22 06	14 22	21 01	21 50	23 37	22 00	01 33
05	16 58	20 52	07 28	21 59	14 33	20 58	21 50	23 37	22 00	01 34
06	16 42	16 33	06 52	21 51	14 44	20 56	21 51	23 37	22 00	01 35
07	16 25	11 24	06 16	21 42	14 55	20 54	21 52	23 38	22 00	01 35
08	16 08	05 S 39	05 41	21 32	15 05	20 51	21 52	23 38	22 01	01 36
09	15 51	00 N 27	05 06	21 22	15 16	20 49	21 53	23 38	22 01	01 37
10	15 34	06 36	04 32	21 11	15 26	20 46	21 54	23 38	22 01	01 37
11	15 16	12 31	04 00	21 00	15 36	20 43	21 54	23 38	22 01	01 38
12	14 58	17 51	03 28	20 48	15 46	20 41	21 55	23 38	22 01	01 39
13	14 40	22 15	02 57	20 36	15 56	20 38	21 55	23 38	22 01	01 39
14	14 21	25 21	02 27	20 22	16 06	20 36	21 56	23 38	22 01	01 40
15	14 03	26 50	01 58	20 09	16 15	20 33	21 57	23 38	22 02	01 41
16	13 44	26 33	01 31	19 54	16 25	20 31	21 57	23 38	22 02	01 42
17	13 25	24 32	01 05	19 40	16 34	20 28	21 58	23 38	22 02	01 42
18	13 06	21 01	00 41	19 24	16 43	20 25	21 58	23 38	22 02	01 43
19	12 46	16 20	00 N 19	19 08	16 52	20 23	21 59	23 38	22 02	01 44
20	12 26	10 54	00 S 01	18 52	17 01	20 20	21 59	23 38	22 02	01 45
21	12 06	05 N 04	00 20	18 34	17 09	20 18	22 00	23 38	22 02	01 45
22	11 46	00 S 51	00 36	18 17	17 18	20 15	22 00	23 38	22 02	01 46
23	11 26	06 36	00 50	17 59	17 26	20 12	22 01	23 39	22 03	01 47
24	11 06	11 56	01 01	17 40	17 34	20 10	22 01	23 39	22 03	01 48
25	10 45	16 42	01 09	17 21	17 42	20 07	22 02	23 39	22 03	01 48
26	10 24	20 44	01 15	17 01	17 50	20 04	22 02	23 39	22 03	01 49
27	10 03	23 51	01 17	16 41	17 58	20 02	22 03	23 39	22 03	01 50
28	09 42	25 56	01 16	16 20	18 05	19 59	22 03	23 39	22 03	01 51
29	09 21	26 51	01 12	15 59	18 13	19 57	22 04	23 39	22 03	01 52
30	09 00	26 32	01 04	15 38	18 20	19 54	22 04	23 39	22 03	01 52
31	08 N 38	24 S 56	00 S 53	15 N 16	18 N 27	19 N 51	22 S 04	23 S 39	22 S 03	01 S 53

ZODIAC SIGN ENTRIES

Date	h	m	Planets
03	02	09	☽ ♑
05	12	19	☽ ♒
07	19	54	☽ ♓
10	01	13	☽ ♈
12	04	55	☽ ♉
13	22	05	♀ ♌
14	07	41	☽ ♊
16	10	12	☽ ♋
18	07	30	♃ ♌
18	13	18	☽ ♌
20	17	33	☽ ♍
23	00	17	☽ ♎
23	09	21	☉ ♍
25	09	56	☽ ♏
27	21	57	☽ ♐
30	10	23	☽ ♑
31	11	40	♂ ♊

LATITUDES

Date	Mercury ☿ ° '	Venus ♀ ° '	Mars ♂ ° '	Jupiter ♃ ° '	Saturn ♄ ° '	Uranus ♅ ° '	Neptune ♆ ° '	Pluto ♇ ° '
01	00 N 01	00 S 11	02 S 01	00 N 15	00 N 03	00 S 20	00 N 52	15 N 27
04	00 S 28	00 S 03	02 00	00 16	00 03	00 20	00 52	15 25
07	00 58	00 N 05	01 59	00 16	00 02	00 20	00 52	15 24
10	01 30	00 13	01 58	00 16	00 02	00 20	00 52	15 22
13	02 03	00 20	01 56	00 17	00 02	00 20	00 52	15 20
16	02 36	00 28	01 55	00 17	00 02	00 20	00 52	15 19
19	03 08	00 35	01 53	00 17	00 01	00 20	00 52	15 17
22	03 37	00 42	01 51	00 18	00 01	00 20	00 51	15 15
25	04 02	00 48	01 49	00 18	00 01	00 20	00 51	15 14
28	04 21	00 54	01 47	00 18	00 01	00 20	00 51	15 12
31	04 S 29	01 N 00	01 S 44	00 N 19	00 N 00	00 S 20	00 N 51	15 N 10

DATA

Julian Date	2448105
Delta T	+57 seconds
Ayanamsa	23° 43' 46"
Synetic vernal point	05° ♓ 23' 13"
True obliquity of ecliptic	23° 26' 31"

MOON'S PHASES, APSIDES AND POSITIONS ☽

Date	h	m	Phase	Longitude ° '	Eclipse Indicator
06	14	19	○	13 ♒ 52	partial
13	15	54	☾	20 ♉ 38	
20	12	39	●	27 ♌ 15	
28	07	34	☽	04 ♐ 45	

Day	h	m		
15	09	58	Perigee	
28	02	58	Apogee	
02	09	38	Max dec	26° S 57'
09	10	15	0N	
15	20	15	Max dec	26° N 56'
22	08	30	0S	
29	17	50	Max dec	26° S 54'

LONGITUDES

Date	Chiron ⚷ ° '	Ceres ⚳ ° '	Pallas ⚴ ° '	Juno ⚵ ° '	Vesta ⚶ ° '	Black Moon Lilith ⚸ ° '
01	21 ♋ 02	18 ♌ 39	13 ♋ 46	11 ♏ 17	21 ♉ 34	00 ♐ 13
11	22 ♋ 04	23 ♌ 11	19 ♋ 42	12 ♏ 47	24 ♉ 22	01 ♐ 20
21	23 ♋ 03	27 ♌ 45	25 ♋ 36	14 ♏ 38	26 ♉ 47	02 ♐ 27
31	23 ♋ 58	02 ♍ 19	01 ♌ 26	16 ♏ 47	28 ♉ 44	03 ♐ 34

ASPECTARIAN

h m	Aspects
01 Wednesday	
01 39	☽ ∠ ♄
02 52	☽ ⊻ ♅
03 07	☽ ⊥ ♆
07 30	☽ ± ♀
07 48	☽ △ ☉
12 54	☽ bQ ♃
13 33	♀ △ ♇
15 11	☽ ⊻ ♆
16 22	☽ ⊼ ♂
19 40	☽ ⊥ ♄
20 12	☽ ⊻ ♇
20 56	☽ ⊼ ♀
02 Thursday	
05 04	☽ ± ♂
07 18	☽ ± ♃
07 37	☽ ⊻ ♄
08 13	☽ ⊥ ♇
13 34	☿ bQ ♄
16 28	☽ bQ ☉
19 28	☽ ⊼ ♃
23 58	☽ bQ ♂
03 Friday	
01 26	☿ △ ♅
02 07	☽ ∠ ♇
11 48	☽ ± ☉
14 33	☽ ☌ ♅
16 08	☽ △ ☿
04 Saturday	
00 32	☽ ⊼ ☉
02 33	☽ ☌ ♆
06 57	☽ △ ♂
07 32	☽ ⚹ ♇
14 39	☽ ☍ ♀
16 17	☽ ∥ ♅
18 12	☽ ☌ ♄
18 39	♂ ☍ ♇
22 00	☉ ± ♅
23 06	♂ Q ♃
05 Sunday	
00 29	☽ bQ ☿
01 51	☉ ⊼ ♆
04 22	☽ # ♀
04 30	☽ ∥ ♆
05 38	☽ ∥ ♄
06 37	☽ Q ♇
06 58	☽ ☍ ♃
07 57	♀ # ♆
11 17	☽ # ♃
19 24	☽ ± ☿
06 Monday	
00 04	☽ ⊻ ♅
01 29	♀ ☍ ♄
07 48	☽ ⊼ ☿
10 45	♀ # ♄
11 15	☽ ⊥ ♅
11 15	☽ # ☉
11 34	☽ ⊻ ♆
14 19	☽ ☍ ☉
16 26	☽ □ ♇
18 41	☽ □ ♂
20 35	☽ # ♂
22 39	☽ ⊥ ♆
07 Tuesday	
02 20	☽ ⊻ ♄
03 50	☽ ∠ ♅
05 01	☽ ⊼ ♀
13 15	☽ ⊥ ♄
15 06	☽ ∠ ♆
15 44	☽ ⊼ ♃
17 04	☽ ± ♀
19 08	☉ □ ♇
08 Wednesday	
01 15	☿ △ ♆
02 46	☽ ± ♃
05 05	☽ Q ♂
05 28	☽ ∠ ♄
07 01	☽ ⚹ ♅
10 56	☽ bQ ♀
11 52	☽ # ☿
13 14	☿ ∠ ♃
18 03	☽ ⚹ ♆
19 09	☽ bQ ♃
19 34	☽ ☍ ☿
22 51	☽ △ ♇
09 Thursday	
00 57	☽ ⊼ ☉
03 30	☽ ⚹ ♂
04 00	☽ ∥ ♇
04 21	☽ Q ♅
08 03	☽ ⚹ ♄
12 21	☽ ± ☉
15 15	☽ Q ♆
16 09	☽ △ ♀
16 31	☽ # ♇
22 03	☽ △ ♃
10 Friday	
01 16	☽ bQ ♇
04 35	☽ ∥ ☿
05 00	☽ Q ♄
05 17	☽ bQ ☉
07 01	☽ ∠ ♂
11 50	☽ □ ♅
14 09	☿ ⚹ ♇
16 55	☽ ± ♇
18 26	☉ ∥ ♂
22 32	☽ □ ♆
23 19	☽ ⊥ ♂
11 Saturday	
03 17	☽ ⊼ ♇
04 17	☽ ⊼ ☿
04 50	☉ ± ♆
09 09	☽ △ ☉
10 06	☽ ⊻ ♂
11 58	☽ □ ♄
15 22	☽ ± ☿
23 18	☽ ∥ ☉
12 Sunday	
01 00	☽ □ ♀
01 54	☽ ∥ ♂
02 38	☽ □ ♃
07 50	☽ bQ ☿
15 11	☽ △ ♅
19 27	☉ □ ♂
23 37	♀ ☌ ♃
13 Monday	
00 01	☉ ⊼ ♄
01 42	☽ △ ♆
02 34	☽ ∥ ♃
02 41	☽ ∥ ♀
02 41	♂ △ ♄
05 29	♀ ∥ ♃
06 29	☽ ☍ ♇
09 40	☽ Q ♃
09 59	☽ # ♄
10 27	☽ Q ♀
10 35	☽ # ♆
11 00	☽ △ ☿
14 46	☽ △ ♄
15 21	☽ ☌ ♂
15 54	☽ □ ☉
16 32	☽ bQ ♅
21 24	☽ # ♅
14 Tuesday	
01 03	☉ bQ ♅
03 00	☽ bQ ♆
06 14	☽ ⚹ ♃
07 41	☽ ± ♅
08 35	☽ ⚹ ♀
15 57	☽ bQ ♄
17 46	☽ ⊼ ♅
18 09	☽ ± ♆
19 28	♂ bQ ♅
15 Wednesday	
00 26	☽ Q ☉
04 14	☽ ⊼ ♆
07 02	☽ ± ♄
07 52	☽ ∠ ♃
09 05	☽ ⊼ ♇
12 12	☽ ∠ ♀
16 35	☽ □ ☿
17 06	☽ ⊼ ♄
19 11	☽ ± ♇
20 00	☽ ⊻ ♂
22 05	☽ ⚹ ☉
23 15	☽ ⊥ ♃
16 Thursday	
01 51	☿ △ ♄
02 58	☿ # ♇
04 47	☽ ⊥ ♀
06 32	☽ ⊥ ♂
09 31	☽ ⊻ ♃
10 22	☽ bQ ♇
15 51	☽ ⊻ ♀
20 14	☽ ☍ ♅
22 22	☽ ∠ ♂
17 Friday	
00 33	☽ Q ☿
01 14	☽ ∠ ☉
06 46	☽ ☍ ♆
11 46	☽ △ ♇
17 37	☽ ⊥ ☉
17 41	♀ Q ♂
19 06	☽ # ♅
19 38	☽ ☍ ♄
21 54	☽ ⚹ ☿
18 Saturday	
00 55	☽ ⚹ ♂
04 37	☽ ⊻ ☉
05 55	☽ # ♆
06 17	☽ # ♄
13 16	☽ ☌ ♃
15 22	☽ ∥ ♃
17 55	♀ ⊼ ♅
21 24	☽ ∥ ♀
21 39	☉ ± ♄
22 22	☽ Q ♂
23 20	☽ ⊼ ♅
23 52	☽ ☌ ♀
19 Sunday	
00 44	☽ ∠ ☿
09 35	☽ ∥ ♂
09 42	☽ ± ♅
10 07	☽ ⊼ ♆
15 23	☽ □ ♇
16 21	☿ ∠ ♀
16 58	☽ ⊥ ☿
20 34	☽ ± ♆
23 13	☽ ⊼ ♄
20 Monday	
01 24	☽ bQ ♅
03 52	☽ ⊻ ☿
05 01	☽ ∥ ☉
07 05	☽ □ ♂
09 15	☉ bQ ♆
09 45	☽ ± ♄
12 24	☽ bQ ♆
12 39	☽ ☌ ☉
18 29	☽ ⊻ ♃
23 11	☽ Q ♇
21 Tuesday	
01 42	☽ bQ ♄
04 00	☽ △ ♅
05 23	☽ ⊥ ♃
09 59	☽ ⊻ ♀
15 16	☽ △ ♆
20 53	☽ ⚹ ♇
21 57	☽ ∠ ♃
21 59	☽ ⊥ ♀
22 Wednesday	
01 21	☽ # ♇
04 49	☽ △ ♄
06 19	☽ # ☿
10 54	☽ ∥ ☿
11 29	☽ ☌ ☿
15 41	☽ △ ♂
15 45	☽ ∥ ♇
16 17	☽ ∠ ♀
23 32	☽ ⊻ ☉
23 Thursday	
00 38	☽ ∠ ♇
02 10	☽ ⚹ ♃
11 12	☽ □ ♅
11 45	☽ ⊥ ☉
13 36	♀ ± ♅
17 39	☽ ⊥ ♇
18 25	♀ ⊼ ♆
21 11	☽ bQ ♂
23 05	☽ □ ♆
23 35	☽ ⚹ ♀
24 Friday	
01 18	☽ Q ♃
05 09	☽ ⊻ ♇
06 22	☽ ∠ ☉
08 18	☽ # ☉
13 14	☽ □ ♄
15 17	☽ ± ♂
17 14	♀ ∥ ♂
19 18	☉ ⊻ ♃
21 17	☽ ⊻ ☿
21 45	☽ Q ♅
25 Saturday	
01 22	☽ Q ♀
03 31	☽ ⊼ ♂
09 05	☽ ⊥ ☿
09 59	☽ Q ♆
12 55	☽ □ ♃
14 07	☽ ⚹ ☉
14 08	☿ St R
15 17	☽ # ♀
16 30	♂ bQ ♆
17 42	☽ # ♂
21 20	☽ ⚹ ♅
26 Sunday	
00 27	☽ Q ♄
02 52	☽ ∠ ☿
07 47	☽ # ♃
08 55	♀ □ ♇
09 46	☽ ⚹ ♆
16 04	☽ Q ☉
16 15	☽ ☌ ♇
17 05	☽ □ ♀
18 17	☉ Q ♇
21 12	☽ ∥ ♄
21 15	☽ ∥ ♆
27 Monday	
00 22	☽ ⚹ ♄
03 17	☽ ∠ ♅
08 40	☽ ⚹ ☿
10 07	☽ ∥ ♅
15 53	☽ ∠ ♆
18 07	☽ ☍ ♂
19 20	☉ bQ ♄
21 25	☽ ⊥ ♅
28 Tuesday	
00 37	☿ ⊥ ♀
02 05	☽ △ ♃
04 06	♄ ∥ ♆
06 32	☽ ∠ ♄
07 34	☽ □ ☉
08 28	☽ Q ☿
09 32	☽ ⊻ ♅
10 03	☽ ⊥ ♆
14 17	♀ ± ♆
22 10	☽ ⊻ ♆
29 Wednesday	
00 40	☽ ⊥ ♄
04 53	☽ ⊻ ♇
07 25	☉ △ ♅
08 55	☽ bQ ♃
12 35	☽ △ ♀
12 44	☽ ⊻ ♄
13 55	♀ ⊼ ♄
16 59	☽ ⊥ ♇
19 45	☽ □ ☿
30 Thursday	
03 19	☽ ± ♃
09 17	☽ ⊼ ♂
11 01	☽ ∠ ♇
15 29	☽ ⊼ ♃
19 14	♀ bQ ♅
20 55	♂ ± ♅
21 41	☽ ☌ ♅
21 43	☽ ± ♂
21 58	☽ bQ ♀
31 Friday	
01 02	☽ △ ☉
10 00	☽ ☌ ♆
15 44	☿ ⊻ ♀
16 09	☽ bQ ♂
16 39	☽ ⚹ ♇
17 37	☽ ± ♀
23 57	☽ ☌ ♄

All ephemeris data is given at 12.00 UT and the Moon's longitude is additionally given for 24.00 UT

SEPTEMBER 1990

LONGITUDES

Date	Sidereal time h m s	Sun ☉ ° ' "	Moon ☽ ° ' "	Moon ☽ 24.00 ° ' "	Mercury ☿ ° '	Venus ♀ ° '	Mars ♂ ° '	Jupiter ♃ ° '	Saturn ♄ ° '	Uranus ♅ ° '	Neptune ♆ ° '	Pluto ♇ ° '
01	10 41 33	08 ♍ 47 54	25 ♑ 21 30	01 ♒ 39 49	21 ♍ 12	22 ♌ 47	00 ♊ 30	02 ♌ 56	19 ♑ 05	05 ♑ 40	11 ♑ 56	15 ♏ 22
02	10 45 30	09 45 58	08 ♒ 02 56	14 ♒ 31 04	20 R 29	24 01	01 00	03 08	19 R 03	05 R 40	11 R 55	15 23
03	10 49 26	10 44 03	21 ♒ 04 23	27 ♒ 42 53	19 40	25 15	01 29	03 20	19 01	05 39	11 54	15 24
04	10 53 23	11 42 10	04 ♓ 26 32	11 ♓ 15 09	18 48	26 29	01 57	03 32	18 59	05 39	11 54	15 26
05	10 57 20	12 40 18	18 ♓ 08 26	25 ♓ 05 59	17 52	27 43	02 26	03 44	18 58	05 38	11 53	15 27
06	11 01 16	13 38 28	02 ♈ 07 20	09 ♈ 11 54	16 54	28 ♌ 57	02 53	03 55	18 56	05 38	11 53	15 28
07	11 05 13	14 36 39	16 ♈ 19 06	23 ♈ 28 15	15 55	00 ♍ 11	03 21	04 07	18 54	05 37	11 52	15 30
08	11 09 09	15 34 53	00 ♉ 38 44	07 ♉ 49 52	14 56	01 25	03 48	04 19	18 53	05 37	11 52	15 31
09	11 13 06	16 33 09	15 ♉ 01 04	22 ♉ 11 48	13 58	02 40	04 15	04 30	18 51	05 37	11 51	15 33
10	11 17 02	17 31 26	29 ♉ 21 33	06 ♊ 29 56	13 03	03 54	04 41	04 42	18 50	05 36	11 51	15 34
11	11 20 59	18 29 46	13 ♊ 36 38	20 ♊ 41 21	12 13	05 08	05 07	04 53	18 49	05 36	11 50	15 36
12	11 24 55	19 28 08	27 ♊ 43 57	04 ♋ 44 15	11 27	06 22	05 33	05 04	18 48	05 36	11 50	15 37
13	11 28 52	20 26 33	11 ♋ 42 11	18 ♋ 37 40	10 49	07 37	05 58	05 16	18 47	05 36	11 50	15 39
14	11 32 49	21 24 59	25 ♋ 30 40	02 ♌ 21 08	10 17	08 51	06 22	05 27	18 46	05 36	11 49	15 40
15	11 36 45	22 23 28	09 ♌ 09 02	15 ♌ 54 17	09 54	10 05	06 47	05 38	18 45	05 D 36	11 49	15 42
16	11 40 42	23 21 58	22 ♌ 36 50	29 ♌ 16 34	09 40	11 20	07 10	05 49	18 44	05 36	11 49	15 44
17	11 44 38	24 20 30	05 ♍ 53 25	12 ♍ 27 14	09 35	12 34	07 34	06 00	18 44	05 36	11 49	15 45
18	11 48 35	25 19 05	18 ♍ 57 57	25 ♍ 25 26	09 D 40	13 49	07 56	06 11	18 43	05 36	11 48	15 47
19	11 52 31	26 17 41	01 ♎ 49 37	08 ♎ 10 27	09 55	15 03	08 18	06 21	18 43	05 37	11 48	15 49
20	11 56 28	27 16 19	14 ♎ 27 54	20 ♎ 42 02	10 18	16 18	08 40	06 32	18 43	05 37	11 48	15 51
21	12 00 24	28 14 59	26 ♎ 52 54	03 ♏ 00 39	10 51	17 32	09 01	06 43	18 42	05 37	11 48	15 52
22	12 04 21	29 ♍ 13 41	09 ♏ 05 30	15 ♏ 07 42	11 33	18 47	09 22	06 53	18 42	05 38	11 48	15 54
23	12 08 18	00 ♎ 12 25	21 ♏ 07 36	27 ♏ 05 33	12 24	20 02	09 42	07 03	18 D 42	05 38	11 48	15 56
24	12 12 14	01 11 11	03 ♐ 02 01	08 ♐ 57 29	13 22	21 16	10 01	07 14	18 42	05 38	11 D 48	15 58
25	12 16 11	02 09 58	14 ♐ 52 30	20 ♐ 47 39	14 27	22 31	10 20	07 24	18 43	05 39	11 48	16 00
26	12 20 07	03 08 47	26 ♐ 43 32	02 ♑ 40 48	15 39	23 46	10 39	07 34	18 43	05 40	11 48	16 02
27	12 24 04	04 07 38	08 ♑ 40 06	14 ♑ 42 07	16 57	25 00	10 57	07 44	18 43	05 40	11 48	16 03
28	12 28 00	05 06 31	20 ♑ 47 29	26 ♑ 56 51	18 19	26 15	11 14	07 54	18 44	05 41	11 48	16 05
29	12 31 57	06 05 25	03 ♒ 10 50	09 ♒ 30 00	19 47	27 30	11 30	08 04	18 44	05 42	11 48	16 07
30	12 35 53	07 ♎ 04 21	15 ♒ 54 50	22 ♒ 25 45	21 ♍ 18	28 ♍ 45	11 ♊ 46	08 ♌ 13	18 ♑ 45	05 ♑ 42	11 ♑ 49	16 ♏ 09

Moon True ☊, Moon Mean ☊, Moon ☽ Latitude

Date	Moon True ☊ ° '	Moon Mean ☊ ° '	Moon ☽ Latitude ° '
01	07 ♒ 02	05 ♒ 34	01 S 03
02	07 R 03	05 31	00 N 05
03	07 02	05 27	01 16
04	06 59	05 24	02 24
05	06 55	05 21	03 25
06	06 50	05 18	04 15
07	06 44	05 15	04 50
08	06 39	05 12	05 08
09	06 34	05 08	05 06
10	06 31	05 05	04 46
11	06 29	05 02	04 07
12	06 D 30	04 59	03 14
13	06 31	04 56	02 10
14	06 32	04 53	00 N 59
15	06 R 32	04 49	00 S 14
16	06 31	04 46	01 26
17	06 28	04 43	02 31
18	06 22	04 40	03 28
19	06 14	04 37	04 13
20	06 05	04 33	04 44
21	05 55	04 30	05 02
22	05 46	04 27	05 05
23	05 38	04 24	04 55
24	05 32	04 21	04 31
25	05 28	04 18	03 57
26	05 26	04 14	03 12
27	05 D 26	04 11	02 18
28	05 27	04 08	01 18
29	05 27	04 05	00 12
30	05 ♒ 27	04 ♒ 02	00 N 56

DECLINATIONS

Date	Sun ☉ ° '	Moon ☽ ° '	Mercury ☿ ° '	Venus ♀ ° '	Mars ♂ ° '	Jupiter ♃ ° '	Saturn ♄ ° '	Uranus ♅ ° '	Neptune ♆ ° '	Pluto ♇ ° '
01	08 N 16	22 S 07	00 S 37	14 N 54	18 N 34	19 N 49	22 S 05	23 S 39	22 S 03	01 S 54
02	07 54	18 10	00 S 19	14 31	18 41	19 46	22 05	23 39	22 04	01 55
03	07 33	13 17	00 N 04	14 08	18 48	19 43	22 05	23 39	22 04	01 56
04	07 10	07 39	00 30	13 44	18 55	19 41	22 06	23 39	22 04	01 57
05	06 48	01 S 33	00 58	13 20	19 01	19 38	22 06	23 39	22 04	01 58
06	06 26	04 N 45	01 30	12 56	19 07	19 35	22 06	23 39	22 04	01 58
07	06 04	10 53	02 03	12 31	19 14	19 33	22 07	23 39	22 04	01 59
08	05 41	16 30	02 38	12 06	19 20	19 30	22 07	23 39	22 04	02 00
09	05 19	21 13	03 14	11 40	19 26	19 28	22 07	23 39	22 04	02 01
10	04 56	24 40	03 50	11 15	19 32	19 25	22 08	23 39	22 04	02 02
11	04 33	26 32	04 25	10 49	19 37	19 22	22 08	23 39	22 04	02 03
12	04 10	26 40	05 00	10 22	19 43	19 20	22 08	23 39	22 04	02 04
13	03 47	25 05	05 32	09 56	19 48	19 17	22 08	23 39	22 04	02 04
14	03 24	22 01	06 02	09 29	19 54	19 15	22 09	23 39	22 05	02 05
15	03 01	17 45	06 29	09 01	19 59	19 12	22 09	23 39	22 05	02 06
16	02 38	12 38	06 52	08 34	20 04	19 09	22 09	23 39	22 05	02 07
17	02 15	07 00	07 12	08 06	20 09	19 07	22 09	23 39	22 05	02 08
18	01 52	01 N 11	07 28	07 38	20 14	19 04	22 09	23 39	22 05	02 09
19	01 28	04 S 36	07 39	07 10	20 19	19 02	22 09	23 39	22 05	02 10
20	01 05	10 04	07 45	06 42	20 24	18 59	22 09	23 39	22 05	02 11
21	00 42	15 03	07 47	06 13	20 28	18 57	22 10	23 39	22 05	02 11
22	00 N 18	19 20	07 45	05 44	20 33	18 54	22 10	23 39	22 05	02 12
23	00 S 05	22 46	07 38	05 16	20 37	18 52	22 10	23 39	22 05	02 13
24	00 28	25 12	07 27	04 46	20 42	18 49	22 10	23 39	22 05	02 14
25	00 52	26 30	07 12	04 17	20 46	18 47	22 10	23 39	22 05	02 15
26	01 15	26 36	06 53	03 48	20 50	18 44	22 10	23 39	22 05	02 16
27	01 38	25 27	06 31	03 18	20 54	18 42	22 10	23 39	22 05	02 17
28	02 02	23 07	06 05	02 49	20 58	18 39	22 10	23 39	22 05	02 18
29	02 25	19 39	05 36	02 19	21 02	18 37	22 10	23 39	22 05	02 18
30	02 S 48	15 S 11	05 N 04	01 N 49	21 N 06	18 N 35	22 S 10	23 S 38	22 S 05	02 S 19

ZODIAC SIGN ENTRIES

Date	h	m	Planets
01	20	51	☽ ♒
04	04	06	☽ ♓
06	08	23	☽ ♈
07	08	21	♀ ♍
08	10	55	☽ ♉
10	13	05	☽ ♊
12	15	53	☽ ♋
14	19	52	☽ ♌
17	01	19	☽ ♍
19	08	34	☽ ♎
21	18	06	☽ ♏
23	06	56	☉ ♎
24	05	52	☽ ♐
26	18	36	☽ ♑
29	05	54	☽ ♒

LATITUDES

Date	Mercury ☿ ° '	Venus ♀ ° '	Mars ♂ ° '	Jupiter ♃ ° '	Saturn ♄ ° '	Uranus ♅ ° '	Neptune ♆ ° '	Pluto ♇ ° '
01	04 S 28	01 N 02	01 S 43	00 N 19	00 00	00 S 20	00 N 51	15 N 10
04	04 17	01 07	01 41	00 19	00 00	00 20	00 51	15 08
07	03 48	01 11	01 38	00 20	00 00	00 20	00 51	15 07
10	03 04	01 15	01 35	00 20	00 S 01	00 20	00 51	15 05
13	02 08	01 18	01 31	00 20	00 01	00 19	00 51	15 04
16	01 09	01 21	01 28	00 21	00 01	00 19	00 51	15 03
19	00 S 14	01 23	01 24	00 21	00 01	00 19	00 50	15 01
22	00 N 34	01 25	01 19	00 22	00 02	00 19	00 50	15 00
25	01 10	01 26	01 15	00 22	00 02	00 19	00 50	14 59
28	01 35	01 26	01 10	00 23	00 02	00 19	00 50	14 58
31	01 N 50	01 N 26	01 S 05	00 N 23	00 S 02	00 S 19	00 N 50	14 N 57

DATA

Julian Date	2448136
Delta T	+57 seconds
Ayanamsa	23° 43' 51"
Synetic vernal point	05° ♓ 23' 09"
True obliquity of ecliptic	23° 26' 32"

MOON'S PHASES, APSIDES AND POSITIONS ☽

Date	h	m	Phase	Longitude ° '	Eclipse Indicator
05	01	46	○	12 ♓ 15	
11	20	53	☾	18 ♊ 51	
19	00	46	●	25 ♍ 50	
27	02	06	☽	03 ♑ 43	

Day	h	m		
09	11	24	Perigee	
24	22	16	Apogee	
05	17	55	0N	
12	01	49	Max dec	26° N 49'
18	16	51	0S	
26	01	50	Max dec	26° S 42'

LONGITUDES

Date	Chiron ⚷ ° '	Ceres ⚳ ° '	Pallas ⚴ ° '	Juno ⚵ ° '	Vesta ⚶ ° '	Black Moon Lilith ⚸ ° '
01	24 ♋ 03	02 ♍ 47	02 ♌ 01	17 ♏ 01	28 ♉ 55	03 ♐ 41
11	24 ♋ 53	07 ♍ 21	07 ♌ 45	19 ♏ 27	00 ♊ 17	04 ♐ 48
21	25 ♋ 37	11 ♍ 54	13 ♌ 22	22 ♏ 06	01 ♊ 03	05 ♐ 55
31	26 ♋ 14	16 ♍ 26	18 ♌ 49	24 ♏ 57	01 ♊ 07	07 ♐ 02

All ephemeris data is given at 12.00 UT and the Moon's longitude is additionally given for 24.00 UT

ASPECTARIAN

01 Saturday
00 11 ☽ ∥ ♅
04 25 ☽ △ ☿
06 31 ☽ ⊼ ♀
08 45 ☽ bQ ☉
12 12 ☽ ∥ ♄
12 21 ☽ ∥ ♆
15 50 ☽ Q ♇
16 17 ☉ ⊥ ♃
22 12 ☽ △ ♂

02 Sunday
02 38 ☽ ☍ ♃
02 54 ☽ # ♃
03 18 ☽ ± ☉
07 28 ☽ bQ ☿
07 32 ☽ ⊻ ♅
09 14 ☽ # ♂
15 28 ☽ ⊼ ☉
18 43 ☽ ⊥ ♅
19 12 ☽ ⊻ ♆
23 15 ☽ ± ☿

03 Monday
01 37 ☽ □ ♇
06 14 ☽ ⊥ ♆
07 36 ♀ ± ♄
07 48 ☽ # ♀
08 16 ☽ ⊻ ♄
09 36 ☽ ⊼ ☿
11 14 ☽ ∠ ♅
19 08 ☽ ⊥ ♄
20 20 ☽ ☍ ♀
22 33 ☽ ∠ ♆

04 Tuesday
06 48 ☿ △ ♄
07 25 ☽ □ ♂
10 22 ☽ ⊼ ♃
11 12 ☽ ∠ ♄
14 03 ☽ # ☉
14 08 ☽ ⚹ ♅
16 47 ☉ △ ♆
17 49 ☿ ∠ ♃
19 59 ♀ bQ ♆
21 07 ☽ ± ♃

05 Wednesday
01 07 ☽ ⚹ ♆
01 46 ☽ ☍ ☉
07 19 ☽ △ ♇
10 25 ☽ ∥ ♇
11 08 ☽ Q ♅
11 34 ☽ ☍ ☿
13 02 ☽ bQ ♃
13 25 ☽ ⚹ ♄
14 02 ☽ # ☿
16 05 ☽ Q ♂
21 55 ☽ Q ♆
22 28 ☽ ∥ ☿

06 Thursday
01 24 ☽ # ♇
06 05 ☽ ⊼ ♀
09 11 ☽ bQ ♇
09 59 ☽ Q ♄
13 21 ☽ ⚹ ♂
15 06 ☽ △ ♃
17 16 ☽ ± ♀
17 57 ☽ □ ♅
18 07 ☽ ∥ ☉

07 Friday
00 29 ☽ ± ♇
04 31 ☽ □ ♆
08 55 ☽ ⊼ ☉
09 17 ☿ # ♇
09 55 ☽ bQ ♀
10 37 ☽ ⊼ ♇
11 22 ☽ ⊼ ☿
15 32 ☽ ∠ ♂
16 20 ☽ □ ♄
18 14 ☽ ∥ ♀
19 44 ☽ ± ☉
20 47 ☽ ± ☿
21 56 ☿ ⚹ ♇

08 Saturday
04 00 ☉ ☌ ☿
07 06 ☽ ⊥ ♂
10 27 ☉ ⚹ ♇
10 53 ☽ bQ ☿
11 53 ☽ bQ ☉
13 25 ☽ △ ♀
15 58 ♂ bQ ♄
17 27 ☽ ⊻ ♂
18 12 ☽ □ ♃
20 18 ☽ △ ♅

09 Sunday
02 00 ☽ ∥ ♂
02 27 ☽ ∥ ♃
06 43 ☽ △ ♆
10 21 ☽ △ ☿
12 53 ☽ ☍ ♇
14 45 ☽ △ ☉
16 58 ♂ ∥ ♃
17 06 ☽ # ♆
17 26 ☽ # ♄
18 24 ☽ △ ♄
21 21 ☽ bQ ♅

10 Monday
00 41 ☽ Q ♃
03 48 ☽ # ♅
05 31 ♀ Q ♇
07 47 ☽ bQ ♆
10 51 ♀ bQ ♄
12 25 ☽ ± ♅
12 39 ♂ ⚹ ♃
19 31 ☽ bQ ♄
20 21 ☽ □ ♀
21 05 ☽ ⚹ ♃
21 14 ☽ ☌ ♂
22 30 ☽ ⊼ ♅
22 54 ☽ ± ♆

11 Tuesday
06 17 ♀ ⊻ ♃
09 00 ☽ ⊼ ♆
09 46 ☽ □ ☿
10 40 ☽ ± ♄
11 35 ♀ □ ♂
15 11 ☉ ∥ ☿
15 22 ☽ ⊼ ♇
19 46 ☉ △ ♄
20 49 ☽ ⊼ ♄
20 53 ☽ □ ☉
21 06 ♀ △ ♅
22 47 ☽ ∠ ♃
23 33 ☿ △ ♆

12 Wednesday
01 34 ☽ ± ♇
05 43 ☽ Q ♀
14 19 ☽ ⊥ ♃
14 48 ☽ Q ☿
15 12 ♂ ⊼ ♅
16 57 ☽ bQ ♇
22 35 ☿ ⊥ ♃

13 Thursday
00 45 ☽ ⊻ ♃
01 29 ☽ ☍ ♅
01 48 ☽ ⊻ ♂
04 09 ♂ ± ♆
04 15 ☽ ⚹ ♀
05 57 ☽ Q ☉
06 28 ☉ ∠ ♃
10 31 ☽ ⚹ ☿
12 13 ☽ ☍ ♆
12 28 ☽ ⊥ ♂
18 50 ☽ △ ♇

14 Friday
00 15 ☽ ☍ ♄
00 34 ☽ # ♅
04 19 ☽ ⚹ ☉
04 34 ☽ ∠ ♂
08 49 ☽ ∠ ♀
11 10 ☽ # ♄
11 36 ☽ # ♆
11 37 ☽ ∠ ☿
18 31 ♅ St D

15 Saturday
00 21 ☽ ∥ ♂
02 10 ☽ ⊥ ♀
02 58 ☽ ⊥ ☿
04 21 ☽ ∥ ♃
05 42 ☽ ☌ ♃
05 43 ☽ ⊼ ♅
07 40 ☽ ⚹ ♂
07 55 ♃ ⊼ ♅
08 39 ☽ ∠ ☉
09 08 ☿ ☌ ♀
13 18 ☽ ⊻ ☿
13 50 ☽ ⊻ ♀
16 21 ☽ ± ♅
16 44 ☽ ⊼ ♆
23 40 ☽ □ ♇

16 Sunday
01 52 ☽ ⊥ ☉
03 24 ☽ ± ♆
05 04 ☽ ⊼ ♄
05 39 ☽ Q ♂
08 24 ☽ bQ ♅
13 27 ☽ ⊻ ☉
15 49 ☽ ± ♄
19 33 ☽ bQ ♆
21 16 ♀ △ ♆
22 58 ♀ ⊥ ♃

17 Monday
07 01 ☽ ∥ ♀
08 05 ☽ bQ ♄
08 07 ☽ Q ♇
11 13 ☽ ∥ ☿
11 29 ☽ △ ♅
12 06 ☿ St D
12 12 ☽ ⊻ ♃
15 08 ☽ □ ♂
18 46 ☽ ☌ ☿
18 58 ☉ # ♇
22 49 ☽ △ ♆
23 19 ☽ ⊥ ♃

18 Tuesday
01 30 ☽ ☌ ♀
06 07 ☽ ⚹ ♇
08 02 ☽ # ♇
08 59 ☽ ∥ ☉
11 33 ☽ △ ♄
16 09 ☽ ∠ ♃
18 21 ☿ ∥ ♀
23 45 ☽ # ☉

19 Wednesday
00 46 ☽ ☌ ☉
01 46 ☽ ∥ ♇
10 05 ☽ ∠ ♇
19 09 ☽ □ ♅
20 41 ☽ ⚹ ♃
22 12 ☽ # ♀

20 Thursday
00 37 ☽ △ ♂
01 28 ☽ # ☿
02 59 ♀ ⚹ ♇
03 09 ☽ ⊥ ♇
03 46 ☽ ⊻ ☿
06 54 ☽ □ ♆
14 39 ☽ ⊻ ♇
15 40 ☽ ⊥ ☿
15 54 ☽ ⊻ ♀
19 56 ☽ Q ♃
20 09 ☽ □ ♄

21 Friday
04 46 ☽ ⊥ ♀
05 39 ☽ Q ♅
06 16 ☽ bQ ♂
09 54 ☽ ∠ ☿
14 54 ☽ ⊻ ☉
17 42 ☽ Q ♆

22 Saturday
00 20 ☽ ∠ ♀
00 22 ☽ ± ♂
03 42 ☽ ⊥ ☉
05 09 ☽ ⚹ ♅
07 17 ☽ Q ♄
07 34 ☽ □ ♃
09 22 ☽ # ♃
10 29 ♀ △ ♄
12 33 ☽ ⊼ ♂
17 13 ☽ ⚹ ☿
17 22 ☽ ⚹ ♆
19 23 ☿ △ ♆
19 54 ☽ # ♂
23 06 ☽ ∠ ☉

23 Sunday
01 35 ☽ ☌ ♇
05 12 ♄ St D
06 35 ☽ ∥ ♆
07 09 ☽ ⚹ ♄
07 12 ☽ ∥ ♄
09 32 ☽ ⚹ ♀
11 01 ☽ ∠ ♅
18 36 ♆ St D
19 07 ☽ Q ☿
19 29 ☽ ∥ ♅
23 24 ☽ ∠ ♆

24 Monday
05 08 ☽ ⊥ ♅
06 21 ☉ ∠ ♇
07 56 ☽ ⚹ ☉
08 18 ☿ ⊥ ♃
12 32 ☽ Q ♀
13 22 ☽ ∠ ♄
17 17 ☽ ⊻ ♅
17 36 ☽ ⊥ ♆
20 37 ☽ △ ♃

25 Tuesday
02 33 ☽ ☍ ♂
05 46 ☽ ⊻ ♆
07 36 ☽ ⊥ ♄
09 25 ♀ ∠ ♃
10 26 ☽ Q ☉
11 03 ☽ □ ☿
14 17 ☽ ⊻ ♇
19 47 ☽ ⊻ ♄

26 Wednesday
02 28 ☽ ⊥ ♇
03 28 ☽ bQ ♃
05 18 ☽ □ ♀
19 24 ☿ ⚹ ♇
20 42 ☽ ∠ ♇
21 54 ☽ ± ♃

27 Thursday
02 06 ☽ □ ☉
06 00 ☽ ☌ ♅
10 06 ☽ ⊼ ♃
16 39 ☽ ⊼ ♂
18 15 ☽ ☌ ♆

28 Friday
02 43 ☽ ⚹ ♇
04 50 ☽ ± ♂
06 31 ☽ △ ☿
07 29 ☽ ∥ ♅
07 57 ☽ ☌ ♄
18 48 ☿ △ ♄
19 21 ☽ ∥ ♄
19 56 ☽ ∥ ♆
22 51 ☽ bQ ♂
23 50 ☽ △ ♀

29 Saturday
02 10 ☉ □ ♅
02 15 ☽ Q ♇
03 24 ☽ # ♂
04 50 ☉ ∥ ♇
09 15 ☉ # ♀
12 29 ♀ # ♇
15 28 ☽ bQ ☿
16 48 ☽ ⊻ ♅
18 01 ☽ △ ☉
18 03 ☽ # ♃
21 24 ☽ ☍ ♃

30 Sunday
04 06 ☽ △ ♂
04 08 ☽ ⊥ ♅
04 20 ☽ ⊻ ♆
07 31 ☽ bQ ♀
10 43 ☽ ± ☿
12 27 ☽ □ ♇
15 31 ☽ ⊥ ♆
15 49 ♂ ⊼ ♆
17 15 ☽ ⊻ ♄
20 51 ☽ ∠ ♅
23 18 ☽ ⊼ ☿

OCTOBER 1990

Date	Sidereal time h m s	Sun ☉ ° ′ ″	Moon ☽ ° ′ ″	Moon ☽ 24.00 ° ′ ″	Mercury ☿ ° ′	Venus ♀ ° ′	Mars ♂ ° ′	Jupiter ♃ ° ′	Saturn ♄ ° ′	Uranus ♅ ° ′	Neptune ♆ ° ′	Pluto ♇ ° ′
01	12 39 50	08 ♎ 03 19	29 ♒ 03 02	05 ♓ 46 53	22 ♍ 53	29 ♍ 59	12 ♊ 01	08 ♌ 23	18 ♑ 46	05 ♑ 43	11 ♑ 49	16 ♏ 11
02	12 43 47	09 02 18	12 ♓ 37 18	19 ♓ 34 10	24 31	01 ♎ 14	12 16	08 32	18 47	05 44	11 49	16 13
03	12 47 43	10 01 20	26 ♓ 37 09	03 ♈ 45 46	26 11	02 29	12 30	08 42	18 48	05 45	11 49	16 15
04	12 51 40	11 00 23	10 ♈ 59 21	18 ♈ 17 06	27 52	03 44	12 43	08 51	18 49	05 46	11 50	16 17
05	12 55 36	11 59 28	25 ♈ 38 03	03 ♉ 01 13	29 ♍ 35	04 59	12 56	09 00	18 50	05 47	11 50	16 20
06	12 59 33	12 58 36	10 ♉ 25 31	17 ♉ 49 53	01 ♎ 19	06 14	13 08	09 09	18 51	05 48	11 51	16 22
07	13 03 29	13 57 46	25 ♉ 13 19	02 ♊ 34 55	03 04	07 28	13 19	09 18	18 52	05 49	11 51	16 24
08	13 07 26	14 56 58	09 ♊ 53 52	17 ♊ 09 31	04 50	08 43	13 30	09 27	18 54	05 50	11 51	16 26
09	13 11 22	15 56 12	24 ♊ 21 24	01 ♋ 29 10	06 35	09 58	13 39	09 36	18 55	05 52	11 52	16 28
10	13 15 19	16 55 29	08 ♋ 32 36	15 ♋ 31 39	08 21	11 13	13 48	09 44	18 57	05 53	11 52	16 30
11	13 19 16	17 54 48	22 ♋ 26 20	29 ♋ 16 45	10 07	12 28	13 57	09 53	18 59	05 54	11 53	16 32
12	13 23 12	18 54 10	06 ♌ 03 05	12 ♌ 45 33	11 52	13 43	14 04	10 01	19 01	05 56	11 54	16 35
13	13 27 09	19 53 33	19 ♌ 24 20	25 ♌ 59 42	13 38	14 58	14 11	10 09	19 03	05 57	11 54	16 37
14	13 31 05	20 52 59	02 ♍ 31 49	09 ♍ 00 54	15 23	16 13	14 17	10 17	19 05	05 59	11 55	16 39
15	13 35 02	21 52 28	15 ♍ 27 06	21 ♍ 50 32	17 07	17 28	14 22	10 25	19 07	06 00	11 56	16 41
16	13 38 58	22 51 58	28 ♍ 11 18	04 ♎ 29 29	18 51	18 44	14 26	10 33	19 09	06 02	11 56	16 43
17	13 42 55	23 51 30	10 ♎ 45 07	16 ♎ 58 16	20 34	19 59	14 29	10 40	19 11	06 03	11 57	16 46
18	13 46 51	24 51 05	23 ♎ 08 58	29 ♎ 17 14	22 17	21 14	14 31	10 48	19 14	06 05	11 58	16 48
19	13 50 48	25 50 42	05 ♏ 23 10	11 ♏ 26 50	23 59	22 29	14 33	10 55	19 16	06 07	11 59	16 50
20	13 54 45	26 50 20	17 ♏ 28 21	23 ♏ 27 54	25 41	23 44	14 34	11 02	19 19	06 08	12 00	16 53
21	13 58 41	27 50 01	29 ♏ 25 42	05 ♐ 22 00	27 22	24 59	14 R 34	11 09	19 22	06 10	12 01	16 55
22	14 02 38	28 49 43	11 ♐ 17 08	17 ♐ 11 28	29 ♎ 03	26 14	14 33	11 16	19 24	06 12	12 01	16 57
23	14 06 34	29 ♎ 49 28	23 ♐ 05 26	28 ♐ 59 31	00 ♏ 43	27 29	14 31	11 23	19 27	06 14	12 02	17 00
24	14 10 31	00 ♏ 49 14	04 ♑ 54 16	10 ♑ 50 14	02 22	28 ♎ 45	14 28	11 30	19 30	06 16	12 03	17 02
25	14 14 27	01 49 02	16 ♑ 48 25	22 ♑ 48 25	04 01	00 ♏ 00	14 24	11 36	19 33	06 18	12 04	17 04
26	14 18 24	02 48 51	28 ♑ 51 56	04 ♒ 59 20	05 39	01 15	14 19	11 43	19 36	06 20	12 06	17 07
27	14 22 20	03 48 42	11 ♒ 11 16	17 ♒ 28 26	07 16	02 30	14 14	11 49	19 40	06 22	12 07	17 09
28	14 26 17	04 48 35	23 ♒ 51 27	00 ♓ 20 52	08 53	03 45	14 08	11 55	19 43	06 24	12 08	17 11
29	14 30 14	05 48 29	06 ♓ 57 11	13 ♓ 40 45	10 30	05 01	14 00	12 01	19 46	06 26	12 09	17 14
30	14 34 10	06 48 25	20 ♓ 31 47	27 ♓ 30 21	12 06	06 16	13 52	12 06	19 50	06 28	12 10	17 16
31	14 38 07	07 ♏ 48 23	04 ♈ 36 18	11 ♈ 49 14	13 ♏ 42	07 ♏ 31	13 ♊ 43	12 ♌ 12	19 ♑ 54	06 ♑ 31	12 ♑ 11	17 ♏ 19

Date	Moon True ☊ ° ′	Moon Mean ☊ ° ′	Moon Latitude ° ′
01	05 ♒ 25	03 ♒ 59	02 N 02
02	05 R 20	03 55	03 04
03	05 12	03 52	03 57
04	05 03	03 49	04 36
05	04 53	03 46	04 58
06	04 43	03 43	05 01
07	04 34	03 39	04 43
08	04 27	03 36	04 07
09	04 23	03 33	03 15
10	04 22	03 30	02 12
11	04 D 21	03 27	01 N 03
12	04 R 22	03 24	00 S 09
13	04 21	03 20	01 19
14	04 18	03 17	02 23
15	04 12	03 14	03 19
16	04 04	03 11	04 04
17	03 52	03 08	04 36
18	03 39	03 05	04 55
19	03 25	03 01	04 59
20	03 11	02 58	04 51
21	02 58	02 55	04 29
22	02 48	02 52	03 56
23	02 41	02 49	03 12
24	02 37	02 45	02 21
25	02 35	02 42	01 22
26	02 D 34	02 39	00 S 20
27	02 R 34	02 36	00 N 45
28	02 33	02 33	01 50
29	02 31	02 30	02 51
30	02 25	02 26	03 45
31	02 ♒ 17	02 ♒ 23	04 N 27

DECLINATIONS

Date	Sun ☉ ° ′	Moon ☽ ° ′	Mercury ☿ ° ′	Venus ♀ ° ′	Mars ♂ ° ′	Jupiter ♃ ° ′	Saturn ♄ ° ′	Uranus ♅ ° ′	Neptune ♆ ° ′	Pluto ♇ ° ′
01	03 S 12	09 S 54	04 N 30	01 N 19	21 N 09	18 N 32	22 S 10	23 S 38	22 S 05	02 S 20
02	03 35	03 S 59	03 54	00 50	21 13	18 30	22 10	23 38	22 05	02 21
03	03 58	02 N 17	03 16	00 N 20	21 17	18 28	22 10	23 38	22 05	02 22
04	04 21	08 35	02 36	00 S 10	21 20	18 26	22 10	23 38	22 05	02 23
05	04 44	14 32	01 55	00 40	21 24	18 23	22 10	23 38	22 05	02 24
06	05 08	19 42	01 13	01 11	21 27	18 21	22 10	23 38	22 05	02 25
07	05 30	23 39	00 N 30	01 41	21 30	18 19	22 09	23 38	22 05	02 25
08	05 53	26 00	00 S 14	02 11	21 34	18 17	22 09	23 38	22 05	02 26
09	06 16	26 34	00 58	02 41	21 37	18 14	22 09	23 38	22 05	02 27
10	06 39	25 22	01 43	03 11	21 40	18 12	22 09	23 38	22 05	02 28
11	07 02	22 36	02 27	03 41	21 43	18 10	22 09	23 38	22 05	02 29
12	07 24	18 37	03 12	04 10	21 46	18 08	22 09	23 38	22 05	02 30
13	07 47	13 45	03 57	04 40	21 49	18 06	22 08	23 38	22 05	02 31
14	08 09	08 21	04 42	05 10	21 52	18 04	22 08	23 38	22 05	02 31
15	08 31	02 N 41	05 27	05 40	21 55	18 02	22 08	23 38	22 05	02 32
16	08 54	03 S 00	06 12	06 09	21 58	18 00	22 08	23 37	22 05	02 33
17	09 16	08 29	06 56	06 39	22 00	17 58	22 07	23 37	22 05	02 34
18	09 37	13 33	07 40	07 08	22 03	17 56	22 07	23 37	22 05	02 35
19	09 59	18 01	08 23	07 37	22 06	17 55	22 07	23 37	22 05	02 36
20	10 21	21 41	09 06	08 06	22 08	17 53	22 07	23 37	22 05	02 36
21	10 42	24 24	09 48	08 35	22 11	17 51	22 06	23 37	22 05	02 37
22	11 04	26 01	10 30	09 03	22 13	17 49	22 06	23 37	22 05	02 38
23	11 25	26 28	11 11	09 32	22 15	17 48	22 06	23 37	22 05	02 39
24	11 46	25 42	11 52	10 00	22 18	17 46	22 05	23 37	22 05	02 40
25	12 06	23 45	12 32	10 28	22 20	17 44	22 05	23 37	22 05	02 40
26	12 27	20 43	13 11	10 55	22 22	17 43	22 05	23 37	22 05	02 41
27	12 47	16 42	13 49	11 23	22 24	17 41	22 04	23 36	22 05	02 42
28	13 08	11 50	14 27	11 50	22 26	17 40	22 04	23 36	22 05	02 43
29	13 28	06 19	15 04	12 17	22 28	17 38	22 03	23 36	22 04	02 44
30	13 47	00 19	15 40	12 43	22 30	17 37	22 03	23 36	22 04	02 44
31	14 S 07	05 N 54	16 S 16	13 S 10	22 N 32	17 N 35	22 S 02	23 S 36	22 S 04	02 S 45

ZODIAC SIGN ENTRIES

Date	h	m	Planets
01	12	13	♀ ♎
01	13	42	☽ ♓
03	17	42	☽ ♈
05	17	44	☿ ♎
05	19	06	☽ ♉
07	19	47	☽ ♊
09	21	29	☽ ♋
12	01	16	☽ ♌
14	07	21	☽ ♍
16	15	26	☽ ♎
19	01	24	☽ ♏
21	13	09	☽ ♐
23	01	46	☿ ♏
23	16	14	☉ ♏
24	02	03	☽ ♑
25	12	03	♀ ♏
26	14	14	☽ ♒
28	23	22	☽ ♓
31	04	14	☽ ♈

LATITUDES

Date	Mercury ☿ ° ′	Venus ♀ ° ′	Mars ♂ ° ′	Jupiter ♃ ° ′	Saturn ♄ ° ′	Uranus ♅ ° ′	Neptune ♆ ° ′	Pluto ♇ ° ′
01	01 N 50	01 N 26	01 S 05	00 N 23	00 S 02	00 S 19	00 N 50	14 N 57
04	01 55	01 26	01 00	00 23	00 02	00 19	00 50	14 56
07	01 53	01 24	00 54	00 24	00 03	00 19	00 50	14 55
10	01 45	01 22	00 48	00 24	00 03	00 19	00 50	14 54
13	01 33	01 20	00 42	00 25	00 03	00 19	00 50	14 53
16	01 17	01 17	00 35	00 25	00 03	00 19	00 49	14 52
19	01 00	01 13	00 28	00 26	00 04	00 19	00 49	14 51
22	00 41	01 09	00 20	00 26	00 04	00 19	00 49	14 51
25	00 21	01 05	00 12	00 27	00 04	00 19	00 49	14 50
28	00 N 01	01 00	00 S 04	00 28	00 04	00 19	00 49	14 50
31	00 S 19	00 N 55	00 N 05	00 N 28	00 S 04	00 S 19	00 N 49	14 N 49

DATA

Julian Date	2448166
Delta T	+57 seconds
Ayanamsa	23° 43′ 54″
Synetic vernal point	05° ♓ 23′ 05″
True obliquity of ecliptic	23° 26′ 32″

MOON'S PHASES, APSIDES AND POSITIONS ☽

Date	h	m	Phase	Longitude ° ′	Eclipse Indicator
04	12	02	○	11 ♈ 00	
11	03	31	☾	17 ♋ 34	
18	15	37	●	25 ♎ 00	
26	20	26	☽	03 ♒ 10	

Day	h	m		
06	18	38	Perigee	
22	16	04	Apogee	
03	03	21	0N	
09	07	29	Max dec	26° N 36′
15	23	15	0S	
23	08	43	Max dec	26° S 28′
30	13	13	0N	

LONGITUDES

Date	Chiron ⚷ ° ′	Ceres ⚳ ° ′	Pallas ⚴ ° ′	Juno ⚵ ° ′	Vesta ⚶ ° ′	Black Moon Lilith ⚸ ° ′
01	26 ♋ 14	16 ♍ 26	18 ♌ 49	24 ♏ 57	01 ♊ 07	07 ♐ 02
11	26 ♋ 44	20 ♍ 55	24 ♌ 05	27 ♏ 58	01 ♊ 27	08 ♐ 09
21	27 ♋ 05	25 ♍ 21	29 ♌ 09	01 ♐ 07	29 ♉ 03	09 ♐ 16
31	27 ♋ 17	29 ♍ 43	03 ♍ 56	04 ♐ 22	27 ♉ 02	10 ♐ 23

ASPECTARIAN

h m Aspects

01 Monday
00 16 ☽ ᑫ ☉
01 53 ☽ ± ♀
04 14 ☽ ⊥ ♄
07 58 ☽ ∠ ♆
13 51 ☽ ⊼ ♀
17 48 ☽ ± ☉
20 12 ☿ ∠ ♃
20 26 ☽ ∠ ♄
21 32 ☉ ✶ ♃
23 54 ☽ ✶ ♅

02 Tuesday
04 47 ☽ ⊼ ♃
05 15 ☽ ⊼ ☉
10 36 ☽ ✶ ♆
11 22 ☽ □ ♂
11 45 ♀ ∠ ♇
12 22 ☽ # ☿
13 28 ☽ ∥ ☉
15 22 ☽ ± ♃
18 15 ☽ △ ♇
18 20 ☽ ∥ ♇
19 30 ☉ # ☿
20 51 ☽ Q ♅
22 39 ☽ ✶ ♄

03 Wednesday
01 15 ☽ # ♀
05 08 ☽ ∥ ♀
06 59 ☽ ᑫ ♃
07 16 ☽ Q ♆
11 09 ☽ ☍ ☿
12 19 ☽ # ♇
15 22 ☽ ∥ ☿
17 56 ☉ ⊥ ♇
18 39 ☽ Q ♂
18 48 ☽ # ☉
19 02 ☽ Q ♄
19 50 ☽ ᑫ ♇
22 48 ☽ ☍ ♀

04 Thursday
03 20 ☽ □ ♅
08 25 ☽ △ ♃
10 51 ☽ ± ♇
12 02 ☽ ☍ ☉
13 23 ☽ □ ♆
14 54 ☽ ✶ ♂
19 46 ☿ # ♇
20 45 ☽ ⊼ ♇
22 35 ♂ ± ♄

05 Friday
00 53 ☽ □ ♄
08 11 ☉ □ ♆
15 48 ☽ ∠ ♂
19 17 ☽ ⊼ ☿

06 Saturday
03 43 ♀ □ ♅
04 30 ☽ △ ♅
04 34 ☽ ⊼ ♀
05 17 ☽ ∥ ♃
06 18 ☽ ± ☿
06 36 ☽ ⊥ ♂
09 55 ☽ □ ♃
12 32 ☿ ∠ ♇
12 51 ☿ # ♀
14 18 ☽ △ ♆
15 11 ☽ ± ♀
16 26 ☽ ⊼ ☉
16 27 ☽ ⚺ ♂
16 44 ☉ △ ♂
21 38 ☽ ☍ ♇
21 44 ☽ ∥ ♂
22 50 ☽ ᑫ ☿

07 Sunday
01 26 ☽ # ♆
01 40 ☽ △ ♄
01 52 ☽ # ♄
02 51 ☽ ± ☉
04 50 ☽ ᑫ ♅
07 07 ☽ ᑫ ♀
11 55 ☽ # ♅
14 39 ☽ ᑫ ♆
15 25 ☽ Q ♃
18 32 ☽ ᑫ ☉
19 30 ☽ ± ♅

08 Monday
02 08 ☽ ᑫ ♄
02 32 ☽ △ ☿
05 20 ☽ ⊼ ♅
05 22 ☽ ± ♆
09 53 ☽ △ ♀
11 15 ☽ ✶ ♃
15 14 ☽ ⊼ ♆
16 57 ☽ ± ♄
18 00 ☽ ☌ ♂
20 57 ☽ △ ☉
22 49 ☽ ⊼ ♇

09 Tuesday
00 52 ♀ ∥ ♇
01 58 ☿ □ ♅
02 55 ☽ ⊼ ♄
03 45 ♀ ✶ ♃
08 50 ☽ ± ♇
12 24 ☽ ∠ ♃
21 47 ♀ ⊥ ♇

10 Wednesday
00 00 ☽ ᑫ ♇
01 22 ☉ ⚺ ♇
03 44 ☽ ⊥ ♃
07 27 ☽ ☍ ♅
11 38 ☽ □ ☿
14 04 ☽ ⚺ ♃
17 02 ☽ □ ♀
17 43 ☽ ☍ ♆
21 08 ☽ ⚺ ♂

11 Thursday
00 38 ♀ □ ♆
01 43 ☽ △ ♇
03 31 ☽ □ ☉
04 24 ☽ # ♅
05 58 ☽ ☍ ♄
07 37 ☽ ⊥ ♂
08 28 ☿ ✶ ♃
12 48 ☿ ∥ ♇
15 08 ☽ # ♄
15 32 ☽ # ♆
17 52 ☽ ∥ ♂
17 55 ☿ ⊥ ♇
23 25 ☽ Q ☿
23 31 ☽ ∠ ♂

12 Friday
03 32 ☽ Q ♀
11 47 ☽ ⊼ ♅
12 17 ☿ □ ♆
13 38 ☽ Q ☉
14 34 ☽ ∥ ♃
14 43 ☉ □ ♄
19 09 ☽ ☌ ♃
19 19 ♀ △ ♂
22 27 ☽ ⊼ ♆
22 31 ☽ ± ♅
23 59 ☽ ✶ ☿

13 Saturday
02 29 ☽ ✶ ♂
03 09 ☽ ✶ ♀
06 56 ☽ □ ♇
09 17 ☽ ± ♆
11 21 ☽ ⊼ ♄
12 57 ☽ ✶ ☉
14 49 ☽ ᑫ ♅
20 02 ☿ △ ♂
22 17 ☽ ± ♄

14 Sunday
00 26 ☽ Q ♂
01 41 ☽ ᑫ ♆
07 25 ☽ ∠ ☿
09 20 ☽ ∠ ♀
12 47 ☽ # ☉
14 52 ☽ ᑫ ♄
15 55 ☽ Q ♇
18 23 ☽ △ ♅
18 42 ☽ ∠ ☉
20 26 ♀ ⚺ ♇

15 Monday
00 27 ☽ # ♀
01 40 ☽ # ☿
02 30 ☽ ⚺ ♃
02 39 ☽ ⊥ ☿
03 47 ☽ ⊥ ♀
05 25 ☽ △ ♆
05 57 ☿ ⚺ ♇
09 57 ☽ □ ♂
12 35 ☽ # ♇
12 51 ☽ ⊥ ☉
13 49 ☽ ⊥ ♃
14 19 ☽ ✶ ♇
15 36 ☽ ⚺ ☿
16 12 ☽ ⚺ ♀
18 53 ☽ △ ♄

16 Tuesday
01 05 ☽ ⚺ ☉
03 04 ☉ Q ♃
05 51 ☿ ☌ ♀
06 56 ☽ ∠ ♃
08 03 ☿ ∥ ♀
10 03 ☽ ∥ ♇
16 17 ☿ □ ♄
18 45 ☽ ∠ ♇
20 24 ♀ □ ♄

17 Wednesday
02 57 ☽ ∥ ♀
02 58 ☽ □ ♅
03 57 ☽ ∥ ☿
11 51 ☽ ✶ ♃
12 01 ☽ ⊥ ♇
14 19 ☽ □ ♆
15 47 ☽ ∥ ☉
16 51 ☉ Q ♅
19 13 ☽ △ ♂
23 38 ☽ ⚺ ♇

18 Thursday
04 21 ☽ □ ♄
07 50 ☽ ☌ ♀
10 03 ☽ ☌ ☿
11 18 ☽ Q ♃
13 49 ☽ Q ♅
15 37 ☽ ☌ ☉
19 45 ☿ Q ♃

19 Friday
00 30 ☽ ᑫ ♂
01 21 ☽ Q ♆
07 19 ♂ # ♆
11 22 ☽ # ♃
13 26 ☽ ✶ ♅
13 43 ☿ Q ♅
15 44 ☽ Q ♄
18 16 ☽ ± ♂
21 18 ♀ Q ♃
23 04 ☽ □ ♃

20 Saturday
00 05 ♂ # ♄
01 04 ☽ ✶ ♆
06 12 ☽ ⊼ ♂
10 48 ☽ ☌ ♇
14 59 ☽ ∥ ♆
15 12 ☽ ∥ ♄
15 26 ☽ # ♂
15 42 ☽ ✶ ♄
19 21 ☽ ∠ ♅
19 29 ♂ St R
19 58 ♀ Q ♅

21 Sunday
02 00 ☽ ⚺ ♀
04 02 ☽ ∥ ♅
07 07 ☽ ∠ ♆
07 10 ☽ ⚺ ☿
08 30 ☽ ⚺ ☉
13 30 ☽ ⊥ ♅
15 31 ☽ ⊥ ♀
21 16 ☽ ⊥ ☿
21 43 ☽ ⊥ ☉
22 00 ☽ ∠ ♄

22 Monday
01 19 ☽ ⊥ ♆
01 40 ☽ ⚺ ♅
04 21 ☉ ☌ ☿
11 54 ☽ ∠ ♀
11 58 ☽ △ ♃
13 30 ☽ ⚺ ♆
16 19 ☽ ⊥ ♄
17 38 ☽ ∠ ☉
18 31 ☽ ∠ ☿
18 36 ☽ ☍ ♂
19 03 ☿ ᑫ ♂
23 33 ☽ ⚺ ♇

23 Tuesday
02 15 ☿ Q ♆
04 34 ☽ ⚺ ♄
04 42 ☉ ᑫ ♂
11 48 ☽ ⊥ ♇
17 18 ☉ Q ♆
18 46 ☽ ᑫ ♃
22 01 ☽ ✶ ♀

24 Wednesday
02 57 ☽ ✶ ☉
04 27 ☉ ∥ ☿
06 01 ☽ ✶ ☿
06 09 ☽ ∠ ♇
13 13 ☽ ± ♃
14 46 ☽ ☌ ♅

25 Thursday
01 09 ♀ ᑫ ♂
01 12 ☽ Q ♀
01 27 ☽ ⊼ ♃
02 29 ☽ ☌ ♆
05 27 ☽ Q ☉
07 12 ☽ ⊼ ♂
10 09 ☽ Q ☿
12 33 ☽ ✶ ♇
13 18 ☽ ∥ ♅
13 30 ♀ Q ♆
17 32 ☽ ☌ ♄
19 10 ☽ ± ♂

26 Friday
00 04 ☽ # ♂
02 10 ☽ ∥ ♄
02 11 ☽ ∥ ♆
05 39 ♄ ∥ ♆
12 29 ☽ Q ♇
12 54 ☽ ᑫ ♂
17 14 ☽ □ ♀
20 26 ☽ □ ☉
22 19 ☿ ✶ ♅

27 Saturday
02 39 ☽ ⚺ ♅
03 18 ☽ □ ☿
06 30 ☽ # ♃
13 13 ☽ ☍ ♃
13 46 ☽ ⚺ ♆
14 16 ☽ ⊥ ♅
17 47 ☽ △ ♂
17 59 ☿ Q ♄
23 25 ☽ □ ♇

28 Sunday
00 58 ☽ ∥ ☿
01 13 ☽ ⊥ ♆
01 23 ☿ ± ♂
04 12 ☽ ⚺ ♄
06 21 ☽ ∥ ☉
07 23 ☽ ∠ ♅
12 02 ☽ ∥ ♀
15 29 ☽ ⊥ ♄
18 05 ☽ ∠ ♆

29 Monday
08 03 ☽ ∠ ♄
08 08 ☽ △ ♀
09 46 ☽ △ ☉
11 04 ☽ ✶ ♅
19 13 ☽ △ ☿
21 07 ☽ ⊼ ♃
21 18 ☽ ✶ ♆

30 Tuesday
00 27 ☽ □ ♂
02 29 ☽ ∥ ♇
03 40 ☉ ✶ ♅
06 18 ☽ △ ♇
07 45 ☽ ± ♃
08 25 ☽ Q ♅
10 47 ☽ ✶ ♄
12 06 ☿ □ ♃
13 01 ☿ ✶ ♆
13 24 ☽ ᑫ ♀
14 23 ☽ ᑫ ☉
16 07 ♀ ✶ ♅
16 24 ♂ ± ♄
18 18 ☽ Q ♆
23 24 ☽ ᑫ ♃
23 50 ☽ # ♇

31 Wednesday
00 45 ☽ ᑫ ☿
06 18 ☽ ± ♀
06 56 ☽ ± ☉
07 12 ☽ Q ♂
07 25 ☽ Q ♄
08 04 ♃ ⊼ ♆
08 08 ☽ ᑫ ♇
10 12 ☉ ± ♂
12 23 ☿ ⊼ ♂
14 12 ☉ Q ♄
15 12 ☽ □ ♅
15 26 ♀ ± ♂
17 20 ☽ ⊼ ♀
17 45 ☽ ⊼ ☉
17 48 ☽ ± ☿
19 32 ♀ Q ♄
23 11 ☽ ± ♇

All ephemeris data is given at 12.00 UT and the Moon's longitude is additionally given for 24.00 UT

NOVEMBER 1990

LONGITUDES

Date	Sidereal time h m s	Sun ☉ ° ' "	Moon ☽ ° ' "	Moon ☽ 24.00 ° ' "	Mercury ☿ ° '	Venus ♀ ° '	Mars ♂ ° '	Jupiter ♃ ° '	Saturn ♄ ° '	Uranus ♅ ° '	Neptune ♆ ° '	Pluto ♇ ° '
01	14 42 03	08 ♏ 48 22	19 ♈ 08 36	26 ♈ 33 32	15 ♏ 17	08 ♏ 46	13 ♊ 33	12 ♌ 17	19 ♑ 57	06 ♑ 33	12 ♑ 12	17 ♏ 21
02	14 46 00	09 48 24	04 ♉ 03 04	11 ♉ 35 58	16 51	10 02	13 R 23	12 23	20 01	06 35	12 14	17 23
03	14 49 56	10 48 27	19 ♉ 10 57	26 ♉ 46 40	18 25	11 17	13 11	12 28	20 05	06 38	12 15	17 26
04	14 53 53	11 48 32	04 ♊ 21 44	11 ♊ 54 54	19 59	12 32	12 59	12 32	20 09	06 40	12 16	17 28
05	14 57 49	12 48 39	19 ♊ 24 59	26 ♊ 51 00	21 33	13 47	12 45	12 37	20 13	06 42	12 18	17 31
06	15 01 46	13 48 48	04 ♋ 12 10	11 ♋ 27 53	23 06	15 03	12 31	12 42	20 17	06 45	12 19	17 33
07	15 05 43	14 48 59	18 ♋ 37 47	25 ♋ 41 39	24 38	16 18	12 17	12 46	20 21	06 47	12 20	17 35
08	15 09 39	15 49 12	02 ♌ 39 27	09 ♌ 31 19	26 11	17 33	12 01	12 50	20 25	06 50	12 22	17 38
09	15 13 36	16 49 27	16 ♌ 17 27	22 ♌ 58 09	27 42	18 48	11 45	12 54	20 30	06 53	12 23	17 40
10	15 17 32	17 49 44	29 ♌ 33 46	06 ♍ 04 42	29 ♏ 14	20 04	11 28	12 58	20 34	06 55	12 25	17 43
11	15 21 29	18 50 04	12 ♍ 31 21	18 ♍ 54 06	00 ♐ 45	21 19	11 10	13 02	20 38	06 58	12 26	17 45
12	15 25 25	19 50 25	25 ♍ 13 19	01 ♎ 29 21	02 16	22 34	10 52	13 05	20 43	07 00	12 28	17 48
13	15 29 22	20 50 47	07 ♎ 42 32	13 ♎ 53 08	03 47	23 50	10 33	13 09	20 48	07 03	12 29	17 50
14	15 33 18	21 51 12	20 ♎ 01 24	26 ♎ 07 32	05 17	25 05	10 13	13 12	20 52	07 06	12 31	17 52
15	15 37 15	22 51 39	02 ♏ 11 45	08 ♏ 14 11	06 47	26 20	09 53	13 15	20 57	07 09	12 33	17 55
16	15 41 12	23 52 07	14 ♏ 15 01	20 ♏ 14 21	08 16	27 36	09 33	13 17	21 02	07 12	12 34	17 57
17	15 45 08	24 52 37	26 ♏ 12 22	02 ♐ 09 11	09 45	28 ♏ 51	09 12	13 20	21 07	07 14	12 36	18 00
18	15 49 05	25 53 09	08 ♐ 05 00	13 ♐ 59 59	11 14	00 ♐ 06	08 50	13 22	21 12	07 17	12 38	18 02
19	15 53 01	26 53 42	19 ♐ 54 24	25 ♐ 48 29	12 42	01 22	08 29	13 25	21 17	07 20	12 39	18 04
20	15 56 58	27 54 17	01 ♑ 42 34	07 ♑ 36 59	14 10	02 37	08 07	13 27	21 22	07 23	12 41	18 07
21	16 00 54	28 54 53	13 ♑ 32 09	19 ♑ 28 30	15 37	03 52	07 44	13 28	21 27	07 26	12 43	18 09
22	16 04 51	29 ♏ 55 30	25 ♑ 26 33	01 ♒ 26 49	17 04	05 08	07 22	13 30	21 32	07 29	12 44	18 12
23	16 08 47	00 ♐ 56 08	07 ♒ 29 53	13 ♒ 36 21	18 30	06 23	06 59	13 31	21 38	07 32	12 46	18 14
24	16 12 44	01 56 48	19 ♒ 46 50	26 ♒ 01 58	19 55	07 38	06 37	13 33	21 43	07 35	12 48	18 16
25	16 16 41	02 57 28	02 ♓ 22 23	08 ♓ 48 41	21 19	08 54	06 14	13 34	21 48	07 38	12 50	18 19
26	16 20 37	03 58 10	15 ♓ 21 23	22 ♓ 01 00	22 43	10 09	05 51	13 34	21 54	07 42	12 52	18 21
27	16 24 34	04 58 53	28 ♓ 47 53	05 ♈ 42 16	24 05	11 25	05 29	13 35	21 59	07 45	12 54	18 24
28	16 28 30	05 59 36	12 ♈ 44 13	19 ♈ 53 37	25 27	12 40	05 06	13 35	22 05	07 48	12 56	18 26
29	16 32 27	07 00 21	27 ♈ 10 09	04 ♉ 33 13	26 47	13 55	04 44	13 36	22 11	07 51	12 57	18 28
30	16 36 23	08 ♐ 01 07	12 ♉ 02 03	19 ♉ 35 36	28 ♐ 05	15 ♐ 11	04 ♊ 22	13 ♌ 36	22 ♑ 16	07 ♑ 54	12 ♑ 59	18 ♏ 31

Date	Moon True ☊ ° '	Moon Mean ☊ ° '	Moon ☽ Latitude ° '
01	02 ♒ 07	02 ♒ 20	04 N 53
02	01 R 56	02 17	05 00
03	01 44	02 14	04 47
04	01 34	02 10	04 13
05	01 27	02 07	03 22
06	01 22	02 04	02 18
07	01 19	02 01	01 N 07
08	01 D 19	01 58	00 S 07
09	01 19	01 55	01 19
10	01 R 19	01 51	02 24
11	01 16	01 48	03 20
12	01 11	01 45	04 05
13	01 04	01 42	04 37
14	00 53	01 39	04 56
15	00 41	01 36	05 02
16	00 28	01 32	04 53
17	00 15	01 29	04 32
18	00 03	01 26	03 59
19	29 ♑ 54	01 23	03 15
20	29 47	01 20	02 24
21	29 43	01 16	01 25
22	29 41	01 13	00 S 23
23	29 D 41	01 10	00 N 42
24	29 42	01 07	01 46
25	29 43	01 04	02 46
26	29 R 42	01 01	03 40
27	29 39	00 57	04 24
28	29 35	00 54	04 54
29	29 28	00 51	05 06
30	29 ♑ 20	00 ♒ 48	04 N 59

DECLINATIONS

Date	Sun ☉ ° '	Moon ☽ ° '	Mercury ☿ ° '	Venus ♀ ° '	Mars ♂ ° '	Jupiter ♃ ° '	Saturn ♄ ° '	Uranus ♅ ° '	Neptune ♆ ° '	Pluto ♇ ° '
01	14 S 26	12 N 00	16 S 50	13 S 36	22 N 34	17 N 34	22 S 02	23 S 36	22 S 04	02 S 46
02	14 45	17 34	17 24	14 01	22 35	17 33	22 01	23 36	22 04	02 47
03	15 04	22 07	17 57	14 26	22 37	17 32	22 01	23 36	22 04	02 47
04	15 23	25 10	18 29	14 51	22 38	17 30	22 00	23 36	22 04	02 48
05	15 41	26 23	19 00	15 16	22 40	17 29	22 00	23 35	22 04	02 49
06	15 59	25 41	19 30	15 40	22 41	17 28	21 59	23 35	22 04	02 49
07	16 17	23 15	19 59	16 04	22 42	17 27	21 59	23 35	22 04	02 50
08	16 35	19 27	20 27	16 27	22 43	17 26	21 58	23 35	22 04	02 51
09	16 52	14 42	20 54	16 50	22 44	17 25	21 58	23 35	22 03	02 51
10	17 09	09 23	21 20	17 12	22 45	17 24	21 57	23 35	22 03	02 52
11	17 26	03 N 47	21 45	17 34	22 45	17 23	21 57	23 35	22 03	02 53
12	17 42	01 S 51	22 09	17 56	22 46	17 23	21 56	23 34	22 03	02 53
13	17 58	07 18	22 32	18 17	22 46	17 22	21 55	23 34	22 03	02 54
14	18 14	12 24	22 54	18 38	22 47	17 21	21 55	23 34	22 03	02 55
15	18 29	16 57	23 14	18 58	22 47	17 20	21 54	23 34	22 03	02 55
16	18 45	20 47	23 34	19 17	22 47	17 20	21 53	23 34	22 03	02 56
17	18 59	23 42	23 52	19 36	22 47	17 19	21 53	23 34	22 03	02 57
18	19 14	25 35	24 09	19 55	22 46	17 19	21 52	23 33	22 02	02 57
19	19 28	26 18	24 25	20 13	22 46	17 18	21 51	23 33	22 02	02 58
20	19 42	25 49	24 40	20 30	22 45	17 18	21 50	23 33	22 02	02 58
21	19 55	24 10	24 53	20 47	22 45	17 18	21 50	23 33	22 02	02 59
22	20 08	21 26	25 05	21 03	22 44	17 18	21 49	23 33	22 02	03 00
23	20 21	17 43	25 16	21 18	22 43	17 17	21 48	23 33	22 02	03 00
24	20 33	13 13	25 25	21 33	22 42	17 17	21 47	23 32	22 02	03 01
25	20 45	08 03	25 33	21 48	22 41	17 17	21 47	23 32	22 01	03 01
26	20 57	02 S 23	25 40	22 01	22 40	17 17	21 46	23 32	22 01	03 02
27	21 08	03 N 33	25 45	22 14	22 38	17 17	21 45	23 32	22 01	03 02
28	21 19	09 32	25 49	22 27	22 37	17 17	21 44	23 32	22 01	03 03
29	21 29	15 13	25 51	22 39	22 35	17 17	21 43	23 32	22 01	03 03
30	21 S 39	20 N 11	25 S 52	22 S 50	22 N 34	17 N 17	21 S 42	23 S 31	22 S 01	03 S 04

ZODIAC SIGN ENTRIES

Date	h	m	Planets
02	05	31	☽ ♉
04	05	06	☽ ♊
06	05	07	☽ ♋
08	07	24	☽ ♌
10	12	48	☽ ♍
11	00	06	☿ ♐
12	21	08	☽ ♎
15	07	39	☽ ♏
17	19	39	☽ ♐
18	09	58	♀ ♐
20	08	31	☽ ♑
22	13	47	☉ ♐
22	21	07	☽ ♒
25	07	32	☽ ♓
27	14	06	☽ ♈
29	16	37	☽ ♉

LATITUDES

Date	Mercury ☿ ° '	Venus ♀ ° '	Mars ♂ ° '	Jupiter ♃ ° '	Saturn ♄ ° '	Uranus ♅ ° '	Neptune ♆ ° '	Pluto ♇ ° '
01	00 S 26	00 N 53	00 N 08	00 N 28	00 S 04	00 S 19	00 N 49	14 N 49
04	00 46	00 47	00 17	00 29	00 05	00 19	00 49	14 49
07	01 05	00 41	00 26	00 29	00 05	00 19	00 49	14 49
10	01 23	00 34	00 35	00 30	00 05	00 19	00 49	14 48
13	01 39	00 28	00 45	00 31	00 05	00 19	00 48	14 48
16	01 54	00 21	00 54	00 31	00 06	00 19	00 48	14 48
19	02 07	00 14	01 03	00 32	00 06	00 19	00 48	14 48
22	02 17	00 N 07	01 12	00 33	00 06	00 19	00 48	14 48
25	02 24	00 S 01	01 21	00 33	00 06	00 19	00 48	14 49
28	02 27	00 08	01 29	00 34	00 06	00 19	00 48	14 49
31	02 S 25	00 S 15	01 N 37	00 N 34	00 S 07	00 S 19	00 N 48	14 N 49

DATA

Julian Date	2448197
Delta T	+57 seconds
Ayanamsa	23° 43' 57"
Synetic vernal point	05° ♓ 23' 02"
True obliquity of ecliptic	23° 26' 31"

MOON'S PHASES, APSIDES AND POSITIONS ☽

Date	h	m	Phase	Longitude ° '	Eclipse Indicator
02	21	48	○	10 ♉ 13	
09	13	02	☾	16 ♌ 52	
17	09	05	●	24 ♏ 45	
25	13	12	☽	03 ♓ 00	

Day	h	m		
03	23	15	Perigee	
19	03	09	Apogee	
05	15	05	Max dec	26° N 24'
12	04	04	0S	
19	14	24	Max dec	26° S 19'
26	21	45	0N	

LONGITUDES

Date	Chiron ⚷ ° '	Ceres ⚳ ° '	Pallas ⚴ ° '	Juno ⚵ ° '	Vesta ⚶ ° '	Black Moon Lilith ⚸ ° '
01	27 ♋ 18	00 ♎ 08	04 ♍ 24	04 ♐ 42	26 ♉ 48	10 ♐ 29
11	27 ♋ 19	04 ♎ 23	08 ♍ 51	08 ♐ 03	24 ♉ 18	11 ♐ 36
21	27 ♋ 12	08 ♎ 31	12 ♍ 56	11 ♐ 29	21 ♉ 42	12 ♐ 43
31	26 ♋ 55	12 ♎ 30	16 ♍ 35	14 ♐ 58	19 ♉ 18	13 ♐ 50

All ephemeris data is given at 12.00 UT and the Moon's longitude is additionally given for 24.00 UT

ASPECTARIAN

h m Aspects

01 Thursday
00 37 ☽ □ ♆
00 42 ☽ △ ♃
02 58 ☽ ⚹ ♂
04 55 ☽ ⊼ ☿
09 04 ☽ ⊼ ♇
13 19 ☽ □ ♄
15 15 ☉ ☌ ♀
19 03 ☽ # ♀
22 41 ☽ # ☉

02 Friday
03 02 ☽ ∠ ♂
11 05 ☽ # ☿
11 53 ☽ ∥ ♃
16 03 ☽ △ ♅
17 14 ☽ ⊥ ♂
18 15 ☿ # ♃
20 23 ☿ ☌ ♇
21 48 ☽ ☍ ☉
22 22 ☽ ☍ ♀

03 Saturday
01 01 ☽ △ ♆
01 18 ☽ □ ♃
02 38 ☽ ⊻ ♂
09 13 ☽ ☍ ♇
10 40 ☽ ☍ ☿
11 23 ☽ # ♄
11 42 ☽ # ♆
13 25 ☽ △ ♄
15 12 ☽ ∥ ♂
15 52 ☽ bQ ♅
22 01 ☽ # ♅

04 Sunday
00 46 ☽ bQ ♆
05 55 ☽ Q ♃
06 08 ☽ ± ♅
06 53 ♀ ⚹ ♆
12 08 ♀ □ ♃
13 15 ☽ bQ ♄
14 32 ☿ ⚹ ♄
15 02 ☽ ± ♆
15 40 ☽ ⊼ ♅
19 15 ♀ ⊼ ♂
23 20 ☉ ⚹ ♆

05 Monday
00 35 ☽ ⊼ ♆
00 40 ☽ ⊼ ☉
01 04 ☽ ⚹ ♃
01 30 ☽ ☌ ♂
02 10 ☽ ⊼ ♀
03 38 ☽ ± ♄
07 02 ☉ □ ♃
08 56 ☽ ⊼ ♇
10 57 ☽ ± ☉
10 58 ☉ ⊼ ♂
12 39 ☽ ± ♀
13 17 ☽ ⊼ ♄
14 35 ☿ ∠ ♅
15 49 ☽ ⊼ ☿
18 37 ☽ ± ♇
22 52 ♂ ⚹ ♃

06 Tuesday
01 19 ☽ ∠ ♃
02 33 ☽ bQ ☉
02 40 ☽ ± ☿
04 34 ☽ bQ ♀
09 17 ☽ bQ ♇
16 07 ☽ ⊥ ♃
16 12 ☽ ☍ ♅
19 10 ☽ bQ ☿

07 Wednesday
01 26 ☽ ☍ ♆
01 32 ☽ ⊻ ♂
02 07 ☽ ⊻ ♃
05 07 ☽ △ ☉
06 36 ♂ ⊼ ♆
07 42 ☽ △ ♀
09 23 ☽ # ♅
10 15 ☽ △ ♇
11 25 ☽ ⊥ ♂
14 55 ☽ ☍ ♄
15 58 ☽ ∥ ♂
20 21 ☽ # ♆
20 54 ☽ # ♄
23 27 ☽ △ ☿

08 Thursday
02 27 ☽ ∠ ♂
06 23 ♃ ± ♅
06 55 ☽ # ☿
13 33 ♀ ☌ ♇
19 18 ☽ ⊼ ♅
22 43 ☽ ∥ ♃

09 Friday
02 08 ☽ # ☉
02 27 ☽ # ♀
04 05 ☽ ⚹ ♂
05 03 ☽ ⊼ ♆
05 54 ☽ ± ♅
05 57 ☽ ☌ ♃
06 55 ☿ ∠ ♆
13 02 ☽ □ ☉
14 29 ☽ □ ♇
15 46 ☽ ± ♆
16 58 ☽ □ ♀
19 34 ☽ ⊼ ♄
20 06 ☉ ∥ ♀
22 03 ☽ bQ ♅

10 Saturday
01 08 ☽ Q ♂
06 30 ☽ ± ♄
08 04 ☽ bQ ♆
09 05 ☉ ☌ ♇
11 19 ☽ □ ☿
22 13 ♀ ⚹ ♄
23 07 ☽ bQ ♄
23 21 ☽ Q ♇

11 Sunday
00 14 ♀ # ♃
00 31 ☽ Q ☉
01 36 ☽ △ ♅
05 22 ☽ Q ♀
08 55 ☉ # ♃
09 32 ☽ □ ♂
11 50 ☽ △ ♆
12 57 ☽ ⊻ ♃
15 26 ☿ ⊥ ♅
15 49 ☽ # ♇
21 52 ☽ ⚹ ♇
23 01 ☿ ∥ ♄

12 Monday
00 18 ☽ ⊥ ♃
00 48 ♀ ∠ ♅
00 54 ☽ ⚹ ☉
01 18 ☽ Q ☿
03 23 ☽ △ ♄
06 01 ☿ ∥ ♆
06 24 ☽ ⚹ ♀
16 31 ☽ ∥ ♇
17 30 ☽ ∠ ♃

13 Tuesday
02 33 ☽ ∠ ♇
03 21 ☽ ⚹ ☿
08 05 ☽ ∠ ☉
10 36 ☉ ⚹ ♄
10 44 ☽ □ ♅
14 25 ☽ ∠ ♀
17 22 ☽ △ ♂
20 02 ☽ ⊥ ♇
21 18 ☽ □ ♆
22 36 ☽ ⚹ ♃

14 Wednesday
03 07 ☽ ⊥ ☉
03 51 ☿ # ♂
07 46 ☽ ⊻ ♇
09 57 ☽ ⊥ ♀
12 34 ☽ ∠ ☿
13 40 ☽ □ ♄
15 55 ☽ ⊻ ☉
18 09 ☉ ∠ ♅
21 57 ☽ bQ ♂
22 01 ☿ ∠ ♄
22 01 ☽ Q ♅
22 13 ☽ Q ♃
23 05 ☽ ⊻ ♀

15 Thursday
08 11 ☿ ⊥ ♆
08 43 ☽ Q ♆
08 47 ☽ ⊥ ☿
14 14 ☽ # ♃
15 16 ☽ ± ♂
18 09 ☿ ⊻ ♅
21 41 ☽ ∥ ☉
21 52 ☽ ⚹ ♅
22 23 ☽ ⊻ ☿

16 Friday
01 06 ☽ ∥ ♀
01 31 ☽ Q ♄
02 52 ☽ ⊼ ♂
08 38 ☽ ⚹ ♆
10 04 ☽ □ ♃
11 30 ♀ ∠ ♆
11 58 ☿ ∥ ♅
19 26 ☽ ☌ ♇
20 10 ☽ ∥ ♄
21 26 ☽ ∥ ♆

17 Saturday
01 41 ☽ ⚹ ♄
03 27 ☽ # ♂
03 59 ☽ ∠ ♅
04 46 ☿ ☍ ♂
09 05 ☽ ☌ ☉
10 35 ☽ ∥ ♅
13 52 ☽ ∥ ☿
14 49 ☽ ∠ ♆
17 58 ☽ ☌ ♀
22 12 ☽ ⊥ ♅

18 Sunday
08 09 ☽ ∠ ♄
09 02 ☽ ⊥ ♆
10 23 ☽ ⊻ ♅
13 29 ☽ ☍ ♂
19 17 ☽ ☌ ☿
21 14 ☽ ⊻ ♆
22 46 ☽ △ ♃

19 Monday
02 32 ☽ ⊥ ♄
08 16 ☽ ⊻ ♇
11 15 ☿ ⊻ ♆
11 31 ♀ ⊥ ♅
14 49 ☽ ⊻ ♄
20 30 ☽ ⊥ ♇
23 56 ☿ △ ♃

20 Tuesday
03 32 ☽ ⊻ ☉
05 20 ☽ bQ ♃
06 34 ☉ ∠ ♆
14 04 ☽ ⊻ ♀
14 52 ☽ ∠ ♇
16 53 ☽ ⊥ ☉
23 35 ☽ ☌ ♅
23 41 ☽ ± ♃

21 Wednesday
00 37 ☽ ⊼ ♂
03 42 ☽ ⊥ ♀
04 24 ☽ ∥ ☿
09 07 ☿ ⊥ ♄
10 20 ☽ ☌ ♆
11 52 ☽ ⊼ ♃
12 24 ☽ ± ♂
12 50 ☽ ∠ ☉
16 47 ☽ ⊻ ☿
18 18 ☽ ∥ ♅
21 22 ☽ ⚹ ♇

22 Thursday
00 04 ☽ ∠ ♀
01 37 ☽ # ♂
04 06 ☽ ☌ ♄
05 11 ♂ ⊼ ♅
06 01 ☽ bQ ♂
06 33 ☽ ⊥ ☿
07 23 ☽ ∥ ♆
09 03 ☽ ∥ ♄
14 33 ☽ ∥ ♀
20 32 ☽ ∥ ☉
21 32 ☽ Q ♇
21 47 ☽ ⚹ ☉
23 00 ♃ ± ♅

23 Friday
03 00 ☽ ∠ ☿
07 33 ☿ ⊻ ♇
09 33 ☽ ⚹ ♀
11 02 ☽ △ ♂
12 05 ☽ ⊻ ♅
14 31 ☽ # ♃
16 58 ♀ ∠ ♄
19 33 ♀ ⊥ ♆
20 52 ♀ ☍ ♂
22 24 ☽ ⊻ ♆
23 39 ☽ Q ☉
23 51 ☽ ☍ ♃
23 55 ☽ ⊥ ♅

24 Saturday
00 50 ♂ ± ♆
03 03 ☉ ⊥ ♅
06 36 ♂ bQ ♄
09 05 ☽ □ ♇
10 06 ☽ ⊥ ♆
10 57 ♀ ⊻ ♅
11 42 ☽ Q ♀
12 17 ☽ ⚹ ☿
15 46 ☽ ⊻ ♄
17 26 ☽ ∠ ♅

25 Sunday
03 19 ☽ ⊥ ♄
03 24 ☽ ∠ ♆
10 08 ♀ ∥ ♄
13 12 ☽ □ ☉
14 00 ☽ Q ☿
19 01 ☽ □ ♂
20 21 ☽ ∠ ♄
20 55 ☿ ⊻ ♄
21 52 ☽ ⚹ ♅

26 Monday
01 28 ☽ □ ♀
07 26 ☽ ⚹ ♆
08 45 ☽ ⊼ ♃
09 22 ☽ ∥ ♇
11 48 ♀ ∥ ♆
17 26 ☽ △ ♇
19 38 ☽ ± ♃
19 52 ☽ Q ♅
23 52 ☽ ⚹ ♄

27 Tuesday
02 45 ☽ □ ☿
02 51 ☽ Q ♂
05 06 ☽ Q ♆
09 57 ☽ # ♇
11 37 ☽ bQ ♃
17 28 ☿ ⊥ ♇
20 02 ☽ bQ ♇
20 33 ☉ ☍ ♂
21 06 ☽ Q ♄
23 18 ☽ ⚹ ♂
23 36 ☽ △ ☉

28 Wednesday
03 34 ☽ □ ♅
11 29 ☽ ± ♇
11 52 ☽ △ ♀
12 19 ☽ □ ♆
13 27 ☽ △ ♃
17 06 ♀ ⊻ ♆
21 35 ☽ ⊼ ♇

29 Thursday
00 02 ☽ ∠ ♂
02 52 ☽ bQ ☉
03 44 ☽ □ ♄
05 46 ♀ △ ♃
06 13 ♀ # ♂
10 49 ☉ ⊥ ♆
11 18 ☽ △ ☿
14 29 ☽ ⊥ ♂
15 08 ☽ bQ ♀
16 32 ☉ ∠ ♄
18 43 ☽ ± ☉
21 26 ☽ ∥ ♃
23 59 ☽ ⊻ ♂

30 Friday
05 06 ♃ St R
05 07 ☽ ⊼ ☉
05 22 ☽ △ ♅
07 01 ☽ ± ♀
09 09 ☉ ⊻ ♅
13 32 ☽ △ ♆
13 50 ☽ bQ ☿
14 29 ☽ □ ♃
17 27 ☽ ⊼ ♀
20 04 ☉ ∥ ♄
20 38 ☽ # ♄
20 39 ☽ # ☉
21 30 ☿ bQ ♃
22 19 ☽ ☍ ♇
22 31 ☽ # ♆

DECEMBER 1990

LONGITUDES

Date	Sidereal time h m s	Sun ☉ ° ' "	Moon ☽ ° ' "	Moon ☽ 24.00 ° ' "	Mercury ☿ ° '	Venus ♀ ° '	Mars ♂ ° '	Jupiter ♃ ° '	Saturn ♄ ° '	Uranus ♅ ° '	Neptune ♆ ° '	Pluto ♇ ° '
01	16 40 20	09 ♐ 01 54	27 ♉ 12 42	04 ♊ 52 00	29 ♐ 22	16 ♐ 26	04 ♊ 00	13 ♌ 36	22 ♑ 22	07 ♑ 58	13 ♑ 01	18 ♏ 33
02	16 44 16	10 02 42	12 ♊ 32 05	20 ♊ 11 31	00 ♑ 36	17 41	03 R 38	13 R 35	22 28	08 01	13 03	18 35
03	16 48 13	11 03 31	27 ♊ 48 55	05 ♋ 23 00	01 49	18 57	03 17	13 35	22 34	08 04	13 05	18 37
04	16 52 10	12 04 22	12 ♋ 52 41	20 ♋ 17 03	02 58	20 12	02 56	13 34	22 40	08 07	13 07	18 40
05	16 56 06	13 05 14	27 ♋ 35 22	04 ♌ 47 10	04 04	21 27	02 36	13 33	22 46	08 11	13 09	18 42
06	17 00 03	14 06 07	11 ♌ 52 09	18 ♌ 50 14	05 07	22 43	02 16	13 32	22 52	08 14	13 11	18 44
07	17 03 59	15 07 01	25 ♌ 41 29	02 ♍ 26 05	06 05	23 58	01 56	13 31	22 58	08 17	13 13	18 46
08	17 07 56	16 07 56	09 ♍ 04 21	15 ♍ 36 41	06 59	25 13	01 37	13 29	23 04	08 21	13 15	18 49
09	17 11 52	17 08 53	22 ♍ 03 30	28 ♍ 25 17	07 47	26 29	01 19	13 27	23 10	08 24	13 18	18 51
10	17 15 49	18 09 51	04 ♎ 42 33	10 ♎ 55 46	08 30	27 44	01 01	13 25	23 17	08 28	13 20	18 53
11	17 19 45	19 10 50	17 ♎ 05 25	23 ♎ 11 59	09 05	28 ♐ 59	00 44	13 23	23 23	08 31	13 22	18 55
12	17 23 42	20 11 50	29 ♎ 15 53	05 ♏ 17 33	09 32	00 ♑ 15	00 28	13 21	23 29	08 35	13 24	18 57
13	17 27 39	21 12 52	11 ♏ 17 20	17 ♏ 15 35	09 51	01 30	00 ♊ 12	13 18	23 36	08 38	13 26	19 00
14	17 31 35	22 13 54	23 ♏ 12 37	29 ♏ 08 43	10 00	02 45	29 ♉ 57	13 16	23 42	08 41	13 28	19 02
15	17 35 32	23 14 57	05 ♐ 04 08	10 ♐ 59 07	09 R 58	04 01	29 43	13 13	23 48	08 45	13 30	19 04
16	17 39 28	24 16 01	16 ♐ 53 54	22 ♐ 48 40	09 46	05 16	29 30	13 10	23 55	08 48	13 32	19 06
17	17 43 25	25 17 06	28 ♐ 43 40	04 ♑ 39 07	09 22	06 31	29 17	13 06	24 02	08 52	13 35	19 08
18	17 47 21	26 18 11	10 ♑ 35 14	16 ♑ 32 17	08 47	07 47	29 05	13 03	24 08	08 56	13 37	19 10
19	17 51 18	27 19 17	22 ♑ 30 32	28 ♑ 30 16	08 00	09 02	28 54	12 59	24 15	08 59	13 39	19 12
20	17 55 14	28 20 24	04 ♒ 31 50	10 ♒ 35 35	07 02	10 18	28 44	12 55	24 21	09 03	13 41	19 14
21	17 59 11	29 ♐ 21 30	16 ♒ 41 54	22 ♒ 51 12	05 55	11 33	28 34	12 51	24 28	09 06	13 43	19 16
22	18 03 08	00 ♑ 22 37	29 ♒ 03 55	05 ♓ 20 31	04 41	12 48	28 26	12 47	24 35	09 10	13 46	19 18
23	18 07 04	01 23 44	11 ♓ 41 27	18 ♓ 07 12	03 21	14 03	28 18	12 43	24 41	09 13	13 48	19 20
24	18 11 01	02 24 52	24 ♓ 38 11	01 ♈ 14 50	01 59	15 19	28 11	12 38	24 48	09 17	13 50	19 22
25	18 14 57	03 25 59	07 ♈ 57 30	14 ♈ 46 27	00 ♑ 37	16 34	28 05	12 34	24 55	09 20	13 52	19 24
26	18 18 54	04 27 07	21 ♈ 41 51	28 ♈ 43 46	29 ♐ 17	17 49	28 00	12 29	25 02	09 24	13 55	19 26
27	18 22 50	05 28 14	05 ♉ 52 06	13 ♉ 06 34	28 03	19 05	27 55	12 24	25 09	09 28	13 57	19 27
28	18 26 47	06 29 22	20 ♉ 26 43	27 ♉ 51 56	26 57	20 20	27 52	12 18	25 16	09 31	13 59	19 29
29	18 30 43	07 30 29	05 ♊ 21 22	12 ♊ 54 03	25 59	21 35	27 49	12 13	25 22	09 35	14 01	19 31
30	18 34 40	08 31 37	20 ♊ 28 53	28 ♊ 04 40	25 11	22 50	27 47	12 08	25 29	09 38	14 04	19 33
31	18 38 36	09 ♑ 32 45	05 ♋ 40 07	13 ♋ 14 02	24 ♐ 34	24 ♑ 06	27 ♉ 46	12 ♌ 02	25 ♑ 36	09 ♑ 42	14 ♑ 06	19 ♏ 35

Date	Moon True ☊ ° '	Moon Mean ☊ ° '	Moon ☽ Latitude ° '
01	29 ♑ 12	00 ♒ 45	04 N 31
02	29 R 04	00 42	03 43
03	28 59	00 38	02 40
04	28 56	00 35	01 26
05	28 55	00 32	00 N 07
06	28 D 55	00 29	01 S 10
07	28 57	00 26	02 20
08	28 58	00 22	03 20
09	28 R 58	00 19	04 08
10	28 56	00 16	04 43
11	28 53	00 13	05 04
12	28 48	00 10	05 10
13	28 41	00 07	05 03
14	28 34	00 03	04 42
15	28 27	00 ♒ 00	04 10
16	28 20	29 ♑ 57	03 26
17	28 15	29 54	02 34
18	28 12	29 51	01 35
19	28 11	29 48	00 S 31
20	28 D 11	29 44	00 N 35
21	28 12	29 41	01 40
22	28 14	29 38	02 42
23	28 15	29 35	03 37
24	28 16	29 32	04 22
25	28 R 17	29 28	04 56
26	28 16	29 25	05 13
27	28 13	29 22	05 12
28	28 11	29 19	04 52
29	28 08	29 16	04 11
30	28 05	29 13	03 13
31	28 ♑ 03	29 ♑ 09	02 N 01

DECLINATIONS

Date	Sun ☉ ° '	Moon ☽ ° '	Mercury ☿ ° '	Venus ♀ ° '	Mars ♂ ° '	Jupiter ♃ ° '	Saturn ♄ ° '	Uranus ♅ ° '	Neptune ♆ ° '	Pluto ♇ ° '
01	21 S 48	23 N 56	25 S 51	23 S 00	22 N 32	17 N 18	21 S 42	23 S 31	22 S 01	03 S 04
02	21 57	26 00	25 50	23 10	22 30	17 18	21 41	23 31	22 00	03 05
03	22 06	26 05	25 46	23 19	22 29	17 18	21 40	23 31	22 00	03 05
04	22 14	24 14	25 42	23 27	22 27	17 19	21 39	23 31	22 00	03 06
05	22 22	20 46	25 36	23 34	22 25	17 19	21 38	23 30	22 00	03 06
06	22 30	16 07	25 28	23 41	22 23	17 20	21 37	23 30	22 00	03 06
07	22 37	10 46	25 20	23 47	22 21	17 20	21 36	23 30	21 59	03 07
08	22 43	05 N 04	25 10	23 53	22 20	17 21	21 35	23 30	21 59	03 07
09	22 49	00 S 39	24 59	23 58	22 18	17 22	21 34	23 30	21 59	03 08
10	22 55	06 12	24 47	24 01	22 16	17 23	21 33	23 29	21 59	03 08
11	23 00	11 23	24 34	24 05	22 14	17 23	21 32	23 29	21 59	03 08
12	23 05	16 02	24 20	24 07	22 12	17 24	21 31	23 29	21 59	03 09
13	23 09	20 01	24 06	24 09	22 11	17 25	21 30	23 29	21 58	03 09
14	23 13	23 08	23 51	24 10	22 09	17 26	21 29	23 28	21 58	03 09
15	23 16	25 14	23 35	24 10	22 07	17 27	21 28	23 28	21 58	03 10
16	23 19	26 13	23 19	24 09	22 06	17 28	21 27	23 28	21 58	03 10
17	23 21	26 00	23 02	24 08	22 04	17 29	21 26	23 28	21 58	03 10
18	23 23	24 36	22 46	24 06	22 03	17 31	21 25	23 27	21 57	03 11
19	23 25	22 04	22 29	24 03	22 02	17 32	21 24	23 27	21 57	03 11
20	23 26	18 34	22 12	24 00	22 01	17 33	21 23	23 27	21 57	03 11
21	23 26	14 15	21 56	23 55	22 00	17 34	21 21	23 27	21 57	03 11
22	23 26	09 16	21 40	23 50	21 59	17 36	21 20	23 27	21 57	03 12
23	23 26	03 S 50	21 24	23 45	21 58	17 37	21 19	23 26	21 56	03 12
24	23 25	01 N 53	21 09	23 38	21 57	17 39	21 18	23 26	21 56	03 12
25	23 24	07 41	20 55	23 31	21 57	17 40	21 17	23 26	21 56	03 12
26	23 22	13 17	20 43	23 23	21 56	17 42	21 16	23 26	21 56	03 12
27	23 20	18 23	20 32	23 14	21 56	17 44	21 15	23 25	21 55	03 13
28	23 17	22 33	20 24	23 05	21 56	17 45	21 13	23 25	21 55	03 13
29	23 14	25 19	20 17	22 55	21 56	17 47	21 12	23 25	21 55	03 13
30	23 10	26 18	20 13	22 44	21 56	17 49	21 11	23 25	21 55	03 13
31	23 S 06	25 N 20	20 S 11	22 S 33	21 N 56	17 N 50	21 S 10	23 S 24	21 S 55	03 S 13

ZODIAC SIGN ENTRIES

Date	h m	Planets
01	16 23	☽ ♊
02	00 13	☿ ♑
03	15 27	☽ ♋
05	16 00	☽ ♌
07	19 39	☽ ♍
10	03 00	☽ ♎
12	07 18	♀ ♑
12	13 28	☽ ♏
14	07 46	♂ ♉
15	01 44	☽ ♐
17	14 35	☽ ♑
20	02 59	☽ ♒
22	03 07	☉ ♑
22	13 48	☽ ♓
24	21 45	☽ ♈
25	22 57	☿ ♐
27	02 09	☽ ♉
29	03 26	☽ ♊
31	03 02	☽ ♋

LATITUDES

Date	Mercury ☿ ° '	Venus ♀ ° '	Mars ♂ ° '	Jupiter ♃ ° '	Saturn ♄ ° '	Uranus ♅ ° '	Neptune ♆ ° '	Pluto ♇ ° '
01	02 S 25	00 S 15	01 N 37	00 N 34	00 S 07	00 S 19	00 N 48	14 N 49
04	02 17	00 22	01 44	00 35	00 07	00 19	00 48	14 50
07	02 02	00 29	01 51	00 36	00 07	00 19	00 48	14 50
10	01 37	00 36	01 56	00 36	00 07	00 19	00 48	14 51
13	01 01	00 43	02 02	00 37	00 07	00 19	00 48	14 51
16	00 S 14	00 49	02 06	00 38	00 08	00 19	00 48	14 52
19	00 N 43	00 55	02 10	00 38	00 08	00 19	00 48	14 53
22	01 42	01 01	02 13	00 39	00 08	00 19	00 48	14 53
25	02 31	01 07	02 16	00 40	00 08	00 19	00 48	14 54
28	03 01	01 12	02 18	00 40	00 08	00 19	00 47	14 55
31	03 N 09	01 S 16	02 N 20	00 N 41	00 S 09	00 S 19	00 N 47	14 N 56

DATA

Julian Date	2448227
Delta T	+57 seconds
Ayanamsa	23° 44' 02"
Synetic vernal point	05° ♓ 22' 57"
True obliquity of ecliptic	23° 26' 30"

MOON'S PHASES, APSIDES AND POSITIONS ☽

Date	h m	Phase	Longitude ° '	Eclipse Indicator
02	07 50	○	09 ♊ 52	
09	02 04	☾	16 ♍ 44	
17	04 22	●	24 ♐ 58	
25	03 16	☽	03 ♈ 04	
31	18 35	○	09 ♋ 50	

Day	h m		
02	10 56	Perigee	
16	03 49	Apogee	
30	23 55	Perigee	
03	01 05	Max dec	26° N 18'
09	09 14	0S	
16	19 43	Max dec	26° S 17'
24	04 10	0N	
30	12 08	Max dec	26° N 18'

LONGITUDES

Date	Chiron ⚷ ° '	Ceres ⚳ ° '	Pallas ⚴ ° '	Juno ⚵ ° '	Vesta ⚶ ° '	Black Moon Lilith ⚸ ° '
01	26 ♋ 55	12 ♎ 30	16 ♍ 35	14 ♐ 58	19 ♉ 18	13 ♐ 50
11	26 ♋ 29	16 ♎ 17	19 ♍ 44	18 ♐ 29	17 ♉ 22	14 ♐ 57
21	25 ♋ 57	19 ♎ 51	22 ♍ 17	22 ♐ 01	16 ♉ 06	16 ♐ 04
31	25 ♋ 18	23 ♎ 09	24 ♍ 09	25 ♐ 33	15 ♉ 34	17 ♐ 11

All ephemeris data is given at 12.00 UT and the Moon's longitude is additionally given for 24.00 UT

ASPECTARIAN

h m Aspects

01 Saturday
01 58 ☽ ∥ ♂
04 20 ☽ △ ♄
04 40 ☽ # ♀
05 17 ☽ ⚼ ♅
05 24 ☽ ± ☿
08 47 ☽ # ♅
10 44 ♀ ⊥ ♄
13 16 ☽ ⚼ ♆
15 41 ☽ ⊼ ☿
18 52 ☽ Q ♃
19 28 ☽ ± ♅
22 23 ☽ ☌ ♂

02 Sunday
03 24 ☽ ± ♆
04 01 ☽ ⚼ ♄
04 54 ☽ ⊼ ♅
07 50 ☽ ☍ ☉
08 52 ☽ # ☿
12 49 ☽ ⊼ ♆
13 39 ☽ ⚹ ♃
18 12 ☽ ± ♄
19 35 ☉ ∥ ♆
20 48 ☽ ☍ ♀
21 30 ☽ ⊼ ♇

03 Monday
03 41 ☽ ⊼ ♄
05 42 ♀ ⊻ ♇
06 57 ☽ ± ♇
13 12 ☽ ∠ ♃
18 45 ☽ # ☿
18 51 ☽ ☍ ☿
20 28 ☽ ⊻ ♂
21 13 ☽ ⚼ ♇

04 Tuesday
03 29 ☽ ⊥ ♃
04 21 ☽ ☍ ♅
05 49 ☽ ⊥ ♂
10 37 ☽ ⊼ ☉
11 30 ☿ ⊼ ♂
12 24 ☽ ☍ ♆
13 06 ☽ ⊻ ♃
17 58 ☽ # ♅
18 10 ☽ # ♀
19 59 ☽ ∠ ♂
21 01 ☽ ± ☉
21 23 ☽ △ ♇
22 58 ♀ ∥ ♅

05 Wednesday
00 58 ☽ ⊼ ♀
01 39 ☽ ∥ ♂
02 24 ☽ # ☉
03 30 ☿ ∠ ♇
04 00 ☽ ☍ ♄
04 29 ☽ # ♆
06 46 ☽ # ♄
11 45 ☽ ± ♀
12 53 ☽ ⚼ ☉
13 39 ☉ ⊻ ♆
19 11 ☉ # ♂
20 08 ☽ ⚹ ♂
22 46 ☉ △ ♃
23 40 ☽ ⊼ ☿

06 Thursday
04 15 ☽ ⚼ ♀
05 48 ☽ ⊼ ♅
06 10 ☽ ∥ ♃
10 37 ☽ ± ☿
13 47 ♂ ± ♅
14 16 ☽ ⊼ ♆
14 50 ☽ ☌ ♃
15 14 ♀ ⊻ ♄
16 00 ☽ Q ♂
16 04 ☽ ± ♅
16 07 ☽ △ ☉
23 51 ☽ □ ♇

07 Friday
00 39 ☽ ± ♆
03 19 ☽ ⚼ ☿
07 10 ☽ ⊼ ♄
07 45 ☽ ⚼ ♅
08 39 ☽ △ ♀
16 30 ☽ ⚼ ♆
17 51 ☽ ± ♄
22 51 ☽ □ ♂

08 Saturday
03 54 ♀ ⊥ ♇
07 53 ☽ Q ♇
07 57 ☽ △ ☿
10 10 ☽ ⚼ ♄
10 40 ☽ △ ♅
19 41 ☽ △ ♆
20 03 ☽ ⊻ ♃
20 08 ☽ # ♇

09 Sunday
00 09 ♂ Q ♃
01 32 ☿ ± ♂
02 00 ☿ ± ♃
02 04 ☽ □ ☉
05 59 ☽ ⚹ ♇
07 09 ☽ ⊥ ♃
12 40 ☉ ⊥ ♄
14 07 ☽ △ ♄
21 14 ☽ □ ♀
22 34 ☽ ∥ ♇

10 Monday
00 02 ☽ ∠ ♃
05 07 ☽ △ ♂
10 25 ☽ ∠ ♇
10 41 ☿ ☌ ♅
15 03 ☽ Q ☉
19 15 ☽ □ ♅
19 40 ☽ □ ☿

11 Tuesday
00 48 ♀ ⚼ ♃
03 51 ☽ ⊥ ♇
04 42 ☽ □ ♆
04 48 ☽ ⚹ ♃
05 38 ☉ ⊻ ♇
09 25 ☽ ⚼ ♂
11 47 ☽ Q ♀
15 36 ☽ ⊻ ♇
16 28 ☽ ⚹ ☉
20 13 ♃ ⊼ ♆

12 Wednesday
00 28 ☽ □ ♄
02 42 ☽ ± ♂
04 16 ☽ Q ♃
06 39 ☽ Q ♅
08 27 ☽ Q ☿
14 10 ☽ ⚹ ♀
14 20 ☽ ⊼ ♂
15 30 ♀ ⊼ ♂
16 15 ☽ Q ♆
19 46 ☽ # ♃

13 Thursday
00 54 ☽ ∠ ☉
06 39 ☽ ⚹ ♅
07 44 ☿ ∥ ♀
09 03 ☽ ⚹ ☿
12 37 ☽ Q ♄
16 02 ☽ □ ♃
16 19 ☽ ⚹ ♆
20 37 ☽ ⊥ ☉
22 33 ☽ ∥ ♄
23 42 ☽ ∠ ♀

14 Friday
02 14 ☽ ∥ ♆
03 32 ☽ ☌ ♇
03 43 ☽ # ♂
09 50 ☽ ⊻ ☉
12 50 ☽ ∥ ☉
12 59 ☽ ∠ ♅
13 00 ☽ ⚹ ♄
15 15 ☽ ∥ ♅
15 37 ☽ ∠ ☿
18 15 ☽ ∥ ☿
20 01 ☽ ⊥ ♀
21 08 ☿ St R
22 19 ☽ ∥ ♀
22 40 ☽ ∠ ♆

15 Saturday
01 22 ☽ ☍ ♂
07 17 ☽ ⊥ ♅
09 36 ☽ ⊻ ♀
09 48 ☽ ⊥ ☿
12 59 ♀ ∠ ♇
16 57 ☽ ⊥ ♆
19 30 ☽ ⊻ ♅
19 39 ☽ ∠ ♄
21 49 ☽ ⊻ ☿
22 23 ☿ ∥ ♅

16 Sunday
02 45 ☉ ⊻ ♄
04 27 ☽ △ ♃
05 10 ☽ ⊻ ♆
11 48 ☉ ∥ ☿
14 05 ☽ ⊥ ♄
15 42 ♀ ± ♂
16 29 ☽ ⊻ ♇

17 Monday
02 22 ☽ ⊻ ♄
04 22 ☽ ☌ ☉
04 41 ☽ ⊥ ♇
08 16 ☉ ⊥ ♇
10 45 ☽ ⚼ ♃
13 07 ☽ ⊼ ♂
22 38 ♀ ± ♃
22 59 ☽ ∠ ♇

18 Tuesday
01 04 ☽ ± ♂
04 53 ☽ ± ♃
05 39 ☽ ☌ ♀
07 11 ☿ ☌ ♅
08 33 ☽ ☌ ☿
08 38 ☽ ☌ ♅
16 57 ☽ ⊼ ♃
17 43 ☽ ∥ ♀
18 07 ☽ ☌ ♆
18 57 ☽ ⚼ ♂

19 Wednesday
00 02 ☿ ☌ ♀
00 05 ☽ ∥ ♅
00 35 ☽ ∥ ☉
05 20 ☽ ⚹ ♇
08 19 ☽ ∥ ☿
10 57 ♀ ☌ ♅
12 20 ☽ # ♂
12 56 ☽ ∥ ♆
15 31 ☽ ☌ ♄
17 12 ☽ ∥ ♄
22 32 ☽ ⊻ ☉

20 Thursday
00 37 ☽ △ ♂
02 46 ☉ ⚼ ♃
05 25 ☽ Q ♇
11 35 ☽ ⊥ ☉
14 47 ☿ ± ♃
16 34 ☽ ⊻ ☿
17 18 ♂ ⚼ ♆
17 59 ☽ # ♃
19 57 ☉ ⊼ ♂
20 59 ☽ ⊻ ♅

21 Friday
00 43 ☽ ⊻ ♀
03 26 ☽ ⊥ ☿
04 30 ☽ ☍ ♃
06 07 ☿ # ♂
06 09 ☽ ⊻ ♆
07 00 ☽ ∠ ☉
08 51 ☽ ⊥ ♅
10 35 ☿ ∥ ♆
13 51 ☽ ⊥ ♀
17 02 ☽ □ ♇
17 56 ☽ ⊥ ♆
19 31 ☽ ∠ ☿

22 Saturday
02 30 ☽ ∠ ♅
03 16 ☽ ⊻ ♄
09 18 ☽ ∠ ♀
10 47 ☽ □ ♂
11 25 ☽ ∠ ♆
11 42 ♀ ⊼ ♃
14 44 ☽ ⚹ ☉
14 56 ☽ ⊥ ♄
17 10 ☿ ± ♂
18 51 ☿ ∠ ♇
21 44 ☽ ⚹ ☿
22 50 ♀ ⚼ ♂

23 Sunday
06 52 ♀ ☌ ♆
07 20 ☽ ⚹ ♅
08 12 ☽ ∠ ♄
13 54 ☽ ⊼ ♃
14 43 ☽ ∥ ♇
15 28 ☽ Q ☉
15 58 ☽ ⚹ ♆
16 55 ☽ ⚹ ♀
18 12 ☽ Q ☿
19 50 ☿ ∥ ♄
20 32 ☽ Q ♂

24 Monday
01 01 ☽ ± ♃
02 17 ☽ △ ♇
05 49 ☽ Q ♅
07 39 ☉ ☌ ☿
12 18 ☽ ⚹ ♄
14 12 ☽ Q ♆
17 24 ☽ Q ♀
17 26 ☽ ⚼ ♃
17 26 ☽ # ♇
18 25 ☽ ⚹ ♂

25 Tuesday
00 04 ☽ □ ☿
03 16 ☽ □ ☉
05 38 ☽ ⚼ ♇
10 08 ☽ Q ♄
14 28 ☽ □ ♅
20 04 ☽ △ ♃
20 59 ☽ ∠ ♂
21 37 ☽ ± ♇
22 27 ☽ □ ♆

26 Wednesday
02 03 ☉ ± ♂
04 25 ♀ ∥ ♅
04 38 ☽ □ ♀
08 05 ☽ ⊼ ♇
11 23 ☉ ∠ ♇
12 30 ☽ ⊥ ♂
16 04 ☉ ∥ ♀
17 45 ☽ □ ♄
22 42 ☽ ⊻ ♂
23 54 ☽ △ ☿

27 Thursday
08 41 ☽ ∥ ♃
11 17 ☽ △ ☉
14 59 ☿ ⊼ ♂
18 00 ☽ △ ♅
19 27 ♀ ⚹ ♇
22 45 ☽ □ ♃
23 03 ☽ ⚼ ☿
23 11 ☽ # ☿

28 Friday
01 25 ☽ △ ♆
03 12 ☿ ⚼ ♃
03 39 ☽ # ♄
07 55 ☽ # ♆
07 59 ☽ ∥ ♂
08 02 ☉ ± ♃
10 26 ☽ ☍ ♇
11 48 ☽ △ ♀
12 46 ☽ ± ☿
13 49 ☽ ⚼ ☉
15 37 ☽ # ♀
17 10 ☽ # ☉
18 15 ☽ # ♅
18 38 ☽ ⚼ ♅
19 52 ☽ △ ♄
21 51 ☽ ⊼ ☿
23 57 ☽ ☌ ♂

29 Saturday
01 50 ☽ ⚼ ♆
03 50 ☽ Q ♃
05 24 ☽ ± ☉
09 09 ☽ ± ♅
14 08 ☽ ⚼ ♀
15 41 ☽ ⊼ ☉
16 16 ☽ ± ♆
18 45 ☽ ⊼ ♅
20 03 ☽ ⚼ ♄
22 51 ☽ ⚹ ♃

30 Sunday
00 54 ☿ ⊥ ♇
01 49 ☽ ⊼ ♆
03 31 ☿ ⊻ ♄
05 43 ☽ ± ♀
10 25 ☽ ± ♄
10 31 ☽ ⊼ ♇
16 04 ☽ ⊼ ♀
19 07 ☽ ☍ ☿
19 58 ☽ ⊼ ♄
20 01 ☽ ± ♇
22 26 ☽ ∠ ♃
23 31 ☽ ⊻ ♂

31 Monday
08 59 ☽ ⊥ ♂
10 16 ☽ ⚼ ♇
12 34 ☽ ⊥ ♃
15 51 ☉ ☌ ♅
18 18 ☿ ⊻ ♀
18 25 ☽ ☍ ♅
18 35 ☽ ☍ ☉
22 01 ☽ ⊻ ♃
23 14 ☽ ∠ ♂

JANUARY 1991

LONGITUDES

Date	Sidereal time h m s	Sun ☉ ° ' "	Moon ☽ ° ' "	Moon ☽ 24.00 ° ' "	Mercury ☿ ° '	Venus ♀ ° '	Mars ♂ ° '	Jupiter ♃ ° '	Saturn ♄ ° '	Uranus ♅ ° '	Neptune ♆ ° '	Pluto ♇ ° '
01	18 42 33	10 ♑ 33 53	20 ♋ 45 11	28 ♋ 12 31	24 ♐ 06	25 ♑ 21	27 ♉ 45	11 ♌ 56	25 ♑ 43	09 ♑ 46	14 ♑ 08	19 ♏ 36
02	18 46 30	11 35 01	05 ♌ 35 04	12 ♌ 52 04	23 R 50	26 36	27 D 46	11 R 50	25 50	09 49	14 10	19 38
03	18 50 26	12 36 09	20 ♌ 02 55	27 ♌ 07 11	23 43	27 51	27 47	11 44	25 57	09 53	14 13	19 40
04	18 54 23	13 37 17	04 ♍ 07 37	10 ♍ 55 10	23 D 45	29 ♑ 07	27 49	11 38	26 04	09 56	14 15	19 41
05	18 58 19	14 38 25	17 ♍ 38 54	24 ♍ 16 01	23 56	00 ♒ 22	27 51	11 31	26 11	10 00	14 17	19 43
06	19 02 16	15 39 34	00 ♎ 46 50	07 ♎ 11 44	24 15	01 37	27 55	11 25	26 18	10 04	14 20	19 45
07	19 06 12	16 40 43	13 ♎ 31 12	19 ♎ 45 44	24 40	02 52	27 59	11 18	26 25	10 07	14 22	19 46
08	19 10 09	17 41 52	25 ♎ 55 53	02 ♏ 02 11	25 13	04 07	28 04	11 11	26 32	10 11	14 24	19 48
09	19 14 05	18 43 01	08 ♏ 05 12	14 ♏ 05 29	25 51	05 22	28 09	11 05	26 40	10 14	14 26	19 49
10	19 18 02	19 44 10	20 ♏ 03 36	26 ♏ 00 02	26 34	06 38	28 15	10 58	26 47	10 18	14 29	19 51
11	19 21 59	20 45 19	01 ♐ 55 17	07 ♐ 49 49	27 22	07 53	28 22	10 51	26 54	10 21	14 31	19 52
12	19 25 55	21 46 28	13 ♐ 44 04	19 ♐ 38 25	28 15	09 08	28 30	10 43	27 01	10 25	14 33	19 54
13	19 29 52	22 47 37	25 ♐ 33 16	01 ♑ 28 54	29 ♐ 11	10 23	28 38	10 36	27 08	10 28	14 35	19 55
14	19 33 48	23 48 46	07 ♑ 25 40	13 ♑ 23 49	00 ♑ 10	11 38	28 47	10 29	27 15	10 32	14 38	19 56
15	19 37 45	24 49 54	19 ♑ 23 36	25 ♑ 25 15	01 12	12 53	28 56	10 21	27 22	10 36	14 40	19 58
16	19 41 41	25 51 02	01 ♒ 28 58	07 ♒ 34 57	02 18	14 08	29 06	10 14	27 29	10 39	14 42	19 59
17	19 45 38	26 52 10	13 ♒ 43 23	19 ♒ 54 28	03 25	15 23	29 17	10 06	27 37	10 43	14 44	20 00
18	19 49 34	27 53 16	26 ♒ 08 22	02 ♓ 25 16	04 35	16 38	29 28	09 58	27 44	10 46	14 47	20 01
19	19 53 31	28 54 22	08 ♓ 45 22	15 ♓ 08 49	05 46	17 53	29 40	09 51	27 51	10 50	14 49	20 03
20	19 57 28	29 ♑ 55 27	21 ♓ 35 51	28 ♓ 06 38	07 00	19 09	29 ♉ 53	09 43	27 58	10 53	14 51	20 04
21	20 01 24	00 ♒ 56 32	04 ♈ 41 23	11 ♈ 20 15	08 15	20 23	00 ♊ 06	09 35	28 05	10 56	14 53	20 05
22	20 05 21	01 57 35	18 ♈ 03 25	24 ♈ 50 59	09 32	21 38	00 20	09 27	28 12	11 00	14 55	20 06
23	20 09 17	02 58 38	01 ♉ 43 05	08 ♉ 39 43	10 50	22 53	00 34	09 19	28 19	11 03	14 58	20 07
24	20 13 14	03 59 39	15 ♉ 40 52	22 ♉ 46 25	12 09	24 08	00 48	09 11	28 27	11 07	15 00	20 08
25	20 17 10	05 00 40	29 ♉ 56 08	07 ♊ 09 45	13 30	25 23	01 04	09 03	28 34	11 10	15 02	20 09
26	20 21 07	06 01 39	14 ♊ 26 47	21 ♊ 46 44	14 52	26 38	01 19	08 55	28 41	11 13	15 04	20 10
27	20 25 03	07 02 37	29 ♊ 08 57	06 ♋ 32 39	16 14	27 53	01 35	08 47	28 48	11 17	15 06	20 11
28	20 29 00	08 03 35	13 ♋ 57 03	21 ♋ 21 13	17 38	29 ♒ 08	01 52	08 39	28 55	11 20	15 08	20 12
29	20 32 57	09 04 31	28 ♋ 44 16	06 ♌ 05 15	19 03	00 ♓ 23	02 09	08 31	29 02	11 23	15 11	20 13
30	20 36 53	10 05 26	13 ♌ 23 18	20 ♌ 37 36	20 29	01 37	02 27	08 23	29 09	11 27	15 13	20 13
31	20 40 50	11 ♒ 06 20	27 ♌ 47 23	04 ♍ 52 02	21 ♑ 55	02 ♓ 52	02 ♊ 45	08 ♌ 15	29 ♑ 16	11 ♑ 30	15 ♑ 15	20 ♏ 14

Date	Moon ☽ True ☊ ° '	Moon ☽ Mean ☊ ° '	Moon ☽ Latitude ° '
01	28 ♑ 02	29 ♑ 06	00 N 40
02	28 D 02	29 03	00 S 42
03	28 03	29 00	01 59
04	28 04	28 57	03 07
05	28 05	28 54	04 02
06	28 06	28 50	04 42
07	28 07	28 47	05 07
08	28 R 07	28 44	05 17
09	28 06	28 41	05 12
10	28 05	28 38	04 54
11	28 04	28 34	04 24
12	28 03	28 31	03 42
13	28 02	28 28	02 51
14	28 02	28 25	01 52
15	28 01	28 22	00 S 48
16	28 D 01	28 19	00 N 19
17	28 01	28 15	01 26
18	28 01	28 12	02 30
19	28 R 01	28 09	03 28
20	28 01	28 06	04 16
21	28 01	28 03	04 52
22	28 01	27 59	05 13
23	28 01	27 56	05 17
24	28 D 01	27 53	05 03
25	28 01	27 50	04 30
26	28 02	27 47	03 39
27	28 03	27 44	02 34
28	28 03	27 40	01 N 18
29	28 R 03	27 37	00 S 04
30	28 03	27 34	01 24
31	28 ♑ 02	27 ♑ 31	02 S 37

DECLINATIONS

Date	Sun ☉ ° '	Moon ☽ ° '	Mercury ☿ ° '	Venus ♀ ° '	Mars ♂ ° '	Jupiter ♃ ° '	Saturn ♄ ° '	Uranus ♅ ° '	Neptune ♆ ° '	Pluto ♇ ° '
01	23 S 01	22 N 30	20 S 11	22 S 21	21 N 57	17 N 52	21 S 09	23 S 24	21 S 54	03 S 13
02	22 56	18 12	20 13	22 08	21 57	17 54	21 07	23 24	21 54	03 13
03	22 51	12 55	20 17	21 54	21 58	17 56	21 06	23 24	21 54	03 13
04	22 45	07 07	20 22	21 40	21 58	17 58	21 05	23 23	21 54	03 13
05	22 38	01 N 10	20 29	21 26	21 59	18 00	21 04	23 23	21 53	03 13
06	22 31	04 S 37	20 37	21 10	22 00	18 02	21 02	23 23	21 53	03 14
07	22 24	10 03	20 46	20 54	22 02	18 04	21 01	23 22	21 53	03 14
08	22 16	14 56	20 56	20 37	22 03	18 06	21 00	23 22	21 53	03 14
09	22 08	19 07	21 06	20 20	22 04	18 08	20 59	23 22	21 52	03 14
10	21 59	22 29	21 16	20 02	22 06	18 10	20 57	23 22	21 52	03 14
11	21 50	24 51	21 26	19 44	22 07	18 12	20 56	23 21	21 52	03 14
12	21 41	26 07	21 36	19 25	22 09	18 14	20 55	23 21	21 52	03 14
13	21 31	26 13	21 46	19 05	22 11	18 16	20 53	23 21	21 52	03 14
14	21 21	25 06	21 56	18 45	22 13	18 19	20 52	23 21	21 51	03 14
15	21 10	22 50	22 05	18 25	22 15	18 21	20 51	23 20	21 51	03 13
16	20 59	19 31	22 13	18 04	22 18	18 23	20 49	23 20	21 51	03 13
17	20 47	15 17	22 21	17 42	22 20	18 25	20 48	23 20	21 51	03 13
18	20 35	10 27	22 29	17 20	22 22	18 27	20 47	23 20	21 50	03 13
19	20 23	05 S 05	22 35	16 57	22 25	18 30	20 45	23 19	21 50	03 13
20	20 10	00 N 35	22 40	16 34	22 28	18 32	20 44	23 19	21 50	03 13
21	19 57	06 20	22 45	16 11	22 30	18 34	20 43	23 19	21 50	03 13
22	19 44	11 54	22 48	15 47	22 33	18 36	20 41	23 18	21 49	03 13
23	19 30	17 01	22 51	15 22	22 36	18 39	20 40	23 18	21 49	03 12
24	19 15	21 21	22 52	14 57	22 39	18 41	20 39	23 18	21 49	03 12
25	19 01	24 32	22 53	14 32	22 42	18 43	20 37	23 18	21 49	03 12
26	18 46	26 10	22 52	14 06	22 45	18 45	20 36	23 17	21 48	03 12
27	18 31	26 00	22 50	13 40	22 48	18 48	20 34	23 17	21 48	03 12
28	18 15	24 00	22 47	13 14	22 51	18 50	20 33	23 17	21 48	03 11
29	17 59	20 21	22 43	12 47	22 54	18 52	20 32	23 17	21 48	03 11
30	17 43	15 28	22 37	12 20	22 58	18 54	20 30	23 16	21 47	03 11
31	17 S 27	09 N 47	22 S 31	11 S 53	23 N 01	18 N 56	20 S 29	23 S 16	21 S 47	03 S 11

ZODIAC SIGN ENTRIES

Date	h m	Planets
02	02 54	☽ ♌
04	04 57	☽ ♍
05	05 03	♀ ♒
06	10 33	☽ ♎
08	19 59	☽ ♏
11	08 06	☽ ♐
13	21 00	☽ ♑
14	08 02	☿ ♑
16	09 04	☽ ♒
18	19 23	☽ ♓
20	13 47	☉ ♒
21	01 15	♂ ♊
21	03 28	☽ ♈
23	09 01	☽ ♉
25	12 06	☽ ♊
27	13 23	☽ ♋
29	04 44	♀ ♓
29	14 03	☽ ♌
31	15 44	☽ ♍

LATITUDES

Date	Mercury ☿ ° '	Venus ♀ ° '	Mars ♂ ° '	Jupiter ♃ ° '	Saturn ♄ ° '	Uranus ♅ ° '	Neptune ♆ ° '	Pluto ♇ ° '
01	03 N 08	01 S 18	02 N 21	00 N 41	00 S 09	00 S 19	00 N 47	14 N 57
04	02 55	01 22	02 22	00 42	00 09	00 19	00 47	14 58
07	02 34	01 25	02 22	00 42	00 09	00 19	00 47	14 59
10	02 08	01 28	02 23	00 43	00 09	00 19	00 47	15 00
13	01 40	01 31	02 24	00 43	00 09	00 19	00 47	15 01
16	01 12	01 32	02 24	00 44	00 10	00 19	00 47	15 02
19	00 44	01 33	02 23	00 44	00 10	00 19	00 47	15 03
22	00 N 18	01 34	02 23	00 45	00 10	00 19	00 47	15 05
25	00 S 08	01 34	02 22	00 45	00 10	00 19	00 47	15 06
28	00 31	01 33	02 22	00 45	00 11	00 19	00 47	15 07
31	00 S 52	01 S 32	02 N 21	00 N 46	00 S 11	00 S 20	00 N 47	15 N 09

DATA

Julian Date	2448258
Delta T	+58 seconds
Ayanamsa	23° 44' 08"
Synetic vernal point	05° ♓ 22' 51"
True obliquity of ecliptic	23° 26' 30"

MOON'S PHASES, APSIDES AND POSITIONS ☽

Date	h	m	Phase	Longitude ° '	Eclipse Indicator
07	18	35	☾	16 ♎ 58	
15	23	50	●	25 ♑ 20	Annular
23	14	22	☽	03 ♉ 05	
30	06	10	○	09 ♌ 51	

Day	h	m		
12	11	09	Apogee	
28	08	40	Perigee	
05	16	47	0S	
13	01	47	Max dec	26° S 19'
20	09	33	0N	
26	21	56	Max dec	26° N 19'

LONGITUDES

Date	Chiron ⚷ ° '	Ceres ⚳ ° '	Pallas ⚴ ° '	Juno ⚵ ° '	Vesta ⚶ ° '	Black Moon Lilith ⚸ ° '
01	25 ♋ 14	23 ♎ 28	24 ♍ 17	25 ♐ 55	15 ♉ 34	17 ♐ 17
11	24 ♋ 33	26 ♎ 25	25 ♍ 15	29 ♐ 25	15 ♉ 50	18 ♐ 24
21	23 ♋ 50	28 ♎ 58	25 ♍ 17	02 ♑ 54	16 ♉ 47	19 ♐ 31
31	23 ♋ 08	01 ♏ 04	24 ♍ 20	06 ♑ 18	18 ♉ 20	20 ♐ 38

All ephemeris data is given at 12.00 UT and the Moon's longitude is additionally given for 24.00 UT

ASPECTARIAN

h m	Aspects
01 Tuesday	
01 25	☽ ☍ ♆
05 43	☽ # ♅
08 23	☽ # ☉
10 10	☽ △ ♇
12 50	♂ St D
13 05	☽ # ♀
15 34	☽ ∥ ♂
15 49	☽ # ♆
17 16	☽ ⊼ ☿
19 51	♀ ☌ ♄
20 03	☽ ☍ ♄
20 04	☽ ☍ ♀
20 25	☽ # ♄
23 16	☽ ⚹ ♂
02 Wednesday	
01 42	☽ # ☿
02 46	☽ ± ☿
13 28	☽ ∥ ♃
17 16	☽ bQ ☿
17 22	☉ ⊼ ♃
18 52	☽ Q ♂
18 59	☽ ⊼ ♅
22 13	☽ ☌ ♃
22 37	☽ ⊼ ☉
03 Thursday	
02 12	☽ ⊼ ♆
04 59	☽ ± ♅
06 29	♀ # ♂
09 23	☽ ± ☉
10 28	♀ △ ♂
11 21	☽ □ ♇
12 17	☽ ± ♆
12 56	♀ ∥ ♆
16 15	☉ bQ ♂
17 53	☿ St D
18 11	☽ △ ☿
20 13	☽ bQ ♅
22 06	☽ ⊼ ♄
04 Friday	
01 09	☽ □ ♂
01 50	☽ bQ ☉
02 34	☽ ⊼ ♀
03 38	☽ bQ ♆
08 29	☽ ± ♄
13 59	☽ ± ♀
18 19	☽ Q ♇
22 19	☽ △ ♅
05 Saturday	
00 22	☽ bQ ♄
01 09	☽ ⊻ ♃
01 57	☿ ⊥ ♀
03 22	☉ ☌ ♆
03 42	☽ # ♇
05 58	☽ △ ♆
06 10	☽ △ ☉
07 29	☽ bQ ♀
11 46	☽ ⊥ ♃
15 44	☽ ⚹ ♇
23 38	☽ □ ☿
06 Sunday	
03 40	☽ △ ♄
04 00	☽ ∠ ♃
06 07	☽ ∥ ♇
06 40	☽ △ ♂
13 43	☽ △ ♀
14 29	♀ Q ♇
19 24	☽ ∠ ♇
07 Monday	
00 45	♀ ∥ ♄
05 30	☽ □ ♅
07 49	☽ ⚹ ♃
10 19	☽ Q ☿
10 57	☽ bQ ♂
12 29	☽ ⊥ ♇
13 37	☽ □ ♆
18 35	☽ □ ☉
19 24	☿ ∥ ♀
08 Tuesday	
00 02	☽ ⊻ ♇
04 24	☽ ± ♂
06 42	☽ Q ♃
10 32	☽ ⚹ ☿
13 12	☽ □ ♄
16 12	☽ ⊼ ♂
16 25	☽ Q ♅
20 58	☿ ∥ ♄
09 Wednesday	
00 46	☽ Q ♆
05 51	☽ # ♃
05 59	☽ □ ♀
09 01	☽ Q ☉
10 55	☿ ⊥ ♇
16 19	☽ ⚹ ♅
17 51	☽ ∠ ☿
17 54	☽ □ ♃
18 45	☿ bQ ♃
19 21	☽ ∥ ♀
20 48	☉ # ♂
10 Thursday	
00 22	☽ ∥ ♄
00 44	☽ ⚹ ♆
01 16	☽ Q ♄
02 01	☽ ∥ ☿
07 08	☽ ∥ ♆
08 15	☽ ∥ ☉
08 54	☽ # ♂
11 17	☽ ⚹ ☉
11 34	☽ ☌ ♇
13 06	☽ ⊥ ☿
14 36	☉ ⚹ ♇
19 30	☿ ⊻ ♄
19 47	☽ ∥ ♅
22 19	☽ Q ♀
22 38	☽ ∠ ♅
11 Friday	
01 43	☽ ⚹ ♄
02 05	☽ ⊻ ☿
04 43	☽ ☍ ♂
07 06	☽ ∠ ♆
07 34	☉ ∥ ♆
16 58	☽ ⊥ ♅
20 31	☽ ∠ ☉
12 Saturday	
01 26	☽ ⊥ ♆
01 32	☽ ⚹ ♀
05 13	☽ ⊻ ♅
05 56	☽ △ ♃
08 28	☽ ∠ ♄
13 40	☽ ⊻ ♆
16 32	☽ ⊥ ☉
17 33	☉ ∥ ☿
19 40	☿ ⊼ ♂
13 Sunday	
00 32	☽ ⊻ ♇
02 56	☽ ⊥ ♄
05 52	☽ ⊻ ☉
09 19	♂ Q ♃
11 37	☽ ∠ ♀
12 06	☽ bQ ♃
12 44	☽ ⊥ ♇
13 50	♀ ⊻ ♅
15 14	☽ ⊻ ♄
15 47	♀ ☍ ♃
18 18	☽ ⊼ ♂
20 00	☽ ☌ ☿
14 Monday	
00 52	☿ ∥ ♆
04 40	♃ ⊼ ♅
06 07	☽ ± ♃
06 35	☽ ± ♂
06 58	☽ ∠ ♇
07 58	☽ ⊥ ♀
18 05	☽ ⊼ ♃
18 17	☽ ☌ ♅
21 27	☽ ⊻ ♀
15 Tuesday	
00 56	☽ bQ ♂
02 31	☽ ☌ ♆
07 28	☽ ∥ ♅
13 08	☽ ⚹ ♇
16 10	♀ # ♃
16 38	☽ # ♂
17 48	☽ ∥ ☿
19 52	☽ ∥ ♆
23 50	☽ ☌ ☉
16 Wednesday	
01 41	☽ ∥ ☉
03 17	☽ ∥ ♄
04 02	☽ ☌ ♄
07 14	☽ △ ♂
12 59	☽ Q ♇
13 45	☽ ⊻ ☿
18 56	☽ # ♃
21 45	☽ ∥ ♀
23 08	♀ ⊻ ♆
17 Thursday	
02 44	☽ ⊥ ☿
05 00	☽ ☍ ♃
05 50	☿ # ♂
06 06	☽ ⊻ ♅
09 55	☉ ∥ ♄
13 59	☽ ⊻ ♆
15 37	☽ ☌ ♀
17 50	☽ ⊥ ♅
22 03	☽ ∠ ☿
18 Friday	
00 12	☽ □ ♇
00 49	☿ ± ♃
01 39	☽ ⊥ ♆
07 45	☉ ☌ ♄
11 17	☽ ∠ ♅
14 33	♀ ⊥ ♅
15 04	☽ ⊻ ♄
15 39	☽ ⊻ ☉
18 29	☽ □ ♂
18 59	☽ ∠ ♆
21 08	☿ ∠ ♇
19 Saturday	
02 37	☽ ⊥ ♄
04 05	☽ ⊥ ☉
05 46	☽ ⚹ ☿
09 34	☿ ± ♂
14 02	☽ ⊼ ♃
15 55	☽ ⚹ ♅
19 46	☽ ∠ ♄
19 59	☽ ∥ ♇
22 31	☽ ∠ ☉
23 25	☽ ⚹ ♆
20 Sunday	
01 10	☽ ± ♃
04 59	☽ Q ♂
06 40	☽ Q ☿
06 57	☽ ⊻ ♀
07 55	♂ bQ ♆
09 09	☽ △ ♇
10 42	☉ △ ♂
14 23	☽ Q ♅
17 42	☽ bQ ♃
19 14	☽ ⊥ ♀
21 43	☽ Q ♆
22 58	☽ # ♇
23 51	☽ ⚹ ♄
21 Monday	
03 30	☽ ⚹ ♂
04 36	☽ ⚹ ☉
05 56	♀ □ ♇
12 43	☽ bQ ♇
13 24	☽ ∠ ♀
19 08	☽ □ ☿
20 46	☽ △ ♃
21 49	♀ ⊥ ♆
21 50	☽ Q ♄
23 20	☽ □ ♅
22 Tuesday	
04 06	☽ Q ☉
04 56	☽ ± ♇
06 24	☽ □ ♆
07 03	☽ ∠ ♂
10 41	☿ ⊼ ♃
15 20	☉ Q ♇
15 38	☽ ⊼ ♇
18 59	☽ ⚹ ♀
23 16	☽ ⊥ ♂
23 Wednesday	
04 33	☽ # ♀
06 02	☽ □ ♄
09 57	☽ ⊻ ♂
14 22	☽ □ ☉
16 13	☿ ☌ ♅
18 03	☽ Q ♀
20 27	☽ ∥ ♃
24 Thursday	
00 21	☽ # ☉
01 01	☽ □ ♃
04 10	☽ △ ♅
05 22	☽ △ ☿
07 39	☽ # ♄
10 50	☽ △ ♆
14 55	☽ # ♆
19 33	☽ ☍ ♇
20 43	☽ ∥ ♂
22 16	☽ # ☿
25 Friday	
01 24	☽ # ♅
03 40	☽ □ ♀
05 40	☽ bQ ♅
07 14	☽ Q ♃
09 21	☽ bQ ☿
09 41	☽ △ ♄
12 10	☽ bQ ♆
13 54	☽ ☌ ♂
20 44	☽ ± ♅
21 05	☽ △ ☉
26 Saturday	
01 51	☽ ± ☿
02 59	☽ ⚹ ♃
03 08	☽ ± ♆
03 41	♀ ∠ ♅
06 41	☽ ⊼ ♅
10 44	☽ bQ ♄
12 45	☽ ⊼ ☿
13 02	☽ ⊼ ♆
13 03	☉ # ♃
15 45	☿ ☌ ♆
21 23	☽ ⊼ ♇
23 34	☽ bQ ☉
27 Sunday	
01 35	☽ ± ♄
03 21	☽ ∠ ♃
07 10	☽ ± ♇
09 45	☽ △ ♀
11 25	☽ ⊼ ♄
15 18	☽ ± ☉
16 02	☽ ⊻ ♂
17 51	☽ ⊥ ♃
19 29	☿ bQ ♂
21 21	☿ # ♂
21 48	☽ bQ ♇
28 Monday	
01 45	☽ ⊼ ☉
01 57	☽ ⊥ ♂
03 30	☽ ⊻ ♃
07 26	♀ ⊻ ♄
07 45	☽ ☍ ♅
12 19	☽ bQ ♀
13 56	☽ ☍ ♆
16 49	☽ ∠ ♂
17 36	☽ # ♅
18 36	☽ ☍ ☿
20 32	☽ ∥ ♂
21 17	☽ # ☿
22 08	☽ △ ♇
29 Tuesday	
00 26	☉ ☍ ♃
03 35	☽ # ♆
04 15	☽ ± ♀
08 02	♀ ∠ ♆
11 02	☽ # ♄
12 29	☽ ☍ ♄
14 55	☽ ⊼ ♀
17 41	☽ ⚹ ♂
19 48	☽ ∥ ♃
30 Wednesday	
00 56	☽ # ☉
03 51	☽ ☌ ♃
06 10	☽ ☍ ☉
07 44	☿ ⚹ ♇
08 47	☽ ⊼ ♅
13 47	☽ Q ♂
15 01	☽ ⊼ ♆
18 44	☽ ± ♅
23 21	☽ □ ♇
31 Thursday	
01 00	☽ ± ♆
01 03	☽ ⊼ ☿
02 38	☽ # ♀
08 46	♀ □ ♂
09 49	☽ bQ ♅
12 15	☽ ± ☿
14 31	☽ ⊼ ♄
16 10	☽ bQ ♆
20 34	☽ □ ♂
21 26	☽ ☍ ♀
21 47	☉ ⊻ ♅

FEBRUARY 1991

LONGITUDES

Date	Sidereal time h m s	Sun ☉ ° ' "	Moon ☽ ° ' "	Moon ☽ 24.00 ° ' "	Mercury ☿ ° '	Venus ♀ ° '	Mars ♂ ° '	Jupiter ♃ ° '	Saturn ♄ ° '	Uranus ♅ ° '	Neptune ♆ ° '	Pluto ♇ ° '
01	20 44 46	12 ♒ 07 13	11 ♍ 51 05	18 ♍ 44 09	23 ♑ 23	04 ♓ 07	03 ♊ 03	08 ♌ 07	29 ♑ 23	11 ♑ 33	15 ♑ 17	20 ♏ 15
02	20 48 43	13 08 05	25 ♍ 31 03	02 ♎ 11 42	24 51	05 22	03 22	07 R 59	29 30	11 36	15 19	20 16
03	20 52 39	14 08 57	08 ♎ 46 09	15 ♎ 14 36	26 20	06 36	03 41	07 51	29 37	11 39	15 21	20 16
04	20 56 36	15 09 47	21 ♎ 37 20	27 ♎ 54 43	27 50	07 51	04 00	07 43	29 44	11 43	15 23	20 17
05	21 00 32	16 10 37	04 ♏ 07 14	10 ♏ 15 22	29 ♑ 21	09 06	04 20	07 36	29 51	11 46	15 25	20 18
06	21 04 29	17 11 25	16 ♏ 19 41	22 ♏ 20 47	00 ♒ 52	10 20	04 41	07 28	29 ♑ 58	11 49	15 27	20 18
07	21 08 26	18 12 13	28 ♏ 19 16	04 ♐ 15 45	02 25	11 35	05 01	07 20	00 ♒ 05	11 52	15 29	20 19
08	21 12 22	19 13 00	10 ♐ 10 52	16 ♐ 05 12	03 58	12 49	05 23	07 12	00 12	11 55	15 31	20 19
09	21 16 19	20 13 46	21 ♐ 59 22	27 ♐ 53 55	05 32	14 04	05 44	07 05	00 19	11 58	15 33	20 20
10	21 20 15	21 14 31	03 ♑ 49 23	09 ♑ 46 16	07 06	15 18	06 06	06 57	00 26	12 01	15 35	20 20
11	21 24 12	22 15 15	15 ♑ 45 02	21 ♑ 46 04	08 42	16 33	06 28	06 49	00 32	12 04	15 37	20 21
12	21 28 08	23 15 57	27 ♑ 49 47	03 ♒ 56 26	10 18	17 47	06 50	06 42	00 39	12 07	15 39	20 21
13	21 32 05	24 16 39	10 ♒ 06 19	16 ♒ 19 36	11 55	19 01	07 13	06 35	00 46	12 10	15 41	20 21
14	21 36 01	25 17 18	22 ♒ 36 27	28 ♒ 56 56	13 33	20 16	07 36	06 27	00 53	12 13	15 43	20 22
15	21 39 58	26 17 57	05 ♓ 21 05	11 ♓ 48 54	15 12	21 30	07 59	06 20	00 59	12 16	15 44	20 22
16	21 43 55	27 18 34	18 ♓ 20 19	24 ♓ 55 14	16 52	22 44	08 23	06 13	01 06	12 18	15 46	20 22
17	21 47 51	28 19 09	01 ♈ 33 33	08 ♈ 15 05	18 32	23 59	08 47	06 06	01 13	12 21	15 48	20 22
18	21 51 48	29 ♒ 19 43	14 ♈ 59 43	21 ♈ 47 15	20 14	25 13	09 11	05 59	01 19	12 24	15 50	20 22
19	21 55 44	00 ♓ 20 15	28 ♈ 37 34	05 ♉ 30 28	21 56	26 27	09 36	05 53	01 26	12 27	15 52	20 23
20	21 59 41	01 20 45	12 ♉ 25 49	19 ♉ 23 28	23 39	27 41	10 01	05 46	01 32	12 29	15 53	20 23
21	22 03 37	02 21 13	26 ♉ 23 17	03 ♊ 25 07	25 24	28 ♓ 55	10 26	05 39	01 39	12 32	15 55	20 23
22	22 07 34	03 21 40	10 ♊ 28 49	17 ♊ 34 14	27 09	00 ♈ 09	10 51	05 33	01 45	12 34	15 57	20 R 23
23	22 11 30	04 22 04	24 ♊ 41 09	01 ♋ 49 19	28 ♒ 55	01 23	11 17	05 27	01 52	12 37	15 58	20 23
24	22 15 27	05 22 27	08 ♋ 58 30	16 ♋ 08 19	00 ♓ 42	02 37	11 42	05 21	01 58	12 40	16 00	20 23
25	22 19 24	06 22 47	23 ♋ 18 25	00 ♌ 28 19	02 30	03 51	12 08	05 15	02 04	12 42	16 02	20 22
26	22 23 20	07 23 06	07 ♌ 37 33	14 ♌ 45 33	04 19	05 05	12 35	05 09	02 11	12 44	16 03	20 22
27	22 27 17	08 23 23	21 ♌ 51 44	28 ♌ 55 32	06 09	06 19	13 01	05 03	02 17	12 47	16 05	20 22
28	22 31 13	09 ♓ 23 38	05 ♍ 56 21	12 ♍ 53 38	08 ♓ 00	07 ♈ 32	13 ♊ 28	04 ♌ 58	02 ♒ 23	12 ♑ 49	16 ♑ 06	20 ♏ 22

MOON NODES AND LATITUDE

Date	Moon True ☊ ° '	Moon Mean ☊ ° '	Moon ☽ Latitude ° '
01	28 ♑ 00	27 ♑ 28	03 S 40
02	27 R 58	27 25	04 27
03	27 56	27 21	04 59
04	27 54	27 18	05 14
05	27 53	27 15	05 14
06	27 52	27 12	05 00
07	27 D 53	27 09	04 33
08	27 54	27 05	03 54
09	27 55	27 02	03 06
10	27 57	26 59	02 09
11	27 59	26 56	01 07
12	27 59	26 53	00 S 01
13	27 R 59	26 50	01 N 06
14	27 57	26 46	02 11
15	27 53	26 43	03 11
16	27 49	26 40	04 02
17	27 43	26 37	04 40
18	27 38	26 34	05 05
19	27 34	26 31	05 12
20	27 31	26 27	05 02
21	27 30	26 24	04 34
22	27 D 30	26 21	03 49
23	27 31	26 18	02 50
24	27 32	26 15	01 40
25	27 33	26 11	00 N 23
26	27 R 33	26 08	00 S 55
27	27 31	26 05	02 08
28	27 ♑ 26	26 ♑ 02	03 S 13

DECLINATIONS

Date	Sun ☉ ° '	Moon ☽ ° '	Mercury ☿ ° '	Venus ♀ ° '	Mars ♂ ° '	Jupiter ♃ ° '	Saturn ♄ ° '	Uranus ♅ ° '	Neptune ♆ ° '	Pluto ♇ ° '
01	17 S 10	03 N 44	22 S 23	11 S 25	23 N 04	18 N 59	20 S 27	23 S 16	21 S 47	03 S 11
02	16 53	02 S 18	22 13	10 57	23 08	19 01	20 26	23 16	21 47	03 10
03	16 35	08 03	22 03	10 29	23 11	19 03	20 25	23 15	21 47	03 10
04	16 17	13 17	21 51	10 00	23 14	19 05	20 23	23 15	21 46	03 10
05	15 59	17 49	21 38	09 32	23 18	19 07	20 22	23 15	21 46	03 09
06	15 41	21 30	21 23	09 02	23 21	19 09	20 21	23 15	21 46	03 09
07	15 23	24 13	21 07	08 33	23 25	19 11	20 19	23 14	21 46	03 09
08	15 04	25 50	20 50	08 04	23 28	19 13	20 18	23 14	21 45	03 08
09	14 45	26 17	20 32	07 34	23 32	19 15	20 16	23 14	21 45	03 08
10	14 25	25 32	20 12	07 04	23 35	19 17	20 15	23 14	21 45	03 08
11	14 06	23 37	19 51	06 34	23 39	19 19	20 14	23 13	21 45	03 07
12	13 46	20 37	19 28	06 04	23 42	19 21	20 12	23 13	21 44	03 07
13	13 26	16 39	19 04	05 33	23 46	19 23	20 11	23 13	21 44	03 07
14	13 06	11 55	18 39	05 03	23 49	19 25	20 09	23 13	21 44	03 06
15	12 45	06 35	18 12	04 32	23 53	19 27	20 08	23 13	21 44	03 06
16	12 24	00 S 54	17 44	04 01	23 56	19 29	20 07	23 12	21 44	03 05
17	12 04	04 N 54	17 14	03 30	24 00	19 31	20 05	23 12	21 43	03 05
18	11 42	10 35	16 44	02 59	24 03	19 32	20 04	23 12	21 43	03 05
19	11 21	15 50	16 11	02 28	24 06	19 34	20 03	23 11	21 43	03 04
20	11 00	20 21	15 38	01 57	24 10	19 36	20 01	23 11	21 43	03 04
21	10 38	23 47	15 03	01 25	24 13	19 37	20 00	23 11	21 42	03 03
22	10 16	25 48	14 26	00 54	24 16	19 39	19 58	23 11	21 42	03 03
23	09 55	26 09	13 49	00 S 23	24 19	19 41	19 57	23 10	21 42	03 02
24	09 33	24 48	13 10	00 N 08	24 23	19 42	19 56	23 10	21 42	03 02
25	09 10	21 49	12 29	00 40	24 26	19 44	19 54	23 10	21 42	03 01
26	08 48	17 29	11 47	01 11	24 29	19 45	19 53	23 10	21 41	03 01
27	08 25	12 12	11 04	01 42	24 32	19 46	19 52	23 10	21 41	03 01
28	08 S 03	06 N 20	10 S 20	02 N 14	24 N 35	19 N 48	19 S 50	23 S 09	21 S 41	03 S 00

ZODIAC SIGN ENTRIES

Date	h	m	Planets
02	20	02	☽ ♎
05	04	01	☽ ♏
05	22	20	☿ ♒
06	18	51	♄ ♒
07	15	23	☽ ♐
10	04	16	☽ ♑
12	16	16	☽ ♒
15	01	59	☽ ♓
17	09	11	☽ ♈
18	03	58	☉ ♓
19	14	24	☽ ♉
21	18	10	☽ ♊
22	09	02	♀ ♈
23	20	56	☽ ♋
24	02	35	☿ ♓
25	23	13	☽ ♌
28	01	50	☽ ♍

LATITUDES

Date	Mercury ☿ ° '	Venus ♀ ° '	Mars ♂ ° '	Jupiter ♃ ° '	Saturn ♄ ° '	Uranus ♅ ° '	Neptune ♆ ° '	Pluto ♇ ° '
01	00 S 59	01 S 31	02 N 21	00 N 46	00 S 11	00 S 20	00 N 47	15 N 09
04	01 17	01 29	02 20	00 46	00 11	00 20	00 47	15 11
07	01 32	01 26	02 19	00 46	00 11	00 20	00 47	15 12
10	01 45	01 23	02 18	00 47	00 12	00 20	00 47	15 13
13	01 56	01 19	02 17	00 47	00 12	00 20	00 47	15 14
16	02 03	01 14	02 16	00 47	00 12	00 20	00 47	15 16
19	02 06	01 09	02 14	00 47	00 12	00 20	00 47	15 18
22	02 06	01 03	02 13	00 47	00 13	00 20	00 47	15 19
25	02 02	00 57	02 12	00 47	00 13	00 20	00 47	15 20
28	01 54	00 50	02 11	00 47	00 13	00 20	00 47	15 22
31	01 S 41	00 S 42	02 N 09	00 N 48	00 S 13	00 S 20	00 N 48	15 N 23

DATA

Julian Date	2448289
Delta T	+58 seconds
Ayanamsa	23° 44' 14"
Synetic vernal point	05° ♓ 22' 46"
True obliquity of ecliptic	23° 26' 30"

MOON'S PHASES, APSIDES AND POSITIONS ☽

Date	h	m	Phase	Longitude ° '	Eclipse Indicator
06	13	52	☾	17 ♏ 16	
14	17	32	●	25 ♒ 31	
21	22	58	☽	02 ♊ 49	
28	18	25	○	09 ♍ 40	

Day	h	m		
09	04	25	Apogee	
25	01	21	Perigee	
02	02	46	0S	
09	09	03	Max dec	26° S 18'
16	15	45	0N	
23	05	06	Max dec	26° N 14'

LONGITUDES

Date	Chiron ⚷ ° '	Ceres ⚳ ° '	Pallas ⚴ ° '	Juno ⚵ ° '	Vesta ⚶ ° '	Black Moon Lilith ⚸ ° '
01	23 ♋ 04	01 ♏ 15	24 ♍ 11	06 ♑ 38	18 ♉ 31	20 ♐ 44
11	22 ♋ 27	02 ♏ 45	22 ♍ 11	09 ♑ 58	20 ♉ 37	21 ♐ 51
21	21 ♋ 55	03 ♏ 39	19 ♍ 25	13 ♑ 10	23 ♉ 08	21 ♐ 58
31	21 ♋ 31	03 ♏ 52	16 ♍ 12	16 ♑ 15	26 ♉ 01	24 ♐ 04

ASPECTARIAN

h m Aspects

01 Friday
00 47 ☽ ± ♄
05 19 ☽ ⚼ ☿
05 38 ☽ ⊻ ♃
05 47 ☽ Q ♇
11 29 ☽ △ ♅
12 30 ☽ ⊼ ☉
14 12 ☽ # ♇
15 54 ☽ ⊥ ♃
16 26 ☽ ⚼ ♄
17 58 ☽ △ ♆
23 48 ☽ ± ☉

02 Saturday
02 41 ☽ ⚹ ♇
07 33 ☽ ∠ ♃
10 40 ☽ △ ☿
15 02 ♀ ⊥ ♄
15 31 ☽ ∥ ♇
17 04 ☽ ⚼ ☉
19 12 ☽ △ ♄

03 Sunday
02 28 ☽ △ ♂
05 36 ☽ ∠ ♇
07 37 ☽ ⊼ ♀
10 20 ☽ ⚹ ♃
17 21 ☽ □ ♅
19 50 ☽ ± ♀
21 55 ☽ ∥ ♀
22 12 ☽ ⊥ ♇
22 49 ☽ △ ☉

04 Monday
00 14 ☽ □ ♆
06 56 ☽ ⚼ ♂
08 27 ☽ Q ♃
09 28 ☽ ⊻ ♇
09 50 ♀ ⊼ ♃
14 35 ☽ ⚼ ♀
15 42 ♂ # ♅
17 25 ☉ ⊻ ♆
21 01 ☿ ∥ ♆

05 Tuesday
00 31 ☽ ± ♂
01 29 ☽ □ ☿
02 28 ☽ ∥ ☉
03 31 ☽ Q ♅
03 39 ☽ □ ♄
10 38 ☽ Q ♆
12 26 ☽ ⊼ ♂
18 42 ☽ □ ♃
19 54 ☽ # ♃
20 39 ☿ ☌ ♄
22 49 ☽ △ ♀

06 Wednesday
03 02 ☽ ⚹ ♅
03 48 ☽ ∥ ♄
10 15 ☽ ⚹ ♆
11 12 ☽ ∥ ☿
13 52 ☽ □ ☉
13 56 ☽ ∥ ♆
15 17 ☽ Q ♄
17 48 ☽ Q ☿
19 55 ☽ ☌ ♇

07 Thursday
02 14 ☽ ∥ ♅
03 33 ☉ ⊥ ♅
03 40 ☽ # ♂
09 03 ☽ ∠ ♅
10 31 ☿ Q ♇
15 35 ☽ ⚹ ♄
16 23 ☽ ∠ ♆
17 48 ♀ ⚹ ♅
21 29 ☽ ⚹ ☿

08 Friday
01 57 ☽ ☍ ♂
03 19 ☽ ⊥ ♅
05 25 ☽ Q ☉
06 01 ☽ △ ♃
10 39 ☽ ⊥ ♆
15 32 ☽ ⊻ ♅
17 59 ☽ □ ♀
18 44 ♀ ± ♃
22 18 ☽ ∠ ♄
22 52 ☽ ⊻ ♆

09 Saturday
08 05 ☽ ⚹ ☉
08 34 ☽ ∠ ☿
08 37 ☽ ⊻ ♇
12 10 ☽ ⚼ ♃
14 24 ☉ □ ♇
16 06 ☿ △ ♂
16 46 ☽ ⊥ ♄
20 49 ☽ ⊥ ♇

10 Sunday
05 03 ☽ ⊻ ♄
05 39 ☽ ⊥ ☿
06 01 ♂ ± ♅
06 15 ☽ ± ♃
08 09 ☿ ∥ ♄
09 51 ☿ ☍ ♃
10 50 ☽ Q ♀
14 37 ♀ ∠ ♄
15 04 ☽ ∠ ♇
16 44 ☽ ⊼ ♂
17 20 ☽ ∠ ☉
17 34 ♀ ⚹ ♆
18 15 ☽ ⊼ ♃
19 39 ☽ ⊻ ☿
20 21 ☉ ⊥ ♆

11 Monday
04 35 ☽ ☌ ♅
05 12 ☽ ± ♂
11 44 ☽ ☌ ♆
11 44 ☽ # ♂
13 06 ☽ ⊥ ☉
13 46 ☽ ⚹ ♀
15 44 ☽ ∥ ♅
21 10 ☽ ⚹ ♇
23 45 ☽ ⚼ ♂

12 Tuesday
02 09 ☽ ⊻ ☉
03 54 ☽ ∥ ♆
05 26 ♂ ⚹ ♃
14 46 ☽ ∥ ♄
17 37 ☽ ☌ ♄
18 25 ☿ # ♃
20 09 ☽ # ♃
20 22 ☽ ∥ ☿
20 53 ☽ Q ♇
22 50 ☽ ∠ ♀

13 Wednesday
05 13 ☽ ☍ ♃
06 13 ☽ △ ♂
15 42 ☿ ⊻ ♅
16 00 ☽ ⊻ ♅
16 03 ☽ ☌ ☿
18 16 ☽ ⊥ ♀
22 47 ☽ ⊻ ♆

14 Thursday
03 35 ☽ ⊥ ♅
05 54 ☽ ∥ ☉
07 03 ☽ ⊻ ♀
07 43 ☽ □ ♇
10 17 ☽ ⊥ ♆
13 54 ♀ △ ♇
17 32 ☽ ☌ ☉
20 45 ☽ ∠ ♅

15 Friday
03 21 ☽ ∠ ♆
03 46 ☽ ⊻ ♄
09 06 ♀ ⚼ ♃
13 49 ☽ ⊼ ♃
15 05 ☽ ⊥ ♄
17 04 ☽ □ ♂
19 58 ☿ ⊻ ♆
21 42 ☽ ∥ ♀

16 Saturday
00 51 ☽ ± ♃
00 52 ☽ ⚹ ♅
02 53 ☽ ∥ ♇
05 06 ☿ ⊥ ♀
07 17 ☽ ⚹ ♆
07 52 ☽ ∠ ♄
08 54 ☽ ⊻ ☿
11 54 ☉ ∠ ♅
15 43 ☽ △ ♇
17 13 ☽ ⚼ ♃
20 52 ☽ ☌ ♀
21 27 ☽ ⊥ ☿
22 55 ☽ Q ♅

17 Sunday
03 07 ☽ Q ♂
04 28 ☽ # ♇
05 12 ☽ Q ♆
05 40 ☽ ⊻ ☉
06 39 ☽ # ♀
09 15 ☿ ⊥ ♅
11 22 ☽ ⚹ ♄
16 04 ☽ ∠ ☿
17 22 ☽ ⊥ ☉
18 51 ☽ ⚼ ♇
19 33 ♀ Q ♅
20 05 ☽ △ ♃

18 Monday
01 21 ☽ ⚹ ♂
07 22 ☽ □ ♅
07 37 ♀ ∥ ♇
09 00 ☽ Q ♄
10 43 ☽ ∠ ☉
10 54 ☽ ± ♇
13 29 ☽ □ ♆
14 02 ☿ □ ♇
16 37 ☽ # ☉
21 31 ☽ ⊼ ♇
22 35 ☽ ⚹ ☿

19 Tuesday
04 43 ☽ ∠ ♂
07 49 ☽ ⊻ ♀
10 55 ☿ ⊥ ♆
13 32 ☽ # ☿
15 14 ☽ ⚹ ☉
16 56 ☽ □ ♄
19 20 ☽ ⊥ ♀
20 57 ☽ ⊥ ♂
22 35 ☽ Q ☿

20 Wednesday
00 32 ☽ □ ♃
00 48 ☉ ∠ ♆
04 26 ♂ ± ♆
07 34 ☽ ∥ ♃
07 41 ☽ ⊻ ♂
10 03 ☽ # ♄
12 06 ☽ △ ♅
12 29 ☽ ∠ ♀
13 42 ☽ Q ☉
16 04 ♀ Q ♆
17 11 ☉ ⊻ ♄
17 59 ☽ △ ♆
20 31 ☽ # ♆
21 34 ♀ Q ♂

21 Thursday
01 42 ☽ ☍ ♇
07 06 ☽ # ♅
07 22 ☽ Q ♃
10 04 ☽ □ ☿
13 58 ☽ ⚼ ♅
16 04 ☽ ∥ ♂
16 45 ☽ ⚹ ♀
19 45 ☽ ⚼ ♆
21 03 ☽ △ ♄
22 58 ☽ □ ☉

22 Friday
01 30 ♇ St R
03 41 ☽ ⚹ ♃
05 21 ☽ ± ♅
11 05 ☽ ± ♆
12 39 ☽ ☌ ♂
15 06 ☽ Q ♀
15 34 ☽ ⊼ ♅
17 56 ☿ ∠ ♅
21 16 ☽ ⊼ ♆
22 42 ☽ ⚼ ♄

23 Saturday
04 44 ☽ ⊼ ♇
04 54 ☽ ∠ ♃
14 00 ☽ ± ♄
14 51 ☽ ± ♇
19 57 ☽ ⊥ ♃
20 08 ☽ △ ☿
22 10 ♀ ⚹ ♄

24 Sunday
00 10 ☽ ⊼ ♄
00 20 ☽ □ ♀
05 30 ☽ △ ☉
05 57 ☽ ⊻ ♃
05 58 ☽ ⚼ ♇
11 23 ☉ ⊼ ♃
16 04 ☿ ∠ ♆
16 10 ☽ ∥ ♂
16 43 ☽ ⊻ ♂
18 11 ☽ ☍ ♅
23 48 ☽ ☍ ♆

25 Monday
00 53 ☽ ⚼ ☿
02 30 ☽ # ♅
03 05 ☽ ⊥ ♂
05 58 ☿ ⊻ ♄
07 05 ☽ △ ♇
08 32 ☽ ⚼ ☉
12 44 ☽ # ♆
18 08 ☽ ± ☿
18 37 ☽ ∠ ♂
23 26 ☽ # ♄

26 Tuesday
00 18 ☽ ∥ ♃
00 44 ☽ ± ☉
02 48 ☽ ☍ ♄
05 38 ☽ ⊼ ☿
07 19 ☽ △ ♀
07 52 ☽ ☌ ♃
11 34 ☽ ⊼ ☉
13 15 ♀ △ ♃
17 41 ♀ ⚼ ♇
20 35 ☽ ⚹ ♂
20 37 ☽ ⊼ ♅
21 46 ♂ ⊼ ♅
22 21 ☿ ⊼ ♃

27 Wednesday
02 13 ☽ ⊼ ♆
06 46 ☽ ± ♅
09 10 ☉ ⊥ ♄
09 28 ☽ □ ♇
10 59 ☽ ⚼ ♀
12 22 ☽ ± ♆
17 27 ☽ # ☿
17 32 ☽ Q ♂
18 07 ☿ ⊻ ♀
22 05 ☽ ⚼ ♅

28 Thursday
03 42 ☽ ⚼ ♆
03 44 ☽ ± ♀
04 41 ☽ # ☉
05 52 ☽ ⊼ ♄
07 06 ♄ Q ♇
10 20 ☽ ⊻ ♃
15 01 ☽ ⊼ ♀
16 06 ☽ ☍ ☿
16 10 ☽ Q ♇
16 15 ☽ ± ♄
17 16 ☿ ⊥ ♄
18 25 ☽ ☍ ☉
20 36 ☽ ⊥ ♃
23 54 ☽ △ ♅

All ephemeris data is given at 12.00 UT and the Moon's longitude is additionally given for 24.00 UT

MARCH 1991

Date	Sidereal time h m s	Sun ☉ ° ' "	Moon ☽ ° ' "	Moon ☽ 24.00 ° ' "	Mercury ☿ ° '	Venus ♀ ° '	Mars ♂ ° '	Jupiter ♃ ° '	Saturn ♄ ° '	Uranus ♅ ° '	Neptune ♆ ° '	Pluto ♇ ° '
01	22 35 10	10 ♓ 23 51	19 ♍ 46 54	26 ♍ 35 40	09 ♓ 52	08 ♈ 46	13 ♊ 55	04 ♌ 52	02 ♒ 29	12 ♑ 52	16 ♑ 08	20 ♏ 22
02	22 39 06	11 24 02	03 ♎ 19 38	09 ♎ 58 32	11 45	10 00	14 22	04 R 47	02 35	12 54	16 09	20 R 21
03	22 43 03	12 24 12	16 ♎ 32 13	23 ♎ 00 40	13 38	11 13	14 49	04 42	02 42	12 56	16 11	20 21
04	22 46 59	13 24 20	29 ♎ 23 57	05 ♏ 42 16	15 33	12 27	15 17	04 37	02 48	12 58	16 12	20 21
05	22 50 56	14 24 26	11 ♏ 55 52	18 ♏ 05 10	17 28	13 40	15 44	04 32	02 53	13 00	16 14	20 20
06	22 54 53	15 24 31	24 ♏ 10 34	00 ♐ 12 36	19 24	14 54	16 12	04 28	02 59	13 03	16 15	20 20
07	22 58 49	16 24 34	06 ♐ 11 50	12 ♐ 08 53	21 20	16 07	16 40	04 23	03 05	13 05	16 17	20 19
08	23 02 46	17 24 36	18 ♐ 04 21	23 ♐ 58 56	23 17	17 20	17 09	04 19	03 11	13 07	16 18	20 19
09	23 06 42	18 24 36	29 ♐ 53 17	05 ♑ 48 04	25 14	18 34	17 37	04 15	03 17	13 09	16 19	20 18
10	23 10 39	19 24 35	11 ♑ 43 56	17 ♑ 41 33	27 11	19 47	18 06	04 11	03 22	13 11	16 20	20 18
11	23 14 35	20 24 32	23 ♑ 41 31	29 ♑ 44 25	29 ♓ 08	21 00	18 34	04 08	03 28	13 13	16 22	20 17
12	23 18 32	21 24 27	05 ♒ 50 44	12 ♒ 00 58	01 ♈ 05	22 13	19 03	04 04	03 34	13 14	16 23	20 17
13	23 22 28	22 24 20	18 ♒ 15 30	24 ♒ 34 38	03 00	23 26	19 32	04 01	03 39	13 16	16 24	20 16
14	23 26 25	23 24 12	00 ♓ 58 35	07 ♓ 27 28	04 55	24 39	20 02	03 58	03 45	13 18	16 25	20 15
15	23 30 22	24 24 01	14 ♓ 01 18	20 ♓ 40 00	06 48	25 52	20 31	03 55	03 50	13 20	16 26	20 15
16	23 34 18	25 23 49	27 ♓ 23 21	04 ♈ 11 03	08 40	27 05	21 01	03 52	03 55	13 21	16 27	20 14
17	23 38 15	26 23 35	11 ♈ 02 45	17 ♈ 57 57	10 29	28 18	21 30	03 49	04 01	13 23	16 29	20 13
18	23 42 11	27 23 18	24 ♈ 56 10	01 ♉ 56 52	12 15	29 ♈ 30	22 00	03 47	04 06	13 25	16 30	20 12
19	23 46 08	28 23 00	08 ♉ 59 29	16 ♉ 03 30	13 59	00 ♉ 43	22 30	03 45	04 11	13 26	16 31	20 11
20	23 50 04	29 ♓ 22 39	23 ♉ 08 26	00 ♊ 13 50	15 38	01 56	23 00	03 43	04 16	13 28	16 32	20 11
21	23 54 01	00 ♈ 22 16	07 ♊ 19 18	14 ♊ 24 32	17 14	03 08	23 31	03 41	04 21	13 29	16 33	20 10
22	23 57 57	01 21 51	21 ♊ 29 16	28 ♊ 33 18	18 45	04 21	24 01	03 39	04 26	13 30	16 33	20 09
23	00 01 54	02 21 24	05 ♋ 36 31	12 ♋ 38 46	20 11	05 33	24 32	03 38	04 31	13 32	16 34	20 08
24	00 05 51	03 20 54	19 ♋ 39 59	26 ♋ 40 05	21 31	06 46	25 02	03 36	04 36	13 33	16 35	20 07
25	00 09 47	04 20 22	03 ♌ 38 59	10 ♌ 36 32	22 46	07 58	25 33	03 35	04 40	13 34	16 36	20 06
26	00 13 44	05 19 48	17 ♌ 32 38	24 ♌ 27 04	23 55	09 10	26 04	03 35	04 45	13 36	16 37	20 05
27	00 17 40	06 19 11	01 ♍ 19 39	08 ♍ 10 06	24 57	10 22	26 35	03 34	04 50	13 37	16 38	20 04
28	00 21 37	07 18 32	14 ♍ 58 09	21 ♍ 43 29	25 52	11 34	27 06	03 33	04 54	13 38	16 38	20 03
29	00 25 33	08 17 50	28 ♍ 25 49	05 ♎ 04 49	26 41	12 46	27 37	03 33	04 59	13 39	16 39	20 02
30	00 29 30	09 17 07	11 ♎ 40 16	18 ♎ 11 53	27 22	13 58	28 08	03 33	05 03	13 40	16 39	20 00
31	00 33 26	10 ♈ 16 22	24 ♎ 39 33	01 ♏ 03 07	27 ♈ 57	15 ♉ 10	28 ♊ 40	03 ♌ 33	05 ♒ 07	13 ♑ 41	16 ♑ 40	19 ♏ 59

Date	Moon True ☊ ° '	Moon Mean ☊ ° '	Moon ☽ Latitude ° '
01	27 ♑ 20	25 ♑ 59	04 S 05
02	27 R 13	25 56	04 42
03	27 05	25 52	05 03
04	26 57	25 49	05 08
05	26 51	25 46	04 58
06	26 46	25 43	04 34
07	26 44	25 40	03 59
08	26 D 43	25 37	03 13
09	26 43	25 33	02 20
10	26 45	25 30	01 20
11	26 45	25 27	00 S 17
12	26 R 45	25 24	00 N 49
13	26 43	25 21	01 53
14	26 38	25 17	02 53
15	26 31	25 14	03 45
16	26 21	25 11	04 26
17	26 11	25 08	04 53
18	26 01	25 05	05 04
19	25 52	25 02	04 56
20	25 45	24 58	04 30
21	25 40	24 55	03 48
22	25 38	24 52	02 51
23	25 D 38	24 49	01 45
24	25 38	24 46	00 N 32
25	25 R 38	24 43	00 S 43
26	25 36	24 39	01 54
27	25 32	24 36	02 57
28	25 24	24 33	03 50
29	25 15	24 30	04 29
30	25 03	24 27	04 53
31	24 ♑ 50	24 ♑ 23	05 S 01

DECLINATIONS

Date	Sun ☉	Moon ☽	Mercury ☿	Venus ♀	Mars ♂	Jupiter ♃	Saturn ♄	Uranus ♅	Neptune ♆	Pluto ♇
01	07 S 40	00 N 17	09 S 34	02 N 45	24 N 37	19 N 49	19 S 49	23 S 09	21 S 41	03 S 00
02	07 17	05 S 38	08 47	03 16	24 40	19 50	19 48	23 09	21 41	02 59
03	06 54	11 10	07 59	03 47	24 43	19 52	19 47	23 09	21 40	02 59
04	06 31	16 03	07 10	04 18	24 46	19 53	19 45	23 09	21 40	02 58
05	06 08	20 08	06 19	04 49	24 48	19 54	19 44	23 08	21 40	02 58
06	05 45	23 14	05 28	05 20	24 51	19 55	19 43	23 08	21 40	02 57
07	05 22	25 16	04 35	05 51	24 53	19 56	19 41	23 08	21 40	02 56
08	04 58	26 07	03 42	06 22	24 56	19 57	19 40	23 08	21 40	02 56
09	04 35	25 46	02 48	06 52	24 58	19 58	19 39	23 08	21 39	02 55
10	04 12	24 15	01 53	07 22	25 00	19 59	19 38	23 08	21 39	02 55
11	03 48	21 38	00 58	07 53	25 02	20 00	19 36	23 07	21 39	02 54
12	03 24	18 02	00 S 02	08 23	25 04	20 01	19 35	23 07	21 39	02 54
13	03 01	13 34	00 N 54	08 52	25 06	20 01	19 34	23 07	21 39	02 53
14	02 37	08 26	01 49	09 22	25 08	20 02	19 33	23 07	21 39	02 53
15	02 13	02 S 49	02 45	09 51	25 10	20 03	19 32	23 07	21 38	02 52
16	01 50	03 N 02	03 40	10 21	25 11	20 03	19 31	23 07	21 38	02 52
17	01 26	08 52	04 34	10 50	25 13	20 04	19 29	23 06	21 38	02 51
18	01 02	14 22	05 27	11 18	25 14	20 05	19 28	23 06	21 38	02 50
19	00 39	19 10	06 19	11 47	25 15	20 05	19 27	23 06	21 38	02 50
20	00 S 15	22 55	07 09	12 15	25 17	20 06	19 26	23 06	21 38	02 49
21	00 N 09	25 17	07 57	12 43	25 18	20 06	19 25	23 06	21 38	02 49
22	00 33	26 01	08 44	13 10	25 19	20 06	19 24	23 06	21 37	02 48
23	00 56	25 04	09 28	13 38	25 20	20 07	19 23	23 06	21 37	02 48
24	01 20	22 31	10 09	14 05	25 20	20 07	19 22	23 06	21 37	02 47
25	01 43	18 39	10 48	14 31	25 21	20 07	19 21	23 06	21 37	02 46
26	02 07	13 46	11 24	14 58	25 22	20 07	19 20	23 05	21 37	02 46
27	02 31	08 14	11 56	15 24	25 22	20 07	19 19	23 05	21 37	02 45
28	02 54	02 N 23	12 26	15 49	25 22	20 07	19 18	23 05	21 37	02 45
29	03 17	03 S 30	12 52	16 14	25 22	20 07	19 17	23 05	21 37	02 44
30	03 41	09 07	13 14	16 39	25 22	20 07	19 16	23 05	21 37	02 43
31	04 N 04	14 S 13	13 N 33	17 N 03	25 N 22	20 N 07	19 S 15	23 S 05	21 S 37	02 S 43

ZODIAC SIGN ENTRIES

Date	h	m	Planets
02	06	03	☽ ♎
04	13	08	☽ ♏
06	23	35	☽ ♐
09	12	14	☽ ♑
11	22	40	☿ ♈
12	00	31	☽ ♒
14	10	11	☽ ♓
16	16	38	☽ ♈
18	20	40	☽ ♉
18	21	45	♀ ♉
20	23	37	☽ ♊
21	03	02	☉ ♈
23	02	27	☽ ♋
25	05	43	☽ ♌
27	09	41	☽ ♍
29	14	49	☽ ♎
31	22	01	☽ ♏

LATITUDES

Date	Mercury ☿ ° '	Venus ♀ ° '	Mars ♂ ° '	Jupiter ♃ ° '	Saturn ♄ ° '	Uranus ♅ ° '	Neptune ♆ ° '	Pluto ♇ ° '
01	01 S 50	00 S 47	02 N 10	00 N 47	00 S 13	00 S 20	00 N 47	15 N 22
04	01 35	00 40	02 09	00 48	00 13	00 20	00 48	15 24
07	01 15	00 32	02 07	00 48	00 14	00 20	00 48	15 25
10	00 50	00 24	02 06	00 48	00 14	00 20	00 48	15 26
13	00 S 20	00 15	02 05	00 48	00 14	00 20	00 48	15 27
16	00 N 15	00 S 06	02 03	00 47	00 14	00 20	00 48	15 29
19	00 52	00 N 03	02 02	00 47	00 15	00 21	00 48	15 30
22	01 30	00 13	02 00	00 47	00 15	00 21	00 48	15 31
25	02 06	00 22	01 59	00 47	00 15	00 21	00 48	15 32
28	02 37	00 32	01 58	00 47	00 15	00 21	00 48	15 33
31	03 N 00	00 N 42	01 N 56	00 N 47	00 S 16	00 S 21	00 N 48	15 N 34

DATA

Julian Date	2448317
Delta T	+58 seconds
Ayanamsa	23° 44' 17"
Synetic vernal point	05° ♓ 22' 42"
True obliquity of ecliptic	23° 26' 30"

MOON'S PHASES, APSIDES AND POSITIONS ☽

Date	h	m	Phase	Longitude ° '	Eclipse Indicator
08	10	32	☾	17 ♐ 21	
16	08	10	●	25 ♓ 14	
23	06	03	☽	02 ♋ 07	
30	07	17	○	09 ♎ 05	

Day	h	m		
09	00	57	Apogee	
22	04	33	Perigee	
01	13	09	0S	
08	17	07	Max dec	26° S 08'
15	23	37	0N	
22	10	28	Max dec	26° N 01'
28	21	40	0S	

LONGITUDES

Date	Chiron ⚷ ° '	Ceres ⚳ ° '	Pallas ⚴ ° '	Juno ⚵ ° '	Vesta ⚶ ° '	Black Moon Lilith ⚸ ° '
01	21 ♋ 35	03 ♏ 52	16 ♍ 51	15 ♑ 38	25 ♉ 25	23 ♐ 51
11	21 ♋ 18	03 ♏ 32	13 ♍ 32	18 ♑ 35	28 ♉ 31	24 ♐ 58
21	21 ♋ 10	02 ♏ 29	10 ♍ 38	21 ♑ 20	01 ♊ 53	26 ♐ 04
31	21 ♋ 11	00 ♏ 51	08 ♍ 20	23 ♑ 51	05 ♊ 27	27 ♐ 11

All ephemeris data is given at 12.00 UT and the Moon's longitude is additionally given for 24.00 UT

ASPECTARIAN

h m	Aspects
01 Friday	
01 16	☽ # ♇
01 25	☽ □ ♂
03 01	☽ ∥ ♀
05 37	☽ △ ♆
07 58	☽ bQ ♄
12 10	☽ ∠ ♃
12 55	♃ # ♄
13 01	☽ ✱ ♇
22 27	☉ ± ♃
23 00	♀ # ♇
02 Saturday	
00 18	☿ ± ♃
01 07	☽ ∥ ♇
01 20	☽ # ♀
02 35	☉ ☌ ☿
10 40	☽ △ ♄
14 36	☽ ✱ ♃
15 39	☽ ∠ ♇
18 30	☽ ∥ ☉
23 41	☽ ∥ ☿
03 Sunday	
01 16	☽ ☍ ♀
02 54	☿ ✱ ♅
03 48	☽ ⊼ ☉
05 22	☽ □ ♅
05 47	☽ ⊼ ☿
07 59	☽ ⊥ ♇
08 44	☽ △ ♂
11 21	☽ □ ♆
12 18	☽ Q ♃
15 44	☽ ± ☉
18 43	☽ ± ☿
19 03	☽ ⊻ ♇
04 Monday	
01 13	☉ ✱ ♅
07 33	☿ □ ♂
09 58	☽ bQ ☉
13 44	☽ bQ ♂
14 34	☽ bQ ☿
14 59	☽ Q ♅
18 29	☽ □ ♄
20 21	☿ ✱ ♆
21 09	☽ Q ♆
21 52	☽ □ ♃
22 37	♀ □ ♅
05 Tuesday	
07 36	☽ ± ♂
09 23	☽ ∥ ♄
10 26	☽ # ♃
14 06	☽ ✱ ♅
15 45	☽ ⊼ ♀
17 14	☽ △ ☉
17 34	☿ ∠ ♄
19 42	☽ ⊼ ♂
20 23	☽ ✱ ♆
21 11	☉ ∥ ☿
22 51	☽ ∥ ♆
06 Wednesday	
00 48	☽ △ ☿
01 02	♀ ± ♇
04 25	☽ ☌ ♇
04 48	☽ ± ♀
05 39	☽ Q ♄
11 03	☽ ∥ ♅
12 48	☿ bQ ♃
14 02	♀ Q ♄
14 07	☿ # ♀
14 36	♂ ⊼ ♆
19 42	☽ ∠ ♅
23 03	☉ # ♀
23 31	☿ △ ♇
07 Thursday	
00 39	☽ bQ ♀
02 07	☽ ∠ ♆
05 42	☽ ✱ ♄
06 13	☽ # ♂
08 24	☽ △ ♃
08 42	☉ ✱ ♆
13 47	☽ ⊥ ♅
15 10	♀ □ ♆
20 14	☽ ⊥ ♆
23 52	☉ □ ♂
08 Friday	
01 55	☽ ⊻ ♅
05 43	♀ ✱ ♂
08 24	☽ ⊻ ♆
10 02	☽ ☍ ♂
10 20	☽ △ ♀
10 32	☽ □ ☉
12 14	☽ ∠ ♄
14 31	☽ bQ ♃
16 33	☽ ⊻ ♇
19 43	☉ ⊻ ♀
09 Saturday	
00 41	☽ □ ☿
04 44	☽ ⊥ ♇
06 39	☽ ⊥ ♄
08 31	☿ ∥ ♇
08 32	☉ ∠ ♄
08 42	☽ ± ♃
10 52	☿ Q ♅
18 57	☽ ⊻ ♄
20 49	☽ ⊼ ♃
22 59	☽ ∠ ♇
10 Sunday	
02 27	☽ Q ☉
02 33	☽ # ♂
07 01	☉ bQ ♃
14 55	☽ ☌ ♅
20 20	☽ Q ☿
21 18	☽ ☌ ♆
22 07	♀ ⊼ ♇
23 33	☽ ∥ ♅
11 Monday	
01 20	☽ ⊼ ♂
02 21	☿ Q ♆
04 51	☽ ✱ ☉
05 13	☽ ✱ ♇
05 33	♂ bQ ♄
06 01	☽ □ ♀
09 08	☉ △ ♇
11 52	☽ ∥ ♆
13 50	☽ ± ♂
23 37	☽ # ♃
12 Tuesday	
00 52	☽ ✱ ☿
02 18	☽ ∥ ♄
05 01	☽ Q ♇
07 29	☽ ☌ ♄
08 21	☽ bQ ♂
08 32	☽ ☍ ♃
12 36	♂ ∠ ♃
13 12	☽ ∠ ☉
21 27	☽ Q ♀
13 Wednesday	
02 24	☽ ⊻ ♅
08 09	☽ ⊥ ☉
08 26	☽ ⊻ ♆
11 26	☽ ∠ ☿
13 56	☽ ⊥ ♅
14 33	☽ △ ♂
15 50	☽ □ ♇
19 54	☽ ⊥ ♆
19 55	☉ ∥ ♇
20 29	☿ ✱ ♄
20 34	☽ ⊻ ☉
22 53	☽ ✱ ♀
14 Thursday	
00 15	☿ △ ♃
06 59	☽ ∠ ♅
07 29	☽ ⊥ ☿
08 11	☽ # ♀
12 50	☽ ∠ ♆
16 14	☿ bQ ♇
17 11	☽ ⊻ ♄
17 31	☽ ⊼ ♃
20 34	☽ ⊻ ☿
22 58	♂ ⊼ ♇
15 Friday	
02 28	☉ # ☿
04 19	☽ ⊥ ♄
04 32	☽ ± ♃
05 40	☽ ∠ ♀
10 44	☽ ✱ ♅
11 49	☽ ∥ ♇
12 16	☽ # ☿
14 40	☽ ∥ ☉
15 06	☿ # ♇
16 23	☽ ✱ ♆
20 46	☽ ∠ ♄
20 49	☽ bQ ♃
23 14	☽ △ ♇
23 37	☽ ⊥ ♀
16 Saturday	
00 11	☽ □ ♂
01 26	♃ ☍ ♄
07 24	☽ # ☉
08 10	☽ ☌ ☉
08 23	☽ Q ♅
11 00	☉ Q ♅
11 17	☽ # ♇
11 24	☽ ⊻ ♀
13 54	☽ Q ♆
15 01	☽ ∥ ☿
18 11	☿ Q ♂
23 24	☽ △ ♃
23 37	☽ ✱ ♄
17 Sunday	
01 50	☽ bQ ♇
09 13	☽ Q ♂
10 52	☽ ☌ ☿
16 04	☽ □ ♅
17 31	☽ ± ♇
20 40	☽ Q ♄
21 06	☽ ∥ ♀
21 26	☽ □ ♆
18 Monday	
03 52	☽ ⊼ ♇
06 46	☽ ✱ ♂
16 32	☽ ⊻ ☉
20 34	☽ ☌ ♀
19 Tuesday	
03 05	☽ □ ♃
03 34	☽ ⊥ ☉
03 46	☽ □ ♄
04 13	☿ □ ♅
09 22	☽ ∠ ♂
13 36	☽ # ♄
14 59	☿ ± ♇
15 07	☉ Q ♆
17 15	☽ ∥ ♃
19 34	☽ △ ♅
20 02	☽ ∠ ☉
21 37	☽ ⊻ ☿
20 Wednesday	
00 47	☽ △ ♆
01 13	☽ ⊥ ♂
02 51	☽ # ♆
06 59	☽ ☍ ♇
09 08	☽ ⊥ ☿
09 35	☽ Q ♃
11 46	☽ ⊻ ♂
13 28	☽ # ♅
21 01	☽ bQ ♅
21 51	☿ Q ♄
23 21	☽ ✱ ☉
21 Thursday	
01 22	☿ □ ♆
02 13	☽ bQ ♆
02 19	☽ ∠ ☿
04 16	☽ ⊻ ♀
05 51	☽ ✱ ♃
06 57	☽ △ ♄
12 17	☽ ± ♅
12 18	☽ ∥ ♂
15 22	☽ ⊥ ♀
17 27	☽ ± ♆
21 12	☽ Q ☉
22 27	☽ ⊼ ♅
22 30	♀ □ ♃
22 Friday	
03 38	☽ ⊼ ♆
06 49	☽ ✱ ☿
07 12	☽ ∠ ♃
08 02	☽ ∠ ♀
08 30	☽ bQ ♄
09 44	☽ ⊼ ♇
13 52	♀ □ ♄
16 27	☽ ☌ ♂
19 54	☽ ± ♇
22 27	☽ ⊥ ♃
23 52	☽ ± ♄
23 Saturday	
05 31	☽ Q ☿
06 03	☽ □ ☉
08 12	☽ ∥ ♂
08 38	☽ ⊻ ♃
10 08	☽ ⊼ ♄
11 09	☿ ⊼ ♇
11 11	☽ bQ ♇
11 54	☽ ✱ ♀
24 Sunday	
01 32	☽ ☍ ♅
06 44	☽ ☍ ♆
07 38	☽ # ♅
10 18	☽ Q ♀
12 46	☽ △ ♇
15 30	☽ □ ☿
18 09	☉ △ ♃
18 19	☽ # ♆
21 33	☽ ⊻ ♂
25 Monday	
03 43	☽ ∥ ♃
08 09	☽ # ♄
08 15	☽ ⊥ ♂
11 54	☽ ☌ ♃
13 17	☽ △ ☉
13 46	☽ ☍ ♄
20 08	☽ □ ♀
20 48	☉ ✱ ♄
26 Tuesday	
00 21	☽ ∠ ♂
05 09	☽ ⊼ ♅
06 04	☉ bQ ♇
06 56	☽ ∥ ♀
10 23	☽ ⊼ ♆
12 45	♂ ± ♇
15 33	☽ ± ♅
16 24	☽ □ ♇
17 12	☽ bQ ☉
20 49	☽ ± ♆
21 36	☽ ∥ ☿
23 59	☽ △ ☿
27 Wednesday	
03 23	☽ ✱ ♂
07 15	☽ bQ ♅
10 06	☽ ± ☉
12 31	☽ bQ ♆
15 55	☽ ⊻ ♃
18 10	☽ ⊼ ♄
21 26	☽ ⊼ ☉
23 48	☽ Q ♇
28 Thursday	
01 14	☽ Q ♂
02 27	☽ ⊥ ♃
02 36	☉ # ♇
04 15	☽ bQ ☿
04 47	☽ ± ♄
05 24	☽ △ ♀
09 38	☽ △ ♅
10 01	☽ ∥ ☉
10 31	☽ # ♇
14 58	☽ △ ♆
18 21	☽ ∠ ♃
20 48	☽ bQ ♄
21 00	☽ ✱ ♇
21 17	☽ ± ☿
29 Friday	
06 48	♀ ∠ ♂
08 40	☽ ⊼ ☿
08 52	♂ ⊥ ♃
08 52	☽ ∥ ♇
10 29	☽ □ ♂
10 41	☽ bQ ♀
11 06	☽ # ☉
21 14	☽ ✱ ♃
23 53	☽ △ ♄
23 53	☽ ∠ ♇
30 Saturday	
04 34	☽ ± ♀
05 53	♀ △ ♅
07 17	☽ ☍ ☉
13 14	♃ St D
15 39	☽ □ ♅
16 17	☽ ⊥ ♇
16 37	☽ ⊼ ♀
19 07	☽ Q ♃
21 10	☽ □ ♆
31 Sunday	
03 19	☽ ⊻ ♇
08 27	☽ # ☿
18 24	☽ ☍ ☿
19 50	☽ △ ♂

APRIL 1991

Date	Sidereal time	Sun ☉	Moon ☽	Moon ☽ 24.00	Mercury ☿	Venus ♀	Mars ♂	Jupiter ♃	Saturn ♄	Uranus ♅	Neptune ♆	Pluto ♇
	h m s	° ' "	° ' "	° ' "	° '	° '	° '	° '	° '	° '	° '	° '
01	00 37 23	11 ♈ 15 34	07 ♏ 22 35	13 ♏ 38 00	28 ♈ 24	16 ♉ 21	29 ♊ 11	03 ♌ 33	05 ♒ 11	13 ♑ 42	16 ♑ 41	19 ♏ 58
02	00 41 20	12 14 45	19 ♏ 49 31	25 ♏ 57 19	28 43	17 33	29 ♊ 43	03 D 34	05 16	13 42	16 41	19 R 57
03	00 45 16	13 13 54	02 ♐ 01 45	08 ♐ 03 09	28 56	18 44	00 ♋ 15	03 34	05 20	13 43	16 42	19 56
04	00 49 13	14 13 01	14 ♐ 02 00	19 ♐ 58 47	29 01	19 56	00 47	03 35	05 24	13 44	16 42	19 54
05	00 53 09	15 12 07	25 ♐ 54 05	01 ♑ 48 29	28 R 59	21 07	01 19	03 36	05 28	13 45	16 43	19 53
06	00 57 06	16 11 10	07 ♑ 42 39	13 ♑ 37 15	28 51	22 18	01 51	03 38	05 31	13 45	16 43	19 52
07	01 01 02	17 10 12	19 ♑ 32 57	25 ♑ 30 28	28 36	23 30	02 23	03 39	05 35	13 46	16 44	19 50
08	01 04 59	18 09 12	01 ♒ 30 30	07 ♒ 33 42	28 16	24 41	02 55	03 41	05 39	13 46	16 44	19 49
09	01 08 55	19 08 11	13 ♒ 40 44	19 ♒ 52 12	27 49	25 52	03 27	03 42	05 42	13 47	16 44	19 48
10	01 12 52	20 07 07	26 ♒ 08 37	02 ♓ 30 28	27 18	27 03	04 00	03 44	05 46	13 47	16 45	19 46
11	01 16 49	21 06 02	08 ♓ 58 07	15 ♓ 31 50	26 43	28 13	04 32	03 46	05 49	13 48	16 45	19 45
12	01 20 45	22 04 54	22 ♓ 11 42	28 ♓ 57 43	26 04	29 ♉ 24	05 05	03 49	05 53	13 48	16 45	19 44
13	01 24 42	23 03 45	05 ♈ 49 43	12 ♈ 47 21	25 23	00 ♊ 35	05 37	03 51	05 56	13 48	16 45	19 42
14	01 28 38	24 02 34	19 ♈ 50 09	26 ♈ 57 28	24 40	01 45	06 10	03 54	05 59	13 49	16 46	19 41
15	01 32 35	25 01 22	04 ♉ 08 35	11 ♉ 22 40	23 56	02 56	06 43	03 57	06 02	13 49	16 46	19 39
16	01 36 31	26 00 07	18 ♉ 38 50	25 ♉ 56 12	23 11	04 06	07 15	04 00	06 05	13 49	16 46	19 38
17	01 40 28	26 58 50	03 ♊ 13 53	10 ♊ 31 07	22 28	05 16	07 48	04 03	06 08	13 49	16 46	19 36
18	01 44 24	27 57 31	17 ♊ 47 09	25 ♊ 01 23	21 45	06 26	08 21	04 06	06 11	13 R 49	16 46	19 35
19	01 48 21	28 56 10	02 ♋ 13 20	09 ♋ 22 37	21 05	07 36	08 55	04 10	06 13	13 49	16 46	19 33
20	01 52 18	29 ♈ 54 47	16 ♋ 29 00	23 ♋ 32 18	20 28	08 46	09 28	04 14	06 16	13 49	16 R 46	19 32
21	01 56 14	00 ♉ 53 21	00 ♌ 32 27	07 ♌ 29 26	19 54	09 56	10 01	04 18	06 18	13 49	16 46	19 30
22	02 00 11	01 51 53	14 ♌ 23 19	21 ♌ 14 09	19 24	11 06	10 34	04 22	06 21	13 49	16 46	19 28
23	02 04 07	02 50 23	28 ♌ 02 00	04 ♍ 46 57	18 58	12 15	11 08	04 26	06 23	13 48	16 46	19 27
24	02 08 04	03 48 51	11 ♍ 29 04	18 ♍ 08 21	18 37	13 25	11 41	04 31	06 25	13 48	16 45	19 25
25	02 12 00	04 47 16	24 ♍ 44 50	01 ♎ 18 30	18 20	14 34	12 15	04 35	06 28	13 48	16 45	19 24
26	02 15 57	05 45 39	07 ♎ 49 18	14 ♎ 17 12	18 08	15 43	12 48	04 40	06 30	13 47	16 45	19 22
27	02 19 53	06 44 01	20 ♎ 42 07	27 ♎ 04 01	18 01	16 52	13 22	04 45	06 32	13 47	16 45	19 20
28	02 23 50	07 42 21	03 ♏ 22 51	09 ♏ 38 34	17 D 59	18 01	13 55	04 50	06 33	13 46	16 44	19 19
29	02 27 47	08 40 38	15 ♏ 51 13	22 ♏ 00 50	18 02	19 10	14 29	04 55	06 35	13 46	16 44	19 17
30	02 31 43	09 ♉ 38 55	28 ♏ 07 31	04 ♐ 11 25	18 ♈ 10	20 ♊ 18	15 ♋ 03	05 ♌ 00	06 ♒ 37	13 ♑ 45	16 ♑ 44	19 ♏ 16

Date	Moon True ☊	Moon Mean ☊	Moon ☽ Latitude
	° '	° '	° '
01	24 ♑ 38	24 ♑ 20	04 S 54
02	24 R 27	24 17	04 33
03	24 18	24 14	03 59
04	24 12	24 11	03 15
05	24 09	24 08	02 24
06	24 07	24 04	01 26
07	24 D 07	24 01	00 S 24
08	24 R 07	23 58	00 N 39
09	24 06	23 55	01 42
10	24 03	23 52	02 41
11	23 57	23 48	03 34
12	23 49	23 45	04 16
13	23 38	23 42	04 46
14	23 26	23 39	05 00
15	23 14	23 36	04 55
16	23 03	23 33	04 31
17	22 55	23 29	03 50
18	22 49	23 26	02 53
19	22 46	23 23	01 46
20	22 45	23 20	00 N 33
21	22 D 45	23 17	00 S 41
22	22 R 44	23 14	01 52
23	22 42	23 10	02 55
24	22 38	23 07	03 48
25	22 31	23 04	04 27
26	22 21	23 01	04 52
27	22 09	22 58	05 01
28	21 56	22 54	04 55
29	21 44	22 51	04 35
30	21 ♑ 33	22 ♑ 48	04 S 03

DECLINATIONS

Date	Sun ☉	Moon ☽	Mercury ☿	Venus ♀	Mars ♂	Jupiter ♃	Saturn ♄	Uranus ♅	Neptune ♆	Pluto ♇
	° '	° '	° '	° '	° '	° '	° '	° '	° '	° '
01	04 N 27	18 S 36	13 N 48	17 N 27	25 N 22	20 N 07	19 S 14	23 S 05	21 S 36	02 S 42
02	04 50	22 04	13 59	17 51	25 22	20 07	19 13	23 05	21 36	02 42
03	05 13	24 28	14 06	18 14	25 21	20 07	19 12	23 05	21 36	02 41
04	05 36	25 43	14 10	18 37	25 21	20 07	19 11	23 05	21 36	02 41
05	05 59	25 46	14 10	18 59	25 20	20 06	19 10	23 05	21 36	02 40
06	06 22	24 39	14 05	19 21	25 19	20 06	19 09	23 05	21 36	02 39
07	06 45	22 25	13 58	19 42	25 18	20 06	19 09	23 05	21 36	02 39
08	07 07	19 11	13 46	20 03	25 17	20 05	19 08	23 05	21 36	02 38
09	07 30	15 06	13 31	20 23	25 16	20 05	19 07	23 05	21 36	02 38
10	07 52	10 17	13 13	20 42	25 14	20 04	19 06	23 05	21 36	02 37
11	08 14	04 S 54	12 51	21 02	25 13	20 04	19 06	23 05	21 36	02 37
12	08 36	00 N 50	12 28	21 20	25 11	20 03	19 05	23 05	21 36	02 36
13	08 58	06 42	12 01	21 38	25 09	20 03	19 04	23 05	21 36	02 36
14	09 20	12 23	11 33	21 56	25 07	20 02	19 03	23 05	21 36	02 35
15	09 41	17 31	11 04	22 13	25 05	20 01	19 03	23 05	21 36	02 34
16	10 03	21 43	10 34	22 30	25 03	20 00	19 02	23 05	21 36	02 34
17	10 24	24 34	10 03	22 45	25 01	20 00	19 02	23 05	21 36	02 33
18	10 45	25 46	09 32	23 01	24 58	19 59	19 01	23 05	21 36	02 33
19	11 06	25 12	09 01	23 15	24 56	19 58	19 00	23 05	21 35	02 32
20	11 27	22 58	08 32	23 29	24 53	19 57	19 00	23 05	21 35	02 32
21	11 47	19 22	08 03	23 43	24 50	19 56	18 59	23 05	21 35	02 31
22	12 07	14 44	07 36	23 56	24 47	19 55	18 59	23 05	21 35	02 31
23	12 27	09 25	07 11	24 08	24 44	19 54	18 58	23 05	21 35	02 30
24	12 47	03 N 45	06 48	24 19	24 40	19 53	18 58	23 05	21 35	02 30
25	13 07	02 S 00	06 28	24 30	24 37	19 52	18 57	23 05	21 35	02 29
26	13 27	07 34	06 09	24 41	24 33	19 50	18 57	23 05	21 35	02 29
27	13 46	12 44	05 53	24 50	24 29	19 49	18 57	23 05	21 36	02 28
28	14 05	17 16	05 40	24 59	24 25	19 48	18 56	23 05	21 36	02 28
29	14 24	20 58	05 30	25 07	24 21	19 47	18 56	23 05	21 36	02 27
30	14 N 42	23 S 41	05 N 21	25 N 15	24 N 17	19 N 45	18 S 56	23 S 05	21 S 36	02 S 27

ZODIAC SIGN ENTRIES

Date	h m	Planets
03	00 49	♂ ♋
03	07 59	☽ ♐
05	20 19	☽ ♑
08	09 00	☽ ♒
10	19 18	☽ ♓
13	00 10	♀ ♊
13	01 50	☽ ♈
15	05 06	☽ ♉
17	06 41	☽ ♊
19	08 17	☽ ♋
20	14 08	☉ ♉
21	11 04	☽ ♌
23	15 29	☽ ♍
25	21 36	☽ ♎
28	05 34	☽ ♏
30	15 42	☽ ♐

LATITUDES

Date	Mercury ☿	Venus ♀	Mars ♂	Jupiter ♃	Saturn ♄	Uranus ♅	Neptune ♆	Pluto ♇
	° '	° '	° '	° '	° '	° '	° '	° '
01	03 N 06	00 N 45	01 N 56	00 N 47	00 S 16	00 S 21	00 N 48	15 N 34
04	03 15	00 55	01 54	00 47	00 16	00 21	00 48	15 35
07	03 11	01 05	01 53	00 47	00 17	00 21	00 48	15 36
10	02 54	01 15	01 52	00 47	00 17	00 21	00 48	15 36
13	02 22	01 24	01 50	00 46	00 17	00 21	00 48	15 37
16	01 40	01 33	01 49	00 46	00 18	00 21	00 48	15 38
19	00 51	01 42	01 47	00 46	00 18	00 21	00 48	15 38
22	00 N 01	01 50	01 46	00 46	00 18	00 21	00 48	15 38
25	00 S 47	01 58	01 44	00 46	00 18	00 21	00 48	15 39
28	01 30	02 06	01 43	00 46	00 19	00 22	00 48	15 39
31	02 S 06	02 N 12	01 N 42	00 N 46	00 S 19	00 S 22	00 N 48	15 N 39

DATA

Julian Date	2448348
Delta T	+58 seconds
Ayanamsa	23° 44' 20"
Synetic vernal point	05° ♓ 22' 39"
True obliquity of ecliptic	23° 26' 30"

MOON'S PHASES, APSIDES AND POSITIONS ☽

Date	h m	Phase	Longitude ° '	Eclipse Indicator
07	06 45	☾	16 ♑ 57	
14	19 38	●	24 ♈ 21	
21	12 39	☽	00 ♌ 55	
28	20 59	○	08 ♏ 04	

Day	h m		
05	21 03	Apogee	
17	17 00	Perigee	
05	01 00	Max dec	25° S 54'
12	08 35	0N	
18	16 16	Max dec	25° N 47'
25	03 37	0S	

LONGITUDES

Date	Chiron ⚷	Ceres ⚳	Pallas ⚴	Juno ⚵	Vesta ⚶	Black Moon Lilith ⚸
	° '	° '	° '	° '	° '	° '
01	21 ♋ 12	00 ♏ 39	08 ♍ 09	24 ♑ 05	05 ♊ 49	27 ♐ 18
11	21 ♋ 24	28 ♎ 33	06 ♍ 47	26 ♑ 18	09 ♊ 35	28 ♐ 24
21	21 ♋ 46	26 ♎ 17	06 ♍ 16	28 ♑ 10	13 ♊ 28	29 ♐ 31
31	22 ♋ 16	24 ♎ 08	06 ♍ 34	29 ♑ 40	17 ♊ 29	00 ♑ 37

All ephemeris data is given at 12.00 UT and the Moon's longitude is additionally given for 24.00 UT

ASPECTARIAN

01 Monday
- 01 12 ☽ Q ♅
- 04 35 ☽ ⋕ ♀
- 04 44 ☽ □ ♃
- 06 52 ☽ Q ♆
- 07 49 ☽ □ ♄
- 12 03 ♂ ± ♄
- 15 54 ☽ ∥ ♄
- 18 37 ♀ △ ♆
- 20 04 ☽ ⊼ ☉
- 21 43 ☽ ⋕ ♃

02 Tuesday
- 00 08 ☽ ✶ ♅
- 01 39 ☽ Ꝑ ♂
- 05 54 ☽ ✶ ♆
- 07 06 ☽ ☍ ♀
- 08 22 ☽ ∥ ♆
- 08 40 ☽ ± ☉
- 12 14 ☽ ☌ ♇
- 18 45 ☽ Q ♄
- 19 57 ☽ ± ♂
- 20 50 ☽ ∥ ♅

03 Wednesday
- 03 49 ☽ Ꝑ ☉
- 05 27 ☽ ∠ ♅
- 05 47 ☽ ⊼ ☿
- 08 18 ☽ ⊼ ♂
- 11 21 ☽ ∠ ♆
- 15 04 ☽ △ ♃
- 17 50 ☽ ± ☿
- 18 36 ☽ ✶ ♄
- 23 21 ☽ ⊥ ♅

04 Thursday
- 00 04 ☉ □ ♅
- 02 07 ☽ ⋕ ♂
- 04 35 ☉ ± ♇
- 05 19 ☽ ⊥ ♆
- 11 24 ☽ ⊻ ♅
- 11 32 ♀ ☍ ♇
- 11 58 ☽ Ꝑ ☿
- 12 24 ☽ △ ☉
- 17 24 ☽ ⊻ ♆
- 18 09 ☿ St R
- 21 12 ☽ Ꝑ ♃
- 23 50 ☽ ⊻ ♇

05 Friday
- 00 55 ☽ ∠ ♄
- 01 14 ☽ ⊼ ♀
- 11 58 ☽ ⊥ ♇
- 14 45 ☽ ± ♀
- 15 28 ☽ ± ♃
- 18 13 ☽ △ ☿
- 19 16 ☽ ⊥ ♄
- 21 42 ☉ ⊥ ♀
- 21 58 ♀ Q ♃
- 23 30 ☽ ☍ ♂

06 Saturday
- 00 08 ♀ ⋕ ♄
- 00 32 ☽ ⋕ ♂
- 03 41 ☽ ⊼ ♃
- 06 13 ☽ ∠ ♇
- 07 32 ☽ ⊻ ♄
- 11 05 ☽ Ꝑ ♀

07 Sunday
- 00 17 ☽ ☌ ♅
- 01 10 ☉ □ ♆
- 05 56 ☽ ∥ ♅
- 06 18 ☽ ☌ ♆
- 06 45 ☽ □ ☉
- 12 35 ☽ ✶ ♇
- 18 48 ☽ ∥ ♆
- 20 50 ☽ △ ♀
- 22 49 ☉ Q ♄

08 Monday
- 05 43 ☽ □ ☿
- 05 57 ☽ ⋕ ♃
- 06 48 ☽ ⋕ ♀
- 12 23 ☽ ∥ ♄
- 12 37 ☽ Q ♇
- 14 56 ☽ ⊼ ♂
- 15 11 ♀ ∥ ♃
- 16 19 ☽ ☍ ♃
- 20 15 ☽ ☌ ♄
- 22 02 ☽ Q ☉

09 Tuesday
- 03 20 ☽ ± ♂
- 12 12 ☽ ⊻ ♅
- 16 01 ☽ Q ☿
- 17 57 ☽ ⊻ ♆
- 20 47 ☽ ⋕ ☿
- 21 41 ☽ Ꝑ ♂
- 23 30 ☽ ✶ ☉
- 23 50 ☽ □ ♇
- 23 50 ☽ ⊥ ♅
- 23 58 ♂ ⊻ ♃

10 Wednesday
- 03 45 ☉ ⊼ ♇
- 05 31 ☽ ⊥ ♆
- 13 53 ☽ □ ♀
- 14 07 ☽ ✶ ☿
- 15 37 ☿ ⊻ ♀
- 17 01 ☽ ∠ ♅
- 22 20 ☽ ⋕ ☉
- 22 35 ☽ ∠ ♆

11 Thursday
- 02 20 ☽ ⊼ ♃
- 03 25 ☽ △ ♂
- 06 09 ☽ ⊻ ♄
- 06 16 ☽ ∠ ☉
- 13 29 ☽ ± ♃
- 16 49 ☽ ∠ ☿
- 17 16 ☽ ⊥ ♄
- 20 51 ☽ ✶ ♅
- 21 13 ♂ Ꝑ ♇
- 21 45 ☽ ∥ ♇
- 23 38 ♀ ⊥ ♂
- 23 41 ♀ Ꝑ ♅

12 Friday
- 00 07 ☽ ⊥ ☉
- 02 13 ☽ ✶ ♆
- 02 33 ☽ Q ♀
- 05 55 ☽ Ꝑ ♃
- 07 35 ☽ △ ♇
- 08 23 ☽ ⊥ ☿
- 09 38 ☽ ∠ ♄
- 11 47 ☽ ⊻ ☉
- 18 25 ☽ Q ♅
- 18 34 ☽ ⊻ ☿
- 19 14 ☽ ⋕ ♇
- 23 38 ☽ Q ♆

13 Saturday
- 01 59 ☽ ✶ ♀
- 08 12 ♀ ⋕ ♆
- 08 33 ☽ △ ♃
- 10 03 ☽ Ꝑ ♇
- 11 37 ☽ □ ♂
- 12 11 ☽ ✶ ♄
- 22 04 ☽ ∥ ☉
- 22 13 ☿ ⊥ ♀

14 Sunday
- 01 32 ☽ ± ♇
- 01 45 ☽ □ ♅
- 03 10 ♂ ⊼ ♄
- 06 18 ☽ ∠ ♀
- 06 47 ☽ □ ♆
- 08 43 ☽ ∥ ☿
- 08 51 ☽ Q ♄
- 11 44 ☽ ⊼ ♇
- 12 05 ♀ Ꝑ ♆
- 18 42 ☉ Q ♂
- 19 36 ☽ Q ♂
- 19 38 ☽ ☌ ☉
- 19 45 ☽ ☌ ☿
- 20 42 ☉ ☌ ☿
- 21 23 ☿ Q ♂
- 22 53 ☽ ⊥ ♀

15 Monday
- 09 48 ☽ ⊻ ♀
- 11 40 ☽ □ ♃
- 15 09 ☽ □ ♄
- 16 26 ☽ ✶ ♂
- 19 59 ☽ ⋕ ♄

16 Tuesday
- 01 25 ☽ ∥ ♃
- 04 02 ☽ △ ♅
- 08 54 ☽ △ ♆
- 09 46 ♀ ✶ ♃
- 11 10 ☽ ⋕ ♆
- 13 37 ☽ ☍ ♇
- 17 32 ☽ Q ♃
- 17 55 ☽ ∥ ♀
- 18 11 ☽ ∠ ♂
- 19 07 ☽ ⊻ ☿
- 21 55 ☽ ⋕ ♅

17 Wednesday
- 00 59 ☽ ⊻ ☉
- 02 14 ☉ ∥ ☿
- 04 31 ☽ ⊥ ☿
- 04 44 ☽ Ꝑ ♅
- 09 34 ☽ ⊥ ♂
- 09 35 ☽ Ꝑ ♆
- 11 33 ☽ ⊥ ☉
- 13 21 ☽ ✶ ♃
- 15 39 ☽ ☌ ♀
- 16 47 ☽ △ ♄
- 17 41 ☽ ∥ ♂
- 18 38 ☽ ∠ ☿
- 19 33 ☽ ± ♅
- 19 50 ☽ ⊻ ♂

18 Thursday
- 00 24 ☽ ± ♆
- 03 27 ☽ ∠ ☉
- 05 26 ☽ ⊼ ♅
- 06 22 ♀ △ ♄
- 10 19 ☽ ⊼ ♆
- 10 33 ♅ St R
- 14 12 ☽ ∠ ♃
- 14 58 ☽ ⊼ ♇
- 16 06 ☿ ∠ ♀
- 17 38 ☽ Ꝑ ♄
- 18 17 ☽ ✶ ☿
- 18 31 ♀ ⋕ ♅

19 Friday
- 00 11 ♆ St R
- 00 54 ☽ ± ♇
- 05 12 ☽ ⊥ ♃
- 06 07 ☽ ✶ ☉
- 08 39 ☽ ± ♄
- 13 23 ☽ Q ☿
- 15 16 ☽ ⊻ ♃
- 15 54 ☽ Ꝑ ♇
- 16 12 ☽ ∥ ♂
- 16 17 ♀ ± ♅
- 18 43 ☽ ⊼ ♄
- 21 50 ☽ ⊻ ♀
- 23 40 ☽ ☌ ♂

20 Saturday
- 03 42 ☽ Q ☉
- 07 29 ☽ ☍ ♅
- 07 58 ☽ ∥ ♀
- 08 51 ☽ ⊥ ♀
- 11 08 ☽ ⋕ ♅
- 12 29 ☽ ☍ ♆
- 17 09 ☽ △ ♇
- 18 30 ☽ □ ☿
- 22 11 ☽ ⋕ ♆

21 Sunday
- 01 31 ☽ ∠ ♀
- 08 41 ☽ ∥ ♃
- 12 39 ☽ □ ☉
- 14 09 ☽ ⋕ ♄
- 15 11 ♀ ⊻ ♂
- 18 30 ☽ ☌ ♃
- 21 59 ☽ ☍ ♄

22 Monday
- 05 04 ☽ ⊻ ♂
- 05 05 ♀ ± ♆
- 05 44 ☽ ✶ ♀
- 08 01 ☿ ⊼ ♇
- 10 59 ☽ ⊼ ♅
- 15 59 ☽ ⊥ ♂
- 16 09 ☽ ⊼ ♆
- 20 30 ☽ △ ☿
- 20 53 ☽ □ ♇
- 21 29 ☽ ± ♅
- 23 21 ☽ ∥ ☉

23 Tuesday
- 02 41 ☽ ± ♆
- 04 42 ☽ Q ♀
- 08 29 ☽ ∠ ♂
- 13 22 ☽ Ꝑ ♅
- 18 37 ☽ Ꝑ ♆
- 21 12 ☽ △ ☉
- 22 16 ☽ Ꝑ ☿
- 22 16 ☽ ∥ ☿
- 23 26 ☽ ⊻ ♃

24 Wednesday
- 02 54 ☽ ⊼ ♄
- 04 44 ☽ Q ♇
- 10 14 ☽ ⊥ ♃
- 12 23 ☽ ✶ ♂
- 13 42 ☽ ± ♄
- 13 59 ☽ ± ☿
- 15 48 ☽ □ ♀
- 16 10 ☽ △ ♅
- 17 13 ☽ ⋕ ♇
- 20 01 ♀ ⊼ ♅
- 21 30 ☽ △ ♆

25 Thursday
- 00 34 ☽ ⊼ ☿
- 01 29 ☿ Q ♄
- 02 16 ☽ Ꝑ ☉
- 02 18 ☽ ✶ ♇
- 02 34 ☽ ∠ ♃
- 06 00 ☽ Ꝑ ♄
- 06 34 ☉ □ ♃
- 11 02 ☽ Q ♂
- 14 04 ☽ ∥ ♇
- 19 58 ☽ ± ☉
- 22 56 ♀ ∥ ♂

26 Friday
- 05 38 ☽ ∠ ♇
- 06 06 ☽ ⋕ ☿
- 06 08 ☽ ✶ ♃
- 07 53 ☽ ⊼ ☉
- 09 32 ☽ △ ♄
- 21 39 ☽ □ ♂
- 22 16 ☽ ⊥ ♇
- 23 04 ☽ □ ♅

27 Saturday
- 04 07 ☽ △ ♀
- 04 32 ☽ Q ♃
- 04 35 ☽ □ ♆
- 06 43 ☉ □ ♄
- 07 00 ☽ ☍ ☿
- 09 24 ♀ ⊼ ♆
- 09 27 ☽ ⊻ ♇
- 17 33 ☽ ⋕ ☉

28 Sunday
- 05 41 ♂ ☍ ♅
- 08 56 ☽ Q ♅
- 09 49 ☿ St D
- 11 14 ☽ Ꝑ ♀
- 11 16 ☿ ✶ ♀
- 14 36 ☽ Q ♆
- 14 47 ☽ □ ♃
- 18 05 ☽ □ ♄
- 20 59 ☽ ☍ ☉
- 22 07 ☽ ∥ ♄

29 Monday
- 03 38 ☽ ⋕ ♃
- 06 16 ☽ ± ♀
- 07 57 ☽ ✶ ♅
- 09 13 ☽ △ ♂
- 13 43 ☽ ✶ ♆
- 14 30 ♀ ⊼ ♇
- 16 16 ☽ ⊼ ☿
- 16 45 ☽ ∥ ♆
- 18 40 ☽ ☌ ♇
- 19 06 ☽ ⊼ ♀

30 Tuesday
- 04 06 ☽ ± ☿
- 05 05 ☽ Q ♄
- 05 07 ♀ ∠ ♃
- 05 48 ☽ ∥ ♅
- 13 14 ☽ ∠ ♅
- 15 59 ☽ Ꝑ ♂
- 18 55 ☽ ⋕ ♂
- 19 07 ☽ ∠ ♆
- 22 07 ☽ Ꝑ ☿

MAY 1991

LONGITUDES

Date	Sidereal time h m s	Sun ☉ ° ' "	Moon ☽ ° ' "	Moon ☽ 24.00 ° ' "	Mercury ☿ ° '	Venus ♀ ° '	Mars ♂ ° '	Jupiter ♃ ° '	Saturn ♄ ° '	Uranus ♅ ° '	Neptune ♆ ° '	Pluto ♇ ° '
01	02 35 40	10 ♉ 37 09	10 ♐ 12 46	16 ♐ 11 50	18 ♈ 22	21 ♊ 27	15 ♋ 37	05 ♌ 06	06 ♒ 38	13 ♑ 45	16 ♑ 43	19 ♏ 14
02	02 39 36	11 35 22	22 ♐ 08 56	28 ♐ 04 28	18 39	22 35	16 11	05 12	06 40	13 R 44	16 R 43	19 R 12
03	02 43 33	12 33 33	03 ♑ 58 53	09 ♑ 52 41	19 01	23 43	16 45	05 17	06 41	13 43	16 42	19 11
04	02 47 29	13 31 43	15 ♑ 46 24	21 ♑ 40 38	19 27	24 52	17 19	05 23	06 43	13 42	16 42	19 09
05	02 51 26	14 29 51	27 ♑ 36 00	03 ♒ 33 09	19 57	25 59	17 53	05 30	06 44	13 42	16 41	19 07
06	02 55 22	15 27 58	09 ♒ 32 45	15 ♒ 35 29	20 32	27 07	18 27	05 36	06 45	13 41	16 41	19 06
07	02 59 19	16 26 03	21 ♒ 42 01	27 ♒ 53 01	21 10	28 15	19 01	05 42	06 46	13 40	16 40	19 04
08	03 03 16	17 24 07	04 ♓ 09 05	10 ♓ 30 49	21 52	29 ♊ 22	19 35	05 49	06 47	13 39	16 40	19 02
09	03 07 12	18 22 10	16 ♓ 58 40	23 ♓ 33 05	22 38	00 ♋ 29	20 09	05 55	06 48	13 38	16 39	19 01
10	03 11 09	19 20 11	00 ♈ 14 18	07 ♈ 02 29	23 27	01 37	20 44	06 02	06 48	13 37	16 38	18 59
11	03 15 05	20 18 11	13 ♈ 57 36	20 ♈ 59 26	24 20	02 44	21 18	06 09	06 49	13 36	16 38	18 57
12	03 19 02	21 16 09	28 ♈ 07 37	05 ♉ 21 34	25 15	03 50	21 53	06 16	06 49	13 35	16 37	18 55
13	03 22 58	22 14 06	12 ♉ 40 31	20 ♉ 03 34	26 14	04 57	22 27	06 24	06 50	13 33	16 36	18 54
14	03 26 55	23 12 02	27 ♉ 29 42	04 ♊ 57 48	27 16	06 03	23 02	06 31	06 50	13 32	16 36	18 52
15	03 30 51	24 09 56	12 ♊ 26 46	19 ♊ 55 27	28 21	07 10	23 36	06 39	06 50	13 31	16 35	18 50
16	03 34 48	25 07 49	27 ♊ 22 51	04 ♋ 47 59	29 ♈ 29	08 16	24 11	06 46	06 51	13 30	16 34	18 49
17	03 38 45	26 05 40	12 ♋ 10 04	19 ♋ 28 26	00 ♉ 39	09 21	24 45	06 54	06 R 51	13 28	16 33	18 47
18	03 42 41	27 03 29	26 ♋ 42 35	03 ♌ 52 09	01 52	10 27	25 20	07 02	06 50	13 27	16 32	18 45
19	03 46 38	28 01 17	10 ♌ 56 56	17 ♌ 56 48	03 08	11 33	25 55	07 10	06 50	13 26	16 31	18 44
20	03 50 34	28 59 03	24 ♌ 51 48	01 ♍ 42 00	04 26	12 38	26 30	07 18	06 50	13 24	16 30	18 42
21	03 54 31	29 ♉ 56 47	08 ♍ 27 32	15 ♍ 08 38	05 47	13 43	27 05	07 26	06 50	13 23	16 29	18 40
22	03 58 27	00 ♊ 54 29	21 ♍ 45 28	28 ♍ 18 17	07 10	14 48	27 40	07 35	06 49	13 21	16 28	18 39
23	04 02 24	01 52 10	04 ♎ 47 19	11 ♎ 12 47	08 36	15 52	28 14	07 43	06 49	13 19	16 27	18 37
24	04 06 20	02 49 50	17 ♎ 34 53	23 ♎ 53 48	10 06	16 56	28 49	07 52	06 48	13 18	16 26	18 36
25	04 10 17	03 47 28	00 ♏ 09 43	06 ♏ 22 49	11 35	18 00	29 ♋ 24	08 00	06 47	13 16	16 25	18 34
26	04 14 14	04 45 04	12 ♏ 33 13	18 ♏ 41 05	13 08	19 04	00 ♌ 00	08 09	06 46	13 14	16 24	18 32
27	04 18 10	05 42 39	24 ♏ 46 34	00 ♐ 49 49	14 43	20 08	00 35	08 18	06 45	13 13	16 23	18 31
28	04 22 07	06 40 14	06 ♐ 51 01	12 ♐ 50 19	16 21	21 11	01 10	08 27	06 44	13 11	16 22	18 29
29	04 26 03	07 37 47	18 ♐ 47 57	24 ♐ 44 10	18 01	22 14	01 45	08 36	06 43	13 09	16 21	18 28
30	04 30 00	08 35 18	00 ♑ 39 22	06 ♑ 33 24	19 43	23 17	02 20	08 46	06 42	13 07	16 20	18 26
31	04 33 56	09 ♊ 32 49	12 ♑ 27 05	18 ♑ 20 38	21 ♉ 28	24 ♋ 19	02 ♌ 55	08 ♌ 55	06 ♒ 41	13 ♑ 05	16 ♑ 18	18 ♏ 25

Date	Moon True ☊ ° '	Moon Mean ☊ ° '	Moon ☽ Latitude ° '
01	21 ♑ 24	22 ♑ 45	03 S 20
02	21 R 18	22 42	02 28
03	21 14	22 39	01 30
04	21 12	22 35	00 S 29
05	21 D 12	22 32	00 N 34
06	21 13	22 29	01 36
07	21 R 13	22 26	02 35
08	21 12	22 23	03 28
09	21 08	22 20	04 13
10	21 02	22 16	04 45
11	20 55	22 13	05 03
12	20 46	22 10	05 02
13	20 36	22 07	04 43
14	20 28	22 04	04 04
15	20 21	22 00	03 08
16	20 17	21 57	02 00
17	20 15	21 54	00 N 43
18	20 D 15	21 51	00 S 35
19	20 16	21 48	01 49
20	20 17	21 45	02 55
21	20 R 16	21 41	03 50
22	20 14	21 38	04 31
23	20 10	21 35	04 57
24	20 04	21 32	05 08
25	19 57	21 29	05 03
26	19 49	21 26	04 44
27	19 41	21 22	04 13
28	19 34	21 19	03 30
29	19 28	21 16	02 38
30	19 24	21 13	01 40
31	19 ♑ 22	21 ♑ 10	00 S 38

DECLINATIONS

Date	Sun ☉ ° '	Moon ☽ ° '	Mercury ☿ ° '	Venus ♀ ° '	Mars ♂ ° '	Jupiter ♃ ° '	Saturn ♄ ° '	Uranus ♅ ° '	Neptune ♆ ° '	Pluto ♇ ° '
01	15 N 01	25 S 16	05 N 16	25 N 22	24 N 13	19 N 44	18 S 55	23 S 05	21 S 36	02 S 27
02	15 19	25 40	05 13	25 28	24 08	19 42	18 55	23 06	21 36	02 26
03	15 37	24 53	05 13	25 34	24 03	19 41	18 55	23 06	21 36	02 26
04	15 54	22 59	05 15	25 39	23 59	19 39	18 55	23 06	21 36	02 25
05	16 11	20 05	05 19	25 43	23 54	19 38	18 54	23 06	21 36	02 25
06	16 28	16 19	05 25	25 47	23 49	19 36	18 54	23 06	21 36	02 24
07	16 45	11 49	05 34	25 50	23 43	19 35	18 54	23 06	21 36	02 24
08	17 02	06 45	05 45	25 52	23 38	19 33	18 54	23 06	21 36	02 24
09	17 18	01 S 16	05 58	25 54	23 33	19 31	18 54	23 06	21 36	02 23
10	17 34	04 N 27	06 12	25 55	23 27	19 30	18 54	23 06	21 36	02 23
11	17 49	10 09	06 29	25 55	23 21	19 28	18 54	23 07	21 36	02 22
12	18 05	15 30	06 47	25 55	23 15	19 26	18 54	23 07	21 36	02 22
13	18 20	20 08	07 07	25 54	23 09	19 24	18 54	23 07	21 36	02 22
14	18 34	23 34	07 29	25 53	23 03	19 22	18 54	23 07	21 36	02 21
15	18 49	25 24	07 52	25 50	22 56	19 20	18 54	23 07	21 36	02 21
16	19 03	25 24	08 16	25 47	22 50	19 18	18 54	23 07	21 37	02 21
17	19 17	23 36	08 42	25 43	22 43	19 16	18 54	23 07	21 37	02 20
18	19 30	20 15	09 09	25 39	22 36	19 14	18 54	23 08	21 37	02 20
19	19 43	15 44	09 37	25 34	22 30	19 12	18 54	23 08	21 37	02 20
20	19 56	10 29	10 07	25 28	22 22	19 10	18 55	23 08	21 37	02 20
21	20 08	04 N 51	10 37	25 22	22 15	19 08	18 55	23 08	21 37	02 19
22	20 21	00 S 53	11 08	25 15	22 08	19 06	18 55	23 08	21 37	02 19
23	20 32	06 27	11 40	25 08	22 00	19 04	18 55	23 09	21 37	02 19
24	20 44	11 38	12 14	25 00	21 53	19 01	18 56	23 09	21 37	02 18
25	20 55	16 15	12 47	24 52	21 45	18 59	18 56	23 09	21 38	02 18
26	21 05	20 07	13 22	24 42	21 37	18 57	18 56	23 09	21 38	02 18
27	21 16	23 03	13 57	24 33	21 29	18 54	18 57	23 09	21 38	02 18
28	21 26	24 54	14 32	24 23	21 21	18 52	18 57	23 09	21 38	02 18
29	21 35	25 36	15 08	24 12	21 13	18 50	18 57	23 10	21 38	02 18
30	21 44	25 06	15 44	24 00	21 04	18 47	18 58	23 10	21 38	02 17
31	21 N 53	23 S 29	16 N 20	23 N 49	20 N 56	18 N 45	18 S 58	23 S 10	21 S 38	02 S 17

ZODIAC SIGN ENTRIES

Date	h	m	Planets
03	03	55	☽ ♑
05	16	51	☽ ♒
08	04	04	☽ ♓
09	01	28	♀ ♋
10	11	35	☽ ♈
12	15	07	☽ ♉
14	16	02	☽ ♊
16	16	14	☽ ♋
16	22	45	☿ ♉
18	17	30	☽ ♌
20	21	00	☽ ♍
21	13	20	☉ ♊
23	03	08	☽ ♎
25	11	41	☽ ♏
26	12	19	♂ ♌
27	22	21	☽ ♐
30	10	40	☽ ♑

LATITUDES

Date	Mercury ☿ ° '	Venus ♀ ° '	Mars ♂ ° '	Jupiter ♃ ° '	Saturn ♄ ° '	Uranus ♅ ° '	Neptune ♆ ° '	Pluto ♇ ° '
01	02 S 06	02 N 12	01 N 42	00 N 46	00 S 19	00 S 22	00 N 48	15 N 39
04	02 34	02 19	01 40	00 45	00 19	00 22	00 49	15 39
07	02 54	02 24	01 39	00 45	00 20	00 22	00 49	15 39
10	03 07	02 29	01 37	00 45	00 20	00 22	00 49	15 39
13	03 14	02 33	01 36	00 45	00 20	00 22	00 49	15 38
16	03 13	02 36	01 35	00 45	00 21	00 22	00 49	15 38
19	03 07	02 38	01 33	00 45	00 21	00 22	00 49	15 38
22	02 56	02 39	01 32	00 45	00 22	00 22	00 49	15 38
25	02 39	02 39	01 30	00 45	00 22	00 22	00 49	15 37
28	02 18	02 38	01 29	00 45	00 22	00 22	00 49	15 36
31	01 S 52	02 N 36	01 N 28	00 N 44	00 S 23	00 S 22	00 N 49	15 N 36

DATA

Julian Date	2448378
Delta T	+58 seconds
Ayanamsa	23° 44' 24"
Synetic vernal point	05° ♓ 22' 36"
True obliquity of ecliptic	23° 26' 29"

MOON'S PHASES, APSIDES AND POSITIONS ☽

Date	h	m	Phase	Longitude ° '	Eclipse Indicator
07	00	46	☾	15 ♒ 59	
14	04	36	●	22 ♉ 54	
20	19	46	☽	29 ♌ 18	
28	11	37	○	06 ♐ 39	

Day	h	m		
03	14	29	Apogee	
15	16	35	Perigee	
31	02	38	Apogee	
02	08	00	Max dec	25° S 41'
09	17	21	ON	
16	00	05	Max dec	25° N 38'
22	08	17	OS	
29	14	00	Max dec	25° S 36'

LONGITUDES

Date	Chiron ⚷ ° '	Ceres ⚳ ° '	Pallas ⚴ ° '	Juno ⚵ ° '	Vesta ⚶ ° '	Black Moon Lilith ⚸ ° '
01	22 ♋ 16	24 ♎ 08	06 ♍ 34	29 ♑ 40	17 ♊ 29	00 ♑ 37
11	22 ♋ 56	22 ♎ 22	07 ♍ 34	00 ♒ 42	21 ♊ 01	01 ♑ 44
21	23 ♋ 39	21 ♎ 08	09 ♍ 11	01 ♒ 14	25 ♊ 46	02 ♑ 50
31	24 ♋ 31	20 ♎ 33	11 ♍ 18	01 ♒ 13	00 ♋ 00	03 ♑ 57

All ephemeris data is given at 12.00 UT and the Moon's longitude is additionally given for 24.00 UT

ASPECTARIAN

h m	Aspects
01 Wednesday	
01 44	☽ △ ♃
04 52	☽ ⚹ ♄
07 05	☽ ⊥ ♅
10 44	☽ ± ♂
12 53	☽ ⊼ ☉
13 01	☽ ⊥ ♆
14 47	☽ # ♀
16 08	♀ bQ ♄
19 04	☽ ⊻ ♅
23 22	☽ ⊼ ♂
02 Thursday	
01 03	☽ ⊻ ♆
01 59	☽ ± ☉
04 46	☽ △ ☿
06 04	☽ ⊻ ♇
08 01	☽ bQ ♃
11 01	☽ ∠ ♄
12 45	☿ Q ♄
12 59	☽ ☍ ♀
18 10	☽ ⊥ ♇
21 08	☽ # ♀
21 47	☽ bQ ☉
03 Friday	
02 23	☽ ± ♃
05 18	☽ ⊥ ♄
10 32	♂ ☍ ♆
12 24	☽ ∠ ♇
14 41	☽ ⊼ ♃
17 31	☽ ⊻ ♄
20 37	☿ ⊼ ♇
04 Saturday	
00 41	☽ # ♂
07 01	☽ △ ☉
07 48	☽ ☌ ♅
10 56	☽ ∥ ♅
13 53	☽ ☌ ♆
15 17	☽ ☍ ♂
16 23	☉ △ ♅
17 59	♀ ± ♇
18 51	☽ ⚹ ♇
19 48	☽ □ ☿
05 Sunday	
00 29	☽ ∥ ♆
08 24	☽ ⊼ ♀
15 13	☽ # ♃
19 05	☽ Q ♇
20 05	☽ ∥ ♄
21 47	☽ ± ♀
06 Monday	
04 02	☽ ☍ ♃
06 24	☽ ☌ ♄
09 52	☽ Q ☿
11 09	☽ # ☉
17 39	☽ bQ ♀
20 13	☽ ⊻ ♅
07 Tuesday	
00 46	☽ □ ☉
02 09	☽ ⊻ ♆
06 29	☽ ⊼ ♂
06 51	☽ □ ♇
08 01	☽ ⊥ ♅
10 54	☽ ⚹ ☿
13 54	☽ ⊥ ♆
13 58	♂ △ ♇
17 51	☉ △ ♆
18 46	☽ ± ♂
08 Wednesday	
01 29	☽ ∠ ♅
01 57	☽ △ ♀
07 16	☽ ∠ ♆
12 52	☽ bQ ♂
14 34	☽ Q ☉
15 11	☽ ⊼ ♃
16 20	☽ # ☿
16 59	☽ ⊻ ♄
17 28	☽ ∠ ☿
22 31	♀ ⊥ ♃
09 Thursday	
02 33	☽ ± ♃
04 15	☽ ⊥ ♄
05 49	☽ ⚹ ♅
07 11	☽ ∥ ♇
11 19	☽ ⊥ ☿
11 24	☽ ⚹ ♆
14 46	☽ ⚹ ☉
15 43	☽ △ ♇
18 06	☽ △ ♂
18 34	♀ ± ♄
19 18	☽ bQ ♃
20 49	☽ ∠ ♄
23 00	☽ ⊻ ☿
10 Friday	
03 21	☽ # ♇
03 25	☉ ☍ ♇
03 44	☽ Q ♅
09 09	☽ Q ♆
14 39	☽ □ ♀
18 37	☽ bQ ♇
19 41	☽ ∥ ☿
19 48	☽ ∠ ☉
22 20	☽ △ ♃
23 36	☽ ⚹ ♄
11 Saturday	
10 16	☽ ± ♇
11 23	☽ □ ♅
12 38	☽ ⊥ ☉
16 35	☽ □ ♆
20 19	☽ Q ♄
20 31	☽ ⊼ ♇
23 38	☽ ⊻ ☉
12 Sunday	
00 32	☽ Q ♀
01 03	☽ □ ♂
06 51	☽ ☌ ☿
13 49	♀ bQ ♇
22 16	☽ ⚹ ♀
13 Monday	
01 28	☽ ∥ ☉
01 37	☽ □ ♃
02 25	☽ □ ♄
05 05	☽ # ♄
07 53	☽ ∥ ♃
08 13	☽ Q ♂
13 26	☽ △ ♅
18 24	☽ △ ♆
20 11	♂ # ♅
21 11	☽ # ♆
22 06	☽ ☍ ♇
14 Tuesday	
00 46	☽ ∠ ♀
01 16	☉ ⚹ ♂
04 30	☽ ⚹ ♂
04 36	☽ ☌ ☉
07 10	☽ Q ♃
07 48	☽ ∥ ♂
08 13	☽ # ♅
11 37	☽ ⊻ ☿
13 41	☽ bQ ♅
16 27	☽ ⊥ ♀
18 35	☽ bQ ♆
22 00	☽ ⊥ ☿
23 19	♀ ⊻ ♃
15 Wednesday	
02 37	☽ ⚹ ♃
02 51	☽ ⊻ ♀
03 00	☽ △ ♄
04 06	☽ ± ♅
05 02	♀ ⊼ ♄
05 35	☽ ∠ ♂
09 01	☽ ± ♆
13 34	☽ ∠ ☿
13 43	☽ ⊼ ♅
18 37	☽ ⊼ ♆
20 35	☉ # ♄
20 36	☽ ⊥ ♂
22 14	☽ ⊼ ♇
16 Thursday	
01 40	☉ Q ♃
02 53	☽ ∠ ♃
03 05	☽ bQ ♄
06 38	☽ ⊻ ♂
07 52	☽ ± ♇
08 07	☽ ⊻ ☉
15 40	☽ ⚹ ☿
17 31	☽ ⊥ ♃
17 35	☽ ± ♄
18 29	☽ ⊥ ☉
22 23	☽ bQ ♇
17 Friday	
01 24	♃ ☍ ♄
03 19	☽ ⊼ ♄
03 20	☽ ⊻ ♃
04 05	♄ St R
07 03	☽ ☌ ♀
10 07	☽ ∠ ☉
11 28	☉ ∥ ♃
12 52	☽ Q ☿
14 08	☽ ☍ ♅
16 08	☽ # ♅
19 11	☽ ☍ ♆
19 39	☽ ∥ ♂
22 51	☽ △ ♇
18 Saturday	
03 21	☽ # ♆
09 37	☽ ☌ ♂
12 37	☽ ⚹ ☉
16 08	☽ ∥ ☉
17 54	☽ ∥ ♃
19 42	☽ # ♄
21 28	☽ □ ☿
19 Sunday	
05 01	☽ ☍ ♄
05 30	☽ ☌ ♃
10 18	☽ Q ☉
13 06	☽ ⊻ ♀
16 13	☽ ⊼ ♅
21 32	☽ ⊼ ♆
21 49	☉ bQ ♅
20 Monday	
00 16	☽ ⊥ ♀
01 19	☽ □ ♇
02 32	☽ ± ♅
07 54	☽ ± ♆
13 29	☽ ∥ ☿
14 59	☽ ⊻ ♂
17 15	☽ ∠ ♀
18 11	☽ bQ ♅
19 46	☽ □ ☉
23 39	☽ bQ ♆
21 Tuesday	
02 00	☽ ⊥ ♂
04 42	♀ ☍ ♅
06 42	☽ △ ☿
08 50	☽ Q ♇
09 05	☽ ⊼ ♄
10 10	☽ ⊻ ♃
18 46	☽ ∠ ♂
19 49	☽ ± ♄
20 48	☽ △ ♅
21 01	☽ ⊥ ♃
22 15	☽ ⚹ ♀
22 33	☽ # ♇
22 Wednesday	
02 25	☽ △ ♆
05 58	☿ □ ♄
06 21	☽ ⚹ ♇
12 07	☽ bQ ♄
12 51	☽ bQ ☿
13 31	☽ ∠ ♃
18 05	☽ ∥ ♇
19 36	☿ □ ♃
22 03	☽ Q ♀
23 19	☽ ⚹ ♂
23 Thursday	
01 51	☉ bQ ♆
06 09	☽ △ ☉
07 26	☽ ± ☿
09 50	☽ ∠ ♇
15 46	☽ △ ♄
17 31	☽ ⚹ ♃
20 01	☽ ⊼ ☿
22 40	☽ Q ♂
24 Friday	
00 58	♀ ☍ ♆
02 37	☽ ⊥ ♇
03 56	☽ □ ♅
09 51	☽ □ ♆
10 40	☽ □ ♀
12 31	☽ bQ ☉
13 55	☽ ⊻ ♇
15 16	☽ # ☿
16 22	☽ Q ♃
25 Saturday	
07 04	☽ ± ☉
10 29	☽ □ ♂
14 07	☽ Q ♅
19 35	☽ ⊼ ☉
20 12	☽ Q ♆
26 Sunday	
00 20	♀ △ ♇
00 46	☽ □ ♄
03 20	☽ □ ♃
04 05	☽ ∥ ♄
04 14	☽ # ♃
10 32	♂ # ♆
13 18	☽ ☍ ☿
13 20	☽ ⚹ ♅
13 39	☿ △ ♅
17 02	♃ # ♄
19 31	☽ ⚹ ♆
19 33	☽ # ☉
22 49	☽ # ♂
23 22	☽ ∥ ♆
23 41	☽ ☌ ♇
27 Monday	
01 58	☽ △ ♀
11 58	☽ Q ♄
13 06	☽ ∥ ♅
18 47	☽ ∠ ♅
28 Tuesday	
00 05	☽ △ ♂
01 05	☽ ∠ ♆
04 19	☽ # ♀
05 56	☉ ∥ ♂
10 32	☽ bQ ♀
11 37	☽ ☍ ☉
11 47	☽ ⚹ ♄
12 15	☿ △ ♆
12 40	☽ ⊥ ♅
13 39	☉ △ ♄
15 15	☽ △ ♃
19 02	☽ ⊥ ♆
29 Wednesday	
00 25	☉ ± ♅
00 40	☽ ⊻ ♅
06 20	☽ ± ♀
07 04	☽ ⊻ ♆
07 39	☽ bQ ♂
10 10	☽ ⊼ ☿
11 19	☽ ⊻ ♇
17 53	☽ ∠ ♄
18 12	☿ ☍ ♇
19 36	☽ ⊼ ♀
19 42	☉ # ♆
21 51	☽ bQ ♃
23 25	☽ ⊥ ♇
30 Thursday	
00 18	☽ ± ☿
02 47	☽ ± ♂
12 05	☽ ⊥ ♄
15 36	☽ ⊼ ♂
16 20	☽ ± ♃
17 11	☉ ⚹ ♃
17 38	☽ ∠ ♇
21 42	☽ bQ ☿
31 Friday	
00 16	☽ ⊻ ♄
00 47	☿ Q ♂
04 43	☽ ⊼ ♃
05 34	☽ ⊼ ☉
07 47	☽ # ♀
13 18	☽ ☌ ♅
15 23	☽ ∥ ♅
18 52	☽ ± ☉
19 50	☽ ☌ ♆

JUNE 1991

LONGITUDES

Date	Sidereal time h m s	Sun ☉ ° ' "	Moon ☽ ° ' "	Moon ☽ 24.00 ° ' "	Mercury ☿ ° '	Venus ♀ ° '	Mars ♂ ° '	Jupiter ♃ ° '	Saturn ♄ ° '	Uranus ♅ ° '	Neptune ♆ ° '	Pluto ⯓ ° '
01	04 37 53	10 ♊ 30 19	24 ♑ 14 30	00 ♒ 09 07	23 ♉ 15	25 ♋ 21	03 ♌ 31	09 ♌ 05	06 ♒ 39	13 ♑ 04	16 ♑ 17	18 ♏ 23
02	04 41 49	11 27 48	06 ♒ 05 00	12 ♒ 02 40	25 05	26 23	04 06	09 14	06 R 38	13 R 02	16 R 16	18 R 22
03	04 45 46	12 25 17	18 ♒ 02 42	24 ♒ 05 40	26 57	27 24	04 41	09 24	06 36	13 00	16 15	18 20
04	04 49 43	13 22 44	00 ♓ 12 09	06 ♓ 22 47	28 ♉ 51	28 26	05 17	09 33	06 34	12 58	16 13	18 19
05	04 53 39	14 20 11	12 ♓ 38 07	18 ♓ 58 45	00 ♊ 47	29 ♋ 26	05 52	09 43	06 33	12 56	16 12	18 18
06	04 57 36	15 17 37	25 ♓ 25 11	01 ♈ 57 53	02 45	00 ♌ 27	06 28	09 53	06 31	12 54	16 11	18 16
07	05 01 32	16 15 02	08 ♈ 37 15	15 ♈ 23 31	04 46	01 27	07 03	10 03	06 29	12 52	16 09	18 14
08	05 05 29	17 12 27	22 ♈ 16 52	29 ♈ 17 15	06 48	02 27	07 39	10 13	06 27	12 50	16 08	18 13
09	05 09 25	18 09 52	06 ♉ 24 29	13 ♉ 38 14	08 53	03 26	08 14	10 24	06 25	12 47	16 07	18 11
10	05 13 22	19 07 15	20 ♉ 57 53	28 ♉ 22 44	10 59	04 26	08 50	10 34	06 22	12 45	16 05	18 10
11	05 17 18	20 04 39	05 ♊ 51 50	13 ♊ 24 09	13 06	05 24	09 26	10 44	06 20	12 43	16 04	18 09
12	05 21 15	21 02 01	20 ♊ 58 31	28 ♊ 33 43	15 15	06 22	10 01	10 55	06 18	12 41	16 03	18 07
13	05 25 12	21 59 23	06 ♋ 08 33	13 ♋ 41 49	17 25	07 20	10 37	11 05	06 15	12 39	16 01	18 06
14	05 29 08	22 56 44	21 ♋ 12 28	28 ♋ 39 29	19 36	08 18	11 13	11 16	06 13	12 37	16 00	18 05
15	05 33 05	23 54 04	06 ♌ 02 06	13 ♌ 19 38	21 48	09 15	11 49	11 27	06 10	12 34	15 58	18 03
16	05 37 01	24 51 24	20 ♌ 31 36	27 ♌ 37 40	24 00	10 11	12 25	11 38	06 07	12 32	15 57	18 02
17	05 40 58	25 48 42	04 ♍ 37 41	11 ♍ 31 35	26 12	11 07	13 00	11 48	06 04	12 30	15 55	18 01
18	05 44 54	26 46 00	18 ♍ 19 28	25 ♍ 01 31	28 ♊ 23	12 03	13 36	11 59	06 02	12 27	15 54	18 00
19	05 48 51	27 43 16	01 ♎ 37 57	08 ♎ 09 06	00 ♋ 35	12 58	14 12	12 10	05 59	12 25	15 52	17 58
20	05 52 47	28 40 32	14 ♎ 35 17	20 ♎ 56 55	02 45	13 52	14 48	12 22	05 56	12 23	15 51	17 57
21	05 56 44	29 ♊ 37 47	27 ♎ 14 20	03 ♏ 27 57	04 55	14 46	15 24	12 33	05 53	12 20	15 49	17 56
22	06 00 41	00 ♋ 35 02	09 ♏ 38 09	15 ♏ 45 17	07 03	15 39	16 00	12 44	05 49	12 18	15 48	17 55
23	06 04 37	01 32 16	21 ♏ 49 43	27 ♏ 51 47	09 10	16 32	16 37	12 55	05 46	12 16	15 46	17 54
24	06 08 34	02 29 29	03 ♐ 51 50	09 ♐ 50 08	11 15	17 24	17 13	13 07	05 43	12 13	15 44	17 53
25	06 12 30	03 26 42	15 ♐ 47 01	21 ♐ 42 45	13 19	18 16	17 49	13 18	05 40	12 11	15 43	17 52
26	06 16 27	04 23 55	27 ♐ 37 36	03 ♑ 31 52	15 21	19 07	18 25	13 30	05 36	12 09	15 41	17 51
27	06 20 23	05 21 07	09 ♑ 25 49	15 ♑ 19 43	17 21	19 57	19 01	13 41	05 33	12 06	15 40	17 50
28	06 24 20	06 18 19	21 ♑ 13 52	27 ♑ 08 35	19 19	20 46	19 37	13 53	05 29	12 04	15 38	17 49
29	06 28 16	07 15 31	03 ♒ 04 10	09 ♒ 00 58	21 16	21 35	20 14	14 04	05 26	12 01	15 37	17 48
30	06 32 13	08 ♋ 12 42	14 ♒ 59 20	20 ♒ 59 39	23 ♋ 10	22 ♌ 23	20 ♌ 50	14 ♌ 16	05 ♒ 22	11 ♑ 59	15 ♑ 35	17 ♏ 47

Date	Moon True ☊ ° '	Moon Mean ☊ ° '	Moon ☽ Latitude ° '
01	19 ♑ 22	21 ♑ 06	00 N 27
02	19 D 23	21 03	01 30
03	19 25	21 00	02 30
04	19 26	20 57	03 25
05	19 27	20 54	04 11
06	19 R 27	20 51	04 46
07	19 25	20 47	05 07
08	19 22	20 44	05 12
09	19 18	20 41	04 59
10	19 13	20 38	04 26
11	19 09	20 35	03 35
12	19 06	20 32	02 28
13	19 04	20 28	01 N 10
14	19 D 04	20 25	00 S 12
15	19 05	20 22	01 32
16	19 06	20 19	02 45
17	19 07	20 16	03 45
18	19 08	20 12	04 31
19	19 R 08	20 09	05 01
20	19 08	20 06	05 14
21	19 06	20 03	05 12
22	19 04	20 00	04 55
23	19 01	19 57	04 25
24	18 58	19 53	03 44
25	18 56	19 50	02 53
26	18 55	19 47	01 55
27	18 54	19 44	00 S 52
28	18 D 54	19 41	00 N 13
29	18 54	19 37	01 18
30	18 ♑ 55	19 ♑ 34	02 N 20

DECLINATIONS

Date	Sun ☉ ° '	Moon ☽ ° '	Mercury ☿ ° '	Venus ♀ ° '	Mars ♂ ° '	Jupiter ♃ ° '	Saturn ♄ ° '	Uranus ♅ ° '	Neptune ♆ ° '	Pluto ⯓ ° '
01	22 N 01	20 S 50	16 N 56	23 N 36	20 N 47	18 N 42	18 S 59	23 S 10	21 S 38	02 S 17
02	22 10	17 18	17 32	23 23	20 38	18 40	18 59	23 10	21 39	02 17
03	22 17	13 02	18 08	23 10	20 29	18 37	19 00	23 11	21 39	02 17
04	22 24	08 12	18 43	22 56	20 20	18 34	19 00	23 11	21 39	02 17
05	22 31	02 S 57	19 18	22 42	20 11	18 32	19 01	23 11	21 39	02 17
06	22 38	02 N 33	19 52	22 27	20 02	18 29	19 01	23 11	21 39	02 17
07	22 44	08 07	20 25	22 12	19 52	18 26	19 02	23 12	21 39	02 16
08	22 50	13 30	20 58	21 56	19 43	18 24	19 03	23 12	21 39	02 16
09	22 55	18 21	21 28	21 40	19 33	18 21	19 03	23 12	21 40	02 16
10	23 00	22 17	21 58	21 24	19 23	18 18	19 04	23 12	21 40	02 16
11	23 04	24 49	22 25	21 07	19 13	18 15	19 05	23 12	21 40	02 16
12	23 08	25 36	22 51	20 50	19 03	18 12	19 05	23 13	21 40	02 16
13	23 12	24 28	23 15	20 32	18 53	18 09	19 06	23 13	21 40	02 16
14	23 15	21 35	23 36	20 14	18 43	18 06	19 07	23 13	21 40	02 16
15	23 18	17 17	23 55	19 56	18 32	18 03	19 08	23 13	21 41	02 16
16	23 20	12 03	24 12	19 38	18 22	18 00	19 08	23 14	21 41	02 16
17	23 22	06 19	24 25	19 19	18 11	17 57	19 09	23 14	21 41	02 16
18	23 24	00 N 27	24 36	19 00	18 00	17 54	19 10	23 14	21 41	02 16
19	23 25	05 S 15	24 44	18 40	17 49	17 51	19 11	23 14	21 41	02 17
20	23 26	10 35	24 50	18 21	17 38	17 48	19 12	23 14	21 42	02 17
21	23 26	15 20	24 52	18 01	17 27	17 45	19 13	23 15	21 42	02 17
22	23 26	19 22	24 52	17 41	17 16	17 42	19 13	23 15	21 42	02 17
23	23 26	22 30	24 49	17 20	17 05	17 38	19 14	23 15	21 42	02 17
24	23 25	24 35	24 43	17 00	16 53	17 35	19 15	23 15	21 42	02 17
25	23 24	25 33	24 35	16 39	16 42	17 32	19 16	23 16	21 42	02 17
26	23 22	25 20	24 24	16 18	16 30	17 29	19 17	23 16	21 43	02 17
27	23 20	23 58	24 11	15 57	16 18	17 25	19 18	23 16	21 43	02 18
28	23 17	21 33	23 56	15 36	16 07	17 22	19 19	23 16	21 43	02 18
29	23 15	18 13	23 39	15 15	15 55	17 19	19 20	23 17	21 43	02 18
30	23 N 11	14 S 07	23 N 20	14 N 53	15 N 43	17 N 15	19 S 21	23 S 17	21 S 43	02 S 18

ZODIAC SIGN ENTRIES

Date	h	m	Planets
01	23	42	☽ ♒
04	11	36	☽ ♓
05	02	24	☿ ♊
06	01	16	♀ ♌
06	20	25	☽ ♈
09	01	13	☽ ♉
11	02	36	☽ ♊
13	02	16	☽ ♋
15	02	10	☽ ♌
17	04	03	☽ ♍
19	05	40	☿ ♋
19	09	01	☽ ♎
21	17	18	☽ ♏
21	21	19	☉ ♋
24	04	16	☽ ♐
26	16	49	☽ ♑
29	05	47	☽ ♒

LATITUDES

Date	Mercury ☿ ° '	Venus ♀ ° '	Mars ♂ ° '	Jupiter ♃ ° '	Saturn ♄ ° '	Uranus ♅ ° '	Neptune ♆ ° '	Pluto ⯓ ° '
01	01 S 43	02 N 35	01 N 27	00 N 44	00 S 23	00 S 22	00 N 49	15 N 35
04	01 13	02 31	01 26	00 44	00 23	00 22	00 49	15 35
07	00 41	02 25	01 24	00 44	00 23	00 22	00 49	15 34
10	00 S 08	02 19	01 23	00 44	00 24	00 23	00 49	15 33
13	00 N 24	02 10	01 22	00 44	00 24	00 23	00 49	15 32
16	00 53	02 01	01 20	00 44	00 24	00 23	00 49	15 31
19	01 18	01 49	01 19	00 44	00 25	00 23	00 49	15 30
22	01 37	01 36	01 17	00 44	00 25	00 23	00 49	15 28
25	01 49	01 22	01 16	00 44	00 25	00 23	00 49	15 27
28	01 54	01 05	01 14	00 44	00 26	00 23	00 49	15 26
31	01 N 54	00 N 47	01 N 13	00 N 44	00 S 26	00 S 23	00 N 49	15 N 24

DATA

Julian Date	2448409
Delta T	+58 seconds
Ayanamsa	23° 44' 29"
Synetic vernal point	05° ♓ 22' 30"
True obliquity of ecliptic	23° 26' 28"

MOON'S PHASES, APSIDES AND POSITIONS ☽

Date	h	m	Phase	Longitude ° '	Eclipse Indicator
05	15	30	☾	14 ♓ 29	
12	12	06	●	21 ♊ 02	
19	04	19	☽	27 ♍ 25	
27	02	58	○	05 ♑ 00	

Day	h	m		
13	00	18	Perigee	
27	07	06	Apogee	
06	00	58	0N	
12	09	53	Max dec	25° N 36'
18	13	53	0S	
25	19	37	Max dec	25° S 37'

LONGITUDES

Date	Chiron ⚷ ° '	Ceres ⚳ ° '	Pallas ⚴ ° '	Juno ⚵ ° '	Vesta ⚶ ° '	Black Moon Lilith ⚸ ° '
01	24 ♋ 37	20 ♎ 32	11 ♍ 32	01 ♒ 10	00 ♋ 26	04 ♑ 04
11	25 ♋ 34	20 ♎ 40	14 ♍ 06	00 ♒ 29	03 ♋ 44	05 ♑ 10
21	26 ♋ 36	21 ♎ 25	17 ♍ 01	29 ♑ 12	09 ♋ 04	06 ♑ 17
31	27 ♋ 42	22 ♎ 43	20 ♍ 12	27 ♑ 24	13 ♋ 27	07 ♑ 23

ASPECTARIAN

h m Aspects

01 Saturday
00 06 ☽ ⚹ ⯓
02 43 ☽ # ☉
05 32 ☽ ∥ ♆
06 38 ☉ ± ♆
09 38 ☽ △ ☿
12 24 ☽ # ♂
14 28 ☽ ☍ ♀
14 48 ☽ ♇ ☉

02 Sunday
00 27 ☽ Q ⯓
01 15 ☽ ∥ ♄
03 17 ☽ # ♃
07 47 ☽ ☍ ♂
10 45 ☽ # ☿
13 06 ☽ ☌ ♄
18 26 ☽ ☍ ♃
23 47 ☽ △ ☉

03 Monday
01 56 ☽ ⊻ ♅
08 25 ☽ ⊻ ♆
10 57 ♀ # ♅
12 34 ☽ □ ⯓
13 53 ☽ ⊥ ♅
17 04 ☉ ∠ ♀
18 19 ☿ Q ♃
20 20 ☽ ⊥ ♆

04 Tuesday
00 50 ☿ ⚹ ♀
01 08 ☿ ♇ ♅
01 54 ☉ ⊼ ♅
06 22 ☿ ∥ ♃
07 37 ☽ ∠ ♅
08 13 ☽ ⊼ ♀
08 51 ☽ □ ☿
13 59 ☽ ∠ ♆
20 57 ☽ ± ♀
22 22 ☽ ⊼ ♂
23 49 ☿ # ♄

05 Wednesday
00 21 ☽ ⊻ ♄
06 21 ☽ ⊼ ♃
10 28 ☽ ± ♂
11 49 ☽ ⊥ ♄
12 33 ☽ ⚹ ♅
15 00 ☽ ∥ ⯓
15 30 ☽ □ ☉
15 44 ☽ ♇ ♀
17 07 ☿ ♇ ♆
17 56 ☽ ± ♃
18 45 ☽ ⚹ ♆
22 40 ☽ △ ⯓

06 Thursday
00 03 ☉ ∥ ♀
01 44 ☽ Q ☿
04 18 ☽ ♇ ♂
04 46 ☽ ∠ ♄
10 48 ☽ # ⯓
11 00 ☽ ♇ ♃
11 02 ☽ Q ♅
13 56 ♂ ☍ ♄
17 05 ☽ Q ♆
17 15 ☿ ∥ ♂
22 00 ☽ △ ♀

07 Friday
02 20 ☽ ♇ ⯓
03 32 ☽ Q ☉
03 49 ☽ ⚹ ☿
08 10 ☽ ⚹ ♄
09 04 ☽ △ ♂
09 43 ☉ ⊼ ♆
14 36 ☽ △ ♃
18 25 ☽ ± ⯓
19 31 ☽ □ ♅

08 Saturday
01 19 ☽ □ ♆
02 31 ☽ ⚹ ☉
04 57 ☽ ⊼ ⯓
05 22 ☽ Q ♄
07 53 ☿ △ ♄
11 02 ☽ ∠ ☿
12 15 ☿ ± ♅

09 Sunday
01 43 ☿ ⚹ ♂
05 03 ☽ ⊥ ☿
06 10 ☽ ∠ ☉
06 39 ☽ □ ♀
11 56 ☽ ∥ ♃
12 00 ☽ □ ♄
12 36 ☉ ⊼ ⯓
13 09 ♀ # ♆
15 11 ☽ □ ♂
15 49 ☽ # ♄
16 49 ☽ ⊻ ☿
18 12 ☿ ∥ ♀
18 22 ☽ ∥ ♂
18 43 ☽ □ ♃
21 01 ☿ # ♆
22 15 ☽ ⊥ ☉
22 35 ☽ △ ♅

10 Monday
02 00 ☿ ± ♆
04 03 ☽ △ ♆
06 22 ☽ ∥ ♀
06 55 ☿ ⚹ ♃
07 26 ☽ ☍ ⯓
07 41 ☽ # ♆
08 47 ☽ ⊻ ☉
09 24 ☽ ∥ ☿
14 33 ☽ Q ♀
17 38 ☽ ∥ ☉
19 12 ☽ # ♅
21 54 ☽ Q ♂
22 58 ☽ ♇ ♅

11 Tuesday
00 27 ☽ Q ♃
04 20 ☽ ♇ ♆
07 44 ☿ ⊼ ♅
11 13 ☽ ⚹ ♀
12 45 ☽ △ ♄
13 22 ☽ ± ♅
17 55 ☽ ⚹ ♂
18 41 ☽ ± ♆
19 52 ☽ ⚹ ♃
22 53 ☽ ⊼ ♅

12 Wednesday
01 26 ☽ ☌ ☿
04 12 ☽ ⊼ ♆
07 12 ♂ # ♄
07 29 ☽ ⊼ ⯓
10 05 ♀ ☍ ♄
12 06 ☽ ☌ ☉
12 30 ☽ ♇ ♄
12 40 ☽ ∠ ♀
16 58 ☽ ± ⯓
18 16 ☉ ♇ ♄
18 40 ☽ ∠ ♂
19 54 ☽ ∠ ♃
20 39 ☿ ⊼ ♆

13 Thursday
02 42 ☽ ± ♄
03 53 ☽ ⊥ ♀
07 11 ☽ ♇ ⯓
08 25 ☉ ∥ ☿
09 29 ☽ ⊥ ♂
09 54 ☿ # ♅
10 19 ☽ ⊥ ♃
12 10 ☽ ⊼ ♄
14 02 ☽ ⊻ ♀
18 55 ☉ # ♅
19 23 ☿ ⊼ ⯓
19 24 ☽ ⊻ ♂
19 57 ☽ ⊻ ♃
22 18 ☽ ☍ ♅
22 29 ☽ ∥ ☿
23 59 ☽ ∥ ☉

14 Friday
00 05 ☽ # ♅
03 40 ☽ ☍ ♆
07 00 ☽ △ ⯓
09 00 ☽ ⊻ ☿
11 21 ☽ # ♆
14 59 ☽ ⊻ ☉
15 01 ♂ ☌ ♃
20 17 ☽ ⊥ ☿
20 51 ☽ ∥ ♀

15 Saturday
01 19 ☽ ⊥ ☉
02 29 ☽ # ♄
05 15 ☿ ♇ ♄
05 20 ☽ ∥ ♂
08 02 ☽ ∥ ♃
12 13 ☽ ☍ ♄
13 28 ☽ ∠ ☿
15 48 ☉ ± ⯓
17 02 ☽ ∠ ☉
17 38 ☽ ☌ ♀
21 00 ☽ ☌ ♃
21 54 ☽ ☌ ♂
22 43 ☽ ⊼ ♅

16 Sunday
04 22 ☽ ⊼ ♆
07 50 ☽ □ ⯓
08 40 ☽ ± ♅
12 26 ☿ ± ⯓
14 23 ☽ ± ♆
16 40 ♂ ⊼ ♅
18 55 ☽ ⚹ ☿
19 49 ☽ ⚹ ☉
23 48 ☽ ♇ ♅

17 Monday
04 38 ☉ ☌ ☿
05 38 ☽ ♇ ♆
10 37 ☿ ∠ ♀
14 24 ☽ Q ⯓
14 30 ☽ ⊼ ♄
17 56 ☽ Q ☉
19 19 ☿ ∠ ♃
19 21 ☽ Q ☿
23 43 ♀ # ♄

18 Tuesday
00 07 ☽ ⊻ ♀
00 40 ☽ ⊻ ♃
00 55 ☽ ± ♄
01 40 ☽ △ ♅
03 16 ☽ ⊻ ♂
04 32 ☽ # ⯓
07 42 ☽ △ ♆
10 11 ♀ ☌ ♃
11 24 ☽ ⊥ ♃
11 25 ☽ ⚹ ⯓
11 28 ☽ ⊥ ♀
14 23 ☽ ⊥ ♂
15 17 ☿ ∠ ♂
16 48 ☽ ♇ ♄
18 58 ☉ ∠ ♃
22 17 ♀ ⊼ ♅
23 21 ☽ ∥ ⯓

19 Wednesday
03 46 ☽ ∠ ♃
04 19 ☽ □ ☉
04 49 ☽ ∠ ♀
05 34 ☿ ± ♄
06 37 ♂ ∥ ♃
07 22 ☽ ∠ ♂
09 41 ☽ □ ☿
14 27 ☽ ∠ ⯓
19 57 ☽ △ ♄

20 Thursday
07 05 ☽ ⊥ ⯓
07 46 ☽ ⚹ ♃
07 53 ☽ □ ♅
10 33 ☽ ⚹ ♀
12 26 ☽ ⚹ ♂
14 10 ♃ ⊼ ♅
14 15 ☿ ♇ ⯓
14 21 ☽ □ ♆
18 19 ☽ ⊻ ⯓

21 Friday
06 46 ☽ Q ♃
11 02 ☽ Q ♀
12 20 ☽ Q ♂
16 58 ☽ △ ☉
17 53 ☉ ± ♄
17 57 ☽ Q ♅
22 34 ☿ ⊼ ♄
23 28 ☽ # ♂

22 Saturday
00 39 ☽ Q ♆
01 35 ☽ # ♃
02 10 ☽ # ♀
03 49 ♂ ⊼ ♆
04 36 ☽ □ ♄
05 54 ☽ △ ☿
08 07 ☿ ⊥ ♃
10 24 ♀ ∥ ♃
11 03 ☽ ∥ ♄
15 34 ♀ ⊼ ♆
17 12 ☽ ⚹ ♅
18 09 ☽ □ ♃

23 Sunday
00 03 ☽ ⚹ ♆
00 39 ☽ ♇ ☉
00 44 ☽ □ ♀
01 09 ☽ □ ♂
04 14 ☽ ☌ ⯓
05 09 ☽ ∥ ♆
11 49 ☉ ∠ ♀
15 50 ☽ Q ♄
16 52 ☉ ∠ ♂
17 37 ☽ ♇ ☿
18 23 ♀ ☌ ♂
19 22 ☽ ∥ ♅
20 00 ☽ ± ☉
21 13 ☽ # ☉
22 46 ☽ ∠ ♅

24 Monday
05 46 ☽ ∠ ♆
09 01 ☽ ⊼ ☉
11 18 ☿ ⊥ ♂
13 54 ☽ # ☿
15 00 ☿ ⊥ ♀
15 22 ☽ ± ☿
15 42 ☽ ⚹ ♄
16 43 ☽ ⊥ ♅
21 37 ☉ ♇ ⯓
23 01 ☿ ☍ ♅
23 47 ☽ ⊥ ♆

25 Tuesday
00 59 ♀ □ ⯓
04 26 ♀ ∥ ♂
04 45 ☽ ⊻ ♅
05 59 ☽ ⊼ ☿
06 54 ☽ △ ♃
09 50 ♀ ± ♅
11 48 ☿ ⊻ ♃
11 52 ☽ ⊻ ♆
13 55 ♂ □ ⯓
16 12 ☽ ⊻ ⯓
16 19 ☽ △ ♂
17 24 ☽ △ ♀
21 49 ☽ ∠ ♄

26 Wednesday
01 47 ♂ ± ♅
04 20 ☽ ⊥ ⯓
13 47 ☽ ♇ ♃
15 59 ☿ ☍ ♆
16 00 ☽ ⊥ ♄
22 36 ☽ ∠ ⯓

27 Thursday
00 24 ☽ ♇ ♂
02 11 ☽ ♇ ♀
02 58 ☽ ☍ ☉
04 08 ☽ ⊻ ♄
08 24 ☽ ± ♃
08 48 ☽ # ☿
16 34 ☉ ⊼ ♄
17 25 ☽ ☌ ♅
17 44 ☿ △ ⯓
19 33 ☽ # ☉
19 42 ☽ ± ♂
20 06 ☽ ∥ ♅
20 48 ☽ ⊼ ♃
21 53 ☽ ± ♀

28 Friday
00 39 ☽ ☌ ♆
05 04 ☽ ⚹ ⯓
07 21 ☽ ☍ ☿
08 34 ☽ ⊼ ♂
10 38 ☽ ∥ ♆
10 59 ☽ ⊼ ♀
17 24 ☿ ⊻ ♂
21 15 ☉ # ♅

29 Saturday
04 36 ☽ ∥ ♄
05 23 ☽ Q ⯓
12 51 ♀ ± ♆
16 44 ☽ ☌ ♄
17 43 ☽ # ♃
18 56 ☿ ⊻ ♀
21 12 ☽ ⊼ ☉

30 Sunday
02 38 ☽ # ♂
05 59 ☽ ⊻ ♅
07 21 ☽ # ♀
10 18 ☽ ± ☉
10 32 ☽ ☍ ♃
13 11 ☽ ⊻ ♆
13 52 ☉ ⊥ ♃
16 00 ☿ # ♅
17 35 ☽ □ ⯓
17 58 ☽ ⊥ ♅

All ephemeris data is given at 12.00 UT and the Moon's longitude is additionally given for 24.00 UT

JULY 1991

LONGITUDES

Date	Sidereal time h m s	Sun ☉ ° ' "	Moon ☽ ° ' "	Moon ☽ 24.00 ° ' "	Mercury ☿ ° '	Venus ♀ ° '	Mars ♂ ° '	Jupiter ♃ ° '	Saturn ♄ ° '	Uranus ♅ ° '	Neptune ♆ ° '	Pluto ⯓ ° '
01	06 36 10	09 ♋ 09 54	27 ♒ 02 18	03 ♓ 07 45	25 ♋ 02	23 ♌ 10	21 ♌ 26	14 ♌ 28	05 ♒ 18	11 ♑ 57	15 ♑ 33	17 ♏ 46
02	06 40 06	10 07 06	09 ♓ 16 24	15 ♓ 28 42	26 52	23 56	22 03	14 40	05 R 14	11 R 54	15 R 32	17 R 45
03	06 44 03	11 04 18	21 ♓ 45 07	28 ♓ 06 06	28 ♋ 40	24 42	22 39	14 52	05 11	11 52	15 30	17 44
04	06 47 59	12 01 30	04 ♈ 32 05	11 ♈ 03 28	00 ♌ 26	25 26	23 16	15 04	05 07	11 49	15 29	17 44
05	06 51 56	12 58 42	17 ♈ 40 38	24 ♈ 23 51	02 10	26 10	23 52	15 16	05 03	11 47	15 27	17 43
06	06 55 52	13 55 54	01 ♉ 13 22	08 ♉ 09 16	03 51	26 53	24 29	15 28	04 59	11 44	15 25	17 42
07	06 59 49	14 53 07	15 ♉ 11 35	22 ♉ 20 08	05 31	27 34	25 05	15 40	04 55	11 42	15 24	17 41
08	07 03 45	15 50 21	29 ♉ 34 39	06 ♊ 54 39	07 09	28 15	25 42	15 52	04 51	11 40	15 22	17 41
09	07 07 42	16 47 35	14 ♊ 19 30	21 ♊ 48 25	08 44	28 55	26 19	16 04	04 47	11 37	15 20	17 40
10	07 11 39	17 44 49	29 ♊ 20 29	06 ♋ 54 37	10 17	29 ♌ 33	26 55	16 17	04 42	11 35	15 19	17 39
11	07 15 35	18 42 03	14 ♋ 29 42	22 ♋ 04 34	11 49	00 ♍ 11	27 32	16 29	04 38	11 32	15 17	17 39
12	07 19 32	19 39 18	29 ♋ 38 02	07 ♌ 08 58	13 18	00 47	28 09	16 41	04 34	11 30	15 16	17 38
13	07 23 28	20 36 33	14 ♌ 36 20	21 ♌ 59 13	14 45	01 22	28 45	16 54	04 30	11 28	15 14	17 38
14	07 27 25	21 33 47	29 ♌ 16 50	06 ♍ 28 34	16 09	01 56	29 22	17 06	04 26	11 25	15 12	17 37
15	07 31 21	22 31 02	13 ♍ 34 00	20 ♍ 32 50	17 32	02 28	29 ♌ 59	17 19	04 21	11 23	15 11	17 37
16	07 35 18	23 28 17	27 ♍ 24 58	04 ♎ 10 25	18 52	02 59	00 ♍ 36	17 31	04 17	11 20	15 09	17 36
17	07 39 14	24 25 32	10 ♎ 49 22	17 ♎ 22 03	20 10	03 29	01 13	17 44	04 13	11 18	15 08	17 36
18	07 43 11	25 22 47	23 ♎ 48 51	00 ♏ 10 10	21 26	03 57	01 50	17 56	04 08	11 16	15 06	17 36
19	07 47 08	26 20 02	06 ♏ 26 28	12 ♏ 38 16	22 39	04 23	02 27	18 09	04 04	11 13	15 04	17 35
20	07 51 04	27 17 18	18 ♏ 46 04	24 ♏ 50 24	23 50	04 48	03 04	18 22	03 59	11 11	15 03	17 35
21	07 55 01	28 14 34	00 ♐ 51 48	06 ♐ 50 46	24 58	05 11	03 41	18 34	03 55	11 09	15 01	17 35
22	07 58 57	29 ♋ 11 50	12 ♐ 47 47	18 ♐ 43 21	26 04	05 33	04 18	18 47	03 51	11 07	15 00	17 35
23	08 02 54	00 ♌ 09 06	24 ♐ 37 53	00 ♑ 31 50	27 05	05 52	04 55	19 00	03 46	11 04	14 58	17 34
24	08 06 50	01 06 23	06 ♑ 25 35	12 ♑ 19 30	28 07	06 10	05 32	19 13	03 42	11 02	14 57	17 34
25	08 10 47	02 03 40	18 ♑ 13 55	24 ♑ 09 09	29 04	06 26	06 09	19 25	03 37	11 00	14 55	17 34
26	08 14 43	03 00 58	00 ♒ 05 29	06 ♒ 03 13	29 ♌ 58	06 40	06 47	19 38	03 33	10 58	14 53	17 34
27	08 18 40	03 58 17	12 ♒ 02 35	18 ♒ 03 50	00 ♍ 49	06 52	07 24	19 51	03 28	10 55	14 52	17 34
28	08 22 37	04 55 36	24 ♒ 07 13	00 ♓ 12 56	01 36	07 02	08 01	20 04	03 24	10 53	14 50	17 34
29	08 26 33	05 52 56	06 ♓ 21 15	12 ♓ 32 21	02 21	07 10	08 38	20 17	03 19	10 51	14 49	17 D 34
30	08 30 30	06 50 17	18 ♓ 46 31	25 ♓ 03 57	03 01	07 15	09 16	20 30	03 15	10 49	14 47	17 34
31	08 34 26	07 ♌ 47 39	01 ♈ 24 55	07 ♈ 49 38	03 ♍ 38	07 ♍ 18	09 ♍ 53	20 ♌ 43	03 ♒ 10	10 ♑ 47	14 ♑ 46	17 ♏ 34

Moon

Date	Moon True ☊ ° '	Moon Mean ☊ ° '	Moon ☽ Latitude ° '
01	18 ♑ 56	19 ♑ 31	03 N 16
02	18 D 57	19 28	04 05
03	18 57	19 25	04 43
04	18 58	19 22	05 08
05	18 R 58	19 18	05 18
06	18 57	19 15	05 10
07	18 57	19 12	04 45
08	18 57	19 09	04 01
09	18 56	19 06	03 01
10	18 56	19 03	01 47
11	18 D 56	18 59	00 N 25
12	18 R 56	18 56	00 S 59
13	18 56	18 53	02 18
14	18 56	18 50	03 26
15	18 56	18 47	04 19
16	18 55	18 43	04 56
17	18 55	18 40	05 14
18	18 55	18 37	05 16
19	18 D 55	18 34	05 03
20	18 55	18 31	04 36
21	18 56	18 28	03 57
22	18 57	18 24	03 08
23	18 58	18 21	02 11
24	18 59	18 18	01 09
25	18 59	18 15	00 S 04
26	18 R 59	18 12	01 N 01
27	18 58	18 09	02 04
28	18 56	18 05	03 02
29	18 54	18 02	03 53
30	18 51	17 59	04 33
31	18 ♑ 49	17 ♑ 56	05 N 01

DECLINATIONS

Date	Sun ☉ ° '	Moon ☽ ° '	Mercury ☿ ° '	Venus ♀ ° '	Mars ♂ ° '	Jupiter ♃ ° '	Saturn ♄ ° '	Uranus ♅ ° '	Neptune ♆ ° '	Pluto ⯓ ° '
01	23 N 07	09 S 25	22 N 59	14 N 32	15 N 31	17 N 12	19 S 22	23 S 17	21 S 44	02 S 18
02	23 03	04 S 19	22 37	14 10	15 18	17 08	19 23	23 17	21 44	02 19
03	22 59	01 N 04	22 13	13 49	15 06	17 05	19 24	23 17	21 44	02 19
04	22 54	06 30	21 48	13 27	14 54	17 01	19 25	23 18	21 44	02 19
05	22 48	11 49	21 21	13 06	14 41	16 58	19 26	23 18	21 44	02 19
06	22 43	16 45	20 54	12 44	14 28	16 54	19 27	23 18	21 45	02 20
07	22 37	20 56	20 25	12 22	14 16	16 51	19 28	23 18	21 45	02 20
08	22 30	23 59	19 55	12 01	14 03	16 47	19 29	23 19	21 45	02 20
09	22 23	25 31	19 25	11 39	13 50	16 43	19 30	23 19	21 45	02 21
10	22 16	25 13	18 54	11 18	13 37	16 40	19 32	23 19	21 45	02 21
11	22 08	23 04	18 21	10 57	13 24	16 36	19 33	23 19	21 45	02 21
12	22 00	19 16	17 50	10 36	13 11	16 32	19 34	23 19	21 46	02 22
13	21 52	14 15	17 17	10 15	12 57	16 29	19 35	23 20	21 46	02 22
14	21 43	08 30	16 45	09 54	12 44	16 25	19 36	23 20	21 46	02 22
15	21 34	02 N 28	16 11	09 33	12 31	16 21	19 37	23 20	21 46	02 23
16	21 24	03 S 30	15 38	09 13	12 17	16 17	19 38	23 20	21 46	02 23
17	21 14	09 06	15 05	08 53	12 04	16 13	19 39	23 20	21 47	02 24
18	21 04	14 08	14 31	08 33	11 50	16 10	19 41	23 21	21 47	02 24
19	20 53	18 26	13 58	08 13	11 36	16 06	19 42	23 21	21 47	02 25
20	20 42	21 49	13 25	07 54	11 22	16 02	19 43	23 21	21 47	02 25
21	20 31	24 11	12 52	07 35	11 09	15 58	19 44	23 21	21 47	02 25
22	20 19	25 26	12 20	07 17	10 55	15 54	19 45	23 21	21 48	02 26
23	20 07	25 31	11 48	06 59	10 41	15 50	19 46	23 22	21 48	02 26
24	19 55	24 26	11 16	06 42	10 27	15 46	19 47	23 22	21 48	02 27
25	19 42	22 16	10 46	06 25	10 12	15 42	19 48	23 22	21 48	02 27
26	19 29	19 08	10 15	06 08	09 58	15 38	19 50	23 22	21 48	02 28
27	19 16	15 12	09 46	05 52	09 44	15 34	19 51	23 23	21 49	02 28
28	19 02	10 37	09 18	05 37	09 29	15 30	19 52	23 23	21 49	02 29
29	18 48	05 34	08 50	05 22	09 15	15 26	19 53	23 23	21 49	02 29
30	18 34	00 15	08 24	05 08	09 01	15 22	19 54	23 23	21 49	02 30
31	18 N 19	05 N 10	07 N 59	04 N 55	08 N 46	15 N 18	19 S 55	23 S 23	21 S 49	02 S 31

ZODIAC SIGN ENTRIES

Date	h	m	Planets
01	17	51	☽ ♓
04	03	33	☽ ♈
04	06	05	☿ ♌
06	09	52	☽ ♉
08	12	42	☽ ♊
10	13	03	☽ ♋
11	05	06	♀ ♍
12	12	35	☽ ♌
14	13	12	☽ ♍
15	12	32	♂ ♍
16	16	34	☽ ♎
18	23	41	☽ ♏
21	10	16	☽ ♐
23	08	11	☉ ♌
23	22	55	☽ ♑
26	11	49	☽ ♒
26	13	00	☿ ♍
28	23	35	☽ ♓
31	09	20	☽ ♈

LATITUDES

Date	Mercury ☿ ° '	Venus ♀ ° '	Mars ♂ ° '	Jupiter ♃ ° '	Saturn ♄ ° '	Uranus ♅ ° '	Neptune ♆ ° '	Pluto ⯓ ° '
01	01 N 54	00 N 47	01 N 13	00 N 44	00 S 26	00 S 23	00 N 49	15 N 24
04	01 47	00 26	01 12	00 44	00 26	00 23	00 49	15 23
07	01 34	00 N 04	01 10	00 44	00 27	00 23	00 49	15 21
10	01 17	00 S 21	01 09	00 44	00 27	00 23	00 49	15 20
13	00 55	00 48	01 07	00 44	00 27	00 23	00 49	15 18
16	00 N 29	01 17	01 06	00 44	00 28	00 23	00 49	15 17
19	00 00	01 48	01 04	00 44	00 28	00 23	00 49	15 15
22	00 S 32	02 21	01 03	00 45	00 28	00 23	00 49	15 13
25	01 07	02 57	01 01	00 45	00 29	00 23	00 49	15 12
28	01 43	03 34	01 00	00 45	00 29	00 23	00 48	15 10
31	02 S 21	04 S 13	00 N 58	00 N 45	00 S 29	00 S 23	00 N 48	15 N 08

DATA

Julian Date	2448439
Delta T	+58 seconds
Ayanamsa	23° 44' 34"
Synetic vernal point	05° ♓ 22' 25"
True obliquity of ecliptic	23° 26' 28"

MOON'S PHASES, APSIDES AND POSITIONS ☽

Date	h	m	Phase	Longitude ° '	Eclipse Indicator
05	02	50	☾	12 ♈ 37	
11	19	06	●	18 ♋ 59	Total
18	15	11	☽	25 ♎ 30	
26	18	24	○	03 ♒ 16	

Day	h	m		
11	09	57	Perigee	
24	10	59	Apogee	
03	07	19	0N	
09	20	18	Max dec	25° N 38'
15	21	50	0S	
23	01	37	Max dec	25° S 37'
30	13	06	0N	

LONGITUDES

Date	Chiron ⚷ ° '	Ceres ⚳ ° '	Pallas ⚴ ° '	Juno ⚵ ° '	Vesta ⚶ ° '	Black Moon Lilith ⚸ ° '
01	27 ♋ 42	22 ♎ 43	20 ♍ 12	27 ♑ 24	13 ♋ 27	07 ♑ 23
11	28 ♋ 50	24 ♎ 30	23 ♍ 37	25 ♑ 13	17 ♋ 50	08 ♑ 30
21	29 ♋ 59	26 ♎ 43	27 ♍ 14	22 ♑ 51	22 ♋ 04	09 ♑ 36
31	01 ♌ 08	29 ♎ 17	01 ♎ 00	20 ♑ 32	26 ♋ 39	10 ♑ 43

All ephemeris data is given at 12.00 UT and the Moon's longitude is additionally given for 24.00 UT

ASPECTARIAN

h m	Aspects
01 Monday	
00 18	☽ ☍ ♂
00 48	☉ ∥ ☿
01 09	☽ ⊥ ♆
03 47	☽ ☍ ♀
05 49	☽ ♇ ☉
07 18	☽ ⊼ ☿
11 49	☽ ∠ ♅
16 23	♂ ± ♆
18 56	☽ ∠ ♆
21 17	☽ ± ☿
02 Tuesday	
04 10	☽ ⊻ ♄
13 47	☽ △ ☉
15 48	☽ ⊥ ♄
17 05	☽ ⚹ ♅
17 53	☽ ♇ ☿
21 02	☽ ∥ ⯓
22 36	☽ ⊼ ♃
03 Wednesday	
00 04	☽ ⚹ ♆
04 21	☽ △ ⯓
09 01	☽ ∠ ♄
10 17	☽ ± ♃
13 48	☽ ⊼ ♂
15 59	☽ Q ♅
17 32	☽ # ⯓
17 56	☽ ⊼ ♀
22 51	☽ Q ♆
04 Thursday	
01 41	☽ ± ♂
03 09	☽ △ ☿
03 33	☽ ♇ ♃
05 53	☽ ± ♀
07 04	☉ ☍ ♅
08 39	☽ ♇ ⯓
13 04	☽ ⚹ ♄
15 21	☿ # ♆
19 13	☽ ♇ ♂
23 30	☽ ♇ ♀
05 Friday	
01 12	☽ ± ⯓
01 21	☽ □ ♅
02 50	☽ □ ☉
07 35	☽ △ ♃
07 59	☽ □ ♆
10 52	☽ Q ♄
12 04	☽ ⊼ ⯓
17 33	☽ ∥ ♀
23 35	☽ △ ♂
06 Saturday	
01 03	☽ ∥ ♂
03 58	☽ △ ♀
07 28	♃ ⊼ ♆
07 39	♀ ♇ ♅
12 49	☽ ∥ ♃
13 20	☽ Q ☉
17 13	☽ □ ☿
18 30	☽ □ ♄
07 Sunday	
02 58	☽ # ♄
03 31	☿ ☍ ♄
06 05	☽ △ ♅
09 05	☽ ∥ ☿
11 27	☽ ⚹ ☉
12 20	☽ △ ♆
12 49	☽ □ ♃
16 13	☽ ☍ ⯓
17 29	☽ # ♆
23 28	☽ ∥ ☉
08 Monday	
00 27	☉ ☍ ♆
03 45	☽ Q ☿
05 19	☽ □ ♂
05 36	☽ # ♅
07 12	☽ ♇ ♅
09 42	☽ □ ♀
12 58	☉ ⊻ ♃
13 18	☽ ♇ ♆
14 13	☽ ∠ ☉
19 08	☽ Q ♃
20 36	☽ △ ♄
21 56	☽ ± ♅
09 Tuesday	
01 53	☽ ⚹ ☿
03 58	☽ ± ♆
05 54	☽ ⊥ ☉
07 39	☽ ⊼ ♅
07 51	☿ # ♄
11 58	☽ Q ♂
13 38	☽ ⊼ ♆
14 51	☽ ⚹ ♃
16 14	☽ ⊻ ☉
16 21	☽ Q ♀
17 22	☽ ⊼ ⯓
20 43	☽ ♇ ♄
23 26	♂ ♇ ♅
10 Wednesday	
02 58	☽ ± ⯓
04 49	☽ ∠ ☿
07 59	☽ ⚹ ♂
09 46	☉ △ ⯓
11 00	☽ ± ♄
12 21	☽ ⚹ ♀
15 07	☽ ∠ ♃
17 16	☽ ♇ ⯓
20 28	☽ ⊼ ♄
20 44	☽ ⊥ ☿
11 Thursday	
05 34	☽ ⊥ ♃
07 17	☽ ⊻ ☿
07 20	☽ ☍ ♅
07 45	☿ ⊼ ♅
08 46	☽ ∠ ♂
09 55	☽ # ♅
13 07	☽ ∠ ♀
13 15	☽ ☍ ♆
15 11	☽ ⊻ ♃
16 06	♀ ♇ ♆
16 59	☽ △ ⯓
19 04	☽ ∥ ☉
19 06	☽ ☌ ☉
21 20	☽ # ♆
23 36	☽ ⊥ ♂
12 Friday	
03 58	☽ ⊥ ♀
09 32	☽ ⊻ ♂
10 24	☽ # ♄
13 54	☽ ⊻ ♀
19 50	☽ ☍ ♄
20 17	☽ ∥ ☿
13 Saturday	
01 47	☽ ∥ ♃
06 56	☽ ⊼ ♅
12 15	☽ ☌ ☿
13 01	☽ ⊼ ♆
15 46	☽ ☌ ♃
16 37	☽ ± ♅
16 54	☽ □ ⯓
17 50	☽ ∥ ♂
20 04	☿ ⊼ ♆
22 26	☽ ⊻ ☉
22 45	☽ ± ♆
14 Sunday	
03 25	☉ # ♆
06 00	☽ ∥ ♀
07 17	☽ ♇ ♅
08 58	☽ ⊥ ☉
12 09	☽ ☌ ♂
13 32	☽ ♇ ♆
16 34	☽ ☌ ♀
20 31	☽ ⊼ ♄
22 34	☽ Q ⯓
15 Monday	
01 01	☽ ∠ ☉
04 03	☿ ∥ ♃
06 34	☽ ± ♄
07 23	☿ ☌ ♃
08 17	☽ △ ♅
09 22	☿ ± ♅
12 21	☽ # ⯓
13 25	☿ □ ⯓
14 45	☽ △ ♆
18 31	☽ ⊻ ♃
18 34	♃ ± ♅
18 56	☽ ⚹ ⯓
19 16	♂ ♇ ♆
19 31	☽ ⊻ ☿
21 53	☽ ♇ ♄
16 Tuesday	
04 34	☽ ⚹ ☉
05 04	☽ ⊥ ♃
07 03	☽ ⊥ ☿
07 28	☽ ∥ ⯓
17 54	☽ ⊻ ♂
21 11	☽ ∠ ♃
21 12	☽ ∠ ⯓
21 37	♃ □ ⯓
22 15	☽ ⊻ ♀
17 Wednesday	
00 08	☽ △ ♄
00 42	☽ ∠ ☿
03 26	☽ Q ☉
05 09	☽ ⊥ ♂
09 28	☽ ⊥ ♀
11 04	☽ # ♀
12 52	☽ □ ♅
13 25	☽ ⊥ ⯓
19 51	☽ □ ♆
22 22	☽ ∠ ♂
18 Thursday	
00 25	☽ ⊻ ⯓
00 53	☽ ⚹ ♃
01 07	☽ # ♂
02 35	☽ ∠ ♀
05 44	☿ ± ♆
07 04	☽ ⚹ ☿
13 45	☽ # ☿
15 11	☽ □ ☉
20 49	♀ ⊼ ♄
22 15	☽ Q ♅
22 35	☽ # ♃
23 45	☽ Q ♃
19 Friday	
03 57	☽ ⚹ ♂
05 33	☽ Q ♆
07 28	☽ □ ♄
07 55	☽ ⚹ ♀
08 12	☽ Q ☿
19 59	☉ ⊥ ♂
20 14	☽ ∥ ♄
21 13	☽ ⚹ ♅
20 Saturday	
03 46	☽ # ☉
04 21	☽ Q ♂
04 43	☽ ⚹ ♆
08 00	☽ Q ♀
09 40	☽ ☌ ⯓
11 11	☽ □ ♃
11 43	☽ ∥ ♆
18 19	☽ Q ♄
23 03	☽ □ ☿
21 Sunday	
02 12	☽ ∥ ♅
02 37	☽ ∠ ♅
06 19	☽ △ ☉
10 19	☽ ∠ ♆
17 57	☽ □ ♂
18 05	☽ ⚹ ♄
20 12	♂ ⊼ ♄
20 33	☽ ⊥ ♅
20 57	☽ □ ♀
22 Monday	
04 20	☽ ⊥ ♆
08 36	☽ ⊻ ♅
13 02	☿ ♇ ♅
13 59	♀ Q ⯓
15 05	☽ ♇ ☉
16 26	☽ ⊻ ♆
21 40	☽ ⊻ ⯓
23 Tuesday	
00 10	☽ ∠ ♄
00 21	☽ △ ♃
01 31	☉ ⊥ ♀
09 51	☽ ⊥ ⯓
10 56	☽ ± ☉
17 31	☽ △ ☿
18 20	☽ ⊥ ♄
24 Wednesday	
00 13	☽ ⊼ ☉
04 09	☽ ∠ ⯓
06 28	☽ ⊻ ♄
07 24	☽ ♇ ♃
10 05	☽ △ ♂
11 28	☽ △ ♀
13 17	♂ Q ⯓
21 21	☽ ☌ ♅
25 Thursday	
01 08	☉ # ♄
01 16	☽ ∥ ♅
02 03	☽ ± ♃
02 48	☽ ♇ ☿
05 17	☽ ☌ ♆
10 39	☽ ⚹ ⯓
14 28	☽ ⊼ ♃
16 03	☽ ∥ ♆
18 15	☽ ♇ ♂
18 38	☽ ♇ ♀
22 37	☽ ± ☿
26 Friday	
05 39	♀ ☌ ♂
07 16	☽ ∥ ♄
09 28	☽ # ☉
10 04	☿ ♇ ♆
10 56	☽ Q ⯓
11 43	☽ ⊼ ☿
13 11	☽ ± ♀
13 27	☽ ± ♂
18 24	☽ ☍ ☉
18 55	☽ ☌ ♄
27 Saturday	
00 21	☉ ☍ ♄
01 28	☽ ⊼ ♀
02 11	☽ ⊼ ♂
09 46	☽ ⊻ ♅
09 51	☽ # ♃
15 53	☿ ∥ ♂
17 37	☽ ⊻ ♆
21 42	☽ ⊥ ♅
23 00	☽ □ ⯓
28 Sunday	
03 50	☽ ☍ ♃
05 31	☽ ⊥ ♆
15 29	☽ ∠ ♅
17 47	☽ # ♂
19 06	☽ # ☿
23 14	☽ ∠ ♆
23 51	⯓ St D
29 Monday	
03 41	☽ ☍ ☿
06 07	☽ ⊻ ♄
11 00	☽ ⊼ ☉
12 58	☽ # ♀
13 35	☽ ☍ ♀
16 41	☽ ☍ ♂
17 44	☽ ⊥ ♄
20 43	☽ ⚹ ♅
23 37	☽ ± ☉
30 Tuesday	
01 58	☽ ∥ ⯓
04 22	☽ ⚹ ♆
09 41	☽ △ ⯓
11 00	☽ ∠ ♄
11 34	♂ ± ♄
15 21	☽ ⊼ ♃
18 20	☽ ♇ ☉
19 42	☽ Q ♅
19 51	☿ ⊼ ♄
23 09	☉ ⊻ ♀
31 Wednesday	
00 12	☽ # ⯓
02 58	☽ ± ♃
03 15	☽ Q ♆
10 56	☽ ∥ ♀
14 10	☽ ♇ ⯓
15 17	☽ ⚹ ♄
16 21	☽ ⊼ ☿
17 22	♃ ± ♆
20 12	☽ ♇ ♃
23 03	☽ ⊼ ♀
23 47	☽ ∥ ☿

AUGUST 1991

LONGITUDES

Date	Sidereal time h m s	Sun ☉ ° ' "	Moon ☽ ° ' "	Moon ☽ 24.00 ° ' "	Mercury ☿ ° '	Venus ♀ ° '	Mars ♂ ° '	Jupiter ♃ ° '	Saturn ♄ ° '	Uranus ♅ ° '	Neptune ♆ ° '	Pluto ⯓ ° '
01	08 38 23	08 ♌ 45 02	14 ♈ 18 23	20 ♈ 51 23	04 ♍ 11	07 ♍ 19	10 ♍ 30	20 ♌ 56	03 ♒ 06	10 ♑ 45	14 ♑ 45	17 ♏ 34
02	08 42 19	09 42 26	27 ♈ 28 52	04 ♉ 11 03	04 39	07 R 18	11 08	21 09	03 R 02	10 R 43	14 R 43	17 34
03	08 46 16	10 39 51	10 ♉ 58 06	17 ♉ 50 08	05 04	07 15	11 45	21 22	02 57	10 41	14 42	17 34
04	08 50 12	11 37 18	24 ♉ 47 14	01 ♊ 49 23	05 23	07 09	12 23	21 35	02 53	10 39	14 40	17 35
05	08 54 09	12 34 45	08 ♊ 56 30	16 ♊ 08 21	05 38	07 00	13 00	21 48	02 48	10 37	14 39	17 35
06	08 58 06	13 32 15	23 ♊ 24 37	00 ♋ 44 50	05 49	06 49	13 38	22 01	02 44	10 35	14 38	17 35
07	09 02 02	14 29 45	08 ♋ 08 25	15 ♋ 34 39	05 54	06 36	14 16	22 14	02 40	10 33	14 36	17 35
08	09 05 59	15 27 17	23 ♋ 02 41	00 ♌ 31 35	05 R 54	06 21	14 53	22 27	02 35	10 31	14 35	17 36
09	09 09 55	16 24 49	08 ♌ 00 21	15 ♌ 27 56	05 49	06 03	15 31	22 40	02 31	10 29	14 35	17 36
10	09 13 52	17 22 23	22 ♌ 53 17	00 ♍ 15 26	05 38	05 43	16 09	22 53	02 27	10 28	14 32	17 36
11	09 17 48	18 19 58	07 ♍ 33 25	14 ♍ 46 28	05 22	05 21	16 47	23 06	02 23	10 26	14 31	17 37
12	09 21 45	19 17 34	21 ♍ 53 53	28 ♍ 55 11	05 00	04 57	17 24	23 19	02 19	10 24	14 30	17 37
13	09 25 41	20 15 10	05 ♎ 49 59	12 ♎ 38 09	04 33	04 31	18 02	23 33	02 14	10 23	14 28	17 38
14	09 29 38	21 12 48	19 ♎ 19 37	25 ♎ 54 31	04 01	04 02	18 40	23 46	02 10	10 21	14 27	17 38
15	09 33 35	22 10 27	02 ♏ 23 06	08 ♏ 45 43	03 25	03 32	19 18	23 59	02 06	10 19	14 26	17 39
16	09 37 31	23 08 07	15 ♏ 02 49	21 ♏ 14 55	02 44	03 01	19 56	24 12	02 02	10 18	14 25	17 40
17	09 41 28	24 05 47	27 ♏ 22 34	03 ♐ 26 23	01 59	02 28	20 34	24 25	01 58	10 16	14 24	17 40
18	09 45 24	25 03 29	09 ♐ 26 58	15 ♐ 24 58	01 11	01 53	21 12	24 38	01 54	10 15	14 22	17 41
19	09 49 21	26 01 12	21 ♐ 21 00	27 ♐ 15 39	00 ♍ 21	01 18	21 50	24 51	01 50	10 13	14 21	17 42
20	09 53 17	26 58 56	03 ♑ 09 33	09 ♑ 03 13	29 ♌ 29	00 42	22 28	25 04	01 47	10 12	14 20	17 42
21	09 57 14	27 56 41	14 ♑ 57 13	20 ♑ 52 02	28 37	00 ♍ 05	23 07	25 18	01 43	10 10	14 19	17 43
22	10 01 10	28 54 28	26 ♑ 48 06	02 ♒ 45 49	27 45	29 ♌ 28	23 45	25 31	01 39	10 09	14 18	17 44
23	10 05 07	29 ♌ 52 16	08 ♒ 45 34	14 ♒ 47 38	26 54	28 50	24 23	25 44	01 35	10 08	14 17	17 45
24	10 09 04	00 ♍ 50 04	20 ♒ 52 18	26 ♒ 59 45	26 07	28 13	25 01	25 57	01 32	10 07	14 16	17 46
25	10 13 00	01 47 55	03 ♓ 10 10	09 ♓ 23 39	25 22	27 36	25 39	26 10	01 28	10 05	14 15	17 47
26	10 16 57	02 45 47	15 ♓ 40 19	22 ♓ 00 10	24 43	27 00	26 18	26 23	01 25	10 04	14 14	17 48
27	10 20 53	03 43 40	28 ♓ 23 16	04 ♈ 49 35	24 09	26 24	26 56	26 36	01 21	10 03	14 13	17 49
28	10 24 50	04 41 35	11 ♈ 19 08	17 ♈ 51 53	23 42	25 50	27 35	26 49	01 18	10 02	14 12	17 50
29	10 28 46	05 39 32	24 ♈ 27 50	01 ♉ 06 57	23 21	25 17	28 13	27 02	01 15	10 01	14 12	17 51
30	10 32 43	06 37 30	07 ♉ 49 15	14 ♉ 34 44	23 08	24 45	28 52	27 15	01 12	10 00	14 11	17 52
31	10 36 39	07 ♍ 35 31	21 ♉ 23 25	28 ♉ 15 18	23 ♌ 03	24 ♌ 15	29 ♍ 30	27 ♌ 28	01 ♒ 08	09 ♑ 59	14 ♑ 10	17 ♏ 53

Date	Moon True ☊ ° '	Moon Mean ☊ ° '	Moon ☽ Latitude ° '
01	18 ♑ 47	17 ♑ 53	05 N 14
02	18 R 45	17 49	05 12
03	18 D 45	17 46	04 52
04	18 45	17 43	04 15
05	18 46	17 40	03 22
06	18 48	17 37	02 15
07	18 49	17 34	00 N 58
08	18 R 49	17 30	00 S 23
09	18 48	17 27	01 44
10	18 46	17 24	02 56
11	18 42	17 21	03 56
12	18 38	17 18	04 40
13	18 33	17 15	05 05
14	18 29	17 11	05 13
15	18 26	17 08	05 04
16	18 24	17 05	04 40
17	18 D 24	17 02	04 03
18	18 25	16 59	03 17
19	18 27	16 55	02 23
20	18 28	16 52	01 23
21	18 29	16 49	00 S 19
22	18 R 29	16 46	00 N 45
23	18 27	16 43	01 48
24	18 23	16 40	02 47
25	18 17	16 36	03 38
26	18 10	16 33	04 20
27	18 02	16 30	04 50
28	17 54	16 27	05 06
29	17 47	16 24	05 06
30	17 42	16 21	04 49
31	17 ♑ 38	16 ♑ 17	04 N 16

DECLINATIONS

Date	Sun ☉ ° '	Moon ☽ ° '	Mercury ☿ ° '	Venus ♀ ° '	Mars ♂ ° '	Jupiter ♃ ° '	Saturn ♄ ° '	Uranus ♅ ° '	Neptune ♆ ° '	Pluto ⯓ ° '
01	18 N 04	10 N 28	07 N 36	04 N 42	08 N 31	15 N 14	19 S 56	23 S 23	21 S 49	02 S 31
02	17 49	15 25	07 14	04 31	08 17	15 10	19 57	23 23	21 50	02 32
03	17 34	19 44	06 54	04 20	08 02	15 06	19 58	23 24	21 50	02 32
04	17 18	23 05	06 35	04 09	07 47	15 01	20 00	23 24	21 50	02 33
05	17 02	25 07	06 18	04 00	07 32	14 57	20 01	23 24	21 50	02 34
06	16 46	25 32	06 04	03 52	07 17	14 53	20 02	23 24	21 50	02 34
07	16 29	24 10	05 52	03 44	07 02	14 49	20 03	23 24	21 51	02 35
08	16 12	21 05	05 42	03 38	06 47	14 45	20 04	23 24	21 51	02 35
09	15 55	16 36	05 34	03 33	06 32	14 40	20 05	23 24	21 51	02 36
10	15 38	11 07	05 30	03 28	06 17	14 36	20 06	23 25	21 51	02 37
11	15 20	05 N 05	05 28	03 25	06 02	14 32	20 07	23 25	21 51	02 37
12	15 02	01 S 04	05 29	03 23	05 47	14 28	20 08	23 25	21 51	02 38
13	14 44	06 59	05 33	03 22	05 32	14 23	20 09	23 25	21 52	02 39
14	14 26	12 23	05 40	03 22	05 16	14 19	20 10	23 25	21 52	02 39
15	14 07	17 03	05 50	03 23	05 01	14 15	20 11	23 25	21 52	02 40
16	13 48	20 48	06 02	03 25	04 46	14 11	20 12	23 25	21 52	02 41
17	13 29	23 31	06 18	03 28	04 30	14 06	20 13	23 25	21 52	02 41
18	13 10	25 07	06 37	03 32	04 15	14 02	20 14	23 26	21 52	02 42
19	12 51	25 32	06 57	03 37	03 59	13 58	20 15	23 26	21 52	02 43
20	12 31	24 47	07 21	03 43	03 44	13 53	20 16	23 26	21 53	02 44
21	12 11	22 55	07 45	03 50	03 28	13 49	20 17	23 26	21 53	02 44
22	11 51	20 04	08 12	03 58	03 13	13 45	20 17	23 26	21 53	02 45
23	11 31	16 20	08 39	04 07	02 57	13 40	20 18	23 26	21 53	02 46
24	11 11	11 54	09 07	04 16	02 41	13 36	20 19	23 26	21 53	02 47
25	10 50	06 57	09 35	04 26	02 26	13 31	20 20	23 26	21 53	02 47
26	10 29	01 S 39	10 02	04 37	02 10	13 27	20 21	23 26	21 53	02 48
27	10 09	03 N 48	10 29	04 48	01 54	13 23	20 22	23 26	21 54	02 49
28	09 47	09 10	10 55	04 59	01 39	13 18	20 22	23 26	21 54	02 50
29	09 26	14 13	11 18	05 11	01 23	13 14	20 23	23 27	21 54	02 50
30	09 05	18 40	11 40	05 23	01 07	13 09	20 24	23 27	21 54	02 51
31	08 N 43	22 N 14	11 N 59	05 N 35	00 N 51	13 N 05	20 S 25	23 S 27	21 S 54	02 S 52

ZODIAC SIGN ENTRIES

Date	h	m	Planets
02	16	32	☽ ♉
04	20	54	☽ ♊
06	22	47	☽ ♋
08	23	09	☽ ♌
10	23	35	☽ ♍
13	01	52	☽ ♎
15	07	34	☽ ♏
17	17	11	☽ ♐
19	21	40	☿ ♌
20	05	34	☽ ♑
21	15	06	♀ ♌
22	18	27	☽ ♒
23	15	13	☉ ♍
25	05	51	☽ ♓
27	15	01	☽ ♈
29	22	00	☽ ♉

LATITUDES

Date	Mercury ☿ ° '	Venus ♀ ° '	Mars ♂ ° '	Jupiter ♃ ° '	Saturn ♄ ° '	Uranus ♅ ° '	Neptune ♆ ° '	Pluto ⯓ ° '
01	02 S 33	04 S 26	00 N 58	00 N 45	00 S 29	00 S 23	00 N 48	15 N 08
04	03 10	05 06	00 57	00 45	00 29	00 23	00 48	15 06
07	03 45	05 45	00 55	00 45	00 30	00 23	00 48	15 04
10	04 15	06 23	00 54	00 45	00 30	00 23	00 48	15 03
13	04 37	06 58	00 52	00 46	00 30	00 23	00 48	15 01
16	04 47	07 29	00 51	00 46	00 30	00 23	00 48	14 59
19	04 41	07 54	00 49	00 46	00 30	00 23	00 48	14 58
22	04 19	08 12	00 47	00 46	00 31	00 23	00 48	14 56
25	03 42	08 22	00 46	00 46	00 31	00 23	00 48	14 54
28	02 52	08 24	00 44	00 47	00 31	00 23	00 48	14 53
31	01 S 57	08 S 18	00 N 43	00 N 47	00 S 31	00 S 23	00 N 48	14 N 51

DATA

Julian Date	2448470
Delta T	+58 seconds
Ayanamsa	23° 44' 39"
Synetic vernal point	05° ♓ 22' 20"
True obliquity of ecliptic	23° 26' 28"

MOON'S PHASES, APSIDES AND POSITIONS ☽

Date	h	m	Phase	Longitude ° '	Eclipse Indicator
03	11	26	☾	10 ♉ 38	
10	02	28	●	17 ♌ 00	
17	05	01	☽	23 ♏ 49	
25	09	07	○	01 ♓ 41	

Day	h	m	
08	18	04	Perigee
20	22	27	Apogee
06	05	39	Max dec 25° N 36'
12	07	48	0S
19	08	29	Max dec 25° S 33'
26	19	18	0N

LONGITUDES

Date	Chiron ⚷ ° '	Ceres ⚳ ° '	Pallas ⚴ ° '	Juno ⚵ ° '	Vesta ⚶ ° '	Black Moon Lilith ⚸ ° '
01	01 ♌ 15	29 ♎ 33	01 ♎ 23	20 ♑ 19	27 ♋ 05	10 ♑ 49
11	02 ♌ 24	02 ♏ 27	05 ♎ 18	18 ♑ 21	01 ♌ 29	11 ♑ 56
21	03 ♌ 32	05 ♏ 35	09 ♎ 20	16 ♑ 53	05 ♌ 53	13 ♑ 02
31	04 ♌ 36	08 ♏ 56	13 ♎ 28	16 ♑ 01	10 ♌ 15	14 ♑ 09

ASPECTARIAN

h m Aspects

01 Thursday
00 54 ☽ △ ☉
03 28 ☽ ∥ ♂
04 03 ☽ ± ☿
04 37 ☽ ⊼ ♂
05 27 ☽ □ ♅
06 56 ☽ ± ⯓
10 11 ☽ ± ♀
10 35 ♀ St R
12 48 ☽ □ ♆
13 27 ☽ Q ♄
16 15 ☽ ± ♂
17 59 ☽ ⊼ ⯓
20 47 ♂ △ ♅
21 17 ☽ ♇ ☿

02 Friday
00 20 ☽ △ ♃
02 39 ☽ ♇ ♀
09 27 ☽ ♇ ♂
10 42 ☽ ∥ ♃
21 53 ☽ □ ♄

03 Saturday
00 04 ☽ ∥ ☉
01 15 ☽ △ ☿
05 28 ☽ △ ♀
11 26 ☽ □ ☉
11 30 ☽ △ ♅
12 25 ☉ ⊼ ♅
13 27 ☽ △ ♂
13 28 ☽ ⋕ ♄
18 31 ☽ △ ♆
23 33 ☽ ☍ ⯓

04 Sunday
01 59 ☽ ⋕ ♆
06 24 ☽ □ ♃
13 28 ☽ ♇ ♅
14 44 ☽ ⋕ ♅
20 20 ☽ ♇ ♆
20 52 ☽ Q ☉

05 Monday
01 43 ☽ △ ♄
04 44 ☽ ± ♅
05 14 ☿ Q ⯓
06 22 ☽ □ ☿
08 47 ☽ □ ♀
11 31 ☽ ± ♆
13 27 ☽ Q ♃
14 48 ☽ ⊼ ♅
18 31 ☽ ✶ ☉
19 06 ☽ □ ♂
21 31 ☽ ⊼ ♆

06 Tuesday
02 23 ☽ ⊼ ⯓
02 41 ☽ ♇ ♄
09 40 ☽ ✶ ♃
12 17 ☽ ± ⯓
12 40 ☽ Q ☿
14 17 ☽ Q ♀
17 25 ☽ ± ♄
19 03 ☉ ⊻ ♂
20 59 ☽ ∠ ☉

07 Wednesday
02 03 ☽ Q ♂
03 00 ☽ ♇ ⯓
03 10 ☽ ⊼ ♄
08 21 ☽ ✶ ☿
09 33 ☽ ✶ ♀
10 31 ☽ ∠ ♃
12 37 ☽ ⊥ ☉
14 37 ☉ ⊼ ♆
15 53 ☽ ☍ ♅
19 16 ☽ ⋕ ♅
22 19 ☽ ✶ ♂
22 25 ☽ ☍ ♆
22 58 ☽ ⊻ ☉
23 57 ☿ St R

08 Thursday
00 34 ♂ △ ♆
01 15 ☽ ⊥ ♃
03 15 ☽ △ ⯓
07 07 ☽ ⋕ ♆
08 34 ☽ ∠ ☿
09 20 ☽ ∠ ♀
11 02 ☽ ⊻ ♃
18 06 ☽ ⋕ ♄
22 57 ☽ ⊥ ☿
23 28 ☽ ∠ ♂
23 30 ☽ ⊥ ♀

09 Friday
03 14 ☽ ☍ ♄
08 30 ☽ ⊻ ☿
08 56 ☽ ⊻ ♀
13 53 ☉ ± ♅
14 32 ☽ ⊥ ♂
15 22 ☽ ∥ ☉
15 59 ☽ ⊼ ♅
20 56 ☽ ∥ ♃
22 31 ☽ ⊼ ♆

10 Saturday
00 37 ☽ ⊻ ♂
01 38 ☽ ± ♅
02 28 ☽ ☌ ☉
03 27 ☽ □ ⯓
08 12 ☽ ± ♆
12 00 ☽ ☌ ♃
14 16 ☿ Q ⯓
16 10 ☽ ♇ ♅
17 55 ☉ □ ⯓
19 32 ♀ Q ⯓
22 48 ☽ ♇ ♆

11 Sunday
03 31 ☽ ⊼ ♄
08 07 ☽ ∥ ♂
08 27 ☽ ☌ ☿
08 27 ☽ ☌ ♀
08 48 ☽ Q ⯓
09 24 ☿ ☌ ♀
10 30 ☽ ∥ ☿
13 21 ☽ ± ♄
16 45 ☽ △ ♅
18 31 ☽ ∥ ♀
21 33 ☽ ⋕ ⯓
23 33 ☽ △ ♆

12 Monday
04 04 ☽ ☌ ♂
04 17 ☽ ♇ ♄
04 47 ☽ ✶ ⯓
07 16 ☽ ⊻ ☉
08 37 ♂ ♇ ♄
14 28 ☽ ⊻ ♃
18 12 ☽ ⊥ ☉
18 13 ☽ ∥ ⯓
20 18 ♂ ✶ ⯓
21 10 ☽ ⋕ ♀

13 Tuesday
00 54 ☽ ⊥ ♃
05 46 ☽ △ ♄
05 55 ☽ ⋕ ☿
06 11 ☽ ⋕ ♂
06 25 ☽ ∠ ⯓
09 46 ☽ ⊻ ♀
09 51 ☽ ⊻ ☿
10 43 ☿ ∥ ♂
10 55 ☽ ∠ ☉
16 50 ☽ ∠ ♃
17 23 ☉ ± ♆
19 57 ☽ ⊥ ♀
19 59 ☽ □ ♅
20 01 ☽ ⊥ ☿
22 13 ☽ ⊥ ⯓

14 Wednesday
03 15 ☽ □ ♆
07 50 ☿ ☌ ♀
08 57 ☽ ⊻ ⯓
10 45 ☽ ⊻ ♂
11 29 ☽ ∠ ☿
11 30 ☽ ∠ ♀
15 41 ☽ ✶ ☉
20 12 ☽ ✶ ♃
21 19 ☽ ⋕ ♃
21 25 ☽ ⋕ ☉
22 13 ☽ ⊥ ♂
23 15 ☉ ∥ ♃

15 Thursday
04 28 ☽ Q ♅
11 29 ☽ □ ♄
12 05 ☽ Q ♆
13 50 ☽ ✶ ☿
14 04 ☽ ✶ ♀
15 37 ☽ Q ☉
15 47 ☽ ∠ ♂
18 52 ☽ Q ♃

16 Friday
02 56 ☽ ✶ ♅
07 39 ☽ ∥ ♄
10 47 ☽ ✶ ♆
11 26 ☽ Q ☿
11 56 ☽ Q ♀
17 03 ☽ ☌ ⯓
20 18 ☽ ∥ ♆
21 36 ☽ Q ♄
21 57 ☽ ✶ ♂

17 Saturday
05 01 ☽ □ ☉
06 05 ☽ □ ♃
07 52 ☽ ∠ ♅
10 55 ☽ ∥ ♅
12 28 ☿ ⊼ ♄
15 58 ☽ ∠ ♆
20 33 ☽ □ ☿
21 02 ☽ ✶ ♄
21 36 ☽ □ ♀
22 24 ☉ ☌ ♃
22 50 ☽ Q ♂

18 Sunday
01 37 ☽ ⊥ ♅
09 51 ☽ ⊥ ♆
11 12 ♀ ⊼ ♄
13 35 ☽ ⊻ ♅
16 32 ☉ ♇ ♅
21 53 ☽ ⊻ ♆

19 Monday
02 56 ☽ ∠ ♄
04 36 ☽ ⊻ ⯓
13 03 ☽ □ ♂
16 46 ☽ ⊥ ⯓
19 15 ☽ △ ♃
21 04 ☽ ⊥ ♄
22 19 ☽ △ ☉

20 Tuesday
05 02 ☽ △ ☿
07 14 ☽ △ ♀
09 12 ☽ ⊻ ♄
11 05 ☽ ∠ ⯓
12 38 ♀ ∥ ♂
16 04 ☿ ♇ ♆

21 Wednesday
00 06 ♃ ♇ ♅
02 18 ☽ ☌ ♅
02 21 ☽ ♇ ♃
06 37 ☽ ∥ ♅
07 33 ☽ ♇ ☉
09 27 ☽ ♇ ☿
10 43 ☽ ☌ ♆
12 15 ☽ ♇ ♀
17 38 ☽ ✶ ⯓
20 42 ☉ ☌ ☿
20 59 ☽ ± ♃
21 41 ☽ ∥ ♆

22 Thursday
02 28 ☽ ± ☿
03 26 ☽ ± ☉
05 29 ☽ △ ♂
05 35 ☽ ± ♀
09 21 ☽ ⊼ ♃
10 19 ☽ ∥ ♄
13 46 ☽ ⊼ ☿
16 37 ☽ ⊼ ☉
17 05 ☽ ⊼ ♀
17 55 ☽ Q ⯓
18 13 ♀ ♇ ♆
20 21 ☉ ☌ ♀
21 40 ☉ ♇ ♆
21 43 ☽ ☌ ♄

23 Friday
13 19 ☽ ♇ ♂
14 44 ☽ ⊻ ♅
22 59 ☽ ⊻ ♆

24 Saturday
02 37 ☽ ⊥ ♅
03 05 ☽ ⋕ ♃
04 18 ♂ ⋕ ⯓
05 52 ☽ □ ⯓
08 09 ☽ ± ♂
10 49 ☽ ⊥ ♆
15 55 ☿ ☌ ♃
15 57 ☽ ⋕ ☉
20 18 ☽ ∠ ♅
20 35 ☽ ⊼ ♂
21 40 ☽ ☍ ☿
22 08 ☽ ☍ ♃

25 Sunday
00 35 ☽ ⋕ ☿
01 42 ☽ ☍ ♀
04 20 ☉ ⊼ ♄
04 24 ☽ ∠ ♆
06 57 ☿ ⊻ ♂
08 44 ☽ ⊻ ♄
09 07 ☽ ☍ ☉
20 16 ☽ ⊥ ♄
22 22 ☿ ♇ ♅
23 10 ☽ ⋕ ♀

26 Monday
01 19 ☽ ✶ ♅
06 52 ☽ ∥ ⯓
09 16 ☽ ✶ ♆
09 35 ☽ ⋕ ♂
13 24 ☽ ∠ ♄
16 02 ☽ △ ⯓
17 04 ♂ ⊻ ♃

27 Tuesday
00 06 ☽ Q ♅
01 31 ☉ ∥ ☿
01 36 ♀ ⊻ ♂
04 04 ☽ ∥ ♂
04 22 ☽ ⊼ ☿
06 03 ♀ ☌ ♃
07 41 ☽ ⋕ ⯓
07 57 ☽ Q ♆
08 27 ☽ ⊼ ♀
08 36 ☽ ⊼ ♃
09 08 ☽ ☍ ♂
15 10 ☽ ± ☿
16 34 ☽ ∥ ♀
17 31 ☽ ✶ ♄
19 11 ☽ ± ♀
20 00 ☽ ± ♃
20 16 ☽ ♇ ⯓
22 46 ☽ ⊼ ☉

28 Wednesday
07 19 ☽ ♇ ☿
09 38 ☽ □ ♅
10 45 ☽ ± ☉
11 08 ☽ ♇ ♀
12 56 ☽ ± ⯓
12 56 ☽ ♇ ♃
14 41 ☽ ∥ ☉
15 38 ☽ Q ♄
17 18 ☽ □ ♆
20 46 ☽ ∥ ☿
23 57 ☽ ⊼ ⯓

29 Thursday
04 33 ☽ ♇ ☉
07 13 ☽ ∥ ♃
10 02 ☽ △ ☿
13 25 ☽ △ ♀
16 41 ☉ Q ⯓
16 44 ☽ △ ♃
19 08 ☽ ⊼ ♂

30 Friday
00 11 ☽ □ ♄
00 13 ♀ ♇ ♅
06 27 ☽ ± ♂
09 42 ☽ △ ☉
15 52 ☽ △ ♅
20 39 ☿ ⊥ ♂
22 48 ☽ ⋕ ♄
23 16 ☽ ♇ ♂
23 17 ☽ △ ♆

31 Saturday
01 22 ☉ ± ♄
05 49 ☽ ☍ ⯓
09 24 ☽ ⋕ ♆
14 34 ☿ St D
14 55 ☽ □ ☿
16 50 ☽ □ ♀
18 17 ☽ ♇ ♅
22 42 ☽ ⋕ ♅
22 49 ☽ □ ♃

All ephemeris data is given at 12.00 UT and the Moon's longitude is additionally given for 24.00 UT

SEPTEMBER 1991

LONGITUDES

Date	Sidereal time h m s	Sun ☉ ° ' "	Moon ☽ ° ' "	Moon ☽ 24.00 ° ' "	Mercury ☿ ° '	Venus ♀ ° '	Mars ♂ ° '	Jupiter ♃ ° '	Saturn ♄ ° '	Uranus ♅ ° '	Neptune ♆ ° '	Pluto ♇ ° '
01	10 40 36	08 ♍ 33 33	05 ♊ 10 26	12 ♊ 08 47	23 ♌ 06	23 ♌ 47	00 ♎ 09	27 ♌ 41	01 ♒ 05	09 ♑ 58	14 ♑ 09	17 ♏ 54
02	10 44 33	09 31 38	19 ♊ 10 19	26 ♊ 14 59	23 D 18	23 R 20	00 47	27 54	01 R 02	09 R 57	14 R 08	17 55
03	10 48 29	10 29 44	03 ♋ 22 39	10 ♋ 33 06	23 38	22 56	01 26	28 07	00 59	09 56	14 08	17 56
04	10 52 26	11 27 52	17 ♋ 46 03	25 ♋ 01 06	24 07	22 34	02 05	28 20	00 57	09 56	14 07	17 58
05	10 56 22	12 26 03	02 ♌ 17 48	09 ♌ 35 31	24 44	22 14	02 43	28 33	00 54	09 55	14 06	17 59
06	11 00 19	13 24 15	16 ♌ 53 36	24 ♌ 11 17	25 29	21 57	03 22	28 46	00 51	09 54	14 06	18 00
07	11 04 15	14 22 29	01 ♍ 27 45	08 ♍ 42 10	26 22	21 41	04 01	28 59	00 48	09 54	14 05	18 01
08	11 08 12	15 20 45	15 ♍ 53 42	23 ♍ 01 32	27 22	21 29	04 40	29 12	00 46	09 53	14 04	18 03
09	11 12 08	16 19 02	00 ♎ 05 00	07 ♎ 03 27	28 29	21 18	05 19	29 25	00 43	09 53	14 04	18 04
10	11 16 05	17 17 21	13 ♎ 56 24	20 ♎ 43 30	29 ♌ 43	21 10	05 58	29 38	00 41	09 52	14 03	18 06
11	11 20 02	18 15 42	27 ♎ 24 34	03 ♏ 59 31	01 ♍ 03	21 05	06 37	29 ♌ 50	00 39	09 52	14 03	18 07
12	11 23 58	19 14 05	10 ♏ 28 27	16 ♏ 51 35	02 28	21 01	07 16	00 ♍ 03	00 37	09 51	14 02	18 08
13	11 27 55	20 12 29	23 ♏ 09 14	29 ♏ 21 49	03 58	21 D 00	07 55	00 16	00 34	09 51	14 02	18 10
14	11 31 51	21 10 55	05 ♐ 29 52	11 ♐ 33 56	05 32	21 02	08 34	00 29	00 32	09 51	14 02	18 11
15	11 35 48	22 09 23	17 ♐ 34 38	23 ♐ 32 38	07 10	21 06	09 13	00 41	00 30	09 51	14 01	18 13
16	11 39 44	23 07 52	29 ♐ 28 37	05 ♑ 23 15	08 51	21 12	09 52	00 54	00 29	09 50	14 01	18 15
17	11 43 41	24 06 22	11 ♑ 17 13	17 ♑ 11 11	10 34	21 20	10 31	01 06	00 27	09 50	14 01	18 16
18	11 47 37	25 04 55	23 ♑ 05 49	29 ♑ 01 42	12 20	21 30	11 11	01 19	00 25	09 50	14 00	18 18
19	11 51 34	26 03 29	04 ♒ 59 27	10 ♒ 59 33	14 07	21 43	11 50	01 32	00 24	09 D 50	14 00	18 19
20	11 55 31	27 02 05	17 ♒ 02 31	23 ♒ 08 44	15 56	21 57	12 29	01 44	00 22	09 50	14 00	18 21
21	11 59 27	28 00 42	29 ♒ 18 32	05 ♓ 32 11	17 46	22 14	13 09	01 56	00 21	09 50	14 00	18 23
22	12 03 24	28 59 21	11 ♓ 49 52	18 ♓ 11 41	19 36	22 32	13 48	02 09	00 20	09 50	14 00	18 25
23	12 07 20	29 ♍ 58 02	24 ♓ 37 39	01 ♈ 07 41	21 27	22 53	14 28	02 21	00 18	09 51	13 59	18 26
24	12 11 17	00 ♎ 56 45	07 ♈ 41 40	14 ♈ 19 22	23 18	23 15	15 07	02 33	00 17	09 51	13 59	18 28
25	12 15 13	01 55 30	21 ♈ 00 34	27 ♈ 44 58	25 09	23 39	15 47	02 46	00 16	09 51	13 59	18 30
26	12 19 10	02 54 18	04 ♉ 32 15	11 ♉ 22 06	26 59	24 05	16 26	02 58	00 15	09 51	13 D 59	18 32
27	12 23 06	03 53 07	18 ♉ 14 12	25 ♉ 08 17	28 ♍ 49	24 32	17 06	03 10	00 15	09 52	13 59	18 34
28	12 27 03	04 51 59	02 ♊ 04 07	09 ♊ 01 27	00 ♎ 39	25 01	17 46	03 22	00 14	09 52	13 59	18 35
29	12 31 00	05 50 52	16 ♊ 00 09	23 ♊ 00 04	02 28	25 32	18 25	03 34	00 13	09 53	13 59	18 37
30	12 34 56	06 ♎ 49 49	00 ♋ 01 05	07 ♋ 03 09	04 ♎ 17	26 ♌ 04	19 ♎ 05	03 ♍ 46	00 ♒ 13	09 ♑ 53	14 ♑ 00	18 ♏ 39

Date	Moon True ☊ ° '	Moon Mean ☊ ° '	Moon ☽ Latitude ° '
01	17 ♑ 37	16 ♑ 14	03 N 28
02	17 D 37	16 11	02 27
03	17 38	16 08	01 N 16
04	17 R 39	16 05	00 S 01
05	17 38	16 01	01 18
06	17 35	15 58	02 30
07	17 29	15 55	03 32
08	17 22	15 52	04 20
09	17 12	15 49	04 51
10	17 03	15 46	05 04
11	16 53	15 42	04 59
12	16 46	15 39	04 39
13	16 40	15 36	04 06
14	16 37	15 33	03 21
15	16 35	15 30	02 28
16	16 D 35	15 26	01 30
17	16 36	15 23	00 S 28
18	16 R 36	15 20	00 N 35
19	16 34	15 17	01 36
20	16 30	15 14	02 34
21	16 24	15 11	03 26
22	16 14	15 07	04 09
23	16 03	15 04	04 41
24	15 51	15 01	04 58
25	15 39	14 58	05 00
26	15 28	14 55	04 45
27	15 19	14 52	04 13
28	15 13	14 48	03 27
29	15 10	14 45	02 28
30	15 ♑ 09	14 ♑ 42	01 N 19

DECLINATIONS

Date	Sun ☉ ° '	Moon ☽ ° '	Mercury ☿ ° '	Venus ♀ ° '	Mars ♂ ° '	Jupiter ♃ ° '	Saturn ♄ ° '	Uranus ♅ ° '	Neptune ♆ ° '	Pluto ♇ ° '
01	08 N 22	24 N 35	12 N 16	05 N 48	00 N 35	13 N 01	20 S 25	23 S 27	21 S 54	02 S 53
02	08 00	25 27	12 30	06 00	00 20	12 56	20 26	23 27	21 54	02 54
03	07 38	24 40	12 40	06 13	00 N 04	12 52	20 27	23 27	21 54	02 54
04	07 16	22 15	12 48	06 25	00 S 12	12 47	20 28	23 27	21 54	02 55
05	06 54	18 23	12 51	06 37	00 28	12 43	20 28	23 27	21 55	02 56
06	06 31	13 23	12 52	06 49	00 44	12 39	20 29	23 27	21 55	02 57
07	06 09	07 39	12 48	07 00	01 00	12 34	20 29	23 27	21 55	02 58
08	05 47	01 N 34	12 41	07 12	01 16	12 30	20 30	23 27	21 55	02 59
09	05 24	04 S 29	12 31	07 23	01 32	12 25	20 31	23 27	21 55	02 59
10	05 01	10 10	12 17	07 33	01 48	12 21	20 31	23 27	21 55	03 00
11	04 39	15 12	11 59	07 43	02 03	12 17	20 32	23 27	21 55	03 01
12	04 16	19 23	11 38	07 53	02 19	12 12	20 32	23 27	21 55	03 02
13	03 53	22 31	11 14	08 02	02 35	12 08	20 33	23 27	21 55	03 03
14	03 30	24 31	10 47	08 11	02 51	12 03	20 33	23 27	21 55	03 04
15	03 07	25 19	10 17	08 19	03 07	11 59	20 34	23 27	21 55	03 04
16	02 44	24 57	09 44	08 27	03 23	11 55	20 34	23 27	21 55	03 05
17	02 20	23 26	09 10	08 34	03 39	11 50	20 34	23 27	21 56	03 06
18	01 57	20 54	08 33	08 40	03 55	11 46	20 35	23 27	21 56	03 07
19	01 34	17 28	07 54	08 46	04 11	11 42	20 35	23 27	21 56	03 08
20	01 11	13 16	07 14	08 51	04 26	11 37	20 36	23 27	21 56	03 09
21	00 47	08 29	06 32	08 56	04 42	11 33	20 36	23 27	21 56	03 10
22	00 24	03 S 17	05 49	09 00	04 58	11 29	20 36	23 27	21 56	03 10
23	00 N 01	02 N 10	05 05	09 03	05 14	11 24	20 37	23 27	21 56	03 11
24	00 S 23	07 37	04 20	09 06	05 30	11 20	20 37	23 27	21 56	03 12
25	00 46	12 50	03 34	09 08	05 45	11 16	20 37	23 27	21 56	03 13
26	01 09	17 30	02 48	09 09	06 01	11 11	20 37	23 27	21 56	03 14
27	01 33	21 19	02 02	09 10	06 17	11 07	20 37	23 27	21 56	03 15
28	01 56	23 57	01 15	09 09	06 32	11 03	20 38	23 27	21 56	03 16
29	02 19	25 09	00 28	09 09	06 48	10 59	20 38	23 27	21 56	03 16
30	02 S 43	24 N 46	00 S 19	09 N 08	07 S 04	10 N 54	20 S 38	23 S 27	21 S 56	03 S 17

ZODIAC SIGN ENTRIES

Date	h	m	Planets
01	03	02	☽ ♊
01	06	38	♂ ♎
03	06	19	☽ ♋
05	08	13	☽ ♌
07	09	35	☽ ♍
09	11	51	☽ ♎
10	17	14	☿ ♍
11	16	42	☽ ♏
12	06	00	♃ ♍
14	01	14	☽ ♐
16	13	04	☽ ♑
19	01	58	☽ ♒
21	13	20	☽ ♓
23	12	48	☉ ♎
23	21	56	☽ ♈
26	03	59	☽ ♉
28	03	26	☿ ♎
28	08	25	☽ ♊
30	11	58	☽ ♋

LATITUDES

Date	Mercury ☿ ° '	Venus ♀ ° '	Mars ♂ ° '	Jupiter ♃ ° '	Saturn ♄ ° '	Uranus ♅ ° '	Neptune ♆ ° '	Pluto ♇ ° '
01	01 S 38	08 S 15	00 N 42	00 N 47	00 S 31	00 S 23	00 N 48	14 N 51
04	00 S 44	08 00	00 41	00 47	00 31	00 23	00 47	14 49
07	00 N 05	07 40	00 39	00 48	00 31	00 23	00 47	14 48
10	00 45	07 16	00 38	00 48	00 32	00 23	00 47	14 46
13	01 16	06 48	00 36	00 48	00 32	00 23	00 47	14 45
16	01 36	06 19	00 34	00 48	00 32	00 23	00 47	14 43
19	01 47	05 48	00 33	00 49	00 32	00 23	00 47	14 42
22	01 51	05 17	00 31	00 49	00 32	00 23	00 47	14 41
25	01 47	04 46	00 30	00 50	00 32	00 22	00 47	14 39
28	01 39	04 16	00 28	00 50	00 32	00 22	00 47	14 38
31	01 N 26	03 S 46	00 N 26	00 N 50	00 S 32	00 S 22	00 N 46	14 N 37

DATA

Julian Date	2448501
Delta T	+58 seconds
Ayanamsa	23° 44' 43"
Synetic vernal point	05° ♓ 22' 16"
True obliquity of ecliptic	23° 26' 28"

MOON'S PHASES, APSIDES AND POSITIONS ☽

Date	h	m	Phase	Longitude ° '	Eclipse Indicator
01	18	16	☾	08 ♊ 49	
08	11	01	●	15 ♍ 18	
15	22	01	☽	22 ♐ 34	
23	22	40	○	00 ♈ 24	

Day	h	m		
05	19	18	Perigee	
17	15	20	Apogee	
02	12	49	Max dec	25° N 27'
08	18	08	0S	
15	16	09	Max dec	25° S 21'
23	02	32	0N	
29	18	11	Max dec	25° N 12'

LONGITUDES

Date	Chiron ⚷ ° '	Ceres ⚳ ° '	Pallas ⚴ ° '	Juno ⚵ ° '	Vesta ⚶ ° '	Black Moon Lilith ⚸ ° '
01	04 ♌ 43	09 ♏ 16	13 ♎ 53	15 ♑ 58	10 ♌ 41	14 ♑ 15
11	05 ♌ 43	12 ♏ 48	18 ♎ 06	15 ♑ 50	15 ♌ 01	15 ♑ 22
21	06 ♌ 39	16 ♏ 29	22 ♎ 22	16 ♑ 05	19 ♌ 17	16 ♑ 28
31	07 ♌ 29	20 ♏ 17	26 ♎ 42	17 ♑ 28	23 ♌ 30	17 ♑ 35

ASPECTARIAN

01 Sunday
01 34 ☽ ⚼ ♆
02 52 ☽ △ ♂
04 04 ♀ ⊥ ♂
04 57 ☽ △ ♄
09 55 ☽ ± ♅
17 08 ☽ ± ♆
18 16 ☽ □ ☉
20 15 ☽ ⚻ ♅
22 20 ☽ Q ☿
23 01 ☽ Q ♀

02 Monday
03 25 ☽ ⚻ ♆
06 21 ☽ Q ♃
06 41 ☽ ⚼ ♄
09 52 ☽ ⚻ ♇
13 19 ☿ ☌ ♀
18 53 ☽ ⚹ ♀
19 10 ☽ ⚹ ☿
20 04 ☽ ± ♇
20 42 ♂ △ ♄
21 55 ☽ ± ♄
22 26 ☉ △ ♅

03 Tuesday
03 02 ☽ ⚹ ♃
03 11 ☽ Q ☉
08 00 ☽ ⚻ ♄
08 35 ☽ □ ♂
11 16 ☽ ⚼ ♇
19 26 ☽ ∠ ♀
21 05 ☽ ∠ ☿
22 58 ☽ ☍ ♅

04 Wednesday
00 46 ☽ ⚹ ☉
02 00 ☽ ⋕ ♅
04 32 ☽ ∠ ♃
05 56 ☽ ☍ ♆
10 03 ☽ ⊥ ♀
11 44 ☿ ∥ ♃
12 19 ☽ △ ♇
12 36 ☽ ⊥ ☿
14 32 ☽ ⋕ ♆
16 00 ☽ Q ♂
19 41 ☽ ⊥ ♃
19 46 ☽ ⊻ ♀
22 57 ☽ ⊻ ☿

05 Thursday
00 06 ☽ ⋕ ♄
03 25 ☽ ∠ ☉
05 45 ☽ ⊻ ♃
09 42 ☽ ☍ ♄
12 44 ☽ ⚹ ♂
18 06 ☿ ⚼ ♅
19 18 ☽ ⊥ ☉
21 54 ♂ ∠ ♇
23 49 ☉ ∥ ♀

06 Friday
00 31 ☽ ⚻ ♅
05 51 ☽ ⊻ ☉
07 24 ☽ ⚻ ♆
10 23 ☽ ± ♅
13 49 ☽ □ ♇
14 19 ☽ ∥ ☿
14 32 ☽ ∠ ♂
15 17 ☽ ∥ ♃
17 15 ☽ ± ♆
20 09 ☽ ☌ ♀

07 Saturday
01 10 ☽ ⚼ ♅
03 01 ☽ ☌ ☿
04 52 ☉ △ ♆
06 03 ☽ ⊥ ♂
07 51 ☽ ☌ ♃
08 04 ☽ ⚼ ♆
10 55 ☽ ⚻ ♄
14 28 ☽ ∥ ♀
16 25 ☽ ⊻ ♂
18 21 ☽ ∥ ☉
19 34 ☽ Q ♇
20 50 ☽ ± ♄
22 27 ♃ ⚼ ♆

08 Sunday
01 58 ☽ △ ♅
06 28 ☽ ⋕ ♇
08 57 ☽ △ ♆
11 01 ☽ ☌ ☉
11 47 ☽ ⚼ ♄
13 07 ☽ ⋕ ♂
15 37 ☽ ⚹ ♇
21 16 ☽ ⊻ ♀
21 56 ☉ ⚼ ♄
23 37 ☽ ∥ ♂

09 Monday
05 58 ☽ ∥ ♇
07 19 ☽ ⊥ ♀
09 02 ☽ ⊻ ☿
10 50 ☽ ⊻ ♃
13 06 ☽ △ ♄
15 30 ☽ ⋕ ☉
17 08 ☽ ∠ ♇
17 37 ♂ ⊥ ♃
20 17 ☽ ⊥ ☿
21 18 ☽ ⊥ ♃
21 25 ☽ ☌ ♂
22 35 ☽ ∠ ♀
23 23 ☿ ⚼ ♆

10 Tuesday
00 20 ☽ ⋕ ♀
02 08 ☿ ∥ ♃
04 53 ☽ □ ♅
08 45 ☽ ⊥ ♇
09 56 ☿ ☌ ♃
12 12 ☽ □ ♆
13 14 ☽ ∠ ♃
13 31 ☽ ∠ ☿
18 21 ☽ ⊻ ☉
18 36 ♀ ∠ ♂
19 20 ☽ ⊻ ♇
20 59 ☿ ⊥ ♂
21 08 ☽ ⋕ ☿
21 49 ☽ ⋕ ♃

11 Wednesday
00 42 ☽ ⚹ ♀
05 08 ☿ ⚻ ♄
05 53 ☽ ⊥ ☉
08 19 ☉ ⚹ ♇
12 49 ☽ Q ♅
16 29 ☽ ⚹ ♃
17 52 ☽ □ ♄
19 23 ☽ ⚹ ☿
20 26 ☽ Q ♆
22 16 ☽ Q ♀
23 31 ☽ ∠ ☉

12 Thursday
05 43 ☽ ⊻ ♂
10 51 ☽ ⚹ ♅
15 00 ☽ Q ♃
17 30 ☽ ⊥ ♂
18 40 ☽ ⚹ ♆
19 57 ☽ ∥ ♄
20 27 ☽ Q ☿

13 Friday
02 27 ☽ ☌ ♇
03 17 ☽ Q ♄
05 53 ☽ ⚹ ☉
06 42 ☽ ∥ ♆
07 53 ☽ □ ♀
08 56 ♀ St D
11 31 ☽ ∠ ♂
15 16 ☽ ∠ ♅
21 30 ☽ ∥ ♅
23 21 ☽ ∠ ♆

14 Saturday
02 00 ☽ □ ♃
02 19 ☽ ⚹ ♄
07 04 ☽ Q ☉
08 08 ☉ ⊻ ♀
08 45 ☽ ⊥ ♅
12 04 ☽ □ ☿
16 59 ☽ ⊥ ♆
18 17 ♃ ⚻ ♄
18 24 ☽ ⚹ ♂
20 35 ☽ ⊻ ♅
22 00 ☿ Q ♇

15 Sunday
02 41 ☿ ± ♄
04 53 ☽ ⊻ ♆
07 50 ♂ ∥ ♇
07 52 ☽ ∠ ♄
11 48 ☉ ⋕ ♂
13 17 ☽ ⊻ ♇
14 18 ☉ ⋕ ♇
19 07 ☽ △ ♀
19 44 ☽ Q ♂
22 01 ☽ □ ☉

16 Monday
01 23 ☽ ⊥ ♇
01 55 ☽ ⊥ ♄
10 52 ♂ □ ♅
14 01 ☽ ⊻ ♄
14 56 ☽ △ ♃
19 40 ☽ ∠ ♇

17 Tuesday
01 48 ☽ ⚼ ♀
01 54 ☿ △ ♅
09 03 ☽ ☌ ♅
10 17 ☽ △ ☿
10 21 ☽ □ ♂
11 00 ☿ ⊻ ♂
11 46 ☽ ∥ ♅
17 32 ☽ ☌ ♆
20 21 ☽ ± ♀
21 59 ☽ ⚼ ♃

18 Wednesday
02 14 ☽ ⚹ ♇
03 18 ☽ ∥ ♆
08 02 ☿ ∥ ♀
08 43 ☽ ⚻ ♀
14 28 ☽ ∥ ♄
16 23 ☽ △ ☉
16 35 ☽ ± ♃
22 05 ☽ ⚼ ☿

19 Thursday
02 35 ☽ Q ♇
02 47 ☽ ☌ ♄
04 55 ☽ ⚻ ♃
08 35 ♅ St D
10 22 ☿ △ ♆
19 23 ☽ ± ☿
21 42 ☽ ⊻ ♅

20 Friday
01 12 ☽ ⚼ ☉
02 28 ☽ △ ♂
04 37 ☿ ⚼ ♄
05 59 ☽ ⊻ ♆
06 44 ♂ ⊥ ♇
09 26 ☽ ⚻ ☿
09 37 ☽ ⊥ ♅
14 35 ☽ □ ♇
17 49 ☽ ⊥ ♆
20 17 ♄ Q ♇
20 33 ☽ ± ☉
20 44 ☽ ⋕ ♃
21 53 ☽ ☍ ♀

21 Saturday
03 18 ☽ ∠ ♅
09 16 ☽ ⚻ ☉
09 37 ☽ ⚼ ♂
09 56 ☽ ⋕ ♀
11 23 ☽ ∠ ♆
14 00 ☽ ⊻ ♄
17 10 ☽ ☍ ♃
19 48 ☉ ⊥ ♀
20 10 ☿ ⚹ ♇
22 41 ☽ ⋕ ☿

22 Sunday
01 32 ☽ ⊥ ♄
03 55 ☽ ± ♂
04 45 ☽ ∥ ♂
08 13 ☽ ⚹ ♅
12 29 ☽ ∥ ♇
15 56 ☽ ⚻ ♂
16 05 ☽ ⚹ ♆
18 36 ☽ ∠ ♄
18 52 ♂ □ ♆

23 Monday
00 26 ☽ △ ♇
01 45 ☽ ⋕ ☉
03 13 ☽ ∥ ☉
05 06 ☽ ☍ ☿
06 49 ☽ Q ♅
08 21 ☿ ⋕ ♂
08 40 ☽ ⚻ ♀
14 32 ☽ Q ♆
16 30 ☽ ⋕ ♇
20 05 ☽ ± ♀
20 08 ☉ △ ♄
22 28 ☽ ⚹ ♄
22 40 ☽ ☍ ☉
23 14 ☽ ∥ ☿

24 Tuesday
02 07 ☽ ⋕ ♂
02 28 ☽ ⚻ ♃
04 16 ☽ ⚼ ♇
11 15 ☿ ⊻ ♀
13 03 ☽ ⚼ ♀
13 35 ☽ ± ♃
15 55 ☽ □ ♅
18 41 ☽ ∥ ♀
20 19 ☽ Q ♄
20 40 ☽ ± ♇
23 24 ☽ □ ♆

25 Wednesday
02 08 ☽ ☍ ♂
04 42 ☽ ∥ ♃
06 05 ☽ ⚼ ♃
07 30 ☽ ⚻ ♇
16 52 ☽ △ ♀
20 32 ☽ ⚻ ☿
22 52 ☿ ⋕ ♇

26 Thursday
04 27 ☽ □ ♄
07 11 ♆ St D
08 50 ☽ ± ☿
08 54 ☽ ⚻ ☉
09 11 ☽ △ ♃
13 49 ☉ ⊻ ♃
20 16 ☽ ± ☉
21 21 ☽ △ ♅

27 Friday
03 06 ☽ ⚼ ☿
03 45 ☉ ∠ ♇
04 35 ☽ △ ♆
07 07 ☽ ⋕ ♄
09 55 ☽ ⚻ ♂
12 34 ☽ ☍ ♇
13 13 ☽ ⚼ ☉
16 42 ☽ ⋕ ♆
20 53 ☽ ± ♂
21 55 ☉ ⋕ ☿
23 21 ☽ □ ♀
23 32 ☽ ⚼ ♅

28 Saturday
04 35 ♀ ⚼ ♅
06 16 ☽ ⋕ ♅
06 30 ☿ △ ♄
06 40 ☽ ⚼ ♆
08 49 ☽ △ ♄
09 11 ☽ △ ☿
13 15 ☽ ⚼ ♂
14 17 ☽ □ ♃
15 07 ☽ ± ♅
17 12 ☽ △ ☉
18 42 ☿ ⊥ ♀
22 13 ☽ ± ♆

29 Sunday
01 28 ☽ ⚻ ♅
07 35 ☽ Q ♀
08 33 ☽ ⚻ ♆
10 39 ☽ ⚼ ♄
16 22 ☽ △ ♂
16 30 ☽ ⚻ ♇
19 29 ♂ ∠ ♃
19 31 ♂ ⊻ ♇
21 41 ☽ Q ♃

30 Monday
02 05 ☽ ± ♄
02 48 ☽ ± ♇
03 33 ☿ ∠ ♇
04 23 ☿ ⊻ ♃
04 58 ☽ ⚹ ♀
12 20 ☽ ⚻ ♄
18 13 ☽ ⚼ ♇
18 30 ☽ ⚹ ♃
20 20 ☽ □ ☿

All ephemeris data is given at 12.00 UT and the Moon's longitude is additionally given for 24.00 UT

OCTOBER 1991

LONGITUDES

Date	Sidereal time h m s	Sun ☉ ° ′ ″	Moon ☽ ° ′ ″	Moon ☽ 24.00 ° ′ ″	Mercury ☿ ° ′	Venus ♀ ° ′	Mars ♂ ° ′	Jupiter ♃ ° ′	Saturn ♄ ° ′	Uranus ♅ ° ′	Neptune ♆ ° ′	Pluto ♇ ° ′
01	12 38 53	07 ♎ 48 47	14 ♋ 06 10	21 ♋ 10 02	06 ♎ 04	26 ♌ 37	19 ♎ 45	03 ♍ 58	00 ♒ 12	09 ♑ 54	14 ♑ 00	18 ♏ 41
02	12 42 49	08 47 48	28 ♋ 14 40	05 ♌ 19 52	07 51	27 12	20 25	04 10	00 R 12	09 55	14 00	18 43
03	12 46 46	09 46 51	12 ♌ 25 28	19 ♌ 31 09	09 38	27 48	21 05	04 22	00 12	09 55	14 00	18 45
04	12 50 42	10 45 57	26 ♌ 36 36	03 ♍ 41 23	11 23	28 26	21 45	04 33	00 12	09 56	14 00	18 47
05	12 54 39	11 45 04	10 ♍ 45 02	17 ♍ 47 01	13 08	29 04	22 25	04 45	00 D 12	09 57	14 01	18 49
06	12 58 35	12 44 14	24 ♍ 46 47	01 ♎ 43 47	14 52	29 ♌ 44	23 05	04 57	00 12	09 58	14 01	18 51
07	13 02 32	13 43 26	08 ♎ 37 27	15 ♎ 27 17	16 35	00 ♍ 25	23 45	05 08	00 12	09 59	14 01	18 53
08	13 06 29	14 42 40	22 ♎ 12 50	28 ♎ 53 46	18 17	01 07	24 25	05 20	00 12	09 59	14 02	18 55
09	13 10 25	15 41 56	05 ♏ 29 48	12 ♏ 00 46	19 59	01 51	25 05	05 31	00 13	10 00	14 02	18 57
10	13 14 22	16 41 14	18 ♏ 26 39	24 ♏ 47 32	21 39	02 35	25 46	05 43	00 13	10 02	14 03	18 59
11	13 18 18	17 40 34	01 ♐ 03 34	07 ♐ 15 03	23 19	03 20	26 26	05 54	00 14	10 03	14 03	19 02
12	13 22 15	18 39 56	13 ♐ 22 22	19 ♐ 25 57	24 59	04 06	27 06	06 05	00 14	10 04	14 04	19 04
13	13 26 11	19 39 20	25 ♐ 26 22	01 ♑ 24 10	26 37	04 53	27 47	06 16	00 15	10 05	14 04	19 06
14	13 30 08	20 38 45	07 ♑ 19 59	13 ♑ 14 29	28 15	05 41	28 27	06 27	00 16	10 06	14 05	19 08
15	13 34 04	21 38 13	19 ♑ 08 22	25 ♑ 02 19	29 ♎ 52	06 30	29 08	06 38	00 17	10 08	14 05	19 10
16	13 38 01	22 37 42	00 ♒ 57 02	06 ♒ 53 12	01 ♏ 28	07 19	29 ♎ 48	06 49	00 18	10 09	14 06	19 12
17	13 41 58	23 37 12	12 ♒ 51 31	18 ♒ 52 35	03 04	08 09	00 ♏ 29	07 00	00 19	10 10	14 07	19 15
18	13 45 54	24 36 45	24 ♒ 57 02	01 ♓ 05 23	04 39	09 00	01 09	07 11	00 21	10 12	14 08	19 17
19	13 49 51	25 36 19	07 ♓ 18 07	13 ♓ 35 37	06 13	09 52	01 50	07 22	00 22	10 13	14 08	19 19
20	13 53 47	26 35 55	19 ♓ 58 12	26 ♓ 26 02	07 47	10 45	02 31	07 32	00 23	10 15	14 09	19 21
21	13 57 44	27 35 33	02 ♈ 59 12	09 ♈ 37 40	09 20	11 38	03 11	07 43	00 25	10 16	14 10	19 24
22	14 01 40	28 35 12	16 ♈ 21 16	23 ♈ 09 42	10 53	12 32	03 52	07 53	00 27	10 18	14 11	19 26
23	14 05 37	29 ♎ 34 54	00 ♉ 02 37	06 ♉ 59 30	12 25	13 26	04 33	08 03	00 29	10 20	14 12	19 28
24	14 09 33	00 ♏ 34 38	13 ♉ 59 48	21 ♉ 02 56	13 57	14 22	05 14	08 14	00 30	10 21	14 13	19 31
25	14 13 30	01 34 24	28 ♉ 08 16	05 ♊ 15 11	15 27	15 17	05 55	08 24	00 32	10 23	14 13	19 33
26	14 17 27	02 34 12	12 ♊ 23 05	19 ♊ 31 25	16 58	16 14	06 36	08 34	00 34	10 25	14 14	19 35
27	14 21 23	03 34 02	26 ♊ 39 44	03 ♋ 47 36	18 28	17 11	07 17	08 44	00 37	10 27	14 15	19 38
28	14 25 20	04 33 54	10 ♋ 54 42	18 ♋ 00 47	19 57	18 08	07 58	08 54	00 39	10 29	14 16	19 40
29	14 29 16	05 33 49	25 ♋ 05 39	02 ♌ 09 09	21 25	19 06	08 39	09 03	00 41	10 31	14 18	19 42
30	14 33 13	06 33 45	09 ♌ 11 12	16 ♌ 11 42	22 54	20 05	09 20	09 13	00 44	10 33	14 19	19 45
31	14 37 09	07 ♏ 33 44	23 ♌ 10 35	00 ♍ 07 46	24 ♏ 21	21 ♍ 04	10 ♏ 01	09 ♍ 23	00 ♒ 46	10 ♑ 35	14 ♑ 20	19 ♏ 47

Date	Moon True ☊ ° ′	Moon Mean ☊ ° ′	Moon ☽ Latitude ° ′
01	15 ♑ 09	14 ♑ 39	00 N 06
02	15 R 09	14 36	01 S 09
03	15 07	14 32	02 19
04	15 02	14 29	03 20
05	14 55	14 26	04 09
06	14 45	14 23	04 43
07	14 33	14 20	04 59
08	14 20	14 17	04 57
09	14 07	14 13	04 40
10	13 57	14 10	04 08
11	13 48	14 07	03 25
12	13 43	14 04	02 33
13	13 40	14 01	01 35
14	13 39	13 58	00 S 33
15	13 D 39	13 54	00 N 29
16	13 R 38	13 51	01 31
17	13 37	13 48	02 28
18	13 34	13 45	03 20
19	13 27	13 42	04 04
20	13 19	13 38	04 37
21	13 08	13 35	04 57
22	12 56	13 32	05 01
23	12 44	13 29	04 48
24	12 33	13 26	04 18
25	12 24	13 23	03 32
26	12 18	13 19	02 32
27	12 15	13 16	01 22
28	12 14	13 13	00 N 07
29	12 D 15	13 10	01 S 08
30	12 R 15	13 07	02 18
31	12 ♑ 14	13 ♑ 04	03 S 20

DECLINATIONS

Date	Sun ☉ ° ′	Moon ☽ ° ′	Mercury ☿ ° ′	Venus ♀ ° ′	Mars ♂ ° ′	Jupiter ♃ ° ′	Saturn ♄ ° ′	Uranus ♅ ° ′	Neptune ♆ ° ′	Pluto ♇ ° ′
01	03 S 06	22 N 47	01 S 06	09 N 06	07 S 19	10 N 50	20 S 38	23 S 27	21 S 56	03 S 18
02	03 29	19 24	01 53	09 04	07 35	10 46	20 38	23 27	21 56	03 19
03	03 52	14 51	02 39	09 01	07 50	10 42	20 38	23 27	21 56	03 20
04	04 16	09 30	03 25	08 57	08 06	10 38	20 38	23 26	21 56	03 21
05	04 39	03 N 41	04 11	08 52	08 21	10 33	20 38	23 26	21 56	03 22
06	05 02	02 S 15	04 57	08 47	08 36	10 29	20 38	23 26	21 56	03 22
07	05 25	08 00	05 42	08 42	08 52	10 25	20 38	23 26	21 56	03 23
08	05 48	13 15	06 27	08 36	09 07	10 21	20 38	23 26	21 56	03 24
09	06 11	17 45	07 12	08 29	09 22	10 17	20 38	23 26	21 56	03 25
10	06 34	21 18	07 55	08 22	09 37	10 13	20 38	23 26	21 56	03 26
11	06 56	23 43	08 39	08 14	09 53	10 09	20 38	23 26	21 56	03 27
12	07 19	24 56	09 21	08 05	10 08	10 05	20 38	23 26	21 56	03 27
13	07 41	24 57	10 03	07 56	10 23	10 01	20 38	23 26	21 56	03 28
14	08 04	23 48	10 45	07 46	10 38	09 57	20 38	23 26	21 56	03 29
15	08 26	21 36	11 26	07 36	10 53	09 53	20 37	23 25	21 56	03 30
16	08 48	18 29	12 06	07 26	11 07	09 49	20 37	23 25	21 56	03 31
17	09 10	14 35	12 45	07 14	11 22	09 45	20 37	23 25	21 56	03 32
18	09 32	10 03	13 24	07 03	11 37	09 42	20 37	23 25	21 56	03 32
19	09 54	05 S 03	14 02	06 50	11 51	09 38	20 36	23 25	21 56	03 33
20	10 16	00 N 17	14 39	06 37	12 06	09 34	20 36	23 25	21 56	03 34
21	10 37	05 44	15 15	06 24	12 20	09 30	20 36	23 25	21 56	03 35
22	10 58	11 03	15 51	06 10	12 35	09 26	20 35	23 25	21 56	03 36
23	11 20	15 59	16 26	05 56	12 49	09 23	20 35	23 24	21 56	03 36
24	11 40	20 08	17 00	05 41	13 03	09 19	20 35	23 24	21 56	03 37
25	12 01	23 11	17 33	05 26	13 17	09 16	20 34	23 24	21 56	03 38
26	12 22	24 47	18 05	05 11	13 32	09 12	20 34	23 24	21 55	03 39
27	12 42	24 46	18 37	04 55	13 45	09 08	20 33	23 24	21 55	03 39
28	13 03	23 07	19 07	04 38	13 59	09 05	20 33	23 24	21 55	03 40
29	13 23	20 00	19 37	04 21	14 13	09 01	20 32	23 24	21 55	03 41
30	13 42	15 44	20 05	04 04	14 27	08 58	20 32	23 23	21 55	03 42
31	14 S 02	10 N 39	20 S 33	03 N 46	14 S 40	08 N 54	20 S 31	23 S 23	21 S 55	03 S 42

ZODIAC SIGN ENTRIES

Date	h	m	Planets
02	14	58	☽ ♌
04	17	45	☽ ♍
06	21	00	☽ ♎
06	21	15	♀ ♍
09	02	00	☽ ♏
11	09	58	☽ ♐
13	21	10	☽ ♑
15	14	01	☿ ♏
16	10	04	☽ ♒
16	19	05	♂ ♏
18	21	53	☽ ♓
21	06	33	☽ ♈
23	11	55	☽ ♉
23	22	05	☉ ♏
25	15	09	☽ ♊
27	17	37	☽ ♋
29	20	20	☽ ♌
31	23	47	☽ ♍

LATITUDES

Date	Mercury ☿ ° ′	Venus ♀ ° ′	Mars ♂ ° ′	Jupiter ♃ ° ′	Saturn ♄ ° ′	Uranus ♅ ° ′	Neptune ♆ ° ′	Pluto ♇ ° ′
01	01 N 26	03 S 46	00 N 26	00 N 50	00 S 32	00 S 22	00 N 46	14 N 37
04	01 10	03 16	00 25	00 51	00 32	00 22	00 46	14 36
07	00 53	02 48	00 23	00 51	00 32	00 22	00 46	14 35
10	00 34	02 21	00 21	00 52	00 33	00 22	00 46	14 34
13	00 N 14	01 55	00 20	00 52	00 33	00 22	00 46	14 33
16	00 S 07	01 30	00 18	00 53	00 33	00 22	00 46	14 32
19	00 28	01 07	00 16	00 53	00 33	00 22	00 46	14 31
22	00 48	00 44	00 15	00 54	00 33	00 22	00 46	14 31
25	01 08	00 23	00 13	00 54	00 33	00 22	00 46	14 30
28	01 27	00 S 04	00 11	00 55	00 33	00 22	00 46	14 30
31	01 S 44	00 N 15	00 N 09	00 N 55	00 S 33	00 S 22	00 N 45	14 N 29

DATA

Julian Date	2448531
Delta T	+58 seconds
Ayanamsa	23° 44' 47"
Synetic vernal point	05° ♓ 22' 13"
True obliquity of ecliptic	23° 26' 28"

LONGITUDES

Date	Chiron ⚷ ° ′	Ceres ⚳ ° ′	Pallas ⚴ ° ′	Juno ⚵ ° ′	Vesta ⚶ ° ′	Black Moon Lilith ⚸ ° ′
01	07 ♌ 29	20 ♏ 17	26 ♎ 42	17 ♑ 28	23 ♌ 30	17 ♑ 35
11	08 ♌ 12	24 ♏ 11	01 ♏ 04	19 ♑ 07	27 ♌ 37	18 ♑ 41
21	08 ♌ 47	28 ♏ 10	05 ♏ 28	21 ♑ 16	01 ♍ 38	19 ♑ 47
31	09 ♌ 13	02 ♐ 13	09 ♏ 52	23 ♑ 50	05 ♍ 31	20 ♑ 54

MOON'S PHASES, APSIDES AND POSITIONS ☽

Date	h	m	Phase	Longitude ° ′	Eclipse Indicator
01	00	30	☾	07 ♋ 21	
07	21	39	●	14 ♎ 07	
15	17	33	☽	21 ♑ 52	
23	11	08	○	29 ♈ 33	
30	07	11	☾	06 ♌ 22	

Day	h	m		
02	17	51	Perigee	
15	11	01	Apogee	
27	15	53	Perigee	
06	02	53	0S	
13	00	06	Max dec	25° S 05'
20	10	46	0N	
26	23	40	Max dec	24° N 59'

ASPECTARIAN

01 Tuesday
00 30 ☽ □ ☉
04 50 ☽ ☍ ♅
05 46 ☽ # ♅
07 36 ☽ ∠ ♀
11 49 ☽ ☍ ♆
18 59 ☽ # ♆
19 48 ☽ △ ♇
20 23 ☽ ∠ ♃
22 04 ☽ □ ♂
23 33 ☽ ⊥ ♀

02 Wednesday
00 57 ☉ ∥ ♇
04 12 ☽ # ♄
07 23 ☽ Q ☿
09 22 ☽ Q ☉
10 09 ☽ ⊻ ♀
11 52 ☽ ⊥ ♃
15 19 ☽ ☍ ♄
22 10 ☽ ⊻ ♃

03 Thursday
06 04 ☽ Q ♂
06 36 ☽ ⚹ ☿
07 12 ☽ ⚹ ☉
07 46 ☽ ⊼ ♅
14 40 ☽ ⊼ ♆
15 27 ☉ □ ♅
15 59 ☿ □ ♅
16 40 ☉ ☌ ☿
17 55 ☽ ± ♅
22 44 ☽ □ ♇
23 14 ☿ ⊥ ♃

04 Friday
00 49 ☽ ± ♆
03 22 ☽ ⚹ ♂
05 41 ☉ ⊥ ♃
07 07 ☽ ∥ ♃
09 09 ☽ ⚻ ♅
09 27 ☿ ∥ ♇
10 28 ☽ ∠ ☉
11 34 ☽ ∠ ☿
14 23 ☽ ∥ ♀
15 14 ☽ ☌ ♀
16 04 ☽ ⚻ ♆
17 41 ☽ # ♂
18 04 ☽ ⊼ ♄

05 Saturday
01 40 ☽ ☌ ♃
02 52 ☽ ⊥ ☉
03 57 ♄ St D
04 15 ☽ ± ♄
04 59 ☽ ⊥ ☿
05 18 ☽ Q ♇
06 03 ☽ ∠ ♂
07 35 ☿ ⊥ ♇
08 21 ☽ # ☉
09 42 ♀ ⚻ ♆
10 12 ☽ # ☿
10 38 ☽ △ ♅
13 20 ☽ # ♇
13 50 ☽ ⊻ ☉
16 38 ☽ ⊻ ☿
17 34 ☽ △ ♆
19 34 ☽ ⚻ ♄
22 08 ☽ ⊥ ♂

06 Sunday
00 12 ☿ □ ♆
01 48 ☽ ⚹ ♇
08 56 ☽ ⊻ ♂
09 09 ☿ ∠ ♀
14 54 ☉ ⊥ ♇
16 36 ☽ ∥ ♇
17 07 ☉ ∥ ☿
21 00 ☽ ⊻ ♀
21 21 ☽ △ ♄

07 Monday
00 16 ☽ ∥ ☉
00 44 ♀ # ♂
00 48 ☽ ∥ ☿
03 43 ☽ ∠ ♇
04 09 ♀ ⊼ ♄
05 50 ☽ ⊻ ♃
07 57 ☽ ⊥ ♀
14 22 ☽ □ ♅
15 01 ☽ # ♀
15 59 ☽ ∥ ♂
16 28 ☽ ⊥ ♃
19 19 ☉ □ ♆
19 30 ☽ ⊥ ♇
21 29 ☽ □ ♆
21 39 ☽ ☌ ☉
22 36 ☽ # ♃

08 Tuesday
00 35 ☽ ∠ ♀
04 00 ☽ ☌ ☿
06 07 ☽ ⊻ ♇
08 36 ☽ ∠ ♃
16 09 ☽ ☌ ♂
21 10 ☿ ⊻ ♇
22 23 ☽ Q ♅

09 Wednesday
02 22 ☽ □ ♄
04 57 ☽ ⚹ ♀
05 41 ☽ Q ♆
12 03 ☽ ⚹ ♃
20 18 ☽ ⚹ ♅
20 45 ☿ ∠ ♃

10 Thursday
03 46 ☽ ⚹ ♆
04 20 ☽ Q ♀
06 57 ☽ ∥ ♄
08 26 ☽ ⊻ ☉
10 36 ☽ Q ♃
11 34 ☽ Q ♄
13 02 ☽ ☌ ♇
17 22 ☽ ∥ ♆
18 58 ☽ ⊻ ☿
20 41 ☽ ⊥ ☉

11 Friday
00 19 ☿ # ♀
00 28 ☽ ∠ ♅
02 37 ☽ ⊻ ♂
08 08 ☽ ∠ ♆
08 09 ☽ ⊥ ☿
08 27 ☽ ∥ ♅
10 24 ☽ ⚹ ♄
14 48 ☽ ⊥ ♂
15 23 ☽ ∠ ☉
16 40 ☽ □ ♀
17 46 ☽ ⊥ ♅
21 31 ☽ □ ♃

12 Saturday
01 34 ☽ ⊥ ♆
04 17 ☽ ∠ ☿
05 29 ☽ ⊻ ♅
08 39 ♂ # ♃
09 21 ☽ ∠ ♂
13 22 ☽ ⊻ ♆
15 41 ☽ ∠ ♄
21 59 ☉ ⊻ ♇
23 18 ☽ ⊻ ♇
23 25 ☽ ⚹ ☉

13 Sunday
09 05 ♃ ± ♄
09 37 ☽ ⊥ ♄
10 48 ☿ # ♃
11 19 ☽ ⊥ ♇
14 44 ☽ ⚹ ☿
16 59 ☽ ⚹ ♂
21 41 ☽ ⊻ ♄
23 03 ☉ # ♀
23 14 ♂ Q ♅

14 Monday
01 39 ☽ Q ☉
05 28 ☿ ∥ ♂
05 30 ☽ ∠ ♇
08 24 ☽ △ ♀
09 51 ☿ Q ♅
10 12 ☽ △ ♃
16 30 ☉ ∠ ♀
16 54 ☽ ∥ ♅
17 10 ☿ ☌ ♂
17 38 ☽ ☌ ♅
18 43 ☽ Q ♂
18 51 ☽ Q ☿

15 Tuesday
01 43 ☽ ☌ ♆
05 44 ♀ ± ♄
08 53 ☽ ∥ ♆
12 04 ☽ ⚹ ♇
12 06 ☉ ∠ ♃
17 09 ☽ ⚻ ♀
17 10 ☽ ⚻ ♃
17 33 ☽ □ ☉
17 34 ♀ ☌ ♃
18 19 ☿ □ ♄
20 13 ☽ ∥ ♄

16 Wednesday
08 41 ♀ Q ♇
09 32 ☽ □ ♂
10 41 ☽ ☌ ♄
11 44 ☽ ± ♃
12 31 ☽ Q ♇
12 48 ☽ ± ♀
13 13 ☽ □ ☿
21 32 ☿ Q ♆

17 Thursday
00 03 ☽ ⊼ ♃
01 50 ☽ ⊼ ♀
06 19 ♂ □ ♄
06 36 ☽ ⊻ ♅
14 31 ☽ ⊻ ♆
18 38 ☽ ⊥ ♅
20 47 ☽ ∥ ☿

18 Friday
00 46 ☽ □ ♇
02 28 ☽ ⊥ ♆
04 28 ☽ ∥ ♂
11 17 ☽ △ ☉
12 29 ☽ ∠ ♅
13 50 ☽ # ♃
14 24 ☽ ∥ ☉
20 11 ☽ ∠ ♆
20 47 ☉ # ♃
22 34 ☽ ⊻ ♄

19 Saturday
00 50 ☽ △ ♂
03 17 ☽ # ♀
05 07 ♃ Q ♇
09 38 ☽ △ ☿
10 12 ☽ ⊥ ♄
12 07 ☽ ☍ ♃
17 17 ☽ ☍ ♀
17 36 ☽ ⚹ ♅
18 50 ☽ ∥ ♇
18 52 ☽ ⚻ ☉
21 53 ♀ △ ♅
23 05 ♂ Q ♆

20 Sunday
01 03 ☽ ⚹ ♆
03 23 ☽ ∠ ♄
07 08 ☽ ⚻ ♂
07 38 ☿ ⚹ ♃
10 51 ☽ △ ♇
13 16 ☽ ± ☉
16 15 ☽ Q ♅
17 58 ☽ ⚻ ☿
23 29 ☽ Q ♆

21 Monday
00 48 ☽ ± ♂
01 19 ☽ ⊼ ☉
02 31 ☽ # ♇
07 18 ☽ ⚹ ♄
12 23 ☽ ⊼ ♂
12 44 ☽ ± ☿
14 34 ☽ ⚻ ♇
14 52 ☽ ∥ ♀
20 40 ☽ ⊼ ♃

22 Tuesday
00 59 ☽ ⊼ ☿
01 11 ☽ □ ♅
02 43 ☿ ⚹ ♅
04 41 ☽ ∥ ♃
04 43 ☽ ⊼ ♀
04 52 ☉ Q ♅
05 02 ☽ Q ♄
06 48 ☽ ± ♇
07 33 ☽ ± ♃
08 08 ☽ □ ♆
11 35 ☽ # ☉
16 08 ☽ ± ♀
17 28 ☽ ⊼ ♇
19 30 ☽ # ♂
23 40 ☽ ⚻ ♃

23 Wednesday
09 01 ☽ ⚻ ♀
11 08 ☽ ☍ ☉
12 45 ☽ □ ♄
14 44 ☽ # ☿
20 12 ☽ ☍ ♂

24 Thursday
02 00 ☽ △ ♃
05 46 ☽ △ ♅
08 00 ♀ △ ♆
10 15 ☉ □ ♄
11 54 ☽ ☍ ☿
12 22 ☽ △ ♆
12 40 ☽ △ ♀
14 56 ☽ # ♄
16 14 ☿ ⚹ ♆
21 25 ☽ ☍ ♇

25 Friday
00 49 ☽ # ♆
05 00 ☿ ⚹ ♀
07 20 ☽ ⚻ ♅
13 50 ☽ ⚻ ♆
14 18 ☽ # ♅
16 04 ☽ △ ♄
18 14 ☽ ⊼ ☉
18 41 ♀ ⚻ ♄
22 34 ☽ ± ♅

26 Saturday
01 46 ☽ ⊼ ♂
03 56 ☉ Q ♆
05 01 ☽ ± ♆
05 06 ☽ ± ☉
05 30 ☽ □ ♃
08 41 ☽ ⊼ ♅
12 22 ☽ ± ♂
15 07 ☽ ⊼ ♆
17 23 ☽ ⚻ ♄
18 55 ☽ □ ♀
20 36 ☽ ⊼ ☿
21 22 ☽ ⚻ ☉

27 Sunday
00 08 ☽ ⊼ ♇
04 15 ☽ ⚻ ♂
07 52 ☽ ± ☿
08 33 ☽ ± ♄
10 15 ☽ ± ♇
12 07 ☽ Q ♃
14 30 ☿ Q ♄
18 40 ☽ ⊼ ♄

28 Monday
00 30 ☽ △ ☉
00 46 ☽ ⚻ ☿
01 26 ☽ ⚻ ♇
03 22 ☽ Q ♀
06 46 ☽ △ ♂
07 20 ☿ ☌ ♇
08 33 ☽ ⚹ ♃
09 04 ☽ # ♅
11 16 ☽ ☍ ♅
17 41 ☽ ☍ ♆
22 26 ☽ # ♆

29 Tuesday
01 06 ☽ ⚹ ♀
02 50 ☽ △ ♇
05 03 ☽ △ ☿
05 16 ☿ Q ♃
08 27 ☽ # ♄
10 13 ☽ ∠ ♃
14 12 ☽ # ☿
21 32 ☽ ☍ ♄

30 Wednesday
01 42 ☽ ⊥ ♃
03 32 ♀ ⚹ ♇
04 27 ☽ ∠ ♀
06 40 ♂ ⚹ ♃
07 11 ☽ □ ☉
12 03 ☽ ⊻ ♃
12 16 ☽ □ ♂
14 20 ☽ ⊼ ♅
18 06 ☽ # ♂
20 47 ☽ ⊼ ♆
21 00 ☽ ⊥ ♀
21 19 ☽ # ☉

31 Thursday
00 38 ☽ ± ♅
06 09 ☽ □ ♇
07 06 ☽ ± ♆
08 05 ☽ ⊻ ♀
11 00 ☿ ∥ ♄
14 15 ☽ □ ☿
16 09 ☽ ⚻ ♅
16 26 ☽ Q ☉
16 51 ♀ # ♇
19 42 ☽ ∥ ♃
20 47 ☽ Q ♂
22 38 ☽ ⚻ ♆

All ephemeris data is given at 12.00 UT and the Moon's longitude is additionally given for 24.00 UT

NOVEMBER 1991

LONGITUDES

Date	Sidereal time h m s	Sun ☉ ° ' "	Moon ☽ ° ' "	Moon ☽ 24.00 ° ' "	Mercury ☿ ° '	Venus ♀ ° '	Mars ♂ ° '	Jupiter ♃ ° '	Saturn ♄ ° '	Uranus ♅ ° '	Neptune ♆ ° '	Pluto ♇ ° '
01	14 41 06	08 ♏ 33 45	07 ♍ 03 09	13 ♍ 56 36	25 ♏ 48	22 ♍ 03	10 ♏ 42	09 ♍ 32	00 ♒ 49	10 ♑ 37	14 ♑ 21	19 ♏ 49
02	14 45 02	09 33 48	20 ♍ 47 59	27 ♍ 37 06	27 14	23 03	11 24	09 41	00 52	10 39	14 22	19 52
03	14 48 59	10 33 54	04 ♎ 23 44	11 ♎ 07 39	28 ♏ 40	24 03	12 05	09 51	00 55	10 41	14 23	19 54
04	14 52 56	11 34 01	17 ♎ 48 37	24 ♎ 26 24	00 ♐ 05	25 04	12 47	10 00	00 57	10 43	14 25	19 57
05	14 56 52	12 34 10	01 ♏ 00 46	07 ♏ 31 31	01 29	26 05	13 28	10 09	01 00	10 46	14 26	19 59
06	15 00 49	13 34 21	13 ♏ 58 31	20 ♏ 21 38	02 52	27 07	14 10	10 18	01 04	10 48	14 27	20 01
07	15 04 45	14 34 34	26 ♏ 40 52	02 ♐ 56 13	04 15	28 09	14 51	10 26	01 07	10 50	14 28	20 04
08	15 08 42	15 34 49	09 ♐ 07 48	15 ♐ 15 48	05 36	29 ♍ 11	15 33	10 35	01 10	10 52	14 30	20 06
09	15 12 38	16 35 05	21 ♐ 20 26	27 ♐ 22 03	06 57	00 ♎ 14	16 14	10 43	01 13	10 55	14 31	20 09
10	15 16 35	17 35 24	03 ♑ 21 02	09 ♑ 17 50	08 16	01 17	16 56	10 52	01 17	10 57	14 33	20 11
11	15 20 31	18 35 43	15 ♑ 12 55	21 ♑ 06 53	09 35	02 21	17 38	11 00	01 20	11 00	14 34	20 13
12	15 24 28	19 36 04	27 ♑ 00 17	02 ♒ 53 47	10 51	03 24	18 20	11 08	01 24	11 02	14 36	20 16
13	15 28 25	20 36 27	08 ♒ 48 00	14 ♒ 43 37	12 07	04 29	19 01	11 16	01 28	11 05	14 37	20 18
14	15 32 21	21 36 50	20 ♒ 41 18	26 ♒ 41 44	13 20	05 33	19 43	11 24	01 32	11 08	14 39	20 21
15	15 36 18	22 37 15	02 ♓ 45 34	08 ♓ 53 26	14 32	06 38	20 25	11 32	01 35	11 10	14 40	20 23
16	15 40 14	23 37 42	15 ♓ 05 56	21 ♓ 23 35	15 41	07 43	21 07	11 40	01 39	11 13	14 42	20 26
17	15 44 11	24 38 10	27 ♓ 46 50	04 ♈ 16 04	16 48	08 48	21 49	11 47	01 44	11 16	14 43	20 28
18	15 48 07	25 38 39	10 ♈ 51 32	17 ♈ 33 20	17 52	09 53	22 31	11 54	01 48	11 18	14 45	20 30
19	15 52 04	26 39 09	24 ♈ 21 30	01 ♉ 15 50	18 53	10 59	23 13	12 02	01 52	11 21	14 47	20 33
20	15 56 00	27 39 41	08 ♉ 16 01	15 ♉ 21 37	19 51	12 05	23 56	12 09	01 56	11 24	14 48	20 35
21	15 59 57	28 40 14	22 ♉ 31 59	29 ♉ 46 25	20 44	13 12	24 38	12 16	02 01	11 27	14 50	20 38
22	16 03 54	29 ♏ 40 49	07 ♊ 04 06	14 ♊ 24 07	21 33	14 18	25 20	12 22	02 05	11 30	14 52	20 40
23	16 07 50	00 ♐ 41 26	21 ♊ 45 36	29 ♊ 07 37	22 16	15 25	26 02	12 29	02 10	11 32	14 53	20 42
24	16 11 47	01 42 04	06 ♋ 29 18	13 ♋ 49 53	22 54	16 32	26 45	12 36	02 14	11 35	14 55	20 45
25	16 15 43	02 42 43	21 ♋ 08 39	28 ♋ 24 59	23 25	17 40	27 27	12 42	02 19	11 38	14 57	20 47
26	16 19 40	03 43 24	05 ♌ 38 25	12 ♌ 48 35	23 49	18 47	28 10	12 48	02 24	11 41	14 59	20 50
27	16 23 36	04 44 07	19 ♌ 55 10	26 ♌ 58 01	24 05	19 55	28 52	12 54	02 28	11 44	15 01	20 52
28	16 27 33	05 44 51	03 ♍ 57 00	10 ♍ 52 07	24 11	21 03	29 ♏ 35	13 00	02 33	11 47	15 02	20 54
29	16 31 29	06 45 36	17 ♍ 43 21	24 ♍ 30 45	24 R 08	22 11	00 ♐ 17	13 06	02 38	11 50	15 04	20 57
30	16 35 26	07 ♐ 46 24	01 ♎ 14 24	07 ♎ 54 23	23 ♐ 55	23 ♎ 19	01 ♐ 00	13 ♍ 11	02 ♒ 43	11 ♑ 54	15 ♑ 06	20 ♏ 59

Date	Moon True ☊ ° '	Moon Mean ☊ ° '	Moon ☽ Latitude ° '
01	12 ♑ 11	13 ♑ 00	04 S 09
02	12 R 05	12 57	04 44
03	11 56	12 54	05 01
04	11 46	12 51	05 02
05	11 35	12 48	04 47
06	11 25	12 44	04 17
07	11 16	12 41	03 35
08	11 09	12 38	02 43
09	11 04	12 35	01 44
10	11 02	12 32	00 S 41
11	11 D 02	12 29	00 N 23
12	11 02	12 25	01 25
13	11 04	12 22	02 24
14	11 04	12 19	03 18
15	11 R 04	12 16	04 03
16	11 01	12 13	04 38
17	10 57	12 10	05 01
18	10 51	12 06	05 09
19	10 43	12 03	05 00
20	10 36	12 00	04 34
21	10 29	11 57	03 50
22	10 24	11 54	02 51
23	10 20	11 50	01 39
24	10 19	11 47	00 N 21
25	10 D 19	11 44	00 S 59
26	10 21	11 41	02 14
27	10 22	11 38	03 19
28	10 23	11 35	04 12
29	10 R 22	11 31	04 49
30	10 ♑ 20	11 ♑ 28	05 S 09

DECLINATIONS

Date	Sun ☉	Moon ☽	Mercury ☿	Venus ♀	Mars ♂	Jupiter ♃	Saturn ♄	Uranus ♅	Neptune ♆	Pluto ♇
01	14 S 21	05 N 04	20 S 59	03 N 28	14 S 54	08 N 51	20 S 31	23 S 23	21 S 55	03 S 43
02	14 41	00 S 42	21 25	03 10	15 07	08 48	20 30	23 23	21 55	03 44
03	15 00	06 21	21 49	02 51	15 21	08 44	20 30	23 23	21 55	03 45
04	15 18	11 39	22 12	02 32	15 34	08 41	20 29	23 22	21 55	03 45
05	15 37	16 18	22 35	02 12	15 47	08 38	20 28	23 22	21 55	03 46
06	15 55	20 07	22 56	01 53	16 00	08 35	20 28	23 22	21 54	03 47
07	16 13	22 54	23 16	01 33	16 13	08 32	20 27	23 22	21 54	03 47
08	16 30	24 30	23 34	01 12	16 25	08 29	20 26	23 22	21 54	03 48
09	16 48	24 53	23 52	00 52	16 38	08 26	20 26	23 21	21 54	03 49
10	17 05	24 05	24 08	00 31	16 50	08 23	20 25	23 21	21 54	03 50
11	17 22	22 12	24 23	00 N 09	17 03	08 20	20 24	23 21	21 54	03 50
12	17 38	19 22	24 37	00 S 12	17 15	08 17	20 23	23 21	21 54	03 51
13	17 54	15 44	24 49	00 34	17 27	08 14	20 23	23 21	21 54	03 51
14	18 10	11 28	25 00	00 55	17 39	08 11	20 22	23 20	21 53	03 52
15	18 26	06 43	25 09	01 17	17 51	08 08	20 21	23 20	21 53	03 52
16	18 41	01 S 36	25 17	01 40	18 02	08 06	20 20	23 20	21 53	03 53
17	18 56	03 N 43	25 24	02 02	18 14	08 03	20 20	23 20	21 53	03 54
18	19 10	09 02	25 29	02 25	18 25	08 00	20 18	23 19	21 53	03 55
19	19 25	14 06	25 33	02 47	18 37	07 58	20 17	23 19	21 53	03 55
20	19 38	18 35	25 35	03 10	18 48	07 55	20 16	23 19	21 53	03 56
21	19 52	22 07	25 35	03 33	18 59	07 53	20 16	23 19	21 52	03 56
22	20 05	24 18	25 34	03 56	19 09	07 50	20 15	23 18	21 52	03 57
23	20 18	24 50	25 31	04 20	19 20	07 48	20 14	23 18	21 52	03 58
24	20 30	23 38	25 27	04 43	19 31	07 46	20 13	23 18	21 52	03 58
25	20 42	20 49	25 21	05 06	19 41	07 44	20 12	23 18	21 52	03 59
26	20 54	16 42	25 13	05 30	19 51	07 42	20 10	23 17	21 52	03 59
27	21 05	11 41	25 03	05 53	20 01	07 39	20 09	23 17	21 51	04 00
28	21 16	06 09	24 52	06 17	20 11	07 37	20 08	23 17	21 51	04 00
29	21 26	00 24	24 38	06 40	20 20	07 35	20 07	23 17	21 51	04 01
30	21 S 36	05 S 13	24 S 23	07 S 04	20 S 30	07 N 34	20 S 06	23 S 16	21 S 51	04 S 01

ZODIAC SIGN ENTRIES

Date	h	m	Planets
03	04	13	☽ ♎
04	10	41	☿ ♐
05	10	09	☽ ♏
07	18	21	☽ ♐
08	06	37	♀ ♎
10	05	16	☽ ♑
12	18	06	☽ ♒
15	06	33	☽ ♓
17	16	08	☽ ♈
19	21	49	☽ ♉
22	00	22	☽ ♊
22	19	36	☉ ♐
24	01	25	☽ ♋
26	02	37	☽ ♌
28	05	12	☽ ♍
29	02	19	♂ ♐
30	09	47	☽ ♎

LATITUDES

Date	Mercury ☿	Venus ♀	Mars ♂	Jupiter ♃	Saturn ♄	Uranus ♅	Neptune ♆	Pluto ♇
01	01 S 50	00 N 21	00 N 09	00 N 55	00 S 33	00 S 22	00 N 45	14 N 29
04	02 05	00 37	00 07	00 56	00 33	00 22	00 45	14 28
07	02 18	00 53	00 05	00 57	00 33	00 22	00 45	14 28
10	02 29	01 07	00 04	00 57	00 33	00 22	00 45	14 28
13	02 35	01 20	00 N 02	00 58	00 33	00 22	00 45	14 28
16	02 38	01 31	00 00	00 59	00 33	00 22	00 45	14 28
19	02 35	01 42	00 S 02	00 59	00 33	00 22	00 45	14 27
22	02 24	01 51	00 04	01 00	00 34	00 22	00 45	14 28
25	02 04	01 59	00 05	01 01	00 34	00 22	00 45	14 28
28	01 33	02 05	00 07	01 01	00 34	00 22	00 45	14 28
31	00 S 49	02 N 11	00 S 09	01 N 02	00 S 34	00 S 22	00 N 45	14 N 28

DATA

Julian Date	2448562
Delta T	+58 seconds
Ayanamsa	23° 44' 50"
Synetic vernal point	05° ♓ 22' 09"
True obliquity of ecliptic	23° 26' 28"

MOON'S PHASES, APSIDES AND POSITIONS ☽

Date	h	m	Phase	Longitude ° '	Eclipse Indicator
06	11	11	●	13 ♏ 32	
14	14	02	☽	21 ♒ 42	
21	22	56	○	29 ♉ 08	
28	15	21	☾	05 ♍ 53	

Day	h	m		
12	07	34	Apogee	
24	02	33	Perigee	
02	09	05	0S	
09	07	38	Max dec	24° S 54'
16	19	15	0N	
23	07	25	Max dec	24° N 52'
29	13	46	0S	

LONGITUDES

Date	Chiron ⚷ ° '	Ceres ⚳ ° '	Pallas ⚴ ° '	Juno ⚵ ° '	Vesta ⚶ ° '	Black Moon Lilith ⚸ ° '
01	09 ♌ 15	02 ♐ 37	10 ♍ 19	24 ♑ 07	05 ♍ 54	21 ♑ 01
11	09 ♌ 30	06 ♐ 43	14 ♍ 43	27 ♑ 05	09 ♍ 35	22 ♑ 07
21	09 ♌ 36	10 ♐ 50	19 ♍ 07	00 ♒ 23	13 ♍ 03	23 ♑ 13
31	09 ♌ 32	14 ♐ 58	23 ♍ 29	03 ♒ 57	16 ♍ 15	24 ♑ 20

All ephemeris data is given at 12.00 UT and the Moon's longitude is additionally given for 24.00 UT

ASPECTARIAN

h m	Aspects
01 Friday	
01 09	☽ ⚻ ♄
08 30	♂ ✶ ♅
08 50	☿ ∠ ♅
11 35	☽ ± ♄
13 21	☽ Q ♇
14 50	☽ ✶ ☉
16 22	☽ ☌ ♃
17 36	☽ # ♇
18 13	☽ △ ♅
18 42	☽ ✶ ♂
19 02	☽ ∥ ♀
02 Saturday	
00 44	☽ △ ♆
01 07	☽ Q ☿
03 19	☽ ⚼ ♄
10 21	☽ ✶ ♇
15 32	☉ ✶ ♃
16 16	☽ ☌ ♀
19 08	☽ ∠ ☉
21 47	☽ # ♀
22 22	☽ ∠ ♂
03 Sunday	
00 39	☽ ✶ ☿
00 46	☽ ∥ ♇
05 48	☽ △ ♄
12 20	☽ ⊥ ☉
12 54	☽ ∠ ♇
14 58	☉ ✶ ♅
15 10	☽ ⊥ ♂
17 48	☿ ∥ ♆
21 49	☽ ⚺ ♃
22 26	☽ # ♃
23 14	☽ □ ♅
23 53	☽ ⚺ ☉
04 Monday	
00 28	☿ ∠ ♆
02 28	☽ ⚺ ♂
05 01	☽ ⊥ ♇
05 52	☽ □ ♆
06 30	☽ ∠ ☿
08 41	☽ ⊥ ♃
15 52	☽ ⚺ ♇
05 Tuesday	
00 42	☽ ⊥ ☿
01 09	☽ ∠ ♃
02 15	☽ ⚺ ♀
03 35	☿ ✶ ♄
07 52	☽ Q ♅
07 52	☽ ∥ ☉
08 56	☽ ∥ ♂
11 59	☽ □ ♄
12 58	☽ ⚺ ☿
14 09	☽ ⊥ ♀
14 36	☽ Q ♆
06 Wednesday	
05 03	☽ ✶ ♃
06 03	☽ ✶ ♅
08 14	☽ ∠ ♀
11 11	☽ ☌ ☉
12 22	☽ ☌ ♂
12 54	☽ ✶ ♆
14 31	☽ ∥ ♄
21 35	☽ Q ♄
22 30	♂ ✶ ♆
23 24	☽ ☌ ♇
07 Thursday	
02 18	☽ ∥ ♆
03 50	☽ Q ♃
09 31	☉ ✶ ♆
10 23	☽ ∠ ♅
11 05	☉ ∥ ♂
15 04	☽ ✶ ♀
16 59	☽ ∥ ☿
17 21	☽ ∠ ♆
17 28	☽ ∥ ♅
19 54	☿ ∥ ♅
20 31	☽ ✶ ♄
22 42	☿ ⊥ ♅
08 Friday	
03 43	☽ ⊥ ♅
04 19	☽ ☌ ☿
09 16	☉ ☌ ♂
10 46	☽ ⊥ ♆
14 52	☽ □ ♃
15 25	☽ ⚺ ♅
16 23	☽ Q ♀
22 31	☽ ⚺ ♆
09 Saturday	
01 19	☽ ⚺ ♂
01 45	☽ ⚺ ☉
01 51	☽ ∠ ♄
09 37	☽ ⚺ ♇
13 53	☽ ⊥ ♂
14 42	☽ ⊥ ☉
19 45	☽ ⊥ ♄
21 35	☽ ⊥ ♇
10 Sunday	
07 27	☽ □ ♀
07 49	☽ ⚺ ♄
08 59	☽ ∠ ♂
10 20	☽ ∠ ☉
11 18	☽ ∥ ☿
11 52	♀ △ ♄
15 42	☽ ∠ ♇
17 04	☿ ⊥ ♆
22 56	☽ ∥ ♅
23 09	☽ ⚺ ☿
11 Monday	
03 21	☽ △ ♃
03 25	☽ ☌ ♅
10 41	☽ ☌ ♆
10 51	♃ △ ♅
12 49	☽ ⊥ ☿
14 57	☽ ∥ ♆
17 13	☽ ✶ ♂
19 31	☽ ✶ ☉
22 13	☽ ✶ ♇
12 Tuesday	
04 05	☽ ∥ ♄
06 44	♀ ∠ ♂
06 55	☉ Q ♄
09 22	☽ ∠ ☿
10 13	☽ ⚼ ♃
15 37	☿ ⚺ ♅
18 00	☿ □ ♃
19 12	☽ Q ♂
21 00	☽ ☌ ♄
22 14	☽ Q ☉
22 45	☽ Q ♇
23 12	☽ ∥ ☉
13 Wednesday	
01 49	☽ ∥ ♂
02 20	☽ △ ♀
04 30	☉ ☌ ♇
04 45	☽ ± ♃
16 39	☽ ⚺ ♅
17 04	☽ ⚻ ♃
19 29	☽ ✶ ☿
23 48	☽ ⚺ ♆
14 Thursday	
04 38	♂ Q ♄
04 49	☽ ⊥ ♅
07 18	♀ ∠ ♇
09 56	☽ □ ♂
11 19	☽ □ ♇
11 41	☽ ⚼ ♀
11 55	☽ ⊥ ♆
14 02	☽ □ ☉
22 19	☽ Q ☿
22 54	☽ ∠ ♅
15 Friday	
04 57	☽ # ♃
05 53	☽ ∠ ♆
07 23	☽ ± ♀
09 41	☽ ⚺ ♄
10 45	♂ ☌ ♇
14 55	☿ ⚺ ♆
20 19	☽ ⚻ ♀
21 31	☽ ⊥ ♄
16 Saturday	
01 24	☽ ∥ ♇
04 29	☽ ✶ ♅
05 18	☽ ☍ ♃
11 13	☽ ✶ ♆
11 43	☽ ∥ ♀
12 51	☉ Q ♃
13 14	☽ □ ☿
15 00	☽ ∠ ♄
22 12	☽ △ ♇
17 Sunday	
00 09	☽ △ ♂
03 30	☽ Q ♅
03 51	☽ # ♀
05 37	☽ △ ☉
10 01	☽ Q ♆
10 12	☿ ∠ ♄
12 49	☽ # ♇
19 21	☽ ✶ ♄
18 Monday	
02 14	☽ ⚼ ♇
05 36	☽ ⚼ ♂
07 21	☽ ∥ ♃
10 05	☽ ☍ ♀
11 35	☽ ⚼ ☉
12 48	☽ □ ♅
13 54	☽ ⚻ ♃
17 19	☽ Q ♄
18 35	☽ ± ♇
19 00	☽ □ ♆
22 43	☽ ± ♂
19 Tuesday	
00 44	☽ ± ♃
01 36	☽ △ ☿
04 28	☉ ∠ ♅
04 58	☽ ± ☉
05 17	☽ ⚻ ♇
09 54	☽ ⚻ ♂
16 19	☽ ⚻ ☉
16 42	☽ ⚼ ♃
20 17	♀ □ ♅
20 Wednesday	
01 06	☽ □ ♄
05 44	☽ ⚼ ☿
13 19	☽ # ♂
13 22	♀ ⚺ ♃
17 20	☽ △ ♅
18 38	☽ △ ♃
18 51	☽ # ☉
19 02	☽ ⚻ ♀
20 58	♂ Q ♃
22 05	☽ ± ☿
22 30	☽ # ♄
23 05	☽ △ ♆
21 Thursday	
05 58	☽ ± ♀
08 49	☽ ⚻ ☿
08 49	☽ ☍ ♇
08 54	☿ ⚺ ♇
10 04	☽ # ♆
15 40	☽ ☍ ♂
18 31	☽ ⚼ ♅
22 10	☽ ⚼ ♀
22 56	☽ ☍ ☉
23 05	☽ # ♅
22 Friday	
00 07	☽ ⚼ ♆
03 46	☽ △ ♄
09 24	☽ ± ♅
12 40	♀ ∥ ♇
14 57	☽ ± ♆
16 24	☉ ∠ ♆
19 16	☽ ⚻ ♅
20 07	♀ ⊥ ♇
20 45	☽ □ ♃
23 Saturday	
00 18	♀ □ ♆
00 46	☽ ⚻ ♆
00 49	☽ △ ♀
04 28	☽ ⚼ ♄
04 33	☉ ∥ ♄
10 17	☽ ⚻ ♇
12 52	☽ ☍ ☿
19 12	☽ ± ♄
19 19	☽ ⚻ ♂
20 05	☽ ± ♇
24 Sunday	
02 19	☽ Q ♃
03 37	☽ ⚻ ☉
05 02	☽ ⚻ ♄
05 35	☽ ± ♂
06 20	♂ ∠ ♅
10 47	☽ ⚼ ♇
14 08	☽ ± ☉
15 38	☽ # ♅
20 22	☽ ☍ ♅
21 01	☽ ⚼ ♂
22 03	☽ ✶ ♃
25 Monday	
01 44	☉ ✶ ♄
01 49	☽ ☍ ♆
04 23	☽ # ♆
05 48	☽ □ ♀
05 57	☽ ⚼ ☉
11 25	☽ △ ♇
12 41	☽ # ☉
15 52	☽ ⚻ ☿
16 06	☽ # ♄
19 00	☽ # ♂
22 53	☽ ∠ ♃
22 56	☽ △ ♂
23 12	☉ ∠ ♀
26 Tuesday	
02 04	☽ ± ☿
06 34	☽ ☍ ♄
08 34	☽ △ ☉
12 25	♀ ⊥ ♃
13 57	☽ ⊥ ♃
14 04	☽ Q ♀
17 25	☽ ⚼ ☿
22 09	☽ ⚻ ♅
27 Wednesday	
00 04	☽ ⚺ ♃
03 41	☽ ⚻ ♆
08 18	☽ ± ♅
11 59	☽ ✶ ♀
13 37	☽ □ ♇
13 51	☽ ± ♆
19 09	☽ △ ☿
23 39	☽ ⚼ ♅
28 Thursday	
04 04	☽ □ ♂
05 15	☽ ⚼ ♆
05 42	☽ ∥ ♃
06 33	♂ ∥ ♄
08 54	♀ ⚺ ♇
09 35	☽ ⚻ ♄
11 29	☽ # ♀
13 03	☉ ⊥ ♅
15 21	☽ □ ☉
15 57	☽ ∠ ♀
17 00	☿ St R
20 01	☽ ± ♄
20 36	☽ Q ♇
21 00	☽ # ♇
29 Friday	
01 39	☽ △ ♅
03 50	☽ ☌ ♃
04 22	♂ ∠ ♆
07 20	☽ △ ♆
09 03	☽ ⊥ ♀
11 51	☽ ⚼ ♄
13 03	☽ Q ♂
17 42	☽ ✶ ♇
20 36	☽ ⚺ ♀
23 12	☽ □ ☿
30 Saturday	
01 27	☽ Q ☉
06 49	☽ ∥ ♇
11 32	☽ ✶ ♂
14 40	☽ △ ♄
20 33	☽ ∠ ♇
20 48	☽ ∥ ♀
21 36	☿ ✶ ♀
22 19	☽ # ♃

DECEMBER 1991

LONGITUDES

Date	Sidereal time h m s	Sun ☉ ° ' "	Moon ☽ ° ' "	Moon ☽ 24.00 ° ' "	Mercury ☿ ° '	Venus ♀ ° '	Mars ♂ ° '	Jupiter ♃ ° '	Saturn ♄ ° '	Uranus ♅ ° '	Neptune ♆ ° '	Pluto ♇ ° '
01	16 39 23	08 ♐ 47 12	14 ♎ 30 46	21 ♎ 03 41	23 ♐ 31	24 ♎ 28	01 ♐ 43	13 ♍ 17	02 ♒ 48	11 ♑ 57	15 ♑ 08	21 ♏ 01
02	16 43 19	09 48 03	27 ♎ 33 10	03 ♏ 59 20	22 R 56	25 37	02 25	13 22	02 53	12 00	15 10	21 04
03	16 47 16	10 48 54	10 ♏ 22 16	16 ♏ 42 00	22 10	26 46	03 08	13 27	02 59	12 03	15 12	21 06
04	16 51 12	11 49 47	22 ♏ 58 39	29 ♏ 12 18	21 13	27 55	03 51	13 32	03 04	12 06	15 14	21 08
05	16 55 09	12 50 41	05 ♐ 23 02	11 ♐ 30 58	20 07	29 ♎ 04	04 34	13 37	03 09	12 09	15 16	21 11
06	16 59 05	13 51 37	17 ♐ 36 15	23 ♐ 39 03	18 53	00 ♏ 13	05 17	13 42	03 15	12 13	15 18	21 13
07	17 03 02	14 52 33	29 ♐ 39 34	05 ♑ 38 02	17 34	01 23	06 00	13 46	03 20	12 16	15 20	21 15
08	17 06 58	15 53 30	11 ♑ 34 44	17 ♑ 29 58	16 11	02 33	06 43	13 50	03 26	12 19	15 22	21 17
09	17 10 55	16 54 28	23 ♑ 24 07	29 ♑ 17 33	14 48	03 43	07 26	13 54	03 32	12 23	15 24	21 20
10	17 14 52	17 55 27	05 ♒ 10 43	11 ♒ 04 05	13 28	04 53	08 09	13 58	03 37	12 26	15 26	21 22
11	17 18 48	18 56 26	16 ♒ 58 11	22 ♒ 53 33	12 13	06 03	08 52	14 02	03 43	12 29	15 28	21 24
12	17 22 45	19 57 27	28 ♒ 50 44	04 ♓ 50 21	11 05	07 13	09 35	14 05	03 49	12 33	15 30	21 26
13	17 26 41	20 58 27	10 ♓ 52 59	16 ♓ 59 16	10 07	08 24	10 19	14 09	03 55	12 36	15 32	21 28
14	17 30 38	21 59 28	23 ♓ 09 46	29 ♓ 25 05	09 18	09 34	11 02	14 12	04 00	12 39	15 34	21 31
15	17 34 34	23 00 30	05 ♈ 45 45	12 ♈ 12 16	08 41	10 45	11 45	14 15	04 06	12 43	15 37	21 33
16	17 38 31	24 01 32	18 ♈ 45 04	25 ♈ 24 28	08 15	11 56	12 29	14 18	04 12	12 46	15 39	21 35
17	17 42 27	25 02 34	02 ♉ 10 42	09 ♉ 03 51	07 59	13 07	13 12	14 20	04 18	12 50	15 41	21 37
18	17 46 24	26 03 37	16 ♉ 03 51	23 ♉ 10 29	07 D 55	14 18	13 56	14 23	04 25	12 53	15 43	21 39
19	17 50 21	27 04 41	00 ♊ 23 22	07 ♊ 41 54	08 00	15 29	14 39	14 25	04 31	12 57	15 45	21 41
20	17 54 17	28 05 45	15 ♊ 05 23	22 ♊ 32 53	08 14	16 40	15 23	14 27	04 37	13 00	15 47	21 43
21	17 58 14	29 ♐ 06 49	00 ♋ 03 26	07 ♋ 35 55	08 37	17 51	16 06	14 29	04 43	13 04	15 50	21 45
22	18 02 10	00 ♑ 07 54	15 ♋ 09 11	22 ♋ 42 05	09 07	19 03	16 50	14 31	04 49	13 07	15 52	21 47
23	18 06 07	01 09 00	00 ♌ 13 32	07 ♌ 42 29	09 44	20 14	17 34	14 32	04 56	13 11	15 54	21 49
24	18 10 03	02 10 06	15 ♌ 08 01	22 ♌ 29 21	10 27	21 26	18 18	14 34	05 02	13 14	15 56	21 51
25	18 14 00	03 11 12	29 ♌ 45 51	06 ♍ 57 02	11 15	22 38	19 01	14 35	05 09	13 18	15 58	21 53
26	18 17 56	04 12 19	14 ♍ 02 35	21 ♍ 02 19	12 08	23 49	19 45	14 36	05 15	13 21	16 01	21 55
27	18 21 53	05 13 27	27 ♍ 56 11	04 ♎ 44 15	13 05	25 01	20 29	14 37	05 22	13 25	16 03	21 57
28	18 25 50	06 14 36	11 ♎ 26 39	18 ♎ 03 37	14 06	26 13	21 13	14 37	05 28	13 28	16 05	21 59
29	18 29 46	07 15 44	24 ♎ 35 28	01 ♏ 02 29	15 10	27 26	21 57	14 38	05 35	13 32	16 07	22 01
30	18 33 43	08 16 54	07 ♏ 25 02	13 ♏ 43 29	16 18	28 38	22 41	14 38	05 41	13 35	16 10	22 03
31	18 37 39	09 ♑ 18 04	19 ♏ 58 11	26 ♏ 09 30	17 ♐ 27	29 ♏ 50	23 ♐ 25	14 ♍ 38	05 ♒ 48	13 ♑ 39	16 ♑ 12	22 ♏ 05

Date	Moon True ☊ ° '	Moon Mean ☊ ° '	Moon ☽ Latitude ° '
01	10 ♑ 17	11 ♑ 25	05 S 12
02	10 R 12	11 22	04 59
03	10 07	11 19	04 30
04	10 02	11 15	03 50
05	09 57	11 12	02 58
06	09 54	11 09	02 00
07	09 52	11 06	00 S 56
08	09 D 52	11 03	00 N 09
09	09 53	11 00	01 14
10	09 54	10 56	02 15
11	09 56	10 53	03 11
12	09 58	10 50	03 59
13	09 59	10 47	04 37
14	09 R 59	10 44	05 03
15	09 58	10 41	05 16
16	09 57	10 37	05 13
17	09 55	10 34	04 53
18	09 53	10 31	04 16
19	09 52	10 28	03 22
20	09 50	10 25	02 13
21	09 50	10 21	00 N 54
22	09 D 50	10 18	00 S 30
23	09 50	10 15	01 51
24	09 51	10 12	03 04
25	09 51	10 09	04 04
26	09 52	10 06	04 46
27	09 52	10 02	05 11
28	09 R 52	09 59	05 18
29	09 52	09 56	05 08
30	09 52	09 53	04 42
31	09 ♑ 52	09 ♑ 50	04 S 04

DECLINATIONS

Date	Sun ☉ ° '	Moon ☽ ° '	Mercury ☿ ° '	Venus ♀ ° '	Mars ♂ ° '	Jupiter ♃ ° '	Saturn ♄ ° '	Uranus ♅ ° '	Neptune ♆ ° '	Pluto ♇ ° '
01	21 S 46	10 S 30	24 S 06	07 S 27	20 S 39	07 N 32	20 S 05	23 S 16	21 S 51	04 S 02
02	21 55	15 14	23 46	07 51	20 48	07 30	20 04	23 16	21 50	04 02
03	22 04	19 12	23 25	08 14	20 57	07 28	20 03	23 16	21 50	04 03
04	22 12	22 13	23 02	08 38	21 06	07 26	20 01	23 15	21 50	04 03
05	22 20	24 08	22 37	09 01	21 15	07 25	20 00	23 15	21 50	04 03
06	22 28	24 51	22 11	09 24	21 23	07 23	19 59	23 15	21 50	04 04
07	22 35	24 22	21 44	09 48	21 31	07 22	19 58	23 14	21 49	04 04
08	22 42	22 47	21 17	10 11	21 39	07 20	19 56	23 14	21 49	04 05
09	22 48	20 12	20 50	10 34	21 47	07 19	19 55	23 14	21 49	04 05
10	22 54	16 47	20 25	10 57	21 54	07 18	19 54	23 13	21 49	04 06
11	22 59	12 43	20 01	11 19	22 02	07 17	19 53	23 13	21 49	04 06
12	23 04	08 08	19 40	11 42	22 09	07 16	19 51	23 13	21 48	04 06
13	23 08	03 S 13	19 23	12 04	22 16	07 15	19 50	23 12	21 48	04 07
14	23 12	01 N 56	19 08	12 26	22 23	07 14	19 49	23 12	21 48	04 07
15	23 15	07 07	18 58	12 48	22 29	07 13	19 47	23 12	21 48	04 07
16	23 18	12 10	18 51	13 10	22 36	07 12	19 46	23 11	21 48	04 08
17	23 21	16 49	18 47	13 32	22 42	07 11	19 44	23 11	21 47	04 08
18	23 23	20 44	18 47	13 53	22 48	07 10	19 43	23 11	21 47	04 08
19	23 24	23 31	18 50	14 14	22 53	07 10	19 42	23 10	21 47	04 08
20	23 26	24 49	18 55	14 35	22 59	07 09	19 40	23 10	21 47	04 09
21	23 26	24 20	19 03	14 55	23 04	07 09	19 39	23 10	21 46	04 09
22	23 26	22 05	19 12	15 15	23 09	07 08	19 37	23 09	21 46	04 09
23	23 26	18 18	19 24	15 35	23 14	07 08	19 36	23 09	21 46	04 09
24	23 25	13 22	19 36	15 55	23 19	07 08	19 34	23 09	21 46	04 10
25	23 24	07 45	19 50	16 14	23 23	07 07	19 33	23 08	21 45	04 10
26	23 22	01 N 52	20 04	16 33	23 27	07 07	19 31	23 08	21 45	04 10
27	23 20	03 S 56	20 19	16 52	23 31	07 07	19 30	23 08	21 45	04 10
28	23 18	09 24	20 35	17 10	23 35	07 07	19 28	23 07	21 45	04 10
29	23 15	14 18	20 50	17 27	23 38	07 07	19 26	23 07	21 44	04 11
30	23 11	18 26	21 05	17 45	23 41	07 08	19 25	23 07	21 44	04 11
31	23 S 07	21 S 38	21 S 21	18 S 02	23 S 44	07 N 08	19 S 23	23 S 06	21 S 44	04 S 11

ZODIAC SIGN ENTRIES

Date	h	m	Planets
02	16	33	☽ ♏
05	01	32	☽ ♐
06	07	21	♀ ♏
07	12	41	☽ ♑
10	01	27	☽ ♒
12	14	19	☽ ♓
15	01	06	☽ ♈
17	08	10	☽ ♉
19	11	21	☽ ♊
21	11	55	☽ ♋
22	08	54	☉ ♑
23	11	38	☽ ♌
25	12	23	☽ ♍
27	15	37	☽ ♎
29	22	03	☽ ♏
31	15	19	♀ ♐

LATITUDES

Date	Mercury ☿ ° '	Venus ♀ ° '	Mars ♂ ° '	Jupiter ♃ ° '	Saturn ♄ ° '	Uranus ♅ ° '	Neptune ♆ ° '	Pluto ♇ ° '
01	00 S 49	02 N 11	00 S 09	01 N 02	00 S 34	00 S 22	00 N 45	14 N 28
04	00 N 07	02 15	00 11	01 03	00 34	00 22	00 44	14 28
07	01 08	02 18	00 13	01 04	00 34	00 22	00 44	14 29
10	02 01	02 20	00 15	01 04	00 34	00 22	00 44	14 29
13	02 37	02 21	00 16	01 05	00 34	00 22	00 44	14 30
16	02 53	02 22	00 18	01 06	00 34	00 22	00 44	14 30
19	02 51	02 21	00 20	01 07	00 34	00 22	00 44	14 31
22	02 39	02 19	00 22	01 08	00 35	00 22	00 44	14 32
25	02 19	02 16	00 24	01 09	00 35	00 22	00 44	14 32
28	01 56	02 13	00 26	01 09	00 35	00 22	00 44	14 33
31	01 N 31	02 N 08	00 S 27	01 N 10	00 S 35	00 S 22	00 N 44	14 N 34

DATA

Julian Date	2448592
Delta T	+58 seconds
Ayanamsa	23° 44' 55"
Synetic vernal point	05° ♓ 22' 05"
True obliquity of ecliptic	23° 26' 27"

MOON'S PHASES, APSIDES AND POSITIONS ☽

Date	h	m	Phase	Longitude ° '	Eclipse Indicator
06	03	56	●	13 ♐ 31	
14	09	32	☽	21 ♓ 53	
21	10	23	○	29 ♊ 03	partial
28	01	55	☾	05 ♎ 49	

Day	h	m		
10	01	56	Apogee	
22	09	32	Perigee	
06	14	22	Max dec	24° S 51'
14	03	04	0N	
20	17	48	Max dec	24° N 52'
26	19	38	0S	

LONGITUDES

Date	Chiron ⚷ ° '	Ceres ⚳ ° '	Pallas ⚴ ° '	Juno ⚵ ° '	Vesta ⚶ ° '	Black Moon Lilith ⚸ ° '
01	09 ♌ 32	14 ♐ 58	23 ♏ 29	03 ♒ 57	16 ♍ 15	24 ♑ 20
11	09 ♌ 17	19 ♐ 07	27 ♏ 49	07 ♒ 46	19 ♍ 07	25 ♑ 26
21	08 ♌ 53	23 ♐ 14	02 ♐ 05	11 ♒ 47	21 ♍ 35	26 ♑ 33
31	08 ♌ 21	27 ♐ 20	06 ♐ 16	16 ♒ 00	23 ♍ 34	27 ♑ 39

All ephemeris data is given at 12.00 UT and the Moon's longitude is additionally given for 24.00 UT

ASPECTARIAN

h m Aspects

01 Sunday
00 44 ☽ ⚹ ☉
06 44 ☽ Q ☿
07 18 ☽ □ ♅
09 44 ☽ ⊻ ♃
12 56 ☽ ⊥ ♇
13 08 ☽ □ ♆
16 06 ♀ # ♃
16 15 ☽ ∠ ♂
20 29 ☉ ⊥ ♆
20 47 ☽ ⊥ ♃
23 38 ☉ ∥ ♆
23 58 ☽ ⊻ ♇

02 Monday
03 51 ☽ ⚹ ☿
06 28 ☽ ∠ ☉
08 03 ☽ ☌ ♀
09 47 ☽ ⊥ ♂
13 32 ☽ ∠ ♃
16 34 ☽ Q ♅
21 36 ☽ ⊻ ♂
22 01 ☽ □ ♄
22 29 ☽ Q ♆

03 Tuesday
00 39 ☽ ⊥ ☉
06 00 ♂ ⚹ ♄
06 19 ☽ ∠ ☿
12 55 ☽ ⊻ ☉
15 11 ☽ ⚹ ♅
17 52 ☽ ⚹ ♃
17 52 ☽ ∥ ♄
21 10 ☽ ⚹ ♆
22 15 ☽ ⊥ ☿
22 24 ☿ ∥ ♅

04 Wednesday
01 34 ☽ ∥ ♂
08 19 ☽ Q ♄
08 25 ☽ ∥ ♆
08 28 ☽ ☌ ♇
08 32 ♀ ⊥ ♂
08 52 ☽ ⊻ ☿
11 54 ☽ ∥ ☉
13 43 ☿ ⊻ ♇
16 57 ☽ Q ♃
18 50 ☉ ⊻ ♅
19 13 ☽ ∥ ☿
19 58 ☽ ∠ ♅
22 28 ☽ ⊻ ♀
23 12 ☽ ∥ ♅

05 Thursday
01 56 ♀ ∠ ♃
02 01 ☽ ∠ ♆
07 38 ☽ ⚹ ♄
10 18 ☽ ☌ ♂
11 19 ☽ ⊥ ♀
13 31 ☽ ⊥ ♅
19 36 ☽ ⊥ ♆
21 13 ♄ Q ♇

06 Friday
00 07 ☉ ∥ ☿
01 19 ☽ ⊻ ♅
03 56 ☽ ☌ ☉
04 14 ☽ □ ♃
06 48 ☽ ∠ ♀
07 26 ☽ ⊻ ♆
07 42 ☉ □ ♃
11 43 ♀ Q ♅
13 17 ☽ ∠ ♄
14 18 ☽ ☌ ☿
19 11 ☽ ⊻ ♇
22 59 ☿ ∠ ♄

07 Saturday
07 10 ☽ ⊥ ♇
07 19 ☽ ⊥ ♄
07 20 ☿ ∥ ♆
15 50 ☽ ⚹ ♀
19 26 ☽ ⊻ ♄
20 59 ☿ ∥ ♂
21 45 ♂ ⊥ ♅
23 08 ☉ ⊻ ♆
23 11 ☿ ∠ ♀

08 Sunday
01 17 ☽ ∠ ♇
01 33 ☽ ⊻ ♂
06 32 ☽ ∥ ♅
12 55 ☽ ∥ ☉
13 30 ☽ ☌ ♅
14 27 ☽ ⊥ ♂
14 58 ☉ ☌ ☿
16 36 ☽ △ ♃
18 40 ☽ Q ♀
19 41 ☽ ☌ ♆
20 22 ☽ ⊻ ☿
21 33 ☽ ⊻ ☉
21 57 ☽ ∥ ♆
22 59 ☽ ∥ ♂

09 Monday
01 56 ☿ ⊻ ♆
05 22 ♀ Q ♆
05 46 ☽ ∥ ☿
07 16 ☽ ⊥ ☿
07 46 ☽ ⚹ ♇
07 50 ♀ □ ♄
09 54 ☽ ∠ ♂
10 54 ☽ ⊥ ☉
14 11 ☽ ∥ ♄
18 48 ♂ ∥ ♆
23 16 ☽ bQ ♃
23 42 ☽ ∠ ☿

10 Tuesday
03 19 ☿ □ ♃
06 58 ☽ ∠ ☉
08 17 ☽ Q ♇
08 48 ☽ ☌ ♄
11 19 ☽ □ ♀
17 43 ☽ ± ♃
18 27 ☽ ⚹ ♂

11 Wednesday
02 51 ☽ ⊻ ♅
03 15 ☽ ⚹ ☿
06 00 ☽ ⊼ ♃
06 07 ☉ ∠ ♄
06 54 ☿ ⊻ ♅
08 56 ☽ ⊻ ♆
13 43 ☿ ⊥ ♀
15 06 ☽ ⊥ ♅
16 22 ☽ ⚹ ☉
18 58 ☽ ∥ ♀
20 25 ☽ Q ♂
21 01 ☽ □ ♇
21 09 ☽ ⊥ ♆
22 20 ☿ ∥ ♄

12 Thursday
01 22 ☽ Q ☿
08 55 ♂ ⊥ ♆
09 22 ☽ ∠ ♅
15 20 ☽ ∠ ♆
16 25 ☽ # ♃
18 49 ☽ Q ☉
22 02 ☽ ⊻ ♄

13 Friday
06 32 ☽ △ ♀
07 44 ☽ ∥ ♇
09 01 ☿ ☌ ♂
10 03 ☽ ⊥ ♄
10 35 ☽ □ ☿
10 48 ☽ □ ♂
15 24 ☽ ⚹ ♅
18 28 ☽ ☍ ♃
21 11 ☽ ⚹ ♆

14 Saturday
00 15 ☉ ⊻ ♇
03 44 ☿ ⊥ ♆
03 53 ☽ ∠ ♄
08 39 ☿ ⊻ ♀
08 48 ☽ △ ♇
09 32 ☽ □ ☉
13 19 ☉ ∥ ♅
14 54 ☽ Q ♅
15 00 ☽ bQ ♀
20 30 ☽ Q ♆
22 06 ☽ # ♇

15 Sunday
08 52 ☽ ⚹ ♄
09 54 ☽ ± ♀
12 25 ☽ ∥ ♃
13 28 ☽ bQ ♇
17 15 ☽ △ ☿
22 14 ☽ ⊼ ♀
23 50 ☽ △ ♂

16 Monday
01 00 ☽ □ ♅
03 50 ☽ ⊼ ♃
06 12 ☽ ± ♇
06 19 ☽ □ ♆
07 19 ☽ Q ♄
14 49 ☽ ± ♃
17 09 ☽ ⊼ ♇
17 21 ☽ # ♀
19 56 ☽ bQ ☿
22 18 ☽ △ ☉
22 28 ♂ ⊻ ♅

17 Tuesday
04 35 ☽ bQ ♂
06 01 ♀ ⚹ ♅
06 59 ☽ bQ ♃
11 40 ☽ ± ☿
15 46 ☽ □ ♄
16 59 ♀ ⊻ ♂
21 16 ☽ ± ♂
22 03 ☽ ⊼ ☿
23 20 ☽ # ☿

18 Wednesday
02 46 ☽ bQ ☉
05 15 ☽ # ♄
06 33 ☽ △ ♅
08 10 ☽ ⊼ ♂
08 42 ☽ ☍ ♀
09 08 ☽ △ ♃
11 13 ☿ St D
11 24 ☽ △ ♆
13 53 ♀ ⚹ ♃
19 17 ☽ ± ☉
19 50 ☽ # ♆
21 28 ☽ ☍ ♇

19 Thursday
03 51 ♂ □ ♃
05 10 ☽ # ♂
06 06 ☽ ⊼ ☉
07 56 ☽ bQ ♅
08 15 ☽ # ♅
10 43 ☽ # ☉
12 36 ☽ bQ ♆
17 44 ♀ ⚹ ♆
18 50 ☽ △ ♄
22 49 ☽ ± ♅

20 Friday
00 40 ☽ ☍ ☿
02 52 ☉ ⊥ ♇
03 23 ☽ ± ♆
08 37 ☽ ⊼ ♅
10 58 ☽ □ ♃
12 30 ☽ ☍ ♂
13 08 ☽ ⊼ ♆
14 46 ☽ ⊼ ♀
19 21 ☽ bQ ♄
22 42 ☽ ⊼ ♇

21 Saturday
01 14 ☽ ± ♀
01 39 ☉ ⊥ ♄
02 11 ♂ ⊻ ♆
08 19 ☽ ± ♇
09 51 ☽ ± ♄
10 23 ☽ ☍ ☉
15 53 ☽ Q ♃
16 50 ☽ bQ ♀
19 28 ☽ ⊼ ♄
22 41 ☽ bQ ♇
23 53 ☽ # ☉

22 Sunday
02 03 ☽ ⊼ ☿
02 44 ☽ # ♅
03 06 ☽ # ♂
08 45 ☽ ☍ ♅
10 59 ☽ ⚹ ♃
11 56 ☽ ± ☿
13 08 ☽ ☍ ♆
13 23 ♂ ∥ ♅
14 27 ☽ # ♆
14 49 ☽ ⊼ ♂
18 43 ☽ △ ♀
22 34 ☽ △ ♇

23 Monday
00 50 ☽ ± ♂
02 50 ☽ bQ ☿
04 38 ☽ # ♄
06 08 ☽ # ☿
10 54 ☽ ∠ ♃
13 35 ☽ ⊼ ☉
15 56 ☽ bQ ♂
18 20 ☿ ⊥ ♆
19 35 ☽ ☍ ♄
23 55 ☽ ± ☉

24 Tuesday
00 52 ☽ # ♀
01 22 ☽ ⊥ ♃
04 01 ☽ △ ☿
08 27 ☿ ∥ ♄
08 55 ☽ ⊼ ♅
11 04 ☽ ⊻ ♃
13 18 ☽ ⊼ ♆
15 33 ☽ bQ ☉
17 25 ☽ △ ♂
18 42 ☽ ± ♅
20 45 ♀ ☌ ♇
22 59 ☽ □ ♇
23 07 ☽ ± ♆
23 11 ☽ □ ♀

25 Wednesday
09 33 ☽ bQ ♅
14 01 ☽ bQ ♆
14 35 ☽ ∥ ♃
17 15 ☉ ∥ ♂
18 08 ☽ △ ☉
21 02 ☽ ⊼ ♄
23 22 ♀ Q ♄

26 Thursday
02 40 ☽ # ♇
04 59 ☽ Q ♇
07 13 ☽ ± ♄
07 53 ☽ Q ♀
08 32 ☽ □ ☿
10 49 ☽ △ ♅
12 57 ☽ ☌ ♃
15 22 ☽ △ ♆
22 20 ☽ □ ♂
22 43 ☽ bQ ♄

27 Friday
01 33 ☽ ⚹ ♇
06 26 ☽ ⚹ ♀
07 06 ♂ ∠ ♄
13 00 ☽ ∥ ♇
15 34 ☉ ⊻ ♄
17 59 ☽ Q ☿
20 19 ☿ ⊻ ♅

28 Saturday
01 13 ☽ △ ♄
01 45 ☽ # ♃
01 55 ☽ □ ☉
03 59 ☽ ∠ ♇
07 46 ☽ Q ♂
11 34 ☽ ∠ ♀
15 41 ☽ □ ♅
17 13 ☽ ⚹ ☿
17 45 ☽ ⊻ ♃
19 57 ♀ Q ♃
20 14 ☽ ⊥ ♇
20 25 ☽ □ ♆
23 47 ☿ □ ♃

29 Sunday
04 42 ☽ ⊥ ♃
05 34 ☽ ⊥ ♀
05 59 ☉ ∠ ♇
06 51 ☽ ⚹ ♂
07 14 ☽ ⊻ ♇
13 21 ☽ Q ☉
14 07 ♂ ⊻ ♇
17 48 ☽ ⊻ ♀
21 21 ☽ ∠ ♃
23 21 ☽ ∠ ☿

30 Monday
00 59 ☽ Q ♅
05 50 ☽ Q ♆
07 20 ☽ ∥ ♀
08 42 ☽ □ ♄
09 06 ☿ ⊻ ♆
11 12 ♀ ∠ ♅
12 33 ☽ ∠ ♂
13 47 ☽ ⚹ ☉
18 00 ☽ ⊥ ☿
18 35 ☽ ∥ ♄
21 35 ♃ St R
23 48 ☽ ⚹ ♅

31 Tuesday
01 44 ☽ ⚹ ♃
04 43 ☽ ⚹ ♆
06 39 ☽ ⊻ ☿
06 47 ☽ ⊥ ♂
09 10 ☽ ∥ ☿
12 49 ☽ ∥ ♆
15 26 ☉ ∥ ♅
16 05 ☽ ☌ ♇
19 06 ☽ ⊻ ♂
19 29 ☽ Q ♄
21 09 ☽ ∠ ☉

JANUARY 1992

LONGITUDES

Date	Sidereal time h m s	Sun ☉ ° ' "	Moon ☽ ° ' "	Moon ☽ 24.00 ° ' "	Mercury ☿ ° '	Venus ♀ ° '	Mars ♂ ° '	Jupiter ♃ ° '	Saturn ♄ ° '	Uranus ♅ ° '	Neptune ♆ ° '	Pluto ♇ ° '
01	18 41 36	10 ♑ 19 14	02 ♐ 17 45	08 ♐ 23 18	18 ♐ 39	01 ♐ 02	24 ♐ 09	14 ♍ 38	05 ♒ 55	13 ♑ 43	16 ♑ 14	22 ♏ 06
02	18 45 32	11 20 25	14 ♐ 26 27	20 ♐ 27 29	19 53	02 15	24 54	14 R 37	06 01	13 46	16 16	22 08
03	18 49 29	12 21 35	26 ♐ 26 42	02 ♑ 24 22	21 09	03 27	25 38	14 36	06 08	13 50	16 19	22 10
04	18 53 25	13 22 46	08 ♑ 20 45	14 ♑ 16 04	22 26	04 40	26 22	14 36	06 15	13 53	16 21	22 12
05	18 57 22	14 23 57	20 ♑ 10 37	26 ♑ 04 37	23 45	05 52	27 06	14 35	06 22	13 57	16 23	22 13
06	19 01 19	15 25 08	01 ♒ 58 22	07 ♒ 52 07	25 05	07 05	27 51	14 33	06 29	14 01	16 25	22 15
07	19 05 15	16 26 18	13 ♒ 46 10	19 ♒ 40 49	26 26	08 18	28 35	14 32	06 36	14 04	16 28	22 17
08	19 09 12	17 27 28	25 ♒ 36 26	01 ♓ 33 21	27 49	09 31	29 ♐ 20	14 31	06 42	14 08	16 30	22 18
09	19 13 08	18 28 38	07 ♓ 31 59	13 ♓ 32 43	29 ♐ 12	10 43	00 ♑ 04	14 29	06 49	14 11	16 32	22 20
10	19 17 05	19 29 48	19 ♓ 36 01	25 ♓ 42 20	00 ♑ 36	11 56	00 49	14 27	06 56	14 15	16 35	22 21
11	19 21 01	20 30 57	01 ♈ 52 09	08 ♈ 05 58	02 01	13 09	01 33	14 25	07 03	14 18	16 37	22 23
12	19 24 58	21 32 05	14 ♈ 24 15	20 ♈ 47 31	03 27	14 22	02 18	14 22	07 10	14 22	16 39	22 24
13	19 28 54	22 33 13	27 ♈ 16 14	03 ♉ 50 48	04 53	15 35	03 02	14 20	07 17	14 26	16 41	22 26
14	19 32 51	23 34 20	10 ♉ 31 36	17 ♉ 18 55	06 21	16 48	03 47	14 17	07 24	14 29	16 44	22 27
15	19 36 48	24 35 26	24 ♉ 12 58	01 ♊ 13 48	07 48	18 01	04 32	14 14	07 31	14 33	16 46	22 29
16	19 40 44	25 36 32	08 ♊ 21 27	15 ♊ 35 25	09 17	19 14	05 17	14 11	07 38	14 36	16 48	22 30
17	19 44 41	26 37 38	22 ♊ 55 33	00 ♋ 21 10	10 46	20 28	06 01	14 08	07 45	14 40	16 50	22 31
18	19 48 37	27 38 42	07 ♋ 51 29	15 ♋ 25 32	12 16	21 41	06 46	14 04	07 53	14 43	16 53	22 33
19	19 52 34	28 39 46	23 ♋ 02 14	00 ♌ 40 20	13 46	22 54	07 31	14 01	08 00	14 47	16 55	22 34
20	19 56 30	29 ♑ 40 49	08 ♌ 18 34	15 ♌ 55 37	15 17	24 07	08 16	13 57	08 07	14 50	16 57	22 35
21	20 00 27	00 ♒ 41 52	23 ♌ 30 14	01 ♍ 01 13	16 49	25 21	09 01	13 53	08 14	14 54	16 59	22 36
22	20 04 23	01 42 53	08 ♍ 27 31	15 ♍ 48 15	18 21	26 34	09 46	13 49	08 21	14 57	17 02	22 38
23	20 08 20	02 43 55	23 ♍ 02 45	00 ♎ 10 30	19 54	27 48	10 31	13 44	08 28	15 01	17 04	22 39
24	20 12 17	03 44 56	07 ♎ 11 13	14 ♎ 04 46	21 27	29 ♐ 01	11 16	13 40	08 35	15 04	17 06	22 40
25	20 16 13	04 45 56	20 ♎ 51 15	27 ♎ 30 49	23 01	00 ♑ 15	12 01	13 35	08 43	15 08	17 08	22 41
26	20 20 10	05 46 56	04 ♏ 03 49	10 ♏ 30 39	24 35	01 28	12 46	13 30	08 50	15 11	17 10	22 42
27	20 24 06	06 47 56	16 ♏ 51 49	23 ♏ 07 49	26 10	02 42	13 31	13 25	08 57	15 14	17 13	22 43
28	20 28 03	07 48 55	29 ♏ 19 13	05 ♐ 26 35	27 46	03 55	14 17	13 20	09 04	15 18	17 15	22 44
29	20 31 59	08 49 53	11 ♐ 30 28	17 ♐ 31 27	29 ♑ 23	05 09	15 02	13 15	09 11	15 21	17 17	22 45
30	20 35 56	09 50 51	23 ♐ 30 01	29 ♐ 26 42	01 ♒ 00	06 23	15 47	13 10	09 18	15 24	17 19	22 46
31	20 39 52	10 ♒ 51 47	05 ♑ 21 58	11 ♑ 16 14	02 ♒ 38	07 ♑ 36	16 ♑ 33	13 ♍ 04	09 ♒ 26	15 ♑ 28	17 ♑ 21	22 ♏ 47

DECLINATIONS

Date	Moon True ☊ ° '	Moon Mean ☊ ° '	Moon ☽ Latitude ° '	Sun ☉ ° '	Moon ☽ ° '	Mercury ☿ ° '	Venus ♀ ° '	Mars ♂ ° '	Jupiter ♃ ° '	Saturn ♄ ° '	Uranus ♅ ° '	Neptune ♆ ° '	Pluto ♇ ° '
01	09 ♑ 52	09 ♑ 47	03 S 14	23 S 02	23 S 48	21 S 35	18 S 18	23 S 47	07 N 08	19 S 22	23 S 06	21 S 44	04 S 11
02	09 D 52	09 43	02 17	22 57	24 48	21 50	18 34	23 49	07 09	19 20	23 06	21 43	04 11
03	09 52	09 40	01 14	22 52	24 38	22 04	18 51	23 51	07 09	19 18	23 05	21 43	04 11
04	09 52	09 37	00 S 08	22 46	23 19	22 17	19 05	23 53	07 10	19 17	23 05	21 43	04 11
05	09 R 52	09 34	00 N 57	22 40	20 59	22 29	19 20	23 55	07 10	19 15	23 05	21 43	04 11
06	09 52	09 31	02 00	22 33	17 46	22 41	19 34	23 56	07 11	19 13	23 04	21 42	04 12
07	09 51	09 27	02 58	22 26	13 51	22 52	19 47	23 58	07 12	19 12	23 04	21 42	04 12
08	09 50	09 24	03 48	22 18	09 24	23 02	20 01	23 59	07 13	19 10	23 03	21 42	04 12
09	09 49	09 21	04 29	22 10	04 S 35	23 11	20 13	23 59	07 14	19 08	23 03	21 41	04 12
10	09 47	09 18	04 58	22 01	00 N 27	23 19	20 25	24 00	07 15	19 07	23 03	21 41	04 12
11	09 46	09 15	05 14	21 52	05 33	23 26	20 37	24 00	07 16	19 05	23 02	21 41	04 12
12	09 46	09 12	05 16	21 43	10 32	23 32	20 48	24 00	07 17	19 03	23 02	21 41	04 12
13	09 D 46	09 08	05 03	21 33	15 12	23 36	20 58	24 00	07 18	19 01	23 02	21 40	04 12
14	09 46	09 05	04 33	21 23	19 18	23 40	21 08	23 59	07 19	19 00	23 01	21 40	04 12
15	09 47	09 02	03 47	21 12	22 29	23 42	21 17	23 58	07 21	18 58	23 01	21 40	04 12
16	09 48	08 59	02 46	21 01	24 26	23 44	21 26	23 57	07 22	18 56	23 00	21 40	04 12
17	09 49	08 56	01 32	20 50	24 47	23 44	21 34	23 56	07 24	18 54	23 00	21 39	04 12
18	09 50	08 53	00 N 11	20 38	23 23	23 42	21 42	23 54	07 25	18 53	23 00	21 39	04 12
19	09 R 50	08 49	01 S 12	20 26	20 17	23 40	21 49	23 53	07 27	18 51	22 59	21 39	04 12
20	09 48	08 46	02 31	20 13	15 45	23 36	21 55	23 50	07 29	18 49	22 59	21 38	04 11
21	09 46	08 43	03 38	20 00	10 15	23 31	22 01	23 48	07 30	18 47	22 59	21 38	04 11
22	09 43	08 40	04 30	19 47	04 N 14	23 24	22 06	23 46	07 32	18 46	22 58	21 38	04 11
23	09 40	08 37	05 02	19 33	01 S 52	23 17	22 10	23 43	07 34	18 44	22 58	21 38	04 11
24	09 37	08 33	05 15	19 19	07 40	23 07	22 14	23 40	07 36	18 42	22 57	21 37	04 11
25	09 35	08 30	05 09	19 04	12 55	22 57	22 17	23 36	07 38	18 40	22 57	21 37	04 11
26	09 34	08 27	04 47	18 50	17 22	22 45	22 19	23 33	07 40	18 38	22 57	21 37	04 10
27	09 D 35	08 24	04 11	18 34	20 53	22 32	22 21	23 29	07 42	18 36	22 56	21 37	04 10
28	09 36	08 21	03 24	18 19	23 19	22 17	22 22	23 25	07 45	18 35	22 56	21 36	04 10
29	09 38	08 18	02 29	18 03	24 37	22 01	22 23	23 20	07 47	18 33	22 56	21 36	04 10
30	09 40	08 14	01 28	17 47	24 44	21 43	22 23	23 16	07 49	18 31	22 55	21 36	04 10
31	09 ♑ 41	08 ♑ 11	00 S 24	17 S 31	23 S 44	21 S 24	22 S 22	23 S 11	07 N 51	18 S 29	22 S 55	21 S 35	04 S 09

ZODIAC SIGN ENTRIES

Date	h	m	Planets
01	07	30	☽ ♐
03	19	09	☽ ♑
06	07	59	☽ ♒
08	20	52	☽ ♓
09	09	47	♂ ♑
10	01	46	☿ ♑
11	08	22	☽ ♈
13	17	00	☽ ♉
15	21	55	☽ ♊
17	23	26	☽ ♋
19	22	57	☽ ♌
20	19	32	☉ ♒
21	22	22	☽ ♍
23	23	42	☽ ♎
25	07	14	♀ ♑
26	04	32	☽ ♏
28	13	20	☽ ♐
29	21	15	☿ ♒
31	01	07	☽ ♑

LATITUDES

Date	Mercury ☿ ° '	Venus ♀ ° '	Mars ♂ ° '	Jupiter ♃ ° '	Saturn ♄ ° '	Uranus ♅ ° '	Neptune ♆ ° '	Pluto ♇ ° '
01	01 N 22	02 N 07	00 S 28	01 N 10	00 S 35	00 S 22	00 N 44	14 N 34
04	00 57	02 01	00 30	01 11	00 35	00 22	00 44	14 35
07	00 32	01 55	00 32	01 12	00 35	00 22	00 44	14 36
10	00 N 07	01 49	00 34	01 13	00 35	00 22	00 44	14 37
13	00 S 15	01 42	00 35	01 14	00 36	00 22	00 44	14 39
16	00 37	01 34	00 37	01 15	00 36	00 22	00 44	14 40
19	00 56	01 26	00 39	01 15	00 36	00 22	00 44	14 41
22	01 14	01 18	00 41	01 16	00 36	00 22	00 44	14 42
25	01 29	01 10	00 43	01 16	00 36	00 22	00 44	14 43
28	01 42	01 01	00 44	01 17	00 37	00 22	00 44	14 45
31	01 S 53	00 N 52	00 S 46	01 N 18	00 S 37	00 S 22	00 N 44	14 N 46

DATA

Julian Date	2448623
Delta T	+58 seconds
Ayanamsa	23° 45' 00"
Synetic vernal point	05° ♓ 21' 59"
True obliquity of ecliptic	23° 26' 26"

MOON'S PHASES, APSIDES AND POSITIONS ☽

Date	h	m	Phase	Longitude ° '	Eclipse Indicator
04	23	10	●	13 ♑ 51	Annular
13	02	32	◑	22 ♈ 09	
19	21	28	○	29 ♋ 04	
26	15	27	☾	05 ♏ 56	

Day	h	m		
06	11	54	Apogee	
19	22	20	Perigee	
02	20	26	Max dec	24° S 52'
10	09	53	0N	
17	05	06	Max dec	24° N 52'
23	04	36	0S	
30	02	30	Max dec	24° S 50'

LONGITUDES

Date	Chiron ⚷ ° '	Ceres ⚳ ° '	Pallas ⚴ ° '	Juno ⚵ ° '	Vesta ⚶ ° '	Black Moon Lilith ⚸ ° '
01	08 ♌ 18	27 ♐ 45	06 ♐ 41	16 ♒ 26	23 ♍ 44	27 ♑ 46
11	07 ♌ 39	01 ♑ 48	10 ♐ 45	20 ♒ 50	25 ♍ 05	28 ♑ 53
21	06 ♌ 56	05 ♑ 47	14 ♐ 20	25 ♒ 05	25 ♍ 44	29 ♑ 59
31	06 ♌ 11	09 ♑ 41	18 ♐ 28	00 ♓ 02	25 ♍ 38	01 ♒ 06

All ephemeris data is given at 12.00 UT and the Moon's longitude is additionally given for 24.00 UT

ASPECTARIAN

h m Aspects

01 Wednesday
00 55 ☽ Q ♃
02 27 ☽ ∥ ☉
02 49 ☽ ∥ ♅
04 57 ☽ ∠ ♅
09 16 ☽ ☌ ♀
09 55 ☽ ∠ ♆
11 46 ☽ ∥ ♂
16 02 ♀ ∠ ♆
16 20 ☽ ⊥ ☉
19 11 ☽ ⚹ ♄
22 43 ☽ ⊥ ♅

02 Thursday
01 18 ☿ ∥ ♆
03 42 ☽ ⊥ ♆
05 17 ☽ ⊻ ☉
10 40 ☽ ⊻ ♅
12 21 ☽ □ ♃
15 39 ☽ ⊻ ♆

03 Friday
00 07 ☽ ☌ ☿
01 15 ☽ ∠ ♄
03 24 ☽ ⊻ ♇
10 15 ☽ ☌ ♂
11 46 ☿ ∠ ♄
15 28 ☽ ⊥ ♇
19 30 ☽ ⊥ ♄

04 Saturday
03 43 ☽ ⊻ ♀
03 57 ☽ ∥ ♂
07 24 ☿ ⊻ ♇
07 43 ☽ ⊻ ♄
09 40 ☽ ∠ ♇
15 00 ☽ ∥ ♅
17 13 ☽ ⊥ ♀
19 00 ☽ ∥ ☉
22 54 ☽ ∥ ☿
23 10 ☽ ☌ ☉
23 17 ☽ ☌ ♅

05 Sunday
00 39 ☽ △ ♃
00 45 ☉ ☌ ♅
04 17 ☽ ☌ ♆
05 17 ♀ ∥ ♄
05 30 ☽ ∥ ♆
13 35 ☽ ∠ ♀
16 08 ☉ △ ♃
16 10 ☽ ⚹ ♇
20 11 ☽ ⊻ ☿
22 44 ♀ ⚹ ♄

06 Monday
00 14 ☽ ∥ ♀
01 23 ☉ ∥ ☿
01 48 ☽ ∥ ♄
03 02 ☽ ⊻ ♂
07 06 ☽ ⅊ ♃
09 57 ☽ ⊥ ☿
16 04 ☽ ⊥ ♂
16 39 ☽ Q ♇
21 15 ☽ ☌ ♄
23 36 ☽ ⚹ ♀

07 Tuesday
01 23 ☽ ± ♃
01 33 ♂ ⊥ ♇
06 39 ☽ ∠ ☿
07 17 ♀ ⊥ ♅
11 36 ☽ ∠ ♂
12 35 ☉ ☌ ♆
12 37 ☽ ⊻ ♅
13 33 ☽ ⊼ ♃
17 29 ☽ ⊻ ♆
17 56 ☽ ⊻ ☉

08 Wednesday
00 51 ☽ ⊥ ♅
02 46 ☽ Q ♀
05 18 ☽ □ ♇
05 42 ☽ ⊥ ♆
07 15 ☽ ⊥ ☉
15 36 ☿ ∥ ♅
17 02 ☽ ⚹ ☿
19 09 ☽ ∠ ♅
20 01 ☽ ⚹ ♂
20 44 ☿ ⊥ ♇
23 05 ☽ # ♃
23 56 ☽ ∠ ♆

09 Thursday
03 06 ☽ ∠ ☉
08 15 ♀ ⊥ ♆
10 34 ☽ ⊻ ♄
13 54 ☽ ∥ ♇
19 05 ☽ □ ♀
20 17 ☽ Q ☿
21 39 ☽ Q ♂
22 40 ☽ ⊥ ♄

10 Friday
01 21 ☽ ⚹ ♅
01 49 ☽ ☍ ♃
06 00 ☽ ⚹ ♆
11 47 ☽ ⚹ ☉
16 39 ☽ ∠ ♄
16 53 ♂ ⊥ ♄
17 26 ☽ △ ♇
18 15 ☿ ⊥ ♄
19 33 ☿ ☌ ♂

11 Saturday
01 07 ☽ Q ♅
05 37 ☽ # ♇
05 39 ☽ Q ♆
11 21 ☽ □ ♂
12 19 ☽ □ ☿
13 22 ☽ Q ☉
20 11 ☽ ∥ ♃
22 05 ☽ ⚹ ♄
22 39 ☽ ⅊ ♇

12 Sunday
11 55 ☽ △ ♀
11 56 ☽ □ ♅
11 56 ☽ ⊼ ♃
11 58 ♀ ⊻ ♅
12 04 ♀ □ ♃
13 07 ♃ △ ♅
15 47 ☽ ± ♇
16 15 ☽ □ ♆
18 03 ☉ ∥ ♆
21 03 ☽ Q ♄
23 11 ☽ ± ♃

13 Monday
02 32 ☽ □ ☉
03 02 ☽ ⊼ ♇
09 01 ☉ ⚹ ♇
15 46 ☽ ⅊ ♃
18 41 ☽ ⅊ ♀
23 10 ☽ △ ♂

14 Tuesday
03 35 ☽ △ ☿
06 22 ☽ □ ♄
10 06 ☽ # ♄
10 28 ♀ ⊻ ♆
12 32 ☽ ± ♀
18 39 ☽ △ ♃
19 03 ☽ △ ♅
23 00 ☽ △ ♆

15 Wednesday
00 11 ☽ ⊼ ♀
01 31 ☽ # ♀
01 56 ☽ # ☉
03 24 ☽ ⅊ ♂
04 58 ☽ # ♆
05 49 ☉ ∥ ♀
06 31 ☿ ∠ ♇
06 57 ☿ ⊻ ♄
08 59 ☽ ☍ ♇
09 17 ☽ ⅊ ☿
12 42 ☽ △ ☉
17 03 ☽ # ♅
19 49 ☽ ± ♂
21 10 ☽ ⅊ ♅

16 Thursday
00 57 ☽ ⅊ ♆
01 03 ☽ # ☿
02 29 ☽ ± ☿
04 15 ☽ # ♂
06 33 ☽ ⊼ ♂
10 48 ☽ △ ♄
12 25 ☽ ± ♅
13 43 ☽ ⊼ ☿
16 02 ☽ ⅊ ☉
16 05 ☽ ± ♆
21 39 ☽ □ ♃
22 25 ☽ ⊼ ♅

17 Friday
02 02 ☽ ⊼ ♆
07 37 ☽ ☍ ♀
07 59 ☽ ± ☉
11 21 ☽ ⊼ ♇
11 43 ☽ ⅊ ♄
18 26 ☽ ⊼ ☉
21 04 ☽ ± ♇

18 Saturday
02 22 ☽ ± ♄
02 48 ☽ Q ♃
03 20 ♀ ∥ ♆
05 54 ☽ # ♂
07 45 ☉ ⊥ ♀
08 26 ☽ # ☿
10 11 ☽ ☍ ♂
11 30 ☽ ⅊ ♇
12 02 ☽ ⊼ ♄
15 55 ☽ # ♅
19 46 ☽ ☍ ☿
21 49 ☽ ⚹ ♃
22 56 ☽ ☍ ♅

19 Sunday
02 03 ☽ # ♀
02 20 ☽ ☍ ♆
02 54 ☽ # ♆
05 16 ♀ ⊻ ♇
11 02 ☽ # ☉
11 15 ☽ △ ♇
11 46 ☽ ⊼ ♀
13 34 ♂ ∠ ♇
14 01 ♀ ∠ ♄
14 10 ♃ ± ♄
15 41 ☿ △ ♃
19 45 ☉ ⅊ ♃
20 24 ☽ # ♄
21 21 ☽ ∠ ♃
21 28 ☽ ☍ ☉
22 01 ☽ ± ♀

20 Monday
04 39 ☿ ☌ ♅
06 10 ♂ ⊻ ♄
11 26 ☽ ⊥ ♃
11 41 ☽ ☍ ♄
11 56 ☽ ⊼ ♂
13 24 ☽ ⅊ ♀
20 50 ☽ ⊻ ♃
21 52 ☽ ± ♂
22 19 ☽ ⊼ ♅

21 Tuesday
00 13 ☽ ⊼ ☿
01 39 ☽ ⊼ ♆
07 51 ☽ ± ♅
10 34 ☽ □ ♇
10 47 ☽ ± ☿
11 11 ☽ ± ♆
12 51 ☽ ⅊ ♂
14 52 ☿ ☌ ♆
15 11 ☽ △ ♀
22 14 ☽ ⅊ ♅
23 00 ☽ ∥ ♃

22 Wednesday
00 19 ☽ ⊼ ☉
01 35 ☽ ⅊ ♆
02 47 ☽ ⅊ ☿
10 42 ☽ ± ☉
11 49 ☽ ⊼ ♄
12 10 ☽ # ♇
14 14 ☽ △ ♂
15 32 ☽ Q ♇
20 41 ☽ ☌ ♃
21 42 ☽ ± ♄
22 39 ☽ △ ♅

23 Thursday
02 03 ☽ △ ♆
02 31 ☽ ⅊ ☉
06 07 ☽ △ ☿
11 20 ☽ ⚹ ♇
12 43 ☽ ⅊ ♄
20 43 ☽ □ ♀
21 22 ☽ ∥ ♇

24 Friday
04 58 ♀ ⊥ ♇
05 38 ☽ △ ☉
11 42 ☽ # ♃
12 50 ☽ ∠ ♇
14 27 ☽ △ ♄
19 29 ☽ □ ♂
23 13 ☽ ⊻ ♃

25 Saturday
01 48 ☽ □ ♅
04 35 ☽ ⊥ ♇
05 22 ☽ □ ♆
06 53 ☽ Q ♀
06 54 ☿ ⚹ ♇
09 45 ☽ ⊥ ♃
10 00 ☉ Q ♇
11 28 ☿ ∥ ♅
15 17 ☽ ⊻ ♇
16 23 ☽ □ ☿

26 Sunday
01 52 ☽ ∠ ♃
05 35 ☽ Q ♂
06 44 ☽ ⚹ ♀
10 22 ☽ Q ♅
14 03 ☽ Q ♆
15 27 ☽ □ ☉
19 51 ☽ ∥ ♄
20 34 ☽ ∥ ☉
20 56 ☽ □ ♄

27 Monday
05 16 ☽ ⚹ ♂
05 31 ☽ ⚹ ♃
06 10 ☽ Q ☿
08 25 ☉ ∥ ♄
08 54 ☽ ⚹ ♅
09 06 ♂ △ ♃
12 40 ☽ ⚹ ♆
13 45 ☽ ∠ ♀
17 30 ♀ ⊥ ♄
18 06 ☽ ∥ ♆
23 13 ☽ ☌ ♇

28 Tuesday
01 11 ☽ ∥ ♀
01 29 ☽ ∥ ☿
01 37 ☉ ± ♃
04 07 ☿ ∥ ♀
04 19 ☽ Q ♃
04 35 ☽ Q ☉
07 17 ☽ ∥ ♅
07 35 ☽ Q ♄
08 32 ☽ ⚹ ☿
08 59 ☽ ⊥ ♀
11 55 ☽ ∠ ♂
13 04 ☽ ∥ ♂
13 55 ☽ ∠ ♅
17 44 ☽ ∠ ♆
20 03 ☿ ⅊ ♃
22 01 ☽ ⊻ ♀

29 Wednesday
06 12 ☽ ⚹ ☉
06 46 ☽ ⊥ ♂
07 21 ☽ ⚹ ♄
07 42 ☽ ⊥ ♅
11 33 ☽ ⊥ ♆
15 26 ☽ □ ♃
18 36 ☽ ∠ ☿
19 29 ☽ ⊻ ♂
19 42 ☽ ⊻ ♅
21 33 ☉ ☌ ♄
22 58 ♂ ☌ ♅
23 33 ☽ ⊻ ♆

30 Thursday
10 31 ☽ ⊻ ♇
13 39 ☽ ∠ ♄
14 58 ☽ ∠ ☉
15 29 ☽ ⊥ ☿
21 47 ☿ ∥ ♆
22 38 ☽ ⊥ ♇

31 Friday
05 33 ☽ ⊻ ☿
08 02 ☽ ⊥ ♄
10 53 ☽ ⊥ ☉
15 27 ♀ ∠ ♇
16 54 ☽ ∠ ♇
17 04 ☽ ☌ ♀
19 59 ☽ ∥ ♂
20 20 ☽ ⊻ ♄
22 54 ☽ ∥ ♅

FEBRUARY 1992

LONGITUDES

Date	Sidereal time h m s	Sun ☉ ° ' "	Moon ☽ ° ' "	Moon ☽ 24.00 ° ' "	Mercury ☿ ° '	Venus ♀ ° '	Mars ♂ ° '	Jupiter ♃ ° '	Saturn ♄ ° '	Uranus ♅ ° '	Neptune ♆ ° '	Pluto ♇ ° '
01	20 43 49	11 ♒ 52 43	17 ♑ 09 54	23 ♑ 03 19	04 ♒ 16	08 ♑ 50	17 ♑ 18	12 ♍ 58	09 ♒ 33	15 ♑ 31	17 ♑ 23	22 ♏ 48
02	20 47 46	12 53 39	28 ♑ 56 50	04 ♒ 50 44	05 55	10 04	18 03	12 R 52	09 40	15 34	17 25	22 48
03	20 51 42	13 54 33	10 ♒ 45 16	16 ♒ 40 40	07 35	11 17	18 49	12 46	09 47	15 38	17 27	22 49
04	20 55 39	14 55 25	22 ♒ 37 11	28 ♒ 35 01	09 16	12 31	19 34	12 40	09 54	15 41	17 30	22 50
05	20 59 35	15 56 17	04 ♓ 34 21	10 ♓ 35 23	10 57	13 45	20 20	12 34	10 02	15 44	17 32	22 51
06	21 03 32	16 57 07	16 ♓ 38 20	22 ♓ 43 25	12 39	14 59	21 05	12 27	10 09	15 47	17 34	22 51
07	21 07 28	17 57 56	28 ♓ 50 51	05 ♈ 00 54	14 22	16 13	21 51	12 21	10 16	15 51	17 36	22 52
08	21 11 25	18 58 44	11 ♈ 13 50	17 ♈ 29 57	16 06	17 27	22 37	12 14	10 23	15 54	17 38	22 53
09	21 15 21	19 59 30	23 ♈ 49 34	00 ♉ 13 02	17 50	18 40	23 22	12 07	10 30	15 57	17 40	22 53
10	21 19 18	21 00 15	06 ♉ 40 41	13 ♉ 12 53	19 35	19 54	24 08	12 01	10 37	16 00	17 42	22 54
11	21 23 15	22 00 58	19 ♉ 49 59	26 ♉ 32 19	21 21	21 08	24 54	11 54	10 44	16 03	17 44	22 54
12	21 27 11	23 01 40	03 ♊ 20 09	10 ♊ 13 45	23 08	22 22	25 39	11 47	10 51	16 06	17 46	22 55
13	21 31 08	24 02 20	17 ♊ 13 14	24 ♊ 18 39	24 56	23 36	26 25	11 39	10 58	16 09	17 48	22 55
14	21 35 04	25 02 58	01 ♋ 29 54	08 ♋ 46 45	26 44	24 50	27 11	11 32	11 05	16 12	17 49	22 55
15	21 39 01	26 03 35	16 ♋ 08 47	23 ♋ 35 23	28 ♒ 33	26 04	27 57	11 25	11 13	16 15	17 51	22 56
16	21 42 57	27 04 09	01 ♌ 05 47	08 ♌ 39 01	00 ♓ 23	27 18	28 42	11 18	11 20	16 18	17 53	22 56
17	21 46 54	28 04 43	16 ♌ 13 59	23 ♌ 49 27	02 13	28 32	29 ♑ 28	11 10	11 27	16 21	17 55	22 56
18	21 50 50	29 ♒ 05 14	01 ♍ 24 07	08 ♍ 56 43	04 04	29 ♑ 46	00 ♒ 14	11 03	11 33	16 24	17 57	22 57
19	21 54 47	00 ♓ 05 44	16 ♍ 25 59	23 ♍ 50 46	05 55	01 ♒ 00	01 00	10 55	11 40	16 27	17 59	22 57
20	21 58 44	01 06 13	01 ♎ 10 06	08 ♎ 23 10	07 47	02 14	01 46	10 47	11 47	16 29	18 01	22 57
21	22 02 40	02 06 40	15 ♎ 29 22	22 ♎ 28 18	09 39	03 28	02 32	10 40	11 54	16 32	18 02	22 57
22	22 06 37	03 07 05	29 ♎ 19 47	06 ♏ 03 51	11 31	04 42	03 18	10 32	12 01	16 35	18 04	22 57
23	22 10 33	04 07 30	12 ♏ 40 41	19 ♏ 10 35	13 24	05 56	04 04	10 24	12 08	16 38	18 06	22 57
24	22 14 30	05 07 53	25 ♏ 34 02	01 ♐ 51 32	15 15	07 10	04 50	10 16	12 15	16 40	18 08	22 57
25	22 18 26	06 08 15	08 ♐ 03 42	14 ♐ 11 11	17 07	08 24	05 36	10 09	12 22	16 43	18 09	22 R 57
26	22 22 23	07 08 35	20 ♐ 14 39	26 ♐ 14 47	18 57	09 38	06 22	10 01	12 28	16 46	18 11	22 57
27	22 26 19	08 08 54	02 ♑ 12 14	08 ♑ 07 39	20 47	10 52	07 08	09 53	12 35	16 48	18 12	22 57
28	22 30 16	09 09 11	14 ♑ 01 41	19 ♑ 54 55	22 34	12 06	07 54	09 45	12 42	16 51	18 14	22 57
29	22 34 13	10 ♓ 09 27	25 ♑ 47 53	01 ♒ 41 06	24 ♓ 20	13 ♒ 20	08 ♒ 41	09 ♍ 37	12 ♒ 48	16 ♑ 53	18 ♑ 16	22 ♏ 57

Date	Moon True ☊ ° '	Moon Mean ☊ ° '	Moon ☽ Latitude ° '
01	09 ♑ 40	08 ♑ 08	00 N 41
02	09 R 39	08 05	01 44
03	09 35	08 02	02 42
04	09 30	07 59	03 33
05	09 23	07 55	04 15
06	09 16	07 52	04 47
07	09 08	07 49	05 05
08	09 02	07 46	05 10
09	08 57	07 43	05 00
10	08 54	07 39	04 35
11	08 53	07 36	03 55
12	08 D 53	07 33	03 01
13	08 55	07 30	01 55
14	08 56	07 27	00 N 40
15	08 R 56	07 24	00 S 39
16	08 54	07 20	01 57
17	08 50	07 17	03 07
18	08 43	07 14	04 05
19	08 35	07 11	04 44
20	08 27	07 08	05 04
21	08 19	07 04	05 04
22	08 12	07 01	04 46
23	08 08	06 58	04 13
24	08 05	06 55	03 28
25	08 D 05	06 52	02 34
26	08 05	06 49	01 34
27	08 06	06 45	00 S 32
28	08 R 06	06 42	00 N 32
29	08 ♑ 04	06 ♑ 39	01 N 33

DECLINATIONS

Date	Sun ☉ ° '	Moon ☽ ° '	Mercury ☿ ° '	Venus ♀ ° '	Mars ♂ ° '	Jupiter ♃ ° '	Saturn ♄ ° '	Uranus ♅ ° '	Neptune ♆ ° '	Pluto ♇ ° '
01	17 S 14	21 S 40	21 S 04	22 S 20	23 S 05	07 N 54	18 S 27	22 S 54	21 S 35	04 S 09
02	16 57	18 41	20 42	22 18	23 00	07 56	18 25	22 54	21 35	04 09
03	16 39	14 57	20 19	22 15	22 54	07 59	18 24	22 54	21 35	04 09
04	16 22	10 37	19 54	22 12	22 49	08 01	18 22	22 53	21 34	04 09
05	16 04	05 52	19 28	22 07	22 42	08 04	18 20	22 53	21 34	04 08
06	15 45	00 S 52	19 00	22 03	22 36	08 07	18 18	22 53	21 34	04 08
07	15 27	04 N 12	18 31	21 57	22 29	08 09	18 16	22 52	21 34	04 08
08	15 08	09 12	18 00	21 51	22 23	08 12	18 14	22 52	21 33	04 07
09	14 49	13 53	17 28	21 44	22 15	08 15	18 12	22 52	21 33	04 07
10	14 30	18 04	16 55	21 37	22 08	08 17	18 11	22 51	21 33	04 07
11	14 10	21 28	16 20	21 29	22 01	08 20	18 09	22 51	21 33	04 06
12	13 51	23 47	15 43	21 20	21 53	08 23	18 07	22 51	21 32	04 06
13	13 31	24 44	15 05	21 11	21 45	08 26	18 05	22 50	21 32	04 06
14	13 10	24 06	14 26	21 01	21 36	08 29	18 03	22 50	21 32	04 05
15	12 50	21 49	13 45	20 50	21 28	08 32	18 01	22 50	21 32	04 05
16	12 29	18 01	13 03	20 39	21 19	08 35	17 59	22 49	21 31	04 05
17	12 09	12 59	12 20	20 27	21 10	08 38	17 58	22 49	21 31	04 04
18	11 48	07 10	11 35	20 15	21 01	08 41	17 56	22 49	21 31	04 04
19	11 26	00 N 59	10 49	20 02	20 51	08 44	17 54	22 48	21 31	04 04
20	11 05	05 S 07	10 02	19 48	20 42	08 47	17 52	22 48	21 30	04 03
21	10 43	10 46	09 14	19 34	20 32	08 50	17 50	22 48	21 30	04 03
22	10 22	15 42	08 25	19 19	20 22	08 53	17 48	22 47	21 30	04 02
23	10 00	19 39	07 34	19 04	20 11	08 56	17 47	22 47	21 30	04 02
24	09 38	22 31	06 43	18 48	20 01	08 59	17 45	22 47	21 29	04 02
25	09 16	24 11	05 52	18 31	19 50	09 02	17 43	22 46	21 29	04 01
26	08 53	24 39	04 59	18 14	19 39	09 05	17 41	22 46	21 29	04 01
27	08 31	23 57	04 07	17 57	19 28	09 08	17 39	22 46	21 29	04 00
28	08 08	22 11	03 14	17 39	19 17	09 11	17 37	22 45	21 28	04 00
29	07 S 46	19 S 28	02 S 22	17 S 21	19 S 05	09 N 14	17 S 36	22 S 45	21 S 28	03 S 59

ZODIAC SIGN ENTRIES

Date	h	m	Planets
02	14	09	☽ ♒
05	02	51	☽ ♓
07	14	15	☽ ♈
09	23	36	☽ ♉
12	06	08	☽ ♊
14	09	31	☽ ♋
16	07	04	☿ ♓
16	10	15	☽ ♌
18	04	38	♂ ♒
18	09	47	☽ ♍
18	16	40	♀ ♒
19	09	43	☉ ♓
20	10	04	☽ ♎
22	13	11	☽ ♏
24	20	26	☽ ♐
27	07	33	☽ ♑
29	20	34	☽ ♒

LATITUDES

Date	Mercury ☿ ° '	Venus ♀ ° '	Mars ♂ ° '	Jupiter ♃ ° '	Saturn ♄ ° '	Uranus ♅ ° '	Neptune ♆ ° '	Pluto ♇ ° '
01	01 S 56	00 N 49	00 S 47	01 N 18	00 S 37	00 S 22	00 N 44	14 N 46
04	02 02	00 40	00 48	01 19	00 37	00 22	00 44	14 48
07	02 05	00 31	00 50	01 20	00 37	00 22	00 44	14 49
10	02 04	00 21	00 52	01 20	00 38	00 22	00 44	14 50
13	02 00	00 12	00 53	01 21	00 38	00 23	00 44	14 52
16	01 50	00 N 03	00 55	01 21	00 38	00 23	00 44	14 53
19	01 36	00 S 05	00 57	01 21	00 38	00 23	00 44	14 54
22	01 16	00 14	00 58	01 22	00 39	00 23	00 44	14 56
25	00 50	00 22	01 00	01 22	00 39	00 23	00 44	14 57
28	00 S 19	00 30	01 01	01 22	00 39	00 23	00 44	14 58
31	00 N 17	00 S 38	01 S 03	01 N 23	00 S 39	00 S 23	00 N 44	15 N 00

DATA

Julian Date	2448654
Delta T	+58 seconds
Ayanamsa	23° 45' 06"
Synetic vernal point	05° ♓ 21' 53"
True obliquity of ecliptic	23° 26' 26"

MOON'S PHASES, APSIDES AND POSITIONS ☽

Date	h	m	Phase	Longitude ° '	Eclipse Indicator
03	19	00	●	14 ♒ 12	
11	16	15	☽	22 ♉ 12	
18	08	04	○	28 ♌ 55	
25	07	56	☾	05 ♐ 58	

Day	h	m		
02	12	06	Apogee	
17	10	56	Perigee	
29	21	04	Apogee	
06	16	08	0N	
13	14	46	Max dec	24° N 45'
19	15	49	0S	
26	09	21	Max dec	24° S 39'

LONGITUDES

Date	Chiron ⚷ ° '	Ceres ⚳ ° '	Pallas ⚴ ° '	Juno ⚵ ° '	Vesta ⚶ ° '	Black Moon Lilith ⚸ ° '
01	06 ♌ 07	10 ♑ 04	18 ♐ 50	00 ♓ 30	25 ♍ 35	01 ♒ 12
11	05 ♌ 23	13 ♑ 52	22 ♐ 23	05 ♓ 17	24 ♍ 36	02 ♒ 19
21	04 ♌ 44	17 ♑ 32	25 ♐ 43	10 ♓ 11	22 ♍ 53	03 ♒ 26
31	04 ♌ 10	21 ♑ 03	28 ♐ 45	15 ♓ 09	20 ♍ 35	04 ♒ 32

ASPECTARIAN

h m	Aspects
01 Saturday	
00 13	☽ ⊻ ☉
03 32	☽ △ ♃
05 14	☽ ∥ ♀
08 38	☽ ☌ ♅
12 18	☽ ☌ ♂
12 27	☽ ☌ ♆
12 42	☽ ∥ ♆
14 56	♂ ☌ ♆
18 13	☽ ∥ ☿
19 44	☿ Q ♇
23 29	☽ ⚹ ♇
02 Sunday	
03 27	♀ ⊻ ♄
09 50	☽ P ♃
11 31	☉ ⊼ ♃
13 49	☽ ∥ ♄
23 56	☽ Q ♇
03 Monday	
00 40	☽ ∥ ☉
00 58	☿ ± ♃
03 59	☽ ± ♃
04 30	☽ ☌ ☿
10 01	☽ ☌ ♄
13 13	☽ ⊻ ♀
15 15	♂ ∥ ♅
16 03	☽ ⊼ ♃
19 00	☽ ☌ ☉
21 55	☽ ⊻ ♅
04 Tuesday	
01 37	☽ ⊻ ♆
02 46	☽ ⊥ ♀
05 26	☽ ⊻ ♂
10 06	☽ ⊥ ♅
12 26	☽ □ ♇
13 46	☽ ⊥ ♆
14 39	♀ △ ♃
18 21	☽ ⊥ ♂
21 53	☿ ☌ ♄
23 00	☽ ∠ ♀
05 Wednesday	
01 12	☽ # ♃
04 17	☽ ∠ ♅
06 58	☉ ⊻ ♅
07 54	☽ ∠ ♆
13 37	☽ ∠ ♂
20 25	☽ ∥ ♇
22 59	☽ ⊻ ♄
06 Thursday	
02 48	☽ ⊻ ☿
03 47	☽ ☍ ♃
08 21	☽ ⚹ ♀
09 26	☿ ⊼ ♃
10 19	☽ ⚹ ♅
11 01	☽ ⊥ ♄
12 41	☽ ⊻ ☉
13 50	☽ ⚹ ♆
16 38	☽ ⊥ ☿
21 22	☽ ⚹ ♂
07 Friday	
00 16	☽ △ ♇
01 35	☽ ⊥ ☉
02 56	☉ ⊻ ♆
04 29	♀ ☌ ♅
04 55	☽ ∠ ♄
10 02	☽ Q ♅
10 37	☽ Q ♀
11 38	☽ # ♇
13 11	☽ ∠ ☿
13 28	☽ Q ♆
20 44	☽ ∠ ☉
22 23	☽ Q ♂
08 Saturday	
00 26	☿ ∥ ♄
05 32	☽ P ♇
07 06	☽ ∥ ♃
09 11	☿ ⊻ ♅
10 21	☽ ⚹ ♄
13 55	☽ ⊼ ♃
15 45	♀ ☌ ♆
20 34	♂ ⚹ ♇
20 59	☽ □ ♅
22 49	☽ ⚹ ☿
22 49	☽ ± ♇
09 Sunday	
00 17	☽ □ ♆
01 11	☽ □ ♀
01 17	☽ ± ♃
04 07	☽ ⚹ ☉
09 28	☽ Q ♄
09 36	☿ ⊻ ♆
10 13	☽ ⊼ ♇
11 05	☽ □ ♂
16 43	☽ # ☉
18 09	☽ P ♃
10 Monday	
01 04	☽ Q ☿
04 37	☽ Q ☉
05 51	☽ # ☿
12 42	☽ # ♄
19 19	☽ □ ♄
21 43	☽ △ ♃
11 Tuesday	
00 11	♀ ∥ ♆
02 16	☿ ⊻ ♀
05 08	☽ △ ♅
08 11	☽ △ ♆
12 08	☽ # ♀
12 41	☽ # ♆
12 51	☉ ⊥ ♅
14 35	☽ △ ♀
15 09	☽ □ ☿
16 15	☽ □ ☉
16 29	☽ # ♂
17 31	☽ ☍ ♇
21 37	☽ △ ♂
21 42	☿ ⊥ ♅
12 Wednesday	
00 45	☽ # ♅
08 04	☽ P ♅
08 43	☉ ☌ ☿
09 00	☿ □ ♇
09 12	☉ □ ♇
10 59	☽ P ♆
19 44	☽ P ♀
20 34	☿ ⊥ ♆
22 40	♀ ⚹ ♇
23 49	☽ ± ♅
13 Thursday	
01 12	☽ △ ♄
01 28	☽ P ♂
02 33	☽ □ ♃
02 41	☽ ± ♆
05 58	☉ ⊥ ♆
10 10	☽ ⊼ ♅
12 42	☽ ± ♀
12 59	☽ ⊼ ♆
17 44	☽ ± ♂
18 34	♂ P ♃
21 40	☽ ⊼ ♇
23 49	☽ ⊼ ♀
14 Friday	
00 25	☽ △ ☉
02 55	☽ △ ☿
02 55	☽ P ♄
04 24	☽ ⊼ ♂
07 43	☽ ± ♇
08 46	☽ Q ♃
17 59	☽ ± ♄
22 14	☿ ⊻ ♂
22 36	☽ P ♇
15 Saturday	
01 11	♂ ∥ ♆
03 07	☽ P ☉
03 22	☽ # ♅
03 54	☽ ⊼ ♄
04 22	☽ ⚹ ♃
07 11	☽ P ☿
11 40	☉ ⊻ ♀
12 10	☽ ☍ ♅
14 15	☽ # ♆
14 46	☽ ☍ ♆
14 50	☽ # ♂
18 15	♀ P ♃
18 47	☽ ± ☉
19 33	☽ # ♀
22 57	☽ △ ♇
23 46	☽ ± ☿
16 Sunday	
04 24	☽ ∠ ♃
05 07	☽ ⊼ ☉
05 23	☽ ☍ ♀
07 59	☽ ☍ ♂
08 43	♃ ⊼ ♄
10 42	☽ ⊼ ☿
12 08	☽ # ♄
18 37	☽ ⊥ ♃
17 Monday	
00 23	☿ ∠ ♅
04 03	☽ ⊻ ♃
04 22	☽ ☍ ♄
12 11	☽ ⊼ ♅
14 40	☽ ⊼ ♆
15 17	☽ # ☿
15 54	☽ # ☉
21 17	☿ ∠ ♆
21 42	☽ ± ♅
22 36	☽ □ ♇
23 45	☉ ∥ ☿
18 Tuesday	
00 10	☽ ± ♆
06 01	☽ ∥ ♃
08 04	☽ ☍ ☉
09 10	☽ ⊼ ♀
10 03	☽ ⊼ ♂
11 59	☽ P ♅
14 28	☽ P ♆
16 49	☽ ☍ ☿
19 32	☽ ± ♀
20 05	☽ ± ♂
19 Wednesday	
00 07	☽ # ♇
03 12	☽ Q ♇
03 13	☽ ☌ ♃
04 18	☽ ⊼ ♄
06 13	♃ Q ♇
11 14	☽ P ♀
11 16	☽ P ♂
12 01	☽ △ ♅
12 22	♀ ☌ ♂
14 01	☽ ± ♄
14 30	☽ △ ♆
22 32	☽ ⚹ ♇
20 Thursday	
04 45	☽ P ♄
07 44	☽ ∥ ♇
11 34	☿ ⊥ ♂
11 53	☽ ⊼ ☉
13 02	☽ △ ♂
13 55	☽ △ ♀
21 39	☉ ∠ ♅
22 36	☽ ± ☉
23 16	☽ ∠ ♇
21 Friday	
00 39	☽ ⊼ ☿
03 23	☽ # ♃
03 54	☽ ⊻ ♃
04 40	☿ ⊥ ♀
05 52	☽ △ ♄
06 04	☽ ∥ ☿
11 48	☽ ∥ ☉
12 19	☽ ± ☿
13 48	☽ □ ♅
13 59	☽ ⊥ ♃
14 30	☽ ⊥ ♇
14 59	☽ P ☉
16 22	☽ □ ♆
23 09	☿ # ♃
22 Saturday	
00 06	☿ ☍ ♃
00 50	☽ ⊻ ♇
05 23	☽ ∠ ♃
06 17	☽ P ☿
09 10	♀ ± ♃
10 46	☉ ∠ ♆
17 06	♀ Q ♇
18 46	☿ ⊻ ♄
19 16	☽ △ ☉
19 28	☽ □ ♂
21 22	☽ Q ♅
22 30	☽ □ ♀
23 55	☽ ∥ ♄
23 Sunday	
00 02	☽ Q ♆
06 01	☉ ⊻ ♂
07 53	☽ ⚹ ♃
08 13	☽ ∥ ♀
11 00	☽ □ ♄
13 32	☽ △ ☿
15 37	☽ ∥ ♂
19 18	☽ ⚹ ♅
21 02	♂ ± ♃
22 01	☽ ⚹ ♆
24 Monday	
02 13	☽ ∥ ♆
05 51	☽ Q ♃
06 31	☽ Q ♂
07 04	☽ ☌ ♇
11 09	☽ Q ♀
14 50	☽ ∥ ♅
15 52	♂ Q ♇
20 59	☽ Q ♄
21 35	♇ St R
23 41	☽ ∠ ♅
25 Tuesday	
02 28	☽ ∠ ♆
06 44	☿ ⚹ ♅
06 54	☽ ⚹ ♂
07 56	☽ □ ☉
12 43	☽ ⚹ ♀
16 01	☽ □ ♃
17 12	☽ ⊥ ♅
20 01	☽ ⊥ ♆
20 29	☽ ⚹ ♄
26 Wednesday	
00 58	☉ # ♃
01 44	☿ ⚹ ♆
05 03	☽ ⊻ ♅
05 17	☿ ⊥ ♄
07 53	☽ ⊻ ♆
08 59	☽ □ ☿
14 24	☽ ∠ ♂
17 25	☽ ⊻ ♇
18 44	♀ ⊼ ♃
21 46	☽ ∠ ♀
22 41	☽ Q ☉
27 Thursday	
02 36	☽ ∠ ♄
05 27	☽ ⊥ ♇
09 42	☽ ⊥ ♂
15 07	☿ ∥ ♇
18 00	☽ ⊥ ♀
20 57	☽ ⊥ ♄
22 41	☽ ⊻ ♂
23 39	☽ ∠ ♇
28 Friday	
01 10	☽ ⚹ ☉
03 23	☽ △ ♃
03 43	☽ Q ☿
05 28	☽ ∥ ♅
07 37	☽ ⊻ ♀
09 16	☽ ⊻ ♄
14 14	♀ ∥ ♄
17 05	☿ △ ♇
17 46	☽ ☌ ♅
19 01	☽ ∥ ♆
19 54	☿ ∠ ♂
20 36	☽ ☌ ♆
29 Saturday	
00 37	☉ ☍ ♃
00 48	♀ ☌ ♄
06 12	☽ ⚹ ♇
08 30	☽ ⚹ ☿
09 37	☽ P ♃
10 34	☽ ∠ ☉
15 01	☽ ∥ ♂

All ephemeris data is given at 12.00 UT and the Moon's longitude is additionally given for 24.00 UT

MARCH 1992

LONGITUDES

Date	Sidereal time h m s	Sun ☉ ° ' "	Moon ☽ ° ' "	Moon ☽ 24.00 ° ' "	Mercury ☿ ° '	Venus ♀ ° '	Mars ♂ ° '	Jupiter ♃ ° '	Saturn ♄ ° '	Uranus ♅ ° '	Neptune ♆ ° '	Pluto ⯓ ° '
01	22 38 09	11 ♓ 09 41	07 ♒ 35 01	13 ♒ 30 01	26 ♓ 04	14 ♒ 34	09 ♒ 27	09 ♍ 29	12 ♒ 55	16 ♑ 56	18 ♑ 17	22 ♏ 57
02	22 42 06	12 09 54	19 ♒ 26 28	25 ♒ 24 40	27 44	15 48	10 13	09 R 21	13 02	16 58	18 19	22 R 57
03	22 46 02	13 10 05	01 ♓ 24 51	07 ♓ 27 12	29 ♓ 22	17 02	10 59	09 14	13 08	17 00	18 20	22 56
04	22 49 59	14 10 14	13 ♓ 31 54	19 ♓ 39 02	00 ♈ 55	18 16	11 45	09 06	13 15	17 03	18 22	22 56
05	22 53 55	15 10 21	25 ♓ 48 42	02 ♈ 00 57	02 24	19 30	12 32	08 58	13 21	17 05	18 23	22 56
06	22 57 52	16 10 26	08 ♈ 15 51	14 ♈ 33 26	03 47	20 44	13 18	08 50	13 28	17 07	18 25	22 55
07	23 01 48	17 10 29	20 ♈ 53 45	27 ♈ 16 52	05 05	21 59	14 04	08 42	13 34	17 09	18 26	22 55
08	23 05 45	18 10 31	03 ♉ 42 51	10 ♉ 11 50	06 17	23 13	14 51	08 35	13 40	17 12	18 27	22 55
09	23 09 42	19 10 30	16 ♉ 43 55	23 ♉ 19 18	07 22	24 27	15 37	08 27	13 47	17 14	18 29	22 54
10	23 13 38	20 10 27	29 ♉ 58 08	06 ♊ 40 37	08 19	25 41	16 23	08 19	13 53	17 16	18 30	22 54
11	23 17 35	21 10 22	13 ♊ 26 58	20 ♊ 17 23	09 10	26 55	17 10	08 12	13 59	17 18	18 31	22 53
12	23 21 31	22 10 15	27 ♊ 12 01	04 ♋ 10 59	09 52	28 09	17 56	08 04	14 05	17 20	18 33	22 53
13	23 25 28	23 10 05	11 ♋ 14 20	18 ♋ 22 02	10 26	29 ♒ 23	18 42	07 57	14 12	17 22	18 34	22 52
14	23 29 24	24 09 54	25 ♋ 33 54	02 ♌ 49 40	10 52	00 ♓ 37	19 29	07 50	14 18	17 24	18 35	22 51
15	23 33 21	25 09 39	10 ♌ 08 52	17 ♌ 30 53	11 09	01 51	20 15	07 42	14 24	17 26	18 36	22 51
16	23 37 17	26 09 23	24 ♌ 55 01	02 ♍ 20 20	11 18	03 05	21 02	07 35	14 30	17 28	18 37	22 50
17	23 41 14	27 09 05	09 ♍ 45 51	17 ♍ 10 31	11 R 18	04 19	21 48	07 28	14 36	17 29	18 39	22 49
18	23 45 11	28 08 44	24 ♍ 33 14	01 ♎ 52 55	11 10	05 34	22 34	07 21	14 42	17 31	18 40	22 49
19	23 49 07	29 ♓ 08 21	09 ♎ 08 34	16 ♎ 19 19	10 54	06 48	23 21	07 14	14 47	17 33	18 41	22 48
20	23 53 04	00 ♈ 07 56	23 ♎ 24 24	00 ♏ 23 16	10 31	08 02	24 07	07 08	14 53	17 34	18 42	22 47
21	23 57 00	01 07 30	07 ♏ 15 33	14 ♏ 01 01	10 01	09 16	24 54	07 01	14 59	17 36	18 43	22 46
22	00 00 57	02 07 02	20 ♏ 39 40	27 ♏ 11 39	09 25	10 30	25 40	06 54	15 05	17 38	18 44	22 45
23	00 04 53	03 06 32	03 ♐ 37 14	09 ♐ 56 50	08 43	11 44	26 27	06 48	15 10	17 39	18 45	22 44
24	00 08 50	04 06 00	16 ♐ 10 57	22 ♐ 20 10	07 58	12 58	27 13	06 42	15 16	17 40	18 45	22 43
25	00 12 46	05 05 26	28 ♐ 25 06	04 ♑ 26 27	07 09	14 12	28 00	06 35	15 21	17 42	18 46	22 42
26	00 16 43	06 04 51	10 ♑ 24 50	16 ♑ 21 09	06 18	15 26	28 46	06 29	15 27	17 43	18 47	22 42
27	00 20 40	07 04 14	22 ♑ 15 53	28 ♑ 09 47	05 26	16 40	29 ♒ 33	06 23	15 32	17 45	18 48	22 41
28	00 24 36	08 03 35	04 ♒ 03 31	09 ♒ 57 41	04 35	17 54	00 ♓ 19	06 18	15 37	17 46	18 49	22 39
29	00 28 33	09 02 54	15 ♒ 52 52	21 ♒ 49 35	03 44	19 08	01 06	06 12	15 42	17 47	18 49	22 38
30	00 32 29	10 02 11	27 ♒ 48 19	03 ♓ 49 28	02 55	20 22	01 52	06 06	15 48	17 48	18 50	22 37
31	00 36 26	11 ♈ 01 27	09 ♓ 53 23	16 ♓ 00 19	02 ♈ 09	21 ♓ 36	02 ♓ 39	06 ♍ 01	15 ♒ 53	17 ♑ 49	18 ♑ 51	22 ♏ 36

DECLINATIONS

Date	Moon True ☊ ° '	Moon Mean ☊ ° '	Moon ☽ Latitude ° '	Sun ☉ ° '	Moon ☽ ° '	Mercury ☿ ° '	Venus ♀ ° '	Mars ♂ ° '	Jupiter ♃ ° '	Saturn ♄ ° '	Uranus ♅ ° '	Neptune ♆ ° '	Pluto ⯓ ° '
01	08 ♑ 00	06 ♑ 36	02 N 31	07 S 23	15 S 57	01 S 30	17 S 02	18 S 53	09 N 17	17 S 34	22 S 45	21 S 28	03 S 59
02	07 R 53	06 33	03 22	07 00	11 48	00 S 38	16 42	18 41	09 20	17 32	22 45	21 28	03 58
03	07 43	06 30	04 04	06 37	07 10	00 N 12	16 22	18 29	09 23	17 30	22 44	21 28	03 58
04	07 32	06 26	04 36	06 14	02 S 13	01 02	16 02	18 17	09 26	17 29	22 44	21 27	03 57
05	07 19	06 23	04 56	05 51	02 N 52	01 49	15 41	18 04	09 29	17 27	22 44	21 27	03 57
06	07 06	06 20	05 02	05 27	07 54	02 35	15 20	17 51	09 32	17 25	22 43	21 27	03 56
07	06 55	06 17	04 53	05 04	12 41	03 19	14 58	17 39	09 35	17 23	22 43	21 27	03 56
08	06 45	06 14	04 30	04 41	16 59	04 00	14 36	17 25	09 38	17 22	22 43	21 27	03 55
09	06 38	06 10	03 52	04 17	20 32	04 38	14 13	17 12	09 41	17 20	22 43	21 26	03 55
10	06 34	06 07	03 01	03 54	23 06	05 13	13 50	16 59	09 44	17 18	22 42	21 26	03 54
11	06 33	06 04	01 59	03 30	24 23	05 45	13 27	16 45	09 47	17 16	22 42	21 26	03 54
12	06 D 33	06 01	00 N 49	03 06	24 14	06 13	13 03	16 31	09 49	17 15	22 42	21 26	03 53
13	06 R 33	05 58	00 S 25	02 43	22 33	06 37	12 39	16 17	09 52	17 13	22 42	21 26	03 53
14	06 32	05 55	01 39	02 19	19 24	06 57	12 15	16 03	09 55	17 11	22 42	21 25	03 52
15	06 28	05 51	02 48	01 55	15 00	07 12	11 50	15 49	09 58	17 10	22 41	21 25	03 52
16	06 22	05 48	03 46	01 32	09 40	07 23	11 25	15 35	10 00	17 08	22 41	21 25	03 51
17	06 13	05 45	04 30	01 08	03 N 45	07 30	11 00	15 20	10 03	17 07	22 41	21 25	03 51
18	06 02	05 42	04 55	00 44	02 S 21	07 32	10 34	15 05	10 05	17 05	22 41	21 25	03 50
19	05 51	05 39	05 00	00 S 21	08 13	07 30	10 08	14 50	10 08	17 03	22 41	21 25	03 50
20	05 39	05 36	04 47	00 N 03	13 31	07 23	09 42	14 35	10 10	17 02	22 40	21 24	03 49
21	05 29	05 32	04 16	00 27	17 58	07 11	09 16	14 20	10 13	17 00	22 40	21 24	03 49
22	05 22	05 29	03 32	00 51	21 20	06 56	08 49	14 05	10 15	16 59	22 40	21 24	03 48
23	05 17	05 26	02 39	01 14	23 29	06 37	08 22	13 49	10 18	16 57	22 40	21 24	03 47
24	05 14	05 23	01 39	01 38	24 22	06 15	07 55	13 34	10 20	16 56	22 40	21 24	03 47
25	05 D 14	05 20	00 S 36	02 01	24 02	05 49	07 27	13 18	10 22	16 54	22 40	21 24	03 46
26	05 R 14	05 16	00 N 27	02 25	22 35	05 22	07 00	13 02	10 24	16 53	22 39	21 24	03 46
27	05 13	05 13	01 29	02 48	20 08	04 52	06 32	12 46	10 26	16 51	22 39	21 24	03 45
28	05 11	05 10	02 26	03 12	16 52	04 21	06 04	12 30	10 28	16 50	22 39	21 23	03 45
29	05 06	05 07	03 17	03 35	12 56	03 49	05 36	12 14	10 30	16 48	22 39	21 23	03 44
30	04 59	05 04	04 00	03 59	08 29	03 17	05 08	11 58	10 32	16 47	22 39	21 23	03 44
31	04 ♑ 49	05 ♑ 01	04 N 33	04 N 22	03 S 39	02 N 46	04 S 39	11 S 41	10 N 34	16 S 45	22 S 39	21 S 23	03 S 43

ZODIAC SIGN ENTRIES

Date	h m	Planets
03	09 11	☽ ♓
03	21 45	☿ ♈
05	20 07	☽ ♈
08	05 05	☽ ♉
10	12 03	☽ ♊
12	16 50	☽ ♋
13	23 57	♀ ♓
14	19 20	☽ ♌
16	20 13	☽ ♍
18	20 55	☽ ♎
20	08 48	☉ ♈
20	23 20	☽ ♏
23	05 13	☽ ♐
25	15 08	☽ ♑
28	02 04	♂ ♓
28	03 44	☽ ♒
30	16 23	☽ ♓

LATITUDES

Date	Mercury ☿ ° '	Venus ♀ ° '	Mars ♂ ° '	Jupiter ♃ ° '	Saturn ♄ ° '	Uranus ♅ ° '	Neptune ♆ ° '	Pluto ⯓ ° '
01	00 N 05	00 S 35	01 S 02	01 N 23	00 S 39	00 S 23	00 N 44	14 N 59
04	00 44	00 43	01 04	01 23	00 40	00 23	00 44	15 01
07	01 25	00 50	01 05	01 23	00 40	00 23	00 44	15 02
10	02 05	00 56	01 06	01 23	00 40	00 23	00 44	15 03
13	02 42	01 02	01 08	01 23	00 41	00 23	00 44	15 04
16	03 11	01 07	01 09	01 23	00 41	00 23	00 44	15 05
19	03 27	01 12	01 10	01 23	00 41	00 23	00 44	15 07
22	03 29	01 17	01 11	01 22	00 42	00 23	00 44	15 08
25	03 15	01 21	01 12	01 22	00 42	00 24	00 44	15 09
28	02 45	01 24	01 13	01 22	00 42	00 24	00 44	15 10
31	02 N 05	01 S 26	01 S 14	01 N 22	00 S 43	00 S 24	00 N 44	15 N 11

DATA

Julian Date	2448683
Delta T	+58 seconds
Ayanamsa	23° 45' 10"
Synetic vernal point	05° ♓ 21' 50"
True obliquity of ecliptic	23° 26' 27"

MOON'S PHASES, APSIDES AND POSITIONS ☽

Date	h m	Phase	Longitude ° '	Eclipse Indicator
04	13 22	●	14 ♓ 14	
12	02 36	☽	21 ♊ 47	
18	18 18	○	28 ♍ 24	
26	02 30	☾	05 ♑ 41	

Day	h m		
16	17 46	Perigee	
28	14 20	Apogee	
04	22 31	0N	
11	21 36	Max dec	24° N 30'
18	02 44	0S	
24	17 13	Max dec	24° S 24'

LONGITUDES

Date	Chiron ⚷ ° '	Ceres ⚳ ° '	Pallas ⚴ ° '	Juno ⚵ ° '	Vesta ⚶ ° '	Black Moon Lilith ⚸ ° '
01	04 ♌ 13	20 ♑ 42	28 ♐ 28	14 ♓ 39	20 ♍ 50	04 ♒ 26
11	03 ♌ 45	24 ♑ 05	01 ♑ 12	19 ♓ 41	18 ♍ 15	05 ♒ 32
21	03 ♌ 27	27 ♑ 15	03 ♑ 34	24 ♓ 48	15 ♍ 43	06 ♒ 39
31	03 ♌ 18	00 ♒ 10	05 ♑ 27	29 ♓ 58	13 ♍ 34	07 ♒ 45

All ephemeris data is given at 12.00 UT and the Moon's longitude is additionally given for 24.00 UT

ASPECTARIAN

h m Aspects

01 Sunday
01 27 ☽ ∥ ♄
03 46 ☽ ± ♃
04 28 ☽ ∥ ♀
06 37 ☽ ⊥ ☉
06 39 ☽ Q ⯓
13 06 ♂ ⊼ ♃
15 49 ☽ ⊼ ♃
16 03 ☽ ☌ ♂
19 56 ☽ ⊻ ☉
20 15 ☽ ∠ ☿
22 55 ☽ ☌ ♄

02 Monday
03 48 ☽ ☌ ♀
07 00 ☽ ⊻ ♅
09 43 ☽ ⊻ ♆
16 30 ☿ ∠ ♄
17 22 ☽ ⊥ ☿
19 03 ☽ □ ⯓
19 07 ☽ ⊥ ♅
21 49 ☽ ⊥ ♆

03 Tuesday
00 51 ☽ # ♃
06 32 ☿ Q ♅
07 17 ☽ ⊻ ☿
11 12 ☉ ⊻ ♄
11 24 ♀ ⊻ ♅
13 11 ☽ ∠ ♅
14 58 ☽ ∥ ☉
15 50 ☽ ∠ ♆

04 Wednesday
03 13 ☿ Q ♆
03 20 ☽ ☍ ♃
03 41 ☽ ∥ ⯓
08 16 ☽ ⊻ ♂
11 26 ☽ ⊻ ♄
13 22 ☽ ☌ ☉
13 48 ♀ ⊻ ♆
16 52 ☽ # ☿
18 55 ☽ ✶ ♅
20 51 ☽ ⊥ ♂
21 30 ☽ ✶ ♆
22 21 ☽ ⊻ ♀
23 19 ☽ ⊥ ♄

05 Thursday
06 13 ☽ ∥ ☿
06 24 ☽ △ ⯓
11 21 ☽ ⊥ ♀
15 33 ☽ ∠ ♂
16 58 ☽ ∠ ♄
17 07 ☽ # ⯓
18 22 ☽ Q ♅
20 52 ☽ Q ♆

06 Friday
01 06 ☽ # ☉
02 21 ☽ ☌ ☿
06 38 ☽ ∠ ♀
11 21 ☽ ♇ ⯓
13 05 ☽ ⊼ ♃
17 52 ♂ ☌ ♄
20 06 ☽ ∥ ♃
22 00 ☽ ✶ ♄
22 14 ☽ ✶ ♂

07 Saturday
00 24 ☽ ± ♃
04 22 ☽ ⊻ ☉
04 29 ☽ ± ⯓
04 55 ☽ □ ♅
07 20 ☽ □ ♆
11 35 ☉ ✶ ♅
14 15 ☽ ✶ ♀
15 48 ☽ ⊼ ⯓
16 40 ☽ ⊥ ☉
17 14 ☽ ♇ ♃
20 52 ☽ Q ♄
22 21 ☽ Q ♂
23 22 ☽ # ♀

08 Sunday
06 10 ♀ □ ⯓
09 16 ☿ # ⯓
10 55 ☽ ∠ ☉
11 40 ♀ ⊥ ♅
14 19 ☽ # ♄
14 35 ☽ # ♂
15 04 ☽ Q ♀
17 13 ☽ ⊻ ☿
18 55 ☉ ✶ ♆
20 00 ♂ ∥ ♄
20 55 ☽ △ ♃

09 Monday
03 40 ☉ # ☿
05 17 ☽ ⊥ ☿
06 33 ☽ □ ♄
09 50 ☽ □ ♂
12 40 ♀ ⊥ ♆
12 55 ☽ △ ♅
15 12 ☽ △ ♆
16 50 ☽ ✶ ☉
19 21 ☽ # ♆
23 06 ☽ ∠ ☿
23 14 ☽ ☍ ⯓

10 Tuesday
00 59 ☿ ♇ ⯓
03 29 ☽ □ ♀
04 14 ☉ ⊥ ♄
07 35 ☽ # ♅
11 09 ☉ ∥ ⯓
11 59 ☿ ⊼ ♃
16 08 ☽ ♇ ♅
16 17 ☽ Q ☉
18 21 ☽ ♇ ♆

11 Wednesday
02 47 ☽ □ ♃
03 57 ☽ ✶ ☿
08 12 ☽ ± ♅
10 22 ☽ ± ♆
12 57 ☽ △ ♄
16 30 ♂ ⊻ ♅
18 47 ☽ ⊼ ♅
18 55 ☽ △ ♂
20 56 ☽ ⊼ ♆

12 Thursday
02 17 ☽ Q ☿
02 36 ☽ □ ☉
04 31 ☽ ⊼ ⯓
10 04 ☽ Q ♃
13 48 ☽ △ ♀
14 53 ☽ ± ⯓
15 17 ☽ ♇ ♄
22 26 ☽ ♇ ♂

13 Friday
04 48 ☉ △ ⯓
06 17 ☽ ♇ ⯓
06 28 ☽ ✶ ♃
06 48 ☽ ± ♄
07 30 ♂ ⊻ ♆
10 33 ☽ # ♅
10 35 ☽ □ ☿
14 37 ☽ ± ♂
17 01 ☽ ⊼ ♄
17 49 ☽ ♇ ♀
21 53 ☽ # ♆
22 21 ☽ ☍ ♅

14 Saturday
00 21 ☽ ☍ ♆
01 17 ☽ ⊼ ♂
07 29 ☽ ∠ ♃
07 30 ☽ △ ⯓
09 30 ☽ △ ☉
10 17 ☽ ± ♀
21 08 ☽ ⊼ ♀
22 16 ☽ ⊥ ♃

15 Sunday
00 56 ☽ # ♄
07 47 ☽ # ♂
08 03 ☽ ⊻ ♃
12 01 ☽ ♇ ☉
13 40 ☽ △ ☿
18 59 ☽ ☍ ♄
23 26 ♀ ∠ ♅
23 53 ☽ ⊼ ♅

16 Monday
01 47 ☽ ⊼ ♆
03 44 ☽ ± ☉
03 50 ☽ # ♀
05 21 ☽ ☍ ♂
08 38 ☽ □ ⯓
09 38 ☽ ± ♅
10 34 ☽ ∥ ♃
11 31 ☽ ± ♆
14 09 ☽ ⊼ ☉
14 14 ☽ ♇ ☿
21 12 ☽ ∥ ☿
22 32 ♀ ∠ ♆

17 Tuesday
00 13 ☽ ♇ ♅
00 33 ☿ St R
02 06 ☽ ♇ ♆
02 25 ☽ ☍ ♀
04 48 ☽ ± ☿
08 19 ☽ ☌ ♃
11 36 ☽ # ⯓
13 43 ☽ Q ⯓
14 28 ☽ ⊼ ☿
19 52 ☽ ⊼ ♄
23 00 ☽ # ☉

18 Wednesday
00 32 ☽ △ ♅
02 24 ☽ △ ♆
05 17 ☿ ⊥ ♀
05 40 ☽ ± ♄
06 02 ☽ ∥ ☉
08 36 ☽ ⊼ ♂
09 10 ☽ ✶ ⯓
17 56 ☽ ∥ ⯓
18 18 ☽ ☍ ☉
18 56 ☽ ± ♂
19 11 ♂ □ ⯓
20 28 ☽ ♇ ♄

19 Thursday
07 44 ☽ ⊼ ♀
08 52 ☽ ⊻ ♃
08 58 ☽ # ☿
09 46 ☽ ∠ ⯓
10 36 ☽ ♇ ♂
12 23 ♀ # ♃
14 51 ☽ ☍ ☿
18 22 ♂ ⊥ ♅
18 39 ☽ ± ♀
18 47 ☽ ⊥ ♃
19 42 ☽ ∥ ♀
19 57 ♀ ☍ ♃
20 21 ☽ # ♃
21 29 ☽ △ ♄
22 05 ☉ Q ♅

20 Friday
00 47 ☽ ⊥ ⯓
02 05 ☽ □ ♅
03 59 ☽ □ ♆
05 24 ☉ ∠ ♄
09 50 ☽ ∠ ♃
10 49 ☉ ⊥ ♂
10 56 ☽ ⊻ ⯓
11 17 ☽ ♇ ♀
13 18 ☽ △ ♂
17 02 ☽ ∥ ♂

21 Saturday
00 27 ☽ ⊼ ☉
01 49 ☉ Q ♆
06 09 ♂ ⊥ ♆
06 20 ☽ ∥ ♄
09 05 ☽ Q ♅
11 02 ☽ Q ♆
11 34 ☽ ✶ ♃
11 45 ☽ ± ☉
14 04 ☿ ∠ ♂
15 53 ☽ △ ♀
16 40 ☽ ⊼ ☿
21 56 ☿ ⊻ ♀

22 Sunday
01 50 ☽ □ ♄
02 56 ☽ ± ☿
05 03 ☽ ♇ ☉
06 29 ☽ ✶ ♅
08 29 ☽ ✶ ♆
08 50 ☽ Q ♃
12 38 ☽ ∥ ♆
15 49 ☽ ☌ ⯓
18 32 ☽ ♇ ☿
21 46 ☽ □ ♂

23 Monday
01 11 ☽ ∥ ♅
10 10 ☽ ∠ ♅
10 57 ☽ △ ☉
11 09 ☽ Q ♄
12 14 ☽ ∠ ♆
17 57 ☽ □ ♃
21 08 ☽ △ ☿

24 Tuesday
03 17 ☽ ⊥ ♅
05 06 ☽ □ ♀
05 23 ☽ ⊥ ♆
10 01 ☽ Q ♂
10 12 ☽ ✶ ♄
14 54 ☽ ⊻ ♅
17 01 ☽ ⊻ ♆
19 23 ☿ ♇ ⯓

25 Wednesday
00 45 ☽ ⊻ ⯓
11 06 ☽ ✶ ♂
12 34 ☽ ⊥ ⯓
15 52 ☽ ∠ ♄
20 23 ☽ Q ♀

26 Thursday
02 30 ☽ □ ☉
04 10 ☽ △ ♃
04 17 ☽ □ ☿
06 12 ☿ ⊼ ♃
06 32 ☽ ∠ ⯓
10 02 ☽ ⊥ ♄
11 00 ☽ ∥ ♅
12 11 ♀ ⊻ ♄
14 55 ☉ ☌ ☿
19 15 ☽ ∠ ♂
20 59 ☉ ⊼ ♃
22 14 ☽ ⊻ ♄
23 19 ☽ ✶ ♀

27 Friday
00 45 ☽ ∥ ♆
02 48 ☽ ☌ ♅
04 57 ☽ ☌ ♆
10 14 ☽ ♇ ♃
10 29 ☿ ⊥ ♂
12 50 ☽ ✶ ⯓
14 14 ☽ Q ☿
14 47 ☽ ⊥ ♂
18 14 ☽ Q ☉

28 Saturday
02 25 ☉ ♇ ⯓
03 51 ☽ ⊻ ♂
04 24 ☽ ± ♃
09 14 ♀ ✶ ♅
09 22 ☽ ∠ ♀
12 18 ☽ ∥ ♄
12 59 ☽ ✶ ☿
13 13 ☽ Q ⯓
16 31 ☽ ⊼ ♃
20 53 ☽ ✶ ☉

29 Sunday
05 48 ☽ ⊥ ♀
05 51 ♀ ✶ ♆
11 39 ☽ ☌ ♄
15 51 ☽ ⊻ ♅
15 52 ☿ # ⯓
16 13 ☽ ∥ ♂
17 23 ☽ ∠ ☿
17 57 ☽ ⊻ ♆
18 04 ☉ ∥ ☿
19 21 ☽ ⊻ ♀
20 57 ☉ # ⯓

30 Monday
01 15 ☽ # ♃
01 37 ☽ □ ⯓
03 58 ☽ ⊥ ♅
05 57 ☽ ∠ ☉
06 03 ☽ ⊥ ♆
10 20 ☽ ⊥ ☿
20 40 ☽ ☌ ♂
21 35 ☽ ⊻ ☿
21 59 ☽ ∠ ♅

31 Tuesday
00 02 ☽ ∠ ♆
01 32 ☽ ⊥ ☉
04 11 ☿ ⊻ ♂
04 24 ☽ ☍ ♃
06 36 ☽ ∥ ♀
08 47 ☽ # ☉
11 40 ☽ ∥ ⯓
14 26 ☽ ⊻ ☉
16 47 ☽ # ☿
17 32 ♂ ∠ ♅
17 42 ♀ ⊥ ♄
20 02 ☉ # ♀
23 50 ☽ ⊻ ♄

APRIL 1992

Date	Sidereal time h m s	LONGITUDES Sun ☉ ° ' "	Moon ☽ ° ' "	Moon ☽ 24.00 ° ' "	Mercury ☿ ° '	Venus ♀ ° '	Mars ♂ ° '	Jupiter ♃ ° '	Saturn ♄ ° '	Uranus ♅ ° '	Neptune ♆ ° '	Pluto ♇ ° '
01	00 40 22	12 ♈ 00 40	22 ♓ 10 30	28 ♓ 24 04	01 ♈ 27	22 ♓ 50	03 ♓ 25	05 ♍ 56	15 ♒ 58	17 ♑ 50	18 ♑ 52	22 ♏ 35
02	00 44 19	12 59 52	04 ♈ 41 04	11 ♈ 01 31	00 R 48	24 04	04 12	05 R 51	16 03	17 51	18 52	22 R 34
03	00 48 15	13 59 01	17 ♈ 25 23	23 ♈ 52 34	00 ♈ 15	25 19	04 58	05 46	16 08	17 52	18 53	22 33
04	00 52 12	14 58 09	00 ♉ 22 57	06 ♉ 56 22	29 ♓ 46	26 33	05 45	05 41	16 13	17 53	18 53	22 32
05	00 56 09	15 57 14	13 ♉ 32 42	20 ♉ 11 47	29 23	27 47	06 31	05 36	16 17	17 54	18 54	22 30
06	01 00 05	16 56 18	26 ♉ 53 30	03 ♊ 37 44	29 05	29 ♓ 01	07 18	05 32	16 22	17 55	18 54	22 29
07	01 04 02	17 55 19	10 ♊ 24 24	17 ♊ 13 28	28 53	00 ♈ 15	08 04	05 28	16 27	17 56	18 55	22 28
08	01 07 58	18 54 18	24 ♊ 04 53	00 ♋ 58 40	28 46	01 29	08 51	05 24	16 31	17 56	18 55	22 26
09	01 11 55	19 53 14	07 ♋ 54 50	14 ♋ 53 23	28 D 44	02 43	09 37	05 20	16 36	17 57	18 55	22 25
10	01 15 51	20 52 09	21 ♋ 54 18	28 ♋ 57 33	28 48	03 57	10 24	05 16	16 40	17 58	18 56	22 24
11	01 19 48	21 51 01	06 ♌ 03 02	13 ♌ 10 35	28 57	05 11	11 10	05 12	16 44	17 58	18 56	22 22
12	01 23 44	22 49 50	20 ♌ 19 57	27 ♌ 30 47	29 11	06 25	11 57	05 09	16 49	17 59	18 56	22 21
13	01 27 41	23 48 37	04 ♍ 42 40	11 ♍ 55 02	29 30	07 38	12 43	05 06	16 53	17 59	18 57	22 20
14	01 31 38	24 47 22	19 ♍ 07 17	26 ♍ 18 43	29 ♓ 54	08 52	13 30	05 02	16 57	18 00	18 57	22 18
15	01 35 34	25 46 05	03 ♎ 28 36	10 ♎ 36 12	00 ♈ 22	10 06	14 16	05 00	17 01	18 00	18 57	22 17
16	01 39 31	26 44 46	17 ♎ 40 46	24 ♎ 41 38	00 54	11 20	15 03	04 57	17 05	18 00	18 57	22 15
17	01 43 27	27 43 24	01 ♏ 38 11	08 ♏ 29 54	01 31	12 34	15 49	04 54	17 09	18 00	18 57	22 14
18	01 47 24	28 42 01	15 ♏ 16 24	21 ♏ 57 26	02 11	13 48	16 35	04 52	17 12	18 01	18 57	22 12
19	01 51 20	29 ♈ 40 36	28 ♏ 32 51	05 ♐ 02 39	02 55	15 02	17 22	04 50	17 16	18 01	18 57	22 11
20	01 55 17	00 ♉ 39 10	11 ♐ 26 59	17 ♐ 46 04	03 42	16 16	18 08	04 48	17 20	18 01	18 57	22 09
21	01 59 13	01 37 41	24 ♐ 00 15	00 ♑ 09 58	04 33	17 30	18 54	04 46	17 23	18 01	18 R 57	22 08
22	02 03 10	02 36 11	06 ♑ 15 42	12 ♑ 18 01	05 27	18 44	19 41	04 44	17 27	18 R 01	18 57	22 06
23	02 07 07	03 34 39	18 ♑ 17 32	24 ♑ 14 52	06 24	19 58	20 27	04 43	17 30	18 01	18 57	22 05
24	02 11 03	04 33 06	00 ♒ 10 41	06 ♒ 05 39	07 24	21 12	21 13	04 42	17 33	18 01	18 57	22 03
25	02 15 00	05 31 31	12 ♒ 00 27	17 ♒ 55 42	08 27	22 25	22 00	04 41	17 36	18 01	18 57	22 02
26	02 18 56	06 29 54	23 ♒ 52 04	29 ♒ 50 09	09 32	23 39	22 46	04 40	17 40	18 00	18 57	22 00
27	02 22 53	07 28 16	05 ♓ 50 32	11 ♓ 53 42	10 41	24 53	23 32	04 39	17 43	18 00	18 57	21 58
28	02 26 49	08 26 36	18 ♓ 00 09	24 ♓ 10 15	11 51	26 07	24 19	04 38	17 45	18 00	18 56	21 57
29	02 30 46	09 24 54	00 ♈ 24 22	06 ♈ 42 42	13 04	27 21	25 05	04 38	17 48	18 00	18 56	21 55
30	02 34 42	10 ♉ 23 11	13 ♈ 05 27	19 ♈ 32 41	14 ♈ 20	28 ♈ 35	25 ♓ 51	04 ♍ 38	17 ♒ 51	17 ♑ 59	18 ♑ 56	21 ♏ 54

Date	Moon True ☊ ° '	Moon Mean ☊ ° '	Moon ☽ Latitude ° '
01	04 ♑ 37	04 ♑ 57	04 N 53
02	04 R 24	04 54	05 00
03	04 10	04 51	04 52
04	03 58	04 48	04 29
05	03 48	04 45	03 52
06	03 41	04 42	03 01
07	03 36	04 38	01 59
08	03 34	04 35	00 N 50
09	03 D 34	04 32	00 S 23
10	03 R 34	04 29	01 36
11	03 34	04 26	02 43
12	03 31	04 22	03 41
13	03 26	04 19	04 26
14	03 18	04 16	04 53
15	03 09	04 13	05 03
16	02 58	04 10	04 53
17	02 48	04 07	04 26
18	02 39	04 03	03 44
19	02 32	04 00	02 50
20	02 27	03 57	01 50
21	02 25	03 54	00 S 45
22	02 D 25	03 51	00 N 21
23	02 25	03 47	01 24
24	02 26	03 44	02 23
25	02 R 26	03 41	03 16
26	02 24	03 38	04 00
27	02 20	03 35	04 34
28	02 14	03 32	04 57
29	02 07	03 28	05 06
30	01 ♑ 58	03 ♑ 25	05 N 01

DECLINATIONS

Date	Sun ☉	Moon ☽	Mercury ☿	Venus ♀	Mars ♂	Jupiter ♃	Saturn ♄	Uranus ♅	Neptune ♆	Pluto ♇
01	04 N 45	01 N 23	02 N 15	04 S 10	11 S 25	10 N 36	16 S 44	22 S 39	21 S 23	03 S 42
02	05 08	06 27	01 45	03 42	11 08	10 38	16 43	22 38	21 23	03 42
03	05 31	11 20	01 17	03 13	10 51	10 40	16 42	22 38	21 23	03 41
04	05 54	15 48	00 51	02 44	10 34	10 41	16 40	22 38	21 23	03 41
05	06 17	19 35	00 27	02 15	10 17	10 43	16 39	22 38	21 23	03 40
06	06 39	22 24	00 N 05	01 46	10 00	10 44	16 37	22 38	21 22	03 40
07	07 02	23 59	00 S 14	01 17	09 43	10 46	16 36	22 38	21 22	03 39
08	07 24	24 09	00 31	00 48	09 26	10 47	16 35	22 38	21 22	03 39
09	07 47	22 49	00 45	00 S 18	09 09	10 49	16 34	22 38	21 22	03 38
10	08 09	20 05	00 56	00 N 11	08 51	10 50	16 33	22 38	21 22	03 38
11	08 31	16 07	01 05	00 40	08 34	10 51	16 31	22 38	21 22	03 37
12	08 53	11 12	01 11	01 09	08 16	10 52	16 30	22 38	21 22	03 37
13	09 14	05 N 39	01 15	01 39	07 59	10 53	16 29	22 38	21 22	03 36
14	09 36	00 S 12	01 17	02 08	07 41	10 54	16 28	22 38	21 22	03 35
15	09 57	06 00	01 15	02 37	07 24	10 55	16 27	22 38	21 22	03 35
16	10 19	11 27	01 12	03 06	07 06	10 56	16 26	22 38	21 22	03 34
17	10 40	16 11	01 06	03 35	06 48	10 57	16 25	22 38	21 22	03 34
18	11 01	19 59	00 58	04 05	06 30	10 58	16 24	22 38	21 22	03 33
19	11 22	22 36	00 49	04 34	06 12	10 58	16 23	22 38	21 22	03 33
20	11 42	23 58	00 37	05 02	05 54	10 59	16 22	22 38	21 22	03 32
21	12 02	24 03	00 23	05 31	05 36	10 59	16 21	22 38	21 22	03 32
22	12 23	22 57	00 S 07	06 00	05 18	11 00	16 20	22 38	21 22	03 31
23	12 43	20 48	00 N 11	06 28	05 00	11 00	16 19	22 38	21 22	03 31
24	13 02	17 47	00 30	06 57	04 42	11 00	16 18	22 38	21 22	03 30
25	13 22	14 03	00 51	07 25	04 24	11 01	16 17	22 38	21 22	03 30
26	13 41	09 47	01 13	07 53	04 05	11 01	16 17	22 38	21 22	03 30
27	14 00	05 07	01 37	08 21	03 47	11 01	16 16	22 38	21 22	03 29
28	14 19	00 S 11	02 03	08 49	03 29	11 01	16 15	22 38	21 22	03 29
29	14 38	04 N 50	02 30	09 17	03 11	11 01	16 14	22 38	21 22	03 28
30	14 N 56	09 N 47	02 N 58	09 N 44	02 S 53	11 N 01	16 S 14	22 S 38	21 S 22	03 S 28

ZODIAC SIGN ENTRIES

Date	h m	Planets
02	03 04	☽ ♈
03	23 52	☿ ♓
04	11 18	☽ ♉
06	17 33	☽ ♊
07	07 16	♀ ♈
08	22 18	☽ ♋
11	01 46	☽ ♌
13	04 09	☽ ♍
14	17 35	☿ ♈
15	06 10	☽ ♎
17	09 10	☽ ♏
19	14 40	☽ ♐
19	19 57	☉ ♉
21	23 40	☽ ♑
24	11 38	☽ ♒
27	00 20	☽ ♓
29	11 13	☽ ♈

LATITUDES

Date	Mercury ☿	Venus ♀	Mars ♂	Jupiter ♃	Saturn ♄	Uranus ♅	Neptune ♆	Pluto ♇
01	01 N 49	01 S 27	01 S 14	01 N 22	00 S 43	00 S 24	00 N 44	15 N 11
04	01 02	01 29	01 15	01 21	00 43	00 24	00 44	15 12
07	00 N 14	01 30	01 16	01 21	00 44	00 24	00 44	15 12
10	00 S 30	01 31	01 17	01 21	00 44	00 24	00 44	15 13
13	01 09	01 31	01 18	01 20	00 44	00 24	00 44	15 14
16	01 42	01 30	01 18	01 20	00 45	00 24	00 45	15 14
19	02 09	01 29	01 19	01 19	00 45	00 24	00 45	15 15
22	02 29	01 27	01 19	01 19	00 46	00 24	00 45	15 15
25	02 43	01 25	01 20	01 18	00 46	00 24	00 45	15 16
28	02 52	01 22	01 20	01 18	00 47	00 24	00 45	15 16
31	02 S 55	01 S 18	01 S 20	01 N 17	00 S 47	00 S 25	00 N 45	15 N 16

DATA

Julian Date	2448714
Delta T	+58 seconds
Ayanamsa	23° 45' 12"
Synetic vernal point	05° ♓ 21' 47"
True obliquity of ecliptic	23° 26' 27"

MOON'S PHASES, APSIDES AND POSITIONS ☽

Date	h m	Phase	Longitude ° '	Eclipse Indicator
03	05 01	●	13 ♈ 42	
10	10 06	☽	20 ♋ 47	
17	04 42	○	27 ♎ 26	
24	21 40	☾	04 ♒ 57	

Day	h m		
13	07 00	Perigee	
25	09 49	Apogee	
01	05 26	0N	
08	02 40	Max dec	24° N 15'
14	11 12	0S	
21	01 37	Max dec	24° S 10'
28	12 53	0N	

LONGITUDES

Date	Chiron ⚷ ° '	Ceres ⚳ ° '	Pallas ⚴ ° '	Juno ⚵ ° '	Vesta ⚶ ° '	Black Moon Lilith ⚸ ° '
01	03 ♌ 18	00 ♒ 27	05 ♑ 37	00 ♈ 30	13 ♍ 22	07 ♒ 52
11	03 ♌ 21	03 ♒ 05	06 ♑ 54	05 ♈ 44	11 ♍ 56	08 ♒ 59
21	03 ♌ 33	05 ♒ 23	07 ♑ 34	11 ♈ 00	11 ♍ 17	10 ♒ 06
31	03 ♌ 56	07 ♒ 19	07 ♑ 32	16 ♈ 20	11 ♍ 27	11 ♒ 12

All ephemeris data is given at 12.00 UT and the Moon's longitude is additionally given for 24.00 UT

ASPECTARIAN

h m Aspects

01 Wednesday
03 34 ☽ ✱ ♅
05 33 ☽ ✱ ♆
07 06 ♀ △ ♇
10 10 ☉ ± ♃
11 35 ☽ ⊥ ♄
12 47 ☽ △ ♇
13 26 ☽ ☌ ♀
15 41 ☽ ∥ ☿
22 56 ☽ # ♇

02 Thursday
00 02 ☽ # ♀
01 41 ♂ ∠ ♆
02 47 ☽ Q ♅
03 43 ☿ ∠ ♄
04 43 ☽ Q ♆
04 57 ☽ ☌ ☿
05 01 ☽ ∠ ♄
05 12 ☽ ∥ ☉
09 33 ☿ Q ♆
11 01 ☽ ⊻ ♂
11 53 ♀ ∥ ♇
14 11 ☽ ⊼ ♃
17 27 ☽ ꝑ ♇
23 07 ☽ ⊥ ♂

03 Friday
01 27 ☽ ± ♃
05 01 ☽ ☌ ☉
08 33 ☽ ∥ ♃
09 34 ☽ ✱ ♄
09 40 ☽ # ♂
10 22 ☽ ± ♇
12 50 ☽ □ ♅
14 43 ☽ □ ♆
17 03 ☽ ∠ ♂
18 11 ☽ ꝑ ♃
21 31 ☽ ⊼ ♇

04 Saturday
02 55 ♂ # ♃
04 11 ☽ ⊻ ♀
05 52 ☿ Q ♅
07 59 ☽ Q ♄
10 09 ♂ ☍ ♃
10 55 ☽ ⊻ ☿
16 23 ☽ ⊥ ♀
17 00 ☽ # ♄
21 34 ☽ ⊥ ☿
21 39 ☽ △ ♃
22 27 ☽ ✱ ♂

05 Sunday
10 28 ☽ ∠ ♀
13 29 ☽ ∠ ☿
16 42 ☽ ⊻ ☉
16 59 ☽ □ ♄
19 53 ☽ △ ♅
20 51 ☉ ✱ ♄
21 33 ☽ Q ♂
21 40 ☽ △ ♆

06 Monday
01 09 ☉ ± ♇
02 06 ☽ # ♆
04 07 ☽ ☍ ♇
04 22 ☽ ⊥ ☉
13 12 ☿ ☌ ♀
14 37 ☽ # ♅
15 51 ☽ ✱ ☿
16 10 ☽ ✱ ♀
21 42 ☽ ∠ ☉
22 45 ☽ ꝑ ♅

07 Tuesday
00 30 ☽ ꝑ ♆
03 18 ☽ □ ♃
05 48 ♀ Q ♅
07 38 ☽ □ ♂
12 09 ☉ □ ♅
12 49 ☽ Q ☿
14 41 ☽ ± ♅
15 34 ☽ Q ♀
16 25 ☽ ± ♆
22 41 ☽ △ ♄

08 Wednesday
01 04 ♀ Q ♆
01 15 ☽ ⊼ ♅
02 15 ☽ ✱ ☉
02 58 ☽ ⊼ ♆
09 08 ☽ ⊼ ♇
10 48 ☽ Q ♃
12 20 ☉ □ ♆
12 52 ♀ ∠ ♄
19 35 ☽ ± ♇
20 07 ☽ □ ☿
21 15 ☿ ∥ ♀

09 Thursday
00 47 ☽ Q ☉
01 01 ☽ ꝑ ♄
02 07 ☽ □ ♀
06 26 ☿ St D
07 33 ☽ ✱ ♃
11 09 ☽ ꝑ ♇
14 08 ☽ # ♅
15 07 ☽ △ ♂
16 39 ☽ ± ♄
22 04 ☉ ꝑ ♃

10 Friday
02 08 ☽ # ♆
03 00 ☽ ⊼ ♄
05 15 ☽ ☍ ♅
06 55 ☽ ☍ ♆
09 13 ☽ ∠ ♃
10 06 ☽ □ ☉
12 50 ☽ △ ♇
18 18 ☽ ꝑ ♂
23 51 ☽ △ ☿

11 Saturday
00 28 ☽ ⊥ ♃
09 50 ☽ # ♄
10 23 ☽ △ ♀
10 26 ☽ ± ♂
10 35 ☽ ⊻ ♃
12 30 ♀ ⊼ ♃
13 53 ☉ # ♂
21 08 ☽ ⊼ ♂

12 Sunday
00 31 ☉ ⊼ ♇
01 31 ☽ ꝑ ☿
06 04 ☽ ☍ ♄
08 03 ☽ ⊼ ♅
09 40 ☽ ⊼ ♆
13 31 ☽ ∥ ♃
13 57 ☿ # ♀
13 58 ☽ ꝑ ♀
15 22 ☽ □ ♇
16 29 ☽ △ ☉
16 52 ☽ ± ☿
18 06 ☽ ± ♅
19 42 ☽ ± ♆
21 41 ☽ ∥ ☉

13 Monday
01 39 ☽ # ♂
03 07 ☽ ⊼ ☿
06 00 ♀ ꝑ ♇
06 24 ☽ ± ♀
09 07 ☽ ꝑ ♅
10 43 ☽ ꝑ ♆
12 38 ☽ ☌ ♃
17 20 ☽ ⊼ ♀
19 19 ☽ ꝑ ☉
20 31 ☽ # ♇
21 20 ☽ Q ♇

14 Tuesday
02 06 ☽ ☍ ♂
03 14 ☽ ∥ ♀
06 01 ☽ # ☿
08 22 ☽ ⊼ ♄
10 07 ☽ △ ♅
11 24 ☽ ± ☉
11 43 ☽ △ ♆
16 24 ☽ ∥ ☿
17 16 ☿ Q ♅
17 18 ☽ ✱ ♇
18 25 ☽ ± ♄
20 38 ☽ # ♀
22 09 ☽ ⊼ ☉

15 Wednesday
01 53 ☽ ∥ ♇
06 36 ☽ ☍ ☿
09 32 ☽ ꝑ ♄
14 32 ☽ ⊻ ♃
17 36 ☽ ∥ ♂
18 23 ☽ ∠ ♇

16 Thursday
00 13 ☽ ☍ ♀
00 37 ☽ ⊥ ♃
04 39 ♀ ± ♃
06 25 ☽ # ☉
07 15 ☽ ⊼ ♂
09 35 ☽ ⊥ ♇
09 38 ☽ # ♃
10 58 ☽ △ ♄
12 33 ☽ □ ♅
14 06 ☿ Q ♆
14 10 ☽ □ ♆
15 51 ☽ ∠ ♃
18 04 ☽ ± ♂
19 48 ☽ ⊻ ♇

17 Friday
04 42 ☽ ☍ ☉
10 29 ☽ ꝑ ♂
10 44 ♀ # ♇
11 46 ☽ ⊼ ☿
13 14 ☽ ∥ ♄
17 41 ☽ ✱ ♃
19 38 ☽ Q ♅
21 17 ☽ Q ♆
22 47 ☽ ± ☿

18 Saturday
08 09 ☉ ∥ ♃
09 07 ☽ ⊼ ♀
13 01 ☿ ∠ ♄
14 30 ☽ △ ♂
14 50 ☽ Q ♃
15 28 ☽ □ ♄
15 36 ☽ ꝑ ☿
16 54 ☽ ✱ ♅
18 36 ☽ ✱ ♆
20 56 ☽ ± ♀
23 16 ☽ ∥ ♆

19 Sunday
00 26 ☽ ☌ ♇
01 17 ☉ Q ♄
08 50 ♂ ⊻ ♄
12 17 ☽ ∥ ♅
14 14 ☽ ⊼ ☉
15 01 ☽ ꝑ ♀
20 14 ☽ ∠ ♅
20 33 ☽ △ ☿
21 59 ☽ ∠ ♆
23 34 ☽ □ ♃

20 Monday
00 28 ☽ Q ♄
02 15 ☽ ± ☉
08 17 ♂ ✱ ♅
09 54 ♀ ± ♇
12 14 ♆ St R
13 04 ☽ ⊥ ♅
14 51 ☽ ⊥ ♆
20 38 ☽ ꝑ ☉
22 07 ☽ △ ♀
23 13 ☽ ✱ ♄

21 Tuesday
00 28 ☽ ⊻ ♅
01 32 ☽ □ ♂
02 17 ☽ ⊻ ♆
08 23 ☽ ⊻ ♇
09 44 ♀ ✱ ♄
13 33 ♂ ✱ ♆
14 31 ♀ # ♂
17 43 ☿ ⊼ ♃
20 00 ☽ ⊥ ♇
22 06 ♀ □ ♅
23 20 ♅ St R

22 Wednesday
04 09 ☽ △ ☉
04 26 ☽ ∠ ♄
09 00 ☽ △ ♃
10 16 ☽ □ ☿
13 40 ☽ ∠ ♇
15 00 ☽ Q ♂
16 23 ☽ ∥ ♅
16 25 ♀ □ ♆
22 20 ☽ ⊥ ♄

23 Thursday
06 39 ☽ ∥ ♆
07 17 ♀ ꝑ ♃
10 24 ☽ ⊻ ♄
11 27 ☽ ☌ ♅
13 20 ☽ ☌ ♆
14 51 ☽ ꝑ ♃
15 45 ☽ □ ♀
16 39 ☽ ✱ ♂
19 36 ☽ ✱ ♇

24 Friday
01 26 ☽ Q ☿
03 58 ☿ ꝑ ♇
09 00 ☽ ± ♃
13 39 ♀ ⊻ ♂
15 26 ☉ △ ♃
19 50 ☽ Q ♇
21 09 ☽ ⊼ ♃
21 40 ☽ □ ☉
22 03 ☽ ∥ ♄

25 Saturday
01 07 ☽ ∠ ♂
04 04 ☽ ✱ ☿
04 25 ♀ ⊼ ♇
08 25 ☽ Q ♀
12 55 ♂ △ ♇
15 47 ☽ # ☉
20 39 ☽ ⊥ ♂
23 24 ☽ ☌ ♄

26 Sunday
00 10 ☽ ⊻ ♅
02 04 ☽ ⊻ ♆
05 22 ☽ # ♃
08 14 ☽ □ ♇
09 38 ☽ ⊻ ♂
11 31 ☽ ✱ ♀
12 17 ☽ ⊥ ♅
13 23 ☽ Q ☉
13 30 ☽ ∠ ☿
14 11 ☽ ⊥ ♆
21 02 ☽ # ♀

27 Monday
06 20 ☽ ∠ ♅
08 13 ☽ ∠ ♆
09 26 ☽ ⊥ ☿
09 37 ☽ ☍ ♃
11 25 ☿ ± ♃
15 31 ☽ ✱ ☉
17 41 ♂ ⊥ ♄
18 59 ☽ ∥ ♂
20 03 ☽ ∥ ♇
20 56 ☽ ∠ ♀
22 37 ☽ ⊻ ☿

28 Tuesday
03 44 ☽ # ☿
11 31 ☽ ⊻ ♄
11 59 ☽ ✱ ♅
12 34 ♂ ∥ ♇
13 50 ☽ ✱ ♆
16 35 ☽ ⊥ ♀
19 40 ☽ △ ♇
23 15 ☽ ⊥ ♄
23 30 ☽ ∠ ☉
23 43 ☽ ∥ ☿

29 Wednesday
01 05 ☽ ☌ ♂
04 31 ☽ # ♂
05 28 ☽ # ♇
05 29 ☽ ⊻ ♀
11 12 ☽ Q ♅
13 01 ☽ Q ♆
16 36 ☽ ∠ ♄
18 13 ☽ ⊥ ☉
20 04 ☽ ⊼ ♃

30 Thursday
00 22 ☽ ꝑ ♇
06 31 ☽ ⊻ ☉
07 23 ☽ ± ♃
09 06 ☿ # ♂
11 45 ☽ ∥ ♀
14 34 ☽ ☌ ☿
17 13 ☽ ± ♇
18 11 ☽ ∥ ♃
19 40 ♃ St D
20 54 ☽ ✱ ♄
21 06 ☽ □ ♅
22 52 ☽ □ ♆

MAY 1992

LONGITUDES

Date	Sidereal time h m s	Sun ☉ ° ' "	Moon ☽ ° ' "	Moon ☽ 24.00 ° ' "	Mercury ☿ ° '	Venus ♀ ° '	Mars ♂ ° '	Jupiter ♃ ° '	Saturn ♄ ° '	Uranus ♅ ° '	Neptune ♆ ° '	Pluto ♇ ° '
01	02 38 39	11 ♉ 21 27	26 ♈ 04 23	02 ♉ 40 27	15 ♈ 38	29 ♈ 49	26 ♓ 37	04 ♍ 38	17 ♒ 54	17 ♑ 59	18 ♑ 55	21 ♏ 52
02	02 42 36	12 19 40	09 ♉ 20 42	16 ♉ 04 54	16 58	01 ♉ 02	27 23	04 D 38	17 56	17 R 58	18 R 55	21 R 50
03	02 46 32	13 17 52	22 ♉ 52 44	29 ♉ 43 53	18 20	02 16	28 09	04 39	17 59	17 58	18 55	21 49
04	02 50 29	14 16 02	06 ♊ 37 57	13 ♊ 34 34	19 44	03 30	28 56	04 39	18 01	17 57	18 54	21 47
05	02 54 25	15 14 11	20 ♊ 33 21	27 ♊ 33 56	21 11	04 44	29 ♓ 42	04 40	18 03	17 56	18 54	21 45
06	02 58 22	16 12 18	04 ♋ 35 58	11 ♋ 39 09	22 39	05 58	00 ♈ 28	04 41	18 06	17 56	18 53	21 44
07	03 02 18	17 10 22	18 ♋ 43 13	25 ♋ 47 53	24 10	07 12	01 14	04 42	18 08	17 55	18 53	21 42
08	03 06 15	18 08 25	02 ♌ 52 57	09 ♌ 58 13	25 42	08 25	02 00	04 43	18 10	17 54	18 52	21 40
09	03 10 11	19 06 26	17 ♌ 03 27	24 ♌ 08 29	27 17	09 39	02 46	04 45	18 12	17 53	18 52	21 39
10	03 14 08	20 04 25	01 ♍ 13 05	08 ♍ 17 01	28 ♈ 54	10 53	03 31	04 47	18 13	17 52	18 51	21 37
11	03 18 05	21 02 22	15 ♍ 20 02	22 ♍ 21 52	00 ♉ 32	12 07	04 17	04 48	18 15	17 51	18 50	21 35
12	03 22 01	22 00 17	29 ♍ 22 10	06 ♎ 20 38	02 13	13 21	05 03	04 50	18 17	17 51	18 50	21 34
13	03 25 58	22 58 10	13 ♎ 16 53	20 ♎ 10 35	03 56	14 34	05 49	04 53	18 18	17 49	18 49	21 32
14	03 29 54	23 56 02	27 ♎ 01 21	03 ♏ 48 51	05 41	15 48	06 35	04 55	18 20	17 48	18 48	21 30
15	03 33 51	24 53 52	10 ♏ 32 47	17 ♏ 12 52	07 27	17 02	07 21	04 58	18 21	17 47	18 47	21 29
16	03 37 47	25 51 40	23 ♏ 48 55	00 ♐ 20 46	09 16	18 16	08 06	05 00	18 22	17 46	18 47	21 27
17	03 41 44	26 49 27	06 ♐ 48 20	13 ♐ 11 37	11 07	19 29	08 52	05 03	18 23	17 45	18 46	21 25
18	03 45 40	27 47 13	19 ♐ 30 41	25 ♐ 45 41	13 00	20 43	09 38	05 06	18 24	17 44	18 45	21 24
19	03 49 37	28 44 58	01 ♑ 56 51	08 ♑ 04 28	14 55	21 57	10 23	05 10	18 25	17 42	18 44	21 22
20	03 53 34	29 ♉ 42 41	14 ♑ 08 52	20 ♑ 10 28	16 51	23 11	11 09	05 13	18 26	17 41	18 43	21 20
21	03 57 30	00 ♊ 40 23	26 ♑ 09 45	02 ♒ 07 11	18 50	24 24	11 54	05 17	18 27	17 40	18 42	21 19
22	04 01 27	01 38 04	08 ♒ 03 20	13 ♒ 58 46	20 51	25 38	12 40	05 20	18 28	17 38	18 41	21 17
23	04 05 23	02 35 44	19 ♒ 54 03	25 ♒ 49 49	22 53	26 52	13 25	05 24	18 28	17 37	18 40	21 15
24	04 09 20	03 33 23	01 ♓ 46 41	07 ♓ 45 14	24 57	28 05	14 11	05 28	18 29	17 35	18 39	21 14
25	04 13 16	04 31 00	13 ♓ 46 05	19 ♓ 49 48	27 03	29 ♉ 19	14 56	05 33	18 29	17 34	18 38	21 12
26	04 17 13	05 28 37	25 ♓ 56 57	02 ♈ 08 02	29 ♉ 10	00 ♊ 33	15 41	05 37	18 29	17 32	18 37	21 10
27	04 21 09	06 26 13	08 ♈ 23 30	14 ♈ 43 45	01 ♊ 19	01 47	16 27	05 41	18 29	17 31	18 36	21 09
28	04 25 06	07 23 48	21 ♈ 09 05	27 ♈ 39 45	03 29	03 00	17 12	05 46	18 29	17 29	18 35	21 07
29	04 29 03	08 21 22	04 ♉ 15 52	10 ♉ 57 27	05 40	04 14	17 57	05 51	18 R 29	17 27	18 34	21 06
30	04 32 59	09 18 56	17 ♉ 44 27	24 ♉ 36 37	07 51	05 28	18 42	05 56	18 29	17 26	18 33	21 04
31	04 36 56	10 ♊ 16 28	01 ♊ 33 42	08 ♊ 35 15	10 ♊ 03	06 ♊ 42	19 ♈ 27	06 ♍ 01	18 ♒ 29	17 ♑ 24	18 ♑ 32	21 ♏ 02

DECLINATIONS

Date	Moon True ☊ ° '	Moon Mean ☊ ° '	Moon ☽ Latitude ° '	Sun ☉ ° '	Moon ☽ ° '	Mercury ☿ ° '	Venus ♀ ° '	Mars ♂ ° '	Jupiter ♃ ° '	Saturn ♄ ° '	Uranus ♅ ° '	Neptune ♆ ° '	Pluto ♇ ° '
01	01 ♑ 49	03 ♑ 22	04 N 40	15 N 14	14 N 24	03 N 28	10 N 11	02 S 34	11 N 01	16 S 13	22 S 38	21 S 22	03 S 27
02	01 R 40	03 19	04 03	15 32	18 27	03 59	10 38	02 16	11 00	16 12	22 38	21 22	03 27
03	01 34	03 16	03 13	15 50	21 36	04 30	11 05	01 58	11 00	16 12	22 39	21 22	03 26
04	01 29	03 13	02 10	16 07	23 33	05 04	11 31	01 39	11 00	16 11	22 39	21 22	03 26
05	01 27	03 09	00 N 58	16 24	24 05	05 38	11 57	01 21	10 59	16 11	22 39	21 22	03 26
06	01 D 26	03 06	00 S 17	16 41	23 05	06 13	12 23	01 03	10 59	16 10	22 39	21 22	03 25
07	01 27	03 03	01 32	16 58	20 37	06 49	12 49	00 44	10 58	16 10	22 39	21 22	03 25
08	01 28	03 00	02 42	17 14	16 53	07 26	13 14	00 26	10 58	16 09	22 39	21 22	03 24
09	01 29	02 57	03 42	17 30	12 12	08 04	13 39	00 S 08	10 57	16 09	22 39	21 22	03 24
10	01 R 28	02 53	04 28	17 46	06 52	08 43	14 04	00 N 10	10 56	16 08	22 39	21 22	03 24
11	01 26	02 50	04 58	18 01	01 N 12	09 22	14 28	00 28	10 55	16 08	22 40	21 23	03 23
12	01 22	02 47	05 10	18 16	04 S 29	10 02	14 52	00 47	10 54	16 08	22 40	21 23	03 23
13	01 17	02 44	05 03	18 31	09 54	10 43	15 16	01 05	10 53	16 07	22 40	21 23	03 23
14	01 11	02 41	04 39	18 45	14 45	11 24	15 39	01 23	10 52	16 07	22 40	21 23	03 22
15	01 05	02 38	04 00	19 00	18 47	12 06	16 02	01 41	10 51	16 07	22 40	21 23	03 22
16	01 00	02 34	03 08	19 13	21 46	12 48	16 24	01 59	10 50	16 06	22 40	21 23	03 22
17	00 57	02 31	02 08	19 27	23 33	13 30	16 46	02 17	10 49	16 06	22 41	21 23	03 21
18	00 54	02 28	01 S 02	19 40	24 03	14 12	17 08	02 35	10 47	16 06	22 41	21 23	03 21
19	00 D 54	02 25	00 N 06	19 53	23 20	14 55	17 29	02 53	10 46	16 06	22 41	21 23	03 21
20	00 54	02 22	01 12	20 05	21 30	15 37	17 49	03 11	10 45	16 06	22 41	21 23	03 21
21	00 56	02 19	02 14	20 18	18 43	16 20	18 10	03 29	10 43	16 06	22 41	21 24	03 20
22	00 58	02 15	03 10	20 29	15 12	17 01	18 29	03 47	10 42	16 06	22 42	21 24	03 20
23	00 59	02 12	03 57	20 41	11 06	17 43	18 49	04 05	10 40	16 06	22 42	21 24	03 20
24	01 00	02 09	04 34	20 52	06 34	18 23	19 07	04 22	10 38	16 06	22 42	21 24	03 20
25	00 R 59	02 06	05 00	21 03	01 S 46	19 03	19 26	04 40	10 37	16 06	22 42	21 24	03 19
26	00 58	02 03	05 13	21 13	03 N 11	19 42	19 43	04 58	10 35	16 06	22 42	21 24	03 19
27	00 55	01 59	05 12	21 23	08 06	20 19	20 01	05 15	10 33	16 06	22 43	21 24	03 19
28	00 52	01 56	04 55	21 33	12 48	20 55	20 17	05 33	10 31	16 06	22 43	21 24	03 19
29	00 49	01 53	04 23	21 42	17 03	21 30	20 33	05 50	10 29	16 06	22 43	21 25	03 19
30	00 46	01 50	03 35	21 51	20 34	22 02	20 49	06 07	10 27	16 06	22 43	21 25	03 18
31	00 ♑ 43	01 ♑ 47	02 N 34	21 N 59	22 N 59	22 N 32	21 N 04	06 N 25	10 N 25	16 S 07	22 S 44	21 S 25	03 S 18

ZODIAC SIGN ENTRIES

Date	h	m	Planets
01	15	41	♀ ♉
01	19	09	☽ ♉
04	00	28	☽ ♊
05	21	36	♂ ♈
06	04	09	☽ ♋
08	07	07	☽ ♌
10	09	56	☽ ♍
11	04	10	☿ ♉
12	13	05	☽ ♎
14	17	15	☽ ♏
16	23	22	☽ ♐
19	08	13	☽ ♑
20	19	12	☉ ♊
21	19	43	☽ ♒
24	08	25	☽ ♓
26	01	18	♀ ♊
26	19	52	☽ ♈
26	21	16	☿ ♊
29	04	16	☽ ♉
31	09	19	☽ ♊

LATITUDES

Date	Mercury ☿ ° '	Venus ♀ ° '	Mars ♂ ° '	Jupiter ♃ ° '	Saturn ♄ ° '	Uranus ♅ ° '	Neptune ♆ ° '	Pluto ♇ ° '
01	02 S 55	01 S 18	01 S 20	01 N 17	00 S 47	00 S 25	00 N 45	15 N 16
04	02 52	01 14	01 20	01 17	00 48	00 25	00 45	15 16
07	02 45	01 10	01 20	01 16	00 48	00 25	00 45	15 16
10	02 32	01 05	01 20	01 16	00 48	00 25	00 45	15 16
13	02 15	01 00	01 20	01 15	00 49	00 25	00 45	15 16
16	01 53	00 54	01 20	01 15	00 49	00 25	00 45	15 16
19	01 27	00 48	01 20	01 14	00 50	00 25	00 45	15 15
22	00 59	00 42	01 20	01 14	00 50	00 25	00 45	15 15
25	00 S 28	00 36	01 19	01 13	00 51	00 25	00 45	15 14
28	00 N 04	00 29	01 19	01 13	00 51	00 25	00 45	15 14
31	00 N 35	00 S 22	01 S 18	01 N 12	00 S 52	00 S 25	00 N 45	15 N 13

DATA

Julian Date	2448744
Delta T	+58 seconds
Ayanamsa	23° 45' 16"
Synetic vernal point	05° ♓ 21' 44"
True obliquity of ecliptic	23° 26' 26"

MOON'S PHASES, APSIDES AND POSITIONS ☽

Date	h	m	Phase	Longitude ° '	Eclipse Indicator
02	17	44	●	12 ♉ 34	
09	15	44	☽	19 ♌ 15	
16	16	03	○	26 ♏ 01	
24	15	53	☾	03 ♓ 43	

Day	h	m		
08	11	37	Perigee	
23	04	49	Apogee	
05	08	21	Max dec	24° N 06'
11	17	01	0S	
18	09	40	Max dec	24° S 04'
25	20	36	0N	

LONGITUDES

Date	Chiron ⚷ ° '	Ceres ⚳ ° '	Pallas ⚴ ° '	Juno ⚵ ° '	Vesta ⚶ ° '	Black Moon Lilith ⚸ ° '
01	03 ♌ 56	07 ♒ 19	07 ♑ 32	16 ♈ 20	11 ♍ 27	11 ♒ 12
11	04 ♌ 28	08 ♒ 49	06 ♑ 45	21 ♈ 41	12 ♍ 23	12 ♒ 19
21	05 ♌ 08	09 ♒ 50	05 ♑ 14	27 ♈ 03	14 ♍ 00	13 ♒ 26
31	05 ♌ 56	10 ♒ 18	03 ♑ 02	02 ♉ 27	16 ♍ 13	14 ♒ 33

All ephemeris data is given at 12.00 UT and the Moon's longitude is additionally given for 24.00 UT

ASPECTARIAN

01 Friday
00 10 ☽ ⚼ ♃
04 18 ☽ ⊼ ♇
11 39 ☿ # ♇
13 04 ☽ ⊻ ♂
13 42 ♀ Q ♄
16 14 ☿ ± ♇
16 58 ☽ ∥ ☉
18 59 ☽ Q ♄
19 31 ☽ ☌ ♀
22 12 ☽ # ♄

02 Saturday
00 38 ☽ ⊥ ♂
03 32 ☽ △ ♃
17 44 ☽ ☌ ☉
17 46 ☽ ∠ ♂
19 11 ☉ ∠ ♂

03 Sunday
03 04 ☽ ⊻ ☿
03 19 ♄ ⊻ ♅
03 20 ☽ △ ♅
03 20 ☽ □ ♄
05 01 ☽ △ ♆
05 37 ☿ □ ♅
05 42 ☿ ✱ ♄
07 49 ♀ ∥ ♃
09 56 ☽ # ♆
10 07 ☽ ☍ ♇
14 50 ☽ ⊥ ☿
21 48 ☽ ✱ ♂
21 56 ☿ □ ♆
22 51 ☽ # ♅

04 Monday
05 37 ☽ ⚼ ♅
06 02 ☽ ⊻ ♀
07 16 ☽ ⚼ ♆
08 20 ☽ ∠ ☿
08 34 ☽ □ ♃
10 35 ☿ ⚼ ♃
17 15 ☉ # ♄
17 27 ☽ ⊥ ♀
19 52 ☽ Q ♂
21 11 ☽ ± ♅
22 50 ☽ ± ♆

05 Tuesday
02 11 ☽ ⊻ ☉
07 31 ☽ ⊼ ♅
07 42 ☽ △ ♄
09 09 ☽ ⊼ ♆
10 27 ☽ ∠ ♀
10 40 ♀ △ ♃
13 11 ☽ ✱ ☿
13 15 ☽ ⊥ ☉
14 03 ☽ ⊼ ♇
15 37 ☽ Q ♃
19 35 ♂ Q ♅
21 17 ☿ ⊼ ♇

06 Wednesday
00 18 ☽ ± ♇
04 32 ☽ □ ♂
05 47 ☽ ∠ ☉
09 26 ☽ ⚼ ♄
12 06 ☽ Q ☿
12 08 ☽ ✱ ♃
14 33 ☽ ✱ ♀
15 37 ☽ ⚼ ♇
17 20 ☽ # ♅

07 Thursday
00 47 ☽ ± ♄
01 15 ♂ Q ♆
05 52 ☽ # ♆
09 11 ☽ ✱ ☉
10 38 ☽ ☍ ♅
10 59 ☽ ⊼ ♄
12 16 ☽ ☍ ♆
12 53 ☽ Q ♀
13 40 ☽ ∠ ♃
13 43 ♀ ⊥ ♂
17 03 ☽ △ ♇
22 21 ☽ □ ☿

08 Friday
04 57 ☽ ⊥ ♃
06 10 ☉ △ ♅
07 01 ☽ Q ☉
10 09 ☽ ∥ ☉
10 24 ☽ △ ♂
12 31 ☉ □ ♄
15 07 ☽ ⊻ ♃
16 04 ☽ # ♄
22 16 ☽ □ ♀

09 Saturday
05 31 ☽ ∥ ♀
05 56 ☉ △ ♆
13 15 ☽ ⚼ ♂
13 24 ☽ ⊼ ♅
13 56 ☽ ☍ ♄
15 03 ☽ ⊼ ♆
15 44 ☽ □ ☉
17 51 ☽ ∥ ♃
19 45 ☽ □ ♇
23 34 ☽ ± ♅

10 Sunday
01 13 ☽ ± ♆
02 08 ♂ ∠ ♄
04 47 ☽ ∥ ☿
05 23 ☽ ± ♂
07 33 ☽ △ ☿
14 48 ☽ ⚼ ♅
16 08 ☽ ⊼ ♂
16 28 ☽ ⚼ ♆
18 03 ☽ ☌ ♃

11 Monday
02 14 ☽ Q ♇
02 47 ☽ # ♇
05 59 ☽ △ ♀
07 45 ☿ Q ♄
12 24 ☽ ⚼ ☿
14 53 ☽ ∥ ♂
16 18 ☽ △ ♅
16 59 ☽ ⊼ ♄
17 58 ☽ △ ♆
19 24 ☽ # ♂
22 27 ☽ △ ☉
22 39 ☽ ✱ ♇

12 Tuesday
01 14 ☉ ☍ ♇
03 15 ☽ ± ♄
05 01 ♂ ⊼ ♃
05 51 ☽ ± ☿
07 18 ☽ ∥ ♇
10 04 ☽ ⚼ ♀
17 34 ☽ ⊼ ☿
18 44 ☽ ⚼ ♄
21 26 ☽ ⊻ ♃
22 20 ☽ ☍ ♂

13 Wednesday
00 21 ☽ ∠ ♇
02 07 ☽ ⚼ ☉
03 03 ☽ ± ♀
07 49 ☽ ⊥ ♃
14 28 ☽ ⊼ ♀
15 54 ☽ ⊥ ♇
16 25 ☽ # ☿
16 37 ☽ # ♃
17 58 ☿ ∥ ♃
18 53 ☽ ± ☉
19 53 ☽ □ ♅
20 45 ☽ △ ♄
21 37 ☽ □ ♆
23 31 ☽ ∠ ♃

14 Thursday
01 23 ☿ △ ♃
02 20 ☽ ⊻ ♇
06 10 ☽ ⊼ ☉
09 42 ♂ ⚼ ♇
17 21 ☽ # ♀
19 31 ☽ ∥ ♄

15 Friday
02 00 ☽ ✱ ♃
03 31 ☽ Q ♅
05 18 ☽ Q ♆
05 38 ☽ ☍ ☿
05 56 ☽ ⊼ ♂
09 21 ☿ ⊻ ♂
13 32 ☽ # ☉
17 19 ☽ ± ♂
17 24 ♀ # ♄
23 35 ☽ Q ♃

16 Saturday
00 51 ☽ ☍ ♀
01 01 ☽ ✱ ♅
02 05 ☽ □ ♄
02 35 ♀ △ ♅
02 50 ☽ ✱ ♆
07 41 ☽ ☌ ♇
08 18 ☽ ∥ ♆
10 37 ☽ ⚼ ♂
14 11 ♀ □ ♄
16 03 ☽ ☍ ☉
22 01 ♀ △ ♆
22 11 ☽ ∥ ♅

17 Sunday
04 28 ☽ ∠ ♅
06 20 ☽ ∠ ♆
08 43 ☽ □ ♃
11 13 ☽ Q ♄
16 06 ☽ △ ♂
21 16 ☽ ⊥ ♅
21 28 ☽ ⊼ ☿
23 11 ☽ ⊥ ♆

18 Monday
08 36 ☽ ⊻ ♅
09 53 ☽ ✱ ♄
10 33 ☽ ⊻ ♆
10 50 ☽ ± ☿
14 33 ☽ ⊼ ♀
15 35 ☽ ⊻ ♇

19 Tuesday
00 53 ♀ ☍ ♇
03 07 ☽ ⊥ ♇
03 22 ☽ ± ♀
05 15 ☽ ⊼ ☉
07 18 ☽ ⚼ ☿
14 53 ☽ ∠ ♄
17 56 ☽ ± ☉
18 18 ☽ △ ♃
20 37 ☽ ∠ ♇
22 07 ☽ ∥ ♅
22 53 ☽ ⚼ ♀

20 Wednesday
05 40 ☽ □ ♂
08 36 ☽ ⊥ ♄
13 05 ☽ ∥ ♆
13 13 ☽ ⚼ ☉
14 27 ♂ ± ♃
17 45 ☿ ⊥ ♂
18 26 ☽ △ ☿
19 01 ☽ ☌ ♅
20 32 ☽ ⊻ ♄
21 05 ☽ ☌ ♆
22 00 ☿ △ ♅

21 Thursday
00 09 ☽ ⚼ ♃
00 12 ☽ # ☉
00 29 ♂ # ♇
02 17 ☽ ✱ ♇
04 08 ☿ # ♄
07 19 ☿ □ ♄
08 04 ☽ △ ♀
10 28 ☿ △ ♆
15 47 ☽ # ♀
18 18 ☽ ± ♃
20 03 ☽ Q ♂
21 53 ☽ △ ☉

22 Friday
02 04 ☽ # ☿
02 22 ☽ Q ♇
06 17 ☽ ∥ ♄
06 28 ☽ ⊼ ♃
17 06 ☿ ☍ ♇
21 58 ☽ ✱ ♂

23 Saturday
07 23 ☽ ⊻ ♅
09 06 ☽ ☌ ♄
09 31 ☽ ⊻ ♆
12 29 ☉ ⚼ ♅
14 21 ☽ # ♃
14 44 ☽ □ ♇
19 19 ☽ □ ☿
19 30 ☽ ⊥ ♅
21 39 ☽ ⊥ ♆

24 Sunday
03 43 ☽ □ ♀
06 24 ☽ ∠ ♂
13 38 ☽ ∠ ♅
14 29 ☉ ⚼ ♆
15 47 ☽ ∠ ♆
15 53 ☽ □ ☉
19 28 ☽ ☍ ♃
22 27 ☽ # ♂

25 Monday
01 43 ☽ ⊥ ♂
04 18 ☽ ∥ ♇
14 28 ☽ ⊻ ♂
15 05 ☽ Q ☿
19 31 ☽ ✱ ♅
19 50 ☽ Q ♀
20 13 ♂ ± ♇
21 21 ☽ ⊻ ♄
21 38 ☽ ✱ ♆

26 Tuesday
02 40 ☽ △ ♇
06 45 ☽ Q ☉
09 08 ☽ ⊥ ♄
12 41 ☽ # ♇
13 45 ☿ ∥ ♀
15 44 ☉ □ ♃
18 58 ☽ Q ♅
19 05 ♀ ∠ ♂
19 35 ☽ ✱ ☿
21 04 ☽ Q ♆
21 11 ☽ ∥ ♂
21 55 ☽ ✱ ♀

27 Wednesday
02 36 ☽ ∠ ♄
06 48 ☽ ⊼ ♃
07 43 ☽ ⚼ ♇
07 57 ☽ ✱ ☉
14 08 ☿ ∠ ♂
15 02 ☉ # ♆
18 18 ☽ ± ♃
23 50 ☿ ☌ ♀

28 Thursday
00 14 ☽ ∥ ♃
00 46 ☽ ± ♇
01 07 ☿ ⚼ ♅
02 04 ♀ ⚼ ♅
04 10 ☽ ☌ ♂
05 11 ☽ □ ♅
05 31 ☽ ∠ ♀
06 01 ☽ ∠ ☿
07 03 ☽ ✱ ♄
07 14 ☽ □ ♆
11 17 ☽ ⚼ ♃
11 57 ☽ ⊼ ♇
13 09 ☿ ⚼ ♆
13 36 ♄ St R
14 30 ☽ ∠ ☉
20 53 ♂ □ ♅
23 11 ♀ ⚼ ♆
23 55 ☽ ⊥ ♀

29 Friday
01 59 ☽ ⊥ ☿
05 09 ☽ Q ♄
06 18 ☽ # ♄
08 17 ☽ ⊥ ☉
08 23 ☿ # ♆
11 56 ☽ ⊻ ♀
14 10 ☿ □ ♃
14 53 ☽ △ ♃
15 01 ☽ ⊻ ☿
19 55 ☽ ⊻ ☉

30 Saturday
00 25 ☉ ∥ ☿
05 07 ♂ ✱ ♄
07 14 ♂ □ ♆
11 27 ☽ △ ♅
13 19 ☽ □ ♄
13 25 ☽ △ ♆
13 47 ☽ ⊻ ♂
14 17 ☽ ∥ ♀
17 49 ☽ ☍ ♇
19 11 ☽ # ♆
21 53 ♀ □ ♃
23 56 ☽ ∥ ☉

31 Sunday
00 51 ☽ ⊥ ♂
04 51 ☽ ∥ ☿
08 42 ☽ # ♅
13 26 ☽ ⚼ ♅
15 22 ☽ ⚼ ♆
16 21 ☉ ☌ ☿
17 14 ☽ ∠ ♂
19 41 ☽ □ ♃
21 24 ☿ # ♅
21 37 ☽ ☌ ♀

JUNE 1992

LONGITUDES

Date	Sidereal time h m s	Sun ☉ ° ' "	Moon ☽ ° ' "	Moon ☽ 24.00 ° ' "	Mercury ☿ ° '	Venus ♀ ° '	Mars ♂ ° '	Jupiter ♃ ° '	Saturn ♄ ° '	Uranus ♅ ° '	Neptune ♆ ° '	Pluto ♇ ° '
01	04 40 52	11 ♊ 13 59	15 ♊ 40 48	22 ♊ 49 45	12 ♊ 15	07 ♊ 55	20 ♈ 12	06 ♍ 07	18 ♒ 29	17 ♑ 22	18 ♑ 31	21 ♏ 01
02	04 44 49	12 11 30	00 ♋ 01 29	07 ♋ 15 19	14 27	09 09	20 57	06 12	18 R 28	17 R 20	18 R 29	20 R 59
03	04 48 45	13 08 59	14 ♋ 30 34	21 ♋ 46 33	16 39	10 23	21 42	06 18	18 28	17 19	18 28	20 58
04	04 52 42	14 06 27	29 ♋ 02 37	06 ♌ 18 09	18 50	11 36	22 27	06 23	18 27	17 17	18 27	20 56
05	04 56 38	15 03 54	13 ♌ 32 35	20 ♌ 45 24	21 00	12 50	23 12	06 29	18 26	17 15	18 26	20 55
06	05 00 35	16 01 19	27 ♌ 56 11	05 ♍ 04 32	23 09	14 04	23 57	06 35	18 25	17 13	18 24	20 53
07	05 04 32	16 58 44	12 ♍ 10 10	19 ♍ 12 50	25 16	15 18	24 41	06 42	18 25	17 11	18 23	20 52
08	05 08 28	17 56 07	26 ♍ 12 20	03 ♎ 08 32	27 23	16 31	25 26	06 48	18 23	17 09	18 22	20 50
09	05 12 25	18 53 29	10 ♎ 01 19	16 ♎ 50 37	29 ♊ 27	17 45	26 11	06 54	18 22	17 07	18 20	20 49
10	05 16 21	19 50 50	23 ♎ 36 23	00 ♏ 18 35	01 ♋ 29	18 59	26 55	07 01	18 21	17 05	18 19	20 48
11	05 20 18	20 48 10	06 ♏ 57 14	13 ♏ 32 19	03 30	20 13	27 40	07 08	18 20	17 03	18 18	20 46
12	05 24 14	21 45 29	20 ♏ 03 52	26 ♏ 31 54	05 28	21 26	28 24	07 15	18 18	17 01	18 16	20 45
13	05 28 11	22 42 48	02 ♐ 56 30	09 ♐ 17 42	07 24	22 40	29 08	07 22	18 17	16 59	18 15	20 44
14	05 32 07	23 40 05	15 ♐ 35 37	21 ♐ 50 20	09 18	23 53	29 ♈ 53	07 29	18 15	16 57	18 13	20 42
15	05 36 04	24 37 22	28 ♐ 01 59	04 ♑ 10 45	11 10	25 07	00 ♉ 37	07 36	18 14	16 54	18 12	20 41
16	05 40 01	25 34 39	10 ♑ 16 48	16 ♑ 20 22	12 59	26 21	01 21	07 43	18 12	16 52	18 10	20 40
17	05 43 57	26 31 55	22 ♑ 21 41	28 ♑ 21 05	14 46	27 35	02 05	07 51	18 10	16 50	18 09	20 38
18	05 47 54	27 29 10	04 ♒ 18 52	10 ♒ 15 24	16 31	28 ♊ 48	02 49	07 59	18 08	16 48	18 08	20 37
19	05 51 50	28 26 25	16 ♒ 11 05	22 ♒ 06 20	18 13	00 ♋ 02	03 33	08 06	18 06	16 46	18 06	20 36
20	05 55 47	29 ♊ 23 40	28 ♒ 01 38	03 ♓ 57 28	19 53	01 16	04 17	08 14	18 04	16 43	18 05	20 35
21	05 59 43	00 ♋ 20 54	09 ♓ 54 21	15 ♓ 52 48	21 30	02 29	05 01	08 22	18 02	16 41	18 03	20 33
22	06 03 40	01 18 08	21 ♓ 53 24	27 ♓ 56 40	23 05	03 43	05 45	08 30	18 00	16 39	18 02	20 32
23	06 07 36	02 15 22	04 ♈ 03 11	10 ♈ 13 30	24 38	04 57	06 29	08 39	17 57	16 36	18 00	20 31
24	06 11 33	03 12 36	16 ♈ 28 09	22 ♈ 47 36	26 08	06 11	07 12	08 47	17 55	16 34	17 58	20 30
25	06 15 30	04 09 51	29 ♈ 12 21	05 ♉ 42 44	27 36	07 24	07 56	08 55	17 52	16 32	17 57	20 29
26	06 19 26	05 07 05	12 ♉ 19 07	19 ♉ 01 41	29 ♋ 01	08 38	08 40	09 04	17 50	16 29	17 55	20 28
27	06 23 23	06 04 19	25 ♉ 50 35	02 ♊ 45 45	00 ♌ 23	09 52	09 23	09 13	17 47	16 27	17 54	20 27
28	06 27 19	07 01 33	09 ♊ 47 05	16 ♊ 54 15	01 43	11 05	10 06	09 21	17 44	16 25	17 52	20 26
29	06 31 16	07 58 47	24 ♊ 06 49	01 ♋ 24 12	03 01	12 19	10 50	09 30	17 41	16 22	17 51	20 25
30	06 35 12	08 ♋ 56 01	08 ♋ 45 37	16 ♋ 10 18	04 ♌ 15	13 ♋ 33	11 ♉ 33	09 ♍ 39	17 ♒ 38	16 ♑ 20	17 ♑ 49	20 ♏ 24

Date	Moon True ☊ ° '	Moon Mean ☊ ° '	Moon ☽ Latitude ° '
01	00 ♑ 42	01 ♑ 44	01 N 22
02	00 R 41	01 40	00 N 04
03	00 D 42	01 37	01 S 16
04	00 43	01 34	02 30
05	00 44	01 31	03 35
06	00 45	01 28	04 26
07	00 45	01 25	05 00
08	00 R 45	01 21	05 15
09	00 44	01 18	05 12
10	00 43	01 15	04 52
11	00 42	01 12	04 15
12	00 41	01 09	03 26
13	00 40	01 05	02 28
14	00 39	01 02	01 23
15	00 39	00 59	00 S 15
16	00 D 39	00 56	00 N 53
17	00 40	00 53	01 58
18	00 40	00 50	02 56
19	00 40	00 46	03 47
20	00 40	00 43	04 27
21	00 41	00 40	04 57
22	00 R 41	00 37	05 14
23	00 D 40	00 34	05 17
24	00 41	00 31	05 05
25	00 41	00 27	04 39
26	00 41	00 24	03 57
27	00 42	00 21	03 02
28	00 42	00 18	01 54
29	00 42	00 15	00 37
30	00 ♑ 42	00 ♑ 11	00 S 45

DECLINATIONS

Date	Sun ☉ ° '	Moon ☽ ° '	Mercury ☿ ° '	Venus ♀ ° '	Mars ♂ ° '	Jupiter ♃ ° '	Saturn ♄ ° '	Uranus ♅ ° '	Neptune ♆ ° '	Pluto ♇ ° '
01	22 N 08	24 N 02	23 N 01	21 N 18	06 N 42	10 N 23	16 S 07	22 S 44	21 S 25	03 S 18
02	22 15	23 30	23 26	21 32	06 59	10 21	16 07	22 44	21 25	03 18
03	22 23	21 24	23 49	21 46	07 16	10 19	16 08	22 44	21 25	03 18
04	22 30	17 54	24 10	21 58	07 33	10 16	16 08	22 45	21 26	03 18
05	22 36	13 19	24 28	22 10	07 49	10 14	16 08	22 45	21 26	03 18
06	22 42	08 01	24 43	22 22	08 06	10 12	16 09	22 45	21 26	03 18
07	22 48	02 N 22	24 55	22 32	08 23	10 09	16 09	22 45	21 26	03 18
08	22 54	03 S 19	25 05	22 42	08 39	10 07	16 10	22 46	21 26	03 17
09	22 59	08 45	25 11	22 52	08 56	10 04	16 10	22 46	21 26	03 17
10	23 03	13 41	25 15	23 01	09 12	10 02	16 11	22 46	21 27	03 17
11	23 07	17 51	25 17	23 09	09 28	09 59	16 11	22 47	21 27	03 17
12	23 11	21 04	25 16	23 16	09 44	09 56	16 12	22 47	21 27	03 17
13	23 14	23 10	25 12	23 23	10 00	09 53	16 12	22 47	21 27	03 17
14	23 17	24 02	25 07	23 29	10 16	09 51	16 13	22 47	21 27	03 17
15	23 20	23 40	24 59	23 35	10 32	09 48	16 14	22 48	21 28	03 17
16	23 22	22 10	24 49	23 39	10 48	09 45	16 14	22 48	21 28	03 17
17	23 24	19 39	24 37	23 43	11 03	09 42	16 15	22 48	21 28	03 17
18	23 25	16 20	24 23	23 47	11 19	09 39	16 16	22 49	21 28	03 17
19	23 26	12 23	24 07	23 49	11 34	09 36	16 17	22 49	21 28	03 17
20	23 26	07 58	23 50	23 51	11 49	09 33	16 17	22 49	21 28	03 18
21	23 26	03 S 16	23 32	23 52	12 04	09 30	16 18	22 49	21 29	03 18
22	23 26	01 N 35	23 12	23 53	12 19	09 27	16 19	22 50	21 29	03 18
23	23 25	06 27	22 51	23 53	12 34	09 23	16 20	22 50	21 29	03 18
24	23 24	11 10	22 29	23 52	12 49	09 20	16 21	22 50	21 29	03 18
25	23 22	15 32	22 07	23 50	13 03	09 17	16 22	22 51	21 30	03 18
26	23 20	19 18	21 43	23 48	13 18	09 13	16 23	22 51	21 30	03 18
27	23 18	22 09	21 18	23 45	13 32	09 10	16 24	22 51	21 30	03 18
28	23 15	23 47	20 53	23 41	13 46	09 07	16 25	22 52	21 30	03 19
29	23 12	23 55	20 27	23 37	14 00	09 03	16 26	22 52	21 30	03 19
30	23 N 08	22 N 25	20 N 01	23 N 32	14 N 14	09 N 00	16 S 27	22 S 52	21 S 31	03 S 19

ZODIAC SIGN ENTRIES

Date	h	m	Planets
02	11	58	☽ ♋
04	13	35	☽ ♌
06	15	28	☽ ♍
08	18	33	☽ ♎
09	18	27	☿ ♋
10	23	27	☽ ♏
13	06	29	☽ ♐
14	15	56	♂ ♉
15	15	50	☽ ♑
18	03	19	☽ ♒
19	11	22	♀ ♋
20	16	00	☽ ♓
21	03	14	☉ ♋
23	04	03	☽ ♈
25	13	28	☽ ♉
27	05	11	☿ ♌
27	19	14	☽ ♊
29	21	42	☽ ♋

LATITUDES

Date	Mercury ☿ ° '	Venus ♀ ° '	Mars ♂ ° '	Jupiter ♃ ° '	Saturn ♄ ° '	Uranus ♅ ° '	Neptune ♆ ° '	Pluto ♇ ° '
01	00 N 45	00 S 20	01 S 18	01 N 12	00 S 52	00 S 25	00 N 45	15 N 13
04	01 12	00 13	01 17	01 11	00 53	00 25	00 45	15 12
07	01 34	00 S 05	01 16	01 11	00 53	00 25	00 45	15 11
10	01 50	00 N 02	01 15	01 10	00 54	00 26	00 45	15 10
13	01 59	00 09	01 14	01 10	00 54	00 26	00 45	15 09
16	02 01	00 16	01 13	01 10	00 54	00 26	00 45	15 08
19	01 57	00 23	01 12	01 09	00 55	00 26	00 45	15 07
22	01 46	00 30	01 11	01 09	00 55	00 26	00 45	15 06
25	01 30	00 36	01 10	01 08	00 56	00 26	00 45	15 05
28	01 08	00 43	01 08	01 08	00 56	00 26	00 45	15 03
31	00 N 41	00 N 49	01 S 07	01 N 08	00 S 57	00 S 26	00 N 45	15 N 02

DATA

Julian Date	2448775
Delta T	+58 seconds
Ayanamsa	23° 45' 20"
Synetic vernal point	05° ♓ 21' 39"
True obliquity of ecliptic	23° 26' 25"

MOON'S PHASES, APSIDES AND POSITIONS ☽

Date	h	m	Phase	Longitude ° '	Eclipse Indicator
01	03	57	●	10 ♊ 55	
07	20	47	☽	17 ♍ 20	
15	04	50	○	24 ♐ 20	partial
23	08	11	☾	02 ♈ 06	
30	12	18	●	08 ♋ 57	Total

Day	h	m		
04	01	54	Perigee	
19	21	47	Apogee	
01	16	09	Max dec	24° N 03'
07	21	57	0S	
14	16	49	Max dec	24° S 03'
22	04	12	0N	
29	01	58	Max dec	24° N 04'

LONGITUDES

Date	Chiron ⚷ ° '	Ceres ⚳ ° '	Pallas ⚴ ° '	Juno ⚵ ° '	Vesta ⚶ ° '	Black Moon Lilith ⚸ ° '
01	06 ♌ 02	10 ♒ 19	02 ♑ 47	02 ♉ 59	16 ♍ 28	14 ♒ 39
11	06 ♌ 57	10 ♒ 09	00 ♑ 05	08 ♉ 35	19 ♍ 14	15 ♒ 46
21	07 ♌ 59	09 ♒ 23	27 ♐ 11	13 ♉ 47	22 ♍ 26	16 ♒ 53
31	09 ♌ 05	08 ♒ 04	24 ♐ 25	19 ♉ 10	25 ♍ 59	18 ♒ 00

ASPECTARIAN

h m Aspects

01 Monday
02 32 ☿ ± ♅
03 57 ☽ ☌ ☉
04 44 ☽ ± ♅
05 09 ☽ ☌ ☿
06 39 ☽ ± ♆
14 48 ☿ ± ♆
14 50 ☽ ⊼ ♅
15 21 ☉ ± ♅
16 42 ☽ △ ♄
16 45 ☽ ⊼ ♆
20 01 ☽ ⚹ ♂
20 57 ☽ ⊼ ♇
23 24 ♀ ⋕ ♆

02 Tuesday
02 14 ☽ Q ♃
06 57 ☽ ± ♇
12 52 ☽ ∥ ☿
13 07 ♂ ⊼ ♇
17 08 ☽ Q ♂
17 43 ☽ bQ ♄
19 18 ☉ ± ♆
20 59 ♂ bQ ♃
21 53 ☽ bQ ♇
22 19 ☽ ⚹ ♃
22 52 ☽ ⋕ ♅

03 Wednesday
03 18 ☽ ∥ ☉
04 32 ☽ ⊻ ♀
08 37 ☽ ± ♄
09 05 ☽ ∥ ♀
09 36 ☽ ⊻ ☉
11 47 ☽ ⋕ ♆
15 22 ☽ ⊥ ♀
16 09 ☽ ⊻ ☿
16 37 ☽ ☍ ♅
18 31 ☽ ⊼ ♄
18 32 ☽ ☍ ♆
19 12 ☿ ⊼ ♅
20 12 ☽ ⊥ ☉
22 39 ☽ △ ♇
23 17 ☽ ∠ ♃

04 Thursday
00 31 ☽ □ ♂
03 49 ☽ ⊥ ☿
05 46 ♀ ± ♅
07 36 ☽ ∠ ♀
07 51 ☿ △ ♄
07 51 ☿ ⊼ ♆
08 16 ♄ ⊻ ♆
12 07 ☽ ∠ ☉
14 14 ☽ ⊥ ♃
21 18 ☽ ∠ ☿
21 49 ☽ ⋕ ♄

05 Friday
00 14 ☽ ⊻ ♃
04 10 ♀ ± ♆
10 43 ☽ ⚹ ♀
11 05 ☿ ⊼ ♇
14 42 ☽ ⚹ ☉
18 09 ☽ ⊼ ♅
20 06 ☽ ⊼ ♆
20 08 ☽ ☍ ♄

06 Saturday
00 15 ☽ □ ♇
02 23 ☽ ∥ ♃
02 35 ☽ ⚹ ☿
04 07 ☽ ± ♅
04 57 ☽ △ ♂
06 06 ☽ ± ♆
08 34 ☽ Q ♀
11 40 ☽ ∥ ♂
12 09 ☽ Q ☉
19 10 ☽ bQ ♅
21 10 ☽ bQ ♆
23 54 ☉ ⋕ ♅

07 Sunday
01 47 ☿ ⚹ ♂
02 16 ☽ Q ☿
02 40 ☽ ☌ ♃
05 04 ☿ Q ♃
06 25 ☽ Q ♇
07 34 ☽ bQ ♂
08 08 ☽ ⋕ ♇
16 57 ☉ ⊼ ♅
17 49 ☽ □ ♀
20 31 ☽ △ ♅
20 47 ☽ □ ☉
22 34 ☽ △ ♆
22 37 ☽ ⊼ ♄
23 44 ☽ ± ♂

08 Monday
02 48 ☽ ⚹ ♇
05 56 ☿ ± ♇
08 53 ☽ ± ♄
10 36 ☽ ⊼ ♂
11 53 ☽ ∥ ♇
14 22 ☽ □ ☿
20 30 ♀ ⋕ ♅
22 27 ☉ ⊼ ♆
23 14 ☉ △ ♄
23 58 ♀ ⊼ ♅

09 Tuesday
00 25 ☽ bQ ♄
04 40 ☽ ∠ ♇
06 31 ☽ ⊻ ♃
12 52 ☽ ⋕ ♂
17 06 ☽ ⊥ ♃
18 05 ☽ ⋕ ♃
20 25 ☽ ⊥ ♇
23 20 ♀ ⊼ ♆

10 Wednesday
00 00 ♀ △ ♄
00 27 ☽ □ ♅
02 37 ☽ □ ♆
02 41 ☽ △ ♄
02 57 ☽ △ ♀
04 49 ☽ △ ☉
07 00 ☽ ⊻ ♇
09 09 ☽ ∠ ♃
18 16 ☽ ☍ ♂

11 Thursday
01 47 ☽ ∥ ♄
02 44 ☉ ∥ ♀
04 38 ☽ △ ☿
08 30 ☽ bQ ♀
08 33 ☽ Q ♅
09 45 ☽ bQ ☉
10 01 ☿ bQ ♄
10 48 ☽ Q ♆
11 13 ☉ ⊼ ♇
12 19 ☽ ⚹ ♃
22 51 ♀ ⊼ ♇

12 Friday
02 35 ☽ ± ♀
03 27 ☽ ± ☉
06 23 ☽ ⚹ ♅
08 42 ☽ ⚹ ♆
08 46 ☽ □ ♄
10 28 ☽ Q ♃
12 53 ☽ bQ ☿
13 16 ☽ ☌ ♇
14 48 ☽ ⊼ ♀
15 22 ☿ bQ ♇
15 23 ☽ ⊼ ☉
15 30 ☽ ∥ ♆

13 Saturday
02 56 ♂ ∥ ♃
04 26 ☽ ⊼ ♂
06 27 ☽ ∥ ♅
08 37 ☽ ± ☿
10 12 ☽ ∠ ♅
11 22 ☿ ⚹ ♃
12 34 ☽ ∠ ♆
13 20 ☽ ⋕ ☉
16 11 ☽ ⋕ ♀
16 24 ☽ ± ♂
16 30 ☉ ☌ ♀
18 17 ☽ Q ♄
20 25 ☽ □ ♃
21 55 ☽ ⊼ ☿

14 Sunday
03 09 ☽ ⊥ ♅
05 35 ☽ ⊥ ♆
10 33 ☽ bQ ♂
14 34 ☽ ⊻ ♅
17 02 ☽ ⊻ ♆
17 06 ☽ ⚹ ♄
21 48 ☽ ⊻ ♇
23 52 ♂ Q ♄

15 Monday
04 50 ☽ ☍ ☉
05 43 ☽ ☍ ♀
09 23 ☽ ⊥ ♇
14 02 ☽ ⋕ ♀
17 21 ☽ △ ♂
18 58 ☽ ⋕ ☉
22 07 ☽ ∠ ♄
22 28 ♀ Q ♃

16 Tuesday
01 42 ☿ ± ♄
02 55 ☽ ∠ ♇
03 45 ☽ ∥ ♅
06 55 ☽ △ ♃
15 47 ☽ ⊥ ♄
16 14 ☉ Q ♃
18 00 ♀ ± ♇
18 17 ☽ ☍ ☿
19 37 ☽ ∥ ♆
20 14 ☿ Q ♂

17 Wednesday
01 01 ☽ ☌ ♅
03 37 ☽ ☌ ♆
03 40 ☽ ⊻ ♄
08 34 ☽ ⚹ ♇
12 59 ☽ bQ ♃
14 37 ☉ ± ♇
21 04 ☽ ⊼ ☉
23 38 ☽ ⊼ ♀

18 Thursday
07 14 ☽ ± ♃
08 35 ☽ Q ♇
08 48 ☽ □ ♂
10 11 ☽ ± ☉
12 25 ☽ ∥ ♄
13 06 ☽ ± ♀
15 49 ☿ ☍ ♅
19 28 ☽ ⊼ ♃

19 Friday
05 57 ☽ bQ ☉
09 24 ☽ bQ ♀
10 20 ☿ ☍ ♆
10 23 ☿ ⊼ ♄
13 10 ☽ ⊻ ♅
15 52 ☽ ⊻ ♆
15 53 ☽ ☌ ♄
16 19 ☽ ⋕ ♂
16 48 ☽ ⊼ ☿
19 35 ♄ ⊻ ♆
20 56 ☽ □ ♇
23 36 ☽ Q ♂

20 Saturday
01 17 ☽ ⊥ ♅
03 33 ☽ ⋕ ♃
04 01 ☽ ⊥ ♆
06 57 ☽ ± ☿
11 09 ☿ ∥ ♀
15 01 ☽ △ ☉
19 18 ☽ △ ♀
19 27 ☽ ∠ ♅
22 04 ☿ △ ♇
22 12 ☽ ∠ ♆

21 Sunday
01 30 ☽ ⚹ ♂
04 04 ☽ bQ ☿
08 52 ☽ ☍ ♃
11 53 ☽ ∥ ♇
19 10 ☉ ∥ ☿
22 18 ♀ bQ ♄

22 Monday
01 34 ☽ ⚹ ♅
04 15 ☽ ⊻ ♄
04 19 ☽ ⚹ ♆
09 19 ☽ △ ♇
09 35 ☽ ∠ ♂
14 44 ☽ △ ☿
16 10 ☽ ⊥ ♄
16 49 ♃ Q ♇
19 03 ☿ ∠ ♃
20 24 ☽ ⋕ ♇

23 Tuesday
01 20 ☽ Q ♅
04 04 ☽ Q ♆
04 33 ☽ ⊥ ♂
08 11 ☽ □ ☉
09 51 ☽ ∠ ♄
13 33 ☿ ⋕ ♅
13 56 ☽ □ ♀
14 51 ☽ bQ ♇
17 02 ☽ ⊻ ♂
21 02 ☽ ⊼ ♃
23 00 ♀ bQ ♇

24 Wednesday
02 37 ☽ ∥ ♃
04 52 ☉ bQ ♄
08 14 ☽ ± ♇
08 44 ☽ ± ♃
12 11 ☽ □ ♅
14 45 ☽ ⚹ ♄
14 52 ☽ □ ♆
19 39 ☽ ⊼ ♇
21 15 ☽ ∥ ♂
21 45 ☽ Q ☉

25 Thursday
02 01 ☽ bQ ♃
04 09 ☽ Q ♀
08 37 ☽ □ ☿
13 14 ☽ Q ♄
16 55 ☽ ⋕ ♄
21 53 ☽ ⚹ ☉

26 Friday
04 39 ☽ ⚹ ♀
05 00 ☽ ☌ ♂
06 03 ☽ △ ♃
13 17 ♀ ⚹ ♂
19 28 ☽ △ ♅
20 32 ☉ bQ ♇
21 23 ☽ Q ☿
21 35 ♀ ⚹ ♃
21 50 ☽ □ ♄
22 01 ☽ △ ♆

27 Saturday
00 38 ☿ ⋕ ♆
02 31 ☽ ☍ ♇
02 59 ☽ ∠ ☉
04 48 ♂ △ ♃
04 53 ☽ ∥ ☿
05 34 ☽ ⋕ ♆
10 07 ☽ ∠ ♀
19 54 ☽ ⊥ ☉
20 02 ☽ ⋕ ♅
20 45 ☽ ⚹ ☿
21 43 ☽ bQ ♅

28 Sunday
00 12 ☽ bQ ♆
01 56 ☽ ∥ ☉
03 14 ☽ ⊥ ♀
06 58 ☽ ⊻ ☉
09 39 ☽ ∥ ♀
11 16 ☽ □ ♃
12 35 ☽ ⊻ ♂
13 03 ☽ ± ♅
14 25 ☽ ⊻ ♀
15 31 ☽ ± ♆
23 08 ☽ ⊼ ♅
23 14 ☽ ⊥ ♂

29 Monday
00 09 ♀ ± ♄
00 51 ☽ ∠ ☿
01 21 ☽ △ ♄
01 35 ☽ ⊼ ♆
05 52 ☽ ⊼ ♇
14 59 ☽ ∠ ♂
15 48 ☽ ± ♇
17 14 ☽ ⊥ ☿
17 39 ☽ Q ♃
20 16 ☽ ∥ ♀
22 45 ☿ ⊥ ♃

30 Tuesday
02 04 ☽ bQ ♄
03 04 ☽ ∥ ☉
04 00 ☽ ⊻ ☿
06 32 ☽ bQ ♇
06 53 ☽ ⋕ ♅
12 18 ☽ ☌ ☉
13 28 ☽ ⚹ ♃
16 39 ☽ ± ♄
16 46 ☽ ⚹ ♂
20 20 ☽ ⋕ ♆
20 28 ☽ ☌ ♀

All ephemeris data is given at 12.00 UT and the Moon's longitude is additionally given for 24.00 UT

JULY 1992

LONGITUDES

Date	Sidereal time h m s	Sun ☉ ° ' "	Moon ☽ ° ' "	Moon ☽ 24.00 ° ' "	Mercury ☿ ° '	Venus ♀ ° '	Mars ♂ ° '	Jupiter ♃ ° '	Saturn ♄ ° '	Uranus ♅ ° '	Neptune ♆ ° '	Pluto ♇ ° '
01	06 39 09	09 ♋ 53 15	23 ♋ 37 14	01 ♌ 05 26	05 ♌ 27	14 ♋ 47	12 ♉ 16	09 ♍ 49	17 ♒ 35	16 ♑ 17	17 ♑ 47	20 ♏ 23
02	06 43 05	10 50 28	08 ♌ 33 53	16 ♌ 01 33	06 37	16 01	12 59	09 58	17 R 32	16 R 15	17 R 46	20 R 22
03	06 47 02	11 47 42	23 ♌ 27 29	00 ♍ 50 46	07 43	17 14	13 42	10 07	17 29	16 13	17 44	20 21
04	06 50 59	12 44 55	08 ♍ 10 39	15 ♍ 26 28	08 46	18 28	14 25	10 16	17 26	16 10	17 43	20 20
05	06 54 55	13 42 07	22 ♍ 37 42	29 ♍ 43 58	09 47	19 42	15 08	10 26	17 23	16 08	17 41	20 19
06	06 58 52	14 39 20	06 ♎ 45 02	13 ♎ 40 47	10 44	20 56	15 51	10 36	17 19	16 05	17 39	20 19
07	07 02 48	15 36 32	20 ♎ 31 12	27 ♎ 16 22	11 38	22 09	16 34	10 45	17 16	16 03	17 38	20 18
08	07 06 45	16 33 44	03 ♏ 56 27	10 ♏ 31 41	12 29	23 23	17 16	10 55	17 13	16 01	17 36	20 17
09	07 10 41	17 30 56	17 ♏ 02 18	23 ♏ 28 37	13 16	24 37	17 59	11 05	17 09	15 58	17 34	20 16
10	07 14 38	18 28 08	29 ♏ 50 55	06 ♐ 09 32	14 00	25 51	18 41	11 15	17 05	15 56	17 33	20 16
11	07 18 34	19 25 20	12 ♐ 24 46	18 ♐ 36 55	14 40	27 04	19 23	11 25	17 02	15 53	17 31	20 15
12	07 22 31	20 22 32	24 ♐ 46 15	00 ♑ 53 05	15 16	28 18	20 06	11 35	16 58	15 51	17 30	20 14
13	07 26 28	21 19 44	06 ♑ 57 39	13 ♑ 00 12	15 48	29 ♋ 32	20 48	11 45	16 54	15 48	17 28	20 14
14	07 30 24	22 16 57	19 ♑ 01 00	25 ♑ 00 16	16 16	00 ♌ 46	21 30	11 56	16 50	15 46	17 26	20 13
15	07 34 21	23 14 09	00 ♒ 58 14	06 ♒ 55 09	16 39	02 00	22 12	12 06	16 47	15 44	17 25	20 13
16	07 38 17	24 11 22	12 ♒ 51 16	18 ♒ 46 49	16 58	03 13	22 54	12 17	16 43	15 41	17 23	20 12
17	07 42 14	25 08 36	24 ♒ 42 06	00 ♓ 37 25	17 13	04 27	23 36	12 27	16 39	15 39	17 22	20 12
18	07 46 10	26 05 50	06 ♓ 33 04	12 ♓ 29 25	17 23	05 41	24 17	12 38	16 35	15 37	17 20	20 11
19	07 50 07	27 03 04	18 ♓ 26 49	24 ♓ 25 43	17 28	06 55	24 59	12 48	16 31	15 34	17 18	20 11
20	07 54 03	28 00 20	00 ♈ 26 31	06 ♈ 29 41	17 R 28	08 08	25 41	12 59	16 26	15 32	17 17	20 11
21	07 58 00	28 57 36	12 ♈ 35 43	18 ♈ 45 06	17 23	09 22	26 22	13 10	16 22	15 29	17 15	20 10
22	08 01 57	29 ♋ 54 52	24 ♈ 58 22	01 ♉ 16 01	17 14	10 36	27 04	13 21	16 18	15 27	17 14	20 10
23	08 05 53	00 ♌ 52 10	07 ♉ 38 35	14 ♉ 06 31	16 59	11 50	27 45	13 32	16 14	15 25	17 12	20 10
24	08 09 50	01 49 29	20 ♉ 40 18	27 ♉ 20 17	16 40	13 04	28 26	13 43	16 10	15 23	17 10	20 10
25	08 13 46	02 46 48	04 ♊ 06 48	11 ♊ 00 03	16 16	14 17	29 07	13 54	16 05	15 20	17 09	20 09
26	08 17 43	03 44 09	18 ♊ 00 06	25 ♊ 06 54	15 47	15 31	29 ♉ 48	14 05	16 01	15 18	17 07	20 09
27	08 21 39	04 41 30	02 ♋ 20 10	09 ♋ 39 30	15 15	16 45	00 ♊ 29	14 17	15 57	15 16	17 06	20 09
28	08 25 36	05 38 53	17 ♋ 04 16	24 ♋ 33 39	14 39	17 59	01 10	14 28	15 52	15 14	17 04	20 09
29	08 29 32	06 36 16	02 ♌ 06 40	09 ♌ 42 09	14 00	19 13	01 50	14 39	15 48	15 11	17 03	20 09
30	08 33 29	07 33 40	17 ♌ 18 54	24 ♌ 55 34	13 18	20 27	02 31	14 51	15 44	15 09	17 01	20 09
31	08 37 26	08 ♌ 31 05	02 ♍ 30 53	10 ♍ 03 35	12 ♌ 34	21 ♌ 40	03 ♊ 11	15 ♍ 02	15 ♒ 39	15 ♑ 07	17 ♑ 00	20 ♏ 09

Date	Moon True ☊ ° '	Moon Mean ☊ ° '	Moon ☽ Latitude ° '
01	00 ♑ 42	00 ♑ 08	02 S 04
02	00 R 41	00 05	03 15
03	00 40	00 ♑ 02	04 12
04	00 39	29 ♐ 59	04 53
05	00 37	29 56	05 13
06	00 37	29 52	05 15
07	00 D 37	29 49	04 58
08	00 37	29 46	04 25
09	00 38	29 43	03 38
10	00 40	29 40	02 42
11	00 41	29 36	01 39
12	00 42	29 33	00 S 33
13	00 R 42	29 30	00 N 34
14	00 41	29 27	01 39
15	00 39	29 24	02 39
16	00 35	29 21	03 31
17	00 32	29 17	04 15
18	00 27	29 14	04 47
19	00 23	29 11	05 06
20	00 20	29 08	05 13
21	00 17	29 05	05 06
22	00 16	29 02	04 45
23	00 D 16	28 58	04 09
24	00 17	28 55	03 20
25	00 19	28 52	02 18
26	00 20	28 49	01 N 07
27	00 R 21	28 46	00 S 11
28	00 20	28 42	01 30
29	00 17	28 39	02 44
30	00 13	28 36	03 48
31	00 ♑ 08	28 ♐ 33	04 S 35

DECLINATIONS

Date	Sun ☉ ° '	Moon ☽ ° '	Mercury ☿ ° '	Venus ♀ ° '	Mars ♂ ° '	Jupiter ♃ ° '	Saturn ♄ ° '	Uranus ♅ ° '	Neptune ♆ ° '	Pluto ♇ ° '
01	23 N 04	19 N 21	19 N 34	23 N 26	14 N 28	08 N 56	16 S 28	22 S 52	21 S 31	03 S 19
02	23 00	14 59	19 08	23 19	14 41	08 52	16 29	22 53	21 31	03 19
03	22 55	09 44	18 41	23 12	14 54	08 49	16 30	22 53	21 31	03 20
04	22 50	03 N 59	18 14	23 04	15 08	08 45	16 31	22 53	21 31	03 20
05	22 44	01 S 52	17 47	22 56	15 21	08 41	16 32	22 54	21 32	03 20
06	22 38	07 30	17 20	22 46	15 34	08 38	16 33	22 54	21 32	03 20
07	22 32	12 36	16 54	22 36	15 46	08 34	16 35	22 54	21 32	03 21
08	22 25	16 59	16 27	22 26	15 59	08 30	16 36	22 55	21 32	03 21
09	22 18	20 25	16 02	22 15	16 11	08 26	16 37	22 55	21 33	03 21
10	22 10	22 46	15 37	22 03	16 24	08 22	16 38	22 55	21 33	03 22
11	22 02	23 55	15 12	21 50	16 36	08 18	16 39	22 55	21 33	03 22
12	21 54	23 53	14 49	21 37	16 48	08 14	16 41	22 56	21 33	03 22
13	21 45	22 41	14 26	21 23	16 59	08 10	16 42	22 56	21 33	03 23
14	21 36	20 27	14 04	21 09	17 11	08 06	16 43	22 56	21 34	03 23
15	21 26	17 21	13 44	20 54	17 22	08 02	16 44	22 57	21 34	03 23
16	21 17	13 34	13 24	20 38	17 34	07 58	16 46	22 57	21 34	03 24
17	21 06	09 17	13 07	20 22	17 45	07 54	16 47	22 57	21 34	03 24
18	20 56	04 S 40	12 50	20 05	17 56	07 50	16 48	22 57	21 35	03 25
19	20 45	00 N 08	12 35	19 48	18 06	07 46	16 50	22 58	21 35	03 25
20	20 34	04 58	12 22	19 30	18 17	07 42	16 51	22 58	21 35	03 25
21	20 22	09 40	12 11	19 12	18 27	07 37	16 52	22 58	21 35	03 26
22	20 10	14 05	12 02	18 53	18 37	07 33	16 54	22 59	21 35	03 26
23	19 58	17 59	11 55	18 33	18 48	07 29	16 55	22 59	21 36	03 27
24	19 45	21 08	11 50	18 13	18 57	07 24	16 57	22 59	21 36	03 27
25	19 32	23 14	11 47	17 53	19 07	07 20	16 58	22 59	21 36	03 28
26	19 19	24 01	11 47	17 32	19 17	07 16	16 59	23 00	21 36	03 28
27	19 05	23 14	11 48	17 10	19 26	07 11	17 01	23 00	21 36	03 29
28	18 52	20 52	11 52	16 48	19 35	07 07	17 02	23 00	21 37	03 29
29	18 37	17 01	11 59	16 26	19 44	07 02	17 04	23 00	21 37	03 30
30	18 23	12 01	12 07	16 03	19 53	06 58	17 05	23 01	21 37	03 30
31	18 N 08	06 N 18	12 N 17	15 N 39	20 N 01	06 N 53	17 S 06	23 S 01	21 S 37	03 S 31

ZODIAC SIGN ENTRIES

Date	h	m	Planets
01	22	15	☽ ♌
03	22	37	☽ ♍
06	00	27	☽ ♎
08	04	53	☽ ♏
10	12	17	☽ ♐
12	22	16	☽ ♑
13	21	07	♀ ♌
15	10	03	☽ ♒
17	22	44	☽ ♓
20	11	07	☽ ♈
22	14	09	☉ ♌
22	21	36	☽ ♉
25	04	44	☽ ♊
26	18	59	♂ ♊
27	08	08	☽ ♋
29	08	39	☽ ♌
31	08	01	☽ ♍

LATITUDES

Date	Mercury ☿ ° '	Venus ♀ ° '	Mars ♂ ° '	Jupiter ♃ ° '	Saturn ♄ ° '	Uranus ♅ ° '	Neptune ♆ ° '	Pluto ♇ ° '
01	00 N 41	00 N 49	01 S 07	01 N 08	00 S 57	00 S 26	00 N 45	15 N 02
04	00 N 10	00 55	01 05	01 07	00 57	00 26	00 45	15 01
07	00 S 25	01 00	01 04	01 07	00 58	00 26	00 45	14 59
10	01 04	01 05	01 02	01 07	00 58	00 26	00 45	14 58
13	01 45	01 10	01 00	01 06	00 58	00 26	00 45	14 56
16	02 27	01 14	00 58	01 06	00 59	00 26	00 45	14 55
19	03 09	01 18	00 56	01 06	00 59	00 26	00 45	14 53
22	03 49	01 21	00 54	01 05	00 59	00 26	00 45	14 52
25	04 22	01 24	00 52	01 05	01 00	00 26	00 45	14 50
28	04 46	01 26	00 50	01 05	01 00	00 26	00 45	14 48
31	04 S 57	01 N 27	00 S 47	01 N 05	01 S 00	00 S 26	00 N 45	14 N 47

DATA

Julian Date	2448805
Delta T	+58 seconds
Ayanamsa	23° 45' 26"
Synetic vernal point	05° ♓ 21' 33"
True obliquity of ecliptic	23° 26' 24"

MOON'S PHASES, APSIDES AND POSITIONS ☽

Date	h	m	Phase	Longitude ° '	Eclipse Indicator
07	02	43	☽	15 ♎ 14	
14	19	06	○	22 ♑ 34	
22	22	12	☾	00 ♉ 19	
29	19	35	●	06 ♌ 54	

Day	h	m		
02	00	27	Perigee	
17	10	24	Apogee	
30	07	38	Perigee	
05	04	16	0S	
11	23	05	Max dec	24° S 03'
19	11	20	0N	
26	12	22	Max dec	24° N 01'

LONGITUDES

Date	Chiron ⚷ ° '	Ceres ⚳ ° '	Pallas ⚴ ° '	Juno ⚵ ° '	Vesta ⚶ ° '	Black Moon Lilith ⚸ ° '
01	09 ♌ 05	08 ♒ 04	24 ♐ 25	19 ♉ 10	25 ♍ 59	18 ♒ 00
11	10 ♌ 15	06 ♒ 16	22 ♐ 03	24 ♉ 30	29 ♍ 50	19 ♒ 07
21	11 ♌ 27	04 ♒ 10	20 ♐ 17	29 ♉ 48	03 ♎ 57	20 ♒ 14
31	12 ♌ 42	01 ♒ 59	19 ♐ 12	05 ♊ 00	08 ♎ 16	21 ♒ 21

ASPECTARIAN

h m	Aspects
01 Wednesday	
00 13	☽ ☍ ♅
02 19	☽ ⚻ ♄
02 38	☽ ☍ ♆
06 48	☽ △ ♇
09 38	☉ ✶ ♃
10 19	☽ ∥ ☿
13 06	☽ Q ♂
13 56	☽ ∠ ♃
02 Thursday	
04 27	☽ ⋕ ♄
04 32	☽ ⊥ ♃
08 36	☽ ☌ ☿
13 25	☽ ∥ ♂
14 16	☽ ⚺ ♃
15 54	☽ ⚺ ☉
16 34	♀ ☍ ♅
19 28	☽ □ ♂
03 Friday	
00 20	☽ ⚻ ♅
01 03	☽ ⚺ ♀
02 13	☽ ⊥ ☉
02 23	☽ ☍ ♄
02 46	☽ ⚻ ♆
04 41	☉ ± ♄
06 59	☽ □ ♇
09 59	☽ ± ♅
11 37	☽ ⊥ ♀
12 27	☽ ± ♆
15 57	☽ ∥ ♃
16 40	♀ ⚻ ♄
17 47	☽ ∠ ☉
20 33	☉ ⋕ ♅
21 30	♀ ☍ ♆
04 Saturday	
00 34	☽ ⚼ ♅
03 03	☽ ⚼ ♆
03 34	☽ ∠ ♀
12 16	☽ Q ♇
13 03	☽ ⚺ ☿
14 38	☽ ⋕ ♇
15 29	☽ ☌ ♃
20 04	☽ ✶ ☉
22 51	☽ △ ♂
23 43	☽ ⊥ ☿
05 Sunday	
01 10	☽ △ ♅
03 15	☽ ⚻ ♄
03 45	☽ △ ♆
06 38	☽ ✶ ♀
08 08	☽ ✶ ♇
13 15	☽ ± ♄
15 53	☽ ∠ ☿
16 48	♀ ⋕ ♅
17 33	☽ Q ☉
18 06	☽ ∥ ♇
06 Monday	
00 06	♀ △ ♇
01 22	☽ ⚼ ♂
04 27	☽ ⚼ ♄
04 49	☽ Q ♀
07 35	☿ ⚺ ♃
09 32	☽ ∠ ♇
17 02	☽ ⋕ ♃
17 38	☽ ± ♂
18 43	☽ ⚺ ♃
19 22	☽ ✶ ☿
19 43	♂ △ ♅
07 Tuesday	
01 05	☽ ⊥ ♇
02 43	☽ □ ☉
04 10	☽ □ ♅
04 39	☽ ⚻ ♂
05 18	☽ ⊥ ♃
06 18	☽ △ ♄
06 55	☽ □ ♆
11 36	☽ ⚺ ♇
15 11	☽ □ ♀
17 54	☽ Q ☿
21 24	☽ ∠ ♃
22 38	☉ ☍ ♅
08 Wednesday	
04 42	☿ ⋕ ♄
05 47	☽ ⋕ ♂
09 11	☽ ⋕ ☿
09 41	☽ ∥ ♄
10 07	♂ □ ♄
12 07	☽ Q ♅
15 00	☽ Q ♆
17 52	☉ ∥ ♀
22 49	♂ △ ♆
09 Thursday	
00 53	☽ ✶ ♃
03 20	☉ ⚻ ♄
04 36	☽ □ ☿
05 58	☿ ∥ ♂
10 01	☽ ✶ ♅
12 12	☽ □ ♄
12 57	☽ △ ☉
12 59	☽ ✶ ♆
13 25	☉ ☍ ♆
13 51	☽ ☍ ♂
18 01	☽ ☌ ♇
22 08	☽ ∥ ♆
23 25	☽ Q ♃
10 Friday	
03 38	☽ △ ♀
04 03	☽ ⋕ ♀
05 06	☽ ⋕ ☉
14 02	☽ ∠ ♅
14 14	☽ ∥ ♅
17 06	☽ ∠ ♆
19 26	☽ ⚼ ☉
21 08	♀ ∠ ♃
21 55	☽ Q ♄
11 Saturday	
07 09	☽ ⊥ ♅
08 57	☉ ✶ ♂
10 03	☽ □ ♃
10 17	☽ ⊥ ♆
11 17	☽ ⚼ ♀
14 06	☽ ± ☉
16 34	☽ △ ☿
18 41	☽ ⚺ ♅
20 21	♂ ⋕ ♄
20 53	☽ ✶ ♄
21 51	☽ ⚺ ♆
12 Sunday	
02 20	☽ ⚻ ♂
02 42	☽ ⚻ ☉
03 10	☽ ⚺ ♇
06 39	☽ ± ♀
08 40	☉ △ ♇
14 45	☽ ± ♂
14 53	☽ ⊥ ♇
16 55	♂ ☍ ♇
18 32	♀ ⋕ ♆
19 42	☽ ⚻ ♀
23 17	☽ ⚼ ☿
13 Monday	
02 04	☽ ∠ ♄
08 23	☽ ∥ ♅
08 35	☽ ∠ ♇
09 33	☽ ⚼ ♂
12 30	☿ ⚻ ♅
17 52	☽ ± ☿
19 47	☽ ⊥ ♄
21 39	☽ △ ♃
14 Tuesday	
00 09	☽ ⋕ ☉
01 22	☽ ∥ ♆
04 49	☽ ⋕ ♀
05 32	☽ ☌ ♅
06 17	☽ ⚻ ☿
07 40	☽ ⚺ ♄
08 51	☽ ☌ ♆
14 25	☽ ✶ ♇
17 17	☽ △ ♂
17 35	☉ ⋕ ♆
19 06	☽ ☍ ☉
15 Wednesday	
04 06	☽ ⚼ ♃
11 53	☽ ⋕ ♂
14 18	☽ ☍ ♀
14 30	☽ Q ♇
16 09	☽ ∥ ♄
19 08	☿ ☍ ♄
22 30	☽ ± ♃
16 Thursday	
10 49	☽ ⚻ ♃
13 03	☽ ⋕ ☿
17 43	☽ ⚺ ♅
19 46	☽ ☌ ♄
20 32	☽ ☍ ☿
21 09	☽ ⚺ ♆
17 Friday	
02 53	☽ □ ♇
05 50	☽ ⊥ ♅
09 17	☽ ⊥ ♆
09 37	☽ □ ♂
12 58	☽ ⚻ ☉
19 27	☽ ⋕ ♃
18 Saturday	
00 01	☽ ∠ ♅
02 12	☽ ± ☉
03 29	☽ ∠ ♆
04 58	☿ ⚻ ♆
10 02	☽ ⚻ ♀
18 20	☽ ∥ ♇
21 59	☽ ⚼ ☉
23 34	☽ ± ♀
19 Sunday	
00 19	☽ Q ♂
00 28	☽ ☍ ♃
06 14	☽ ✶ ♅
08 07	☽ ⚺ ♄
09 38	♀ ⊥ ♃
09 42	☽ ✶ ♆
10 01	☽ ⚻ ☿
15 20	♀ Q ♂
15 29	☽ △ ♇
19 45	☽ ⚼ ♀
20 07	☽ ⊥ ♄
22 05	☽ ± ☿
20 Monday	
00 54	☿ St R
01 55	☽ ✶ ♂
04 18	☽ ⋕ ♇
06 13	☽ Q ♅
06 44	☽ △ ☉
09 41	☽ Q ♆
11 26	☉ ∠ ♃
13 59	☽ ∠ ♄
16 01	☽ ⚼ ☿
21 24	☽ ⚼ ♇
21 Tuesday	
01 35	☽ ∥ ♃
04 58	☽ △ ♀
09 27	☽ ∠ ♂
13 08	☽ ⚻ ♃
15 05	☽ ± ♇
17 38	☽ □ ♅
19 20	☽ ✶ ♄
21 04	☽ □ ♆
21 15	☽ △ ☿
22 Wednesday	
00 58	☽ ∥ ☿
01 00	☽ ± ♃
02 45	☽ ⚻ ♇
04 01	☽ ⊥ ♂
12 06	☿ ⚻ ♆
16 14	☽ ⚺ ♂
18 20	☽ Q ♄
18 33	☽ ⚼ ♃
22 12	☽ □ ☉
23 Thursday	
00 20	♀ ∥ ♂
05 01	☽ ⋕ ♄
15 32	☽ ∥ ♀
17 51	☽ ∥ ♂
20 37	☽ □ ♀
23 06	☽ △ ♃
24 Friday	
01 22	☽ ∥ ☉
02 22	☽ △ ♅
03 49	☽ □ ♄
04 53	☽ □ ☿
05 38	☽ △ ♆
10 21	☽ Q ☉
11 04	☽ ☍ ♇
16 19	☽ ⋕ ♆
25 Saturday	
02 42	☽ ☌ ♂
03 06	♀ ⚺ ♃
05 21	☽ ⚼ ♅
08 14	☽ ⋕ ♅
08 28	☽ Q ♀
08 33	☽ ⚼ ♆
09 29	☽ ✶ ☉
12 15	☽ Q ☿
21 06	☽ ± ♅
22 50	☿ ☍ ♄
26 Sunday	
00 14	☽ ± ♆
05 14	☽ □ ♃
07 22	☽ ✶ ♀
07 24	☽ ⚻ ♅
07 48	♀ ⚻ ♅
08 22	☽ ✶ ☿
08 38	☽ △ ♄
10 30	☽ ⚻ ♆
13 20	☽ ∠ ☉
14 41	☉ ∥ ♂
15 39	☽ ⚻ ♇
15 41	☿ ☌ ♀
21 08	♀ ☍ ♄
27 Monday	
01 44	☽ ± ♇
04 38	♂ ⚼ ♅
05 32	☽ ⊥ ☉
08 41	☽ ∠ ☿
08 47	☽ ⚺ ♂
09 43	☽ ⚼ ♄
10 57	☽ ∠ ♀
11 22	☿ ⚻ ♅
11 54	☽ Q ♃
15 24	☽ ⋕ ♅
16 09	☽ ⚺ ☉
16 38	☽ ⚼ ♇
18 35	♀ ⚻ ♆
19 09	☽ ⊥ ♂
21 34	♀ ⋕ ♄
22 54	☽ ⊥ ☿
28 Tuesday	
00 24	☽ ± ♄
03 02	☽ ⊥ ♀
05 56	☽ ⋕ ♆
07 45	☽ ✶ ♃
08 15	☽ ⚺ ☿
09 02	☽ ☍ ♅
10 05	☽ ⚻ ♄
10 28	☽ ∠ ♂
12 00	☽ ☍ ♆
13 36	☽ ⚺ ♀
16 57	☽ △ ♇
17 12	☿ ⚺ ♃
20 37	☽ ∥ ♂
29 Wednesday	
02 23	☽ ∥ ☉
08 04	☽ ∠ ♃
11 33	☽ ✶ ♂
11 47	☽ ⋕ ♄
14 41	☿ Q ♂
15 24	☽ ∥ ♀
19 01	♂ ⚼ ♆
19 35	☽ ☌ ☉
22 29	☽ ⊥ ♃
30 Thursday	
05 57	☽ ☌ ☿
06 15	♀ □ ♇
07 23	☽ Q ♂
08 04	☽ ⚺ ♃
08 36	☽ ⚻ ♅
09 30	☽ ☍ ♄
11 32	☽ ⚻ ♆
11 37	☽ ∥ ☿
16 28	☽ □ ♇
17 22	☽ ☌ ♀
18 02	☽ ± ♅
20 59	☽ ± ♆
21 34	♇ St D
31 Friday	
01 26	♀ ± ♅
08 13	☽ ⚼ ♅
09 34	☽ ∥ ♃
11 11	☽ ⚼ ♆
13 07	☽ □ ♂
19 49	♃ △ ♅
20 57	☽ Q ♇
22 11	☽ ⚺ ☉
23 07	☽ ⋕ ♇

All ephemeris data is given at 12.00 UT and the Moon's longitude is additionally given for 24.00 UT

AUGUST 1992

LONGITUDES

Date	Sidereal time h m s	Sun ☉ ° ' "	Moon ☽ ° ' "	Moon ☽ 24.00 ° ' "	Mercury ☿ ° '	Venus ♀ ° '	Mars ♂ ° '	Jupiter ♃ ° '	Saturn ♄ ° '	Uranus ♅ ° '	Neptune ♆ ° '	Pluto ♇ ° '
01	08 41 22	09 ♌ 28 30	17 ♍ 32 31	24 ♍ 56 42	11 ♌ 49	22 ♌ 54	03 ♊ 52	15 ♍ 14	15 ♒ 35	15 ♑ 05	16 ♑ 58	20 ♏ 09
02	08 45 19	10 25 56	02 ♎ 15 19	09 ♎ 27 47	11 R 03	24 08	04 32	15 26	15 R 30	15 R 03	16 R 57	20 D 09
03	08 49 15	11 23 23	16 ♎ 33 39	23 ♎ 32 45	10 18	25 22	05 12	15 37	15 26	15 01	16 55	20 09
04	08 53 12	12 20 50	00 ♏ 25 01	07 ♏ 10 34	09 34	26 36	05 52	15 49	15 21	14 59	16 54	20 09
05	08 57 08	13 18 18	13 ♏ 49 40	20 ♏ 22 39	08 52	27 50	06 32	16 01	15 17	14 57	16 53	20 09
06	09 01 05	14 15 47	26 ♏ 49 56	03 ♐ 12 01	08 13	29 ♌ 03	07 12	16 13	15 12	14 54	16 51	20 10
07	09 05 01	15 13 17	09 ♐ 29 23	15 ♐ 42 34	07 38	00 ♍ 17	07 51	16 25	15 08	14 53	16 50	20 10
08	09 08 58	16 10 47	21 ♐ 52 04	27 ♐ 58 23	07 07	01 31	08 31	16 37	15 03	14 51	16 48	20 10
09	09 12 55	17 08 19	04 ♑ 02 00	10 ♑ 03 23	06 41	02 45	09 10	16 49	14 59	14 49	16 47	20 10
10	09 16 51	18 05 51	16 ♑ 02 55	22 ♑ 01 01	06 21	03 59	09 50	17 01	14 54	14 47	16 46	20 11
11	09 20 48	19 03 24	27 ♑ 58 00	03 ♒ 54 11	06 07	05 12	10 29	17 13	14 50	14 45	16 44	20 11
12	09 24 44	20 00 58	09 ♒ 49 52	15 ♒ 45 17	05 59	06 26	11 08	17 25	14 45	14 43	16 43	20 12
13	09 28 41	20 58 33	21 ♒ 40 40	27 ♒ 36 14	05 D 59	07 40	11 47	17 37	14 41	14 41	16 42	20 12
14	09 32 37	21 56 10	03 ♓ 32 11	09 ♓ 28 44	06 05	08 54	12 26	17 50	14 37	14 39	16 41	20 13
15	09 36 34	22 53 48	15 ♓ 26 05	21 ♓ 24 26	06 18	10 07	13 04	18 02	14 32	14 38	16 39	20 13
16	09 40 30	23 51 27	27 ♓ 24 03	03 ♈ 25 10	06 39	11 21	13 43	18 14	14 28	14 36	16 38	20 14
17	09 44 27	24 49 07	09 ♈ 28 04	15 ♈ 33 05	07 07	12 35	14 21	18 27	14 23	14 34	16 37	20 14
18	09 48 23	25 46 49	21 ♈ 40 34	27 ♈ 50 52	07 43	13 49	15 00	18 39	14 19	14 33	16 36	20 15
19	09 52 20	26 44 33	04 ♉ 04 26	10 ♉ 21 41	08 25	15 02	15 38	18 51	14 15	14 31	16 35	20 15
20	09 56 17	27 42 18	16 ♉ 43 05	23 ♉ 09 07	09 15	16 16	16 16	19 04	14 10	14 30	16 33	20 16
21	10 00 13	28 40 06	29 ♉ 40 13	06 ♊ 16 51	10 11	17 30	16 54	19 16	14 06	14 28	16 32	20 17
22	10 04 10	29 ♌ 37 54	12 ♊ 59 25	19 ♊ 48 15	11 15	18 44	17 32	19 29	14 02	14 27	16 31	20 17
23	10 08 06	00 ♍ 35 45	26 ♊ 43 37	03 ♋ 45 39	12 24	19 57	18 09	19 41	13 58	14 25	16 30	20 18
24	10 12 03	01 33 37	10 ♋ 54 20	18 ♋ 09 28	13 40	21 11	18 47	19 54	13 53	14 24	16 29	20 19
25	10 15 59	02 31 31	25 ♋ 30 42	02 ♌ 57 23	15 02	22 25	19 24	20 07	13 49	14 22	16 28	20 20
26	10 19 56	03 29 27	10 ♌ 28 45	18 ♌ 03 45	16 28	23 39	20 01	20 19	13 45	14 21	16 27	20 21
27	10 23 52	04 27 24	25 ♌ 41 13	03 ♍ 19 48	18 00	24 52	20 38	20 32	13 41	14 20	16 26	20 22
28	10 27 49	05 25 23	10 ♍ 58 05	18 ♍ 34 41	19 36	26 06	21 15	20 45	13 37	14 19	16 25	20 23
29	10 31 46	06 23 23	26 ♍ 08 13	03 ♎ 37 27	21 16	27 20	21 52	20 58	13 33	14 17	16 24	20 24
30	10 35 42	07 21 25	11 ♎ 01 18	18 ♎ 18 54	22 59	28 34	22 29	21 10	13 29	14 16	16 24	20 25
31	10 39 39	08 ♍ 19 29	25 ♎ 29 37	02 ♏ 33 03	24 ♌ 46	29 ♍ 47	23 ♊ 05	21 ♍ 23	13 ♒ 25	14 ♑ 15	16 ♑ 23	20 ♏ 26

Date	Moon True ☊ ° '	Moon Mean ☊ ° '	Moon ☽ Latitude ° '
01	00 ♑ 02	28 ♐ 30	05 S 03
02	29 ♐ 57	28 27	05 10
03	29 53	28 23	04 57
04	29 51	28 20	04 27
05	29 D 50	28 17	03 44
06	29 51	28 14	02 49
07	29 52	28 11	01 48
08	29 53	28 08	00 S 43
09	29 R 53	28 04	00 N 22
10	29 52	28 01	01 26
11	29 48	27 58	02 25
12	29 42	27 55	03 18
13	29 35	27 52	04 02
14	29 25	27 48	04 35
15	29 15	27 45	04 56
16	29 05	27 42	05 05
17	28 56	27 39	05 00
18	28 49	27 36	04 41
19	28 45	27 33	04 09
20	28 42	27 29	03 25
21	28 D 41	27 26	02 29
22	28 42	27 23	01 23
23	28 42	27 20	00 N 11
24	28 R 42	27 17	01 S 05
25	28 39	27 14	02 18
26	28 34	27 10	03 23
27	28 27	27 07	04 15
28	28 18	27 04	04 49
29	28 08	27 01	05 02
30	27 58	26 58	04 55
31	27 ♐ 50	26 ♐ 54	04 S 28

DECLINATIONS

Date	Sun ☉ ° '	Moon ☽ ° '	Mercury ☿ ° '	Venus ♀ ° '	Mars ♂ ° '	Jupiter ♃ ° '	Saturn ♄ ° '	Uranus ♅ ° '	Neptune ♆ ° '	Pluto ♇ ° '
01	17 N 53	00 N 16	12 N 29	15 N 16	20 N 10	06 N 49	17 S 08	23 S 01	21 S 37	03 S 31
02	17 38	05 S 38	12 43	14 52	20 18	06 44	17 09	23 01	21 38	03 32
03	17 22	11 05	12 58	14 27	20 26	06 40	17 11	23 02	21 38	03 32
04	17 06	15 47	13 15	14 02	20 34	06 35	17 12	23 02	21 38	03 33
05	16 50	19 33	13 33	13 37	20 41	06 30	17 13	23 02	21 38	03 34
06	16 33	22 12	13 51	13 11	20 49	06 26	17 15	23 02	21 38	03 34
07	16 16	23 39	14 10	12 45	20 56	06 21	17 16	23 02	21 38	03 35
08	15 59	23 55	14 29	12 19	21 03	06 16	17 18	23 03	21 39	03 35
09	15 42	23 00	14 48	11 52	21 10	06 12	17 19	23 03	21 39	03 36
10	15 24	21 03	15 06	11 25	21 17	06 07	17 20	23 03	21 39	03 37
11	15 07	18 12	15 24	10 58	21 24	06 02	17 22	23 03	21 39	03 37
12	14 49	14 36	15 42	10 30	21 30	05 57	17 23	23 03	21 40	03 38
13	14 30	10 28	15 58	10 02	21 36	05 52	17 25	23 04	21 40	03 39
14	14 12	05 56	16 13	09 34	21 42	05 48	17 26	23 04	21 40	03 39
15	13 53	01 S 11	16 26	09 06	21 48	05 43	17 27	23 04	21 40	03 40
16	13 34	03 N 38	16 38	08 37	21 54	05 38	17 29	23 04	21 40	03 41
17	13 15	08 21	16 48	08 09	21 59	05 33	17 30	23 04	21 41	03 41
18	12 56	12 48	16 55	07 39	22 04	05 28	17 31	23 05	21 41	03 42
19	12 36	16 47	17 01	07 10	22 10	05 23	17 33	23 05	21 41	03 43
20	12 16	20 06	17 04	06 41	22 14	05 18	17 34	23 05	21 41	03 43
21	11 56	22 30	17 04	06 11	22 19	05 13	17 35	23 05	21 41	03 44
22	11 36	23 44	17 02	05 41	22 24	05 08	17 37	23 05	21 41	03 45
23	11 16	23 35	16 57	05 11	22 28	05 04	17 38	23 05	21 41	03 45
24	10 55	21 55	16 49	04 41	22 32	04 59	17 39	23 05	21 42	03 46
25	10 35	18 47	16 38	04 11	22 37	04 54	17 40	23 06	21 42	03 47
26	10 14	14 21	16 25	03 41	22 40	04 49	17 42	23 06	21 42	03 48
27	09 53	08 57	16 08	03 10	22 44	04 44	17 43	23 06	21 42	03 48
28	09 31	03 N 00	15 48	02 40	22 48	04 39	17 44	23 06	21 42	03 49
29	09 10	03 S 05	15 25	02 09	22 51	04 33	17 45	23 06	21 42	03 50
30	08 49	08 53	15 00	01 38	22 54	04 28	17 46	23 06	21 43	03 51
31	08 N 27	14 S 01	14 N 32	01 N 07	22 N 57	04 N 23	17 S 47	23 S 06	21 S 43	03 S 51

ZODIAC SIGN ENTRIES

Date	h	m	Planets
02	08	17	☽ ♎
04	11	16	☽ ♏
06	17	57	☽ ♐
07	06	26	♀ ♍
09	04	00	☽ ♑
11	16	06	☽ ♒
14	04	51	☽ ♓
16	17	11	☽ ♈
19	04	10	☽ ♉
21	12	36	☽ ♊
22	21	10	☉ ♍
23	17	36	☽ ♋
25	19	15	☽ ♌
27	18	46	☽ ♍
29	18	10	☽ ♎
31	16	09	♀ ♎
31	19	38	☽ ♏

LATITUDES

Date	Mercury ☿ ° '	Venus ♀ ° '	Mars ♂ ° '	Jupiter ♃ ° '	Saturn ♄ ° '	Uranus ♅ ° '	Neptune ♆ ° '	Pluto ♇ ° '
01	04 S 57	01 N 28	00 S 47	01 N 05	01 S 01	00 S 26	00 N 45	14 N 46
04	04 47	01 28	00 44	01 04	01 01	00 26	00 45	14 44
07	04 20	01 29	00 42	01 04	01 01	00 26	00 44	14 43
10	03 42	01 28	00 39	01 04	01 01	00 26	00 44	14 41
13	02 54	01 27	00 36	01 04	01 02	00 26	00 44	14 39
16	02 03	01 25	00 33	01 04	01 02	00 26	00 44	14 38
19	01 11	01 23	00 31	01 04	01 02	00 26	00 44	14 36
22	00 S 23	01 20	00 28	01 04	01 02	00 26	00 44	14 34
25	00 N 20	01 17	00 25	01 04	01 02	00 26	00 44	14 33
28	00 54	01 13	00 21	01 04	01 02	00 26	00 44	14 31
31	01 N 20	01 N 08	00 S 18	01 N 04	01 S 02	00 S 26	00 N 44	14 N 30

DATA

Julian Date	2448836
Delta T	+58 seconds
Ayanamsa	23° 45' 31"
Synetic vernal point	05° ♓ 21' 28"
True obliquity of ecliptic	23° 26' 25"

MOON'S PHASES, APSIDES AND POSITIONS ☽

Date	h	m	Phase	Longitude ° '	Eclipse Indicator
05	10	59	☽	13 ♏ 16	
13	10	27	○	20 ♒ 55	
21	10	01	☾	28 ♉ 35	
28	02	42	●	05 ♍ 03	

Day	h	m		
13	15	21	Apogee	
27	17	31	Perigee	
01	13	05	0S	
08	05	05	Max dec	23° S 58'
15	17	54	0N	
22	21	33	Max dec	23° N 51'
28	23	46	0S	

LONGITUDES

Date	Chiron ⚷ ° '	Ceres ⚳ ° '	Pallas ⚴ ° '	Juno ⚵ ° '	Vesta ⚶ ° '	Black Moon Lilith ⚸ ° '
01	12 ♌ 49	01 ♒ 46	19 ♐ 08	05 ♊ 31	08 ♎ 43	21 ♒ 27
11	14 ♌ 04	29 ♑ 44	18 ♐ 50	10 ♊ 37	13 ♎ 15	22 ♒ 34
21	15 ♌ 19	28 ♑ 03	19 ♐ 11	15 ♊ 34	17 ♎ 56	23 ♒ 41
31	16 ♌ 32	26 ♑ 52	20 ♐ 08	20 ♊ 20	22 ♎ 45	24 ♒ 48

All ephemeris data is given at 12.00 UT and the Moon's longitude is additionally given for 24.00 UT

ASPECTARIAN

h m	Aspects
01 Saturday	
03 15	☽ ⊻ ☿
04 27	☉ ⊥ ♃
08 03	☽ △ ♅
08 14	☽ ☌ ♃
08 27	☽ ⊥ ☉
08 51	☽ ⚻ ♄
11 05	☽ △ ♆
12 25	☽ ⊥ ☿
13 17	♀ ± ♆
16 13	☽ ✶ ♇
18 30	☽ ± ♄
21 28	☽ ⊻ ♀
02 Sunday	
00 01	☽ ∠ ☉
02 19	☽ ∠ ☿
03 18	☽ ∥ ♇
08 11	☽ ⊥ ♀
09 07	☽ ⚼ ♄
15 57	☽ △ ♂
16 36	☽ ⋕ ♃
16 48	☽ ∠ ♇
18 39	♃ ⚻ ♄
20 43	☉ ☌ ☿
03 Monday	
00 32	☽ ∠ ♀
01 56	☽ ✶ ☿
02 36	☽ ✶ ☉
07 54	☽ ⊥ ♇
09 22	☽ □ ♅
10 05	☽ △ ♄
10 23	☽ ⊻ ♃
12 37	☽ □ ♆
18 09	☽ ⊻ ♇
18 33	☽ ⚼ ♂
20 48	☽ ⊥ ♃
21 21	☽ Q ☿
21 41	☽ ⋕ ☿
04 Tuesday	
00 36	☽ Q ☉
03 20	☽ ⋕ ♀
03 28	☉ ⋕ ♄
04 39	☽ ✶ ♀
05 29	☿ ⊥ ♃
10 59	☽ ± ♂
12 43	☽ ∠ ♃
16 31	☽ Q ♅
19 10	☽ ⋕ ☉
19 55	☽ Q ♆
20 21	☽ ∥ ♄
22 10	☽ ⚻ ♂
05 Wednesday	
03 29	☽ □ ☿
04 01	☽ Q ♀
10 59	☽ □ ☉
14 01	☽ ✶ ♅
14 21	☿ ∥ ♀
14 38	☽ □ ♄
16 03	☽ ✶ ♃
17 33	☽ ✶ ♆
21 35	☽ ⋕ ♂
23 36	☽ ☌ ♇
06 Thursday	
06 00	☽ ∥ ♆
14 38	☽ Q ♃
16 37	☽ □ ♀
17 46	☽ ∠ ♅
21 26	☽ ∠ ♆
23 23	☽ ∥ ♅
23 56	☽ Q ♄
07 Friday	
03 37	☉ ⚻ ♅
04 12	♀ ⚼ ♅
07 39	☿ ✶ ♂
08 36	☽ △ ☿
08 42	☽ ☍ ♂
09 53	☉ ☍ ♄
10 49	☽ ⊥ ♅
14 34	☽ ⊥ ♆
22 21	☽ ⊻ ♅
22 49	☽ ✶ ♄
23 59	☽ △ ☉
08 Saturday	
01 35	☽ □ ♃
02 09	☽ ⊻ ♆
08 41	☽ ⊻ ♇
12 29	☽ ⚼ ☿
17 34	♀ ⚼ ♆
20 27	☽ ⊥ ♇
23 20	♂ ± ♅
09 Sunday	
01 42	☉ ⊻ ♃
03 20	☉ ⚻ ♆
04 01	☽ ∠ ♄
05 35	☽ ± ☿
07 55	☽ ⚼ ☉
08 51	♃ △ ♆
09 09	☽ △ ♀
11 18	☽ ∥ ♅
14 16	☽ ∠ ♇
17 08	☽ ⚻ ☿
21 47	☽ ⊥ ♄
22 50	☽ ⚻ ♂
10 Monday	
03 24	☽ ± ☉
05 44	☽ ∥ ♆
09 28	☽ ☌ ♅
09 43	☽ ⊻ ♄
09 46	☽ ⋕ ♂
11 32	☽ ± ♂
13 26	☽ ☌ ♆
13 58	☽ △ ♃
16 28	☽ ⚻ ☉
18 33	☽ ⚼ ♀
20 18	☽ ✶ ♇
11 Tuesday	
00 05	☉ ∥ ☿
06 42	☽ ⚼ ♂
14 47	☽ ± ♀
17 55	☽ ∥ ♄
20 32	☽ Q ♇
20 44	☽ ⚼ ♃
21 11	♂ ± ♆
12 Wednesday	
03 54	☿ ⊻ ♀
04 17	☽ ☍ ☿
04 20	☽ ⚻ ♀
05 38	☽ ⋕ ☿
10 38	☽ ⋕ ☉
14 47	☽ △ ♂
15 16	☽ ± ♃
16 28	☉ □ ♇
21 52	☽ ⊻ ♅
21 55	☽ ☌ ♄
13 Thursday	
01 56	☽ ⊻ ♆
02 53	☿ St D
03 38	☽ ⚻ ♃
05 00	☉ ± ♅
09 00	☽ □ ♇
09 29	♄ ⊻ ♅
10 00	☽ ⊥ ♅
10 27	☽ ☍ ☉
14 04	☽ ⊥ ♆
14 35	☽ ⋕ ♀
22 32	♀ Q ♇
14 Friday	
03 07	♂ ⋕ ♆
04 11	☽ ∠ ♅
08 15	☽ ∠ ♆
12 44	☽ ⋕ ♃
17 13	☽ ⚻ ☿
23 35	☽ ∥ ♇
15 Saturday	
00 04	☽ ☍ ♀
05 33	☽ ± ☿
06 06	☉ ± ♆
06 59	☽ □ ♂
10 12	☽ ⊻ ♄
10 23	☽ ✶ ♅
14 27	☽ ✶ ♆
17 19	☽ ☍ ♃
21 37	☽ △ ♇
22 11	☽ ⊥ ♄
16 Sunday	
00 07	☽ ⚼ ☿
04 18	☽ ⚻ ☉
10 24	☽ Q ♅
12 14	☽ ⋕ ♇
14 28	☽ Q ♆
16 05	☽ ∠ ♄
17 20	☽ ± ☉
21 06	☽ Q ♂
21 55	☽ ∥ ♃
17 Monday	
03 36	☽ ⚼ ♇
07 08	☽ △ ☿
11 02	☽ ∥ ♀
12 45	☽ ⚼ ☉
13 02	♂ △ ♄
18 51	☽ ⚻ ♀
19 44	♂ ⚻ ♅
21 25	☽ ± ♇
21 39	☽ ✶ ♄
22 03	☽ □ ♅
22 11	☽ ✶ ♂
18 Tuesday	
02 04	☽ □ ♆
05 59	☽ ⚻ ♃
07 48	☿ ⊥ ♀
07 57	☽ ± ♀
09 12	☽ ⚻ ♇
12 41	☽ ∥ ☉
17 53	☽ ± ♃
20 40	☽ △ ☉
20 59	☽ Q ♄
21 17	♀ ⚻ ♄
19 Wednesday	
02 01	♀ △ ♅
03 23	☽ ⚼ ♀
05 01	☽ ∠ ♂
11 34	☽ ⚼ ♃
13 31	☽ ∥ ☿
17 03	☽ ⋕ ♄
20 52	☽ □ ☿
23 11	☽ ⊥ ♂
20 Thursday	
07 14	☽ □ ♄
07 49	☽ △ ♅
11 04	☽ △ ♀
11 06	☽ ⊻ ♂
11 42	☽ △ ♆
11 50	♀ □ ♂
16 28	☽ △ ♃
17 31	♀ △ ♆
18 39	☽ ☍ ♇
22 44	♂ ⚻ ♆
21 Friday	
02 35	☽ ⋕ ♆
09 04	☽ Q ☿
09 39	☽ ∥ ♂
10 01	☽ □ ☉
11 38	☽ ⚼ ♅
15 24	☽ ⚼ ♆
20 33	☽ ⋕ ♅
22 Saturday	
03 54	☽ ± ♅
04 33	☉ Q ♂
07 25	☉ ⚼ ♅
07 37	☽ ± ♆
08 37	☽ ✶ ☿
13 50	☽ △ ♄
14 34	☽ ⚻ ♅
18 14	☽ ⚻ ♆
20 24	☽ ☌ ♂
20 49	☽ Q ☉
23 07	☽ □ ♀
23 37	☽ □ ♃
23 Sunday	
00 52	☽ ⚻ ♇
05 43	♀ ☌ ♃
11 17	☽ ± ♇
12 01	♀ ± ♄
13 17	☽ ∠ ☿
15 49	☽ ⚼ ♄
18 52	♀ ✶ ♇
19 07	☽ ✶ ☉
19 28	♀ ∥ ♃
21 42	☽ ⋕ ♅
24 Monday	
02 37	☽ ⚼ ♇
05 27	☽ ∥ ♂
06 03	☽ ⊥ ☿
06 54	☽ Q ♃
06 59	☽ ± ♄
08 52	☽ Q ♀
10 11	☉ ⚼ ♆
10 58	♃ ± ♄
14 10	☽ ⋕ ♆
15 49	☿ ☍ ♄
16 56	☽ ⚻ ♄
17 02	☿ ⊥ ♃
17 03	☽ ⊻ ☿
17 47	☽ ☍ ♅
21 14	☽ ☍ ♆
22 02	☽ ∠ ☉
25 Tuesday	
00 51	☿ ⚻ ♅
01 36	☽ ⊻ ♂
03 05	☽ ✶ ♃
03 34	☽ △ ♇
06 31	☽ ✶ ♀
11 49	☽ ⊥ ♂
13 45	☽ ⊥ ☉
18 37	☽ ⋕ ♄
26 Wednesday	
00 05	☽ ⊻ ☉
01 01	☽ ∥ ☿
02 56	☽ ∠ ♂
03 40	☽ ∠ ♃
06 36	♀ ⋕ ♇
08 50	☽ ∠ ♀
11 43	☿ ⚻ ♆
15 08	♃ ✶ ♇
17 10	☽ ☍ ♄
18 08	☽ ⚻ ♅
18 11	☽ ⊥ ♃
21 27	☽ ⚻ ♆
22 32	☽ ☌ ☿
27 Thursday	
00 20	☽ ⊥ ♀
01 00	♂ ⚻ ♇
03 35	☽ ± ♅
03 37	☽ □ ♇
03 44	☽ ✶ ♂
03 47	☽ ⊻ ♃
05 46	♂ □ ♃
06 54	☽ ± ♆
07 51	☽ ∥ ☉
10 37	☽ ⊻ ♀
17 43	☽ ⚼ ♅
21 01	☽ ⚼ ♆
23 22	☽ Q ♂
28 Friday	
02 42	☽ ☌ ☉
05 23	☽ ∥ ♃
07 56	☽ Q ♇
08 45	☽ ⋕ ♇
13 26	☽ ∥ ♀
16 09	☽ ⚻ ♄
17 15	☽ △ ♅
20 35	☽ △ ♆
22 13	☿ ± ♅
23 28	☿ □ ♇
29 Saturday	
01 35	☽ ± ♄
02 52	☽ ✶ ♇
03 17	☽ ⊻ ☿
03 39	☽ ☌ ♃
04 55	☽ □ ♂
07 02	☿ ⊻ ♃
08 33	☽ ⋕ ♀
14 02	☽ ⊥ ☿
14 04	☽ ☌ ♀
14 59	☽ ∥ ♇
15 21	☿ ⊥ ♀
15 50	☽ ⚼ ♄
17 50	☽ ⋕ ♃
30 Sunday	
01 05	☿ ✶ ♂
02 53	☽ ∠ ♇
03 50	☿ ± ♆
05 37	☽ ⊻ ☉
06 24	☽ ∠ ☿
10 38	♀ ⚼ ♄
11 43	☽ ⋕ ☉
16 01	☽ △ ♄
16 06	☽ ⊥ ☉
17 19	☽ □ ♅
17 34	☽ ⊥ ♇
20 49	☽ □ ♆
31 Monday	
03 30	☽ ⊻ ♇
05 00	☽ ⊻ ♃
07 46	☽ △ ♂
08 05	☽ ∠ ☉
10 36	☽ ✶ ☿
14 21	☽ ⋕ ☿
14 39	☉ Q ♇
15 15	☽ ⊥ ♃
19 58	☽ ⊻ ♀
23 28	☽ Q ♅

SEPTEMBER 1992

Date	Sidereal time h m s	Sun ☉ ° ' "	Moon ☽ ° ' "	Moon ☽ 24.00 ° ' "	Mercury ☿ ° '	Venus ♀ ° '	Mars ♂ ° '	Jupiter ♃ ° '	Saturn ♄ ° '	Uranus ♅ ° '	Neptune ♆ ° '	Pluto ♇ ° '
01	10 43 35	09 ♍ 17 33	09 ♏ 28 59	16 ♏ 17 27	26 ♌ 35	01 ♎ 01	23 ♊ 41	21 ♍ 36	13 ♒ 21	14 ♑ 14	16 ♑ 22	20 ♏ 27
02	10 47 32	10 15 40	22 58 38	29 ♏ 32 53	28 ♌ 26	02 15	24 17	21 49	13 R 18	14 R 13	16 R 21	20 28
03	10 51 28	11 13 47	06 ♐ 00 38	12 ♐ 22 25	00 ♍ 19	03 28	24 53	22 02	13 14	14 12	16 20	20 29
04	10 55 25	12 11 56	18 ♐ 38 48	24 ♐ 50 27	02 13	04 42	25 29	22 14	13 10	14 11	16 20	20 30
05	10 59 21	13 10 07	00 ♑ 57 58	07 ♑ 02 00	04 08	05 56	26 04	22 27	13 07	14 11	16 19	20 31
06	11 03 18	14 08 19	13 ♑ 03 10	19 ♑ 02 03	06 03	07 09	26 40	22 40	13 03	14 09	16 18	20 33
07	11 07 15	15 06 32	24 ♑ 59 13	00 ♒ 55 11	07 59	08 23	27 15	22 53	13 00	14 09	16 17	20 34
08	11 11 11	16 04 47	06 ♒ 50 24	12 ♒ 45 19	09 54	09 36	27 50	23 06	12 56	14 08	16 17	20 35
09	11 15 08	17 03 04	18 ♒ 40 38	24 ♒ 35 40	11 50	10 50	28 25	23 19	12 53	14 07	16 16	20 37
10	11 19 04	18 01 22	00 ♓ 31 42	06 ♓ 28 39	13 45	12 04	29 00	23 32	12 50	14 07	16 16	20 38
11	11 23 01	18 59 42	12 ♓ 26 42	18 ♓ 26 03	15 40	13 17	29 ♊ 34	23 45	12 47	14 06	16 15	20 39
12	11 26 57	19 58 04	24 ♓ 26 50	00 ♈ 29 11	17 34	14 31	00 ♋ 08	23 58	12 44	14 05	16 15	20 41
13	11 30 54	20 56 27	06 ♈ 33 13	12 ♈ 39 04	19 27	15 44	00 43	24 11	12 40	14 05	16 14	20 42
14	11 34 50	21 54 53	18 ♈ 46 52	24 ♈ 56 47	21 20	16 58	01 16	24 24	12 38	14 04	16 14	20 44
15	11 38 47	22 53 20	01 ♉ 08 59	07 ♉ 23 42	23 11	18 11	01 50	24 37	12 35	14 04	16 13	20 45
16	11 42 44	23 51 50	13 ♉ 41 08	20 ♉ 01 36	25 02	19 25	02 24	24 50	12 32	14 04	16 13	20 47
17	11 46 40	24 50 22	26 ♉ 25 24	02 ♊ 52 52	26 52	20 38	02 57	25 03	12 29	14 03	16 12	20 48
18	11 50 37	25 48 56	09 ♊ 24 21	16 ♊ 00 14	28 ♍ 41	21 52	03 30	25 16	12 26	14 03	16 12	20 50
19	11 54 33	26 47 32	22 ♊ 40 52	29 ♊ 26 35	00 ♎ 28	23 05	04 03	25 29	12 24	14 03	16 12	20 51
20	11 58 30	27 46 11	06 ♋ 17 40	13 ♋ 14 21	02 15	24 19	04 36	25 42	12 21	14 03	16 12	20 53
21	12 02 26	28 44 51	20 ♋ 16 43	27 ♋ 24 47	04 01	25 32	05 08	25 55	12 19	14 03	16 11	20 55
22	12 06 23	29 ♍ 43 34	04 ♌ 38 21	11 ♌ 57 05	05 45	26 46	05 41	26 08	12 17	14 03	16 11	20 56
23	12 10 19	00 ♎ 42 19	19 ♌ 20 26	26 ♌ 47 41	07 29	27 59	06 13	26 21	12 14	14 D 03	16 11	20 58
24	12 14 16	01 41 07	04 ♍ 17 55	11 ♍ 50 01	09 12	29 ♎ 13	06 44	26 34	12 12	14 03	16 11	21 00
25	12 18 13	02 39 56	19 ♍ 22 49	26 ♍ 55 01	10 54	00 ♏ 26	07 16	26 47	12 10	14 03	16 11	21 01
26	12 22 09	03 38 47	04 ♎ 25 20	11 ♎ 52 31	12 35	01 39	07 47	27 00	12 08	14 03	16 11	21 03
27	12 26 06	04 37 41	19 ♎ 15 25	26 ♎ 33 04	14 15	02 53	08 18	27 12	12 06	14 03	16 11	21 05
28	12 30 02	05 36 36	03 ♏ 44 39	10 ♏ 49 34	15 54	04 06	08 49	27 25	12 05	14 03	16 D 11	21 07
29	12 33 59	06 35 34	17 ♏ 47 26	24 ♏ 38 06	17 32	05 19	09 19	27 38	12 03	14 04	16 11	21 09
30	12 37 55	07 ♎ 34 33	01 ♐ 21 34	07 ♐ 58 01	19 ♎ 09	06 ♏ 33	09 ♋ 50	27 ♍ 51	12 ♒ 01	14 ♑ 04	16 ♑ 11	21 ♏ 11

Date	Moon True ☊ ° '	Moon Mean ☊ ° '	Moon ☽ Latitude ° '
01	27 ♐ 45	26 ♐ 51	03 S 47
02	27 R 42	26 48	02 53
03	27 41	26 45	01 52
04	27 D 41	26 42	00 S 48
05	27 R 41	26 39	00 N 17
06	27 40	26 35	01 21
07	27 37	26 32	02 19
08	27 31	26 29	03 11
09	27 23	26 26	03 55
10	27 12	26 23	04 28
11	26 59	26 20	04 50
12	26 45	26 16	04 59
13	26 31	26 13	04 55
14	26 19	26 10	04 37
15	26 09	26 07	04 06
16	26 01	26 04	03 23
17	25 57	26 00	02 29
18	25 55	25 57	01 26
19	25 D 54	25 54	00 N 17
20	25 R 54	25 51	00 S 55
21	25 53	25 48	02 05
22	25 50	25 45	03 09
23	25 45	25 41	04 03
24	25 36	25 38	04 40
25	25 26	25 35	04 59
26	25 15	25 32	04 57
27	25 04	25 29	04 34
28	24 54	25 25	03 55
29	24 47	25 22	03 02
30	24 ♐ 43	25 ♐ 19	02 S 00

DECLINATIONS

Date	Sun ☉ ° '	Moon ☽ ° '	Mercury ☿ ° '	Venus ♀ ° '	Mars ♂ ° '	Jupiter ♃ ° '	Saturn ♄ ° '	Uranus ♅ ° '	Neptune ♆ ° '	Pluto ♇ ° '
01	08 N 05	18 S 14	14 N 01	00 N 37	23 N 00	04 N 18	17 S 49	23 S 06	21 S 43	03 S 52
02	07 43	21 19	13 28	00 N 06	23 03	04 13	17 50	23 07	21 43	03 53
03	07 21	23 09	12 53	00 S 25	23 06	04 08	17 51	23 07	21 43	03 54
04	06 59	23 45	12 16	00 56	23 08	04 03	17 52	23 07	21 43	03 54
05	06 37	23 09	11 37	01 27	23 10	03 58	17 53	23 07	21 43	03 55
06	06 15	21 28	10 56	01 58	23 13	03 53	17 54	23 07	21 43	03 56
07	05 52	18 51	10 14	02 29	23 15	03 48	17 55	23 07	21 43	03 57
08	05 29	15 28	09 31	03 00	23 16	03 43	17 56	23 07	21 44	03 58
09	05 07	11 30	08 47	03 30	23 18	03 38	17 57	23 07	21 44	03 58
10	04 44	07 06	08 02	04 01	23 20	03 33	17 58	23 07	21 44	03 59
11	04 21	02 S 25	07 16	04 32	23 21	03 27	17 59	23 07	21 44	04 00
12	03 58	02 N 23	06 30	05 02	23 22	03 22	18 00	23 07	21 44	04 01
13	03 35	07 07	05 43	05 33	23 23	03 17	18 01	23 07	21 44	04 02
14	03 12	11 38	04 56	06 03	23 24	03 12	18 01	23 07	21 44	04 03
15	02 49	15 43	04 08	06 34	23 25	03 07	18 02	23 07	21 44	04 03
16	02 26	19 10	03 21	07 04	23 26	03 02	18 03	23 07	21 44	04 04
17	02 03	21 46	02 33	07 34	23 27	02 57	18 04	23 07	21 44	04 05
18	01 40	23 17	01 45	08 04	23 27	02 52	18 04	23 07	21 44	04 06
19	01 17	23 31	00 58	08 33	23 28	02 46	18 05	23 07	21 45	04 07
20	00 53	22 23	00 N 11	09 03	23 28	02 41	18 06	23 07	21 45	04 07
21	00 30	19 51	00 S 37	09 32	23 28	02 36	18 07	23 07	21 45	04 08
22	00 N 07	16 02	01 23	10 01	23 28	02 31	18 07	23 07	21 45	04 09
23	00 S 17	11 11	02 10	10 30	23 28	02 26	18 08	23 07	21 45	04 10
24	00 40	05 N 35	02 56	10 59	23 28	02 21	18 08	23 07	21 45	04 11
25	01 04	00 S 23	03 42	11 27	23 28	02 16	18 09	23 07	21 45	04 12
26	01 27	06 18	04 27	11 55	23 27	02 11	18 10	23 07	21 45	04 12
27	01 50	11 46	05 12	12 23	23 27	02 05	18 10	23 07	21 45	04 13
28	02 14	16 27	05 57	12 51	23 26	02 00	18 11	23 07	21 45	04 14
29	02 37	20 03	06 41	13 18	23 26	01 55	18 11	23 07	21 45	04 15
30	03 S 00	22 S 24	07 S 24	13 S 45	23 N 25	01 N 50	18 S 11	23 S 07	21 S 45	04 S 16

ZODIAC SIGN ENTRIES

Date	h m	Planets
03	00 50	☽ ♐
03	08 03	☿ ♍
05	10 06	☽ ♑
07	22 08	☽ ♒
10	10 56	☽ ♓
12	06 05	♂ ♋
12	23 02	☽ ♈
15	09 47	☽ ♉
17	18 30	☽ ♊
19	05 41	☿ ♎
20	00 59	☽ ♋
22	04 19	☽ ♌
22	18 43	☉ ♎
24	05 08	☽ ♍
25	03 31	♀ ♏
26	04 55	☽ ♎
28	05 44	☽ ♏
30	09 33	☽ ♐

LATITUDES

Date	Mercury ☿ ° '	Venus ♀ ° '	Mars ♂ ° '	Jupiter ♃ ° '	Saturn ♄ ° '	Uranus ♅ ° '	Neptune ♆ ° '	Pluto ♇ ° '
01	01 N 27	01 N 06	00 S 17	01 N 04	01 S 02	00 S 26	00 N 44	14 N 29
04	01 41	01 01	00 14	01 04	01 03	00 26	00 44	14 28
07	01 47	00 55	00 10	01 04	01 03	00 26	00 44	14 26
10	01 47	00 49	00 07	01 04	01 03	00 26	00 44	14 25
13	01 40	00 42	00 S 03	01 04	01 03	00 26	00 44	14 23
16	01 30	00 35	00 N 01	01 04	01 03	00 26	00 43	14 22
19	01 15	00 27	00 05	01 04	01 03	00 26	00 43	14 20
22	00 59	00 19	00 09	01 04	01 03	00 26	00 43	14 19
25	00 40	00 11	00 13	01 04	01 03	00 25	00 43	14 18
28	00 N 20	00 N 02	00 18	01 04	01 03	00 25	00 43	14 17
31	00 S 01	00 S 06	00 N 22	01 N 04	01 S 03	00 S 25	00 N 43	14 N 15

DATA

Julian Date	2448867
Delta T	+58 seconds
Ayanamsa	23° 45' 35"
Synetic vernal point	05° ♓ 21' 25"
True obliquity of ecliptic	23° 26' 25"

MOON'S PHASES, APSIDES AND POSITIONS ☽

Date	h	m	Phase	Longitude ° '	Eclipse Indicator
03	22	39	☽	11 ♐ 40	
12	02	17	○	19 ♓ 34	
19	19	53	☾	27 ♊ 07	
26	10	40	●	03 ♎ 36	

Day	h	m		
09	18	26	Apogee	
25	02	27	Perigee	
04	11	37	Max dec	23° S 45'
12	00	07	ON	
19	04	22	Max dec	23° N 35'
25	10	29	OS	

LONGITUDES

Date	Chiron ⚷ ° '	Ceres ⚳ ° '	Pallas ⚴ ° '	Juno ⚵ ° '	Vesta ⚶ ° '	Black Moon Lilith ⚸ ° '
01	16 ♌ 40	26 ♑ 47	20 ♐ 15	20 ♊ 48	23 ♎ 15	24 ♒ 55
11	17 ♌ 50	26 ♑ 15	21 ♐ 45	25 ♊ 18	28 ♎ 12	26 ♒ 02
21	18 ♌ 57	26 ♑ 16	23 ♐ 41	29 ♊ 29	03 ♏ 15	27 ♒ 09
31	19 ♌ 59	26 ♑ 57	25 ♐ 58	03 ♋ 17	08 ♏ 24	28 ♒ 16

All ephemeris data is given at 12.00 UT and the Moon's longitude is additionally given for 24.00 UT

ASPECTARIAN

01 Tuesday
03 08 ☽ Q ♆
06 54 ☽ ∠ ♃
07 17 ☽ ⊥ ♀
09 18 ☽ ∥ ♄
10 11 ☽ Q ☿
10 33 ☽ ꝑ ♂
11 39 ☽ ⚹ ☉
18 46 ☽ □ ♄
20 20 ☽ ⚹ ♅

02 Wednesday
00 07 ☽ ⚹ ♆
00 40 ☽ ∠ ♀
03 10 ☽ ± ♂
07 28 ☽ ☌ ♇
09 52 ☽ ⚹ ♃
10 36 ☽ Q ☉
14 29 ☽ ⊼ ♂
16 07 ☽ ∥ ♆
21 58 ☿ ꝑ ♅
23 23 ☽ ∠ ♅
23 37 ☽ □ ☿

03 Thursday
03 09 ☽ Q ♄
03 19 ☽ ∠ ♆
06 46 ☽ ⚹ ♀
08 14 ☽ Q ♃
10 48 ☽ # ♂
11 08 ☽ ∥ ♅
16 06 ☽ ⊥ ♅
20 08 ☽ ⊥ ♆
21 06 ♂ # ♅
22 39 ☽ □ ☉

04 Friday
00 54 ☿ ꝑ ♆
01 34 ☽ ⚹ ♄
03 28 ☽ ⊻ ♅
07 33 ☽ ⊻ ♆
07 51 ☽ Q ♀
15 35 ☽ ⊻ ♇
19 04 ☽ □ ♃

05 Saturday
01 55 ☽ ☍ ♂
03 16 ☽ ⊥ ♇
04 00 ♀ ∠ ♇
06 25 ☽ ∠ ♄
10 42 ☉ ⊼ ♄
11 26 ☽ # ♂
12 40 ☽ ∥ ♅
19 25 ☽ △ ☿
21 01 ☽ ∠ ♇
22 55 ☽ □ ♀
23 14 ♃ # ♇

06 Sunday
00 06 ☽ ⊥ ♄
06 58 ♂ ± ♇
09 03 ☽ ∥ ♆
12 00 ☽ ⊻ ♄
12 27 ☉ △ ♅
14 12 ☽ ☌ ♅
14 22 ☽ △ ☉
18 30 ☽ ☌ ♆

07 Monday
03 04 ☽ ⚹ ♇
07 10 ☽ ꝑ ☿
07 41 ☽ △ ♃
16 49 ☽ ⊼ ♂
19 07 ☽ ∥ ♄
19 23 ☿ Q ♇
23 17 ☽ ꝑ ☉

08 Tuesday
01 44 ☿ ⊻ ♀
03 22 ☽ Q ♇
04 54 ☽ ± ☿
05 35 ☽ ± ♂
10 42 ☿ Q ♂
14 36 ☽ ꝑ ♃
15 57 ♂ ꝑ ♄
16 55 ☉ △ ♆
18 16 ☽ △ ♀
19 10 ☽ ± ☉
19 26 ☽ ⊼ ☿

09 Wednesday
00 19 ☽ ☌ ♄
00 47 ☽ ꝑ ♂
02 47 ☽ ⊻ ♅
07 08 ☽ ⊻ ♆
08 25 ☽ ⊼ ☉
09 12 ☽ ± ♃
14 56 ☽ ⊥ ♅
15 56 ☽ □ ♇
16 46 ♀ # ♃
19 17 ☽ ⊥ ♆
21 35 ☽ ⊼ ♃

10 Thursday
00 45 ☿ ⊼ ♄
04 11 ☽ ꝑ ♀
06 05 ☽ # ☿
08 44 ☽ △ ♂
09 08 ☽ ∠ ♅
10 26 ♀ ∥ ♇
13 29 ☽ ∠ ♆
16 25 ☿ △ ♅

11 Friday
00 26 ☽ ± ♀
01 19 ☽ # ☉
02 20 ☽ ∥ ♀
02 26 ♀ △ ♄
04 00 ☽ ∥ ♇
06 39 ☽ # ♃
06 54 ☉ ± ♄
07 15 ☉ # ♀
12 40 ☽ ⊻ ♄
13 53 ☽ ⊼ ♀
15 19 ☽ ⚹ ♅
19 21 ☿ △ ♆
19 38 ☽ ⚹ ♆
19 41 ☽ ☍ ☿

12 Saturday
00 38 ☽ ⊥ ♄
02 17 ☽ ☍ ☉
03 47 ♀ □ ♅
04 28 ☽ △ ♇
09 32 ☉ # ♇
11 01 ☽ ☍ ♃
15 16 ☽ Q ♅
15 18 ♀ ⊥ ♇
16 54 ☽ ∥ ♃
18 30 ☽ ∠ ♄
19 25 ☽ ∥ ☉
19 33 ☽ Q ♆
20 14 ☽ # ♇
23 52 ☽ □ ♂

13 Sunday
02 19 ☿ ± ♄
03 01 ☽ # ♀
05 50 ☽ ∥ ☿
05 57 ☉ ⚹ ♇
10 19 ☽ ꝑ ♇
15 02 ☿ # ♀
21 41 ♀ □ ♆

14 Monday
00 00 ☽ ⚹ ♄
02 48 ☽ □ ♅
04 03 ☽ ± ♇
04 09 ☿ ⚹ ♇
07 01 ☽ □ ♆
08 03 ☽ ☍ ♀
12 35 ☉ ∥ ♃
13 01 ☽ Q ♂
15 48 ☽ ⊼ ♇
17 51 ☽ ⊼ ☿
18 38 ☽ ⊼ ☉
23 08 ☽ ⊼ ♃
23 20 ☽ Q ♄

15 Tuesday
03 48 ☉ ☌ ☿
07 16 ☽ ± ☉
07 33 ☽ ± ☿
10 57 ☽ ± ♃
13 23 ☽ ⚹ ♂
14 25 ☿ # ♇

16 Wednesday
02 02 ☽ ꝑ ☉
03 33 ☽ # ♄
03 51 ☽ ꝑ ☿
04 32 ☽ ꝑ ♃
08 58 ☿ ☌ ♃
09 49 ☽ □ ♄
12 43 ☽ △ ♅
16 48 ☽ △ ♆
19 21 ☽ ∠ ♂
22 39 ☿ ∥ ♃

17 Thursday
00 00 ☽ ⊼ ♀
01 26 ☽ ☍ ♇
08 48 ☽ △ ☉
09 23 ☽ △ ♃
11 41 ☽ # ♆
12 27 ☽ ± ♀
12 57 ☽ △ ☿
13 02 ☽ ⊥ ♂
15 16 ♀ ⊻ ♇
16 54 ☽ ꝑ ♅
18 31 ☉ ☌ ♃
20 00 ☿ ꝑ ♄
20 54 ☽ ꝑ ♆

18 Friday
00 40 ☽ ⊻ ♂
06 51 ☽ ꝑ ♀
08 15 ☽ # ♅
09 31 ☽ ± ♅
13 28 ☽ ± ♆
17 31 ☽ △ ♄
17 34 ☉ ∥ ☿
17 37 ☽ ∥ ♂
20 28 ☽ ⊼ ♅

19 Saturday
00 21 ☽ ⊼ ♆
08 44 ☽ ⊼ ♇
12 48 ☽ △ ♀
14 48 ☽ ∥ ♂
17 04 ☽ □ ♃
19 26 ☽ ± ♇
19 53 ☽ □ ☉
20 22 ☽ ꝑ ♄

20 Sunday
00 06 ☽ # ♅
02 16 ☉ ꝑ ♄
03 53 ☽ □ ☿
08 55 ☽ ☌ ♂
11 17 ☽ ꝑ ♇
12 06 ☽ ± ♄
19 26 ☽ # ♆
22 27 ☽ ⊼ ♄

21 Monday
00 59 ☽ Q ♃
01 23 ☽ ☍ ♅
05 03 ☽ ☍ ♆
05 33 ☽ Q ☉
09 44 ☉ # ☿
13 04 ☽ △ ♇
15 21 ☽ Q ☿
20 55 ♀ ⊻ ♃
21 38 ☽ ⚹ ♃
21 41 ☽ □ ♀
23 51 ☽ # ♄

22 Tuesday
03 16 ☽ ⚹ ☉
10 21 ☿ □ ♂
13 46 ☽ ⊻ ♂
14 06 ☽ ⚹ ☿
14 31 ☿ ∠ ♇
22 49 ☽ ∠ ♃
23 44 ♅ St D
23 59 ☽ ⊥ ♂

23 Wednesday
00 25 ♂ ꝑ ♇
00 30 ☽ ☍ ♄
03 25 ☽ ⊼ ♅
05 42 ☽ ∠ ☉
06 04 ☽ Q ♀
06 54 ☽ ⊼ ♆
13 08 ☽ ± ♅
13 20 ♂ ± ♄
13 39 ☽ ⊥ ♃
14 38 ☽ □ ♇
14 48 ☽ # ♀
15 08 ☽ ∠ ♂
16 35 ☽ ± ♆
17 44 ☽ ∠ ☿
19 25 ☿ # ♃
21 15 ☽ ⊥ ☉
23 27 ☽ ⊻ ♃

24 Thursday
03 09 ☽ ⚹ ♀
03 36 ☽ ꝑ ♅
07 02 ☽ ꝑ ♆
07 32 ☽ ⊻ ☉
10 01 ☽ ⊥ ☿
16 02 ☽ ⚹ ♂
17 42 ☽ # ♇
19 30 ☽ Q ♇
20 48 ☽ ⊻ ☿
21 31 ☽ # ☿

25 Friday
00 34 ☽ ⊼ ♄
01 18 ☽ ∥ ♃
03 31 ☽ △ ♅
05 10 ☽ ∠ ♀
06 36 ☽ # ☉
06 55 ☽ △ ♆
10 05 ☽ ± ♄
11 49 ☽ Q ♂
14 37 ☽ ⚹ ♇
14 55 ☽ ∥ ☉
19 26 ☽ # ♃
20 45 ☽ ⊥ ♀
23 57 ☽ ☌ ♃

26 Saturday
00 23 ☽ ꝑ ♄
03 19 ☽ ∥ ☿
03 23 ☽ ∥ ♇
03 52 ☿ ∥ ♇
05 48 ☿ △ ♄
07 10 ☽ ⊻ ♀
10 40 ☽ ☌ ☉
14 37 ☽ ∠ ♇
17 36 ☽ □ ♂
19 45 ♀ Q ♅

27 Sunday
00 24 ☽ △ ♄
02 07 ♃ ꝑ ♄
02 48 ☽ ☌ ☿
03 31 ☽ □ ♅
05 11 ☽ ⊥ ♇
06 59 ☽ □ ♆
09 12 ☿ □ ♅
15 00 ☽ ⊻ ♇
15 14 ☽ ∥ ♀
18 35 ♆ St D
19 51 ♀ ⊥ ♃

28 Monday
00 23 ☿ ⊥ ♇
00 42 ☉ # ♃
01 17 ☽ ⊻ ♃
09 10 ☽ Q ♅
11 27 ☽ ⊥ ♃
12 39 ☽ ☌ ♀
12 44 ☽ Q ♆
13 34 ♀ Q ♆
15 23 ☽ ⊻ ☉
16 10 ☿ □ ♆
20 54 ☽ △ ♂
22 35 ☽ ∥ ♄

29 Tuesday
00 42 ☉ ∠ ♇
02 07 ☽ □ ♄
02 21 ☽ ⊥ ☉
02 58 ☽ ∠ ♃
05 33 ☽ ⚹ ♅
09 13 ☽ ⚹ ♆
11 29 ☽ ⊻ ☿
17 52 ☽ ☌ ♇
19 09 ☽ ∠ ☉
23 25 ☽ ⊥ ☿
23 54 ☽ ꝑ ♂

30 Wednesday
03 55 ☽ ∥ ♆
05 37 ☽ ⚹ ♃
07 53 ☽ ∠ ♅
09 36 ☽ Q ♄
11 41 ☽ ∠ ♆
16 38 ☽ ± ♂
17 45 ☽ ∠ ☿
22 22 ☽ ⊻ ♀

OCTOBER 1992

LONGITUDES

Date	Sidereal time h m s	Sun ☉ ° ' "	Moon ☽ ° ' "	Moon ☽ 24.00 ° ' "	Mercury ☿ ° '	Venus ♀ ° '	Mars ♂ ° '	Jupiter ♃ ° '	Saturn ♄ ° '	Uranus ♅ ° '	Neptune ♆ ° '	Pluto ♇ ° '
01	12 41 52	08 ♎ 33 34	14 ♐ 27 47	20 ♐ 51 19	20 ♎ 46	07 ♏ 46	10 ♋ 20	28 ♍ 04	12 ♒ 00	14 ♑ 05	16 ♑ 11	21 ♏ 12
02	12 45 48	09 32 36	27 ♐ 09 08	03 ♑ 21 50	22 21	08 59	10 49	28 17	11 R 58	14 05	16 11	21 14
03	12 49 45	10 31 41	09 ♑ 30 05	15 ♑ 34 31	23 56	10 13	11 19	28 30	11 57	14 05	16 11	21 16
04	12 53 42	11 30 47	21 ♑ 35 48	27 ♑ 34 37	25 30	11 26	11 48	28 43	11 56	14 06	16 12	21 18
05	12 57 38	12 29 55	03 ♒ 31 37	09 ♒ 27 23	27 03	12 39	12 17	28 56	11 55	14 07	16 12	21 20
06	13 01 35	13 29 05	15 ♒ 22 32	21 ♒ 17 36	28 ♎ 35	13 52	12 46	29 08	11 54	14 07	16 12	21 22
07	13 05 31	14 28 17	27 ♒ 13 03	03 ♓ 09 21	00 ♏ 07	15 06	13 14	29 21	11 53	14 08	16 12	21 24
08	13 09 28	15 27 30	09 ♓ 06 52	15 ♓ 05 58	01 38	16 19	13 42	29 34	11 52	14 09	16 13	21 26
09	13 13 24	16 26 45	21 ♓ 06 53	27 ♓ 09 53	03 07	17 32	14 10	29 47	11 51	14 10	16 13	21 28
10	13 17 21	17 26 02	03 ♈ 15 07	09 ♈ 22 43	04 37	18 45	14 37	29 ♍ 59	11 51	14 11	16 14	21 30
11	13 21 17	18 25 21	15 ♈ 32 47	21 ♈ 45 22	06 05	19 58	15 04	00 ♎ 12	11 50	14 11	16 14	21 32
12	13 25 14	19 24 43	28 ♈ 00 30	04 ♉ 18 14	07 33	21 11	15 31	00 25	11 50	14 12	16 14	21 35
13	13 29 11	20 24 06	10 ♉ 38 34	17 ♉ 01 33	08 59	22 24	15 57	00 37	11 50	14 13	16 15	21 37
14	13 33 07	21 23 31	23 ♉ 27 11	29 ♉ 55 34	10 25	23 37	16 23	00 50	11 49	14 14	16 16	21 39
15	13 37 04	22 22 59	06 ♊ 26 46	13 ♊ 00 54	11 51	24 50	16 49	01 02	11 49	14 16	16 16	21 41
16	13 41 00	23 22 29	19 ♊ 38 06	26 ♊ 18 32	13 15	26 04	17 14	01 15	11 D 49	14 17	16 17	21 43
17	13 44 57	24 22 01	03 ♋ 02 20	09 ♋ 49 42	14 38	27 17	17 39	01 27	11 49	14 18	16 17	21 45
18	13 48 53	25 21 36	16 ♋ 40 46	23 ♋ 35 39	16 01	28 29	18 04	01 40	11 49	14 19	16 18	21 48
19	13 52 50	26 21 13	00 ♌ 34 27	07 ♌ 37 09	17 22	29 ♏ 42	18 28	01 52	11 50	14 21	16 19	21 50
20	13 56 46	27 20 52	14 ♌ 43 39	21 ♌ 53 46	18 43	00 ♐ 55	18 52	02 04	11 50	14 22	16 19	21 52
21	14 00 43	28 20 33	29 ♌ 07 09	06 ♍ 23 22	20 02	02 08	19 16	02 17	11 51	14 23	16 20	21 54
22	14 04 40	29 ♎ 20 17	13 ♍ 41 49	21 ♍ 01 47	21 20	03 21	19 39	02 29	11 51	14 25	16 21	21 57
23	14 08 36	00 ♏ 20 03	28 ♍ 22 27	05 ♎ 42 55	22 37	04 34	20 01	02 41	11 52	14 26	16 22	21 59
24	14 12 33	01 19 51	13 ♎ 02 13	20 ♎ 19 24	23 53	05 47	20 24	02 53	11 53	14 28	16 23	22 01
25	14 16 29	02 19 41	27 ♎ 33 34	04 ♏ 43 50	25 07	07 00	20 46	03 06	11 54	14 29	16 24	22 03
26	14 20 26	03 19 33	11 ♏ 49 29	18 ♏ 49 54	26 20	08 13	21 07	03 18	11 55	14 31	16 25	22 06
27	14 24 22	04 19 27	25 ♏ 44 36	02 ♐ 33 17	27 31	09 25	21 28	03 30	11 56	14 33	16 26	22 08
28	14 28 19	05 19 23	09 ♐ 15 48	15 ♐ 52 09	28 39	10 38	21 48	03 42	11 57	14 35	16 27	22 10
29	14 32 15	06 19 20	22 ♐ 22 28	28 ♐ 47 01	29 ♏ 46	11 51	22 08	03 54	11 58	14 36	16 28	22 13
30	14 36 12	07 19 20	05 ♑ 06 08	11 ♑ 20 18	00 ♐ 51	13 04	22 28	04 05	12 00	14 38	16 29	22 15
31	14 40 09	08 ♏ 19 21	17 ♑ 30 00	23 ♑ 35 49	01 ♐ 52	14 ♐ 16	22 ♋ 47	04 ♎ 17	12 ♒ 01	14 ♑ 40	16 ♑ 30	22 ♏ 17

Date	Moon True ☊ ° '	Moon Mean ☊ ° '	Moon ☽ Latitude ° '
01	24 ♐ 41	25 ♐ 16	00 S 54
02	24 D 41	25 13	00 N 13
03	24 41	25 10	01 18
04	24 R 41	25 06	02 18
05	24 39	25 03	03 11
06	24 34	25 00	03 55
07	24 27	24 57	04 29
08	24 18	24 54	04 52
09	24 06	24 51	05 02
10	23 54	24 47	04 58
11	23 41	24 44	04 41
12	23 30	24 41	04 10
13	23 21	24 38	03 26
14	23 15	24 35	02 31
15	23 11	24 31	01 28
16	23 10	24 28	00 N 19
17	23 D 10	24 25	00 S 53
18	23 11	24 22	02 02
19	23 R 11	24 19	03 06
20	23 10	24 16	04 00
21	23 06	24 12	04 40
22	23 00	24 09	05 02
23	22 53	24 06	05 05
24	22 44	24 03	04 47
25	22 36	24 00	04 11
26	22 29	23 57	03 20
27	22 24	23 53	02 18
28	22 21	23 50	01 S 10
29	22 D 20	23 47	00 00
30	22 21	23 44	01 N 09
31	22 ♐ 22	23 ♐ 41	02 N 12

DECLINATIONS

Date	Sun ☉ ° '	Moon ☽ ° '	Mercury ☿ ° '	Venus ♀ ° '	Mars ♂ ° '	Jupiter ♃ ° '	Saturn ♄ ° '	Uranus ♅ ° '	Neptune ♆ ° '	Pluto ♇ ° '
01	03 S 24	23 S 26	08 S 07	14 S 12	23 N 24	01 N 45	18 S 12	23 S 07	21 S 45	04 S 17
02	03 47	23 11	08 50	14 38	23 23	01 40	18 12	23 07	21 45	04 17
03	04 10	21 48	09 31	15 04	23 23	01 35	18 13	23 07	21 45	04 18
04	04 33	19 26	10 12	15 30	23 22	01 30	18 13	23 07	21 45	04 19
05	04 56	16 16	10 53	15 55	23 20	01 25	18 13	23 07	21 45	04 20
06	05 19	12 29	11 33	16 20	23 19	01 20	18 13	23 07	21 45	04 21
07	05 42	08 13	12 12	16 45	23 18	01 15	18 14	23 07	21 45	04 22
08	06 05	03 S 39	12 50	17 09	23 17	01 10	18 14	23 06	21 45	04 22
09	06 28	01 N 06	13 28	17 33	23 16	01 05	18 14	23 06	21 45	04 23
10	06 51	05 51	14 04	17 56	23 14	01 00	18 14	23 06	21 45	04 24
11	07 13	10 26	14 41	18 19	23 13	00 55	18 14	23 06	21 45	04 25
12	07 36	14 39	15 16	18 41	23 12	00 50	18 14	23 06	21 45	04 26
13	07 58	18 17	15 50	19 03	23 10	00 45	18 15	23 06	21 45	04 26
14	08 21	21 05	16 24	19 24	23 09	00 40	18 15	23 06	21 45	04 27
15	08 43	22 50	16 57	19 45	23 08	00 35	18 15	23 06	21 45	04 28
16	09 05	23 21	17 29	20 06	23 06	00 30	18 15	23 06	21 45	04 29
17	09 27	22 32	18 00	20 26	23 05	00 25	18 14	23 05	21 45	04 30
18	09 49	20 22	18 30	20 45	23 03	00 20	18 14	23 05	21 45	04 30
19	10 10	17 00	18 59	21 04	23 02	00 16	18 14	23 05	21 45	04 31
20	10 32	12 35	19 27	21 22	23 00	00 11	18 14	23 05	21 45	04 32
21	10 53	07 24	19 54	21 40	22 59	00 06	18 14	23 05	21 45	04 33
22	11 14	01 N 46	20 20	21 57	22 57	00 N 01	18 14	23 05	21 45	04 33
23	11 35	04 S 01	20 45	22 13	22 56	00 S 04	18 14	23 04	21 44	04 34
24	11 56	09 33	21 09	22 29	22 55	00 08	18 13	23 04	21 44	04 35
25	12 17	14 31	21 31	22 45	22 53	00 13	18 13	23 04	21 44	04 36
26	12 37	18 33	21 52	22 59	22 52	00 18	18 13	23 04	21 44	04 37
27	12 58	21 26	22 12	23 13	22 51	00 22	18 12	23 04	21 44	04 37
28	13 18	23 00	22 31	23 27	22 49	00 27	18 12	23 03	21 44	04 38
29	13 38	23 13	22 49	23 39	22 48	00 32	18 12	23 03	21 44	04 39
30	13 57	22 12	23 05	23 52	22 47	00 36	18 11	23 03	21 44	04 39
31	14 S 17	20 S 07	23 S 19	24 S 03	22 N 46	00 S 41	18 S 11	23 S 03	21 S 44	04 S 40

ZODIAC SIGN ENTRIES

Date	h m	Planets
02	17 29	☽ ♑
05	04 53	☽ ♒
07	10 13	☿ ♏
07	17 38	☽ ♓
10	05 36	☽ ♈
10	13 26	♃ ♎
12	15 48	☽ ♉
15	00 08	☽ ♊
17	06 36	☽ ♋
19	11 01	☽ ♌
19	17 47	♀ ♐
21	13 27	☽ ♍
23	03 57	☉ ♏
23	14 39	☽ ♎
25	16 04	☽ ♏
27	19 29	☽ ♐
29	17 02	☿ ♐
30	02 18	☽ ♑

LATITUDES

Date	Mercury ☿ ° '	Venus ♀ ° '	Mars ♂ ° '	Jupiter ♃ ° '	Saturn ♄ ° '	Uranus ♅ ° '	Neptune ♆ ° '	Pluto ♇ ° '
01	00 S 01	00 S 06	00 N 22	01 N 04	01 S 03	00 S 25	00 N 43	14 N 15
04	00 22	00 15	00 27	01 04	01 03	00 25	00 43	14 14
07	00 44	00 24	00 31	01 05	01 03	00 25	00 43	14 13
10	01 05	00 33	00 36	01 05	01 02	00 25	00 43	14 12
13	01 25	00 42	00 41	01 05	01 02	00 25	00 42	14 11
16	01 45	00 51	00 47	01 05	01 02	00 25	00 42	14 10
19	02 03	01 00	00 52	01 06	01 02	00 25	00 42	14 10
22	02 19	01 08	00 58	01 06	01 02	00 25	00 42	14 09
25	02 33	01 17	01 04	01 06	01 02	00 25	00 42	14 08
28	02 44	01 25	01 10	01 06	01 02	00 25	00 42	14 08
31	02 S 50	01 S 33	01 N 16	01 N 07	01 S 02	00 S 25	00 N 42	14 N 07

DATA

Julian Date	2448897
Delta T	+58 seconds
Ayanamsa	23° 45' 37"
Synetic vernal point	05° ♓ 21' 22"
True obliquity of ecliptic	23° 26' 25"

MOON'S PHASES, APSIDES AND POSITIONS ☽

Date	h m	Phase	Longitude ° '	Eclipse Indicator
03	14 12	☽	10 ♑ 37	
11	18 03	○	18 ♈ 40	
19	04 12	☾	26 ♋ 02	
25	20 34	●	02 ♏ 41	

Day	h m		
07	05 41	Apogee	
23	04 33	Perigee	
01	19 18	Max dec	23° S 29'
09	06 29	ON	
16	09 26	Max dec	23° N 21'
22	19 19	OS	
29	04 04	Max dec	23° S 17'

LONGITUDES

Date	Chiron ⚷ ° '	Ceres ⚳ ° '	Pallas ⚴ ° '	Juno ⚵ ° '	Vesta ⚶ ° '	Black Moon Lilith ⚸ ° '
01	19 ♌ 59	26 ♑ 57	25 ♐ 58	03 ♋ 17	08 ♏ 24	28 ♒ 16
11	20 ♌ 56	28 ♑ 07	28 ♐ 32	06 ♋ 36	13 ♏ 37	29 ♒ 23
21	21 ♌ 45	29 ♑ 45	01 ♑ 22	09 ♋ 20	18 ♏ 53	00 ♓ 30
31	22 ♌ 27	01 ♒ 47	04 ♑ 23	11 ♋ 22	24 ♏ 13	01 ♓ 37

All ephemeris data is given at 12.00 UT and the Moon's longitude is additionally given for 24.00 UT

ASPECTARIAN

h m Aspects

01 Thursday
00 12 ☽ ⊥ ♅
00 12 ☽ ✶ ☉
01 01 ☽ ∥ ♅
03 43 ☽ Q ♃
04 02 ☽ ⊼ ♂
04 04 ☽ ⊥ ♆
07 26 ☽ ✶ ♄
10 34 ☽ ⊥ ♀
10 40 ☽ ⋕ ♂
11 17 ☽ ⊻ ♅
15 13 ☽ ⊻ ♆
18 51 ☿ ⊻ ♇

02 Friday
00 24 ☽ Q ☉
00 42 ☽ ⊻ ♇
01 31 ☽ ✶ ☿
04 32 ☽ ⋕ ♂
05 18 ☽ ∠ ♀
11 40 ☽ ∠ ♄
12 10 ☽ ⊥ ♇
14 02 ☽ ∥ ♅
14 13 ☽ □ ♃

03 Saturday
03 59 ☽ Q ☿
05 03 ☽ ⊥ ♄
05 39 ☽ ∠ ♇
12 41 ☽ ∥ ♆
13 33 ☽ ✶ ♀
14 12 ☽ □ ☉
15 43 ☽ ☍ ♂
16 49 ☽ ⊻ ♄
20 44 ☉ ∥ ♇
21 04 ☽ ☌ ♅

04 Sunday
01 13 ☽ ☌ ♆
11 25 ☽ ✶ ♇
16 06 ☽ Q ♀
18 15 ♂ ⊼ ♄
20 14 ☉ ⊻ ♀
20 59 ☽ □ ☿
21 41 ♀ □ ♄
21 56 ☽ ∥ ♄
22 01 ☉ △ ♄

05 Monday
00 00 ♀ △ ♂
01 46 ☉ □ ♂
02 33 ☽ △ ♃
11 37 ☽ Q ♇
14 08 ☽ ∥ ♀

06 Tuesday
04 57 ☽ ☌ ♄
06 28 ☽ ⊼ ♂
07 49 ☽ △ ☉
08 36 ☽ □ ♀
09 27 ☽ bQ ♃
09 27 ☽ ⊻ ♅
13 41 ☽ ⊻ ♆
16 45 ☽ ∥ ☿
16 56 ♀ ✶ ♅
18 19 ♀ ∠ ♃
19 09 ☽ ± ♂
21 38 ☽ ⊥ ♅
22 03 ☿ ⊻ ♃

07 Wednesday
00 11 ☽ □ ♇
01 51 ☽ ⊥ ♆
03 42 ☉ □ ♅
04 02 ☽ ± ♃
14 08 ☽ bQ ♂
15 53 ☽ ∠ ♅
16 24 ☽ ⊼ ♃
16 58 ☽ bQ ☉
18 43 ☽ △ ☿
20 04 ☽ ∠ ♆

08 Thursday
00 20 ☽ ∥ ☉
08 16 ☽ ∥ ♇
10 01 ♀ ✶ ♆
11 28 ☉ ⊥ ♇
12 45 ☽ ± ☉
17 31 ☽ ⊻ ♄
20 24 ☿ Q ♅
21 34 ☽ △ ♂
22 06 ☽ ✶ ♅

09 Friday
00 50 ☽ ⋕ ♃
01 51 ☽ ⊼ ☉
02 14 ☽ ✶ ♆
04 03 ☽ △ ♀
05 12 ☽ bQ ☿
05 31 ☽ ⊥ ♄
06 28 ☉ □ ♆
11 54 ☽ ∥ ♃
12 04 ♂ ☍ ♅
12 43 ☽ △ ♇
22 02 ☽ Q ♅
23 23 ☽ ∠ ♄

10 Saturday
01 35 ☽ ± ☿
02 06 ☽ Q ♆
04 36 ☽ ⋕ ♇
05 28 ☽ ☍ ♃
05 44 ☿ Q ♆
13 05 ☽ bQ ♀
15 02 ☽ ⊼ ☿
17 33 ☽ ⋕ ☉
18 24 ☽ bQ ♇

11 Sunday
04 48 ☽ ✶ ♄
07 21 ♀ ∥ ♄
08 36 ☽ ± ♀
09 22 ☽ □ ♅
11 02 ☽ □ ♂
11 59 ☽ ± ♇
13 20 ☽ □ ♆
14 11 ☿ ⊥ ♃
18 03 ☽ ☍ ☉
21 29 ☽ ⊼ ♀
23 37 ☽ ⊼ ♇

12 Monday
04 00 ☽ Q ♄
16 26 ☽ ⋕ ☿
16 40 ☽ ⊼ ♃
19 52 ♀ ☌ ♇
22 53 ☽ Q ♂

13 Tuesday
04 16 ☽ ± ♃
08 29 ☽ ☍ ☿
11 44 ☽ ⋕ ♄
14 14 ☽ □ ♄
18 40 ☽ ⋕ ♀
18 45 ☽ △ ♅
21 31 ☽ bQ ♃
22 20 ☽ ✶ ♂
22 33 ☽ △ ♆

14 Wednesday
04 43 ♂ ☍ ♆
07 50 ☽ ⊼ ☉
08 38 ☽ ☍ ♇
12 21 ☽ ☍ ♀
18 25 ☉ ⊻ ♇
19 29 ☽ ⋕ ♆
19 55 ☽ ± ☉
22 45 ☽ bQ ♅

15 Thursday
01 53 ☽ △ ♃
02 28 ☽ bQ ♆
03 12 ☽ ∠ ♂
11 37 ☿ □ ♄
13 51 ☽ bQ ☉
15 20 ☽ ± ♅
18 19 ☽ ⋕ ♅
19 00 ☽ ± ♆
19 04 ☽ ∥ ♂
20 16 ☽ ⊥ ♂
21 49 ☽ △ ♄
23 03 ☽ ⊼ ☿

16 Friday
02 07 ♄ St D
02 17 ☽ ⊼ ♅
05 55 ☽ ⊼ ♆
07 32 ☽ ⊻ ♂
11 13 ☽ ± ☿
15 46 ☽ ⊼ ♇
19 17 ☽ △ ☉
22 44 ♂ ⋕ ♅

17 Saturday
00 30 ☽ ⋕ ♅
00 33 ☽ ∥ ♂
00 42 ☽ ⊼ ♀
00 55 ☽ bQ ♄
02 34 ☽ ± ♇
05 15 ☽ bQ ☿
06 03 ☿ ✶ ♅
09 08 ☽ □ ♃
12 28 ☽ ± ♀
16 56 ☽ ± ♄
18 36 ☽ bQ ♇
22 29 ☽ ⋕ ♆
23 41 ☿ ∥ ♄

18 Sunday
03 30 ☽ ⊼ ♄
05 53 ☽ bQ ♀
07 53 ☽ ☍ ♅
09 02 ☽ ⋕ ♀
10 42 ☽ △ ☿
11 20 ☽ ☍ ♆
14 29 ☽ ☌ ♂
17 07 ☿ ✶ ♆
17 16 ☽ Q ♃
20 54 ☽ △ ♇

19 Monday
00 32 ☽ ⋕ ☿
01 29 ☿ ∠ ♃
04 03 ☽ ⋕ ♄
04 12 ☽ □ ☉
04 40 ♀ ∠ ♅
10 22 ☽ △ ♀
14 15 ☽ ✶ ♃
14 26 ♀ Q ♄

20 Tuesday
07 08 ☽ ☍ ♄
11 23 ☽ ⊼ ♅
13 07 ☽ Q ☉
14 41 ☽ ⊼ ♆
16 00 ☽ ∠ ♃
16 03 ☿ △ ♂
19 09 ☽ ⊻ ♂
19 22 ☽ □ ☿
20 01 ♀ ∠ ♆
21 14 ☽ ⋕ ☉
21 28 ☽ ± ♅
23 59 ☽ □ ♇

21 Wednesday
00 44 ☽ ± ♆
05 25 ☽ ⊥ ♂
07 13 ☽ ⊥ ♃
10 37 ☽ ✶ ☉
12 27 ☽ bQ ♅
15 21 ♀ ✶ ♃
15 40 ☽ bQ ♆
17 18 ☽ ⊻ ♃
17 27 ☽ □ ♀
18 47 ♀ ∥ ♆
20 44 ☽ ∠ ♂

22 Thursday
00 18 ☽ ⋕ ♇
04 09 ☽ Q ☿
05 49 ☽ Q ♇
08 59 ☽ ⊼ ♄
13 08 ☽ ∠ ☉
13 11 ☽ △ ♅
16 21 ☽ △ ♆
18 49 ☽ ± ♄
19 18 ☽ ⋕ ♃
19 20 ☽ ∥ ♃
22 00 ☽ ✶ ♂
23 33 ☿ ☌ ♇

23 Friday
01 32 ☽ ✶ ♇
01 40 ☽ Q ♀
01 43 ☽ ✶ ☿
04 55 ☽ ⊥ ☉
09 32 ☽ bQ ♄
14 22 ☽ ∥ ♇
15 26 ☽ ⊻ ☉
18 07 ☽ Q ♂
19 09 ☽ ☌ ♃
23 02 ☽ ✶ ♀

24 Saturday
01 00 ♀ bQ ♂
02 06 ☽ ∠ ♇
04 33 ☽ ∠ ☿
10 06 ☽ △ ♄
14 21 ☽ □ ♅
16 55 ☽ ⊥ ♇
17 30 ☽ □ ♆
20 43 ☽ ⊥ ☿
23 54 ☽ ∥ ☉

25 Sunday
00 26 ☽ □ ♂
01 55 ☽ ∠ ♀
02 50 ☽ ⊻ ♇
07 34 ☽ ⊻ ☿
16 02 ☉ Q ♅
18 16 ☽ ⊥ ♀
20 15 ☽ Q ♅
20 34 ☽ ☌ ☉
21 23 ☽ ⊻ ♃
23 27 ☽ Q ♆

26 Monday
00 55 ♀ ⋕ ♂
02 43 ☿ ∥ ♆
05 18 ☽ ⊻ ♀
07 38 ☽ ⊥ ♃
09 40 ☽ ∥ ♄
11 03 ☉ ⊻ ♃
12 09 ☽ □ ♄
16 36 ☽ ✶ ♅
18 16 ♀ ⊥ ♅
19 39 ♀ ∥ ♅
19 51 ☽ ✶ ♆
23 14 ☽ ∠ ♃

27 Tuesday
04 22 ☽ △ ♂
05 42 ☽ ☌ ♇
14 29 ☉ Q ♆
15 23 ☽ ☌ ☿
15 23 ☽ ∥ ♆
17 16 ♂ Q ♃
18 42 ☽ ∠ ♅
19 22 ☽ Q ♄
22 01 ☽ ∠ ♆
23 41 ☽ ∥ ☿

28 Wednesday
01 53 ☽ ✶ ♃
04 22 ☽ ⊻ ☉
07 28 ☽ bQ ♂
07 59 ☽ ⋕ ♂
08 07 ♀ ⊥ ♆
10 46 ☽ ⊥ ♅
13 50 ☽ ∥ ♅
14 08 ☽ ⊥ ♆
14 44 ☽ ☌ ♀
16 02 ☽ ⊥ ☉
16 52 ☽ ✶ ♄
21 39 ☽ ⊻ ♅
23 52 ☽ Q ♃

29 Thursday
00 12 ☽ ± ♂
01 04 ☽ ⊻ ♆
08 17 ☿ ∠ ♅
09 53 ☽ ∠ ☉
11 32 ☿ ⋕ ♂
11 33 ☽ ⊼ ♂
11 42 ☽ ⊻ ♇
14 31 ♀ ✶ ♄
16 32 ☿ Q ♄
18 05 ♂ △ ♇
18 39 ☽ ∥ ♅
20 36 ☽ ∠ ♄
22 14 ☽ ∥ ☿
22 57 ☽ ⊥ ♇

30 Friday
01 22 ☽ ⋕ ♂
03 09 ☽ ⊻ ☿
09 38 ☿ ∥ ♅
10 02 ☽ □ ♃
13 43 ☽ ⊥ ♄
15 39 ☽ ⊥ ☿
16 08 ☽ ∠ ♇
16 37 ☽ ✶ ☉
18 31 ☽ ∥ ♆

31 Saturday
01 18 ☽ ⊻ ♄
02 52 ☿ ∠ ♆
05 00 ☽ ⊻ ♀
06 27 ☽ ☌ ♅
10 02 ☽ ☌ ♆
10 40 ☽ ∠ ☿
17 54 ☉ ⊥ ♀
18 02 ☽ Q ☉
18 02 ☽ ⊥ ♀
20 05 ♀ ⊻ ♅
21 27 ☽ ✶ ♇
22 39 ☽ ☍ ♂

NOVEMBER 1992

Date	Sidereal time h m s	Sun ☉ ° ' "	Moon ☽ ° ' "	Moon ☽ 24.00 ° ' "	Mercury ☿ ° '	Venus ♀ ° '	Mars ♂ ° '	Jupiter ♃ ° '	Saturn ♄ ° '	Uranus ♅ ° '	Neptune ♆ ° '	Pluto ♇ ° '
01	14 44 05	09 ♏ 19 23	29 ♑ 38 22	05 ♒ 38 16	02 ♐ 51	15 ♐ 29	23 ♋ 05	04 ♎ 29	12 ♒ 03	14 ♑ 42	16 ♑ 31	22 ♏ 20
02	14 48 02	10 19 27	11 ♒ 36 10	17 ♒ 32 41	03 47	16 41	23 23	04 41	12 05	14 44	16 32	22 22
03	14 51 58	11 19 33	23 ♒ 28 29	29 ♒ 24 09	04 39	17 54	23 41	04 52	12 06	14 46	16 33	22 24
04	14 55 55	12 19 40	05 ♓ 20 17	11 ♓ 17 26	05 27	19 07	23 58	05 04	12 08	14 48	16 34	22 27
05	14 59 51	13 19 49	17 ♓ 16 07	23 ♓ 16 47	06 10	20 19	24 14	05 15	12 10	14 50	16 36	22 29
06	15 03 48	14 20 00	29 ♓ 19 52	05 ♈ 25 43	06 49	21 32	24 30	05 27	12 12	14 52	16 37	22 32
07	15 07 44	15 20 12	11 ♈ 34 38	17 ♈ 46 50	07 21	22 44	24 45	05 38	12 15	14 54	16 38	22 34
08	15 11 41	16 20 25	24 ♈ 02 31	00 ♉ 21 45	07 48	23 56	25 00	05 50	12 17	14 57	16 39	22 36
09	15 15 38	17 20 40	06 ♉ 44 38	13 ♉ 11 07	08 07	25 09	25 14	06 01	12 19	14 59	16 41	22 39
10	15 19 34	18 20 57	19 ♉ 41 11	26 ♉ 14 42	08 19	26 21	25 28	06 12	12 22	15 01	16 42	22 41
11	15 23 31	19 21 16	02 ♊ 51 34	09 ♊ 31 37	08 R 23	27 33	25 41	06 23	12 24	15 04	16 44	22 44
12	15 27 27	20 21 37	16 ♊ 14 41	23 ♊ 00 35	08 17	28 45	25 53	06 34	12 27	15 06	16 45	22 46
13	15 31 24	21 21 59	29 ♊ 49 10	06 ♋ 40 15	08 02	29 ♐ 58	26 05	06 45	12 30	15 08	16 46	22 48
14	15 35 20	22 22 23	13 ♋ 33 41	20 ♋ 29 19	07 37	01 ♑ 10	26 16	06 56	12 33	15 11	16 48	22 51
15	15 39 17	23 22 49	27 ♋ 27 00	04 ♌ 26 35	07 02	02 22	26 27	07 07	12 36	15 13	16 49	22 53
16	15 43 13	24 23 17	11 ♌ 27 56	18 ♌ 30 54	06 16	03 34	26 37	07 17	12 39	15 16	16 51	22 56
17	15 47 10	25 23 47	25 ♌ 35 17	02 ♍ 40 53	05 21	04 46	26 46	07 28	12 42	15 18	16 53	22 58
18	15 51 07	26 24 19	09 ♍ 47 26	16 ♍ 54 40	04 17	05 58	26 54	07 39	12 45	15 21	16 54	23 01
19	15 55 03	27 24 52	24 ♍ 02 13	01 ♎ 09 42	03 05	07 10	27 02	07 49	12 49	15 24	16 56	23 03
20	15 59 00	28 25 28	08 ♎ 16 41	15 ♎ 22 41	01 47	08 21	27 09	07 59	12 52	15 26	16 57	23 05
21	16 02 56	29 ♏ 26 04	22 ♎ 27 12	29 ♎ 29 44	00 ♐ 26	09 33	27 15	08 10	12 55	15 29	16 59	23 08
22	16 06 53	00 ♐ 26 42	06 ♏ 29 45	13 ♏ 26 47	29 ♏ 05	10 45	27 21	08 20	12 59	15 32	17 01	23 10
23	16 10 49	01 27 23	20 ♏ 20 21	27 ♏ 10 04	27 45	11 57	27 25	08 30	13 03	15 35	17 02	23 13
24	16 14 46	02 28 04	03 ♐ 55 35	10 ♐ 36 38	26 30	13 08	27 29	08 40	13 07	15 37	17 04	23 15
25	16 18 42	03 28 47	17 ♐ 13 04	23 ♐ 44 46	25 22	14 20	27 32	08 50	13 10	15 40	17 06	23 17
26	16 22 39	04 29 32	00 ♑ 11 44	06 ♑ 34 04	24 23	15 31	27 35	09 00	13 14	15 43	17 08	23 20
27	16 26 36	05 30 17	12 ♑ 51 57	19 ♑ 05 37	23 34	16 43	27 36	09 09	13 18	15 46	17 10	23 22
28	16 30 32	06 31 04	25 ♑ 15 23	01 ♒ 21 40	22 56	17 54	27 37	09 19	13 22	15 49	17 11	23 24
29	16 34 29	07 31 51	07 ♒ 24 52	13 ♒ 25 31	22 30	19 05	27 R 37	09 28	13 27	15 52	17 13	23 27
30	16 38 25	08 ♐ 32 40	19 ♒ 24 07	25 ♒ 21 15	22 ♏ 16	20 ♑ 17	27 ♋ 36	09 ♎ 38	13 ♒ 31	15 ♑ 55	17 ♑ 15	23 ♏ 29

Date	Moon True ☊ ° '	Moon Mean ☊ ° '	Moon ☽ Latitude ° '
01	22 ♐ 24	23 ♐ 37	03 N 08
02	22 R 24	23 34	03 55
03	22 23	23 31	04 32
04	22 20	23 28	04 57
05	22 16	23 25	05 09
06	22 10	23 22	05 08
07	22 03	23 18	04 53
08	21 57	23 15	04 23
09	21 51	23 12	03 41
10	21 46	23 09	02 46
11	21 43	23 06	01 41
12	21 41	23 03	00 N 30
13	21 D 41	22 59	00 S 44
14	21 43	22 56	01 57
15	21 44	22 53	03 04
16	21 46	22 50	04 00
17	21 R 46	22 47	04 42
18	21 45	22 43	05 07
19	21 44	22 40	05 14
20	21 41	22 37	05 01
21	21 38	22 34	04 30
22	21 34	22 31	03 43
23	21 32	22 28	02 43
24	21 30	22 24	01 35
25	21 29	22 21	00 S 24
26	21 D 29	22 18	00 N 48
27	21 30	22 15	01 55
28	21 31	22 12	02 56
29	21 33	22 09	03 48
30	21 ♐ 34	22 ♐ 05	04 N 28

DECLINATIONS

Date	Sun ☉ ° '	Moon ☽ ° '	Mercury ☿ ° '	Venus ♀ ° '	Mars ♂ ° '	Jupiter ♃ ° '	Saturn ♄ ° '	Uranus ♅ ° '	Neptune ♆ ° '	Pluto ♇ ° '
01	14 S 36	17 S 10	23 S 32	24 S 14	22 N 45	00 S 45	18 S 10	23 S 03	21 S 44	04 S 41
02	14 55	13 32	23 43	24 24	22 44	00 50	18 10	23 02	21 44	04 42
03	15 14	09 25	23 53	24 33	22 43	00 55	18 09	23 02	21 43	04 42
04	15 32	04 57	24 01	24 42	22 43	00 59	18 09	23 02	21 43	04 43
05	15 50	00 S 17	24 07	24 50	22 42	01 03	18 08	23 02	21 43	04 44
06	16 08	04 N 26	24 11	24 57	22 41	01 08	18 07	23 01	21 43	04 44
07	16 26	09 04	24 12	25 04	22 41	01 12	18 07	23 01	21 43	04 45
08	16 43	13 24	24 12	25 10	22 40	01 17	18 06	23 01	21 43	04 46
09	17 01	17 14	24 09	25 15	22 40	01 21	18 05	23 01	21 43	04 46
10	17 17	20 19	24 04	25 19	22 40	01 25	18 05	23 00	21 43	04 47
11	17 34	22 23	23 56	25 23	22 40	01 29	18 04	23 00	21 42	04 48
12	17 50	23 13	23 45	25 26	22 40	01 34	18 03	23 00	21 42	04 48
13	18 06	22 42	23 30	25 28	22 40	01 38	18 02	23 00	21 42	04 49
14	18 22	20 49	23 13	25 29	22 40	01 42	18 01	22 59	21 42	04 50
15	18 37	17 40	22 52	25 30	22 41	01 46	18 01	22 59	21 42	04 50
16	18 52	13 30	22 28	25 30	22 42	01 50	18 00	22 59	21 42	04 51
17	19 07	08 34	22 01	25 29	22 42	01 54	17 59	22 58	21 41	04 52
18	19 21	03 N 09	21 30	25 28	22 43	01 58	17 58	22 58	21 41	04 52
19	19 35	02 S 26	20 57	25 25	22 44	02 02	17 57	22 58	21 41	04 53
20	19 49	07 54	20 22	25 22	22 45	02 06	17 56	22 57	21 41	04 53
21	20 02	12 55	19 45	25 19	22 47	02 10	17 55	22 57	21 41	04 54
22	20 15	17 11	19 08	25 14	22 48	02 14	17 54	22 57	21 41	04 54
23	20 27	20 27	18 32	25 09	22 50	02 18	17 53	22 56	21 40	04 55
24	20 39	22 30	17 57	25 03	22 52	02 22	17 52	22 56	21 40	04 56
25	20 51	23 13	17 26	24 57	22 54	02 26	17 50	22 56	21 40	04 56
26	21 02	22 38	16 58	24 49	22 56	02 29	17 49	22 55	21 40	04 57
27	21 13	20 54	16 35	24 41	22 59	02 33	17 48	22 55	21 40	04 57
28	21 24	18 12	16 16	24 32	23 01	02 36	17 47	22 55	21 39	04 58
29	21 34	14 45	16 02	24 23	23 04	02 40	17 46	22 54	21 39	04 58
30	21 S 44	10 S 45	15 S 53	24 S 13	23 N 07	02 S 43	17 S 44	22 S 54	21 S 39	04 S 59

ZODIAC SIGN ENTRIES

Date	h	m	Planets
01	12	43	☽ ♒
04	01	13	☽ ♓
06	13	19	☽ ♈
08	23	19	☽ ♉
11	06	49	☽ ♊
13	12	19	☽ ♋
13	12	48	♀ ♑
15	16	23	☽ ♌
17	19	28	☽ ♍
19	22	03	☽ ♎
21	19	44	☿ ♏
22	00	52	☽ ♏
22	01	26	☉ ♐
24	05	01	☽ ♐
26	11	38	☽ ♑
28	21	19	☽ ♒

LATITUDES

Date	Mercury ☿ ° '	Venus ♀ ° '	Mars ♂ ° '	Jupiter ♃ ° '	Saturn ♄ ° '	Uranus ♅ ° '	Neptune ♆ ° '	Pluto ♇ ° '
01	02 S 51	01 S 35	01 N 18	01 N 07	01 S 02	00 S 25	00 N 42	14 N 07
04	02 51	01 43	01 25	01 07	01 02	00 25	00 42	14 06
07	02 42	01 49	01 32	01 08	01 02	00 25	00 42	14 06
10	02 24	01 56	01 39	01 08	01 02	00 25	00 42	14 06
13	01 53	02 01	01 46	01 09	01 01	00 25	00 41	14 05
16	01 08	02 06	01 54	01 09	01 01	00 25	00 41	14 05
19	00 S 11	02 11	02 01	01 09	01 01	00 25	00 41	14 05
22	00 N 50	02 14	02 09	01 10	01 01	00 25	00 41	14 05
25	01 43	02 17	02 18	01 10	01 01	00 25	00 41	14 05
28	02 19	02 19	02 26	01 11	01 01	00 25	00 41	14 05
31	02 N 36	02 S 20	02 N 34	01 N 12	01 S 01	00 S 25	00 N 41	14 N 05

DATA

Julian Date	2448928
Delta T	+58 seconds
Ayanamsa	23° 45' 41"
Synetic vernal point	05° ♓ 21' 18"
True obliquity of ecliptic	23° 26' 24"

MOON'S PHASES, APSIDES AND POSITIONS ☽

Date	h	m	Phase	Longitude ° '	Eclipse Indicator
02	09	11	☽	10 ♒ 12	
10	09	20	○	18 ♉ 14	
17	11	39	☾	25 ♌ 23	
24	09	11	●	02 ♐ 21	

Day	h	m		
03	23	24	Apogee	
18	23	51	Perigee	
05	13	27	ON	
12	14	57	Max dec	23° N 14'
19	01	33	OS	
25	13	05	Max dec	23° S 13'

LONGITUDES

Date	Chiron ⚷ ° '	Ceres ⚳ ° '	Pallas ⚴ ° '	Juno ⚵ ° '	Vesta ⚶ ° '	Black Moon Lilith ⚸ ° '
01	22 ♌ 30	02 ♒ 01	04 ♑ 42	11 ♋ 31	24 ♏ 45	01 ♓ 44
11	23 ♌ 02	04 ♒ 27	07 ♑ 54	12 ♋ 38	00 ♐ 07	02 ♓ 51
21	23 ♌ 23	07 ♒ 11	11 ♑ 14	12 ♋ 49	05 ♐ 31	03 ♓ 58
31	23 ♌ 34	10 ♒ 10	14 ♑ 40	12 ♋ 02	10 ♐ 56	05 ♓ 06

All ephemeris data is given at 12.00 UT and the Moon's longitude is additionally given for 24.00 UT

ASPECTARIAN

h m Aspects

01 Sunday
04 26 ☽ ∥ ♄
13 52 ☽ ∠ ♀
18 59 ☽ ⚹ ☿
21 24 ☽ Q ♇
21 51 ☽ △ ♃

02 Monday
03 58 ☽ ∥ ☉
08 48 ♀ ⊻ ♆
09 11 ☽ □ ☉
11 43 ♀ Q ♃
12 58 ☽ ☌ ♄
18 20 ☽ ⊻ ♅
21 07 ☽ Q ☿
21 58 ☽ ⊻ ♆
22 34 ☉ ⊥ ♃
23 26 ☽ ⚹ ♀

03 Tuesday
04 35 ☽ ⚼ ♃
06 15 ♀ ± ♂
06 30 ☽ ⊥ ♅
09 50 ☽ □ ♇
10 08 ☽ ⊥ ♆
12 26 ☽ ⊼ ♂
20 33 ☿ ⚹ ♃
23 07 ☽ ± ♃

04 Wednesday
00 46 ☽ ∠ ♅
00 52 ☽ ± ♂
02 29 ☽ Q ♀
04 23 ☽ ∠ ♆
07 20 ☉ □ ♄
11 27 ☽ ⊼ ♃
12 14 ☽ □ ☿
13 13 ☽ ∥ ♇
19 29 ☽ ⚼ ♂

05 Thursday
01 45 ☽ ⊻ ♄
03 23 ☽ △ ☉
07 07 ☽ ⚹ ♅
08 08 ☽ ∥ ♃
10 39 ☽ ⚹ ♆
13 49 ☽ ⊥ ♄
18 47 ☽ □ ♀
18 55 ☽ # ♃
22 27 ☽ △ ♇

06 Friday
02 13 ☽ △ ♂
07 07 ☽ Q ♅
07 47 ☽ ∠ ♄
10 35 ☽ Q ♆
12 00 ☽ ⚼ ☉
13 33 ☽ # ♇

07 Saturday
00 14 ☽ ☍ ♃
01 21 ☉ ⚹ ♅
03 25 ☽ △ ☿
04 10 ☽ ⚼ ♇
07 15 ☽ ± ☉
08 35 ♀ ⊻ ♇
13 18 ☽ ⚹ ♄
18 28 ☽ □ ♅
19 56 ☽ ⊼ ☉
21 41 ☽ ± ♇
21 49 ☽ □ ♆

08 Sunday
09 15 ☽ ⊼ ♇
09 33 ☽ ⚼ ☿
11 47 ☽ △ ♀
12 28 ☽ Q ♄
13 52 ☽ □ ♂
19 45 ☉ ⚹ ♆

09 Monday
03 07 ☽ ± ☿
10 20 ☽ # ☉
10 37 ☽ ⊼ ♃
14 22 ♀ ⊼ ♂
14 38 ☽ ⊼ ☿
18 00 ☽ # ♄
19 00 ☽ ⚼ ♀
21 58 ☽ ± ♃
22 26 ☽ □ ♄

10 Tuesday
00 19 ☽ Q ♂
03 22 ☽ △ ♅
06 30 ☽ △ ♆
09 20 ☽ ☍ ☉
13 21 ☽ ± ♀
14 49 ☽ ⚼ ♃
17 31 ☽ ☍ ♇
22 46 ☽ ⚹ ♂

11 Wednesday
01 25 ☽ ⊼ ♀
02 35 ☽ # ♆
06 55 ☽ ⚼ ♅
08 59 ♀ ∠ ♄
09 48 ☿ St R
09 57 ☽ ⚼ ♆
16 58 ☽ ∥ ♂
18 27 ☽ △ ♃
21 55 ☽ ☍ ☿
23 12 ☽ ± ♅

12 Thursday
00 54 ☽ # ♅
02 10 ☽ ± ♆
02 18 ☽ ∠ ♂
05 13 ☽ △ ♄
09 57 ☽ ⊼ ♅
12 12 ♀ ⊥ ♇
12 54 ☽ ⊼ ♆
18 34 ☽ ⊥ ♂
19 54 ☽ ⊼ ☉
23 36 ☽ ⊼ ♇

13 Friday
05 04 ☽ # ♅
05 20 ☽ ⊻ ♂
06 19 ☉ ∥ ♄
07 21 ☽ ± ☉
07 54 ☽ ⚼ ♄
10 13 ☽ ± ♇
12 16 ☽ ☍ ♀
12 36 ☽ ∥ ♂
23 12 ☉ ∠ ♃
23 44 ☽ ± ♄

14 Saturday
00 18 ☽ □ ♃
00 22 ☽ ⚼ ☉
02 00 ☽ ⊼ ☿
02 01 ☽ ⚼ ♇
02 41 ☽ # ♆
10 14 ☽ ⊼ ♄
12 06 ☽ ± ☿
14 49 ☽ ☍ ♅
17 38 ☽ ☍ ♆
23 47 ☉ ☌ ♇

15 Sunday
03 05 ☽ ⚼ ☿
04 07 ☽ △ ♇
04 27 ☽ △ ☉
04 45 ☿ ∥ ♅
06 02 ☽ # ☉
07 55 ☽ Q ♃
09 45 ☿ ⚹ ♃
09 45 ☽ # ♄
10 15 ☽ ☌ ♂
21 14 ☽ ⊼ ♀
23 35 ☿ # ♂

16 Monday
03 37 ☽ △ ☿
04 47 ☽ ⚹ ♃
08 27 ☽ ± ♀
14 01 ☽ ☍ ♄
18 30 ☽ ⊼ ♅
21 11 ☽ ⊼ ♆

17 Tuesday
01 12 ☽ ⚼ ♀
04 43 ☽ ± ♅
06 39 ☽ ∠ ♃
07 24 ☽ ± ♆
07 33 ☽ □ ♇
11 39 ☽ □ ☉
14 00 ☽ ⊻ ♂
18 21 ☿ ⊻ ♀
20 01 ☽ ⚼ ♅
22 04 ☽ ⊥ ♃
22 40 ☽ ⚼ ♆

18 Wednesday
00 16 ☽ ⊥ ♂
03 23 ☽ □ ☿
03 33 ☿ ∥ ♆
04 32 ☽ # ♇
04 57 ☽ △ ♀
08 20 ☽ ⊻ ♃
14 04 ☽ Q ♇
15 36 ☽ ∠ ♂
17 01 ☽ ⊼ ♄
17 02 ☽ # ♃
20 22 ☽ Q ☉
21 24 ☽ △ ♅

19 Thursday
00 00 ☽ △ ♆
01 35 ☉ △ ♂
03 10 ☽ ± ♄
04 38 ♀ ⊥ ♄
07 25 ☽ Q ☿
10 16 ☽ ∥ ♃
10 20 ☽ ⚹ ♇
17 05 ☽ ⚹ ♂
18 07 ☽ ⚹ ☉
18 23 ☽ ⚼ ♄
22 37 ☽ ∥ ♇

20 Friday
01 59 ☽ ⚹ ☿
03 25 ♀ □ ♃
06 26 ♀ ∠ ♇
06 34 ☿ ⊥ ♀
09 00 ☿ ∠ ♆
11 31 ☽ ☌ ♃
11 41 ☽ ∠ ♇
12 09 ☽ □ ♀
13 29 ☽ Q ♂
19 47 ☽ △ ♄
21 22 ☽ ∠ ☉

21 Saturday
00 09 ☽ □ ♅
01 09 ☽ ∠ ☿
02 42 ☽ □ ♆
02 56 ☽ ⊥ ♇
03 47 ☿ Q ♄
03 56 ☉ ∥ ☿
05 57 ♃ ∠ ♇
11 13 ☿ ∠ ♅
13 09 ☽ ⊻ ♇
13 48 ☽ ⊥ ☉
15 05 ☽ ⊥ ☿
20 13 ☽ □ ♂
21 29 ☽ Q ♀
22 09 ☉ ☌ ☿

22 Sunday
00 24 ☽ ⊻ ☿
00 49 ☽ ⊻ ☉
06 53 ☽ Q ♅
09 27 ☽ Q ♆
14 07 ☉ ∠ ♅
15 12 ☽ ⊻ ♃
16 32 ☽ ∥ ♄
20 01 ☽ ⚹ ♀
23 10 ☽ ∥ ☿
23 15 ☽ □ ♄

23 Monday
01 38 ☉ Q ♄
01 42 ☽ ⊥ ♃
03 40 ☽ ⚹ ♅
06 14 ☽ ⚹ ♆
11 59 ☽ ∥ ☉
17 03 ☽ ☌ ♇
17 37 ☽ ∠ ♃
17 49 ☿ △ ♂
19 49 ☿ ∠ ♀
23 55 ☽ ☌ ☿

24 Tuesday
00 25 ☽ ∥ ♆
00 31 ☽ △ ♂
00 43 ☽ ∠ ♀
02 16 ☉ ∠ ♆
06 06 ☽ ∠ ♅
06 58 ☽ Q ♄
08 41 ☽ ∠ ♆
09 11 ☽ ☌ ☉
11 25 ♀ ⊻ ♄
16 26 ☿ ∥ ♄
18 19 ☽ ⊥ ♀
19 50 ☽ # ♂
20 36 ☽ ⚹ ♃
21 17 ☽ ∥ ♅
22 15 ☽ ⊥ ♅

25 Wednesday
00 51 ☽ ⊥ ♆
03 28 ☽ ⚼ ♂
04 36 ☽ ⚹ ♄
06 13 ☽ ⊻ ♀
09 10 ☽ ⊻ ♅
11 47 ☽ ⊻ ♆
18 43 ☽ Q ♃
19 57 ☽ ± ♂
23 11 ☽ ⊻ ♇

26 Thursday
01 54 ☽ ⊻ ☿
03 59 ♂ # ♅
05 15 ☽ # ♂
05 19 ☽ ∥ ♅
07 06 ☽ ⊼ ♂
08 20 ☽ ∠ ♄
10 22 ☽ ⊥ ♇
12 19 ☽ ⊥ ☿
16 10 ♀ ☌ ♅
20 46 ☽ ⊻ ☉
20 58 ☿ ∠ ♃

27 Friday
01 21 ☽ ⊥ ♄
03 13 ☽ ∥ ♆
03 23 ☽ ∠ ♇
04 15 ☽ ∠ ☿
04 49 ☽ □ ♃
08 46 ☽ ∥ ☉
09 10 ☽ ⊥ ☉
12 51 ☽ ⊻ ♄
17 36 ☽ ☌ ♅
18 29 ☿ ☌ ♇
20 11 ☽ ☌ ♀
20 17 ☽ ☌ ♆
21 15 ♀ ☌ ♆

28 Saturday
04 03 ☽ ∠ ☉
07 40 ☽ ⚹ ☿
08 23 ☽ ⚹ ♇
15 11 ☽ ∥ ♄
16 38 ☽ ☍ ♂
23 30 ♂ St R

29 Sunday
03 03 ☽ ∥ ☿
06 23 ☽ Q ☿
08 05 ☽ Q ♇
12 15 ☽ ⚹ ☉
16 09 ☽ △ ♃

30 Monday
00 07 ☽ ☌ ♄
00 23 ☉ ∥ ♆
04 58 ☽ ⊻ ♅
07 40 ☽ ⊻ ♆
08 41 ♂ Q ♃
13 57 ☽ ⊻ ♀
14 31 ☽ Q ☉
17 05 ☽ ⊥ ♅
17 42 ☽ □ ☿
19 46 ☽ ⊥ ♆
20 15 ☽ □ ♇
22 41 ☽ ⚼ ♃

Date	Sidereal time h m s	Sun ☉ ° ' "	Moon ☽ ° ' "	Moon ☽ 24.00 ° ' "	Mercury ☿ ° '	Venus ♀ ° '	Mars ♂ ° '	Jupiter ♃ ° '	Saturn ♄ ° '	Uranus ♅ ° '	Neptune ♆ ° '	Pluto ♇ ° '
01	16 42 22	09 ♐ 33 30	01 ♓ 17 28	07 ♓ 13 23	22 ♏ 12	21 ♑ 28	27 ♋ 35	09 ♎ 47	13 ♒ 35	15 ♑ 58	17 ♑ 17	23 ♏ 32
02	16 46 18	10 34 20	13 ♓ 09 37	19 ♓ 06 44	22 D 20	22 39	27 R 32	09 56	13 40	16 01	17 19	23 34
03	16 50 15	11 35 11	25 ♓ 05 20	01 ♈ 06 00	22 36	23 50	27 29	10 05	13 44	16 04	17 21	23 36
04	16 54 11	12 36 03	07 ♈ 09 17	13 ♈ 15 41	23 02	25 00	27 25	10 14	13 49	16 07	17 23	23 39
05	16 58 08	13 36 56	19 ♈ 25 41	25 ♈ 39 41	23 36	26 11	27 20	10 23	13 53	16 10	17 25	23 41
06	17 02 05	14 37 50	01 ♉ 58 02	08 ♉ 21 01	24 17	27 22	27 14	10 32	13 58	16 14	17 27	23 43
07	17 06 01	15 38 44	14 ♉ 48 51	21 ♉ 21 37	25 04	28 32	27 07	10 40	14 03	16 17	17 29	23 45
08	17 09 58	16 39 40	27 ♉ 59 23	04 ♊ 42 03	25 57	29 ♑ 43	27 00	10 49	14 08	16 20	17 31	23 48
09	17 13 54	17 40 36	11 ♊ 29 27	18 ♊ 21 22	26 55	00 ♒ 53	26 51	10 57	14 13	16 23	17 33	23 50
10	17 17 51	18 41 33	25 ♊ 17 25	02 ♋ 17 13	27 57	02 02	26 42	11 05	14 18	16 26	17 35	23 52
11	17 21 47	19 42 31	09 ♋ 20 17	16 ♋ 26 04	29 ♏ 02	03 14	26 32	11 14	14 23	16 30	17 37	23 54
12	17 25 44	20 43 30	23 ♋ 34 02	00 ♌ 43 36	00 ♐ 11	04 24	26 21	11 22	14 28	16 33	17 39	23 57
13	17 29 40	21 44 30	07 ♌ 54 11	15 ♌ 05 14	01 23	05 34	26 09	11 29	14 33	16 36	17 41	23 59
14	17 33 37	22 45 31	22 ♌ 16 13	29 ♌ 26 39	02 38	06 43	25 57	11 37	14 39	16 40	17 43	24 01
15	17 37 34	23 46 33	06 ♍ 36 05	13 ♍ 44 09	03 54	07 53	25 43	11 45	14 44	16 43	17 45	24 03
16	17 41 30	24 47 36	20 ♍ 50 30	27 ♍ 54 52	05 13	09 03	25 29	11 52	14 49	16 46	17 47	24 05
17	17 45 27	25 48 40	04 ♎ 57 01	11 ♎ 56 44	06 32	10 12	25 14	11 59	14 55	16 50	17 49	24 07
18	17 49 23	26 49 45	18 ♎ 53 53	25 ♎ 48 20	07 54	11 21	24 59	12 07	15 00	16 53	17 51	24 10
19	17 53 20	27 50 50	02 ♏ 39 57	09 ♏ 28 40	09 17	12 30	24 42	12 14	15 06	16 56	17 54	24 12
20	17 57 16	28 51 57	16 ♏ 14 24	22 ♏ 57 03	10 40	13 39	24 25	12 21	15 12	17 00	17 56	24 14
21	18 01 13	29 ♐ 53 04	29 ♏ 36 33	06 ♐ 12 51	12 05	14 48	24 07	12 27	15 18	17 03	17 58	24 16
22	18 05 09	00 ♑ 54 12	12 ♐ 45 53	19 ♐ 15 36	13 31	15 57	23 48	12 34	15 23	17 07	18 00	24 18
23	18 09 06	01 55 21	25 ♐ 42 00	02 ♑ 05 02	14 57	17 06	23 29	12 40	15 29	17 10	18 02	24 20
24	18 13 03	02 56 30	08 ♑ 24 46	14 ♑ 41 13	16 24	18 14	23 09	12 47	15 35	17 14	18 05	24 22
25	18 16 59	03 57 39	20 ♑ 54 29	27 ♑ 04 41	17 51	19 22	22 49	12 53	15 41	17 18	18 07	24 24
26	18 20 56	04 58 48	03 ♒ 11 59	09 ♒ 16 37	19 20	20 30	22 28	12 59	15 47	17 21	18 09	24 26
27	18 24 52	05 59 58	15 ♒ 18 49	21 ♒ 18 55	20 48	21 38	22 07	13 05	15 53	17 24	18 11	24 28
28	18 28 49	07 01 07	27 ♒ 17 16	03 ♓ 14 15	22 17	22 46	21 45	13 10	15 59	17 28	18 13	24 30
29	18 32 45	08 02 17	09 ♓ 10 20	15 ♓ 05 59	23 47	23 53	21 22	13 16	16 05	17 31	18 16	24 32
30	18 36 42	09 03 27	21 ♓ 01 43	26 ♓ 58 05	25 17	25 00	21 00	13 21	16 12	17 35	18 18	24 34
31	18 40 38	10 ♑ 04 36	02 ♈ 55 40	08 ♈ 55 02	26 ♐ 47	26 ♒ 07	20 ♋ 37	13 ♎ 27	16 ♒ 18	17 ♑ 38	18 ♑ 20	24 ♏ 35

Date	Moon True ☊ ° '	Moon Mean ☊ ° '	Moon ☽ Latitude ° '
01	21 ♐ 35	22 ♐ 02	04 N 57
02	21 R 35	21 59	05 13
03	21 34	21 56	05 16
04	21 34	21 53	05 05
05	21 33	21 49	04 40
06	21 32	21 46	04 01
07	21 31	21 43	03 10
08	21 30	21 40	02 07
09	21 30	21 37	00 N 55
10	21 D 30	21 34	00 S 21
11	21 30	21 30	01 37
12	21 30	21 27	02 49
13	21 R 30	21 24	03 50
14	21 30	21 21	04 37
15	21 30	21 18	05 07
16	21 30	21 14	05 17
17	21 D 30	21 11	05 09
18	21 30	21 08	04 42
19	21 31	21 05	03 59
20	21 31	21 02	03 03
21	21 32	20 59	01 59
22	21 33	20 55	00 S 48
23	21 R 33	20 52	00 N 23
24	21 33	20 49	01 32
25	21 31	20 46	02 35
26	21 29	20 43	03 30
27	21 27	20 40	04 15
28	21 24	20 36	04 48
29	21 22	20 33	05 08
30	21 20	20 30	05 15
31	21 ♐ 19	20 ♐ 27	05 N 09

DECLINATIONS

Date	Sun ☉ ° '	Moon ☽ ° '	Mercury ☿ ° '	Venus ♀ ° '	Mars ♂ ° '	Jupiter ♃ ° '	Saturn ♄ ° '	Uranus ♅ ° '	Neptune ♆ ° '	Pluto ♇ ° '
01	21 S 53	06 S 23	15 S 48	24 S 02	23 N 10	02 S 47	17 S 43	22 S 54	21 S 39	04 S 59
02	22 02	01 S 47	15 48	23 51	23 13	02 50	17 42	22 53	21 39	05 00
03	22 10	02 N 53	15 52	23 39	23 17	02 54	17 41	22 53	21 38	05 00
04	22 18	07 31	15 59	23 26	23 21	02 57	17 39	22 53	21 38	05 00
05	22 26	11 55	16 09	23 12	23 24	03 00	17 38	22 52	21 38	05 01
06	22 33	15 56	16 22	22 58	23 28	03 03	17 36	22 52	21 38	05 01
07	22 40	19 18	16 37	22 44	23 32	03 07	17 35	22 51	21 37	05 02
08	22 46	21 46	16 54	22 29	23 37	03 10	17 34	22 51	21 37	05 02
09	22 52	23 04	17 13	22 13	23 41	03 13	17 32	22 51	21 37	05 02
10	22 58	23 00	17 32	21 56	23 46	03 16	17 31	22 50	21 37	05 03
11	23 02	21 29	17 53	21 39	23 51	03 19	17 29	22 50	21 36	05 03
12	23 07	18 37	18 14	21 22	23 55	03 22	17 28	22 49	21 36	05 04
13	23 11	14 35	18 35	21 04	24 00	03 25	17 26	22 49	21 36	05 04
14	23 15	09 43	18 57	20 45	24 06	03 27	17 25	22 49	21 36	05 04
15	23 18	04 N 20	19 19	20 26	24 11	03 30	17 23	22 48	21 35	05 05
16	23 20	01 S 14	19 41	20 06	24 16	03 33	17 21	22 48	21 35	05 05
17	23 22	06 41	20 02	19 46	24 22	03 36	17 20	22 47	21 35	05 05
18	23 24	11 45	20 23	19 25	24 27	03 38	17 18	22 47	21 35	05 06
19	23 25	16 08	20 43	19 04	24 33	03 41	17 17	22 46	21 34	05 06
20	23 26	19 37	21 03	18 42	24 38	03 43	17 15	22 46	21 34	05 06
21	23 26	22 00	21 23	18 20	24 44	03 46	17 13	22 46	21 34	05 06
22	23 26	23 08	21 41	17 58	24 50	03 48	17 11	22 45	21 34	05 07
23	23 26	22 59	21 59	17 35	24 56	03 50	17 10	22 45	21 33	05 07
24	23 24	21 38	22 16	17 11	25 01	03 52	17 08	22 44	21 33	05 07
25	23 23	19 15	22 32	16 48	25 07	03 55	17 06	22 44	21 33	05 07
26	23 21	16 02	22 47	16 24	25 13	03 57	17 04	22 43	21 32	05 07
27	23 18	12 11	23 01	15 59	25 18	03 59	17 03	22 43	21 32	05 08
28	23 15	07 54	23 14	15 34	25 24	04 01	17 01	22 42	21 32	05 08
29	23 12	03 S 22	23 26	15 09	25 30	04 03	16 59	22 42	21 32	05 08
30	23 08	01 N 17	23 37	14 43	25 35	04 05	16 57	22 42	21 31	05 08
31	23 S 03	05 N 53	23 S 47	14 S 18	25 N 40	04 S 06	16 S 55	22 S 41	21 S 31	05 S 08

ZODIAC SIGN ENTRIES

Date	h m	Planets
01	09 23	☽ ♓
03	21 49	☽ ♈
06	08 16	☽ ♉
08	15 37	☽ ♊
08	17 49	♀ ♒
10	20 05	☽ ♋
12	08 05	☿ ♐
12	22 47	☽ ♌
15	00 56	☽ ♍
17	03 33	☽ ♎
19	07 20	☽ ♏
21	12 42	☽ ♐
21	14 43	☉ ♑
23	20 04	☽ ♑
26	05 43	☽ ♒
28	17 28	☽ ♓
31	06 07	☽ ♈

LATITUDES

Date	Mercury ☿ ° '	Venus ♀ ° '	Mars ♂ ° '	Jupiter ♃ ° '	Saturn ♄ ° '	Uranus ♅ ° '	Neptune ♆ ° '	Pluto ♇ ° '
01	02 N 36	02 S 20	02 N 34	01 N 12	01 S 01	00 S 25	00 N 41	14 N 05
04	02 38	02 20	02 43	01 12	01 01	00 25	00 41	14 06
07	02 29	02 20	02 52	01 13	01 01	00 25	00 41	14 06
10	02 13	02 18	03 00	01 13	01 01	00 25	00 41	14 06
13	01 53	02 15	03 09	01 14	01 01	00 25	00 41	14 07
16	01 31	02 11	03 17	01 15	01 01	00 25	00 41	14 07
19	01 08	02 06	03 25	01 15	01 01	00 25	00 41	14 08
22	00 44	02 00	03 33	01 16	01 01	00 25	00 41	14 09
25	00 N 21	01 53	03 40	01 17	01 01	00 25	00 40	14 09
28	00 S 01	01 44	03 46	01 17	01 01	00 25	00 40	14 10
31	00 S 23	01 S 35	03 N 52	01 N 18	01 S 01	00 S 25	00 N 40	14 N 11

DATA

Julian Date	2448958
Delta T	+58 seconds
Ayanamsa	23° 45' 46"
Synetic vernal point	05° ♓ 21' 14"
True obliquity of ecliptic	23° 26' 23"

MOON'S PHASES, APSIDES AND POSITIONS ☽

Date	h m	Phase	Longitude ° '	Eclipse Indicator
02	06 17	☽	10 ♓ 20	
09	23 41	○	18 ♊ 10	total
16	19 13	☾	25 ♍ 06	
24	00 43	●	02 ♑ 28	Partial

Day	h m		
01	20 10	Apogee	
13	21 17	Perigee	
29	17 07	Apogee	
02	21 12	0N	
09	22 55	Max dec	23° N 13'
16	06 42	0S	
22	21 14	Max dec	23° S 13'
30	05 24	0N	

LONGITUDES

Date	Chiron ⚷ ° '	Ceres ⚳ ° '	Pallas ⚴ ° '	Juno ⚵ ° '	Vesta ⚶ ° '	Black Moon Lilith ⚸ ° '
01	23 ♌ 34	10 ♒ 10	14 ♑ 40	12 ♋ 02	10 ♐ 56	05 ♓ 06
11	23 ♌ 34	13 ♒ 23	18 ♑ 11	10 ♋ 25	16 ♐ 21	06 ♓ 13
21	23 ♌ 23	16 ♒ 45	21 ♑ 45	08 ♋ 11	21 ♐ 47	07 ♓ 20
31	23 ♌ 02	20 ♒ 17	25 ♑ 20	05 ♋ 45	27 ♐ 11	08 ♓ 27

ASPECTARIAN

h m Aspects

01 Tuesday
03 24 ☽ ⊥ ♀
04 31 ☽ ⊼ ♂
07 31 ☿ St D
11 20 ☽ ∠ ♅
14 01 ☽ ∠ ♆
16 37 ☽ ± ♂
17 07 ☽ ± ♃
18 20 ☉ ⚹ ♃
19 23 ☽ ∥ ♇
22 10 ☉ ⊥ ♅
23 37 ☽ ∠ ♀

02 Wednesday
04 26 ☿ ⚹ ♀
05 24 ☽ ⊼ ♃
06 17 ☽ □ ☉
06 40 ☽ ∥ ♃
10 45 ☽ ꝑ ♂
13 01 ☽ ⊻ ♄
17 47 ☽ ⚹ ♅
20 24 ☽ ⚹ ♆

03 Thursday
01 11 ☽ ⊥ ♄
06 07 ☉ ⊥ ♆
06 53 ☽ △ ☿
07 18 ♀ ⚹ ♇
09 01 ☽ △ ♇
09 12 ☽ ⚹ ♀
12 02 ☽ # ♃
16 46 ☽ △ ♂
17 59 ☽ Q ♅
19 20 ☽ ∠ ♄
20 32 ☽ Q ♆
22 55 ☽ # ♇

04 Friday
07 51 ☉ ꝑ ♂
11 41 ☽ Q ♀
13 49 ☽ ꝑ ☿
14 56 ☽ ꝑ ♇
18 09 ☽ ☍ ♃
19 33 ♀ # ♂
23 40 ☽ △ ☉

05 Saturday
01 10 ☽ ⚹ ♄
05 39 ☽ □ ♅
08 05 ☽ □ ♆
08 16 ☽ ± ☿
08 36 ☽ ± ♇
15 18 ☿ ☌ ♇
19 04 ☉ ⚹ ♄
20 14 ☽ ⊼ ♇
20 29 ☽ ⊼ ☿

06 Sunday
00 31 ☽ Q ♄
02 21 ☽ □ ♀
03 04 ☽ □ ♂
07 11 ☽ ꝑ ☉
09 30 ♀ ☍ ♂
15 04 ☽ # ☿
23 12 ☽ # ♄
23 26 ♀ ∥ ♅

07 Monday
01 36 ☽ ± ☉
04 15 ☽ ⊼ ♃
10 35 ☽ □ ♄
12 33 ☽ Q ♂
13 40 ☽ ⊼ ☉
14 42 ☽ △ ♅
15 28 ☽ ± ♃
16 18 ☉ ∥ ♀
16 55 ☽ △ ♆

08 Tuesday
03 47 ☉ ⊻ ♅
04 24 ☽ ☍ ♇
07 53 ☿ ∠ ♃
08 02 ☽ ꝑ ♃
08 03 ☽ ☍ ☿
10 10 ☽ # ♆
10 13 ☽ ⚹ ♂
15 24 ☽ △ ♀
18 01 ☽ ꝑ ♅
20 07 ☽ ꝑ ♆
20 35 ☽ # ♀

09 Wednesday
05 21 ☽ # ☉
05 24 ☽ # ♅
05 54 ☉ ∥ ♅
08 44 ☉ ⊻ ♆
10 03 ☽ ± ♅
10 49 ☿ △ ♂
11 03 ☽ △ ♃
12 06 ☽ ± ♆
12 38 ☽ ∠ ♂
16 48 ☽ △ ♄
20 25 ☽ ꝑ ♀
20 36 ☽ ⊼ ♅
22 37 ☽ ⊼ ♆
23 41 ☽ ☍ ☉

10 Thursday
04 10 ☽ ⊥ ♂
09 33 ☽ ⊼ ♇
10 11 ☿ ∥ ♄
13 14 ☽ # ☉
13 27 ☽ ± ♀
14 24 ☽ ⊻ ♂
16 25 ☽ # ♅
16 57 ☽ ⊼ ☿
18 56 ☽ ꝑ ♄
19 53 ☽ ± ♇

11 Friday
00 40 ☽ ⊼ ♀
04 04 ☽ ± ☿
09 54 ☽ # ♀
10 22 ☽ ± ♄
10 43 ☽ # ♆
11 16 ☽ ꝑ ♇
15 14 ☽ □ ♃
16 02 ♀ ∥ ♆
20 35 ☽ ⊼ ♄
20 39 ☽ ꝑ ☿

12 Saturday
00 09 ☽ ☍ ♅
02 01 ☽ ☍ ♆
04 32 ☉ ± ♂
06 52 ☽ ⊼ ☉
12 38 ☽ △ ♇
14 20 ☽ # ☿
16 37 ☽ ☌ ♂
17 42 ☽ ± ☉
19 34 ☽ # ♄
21 48 ☽ Q ♃

13 Sunday
00 06 ☽ △ ☿
07 44 ☽ ☍ ♀
09 55 ☽ ꝑ ☉
16 25 ☿ ∠ ♅
18 03 ☽ ⚹ ♃
20 57 ♀ Q ♇
23 11 ☽ ☍ ♄

14 Monday
02 35 ☽ ⊼ ♅
04 22 ☽ ⊼ ♆
12 20 ☿ Q ♄
12 39 ☽ ± ♅
12 53 ☽ △ ☉
13 43 ☿ ∠ ♆
14 25 ☽ ± ♆
14 56 ☽ □ ♇
18 03 ☽ ⊻ ♂
19 20 ☽ ∠ ♃

15 Tuesday
03 46 ☽ ꝑ ♅
03 57 ☽ ⊥ ♂
05 31 ☽ ꝑ ♆
07 01 ☽ □ ☿
08 47 ☽ # ♇
10 33 ☽ ⊥ ♃
11 10 ☉ Q ♃
14 21 ☽ ⊼ ♀
15 35 ☽ # ♃
18 47 ☉ ⊻ ♇
18 49 ☽ ∠ ♂
20 43 ☽ ⊻ ♃
21 11 ☽ Q ♇

16 Wednesday
01 20 ☽ ± ♀
01 46 ☽ ⊼ ♄
05 06 ☽ △ ♅
06 49 ☽ △ ♆
11 58 ☽ ± ♄
16 25 ☽ Q ☿
17 31 ☽ ⚹ ♇
17 54 ☽ ꝑ ♀
19 13 ☽ □ ☉
19 44 ☽ ⚹ ♂
22 09 ☽ ∥ ♃

17 Thursday
01 10 ☉ ⊼ ♂
02 45 ☿ ∥ ♀
03 21 ☽ ꝑ ♄
04 50 ☽ ∥ ♇
15 01 ☽ ⚹ ☿
15 51 ☽ Q ♂
19 10 ☽ ∠ ♇
21 49 ☽ △ ♀

18 Friday
00 11 ☽ ☌ ♃
04 25 ☽ Q ☉
05 14 ☽ △ ♄
08 30 ☽ □ ♅
10 12 ☽ □ ♆
10 43 ☽ ⊥ ♇
19 42 ☽ ∠ ☿
21 09 ☽ ⊻ ♇
22 21 ☽ □ ♂

19 Saturday
02 53 ☽ ⚹ ☉
05 32 ♀ △ ♃
13 11 ☽ ⊥ ☿
16 01 ☽ Q ♅
17 41 ☽ Q ♆
18 07 ☿ ꝑ ♂
19 04 ☽ ∥ ♄

20 Sunday
00 58 ☽ ⊻ ☿
05 01 ☽ ⊻ ♃
05 39 ☽ ∥ ♀
06 59 ☽ □ ♀
07 26 ☽ ∠ ☉
10 08 ☽ □ ♄
13 21 ☽ ⚹ ♅
15 01 ☽ ⚹ ♆
15 47 ☽ ⊥ ♃
17 49 ☿ ⊥ ♅

21 Monday
00 49 ☽ ⊥ ☉
01 24 ♂ △ ♇
02 19 ☽ △ ♂
02 20 ☽ ☌ ♇
03 09 ☽ ∥ ☿
06 33 ☽ ∥ ♆
08 05 ☽ ∠ ♃
09 58 ☿ ⊥ ♆
12 32 ☽ ⊻ ☉
16 27 ☽ ∠ ♅
18 06 ☽ ∠ ♆
18 21 ☽ Q ♀
18 44 ☽ Q ♄
18 50 ☿ ⚹ ♃
21 16 ☉ ⊥ ♇
22 36 ☉ ∠ ♄
23 09 ♀ ☌ ♄

22 Tuesday
00 41 ☽ ∥ ♅
02 02 ☿ ∥ ♆
03 12 ☉ ∠ ♀
04 54 ☽ ꝑ ♂
08 57 ☽ ⊥ ♅
10 36 ☽ ⊥ ♆
11 38 ☽ ⚹ ♃
13 32 ☽ ☌ ☿
16 52 ☽ ⚹ ♄
18 26 ☽ ⚹ ♀
20 03 ☽ ⊻ ♅
21 05 ☽ ± ♂
21 42 ☽ ⊻ ♆

23 Wednesday
07 58 ☽ ⊼ ♂
09 26 ☽ ⊻ ♇
10 04 ☽ Q ♃
13 45 ♀ ⊻ ♅
18 20 ☽ ∥ ♅
20 43 ☽ ⊥ ♇
21 03 ☽ ∠ ♄
21 36 ☿ ⚹ ♄

24 Thursday
00 43 ☽ ☌ ☉
01 12 ☽ ∠ ♀
04 32 ☽ ∥ ☿
08 37 ♀ ⊻ ♆
13 08 ☽ ∥ ♆
13 49 ☽ ∠ ♇
14 15 ☽ ⊥ ♄
15 52 ♀ ∥ ♄
20 01 ☽ ⊥ ♀
20 25 ☽ □ ♃
22 10 ☿ ± ♂

25 Friday
01 50 ☽ ⊻ ♄
02 16 ☿ ⊻ ♅
04 58 ☽ ☌ ♅
05 19 ☽ ⊻ ☿
06 35 ☽ ☌ ♆
08 43 ☽ ⊻ ♀
15 36 ☽ ☍ ♂
16 18 ☿ ⊻ ♆
18 30 ☽ ⊥ ☿
18 48 ☽ ⚹ ♇

26 Saturday
04 43 ☽ ∥ ♄
06 07 ☿ ∥ ♅
09 12 ☽ ∥ ♀
14 31 ☽ ∠ ☿
15 50 ☽ ⊻ ☉
18 23 ☽ Q ♇

27 Sunday
04 48 ☽ ⊥ ☉
07 31 ☽ △ ♃
13 09 ☽ ☌ ♄
16 12 ☽ ⊻ ♅
17 45 ☽ ⊻ ♆
19 41 ♀ ⊼ ♂

28 Monday
00 25 ☽ ∠ ☉
00 32 ☽ ⚹ ☿
01 12 ☽ ⊼ ♂
01 57 ☽ ☌ ♀
04 16 ☽ ⊥ ♅
04 58 ☿ ⊼ ♂
05 49 ☽ ⊥ ♆
06 22 ☽ □ ♇
12 54 ☽ ± ♂
13 38 ☉ ∥ ☿
13 48 ☽ ꝑ ♃
22 29 ☽ ∠ ♅

29 Tuesday
00 01 ☽ ∠ ♆
02 45 ☽ ∥ ♇
03 49 ♀ ⊥ ♅
04 09 ☽ Q ☿
06 31 ☽ ꝑ ♂
08 07 ☽ ± ♃
08 28 ☽ ∥ ♃
09 29 ☽ ⚹ ☉
18 34 ☿ ⚹ ♀
20 20 ♀ ⊥ ♆
20 21 ☽ ⊼ ♃

30 Wednesday
00 12 ☿ ⊻ ♇
02 08 ☽ ⊻ ♄
02 10 ♀ □ ♇
04 59 ☽ ⚹ ♅
06 28 ☽ ⚹ ♆
11 56 ☽ △ ♂
12 04 ☽ Q ☉
13 14 ☿ Q ♃
14 23 ☽ ⊥ ♄
19 09 ☽ △ ♇
20 52 ☽ ⊻ ♀
21 51 ☽ □ ☿

31 Thursday
00 12 ☉ ∠ ♇
02 36 ☽ # ♃
05 21 ☽ Q ♅
06 46 ☽ Q ♆
08 03 ☽ # ♇
08 42 ☽ ∠ ♄
10 13 ☽ ⊥ ♀
17 49 ☉ ⊥ ♄
19 49 ♀ ± ♂

All ephemeris data is given at 12.00 UT and the Moon's longitude is additionally given for 24.00 UT

JANUARY 1993

LONGITUDES

Date	Sidereal time h m s	Sun ☉ ° ' "	Moon ☽ ° ' "	Moon ☽ 24.00 ° ' "	Mercury ☿ ° '	Venus ♀ ° '	Mars ♂ ° '	Jupiter ♃ ° '	Saturn ♄ ° '	Uranus ♅ ° '	Neptune ♆ ° '	Pluto ♇ ° '
01	18 44 35	11 ♑ 05 45	14 ♈ 56 47	21 ♈ 01 31	28 ♐ 18	27 ♒ 14	20 ♋ 13	13 ♎ 32	16 ♒ 24	17 ♑ 42	18 ♑ 22	24 ♏ 37
02	18 48 32	12 06 54	27 ♈ 09 51	03 ♉ 22 20	29 ♐ 49	28 21	19 R 50	13 36	16 31	17 45	18 25	24 39
03	18 52 28	13 08 03	09 ♉ 39 30	16 ♉ 01 52	01 ♑ 21	29 ♒ 27	19 26	13 41	16 37	17 49	18 27	24 41
04	18 56 25	14 09 12	22 ♉ 29 50	29 ♉ 03 46	02 53	00 ♓ 33	19 02	13 46	16 43	17 53	18 29	24 43
05	19 00 21	15 10 20	05 ♊ 43 54	12 ♊ 30 23	04 25	01 39	18 38	13 50	16 50	17 56	18 32	24 44
06	19 04 18	16 11 28	19 ♊ 23 12	26 ♊ 22 14	05 58	02 45	18 15	13 54	16 56	18 00	18 34	24 46
07	19 08 14	17 12 36	03 ♋ 27 09	10 ♋ 37 30	07 31	03 50	17 51	13 58	17 03	18 03	18 36	24 48
08	19 12 11	18 13 44	17 ♋ 52 41	25 ♋ 11 55	09 05	04 55	17 27	14 02	17 10	18 07	18 38	24 49
09	19 16 07	19 14 51	02 ♌ 34 20	09 ♌ 58 57	10 39	06 00	17 03	14 06	17 16	18 10	18 41	24 51
10	19 20 04	20 15 58	17 ♌ 24 44	24 ♌ 50 40	12 13	07 04	16 39	14 10	17 23	18 14	18 43	24 53
11	19 24 01	21 17 05	02 ♍ 15 43	09 ♍ 38 56	13 48	08 09	16 16	14 13	17 30	18 18	18 45	24 54
12	19 27 57	22 18 12	16 ♍ 59 29	24 ♍ 16 38	15 23	09 12	15 52	14 16	17 36	18 21	18 47	24 56
13	19 31 54	23 19 19	01 ♎ 29 48	08 ♎ 38 34	16 59	10 16	15 30	14 19	17 43	18 25	18 50	24 57
14	19 35 50	24 20 26	15 ♎ 42 38	22 ♎ 41 50	18 35	11 19	15 07	14 22	17 50	18 28	18 52	24 59
15	19 39 47	25 21 32	29 ♎ 36 10	06 ♏ 25 39	20 11	12 22	14 45	14 24	17 57	18 32	18 54	25 00
16	19 43 43	26 22 39	13 ♏ 10 28	19 ♏ 50 48	21 49	13 24	14 23	14 27	18 04	18 35	18 57	25 02
17	19 47 40	27 23 45	26 ♏ 26 53	02 ♐ 59 00	23 26	14 27	14 01	14 29	18 11	18 39	18 59	25 03
18	19 51 36	28 24 51	09 ♐ 27 23	15 ♐ 52 21	25 04	15 28	13 40	14 31	18 18	18 42	19 01	25 04
19	19 55 33	29 ♑ 25 57	22 ♐ 14 07	28 ♐ 32 55	26 43	16 30	13 20	14 33	18 25	18 46	19 03	25 06
20	19 59 30	00 ♒ 27 02	04 ♑ 48 59	11 ♑ 02 30	28 ♑ 23	17 31	13 00	14 35	18 32	18 50	19 06	25 07
21	20 03 26	01 28 06	17 ♑ 13 38	23 ♑ 22 31	00 ♒ 02	18 31	12 41	14 36	18 39	18 53	19 08	25 08
22	20 07 23	02 29 10	29 ♑ 29 19	05 ♒ 34 08	01 43	19 31	12 22	14 38	18 46	18 57	19 10	25 10
23	20 11 19	03 30 14	11 ♒ 37 08	17 ♒ 38 25	03 24	20 31	12 04	14 39	18 53	19 00	19 12	25 11
24	20 15 16	04 31 16	23 ♒ 38 11	29 ♒ 36 35	05 06	21 30	11 47	14 40	19 00	19 04	19 14	25 12
25	20 19 12	05 32 18	05 ♓ 33 49	11 ♓ 30 10	06 48	22 29	11 30	14 41	19 07	19 07	19 17	25 13
26	20 23 09	06 33 18	17 ♓ 25 52	23 ♓ 21 16	08 31	23 27	11 14	14 41	19 14	19 10	19 19	25 14
27	20 27 05	07 34 18	29 ♓ 16 44	05 ♈ 12 41	10 14	24 25	10 59	14 42	19 21	19 14	19 21	25 15
28	20 31 02	08 35 16	11 ♈ 09 34	17 ♈ 07 52	11 58	25 22	10 45	14 42	19 28	19 17	19 23	25 16
29	20 34 59	09 36 13	23 ♈ 08 09	29 ♈ 10 59	13 43	26 18	10 31	14 R 42	19 35	19 21	19 25	25 17
30	20 38 55	10 37 10	05 ♉ 16 56	11 ♉ 26 39	15 28	27 14	10 18	14 42	19 42	19 24	19 28	25 18
31	20 42 52	11 ♒ 38 04	17 ♉ 40 45	23 ♉ 59 49	17 ♒ 14	28 ♓ 10	10 ♋ 06	14 ♎ 41	19 ♒ 50	19 ♑ 27	19 ♑ 30	25 ♏ 19

Date	Moon True ☊ ° '	Moon Mean ☊ ° '	Moon ☽ Latitude ° '
01	21 ♐ 19	20 ♐ 24	04 N 49
02	21 D 20	20 20	04 16
03	21 21	20 17	03 30
04	21 23	20 14	02 33
05	21 24	20 11	01 25
06	21 25	20 08	00 N 11
07	21 R 25	20 05	01 S 06
08	21 23	20 01	02 20
09	21 20	19 58	03 27
10	21 16	19 55	04 20
11	21 11	19 52	04 56
12	21 07	19 49	05 12
13	21 04	19 46	05 07
14	21 02	19 42	04 44
15	21 D 02	19 39	04 05
16	21 03	19 36	03 12
17	21 04	19 33	02 10
18	21 06	19 30	01 S 03
19	21 R 06	19 26	00 N 06
20	21 05	19 23	01 14
21	21 02	19 20	02 18
22	20 57	19 17	03 13
23	20 50	19 14	03 59
24	20 41	19 11	04 34
25	20 32	19 07	04 57
26	20 23	19 04	05 07
27	20 16	19 01	05 04
28	20 10	18 58	04 47
29	20 06	18 55	04 18
30	20 04	18 52	03 37
31	20 ♐ 04	18 ♐ 48	02 N 45

DECLINATIONS

Date	Sun ☉ ° '	Moon ☽ ° '	Mercury ☿ ° '	Venus ♀ ° '	Mars ♂ ° '	Jupiter ♃ ° '	Saturn ♄ ° '	Uranus ♅ ° '	Neptune ♆ ° '	Pluto ♇ ° '
01	22 S 59	10 N 20	23 S 55	13 S 51	25 N 46	04 S 08	16 S 53	22 S 41	21 S 31	05 S 08
02	22 53	14 26	24 02	13 25	25 51	04 10	16 51	22 40	21 30	05 09
03	22 47	18 01	24 09	12 58	25 56	04 11	16 49	22 40	21 30	05 09
04	22 41	20 51	24 13	12 31	26 01	04 13	16 48	22 39	21 30	05 09
05	22 35	22 40	24 17	12 04	26 05	04 14	16 46	22 39	21 30	05 09
06	22 27	23 12	24 19	11 37	26 10	04 16	16 44	22 38	21 29	05 09
07	22 20	22 18	24 20	11 09	26 14	04 17	16 42	22 38	21 29	05 09
08	22 12	19 56	24 20	10 41	26 18	04 18	16 40	22 37	21 29	05 09
09	22 04	16 14	24 18	10 13	26 22	04 20	16 38	22 37	21 29	05 09
10	21 55	11 29	24 15	09 45	26 26	04 21	16 36	22 36	21 28	05 09
11	21 45	06 04	24 11	09 17	26 30	04 22	16 34	22 36	21 28	05 09
12	21 36	00 N 21	24 05	08 48	26 33	04 23	16 32	22 35	21 27	05 09
13	21 26	05 S 18	23 57	08 19	26 37	04 23	16 29	22 35	21 27	05 09
14	21 15	10 33	23 49	07 51	26 40	04 24	16 27	22 35	21 27	05 09
15	21 04	15 09	23 39	07 22	26 42	04 25	16 25	22 34	21 27	05 09
16	20 53	18 51	23 27	06 53	26 45	04 26	16 23	22 34	21 26	05 09
17	20 41	21 28	23 14	06 24	26 47	04 27	16 21	22 33	21 26	05 09
18	20 29	22 54	22 59	05 54	26 50	04 27	16 19	22 33	21 26	05 09
19	20 16	23 07	22 43	05 25	26 52	04 28	16 17	22 32	21 25	05 09
20	20 03	22 07	22 26	04 56	26 53	04 28	16 15	22 32	21 25	05 09
21	19 50	20 04	22 07	04 27	26 55	04 28	16 13	22 31	21 25	05 09
22	19 36	17 07	21 46	03 57	26 56	04 29	16 10	22 31	21 24	05 09
23	19 22	13 28	21 24	03 28	26 58	04 29	16 08	22 30	21 24	05 09
24	19 08	09 19	21 01	02 59	26 58	04 29	16 06	22 30	21 24	05 09
25	18 53	04 52	20 35	02 30	26 59	04 29	16 04	22 29	21 23	05 08
26	18 38	00 S 15	20 09	02 00	27 00	04 29	16 02	22 29	21 23	05 08
27	18 23	04 N 21	19 41	01 31	27 01	04 29	16 00	22 28	21 23	05 08
28	18 07	08 49	19 11	01 02	27 01	04 29	15 57	22 28	21 23	05 08
29	17 51	12 59	18 40	00 33	27 01	04 28	15 55	22 27	21 22	05 08
30	17 34	16 41	18 07	00 04	27 01	04 28	15 53	22 27	21 22	05 08
31	17 S 18	19 N 45	17 S 33	00 N 25	27 N 01	04 S 28	15 S 51	22 S 26	21 S 22	05 S 07

ZODIAC SIGN ENTRIES

Date	h	m	Planets
02	14	47	☽ ♉
02	17	30	☿ ♑
03	23	54	♀ ♓
05	01	42	☽ ♊
07	06	10	☽ ♋
09	07	49	☽ ♌
11	08	20	☽ ♍
13	09	30	☽ ♎
15	12	42	☽ ♏
17	18	30	☽ ♐
20	01	23	☉ ♒
20	02	46	☽ ♑
21	11	25	☿ ♒
22	13	00	☽ ♒
25	00	47	☽ ♓
27	13	28	☽ ♈
30	01	37	☽ ♉

LATITUDES

Date	Mercury ☿ ° '	Venus ♀ ° '	Mars ♂ ° '	Jupiter ♃ ° '	Saturn ♄ ° '	Uranus ♅ ° '	Neptune ♆ ° '	Pluto ♇ ° '
01	00 S 29	01 S 31	03 N 53	01 N 18	01 S 01	00 S 25	00 N 40	14 N 11
04	00 49	01 20	03 58	01 19	01 01	00 25	00 40	14 12
07	01 07	01 08	04 01	01 20	01 01	00 25	00 40	14 13
10	01 23	00 54	04 04	01 21	01 01	00 25	00 40	14 14
13	01 37	00 39	04 06	01 21	01 01	00 25	00 40	14 15
16	01 48	00 23	04 07	01 22	01 01	00 25	00 40	14 16
19	01 57	00 S 06	04 07	01 23	01 01	00 25	00 40	14 17
22	02 03	00 N 12	04 06	01 24	01 01	00 25	00 40	14 18
25	02 05	00 32	04 04	01 25	01 01	00 25	00 40	14 19
28	02 04	00 53	04 02	01 25	01 02	00 25	00 40	14 21
31	01 S 58	01 N 15	03 N 59	01 N 26	01 S 02	00 S 25	00 N 40	14 N 22

DATA

Julian Date	2448989
Delta T	+59 seconds
Ayanamsa	23° 45' 51"
Synetic vernal point	05° ♓ 21' 08"
True obliquity of ecliptic	23° 26' 23"

MOON'S PHASES, APSIDES AND POSITIONS ☽

Date	h	m	Phase	Longitude ° '	Eclipse Indicator
01	03	38	☽	10 ♈ 44	
08	12	37	○	18 ♋ 15	
15	04	01	☾	25 ♎ 11	
22	18	27	●	02 ♒ 46	
30	23	20	☽	11 ♉ 06	

Day	h	m		
10	12	19	Perigee	
26	10	25	Apogee	
06	09	18	Max dec	23° N 13'
12	13	28	0S	
19	03	56	Max dec	23° S 11'
26	13	19	0N	

LONGITUDES

Date	Chiron ⚷ ° '	Ceres ⚳ ° '	Pallas ⚴ ° '	Juno ⚵ ° '	Vesta ⚶ ° '	Black Moon Lilith ⚸ ° '
01	22 ♌ 59	20 ♒ 39	25 ♑ 42	05 ♋ 31	27 ♐ 43	08 ♓ 34
11	22 ♌ 28	24 ♒ 19	29 ♑ 18	03 ♋ 23	03 ♑ 05	09 ♓ 41
21	21 ♌ 50	28 ♒ 04	02 ♒ 54	01 ♋ 51	08 ♑ 15	10 ♓ 48
31	21 ♌ 07	01 ♓ 54	06 ♒ 29	01 ♋ 07	13 ♑ 41	11 ♓ 56

ASPECTARIAN

01 Friday
01 23 ☽ ⚼ ♇ · 03 38 ☽ □ ☉ · 06 04 ☽ ∠ ♀ · 09 10 ☽ ☍ ♃ · 14 55 ☽ ⚹ ♄ · 17 28 ☽ □ ♅ · 18 48 ☽ □ ♆ · 19 17 ☽ ± ♇ · 22 06 ☽ □ ♂

02 Saturday
06 24 ☽ # ♀ · 07 05 ☽ ⚻ ♇ · 14 31 ☽ ⚹ ♀ · 14 38 ☽ Q ♄ · 17 53 ☽ △ ☿ · 18 05 ♀ ⚼ ♃

03 Sunday
01 18 ☿ ⊥ ♇ · 03 33 ☽ # ♄ · 07 54 ☽ Q ♂ · 15 43 ☽ Q ♀ · 16 30 ☿ ∠ ♄ · 19 08 ☽ △ ☉ · 19 39 ☽ ⚻ ♃

04 Monday
01 12 ☽ □ ♄ · 02 04 ☉ □ ♃ · 02 18 ☽ ⚼ ☿ · 03 24 ☽ △ ♅ · 04 34 ☽ △ ♆ · 05 48 ☽ ⚹ ♂ · 06 55 ☽ ± ♃ · 16 04 ☽ ☍ ♇ · 19 00 ☽ # ♆ · 20 02 ☉ ∥ ♅ · 21 06 ☽ ± ☿ · 23 31 ☽ ⚼ ♃

05 Tuesday
01 11 ☽ ⚼ ☉ · 04 02 ☽ □ ♀ · 06 58 ☽ ⚼ ♅ · 08 02 ☽ ⚼ ♆ · 08 22 ☽ ∠ ♂ · 09 21 ☽ ⚻ ☿ · 10 31 ☽ # ☉ · 11 43 ☽ # ♅ · 18 20 ♂ ☍ ♆ · 18 37 ☽ ± ☉ · 23 03 ☽ ± ♅ · 23 53 ☽ ⊥ ♂

06 Wednesday
00 04 ☽ ± ♆ · 02 24 ☽ △ ♃ · 06 00 ☽ ⚻ ☉ · 07 43 ☽ △ ♄ · 09 35 ☽ ⚻ ♅ · 10 04 ☽ ⚺ ♂ · 10 34 ☽ ⚻ ♆ · 17 48 ♀ ∠ ♅ · 19 59 ♀ ⚼ ♂ · 21 17 ☽ ⚻ ♇

07 Thursday
00 51 ♂ ☍ ♅ · 06 37 ☽ # ♅ · 06 39 ♀ ∠ ♆ · 07 30 ☽ ± ♇ · 07 47 ☉ ⚺ ♄ · 09 37 ☽ ⚼ ♄ · 11 30 ☽ # ☉ · 12 42 ☽ △ ♀ · 19 39 ☽ ☍ ☿ · 22 04 ☽ # ♆ · 22 38 ☽ ⚼ ♇ · 22 42 ☉ ☍ ♂

08 Friday
00 48 ☽ ± ♄ · 05 38 ☽ □ ♃ · 09 09 ☉ ☌ ♅ · 10 49 ☽ ⚻ ♄ · 11 18 ☽ ☌ ♂ · 12 23 ☽ ☍ ♅ · 12 37 ☽ ☍ ☉ · 13 15 ☽ ☍ ♆ · 15 38 ☽ ⚼ ♀ · 22 03 ☉ ☌ ♆ · 23 24 ☽ △ ♇ · 23 40 ☿ ∠ ♇

09 Saturday
01 24 ♂ ⚻ ♄ · 07 30 ☽ ± ♀ · 09 44 ☽ # ♄ · 11 14 ☽ Q ♃ · 17 59 ☽ ⚻ ♀ · 22 21 ☿ ⊥ ♄

10 Sunday
02 37 ☽ ⚻ ☿ · 06 44 ☽ ⚹ ♃ · 10 48 ☽ ⚺ ♂ · 11 57 ☽ ☍ ♄ · 13 20 ☽ ⚻ ♅ · 13 27 ☽ ± ☿ · 14 07 ☽ ⚻ ♆ · 16 57 ☽ ⚻ ☉ · 20 15 ☽ ⊥ ♂ · 20 42 ☽ # ♀ · 23 03 ☽ ± ♅ · 23 49 ☽ ± ♆

11 Monday
00 04 ☽ □ ♇ · 03 21 ☽ ± ☉ · 05 43 ☽ ⚼ ☿ · 07 03 ☽ ∠ ♃ · 10 25 ☽ ∠ ♂ · 13 41 ☽ ⚼ ♅ · 13 42 ♀ ± ♃ · 14 25 ☽ ⚼ ♆ · 15 52 ☽ # ♇ · 18 36 ☿ □ ♃ · 19 01 ☽ ⚼ ☉ · 19 11 ☽ # ♃ · 21 42 ☽ ⊥ ♃ · 22 18 ☽ ☍ ♀

12 Tuesday
05 20 ☽ Q ♇ · 07 31 ☽ ⚺ ♃ · 09 02 ☽ △ ☿ · 10 13 ☽ ⚹ ♂ · 13 01 ☽ ⚻ ♄ · 14 15 ☽ △ ♅ · 14 58 ☽ △ ♆ · 18 00 ☿ ☍ ♂ · 21 24 ☽ △ ☉ · 22 59 ☽ ± ♄

13 Wednesday
01 06 ☽ ⚹ ♇ · 05 30 ☽ Q ♂ · 08 05 ☉ ∥ ♆ · 08 06 ☽ ∥ ♃ · 11 24 ☽ ∥ ♇ · 14 04 ☽ ⚼ ♄

14 Thursday
00 01 ☿ ⚺ ♄ · 00 24 ☽ ∥ ♀ · 02 14 ☽ ∠ ♇ · 03 56 ☽ ⚻ ♀ · 09 42 ☽ ☌ ♃ · 10 21 ☿ ☌ ♅ · 11 01 ☽ □ ♂ · 14 58 ☽ ± ♀ · 15 40 ☽ △ ♄ · 16 25 ☿ ☌ ♆ · 16 45 ☽ □ ♅ · 17 25 ☽ □ ♆ · 17 33 ☽ □ ☿ · 17 36 ☽ ⊥ ♇

15 Friday
03 24 ☉ ⚹ ♇ · 03 59 ☽ ⚺ ♇ · 04 01 ☽ □ ☉ · 07 47 ☽ ⚼ ♀ · 19 33 ☽ ∥ ♄

16 Saturday
00 14 ☽ Q ♅ · 00 53 ☽ Q ♆ · 05 11 ☽ Q ☿ · 07 46 ♂ □ ♃ · 12 27 ☽ △ ♀ · 14 06 ☽ △ ♂ · 14 17 ☽ ⚺ ♃ · 14 20 ☽ Q ☉ · 20 51 ☽ □ ♄ · 21 46 ☽ ⚹ ♅ · 22 24 ☽ ⚹ ♆

17 Sunday
01 08 ☽ ⊥ ♃ · 04 13 ☽ ∥ ☉ · 04 42 ♀ △ ♂ · 05 44 ☽ ⚹ ☿ · 09 27 ☽ ☌ ♇ · 11 32 ☽ ∥ ♆ · 13 02 ♀ ⚻ ♃ · 13 53 ☽ ⚹ ☉ · 16 35 ☽ ⚼ ♂ · 17 35 ☽ ∠ ♃

18 Monday
01 17 ☽ ∠ ♅ · 01 53 ☽ ∠ ♆ · 03 37 ☽ ∥ ♅ · 06 04 ☽ Q ♄ · 08 46 ☽ ± ♂ · 11 59 ☿ ⚹ ♇ · 13 19 ☽ ∠ ☿ · 13 56 ☽ ∥ ☿ · 18 06 ☽ ⊥ ♅ · 18 40 ☽ ⊥ ♆ · 19 40 ☽ ⚻ ♂ · 20 02 ☽ ∠ ☉ · 21 29 ☽ ⚹ ♃

19 Tuesday
00 14 ☽ □ ♀ · 04 42 ☽ ⚹ ♄ · 05 25 ☽ ⚺ ♅ · 05 58 ☽ ⚺ ♆ · 08 42 ☽ ⊥ ☿ · 14 28 ☽ ⊥ ☉ · 17 26 ☽ ⚺ ♇ · 20 13 ☽ Q ♃ · 21 48 ☽ ⚺ ☿

20 Wednesday
01 21 ♀ ∥ ♇ · 02 54 ☽ ⚺ ☉ · 04 04 ☿ ∥ ♅ · 04 54 ☽ ⊥ ♇ · 04 57 ☽ ∥ ♅ · 05 13 ☽ ∥ ☿ · 09 30 ☽ ∠ ♄ · 13 27 ☽ Q ♀ · 21 39 ☽ ∥ ♆ · 22 14 ☽ ∠ ♇

21 Thursday
03 01 ☽ ⊥ ♄ · 03 23 ☽ ☍ ♂ · 06 54 ☽ □ ♃ · 10 37 ♀ ∥ ♃ · 14 04 ♂ ± ♄ · 14 22 ☽ ∥ ☉ · 14 45 ☽ ⚹ ♀ · 14 47 ☽ ⚺ ♄ · 15 15 ☽ ☌ ♅ · 15 17 ♀ ⚺ ♄ · 15 43 ☽ ☌ ♆ · 21 13 ♀ ⚹ ♅

22 Friday
03 08 ♀ ⚹ ♆ · 03 29 ☽ ⚹ ♇ · 17 05 ☽ ☌ ☿ · 18 27 ☽ ☌ ☉ · 18 39 ☽ ∥ ♄ · 22 49 ☽ ∠ ♀

23 Saturday
03 10 ☽ Q ♇ · 12 06 ☿ ∥ ♆ · 12 52 ☽ ⚻ ♂ · 15 42 ☉ ☌ ☿ · 18 02 ☽ △ ♃ · 18 17 ☽ ⊥ ♀

24 Sunday
00 33 ☽ ± ♂ · 02 37 ☽ ☌ ♄ · 02 47 ☽ ⚺ ♅ · 03 10 ☽ ⚺ ♆ · 07 21 ☽ ⚺ ♀ · 14 52 ☽ ⊥ ♅ · 15 08 ☽ □ ♇ · 15 14 ☽ ⊥ ♆ · 18 10 ☽ ⚼ ♂

25 Monday
00 07 ☽ ⚼ ♃ · 09 04 ☽ ∠ ♅ · 09 24 ☽ ∠ ♆ · 10 32 ☽ ∥ ♇ · 11 57 ☽ ⚺ ☉ · 13 18 ♄ ⚺ ♅ · 13 59 ☽ ∥ ♃ · 14 55 ☽ ⚺ ☿ · 17 55 ☿ Q ♇ · 18 18 ☽ ± ♃ · 23 44 ☽ △ ♂

26 Tuesday
01 12 ☽ ⊥ ☉ · 01 51 ☽ ∥ ♀ · 05 06 ☽ ⊥ ☿ · 06 26 ☽ ⚻ ♃ · 10 01 ☿ ∠ ♀ · 15 33 ☽ ⚹ ♅ · 15 41 ☽ ⚺ ♄ · 15 50 ☽ ⚹ ♆ · 21 08 ☽ ∠ ☉ · 22 36 ☽ # ♀

27 Wednesday
01 17 ☽ ☌ ♀ · 02 25 ☽ ∠ ☿ · 03 50 ☽ △ ♇ · 03 58 ☽ ⊥ ♄ · 04 21 ☉ Q ♇ · 12 16 ♄ ⚺ ♆ · 12 40 ☽ # ♃ · 15 58 ☽ Q ♅ · 16 07 ☽ # ♇ · 16 12 ☽ Q ♆ · 21 04 ☿ ⚻ ♂ · 22 22 ☽ ∠ ♄

28 Thursday
06 20 ☽ ⚹ ☉ · 09 34 ♀ △ ♇ · 10 12 ☽ ⚼ ♇ · 11 11 ☽ □ ♂ · 13 55 ☽ ⚹ ☿ · 15 01 ♀ ⊥ ♄ · 19 07 ☽ ☍ ♃ · 23 09 ♃ St R

29 Friday
04 18 ☽ ± ♇ · 04 24 ☽ □ ♅ · 04 34 ☽ □ ♆ · 04 51 ☽ ⚹ ♄ · 08 40 ☽ Q ☉ · 16 17 ☽ ⚻ ♇ · 18 00 ☽ Q ☿ · 18 50 ☽ ⚺ ♀ · 22 30 ☽ Q ♂

30 Saturday
01 25 ☿ △ ♃ · 04 55 ☽ Q ♄ · 05 47 ☉ ⚻ ♂ · 06 30 ☽ # ♄ · 07 40 ☽ ⊥ ♀ · 11 47 ♂ ⚼ ♇ · 17 52 ☽ # ☉ · 20 52 ☽ # ☿ · 21 38 ☽ ⚹ ♂ · 22 12 ☿ ± ♂ · 23 20 ☽ □ ☉

31 Sunday
02 38 ☽ ∠ ♀ · 06 16 ☽ ⚻ ♃ · 11 00 ☽ □ ☿ · 15 25 ☽ △ ♅ · 15 29 ☽ △ ♆ · 16 08 ☽ □ ♄ · 17 44 ☽ ± ♃

All ephemeris data is given at 12.00 UT and the Moon's longitude is additionally given for 24.00 UT

FEBRUARY 1993

Date	Sidereal time h m s	Sun ☉ ° ′ ″	Moon ☽ ° ′ ″	Moon ☽ 24.00 ° ′ ″	Mercury ☿ ° ′	Venus ♀ ° ′	Mars ♂ ° ′	Jupiter ♃ ° ′	Saturn ♄ ° ′	Uranus ♅ ° ′	Neptune ♆ ° ′	Pluto ♇ ° ′
01	20 46 48	12 ♒ 38 58	00 ♊ 24 28	06 ♊ 55 13	19 ♒ 00	29 ♓ 05	09 ♋ 55	14 ♎ 41	19 ♒ 57	19 ♑ 31	19 ♑ 32	25 ♏ 20
02	20 50 45	13 39 50	13 ♊ 32 34	20 ♊ 16 53	20 46	29 ♓ 59	09 R 44	14 R 40	20 04	19 34	19 34	25 21
03	20 54 41	14 40 41	27 ♊ 08 26	04 ♋ 07 19	22 33	00 ♈ 52	09 35	14 39	20 11	19 38	19 36	25 22
04	20 58 38	15 41 31	11 ♋ 13 29	18 ♋ 26 38	24 20	01 45	09 26	14 38	20 18	19 41	19 38	25 23
05	21 02 34	16 42 19	25 ♋ 46 18	03 ♌ 11 46	26 07	02 37	09 18	14 36	20 26	19 44	19 40	25 23
06	21 06 31	17 43 06	10 ♌ 42 06	18 ♌ 16 12	27 54	03 28	09 10	14 35	20 33	19 47	19 42	25 24
07	21 10 28	18 43 51	25 ♌ 52 46	03 ♍ 30 27	29 ♒ 41	04 18	09 04	14 33	20 40	19 51	19 44	25 25
08	21 14 24	19 44 36	11 ♍ 07 50	18 ♍ 43 31	01 ♓ 27	05 08	08 58	14 31	20 47	19 54	19 46	25 25
09	21 18 21	20 45 19	26 ♍ 16 15	03 ♎ 44 51	03 13	05 56	08 53	14 29	20 55	19 57	19 48	25 26
10	21 22 17	21 46 01	11 ♎ 08 24	18 ♎ 26 09	04 57	06 44	08 49	14 27	21 02	20 00	19 50	25 27
11	21 26 14	22 46 42	25 ♎ 37 35	02 ♏ 42 24	06 41	07 31	08 46	14 24	21 09	20 03	19 52	25 27
12	21 30 10	23 47 22	09 ♏ 40 30	16 ♏ 31 57	08 23	08 17	08 44	14 22	21 16	20 06	19 54	25 28
13	21 34 07	24 48 00	23 ♏ 16 58	29 ♏ 55 52	10 02	09 02	08 42	14 19	21 24	20 10	19 56	25 28
14	21 38 03	25 48 38	06 ♐ 29 03	12 ♐ 56 58	11 39	09 46	08 41	14 16	21 31	20 13	19 58	25 29
15	21 42 00	26 49 15	19 ♐ 20 08	25 ♐ 39 02	13 13	10 29	08 D 41	14 13	21 38	20 16	20 00	25 29
16	21 45 57	27 49 50	01 ♑ 54 09	08 ♑ 05 58	14 43	11 10	08 41	14 09	21 45	20 19	20 02	25 30
17	21 49 53	28 50 24	14 ♑ 14 54	20 ♑ 21 22	16 09	11 51	08 42	14 06	21 52	20 22	20 04	25 30
18	21 53 50	29 ♒ 50 57	26 ♑ 25 44	02 ♒ 28 18	17 30	12 30	08 44	14 02	22 00	20 25	20 06	25 30
19	21 57 46	00 ♓ 51 29	08 ♒ 29 22	14 ♒ 29 10	18 46	13 08	08 47	13 58	22 07	20 28	20 08	25 30
20	22 01 43	01 51 59	20 ♒ 27 54	26 ♒ 25 46	19 55	13 45	08 50	13 54	22 14	20 31	20 10	25 31
21	22 05 39	02 52 27	02 ♓ 22 55	08 ♓ 19 31	20 57	14 21	08 54	13 50	22 21	20 33	20 11	25 31
22	22 09 36	03 52 53	14 ♓ 15 42	20 ♓ 11 39	21 51	14 55	08 59	13 46	22 28	20 36	20 13	25 31
23	22 13 32	04 53 18	26 ♓ 07 31	02 ♈ 03 31	22 38	15 27	09 05	13 41	22 35	20 39	20 15	25 31
24	22 17 29	05 53 42	07 ♈ 59 52	13 ♈ 56 50	23 15	15 59	09 11	13 36	22 43	20 42	20 17	25 31
25	22 21 26	06 54 03	19 ♈ 54 45	25 ♈ 53 57	23 44	16 28	09 18	13 31	22 50	20 45	20 18	25 31
26	22 25 22	07 54 22	01 ♉ 54 50	07 ♉ 57 51	24 03	16 56	09 25	13 26	22 57	20 47	20 20	25 31
27	22 29 19	08 54 40	14 ♉ 03 31	20 ♉ 12 19	24 12	17 22	09 33	13 21	23 04	20 50	20 22	25 R 31
28	22 33 15	09 ♓ 54 56	26 ♉ 24 51	02 ♊ 41 41	24 ♓ 12	17 ♈ 47	09 ♋ 42	13 ♎ 16	23 ♒ 11	20 ♑ 53	20 ♑ 23	25 ♏ 31

Date	Moon True ☊ ° ′	Moon Mean ☊ ° ′	Moon ☽ Latitude ° ′
01	20 ♐ 05	18 ♐ 45	01 N 44
02	20 D 06	18 42	00 N 35
03	20 R 06	18 39	00 S 38
04	20 04	18 36	01 51
05	20 00	18 32	02 59
06	19 53	18 29	03 56
07	19 44	18 26	04 38
08	19 35	18 23	05 01
09	19 26	18 20	05 02
10	19 18	18 17	04 42
11	19 13	18 13	04 05
12	19 10	18 10	03 14
13	19 09	18 07	02 13
14	19 D 09	18 04	01 S 07
15	19 R 09	18 01	00 N 01
16	19 09	17 58	01 07
17	19 06	17 54	02 09
18	19 00	17 51	03 04
19	18 51	17 48	03 50
20	18 40	17 45	04 25
21	18 27	17 42	04 49
22	18 12	17 38	04 59
23	17 58	17 35	04 57
24	17 45	17 32	04 42
25	17 34	17 29	04 14
26	17 27	17 26	03 35
27	17 22	17 23	02 46
28	17 ♐ 20	17 ♐ 19	01 N 48

DECLINATIONS

Sun ☉ ° ′	Moon ☽ ° ′	Mercury ☿ ° ′	Venus ♀ ° ′	Mars ♂ ° ′	Jupiter ♃ ° ′	Saturn ♄ ° ′	Uranus ♅ ° ′	Neptune ♆ ° ′	Pluto ♇ ° ′
17 S 01	21 N 56	16 S 57	00 N 53	27 N 01	04 S 27	15 S 49	22 S 26	21 S 21	05 S 07
16 43	23 00	16 20	01 22	27 01	04 27	15 46	22 26	21 21	05 07
16 26	22 47	15 41	01 50	27 00	04 26	15 44	22 25	21 21	05 07
16 08	21 08	15 02	02 19	27 00	04 25	15 42	22 25	21 20	05 06
15 50	18 04	14 20	02 47	26 59	04 25	15 40	22 24	21 20	05 06
15 31	13 45	13 38	03 14	26 58	04 24	15 37	22 24	21 20	05 06
15 13	08 31	12 54	03 42	26 57	04 23	15 35	22 23	21 20	05 06
14 54	02 N 45	12 10	04 09	26 56	04 22	15 33	22 23	21 19	05 05
14 35	03 S 08	11 24	04 36	26 55	04 21	15 31	22 22	21 19	05 05
14 15	08 44	10 38	05 03	26 54	04 20	15 28	22 22	21 19	05 05
13 55	13 42	09 51	05 30	26 53	04 19	15 26	22 21	21 18	05 05
13 35	17 47	09 04	05 56	26 51	04 17	15 24	22 21	21 18	05 04
13 15	20 45	08 16	06 22	26 50	04 16	15 22	22 21	21 18	05 04
12 55	22 30	07 29	06 48	26 49	04 15	15 19	22 20	21 17	05 04
12 34	23 00	06 42	07 13	26 47	04 13	15 17	22 20	21 17	05 03
12 14	22 18	05 55	07 38	26 46	04 12	15 15	22 19	21 17	05 03
11 53	20 32	05 09	08 02	26 44	04 10	15 13	22 19	21 17	05 03
11 32	17 51	04 25	08 26	26 42	04 09	15 10	22 18	21 16	05 02
11 10	14 26	03 42	08 50	26 40	04 07	15 08	22 18	21 16	05 02
10 49	10 28	03 01	09 13	26 39	04 05	15 06	22 18	21 16	05 02
10 27	06 08	02 23	09 36	26 37	04 03	15 03	22 17	21 16	05 01
10 05	01 S 35	01 48	09 58	26 35	04 01	15 01	22 17	21 15	05 01
09 43	03 N 00	01 16	10 19	26 33	03 59	14 59	22 16	21 15	05 00
09 21	07 29	00 47	10 40	26 31	03 57	14 57	22 16	21 15	05 00
08 59	11 42	00 22	11 01	26 29	03 55	14 54	22 16	21 14	05 00
08 36	15 30	00 S 02	11 21	26 27	03 53	14 52	22 15	21 14	04 59
08 14	18 42	00 N 14	11 40	26 24	03 51	14 50	22 15	21 14	04 59
07 S 51	21 N 06	00 N 25	11 N 58	26 N 22	03 S 49	14 S 48	22 S 14	21 S 14	04 S 58

ZODIAC SIGN ENTRIES

Date	h	m	Planets
01	11	15	☽ ♊
02	12	37	♀ ♈
03	16	56	☽ ♋
05	18	51	☽ ♌
07	16	19	☿ ♓
07	18	29	☽ ♍
09	17	58	☽ ♎
11	19	23	☽ ♏
14	00	08	☽ ♐
16	08	20	☽ ♑
18	15	35	☉ ♓
18	19	05	☽ ♒
21	07	12	☽ ♓
23	19	50	☽ ♈
26	08	11	☽ ♉
28	18	52	☽ ♊

LATITUDES

Date	Mercury ☿ ° ′	Venus ♀ ° ′	Mars ♂ ° ′	Jupiter ♃ ° ′	Saturn ♄ ° ′	Uranus ♅ ° ′	Neptune ♆ ° ′	Pluto ♇ ° ′
01	01 S 55	01 N 22	03 N 58	01 N 26	01 S 02	00 S 25	00 N 40	14 N 22
04	01 43	01 46	03 54	01 27	01 02	00 25	00 40	14 23
07	01 25	02 10	03 50	01 28	01 02	00 25	00 40	14 25
10	01 01	02 36	03 46	01 29	01 02	00 25	00 40	14 26
13	00 S 30	03 02	03 41	01 29	01 03	00 25	00 40	14 27
16	00 N 07	03 29	03 37	01 30	01 03	00 25	00 40	14 29
19	00 49	03 57	03 32	01 31	01 03	00 25	00 40	14 30
22	01 34	04 26	03 27	01 32	01 03	00 25	00 40	14 31
25	02 18	04 55	03 22	01 32	01 04	00 25	00 40	14 33
28	02 58	05 24	03 18	01 33	01 04	00 26	00 40	14 34
31	03 N 27	05 N 54	03 N 13	01 N 33	01 S 04	00 S 26	00 N 40	14 N 35

DATA

Julian Date	2449020
Delta T	+59 seconds
Ayanamsa	23° 45′ 56″
Synetic vernal point	05° ♓ 21′ 03″
True obliquity of ecliptic	23° 26′ 23″

LONGITUDES

Date	Chiron ⚷ ° ′	Ceres ⚳ ° ′	Pallas ⚴ ° ′	Juno ⚵ ° ′	Vesta ⚶ ° ′	Black Moon Lilith ⚸ ° ′
01	21 ♌ 02	02 ♓ 17	06 ♒ 50	01 ♋ 06	14 ♑ 13	12 ♓ 02
11	20 ♌ 16	06 ♓ 10	10 ♒ 21	01 ♋ 19	19 ♑ 25	13 ♓ 09
21	19 ♌ 31	10 ♓ 05	13 ♒ 49	02 ♋ 19	24 ♑ 32	14 ♓ 17
31	18 ♌ 49	14 ♓ 02	17 ♒ 11	04 ♋ 01	29 ♑ 32	15 ♓ 24

MOON'S PHASES, APSIDES AND POSITIONS ☽

Date	h	m	Phase	Longitude ° ′	Eclipse Indicator
06	23	55	○	18 ♌ 13	
13	14	57	☾	24 ♏ 55	
21	13	05	●	02 ♓ 55	

Day	h	m		
07	20	28	Perigee	
22	18	10	Apogee	
02	20	07	Max dec	23° N 05′
08	23	11	OS	
15	09	52	Max dec	23° S 00′
22	20	17	ON	

ASPECTARIAN

h m	Aspects
01 Monday	
01 52	☽ ∠ ♂
02 30	☽ ☍ ♇
04 32	☽ # ♆
07 22	☉ ∥ ☿
09 20	☽ ✶ ♀
10 38	☽ ⚼ ♃
18 24	☽ ⊥ ♂
19 14	☿ ⊻ ♅
19 23	☿ ⊻ ♆
19 37	☽ ⚼ ♅
19 38	☽ ⚼ ♆
20 09	☽ # ♅
02 Tuesday	
01 48	☿ ☌ ♄
05 13	☽ ⊻ ♂
08 06	♅ ☌ ♆
08 59	☽ Q ♀
12 03	☽ ± ♆
12 03	☽ ± ♅
12 14	☽ △ ☉
14 00	☽ △ ♃
22 46	☽ ⊼ ♆
22 47	☽ ⊼ ♅
23 44	☽ △ ♄
03 Wednesday	
02 48	☽ △ ☿
08 54	☽ ⊼ ♇
10 16	☿ ∥ ♄
11 17	☉ △ ♃
16 44	☽ ⚼ ☉
18 52	☽ □ ♀
19 17	☽ ± ♇
19 35	☽ # ♅
04 Thursday	
01 56	☽ ⚼ ♄
06 33	☉ ± ♂
08 22	☽ ⚼ ☿
08 55	♀ Q ♆
09 01	☽ ☌ ♂
09 14	☽ ± ☉
09 47	☽ # ♆
10 07	♀ Q ♅
10 35	☽ ⚼ ♇
13 12	☿ ⚼ ♂
17 11	☽ ± ♄
17 40	☽ □ ♃
20 01	☽ ⊼ ☉
05 Friday	
01 28	☽ ± ☿
02 00	☽ ☍ ♆
02 05	☽ ☍ ♅
02 09	☿ □ ♇
03 12	☽ ⊼ ♄
05 53	☿ ⊥ ♆
06 42	☿ ⊥ ♅
11 23	☽ △ ♇
12 38	☽ ⊼ ☿
23 02	☽ Q ♃
23 44	☽ △ ♀
06 Saturday	
00 44	☿ ⊥ ♀
02 12	☽ # ☉
02 15	☽ # ♄
03 10	☉ ∥ ♄
09 35	☽ ⊻ ♂
12 43	☽ # ☿
18 09	☽ ✶ ♃
19 03	☽ ⊥ ♂
23 55	☽ ☍ ☉
07 Sunday	
01 02	☽ ⚼ ♀
02 18	☽ ⊼ ♆
02 27	☽ ⊼ ♅
03 43	☽ ☍ ♄
09 10	☽ ∠ ♂
10 18	☿ ⚼ ♃
11 16	☽ □ ♇
11 47	☽ ± ♆
11 57	☽ ± ♅
16 02	☽ ± ♀
17 46	☽ ∠ ♃
18 46	☽ ☍ ☿
08 Monday	
01 58	☽ ⚼ ♆
02 01	☽ ⊼ ♀
02 09	☽ ⚼ ♅
02 25	☽ # ♇
05 23	☽ # ♃
06 41	☽ ∥ ♀
07 54	☽ ⊥ ♃
08 37	☽ ✶ ♂
12 46	☉ ⊻ ♆
15 37	☽ Q ♇
15 52	☉ ⊻ ♅
17 20	☽ ⊻ ♃
22 52	♀ # ♃
09 Tuesday	
01 42	☽ △ ♆
01 54	☽ △ ♅
02 35	☽ ⊼ ☉
03 24	☽ ⊼ ♄
03 29	☽ Q ♂
10 40	☽ ✶ ♇
11 00	♀ ∠ ♄
12 50	☽ ± ☉
13 02	☽ ± ♄
16 10	☉ ☌ ♄
17 03	☽ ∥ ♃
18 39	☽ # ♀
20 09	☽ ∥ ♇
10 Wednesday	
00 37	☽ ⊼ ☿
03 37	☽ ⚼ ♄
04 22	☽ ⚼ ☉
04 26	☽ ☍ ♀
08 14	☽ □ ♂
08 33	☉ ∠ ♀
10 22	☿ ∠ ♆
10 52	☽ ∠ ♇
11 40	☽ ± ☿
12 39	☿ ∠ ♅
13 29	♀ # ♇
17 24	☽ ☌ ♃
19 35	☽ ∥ ☿
11 Thursday	
01 41	☽ ⊥ ♇
02 22	☽ □ ♆
02 39	☽ □ ♅
04 27	☽ △ ♄
04 30	☽ ⚼ ☿
06 52	☽ △ ☉
11 43	☽ ⊻ ♇
13 04	☽ ∥ ☉
21 23	☽ ∥ ♄
12 Friday	
08 56	☽ Q ♆
09 16	☽ Q ♅
09 26	☽ △ ☿
09 27	☽ ⊼ ♀
09 32	☿ ⊻ ♀
10 21	☽ △ ♂
10 32	☉ ⚼ ♂
11 49	☿ ± ♃
16 54	☿ △ ♂
20 09	☽ ⊻ ♃
20 31	☽ ± ♀
13 Saturday	
01 35	♀ □ ♂
06 01	☽ ✶ ♆
06 24	☽ ✶ ♅
06 43	☽ ⊥ ♃
08 35	☽ □ ♄
12 44	☽ ⚼ ♂
13 25	☽ ⚼ ♀
14 57	☽ □ ☉
15 56	☽ ☌ ♇
17 59	☽ ∥ ♆
22 50	☽ ∠ ♃
14 Sunday	
04 04	☉ □ ♇
05 01	☽ ± ♂
08 54	☽ ∥ ♅
09 13	☽ ∠ ♆
09 39	☽ ∠ ♅
15 59	☉ ⊥ ♆
16 03	☽ ⊼ ♂
17 39	☽ Q ♄
18 25	☽ △ ♀
22 01	☉ ⊥ ♅
22 55	☽ □ ☿
15 Monday	
01 37	☿ # ♀
01 57	☽ ⊥ ♆
02 24	☽ ✶ ♃
02 25	☽ ⊥ ♅
02 46	☽ Q ☉
07 43	♂ St D
12 20	♀ ⚼ ♇
13 16	☽ ⊻ ♆
13 46	☽ ⊻ ♅
16 24	☽ ✶ ♄
23 41	☽ ⊻ ♇
16 Tuesday	
01 01	☽ Q ♃
03 10	☿ ⊼ ♃
03 29	☽ ✶ ☉
11 13	☽ ⊥ ♇
11 43	☽ ∥ ♅
13 48	☽ Q ☿
21 29	☽ ∠ ♄
17 Wednesday	
01 09	☽ ☍ ♂
03 29	☽ ∥ ♆
04 40	☽ ∠ ♇
07 02	☽ □ ♀
11 08	☽ ∠ ☉
11 42	☽ □ ♃
15 13	☽ ⊥ ♄
15 33	☿ ∥ ♇
16 13	☽ ✶ ☿
17 46	☉ ⚼ ♃
23 28	☽ ☌ ♆
18 Thursday	
00 04	☽ ☌ ♅
03 08	☽ ⊻ ♄
06 26	☽ ⊥ ☉
10 10	☽ ✶ ♇
19 24	☽ ⊻ ☉
20 32	☽ Q ♀
21 22	☿ ∥ ♃
19 Friday	
01 30	☽ ∠ ☿
07 25	☽ ∥ ♄
10 02	☽ Q ♇
12 35	☽ ⊼ ♂
21 29	☽ ⊥ ☿
21 49	☽ ✶ ♀
22 54	☽ △ ♃
20 Saturday	
00 39	☽ ± ♂
09 52	☽ ∥ ☉
10 47	☽ ⊻ ☿
11 23	☽ ⊻ ♆
12 05	☽ ⊻ ♅
15 35	☽ ☌ ♄
17 20	♀ ☍ ♃
17 39	☿ ✶ ♆
18 32	☽ # ♀
18 49	☽ ⚼ ♂
22 09	☽ □ ♇
23 29	☽ ⊥ ♆
21 Sunday	
00 13	☽ ⊥ ♅
02 08	☿ ✶ ♅
04 53	☽ ⚼ ♃
05 34	☽ ∠ ♀
13 05	☽ ☌ ☉
17 41	☽ ∠ ♆
17 57	☽ ∥ ♇
18 26	☽ ∠ ♅
22 56	☽ ± ♃
23 08	☽ ∥ ♃
22 Monday	
00 39	☽ ⊥ ♀
01 16	☽ △ ♂
10 45	☽ ∥ ☿
10 59	☽ ⊼ ♃
13 23	☽ ⊻ ♀
16 04	☉ # ♀
23 Tuesday	
00 05	☽ ✶ ♆
00 53	☽ ✶ ♅
03 47	☽ # ☿
04 28	☽ ☌ ☿
04 47	☽ ⊻ ♄
10 25	☿ ⊻ ♄
10 47	☽ △ ♇
17 02	☽ ⊥ ♄
17 10	☽ # ♃
20 52	☉ ∠ ♆
22 35	☽ # ♇
24 Wednesday	
00 25	☽ Q ♆
01 15	☽ Q ♅
07 04	☉ ∠ ♅
07 22	☽ ⊻ ☉
11 25	☽ ∠ ♄
14 25	☽ □ ♂
17 06	☽ ⚼ ♇
20 35	☽ ⊥ ☉
21 31	☽ # ☉
23 14	☽ ☍ ♃
25 Thursday	
04 47	☽ ☌ ♀
07 32	☽ ∥ ♀
11 13	☽ ± ♇
12 48	☽ □ ♆
13 40	☽ □ ♅
16 21	☽ ∠ ☉
17 54	☽ ✶ ♄
19 54	☽ ⊻ ☿
23 15	☽ ⊼ ♇
26 Friday	
01 43	☉ ± ♃
02 56	☽ Q ♂
07 48	☽ # ♄
08 12	☽ ⊥ ☿
14 31	♇ St R
18 05	☽ Q ♄
27 Saturday	
00 57	☽ ✶ ☉
02 21	☽ ∠ ☿
03 02	☽ ✶ ♂
10 38	☽ ⊼ ♃
18 43	☽ ⊻ ♀
22 16	☽ ± ♃
22 54	☿ St R
28 Sunday	
00 20	☽ △ ♆
01 16	☽ △ ♅
02 33	☽ Q ☉
05 42	☽ □ ♄
05 51	☉ △ ♂
06 46	☽ ⊥ ♀
07 45	☽ ✶ ☿
08 39	☽ ∠ ♂
10 17	☽ ☍ ♇
13 34	☽ # ♆
15 32	☽ ⚼ ♃

All ephemeris data is given at 12.00 UT and the Moon's longitude is additionally given for 24.00 UT

MARCH 1993

LONGITUDES

Date	Sidereal time h m s	Sun ☉ ° ' "	Moon ☽ ° ' "	Moon ☽ 24.00 ° ' "	Mercury ☿ ° '	Venus ♀ ° '	Mars ♂ ° '	Jupiter ♃ ° '	Saturn ♄ ° '	Uranus ♅ ° '	Neptune ♆ ° '	Pluto ♇ ° '
01	22 37 12	10 ♓ 55 09	09 ♊ 03 25	15 ♊ 30 37	24 ♓ 02	18 ♈ 09	09 ♋ 51	13 ♎ 10	23 ♒ 18	20 ♑ 55	20 ♑ 25	25 ♏ 31
02	22 41 08	11 55 21	22 ♊ 03 52	28 ♊ 43 39	23 R 42	18 30	10 01	13 R 05	23 25	20 58	20 27	25 R 31
03	22 45 05	12 55 31	05 ♋ 30 24	12 ♋ 24 25	23 14	18 49	10 11	12 59	23 32	21 00	20 28	25 31
04	22 49 01	13 55 38	19 ♋ 25 52	26 ♋ 34 45	22 38	19 06	10 22	12 53	23 39	21 03	20 30	25 31
05	22 52 58	14 55 43	03 ♌ 50 49	11 ♌ 13 38	21 55	19 20	10 34	12 47	23 46	21 05	20 31	25 31
06	22 56 55	15 55 47	18 ♌ 42 28	26 ♌ 16 24	21 06	19 33	10 46	12 41	23 53	21 08	20 33	25 30
07	23 00 51	16 55 48	03 ♍ 54 15	11 ♍ 34 40	20 12	19 43	10 59	12 35	23 59	21 10	20 34	25 30
08	23 04 48	17 55 47	19 ♍ 16 10	26 ♍ 57 15	19 15	19 51	11 12	12 28	24 06	21 12	20 36	25 30
09	23 08 44	18 55 43	04 ♎ 36 23	12 ♎ 12 09	18 16	19 57	11 25	12 22	24 13	21 15	20 37	25 29
10	23 12 41	19 55 39	19 ♎ 43 17	27 ♎ 08 46	17 16	20 00	11 39	12 15	24 20	21 17	20 38	25 29
11	23 16 37	20 55 33	04 ♏ 27 45	11 ♏ 39 40	16 17	20 R 01	11 54	12 09	24 27	21 19	20 40	25 28
12	23 20 34	21 55 25	18 ♏ 44 12	25 ♏ 41 14	15 20	20 00	12 09	12 02	24 33	21 21	20 41	25 28
13	23 24 30	22 55 16	02 ♐ 30 51	09 ♐ 13 18	14 26	19 56	12 25	11 55	24 40	21 24	20 42	25 28
14	23 28 27	23 55 04	15 ♐ 48 58	22 ♐ 18 19	13 36	19 50	12 41	11 48	24 47	21 26	20 44	25 27
15	23 32 24	24 54 52	28 ♐ 41 53	05 ♑ 00 16	12 50	19 41	12 57	11 41	24 53	21 28	20 45	25 26
16	23 36 20	25 54 37	11 ♑ 14 03	17 ♑ 23 50	12 10	19 29	13 14	11 34	25 00	21 30	20 46	25 26
17	23 40 17	26 54 21	23 ♑ 30 12	29 ♑ 33 43	11 36	19 15	13 31	11 27	25 06	21 32	20 47	25 25
18	23 44 13	27 54 03	05 ♒ 34 53	11 ♒ 34 11	11 08	18 59	13 49	11 19	25 13	21 34	20 49	25 25
19	23 48 10	28 53 43	17 ♒ 32 05	23 ♒ 28 56	10 46	18 40	14 07	11 12	25 19	21 36	20 50	25 24
20	23 52 06	29 ♓ 53 21	29 ♒ 25 06	05 ♓ 20 53	10 30	18 19	14 25	11 04	25 26	21 37	20 51	25 23
21	23 56 03	00 ♈ 52 57	11 ♓ 16 33	17 ♓ 12 18	10 21	17 56	14 44	10 57	25 32	21 39	20 52	25 22
22	23 59 59	01 52 32	23 ♓ 08 21	29 ♓ 04 51	10 18	17 31	15 03	10 49	25 38	21 41	20 53	25 22
23	00 03 56	02 52 04	05 ♈ 02 00	10 ♈ 59 56	10 D 20	17 03	15 23	10 42	25 45	21 43	20 54	25 21
24	00 07 53	03 51 34	16 ♈ 58 48	22 ♈ 58 48	10 28	16 34	15 43	10 34	25 51	21 44	20 55	25 20
25	00 11 49	04 51 03	29 ♈ 00 06	05 ♉ 02 56	10 42	16 02	16 03	10 27	25 57	21 46	20 56	25 19
26	00 15 46	05 50 29	11 ♉ 07 33	17 ♉ 14 15	11 00	15 30	16 24	10 19	26 03	21 47	20 57	25 18
27	00 19 42	06 49 53	23 ♉ 23 20	29 ♉ 35 11	11 24	14 55	16 45	10 11	26 09	21 49	20 58	25 17
28	00 23 39	07 49 15	05 ♊ 50 12	12 ♊ 08 49	11 52	14 20	17 06	10 03	26 15	21 50	20 58	25 16
29	00 27 35	08 48 34	18 ♊ 31 29	24 ♊ 58 42	12 25	13 44	17 28	09 56	26 21	21 52	20 59	25 15
30	00 31 32	09 47 52	01 ♋ 30 53	08 ♋ 08 31	13 02	13 07	17 50	09 48	26 27	21 53	21 00	25 14
31	00 35 28	10 ♈ 47 06	14 ♋ 51 59	21 ♋ 41 37	13 ♓ 43	12 ♈ 29	18 ♋ 12	09 ♎ 40	26 ♒ 33	21 ♑ 55	21 ♑ 01	25 ♏ 13

Date	Moon True ☊ ° '	Moon Mean ☊ ° '	Moon ☽ Latitude ° '
01	17 ♐ 19	17 ♐ 16	00 N 44
02	17 R 19	17 13	00 S 25
03	17 18	17 10	01 35
04	17 16	17 07	02 41
05	17 11	17 03	03 39
06	17 03	17 00	04 25
07	16 52	16 57	04 52
08	16 41	16 54	05 00
09	16 30	16 51	04 45
10	16 20	16 48	04 11
11	16 13	16 44	03 21
12	16 08	16 41	02 19
13	16 06	16 38	01 11
14	16 D 06	16 35	00 S 01
15	16 R 06	16 32	01 N 06
16	16 05	16 29	02 09
17	16 02	16 25	03 04
18	15 57	16 22	03 50
19	15 49	16 19	04 25
20	15 38	16 16	04 49
21	15 25	16 13	05 00
22	15 11	16 09	04 58
23	14 57	16 06	04 43
24	14 45	16 03	04 15
25	14 35	16 00	03 36
26	14 27	15 57	02 47
27	14 22	15 54	01 49
28	14 20	15 50	00 N 45
29	14 D 20	15 47	00 S 22
30	14 21	15 44	01 30
31	14 ♐ 21	15 ♐ 41	02 S 35

DECLINATIONS

Date	Sun ☉ ° '	Moon ☽ ° '	Mercury ☿ ° '	Venus ♀ ° '	Mars ♂ ° '	Jupiter ♃ ° '	Saturn ♄ ° '	Uranus ♅ ° '	Neptune ♆ ° '	Pluto ♇ ° '
01	07 S 28	22 N 32	00 N 31	12 N 16	26 N 20	03 S 46	14 S 46	22 S 14	21 S 13	04 S 58
02	07 05	22 47	00 32	12 33	26 18	03 44	14 43	22 14	21 13	04 57
03	06 42	21 45	00 29	12 49	26 15	03 42	14 41	22 13	21 13	04 57
04	06 19	19 23	00 20	13 04	26 13	03 39	14 39	22 13	21 13	04 57
05	05 56	15 44	00 N 08	13 19	26 10	03 37	14 37	22 13	21 13	04 56
06	05 33	11 01	00 S 09	13 32	26 08	03 34	14 34	22 12	21 12	04 56
07	05 10	05 N 32	00 30	13 45	26 05	03 32	14 32	22 12	21 12	04 55
08	04 46	00 S 21	00 54	13 56	26 03	03 29	14 30	22 11	21 12	04 55
09	04 23	06 12	01 21	14 06	26 00	03 27	14 28	22 11	21 12	04 54
10	03 59	11 35	01 49	14 16	25 57	03 24	14 26	22 11	21 11	04 54
11	03 36	16 09	02 19	14 24	25 54	03 21	14 24	22 11	21 11	04 53
12	03 12	19 38	02 50	14 31	25 51	03 18	14 22	22 10	21 11	04 53
13	02 49	21 50	03 20	14 36	25 48	03 15	14 19	22 10	21 11	04 52
14	02 25	22 43	03 51	14 41	25 45	03 13	14 17	22 10	21 11	04 52
15	02 01	22 20	04 20	14 44	25 42	03 10	14 15	22 09	21 10	04 51
16	01 38	20 50	04 48	14 45	25 39	03 07	14 13	22 09	21 10	04 51
17	01 14	18 22	05 14	14 46	25 36	03 04	14 11	22 09	21 10	04 50
18	00 50	15 09	05 38	14 44	25 33	03 01	14 09	22 08	21 10	04 50
19	00 26	11 22	06 01	14 42	25 29	02 58	14 07	22 08	21 10	04 49
20	00 S 03	07 10	06 21	14 38	25 26	02 55	14 05	22 08	21 09	04 49
21	00 N 21	02 S 43	06 38	14 32	25 22	02 52	14 03	22 08	21 09	04 48
22	00 45	01 N 50	06 53	14 25	25 19	02 49	14 01	22 07	21 09	04 48
23	01 08	06 19	07 06	14 16	25 15	02 46	13 59	22 07	21 09	04 47
24	01 32	10 36	07 16	14 06	25 11	02 43	13 57	22 07	21 09	04 47
25	01 56	14 29	07 23	13 54	25 08	02 40	13 55	22 07	21 09	04 46
26	02 19	17 48	07 29	13 41	25 04	02 37	13 53	22 06	21 08	04 46
27	02 43	20 23	07 32	13 26	25 00	02 34	13 51	22 06	21 08	04 45
28	03 06	22 01	07 32	13 10	24 56	02 31	13 49	22 06	21 08	04 45
29	03 30	22 34	07 31	12 53	24 52	02 28	13 47	22 06	21 08	04 44
30	03 53	21 56	07 27	12 35	24 47	02 25	13 45	22 06	21 08	04 44
31	04 N 16	20 N 02	07 S 22	12 N 16	24 N 43	02 S 22	13 S 43	22 S 05	21 S 08	04 S 43

ZODIAC SIGN ENTRIES

Date	h	m	Planets
03	02	16	☽ ♋
05	05	40	☽ ♌
07	05	52	☽ ♍
09	04	46	☽ ♎
11	04	40	☽ ♏
13	07	33	☽ ♐
15	14	28	☽ ♑
18	00	52	☽ ♒
20	13	11	☽ ♓
20	14	41	☉ ♈
23	01	51	☽ ♈
25	13	59	☽ ♉
28	00	48	☽ ♊
30	09	14	☽ ♋

LATITUDES

Date	Mercury ☿ ° '	Venus ♀ ° '	Mars ♂ ° '	Jupiter ♃ ° '	Saturn ♄ ° '	Uranus ♅ ° '	Neptune ♆ ° '	Pluto ♇ ° '
01	03 N 09	05 N 34	03 N 16	01 N 33	01 S 04	00 S 26	00 N 40	14 N 34
04	03 33	06 03	03 11	01 33	01 04	00 26	00 40	14 35
07	03 40	06 31	03 07	01 34	01 04	00 26	00 40	14 37
10	03 29	06 58	03 02	01 34	01 05	00 26	00 40	14 38
13	03 01	07 22	02 58	01 35	01 05	00 26	00 40	14 39
16	02 23	07 43	02 53	01 35	01 05	00 26	00 40	14 40
19	01 38	08 00	02 49	01 35	01 06	00 26	00 40	14 41
22	00 53	08 10	02 45	01 36	01 06	00 26	00 40	14 42
25	00 N 11	08 14	02 40	01 36	01 07	00 26	00 40	14 44
28	00 S 28	08 14	02 36	01 36	01 07	00 26	00 41	14 44
31	01 S 02	07 N 58	02 N 32	01 N 36	01 S 07	00 S 26	00 N 41	14 N 45

DATA

Julian Date	2449048
Delta T	+59 seconds
Ayanamsa	23° 45' 59"
Synetic vernal point	05° ♓ 21' 00"
True obliquity of ecliptic	23° 26' 23"

MOON'S PHASES, APSIDES AND POSITIONS ☽

Date	h	m	Phase	Longitude ° '	Eclipse Indicator
01	15	47	☽	11 ♊ 05	
08	09	46	○	17 ♍ 50	
15	04	17	☾	24 ♐ 36	
23	07	14	●	02 ♈ 40	
31	04	10	☽	10 ♋ 28	

Day	h	m		
08	08	40	Perigee	
21	19	15	Apogee	
02	05	02	Max dec	22° N 50'
08	10	35	0S	
14	16	30	Max dec	22° S 44'
22	02	20	0N	
29	11	19	Max dec	22° N 34'

LONGITUDES

Date	Chiron ⚷ ° '	Ceres ⚳ ° '	Pallas ⚴ ° '	Juno ⚵ ° '	Vesta ⚶ ° '	Black Moon Lilith ⚸ ° '
01	18 ♌ 57	13 ♓ 14	16 ♒ 33	03 ♋ 37	28 ♑ 33	15 ♓ 11
11	18 ♌ 19	17 ♓ 11	19 ♒ 49	05 ♋ 47	03 ♒ 28	16 ♓ 18
21	17 ♌ 48	21 ♓ 07	22 ♒ 59	08 ♋ 56	08 ♒ 15	17 ♓ 25
31	17 ♌ 26	25 ♓ 01	26 ♒ 01	11 ♋ 29	12 ♒ 53	18 ♓ 32

All ephemeris data is given at 12.00 UT and the Moon's longitude is additionally given for 24.00 UT

ASPECTARIAN

h m	Aspects
01 Monday	
00 33	☽ ∠ ♀
02 04	☽ ⊥ ♂
05 08	☽ bQ ♆
05 13	☽ ⧺ ♅
06 05	☽ bQ ♅
06 24	☽ Q ☿
13 30	☽ ⊻ ♂
15 47	☽ □ ☉
19 37	☽ △ ♃
22 00	☽ ± ♆
22 57	☽ ± ♅
02 Tuesday	
05 20	☽ ✶ ♀
09 03	☽ ⊼ ♆
09 59	☽ ⊼ ♅
14 28	☽ △ ♄
14 53	☽ □ ☿
18 15	☽ ⊼ ♇
03 Wednesday	
00 40	☿ ⊻ ♄
03 32	☽ Q ♀
04 25	☽ ⧺ ♅
04 58	☽ ± ♇
08 18	☉ ⊥ ♀
13 17	☉ ⊼ ♃
17 20	☽ bQ ♄
18 47	☽ ⧺ ♆
20 16	☽ ☌ ♂
20 44	☽ bQ ♇
04 Thursday	
00 54	☽ □ ♃
01 53	☽ △ ☉
08 57	☽ ± ♄
11 25	☽ □ ♀
13 48	☽ ☍ ♆
14 44	☽ ☍ ♅
17 10	☽ △ ☿
19 10	☽ ⊼ ♄
22 13	☽ △ ♇
05 Friday	
05 05	☽ bQ ☉
05 18	♂ bQ ♇
07 00	☽ Q ♃
16 46	☽ bQ ☿
18 15	☽ ⧺ ♄
20 53	☽ ± ☉
23 05	☽ ⊻ ♂
06 Saturday	
00 22	☽ ∥ ♀
02 25	☽ ✶ ♃
06 33	☽ ± ☿
06 35	♀ ± ♇
07 15	☽ ⊼ ☉
08 51	☽ ⊥ ♂
11 19	☿ ✶ ♅
13 22	☽ △ ♀
14 56	☽ ⊼ ♆
15 36	☽ ⊼ ☿
15 52	☽ ⊼ ♅
20 16	☽ ☍ ♄
22 47	☽ □ ♇
23 21	☽ ∠ ♂
07 Sunday	
00 27	☽ ± ♆
01 23	☽ ± ♅
02 08	☽ ∠ ♃
02 43	☿ ✶ ♆
13 18	☽ bQ ♀
13 39	☽ ⧺ ☉
14 33	☽ ⧺ ♇
14 37	☽ bQ ♆
15 33	☽ bQ ♅
16 10	☽ ⊥ ♃
20 20	☽ ⧺ ♃
22 44	☿ ⊻ ♀
23 13	☽ ✶ ♂
08 Monday	
01 28	☽ ⊻ ♃
03 00	☽ Q ♇
03 11	☉ ∥ ♇
03 29	☽ ± ♀
07 17	☽ ⧺ ☿
09 46	☽ ☍ ☉
11 59	☽ ☍ ☿
12 55	☽ ⊼ ♀
14 04	☽ △ ♆
14 24	☽ ∥ ☿
15 02	☽ △ ♅
18 13	☽ Q ♂
19 36	☽ ⊼ ♄
21 43	☽ ✶ ♇
09 Tuesday	
00 39	☽ ∥ ♃
04 01	☉ ☌ ☿
04 55	☽ ∥ ☉
05 04	☽ ± ♄
06 36	☽ ∥ ♇
19 20	☽ bQ ♄
21 17	☽ ∠ ♇
22 56	☽ □ ♂
10 Wednesday	
00 10	☽ ☌ ♃
08 19	☽ ⊼ ☿
11 37	☽ ⊥ ♇
12 21	☽ ⊼ ☉
12 28	☽ ☍ ♀
13 29	☽ □ ♆
13 58	☉ ⊻ ♀
14 31	☽ □ ♅
17 21	☽ ± ☿
19 29	☽ △ ♄
21 17	☽ ⊻ ♇
22 45	☽ ± ☉
11 Thursday	
01 50	☽ ⧺ ♀
02 12	☽ ∥ ♄
05 33	☉ ✶ ♆
07 05	☽ bQ ☿
09 28	♀ St R
11 41	♀ ⧺ ♄
14 36	☽ bQ ☉
18 59	☽ Q ♆
20 06	☽ Q ♅
21 51	☉ ✶ ♅
12 Friday	
00 37	☽ △ ♂
00 43	☽ ⊻ ♃
04 09	♂ □ ♃
05 05	☉ ∥ ♃
06 33	☽ △ ☿
10 48	☽ ⊥ ♃
14 09	☽ ⊼ ♀
15 21	☽ ✶ ♆
16 31	☽ ✶ ♅
17 54	☽ △ ☉
21 58	☉ ∥ ☿
22 07	☽ □ ♄
23 37	☽ ☌ ♇
13 Saturday	
00 29	☽ ± ♀
02 14	☽ ∠ ♃
02 50	☽ bQ ♂
03 18	☽ ∥ ♆
08 28	☿ ∥ ♃
16 16	☽ bQ ♀
17 42	☽ ∠ ♆
17 52	☽ ∥ ♅
18 56	☽ ∠ ♅
19 05	☽ ± ♂
14 Sunday	
03 53	☿ ⊥ ♀
04 44	☽ ✶ ♃
06 08	☽ ⊼ ♂
06 24	☽ Q ♄
08 10	☽ □ ☿
10 00	☽ ⊥ ♆
11 17	☽ ⊥ ♅
19 19	☽ △ ♀
21 05	☽ ⊻ ♆
22 24	☽ ⊻ ♅
15 Monday	
02 39	☽ Q ♃
04 17	☽ □ ☉
04 46	☽ ✶ ♄
05 52	☽ ⊻ ♇
09 17	☿ △ ♂
11 18	☉ ⊻ ♄
15 50	☽ Q ☿
16 00	☽ ∥ ♅
17 12	☽ ⊥ ♇
16 Tuesday	
00 34	☉ △ ♇
07 43	☽ ∥ ♆
09 35	☽ ∠ ♄
10 27	☽ ∠ ♇
12 38	☽ □ ♃
13 44	☽ ✶ ☿
14 34	☿ ∥ ♇
15 58	☽ ☍ ♂
17 39	☽ Q ☉
18 32	♀ ± ♇
17 Wednesday	
03 16	☽ ⊥ ♄
03 48	☽ □ ♀
06 39	☽ ☌ ♆
08 06	☽ ☌ ♅
15 12	☽ ⊻ ♄
15 47	☽ ✶ ♇
17 53	☽ ∠ ☿
19 20	☽ ✶ ☉
22 05	☿ ⊼ ♃
18 Thursday	
11 08	☽ ⊥ ☿
14 44	☽ Q ♀
14 50	☽ ⧺ ♀
15 39	☽ Q ♇
18 46	☽ ∥ ♄
22 46	☽ ⊻ ☿
23 23	☽ △ ♃
19 Friday	
04 00	☽ ∠ ☉
04 56	☽ ⊼ ♂
14 14	☽ ✶ ♀
17 20	☽ ± ♂
18 39	☽ ⊻ ♆
20 12	☽ ⊻ ♅
23 48	☽ ⊥ ☉
20 Saturday	
03 09	♄ □ ♇
03 51	☽ □ ♇
03 52	☽ ☌ ♄
05 18	☽ bQ ♃
06 47	☽ ⊥ ♆
08 22	☽ ⊥ ♅
12 00	☽ bQ ♂
13 02	☽ ⊻ ☉
16 13	☽ ∥ ☿
19 39	☽ ∠ ♀
23 19	☽ ± ♃
21 Sunday	
00 49	☽ ∥ ♇
01 02	☽ ∠ ♆
02 37	☽ ∠ ♅
10 08	☽ ☌ ☿
11 10	☽ ∥ ♃
11 21	☽ ⊼ ♃
13 17	☽ ⊥ ♀
19 12	☽ △ ♂
23 29	☽ ⧺ ☉
22 Monday	
01 01	☽ ⊻ ♀
05 40	☉ ⊥ ♄
05 44	☽ ∥ ☉
07 26	☽ ✶ ♆
09 03	☽ ✶ ♅
13 44	☿ St D
16 29	☽ △ ♇
17 06	☽ ⊻ ♄
17 10	☽ ⧺ ♃
23 Tuesday	
03 44	☽ ⧺ ♇
05 19	☽ ⊥ ♄
06 50	♃ bQ ♄
07 14	☽ ☌ ☉
07 42	☽ Q ♆
09 20	☽ Q ♅
12 45	☉ Q ♆
16 25	☽ ⧺ ☿
22 41	☽ bQ ♇
22 46	☽ ⊻ ☿
23 16	☽ ☍ ♃
23 36	☽ ∠ ♄
24 Wednesday	
08 58	☉ Q ♅
09 23	☽ □ ♂
10 58	☽ ⊥ ☿
11 11	☽ ☌ ♀
15 07	☿ ⊥ ♀
16 42	☽ ± ♇
19 27	☿ ⊼ ♃
19 53	☽ □ ♆
21 32	☽ □ ♅
25 Thursday	
04 40	☽ ⊼ ♇
05 16	☽ ∠ ☿
05 53	☽ ✶ ♄
08 19	☽ ⧺ ♄
08 23	☽ ∥ ♀
09 54	♀ ⧺ ♄
11 40	♀ □ ♂
22 19	☽ Q ♂
26 Friday	
00 38	☽ ⊻ ☉
05 54	☽ Q ♄
10 25	☽ ⊼ ♃
11 46	☽ ✶ ☿
13 32	☽ ⊥ ☉
14 09	♃ ∠ ♇
20 12	☽ ⊻ ♀
22 05	☽ ± ♃
22 39	☽ ✶ ♂
27 Saturday	
04 01	☉ ⧺ ♃
07 16	☽ △ ♆
07 25	☽ ⊥ ♀
08 42	☽ ∠ ☉
08 56	☽ △ ♅
12 02	☽ Q ☿
15 27	☽ bQ ♃
15 41	☽ ☍ ♇
17 25	☽ □ ♄
21 21	☽ ⧺ ♆
28 Sunday	
00 05	☽ ∠ ♀
04 38	☽ ∠ ♂
12 16	☽ bQ ♆
13 47	☽ ⧺ ♅
13 55	☽ bQ ♅
16 07	☽ ✶ ☉
19 57	☽ △ ♃
22 19	☽ ⊥ ♂
23 59	☽ □ ☿
29 Monday	
03 24	☽ ✶ ♀
05 21	☽ ± ♆
07 00	☽ ± ♅
09 57	☽ ⊻ ♂
16 36	☽ ⊼ ♆
16 37	☽ Q ☉
18 14	☽ ⊼ ♅
30 Tuesday	
00 30	☽ ⊼ ♇
00 47	☽ Q ♀
02 39	☽ △ ♄
08 35	☽ ⧺ ♅
11 30	☽ ± ♇
12 01	☉ ☍ ♃
13 29	☿ ⊻ ♀
22 32	☉ bQ ♇
31 Wednesday	
00 05	☽ ⧺ ♆
02 50	☽ □ ♃
03 44	☽ bQ ♇
04 10	☽ □ ☉
06 04	☽ bQ ♄
07 57	☽ □ ♀
09 50	☽ △ ☿
18 03	☽ ☌ ♂
22 05	☽ ± ♄
22 49	☽ ☍ ♆

APRIL 1993

Date	Sidereal time h m s	Sun ☉	Moon ☽	Moon ☽ 24.00	Mercury ☿	Venus ♀	Mars ♂	Jupiter ♃	Saturn ♄	Uranus ♅	Neptune ♆	Pluto ♇
		° ′ ″	° ′ ″	° ′ ″	° ′	° ′	° ′	° ′	° ′	° ′	° ′	° ′
01	00 39 25	11 ♈ 46 19	28 ♋ 37 39	05 ♌ 40 11	14 ♓ 27	11 ♈ 51	18 ♋ 35	09 ♎ 32	26 ♒ 39	21 ♑ 56	21 ♑ 02	25 ♏ 12
02	00 43 22	12 45 29	12 ♌ 49 11	20 ♌ 04 24	15 15	11 R 13	18 57	09 R 25	26 45	21 57	21 02	25 R 11
03	00 47 18	13 44 37	27 ♌ 25 24	04 ♍ 51 33	16 06	10 36	19 20	09 17	26 51	21 58	21 03	25 10
04	00 51 15	14 43 42	12 ♍ 22 00	19 ♍ 55 42	17 01	09 59	19 44	09 09	26 56	21 59	21 04	25 09
05	00 55 11	15 42 46	27 ♍ 31 26	05 ♎ 07 55	17 58	09 22	20 08	09 02	27 02	22 00	21 04	25 08
06	00 59 08	16 41 47	12 ♎ 43 46	20 ♎ 17 39	18 58	08 47	20 32	08 54	27 07	22 01	21 05	25 07
07	01 03 04	17 40 46	27 ♎ 48 16	05 ♏ 14 30	20 01	08 13	20 56	08 46	27 13	22 02	21 05	25 05
08	01 07 01	18 39 43	12 ♏ 35 22	19 ♏ 50 06	21 07	07 40	21 20	08 39	27 18	22 03	21 06	25 04
09	01 10 57	19 38 38	26 ♏ 58 07	03 ♐ 59 04	22 15	07 09	21 45	08 31	27 24	22 04	21 06	25 03
10	01 14 54	20 37 32	10 ♐ 52 50	17 ♐ 39 27	23 25	06 40	22 10	08 24	27 29	22 05	21 07	25 02
11	01 18 51	21 36 24	24 ♐ 19 04	00 ♑ 52 04	24 37	06 13	22 35	08 17	27 34	22 06	21 07	25 00
12	01 22 47	22 35 14	07 ♑ 18 49	13 ♑ 39 52	25 52	05 47	23 00	08 09	27 39	22 07	21 07	24 59
13	01 26 44	23 34 02	19 ♑ 55 45	26 ♑ 07 03	27 09	05 24	23 26	08 02	27 44	22 07	21 08	24 58
14	01 30 40	24 32 49	02 ♒ 14 23	08 ♒ 18 22	28 28	05 04	23 52	07 55	27 49	22 08	21 08	24 56
15	01 34 37	25 31 34	14 ♒ 19 35	20 ♒ 18 36	29 ♓ 49	04 45	24 18	07 48	27 54	22 08	21 08	24 55
16	01 38 33	26 30 17	26 ♒ 16 00	02 ♓ 12 16	01 ♈ 12	04 29	24 44	07 41	27 59	22 09	21 08	24 53
17	01 42 30	27 28 58	08 ♓ 07 53	14 ♓ 03 17	02 36	04 16	25 11	07 34	28 04	22 09	21 09	24 52
18	01 46 26	28 27 38	19 ♓ 58 53	25 ♓ 55 01	04 03	04 05	25 37	07 27	28 09	22 10	21 09	24 51
19	01 50 23	29 ♈ 26 16	01 ♈ 51 59	07 ♈ 50 04	05 31	03 56	26 04	07 20	28 14	22 10	21 09	24 49
20	01 54 20	00 ♉ 24 52	13 ♈ 49 30	19 ♈ 50 30	07 01	03 50	26 31	07 13	28 18	22 11	21 09	24 48
21	01 58 16	01 23 26	25 ♈ 53 14	01 ♉ 57 51	08 33	03 46	26 59	07 07	28 23	22 11	21 09	24 46
22	02 02 13	02 21 58	08 ♉ 04 32	14 ♉ 13 23	10 07	03 44	27 26	07 00	28 27	22 11	21 09	24 45
23	02 06 09	03 20 29	20 ♉ 24 35	26 ♉ 38 15	11 42	03 D 45	27 54	06 54	28 32	22 11	21 R 09	24 43
24	02 10 06	04 18 57	02 ♊ 54 34	09 ♊ 13 43	13 19	03 49	28 22	06 48	28 36	22 11	21 09	24 42
25	02 14 02	05 17 24	15 ♊ 35 53	22 ♊ 01 17	14 58	03 54	28 50	06 42	28 40	22 11	21 09	24 40
26	02 17 59	06 15 48	28 ♊ 30 09	05 ♋ 02 43	16 39	04 02	29 18	06 36	28 44	22 R 11	21 09	24 39
27	02 21 55	07 14 11	11 ♋ 39 14	18 ♋ 19 56	18 22	04 12	29 ♋ 46	06 30	28 48	22 11	21 09	24 37
28	02 25 52	08 12 31	25 ♋ 05 01	01 ♌ 54 40	20 06	04 24	00 ♌ 15	06 24	28 52	22 11	21 09	24 35
29	02 29 49	09 10 49	08 ♌ 49 01	15 ♌ 48 05	21 52	04 38	00 43	06 19	28 56	22 11	21 08	24 34
30	02 33 45	10 ♉ 09 06	22 ♌ 51 50	00 ♍ 00 06	23 ♈ 39	04 ♈ 54	01 ♌ 12	06 ♎ 13	29 ♒ 00	22 ♑ 11	21 ♑ 08	24 ♏ 32

Date	Moon True ☊ ° ′	Moon Mean ☊ ° ′	Moon ☽ Latitude ° ′
01	14 ♐ 19	15 ♐ 38	03 S 33
02	14 R 16	15 35	04 20
03	14 10	15 31	04 52
04	14 03	15 28	05 04
05	13 54	15 25	04 56
06	13 45	15 22	04 27
07	13 38	15 19	03 39
08	13 32	15 15	02 38
09	13 28	15 12	01 27
10	13 27	15 09	00 S 14
11	13 D 28	15 06	00 N 58
12	13 29	15 03	02 05
13	13 30	15 00	03 03
14	13 R 29	14 56	03 52
15	13 27	14 53	04 29
16	13 23	14 50	04 54
17	13 17	14 47	05 06
18	13 10	14 44	05 06
19	13 02	14 41	04 51
20	12 53	14 37	04 25
21	12 46	14 34	03 46
22	12 40	14 31	02 56
23	12 36	14 28	01 57
24	12 33	14 25	00 N 52
25	12 D 33	14 21	00 S 17
26	12 34	14 18	01 26
27	12 35	14 15	02 32
28	12 37	14 12	03 31
29	12 37	14 09	04 20
30	12 ♐ 37	14 ♐ 06	04 S 54

DECLINATIONS

Date	Sun ☉ ° ′	Moon ☽ ° ′	Mercury ☿ ° ′	Venus ♀ ° ′	Mars ♂ ° ′	Jupiter ♃ ° ′	Saturn ♄ ° ′	Uranus ♅ ° ′	Neptune ♆ ° ′	Pluto ♇ ° ′
01	04 N 39	16 N 57	07 S 14	11 N 55	24 N 39	02 S 19	13 S 41	22 S 05	21 S 08	04 S 43
02	05 02	12 48	07 04	11 34	24 34	02 16	13 40	22 05	21 08	04 42
03	05 25	07 48	06 53	11 12	24 30	02 13	13 38	22 05	21 07	04 42
04	05 48	02 N 14	06 39	10 50	24 25	02 10	13 36	22 05	21 07	04 41
05	06 11	03 S 32	06 24	10 27	24 20	02 07	13 34	22 05	21 07	04 40
06	06 34	09 07	06 07	10 04	24 15	02 04	13 32	22 05	21 07	04 40
07	06 56	14 06	05 49	09 41	24 10	02 01	13 31	22 04	21 07	04 39
08	07 19	18 07	05 29	09 18	24 05	01 58	13 29	22 04	21 07	04 39
09	07 41	20 54	05 07	08 55	24 00	01 55	13 27	22 04	21 07	04 38
10	08 03	22 18	04 44	08 32	23 55	01 52	13 26	22 04	21 07	04 38
11	08 25	22 21	04 19	08 10	23 50	01 49	13 24	22 04	21 07	04 37
12	08 47	21 10	03 53	07 48	23 44	01 46	13 22	22 04	21 07	04 37
13	09 09	18 56	03 26	07 27	23 39	01 44	13 21	22 04	21 07	04 36
14	09 31	15 54	02 57	07 06	23 33	01 41	13 19	22 04	21 07	04 36
15	09 52	12 14	02 27	06 46	23 27	01 38	13 17	22 04	21 06	04 35
16	10 14	08 09	01 56	06 27	23 21	01 35	13 16	22 04	21 06	04 35
17	10 35	03 S 47	01 23	06 09	23 15	01 33	13 14	22 04	21 06	04 34
18	10 56	00 N 43	00 50	05 52	23 09	01 30	13 13	22 04	21 06	04 34
19	11 16	05 12	00 S 15	05 35	23 03	01 28	13 11	22 04	21 06	04 33
20	11 37	09 31	00 N 21	05 20	22 57	01 25	13 10	22 03	21 06	04 33
21	11 57	13 30	00 58	05 06	22 51	01 23	13 09	22 03	21 06	04 32
22	12 18	16 58	01 36	04 53	22 44	01 20	13 07	22 03	21 06	04 32
23	12 38	19 44	02 15	04 41	22 37	01 18	13 06	22 03	21 06	04 32
24	12 58	21 36	02 55	04 30	22 31	01 16	13 04	22 03	21 06	04 31
25	13 17	22 23	03 36	04 20	22 24	01 13	13 03	22 03	21 06	04 31
26	13 37	22 00	04 18	04 11	22 17	01 11	13 02	22 04	21 06	04 30
27	13 56	20 24	05 01	04 03	22 10	01 09	13 01	22 04	21 06	04 30
28	14 15	17 39	05 44	03 57	22 03	01 07	12 59	22 04	21 06	04 29
29	14 33	13 52	06 28	03 51	21 55	01 05	12 58	22 04	21 06	04 29
30	14 N 52	09 N 15	07 N 13	03 N 46	21 N 48	01 S 03	12 S 57	22 S 04	21 S 06	04 S 28

ZODIAC SIGN ENTRIES

Date	h m	Planets
01	14 21	☽ ♌
03	16 10	☽ ♍
05	15 54	☽ ♎
07	15 32	☽ ♏
09	17 10	☽ ♐
11	22 24	☽ ♑
14	07 36	☽ ♒
15	15 18	☿ ♈
16	19 33	☽ ♓
19	08 14	☽ ♈
20	01 49	☉ ♉
21	20 08	☽ ♉
24	06 27	☽ ♊
26	14 45	☽ ♋
27	23 40	♂ ♌
28	20 39	☽ ♌

LATITUDES

Date	Mercury ☿ ° ′	Venus ♀ ° ′	Mars ♂ ° ′	Jupiter ♃ ° ′	Saturn ♄ ° ′	Uranus ♅ ° ′	Neptune ♆ ° ′	Pluto ♇ ° ′
01	01 S 12	07 N 52	02 N 31	01 N 36	01 S 08	00 S 26	00 N 41	14 N 46
04	01 39	07 30	02 27	01 36	01 08	00 26	00 41	14 47
07	02 01	07 00	02 23	01 36	01 08	00 27	00 41	14 47
10	02 18	06 25	02 20	01 36	01 09	00 27	00 41	14 48
13	02 30	05 46	02 16	01 35	01 09	00 27	00 41	14 49
16	02 37	05 05	02 13	01 35	01 10	00 27	00 41	14 49
19	02 40	04 23	02 09	01 35	01 10	00 27	00 41	14 50
22	02 37	03 42	02 06	01 34	01 11	00 27	00 41	14 50
25	02 30	03 02	02 03	01 34	01 11	00 27	00 41	14 51
28	02 17	02 23	02 00	01 33	01 12	00 27	00 41	14 51
31	02 S 01	01 N 47	01 N 57	01 N 33	01 S 12	00 S 27	00 N 41	14 N 51

DATA

Julian Date	2449079
Delta T	+59 seconds
Ayanamsa	23° 46′ 02″
Synetic vernal point	05° ♓ 20′ 57″
True obliquity of ecliptic	23° 26′ 23″

MOON'S PHASES, APSIDES AND POSITIONS ☽

Date	h m	Phase	Longitude ° ′	Eclipse Indicator
06	18 43	○	16 ♎ 58	
13	19 39	☾	23 ♑ 53	
21	23 49	●	01 ♉ 52	
29	12 41	☽	09 ♌ 12	

Day	h m		
05	19 35	Perigee	
18	05 16	Apogee	
04	21 18	0S	
11	00 46	Max dec	22° S 30′
18	08 11	0N	
25	16 20	Max dec	22° N 24′

LONGITUDES

Date	Chiron ⚷ ° ′	Ceres ⚳ ° ′	Pallas ⚴ ° ′	Juno ⚵ ° ′	Vesta ⚶ ° ′	Black Moon Lilith ⚸ ° ′
01	17 ♌ 24	25 ♓ 25	26 ♒ 19	11 ♋ 48	13 ♒ 21	18 ♓ 39
11	17 ♌ 13	29 ♓ 16	29 ♒ 09	15 ♋ 10	17 ♒ 47	19 ♓ 46
21	17 ♌ 12	03 ♈ 05	01 ♓ 48	18 ♋ 47	22 ♒ 00	20 ♓ 53
31	17 ♌ 23	06 ♈ 49	04 ♓ 12	22 ♋ 34	25 ♒ 58	22 ♓ 01

ASPECTARIAN

h m	Aspects
01 Thursday	
00 24	☽ ☍ ♅
06 06	☽ △ ♇
08 35	☽ ⚻ ♄
08 45	☉ ∠ ♄
10 09	☽ Q ♃
13 11	☉ ☌ ♀
13 30	☽ bQ ☿
15 19	☉ # ♇
18 37	♀ ∠ ♄
02 Friday	
05 40	☽ ± ☿
06 21	☽ ✶ ♃
07 27	☽ # ♄
09 27	☽ △ ♀
11 53	☽ △ ☉
16 17	☽ ⚻ ☿
18 45	☽ ∥ ♀
22 26	☽ ⊻ ♂
03 Saturday	
01 36	☽ ⚻ ♆
03 06	☽ ⚻ ♅
06 56	☽ ∠ ♃
08 20	☽ □ ♇
08 31	☽ ⊥ ♂
09 09	☽ bQ ♀
11 03	☽ ☍ ♄
11 23	☽ ± ♆
12 53	☽ ± ♅
14 18	☽ bQ ☉
16 16	☽ # ☿
21 23	☽ ⊥ ♃
21 47	☽ ∥ ☉
23 07	☽ ± ♀
23 28	☽ ∠ ♂
04 Sunday	
01 37	☽ # ♇
01 55	☽ bQ ♆
03 24	☽ bQ ♅
05 06	♀ bQ ♇
05 47	☽ ± ☉
06 55	☽ ⊻ ♃
08 20	☽ ⚻ ♀
12 18	☽ # ♃
13 15	☽ Q ♇
16 01	☽ ⚻ ☉
19 52	☽ ☍ ☿
05 Monday	
00 00	☽ ✶ ♂
01 48	☽ △ ♆
03 17	☽ △ ♅
06 06	☽ ∥ ♃
08 13	☽ ✶ ♇
11 13	☽ ⚻ ♄
16 45	☽ ∥ ♇
19 27	☽ Q ♂
20 05	☉ # ☿
20 44	☽ ± ♄
23 05	♄ ⊥ ♆
23 32	☽ ∥ ☿
23 58	☽ # ☉
06 Tuesday	
05 49	♀ ☍ ♃
06 00	☽ ☍ ♀
06 00	☽ ☌ ♃
07 52	☽ ∠ ♇
11 02	☽ bQ ♄
16 02	☽ # ♀
18 43	☽ ☍ ☉
22 06	☽ ⊥ ♇
22 38	☽ ⚻ ☿
07 Wednesday	
00 42	☽ □ ♂
01 15	☽ □ ♆
02 46	☽ □ ♅
07 39	☽ ⊻ ♇
08 55	☽ ± ☿
08 57	☽ ∥ ♄
11 03	☽ △ ♄
21 33	♂ ☍ ♆
08 Thursday	
00 34	☽ bQ ☿
04 14	☽ ⚻ ♀
05 36	☽ ⊻ ♃
06 16	☽ Q ♆
07 50	☽ Q ♅
09 37	♂ ± ♄
11 38	☿ ✶ ♆
13 43	☽ ± ♀
15 22	☽ ⊥ ♃
19 36	☿ △ ♂
21 43	☉ ± ♇
22 47	☽ ⚻ ☉
09 Friday	
02 07	☽ ✶ ♆
02 56	☽ △ ♂
03 20	☽ △ ☿
03 44	☽ ✶ ♅
04 09	☽ bQ ♀
06 14	☽ ∠ ♃
08 21	☿ ✶ ♅
08 45	☽ ☌ ♇
09 35	☽ ± ☉
12 44	☽ □ ♄
14 35	☽ ∥ ♆
10 Saturday	
02 08	☽ bQ ☉
03 40	☽ ∠ ♆
04 53	☽ △ ♀
05 18	☽ bQ ♂
05 22	☽ ∠ ♄
05 41	☽ ∥ ♄
07 22	♂ ☍ ♅
07 42	☽ ✶ ♃
18 13	☿ ∥ ♇
19 28	☽ ⊥ ♆
20 11	☽ Q ♄
21 12	☽ ⊥ ♅
21 38	☽ ± ♂
23 56	☉ □ ♆
11 Sunday	
03 36	☉ ∥ ♀
04 46	☽ Q ♃
06 12	☽ ⊻ ♆
06 42	☽ △ ☉
07 58	☽ ⊻ ♅
08 45	☽ ⚻ ♂
12 37	☽ □ ☿
13 15	☽ ⊻ ♇
17 58	☽ ✶ ♄
19 17	☿ △ ♇
20 24	☽ ∥ ♅
12 Monday	
00 10	☉ □ ♅
00 14	☽ ⊥ ♇
09 14	☽ □ ♀
12 42	☽ ∥ ♆
13 34	☽ □ ♃
17 01	☽ ∠ ♇
22 09	☽ ∠ ♄
13 Tuesday	
01 47	☽ Q ☿
06 07	☉ □ ♂
14 19	☽ ☌ ♆
15 31	☽ ⊥ ♄
16 14	☽ ☌ ♅
18 32	☽ Q ♀
19 01	☽ ☍ ♂
19 39	☽ □ ☉
21 44	☽ ✶ ♇
23 35	☿ ⊻ ♄
14 Wednesday	
03 16	☽ ⊻ ♄
03 41	☽ ✶ ☿
17 26	☽ ✶ ♀
21 16	☽ Q ♇
21 21	☉ ⚻ ♇
23 07	☽ △ ♃
15 Thursday	
05 22	☽ ∥ ♄
10 15	☽ Q ☉
13 06	☽ ∠ ☿
22 38	☽ ∠ ♀
16 Friday	
01 00	☽ # ☉
01 40	☽ ⊻ ♆
03 42	☽ ⊻ ♅
04 50	☽ bQ ♃
08 47	☽ ⚻ ♂
09 14	☽ □ ♇
09 33	☽ ⊥ ☿
12 31	☽ ✶ ☉
13 46	☽ ⊥ ♆
15 30	☽ ☌ ♄
15 48	☽ ⊥ ♅
16 24	☽ ⊥ ♀
20 05	♂ △ ♇
21 23	☽ ± ♂
22 09	☽ # ♀
22 50	☽ ± ♃
23 17	☽ ⊻ ☿
17 Saturday	
04 18	☽ ⊻ ♀
04 36	☿ ∥ ♃
07 42	☽ ∥ ♇
07 58	☽ ∠ ♆
10 02	☽ ∠ ♅
10 51	☽ ⚻ ♃
16 18	☽ bQ ♂
21 05	☿ Q ♆
21 37	☽ ∠ ☉
18 Sunday	
00 04	☽ ∥ ♃
00 17	♀ Q ♅
02 35	☽ ∥ ☿
03 38	☉ ✶ ♄
04 33	♂ Q ♃
05 11	♀ ⊥ ♄
12 27	☿ ☌ ♀
12 32	☽ # ☿
13 46	☿ ⊥ ♄
13 59	☿ Q ♅
14 22	☽ ✶ ♆
16 09	☽ # ♃
16 25	☽ ✶ ♅
17 28	☽ ⊥ ☉
17 57	♄ ⊥ ♅
21 49	☽ △ ♇
23 51	☽ △ ♂
19 Monday	
04 37	☽ ⊻ ♄
06 40	☽ ⊻ ☉
08 32	☽ # ♇
14 01	☽ ∥ ♀
14 35	☽ Q ♆
16 07	☽ ☌ ♀
16 39	☽ Q ♅
16 47	☽ ⊥ ♄
20 24	☽ ☌ ☿
22 54	☽ ☍ ♃
20 Tuesday	
03 57	☽ bQ ♇
10 57	☽ bQ ♄
14 59	☽ ☍ ♃
21 54	☽ ± ♇
21 Wednesday	
01 30	☽ ∥ ☉
02 36	☽ □ ♆
04 39	☽ □ ♅
09 44	☽ # ♄
09 48	☽ ⚻ ♇
14 14	☽ □ ♂
16 57	☽ ✶ ♄
23 49	☽ ☌ ☉
22 Thursday	
02 45	☿ # ♃
03 30	☽ ⊻ ♀
06 27	☿ bQ ♇
09 55	☽ ⚻ ♃
14 14	♀ St D
15 15	☽ ⊥ ♀
16 35	☽ ⊻ ☿
16 41	☽ Q ♄
21 33	☽ ± ♃
22 35	♆ St R
23 Friday	
02 55	☽ Q ♂
05 59	☽ ⊥ ☿
08 48	☽ ∠ ♀
13 26	☽ △ ♆
14 51	☽ bQ ♃
15 26	☽ △ ♅
20 18	☽ ☍ ♇
22 43	☉ ⊻ ♀
24 Saturday	
01 56	☽ ∠ ☿
02 58	☽ ✶ ♂
03 43	☽ □ ♄
04 15	☽ # ♆
09 10	♀ # ♇
13 44	☽ ✶ ♀
14 54	☽ ⊻ ☉
16 10	☿ ∠ ♄
18 10	☽ bQ ♆
19 20	☽ △ ♃
19 55	☉ # ♄
20 08	☽ bQ ♅
21 54	☽ # ♅
25 Sunday	
02 27	♂ ⚻ ♄
03 13	☽ ⊥ ☉
08 33	☽ ∠ ♂
10 39	☽ ✶ ☿
11 10	☽ ± ♆
12 35	☽ Q ♀
12 55	☽ ∥ ♂
13 07	☽ ± ♅
21 30	☽ ∠ ☉
22 23	☽ ⚻ ♆
26 Monday	
00 19	☽ ⚻ ♅
00 21	☽ ∥ ♂
02 00	☽ ⊥ ♂
04 53	☽ ⚻ ♇
08 47	☿ ∥ ♀
10 04	♅ St R
10 32	☽ # ♅
12 19	☽ Q ☿
12 26	☽ △ ♄
13 31	☽ ⊻ ♂
15 56	☽ ± ♇
18 53	☿ # ♇
19 28	☉ ⚻ ♃
22 16	☽ □ ♀
27 Tuesday	
02 43	☽ □ ♃
03 21	☽ ✶ ☉
03 33	☽ # ♆
08 19	☽ bQ ♇
10 57	♀ Q ♅
15 33	☿ ± ♇
15 54	☽ bQ ♄
28 Wednesday	
01 50	☽ □ ☿
02 41	☽ Q ☉
05 01	☽ ☍ ♆
06 52	☽ ☍ ♅
08 04	☽ ± ♄
09 12	♂ # ♅
10 48	☽ Q ♃
11 08	☽ △ ♇
18 42	☽ ⚻ ♄
21 25	☽ ☌ ♂
29 Thursday	
02 18	☿ □ ♆
04 37	☽ △ ♀
07 41	☽ ✶ ♃
08 21	☽ ∥ ☉
12 41	☽ □ ☉
16 24	☿ □ ♅
17 03	☽ # ♄
30 Friday	
06 53	☽ bQ ♀
09 05	☽ ⚻ ♆
09 14	☽ ∠ ♃
10 51	☽ ⚻ ♅
13 32	☽ △ ☿
14 49	☽ □ ♇
19 12	☽ ± ♆
20 25	☽ ∥ ☿
20 57	☽ ± ♅
22 20	♀ ⊥ ♄
22 22	☽ ☍ ♄
22 22	☽ ± ♀
23 30	☽ ⚻ ♃

All ephemeris data is given at 12.00 UT and the Moon's longitude is additionally given for 24.00 UT

MAY 1993

LONGITUDES

Date	Sidereal time h m s	Sun ☉ ° ' "	Moon ☽ ° ' "	Moon ☽ 24.00 ° ' "	Mercury ☿ ° '	Venus ♀ ° '	Mars ♂ ° '	Jupiter ♃ ° '	Saturn ♄ ° '	Uranus ♅ ° '	Neptune ♆ ° '	Pluto ♇ ° '
01	02 37 42	11 ♉ 07 20	07 ♍ 12 37	14 ♍ 28 58	25 ♈ 29	05 ♈ 12	01 ♌ 41	06 ♎ 08	29 ♒ 04	22 ♑ 11	21 ♑ 08	24 ♏ 31
02	02 41 38	12 05 32	21 ♍ 48 34	29 ♍ 10 46	27 20	05 32	02 10	06 R 03	29 07	22 R 11	21 R 08	24 R 29
03	02 45 35	13 03 42	06 ♎ 34 44	13 ♎ 59 34	29 ♈ 13	05 54	02 40	05 58	29 11	22 10	21 07	24 27
04	02 49 31	14 01 50	21 ♎ 24 18	28 ♎ 47 58	01 ♉ 08	06 17	03 09	05 53	29 15	22 10	21 07	24 26
05	02 53 28	14 59 56	06 ♏ 09 32	13 ♏ 28 06	03 04	06 42	03 39	05 48	29 18	22 09	21 07	24 24
06	02 57 24	15 58 00	20 ♏ 42 48	27 ♏ 52 54	05 02	07 09	04 09	05 44	29 21	22 09	21 06	24 22
07	03 01 21	16 56 03	04 ♐ 57 46	11 ♐ 56 57	07 02	07 37	04 38	05 39	29 25	22 08	21 06	24 21
08	03 05 18	17 54 05	18 ♐ 50 09	25 ♐ 37 11	09 04	08 07	05 08	05 35	29 28	22 08	21 05	24 19
09	03 09 14	18 52 05	02 ♑ 18 02	08 ♑ 52 49	11 07	08 38	05 39	05 31	29 31	22 07	21 05	24 18
10	03 13 11	19 50 04	15 ♑ 21 44	21 ♑ 45 07	13 12	09 11	06 09	05 27	29 34	22 07	21 04	24 16
11	03 17 07	20 48 01	28 ♑ 03 22	04 ♒ 16 55	15 18	09 45	06 39	05 23	29 37	22 06	21 04	24 14
12	03 21 04	21 45 57	10 ♒ 26 19	16 ♒ 32 04	17 26	10 20	07 10	05 20	29 39	22 05	21 03	24 13
13	03 25 00	22 43 52	22 ♒ 34 46	28 ♒ 35 00	19 34	10 57	07 40	05 17	29 42	22 04	21 02	24 11
14	03 28 57	23 41 45	04 ♓ 33 19	10 ♓ 30 19	21 44	11 35	08 11	05 13	29 45	22 03	21 02	24 09
15	03 32 53	24 39 37	16 ♓ 26 33	22 ♓ 22 34	23 54	12 14	08 42	05 10	29 47	22 03	21 01	24 08
16	03 36 50	25 37 28	28 ♓ 18 52	04 ♈ 15 56	26 05	12 54	09 13	05 07	29 50	22 02	21 00	24 06
17	03 40 47	26 35 17	10 ♈ 14 13	16 ♈ 14 08	28 ♉ 17	13 35	09 44	05 05	29 52	22 01	21 00	24 04
18	03 44 43	27 33 06	22 ♈ 16 03	28 ♈ 20 16	00 ♊ 28	14 17	10 16	05 02	29 54	22 00	20 59	24 03
19	03 48 40	28 30 53	04 ♉ 27 05	10 ♉ 36 45	02 39	15 00	10 47	05 00	29 57	21 58	20 58	24 01
20	03 52 36	29 ♉ 28 39	16 ♉ 49 26	23 ♉ 05 19	04 50	15 44	11 19	04 58	29 ♒ 59	21 57	20 57	23 59
21	03 56 33	00 ♊ 26 24	29 ♉ 24 29	05 ♊ 47 02	06 59	16 29	11 50	04 56	00 ♓ 01	21 56	20 56	23 58
22	04 00 29	01 24 07	12 ♊ 13 01	18 ♊ 42 26	09 08	17 15	12 22	04 54	00 02	21 55	20 55	23 56
23	04 04 26	02 21 49	25 ♊ 15 17	01 ♋ 51 34	11 16	18 01	12 54	04 52	00 04	21 54	20 55	23 54
24	04 08 22	03 19 30	08 ♋ 31 12	15 ♋ 14 10	13 21	18 49	13 26	04 51	00 06	21 52	20 54	23 53
25	04 12 19	04 17 09	22 ♋ 00 58	28 ♋ 49 45	15 25	19 37	13 58	04 49	00 08	21 51	20 53	23 51
26	04 16 16	05 14 47	05 ♌ 42 11	12 ♌ 37 35	17 27	20 26	14 30	04 48	00 09	21 50	20 52	23 49
27	04 20 12	06 12 23	19 ♌ 35 49	26 ♌ 36 42	19 27	21 15	15 02	04 47	00 10	21 48	20 51	23 48
28	04 24 09	07 09 58	03 ♍ 40 03	10 ♍ 45 38	21 25	22 05	15 34	04 47	00 12	21 47	20 50	23 46
29	04 28 05	08 07 32	17 ♍ 53 11	25 ♍ 02 22	23 20	22 56	16 07	04 46	00 13	21 45	20 49	23 44
30	04 32 02	09 05 04	02 ♎ 12 50	09 ♎ 24 08	25 13	23 48	16 39	04 46	00 14	21 44	20 47	23 43
31	04 35 58	10 ♊ 02 34	16 ♎ 35 51	23 ♎ 47 26	27 ♊ 03	24 ♈ 40	17 ♌ 12	04 ♎ 45	00 ♓ 15	21 ♑ 42	20 ♑ 46	23 ♏ 41

DECLINATIONS

Date	Moon True ☊ ° '	Moon Mean ☊ ° '	Moon Latitude ° '	Sun ☉ ° '	Moon ☽ ° '	Mercury ☿ ° '	Venus ♀ ° '	Mars ♂ ° '	Jupiter ♃ ° '	Saturn ♄ ° '	Uranus ♅ ° '	Neptune ♆ ° '	Pluto ♇ ° '
01	12 ♐ 35	14 ♐ 02	05 S 11	15 N 10	04 N 03	07 N 59	03 N 43	21 N 41	01 S 01	12 S 56	22 S 04	21 S 06	04 S 28
02	12 R 31	13 59	05 08	15 28	01 S 28	08 45	03 40	21 33	00 59	12 55	22 04	21 06	04 28
03	12 28	13 56	04 45	15 46	06 59	09 32	03 38	21 25	00 57	12 54	22 04	21 06	04 27
04	12 23	13 53	04 03	16 03	12 06	10 19	03 37	21 18	00 55	12 53	22 04	21 06	04 27
05	12 20	13 50	03 05	16 20	16 29	11 06	03 37	21 10	00 54	12 52	22 04	21 06	04 26
06	12 17	13 47	01 56	16 37	19 47	11 54	03 38	21 02	00 52	12 51	22 04	21 07	04 26
07	12 16	13 43	00 S 40	16 54	21 47	12 41	03 40	20 53	00 51	12 50	22 04	21 07	04 26
08	12 D 16	13 40	00 N 36	17 10	22 22	13 29	03 43	20 45	00 49	12 49	22 04	21 07	04 25
09	12 17	13 37	01 48	17 26	21 37	14 17	03 47	20 37	00 48	12 48	22 05	21 07	04 25
10	12 18	13 34	02 52	17 42	19 42	15 04	03 51	20 28	00 46	12 47	22 05	21 07	04 24
11	12 20	13 31	03 46	17 57	16 52	15 51	03 56	20 20	00 45	12 46	22 05	21 07	04 24
12	12 21	13 27	04 28	18 12	13 19	16 37	04 02	20 11	00 44	12 45	22 05	21 07	04 24
13	12 22	13 24	04 57	18 27	09 18	17 23	04 09	20 02	00 43	12 44	22 05	21 07	04 23
14	12 R 21	13 21	05 12	18 42	04 59	18 07	04 16	19 53	00 42	12 43	22 05	21 07	04 23
15	12 20	13 18	05 14	18 56	00 S 31	18 50	04 24	19 44	00 41	12 43	22 06	21 07	04 23
16	12 18	13 15	05 03	19 10	03 N 58	19 32	04 32	19 35	00 40	12 42	22 06	21 07	04 22
17	12 16	13 12	04 39	19 24	08 19	20 12	04 41	19 26	00 39	12 42	22 06	21 07	04 22
18	12 13	13 08	04 02	19 37	12 24	20 51	04 51	19 17	00 38	12 41	22 06	21 08	04 22
19	12 11	13 05	03 14	19 50	16 02	21 27	05 01	19 07	00 37	12 41	22 06	21 08	04 22
20	12 10	13 02	02 15	20 02	19 02	22 02	05 12	18 58	00 37	12 40	22 06	21 08	04 21
21	12 09	12 59	01 N 10	20 15	21 10	22 33	05 23	18 48	00 36	12 39	22 07	21 08	04 21
22	12 D 09	12 56	00 00	20 26	22 15	23 03	05 35	18 38	00 36	12 39	22 07	21 08	04 21
23	12 09	12 52	01 S 12	20 38	22 09	23 30	05 47	18 28	00 35	12 39	22 07	21 08	04 21
24	12 10	12 49	02 21	20 49	20 49	23 54	06 00	18 18	00 35	12 38	22 07	21 08	04 20
25	12 10	12 46	03 23	21 00	18 18	24 16	06 13	18 08	00 35	12 38	22 08	21 08	04 20
26	12 11	12 43	04 15	21 11	14 43	24 35	06 27	17 58	00 34	12 37	22 08	21 09	04 20
27	12 11	12 40	04 53	21 21	10 18	24 51	06 41	17 47	00 34	12 37	22 08	21 09	04 20
28	12 12	12 37	05 13	21 30	05 N 17	25 04	06 55	17 37	00 34	12 37	22 08	21 09	04 19
29	12 R 12	12 33	05 15	21 40	00 S 03	25 15	07 09	17 27	00 34	12 37	22 09	21 09	04 19
30	12 11	12 30	04 58	21 49	05 26	25 24	07 24	17 16	00 34	12 36	22 09	21 09	04 19
31	12 ♐ 11	12 ♐ 27	04 S 22	21 N 57	10 S 33	25 N 30	07 N 39	17 N 05	00 S 34	12 S 36	22 S 09	21 S 09	04 S 19

ZODIAC SIGN ENTRIES

Date	h	m	Planets
01	00	00	☽ ♍
03	01	20	☽ ♎
03	21	54	☿ ♉
05	01	57	☽ ♏
07	03	34	☽ ♐
09	07	51	☽ ♑
11	15	44	☽ ♒
14	02	50	☽ ♓
16	15	24	☽ ♈
18	06	53	☿ ♊
19	03	16	☽ ♉
21	01	02	☉ ♊
21	04	58	♄ ♓
21	13	07	☽ ♊
23	20	38	☽ ♋
26	02	03	☽ ♌
28	05	46	☽ ♍
30	08	18	☽ ♎

LATITUDES

Date	Mercury ☿ ° '	Venus ♀ ° '	Mars ♂ ° '	Jupiter ♃ ° '	Saturn ♄ ° '	Uranus ♅ ° '	Neptune ♆ ° '	Pluto ♇ ° '
01	02 S 01	01 N 47	01 N 57	01 N 33	01 S 12	00 S 27	00 N 41	14 N 51
04	01 40	01 14	01 53	01 32	01 13	00 27	00 41	14 51
07	01 15	00 42	01 51	01 32	01 14	00 28	00 41	14 51
10	00 46	00 N 14	01 48	01 31	01 14	00 28	00 41	14 51
13	00 S 15	00 S 12	01 45	01 30	01 15	00 28	00 41	14 51
16	00 N 16	00 36	01 42	01 30	01 15	00 28	00 41	14 51
19	00 47	00 58	01 39	01 29	01 16	00 28	00 41	14 51
22	01 15	01 17	01 36	01 28	01 16	00 28	00 41	14 50
25	01 38	01 34	01 34	01 28	01 17	00 28	00 41	14 50
28	01 55	01 49	01 31	01 27	01 18	00 28	00 41	14 50
31	02 N 05	02 S 02	01 N 29	01 N 26	01 S 18	00 S 28	00 N 41	14 N 49

DATA

Julian Date	2449109
Delta T	+59 seconds
Ayanamsa	23° 46' 06"
Synetic vernal point	05° ♓ 20' 54"
True obliquity of ecliptic	23° 26' 22"

MOON'S PHASES, APSIDES AND POSITIONS ☽

Date	h	m	Phase	Longitude ° '	Eclipse Indicator
06	03	34	○	15 ♏ 38	
13	12	20	☾	22 ♒ 45	
21	14	07	●	00 ♊ 31	Partial
28	18	21	☽	07 ♍ 25	

Day	h	m		
04	00	18	Perigee	
15	21	52	Apogee	
31	11	16	Perigee	
02	05	39	OS	
08	10	13	Max dec	22° S 23'
15	14	46	ON	
22	22	10	Max dec	22° N 22'
29	11	46	OS	

LONGITUDES

Date	Chiron ⚷ ° '	Ceres ⚳ ° '	Pallas ⚴ ° '	Juno ⚵ ° '	Vesta ⚶ ° '	Black Moon Lilith ⚸ ° '
01	17 ♌ 23	06 ♈ 49	04 ♓ 12	22 ♋ 34	25 ♒ 58	22 ♓ 01
11	17 ♌ 43	10 ♈ 28	06 ♓ 20	26 ♋ 30	29 ♒ 39	23 ♓ 08
21	18 ♌ 14	14 ♈ 00	08 ♓ 30	00 ♌ 33	03 ♓ 00	24 ♓ 15
31	18 ♌ 54	17 ♈ 25	09 ♓ 36	04 ♌ 40	05 ♓ 56	25 ♓ 22

All ephemeris data is given at 12.00 UT and the Moon's longitude is additionally given for 24.00 UT

ASPECTARIAN

h m	Aspects
01 Saturday	
00 17	☽ ⊥ ♃
02 30	☽ ⊻ ♂
08 36	☽ ⊼ ♀
10 08	☽ ⋕ ♇
10 13	☽ bQ ♆
10 13	☽ ⊻ ♃
10 26	☉ Q ♄
11 57	☽ bQ ♅
12 49	☽ ⊥ ♂
13 31	☽ ∥ ♀
14 57	☉ ⊥ ♀
18 11	☽ bQ ☿
18 56	☽ △ ☉
20 44	☽ Q ♇
02 Sunday	
01 20	☽ ⋕ ♃
04 09	☽ ∠ ♂
09 54	☽ ∥ ♃
10 53	☽ △ ♆
10 59	☉ ± ♃
11 06	☽ ± ☿
12 36	☽ △ ♅
16 21	☽ ⚹ ♇
21 13	☽ bQ ☉
21 24	☽ ⋕ ♀
22 18	☽ ⊼ ☿
23 58	☽ ⊼ ♄
03 Monday	
00 54	☽ ∥ ♇
05 26	☽ ⚹ ♂
09 44	☽ ± ♄
10 52	☽ ☍ ♀
11 01	☽ ☌ ♃
11 35	☿ ⚹ ♄
12 50	☽ ± ☉
15 39	♀ ☍ ♃
16 39	☽ ∠ ♇
23 14	☽ ⊼ ☉
04 Tuesday	
00 21	☽ bQ ♄
01 32	☽ Q ♂
01 42	☽ ⋕ ☿
07 11	☽ ⊥ ♇
11 32	☽ □ ♆
13 14	☽ □ ♅
15 54	☽ ∥ ♄
16 54	☽ ⊻ ♇
05 Wednesday	
00 46	☽ △ ♄
06 11	☽ ☍ ☿
07 45	☽ □ ♂
11 02	☽ ⋕ ☉
11 25	☽ ⊻ ♃
12 55	☽ ⊼ ♀
16 50	☽ Q ♆
18 33	☽ Q ♅
21 13	☽ ⊥ ♃
21 23	♂ ⋕ ♆
21 27	☿ □ ♂
23 04	☽ ± ♀
06 Thursday	
03 34	☽ ☍ ☉
12 02	☽ ∠ ♃
12 39	☽ ⚹ ♆
14 23	☽ ⚹ ♅
14 28	☽ bQ ♀
18 06	☽ ☌ ♇
20 01	☿ ⊼ ♃
07 Friday	
00 01	☽ ⋕ ♂
01 52	☽ ∥ ♆
02 32	☽ □ ♄
11 26	☽ △ ♂
13 11	☽ ⚹ ♃
13 56	☽ ∠ ♆
15 43	☽ ∠ ♅
16 04	☿ ⋕ ♄
16 09	☽ ⊼ ☿
16 43	☽ △ ♀
18 20	☽ ∥ ♅
21 07	☿ ⊻ ♀
08 Saturday	
04 16	☽ ± ☿
05 27	☽ ⊥ ♆
07 16	☽ ⊥ ♅
09 35	☽ Q ♄
09 49	☽ Q ♃
10 14	☽ ⊼ ☉
14 23	☽ bQ ♂
15 58	☽ ⊻ ♆
17 48	☽ ⊻ ♅
21 38	☽ ± ☉
21 40	☽ ⊻ ♇
22 53	☽ bQ ☿
09 Sunday	
02 19	☽ ∥ ♅
06 45	♂ ⚹ ♃
06 57	☽ ⚹ ♄
07 01	☽ ± ♂
08 23	☽ ⊥ ♇
13 15	♃ ± ♄
15 04	☽ bQ ☉
16 28	☿ ± ♃
16 39	☿ Q ♄
17 49	☽ □ ♃
18 19	☽ ⊼ ♂
19 50	☽ ∥ ♆
10 Monday	
00 03	☽ □ ♀
00 44	☽ ∠ ♇
03 15	☽ ⋕ ♂
07 12	☽ △ ☿
10 30	☽ ∠ ♄
15 18	♀ bQ ♇
21 04	☽ △ ☉
22 42	☽ ☌ ♆
11 Tuesday	
00 40	☽ ☌ ♅
02 27	☉ bQ ♃
03 29	☽ ⊥ ♄
04 12	☽ ⋕ ☉
04 44	☽ ⚹ ♇
11 23	☽ Q ♀
15 00	☽ ⊻ ♄
17 56	☽ ⋕ ☿
18 24	☉ △ ♆
18 59	☿ ⊥ ♀
12 Wednesday	
02 05	☽ △ ♃
03 46	☽ Q ♇
05 20	☽ ☍ ♂
11 48	☽ ⚹ ♀
15 34	☽ ∥ ♄
19 50	☉ △ ♅
13 Thursday	
04 44	☽ □ ☿
07 26	☽ bQ ♃
08 56	☽ ⊻ ♆
10 59	☽ ⊻ ♅
12 20	☽ □ ☉
15 11	☽ □ ♇
19 06	☽ ∠ ♀
19 37	☿ bQ ♃
20 54	☽ ⊥ ♆
22 58	☽ ⊥ ♅
14 Friday	
01 20	☽ ± ♃
02 18	☽ ☌ ♄
04 14	☿ △ ♆
13 20	☽ ⊼ ♃
14 11	☽ ⊥ ♀
14 58	☽ ∠ ♆
15 17	☽ ∥ ♇
15 33	☿ △ ♅
15 49	☽ ⋕ ♀
17 02	☽ ∠ ♅
19 39	☽ ⊼ ♂
23 04	☉ ☍ ♇
15 Saturday	
00 47	☽ Q ☿
02 59	☽ ⊻ ♀
03 41	☽ Q ☉
08 19	☽ ± ♂
09 25	♀ ⋕ ♇
11 09	☽ ∥ ♃
14 22	☿ ☍ ♇
16 46	☉ ∥ ☿
18 22	☽ ⋕ ♃
21 15	☽ ⚹ ♆
23 19	☽ ⚹ ♅
16 Sunday	
02 51	☉ ☌ ☿
03 21	☽ bQ ♂
03 30	☽ △ ♇
06 05	☽ ⚹ ☉
06 30	☽ ⚹ ☿
13 24	☿ ∥ ♂
14 13	☽ ⋕ ♇
15 04	☽ ⊻ ♄
15 13	☽ ∥ ♀
21 27	☽ Q ♆
23 30	☽ Q ♅
17 Monday	
01 40	☽ ☍ ♃
03 12	☽ ⊥ ♄
04 15	☿ Q ♂
09 40	☽ bQ ♇
10 57	☽ △ ♂
14 30	☉ ∥ ♂
14 57	☽ ∠ ☉
16 52	☿ ∠ ♀
19 06	☽ ☌ ♀
19 27	☽ ∠ ☿
21 18	☽ ∠ ♄
18 Tuesday	
03 37	☽ ± ♇
05 45	☿ □ ♄
09 27	☽ □ ♆
10 27	☽ ⊥ ☉
11 27	☽ □ ♅
13 44	☽ ⋕ ♄
15 30	☽ ⊼ ♇
17 19	☽ ⊥ ☿
22 49	☿ ⋕ ♆
23 21	☽ ⊻ ☉
19 Wednesday	
03 08	☽ ⚹ ♄
07 43	☽ ⊻ ☿
09 59	♀ ∠ ♄
13 04	☽ ⊼ ♃
20 Thursday	
00 42	☽ ± ♃
00 53	☽ □ ♂
02 37	☽ Q ♄
02 44	☉ Q ♂
09 46	☽ ⊻ ♀
11 25	☽ ∥ ♂
13 26	☿ △ ♃
15 38	☿ ⋕ ♅
18 00	☽ bQ ♃
19 55	☽ △ ♆
21 49	☽ △ ♅
22 00	☽ ⊥ ♀
23 07	☽ ∥ ☉
21 Friday	
00 23	☿ bQ ♆
00 53	☉ □ ♄
01 41	☽ ☍ ♇
11 24	☿ bQ ♅
11 34	☽ ⋕ ♆
12 51	☽ Q ♂
13 08	☽ □ ♄
14 07	☽ ☌ ☉
16 10	☽ ∠ ♀
22 22	☽ △ ♃
22 Saturday	
00 17	☽ bQ ♆
02 08	☽ bQ ♅
05 07	☽ ☌ ☿
06 53	☽ ⋕ ♅
12 17	☽ ⚹ ♂
17 01	☽ ± ♆
18 50	☽ ± ♅
21 53	☽ ⚹ ♀
23 Sunday	
04 03	☽ ⊼ ♆
05 52	☽ ⊼ ♅
08 32	♀ ± ♇
09 32	☽ ⊼ ♇
13 10	☽ ⋕ ♅
17 01	☽ ∠ ♂
20 26	☽ ± ♇
20 47	☽ △ ♄
21 13	☽ Q ♀
24 Monday	
01 55	☽ ⊻ ☉
05 24	☽ □ ♃
07 53	☽ ⋕ ♆
09 57	☽ ⊥ ♂
12 01	☽ ∥ ☉
12 38	☽ bQ ♇
13 07	☿ ⚹ ♂
13 33	☽ ⊥ ☉
21 08	☽ ⊻ ♂
22 14	☽ ⊻ ☿
23 47	☽ bQ ♄
25 Tuesday	
05 41	☿ ± ♆
06 49	☽ ∠ ☉
07 30	☽ □ ♀
10 00	☽ ☍ ♆
10 47	☽ ⊥ ☿
11 44	☽ ☍ ♅
13 20	☽ ∥ ♂
13 26	☽ Q ♃
15 15	☽ △ ♇
15 45	☽ ± ♄
16 58	☿ ± ♅
26 Wednesday	
01 09	☉ △ ♃
02 18	☽ ⊼ ♄
05 22	☽ ∠ ☿
07 20	☉ ⋕ ♆
10 26	☽ ⚹ ♃
11 09	☽ ⚹ ☉
23 54	☽ ⋕ ♄
27 Thursday	
00 24	♀ □ ♆
03 05	☉ bQ ♆
03 50	☽ ☌ ♂
09 26	☽ Q ☉
11 43	☽ ⚹ ☿
12 20	☽ ∠ ♃
14 08	☽ ⊼ ♆
15 01	☽ △ ♀
15 47	☽ ⊼ ♅
19 11	☽ □ ♇
20 49	☉ ∠ ♀
28 Friday	
00 23	☽ ± ♆
02 01	☽ ± ♅
02 36	☉ bQ ♅
03 25	♀ □ ♅
03 42	☽ ⊥ ♃
04 48	☽ ∥ ♀
04 48	☿ ⊼ ♆
06 06	☽ ☍ ♄
11 30	☽ Q ☿
13 53	☽ ⊻ ♃
15 39	☽ bQ ♆
16 25	☽ ⋕ ♇
16 28	☿ ⊼ ♅
17 16	☽ bQ ♅
18 10	☽ bQ ♀
18 21	☽ □ ☉
29 Saturday	
01 40	☽ Q ♇
02 59	☿ ⚹ ♀
08 54	☽ ⊻ ♂
09 14	☽ ⋕ ♃
10 18	☽ ± ♀
14 18	☽ ∥ ♃
16 54	☽ △ ♆
17 04	☿ ⊼ ♇
18 29	☽ △ ♅
19 23	☽ ⊥ ♂
21 01	☽ ⊼ ♀
21 48	☽ ⚹ ♇
22 32	☽ □ ☿
30 Sunday	
06 59	☽ ∥ ♇
08 41	☽ ⊼ ♄
09 46	♀ ⊼ ♇
11 02	☽ ∠ ♂
16 15	☽ ☌ ♃
18 43	☽ ± ♄
21 29	☽ ⋕ ♀
22 50	☽ ∠ ♇
31 Monday	
00 17	☽ △ ☉
09 45	☽ bQ ♄
13 03	☽ ⚹ ♂
13 49	☽ ⊥ ♇
18 57	☽ □ ♆
20 30	☽ □ ♅
22 25	☽ ∥ ♄
23 48	☽ ⊻ ♇

LONGITUDES

Date	Sidereal time h m s	Sun ☉ ° ′ ″	Moon ☽ ° ′ ″	Moon ☽ 24.00 ° ′ ″	Mercury ☿ ° ′	Venus ♀ ° ′	Mars ♂ ° ′	Jupiter ♃ ° ′	Saturn ♄ ° ′	Uranus ♅ ° ′	Neptune ♆ ° ′	Pluto ♇ ° ′
01	04 39 55	11 ♊ 00 03	00 ♏ 58 22	08 ♏ 08 05	28 ♊ 50	25 ♈ 33	17 ♌ 45	04 ♎ 45	00 ♓ 16	21 ♑ 41	20 ♑ 45	23 ♏ 40
02	04 43 51	11 57 31	15 ♏ 16 01	22 ♏ 21 36	00 ♋ 35	26 26	18 18	04 D 46	00 17	21 R 39	20 R 44	23 R 38
03	04 47 48	12 54 58	29 ♏ 24 19	06 ♐ 23 39	02 17	27 20	18 51	04 46	00 17	21 37	20 43	23 37
04	04 51 45	13 52 24	13 ♐ 19 10	20 ♐ 10 30	03 56	28 14	19 23	04 47	00 18	21 36	20 42	23 35
05	04 55 41	14 49 49	26 ♐ 57 19	03 ♑ 39 25	05 33	29 ♈ 09	19 57	04 47	00 19	21 34	20 40	23 33
06	04 59 38	15 47 14	10 ♑ 16 39	16 ♑ 48 59	07 06	00 ♉ 05	20 30	04 48	00 19	21 32	20 39	23 32
07	05 03 34	16 44 37	23 ♑ 16 28	29 ♑ 39 14	08 37	01 00	21 03	04 49	00 19	21 30	20 38	23 30
08	05 07 31	17 42 00	05 ♒ 57 28	12 ♒ 11 29	10 05	01 57	21 36	04 50	00 20	21 28	20 37	23 29
09	05 11 27	18 39 22	18 ♒ 21 36	24 ♒ 28 14	11 30	02 54	22 10	04 52	00 20	21 27	20 35	23 28
10	05 15 24	19 36 43	00 ♓ 31 51	06 ♓ 32 56	12 52	03 51	22 43	04 53	00 R 20	21 25	20 34	23 26
11	05 19 20	20 34 04	12 ♓ 32 01	18 ♓ 29 40	14 11	04 48	23 17	04 55	00 20	21 23	20 33	23 25
12	05 23 17	21 31 24	24 ♓ 26 26	00 ♈ 22 53	15 27	05 46	23 51	04 57	00 19	21 21	20 31	23 23
13	05 27 14	22 28 44	06 ♈ 19 38	12 ♈ 17 14	16 40	06 45	24 24	04 59	00 19	21 19	20 30	23 22
14	05 31 10	23 26 04	18 ♈ 16 14	24 ♈ 17 12	17 50	07 44	24 58	05 02	00 19	21 17	20 29	23 20
15	05 35 07	24 23 23	00 ♉ 20 38	06 ♉ 26 59	18 56	08 43	25 32	05 04	00 18	21 15	20 27	23 19
16	05 39 03	25 20 42	12 ♉ 36 43	18 ♉ 50 12	20 00	09 42	26 06	05 07	00 18	21 13	20 26	23 18
17	05 43 00	26 18 00	25 ♉ 07 45	01 ♊ 29 37	21 00	10 42	26 40	05 09	00 17	21 10	20 24	23 16
18	05 46 56	27 15 18	07 ♊ 55 59	14 ♊ 26 59	21 56	11 42	27 14	05 12	00 16	21 08	20 23	23 15
19	05 50 53	28 12 36	21 ♊ 02 36	27 ♊ 42 48	22 49	12 43	27 48	05 16	00 15	21 06	20 21	23 14
20	05 54 49	29 ♊ 09 53	04 ♋ 27 27	11 ♋ 16 18	23 39	13 44	28 23	05 19	00 14	21 04	20 20	23 12
21	05 58 46	00 ♋ 07 10	18 ♋ 09 04	25 ♋ 05 24	24 25	14 45	28 57	05 22	00 13	21 02	20 19	23 11
22	06 02 43	01 04 26	02 ♌ 04 53	09 ♌ 07 03	25 07	15 46	29 ♌ 32	05 26	00 12	21 00	20 17	23 10
23	06 06 39	02 01 42	16 ♌ 11 26	23 ♌ 17 32	25 45	16 48	00 ♍ 06	05 30	00 11	20 57	20 16	23 09
24	06 10 36	02 58 57	00 ♍ 24 51	07 ♍ 32 56	26 19	17 50	00 41	05 34	00 10	20 55	20 14	23 08
25	06 14 32	03 56 11	14 ♍ 41 20	21 ♍ 49 36	26 49	18 52	01 15	05 38	00 08	20 53	20 12	23 06
26	06 18 29	04 53 25	28 ♍ 57 24	06 ♎ 04 23	27 14	19 54	01 50	05 42	00 07	20 51	20 11	23 05
27	06 22 25	05 50 38	13 ♎ 10 15	20 ♎ 14 44	27 36	20 57	02 25	05 47	00 05	20 48	20 09	23 04
28	06 26 22	06 47 51	27 ♎ 17 37	04 ♏ 18 41	27 53	22 00	03 00	05 51	00 03	20 46	20 08	23 03
29	06 30 18	07 45 03	11 ♏ 17 44	18 ♏ 14 36	28 05	23 03	03 35	05 56	00 02	20 44	20 06	23 02
30	06 34 15	08 ♋ 42 15	25 ♏ 09 05	02 ♐ 01 03	28 ♋ 12	24 ♉ 07	04 ♍ 10	06 ♎ 01	00 ♓ 00	20 ♑ 41	20 ♑ 05	23 ♏ 01

Date	Moon True ☊ ° ′	Moon Mean ☊ ° ′	Moon ☽ Latitude ° ′
01	12 ♐ 11	12 ♐ 24	03 S 29
02	12 R 11	12 21	02 24
03	12 D 11	12 18	01 S 10
04	12 R 11	12 14	00 N 06
05	12 11	12 11	01 21
06	12 11	12 08	02 30
07	12 10	12 05	03 29
08	12 10	12 02	04 16
09	12 09	11 58	04 50
10	12 08	11 55	05 10
11	12 08	11 52	05 17
12	12 D 08	11 49	05 09
13	12 08	11 46	04 49
14	12 09	11 43	04 16
15	12 10	11 39	03 31
16	12 11	11 36	02 36
17	12 12	11 33	01 33
18	12 12	11 30	00 N 24
19	12 R 12	11 27	00 S 49
20	12 11	11 24	02 00
21	12 10	11 20	03 05
22	12 07	11 17	04 01
23	12 05	11 14	04 43
24	12 02	11 11	05 08
25	12 00	11 08	05 14
26	12 00	11 04	05 01
27	12 D 00	11 01	04 29
28	12 00	10 58	03 42
29	12 02	10 55	02 41
30	12 ♐ 03	10 ♐ 52	01 S 32

DECLINATIONS

Date	Sun ☉ ° ′	Moon ☽ ° ′	Mercury ☿ ° ′	Venus ♀ ° ′	Mars ♂ ° ′	Jupiter ♃ ° ′	Saturn ♄ ° ′	Uranus ♅ ° ′	Neptune ♆ ° ′	Pluto ♇ ° ′
01	22 N 06	15 S 05	25 N 33	07 N 55	16 N 54	00 S 35	12 S 36	22 S 09	21 S 09	04 S 19
02	22 13	18 42	25 35	08 11	16 43	00 35	12 36	22 10	21 10	04 19
03	22 21	21 10	25 34	08 26	16 32	00 35	12 36	22 10	21 10	04 19
04	22 28	22 18	25 31	08 43	16 21	00 36	12 36	22 10	21 10	04 18
05	22 35	22 03	25 27	08 59	16 10	00 36	12 36	22 11	21 10	04 18
06	22 41	20 33	25 20	09 16	15 59	00 37	12 36	22 11	21 10	04 18
07	22 47	18 00	25 12	09 32	15 47	00 38	12 36	22 11	21 11	04 18
08	22 52	14 39	25 02	09 49	15 36	00 38	12 36	22 12	21 11	04 18
09	22 57	10 43	24 51	10 06	15 24	00 39	12 36	22 12	21 11	04 18
10	23 02	06 27	24 38	10 23	15 12	00 40	12 37	22 12	21 11	04 18
11	23 06	01 S 59	24 25	10 40	15 00	00 41	12 37	22 13	21 11	04 18
12	23 10	02 N 32	24 10	10 58	14 49	00 42	12 37	22 13	21 12	04 18
13	23 14	06 56	23 54	11 15	14 37	00 43	12 37	22 13	21 12	04 18
14	23 17	11 06	23 37	11 32	14 24	00 44	12 38	22 13	21 12	04 18
15	23 19	14 53	23 20	11 50	14 12	00 45	12 38	22 14	21 12	04 18
16	23 21	18 06	23 01	12 07	14 00	00 47	12 38	22 14	21 12	04 18
17	23 23	20 33	22 43	12 25	13 48	00 48	12 39	22 15	21 13	04 18
18	23 25	22 01	22 23	12 42	13 35	00 49	12 39	22 15	21 13	04 18
19	23 26	22 20	22 04	13 00	13 23	00 51	12 40	22 15	21 13	04 18
20	23 26	21 22	21 44	13 17	13 10	00 52	12 40	22 16	21 13	04 18
21	23 26	19 09	21 24	13 34	12 57	00 54	12 41	22 16	21 13	04 18
22	23 26	15 47	21 04	13 51	12 44	00 56	12 41	22 16	21 14	04 18
23	23 25	11 29	20 44	14 08	12 32	00 57	12 42	22 17	21 14	04 18
24	23 24	06 31	20 24	14 25	12 19	00 59	12 43	22 17	21 14	04 18
25	23 23	01 N 12	20 04	14 42	12 05	01 01	12 43	22 17	21 14	04 18
26	23 21	04 S 11	19 45	14 59	11 52	01 03	12 44	22 18	21 15	04 19
27	23 19	09 20	19 26	15 16	11 39	01 05	12 45	22 18	21 15	04 19
28	23 16	13 57	19 07	15 32	11 26	01 07	12 46	22 18	21 15	04 19
29	23 13	17 46	18 50	15 48	11 12	01 09	12 46	22 19	21 15	04 19
30	23 N 09	20 S 32	18 N 33	16 N 04	10 N 59	01 S 11	12 S 47	22 S 19	21 S 16	04 S 19

ZODIAC SIGN ENTRIES

Date	h	m	Planets
01	10	22	☽ ♏
02	03	54	☿ ♋
03	13	01	☽ ♐
05	17	26	☽ ♑
06	10	03	♀ ♉
08	00	39	☽ ♒
10	10	57	☽ ♓
12	23	14	☽ ♈
15	11	19	☽ ♉
17	21	12	☽ ♊
20	04	05	☽ ♋
21	09	00	☉ ♋
22	08	26	☽ ♌
23	07	42	♂ ♍
24	11	18	☽ ♍
26	13	45	☽ ♎
28	16	37	☽ ♏
30	08	29	♄ ♒
30	20	28	☽ ♐

LATITUDES

Date	Mercury ☿ ° ′	Venus ♀ ° ′	Mars ♂ ° ′	Jupiter ♃ ° ′	Saturn ♄ ° ′	Uranus ♅ ° ′	Neptune ♆ ° ′	Pluto ♇ ° ′
01	02 N 07	02 S 06	01 N 28	01 N 26	01 S 18	00 S 28	00 N 41	14 N 49
04	02 09	02 17	01 25	01 25	01 19	00 28	00 41	14 48
07	02 03	02 26	01 23	01 24	01 20	00 28	00 41	14 47
10	01 50	02 34	01 20	01 23	01 20	00 28	00 41	14 46
13	01 31	02 40	01 18	01 23	01 21	00 28	00 41	14 46
16	01 05	02 44	01 15	01 22	01 21	00 29	00 41	14 45
19	00 N 34	02 48	01 13	01 21	01 22	00 29	00 41	14 44
22	00 S 03	02 49	01 11	01 20	01 22	00 29	00 41	14 42
25	00 45	02 50	01 08	01 20	01 23	00 29	00 41	14 41
28	01 30	02 50	01 06	01 19	01 24	00 29	00 41	14 40
31	02 S 17	02 S 48	01 N 04	01 N 18	01 S 24	00 S 29	00 N 41	14 N 39

DATA

Julian Date	2449140
Delta T	+59 seconds
Ayanamsa	23° 46′ 10″
Synetic vernal point	05° ♓ 20′ 49″
True obliquity of ecliptic	23° 26′ 21″

MOON'S PHASES, APSIDES AND POSITIONS ☽

Date	h	m	Phase	Longitude ° ′	Eclipse Indicator
04	13	02	○	13 ♐ 55	total
12	05	36	☾	21 ♓ 16	
20	01	52	●	28 ♊ 46	
26	22	43	☽	05 ♎ 19	

Day	h	m		
12	16	20	Apogee	
25	17	23	Perigee	
04	19	40	Max dec	22° S 22′
11	22	31	0N	
19	06	01	Max dec	22° N 22′
25	17	19	0S	

LONGITUDES

Date	Chiron ⚷ ° ′	Ceres ⚳ ° ′	Pallas ⚴ ° ′	Juno ⚵ ° ′	Vesta ⚶ ° ′	Black Moon Lilith ⚸ ° ′
01	18 ♌ 59	17 ♈ 45	09 ♓ 43	05 ♌ 05	06 ♓ 12	25 ♓ 29
11	19 ♌ 48	20 ♈ 59	10 ♓ 42	09 ♌ 16	08 ♓ 37	26 ♓ 36
21	20 ♌ 45	24 ♈ 02	11 ♓ 13	13 ♌ 29	10 ♓ 29	27 ♓ 43
31	21 ♌ 48	26 ♈ 52	11 ♓ 11	17 ♌ 43	11 ♓ 42	28 ♓ 51

ASPECTARIAN

h m Aspects

01 Tuesday
01 11 ♃ St D
02 20 ☽ ☍ ♀
03 06 ☽ ⚼ ☉
07 56 ☽ △ ☿
09 52 ☽ Q ♂
10 49 ☽ △ ♄
18 20 ☽ ⊻ ♃
19 14 ☽ ± ☉
22 42 ☽ # ♂
23 04 ☿ ± ♇

02 Wednesday
00 15 ☉ # ♅
01 01 ☽ Q ♆
02 34 ☽ Q ♅
04 25 ☽ ⊥ ♃
06 02 ☽ ⊼ ☉
07 43 ☿ △ ♄
12 37 ☽ ⚼ ☿
17 19 ☽ □ ♂
19 36 ☽ ∠ ♃
21 14 ☽ ⚹ ♆
22 46 ☽ ⚹ ♅

03 Thursday
02 08 ☽ ☌ ♇
05 57 ☽ ± ☿
08 13 ☽ ⊼ ♀
11 58 ☽ ∥ ♆
13 31 ☽ □ ♄
17 36 ☽ ⊼ ☿
19 12 ☽ ± ♀
21 12 ☽ ⚹ ♃
22 49 ☽ ∠ ♆

04 Friday
00 22 ☽ ∠ ♅
06 57 ☽ ∥ ♅
11 51 ☽ ⚼ ♀
13 02 ☽ ☍ ☉
14 24 ☽ ⊥ ♆
15 57 ☽ ⊥ ♅
18 02 ☽ Q ♃
20 43 ☽ Q ♄
22 11 ☿ ∠ ♂
23 04 ☽ △ ♂

05 Saturday
00 30 ☿ □ ♃
00 54 ☽ ⊻ ♆
02 28 ☽ ⊻ ♅
05 04 ♂ ∠ ♃
05 59 ☽ ⊻ ♇
08 10 ☉ ± ♆
08 20 ☽ ∥ ♅
16 12 ☽ △ ♀
16 38 ☽ ⊥ ♇
17 59 ☽ ⚹ ♄

06 Sunday
02 03 ☽ □ ♃
02 56 ☽ ⚼ ♂
04 10 ☽ ∥ ♆
05 28 ☽ ☍ ☿
05 51 ☉ ± ♅
08 50 ☽ ∠ ♇
18 16 ♀ ⚹ ♄
18 36 ♂ ⊼ ♆
20 04 ☽ ± ♂
21 14 ☽ ∠ ♄
22 54 ☽ ⊼ ☉

07 Monday
07 05 ☽ ☌ ♆
07 40 ☽ ⊼ ♂
08 42 ☽ ☌ ♅
10 16 ☿ ⚼ ♇
10 56 ☽ ± ☉
12 26 ☽ ⚹ ♇
13 58 ☽ ⊥ ♄

08 Tuesday
01 16 ☽ ⊻ ♄
03 44 ☽ □ ♀
05 16 ☽ ⚼ ☉
05 20 ☽ # ♂
06 36 ♂ ⊼ ♅
09 52 ☽ △ ♃
11 06 ☽ Q ♇
20 57 ☽ ⊼ ☿

09 Wednesday
00 51 ☽ ∥ ♄
10 06 ☽ ± ☿
12 38 ☽ △ ☉
14 57 ☽ ⚼ ♃
15 22 ☽ # ♀
16 22 ☽ ⊻ ♆
17 23 ☽ Q ♀
18 01 ☽ ⊻ ♅
19 49 ☽ ☍ ♂
21 59 ☽ □ ♇

10 Thursday
04 09 ☽ ⊥ ♆
05 29 ♄ St R
05 50 ☽ ⊥ ♅
06 03 ☽ ⚼ ☿
08 44 ☽ ± ♃
11 36 ☽ ☌ ♄
19 10 ☽ ⚹ ♀
20 42 ☽ ⊼ ♃
22 01 ☽ ∠ ♆
23 36 ☽ ∥ ♇
23 42 ☽ ∠ ♅

11 Friday
11 28 ☉ ⊼ ♆
14 55 ♀ ⊼ ♃
15 43 ☽ △ ☿
17 19 ♂ □ ♇
18 52 ☽ ∥ ♃

12 Saturday
02 12 ☽ # ♃
03 56 ☽ ∠ ♀
04 06 ☽ ⚹ ♆
05 36 ☽ □ ☉
05 46 ☽ ⚹ ♅
07 41 ☉ ⊼ ♅
09 37 ☿ ⚼ ♄
09 53 ☽ △ ♇
10 44 ☽ ⊼ ♂
21 33 ☽ # ♇
23 27 ☽ ± ♂
23 44 ☽ ⊥ ♀
23 53 ☽ ⊻ ♄

13 Sunday
04 18 ☽ Q ♆
05 56 ☽ Q ♅
09 17 ☽ ☍ ♃
11 59 ☽ ⊥ ♄
12 55 ☽ ⊻ ♀
16 06 ☽ ⚼ ♇
18 30 ☽ ⚼ ♂
21 06 ☽ Q ☉

14 Monday
06 05 ☽ ∠ ♄
09 41 ☉ ⊼ ♇
10 08 ☽ ± ♇
11 01 ☽ □ ☿
14 50 ☽ ∥ ♀
16 24 ☽ □ ♆
18 00 ☽ □ ♅
21 19 ☽ # ♄
22 06 ☽ ⊼ ♇
23 11 ☽ ⚹ ☉

15 Tuesday
02 01 ☽ △ ♂
07 40 ☽ ∥ ♂
11 55 ☽ ⚹ ♄
12 35 ☉ ∥ ☿
21 19 ☽ ⊼ ♃

16 Wednesday
02 11 ☽ Q ☿
05 52 ☽ ☌ ♀
07 14 ☽ ∠ ☉
09 05 ☽ ± ♃
11 23 ☽ Q ♄
17 35 ☿ ⊥ ♂
22 03 ☿ ☍ ♆

17 Thursday
01 24 ♂ ± ♆
02 03 ☽ ⊥ ☉
02 30 ☽ ⚼ ♃
03 01 ☽ △ ♆
03 28 ☽ ⚹ ☿
04 30 ☽ △ ♅
08 29 ☽ ☍ ♇
14 24 ☽ ⊻ ☉
15 03 ☽ □ ♂
16 18 ☿ ☍ ♅
20 45 ☽ # ♆
21 43 ☽ □ ♄
23 26 ♀ ± ♃

18 Friday
06 55 ☽ △ ♃
07 16 ☽ ⚼ ♆
07 50 ♀ # ♄
08 08 ♂ ± ♅
08 41 ☽ ⚼ ♅
10 01 ☽ ∠ ☿
10 53 ☉ ⚹ ♂
19 33 ☽ ⊻ ♀
19 35 ☽ # ♅
20 40 ☽ ∥ ☿
22 17 ☿ # ♅
23 51 ☽ ± ♆

19 Saturday
01 14 ☽ ± ♅
01 16 ♀ Q ♄
02 03 ☽ Q ♂
03 48 ☽ ⊥ ☿
07 26 ☽ ⊥ ♀
10 46 ☽ ⊼ ♆
12 07 ☽ ⊼ ♅
15 26 ☽ ⊻ ☿
15 56 ☽ ⊼ ♇
16 09 ☽ # ♅
23 14 ☿ △ ♇

20 Sunday
00 43 ☽ ⚹ ♂
00 59 ☽ ∠ ♀
01 18 ☿ Q ♃
01 52 ☽ ☌ ☉
02 41 ☽ ± ♇
03 44 ☽ ∥ ☿
04 31 ☽ △ ♄
06 26 ♀ ∥ ♂
13 03 ☉ ± ♇
13 31 ☽ □ ♃
14 09 ☽ # ♆
18 37 ☽ ⚼ ♇

21 Monday
04 22 ☽ ∠ ♂
05 36 ☽ ⚹ ♀
05 59 ☿ ± ♄
06 55 ☽ ⚼ ♄
14 34 ☉ △ ♄
15 44 ☽ ☍ ♆
16 59 ☽ ☍ ♅
20 40 ☽ ⊥ ♂
20 42 ☽ △ ♇
21 05 ☽ Q ♃
22 30 ☽ ± ♄
23 25 ☽ ☌ ☿

22 Tuesday
00 10 ☿ # ♆
04 02 ☽ Q ♀
07 26 ☽ ⊻ ♂
07 36 ♂ ⊥ ♃
08 47 ☽ ⊼ ♄
10 09 ☽ ⊻ ☉
17 22 ♂ # ♄
17 45 ☽ ⚹ ♃
21 08 ☽ ⊥ ☉
22 34 ☽ ∥ ♀

23 Wednesday
05 36 ☽ # ♄
06 15 ☽ ∥ ♂
13 07 ☽ □ ♀
13 31 ☽ ∠ ☉
15 15 ♂ ☍ ♄
18 52 ☽ ⊼ ♆
19 19 ☽ ∠ ♃
20 02 ☽ ⊼ ♅
23 44 ☽ □ ♇

24 Thursday
04 49 ☽ ⊻ ☿
04 58 ☽ ± ♆
06 08 ☽ ± ♅
10 34 ☽ ⊥ ♃
11 35 ☽ ☍ ♄
12 28 ☽ ☌ ♂
15 19 ☽ ⊥ ☿
16 38 ☽ ⚹ ☉
20 06 ☽ ⚼ ♆
20 42 ☽ ⊻ ♃
21 14 ☽ ⚼ ♅
22 06 ☽ # ♇

25 Friday
05 59 ☽ Q ♇
07 00 ☽ ∠ ☿
12 47 ☽ # ♃
14 15 ☽ Q ☉
19 34 ☽ △ ♀
21 16 ☽ △ ♆
21 53 ☽ ∥ ♃
22 23 ☽ △ ♅

26 Saturday
02 08 ☽ ⚹ ♇
07 33 ☉ ∠ ♀
09 02 ☽ ⚹ ☿
12 34 ☽ ∥ ♇
13 57 ☽ ⊼ ♄
17 04 ☽ ⊻ ♂
18 11 ♀ △ ♆
22 43 ☽ □ ☉
22 50 ☽ ⚼ ♀
23 26 ☽ ☌ ♃

27 Sunday
00 03 ☽ ± ♄
03 23 ☽ ∠ ♇
03 37 ☽ ⊥ ♂
05 48 ☽ Q ☿
07 42 ♀ ⚼ ♃
08 47 ♀ △ ♅
10 11 ☉ □ ♃
15 14 ☽ ⚼ ♄
15 15 ☽ ± ♀
18 36 ☽ ⊥ ♇
19 30 ☽ ∠ ♂
23 06 ☽ # ♂
23 50 ☽ □ ♆

28 Monday
00 55 ☽ □ ♅
02 15 ☽ ⊼ ♀
04 47 ☽ ⊻ ♇
05 26 ☽ ∥ ♄
13 01 ☽ □ ☿
16 43 ☽ △ ♄
21 58 ☽ # ♀
22 11 ☽ ⚹ ♂

29 Tuesday
02 44 ☽ ⊻ ♃
05 27 ☽ △ ☉
06 31 ☽ Q ♆
07 36 ☽ Q ♅
11 35 ♀ ☍ ♇
13 06 ☽ ⊥ ♃
19 00 ☉ ⚼ ♇
19 23 ☽ # ☿
19 43 ☽ Q ♂

30 Wednesday
03 12 ☽ ⚹ ♆
04 16 ☽ ⚹ ♅
04 46 ☽ ∠ ♃
07 53 ♃ ± ♄
08 17 ☽ ☌ ♇
09 18 ☽ ⚼ ☉
10 02 ☽ ☍ ♀
13 57 ☿ ⊥ ♂
17 22 ☽ △ ☿
20 26 ☽ □ ♄
21 01 ☽ ∥ ♆

All ephemeris data is given at 12.00 UT and the Moon's longitude is additionally given for 24.00 UT

JULY 1993

LONGITUDES

Date	Sidereal time h m s	Sun ☉ ° ′ ″	Moon ☽ ° ′ ″	Moon ☽ 24.00 ° ′ ″	Mercury ☿ ° ′	Venus ♀ ° ′	Mars ♂ ° ′	Jupiter ♃ ° ′	Saturn ♄ ° ′	Uranus ♅ ° ′	Neptune ♆ ° ′	Pluto ♇ ° ′
01	06 38 12	09 ♋ 39 26	08 ♐ 50 17	15 ♐ 36 38	28 ♋ 15	25 ♉ 10	04 ♍ 45	06 ♎ 06	29 ♒ 58	20 ♑ 39	20 ♑ 03	23 ♏ 00
02	06 42 08	10 36 38	22 ♐ 19 56	29 ♐ 00 00	28 R 14	26 14	05 20	06 11	29 R 56	20 R 37	20 R 01	22 R 59
03	06 46 05	11 33 49	05 ♑ 36 42	12 ♑ 09 54	28 07	27 18	05 55	06 17	29 54	20 34	20 00	22 58
04	06 50 01	12 31 00	18 ♑ 39 31	25 ♑ 05 29	27 57	28 22	06 31	06 22	29 51	20 32	19 58	22 57
05	06 53 58	13 28 11	01 ♒ 27 47	07 ♒ 46 26	27 41	29 ♉ 27	07 06	06 28	29 49	20 30	19 57	22 56
06	06 57 54	14 25 22	14 ♒ 01 31	20 ♒ 13 11	27 22	00 ♊ 31	07 42	06 33	29 47	20 27	19 55	22 55
07	07 01 51	15 22 33	26 ♒ 21 38	02 ♓ 27 08	26 59	01 36	08 17	06 39	29 44	20 25	19 53	22 55
08	07 05 47	16 19 45	08 ♓ 29 58	14 ♓ 30 32	26 31	02 41	08 53	06 45	29 41	20 22	19 52	22 54
09	07 09 44	17 16 57	20 ♓ 29 14	26 ♓ 26 33	26 01	03 47	09 28	06 52	29 39	20 20	19 50	22 53
10	07 13 41	18 14 09	02 ♈ 23 00	08 ♈ 19 05	25 27	04 52	10 04	06 58	29 36	20 18	19 49	22 52
11	07 17 37	19 11 21	14 ♈ 15 26	20 ♈ 12 36	24 52	05 58	10 40	07 05	29 33	20 15	19 47	22 51
12	07 21 34	20 08 34	26 ♈ 11 13	02 ♉ 11 54	24 14	07 03	11 15	07 11	29 30	20 13	19 45	22 51
13	07 25 30	21 05 48	08 ♉ 15 16	14 ♉ 21 54	23 35	08 09	11 51	07 18	29 27	20 10	19 44	22 50
14	07 29 27	22 03 02	20 ♉ 32 24	26 ♉ 47 17	22 56	09 15	12 27	07 25	29 24	20 08	19 42	22 49
15	07 33 23	23 00 17	03 ♊ 07 03	09 ♊ 32 07	22 17	10 22	13 03	07 32	29 21	20 05	19 40	22 49
16	07 37 20	23 57 33	16 ♊ 02 48	22 ♊ 39 21	21 38	11 28	13 39	07 39	29 18	20 03	19 39	22 48
17	07 41 16	24 54 49	29 ♊ 21 52	06 ♋ 10 21	21 01	12 35	14 16	07 46	29 15	20 01	19 37	22 48
18	07 45 13	25 52 05	13 ♋ 04 38	20 ♋ 04 25	20 26	13 41	14 52	07 54	29 11	19 58	19 36	22 47
19	07 49 10	26 49 22	27 ♋ 09 16	04 ♌ 18 35	19 54	14 48	15 28	08 01	29 08	19 56	19 34	22 47
20	07 53 06	27 46 41	11 ♌ 31 41	18 ♌ 47 44	19 25	15 55	16 04	08 09	29 05	19 53	19 32	22 46
21	07 57 03	28 43 58	26 ♌ 05 53	03 ♍ 25 14	19 01	17 02	16 41	08 17	29 01	19 51	19 31	22 46
22	08 00 59	29 ♋ 41 16	10 ♍ 44 53	18 ♍ 03 58	18 40	18 10	17 17	08 25	28 57	19 49	19 29	22 45
23	08 04 56	00 ♌ 38 34	25 ♍ 21 42	02 ♎ 37 25	18 25	19 17	17 54	08 33	28 54	19 46	19 28	22 45
24	08 08 52	01 35 53	09 ♎ 50 31	17 ♎ 00 33	18 14	20 25	18 30	08 41	28 50	19 44	19 26	22 44
25	08 12 49	02 33 12	24 ♎ 07 12	01 ♏ 10 15	18 09	21 32	19 07	08 49	28 46	19 42	19 24	22 44
26	08 16 45	03 30 32	08 ♏ 09 33	15 ♏ 05 07	18 D 10	22 40	19 44	08 57	28 43	19 39	19 23	22 44
27	08 20 42	04 27 52	21 ♏ 56 56	28 ♏ 45 08	18 17	23 48	20 21	09 06	28 39	19 37	19 21	22 44
28	08 24 39	05 25 12	05 ♐ 29 48	12 ♐ 11 05	18 29	24 56	20 57	09 15	28 35	19 35	19 20	22 44
29	08 28 35	06 22 33	18 ♐ 49 06	25 ♐ 24 00	18 48	26 04	21 34	09 23	28 31	19 32	19 18	22 44
30	08 32 32	07 19 55	01 ♑ 55 52	08 ♑ 24 48	19 13	27 13	22 11	09 32	28 27	19 30	19 17	22 43
31	08 36 28	08 ♌ 17 17	14 ♑ 50 52	21 ♑ 14 08	19 ♋ 45	28 ♊ 21	22 ♍ 48	09 ♎ 41	28 ♒ 23	19 ♑ 28	19 ♑ 15	22 ♏ 43

DECLINATIONS

Date	Moon True ☊ ° ′	Moon Mean ☊ ° ′	Moon ☽ Latitude ° ′	Sun ☉ ° ′	Moon ☽ ° ′	Mercury ☿ ° ′	Venus ♀ ° ′	Mars ♂ ° ′	Jupiter ♃ ° ′	Saturn ♄ ° ′	Uranus ♅ ° ′	Neptune ♆ ° ′	Pluto ♇ ° ′
01	12 ♐ 04	10 ♐ 49	00 S 18	23 N 05	22 S 04	18 N 17	16 N 20	10 N 45	01 S 13	12 S 48	22 S 20	21 S 16	04 S 19
02	12 R 04	10 45	00 N 56	23 01	22 17	18 02	16 36	10 32	01 16	12 49	22 20	21 16	04 20
03	12 02	10 42	02 05	22 56	21 14	17 48	16 51	10 18	01 18	12 50	22 20	21 16	04 20
04	11 59	10 39	03 07	22 51	19 03	17 35	17 07	10 04	01 20	12 51	22 21	21 16	04 20
05	11 54	10 36	03 57	22 45	15 58	17 23	17 21	09 51	01 23	12 52	22 21	21 17	04 20
06	11 49	10 33	04 36	22 40	12 13	17 13	17 36	09 37	01 25	12 53	22 21	21 17	04 20
07	11 43	10 30	05 00	22 33	08 01	17 04	17 50	09 23	01 28	12 54	22 22	21 17	04 21
08	11 38	10 26	05 11	22 26	03 S 35	16 56	18 05	09 09	01 31	12 55	22 22	21 17	04 21
09	11 34	10 23	05 07	22 19	00 N 57	16 50	18 18	08 55	01 33	12 56	22 23	21 18	04 21
10	11 31	10 20	04 51	22 12	05 24	16 46	18 32	08 41	01 36	12 57	22 23	21 18	04 22
11	11 30	10 17	04 22	22 04	09 39	16 43	18 45	08 26	01 39	12 59	22 23	21 18	04 22
12	11 D 30	10 14	03 42	21 56	13 33	16 41	18 57	08 12	01 42	13 00	22 24	21 18	04 22
13	11 31	10 10	02 51	21 47	16 57	16 41	19 10	07 58	01 44	13 01	22 24	21 19	04 22
14	11 32	10 07	01 52	21 38	19 41	16 42	19 22	07 43	01 47	13 02	22 24	21 19	04 23
15	11 34	10 04	00 N 46	21 29	21 32	16 45	19 34	07 29	01 50	13 03	22 25	21 19	04 23
16	11 R 34	10 01	00 S 24	21 19	22 18	16 49	19 45	07 14	01 53	13 05	22 25	21 19	04 23
17	11 32	09 58	01 35	21 09	21 51	16 55	19 56	07 00	01 56	13 06	22 25	21 20	04 24
18	11 29	09 55	02 42	20 58	20 07	17 02	20 06	06 45	01 59	13 07	22 26	21 20	04 24
19	11 24	09 51	03 41	20 48	17 07	17 09	20 16	06 30	02 03	13 09	22 26	21 20	04 25
20	11 17	09 48	04 27	20 36	13 03	17 18	20 25	06 16	02 06	13 10	22 27	21 20	04 25
21	11 10	09 45	04 57	20 25	08 10	17 28	20 35	06 01	02 09	13 11	22 27	21 21	04 25
22	11 03	09 42	05 07	20 13	02 N 43	17 38	20 43	05 46	02 12	13 13	22 27	21 21	04 26
23	10 57	09 39	04 58	20 01	02 S 43	17 49	20 51	05 31	02 16	13 14	22 28	21 21	04 26
24	10 53	09 36	04 29	19 48	08 01	18 01	20 59	05 16	02 19	13 16	22 28	21 22	04 27
25	10 52	09 32	03 44	19 35	12 50	18 13	21 06	05 01	02 23	13 17	22 28	21 22	04 27
26	10 D 51	09 29	02 46	19 22	16 51	18 25	21 13	04 46	02 26	13 19	22 29	21 22	04 28
27	10 52	09 26	01 40	19 09	19 52	18 37	21 19	04 31	02 30	13 20	22 29	21 22	04 28
28	10 53	09 23	00 S 29	18 55	21 42	18 49	21 25	04 16	02 33	13 22	22 29	21 22	04 28
29	10 R 53	09 20	00 N 42	18 41	22 16	19 01	21 30	04 01	02 37	13 23	22 29	21 23	04 29
30	10 51	09 16	01 50	18 26	21 35	19 13	21 35	03 45	02 40	13 25	22 30	21 23	04 29
31	10 ♐ 46	09 ♐ 13	02 N 51	18 N 12	19 S 47	19 N 23	21 N 39	03 N 30	02 S 44	13 S 26	22 S 30	21 S 23	04 S 30

ZODIAC SIGN ENTRIES

Date	h	m	Planets
03	01	49	☽ ♑
05	09	14	☽ ♒
06	00	21	♀ ♊
07	19	09	☽ ♓
10	07	11	☽ ♈
12	19	37	☽ ♉
15	06	07	☽ ♊
17	13	08	☽ ♋
19	16	47	☽ ♌
21	18	24	☽ ♍
22	09	51	☉ ♌
23	19	39	☽ ♎
25	22	00	☽ ♏
28	02	13	☽ ♐
30	08	27	☽ ♑

LATITUDES

Date	Mercury ☿ ° ′	Venus ♀ ° ′	Mars ♂ ° ′	Jupiter ♃ ° ′	Saturn ♄ ° ′	Uranus ♅ ° ′	Neptune ♆ ° ′	Pluto ♇ ° ′
01	02 S 17	02 S 48	01 N 04	01 N 18	01 S 24	00 S 29	00 N 41	14 N 39
04	03 03	02 45	01 02	01 18	01 25	00 29	00 41	14 37
07	03 46	02 42	00 59	01 17	01 26	00 29	00 41	14 36
10	04 21	02 37	00 57	01 16	01 26	00 29	00 41	14 35
13	04 46	02 32	00 55	01 16	01 27	00 29	00 41	14 33
16	04 56	02 26	00 53	01 15	01 27	00 29	00 41	14 32
19	04 51	02 19	00 51	01 14	01 28	00 29	00 41	14 30
22	04 33	02 12	00 48	01 13	01 28	00 29	00 41	14 29
25	04 06	02 04	00 46	01 13	01 29	00 29	00 41	14 27
28	03 22	01 56	00 44	01 13	01 29	00 29	00 41	14 25
31	02 S 38	01 S 47	00 N 42	01 N 12	01 S 29	00 S 29	00 N 41	14 N 24

DATA

Julian Date	2449170
Delta T	+59 seconds
Ayanamsa	23° 46′ 15″
Synetic vernal point	05° ♓ 20′ 44″
True obliquity of ecliptic	23° 26′ 21″

MOON'S PHASES, APSIDES AND POSITIONS ☽

Date	h	m	Phase	Longitude ° ′	Eclipse Indicator
03	23	45	○	12 ♑ 02	
11	22	49	☾	19 ♈ 37	
19	11	24	●	26 ♋ 48	
26	03	25	☽	03 ♏ 10	

Day	h	m	
10	10	46	Apogee
22	08	22	Perigee

Day	h	m		
02	04	00	Max dec	22° S 21′
09	07	00	0N	
16	15	32	Max dec	22° N 19′
23	00	10	0S	
29	10	47	Max dec	22° S 16′

LONGITUDES

Date	Chiron ⚷ ° ′	Ceres ⚳ ° ′	Pallas ⚴ ° ′	Juno ⚵ ° ′	Vesta ⚶ ° ′	Black Moon Lilith ☾ ° ′
01	21 ♌ 48	26 ♈ 52	11 ♓ 11	17 ♌ 43	11 ♓ 42	28 ♓ 51
11	22 ♌ 57	29 ♈ 26	10 ♓ 36	21 ♌ 58	12 ♓ 12	29 ♓ 58
21	24 ♌ 10	01 ♉ 42	09 ♓ 26	26 ♌ 21	11 ♓ 54	01 ♈ 05
31	25 ♌ 27	03 ♉ 36	07 ♓ 43	00 ♍ 28	10 ♓ 50	02 ♈ 12

ASPECTARIAN

01 Thursday
02 12 ☽ ± ☉
04 29 ☽ □ ♂
05 20 ☽ ∠ ♆
06 24 ☽ ∠ ♅
07 08 ☽ ⚹ ♃
13 33 ☽ ⊼ ☉
15 29 ☿ St R
19 49 ☽ ⚼ ☿
21 13 ☽ ⊥ ♆
22 16 ☽ ⊥ ♅
22 54 ☽ ∥ ♅
23 45 ♂ ⚼ ♆

02 Friday
04 09 ☽ Q ♄
04 32 ☽ Q ♃
07 53 ☽ ⊻ ♆
08 51 ☽ ∥ ♅
08 56 ☽ ⊻ ♅
11 49 ☽ ± ☿
13 10 ☽ ⊻ ♇
19 37 ☽ ⊼ ♀
22 30 ♂ ⚼ ♅
22 32 ☽ ⊼ ☿
23 57 ☽ ⊥ ♇

03 Saturday
01 38 ☽ ⚹ ♄
07 25 ☽ ± ♀
11 27 ☽ ∥ ♆
12 36 ☽ △ ♂
13 13 ☽ □ ♃
16 18 ☽ ∠ ♇
23 45 ☽ ☍ ☉

04 Sunday
01 21 ☽ ⚼ ♀
03 59 ☿ ⚹ ♀
04 58 ♂ ⊻ ♃
04 59 ☽ ∠ ♄
14 26 ☽ ☌ ♆
15 28 ☽ ☌ ♅
17 34 ☽ ⚼ ♂
19 59 ☽ ⚹ ♇
21 39 ☽ ⊥ ♄

05 Monday
01 04 ☽ # ☿
02 41 ☽ # ♀
05 02 ☽ ☍ ☿
07 50 ☽ △ ♀
08 54 ☽ ⊻ ♄
11 17 ☽ ± ♂
13 42 ☿ ∥ ♀
18 35 ☽ Q ♇
19 58 ♀ □ ♄
21 34 ☽ △ ♃
23 15 ☽ ⊼ ♂

06 Tuesday
08 00 ☽ ∥ ♄
12 50 ☽ ⊼ ☉
20 31 ☉ ⚼ ♄
23 23 ☽ ⊻ ♆

07 Wednesday
00 25 ☽ ⊻ ♅
01 26 ☽ ± ☉
02 43 ☽ ⚼ ♃
03 59 ☽ # ♂
05 15 ☽ □ ♇
11 05 ☽ ⊥ ♆
12 06 ☽ ⊥ ♅
13 10 ☽ ⊼ ☿
18 37 ☽ ☌ ♄
20 31 ☽ ± ♃
20 34 ☽ ⚼ ☉
23 20 ☽ □ ♀

08 Thursday
00 35 ☽ ± ☿
04 48 ☽ ∠ ♆
05 48 ☽ ∠ ♅
07 53 ☽ ∥ ♇
08 30 ☽ ⊼ ♃
12 47 ☽ ☍ ♂
17 48 ☽ ⚼ ☿
22 53 ☽ ∥ ♃

09 Friday
01 57 ☉ # ♅
05 00 ☽ △ ☉
10 42 ☽ ⚹ ♆
11 41 ☽ ⚹ ♅
14 51 ☽ Q ♀
15 18 ☽ # ♃
16 49 ☽ △ ♇
22 39 ☽ △ ☿

10 Saturday
06 20 ☽ # ♇
06 24 ☽ ⊻ ♄
10 48 ♀ ⚼ ♆
10 51 ☽ Q ♆
11 49 ☽ Q ♅
17 32 ☽ ⚹ ♀
18 29 ☽ ⊥ ♄
19 59 ☿ ∠ ♂
21 02 ♀ ⚼ ♅
21 21 ☽ ☍ ♃
23 05 ☽ ⚼ ♇

11 Sunday
04 21 ☽ ⊼ ♂
04 50 ☿ Q ♃
05 24 ☽ ∥ ♂
12 36 ☽ ∠ ♄
17 06 ☽ ± ♂
17 14 ☽ ± ♇
19 43 ♂ Q ♇
22 49 ☽ □ ☉
23 07 ☽ □ ♆

12 Monday
00 03 ☽ □ ♅
02 31 ☉ ☍ ♆
02 52 ☽ ∠ ♀
05 18 ☽ ⊼ ♇
08 17 ☽ □ ☿
08 25 ☽ # ♄
12 09 ☽ ⚼ ♂
13 39 ☉ ☍ ♅
15 11 ♀ △ ♃
18 37 ☽ ⚹ ♄
22 42 ☽ ⊥ ♀

13 Tuesday
09 56 ☽ ∥ ☿
10 06 ☽ ⊼ ♃
11 47 ☽ ⊻ ♀
13 48 ☽ Q ☉
17 13 ☿ ± ♄
17 56 ☿ ∠ ♀
18 14 ☽ Q ☿
18 17 ☽ Q ♄
19 27 ☽ △ ♂
22 01 ☽ ± ♃

14 Wednesday
08 33 ☽ ∥ ♀
10 23 ☽ △ ♆
11 13 ☽ △ ♅
15 09 ☽ ⚹ ☉
15 39 ☽ ⚼ ♃
16 02 ☿ △ ♇
16 23 ☽ ⚹ ☿
16 24 ☽ ☍ ♇

15 Thursday
01 08 ☉ ☌ ☿
04 55 ☽ □ ♄
07 13 ☉ △ ♇
08 28 ☽ # ♆
11 13 ☽ ∥ ☉
14 55 ☽ ⚼ ♆
15 42 ☽ ⚼ ♅
19 25 ☽ ∠ ☿
20 19 ☉ ± ♄
20 21 ☽ △ ♃
21 53 ☽ ∠ ☉

16 Friday
02 48 ☽ ☌ ♀
07 24 ☽ □ ♂
07 37 ☽ ± ♆
08 21 ☽ ± ♅
10 51 ☉ # ♆
11 17 ☽ ⊥ ☿
15 46 ☽ ⊥ ☉
18 33 ☽ ⊼ ♆
19 16 ☽ ⊼ ♅
21 42 ☽ ⊻ ☿

17 Saturday
00 15 ☽ ⊼ ♇
03 27 ☽ ⊻ ☉
10 59 ☽ ± ♇
11 47 ☽ △ ♄
16 03 ♃ ∠ ♇
17 22 ☽ Q ♂
21 22 ☽ # ♆

18 Sunday
01 16 ☽ ∥ ☉
02 49 ☽ ⚼ ♇
02 56 ☽ □ ♃
09 58 ♀ ± ♆
12 06 ☽ ∥ ♀
12 48 ☉ Q ♃
13 09 ☽ ⊻ ♀
13 55 ☽ ⚼ ♄
15 13 ☽ ⚹ ♂
17 50 ♀ ± ♅
22 48 ☿ ⊥ ♀
23 10 ☽ ☍ ♆
23 47 ☽ ☍ ♅

19 Monday
00 09 ☽ ☌ ☿
00 19 ☽ ⊥ ♀
04 36 ☽ △ ♇
05 14 ☽ ± ♄
10 04 ☽ Q ♃
10 36 ☿ ☍ ♅
11 24 ☽ ☌ ☉
11 44 ☽ ∥ ☿
15 19 ☽ ⊼ ♄
16 50 ☽ ∠ ♀
17 49 ☽ ∠ ♂

20 Tuesday
05 32 ☿ ☍ ♆
06 21 ☽ ⚹ ♃
09 29 ☽ ⊥ ♂
11 20 ☽ # ♄
19 09 ♀ □ ♂
19 50 ☽ ⊻ ♂
19 52 ☽ ⚹ ♀

21 Wednesday
00 38 ☉ ∥ ♀
00 40 ☽ ⊻ ☿
01 12 ☽ ⊼ ♆
01 46 ☽ ⊼ ♅
06 32 ☽ □ ♇
07 20 ☽ ∠ ♃
10 16 ☽ ⊥ ☿
11 03 ☽ ± ♆
11 36 ☽ ± ♅
16 37 ☽ ⊻ ☉
16 46 ☽ ☍ ♄
17 13 ☽ Q ♀
18 45 ☉ ⊼ ♄
22 13 ☽ ⊥ ♃
22 15 ☽ ∥ ♂

22 Thursday
00 39 ☽ ∠ ☿
01 46 ☽ ⚼ ♆
02 18 ☽ ⚼ ♅
03 08 ☽ ⊥ ☉
04 50 ☽ # ♇
08 08 ☽ ⊻ ♃
12 01 ☽ Q ♇
14 32 ☽ # ♃
18 54 ☽ ∠ ☉
20 39 ☿ ⊻ ♀
23 11 ☽ ☌ ♂

23 Friday
00 45 ☽ ⚹ ☿
01 10 ☽ □ ♀
02 18 ☽ △ ♆
02 50 ☽ △ ♅
07 42 ☽ ⚹ ♇
10 01 ☽ ∥ ♃
15 38 ♀ ⊼ ♆
17 49 ☽ ⊼ ♄
19 39 ☽ ∥ ♇
20 13 ☽ Q ☿
21 20 ☽ ⚹ ☉
22 01 ♀ ⊼ ♅
23 55 ☽ # ♂

24 Saturday
03 23 ☿ ⚹ ♂
03 42 ☽ ± ♄
08 30 ☽ ∠ ♇
10 03 ☽ ☌ ♃
18 39 ☽ ⚼ ♄
18 44 ☽ Q ☉
23 33 ☽ ⊥ ♇

25 Sunday
01 58 ☽ □ ☿
03 10 ☽ ⊻ ♂
04 03 ☽ □ ♆
04 32 ☽ □ ♅
07 15 ☽ △ ♀
09 40 ☽ ⊻ ♇
13 46 ☽ ⊥ ♂
14 30 ☽ ∥ ♄
19 52 ☽ △ ♄
20 49 ☿ St D
22 53 ♂ △ ♆

26 Monday
03 25 ☽ □ ☉
05 50 ☽ ∠ ♂
09 13 ♂ △ ♅
10 40 ☽ Q ♆
11 05 ☽ ⚼ ♀
11 08 ☽ Q ♅
13 23 ☽ ⊻ ♃
13 24 ♀ ⊼ ♇
23 54 ☽ ⊥ ♃

27 Tuesday
00 12 ☽ # ☿
04 05 ☽ ± ♀
05 30 ☽ △ ☿
05 54 ☽ # ☉
07 28 ☽ ⚹ ♆
07 55 ☽ ⚹ ♅
09 03 ☽ ⚹ ♂
13 23 ☽ ☌ ♇
15 33 ☽ ⊼ ♀
15 49 ☽ ∠ ♃
16 20 ♂ # ♇
23 45 ☽ □ ♄

28 Wednesday
00 26 ♀ # ♆
06 17 ☽ ∥ ♆
06 39 ☽ # ♀
07 15 ☽ Q ♂
08 21 ☽ ⚼ ☿
09 55 ☽ ∠ ♆
10 22 ☽ ∠ ♅
11 51 ☽ △ ☉
17 09 ☉ ∥ ☿
18 47 ☽ ⚹ ♃

29 Thursday
00 50 ☽ ± ☿
02 02 ☽ ⊥ ♆
02 28 ☽ ⊥ ♅
07 50 ☽ Q ♄
11 59 ☽ ⊼ ☿
12 53 ☽ ⊻ ♆
13 18 ☽ ⊻ ♅
16 44 ☽ Q ♃
17 01 ☽ ⚼ ☉
17 15 ☽ □ ♂
19 07 ☽ ⊻ ♇

30 Friday
01 51 ☉ ∠ ♂
02 29 ☽ ☍ ♀
05 37 ☽ ⚹ ♄
06 06 ☽ ⊥ ♇
10 48 ☽ ± ☉
12 10 ☽ # ♀
14 42 ☿ ☍ ♆
15 40 ☽ ∥ ♆
22 43 ☽ ∠ ♇
22 47 ☽ ⊼ ☉

31 Saturday
00 35 ☿ ☍ ♅
02 15 ☽ □ ♃
08 47 ♂ ⚹ ♇
09 16 ☽ ∠ ♄
12 31 ♀ △ ♄
15 37 ☽ # ☿
19 48 ♀ ± ♇
20 15 ☽ ☌ ♆
20 38 ☽ ☌ ♅
21 38 ☽ ☍ ☿

All ephemeris data is given at 12.00 UT and the Moon's longitude is additionally given for 24.00 UT

AUGUST 1993

LONGITUDES

Date	Sidereal time h m s	Sun ☉ ° ' "	Moon ☽ ° ' "	Moon ☽ 24.00 ° ' "	Mercury ☿ ° '	Venus ♀ ° '	Mars ♂ ° '	Jupiter ♃ ° '	Saturn ♄ ° '	Uranus ♅ ° '	Neptune ♆ ° '	Pluto ♇ ° '
01	08 40 25	09 ♌ 14 40	27 ♑ 34 38	03 ♒ 52 23	20 ♋ 22	29 ♊ 30	23 ♍ 25	09 ♎ 50	28 ♒ 19	19 ♑ 26	19 ♑ 14	22 ♏ 43
02	08 44 21	10 12 04	10 ♒ 07 27	16 ♒ 19 51	21 06	00 ♋ 38	24 03	09 59	28 R 14	19 R 23	19 R 12	22 R 43
03	08 48 18	11 09 28	22 ♒ 29 39	28 ♒ 36 56	21 55	01 47	24 40	10 08	28 10	19 21	19 11	22 D 43
04	08 52 14	12 06 54	04 ♓ 41 49	10 ♓ 44 27	22 51	02 56	25 17	10 18	28 06	19 19	19 09	22 43
05	08 56 11	13 04 20	16 ♓ 45 03	22 ♓ 43 51	23 53	04 05	25 54	10 27	28 02	19 17	19 08	22 43
06	09 00 08	14 01 48	28 ♓ 41 10	04 ♈ 37 21	25 00	05 14	26 32	10 37	27 57	19 15	19 06	22 43
07	09 04 04	14 59 17	10 ♈ 32 49	16 ♈ 28 00	26 13	06 23	27 09	10 46	27 53	19 13	19 05	22 44
08	09 08 01	15 56 47	22 ♈ 23 25	28 ♈ 19 38	27 31	07 33	27 46	10 56	27 49	19 11	19 04	22 44
09	09 11 57	16 54 18	04 ♉ 17 12	10 ♉ 16 46	28 ♋ 55	08 42	28 24	11 06	27 44	19 09	19 02	22 44
10	09 15 54	17 51 51	16 ♉ 18 58	22 ♉ 24 26	00 ♌ 23	09 52	29 02	11 16	27 40	19 07	19 01	22 44
11	09 19 50	18 49 26	28 ♉ 33 51	04 ♊ 47 50	01 56	11 01	29 ♍ 39	11 26	27 35	19 05	18 59	22 45
12	09 23 47	19 47 01	11 ♊ 07 01	17 ♊ 31 58	03 34	12 11	00 ♎ 17	11 36	27 31	19 03	18 58	22 45
13	09 27 43	20 44 39	24 ♊ 03 10	00 ♋ 41 02	05 15	13 21	00 55	11 46	27 27	19 01	18 57	22 45
14	09 31 40	21 42 17	07 ♋ 25 51	14 ♋ 17 44	07 00	14 31	01 33	11 56	27 22	18 59	18 55	22 46
15	09 35 37	22 39 57	21 ♋ 16 41	28 ♋ 22 27	08 48	15 41	02 11	12 07	27 18	18 57	18 54	22 46
16	09 39 33	23 37 39	05 ♌ 34 38	12 ♌ 52 35	10 40	16 51	02 49	12 17	27 13	18 55	18 53	22 46
17	09 43 30	24 35 22	20 ♌ 15 30	27 ♌ 42 22	12 33	18 01	03 27	12 27	27 08	18 53	18 52	22 47
18	09 47 26	25 33 06	05 ♍ 12 03	12 ♍ 43 20	14 29	19 12	04 05	12 38	27 04	18 51	18 50	22 47
19	09 51 23	26 30 51	20 ♍ 14 58	27 ♍ 45 43	16 26	20 22	04 43	12 49	26 59	18 50	18 49	22 48
20	09 55 19	27 28 38	05 ♎ 14 25	12 ♎ 40 03	18 25	21 33	05 21	12 59	26 55	18 48	18 48	22 49
21	09 59 16	28 26 26	20 ♎ 01 47	27 ♎ 18 55	20 24	22 43	05 59	13 10	26 50	18 46	18 47	22 49
22	10 03 12	29 ♌ 24 15	04 ♏ 30 58	11 ♏ 37 37	22 24	23 54	06 38	13 21	26 46	18 45	18 46	22 50
23	10 07 09	00 ♍ 22 05	18 ♏ 38 44	25 ♏ 34 19	24 25	25 05	07 16	13 32	26 41	18 43	18 45	22 51
24	10 11 06	01 19 56	02 ♐ 24 30	09 ♐ 09 30	26 25	26 16	07 54	13 43	26 37	18 41	18 44	22 51
25	10 15 02	02 17 49	15 ♐ 49 36	22 ♐ 25 07	28 ♌ 25	27 27	08 33	13 54	26 32	18 40	18 42	22 52
26	10 18 59	03 15 43	28 ♐ 56 23	05 ♑ 23 46	00 ♍ 24	28 38	09 12	14 06	26 28	18 38	18 41	22 53
27	10 22 55	04 13 38	11 ♑ 47 36	18 ♑ 08 11	02 23	29 ♋ 49	09 50	14 17	26 23	18 37	18 40	22 54
28	10 26 52	05 11 34	24 ♑ 25 49	00 ♒ 40 44	04 22	01 ♌ 00	10 29	14 28	26 19	18 36	18 39	22 54
29	10 30 48	06 09 32	06 ♒ 53 11	13 ♒ 03 19	06 19	02 11	11 08	14 40	26 14	18 34	18 38	22 55
30	10 34 45	07 07 31	19 ♒ 11 20	25 ♒ 17 22	08 15	03 23	11 46	14 51	26 10	18 33	18 37	22 56
31	10 38 41	08 ♍ 05 31	01 ♓ 21 31	07 ♓ 23 56	10 ♍ 11	04 ♌ 34	12 ♎ 25	15 ♎ 02	26 ♒ 06	18 ♑ 31	18 ♑ 37	22 ♏ 57

DECLINATIONS

Date	Moon True ☊ ° '	Moon Mean ☊ ° '	Moon ☽ Latitude ° '	Sun ☉ ° '	Moon ☽ ° '	Mercury ☿ ° '	Venus ♀ ° '	Mars ♂ ° '	Jupiter ♃ ° '	Saturn ♄ ° '	Uranus ♅ ° '	Neptune ♆ ° '	Pluto ♇ ° '
01	10 ♐ 40	09 ♐ 10	03 N 43	17 N 57	17 S 00	19 N 33	21 N 42	03 N 15	02 S 48	13 S 28	22 S 31	21 S 23	04 S 30
02	10 R 30	09 07	04 22	17 41	13 30	19 42	21 45	02 59	02 51	13 29	22 31	21 23	04 31
03	10 20	09 04	04 49	17 26	09 27	19 50	21 48	02 44	02 55	13 31	22 31	21 24	04 32
04	10 09	09 01	05 02	17 10	05 06	19 57	21 50	02 29	02 59	13 32	22 31	21 24	04 32
05	09 58	08 57	05 01	16 54	00 S 36	20 02	21 51	02 13	03 03	13 34	22 32	21 24	04 33
06	09 48	08 54	04 47	16 37	03 N 52	20 05	21 52	01 58	03 07	13 36	22 32	21 24	04 33
07	09 41	08 51	04 21	16 20	08 10	20 07	21 52	01 42	03 11	13 37	22 32	21 24	04 34
08	09 36	08 48	03 43	16 03	12 10	20 06	21 52	01 27	03 15	13 39	22 33	21 25	04 34
09	09 33	08 45	02 56	15 46	15 42	20 04	21 51	01 11	03 19	13 41	22 33	21 25	04 35
10	09 33	08 41	02 00	15 29	18 38	19 58	21 49	00 56	03 23	13 42	22 33	21 25	04 35
11	09 D 33	08 38	00 N 58	15 11	20 47	19 51	21 47	00 40	03 27	13 44	22 34	21 25	04 36
12	09 R 33	08 35	00 S 08	14 53	21 58	19 41	21 45	00 24	03 31	13 45	22 34	21 26	04 37
13	09 32	08 32	01 16	14 35	22 02	19 28	21 41	00 N 09	03 35	13 47	22 34	21 26	04 37
14	09 29	08 29	02 22	14 16	20 52	19 12	21 37	00 S 07	03 39	13 49	22 34	21 26	04 38
15	09 24	08 26	03 22	13 58	18 26	18 54	21 33	00 23	03 43	13 50	22 35	21 26	04 39
16	09 16	08 22	04 11	13 39	14 49	18 33	21 28	00 38	03 47	13 52	22 35	21 26	04 39
17	09 07	08 19	04 45	13 20	10 13	18 10	21 22	00 54	03 52	13 54	22 35	21 27	04 40
18	08 56	08 16	05 00	13 00	04 N 56	17 44	21 16	01 10	03 56	13 55	22 35	21 27	04 40
19	08 46	08 13	04 55	12 41	00 S 40	17 15	21 09	01 26	04 00	13 57	22 36	21 27	04 41
20	08 37	08 10	04 30	12 21	06 12	16 44	21 02	01 41	04 04	13 58	22 36	21 27	04 42
21	08 30	08 07	03 46	12 01	11 19	16 11	20 54	01 57	04 09	14 00	22 36	21 27	04 42
22	08 26	08 03	02 49	11 41	15 41	15 36	20 45	02 13	04 13	14 02	22 36	21 28	04 43
23	08 24	08 00	01 42	11 21	19 01	14 59	20 36	02 29	04 18	14 03	22 36	21 28	04 44
24	08 D 24	07 57	00 S 32	11 00	21 10	14 20	20 26	02 45	04 22	14 05	22 37	21 28	04 45
25	08 R 24	07 54	00 N 39	10 39	22 02	13 40	20 16	03 00	04 26	14 07	22 37	21 28	04 45
26	08 23	07 51	01 46	10 19	21 40	12 59	20 05	03 16	04 31	14 08	22 37	21 28	04 46
27	08 20	07 47	02 47	09 58	20 09	12 16	19 53	03 32	04 35	14 10	22 37	21 28	04 47
28	08 14	07 44	03 37	09 36	17 40	11 33	19 41	03 48	04 40	14 11	22 37	21 29	04 47
29	08 05	07 41	04 17	09 15	14 24	10 48	19 29	04 03	04 44	14 13	22 38	21 29	04 48
30	07 54	07 38	04 44	08 54	10 34	10 03	19 16	04 19	04 49	14 14	22 38	21 29	04 49
31	07 ♐ 40	07 ♐ 35	04 N 58	08 N 32	06 S 21	09 N 18	19 N 02	04 S 35	04 S 53	14 S 16	22 S 38	21 S 29	04 S 50

ZODIAC SIGN ENTRIES

Date	h m	Planets
01	16 36	☽ ♒
01	22 38	♀ ♋
04	02 44	☽ ♓
06	14 39	☽ ♈
09	03 22	☽ ♉
10	05 51	☿ ♌
11	14 47	☽ ♊
12	01 10	♂ ♎
13	22 46	☽ ♋
16	02 43	☽ ♌
18	03 41	☽ ♍
20	03 35	☽ ♎
22	04 27	☽ ♏
23	02 50	☉ ♍
24	07 45	☽ ♐
26	07 06	☿ ♍
26	13 58	☽ ♑
27	15 48	♀ ♌
28	22 42	☽ ♒
31	09 19	☽ ♓

LATITUDES

Date	Mercury ☿ ° '	Venus ♀ ° '	Mars ♂ ° '	Jupiter ♃ ° '	Saturn ♄ ° '	Uranus ♅ ° '	Neptune ♆ ° '	Pluto ♇ ° '
01	02 S 22	01 S 44	00 N 41	01 N 12	01 S 30	00 S 29	00 N 41	14 N 23
04	01 35	01 35	00 39	01 11	01 30	00 29	00 41	14 22
07	00 49	01 25	00 37	01 11	01 30	00 29	00 41	14 20
10	00 S 06	01 15	00 35	01 10	01 31	00 29	00 41	14 18
13	00 N 32	01 05	00 33	01 10	01 31	00 29	00 41	14 17
16	01 02	00 55	00 31	01 09	01 31	00 29	00 40	14 15
19	01 24	00 45	00 29	01 09	01 31	00 29	00 40	14 13
22	01 39	00 35	00 27	01 08	01 32	00 29	00 40	14 12
25	01 45	00 25	00 25	01 08	01 32	00 29	00 40	14 10
28	01 45	00 15	00 23	01 08	01 32	00 29	00 40	14 09
31	01 N 40	00 S 05	00 N 21	01 N 07	01 S 32	00 S 29	00 N 40	14 N 07

DATA

Julian Date	2449201
Delta T	+59 seconds
Ayanamsa	23° 46' 20"
Synetic vernal point	05° ♓ 20' 39"
True obliquity of ecliptic	23° 26' 21"

MOON'S PHASES, APSIDES AND POSITIONS ☽

Date	h m	Phase	Longitude ° '	Eclipse Indicator
02	12 10	○	10 ♒ 12	
10	15 19	☾	18 ♉ 00	
17	19 28	●	24 ♌ 53	
24	09 57	☽	01 ♐ 15	

Day	h m		
07	03 49	Apogee	
19	06 46	Perigee	
05	15 13	ON	
13	01 22	Max dec	22° N 09'
19	09 11	OS	
25	16 38	Max dec	22° S 04'

LONGITUDES

Date	Chiron ⚷ ° '	Ceres ⚳ ° '	Pallas ⚴ ° '	Juno ⚵ ° '	Vesta ⚶ ° '	Black Moon Lilith ⚸ ° '
01	25 ♌ 34	03 ♉ 47	07 ♓ 31	00 ♍ 53	10 ♓ 41	02 ♈ 19
11	26 ♌ 53	05 ♉ 13	05 ♓ 19	05 ♍ 17	08 ♓ 51	03 ♈ 26
21	28 ♌ 14	06 ♉ 09	02 ♓ 50	09 ♍ 19	06 ♓ 32	04 ♈ 33
31	29 ♌ 34	06 ♉ 32	00 ♓ 17	13 ♍ 29	04 ♓ 02	05 ♈ 40

ASPECTARIAN

h m	Aspects
01 Sunday	
02 05	☽ ⊥ ♄
02 48	☽ ⚹ ♇
03 44	☽ △ ♂
03 58	☽ # ☉
13 23	☽ ⊻ ♄
16 00	☽ ⊼ ♀
02 Monday	
01 37	☽ Q ♇
04 37	☽ ± ♀
05 34	☉ ⚹ ♃
09 48	☽ ⚼ ♂
11 44	☽ △ ♃
12 02	☽ ∥ ♄
12 10	☽ ☍ ☉
19 29	♇ St D
21 57	♂ # ♃
23 45	☽ ⚼ ♀
03 Tuesday	
04 08	☽ ± ♂
05 33	☽ ⊻ ♆
05 54	☽ ⊻ ♅
10 48	☽ ⊼ ☿
12 27	☽ □ ♇
16 28	☽ ⊼ ♂
17 15	☽ ⚼ ♃
17 15	☽ ⊥ ♆
17 35	☽ ⊥ ♅
18 10	☿ ± ♄
23 04	☽ ☌ ♄
23 29	☽ ± ☿
04 Wednesday	
08 09	☽ △ ♀
08 48	☿ △ ♇
10 56	☽ ∠ ♆
11 12	☽ ± ♃
11 15	☽ ∠ ♅
15 02	☽ ∥ ♇
18 49	☽ ⚼ ☿
23 11	☽ ∥ ♃
23 15	☽ ⊼ ♃
05 Thursday	
02 54	☽ # ♂
04 01	☽ ⊼ ☉
16 46	☽ ⚹ ♆
17 03	☽ ± ☉
17 04	☽ ⚹ ♅
23 59	☽ △ ♇
06 Friday	
02 16	☽ ∥ ♂
03 46	☽ △ ☿
07 24	☽ ☍ ♂
07 51	☽ # ♃
10 32	☽ ⊻ ♄
12 45	☽ ⚼ ☉
15 45	☽ # ♇
16 53	☽ Q ♆
17 09	☽ Q ♅
22 35	☽ ⊥ ♄
07 Saturday	
02 40	☽ □ ♀
06 17	☽ ⚼ ♇
12 28	☽ ☍ ♃
16 43	☽ ∠ ♄
21 48	☽ △ ☉
08 Sunday	
00 32	☽ ± ♇
05 16	☽ □ ♆
05 31	☽ □ ♅
12 41	☽ ⊼ ♇
13 15	♂ ⊼ ♄
15 54	♀ ⚼ ♇
16 53	☿ ⊼ ♄
19 04	☽ Q ♀
20 17	☿ ⚹ ♂
21 41	☽ # ♄
22 54	☽ ⚹ ♄
23 29	☽ ⊼ ♂
23 43	☽ □ ☿
09 Monday	
12 14	☽ ± ♂
12 25	☽ ∥ ☉
15 28	☿ Q ♃
21 47	☽ ⚹ ♀
22 51	☽ Q ♄
10 Tuesday	
01 49	☽ ⊼ ♃
07 13	☽ ⚼ ♂
13 54	☽ ± ♃
15 19	☽ □ ☉
16 40	☽ Q ☿
17 19	☽ △ ♆
17 30	☽ △ ♅
11 Wednesday	
00 39	☽ ☍ ♇
01 01	☽ ∥ ☿
06 33	☽ ∠ ♀
07 48	☽ ⚼ ♃
10 07	☽ □ ♄
14 13	☽ △ ♂
16 04	☉ ⊼ ♆
18 06	☉ ⊼ ♅
19 28	☽ ⚹ ☿
21 50	♀ □ ♃
22 22	☽ # ♆
22 26	☽ ⚼ ♆
22 36	☽ ⚼ ♅
12 Thursday	
01 42	☽ ⊥ ♀
05 10	☽ Q ☉
05 31	☽ ∥ ♀
12 55	☽ △ ♃
14 12	☽ ⊻ ♀
15 28	☽ ± ♆
15 37	☽ ± ♅
18 28	♀ ⚼ ♄
13 Friday	
02 38	☽ ⊼ ♆
02 46	☽ ⊼ ♅
03 58	☽ ∠ ☿
05 27	☽ ⚹ ☉
09 37	☽ ⊼ ♇
18 07	☽ △ ♄
20 32	☽ ± ♇
22 50	☽ ⊥ ☿
22 57	☽ ∥ ♀
14 Saturday	
01 02	☽ □ ♂
03 21	☽ # ♆
10 37	☽ ∠ ☉
11 08	☽ ⊻ ☿
12 35	☽ ⚼ ♇
20 00	☽ □ ♃
20 36	☽ ⚼ ♄
15 Sunday	
01 31	☽ ☌ ♀
03 31	☽ ⊥ ☉
07 40	☽ ∥ ☿
07 57	☽ ☍ ♆
08 02	☽ ☍ ♅
10 02	☽ Q ♂
12 01	☽ ± ♄
14 32	☉ □ ♇
14 32	☽ △ ♇
14 32	☽ ⊻ ☉
20 30	☉ # ♄
22 08	☽ ⊼ ♄
16 Monday	
03 05	☽ Q ♃
05 41	♃ ⚼ ♄
07 12	☽ ⚹ ♂
17 19	☽ # ♄
17 48	♀ # ♆
19 04	☽ ∥ ☉
21 37	☽ ☌ ☿
23 10	☽ ⚹ ♃
17 Tuesday	
01 49	♂ ± ♄
08 04	☽ ⊻ ♀
08 56	☽ ∠ ♂
09 44	☽ ⊼ ♆
09 47	☽ ⊼ ♅
10 39	☿ ⚹ ♃
16 05	☽ □ ♇
18 36	☽ ⊥ ♀
18 38	☉ ± ♆
19 11	☉ ± ♅
19 25	☽ ± ♆
19 27	☽ ± ♅
19 28	☽ ☌ ☉
23 02	☽ ☍ ♄
23 44	☽ ∠ ♃
18 Wednesday	
00 05	☽ ⊥ ♂
04 53	♀ ☍ ♆
05 16	♀ ☍ ♅
09 50	☽ ⚼ ♆
09 51	☽ ⚼ ♅
10 08	☽ ⊻ ♂
10 15	☽ ∠ ♀
13 09	☽ # ♇
14 19	☽ ⊥ ♃
16 19	☽ # ♃
20 56	☽ Q ♇
19 Thursday	
00 00	☽ ⊻ ♃
03 30	☽ # ♂
05 01	☽ ⊻ ☿
09 43	☽ △ ♆
09 44	☽ △ ♅
12 12	☽ ⚹ ♀
15 26	☽ ∥ ♂
16 01	☽ ⊥ ☿
16 04	☽ ⚹ ♇
22 41	☽ ⊻ ☉
22 43	☽ ⊼ ♄
23 01	☉ ☍ ♄
23 57	♀ ± ♄
20 Friday	
02 31	☽ ∥ ♃
04 30	☉ ⊥ ♀
05 20	☽ ∥ ♇
07 55	♅ ☌ ♆
08 17	☽ ± ♄
08 37	☽ ∠ ☿
08 58	☽ ⊥ ☉
09 02	☽ Q ♀
12 11	☽ ☌ ♂
16 09	☽ ∠ ♇
16 36	☿ ⊼ ♅
16 38	☿ ⊼ ♆
22 43	☽ ⚼ ♄
21 Saturday	
00 30	☽ ∠ ☉
00 41	☽ ☌ ♃
03 46	☉ ∠ ♃
06 45	☽ ⊥ ♇
09 57	☽ □ ♅
09 58	☽ □ ♆
12 43	☽ ⚹ ☿
14 02	♀ △ ♇
15 18	☽ # ☉
16 35	☽ ⊻ ♇
16 48	☽ □ ♀
22 19	☿ ∠ ♂
23 09	☽ △ ♄
22 Sunday	
02 16	☽ ∥ ♄
02 51	☽ ⚹ ☉
11 35	☽ # ☿
11 47	☽ Q ☿
15 43	☽ ⊻ ♂
15 44	☽ Q ♅
15 46	☽ Q ♆
17 08	☿ □ ♇
23 Monday	
00 28	☽ Q ☉
02 21	☽ ⊥ ♂
03 08	☽ ⊻ ♃
12 07	☽ ⚹ ♅
12 10	☽ ⚹ ♆
13 33	☽ ⊥ ♃
15 38	☿ ± ♅
15 58	☿ ± ♆
18 33	☽ ∠ ♂
19 15	☽ ☌ ♇
20 22	♀ Q ♂
23 40	☽ □ ☿
24 Tuesday	
00 11	☽ △ ♀
01 52	☽ □ ♄
02 57	☽ # ♀
05 25	☽ ∠ ♃
07 32	☿ ⊻ ♀
09 56	♂ ∠ ♇
09 57	☽ □ ☉
14 16	☽ ∠ ♅
14 19	☿ ☍ ♄
14 20	☽ ∠ ♆
17 22	☽ ∥ ♆
18 45	♀ ⊼ ♄
21 00	☿ # ♄
22 15	☽ ⚹ ♂
25 Wednesday	
05 18	☽ ⚼ ♀
06 18	☽ ⊥ ♅
06 23	☽ ⊥ ♆
08 29	☽ ⚹ ♃
09 41	☽ Q ♄
17 08	☽ ⊻ ♅
17 13	☽ ⊻ ♆
18 32	☿ ∠ ♃
21 01	☽ Q ♂
23 14	☽ ± ♀
26 Thursday	
00 50	☽ ⊻ ♇
06 40	☽ Q ♃
06 48	☉ ⊥ ♂
07 27	☽ ⚹ ♄
11 22	☽ ⊼ ♀
11 53	☽ ⊥ ♇
15 12	☽ △ ☿
16 18	☽ ∥ ♆
20 40	☽ △ ☉
21 10	☉ ⚼ ♅
22 27	☉ ⚼ ♆
27 Friday	
04 40	☽ ∠ ♇
08 07	☽ □ ♂
11 15	☽ ∠ ♄
15 17	☽ # ♀
16 46	☽ □ ♃
28 Saturday	
00 32	☽ ⚼ ☿
00 53	☽ ☌ ♅
01 00	☽ ☌ ♆
02 45	☿ ⚼ ♅
03 14	☽ ⚼ ☉
03 29	☿ ⚼ ♆
04 11	☽ ⊥ ♄
09 05	☽ ⚹ ♇
14 13	☿ ⊥ ♂
15 35	☽ ⊻ ♄
20 56	☽ ± ☿
21 54	☽ ± ☉
29 Sunday	
01 57	☽ ☍ ♀
08 11	☽ Q ♇
08 12	☉ ☌ ☿
10 28	☽ ⊼ ☉
10 41	☽ ⊼ ☿
13 15	☽ ∥ ♄
15 51	♂ ⚼ ♄
20 42	☽ △ ♂
23 19	♀ Q ♃
30 Monday	
03 22	☽ △ ♃
10 45	☽ ⊻ ♅
10 54	☽ ⊻ ♆
11 45	♃ ∥ ♇
15 41	☽ # ☿
19 22	☽ □ ♇
20 13	☿ ⊥ ♃
22 31	☽ ⊥ ♅
22 37	☽ # ☉
22 41	☽ ⊥ ♆
31 Tuesday	
01 39	☽ ☌ ♄
03 46	☽ ⚼ ♂
09 21	☽ ⚼ ♃
16 17	☽ ∠ ♅
16 28	☽ ∠ ♆
19 04	☽ ⊼ ♀
19 52	☽ ∥ ♃
20 20	☽ ∥ ♇
21 08	☽ ∥ ♂
21 50	☿ Q ♇
22 37	☽ ± ♂

All ephemeris data is given at 12.00 UT and the Moon's longitude is additionally given for 24.00 UT

SEPTEMBER 1993

LONGITUDES

Date	Sidereal time h m s	Sun ☉ ° ' "	Moon ☽ ° ' "	Moon ☽ 24.00 ° ' "	Mercury ☿ ° '	Venus ♀ ° '	Mars ♂ ° '	Jupiter ♃ ° '	Saturn ♄ ° '	Uranus ♅ ° '	Neptune ♆ ° '	Pluto ♇ ° '
01	10 42 38	09 ♍ 03 34	13 ♓ 24 44	19 ♓ 24 02	12 ♍ 05	05 ♌ 46	13 ♎ 04	15 ♎ 14	26 ♒ 01	18 ♑ 30	18 ♑ 36	22 ♏ 58
02	10 46 35	10 01 37	25 ♓ 22 00	01 ♈ 18 49	13 58	06 57	13 43	15 26	25 R 57	18 R 29	18 R 35	22 59
03	10 50 31	10 59 43	07 ♈ 14 42	13 ♈ 09 54	15 50	08 09	14 22	15 37	25 53	18 28	18 34	23 00
04	10 54 28	11 57 50	19 ♈ 04 43	24 ♈ 59 30	17 41	09 21	15 01	15 49	25 48	18 27	18 33	23 01
05	10 58 24	12 56 00	00 ♉ 54 38	06 ♉ 50 35	19 30	10 33	15 40	16 01	25 44	18 26	18 32	23 02
06	11 02 21	13 54 11	12 ♉ 47 50	18 ♉ 46 55	21 19	11 45	16 19	16 13	25 40	18 25	18 32	23 03
07	11 06 17	14 52 24	24 ♉ 48 25	00 ♊ 52 55	23 06	12 57	16 59	16 25	25 36	18 24	18 31	23 05
08	11 10 14	15 50 40	07 ♊ 01 05	13 ♊ 13 33	24 52	14 09	17 38	16 37	25 32	18 23	18 30	23 06
09	11 14 10	16 48 57	19 ♊ 30 58	25 ♊ 53 56	26 37	15 21	18 17	16 49	25 28	18 22	18 29	23 07
10	11 18 07	17 47 16	02 ♋ 23 03	08 ♋ 58 51	28 ♍ 20	16 33	18 57	17 01	25 24	18 21	18 29	23 08
11	11 22 04	18 45 38	15 ♋ 41 44	22 ♋ 32 01	00 ♎ 03	17 46	19 36	17 13	25 20	18 20	18 28	23 10
12	11 26 00	19 44 02	29 ♋ 29 51	06 ♌ 35 12	01 44	18 58	20 16	17 25	25 16	18 19	18 28	23 11
13	11 29 57	20 42 27	13 ♌ 47 48	21 ♌ 07 12	03 25	20 10	20 56	17 37	25 12	18 19	18 27	23 12
14	11 33 53	21 40 55	28 ♌ 32 41	06 ♍ 03 18	05 04	21 23	21 35	17 49	25 08	18 18	18 27	23 14
15	11 37 50	22 39 25	13 ♍ 37 55	21 ♍ 15 15	06 42	22 36	22 15	18 02	25 04	18 17	18 26	23 15
16	11 41 46	23 37 56	28 ♍ 53 54	06 ♎ 32 25	08 19	23 48	22 55	18 14	25 01	18 17	18 26	23 16
17	11 45 43	24 36 30	14 ♎ 09 24	21 ♎ 43 33	09 55	25 01	23 35	18 27	24 57	18 16	18 25	23 18
18	11 49 39	25 35 05	29 ♎ 13 42	06 ♏ 38 54	11 30	26 14	24 15	18 39	24 53	18 16	18 25	23 19
19	11 53 36	26 33 42	13 ♏ 58 24	21 ♏ 11 39	13 04	27 27	24 54	18 51	24 50	18 15	18 24	23 21
20	11 57 33	27 32 21	28 ♏ 18 21	05 ♐ 18 23	14 37	28 40	25 35	19 04	24 46	18 15	18 24	23 22
21	12 01 29	28 31 01	12 ♐ 11 46	18 ♐ 58 42	16 09	29 ♌ 53	26 15	19 16	24 43	18 15	18 24	23 24
22	12 05 26	29 ♍ 29 43	25 ♐ 39 29	02 ♑ 14 29	17 40	01 ♍ 06	26 55	19 29	24 40	18 14	18 24	23 26
23	12 09 22	00 ♎ 28 27	08 ♑ 44 08	15 ♑ 08 54	19 10	02 19	27 35	19 42	24 37	18 14	18 23	23 27
24	12 13 19	01 27 13	21 ♑ 29 14	27 ♑ 45 38	20 39	03 32	28 15	19 54	24 33	18 14	18 23	23 29
25	12 17 15	02 26 00	03 ♒ 58 31	10 ♒ 08 20	22 07	04 45	28 56	20 07	24 30	18 14	18 23	23 31
26	12 21 12	03 24 49	16 ♒ 15 29	22 ♒ 20 19	23 34	05 59	29 ♎ 36	20 20	24 27	18 14	18 23	23 32
27	12 25 08	04 23 39	28 ♒ 23 10	04 ♓ 24 18	25 00	07 12	00 ♏ 16	20 32	24 24	18 14	18 23	23 34
28	12 29 05	05 22 32	10 ♓ 24 01	16 ♓ 22 32	26 25	08 26	00 57	20 45	24 21	18 D 14	18 23	23 36
29	12 33 02	06 21 26	22 ♓ 20 02	28 ♓ 16 45	27 49	09 39	01 37	20 58	24 19	18 14	18 23	23 38
30	12 36 58	07 ♎ 20 22	04 ♈ 12 51	10 ♈ 08 31	29 ♎ 12	10 ♍ 53	02 ♏ 18	21 ♎ 11	24 ♒ 16	18 ♑ 14	18 ♑ 23	23 ♏ 39

DECLINATIONS

Date	Moon True ☊ ° '	Moon Mean ☊ ° '	Moon ☽ Latitude ° '	Sun ☉ ° '	Moon ☽ ° '	Mercury ☿ ° '	Venus ♀ ° '	Mars ♂ ° '	Jupiter ♃ ° '	Saturn ♄ ° '	Uranus ♅ ° '	Neptune ♆ ° '	Pluto ♇ ° '
01	07 ♐ 26	07 ♐ 32	04 N 58	08 N 10	01 S 56	08 N 31	18 N 48	04 S 51	04 S 58	14 S 17	22 S 38	21 S 29	04 S 50
02	07 R 13	07 28	04 45	07 49	02 N 31	07 45	18 33	05 06	05 03	14 19	22 38	21 29	04 51
03	07 01	07 25	04 20	07 27	06 51	06 58	18 18	05 22	05 07	14 20	22 38	21 30	04 52
04	06 51	07 22	03 43	07 04	10 55	06 11	18 02	05 38	05 12	14 22	22 39	21 30	04 53
05	06 44	07 19	02 57	06 42	14 33	05 24	17 46	05 54	05 16	14 23	22 39	21 30	04 53
06	06 40	07 16	02 03	06 20	17 37	04 36	17 29	06 09	05 21	14 25	22 39	21 30	04 54
07	06 38	07 13	01 N 02	05 57	19 58	03 49	17 11	06 25	05 26	14 26	22 39	21 30	04 55
08	06 D 37	07 09	00 S 02	05 35	21 27	03 02	16 54	06 41	05 30	14 28	22 39	21 30	04 56
09	06 R 37	07 06	01 08	05 12	21 54	02 15	16 35	06 56	05 35	14 29	22 39	21 30	04 56
10	06 37	07 03	02 12	04 50	21 13	01 28	16 16	07 12	05 40	14 30	22 39	21 30	04 57
11	06 34	07 00	03 11	04 27	19 21	00 N 42	15 57	07 27	05 45	14 32	22 39	21 31	04 58
12	06 29	06 57	04 02	04 04	16 19	00 S 05	15 37	07 43	05 49	14 33	22 39	21 31	04 59
13	06 22	06 53	04 39	03 41	12 14	00 51	15 17	07 58	05 54	14 34	22 40	21 31	04 59
14	06 13	06 50	04 59	03 18	07 18	01 36	14 57	08 14	05 59	14 36	22 40	21 31	05 00
15	06 02	06 47	04 59	02 55	01 N 50	02 22	14 36	08 29	06 03	14 37	22 40	21 31	05 01
16	05 52	06 44	04 38	02 32	03 S 49	03 06	14 14	08 45	06 08	14 38	22 40	21 31	05 02
17	05 43	06 41	03 57	02 09	09 13	03 51	13 52	09 00	06 13	14 39	22 40	21 31	05 03
18	05 36	06 38	03 00	01 45	14 00	04 35	13 30	09 15	06 18	14 41	22 40	21 31	05 03
19	05 32	06 34	01 52	01 22	17 49	05 18	13 08	09 30	06 23	14 42	22 40	21 31	05 04
20	05 31	06 31	00 S 38	00 59	20 24	06 01	12 44	09 46	06 27	14 43	22 40	21 31	05 05
21	05 D 30	06 28	00 N 36	00 35	21 40	06 44	12 21	10 01	06 32	14 44	22 40	21 31	05 06
22	05 31	06 25	01 45	00 N 12	21 37	07 26	11 57	10 16	06 37	14 45	22 40	21 32	05 07
23	05 R 31	06 22	02 47	00 S 11	20 22	08 07	11 33	10 31	06 42	14 46	22 40	21 32	05 07
24	05 29	06 19	03 39	00 35	18 07	08 47	11 09	10 46	06 47	14 47	22 40	21 32	05 08
25	05 24	06 15	04 20	00 58	15 03	09 27	10 44	11 01	06 52	14 48	22 40	21 32	05 09
26	05 18	06 12	04 47	01 21	11 23	10 07	10 19	11 16	06 56	14 49	22 40	21 32	05 10
27	05 08	06 09	05 02	01 45	07 19	10 45	09 54	11 31	07 01	14 50	22 40	21 32	05 11
28	04 58	06 06	05 02	02 08	03 S 00	11 23	09 28	11 45	07 06	14 51	22 40	21 32	05 11
29	04 46	06 03	04 50	02 31	01 N 24	12 00	09 02	12 00	07 11	14 52	22 40	21 32	05 12
30	04 ♐ 35	05 ♐ 59	04 N 25	02 S 55	05 N 43	12 S 37	08 N 36	12 S 15	07 S 16	14 S 53	22 S 40	21 S 32	05 S 13

ZODIAC SIGN ENTRIES

Date	h	m	Planets
02	21	21	☽ ♈
05	10	09	☽ ♉
07	22	16	☽ ♊
10	07	37	☽ ♋
11	11	18	☿ ♎
12	12	51	☽ ♌
14	14	20	☽ ♍
16	13	44	☽ ♎
18	13	14	☽ ♏
20	14	53	☽ ♐
21	14	22	♀ ♍
22	19	54	☽ ♑
23	00	22	☉ ♎
25	04	19	☽ ♒
27	02	15	♂ ♏
27	15	13	☽ ♓
30	03	29	☽ ♈

LATITUDES

Date	Mercury ☿ ° '	Venus ♀ ° '	Mars ♂ ° '	Jupiter ♃ ° '	Saturn ♄ ° '	Uranus ♅ ° '	Neptune ♆ ° '	Pluto ♇ ° '
01	01 N 37	00 S 02	00 N 20	01 N 07	01 S 32	00 S 29	00 N 40	14 N 06
04	01 25	00 N 07	00 19	01 07	01 32	00 29	00 40	14 05
07	01 11	00 16	00 17	01 07	01 32	00 29	00 40	14 03
10	00 53	00 25	00 15	01 06	01 32	00 29	00 40	14 02
13	00 34	00 33	00 13	01 06	01 32	00 29	00 40	14 00
16	00 N 13	00 41	00 11	01 06	01 32	00 28	00 40	13 59
19	00 S 09	00 49	00 09	01 05	01 32	00 28	00 40	13 58
22	00 32	00 56	00 07	01 05	01 32	00 28	00 39	13 56
25	00 54	01 02	00 05	01 05	01 32	00 28	00 39	13 55
28	01 17	01 08	00 03	01 05	01 32	00 28	00 39	13 54
31	01 S 39	01 N 14	00 N 01	01 N 05	01 S 32	00 S 28	00 N 39	13 N 53

DATA

Julian Date	2449232
Delta T	+59 seconds
Ayanamsa	23° 46' 24"
Synetic vernal point	05° ♓ 20' 35"
True obliquity of ecliptic	23° 26' 21"

MOON'S PHASES, APSIDES AND POSITIONS ☽

Date	h	m	Phase	Longitude ° '	Eclipse Indicator
01	02	33	○	08 ♓ 41	
09	06	26	☾	16 ♊ 35	
16	03	10	●	23 ♍ 16	
22	19	32	☽	29 ♐ 48	
30	18	54	○	07 ♈ 37	

Day	h	m		
03	16	47	Apogee	
16	14	34	Perigee	
30	21	07	Apogee	
01	22	23	ON	
09	09	52	Max dec	21° N 54'
15	19	47	OS	
21	22	58	Max dec	21° S 48'
29	04	22	ON	

LONGITUDES

Date	Chiron ⚷ ° '	Ceres ⚳ ° '	Pallas ⚴ ° '	Juno ⚵ ° '	Vesta ⚶ ° '	Black Moon Lilith ⚸ ° '
01	29 ♌ 42	06 ♉ 33	00 ♓ 02	13 ♍ 54	03 ♓ 47	05 ♈ 47
11	01 ♍ 02	06 ♉ 16	27 ♒ 39	18 ♍ 01	01 ♓ 28	06 ♈ 54
21	02 ♍ 19	05 ♉ 22	25 ♒ 37	22 ♍ 04	29 ♒ 38	08 ♈ 01
31	03 ♍ 33	03 ♉ 53	24 ♒ 06	26 ♍ 04	28 ♒ 29	09 ♈ 08

All ephemeris data is given at 12.00 UT and the Moon's longitude is additionally given for 24.00 UT

ASPECTARIAN

h m	Aspects
01 Wednesday	
01 08	☿ ⊥ ♀
02 33	☽ ☍ ☉
03 31	☽ ± ♃
08 20	☽ ± ♀
08 50	☽ ☍ ☿
11 16	☽ ⊼ ♂
11 16	♂ ∥ ♇
15 42	☽ ⊼ ♃
17 25	☉ ⊥ ♃
22 11	☽ ✱ ♅
22 22	☽ ✱ ♆
02 Thursday	
03 33	♂ ∥ ♃
04 22	☽ ₽ ♀
07 09	☿ ⊻ ♂
07 12	☽ △ ♇
08 19	☉ ∥ ☿
13 10	☽ ⊻ ♄
22 18	☽ Q ♅
22 30	☽ Q ♆
03 Friday	
00 49	☽ # ♇
01 12	☽ ⊥ ♄
02 06	☽ # ♃
03 08	☽ # ♂
08 59	☿ ⊻ ♃
12 12	☉ Q ♇
12 33	☽ ∥ ☿
13 32	☽ ₽ ♇
14 02	☽ △ ♀
15 07	☽ ∥ ☉
19 19	☽ ∠ ♄
20 17	☽ ⊼ ☉
04 Saturday	
03 17	☽ ☍ ♂
05 16	☽ ☍ ♃
07 49	☽ ± ♇
08 38	☽ ⊼ ☿
09 32	☽ ± ☉
10 43	☽ □ ♅
10 56	☽ □ ♆
20 01	☽ ⊼ ♇
21 58	☿ △ ♅
23 03	☽ ± ☿
23 23	☿ △ ♆
05 Sunday	
00 34	☿ # ♂
01 34	☽ ✱ ♄
05 26	☽ ₽ ☉
10 51	☽ # ♄
15 20	☿ # ♃
20 35	☽ ₽ ☿
06 Monday	
01 43	☽ Q ♄
03 11	☿ # ♇
06 08	♂ ☌ ♃
09 39	☽ □ ♀
10 51	☽ ∥ ♀
14 25	☽ △ ☉
18 43	☉ # ♂
18 58	☽ ⊼ ♃
19 29	☽ ⊼ ♂
23 14	☽ △ ♅
23 29	☽ △ ♆
07 Tuesday	
07 10	☽ ± ♃
08 01	☽ △ ☿
08 10	☽ ± ♂
08 34	☽ ☍ ♇
11 44	☿ ✱ ♇
13 34	☽ □ ♄
15 45	♂ ⊥ ♇
08 Wednesday	
01 15	☽ ₽ ♃
01 27	☽ Q ♀
02 57	☽ ₽ ♂
04 54	☽ ₽ ♅
05 09	☽ ₽ ♆
13 25	☽ # ♆
16 01	☉ # ♃
20 46	☿ ⊼ ♄
22 21	☽ ± ♅
22 36	☽ ± ♆
09 Thursday	
03 14	☽ ✱ ♀
06 26	☽ □ ☉
06 47	☽ △ ♃
09 33	☽ △ ♂
09 49	☽ ⊼ ♅
10 04	☽ ⊼ ♆
11 51	☉ ⊻ ♃
14 36	♂ □ ♅
18 48	☽ ⊼ ♇
19 15	♂ □ ♆
23 07	☽ △ ♄
10 Friday	
03 24	☽ □ ☿
04 23	☉ # ♇
05 48	☽ # ♆
06 01	☽ ± ♇
10 19	☽ ∠ ♀
18 43	☽ Q ☉
22 30	☽ ₽ ♇
22 59	♀ ✱ ♃
11 Saturday	
01 38	☉ △ ♅
02 28	☽ ₽ ♄
04 18	☽ ⊥ ♀
04 51	♃ ⊥ ♇
04 55	☉ △ ♆
14 43	☽ □ ♃
16 00	☽ ⊻ ♀
16 39	☽ ☍ ♅
16 45	☽ Q ☿
16 54	☽ ☍ ♆
17 49	☽ ✱ ☉
18 22	☽ ± ♄
19 14	☽ □ ♂
23 20	♀ ⊼ ♅
12 Sunday	
01 07	☽ △ ♇
02 02	♀ ⊼ ♆
04 46	☽ ⊼ ♄
05 27	☿ ± ♄
16 20	☽ ✱ ☿
16 56	☽ ∥ ♀
21 32	☽ ∠ ☉
22 11	☽ Q ♃
23 01	☽ # ♄
13 Monday	
03 31	☽ Q ♂
13 37	☽ ⊥ ☉
18 23	☽ ✱ ♃
19 25	☽ ⊼ ♅
19 39	☽ ⊼ ♆
20 33	☽ ∠ ☿
23 24	☽ ☌ ♀
14 Tuesday	
00 08	☽ ⊻ ☉
00 14	☽ ✱ ♂
03 25	☽ □ ♇
04 40	☉ ⊻ ♂
05 10	☽ ± ♅
05 24	☽ ± ♆
06 32	☽ ☍ ♄
07 56	☽ # ♂
12 57	☽ ⊥ ☿
17 53	☽ # ♃
18 57	☽ ∠ ♃
19 36	☽ ₽ ♅
19 50	☽ ₽ ♆
20 53	♀ ✱ ♂
22 16	☽ # ♇
23 42	☽ ⊻ ☿
15 Wednesday	
01 26	☽ ∠ ♂
07 01	☽ ∥ ☉
08 14	☽ Q ♇
09 26	☽ ⊥ ♃
10 01	☽ # ☿
10 44	♀ # ♄
16 19	☽ ⊥ ♂
18 24	☉ ⊻ ♀
19 01	☽ ⊻ ♃
19 20	☽ △ ♅
19 34	☽ △ ♆
23 43	☉ # ☿
16 Thursday	
01 17	♀ □ ♇
02 11	☽ ⊻ ♂
02 59	☉ ✱ ♇
03 10	☽ ✱ ♇
03 10	☽ ☌ ☉
03 19	☽ ⊻ ♀
05 55	☽ ⊼ ♄
06 52	☽ # ☉
08 32	☽ ∥ ☿
11 18	☿ ∠ ♇
13 33	☽ ⊥ ♀
15 18	☽ ± ♄
16 47	♃ □ ♅
17 16	☽ ∥ ♇
21 18	♀ ± ♅
22 14	☽ ∥ ♃
17 Friday	
00 14	♀ ± ♆
01 36	♂ ⊻ ♇
02 45	☽ ∠ ♇
04 33	☽ ☌ ☿
04 54	☽ ∠ ♀
05 23	☽ ₽ ♄
09 30	♃ □ ♆
10 42	♀ ☍ ♄
10 54	☽ ∥ ♂
12 23	☿ ₽ ♄
16 59	☽ ⊥ ♇
18 04	☿ ∠ ♀
18 30	☽ □ ♅
18 45	☽ □ ♆
18 53	☽ ☌ ♃
19 54	☉ ⊼ ♄
18 Saturday	
02 32	☽ ⊻ ♇
03 38	☽ ☌ ♂
05 04	☽ △ ♄
05 45	☽ ⊻ ☉
06 46	☽ ✱ ♀
09 29	☽ # ♀
15 48	☽ ∥ ♄
16 04	☽ ⊥ ☉
23 22	☽ Q ♅
23 37	☽ Q ♆
19 Sunday	
03 54	☽ Q ♀
04 00	☿ ∥ ♇
07 45	☽ ∠ ☉
09 27	♂ △ ♄
10 20	☽ ⊻ ☿
19 05	☽ ✱ ♅
19 21	☽ ✱ ♆
20 13	☽ ⊻ ♃
21 29	☽ ⊥ ☿
20 Monday	
03 39	☽ ☌ ♇
06 02	☽ □ ♄
06 26	☽ ⊥ ♃
07 09	☽ ⊻ ♂
10 36	☽ ✱ ☉
12 40	☽ □ ♀
14 31	☽ ∠ ☿
17 52	☽ ⊥ ♂
20 27	☽ ∠ ♅
20 43	☽ ∠ ♆
22 01	☽ ∠ ♃
21 Tuesday	
04 40	☿ ∥ ♃
07 15	☽ ∥ ♆
08 50	☽ Q ☉
10 15	☽ ∠ ♂
12 05	☽ ⊥ ♅
12 21	☽ ⊥ ♆
12 55	☽ Q ♄
19 52	☽ ✱ ☿
22 41	☽ ⊻ ♅
22 58	☽ ⊻ ♆
22 Wednesday	
00 44	☽ ✱ ♃
07 58	☽ ⊻ ♇
08 01	☿ ⊥ ♇
10 12	☽ ✱ ♄
14 24	☽ ✱ ♂
14 54	☽ ∥ ♆
18 52	☽ ⊥ ♇
19 32	☽ □ ☉
20 15	☽ Q ☿
20 58	☿ □ ♅
22 47	☽ Q ♃
22 55	☽ △ ♀
23 26	☿ □ ♆
23 Thursday	
11 29	☽ ∠ ♇
13 37	☽ ∠ ♄
13 40	☽ Q ♂
15 07	☉ ± ♄
21 47	☿ ☌ ♃
24 Friday	
05 48	☽ ₽ ♀
05 49	☽ ☌ ♅
06 01	♀ ₽ ♅
06 07	☽ ☌ ♆
06 27	☽ ⊥ ♄
08 56	☽ □ ♃
09 02	♀ ₽ ♆
10 13	☽ □ ☿
15 49	☽ ✱ ♇
17 50	☽ ⊻ ♄
25 Saturday	
00 49	☽ ± ♀
01 41	☽ □ ♂
01 50	♀ # ♂
08 45	☽ △ ☉
13 41	☽ ⊼ ♀
13 44	☽ ∥ ♄
14 59	☽ Q ♇
20 32	♀ ∠ ♃
26 Sunday	
11 25	☿ ⊻ ♇
12 43	☽ ∥ ♂
15 53	☽ ⊻ ♅
16 11	☽ ⊻ ♆
16 37	☽ ₽ ☉
16 37	☿ # ♀
18 40	☽ ∥ ☿
19 16	☽ # ♀
20 10	☽ △ ♃
27 Monday	
02 15	☿ △ ♄
02 25	☽ □ ♇
03 45	☽ ⊥ ♅
04 02	☽ ⊥ ♆
04 08	☽ ☌ ♄
04 23	☽ △ ☿
12 01	☽ ± ☉
12 27	♅ St D
13 38	☽ ∥ ♃
15 59	☽ △ ♂
21 39	☽ ∠ ♅
21 57	☽ ∠ ♆
23 58	☽ ∥ ♇
28 Tuesday	
01 03	☽ ⊼ ☉
01 58	♄ ⊥ ♆
02 32	☽ ₽ ♃
07 36	☽ ☍ ♀
14 19	☽ ₽ ☿
16 20	☽ ∥ ☉
20 54	☽ ± ♃
23 49	☽ ₽ ♂
29 Wednesday	
03 44	☽ ✱ ♅
04 02	☽ ✱ ♆
09 11	☽ ⊼ ♃
10 49	☽ ± ☿
11 38	☿ ∥ ♂
14 37	☽ △ ♇
15 58	☽ ⊻ ♄
18 47	☽ # ☉
19 02	☽ ± ♂
30 Thursday	
00 31	☽ ⊼ ☿
03 57	☽ Q ♅
04 03	☽ ⊥ ♄
04 14	☽ Q ♆
06 05	♆ St D
07 54	☽ ⊼ ♂
09 07	☽ # ♇
18 54	☽ ☍ ☉
20 58	☽ # ♃
21 01	☽ ₽ ♇
22 11	☽ ∠ ♄

OCTOBER 1993

Date	Sidereal time h m s	Sun ☉ ° ' "	Moon ☽ ° ' "	Moon ☽ 24.00 ° ' "	Mercury ☿ ° '	Venus ♀ ° '	Mars ♂ ° '	Jupiter ♃ ° '	Saturn ♄ ° '	Uranus ♅ ° '	Neptune ♆ ° '	Pluto ♇ ° '
01	12 40 55	08 ♎ 19 21	16 ♈ 03 56	21 ♈ 59 20	00 ♏ 33	12 ♍ 06	02 ♏ 59	21 ♎ 24	24 ♒ 13	18 ♑ 14	18 ♑ 23	23 ♏ 41
02	12 44 51	09 18 21	27 ♈ 54 54	03 ♉ 50 56	01 54	13 20	03 40	21 36	24 R 11	18 14	18 D 23	23 43
03	12 48 48	10 17 23	09 ♉ 47 41	15 ♉ 45 29	03 13	14 34	04 20	21 49	24 08	18 15	18 23	23 45
04	12 52 44	11 16 28	21 ♉ 44 41	27 ♉ 45 42	04 31	15 47	05 01	22 02	24 06	18 15	18 23	23 47
05	12 56 41	12 15 35	03 ♊ 48 57	09 ♊ 54 55	05 47	17 01	05 42	22 15	24 04	18 15	18 23	23 49
06	13 00 37	13 14 44	16 ♊ 04 07	22 ♊ 17 04	07 03	18 15	06 23	22 28	24 02	18 16	18 23	23 51
07	13 04 34	14 13 56	28 ♊ 34 18	04 ♋ 56 23	08 16	19 29	07 04	22 41	24 00	18 16	18 23	23 53
08	13 08 31	15 13 10	11 ♋ 23 51	17 ♋ 57 11	09 28	20 43	07 45	22 54	23 58	18 17	18 24	23 55
09	13 12 27	16 12 26	24 ♋ 36 51	01 ♌ 23 11	10 39	21 57	08 26	23 07	23 56	18 17	18 24	23 57
10	13 16 24	17 11 44	08 ♌ 16 28	15 ♌ 16 48	11 47	23 11	09 08	23 20	23 54	18 18	18 24	23 59
11	13 20 20	18 11 05	22 ♌ 24 08	29 ♌ 38 14	12 54	24 25	09 49	23 33	23 52	18 19	18 25	24 01
12	13 24 17	19 10 28	06 ♍ 58 38	14 ♍ 24 41	13 58	25 39	10 30	23 46	23 50	18 19	18 25	24 03
13	13 28 13	20 09 54	21 ♍ 55 30	29 ♍ 30 01	15 00	26 54	11 12	23 59	23 49	18 20	18 26	24 05
14	13 32 10	21 09 21	07 ♎ 07 01	14 ♎ 45 10	16 00	28 08	11 53	24 12	23 48	18 21	18 26	24 07
15	13 36 06	22 08 51	22 ♎ 23 08	29 ♎ 59 31	16 57	29 ♍ 22	12 35	24 25	23 46	18 22	18 26	24 09
16	13 40 03	23 08 22	07 ♏ 33 04	15 ♏ 02 36	17 50	00 ♎ 37	13 16	24 38	23 45	18 23	18 27	24 11
17	13 44 00	24 07 56	22 ♏ 27 10	29 ♏ 45 56	18 40	01 51	13 58	24 51	23 44	18 24	18 28	24 13
18	13 47 56	25 07 32	06 ♐ 58 21	14 ♐ 04 00	19 27	03 05	14 40	25 04	23 43	18 25	18 28	24 15
19	13 51 53	26 07 09	21 ♐ 02 43	27 ♐ 54 29	20 09	04 20	15 21	25 17	23 42	18 26	18 29	24 17
20	13 55 49	27 06 48	04 ♑ 39 24	11 ♑ 17 46	20 47	05 34	16 03	25 30	23 41	18 27	18 29	24 20
21	13 59 46	28 06 29	17 ♑ 49 56	24 ♑ 16 19	21 20	06 49	16 45	25 43	23 40	18 28	18 30	24 22
22	14 03 42	29 ♎ 06 11	00 ♒ 37 24	06 ♒ 53 42	21 47	08 03	17 27	25 56	23 40	18 30	18 31	24 24
23	14 07 39	00 ♏ 05 56	13 ♒ 05 45	19 ♒ 14 06	22 08	09 18	18 09	26 09	23 39	18 31	18 32	24 26
24	14 11 35	01 05 41	25 ♒ 19 15	01 ♓ 21 43	22 23	10 33	18 51	26 22	23 39	18 32	18 32	24 29
25	14 15 32	02 05 29	07 ♓ 21 59	13 ♓ 20 30	22 30	11 47	19 33	26 35	23 38	18 34	18 33	24 31
26	14 19 29	03 05 18	19 ♓ 17 41	25 ♓ 13 56	22 R 29	13 02	20 15	26 48	23 38	18 35	18 34	24 33
27	14 23 25	04 05 09	01 ♈ 09 36	07 ♈ 05 01	22 21	14 17	20 57	27 01	23 38	18 37	18 35	24 35
28	14 27 22	05 05 02	13 ♈ 00 28	18 ♈ 56 13	22 03	15 32	21 40	27 14	23 D 38	18 38	18 36	24 38
29	14 31 18	06 04 57	24 ♈ 52 31	00 ♉ 49 36	21 36	16 46	22 22	27 27	23 38	18 40	18 37	24 40
30	14 35 15	07 04 53	06 ♉ 47 42	12 ♉ 47 01	21 00	18 01	23 04	27 40	23 38	18 41	18 38	24 42
31	14 39 11	08 ♏ 04 52	18 ♉ 47 46	24 ♉ 50 11	20 ♏ 14	19 ♎ 16	23 ♏ 47	27 ♎ 53	23 ♒ 39	18 ♑ 43	18 ♑ 39	24 ♏ 45

Date	Moon True ☊ ° '	Moon Mean ☊ ° '	Moon ☽ Latitude ° '
01	04 ♐ 25	05 ♐ 56	03 N 48
02	04 R 17	05 53	03 02
03	04 12	05 50	02 07
04	04 09	05 47	01 06
05	04 08	05 44	00 N 02
06	04 D 08	05 40	01 S 04
07	04 10	05 37	02 08
08	04 11	05 34	03 07
09	04 R 10	05 31	03 59
10	04 08	05 28	04 38
11	04 04	05 25	05 03
12	03 59	05 21	05 08
13	03 52	05 18	04 54
14	03 45	05 15	04 18
15	03 40	05 12	03 25
16	03 36	05 09	02 16
17	03 33	05 05	01 S 00
18	03 D 33	05 02	00 N 19
19	03 34	04 59	01 34
20	03 35	04 56	02 41
21	03 37	04 53	03 38
22	03 R 37	04 50	04 22
23	03 36	04 46	04 53
24	03 34	04 43	05 09
25	03 30	04 40	05 12
26	03 25	04 37	05 00
27	03 20	04 34	04 37
28	03 14	04 30	04 01
29	03 10	04 27	03 14
30	03 06	04 24	02 19
31	03 ♐ 04	04 ♐ 21	01 N 18

DECLINATIONS

Date	Sun ☉ ° '	Moon ☽ ° '	Mercury ☿ ° '	Venus ♀ ° '	Mars ♂ ° '	Jupiter ♃ ° '	Saturn ♄ ° '	Uranus ♅ ° '	Neptune ♆ ° '	Pluto ♇ ° '
01	03 S 18	09 N 50	13 S 12	08 N 09	12 S 29	07 S 21	14 S 54	22 S 40	21 S 32	05 S 14
02	03 41	13 33	13 47	07 43	12 44	07 25	14 54	22 40	21 32	05 15
03	04 04	16 45	14 21	07 16	12 58	07 30	14 55	22 40	21 32	05 15
04	04 28	19 16	14 54	06 49	13 12	07 35	14 56	22 40	21 32	05 16
05	04 51	20 56	15 26	06 21	13 27	07 40	14 57	22 40	21 32	05 17
06	05 14	21 39	15 57	05 54	13 41	07 45	14 57	22 40	21 32	05 18
07	05 37	21 18	16 27	05 26	13 55	07 50	14 58	22 39	21 32	05 19
08	06 00	19 50	16 56	04 58	14 09	07 54	14 58	22 39	21 32	05 19
09	06 22	17 17	17 24	04 30	14 23	07 59	14 59	22 39	21 32	05 20
10	06 45	13 43	17 51	04 01	14 36	08 04	15 00	22 39	21 32	05 21
11	07 08	09 16	18 17	03 33	14 50	08 09	15 00	22 39	21 32	05 22
12	07 30	04 N 10	18 41	03 04	15 04	08 14	15 01	22 39	21 32	05 22
13	07 53	01 S 18	19 04	02 36	15 17	08 19	15 01	22 39	21 32	05 23
14	08 15	06 47	19 25	02 07	15 31	08 23	15 01	22 39	21 32	05 24
15	08 37	11 53	19 46	01 38	15 44	08 28	15 02	22 39	21 32	05 25
16	09 00	16 11	20 04	01 09	15 57	08 33	15 02	22 38	21 32	05 26
17	09 22	19 21	20 21	00 40	16 10	08 38	15 02	22 38	21 32	05 26
18	09 43	21 10	20 36	00 N 11	16 23	08 43	15 03	22 38	21 32	05 27
19	10 05	21 34	20 48	00 S 18	16 36	08 47	15 03	22 38	21 32	05 28
20	10 27	20 30	20 59	00 48	16 49	08 52	15 03	22 38	21 32	05 28
21	10 48	18 39	21 08	01 17	17 01	08 57	15 03	22 38	21 32	05 29
22	11 09	15 45	21 14	01 46	17 14	09 02	15 03	22 37	21 31	05 30
23	11 30	12 12	21 18	02 15	17 26	09 06	15 04	22 37	21 31	05 31
24	11 51	08 13	21 18	02 44	17 38	09 11	15 04	22 37	21 31	05 32
25	12 12	03 S 59	21 16	03 14	17 50	09 16	15 04	22 37	21 31	05 32
26	12 33	00 N 22	21 10	03 43	18 02	09 20	15 04	22 37	21 31	05 33
27	12 53	04 41	21 01	04 12	18 14	09 25	15 04	22 36	21 31	05 34
28	13 13	08 50	20 48	04 41	18 26	09 30	15 04	22 36	21 31	05 35
29	13 33	12 39	20 31	05 10	18 37	09 34	15 03	22 36	21 31	05 35
30	13 53	15 59	20 10	05 39	18 49	09 39	15 03	22 36	21 31	05 36
31	14 S 12	18 N 40	19 S 44	06 S 07	19 S 00	09 S 44	15 S 03	22 S 35	21 S 31	05 S 37

ZODIAC SIGN ENTRIES

Date	h	m	Planets
01	02	09	☿ ♏
02	16	13	☽ ♉
05	04	27	☽ ♊
07	14	42	☽ ♋
09	21	34	☽ ♌
12	00	36	☽ ♍
14	00	47	☽ ♎
16	00	01	☽ ♏
16	00	13	♀ ♎
18	00	23	☽ ♐
20	03	42	☽ ♑
22	10	49	☽ ♒
23	09	37	☉ ♏
24	21	17	☽ ♓
27	09	39	☽ ♈
29	22	20	☽ ♉

LATITUDES

Date	Mercury ☿ ° '	Venus ♀ ° '	Mars ♂ ° '	Jupiter ♃ ° '	Saturn ♄ ° '	Uranus ♅ ° '	Neptune ♆ ° '	Pluto ♇ ° '
01	01 S 39	01 N 14	00 N 01	01 N 05	01 S 32	00 S 28	00 N 39	13 N 53
04	02 00	01 18	00 01	01 05	01 32	00 28	00 39	13 51
07	02 19	01 22	00 03	01 04	01 32	00 28	00 39	13 50
10	02 37	01 26	00 04	01 04	01 31	00 28	00 39	13 49
13	02 51	01 29	00 06	01 04	01 31	00 28	00 39	13 48
16	03 02	01 31	00 08	01 04	01 31	00 28	00 39	13 47
19	03 09	01 32	00 10	01 04	01 31	00 28	00 39	13 47
22	03 08	01 33	00 12	01 04	01 31	00 28	00 38	13 46
25	02 58	01 34	00 14	01 04	01 30	00 28	00 38	13 45
28	02 37	01 33	00 15	01 04	01 30	00 28	00 38	13 44
31	02 S 00	01 N 32	00 S 17	01 N 04	01 S 30	00 S 28	00 N 38	13 N 44

DATA

Julian Date	2449262
Delta T	+59 seconds
Ayanamsa	23° 46' 26"
Synetic vernal point	05° ♓ 20' 33"
True obliquity of ecliptic	23° 26' 21"

MOON'S PHASES, APSIDES AND POSITIONS ☽

Date	h	m	Phase	Longitude ° '	Eclipse Indicator
08	19	35	☾	15 ♋ 32	
15	11	36	●	22 ♎ 08	
22	08	52	☽	28 ♑ 58	
30	12	38	○	07 ♉ 06	

Day	h	m	
15	01	40	Perigee
27	23	46	Apogee
06	16	16	Max dec 21° N 40'
13	06	22	0S
19	07	04	Max dec 21° S 36'
26	09	57	0N

LONGITUDES

Date	Chiron ⚷ ° '	Ceres ⚳ ° '	Pallas ⚴ ° '	Juno ⚵ ° '	Vesta ⚶ ° '	Black Moon Lilith ° '
01	03 ♍ 33	03 ♉ 53	24 ♒ 06	26 ♍ 04	28 ♒ 29	09 ♈ 08
11	04 ♍ 43	01 ♉ 56	23 ♒ 09	29 ♍ 59	28 ♒ 06	10 ♈ 15
21	05 ♍ 48	29 ♈ 42	22 ♒ 48	03 ♎ 49	28 ♒ 29	11 ♈ 22
31	06 ♍ 46	27 ♈ 25	23 ♒ 01	07 ♎ 32	29 ♒ 34	12 ♈ 29

All ephemeris data is given at 12.00 UT and the Moon's longitude is additionally given for 24.00 UT

ASPECTARIAN

h m	Aspects
01 Friday	
02 59	☽ ∥ ♀
03 03	☽ ⊼ ♀
03 38	♀ Q ♇
05 52	♄ ⊥ ♅
15 18	☽ ± ♇
16 24	☽ □ ♅
16 36	☽ ± ♀
16 41	☽ □ ♆
21 10	☉ ∠ ♇
22 59	☽ ☍ ♃
02 Saturday	
03 29	☽ ⊼ ♇
04 28	☽ ⚹ ♄
05 59	☽ # ♂
09 04	☉ ⚼ ♄
12 56	☽ ⚼ ♀
13 53	☽ # ☿
21 04	☽ ☍ ☿
21 37	☽ # ♄
03 Sunday	
00 19	☽ ☍ ♂
01 01	♀ # ♃
04 39	☽ Q ♄
13 05	☽ ⊼ ☉
22 41	☽ △ ♀
04 Monday	
02 14	☽ ± ☉
05 00	☽ △ ♅
05 16	☽ △ ♆
12 36	☽ ⊼ ♃
13 21	☿ ∥ ♄
16 05	☽ ☍ ♇
16 42	☽ □ ♄
17 52	♀ ⊥ ♃
21 51	☽ ⚼ ☉
05 Tuesday	
00 46	☽ ± ♃
08 18	☿ ☌ ♂
10 53	☽ ⚼ ♅
11 09	☽ ⚼ ♆
15 56	☽ ⊼ ♂
16 21	☽ ⊼ ☿
18 54	☽ ⚼ ♃
20 52	☿ Q ♅
23 19	☿ Q ♆
06 Wednesday	
04 14	☽ # ♆
04 24	☽ ± ♂
04 35	☽ ± ♅
04 50	☽ ± ♆
05 27	☽ ± ☿
06 02	☽ △ ☉
07 40	♂ Q ♅
12 06	♂ Q ♆
12 14	♀ △ ♅
14 40	♀ △ ♆
16 15	☽ ⊼ ♅
16 17	☉ ∥ ♇
16 30	☽ ⊼ ♆
16 42	☽ □ ♀
22 52	☽ ⚼ ♂
07 Thursday	
00 34	☽ △ ♃
00 48	☽ ⚼ ☿
03 02	☽ ⊼ ♇
03 18	☽ △ ♄
04 12	☽ # ♆
06 48	☉ # ♀
14 29	☽ ± ♇
18 01	♀ # ♇
08 Friday	
04 52	☽ △ ♂
06 31	☽ Q ♀
07 23	☽ ⚼ ♇
07 30	☽ ⚼ ♄
08 05	☽ △ ☿
19 35	☽ □ ☉
23 59	☽ ± ♄
09 Saturday	
00 36	☽ ☍ ♅
00 48	☽ ☍ ♆
06 23	♄ □ ♇
06 44	☽ ⚹ ♀
09 16	☽ □ ♃
10 46	☽ ⊼ ♄
10 48	☽ △ ♇
11 11	☽ # ☿
10 Sunday	
04 08	☽ # ♄
06 15	☽ Q ☉
06 52	☽ # ♂
11 50	☽ ∠ ♀
13 33	☽ □ ♂
15 29	♀ ⊻ ♃
17 21	☽ Q ♃
18 34	☽ □ ☿
11 Monday	
01 32	♀ ⊼ ♄
03 49	♀ ⚹ ♇
04 23	☽ ⚹ ☉
04 41	☽ ⊥ ♀
05 08	☽ ⊼ ♅
05 18	☽ ⊼ ♆
07 36	☉ ⊥ ♇
13 57	☽ ⚹ ♃
14 26	☽ ☍ ♄
14 41	☽ □ ♇
15 06	☉ □ ♅
15 11	☽ ± ♅
15 21	☽ ± ♆
15 41	☽ ⊻ ♀
17 25	☽ # ♃
17 32	☉ □ ♆
21 26	☽ Q ♂
21 40	☽ # ☉
12 Tuesday	
03 12	☽ Q ☿
05 18	♀ ∠ ♂
06 03	☽ ⚼ ♅
06 12	☽ ⚼ ♆
06 15	♂ ∥ ♄
06 34	☽ # ♇
07 06	☽ ∠ ☉
14 57	☽ ∠ ♃
17 23	☽ ∥ ♀
17 59	☽ ⚹ ♂
19 28	♃ △ ♄
20 13	☽ Q ♇
22 43	☽ ⊥ ☉
13 Wednesday	
00 09	☽ ⚹ ☿
05 38	☽ ⊥ ♃
06 17	☽ △ ♅
06 25	☽ △ ♆
09 00	☽ ⊻ ☉
15 00	☽ ⊼ ♄
15 19	☽ ⊻ ♃
15 26	☽ ⚹ ♇
17 10	☽ # ♀
19 06	☽ ∠ ♂
20 35	☽ ☌ ♀
14 Thursday	
00 29	☽ ± ♄
00 35	♃ ⊻ ♇
01 42	☽ ∠ ☿
05 51	☽ ∥ ♇
09 58	☽ ⊥ ♂
14 38	☽ ⚼ ♄
15 09	☽ ∠ ♇
16 50	☽ ⊥ ☿
19 11	☽ ∥ ☉
19 25	☽ ∥ ♃
19 51	☽ ⊻ ♂
23 13	☉ ∥ ♃
15 Friday	
02 53	☽ ⊻ ☿
05 19	☽ ⊥ ♇
05 40	☽ □ ♅
05 48	☽ □ ♆
11 36	☽ ☌ ☉
14 11	☽ △ ♄
14 47	☽ ⊻ ♇
15 15	☽ ☌ ♃
19 37	♀ ± ♄
16 Saturday	
00 00	☽ ⊻ ♀
05 02	☽ ∥ ♄
10 08	☽ Q ♅
10 15	☽ Q ♆
10 22	☽ ⊥ ♀
10 29	☽ ∥ ♂
21 36	☽ ☌ ♂
17 Sunday	
02 05	☽ ∠ ♀
02 27	☉ △ ♄
03 45	☿ ⚹ ♅
05 24	☽ ⚹ ♅
05 30	☽ ☌ ☿
05 31	☽ ⚹ ♆
05 38	☿ ⚹ ♆
14 05	☽ □ ♄
14 11	☉ ⊻ ♇
14 53	☽ ☌ ♇
14 56	☽ ⊻ ☉
15 59	☽ ⊻ ♃
18 Monday	
00 34	☽ ∥ ☿
01 32	☽ ⊥ ☉
02 00	☽ ⊥ ♃
04 54	☽ ⚹ ♀
06 03	☽ ∠ ♅
06 08	☽ ∠ ♆
10 16	☉ ☌ ♃
17 18	☽ ∠ ♃
17 43	☽ ∠ ☉
19 59	☽ Q ♄
21 12	☽ ⊥ ♅
21 17	☽ ⊥ ♆
23 05	☽ ∥ ♆
19 Tuesday	
01 41	☽ ⊻ ♂
03 05	☽ Q ♀
07 29	☽ ⊻ ♅
07 34	☽ ⊻ ♆
10 22	☽ ⊻ ☿
12 34	☽ ⊥ ♂
15 08	☽ ∥ ♆
16 37	☽ ⚹ ♄
17 40	☽ ⊻ ♇
19 31	☽ ⚹ ♃
21 21	☽ ⊥ ☿
21 33	☽ ⚹ ☉
20 Wednesday	
04 16	☽ ⊥ ♇
05 13	☽ ∠ ♂
07 02	☽ ∥ ☿
13 49	☽ □ ♀
14 07	☽ ∠ ☿
17 13	☽ Q ♃
19 15	☽ ∠ ♄
19 33	☿ ∠ ♀
20 27	☽ ∠ ♇
20 41	☽ Q ☉
21 Thursday	
09 53	☽ ⚹ ♂
11 42	☽ ⊥ ♄
13 11	☽ ☌ ♅
13 14	☽ ☌ ♆
18 45	☽ ⚹ ☿
22 52	☽ ⊻ ♄
22 Friday	
00 13	☽ ⚹ ♇
01 17	☽ ∥ ♂
02 59	☽ □ ♃
08 52	☽ □ ☉
09 38	☽ Q ♂
16 58	☽ ∥ ♄
18 13	☽ Q ☿
23 05	☽ Q ♇
23 34	♀ ⚼ ♄
23 Saturday	
03 49	☽ △ ♀
14 45	♀ ∠ ♇
16 01	☽ ∥ ☉
22 28	☽ □ ♂
22 36	☽ ⊻ ♅
22 37	☽ ⊻ ♆
24 Sunday	
00 56	♂ ⚹ ♅
01 12	♂ ⚹ ♆
06 05	☽ □ ☿
06 30	☽ ∥ ♃
08 41	☽ ☌ ♄
10 19	☽ □ ♇
10 27	☽ ⊥ ♅
10 27	☽ ⊥ ♆
12 30	☽ ⚼ ♀
14 07	☽ △ ♃
20 07	♅ ☌ ♆
25 Monday	
00 30	☽ △ ☉
03 19	☽ ∥ ♇
04 22	☽ ∠ ♆
04 22	☽ ∠ ♅
08 29	☽ ± ♀
15 45	☽ ∥ ♀
20 38	☽ ⚼ ♃
21 55	☽ ⊼ ♀
22 39	☿ St R
26 Tuesday	
09 21	☽ ⚼ ☉
10 32	☽ ⚹ ♆
10 34	☽ ⚹ ♅
14 04	☽ △ ♂
15 07	☽ ± ♃
18 24	☽ △ ☿
20 46	☽ ⊻ ♄
22 39	☽ △ ♇
27 Wednesday	
03 28	☽ ⊼ ♃
05 12	☽ ± ☉
08 52	☽ # ♀
08 55	☽ ⊥ ♄
10 50	☽ Q ♆
10 53	☽ Q ♅
16 59	☽ # ♇
18 28	☽ ⊼ ☉
22 20	☽ ⚼ ♂
28 Thursday	
00 16	☽ ⚼ ☿
03 08	☽ ∠ ♄
03 39	♄ St D
05 08	☽ ⚼ ♇
16 06	☽ # ♃
17 42	☽ ☍ ♀
17 43	☽ ± ♂
17 58	☽ ± ☿
18 05	♀ ⊥ ♂
19 40	☿ ⊥ ♀
20 26	☿ ☌ ♂
23 20	☽ □ ♆
23 25	☽ ± ♇
23 25	☽ □ ♅
29 Friday	
05 40	☽ ⊼ ☿
06 37	☽ ⊼ ♂
09 30	☽ ⚹ ♄
11 35	☽ ⊼ ♇
17 18	☽ ☍ ♃
18 46	☽ # ☉
30 Saturday	
00 58	☉ Q ♆
02 16	☉ Q ♅
04 56	☽ # ♄
09 41	☽ Q ♄
09 49	♀ ∥ ♇
12 38	☽ ☍ ☉
23 55	♀ □ ♆
31 Sunday	
01 09	♀ □ ♅
01 36	♀ ⊥ ♇
07 15	♂ □ ♄
11 42	☽ △ ♆
11 50	☽ △ ♅
13 03	☽ ⊼ ♀
14 41	☽ ☍ ☿
15 57	☽ # ♂
21 38	☽ □ ♄
21 54	☽ # ☿
22 32	☽ ☍ ♂
23 03	☿ ⊻ ♀
23 51	☽ ☍ ♇

NOVEMBER 1993

Date	Sidereal time h m s	Sun ☉ ° ′ ″	Moon ☽ ° ′ ″	Moon ☽ 24.00 ° ′ ″	Mercury ☿ ° ′	Venus ♀ ° ′	Mars ♂ ° ′	Jupiter ♃ ° ′	Saturn ♄ ° ′	Uranus ♅ ° ′	Neptune ♆ ° ′	Pluto ♇ ° ′
01	14 43 08	09 ♏ 04 52	00 ♊ 54 28	07 ♊ 00 54	19 ♏ 20	20 ♎ 31	24 ♏ 29	28 ♎ 06	23 ♒ 39	18 ♑ 45	18 ♑ 40	24 ♏ 47
02	14 47 04	10 04 55	13 ♊ 09 42	19 ♊ 21 12	19 R 18	21 46	25 12	28 19	23 39	18 46	18 41	24 49
03	14 51 01	11 04 59	25 ♊ 35 39	01 ♋ 53 24	17 08	23 01	25 55	28 32	23 40	18 48	18 42	24 52
04	14 54 58	12 05 06	08 ♋ 14 46	14 ♋ 40 06	15 53	24 16	26 37	28 45	23 41	18 50	18 43	24 54
05	14 58 54	13 05 14	21 ♋ 09 43	27 ♋ 43 58	14 35	25 31	27 20	28 58	23 42	18 52	18 44	24 56
06	15 02 51	14 05 25	04 ♌ 23 08	11 ♌ 07 29	13 16	26 46	28 03	29 11	23 43	18 54	18 46	24 59
07	15 06 47	15 05 37	17 ♌ 57 13	24 ♌ 52 27	11 59	28 01	28 46	29 24	23 44	18 56	18 47	25 01
08	15 10 44	16 05 52	01 ♍ 53 13	08 ♍ 59 26	10 46	29 ♎ 16	29 ♏ 29	29 ♎ 36	23 45	18 58	18 48	25 03
09	15 14 40	17 06 09	16 ♍ 10 52	23 ♍ 27 08	09 39	00 ♏ 31	00 ♐ 12	29 ♎ 49	23 46	19 00	18 49	25 06
10	15 18 37	18 06 27	00 ♎ 47 42	08 ♎ 11 56	08 41	01 46	00 55	00 ♏ 02	23 47	19 02	18 51	25 08
11	15 22 33	19 06 48	15 ♎ 38 59	23 ♎ 07 56	07 52	03 01	01 38	00 15	23 49	19 04	18 52	25 11
12	15 26 30	20 07 11	00 ♏ 37 45	08 ♏ 07 22	07 15	04 16	02 21	00 27	23 50	19 06	18 53	25 13
13	15 30 27	21 07 35	15 ♏ 35 42	23 ♏ 01 39	06 49	05 32	03 04	00 40	23 52	19 09	18 55	25 15
14	15 34 23	22 08 01	00 ♐ 24 16	07 ♐ 42 38	06 35	06 47	03 47	00 53	23 54	19 11	18 56	25 18
15	15 38 20	23 08 29	14 ♐ 55 59	22 ♐ 03 44	06 D 32	08 02	04 31	01 05	23 55	19 13	18 58	25 20
16	15 42 16	24 08 58	29 ♐ 05 23	06 ♑ 00 41	06 41	09 17	05 14	01 18	23 57	19 16	18 59	25 23
17	15 46 13	25 09 29	12 ♑ 49 28	19 ♑ 31 46	06 59	10 32	05 58	01 30	23 59	19 18	19 01	25 25
18	15 50 09	26 10 01	26 ♑ 07 42	02 ♒ 37 32	07 27	11 48	06 41	01 43	24 01	19 20	19 02	25 27
19	15 54 06	27 10 34	09 ♒ 01 36	15 ♒ 20 28	08 04	13 03	07 25	01 55	24 04	19 23	19 04	25 30
20	15 58 02	28 11 08	21 ♒ 34 12	27 ♒ 43 45	08 48	14 18	08 08	02 08	24 06	19 25	19 05	25 32
21	16 01 59	29 ♏ 11 44	03 ♓ 49 32	09 ♓ 52 07	09 39	15 34	08 52	02 20	24 08	19 28	19 07	25 35
22	16 05 56	00 ♐ 12 20	15 ♓ 52 04	21 ♓ 49 58	10 36	16 49	09 36	02 32	24 11	19 31	19 08	25 37
23	16 09 52	01 12 58	27 ♓ 45 50	03 ♈ 41 50	11 38	18 04	10 20	02 45	24 14	19 33	19 10	25 39
24	16 13 49	02 13 37	09 ♈ 36 52	15 ♈ 31 58	12 44	19 20	11 03	02 57	24 16	19 36	19 12	25 42
25	16 17 45	03 14 17	21 ♈ 27 34	27 ♈ 24 07	13 55	20 35	11 47	03 09	24 19	19 39	19 13	25 44
26	16 21 42	04 14 59	03 ♉ 21 58	09 ♉ 21 30	15 09	21 50	12 31	03 21	24 22	19 41	19 15	25 47
27	16 25 38	05 15 41	15 ♉ 23 00	21 ♉ 26 45	16 25	23 06	13 15	03 33	24 25	19 44	19 17	25 49
28	16 29 35	06 16 25	27 ♉ 32 59	03 ♊ 41 53	17 44	24 21	13 59	03 45	24 28	19 47	19 19	25 51
29	16 33 31	07 17 10	09 ♊ 53 38	16 ♊ 08 21	19 06	25 36	14 43	03 57	24 31	19 50	19 20	25 54
30	16 37 28	08 ♐ 17 57	22 ♊ 26 09	28 ♊ 47 08	20 ♏ 29	26 ♏ 52	15 ♐ 27	04 ♏ 09	24 ♒ 35	19 ♑ 52	19 ♑ 22	25 ♏ 56

DECLINATIONS

Date	Moon True ☊ ° ′	Moon Mean ☊ ° ′	Moon ☽ Latitude ° ′	Sun ☉ ° ′	Moon ☽ ° ′	Mercury ☿ ° ′	Venus ♀ ° ′	Mars ♂ ° ′	Jupiter ♃ ° ′	Saturn ♄ ° ′	Uranus ♅ ° ′	Neptune ♆ ° ′	Pluto ♇ ° ′
01	03 ♐ 03	04 ♐ 18	00 N 12	14 S 31	20 N 32	19 S 14	06 S 36	19 S 11	09 S 48	15 S 03	22 S 35	21 S 31	05 S 37
02	03 D 03	04 15	00 S 55	14 50	21 28	18 41	07 04	19 22	09 53	15 03	22 35	21 30	05 38
03	03 05	04 11	02 01	15 09	21 21	18 03	07 33	19 32	09 57	15 02	22 35	21 30	05 39
04	03 06	04 08	03 02	15 28	20 09	17 23	08 01	19 43	10 02	15 02	22 34	21 30	05 40
05	03 08	04 05	03 55	15 46	17 54	16 41	08 29	19 53	10 07	15 02	22 34	21 30	05 40
06	03 09	04 02	04 37	16 04	14 40	15 57	08 57	20 04	10 11	15 01	22 34	21 30	05 41
07	03 R 09	03 59	05 05	16 22	10 36	15 14	09 25	20 14	10 16	15 01	22 33	21 30	05 42
08	03 08	03 56	05 16	16 39	05 53	14 33	09 52	20 24	10 20	15 00	22 33	21 30	05 42
09	03 07	03 52	05 08	16 56	00 N 43	13 54	10 19	20 33	10 25	15 00	22 33	21 29	05 43
10	03 05	03 49	04 40	17 13	04 S 36	13 19	10 46	20 43	10 29	14 59	22 33	21 29	05 44
11	03 03	03 46	03 53	17 30	09 45	12 48	11 13	20 52	10 33	14 59	22 32	21 29	05 44
12	03 02	03 43	02 50	17 46	14 21	12 23	11 40	21 02	10 38	14 58	22 32	21 29	05 45
13	03 01	03 40	01 35	18 02	18 03	12 03	12 06	21 11	10 42	14 58	22 32	21 29	05 45
14	03 00	03 36	00 S 14	18 18	20 28	11 48	12 32	21 19	10 47	14 57	22 31	21 29	05 46
15	03 D 00	03 33	01 N 06	18 33	21 30	11 40	12 58	21 28	10 51	14 56	22 31	21 29	05 47
16	03 01	03 30	02 20	18 49	21 06	11 37	13 23	21 36	10 55	14 56	22 31	21 28	05 47
17	03 02	03 27	03 24	19 03	19 27	11 38	13 48	21 45	10 59	14 55	22 30	21 28	05 48
18	03 03	03 24	04 14	19 18	16 46	11 44	14 13	21 53	11 04	14 54	22 30	21 28	05 48
19	03 03	03 21	04 50	19 32	13 20	11 54	14 37	22 01	11 08	14 53	22 30	21 28	05 49
20	03 03	03 17	05 11	19 45	09 24	12 08	15 01	22 08	11 12	14 52	22 29	21 28	05 50
21	03 R 03	03 14	05 18	19 59	05 10	12 25	15 24	22 16	11 16	14 52	22 29	21 27	05 50
22	03 03	03 11	05 10	20 12	00 S 49	12 44	15 48	22 23	11 21	14 51	22 28	21 27	05 51
23	03 03	03 08	04 49	20 24	03 N 32	13 06	16 10	22 30	11 25	14 50	22 28	21 27	05 51
24	03 03	03 05	04 15	20 36	07 43	13 29	16 33	22 37	11 29	14 49	22 28	21 27	05 52
25	03 D 03	03 02	03 31	20 48	11 38	13 54	16 55	22 43	11 33	14 48	22 27	21 27	05 52
26	03 03	02 58	02 38	21 00	15 06	14 20	17 16	22 50	11 37	14 47	22 27	21 26	05 53
27	03 03	02 55	01 37	21 11	17 59	14 47	17 37	22 56	11 41	14 46	22 26	21 26	05 53
28	03 02	02 52	00 N 31	21 21	20 07	15 15	17 58	23 02	11 45	14 44	22 26	21 26	05 54
29	03 R 03	02 49	00 S 38	21 32	21 19	15 43	18 18	23 07	11 49	14 43	22 26	21 26	05 54
30	03 ♐ 03	02 ♐ 46	01 S 46	21 S 41	21 N 28	16 S 11	18 S 37	23 S 13	11 S 53	14 S 42	22 S 25	21 S 25	05 S 55

ZODIAC SIGN ENTRIES

Date	h	m	Planets
01	10	13	☽ ♊
03	20	25	☽ ♋
06	04	06	☽ ♌
08	08	47	☽ ♍
09	02	07	♀ ♏
09	05	29	♂ ♐
10	08	15	♃ ♏
10	10	42	☽ ♎
12	11	00	☽ ♏
14	11	20	☽ ♐
16	13	34	☽ ♑
18	19	08	☽ ♒
21	04	27	☽ ♓
22	07	07	☉ ♐
23	16	30	☽ ♈
26	05	14	☽ ♉
28	16	48	☽ ♊

LATITUDES

Date	Mercury ☿ ° ′	Venus ♀ ° ′	Mars ♂ ° ′	Jupiter ♃ ° ′	Saturn ♄ ° ′	Uranus ♅ ° ′	Neptune ♆ ° ′	Pluto ♇ ° ′
01	01 S 45	01 N 32	00 S 18	01 N 04	01 S 30	00 S 28	00 N 38	13 N 43
04	00 S 49	01 30	00 20	01 04	01 30	00 28	00 38	13 43
07	00 N 12	01 27	00 21	01 04	01 29	00 28	00 38	13 43
10	01 09	01 24	00 23	01 04	01 29	00 28	00 38	13 42
13	01 51	01 20	00 25	01 04	01 29	00 28	00 38	13 42
16	02 16	01 16	00 26	01 04	01 28	00 28	00 38	13 42
19	02 25	01 12	00 28	01 04	01 28	00 28	00 38	13 41
22	02 23	01 07	00 30	01 05	01 28	00 27	00 38	13 41
25	02 13	01 01	00 31	01 05	01 28	00 27	00 37	13 41
28	01 57	00 55	00 33	01 05	01 27	00 27	00 37	13 41
31	01 N 38	00 N 49	00 S 35	01 N 05	01 S 27	00 S 27	00 N 37	13 N 41

DATA

Julian Date	2449293
Delta T	+59 seconds
Ayanamsa	23° 46′ 29″
Synetic vernal point	05° ♓ 20′ 30″
True obliquity of ecliptic	23° 26′ 20″

MOON'S PHASES, APSIDES AND POSITIONS ☽

Date	h	m	Phase	Longitude ° ′	Eclipse Indicator
07	06	36	☾	14 ♌ 52	
13	21	34	●	21 ♏ 32	Partial
21	02	03	☽	28 ♒ 47	
29	06	31	○	07 ♊ 03	total

Day	h	m		
12	11	53	Perigee	
24	12	29	Apogee	
02	21	29	Max dec	21° N 33′
09	15	15	OS	
15	17	08	Max dec	21° S 32′
22	16	28	ON	
30	03	30	Max dec	21° N 32′

LONGITUDES

Date	Chiron ⚷ ° ′	Ceres ⚳ ° ′	Pallas ⚴ ° ′	Juno ⚵ ° ′	Vesta ⚶ ° ′	Black Moon Lilith ☾ ° ′
01	06 ♍ 51	27 ♈ 12	23 ♒ 04	07 ♎ 54	29 ♒ 43	12 ♈ 35
11	07 ♍ 40	25 ♈ 11	23 ♒ 52	11 ♎ 29	01 ♓ 29	13 ♈ 42
21	08 ♍ 20	23 ♈ 36	25 ♒ 07	14 ♎ 55	03 ♓ 46	14 ♈ 49
31	08 ♍ 50	22 ♈ 38	26 ♒ 47	18 ♎ 10	06 ♓ 30	15 ♈ 56

All ephemeris data is given at 12.00 UT and the Moon's longitude is additionally given for 24.00 UT

ASPECTARIAN

h m	Aspects
01 Monday	
02 20	☽ ± ♀
06 22	☽ ⊼ ♃
13 57	☿ ∥ ♂
17 26	☽ ⚼ ♆
17 36	☽ ⚼ ♅
18 24	☽ ± ♃
22 05	☽ ⚼ ♀
22 25	♂ ☌ ♇
02 Tuesday	
01 36	☿ ⚹ ♅
03 31	☿ ⚹ ♆
05 28	☽ ⊼ ☉
11 04	☽ ± ♆
11 14	☽ ± ♅
12 19	☽ ⚼ ♃
15 10	☽ # ♆
18 10	☽ ± ☉
21 07	☽ ⊼ ☿
22 43	☽ ⊼ ♆
22 54	☽ ⊼ ♅
03 Wednesday	
03 28	☉ ∥ ♄
03 53	☽ # ♆
06 30	☽ △ ♀
07 42	☽ ± ☿
08 18	☽ △ ♄
10 35	☽ ⊼ ♇
12 39	☽ ⊼ ♂
13 01	☽ ⚼ ☉
17 43	☽ △ ♃
22 04	☽ ± ♇
23 22	☽ ⚼ ☿
04 Thursday	
00 43	♀ △ ♄
00 45	☽ ± ♂
12 49	☽ ⚼ ♄
15 07	☽ ⚼ ♇
17 09	☽ # ♂
18 42	☽ ⚼ ♂
19 48	☽ △ ☉
05 Friday	
00 38	♀ ⊻ ♇
00 58	☽ △ ☿
05 36	☽ ± ♄
07 32	☽ ☍ ♆
07 46	☽ ☍ ♅
16 39	☽ ⊼ ♄
18 56	☽ △ ♇
20 48	☽ □ ♀
23 55	☽ △ ♂
06 Saturday	
00 46	☽ # ☿
02 28	☽ □ ♃
03 13	☽ # ☉
03 33	☉ ☌ ☿
09 23	☉ ∥ ☿
09 40	☽ # ♄
07 Sunday	
02 25	☽ □ ☿
06 36	☽ □ ☉
08 16	☽ Q ♀
11 00	☽ Q ♃
13 27	☽ ⊼ ♆
13 43	☽ ⊼ ♅
13 49	☽ # ♃
17 46	☽ # ♀
19 38	☿ ∥ ♄
22 02	☽ ☍ ♄
23 51	☽ ± ♆
08 Monday	
00 08	☽ ± ♅
00 17	☽ □ ♇
07 06	☽ Q ☿
07 06	☽ ⚹ ♀
07 41	☽ □ ♂
08 03	☽ ⚹ ♃
12 50	☽ # ♇
15 15	☽ ⚼ ♆
15 32	☽ ⚼ ♅
16 02	☽ Q ☉
18 11	♂ ⊻ ♃
19 56	♀ ☌ ♃
21 35	♀ ⊻ ♂
09 Tuesday	
01 52	☽ ⚹ ☿
06 51	☽ Q ♇
09 42	☽ ∠ ♃
10 47	☽ ∠ ♀
13 39	☽ ⚹ ☉
15 31	☽ Q ♂
16 23	☽ △ ♆
16 41	☽ △ ♅
17 26	♀ ∥ ♃
10 Wednesday	
00 32	☽ ⊼ ♄
00 47	☽ ⊥ ♃
01 04	☽ ∠ ☿
02 44	☽ ⚹ ♇
03 02	☽ ⊥ ♀
10 21	☽ ± ♄
10 44	☽ ⊻ ♃
12 12	☽ ⚹ ♂
13 44	☽ ⊻ ♀
14 53	☽ ⊥ ☿
16 02	☽ ∠ ☉
17 08	☽ ∥ ♇
11 Thursday	
00 05	☽ ⊻ ☿
00 58	☽ ⚼ ♄
03 10	☽ ∠ ♇
05 57	☉ ⚹ ♆
07 38	☽ ⊥ ☉
10 56	☉ ⚹ ♅
13 39	☽ ∠ ♂
16 03	☽ ∥ ♃
17 10	☽ □ ♆
17 30	☽ □ ♅
17 41	☽ ⊥ ♇
17 58	☽ ⊻ ☉
20 03	☽ ∥ ♀
12 Friday	
01 06	☽ △ ♄
02 07	☽ ∥ ☿
03 19	☽ ⊻ ♇
04 48	☽ ⊥ ♂
11 43	☽ ☌ ♃
14 54	☽ ⊻ ♂
15 38	☽ ∥ ♄
18 18	☿ Q ♅
18 22	☽ ☌ ♀
22 02	☽ Q ♆
22 16	☽ ☌ ☿
22 24	☽ Q ♅
13 Saturday	
06 07	☿ Q ♆
10 12	☿ ∥ ♀
12 08	☽ ∥ ☉
17 21	☽ ⚹ ♆
17 44	☽ ⚹ ♅
21 34	☽ ☌ ☉
14 Sunday	
01 23	☽ □ ♄
03 39	☽ ☌ ♇
08 38	☿ ☌ ♀
12 47	☽ ⊻ ♃
15 02	♀ Q ♆
16 55	♂ ∠ ♆
17 48	☽ ∠ ♆
17 50	☽ ☌ ♂
18 12	☽ ∠ ♅
19 57	♀ Q ♅
22 04	☽ ⊻ ☿
22 47	☽ ⊥ ♃
23 27	☽ ⊻ ♀
15 Monday	
01 43	♂ ∠ ♅
05 40	☿ St D
06 58	☽ Q ♄
08 00	☽ ⊥ ☿
08 42	☽ ⊥ ♆
08 58	☽ ∥ ♂
09 08	☽ ⊥ ♅
10 21	☽ ⊥ ♀
10 23	☽ ∥ ♆
13 18	♂ ∥ ♆
13 58	☽ ∠ ♃
18 46	☽ ⊻ ♆
19 13	☽ ⊻ ♅
20 30	☽ ∥ ♂
23 11	☽ ∠ ☿
16 Tuesday	
00 01	☽ ∥ ♆
02 54	☽ ⊻ ☉
02 59	☽ ∠ ♀
03 12	☽ ⚹ ♄
05 37	☽ ⊻ ♇
07 13	☉ □ ♄
13 58	☽ ⊥ ☉
15 52	☽ ⚹ ♃
15 57	☽ ⊥ ♇
23 14	☽ ⊻ ♂
17 Wednesday	
01 26	☽ ⚹ ☿
05 12	☽ ∠ ♄
06 54	☽ ∠ ☉
07 33	☽ ⚹ ♀
07 43	☽ ∠ ♇
10 23	☽ ⊥ ♂
13 14	☽ Q ♃
13 29	☿ Q ♆
15 41	☽ ∥ ☉
18 25	☉ ☌ ♇
21 15	☽ ⊥ ♄
23 05	☽ ☌ ♆
23 23	☽ Q ☿
23 37	☽ ☌ ♅
18 Thursday	
03 26	☽ ∠ ♂
06 29	☿ Q ♅
07 17	☽ Q ♀
08 09	☽ ⊻ ♄
10 46	☽ ⚹ ♇
12 05	☽ ⚹ ☉
22 28	☽ □ ♃
19 Friday	
01 35	☽ ∥ ♄
04 22	☽ ∥ ♀
08 46	☽ ⚹ ♂
09 06	☽ Q ♇
10 05	☽ □ ☿
12 18	☽ Q ☉
20 29	☽ □ ♀
20 31	☽ ∥ ☿
20 Saturday	
01 27	☽ ∥ ♃
03 48	♀ ∥ ♄
07 11	☽ ⊻ ♆
07 50	☽ ⊻ ♅
09 04	☽ Q ♂
11 28	♂ ⊥ ♃
16 56	☽ ☌ ♄
18 51	☽ ⊥ ♆
19 31	☽ ⊥ ♅
19 44	☽ □ ♇
21 Sunday	
02 03	☽ □ ☉
08 17	☽ ∥ ♇
09 00	☽ △ ♃
12 34	☽ ∠ ♆
13 16	☽ ∠ ♅
22 39	☽ □ ♂
22 Monday	
00 31	☽ △ ☿
14 08	☽ △ ♀
15 25	☽ ⚼ ♃
18 35	☽ ⚹ ♆
19 21	☽ ⚹ ♅
23 Tuesday	
04 48	☽ ⊻ ♄
05 55	♂ ∥ ♅
07 42	☽ △ ♇
09 27	☽ ⚼ ☿
09 53	☽ ± ♃
16 59	☽ ⊥ ♄
18 53	☽ Q ♆
19 37	☽ △ ☉
19 41	☽ Q ♅
22 15	☽ ⊼ ♃
24 Wednesday	
00 00	☽ ⚼ ♀
01 12	☽ # ♇
05 33	☽ ± ☿
09 26	♀ ⚹ ♆
11 18	☽ ∠ ♄
14 12	☽ ⚼ ♇
15 07	☽ △ ♂
17 22	♀ ⚹ ♅
19 01	☽ ⊼ ☿
20 25	☽ ± ♀
25 Thursday	
04 52	☽ ⚼ ☉
07 28	☽ □ ♆
08 19	☽ □ ♅
08 30	☽ ± ♇
09 22	☉ ⊻ ♃
10 01	☽ ⊼ ♀
11 28	☽ # ♃
17 48	☽ ⚹ ♄
20 40	☽ ⊼ ♇
23 28	☽ ⚼ ♂
26 Friday	
00 45	☽ ± ☉
05 31	☽ # ☿
06 41	♂ Q ♄
09 35	☽ # ♄
11 58	☽ ☍ ♃
12 04	☉ ∠ ♆
13 56	☽ ⊼ ☉
18 03	☽ Q ♄
18 44	☽ ± ♂
22 53	☉ ∠ ♅
27 Saturday	
07 29	☽ ⊼ ♂
08 03	☽ # ♀
10 36	☿ ∥ ♄
12 59	♂ ⊥ ♆
14 18	☽ ☍ ☿
19 45	☽ △ ♆
20 39	☽ △ ♅
28 Sunday	
04 48	♂ ⊥ ♅
05 00	☽ ☍ ♀
05 56	☽ □ ♄
08 40	☽ ☍ ♇
14 23	♀ □ ♄
22 21	☉ ∥ ♆
29 Monday	
00 18	☽ ⊼ ♃
01 13	☽ ⚼ ♆
02 09	☽ ⚼ ♅
06 31	☽ ☍ ☉
12 07	☽ ± ♃
16 26	☿ ⚹ ♆
16 49	☽ # ♆
17 42	♀ ☌ ♇
18 39	☽ ± ♆
19 36	☽ ± ♅
21 52	☽ ☍ ♂
30 Tuesday	
01 15	☿ ⚹ ♅
05 39	☽ ⚼ ♃
06 09	☽ ⊼ ♆
07 07	☽ ⊼ ♅
07 49	☽ ⊼ ☿
14 19	☽ # ♆
16 05	☽ △ ♄
18 39	☽ ⊼ ♇
20 36	☽ ± ☿
21 18	☽ ⊼ ♀

DECEMBER 1993

LONGITUDES

Date	Sidereal time h m s	Sun ☉ ° ' "	Moon ☽ ° ' "	Moon ☽ 24.00 ° ' "	Mercury ☿ ° '	Venus ♀ ° '	Mars ♂ ° '	Jupiter ♃ ° '	Saturn ♄ ° '	Uranus ♅ ° '	Neptune ♆ ° '	Pluto ♇ ° '
01	16 41 25	09 ♐ 18 44	05 ♋ 11 21	11 ♋ 38 52	21 ♏ 53	28 ♏ 07	16 ♐ 12	04 ♏ 21	24 ♒ 38	19 ♑ 55	19 ♑ 24	25 ♏ 58
02	16 45 21	10 19 33	18 ♋ 09 44	24 ♋ 43 58	23 19	29 ♏ 23	16 56	04 33	24 42	19 58	19 26	26 01
03	16 49 18	11 20 24	01 ♌ 21 36	08 ♌ 02 39	24 46	00 ♐ 38	17 40	04 44	24 45	20 01	19 28	26 03
04	16 53 14	12 21 15	14 ♌ 47 08	21 ♌ 35 04	26 13	01 53	18 25	04 56	24 49	20 04	19 30	26 06
05	16 57 11	13 22 08	28 ♌ 26 24	05 ♍ 21 07	27 42	03 09	19 09	05 08	24 52	20 07	19 32	26 08
06	17 01 07	14 23 02	12 ♍ 19 09	19 ♍ 20 24	29 ♏ 11	04 24	19 53	05 19	24 56	20 10	19 34	26 10
07	17 05 04	15 23 57	26 ♍ 21 36	03 ♎ 31 53	00 ♐ 41	05 40	20 38	05 31	25 00	20 13	19 35	26 12
08	17 09 00	16 24 54	10 ♎ 41 38	17 ♎ 53 38	02 11	06 55	21 23	05 42	25 04	20 16	19 37	26 15
09	17 12 57	17 25 52	25 ♎ 07 27	02 ♏ 22 38	03 42	08 11	22 07	05 53	25 08	20 19	19 39	26 17
10	17 16 54	18 26 51	09 ♏ 38 37	16 ♏ 54 45	05 13	09 26	22 52	06 05	25 12	20 23	19 41	26 19
11	17 20 50	19 27 52	24 ♏ 10 23	01 ♐ 24 50	06 44	10 42	23 37	06 16	25 17	20 26	19 43	26 22
12	17 24 47	20 28 53	08 ♐ 37 21	15 ♐ 47 14	08 16	11 57	24 21	06 27	25 21	20 29	19 45	26 24
13	17 28 43	21 29 55	22 ♐ 53 50	29 ♐ 56 30	09 47	13 13	25 06	06 38	25 25	20 32	19 48	26 26
14	17 32 40	22 30 58	06 ♑ 54 43	13 ♑ 48 01	11 19	14 28	25 51	06 49	25 30	20 35	19 50	26 28
15	17 36 36	23 32 02	20 ♑ 36 05	27 ♑ 18 40	12 51	15 44	26 36	07 00	25 34	20 39	19 52	26 31
16	17 40 33	24 33 06	03 ♒ 55 39	10 ♒ 27 05	14 24	16 59	27 21	07 10	25 39	20 42	19 54	26 33
17	17 44 29	25 34 11	16 ♒ 53 02	23 ♒ 13 45	15 56	18 15	28 06	07 21	25 44	20 45	19 56	26 35
18	17 48 26	26 35 16	29 ♒ 29 33	05 ♓ 40 49	17 29	19 30	28 51	07 32	25 49	20 49	19 58	26 37
19	17 52 23	27 36 21	11 ♓ 48 02	17 ♓ 51 41	19 02	20 46	29 ♐ 36	07 42	25 53	20 52	20 00	26 39
20	17 56 19	28 37 26	23 ♓ 52 22	29 ♓ 50 39	20 35	22 01	00 ♑ 22	07 53	25 58	20 55	20 02	26 41
21	18 00 16	29 ♐ 38 32	05 ♈ 47 10	11 ♈ 42 32	22 08	23 17	01 07	08 03	26 03	20 59	20 04	26 43
22	18 04 12	00 ♑ 39 38	17 ♈ 37 23	23 ♈ 32 22	23 41	24 32	01 52	08 13	26 08	21 02	20 07	26 46
23	18 08 09	01 40 45	29 ♈ 28 03	05 ♉ 25 03	25 15	25 48	02 37	08 23	26 14	21 05	20 09	26 48
24	18 12 05	02 41 51	11 ♉ 23 54	17 ♉ 25 08	26 49	27 03	03 23	08 33	26 19	21 09	20 11	26 50
25	18 16 02	03 42 58	23 ♉ 29 14	29 ♉ 36 35	28 23	28 19	04 08	08 43	26 24	21 12	20 13	26 52
26	18 19 58	04 44 05	05 ♊ 47 34	12 ♊ 02 28	29 ♐ 57	29 ♐ 34	04 54	08 53	26 29	21 16	20 15	26 54
27	18 23 55	05 45 12	18 ♊ 21 30	24 ♊ 44 49	01 ♑ 31	00 ♑ 50	05 39	09 03	26 35	21 19	20 18	26 56
28	18 27 52	06 46 19	01 ♋ 12 28	07 ♋ 44 26	03 06	02 05	06 25	09 13	26 40	21 23	20 20	26 58
29	18 31 48	07 47 26	14 ♋ 20 38	21 ♋ 00 52	04 41	03 21	07 10	09 22	26 46	21 26	20 22	27 00
30	18 35 45	08 48 34	27 ♋ 44 56	04 ♌ 32 31	06 16	04 36	07 56	09 32	26 52	21 30	20 24	27 02
31	18 39 41	09 ♑ 49 42	11 ♌ 23 18	18 ♌ 16 54	07 ♑ 52	05 ♑ 52	08 ♑ 41	09 ♏ 41	26 ♒ 57	21 ♑ 33	20 ♑ 26	27 ♏ 04

DECLINATIONS

Date	Moon True ☊ ° '	Moon Mean ☊ ° '	Moon ☽ Latitude ° '	Sun ☉ ° '	Moon ☽ ° '	Mercury ☿ ° '	Venus ♀ ° '	Mars ♂ ° '	Jupiter ♃ ° '	Saturn ♄ ° '	Uranus ♅ ° '	Neptune ♆ ° '	Pluto ♇ ° '
01	03 ♐ 02	02 ♐ 42	02 S 49	21 S 51	20 N 31	16 S 39	18 S 56	23 S 18	11 S 57	14 S 41	22 S 25	21 S 25	05 S 55
02	03 R 01	02 39	03 45	22 00	18 29	17 07	19 15	23 23	12 01	14 40	22 24	21 25	05 56
03	03 00	02 36	04 30	22 08	15 28	17 35	19 33	23 28	12 05	14 38	22 24	21 25	05 56
04	02 59	02 33	05 01	22 17	11 36	18 03	19 50	23 32	12 08	14 37	22 23	21 24	05 57
05	02 59	02 30	05 16	22 24	07 04	18 30	20 07	23 36	12 12	14 36	22 23	21 24	05 57
06	02 58	02 27	05 13	22 32	02 N 07	18 57	20 24	23 40	12 16	14 34	22 22	21 24	05 58
07	02 D 58	02 23	04 51	22 38	03 S 01	19 23	20 39	23 44	12 20	14 33	22 22	21 24	05 58
08	02 59	02 20	04 11	22 45	08 05	19 48	20 54	23 48	12 23	14 32	22 22	21 23	05 58
09	03 00	02 17	03 15	22 51	12 45	20 13	21 09	23 51	12 27	14 30	22 21	21 23	05 59
10	03 01	02 14	02 06	22 56	16 41	20 37	21 23	23 54	12 31	14 29	22 21	21 23	05 59
11	03 02	02 11	00 S 49	23 01	19 36	21 00	21 36	23 57	12 34	14 27	22 20	21 23	06 00
12	03 R 02	02 08	00 N 31	23 06	21 14	21 22	21 49	23 59	12 38	14 26	22 20	21 22	06 00
13	03 01	02 04	01 48	23 10	21 27	21 43	22 01	24 02	12 41	14 24	22 19	21 22	06 00
14	03 00	02 01	02 56	23 14	20 19	22 03	22 12	24 04	12 45	14 23	22 19	21 22	06 01
15	02 57	01 58	03 53	23 17	18 01	22 22	22 23	24 06	12 48	14 21	22 18	21 22	06 01
16	02 53	01 55	04 36	23 20	14 48	22 41	22 33	24 07	12 52	14 19	22 18	21 21	06 01
17	02 50	01 52	05 03	23 22	10 58	22 58	22 42	24 08	12 55	14 18	22 17	21 21	06 02
18	02 47	01 48	05 14	23 24	06 44	23 14	22 51	24 09	12 58	14 16	22 17	21 21	06 02
19	02 45	01 45	05 11	23 25	02 S 21	23 28	22 59	24 10	13 02	14 14	22 16	21 20	06 02
20	02 44	01 42	04 54	23 26	02 N 04	23 42	23 06	24 11	13 05	14 13	22 16	21 20	06 03
21	02 D 44	01 39	04 24	23 26	06 20	23 54	23 13	24 11	13 08	14 11	22 15	21 20	06 03
22	02 45	01 36	03 43	23 26	10 21	24 06	23 19	24 11	13 11	14 09	22 15	21 19	06 03
23	02 46	01 33	02 53	23 26	13 59	24 16	23 24	24 10	13 14	14 07	22 14	21 19	06 03
24	02 48	01 29	01 55	23 25	17 04	24 24	23 29	24 10	13 18	14 05	22 14	21 19	06 04
25	02 50	01 26	00 N 51	23 23	19 28	24 32	23 32	24 09	13 21	14 04	22 13	21 19	06 04
26	02 R 50	01 23	00 S 16	23 21	21 00	24 38	23 35	24 08	13 24	14 02	22 12	21 18	06 04
27	02 49	01 20	01 24	23 19	21 32	24 43	23 38	24 07	13 27	14 00	22 12	21 18	06 04
28	02 46	01 17	02 29	23 16	20 57	24 46	23 39	24 05	13 30	13 58	22 11	21 18	06 04
29	02 42	01 14	03 27	23 13	19 14	24 48	23 40	24 03	13 32	13 56	22 11	21 17	06 05
30	02 36	01 10	04 15	23 09	16 26	24 49	23 40	24 01	13 35	13 54	22 10	21 17	06 05
31	02 ♐ 29	01 ♐ 07	04 S 50	23 S 04	12 N 43	24 S 48	23 S 39	23 S 58	13 S 38	13 S 52	22 S 10	21 S 17	06 S 05

ZODIAC SIGN ENTRIES

Date	h	m	Planets
01	02	17	☽ ♋
02	23	54	♀ ♐
03	09	33	☽ ♌
05	14	43	☽ ♍
07	01	04	☿ ♐
07	18	03	☽ ♎
09	20	04	☽ ♏
11	21	39	☽ ♐
14	00	06	☽ ♑
16	04	51	☽ ♒
18	12	59	☽ ♓
20	00	34	♂ ♑
21	00	19	☽ ♈
21	20	26	☉ ♑
23	13	05	☽ ♉
26	00	46	☽ ♊
26	12	47	☿ ♑
26	20	09	♀ ♑
28	09	46	☽ ♋
30	15	59	☽ ♌

LATITUDES

Date	Mercury ☿ ° '	Venus ♀ ° '	Mars ♂ ° '	Jupiter ♃ ° '	Saturn ♄ ° '	Uranus ♅ ° '	Neptune ♆ ° '	Pluto ♇ ° '
01	01 N 38	00 N 49	00 S 35	01 N 05	01 S 27	00 S 27	00 N 37	13 N 41
04	01 18	00 43	00 36	01 05	01 27	00 27	00 37	13 41
07	00 56	00 36	00 38	01 06	01 27	00 27	00 37	13 42
10	00 34	00 29	00 39	01 06	01 27	00 27	00 37	13 42
13	00 N 12	00 22	00 41	01 06	01 26	00 27	00 37	13 42
16	00 S 09	00 15	00 42	01 06	01 26	00 27	00 37	13 43
19	00 29	00 08	00 44	01 07	01 26	00 27	00 37	13 43
22	00 48	00 N 01	00 45	01 07	01 26	00 27	00 37	13 44
25	01 06	00 S 07	00 47	01 07	01 26	00 27	00 37	13 44
28	01 22	00 14	00 48	01 08	01 26	00 27	00 37	13 45
31	01 S 36	00 S 21	00 S 49	01 N 08	01 S 25	00 S 27	00 N 37	13 N 46

DATA

Julian Date	2449323
Delta T	+59 seconds
Ayanamsa	23° 46' 34"
Synetic vernal point	05° ♓ 20' 25"
True obliquity of ecliptic	23° 26' 20"

MOON'S PHASES, APSIDES AND POSITIONS ☽

Date	h	m	Phase	Longitude ° '	Eclipse Indicator
06	15	49	☾	14 ♍ 33	
13	09	27	●	21 ♐ 23	
20	22	26	☽	29 ♓ 04	
28	23	05	○	07 ♋ 15	

Day	h	m		
10	13	56	Perigee	
22	07	48	Apogee	
06	21	55	0S	
13	03	48	Max dec	21° S 32'
20	00	44	0N	
27	11	40	Max dec	21° N 32'

LONGITUDES

Date	Chiron ⚷ ° '	Ceres ⚳ ° '	Pallas ⚴ ° '	Juno ⚵ ° '	Vesta ⚶ ° '	Black Moon Lilith ⚸ ° '
01	08 ♍ 50	22 ♈ 38	26 ♒ 47	18 ♎ 10	06 ♓ 30	15 ♈ 56
11	09 ♍ 10	22 ♈ 17	28 ♒ 48	21 ♎ 12	09 ♓ 36	17 ♈ 03
21	09 ♍ 18	22 ♈ 37	01 ♓ 07	24 ♎ 00	13 ♓ 00	18 ♈ 10
31	09 ♍ 14	23 ♈ 32	03 ♓ 42	26 ♎ 30	16 ♓ 40	19 ♈ 17

All ephemeris data is given at 12.00 UT and the Moon's longitude is additionally given for 24.00 UT

ASPECTARIAN

h m	Aspects
01 Wednesday	
05 58	☽ ± ♇
09 47	☽ ± ♀
10 24	☽ △ ♃
15 33	☽ ⚼ ☿
20 18	☽ ⚼ ♄
20 19	☽ ⊼ ☉
22 47	☽ ⚼ ♇
02 Thursday	
04 18	☽ ⚼ ♀
05 17	☽ # ♀
08 21	☽ ± ☉
09 36	☽ ⊼ ♂
12 59	☽ ± ♄
14 20	☽ ☍ ♆
15 20	☽ ☍ ♅
18 23	☉ ⊥ ♃
21 14	☽ ± ♂
22 10	☽ # ☿
22 35	☽ △ ☿
23 59	☽ ⊼ ♄
03 Friday	
02 10	☽ ⚼ ☉
02 22	☽ △ ♇
10 33	☽ △ ♀
11 49	☿ □ ♄
14 30	☽ ⚼ ♂
17 34	☽ # ♄
18 10	☽ □ ♃
04 Saturday	
07 20	☽ △ ☉
08 55	☽ # ♃
09 47	☿ ☌ ♇
18 47	☽ △ ♂
20 21	☽ ⊼ ♆
21 22	☽ ⊼ ♅
23 32	☉ Q ♄
05 Sunday	
02 34	☽ Q ♃
05 45	☽ ☍ ♄
06 54	☽ ± ♆
07 56	☽ ± ♅
07 57	☽ □ ♇
08 05	☉ ∥ ♅
10 33	☽ □ ☿
15 52	☉ ⊥ ♆
17 33	☽ # ♇
21 00	☽ □ ♀
22 36	☽ ⚼ ♆
23 38	☽ ⚼ ♅
23 46	☽ ⚹ ♃
06 Monday	
00 45	♂ ⊻ ♆
06 42	☉ ⊥ ♅
14 59	♀ ∠ ♆
15 11	☽ Q ♇
15 49	☽ □ ☉
21 19	☽ Q ☿
21 42	♂ ⊻ ♅
07 Tuesday	
00 24	☽ △ ♆
01 28	☽ △ ♅
01 39	☽ □ ♂
01 51	☽ ∠ ♃
03 12	♀ ∠ ♅
06 36	♂ ∠ ♃
06 54	☽ Q ♀
08 31	♀ ⊻ ♃
09 36	☽ ⊼ ♄
11 39	☽ ⚹ ♇
17 18	☽ ⊥ ♃
19 47	☽ ± ♄
20 03	☽ ⚹ ☿
08 Wednesday	
00 41	☽ Q ☉
01 51	☽ ∥ ♇
03 32	☽ ⊻ ♃
05 05	☽ ⚹ ♀
09 41	☽ Q ♂
10 57	☽ ⚼ ♄
12 56	☽ ∠ ♇
22 16	☽ ⚹ ☉
09 Thursday	
00 05	☽ ∠ ☿
02 55	☽ □ ♆
03 57	☽ ⊥ ♇
04 01	☽ □ ♅
06 45	☽ ⚹ ♂
08 28	☽ ∠ ♀
10 22	☽ ∥ ♃
12 01	☽ △ ♄
13 56	☽ ⊻ ♇
16 45	☽ ⊥ ☿
21 59	☽ ∥ ♄
10 Friday	
00 47	☽ ⊥ ♀
01 00	☽ ∠ ☉
03 33	☿ ∠ ♆
03 50	☽ ⊻ ☿
06 02	☽ ☌ ♃
08 46	☽ Q ♆
08 54	☽ ∠ ♂
09 54	☽ Q ♅
11 38	☽ ⊻ ♀
11 54	♀ ∥ ♆
14 41	☿ ∠ ♅
16 59	☽ ⊥ ☉
11 Saturday	
00 34	☽ ⊥ ♂
03 31	☿ ⊻ ♃
03 38	☽ ⊻ ☉
04 38	☽ ⚹ ♆
05 47	☽ ⚹ ♅
11 01	☽ ⊻ ♂
13 50	☽ □ ♄
15 38	☽ ☌ ♇
18 21	☉ ⊻ ♆
12 Sunday	
05 33	☽ ∠ ♆
06 45	☽ ∠ ♅
08 19	☽ ⊻ ♃
11 19	☽ ☌ ☿
12 02	☉ ⊻ ♅
12 31	☿ ∥ ♆
16 06	☽ ∥ ♆
18 06	☽ ☌ ♀
18 29	☽ ⊥ ♃
19 57	☽ Q ♄
20 37	☽ ⊥ ♆
21 51	☽ ⊥ ♅
22 59	♀ ⊥ ♃
13 Monday	
06 44	☽ ⊻ ♆
07 59	☽ ⊻ ♅
09 27	☽ ☌ ☉
09 50	☽ ∠ ♃
15 45	☽ ∥ ♆
15 47	☉ ∠ ♃
15 57	☽ ☌ ♂
16 14	♀ Q ♄
16 19	☽ ⚹ ♄
18 02	☽ ⊻ ♇
23 18	♂ ⚹ ♄
23 22	♀ ⊥ ♆
14 Tuesday	
04 19	☽ ⊥ ♇
11 50	☽ ⚹ ♃
14 22	♀ ⊥ ♅
18 16	☽ ∠ ♄
19 57	☽ ∠ ♇
20 38	☽ ⊻ ☿
15 Wednesday	
01 17	♀ ∥ ♅
02 31	☽ ⊻ ♀
06 42	☿ ∥ ♅
08 31	☽ ⊥ ☿
08 51	♂ ⊻ ♇
09 07	☽ Q ♃
10 10	☽ ⊥ ♄
10 41	☽ ☌ ♆
12 05	☽ ☌ ♅
13 56	☿ ∥ ♀
14 13	☽ ⊥ ♀
14 24	☿ ⊥ ♃
17 39	☽ ⊻ ☉
20 56	☽ ⊻ ♄
22 35	☽ ⚹ ♇
23 22	☽ ⊻ ♂
23 45	☿ Q ♄
16 Thursday	
02 41	☽ ∠ ☿
04 01	☿ ⊥ ♆
05 21	☽ ⊥ ☉
08 06	☽ ∠ ♀
10 53	☽ ⊥ ♂
15 13	☽ ∥ ♄
16 53	☿ ⊥ ♅
18 02	☽ □ ♃
20 30	☽ Q ♇
23 13	☽ ∠ ☉
17 Friday	
00 21	☽ ∥ ♃
04 29	☽ ∠ ♂
09 59	☽ ⚹ ☿
14 51	☽ ⚹ ♀
16 04	☉ ⚹ ♄
17 46	☽ ⊻ ♆
19 20	☽ ⊻ ♅
18 Saturday	
04 53	☽ ☌ ♄
05 12	☽ ⊥ ♆
05 55	☽ ⚹ ☉
06 27	☽ □ ♇
06 49	☽ ⊥ ♅
10 41	☽ ⚹ ♂
11 59	☽ Q ☿
12 44	☉ ⊻ ♇
15 54	☽ ∥ ♇
16 20	☽ Q ♀
21 03	♀ ⊻ ♆
22 38	☽ ∠ ♆
19 Sunday	
00 18	☽ ∠ ♅
03 50	☽ △ ♃
06 00	☉ ∥ ☿
07 17	☽ Q ☉
11 35	☽ Q ♂
14 01	♀ ⊻ ♅
20 Monday	
03 25	☿ ⊻ ♆
04 19	☽ ⚹ ♆
04 27	☽ □ ☿
06 04	☽ ⚹ ♅
07 52	☽ □ ♀
09 59	☽ ⚼ ♃
16 14	☽ ⊻ ♄
17 27	☿ ⊻ ♅
17 40	☽ △ ♇
22 26	☽ □ ☉
21 Tuesday	
01 55	☽ □ ♂
04 20	☽ ± ♃
04 24	☽ ⊥ ♄
04 28	☽ Q ♆
06 18	☽ Q ♅
06 53	♀ ∠ ♃
10 20	☽ # ♇
16 24	♃ Q ♆
16 39	☽ ⊼ ♃
21 07	♄ ⊥ ♆
22 45	☽ ∠ ♄
22 Wednesday	
00 04	☽ ⚼ ♇
03 52	☿ ∠ ♃
17 04	☽ □ ♆
18 23	☽ ± ♇
18 57	☽ □ ♅
23 49	☿ ∥ ♂
23 Thursday	
02 10	☽ △ ☿
03 42	☽ △ ♀
05 24	☽ ⚹ ♄
06 35	☽ ⊼ ♇
06 46	☽ # ♃
13 00	☽ # ♄
16 53	☽ △ ☉
17 43	♂ ⊥ ♇
18 39	☉ ∥ ♀
18 48	☽ △ ♂
20 47	♀ ⚹ ♄
24 Friday	
03 53	☿ ⚹ ♄
05 46	☽ Q ♄
06 14	☽ ☍ ♃
07 32	♀ ⊻ ♇
12 15	☿ ⊻ ♇
12 57	☽ ⚼ ☿
13 28	☽ ⚼ ♀
15 11	☉ ⊥ ♇
25 Saturday	
01 42	☽ ⚼ ☉
02 50	☽ ⚼ ♂
05 32	☽ △ ♆
07 01	☿ ☌ ♀
07 28	☽ △ ♅
09 25	☽ ± ♀
09 30	☽ ± ☿
17 46	☽ □ ♄
18 39	☽ ☍ ♇
21 03	☽ ± ☉
21 43	☽ ± ♂
22 33	☽ ⊼ ♀
23 00	☽ ⊼ ☿
26 Sunday	
09 46	☽ ⊼ ☉
10 09	☽ ⊼ ♂
10 58	☽ ⚼ ♆
12 54	☽ ⚼ ♅
18 02	☽ ⊼ ♃
20 03	☽ # ♆
27 Monday	
02 28	☉ ☌ ♂
04 16	☽ ± ♆
05 39	☽ ± ♃
06 13	☽ ± ♅
15 40	☽ ⊼ ♆
17 36	☽ ⊼ ♅
22 50	☽ ⚼ ♃
28 Tuesday	
03 21	☽ # ♆
03 32	☽ △ ♄
04 07	☽ ⊼ ♇
09 50	☿ ⊥ ♇
13 48	☽ ☍ ♀
15 15	☽ ± ♇
15 58	☽ ☍ ☿
22 09	☽ ☍ ♂
23 05	☽ ☍ ☉
29 Wednesday	
02 52	☽ △ ♃
05 05	♀ ⊥ ♇
07 18	☽ ⚼ ♄
07 44	☽ ⚼ ♇
22 52	☽ ☍ ♆
23 38	☽ ± ♄
30 Thursday	
00 48	☽ ☍ ♅
03 39	♃ Q ♅
10 25	☽ ⊼ ♄
10 43	☽ △ ♇
31 Friday	
01 21	☽ ⊼ ♀
05 01	☽ ⊼ ☿
05 05	☽ # ♄
06 36	☽ # ♃
07 00	☽ ⊼ ♂
07 56	☉ ⚹ ♃
08 59	☽ □ ♃
09 03	☽ ⊼ ☉
12 55	☽ ± ♀
16 53	☽ ± ☿
18 06	☽ ± ♂
20 21	☽ ± ☉

JANUARY 1994

LONGITUDES

Date	Sidereal time h m s	Sun ☉ ° ' "	Moon ☽ ° ' "	Moon ☽ 24.00 ° ' "	Mercury ☿ ° '	Venus ♀ ° '	Mars ♂ ° '	Jupiter ♃ ° '	Saturn ♄ ° '	Uranus ♅ ° '	Neptune ♆ ° '	Pluto ♇ ° '
01	18 43 38	10 ♑ 50 50	25 ♌ 12 56	02 ♍ 11 03	09 ♑ 28	07 ♑ 07	09 ♑ 27	09 ♏ 50	27 ♒ 03	21 ♑ 37	20 ♑ 29	27 ♏ 05
02	18 47 34	11 51 58	09 ♍ 10 51	16 ♍ 11 01	11 04	08 23	10 13	09 59	27 09	21 40	20 31	27 07
03	18 51 31	12 53 07	23 ♍ 14 14	00 ♎ 17 12	12 41	09 38	10 59	10 08	27 15	21 44	20 33	27 09
04	18 55 27	13 54 15	07 ♎ 20 43	14 ♎ 24 33	14 18	10 54	11 44	10 17	27 20	21 47	20 36	27 11
05	18 59 24	14 55 24	21 ♎ 28 31	28 ♎ 32 27	15 55	12 09	12 30	10 26	27 26	21 51	20 38	27 13
06	19 03 21	15 56 34	05 ♏ 36 12	12 ♏ 39 36	17 33	13 25	13 16	10 35	27 32	21 54	20 40	27 14
07	19 07 17	16 57 44	19 ♏ 42 28	26 ♏ 44 35	19 11	14 40	14 02	10 43	27 39	21 58	20 42	27 16
08	19 11 14	17 58 53	03 ♐ 45 42	10 ♐ 45 34	20 49	15 56	14 48	10 52	27 45	22 01	20 45	27 18
09	19 15 10	19 00 03	17 ♐ 43 50	24 ♐ 40 10	22 28	17 11	15 34	11 00	27 51	22 05	20 47	27 20
10	19 19 07	20 01 13	01 ♑ 34 12	08 ♑ 25 33	24 08	18 27	16 20	11 08	27 57	22 08	20 49	27 21
11	19 23 03	21 02 23	15 ♑ 13 49	21 ♑ 58 39	25 47	19 42	17 06	11 16	28 03	22 12	20 51	27 23
12	19 27 00	22 03 32	28 ♑ 39 42	05 ♒ 16 42	27 27	20 58	17 53	11 24	28 10	22 15	20 54	27 25
13	19 30 56	23 04 41	11 ♒ 49 27	18 ♒ 17 46	29 ♑ 08	22 13	18 39	11 32	28 16	22 19	20 56	27 26
14	19 34 53	24 05 49	24 ♒ 41 38	01 ♓ 01 04	00 ♒ 49	23 29	19 25	11 39	28 23	22 23	20 58	27 28
15	19 38 50	25 06 57	07 ♓ 16 11	13 ♓ 27 12	02 30	24 44	20 11	11 47	28 29	22 26	21 01	27 29
16	19 42 46	26 08 04	19 ♓ 34 26	25 ♓ 38 14	04 11	26 00	20 58	11 54	28 36	22 30	21 03	27 31
17	19 46 43	27 09 10	01 ♈ 39 03	07 ♈ 37 24	05 53	27 15	21 44	12 01	28 42	22 33	21 05	27 32
18	19 50 39	28 10 16	13 ♈ 33 51	19 ♈ 28 59	07 35	28 31	22 30	12 08	28 49	22 37	21 07	27 34
19	19 54 36	29 ♑ 11 20	25 ♈ 23 27	01 ♉ 17 55	09 17	29 ♑ 46	23 17	12 15	28 55	22 40	21 10	27 35
20	19 58 32	00 ♒ 12 24	07 ♉ 13 04	13 ♉ 09 30	10 59	01 ♒ 01	24 03	12 22	29 02	22 44	21 12	27 36
21	20 02 29	01 13 27	19 ♉ 08 07	25 ♉ 09 23	12 41	02 17	24 49	12 29	29 09	22 47	21 14	27 38
22	20 06 25	02 14 30	01 ♊ 13 58	07 ♊ 22 29	14 22	03 32	25 36	12 35	29 15	22 51	21 16	27 39
23	20 10 22	03 15 31	13 ♊ 35 29	19 ♊ 53 24	16 03	04 48	26 22	12 42	29 22	22 54	21 19	27 40
24	20 14 19	04 16 31	26 ♊ 16 39	02 ♋ 45 29	17 44	06 03	27 09	12 48	29 29	22 58	21 21	27 42
25	20 18 15	05 17 30	09 ♋ 20 05	16 ♋ 00 28	19 23	07 18	27 56	12 54	29 36	23 01	21 23	27 43
26	20 22 12	06 18 29	22 ♋ 46 33	29 ♋ 38 05	21 02	08 34	28 42	13 00	29 43	23 05	21 25	27 44
27	20 26 08	07 19 26	06 ♌ 34 42	13 ♌ 35 52	22 38	09 49	29 ♑ 29	13 06	29 50	23 08	21 28	27 45
28	20 30 05	08 20 23	20 ♌ 40 58	27 ♌ 49 18	24 13	11 04	00 ♒ 15	13 11	29 ♒ 57	23 12	21 30	27 46
29	20 34 01	09 21 18	05 ♍ 00 05	12 ♍ 12 31	25 46	12 20	01 02	13 17	00 ♓ 04	23 15	21 32	27 47
30	20 37 58	10 22 13	19 ♍ 25 50	26 ♍ 39 17	27 16	13 35	01 49	13 22	00 11	23 19	21 34	27 49
31	20 41 54	11 ♒ 23 07	03 ♎ 52 10	11 ♎ 03 56	28 ♒ 43	14 ♒ 50	02 ♒ 35	13 ♏ 27	00 ♓ 18	23 ♑ 22	21 ♑ 36	27 ♏ 50

Moon Nodes and Latitude

Date	Moon True ☊ ° '	Moon Mean ☊ ° '	Moon ☽ Latitude ° '
01	02 ♐ 23	01 ♐ 04	05 S 08
02	02 R 18	01 05	05 08
03	02 14	00 58	04 50
04	02 13	00 54	04 14
05	02 D 13	00 51	03 23
06	02 14	00 48	02 19
07	02 15	00 45	01 S 07
08	02 R 16	00 42	00 N 08
09	02 16	00 39	01 23
10	02 11	00 35	02 31
11	02 05	00 32	03 30
12	01 57	00 29	04 17
13	01 48	00 26	04 48
14	01 38	00 23	05 04
15	01 28	00 19	05 04
16	01 21	00 16	04 51
17	01 15	00 13	04 24
18	01 11	00 10	03 47
19	01 10	00 07	03 00
20	01 D 10	00 04	02 05
21	01 11	00 ♐ 00	01 N 04
22	01 R 11	29 ♏ 57	00 00
23	01 10	29 54	01 S 06
24	01 07	29 51	02 10
25	01 02	29 48	03 09
26	00 54	29 45	03 59
27	00 44	29 41	04 36
28	00 32	29 38	04 58
29	00 21	29 35	05 01
30	00 11	29 32	04 46
31	00 ♐ 04	29 ♏ 29	04 S 12

DECLINATIONS

Date	Sun ☉ ° '	Moon ☽ ° '	Mercury ☿ ° '	Venus ♀ ° '	Mars ♂ ° '	Jupiter ♃ ° '	Saturn ♄ ° '	Uranus ♅ ° '	Neptune ♆ ° '	Pluto ♇ ° '
01	23 S 00	08 N 17	24 S 46	23 S 38	23 S 56	13 S 41	13 S 50	22 S 09	21 S 16	06 S 05
02	22 55	03 N 22	24 42	23 36	23 53	13 44	13 48	22 09	21 16	06 05
03	22 49	01 S 45	24 37	23 33	23 50	13 46	13 46	22 08	21 16	06 05
04	22 43	06 48	24 31	23 29	23 46	13 49	13 44	22 08	21 15	06 05
05	22 36	11 30	24 23	23 25	23 42	13 52	13 42	22 07	21 15	06 06
06	22 29	15 35	24 13	23 20	23 38	13 54	13 40	22 06	21 15	06 06
07	22 22	18 45	24 02	23 14	23 34	13 57	13 37	22 06	21 14	06 06
08	22 14	20 46	23 50	23 08	23 29	13 59	13 35	22 05	21 14	06 06
09	22 06	21 30	23 36	23 01	23 25	14 02	13 33	22 05	21 14	06 06
10	21 57	20 54	23 20	22 53	23 19	14 04	13 31	22 04	21 13	06 06
11	21 48	19 05	23 03	22 44	23 14	14 06	13 29	22 04	21 13	06 06
12	21 38	16 14	22 44	22 35	23 08	14 09	13 27	22 03	21 13	06 06
13	21 28	12 38	22 24	22 25	23 03	14 11	13 24	22 02	21 12	06 06
14	21 17	08 31	22 02	22 14	22 56	14 13	13 22	22 02	21 12	06 06
15	21 07	04 S 08	21 38	22 03	22 50	14 15	13 20	22 01	21 12	06 06
16	20 55	00 N 20	21 13	21 51	22 43	14 17	13 17	22 01	21 11	06 06
17	20 44	04 42	20 47	21 38	22 36	14 19	13 15	22 00	21 11	06 06
18	20 32	08 50	20 19	21 25	22 29	14 21	13 13	22 00	21 11	06 06
19	20 19	12 36	19 49	21 11	22 22	14 23	13 11	21 59	21 10	06 06
20	20 06	15 53	19 18	20 56	22 14	14 25	13 08	21 58	21 10	06 06
21	19 53	18 32	18 46	20 41	22 06	14 27	13 06	21 58	21 10	06 06
22	19 40	20 24	18 12	20 25	21 58	14 29	13 03	21 57	21 09	06 06
23	19 26	21 20	17 37	20 09	21 50	14 31	13 01	21 57	21 09	06 06
24	19 11	21 13	17 01	19 52	21 41	14 32	12 59	21 56	21 09	06 06
25	18 57	19 58	16 24	19 34	21 32	14 34	12 56	21 56	21 08	06 06
26	18 42	17 35	15 45	19 16	21 23	14 36	12 54	21 55	21 08	06 05
27	18 26	14 10	15 06	18 57	21 14	14 37	12 51	21 54	21 08	06 05
28	18 11	09 53	14 26	18 38	21 04	14 39	12 49	21 54	21 07	06 05
29	17 55	05 N 00	13 46	18 18	20 54	14 40	12 47	21 53	21 07	06 05
30	17 38	00 S 12	13 05	17 58	20 44	14 42	12 44	21 53	21 07	06 05
31	17 S 22	05 S 23	12 S 24	17 S 37	20 S 34	14 S 43	12 S 42	21 S 52	21 S 06	06 S 05

ZODIAC SIGN ENTRIES

Date	h	m	Planets
01	20	15	☽ ♍
03	23	31	☽ ♎
06	02	29	☽ ♏
08	05	34	☽ ♐
10	09	16	☽ ♑
12	14	25	☽ ♒
14	00	25	☿ ♒
14	22	04	☽ ♓
17	08	42	☽ ♈
19	16	28	♀ ♒
19	21	52	☽ ♉
20	07	07	☉ ♒
22	09	35	☽ ♊
24	18	55	☽ ♋
27	00	38	☽ ♌
28	04	05	♂ ♒
28	23	43	♄ ♓
29	03	00	☽ ♍
31	05	34	☽ ♎

LATITUDES

Date	Mercury ☿ ° '	Venus ♀ ° '	Mars ♂ ° '	Jupiter ♃ ° '	Saturn ♄ ° '	Uranus ♅ ° '	Neptune ♆ ° '	Pluto ♇ ° '
01	01 S 40	00 S 23	00 S 50	01 N 08	01 S 25	00 S 27	00 N 37	13 N 46
04	01 51	00 30	00 51	01 09	01 25	00 27	00 37	13 47
07	01 59	00 37	00 52	01 09	01 25	00 27	00 37	13 48
10	02 05	00 43	00 53	01 10	01 25	00 27	00 37	13 49
13	02 07	00 49	00 55	01 10	01 25	00 27	00 37	13 49
16	02 04	00 55	00 56	01 10	01 25	00 27	00 37	13 50
19	01 58	01 00	00 57	01 11	01 25	00 27	00 37	13 51
22	01 46	01 05	00 58	01 11	01 25	00 27	00 37	13 53
25	01 28	01 10	00 59	01 12	01 25	00 27	00 37	13 54
28	01 03	01 14	01 00	01 12	01 25	00 27	00 37	13 55
31	00 S 31	01 S 17	01 S 01	01 N 13	01 S 25	00 S 28	00 N 37	13 N 56

DATA

Julian Date	2449354
Delta T	+60 seconds
Ayanamsa	23° 46' 40"
Synetic vernal point	05° ♓ 20' 20"
True obliquity of ecliptic	23° 26' 19"

MOON'S PHASES, APSIDES AND POSITIONS ☽

Date	h	m	Phase	Longitude ° '	Eclipse Indicator
05	00	01	☾	14 ♎ 25	
11	23	10	●	21 ♑ 31	
19	20	27	☽	29 ♈ 33	
27	13	23	○	07 ♌ 23	

Day	h	m		
06	01	08	Perigee	
19	05	16	Apogee	
31	03	55	Perigee	
03	03	49	0S	
09	13	07	Max dec	21° S 30'
16	10	12	0N	
23	21	30	Max dec	21° N 26'
30	11	06	0S	

LONGITUDES

Date	Chiron ⚷ ° '	Ceres ⚳ ° '	Pallas ⚴ ° '	Juno ⚵ ° '	Vesta ⚶ ° '	Black Moon Lilith ⚸ ° '
01	09 ♍ 13	23 ♈ 40	03 ♓ 58	26 ♎ 44	17 ♓ 02	19 ♈ 23
11	08 ♍ 57	25 ♈ 11	06 ♓ 47	28 ♎ 52	20 ♓ 55	20 ♈ 30
21	08 ♍ 31	27 ♈ 11	09 ♓ 47	00 ♏ 36	24 ♓ 47	21 ♈ 37
31	07 ♍ 56	29 ♈ 35	12 ♓ 55	01 ♏ 52	29 ♓ 08	22 ♈ 44

ASPECTARIAN

01 Saturday
03 47 ☽ ⊼ ♆
05 44 ☽ ⊼ ♅
06 07 ☽ ⚼ ♀
10 32 ☽ ⚼ ☿
10 36 ☽ ⚼ ♂
11 37 ☿ ☌ ♂
13 11 ☽ ⚼ ☉
14 11 ☽ ± ♆
15 11 ☽ ☍ ♄
15 14 ☽ □ ♇
16 09 ☽ ± ♅
16 34 ☽ Q ♃
18 09 ☿ ⚹ ♃
22 55 ☽ # ♇

02 Sunday
03 08 ♂ ⚹ ♃
03 11 ♄ □ ♇
05 42 ☽ ⚼ ♆
07 41 ☽ ⚼ ♅
10 30 ☽ △ ♀
13 24 ☽ ⚹ ♃
13 52 ☽ △ ♂
15 39 ☽ △ ☿
16 57 ☽ △ ☉
18 12 ☉ ∠ ♇
19 16 ☉ ∠ ♄
22 11 ☽ Q ♇

03 Monday
04 00 ☿ ∠ ♇
05 06 ☿ ∠ ♄
07 25 ☽ △ ♆
09 25 ☽ △ ♅
09 31 ♃ ∥ ♄
15 16 ☽ ∠ ♃
18 41 ☽ ⚹ ♇
18 52 ☽ ⊼ ♄
20 20 ☉ ☌ ☿
22 47 ♀ ⚹ ♃

04 Tuesday
05 09 ☽ ± ♄
06 45 ☽ ⊥ ♃
08 33 ☽ ∥ ♇
17 03 ☽ ⊻ ♃
18 37 ☽ □ ♀
19 54 ☽ □ ♂
20 14 ☽ ∠ ♇
20 33 ☽ ⚼ ♄

05 Wednesday
00 01 ☽ □ ☉
01 20 ☽ □ ☿
02 26 ♂ ∠ ♇
09 41 ♂ ∠ ♄
10 34 ☽ □ ♆
11 33 ☽ ⊥ ♇
12 38 ☽ □ ♅
13 06 ♀ ∠ ♇
17 55 ♀ ∠ ♄
21 46 ☽ ⊻ ♇
22 12 ☽ △ ♄

06 Thursday
00 13 ☽ ∥ ♄
01 27 ☽ ∥ ♃
04 11 ☽ Q ♀
04 13 ☽ Q ♂
05 01 ♀ ☌ ♂
08 57 ☽ Q ☉
11 53 ☽ Q ☿
17 13 ☽ Q ♆
19 20 ☽ Q ♅
20 32 ☽ ☌ ♃

07 Friday
01 47 ☽ ⚹ ♂
02 35 ☽ ⚹ ♀
06 57 ☽ ⚹ ☉
10 59 ☽ ⚹ ☿
13 42 ☽ ⚹ ♆
15 51 ☽ ⚹ ♅

08 Saturday
00 56 ☽ ☌ ♇
01 38 ☽ □ ♄
04 50 ☽ ∠ ♂
06 41 ☽ ∠ ♀
10 34 ☽ ∠ ☉
10 49 ☿ ☌ ♆
15 24 ☽ ∠ ♆
16 00 ☽ ∠ ☿
17 36 ☽ ∠ ♅
21 09 ☽ ⊥ ♂
21 52 ☽ ∥ ♆
23 37 ☽ ⊥ ♀

09 Sunday
00 18 ☽ ⊻ ♃
02 20 ☿ ⊥ ♄
03 13 ☽ ⊥ ☉
06 07 ☿ ☌ ♅
06 54 ☽ ⊥ ♆
08 04 ☽ ⊻ ♂
08 44 ☽ Q ♄
09 09 ☽ ⊥ ♅
09 32 ☽ ⊥ ☿
10 43 ☽ ⊥ ♃
10 58 ☽ ⊻ ♀
14 22 ☽ ⊻ ☉
14 28 ☉ ∥ ♅
17 17 ☽ ⊻ ♆
19 33 ☽ ⊻ ♅
20 19 ☿ Q ♃
21 18 ☽ ⊻ ☿

10 Monday
02 27 ☽ ∠ ♃
04 39 ☽ ⊻ ♇
04 41 ☽ ∥ ♆
05 39 ☽ ⚹ ♄
13 15 ☿ ∥ ♂
15 07 ☽ ⊥ ♇

11 Tuesday
04 56 ☽ ⚹ ♃
06 57 ☽ ∠ ♇
07 32 ☉ ☌ ♆
08 08 ☽ ∠ ♄
15 32 ☽ ☌ ♂
20 46 ☽ ☌ ♀
22 02 ☽ ☌ ♆
23 10 ☽ ☌ ☉

12 Wednesday
00 14 ☽ ⊥ ♄
00 27 ☽ ☌ ♅
02 27 ☽ Q ♃
09 31 ☽ ☌ ☿
09 44 ☽ ⚹ ♇
10 40 ♀ ☌ ♆
11 05 ☽ ⊻ ♄
11 19 ☿ ⚹ ♇
14 43 ☉ ⊥ ♄
16 59 ☉ ☌ ♅
22 48 ☿ ⊻ ♄

13 Thursday
02 17 ☽ ∥ ♃
07 07 ☽ ∥ ♄
07 36 ☽ Q ♇
09 46 ☿ ∥ ♀
11 27 ☽ □ ♃
12 59 ♀ ⊥ ♄
13 56 ♀ ☌ ♅

14 Friday
00 06 ☉ Q ♃
01 28 ☽ ⊻ ♂
04 59 ☽ ⊻ ♆
07 37 ☽ ⊻ ♅
09 28 ☽ ⊻ ♀
10 47 ☽ ⊻ ☉
12 03 ☿ ∥ ♅
12 44 ♄ ⊥ ♅
13 27 ☽ ⊥ ♂
15 44 ♀ Q ♃
16 19 ☽ ⊥ ♆
17 15 ☽ □ ♇
19 00 ☽ ⊥ ♅
19 02 ☽ ☌ ♄
22 04 ☽ ⊥ ♀
23 09 ☽ ⊥ ☉

15 Saturday
00 31 ☉ ∥ ♆
01 18 ☽ ∥ ♇
01 24 ☽ ⊻ ☿
07 43 ☽ ∠ ♂
09 34 ☽ ∠ ♆
12 19 ☽ ∠ ♅
14 45 ☽ ⊥ ☿
15 09 ♀ ∥ ♅
17 19 ☽ ∠ ♀
18 00 ☽ ∠ ☉
20 50 ☽ △ ♃

16 Sunday
11 07 ☽ ∠ ☿
14 02 ☿ ∥ ♆
14 53 ♂ ☌ ♆
14 55 ☽ ⚹ ♆
14 55 ☽ ⚹ ♂
17 48 ☽ ⚹ ♅

17 Monday
02 04 ☉ ☌ ♀
02 11 ☽ ⚹ ☉
02 11 ☽ ⚹ ♀
02 40 ☽ ⚼ ♃
03 46 ☽ △ ♇
06 03 ☽ ⊻ ♄
14 53 ☽ Q ♆
16 28 ☽ Q ♂
17 08 ☉ ∥ ☿
17 33 ♀ ⚹ ♇
17 51 ☽ Q ♅
18 11 ☽ ⊥ ♄
19 57 ☽ # ♇
20 52 ☽ ± ♃
21 16 ☉ ⚹ ♇
21 54 ☽ ⚹ ☿

18 Tuesday
04 30 ☽ Q ☉
05 05 ☽ Q ♀
09 05 ☽ ⊼ ♃
09 58 ☽ ⚼ ♇
12 30 ☽ ∠ ♄
15 43 ♂ ☌ ♅
18 19 ♀ ⊻ ♄
23 08 ♂ ⊥ ♄

19 Wednesday
02 15 ☽ Q ☿
03 23 ☽ □ ♆
04 15 ☽ ± ♇
04 55 ☉ ⊻ ♄
06 27 ☽ □ ♅
07 24 ☽ □ ♂
12 43 ♀ ∥ ♆
15 51 ☽ # ♄
16 22 ☿ Q ♇
16 28 ☽ ⊼ ♇
19 14 ☽ ⚹ ♄
20 27 ☽ □ ☉
21 57 ☽ □ ♀

20 Thursday
00 39 ☽ # ♃
19 47 ☽ Q ♄
20 52 ☽ □ ☿
22 30 ☽ ☍ ♃
23 35 ♂ Q ♃

21 Friday
09 03 ☿ □ ♃
14 00 ☽ # ☿
16 13 ☽ △ ♆
19 20 ☽ △ ♅

22 Saturday
00 07 ☽ △ ♂
02 19 ☽ # ☉
04 56 ☽ ☍ ♇
08 05 ☽ □ ♄
12 13 ☽ # ♀
14 10 ☽ △ ☉
14 48 ♂ ∥ ♅
17 02 ☽ △ ♀
21 53 ☽ ⚼ ♆

23 Sunday
00 59 ☽ ⚼ ♅
04 20 ☽ # ♆
07 27 ☽ ⚼ ♂
10 16 ☽ ⊼ ♃
15 18 ☽ ± ♆
17 26 ☽ △ ☿
18 22 ☽ ± ♅
21 41 ☽ ⚼ ☉
21 49 ☽ ± ♃

24 Monday
01 07 ☽ ⚼ ♀
01 45 ☽ ± ♂
02 44 ☽ ⊼ ♆
05 46 ☽ ⊼ ♅
13 44 ☽ ⊼ ♂
14 39 ☽ ⊼ ♇
14 39 ☽ # ♆
14 51 ☽ ⚼ ♃
16 02 ☽ ± ☉
18 01 ☽ △ ♄
19 45 ☽ ± ♀

25 Tuesday
01 41 ☽ ⚼ ☿
01 44 ☽ ± ♇
04 02 ☽ ⊼ ☉
05 19 ♂ ⚹ ♇
07 56 ☽ ⊼ ♀
17 52 ☽ # ♀
18 07 ☽ ⚼ ♇
18 29 ☽ △ ♃
20 20 ☽ ± ☿
21 34 ☽ ⚼ ♄

26 Wednesday
01 06 ☽ # ☉
08 29 ☽ ⊼ ☿
09 36 ☽ ☍ ♆
12 33 ☽ ☍ ♅
13 40 ☽ ± ♄
17 59 ☿ ⊻ ♆
20 42 ☽ △ ♇
23 00 ☽ ☍ ♂

27 Thursday
00 14 ☽ ⊼ ♄
04 47 ☽ # ☿
09 11 ☽ # ♃
10 45 ♀ Q ♇
13 23 ☽ ☍ ☉
18 06 ☽ ☍ ♀
19 49 ☿ ⊻ ♅
19 53 ☽ # ♄
23 13 ☽ □ ♃

28 Friday
00 39 ♂ ⊻ ♄
03 50 ♂ ∥ ♆
04 42 ☿ ∥ ♃
13 22 ☽ ⊼ ♆
16 15 ☽ ⊼ ♅
18 42 ☽ ☍ ☿
23 29 ☽ ± ♆
23 56 ☽ □ ♇

29 Saturday
02 22 ☽ ± ♅
03 41 ☽ ☍ ♄
05 00 ☽ ⊼ ♂
05 45 ☽ Q ♃
06 51 ☽ # ♇
14 33 ☽ ⚼ ♆
15 35 ☽ ± ♂
17 27 ☽ ⚼ ♅
19 48 ☽ ⊼ ☉
22 30 ☉ Q ♇

30 Sunday
01 22 ☽ ⊼ ♀
01 52 ☽ ⚹ ♃
05 59 ☽ Q ♇
06 32 ☽ ± ☉
07 24 ☽ ⚼ ♂
07 29 ♀ □ ♃
12 17 ☽ ± ♀
15 34 ☽ △ ♆
17 03 ☿ ⊥ ♆
18 29 ☽ △ ♅
21 01 ☿ □ ♇
22 37 ☽ ⚼ ☉

31 Monday
01 00 ☿ ∥ ♄
01 56 ☽ ⚹ ♇
02 29 ☽ ⊼ ☿
02 56 ☽ ∠ ♃
04 39 ☽ ⚼ ♀
06 00 ☽ ⊼ ♄
09 45 ☽ △ ♂
13 33 ☽ ± ☿
15 16 ☽ ∥ ♇
16 04 ☽ ± ♄
18 00 ☽ ⊥ ♃
23 52 ☿ ⊥ ♅

All ephemeris data is given at 12.00 UT and the Moon's longitude is additionally given for 24.00 UT

FEBRUARY 1994

Date	Sidereal time h m s	Sun ☉ ° ' "	Moon ☽ ° ' "	Moon ☽ 24.00 ° ' "	Mercury ☿ ° '	Venus ♀ ° '	Mars ♂ ° '	Jupiter ♃ ° '	Saturn ♄ ° '	Uranus ♅ ° '	Neptune ♆ ° '	Pluto ♇ ° '
01	20 45 51	12 ♒ 24 00	18 ♎ 14 03	25 ♎ 22 11	00 ♓ 05	16 ♒ 06	03 ♒ 22	13 ♏ 32	00 ♓ 25	23 ♑ 26	21 ♑ 38	27 ♏ 51
02	20 49 48	13 24 52	02 ♏ 28 02	09 ♏ 31 25	01 23	17 21	04 09	13 37	00 32	23 29	21 41	27 52
03	20 53 44	14 25 44	16 ♏ 32 16	23 ♏ 30 33	02 36	18 36	04 56	13 41	00 39	23 32	21 43	27 52
04	20 57 41	15 26 35	00 ♐ 26 15	07 ♐ 19 27	03 42	19 52	05 43	13 46	00 46	23 36	21 45	27 53
05	21 01 37	16 27 25	14 ♐ 10 10	20 ♐ 58 27	04 42	21 07	06 29	13 50	00 53	23 39	21 47	27 54
06	21 05 34	17 28 14	27 ♐ 44 20	04 ♑ 27 48	05 34	22 22	07 16	13 54	01 00	23 43	21 49	27 55
07	21 09 30	18 29 02	11 ♑ 08 49	17 ♑ 47 20	06 17	23 37	08 03	13 58	01 07	23 46	21 51	27 56
08	21 13 27	19 29 50	24 ♑ 23 14	00 ♒ 56 23	06 52	24 53	08 50	14 02	01 15	23 49	21 53	27 57
09	21 17 23	20 30 36	07 ♒ 26 40	13 ♒ 53 57	07 16	26 08	09 37	14 06	01 22	23 52	21 55	27 57
10	21 21 20	21 31 20	20 ♒ 18 05	26 ♒ 38 59	07 31	27 23	10 24	14 09	01 29	23 56	21 57	27 58
11	21 25 17	22 32 04	02 ♓ 56 35	09 ♓ 10 51	07 R 35	28 38	11 11	14 12	01 36	23 59	21 59	27 59
12	21 29 13	23 32 46	15 ♓ 21 48	21 ♓ 29 34	07 28	29 ♒ 54	11 58	14 15	01 43	24 02	22 01	27 59
13	21 33 10	24 33 26	27 ♓ 34 18	03 ♈ 36 14	07 10	01 ♓ 09	12 45	14 18	01 51	24 05	22 03	28 00
14	21 37 06	25 34 05	09 ♈ 35 39	15 ♈ 32 57	06 42	02 24	13 32	14 21	01 58	24 09	22 05	28 00
15	21 41 03	26 34 42	21 ♈ 28 33	27 ♈ 22 57	06 05	03 39	14 19	14 23	02 05	24 12	22 07	28 01
16	21 44 59	27 35 18	03 ♉ 16 42	09 ♉ 10 24	05 19	04 54	15 06	14 26	02 13	24 15	22 09	28 01
17	21 48 56	28 35 52	15 ♉ 04 40	21 ♉ 00 12	04 26	06 09	15 53	14 28	02 20	24 18	22 11	28 02
18	21 52 52	29 ♒ 36 24	26 ♉ 57 39	02 ♊ 57 46	03 26	07 24	16 40	14 30	02 27	24 21	22 13	28 02
19	21 56 49	00 ♓ 36 54	09 ♊ 01 13	15 ♊ 08 43	02 23	08 39	17 27	14 32	02 34	24 24	22 15	28 03
20	22 00 46	01 37 23	21 ♊ 20 55	27 ♊ 38 26	01 16	09 54	18 15	14 33	02 42	24 27	22 17	28 03
21	22 04 42	02 37 50	04 ♋ 01 50	10 ♋ 31 34	00 ♓ 09	11 09	19 02	14 35	02 49	24 30	22 19	28 03
22	22 08 39	03 38 14	17 ♋ 07 59	23 ♋ 51 18	29 ♒ 03	12 24	19 49	14 36	02 56	24 33	22 21	28 04
23	22 12 35	04 38 37	00 ♌ 41 34	07 ♌ 38 41	27 59	13 39	20 36	14 37	03 04	24 36	22 22	28 04
24	22 16 32	05 38 59	14 ♌ 42 19	21 ♌ 52 00	26 59	14 54	21 23	14 38	03 11	24 39	22 24	28 04
25	22 20 28	06 39 18	29 ♌ 07 02	06 ♍ 26 34	26 03	16 09	22 10	14 38	03 18	24 42	22 26	28 04
26	22 24 25	07 39 35	13 ♍ 49 36	21 ♍ 15 04	25 14	17 24	22 57	14 39	03 25	24 45	22 28	28 04
27	22 28 21	08 39 51	28 ♍ 41 50	06 ♎ 08 47	24 31	18 39	23 44	14 39	03 33	24 47	22 29	28 04
28	22 32 18	09 ♓ 40 05	13 ♎ 34 50	20 ♎ 59 01	23 ♒ 54	19 ♓ 54	24 ♒ 32	14 ♏ 39	03 ♓ 40	24 ♑ 50	22 ♑ 31	28 ♏ 04

Date	Moon True ☊ ° '	Moon Mean ☊ ° '	Moon ☽ Latitude ° '
01	29 ♏ 59	29 ♏ 25	03 S 22
02	29 R 57	29 22	02 20
03	29 D 57	29 19	01 S 11
04	29 R 57	29 16	00 N 03
05	29 56	29 13	01 15
06	29 54	29 10	02 21
07	29 48	29 06	03 20
08	29 40	29 03	04 06
09	29 28	29 00	04 39
10	29 15	28 57	04 56
11	29 00	28 54	04 59
12	28 47	28 51	04 48
13	28 35	28 47	04 23
14	28 25	28 44	03 47
15	28 18	28 41	03 01
16	28 14	28 38	02 08
17	28 13	28 35	01 09
18	28 D 12	28 31	00 N 07
19	28 R 12	28 28	00 S 57
20	28 11	28 25	01 59
21	28 08	28 22	02 57
22	28 02	28 19	03 48
23	27 54	28 16	04 28
24	27 44	28 12	04 53
25	27 32	28 09	05 01
26	27 20	28 06	04 49
27	27 09	28 03	04 17
28	27 ♏ 01	28 ♏ 00	03 S 28

DECLINATIONS

Date	Sun ☉ ° '	Moon ☽ ° '	Mercury ☿ ° '	Venus ♀ ° '	Mars ♂ ° '	Jupiter ♃ ° '	Saturn ♄ ° '	Uranus ♅ ° '	Neptune ♆ ° '	Pluto ♇ ° '
01	17 S 05	10 S 16	11 S 44	17 S 15	20 S 23	14 S 44	12 S 39	21 S 52	21 S 06	06 S 05
02	16 48	14 31	11 04	16 54	20 13	14 45	12 37	21 51	21 06	06 04
03	16 30	17 54	10 25	16 31	20 02	14 47	12 34	21 50	21 05	06 04
04	16 12	20 12	09 47	16 09	19 50	14 48	12 32	21 50	21 05	06 04
05	15 54	21 16	09 11	15 45	19 39	14 49	12 29	21 49	21 05	06 04
06	15 36	21 04	08 38	15 22	19 27	14 50	12 27	21 49	21 04	06 04
07	15 17	19 39	08 07	14 58	19 15	14 51	12 24	21 48	21 04	06 03
08	14 58	17 12	07 39	14 33	19 03	14 52	12 22	21 48	21 04	06 03
09	14 39	13 55	07 14	14 08	18 51	14 53	12 19	21 47	21 03	06 03
10	14 20	10 02	06 53	13 43	18 39	14 54	12 16	21 46	21 03	06 03
11	14 00	05 46	06 37	13 17	18 26	14 54	12 14	21 46	21 03	06 02
12	13 40	01 S 21	06 25	12 51	18 13	14 55	12 11	21 45	21 02	06 02
13	13 20	03 N 03	06 18	12 25	18 00	14 56	12 09	21 45	21 02	06 02
14	13 00	07 17	06 16	11 58	17 47	14 56	12 06	21 44	21 02	06 01
15	12 39	11 10	06 18	11 31	17 33	14 57	12 04	21 44	21 01	06 01
16	12 19	14 37	06 25	11 04	17 20	14 57	12 01	21 43	21 01	06 01
17	11 58	17 27	06 36	10 36	17 06	14 58	11 58	21 43	21 01	06 01
18	11 37	19 35	06 52	10 09	16 52	14 58	11 56	21 42	21 00	06 00
19	11 15	20 52	07 11	09 40	16 38	14 59	11 53	21 42	21 00	06 00
20	10 54	21 10	07 32	09 12	16 24	14 59	11 51	21 41	21 00	06 00
21	10 32	20 25	07 56	08 43	16 09	14 59	11 48	21 41	20 59	05 59
22	10 10	18 34	08 22	08 15	15 54	14 59	11 46	21 40	20 59	05 59
23	09 48	15 38	08 48	07 45	15 40	15 00	11 43	21 40	20 59	05 58
24	09 26	11 45	09 15	07 16	15 25	15 00	11 40	21 39	20 59	05 58
25	09 04	07 05	09 41	06 47	15 09	15 00	11 38	21 39	20 58	05 58
26	08 42	01 N 55	10 07	06 17	14 54	15 00	11 35	21 38	20 58	05 57
27	08 19	03 S 25	10 32	05 47	14 39	14 59	11 33	21 38	20 58	05 57
28	07 S 57	08 S 33	10 S 55	05 S 17	14 S 23	14 S 59	11 S 30	21 S 37	20 S 57	05 S 57

ZODIAC SIGN ENTRIES

Date	h	m	Planets
01	10	28	☿ ♓
02	07	49	☽ ♏
04	11	14	☽ ♐
06	16	02	☽ ♑
08	22	16	☽ ♒
11	06	22	☽ ♓
12	14	04	♀ ♓
13	16	49	☽ ♈
16	05	20	☽ ♉
18	18	05	☽ ♊
18	21	22	☉ ♓
21	04	27	☽ ♋
21	15	15	☿ ♒
23	10	48	☽ ♌
25	13	27	☽ ♍
27	14	06	☽ ♎

LATITUDES

Date	Mercury ☿ ° '	Venus ♀ ° '	Mars ♂ ° '	Jupiter ♃ ° '	Saturn ♄ ° '	Uranus ♅ ° '	Neptune ♆ ° '	Pluto ♇ ° '
01	00 S 18	01 S 18	01 S 01	01 N 13	01 S 25	00 S 28	00 N 37	13 N 56
04	00 N 23	01 21	01 02	01 14	01 25	00 28	00 37	13 57
07	01 11	01 24	01 02	01 14	01 25	00 28	00 36	13 59
10	02 00	01 25	01 03	01 15	01 25	00 28	00 36	14 00
13	02 47	01 27	01 04	01 15	01 25	00 28	00 36	14 01
16	03 23	01 27	01 04	01 16	01 25	00 28	00 36	14 02
19	03 42	01 27	01 05	01 17	01 26	00 28	00 37	14 04
22	03 40	01 27	01 05	01 17	01 26	00 28	00 37	14 05
25	03 20	01 26	01 06	01 18	01 26	00 28	00 37	14 06
28	02 47	01 24	01 06	01 18	01 26	00 28	00 37	14 07
31	02 N 08	01 S 22	01 S 06	01 N 19	01 S 26	00 S 28	00 N 37	14 N 09

DATA

Julian Date	2449385
Delta T	+60 seconds
Ayanamsa	23° 46' 44"
Synetic vernal point	05° ♓ 20' 15"
True obliquity of ecliptic	23° 26' 20"

MOON'S PHASES, APSIDES AND POSITIONS ☽

Date	h	m	Phase	Longitude ° '	Eclipse Indicator
03	08	06	☾	14 ♏ 16	
10	14	30	●	21 ♒ 38	
18	17	47	☽	29 ♉ 51	
26	01	15	○	07 ♍ 13	

Day	h	m	
16	01	37	Apogee
27	22	17	Perigee

05	20	06	Max dec	21° S 20'
12	19	16	0N	
20	07	14	Max dec	21° N 12'
26	20	39	0S	

LONGITUDES

Date	Chiron ⚷ ° '	Ceres ⚳ ° '	Pallas ⚴ ° '	Juno ⚵ ° '	Vesta ⚶ ° '	Black Moon Lilith ⚸ ° '
01	07 ♍ 53	29 ♈ 50	13 ♓ 15	01 ♏ 58	29 ♓ 34	22 ♈ 50
11	07 ♍ 11	02 ♉ 36	16 ♓ 32	02 ♏ 41	03 ♈ 51	23 ♈ 57
21	06 ♍ 25	05 ♉ 40	19 ♓ 55	02 ♏ 50	08 ♈ 13	25 ♈ 04
31	05 ♍ 38	08 ♉ 57	23 ♓ 23	02 ♏ 22	12 ♈ 38	26 ♈ 10

All ephemeris data is given at 12.00 UT and the Moon's longitude is additionally given for 24.00 UT

ASPECTARIAN

h m	Aspects
01 Tuesday	
01 29	☽ △ ☉
02 58	☽ ∠ ♇
04 05	☽ ⊻ ♃
06 11	☽ ⅊ ☿
07 14	☽ ⅊ ♄
08 04	☽ △ ♀
17 44	☽ □ ♆
18 04	☽ ⊥ ♇
18 26	☿ ☌ ♄
18 46	☽ ∥ ☿
20 46	☽ □ ♅
02 Wednesday	
00 48	☽ ∥ ♄
04 12	☽ ⊻ ♇
08 41	☽ △ ♄
09 59	☽ △ ☿
13 28	☽ ∥ ♃
15 01	☽ □ ♂
17 05	☉ □ ♃
03 Thursday	
00 17	☽ Q ♆
02 02	☽ ∥ ☉
02 23	☽ ∥ ♀
03 25	☽ Q ♅
07 06	☽ ☌ ♃
08 06	☽ □ ☉
15 54	☽ □ ♀
17 50	☉ ∥ ♀
20 55	☽ ⚹ ♆
23 39	☽ Q ♂
04 Friday	
00 06	☽ ⚹ ♅
07 34	☽ ☌ ♇
07 40	☽ ∥ ♂
12 34	☽ □ ♄
17 39	☽ Q ☉
18 09	☽ □ ☿
21 44	☽ ⚹ ♂
23 01	☽ ∠ ♆
05 Saturday	
02 14	☽ Q ♀
02 17	☽ ∠ ♅
04 47	☽ ∥ ♆
11 25	☽ ⊻ ♃
14 51	☽ ⊥ ♆
16 21	☽ ⚹ ☉
18 10	☽ ⊥ ♅
20 23	☽ Q ♄
22 02	☽ ⊥ ♃
06 Sunday	
01 09	♀ ⊻ ♆
01 28	☽ ⊻ ♆
01 30	☽ ⚹ ♀
01 42	☽ ∠ ♂
04 06	☽ Q ☿
04 49	☽ ⊻ ♅
11 45	☽ ∥ ♆
12 19	☽ ⊻ ♇
14 05	☽ ∠ ♃
17 52	☽ ⚹ ♄
18 41	☽ ⊥ ♂
21 08	☽ ∠ ☉
23 02	☽ ⊥ ♇
07 Monday	
02 48	☽ ⚹ ☿
06 06	☽ ⊻ ♂
06 59	☽ ∠ ♀
14 36	☽ ⊥ ☉
14 49	♀ ⊻ ♅
15 13	☽ ∠ ♇
17 07	☽ ⚹ ♃
17 07	☽ ∥ ♂
18 19	♀ ∥ ♃
21 04	☽ ∠ ♄
08 Tuesday	
00 56	☽ ⊥ ♀
02 22	☽ ⊻ ☉
07 13	☽ ∠ ☿
07 26	☽ ☌ ♆
10 58	☽ ☌ ♅
12 59	☽ ⊻ ♀
13 20	☿ ∠ ♆
13 35	☽ ⊥ ♄
15 01	☽ Q ♃
18 31	☽ ⚹ ♇
19 42	☉ ∥ ♃
09 Wednesday	
00 17	☽ ⊥ ☿
00 40	☽ ⊻ ♄
05 30	☽ ∥ ♃
06 33	☽ ∥ ☉
10 20	☽ ∥ ♀
11 40	☽ ⊻ ☿
16 18	☽ ☌ ♂
16 40	☽ Q ♇
22 21	☽ ∥ ♄
22 29	♂ Q ♇
10 Thursday	
00 25	☽ □ ♃
14 30	☽ ☌ ☉
15 08	☽ ⊻ ♆
18 52	☽ ⊻ ♅
22 39	☉ ⊻ ♆
23 14	♀ ⊥ ♆
23 15	♀ □ ♇
11 Friday	
02 31	☽ □ ♇
02 31	☽ ⊥ ♆
02 52	☽ ☌ ♀
06 19	☽ ⊥ ♅
07 03	☽ ∥ ☿
08 30	☿ St R
09 25	☽ ☌ ♄
10 30	☽ ∥ ♇
19 48	☽ ∠ ♆
20 52	☽ ☌ ☿
23 40	☽ ∠ ♅
12 Saturday	
04 57	☽ ⊻ ♂
09 50	☽ △ ♃
14 53	♀ ⊥ ♅
17 26	☽ ⊥ ♂
13 Sunday	
00 17	☉ ⊻ ♅
01 05	☽ ⚹ ♆
05 05	☽ ⚹ ♅
05 30	☽ ⊻ ☉
12 23	☽ ∠ ♂
12 51	☽ △ ♇
15 27	☽ ⅊ ♃
18 08	☿ ∠ ♆
18 28	☽ ⊥ ☉
19 55	☽ ⊻ ♀
20 35	☽ ⊻ ♄
14 Monday	
00 56	☽ Q ♆
02 51	♀ ☌ ♄
04 12	♀ ∥ ♄
04 43	☽ ⋕ ♇
05 03	☽ Q ♅
06 04	☽ ⋕ ☿
06 27	☽ ⊻ ☿
08 42	☽ ⊥ ♄
09 19	☽ ⊥ ♀
09 29	☽ ± ♃
14 08	☽ ∠ ☉
17 59	☽ ⊥ ☿
18 52	☽ ⅊ ♇
20 30	☽ ⚹ ♂
21 37	☽ ⊼ ♃
15 Tuesday	
03 01	☽ ∠ ♄
05 36	☽ ∠ ♀
11 15	☽ ∠ ☿
13 06	☽ ± ♇
13 19	☽ □ ♆
14 02	☽ ⋕ ♀
14 17	♂ □ ♃
17 33	☽ □ ♅
17 48	☽ ⋕ ♄
21 01	☽ ⋕ ☉
22 32	☽ Q ♂
23 20	☽ ⚹ ☉
16 Wednesday	
01 18	☽ ⊼ ♇
09 48	☽ ⚹ ♄
14 41	☽ ⋕ ♃
15 42	☽ ⚹ ♀
15 52	☽ ⚹ ☿
16 44	☿ ☌ ♀
22 26	☉ □ ♇
17 Thursday	
01 55	☉ ⊥ ♆
02 03	☽ Q ☉
08 56	☽ ⋕ ♂
10 28	☽ Q ♄
10 45	☽ ☍ ♃
10 59	☉ ∥ ♄
13 46	☽ □ ♂
14 32	☽ Q ☿
18 58	☽ Q ♀
18 Friday	
02 26	☽ △ ♆
06 44	☽ △ ♅
08 21	♀ ∠ ♆
14 10	☽ ☍ ♇
17 47	☽ □ ☉
23 06	☽ □ ♄
23 55	☽ □ ☿
19 Saturday	
06 37	☉ ⊥ ♅
08 03	☿ ☌ ♄
08 30	☽ ⅊ ♆
11 12	☽ □ ♀
12 45	☽ ⅊ ♅
16 35	☽ ⋕ ♆
22 49	☽ ⊼ ♃
20 Sunday	
02 11	☽ ± ♆
02 53	♀ ∠ ♅
05 36	☽ △ ♂
06 24	☽ ± ♅
08 01	☉ ☌ ☿
10 28	☽ ± ♃
13 48	☽ ⊼ ♆
17 58	☽ ⊼ ♅
21 54	☽ ⋕ ♆
21 Monday	
00 47	☽ ⊼ ♇
03 39	☽ ⅊ ♃
04 49	☿ ⊥ ♅
05 19	☽ △ ☿
09 10	☽ △ ☉
09 43	☽ △ ♄
12 00	☽ ⅊ ♂
12 03	☽ ± ♇
17 03	☉ ☌ ♄
22 Tuesday	
02 32	☽ △ ♀
04 37	☽ ⅊ ♇
05 37	☽ ± ♂
06 51	☽ ⅊ ☿
07 25	☽ △ ♃
08 52	☿ ∥ ♀
13 28	☽ ⅊ ♄
14 55	☽ ⅊ ☉
17 06	☽ ⊼ ♂
21 20	☽ ☍ ♆
21 47	☽ ± ☿
23 Wednesday	
01 17	☽ ☍ ♅
03 15	☿ ⊥ ♆
05 36	☽ ± ♄
07 24	☽ △ ♇
07 36	☽ ⊼ ☿
08 06	☽ ⅊ ♀
08 09	☽ ± ☉
10 06	☿ □ ♇
11 50	☽ ⋕ ♂
16 09	☽ ⊼ ♄
16 24	☽ ⋕ ♃
19 23	☽ ⊼ ☉
24 Thursday	
01 12	☽ ± ♀
06 38	♀ △ ♃
11 52	☽ □ ♃
12 22	☽ ⊼ ♀
12 24	☽ ⋕ ♄
17 39	☉ ∥ ☿
23 50	☽ ☍ ♂
25 Friday	
00 06	☽ ⋕ ☿
00 55	☽ ⊼ ♆
01 19	☽ ⋕ ☉
04 40	☽ ⊼ ♅
07 14	☽ ☍ ☿
10 16	☽ □ ♇
10 52	☽ ± ♆
13 37	☽ ⋕ ♀
14 36	☽ ± ♅
17 21	☽ ⋕ ♇
17 47	☽ Q ♃
18 56	☽ ☍ ♄
20 22	♂ ⊻ ♆
26 Saturday	
01 15	☽ ☍ ☉
01 38	☽ ⅊ ♆
03 30	♂ ∥ ♃
05 21	☽ ⅊ ♅
07 08	☉ ∠ ♆
13 20	☽ ⚹ ♃
15 38	☽ Q ♇
18 19	☽ ☍ ♀
27 Sunday	
01 59	☽ △ ♆
02 48	☿ ⊻ ♅
03 34	☽ ⊼ ♂
04 07	♀ ∥ ♇
05 33	☽ ⊼ ☿
05 41	☽ △ ♅
11 00	☽ ⚹ ♇
13 33	☽ ∠ ♃
13 46	☽ ± ♂
14 48	☽ ± ☿
19 52	☽ ⊼ ♄
21 56	☽ ∥ ♀
23 37	☽ ∥ ♇
28 Monday	
00 59	☿ ☌ ♂
04 03	☽ ⊥ ♃
04 43	☽ ⅊ ☿
05 05	☽ ⅊ ♂
05 13	☽ ⊼ ☉
05 38	☽ ± ♄
09 16	☽ ∥ ☉
11 11	☽ ∠ ♇
13 44	☽ ⊻ ♃
13 49	♃ St R
15 37	☽ ± ☉
16 12	☉ ∠ ♅
20 18	☽ ⅊ ♄
22 01	♂ ⊻ ♅
23 11	☽ ⊼ ♀

MARCH 1994

LONGITUDES

Date	Sidereal time h m s	Sun ☉ ° ' "	Moon ☽ ° ' "	Moon ☽ 24.00 ° ' "	Mercury ☿ ° '	Venus ♀ ° '	Mars ♂ ° '	Jupiter ♃ ° '	Saturn ♄ ° '	Uranus ♅ ° '	Neptune ♆ ° '	Pluto ♇ ° '
01	22 36 15	10 ♓ 40 18	28 ♎ 20 30	05 ♏ 38 35	23 ♒ 25	21 ♓ 09	25 ♒ 19	14 ♏ 39	03 ♓ 47	24 ♑ 53	22 ♑ 33	28 ♏ 04
02	22 40 11	11 40 29	12 ♏ 52 44	20 ♏ 02 36	23 R 03	22 24	26 06	14 R 39	03 55	24 56	22 34	28 R 04
03	22 44 08	12 40 39	27 ♏ 07 56	04 ♐ 08 40	22 48	23 39	26 53	14 39	04 02	24 58	22 36	28 04
04	22 48 04	13 40 47	11 ♐ 04 48	17 ♐ 56 27	22 39	24 53	27 40	14 38	04 09	25 01	22 38	28 04
05	22 52 01	14 40 54	24 ♐ 43 47	01 ♑ 26 59	22 D 38	26 08	28 28	14 37	04 16	25 04	22 39	28 04
06	22 55 57	15 40 59	08 ♑ 06 18	14 ♑ 41 56	22 43	27 23	29 ♒ 15	14 36	04 23	25 06	22 41	28 04
07	22 59 54	16 41 03	21 ♑ 14 07	27 ♑ 43 03	22 53	28 38	00 ♓ 02	14 35	04 31	25 09	22 42	28 04
08	23 03 50	17 41 05	04 ♒ 08 55	10 ♒ 31 50	23 10	29 ♓ 52	00 49	14 33	04 38	25 11	22 44	28 04
09	23 07 47	18 41 05	16 ♒ 51 56	23 ♒ 09 19	23 32	01 ♈ 07	01 36	14 32	04 45	25 14	22 45	28 03
10	23 11 44	19 41 03	29 ♒ 24 04	05 ♓ 36 15	23 59	02 22	02 24	14 30	04 52	25 16	22 47	28 03
11	23 15 40	20 41 00	11 ♓ 45 55	17 ♓ 53 10	24 30	03 36	03 11	14 28	04 59	25 18	22 48	28 03
12	23 19 37	21 40 55	23 ♓ 58 03	00 ♈ 00 41	25 06	04 51	03 58	14 26	05 06	25 21	22 50	28 02
13	23 23 33	22 40 47	06 ♈ 01 14	11 ♈ 59 51	25 46	06 06	04 45	14 24	05 14	25 23	22 51	28 02
14	23 27 30	23 40 38	17 ♈ 56 45	23 ♈ 52 13	26 30	07 20	05 32	14 21	05 21	25 25	22 52	28 01
15	23 31 26	24 40 27	29 ♈ 46 33	05 ♉ 40 09	27 18	08 35	06 19	14 18	05 28	25 28	22 54	28 01
16	23 35 23	25 40 14	11 ♉ 33 24	17 ♉ 26 47	28 09	09 49	07 07	14 16	05 35	25 30	22 55	28 00
17	23 39 19	26 39 58	23 ♉ 20 49	29 ♉ 16 04	29 ♒ 03	11 04	07 54	14 13	05 42	25 32	22 56	28 00
18	23 43 16	27 39 40	05 ♊ 13 06	11 ♊ 12 35	00 ♓ 00	12 18	08 41	14 09	05 49	25 34	22 57	27 59
19	23 47 13	28 39 20	17 ♊ 15 08	23 ♊ 21 26	01 00	13 33	09 28	14 06	05 56	25 36	22 59	27 59
20	23 51 09	29 ♓ 38 58	29 ♊ 32 09	05 ♋ 47 54	02 02	14 47	10 15	14 02	06 02	25 38	23 00	27 58
21	23 55 06	00 ♈ 38 34	12 ♋ 09 20	18 ♋ 36 59	03 07	16 01	11 02	13 59	06 09	25 40	23 01	27 57
22	23 59 02	01 38 07	25 ♋ 11 20	01 ♌ 52 46	04 14	17 16	11 49	13 55	06 16	25 42	23 02	27 57
23	00 02 59	02 37 38	08 ♌ 41 32	15 ♌ 37 42	05 24	18 30	12 36	13 51	06 23	25 44	23 03	27 56
24	00 06 55	03 37 07	22 ♌ 41 11	29 ♌ 51 43	06 35	19 44	13 24	13 46	06 30	25 46	23 04	27 55
25	00 10 52	04 36 33	07 ♍ 08 45	14 ♍ 31 36	07 49	20 59	14 11	13 42	06 37	25 48	23 05	27 55
26	00 14 48	05 35 57	21 ♍ 59 21	29 ♍ 30 56	09 05	22 13	14 58	13 37	06 43	25 49	23 06	27 54
27	00 18 45	06 35 19	07 ♎ 05 08	14 ♎ 40 41	10 22	23 27	15 45	13 33	06 50	25 51	23 07	27 53
28	00 22 41	07 34 39	22 ♎ 16 18	29 ♎ 50 42	11 42	24 41	16 32	13 28	06 57	25 53	23 08	27 52
29	00 26 38	08 33 57	07 ♏ 22 44	14 ♏ 51 22	13 03	25 55	17 19	13 23	07 03	25 54	23 09	27 51
30	00 30 35	09 33 14	22 ♏ 15 44	29 ♏ 35 08	14 26	27 09	18 06	13 18	07 10	25 56	23 10	27 50
31	00 34 31	10 ♈ 32 28	06 ♐ 49 03	13 ♐ 57 10	15 ♓ 50	28 ♈ 23	18 ♓ 53	13 ♏ 12	07 ♓ 16	25 ♑ 57	23 ♑ 11	27 ♏ 49

Date	Moon True ☊ ° '	Moon Mean ☊ ° '	Moon ☽ Latitude ° '
01	26 ♏ 56	27 ♏ 57	02 S 25
02	26 R 53	27 53	01 S 14
03	26 D 53	27 50	00 N 01
04	26 53	27 47	01 15
05	26 R 53	27 44	02 22
06	26 51	27 41	03 20
07	26 46	27 37	04 07
08	26 39	27 34	04 40
09	26 29	27 31	04 58
10	26 17	27 28	05 02
11	26 04	27 25	04 51
12	25 52	27 22	04 27
13	25 41	27 18	03 52
14	25 33	27 15	03 06
15	25 27	27 12	02 12
16	25 23	27 09	01 13
17	25 22	27 06	00 N 11
18	25 D 22	27 03	00 S 53
19	25 23	26 59	01 55
20	25 24	26 56	02 53
21	25 R 23	26 53	03 44
22	25 20	26 50	04 26
23	25 16	26 47	04 55
24	25 09	26 43	05 07
25	25 01	26 40	05 00
26	24 54	26 37	04 34
27	24 46	26 34	03 48
28	24 41	26 31	02 46
29	24 38	26 28	01 32
30	24 37	26 24	00 13
31	24 ♏ 37	26 ♏ 21	01 N 06

DECLINATIONS

Date	Sun ☉ ° '	Moon ☽ ° '	Mercury ☿ ° '	Venus ♀ ° '	Mars ♂ ° '	Jupiter ♃ ° '	Saturn ♄ ° '	Uranus ♅ ° '	Neptune ♆ ° '	Pluto ♇ ° '
01	07 S 34	13 S 09	11 S 17	04 S 47	14 S 07	14 S 59	11 S 27	21 S 37	20 S 57	05 S 56
02	07 11	16 52	11 36	04 17	13 51	14 59	11 25	21 36	20 57	05 56
03	06 48	19 30	11 54	03 47	13 35	14 59	11 22	21 36	20 57	05 55
04	06 25	20 52	12 10	03 16	13 19	14 58	11 20	21 35	20 56	05 55
05	06 02	20 58	12 24	02 46	13 03	14 58	11 17	21 35	20 56	05 55
06	05 39	19 51	12 35	02 15	12 47	14 57	11 15	21 34	20 56	05 54
07	05 15	17 42	12 45	01 44	12 30	14 57	11 12	21 34	20 56	05 54
08	04 52	14 41	12 52	01 14	12 13	14 56	11 10	21 34	20 55	05 53
09	04 29	11 02	12 57	00 43	11 57	14 56	11 07	21 33	20 55	05 53
10	04 05	06 58	13 01	00 S 12	11 40	14 55	11 04	21 33	20 55	05 52
11	03 42	02 S 39	13 02	00 N 19	11 23	14 54	11 02	21 32	20 55	05 52
12	03 18	01 N 42	13 02	00 50	11 06	14 53	10 59	21 32	20 54	05 51
13	02 54	05 56	12 59	01 21	10 48	14 53	10 57	21 32	20 54	05 51
14	02 31	09 54	12 55	01 51	10 31	14 52	10 54	21 31	20 54	05 51
15	02 07	13 28	12 49	02 22	10 14	14 51	10 52	21 31	20 54	05 50
16	01 43	16 28	12 41	02 53	09 56	14 50	10 49	21 30	20 53	05 50
17	01 20	18 47	12 32	03 24	09 39	14 49	10 47	21 30	20 53	05 49
18	00 56	20 18	12 21	03 54	09 21	14 48	10 44	21 30	20 53	05 49
19	00 32	20 55	12 08	04 25	09 03	14 46	10 42	21 29	20 53	05 48
20	00 S 08	20 33	11 54	04 55	08 45	14 45	10 39	21 29	20 53	05 48
21	00 N 15	19 09	11 38	05 26	08 27	14 44	10 37	21 29	20 52	05 47
22	00 39	16 44	11 21	05 56	08 09	14 43	10 35	21 28	20 52	05 47
23	01 03	13 21	11 02	06 26	07 51	14 41	10 32	21 28	20 52	05 46
24	01 26	09 07	10 42	06 56	07 33	14 40	10 30	21 28	20 52	05 46
25	01 50	04 N 14	10 20	07 26	07 15	14 39	10 27	21 27	20 52	05 45
26	02 13	01 S 01	09 57	07 55	06 57	14 37	10 25	21 27	20 51	05 45
27	02 37	06 18	09 33	08 25	06 38	14 36	10 23	21 27	20 51	05 44
28	03 00	11 14	09 07	08 54	06 20	14 34	10 20	21 27	20 51	05 44
29	03 24	15 25	08 40	09 23	06 02	14 32	10 18	21 26	20 51	05 43
30	03 47	18 32	08 12	09 52	05 43	14 31	10 16	21 26	20 51	05 43
31	04 N 10	20 S 22	07 S 42	10 N 20	05 S 25	14 S 29	10 S 13	21 S 26	20 S 51	05 S 42

ZODIAC SIGN ENTRIES

Date	h	m	Planets
01	14	43	☽ ♏
03	16	54	☽ ♐
05	21	24	☽ ♑
07	11	01	♂ ♓
08	04	15	☽ ♒
08	14	28	♀ ♈
10	13	09	☽ ♓
12	23	59	☽ ♈
15	12	27	☽ ♉
18	01	29	☽ ♊
18	12	03	☿ ♓
20	12	54	☽ ♋
20	20	28	☉ ♈
22	20	39	☽ ♌
25	00	14	☽ ♍
27	00	46	☽ ♎
29	00	15	☽ ♏
31	00	41	☽ ♐

LATITUDES

Date	Mercury ☿ ° '	Venus ♀ ° '	Mars ♂ ° '	Jupiter ♃ ° '	Saturn ♄ ° '	Uranus ♅ ° '	Neptune ♆ ° '	Pluto ♇ ° '
01	02 N 35	01 S 23	01 S 06	01 N 18	01 S 26	00 S 28	00 N 37	14 N 08
04	01 54	01 21	01 06	01 19	01 26	00 28	00 37	14 09
07	01 12	01 18	01 07	01 19	01 27	00 28	00 37	14 10
10	00 N 33	01 14	01 07	01 20	01 27	00 28	00 37	14 11
13	00 S 04	01 10	01 07	01 20	01 27	00 28	00 37	14 12
16	00 37	01 06	01 07	01 21	01 28	00 28	00 37	14 14
19	01 05	01 01	01 07	01 21	01 28	00 29	00 37	14 15
22	01 30	00 55	01 07	01 22	01 28	00 29	00 37	14 16
25	01 50	00 49	01 06	01 22	01 29	00 29	00 37	14 17
28	02 06	00 43	01 06	01 23	01 29	00 29	00 37	14 18
31	02 S 18	00 S 36	01 S 06	01 N 23	01 S 29	00 S 29	00 N 37	14 N 19

DATA

Julian Date	2449413
Delta T	+60 seconds
Ayanamsa	23° 46' 47"
Synetic vernal point	05° ♓ 20' 12"
True obliquity of ecliptic	23° 26' 20"

MOON'S PHASES, APSIDES AND POSITIONS ☽

Date	h	m	Phase	Longitude ° '	Eclipse Indicator
04	16	53	☾	13 ♐ 53	
12	07	05	●	21 ♓ 29	
20	12	14	☽	29 ♊ 40	
27	11	10	○	06 ♎ 33	

Day	h	m	
15	17	20	Apogee
28	06	20	Perigee

Day	h	m		
05	01	47	Max dec	21° S 05'
12	02	37	0N	
19	15	15	Max dec	20° N 56'
26	07	25	0S	

LONGITUDES

Date	Chiron ⚷ ° '	Ceres ⚳ ° '	Pallas ⚴ ° '	Juno ⚵ ° '	Vesta ⚶ ° '	Black Moon Lilith ⚸ ° '
01	05 ♍ 47	08 ♉ 17	22 ♓ 41	02 ♏ 31	11 ♈ 45	25 ♈ 57
11	05 ♍ 01	11 ♉ 44	26 ♓ 12	01 ♏ 35	16 ♈ 13	27 ♈ 04
21	04 ♍ 19	15 ♉ 21	29 ♓ 46	00 ♏ 05	20 ♈ 42	28 ♈ 10
31	03 ♍ 43	19 ♉ 07	03 ♈ 22	28 ♎ 08	25 ♈ 12	29 ♈ 17

All ephemeris data is given at 12.00 UT and the Moon's longitude is additionally given for 24.00 UT

ASPECTARIAN

h m	Aspects
01 Tuesday	
00 57	☽ ∥ ☿
01 46	☽ ⊥ ♇
02 31	☽ □ ♆
02 50	☽ ∥ ♄
04 12	☽ △ ☿
06 20	☽ □ ♅
06 46	☽ △ ♂
07 19	☽ ⚼ ☉
09 52	☽ ± ♀
10 06	♇ St R
11 34	☽ ⚺ ♇
17 23	☽ ∥ ♂
21 01	☽ △ ♄
23 03	☽ ∥ ♃
23 19	☿ ∥ ♄
02 Wednesday	
02 02	☽ ⚼ ♀
08 10	☽ Q ♆
09 51	☽ △ ☉
12 05	☽ Q ♅
14 57	☽ ☌ ♃
15 32	♀ ⚹ ♆
22 10	☿ ⚺ ♀
03 Thursday	
04 18	☽ ⚹ ♆
04 45	☽ □ ☿
05 30	☽ △ ♀
08 19	☽ ⚹ ♅
11 33	☽ □ ♂
13 36	☽ ☌ ♇
23 54	☽ □ ♄
04 Friday	
06 00	☽ ∠ ♆
10 09	☽ ∠ ♅
11 16	☽ Q ☿
14 25	☽ ∥ ♆
14 32	♀ ⚹ ♅
16 53	☽ □ ☉
18 11	☽ ⚺ ♃
19 05	☿ ⚺ ♆
20 30	☽ Q ♂
21 43	☽ ⊥ ♆
05 Saturday	
00 07	♂ □ ♇
01 56	☽ ⊥ ♅
04 44	☽ ⊥ ♃
05 49	☿ St D
07 36	☽ Q ♄
08 16	☽ ⚹ ☿
08 19	☽ ⚺ ♆
10 30	☉ △ ♃
12 35	☽ ⚺ ♅
13 24	☽ ∥ ♆
14 45	☽ □ ♀
17 57	☽ ⚺ ♇
18 12	♂ ⊥ ♆
19 03	☽ ⚹ ♂
19 47	☉ ∥ ♇
20 42	☽ ∠ ♃
06 Sunday	
03 22	☽ Q ☉
04 30	☿ ⚺ ♆
04 42	☽ ⊥ ♇
05 14	☽ ⚹ ♄
11 17	☽ ∠ ☿
21 01	☽ ∠ ♇
22 14	☿ ∥ ♂
23 48	☽ ⚹ ♃
23 53	☽ ∠ ♂
07 Monday	
01 09	♀ △ ♇
02 38	☽ Q ♀
02 56	☽ ⚹ ☉
03 53	☽ ⊥ ☿
08 48	☽ ∠ ♄
14 43	☽ ☌ ♆
15 06	☽ ⚺ ☿
17 30	☽ ⊥ ♂
18 16	☿ ⊥ ♀
19 15	☽ ☌ ♅
21 52	☽ Q ♃
08 Tuesday	
00 38	☽ ⚹ ♇
01 36	☽ ⊥ ♄
03 09	☽ ⚹ ♀
05 22	☽ ⚺ ♂
06 03	♀ ⚼ ♃
09 02	☽ ∠ ☉
10 12	☽ ∥ ♃
12 55	☽ ⚺ ♄
23 06	☽ Q ♇
23 50	♂ ⊥ ♅
09 Wednesday	
00 05	☽ ∥ ☿
03 24	☽ ⊥ ☉
05 54	☽ ∥ ♂
07 35	☽ □ ♃
10 25	☽ ∠ ♀
11 31	☽ ∥ ♄
15 46	☽ ⚺ ☉
23 15	☽ ⚺ ♆
10 Thursday	
01 10	☽ ☌ ☿
04 01	☽ ⚺ ♅
05 30	☽ ⊥ ♀
09 24	☽ □ ♇
10 48	☽ ⊥ ♆
13 34	♀ ⚺ ♂
15 37	☽ ⊥ ♅
18 10	☽ ☌ ♂
18 10	☽ ∥ ♇
18 21	☽ ⚺ ♀
22 41	☽ ☌ ♄
11 Friday	
04 16	☽ ∠ ♆
05 44	☽ ∥ ☉
09 09	☽ ∠ ♅
17 17	☽ △ ♃
23 31	☽ # ♀
12 Saturday	
06 33	☽ ∥ ♀
07 05	☽ ☌ ☉
09 44	☽ ⚹ ♆
11 34	♀ Q ♆
14 23	☽ ⚺ ☿
14 44	☽ ⚹ ♅
17 29	♀ ⚺ ♄
20 04	☽ △ ♇
20 11	☽ # ☉
21 33	☿ ⚺ ♅
22 06	♂ ∥ ♄
22 49	☽ ⚼ ♃
13 Sunday	
03 00	☽ ⊥ ☿
09 17	☽ ⚺ ♂
09 39	☽ Q ♆
10 24	☽ ⚺ ♄
11 30	☽ # ♇
12 10	☽ ☌ ♀
14 45	☽ Q ♅
16 12	☉ ⚹ ♆
16 45	☽ ± ♃
22 09	☽ ∠ ☿
22 10	☽ ⊥ ♂
22 33	☽ ⊥ ♄
14 Monday	
02 04	☽ ⚼ ♇
04 46	☽ ⚻ ♃
05 01	♂ ☌ ♄
13 44	♀ Q ♅
15 38	☽ # ♂
16 54	☽ ∠ ♄
17 37	☽ ∠ ♂
18 23	☽ # ♄
20 15	☽ ± ♇
22 00	☽ □ ♆
15 Tuesday	
00 41	☽ ⚺ ☉
03 12	☽ □ ♅
05 16	☉ # ♀
06 35	☽ ⚹ ☿
06 58	♀ ± ♃
07 34	☽ # ☿
08 25	☽ ⚻ ♇
14 00	☽ ⊥ ☉
22 25	☽ # ♃
23 41	☽ ⚹ ♄
16 Wednesday	
02 17	☽ ⚹ ♂
07 39	☉ ⚹ ♅
08 03	☽ ⚺ ♀
08 13	☿ □ ♇
08 54	☽ Q ☿
10 02	☽ ∠ ☉
17 29	☽ ☍ ♃
21 43	☽ ⊥ ♀
17 Thursday	
00 23	☽ Q ♄
04 29	☽ Q ♂
09 04	☿ ⊥ ♆
11 10	☽ △ ♆
13 18	♂ ∠ ♆
16 27	☽ △ ♅
18 09	☽ ∠ ♀
19 21	☽ ⚹ ☉
21 25	☽ ☍ ♇
18 Friday	
00 33	☽ □ ☿
01 29	♀ ⊥ ♄
13 12	☽ □ ♄
17 30	☽ ⚼ ♆
19 26	☽ □ ♂
19 50	☉ △ ♇
21 43	☽ Q ☉
22 45	☽ ⚼ ♅
19 Saturday	
01 10	♀ ⚼ ♇
03 48	☽ ⚹ ♀
05 47	☽ ⚻ ♃
07 29	☽ # ♆
11 27	☽ ± ♆
16 39	☽ ± ♅
17 35	☽ ± ♃
22 06	☉ ⚼ ♃
22 16	♀ ⚻ ♃
23 09	☽ # ♆
23 17	☽ ⚻ ♆
20 Sunday	
02 37	☿ ⊥ ♅
04 25	☽ ⚻ ♅
06 05	☽ Q ♀
08 58	☽ ⚻ ♇
11 03	☽ ⚼ ♃
12 14	☽ □ ☉
17 15	☽ △ ☿
20 30	☽ ± ♇
21 Monday	
00 13	♂ ∠ ♅
00 35	☽ △ ♄
09 46	☽ △ ♂
13 30	☽ ⚼ ♇
15 23	☽ △ ♃
19 58	☽ □ ♀
22 Tuesday	
00 07	☽ ⚼ ☿
04 48	☽ ⚼ ♄
04 55	♀ # ♇
08 05	☽ ☍ ♆
12 56	☽ ☍ ♅
15 08	☽ ⚼ ♂
16 58	☽ △ ♇
18 01	☽ ± ☿
21 13	☽ ± ♄
23 Wednesday	
00 29	☽ △ ☉
03 10	☽ # ♃
05 41	☽ ⚻ ☿
07 55	☽ ⚻ ♄
08 08	☽ ± ♂
17 41	♀ ⊥ ♂
19 12	☽ ⚻ ♂
20 53	☽ □ ♃
24 Thursday	
02 46	☽ # ☿
04 32	☽ # ♄
04 35	☽ ⚼ ☉
06 32	☽ △ ♀
09 57	☿ ☌ ♄
12 39	☽ ⚻ ♆
14 20	♃ Q ♅
17 11	☽ ⚻ ♅
20 30	☽ # ♂
20 46	☽ □ ♇
20 53	☽ ± ☉
22 02	☽ ∥ ♀
22 40	♂ △ ♃
22 42	☽ ± ♆
25 Friday	
03 05	☽ Q ♃
03 10	☿ ∥ ♄
03 11	☽ ± ♅
04 46	☽ # ♇
06 39	♀ # ♂
07 32	☽ ⚻ ☉
09 54	☽ ⚼ ♀
11 07	☽ ☍ ♄
13 12	☽ ☍ ☿
13 32	☽ ⚼ ♆
17 15	☿ ∠ ♆
17 58	☽ ⚼ ♅
22 20	☽ ∥ ☉
22 37	☽ ⚹ ♃
23 48	☉ Q ♆
26 Saturday	
00 04	☽ ☍ ♂
01 31	♀ ∠ ♄
01 53	☽ ± ♀
02 13	☽ Q ♇
05 54	♀ ± ♇
12 23	☽ ⚻ ♀
13 47	☽ △ ♆
17 52	☽ # ☉
18 08	☽ △ ♅
21 25	☽ ⚹ ♇
22 32	☽ ∠ ♃
27 Sunday	
05 32	♀ □ ♆
09 24	☽ ∥ ♇
11 10	☽ ☍ ☉
11 36	☽ ⚻ ♄
12 43	☽ ⊥ ♃
13 29	☽ ∥ ♂
17 41	☽ ⚻ ☿
18 40	☉ ⚺ ♄
20 59	☿ ∠ ♅
21 09	☽ ± ♄
21 09	☽ ∠ ♇
22 09	☽ ⚺ ♃
23 02	☽ # ♀
28 Monday	
02 14	☽ ∥ ☿
02 26	☽ ⚻ ♂
04 04	☽ ± ☿
07 28	☽ ∥ ♄
09 29	☉ ± ♃
11 22	☽ ⊥ ♇
11 29	☽ ⚼ ♄
12 26	☽ ± ♂
13 22	☽ □ ♆
16 09	☽ ☍ ♀
17 38	☿ # ♀
17 43	☽ □ ♅
19 32	☉ Q ♅
19 41	☽ ⚼ ☿
20 51	☽ ⚺ ♇
29 Tuesday	
03 29	☽ ⚼ ♂
06 30	☽ ∥ ♃
11 29	☽ △ ♄
11 43	♀ □ ♅
14 02	☽ ⚻ ☉
17 33	☿ △ ♃
18 03	☽ Q ♆
21 34	☽ ☌ ♃
22 00	☽ △ ☿
22 29	☽ Q ♅
30 Wednesday	
00 21	☽ ± ☉
04 51	☽ △ ♂
12 25	♂ ∥ ♇
13 28	☽ ⚹ ♆
16 01	☽ ⚼ ☉
18 00	☽ ⚹ ♅
20 44	☽ ⚻ ♀
21 07	☽ ☌ ♇
31 Thursday	
01 04	♀ ⚻ ♇
06 30	♀ # ♄
07 34	☽ ± ♀
12 46	☽ □ ♄
14 17	☽ ∠ ♆
18 42	☽ △ ☉
18 57	☽ ∠ ♅
22 40	☽ ⚺ ♃

APRIL 1994

LONGITUDES

Date	Sidereal time h m s	Sun ☉ ° ' "	Moon ☽ ° ' "	Moon ☽ 24.00 ° ' "	Mercury ☿ ° '	Venus ♀ ° '	Mars ♂ ° '	Jupiter ♃ ° '	Saturn ♄ ° '	Uranus ♅ ° '	Neptune ♆ ° '	Pluto ⯓ ° '
01	00 38 28	11 ♈ 31 41	20 ♐ 59 19	27 ♐ 55 28	17 ♓ 16	29 ♈ 37	19 ♓ 40	13 ♏ 07	07 ♓ 23	25 ♑ 59	23 ♑ 12	27 ♏ 48
02	00 42 24	12 30 52	04 ♑ 45 43	11 ♑ 30 15	18 44	00 ♉ 51	20 27	13 R 01	07 29	26 00	23 12	27 R 47
03	00 46 21	13 30 01	18 ♑ 09 20	24 ♑ 43 16	20 13	02 05	21 14	12 56	07 36	26 02	23 13	27 46
04	00 50 17	14 29 09	01 ♒ 12 25	07 ♒ 37 07	21 44	03 19	22 00	12 50	07 42	26 03	23 14	27 45
05	00 54 14	15 28 14	13 ♒ 57 45	20 ♒ 14 40	23 16	04 33	22 47	12 44	07 48	26 04	23 14	27 44
06	00 58 10	16 27 18	26 ♒ 28 12	02 ♓ 38 42	24 50	05 47	23 34	12 38	07 55	26 06	23 15	27 43
07	01 02 07	17 26 20	08 ♓ 46 26	14 ♓ 51 41	26 26	07 01	24 21	12 32	08 01	26 07	23 16	27 42
08	01 06 04	18 25 20	20 ♓ 54 44	26 ♓ 55 49	28 03	08 15	25 08	12 25	08 07	26 08	23 16	27 41
09	01 10 00	19 24 19	02 ♈ 55 10	08 ♈ 53 00	29 ♓ 41	09 28	25 55	12 19	08 13	26 09	23 17	27 39
10	01 13 57	20 23 15	14 ♈ 49 31	20 ♈ 44 59	01 ♈ 21	10 42	26 41	12 12	08 19	26 10	23 17	27 38
11	01 17 53	21 22 09	26 ♈ 39 35	02 ♉ 33 35	03 03	11 56	27 28	12 06	08 25	26 11	23 18	27 37
12	01 21 50	22 21 02	08 ♉ 27 15	14 ♉ 20 52	04 46	13 10	28 15	11 59	08 31	26 12	23 18	27 36
13	01 25 46	23 19 52	20 ♉ 14 45	26 ♉ 09 15	06 31	14 23	29 02	11 52	08 37	26 13	23 19	27 34
14	01 29 43	24 18 40	02 ♊ 04 44	08 ♊ 01 38	08 17	15 37	29 ♓ 48	11 45	08 43	26 14	23 19	27 33
15	01 33 39	25 17 26	14 ♊ 00 22	20 ♊ 01 25	10 04	16 50	00 ♈ 35	11 38	08 49	26 15	23 19	27 32
16	01 37 36	26 16 10	26 ♊ 05 18	02 ♋ 12 32	11 54	18 04	01 22	11 31	08 54	26 15	23 20	27 30
17	01 41 33	27 14 52	08 ♋ 23 39	14 ♋ 39 12	13 44	19 17	02 08	11 24	09 00	26 16	23 20	27 29
18	01 45 29	28 13 32	20 ♋ 59 43	27 ♋ 25 42	15 37	20 31	02 55	11 17	09 06	26 17	23 20	27 28
19	01 49 26	29 ♈ 12 09	03 ♌ 57 37	10 ♌ 35 54	17 31	21 44	03 41	11 09	09 11	26 17	23 20	27 26
20	01 53 22	00 ♉ 10 44	17 ♌ 20 51	24 ♌ 12 41	19 26	22 58	04 28	11 02	09 17	26 18	23 21	27 25
21	01 57 19	01 09 17	01 ♍ 11 29	08 ♍ 17 12	21 23	24 11	05 14	10 54	09 22	26 18	23 21	27 23
22	02 01 15	02 07 48	15 ♍ 29 34	22 ♍ 48 09	23 22	25 24	06 01	10 47	09 28	26 19	23 21	27 22
23	02 05 12	03 06 16	00 ♎ 12 20	07 ♎ 41 19	25 22	26 38	06 47	10 39	09 33	26 19	23 21	27 21
24	02 09 08	04 04 43	15 ♎ 14 07	22 ♎ 49 37	27 24	27 51	07 33	10 32	09 38	26 20	23 21	27 19
25	02 13 05	05 03 07	00 ♏ 26 38	08 ♏ 03 53	29 ♈ 27	29 ♉ 04	08 20	10 24	09 43	26 20	23 R 21	27 18
26	02 17 02	06 01 30	15 ♏ 38 45	23 ♏ 14 10	01 ♉ 31	00 ♊ 17	09 06	10 17	09 48	26 20	23 21	27 16
27	02 20 58	06 59 51	00 ♐ 44 51	08 ♐ 11 15	03 37	01 30	09 52	10 09	09 53	26 20	23 21	27 15
28	02 24 55	07 58 10	15 ♐ 32 30	22 ♐ 47 59	05 43	02 43	10 38	10 02	09 58	26 20	23 21	27 13
29	02 28 51	08 56 28	29 ♐ 57 13	06 ♑ 59 55	07 51	03 56	11 24	09 54	10 03	26 21	23 21	27 11
30	02 32 48	09 ♉ 54 44	13 ♑ 55 56	20 ♑ 45 19	10 ♉ 00	05 ♊ 09	12 ♈ 11	09 ♏ 46	10 ♓ 08	26 ♑ 21	23 ♑ 21	27 ♏ 10

Date	Moon True ☊ ° '	Moon Mean ☊ ° '	Moon ☽ Latitude ° '
01	24 ♏ 38	26 ♏ 18	02 N 18
02	24 D 39	26 15	03 20
03	24 R 40	26 12	04 10
04	24 38	26 08	04 45
05	24 35	26 05	05 05
06	24 30	26 02	05 10
07	24 24	25 59	05 01
08	24 17	25 56	04 38
09	24 11	25 53	04 03
10	24 05	25 49	03 18
11	24 01	25 46	02 24
12	23 58	25 43	01 24
13	23 57	25 40	00 N 20
14	23 D 57	25 37	00 S 45
15	23 58	25 34	01 48
16	24 00	25 30	02 48
17	24 01	25 27	03 41
18	24 02	25 24	04 24
19	24 R 03	25 21	04 56
20	24 02	25 18	05 13
21	24 00	25 14	05 13
22	23 57	25 11	04 53
23	23 54	25 08	04 14
24	23 52	25 05	03 17
25	23 50	25 02	02 06
26	23 49	24 59	00 S 45
27	23 D 49	24 55	00 N 38
28	23 49	24 52	01 58
29	23 51	24 49	03 07
30	23 ♏ 52	24 ♏ 46	04 N 03

DECLINATIONS

Date	Sun ☉ ° '	Moon ☽ ° '	Mercury ☿ ° '	Venus ♀ ° '	Mars ♂ ° '	Jupiter ♃ ° '	Saturn ♄ ° '	Uranus ♅ ° '	Neptune ♆ ° '	Pluto ⯓ ° '
01	04 N 34	20 S 50	07 S 11	10 N 49	05 S 06	14 S 27	10 S 11	21 S 25	20 S 51	05 S 42
02	04 57	20 01	06 39	11 17	04 48	14 26	10 09	21 25	20 50	05 41
03	05 20	18 05	06 06	11 45	04 29	14 24	10 07	21 25	20 50	05 41
04	05 43	15 15	05 31	12 12	04 10	14 22	10 04	21 25	20 50	05 40
05	06 05	11 46	04 56	12 40	03 52	14 20	10 02	21 25	20 50	05 40
06	06 28	07 49	04 19	13 07	03 33	14 18	10 00	21 24	20 50	05 39
07	06 51	03 S 38	03 41	13 33	03 14	14 16	09 58	21 24	20 50	05 39
08	07 13	00 N 40	03 02	14 00	02 55	14 14	09 55	21 24	20 50	05 38
09	07 36	04 53	02 22	14 26	02 37	14 12	09 53	21 24	20 50	05 38
10	07 58	08 53	01 41	14 51	02 18	14 10	09 51	21 24	20 49	05 37
11	08 20	12 31	00 59	15 16	01 59	14 08	09 49	21 23	20 49	05 37
12	08 42	15 39	00 S 15	15 41	01 40	14 06	09 47	21 23	20 49	05 36
13	09 04	18 08	00 N 29	16 06	01 22	14 04	09 45	21 23	20 49	05 36
14	09 25	19 51	01 14	16 30	01 03	14 02	09 43	21 23	20 49	05 35
15	09 47	20 42	02 00	16 54	00 44	13 59	09 41	21 23	20 49	05 35
16	10 08	20 35	02 47	17 17	00 25	13 57	09 39	21 23	20 49	05 35
17	10 30	19 30	03 34	17 40	00 S 07	13 55	09 37	21 23	20 49	05 34
18	10 51	17 27	04 23	18 02	00 N 12	13 53	09 35	21 23	20 49	05 34
19	11 11	14 28	05 12	18 24	00 31	13 51	09 33	21 23	20 49	05 33
20	11 32	10 39	06 01	18 45	00 50	13 48	09 31	21 22	20 49	05 33
21	11 53	06 10	06 52	19 06	01 08	13 46	09 29	21 22	20 49	05 32
22	12 13	01 N 12	07 43	19 27	01 27	13 44	09 27	21 22	20 49	05 32
23	12 33	03 S 58	08 34	19 47	01 45	13 41	09 25	21 22	20 49	05 31
24	12 53	09 02	09 26	20 06	02 04	13 39	09 24	21 22	20 49	05 31
25	13 12	13 35	10 18	20 25	02 23	13 37	09 22	21 22	20 49	05 30
26	13 32	17 15	11 10	20 43	02 41	13 34	09 20	21 22	20 49	05 30
27	13 51	19 41	12 01	21 01	02 59	13 32	09 18	21 22	20 49	05 30
28	14 10	20 42	12 53	21 18	03 18	13 30	09 17	21 22	20 49	05 29
29	14 29	20 19	13 44	21 35	03 36	13 27	09 15	21 22	20 49	05 29
30	14 N 47	18 S 40	14 N 35	21 N 51	03 N 55	13 S 25	09 S 13	21 S 22	20 S 49	05 S 28

ZODIAC SIGN ENTRIES

Date	h m	Planets
01	19 20	♀ ♉
02	03 37	☽ ♑
04	09 45	☽ ♒
06	18 51	☽ ♓
09	06 09	☽ ♈
09	16 30	☿ ♈
11	18 48	☽ ♉
14	07 48	☽ ♊
14	18 02	♂ ♈
16	19 41	☽ ♋
19	04 45	☽ ♌
20	07 36	☉ ♉
21	09 58	☽ ♍
23	11 40	☽ ♎
25	11 18	☽ ♏
25	18 27	☿ ♉
26	06 24	♀ ♊
27	10 48	☽ ♐
29	12 05	☽ ♑

LATITUDES

Date	Mercury ☿ ° '	Venus ♀ ° '	Mars ♂ ° '	Jupiter ♃ ° '	Saturn ♄ ° '	Uranus ♅ ° '	Neptune ♆ ° '	Pluto ⯓ ° '
01	02 S 20	00 S 34	01 S 06	01 N 23	01 S 29	00 S 29	00 N 37	14 N 19
04	02 26	00 27	01 05	01 23	01 30	00 29	00 37	14 20
07	02 28	00 19	01 05	01 23	01 30	00 29	00 37	14 21
10	02 25	00 12	01 04	01 24	01 31	00 29	00 37	14 21
13	02 18	00 S 04	01 04	01 24	01 31	00 29	00 37	14 22
16	02 06	00 N 04	01 03	01 24	01 32	00 29	00 37	14 23
19	01 49	00 12	01 02	01 24	01 32	00 29	00 37	14 23
22	01 28	00 20	01 01	01 24	01 33	00 30	00 37	14 24
25	01 03	00 29	01 01	01 24	01 33	00 30	00 37	14 24
28	00 35	00 37	01 00	01 24	01 34	00 30	00 37	14 24
31	00 S 04	00 N 45	00 S 58	01 N 24	01 S 34	00 S 30	00 N 37	14 N 25

DATA

Julian Date	2449444
Delta T	+60 seconds
Ayanamsa	23° 46' 50"
Synetic vernal point	05° ♓ 20' 09"
True obliquity of ecliptic	23° 26' 20"

MOON'S PHASES, APSIDES AND POSITIONS ☽

Date	h m	Phase	Longitude ° '	Eclipse Indicator
03	02 55	☾	13 ♑ 08	
11	00 17	●	20 ♈ 53	
19	02 34	☽	28 ♋ 49	
25	19 45	○	05 ♏ 22	

Day	h m	
12	00 04	Apogee
25	17 23	Perigee

Day	h m		
01	08 22	Max dec	20° S 51'
08	08 18	0N	
15	21 27	Max dec	20° N 46'
22	17 38	0S	
28	17 11	Max dec	20° S 44'

LONGITUDES

Date	Chiron ⚷ ° '	Ceres ⚳ ° '	Pallas ⚴ ° '	Juno ⚵ ° '	Vesta ⚶ ° '	Black Moon Lilith ⚸ ° '
01	03 ♍ 40	19 ♉ 30	03 ♈ 44	27 ♎ 56	25 ♈ 39	29 ♈ 24
11	03 ♍ 13	23 ♉ 23	07 ♈ 22	25 ♎ 40	00 ♉ 30	00 ♉ 30
21	02 ♍ 57	27 ♉ 22	11 ♈ 01	23 ♎ 20	04 ♉ 40	01 ♉ 37
31	02 ♍ 51	01 ♊ 25	14 ♈ 40	21 ♎ 13	09 ♉ 09	02 ♉ 43

ASPECTARIAN

h m	Aspects
01 Friday	
00 06	☽ ♇ ♀
04 54	☽ □ ☿
05 30	☽ ⊥ ♆
05 41	☽ ∥ ♆
08 49	☽ ⊥ ♃
09 35	☽ □ ♂
10 16	☽ ⊥ ♅
11 07	☽ ∥ ♆
15 48	☽ ⊻ ♆
19 38	☽ Q ♄
20 38	☽ ⊻ ♅
23 47	☽ ⊻ ⯓
02 Saturday	
00 15	☽ ∠ ♃
04 27	☽ △ ♀
06 43	☉ # ♂
10 17	☽ ⊥ ⯓
15 55	☽ Q ☿
16 52	☽ ⚹ ♄
18 31	☉ ♇ ⯓
18 56	☽ Q ♂
23 18	☉ ⊼ ♃
03 Sunday	
02 17	☽ ∠ ⯓
02 37	☽ ⚹ ♃
02 55	☽ □ ☉
14 33	☉ ⊥ ♄
16 15	☽ ⚹ ☿
17 57	☽ ⚹ ♂
20 10	☽ ∠ ♄
21 15	☽ ☌ ♆
04 Monday	
00 17	☽ Q ♃
02 26	☽ ☌ ♅
05 35	☿ ∥ ⯓
05 36	☽ ⚹ ⯓
07 15	☉ # ☿
09 41	☉ # ⯓
12 56	☽ ⊥ ♄
14 35	☽ Q ☉
16 22	☽ □ ♀
18 32	☽ ∥ ♃
20 43	☿ ☌ ♂
23 33	☽ ∠ ♂
23 45	☽ ∠ ☿
05 Tuesday	
00 15	☽ ⊻ ♄
04 00	☽ Q ⯓
06 48	☽ # ♀
09 41	☽ □ ♃
11 28	☿ ⚹ ♆
15 07	☽ ⚹ ☉
17 45	☽ ⊥ ♂
19 12	☽ ⊥ ☿
22 53	☽ ∥ ♄
06 Wednesday	
02 04	♂ ⚹ ♆
05 47	☽ ⊻ ♆
06 01	☽ ⊻ ♂
06 14	☽ Q ♀
08 23	☽ ⊻ ☿
11 16	☽ ⊻ ♅
14 25	☽ □ ⯓
17 24	☽ ⊥ ♆
19 12	☽ # ☉
22 31	☽ ∠ ☉
22 57	☽ ⊥ ♅
07 Thursday	
00 32	☽ ∥ ⯓
07 11	☿ ⚹ ♅
08 10	☽ ⚹ ♀
10 30	☽ ☌ ♄
11 00	☽ ∠ ♆
11 38	☽ ∥ ☿
14 22	☽ ∥ ♂
16 37	☽ ∠ ♅
17 42	☽ ⊥ ☉
19 19	☽ △ ♃
08 Friday	
03 18	☿ ♇ ♃
06 35	☿ △ ⯓
06 37	☽ ⊻ ☉
09 15	♀ ⚹ ♄
16 42	☽ ⚹ ♆
17 10	☽ ∠ ♀
19 21	☿ ∥ ♂
20 59	☽ ☌ ♂
22 25	☽ ⚹ ♅
23 34	☽ # ☿
23 54	☽ # ♂
09 Saturday	
00 32	♀ # ♃
00 52	☽ ♇ ♃
01 28	☽ △ ⯓
04 29	☽ ☌ ☿
13 15	☽ ⊥ ♀
16 24	☽ # ⯓
16 45	☽ Q ♆
18 46	☽ ± ♃
19 35	♂ ⚹ ♅
22 33	☽ Q ♅
22 45	☽ ⊻ ♄
10 Sunday	
02 43	☽ ⊻ ♀
04 12	♄ ∠ ♆
05 47	☽ ∥ ☉
06 45	☽ ⊼ ♃
07 35	☽ ♇ ⯓
10 58	☽ ⊥ ♄
18 07	☽ # ♄
11 Monday	
00 17	☽ ☌ ☉
01 47	☽ ± ⯓
01 50	♂ ♇ ♃
05 10	☽ □ ♆
05 22	☽ ∠ ♄
11 02	☽ □ ♅
13 46	☽ ⊻ ♂
13 56	☽ ⊼ ⯓
14 52	♀ ☍ ♃
16 23	♂ △ ⯓
17 55	☉ ± ⯓
23 44	☽ # ♃
12 Tuesday	
02 50	☽ ⊥ ♂
03 12	☽ ⊻ ☿
12 08	☽ ⚹ ♄
12 26	☽ ∥ ♀
16 44	♀ ∠ ♂
17 31	☽ ⊥ ☿
19 07	☽ ☍ ♃
19 28	☿ Q ♆
22 27	☽ ∠ ♂
22 42	☽ ☌ ♀
13 Wednesday	
03 44	☿ ± ♃
11 30	☉ □ ♆
12 46	☽ Q ♄
15 01	☽ ∠ ☿
18 14	☽ △ ♆
18 50	☽ ⊻ ☉
19 46	☉ ∠ ♄
14 Thursday	
00 09	☽ △ ♅
02 51	☽ ☍ ⯓
07 05	☽ ⚹ ♂
07 55	☿ # ♂
08 06	☽ ⊥ ☉
11 22	☿ Q ♅
18 12	☿ ⊻ ♄
15 Friday	
00 35	☽ ♇ ♆
01 30	☽ □ ♄
02 43	☽ ⚹ ☿
03 53	☽ ∠ ☉
05 40	☉ # ♄
06 28	☽ ♇ ♅
07 18	☽ ⊼ ♃
08 57	☽ Q ♂
18 18	☽ ⊻ ♀
18 38	☽ ± ♆
19 10	☽ ± ♃
16 Saturday	
00 27	☽ ± ♅
06 33	☽ ⊼ ♆
06 54	☽ Q ☿
07 21	☿ ⊼ ♃
07 34	☽ ⊥ ♀
11 41	☉ □ ♅
12 20	☽ ⊼ ♅
12 23	☽ ⚹ ☉
12 50	☽ ♇ ♃
14 47	☽ ⊼ ⯓
18 47	☿ ⊥ ♀
19 56	☿ ♇ ⯓
23 02	☽ □ ♂
17 Sunday	
02 30	☽ ± ⯓
03 11	☽ ∠ ♀
13 11	☽ △ ♄
13 47	☽ Q ☉
17 42	☉ ⊼ ⯓
17 43	☽ △ ♃
19 51	☽ ♇ ⯓
18 Monday	
00 03	☽ □ ☿
00 11	♃ Q ♆
05 03	☿ ⊥ ♄
06 56	☽ ∥ ♀
11 00	☽ ⚹ ♀
16 23	☽ ☍ ♆
17 51	☽ ♇ ♄
21 53	☽ ☍ ♅
19 Tuesday	
00 02	☽ △ ⯓
00 18	♀ Q ♄
02 34	☽ □ ☉
10 35	☽ ± ♄
11 28	☽ △ ♂
11 33	☽ Q ♀
16 19	☽ # ♃
21 32	☽ ⊼ ♄
22 17	☿ # ⯓
20 Wednesday	
00 53	☽ □ ♃
07 16	☽ ∥ ☉
15 56	☽ ♇ ♂
16 17	☽ △ ☿
18 28	☽ # ♄
19 30	♀ △ ♆
22 30	☽ ⊼ ♆
22 47	☽ □ ♀
21 Thursday	
03 15	♂ ± ♃
03 37	☽ ⊼ ♅
05 30	☽ □ ⯓
08 08	☽ Q ♃
08 28	☽ ± ♂
08 51	☽ ± ♆
09 02	☽ ∥ ☿
11 56	☽ △ ☉
12 02	☿ ± ⯓
13 54	☽ ± ♅
15 11	☽ # ⯓
15 27	♂ Q ♆
19 15	☽ ⊼ ♂
22 13	☽ ♇ ☿
22 Friday	
00 06	☽ ♇ ♆
01 54	☽ ☍ ♄
04 15	☽ ⚹ ♃
05 04	☽ ♇ ♅
10 56	☽ ∥ ♂
11 47	☿ □ ♆
11 48	☽ Q ⯓
14 54	☽ ♇ ☉
15 35	☽ ± ☿
23 Saturday	
00 53	☽ △ ♆
01 07	☽ # ♂
01 45	☿ ∠ ♄
02 57	☽ ⊼ ☿
04 42	☽ ∠ ♃
05 42	☽ △ ♀
05 43	☽ △ ♅
05 58	♀ △ ♅
06 38	☽ ± ☉
07 23	☽ ⚹ ⯓
16 59	☽ ⊼ ☉
19 06	☽ ⊥ ♃
19 12	☽ ∥ ⯓
23 07	☽ ☍ ♂
23 21	☿ □ ♅
24 Sunday	
01 49	♀ ☍ ⯓
03 03	☽ ⊼ ♄
04 36	☽ ⊻ ♃
07 23	☽ ∠ ⯓
07 53	☽ ♇ ♀
11 03	☿ # ♄
11 07	☿ ⊼ ⯓
12 38	☽ ± ♄
13 46	☽ ∥ ♄
14 23	☽ # ☿
21 36	☽ ⊥ ⯓
23 22	☽ ± ♀
25 Monday	
00 49	☽ □ ♆
01 08	☿ ⊻ ♀
02 56	☽ ♇ ♄
05 31	☽ □ ♅
07 03	☽ ⊻ ⯓
09 38	☽ # ☉
09 39	☽ ⊼ ♀
10 11	☽ ☍ ☿
10 38	♆ St R
12 08	☽ ∥ ♃
12 12	♂ Q ♅
19 45	☽ ☍ ☉
26 Tuesday	
01 04	☽ ⊼ ♂
02 42	☽ △ ♄
03 34	☽ ☌ ♃
05 11	☽ Q ♆
09 54	☽ Q ♅
11 03	☽ ± ♂
14 55	☉ # ♃
19 18	♀ # ♆
27 Wednesday	
00 11	☽ ⚹ ♆
02 05	☽ ♇ ♂
04 57	☽ ⚹ ♅
06 24	☽ ☌ ⯓
12 49	♂ ⊻ ♄
13 19	☽ ☍ ♀
17 22	☽ ⊼ ☿
19 38	♂ ⊼ ♃
22 47	☽ ⊼ ☉
28 Thursday	
00 16	☽ ∠ ♆
02 51	☽ □ ♄
03 04	☽ ⊻ ♃
03 32	☽ △ ♂
04 42	☽ ± ☿
05 08	☽ ∠ ♅
09 14	☽ ± ☉
12 47	☽ ⊥ ♃
14 58	☽ ⊥ ♆
17 50	♀ # ♅
17 56	♃ △ ♄
19 55	☽ ⊥ ♅
22 01	☽ ♇ ☿
29 Friday	
00 55	☽ ⊻ ♆
01 10	☽ ♇ ☉
03 35	☽ ∠ ♃
04 19	☿ # ♃
05 55	☽ ⊻ ♅
07 21	☽ ⊻ ⯓
08 47	☽ Q ♄
17 29	☽ ⊥ ⯓
19 24	☽ ⊼ ♀
30 Saturday	
03 55	☽ △ ☿
04 30	☽ △ ☉
04 50	☽ ⚹ ♃
05 22	☽ ⚹ ♄
06 42	☽ ± ♀
08 46	☽ □ ♂
08 55	☉ ☍ ♃
08 56	☽ ∠ ⯓
09 39	☿ ☍ ♃
10 21	☉ ☌ ☿
11 41	♂ ♇ ⯓
13 40	☿ ⚹ ♄
18 03	☉ ⚹ ♄
20 55	☉ ∥ ☿
22 18	♅ St R

All ephemeris data is given at 12.00 UT and the Moon's longitude is additionally given for 24.00 UT

MAY 1994

LONGITUDES

Date	Sidereal time h m s	Sun ☉ ° ' "	Moon ☽ ° ' "	Moon ☽ 24.00 ° ' "	Mercury ☿ ° '	Venus ♀ ° '	Mars ♂ ° '	Jupiter ♃ ° '	Saturn ♄ ° '	Uranus ♅ ° '	Neptune ♆ ° '	Pluto ♇ ° '
01	02 36 44	10 ♉ 52 59	27 ♑ 28 10	04 ♒ 04 45	12 ♉ 09	06 ♊ 22	12 ♈ 57	09 ♏ 39	10 ♓ 13	26 ♑ 21	23 ♑ 20	27 ♏ 08
02	02 40 41	11 51 12	10 ♒ 35 23	17 ♒ 00 29	14 18	07 35	13 43	09 R 31	10 18	26 R 21	23 R 20	27 R 07
03	02 44 37	12 49 24	23 ♒ 20 27	29 ♒ 35 47	16 28	08 48	14 29	09 23	10 22	26 20	23 20	27 05
04	02 48 34	13 47 34	05 ♓ 46 56	11 ♓ 54 25	18 37	10 01	15 15	09 16	10 27	26 20	23 20	27 04
05	02 52 31	14 45 42	17 ♓ 58 41	24 ♓ 00 12	20 47	11 13	16 01	09 08	10 31	26 20	23 19	27 02
06	02 56 27	15 43 50	29 ♓ 59 26	05 ♈ 56 48	22 55	12 26	16 47	09 00	10 36	26 20	23 19	27 00
07	03 00 24	16 41 55	11 ♈ 52 42	17 ♈ 47 32	25 02	13 39	17 32	08 53	10 40	26 20	23 19	26 59
08	03 04 20	17 40 00	23 ♈ 41 38	29 ♈ 35 22	27 09	14 51	18 18	08 45	10 44	26 19	23 18	26 57
09	03 08 17	18 38 02	05 ♉ 29 01	11 ♉ 22 53	29 ♉ 13	16 04	19 04	08 38	10 48	26 19	23 18	26 55
10	03 12 13	19 36 04	17 ♉ 17 17	23 ♉ 12 28	01 ♊ 16	17 17	19 50	08 30	10 53	26 18	23 17	26 54
11	03 16 10	20 34 03	29 ♉ 08 44	05 ♊ 06 19	03 16	18 29	20 36	08 23	10 57	26 18	23 17	26 52
12	03 20 06	21 32 01	11 ♊ 05 31	17 ♊ 06 37	05 15	19 42	21 21	08 16	11 01	26 17	23 16	26 50
13	03 24 03	22 29 58	23 ♊ 09 53	29 ♊ 15 39	07 10	20 54	22 07	08 09	11 04	26 17	23 16	26 49
14	03 28 00	23 27 53	05 ♋ 24 12	11 ♋ 35 53	09 04	22 06	22 52	08 01	11 08	26 16	23 15	26 47
15	03 31 56	24 25 46	17 ♋ 51 00	24 ♋ 09 56	10 54	23 19	23 38	07 54	11 12	26 15	23 15	26 46
16	03 35 53	25 23 38	00 ♌ 33 00	07 ♌ 00 33	12 41	24 31	24 23	07 47	11 15	26 15	23 14	26 44
17	03 39 49	26 21 27	13 ♌ 32 54	20 ♌ 10 20	14 25	25 43	25 09	07 40	11 19	26 14	23 13	26 42
18	03 43 46	27 19 15	26 ♌ 53 08	03 ♍ 41 30	16 07	26 55	25 54	07 33	11 22	26 13	23 13	26 41
19	03 47 42	28 17 02	10 ♍ 35 32	17 ♍ 35 17	17 44	28 08	26 39	07 27	11 26	26 12	23 12	26 39
20	03 51 39	29 ♉ 14 46	24 ♍ 40 41	01 ♎ 51 31	19 19	29 ♊ 20	27 25	07 20	11 29	26 11	23 11	26 37
21	03 55 35	00 ♊ 12 29	09 ♎ 07 29	16 ♎ 28 06	20 50	00 ♋ 32	28 10	07 13	11 32	26 10	23 11	26 36
22	03 59 32	01 10 10	23 ♎ 52 44	01 ♏ 20 38	22 18	01 44	28 55	07 07	11 35	26 09	23 09	26 34
23	04 03 29	02 07 50	08 ♏ 50 55	16 ♏ 22 36	23 43	02 56	29 ♈ 40	07 00	11 38	26 08	23 09	26 32
24	04 07 25	03 05 29	23 ♏ 54 38	01 ♐ 25 54	25 04	04 07	00 ♉ 25	06 54	11 41	26 07	23 08	26 31
25	04 11 22	04 03 06	08 ♐ 55 19	16 ♐ 21 51	26 21	05 19	01 10	06 48	11 44	26 06	23 07	26 29
26	04 15 18	05 00 42	23 ♐ 44 31	01 ♑ 02 29	27 35	06 31	01 55	06 42	11 47	26 05	23 06	26 27
27	04 19 15	05 58 17	08 ♑ 15 00	15 ♑ 21 31	28 46	07 43	02 40	06 36	11 49	26 04	23 05	26 26
28	04 23 11	06 55 51	22 ♑ 21 39	29 ♑ 15 08	29 ♊ 52	08 54	03 25	06 30	11 52	26 02	23 04	26 24
29	04 27 08	07 53 24	06 ♒ 01 55	12 ♒ 42 02	00 ♋ 55	10 06	04 10	06 25	11 54	26 01	23 03	26 22
30	04 31 04	08 50 56	19 ♒ 15 41	25 ♒ 43 11	01 54	11 17	04 54	06 19	11 57	26 00	23 02	26 21
31	04 35 01	09 ♊ 48 28	02 ♓ 04 54	08 ♓ 21 17	02 ♋ 50	12 ♋ 29	05 ♉ 39	06 ♏ 14	11 ♓ 59	25 ♑ 58	23 ♑ 01	26 ♏ 19

MOON NODES AND LATITUDE; DECLINATIONS

Date	Moon True ☊ ° '	Moon Mean ☊ ° '	Moon ☽ Latitude ° '	Sun ☉ ° '	Moon ☽ ° '	Mercury ☿ ° '	Venus ♀ ° '	Mars ♂ ° '	Jupiter ♃ ° '	Saturn ♄ ° '	Uranus ♅ ° '	Neptune ♆ ° '	Pluto ♇ ° '
01	23 ♏ 52	24 ♏ 43	04 N 44	15 N 05	16 S 01	15 N 25	22 N 06	04 N 13	13 S 23	09 S 12	21 S 22	20 S 49	05 S 28
02	23 D 53	24 40	05 09	15 23	12 37	16 14	22 21	04 31	13 20	09 10	21 22	20 49	05 27
03	23 R 53	24 36	05 17	15 41	08 44	17 02	22 35	04 49	13 18	09 08	21 22	20 49	05 27
04	23 52	24 33	05 10	15 59	04 35	17 49	22 49	05 07	13 16	09 07	21 22	20 49	05 27
05	23 51	24 30	04 50	16 16	00 S 18	18 34	23 02	05 25	13 13	09 05	21 23	20 49	05 26
06	23 50	24 27	04 16	16 33	03 N 55	19 17	23 14	05 43	13 11	09 04	21 23	20 49	05 26
07	23 49	24 24	03 32	16 50	07 57	19 59	23 26	06 01	13 09	09 02	21 23	20 49	05 26
08	23 48	24 20	02 40	17 06	11 40	20 38	23 37	06 19	13 06	09 01	21 23	20 49	05 25
09	23 47	24 17	01 40	17 22	14 55	21 15	23 47	06 37	13 04	09 00	21 23	20 49	05 25
10	23 47	24 14	00 N 36	17 38	17 34	21 50	23 57	06 54	13 02	08 58	21 23	20 49	05 24
11	23 D 47	24 11	00 S 30	17 54	19 29	22 22	24 06	07 12	13 00	08 57	21 23	20 49	05 24
12	23 47	24 08	01 35	18 09	20 32	22 52	24 14	07 29	12 57	08 56	21 23	20 49	05 24
13	23 47	24 05	02 36	18 24	20 40	23 19	24 22	07 47	12 55	08 54	21 23	20 49	05 23
14	23 47	24 01	03 31	18 38	19 48	23 43	24 29	08 04	12 53	08 53	21 24	20 49	05 23
15	23 R 47	23 58	04 18	18 53	17 59	24 05	24 35	08 21	12 51	08 52	21 24	20 50	05 23
16	23 47	23 55	04 52	19 07	15 16	24 25	24 41	08 38	12 49	08 51	21 24	20 50	05 23
17	23 47	23 52	05 13	19 20	11 45	24 42	24 45	08 55	12 47	08 50	21 24	20 50	05 22
18	23 47	23 49	05 17	19 34	07 34	24 56	24 49	09 12	12 44	08 48	21 24	20 50	05 22
19	23 D 47	23 46	05 04	19 47	02 N 54	25 09	24 53	09 29	12 42	08 47	21 24	20 50	05 22
20	23 47	23 42	04 33	19 59	02 S 04	25 18	24 56	09 46	12 40	08 46	21 25	20 50	05 21
21	23 48	23 39	03 44	20 12	07 03	25 26	24 58	10 02	12 38	08 45	21 25	20 50	05 21
22	23 49	23 36	02 39	20 24	11 44	25 32	24 59	10 19	12 37	08 44	21 25	20 50	05 21
23	23 49	23 33	01 S 22	20 35	15 45	25 35	24 59	10 35	12 35	08 43	21 25	20 50	05 21
24	23 49	23 30	00 00	20 46	18 44	25 37	24 59	10 52	12 33	08 43	21 26	20 51	05 20
25	23 R 49	23 26	01 N 23	20 57	20 25	25 37	24 58	11 08	12 31	08 42	21 26	20 51	05 20
26	23 48	23 23	02 39	21 08	20 39	25 35	24 57	11 24	12 29	08 41	21 26	20 51	05 20
27	23 47	23 20	03 42	21 18	19 29	25 32	24 54	11 40	12 27	08 40	21 26	20 51	05 20
28	23 45	23 17	04 31	21 28	17 08	25 27	24 51	11 55	12 26	08 39	21 27	20 51	05 20
29	23 44	23 14	05 02	21 37	13 53	25 21	24 48	12 11	12 24	08 39	21 27	20 51	05 19
30	23 42	23 11	05 16	21 47	10 02	25 13	24 43	12 27	12 22	08 38	21 27	20 52	05 19
31	23 ♏ 42	23 ♏ 07	05 N 13	21 N 55	05 S 51	25 N 05	24 N 38	12 N 42	12 S 21	08 S 37	21 S 27	20 S 52	05 S 19

ZODIAC SIGN ENTRIES

Date	h	m	Planets
01	16	34	☽ ♒
04	00	47	☽ ♓
06	12	01	☽ ♈
09	00	50	☽ ♉
09	21	08	☿ ♊
11	13	43	☽ ♊
14	01	27	☽ ♋
16	10	58	☽ ♌
18	17	31	☽ ♍
20	20	54	☽ ♎
21	01	26	♀ ♋
21	06	48	☉ ♊
22	21	51	☽ ♏
23	22	37	♂ ♉
24	21	43	☽ ♐
26	22	17	☽ ♑
28	14	52	☿ ♋
29	01	19	☽ ♒
31	08	03	☽ ♓

LATITUDES

Date	Mercury ☿ ° '	Venus ♀ ° '	Mars ♂ ° '	Jupiter ♃ ° '	Saturn ♄ ° '	Uranus ♅ ° '	Neptune ♆ ° '	Pluto ♇ ° '
01	00 S 04	00 N 45	00 S 58	01 N 24	01 S 34	00 S 30	00 N 37	14 N 25
04	00 N 28	00 52	00 57	01 23	01 35	00 30	00 37	14 25
07	00 59	01 00	00 56	01 23	01 36	00 30	00 37	14 25
10	01 27	01 07	00 55	01 23	01 36	00 30	00 37	14 25
13	01 50	01 14	00 54	01 23	01 37	00 30	00 37	14 25
16	02 07	01 21	00 52	01 22	01 37	00 30	00 37	14 25
19	02 17	01 27	00 51	01 22	01 38	00 30	00 37	14 25
22	02 19	01 33	00 50	01 21	01 39	00 30	00 37	14 24
25	02 14	01 39	00 48	01 21	01 39	00 31	00 37	14 24
28	02 01	01 43	00 47	01 20	01 40	00 31	00 37	14 24
31	01 N 40	01 N 48	00 S 45	01 N 19	01 S 41	00 S 31	00 N 37	14 N 23

DATA

Julian Date	2449474
Delta T	+60 seconds
Ayanamsa	23° 46' 54"
Synetic vernal point	05° ♓ 20' 06"
True obliquity of ecliptic	23° 26' 19"

MOON'S PHASES, APSIDES AND POSITIONS ☽

Date	h	m	Phase	Longitude ° '	Eclipse Indicator
02	14	32	☾	11 ♒ 57	
10	17	07	●	19 ♉ 48	Annular
18	12	50	☽	27 ♌ 21	
25	03	39	○	03 ♐ 43	partial

Day	h	m		
09	02	34	Apogee	
24	03	01	Perigee	
05	13	44	ON	
13	03	02	Max dec	20° N 44'
20	02	07	OS	
26	03	45	Max dec	20° S 44'

LONGITUDES

Date	Chiron ⚷ ° '	Ceres ⚳ ° '	Pallas ⚴ ° '	Juno ⚵ ° '	Vesta ⚶ ° '	Black Moon Lilith ⚸ ° '
01	02 ♍ 51	01 ♊ 25	14 ♈ 40	21 ♎ 13	09 ♉ 54	02 ♉ 43
11	02 ♍ 56	05 ♊ 32	18 ♈ 18	19 ♎ 24	13 ♉ 37	03 ♉ 50
21	03 ♍ 13	09 ♊ 42	21 ♈ 55	18 ♎ 04	18 ♉ 03	04 ♉ 56
31	03 ♍ 40	13 ♊ 53	25 ♈ 29	17 ♎ 17	22 ♉ 25	06 ♉ 03

All ephemeris data is given at 12.00 UT and the Moon's longitude is additionally given for 24.00 UT

ASPECTARIAN

h m Aspects

01 Sunday
00 01 ☽ ⚼ ♀
01 40 ☽ Q ♃
04 36 ☽ ☌ ♆
07 56 ☽ ∠ ♄
09 58 ☽ ☌ ♅
11 24 ☽ ⚹ ♇
15 37 ☽ # ☿
18 26 ☽ # ☉
18 40 ☽ Q ♂

02 Monday
00 19 ☽ ⊥ ♄
01 48 ☿ ⊻ ♂
05 52 ☽ △ ♀
07 14 ☽ ∥ ♃
09 16 ☽ Q ♇
10 02 ☽ □ ♃
11 27 ☽ ⊻ ♄
14 32 ☽ □ ☉
18 11 ☽ ⚹ ♂
20 20 ☽ □ ☿

03 Tuesday
02 51 ♀ ⚼ ♆
09 37 ☽ ∥ ♄
11 59 ☽ ⊻ ♆
17 44 ☽ ⊻ ♅
19 09 ☽ □ ♇
22 35 ♀ ⊼ ♃
23 29 ☽ ⊥ ♆

04 Wednesday
00 33 ☽ ∠ ♂
03 35 ☽ Q ☉
05 18 ☽ ⊥ ♅
07 04 ☽ ∥ ♇
09 07 ☽ # ♂
13 59 ☽ Q ☿
16 58 ☽ ∠ ♆
18 44 ☽ △ ♃
19 13 ☽ ⊥ ♂
21 11 ☽ □ ♀
21 11 ☽ ☌ ♄
21 13 ♀ □ ♄
22 53 ☽ ∠ ♅

05 Thursday
05 05 ☽ ⚹ ☉
07 50 ☽ ⊻ ♂
13 21 ♂ # ♇
14 12 ♀ ⚼ ♅
18 46 ☽ ⚹ ☿
22 38 ☽ ⚹ ♆

06 Friday
00 08 ☽ ⚼ ♃
04 40 ☽ ⚹ ♅
05 43 ♂ ⊥ ♄
06 01 ☽ △ ♇
08 16 ☿ Q ♄
09 33 ☿ ⊥ ♂
13 00 ☽ Q ♀
13 37 ☽ ∠ ☉
16 30 ☿ △ ♆
18 00 ☽ ± ♃
20 50 ☽ # ♇
22 43 ☽ Q ♆
23 23 ☽ ∥ ♂

07 Saturday
04 49 ☽ Q ♅
06 00 ☽ ⊼ ♃
07 28 ☽ ∠ ☿
09 24 ☽ ⊥ ☉
09 32 ☽ ⊻ ♄
12 12 ☽ ⚼ ♇
16 00 ☽ ⚹ ♀
18 44 ☽ # ♄
21 46 ☽ ⊥ ♄
22 39 ☽ ⊻ ☉

08 Sunday
00 17 ☽ ☌ ♂
02 37 ☿ △ ♅
05 41 ☽ ⊥ ☿
06 26 ☽ ± ♇
09 51 ☿ ☍ ♇
10 11 ♀ ± ♃
11 12 ☽ □ ♆
16 11 ☽ ∠ ♄
17 20 ☽ □ ♅
18 37 ☽ ⊼ ♇
18 57 ☿ # ♆
20 31 ☽ ⊻ ☿
22 01 ☽ # ♃

09 Monday
01 59 ☽ ∠ ♀
17 17 ☿ # ♅
18 20 ☽ ☍ ♃
22 24 ☽ ⊥ ♀
22 54 ☽ ⚹ ♄

10 Tuesday
11 59 ☽ ⊻ ♀
12 14 ♀ ± ♆
12 45 ☽ ∥ ☉
17 07 ☽ ☌ ☉
17 31 ☽ ⊻ ♂
23 24 ☽ Q ♄

11 Wednesday
00 09 ☽ △ ♆
06 15 ☽ △ ♅
06 29 ☽ ⊥ ♂
07 25 ☽ ☍ ♇
14 49 ☉ ⊻ ♂
20 27 ♂ ± ♇
21 58 ☽ ☌ ☿

12 Thursday
01 52 ☽ ∠ ♂
06 22 ☽ ⚼ ♆
06 24 ☽ ⊼ ♃
11 50 ☽ □ ♄
12 24 ☽ ⚼ ♅
18 16 ☽ ± ♃
23 43 ♀ ± ♅

13 Friday
00 19 ☽ ± ♆
06 18 ☽ ± ♅
07 02 ☽ ☌ ♀
09 47 ☽ ⚹ ♂
10 34 ☽ ⊻ ☉
10 42 ☿ ∠ ♂
11 57 ☽ ⚼ ♃
12 12 ☽ ⊼ ♆
18 08 ☽ ⊼ ♅
19 11 ☽ ⊼ ♇
23 24 ☽ ⊥ ☉
23 30 ☿ ⊼ ♃

14 Saturday
01 43 ☿ ⚼ ♆
03 15 ☉ Q ♄
06 47 ☉ △ ♆
06 55 ☽ ± ♇
10 54 ☽ Q ♂
17 02 ☽ △ ♃
18 27 ☽ ∠ ☉
20 21 ☽ ⊻ ☿
23 10 ☽ △ ♄
23 54 ♂ □ ♆

15 Sunday
00 20 ☽ ⚼ ♇
02 56 ☽ ∥ ☉
04 35 ♀ ⚼ ♃
09 52 ☽ ⊥ ☿
10 37 ♀ ⊼ ♆
16 06 ☿ □ ♄
16 43 ☿ ⚼ ♅
22 15 ☽ ☍ ♆
23 29 ☽ ⊻ ♀
23 41 ☽ □ ♂

16 Monday
01 32 ☽ ⚹ ☉
03 54 ☽ ⚼ ♄
03 56 ☽ ☍ ♅
04 51 ☽ △ ♇
05 03 ♀ ⚹ ♂
05 46 ☽ ∠ ☿
07 36 ♄ ∠ ♅
11 56 ☽ ⊥ ♀
20 48 ☽ ± ♄

17 Tuesday
01 19 ☽ □ ♃
01 44 ☽ Q ☉
02 09 ☿ ± ♃
04 15 ♂ # ♄
05 28 ☽ # ♃
06 18 ☽ ∠ ♀
07 54 ☽ ⊼ ♄
08 53 ☉ △ ♅
13 50 ☽ ⚹ ☿
19 46 ☿ ∥ ♀
20 22 ☉ ☍ ♇
22 03 ♀ ⊼ ♅

18 Wednesday
03 35 ☽ ∥ ♂
05 13 ☽ # ♄
05 27 ☽ ⊼ ♆
07 09 ♀ ⊼ ♇
09 39 ☽ Q ♃
10 09 ☽ △ ♂
10 49 ☽ ⊼ ♅
11 38 ☽ □ ♇
12 05 ☽ ⚹ ♀
12 50 ☽ □ ☉
14 28 ☽ Q ☿
16 07 ☽ ± ♆
21 24 ☽ ± ♅
21 51 ♂ □ ♅
23 36 ☽ # ♇

19 Thursday
03 58 ☿ ± ♆
04 11 ♂ ∠ ♄
06 35 ☽ ⚹ ♃
07 51 ☽ ⚼ ♆
11 07 ☽ Q ♀
11 45 ♂ ⊼ ♇
13 03 ☽ ⚼ ♅
13 27 ☽ ☍ ♄
13 56 ☽ ⚼ ♂
18 58 ☽ Q ♇

20 Friday
01 49 ☽ □ ☿
03 47 ☉ ⊻ ♀
06 11 ☽ ± ♂
08 05 ☽ ∠ ♃
09 29 ☽ △ ♆
14 32 ☽ △ ♅
15 15 ☽ ⚹ ♇
16 50 ☽ ⊼ ♂
20 12 ☽ △ ☉
20 30 ☽ □ ♀
23 02 ☽ ⊥ ♃

21 Saturday
01 27 ☿ ± ♅
03 47 ☽ ∥ ♇
08 54 ☽ ⊻ ♃
15 58 ☽ ⊼ ♄
16 02 ☽ ∠ ♇
20 29 ☽ ∥ ♄
22 39 ☽ ⚼ ☉

22 Sunday
01 47 ☽ ± ♄
04 01 ☽ # ♂
06 39 ☽ ⊥ ♇
09 01 ☿ ⚼ ♃
09 11 ☽ △ ☿
10 50 ☽ □ ♆
14 13 ☽ ± ☉
15 40 ☽ □ ♅
16 19 ☽ ⊻ ♇
16 23 ☽ ⚼ ♄
16 50 ☽ ∥ ♃
20 32 ☽ ☍ ♂

23 Monday
00 31 ☽ ⊼ ☉
01 43 ☽ △ ♀
02 16 ☿ ⊼ ♆
04 23 ♀ ± ♇
09 05 ☽ ☌ ♃
11 46 ☽ ⚼ ☿
15 39 ☽ Q ♆
16 28 ☽ △ ♄
20 25 ☽ Q ♅

24 Tuesday
03 32 ☽ ± ☿
03 43 ☽ ⚼ ♀
10 45 ☽ ⚹ ♆
14 01 ☽ ⊼ ☿
15 31 ☽ ⚹ ♅
16 08 ☽ ☌ ♇
19 18 ☽ ± ♀
21 08 ☉ # ♆
22 56 ☽ ⊼ ♂

25 Wednesday
03 39 ☽ ☍ ☉
05 43 ☽ ⊼ ♀
07 13 ☿ ⊼ ♅
08 37 ☽ ⊻ ♃
09 02 ☽ ± ♂
10 42 ☽ ∠ ♆
14 21 ☿ ⊼ ♇
15 30 ☽ ∠ ♅
16 32 ☽ □ ♄
18 12 ☽ ⊥ ♃

26 Thursday
00 18 ☽ ⚼ ♂
01 12 ☽ ⊥ ♆
06 02 ☽ ⊥ ♅
08 41 ☽ ∠ ♃
10 57 ☽ ⊻ ♆
15 24 ♀ △ ♃
15 49 ☽ ⊻ ♅
16 26 ☽ ⊻ ♇
18 52 ☽ ☍ ☿
21 57 ☽ Q ♄

27 Friday
02 11 ☽ △ ♂
02 19 ☽ ⊥ ♇
07 55 ☽ ⊼ ☉
09 16 ☽ ⚹ ♃
11 01 ☽ ☍ ♀
17 20 ☽ ∠ ♇
18 01 ☽ ⚹ ♄
18 43 ☽ ± ☉

28 Saturday
02 18 ☉ ⊼ ♃
05 25 ☽ Q ♃
08 25 ☉ # ♅
11 12 ☽ ⚼ ☉
13 13 ☽ ☌ ♆
18 22 ☽ ☌ ♅
19 00 ☽ ⚹ ♇
19 51 ☽ ∠ ♄

29 Sunday
02 12 ☽ ⊼ ☿
08 28 ☽ □ ♂
11 46 ☽ ⊥ ♄
12 40 ☽ □ ♃
13 43 ☽ ± ☿
15 35 ☽ △ ☉
15 56 ☉ ⚼ ♆
16 11 ☽ Q ♇
20 01 ☽ ⊼ ♀
21 41 ☽ ∥ ♃
22 17 ☽ # ♂
22 35 ☽ ⊻ ♄

30 Monday
06 13 ♂ # ♃
07 20 ☽ ⚼ ☿
08 01 ☽ ± ♀
13 06 ♀ ⚼ ♇
18 59 ☽ ⊻ ♆
19 10 ☽ Q ♂
20 15 ☽ ∥ ♄
22 56 ☿ ± ♇

31 Tuesday
00 29 ☽ ⊻ ♅
01 09 ☽ □ ♇
01 37 ♀ △ ♄
02 25 ☽ ⚼ ♀
06 12 ☽ ⊥ ♆
11 47 ☽ ⊥ ♅
13 32 ☽ △ ☿
14 59 ☽ ∥ ♇
19 14 ☽ ⚹ ♂
19 51 ☽ △ ♃
23 20 ☽ ∠ ♆

JUNE 1994

LONGITUDES

Date	Sidereal time h m s	Sun ☉ ° ′ ″	Moon ☽ ° ′ ″	Moon ☽ 24.00 ° ′ ″	Mercury ☿ ° ′	Venus ♀ ° ′	Mars ♂ ° ′	Jupiter ♃ ° ′	Saturn ♄ ° ′	Uranus ♅ ° ′	Neptune ♆ ° ′	Pluto ♇ ° ′
01	04 38 58	10 ♊ 45 58	14 ♓ 32 52	20 ♓ 40 11	03 ♋ 41	13 ♋ 40	06 ♉ 24	06 ♏ 08	12 ♓ 01	25 ♑ 57	23 ♑ 00	26 ♏ 18
02	04 42 54	11 43 28	26 ♓ 43 50	02 ♈ 44 22	04 28	14 52	07 08	06 R 03	12 03	25 R 55	22 R 59	26 R 16
03	04 46 51	12 40 57	08 ♈ 42 23	14 ♈ 38 27	05 12	16 03	07 53	05 58	12 05	25 54	22 58	26 14
04	04 50 47	13 38 25	20 ♈ 33 07	26 ♈ 26 57	05 51	17 14	08 37	05 53	12 07	25 52	22 57	26 13
05	04 54 44	14 35 52	02 ♉ 20 25	08 ♉ 14 01	06 25	18 25	09 22	05 49	12 09	25 51	22 55	26 11
06	04 58 40	15 33 19	14 ♉ 08 12	20 ♉ 03 21	06 56	19 36	10 06	05 44	12 10	25 49	22 54	26 10
07	05 02 37	16 30 45	25 ♉ 59 52	01 ♊ 58 04	07 22	20 47	10 51	05 40	12 12	25 47	22 53	26 08
08	05 06 33	17 28 10	07 ♊ 58 14	14 ♊ 00 40	07 44	21 58	11 35	05 36	12 14	25 46	22 52	26 07
09	05 10 30	18 25 35	20 ♊ 05 33	26 ♊ 13 07	08 01	23 09	12 19	05 32	12 15	25 44	22 50	26 05
10	05 14 27	19 22 58	02 ♋ 23 31	08 ♋ 36 55	08 13	24 20	13 03	05 28	12 16	25 42	22 49	26 04
11	05 18 23	20 20 21	14 ♋ 53 25	21 ♋ 13 09	08 21	25 31	13 47	05 24	12 17	25 40	22 48	26 02
12	05 22 20	21 17 43	27 ♋ 36 13	04 ♌ 02 42	08 25	26 42	14 31	05 21	12 19	25 39	22 47	26 01
13	05 26 16	22 15 04	10 ♌ 32 42	17 ♌ 06 17	08 R 23	27 52	15 15	05 17	12 20	25 37	22 45	25 59
14	05 30 13	23 12 24	23 ♌ 43 32	00 ♍ 24 33	08 18	29 ♋ 03	15 59	05 14	12 20	25 35	22 44	25 58
15	05 34 09	24 09 43	07 ♍ 09 23	13 ♍ 58 06	08 08	00 ♌ 14	16 43	05 11	12 21	25 33	22 43	25 56
16	05 38 06	25 07 02	20 ♍ 50 44	27 ♍ 47 18	07 54	01 24	17 27	05 08	12 22	25 31	22 41	25 55
17	05 42 02	26 04 19	04 ♎ 47 44	11 ♎ 51 59	07 36	02 34	18 10	05 05	12 23	25 29	22 40	25 54
18	05 45 59	27 01 35	18 ♎ 59 53	26 ♎ 11 12	07 14	03 45	18 54	05 03	12 23	25 27	22 38	25 52
19	05 49 56	27 58 51	03 ♏ 25 37	10 ♏ 42 42	06 49	04 55	19 38	05 00	12 24	25 25	22 37	25 51
20	05 53 52	28 56 06	18 ♏ 01 57	25 ♏ 22 46	06 21	06 05	20 21	04 58	12 24	25 23	22 36	25 50
21	05 57 49	29 ♊ 53 20	02 ♐ 44 24	10 ♐ 06 06	05 51	07 15	21 05	04 56	12 24	25 21	22 34	25 48
22	06 01 45	00 ♋ 50 34	17 ♐ 27 00	24 ♐ 46 15	05 18	08 25	21 48	04 54	12 24	25 19	22 33	25 47
23	06 05 42	01 47 48	02 ♑ 03 00	09 ♑ 16 22	04 44	09 35	22 31	04 53	12 R 24	25 17	22 31	25 46
24	06 09 38	02 45 01	16 ♑ 25 38	23 ♑ 30 06	04 09	10 44	23 15	04 51	12 24	25 15	22 30	25 44
25	06 13 35	03 42 13	00 ♒ 29 12	07 ♒ 22 32	03 34	11 54	23 58	04 50	12 24	25 12	22 28	25 43
26	06 17 31	04 39 26	14 ♒ 09 48	20 ♒ 50 53	03 00	13 04	24 41	04 49	12 24	25 10	22 27	25 42
27	06 21 28	05 36 38	27 ♒ 25 45	03 ♓ 54 35	02 26	14 13	25 24	04 48	12 23	25 08	22 25	25 41
28	06 25 25	06 33 50	10 ♓ 17 35	16 ♓ 35 09	01 53	15 23	26 07	04 47	12 23	25 06	22 24	25 40
29	06 29 21	07 31 02	22 ♓ 47 42	28 ♓ 55 45	01 23	16 32	26 50	04 47	12 22	25 03	22 22	25 39
30	06 33 18	08 ♋ 28 15	04 ♈ 59 53	11 ♈ 00 42	00 ♋ 55	17 ♌ 41	27 ♉ 33	04 ♏ 46	12 ♓ 22	25 ♑ 01	22 ♑ 20	25 ♏ 37

Date	Moon True ☊ ° ′	Moon Mean ☊ ° ′	Moon ☽ Latitude ° ′
01	23 ♏ 41	23 ♏ 04	04 N 56
02	23 D 42	23 01	04 26
03	23 43	22 58	03 44
04	23 45	22 55	02 54
05	23 46	22 52	01 56
06	23 47	22 48	00 N 53
07	23 R 48	22 45	00 S 12
08	23 47	22 42	01 17
09	23 45	22 39	02 20
10	23 42	22 36	03 17
11	23 38	22 32	04 05
12	23 34	22 29	04 42
13	23 29	22 26	05 05
14	23 26	22 23	05 13
15	23 23	22 20	05 04
16	23 22	22 17	04 38
17	23 D 22	22 13	03 55
18	23 24	22 10	02 57
19	23 25	22 07	01 47
20	23 26	22 04	00 S 29
21	23 R 26	22 01	00 N 51
22	23 24	21 57	02 07
23	23 20	21 54	03 15
24	23 15	21 51	04 09
25	23 09	21 48	04 46
26	23 03	21 45	05 06
27	22 57	21 42	05 09
28	22 52	21 38	04 56
29	22 49	21 35	04 29
30	22 ♏ 48	21 ♏ 32	03 N 50

DECLINATIONS

Date	Sun ☉ ° ′	Moon ☽ ° ′	Mercury ☿ ° ′	Venus ♀ ° ′	Mars ♂ ° ′	Jupiter ♃ ° ′	Saturn ♄ ° ′	Uranus ♅ ° ′	Neptune ♆ ° ′	Pluto ♇ ° ′
01	22 N 04	01 S 32	24 N 55	24 N 32	12 N 57	12 S 19	08 S 37	21 S 28	20 S 52	05 S 19
02	22 11	02 N 46	24 44	24 26	13 12	12 18	08 36	21 28	20 52	05 19
03	22 19	06 53	24 32	24 19	13 27	12 16	08 35	21 28	20 52	05 19
04	22 26	10 42	24 19	24 11	13 42	12 15	08 35	21 29	20 52	05 18
05	22 33	14 06	24 06	24 02	13 57	12 14	08 34	21 29	20 53	05 18
06	22 39	16 56	23 52	23 53	14 11	12 12	08 34	21 29	20 53	05 18
07	22 45	19 03	23 37	23 43	14 26	12 11	08 34	21 30	20 53	05 18
08	22 51	20 22	23 22	23 33	14 40	12 10	08 33	21 30	20 53	05 18
09	22 56	20 45	23 06	23 22	14 54	12 09	08 33	21 30	20 53	05 18
10	23 01	20 08	22 51	23 10	15 08	12 08	08 33	21 31	20 54	05 18
11	23 05	18 33	22 34	22 57	15 22	12 07	08 32	21 31	20 54	05 18
12	23 09	16 02	22 18	22 44	15 35	12 06	08 32	21 31	20 54	05 18
13	23 13	12 41	22 02	22 31	15 49	12 05	08 32	21 32	20 54	05 18
14	23 16	08 41	21 45	22 16	16 02	12 04	08 32	21 32	20 55	05 18
15	23 19	04 N 10	21 29	22 02	16 15	12 03	08 32	21 32	20 55	05 18
16	23 21	00 S 38	21 13	21 46	16 28	12 03	08 32	21 33	20 55	05 18
17	23 23	05 30	20 58	21 30	16 41	12 02	08 32	21 33	20 55	05 18
18	23 24	10 10	20 42	21 14	16 54	12 01	08 32	21 34	20 55	05 18
19	23 25	14 20	20 27	20 57	17 06	12 01	08 32	21 34	20 56	05 18
20	23 26	17 40	20 13	20 39	17 18	12 00	08 32	21 34	20 56	05 18
21	23 26	19 53	20 00	20 21	17 30	12 00	08 32	21 35	20 56	05 18
22	23 26	20 44	19 47	20 02	17 42	12 00	08 32	21 35	20 56	05 18
23	23 26	20 11	19 35	19 43	17 54	11 59	08 33	21 36	20 57	05 18
24	23 25	18 19	19 24	19 23	18 06	11 59	08 33	21 36	20 57	05 18
25	23 23	15 23	19 15	19 03	18 17	11 59	08 33	21 36	20 57	05 18
26	23 21	11 41	19 06	18 43	18 28	11 59	08 33	21 37	20 57	05 18
27	23 19	07 31	18 59	18 22	18 39	11 59	08 34	21 37	20 58	05 18
28	23 17	03 S 08	18 53	18 00	18 50	11 59	08 34	21 38	20 58	05 19
29	23 14	01 N 15	18 48	17 38	19 01	11 59	08 35	21 38	20 58	05 19
30	23 N 10	05 N 30	18 N 45	17 N 16	19 N 11	11 S 59	08 S 35	21 S 39	20 S 58	05 S 19

ZODIAC SIGN ENTRIES

Date	h m	Planets
02	18 31	☽ ♈
05	07 14	☽ ♉
07	20 03	☽ ♊
10	07 22	☽ ♋
12	16 29	☽ ♌
14	23 16	☽ ♍
15	07 23	♀ ♌
17	03 48	☽ ♎
19	06 20	☽ ♏
21	07 32	☽ ♐
21	14 48	☉ ♋
23	08 37	☽ ♑
25	11 10	☽ ♒
27	16 44	☽ ♓
30	02 07	☽ ♈

LATITUDES

Date	Mercury ☿ ° ′	Venus ♀ ° ′	Mars ♂ ° ′	Jupiter ♃ ° ′	Saturn ♄ ° ′	Uranus ♅ ° ′	Neptune ♆ ° ′	Pluto ♇ ° ′
01	01 N 32	01 N 49	00 S 44	01 N 19	01 S 41	00 S 31	00 N 37	14 N 23
04	01 01	01 52	00 43	01 18	01 42	00 31	00 37	14 22
07	00 N 23	01 55	00 41	01 18	01 42	00 31	00 37	14 22
10	00 S 20	01 57	00 39	01 17	01 43	00 31	00 37	14 21
13	01 09	01 58	00 37	01 16	01 44	00 31	00 37	14 20
16	01 59	01 58	00 36	01 16	01 45	00 31	00 37	14 19
19	02 49	01 58	00 34	01 15	01 45	00 31	00 37	14 18
22	03 33	01 57	00 32	01 14	01 46	00 31	00 37	14 17
25	04 09	01 55	00 30	01 13	01 47	00 31	00 37	14 16
28	04 33	01 52	00 28	01 12	01 47	00 31	00 37	14 15
31	04 S 43	01 N 48	00 S 26	01 N 12	01 S 48	00 S 31	00 N 37	14 N 14

DATA

Julian Date	2449505
Delta T	+60 seconds
Ayanamsa	23° 46′ 58″
Synetic vernal point	05° ♓ 20′ 01″
True obliquity of ecliptic	23° 26′ 18″

MOON'S PHASES, APSIDES AND POSITIONS ☽

Date	h m	Phase	Longitude ° ′	Eclipse Indicator
01	04 02	☾	10 ♓ 27	
09	08 26	●	18 ♊ 17	
16	19 57	☽	25 ♍ 26	
23	11 33	○	01 ♑ 47	
30	19 31	☾	08 ♈ 46	

Day	h m		
05	12 48	Apogee	
21	06 49	Perigee	
01	20 29	0N	
09	09 24	Max dec	20° N 45′
16	08 53	0S	
22	14 26	Max dec	20° S 45′
29	05 05	0N	

LONGITUDES

Date	Chiron ⚷ ° ′	Ceres ⚳ ° ′	Pallas ⚴ ° ′	Juno ⚵ ° ′	Vesta ⚶ ° ′	Black Moon Lilith ⚸ ° ′
01	03 ♍ 43	14 ♊ 19	25 ♈ 50	17 ♎ 14	22 ♉ 52	06 ♉ 09
11	04 ♍ 21	18 ♊ 32	29 ♈ 21	17 ♎ 03	27 ♉ 11	07 ♉ 16
21	05 ♍ 08	22 ♊ 45	02 ♉ 47	17 ♎ 24	01 ♊ 26	08 ♉ 23
31	06 ♍ 04	26 ♊ 59	06 ♉ 07	18 ♎ 14	05 ♊ 37	09 ♉ 29

All ephemeris data is given at 12.00 UT and the Moon's longitude is additionally given for 24.00 UT

ASPECTARIAN

h m	Aspects
01 Wednesday	
04 02	☽ □ ☉
04 36	♂ ☍ ♃
05 01	☽ ∠ ♅
07 04	☽ ☌ ♄
10 07	☽ △ ♀
16 24	☉ bQ ♅
02 Thursday	
00 50	☽ bQ ♃
02 18	☽ ∠ ♂
04 34	☽ ⚹ ♆
10 24	☽ ⚹ ♅
11 05	☽ △ ♇
18 29	☽ Q ☉
18 35	☽ ± ♃
19 37	☉ ± ♃
20 31	☉ □ ♄
21 23	☽ ⊥ ♂
03 Friday	
02 39	☽ # ♇
04 28	☽ Q ♆
04 29	☽ □ ☿
06 32	☽ ⊼ ♃
10 14	☽ ⊻ ♂
10 22	☽ Q ♅
17 06	☽ bQ ♇
18 51	☽ ⊻ ♄
20 44	☽ ⚹ ☉
22 25	☽ # ♄
04 Saturday	
04 31	☽ □ ♀
07 02	☽ ⊥ ♄
11 19	☽ ± ♇
13 38	☿ △ ♃
16 51	☽ □ ♆
19 04	☽ Q ☿
22 26	☽ # ♃
22 48	☽ □ ♅
23 30	☽ ⊼ ♇
05 Sunday	
01 24	☽ ∠ ♄
05 55	☽ ∠ ☉
10 45	☽ ∥ ♂
19 02	☽ ☍ ♃
20 43	☽ ⚹ ☿
21 15	☽ Q ♀
06 Monday	
01 52	☽ ⊥ ☉
03 15	☽ ☌ ♂
05 54	☿ ∥ ♀
08 00	☽ ⚹ ♄
15 08	☽ ⊻ ☉
07 Tuesday	
00 19	☽ ⚹ ♀
04 25	☽ ∠ ☿
05 44	☽ △ ♆
08 22	☽ Q ♄
11 35	☽ △ ♅
12 17	☽ ☍ ♇
21 05	☉ ± ♆
23 09	☽ ⊥ ☿
08 Wednesday	
07 17	☽ ⊼ ♃
09 48	☽ ∠ ♀
11 30	☽ ⊻ ☿
11 47	☽ bQ ♆
17 32	☽ bQ ♅
19 10	☽ ± ♃
19 39	☽ ⊻ ♂
20 29	☽ □ ♄
09 Thursday	
00 04	☉ ⊥ ♂
05 36	☽ ⊥ ♀
05 36	☽ ± ♆
05 42	♀ ☍ ♆
08 17	☽ ⊥ ♂
08 26	☽ ☌ ☉
09 44	♂ ⚹ ♄
11 18	☽ ± ♅
12 51	☽ bQ ♃
17 23	☽ ⊼ ♆
18 39	☽ ⊻ ♀
23 02	☽ ⊼ ♅
23 43	☽ ⊼ ♇
10 Friday	
00 06	☉ ∥ ☿
03 02	☽ ∠ ♂
11 22	☽ ± ♇
17 54	☽ △ ♃
19 49	☉ ± ♅
23 23	☽ ☌ ☿
11 Saturday	
00 53	☉ ∥ ♀
04 39	☽ bQ ♇
07 02	☽ △ ♄
09 46	☽ ⚹ ♂
13 28	☉ bQ ♃
15 06	♀ ☍ ♅
22 19	♀ △ ♇
23 11	☽ ⊻ ☉
12 Sunday	
02 32	♀ Q ♂
02 57	☽ ☍ ♆
08 20	☽ ☍ ♅
09 01	☽ △ ♇
09 51	☽ Q ♂
10 08	☽ ☌ ♀
11 23	☽ ⊥ ☉
11 27	☽ bQ ♄
15 17	☽ ∥ ♂
17 49	☿ St R
13 Monday	
00 40	♀ bQ ♄
02 20	☽ □ ♃
04 13	☽ ± ♄
05 27	☽ ∠ ☉
08 03	☽ ⊻ ☿
15 16	☽ ⊼ ♄
15 54	☽ # ♃
19 01	☽ ⊥ ☿
21 08	☽ □ ♂
14 Tuesday	
00 22	☉ ⊼ ♆
10 13	☽ ⊼ ♆
10 59	☽ ⚹ ☉
11 07	☽ Q ♃
11 14	☽ ∠ ☿
12 49	☽ # ♄
15 20	☽ ⊼ ♅
16 01	☽ □ ♇
20 59	☽ ± ♆
22 29	☽ ⊻ ♀
15 Wednesday	
02 04	☽ ± ♅
05 01	☉ ⊥ ♀
06 12	☽ # ♇
07 19	☿ # ♅
08 31	☽ ⚹ ♃
10 06	☽ Q ☉
10 12	☽ ⊥ ♀
12 59	☽ bQ ♆
13 42	☽ ⚹ ☿
17 59	☽ bQ ♅
21 11	☽ ☍ ♄
23 56	☽ Q ♇
16 Thursday	
03 32	☽ ∠ ♀
05 45	☽ △ ♂
10 23	☽ Q ☿
10 46	☽ ∠ ♃
15 11	☽ △ ♆
19 57	☽ □ ☉
20 04	☽ △ ♅
20 46	☽ ⚹ ♇
21 43	☉ ⊼ ♅
17 Friday	
02 16	☽ ⊥ ♃
07 36	☉ ⊼ ♇
07 41	♀ # ♅
07 52	☽ ⚹ ♀
09 05	☽ bQ ♂
10 58	☽ ∥ ♇
12 30	☽ ⊻ ♃
15 34	☿ # ♆
16 39	☽ □ ☿
22 20	☽ ∠ ♇
18 Saturday	
00 52	☽ ⊼ ♄
01 12	☽ ± ♂
03 21	☽ ∥ ♄
06 03	☽ Q ♀
10 58	☽ ± ♄
11 50	☽ ⊼ ♂
13 27	☽ ⊥ ♇
18 05	☽ □ ♆
22 11	☽ ∥ ♃
22 45	☽ □ ♅
23 27	☽ ⊻ ♇
19 Sunday	
02 00	☽ bQ ♄
02 21	☽ △ ☉
13 20	♀ # ♆
13 52	♀ □ ♃
14 36	☽ ☌ ♃
14 40	☽ □ ♀
17 26	☽ △ ☿
23 49	☽ Q ♆
20 Monday	
02 46	☽ △ ♄
04 24	☽ Q ♅
04 49	☽ bQ ☉
08 42	☽ # ♂
15 54	☿ ⊻ ♀
15 59	☽ ☍ ♂
17 15	☽ bQ ☿
18 31	♀ ± ♄
19 26	☽ ⚹ ♆
20 34	☽ ± ☉
23 59	☽ ⚹ ♅
21 Tuesday	
00 43	☽ ☌ ♇
07 02	☽ ⊼ ☉
07 27	☽ ± ☿
07 30	☿ ∠ ♂
13 36	☽ # ☿
15 34	☽ ⊻ ♃
16 53	☽ ⊼ ☿
18 44	☽ # ♀
19 51	☽ ∠ ♆
19 59	☽ △ ♀
22 Wednesday	
00 22	☽ ∠ ♅
01 20	☽ ⊥ ♃
03 45	☽ □ ♄
10 31	☽ ⊥ ♆
15 02	☽ ⊥ ♅
16 01	☽ ∠ ♃
19 30	☽ ⊼ ♂
20 20	☽ ⊻ ♆
22 37	☽ bQ ♀
23 Thursday	
00 52	☽ ⊻ ♅
01 39	☽ ⊻ ♇
03 56	♄ St R
05 46	☿ △ ♃
05 52	☽ ± ♂
09 17	☽ Q ♄
11 08	☉ ± ♇
11 31	☽ ⊥ ♇
11 33	☽ ☍ ☉
11 52	♂ △ ♆
14 45	☽ ± ♀
16 17	☽ ☍ ☿
16 41	☽ ⚹ ♃
21 24	☽ # ♀
21 34	☽ bQ ♂
22 35	☽ # ☿
24 Friday	
01 37	☽ ⊼ ♀
02 28	☽ ∠ ♇
03 59	☿ ⊥ ♀
05 14	☽ ⚹ ♄
09 46	☿ ∥ ♀
12 43	☽ Q ♃
13 58	☽ # ♂
22 16	☽ ☌ ♆
25 Saturday	
00 11	☽ △ ♂
02 56	☽ ☌ ♅
03 48	☽ ⚹ ♇
06 41	☽ ∠ ♄
09 57	☉ ☌ ☿
17 08	☽ ⊼ ☿
18 00	☽ ⊼ ☉
19 33	☽ □ ♃
22 14	♀ ⊼ ♄
22 17	☽ ⊥ ♄
26 Sunday	
00 35	☽ Q ♇
02 23	♂ Q ♄
03 13	☽ ± ☿
05 19	☽ ± ☉
08 51	☽ ⊻ ♄
09 51	☽ ☍ ♀
10 13	☽ ∥ ♃
15 56	☉ △ ♃
18 34	☽ bQ ☿
22 37	☽ bQ ☉
22 54	♀ ∥ ♂
27 Monday	
02 52	☽ ⊻ ♆
03 25	♂ △ ♅
06 10	☽ ∥ ♄
07 48	☽ ⊻ ♅
08 04	☽ □ ♂
08 48	☽ □ ♇
13 49	☽ ⊥ ♆
18 49	☽ ⊥ ♅
20 52	☽ △ ☿
21 04	♂ ☍ ♇
28 Tuesday	
00 12	☽ ∥ ♇
01 39	☽ △ ♃
04 24	☽ △ ☉
06 32	☽ ∠ ♆
07 29	☿ ⊥ ♂
11 38	☽ ∠ ♅
15 57	☽ ☌ ♄
16 05	☿ ∥ ♂
19 43	☽ Q ♂
22 40	☽ ⊼ ♀
22 45	☿ ± ♇
29 Wednesday	
06 09	☽ bQ ♃
09 46	☿ ∠ ♀
11 10	☽ ⚹ ♆
11 26	☽ ± ♀
16 24	☽ ⚹ ♅
17 33	☽ △ ♇
20 23	☽ ⚹ ♂
23 42	☽ ± ♃
30 Thursday	
04 12	☽ □ ☿
06 56	☽ bQ ♀
10 42	☽ Q ♆
10 53	☽ # ♇
11 33	☽ ⊼ ♃
16 01	☽ Q ♅
19 31	☽ □ ☉
23 12	☽ bQ ♇

JULY 1994

LONGITUDES

Date	Sidereal time h m s	Sun ☉ ° ' "	Moon ☽ ° ' "	Moon ☽ 24.00 ° ' "	Mercury ☿ ° '	Venus ♀ ° '	Mars ♂ ° '	Jupiter ♃ ° '	Saturn ♄ ° '	Uranus ♅ ° '	Neptune ♆ ° '	Pluto ♇ ° '
01	06 37 14	09 ♋ 25 27	16 ♈ 58 50	22 ♈ 54 56	00 ♋ 30	18 ♌ 50	28 ♉ 16	04 ♏ 46	12 ♓ 21	24 ♑ 59	22 ♑ 19	25 ♏ 36
02	06 41 10	10 22 40	28 ♈ 49 39	04 ♉ 43 37	00 ♋ 09	19 59	28 59	04 D 46	12 R 20	24 R 57	22 R 17	25 R 35
03	06 45 07	11 19 53	10 ♉ 37 29	16 ♉ 31 50	29 ♊ 51	21 08	29 ♉ 41	04 46	12 19	24 54	22 16	25 34
04	06 49 04	12 17 06	22 ♉ 27 16	28 ♉ 24 17	29 R 38	22 17	00 ♊ 24	04 47	12 18	24 52	22 14	25 33
05	06 53 00	13 14 19	04 ♊ 23 24	10 ♊ 25 03	29 29	23 26	01 07	04 47	12 17	24 50	22 13	25 32
06	06 56 57	14 11 32	16 ♊ 29 37	22 ♊ 37 25	29 25	24 34	01 49	04 48	12 15	24 47	22 11	25 31
07	07 00 54	15 08 46	28 ♊ 48 43	05 ♋ 03 41	29 D 26	25 43	02 31	04 49	12 14	24 45	22 09	25 30
08	07 04 50	16 06 00	11 ♋ 22 27	17 ♋ 45 05	29 32	26 51	03 14	04 50	12 13	24 43	22 08	25 30
09	07 08 47	17 03 14	24 ♋ 11 33	00 ♌ 41 47	29 43	27 59	03 56	04 51	12 11	24 40	22 06	25 29
10	07 12 43	18 00 28	07 ♌ 15 41	13 ♌ 53 04	29 ♊ 59	29 ♌ 07	04 38	04 52	12 09	24 38	22 05	25 28
11	07 16 40	18 57 42	20 ♌ 33 46	27 ♌ 17 34	00 ♋ 21	00 ♍ 15	05 21	04 54	12 08	24 35	22 03	25 27
12	07 20 36	19 54 56	04 ♍ 04 15	10 ♍ 53 37	00 48	01 23	06 03	04 56	12 06	24 33	22 01	25 26
13	07 24 33	20 52 10	17 ♍ 45 26	24 ♍ 39 33	01 20	02 31	06 45	04 58	12 04	24 31	22 00	25 25
14	07 28 29	21 49 24	01 ♎ 35 49	08 ♎ 34 03	01 57	03 39	07 27	05 00	12 02	24 28	21 58	25 25
15	07 32 26	22 46 38	15 ♎ 34 10	22 ♎ 36 03	02 39	04 46	08 08	05 02	12 00	24 26	21 56	25 24
16	07 36 23	23 43 52	29 ♎ 39 35	06 ♏ 44 38	03 27	05 54	08 50	05 05	11 58	24 23	21 55	25 23
17	07 40 19	24 41 06	13 ♏ 51 04	20 ♏ 58 41	04 19	07 01	09 32	05 07	11 55	24 21	21 53	25 23
18	07 44 16	25 38 21	28 ♏ 07 14	05 ♐ 16 26	05 17	08 08	10 14	05 10	11 53	24 19	21 52	25 22
19	07 48 12	26 35 35	12 ♐ 25 53	19 ♐ 35 11	06 19	09 15	10 55	05 13	11 51	24 16	21 50	25 22
20	07 52 09	27 32 50	26 ♐ 43 48	03 ♑ 51 10	07 27	10 22	11 37	05 16	11 48	24 14	21 48	25 21
21	07 56 05	28 30 05	10 ♑ 56 43	17 ♑ 59 49	08 39	11 28	12 18	05 19	11 45	24 11	21 47	25 21
22	08 00 02	29 ♋ 27 21	24 ♑ 59 52	01 ♒ 56 16	09 56	12 35	12 59	05 23	11 43	24 09	21 45	25 20
23	08 03 58	00 ♌ 24 37	08 ♒ 48 32	15 ♒ 36 11	11 17	13 41	13 41	05 27	11 40	24 07	21 44	25 20
24	08 07 55	01 21 54	22 ♒ 18 54	28 ♒ 56 25	12 44	14 47	14 22	05 30	11 37	24 04	21 42	25 19
25	08 11 52	02 19 11	05 ♓ 28 39	11 ♓ 55 33	14 14	15 53	15 03	05 34	11 34	24 02	21 40	25 19
26	08 15 48	03 16 29	18 ♓ 17 15	24 ♓ 33 58	15 49	16 59	15 44	05 39	11 31	23 59	21 39	25 18
27	08 19 45	04 13 48	00 ♈ 46 01	06 ♈ 53 48	17 27	18 04	16 25	05 43	11 28	23 57	21 37	25 18
28	08 23 41	05 11 08	12 ♈ 57 49	18 ♈ 58 36	19 10	19 10	17 06	05 47	11 25	23 55	21 36	25 18
29	08 27 38	06 08 29	24 ♈ 56 45	00 ♉ 52 55	20 56	20 15	17 47	05 52	11 22	23 52	21 34	25 18
30	08 31 34	07 05 51	06 ♉ 47 45	12 ♉ 41 56	22 45	21 20	18 28	05 57	11 18	23 50	21 33	25 17
31	08 35 31	08 ♌ 03 14	18 ♉ 36 09	24 ♉ 31 05	24 ♋ 37	22 ♍ 25	19 ♊ 08	06 ♏ 02	11 ♓ 15	23 ♑ 48	21 ♑ 32	25 ♏ 17

Date	Moon True ☊ ° '	Moon Mean ☊ ° '	Moon ☽ Latitude ° '
01	22 ♏ 48	21 ♏ 29	03 N 02
02	22 D 49	21 26	02 06
03	22 51	21 23	01 06
04	22 51	21 19	00 N 02
05	22 R 50	21 16	01 S 02
06	22 48	21 13	02 04
07	22 43	21 10	03 01
08	22 36	21 07	03 51
09	22 27	21 03	04 29
10	22 18	21 00	04 55
11	22 08	20 57	05 05
12	22 00	20 54	04 59
13	21 54	20 51	04 35
14	21 50	20 48	03 55
15	21 48	20 44	03 01
16	21 D 48	20 41	01 56
17	21 49	20 38	00 S 42
18	21 R 49	20 35	00 N 34
19	21 47	20 32	01 48
20	21 43	20 29	02 55
21	21 37	20 25	03 50
22	21 28	20 22	04 31
23	21 18	20 19	04 56
24	21 06	20 16	05 02
25	20 56	20 13	04 53
26	20 47	20 09	04 29
27	20 40	20 06	03 52
28	20 36	20 03	03 06
29	20 34	20 00	02 12
30	20 34	19 57	01 13
31	20 ♏ 34	19 ♏ 54	00 N 10

DECLINATIONS

Date	Sun ☉ ° '	Moon ☽ ° '	Mercury ☿ ° '	Venus ♀ ° '	Mars ♂ ° '	Jupiter ♃ ° '	Saturn ♄ ° '	Uranus ♅ ° '	Neptune ♆ ° '	Pluto ♇ ° '
01	23 N 06	09 N 28	18 N 43	16 N 53	19 N 21	11 S 59	08 S 36	21 S 39	20 S 59	05 S 19
02	23 02	13 01	18 43	16 30	19 31	11 59	08 36	21 39	20 59	05 19
03	22 57	16 03	18 44	16 06	19 41	12 00	08 37	21 40	20 59	05 19
04	22 52	18 25	18 46	15 42	19 51	12 00	08 37	21 40	20 59	05 19
05	22 47	20 00	18 50	15 18	20 01	12 00	08 38	21 41	21 00	05 20
06	22 41	20 42	18 55	14 53	20 10	12 01	08 39	21 41	21 00	05 20
07	22 35	20 25	19 01	14 28	20 19	12 02	08 39	21 42	21 00	05 20
08	22 28	19 07	19 09	14 03	20 28	12 02	08 40	21 42	21 00	05 20
09	22 21	16 51	19 17	13 38	20 37	12 03	08 41	21 42	21 01	05 21
10	22 14	13 42	19 26	13 12	20 45	12 04	08 42	21 43	21 01	05 21
11	22 06	09 48	19 37	12 45	20 53	12 04	08 43	21 43	21 01	05 21
12	21 58	05 22	19 47	12 19	21 02	12 05	08 44	21 44	21 02	05 21
13	21 49	00 N 37	19 59	11 52	21 09	12 06	08 45	21 44	21 02	05 22
14	21 40	04 S 14	20 10	11 25	21 17	12 07	08 46	21 45	21 02	05 22
15	21 31	08 55	20 22	10 58	21 25	12 08	08 47	21 45	21 02	05 22
16	21 21	13 09	20 34	10 30	21 32	12 09	08 48	21 45	21 02	05 23
17	21 11	16 40	20 46	10 03	21 39	12 11	08 49	21 46	21 02	05 23
18	21 01	19 12	20 58	09 35	21 46	12 11	08 50	21 46	21 03	05 23
19	20 50	20 30	21 09	09 06	21 53	12 13	08 51	21 47	21 03	05 24
20	20 39	20 29	21 20	08 38	21 59	12 14	08 52	21 47	21 04	05 24
21	20 28	19 10	21 30	08 10	22 05	12 15	08 53	21 48	21 04	05 24
22	20 16	16 41	21 39	07 41	22 11	12 17	08 54	21 48	21 04	05 25
23	20 04	13 18	21 46	07 12	22 17	12 18	08 56	21 48	21 05	05 25
24	19 51	09 18	21 53	06 43	22 23	12 20	08 57	21 49	21 05	05 26
25	19 39	04 57	21 58	06 14	22 28	12 21	08 58	21 49	21 05	05 26
26	19 25	00 S 30	22 00	05 45	22 34	12 23	09 00	21 50	21 05	05 26
27	19 12	03 N 51	22 01	05 16	22 39	12 24	09 01	21 50	21 06	05 27
28	18 58	07 58	22 00	04 46	22 44	12 26	09 02	21 51	21 06	05 27
29	18 44	11 42	21 57	04 17	22 48	12 28	09 04	21 51	21 06	05 28
30	18 30	14 55	21 51	03 47	22 53	12 30	09 05	21 51	21 06	05 28
31	18 N 15	17 N 32	21 N 43	03 N 17	22 N 57	12 S 32	09 S 07	21 S 52	21 S 07	05 S 29

ZODIAC SIGN ENTRIES

Date	h	m	Planets
02	14	23	☽ ♉
02	23	18	☿ ♊
03	22	30	♂ ♊
05	03	12	☽ ♊
07	14	17	☽ ♋
09	22	43	☽ ♌
10	12	41	☿ ♋
11	06	36	♀ ♍
12	04	48	☽ ♍
14	09	15	☽ ♎
16	12	35	☽ ♏
18	15	09	☽ ♐
20	17	30	☽ ♑
22	20	38	☽ ♒
23	01	41	☉ ♌
25	01	56	☽ ♓
27	10	30	☽ ♈
29	22	13	☽ ♉

LATITUDES

Date	Mercury ☿ ° '	Venus ♀ ° '	Mars ♂ ° '	Jupiter ♃ ° '	Saturn ♄ ° '	Uranus ♅ ° '	Neptune ♆ ° '	Pluto ♇ ° '
01	04 S 43	01 N 48	00 S 26	01 N 12	01 S 48	00 S 31	00 N 37	14 N 14
04	04 40	01 43	00 23	01 11	01 49	00 32	00 37	14 12
07	04 25	01 37	00 21	01 10	01 50	00 32	00 37	14 11
10	04 00	01 31	00 19	01 09	01 50	00 32	00 37	14 10
13	03 27	01 23	00 17	01 08	01 51	00 32	00 37	14 08
16	02 49	01 15	00 15	01 08	01 52	00 32	00 37	14 07
19	02 08	01 05	00 12	01 07	01 52	00 32	00 37	14 06
22	01 25	00 55	00 10	01 06	01 53	00 32	00 37	14 04
25	00 43	00 44	00 08	01 05	01 53	00 32	00 37	14 02
28	00 S 04	00 31	00 05	01 04	01 54	00 32	00 37	14 01
31	00 N 31	00 N 18	00 S 03	01 N 04	01 S 55	00 S 32	00 N 37	13 N 59

DATA

Julian Date	2449535
Delta T	+60 seconds
Ayanamsa	23° 47' 03"
Synetic vernal point	05° ♓ 19' 57"
True obliquity of ecliptic	23° 26' 18"

MOON'S PHASES, APSIDES AND POSITIONS ☽

Date	h	m	Phase	Longitude ° '	Eclipse Indicator
08	21	37	●	16 ♋ 29	
16	01	12	☽	23 ♎ 18	
22	20	16	○	29 ♑ 47	
30	12	40	☾	07 ♉ 07	

Day	h	m		
03	04	45	Apogee	
18	17	36	Perigee	
30	22	56	Apogee	
06	17	12	Max dec	20° N 43'
13	15	03	0S	
19	23	39	Max dec	20° S 40'
26	14	44	0N	

LONGITUDES

Date	Chiron ⚷ ° '	Ceres ⚳ ° '	Pallas ⚴ ° '	Juno ⚵ ° '	Vesta ⚶ ° '	Black Moon Lilith ⚸ ° '
01	06 ♍ 04	26 ♊ 59	06 ♉ 07	18 ♎ 14	05 ♊ 37	09 ♉ 29
11	07 ♍ 07	01 ♋ 12	09 ♉ 20	19 ♎ 29	09 ♊ 12	10 ♉ 35
21	08 ♍ 16	05 ♋ 24	12 ♉ 24	21 ♎ 08	13 ♊ 41	11 ♉ 42
31	09 ♍ 30	09 ♋ 34	15 ♉ 16	23 ♎ 05	17 ♊ 32	12 ♉ 48

All ephemeris data is given at 12.00 UT and the Moon's longitude is additionally given for 24.00 UT

ASPECTARIAN

h m	Aspects
01 Friday	
02 41	☽ ⊻ ♄
04 03	☽ ∠ ♂
06 30	☽ ⌗ ♄
14 45	☽ ⊥ ♄
14 58	☽ Q ☿
16 09	☽ △ ♀
17 18	☽ ± ♇
22 46	☽ □ ♆
23 22	☽ ⊥ ♂
02 Saturday	
03 35	♃ St D
04 08	☽ □ ♅
04 40	☽ ⌗ ♃
05 26	☽ ⊼ ♇
08 58	☽ ∠ ♄
11 00	☽ Q ☉
12 19	☽ ⊻ ♂
14 36	☽ ✶ ☿
17 13	☉ ⚼ ♇
03 Sunday	
00 05	☽ ☍ ♃
12 26	☽ ∥ ♀
13 34	☽ ✶ ☉
15 26	☽ ✶ ♄
16 09	☿ ⊻ ♂
20 25	☽ ∠ ☿
04 Monday	
11 05	♀ ⊼ ♆
11 34	☽ △ ♆
11 37	☽ □ ♀
12 18	☉ △ ♄
14 21	☽ ⊥ ☿
15 43	☽ Q ♄
16 31	☽ ∥ ☿
16 51	☽ △ ♅
18 15	☽ ☍ ♇
22 27	♀ Q ♃
22 36	☽ ∠ ☉
05 Tuesday	
02 17	☽ ⊻ ☿
05 01	☽ ☌ ♂
12 06	☽ ∥ ♂
12 47	☽ ⊼ ♃
17 37	☽ ⚼ ♆
18 10	☽ ⊥ ☉
22 48	☽ ⚼ ♅
06 Wednesday	
00 44	☽ ± ♃
03 27	☽ Q ♀
03 39	☽ □ ♄
07 05	☽ ⊻ ☉
11 23	☽ ± ♆
16 27	♀ ⊼ ♅
16 29	☽ ± ♅
18 29	☽ ⚼ ♃
19 42	☿ St D
23 07	☽ ⊼ ♆
07 Thursday	
04 10	☽ ⊼ ♅
05 23	☽ ✶ ♀
05 37	☽ ⊼ ♇
07 46	♀ □ ♇
13 13	☽ ☌ ☿
14 22	☽ ∥ ♂
17 11	☽ ± ♇
19 34	☽ ⊻ ♂
23 32	☽ △ ♃
08 Friday	
07 42	☽ ⊥ ♂
10 20	☽ ⚼ ♇
11 43	☽ ∥ ☿
12 59	☽ ∠ ♀
13 35	☽ △ ♄
21 37	☽ ☌ ☉
09 Saturday	
01 39	☽ ∠ ♂
07 31	☽ ⊥ ♀
08 08	☽ ☍ ♆
12 53	☽ ☍ ♅
14 22	♀ ± ♆
14 23	☽ △ ♇
17 31	☽ ⚼ ♄
19 41	☽ ⊻ ♀
22 24	☽ ⊻ ☿
10 Sunday	
06 57	☽ ✶ ♂
07 38	☽ □ ♃
09 38	☽ ⊥ ☿
10 00	☽ ± ♄
15 48	☽ ∥ ♀
20 16	♂ ⊼ ♃
20 52	☽ ⊼ ♄
22 34	☽ ⌗ ♃
11 Monday	
02 22	☽ ∠ ☿
05 54	☽ Q ♂
08 55	☽ ⊻ ☉
14 39	☽ ⊼ ♆
15 02	☿ ✶ ♀
16 11	☽ Q ♃
18 09	☽ ⌗ ♄
18 49	♀ ± ♅
19 10	☽ ⊼ ♅
20 27	☽ ⊥ ☉
20 43	☽ □ ♇
12 Tuesday	
01 19	☽ ± ♆
05 48	☽ ± ♅
06 00	☽ ✶ ☿
06 50	☽ ☌ ♀
12 05	☽ ⌗ ♇
12 06	♂ ⌗ ♆
13 31	☽ ✶ ♃
13 36	☽ ∠ ☉
15 40	☽ □ ♂
17 11	☽ ⚼ ♆
21 37	☽ ⚼ ♅
23 56	♀ ⌗ ♃
13 Wednesday	
02 05	☽ ☍ ♄
03 56	☽ Q ☿
04 26	☽ Q ♇
15 51	☽ ∠ ♃
15 56	☿ ± ♇
17 49	☽ ✶ ☉
19 22	☽ △ ♆
20 17	♂ ⚼ ♆
23 42	☽ △ ♅
14 Thursday	
00 42	☉ ⌗ ♅
01 19	☽ ✶ ♇
07 30	☽ ⊥ ♃
12 38	☽ □ ☿
15 31	☉ ☍ ♆
15 51	☽ ⊻ ♀
16 07	☽ Q ☉
17 42	☽ ∥ ♇
17 53	☽ ⊻ ♃
22 36	☽ △ ♂
15 Friday	
03 04	☽ ⊥ ♀
03 09	☽ ∠ ♇
05 54	☽ ⊼ ♄
11 16	☽ ∥ ♄
16 08	☽ ± ♄
17 51	♀ ✶ ♃
18 32	☽ ⊥ ♇
19 48	☽ ∠ ♀
20 51	☉ ∥ ♂
22 07	☽ ⌗ ♀
22 51	☽ □ ♆
16 Saturday	
01 12	☽ □ ☉
01 35	☽ ⚼ ♂
03 04	☽ □ ♅
04 45	☽ ⊻ ♇
05 59	☽ ∥ ♃
07 26	☽ ⚼ ♄
17 40	☽ ± ♂
18 49	☽ △ ☿
21 12	☽ ☌ ♃
21 47	☉ ∠ ♂
23 28	☽ ✶ ♀
17 Sunday	
03 54	☉ ☍ ♅
04 20	☽ ⊼ ♂
05 19	☽ Q ♆
06 03	♂ ⚼ ♅
08 45	☽ △ ♄
09 19	♀ ⚼ ♆
09 28	☽ Q ♅
21 26	☽ Q ♀
21 51	☽ ⚼ ☿
18 Monday	
01 30	☽ ✶ ♆
05 16	☉ △ ♇
05 37	☽ ✶ ♅
06 48	☉ ⌗ ♆
07 23	☽ ☌ ♇
07 32	☽ △ ☉
09 11	☿ △ ♃
13 32	♂ ⌗ ♅
14 05	☽ ± ☿
14 58	☉ ∥ ☿
22 54	☿ ⌗ ♆
23 52	☽ ⊻ ♃
19 Tuesday	
00 56	☽ ⊼ ☿
02 38	☽ ∠ ♆
06 13	☽ □ ♀
06 43	☽ ∠ ♅
09 20	☽ ☍ ♂
09 57	☽ ⊥ ♃
10 30	☽ ⚼ ☉
11 01	☽ □ ♄
12 27	♀ ⚼ ♅
17 41	☽ ⊥ ♆
18 01	☉ ⚼ ♄
21 46	☽ ⊥ ♅
23 11	♂ ± ♃
20 Wednesday	
00 43	♀ ⌗ ♄
01 06	☽ ∠ ♃
02 40	☽ ± ☉
03 44	☽ ⊻ ♆
07 48	☽ ⊻ ♅
09 41	☽ ⊻ ♇
13 28	☽ ⊼ ☉
17 09	☽ Q ♄
18 14	♂ □ ♄
19 46	☽ ⊥ ♇
21 Thursday	
02 27	☽ ✶ ♃
07 44	☽ ☍ ☿
10 59	☽ ∠ ♇
12 58	☽ △ ♀
13 22	☽ ✶ ♄
14 25	☽ ⊼ ♂
17 58	♀ ☍ ♄
22 54	☽ Q ♃
22 Friday	
01 10	☽ ± ♂
06 26	☽ ☌ ♆
10 33	☽ ☌ ♅
12 35	☽ ✶ ♇
14 57	☽ ∠ ♄
16 50	☽ ⚼ ♀
17 26	☽ ⚼ ♂
19 12	☿ ⚼ ♇
20 16	☽ ☍ ☉
23 Saturday	
04 18	♀ Q ♇
06 05	☽ □ ♃
06 31	☽ ⊥ ♄
09 24	☽ Q ♇
09 51	☽ ± ♀
11 41	♀ □ ♂
16 52	☽ ⊼ ☿
17 01	☽ ⊻ ♄
18 11	☿ △ ♄
18 14	☽ ∥ ♃
19 31	☿ ⌗ ♅
21 03	☽ △ ♂
21 21	☽ ⊼ ♀
24 Sunday	
04 47	☽ ± ☿
10 54	☽ ⊻ ♆
13 59	☽ ∥ ♄
15 09	☽ ⊻ ♅
17 25	☽ □ ♇
21 43	☽ ⊥ ♆
23 02	☽ ⚼ ☿
25 Monday	
02 01	☽ ⊥ ♅
04 11	☽ ⌗ ♀
05 43	☽ ⊼ ☉
09 25	☽ ∥ ♇
12 11	☽ △ ♃
14 12	☽ ∠ ♆
17 41	☽ ± ☉
18 34	☽ ∠ ♅
23 17	☽ ☌ ♄
26 Tuesday	
06 38	☽ △ ☿
06 53	☽ □ ♂
09 00	♂ ± ♆
09 17	☽ ☍ ♀
10 01	☿ ⊻ ♂
11 58	☽ ⚼ ☉
16 30	☽ ⚼ ♃
18 23	☽ ✶ ♆
22 52	☽ ✶ ♅
27 Wednesday	
01 25	☽ △ ♇
02 56	♀ ⌗ ♇
09 56	☽ ± ♃
17 33	☽ Q ♆
19 09	☽ ∥ ♀
19 20	☽ △ ☉
19 33	☽ Q ♂
21 05	☽ ⌗ ♇
21 44	☽ ⊼ ♃
22 06	☽ Q ♅
28 Thursday	
06 43	☽ ⚼ ♇
08 56	☽ ⊻ ♄
12 04	☿ ✶ ♀
17 29	☉ ± ♄
18 38	☽ ⌗ ♄
20 44	☽ ✶ ♂
20 50	☽ ⊥ ♄
29 Friday	
00 38	☽ ± ♇
01 37	☽ ⊼ ♀
02 30	☽ □ ☿
04 26	☉ □ ♃
05 13	☽ □ ♆
09 51	☽ □ ♅
12 42	☽ ⊼ ♇
14 51	☽ ∠ ♄
14 54	☽ ± ♀
15 07	♂ ± ♅
17 23	☽ ⌗ ♃
20 21	☿ ☍ ♆
30 Saturday	
02 38	♀ ∠ ♃
04 49	☽ ∠ ♂
10 15	☽ ☍ ♃
10 45	☿ ⌗ ♅
10 58	☽ ⚼ ♀
12 40	☽ □ ☉
16 27	♀ △ ♆
21 08	☽ ✶ ♄
21 33	☽ Q ☿
31 Sunday	
00 13	☽ ⊥ ♂
01 40	☿ ☍ ♅
13 09	☽ ⊻ ♂
17 54	☽ △ ♆
19 18	☽ ∥ ☉
20 20	☿ △ ♇
20 31	☽ △ ♀
21 23	☽ Q ♄
22 02	☿ ⊥ ♂
22 30	☽ △ ♅

AUGUST 1994

LONGITUDES

Date	Sidereal time h m s	Sun ☉ ° ' "	Moon ☽ ° ' "	Moon ☽ 24.00 ° ' "	Mercury ☿ ° '	Venus ♀ ° '	Mars ♂ ° '	Jupiter ♃ ° '	Saturn ♄ ° '	Uranus ♅ ° '	Neptune ♆ ° '	Pluto ♇ ° '
01	08 39 27	09 ♌ 00 38	00 ♊ 27 24	06 ♊ 25 45	26 ♋ 32	23 ♍ 30	19 ♊ 49	06 ♏ 07	11 ♓ 12	23 ♑ 45	21 ♑ 29	25 ♏ 17
02	08 43 24	09 58 03	12 ♊ 26 43	18 ♊ 30 54	28 ♋ 30	24 34	20 29	06 12	11 R 08	23 R 43	21 R 28	25 R 17
03	08 47 21	10 55 30	24 ♊ 38 46	00 ♋ 50 46	00 ♌ 29	25 38	21 10	06 17	11 05	23 41	21 26	25 17
04	08 51 17	11 52 58	07 ♋ 07 14	13 ♋ 28 28	02 30	26 42	21 50	06 23	11 01	23 39	21 25	25 17
05	08 55 14	12 50 26	19 ♋ 54 36	26 ♋ 25 42	04 33	27 46	22 30	06 28	10 57	23 36	21 23	25 17
06	08 59 10	13 47 56	03 ♌ 01 43	09 ♌ 42 29	06 36	28 50	23 11	06 34	10 53	23 34	21 22	25 17
07	09 03 07	14 45 27	16 ♌ 27 46	23 ♌ 17 12	08 40	29 ♍ 53	23 51	06 40	10 50	23 32	21 20	25 D 17
08	09 07 03	15 42 58	00 ♍ 10 21	07 ♍ 06 46	10 44	00 ♎ 56	24 31	06 46	10 46	23 30	21 19	25 17
09	09 11 00	16 40 31	14 ♍ 05 53	21 ♍ 07 13	12 48	01 59	25 11	06 53	10 42	23 28	21 18	25 17
10	09 14 56	17 38 05	28 ♍ 10 13	05 ♎ 14 24	14 52	03 02	25 51	06 59	10 38	23 25	21 16	25 17
11	09 18 53	18 35 39	12 ♎ 19 19	19 ♎ 24 36	16 56	04 04	26 30	07 06	10 34	23 23	21 15	25 17
12	09 22 50	19 33 15	26 ♎ 29 55	03 ♏ 35 00	18 59	05 06	27 10	07 12	10 30	23 21	21 13	25 17
13	09 26 46	20 30 51	10 ♏ 39 40	17 ♏ 43 44	21 01	06 08	27 50	07 19	10 26	23 19	21 12	25 18
14	09 30 43	21 28 28	24 ♏ 47 06	01 ♐ 49 40	23 02	07 10	28 29	07 26	10 21	23 17	21 11	25 18
15	09 34 39	22 26 07	08 ♐ 51 20	15 ♐ 51 58	25 02	08 11	29 09	07 33	10 17	23 15	21 09	25 18
16	09 38 36	23 23 46	22 ♐ 51 28	29 ♐ 49 40	27 00	09 12	29 ♊ 48	07 40	10 13	23 13	21 08	25 19
17	09 42 32	24 21 26	06 ♑ 46 20	13 ♑ 41 16	28 ♌ 58	10 12	00 ♋ 27	07 48	10 09	23 11	21 07	25 19
18	09 46 29	25 19 08	20 ♑ 34 08	27 ♑ 24 40	00 ♍ 54	11 12	01 07	07 55	10 04	23 09	21 05	25 19
19	09 50 25	26 16 50	04 ♒ 12 30	10 ♒ 57 19	02 49	12 12	01 46	08 03	10 00	23 07	21 04	25 20
20	09 54 22	27 14 34	17 ♒ 38 48	24 ♒ 16 38	04 43	13 12	02 25	08 11	09 56	23 06	21 03	25 20
21	09 58 19	28 12 19	00 ♓ 50 35	07 ♓ 20 27	06 35	14 11	03 04	08 19	09 51	23 04	21 02	25 21
22	10 02 15	29 ♌ 10 05	13 ♓ 46 06	20 ♓ 07 30	08 26	15 10	03 42	08 27	09 47	23 02	21 00	25 21
23	10 06 12	00 ♍ 07 53	26 ♓ 24 41	02 ♈ 37 47	10 16	16 08	04 21	08 35	09 42	23 00	20 59	25 22
24	10 10 08	01 05 42	08 ♈ 46 59	14 ♈ 52 35	12 04	17 06	05 00	08 43	09 38	22 58	20 58	25 23
25	10 14 05	02 03 33	20 ♈ 54 57	26 ♈ 54 32	13 51	18 04	05 38	08 51	09 33	22 57	20 57	25 23
26	10 18 01	03 01 26	02 ♉ 51 49	08 ♉ 47 20	15 36	19 01	06 17	09 00	09 29	22 55	20 56	25 24
27	10 21 58	03 59 20	14 ♉ 41 42	20 ♉ 35 33	17 21	19 58	06 55	09 08	09 24	22 53	20 55	25 25
28	10 25 54	04 57 17	26 ♉ 29 32	02 ♊ 24 20	19 03	20 54	07 34	09 17	09 20	22 52	20 54	25 25
29	10 29 51	05 55 15	08 ♊ 20 38	14 ♊ 19 07	20 45	21 50	08 12	09 26	09 15	22 50	20 53	25 26
30	10 33 48	06 53 15	20 ♊ 20 28	26 ♊ 25 20	22 25	22 45	08 50	09 35	09 11	22 49	20 52	25 27
31	10 37 44	07 ♍ 51 17	02 ♋ 34 18	08 ♋ 47 56	24 ♍ 04	23 ♎ 40	09 ♋ 28	09 ♏ 44	09 ♓ 06	22 ♑ 47	20 ♑ 51	25 ♏ 28

MOON NODES AND LATITUDE

Date	Moon True ☊ ° '	Moon Mean ☊ ° '	Moon ☽ Latitude ° '
01	20 ♏ 33	19 ♏ 50	00 S 52
02	20 R 32	19 47	01 53
03	20 28	19 44	02 50
04	20 21	19 41	03 40
05	20 12	19 38	04 20
06	20 01	19 35	04 48
07	19 49	19 31	05 00
08	19 37	19 28	04 56
09	19 26	19 25	04 34
10	19 18	19 22	03 55
11	19 12	19 19	03 02
12	19 09	19 15	01 57
13	19 D 08	19 12	00 S 45
14	19 R 08	19 09	00 N 30
15	19 08	19 06	01 42
16	19 05	19 03	02 48
17	19 01	19 00	03 44
18	18 53	18 56	04 25
19	18 43	18 53	04 51
20	18 32	18 50	05 01
21	18 19	18 47	04 54
22	18 07	18 44	04 32
23	17 57	18 41	03 57
24	17 49	18 37	03 11
25	17 44	18 34	02 17
26	17 41	18 31	01 18
27	17 40	18 28	00 N 16
28	17 D 40	18 25	00 S 47
29	17 R 40	18 21	01 48
30	17 39	18 18	02 45
31	17 ♏ 37	18 ♏ 15	03 S 35

DECLINATIONS

Date	Sun ☉ ° '	Moon ☽ ° '	Mercury ☿ ° '	Venus ♀ ° '	Mars ♂ ° '	Jupiter ♃ ° '	Saturn ♄ ° '	Uranus ♅ ° '	Neptune ♆ ° '	Pluto ♇ ° '
01	18 N 00	19 N 24	21 N 32	02 N 48	23 N 01	12 S 34	09 S 08	21 S 52	21 S 07	05 S 29
02	17 45	20 25	21 18	02 18	23 05	12 35	09 10	21 53	21 07	05 30
03	17 29	20 30	21 02	01 48	23 08	12 37	09 11	21 53	21 07	05 30
04	17 14	19 35	20 43	01 18	23 12	12 40	09 13	21 53	21 08	05 31
05	16 57	17 40	20 21	00 48	23 15	12 42	09 14	21 54	21 08	05 31
06	16 41	14 49	19 57	00 N 19	23 18	12 44	09 16	21 54	21 08	05 32
07	16 24	11 07	19 31	00 S 11	23 21	12 46	09 17	21 54	21 08	05 32
08	16 07	06 48	19 02	00 41	23 24	12 48	09 19	21 55	21 09	05 33
09	15 50	02 N 03	18 31	01 11	23 26	12 50	09 21	21 55	21 09	05 33
10	15 33	02 S 52	17 58	01 41	23 28	12 53	09 22	21 56	21 09	05 34
11	15 15	07 39	17 24	02 10	23 30	12 55	09 24	21 56	21 09	05 34
12	14 57	12 02	16 47	02 40	23 32	12 57	09 26	21 56	21 10	05 35
13	14 39	15 44	16 10	03 10	23 34	13 00	09 27	21 57	21 10	05 36
14	14 21	18 29	15 30	03 39	23 35	13 02	09 29	21 57	21 10	05 36
15	14 02	20 05	14 50	04 09	23 36	13 05	09 31	21 57	21 10	05 37
16	13 43	20 27	14 08	04 38	23 37	13 07	09 33	21 58	21 10	05 37
17	13 24	19 33	13 26	05 07	23 38	13 10	09 34	21 58	21 10	05 38
18	13 05	17 29	12 43	05 37	23 39	13 13	09 36	21 58	21 11	05 39
19	12 45	14 29	11 59	06 06	23 39	13 15	09 38	21 59	21 11	05 39
20	12 26	10 46	11 15	06 34	23 40	13 18	09 40	21 59	21 11	05 40
21	12 06	06 35	10 30	07 03	23 40	13 21	09 41	21 59	21 12	05 40
22	11 46	02 S 12	09 45	07 32	23 40	13 24	09 43	21 59	21 12	05 41
23	11 26	02 N 12	08 59	08 00	23 39	13 26	09 45	22 00	21 12	05 42
24	11 05	06 24	08 13	08 28	23 39	13 29	09 47	22 00	21 12	05 42
25	10 44	10 17	07 27	08 56	23 38	13 32	09 48	22 00	21 12	05 43
26	10 24	13 41	06 41	09 24	23 37	13 35	09 50	22 01	21 13	05 44
27	10 03	16 30	05 55	09 52	23 36	13 38	09 52	22 01	21 13	05 44
28	09 42	18 37	05 10	10 19	23 35	13 41	09 54	22 01	21 13	05 45
29	09 20	19 55	04 24	10 46	23 34	13 44	09 56	22 01	21 13	05 46
30	08 59	20 21	03 38	11 13	23 32	13 47	09 57	22 02	21 13	05 46
31	08 N 37	19 N 49	02 N 53	11 S 40	23 N 31	13 S 50	09 S 59	22 S 02	21 S 13	05 S 47

ZODIAC SIGN ENTRIES

Date	h m	Planets
01	11 05	☽ ♊
03	06 09	☿ ♌
03	22 22	☽ ♋
06	06 31	☽ ♌
07	14 36	♀ ♎
08	11 42	☽ ♍
10	15 07	☽ ♎
12	17 56	☽ ♏
14	20 53	☽ ♐
16	19 15	♂ ♋
17	00 18	☽ ♑
18	00 44	☿ ♍
19	04 34	☽ ♒
21	10 27	☽ ♓
23	08 44	☉ ♍
23	18 55	☽ ♈
26	06 13	☽ ♉
28	19 07	☽ ♊
31	07 00	☽ ♋

LATITUDES

Date	Mercury ☿ ° '	Venus ♀ ° '	Mars ♂ ° '	Jupiter ♃ ° '	Saturn ♄ ° '	Uranus ♅ ° '	Neptune ♆ ° '	Pluto ♇ ° '
01	00 N 42	00 N 14	00 S 02	01 N 03	01 S 55	00 S 32	00 N 37	13 N 59
04	01 09	00 00	00 N 01	01 03	01 55	00 32	00 37	13 57
07	01 28	00 S 15	00 03	01 02	01 56	00 32	00 37	13 56
10	01 40	00 31	00 06	01 01	01 56	00 32	00 37	13 54
13	01 46	00 48	00 08	01 01	01 57	00 32	00 37	13 52
16	01 44	01 05	00 11	01 00	01 57	00 32	00 37	13 51
19	01 38	01 23	00 14	00 59	01 57	00 32	00 37	13 49
22	01 27	01 41	00 16	00 59	01 58	00 32	00 36	13 48
25	01 12	02 00	00 19	00 58	01 58	00 32	00 36	13 46
28	00 54	02 20	00 22	00 57	01 58	00 32	00 36	13 44
31	00 N 34	02 S 40	00 N 25	00 N 57	01 S 58	00 S 31	00 N 36	13 N 43

DATA

Julian Date	2449566
Delta T	+60 seconds
Ayanamsa	23° 47' 07"
Synetic vernal point	05° ♓ 19' 52"
True obliquity of ecliptic	23° 26' 18"

MOON'S PHASES, APSIDES AND POSITIONS ☽

Date	h	m	Phase	Longitude ° '	Eclipse Indicator
07	08	45	●	14 ♌ 38	
14	05	57	☽	21 ♏ 14	
21	06	47	○	28 ♒ 00	
29	06	41	☾	05 ♊ 42	

Day	h	m	
12	23	06	Perigee
27	17	49	Apogee

Day	h	m		
03	02	04	Max dec	20° N 35'
09	22	01	0S	
16	06	40	Max dec	20° S 29'
22	23	57	0N	
30	10	58	Max dec	20° N 21'

LONGITUDES

Date	Chiron ⚷ ° '	Ceres ⚳ ° '	Pallas ⚴ ° '	Juno ⚵ ° '	Vesta ⚶ ° '	Black Moon Lilith ⚸ ° '
01	09 ♍ 38	09 ♋ 59	15 ♉ 32	23 ♎ 18	17 ♊ 54	12 ♉ 55
11	10 ♍ 57	14 ♋ 05	18 ♉ 08	25 ♎ 34	21 ♊ 36	14 ♉ 02
21	12 ♍ 19	18 ♋ 08	20 ♉ 23	28 ♎ 04	25 ♊ 06	15 ♉ 08
31	13 ♍ 43	22 ♋ 07	22 ♉ 15	00 ♏ 45	28 ♊ 24	16 ♉ 14

ASPECTARIAN

h m Aspects

01 Monday
01 33 ☽ ☍ ♇
02 34 ☽ ⚹ ☿
04 26 ☽ Q ☉
07 49 ☿ ⅊ ♄
17 40 ♀ △ ♅
23 27 ☽ ⊼ ♃

02 Tuesday
00 06 ☽ ⅊ ♆
04 36 ☽ ⅊ ♅
06 39 ☽ ⚹ ☉
09 24 ☽ □ ♄
11 30 ☽ ± ♃
14 30 ☽ ∠ ☿
17 58 ☽ ± ♆
22 24 ☽ ± ♅

03 Wednesday
03 56 ♀ ⚹ ♇
04 17 ☿ # ♆
04 48 ☽ ☌ ♂
05 24 ☽ ⅊ ♃
05 45 ☽ ⊼ ♆
10 08 ☽ ⊼ ♅
11 38 ☽ ⊥ ☿
13 14 ☽ ⊼ ♇
14 07 ☽ □ ♀
14 42 ☽ ∠ ☉
15 33 ☉ ⊼ ♄
17 08 ♂ ⅊ ♃
21 28 ♂ ⊼ ♆

04 Thursday
00 50 ☽ ± ♇
01 29 ☽ ⊻ ☿
09 27 ☽ ⊥ ☉
10 35 ☽ △ ♃
17 59 ☽ ⅊ ♇
19 20 ☽ △ ♄
21 44 ☽ ⊻ ☉

05 Friday
03 36 ☽ Q ♀
14 44 ☽ ☍ ♆
16 40 ☿ ± ♄
17 04 ☽ ⊻ ♂
17 08 ♇ St D
18 48 ☽ ☍ ♅
19 30 ☽ ∥ ☉
21 54 ☽ △ ♇
23 05 ☽ ⅊ ♄

06 Saturday
03 43 ☽ ⚹ ♀
04 39 ☽ ⊥ ♂
04 39 ☉ ∠ ♀
11 41 ☿ □ ♃
15 21 ☽ ± ♄
18 26 ☽ □ ♃
19 36 ☽ ☌ ☿
21 45 ☽ ∠ ♂

07 Sunday
01 18 ♂ ⊼ ♅
02 00 ☽ # ♃
02 03 ☽ ⊼ ♄
08 45 ☽ ☌ ☉
08 58 ☽ ∠ ♀
15 10 ☿ ∠ ♂
20 34 ☽ ⊼ ♆
22 28 ☽ # ♄

08 Monday
00 24 ☽ ⊼ ♅
01 39 ☽ ⚹ ♂
02 08 ☽ ⊥ ♀
02 32 ☽ Q ♃
03 29 ☽ □ ♇
07 03 ☽ ± ♆
07 51 ♀ ⊥ ♃
10 50 ☽ ± ♅
12 20 ☿ ⊼ ♄
13 26 ☽ ⊻ ♀
18 27 ☽ # ♇
22 37 ☽ ⅊ ♆
23 30 ☽ ⚹ ♃
23 31 ☽ Q ♂

09 Tuesday
02 21 ☽ ⅊ ♅
06 12 ☽ ☍ ♄
09 24 ☽ ⊻ ☿
10 36 ☽ Q ♇
15 40 ♂ ⊼ ♇
15 51 ☽ # ♀
16 44 ☽ ⊻ ☉
21 26 ☽ ⊥ ☿

10 Wednesday
00 16 ☽ △ ♆
01 24 ☽ ∠ ♃
03 43 ☽ ⊥ ☉
03 57 ☽ △ ♅
05 33 ☽ ∥ ♀
07 05 ☽ ⚹ ♇
07 51 ☽ □ ♂
15 23 ☽ ∠ ☿
16 49 ☽ ⊥ ♃
20 08 ☽ ∠ ☉
20 54 ☽ ☌ ♀

11 Thursday
01 23 ☽ ∥ ♇
03 04 ☽ ⊻ ♃
08 33 ☽ ∠ ♇
09 02 ☽ ⊼ ♄
19 09 ☽ ± ♄
21 07 ☽ ⅊ ♂
21 16 ☽ ∥ ♄
23 23 ☽ ⚹ ☉
23 48 ☽ ⊥ ♇

12 Friday
03 05 ☽ □ ♆
06 41 ☽ □ ♅
09 57 ☽ ⊻ ♇
10 19 ☽ ⅊ ♄
13 12 ☽ △ ♂
17 36 ☽ ∥ ♃
20 51 ☽ Q ☿
21 11 ☽ Q ☉

13 Saturday
00 54 ☉ ☌ ☿
03 43 ☽ ⊻ ♀
05 00 ☽ # ☉
06 17 ☽ ☌ ♃
09 31 ☽ Q ♆
11 36 ☽ △ ♄
13 07 ☽ Q ♅
14 14 ☿ ⊼ ♆
14 41 ☽ # ☿
14 42 ☽ ⊥ ♀
14 59 ☿ ∠ ♀
15 52 ☽ ⅊ ♂

14 Sunday
04 44 ☉ ⊼ ♆
05 52 ☽ ⚹ ♆
05 57 ☽ □ ☉
07 11 ☽ ∠ ♀
07 54 ☽ ± ♂
08 31 ☽ □ ☿
09 27 ☽ ⚹ ♅
12 53 ☽ ☌ ♇
15 03 ☿ ⊼ ♅
18 37 ☽ ⊼ ♂
19 21 ♀ ⊻ ♃

15 Monday
07 24 ☽ ∠ ♆
09 45 ☽ ⊻ ♃
10 45 ☽ ⚹ ♀
10 58 ☽ ∠ ♅
14 26 ☽ □ ♄
15 21 ☿ □ ♇
18 46 ☿ Q ♃
20 06 ☽ ⊥ ♃
22 46 ☽ ⊥ ♆

16 Tuesday
02 21 ☽ ⊥ ♅
07 44 ☉ ⊼ ♅
08 55 ☽ Q ♀
09 02 ☽ ⊻ ♆
11 41 ☽ ∠ ♃
12 37 ☽ ⊻ ♅
13 00 ☽ △ ☉
13 30 ☿ ± ♆
16 13 ☽ ⊻ ♇
20 19 ☽ △ ☿
21 11 ☽ Q ♄

17 Wednesday
00 33 ☽ ☍ ♂
02 34 ☽ ⊥ ♇
10 43 ♀ ⊼ ♄
13 47 ☽ ⚹ ♃
13 58 ☉ ∥ ☿
14 39 ☿ ± ♅
14 43 ♀ ∠ ♇
16 49 ☽ ⅊ ☉
17 49 ☽ ⚹ ♄
18 09 ☽ ∠ ♇
18 25 ☽ □ ♀
20 27 ☿ # ♃

18 Thursday
02 32 ☽ ⅊ ☿
03 29 ☉ # ♃
09 39 ☽ ± ☉
10 52 ☽ Q ♃
12 08 ☉ □ ♇
12 55 ☽ ☌ ♆
13 45 ♀ ∥ ♇
15 48 ☿ ⚹ ♂
16 31 ☽ ☌ ♅
19 51 ☽ ∠ ♄
19 58 ♂ ± ♇
20 20 ☽ ⚹ ♇
20 51 ☽ ± ☿
20 57 ☽ ⊼ ☉

19 Friday
05 20 ☉ Q ♃
07 27 ☽ ⊼ ♂
09 09 ☽ ⊼ ☿
11 38 ☽ ⊥ ♄
17 33 ☽ Q ♇
18 38 ☽ ± ♂
18 53 ☽ □ ♃
20 15 ☽ ∥ ♃
22 15 ☽ ⊻ ♄

20 Saturday
00 44 ☽ # ☉
03 22 ☽ △ ♀
07 15 ☉ ± ♆
08 25 ☽ # ☿
11 33 ☽ ⅊ ♂
18 08 ☽ ⊻ ♆
18 31 ☽ ∥ ♄
21 49 ☽ ⊻ ♅

21 Sunday
01 57 ☽ □ ♇
04 50 ☽ ⅊ ♅
05 01 ☽ ⊥ ♅
06 47 ☽ ☍ ♀
08 43 ☽ ⅊ ♀
08 44 ☽ ⊥ ☉
09 41 ☽ ∥ ♀
16 18 ☽ △ ♂
17 03 ☽ ∥ ♇
21 33 ☽ ∠ ♆

22 Monday
00 23 ☽ ☍ ☿
01 19 ☽ ∠ ♅
01 57 ☽ △ ♃
02 41 ☽ ± ♀
04 35 ☽ ☌ ♄
06 46 ☿ ⅊ ♅
08 43 ☉ ± ♅
12 03 ☿ ⚹ ♃
12 44 ☿ # ♄
14 51 ☽ ⊼ ♀

23 Tuesday
01 39 ☽ ⚹ ♆
02 05 ♀ ± ♄
04 55 ☿ ☍ ♄
05 29 ☽ ⚹ ♅
06 31 ☽ ⅊ ♃
08 25 ☿ ⊥ ♀
10 00 ☽ △ ♇
19 46 ☽ ⊼ ☉

24 Wednesday
00 02 ☽ ± ♃
00 41 ☽ Q ♆
04 12 ☽ □ ♂
04 34 ☽ Q ♅
07 08 ☿ # ♀
07 54 ☽ # ♇
08 25 ☽ ± ☉
11 52 ☽ ⊼ ♃
13 39 ☽ ⊻ ♄
15 08 ☽ ⅊ ♇
19 34 ☽ ⊼ ☿
21 09 ☽ ∥ ☿

25 Thursday
01 25 ☽ ⊥ ♄
02 12 ☽ # ♀
03 40 ☽ ⅊ ☉
05 43 ☿ Q ♇
05 50 ☽ ☍ ♀
08 54 ☽ # ♄
08 57 ☽ ± ♇
09 30 ☽ ± ☿
12 04 ☽ □ ♆
14 47 ☽ ∥ ☉
16 03 ☽ □ ♅
17 45 ☽ Q ♂
19 14 ☽ ∠ ♄
20 57 ☽ ⊼ ♇

26 Friday
02 37 ♃ Q ♆
06 40 ☽ ⅊ ☿
11 12 ☽ # ♃
12 21 ☽ △ ☉
19 19 ☽ ⚹ ♂
21 44 ♀ ⊥ ♇

27 Saturday
00 34 ☽ ☍ ♃
01 19 ☽ ⚹ ♄
12 16 ♀ ∥ ♄
17 27 ☉ # ♀
17 41 ☿ # ♇
18 18 ☽ △ ☿
23 15 ☉ # ♄
23 39 ☽ ⊼ ♀

28 Sunday
00 38 ☽ △ ♆
01 34 ☽ Q ♄
03 33 ☽ ∠ ♂
04 38 ☽ △ ♅
09 50 ☽ ☍ ♇
11 48 ♀ □ ♆
12 55 ☽ ± ♀
17 10 ♃ △ ♄
22 52 ☽ ⊥ ♂
23 22 ☿ Q ♂

29 Monday
06 41 ☽ □ ☉
07 02 ☽ ⅊ ♆
08 42 ☽ ⅊ ♀
10 59 ☉ ⅊ ♆
10 59 ☽ ⅊ ♅
11 41 ☽ ⊻ ♂
13 49 ☽ □ ♄
13 50 ☿ △ ♆
14 13 ☽ ⊼ ♃

30 Tuesday
01 06 ☽ ± ♆
02 24 ☽ ± ♃
05 00 ☽ ± ♅
13 02 ☽ ⊼ ♆
13 23 ♀ □ ♅
16 46 ☽ □ ☿
16 53 ☽ ⊼ ♅
17 11 ☽ △ ♀
17 36 ☿ △ ♅
20 29 ☽ ⅊ ♃
21 45 ☽ Q ☉
22 06 ☽ ⊼ ♇
22 59 ☿ ⊻ ♀
23 43 ♂ △ ♄

31 Wednesday
09 51 ☽ ± ♇
10 23 ☉ ± ♇
22 28 ♀ ⅊ ♄
22 40 ☽ ∠ ♃
23 03 ☽ ⚹ ☉

All ephemeris data is given at 12.00 UT and the Moon's longitude is additionally given for 24.00 UT

SEPTEMBER 1994

LONGITUDES

Date	Sidereal time h m s	Sun ☉ ° ' "	Moon ☽ ° ' "	Moon ☽ 24.00 ° ' "	Mercury ☿ ° '	Venus ♀ ° '	Mars ♂ ° '	Jupiter ♃ ° '	Saturn ♄ ° '	Uranus ♅ ° '	Neptune ♆ ° '	Pluto ♇ ° '
01	10 41 41	08 ♍ 49 20	15 ♋ 06 43	21 ♋ 31 03	25 ♍ 42	24 ♎ 35	10 ♋ 06	09 ♏ 53	09 ♓ 02	22 ♑ 46	20 ♑ 50	25 ♏ 29
02	10 45 37	09 47 26	28 ♋ 01 13	04 ♌ 37 23	27 18	25 29	10 44	10 02	08 R 57	22 R 44	20 R 49	25 30
03	10 49 34	10 45 34	11 ♌ 19 36	18 ♌ 07 46	28 ♍ 54	26 22	11 21	10 11	08 52	22 43	20 48	25 31
04	10 53 30	11 43 43	25 ♌ 01 37	02 ♍ 00 47	00 ♎ 28	27 15	11 59	10 21	08 48	22 42	20 47	25 32
05	10 57 27	12 41 54	09 ♍ 04 43	16 ♍ 12 48	02 00	28 07	12 36	10 30	08 43	22 40	20 46	25 33
06	11 01 23	13 40 07	23 ♍ 24 16	00 ♎ 38 21	03 32	28 58	13 14	10 40	08 39	22 39	20 45	25 34
07	11 05 20	14 38 21	07 ♎ 54 13	15 ♎ 11 02	05 02	29 ♎ 49	13 51	10 50	08 34	22 38	20 45	25 35
08	11 09 17	15 36 37	22 ♎ 28 03	29 ♎ 44 31	06 31	00 ♏ 39	14 28	11 00	08 30	22 37	20 44	25 36
09	11 13 13	16 34 55	06 ♏ 59 49	14 ♏ 13 25	07 59	01 29	15 05	11 10	08 25	22 36	20 43	25 37
10	11 17 10	17 33 14	21 ♏ 24 53	28 ♏ 33 53	09 26	02 18	15 42	11 20	08 21	22 35	20 42	25 38
11	11 21 06	18 31 35	05 ♐ 40 10	12 ♐ 43 35	10 51	03 06	16 19	11 30	08 16	22 34	20 42	25 39
12	11 25 03	19 29 57	19 ♐ 44 02	26 ♐ 41 29	12 15	03 53	16 56	11 40	08 12	22 33	20 41	25 41
13	11 28 59	20 28 21	03 ♑ 35 53	10 ♑ 27 15	13 38	04 39	17 33	11 50	08 07	22 32	20 40	25 42
14	11 32 56	21 26 47	17 ♑ 15 36	24 ♑ 00 57	14 59	05 25	18 09	12 01	08 03	22 31	20 40	25 43
15	11 36 52	22 25 14	00 ♒ 43 16	07 ♒ 22 33	16 19	06 10	18 46	12 11	07 59	22 30	20 39	25 44
16	11 40 49	23 23 42	13 ♒ 58 45	20 ♒ 31 51	17 38	06 53	19 22	12 22	07 54	22 29	20 39	25 46
17	11 44 46	24 22 13	27 ♒ 01 48	03 ♓ 28 31	18 55	07 36	19 58	12 33	07 50	22 29	20 38	25 47
18	11 48 42	25 20 45	09 ♓ 52 00	16 ♓ 12 12	20 11	08 18	20 34	12 43	07 46	22 28	20 38	25 49
19	11 52 39	26 19 18	22 ♓ 29 07	28 ♓ 42 47	21 25	08 59	21 10	12 54	07 41	22 27	20 37	25 50
20	11 56 35	27 17 54	04 ♈ 53 17	11 ♈ 00 44	22 37	09 39	21 46	13 05	07 37	22 27	20 37	25 51
21	12 00 32	28 16 32	17 ♈ 05 16	23 ♈ 07 08	23 48	10 17	22 22	13 16	07 33	22 26	20 36	25 53
22	12 04 28	29 ♍ 15 22	29 ♈ 06 36	05 ♉ 03 58	24 56	10 55	22 57	13 27	07 29	22 26	20 36	25 54
23	12 08 25	00 ♎ 13 54	10 ♉ 59 38	16 ♉ 54 01	26 03	11 31	23 33	13 38	07 25	22 25	20 36	25 56
24	12 12 21	01 12 38	22 ♉ 48 56	28 ♉ 40 53	27 08	12 06	24 08	13 50	07 21	22 25	20 36	25 58
25	12 16 18	02 11 24	04 ♊ 34 26	10 ♊ 28 50	28 10	12 40	24 43	14 01	07 17	22 24	20 35	25 59
26	12 20 15	03 10 13	16 ♊ 24 42	22 ♊ 22 41	29 ♎ 10	13 12	25 19	14 12	07 13	22 24	20 35	26 01
27	12 24 11	04 09 03	28 ♊ 23 26	04 ♋ 27 34	00 ♏ 07	13 43	25 53	14 24	07 09	22 24	20 35	26 02
28	12 28 08	05 07 57	10 ♋ 35 44	16 ♋ 48 33	01 02	14 12	26 28	14 35	07 06	22 24	20 35	26 04
29	12 32 04	06 06 52	23 ♋ 06 35	29 ♋ 30 21	01 53	14 40	27 03	14 47	07 02	22 23	20 35	26 06
30	12 36 01	07 ♎ 05 50	06 ♌ 00 16	12 ♌ 36 41	02 ♏ 41	15 ♏ 07	27 ♋ 38	14 ♏ 58	06 ♓ 58	22 ♑ 23	20 ♑ 35	26 ♏ 07

Date	Moon True ☊ ° '	Moon Mean ☊ ° '	Moon ☽ Latitude ° '
01	17 ♏ 32	18 ♏ 18	04 S 17
02	17 R 24	18 09	04 47
03	17 15	18 06	05 02
04	17 04	18 02	05 01
05	16 54	17 59	04 41
06	16 45	17 56	04 04
07	16 37	17 53	03 11
08	16 33	17 50	02 05
09	16 31	17 46	00 S 51
10	16 D 30	17 43	00 N 26
11	16 31	17 40	01 41
12	16 R 31	17 37	02 48
13	16 31	17 34	03 45
14	16 28	17 31	04 28
15	16 23	17 27	04 55
16	16 16	17 24	05 06
17	16 07	17 21	05 01
18	15 58	17 18	04 41
19	15 49	17 15	04 07
20	15 41	17 12	03 21
21	15 36	17 08	02 28
22	15 32	17 05	01 28
23	15 30	17 02	00 N 24
24	15 D 31	16 59	00 S 40
25	15 32	16 56	01 42
26	15 33	16 52	02 40
27	15 34	16 49	03 32
28	15 R 34	16 46	04 16
29	15 33	16 43	04 48
30	15 ♏ 30	16 ♏ 40	05 S 07

DECLINATIONS

Date	Sun ☉ ° '	Moon ☽ ° '	Mercury ☿ ° '	Venus ♀ ° '	Mars ♂ ° '	Jupiter ♃ ° '	Saturn ♄ ° '	Uranus ♅ ° '	Neptune ♆ ° '	Pluto ♇ ° '
01	08 N 16	18 N 20	02 N 07	12 S 06	23 N 29	13 S 53	10 S 01	22 S 02	21 S 14	05 S 48
02	07 54	15 52	01 22	12 33	23 27	13 56	10 03	22 02	21 14	05 49
03	07 32	12 32	00 N 38	12 58	23 25	13 59	10 05	22 02	21 14	05 49
04	07 10	08 27	00 S 07	13 24	23 22	14 02	10 06	22 03	21 14	05 50
05	06 48	03 N 49	00 51	13 49	23 20	14 05	10 08	22 03	21 14	05 51
06	06 25	01 S 07	01 34	14 14	23 17	14 09	10 10	22 03	21 14	05 51
07	06 03	06 04	02 17	14 39	23 14	14 12	10 12	22 03	21 15	05 52
08	05 40	10 41	03 00	15 03	23 11	14 15	10 13	22 03	21 15	05 53
09	05 18	14 39	03 42	15 27	23 08	14 18	10 15	22 04	21 15	05 54
10	04 55	17 41	04 24	15 51	23 05	14 21	10 17	22 04	21 15	05 54
11	04 32	19 36	05 05	16 14	23 02	14 25	10 18	22 04	21 15	05 55
12	04 09	20 15	05 45	16 37	22 58	14 28	10 20	22 04	21 15	05 56
13	03 47	19 38	06 25	17 00	22 54	14 31	10 22	22 04	21 15	05 57
14	03 24	17 54	07 04	17 22	22 51	14 35	10 23	22 04	21 16	05 57
15	03 00	15 11	07 42	17 44	22 47	14 38	10 25	22 04	21 16	05 58
16	02 37	11 44	08 20	18 05	22 43	14 41	10 27	22 05	21 16	05 59
17	02 14	07 47	08 57	18 26	22 38	14 45	10 28	22 05	21 16	06 00
18	01 51	03 S 32	09 33	18 46	22 34	14 48	10 30	22 05	21 16	06 00
19	01 28	00 N 48	10 08	19 06	22 30	14 52	10 32	22 05	21 16	06 01
20	01 04	05 01	10 42	19 25	22 25	14 55	10 33	22 05	21 16	06 02
21	00 41	08 59	11 16	19 44	22 20	14 59	10 35	22 05	21 16	06 03
22	00 N 18	12 31	11 48	20 03	22 15	15 02	10 36	22 05	21 16	06 03
23	00 S 06	15 31	12 19	20 20	22 10	15 06	10 38	22 05	21 16	06 04
24	00 29	17 50	12 49	20 38	22 05	15 09	10 39	22 05	21 16	06 05
25	00 52	19 23	13 18	20 55	22 00	15 13	10 40	22 05	21 17	06 06
26	01 16	20 05	13 46	21 11	21 55	15 16	10 42	22 05	21 17	06 06
27	01 39	19 54	14 12	21 26	21 49	15 20	10 43	22 05	21 17	06 07
28	02 02	18 46	14 37	21 41	21 44	15 23	10 45	22 05	21 17	06 08
29	02 26	16 43	15 01	21 56	21 38	15 27	10 46	22 05	21 17	06 09
30	02 S 49	13 N 48	15 S 22	22 S 09	21 N 33	15 S 30	10 S 47	22 S 05	21 S 17	06 S 10

ZODIAC SIGN ENTRIES

Date	h	m	Planets
02	15	37	☽ ♌
04	04	55	☿ ♎
04	20	33	☽ ♍
06	22	57	☽ ♎
07	17	12	♀ ♏
09	00	26	☽ ♏
11	02	25	☽ ♐
13	05	44	☽ ♑
15	10	42	☽ ♒
17	17	31	☽ ♓
20	02	30	☽ ♈
22	13	47	☽ ♉
23	06	19	☉ ♎
25	02	41	☽ ♊
27	08	51	☿ ♏
27	15	12	☽ ♋
30	00	55	☽ ♌

LATITUDES

Date	Mercury ☿ ° '	Venus ♀ ° '	Mars ♂ ° '	Jupiter ♃ ° '	Saturn ♄ ° '	Uranus ♅ ° '	Neptune ♆ ° '	Pluto ♇ ° '
01	00 N 27	02 S 47	00 N 26	00 N 57	01 S 59	00 S 31	00 N 36	13 N 42
04	00 N 05	03 07	00 29	00 56	01 59	00 31	00 36	13 41
07	00 S 19	03 28	00 32	00 56	01 59	00 31	00 36	13 39
10	00 43	03 49	00 35	00 55	01 59	00 31	00 36	13 38
13	01 07	04 10	00 38	00 54	01 59	00 31	00 36	13 36
16	01 32	04 31	00 41	00 54	01 59	00 31	00 36	13 35
19	01 55	04 52	00 44	00 53	01 59	00 31	00 36	13 33
22	02 18	05 13	00 47	00 53	01 59	00 31	00 36	13 32
25	02 40	05 32	00 50	00 53	01 59	00 31	00 36	13 31
28	02 59	05 51	00 53	00 52	01 59	00 31	00 36	13 30
31	03 S 14	06 S 09	00 N 57	00 N 52	01 S 59	00 S 31	00 N 35	13 N 28

DATA

Julian Date	2449597
Delta T	+60 seconds
Ayanamsa	23° 47' 11"
Synetic vernal point	05° ♓ 19' 48"
True obliquity of ecliptic	23° 26' 18"

MOON'S PHASES, APSIDES AND POSITIONS ☽

Date	h	m	Phase	Longitude ° '	Eclipse Indicator
05	18	33	●	12 ♍ 58	
12	11	34	☽	19 ♐ 29	
19	20	01	○	26 ♓ 39	
28	00	23	☾	04 ♋ 39	

Day	h	m		
08	14	17	Perigee	
24	11	54	Apogee	
06	06	36	0S	
12	12	16	Max dec	20° S 15'
19	07	35	0N	
26	18	52	Max dec	20° N 08'

LONGITUDES

Date	Chiron ⚷ ° '	Ceres ⚳ ° '	Pallas ⚴ ° '	Juno ⚵ ° '	Vesta ⚶ ° '	Black Moon Lilith ⚸ ° '
01	13 ♍ 51	22 ♋ 30	22 ♉ 25	01 ♏ 02	28 ♊ 42	16 ♉ 21
11	15 ♍ 16	26 ♋ 22	23 ♉ 42	03 ♏ 55	01 ♋ 43	17 ♉ 27
21	16 ♍ 41	00 ♌ 05	24 ♉ 22	06 ♏ 56	04 ♋ 25	18 ♉ 34
31	18 ♍ 04	03 ♌ 39	24 ♉ 17	10 ♏ 04	06 ♋ 45	19 ♉ 40

All ephemeris data is given at 12.00 UT and the Moon's longitude is additionally given for 24.00 UT

ASPECTARIAN

h m	Aspects
01 Thursday	
00 30	☽ △ ♄
01 10	♂ △ ♃
01 57	☽ △ ♃
01 59	☽ ☌ ♂
03 12	☽ ⚼ ♇
08 44	☿ ⚹ ♇
08 56	☽ Q ☿
16 40	☉ ☍ ♄
22 42	☽ ☍ ♆
02 Friday	
02 17	☽ ☍ ♅
02 55	♂ ⚼ ♇
04 33	☽ ⚼ ♄
05 35	☽ ∠ ☉
06 59	☽ □ ♀
07 21	☽ △ ♇
10 31	☽ ⚹ ☿
12 30	♀ ⊻ ♇
19 12	☉ ⚹ ♃
20 55	☽ ± ♄
23 19	☽ ⊥ ☉
03 Saturday	
02 27	☽ ⋕ ♃
07 39	☽ ⊼ ♄
09 28	☽ ⋕ ♀
09 57	☽ □ ♃
10 55	☽ ⊻ ☉
12 03	☽ ⊻ ♂
17 09	☽ ∠ ☿
17 45	☽ Q ♀
23 09	☽ ⊥ ♂
04 Sunday	
02 49	☽ ⋕ ♄
04 39	☽ ⊼ ♆
07 58	☽ ⊼ ♅
10 54	☽ ⊥ ☿
12 52	☽ □ ♇
15 02	☽ ± ♆
15 32	☽ ∠ ♂
16 05	☽ ⚹ ♀
17 47	☽ Q ♃
18 19	☽ ± ♅
19 31	☽ ∥ ☉
22 30	☽ ⊻ ☿
05 Monday	
01 48	☽ ⋕ ♇
05 41	☉ ⚹ ♂
06 24	☽ ⚼ ♆
09 38	☽ ⚼ ♅
11 24	☽ ☍ ♄
14 26	☽ ⚹ ♃
18 13	☽ ⚹ ♂
18 33	☽ ☌ ☉
19 14	☽ ∠ ♀
19 32	☽ Q ♇
06 Tuesday	
00 41	☽ ⋕ ☿
05 28	♀ ∥ ♃
07 36	☽ △ ♆
09 17	☉ Q ♇
10 06	♃ Q ♅
10 45	☽ △ ♅
11 14	☽ ⊥ ♀
14 31	☽ ∥ ☿
15 10	☽ Q ♂
15 35	☽ ⚹ ♇
15 48	☽ ∠ ♃
21 49	☽ ⊻ ♀
07 Wednesday	
06 44	☽ ☌ ☿
06 52	☽ ⊥ ♃
08 17	☿ ⊥ ♃
11 03	☽ ∥ ♇
11 57	☽ ⋕ ☉
13 06	☽ ⊼ ♄
16 25	☽ ∠ ♇
16 53	☽ ⊻ ♃
22 15	☽ □ ♂
22 56	☽ ± ♄
23 05	☉ ⋕ ♇
23 54	☽ ⊻ ☉
08 Thursday	
07 16	☽ ⊥ ♇
09 08	☽ □ ♆
09 30	☽ ∥ ♄
10 29	☽ ⊥ ☉
12 15	☽ □ ♅
13 41	☽ ⚼ ♄
17 10	☽ ⊻ ♇
19 28	☉ ∠ ♀
09 Friday	
02 19	☽ ☌ ♀
02 24	☽ ∠ ☉
09 40	☽ ∥ ♃
13 49	☽ ⊻ ☿
14 21	☽ △ ♄
14 51	☽ Q ♆
17 58	☽ Q ♅
18 25	☽ ∥ ♀
18 47	☿ ⊼ ♄
18 59	☽ ☌ ♃
10 Saturday	
00 54	☽ ⊥ ☿
02 03	☽ △ ♂
05 05	☽ ⚹ ☉
10 49	☽ ⚹ ♆
13 57	☽ ⚹ ♅
17 37	☽ ∠ ☿
19 05	☽ ☌ ♇
23 51	☉ ⋕ ☿
11 Sunday	
02 40	☽ Q ☉
04 19	☽ ⚼ ♂
07 23	☽ ⊻ ♀
08 36	☿ ∠ ♇
12 03	☽ ∠ ♆
15 12	☽ ∠ ♅
16 23	☽ □ ♄
18 09	☽ ⊥ ♀
20 16	☽ ± ♂
21 47	☽ ⚹ ☿
22 02	☽ ⊻ ♃
12 Monday	
00 32	☿ ⊻ ♃
03 21	☽ ⊥ ♆
06 32	☽ ⊥ ♅
06 58	☽ ⊼ ♂
08 25	☽ ⊥ ♃
10 27	☽ ∠ ♀
11 34	☽ □ ☉
13 38	☽ ⊻ ♆
16 50	☽ ⊻ ♅
18 41	☿ ∥ ♇
20 39	☽ Q ☿
22 15	☽ ⊻ ♇
23 05	☽ Q ♄
13 Tuesday	
00 07	☽ ∠ ♃
08 41	☽ ⊥ ♇
13 57	☽ ⚹ ♀
16 54	☉ △ ♆
19 52	☽ ⚹ ♄
20 08	☿ ± ♄
14 Wednesday	
00 27	☽ ∠ ♇
02 37	☽ ⚹ ♃
07 33	☽ □ ☿
12 17	☽ Q ♀
13 39	☽ ☍ ♂
16 41	☽ ∥ ♀
18 02	☽ ☌ ♆
20 00	☽ △ ☉
21 19	☽ ☌ ♅
22 13	☽ ∠ ♄
15 Thursday	
00 09	☽ Q ♃
03 04	☽ ⚹ ♇
13 58	☉ △ ♅
14 15	☽ ⊥ ♄
16 06	☽ ∥ ♃
22 23	☽ □ ♀
16 Friday	
00 41	☽ Q ♇
01 01	☽ ⊻ ♄
01 02	☽ ⚼ ☉
09 01	☽ □ ♃
18 45	☽ ± ☉
19 25	☽ △ ☿
20 04	☽ ∥ ♄
22 20	☽ ⊼ ♂
17 Saturday	
00 12	☽ ⊻ ♆
03 35	☽ ⊻ ♅
06 04	☽ ∥ ☿
06 41	☽ ⊼ ☉
09 41	☽ □ ♇
09 56	☽ ± ♂
11 16	☽ ⊥ ♆
14 41	☽ ⊥ ♅
19 02	♀ △ ♄
22 12	☽ ∥ ♇
18 Sunday	
02 14	☽ ⚼ ☿
03 31	☽ ⚼ ♂
04 02	☽ ∠ ♆
04 44	☿ ⊥ ♇
07 29	☽ ∠ ♅
08 03	☽ ☌ ♄
08 53	☽ △ ♀
14 20	♂ ☍ ♆
17 28	☽ △ ♃
20 36	☿ □ ♆
21 03	☽ ± ☿
22 15	☽ ⋕ ☉
23 21	♀ Q ♆
23 40	☉ ⚹ ♇
19 Monday	
02 36	☿ □ ♂
08 26	☽ ⚹ ♆
09 21	☽ △ ♂
09 44	☽ ⊼ ☿
11 56	☽ ⚹ ♅
15 02	☽ ⚼ ♀
15 25	☽ ∥ ☉
18 27	☽ △ ♇
20 01	☽ ☍ ☉
22 35	☽ ⚼ ♃
20 Tuesday	
05 11	☿ ∥ ♄
07 34	☽ Q ♆
08 29	☿ □ ♅
09 26	☽ ± ♀
11 08	☽ Q ♅
12 01	☿ ⚼ ♄
16 22	☽ ± ♃
17 19	☽ ⊻ ♄
17 56	☽ ⋕ ♇
21 50	☽ ⊼ ♀
23 43	☽ ⚼ ♇
21 Wednesday	
04 20	☽ ⊼ ♃
05 03	☽ ⊥ ♄
11 47	☉ ∠ ♃
14 55	♂ ☍ ♅
17 32	♀ Q ♅
17 34	☽ ± ♇
18 53	♂ ⚼ ♄
18 59	☽ □ ♆
22 28	☽ ⋕ ♄
22 38	☽ □ ♅
22 48	☽ ∠ ♄
23 02	☽ □ ♂
22 Thursday	
02 46	☽ ☍ ☿
05 34	☽ ⊼ ♇
05 50	☽ ⋕ ☿
12 19	☽ ⊼ ☉
23 Friday	
01 29	☽ ± ☉
04 48	☽ ⚹ ♄
08 13	☽ ⋕ ♃
09 17	☿ ⊻ ♇
10 54	♄ ∠ ♅
13 07	☽ ☍ ♀
13 11	☽ Q ♂
17 27	☽ ☍ ♃
21 23	☽ ⚼ ☉
24 Saturday	
05 01	☽ Q ♄
07 31	☽ △ ♆
11 13	☽ △ ♅
12 25	♂ ⋕ ♅
14 53	☽ ⚹ ♂
18 28	☽ ☍ ♇
21 42	☽ ⊼ ☿
25 Sunday	
06 42	☽ △ ☉
11 06	☽ ± ☿
14 04	☽ ⚼ ♆
17 29	☽ □ ♄
17 45	☽ ⚼ ♅
23 01	☽ ∠ ♂
26 Monday	
05 12	☽ ⊼ ♀
07 04	☽ ⚼ ☿
07 28	☽ ⊼ ♃
08 19	☽ ± ♆
11 59	☽ ± ♅
17 53	☽ ± ♀
18 08	☽ ⊥ ♂
19 46	☽ ± ♃
20 24	☽ ⊼ ♆
20 57	♀ ∥ ♆
27 Tuesday	
00 02	☽ ⊼ ♅
06 46	☽ ⊻ ♂
07 19	☽ ⊼ ♇
12 40	☽ ⚼ ♀
14 01	☽ ⚼ ♃
15 43	☽ △ ☿
18 24	♂ △ ♇
19 15	☽ ± ♇
28 Wednesday	
00 23	☽ □ ☉
05 12	☽ △ ♄
12 55	☽ ⚼ ♇
15 07	♀ ⋕ ♂
19 16	☽ △ ♀
19 51	☽ △ ♃
29 Thursday	
07 12	☽ ☍ ♆
09 58	☽ ⚼ ♄
10 38	☽ ☍ ♅
14 03	☽ Q ☉
17 38	☽ △ ♇
19 46	☽ ☌ ♂
21 51	♀ ☌ ♃
23 09	☽ ⋕ ♃
30 Friday	
01 15	☽ ⋕ ☿
02 46	☽ ± ♄
05 02	♀ ∥ ♅
05 31	☽ □ ☿
09 05	☉ ⊼ ♄
13 45	☽ ⊼ ♄
14 10	☽ ⚹ ☉
23 08	☿ ∥ ♃

OCTOBER 1994

LONGITUDES

Date	Sidereal time h m s	Sun ☉ ° ' "	Moon ☽ ° ' "	Moon ☽ 24.00 ° ' "	Mercury ☿ ° '	Venus ♀ ° '	Mars ♂ ° '	Jupiter ♃ ° '	Saturn ♄ ° '	Uranus ♅ ° '	Neptune ♆ ° '	Pluto ♇ ° '
01	12 39 57	08 ♎ 04 49	19 ♌ 19 49	26 ♌ 09 44	03 ♏ 26	15 ♏ 32	28 ♋ 12	15 ♏ 10	06 ♓ 55	22 ♑ 23	20 ♑ 34	26 ♏ 09
02	12 43 54	09 03 51	03 ♍ 06 22	10 ♍ 09 28	04 07	15 55	28 46	15 22	06 R 51	22 D 23	20 R 34	26 11
03	12 47 50	10 02 56	17 ♍ 18 37	24 ♍ 33 15	04 43	16 16	29 21	15 34	06 48	22 23	20 D 34	26 13
04	12 51 47	11 02 02	01 ♎ 52 36	09 ♎ 15 50	05 15	16 36	29 ♋ 55	15 46	06 44	22 23	20 34	26 15
05	12 55 44	12 01 11	16 ♎ 41 56	24 ♎ 09 54	05 41	16 54	00 ♌ 28	15 58	06 41	22 24	20 35	26 16
06	12 59 40	13 00 21	01 ♏ 38 38	09 ♏ 07 07	06 03	17 09	01 02	16 10	06 38	22 24	20 35	26 18
07	13 03 37	13 59 34	16 ♏ 34 21	23 ♏ 59 25	06 18	17 23	01 36	16 22	06 35	22 24	20 35	26 20
08	13 07 33	14 58 48	01 ♐ 21 33	08 ♐ 40 05	06 27	17 35	02 09	16 34	06 31	22 24	20 35	26 22
09	13 11 30	15 58 05	15 ♐ 54 29	23 ♐ 04 22	06 R 29	17 44	02 42	16 46	06 28	22 25	20 35	26 24
10	13 15 26	16 57 23	00 ♑ 09 30	07 ♑ 09 42	06 23	17 52	03 16	16 58	06 26	22 25	20 35	26 26
11	13 19 23	17 56 42	14 ♑ 04 56	20 ♑ 55 14	06 10	17 57	03 49	17 11	06 23	22 26	20 36	26 28
12	13 23 19	18 56 04	27 ♑ 40 44	04 ♒ 21 33	05 49	18 00	04 21	17 23	06 20	22 26	20 36	26 30
13	13 27 16	19 55 27	10 ♒ 57 53	17 ♒ 29 58	05 20	18 R 01	04 54	17 35	06 17	22 27	20 36	26 32
14	13 31 13	20 54 52	23 ♒ 58 00	00 ♓ 22 13	04 42	17 59	05 26	17 48	06 15	22 27	20 37	26 34
15	13 35 09	21 54 19	06 ♓ 42 52	13 ♓ 00 08	03 56	17 55	05 59	18 00	06 12	22 28	20 37	26 36
16	13 39 06	22 53 47	19 ♓ 14 16	25 ♓ 25 27	03 02	17 48	06 31	18 13	06 10	22 29	20 38	26 38
17	13 43 02	23 53 18	01 ♈ 33 53	07 ♈ 39 48	02 01	17 39	07 03	18 25	06 07	22 29	20 38	26 40
18	13 46 59	24 52 50	13 ♈ 43 22	19 ♈ 44 47	00 ♏ 54	17 27	07 34	18 38	06 05	22 30	20 39	26 42
19	13 50 55	25 52 24	25 ♈ 44 17	01 ♉ 42 05	29 ♎ 43	17 13	08 06	18 51	06 03	22 31	20 39	26 44
20	13 54 52	26 52 01	07 ♉ 38 26	13 ♉ 33 34	28 29	16 57	08 37	19 03	06 01	22 32	20 40	26 46
21	13 58 48	27 51 39	19 ♉ 27 48	25 ♉ 21 27	27 14	16 38	09 09	19 16	05 59	22 33	20 40	26 49
22	14 02 45	28 51 20	01 ♊ 14 50	07 ♊ 08 21	26 01	16 18	09 40	19 29	05 57	22 34	20 41	26 51
23	14 06 42	29 ♎ 51 02	13 ♊ 02 24	18 ♊ 57 25	24 52	15 54	10 11	19 42	05 55	22 35	20 42	26 53
24	14 10 38	00 ♏ 50 47	24 ♊ 53 52	00 ♋ 52 16	23 48	15 29	10 41	19 54	05 54	22 36	20 42	26 55
25	14 14 35	01 50 34	06 ♋ 53 08	12 ♋ 57 00	22 52	15 02	11 12	20 07	05 52	22 37	20 43	26 57
26	14 18 31	02 50 24	19 ♋ 04 25	25 ♋ 15 57	22 06	14 33	11 42	20 20	05 51	22 39	20 44	27 00
27	14 22 28	03 50 15	01 ♌ 32 08	07 ♌ 53 30	21 30	14 03	12 12	20 33	05 49	22 40	20 45	27 02
28	14 26 24	04 50 09	14 ♌ 20 33	20 ♌ 53 43	21 05	13 31	12 42	20 46	05 48	22 41	20 46	27 04
29	14 30 21	05 50 04	27 ♌ 33 21	04 ♍ 19 44	20 51	12 57	13 11	20 59	05 47	22 42	20 46	27 06
30	14 34 17	06 50 02	11 ♍ 13 01	18 ♍ 13 11	20 D 50	12 23	13 41	21 12	05 46	22 44	20 47	27 09
31	14 38 14	07 ♏ 50 03	25 ♍ 20 08	02 ♎ 33 31	20 ♎ 59	11 ♏ 47	14 ♌ 10	21 ♏ 25	05 ♓ 45	22 ♑ 45	20 ♑ 48	27 ♏ 11

Date	Moon True ☊ ° '	Moon Mean ☊ ° '	Moon ☽ Latitude ° '
01	15 ♏ 25	16 ♏ 37	05 S 11
02	15 R 20	16 33	04 57
03	15 14	16 30	04 24
04	15 09	16 27	03 34
05	15 05	16 24	02 29
06	15 03	16 21	01 S 13
07	15 D 02	16 18	00 N 08
08	15 03	16 14	01 28
09	15 05	16 11	02 42
10	15 06	16 08	03 43
11	15 07	16 05	04 30
12	15 R 07	16 02	05 00
13	15 05	15 58	05 14
14	15 03	15 55	05 11
15	14 59	15 52	04 53
16	14 56	15 49	04 20
17	14 52	15 46	03 37
18	14 49	15 43	02 43
19	14 47	15 39	01 43
20	14 46	15 36	00 N 39
21	14 D 46	15 33	00 S 26
22	14 47	15 30	01 30
23	14 48	15 27	02 31
24	14 49	15 24	03 25
25	14 50	15 20	04 11
26	14 51	15 17	04 46
27	14 52	15 14	05 09
28	14 R 52	15 11	05 18
29	14 51	15 08	05 10
30	14 50	15 04	04 45
31	14 ♏ 49	15 ♏ 01	04 S 02

DECLINATIONS

Date	Sun ☉ ° '	Moon ☽ ° '	Mercury ☿ ° '	Venus ♀ ° '	Mars ♂ ° '	Jupiter ♃ ° '	Saturn ♄ ° '	Uranus ♅ ° '	Neptune ♆ ° '	Pluto ♇ ° '
01	03 S 12	10 N 06	15 S 42	22 S 22	21 N 27	15 S 34	10 S 49	22 S 05	21 S 17	06 S 10
02	03 36	05 45	16 00	22 34	21 21	15 37	10 50	22 05	21 17	06 11
03	03 59	00 N 57	16 16	22 46	21 15	15 41	10 51	22 05	21 17	06 12
04	04 22	04 S 01	16 30	22 57	21 09	15 44	10 52	22 05	21 17	06 13
05	04 45	08 51	16 41	23 07	21 03	15 48	10 53	22 05	21 17	06 13
06	05 08	13 11	16 49	23 16	20 56	15 52	10 55	22 05	21 17	06 14
07	05 31	16 39	16 55	23 24	20 50	15 55	10 56	22 05	21 17	06 15
08	05 54	18 59	16 58	23 31	20 44	15 59	10 57	22 05	21 17	06 16
09	06 17	20 01	16 57	23 38	20 37	16 02	10 58	22 05	21 17	06 16
10	06 40	19 43	16 52	23 43	20 31	16 06	10 59	22 05	21 17	06 17
11	07 02	18 13	16 44	23 47	20 24	16 09	11 00	22 05	21 17	06 18
12	07 25	15 43	16 31	23 51	20 18	16 13	11 01	22 05	21 17	06 19
13	07 47	12 27	16 15	23 53	20 11	16 17	11 02	22 04	21 17	06 20
14	08 10	08 38	15 53	23 54	20 04	16 20	11 03	22 04	21 17	06 20
15	08 32	04 31	15 27	23 54	19 57	16 24	11 03	22 04	21 17	06 21
16	08 54	00 S 16	14 57	23 53	19 51	16 27	11 04	22 04	21 17	06 22
17	09 16	03 N 56	14 22	23 51	19 44	16 31	11 05	22 04	21 17	06 23
18	09 38	07 55	13 44	23 47	19 37	16 34	11 06	22 04	21 17	06 23
19	10 00	11 33	13 02	23 42	19 30	16 38	11 06	22 04	21 17	06 24
20	10 21	14 41	12 18	23 35	19 23	16 42	11 07	22 03	21 17	06 25
21	10 43	17 11	11 32	23 28	19 16	16 45	11 08	22 03	21 17	06 26
22	11 04	18 56	10 46	23 18	19 09	16 49	11 08	22 03	21 16	06 26
23	11 25	19 52	10 02	23 08	19 02	16 52	11 09	22 03	21 16	06 27
24	11 46	19 56	09 19	22 56	18 55	16 56	11 09	22 03	21 16	06 28
25	12 07	19 05	08 40	22 43	18 47	16 59	11 10	22 02	21 16	06 28
26	12 27	17 21	08 06	22 28	18 40	17 03	11 10	22 02	21 16	06 29
27	12 48	14 47	07 36	22 12	18 33	17 06	11 11	22 02	21 16	06 30
28	13 08	11 27	07 12	21 55	18 26	17 10	11 11	22 02	21 16	06 31
29	13 28	07 28	06 54	21 36	18 19	17 13	11 11	22 02	21 16	06 31
30	13 48	02 58	06 42	21 17	18 12	17 17	11 12	22 01	21 16	06 32
31	14 S 07	01 S 51	06 S 35	20 S 56	18 N 05	17 S 20	11 S 12	22 S 01	21 S 16	06 S 33

ZODIAC SIGN ENTRIES

Date	h	m	Planets
02	06	39	☽ ♍
04	08	56	☽ ♎
04	15	48	♂ ♌
06	09	22	☽ ♏
08	09	47	☽ ♐
10	11	44	☽ ♑
12	16	09	☽ ♒
14	23	18	☽ ♓
17	08	56	☽ ♈
19	06	19	☿ ♎
19	20	34	☽ ♉
22	09	28	☽ ♊
23	15	36	☉ ♏
24	22	15	☽ ♋
27	09	05	☽ ♌
29	16	21	☽ ♍
31	19	46	☽ ♎

LATITUDES

Date	Mercury ☿ ° '	Venus ♀ ° '	Mars ♂ ° '	Jupiter ♃ ° '	Saturn ♄ ° '	Uranus ♅ ° '	Neptune ♆ ° '	Pluto ♇ ° '
01	03 S 14	06 S 09	00 N 57	00 N 52	01 S 59	00 S 31	00 N 35	13 N 28
04	03 25	06 25	01 00	00 51	01 58	00 31	00 35	13 27
07	03 30	06 39	01 04	00 51	01 58	00 31	00 35	13 26
10	03 25	06 51	01 07	00 50	01 58	00 31	00 35	13 25
13	03 08	06 59	01 11	00 50	01 58	00 31	00 35	13 24
16	02 35	07 02	01 14	00 50	01 57	00 31	00 35	13 23
19	01 46	07 01	01 18	00 49	01 57	00 31	00 35	13 22
22	00 S 47	06 54	01 22	00 49	01 57	00 31	00 35	13 21
25	00 N 14	06 40	01 26	00 49	01 56	00 30	00 35	13 20
28	01 06	06 19	01 30	00 49	01 56	00 30	00 35	13 20
31	01 N 43	05 S 51	01 N 34	00 N 48	01 S 56	00 S 30	00 N 34	13 N 19

DATA

Julian Date	2449627
Delta T	+60 seconds
Ayanamsa	23° 47' 14"
Synetic vernal point	05° ♓ 19' 46"
True obliquity of ecliptic	23° 26' 18"

MOON'S PHASES, APSIDES AND POSITIONS ☽

Date	h	m	Phase	Longitude ° '	Eclipse Indicator
05	03	55	●	11 ♎ 41	
11	19	17	☽	18 ♑ 15	
19	12	18	○	25 ♈ 53	
27	16	44	☾	04 ♌ 02	

Day	h	m		
06	13	59	Perigee	
22	01	38	Apogee	
03	16	37	0S	
09	18	29	Max dec	20° S 04'
16	13	30	0N	
24	01	31	Max dec	20° N 01'
31	02	53	0S	

LONGITUDES

Date	Chiron ⚷ ° '	Ceres ⚳ ° '	Pallas ⚴ ° '	Juno ⚵ ° '	Vesta ⚶ ° '	Black Moon Lilith ⚸ ° '
01	18 ♍ 04	03 ♌ 39	24 ♉ 17	10 ♏ 04	06 ♋ 45	19 ♉ 40
11	19 ♍ 25	07 ♌ 02	23 ♉ 20	13 ♏ 17	08 ♋ 37	20 ♉ 47
21	20 ♍ 42	10 ♌ 09	21 ♉ 30	16 ♏ 35	09 ♋ 59	21 ♉ 53
31	21 ♍ 54	13 ♌ 00	18 ♉ 50	19 ♏ 56	10 ♋ 44	23 ♉ 00

ASPECTARIAN

01 Saturday
04 29 ☽ □ ♃
05 01 ☽ □ ♀
07 45 ☽ # ♄
14 12 ☽ ⚻ ♆
15 55 ☽ Q ☿
17 24 ☽ ⚻ ♅
19 07 ☽ ∠ ☉

02 Sunday
00 01 ☽ □ ♇
00 43 ☽ ± ♆
01 46 ♅ St D
03 52 ☽ ± ♅
04 13 ☽ ⚺ ♂
09 43 ☽ # ♇
11 55 ☽ ⊥ ☉
12 27 ☽ Q ♃
13 25 ☽ Q ♀
13 48 ☽ ⚹ ☿
14 59 ☽ ⊥ ♂
16 13 ☽ bQ ♆
17 49 ♆ St D
18 22 ☽ ☍ ♄
19 19 ☽ bQ ♅
21 09 ☉ ⊥ ♃
22 10 ☽ # ☉
22 54 ☽ ⚺ ☉

03 Monday
03 57 ♂ # ♆
06 49 ☽ Q ♇
06 50 ☽ ∠ ♂
09 02 ☽ ⚹ ♃
10 13 ☽ ⚹ ♀
16 10 ☽ ∠ ☿
17 25 ☽ △ ♆
20 13 ☉ ⊥ ♀
20 26 ☽ △ ♅

04 Tuesday
02 46 ☽ ⚹ ♇
07 33 ☽ ⊥ ☿
08 40 ☽ ⚹ ♂
10 09 ☽ ∠ ♃
11 32 ☽ ∠ ♀
13 48 ☽ ∥ ☉
17 15 ☉ ∠ ♇
17 40 ☽ ⚺ ☿
19 53 ☽ ⚻ ♄
22 42 ☽ ∥ ♇

05 Wednesday
00 59 ☽ ⊥ ♃
02 27 ☽ ⊥ ♀
03 14 ☽ ∠ ♇
03 55 ☽ ☌ ☉
04 55 ☽ Q ♂
05 33 ☽ ± ♄
10 48 ☽ ⚺ ♃
12 19 ☽ ⚺ ♀
17 46 ☽ ⊥ ♇
18 14 ☽ □ ♆
19 59 ☽ bQ ♄
20 05 ♂ ± ♄
21 09 ☽ □ ♅
22 52 ☽ ∥ ♄

06 Thursday
03 18 ☉ ± ♄
03 25 ☽ ⚺ ♇
10 59 ☽ □ ♂
13 45 ☉ Q ♂
19 12 ☽ ☌ ☿
19 58 ☽ △ ♄
23 08 ☽ Q ♆

07 Friday
02 03 ☽ Q ♅
06 08 ☽ ∥ ♃
07 33 ☽ ⚺ ☉
11 39 ☽ ☌ ♃
13 20 ☽ ☌ ♀
14 14 ☽ ∥ ☿
17 55 ☽ ⊥ ☉
18 29 ☽ ⚹ ♆
21 26 ☽ ⚹ ♅

08 Saturday
03 51 ☽ ☌ ♇
09 35 ☽ ∠ ☉
13 21 ☽ △ ♂
18 55 ☽ ∠ ♆
20 23 ☽ ⚺ ☿
20 26 ☽ □ ♄
21 55 ☽ ∠ ♅

09 Sunday
06 18 ☽ ⊥ ☿
06 43 ☿ St R
08 27 ☿ △ ♄
09 48 ☽ ⊥ ♆
11 32 ☉ ∥ ♇
12 06 ☽ ⚹ ☉
12 50 ☽ ⊥ ♅
13 27 ☽ ⚺ ♃
15 05 ☽ ⚺ ♀
15 07 ☽ bQ ♂
19 49 ☽ ⚺ ♆
21 17 ☽ ∠ ☿
22 54 ☽ ⚺ ♅
23 39 ☽ ⊥ ♃

10 Monday
00 14 ☿ △ ♄
01 15 ☽ ⊥ ♀
02 19 ☽ Q ♄
05 40 ☽ ⚺ ♇
06 52 ☽ ± ♂
09 48 ☽ Q ☉
12 27 ☉ ⚺ ♃
15 08 ☽ ∠ ♃
15 53 ☽ ⊥ ♇
16 39 ☽ ∠ ♀
17 31 ☽ ⚻ ♂
22 32 ☽ ⚹ ☿
22 42 ☽ ⚹ ♄

11 Tuesday
07 26 ☽ ∠ ♇
12 10 ☉ ⚺ ♀
17 30 ☽ ⚹ ♃
18 48 ☽ ⚹ ♀
19 00 ☽ Q ☿
19 17 ☽ □ ☉
23 26 ☽ ☌ ♆

12 Wednesday
00 46 ☽ ∠ ♄
02 40 ☽ ☌ ♅
04 26 ☽ ∥ ☿
07 53 ☽ ∥ ♃
09 53 ☽ ⚹ ♇
15 06 ☽ Q ♃
16 10 ☽ Q ♀
16 44 ☽ ⊥ ♄

13 Thursday
00 30 ☽ ☍ ♂
02 09 ☽ □ ☿
03 31 ☽ ⚺ ♄
05 41 ♀ St R
07 33 ☽ Q ♇
09 54 ☿ ∥ ♃
21 07 ☿ □ ♂
21 12 ☽ ∥ ♄

14 Friday
00 22 ☽ □ ♃
00 55 ☽ □ ♀
03 15 ☉ ⊥ ♇
04 38 ☉ □ ♆
05 46 ☽ ⚺ ♆
05 51 ☽ △ ☉
09 11 ☽ ⚺ ♅
14 36 ☽ ∥ ☉
16 52 ☽ □ ♇
16 57 ☽ ⊥ ♆
19 41 ☉ bQ ♄
20 24 ☽ ⊥ ♅

15 Saturday
01 29 ☽ ∥ ♇
04 13 ♀ ☌ ♃
07 02 ☽ △ ☿
09 55 ☽ ∠ ♆
10 32 ☽ ⚻ ♂
11 02 ☽ ☌ ♄
12 24 ☽ bQ ☉
13 26 ☽ ∠ ♅
21 25 ♂ ⚻ ♄
22 29 ☽ ± ♂

16 Sunday
01 44 ☉ □ ♅
07 06 ☽ ± ☉
09 15 ☽ △ ♀
09 50 ☽ bQ ☿
09 59 ☽ △ ♃
14 41 ☽ ⚹ ♆
16 36 ☽ bQ ♂
18 17 ☽ ⚹ ♅
19 42 ☽ ⚻ ☉

17 Monday
02 00 ☽ ± ☿
02 24 ☽ △ ♇
12 49 ☽ ⚻ ☿
14 06 ☽ bQ ♀
14 06 ☽ Q ♆
15 43 ☽ bQ ♃
17 45 ☽ Q ♅
20 56 ☽ ⚺ ♄
23 16 ☽ △ ♂

18 Tuesday
02 32 ☽ # ♇
07 35 ☽ ± ♀
07 59 ☽ bQ ♇
08 46 ☽ ⊥ ♄
09 48 ☽ ± ♃
19 18 ☽ ⚻ ♀
21 57 ☽ ⚻ ♃

19 Wednesday
00 11 ☽ # ☉
01 48 ☽ □ ♆
01 57 ☽ ± ♇
02 38 ☽ ∠ ♄
05 32 ☽ □ ♅
08 53 ☽ # ♄
12 18 ☽ ☍ ☉
14 01 ☽ ⚻ ♇
19 15 ☽ ☍ ☿
20 48 ☽ # ☿

20 Thursday
08 44 ☽ ⚹ ♄
09 42 ☉ ⚺ ♇
14 05 ☽ □ ♂

21 Friday
05 17 ☉ ☌ ☿
06 25 ☽ ☍ ♀
07 16 ☽ # ♃
09 00 ☽ Q ♄
11 36 ☽ ☍ ♃
14 28 ☽ △ ♆
18 17 ☽ △ ♅
20 01 ☿ ⚺ ♇

22 Saturday
00 39 ☿ ∥ ♄
02 20 ☽ ⚻ ☿
03 00 ☽ ☍ ♇
04 22 ☽ Q ♂
05 40 ☉ ∥ ☿
06 41 ☽ ⚻ ☉
13 25 ☽ ± ☿
15 31 ☽ ∥ ♂
16 53 ☉ ∥ ♄
20 02 ☽ ± ☉
21 03 ☽ bQ ♆
21 34 ☽ □ ♄

23 Sunday
00 53 ☽ bQ ♅
05 55 ☽ ⚹ ♂
06 06 ☽ bQ ☿
15 22 ☽ ± ♆
16 01 ☽ bQ ☉
17 38 ☽ ⚻ ♀
19 12 ☽ ± ♅

24 Monday
01 44 ☽ ⚻ ♃
03 32 ☽ ⚻ ♆
05 22 ☽ ± ♀
07 22 ☽ ⚻ ♅
09 58 ☽ △ ☿
13 40 ☽ ∠ ♂
14 04 ☽ ± ♃
16 05 ☽ ⚻ ♇
22 50 ☽ bQ ♀

25 Tuesday
01 02 ☽ △ ☉
04 09 ☽ ± ♇
06 22 ☿ Q ♂
08 26 ☽ bQ ♃
08 29 ☽ ⊥ ♂
09 59 ☽ △ ♄
17 33 ☽ ∥ ♂
19 00 ☿ □ ♅
20 54 ☽ ⚺ ♂
22 04 ☽ bQ ♇

26 Wednesday
03 30 ☽ △ ♀
14 30 ☽ △ ♃
15 14 ☽ ☍ ♆
15 15 ☽ # ♃
15 27 ☽ bQ ♄
17 34 ☽ □ ☿
18 57 ☽ ☍ ♅

27 Thursday
03 22 ☽ △ ♇
08 45 ☽ ± ♄
16 44 ☽ □ ☉
20 06 ☽ ⚻ ♄

28 Friday
01 35 ☽ # ☉
02 18 ♀ ∥ ♅
02 29 ☽ Q ☿
08 50 ☽ ☌ ♂
10 31 ☽ □ ♀
10 50 ♃ ⚹ ♆
12 50 ☿ ⊥ ♇
13 45 ☽ # ♄
23 46 ☽ ⚻ ♆
23 58 ☽ □ ♃

29 Saturday
00 05 ☽ ⚹ ☿
03 16 ☽ ⚻ ♅
03 51 ☿ ⚺ ♃
04 47 ☽ Q ☉
06 36 ♀ □ ♂
10 36 ☽ ± ♆
10 46 ☉ △ ♄
11 12 ☽ □ ♇
14 03 ☽ ± ♅
15 19 ☽ # ☿
17 13 ☽ # ♇
17 48 ☽ Q ♀
18 10 ♄ ∠ ♆

30 Sunday
02 32 ☽ ☍ ♄
02 33 ☽ bQ ♆
02 36 ☽ ∠ ☿
03 47 ☽ ⚹ ☉
04 06 ☿ St D
05 57 ☽ bQ ♅
08 27 ☽ Q ♃
13 11 ♀ ∥ ♆
13 55 ☽ ⚹ ♀
16 24 ☽ ⚺ ♂
18 15 ☽ ⊥ ☿
18 46 ☽ Q ♇

31 Monday
03 00 ☽ ⊥ ♂
04 22 ☽ △ ♆
04 34 ☽ ⚺ ☿
05 19 ☽ ⚹ ♃
07 29 ☽ ∠ ☉
07 40 ☽ △ ♅
14 20 ☽ ∠ ♀
15 06 ☽ ⚹ ♇
18 36 ☽ ∠ ♂
23 36 ☽ ⊥ ☉

All ephemeris data is given at 12.00 UT and the Moon's longitude is additionally given for 24.00 UT

NOVEMBER 1994

Date	Sidereal time h m s	Sun ☉ ° ' "	Moon ☽ ° ' "	Moon ☽ 24.00 ° ' "	Mercury ☿ ° '	Venus ♀ ° '	Mars ♂ ° '	Jupiter ♃ ° '	Saturn ♄ ° '	Uranus ♅ ° '	Neptune ♆ ° '	Pluto ♇ ° '
01	14 42 11	08 ♏ 50 05	09 ♎ 52 51	17 ♎ 17 27	21 ♎ 18	11 ♏ 12	14 ♌ 39	21 ♏ 38	05 ♓ 44	22 ♑ 47	20 ♑ 49	27 ♏ 13
02	14 46 07	09 50 09	24 ♎ 46 30	02 ♏ 18 59	21 48	10 R 35	15 08	21 51	05 R 43	22 48	20 50	27 16
03	14 50 04	10 50 16	09 ♏ 53 48	17 ♏ 29 46	22 27	09 59	15 36	22 05	05 43	22 50	20 51	27 18
04	14 54 00	11 50 24	25 ♏ 05 40	02 ♐ 40 18	23 13	09 22	16 04	22 18	05 42	22 52	20 52	27 20
05	14 57 57	12 50 34	10 ♐ 12 30	17 ♐ 41 14	24 07	08 46	16 32	22 31	05 42	22 53	20 53	27 22
06	15 01 53	13 50 46	25 ♐ 05 35	02 ♑ 24 48	25 07	08 11	17 00	22 44	05 41	22 55	20 55	27 25
07	15 05 50	14 50 59	09 ♑ 38 18	16 ♑ 45 40	26 13	07 36	17 28	22 57	05 41	22 57	20 56	27 27
08	15 09 46	15 51 14	23 ♑ 46 39	00 ♒ 41 09	27 24	07 03	17 55	23 10	05 41	22 59	20 57	27 30
09	15 13 43	16 51 30	07 ♒ 29 13	14 ♒ 11 02	28 39	06 30	18 22	23 24	05 D 41	23 01	20 58	27 32
10	15 17 40	17 51 47	20 ♒ 46 49	27 ♒ 16 56	29 ♎ 57	05 59	18 48	23 37	05 41	23 02	20 59	27 34
11	15 21 36	18 52 06	03 ♓ 41 44	10 ♓ 01 41	01 ♏ 19	05 30	19 15	23 50	05 41	23 04	21 01	27 37
12	15 25 33	19 52 27	16 ♓ 17 12	22 ♓ 28 45	02 43	05 02	19 41	24 03	05 41	23 06	21 02	27 39
13	15 29 29	20 52 49	28 ♓ 36 47	04 ♈ 41 44	04 09	04 37	20 07	24 17	05 42	23 08	21 03	27 41
14	15 33 26	21 53 12	10 ♈ 44 02	16 ♈ 44 06	05 37	04 13	20 32	24 30	05 42	23 11	21 05	27 44
15	15 37 22	22 53 37	22 ♈ 42 18	28 ♈ 39 01	07 06	03 52	20 58	24 43	05 43	23 13	21 07	27 46
16	15 41 19	23 54 03	04 ♉ 34 34	10 ♉ 29 16	08 37	03 33	21 22	24 57	05 43	23 15	21 08	27 49
17	15 45 15	24 54 31	16 ♉ 23 25	22 ♉ 17 19	10 08	03 17	21 47	25 10	05 44	23 17	21 09	27 51
18	15 49 12	25 55 00	28 ♉ 11 13	04 ♊ 05 22	11 40	03 02	22 11	25 23	05 45	23 19	21 10	27 53
19	15 53 09	26 55 31	10 ♊ 00 04	15 ♊ 55 32	13 13	02 51	22 35	25 36	05 46	23 22	21 12	27 56
20	15 57 05	27 56 03	21 ♊ 52 18	27 ♊ 49 56	14 46	02 42	22 59	25 50	05 47	23 24	21 13	27 58
21	16 01 02	28 56 38	03 ♋ 49 26	09 ♋ 50 51	16 20	02 35	23 22	26 03	05 48	23 26	21 15	28 01
22	16 04 58	29 ♏ 57 13	15 ♋ 54 33	22 ♋ 00 51	17 54	02 31	23 45	26 16	05 50	23 29	21 16	28 03
23	16 08 55	00 ♐ 57 51	28 ♋ 10 09	04 ♌ 22 49	19 28	02 29	24 08	26 30	05 51	23 31	21 18	28 05
24	16 12 51	01 58 30	10 ♌ 39 15	16 ♌ 59 52	21 03	02 D 30	24 30	26 43	05 53	23 34	21 20	28 08
25	16 16 48	02 59 10	23 ♌ 25 05	29 ♌ 55 18	22 37	02 33	24 52	26 56	05 54	23 36	21 21	28 10
26	16 20 44	03 59 53	06 ♍ 30 53	13 ♍ 12 11	24 11	02 38	25 14	27 09	05 56	23 39	21 23	28 13
27	16 24 41	05 00 36	19 ♍ 59 28	26 ♍ 52 56	25 46	02 46	25 35	27 23	05 58	23 41	21 25	28 15
28	16 28 38	06 01 22	03 ♎ 52 42	10 ♎ 58 44	27 20	02 56	25 56	27 36	06 00	23 44	21 26	28 17
29	16 32 34	07 02 09	18 ♎ 10 53	25 ♎ 28 49	28 ♏ 55	03 09	26 16	27 49	06 02	23 47	21 28	28 20
30	16 36 31	08 ♐ 02 57	02 ♏ 52 03	10 ♏ 19 55	00 ♐ 29	03 ♏ 23	26 ♌ 36	28 ♏ 02	06 ♓ 04	23 ♑ 49	21 ♑ 30	28 ♏ 22

Date	Moon True ☊ ° '	Moon Mean ☊ ° '	Moon ☽ Latitude ° '
01	14 ♏ 48	14 ♏ 58	03 S 02
02	14 R 48	14 55	01 49
03	14 48	14 52	00 S 27
04	14 D 48	14 49	00 N 57
05	14 48	14 45	02 17
06	14 48	14 42	03 26
07	14 48	14 39	04 20
08	14 R 48	14 36	04 57
09	14 R 48	14 33	05 16
10	14 D 48	14 30	05 16
11	14 48	14 26	05 01
12	14 49	14 23	04 32
13	14 49	14 20	03 50
14	14 50	14 17	02 59
15	14 51	14 14	02 00
16	14 51	14 10	00 N 57
17	14 R 51	14 07	00 S 09
18	14 51	14 04	01 13
19	14 50	14 01	02 15
20	14 48	13 58	03 11
21	14 46	13 55	03 59
22	14 44	13 51	04 37
23	14 42	13 48	05 03
24	14 40	13 45	05 16
25	14 39	13 42	05 13
26	14 D 38	13 39	04 54
27	14 39	13 35	04 18
28	14 40	13 32	03 27
29	14 42	13 29	02 21
30	14 ♏ 43	13 ♏ 26	01 S 05

DECLINATIONS

Date	Sun ☉ ° '	Moon ☽ ° '	Mercury ☿ ° '	Venus ♀ ° '	Mars ♂ ° '	Jupiter ♃ ° '	Saturn ♄ ° '	Uranus ♅ ° '	Neptune ♆ ° '	Pluto ♇ ° '
01	14 S 27	06 S 42	06 S 34	20 S 34	17 N 57	17 S 24	11 S 12	22 S 01	21 S 15	06 S 33
02	14 46	11 17	06 39	20 12	17 50	17 27	11 12	22 01	21 15	06 34
03	15 05	15 13	06 48	19 49	17 43	17 31	11 12	22 00	21 15	06 35
04	15 23	18 07	07 01	19 25	17 36	17 34	11 12	22 00	21 15	06 35
05	15 42	19 43	07 18	19 01	17 29	17 38	11 12	22 00	21 15	06 36
06	16 00	19 55	07 39	18 36	17 22	17 41	11 12	21 59	21 15	06 37
07	16 17	18 46	08 02	18 11	17 15	17 44	11 12	21 59	21 15	06 37
08	16 35	16 28	08 28	17 46	17 08	17 48	11 12	21 59	21 15	06 38
09	16 52	13 18	08 56	17 22	17 01	17 51	11 12	21 58	21 14	06 39
10	17 09	09 34	09 26	16 57	16 54	17 55	11 12	21 58	21 14	06 39
11	17 26	05 28	09 57	16 33	16 47	17 58	11 12	21 58	21 14	06 40
12	17 42	01 S 14	10 29	16 09	16 41	18 01	11 12	21 57	21 14	06 41
13	17 58	02 N 58	11 02	15 46	16 34	18 05	11 11	21 57	21 14	06 41
14	18 14	06 59	11 35	15 24	16 27	18 08	11 11	21 57	21 14	06 42
15	18 30	10 41	12 09	15 02	16 21	18 11	11 11	21 56	21 13	06 42
16	18 45	13 56	12 44	14 41	16 14	18 14	11 10	21 56	21 13	06 43
17	19 00	16 36	13 18	14 21	16 08	18 18	11 10	21 55	21 13	06 44
18	19 14	18 34	13 52	14 02	16 01	18 21	11 09	21 55	21 13	06 44
19	19 28	19 43	14 26	13 45	15 55	18 24	11 09	21 55	21 13	06 45
20	19 42	20 00	15 00	13 28	15 49	18 27	11 08	21 54	21 12	06 45
21	19 55	19 24	15 33	13 13	15 43	18 31	11 08	21 54	21 12	06 46
22	20 08	17 54	16 06	12 58	15 37	18 34	11 07	21 53	21 12	06 46
23	20 21	15 34	16 38	12 45	15 31	18 37	11 06	21 53	21 12	06 47
24	20 33	12 30	17 10	12 33	15 25	18 40	11 06	21 53	21 11	06 47
25	20 45	08 47	17 41	12 22	15 19	18 43	11 05	21 52	21 11	06 48
26	20 57	04 N 34	18 11	12 13	15 13	18 46	11 04	21 52	21 11	06 48
27	21 08	00 00	18 41	12 04	15 08	18 49	11 03	21 51	21 11	06 49
28	21 19	04 S 42	19 10	11 57	15 02	18 52	11 03	21 51	21 10	06 49
29	21 29	09 18	19 37	11 51	14 57	18 55	11 02	21 50	21 10	06 50
30	21 S 39	13 S 29	20 S 04	11 S 46	14 N 52	18 S 58	11 S 01	21 S 50	21 S 10	06 S 50

ZODIAC SIGN ENTRIES

Date	h	m	Planets	
02	20	19	☽	♏
04	19	46	☽	♐
06	20	02	☽	♑
08	22	48	☽	♒
10	12	46	☿	♏
11	05	04	☽	♓
13	14	44	☽	♈
16	02	44	☽	♉
18	15	41	☽	♊
21	04	01	☽	♋
22	13	06	☉	♐
23	15	33	☽	♌
26	00	09	☽	♍
28	05	22	☽	♎
30	04	38	☿	♐
30	07	21	☽	♏

LATITUDES

Date	Mercury ☿ ° '	Venus ♀ ° '	Mars ♂ ° '	Jupiter ♃ ° '	Saturn ♄ ° '	Uranus ♅ ° '	Neptune ♆ ° '	Pluto ♇ ° '
01	01 N 52	05 S 41	01 N 35	00 N 48	01 S 56	00 S 30	00 N 34	13 N 19
04	02 09	05 04	01 40	00 48	01 55	00 30	00 34	13 18
07	02 14	04 23	01 44	00 48	01 55	00 30	00 34	13 18
10	02 10	03 39	01 48	00 47	01 54	00 30	00 34	13 17
13	01 59	02 53	01 53	00 47	01 54	00 30	00 34	13 17
16	01 44	02 07	01 58	00 47	01 54	00 30	00 34	13 16
19	01 26	01 22	02 03	00 47	01 53	00 30	00 34	13 16
22	01 07	00 40	02 08	00 47	01 53	00 30	00 34	13 16
25	00 46	00 S 01	02 13	00 47	01 52	00 30	00 34	13 16
28	00 25	00 N 35	02 18	00 46	01 52	00 30	00 34	13 16
31	00 N 04	01 N 07	02 N 24	00 N 46	01 S 51	00 S 30	00 N 34	13 N 16

DATA

Julian Date	2449658
Delta T	+60 seconds
Ayanamsa	23° 47' 16"
Synetic vernal point	05° ♓ 19' 43"
True obliquity of ecliptic	23° 26' 18"

MOON'S PHASES, APSIDES AND POSITIONS ☽

Date	h	m	Phase	Longitude ° '	Eclipse Indicator
03	13	36	●	10 ♏ 54	Total
10	06	14	☽	17 ♒ 37	
18	06	57	○	25 ♉ 42	
26	07	04	☾	03 ♍ 47	

Day	h	m		
03	23	33	Perigee	
18	04	50	Apogee	
06	03	14	Max dec	20° S 00'
12	19	01	ON	
20	07	41	Max dec	20° N 01'
27	12	01	OS	

LONGITUDES

Date	Chiron ⚷ ° '	Ceres ⚳ ° '	Pallas ⚴ ° '	Juno ⚵ ° '	Vesta ⚶ ° '	Black Moon Lilith ⚸ ° '
01	22 ♍ 01	13 ♌ 15	18 ♉ 32	20 ♏ 16	10 ♋ 47	23 ♉ 06
11	23 ♍ 07	15 ♌ 42	15 ♉ 17	23 ♏ 40	10 ♋ 46	24 ♉ 13
21	24 ♍ 05	17 ♌ 42	11 ♉ 57	27 ♏ 04	10 ♋ 02	25 ♉ 19
31	24 ♍ 54	19 ♌ 11	09 ♉ 02	00 ♐ 29	08 ♋ 34	26 ♉ 26

All ephemeris data is given at 12.00 UT and the Moon's longitude is additionally given for 24.00 UT

ASPECTARIAN

h m	Aspects
01 Tuesday	
04 38	☽ ⊥ ♀
05 14	☽ ⊼ ♄
06 08	☿ ⊥ ♇
06 37	☽ ∠ ♃
10 10	☽ ⊻ ☉
11 14	☽ ∥ ♇
11 20	☽ ∥ ☿
11 39	☉ Q ♆
14 03	☽ ⊻ ♀
15 01	☽ ± ♄
15 49	☽ ∠ ♇
20 00	☽ ✶ ♂
21 28	☽ ⊥ ♃
02 Wednesday	
03 40	♀ Q ♅
05 31	☽ ⚼ ♄
05 41	☽ □ ♆
06 22	☽ ⊥ ♇
07 04	☽ ☌ ☿
07 16	☽ ⊻ ♃
08 51	☽ □ ♅
11 32	☽ ∥ ♄
15 48	☿ ⊻ ♃
15 53	☽ Q ♂
15 58	☽ ⊻ ♇
23 12	☉ ☌ ♀
03 Thursday	
05 23	☽ △ ♄
10 21	☽ Q ♆
10 59	☽ ∥ ☉
11 53	☉ Q ♅
12 07	☽ ☌ ♀
13 29	☽ Q ♅
13 36	☽ ☌ ☉
21 18	☽ □ ♂
04 Friday	
01 01	☿ □ ♅
05 19	☽ ✶ ♆
06 30	☽ ∥ ♃
07 09	☽ # ♂
07 31	☽ ☌ ♃
08 28	☽ ✶ ♅
08 52	☽ ⊻ ☿
15 33	☽ ☌ ♇
16 19	♂ # ♃
18 55	☽ ⊥ ☿
05 Saturday	
01 29	☽ ∥ ♀
04 48	☽ □ ♄
05 06	☽ ∠ ♆
07 24	♀ Q ♆
08 17	☽ ∠ ♅
09 48	☽ ⊻ ♀
10 09	☽ ∠ ☿
16 31	☽ ⊻ ☉
19 02	☽ ⊥ ♀
19 31	☽ ⊥ ♆
22 29	☽ △ ♂
22 44	☽ ⊥ ♅
06 Sunday	
02 52	☽ ⊥ ☉
05 12	☽ ⊻ ♆
08 06	☽ ⊻ ♃
08 27	☽ ⊻ ♅
09 00	☽ ∠ ♀
09 43	☽ Q ♄
12 03	☽ ✶ ☿
15 48	☽ ⊻ ♇
18 02	☽ ⊥ ♃
18 35	☽ ∠ ☉
23 41	☽ ⚼ ♂
07 Monday	
01 41	☽ ⊥ ♇
05 25	☽ ✶ ♄
08 44	☽ ✶ ♀
09 09	☽ ∠ ♃
09 26	☽ Q ☿
11 09	♃ ✶ ♅
15 09	☽ ± ♂
16 44	☽ ∠ ♇
20 48	☽ ∥ ♀
21 26	☽ ✶ ☉
23 41	☽ ∥ ♃
08 Tuesday	
01 38	☽ ⊼ ♂
04 11	☽ Q ♀
05 41	☽ # ♂
06 41	☽ ∠ ♄
07 08	☽ ☌ ♆
10 37	☽ ☌ ♅
10 47	♀ ∥ ♃
10 57	☽ ✶ ♃
11 04	☽ ∥ ☉
13 51	☿ ⊻ ♇
18 27	☽ ✶ ♇
18 53	☽ □ ☿
19 37	☽ Q ☉
22 14	☽ ⊥ ♄
09 Wednesday	
08 14	☽ Q ♃
08 36	♄ St D
08 48	☽ ⊻ ♄
10 19	☽ □ ♀
15 39	☽ Q ♇
20 52	☉ # ♂
10 Thursday	
01 50	☽ ∥ ♄
04 56	☉ ∥ ♀
06 14	☽ □ ☉
08 16	☽ ☍ ♂
12 23	☽ ⊻ ♆
12 43	☽ ∥ ☿
15 53	♀ # ♂
16 10	☽ ⊻ ♅
17 18	☽ □ ♃
23 28	☽ ⊥ ♆
11 Friday	
00 35	☽ □ ♇
02 51	♀ △ ♄
03 19	☽ ⊥ ♅
05 08	☽ ∥ ♇
06 59	☽ △ ☿
10 16	☿ Q ♂
15 17	☽ △ ♀
15 45	☽ ☌ ♄
16 23	☽ ∠ ♆
20 18	☽ ∠ ♅
12 Saturday	
03 56	☉ □ ♂
15 07	☽ ⚼ ☿
18 48	☽ ⊼ ♂
19 01	☽ ⚼ ♀
19 33	☽ △ ☉
21 12	☽ ✶ ♆
13 Sunday	
01 16	☽ ✶ ♅
03 21	☽ △ ♃
06 55	☽ ± ♂
10 11	☽ △ ♇
10 58	☽ ± ☿
12 00	☽ ± ♀
16 13	☉ ✶ ♆
18 01	☿ ☌ ♀
18 48	☿ ∥ ♄
20 46	☽ Q ♆
23 28	☽ ⊼ ♀
23 47	☉ ∥ ♃
14 Monday	
00 25	☽ ⊼ ☿
00 55	☽ Q ♅
01 18	☽ ⚼ ♂
01 59	☽ ⊻ ♄
03 39	☽ ⚼ ☉
09 30	☽ ⚼ ♃
10 12	☽ # ♇
13 27	☿ △ ♄
13 56	☽ ⊥ ♄
16 00	☽ ⚼ ♇
23 15	☽ ± ☉
15 Tuesday	
03 50	☽ ± ♃
07 59	☽ ∠ ♄
08 21	☽ △ ♂
08 46	☽ □ ♆
10 07	☽ ± ♇
12 25	☽ ⊼ ☉
13 01	☽ □ ♅
15 23	☽ # ♄
16 09	☽ ⊼ ♃
19 51	☉ ✶ ♅
20 31	♂ ⊼ ♆
22 15	☽ ⊼ ♇
16 Wednesday	
00 33	☽ # ☿
09 59	☽ ☍ ♀
14 20	☽ ✶ ♄
17 32	☽ # ♀
18 06	♀ Q ♂
20 13	☿ Q ♆
21 24	☽ ☍ ☿
17 Thursday	
07 27	☽ ∥ ♂
14 45	☽ Q ♄
19 48	☉ ☌ ♃
21 42	☽ △ ♆
23 22	☽ □ ♂
18 Friday	
02 04	☽ △ ♅
06 12	☽ ☍ ♃
06 25	☿ Q ♅
06 57	☽ ☍ ☉
08 45	☽ # ♃
11 24	☽ ☍ ♇
16 46	☿ ∥ ♀
21 42	☽ ⊼ ♀
19 Saturday	
02 54	☽ # ☉
03 24	☽ □ ♄
04 16	☽ ⚼ ♆
08 40	☽ ⚼ ♅
09 41	☽ ± ♀
13 14	☽ Q ♂
19 30	☽ ⊼ ☿
22 33	☽ ± ♆
20 Sunday	
02 57	☽ ± ♅
03 40	☽ ⚼ ♀
09 28	☽ ± ☿
10 42	☽ ⊼ ♆
12 52	☉ ☌ ♇
14 20	☽ ✶ ♂
15 06	☽ ⊼ ♅
20 08	☽ ⊼ ♃
21 Monday	
00 03	☽ # ☉
00 19	☽ ⊼ ♇
01 20	☽ ⊼ ☉
06 17	☽ ⚼ ☿
08 23	☽ ± ♃
09 32	☽ △ ♀
12 22	☽ ± ♇
14 26	☽ ± ☉
15 58	☽ △ ♄
16 29	♂ ⊼ ♅
17 46	☿ # ♂
21 22	☽ ∠ ♂
22 Tuesday	
02 39	☽ ⚼ ♃
03 16	☽ # ♃
06 20	☽ ⚼ ♇
09 57	☽ ⚼ ☉
15 46	☽ ⊥ ♂
16 30	☽ △ ☿
21 42	☽ ⚼ ♄
22 34	☽ ☍ ♆
23 Wednesday	
02 55	☽ ☍ ♅
03 47	☽ # ☿
03 54	☽ ⊻ ♂
08 41	☽ △ ♃
11 51	☽ △ ♇
12 34	☽ ∥ ♂
15 16	☽ ± ♄
16 57	♀ St D
17 54	☽ △ ☉
20 20	☽ □ ♀
24 Thursday	
02 52	☽ ⊼ ♄
11 36	☽ # ♀
16 25	☿ ✶ ♆
21 32	☽ # ♄
25 Friday	
00 50	☉ ⊻ ♀
06 37	☽ Q ♀
08 09	☽ ⊼ ♆
10 18	☽ □ ☿
12 21	☽ ⊼ ♅
14 46	☽ ☌ ♂
18 37	☽ □ ♃
19 18	☽ ± ♆
20 49	☽ □ ♇
23 27	☽ ± ♅
23 36	☽ # ♇
26 Saturday	
03 30	☿ ✶ ♅
04 54	☽ ✶ ♀
07 04	☽ □ ☉
10 57	☽ ☍ ♄
11 45	☽ ⚼ ♆
15 51	☽ ⚼ ♅
23 32	☽ Q ☿
27 Sunday	
03 44	☽ Q ♃
05 23	☽ Q ♇
08 03	☽ ∠ ♀
08 31	☿ □ ♂
14 29	☽ △ ♆
17 42	☽ Q ☉
18 03	☉ ∥ ♆
18 29	☽ △ ♅
19 42	☿ ∥ ♃
22 00	☽ ⊻ ♂
23 21	☽ ✶ ☿
23 56	☽ ⊥ ♀
28 Monday	
01 04	☽ ✶ ♃
02 24	☽ ✶ ♇
08 36	☽ ⊥ ♂
10 22	☽ ⊻ ♀
11 25	☉ □ ♄
15 37	☽ ⊼ ♄
15 55	☽ ✶ ☉
16 40	☿ ☌ ♃
22 07	☉ ∠ ♆
22 55	☽ ∥ ♇
29 Tuesday	
00 13	☽ ∠ ♂
01 44	☽ ± ♄
02 54	☿ ☌ ♇
02 57	☽ ∠ ♃
03 55	☽ ∠ ♇
04 02	☽ ∠ ☿
16 43	☽ ⚼ ♄
17 26	☽ □ ♆
18 06	☽ ⊥ ♃
18 50	☽ ∠ ☉
18 51	☽ ⊥ ♇
20 44	☽ ⊥ ☿
21 15	☽ □ ♅
21 28	☽ ∥ ♄
30 Wednesday	
01 36	☽ ✶ ♂
01 52	☽ ∥ ♀
04 03	☽ ⊻ ♃
04 41	☽ ⊻ ♇
07 41	☽ ⊻ ☿
10 35	☽ ⊥ ☉
12 51	☽ ☌ ♀
17 10	☽ △ ♄
20 48	☽ # ♂
20 57	☽ ⊻ ☉
21 26	☽ Q ♂
22 41	☽ Q ♆

DECEMBER 1994

Date	Sidereal time h m s	Sun ☉ ° ' "	Moon ☽ ° ' "	Moon ☽ 24.00 ° ' "	Mercury ☿ ° '	Venus ♀ ° '	Mars ♂ ° '	Jupiter ♃ ° '	Saturn ♄ ° '	Uranus ♅ ° '	Neptune ♆ ° '	Pluto ♇ ° '
01	16 40 27	09 ♐ 03 47	17 ♏ 51 34	25 ♏ 26 00	02 ♐ 03	03 ♏ 40	26 ♌ 56	28 ♏ 16	06 ♓ 06	23 ♑ 52	21 ♑ 32	28 ♏ 24
02	16 44 24	10 04 39	03 ♐ 02 05	10 ♐ 38 36	03 38	03 58	27 15	28 29	06 09	23 55	21 33	28 27
03	16 48 20	11 05 31	18 ♐ 14 17	25 ♐ 47 52	05 12	04 19	27 34	28 42	06 11	23 58	21 35	28 29
04	16 52 17	12 06 25	03 ♑ 18 09	10 ♑ 44 04	06 46	04 41	27 52	28 55	06 14	24 01	21 37	28 31
05	16 56 13	13 07 20	18 ♑ 04 38	25 ♑ 19 06	08 20	05 06	28 10	29 08	06 16	24 03	21 39	28 34
06	17 00 10	14 08 15	02 ♒ 26 54	09 ♒ 27 40	09 54	05 32	28 27	29 21	06 19	24 06	21 41	28 36
07	17 04 07	15 09 11	16 ♒ 21 11	23 ♒ 07 30	11 29	06 00	28 44	29 34	06 22	24 09	21 43	28 39
08	17 08 03	16 10 08	29 ♒ 46 44	06 ♓ 19 11	13 03	06 29	29 00	29 ♏ 48	06 25	24 12	21 45	28 41
09	17 12 00	17 11 06	12 ♓ 45 16	19 ♓ 05 27	14 37	07 00	29 16	00 ♐ 01	06 28	24 15	21 47	28 43
10	17 15 56	18 12 04	25 ♓ 20 18	01 ♈ 30 23	16 11	07 32	29 31	00 14	06 31	24 18	21 48	28 45
11	17 19 53	19 13 03	07 ♈ 36 19	13 ♈ 38 44	17 45	08 06	29 ♌ 46	00 27	06 34	24 21	21 50	28 48
12	17 23 49	20 14 02	19 ♈ 38 13	25 ♈ 35 23	19 19	08 42	00 ♍ 00	00 40	06 38	24 24	21 52	28 50
13	17 27 46	21 15 02	01 ♉ 30 48	07 ♉ 25 00	20 54	09 19	00 14	00 52	06 41	24 27	21 54	28 52
14	17 31 42	22 16 02	13 ♉ 18 30	19 ♉ 11 45	22 28	09 57	00 27	01 05	06 44	24 30	21 56	28 55
15	17 35 39	23 17 04	25 ♉ 05 10	00 ♊ 59 10	24 03	10 36	00 40	01 18	06 48	24 34	21 59	28 57
16	17 39 36	24 18 05	06 ♊ 54 03	12 ♊ 50 08	25 37	11 17	00 52	01 31	06 52	24 37	22 01	28 59
17	17 43 32	25 19 08	18 ♊ 47 40	24 ♊ 46 52	27 12	11 58	01 04	01 44	06 55	24 40	22 03	29 01
18	17 47 29	26 20 11	00 ♋ 47 58	06 ♋ 51 05	28 ♐ 47	12 41	01 15	01 56	06 59	24 43	22 05	29 03
19	17 51 25	27 21 15	12 ♋ 56 23	19 ♋ 04 01	00 ♑ 22	13 25	01 25	02 09	07 03	24 46	22 07	29 06
20	17 55 22	28 22 19	25 ♋ 14 06	01 ♌ 26 45	01 57	14 10	01 35	02 22	07 07	24 50	22 09	29 08
21	17 59 18	29 ♐ 23 24	07 ♌ 42 07	14 ♌ 00 22	03 33	14 56	01 44	02 34	07 11	24 53	22 11	29 10
22	18 03 15	00 ♑ 24 30	20 ♌ 21 38	26 ♌ 46 07	05 08	15 43	01 52	02 47	07 16	24 56	22 13	29 12
23	18 07 11	01 25 36	03 ♍ 14 02	09 ♍ 45 36	06 44	16 31	02 00	03 00	07 20	25 00	22 15	29 14
24	18 11 08	02 26 43	16 ♍ 21 02	23 ♍ 00 37	08 20	17 20	02 07	03 12	07 24	25 03	22 18	29 16
25	18 15 05	03 27 50	29 ♍ 44 33	06 ♎ 33 04	09 56	18 10	02 14	03 24	07 29	25 06	22 20	29 18
26	18 19 01	04 28 58	13 ♎ 26 22	20 ♎ 24 34	11 33	19 00	02 20	03 37	07 33	25 10	22 22	29 20
27	18 22 58	05 30 07	27 ♎ 27 43	04 ♏ 35 46	13 09	19 52	02 25	03 49	07 38	25 13	22 24	29 22
28	18 26 54	06 31 17	11 ♏ 48 34	19 ♏ 05 48	14 46	20 44	02 29	04 01	07 42	25 16	22 26	29 24
29	18 30 51	07 32 27	26 ♏ 27 01	03 ♐ 51 36	16 23	21 36	02 33	04 14	07 47	25 20	22 29	29 26
30	18 34 47	08 33 37	11 ♐ 18 45	18 ♐ 47 33	18 00	22 30	02 36	04 26	07 52	25 23	22 31	29 28
31	18 38 44	09 ♑ 34 48	26 ♐ 16 59	03 ♑ 45 54	19 ♑ 37	23 ♏ 24	02 ♍ 38	04 ♐ 38	07 ♓ 57	25 ♑ 27	22 ♑ 33	29 ♏ 30

Date	Moon True ☊ ° '	Moon Mean ☊ ° '	Moon ☽ Latitude ° '
01	14 ♏ 43	13 ♏ 23	00 N 17
02	14 R 42	13 20	01 39
03	14 40	13 16	02 54
04	14 36	13 13	03 56
05	14 32	13 10	04 41
06	14 28	13 07	05 07
07	14 24	13 04	05 13
08	14 21	13 01	05 02
09	14 20	12 57	04 36
10	14 D 20	12 54	03 57
11	14 22	12 51	03 08
12	14 23	12 48	02 11
13	14 25	12 45	01 10
14	14 26	12 41	00 N 06
15	14 R 25	12 38	00 S 58
16	14 23	12 35	01 59
17	14 18	12 32	02 56
18	14 11	12 29	03 45
19	14 03	12 26	04 24
20	13 55	12 22	04 52
21	13 46	12 19	05 06
22	13 39	12 16	05 06
23	13 33	12 13	04 50
24	13 30	12 10	04 19
25	13 28	12 07	03 34
26	13 D 28	12 03	02 35
27	13 29	12 00	01 25
28	13 30	11 57	00 S 09
29	13 R 29	11 54	01 N 09
30	13 26	11 51	02 24
31	13 ♏ 21	11 ♏ 47	03 N 29

DECLINATIONS

Date	Sun ☉ ° '	Moon ☽ ° '	Mercury ☿ ° '	Venus ♀ ° '	Mars ♂ ° '	Jupiter ♃ ° '	Saturn ♄ ° '	Uranus ♅ ° '	Neptune ♆ ° '	Pluto ♇ ° '
01	21 S 48	16 S 53	20 S 30	11 S 42	14 N 47	19 S 01	11 S 00	21 S 49	21 S 10	06 S 51
02	21 58	19 08	20 55	11 39	14 42	19 04	10 59	21 49	21 09	06 51
03	22 06	20 02	21 19	11 37	14 38	19 07	10 58	21 48	21 09	06 52
04	22 14	19 28	21 42	11 35	14 33	19 10	10 57	21 48	21 09	06 52
05	22 22	17 34	22 04	11 35	14 29	19 13	10 56	21 47	21 09	06 53
06	22 30	14 38	22 25	11 36	14 24	19 16	10 55	21 47	21 08	06 53
07	22 37	10 57	22 45	11 38	14 20	19 19	10 53	21 46	21 08	06 54
08	22 43	06 49	23 04	11 40	14 16	19 21	10 52	21 46	21 08	06 54
09	22 49	02 S 31	23 21	11 44	14 13	19 24	10 51	21 45	21 08	06 54
10	22 55	01 N 46	23 37	11 48	14 09	19 27	10 50	21 45	21 07	06 55
11	23 00	05 54	23 53	11 52	14 06	19 30	10 48	21 44	21 07	06 55
12	23 05	09 43	24 06	11 58	14 03	19 32	10 47	21 44	21 07	06 56
13	23 09	13 06	24 19	12 04	14 00	19 35	10 46	21 43	21 06	06 56
14	23 13	15 56	24 30	12 11	13 57	19 38	10 44	21 42	21 06	06 56
15	23 16	18 06	24 40	12 18	13 54	19 40	10 43	21 42	21 06	06 57
16	23 19	19 30	24 49	12 26	13 52	19 43	10 41	21 41	21 05	06 57
17	23 21	20 03	24 56	12 34	13 50	19 46	10 40	21 41	21 05	06 57
18	23 23	19 41	25 02	12 43	13 48	19 48	10 38	21 40	21 05	06 58
19	23 25	18 25	25 07	12 52	13 46	19 51	10 37	21 40	21 04	06 58
20	23 26	16 18	25 10	13 02	13 45	19 53	10 35	21 39	21 04	06 58
21	23 26	13 24	25 12	13 12	13 43	19 56	10 33	21 38	21 04	06 58
22	23 26	09 51	25 13	13 22	13 42	19 58	10 32	21 38	21 04	06 59
23	23 26	05 48	25 11	13 33	13 42	20 00	10 30	21 37	21 03	06 59
24	23 25	01 N 24	25 09	13 44	13 41	20 03	10 28	21 37	21 03	06 59
25	23 24	03 S 10	25 05	13 56	13 41	20 05	10 26	21 36	21 03	06 59
26	23 22	07 41	24 59	14 07	13 41	20 08	10 25	21 35	21 02	07 00
27	23 19	11 54	24 52	14 19	13 41	20 10	10 23	21 35	21 02	07 00
28	23 17	15 31	24 44	14 31	13 42	20 12	10 21	21 34	21 02	07 00
29	23 13	18 14	24 34	14 43	13 42	20 15	10 19	21 33	21 01	07 00
30	23 10	19 46	24 22	14 55	13 43	20 17	10 17	21 33	21 01	07 01
31	23 S 06	19 S 55	24 S 09	15 S 08	13 N 45	20 S 19	10 S 15	21 S 32	21 S 00	07 S 01

ZODIAC SIGN ENTRIES

Date	h	m	Planets	
02	07	13	☽	♐
04	06	42	☽	♑
06	07	51	☽	♒
08	12	24	☽	♓
09	10	54	♃	♐
10	21	03	☽	♈
12	11	32	♂	♍
13	08	56	☽	♉
15	22	00	☽	♊
18	10	25	☽	♋
19	06	26	☿	♑
20	21	13	☽	♌
22	02	23	☉	♑
23	06	01	☽	♍
25	12	27	☽	♎
27	16	17	☽	♏
29	17	46	☽	♐
31	17	57	☽	♑

LATITUDES

Date	Mercury ☿ ° '	Venus ♀ ° '	Mars ♂ ° '	Jupiter ♃ ° '	Saturn ♄ ° '	Uranus ♅ ° '	Neptune ♆ ° '	Pluto ♇ ° '
01	00 N 04	01 N 07	02 N 24	00 N 46	01 S 51	00 S 30	00 N 34	13 N 16
04	00 S 16	01 35	02 29	00 46	01 51	00 30	00 34	13 16
07	00 36	02 00	02 35	00 46	01 51	00 30	00 34	13 16
10	00 54	02 22	02 41	00 46	01 50	00 30	00 33	13 16
13	01 12	02 40	02 47	00 46	01 50	00 30	00 33	13 17
16	01 27	02 56	02 53	00 46	01 50	00 30	00 33	13 17
19	01 41	03 08	03 00	00 46	01 49	00 30	00 33	13 17
22	01 52	03 19	03 06	00 46	01 49	00 30	00 33	13 18
25	02 01	03 27	03 13	00 46	01 48	00 30	00 33	13 18
28	02 07	03 33	03 20	00 46	01 48	00 30	00 33	13 19
31	02 S 10	03 N 36	03 N 27	00 N 46	01 S 48	00 S 30	00 N 33	13 N 19

DATA

Julian Date	2449688
Delta T	+60 seconds
Ayanamsa	23° 47' 21"
Synetic vernal point	05° ♓ 19' 39"
True obliquity of ecliptic	23° 26' 17"

MOON'S PHASES, APSIDES AND POSITIONS ☽

Date	h	m	Phase	Longitude ° '	Eclipse Indicator
02	23	54	●	10 ♐ 35	
09	21	06	☽	17 ♓ 34	
18	02	17	○	25 ♊ 55	
25	19	06	☾	03 ♎ 46	

Day	h	m		
02	12	12	Perigee	
15	07	38	Apogee	
30	22	56	Perigee	
03	14	34	Max dec	20° S 02'
10	02	01	ON	
17	14	33	Max dec	20° N 03'
24	19	25	OS	
31	02	28	Max dec	20° S 01'

LONGITUDES

Date	Chiron ⚷ ° '	Ceres ⚳ ° '	Pallas ⚴ ° '	Juno ⚵ ° '	Vesta ⚶ ° '	Black Moon Lilith ⚸ ° '
01	24 ♍ 54	19 ♌ 11	09 ♉ 02	00 ♐ 29	08 ♋ 34	26 ♉ 26
11	25 ♍ 34	20 ♌ 04	06 ♉ 56	03 ♐ 53	06 ♋ 59	27 ♉ 32
21	26 ♍ 03	20 ♌ 17	05 ♉ 53	07 ♐ 14	03 ♋ 59	28 ♉ 39
31	26 ♍ 21	19 ♌ 47	05 ♉ 57	10 ♐ 32	01 ♋ 21	29 ♉ 45

ASPECTARIAN

h m Aspects

01 Thursday
02 26 ☽ Q ♅
07 11 ☉ ∠ ♅
13 55 ☉ ∥ ♅
17 50 ☽ ✱ ♆
21 33 ☽ ✱ ♅

02 Friday
02 40 ☽ □ ♂
04 42 ☽ ☌ ♃
04 44 ☽ ☌ ♇
07 30 ♃ ☌ ♇
08 19 ☉ ⊥ ♀
10 56 ☽ ∥ ♃
13 02 ☽ ☌ ☿
13 31 ☽ ⊻ ♀
16 55 ☽ □ ♄
17 34 ☽ ∠ ♆
18 41 ☿ ⊻ ♀
21 18 ☽ ∠ ♅
23 11 ☽ ⊥ ♀
23 54 ☽ ☌ ☉

03 Saturday
01 48 ☿ ∥ ♆
07 48 ☽ ⊥ ♆
11 34 ☽ ⊥ ♅
13 45 ☽ ∠ ♀
14 23 ♂ ± ♆
17 19 ☽ ⊻ ♆
21 06 ☽ ⊻ ♅
21 28 ☽ Q ♄

04 Sunday
03 07 ☽ △ ♂
03 31 ☿ □ ♄
04 20 ☽ ⊻ ♇
04 52 ☽ ⊻ ♃
09 39 ☿ ∠ ♆
13 58 ☽ ⊥ ♇
14 17 ☽ ✱ ♀
14 38 ☽ ⊥ ♃
16 43 ☽ ✱ ♄
16 54 ☽ ∥ ♃
17 35 ☿ ∥ ♅
18 14 ☽ ⊻ ☿

05 Monday
03 17 ☽ ⊻ ☉
03 47 ☽ ₽ ♂
04 35 ☽ ∠ ♇
05 08 ☽ ⊥ ☿
05 27 ☽ ∠ ♃
10 20 ☽ Q ♀
13 51 ☽ ⊥ ☉
17 17 ☽ ∠ ♄
17 54 ☽ ☌ ♆
18 53 ☽ ± ♂
21 46 ☽ ∠ ☿
21 56 ☽ ☌ ♅
23 21 ☿ ∠ ♅

06 Tuesday
05 06 ☽ ⊼ ♂
05 29 ☽ ✱ ♇
05 58 ☽ ∠ ☉
06 41 ☽ ✱ ♃
08 23 ☽ ⊥ ♄
13 36 ☽ # ♂
17 25 ☽ □ ♀
18 37 ☽ ⊻ ♄
20 12 ☉ ∥ ☿

07 Wednesday
02 01 ☽ Q ♇
02 24 ☽ ✱ ☿
03 26 ♂ □ ♇
03 31 ☽ Q ♃
07 50 ☽ ∥ ♀
09 44 ☽ ✱ ☉
12 20 ☽ ∥ ♄
21 30 ☽ ⊻ ♆
23 25 ☿ ⊥ ♀

08 Thursday
01 37 ☉ ⊥ ♆
01 54 ☽ ⊻ ♅
02 18 ☽ Q ☿
08 18 ☽ ⊥ ♆
08 23 ♀ △ ♄
08 50 ☽ Q ☉
10 00 ☽ □ ♇
10 33 ☽ ☍ ♂
11 35 ☽ ∥ ♇
12 02 ☽ □ ♃
12 46 ☽ ⊥ ♅

09 Friday
00 13 ☽ ☌ ♄
00 48 ☽ △ ♀
00 49 ☽ ∠ ♆
05 25 ☽ ∠ ♅
16 00 ☽ □ ☿
21 06 ☽ □ ☉

10 Saturday
05 11 ☽ ✱ ♆
06 09 ☿ ⊥ ♆
06 21 ☽ ₽ ♀
10 00 ☽ ✱ ♅
14 32 ☉ ⊥ ♅
18 39 ☽ △ ♇
20 17 ☽ ⊼ ♂
21 40 ☽ △ ♃

11 Sunday
00 39 ☽ ± ♀
04 34 ☽ Q ♆
08 18 ☽ ± ♂
09 31 ☽ Q ♅
09 57 ☽ ⊻ ♄
13 03 ☽ ⊼ ♀
18 14 ☽ # ♇
21 29 ☿ ⊥ ♅
21 54 ☽ ⊥ ♄

12 Monday
00 20 ☽ ₽ ♇
02 32 ☽ ₽ ♂
03 53 ☽ ₽ ♃
11 17 ☽ △ ☿
13 19 ☽ △ ☉
16 01 ☽ ∠ ♄
16 31 ☽ □ ♆
18 27 ☽ ± ♇
19 12 ☽ # ♄
21 39 ☽ □ ♅
22 18 ☽ ± ♃

13 Tuesday
04 04 ☽ # ♀
06 37 ☽ ⊼ ♇
09 21 ☽ △ ♂
10 41 ☽ ⊼ ♃
18 59 ☽ ∥ ♂
22 17 ☽ ₽ ☿
22 32 ☽ ₽ ☉
22 33 ☽ ✱ ♄

14 Wednesday
03 16 ☉ ☌ ☿
03 46 ☿ ⊻ ♆
04 03 ☉ ⊻ ♆
04 45 ☽ ☍ ♀
11 50 ♀ Q ♆
18 36 ☽ ± ☉
19 26 ☽ ± ☿
19 53 ♂ ± ♅
23 08 ☽ Q ♄

15 Thursday
05 39 ☽ △ ♆
07 59 ☽ ⊼ ☉
09 33 ☽ ⊼ ☿
10 55 ☽ △ ♅
19 53 ☽ ☍ ♇
20 06 ☿ ⊻ ♅
23 33 ☽ □ ♂
23 57 ☿ Q ♄

16 Friday
00 53 ☽ ☍ ♃
11 55 ☽ □ ♄
12 13 ☽ ₽ ♆
17 31 ☽ ₽ ♅
18 16 ☽ # ♃
19 45 ☉ ⊻ ♅
21 24 ☽ ⊼ ♀

17 Saturday
02 05 ☉ Q ♄
05 47 ☿ ∠ ♀
06 27 ☽ ± ♆
10 15 ☽ ± ♀
11 44 ☽ ± ♅
12 33 ☽ Q ♂
18 32 ☽ ⊼ ♆
23 49 ☽ ⊼ ♅

18 Sunday
02 17 ☽ ☍ ☉
05 25 ☽ ₽ ♀
07 23 ☽ ☍ ☿
08 27 ☽ # ♃
08 31 ☽ ⊼ ♇
12 54 ☽ ✱ ♂
13 04 ♀ Q ♅
14 19 ☽ ⊼ ♃
16 14 ☿ ⊻ ♇
20 28 ☽ ± ♇

19 Monday
00 20 ☽ △ ♄
02 24 ☽ ± ♃
11 41 ♀ Q ♂
13 01 ☽ △ ♀
14 16 ☽ ₽ ♇
18 55 ☽ ∠ ♂
20 24 ☽ ₽ ♃

20 Tuesday
05 39 ☿ △ ♂
05 55 ☽ ₽ ♄
05 59 ☽ ☍ ♆
11 12 ☽ ☍ ♅
12 40 ☽ ⊥ ♂
18 37 ☽ ⊼ ☉
19 07 ☿ ⊻ ♃
19 33 ☽ △ ♇
23 26 ☽ ± ♄

21 Wednesday
00 24 ☽ ⊻ ♂
02 00 ☽ △ ♃
02 53 ☽ ⊼ ☿
06 30 ☉ ⊻ ♇
07 11 ☽ ± ☉
08 36 ♄ ∠ ♆
09 34 ☽ ∥ ♂
11 01 ☽ ⊼ ♄
13 25 ☽ # ♀
16 02 ☽ ± ☿

22 Thursday
01 50 ☽ ₽ ☉
02 40 ☽ □ ♀
07 42 ☽ # ♄
11 31 ☽ ₽ ☿
12 57 ☿ ⊥ ♇
15 30 ☽ ⊼ ♆
20 37 ☽ ⊼ ♅

23 Friday
02 45 ☽ ± ♆
04 34 ☽ □ ♇
05 16 ☽ # ♇
07 50 ☽ ± ♅
08 22 ☽ △ ☉
09 42 ☽ ☌ ♂
11 33 ☽ □ ♃
14 32 ☽ Q ♀
19 21 ☽ △ ☿
19 26 ☽ ₽ ♆
19 35 ☽ ☍ ♄
21 22 ☿ ✱ ♄
23 01 ☉ ∠ ♀

24 Saturday
00 29 ☽ ₽ ♅
03 23 ☉ △ ♂
05 32 ♀ # ♂
13 40 ☽ Q ♇
13 54 ☽ ✱ ♀
20 53 ☽ Q ♃
22 45 ☽ △ ♆

25 Sunday
02 52 ☿ ⊥ ♃
03 43 ☽ △ ♅
10 19 ☉ ⊻ ♃
11 13 ☽ ✱ ♇
16 26 ☽ ⊻ ♂
18 27 ☽ ∠ ♀
18 35 ☽ ✱ ♃
19 06 ☽ □ ☉

26 Monday
01 42 ☽ ⊼ ♄
03 03 ☽ ⊥ ♂
08 17 ☽ □ ☿
08 17 ☽ ∥ ♇
11 12 ☽ ⊥ ♀
12 12 ☽ ± ♄
13 34 ☽ ∠ ♇
18 45 ☽ ∠ ♂
21 03 ☽ ∠ ♃
22 13 ☽ ⊻ ♀

27 Tuesday
03 08 ☽ ∥ ♄
03 23 ☽ □ ♆
03 45 ☽ ₽ ♄
04 45 ☽ Q ☉
05 03 ☽ ⊥ ♇
08 11 ☽ □ ♅
08 52 ☉ ⊥ ♇
12 37 ☽ ⊥ ♃
15 14 ☽ ⊻ ♇
19 01 ☽ Q ☿
20 23 ☽ ✱ ♂
22 51 ☽ ⊻ ♃
23 15 ☽ # ♂

28 Wednesday
02 33 ☽ ✱ ☉
04 22 ☽ ∥ ♀
05 09 ☽ △ ♄
06 32 ☿ ∠ ♇
09 43 ☽ Q ♆
14 26 ☽ Q ♅
16 26 ☽ Q ♂
17 30 ☽ ✱ ☿

29 Thursday
03 36 ☽ ☌ ♀
05 09 ☽ ∠ ☉
05 31 ☽ ✱ ♆
10 10 ☽ ✱ ♅
16 52 ☽ ☌ ♇
18 17 ☉ ✱ ♄
20 52 ☽ ⊥ ☉
20 58 ☽ ∠ ☿
21 55 ☽ □ ♂

30 Friday
00 46 ☽ ☌ ♃
05 50 ☿ ₽ ♂
05 52 ☽ ∠ ♆
06 26 ☽ □ ♄
07 15 ☽ ⊻ ☉
10 30 ☽ ∠ ♅
12 22 ♀ ✱ ♆
13 14 ☽ ⊥ ☿
20 22 ☽ ⊥ ♆

31 Saturday
00 02 ☽ ⊻ ☿
01 00 ☽ ⊥ ♅
06 00 ☽ ⊻ ♆
07 05 ☽ ⊻ ♀
10 39 ☽ ⊻ ♅
11 28 ☽ Q ♄
12 12 ☿ ∠ ♃
17 11 ☽ ⊻ ♇
17 19 ☽ ⊥ ♀
22 12 ☽ △ ♂

All ephemeris data is given at 12.00 UT and the Moon's longitude is additionally given for 24.00 UT

JANUARY 1995

LONGITUDES

Date	Sidereal time h m s	Sun ☉ °	'	"	Moon ☽ °	'	"	Moon ☽ 24.00 °	'	"	Mercury ☿ °	'
01	18 42 40	10 ♑	35	59	11 ♑	13	08	18 ♑	37	34	21 ♑	14
02	18 46 37	11	37	10	25 ♑	58	05	03 ♒	13	45	22	52
03	18 50 34	12	38	21	10 ♒	23	44	17 ♒	27	22	24	29
04	18 54 30	13	39	32	24 ♒	24	15	01 ♓	14	05	26	06
05	18 58 27	14	40	43	07 ♓	56	49	14 ♓	32	33	27	43
06	19 02 23	15	41	53	21 ♓	01	33	27 ♓	24	12	29 ♑	19
07	19 06 20	16	43	03	03 ♈	40	59	09 ♈	52	28	00 ♒	55
08	19 10 16	17	44	12	15 ♈	59	18	22 ♈	02	09	02	30
09	19 14 13	18	45	21	28 ♈	01	41	03 ♉	58	38	04	04
10	19 18 09	19	46	29	09 ♉	53	40	15 ♉	47	28	05	37
11	19 22 06	20	47	37	21 ♉	40	41	27 ♉	33	55	07	09
12	19 26 03	21	48	45	03 ♊	27	46	09 ♊	22	44	08	39
13	19 29 59	22	49	52	15 ♊	19	18	21 ♊	17	52	10	07
14	19 33 56	23	50	58	27 ♊	18	47	03 ♋	22	20	11	32
15	19 37 52	24	52	04	09 ♋	28	45	15 ♋	38	11	12	54
16	19 41 49	25	53	10	21 ♋	50	45	28 ♋	06	28	14	13
17	19 45 45	26	54	15	04 ♌	25	20	10 ♌	47	20	15	27
18	19 49 42	27	55	19	17 ♌	12	22	23 ♌	40	22	16	36
19	19 53 38	28	56	23	00 ♍	11	13	06 ♍	44	52	17	40
20	19 57 35	29 ♑	57	26	13 ♍	21	12	20 ♍	00	12	18	37
21	20 01 32	00 ♒	58	29	26 ♍	41	51	03 ♎	26	08	19	27
22	20 05 28	01	59	32	10 ♎	13	06	17 ♎	02	50	20	08
23	20 09 25	03	00	34	23 ♎	55	24	00 ♏	50	52	20	40
24	20 13 21	04	01	35	07 ♏	49	19	14 ♏	50	46	21	02
25	20 17 18	05	02	37	21 ♏	55	12	29 ♏	02	30	21	14
26	20 21 14	06	03	38	06 ♐	12	30	13 ♐	24	54	21 R	15
27	20 25 11	07	04	38	20 ♐	39	16	27 ♐	55	03	21	04
28	20 29 07	08	05	38	05 ♑	11	38	12 ♑	28	14	20	42
29	20 33 04	09	06	37	19 ♑	44	01	26 ♑	58	07	20	08
30	20 37 01	10	07	35	04 ♒	09	40	11 ♒	17	48	19	25
31	20 40 57	11 ♒	08	32	18 ♒	21	44	25 ♒	20	47	18 ♒	32

Date	Venus ♀ °	'	Mars ♂ °	'	Jupiter ♃ °	'	Saturn ♄ °	'	Uranus ♅ °	'	Neptune ♆ °	'	Pluto ♇ °	'
01	24 ♏	19	02 ♍	39	04 ♐	50	08 ♓	02	25 ♑	30	22 ♑	35	29 ♏	32
02	25	14	02	40	05	02	08	07	25	34	22	37	29	34
03	26	10	02 R	40	05	14	08	12	25	37	22	40	29	36
04	27	07	02	39	05	26	08	17	25	41	22	42	29	38
05	28	04	02	37	05	37	08	23	25	44	22	44	29	40
06	29 ♏	02	02	35	05	49	08	28	25	47	22	46	29	42
07	00 ♐	00	02	32	06	01	08	33	25	51	22	49	29	44
08	00	58	02	28	06	12	08	39	25	55	22	51	29	45
09	01	57	02	23	06	24	08	44	25	58	22	53	29	47
10	02	57	02	17	06	35	08	50	26	02	22	56	29	49
11	03	57	02	11	06	47	08	56	26	05	22	58	29	51
12	04	57	02	04	06	58	09	01	26	09	23	00	29	52
13	05	58	01	56	07	09	09	07	26	12	23	02	29	54
14	07	00	01	47	07	20	09	13	26	16	23	05	29	55
15	08	01	01	37	07	31	09	19	26	19	23	07	29	57
16	09	03	01	27	07	42	09	25	26	23	23	09	29 ♏	59
17	10	05	01	16	07	53	09	31	26	26	23	11	00 ♐	00
18	11	08	01	03	08	04	09	37	26	30	23	14	00	02
19	12	11	00	51	08	15	09	43	26	33	23	16	00	03
20	13	15	00	37	08	25	09	49	26	37	23	18	00	05
21	14	18	00	23	08	36	09	56	26	41	23	21	00	06
22	15	22	00 ♍	08	08	46	10	02	26	44	23	23	00	07
23	16	27	29 ♌	52	08	56	10	08	26	48	23	25	00	09
24	17	31	29	35	09	07	10	15	26	51	23	27	00	10
25	18	36	29	18	09	17	10	21	26	55	23	30	00	11
26	19	41	29	00	09	27	10	28	26	58	23	32	00	13
27	20	46	28	42	09	37	10	34	27	02	23	34	00	14
28	21	52	28	22	09	46	10	41	27	05	23	36	00	15
29	22	58	28	03	09	56	10	47	27	09	23	38	00	16
30	24	04	27	42	10	06	10	54	27	12	23	41	00	18
31	25 ♐	10	27 ♌	22	10 ♐	15	11 ♓	01	27 ♑	16	23 ♑	43	00 ♐	19

Date	Moon True ☊ °	'	Moon Mean ☊ °	'	Moon ☽ Latitude °	'
01	13 ♏	13	11 ♏	44	04 N	19
02	13 R	04	11	41	04	52
03	12	53	11	38	05	04
04	12	44	11	35	04	59
05	12	36	11	32	04	36
06	12	31	11	28	03	59
07	12	28	11	25	03	12
08	12	27	11	22	02	17
09	12 D	27	11	19	01	16
10	12	28	11	16	00 N	14
11	12 R	27	11	13	00 S	49
12	12	25	11	09	01	49
13	12	20	11	06	02	45
14	12	13	11	03	03	35
15	12	02	11	00	04	14
16	11	50	10	57	04	43
17	11	36	10	53	04	59
18	11	23	10	50	04	59
19	11	11	10	47	04	45
20	11	01	10	44	04	15
21	10	54	10	41	03	31
22	10	50	10	38	02	34
23	10	49	10	34	01	28
24	10 D	49	10	31	00 S	16
25	10 R	48	10	28	00 N	58
26	10	47	10	25	02	10
27	10	43	10	22	03	14
28	10	36	10	19	04	06
29	10	26	10	15	04	41
30	10	14	10	12	04	59
31	10 ♏	01	10 ♏	09	04 N	57

DECLINATIONS

Date	Sun ☉ °	'	Moon ☽ °	'	Mercury ☿ °	'	Venus ♀ °	'	Mars ♂ °	'	Jupiter ♃ °	'	Saturn ♄ °	'	Uranus ♅ °	'	Neptune ♆ °	'	Pluto ♇ °	'
01	23 S	01	18 S	40	23 S	54	15 S	20	13 N	46	20 S	21	10 S	13	21 S	32	21 S	00	07 S	01
02	22	56	16	11	23	37	15	33	13	48	20	23	10	11	21	31	21	00	07	01
03	22	50	12	45	23	20	15	45	13	50	20	25	10	09	21	30	20	59	07	01
04	22	44	08	41	23	00	15	58	13	53	20	28	10	07	21	30	20	59	07	01
05	22	38	04 S	19	22	39	16	10	13	56	20	30	10	05	21	29	20	59	07	01
06	22	31	00 N	07	22	17	16	23	13	59	20	32	10	03	21	28	20	58	07	02
07	22	24	04	24	21	53	16	35	14	02	20	34	10	01	21	28	20	58	07	02
08	22	16	08	23	21	28	16	48	14	05	20	36	09	59	21	27	20	58	07	02
09	22	08	11	58	21	01	17	00	14	09	20	38	09	57	21	26	20	57	07	02
10	21	59	15	00	20	33	17	12	14	13	20	39	09	55	21	26	20	57	07	02
11	21	50	17	24	20	04	17	24	14	18	20	41	09	52	21	25	20	57	07	02
12	21	40	19	03	19	34	17	36	14	22	20	43	09	50	21	24	20	56	07	02
13	21	30	19	53	19	03	17	48	14	27	20	45	09	48	21	24	20	56	07	02
14	21	20	19	50	18	32	17	59	14	33	20	47	09	46	21	23	20	55	07	02
15	21	09	18	52	17	59	18	10	14	38	20	49	09	43	21	22	20	55	07	02
16	20	58	17	00	17	27	18	22	14	44	20	50	09	41	21	22	20	55	07	02
17	20	47	14	19	16	54	18	32	14	50	20	52	09	39	21	21	20	54	07	02
18	20	35	10	55	16	21	18	43	14	56	20	54	09	36	21	20	20	54	07	02
19	20	22	06	58	15	49	18	53	15	02	20	56	09	34	21	20	20	54	07	02
20	20	09	02 N	37	15	18	19	03	15	09	20	57	09	31	21	19	20	53	07	02
21	19	56	01 S	55	14	48	19	13	15	16	20	59	09	29	21	18	20	53	07	02
22	19	43	06	25	14	20	19	22	15	23	21	00	09	27	21	18	20	52	07	02
23	19	29	10	39	13	54	19	31	15	30	21	02	09	24	21	17	20	52	07	02
24	19	15	14	22	13	31	19	40	15	38	21	03	09	22	21	16	20	52	07	02
25	19	00	17	18	13	10	19	48	15	45	21	05	09	19	21	16	20	51	07	02
26	18	45	19	13	12	53	19	56	15	53	21	06	09	17	21	15	20	51	07	02
27	18	30	19	53	12	40	20	04	16	01	21	08	09	14	21	14	20	51	07	02
28	18	15	19	15	12	30	20	11	16	09	21	09	09	12	21	14	20	50	07	01
29	17	59	17	21	12	25	20	17	16	17	21	11	09	09	21	13	20	50	07	01
30	17	42	14	23	12	23	20	24	16	25	21	12	09	07	21	12	20	49	07	01
31	17 S	26	10 S	36	12 S	26	20 S	29	16 N	34	21 S	13	09 S	04	21 S	12	20 S	49	07 S	01

ZODIAC SIGN ENTRIES

Date	h	m	Planets
02	18	39	☽ ♒
04	21	49	☽ ♓
06	22	17	☿ ♒
07	04	56	☽ ♈
07	12	07	♀ ♐
09	15	58	☽ ♉
12	04	57	☽ ♊
14	17	20	☽ ♋
17	03	36	☽ ♌
17	09	16	♇ ♐
19	11	39	☽ ♍
20	13	00	☉ ♒
21	17	53	☽ ♎
22	23	48	♂ ♌
23	22	32	☽ ♏
26	01	37	☽ ♐
28	03	26	☽ ♑
30	05	03	☽ ♒

LATITUDES

Date	Mercury ☿ °	'	Venus ♀ °	'	Mars ♂ °	'	Jupiter ♃ °	'	Saturn ♄ °	'	Uranus ♅ °	'	Neptune ♆ °	'	Pluto ♇ °	'
01	02 S	10	03 N	37	03 N	29	00 N	46	01 S	48	00 S	30	00 N	33	13 N	20
04	02	07	03	39	03	36	00	46	01	47	00	30	00	33	13	20
07	01	59	03	39	03	43	00	46	01	47	00	30	00	33	13	21
10	01	45	03	37	03	49	00	46	01	47	00	30	00	33	13	22
13	01	24	03	34	03	56	00	46	01	47	00	30	00	33	13	23
16	00	55	03	29	04	02	00	46	01	46	00	30	00	33	13	24
19	00 S	18	03	24	04	08	00	46	01	46	00	30	00	33	13	25
22	00 N	27	03	17	04	14	00	46	01	46	00	30	00	33	13	26
25	01	19	03	09	04	19	00	46	01	46	00	30	00	33	13	27
28	02	12	03	01	04	23	00	46	01	46	00	30	00	33	13	28
31	02 N	59	02 N	52	04 N	27	00 N	46	01 S	46	00 S	30	00 N	33	13 N	29

DATA

Julian Date	2449719
Delta T	+61 seconds
Ayanamsa	23° 47' 26"
Synetic vernal point	05° ♓ 19' 33"
True obliquity of ecliptic	23° 26' 16"

MOON'S PHASES, APSIDES AND POSITIONS ☽

Date	h	m	Phase	Longitude °	'	Eclipse Indicator
01	10	56	●	10 ♑	33	
08	15	46	☽	17 ♈	54	
16	20	26	○	26 ♋	15	
24	04	58	☾	03 ♏	44	
30	22	48	●	10 ♒	35	

Day	h	m		
11	21	58	Apogee	
27	23	16	Perigee	
06	11	24	ON	
13	22	37	Max dec	19° N 59'
21	01	55	OS	
27	12	24	Max dec	19° S 53'

LONGITUDES

Date	Chiron ⚷ °	'	Ceres ⚳ °	'	Pallas ⚴ °	'	Juno ⚵ °	'	Vesta ⚶ °	'	Black Moon Lilith ⚸ °	'
01	26 ♍	22	19 ♌	42	06 ♉	01	10 ♐	51	01 ♋	05	29 ♉	52
11	26 ♍	27	18 ♌	25	07 ♉	13	14 ♐	04	28 ♊	42	00 ♊	58
21	26 ♍	19	16 ♌	33	09 ♉	19	17 ♐	09	26 ♊	48	02 ♊	05
31	26 ♍	01	14 ♌	19	12 ♉	13	20 ♐	07	25 ♊	35	03 ♊	12

ASPECTARIAN

h m	Aspects
01 Sunday	
01 34	☽ ⊻ ♃
02 50	☽ ⊥ ♇
06 50	☽ ⚹ ♄
08 44	☽ ∠ ♀
10 56	☽ ☌ ☉
11 22	☽ ⊥ ♃
17 23	☽ ∠ ♇
18 48	☉ ⊥ ♃
22 26	☽ ⚼ ♂
02 Monday	
02 09	☽ ∠ ♃
06 16	☽ ☌ ☿
06 30	☽ ☌ ♆
07 18	☽ ∠ ♄
08 25	☿ ☌ ♆
10 43	☽ ⚹ ♀
11 20	☽ ☌ ♅
13 09	☽ ± ♂
16 01	☿ ∠ ♄
16 35	☽ ∥ ♀
17 57	☽ ⚹ ♇
20 49	♀ ⚹ ♅
21 27	♂ St R
22 13	☽ ⊥ ♄
23 04	☽ ⊼ ♂
03 Tuesday	
03 13	☽ ⚹ ♃
04 57	☽ # ♂
08 00	☽ Q ♀
08 17	☽ ⊻ ♄
14 03	☽ Q ♇
16 05	☽ ⊻ ☉
23 47	☽ Q ♃
04 Wednesday	
03 08	☽ ⊥ ☉
03 45	☽ ∥ ♄
05 31	☿ ☌ ♅
09 02	☽ ⊻ ♆
14 14	☽ ⊻ ♅
15 21	☽ ⊻ ☿
17 05	☽ □ ♀
19 33	☽ ⊥ ♆
20 03	☽ ∠ ☉
20 09	☿ ± ♂
21 11	☽ □ ♇
21 16	☽ ∥ ♇
05 Thursday	
00 50	☽ ⊥ ♅
02 29	☽ ☍ ♂
03 22	☽ ⊥ ☿
07 46	☽ □ ♃
11 37	☽ ∠ ♆
11 40	☉ ∠ ♇
12 47	☽ ☌ ♄
14 17	☉ ∥ ☿
17 04	☽ ∠ ♅
21 51	☽ ∠ ☿
06 Friday	
01 13	☿ ⚹ ♀
01 17	☽ ⚹ ☉
15 17	☽ ⚹ ♆
17 49	☿ ⚹ ♇
20 59	☽ ⚹ ♅
07 Saturday	
01 40	☽ Q ☉
04 20	☽ △ ♀
04 24	☽ △ ♇
05 07	♀ ☌ ♇
05 55	☽ ⚹ ☿
09 48	☽ ⊼ ♂
14 11	☽ Q ♆
16 34	☽ △ ♃
20 06	☽ Q ♅
21 20	☽ ± ♂
21 30	☽ ⊻ ♄
08 Sunday	
03 34	☽ # ♇
06 00	☉ ⚼ ♂
08 38	☽ Q ☿
09 20	☽ ⊥ ♄
09 34	☽ ⚼ ♇
11 30	☿ ⊼ ♂
11 58	☽ ⚼ ♀
12 37	☿ ∥ ♅
14 24	☿ ⊥ ♄
14 54	☽ ⚼ ♂
15 46	☽ □ ☉
22 12	☽ # ♄
22 31	☽ ⚼ ♃
09 Monday	
01 40	☽ □ ♆
03 20	☽ ∠ ♄
03 28	☽ ± ♇
07 28	☽ ± ♀
07 51	☽ □ ♅
15 27	☿ ∥ ♆
15 33	☽ ⊼ ♇
16 51	☽ ± ♃
20 38	☽ ⊼ ♀
20 43	☽ △ ♂
21 28	♀ □ ♂
10 Tuesday	
02 02	☽ □ ☿
05 11	☽ ⊼ ♃
05 15	☽ ∥ ♂
07 04	☿ ∥ ♃
09 50	☽ ⚹ ♄
11 Wednesday	
05 16	☿ ⚹ ♃
10 02	☽ △ ☉
10 27	☽ Q ♄
12 05	☽ # ♀
14 38	☽ △ ♆
21 02	☽ △ ♅
12 Thursday	
01 15	♂ ± ♅
04 40	☽ ☍ ♇
09 11	☽ □ ♂
15 19	☽ ☍ ♀
16 24	☉ ∠ ♃
18 28	☿ ⊻ ♄
19 13	☽ ☍ ♃
19 26	☽ ⚼ ☉
19 33	☽ # ☿
21 14	☽ ⚼ ♆
23 22	☽ □ ♄
13 Friday	
00 01	☽ △ ☿
03 39	☽ ⚼ ♅
15 19	☽ ± ☉
15 28	☽ ± ♆
17 06	☉ ☌ ♆
21 09	☽ Q ♂
21 51	☽ ± ♅
14 Saturday	
03 32	☽ ⊼ ♆
04 28	☽ ⊼ ☉
04 28	☉ ∥ ♅
09 54	☽ ⊼ ♅
10 15	☽ ⚼ ☿
17 12	☽ ⊼ ♇
18 51	☿ Q ♇
20 45	☽ ⚹ ♂
21 36	☉ ∠ ♄
21 51	♀ ☌ ♃
15 Sunday	
05 04	☽ ± ♇
05 50	☿ ∥ ♀
06 20	☽ ± ☿
08 06	☽ ⊼ ♃
08 53	☽ ⊼ ♀
11 41	☽ △ ♄
14 20	♀ ∠ ♆
19 30	☽ ⊼ ☿
20 00	☽ ± ♃
21 21	☽ # ♀
21 40	☽ ± ♀
22 42	☽ ⚼ ♇
16 Monday	
01 43	☽ ∠ ♂
03 10	☉ ± ♂
05 35	☽ # ☿
13 40	☽ ⚼ ♃
14 31	☽ ☍ ♆
16 37	☽ ⚼ ♀
16 59	☽ ⚼ ♄
18 49	☽ ⊥ ♂
19 26	☉ ∥ ♆
20 26	☽ ☍ ☉
20 44	☽ ☍ ♅
21 18	♀ □ ♄
17 Tuesday	
00 22	☉ ☌ ♅
01 53	☉ ∥ ♃
03 36	☽ △ ♇
06 05	☽ ⊻ ♂
08 04	☽ ∥ ♂
10 16	☽ ± ♄
18 38	☽ △ ♃
21 41	☽ ⊼ ♄
23 38	☽ △ ♀
18 Wednesday	
10 47	☽ ☍ ☿
12 31	♃ ∥ ♆
20 23	☽ # ♄
20 46	♀ ∠ ♅
23 13	☽ ⊼ ♆
19 Thursday	
05 18	☽ ⊼ ♅
09 31	☽ ⊼ ☉
10 18	☽ ± ♆
11 34	☽ # ♇
11 45	☽ □ ♇
13 11	☽ ☌ ♂
16 20	♃ ∠ ♆
16 22	☽ ± ♅
21 26	☽ ± ☉
20 Friday	
02 49	☽ ⚼ ♆
02 55	☽ □ ♃
05 33	☽ ☍ ♄
08 50	☽ ⚼ ♅
11 47	☽ □ ♀
14 54	☉ ⚹ ♇
15 09	☽ ⚼ ☉
18 02	☿ # ♂
20 33	☽ Q ♇
22 10	☽ ⊼ ☿
21 Saturday	
00 40	☉ ⊼ ♂
05 59	☽ △ ♆
09 37	☽ ± ☿
11 49	☽ Q ♃
11 58	☽ △ ♅
18 05	☽ ⚹ ♇
18 27	☽ ⊻ ♂
20 15	☽ △ ☉
22 51	☽ Q ♀
22 Sunday	
02 34	☽ ⚼ ☿
04 54	☽ ⊥ ♂
09 24	☽ ⚹ ♃
11 40	☽ ⊼ ♄
12 20	♂ □ ♇
15 24	☽ ∥ ♇
20 28	☽ ∠ ♂
20 39	☽ ∠ ♇
21 50	☽ ⚹ ♀
22 18	☽ ± ♄
23 Monday	
04 47	☽ ∥ ♄
06 08	☽ △ ☿
09 45	☉ ∥ ♀
11 07	☽ □ ♆
12 02	☽ ∠ ♃
12 23	☽ ⊥ ♇
14 08	☽ ⚼ ♄
17 00	☽ □ ♅
22 06	☽ ⚹ ♂
22 48	☽ ⊻ ♇
24 Tuesday	
02 07	☽ ∠ ♀
03 48	☽ ⊥ ♃
04 58	☽ □ ☉
06 39	☽ ∥ ☿
10 33	♀ ⊥ ♆
14 14	☽ ⊻ ♃
16 11	☽ △ ♄
17 43	☉ ⊥ ♄
18 14	☽ Q ♆
18 19	☽ Q ♂
18 51	☽ ⊥ ♀
21 49	☽ # ♂
22 00	♂ ± ♆
25 Wednesday	
00 04	☽ Q ♅
05 55	☽ ⊻ ♀
10 50	☽ □ ☿
14 03	☽ Q ☉
14 40	☽ ⚹ ♆
20 27	☽ ⚹ ♅
26 Thursday	
00 11	☽ □ ♂
01 15	☿ St R
01 57	☽ ☌ ♇
05 43	☽ ∥ ☉
11 44	☽ ⚹ ☉
15 53	☽ ∠ ♆
17 02	☽ Q ☿
17 28	☽ ☌ ♃
19 08	☽ □ ♄
21 38	☽ ∠ ♅
27 Friday	
06 53	☽ ⊥ ♆
12 13	☽ ☌ ♀
12 37	☽ ⊥ ♅
12 40	☽ ⚹ ☿
14 32	☽ ∠ ☉
16 50	☽ ⊻ ♆
17 04	☿ ⚹ ♀
17 55	♀ ⊥ ♅
22 34	☽ ⊻ ♅
28 Saturday	
01 00	☽ △ ♂
01 10	☽ Q ♄
03 51	☽ ⊻ ♇
06 30	☽ ⊥ ☉
12 48	☽ ∠ ☿
13 45	☽ ⊥ ♇
17 08	☽ ⊻ ☉
19 38	☽ ⊻ ♃
21 07	☽ ⚹ ♄
29 Sunday	
01 12	☽ ⚼ ♂
03 08	☽ ⊥ ☿
04 36	☽ ∥ ☉
04 37	☽ ∠ ♇
05 39	☽ ⊥ ♃
12 39	☽ ⊻ ☿
15 44	☽ ± ♂
17 47	☽ ⊻ ♀
18 29	☽ ☌ ♆
20 43	☽ ∠ ♃
21 04	☽ # ♂
22 07	☽ ∠ ♄
30 Monday	
00 21	☽ ☌ ♅
01 29	☽ ⊼ ♂
03 19	♀ ⊻ ♆
04 35	☽ ⊥ ♀
05 32	☽ ⚹ ♇
11 09	☉ ⚹ ♃
13 15	☽ ⊥ ♄
15 22	♃ ∥ ♅
20 55	☽ ∠ ♀
22 05	☽ ⚹ ♃
22 48	☽ ☌ ☉
23 25	☽ ⊻ ♄
31 Tuesday	
01 03	☽ ∥ ☿
01 42	☽ Q ♇
08 29	☉ ⊻ ♄
12 16	☽ ☌ ☿
17 45	♂ ⊼ ♅
18 45	☽ Q ♃
21 01	☽ ∥ ♄
21 12	☽ ⊻ ♆

All ephemeris data is given at 12.00 UT and the Moon's longitude is additionally given for 24.00 UT

FEBRUARY 1995

LONGITUDES

Date	Sidereal time h m s	Sun ☉	Moon ☽	Moon ☽ 24.00	Mercury ☿	Venus ♀	Mars ♂	Jupiter ♃	Saturn ♄	Uranus ♅	Neptune ♆	Pluto ♇
01	20 44 54	12 ♒ 09 28	02 ♓ 14 26	09 ♓ 02 15	17 ♒ 31	26 ♐ 17	27 ♌ 00	10 ♐ 25	11 ♓ 07	27 ♑ 19	23 ♑ 45	00 ♐ 20
02	20 48 50	13 10 22	15 ♓ 43 59	22 ♓ 19 34	16 R 24	27 23	26 R 38	10 34	11 14	27 23	23 47	00 21
03	20 52 47	14 11 16	28 ♓ 49 03	05 ♈ 12 37	15 13	28 30	26 16	10 43	11 21	27 26	23 49	00 22
04	20 56 43	15 12 08	11 ♈ 30 36	17 ♈ 43 26	13 59	29 ♐ 37	25 54	10 52	11 28	27 29	23 52	00 23
05	21 00 40	16 12 59	23 ♈ 51 37	29 ♈ 55 44	12 46	00 ♑ 44	25 31	11 01	11 35	27 33	23 54	00 24
06	21 04 36	17 13 48	05 ♉ 56 25	11 ♉ 54 21	11 35	01 52	25 07	11 10	11 42	27 36	23 56	00 25
07	21 08 33	18 14 36	17 ♉ 50 13	23 ♉ 44 43	10 28	02 59	24 44	11 18	11 49	27 40	23 58	00 26
08	21 12 30	19 15 23	29 ♉ 38 35	05 ♊ 32 29	09 26	04 07	24 20	11 27	11 55	27 43	24 00	00 26
09	21 16 26	20 16 08	11 ♊ 27 06	17 ♊ 23 04	08 31	05 15	23 57	11 35	12 02	27 46	24 02	00 27
10	21 20 23	21 16 51	23 ♊ 20 59	29 ♊ 21 23	07 42	06 23	23 33	11 44	12 10	27 50	24 04	00 28
11	21 24 19	22 17 33	05 ♋ 24 41	11 ♋ 31 32	07 02	07 31	23 09	11 52	12 17	27 53	24 06	00 29
12	21 28 16	23 18 14	17 ♋ 42 03	23 ♋ 56 34	06 29	08 39	22 45	12 00	12 24	27 56	24 08	00 29
13	21 32 12	24 18 53	00 ♌ 15 16	06 ♌ 38 14	06 05	09 48	22 21	12 08	12 31	28 00	24 10	00 30
14	21 36 09	25 19 30	13 ♌ 05 27	19 ♌ 36 51	05 48	10 56	21 57	12 16	12 38	28 03	24 12	00 31
15	21 40 05	26 20 06	26 ♌ 12 15	02 ♍ 51 26	05 39	12 05	21 33	12 23	12 45	28 06	24 14	00 31
16	21 44 02	27 20 40	09 ♍ 34 07	16 ♍ 19 59	05 D 38	13 14	21 10	12 31	12 52	28 09	24 16	00 32
17	21 47 59	28 21 13	23 ♍ 08 40	29 ♍ 59 51	05 43	14 23	20 46	12 38	12 59	28 12	24 18	00 32
18	21 51 55	29 ♒ 21 44	06 ♎ 53 11	13 ♎ 48 23	05 55	15 32	20 23	12 45	13 07	28 16	24 20	00 33
19	21 55 52	00 ♓ 22 14	20 ♎ 45 09	27 ♎ 43 17	06 13	16 41	20 00	12 52	13 14	28 19	24 22	00 33
20	21 59 48	01 22 43	04 ♏ 42 34	11 ♏ 42 52	06 37	17 51	19 38	12 59	13 21	28 22	24 24	00 34
21	22 03 45	02 23 10	18 ♏ 44 05	25 ♏ 46 05	07 06	19 00	19 16	13 06	13 28	28 25	24 26	00 34
22	22 07 41	03 23 37	02 ♐ 48 49	09 ♐ 52 09	07 40	20 10	18 54	13 13	13 36	28 28	24 28	00 35
23	22 11 38	04 24 02	16 ♐ 56 00	24 ♐ 00 10	08 18	21 20	18 33	13 19	13 43	28 31	24 30	00 35
24	22 15 34	05 24 25	01 ♑ 04 29	08 ♑ 08 40	09 00	22 29	18 12	13 26	13 50	28 34	24 32	00 35
25	22 19 31	06 24 48	15 ♑ 12 23	22 ♑ 15 16	09 47	23 39	17 52	13 32	13 58	28 37	24 34	00 36
26	22 23 28	07 25 09	29 ♑ 16 52	06 ♒ 16 43	10 37	24 49	17 32	13 38	14 05	28 40	24 35	00 36
27	22 27 24	08 25 28	13 ♒ 14 18	20 ♒ 09 07	11 30	25 59	17 13	13 44	14 12	28 43	24 37	00 36
28	22 31 21	09 ♓ 25 45	27 ♒ 00 42	03 ♓ 48 33	12 ♒ 26	27 ♑ 10	16 ♌ 54	13 ♐ 50	14 ♓ 20	28 ♑ 46	24 ♑ 39	00 ♐ 36

MOON NODES AND LATITUDE

Date	Moon True ☊	Moon Mean ☊	Moon ☽ Latitude
01	09 ♏ 48	10 ♏ 06	04 N 38
02	09 R 38	10 03	04 04
03	09 30	09 59	03 17
04	09 25	09 56	02 22
05	09 22	09 53	01 21
06	09 21	09 50	00 N 18
07	09 D 21	09 47	00 S 45
08	09 R 21	09 44	01 45
09	09 19	09 40	02 41
10	09 15	09 37	03 31
11	09 08	09 34	04 11
12	08 59	09 31	04 41
13	08 47	09 28	04 58
14	08 34	09 24	05 00
15	08 21	09 21	04 47
16	08 10	09 18	04 18
17	08 01	09 15	03 34
18	07 54	09 12	02 37
19	07 51	09 09	01 30
20	07 49	09 05	00 S 17
21	07 D 50	09 02	00 N 58
22	07 50	08 59	02 09
23	07 R 49	08 56	03 13
24	07 47	08 53	04 05
25	07 41	08 50	04 42
26	07 34	08 46	05 01
27	07 24	08 43	05 03
28	07 ♏ 14	08 ♏ 40	04 N 47

DECLINATIONS

Date	Sun ☉	Moon ☽	Mercury ☿	Venus ♀	Mars ♂	Jupiter ♃	Saturn ♄	Uranus ♅	Neptune ♆	Pluto ♇
01	17 S 09	06 S 21	12 S 33	20 S 35	16 N 42	21 S 15	09 S 01	21 S 11	20 S 49	07 S 01
02	16 52	01 S 53	12 43	20 39	16 51	21 16	08 59	21 10	20 48	07 01
03	16 34	02 N 33	12 55	20 44	16 59	21 17	08 56	21 10	20 48	07 01
04	16 17	06 44	13 11	20 48	17 08	21 19	08 53	21 09	20 48	07 00
05	15 59	10 31	13 28	20 51	17 16	21 20	08 51	21 08	20 47	07 00
06	15 40	13 47	13 47	20 54	17 24	21 21	08 48	21 08	20 47	07 00
07	15 22	16 26	14 06	20 56	17 33	21 22	08 45	21 07	20 46	07 00
08	15 03	18 21	14 26	20 58	17 41	21 23	08 43	21 06	20 46	07 00
09	14 44	19 29	14 46	21 00	17 49	21 24	08 40	21 06	20 46	06 59
10	14 24	19 46	15 05	21 01	17 57	21 25	08 37	21 05	20 45	06 59
11	14 05	19 08	15 23	21 01	18 05	21 26	08 35	21 04	20 45	06 59
12	13 45	17 37	15 41	21 01	18 13	21 27	08 32	21 04	20 45	06 59
13	13 25	15 14	15 57	21 00	18 21	21 28	08 29	21 03	20 44	06 59
14	13 05	12 05	16 12	20 58	18 29	21 29	08 26	21 03	20 44	06 58
15	12 44	08 17	16 26	20 56	18 36	21 30	08 24	21 02	20 44	06 58
16	12 24	04 N 00	16 38	20 54	18 43	21 31	08 21	21 01	20 43	06 58
17	12 03	00 S 33	16 49	20 51	18 50	21 32	08 18	21 01	20 43	06 57
18	11 42	05 08	16 58	20 47	18 57	21 33	08 15	21 00	20 43	06 57
19	11 20	09 29	17 06	20 43	19 04	21 34	08 13	20 59	20 42	06 57
20	10 59	13 21	17 13	20 38	19 10	21 35	08 10	20 59	20 42	06 57
21	10 37	16 28	17 17	20 33	19 16	21 35	08 07	20 58	20 42	06 56
22	10 16	18 37	17 20	20 27	19 22	21 36	08 04	20 58	20 41	06 56
23	09 54	19 36	17 22	20 21	19 27	21 37	08 01	20 57	20 41	06 56
24	09 32	19 21	17 22	20 14	19 33	21 38	07 59	20 56	20 41	06 55
25	09 09	17 54	17 21	20 06	19 38	21 38	07 56	20 56	20 40	06 55
26	08 47	15 23	17 19	19 58	19 42	21 39	07 53	20 55	20 40	06 55
27	08 25	12 00	17 14	19 49	19 47	21 40	07 50	20 55	20 40	06 54
28	08 S 02	08 S 01	17 S 09	19 S 40	19 N 51	21 S 40	07 S 47	20 S 54	20 S 39	06 S 54

ZODIAC SIGN ENTRIES

Date	h m	Planets
01	08 05	☽ ♓
03	14 12	☽ ♈
04	20 12	♀ ♑
06	00 08	☽ ♉
08	12 44	☽ ♊
11	01 17	☽ ♋
13	11 31	☽ ♌
15	18 52	☽ ♍
18	00 00	☽ ♎
19	03 11	☉ ♓
20	03 55	☽ ♏
22	07 13	☽ ♐
24	10 11	☽ ♑
26	13 14	☽ ♒
28	17 16	☽ ♓

LATITUDES

Date	Mercury ☿	Venus ♀	Mars ♂	Jupiter ♃	Saturn ♄	Uranus ♅	Neptune ♆	Pluto ♇
01	03 N 11	02 N 49	04 N 28	00 N 46	01 S 46	00 S 30	00 N 33	13 N 29
04	03 36	02 38	04 30	00 46	01 46	00 30	00 33	13 30
07	03 39	02 28	04 32	00 46	01 45	00 30	00 33	13 31
10	03 22	02 17	04 33	00 47	01 45	00 30	00 33	13 32
13	02 53	02 05	04 32	00 47	01 45	00 30	00 33	13 33
16	02 18	01 53	04 31	00 47	01 45	00 30	00 33	13 34
19	01 40	01 41	04 29	00 47	01 45	00 30	00 33	13 36
22	01 03	01 29	04 26	00 47	01 45	00 30	00 33	13 37
25	00 N 28	01 17	04 22	00 47	01 45	00 30	00 33	13 38
28	00 S 05	01 05	04 17	00 47	01 46	00 30	00 33	13 39
31	00 S 34	00 N 53	04 N 12	00 N 48	01 S 46	00 S 30	00 N 33	13 N 41

DATA

Julian Date	2449750
Delta T	+61 seconds
Ayanamsa	23° 47' 31"
Synetic vernal point	05° ♓ 19' 28"
True obliquity of ecliptic	23° 26' 17"

MOON'S PHASES, APSIDES AND POSITIONS ☽

Date	h m	Phase	Longitude	Eclipse Indicator
07	12 54	☽	18 ♉ 17	
15	12 15	○	26 ♌ 21	
22	13 04	☾	03 ♐ 26	

Day	h m		
08	17 59	Apogee	
23	02 05	Perigee	
02	22 07	0N	
10	07 26	Max dec	19° N 47'
17	09 07	0S	
23	19 18	Max dec	19° S 39'

LONGITUDES

Date	Chiron ⚷	Ceres ⚳	Pallas ⚴	Juno ⚵	Vesta ⚶	Black Moon Lilith ⚸
01	25 ♍ 59	14 ♌ 04	12 ♉ 32	20 ♐ 25	25 ♊ 31	03 ♊ 18
11	25 ♍ 29	11 ♌ 44	16 ♉ 08	23 ♐ 11	25 ♊ 07	04 ♊ 25
21	24 ♍ 51	09 ♌ 39	20 ♉ 15	25 ♐ 46	25 ♊ 28	05 ♊ 31
31	24 ♍ 07	08 ♌ 03	24 ♉ 47	28 ♐ 06	26 ♊ 29	06 ♊ 38

ASPECTARIAN

01 Wednesday
00 42 ☽ ⚹ ♀
03 06 ☽ ☍ ♂
03 23 ☽ ⊻ ♅
07 38 ☽ ⊥ ♆
08 21 ☽ ∥ ♇
08 39 ☽ □ ♇
13 54 ☽ ⊥ ♅
16 07 ☉ Q ♇
23 31 ☽ ∠ ♆
23 36 ☽ Q ♀
23 52 ♀ △ ♂

02 Thursday
02 37 ☽ □ ♃
03 51 ☽ ☌ ♄
05 56 ☽ ∠ ♅
07 01 ☽ ⊻ ☉
11 45 ♀ ⊻ ♅
13 06 ☽ ⊻ ☿
13 06 ☉ ⋕ ♂
18 45 ☽ ⊥ ☉
23 08 ☽ ⊥ ☿

03 Friday
02 44 ☽ ⚹ ♆
07 24 ☽ ⊼ ♂
09 25 ☽ ⚹ ♅
11 21 ☽ □ ♀
12 45 ☽ ∠ ☉
14 22 ☽ ∠ ☿
14 53 ☽ △ ♇
18 16 ☽ ± ♂
23 00 ☉ ☌ ☿

04 Saturday
01 12 ☽ Q ♆
05 34 ☿ ∠ ♀
08 07 ☽ Q ♅
08 18 ♀ Q ♄
10 45 ☽ △ ♃
10 51 ♀ ∥ ♆
10 51 ☽ ꝑ ♂
11 55 ☽ ⊻ ♄
13 41 ☽ ⋕ ♇
16 21 ☽ ⚹ ☿
19 28 ☽ ꝑ ♇
19 45 ☽ ⚹ ☉
23 36 ☽ ⊥ ♄

05 Sunday
01 11 ☽ ⋕ ♄
04 34 ♀ ⊻ ♇
12 04 ☽ □ ♆
13 03 ☽ ± ♇
13 38 ☽ Q ☿
15 09 ☽ △ ♂
16 18 ☽ ꝑ ♃
17 25 ☽ ∠ ♄
19 19 ☽ □ ♅
19 26 ☿ Q ♇
21 23 ☽ Q ☉

06 Monday
00 57 ☽ ⊼ ♇
03 00 ☽ △ ♀
10 02 ☿ ⊻ ♄
10 25 ☽ ± ♃
11 57 ☽ ⋕ ☿
19 55 ☿ ⚹ ♃
22 22 ☽ □ ☿
22 38 ☽ ⊼ ♃
23 41 ☽ ⚹ ♄

07 Tuesday
02 40 ☽ ⋕ ☉
12 20 ☽ ꝑ ♀
12 54 ☽ □ ☉

08 Wednesday
00 15 ☽ Q ♄
00 29 ☽ △ ♆
01 33 ☽ □ ♂
01 45 ☽ ∥ ♂
04 19 ☿ ⊥ ♀
08 04 ☽ △ ♅
08 34 ☽ ± ♀
13 37 ☽ ☍ ♇
22 04 ☽ ⊼ ♀

09 Thursday
06 27 ☽ △ ☿
06 49 ♂ ⊼ ♆
07 05 ☽ ꝑ ♆
10 44 ☉ ∥ ☿
12 17 ☽ ☍ ♃
12 58 ☽ Q ♂
13 12 ☽ □ ♄
14 41 ☽ ꝑ ♅
16 22 ☉ ∠ ♀

10 Friday
01 21 ☽ ± ♆
07 28 ☽ △ ☉
08 56 ☽ ± ♅
10 47 ☽ ꝑ ☿
12 23 ☽ ⚹ ♂
13 27 ☽ ⊼ ♆
13 50 ♀ ⊥ ♇
21 00 ☽ ⊼ ♅

11 Saturday
02 13 ☽ ⊼ ♇
03 45 ☽ ± ☿
05 20 ☿ ⊻ ♀
14 06 ☽ ± ♇
15 02 ☽ ⊼ ☿
16 02 ☽ ꝑ ☉
16 34 ☽ ☍ ♀
17 13 ☽ ∠ ♂
21 49 ♀ ꝑ ♂

12 Sunday
00 48 ☽ ⊼ ♃
01 36 ☽ △ ♄
02 31 ☉ ☍ ♂
04 37 ☽ ∥ ♂
07 43 ☽ ꝑ ♇
10 13 ☽ ⊥ ♂
11 10 ☽ ± ☉
12 35 ☽ ± ♃
15 40 ☿ ⊥ ♄
21 25 ☽ ⊻ ♂
23 43 ☽ ⊼ ☉

13 Monday
00 25 ☽ ☍ ♆
06 01 ☽ ꝑ ♃
06 11 ☽ ⋕ ☿
06 45 ☽ ꝑ ♄
06 59 ☉ Q ♃
07 42 ☽ ☍ ♅
08 33 ☉ ⊻ ♆
12 28 ☽ △ ♇
22 41 ☽ ☍ ☿
23 53 ☽ ± ♄

14 Tuesday
04 10 ☽ ⋕ ☉
07 38 ☽ ⊼ ♀
10 27 ☽ △ ♃
11 09 ☽ ⊼ ♄
19 47 ☽ ± ♀

15 Wednesday
03 48 ☽ ☌ ♂
08 26 ☽ ⊼ ♆
11 17 ☽ ⋕ ♄
12 15 ☽ ☍ ☉
13 45 ☽ ꝑ ♀
15 27 ☽ ⊼ ♅
19 05 ♀ ⊻ ♃
19 19 ☽ ± ♆
19 35 ☽ ⋕ ♇
19 48 ☽ □ ♇

16 Thursday
02 17 ☽ ± ♅
03 32 ♀ ⚹ ♄
04 57 ☽ ⊼ ☿
05 07 ☿ St D
11 28 ☽ ꝑ ♆
15 40 ☽ ± ☿
17 17 ☽ □ ♃
17 55 ☽ ☍ ♄
18 24 ☽ ꝑ ♅
19 07 ☽ △ ♀

17 Friday
03 54 ☽ Q ♇
07 41 ☽ ꝑ ☿
07 57 ☽ ⊻ ♂
08 19 ☉ ⊻ ♅
14 03 ☽ △ ♆
18 03 ♀ ± ♂
18 11 ☽ ⊥ ♂
20 54 ☽ △ ♅
21 51 ☽ ⊼ ☉

18 Saturday
00 57 ☽ ⚹ ♇
01 14 ☽ Q ♃
09 08 ☽ ± ☉
09 28 ☽ ∠ ♂
10 17 ☽ △ ☿
12 17 ♀ ∠ ♇
21 48 ☽ ∥ ♇
22 16 ☽ ⚹ ♃
22 53 ☽ ⊼ ♄

19 Sunday
01 59 ☽ ꝑ ☉
03 01 ☽ ∠ ♇
04 21 ☽ □ ♀
04 50 ☽ ∥ ♄
09 21 ☽ ± ♄
10 45 ☽ ⚹ ♂
12 02 ☉ ⊥ ♆
16 29 ☉ □ ♇
16 41 ♀ ∥ ♆
18 15 ☽ □ ♆
18 34 ☽ ⊥ ♇
22 07 ☽ ∥ ☉

20 Monday
00 22 ☽ ∠ ♃
00 59 ☽ ꝑ ♄
01 04 ☽ □ ♅
04 53 ☽ ⊻ ♇
05 50 ☽ △ ☉
06 51 ☽ Q ♂
14 08 ☽ Q ♀
15 22 ☽ □ ☿
15 55 ♄ ∠ ♅
15 57 ☽ ⊥ ♃

21 Tuesday
01 13 ☽ Q ♆
02 18 ☽ ⊻ ♃
02 56 ☽ △ ♄
08 01 ☽ Q ♅
12 30 ☽ ⚹ ♀
12 53 ☽ □ ♂
14 16 ♀ ⊥ ♃
16 03 ♀ ⊼ ♂
20 05 ☽ ∥ ☿
21 45 ☽ ⚹ ♆
23 17 ☽ Q ☿

22 Wednesday
04 34 ☽ ⚹ ♅
08 12 ☽ ☌ ♇
08 46 ☿ ⊥ ♄
13 04 ☽ □ ☉
16 22 ☽ ∠ ♀
20 37 ☽ ⚹ ☿
23 21 ☽ ∠ ♆

23 Thursday
04 43 ☽ ⋕ ♂
05 49 ☽ ☌ ♃
06 11 ☽ ∠ ♅
06 29 ☽ □ ♄
09 02 ☽ ⊥ ♀
14 40 ☽ ⊥ ♆
14 40 ☽ △ ♂
14 57 ☉ ⊥ ♅
20 08 ☽ ⊻ ♀
21 31 ☽ ⊥ ♅
22 00 ☽ Q ☉
23 22 ☽ ∠ ☿

24 Friday
00 52 ☽ ⊻ ♆
06 27 ☽ ⋕ ♂
07 44 ☽ ⊻ ♅
11 10 ☽ ⊻ ♇
13 18 ☽ Q ♄
15 28 ☽ ⊥ ☿
15 31 ☽ ꝑ ♂
19 55 ☽ ⚹ ☉
21 22 ☽ ⊥ ♇

25 Saturday
02 14 ☽ ⊻ ☿
06 27 ☽ ± ♂
09 08 ☽ ⊻ ♃
09 52 ☽ ⚹ ♄
12 39 ☽ ∠ ♇
16 25 ☽ ⊼ ♂
18 17 ☽ ∥ ☿
19 25 ☽ ⊥ ♃
23 23 ☽ ∠ ☉

26 Sunday
03 41 ☽ ☌ ♀
03 58 ☽ ☌ ♆
07 06 ♀ ☌ ♆
10 53 ☽ ∠ ♃
10 57 ☽ ☌ ♅
11 40 ☽ ∠ ♄
14 15 ☽ ⚹ ♇
15 57 ☽ ⊥ ☉

27 Monday
03 03 ☽ ⊻ ☉
03 14 ☽ ⊥ ♄
04 13 ♃ ∠ ♅
08 47 ☽ ☌ ☿
10 54 ☽ Q ♇
12 52 ☽ ⚹ ♃
13 41 ☽ ⊻ ♄
16 23 ♀ ⋕ ♂
18 44 ☽ ☍ ♂

28 Tuesday
07 51 ☽ ⊻ ♆
09 54 ☽ Q ♃
11 52 ☽ ∥ ☉
12 17 ☽ ⊻ ♀
13 17 ☽ ∥ ♄
15 06 ☽ ⊻ ♅
16 08 ☿ Q ♇
17 24 ☉ ∠ ♆
18 19 ☽ ∥ ♇
18 19 ☽ □ ♇
18 25 ☽ ⊥ ♆
23 52 ☽ ⊥ ♀

All ephemeris data is given at 12.00 UT and the Moon's longitude is additionally given for 24.00 UT

MARCH 1995

LONGITUDES

Date	Sidereal time h m s	Sun ☉ ° ' "	Moon ☽ ° ' "	Moon ☽ 24.00 ° ' "	Mercury ☿ ° '	Venus ♀ ° '	Mars ♂ ° '	Jupiter ♃ ° '	Saturn ♄ ° '	Uranus ♅ ° '	Neptune ♆ ° '	Pluto ♇ ° '
01	22 35 17	10 ♓ 26 01	10 ♓ 32 19	17 ♓ 11 40	13 ♒ 25	28 ♑ 20	16 ♌ 36	13 ♐ 55	14 ♓ 27	28 ♑ 49	24 ♑ 41	00 ♐ 36
02	22 39 14	11 26 15	23 ♓ 46 22	00 ♈ 16 17	14 27	29 ♑ 30	16 R 19	14 01	14 34	28 52	24 42	00 36
03	22 43 10	12 26 27	06 ♈ 41 24	13 ♈ 01 45	15 31	00 ♒ 41	16 02	14 06	14 42	28 54	24 44	00 36
04	22 47 07	13 26 38	19 ♈ 17 32	25 ♈ 29 00	16 38	01 51	15 47	14 11	14 49	28 57	24 46	00 R 36
05	22 51 03	14 26 46	01 ♉ 36 30	07 ♉ 40 27	17 46	03 02	15 31	14 16	14 57	29 00	24 47	00 36
06	22 55 00	15 26 52	13 ♉ 41 20	19 ♉ 39 41	18 57	04 12	15 17	14 21	15 04	29 03	24 49	00 36
07	22 58 57	16 26 57	25 ♉ 36 06	01 ♊ 31 11	20 10	05 23	15 03	14 25	15 11	29 05	24 51	00 36
08	23 02 53	17 26 59	07 ♊ 25 37	13 ♊ 20 01	21 24	06 34	14 50	14 30	15 19	29 08	24 52	00 36
09	23 06 50	18 26 59	19 ♊ 15 05	25 ♊ 11 27	22 41	07 44	14 38	14 34	15 26	29 11	24 54	00 36
10	23 10 46	19 26 57	01 ♋ 09 48	07 ♋ 10 44	23 58	08 55	14 27	14 38	15 33	29 13	24 55	00 35
11	23 14 43	20 26 52	13 ♋ 14 50	19 ♋ 22 41	25 18	10 06	14 16	14 42	15 41	29 16	24 57	00 35
12	23 18 39	21 26 46	25 ♋ 34 43	01 ♌ 51 23	26 39	11 17	14 07	14 46	15 48	29 18	24 58	00 35
13	23 22 36	22 26 37	08 ♌ 13 00	14 ♌ 39 49	28 02	12 28	13 58	14 49	15 55	29 21	25 00	00 35
14	23 26 32	23 26 26	21 ♌ 11 57	27 ♌ 49 26	29 ♒ 26	13 39	13 49	14 53	16 03	29 23	25 01	00 34
15	23 30 29	24 26 13	04 ♍ 32 12	11 ♍ 20 01	00 ♓ 52	14 51	13 42	14 56	16 10	29 26	25 02	00 34
16	23 34 26	25 25 58	18 ♍ 12 36	25 ♍ 09 32	02 19	16 02	13 35	14 59	16 17	29 28	25 04	00 34
17	23 38 22	26 25 41	02 ♎ 10 20	09 ♎ 14 26	03 47	17 13	13 29	15 02	16 25	29 30	25 05	00 33
18	23 42 19	27 25 22	16 ♎ 21 16	23 ♎ 30 11	05 17	18 24	13 24	15 05	16 32	29 32	25 06	00 33
19	23 46 15	28 25 01	00 ♏ 40 36	07 ♏ 51 53	06 48	19 36	13 20	15 07	16 39	29 35	25 08	00 32
20	23 50 12	29 ♓ 24 38	15 ♏ 03 30	22 ♏ 14 57	08 20	20 47	13 17	15 09	16 47	29 37	25 09	00 32
21	23 54 08	00 ♈ 24 14	29 ♏ 25 45	06 ♐ 35 31	09 54	21 59	13 14	15 12	16 54	29 39	25 10	00 31
22	23 58 05	01 23 48	13 ♐ 43 57	20 ♐ 50 44	11 28	23 10	13 12	15 14	17 01	29 41	25 11	00 31
23	00 02 01	02 23 20	27 ♐ 55 39	04 ♑ 58 30	13 05	24 22	13 10	15 15	17 08	29 43	25 12	00 30
24	00 05 58	03 22 50	11 ♑ 59 08	18 ♑ 57 24	14 42	25 34	13 10	15 17	17 15	29 45	25 14	00 29
25	00 09 55	04 22 19	25 ♑ 53 10	02 ♒ 46 18	16 21	26 45	13 D 10	15 18	17 23	29 47	25 15	00 29
26	00 13 51	05 21 46	09 ♒ 36 41	16 ♒ 24 11	18 02	27 57	13 11	15 20	17 30	29 49	25 16	00 28
27	00 17 48	06 21 11	23 ♒ 08 41	29 ♒ 50 02	19 43	29 ♒ 09	13 13	15 21	17 37	29 51	25 17	00 27
28	00 21 44	07 20 34	06 ♓ 27 38	13 ♓ 02 52	21 26	00 ♓ 20	13 15	15 22	17 44	29 53	25 18	00 26
29	00 25 41	08 19 55	19 ♓ 34 08	26 ♓ 01 51	23 10	01 32	13 18	15 22	17 51	29 55	25 19	00 25
30	00 29 37	09 19 14	02 ♈ 26 00	08 ♈ 46 34	24 56	02 44	13 22	15 23	17 58	29 57	25 20	00 25
31	00 33 34	10 ♈ 18 32	15 ♈ 03 36	21 ♈ 17 10	26 ♓ 43	03 ♓ 56	13 ♌ 26	15 ♐ 23	18 ♓ 05	29 ♑ 58	25 ♑ 21	00 ♐ 24

Date	Moon True ☊ ° '	Moon Mean ☊ ° '	Moon ☽ Latitude ° '
01	07 ♏ 04	08 ♏ 37	04 N 15
02	06 R 56	08 34	03 29
03	06 49	08 30	02 34
04	06 45	08 27	01 32
05	06 43	08 24	00 N 28
06	06 D 43	08 21	00 S 38
07	06 45	08 18	01 40
08	06 46	08 15	02 38
09	06 R 46	08 11	03 29
10	06 45	08 08	04 12
11	06 43	08 05	04 44
12	06 38	08 02	05 03
13	06 32	07 59	05 09
14	06 24	07 56	04 59
15	06 17	07 52	04 33
16	06 10	07 49	03 50
17	06 05	07 46	02 53
18	06 01	07 43	01 45
19	06 01	07 40	00 S 29
20	06 D 00	07 36	00 N 49
21	06 01	07 33	02 04
22	06 03	07 30	03 11
23	06 04	07 27	04 06
24	06 R 03	07 24	04 46
25	06 02	07 21	05 08
26	05 59	07 17	05 12
27	05 55	07 14	04 59
28	05 50	07 11	04 29
29	05 45	07 08	03 46
30	05 42	07 05	02 52
31	05 ♏ 39	07 ♏ 02	01 N 51

DECLINATIONS

Date	Sun ☉ ° '	Moon ☽ ° '	Mercury ☿ ° '	Venus ♀ ° '	Mars ♂ ° '	Jupiter ♃ ° '	Saturn ♄ ° '	Uranus ♅ ° '	Neptune ♆ ° '	Pluto ♇ ° '
01	07 S 39	03 S 41	17 S 02	19 S 30	19 N 55	21 S 41	07 S 45	20 S 54	20 S 39	06 S 53
02	07 16	00 N 44	16 53	19 20	19 58	21 42	07 42	20 53	20 39	06 53
03	06 54	05 01	16 43	19 09	20 02	21 42	07 39	20 52	20 38	06 53
04	06 30	08 59	16 32	18 57	20 05	21 43	07 36	20 52	20 38	06 52
05	06 07	12 28	16 19	18 45	20 07	21 43	07 33	20 51	20 38	06 52
06	05 44	15 21	16 05	18 33	20 10	21 44	07 31	20 51	20 38	06 52
07	05 21	17 32	15 50	18 20	20 12	21 44	07 28	20 50	20 37	06 51
08	04 58	18 57	15 33	18 06	20 14	21 45	07 25	20 50	20 37	06 51
09	04 34	19 32	15 15	17 52	20 16	21 45	07 22	20 49	20 37	06 50
10	04 11	19 14	14 55	17 38	20 17	21 45	07 19	20 49	20 36	06 50
11	03 47	18 04	14 35	17 22	20 18	21 46	07 17	20 48	20 36	06 50
12	03 24	16 03	14 13	17 07	20 19	21 46	07 14	20 48	20 36	06 49
13	03 00	13 14	13 49	16 51	20 19	21 47	07 11	20 47	20 36	06 49
14	02 36	09 42	13 25	16 34	20 19	21 47	07 08	20 47	20 35	06 48
15	02 13	05 37	12 59	16 17	20 19	21 47	07 05	20 46	20 35	06 48
16	01 49	01 N 08	12 31	16 00	20 19	21 47	07 03	20 46	20 35	06 47
17	01 25	03 S 31	12 03	15 42	20 18	21 48	07 00	20 45	20 35	06 47
18	01 01	08 03	11 33	15 23	20 18	21 48	06 57	20 45	20 34	06 47
19	00 38	12 10	11 02	15 04	20 17	21 48	06 54	20 44	20 34	06 46
20	00 S 14	15 34	10 30	14 45	20 16	21 48	06 52	20 44	20 34	06 46
21	00 N 10	18 00	09 57	14 25	20 14	21 49	06 49	20 44	20 34	06 45
22	00 33	19 17	09 22	14 05	20 12	21 49	06 46	20 43	20 33	06 45
23	00 57	19 19	08 46	13 45	20 11	21 49	06 43	20 43	20 33	06 44
24	01 21	18 09	08 09	13 24	20 08	21 49	06 41	20 42	20 33	06 44
25	01 44	15 55	07 31	13 03	20 06	21 49	06 38	20 42	20 33	06 43
26	02 08	12 50	06 52	12 41	20 04	21 49	06 35	20 42	20 33	06 43
27	02 31	09 06	06 12	12 19	20 01	21 49	06 33	20 41	20 32	06 42
28	02 55	04 58	05 30	11 57	19 58	21 49	06 30	20 41	20 32	06 42
29	03 18	00 S 40	04 47	11 34	19 55	21 49	06 27	20 41	20 32	06 42
30	03 42	03 N 36	04 04	11 11	19 52	21 49	06 24	20 40	20 32	06 41
31	04 N 05	07 N 38	03 S 19	10 S 48	19 N 48	21 S 49	06 S 22	20 S 40	20 S 32	06 S 41

ZODIAC SIGN ENTRIES

Date	h	m	Planets
02	22	10	♀ ♒
02	23	30	☽ ♈
05	08	50	☽ ♉
07	20	55	☽ ♊
10	09	40	☽ ♋
12	20	28	☽ ♌
14	21	35	☿ ♓
15	03	54	☽ ♍
17	08	18	☽ ♎
19	10	52	☽ ♏
21	02	14	☉ ♈
21	12	57	☽ ♐
23	15	31	☽ ♑
25	19	10	☽ ♒
28	00	18	☽ ♓
28	05	10	♀ ♓
30	07	26	☽ ♈

LATITUDES

Date	Mercury ☿ ° '	Venus ♀ ° '	Mars ♂ ° '	Jupiter ♃ ° '	Saturn ♄ ° '	Uranus ♅ ° '	Neptune ♆ ° '	Pluto ♇ ° '
01	00 S 15	01 N 01	04 N 15	00 N 48	01 S 46	00 S 30	00 N 33	13 N 40
04	00 43	00 49	04 10	00 48	01 46	00 30	00 33	13 41
07	01 08	00 37	04 04	00 48	01 46	00 31	00 33	13 42
10	01 29	00 25	03 57	00 48	01 46	00 31	00 33	13 43
13	01 46	00 13	03 51	00 48	01 46	00 31	00 33	13 44
16	02 00	00 N 02	03 44	00 48	01 46	00 31	00 33	13 46
19	02 11	00 S 08	03 37	00 49	01 47	00 31	00 33	13 47
22	02 17	00 19	03 30	00 49	01 47	00 31	00 33	13 48
25	02 19	00 29	03 23	00 49	01 47	00 31	00 33	13 49
28	02 17	00 38	03 16	00 49	01 47	00 31	00 33	13 50
31	02 S 11	00 S 47	03 N 09	00 N 49	01 S 48	00 S 31	00 N 33	13 N 51

DATA

Julian Date	2449778
Delta T	+61 seconds
Ayanamsa	23° 47' 34"
Synetic vernal point	05° ♓ 19' 25"
True obliquity of ecliptic	23° 26' 17"

MOON'S PHASES, APSIDES AND POSITIONS ☽

Date	h	m	Phase	Longitude ° '	Eclipse Indicator
01	11	48	●	10 ♓ 26	
09	10	14	☽	18 ♊ 23	
17	01	26	○	25 ♍ 59	
23	20	10	☾	02 ♑ 44	
31	02	09	●	09 ♈ 54	

Day	h	m		
08	14	59	Apogee	
20	13	23	Perigee	
02	08	00	0N	
09	16	04	Max dec	19° N 32'
16	17	51	0S	
23	00	44	Max dec	19° S 27'
29	15	40	0N	

LONGITUDES

Date	Chiron ⚷ ° '	Ceres ⚳ ° '	Pallas ⚴ ° '	Juno ⚵ ° '	Vesta ⚶ ° '	Black Moon Lilith ⚸ ° '
01	24 ♍ 17	08 ♌ 19	23 ♉ 51	27 ♐ 39	26 ♊ 14	06 ♊ 25
11	23 ♍ 30	07 ♌ 14	28 ♉ 31	29 ♐ 46	27 ♊ 44	07 ♊ 31
21	22 ♍ 42	06 ♌ 52	03 ♊ 46	01 ♑ 33	29 ♊ 46	08 ♊ 38
31	21 ♍ 57	07 ♌ 13	09 ♊ 05	02 ♑ 58	02 ♋ 14	09 ♊ 45

ASPECTARIAN

h m Aspects

01 Wednesday
01 45 ☽ ⊥ ♅
05 33 ☉ ‖ ♄
10 27 ☽ ∠ ♆
11 48 ☽ ☌ ☉
17 30 ☽ ∠ ♀
17 36 ☽ ⊻ ☿
17 54 ☽ ∠ ♅
18 07 ☽ □ ♃
19 06 ☽ ☌ ♄
22 18 ♀ ☌ ♅
22 42 ☽ ⊼ ♂

02 Thursday
00 57 ☿ ⚹ ♃
01 07 ♀ ∠ ♃
05 24 ☽ ⊥ ☿
09 23 ☽ ± ♂
13 37 ♀ ∠ ♄
13 43 ☽ ⚹ ♆
15 13 ☿ ⊻ ♄
21 25 ☽ ⚹ ♅
23 24 ☽ ∠ ☿
23 38 ☽ ⚹ ♀

03 Friday
00 37 ☽ △ ♇
01 39 ☽ ⚼ ♂
10 33 ♀ ⚹ ♇
12 05 ☽ Q ♆
12 48 ☉ ‖ ♇
20 00 ☽ Q ♅
21 10 ☿ ☍ ♂
22 06 ☽ # ☉
22 59 ☽ # ♇
23 49 ☽ ⊻ ☉

04 Saturday
00 30 ☽ Q ♀
02 08 ☽ △ ♃
02 34 ♇ St R
03 20 ☽ ⊻ ♄
03 30 ☽ # ♄
04 55 ☽ ⚼ ♇
05 23 ☽ △ ♂
06 22 ☽ ⚹ ☿
12 19 ☽ ⊥ ☉
14 59 ☽ ⊥ ♄
22 17 ☽ ± ♇
22 37 ☽ □ ♆

05 Sunday
00 47 ☉ ∠ ♅
06 51 ☽ □ ♅
07 19 ☉ □ ♃
07 22 ☽ ⚼ ♃
07 22 ☽ ∠ ☉
08 01 ☽ Q ☿
08 41 ☽ ∠ ♄
10 01 ☽ ⊼ ♇
15 06 ☽ □ ♀

06 Monday
01 16 ☽ ± ♃
01 32 ☉ ☌ ♄
08 48 ☉ ⊼ ♂
13 19 ☽ ⊼ ♃
14 47 ☽ ⚹ ♄
15 08 ☽ □ ♂
15 51 ☽ ⚹ ☉
18 32 ☽ # ☿
23 45 ☽ □ ☿

07 Tuesday
02 49 ♂ ⊼ ♄
10 28 ☽ △ ♆
15 15 ☽ Q ♄
18 18 ☽ Q ☉
19 06 ☽ △ ♅
22 08 ☽ ☍ ♇
22 10 ☽ # ♀

08 Wednesday
02 51 ☽ Q ♂
10 03 ☽ △ ♀
16 58 ☽ ⚼ ♆

09 Thursday
01 40 ☽ ⚼ ♅
02 27 ☽ ☍ ♃
02 48 ☽ ⚹ ♂
04 11 ☽ □ ♄
10 14 ☽ □ ☉
11 17 ☽ ± ♆
18 24 ♂ △ ♃
19 46 ☽ △ ☿
19 50 ☽ ⚼ ♀
19 58 ☽ ± ♅
23 26 ☽ ⊼ ♆

10 Friday
08 05 ☽ ⊼ ♅
08 37 ☽ ∠ ♂
10 52 ☽ ⊼ ♇
15 54 ☽ ± ♀
22 50 ☽ ± ♇

11 Saturday
02 19 ☽ ⊥ ♂
02 21 ♀ ⊥ ♄
05 08 ☽ ⊼ ♀
05 28 ☽ ⚼ ☿
05 28 ☿ ⊻ ♆
08 24 ☉ ± ♂
13 59 ☽ ⊻ ♂
14 52 ☽ ⊼ ♃
16 36 ☽ ⚼ ♇
16 49 ☽ △ ♄
22 56 ☽ # ♀

12 Sunday
01 18 ☽ ± ☿
02 39 ☽ ± ♃
03 19 ☽ △ ☉
10 49 ☽ ☍ ♆
14 01 ☿ Q ♃
14 19 ☽ ⊼ ☿
19 10 ☽ ☍ ♅
20 03 ☽ ⚼ ♃
21 35 ☽ △ ♇
22 05 ☽ ⚼ ♄

13 Monday
06 49 ☽ # ☿
10 26 ☽ ⚼ ☉
14 09 ♀ Q ♇
15 13 ☽ ± ♄
20 44 ☽ ☍ ♀
22 35 ☽ ☌ ♂

14 Tuesday
00 21 ☽ △ ♃
02 28 ☽ ⊼ ♄
04 33 ☽ ± ☉
11 10 ☿ ⊻ ♅
15 00 ♀ ☍ ♂
16 25 ☽ ⊼ ☉
18 57 ☽ ⊼ ♆

15 Wednesday
02 51 ☽ ⊼ ♅
03 34 ☽ # ♄
04 40 ☽ ☍ ☿
04 56 ☽ □ ♇
05 19 ☽ # ♇
05 45 ☽ ± ♆
07 09 ☿ □ ♇
13 35 ☽ ± ♅
13 54 ♀ ⚹ ♃
15 03 ☿ ⊥ ♆
21 45 ☽ ⚼ ♆

16 Thursday
02 52 ☉ ⚹ ♆
04 01 ☽ ⊻ ♂
05 28 ☽ ⚼ ♅
06 22 ☽ □ ♃
07 51 ☽ ⊼ ♀
08 04 ☽ # ☉
08 38 ☽ ☍ ♄
12 37 ☽ Q ♇
14 22 ☽ ⊥ ♂
17 49 ♀ ⊻ ♄
19 14 ☽ ± ♀
23 51 ☽ △ ♆

17 Friday
01 26 ☽ ☍ ☉
02 02 ☽ ‖ ☉
05 45 ☽ ∠ ♂
07 26 ☽ △ ♅
09 14 ☽ ⚹ ♇
12 05 ☽ ⚼ ♀
13 28 ☽ Q ♃
15 04 ☽ ⊼ ☿

18 Saturday
02 26 ☽ ± ☿
05 08 ☽ ‖ ♇
06 07 ☽ ‖ ♄
07 04 ☽ ⚹ ♂
09 51 ☽ ⚹ ♃
10 38 ☽ ∠ ♇
12 18 ☽ ⊼ ♄
15 46 ☽ △ ♀
16 21 ☿ ⊥ ♅
19 22 ☽ ⚼ ☿
22 28 ☽ ± ♄

19 Sunday
01 44 ☽ ⊥ ♇
02 42 ☽ □ ♆
03 07 ☽ Q ♂
05 54 ☽ ‖ ☿
07 57 ☽ ⊼ ☉
10 08 ☉ ⚼ ♂
10 10 ☽ □ ♅
11 04 ☽ ∠ ♃
11 46 ☽ ⊻ ♇
13 39 ☽ ⚼ ♄
18 43 ☽ ± ☉
23 26 ☽ △ ☿

20 Monday
02 08 ☽ ⊥ ♃
06 14 ☽ ‖ ♀
08 49 ☽ Q ♆
09 02 ☽ □ ♂
10 50 ☽ ⚼ ☉
12 10 ☽ ⊻ ♃
14 53 ☽ △ ♄
16 17 ☽ Q ♅
17 08 ☉ ⚹ ♅
22 25 ☽ □ ♀

21 Tuesday
04 52 ☽ ⚹ ♆
12 22 ☽ ⚹ ♅
13 45 ☽ △ ☉
13 49 ☽ ☌ ♇
14 45 ☉ △ ♇
16 16 ☿ ∠ ♆

22 Wednesday
06 02 ☽ ∠ ♆
07 18 ☽ Q ♀
07 43 ☽ □ ☿
11 06 ☽ △ ♂
13 37 ☽ ∠ ♅
14 31 ☽ ☌ ♃
17 35 ☽ □ ♄
21 13 ☽ ⊥ ♆

23 Thursday
01 12 ♄ ‖ ♇
04 51 ☽ ⊥ ♅
05 24 ☽ ⚹ ♀
07 23 ☽ ⊻ ♆
12 25 ☽ ⚼ ♂
13 24 ☿ ⊼ ♂
15 03 ☽ ⊻ ♅
16 22 ☽ ⊻ ♇
18 03 ☽ Q ☿
20 10 ☽ □ ☉

24 Friday
00 23 ☽ Q ♄
02 35 ☽ ⊥ ♇
03 45 ☽ ± ♂
05 13 ♀ ⊻ ♆
09 19 ☽ ∠ ♀
12 44 ☿ ∠ ♅
14 02 ☽ ⊼ ♂
17 18 ☽ ⚹ ☿
17 18 ♂ St D
17 41 ☽ ⊻ ♃
18 01 ☽ ∠ ♇
20 34 ☿ □ ♃
21 09 ☽ ⚹ ♄

25 Saturday
02 16 ☽ ⊥ ♀
04 03 ☽ ⊥ ♃
05 26 ☽ Q ☉
10 53 ☽ ☌ ♆
13 39 ☽ ⊻ ♀
18 48 ☽ ☌ ♅
19 42 ☽ ∠ ♃
19 59 ☽ ⚹ ♇
22 50 ☽ ∠ ☿
23 19 ♀ Q ♃
23 24 ☽ ∠ ♄

26 Sunday
03 51 ☿ ☌ ♄
03 57 ☽ ⚹ ☉
13 06 ☽ ‖ ♀
15 21 ☽ ⊥ ♄
16 51 ☽ ⊥ ☿
17 01 ☽ Q ♇
17 31 ☿ ‖ ♇
18 19 ☽ ☍ ♂
22 06 ☽ ⚹ ♃
22 50 ☿ ‖ ♄

27 Monday
02 04 ☽ ⊻ ♄
04 44 ☿ ± ♂
05 00 ☽ ⊻ ☿
08 33 ☽ ∠ ☉
15 49 ☽ ⊻ ♆
19 32 ☽ Q ♃
23 49 ☽ ☌ ♀

28 Tuesday
00 04 ☽ ⊻ ♅
01 06 ☽ □ ♇
01 58 ☽ ⊥ ☉
02 04 ☽ ‖ ♇
02 35 ♀ ⊻ ♅
02 37 ☽ ⊥ ♆
03 10 ☽ ‖ ♄
08 20 ☽ ‖ ☿
10 51 ☉ Q ♆
10 56 ☽ ⊥ ♅
13 43 ☽ ⊻ ☉
13 56 ♀ □ ♇
18 59 ☽ ∠ ♆
22 31 ☽ # ☉

29 Wednesday
00 25 ☽ ⊼ ♂
03 24 ☽ ∠ ♅
04 15 ☽ □ ♃
07 26 ♀ ⊥ ♆
08 48 ☽ ☌ ♄
11 30 ☽ ± ♂
19 43 ☽ ☌ ☿
22 40 ☽ ⚹ ♆

30 Thursday
04 19 ☽ ⚼ ♂
07 19 ☽ ⚹ ♅
08 12 ☽ △ ♇
12 34 ☽ ‖ ☉
12 38 ☽ ⊻ ♀
14 14 ☽ # ☿
17 27 ☿ ⚹ ♆
19 48 ☉ # ☿
21 16 ☽ Q ♆

31 Friday
01 10 ☽ ⊥ ♀
02 09 ☽ ☌ ☉
04 18 ☽ # ♄
06 05 ☽ Q ♅
06 08 ☽ # ♇
08 52 ☽ △ ♂
12 37 ☽ △ ♃
12 39 ☽ ⚼ ♇
17 52 ☽ ⊻ ♄
20 15 ☽ ∠ ♀

All ephemeris data is given at 12.00 UT and the Moon's longitude is additionally given for 24.00 UT

APRIL 1995

LONGITUDES

Date	Sidereal time h m s	Sun ☉	Moon ☽	Moon ☽ 24.00	Mercury ☿	Venus ♀	Mars ♂	Jupiter ♃	Saturn ♄	Uranus ♅	Neptune ♆	Pluto ♇
01	00 37 30	11 ♈ 17 47	27 ♈ 27 25	03 ♉ 34 31	28 ♓ 31	05 ♓ 08	13 ♌ 31	15 ♐ 23	18 ♓ 12	00 ♒ 00	25 ♑ 21	00 ♐ 23
02	00 41 27	12 17 00	09 ♉ 38 42	15 ♉ 40 17	00 ♈ 21	06 20	13 37	15 R 23	18 19	00 02	25 22	00 R 22
03	00 45 24	13 16 11	21 ♉ 39 35	27 ♉ 36 59	02 12	07 32	13 44	15 23	18 26	00 03	25 23	00 21
04	00 49 20	14 15 20	03 ♊ 32 56	09 ♊ 27 53	04 04	08 44	13 51	15 22	18 33	00 05	25 24	00 20
05	00 53 17	15 14 26	15 ♊ 22 21	21 ♊ 16 52	05 58	09 56	13 58	15 22	18 40	00 06	25 25	00 19
06	00 57 13	16 13 31	27 ♊ 12 01	03 ♋ 08 22	07 54	11 08	14 07	15 21	18 47	00 08	25 25	00 18
07	01 01 10	17 12 33	09 ♋ 06 31	15 ♋ 07 05	09 50	12 20	14 16	15 20	18 53	00 09	25 26	00 17
08	01 05 06	18 11 33	21 ♋ 10 39	27 ♋ 17 49	11 48	13 32	14 25	15 18	19 00	00 10	25 27	00 16
09	01 09 03	19 10 30	03 ♌ 29 09	09 ♌ 45 09	13 48	14 44	14 35	15 17	19 07	00 12	25 27	00 15
10	01 12 59	20 09 25	16 ♌ 06 19	22 ♌ 33 01	15 48	15 56	14 46	15 15	19 14	00 13	25 28	00 13
11	01 16 56	21 08 18	29 ♌ 05 36	05 ♍ 44 18	17 50	17 09	14 57	15 14	19 20	00 14	25 29	00 12
12	01 20 53	22 07 09	12 ♍ 29 12	19 ♍ 20 18	19 53	18 21	15 09	15 12	19 27	00 15	25 29	00 11
13	01 24 49	23 05 57	26 ♍ 17 27	03 ♎ 20 22	21 57	19 33	15 21	15 10	19 34	00 17	25 30	00 10
14	01 28 46	24 04 43	10 ♎ 28 37	17 ♎ 41 39	24 02	20 45	15 34	15 07	19 40	00 18	25 30	00 09
15	01 32 42	25 03 27	24 ♎ 58 47	02 ♏ 19 14	26 08	21 57	15 47	15 05	19 47	00 19	25 30	00 07
16	01 36 39	26 02 10	09 ♏ 42 08	17 ♏ 06 34	28 ♈ 15	23 10	16 01	15 02	19 53	00 20	25 31	00 06
17	01 40 35	27 00 50	24 ♏ 31 38	01 ♐ 56 23	00 ♉ 22	24 22	16 15	14 59	19 59	00 20	25 31	00 05
18	01 44 32	27 59 29	09 ♐ 19 59	16 ♐ 41 38	02 29	25 34	16 30	14 56	20 06	00 21	25 32	00 04
19	01 48 28	28 58 06	24 ♐ 00 36	01 ♑ 16 18	04 36	26 47	16 45	14 53	20 12	00 22	25 32	00 03
20	01 52 25	29 ♈ 56 41	08 ♑ 28 13	15 ♑ 36 00	06 43	27 59	17 01	14 50	20 18	00 23	25 32	00 ♐ 01
21	01 56 22	00 ♉ 55 15	22 ♑ 39 21	29 ♑ 38 07	08 49	29 ♓ 11	17 17	14 46	20 25	00 24	25 32	29 ♏ 59
22	02 00 18	01 53 47	06 ♒ 32 12	13 ♒ 21 37	10 54	00 ♈ 24	17 33	14 42	20 31	00 24	25 33	29 58
23	02 04 15	02 52 17	20 ♒ 06 25	26 ♒ 46 43	12 58	01 36	17 50	14 38	20 37	00 25	25 33	29 57
24	02 08 11	03 50 46	03 ♓ 22 41	09 ♓ 54 28	15 00	02 49	18 08	14 34	20 43	00 26	25 33	29 55
25	02 12 08	04 49 13	16 ♓ 22 18	22 ♓ 46 21	17 00	04 01	18 26	14 30	20 49	00 26	25 33	29 54
26	02 16 04	05 47 38	29 ♓ 06 52	05 ♈ 24 03	18 58	05 14	18 44	14 26	20 55	00 27	25 33	29 52
27	02 20 01	06 46 02	11 ♈ 38 06	17 ♈ 49 14	20 54	06 26	19 02	14 21	21 01	00 27	25 33	29 51
28	02 23 57	07 44 24	23 ♈ 57 40	00 ♉ 03 35	22 47	07 39	19 21	14 17	21 07	00 27	25 R 33	29 49
29	02 27 54	08 42 45	06 ♉ 07 14	12 ♉ 08 48	24 37	08 51	19 41	14 12	21 12	00 28	25 33	29 48
30	02 31 51	09 ♉ 41 03	18 ♉ 08 31	24 ♉ 06 39	26 ♉ 24	10 ♈ 04	20 ♌ 01	14 ♐ 07	21 ♓ 18	00 ♒ 28	25 ♑ 33	29 ♏ 46

Date	Moon True ☊	Moon Mean ☊	Moon ☽ Latitude
01	05 ♏ 37	06 ♏ 58	00 N 45
02	05 D 37	06 55	00 S 22
03	05 38	06 52	01 27
04	05 39	06 49	02 28
05	05 41	06 46	03 22
06	05 43	06 42	04 08
07	05 44	06 39	04 43
08	05 R 44	06 36	05 06
09	05 43	06 33	05 16
10	05 42	06 30	05 11
11	05 40	06 27	04 50
12	05 38	06 23	04 14
13	05 36	06 20	03 21
14	05 35	06 17	02 15
15	05 35	06 14	00 S 58
16	05 D 34	06 11	00 N 23
17	05 35	06 08	01 43
18	05 35	06 04	02 57
19	05 36	06 01	03 58
20	05 36	05 58	04 43
21	05 37	05 55	05 10
22	05 R 37	05 52	05 18
23	05 37	05 48	05 08
24	05 36	05 45	04 41
25	05 36	05 42	04 01
26	05 D 36	05 39	03 09
27	05 36	05 36	02 09
28	05 36	05 33	01 N 04
29	05 R 37	05 29	00 S 03
30	05 ♏ 36	05 ♏ 26	01 S 09

DECLINATIONS

Date	Sun ☉	Moon ☽	Mercury ☿	Venus ♀	Mars ♂	Jupiter ♃	Saturn ♄	Uranus ♅	Neptune ♆	Pluto ♇
01	04 N 28	11 N 16	02 S 33	10 S 24	19 N 44	21 S 49	06 S 19	20 S 40	20 S 32	06 S 40
02	04 51	14 21	01 46	10 00	19 41	21 49	06 17	20 39	20 31	06 40
03	05 14	16 46	00 58	09 36	19 36	21 49	06 14	20 39	20 31	06 39
04	05 37	18 26	00 S 09	09 11	19 32	21 49	06 11	20 39	20 31	06 39
05	06 00	19 17	00 N 40	08 46	19 28	21 49	06 09	20 38	20 31	06 38
06	06 23	19 17	01 31	08 21	19 23	21 49	06 06	20 38	20 31	06 38
07	06 45	18 25	02 23	07 56	19 19	21 49	06 04	20 38	20 31	06 37
08	07 08	16 44	03 15	07 31	19 14	21 48	06 01	20 38	20 31	06 37
09	07 30	14 15	04 08	07 05	19 09	21 48	05 59	20 37	20 30	06 36
10	07 53	11 03	05 01	06 39	19 04	21 48	05 56	20 37	20 30	06 36
11	08 15	07 15	05 55	06 13	18 58	21 48	05 54	20 37	20 30	06 36
12	08 37	02 N 58	06 50	05 46	18 53	21 47	05 51	20 37	20 30	06 35
13	08 59	01 S 36	07 44	05 20	18 47	21 47	05 49	20 36	20 30	06 35
14	09 20	06 13	08 39	04 53	18 41	21 47	05 46	20 36	20 30	06 34
15	09 42	10 35	09 34	04 26	18 35	21 47	05 44	20 36	20 30	06 34
16	10 03	14 21	10 29	03 59	18 29	21 46	05 41	20 36	20 30	06 33
17	10 25	17 14	11 24	03 32	18 23	21 46	05 39	20 36	20 30	06 33
18	10 46	18 56	12 18	03 05	18 17	21 46	05 37	20 35	20 30	06 32
19	11 06	19 21	13 11	02 38	18 10	21 45	05 34	20 35	20 29	06 32
20	11 27	18 28	14 03	02 10	18 03	21 45	05 32	20 35	20 29	06 31
21	11 48	16 27	14 55	01 43	17 57	21 44	05 30	20 35	20 29	06 31
22	12 08	13 31	15 45	01 15	17 50	21 44	05 27	20 35	20 29	06 31
23	12 28	09 55	16 33	00 47	17 43	21 44	05 25	20 35	20 29	06 30
24	12 48	05 54	17 20	00 S 19	17 36	21 43	05 23	20 35	20 29	06 30
25	13 08	01 S 41	18 04	00 N 08	17 28	21 43	05 21	20 35	20 29	06 29
26	13 27	02 N 32	18 47	00 36	17 21	21 42	05 18	20 35	20 29	06 29
27	13 46	06 35	19 28	01 04	17 13	21 42	05 16	20 35	20 29	06 29
28	14 05	10 18	20 06	01 32	17 06	21 41	05 14	20 35	20 29	06 28
29	14 24	13 31	20 41	02 00	16 58	21 41	05 12	20 35	20 29	06 28
30	14 N 43	16 N 08	21 N 15	02 N 28	16 N 50	21 S 40	05 S 10	20 S 34	20 S 29	06 S 27

ZODIAC SIGN ENTRIES

Date	h	m	Planets
01	12	11	♅ ♒
01	16	59	☽ ♉
02	07	29	☿ ♈
04	04	49	☽ ♊
06	17	40	☽ ♋
09	05	16	☽ ♌
11	13	39	☽ ♍
13	18	20	☽ ♎
15	20	13	☽ ♏
17	07	54	☿ ♉
17	20	51	☽ ♐
19	21	53	☽ ♑
20	13	21	☉ ♉
21	02	56	♇ ♏
22	00	38	☽ ♒
22	04	07	♀ ♈
24	05	50	☽ ♓
26	13	41	☽ ♈
28	23	53	☽ ♉

LATITUDES

Date	Mercury ☿	Venus ♀	Mars ♂	Jupiter ♃	Saturn ♄	Uranus ♅	Neptune ♆	Pluto ♇
01	02 S 08	00 S 50	03 N 07	00 N 49	01 S 48	00 S 31	00 N 33	13 N 51
04	01 56	00 58	03 00	00 49	01 48	00 31	00 33	13 52
07	01 39	01 05	02 53	00 50	01 49	00 31	00 33	13 53
10	01 18	01 12	02 47	00 50	01 49	00 32	00 33	13 53
13	00 52	01 18	02 40	00 50	01 49	00 32	00 33	13 54
16	00 S 23	01 23	02 34	00 50	01 50	00 32	00 33	13 55
19	00 N 08	01 28	02 28	00 50	01 50	00 32	00 33	13 55
22	00 41	01 32	02 22	00 50	01 51	00 32	00 33	13 56
25	01 13	01 35	02 17	00 50	01 51	00 32	00 33	13 56
28	01 41	01 38	02 11	00 50	01 52	00 32	00 33	13 57
31	02 N 04	01 S 40	02 N 06	00 N 50	01 S 52	00 S 32	00 N 33	13 N 57

DATA

Julian Date	2449809
Delta T	+61 seconds
Ayanamsa	23° 47' 36"
Synetic vernal point	05° ♓ 19' 23"
True obliquity of ecliptic	23° 26' 17"

MOON'S PHASES, APSIDES AND POSITIONS ☽

Date	h	m	Phase	Longitude	Eclipse Indicator
08	05	35	☽	17 ♋ 56	
15	12	08	○	25 ♎ 04	partial
22	03	18	☾	01 ♒ 33	
29	17	36	●	08 ♉ 56	Annular

Day	h	m	
05	10	11	Apogee
17	08	27	Perigee

Day	h	m		
05	23	52	Max dec	19° N 23'
13	03	41	0S	
19	07	23	Max dec	19° S 22'
25	21	29	0N	

LONGITUDES

Date	Chiron ⚷	Ceres ⚳	Pallas ⚴	Juno ⚵	Vesta ⚶	Black Moon Lilith ⚸
01	21 ♍ 53	07 ♌ 17	09 ♊ 38	03 ♑ 05	02 ♋ 31	09 ♊ 51
11	21 ♍ 12	08 ♌ 22	15 ♊ 08	04 ♑ 01	05 ♋ 24	10 ♊ 58
21	20 ♍ 39	10 ♌ 01	20 ♊ 45	04 ♑ 27	08 ♋ 35	12 ♊ 05
31	20 ♍ 16	12 ♌ 11	26 ♊ 27	04 ♑ 22	12 ♋ 02	13 ♊ 11

ASPECTARIAN

h m	Aspects
01 Saturday	
05 36	☽ ⊥ ♄
06 01	☽ ± ♇
06 35	☽ # ♀
07 54	☽ □ ♆
12 01	♃ St R
12 06	☿ ⅊ ♂
14 26	☽ ⊻ ☿
16 59	☽ □ ♅
17 43	☽ ⊼ ♇
17 44	☽ ⅊ ♃
23 22	☽ ∠ ♄
02 Sunday	
04 18	☽ ⊥ ☿
04 43	☽ ✶ ♀
05 35	☉ Q ♅
05 47	♀ ⊥ ♅
06 37	☉ ⊥ ♀
07 47	☿ ✶ ♅
11 29	☽ ± ♃
12 15	☿ △ ♇
17 43	☽ ⊻ ☉
19 58	☽ □ ♂
23 25	☽ ⊼ ♃
03 Monday	
01 25	☽ ∠ ☿
05 28	☽ ✶ ♄
06 46	☽ ⊥ ☉
07 15	☽ Q ♀
19 31	☽ △ ♆
04 Tuesday	
00 36	☉ △ ♂
02 31	☽ ∠ ☉
04 58	☽ △ ♅
05 30	☽ ☍ ♇
05 52	☽ Q ♄
08 31	☽ Q ♂
13 16	☽ ✶ ☿
23 42	☽ □ ♀
05 Wednesday	
01 55	☽ ⅊ ♆
09 07	☽ ✶ ♂
11 27	☽ ⅊ ♅
11 42	☽ ✶ ☉
11 58	☽ ☍ ♃
13 48	☉ ⅊ ♇
14 50	☉ △ ♃
18 18	☽ Q ☿
18 45	☽ □ ♄
20 08	☉ # ♄
20 13	☽ ± ♆
21 42	♀ ∠ ♆
06 Thursday	
05 46	☽ ± ♅
06 08	☿ Q ♆
08 24	☽ ⊼ ♆
14 16	☽ Q ☉
15 55	☽ ∠ ♂
17 56	☽ ⊼ ♅
18 15	☽ ⊼ ♇
07 Friday	
03 34	☉ # ♇
06 20	☽ ± ♇
10 17	☽ ⊥ ♂
13 45	☽ □ ☿
19 10	☽ △ ♀
22 26	☽ ⊻ ♂
08 Saturday	
00 18	☽ ⅊ ♇
00 24	☽ ⊼ ♃
05 35	☽ □ ☉
07 40	☽ △ ♄
12 15	☽ ± ♃
16 32	☿ Q ♅
20 23	☽ ☍ ♆
23 14	☉ # ♀
09 Sunday	
03 58	☽ ⅊ ♀
05 38	☽ ☍ ♅
05 44	☽ △ ♇
05 49	☽ ⅊ ♃
08 29	♀ ⊼ ♂
10 23	☉ ⊻ ♄
13 14	☽ ⅊ ♄
21 20	♀ ∠ ♅
22 23	☿ △ ♂
22 41	♀ □ ♃
23 08	☽ ± ♀
10 Monday	
05 10	☿ ⅊ ♇
05 34	☿ △ ♃
06 32	☽ ± ♄
09 27	☽ ☌ ♂
10 25	☽ △ ♃
11 20	☽ △ ☿
11 39	☽ ⊼ ♀
14 41	♀ ∥ ♇
15 54	☿ ⊻ ♀
16 39	♅ ✶ ♇
17 53	☽ ⊼ ♄
20 11	☽ △ ☉
11 Tuesday	
05 23	☽ ⊼ ♆
06 31	☽ ∥ ☉
07 23	♃ ∠ ♅
11 16	☿ # ♄
14 01	☽ □ ♇
14 05	☽ ⊼ ♅
15 50	☽ # ♇
16 20	☽ ± ♆
17 12	☿ # ♀
18 19	☽ ∥ ☿
18 43	☽ # ♀
19 55	☽ # ♄
20 01	☽ ⅊ ☿
12 Wednesday	
00 55	☽ ± ♅
01 43	☽ ⅊ ☉
05 37	☿ # ♇
06 35	☿ ⊻ ♄
07 19	♀ ∥ ♄
08 27	☽ ⅊ ♆
14 55	☽ ± ☿
16 45	☽ ⊻ ♂
16 45	☽ □ ♃
16 53	☽ ⅊ ♅
17 04	♂ △ ♃
18 53	☽ ± ☉
21 59	☽ Q ♇
23 15	☽ ☍ ♀
13 Thursday	
00 17	☽ ☍ ♄
03 14	☽ ⊼ ☿
03 22	☽ ⊥ ♂
06 06	☽ ⊼ ☉
10 38	☽ △ ♆
12 14	♀ ☌ ♄
18 36	☽ ✶ ♇
18 49	☽ △ ♅
19 02	☽ ∠ ♂
23 40	☽ Q ♃
14 Friday	
05 39	☽ ∥ ♀
09 41	☽ ∥ ♄
12 49	☉ ☌ ☿
13 11	☿ ± ♇
13 35	☉ ± ♇
13 53	☽ ∥ ♇
19 43	☽ ✶ ♃
19 46	☽ ∠ ♇
20 36	☽ ✶ ♂
15 Saturday	
03 23	☽ ⊼ ♄
04 42	☽ # ☿
04 46	☿ □ ♆
06 30	☽ # ☉
06 35	☽ ⊼ ♀
07 38	☿ ⊥ ♄
07 47	♀ ± ♂
10 36	☽ ⊥ ♇
12 08	☽ ☍ ☉
12 52	☽ □ ♆
13 19	☽ ± ♄
14 13	☽ ☍ ☿
16 40	☽ Q ♂
17 19	☽ ± ♀
17 22	☉ ∥ ☿
20 19	☽ ∠ ♃
20 24	☽ ⊻ ♇
20 44	☽ □ ♅
23 07	☉ □ ♆
16 Sunday	
04 07	☽ ⅊ ♄
07 48	☉ ⊥ ♄
09 17	☽ ⅊ ♀
10 55	☽ ⊥ ♃
18 11	☽ Q ♆
20 37	☽ ⊻ ♃
22 23	☽ □ ♂
17 Monday	
01 59	☽ Q ♅
04 36	☽ △ ♄
07 51	☿ ⅊ ♃
08 51	☿ ⊼ ♇
11 43	☽ △ ♀
11 46	☿ □ ♅
12 07	☿ ⊥ ♀
13 36	☽ ✶ ♆
16 18	☽ ⊼ ☉
20 59	☽ ☌ ♇
21 25	☽ ✶ ♅
23 01	☽ ⊼ ☿
18 Tuesday	
01 12	☽ # ♂
02 43	☽ ± ☉
10 23	☽ ± ☿
11 05	♀ ✶ ♆
13 56	☽ ∠ ♆
18 23	☽ ⅊ ☉
21 06	☽ ☌ ♃
21 49	☽ ∠ ♅
23 53	☽ △ ♂
19 Wednesday	
03 32	☽ ⅊ ☿
04 38	☽ ⊥ ♆
05 42	☽ □ ♄
12 36	☽ ⊥ ♅
14 30	☽ ⊻ ♆
16 58	☽ □ ♀
19 13	☿ ∠ ♄
20 46	☽ △ ☉
21 56	☽ ⊻ ♇
22 31	☽ ⊻ ♅
20 Thursday	
01 02	☽ ⅊ ♂
07 54	☽ ⊥ ♇
08 33	☽ △ ☿
09 16	☉ ⅊ ♃
11 43	☽ Q ♄
13 40	☉ ⊼ ♇
16 21	☽ ± ♂
18 24	☽ # ♂
22 39	☽ ⊻ ♃
22 54	☉ □ ♅
23 00	☽ ∠ ♇
21 Friday	
01 50	☽ Q ♀
02 40	☽ ⊼ ♂
08 08	☽ ✶ ♄
08 47	☽ ⊥ ♃
11 32	☿ ± ♃
16 57	☽ ☌ ♆
22 22	☽ # ☿
22 Saturday	
00 11	☽ ∠ ♃
00 18	☽ ✶ ♀
00 36	☽ ✶ ♇
01 19	☽ ☌ ♅
03 18	☽ □ ☉
03 39	♀ △ ♇
10 12	☽ ∠ ♄
12 11	♀ ✶ ♅
20 47	☽ # ☉
21 01	☽ □ ☿
21 31	☽ Q ♇
23 Sunday	
02 09	☽ ⊥ ♄
02 19	☽ ✶ ♃
05 08	☽ ∠ ♀
07 52	☽ ☍ ♂
12 55	☽ ⊻ ♄
13 29	☽ Q ☉
21 46	☽ ⊻ ♆
22 52	☽ ⊥ ♀
23 42	☽ Q ♃
24 Monday	
05 43	☽ □ ♇
06 37	☽ ⊻ ♅
07 08	☿ ⊼ ♃
08 30	☽ ∥ ♇
08 39	☽ ⊥ ♆
10 52	☽ ⊻ ♀
11 11	☽ Q ☿
12 55	☽ ✶ ☉
14 59	☽ ∥ ♄
17 35	☽ ⊥ ♅
19 15	☿ ∥ ♂
20 21	♀ ⅊ ♂
25 Tuesday	
01 11	☽ ∠ ♆
08 32	☽ □ ♃
10 15	☽ ∠ ♅
13 24	☽ ✶ ☿
15 56	☽ ⊼ ♂
18 59	☽ ∠ ☉
19 50	☽ # ♀
20 23	☽ ☌ ♄
23 33	☽ ∥ ♀
26 Wednesday	
03 29	☽ ± ♂
05 14	☽ ✶ ♆
08 26	☿ □ ♂
13 24	☽ ⊥ ☉
13 26	☽ △ ♇
14 32	☽ ✶ ♅
15 19	☉ ∠ ♄
21 01	☽ ⅊ ♂
22 58	☽ ∠ ☿
27 Thursday	
00 55	☽ ☌ ♀
01 50	☽ ⊻ ☉
04 06	☽ # ♄
04 08	☽ Q ♆
11 20	☽ # ♇
13 30	☿ ✶ ♄
13 34	☽ Q ♅
17 14	☽ △ ♃
18 13	☽ ⅊ ♇
19 28	☽ ⊥ ☿
22 14	♆ St R
28 Friday	
02 45	☽ △ ♂
06 23	☽ ⊻ ♄
06 54	☿ ∠ ♀
09 17	☽ ⊻ ☿
10 09	♀ Q ♆
11 44	☽ ± ♇
15 07	☽ □ ♆
18 14	☽ ⊥ ♄
21 50	☉ ⊻ ♀
22 23	☽ ⅊ ♃
23 30	☽ ⊼ ♇
29 Saturday	
00 13	☉ ± ♃
00 47	☽ □ ♅
03 36	☿ # ♆
07 14	☿ # ♅
12 10	☽ ∠ ♄
16 06	☽ ± ♃
17 36	☽ ☌ ☉
18 02	☽ ⊻ ♀
20 29	☽ ∥ ☉
30 Sunday	
00 31	☿ △ ♆
03 59	☽ ⊼ ♃
07 22	☽ ⊥ ♀
15 51	☽ □ ♂
18 24	☽ ✶ ♄
19 23	☽ ∥ ♂

All ephemeris data is given at 12.00 UT and the Moon's longitude is additionally given for 24.00 UT

MAY 1995

Date	Sidereal time h m s	Sun ☉ ° ' "	Moon ☽ ° ' "	Moon ☽ 24.00 ° ' "	Mercury ☿ ° '	Venus ♀ ° '	Mars ♂ ° '	Jupiter ♃ ° '	Saturn ♄ ° '	Uranus ♅ ° '	Neptune ♆ ° '	Pluto ♇ ° '
01	02 35 47	10 ♉ 39 20	00 ♊ 03 26	05 ♊ 59 09	28 ♉ 07	11 ♈ 16	20 ♌ 21	14 ♐ 02	21 ♓ 24	00 ♒ 28	25 ♑ 33	29 ♏ 45
02	02 39 44	11 37 35	11 ♊ 54 07	17 ♊ 48 40	29 ♉ 47	12 29	20 41	13 R 56	21 29	00 28	25 R 33	29 R 43
03	02 43 40	12 35 48	23 ♊ 43 08	29 ♊ 37 56	01 ♊ 23	13 41	21 02	13 51	21 35	00 28	25 33	29 42
04	02 47 37	13 33 59	05 ♋ 33 28	11 ♋ 30 10	02 55	14 54	21 23	13 45	21 41	00 28	25 32	29 40
05	02 51 33	14 32 09	17 ♋ 28 32	23 ♋ 29 03	04 23	16 06	21 45	13 40	21 46	00 R 28	25 32	29 39
06	02 55 30	15 30 16	29 ♋ 32 14	05 ♌ 38 36	05 47	17 19	22 07	13 34	21 51	00 28	25 32	29 37
07	02 59 26	16 28 21	11 ♌ 48 42	18 ♌ 03 03	07 08	18 32	22 29	13 28	21 57	00 28	25 32	29 35
08	03 03 23	17 26 25	24 ♌ 22 11	00 ♍ 46 34	08 23	19 44	22 51	13 22	22 02	00 28	25 31	29 34
09	03 07 20	18 24 26	07 ♍ 16 41	13 ♍ 52 54	09 35	20 57	23 14	13 16	22 07	00 28	25 31	29 32
10	03 11 16	19 22 26	20 ♍ 35 31	27 ♍ 24 46	10 43	22 10	23 37	13 09	22 12	00 28	25 30	29 31
11	03 15 13	20 20 24	04 ♎ 20 44	11 ♎ 23 23	11 46	23 22	24 00	13 03	22 17	00 28	25 30	29 29
12	03 19 09	21 18 19	18 ♎ 32 09	25 ♎ 47 42	12 44	24 35	24 24	12 56	22 22	00 27	25 30	29 27
13	03 23 06	22 16 14	03 ♏ 08 28	10 ♏ 34 04	13 38	25 48	24 48	12 50	22 27	00 27	25 29	29 26
14	03 27 02	23 14 06	18 ♏ 03 38	25 ♏ 36 09	14 28	27 00	25 12	12 43	22 32	00 26	25 29	29 24
15	03 30 59	24 11 58	03 ♐ 10 29	10 ♐ 45 27	15 13	28 13	25 37	12 36	22 37	00 26	25 28	29 22
16	03 34 55	25 09 47	18 ♐ 19 51	25 ♐ 52 28	15 53	29 ♈ 26	26 02	12 29	22 41	00 25	25 27	29 21
17	03 38 52	26 07 36	03 ♑ 22 14	10 ♑ 48 07	16 28	00 ♉ 38	26 27	12 22	22 46	00 25	25 27	29 19
18	03 42 49	27 05 23	18 ♑ 09 17	25 ♑ 25 04	16 59	01 51	26 52	12 15	22 50	00 24	25 26	29 17
19	03 46 45	28 03 09	02 ♒ 34 56	09 ♒ 38 35	17 25	03 04	27 17	12 08	22 55	00 24	25 26	29 16
20	03 50 42	29 00 54	16 ♒ 35 50	23 ♒ 26 40	17 46	04 17	27 43	12 01	22 59	00 23	25 25	29 14
21	03 54 38	29 ♉ 58 38	00 ♓ 11 13	06 ♓ 49 43	18 02	05 29	28 09	11 54	23 03	00 22	25 24	29 12
22	03 58 35	00 ♊ 56 20	13 ♓ 22 28	19 ♓ 49 50	18 13	06 42	28 35	11 46	23 08	00 21	25 23	29 11
23	04 02 31	01 54 02	26 ♓ 12 14	02 ♈ 30 08	18 20	07 55	29 02	11 39	23 12	00 20	25 23	29 09
24	04 06 28	02 51 43	08 ♈ 43 58	14 ♈ 54 12	18 R 22	09 08	29 28	11 31	23 16	00 20	25 22	29 07
25	04 10 24	03 49 22	21 ♈ 01 15	27 ♈ 05 33	18 19	10 20	29 ♌ 55	11 24	23 20	00 19	25 21	29 06
26	04 14 21	04 47 01	03 ♉ 07 30	09 ♉ 07 29	18 11	11 33	00 ♍ 22	11 16	23 24	00 18	25 20	29 04
27	04 18 18	05 44 38	15 ♉ 05 50	21 ♉ 02 53	18 00	12 46	00 50	11 09	23 28	00 17	25 19	29 02
28	04 22 14	06 42 15	26 ♉ 58 55	02 ♊ 54 13	17 44	13 59	01 17	11 01	23 31	00 15	25 18	29 01
29	04 26 11	07 39 50	08 ♊ 49 02	14 ♊ 43 38	17 25	15 12	01 45	10 54	23 35	00 14	25 17	28 59
30	04 30 07	08 37 24	20 ♊ 38 15	26 ♊ 33 08	17 02	16 25	02 13	10 46	23 39	00 13	25 16	28 58
31	04 34 04	09 ♊ 34 57	02 ♋ 28 31	08 ♋ 24 40	16 ♊ 36	17 ♉ 38	02 ♍ 41	10 ♐ 38	23 ♓ 42	00 ♒ 12	25 ♑ 16	28 ♏ 56

Date	Moon True ☊ ° '	Moon Mean ☊ ° '	Moon ☽ Latitude ° '
01	05 ♏ 36	05 ♏ 23	02 S 12
02	05 R 35	05 20	03 08
03	05 35	05 17	03 57
04	05 34	05 13	04 35
05	05 32	05 10	05 02
06	05 32	05 07	05 15
07	05 31	05 04	05 15
08	05 D 31	05 01	05 00
09	05 32	04 58	04 30
10	05 33	04 54	03 44
11	05 34	04 51	02 45
12	05 35	04 48	01 33
13	05 35	04 45	00 S 14
14	05 R 35	04 42	01 N 08
15	05 34	04 39	02 27
16	05 32	04 35	03 35
17	05 29	04 32	04 27
18	05 27	04 29	05 01
19	05 24	04 26	05 15
20	05 24	04 23	05 09
21	05 D 22	04 19	04 46
22	05 23	04 16	04 09
23	05 24	04 13	03 19
24	05 26	04 10	02 22
25	05 27	04 07	01 19
26	05 28	04 04	00 N 13
27	05 R 28	04 00	00 S 53
28	05 26	03 57	01 55
29	05 22	03 54	02 52
30	05 17	03 51	03 42
31	05 ♏ 12	03 ♏ 48	04 S 22

DECLINATIONS

Date	Sun ☉ ° '	Moon ☽ ° '	Mercury ☿ ° '	Venus ♀ ° '	Mars ♂ ° '	Jupiter ♃ ° '	Saturn ♄ ° '	Uranus ♅ ° '	Neptune ♆ ° '	Pluto ♇ ° '
01	15 N 01	18 N 01	21 N 46	02 N 56	16 N 42	21 S 39	05 S 08	20 S 34	20 S 29	06 S 27
02	15 19	19 06	22 14	03 24	16 33	21 39	05 06	20 34	20 29	06 27
03	15 37	19 21	22 40	03 51	16 25	21 38	05 04	20 34	20 29	06 26
04	15 55	18 45	23 03	04 19	16 17	21 37	05 02	20 34	20 29	06 26
05	16 12	17 19	23 24	04 47	16 08	21 37	05 00	20 35	20 29	06 25
06	16 29	15 06	23 42	05 14	15 59	21 36	04 58	20 35	20 29	06 25
07	16 46	12 12	23 58	05 42	15 51	21 35	04 56	20 35	20 29	06 25
08	17 02	08 41	24 12	06 09	15 42	21 35	04 54	20 35	20 29	06 24
09	17 18	04 40	24 24	06 37	15 33	21 34	04 52	20 35	20 30	06 24
10	17 34	00 N 17	24 33	07 04	15 23	21 33	04 50	20 35	20 30	06 24
11	17 50	04 S 15	24 41	07 31	15 14	21 32	04 49	20 35	20 30	06 23
12	18 05	08 42	24 46	07 58	15 05	21 32	04 47	20 35	20 30	06 23
13	18 20	12 46	24 49	08 24	14 55	21 31	04 45	20 35	20 30	06 23
14	18 35	16 07	24 51	08 51	14 46	21 30	04 43	20 35	20 30	06 22
15	18 49	18 23	24 50	09 17	14 36	21 29	04 42	20 35	20 30	06 22
16	19 03	19 22	24 48	09 44	14 26	21 28	04 40	20 36	20 30	06 22
17	19 17	18 57	24 44	10 10	14 16	21 28	04 38	20 36	20 30	06 21
18	19 30	17 14	24 39	10 35	14 06	21 27	04 37	20 36	20 30	06 21
19	19 44	14 28	24 32	11 01	13 56	21 26	04 35	20 36	20 30	06 21
20	19 56	10 56	24 23	11 26	13 46	21 25	04 34	20 36	20 31	06 21
21	20 09	06 56	24 13	11 51	13 35	21 24	04 32	20 36	20 31	06 20
22	20 21	02 S 42	24 01	12 16	13 25	21 23	04 31	20 37	20 31	06 20
23	20 32	01 N 33	23 49	12 41	13 14	21 22	04 29	20 37	20 31	06 20
24	20 44	05 38	23 35	13 05	13 04	21 21	04 28	20 37	20 31	06 19
25	20 55	09 25	23 19	13 29	12 53	21 20	04 26	20 37	20 31	06 19
26	21 05	12 45	23 03	13 53	12 42	21 19	04 25	20 38	20 31	06 19
27	21 16	15 32	22 46	14 16	12 31	21 18	04 24	20 38	20 32	06 19
28	21 26	17 37	22 28	14 39	12 20	21 17	04 23	20 38	20 32	06 19
29	21 35	18 56	22 09	15 02	12 09	21 17	04 21	20 38	20 32	06 19
30	21 44	19 25	21 49	15 24	11 58	21 16	04 20	20 39	20 32	06 18
31	21 N 53	19 N 03	21 N 29	15 N 46	11 N 46	21 S 15	04 S 19	20 S 39	20 S 32	06 S 18

ZODIAC SIGN ENTRIES

Date	h	m	Planets
01	11	53	☽ ♊
02	15	18	☿ ♊
04	00	45	☽ ♋
06	12	55	☽ ♌
08	22	33	☽ ♍
11	04	30	☽ ♎
13	06	53	☽ ♏
15	06	58	☽ ♐
16	23	22	♀ ♉
17	06	36	☽ ♑
19	07	39	☽ ♒
21	11	40	☽ ♓
21	12	34	☉ ♊
23	19	13	☽ ♈
25	16	09	♂ ♍
26	05	46	☽ ♉
28	18	07	☽ ♊
31	06	59	☽ ♋

LATITUDES

Date	Mercury ☿ ° '	Venus ♀ ° '	Mars ♂ ° '	Jupiter ♃ ° '	Saturn ♄ ° '	Uranus ♅ ° '	Neptune ♆ ° '	Pluto ♇ ° '
01	02 N 04	01 S 40	02 N 06	00 N 50	01 S 52	00 S 32	00 N 33	13 N 57
04	02 21	01 41	02 01	00 50	01 53	00 32	00 33	13 57
07	02 31	01 41	01 56	00 50	01 53	00 32	00 33	13 57
10	02 32	01 41	01 51	00 50	01 54	00 33	00 33	13 57
13	02 24	01 40	01 46	00 50	01 55	00 33	00 33	13 57
16	02 07	01 39	01 42	00 50	01 55	00 33	00 33	13 57
19	01 41	01 37	01 37	00 49	01 56	00 33	00 33	13 57
22	01 07	01 34	01 33	00 49	01 57	00 33	00 33	13 57
25	00 N 24	01 31	01 29	00 49	01 57	00 33	00 33	13 57
28	00 S 25	01 27	01 25	00 49	01 58	00 33	00 33	13 56
31	01 S 17	01 S 23	01 N 21	00 N 48	01 S 59	00 S 33	00 N 33	13 N 56

DATA

Julian Date	2449839
Delta T	+61 seconds
Ayanamsa	23° 47' 39"
Synetic vernal point	05° ♓ 19' 20"
True obliquity of ecliptic	23° 26' 16"

MOON'S PHASES, APSIDES AND POSITIONS ☽

Date	h	m	Phase	Longitude ° '	Eclipse Indicator
07	21	44	☽	16 ♌ 52	
14	20	48	○	23 ♏ 35	
21	11	36	☾	29 ♒ 58	
29	09	27	●	07 ♊ 34	

Day	h	m	
03	00	55	Apogee
15	15	24	Perigee
30	08	01	Apogee

Day	h	m		
03	06	54	Max dec	19° N 22'
10	13	33	0S	
16	16	41	Max dec	19° S 23'
23	03	14	0N	
30	13	39	Max dec	19° N 25'

LONGITUDES

Date	Chiron ⚷ ° '	Ceres ⚳ ° '	Pallas ⚴ ° '	Juno ⚵ ° '	Vesta ⚶ ° '	Black Moon Lilith ⚸ ° '
01	20 ♍ 16	12 ♌ 11	26 ♊ 27	04 ♑ 22	12 ♋ 02	13 ♊ 11
11	20 ♍ 02	14 ♌ 46	02 ♋ 12	03 ♑ 42	15 ♋ 43	14 ♊ 18
21	20 ♍ 00	17 ♌ 43	07 ♋ 57	02 ♑ 30	19 ♋ 34	15 ♊ 25
31	20 ♍ 09	20 ♌ 57	13 ♋ 43	00 ♑ 49	23 ♋ 35	16 ♊ 32

ASPECTARIAN

h m	Aspects
01 Monday	
02 54	☽ △ ♆
03 29	☽ ∠ ♀
07 01	☿ # ♃
07 25	☽ ☌ ☿
11 22	☽ ☍ ♇
12 50	☽ △ ♅
18 49	☽ Q ♄
02 Tuesday	
05 17	☽ Q ♂
09 15	☽ ⚼ ♆
11 12	☿ ☍ ♇
11 23	☽ ⊻ ☉
11 50	♀ Q ♅
13 18	☽ ⚹ ♀
16 06	☽ ☍ ♃
19 15	☽ ⚼ ♅
22 19	☿ △ ♅
03 Wednesday	
00 40	☽ ⊥ ☉
03 31	☽ ± ♆
06 23	☽ ⚹ ♂
07 38	☽ □ ♄
13 32	☽ ± ♅
14 56	♀ △ ♃
15 42	☽ ⊼ ♆
16 27	☽ Q ♀
20 34	☽ ∠ ☉
04 Thursday	
00 06	☽ ⊼ ♇
01 42	☽ ⊼ ♅
05 52	☽ ⊻ ☿
07 32	♀ ⚼ ♇
12 13	☽ ± ♇
13 44	☽ ∠ ♂
16 15	☉ ⊼ ♃
19 45	☽ ⊥ ☿
05 Friday	
01 06	☿ Q ♄
04 24	☽ ⊼ ♃
05 35	☽ ⚹ ☉
06 20	☽ ⚼ ♇
07 47	♅ St R
08 25	☽ ⊥ ♂
08 29	☉ ∥ ♂
08 57	☽ □ ♀
13 50	♂ ⊼ ♄
16 20	☽ ± ♃
16 20	☽ ∠ ☿
20 38	☽ △ ♄
20 48	☽ ⊻ ♂
22 36	♀ # ♄
23 37	☽ ∥ ☉
06 Saturday	
02 42	☽ ∥ ♂
04 04	☽ ☍ ♆
07 38	☽ Q ☉
10 05	☽ ⚼ ♃
12 09	☽ △ ♇
13 51	☽ ☍ ♅
07 Sunday	
01 48	☽ ⚹ ☿
02 28	☽ ⚼ ♄
15 10	☽ △ ♃
20 01	☽ ± ♄
21 44	☽ □ ☉
08 Monday	
02 17	☽ △ ♀
03 38	☽ Q ☿
07 33	☽ ⊼ ♄
09 03	☽ ☌ ♂
14 10	☽ ⊼ ♆
21 43	☽ □ ♇
23 26	☽ ⊼ ♅
09 Tuesday	
01 07	♀ # ♇
01 23	☽ ± ♆
01 49	☽ ∥ ♀
01 54	☽ # ♇
09 19	☽ ⚼ ♀
10 31	☽ ± ♅
10 48	☽ # ♄
16 37	☽ □ ☿
17 54	☽ ⚼ ♆
22 48	☽ □ ♃
10 Wednesday	
02 51	☽ ⚼ ♅
03 19	☽ ± ♀
06 31	☽ Q ♇
07 35	☿ ⚼ ♆
09 40	☽ △ ☉
12 54	♀ ⊻ ♄
14 52	☽ ☍ ♄
15 03	☽ ⊼ ♀
17 30	☽ ⊻ ♂
20 40	☽ △ ♆
11 Thursday	
03 37	☽ ⚹ ♇
04 18	☽ ⊥ ♂
05 18	☽ △ ♅
06 21	☽ Q ♃
13 50	☽ ⚼ ☉
14 10	♀ ± ♇
14 58	☽ ∥ ♄
20 11	☽ ∠ ♂
21 47	☿ Q ♂
23 24	☽ ∥ ♇
12 Friday	
01 35	☽ △ ☿
02 41	☽ ⚹ ♃
05 11	☽ ∠ ♇
06 12	☽ ± ☉
06 44	♀ △ ♂
07 27	☽ # ♀
16 37	☿ ☍ ♃
16 55	☽ ⊼ ☉
18 23	☽ ⊼ ♄
20 08	☽ ⊥ ♇
21 59	☽ ⚹ ♂
22 55	☽ ☍ ♀
23 30	☽ □ ♆
13 Saturday	
03 24	☽ ∠ ♃
04 12	☽ ⚼ ☿
04 19	☽ ± ♄
05 58	☽ ⊻ ♇
05 58	♀ □ ♆
07 37	☽ □ ♅
16 51	☉ ⚹ ♄
17 56	☽ ⊥ ♃
18 06	☽ Q ♂
19 01	☽ ⚼ ♄
19 43	☽ ± ☿
14 Sunday	
01 59	☽ # ♂
03 31	☽ ⊻ ♃
04 40	☽ Q ♆
05 56	☽ ⊼ ☿
12 36	☽ Q ♅
19 09	☽ △ ♄
20 48	☽ ☍ ☉
23 41	☽ □ ♂
23 48	☽ ⚹ ♆
15 Monday	
00 55	♀ ⚼ ♃
03 28	☽ ⊼ ♀
03 40	♂ ⊼ ♆
05 59	☽ ☌ ♇
07 40	☽ ⚹ ♅
13 47	☽ ± ♀
19 26	☿ ⚼ ♅
20 21	♀ ⊥ ♄
20 48	☽ # ☉
23 32	☽ ∠ ♆
16 Tuesday	
02 49	☽ ☌ ♃
05 16	☽ ⚼ ♀
07 23	☽ ∠ ♅
07 57	☽ ☍ ☿
10 25	♀ ⊼ ♇
13 47	☽ ⊥ ♆
18 57	☽ □ ♄
19 17	☉ △ ♆
21 41	☽ ⊥ ♅
23 20	☽ ⊻ ♆
23 37	☽ ⊼ ☉
17 Wednesday	
00 35	☽ △ ♂
04 41	☽ # ☉
05 31	☽ ⊻ ♇
07 14	☽ △ ♀
07 16	☽ ⊻ ♅
07 35	♀ □ ♅
09 52	☽ ± ☉
15 08	☽ ⊥ ♇
18 Thursday	
00 00	☽ Q ♄
01 24	☽ ⚼ ☉
01 25	☽ ⚼ ♂
01 57	☉ □ ♂
02 26	☽ ⊻ ♃
05 41	☽ ∠ ♇
10 01	☽ ⊼ ☿
12 10	☽ ⊥ ♃
16 20	☿ ∠ ♀
16 35	☽ ± ♂
19 46	☽ ⚹ ♄
20 13	☽ ± ☿
19 Friday	
00 01	☽ ☌ ♆
02 51	☽ ⊼ ♂
02 56	☽ ∠ ♃
03 51	☽ △ ☉
06 26	☽ ⚹ ♇
08 19	☽ ☌ ♅
11 43	☽ ⚼ ☿
12 53	☽ □ ♀
16 09	☽ # ♂
21 05	☽ ∠ ♄
20 Saturday	
02 45	☽ Q ♇
04 08	☽ ⚹ ♃
09 10	☽ # ♀
12 41	☽ ⊥ ♄
14 05	☽ △ ☿
17 19	☉ ☍ ♇
22 55	☽ Q ♀
23 15	☽ ⊻ ♄
21 Sunday	
00 54	☽ Q ♃
03 29	☽ ⊻ ♆
08 14	☽ ☍ ♂
10 15	☽ □ ♇
11 36	☽ □ ☉
12 19	☽ ⊻ ♅
14 11	☽ ⊥ ♆
15 26	☽ ∥ ♇
19 17	♀ ± ♃
21 37	☉ △ ♅
22 32	☽ ⚹ ♀
23 09	☽ ⊥ ♅
22 Monday	
01 46	☽ ∥ ♄
06 31	☽ ∠ ♆
09 04	☽ □ ♃
15 40	☽ ∠ ♅
21 06	☽ □ ☿
23 10	☽ Q ☉
23 Tuesday	
05 08	☽ ∠ ♀
06 17	☽ ☌ ♄
08 48	☉ # ♆
10 26	☽ ⚹ ♆
17 34	☽ ⊼ ♂
17 35	☽ △ ♇
17 56	♀ ∠ ♄
18 14	♂ □ ♇
19 51	☽ ⚹ ♅
21 24	☉ # ♅
23 45	☽ ⚹ ☉
24 Wednesday	
00 02	☽ ⊥ ♀
05 02	☽ # ♄
05 28	☽ ± ♂
07 25	☽ Q ☿
09 02	☿ St R
09 21	☽ Q ♆
11 08	♀ ∥ ♂
12 51	☽ ⊻ ♀
16 14	☽ # ♇
17 22	☽ △ ♃
18 58	☽ Q ♅
22 27	☽ ⚼ ♇
23 35	☽ ⚼ ♂
25 Thursday	
06 43	☽ ⚹ ☿
07 19	☽ ∠ ☉
16 05	☽ ± ♇
16 35	☽ ⊻ ♄
20 32	☽ □ ♆
22 31	☽ ⚼ ♃
26 Friday	
02 37	☽ ⊥ ☉
03 56	☽ ⊼ ♇
04 32	☽ ⊥ ♄
04 41	☿ Q ♂
06 18	☽ △ ♂
06 22	☽ □ ♅
06 58	♀ ⊼ ♃
07 51	♂ ⊼ ♅
11 36	☽ ∥ ♂
12 08	☽ ∠ ☿
15 36	☽ ⊻ ☉
16 15	☽ ± ♃
22 33	☽ ∥ ♀
22 36	☽ ∠ ♄
22 59	☿ ⊥ ♀
27 Saturday	
04 09	☽ ⊼ ♃
04 24	☉ Q ♄
05 53	☽ ⊥ ☿
06 47	☽ ☌ ♀
17 44	☽ ⊻ ☿
17 55	☉ # ♃
28 Sunday	
04 58	☽ ⚹ ♄
08 37	☽ △ ♆
12 53	♂ ± ♆
16 06	☽ ☍ ♇
18 38	☽ △ ♅
21 05	☽ □ ♂
29 Monday	
05 24	☽ Q ♄
09 27	☽ ☌ ☉
14 59	☽ ⚼ ♆
16 10	☽ ☍ ♃
30 Tuesday	
01 01	☽ ⚼ ♅
02 26	☽ ⊻ ♀
04 55	☽ ☌ ☿
09 14	☽ ± ♆
11 07	☽ Q ♂
16 01	☽ ⊥ ♀
16 10	☉ ∥ ☿
18 08	☽ □ ♄
19 15	☽ ± ♅
21 06	☿ ⊻ ♀
21 24	☽ ⊼ ♆
31 Wednesday	
04 50	☽ ⊼ ♇
07 24	☽ ⊼ ♅
12 20	☽ ∠ ♀
12 27	☽ ⚹ ♂
16 58	☽ ± ♇

All ephemeris data is given at 12.00 UT and the Moon's longitude is additionally given for 24.00 UT

JUNE 1995

Date	Sidereal time h m s	Sun ☉	Moon ☽	Moon ☽ 24.00	Mercury ☿	Venus ♀	Mars ♂	Jupiter ♃	Saturn ♄	Uranus ♅	Neptune ♆	Pluto ♇
01	04 38 00	10 ♊ 32 29	14 ♋ 21 50	20 ♋ 21 21	16 ♊ 08	18 ♉ 50	03 ♍ 10	10 ♐ 31	23 ♓ 45	00 ♒ 11	25 ♑ 15	28 ♏ 54
02	04 41 57	11 30 00	26 ♋ 20 30	02 ♌ 22 38	15 R 37	20 03	03 38	10 R 23	23 49	00 R 09	25 R 13	28 R 53
03	04 45 53	12 27 29	08 ♌ 27 07	14 ♌ 34 22	15 05	21 16	04 07	10 15	23 52	00 08	25 12	28 51
04	04 49 50	13 24 58	20 ♌ 44 47	26 ♌ 58 48	14 32	22 29	04 36	10 08	23 55	00 07	25 11	28 49
05	04 53 47	14 22 25	03 ♍ 16 54	09 ♍ 39 33	13 58	23 42	05 05	10 00	23 58	00 05	25 10	28 48
06	04 57 43	15 19 50	16 ♍ 07 11	22 ♍ 40 15	13 25	24 55	05 35	09 53	24 01	00 04	25 09	28 46
07	05 01 40	16 17 15	29 ♍ 19 11	06 ♎ 04 18	12 52	26 08	06 04	09 45	24 04	00 02	25 08	28 45
08	05 05 36	17 14 38	12 ♎ 55 53	19 ♎ 54 06	12 21	27 21	06 34	09 37	24 07	00 ♒ 01	25 07	28 43
09	05 09 33	18 12 01	26 ♎ 59 00	04 ♏ 10 28	11 51	28 34	07 04	09 30	24 09	29 ♑ 59	25 06	28 42
10	05 13 29	19 09 22	11 ♏ 28 13	18 ♏ 51 45	11 24	29 ♉ 47	07 34	09 23	24 12	29 58	25 04	28 40
11	05 17 26	20 06 42	26 ♏ 20 23	03 ♐ 53 16	10 59	01 ♊ 00	08 04	09 15	24 14	29 56	25 03	28 39
12	05 21 22	21 04 02	11 ♐ 29 18	19 ♐ 07 17	10 38	02 13	08 34	09 08	24 17	29 54	25 02	28 37
13	05 25 19	22 01 20	26 ♐ 45 54	04 ♑ 23 48	10 20	03 26	09 05	09 00	24 19	29 53	25 01	28 36
14	05 29 16	22 58 38	11 ♑ 59 36	19 ♑ 32 03	10 06	04 39	09 35	08 53	24 21	29 51	24 59	28 34
15	05 33 12	23 55 56	26 ♑ 59 57	04 ♒ 22 22	09 56	05 52	10 06	08 46	24 23	29 49	24 58	28 33
16	05 37 09	24 53 13	11 ♒ 38 28	18 ♒ 47 43	09 50	07 05	10 37	08 39	24 25	29 47	24 57	28 31
17	05 41 05	25 50 29	25 ♒ 49 44	02 ♓ 44 25	09 D 49	08 18	11 08	08 32	24 27	29 46	24 55	28 30
18	05 45 02	26 47 46	09 ♓ 31 46	16 ♓ 12 01	09 52	09 31	11 39	08 25	24 29	29 44	24 54	28 28
19	05 48 58	27 45 02	22 ♓ 45 28	29 ♓ 12 36	10 00	10 44	12 11	08 18	24 31	29 42	24 53	28 27
20	05 52 55	28 42 17	05 ♈ 33 53	11 ♈ 49 55	10 12	11 57	12 42	08 11	24 33	29 40	24 51	28 26
21	05 56 51	29 ♊ 39 33	18 ♈ 01 16	24 ♈ 08 33	10 30	13 11	13 14	08 05	24 34	29 38	24 50	28 24
22	06 00 48	00 ♋ 36 48	00 ♉ 12 22	06 ♉ 13 19	10 51	14 24	13 46	07 58	24 36	29 36	24 48	28 23
23	06 04 45	01 34 03	12 ♉ 11 56	18 ♉ 08 46	11 18	15 37	14 18	07 52	24 37	29 34	24 47	28 22
24	06 08 41	02 31 18	24 ♉ 04 18	29 ♉ 58 58	11 49	16 50	14 50	07 45	24 38	29 32	24 45	28 20
25	06 12 38	03 28 33	05 ♊ 53 11	11 ♊ 47 19	12 24	18 03	15 22	07 39	24 39	29 30	24 44	28 19
26	06 16 34	04 25 48	17 ♊ 41 41	23 ♊ 36 35	13 04	19 17	15 55	07 33	24 40	29 28	24 42	28 18
27	06 20 31	05 23 02	29 ♊ 32 14	05 ♋ 28 52	13 49	20 30	16 27	07 27	24 41	29 26	24 41	28 17
28	06 24 27	06 20 17	11 ♋ 26 40	17 ♋ 25 51	14 37	21 43	17 00	07 21	24 42	29 24	24 39	28 15
29	06 28 24	07 17 31	23 ♋ 26 32	29 ♋ 28 56	15 30	22 56	17 33	07 15	24 43	29 21	24 38	28 14
30	06 32 20	08 ♋ 14 44	05 ♌ 33 11	11 ♌ 39 30	16 ♊ 27	24 ♊ 10	18 ♍ 05	07 ♐ 09	24 ♓ 43	29 ♑ 19	24 ♑ 36	28 ♏ 13

Date	Moon True ☊	Moon Mean ☊	Moon ☽ Latitude
01	05 ♏ 05	03 ♏ 45	04 S 51
02	04 R 59	03 41	05 07
03	04 54	03 38	05 10
04	04 50	03 35	04 59
05	04 48	03 32	04 33
06	04 D 47	03 29	03 54
07	04 48	03 25	03 01
08	04 49	03 22	01 56
09	04 50	03 19	00 S 42
10	04 R 50	03 16	00 N 36
11	04 48	03 13	01 54
12	04 45	03 10	03 05
13	04 39	03 06	04 03
14	04 32	03 03	04 44
15	04 25	03 00	05 05
16	04 19	02 57	05 05
17	04 13	02 54	04 46
18	04 10	02 51	04 11
19	04 09	02 47	03 24
20	04 D 09	02 44	02 28
21	04 09	02 41	01 26
22	04 10	02 38	00 N 21
23	04 R 10	02 35	00 S 43
24	04 08	02 31	01 45
25	04 03	02 28	02 41
26	03 56	02 25	03 31
27	03 47	02 22	04 11
28	03 36	02 19	04 41
29	03 25	02 16	04 58
30	03 ♏ 13	02 ♏ 12	05 S 03

DECLINATIONS

Date	Sun ☉	Moon ☽	Mercury ☿	Venus ♀	Mars ♂	Jupiter ♃	Saturn ♄	Uranus ♅	Neptune ♆	Pluto ♇
01	22 N 02	17 N 51	21 N 09	16 N 08	11 N 35	21 S 14	04 S 18	20 S 39	20 S 32	06 S 18
02	22 10	15 51	20 49	16 29	11 23	21 13	04 17	20 39	20 33	06 18
03	22 17	13 09	20 29	16 50	11 12	21 12	04 16	20 40	20 33	06 18
04	22 24	09 51	20 09	17 10	11 00	21 11	04 15	20 40	20 33	06 18
05	22 31	06 03	19 49	17 30	10 48	21 10	04 14	20 40	20 33	06 18
06	22 38	01 N 53	19 31	17 49	10 36	21 09	04 13	20 41	20 33	06 17
07	22 44	02 S 30	19 13	18 08	10 24	21 08	04 12	20 41	20 34	06 17
08	22 50	06 53	18 56	18 27	10 12	21 07	04 11	20 41	20 34	06 17
09	22 55	11 03	18 41	18 45	10 00	21 06	04 10	20 42	20 34	06 17
10	23 00	14 42	18 27	19 03	09 47	21 05	04 09	20 42	20 34	06 17
11	23 04	17 29	18 14	19 20	09 35	21 04	04 09	20 43	20 34	06 17
12	23 08	19 06	18 03	19 37	09 22	21 03	04 08	20 43	20 35	06 17
13	23 12	19 21	17 54	19 53	09 10	21 02	04 07	20 43	20 35	06 17
14	23 15	18 11	17 47	20 08	08 57	21 01	04 07	20 44	20 35	06 17
15	23 18	15 46	17 42	20 23	08 45	21 00	04 06	20 44	20 35	06 17
16	23 20	12 24	17 39	20 38	08 32	20 59	04 05	20 45	20 36	06 17
17	23 22	08 25	17 37	20 52	08 19	20 58	04 05	20 45	20 36	06 17
18	23 24	04 S 07	17 38	21 05	08 06	20 57	04 04	20 45	20 36	06 17
19	23 25	00 N 15	17 40	21 18	07 53	20 56	04 04	20 46	20 36	06 17
20	23 26	04 28	17 44	21 31	07 40	20 55	04 04	20 46	20 37	06 17
21	23 26	08 23	17 49	21 42	07 27	20 54	04 03	20 47	20 37	06 17
22	23 26	11 53	17 57	21 53	07 13	20 54	04 03	20 47	20 37	06 17
23	23 26	14 49	18 05	22 04	07 00	20 53	04 03	20 48	20 37	06 17
24	23 25	17 06	18 16	22 14	06 47	20 52	04 02	20 48	20 38	06 17
25	23 24	18 38	18 27	22 23	06 33	20 51	04 02	20 48	20 38	06 17
26	23 22	19 22	18 40	22 31	06 20	20 50	04 02	20 49	20 38	06 17
27	23 20	19 15	18 53	22 39	06 06	20 49	04 02	20 49	20 38	06 17
28	23 17	18 17	19 08	22 47	05 52	20 49	04 02	20 50	20 39	06 17
29	23 14	16 30	19 23	22 53	05 39	20 48	04 02	20 50	20 39	06 18
30	23 N 11	13 N 59	19 N 39	22 N 59	05 N 25	20 S 47	04 S 02	20 S 51	20 S 39	06 S 18

ZODIAC SIGN ENTRIES

Date	h	m	Planets
02	19	17	☽ ♌
05	05	46	☽ ♍
07	13	13	☽ ♎
09	01	42	♅ ♑
09	17	03	☽ ♏
10	16	18	♀ ♊
11	17	50	☽ ♐
13	17	05	☽ ♑
15	16	52	☽ ♒
17	19	13	☽ ♓
20	01	29	☽ ♈
21	20	34	☉ ♋
22	11	35	☽ ♉
25	00	02	☽ ♊
27	12	56	☽ ♋
30	01	02	☽ ♌

LATITUDES

Date	Mercury ☿	Venus ♀	Mars ♂	Jupiter ♃	Saturn ♄	Uranus ♅	Neptune ♆	Pluto ♇
01	01 S 34	01 S 21	01 N 19	00 N 48	01 S 59	00 S 33	00 N 33	13 N 56
04	02 25	01 16	01 16	00 48	02 00	00 33	00 33	13 55
07	03 09	01 11	01 12	00 47	02 00	00 33	00 33	13 54
10	03 44	01 05	01 08	00 47	02 01	00 34	00 33	13 54
13	04 08	00 59	01 05	00 47	02 02	00 34	00 33	13 53
16	04 19	00 52	01 01	00 46	02 03	00 34	00 33	13 52
19	04 20	00 45	00 58	00 46	02 03	00 34	00 33	13 51
22	04 10	00 38	00 55	00 45	02 04	00 34	00 33	13 50
25	03 52	00 31	00 51	00 45	02 05	00 34	00 33	13 49
28	03 26	00 24	00 48	00 44	02 06	00 34	00 33	13 48
31	02 S 56	00 S 17	00 N 45	00 N 43	02 S 07	00 S 34	00 N 33	13 N 47

DATA

Julian Date	2449870
Delta T	+61 seconds
Ayanamsa	23° 47' 44"
Synetic vernal point	05° ♓ 19' 15"
True obliquity of ecliptic	23° 26' 15"

MOON'S PHASES, APSIDES AND POSITIONS ☽

Date	h	m	Phase	Longitude	Eclipse Indicator
06	10	26	☽	15 ♍ 16	
13	04	03	○	21 ♐ 42	
19	22	01	☾	28 ♓ 09	
28	00	50	●	05 ♋ 54	

Day	h	m		
13	01	13	Perigee	
26	11	05	Apogee	
06	22	25	0S	
13	04	00	Max dec	19° S 25'
19	10	37	0N	
26	20	38	Max dec	19° N 25'

LONGITUDES

Date	Chiron ⚷	Ceres ⚳	Pallas ⚴	Juno ⚵	Vesta ⚶	Black Moon Lilith ⚸
01	20 ♍ 10	21 ♌ 17	14 ♋ 17	00 ♑ 37	23 ♋ 59	16 ♊ 38
11	20 ♍ 32	24 ♌ 48	20 ♋ 00	28 ♐ 31	28 ♋ 09	17 ♊ 45
21	20 ♍ 04	28 ♌ 31	25 ♋ 41	26 ♐ 15	02 ♌ 25	18 ♊ 52
31	21 ♍ 46	02 ♍ 24	01 ♌ 19	24 ♐ 01	06 ♌ 47	19 ♊ 59

ASPECTARIAN

h m	Aspects
01 Thursday	
03 37	☽ ⊻ ☉
04 19	☽ ⊼ ♃
04 38	☉ ꝓ ♆
06 30	☿ # ♃
11 05	☽ ꝓ ♇
11 22	☉ ☍ ♃
15 24	☽ ⊻ ☿
16 16	☽ ± ♃
16 46	☽ ⊥ ☉
19 57	☽ ∠ ♂
22 01	☽ ⚹ ♀
02 Friday	
02 57	☽ ⊥ ☿
06 15	☽ ∥ ♀
06 56	☽ △ ♄
09 47	☽ ☍ ♆
10 07	☽ ꝓ ♃
12 21	☽ ∠ ☉
14 41	☽ ⊥ ♂
17 02	☽ △ ♇
19 35	☽ ☍ ♅
20 09	☽ ∠ ☿
22 57	☿ # ♅
03 Saturday	
00 38	☽ Q ♀
03 06	☽ ⊻ ♂
07 11	☿ # ♆
09 32	☿ ꝓ ♅
12 49	☽ ꝓ ♄
15 31	☽ △ ♃
20 32	☽ ⚹ ☉
23 04	♃ ∠ ♆
04 Sunday	
00 26	☽ ⚹ ☿
03 37	☽ ∥ ♂
06 29	☽ ± ♄
15 44	☽ □ ♀
18 09	☽ ⊼ ♄
20 33	☽ ⊼ ♆
21 45	☽ Q ☉
22 40	☽ Q ☿
05 Monday	
03 30	☽ □ ♇
05 38	☉ ☌ ☿
05 57	☽ ⊼ ♅
08 00	☽ ± ♆
10 32	☽ # ♇
15 33	☽ ☌ ♂
17 17	☽ ± ♅
17 28	♀ ⚹ ♄
22 44	☽ # ♄
06 Tuesday	
00 31	☽ □ ♃
00 56	☽ ꝓ ♆
05 30	☉ ꝓ ♅
07 12	☽ □ ☿
10 03	☽ ꝓ ♅
10 26	☽ □ ☉
13 12	☽ Q ♇
16 33	♀ △ ♆
07 Wednesday	
02 30	☽ ☍ ♄
04 29	☽ △ ♆
05 42	☽ △ ♀
09 13	☽ Q ♃
10 46	♂ ± ♅
10 58	☽ ⚹ ♇
13 17	☽ △ ♅
21 14	☽ ∥ ♄
08 Thursday	
00 27	☽ ⊻ ♂
06 18	☽ ⚹ ♃
08 41	☽ ∥ ♇
10 54	☽ ꝓ ♀
11 01	☽ △ ☿
11 20	☽ ⊥ ♂
13 22	☽ ∠ ♇
20 00	☽ △ ☉
09 Friday	
03 22	☽ ∠ ♂
03 50	☽ ± ♀
04 46	☽ ⊥ ♇
05 58	☽ # ♂
07 13	☽ ⊼ ♄
07 51	☽ ∠ ♃
08 42	☿ ∥ ♀
08 49	☽ □ ♆
11 47	☽ ꝓ ☿
14 29	♀ ☍ ♇
14 52	☽ ⊻ ♇
14 54	☽ ⊼ ♀
17 02	☽ □ ♅
17 20	☽ ± ♄
22 47	☽ ⊥ ♃
23 07	☽ ꝓ ☉
10 Saturday	
02 19	☽ ± ☿
05 22	☽ ⚹ ♂
08 16	☽ ꝓ ♄
08 36	☽ ⊻ ♃
09 57	☉ ± ♆
11 53	☽ ⊼ ☿
14 37	☽ Q ♆
14 56	☽ ± ☉
15 29	♀ △ ♅
22 32	☽ Q ♅
11 Sunday	
01 20	☽ ⊼ ☉
01 35	☽ Q ♂
08 38	☽ △ ♄
09 57	☽ ⚹ ♆
15 40	☽ ☌ ♇
17 43	☽ ⚹ ♅
20 02	☽ # ☿
20 04	☽ ☍ ♀
12 Monday	
07 14	☽ □ ♂
08 19	☽ ☌ ♃
09 42	☽ ∠ ♆
10 41	☽ ☍ ☿
17 22	☽ ∠ ♅
23 51	☽ ⊥ ♆
13 Tuesday	
04 03	☽ ☍ ☉
07 29	☽ ⊥ ♅
08 09	☽ □ ♄
09 15	☽ ⊻ ♆
09 24	♂ □ ♃
14 52	☽ ⊻ ♇
16 53	☽ ⊻ ♅
23 23	☽ ⊼ ♀
14 Wednesday	
00 17	☽ ⊥ ♇
07 07	☽ ⊻ ♃
08 03	☽ △ ♂
09 02	☽ ⊼ ☿
09 41	☽ ± ♀
12 34	☽ Q ♄
14 30	☽ ∠ ♇
16 33	☽ ⊥ ♃
17 00	☽ # ☿
18 26	☽ ± ☿
15 Thursday	
01 16	☽ ꝓ ♀
04 57	☿ ꝓ ♆
05 49	☿ □ ♂
06 02	♂ ꝓ ♆
06 43	☽ ⊼ ☉
06 50	☽ ∠ ♃
07 47	☽ ⚹ ♄
08 42	☽ ꝓ ☿
08 43	☽ ☌ ♆
08 49	☽ ꝓ ♂
09 17	☉ ± ♅
14 30	☽ ⚹ ♇
16 34	☽ ☌ ♅
17 05	☽ ± ☉
22 37	♀ Q ♄
23 56	☉ □ ♄
23 58	☽ ± ♂
16 Friday	
03 46	☽ △ ♀
07 05	☽ ⚹ ♃
07 49	♀ # ♆
08 19	☽ ∠ ♄
08 53	☽ ꝓ ☉
09 01	☽ △ ☿
10 08	☽ Q ♇
10 14	☽ ⊼ ♂
13 25	☉ ⊼ ♆
23 22	♀ # ♅
23 24	☽ ⊥ ♄
17 Saturday	
03 01	☽ Q ♃
06 58	☿ St D
09 38	☽ ⊻ ♄
10 27	☽ ⊻ ♆
12 01	☽ △ ☉
12 36	☽ # ♂
16 08	♀ ☍ ♃
16 36	☽ □ ♇
18 47	☽ ⊻ ♅
20 48	☽ ⊥ ♆
22 04	♀ # ♃
18 Sunday	
00 02	☽ ∥ ♇
05 16	☽ ⊥ ♅
10 02	☽ □ ♃
11 59	☽ □ ♀
12 13	☽ ∥ ♄
12 37	☽ □ ☿
12 40	☽ ∠ ♆
15 57	☽ ☍ ♂
17 59	☿ ꝓ ♆
19 20	♀ ꝓ ♆
19 30	☿ ☌ ♀
21 19	☽ ∠ ♅
19 Monday	
15 15	☽ ☌ ♄
15 55	☽ ⚹ ♆
21 53	☽ Q ☿
22 01	☽ □ ☉
22 33	☽ △ ♇
20 Tuesday	
00 17	☽ Q ♀
00 53	☽ ⚹ ♅
05 11	☉ ⊼ ♇
09 37	☽ # ♄
14 27	☽ Q ♆
16 58	☽ △ ♃
21 04	☽ ⚹ ☿
22 49	☽ # ♇
23 39	☽ Q ♅
21 Wednesday	
01 34	☽ ⚹ ♀
02 17	☽ ⊼ ♂
03 03	☽ ꝓ ♇
06 20	☽ ∥ ♂
11 14	☽ Q ☉
11 21	☉ ⊼ ♅
13 57	♀ □ ♂
14 28	☽ ± ♂
20 34	☽ ± ♇
21 49	☽ ꝓ ♃
22 Thursday	
00 52	☽ ⊻ ♄
01 20	☽ □ ♆
03 06	☽ ∠ ☿
08 23	☽ ⊼ ♇
09 00	☽ ꝓ ♂
10 13	☽ ∠ ♀
10 48	☽ □ ♅
12 46	☽ ⊥ ♄
12 53	☽ ⚹ ☉
15 28	☽ ± ♃
15 54	♀ ꝓ ♅
21 36	☽ ⊥ ☿
23 Friday	
03 21	☽ ⊼ ♃
06 13	☽ ⊥ ♀
06 48	☽ ∠ ♄
10 06	☽ ⊻ ☿
16 25	☽ △ ♂
19 41	☽ ⊻ ♀
21 35	☽ ∠ ☉
23 24	♂ ꝓ ♅
24 Saturday	
13 09	☽ ⚹ ♄
13 23	☽ △ ♆
17 24	☽ ⊥ ☉
20 39	☽ ☍ ♇
23 03	☽ △ ♅
25 Sunday	
05 50	☉ Q ♂
06 40	☽ ⊻ ☉
07 35	☽ ∥ ☿
13 34	☽ Q ♄
15 33	☽ ☍ ♃
19 48	☽ ꝓ ♆
26 Monday	
01 02	♀ ± ♆
02 01	☽ ☌ ☿
05 27	☽ ꝓ ♅
08 12	☽ □ ♂
08 42	☉ ± ♇
14 03	☽ ± ♆
15 35	☽ ☌ ♀
16 19	♂ # ♇
23 40	☽ ± ♅
27 Tuesday	
02 10	☽ □ ♄
02 12	☽ ⊼ ♆
04 31	♂ Q ♇
09 27	☽ ⊼ ♇
09 32	♄ ⚹ ♆
11 47	☽ ⊼ ♅
13 50	♃ ∥ ♅
21 04	☽ ∥ ☿
21 33	☽ ± ♇
22 24	☽ Q ♂
28 Wednesday	
00 50	☽ ☌ ☉
03 49	☽ ⊼ ♃
05 45	☿ ꝓ ♅
15 38	☽ ꝓ ♇
15 47	☽ ± ♃
18 52	☽ ⊻ ☿
23 40	☽ ⚹ ♂
29 Thursday	
07 49	☽ ⊥ ☿
09 38	☽ ꝓ ♃
10 53	☽ ⊻ ♀
11 01	☉ ⊼ ♃
14 22	☽ ☍ ♆
14 32	☽ △ ♄
19 57	♀ ± ♅
21 31	☽ △ ♇
23 43	☽ ☍ ♅
30 Friday	
00 09	☽ ⊥ ♀
03 12	☽ ∠ ☿
06 55	☽ ∠ ♂
15 08	☽ △ ♃
17 45	☽ ⊻ ☉
19 53	☽ ∠ ♀
20 13	☽ ꝓ ♄
20 34	♀ ⊼ ♆
23 08	♀ □ ♄

All ephemeris data is given at 12.00 UT and the Moon's longitude is additionally given for 24.00 UT

JULY 1995

LONGITUDES

Date	Sidereal time h m s	Sun ☉ ° ' "	Moon ☽ ° ' "	Moon ☽ 24.00 ° ' "	Mercury ☿ ° '	Venus ♀ ° '	Mars ♂ ° '	Jupiter ♃ ° '	Saturn ♄ ° '	Uranus ♅ ° '	Neptune ♆ ° '	Pluto ♇ ° '
01	06 36 17	09 ♋ 11 58	17 ♌ 48 03	23 ♌ 59 06	17 ♊ 29	25 ♊ 23	18 ♍ 39	07 ♐ 04	24 ♓ 44	29 ♑ 17	24 ♑ 35	28 ♏ 12
02	06 40 14	10 09 11	00 ♍ 12 54	06 ♍ 29 43	18 34	26 36	19 12	06 R 58	24 44	29 R 15	24 R 33	28 R 11
03	06 44 10	11 06 24	12 ♍ 49 53	19 ♍ 13 44	19 44	27 50	19 45	06 53	24 45	29 13	24 32	28 10
04	06 48 07	12 03 36	25 ♍ 41 37	02 ♎ 13 56	20 57	29 ♊ 03	20 18	06 48	24 45	29 10	24 30	28 09
05	06 52 03	13 00 48	08 ♎ 51 03	15 ♎ 33 18	22 15	00 ♋ 16	20 52	06 43	24 45	29 08	24 29	28 07
06	06 56 00	13 58 00	22 ♎ 21 01	29 ♎ 14 27	23 36	01 30	21 26	06 38	24 R 45	29 06	24 27	28 06
07	06 59 56	14 55 12	06 ♏ 13 47	13 ♏ 19 04	25 01	02 43	21 59	06 34	24 45	29 03	24 25	28 05
08	07 03 53	15 52 23	20 ♏ 30 13	27 ♏ 47 00	26 31	03 57	22 33	06 29	24 45	29 01	24 24	28 04
09	07 07 49	16 49 34	05 ♐ 09 00	12 ♐ 35 34	28 03	05 10	23 07	06 25	24 45	28 59	24 22	28 04
10	07 11 46	17 46 46	20 ♐ 05 54	27 ♐ 39 00	29 ♊ 40	06 23	23 41	06 21	24 44	28 56	24 21	28 03
11	07 15 43	18 43 57	05 ♑ 13 43	12 ♑ 48 46	01 ♋ 20	07 37	24 16	06 17	24 44	28 54	24 19	28 02
12	07 19 39	19 41 08	20 ♑ 22 50	27 ♑ 54 37	03 03	08 50	24 50	06 13	24 43	28 52	24 17	28 01
13	07 23 36	20 38 20	05 ♒ 22 51	12 ♒ 46 25	04 50	10 04	25 24	06 09	24 43	28 49	24 16	28 00
14	07 27 32	21 35 32	20 ♒ 04 21	27 ♒ 15 53	06 39	11 17	25 59	06 05	24 42	28 47	24 14	27 59
15	07 31 29	22 32 44	04 ♓ 20 28	11 ♓ 17 46	08 32	12 31	26 34	06 02	24 41	28 45	24 13	27 58
16	07 35 25	23 29 57	18 ♓ 07 41	24 ♓ 50 15	10 28	13 45	27 08	05 59	24 40	28 42	24 11	27 58
17	07 39 22	24 27 10	01 ♈ 25 44	07 ♈ 54 29	12 26	14 58	27 43	05 56	24 39	28 40	24 09	27 57
18	07 43 18	25 24 24	14 ♈ 16 59	20 ♈ 33 47	14 26	16 12	28 18	05 53	24 38	28 37	24 08	27 56
19	07 47 15	26 21 39	26 ♈ 45 31	02 ♉ 52 48	16 28	17 25	28 53	05 50	24 37	28 35	24 06	27 56
20	07 51 12	27 18 54	08 ♉ 56 20	14 ♉ 56 46	18 32	18 39	29 ♍ 29	05 48	24 35	28 33	24 04	27 55
21	07 55 08	28 16 11	20 ♉ 54 46	26 ♉ 50 58	20 37	19 53	00 ♎ 04	05 46	24 34	28 30	24 03	27 54
22	07 59 05	29 ♋ 13 28	02 ♊ 45 57	08 ♊ 40 17	22 44	21 06	00 39	05 43	24 32	28 28	24 01	27 54
23	08 03 01	00 ♌ 10 45	14 ♊ 34 31	20 ♊ 29 04	24 50	22 20	01 15	05 41	24 31	28 25	24 00	27 53
24	08 06 58	01 08 04	26 ♊ 24 24	02 ♋ 20 52	26 58	23 34	01 50	05 40	24 29	28 23	23 58	27 53
25	08 10 54	02 05 23	08 ♋ 18 46	14 ♋ 18 22	29 ♋ 05	24 48	02 26	05 38	24 27	28 21	23 56	27 52
26	08 14 51	03 02 43	20 ♋ 19 53	26 ♋ 23 30	01 ♌ 13	26 02	03 02	05 37	24 25	28 18	23 55	27 52
27	08 18 47	04 00 04	02 ♌ 29 20	08 ♌ 37 28	03 19	27 15	03 38	05 35	24 23	28 16	23 53	27 51
28	08 22 44	04 57 26	14 ♌ 48 01	21 ♌ 01 01	05 26	28 29	04 14	05 34	24 21	28 13	23 52	27 51
29	08 26 41	05 54 48	27 ♌ 16 32	03 ♍ 34 38	07 31	29 ♋ 43	04 50	05 33	24 19	28 11	23 50	27 51
30	08 30 37	06 52 11	09 ♍ 55 24	16 ♍ 18 54	09 35	00 ♌ 57	05 26	05 33	24 16	28 09	23 48	27 50
31	08 34 34	07 ♌ 49 34	22 ♍ 45 17	29 ♍ 14 41	11 ♌ 39	02 ♌ 11	06 ♎ 03	05 ♐ 32	24 ♓ 14	28 ♑ 06	23 ♑ 47	27 ♏ 50

DECLINATIONS

Date	Moon True ☊ ° '	Moon Mean ☊ ° '	Moon ☽ Latitude ° '	Sun ☉ ° '	Moon ☽ ° '	Mercury ☿ ° '	Venus ♀ ° '	Mars ♂ ° '	Jupiter ♃ ° '	Saturn ♄ ° '	Uranus ♅ ° '	Neptune ♆ ° '	Pluto ♇ ° '
01	03 ♏ 03	02 ♏ 09	04 S 53	23 N 07	10 N 50	19 N 56	23 N 05	05 N 11	20 S 46	04 S 02	20 S 51	20 S 39	06 S 18
02	02 R 55	02 06	04 30	23 03	07 11	20 13	23 09	04 57	20 46	04 02	20 52	20 40	06 18
03	02 49	02 03	03 53	22 58	03 N 09	20 30	23 13	04 43	20 45	04 02	20 52	20 40	06 18
04	02 46	02 00	03 03	22 53	01 S 06	20 47	23 17	04 29	20 44	04 02	20 53	20 40	06 18
05	02 45	01 56	02 03	22 48	05 24	21 04	23 19	04 15	20 44	04 02	20 53	20 41	06 18
06	02 D 45	01 53	00 S 55	22 42	09 33	21 21	23 21	04 01	20 43	04 03	20 54	20 41	06 19
07	02 R 45	01 50	00 N 18	22 36	13 18	21 37	23 23	03 46	20 43	04 03	20 54	20 41	06 19
08	02 44	01 47	01 33	22 30	16 23	21 53	23 23	03 32	20 42	04 03	20 55	20 41	06 19
09	02 41	01 44	02 43	22 23	18 29	22 08	23 23	03 18	20 41	04 03	20 55	20 42	06 19
10	02 35	01 41	03 43	22 15	19 22	22 21	23 22	03 03	20 41	04 04	20 56	20 42	06 20
11	02 27	01 37	04 28	22 08	18 52	22 34	23 21	02 49	20 40	04 04	20 56	20 42	06 20
12	02 17	01 34	04 55	22 00	17 02	22 45	23 19	02 34	20 40	04 05	20 57	20 43	06 20
13	02 07	01 31	05 01	21 51	14 03	22 54	23 16	02 20	20 40	04 05	20 57	20 43	06 20
14	01 56	01 28	04 47	21 42	10 15	23 02	23 12	02 05	20 39	04 06	20 58	20 43	06 21
15	01 48	01 25	04 15	21 33	05 58	23 08	23 08	01 51	20 39	04 06	20 58	20 44	06 21
16	01 42	01 22	03 29	21 24	01 S 30	23 11	23 03	01 36	20 38	04 07	20 59	20 44	06 21
17	01 38	01 18	02 33	21 14	02 N 54	23 12	22 58	01 21	20 38	04 08	20 59	20 44	06 21
18	01 37	01 15	01 30	21 03	07 01	23 10	22 52	01 07	20 38	04 08	21 00	20 44	06 22
19	01 D 36	01 12	00 N 26	20 53	10 43	23 06	22 45	00 52	20 38	04 09	21 00	20 45	06 22
20	01 R 36	01 09	00 S 39	20 42	13 52	23 00	22 37	00 37	20 37	04 10	21 01	20 45	06 22
21	01 35	01 06	01 40	20 30	16 22	22 50	22 29	00 22	20 37	04 11	21 01	20 45	06 23
22	01 33	01 02	02 37	20 19	18 09	22 38	22 20	00 N 07	20 37	04 11	21 02	20 46	06 23
23	01 27	00 59	03 26	20 07	19 08	22 23	22 10	00 S 08	20 37	04 12	21 02	20 46	06 24
24	01 19	00 56	04 07	19 54	19 17	22 06	22 00	00 23	20 37	04 13	21 03	20 46	06 24
25	01 08	00 53	04 37	19 42	18 34	21 46	21 49	00 38	20 37	04 14	21 03	20 46	06 24
26	00 56	00 50	04 55	19 29	17 02	21 24	21 38	00 53	20 37	04 15	21 04	20 47	06 25
27	00 42	00 47	05 00	19 15	14 44	20 59	21 26	01 08	20 37	04 16	21 04	20 47	06 25
28	00 29	00 43	04 51	19 02	11 45	20 32	21 13	01 23	20 37	04 17	21 04	20 47	06 26
29	00 17	00 40	04 28	18 48	08 13	20 03	20 59	01 38	20 37	04 18	21 05	20 48	06 26
30	00 07	00 37	03 51	18 33	04 16	19 32	20 45	01 53	20 37	04 19	21 05	20 48	06 26
31	00 ♏ 00	00 ♏ 34	03 S 03	18 N 19	00 N 05	18 N 59	20 N 31	02 S 08	20 S 37	04 S 21	21 S 06	20 S 48	06 S 27

ZODIAC SIGN ENTRIES

Date	h	m	Planets
02	11	35	☽ ♍
04	19	55	☽ ♎
05	06	39	♀ ♋
07	01	19	☽ ♏
09	03	37	☽ ♐
10	16	58	☿ ♋
11	03	43	☽ ♑
13	03	21	☽ ♒
15	04	37	☽ ♓
17	09	23	☽ ♈
19	18	20	☽ ♉
21	09	21	♂ ♎
22	06	23	☽ ♊
23	07	30	☉ ♌
24	19	16	☽ ♋
25	22	19	☿ ♌
27	07	07	☽ ♌
29	17	12	☽ ♍
29	17	32	♀ ♌

LATITUDES

Date	Mercury ☿ ° '	Venus ♀ ° '	Mars ♂ ° '	Jupiter ♃ ° '	Saturn ♄ ° '	Uranus ♅ ° '	Neptune ♆ ° '	Pluto ♇ ° '
01	02 S 56	00 S 17	00 N 45	00 N 43	02 S 07	00 S 34	00 N 33	13 N 47
04	02 21	00 09	00 42	00 42	02 07	00 34	00 33	13 46
07	01 43	00 S 02	00 39	00 42	02 08	00 34	00 33	13 45
10	01 05	00 N 05	00 36	00 42	02 09	00 34	00 33	13 43
13	00 S 27	00 13	00 33	00 41	02 10	00 34	00 33	13 42
16	00 N 09	00 20	00 30	00 40	02 11	00 34	00 33	13 41
19	00 41	00 27	00 28	00 40	02 11	00 34	00 33	13 39
22	01 08	00 33	00 25	00 39	02 12	00 34	00 33	13 38
25	01 28	00 40	00 22	00 38	02 13	00 34	00 33	13 36
28	01 40	00 46	00 19	00 38	02 13	00 34	00 33	13 35
31	01 N 46	00 N 52	00 N 17	00 N 37	02 S 14	00 S 34	00 N 33	13 N 33

DATA

Julian Date	2449900
Delta T	+61 seconds
Ayanamsa	23° 47' 49"
Synetic vernal point	05° ♓ 19' 10"
True obliquity of ecliptic	23° 26' 15"

MOON'S PHASES, APSIDES AND POSITIONS ☽

Date	h	m	Phase	Longitude ° '	Eclipse Indicator
05	20	02	☽	13 ♎ 20	
12	10	49	○	19 ♑ 38	
19	11	10	☾	26 ♈ 20	
27	15	13	●	04 ♌ 08	

Day	h	m		
11	10	06	Perigee	
23	20	26	Apogee	
04	05	53	OS	
10	15	26	Max dec	19° S 22'
16	20	03	ON	
24	04	08	Max dec	19° N 19'
31	12	25	OS	

LONGITUDES

Date	Chiron ⚷ ° '	Ceres ⚳ ° '	Pallas ⚴ ° '	Juno ⚵ ° '	Vesta ⚶ ° '	Black Moon Lilith ⚸ ° '
01	21 ♍ 46	02 ♍ 24	01 ♌ 19	24 ♐ 01	06 ♌ 47	19 ♊ 59
11	22 ♍ 37	06 ♍ 26	06 ♌ 53	22 ♐ 01	11 ♌ 15	21 ♊ 08
21	23 ♍ 36	10 ♍ 35	12 ♌ 22	20 ♐ 25	15 ♌ 46	22 ♊ 13
31	24 ♍ 42	14 ♍ 49	17 ♌ 48	19 ♐ 21	20 ♌ 22	23 ♊ 20

ASPECTARIAN

h m Aspects

01 Saturday
01 27 ☽ ⊥ ♂
06 30 ☽ ⊥ ☉
11 19 ☽ ✱ ☿
13 43 ☽ ⊻ ♂
13 49 ☽ ± ♄
18 27 ☉ ∥ ♀

02 Sunday
01 07 ☽ ⊼ ♆
01 27 ☽ ∠ ☉
01 27 ☽ ⊼ ♄
04 18 ☽ ✱ ♀
08 06 ☽ □ ♇
10 09 ☽ ⊼ ♅
11 39 ☿ ± ♆
12 39 ☽ ± ♆
12 45 ☽ Q ☿
17 26 ☽ # ♇
21 36 ☽ ± ♅

03 Monday
00 49 ☽ □ ♃
02 22 ☽ ∥ ♂
05 43 ☽ Q ♀
05 46 ☽ bQ ♆
06 56 ☽ # ♄
08 29 ☽ ✱ ☉
12 44 ☿ □ ♂
14 35 ☽ bQ ♅
18 15 ☽ Q ♇
18 26 ♀ ⊼ ♇

04 Tuesday
01 34 ☽ ☌ ♂
02 12 ☿ # ♆
02 17 ☽ □ ☿
08 08 ☿ # ♃
08 44 ☽ Q ☉
09 48 ☽ △ ♆
10 15 ☽ ☍ ♄
10 22 ☽ Q ♃
14 19 ♀ ⊼ ♅
16 30 ☽ ✱ ♇
18 23 ☽ △ ♅
18 49 ☽ □ ♀
19 49 ☿ # ♅

05 Wednesday
04 22 ☽ ∥ ♄
05 08 ☉ ± ♃
05 53 ☽ # ♂
08 10 ☽ ✱ ♃
14 45 ☉ bQ ♇
17 09 ☽ ∥ ♇
19 39 ☽ ∠ ♇
20 02 ☽ □ ☉

06 Thursday
03 23 ☿ ± ♅
07 46 ♄ St R
08 54 ♂ # ♄
10 19 ☽ ⊻ ♂
10 45 ☽ ∠ ♃
11 34 ☽ ⊥ ♇
14 27 ☽ △ ☿
15 40 ☽ □ ♆
16 12 ☽ ⊼ ♄
21 14 ☽ ⊥ ♂
22 02 ☽ ⊻ ♇
23 43 ☽ □ ♅

07 Friday
02 09 ☿ ⊼ ♆
02 20 ☽ ⊥ ♃
02 37 ☽ ± ♄
05 25 ☽ △ ♀
07 27 ☿ □ ♄
12 34 ☽ ⊻ ♃
13 21 ☽ ∠ ♂
17 59 ☽ bQ ♄
19 11 ☽ bQ ☿
22 29 ☽ Q ♆

08 Saturday
03 44 ☽ △ ☉
06 13 ☽ Q ♅
09 10 ☽ bQ ♀
12 01 ☽ ± ☿
14 33 ♀ ± ♇
15 32 ☽ ✱ ♂
18 25 ☽ ✱ ♆
19 01 ☽ △ ♄
23 04 ☽ ⊼ ☿

09 Sunday
00 28 ☽ ☌ ♇
01 23 ☽ ± ♀
01 59 ☽ ✱ ♅
02 04 ♃ ∥ ♆
06 14 ☽ bQ ☉
11 57 ☽ Q ♂
12 02 ☽ ⊼ ♀
12 04 ☿ ⊼ ♇
14 02 ☽ ☌ ♃
18 48 ☽ ∠ ♆
21 47 ☽ ± ☉

10 Monday
01 37 ☿ ⊼ ♅
02 11 ☽ ∠ ♅
05 03 ☉ ∥ ☿
08 03 ☽ ⊼ ☉
09 12 ☽ ⊥ ♆
11 06 ♀ ⊼ ♃
16 31 ☽ ⊥ ♅
17 56 ☽ □ ♂
18 44 ☽ ⊻ ♆
19 23 ☽ □ ♄

11 Tuesday
00 37 ☽ ⊻ ♇
02 01 ☽ ⊻ ♅
05 03 ☽ ☍ ☿
10 06 ☽ ⊥ ♇
12 32 ♂ Q ♃
13 39 ☽ ⊻ ♃
14 13 ♂ △ ♆
16 06 ☽ ☍ ♀
23 06 ☽ ⊥ ♃
23 52 ☽ Q ♄

12 Wednesday
00 20 ☽ ∠ ♇
07 20 ♂ ☍ ♄
10 49 ☽ ☍ ☉
13 19 ☽ ∠ ♃
18 13 ☽ ☌ ♆
18 54 ☽ ✱ ♄
19 22 ☽ △ ♂

13 Thursday
00 09 ☽ ✱ ♇
01 02 ☿ ± ♇
01 29 ☽ ☌ ♅
10 59 ☽ ⊼ ☿
13 14 ☽ ✱ ♃
19 00 ☽ ∠ ♄
19 28 ☽ Q ♇
20 16 ☽ ⊼ ♀
20 29 ☽ bQ ♂
22 04 ☽ ± ☿

14 Friday
00 07 ☉ bQ ♃
04 53 ☿ ⊼ ♃
06 59 ☽ ± ♀
08 44 ☽ Q ♃
09 44 ☽ ⊥ ♄
11 51 ☽ ± ♂
14 42 ☽ ⊼ ☉
15 01 ☽ bQ ☿
18 55 ☽ ⊻ ♆
19 41 ☽ ⊻ ♄
22 16 ☽ ⊼ ♂
23 20 ☽ bQ ♀

15 Saturday
01 12 ☽ □ ♇
01 27 ☽ ± ☉
02 31 ☽ ⊻ ♅
02 58 ♀ ± ♃
04 59 ☽ ⊥ ♆
09 55 ☽ ∥ ♇
12 41 ☽ ⊥ ♅
14 03 ☿ ∥ ♀
14 54 ☽ □ ♃
17 54 ☽ bQ ☉
20 22 ☽ ∠ ♆
20 22 ☽ △ ☿
20 51 ♀ bQ ♇
21 59 ☽ ∥ ♄

16 Sunday
03 31 ☽ △ ♀
04 14 ☽ ∠ ♅
11 23 ☽ # ♂
22 20 ☽ △ ☉
22 48 ☽ ✱ ♆
23 41 ☽ ☌ ♄

17 Monday
03 53 ☽ ∥ ♂
04 42 ☉ ☍ ♆
04 55 ☽ ☍ ♂
05 39 ☽ △ ♇
06 08 ☿ ± ♃
06 57 ☽ ✱ ♅
16 50 ☉ △ ♄
18 15 ☿ bQ ♇
18 57 ☽ # ♄
20 17 ☽ △ ♃
20 43 ☽ Q ♆
21 10 ♂ ✱ ♇

18 Tuesday
05 07 ☽ Q ♅
08 01 ☽ # ♇
09 27 ☽ bQ ♇
12 20 ☽ □ ☿
16 02 ☽ □ ♀
16 04 ♀ Q ♂
20 17 ☉ # ♅

19 Wednesday
00 17 ♂ △ ♅
00 34 ☽ bQ ♃
02 38 ☽ ± ♇
06 51 ☽ □ ♆
07 50 ☽ ⊻ ♄
11 10 ☽ □ ☉
14 17 ☽ ⊼ ♇
15 33 ☽ □ ♅
16 22 ☽ ⊼ ♂
18 00 ☽ ± ♃
18 52 ☿ Q ♂
19 31 ☽ ⊥ ♄

20 Thursday
04 47 ☽ ± ♂
05 07 ☉ # ♆
05 47 ☽ ⊼ ♃
06 14 ☽ Q ☿
06 57 ☽ Q ♀
13 17 ☽ ∠ ♄
15 19 ☿ ☌ ♀
21 25 ☉ # ♃
23 38 ☽ bQ ♂

21 Friday
01 51 ☽ Q ☉
02 57 ☉ △ ♇
09 41 ☽ ✱ ♀
11 17 ☽ ✱ ☿
13 33 ☿ bQ ♃
17 40 ☉ ☍ ♅
18 19 ☽ △ ♆
19 21 ☽ ✱ ♄

22 Saturday
02 08 ☽ ☍ ♇
03 18 ☽ △ ♅
04 11 ☽ ✱ ☉
04 41 ♀ bQ ♃
07 29 ☽ △ ♂
17 59 ☽ ☍ ♃
19 35 ☽ ∠ ♀
19 39 ☽ Q ♄

23 Sunday
00 16 ☽ ∠ ☿
00 41 ☽ bQ ♆
02 30 ☿ ☍ ♆
08 17 ☿ △ ♄
09 40 ☽ bQ ♅
13 20 ☽ ∠ ☉
16 00 ☽ ⊥ ♀
18 56 ☽ ± ♆
22 34 ☽ ⊥ ☿

24 Monday
03 53 ☽ ± ♅
05 35 ☽ ⊻ ♀
07 04 ☽ ⊼ ♆
08 07 ☽ □ ♄
09 12 ☽ ⊥ ☉
13 22 ☽ ⊻ ☿
14 59 ☽ ⊼ ♇
15 59 ☽ ⊼ ♅
19 39 ♀ ☍ ♆
22 18 ☿ △ ♇
22 23 ☽ ⊻ ☉
23 33 ☽ □ ♂

25 Tuesday
03 05 ☽ ± ♇
03 45 ☿ ☍ ♅
04 17 ☿ ∥ ♀
05 26 ♀ △ ♄
06 38 ☽ ⊼ ♃
18 39 ☽ ± ♃
21 08 ☽ bQ ♇

26 Wednesday
11 09 ☉ ✱ ♂
12 33 ☽ bQ ♃
13 28 ☽ Q ♂
19 05 ☽ ☍ ♆
20 05 ☽ △ ♄

27 Thursday
00 33 ☽ ☌ ♀
02 54 ☽ △ ♇
03 43 ☽ ☍ ♅
07 20 ☿ # ♅
13 59 ☽ ☌ ☿
14 21 ☽ ✱ ♂
15 13 ☽ ☌ ☉
16 54 ☿ ✱ ♂
18 04 ☽ △ ♃
22 42 ☿ # ♆
23 41 ♀ △ ♇

28 Friday
01 27 ☽ bQ ♄
02 09 ☉ ☌ ☿
07 04 ♀ ☍ ♅
07 54 ☿ # ♃
13 39 ☿ △ ♃
18 50 ☽ ± ♄
21 00 ☽ ∠ ♂

29 Saturday
02 35 ♀ # ♅
03 12 ☉ △ ♃
05 25 ☽ ⊼ ♆
06 21 ☽ ⊼ ♄
13 05 ☽ □ ♇
13 44 ☽ ⊼ ♅
15 07 ☽ ⊥ ♂
16 52 ☽ ± ♆
17 10 ☽ ⊻ ♀
23 06 ☽ # ♇

30 Sunday
01 07 ☽ ± ♅
03 06 ☽ ⊻ ♂
03 44 ☽ □ ♃
05 46 ☽ ⊻ ☉
05 47 ☽ ⊥ ♀
08 00 ♀ # ♆
08 24 ☿ bQ ♄
09 54 ☽ bQ ♆
11 15 ☽ ⊻ ☿
11 42 ☽ # ♄
16 15 ♂ ✱ ♃
17 59 ☽ ⊥ ☉
18 03 ☽ bQ ♅
23 06 ☽ Q ♇

31 Monday
00 31 ☽ ∠ ♀
00 41 ☽ ⊥ ☿
00 58 ☽ # ♂
02 08 ♀ # ♃
12 09 ☽ ∠ ☉
13 27 ☽ Q ♃
13 54 ☽ △ ♆
14 44 ☽ ☍ ♄
20 33 ☽ ∠ ☿
21 24 ☽ ✱ ♇
21 52 ☽ △ ♅

All ephemeris data is given at 12.00 UT and the Moon's longitude is additionally given for 24.00 UT

LONGITUDES

Date	Sidereal time h m s	Sun ☉ ° ' "	Moon ☽ ° ' "	Moon ☽ 24.00 ° ' "	Mercury ☿ ° '	Venus ♀ ° '	Mars ♂ ° '	Jupiter ♃ ° '	Saturn ♄ ° '	Uranus ♅ ° '	Neptune ♆ ° '	Pluto ♇ ° '
01	08 38 30	08 ♌ 46 58	05 ♎ 47 16	12 ♎ 23 15	13 ♌ 41	03 ♌ 25	06 ♎ 39	05 ♐ 32	24 ♓ 12	28 ♑ 04	23 ♑ 45	27 ♏ 50
02	08 42 27	09 44 22	19 ♎ 02 51	25 ♎ 46 16	15 41	04 39	07 15	05 R 32	24 R 09	28 R 02	23 R 44	27 R 50
03	08 46 23	10 41 48	02 ♏ 33 45	09 ♏ 25 29	17 41	05 52	07 52	05 D 32	24 06	27 59	23 42	27 49
04	08 50 20	11 39 14	16 ♏ 21 37	23 ♏ 22 13	19 38	07 06	08 29	05 32	24 04	27 57	23 41	27 49
05	08 54 16	12 36 40	00 ♐ 27 19	07 ♐ 36 47	21 35	08 20	09 06	05 33	24 01	27 55	23 39	27 49
06	08 58 13	13 34 07	14 ♐ 50 22	22 ♐ 07 41	23 29	09 34	09 42	05 33	23 58	27 52	23 38	27 49
07	09 02 10	14 31 35	29 ♐ 28 10	06 ♑ 51 08	25 23	10 48	10 19	05 34	23 55	27 50	23 36	27 49
08	09 06 06	15 29 04	14 ♑ 15 45	21 ♑ 41 02	27 14	12 02	10 56	05 35	23 52	27 48	23 35	27 49
09	09 10 03	16 26 34	29 ♑ 05 57	06 ♒ 29 26	29 ♌ 05	13 16	11 34	05 36	23 49	27 46	23 33	27 49
10	09 13 59	17 24 05	13 ♒ 50 25	21 ♒ 07 51	00 ♍ 53	14 31	12 11	05 37	23 46	27 43	23 32	27 D 49
11	09 17 56	18 21 36	28 ♒ 20 51	05 ♓ 28 37	02 40	15 45	12 48	05 39	23 43	27 41	23 30	27 49
12	09 21 52	19 19 09	12 ♓ 30 32	19 ♓ 26 10	04 26	16 59	13 25	05 41	23 39	27 39	23 29	27 49
13	09 25 49	20 16 43	26 ♓ 15 15	02 ♈ 57 42	06 10	18 13	14 03	05 43	23 36	27 37	23 27	27 49
14	09 29 45	21 14 18	09 ♈ 33 34	16 ♈ 03 04	07 52	19 27	14 41	05 45	23 32	27 35	23 26	27 50
15	09 33 42	22 11 55	22 ♈ 26 34	28 ♈ 44 29	09 33	20 41	15 18	05 47	23 29	27 33	23 25	27 50
16	09 37 39	23 09 34	04 ♉ 57 20	11 ♉ 05 43	11 13	21 55	15 56	05 49	23 25	27 31	23 24	27 50
17	09 41 35	24 07 14	17 ♉ 10 15	23 ♉ 11 35	12 51	23 09	16 34	05 52	23 22	27 28	23 23	27 50
18	09 45 32	25 04 55	29 ♉ 10 23	05 ♊ 07 18	14 27	24 24	17 12	05 55	23 18	27 26	23 21	27 51
19	09 49 28	26 02 38	11 ♊ 03 00	16 ♊ 58 07	16 02	25 38	17 50	05 58	23 14	27 24	23 19	27 51
20	09 53 25	27 00 23	22 ♊ 53 14	28 ♊ 48 55	17 36	26 52	18 28	06 01	23 10	27 22	23 18	27 51
21	09 57 21	27 58 09	04 ♋ 45 42	10 ♋ 44 04	19 08	28 06	19 06	06 04	23 06	27 20	23 17	27 52
22	10 01 18	28 55 57	16 ♋ 44 24	22 ♋ 47 05	20 39	29 ♌ 21	19 44	06 07	23 02	27 18	23 16	27 52
23	10 05 14	29 ♌ 53 47	28 ♋ 52 25	05 ♌ 00 39	22 08	00 ♍ 35	20 23	06 11	22 58	27 17	23 14	27 53
24	10 09 11	00 ♍ 51 38	11 ♌ 11 56	17 ♌ 26 24	23 36	01 49	21 01	06 15	22 54	27 15	23 13	27 53
25	10 13 08	01 49 31	23 ♌ 44 08	00 ♍ 05 07	25 03	03 04	21 40	06 19	22 50	27 13	23 12	27 54
26	10 17 04	02 47 25	06 ♍ 29 35	12 ♍ 56 47	26 28	04 18	22 18	06 23	22 46	27 11	23 11	27 54
27	10 21 01	03 45 20	19 ♍ 27 19	26 ♍ 00 52	27 51	05 32	22 57	06 27	22 42	27 09	23 10	27 55
28	10 24 57	04 43 17	02 ♎ 37 20	09 ♎ 16 39	29 ♍ 13	06 47	23 36	06 32	22 38	27 07	23 08	27 56
29	10 28 54	05 41 16	15 ♎ 58 44	22 ♎ 43 30	00 ♎ 33	08 01	24 15	06 36	22 33	27 06	23 07	27 56
30	10 32 50	06 39 16	29 ♎ 30 57	06 ♏ 21 02	01 51	09 15	24 54	06 41	22 29	27 04	23 06	27 57
31	10 36 47	07 ♍ 37 17	13 ♏ 13 44	20 ♏ 09 03	03 ♎ 08	10 ♍ 30	25 ♎ 33	06 ♐ 46	22 ♓ 25	27 ♑ 02	23 ♑ 05	27 ♏ 58

Date	Moon True ☊ ° '	Moon Mean ☊ ° '	Moon ☽ Latitude ° '
01	29 ♎ 56	00 ♏ 31	02 S 04
02	29 R 54	00 28	00 S 57
03	29 D 54	00 24	00 N 14
04	29 R 54	00 21	01 26
05	29 53	00 18	02 34
06	29 50	00 15	03 34
07	29 45	00 12	04 21
08	29 37	00 08	04 51
09	29 27	00 05	05 02
10	29 17	00 ♏ 02	04 52
11	29 06	29 ♎ 59	04 24
12	28 58	29 56	03 40
13	28 52	29 53	02 44
14	28 48	29 49	01 40
15	28 46	29 46	00 N 34
16	28 D 46	29 43	00 S 33
17	28 47	29 40	01 37
18	28 R 47	29 37	02 35
19	28 46	29 34	03 26
20	28 42	29 30	04 08
21	28 37	29 27	04 39
22	28 29	29 24	04 58
23	28 19	29 21	05 04
24	28 09	29 18	04 57
25	27 59	29 14	04 35
26	27 49	29 11	03 59
27	27 42	29 08	03 10
28	27 37	29 05	02 10
29	27 34	29 02	01 S 02
30	27 D 34	28 59	00 N 11
31	27 ♎ 34	28 ♎ 55	01 N 23

DECLINATIONS

Date	Sun ☉ ° '	Moon ☽ ° '	Mercury ☿ ° '	Venus ♀ ° '	Mars ♂ ° '	Jupiter ♃ ° '	Saturn ♄ ° '	Uranus ♅ ° '	Neptune ♆ ° '	Pluto ♇ ° '
01	18 N 04	04 S 12	18 N 25	20 N 16	02 S 24	20 S 37	04 S 22	21 S 06	20 S 48	06 S 27
02	17 49	08 20	17 49	20 00	02 39	20 37	04 23	21 07	20 49	06 28
03	17 33	12 08	17 13	19 44	02 54	20 38	04 24	21 07	20 49	06 28
04	17 17	15 21	16 34	19 27	03 09	20 38	04 25	21 08	20 49	06 29
05	17 01	17 44	15 55	19 09	03 25	20 38	04 27	21 08	20 50	06 29
06	16 45	19 02	15 15	18 51	03 40	20 38	04 28	21 09	20 50	06 30
07	16 28	19 05	14 34	18 33	03 55	20 39	04 29	21 09	20 50	06 30
08	16 12	17 51	13 53	18 14	04 10	20 39	04 31	21 10	20 50	06 31
09	15 54	15 25	13 11	17 54	04 26	20 40	04 32	21 10	20 51	06 31
10	15 37	12 00	12 28	17 34	04 41	20 40	04 34	21 10	20 51	06 32
11	15 19	07 55	11 45	17 13	04 56	20 40	04 35	21 11	20 51	06 32
12	15 02	03 S 29	11 02	16 52	05 12	20 41	04 37	21 11	20 51	06 33
13	14 44	01 N 01	10 18	16 31	05 27	20 42	04 38	21 12	20 52	06 33
14	14 25	05 20	09 35	16 09	05 42	20 42	04 40	21 12	20 52	06 34
15	14 07	09 15	08 51	15 47	05 57	20 43	04 41	21 13	20 52	06 34
16	13 48	12 39	08 07	15 24	06 13	20 43	04 43	21 13	20 52	06 35
17	13 29	15 29	07 23	15 00	06 28	20 44	04 45	21 13	20 53	06 35
18	13 10	17 27	06 39	14 37	06 43	20 45	04 46	21 14	20 53	06 36
19	12 50	18 42	05 56	14 13	06 58	20 45	04 48	21 14	20 53	06 37
20	12 31	19 08	05 12	13 48	07 14	20 46	04 50	21 15	20 53	06 37
21	12 11	18 42	04 29	13 23	07 29	20 47	04 51	21 15	20 54	06 38
22	11 51	17 27	03 46	12 58	07 44	20 48	04 53	21 15	20 54	06 38
23	11 30	15 25	03 03	12 32	07 59	20 48	04 55	21 16	20 54	06 39
24	11 10	12 39	02 21	12 06	08 14	20 49	04 56	21 16	20 54	06 40
25	10 49	09 17	01 39	11 40	08 30	20 50	04 58	21 16	20 55	06 40
26	10 29	05 26	00 57	11 14	08 45	20 51	05 00	21 17	20 55	06 41
27	10 08	01 N 16	00 N 16	10 47	09 00	20 52	05 02	21 17	20 55	06 42
28	09 47	03 S 02	00 S 24	10 20	09 15	20 53	05 04	21 17	20 55	06 42
29	09 25	07 14	01 04	09 52	09 30	20 54	05 05	21 18	20 55	06 43
30	09 04	11 08	01 44	09 24	09 45	20 55	05 07	21 18	20 56	06 43
31	08 N 43	14 S 29	02 S 23	08 N 56	10 S 00	20 S 56	05 S 09	21 S 18	20 S 56	06 S 44

ZODIAC SIGN ENTRIES

Date	h m	Planets
01	01 23	☽ ♎
03	07 29	☽ ♏
05	11 14	☽ ♐
07	12 52	☽ ♑
09	13 28	☽ ♒
10	00 13	☿ ♍
11	14 46	☽ ♓
13	18 41	☽ ♈
16	02 25	☽ ♉
18	13 40	☽ ♊
21	02 24	☽ ♋
23	00 43	♀ ♍
23	14 13	☽ ♌
23	14 35	☉ ♍
25	23 50	☽ ♍
28	07 15	☽ ♎
29	02 07	☿ ♎
30	12 51	☽ ♏

LATITUDES

Date	Mercury ☿ ° '	Venus ♀ ° '	Mars ♂ ° '	Jupiter ♃ ° '	Saturn ♄ ° '	Uranus ♅ ° '	Neptune ♆ ° '	Pluto ♇ ° '
01	01 N 47	00 N 54	00 N 16	00 N 37	02 S 14	00 S 34	00 N 33	13 N 33
04	01 44	00 59	00 13	00 36	02 15	00 34	00 33	13 31
07	01 36	01 04	00 11	00 36	02 16	00 34	00 33	13 30
10	01 24	01 08	00 09	00 35	02 16	00 34	00 33	13 28
13	01 08	01 12	00 06	00 34	02 17	00 34	00 33	13 27
16	00 49	01 16	00 04	00 34	02 17	00 34	00 33	13 25
19	00 28	01 19	00 N 01	00 33	02 18	00 34	00 33	13 24
22	00 N 04	01 21	00 S 01	00 32	02 18	00 34	00 33	13 22
25	00 S 21	01 23	00 03	00 32	02 19	00 34	00 33	13 20
28	00 47	01 24	00 05	00 31	02 19	00 34	00 32	13 19
31	01 S 14	01 N 25	00 S 08	00 N 31	02 S 20	00 S 34	00 N 32	13 N 17

DATA

Julian Date	2449931
Delta T	+61 seconds
Ayanamsa	23° 47' 53"
Synetic vernal point	05° ♓ 19' 06"
True obliquity of ecliptic	23° 26' 15"

MOON'S PHASES, APSIDES AND POSITIONS ☽

Date	h m	Phase	Longitude ° '	Eclipse Indicator
04	03 16	☽	11 ♏ 18	
10	18 16	○	17 ♒ 39	
18	03 04	☾	24 ♉ 43	
26	04 31	●	02 ♍ 29	

Day	h m		
08	13 58	Perigee	
20	11 46	Apogee	
07	01 05	Max dec	19° S 13'
13	06 32	ON	
20	12 05	Max dec	19° N 08'
27	19 05	OS	

LONGITUDES

Date	Chiron ⚷ ° '	Ceres ⚳ ° '	Pallas ⚴ ° '	Juno ⚵ ° '	Vesta ⚶ ° '	Black Moon Lilith ⚸ ° '
01	24 ♍ 49	15 ♍ 15	18 ♌ 20	19 ♐ 17	20 ♌ 50	23 ♊ 26
11	26 ♍ 03	19 ♍ 35	23 ♌ 40	18 ♐ 50	25 ♌ 29	24 ♊ 33
21	27 ♍ 21	23 ♍ 58	28 ♌ 55	18 ♐ 59	00 ♍ 11	25 ♊ 40
31	28 ♍ 43	28 ♍ 25	04 ♍ 05	19 ♐ 40	04 ♍ 56	26 ♊ 47

ASPECTARIAN

h m	Aspects
01 Tuesday	
01 14	☽ ∥ ♂
07 12	☽ ✶ ♀
11 32	☽ ✶ ♃
12 57	☽ ∥ ♄
13 39	☽ ☌ ♂
17 53	☽ ✶ ☉
21 53	☉ ⚼ ♄
02 Wednesday	
00 48	☽ ∠ ♇
00 56	☽ ∥ ♇
04 54	☽ ✶ ☿
07 14	☽ Q ♀
13 03	☉ ∥ ☿
14 39	☽ ∠ ♃
16 41	♃ St D
16 58	☽ ⊥ ♆
17 11	☽ Q ☉
20 21	☽ □ ♆
21 05	☽ ⊼ ♄
03 Thursday	
03 38	☽ ⊻ ♇
03 57	☽ □ ♅
05 18	♀ △ ♃
06 03	☽ Q ☿
06 39	☽ ⊥ ♃
07 41	☽ ± ♄
17 08	☿ ± ♄
17 12	☽ ⊻ ♃
18 23	☽ □ ♀
21 43	☽ ⊻ ♂
23 25	☽ ⚼ ♄
04 Friday	
03 16	☽ □ ☉
03 56	☽ Q ♆
08 37	☽ ⊥ ♂
11 18	☽ Q ♅
18 32	☽ □ ☿
20 48	☽ # ☿
05 Saturday	
00 30	☽ ✶ ♆
00 44	☽ ∠ ♂
01 08	☽ △ ♄
04 38	☽ # ☉
07 33	☽ ☌ ♇
07 43	☽ ✶ ♅
20 33	☽ ☌ ♃
06 Sunday	
00 41	♀ ⚼ ♄
01 43	☽ ∠ ♆
02 27	☽ △ ♀
03 07	☽ ✶ ♂
08 13	☽ # ♀
08 45	☽ ∠ ♅
09 45	☽ △ ☉
13 42	☿ ⊼ ♆
16 36	☽ ⊥ ♆
17 11	♀ ✶ ♂
17 53	☿ ⊼ ♄
23 33	☽ ⊥ ♅
23 48	☽ Q ♂
07 Monday	
02 26	☽ ⊻ ♆
02 58	☽ □ ♄
04 21	☽ △ ☿
05 28	☽ ⚼ ♀
09 18	☽ ⊻ ♇
09 21	☽ ⊻ ♅
10 56	☿ ∠ ♂
12 06	☽ ⚼ ☉
19 04	☽ ⊥ ♇
21 28	☽ ± ♀
21 55	☽ ⊻ ♃
08 Tuesday	
00 26	♅ ✶ ♇
03 44	☽ ± ☉
05 22	☽ # ♀
06 23	☽ □ ♂
07 39	☽ ⊥ ♃
08 05	☽ ⊼ ♀
08 08	☽ Q ♄
08 16	☽ ⚼ ☿
09 40	☽ ∠ ♇
13 34	♇ St D
14 07	☽ ⊼ ☉
19 07	☿ ⊼ ♅
19 30	☿ □ ♇
22 14	☽ ∠ ♃
09 Wednesday	
00 53	☽ ± ☿
03 02	☽ ☌ ♆
03 29	☽ ✶ ♄
07 25	☽ # ☉
09 50	☽ ☌ ♅
09 55	☽ ✶ ♇
11 57	☽ ⊼ ☿
18 13	☿ ± ♆
22 34	☽ ✶ ♃
23 29	♂ ∥ ♄
10 Thursday	
03 44	☽ ∠ ♄
05 25	☽ Q ♇
08 19	☽ # ☿
09 10	☽ △ ♂
13 12	☽ ☍ ♀
18 13	☽ Q ♃
18 16	☽ ☍ ☉
18 25	☽ ⊥ ♄
20 36	☉ ± ♄
11 Friday	
03 57	☽ ⊻ ♆
04 18	☽ ⊻ ♄
10 54	☽ ⊻ ♅
11 03	☽ ⚼ ♂
11 07	☽ □ ♇
12 40	♂ ∠ ♇
13 56	☽ ⊥ ♆
19 35	☽ ∥ ♇
20 17	☽ ☍ ☿
20 57	☽ ⊥ ♅
12 Saturday	
00 19	☽ □ ♃
01 34	☿ ± ♅
02 54	☽ ± ♂
03 21	☽ ∥ ♂
05 07	☽ ∠ ♆
06 00	☽ ∥ ♄
12 15	☽ ∠ ♅
13 39	☽ ⊼ ♂
20 29	☽ ⊼ ♀
13 Sunday	
00 35	♀ ± ♄
00 41	☽ ⊼ ☉
05 35	☿ □ ♃
07 04	☽ ✶ ♆
07 19	☽ ☌ ♄
08 02	☽ ± ♀
12 03	☽ ± ☉
14 25	☽ ✶ ♅
14 47	☽ △ ♇
14 Monday	
01 44	☽ ⚼ ♀
04 29	☽ Q ♆
05 01	☽ △ ♃
05 28	☽ ⚼ ☉
08 10	☽ # ♄
08 27	☽ ⊼ ☿
12 02	☽ Q ♅
14 20	☽ # ♂
18 01	☽ ⚼ ♇
19 17	☽ # ♇
19 55	☿ ⚼ ♆
21 08	☽ ± ☿
21 55	☽ ☍ ♂
15 Tuesday	
06 17	☿ ⊥ ♂
08 19	☽ △ ♀
08 51	☽ ⚼ ♃
09 48	☽ ∥ ☿
10 50	☽ ± ♇
11 30	☽ △ ☉
13 50	☽ □ ♆
13 57	☽ ⊻ ♄
16 37	☽ ⚼ ☿
21 41	☽ □ ♅
22 15	☽ ⊼ ♇
16 Wednesday	
01 21	☽ ⊥ ♄
02 02	☽ ± ♃
13 41	☽ ⊼ ♃
17 36	☉ ⊼ ♆
18 09	☉ ⊼ ♄
18 43	☽ ∠ ♄
20 19	☽ ∥ ☉
17 Thursday	
02 08	☽ △ ☿
06 38	☿ ⚼ ♅
08 10	♄ ✶ ♆
08 33	☽ ∥ ♀
10 44	☽ ⊼ ♂
15 45	♀ ⊼ ♄
16 00	♀ ⊼ ♆
23 20	☽ ± ♂
18 Friday	
00 16	☽ ✶ ♄
00 19	☽ △ ♆
00 20	♂ ∥ ♇
01 18	☽ □ ♀
03 04	☽ □ ☉
08 31	☽ △ ♅
09 19	☽ ☍ ♇
10 28	☿ # ♂
13 49	☿ # ♇
18 26	☽ ⚼ ♂
19 Saturday	
00 17	☽ Q ♄
01 39	☽ ☍ ♃
06 29	☽ ⚼ ♆
09 04	☿ Q ♇
14 44	☽ ⚼ ♅
17 51	☽ Q ♀
18 36	☽ Q ☉
23 40	☽ □ ☿
20 Sunday	
00 42	☽ ± ♆
02 31	☽ △ ♂
08 56	☽ ± ♅
12 34	☽ □ ♄
12 50	☽ ⊼ ♆
20 50	☉ ⊼ ♅
21 00	☽ ✶ ♀
21 03	☽ ⊼ ♅
21 05	☽ ✶ ☉
21 32	♀ ⊼ ♅
22 04	☽ ⊼ ♇
21 Monday	
00 04	☉ ☌ ♀
00 06	☿ # ♄
07 15	♀ □ ♇
09 20	☉ □ ♇
10 11	☽ ± ♇
10 58	☿ ⊻ ♂
14 38	☽ ⊼ ♃
17 29	☽ Q ☿
22 Tuesday	
02 44	☽ ± ♃
04 16	☽ ⚼ ♇
05 55	☽ ∠ ☉
06 40	☽ ∠ ♀
10 23	♀ ± ♆
18 18	☽ □ ♂
19 58	☉ ± ♆
20 45	☽ ⚼ ♃
20 52	☽ ✶ ☿
23 Wednesday	
00 26	☽ △ ♄
00 55	☽ ☍ ♆
01 21	☽ ⊥ ☉
02 36	☽ ⊥ ♀
08 52	☽ ☍ ♅
10 03	☽ △ ♇
14 11	☽ ⊻ ☉
15 44	☽ ⊻ ♀
24 Thursday	
00 59	☿ ☍ ♄
02 21	☽ △ ♃
05 40	☽ ⚼ ♄
05 43	☿ △ ♆
06 19	☽ ∠ ☿
07 33	☽ Q ♂
16 51	☽ ∥ ♀
21 29	♂ ∠ ♃
22 55	☽ ± ♄
23 10	☿ Q ♃
25 Friday	
00 19	☽ ∥ ☉
01 55	☽ ⊥ ☿
07 51	☽ ✶ ♂
10 18	☽ ⊼ ♄
10 59	☽ ⊼ ♆
14 48	☽ ⊻ ☿
14 55	♀ ± ♅
16 49	☽ # ♂
18 34	☽ ⊼ ♅
19 53	☽ □ ♇
20 54	♂ ⊥ ♇
22 19	☽ ± ♆
26 Saturday	
04 29	☽ # ♇
04 31	☽ ☌ ☉
05 50	☽ ± ♅
07 28	☽ ☌ ♀
11 48	☽ □ ♃
13 36	☽ ∠ ♂
14 32	☽ # ♄
15 09	☽ ⚼ ♆
21 30	☉ ± ♅
22 34	☽ ⚼ ♅
27 Sunday	
00 12	☿ △ ♅
03 29	♂ ⊼ ♄
05 29	☽ Q ♇
07 09	☽ ⊥ ♂
13 12	☿ ✶ ♇
17 55	☽ ☍ ♄
18 38	☽ ∥ ☿
18 44	☽ ⊻ ♂
18 47	☽ △ ♆
19 25	☽ # ☿
19 34	♂ □ ♆
21 12	☽ Q ♃
28 Monday	
02 03	☽ △ ♅
03 29	☽ ✶ ♇
05 07	☽ ☌ ☿
06 52	♀ □ ♃
16 05	☽ ⊻ ☉
19 06	☽ ✶ ♃
20 16	☽ ⊻ ♀
23 31	☽ ∥ ♄
29 Tuesday	
03 44	☽ ⊥ ☉
06 34	☽ ∠ ♇
08 08	☽ ⊥ ♀
08 56	☽ ∥ ♇
09 12	☉ # ♂
14 01	♀ ⚼ ♆
21 02	☽ ∠ ☉
22 04	☽ ∠ ♃
22 37	☽ ⊥ ♇
23 38	☽ ⊼ ♄
30 Wednesday	
00 02	☽ # ☉
00 34	♀ # ♂
00 41	☽ □ ♆
01 47	☽ ∠ ♀
02 11	☽ # ♀
02 30	☽ ∥ ♂
03 26	☽ ☌ ♂
07 42	☽ □ ♅
09 14	☽ ⊻ ♇
10 11	☽ ± ♄
12 51	☉ □ ♃
14 04	☽ ⊥ ♃
16 33	☽ ⊻ ☿
31 Thursday	
00 40	☽ ⊻ ♃
01 29	☽ ✶ ☉
01 54	☽ ⚼ ♄
04 09	☽ ⊥ ☿
05 52	♃ ∥ ♆
06 46	☽ ✶ ♀
08 17	☽ Q ♆
13 56	♀ ∠ ♂
15 09	☽ Q ♅
21 22	☽ ∠ ☿
23 22	☉ ⚼ ♆
23 55	☽ Q ☉

All ephemeris data is given at 12.00 UT and the Moon's longitude is additionally given for 24.00 UT

SEPTEMBER 1995

LONGITUDES

Date	Sidereal time h m s	Sun ☉ ° ' "	Moon ☽ ° ' "	Moon ☽ 24.00 ° ' "	Mercury ☿ ° '	Venus ♀ ° '	Mars ♂ ° '	Jupiter ♃ ° '	Saturn ♄ ° '	Uranus ♅ ° '	Neptune ♆ ° '	Pluto ♇ ° '
01	10 40 43	08 ♍ 35 20	27 ♏ 06 58	04 ♐ 07 26	04 ♎ 23	11 ♍ 44	26 ♎ 12	06 ♐ 51	22 ♓ 20	27 ♑ 01	23 ♑ 04	27 ♏ 59
02	10 44 40	09 33 24	11 ♐ 10 21	18 ♐ 15 38	05 37	12 59	26 51	06 57	22 R 16	26 R 59	23 R 03	27 59
03	10 48 37	10 31 29	25 ♐ 23 03	02 ♑ 32 20	06 48	14 13	27 30	07 02	22 11	26 58	23 02	28 00
04	10 52 33	11 29 36	09 ♑ 43 08	16 ♑ 55 01	07 58	15 28	28 10	07 08	22 07	26 56	23 01	28 01
05	10 56 30	12 27 44	24 ♑ 07 27	01 ♒ 19 50	09 05	16 42	28 49	07 13	22 02	26 55	23 00	28 02
06	11 00 26	13 25 54	08 ♒ 31 32	15 ♒ 41 49	10 10	17 57	29 ♎ 29	07 19	21 58	26 53	23 00	28 03
07	11 04 23	14 24 05	22 ♒ 50 00	29 ♒ 55 24	11 13	19 11	00 ♏ 08	07 25	21 53	26 52	22 59	28 04
08	11 08 19	15 22 18	06 ♓ 57 21	13 ♓ 55 17	12 14	20 25	00 48	07 31	21 49	26 51	22 58	28 05
09	11 12 16	16 20 32	20 ♓ 48 41	27 ♓ 37 11	13 12	21 40	01 28	07 38	21 44	26 49	22 57	28 06
10	11 16 12	17 18 48	04 ♈ 20 29	10 ♈ 58 27	14 07	22 54	02 08	07 44	21 39	26 48	22 56	28 07
11	11 20 09	18 17 06	17 ♈ 31 01	23 ♈ 58 16	15 00	24 09	02 47	07 51	21 35	26 47	22 55	28 08
12	11 24 06	19 15 27	00 ♉ 20 22	06 ♉ 37 37	15 49	25 24	03 27	07 58	21 30	26 46	22 55	28 09
13	11 28 02	20 13 49	12 ♉ 50 22	18 ♉ 59 02	16 35	26 38	04 07	08 05	21 26	26 45	22 54	28 10
14	11 31 59	21 12 13	25 ♉ 04 07	01 ♊ 06 10	17 17	27 53	04 48	08 12	21 21	26 44	22 53	28 12
15	11 35 55	22 10 40	07 ♊ 05 45	13 ♊ 03 28	17 56	29 ♍ 07	05 28	08 19	21 16	26 42	22 53	28 13
16	11 39 52	23 09 08	18 ♊ 59 56	24 ♊ 55 45	18 30	00 ♎ 22	06 08	08 26	21 12	26 41	22 52	28 14
17	11 43 48	24 07 39	00 ♋ 51 35	06 ♋ 48 01	19 00	01 36	06 49	08 34	21 07	26 41	22 52	28 15
18	11 47 45	25 06 12	12 ♋ 45 38	18 ♋ 45 00	19 25	02 51	07 29	08 41	21 02	26 40	22 51	28 17
19	11 51 41	26 04 47	24 ♋ 46 39	00 ♌ 51 04	19 45	04 05	08 10	08 49	20 58	26 39	22 50	28 18
20	11 55 38	27 03 24	06 ♌ 58 41	13 ♌ 09 53	20 00	05 20	08 50	08 57	20 53	26 38	22 50	28 19
21	11 59 35	28 02 03	19 ♌ 24 57	25 ♌ 44 09	20 09	06 35	09 31	09 05	20 49	26 37	22 50	28 21
22	12 03 31	29 00 45	02 ♍ 07 38	08 ♍ 35 30	20 R 11	07 49	10 12	09 13	20 44	26 36	22 49	28 22
23	12 07 28	29 ♍ 59 28	15 ♍ 07 45	21 ♍ 44 18	20 07	09 04	10 53	09 21	20 40	26 36	22 49	28 24
24	12 11 24	00 ♎ 58 14	28 ♍ 25 03	05 ♎ 09 46	19 56	10 19	11 34	09 30	20 35	26 35	22 48	28 25
25	12 15 21	01 57 01	11 ♎ 58 12	18 ♎ 50 04	19 38	11 33	12 15	09 38	20 31	26 35	22 48	28 27
26	12 19 17	02 55 51	25 ♎ 45 00	02 ♏ 42 39	19 12	12 48	12 56	09 47	20 26	26 34	22 48	28 28
27	12 23 14	03 54 42	09 ♏ 42 39	16 ♏ 44 37	18 39	14 03	13 37	09 56	20 22	26 34	22 47	28 30
28	12 27 10	04 53 35	23 ♏ 48 12	00 ♐ 53 01	17 59	15 17	14 18	10 05	20 17	26 33	22 47	28 31
29	12 31 07	05 52 30	07 ♐ 58 46	15 ♐ 05 07	17 12	16 32	14 59	10 14	20 13	26 33	22 47	28 33
30	12 35 04	06 ♎ 51 27	22 ♐ 11 45	29 ♐ 18 23	16 ♎ 18	17 ♎ 46	15 ♏ 41	10 ♐ 23	20 ♓ 09	26 ♑ 32	22 ♑ 47	28 ♏ 35

Date	Moon True ☊ ° '	Moon Mean ☊ ° '	Moon ☽ Latitude ° '
01	27 ♎ 35	28 ♎ 52	02 N 32
02	27 R 36	28 49	03 33
03	27 35	28 46	04 21
04	27 32	28 43	04 54
05	27 27	28 40	05 08
06	27 21	28 36	05 03
07	27 14	28 33	04 39
08	27 08	28 30	03 58
09	27 02	28 27	03 03
10	26 58	28 24	02 00
11	26 56	28 20	00 N 51
12	26 D 56	28 17	00 S 19
13	26 57	28 14	01 26
14	26 59	28 11	02 27
15	27 00	28 08	03 22
16	27 01	28 05	04 07
17	27 R 01	28 01	04 41
18	26 59	27 58	05 03
19	26 57	27 55	05 12
20	26 53	27 52	05 08
21	26 48	27 49	04 49
22	26 43	27 45	04 15
23	26 39	27 42	03 28
24	26 36	27 39	02 28
25	26 34	27 36	01 19
26	26 33	27 33	00 S 04
27	26 D 33	27 30	01 N 12
28	26 35	27 26	02 25
29	26 36	27 23	03 29
30	26 ♎ 37	27 ♎ 20	04 N 21

DECLINATIONS

Date	Sun ☉ ° '	Moon ☽ ° '	Mercury ☿ ° '	Venus ♀ ° '	Mars ♂ ° '	Jupiter ♃ ° '	Saturn ♄ ° '	Uranus ♅ ° '	Neptune ♆ ° '	Pluto ♇ ° '
01	08 N 21	17 S 02	03 S 01	08 N 28	10 S 15	20 S 57	05 S 11	21 S 19	20 S 56	06 S 45
02	07 59	18 36	03 38	08 00	10 29	20 58	05 13	21 19	20 56	06 45
03	07 37	19 00	04 15	07 31	10 44	20 59	05 15	21 19	20 56	06 46
04	07 15	18 12	04 51	07 02	10 59	21 00	05 16	21 20	20 57	06 47
05	06 53	16 14	05 25	06 33	11 14	21 02	05 18	21 20	20 57	06 48
06	06 31	13 15	05 59	06 04	11 28	21 03	05 20	21 20	20 57	06 48
07	06 08	09 31	06 32	05 34	11 43	21 04	05 22	21 20	20 57	06 49
08	05 46	05 17	07 04	05 05	11 58	21 05	05 24	21 21	20 57	06 50
09	05 23	00 S 50	07 34	04 35	12 12	21 06	05 26	21 21	20 57	06 50
10	05 01	03 N 33	08 03	04 05	12 26	21 07	05 28	21 21	20 58	06 51
11	04 38	07 40	08 31	03 35	12 41	21 09	05 30	21 21	20 58	06 52
12	04 15	11 18	08 58	03 05	12 55	21 10	05 31	21 21	20 58	06 52
13	03 52	14 20	09 22	02 35	13 09	21 11	05 33	21 22	20 58	06 53
14	03 29	16 39	09 45	02 04	13 24	21 13	05 35	21 22	20 58	06 54
15	03 06	18 11	10 07	01 34	13 38	21 14	05 37	21 22	20 58	06 55
16	02 43	18 53	10 26	01 04	13 52	21 15	05 39	21 22	20 58	06 55
17	02 20	18 45	10 43	00 33	14 06	21 17	05 41	21 22	20 59	06 56
18	01 57	17 48	10 57	00 N 03	14 19	21 18	05 43	21 22	20 59	06 57
19	01 33	16 03	11 09	00 S 28	14 33	21 19	05 45	21 23	20 59	06 57
20	01 10	13 33	11 18	00 59	14 47	21 21	05 46	21 23	20 59	06 58
21	00 47	10 25	11 25	01 29	15 01	21 22	05 48	21 23	20 59	06 59
22	00 N 24	06 44	11 28	02 00	15 14	21 23	05 50	21 23	20 59	07 00
23	00 00	02 N 40	11 27	02 30	15 27	21 25	05 52	21 23	20 59	07 00
24	00 S 23	01 S 38	11 23	03 01	15 41	21 26	05 54	21 23	20 59	07 01
25	00 47	05 57	11 15	03 31	15 54	21 28	05 55	21 23	20 59	07 02
26	01 10	10 01	11 02	04 02	16 07	21 29	05 57	21 23	20 59	07 03
27	01 33	13 35	10 45	04 32	16 20	21 31	05 59	21 23	20 59	07 03
28	01 57	16 23	10 24	05 02	16 33	21 32	06 01	21 24	21 00	07 04
29	02 20	18 12	09 59	05 33	16 46	21 34	06 02	21 24	21 00	07 05
30	02 S 43	18 S 52	09 S 29	06 S 03	16 S 59	21 S 35	06 S 04	21 S 24	21 S 00	07 S 06

ZODIAC SIGN ENTRIES

Date	h	m	Planets
01	16	57	☽ ♐
03	19	45	☽ ♑
05	21	47	☽ ♒
07	07	00	♂ ♏
08	00	08	☽ ♓
10	04	14	☽ ♈
12	11	21	☽ ♉
14	21	48	☽ ♊
16	05	01	♀ ♎
17	10	16	☽ ♋
19	22	19	☽ ♌
22	08	01	☽ ♍
23	12	13	☉ ♎
24	14	50	☽ ♎
26	19	20	☽ ♏
28	22	30	☽ ♐

LATITUDES

Date	Mercury ☿ ° '	Venus ♀ ° '	Mars ♂ ° '	Jupiter ♃ ° '	Saturn ♄ ° '	Uranus ♅ ° '	Neptune ♆ ° '	Pluto ♇ ° '
01	01 S 23	01 N 25	00 S 08	00 N 31	02 S 20	00 S 34	00 N 32	13 N 17
04	01 50	01 25	00 10	00 30	02 20	00 34	00 32	13 15
07	02 16	01 24	00 13	00 29	02 20	00 34	00 32	13 14
10	02 42	01 23	00 15	00 29	02 21	00 34	00 32	13 12
13	03 06	01 21	00 17	00 28	02 21	00 34	00 32	13 11
16	03 26	01 19	00 19	00 28	02 21	00 34	00 32	13 09
19	03 42	01 16	00 21	00 27	02 21	00 34	00 32	13 08
22	03 52	01 12	00 23	00 27	02 21	00 34	00 32	13 07
25	03 51	01 08	00 25	00 26	02 21	00 34	00 32	13 05
28	03 38	01 04	00 26	00 26	02 21	00 34	00 32	13 04
31	03 S 08	00 N 59	00 S 28	00 N 25	02 S 21	00 S 34	00 N 32	13 N 03

DATA

Julian Date	2449962
Delta T	+61 seconds
Ayanamsa	23° 47' 57"
Synetic vernal point	05° ♓ 19' 03"
True obliquity of ecliptic	23° 26' 16"

MOON'S PHASES, APSIDES AND POSITIONS ☽

Date	h	m	Phase	Longitude ° '	Eclipse Indicator
02	09	03	☽	09 ♐ 26	
09	03	37	○	16 ♓ 00	
16	21	09	☾	23 ♊ 31	
24	16	55	●	01 ♎ 10	

Day	h	m		
05	01	21	Perigee	
17	06	14	Apogee	
30	03	31	Perigee	
03	08	05	Max dec	19° S 01'
09	16	29	0N	
16	20	14	Max dec	18° N 56'
24	02	55	0S	
30	13	29	Max dec	18° S 52'

LONGITUDES

Date	Chiron ⚷ ° '	Ceres ⚳ ° '	Pallas ⚴ ° '	Juno ⚵ ° '	Vesta ⚶ ° '	Black Moon Lilith ⚸ ° '
01	28 ♍ 51	28 ♍ 52	04 ♍ 36	19 ♐ 46	05 ♍ 25	26 ♊ 54
11	00 ♎ 16	03 ♎ 21	09 ♍ 40	21 ♐ 00	10 ♍ 11	28 ♊ 01
21	01 ♎ 43	07 ♎ 51	14 ♍ 40	22 ♐ 41	14 ♍ 59	29 ♊ 08
31	03 ♎ 11	12 ♎ 22	19 ♍ 33	24 ♐ 46	19 ♍ 48	00 ♋ 15

ASPECTARIAN

h m Aspects

01 Friday
03 49 ☽ △ ♄
05 03 ☽ ✶ ♆
05 37 ☽ Q ♀
10 21 ☽ ⊻ ♂
11 50 ☽ ✶ ♅
13 29 ☽ ☌ ♇
17 14 ♀ ⚼ ♅
21 08 ☽ ⊥ ♂

02 Saturday
01 39 ☽ ✶ ☿
04 46 ☽ ☌ ♃
06 42 ☽ ∠ ♆
09 03 ☽ □ ☉
13 12 ☽ ∠ ♂
13 23 ☽ ∠ ♅
13 48 ☉ ∥ ♀
15 21 ☽ □ ♀
16 51 ♂ □ ♅
21 57 ☽ ⊥ ♆
23 55 ☽ Q ☿

03 Sunday
04 34 ☽ ⊥ ♅
06 39 ☽ □ ♄
08 04 ☽ ⊻ ♆
14 39 ☽ ⊻ ♅
15 44 ☽ ✶ ♂
16 24 ☽ ⊻ ♇
17 06 ☿ ✶ ♃

04 Monday
02 28 ☽ ⊥ ♇
06 39 ♂ ⊻ ♇
07 39 ☽ ⊻ ♃
08 49 ☽ □ ☿
10 24 ♂ ± ♄
12 39 ☽ Q ♄
12 46 ☽ Q ♂
15 10 ☽ △ ☉
17 31 ☽ ∠ ♇
17 43 ☽ ⊥ ♃
22 29 ☽ △ ♀
22 44 ☉ ⚼ ♅
22 56 ♀ Q ♇

05 Tuesday
00 18 ♀ ⋕ ♇
06 51 ☿ ∥ ♄
08 33 ☽ ✶ ♄
08 49 ☽ ∠ ♃
10 09 ☽ ☌ ♆
16 38 ☽ ☌ ♅
17 46 ☉ ⋕ ♇
17 57 ☽ ⚼ ☉
18 31 ☽ ✶ ♇
20 11 ☽ □ ♂

06 Wednesday
01 48 ☽ ⚼ ♀
09 24 ☽ ∠ ♄
09 58 ☽ ✶ ♃
10 02 ☽ ± ☉
13 45 ☿ ⋕ ♀
14 33 ☽ Q ♇
14 59 ☽ △ ☿
18 15 ☽ ± ♀
20 48 ☽ ⊼ ☉
23 12 ☽ ∥ ♂

07 Thursday
00 23 ☽ ⊥ ♄
01 41 ☉ ⋕ ☿
05 16 ☽ ⊼ ♀
06 13 ☽ Q ♃
10 25 ☽ ⊻ ♄
12 15 ☽ ⊻ ♆
18 10 ☽ ⚼ ☿
18 48 ☽ ⊻ ♅
20 51 ☽ □ ♇
21 18 ♀ ⋕ ♄
22 23 ☽ ⊥ ♆

08 Friday
00 58 ☽ △ ♂
00 59 ☿ ∥ ♇
03 11 ☽ ∥ ☿
03 27 ☽ ∥ ♇
04 59 ☽ ⊥ ♅
09 04 ☽ ⋕ ☉
10 40 ☽ ± ☿
11 19 ☽ ∥ ♄
12 59 ☽ □ ♃
13 13 ☽ ⋕ ♀
13 44 ☽ ∠ ♆
20 24 ☽ ∠ ♅
21 46 ☽ ⊼ ☿

09 Saturday
03 37 ☽ ☍ ☉
04 02 ☽ ⚼ ♂
05 54 ☉ Q ♇
09 25 ☿ ∠ ♇
09 32 ☉ ⋕ ♄
13 13 ♀ ☍ ♄
13 36 ☽ ☌ ♄
13 39 ☽ ☍ ♀
15 45 ☽ ✶ ♆
19 19 ♂ ⊥ ♃
20 36 ☽ ± ♂
21 18 ☉ ∠ ♂
22 34 ☽ ✶ ♅

10 Sunday
00 52 ☽ △ ♇
07 49 ☽ ⊼ ♂
12 33 ♀ △ ♆
13 04 ☽ Q ♆
14 40 ☽ ∥ ♀
18 11 ☽ △ ♃
19 34 ☽ ∥ ☉
20 02 ☽ Q ♅
22 57 ☽ ⋕ ♄

11 Monday
03 56 ☽ ⚼ ♇
07 02 ☽ ☍ ☿
07 07 ☽ ⋕ ♇
13 32 ☽ ⊼ ☉
18 05 ☽ ⋕ ☿
19 29 ☽ ⊻ ♄
20 35 ☽ ± ♇
21 59 ☽ ⚼ ♃
22 02 ☽ □ ♆

12 Tuesday
01 37 ☽ ± ☉
01 40 ☽ ⊼ ♀
02 20 ♃ ∠ ♆
05 15 ☽ □ ♅
06 40 ☽ ⊥ ♄
07 52 ☽ ⊼ ♇
14 13 ☽ ± ♀
15 07 ☽ ± ♃
18 16 ☽ ☍ ♂
20 05 ☽ ⚼ ☉
23 41 ☽ ∠ ♄

13 Wednesday
00 09 ♀ Q ♃
01 12 ☽ ⋕ ♂
02 42 ☽ ⊼ ♃
09 24 ☽ ⚼ ♀
14 05 ♀ △ ♅
19 45 ☽ ⊼ ☿

14 Thursday
03 42 ☽ △ ☉
04 42 ☽ ✶ ♄
07 42 ☽ △ ♆
08 16 ☽ ± ☿
15 17 ☽ △ ♅
15 19 ☉ ☍ ♄
18 13 ☽ △ ♀
18 13 ☽ ☍ ♇
18 14 ♀ ✶ ♇

15 Friday
03 12 ☽ ⚼ ☿
04 23 ☽ Q ♄
08 32 ☽ ⊼ ♂
13 34 ☽ ⚼ ♆
14 29 ☽ ☍ ♃
21 16 ☽ ⚼ ♅
21 19 ☽ ± ♂

16 Saturday
02 31 ♀ ⊥ ♂
05 05 ☉ △ ♆
07 42 ☽ ± ♆
10 57 ☽ △ ☿
13 52 ♂ ⚼ ♄
15 25 ☽ ± ♅
16 25 ☽ □ ♄
16 35 ☽ ⚼ ♂
19 49 ☽ ⊼ ♆
21 09 ☽ □ ☉

17 Sunday
03 33 ☽ ⊼ ♅
06 44 ☽ ⊼ ♇
13 41 ☽ □ ♀
18 53 ☽ ± ♇

18 Monday
00 44 ☽ △ ♂
03 43 ☽ ⊼ ♃
12 45 ☽ Q ☉
13 03 ☽ ⚼ ♇
15 55 ☽ ± ♃

19 Tuesday
01 45 ☽ □ ☿
04 28 ☽ △ ♄
06 03 ☽ Q ♀
08 09 ☽ ☍ ♆
10 05 ☽ ⚼ ♃
14 48 ☽ ✶ ☉
15 42 ☽ ☍ ♅
18 59 ☽ △ ♇

20 Wednesday
01 43 ☉ △ ♅
01 56 ☽ ⋕ ♂
08 26 ☽ ✶ ♀
09 00 ☉ Q ♃
09 53 ☽ ⚼ ♄
14 01 ☽ Q ☿
15 50 ☽ □ ♂
15 53 ☽ △ ♃
17 02 ♂ ⊻ ♃
17 09 ☉ ⋕ ♀
22 42 ☽ ∠ ☉

21 Thursday
03 14 ☽ ± ♄
05 04 ☽ ⋕ ☿
13 24 ☽ ✶ ☿
14 39 ☽ ⊼ ♄
16 34 ☽ ∠ ♀
17 24 ☽ ⊥ ☉
18 29 ☽ ⊼ ♆
19 53 ☉ ✶ ♇

22 Friday
01 39 ☽ ⊼ ♅
03 01 ♃ ∥ ♅
04 13 ☽ Q ♂
04 57 ☽ □ ♇
05 41 ☽ ⊻ ☉
05 48 ☽ ± ♆
09 14 ☿ St R
10 26 ☽ ⋕ ♇
11 22 ☽ ⊥ ♀
12 54 ☽ ± ♅
17 28 ☽ ⋕ ♄
17 41 ☽ ∠ ☿
22 34 ☽ ⚼ ♆
23 42 ☽ ⊻ ♀

23 Saturday
01 18 ☽ □ ♃
03 47 ☽ ✶ ♂
05 32 ☽ ⚼ ♅
09 45 ♂ Q ♆
10 10 ☽ ⊥ ☿
12 47 ☽ ⋕ ♀
14 19 ☽ Q ♇
18 21 ♀ ✶ ♃
20 58 ☽ ⊻ ☿
22 00 ☽ ☍ ♄

24 Sunday
01 42 ☽ ⋕ ☉
01 56 ☽ △ ♆
04 22 ☽ ∥ ☉
08 30 ☽ ∠ ♂
08 43 ☽ △ ♅
10 20 ☽ Q ♃
12 00 ☽ ✶ ♇
16 55 ☽ ☌ ☉
20 36 ☽ ∥ ♀

25 Monday
01 22 ☽ ⊥ ♂
07 52 ☽ ✶ ♃
11 12 ☽ ☌ ♀
11 51 ☽ ∥ ♄
12 30 ☽ ⊻ ♂
14 36 ☽ ∠ ♇
18 12 ☽ ∥ ♇

26 Tuesday
01 01 ☽ ☌ ☿
02 50 ☽ ⊼ ♄
06 19 ☽ ⊥ ♇
06 53 ☽ □ ♆
10 19 ☽ ∠ ♃
13 11 ☽ ± ♄
13 25 ☽ □ ♅
16 43 ☽ ⊻ ♇
17 33 ♀ ⊻ ♂
18 02 ☽ ∥ ☿

27 Wednesday
01 16 ♀ ∠ ♇
01 19 ☽ ⊻ ☉
01 59 ☽ ⊥ ♃
04 35 ☽ ⚼ ♄
12 22 ☽ ⊥ ☉
12 23 ☽ ⊻ ♃
13 51 ☽ Q ♆
19 00 ☽ ☌ ♂
20 07 ☽ ⊻ ♀
20 17 ☽ Q ♅

28 Thursday
02 35 ☽ ⊻ ☿
04 52 ☽ ∠ ☉
06 04 ☽ △ ♄
07 19 ☽ ⊥ ♀
10 16 ☽ ✶ ♆
12 18 ☽ ⊥ ☿
13 53 ☽ ∥ ♂
16 40 ☽ ✶ ♅
20 01 ☽ ☌ ♇
20 43 ♂ Q ♅

29 Friday
00 03 ☽ ∠ ♀
02 46 ☽ ∠ ☿
08 11 ☽ ✶ ☉
11 40 ☽ ∠ ♆
15 50 ☽ ☌ ♃
18 01 ☽ ∠ ♅
19 40 ☿ ☌ ♀

30 Saturday
00 27 ☽ ⊻ ♂
02 40 ☽ ✶ ☿
02 52 ☽ ⊥ ♆
03 49 ☽ ✶ ♀
05 57 ☽ Q ☉
08 33 ☽ □ ♄
09 13 ☽ ⊥ ♅
11 05 ☽ ⊥ ♂
12 59 ☽ ⊻ ♆
13 10 ♀ ∥ ♄
19 20 ☽ ⊻ ♅
21 07 ☿ ⊻ ♂
21 40 ☽ Q ☿
22 48 ☽ ⊻ ♇

All ephemeris data is given at 12.00 UT and the Moon's longitude is additionally given for 24.00 UT

OCTOBER 1995

LONGITUDES

Date	Sidereal time h m s	Sun ☉ ° ' "	Moon ☽ ° ' "	Moon ☽ 24.00 ° ' "	Mercury ☿ ° '	Venus ♀ ° '	Mars ♂ ° '	Jupiter ♃ ° '	Saturn ♄ ° '	Uranus ♅ ° '	Neptune ♆ ° '	Pluto ♇ ° '
01	12 39 00	07 ♎ 50 26	06 ♑ 24 44	13 ♑ 30 32	15 ♎ 19	19 ♎ 01	16 ♏ 22	10 ♐ 32	20 ♓ 04	26 ♑ 32	22 ♑ 47	28 ♏ 36
02	12 42 57	08 49 26	20 ♑ 35 30	27 ♑ 39 22	14 R 15	20 16	17 04	10 41	20 R 00	26 R 32	22 R 47	28 38
03	12 46 53	09 48 28	04 ♒ 41 50	11 ♒ 42 39	13 08	21 30	17 45	10 51	19 56	26 32	22 46	28 40
04	12 50 50	10 47 32	18 ♒ 41 28	25 ♒ 38 02	11 59	22 45	18 27	11 00	19 52	26 32	22 46	28 42
05	12 54 46	11 46 37	02 ♓ 32 03	09 ♓ 23 12	10 49	24 00	19 09	11 10	19 47	26 32	22 D 46	28 43
06	12 58 43	12 45 44	16 ♓ 11 15	22 ♓ 55 55	09 42	25 14	19 51	11 20	19 43	26 32	22 46	28 45
07	13 02 39	13 44 54	29 ♓ 37 01	06 ♈ 14 20	08 39	26 29	20 33	11 30	19 39	26 D 32	22 47	28 47
08	13 06 36	14 44 05	12 ♈ 47 46	19 ♈ 17 12	07 41	27 44	21 15	11 40	19 35	26 32	22 47	28 49
09	13 10 33	15 43 18	25 ♈ 42 38	02 ♉ 04 05	06 50	28 ♎ 58	21 57	11 50	19 31	26 32	22 47	28 51
10	13 14 29	16 42 33	08 ♉ 21 39	14 ♉ 35 28	06 07	00 ♏ 13	22 39	12 00	19 28	26 32	22 47	28 52
11	13 18 26	17 41 51	20 ♉ 45 46	26 ♉ 52 48	05 35	01 28	23 21	12 10	19 24	26 32	22 47	28 54
12	13 22 22	18 41 10	02 ♊ 56 54	08 ♊ 58 27	05 13	02 42	24 03	12 21	19 20	26 32	22 47	28 56
13	13 26 19	19 40 32	14 ♊ 57 51	20 ♊ 55 35	05 01	03 57	24 45	12 31	19 16	26 33	22 48	28 58
14	13 30 15	20 39 57	26 ♊ 52 08	02 ♋ 48 03	05 D 01	05 12	25 28	12 42	19 13	26 33	22 48	29 00
15	13 34 12	21 39 23	08 ♋ 43 51	14 ♋ 40 09	05 11	06 26	26 10	12 53	19 09	26 34	22 48	29 02
16	13 38 08	22 38 52	20 ♋ 37 30	26 ♋ 36 30	05 32	07 41	26 53	13 03	19 06	26 34	22 49	29 04
17	13 42 05	23 38 23	02 ♌ 37 44	08 ♌ 41 47	06 03	08 56	27 36	13 14	19 02	26 35	22 49	29 06
18	13 46 02	24 37 57	14 ♌ 49 11	21 ♌ 00 28	06 43	10 10	28 18	13 25	18 59	26 35	22 49	29 08
19	13 49 58	25 37 32	27 ♌ 16 05	03 ♍ 36 29	07 32	11 25	29 01	13 36	18 56	26 36	22 50	29 10
20	13 53 55	26 37 10	10 ♍ 02 00	16 ♍ 32 54	08 28	12 40	29 ♏ 44	13 47	18 53	26 37	22 50	29 12
21	13 57 51	27 36 50	23 ♍ 09 23	29 ♍ 51 31	09 31	13 54	00 ♐ 27	13 59	18 50	26 37	22 51	29 14
22	14 01 48	28 36 32	06 ♎ 39 16	13 ♎ 32 29	10 40	15 09	01 10	14 10	18 47	26 38	22 51	29 17
23	14 05 44	29 ♎ 36 17	20 ♎ 30 54	27 ♎ 34 06	11 54	16 24	01 53	14 21	18 44	26 39	22 52	29 19
24	14 09 41	00 ♏ 36 03	04 ♏ 41 36	11 ♏ 52 49	13 13	17 39	02 36	14 33	18 41	26 40	22 53	29 21
25	14 13 37	01 35 52	19 ♏ 07 02	26 ♏ 23 30	14 36	18 53	03 19	14 44	18 38	26 41	22 53	29 23
26	14 17 34	02 35 42	03 ♐ 41 28	11 ♐ 00 08	16 02	20 08	04 02	14 56	18 35	26 42	22 54	29 25
27	14 21 31	03 35 34	18 ♐ 18 43	25 ♐ 36 28	17 31	21 23	04 46	15 07	18 33	26 43	22 55	29 27
28	14 25 27	04 35 28	02 ♑ 52 44	10 ♑ 06 54	19 02	22 37	05 29	15 19	18 30	26 44	22 55	29 30
29	14 29 24	05 35 24	17 ♑ 18 28	24 ♑ 27 01	20 34	23 52	06 13	15 31	18 28	26 45	22 56	29 32
30	14 33 20	06 35 21	01 ♒ 32 14	08 ♒ 33 52	22 09	25 07	06 56	15 43	18 26	26 46	22 57	29 34
31	14 37 17	07 ♏ 35 20	15 ♒ 31 46	22 ♒ 25 51	23 ♎ 44	26 ♏ 21	07 ♐ 40	15 ♐ 55	18 ♓ 23	26 ♑ 47	22 ♑ 58	29 ♏ 36

Moon True ☊, Mean ☊, Latitude and DECLINATIONS

Date	Moon True ☊ ° '	Moon Mean ☊ ° '	Moon ☽ Latitude ° '	Sun ☉ ° '	Moon ☽ ° '	Mercury ☿ ° '	Venus ♀ ° '	Mars ♂ ° '	Jupiter ♃ ° '	Saturn ♄ ° '	Uranus ♅ ° '	Neptune ♆ ° '	Pluto ♇ ° '
01	26 ♎ 38	27 ♎ 17	04 N 56	03 S 07	18 S 21	08 S 55	06 S 32	17 S 11	21 S 37	06 S 06	21 S 24	21 S 00	07 S 06
02	26 R 37	27 14	05 14	03 30	16 41	08 18	07 02	17 24	21 38	06 07	21 24	21 00	07 07
03	26 36	27 11	05 13	03 53	14 02	07 37	07 32	17 36	21 40	06 09	21 24	21 00	07 08
04	26 34	27 07	04 53	04 16	10 35	06 54	08 01	17 48	21 41	06 10	21 24	21 00	07 08
05	26 32	27 04	04 16	04 39	06 35	06 10	08 31	18 00	21 42	06 12	21 24	21 00	07 09
06	26 31	27 01	03 25	05 02	02 S 18	05 26	09 00	18 12	21 44	06 14	21 24	21 00	07 10
07	26 29	26 58	02 23	05 25	02 N 02	04 42	09 29	18 24	21 45	06 15	21 24	21 00	07 11
08	26 28	26 55	01 15	05 48	06 12	04 01	09 57	18 36	21 47	06 17	21 24	21 00	07 11
09	26 28	26 51	00 N 04	06 11	10 00	03 22	10 26	18 47	21 48	06 18	21 24	21 00	07 12
10	26 D 28	26 48	01 S 05	06 34	13 15	02 46	10 54	18 58	21 50	06 20	21 23	21 00	07 13
11	26 29	26 45	02 11	06 57	15 50	02 16	11 22	19 10	21 51	06 21	21 23	21 00	07 14
12	26 29	26 42	03 09	07 19	17 39	01 50	11 50	19 21	21 53	06 22	21 23	21 00	07 14
13	26 30	26 39	03 58	07 42	18 39	01 30	12 17	19 32	21 54	06 24	21 23	21 00	07 15
14	26 31	26 36	04 36	08 04	18 48	01 15	12 45	19 42	21 56	06 25	21 23	21 00	07 16
15	26 31	26 32	05 02	08 26	18 07	01 06	13 11	19 53	21 57	06 26	21 23	21 00	07 17
16	26 31	26 29	05 16	08 49	16 39	01 03	13 38	20 04	21 59	06 28	21 23	21 00	07 17
17	26 R 31	26 26	05 16	09 11	14 27	01 05	14 04	20 14	22 00	06 29	21 23	21 00	07 18
18	26 31	26 23	05 01	09 33	11 35	01 12	14 30	20 24	22 02	06 30	21 23	21 00	07 19
19	26 31	26 20	04 33	09 54	08 09	01 23	14 56	20 34	22 03	06 31	21 23	21 00	07 19
20	26 D 31	26 17	03 50	10 16	04 15	01 40	15 21	20 44	22 05	06 32	21 22	21 00	07 20
21	26 31	26 13	02 55	10 37	00 N 02	01 59	15 46	20 53	22 06	06 33	21 22	21 00	07 21
22	26 31	26 10	01 48	10 59	04 S 18	02 23	16 10	21 03	22 08	06 34	21 22	21 00	07 22
23	26 31	26 07	00 S 33	11 20	08 32	02 49	16 34	21 12	22 09	06 36	21 22	20 59	07 22
24	26 R 31	26 04	00 N 45	11 41	12 23	03 18	16 57	21 21	22 11	06 37	21 22	20 59	07 23
25	26 31	26 01	02 02	12 02	15 32	03 50	17 21	21 30	22 12	06 37	21 21	20 59	07 24
26	26 31	25 57	03 12	12 22	17 45	04 23	17 43	21 39	22 14	06 38	21 21	20 59	07 24
27	26 30	25 54	04 09	12 43	18 47	04 58	18 05	21 47	22 15	06 39	21 21	20 59	07 25
28	26 29	25 51	04 51	13 03	18 34	05 34	18 27	21 56	22 16	06 40	21 21	20 59	07 26
29	26 28	25 48	05 13	13 23	17 09	06 11	18 48	22 04	22 18	06 41	21 21	20 59	07 26
30	26 28	25 45	05 16	13 43	14 41	06 49	19 09	22 12	22 19	06 42	21 20	20 59	07 27
31	26 ♎ 28	25 ♎ 42	05 N 00	14 S 02	11 S 24	07 S 28	19 S 29	22 S 19	22 S 21	06 S 42	21 S 20	20 S 59	07 S 28

ZODIAC SIGN ENTRIES

Date	h m	Planets
01	01 10	☽ ♑
03	03 59	☽ ♒
05	07 35	☽ ♓
07	12 41	☽ ♈
09	20 05	☽ ♉
10	07 48	♀ ♏
12	06 10	☽ ♊
14	18 20	☽ ♋
17	06 46	☽ ♌
19	17 11	☽ ♍
20	21 02	♂ ♐
22	00 15	☽ ♎
23	21 32	☉ ♏
24	04 06	☽ ♏
26	05 56	☽ ♐
28	07 15	☽ ♑
30	09 23	☽ ♒

LATITUDES

Date	Mercury ☿ ° '	Venus ♀ ° '	Mars ♂ ° '	Jupiter ♃ ° '	Saturn ♄ ° '	Uranus ♅ ° '	Neptune ♆ ° '	Pluto ♇ ° '
01	03 S 08	00 N 59	00 S 28	00 N 25	02 S 21	00 S 34	00 N 32	13 N 03
04	02 22	00 53	00 30	00 25	02 21	00 33	00 31	13 01
07	01 23	00 48	00 32	00 24	02 21	00 33	00 31	13 00
10	00 S 22	00 41	00 34	00 24	02 20	00 33	00 31	12 59
13	00 N 33	00 35	00 35	00 23	02 20	00 33	00 31	12 58
16	01 15	00 28	00 37	00 23	02 20	00 33	00 31	12 57
19	01 44	00 21	00 39	00 23	02 20	00 33	00 31	12 56
22	02 00	00 13	00 40	00 22	02 19	00 33	00 31	12 55
25	02 05	00 N 06	00 42	00 22	02 19	00 33	00 31	12 54
28	02 02	00 S 02	00 43	00 21	02 18	00 33	00 31	12 54
31	01 N 53	00 S 09	00 S 45	00 N 21	02 S 18	00 S 33	00 N 31	12 N 53

DATA

Julian Date	2449992
Delta T	+61 seconds
Ayanamsa	23° 47' 59"
Synetic vernal point	05° ♓ 19' 00"
True obliquity of ecliptic	23° 26' 16"

MOON'S PHASES, APSIDES AND POSITIONS ☽

Date	h m	Phase	Longitude ° '	Eclipse Indicator
01	14 36	☽	07 ♑ 57	
08	15 52	○	14 ♈ 54	
16	16 26	☾	22 ♋ 50	
24	04 36	●	00 ♏ 18	Total
30	21 17	☽	06 ♒ 59	

Day	h m		
15	02 01	Apogee	
26	20 55	Perigee	
07	00 41	0N	
14	04 18	Max dec	18° N 51'
21	12 13	0S	
27	19 50	Max dec	18° S 51'

LONGITUDES

Date	Chiron ⚷ ° '	Ceres ⚳ ° '	Pallas ⚴ ° '	Juno ⚵ ° '	Vesta ⚶ ° '	Black Moon Lilith ⚸ ° '
01	03 ♎ 11	12 ♎ 22	19 ♍ 33	24 ♐ 46	19 ♍ 48	00 ♋ 15
11	04 ♎ 37	16 ♎ 53	24 ♍ 20	27 ♐ 11	24 ♍ 37	01 ♋ 22
21	06 ♎ 03	21 ♎ 23	29 ♍ 01	29 ♐ 54	29 ♍ 26	02 ♋ 29
31	07 ♎ 25	25 ♎ 52	03 ♎ 35	02 ♑ 53	04 ♎ 14	03 ♋ 36

ASPECTARIAN

h m Aspects

01 Sunday
02 01 ☽ Q ♀
03 03 ☽ ∠ ♂
08 57 ☽ ⊥ ♇
14 36 ☽ □ ☉
14 47 ☽ Q ♄
19 03 ☽ ⊻ ♃

02 Monday
00 11 ☽ ∠ ♇
02 01 ☽ □ ☿
04 25 ☽ ∥ ♂
05 19 ☽ ⊥ ♃
05 43 ☽ ⚹ ♂
07 11 ♀ ⊼ ♄
11 00 ☽ ⚹ ♄
11 23 ☽ □ ♀
15 42 ☽ ☌ ♆
15 56 ♀ ∥ ♇
20 45 ☽ ∠ ♃
22 05 ☽ ☌ ♅

03 Tuesday
00 58 ☿ ∠ ♇
01 42 ☽ ⚹ ♇
03 08 ☽ Q ♂
12 24 ☽ ∠ ♄
13 44 ☿ ∥ ♀
21 24 ☽ △ ☉
22 14 ☽ Q ♇
22 39 ☽ ⚹ ♃

04 Wednesday
01 20 ☽ △ ☿
03 44 ☽ ⊥ ♄
04 18 ☿ ∥ ♇
05 51 ☿ ⊥ ♂
10 49 ♀ ⊥ ♇
11 34 ☽ □ ♂
12 26 ♀ □ ♆
14 00 ☽ ⊻ ♄
18 18 ☉ ⚹ ♃
19 03 ☽ ⊻ ♆
19 32 ☽ Q ♃
19 42 ☽ △ ♀

05 Thursday
01 13 ☽ ⚼ ☉
01 13 ☽ ⚼ ☿
01 17 ☉ ☌ ☿
01 33 ☽ ⊻ ♅
01 58 ☽ ∥ ♀
03 58 ♆ St D
05 21 ☽ □ ♇
05 27 ☽ ⊥ ♆
05 40 ☿ ⚹ ♃
08 45 ☽ ∥ ♇
11 06 ☿ ∥ ♄
11 59 ☽ ⊥ ♅
14 13 ☽ ∥ ♄
14 54 ☽ ∥ ☿
15 42 ☽ ± ☿
18 06 ☽ ± ☉
21 10 ☽ ∠ ♆
22 02 ☽ ∥ ☉

06 Friday
00 27 ☽ ⚼ ♀
01 25 ☽ ⊼ ☿
03 19 ☽ □ ♃
03 46 ☽ ∠ ♅
05 28 ☽ ⊼ ☉
08 11 ♂ △ ♄
12 57 ♅ St D
17 58 ☽ ± ♀
18 15 ☽ ☌ ♄
18 51 ☽ △ ♂
20 25 ☉ ∥ ☿
20 50 ♀ ± ♄
23 43 ☽ ⚹ ♆

07 Saturday
05 47 ☽ ⊼ ♀
06 26 ☽ ⚹ ♅
10 29 ☽ △ ♇
12 17 ♀ ∠ ♃
12 48 ♀ □ ♅
12 50 ☉ ∠ ♇
16 12 ♃ ∠ ♅
21 20 ☽ Q ♆
23 20 ☽ ⚼ ♂

08 Sunday
00 58 ☽ ⋕ ☿
03 14 ☽ ☍ ☿
04 10 ☽ Q ♅
09 23 ☽ ⋕ ☉
09 54 ☽ △ ♃
12 26 ☽ ⋕ ♄
13 52 ☽ ⚼ ♇
15 52 ☽ ☍ ☉
16 46 ☽ ± ♂
17 59 ☽ ⋕ ♇

09 Monday
00 30 ☽ ⊻ ♄
04 32 ☽ ⊼ ♂
06 30 ☽ □ ♆
06 37 ☽ ± ♇
09 26 ♀ ⊻ ♇
10 09 ☿ ∠ ♂
11 39 ☽ ⊥ ♄
13 32 ☽ □ ♅
14 08 ☽ ⚼ ♃
15 24 ☽ ⋕ ♀
17 55 ☽ ⊼ ♇
18 49 ☽ ☍ ♀
19 40 ☉ ∥ ♄

10 Tuesday
04 35 ☽ ∠ ♄
06 37 ☉ ⊥ ♂
07 26 ☽ ± ♃
07 56 ☽ ⊼ ☿
16 42 ♂ ⚹ ♆
18 54 ☽ ± ☿
19 06 ☽ ⊼ ♃

11 Wednesday
05 30 ☽ ⊼ ☉
09 21 ☽ ⚹ ♄
11 39 ☽ ⚼ ☿
15 57 ☽ △ ♆
17 22 ☽ ☍ ♂
18 15 ☽ ± ☉
23 20 ☽ △ ♅

12 Thursday
04 02 ☽ ☍ ♇
06 36 ☉ ∥ ♇
08 49 ☽ Q ♄
11 28 ☽ ⊼ ♀
13 36 ☽ ⚼ ☉
16 24 ☽ △ ☿
21 38 ☽ ⚼ ♆

13 Friday
00 47 ☽ ± ♀
02 49 ☉ ⊼ ♄
05 09 ☽ ⚼ ♅
07 02 ☽ ☍ ♃
15 41 ☽ ± ♆
17 56 ♀ ⚼ ♄
20 38 ☽ □ ♄
20 57 ☽ ⚼ ♀
22 20 ☽ △ ☉
23 14 ☽ ± ♅

14 Saturday
00 48 ☿ St D
03 46 ☽ ⊼ ♆
08 20 ☿ ⊻ ♀
08 59 ☽ ⊼ ♂
11 22 ☽ ⊼ ♅
16 20 ☽ ⊼ ♇
21 53 ☽ ± ♂

15 Sunday
04 30 ☽ ± ♇
04 41 ☽ □ ☿
06 49 ☽ △ ♀
17 15 ☽ ⚼ ♂
20 30 ☽ ⊼ ♃
21 50 ♀ ⊥ ♃
22 45 ☽ ⚼ ♇

16 Monday
01 15 ♂ ⚹ ♅
08 48 ☽ ± ♃
08 56 ☽ △ ♄
15 57 ☉ □ ♆
16 24 ☽ ☍ ♆
16 26 ☽ □ ☉
18 05 ☽ Q ☿
22 34 ☉ ⊥ ♇
23 56 ☽ ☍ ♅

17 Tuesday
01 20 ☽ △ ♂
03 07 ☽ ⚼ ♃
04 58 ☽ △ ♇
14 47 ☽ ⚼ ♄
14 59 ☽ ⋕ ♀
19 09 ☽ ⚹ ☿

18 Wednesday
01 52 ☽ □ ♀
07 21 ☽ Q ☉
08 26 ☽ ± ♄
09 13 ☽ △ ♃
20 03 ☽ ⊼ ♄
20 05 ☉ ± ♄

19 Thursday
00 37 ♀ Q ♆
01 16 ☽ ⋕ ☉
02 16 ☽ ∠ ☿
03 30 ☽ ⊼ ♆
08 36 ☽ ⚹ ☉
10 43 ☽ ⊼ ♅
14 59 ☽ ± ♆
15 31 ☽ □ ♂
15 38 ☽ □ ♇
16 32 ☽ Q ♀
17 14 ☽ ⋕ ♇
17 29 ♂ ☌ ♇
20 42 ☽ ⊥ ☿
22 06 ☽ ± ♅
22 15 ☽ ⋕ ♄

20 Friday
07 55 ☽ ⚼ ♆
08 51 ☽ ⊻ ☿
11 44 ☉ □ ♅
14 55 ☽ ⚼ ♅
15 11 ☽ ∠ ☉
17 23 ☽ ⚹ ♀
19 02 ☽ □ ♃

21 Saturday
01 14 ☽ Q ♇
01 52 ☽ ⋕ ☿
02 58 ☽ Q ♂
04 11 ☽ ☍ ♄
08 59 ☽ ⊥ ☉
11 27 ☽ △ ♆
13 32 ♀ ⊻ ♃
18 14 ☽ △ ♅
20 38 ☽ ⊻ ☉
22 56 ☽ ⚹ ♇
23 21 ☽ ∠ ♀

22 Sunday
00 19 ☽ ∥ ☿
01 46 ☽ ⚹ ♂
01 54 ♀ Q ♅
03 30 ♂ ∥ ♆
03 59 ☽ Q ♃
16 48 ☽ ⊥ ♀
19 41 ☽ ☌ ☿

23 Monday
00 49 ☽ ∥ ♄
01 15 ☽ ⚹ ♃
01 18 ☽ ∠ ♇
04 14 ☽ ⊻ ♀
04 31 ☉ ∠ ♃
04 41 ☉ ⊻ ♇
05 17 ☽ ∥ ♇
05 25 ☽ ∠ ♂
08 57 ☽ ⊼ ♄
16 01 ☽ □ ♆
16 47 ☽ ⊥ ♇
19 10 ☽ ± ♄
21 38 ☽ ⊥ ♂
22 27 ☽ □ ♅

24 Tuesday
02 59 ☽ ⊻ ♇
03 13 ☽ ∠ ♃
04 36 ☽ ☌ ☉
06 56 ☽ ∥ ☉
08 18 ☽ ⊻ ♂
10 18 ☽ ⚼ ♄
13 15 ♂ ∥ ♅
18 32 ☽ ⊥ ♃
22 20 ☽ Q ♆

25 Wednesday
03 43 ☽ ⊻ ☿
04 38 ☽ Q ♅
04 39 ☽ ⊻ ♃
07 18 ♀ △ ♄
08 14 ☿ ∠ ♇
11 12 ☽ △ ♄
11 35 ☽ ☌ ♀
14 43 ☽ ⊥ ☿
14 43 ☿ ⚹ ♃
18 14 ☽ ⚹ ♆

26 Thursday
00 29 ☽ ⚹ ♅
04 58 ☽ ☌ ♇
07 09 ☽ ∠ ☿
10 04 ☽ ⊻ ☉
11 31 ☽ ∥ ♀
12 36 ☽ ☌ ♂
18 55 ☽ ∠ ♆
20 38 ☽ ⊥ ☉

27 Friday
01 09 ☽ ∠ ♅
06 42 ☽ ☌ ♃
09 42 ☽ ⊥ ♆
10 32 ☽ ⚹ ☿
10 56 ☉ ⚼ ♄
12 23 ☽ □ ♄
12 30 ☽ ∠ ☉
15 57 ☽ ⊥ ♅
17 30 ☽ ⊻ ♀
19 34 ☽ ⊻ ♆

28 Saturday
01 50 ☽ ⊻ ♅
04 02 ☿ ⊼ ♄
04 19 ☽ ⊥ ♀
06 24 ☽ ⊻ ♇
08 35 ☽ Q ☿
14 11 ☽ ∥ ♀
15 03 ☽ ⚹ ☉
16 20 ☽ ⊥ ♇
16 32 ☽ ⊻ ♂
17 56 ♀ ⚹ ♆
17 59 ☽ Q ♄
20 36 ☽ ∠ ♀

29 Sunday
03 02 ☽ ⊥ ♂
07 21 ☽ ∠ ♇
08 58 ☽ ⊻ ♃
12 30 ☽ Q ☉
13 56 ☽ ⚹ ♄
18 09 ☽ □ ☿
18 54 ☽ ∠ ♂
19 10 ☽ ⊥ ♃
21 28 ☽ ☌ ♆

30 Monday
00 04 ☽ ⚹ ♀
03 54 ☽ ☌ ♅
06 08 ☿ ∠ ♂
07 03 ☿ ∥ ♄
08 39 ☽ ⚹ ♇
10 35 ☽ ∠ ♃
15 12 ☽ ∠ ♄
18 55 ☽ ∥ ☉
21 17 ☽ □ ☉
21 43 ☽ ⚹ ♂
22 26 ☽ Q ♀

31 Tuesday
00 21 ☿ □ ♆
05 12 ☽ Q ♇
06 36 ☽ ⊥ ♄
10 00 ☿ ⊥ ♇
11 48 ☿ ∥ ♇
12 41 ☽ ⚹ ♃
16 57 ☽ ⊻ ♄
17 08 ♂ ∥ ♃
18 23 ☉ ⊻ ♂
19 34 ☽ Q ♂
20 33 ♀ ⚹ ♅
21 36 ☿ ± ♄
22 18 ♂ ∠ ♆

All ephemeris data is given at 12.00 UT and the Moon's longitude is additionally given for 24.00 UT

NOVEMBER 1995

LONGITUDES

Date	Sidereal time h m s	Sun ☉ ° ' "	Moon ☽ ° ' "	Moon ☽ 24.00 ° ' "	Mercury ☿ ° '	Venus ♀ ° '	Mars ♂ ° '	Jupiter ♃ ° '	Saturn ♄ ° '	Uranus ♅ ° '	Neptune ♆ ° '	Pluto ⯓ ° '
01	14 41 13	08 ♏ 35 20	29 ♒ 16 07	06 ♓ 02 33	25 ♎ 20	27 ♏ 36	08 ♐ 23	16 ♐ 07	18 ♓ 21	26 ♑ 49	22 ♑ 59	29 ♏ 39
02	14 45 10	09 35 22	12 ♓ 45 15	19 ♓ 24 17	26 57	28 ♏ 51	09 07	16 19	18 R 19	26 50	23 00	29 41
03	14 49 06	10 35 25	25 ♓ 59 44	02 ♈ 31 43	28 ♎ 35	00 ♐ 05	09 51	16 31	18 17	26 51	23 01	29 43
04	14 53 03	11 35 31	09 ♈ 00 22	15 ♈ 25 46	00 ♏ 13	01 20	10 35	16 44	18 15	26 53	23 02	29 45
05	14 57 00	12 35 37	21 ♈ 48 01	28 ♈ 07 15	01 51	02 35	11 19	16 56	18 14	26 54	23 03	29 48
06	15 00 56	13 35 46	04 ♉ 23 34	10 ♉ 37 05	03 29	03 49	12 03	17 08	18 12	26 56	23 04	29 50
07	15 04 53	14 35 56	16 ♉ 47 54	22 ♉ 56 10	05 07	05 04	12 47	17 21	18 11	26 57	23 05	29 52
08	15 08 49	15 36 08	29 ♉ 02 01	05 ♊ 05 39	06 46	06 18	13 31	17 33	18 09	26 59	23 06	29 55
09	15 12 46	16 36 23	11 ♊ 07 15	17 ♊ 07 03	08 24	07 33	14 15	17 46	18 08	27 01	23 07	29 57
10	15 16 42	17 36 38	23 ♊ 05 18	29 ♊ 02 19	10 01	08 48	14 59	17 58	18 07	27 02	23 08	29 ♏ 59
11	15 20 39	18 36 56	04 ♋ 58 26	10 ♋ 54 02	11 39	10 02	15 43	18 11	18 05	27 04	23 10	00 ♐ 02
12	15 24 35	19 37 16	16 ♋ 49 32	22 ♋ 45 22	13 17	11 17	16 28	18 24	18 04	27 06	23 11	00 04
13	15 28 32	20 37 38	28 ♋ 42 03	04 ♌ 40 05	14 54	12 32	17 12	18 36	18 03	27 08	23 12	00 06
14	15 32 28	21 38 01	10 ♌ 40 01	16 ♌ 42 25	16 31	13 46	17 57	18 49	18 03	27 10	23 13	00 09
15	15 36 25	22 38 26	22 ♌ 47 53	28 ♌ 57 00	18 07	15 01	18 41	19 02	18 02	27 12	23 15	00 11
16	15 40 22	23 38 54	05 ♍ 10 20	11 ♍ 28 28	19 44	16 15	19 26	19 15	18 01	27 14	23 16	00 13
17	15 44 18	24 39 23	17 ♍ 51 55	24 ♍ 21 09	21 20	17 30	20 10	19 28	18 01	27 16	23 17	00 16
18	15 48 15	25 39 54	00 ♎ 56 36	07 ♎ 38 35	22 56	18 45	20 55	19 41	18 00	27 18	23 19	00 18
19	15 52 11	26 40 27	14 ♎ 27 18	21 ♎ 22 49	24 32	19 59	21 40	19 54	18 00	27 20	23 20	00 21
20	15 56 08	27 41 01	28 ♎ 25 03	05 ♏ 33 45	26 07	21 14	22 25	20 07	18 00	27 22	23 22	00 23
21	16 00 04	28 41 37	12 ♏ 48 27	20 ♏ 08 33	27 42	22 28	23 10	20 20	18 00	27 24	23 23	00 25
22	16 04 01	29 ♏ 42 15	27 ♏ 33 14	05 ♐ 01 33	29 ♏ 17	23 43	23 55	20 33	18 D 00	27 26	23 25	00 28
23	16 07 57	00 ♐ 42 55	12 ♐ 32 23	20 ♐ 04 34	00 ♐ 52	24 57	24 40	20 46	18 00	27 29	23 26	00 30
24	16 11 54	01 43 35	27 ♐ 36 53	05 ♑ 08 07	02 27	26 12	25 25	21 00	18 00	27 31	23 28	00 33
25	16 15 51	02 44 17	12 ♑ 37 09	20 ♑ 02 56	04 01	27 27	26 10	21 13	18 00	27 33	23 29	00 35
26	16 19 47	03 45 00	27 ♑ 24 36	04 ♒ 41 26	05 36	28 41	26 55	21 26	18 01	27 36	23 31	00 37
27	16 23 44	04 45 45	11 ♒ 52 54	18 ♒ 58 39	07 10	29 ♐ 56	27 40	21 39	18 01	27 38	23 33	00 40
28	16 27 40	05 46 30	25 ♒ 58 30	02 ♓ 52 26	08 44	01 ♑ 10	28 26	21 53	18 02	27 41	23 34	00 42
29	16 31 37	06 47 16	09 ♓ 40 33	16 ♓ 23 04	10 18	02 25	29 11	22 06	18 03	27 43	23 36	00 44
30	16 35 33	07 ♐ 48 03	23 ♓ 00 14	29 ♓ 32 27	11 ♐ 52	03 ♑ 39	29 ♐ 56	22 ♐ 20	18 ♓ 04	27 ♑ 46	23 ♑ 38	00 ♐ 47

DECLINATIONS

Date	Moon True ☊ ° '	Moon Mean ☊ ° '	Moon ☽ Latitude ° '	Sun ☉ ° '	Moon ☽ ° '	Mercury ☿ ° '	Venus ♀ ° '	Mars ♂ ° '	Jupiter ♃ ° '	Saturn ♄ ° '	Uranus ♅ ° '	Neptune ♆ ° '	Pluto ⯓ ° '
01	26 ♎ 28	25 ♎ 38	04 N 27	14 S 22	07 S 34	08 S 07	19 S 49	22 S 27	22 S 22	06 S 43	21 S 20	20 S 59	07 S 28
02	26 D 29	25 35	03 39	14 41	03 S 24	08 46	20 08	22 34	22 23	06 44	21 20	20 58	07 29
03	26 31	25 32	02 41	15 00	00 N 52	09 26	20 27	22 41	22 25	06 44	21 19	20 58	07 30
04	26 32	25 29	01 36	15 19	05 02	10 05	20 45	22 48	22 26	06 45	21 19	20 58	07 30
05	26 32	25 26	00 N 26	15 37	08 54	10 44	21 02	22 55	22 27	06 46	21 19	20 58	07 31
06	26 R 32	25 23	00 S 43	15 55	12 18	11 23	21 19	23 01	22 29	06 46	21 18	20 58	07 32
07	26 31	25 19	01 50	16 13	15 06	12 02	21 36	23 07	22 30	06 46	21 18	20 58	07 32
08	26 29	25 16	02 50	16 31	17 07	12 40	21 51	23 13	22 31	06 47	21 18	20 58	07 33
09	26 26	25 13	03 42	16 48	18 27	13 18	22 06	23 19	22 33	06 47	21 17	20 57	07 34
10	26 23	25 10	04 23	17 05	18 53	13 55	22 21	23 25	22 34	06 48	21 17	20 57	07 34
11	26 19	25 07	04 53	17 22	18 28	14 32	22 34	23 30	22 35	06 48	21 17	20 57	07 35
12	26 15	25 03	05 10	17 38	17 15	15 08	22 48	23 35	22 36	06 48	21 16	20 57	07 35
13	26 13	25 00	05 14	17 54	15 18	15 43	23 00	23 40	22 38	06 48	21 16	20 57	07 36
14	26 11	24 57	05 04	18 10	12 41	16 17	23 12	23 45	22 39	06 48	21 15	20 56	07 37
15	26 D 10	24 54	04 41	18 26	09 29	16 51	23 23	23 49	22 40	06 49	21 15	20 56	07 37
16	26 11	24 51	04 04	18 41	05 49	17 24	23 33	23 53	22 41	06 49	21 15	20 56	07 38
17	26 12	24 48	03 15	18 56	01 N 48	17 56	23 43	23 57	22 42	06 49	21 14	20 56	07 38
18	26 14	24 44	02 14	19 10	02 S 26	18 27	23 52	24 00	22 43	06 49	21 14	20 56	07 39
19	26 15	24 41	01 S 04	19 25	06 41	18 58	24 01	24 03	22 44	06 49	21 13	20 55	07 40
20	26 R 16	24 38	00 N 12	19 39	10 44	19 27	24 08	24 07	22 46	06 49	21 13	20 55	07 40
21	26 15	24 35	01 29	19 52	14 16	19 55	24 15	24 09	22 47	06 48	21 12	20 55	07 41
22	26 12	24 32	02 43	20 05	16 58	20 23	24 21	24 12	22 48	06 48	21 12	20 55	07 41
23	26 08	24 29	03 46	20 18	18 34	20 49	24 27	24 14	22 49	06 48	21 12	20 55	07 42
24	26 03	24 25	04 34	20 30	18 51	21 15	24 31	24 16	22 50	06 48	21 11	20 54	07 42
25	25 57	24 22	05 02	20 42	17 49	21 39	24 35	24 18	22 51	06 47	21 11	20 54	07 43
26	25 52	24 19	05 11	20 54	15 36	22 02	24 38	24 20	22 52	06 47	21 10	20 54	07 43
27	25 48	24 16	04 59	21 05	12 26	22 25	24 41	24 21	22 53	06 47	21 10	20 54	07 44
28	25 46	24 13	04 29	21 16	08 38	22 46	24 43	24 22	22 53	06 46	21 09	20 53	07 44
29	25 D 46	24 09	03 43	21 26	04 29	23 06	24 44	24 23	22 54	06 46	21 09	20 53	07 45
30	25 ♎ 46	24 ♎ 06	02 N 49	21 S 37	00 S 12	23 S 24	24 S 44	24 S 23	22 S 55	06 S 45	21 S 08	20 S 53	07 S 45

ZODIAC SIGN ENTRIES

Date	h	m	Planets
01	13	17	☽ ♓
03	10	18	♀ ♐
03	19	21	☽ ♈
04	08	50	☿ ♏
06	03	35	☽ ♉
08	13	55	☽ ♊
10	19	11	⯓ ♐
11	01	57	☽ ♋
13	14	37	☽ ♌
16	02	02	☽ ♍
18	10	18	☽ ♎
20	14	40	☽ ♏
22	15	56	☽ ♐
22	19	01	☉ ♐
22	22	46	☿ ♐
24	15	48	☽ ♑
26	16	15	☽ ♒
27	13	23	♀ ♑
28	18	59	☽ ♓
30	13	58	♂ ♑

LATITUDES

Date	Mercury ☿ ° '	Venus ♀ ° '	Mars ♂ ° '	Jupiter ♃ ° '	Saturn ♄ ° '	Uranus ♅ ° '	Neptune ♆ ° '	Pluto ⯓ ° '
01	01 N 48	00 S 12	00 S 45	00 N 21	02 S 18	00 S 33	00 N 31	12 N 53
04	01 34	00 20	00 47	00 21	02 17	00 33	00 31	12 52
07	01 16	00 27	00 48	00 20	02 17	00 33	00 31	12 52
10	00 57	00 35	00 49	00 20	02 16	00 33	00 30	12 51
13	00 37	00 43	00 51	00 19	02 16	00 33	00 30	12 51
16	00 N 17	00 50	00 52	00 19	02 15	00 33	00 30	12 50
19	00 S 04	00 57	00 53	00 19	02 15	00 32	00 30	12 50
22	00 24	01 04	00 54	00 18	02 14	00 32	00 30	12 50
25	00 43	01 11	00 55	00 18	02 13	00 32	00 30	12 49
28	01 01	01 17	00 56	00 18	02 13	00 32	00 30	12 49
31	01 S 18	01 S 22	00 S 57	00 N 18	02 S 12	00 S 32	00 N 30	12 N 49

DATA

Julian Date	2450023
Delta T	+61 seconds
Ayanamsa	23° 48' 02"
Synetic vernal point	05° ♓ 18' 57"
True obliquity of ecliptic	23° 26' 15"

MOON'S PHASES, APSIDES AND POSITIONS ☽

Date	h	m	Phase	Longitude ° '	Eclipse Indicator
07	07	21	○	14 ♉ 24	
15	11	40	☾	22 ♌ 38	
22	15	43	●	29 ♏ 52	
29	06	28	☽	06 ♓ 33	

Day	h	m	
11	20	53	Apogee
23	22	54	Perigee

Day	h	m		
03	07	05	0N	
10	12	10	Max dec	18° N 53'
17	22	19	0S	
24	05	10	Max dec	18° S 55'
30	13	06	0N	

LONGITUDES

Date	Chiron ⚷ ° '	Ceres ⚳ ° '	Pallas ⚴ ° '	Juno ⚵ ° '	Vesta ⚶ ° '	Black Moon Lilith ⚸ ° '
01	07 ♎ 33	26 ♎ 19	04 ♎ 02	03 ♑ 12	04 ♎ 43	03 ♋ 43
11	08 ♎ 51	00 ♏ 45	08 ♎ 26	06 ♑ 26	09 ♎ 28	04 ♋ 50
21	10 ♎ 03	05 ♏ 08	12 ♎ 42	09 ♑ 51	14 ♎ 11	05 ♋ 57
31	11 ♎ 09	09 ♏ 27	16 ♎ 47	13 ♑ 26	18 ♎ 49	07 ♋ 05

All ephemeris data is given at 12.00 UT and the Moon's longitude is additionally given for 24.00 UT

ASPECTARIAN

01 Wednesday
00 57 ☽ ⊻ ♆
04 10 ☽ △ ☿
07 40 ☽ ⊻ ♅
08 46 ☽ □ ♀
09 08 ☽ ∥ ☿
09 56 ☽ Q ♃
11 30 ☽ ⊥ ♆
12 31 ☽ ∥ ⯓
12 40 ☽ □ ⯓
16 57 ☽ ∥ ♄
18 16 ☽ ⊥ ♅

02 Thursday
03 28 ☽ ∠ ♆
05 07 ☽ □ ♂
05 52 ☽ △ ☉
10 08 ☿ □ ♅
10 21 ☽ ∠ ♅
10 22 ☽ ♇ ☿
18 31 ☽ □ ♃
22 01 ☽ ☌ ♄

03 Friday
04 39 ♀ ☌ ⯓
04 54 ☽ ± ☿
06 33 ☽ ⚹ ♆
09 56 ☉ ⊥ ♃
11 12 ☽ ♇ ☉
13 35 ☽ ⚹ ♅
17 25 ☽ ⊼ ☿
18 51 ☽ △ ⯓
20 18 ☽ △ ♀
22 18 ☉ Q ♆

04 Saturday
04 37 ☽ Q ♆
05 06 ☿ ⊻ ⯓
05 08 ☽ ± ☉
11 46 ☽ Q ♅
15 06 ☽ △ ♂
17 14 ☽ ⊼ ☉
22 24 ☽ # ♄
22 46 ☽ ♇ ⯓

05 Sunday
02 40 ☽ △ ♃
03 07 ☽ # ⯓
03 10 ☽ ♇ ♀
05 17 ☽ ⊻ ♄
06 02 ♀ ∥ ♆
13 21 ☿ ∠ ♃
14 22 ☽ □ ♆
15 47 ☽ ± ⯓
16 35 ☽ ⊥ ♄
21 05 ☽ ♇ ♂
21 42 ☽ □ ♅
22 03 ☽ ± ♀

06 Monday
02 49 ♃ ⊥ ♆
03 14 ☽ ⊼ ⯓
03 29 ☽ # ☿
07 36 ☽ ♇ ♃
07 53 ☿ ♇ ♄
08 13 ♂ ∠ ♅
09 43 ☽ ∠ ♄
10 00 ☽ ☍ ☿
10 42 ♀ ∥ ♅
10 47 ☽ ⊼ ♀
15 22 ☽ ± ♂

07 Tuesday
01 13 ☽ ± ♃
03 41 ☽ ⊼ ♂
07 21 ☽ ☍ ☉
08 23 ☿ ⊻ ♀
13 05 ☽ ⊼ ♃
14 41 ☽ ⚹ ♄
20 48 ☉ Q ♅

08 Wednesday
00 18 ☽ △ ♆
01 44 ☽ # ☉
07 57 ☽ △ ♅
13 44 ☽ ☍ ⯓
14 12 ☽ Q ♄

09 Thursday
04 04 ☽ ☍ ♀
05 42 ☽ ⊼ ☿
06 00 ☽ ♇ ♆
08 08 ☿ ⊥ ♂
13 47 ☽ ♇ ♅
18 40 ☽ ☍ ♂
19 34 ☽ ± ☿
23 08 ♀ ∠ ♆
23 59 ☽ ⊼ ☉

10 Friday
00 01 ☽ ± ♆
01 32 ☽ ☍ ♃
02 00 ☽ □ ♄
07 52 ☽ ± ♅
12 06 ☽ ⊼ ♆
13 09 ☽ ± ☉
16 31 ☽ ♇ ☿
19 59 ☽ ⊼ ♅
22 51 ☉ ⊻ ♃
23 40 ☉ △ ♄

11 Saturday
01 58 ☽ ⊼ ⯓
02 30 ♃ □ ♄
04 39 ☿ Q ♆
09 00 ☽ ♇ ☉
13 15 ♀ ∥ ♃
14 08 ☽ ± ⯓
20 59 ☿ ⊥ ♃
23 27 ☽ ⊼ ♀

12 Sunday
03 40 ☽ △ ☿
06 50 ☽ # ☉
08 26 ☽ ♇ ⯓
11 13 ☽ ⊼ ♂
13 02 ☽ ± ♀
14 31 ☽ △ ♄
15 14 ☽ ⊼ ♃
18 11 ☽ △ ☉

13 Monday
00 10 ☽ ± ♂
00 53 ☽ ☍ ♆
03 35 ☽ ± ♃
04 11 ♀ ∠ ♅
08 28 ☽ # ☿
08 50 ☽ ☍ ♅
09 21 ☽ ♇ ♀
11 56 ♂ ⊥ ♆
14 50 ☽ △ ⯓
15 34 ☿ Q ♅
19 31 ☽ ♇ ♂
20 45 ☽ ♇ ♄
22 03 ☽ ♇ ♃

14 Tuesday
14 44 ☽ ± ♄
15 10 ♂ □ ♄
18 53 ☽ △ ♀

15 Wednesday
01 24 ☽ □ ☿
02 38 ☽ ⊼ ♄
03 23 ☽ △ ♂
04 28 ☽ △ ♃
10 38 ☿ △ ♄
11 40 ☽ □ ☉
12 52 ☽ ⊼ ♆
20 37 ☽ ⊼ ♅

16 Thursday
00 31 ☽ # ⯓
00 36 ☽ ± ♆
02 26 ☽ □ ⯓
02 41 ☉ ⚹ ♆
03 37 ☿ ⊻ ♂
03 39 ☿ ⊻ ♃
03 44 ♂ ☌ ♃
05 46 ☽ # ♄
08 15 ☽ ± ♅
17 36 ☽ Q ☿
17 55 ☽ ♇ ♆

17 Friday
01 23 ☽ Q ☉
01 28 ☽ ♇ ♅
07 52 ♀ ⊥ ♆
11 15 ☽ □ ♀
12 16 ☽ ☍ ♄
12 45 ☽ Q ⯓
15 01 ☽ □ ♃
16 33 ☽ □ ♂
19 21 ☽ ⚹ ☿
21 50 ♀ □ ♄
22 04 ☽ △ ♆

18 Saturday
01 36 ☽ ⚹ ☉
05 22 ☽ △ ♅
10 50 ☽ ⚹ ⯓
17 47 ☿ ⚹ ♆
23 27 ☽ Q ♀

19 Sunday
00 16 ☽ Q ♃
00 44 ♂ ⊥ ♅
02 11 ☽ ∠ ☿
03 06 ☽ Q ♂
06 44 ☽ ∠ ☉
09 54 ♀ ☌ ♃
12 43 ☽ ∥ ♄
13 33 ☽ ∠ ⯓
17 37 ☽ ∥ ⯓
18 10 ☽ ⊼ ♄
20 00 ☽ ⊥ ☿
21 36 ☽ ⚹ ♃
22 32 ☽ ⚹ ♀
23 38 ☽ ⊥ ☉

20 Monday
01 12 ☽ ⚹ ♂
03 09 ♀ ∥ ♂
03 23 ☽ □ ♆
04 11 ☉ ⚹ ♅
04 29 ☽ ± ♄
05 07 ☽ ⊥ ⯓
07 36 ☽ ⊻ ☿
10 13 ☽ □ ♅
10 40 ☽ ⊻ ☉
14 45 ♀ ⊥ ♅
15 20 ☽ ⊻ ⯓
19 43 ☽ ♇ ♄
23 26 ☽ ∠ ♃

21 Tuesday
02 21 ☽ ∠ ♀
03 54 ☽ ∠ ♂
06 30 ☉ ∥ ☿
07 19 ☿ ⚹ ♅
09 39 ☽ Q ♆
14 33 ☽ ⊥ ♃
16 16 ☽ Q ♅
18 34 ☽ ⊥ ♀
19 26 ♂ ⊻ ♆
19 31 ☽ ⊥ ♂
19 48 ♄ St D
20 30 ☽ △ ♄

22 Wednesday
00 30 ☽ ⊻ ♃
05 14 ☽ ⊻ ♀
05 18 ☽ ⚹ ♆
05 48 ☽ ⊻ ♂
05 59 ♀ ⊻ ♆
11 49 ☽ ⚹ ♅
15 08 ☽ ☌ ☿
15 43 ☽ ☌ ☉
16 42 ☽ ☌ ⯓
21 30 ♀ ☌ ♂

23 Thursday
05 26 ☉ ☌ ☿
05 26 ☽ ∠ ♆
06 16 ☿ ☌ ⯓
06 45 ☉ ☌ ⯓
11 54 ☽ ∠ ♅
16 48 ☿ ∥ ♆
19 49 ☽ ⊥ ♆
20 42 ☽ □ ♄

24 Friday
01 18 ☽ ☌ ♃
02 16 ☽ ⊥ ♅
05 23 ☽ ⊻ ♆
08 18 ☽ ☌ ♂
08 34 ☿ ∥ ♅
09 33 ☽ ☌ ♀
11 51 ☽ ⊻ ♅
16 41 ☽ ⊻ ⯓
19 02 ☽ ⊻ ☉
20 37 ☽ ⊻ ☿

25 Saturday
01 23 ☽ Q ♄
02 17 ☽ ⊥ ⯓
05 19 ☽ ⊥ ☉
07 20 ☽ ⊥ ☿
14 15 ♀ ⊻ ♅
16 47 ☽ ∠ ⯓
20 42 ☽ ⚹ ♄
20 52 ☽ ∠ ☉
23 34 ☽ ∠ ☿

26 Sunday
02 06 ☽ ⊻ ♃
05 37 ☽ ☌ ♆
11 09 ☽ ⊻ ♂
11 40 ☉ ∥ ♆
12 02 ☽ ⊥ ♃
12 18 ☽ ☌ ♅
14 17 ☽ ⊻ ♀
17 17 ☽ ⚹ ⯓
21 14 ☽ ∠ ♄
21 34 ☽ ⊥ ♂
23 14 ☽ ⚹ ☉

27 Monday
01 07 ☽ ⊥ ♀
03 08 ☽ ∠ ♃
03 09 ☽ ⚹ ☿
09 23 ♃ ⊥ ♅
10 51 ♂ ⊻ ♅
12 14 ☽ ⊥ ♄
13 19 ☽ Q ⯓
13 24 ☽ ∠ ♂
17 37 ☽ ∠ ♀
20 52 ☽ Q ☉
21 17 ☉ ∥ ♅
22 23 ☽ ⊻ ♄

28 Tuesday
01 52 ☽ Q ☿
02 38 ♀ ⊻ ⯓
04 51 ☽ ⚹ ♃
07 51 ☽ ⊻ ♆
09 26 ☿ ∠ ♆
14 57 ☽ ⊻ ♅
16 29 ☽ ⚹ ♂
17 19 ☽ ∥ ⯓
18 15 ☽ ⊥ ♆
20 14 ☽ □ ⯓
21 35 ☿ ∥ ♃
21 55 ☽ ⚹ ♀
22 58 ☽ ∥ ♄

29 Wednesday
01 27 ☽ ⊥ ♅
01 59 ☽ Q ♃
06 28 ☽ □ ☉
10 05 ☽ ∠ ♆
13 15 ☽ □ ☿
14 50 ☽ Q ♂
17 26 ☽ ∠ ♅
21 19 ☽ Q ♀

30 Thursday
03 01 ☽ ☌ ♄
10 45 ☽ □ ♃
13 08 ☽ ⚹ ♆
20 45 ☽ ⚹ ♅

DECEMBER 1995

LONGITUDES

Date	Sidereal time h m s	Sun ☉ ° ′ ″	Moon ☽ ° ′ ″	Moon ☽ 24.00 ° ′ ″	Mercury ☿ ° ′	Venus ♀ ° ′	Mars ♂ ° ′	Jupiter ♃ ° ′	Saturn ♄ ° ′	Uranus ♅ ° ′	Neptune ♆ ° ′	Pluto ♇ ° ′
01	16 39 30	08 ♐ 48 51	06 ♈ 00 03	12 ♈ 23 28	13 ♐ 26	04 ♑ 54	00 ♑ 42	22 ♐ 33	18 ♓ 05	27 ♑ 48	23 ♑ 39	00 ♐ 49
02	16 43 26	09 49 40	18 ♈ 43 04	24 ♈ 59 15	14 59	06 08	01 27	22 46	18 06	27 51	23 41	00 52
03	16 47 23	10 50 29	01 ♉ 12 24	07 ♉ 22 50	16 33	07 23	02 13	23 00	18 07	27 54	23 43	00 54
04	16 51 20	11 51 20	13 ♉ 30 53	19 ♉ 36 48	18 07	08 37	02 58	23 13	18 08	27 56	23 45	00 56
05	16 55 16	12 52 12	25 ♉ 40 51	01 ♊ 43 15	19 40	09 52	03 44	23 27	18 10	27 59	23 46	00 59
06	16 59 13	13 53 05	07 ♊ 44 12	13 ♊ 43 52	21 14	11 06	04 30	23 41	18 11	28 02	23 48	01 01
07	17 03 09	14 53 59	19 ♊ 42 27	25 ♊ 40 05	22 48	12 20	05 15	23 54	18 13	28 04	23 50	01 03
08	17 07 06	15 54 54	01 ♋ 36 58	07 ♋ 33 16	24 21	13 35	06 01	24 08	18 15	28 07	23 52	01 06
09	17 11 02	16 55 50	13 ♋ 29 11	19 ♋ 24 57	25 55	14 49	06 47	24 21	18 16	28 10	23 54	01 08
10	17 14 59	17 56 47	25 ♋ 20 50	01 ♌ 17 07	27 28	16 03	07 33	24 35	18 18	28 13	23 56	01 10
11	17 18 55	18 57 45	07 ♌ 14 09	13 ♌ 12 18	29 ♐ 02	17 18	08 19	24 49	18 20	28 16	23 58	01 13
12	17 22 52	19 58 44	19 ♌ 12 00	25 ♌ 13 43	00 ♑ 35	18 32	09 05	25 02	18 23	28 19	24 00	01 15
13	17 26 49	20 59 44	01 ♍ 17 57	07 ♍ 25 14	02 09	19 46	09 51	25 16	18 25	28 22	24 02	01 17
14	17 30 45	22 00 45	13 ♍ 36 07	19 ♍ 51 14	03 42	21 01	10 37	25 29	18 27	28 25	24 04	01 19
15	17 34 42	23 01 47	26 ♍ 11 07	02 ♎ 36 23	05 15	22 15	11 23	25 43	18 30	28 28	24 06	01 22
16	17 38 38	24 02 50	09 ♎ 07 35	15 ♎ 45 13	06 48	23 29	12 09	25 57	18 32	28 31	24 08	01 24
17	17 42 35	25 03 55	22 ♎ 29 42	29 ♎ 21 21	08 21	24 44	12 56	26 11	18 35	28 34	24 10	01 26
18	17 46 31	26 05 00	06 ♏ 20 22	13 ♏ 26 46	09 54	25 58	13 42	26 24	18 37	28 37	24 12	01 28
19	17 50 28	27 06 06	20 ♏ 40 21	28 ♏ 00 44	11 26	27 12	14 28	26 38	18 40	28 40	24 14	01 31
20	17 54 24	28 07 13	05 ♐ 27 15	12 ♐ 59 04	12 58	28 26	15 14	26 52	18 43	28 43	24 16	01 33
21	17 58 21	29 ♐ 08 20	20 ♐ 35 03	28 ♐ 13 57	14 30	29 ♑ 40	16 01	27 05	18 46	28 46	24 18	01 35
22	18 02 18	00 ♑ 09 29	05 ♑ 54 20	13 ♑ 34 43	16 01	00 ♒ 54	16 47	27 19	18 49	28 50	24 20	01 37
23	18 06 14	01 10 37	21 ♑ 13 36	28 ♑ 49 35	17 31	02 09	17 34	27 33	18 53	28 53	24 22	01 39
24	18 10 11	02 11 46	06 ♒ 21 22	13 ♒ 47 52	19 00	03 23	18 20	27 46	18 56	28 56	24 24	01 42
25	18 14 07	03 12 55	21 ♒ 08 13	28 ♒ 21 48	20 28	04 37	19 07	28 00	18 59	28 59	24 27	01 44
26	18 18 04	04 14 04	05 ♓ 28 13	12 ♓ 27 19	21 54	05 51	19 53	28 14	19 03	29 03	24 29	01 46
27	18 22 00	05 15 13	19 ♓ 19 06	26 ♓ 03 49	23 19	07 05	20 40	28 27	19 06	29 06	24 31	01 48
28	18 25 57	06 16 22	02 ♈ 41 47	09 ♈ 13 26	24 41	08 19	21 27	28 41	19 10	29 09	24 33	01 50
29	18 29 53	07 17 31	15 ♈ 39 18	21 ♈ 59 55	26 02	09 33	22 13	28 55	19 14	29 13	24 35	01 52
30	18 33 50	08 18 40	28 ♈ 15 53	04 ♉ 27 46	27 19	10 46	23 00	29 08	19 18	29 16	24 37	01 54
31	18 37 47	09 ♑ 19 49	10 ♉ 36 08	16 ♉ 41 33	28 ♑ 33	12 ♒ 00	23 ♑ 47	29 ♐ 22	19 ♓ 22	29 ♑ 19	24 ♑ 40	01 ♐ 56

Moon True ☊, Mean ☊, Latitude

Date	Moon True ☊ ° ′	Moon Mean ☊ ° ′	Moon ☽ Latitude ° ′
01	25 ♎ 48	24 ♎ 03	01 N 45
02	25 D 49	24 00	00 N 38
03	25 R 49	23 57	00 S 29
04	25 47	23 54	01 34
05	25 43	23 50	02 34
06	25 37	23 47	03 26
07	25 29	23 44	04 09
08	25 19	23 41	04 40
09	25 08	23 38	04 59
10	24 58	23 34	05 05
11	24 49	23 31	04 58
12	24 42	23 28	04 38
13	24 37	23 25	04 06
14	24 35	23 22	03 21
15	24 D 35	23 19	02 26
16	24 35	23 15	01 22
17	24 36	23 12	00 S 11
18	24 R 35	23 09	01 N 02
19	24 33	23 06	02 14
20	24 27	23 03	03 20
21	24 19	23 00	04 12
22	24 10	22 56	04 47
23	23 59	22 53	05 02
24	23 49	22 50	04 55
25	23 41	22 47	04 29
26	23 35	22 44	03 46
27	23 32	22 40	02 51
28	23 31	22 37	01 48
29	23 D 31	22 34	00 N 42
30	23 R 31	22 31	00 S 25
31	23 ♎ 30	22 ♎ 28	01 S 29

DECLINATIONS

Date	Sun ☉ ° ′	Moon ☽ ° ′	Mercury ☿ ° ′	Venus ♀ ° ′	Mars ♂ ° ′	Jupiter ♃ ° ′	Saturn ♄ ° ′	Uranus ♅ ° ′	Neptune ♆ ° ′	Pluto ♇ ° ′
01	21 S 46	04 N 00	23 S 42	24 S 43	24 S 23	22 S 56	06 S 45	21 S 08	20 S 52	07 S 46
02	21 55	07 56	23 58	24 42	24 23	22 57	06 44	21 07	20 52	07 46
03	22 04	11 26	24 14	24 40	24 23	22 58	06 43	21 06	20 52	07 47
04	22 12	14 24	24 28	24 37	24 22	22 59	06 43	21 06	20 52	07 47
05	22 20	16 41	24 40	24 33	24 22	22 59	06 42	21 05	20 51	07 47
06	22 28	18 12	24 51	24 29	24 20	23 00	06 41	21 05	20 51	07 48
07	22 35	18 54	25 01	24 24	24 19	23 01	06 40	21 04	20 51	07 48
08	22 42	18 46	25 10	24 18	24 17	23 02	06 40	21 04	20 50	07 49
09	22 48	17 47	25 17	24 12	24 15	23 02	06 39	21 03	20 50	07 49
10	22 53	16 04	25 23	24 05	24 13	23 03	06 38	21 03	20 50	07 49
11	22 59	13 39	25 28	23 57	24 11	23 04	06 37	21 02	20 50	07 50
12	23 04	10 39	25 31	23 48	24 08	23 04	06 36	21 01	20 49	07 50
13	23 08	07 11	25 32	23 38	24 05	23 05	06 35	21 01	20 49	07 51
14	23 12	03 N 21	25 33	23 28	24 02	23 05	06 34	21 00	20 49	07 51
15	23 15	00 S 43	25 31	23 17	23 58	23 06	06 33	20 59	20 48	07 51
16	23 18	04 52	25 28	23 06	23 54	23 06	06 31	20 59	20 48	07 52
17	23 21	08 56	25 24	22 54	23 50	23 07	06 30	20 58	20 48	07 52
18	23 23	12 39	25 18	22 41	23 46	23 07	06 29	20 58	20 47	07 52
19	23 24	15 46	25 11	22 27	23 41	23 08	06 28	20 57	20 47	07 53
20	23 25	17 56	25 02	22 13	23 36	23 08	06 26	20 56	20 47	07 53
21	23 26	18 55	24 52	21 58	23 31	23 09	06 25	20 56	20 46	07 53
22	23 26	18 31	24 40	21 43	23 25	23 09	06 24	20 55	20 46	07 53
23	23 26	16 47	24 27	21 27	23 19	23 09	06 22	20 54	20 45	07 54
24	23 25	13 55	24 12	21 10	23 13	23 10	06 21	20 54	20 45	07 54
25	23 24	10 12	23 56	20 52	23 07	23 10	06 19	20 53	20 45	07 54
26	23 22	06 00	23 38	20 35	23 01	23 10	06 18	20 52	20 44	07 54
27	23 20	01 S 36	23 20	20 16	22 54	23 11	06 16	20 52	20 44	07 55
28	23 17	02 N 44	23 00	19 57	22 47	23 11	06 14	20 51	20 44	07 55
29	23 14	06 48	22 39	19 37	22 39	23 11	06 13	20 50	20 43	07 55
30	23 11	10 28	22 17	19 17	22 32	23 11	06 11	20 49	20 43	07 55
31	23 S 07	13 N 35	21 S 54	18 S 56	22 S 24	23 S 11	06 S 09	20 S 49	20 S 43	07 S 55

ZODIAC SIGN ENTRIES

Date	h	m	Planets
01	00	51	☽ ♈
03	09	40	☽ ♉
05	20	35	☽ ♊
08	08	44	☽ ♋
10	21	24	☽ ♌
12	02	57	☿ ♑
13	09	26	☽ ♍
15	19	09	☽ ♎
18	01	07	☽ ♏
20	03	13	☽ ♐
21	18	23	♀ ♒
22	02	46	☽ ♑
22	08	17	☉ ♑
24	01	52	☽ ♒
26	02	45	☽ ♓
28	07	06	☽ ♈
30	15	21	☽ ♉

LATITUDES

Date	Mercury ☿ ° ′	Venus ♀ ° ′	Mars ♂ ° ′	Jupiter ♃ ° ′	Saturn ♄ ° ′	Uranus ♅ ° ′	Neptune ♆ ° ′	Pluto ♇ ° ′
01	01 S 18	01 S 22	00 S 57	00 N 18	02 S 12	00 S 32	00 N 30	12 N 49
04	01 34	01 28	00 58	00 17	02 12	00 32	00 30	12 49
07	01 47	01 33	00 59	00 17	02 11	00 32	00 30	12 49
10	01 58	01 37	01 00	00 17	02 11	00 32	00 30	12 49
13	02 07	01 41	01 01	00 16	02 10	00 32	00 30	12 50
16	02 13	01 44	01 01	00 16	02 10	00 32	00 30	12 50
19	02 15	01 46	01 02	00 16	02 09	00 32	00 30	12 50
22	02 12	01 48	01 03	00 16	02 09	00 32	00 30	12 50
25	02 05	01 49	01 03	00 15	02 08	00 32	00 29	12 51
28	01 50	01 50	01 04	00 15	02 07	00 32	00 29	12 51
31	01 S 29	01 S 49	01 S 04	00 N 15	02 S 07	00 S 32	00 N 29	12 N 52

DATA

Julian Date	2450053
Delta T	+61 seconds
Ayanamsa	23° 48′ 06″
Synetic vernal point	05° ♓ 18′ 53″
True obliquity of ecliptic	23° 26′ 15″

MOON'S PHASES, APSIDES AND POSITIONS ☽

Date	h	m	Phase	Longitude ° ′	Eclipse Indicator
07	01	27	○	14 ♊ 27	
15	05	31	☾	22 ♍ 45	
22	02	22	●	29 ♐ 45	
28	19	06	☽	06 ♈ 34	

Day	h	m		
09	09	58	Apogee	
22	09	59	Perigee	
07	19	50	Max dec	18° N 57′
15	07	51	0S	
21	17	15	Max dec	18° S 57′
27	20	48	0N	

LONGITUDES

Date	Chiron ⚷ ° ′	Ceres ⚳ ° ′	Pallas ⚴ ° ′	Juno ⚵ ° ′	Vesta ⚶ ° ′	Black Moon Lilith ⚸ ° ′
01	11 ♎ 09	09 ♏ 27	16 ♎ 47	13 ♑ 26	18 ♎ 49	07 ♋ 05
11	12 ♎ 07	13 ♏ 40	20 ♎ 40	17 ♑ 10	23 ♎ 22	08 ♋ 12
21	12 ♎ 55	17 ♏ 47	24 ♎ 19	21 ♑ 01	27 ♎ 48	09 ♋ 19
31	13 ♎ 34	21 ♏ 45	27 ♎ 42	24 ♑ 58	02 ♏ 04	10 ♋ 26

ASPECTARIAN

01 Friday
01 31 ☽ □ ♂
02 10 ☿ ∠ ♅
02 20 ☽ △ ♇
08 07 ☉ ∠ ♆
09 43 ☽ □ ♀
11 21 ☽ Q ♆
16 10 ♂ ⊻ ♇
17 43 ☽ △ ☉
19 09 ☽ Q ♅

02 Saturday
03 55 ☽ △ ☿
04 31 ☽ # ♄
06 33 ☽ bQ ♇
10 49 ☽ ⊻ ♄
11 00 ☽ # ♇
11 13 ♀ Q ♄
19 54 ☽ △ ♃
21 31 ☽ □ ♆
22 18 ☽ ⊥ ♄
23 47 ☽ ± ♇

03 Sunday
00 44 ☽ bQ ☉
02 26 ♀ ⊥ ♇
05 34 ☽ □ ♅
11 24 ☽ ⊼ ♇
12 46 ☽ bQ ☿
14 05 ☽ △ ♂
15 43 ☽ ∠ ♄
19 41 ☽ ± ☉

04 Monday
01 20 ☽ △ ♀
01 27 ☽ bQ ♃
03 27 ☿ ‖ ♂
06 12 ☿ ⊥ ♆
08 27 ☽ ⊼ ☉
08 51 ☽ ± ☿
12 25 ☿ □ ♄
19 26 ☽ ± ♃
21 07 ☽ ⚹ ♄
21 21 ☽ bQ ♂
22 22 ☽ ⊼ ☿

05 Tuesday
02 04 ☿ ‖ ♀
07 30 ☽ ⊼ ♃
08 13 ☽ △ ♆
10 11 ☽ bQ ♀
14 47 ☉ ∠ ♅
16 21 ☽ ± ♂
16 35 ☽ △ ♅
20 55 ☽ Q ♄
22 33 ☽ ☍ ♇

06 Wednesday
05 06 ☽ ⊼ ♂
06 08 ☽ ± ♀
14 08 ☽ bQ ♆
19 30 ☽ ⊼ ♀
22 38 ☽ bQ ♅

07 Thursday
00 37 ☿ ⊥ ♅
01 27 ☽ ☍ ☉
03 48 ♃ ⊻ ♆
08 14 ☽ ± ♆
09 00 ☽ □ ♄
16 47 ☽ ± ♅
19 09 ☽ ☍ ☿
20 20 ☽ ⊼ ♆
20 36 ☽ ☍ ♃

08 Friday
04 22 ☿ ⊻ ♆
04 55 ☽ ⊼ ♅
07 58 ☿ ☌ ♃
10 57 ☽ ⊼ ♇
17 16 ♀ ‖ ♂
19 16 ♂ Q ♄
21 31 ☽ ☍ ♂
23 06 ☽ ± ♇

09 Saturday
15 01 ☽ ☍ ♀
17 22 ☽ bQ ♇
19 37 ☽ ⊼ ☉
21 43 ☽ △ ♄
23 31 ♂ ⊥ ♇

10 Sunday
08 54 ☽ ± ☉
09 08 ☽ ☍ ♆
10 25 ☽ ⊼ ♃
11 36 ☉ ⊥ ♆
14 16 ♀ ∠ ♇
16 57 ☽ ⊼ ☿
17 50 ☽ ☍ ♅
20 47 ☉ □ ♄
22 47 ☽ ± ♃
23 49 ☽ △ ♇
23 52 ☿ ⊻ ♅

11 Monday
04 07 ☽ bQ ♄
04 47 ☽ bQ ☉
06 53 ☽ ± ☿
14 19 ☽ ⊼ ♂
17 17 ☽ bQ ♃
22 18 ☽ ± ♄

12 Tuesday
03 12 ☽ ± ♂
03 42 ☽ bQ ☿
08 48 ♀ ⚹ ♄
10 21 ☽ ⊼ ♄
10 31 ☽ ⊼ ♀
13 42 ☽ △ ☉
15 53 ☉ ‖ ♃
21 35 ☽ ⊼ ♆
22 23 ☽ bQ ♂
22 26 ☿ ⊻ ♇
23 50 ☽ ± ♀
23 50 ☽ △ ♃

13 Wednesday
06 11 ☽ ⊼ ♅
07 39 ☽ # ♇
09 29 ☽ ± ♆
11 59 ☽ □ ♇
13 54 ☽ △ ☿
15 56 ☽ # ♄
18 03 ☽ ± ♅
19 36 ☽ bQ ♀

14 Thursday
03 10 ☽ bQ ♆
05 50 ☽ △ ♂
11 38 ☽ bQ ♅
21 21 ☽ ☍ ♄
21 57 ☉ ⊥ ♅
23 01 ☽ Q ♇

15 Friday
03 45 ☽ △ ♀
05 31 ☽ □ ☉
08 03 ☽ △ ♆
11 06 ☽ □ ♃
15 41 ☉ ‖ ♀
16 18 ☽ △ ♅
21 43 ☽ ⚹ ♇

16 Saturday
07 11 ☽ □ ☿
07 39 ☿ Q ♄
11 03 ♀ ‖ ♃
13 57 ☉ ⊻ ♆
17 45 ☽ Q ☉
17 51 ☽ □ ♂
20 54 ☽ Q ♃
21 23 ☿ ⊥ ♇
21 35 ☽ ‖ ♄

17 Sunday
00 45 ♀ ☌ ♆
01 12 ☽ ∠ ♇
05 02 ☽ ⊼ ♄
05 35 ☽ ‖ ♇
14 56 ☽ □ ♆
15 41 ☽ ± ♄
16 19 ☽ □ ♀
16 53 ☽ ⚹ ☉
17 11 ☽ ⊥ ♇
18 34 ☽ ⚹ ♃
19 38 ☽ Q ☿
22 40 ☽ □ ♅

18 Monday
03 35 ☽ Q ♂
03 38 ☽ ⊻ ♇
07 21 ☽ bQ ♄
18 47 ☽ ⚹ ☿
20 39 ☽ ∠ ☉
20 43 ☽ ∠ ♃
21 43 ☉ ☌ ♃
21 56 ☽ Q ♆
22 29 ♀ ⊻ ♃

19 Tuesday
01 08 ☽ ⚹ ♂
01 18 ☉ ⊻ ♀
02 05 ☽ Q ♀
05 21 ☽ Q ♅
08 41 ☽ △ ♄
11 56 ☽ ⊥ ♃
12 46 ☽ ⊥ ☉
17 51 ☽ ⚹ ♆
21 55 ☽ ⊻ ♃
22 33 ☽ ∠ ☿
23 18 ☽ ⊻ ☉
23 39 ☽ ⚹ ♀

20 Wednesday
01 07 ☽ ⚹ ♅
03 09 ☽ ∠ ♂
05 42 ☽ ☌ ♇
14 42 ☽ ⊥ ☿
17 49 ♀ ☌ ♅
18 06 ☽ ∠ ♆
18 23 ☽ ⊥ ♂

21 Thursday
01 13 ☽ ∠ ♅
01 20 ☽ ⊻ ☿
01 51 ☽ ∠ ♀
02 57 ☉ ⊻ ♅
04 25 ☽ ⊻ ♂
08 24 ☽ ⊥ ♆
09 08 ☽ □ ♄
15 27 ☽ ⊥ ♅
17 17 ☽ ⊥ ♀
17 51 ☽ ⊻ ♆
22 22 ☽ ☌ ♃

22 Friday
00 54 ☽ ⊻ ♅
02 22 ☽ ☌ ☉
03 30 ☽ ⊻ ♀
05 17 ☽ ⊻ ♇
06 33 ♂ ∠ ♇
07 36 ☉ ‖ ♂
13 26 ☽ Q ♄
14 41 ☽ ⊥ ♇
21 57 ☿ ∠ ♇

23 Saturday
02 16 ♀ ⚹ ♇
04 48 ☽ ∠ ♇
05 32 ☽ ☌ ☿
05 56 ☽ ☌ ♂
08 18 ☽ ⚹ ♄
13 41 ☿ ☌ ♂
16 58 ☽ ☌ ♆
22 07 ☽ ⊻ ♃
23 42 ☉ ⊻ ♇

24 Sunday
00 08 ☽ ☌ ♅
04 32 ☽ ⚹ ♇
04 52 ☽ ⊻ ☉
06 49 ☽ ☌ ♀
07 48 ☽ ⊥ ♃
08 06 ☽ ∠ ♄
10 55 ☿ ⚹ ♄
15 10 ☽ ⊥ ☉
21 25 ♀ ⊥ ♃
22 30 ☽ ∠ ♃
22 38 ☽ ⊥ ♄
23 18 ♀ ∠ ♄
23 51 ☽ Q ♇

25 Monday
01 35 ♂ ‖ ♃
06 51 ☽ ∠ ☉
07 49 ♂ ⚹ ♄
08 27 ☽ ⊻ ♄
08 29 ☽ ⊻ ♂
10 46 ☽ ⊻ ☿
11 21 ♀ ‖ ♅
17 29 ☽ ⊻ ♆
18 57 ☽ ⊥ ♂
21 48 ☽ ⊥ ☿
22 39 ♀ ‖ ♆
23 35 ☽ ⚹ ♃

26 Tuesday
01 06 ☽ ⊻ ♅
01 21 ☽ ‖ ♇
03 32 ☽ ⊥ ♆
05 42 ☽ □ ♇
09 44 ☽ ⚹ ☉
10 21 ☽ ‖ ♄
10 57 ☽ ∠ ♂
11 16 ☽ ⊥ ♅
12 42 ☽ ⊻ ♀
14 43 ☽ ∠ ☿
18 53 ☽ ∠ ♆
20 17 ☽ Q ♃

27 Wednesday
00 01 ☽ ⊥ ♀
02 49 ☽ ∠ ♅
08 05 ☽ Q ☉
11 31 ☉ ‖ ☿
11 38 ☽ ☌ ♄
14 32 ☽ ⚹ ♂
17 23 ☽ ∠ ♀
19 54 ☽ ⚹ ☿
21 15 ☽ ⚹ ♆
22 57 ☿ ‖ ♃

28 Thursday
04 35 ☽ □ ♃
05 32 ☽ ⚹ ♅
09 27 ☿ ☌ ♆
10 25 ☽ △ ♇
13 27 ☽ Q ♂
19 05 ☽ Q ♆
19 06 ☽ □ ☉
20 10 ☽ Q ☿
23 24 ☽ ⚹ ♀

29 Friday
03 39 ☽ Q ♅
08 26 ☽ # ♄
10 29 ☉ Q ♄
10 47 ☿ ‖ ♂
14 17 ☽ bQ ♇
18 47 ☽ ⊻ ♄
19 00 ☽ # ♇

30 Saturday
00 20 ☽ Q ♀
01 15 ☽ □ ♂
01 58 ☉ ⊥ ♇
04 59 ☽ □ ♆
06 16 ☽ ⊥ ♄
07 27 ☽ ± ♇
08 26 ☉ ‖ ♃
09 58 ☽ □ ☿
13 43 ☽ △ ♃
13 56 ☽ □ ♅
19 02 ☽ ⊼ ♇
23 44 ☽ ∠ ♄

31 Sunday
06 02 ♃ ⊻ ♅
09 17 ☽ △ ☉
15 04 ☽ □ ♀
19 33 ☽ bQ ♃

All ephemeris data is given at 12.00 UT and the Moon's longitude is additionally given for 24.00 UT

JANUARY 1996

LONGITUDES

Date	Sidereal time h m s	Sun ☉ ° ' "	Moon ☽ ° ' "	Moon ☽ 24.00 ° ' "	Mercury ☿ ° '	Venus ♀ ° '	Mars ♂ ° '	Jupiter ♃ ° '	Saturn ♄ ° '	Uranus ♅ ° '	Neptune ♆ ° '	Pluto ♇ ° '
01	18 41 43	10 ♑ 20 58	22 ♉ 44 29	28 ♉ 45 26	29 ♑ 43	13 ♒ 14	24 ♑ 34	29 ♐ 35	19 ♓ 26	29 ♑ 23	24 ♑ 42	01 ♐ 58
02	18 45 40	11 22 07	04 ♊ 44 47	10 ♊ 42 55	00 ♒ 48	14 28	25 20	29 ♐ 49	19 30	29 26	24 44	02 00
03	18 49 36	12 23 16	16 ♊ 40 10	22 ♊ 36 49	01 48	15 42	26 07	00 ♑ 03	19 34	29 30	24 46	02 02
04	18 53 33	13 24 24	28 ♊ 33 06	04 ♋ 29 13	02 42	16 55	26 54	00 16	19 38	29 33	24 49	02 04
05	18 57 29	14 25 33	10 ♋ 25 22	16 ♋ 21 42	03 29	18 09	27 41	00 30	19 43	29 36	24 51	02 06
06	19 01 26	15 26 41	22 ♋ 18 22	28 ♋ 15 30	04 08	19 23	28 28	00 43	19 47	29 40	24 53	02 08
07	19 05 22	16 27 50	04 ♌ 13 16	10 ♌ 11 50	04 38	20 36	29 ♑ 15	00 57	19 52	29 43	24 55	02 09
08	19 09 19	17 28 58	16 ♌ 11 23	22 ♌ 12 09	04 59	21 50	00 ♒ 02	01 10	19 56	29 47	24 58	02 11
09	19 13 16	18 30 06	28 ♌ 14 24	04 ♍ 18 25	05 09	23 03	00 49	01 24	20 01	29 50	25 00	02 13
10	19 17 12	19 31 14	10 ♍ 24 33	16 ♍ 33 12	05 R 08	24 17	01 36	01 37	20 06	29 54	25 02	02 15
11	19 21 09	20 32 22	22 ♍ 44 47	28 ♍ 59 47	04 55	25 30	02 23	01 50	20 11	29 ♑ 57	25 04	02 17
12	19 25 05	21 33 30	05 ♎ 18 42	11 ♎ 42 02	04 31	26 43	03 10	02 04	20 16	00 ♒ 01	25 07	02 18
13	19 29 02	22 34 38	18 ♎ 10 20	24 ♎ 44 06	03 55	27 56	03 57	02 17	20 21	00 04	25 09	02 20
14	19 32 58	23 35 46	01 ♏ 23 50	08 ♏ 09 56	03 08	29 ♒ 10	04 44	02 30	20 26	00 08	25 11	02 22
15	19 36 55	24 36 54	15 ♏ 02 44	22 ♏ 02 28	02 10	00 ♓ 23	05 32	02 43	20 31	00 11	25 13	02 23
16	19 40 51	25 38 02	29 ♏ 09 11	06 ♐ 22 45	01 ♒ 05	01 36	06 19	02 57	20 36	00 15	25 16	02 25
17	19 44 48	26 39 09	13 ♐ 42 50	21 ♐ 08 51	29 ♑ 53	02 49	07 06	03 10	20 42	00 18	25 18	02 27
18	19 48 45	27 40 16	28 ♐ 39 59	06 ♑ 15 14	28 36	04 02	07 53	03 23	20 47	00 22	25 20	02 28
19	19 52 41	28 41 23	13 ♑ 53 19	21 ♑ 32 54	27 18	05 15	08 40	03 36	20 52	00 25	25 23	02 30
20	19 56 38	29 ♑ 42 30	29 ♑ 12 28	06 ♒ 50 33	26 01	06 28	09 28	03 49	20 58	00 29	25 25	02 31
21	20 00 34	00 ♒ 43 36	14 ♒ 25 42	21 ♒ 56 37	24 46	07 41	10 15	04 02	21 03	00 32	25 27	02 33
22	20 04 31	01 44 41	29 ♒ 22 09	06 ♓ 41 25	23 36	08 53	11 02	04 15	21 09	00 36	25 29	02 34
23	20 08 27	02 45 45	13 ♓ 53 43	20 ♓ 58 38	22 33	10 06	11 50	04 28	21 15	00 40	25 32	02 36
24	20 12 24	03 46 48	27 ♓ 55 58	04 ♈ 45 43	21 37	11 19	12 37	04 41	21 21	00 43	25 34	02 37
25	20 16 20	04 47 50	11 ♈ 28 05	18 ♈ 03 24	20 50	12 31	13 24	04 54	21 26	00 47	25 36	02 39
26	20 20 17	05 48 51	24 ♈ 32 07	00 ♉ 54 47	20 11	13 44	14 12	05 06	21 32	00 50	25 38	02 40
27	20 24 14	06 49 51	07 ♉ 11 59	13 ♉ 24 22	19 42	14 56	14 59	05 19	21 38	00 54	25 41	02 41
28	20 28 10	07 50 49	19 ♉ 32 34	25 ♉ 37 14	19 22	16 08	15 46	05 32	21 44	00 57	25 43	02 43
29	20 32 07	08 51 47	01 ♊ 38 59	07 ♊ 38 26	19 10	17 20	16 34	05 44	21 50	01 01	25 45	02 44
30	20 36 03	09 52 44	13 ♊ 36 07	19 ♊ 32 34	19 D 07	18 32	17 21	05 57	21 56	01 04	25 47	02 45
31	20 40 00	10 ♒ 53 39	25 ♊ 28 16	01 ♋ 23 38	19 ♑ 11	19 ♓ 44	18 ♒ 09	06 ♑ 09	22 ♓ 03	01 ♒ 08	25 ♑ 50	02 ♐ 46

DECLINATIONS

Date	Moon True ☊ ° '	Moon Mean ☊ ° '	Moon ☽ Latitude ° '	Sun ☉ ° '	Moon ☽ ° '	Mercury ☿ ° '	Venus ♀ ° '	Mars ♂ ° '	Jupiter ♃ ° '	Saturn ♄ ° '	Uranus ♅ ° '	Neptune ♆ ° '	Pluto ♇ ° '
01	23 ♎ 26	22 ♎ 25	02 S 28	23 S 02	16 N 04	21 S 30	18 S 35	22 S 16	23 S 11	06 S 08	20 S 48	20 S 42	07 S 56
02	23 R 20	22 21	03 20	22 57	17 49	21 06	18 13	22 07	23 12	06 06	20 47	20 42	07 56
03	23 10	22 18	04 02	22 52	18 46	20 42	17 51	21 59	23 12	06 04	20 47	20 41	07 56
04	22 58	22 15	04 33	22 46	18 53	20 18	17 28	21 50	23 12	06 02	20 46	20 41	07 56
05	22 44	22 12	04 53	22 39	18 10	19 54	17 05	21 41	23 12	06 00	20 45	20 41	07 56
06	22 29	22 09	04 59	22 33	16 40	19 30	16 41	21 32	23 12	05 59	20 44	20 40	07 56
07	22 14	22 06	04 53	22 25	14 27	19 08	16 17	21 22	23 12	05 57	20 44	20 40	07 57
08	22 01	22 02	04 34	22 18	11 38	18 47	15 53	21 12	23 12	05 55	20 43	20 39	07 57
09	21 50	21 59	04 02	22 10	08 18	18 28	15 28	21 02	23 12	05 53	20 42	20 39	07 57
10	21 42	21 56	03 19	22 01	04 36	18 10	15 03	20 52	23 12	05 51	20 42	20 39	07 57
11	21 38	21 53	02 26	21 52	00 N 39	17 55	14 37	20 41	23 12	05 49	20 41	20 38	07 57
12	21 35	21 50	01 25	21 43	03 S 25	17 43	14 11	20 30	23 11	05 47	20 40	20 38	07 57
13	21 D 35	21 46	00 S 18	21 33	07 24	17 33	13 44	20 19	23 11	05 44	20 39	20 37	07 57
14	21 R 35	21 43	00 N 52	21 23	11 09	17 26	13 17	20 08	23 11	05 42	20 39	20 37	07 57
15	21 34	21 40	02 01	21 12	14 26	17 22	12 50	19 57	23 11	05 40	20 38	20 37	07 57
16	21 31	21 37	03 04	21 01	16 58	17 20	12 23	19 45	23 11	05 38	20 37	20 36	07 57
17	21 25	21 34	03 58	20 49	18 30	17 21	11 55	19 33	23 11	05 36	20 36	20 36	07 57
18	21 17	21 31	04 38	20 38	18 48	17 25	11 27	19 21	23 10	05 33	20 36	20 35	07 57
19	21 06	21 27	04 58	20 25	17 46	17 30	10 58	19 09	23 10	05 31	20 35	20 35	07 57
20	20 54	21 24	04 57	20 13	15 28	17 37	10 30	18 56	23 10	05 29	20 34	20 35	07 57
21	20 43	21 21	04 35	20 00	12 06	17 46	10 01	18 43	23 09	05 27	20 33	20 34	07 57
22	20 33	21 18	03 55	19 46	08 01	17 55	09 31	18 30	23 09	05 24	20 33	20 34	07 57
23	20 26	21 15	03 00	19 32	03 S 34	18 05	09 02	18 17	23 09	05 22	20 32	20 33	07 57
24	20 22	21 12	01 56	19 18	00 N 57	18 16	08 32	18 04	23 08	05 20	20 31	20 33	07 57
25	20 20	21 08	00 N 47	19 04	05 15	18 27	08 02	17 50	23 08	05 17	20 30	20 33	07 57
26	20 D 20	21 05	00 S 22	18 49	09 10	18 38	07 32	17 37	23 08	05 15	20 29	20 32	07 57
27	20 R 20	21 02	01 28	18 34	12 31	18 49	07 02	17 23	23 07	05 12	20 29	20 32	07 57
28	20 19	20 59	02 28	18 18	15 14	19 00	06 32	17 09	23 07	05 10	20 28	20 31	07 57
29	20 17	20 56	03 21	18 02	17 13	19 11	06 01	16 54	23 06	05 07	20 27	20 31	07 57
30	20 12	20 52	04 03	17 46	18 24	19 21	05 30	16 40	23 06	05 05	20 26	20 31	07 57
31	20 ♎ 04	20 ♎ 49	04 S 35	17 S 30	18 N 47	19 S 31	05 S 00	16 S 25	23 S 05	05 S 02	20 S 26	20 S 30	07 S 56

ZODIAC SIGN ENTRIES

Date	h m	Planets
01	18 06	☿ ♒
02	02 29	☽ ♊
03	07 22	♃ ♑
04	14 56	☽ ♋
07	03 30	☽ ♌
08	11 02	♂ ♒
09	15 29	☽ ♍
12	01 55	☽ ♎
12	07 13	♅ ♒
14	09 30	☽ ♏
15	04 30	♀ ♓
16	13 25	☽ ♐
17	09 37	☿ ♑
18	14 07	☽ ♑
20	13 15	☽ ♒
20	18 52	☉ ♒
22	13 02	☽ ♓
24	15 37	☽ ♈
26	22 16	☽ ♉
29	08 42	☽ ♊
31	21 11	☽ ♋

LATITUDES

Date	Mercury ☿ ° '	Venus ♀ ° '	Mars ♂ ° '	Jupiter ♃ ° '	Saturn ♄ ° '	Uranus ♅ ° '	Neptune ♆ ° '	Pluto ♇ ° '
01	01 S 19	01 S 49	01 S 04	00 N 15	02 S 07	00 S 32	00 N 29	12 N 52
04	00 46	01 48	01 05	00 14	02 06	00 32	00 29	12 53
07	00 S 02	01 46	01 05	00 14	02 06	00 32	00 29	12 53
10	00 N 50	01 43	01 05	00 14	02 05	00 32	00 29	12 54
13	01 47	01 39	01 05	00 14	02 05	00 32	00 29	12 55
16	02 39	01 34	01 05	00 13	02 05	00 32	00 29	12 56
19	03 15	01 29	01 06	00 13	02 04	00 32	00 29	12 56
22	03 31	01 23	01 05	00 13	02 04	00 32	00 29	12 57
25	03 25	01 17	01 05	00 13	02 03	00 32	00 29	12 58
28	03 04	01 09	01 05	00 13	02 03	00 32	00 29	12 59
31	02 N 35	01 S 01	01 S 05	00 N 12	02 S 03	00 S 32	00 N 29	13 N 00

DATA

Julian Date	2450084
Delta T	+62 seconds
Ayanamsa	23° 48' 11"
Synetic vernal point	05° ♓ 18' 48"
True obliquity of ecliptic	23° 26' 14"

MOON'S PHASES, APSIDES AND POSITIONS ☽

Date	h m	Phase	Longitude ° '	Eclipse Indicator
05	20 51	○	14 ♋ 48	
13	20 45	☾	22 ♎ 57	
20	12 50	●	29 ♑ 45	
27	11 14	☽	06 ♉ 48	

Day	h m		
05	11 15	Apogee	
19	22 59	Perigee	
04	03 21	Max dec	18° N 56'
11	15 50	0S	
18	05 35	Max dec	18° S 51'
24	06 55	0N	
31	10 51	Max dec	18° N 47'

LONGITUDES

Date	Chiron ⚷ ° '	Ceres ⚳ ° '	Pallas ⚴ ° '	Juno ⚵ ° '	Vesta ⚶ ° '	Black Moon Lilith ⚸ ° '
01	13 ♎ 37	22 ♏ 09	28 ♎ 01	25 ♑ 22	02 ♏ 29	10 ♋ 33
11	14 ♎ 03	25 ♏ 56	01 ♏ 02	29 ♑ 25	06 ♏ 34	11 ♋ 40
21	14 ♎ 18	29 ♏ 33	03 ♏ 39	03 ♒ 33	10 ♏ 24	12 ♋ 47
31	14 ♎ 21	02 ♐ 55	05 ♏ 48	07 ♒ 43	13 ♏ 56	13 ♋ 55

All ephemeris data is given at 12.00 UT and the Moon's longitude is additionally given for 24.00 UT

ASPECTARIAN

h m	Aspects
01 Monday	
04 33	☿ ☌ ♅
05 23	☽ ✶ ♄
08 43	☿ ⊻ ♃
13 43	☽ ± ♃
15 52	☽ △ ♂
15 54	☽ △ ♆
15 59	♀ ⊥ ♄
16 25	♂ ☌ ♆
17 41	☽ bQ ☉
02 Tuesday	
01 18	☽ △ ♅
01 56	☽ ⊼ ♃
02 38	♀ Q ♇
03 20	☽ △ ☿
05 27	☽ Q ♄
06 28	☽ ☍ ♇
13 22	☽ ± ☉
18 04	☽ # ♀
20 25	♀ ∠ ♃
22 03	☽ bQ ♆
03 Wednesday	
00 02	☽ bQ ♂
02 34	☽ ⊼ ☉
07 13	☿ ∥ ♅
07 35	☽ bQ ♅
09 48	☽ △ ♀
12 18	☽ bQ ☿
12 34	☿ ∥ ♆
16 15	☽ ± ♆
17 53	☽ □ ♄
17 59	☿ ✶ ♇
19 27	☽ ± ♂
04 Thursday	
01 50	☽ ± ♅
04 25	☽ ⊼ ♆
07 59	☽ ± ☿
08 26	☽ ⊼ ♂
14 02	☽ ⊼ ♅
15 32	☽ ☍ ♃
19 07	☽ ⊼ ♇
19 36	☽ bQ ♀
21 01	☽ ⊼ ☿
05 Friday	
07 17	☽ ± ♇
15 54	☽ ± ♀
20 51	☽ ☍ ☉
06 Saturday	
01 31	☽ bQ ♇
05 25	☽ ⊼ ♀
06 53	☽ △ ♄
11 36	☽ # ♀
17 13	☽ ☍ ♆
20 34	♀ ⊻ ♄
07 Sunday	
01 17	☽ ☍ ♂
02 54	☽ ☍ ♅
05 17	☽ ⊼ ♃
07 50	☽ △ ♇
12 53	☽ ☍ ☿
13 18	☽ bQ ♄
17 35	☽ ± ♃
08 Monday	
03 38	♂ ☌ ♅
04 51	☉ ∠ ♇
06 41	☿ ∠ ♄
07 28	☽ ± ♄
11 57	☽ bQ ♃
14 49	☽ ⊼ ☉
19 32	☽ ⊼ ♄
09 Tuesday	
00 32	☽ ☍ ♀
03 53	☽ ± ☉
05 32	☽ ⊼ ♆
14 22	☽ # ♇
15 11	☽ ⊼ ♅
17 27	☽ ⊼ ♂
17 29	☽ ± ♆
18 22	☽ △ ♃
19 54	☽ □ ♇
21 50	☿ St R
23 22	☽ bQ ☉
10 Wednesday	
01 42	☽ ⊼ ☿
03 05	☽ ± ♅
04 03	☽ # ♄
06 06	☽ ± ♂
11 16	☽ bQ ♆
12 41	♂ ⊻ ♃
13 25	☽ ± ☿
16 15	☿ ∠ ♄
20 49	☽ bQ ♅
11 Thursday	
00 55	☽ bQ ♂
02 47	☉ ✶ ♄
03 22	♀ ⊻ ♆
06 40	☽ bQ ☿
07 00	☽ ☍ ♄
07 13	☽ Q ♇
07 21	☽ △ ☉
08 38	♂ ✶ ♇
12 53	♂ ∥ ♅
16 30	☽ △ ♆
17 52	☽ ⊼ ♀
18 49	♂ ∥ ♆
12 Friday	
01 53	☽ △ ♅
05 44	☽ □ ♃
06 18	☽ ✶ ♇
06 34	☽ ± ♀
07 40	☽ △ ♂
10 33	☽ △ ☿
13 Saturday	
01 18	☽ bQ ♀
01 59	☽ ∥ ♄
10 27	☽ ∠ ♇
11 20	☿ ☌ ♂
15 23	☽ ∥ ♇
15 57	☽ Q ♃
16 01	☽ ⊼ ♄
18 38	♃ ⊻ ♇
20 45	☽ □ ☉
14 Sunday	
00 47	☽ □ ♆
02 56	☽ ⊥ ♇
03 01	☽ ± ♄
07 35	☽ △ ♀
09 43	☽ □ ♅
13 44	☽ ⊻ ♇
14 00	☽ ✶ ♃
14 54	☽ □ ☿
18 19	☽ □ ♂
19 13	☽ bQ ♄
15 Monday	
01 07	☿ ⊻ ♃
01 17	☽ ∥ ♀
06 57	☿ ✶ ♇
07 27	☽ Q ☉
08 00	♀ ⊻ ♅
08 50	☽ Q ♆
11 40	♂ ∠ ♄
16 42	☽ ∠ ♃
17 26	☽ Q ♅
20 12	☽ Q ☿
21 28	☽ △ ♄
16 Tuesday	
02 54	☉ ☌ ♆
03 22	☽ Q ♂
05 08	♀ ⊥ ♆
05 26	☽ ✶ ♆
05 38	☽ ✶ ☉
06 42	☿ ⊻ ♀
08 14	☽ ⊥ ♃
13 50	☽ ✶ ♅
14 59	☽ ✶ ☿
16 10	☉ ⊥ ♀
16 28	☽ □ ♀
16 29	☽ ∥ ☿
17 27	☽ ☌ ♇
18 25	☽ ⊻ ♃
17 Wednesday	
00 34	☽ ✶ ♂
04 02	☿ ☌ ♅
04 31	♀ □ ♇
06 25	☽ ∠ ♆
08 24	☽ ∠ ☉
13 44	☽ ∠ ☿
14 35	☽ ∠ ♅
20 21	♀ ✶ ♃
21 03	☽ ⊥ ♆
23 20	☽ □ ♄
18 Thursday	
00 01	☽ ⊥ ☉
00 29	☽ Q ♀
02 17	☽ ∠ ♂
03 06	☽ ⊥ ☿
05 07	☽ ⊥ ♅
06 41	☽ ⊻ ♆
10 18	☽ ⊻ ☉
11 55	☽ ⊻ ☿
14 42	☽ ⊻ ♅
16 08	☉ ∥ ♅
16 16	☉ ∥ ♆
17 23	☽ ⊥ ♂
17 27	☿ ⊥ ♀
18 02	☽ ⊻ ♇
19 35	☽ ☌ ♃
20 41	♅ ∥ ♆
21 14	☽ ✶ ♀
21 40	☉ ☌ ☿
19 Friday	
03 22	☽ ⊻ ♂
03 31	☽ ⊥ ♇
04 04	☽ Q ♄
15 27	☽ ∥ ☿
17 40	☽ ∠ ♇
22 49	☽ ∠ ♀
23 00	☽ ✶ ♄
20 Saturday	
06 02	☽ ☌ ♆
07 22	☽ ☌ ☿
12 23	♀ ⊥ ♅
12 50	☽ ☌ ☉
14 00	☽ ☌ ♅
14 08	☽ ⊥ ♀
17 13	☽ ✶ ♇
19 21	☽ ⊻ ♃
22 41	☽ ∠ ♄
22 59	☿ ☌ ♆
21 Sunday	
00 23	☽ ⊻ ♀
02 58	♂ ⊥ ♃
04 56	☽ ⊥ ♃
05 01	☽ ☌ ♂
07 21	☉ ☌ ♅
12 11	☽ Q ♇
13 00	☽ ⊥ ♄
19 27	☽ ∠ ♃
22 39	☽ ⊻ ♄
22 Monday	
02 25	☽ ∥ ♀
03 19	☽ ⊻ ☿
05 42	☽ ⊻ ♆
09 02	☿ ∠ ♀
12 21	☽ ⊥ ☿
12 23	☽ ∥ ♇
14 01	☽ ⊻ ♅
15 28	☽ ⊥ ♆
16 10	☽ ⊻ ☉
17 14	☽ □ ♇
20 06	☽ ✶ ♃
23 54	☽ ⊥ ♅
23 Tuesday	
02 07	☽ ∠ ☿
02 21	☽ ∥ ♄
02 47	☽ ⊥ ☉
05 04	☽ ☌ ♀
06 21	☽ ∠ ♆
08 01	☉ ✶ ♇
08 20	☽ ⊻ ♂
14 59	☽ ∠ ♅
16 24	☽ Q ♃
19 02	☽ ⊥ ♂
19 02	☽ ∠ ☉
20 45	♀ ∠ ♆
24 Wednesday	
00 06	☿ ∥ ♂
00 33	☽ ☌ ♄
01 45	☽ ✶ ☿
07 53	☽ ✶ ♆
11 25	☽ ∠ ♂
16 53	☽ ✶ ♅
19 01	☿ ✶ ♄
20 14	☽ △ ♇
21 24	☽ Q ☿
23 06	☽ ✶ ☉
25 Thursday	
00 03	☽ □ ♃
05 02	☽ Q ♆
12 11	☽ # ♄
14 05	☽ ⊻ ♀
14 23	☽ Q ♅
14 54	☉ ⊻ ♃
15 44	☽ ✶ ♂
16 20	♀ ∥ ♇
22 30	☽ Q ☉
23 16	☽ bQ ♇
26 Friday	
02 10	☽ ⊥ ♀
02 52	☽ # ♀
04 14	☽ # ♇
04 17	☽ □ ☿
06 23	☽ ⊻ ♄
14 04	☽ □ ♆
15 19	☽ Q ♂
16 00	☽ ± ♇
17 40	☽ ⊥ ♄
20 41	☽ ∠ ♀
21 53	☉ ∥ ☿
23 54	☽ □ ♅
27 Saturday	
02 44	♂ Q ♇
03 21	☽ ⊼ ♇
06 57	☉ ∠ ♄
08 20	☽ △ ♃
10 55	☽ ∠ ♄
11 14	☽ □ ☉
15 02	♀ ⊻ ♂
28 Sunday	
04 07	☽ □ ♂
04 36	☽ ✶ ♀
08 08	♀ ∠ ♅
10 44	♂ ⊥ ♄
11 39	☽ △ ☿
13 58	☽ bQ ♃
16 21	☽ ✶ ♄
29 Monday	
00 13	☽ △ ♆
06 53	☽ Q ♀
08 07	☽ # ♂
08 07	☽ ± ♃
10 43	☽ △ ♅
14 10	☽ ☍ ♇
16 25	☽ Q ♄
17 00	☽ bQ ☿
20 20	☽ ⊼ ♃
21 40	♀ Q ♃
23 56	☽ # ☉
30 Tuesday	
03 48	☽ △ ☉
06 19	☽ bQ ♆
10 18	☿ St D
11 01	☽ ± ☿
17 00	☽ bQ ♅
20 07	☽ △ ♂
23 06	☽ □ ♀
23 10	☽ ⊼ ☿
23 51	☿ ✶ ♀
31 Wednesday	
00 32	☽ ± ♆
05 00	☽ □ ♄
09 47	♀ ∥ ♄
11 18	☽ ± ♅
12 43	☽ ⊼ ♆
12 56	☽ bQ ☉
23 31	☽ ⊼ ♅

FEBRUARY 1996

LONGITUDES

Date	Sidereal time h m s	Sun ☉ ° ' "	Moon ☽ ° ' "	Moon ☽ 24.00 ° ' "	Mercury ☿ ° '	Venus ♀ ° '	Mars ♂ ° '	Jupiter ♃ ° '	Saturn ♄ ° '	Uranus ♅ ° '	Neptune ♆ ° '	Pluto ♇ ° '
01	20 43 56	11 ♒ 54 33	07 ♋ 19 03	13 ♋ 14 49	19 ♑ 23	20 ♓ 56	18 ♒ 56	06 ♑ 22	22 ♓ 09	01 ♒ 11	25 ♑ 52	02 ♐ 47
02	20 47 53	12 55 26	19 ♋ 11 15	25 ♋ 08 34	19 41	22 08	19 44	06 34	22 15	01 15	25 54	02 49
03	20 51 49	13 56 17	01 ♌ 06 58	07 ♌ 06 38	20 05	23 20	20 31	06 46	22 22	01 18	25 56	02 50
04	20 55 46	14 57 08	13 ♌ 07 41	19 ♌ 10 15	20 36	24 32	21 18	06 59	22 28	01 22	25 58	02 51
05	20 59 43	15 57 57	25 ♌ 14 27	01 ♍ 20 26	21 11	25 43	22 06	07 11	22 34	01 25	26 01	02 52
06	21 03 39	16 58 45	07 ♍ 28 17	13 ♍ 38 10	21 51	26 55	22 53	07 23	22 41	01 28	26 03	02 53
07	21 07 36	17 59 32	19 ♍ 50 16	26 ♍ 04 46	22 35	28 06	23 41	07 35	22 47	01 32	26 05	02 54
08	21 11 32	19 00 18	02 ♎ 21 54	08 ♎ 41 56	23 23	29 ♓ 17	24 28	07 47	22 54	01 35	26 07	02 55
09	21 15 29	20 01 03	15 ♎ 05 10	21 ♎ 31 56	24 15	00 ♈ 28	25 16	07 59	23 01	01 39	26 09	02 56
10	21 19 25	21 01 47	28 ♎ 02 34	04 ♏ 37 24	25 10	01 39	26 03	08 11	23 07	01 42	26 11	02 56
11	21 23 22	22 02 30	11 ♏ 16 49	18 ♏ 01 07	26 08	02 50	26 50	08 22	23 14	01 45	26 13	02 57
12	21 27 18	23 03 12	24 ♏ 50 34	01 ♐ 45 23	27 09	04 01	27 38	08 34	23 21	01 49	26 15	02 58
13	21 31 15	24 03 53	08 ♐ 45 41	15 ♐ 51 27	28 13	05 11	28 25	08 46	23 27	01 52	26 18	02 59
14	21 35 12	25 04 32	23 ♐ 02 31	00 ♑ 18 35	29 ♑ 18	06 22	29 ♒ 13	08 57	23 34	01 55	26 20	03 00
15	21 39 08	26 05 11	07 ♑ 39 09	15 ♑ 03 32	00 ♒ 26	07 32	00 ♓ 00	09 09	23 41	01 59	26 22	03 00
16	21 43 05	27 05 48	22 ♑ 30 54	00 ♒ 00 15	01 37	08 42	00 48	09 20	23 48	02 02	26 24	03 01
17	21 47 01	28 06 25	07 ♒ 30 28	15 ♒ 00 21	02 49	09 53	01 35	09 31	23 55	02 05	26 26	03 02
18	21 50 58	29 ♒ 06 59	22 ♒ 28 44	29 ♒ 54 25	04 02	11 03	02 23	09 43	24 02	02 09	26 28	03 02
19	21 54 54	00 ♓ 07 32	07 ♓ 16 21	14 ♓ 33 32	05 18	12 12	03 10	09 54	24 09	02 12	26 30	03 03
20	21 58 51	01 08 04	21 ♓ 45 13	28 ♓ 50 46	06 35	13 22	03 58	10 05	24 16	02 15	26 32	03 03
21	22 02 47	02 08 33	05 ♈ 49 45	12 ♈ 41 58	07 53	14 32	04 45	10 16	24 23	02 18	26 34	03 04
22	22 06 44	03 09 01	19 ♈ 27 21	26 ♈ 05 59	09 13	15 41	05 32	10 26	24 30	02 21	26 36	03 04
23	22 10 41	04 09 28	02 ♉ 38 09	09 ♉ 04 11	10 34	16 50	06 20	10 37	24 37	02 25	26 37	03 05
24	22 14 37	05 09 52	15 ♉ 24 32	21 ♉ 39 44	11 57	18 00	07 07	10 48	24 44	02 28	26 39	03 05
25	22 18 34	06 10 14	27 ♉ 50 21	03 ♊ 56 59	13 21	19 09	07 55	10 58	24 52	02 31	26 41	03 05
26	22 22 30	07 10 35	10 ♊ 00 15	16 ♊ 00 48	14 46	20 17	08 42	11 09	24 59	02 34	26 43	03 06
27	22 26 27	08 10 53	21 ♊ 59 14	27 ♊ 56 10	16 12	21 26	09 29	11 19	25 06	02 37	26 45	03 06
28	22 30 23	09 11 10	03 ♋ 52 10	09 ♋ 47 46	17 40	22 34	10 17	11 29	25 13	02 40	26 47	03 06
29	22 34 20	10 ♓ 11 24	15 ♋ 43 30	21 ♋ 39 49	19 ♒ 09	23 ♈ 43	11 ♓ 04	11 ♑ 39	25 ♓ 21	02 ♒ 43	26 ♑ 48	03 ♐ 06

Date	Moon True ☊ ° '	Moon Mean ☊ ° '	Moon ☽ Latitude ° '
01	19 ♎ 54	20 ♎ 46	04 S 55
02	19 R 41	20 43	05 02
03	19 28	20 40	04 56
04	19 15	20 37	04 37
05	19 04	20 33	04 05
06	18 54	20 30	03 21
07	18 48	20 27	02 28
08	18 44	20 24	01 26
09	18 43	20 21	00 S 19
10	18 D 43	20 18	00 N 50
11	18 44	20 14	01 58
12	18 R 44	20 11	03 01
13	18 43	20 08	03 55
14	18 40	20 05	04 37
15	18 34	20 02	05 01
16	18 27	19 58	05 06
17	18 19	19 55	04 50
18	18 11	19 52	04 14
19	18 04	19 49	03 21
20	17 59	19 46	02 16
21	17 57	19 43	01 N 05
22	17 D 56	19 39	00 S 08
23	17 57	19 36	01 19
24	17 58	19 33	02 23
25	18 00	19 30	03 19
26	18 R 00	19 27	04 05
27	17 59	19 23	04 39
28	17 56	19 20	05 01
29	17 ♎ 51	19 ♎ 17	05 S 10

DECLINATIONS

Date	Sun ☉ ° '	Moon ☽ ° '	Mercury ☿ ° '	Venus ♀ ° '	Mars ♂ ° '	Jupiter ♃ ° '	Saturn ♄ ° '	Uranus ♅ ° '	Neptune ♆ ° '	Pluto ♇ ° '
01	17 S 13	18 N 20	19 S 40	04 S 29	16 S 11	23 S 05	05 S 00	20 S 25	20 S 30	07 S 56
02	16 56	17 05	19 48	03 58	15 56	23 04	04 57	20 24	20 29	07 56
03	16 39	15 06	19 56	03 26	15 40	23 04	04 55	20 23	20 29	07 56
04	16 21	12 27	20 02	02 55	15 25	23 03	04 52	20 23	20 29	07 56
05	16 03	09 16	20 08	02 24	15 10	23 03	04 49	20 22	20 28	07 56
06	15 45	05 39	20 13	01 53	14 54	23 02	04 47	20 21	20 28	07 56
07	15 26	01 N 46	20 17	01 21	14 38	23 01	04 44	20 20	20 27	07 55
08	15 07	02 S 16	20 20	00 50	14 22	23 01	04 41	20 20	20 27	07 55
09	14 48	06 14	20 22	00 S 18	14 06	23 00	04 39	20 19	20 27	07 55
10	14 29	10 00	20 22	00 N 13	13 50	23 00	04 36	20 18	20 26	07 55
11	14 10	13 21	20 22	00 45	13 34	22 59	04 33	20 17	20 26	07 55
12	13 50	16 03	20 20	01 16	13 17	22 58	04 30	20 17	20 25	07 54
13	13 30	17 53	20 18	01 48	13 01	22 58	04 28	20 16	20 25	07 54
14	13 10	18 39	20 14	02 19	12 44	22 57	04 25	20 15	20 25	07 54
15	12 49	18 13	20 08	02 50	12 27	22 56	04 22	20 15	20 24	07 54
16	12 29	16 32	20 02	03 22	12 10	22 55	04 19	20 14	20 24	07 54
17	12 08	13 43	19 54	03 53	11 53	22 55	04 17	20 13	20 23	07 53
18	11 47	10 01	19 46	04 24	11 36	22 54	04 14	20 12	20 23	07 53
19	11 26	05 43	19 35	04 55	11 19	22 53	04 11	20 12	20 23	07 53
20	11 04	01 S 11	19 24	05 26	11 01	22 53	04 08	20 11	20 22	07 52
21	10 43	03 N 19	19 11	05 57	10 44	22 52	04 05	20 10	20 22	07 52
22	10 21	07 29	18 57	06 28	10 26	22 51	04 02	20 10	20 22	07 52
23	09 59	11 09	18 42	06 58	10 08	22 50	04 00	20 09	20 21	07 52
24	09 37	14 10	18 26	07 29	09 51	22 49	03 57	20 08	20 21	07 51
25	09 15	16 26	18 08	07 59	09 33	22 49	03 54	20 07	20 21	07 51
26	08 53	17 55	17 49	08 29	09 15	22 48	03 51	20 07	20 20	07 51
27	08 30	18 33	17 28	08 59	08 57	22 47	03 48	20 06	20 20	07 50
28	08 08	18 22	17 07	09 29	08 39	22 46	03 45	20 05	20 20	07 50
29	07 S 45	17 N 23	16 S 44	09 N 58	08 S 20	22 S 46	03 S 42	20 S 05	20 S 19	07 S 50

ZODIAC SIGN ENTRIES

Date	h	m	Planets
03	09	46	☽ ♌
05	21	22	☽ ♍
08	07	30	☽ ♎
09	02	30	♀ ♈
10	15	35	☽ ♏
12	20	58	☽ ♐
14	23	29	☽ ♑
15	02	44	☿ ♒
15	11	50	♂ ♓
17	00	00	☽ ♒
19	00	09	☽ ♓
19	09	01	☉ ♓
21	01	58	☽ ♈
23	07	08	☽ ♉
25	16	14	☽ ♊
28	04	10	☽ ♋

LATITUDES

Date	Mercury ☿ ° '	Venus ♀ ° '	Mars ♂ ° '	Jupiter ♃ ° '	Saturn ♄ ° '	Uranus ♅ ° '	Neptune ♆ ° '	Pluto ♇ ° '
01	02 N 24	00 S 58	01 S 05	00 N 12	02 S 03	00 S 32	00 N 29	13 N 01
04	01 50	00 49	01 05	00 12	02 03	00 32	00 29	13 02
07	01 17	00 39	01 04	00 12	02 02	00 32	00 29	13 03
10	00 44	00 28	01 04	00 12	02 02	00 32	00 29	13 04
13	00 N 14	00 17	01 03	00 11	02 02	00 32	00 29	13 05
16	00 S 14	00 S 06	01 03	00 11	02 02	00 32	00 29	13 06
19	00 40	00 N 06	01 02	00 11	02 02	00 32	00 29	13 07
22	01 03	00 19	01 02	00 11	02 01	00 32	00 29	13 08
25	01 23	00 32	01 01	00 10	02 01	00 33	00 29	13 09
28	01 39	00 45	01 00	00 10	02 01	00 33	00 29	13 10
31	01 S 53	00 N 59	00 S 59	00 N 10	02 S 01	00 S 33	00 N 29	13 N 11

DATA

Julian Date	2450115
Delta T	+62 seconds
Ayanamsa	23° 48' 16"
Synetic vernal point	05° ♓ 18' 43"
True obliquity of ecliptic	23° 26' 14"

MOON'S PHASES, APSIDES AND POSITIONS ☽

Date	h	m	Phase	Longitude ° '	Eclipse Indicator
04	15	58	○	15 ♌ 07	
12	08	37	☾	22 ♏ 55	
18	23	30	●	29 ♒ 36	
26	05	52	☽	06 ♊ 55	

Day	h	m		
01	15	35	Apogee	
17	08	33	Perigee	
29	07	03	Apogee	
07	22	33	0S	
14	15	29	Max dec	18° S 40'
20	18	13	0N	
27	18	34	Max dec	18° N 35'

LONGITUDES

Date	Chiron ⚷ ° '	Ceres ⚳ ° '	Pallas ⚴ ° '	Juno ⚵ ° '	Vesta ⚶ ° '	Black Moon Lilith ⚸ ° '
01	14 ♎ 20	03 ♐ 14	06 ♏ 00	08 ♒ 09	14 ♏ 16	14 ♋ 01
11	14 ♎ 10	06 ♐ 18	07 ♏ 32	12 ♒ 22	17 ♏ 25	15 ♋ 09
21	13 ♎ 48	09 ♐ 03	08 ♏ 26	16 ♒ 37	20 ♏ 07	16 ♋ 16
31	13 ♎ 17	11 ♐ 25	08 ♏ 34	20 ♒ 53	22 ♏ 18	17 ♋ 23

ASPECTARIAN

h m	Aspects	h m	Aspects	h m	Aspects
01 Thursday		14 07	☿ ☌ ♆	**20 Tuesday**	
02 49	☽ ⊼ ♇	14 32	♀ △ ♇	04 27	☽ ∠ ♅
04 39	☽ ⚼ ♂	17 16	☽ Q ♆	11 41	☽ ∠ ☿
08 53	☽ ± ☉	17 31	☽ Q ☿	12 33	☽ Q ♃
10 02	☽ ☍ ♃	17 56	☽ ∥ ☉	16 16	☽ ☌ ♄
14 59	☽ ± ♇	**12 Monday**		20 05	☽ ⚹ ♆
22 10	☽ ⊼ ☉	00 47	☽ ⚼ ♀	**21 Wednesday**	
02 Friday		03 08	☽ Q ♅	05 09	☽ ⊻ ☉
00 11	☽ ± ♂	08 37	☽ □ ☉	05 54	☽ ⚹ ♅
01 30	☉ ⊥ ♃	09 21	☽ △ ♄	05 54	☉ ∥ ♂
09 13	☽ ⚼ ♇	09 44	☽ ∠ ♃	07 06	♀ Q ♅
09 45	☿ ⊻ ♂	14 28	☽ ⚹ ♆	07 13	☽ △ ♇
13 02	☽ ☍ ☿	16 21	☽ ⚹ ☿	10 01	☽ ⊻ ♂
13 10	☽ ⊼ ♂	17 09	☽ □ ♂	15 57	☽ ⚹ ☿
14 33	♀ ☌ ♄	19 47	☉ ⊻ ♄	16 04	☉ ⊻ ♅
14 37	☽ # ☉	**13 Tuesday**		16 16	☽ # ♄
18 14	☽ △ ♄	00 09	☽ ⚹ ♅	16 20	☽ ⊥ ☉
18 37	☽ △ ♀	01 35	☽ ⊥ ♃	16 45	☽ Q ♆
03 Saturday		02 06	☽ ☌ ♇	19 49	☽ □ ♃
01 34	☽ ☍ ♆	03 09	☉ ∠ ♃	21 06	☽ ⊥ ♂
05 00	☽ # ♂	05 20	☽ △ ♀	22 15	☉ ⊥ ♆
12 22	☽ ☍ ♅	12 00	☽ ⊻ ♃	**22 Thursday**	
15 26	☽ △ ♇	16 18	☽ ∠ ♆	02 53	☽ Q ♅
23 32	☽ ⊼ ♃	18 02	☽ Q ☉	04 39	☽ ☌ ♀
04 Sunday		20 10	☽ ∠ ☿	05 00	☽ ∥ ♀
00 36	☽ ⚼ ♄	**14 Wednesday**		09 29	☽ ∠ ☉
04 02	☽ ⚼ ♀	01 13	☿ ∥ ♅	09 31	☽ ⚼ ♇
09 26	☉ Q ♇	01 43	☽ Q ♂	10 05	☉ □ ♇
11 42	☽ ± ♃	01 45	☽ ∠ ♅	14 04	☽ ∠ ♂
15 58	☽ ☍ ☉	05 00	☿ ⊻ ♂	14 19	☽ # ♇
18 42	☽ ± ♄	07 28	☽ ⊥ ♆	15 31	☽ Q ☿
23 54	☽ ± ♀	12 29	☽ ⊥ ☿	17 33	☿ ∠ ♄
05 Monday		12 53	☽ □ ♄	21 11	☽ ⊻ ♄
03 33	☽ ⊼ ☿	15 37	☽ ⚹ ☉	**23 Friday**	
05 22	☽ ☍ ♂	16 48	☽ ⊥ ♅	00 56	☽ □ ♆
05 51	☽ ⚼ ♃	17 14	♃ ⊥ ♇	01 47	☽ ± ♇
06 41	☽ ⊼ ♄	17 27	☽ ⊻ ♆	04 40	☽ # ☉
13 03	☽ ⊼ ♀	22 47	☽ ⚹ ♂	05 30	☽ # ♂
13 31	☽ ⊼ ♆	23 13	☽ ⊻ ☿	08 15	☽ ⊥ ♄
15 28	♂ ∠ ♃	**15 Thursday**		11 35	☽ □ ♅
16 02	☽ ± ☿	02 42	☽ ⊻ ♅	12 49	☽ ⊼ ♇
18 03	♀ ⚹ ♆	04 25	☽ ⊻ ♇	12 54	☿ ⊻ ♃
21 08	☽ # ♇	11 48	☽ □ ♀	15 04	☽ ⚹ ☉
06 Tuesday		14 12	☽ ⊥ ♇	19 19	☽ ⚹ ♂
00 12	☽ ⊼ ♅	14 27	☽ ☌ ♃	**24 Saturday**	
01 21	☽ ± ♆	17 59	☽ ∠ ☉	01 10	☽ ∠ ♄
03 00	☽ □ ♇	18 36	☽ Q ♄	03 08	☽ △ ♃
04 04	☉ ⊥ ♄	18 44	☉ ⊻ ♆	04 38	☽ □ ☿
04 43	♂ ⊻ ♄	**16 Friday**		13 54	♀ ⚼ ♇
10 42	☽ ⚼ ☿	00 35	☽ ∠ ♂	15 39	☽ Q ☉
11 50	☽ △ ♃	04 46	☽ ∠ ♇	17 26	☽ ⊻ ♀
12 00	☽ ± ♅	05 23	♀ Q ♆	19 35	☽ Q ♂
17 34	☽ # ♄	09 33	☽ ⊥ ☉	**25 Sunday**	
18 59	☽ ⚼ ♆	14 05	☽ ⚹ ♄	05 46	♀ # ♇
07 Wednesday		15 52	☽ ⊥ ♂	06 08	☽ ⚹ ♄
05 35	☽ ⚼ ♅	18 14	☽ ☌ ♆	06 12	☽ ⊥ ♀
08 07	☽ ⊼ ☉	19 18	☽ Q ♀	08 18	☽ ⚼ ♃
14 03	☽ Q ♇	19 53	☽ ⊻ ☉	09 45	☽ △ ♆
14 49	☽ # ♀	21 00	☿ ☌ ♅	21 12	☽ △ ♅
17 38	☽ △ ☿	**17 Saturday**		22 19	☽ ☍ ♇
17 44	☽ ☍ ♄	02 00	☽ ⊻ ♂	**26 Monday**	
19 21	☿ ⚹ ♄	03 18	☽ ☌ ♅	01 40	☽ ∠ ♀
19 54	☽ ⊼ ♂	03 20	♀ □ ♃	02 13	☽ ± ♃
20 42	☽ ± ☉	03 50	☽ ☌ ☿	05 52	☽ □ ☉
08 Thursday		04 50	☽ ⚹ ♇	05 56	☽ Q ♄
00 02	☽ △ ♆	14 16	☽ ∠ ♄	07 36	♂ ⊥ ♅
04 28	☽ ∥ ♀	15 16	☽ ⊻ ♃	09 13	☽ □ ♂
05 31	☽ ☍ ♀	16 07	☽ ⚹ ♀	10 21	☽ # ☿
08 09	☽ ± ♂	16 20	☿ ⚹ ♇	14 18	☽ ⊼ ♃
09 52	☿ ∥ ♅	**18 Sunday**		15 25	☽ ⚼ ♆
10 31	☽ △ ♅	00 03	☽ Q ♇	17 30	☿ Q ♇
13 02	☽ ⚹ ♇	00 03	☽ ∥ ☉	22 46	☉ # ♀
15 23	☽ ⚼ ☉	00 59	☽ ⊥ ♃	22 48	☽ △ ☿
22 26	☽ □ ♃	01 32	☽ ∥ ♂	**27 Tuesday**	
09 Friday		04 22	♂ ⊻ ♅	03 10	☽ ⚼ ♅
02 21	☽ ⚼ ♂	04 41	♀ # ♄	09 30	☽ ± ♆
02 23	☽ ∥ ♄	04 48	☽ ⊥ ♄	10 46	☽ ⚹ ♀
17 18	☽ ∠ ♇	14 31	☽ ⊻ ♄	10 51	♀ # ♂
21 58	☽ △ ☉	14 38	♂ ⊥ ♆	18 20	☽ □ ♄
22 27	☽ ∥ ♇	15 38	☽ ∠ ♃	21 22	☽ ± ♅
10 Saturday		18 14	☽ ∠ ♀	21 37	☽ ⊼ ♆
02 52	☽ ⊼ ♄	18 26	☽ ⊻ ♆	22 54	☉ ⊥ ♅
06 18	☽ □ ☿	23 30	☽ ☌ ☉	**28 Wednesday**	
08 06	☽ △ ♂	**19 Monday**		08 44	☿ ⊥ ♃
08 32	☽ Q ♃	00 14	☽ ∥ ♇	09 13	☽ ⚼ ☿
08 35	☽ □ ♆	03 41	☽ ⊻ ♅	09 33	☽ ⊼ ♅
09 59	☽ ⊥ ♇	04 11	☽ ⊥ ♆	10 27	☽ ⊼ ♇
13 04	♀ ⚹ ♅	04 55	☽ ☌ ♂	13 34	☽ Q ♀
13 59	☽ ± ♄	05 06	☽ □ ♇	22 36	☽ ± ♇
16 21	♂ ⊻ ♆	08 14	♂ □ ♇	23 46	☽ △ ☉
18 43	☽ □ ♅	08 28	☽ ⊻ ☿	**29 Thursday**	
19 15	☽ ⊼ ♀	10 07	☽ ⊥ ♀	01 54	☽ △ ♂
20 57	☽ ⊻ ♇	13 31	☽ ⊥ ♅	03 39	☽ ☍ ♃
11 Sunday		15 51	☽ # ♀	06 02	☽ ± ☿
06 29	☽ ⚼ ♄	16 21	☽ ⚹ ♃	06 44	☉ ∥ ♇
06 42	☽ ⚹ ♃	18 57	☽ ∠ ♆	15 31	☿ ⊥ ♄
07 11	☽ ± ♀	19 14	☽ ⊥ ☿	16 49	☽ ⚼ ♇
12 34	♀ ⊥ ♂	20 18	☽ ∥ ♄	19 54	☽ ⊼ ☿
13 36	☽ ∥ ♂	20 49	☽ ⊻ ♀		

All ephemeris data is given at 12.00 UT and the Moon's longitude is additionally given for 24.00 UT

MARCH 1996

LONGITUDES

Date	Sidereal time h m s	Sun ☉ ° ' "	Moon ☽ ° ' "	Moon ☽ 24.00 ° ' "	Mercury ☿ ° '	Venus ♀ ° '	Mars ♂ ° '	Jupiter ♃ ° '	Saturn ♄ ° '	Uranus ♅ ° '	Neptune ♆ ° '	Pluto ♇ ° '
01	22 38 16	11 ♓ 11 37	27 ♋ 37 09	03 ♌ 35 52	20 ♒ 38	24 ♈ 51	11 ♓ 51	11 ♑ 49	25 ♓ 28	02 ♒ 46	26 ♑ 50	03 ♐ 07
02	22 42 13	12 11 47	09 ♌ 36 18	15 ♌ 38 44	22 09	25 58	12 39	11 59	25 35	02 49	26 52	03 07
03	22 46 10	13 11 56	21 ♌ 43 24	27 ♌ 50 31	23 41	27 06	13 26	12 09	25 42	02 52	26 54	03 07
04	22 50 06	14 12 02	04 ♍ 00 12	10 ♍ 12 35	25 14	28 14	14 13	12 19	25 50	02 55	26 55	03 07
05	22 54 03	15 12 07	16 ♍ 27 45	22 ♍ 45 47	26 49	29 ♈ 21	15 00	12 28	25 57	02 57	26 57	03 07
06	22 57 59	16 12 09	29 ♍ 06 44	05 ♎ 30 37	28 ♒ 24	00 ♉ 28	15 48	12 38	26 05	03 00	26 59	03 R 07
07	23 01 56	17 12 10	11 ♎ 57 29	18 ♎ 27 22	00 ♓ 00	01 35	16 35	12 47	26 12	03 03	27 00	03 07
08	23 05 52	18 12 10	25 ♎ 00 19	01 ♏ 36 23	01 38	02 41	17 22	12 57	26 19	03 06	27 02	03 07
09	23 09 49	19 12 07	08 ♏ 15 37	14 ♏ 58 05	03 17	03 48	18 09	13 06	26 27	03 08	27 04	03 07
10	23 13 45	20 12 03	21 ♏ 43 51	28 ♏ 32 58	04 57	04 54	18 56	13 15	26 34	03 11	27 05	03 07
11	23 17 42	21 11 57	05 ♐ 25 27	12 ♐ 21 19	06 38	06 00	19 43	13 24	26 42	03 14	27 07	03 06
12	23 21 39	22 11 50	19 ♐ 20 33	26 ♐ 23 02	08 20	07 05	20 30	13 32	26 49	03 16	27 08	03 06
13	23 25 35	23 11 41	03 ♑ 28 37	10 ♑ 37 04	10 03	08 11	21 18	13 41	26 56	03 19	27 10	03 06
14	23 29 32	24 11 30	17 ♑ 48 04	25 ♑ 01 11	11 48	09 16	22 05	13 49	27 04	03 22	27 11	03 06
15	23 33 28	25 11 18	02 ♒ 15 56	09 ♒ 31 42	13 33	10 21	22 52	13 58	27 11	03 24	27 12	03 05
16	23 37 25	26 11 04	16 ♒ 47 49	24 ♒ 03 36	15 20	11 25	23 39	14 06	27 19	03 27	27 14	03 05
17	23 41 21	27 10 48	01 ♓ 18 16	08 ♓ 31 03	17 08	12 30	24 26	14 14	27 26	03 29	27 15	03 05
18	23 45 18	28 10 30	15 ♓ 41 14	22 ♓ 48 07	18 58	13 34	25 13	14 22	27 34	03 31	27 17	03 04
19	23 49 14	29 ♓ 10 10	29 ♓ 51 04	06 ♈ 49 34	20 48	14 38	26 00	14 30	27 41	03 34	27 18	03 04
20	23 53 11	00 ♈ 09 49	13 ♈ 43 09	20 ♈ 31 32	22 40	15 41	26 46	14 38	27 49	03 36	27 19	03 03
21	23 57 08	01 09 25	27 ♈ 14 30	03 ♉ 51 58	24 33	16 44	27 33	14 45	27 56	03 38	27 20	03 03
22	00 01 04	02 08 59	10 ♉ 23 59	16 ♉ 50 40	26 27	17 47	28 20	14 53	28 03	03 41	27 22	03 02
23	00 05 01	03 08 31	23 ♉ 12 16	29 ♉ 29 06	28 ♓ 23	18 50	29 07	15 00	28 11	03 43	27 23	03 02
24	00 08 57	04 08 00	05 ♊ 41 32	11 ♊ 50 02	00 ♈ 19	19 52	29 ♓ 54	15 07	28 18	03 45	27 24	03 01
25	00 12 54	05 07 27	17 ♊ 55 07	23 ♊ 57 17	02 17	20 54	00 ♈ 41	15 14	28 26	03 47	27 25	03 00
26	00 16 50	06 06 53	29 ♊ 57 07	05 ♋ 55 12	04 16	21 55	01 27	15 21	28 33	03 49	27 26	03 00
27	00 20 47	07 06 15	11 ♋ 52 07	17 ♋ 48 27	06 16	22 56	02 14	15 28	28 41	03 51	27 27	02 59
28	00 24 43	08 05 36	23 ♋ 44 47	29 ♋ 41 41	08 17	23 57	03 01	15 34	28 48	03 53	27 28	02 58
29	00 28 40	09 04 54	05 ♌ 39 41	11 ♌ 39 19	10 19	24 57	03 47	15 41	28 55	03 55	27 29	02 58
30	00 32 37	10 04 10	17 ♌ 41 03	23 ♌ 45 20	12 21	25 57	04 34	15 47	29 03	03 57	27 30	02 57
31	00 36 33	11 ♈ 03 23	29 ♌ 52 33	06 ♍ 03 03	14 ♈ 24	26 ♉ 56	05 ♈ 20	15 ♑ 53	29 ♓ 10	03 ♒ 59	27 ♑ 31	02 ♐ 56

Date	Moon True ☊ ° '	Moon Mean ☊ ° '	Moon ☽ Latitude ° '
01	17 ♎ 45	19 ♎ 14	05 S 06
02	17 R 39	19 11	04 48
03	17 32	19 08	04 17
04	17 26	19 04	03 34
05	17 21	19 01	02 40
06	17 18	18 58	01 38
07	17 17	18 55	00 S 29
08	17 D 17	18 52	00 N 42
09	17 18	18 49	01 53
10	17 20	18 45	02 58
11	17 21	18 42	03 54
12	17 22	18 39	04 38
13	17 R 22	18 36	05 05
14	17 20	18 33	05 14
15	17 18	18 29	05 04
16	17 15	18 26	04 34
17	17 12	18 23	03 46
18	17 10	18 20	02 45
19	17 08	18 17	01 34
20	17 07	18 14	00 N 19
21	17 D 08	18 10	00 S 56
22	17 09	18 07	02 05
23	17 10	18 04	03 07
24	17 11	18 01	03 57
25	17 12	17 58	04 36
26	17 13	17 55	05 02
27	17 R 13	17 51	05 15
28	17 12	17 48	05 15
29	17 11	17 45	05 00
30	17 10	17 42	04 33
31	17 ♎ 10	17 ♎ 39	03 S 53

DECLINATIONS

Date	Sun ☉ ° '	Moon ☽ ° '	Mercury ☿ ° '	Venus ♀ ° '	Mars ♂ ° '	Jupiter ♃ ° '	Saturn ♄ ° '	Uranus ♅ ° '	Neptune ♆ ° '	Pluto ♇ ° '
01	07 S 22	15 N 38	16 S 20	10 N 27	08 S 02	22 S 45	03 S 39	20 S 04	20 S 19	07 S 49
02	06 59	13 13	15 54	10 56	07 44	22 44	03 36	20 04	20 18	07 49
03	06 36	10 12	15 27	11 25	07 25	22 43	03 34	20 03	20 18	07 49
04	06 13	06 43	14 59	11 54	07 07	22 42	03 31	20 02	20 18	07 48
05	05 50	02 N 53	14 30	12 22	06 48	22 41	03 28	20 02	20 18	07 48
06	05 27	01 S 09	14 00	12 50	06 30	22 41	03 25	20 01	20 17	07 48
07	05 03	05 10	13 28	13 18	06 11	22 40	03 22	20 00	20 17	07 47
08	04 40	09 01	12 55	13 45	05 52	22 39	03 19	20 00	20 17	07 47
09	04 16	12 29	12 20	14 12	05 34	22 38	03 16	19 59	20 16	07 47
10	03 53	15 20	11 45	14 39	05 15	22 37	03 13	19 59	20 16	07 46
11	03 29	17 22	11 08	15 06	04 56	22 37	03 10	19 58	20 16	07 46
12	03 06	18 24	10 30	15 32	04 37	22 36	03 07	19 57	20 15	07 45
13	02 42	18 18	09 51	15 58	04 18	22 35	03 04	19 57	20 15	07 45
14	02 18	17 04	09 10	16 23	03 59	22 34	03 01	19 56	20 15	07 45
15	01 55	14 43	08 28	16 48	03 41	22 33	02 58	19 56	20 15	07 44
16	01 31	11 27	07 46	17 13	03 22	22 33	02 55	19 55	20 14	07 44
17	01 07	07 29	07 01	17 38	03 03	22 32	02 52	19 55	20 14	07 43
18	00 44	03 S 07	06 16	18 02	02 44	22 31	02 50	19 54	20 14	07 43
19	00 S 20	01 N 23	05 30	18 25	02 25	22 30	02 47	19 54	20 13	07 43
20	00 N 04	05 42	04 42	18 48	02 06	22 30	02 44	19 53	20 13	07 42
21	00 28	09 37	03 54	19 11	01 47	22 29	02 41	19 53	20 13	07 42
22	00 51	12 57	03 04	19 34	01 28	22 28	02 38	19 52	20 13	07 41
23	01 15	15 34	02 13	19 55	01 09	22 27	02 35	19 52	20 12	07 41
24	01 39	17 21	01 21	20 17	00 50	22 27	02 32	19 51	20 12	07 40
25	02 02	18 18	00 S 29	20 38	00 31	22 26	02 29	19 51	20 12	07 40
26	02 26	18 24	00 N 25	20 59	00 S 12	22 25	02 26	19 50	20 12	07 40
27	02 49	17 40	01 19	21 19	00 N 07	22 25	02 23	19 50	20 12	07 39
28	03 13	16 11	02 14	21 38	00 26	22 24	02 20	19 49	20 11	07 39
29	03 36	14 00	03 09	21 57	00 45	22 23	02 18	19 49	20 11	07 38
30	03 59	11 12	04 05	22 16	01 04	22 23	02 15	19 48	20 11	07 38
31	04 N 22	07 N 53	05 N 02	22 N 34	01 N 23	22 S 22	02 S 12	19 S 48	20 S 11	07 S 37

ZODIAC SIGN ENTRIES

Date	h m	Planets
01	16 47	☽ ♌
04	04 13	☽ ♍
06	02 01	♀ ♉
06	13 40	☽ ♎
07	11 53	☿ ♓
08	21 05	☽ ♏
11	02 32	☽ ♐
13	06 08	☽ ♑
15	08 15	☽ ♒
17	09 50	☽ ♓
19	12 15	☽ ♈
20	08 03	☉ ♈
21	16 59	☽ ♉
24	00 59	☽ ♊
24	08 03	☿ ♈
24	15 12	♂ ♈
26	12 06	☽ ♋
29	00 37	☽ ♌
31	12 15	☽ ♍

LATITUDES

Date	Mercury ☿ ° '	Venus ♀ ° '	Mars ♂ ° '	Jupiter ♃ ° '	Saturn ♄ ° '	Uranus ♅ ° '	Neptune ♆ ° '	Pluto ♇ ° '
01	01 S 49	00 N 54	01 S 00	00 N 10	02 S 01	00 S 33	00 N 29	13 N 11
04	02 00	01 08	00 59	00 10	02 01	00 33	00 29	13 12
07	02 08	01 22	00 58	00 10	02 01	00 33	00 29	13 13
10	02 12	01 36	00 57	00 09	02 01	00 33	00 29	13 14
13	02 13	01 50	00 56	00 09	02 01	00 33	00 29	13 15
16	02 09	02 04	00 55	00 09	02 01	00 33	00 29	13 17
19	02 01	02 18	00 54	00 09	02 01	00 33	00 29	13 18
22	01 48	02 32	00 52	00 08	02 01	00 33	00 29	13 19
25	01 31	02 45	00 51	00 08	02 02	00 33	00 29	13 20
28	01 09	02 59	00 50	00 08	02 02	00 33	00 29	13 20
31	00 S 42	03 N 11	00 S 48	00 N 08	02 S 02	00 S 33	00 N 29	13 N 21

DATA

Julian Date	2450144
Delta T	+62 seconds
Ayanamsa	23° 48' 19"
Synetic vernal point	05° ♓ 18' 40"
True obliquity of ecliptic	23° 26' 15"

MOON'S PHASES, APSIDES AND POSITIONS ☽

Date	h m	Phase	Longitude ° '	Eclipse Indicator
05	09 23	○	15 ♍ 06	
12	17 15	☾	22 ♐ 25	
19	10 45	●	29 ♓ 07	
27	01 31	☽	06 ♋ 40	

Day	h m		
16	05 36	Perigee	
28	02 39	Apogee	
06	05 14	0S	
12	22 10	Max dec	18° S 30'
19	04 37	0N	
26	02 43	Max dec	18° N 28'

LONGITUDES

Date	Chiron ⚷ ° '	Ceres ⚳ ° '	Pallas ⚴ ° '	Juno ⚵ ° '	Vesta ⚶ ° '	Black Moon Lilith ⚸ ° '
01	13 ♎ 21	11 ♐ 12	08 ♏ 35	20 ♒ 27	22 ♏ 07	17 ♋ 16
11	12 ♎ 42	13 ♐ 10	08 ♏ 01	24 ♒ 43	23 ♏ 44	18 ♋ 23
21	11 ♎ 58	14 ♐ 38	06 ♏ 36	28 ♒ 59	24 ♏ 39	19 ♋ 31
31	11 ♎ 11	15 ♐ 33	04 ♏ 25	03 ♓ 14	24 ♏ 47	20 ♋ 38

ASPECTARIAN

h m	Aspects
01 Friday	
01 10	☽ # ☿
05 50	☽ □ ♀
07 37	☽ △ ♄
08 52	☽ ⚼ ☉
10 21	☽ ⚼ ♂
10 25	☽ ☍ ♆
10 48	♂ ✶ ♃
11 26	♂ ∠ ♆
22 22	☽ ☍ ♅
23 02	☽ △ ♇
02 Saturday	
02 45	♀ ⚺ ♄
03 52	☉ ∠ ♆
04 35	☽ ± ☉
04 50	♂ ∥ ♇
05 41	☽ ± ♂
06 04	☉ ✶ ♃
13 58	☽ ⚼ ♄
16 49	☽ ⚻ ♃
17 37	☽ ⚻ ☉
18 28	☽ ⚻ ♂
03 Sunday	
04 01	☽ ∥ ♀
04 52	☽ ± ♃
07 28	♀ □ ♆
07 59	☽ ± ♄
12 15	♀ ± ♇
16 25	☽ ☍ ☿
19 54	☽ ⚻ ♄
22 10	☽ ⚻ ♆
22 48	☽ ⚼ ♃
23 37	☽ △ ♀
04 Monday	
04 45	☽ # ♇
09 07	☽ # ♂
09 52	☽ ⚻ ♅
09 54	☽ ± ♆
10 17	☽ □ ♇
14 02	☉ ☌ ♂
15 34	☽ # ☉
21 32	☽ ± ♅
21 49	☿ ⚺ ♄
05 Tuesday	
03 20	☽ ⚼ ♆
04 15	☽ △ ♃
07 33	☽ ⚼ ♀
08 24	☽ # ♄
09 02	☽ ☍ ♂
09 23	☽ ☍ ☉
14 10	☿ ⚺ ♆
14 52	☽ ⚼ ♅
20 19	♇ St R
20 52	☽ Q ♇
23 10	☿ ∠ ♃
06 Wednesday	
02 23	☽ ± ♀
06 13	☽ ☍ ♄
07 58	☽ △ ♆
10 28	☽ ⚻ ☿
14 47	☽ ⚻ ♀
19 20	☽ △ ♅
19 31	☽ ✶ ♇
23 20	☽ ± ☿
07 Thursday	
01 17	☽ ∥ ♄
11 21	☽ ∥ ☉
12 06	♀ ∠ ♂
13 33	☽ □ ♃
15 58	☿ # ♀
17 41	☽ ∥ ♂
18 27	☽ ⚼ ☿
21 06	☽ ⚻ ♂
22 30	☽ ⚻ ☉
23 22	☽ ∠ ♇
08 Friday	
03 05	♀ ⊥ ♄
04 04	☽ ∥ ♇
08 49	☽ ± ♂
09 19	☉ ∠ ♅
10 25	☽ ± ☉
14 25	☽ ⚻ ♄
15 42	☽ □ ♆
15 51	☽ ⊥ ♇
21 05	♅ ✶ ♇
21 13	♀ ⚻ ♇
21 13	♀ □ ♅
22 55	☽ Q ♃
09 Saturday	
01 25	☽ ± ♄
01 45	☽ △ ☿
02 13	☽ ⚼ ♂
02 43	☽ ⚺ ♇
02 45	☽ □ ♅
03 14	☽ ☍ ♀
04 06	☽ ⚼ ☉
08 45	☿ ⊥ ♆
09 34	☿ □ ♇
09 56	☿ ⚺ ♅
11 06	☽ ∥ ☿
11 39	♂ ∠ ♅
17 46	☽ ⚼ ♄
20 45	☽ ✶ ♃
10 Sunday	
00 11	☽ Q ♆
04 27	☽ # ♀
06 45	☽ △ ♂
09 05	☽ △ ☉
10 02	☿ ✶ ♀
11 02	☽ Q ♅
20 36	☽ △ ♄
21 27	☽ ✶ ♆
23 35	☽ ∠ ♃
11 Monday	
07 58	☽ ☌ ♇
08 10	☽ ✶ ♅
13 05	☽ ⚻ ♀
14 23	☽ □ ☿
15 27	☽ ⊥ ♃
23 36	☽ ∠ ♆
12 Tuesday	
00 21	☽ ± ♀
01 56	☽ ⚺ ♃
10 10	☽ ∠ ♅
10 21	☉ ∥ ♄
14 07	☽ □ ♂
15 04	☽ ⊥ ♆
17 05	☽ ⚼ ♀
17 15	☽ □ ☉
13 Wednesday	
00 51	☽ □ ♄
01 18	☽ ⚺ ♆
01 33	☿ ⊥ ♅
01 33	☽ ⊥ ♅
01 33	☽ Q ☿
11 22	☽ ⚺ ♇
11 44	☽ ⚺ ♅
16 16	☉ ∠ ♀
20 34	☽ △ ♀
21 27	☽ ⊥ ♇
22 21	☽ Q ♂
14 Thursday	
00 34	☽ ✶ ☿
01 56	☽ Q ☉
05 18	☽ ☌ ♃
07 24	☽ Q ♄
12 29	☽ ∠ ♇
17 27	☿ ∠ ♆
18 44	☽ # ♀
19 31	☽ ✶ ♂
23 25	☽ ✶ ☉
15 Friday	
03 31	☽ ✶ ♄
03 37	☽ ☌ ♆
05 00	☽ ∠ ☿
13 22	☽ ✶ ♇
13 53	☽ ☌ ♅
16 44	♄ ✶ ♆
18 02	☿ ✶ ♃
21 46	☽ ∠ ♂
21 57	☽ ⊥ ☿
16 Saturday	
02 03	☽ ∠ ☉
02 25	☽ □ ♀
04 32	☽ ∠ ♄
07 31	☽ ⚺ ♃
09 10	☽ Q ♇
09 15	☽ ⚺ ☿
09 42	☉ Q ♃
12 58	☿ ∥ ♇
13 29	☽ ⊥ ♂
17 31	☽ ⊥ ♃
18 00	☽ ⊥ ☉
19 31	☽ ⊥ ♄
23 57	☽ ⚺ ♂
17 Sunday	
04 40	☽ ⚺ ☉
05 17	☽ ⚺ ♆
05 32	☽ ⚺ ♄
08 32	☽ ∠ ♃
10 30	♀ ∠ ♄
10 33	☽ Q ♀
10 38	☽ ∥ ♇
13 50	☉ ✶ ♆
14 57	☽ □ ♇
15 09	☽ ∥ ☿
15 15	☽ ⊥ ♆
15 38	☽ ⚺ ♅
19 04	☉ ☌ ♄
18 Monday	
01 39	☽ ⊥ ♅
03 15	♂ ∥ ♄
06 10	☿ ∠ ♅
06 16	☽ ∠ ♆
08 09	☽ ✶ ♀
09 46	☽ ✶ ♃
13 33	☽ ∥ ♄
14 12	☽ ∥ ♂
16 47	☽ ∠ ♅
18 19	☽ ☌ ☿
19 Tuesday	
01 58	☽ ∥ ☉
05 02	☽ ☌ ♂
06 14	☽ Q ♃
06 50	☽ # ☉
07 38	☽ ✶ ♆
08 16	☽ ☌ ♄
08 48	♀ △ ♃
10 45	☽ ☌ ☉
11 35	☽ ∠ ♀
17 14	☽ # ♂
17 30	☽ △ ♇
18 23	☽ ✶ ♅
19 31	☽ # ♄
20 Wednesday	
04 19	☽ Q ♆
04 22	☽ ⊥ ♀
06 43	♂ Q ♃
07 12	☽ # ☿
13 37	☽ □ ♃
15 19	☽ Q ♅
15 44	☽ ⚺ ♀
19 37	☽ ⚼ ♇
23 50	☽ # ♇
21 Thursday	
05 14	♂ ✶ ♆
06 22	☽ ⚺ ☿
11 39	☽ ± ♇
12 11	☽ □ ♆
12 36	☽ ⚺ ♂
13 15	☽ ⚺ ♄
18 58	☽ ⊥ ☿
19 39	☽ ⚺ ☉
22 30	☽ ⚻ ♇
23 37	☽ □ ♅
22 Friday	
00 09	☽ ⊥ ♂
00 14	☽ ⊥ ♄
01 48	♂ ☌ ♄
07 30	☽ ⊥ ☉
14 17	☽ ∠ ☿
16 59	☽ ∠ ♄
17 44	☿ Q ♃
17 48	☽ ∠ ♂
20 24	☽ △ ♃
23 29	☿ ✶ ♆
23 Saturday	
01 08	☿ ∥ ♄
01 38	☽ ∠ ☉
02 59	☽ ☌ ♀
07 46	♀ # ♅
08 29	☉ # ♂
09 15	☉ △ ♇
09 24	☿ ☌ ♄
19 59	☽ △ ♆
21 35	☽ ✶ ♄
23 41	☽ ✶ ☿
24 Sunday	
00 03	☽ ✶ ♂
01 07	☽ ⚼ ♃
02 21	☉ ✶ ♅
03 17	☿ ☌ ♂
06 34	☉ # ☿
06 46	♀ # ♆
06 49	☽ ☍ ♇
08 13	☽ △ ♅
08 43	☽ ✶ ☉
18 45	☽ ± ♃
21 06	☽ Q ♄
25 Monday	
00 57	☽ Q ♂
01 08	☽ ⚼ ♆
03 26	☽ Q ☿
06 39	☽ ⚻ ♃
10 17	☽ Q ☉
10 39	☿ ∥ ♂
13 43	☽ ⚼ ♅
18 27	☽ ⚺ ♀
18 58	☽ ± ♆
20 43	☿ △ ♇
26 Tuesday	
06 30	☿ ✶ ♅
06 57	☽ ⚻ ♆
07 32	☽ ⊥ ♀
07 43	☽ ± ♅
07 45	☿ # ♂
09 10	☽ □ ♄
12 27	☉ # ♄
15 13	☽ □ ♂
18 06	☽ ⚻ ♇
19 48	☽ ⚻ ♅
22 25	☽ □ ☿
27 Wednesday	
01 31	☽ □ ☉
03 19	☽ ∠ ♀
06 11	☽ ± ♇
19 20	☽ ☍ ♃
28 Thursday	
00 21	☽ ⚼ ♇
07 37	☉ ☌ ☿
10 51	♂ △ ♇
12 26	☽ ✶ ♀
14 42	☿ # ♄
19 32	☽ ☍ ♆
22 18	☽ △ ♄
29 Friday	
02 14	☿ Q ♆
03 34	☿ ∠ ♀
06 35	☽ △ ♇
07 58	☽ △ ♂
08 29	☽ ☍ ♅
14 49	☽ Q ♀
16 18	♂ ✶ ♅
19 28	☽ △ ☉
22 07	☉ Q ♆
23 13	☽ △ ☿
30 Saturday	
04 41	☽ ⚼ ♄
07 31	☉ ∥ ☿
08 12	☽ ⚻ ♃
15 59	☽ ⚼ ♂
20 11	☽ ± ♃
20 14	♀ # ♃
22 42	☽ ± ♄
31 Sunday	
03 52	☽ ⚼ ☉
05 45	☽ □ ♀
07 24	☽ ⚻ ♆
10 36	☽ ⚻ ♄
10 53	☽ ± ♂
10 54	☽ ⚼ ☿
13 42	☽ # ♇
13 59	☽ ⚼ ♃
17 57	☽ □ ♇
19 07	☽ ± ♆
20 01	☽ ⚻ ♅
22 57	☽ ± ☉
23 20	☽ ⚻ ♂

All ephemeris data is given at 12.00 UT and the Moon's longitude is additionally given for 24.00 UT

APRIL 1996

LONGITUDES

Date	Sidereal time h m s	Sun ☉ ° ' "	Moon ☽ ° ' "	Moon ☽ 24.00 ° ' "	Mercury ☿ ° '	Venus ♀ ° '	Mars ♂ ° '	Jupiter ♃ ° '	Saturn ♄ ° '	Uranus ♅ ° '	Neptune ♆ ° '	Pluto ♇ ° '
01	00 40 30	12 ♈ 02 34	12 ♍ 17 07	18 ♍ 35 00	16 ♈ 28	27 ♉ 55	06 ♈ 07	15 ♑ 59	29 ♓ 17	04 ♒ 01	27 ♑ 32	02 ♐ 55
02	00 44 26	13 01 43	24 ♍ 56 50	01 ♎ 22 44	18 32	28 54	06 53	16 05	29 25	04 03	27 33	02 R 54
03	00 48 23	14 00 50	07 ♎ 52 45	14 ♎ 26 51	20 35	29 ♉ 52	07 40	16 11	29 32	04 04	27 34	02 53
04	00 52 19	14 59 55	21 ♎ 04 58	27 ♎ 46 57	22 39	00 ♊ 49	08 26	16 16	29 39	04 06	27 35	02 52
05	00 56 16	15 58 58	04 ♏ 32 37	11 ♏ 21 45	24 42	01 46	09 12	16 21	29 46	04 08	27 36	02 51
06	01 00 12	16 57 59	18 ♏ 14 05	25 ♏ 09 19	26 43	02 43	09 59	16 26	29 ♓ 54	04 09	27 36	02 50
07	01 04 09	17 56 58	02 ♐ 07 07	09 ♐ 07 12	28 ♈ 44	03 39	10 45	16 31	00 ♈ 01	04 11	27 37	02 49
08	01 08 06	18 55 55	16 ♐ 09 12	23 ♐ 12 48	00 ♉ 43	04 34	11 31	16 36	00 08	04 12	27 38	02 48
09	01 12 02	19 54 51	00 ♑ 17 39	07 ♑ 23 27	02 40	05 29	12 17	16 41	00 15	04 14	27 39	02 47
10	01 15 59	20 53 45	14 ♑ 29 52	21 ♑ 36 35	04 35	06 23	13 04	16 45	00 22	04 15	27 39	02 46
11	01 19 55	21 52 37	28 ♑ 43 18	05 ♒ 49 43	06 27	07 17	13 50	16 50	00 30	04 17	27 40	02 45
12	01 23 52	22 51 27	12 ♒ 55 31	20 ♒ 00 24	08 16	08 10	14 36	16 54	00 37	04 18	27 40	02 44
13	01 27 48	23 50 16	27 ♒ 04 03	04 ♓ 06 10	10 01	09 03	15 22	16 58	00 44	04 19	27 41	02 43
14	01 31 45	24 49 03	11 ♓ 06 26	18 ♓ 04 32	11 43	09 54	16 08	17 02	00 51	04 20	27 41	02 42
15	01 35 41	25 47 48	25 ♓ 00 08	01 ♈ 52 57	13 21	10 45	16 54	17 05	00 58	04 22	27 42	02 40
16	01 39 38	26 46 32	08 ♈ 42 42	15 ♈ 29 06	14 55	11 36	17 40	17 09	01 05	04 23	27 42	02 39
17	01 43 35	27 45 13	22 ♈ 11 55	28 ♈ 50 57	16 25	12 25	18 26	17 12	01 12	04 24	27 43	02 38
18	01 47 31	28 43 53	05 ♉ 26 03	11 ♉ 57 07	17 50	13 14	19 11	17 15	01 19	04 25	27 43	02 37
19	01 51 28	29 ♈ 42 30	18 ♉ 24 07	24 ♉ 47 02	19 10	14 02	19 57	17 18	01 26	04 26	27 44	02 35
20	01 55 24	00 ♉ 41 06	01 ♊ 05 58	07 ♊ 21 04	20 25	14 49	20 43	17 21	01 32	04 27	27 44	02 34
21	01 59 21	01 39 40	13 ♊ 32 32	19 ♊ 40 39	21 35	15 35	21 29	17 23	01 39	04 28	27 44	02 33
22	02 03 17	02 38 11	25 ♊ 45 43	01 ♋ 48 08	22 40	16 21	22 14	17 26	01 46	04 28	27 44	02 31
23	02 07 14	03 36 40	07 ♋ 48 20	13 ♋ 46 47	23 40	17 05	23 00	17 28	01 53	04 29	27 45	02 30
24	02 11 10	04 35 08	19 ♋ 43 59	25 ♋ 40 30	24 35	17 49	23 46	17 30	01 59	04 30	27 45	02 29
25	02 15 07	05 33 33	01 ♌ 36 52	07 ♌ 33 41	25 24	18 31	24 31	17 31	02 06	04 31	27 45	02 27
26	02 19 04	06 31 56	13 ♌ 31 32	19 ♌ 31 00	26 08	19 12	25 16	17 33	02 13	04 31	27 45	02 26
27	02 23 00	07 30 17	25 ♌ 32 42	01 ♍ 37 11	26 46	19 53	26 02	17 35	02 19	04 32	27 45	02 24
28	02 26 57	08 28 36	07 ♍ 45 00	13 ♍ 56 39	27 18	20 32	26 47	17 36	02 26	04 32	27 45	02 23
29	02 30 53	09 26 52	20 ♍ 12 36	26 ♍ 33 16	27 45	21 10	27 33	17 37	02 32	04 33	27 R 45	02 21
30	02 34 50	10 ♉ 25 07	02 ♎ 58 59	09 ♎ 29 59	28 ♉ 07	21 ♊ 46	28 ♈ 18	17 ♑ 38	02 ♈ 39	04 ♒ 33	27 ♑ 45	02 ♐ 20

Moon

Date	Moon True ☊ ° '	Moon Mean ☊ ° '	Moon Latitude ° '
01	17 ♎ 09	17 ♎ 35	03 S 01
02	17 R 08	17 32	02 00
03	17 08	17 29	00 S 51
04	17 D 08	17 26	00 N 22
05	17 08	17 23	01 35
06	17 08	17 20	02 44
07	17 R 08	17 16	03 44
08	17 08	17 13	04 32
09	17 08	17 10	05 04
10	17 08	17 07	05 17
11	17 D 08	17 04	05 11
12	17 08	17 01	04 46
13	17 08	16 57	04 03
14	17 09	16 54	03 07
15	17 10	16 51	02 00
16	17 10	16 48	00 N 47
17	17 R 10	16 45	00 S 28
18	17 10	16 41	01 39
19	17 09	16 38	02 44
20	17 07	16 35	03 40
21	17 05	16 32	04 23
22	17 03	16 29	04 54
23	17 01	16 26	05 11
24	16 59	16 22	05 15
25	16 59	16 19	05 05
26	16 D 59	16 16	04 42
27	17 00	16 13	04 07
28	17 01	16 10	03 20
29	17 03	16 07	02 23
30	17 ♎ 04	16 ♎ 03	01 S 17

DECLINATIONS

Date	Sun ☉ ° '	Moon ☽ ° '	Mercury ☿ ° '	Venus ♀ ° '	Mars ♂ ° '	Jupiter ♃ ° '	Saturn ♄ ° '	Uranus ♅ ° '	Neptune ♆ ° '	Pluto ♇ ° '
01	04 N 46	04 N 09	05 N 58	22 N 52	01 N 42	22 S 21	02 S 09	19 S 48	20 S 11	07 S 37
02	05 09	00 N 10	06 55	23 09	02 00	22 21	02 06	19 47	20 10	07 37
03	05 32	03 S 55	07 51	23 26	02 19	22 20	02 03	19 47	20 10	07 36
04	05 54	07 53	08 48	23 42	02 38	22 20	02 00	19 46	20 10	07 36
05	06 17	11 32	09 43	23 58	02 57	22 19	01 58	19 46	20 10	07 35
06	06 40	14 38	10 38	24 13	03 15	22 19	01 55	19 46	20 10	07 35
07	07 02	16 55	11 32	24 28	03 34	22 18	01 52	19 45	20 10	07 34
08	07 25	18 12	12 25	24 42	03 52	22 18	01 49	19 45	20 09	07 34
09	07 47	18 23	13 16	24 55	04 11	22 17	01 46	19 45	20 09	07 34
10	08 09	17 24	14 05	25 08	04 29	22 17	01 44	19 44	20 09	07 33
11	08 31	15 21	14 53	25 21	04 48	22 16	01 41	19 44	20 09	07 33
12	08 53	12 22	15 39	25 33	05 06	22 16	01 38	19 44	20 09	07 32
13	09 15	08 40	16 23	25 45	05 24	22 15	01 36	19 44	20 09	07 32
14	09 37	04 31	17 04	25 55	05 42	22 15	01 33	19 43	20 09	07 31
15	09 58	00 S 09	17 43	26 06	06 00	22 15	01 30	19 43	20 09	07 31
16	10 19	04 N 10	18 20	26 16	06 18	22 14	01 27	19 43	20 09	07 31
17	10 40	08 13	18 53	26 25	06 36	22 14	01 25	19 43	20 08	07 30
18	11 01	11 46	19 25	26 34	06 54	22 14	01 22	19 42	20 08	07 30
19	11 22	14 40	19 53	26 42	07 12	22 13	01 20	19 42	20 08	07 29
20	11 43	16 48	20 19	26 50	07 30	22 13	01 17	19 42	20 08	07 29
21	12 03	18 04	20 43	26 57	07 47	22 13	01 14	19 42	20 08	07 28
22	12 23	18 28	21 04	27 04	08 05	22 13	01 12	19 42	20 08	07 28
23	12 43	18 02	21 22	27 10	08 22	22 13	01 09	19 42	20 08	07 28
24	13 03	16 47	21 37	27 16	08 40	22 12	01 07	19 41	20 08	07 27
25	13 22	14 50	21 50	27 21	08 57	22 12	01 04	19 41	20 08	07 27
26	13 42	12 15	22 00	27 26	09 14	22 12	01 02	19 41	20 08	07 26
27	14 01	09 08	22 07	27 30	09 31	22 12	00 59	19 41	20 08	07 26
28	14 20	05 34	22 12	27 34	09 48	22 12	00 57	19 41	20 08	07 26
29	14 38	01 42	22 15	27 37	10 05	22 12	00 54	19 41	20 08	07 25
30	14 N 57	02 S 22	22 N 15	27 N 40	10 N 21	22 S 12	00 S 52	19 S 41	20 S 08	07 S 25

ZODIAC SIGN ENTRIES

Date	h m	Planets
02	21 26	☽ ♎
03	15 26	♀ ♊
05	03 57	☽ ♏
07	08 21	☽ ♐
07	08 49	♄ ♈
08	03 16	☿ ♉
09	11 30	☽ ♑
11	14 09	☽ ♒
13	17 00	☽ ♓
15	20 42	☽ ♈
18	02 05	☽ ♉
19	19 10	☉ ♉
20	09 54	☽ ♊
22	20 25	☽ ♋
25	08 44	☽ ♌
27	20 49	☽ ♍
30	06 27	☽ ♎

LATITUDES

Date	Mercury ☿ ° '	Venus ♀ ° '	Mars ♂ ° '	Jupiter ♃ ° '	Saturn ♄ ° '	Uranus ♅ ° '	Neptune ♆ ° '	Pluto ♇ ° '
01	00 S 33	03 N 15	00 S 48	00 N 07	02 S 02	00 S 33	00 N 29	13 N 22
04	00 S 01	03 28	00 47	00 07	02 02	00 34	00 29	13 22
07	00 N 33	03 39	00 45	00 07	02 03	00 34	00 29	13 23
10	01 07	03 50	00 44	00 07	02 03	00 34	00 29	13 24
13	01 39	03 59	00 43	00 06	02 03	00 34	00 29	13 25
16	02 07	04 08	00 41	00 06	02 03	00 34	00 29	13 25
19	02 28	04 15	00 39	00 06	02 04	00 34	00 29	13 26
22	02 43	04 21	00 37	00 05	02 04	00 34	00 29	13 26
25	02 48	04 26	00 36	00 05	02 05	00 34	00 29	13 27
28	02 43	04 29	00 34	00 05	02 05	00 34	00 29	13 27
31	02 N 28	04 N 30	00 S 32	00 N 04	02 S 05	00 S 35	00 N 29	13 N 28

DATA

Julian Date	2450175
Delta T	+62 seconds
Ayanamsa	23° 48' 22"
Synetic vernal point	05° ♓ 18' 37"
True obliquity of ecliptic	23° 26' 15"

MOON'S PHASES, APSIDES AND POSITIONS ☽

Date	h m	Phase	Longitude ° '	Eclipse Indicator
04	00 07	○	14 ♎ 31	total
10	23 36	☾	21 ♑ 22	
17	22 49	●	28 ♈ 12	Partial
25	20 40	☽	05 ♌ 55	

Day	h m		
11	02 47	Perigee	
24	22 25	Apogee	
02	13 01	0S	
09	03 34	Max dec	18° S 27'
15	12 50	0N	
22	11 14	Max dec	18° N 28'
29	22 06	0S	

LONGITUDES

Date	Chiron ⚷ ° '	Ceres ⚳ ° '	Pallas ⚴ ° '	Juno ⚵ ° '	Vesta ⚶ ° '	Black Moon Lilith ⚸ ° '
01	11 ♎ 06	15 ♐ 36	04 ♏ 10	03 ♓ 39	24 ♏ 45	20 ♋ 45
11	10 ♎ 20	15 ♐ 50	01 ♏ 22	07 ♓ 51	23 ♏ 58	21 ♋ 52
21	09 ♎ 36	15 ♐ 25	28 ♎ 17	12 ♓ 01	22 ♏ 25	22 ♋ 59
31	08 ♎ 57	14 ♐ 22	25 ♎ 19	16 ♓ 06	20 ♏ 17	24 ♋ 06

All ephemeris data is given at 12.00 UT and the Moon's longitude is additionally given for 24.00 UT

ASPECTARIAN

h m	Aspects
01 Monday	
02 31	♀ △ ♆
02 51	☽ ∥ ☿
06 08	☿ □ ♃
06 39	☿ Q ♅
07 38	☽ ± ♅
07 49	☽ ± ☿
08 35	☽ ∥ ☉
11 30	☽ ⊼ ☉
12 29	☽ ⚻ ♆
19 07	☽ △ ♃
21 32	☽ ⊼ ☿
02 Tuesday	
00 22	☽ # ♄
00 51	☽ ⚻ ♅
01 52	☽ ∥ ♂
04 24	☽ Q ♇
04 47	☿ ⚻ ♇
16 53	☽ △ ♆
18 21	♂ # ♄
19 59	☽ △ ♀
20 25	☽ ☍ ♄
03 Wednesday	
01 13	☽ ∥ ♄
01 52	☽ # ♂
02 36	♀ ⚹ ♄
02 48	☽ ⚹ ♇
04 58	☽ △ ♅
05 32	☿ # ♇
11 34	☽ ☍ ♂
22 38	☽ # ☉
04 Thursday	
00 07	☽ ☍ ☉
01 45	☽ ⚻ ♀
03 15	☽ □ ♃
06 13	☽ ∠ ♇
10 11	☽ ∥ ♇
15 19	☽ ☍ ☿
19 13	☽ ± ♀
19 34	☽ # ☿
22 22	☽ ⊥ ♇
23 39	☽ □ ♆
05 Friday	
00 22	♀ ⚻ ♃
03 28	☽ ⊼ ♄
06 43	☽ ⊼ ♀
09 01	☽ ⊻ ♇
11 16	☽ □ ♅
11 40	☽ Q ♃
14 12	☽ ± ♄
15 37	☉ Q ♅
20 42	☽ ⊼ ♂
21 58	☉ □ ♃
06 Saturday	
00 18	♂ Q ♆
06 08	☽ ⚻ ♄
07 25	☽ Q ♆
07 50	☽ ± ♂
08 51	☽ ⚹ ♃
09 37	☽ ⊼ ☉
11 48	☿ ⊥ ♀
13 22	☿ ± ♇
15 09	♀ ☍ ♇
18 49	☽ Q ♅
20 50	☽ ± ☉
22 35	☿ □ ♆
07 Sunday	
00 23	☽ ⚻ ♂
04 15	☽ ⚹ ♆
05 12	☽ ⊼ ☿
08 21	☽ △ ♄
08 58	☉ ⚻ ♇
10 58	☽ ∠ ♃
13 13	☽ ☌ ♇
13 32	☽ ⚻ ☉
14 49	☽ ☍ ♀
15 33	☽ ⚹ ♅
17 14	☽ ± ☿
08 Monday	
02 10	♀ △ ♅
02 29	☽ ⊥ ♃
03 39	☽ △ ♂
04 28	☿ ⊻ ♄
06 00	☽ ∠ ♆
11 08	☽ ⚻ ☿
12 46	☽ ⊻ ♃
17 05	☽ △ ☉
17 12	☽ ∠ ♅
21 19	☽ ⊥ ♆
21 32	☉ # ♇
09 Tuesday	
07 30	☽ ⊻ ♆
08 30	☽ ⊥ ♅
11 56	☽ □ ♄
13 29	☿ ⊼ ♇
16 13	☽ ⊻ ♇
16 39	☽ △ ☿
18 40	☽ ⊻ ♅
21 23	☽ ⊼ ♀
10 Wednesday	
02 21	☽ ⊥ ♇
07 48	☿ □ ♅
08 12	☽ ± ♀
09 26	☽ □ ♂
15 50	☽ ☌ ♃
17 31	☽ ∠ ♇
18 36	☽ Q ♄
23 36	☽ □ ☉
11 Thursday	
00 25	☽ ⚻ ♀
10 13	☽ ☌ ♆
12 40	☿ ⊥ ♄
15 01	☽ ⚹ ♄
15 14	☽ # ☿
17 33	☽ Q ♂
18 48	☽ ⚹ ♇
21 24	☽ ☌ ♅
12 Friday	
02 58	☽ □ ☿
03 25	☽ △ ♀
08 15	☽ Q ☉
09 34	☿ ⊻ ♀
15 00	☽ ⚹ ♂
15 03	☽ Q ♇
16 35	☽ ∠ ♄
18 46	☽ ⊻ ♃
13 Saturday	
04 59	☽ ⊥ ♃
06 06	☽ ⚹ ☉
07 59	☽ ⊥ ♄
08 46	☽ # ☉
13 03	☽ ⊻ ♆
13 51	☽ Q ☿
17 56	☽ ∠ ♂
18 18	☽ ⊻ ♄
18 49	☽ ∥ ♇
20 23	☽ ∠ ♃
21 37	☽ □ ♇
23 17	☽ ⊥ ♆
14 Sunday	
00 23	☽ ⊻ ♅
05 46	☽ # ♂
09 37	☽ ∠ ☉
09 48	☽ □ ♀
10 14	☽ ⊥ ♂
10 41	☽ ⊥ ♅
13 12	☽ ⚹ ☿
14 43	☽ ∠ ♆
18 42	♂ Q ♅
21 09	☽ ⊻ ♂
22 14	☽ ⚹ ♃
15 Monday	
02 12	☽ ∠ ♅
02 17	☽ ⊥ ☉
04 32	☽ ∥ ♄
04 33	☉ ∠ ♀
13 29	☽ ⊻ ☉
16 42	☽ ⚹ ♆
18 27	♂ □ ♃
18 36	☽ ∠ ☿
18 58	☽ Q ♀
19 09	☽ Q ♃
21 00	☽ # ♄
22 09	♀ ± ♃
22 29	☽ ☌ ♄
16 Tuesday	
01 22	☽ △ ♇
04 22	☽ ⚹ ♅
09 02	☉ ± ♇
11 41	♂ ⚻ ♇
12 25	☽ ⊥ ☿
13 45	☽ Q ♆
17 26	☽ ⚹ ♀
17 Wednesday	
00 23	☽ ⊻ ☿
01 24	☽ ∥ ♂
01 36	☽ Q ♅
03 01	☽ □ ♃
03 50	☽ ⚻ ♇
04 50	☽ ☌ ♂
07 36	☽ # ♇
08 11	☿ ∠ ♄
11 01	☉ □ ♆
19 58	☽ ± ♇
20 42	♀ ⚻ ♆
21 57	☽ □ ♆
22 02	☽ ∠ ♀
22 49	☽ ☌ ☉
18 Thursday	
01 43	☿ △ ♃
04 24	☽ ⊻ ♄
06 00	☽ ∥ ☉
06 51	☽ ⊼ ♇
10 08	☽ □ ♅
14 44	♀ Q ♄
15 28	☽ ⊥ ♄
15 31	☽ ⊥ ♀
19 Friday	
02 22	☿ # ♅
03 19	☽ ⊻ ♀
08 17	☽ ∠ ♄
09 56	☽ △ ♃
13 35	☽ ☌ ☿
15 06	☽ ⊻ ♂
20 Saturday	
01 18	☿ # ♆
03 07	☽ ⊥ ♂
05 35	☽ △ ♆
10 51	♂ # ♇
11 09	☽ ⊻ ☉
12 51	☽ ⚹ ♄
14 23	☽ ⚻ ♃
14 48	☽ ☍ ♇
18 25	☽ △ ♅
21 26	☽ ∠ ♂
23 38	☽ ⊥ ☉
21 Sunday	
05 09	☿ ⊻ ♂
07 48	☽ ± ♃
10 26	☽ ⚻ ♆
11 49	☉ ⊻ ♄
12 13	☽ Q ♄
16 15	☽ ☌ ♀
18 37	☽ ∠ ☉
19 32	☽ ⊼ ♃
23 35	☽ ⚻ ♅
22 Monday	
04 03	☽ ± ♆
04 35	☽ ⚹ ♂
05 19	☽ ⊻ ☿
09 14	☉ ⊼ ♇
15 55	☽ ⊼ ♆
17 23	☽ ± ♅
18 18	☽ ⊥ ☿
23 Tuesday	
00 03	☽ □ ♄
01 25	☽ ⊼ ♇
02 52	☽ ⚹ ☉
05 21	☽ ⊼ ♅
06 00	☽ Q ♂
13 23	☽ ± ♇
13 53	☽ ∠ ☿
24 Wednesday	
01 01	♀ ⊼ ♃
03 35	♃ ∠ ♇
05 05	☽ Q ☉
07 27	☽ ⚻ ♇
07 28	☽ ☍ ♃
07 52	☽ ⊻ ♀
09 51	☉ □ ♅
20 41	☽ □ ♂
20 45	☽ ⊥ ♀
22 32	☽ ⚹ ☿
25 Thursday	
04 11	☽ ☍ ♆
13 00	☽ △ ♄
13 41	☽ △ ♇
16 05	☽ ∠ ♀
17 51	☽ ☍ ♅
20 40	☽ □ ☉
26 Friday	
00 27	☽ Q ☿
00 39	☽ ∥ ☉
19 28	☽ ⚻ ♄
20 05	☽ ⊼ ♃
23 24	♀ ⚻ ♅
27 Saturday	
00 03	☽ ⚹ ♀
08 05	☽ ± ♃
09 24	☽ ∥ ♂
13 02	☽ △ ♂
13 33	☽ ± ♄
14 32	☽ □ ☿
16 22	☽ ⊼ ♆
23 27	♂ ± ♇
23 45	☽ # ♇
28 Sunday	
01 13	☽ Q ♀
01 30	☽ ⊼ ♄
01 31	☽ □ ♇
01 54	☽ ⚻ ♃
02 49	♄ △ ♇
04 11	☽ ± ♆
05 44	☽ ⊼ ♅
09 46	☿ # ♃
10 43	☉ ⊥ ♄
13 32	☽ △ ☉
17 26	☽ ± ♅
20 21	☽ ⚻ ♂
21 42	☽ ⚻ ♆
29 Monday	
07 02	☽ △ ♃
09 55	♆ St R
10 44	☽ ⚻ ♅
11 51	☿ △ ♆
12 17	☽ Q ♇
13 54	☽ □ ♀
14 42	☽ ± ♂
16 46	☽ # ♄
18 41	♂ □ ♆
20 42	☽ ⚻ ☉
30 Tuesday	
01 32	☿ ⊻ ♂
02 15	☽ △ ♆
02 41	☽ △ ☿
02 43	☽ ⊼ ♂
03 18	☽ ∥ ♄
10 47	☽ ⚹ ♇
11 19	♀ ± ♆
11 22	☽ ☍ ♄
14 52	☽ ± ☉
14 55	☽ △ ♅

MAY 1996

Date	Sidereal time h m s	Sun ☉ ° ' "	Moon ☽ ° ' "	Moon ☽ 24.00 ° ' "	Mercury ☿ ° '	Venus ♀ ° '	Mars ♂ ° '	Jupiter ♃ ° '	Saturn ♄ ° '	Uranus ♅ ° '	Neptune ♆ ° '	Pluto ♇ ° '
01	02 38 46	11 ♉ 23 20	16 ♎ 06 26	22 ♎ 48 24	28 ♉ 23	22 ♊ 22	29 ♈ 03	17 ♑ 38	02 ♈ 45	04 ♒ 34	27 ♑ 45	02 ♐ 18
02	02 42 43	12 21 31	29 ♎ 35 48	06 ♏ 28 27	28 33	22 56	29 ♈ 48	17 39	02 51	04 34	27 R 45	02 R 17
03	02 46 39	13 19 40	13 ♏ 26 04	20 ♏ 28 13	28 38	23 28	00 ♉ 33	17 39	02 58	04 34	27 45	02 15
04	02 50 36	14 17 48	27 ♏ 34 23	04 ♐ 43 57	28 R 38	23 59	01 18	17 39	03 04	04 35	27 45	02 14
05	02 54 33	15 15 54	11 ♐ 56 12	19 ♐ 10 26	28 32	24 29	02 03	17 R 39	03 10	04 35	27 45	02 12
06	02 58 29	16 13 58	26 ♐ 25 51	03 ♑ 41 44	28 22	24 57	02 48	17 39	03 16	04 35	27 44	02 11
07	03 02 26	17 12 01	10 ♑ 57 19	18 ♑ 11 58	28 07	25 24	03 33	17 39	03 22	04 35	27 44	02 09
08	03 06 22	18 10 03	25 ♑ 25 05	02 ♒ 36 08	27 48	25 49	04 18	17 38	03 28	04 35	27 44	02 07
09	03 10 19	19 08 03	09 ♒ 44 44	16 ♒ 50 33	27 25	26 12	05 03	17 37	03 34	04 R 35	27 44	02 06
10	03 14 15	20 06 02	23 ♒ 53 21	00 ♓ 52 59	26 59	26 33	05 48	17 36	03 40	04 35	27 43	02 04
11	03 18 12	21 03 59	07 ♓ 49 21	14 ♓ 42 25	26 29	26 53	06 32	17 35	03 46	04 35	27 43	02 03
12	03 22 08	22 01 56	21 ♓ 32 12	28 ♓ 18 44	25 57	27 11	07 17	17 34	03 52	04 35	27 42	02 01
13	03 26 05	22 59 51	05 ♈ 02 03	11 ♈ 42 12	25 24	27 27	08 02	17 32	03 58	04 34	27 42	01 59
14	03 30 02	23 57 44	18 ♈ 19 15	24 ♈ 53 13	24 49	27 40	08 46	17 30	04 03	04 34	27 42	01 58
15	03 33 58	24 55 37	01 ♉ 24 10	07 ♉ 52 05	24 14	27 52	09 31	17 28	04 09	04 34	27 41	01 56
16	03 37 55	25 53 28	14 ♉ 17 01	20 ♉ 38 59	23 38	28 02	10 15	17 26	04 15	04 34	27 41	01 54
17	03 41 51	26 51 18	26 ♉ 57 59	03 ♊ 14 04	23 03	28 09	11 00	17 24	04 20	04 33	27 40	01 53
18	03 45 48	27 49 06	09 ♊ 27 17	15 ♊ 37 43	22 30	28 14	11 44	17 21	04 26	04 33	27 39	01 51
19	03 49 44	28 46 53	21 ♊ 45 29	27 ♊ 50 44	21 58	28 17	12 28	17 19	04 31	04 32	27 39	01 50
20	03 53 41	29 ♉ 44 38	03 ♋ 53 39	09 ♋ 54 28	21 29	28 R 18	13 13	17 16	04 36	04 32	27 38	01 48
21	03 57 37	00 ♊ 42 22	15 ♋ 53 30	21 ♋ 51 05	21 02	28 16	13 57	17 13	04 42	04 31	27 37	01 46
22	04 01 34	01 40 05	27 ♋ 47 35	03 ♌ 43 28	20 39	28 12	14 41	17 09	04 47	04 30	27 37	01 45
23	04 05 31	02 37 46	09 ♌ 39 11	15 ♌ 35 17	20 19	28 05	15 25	17 06	04 52	04 30	27 36	01 43
24	04 09 27	03 35 25	21 ♌ 32 18	27 ♌ 30 49	20 03	27 56	16 09	17 03	04 57	04 29	27 35	01 41
25	04 13 24	04 33 03	03 ♍ 31 28	09 ♍ 34 52	19 51	27 45	16 53	16 59	05 02	04 28	27 34	01 40
26	04 17 20	05 30 39	15 ♍ 41 38	21 ♍ 52 24	19 43	27 31	17 37	16 55	05 07	04 27	27 34	01 38
27	04 21 17	06 28 14	28 ♍ 07 45	04 ♎ 28 16	19 39	27 15	18 21	16 51	05 12	04 27	27 33	01 36
28	04 25 13	07 25 47	10 ♎ 54 26	17 ♎ 26 43	19 D 40	26 56	19 05	16 47	05 16	04 26	27 32	01 35
29	04 29 10	08 23 19	24 ♎ 05 26	00 ♏ 50 48	19 45	26 35	19 48	16 42	05 21	04 25	27 31	01 33
30	04 33 06	09 20 50	07 ♏ 42 55	14 ♏ 41 41	19 55	26 12	20 32	16 38	05 26	04 24	27 30	01 31
31	04 37 03	10 ♊ 18 20	21 ♏ 46 52	28 ♏ 58 00	20 ♉ 10	25 ♊ 47	21 ♉ 16	16 ♑ 33	05 ♈ 30	04 ♒ 23	27 ♑ 29	01 ♐ 30

Date	Moon True ☊ ° '	Moon Mean ☊ ° '	Moon ☽ Latitude ° '	DECLINATIONS Sun ☉ ° '	Moon ☽ ° '	Mercury ☿ ° '	Venus ♀ ° '	Mars ♂ ° '	Jupiter ♃ ° '	Saturn ♄ ° '	Uranus ♅ ° '	Neptune ♆ ° '	Pluto ♇ ° '
01	17 ♎ 04	16 ♎ 00	00 S 05	15 N 15	06 S 25	22 N 12	27 N 42	10 N 38	22 S 12	00 S 50	19 S 41	20 S 08	07 S 25
02	17 R 04	15 57	01 N 08	15 33	10 16	22 07	27 44	10 55	22 12	00 47	19 41	20 08	07 24
03	17 02	15 54	02 20	15 50	13 39	22 00	27 46	11 11	22 12	00 45	19 41	20 08	07 24
04	16 59	15 51	03 24	16 08	16 18	21 50	27 46	11 27	22 12	00 42	19 41	20 08	07 23
05	16 55	15 47	04 16	16 25	17 59	21 38	27 47	11 43	22 12	00 40	19 41	20 08	07 23
06	16 51	15 44	04 53	16 42	18 30	21 24	27 47	11 59	22 13	00 38	19 41	20 08	07 23
07	16 47	15 41	05 11	16 58	17 49	21 08	27 46	12 15	22 13	00 36	19 41	20 08	07 22
08	16 45	15 38	05 09	17 14	15 59	20 50	27 45	12 31	22 13	00 33	19 41	20 08	07 22
09	16 43	15 35	04 48	17 30	13 11	20 30	27 44	12 46	22 13	00 31	19 41	20 08	07 22
10	16 D 43	15 32	04 09	17 46	09 38	20 08	27 42	13 02	22 13	00 29	19 41	20 08	07 21
11	16 43	15 28	03 17	18 01	05 36	19 45	27 39	13 17	22 14	00 27	19 41	20 08	07 21
12	16 45	15 25	02 13	18 16	01 S 19	19 21	27 37	13 32	22 14	00 25	19 41	20 08	07 21
13	16 46	15 22	01 N 04	18 31	02 N 58	18 56	27 33	13 47	22 14	00 23	19 41	20 08	07 21
14	16 R 47	15 19	00 S 08	18 46	07 03	18 31	27 30	14 02	22 15	00 20	19 41	20 09	07 20
15	16 46	15 16	01 19	19 00	10 44	18 05	27 25	14 17	22 15	00 18	19 41	20 09	07 20
16	16 43	15 12	02 24	19 14	13 50	17 40	27 21	14 32	22 15	00 16	19 41	20 09	07 20
17	16 38	15 09	03 21	19 27	16 13	17 15	27 16	14 46	22 16	00 14	19 41	20 09	07 19
18	16 32	15 06	04 07	19 40	17 48	16 50	27 10	15 00	22 16	00 12	19 42	20 09	07 19
19	16 24	15 03	04 41	19 53	18 31	16 26	27 04	15 15	22 17	00 10	19 42	20 09	07 19
20	16 16	15 00	05 01	20 06	18 22	16 04	26 57	15 29	22 17	00 09	19 42	20 09	07 19
21	16 09	14 57	05 08	20 18	17 23	15 43	26 50	15 42	22 18	00 07	19 42	20 09	07 18
22	16 03	14 53	05 02	20 30	15 40	15 24	26 43	15 56	22 18	00 05	19 42	20 09	07 18
23	15 58	14 50	04 43	20 41	13 17	15 07	26 34	16 10	22 19	00 03	19 42	20 10	07 18
24	15 55	14 47	04 11	20 52	10 21	14 51	26 26	16 23	22 19	00 S 01	19 43	20 10	07 18
25	15 54	14 44	03 28	21 03	06 58	14 38	26 17	16 36	22 20	00 N 01	19 43	20 10	07 18
26	15 D 55	14 41	02 36	21 13	03 N 15	14 27	26 07	16 49	22 21	00 02	19 43	20 10	07 17
27	15 56	14 38	01 34	21 23	00 S 42	14 18	25 57	17 02	22 21	00 04	19 43	20 10	07 17
28	15 57	14 34	00 S 27	21 33	04 44	14 12	25 46	17 15	22 22	00 06	19 44	20 10	07 17
29	15 R 57	14 31	00 N 41	21 42	08 40	14 08	25 34	17 27	22 23	00 07	19 44	20 11	07 17
30	15 55	14 28	01 54	21 51	12 17	14 06	25 22	17 39	22 23	00 09	19 44	20 11	07 17
31	15 ♎ 50	14 ♎ 25	03 N 00	22 N 00	15 S 19	14 N 06	25 N 09	17 N 51	22 S 24	00 N 11	19 S 44	20 S 11	07 S 16

ZODIAC SIGN ENTRIES

Date	h m	Planets
02	12 42	☽ ♏
02	18 16	♂ ♉
04	16 05	☽ ♐
06	17 54	☽ ♑
08	19 39	☽ ♒
10	22 29	☽ ♓
13	03 00	☽ ♈
15	09 25	☽ ♉
17	17 48	☽ ♊
20	04 16	☽ ♋
20	18 23	☉ ♊
22	16 28	☽ ♌
25	04 58	☽ ♍
27	15 33	☽ ♎
29	22 30	☽ ♏

LATITUDES

Date	Mercury ☿ ° '	Venus ♀ ° '	Mars ♂ ° '	Jupiter ♃ ° '	Saturn ♄ ° '	Uranus ♅ ° '	Neptune ♆ ° '	Pluto ♇ ° '
01	02 N 28	04 N 30	00 S 32	00 N 04	02 S 05	00 S 35	00 N 29	13 N 28
04	02 02	04 29	00 30	00 04	02 06	00 35	00 29	13 28
07	01 26	04 24	00 29	00 04	02 06	00 35	00 29	13 28
10	00 N 40	04 18	00 27	00 03	02 07	00 35	00 29	13 28
13	00 S 11	04 09	00 25	00 03	02 08	00 35	00 29	13 28
16	01 03	03 55	00 23	00 03	02 08	00 35	00 29	13 28
19	01 53	03 38	00 21	00 02	02 09	00 35	00 29	13 28
22	02 36	03 17	00 19	00 02	02 09	00 35	00 29	13 28
25	03 11	02 52	00 17	00 02	02 10	00 35	00 29	13 28
28	03 35	02 22	00 15	00 01	02 11	00 35	00 29	13 27
31	03 S 49	01 N 47	00 S 14	00 N 01	02 S 11	00 S 36	00 N 29	13 N 27

DATA

Julian Date	2450205
Delta T	+62 seconds
Ayanamsa	23° 48' 24"
Synetic vernal point	05° ♓ 18' 35"
True obliquity of ecliptic	23° 26' 14"

MOON'S PHASES, APSIDES AND POSITIONS ☽

Date	h m	Phase	Longitude ° '	Eclipse Indicator
03	11 48	○	13 ♏ 19	
10	05 04	☾	19 ♒ 49	
17	11 46	●	26 ♉ 51	
25	14 13	☽	04 ♍ 38	

Day	h m		
06	21 55	Perigee	
22	16 23	Apogee	
06	10 25	Max dec	18° S 31'
12	19 18	0N	
19	19 43	Max dec	18° N 34'
27	07 48	0S	

LONGITUDES

Date	Chiron ° '	Ceres ° '	Pallas ⚴ ° '	Juno ⚵ ° '	Vesta ⚶ ° '	Black Moon Lilith ° '
01	08 ♎ 38	14 ♐ 22	25 ♎ 19	16 ♓ 06	20 ♏ 17	24 ♋ 06
11	08 ♎ 27	12 ♐ 45	22 ♎ 48	20 ♓ 07	17 ♏ 52	25 ♋ 14
21	08 ♎ 06	10 ♐ 44	20 ♎ 56	24 ♓ 01	15 ♏ 30	26 ♋ 21
31	07 ♎ 56	08 ♐ 31	19 ♎ 51	27 ♓ 47	13 ♏ 35	27 ♋ 28

ASPECTARIAN

01 Wednesday
02 46 ☽ ⊼ ☉
06 59 ☽ ⚼ ☿
13 34 ☿ # ♃
14 09 ☽ ∠ ♇
14 46 ☽ □ ♃
18 00 ☽ ∥ ♇
23 24 ☽ ± ☿
23 42 ☽ △ ♀

02 Thursday
06 10 ☽ ⊥ ♇
08 45 ☽ □ ♆
10 09 ☽ ⊼ ☿
12 23 ☽ ☍ ♂
16 38 ☽ # ♂
16 41 ☽ ⊻ ♇
17 45 ☽ ⊼ ♄
20 41 ☽ □ ♅
22 34 ☽ Q ♃

03 Friday
03 07 ☽ ⚼ ♀
04 15 ☽ ± ♄
11 48 ☽ ☍ ☉
15 58 ☽ Q ♆
19 10 ☽ ± ♀
19 13 ☽ ✱ ♃
19 48 ☽ ⚼ ♄
22 40 ☿ St R

04 Saturday
03 34 ☽ Q ♅
05 44 ☽ ⊼ ♀
09 51 ☽ # ☉
12 17 ☽ ✱ ♆
13 46 ☽ ☍ ☿
15 37 ♃ St R
18 37 ☽ ⊼ ♂
19 48 ☽ ☌ ♇
20 32 ☽ ∠ ♃
21 17 ☽ △ ♄
23 45 ☽ ✱ ♅

05 Sunday
05 11 ☽ ± ♂
11 32 ☽ ⊥ ♃
13 20 ☽ ∠ ♆
16 31 ♂ ⊼ ♇
17 55 ☽ ⊼ ☉
20 57 ☽ ⚼ ♂
21 29 ☽ ⊻ ♃

06 Monday
00 40 ☽ ∠ ♅
04 15 ☽ ⊥ ♆
04 34 ☽ ± ☉
09 29 ☽ ☍ ♀
14 10 ☽ ⊻ ♆
15 09 ☽ ⊼ ☿
15 33 ☽ ⊥ ♅
20 30 ☽ ⚼ ☉
21 28 ☽ ⊻ ♇
23 06 ☽ △ ♂
23 23 ☽ □ ♄

07 Tuesday
00 54 ☽ ± ☿
01 28 ☽ ⊻ ♅
05 17 ♂ ⊻ ♄
07 22 ☽ ⊥ ♇
15 31 ☽ ⚼ ☿
22 14 ☽ ∠ ♇
22 52 ☉ △ ♃
23 04 ☽ ☌ ♃
23 05 ☽ △ ☉
23 16 ☽ # ☉

08 Wednesday
05 24 ☽ Q ♄
12 41 ☽ ⊼ ♀
15 51 ☽ ☌ ♆
15 53 ☽ △ ☿
16 50 ☿ △ ♆
19 36 ♅ St R
20 30 ☉ ∠ ♄
21 02 ♂ □ ♅
22 59 ☽ ± ♀
23 11 ☽ ✱ ♇

09 Thursday
01 33 ☽ ✱ ♄
03 19 ☽ ☌ ♅
03 40 ☽ □ ♂
14 31 ☽ ⚼ ♀
14 47 ☽ # ♂
19 20 ☽ Q ♇

10 Friday
01 18 ☽ ⊻ ♃
03 03 ☽ ∠ ♄
05 04 ☽ □ ☉
11 31 ☽ ⊥ ♃
11 50 ☽ Q ♂
11 55 ☿ # ♆
16 41 ☽ △ ♀
17 07 ☽ □ ☿
18 32 ☽ ⊥ ♄
18 34 ☽ ⊻ ♆

11 Saturday
00 25 ☿ ⊻ ♀
01 48 ☽ ∥ ♇
02 01 ☽ □ ♇
02 57 ☽ ∠ ♃
04 53 ☽ ⊥ ♆
04 56 ☽ ⊻ ♄
05 15 ☉ ⊥ ♀
06 23 ☽ ⊻ ♅
09 39 ☽ ✱ ♂
14 19 ☽ Q ☉
16 32 ☿ # ♅
16 48 ☽ ⊥ ♅
20 31 ☽ ∠ ♆
23 12 ☽ Q ☿

12 Sunday
05 01 ☽ ✱ ♃
08 33 ☽ ∠ ♅
12 56 ☽ ✱ ☉
13 24 ☽ ∠ ♂
17 03 ☽ ∥ ♄
19 31 ☽ ✱ ☿
21 31 ☽ # ♄
22 12 ☽ □ ♀
22 55 ☽ ✱ ♆

13 Monday
02 11 ☽ Q ♃
06 19 ☽ ⊥ ♂
06 34 ☽ △ ♇
10 04 ☽ ☌ ♄
11 11 ☽ ✱ ♅
17 42 ☽ ⊻ ♂
17 44 ☽ ∠ ☉
20 23 ☽ Q ♆
21 15 ☽ ∠ ☿

14 Tuesday
03 10 ☉ ∥ ☿
07 06 ☽ Q ♀
08 49 ☽ Q ♅
09 32 ☽ ⚼ ♇
10 31 ☽ □ ♃
11 18 ☽ ⊥ ☉
12 52 ☽ ⊥ ☿
13 45 ☽ # ♇
14 14 ♀ ⊼ ♆
23 07 ☽ ⊻ ☉
23 22 ☽ ⊻ ☿

15 Wednesday
01 12 ☉ ☌ ☿
01 57 ☽ ± ♇
05 09 ☽ □ ♆
05 23 ☽ ✱ ♀
12 59 ☽ ⊼ ♇
17 08 ☽ ⊻ ♄
17 51 ☽ □ ♅

16 Thursday
04 00 ☽ ☌ ♂
04 23 ☽ ⊥ ♄
09 37 ☽ ∠ ♀
11 36 ♂ ⊥ ♄
17 55 ☽ △ ♃
18 53 ☽ ∥ ♂
21 25 ☽ ∠ ♄

17 Friday
02 46 ☽ ⊥ ♀
04 53 ☽ ☌ ☿
11 46 ☽ ☌ ☉
13 20 ☽ △ ♆
14 17 ☽ ⊻ ♀
21 23 ☽ ☍ ♇
22 21 ☽ ⚼ ♃
23 02 ☽ ∥ ☿

18 Saturday
02 13 ☽ ✱ ♄
02 32 ☽ △ ♅
08 00 ☉ △ ♆
14 15 ☉ # ♅
15 40 ☽ ± ♃
16 42 ☽ ⊻ ♂
18 12 ☽ ⚼ ♆
22 30 ☿ ⊥ ♀
23 09 ☉ ⊻ ♀

19 Sunday
01 40 ☽ Q ♄
03 19 ☽ ⊼ ♃
05 09 ☽ ⊥ ♂
07 39 ☽ ⚼ ♅
11 47 ☽ ± ♆
12 24 ☽ ⊻ ☿
16 50 ♄ ✱ ♅
23 36 ☽ ⊼ ♆
23 46 ☽ ⊥ ☿

20 Monday
00 00 ☽ ∠ ♂
00 54 ☽ ☌ ♀
01 21 ☽ ± ♅
03 03 ☽ ⊻ ☉
06 08 ♀ St R
07 51 ☽ ⊼ ♇
13 16 ☽ ⊼ ♅
13 26 ☽ □ ♄
14 44 ♀ ∠ ♂
16 00 ☽ ⊥ ☉
16 58 ☽ ∠ ☿
18 57 ☉ # ♆
19 46 ☽ ± ♇

21 Tuesday
07 50 ☽ ✱ ♂
11 36 ☽ ∠ ☉
12 37 ☿ ∥ ♂
13 46 ☽ ⚼ ♇
14 39 ☽ ☍ ♃
22 01 ☽ ✱ ☿

22 Wednesday
09 10 ☽ ∥ ♂
09 37 ☽ Q ♂
11 38 ☽ ☍ ♆
12 49 ☽ ⊻ ♀
13 50 ☉ ☍ ♇
15 32 ☽ ∥ ☿
19 58 ☽ △ ♇
20 32 ☽ ✱ ☉
21 32 ☽ Q ☿
23 34 ☉ ⚼ ♃

23 Thursday
00 51 ☽ ⊥ ♀
01 34 ☽ ☍ ♅
02 14 ☽ △ ♄
18 52 ☽ ∠ ♀
22 57 ☽ Q ☉

24 Friday
00 26 ☽ □ ♂
02 59 ☽ ⊼ ♃
08 47 ☽ ⚼ ♄
09 03 ☽ □ ☿
15 01 ☽ ± ♃
19 22 ☿ ∠ ♄

25 Saturday
00 08 ☽ ⊼ ♆
00 40 ☽ ✱ ♀
02 59 ☽ ± ♄
08 18 ☽ □ ♇
08 56 ☽ ⚼ ♃
09 52 ☽ # ♇
10 01 ☉ △ ♅
12 06 ☽ ± ♆
13 53 ☽ ⊼ ♅
14 13 ☽ □ ☉
14 51 ♂ △ ♃
15 01 ☽ ⊼ ♄

26 Sunday
00 06 ☽ Q ♀
01 08 ☉ ✱ ♄
01 44 ☽ ± ♅
05 52 ☽ ⚼ ♆
07 22 ♀ ⊼ ♆
14 22 ☽ △ ♃
15 59 ☽ △ ♂
19 19 ☽ ⚼ ♅
19 39 ☽ Q ♇
19 46 ☽ △ ☿

27 Monday
07 26 ☽ ∥ ♄
08 11 ☽ # ♄
10 21 ☽ □ ♀
10 53 ☽ △ ♆
18 35 ☽ ✱ ♇
19 02 ☿ St D
22 29 ☽ ⚼ ♂
23 56 ☽ △ ♅

28 Tuesday
00 20 ☽ ⚼ ☿
01 26 ☽ ☍ ♄
05 01 ☽ △ ☉
16 15 ☽ ± ♂
17 07 ☽ ± ☿
22 24 ☽ ∠ ♇
22 43 ☽ □ ♃

29 Wednesday
03 24 ☽ ∥ ♇
03 50 ☽ ⊼ ♂
04 08 ☽ ⊼ ☿
10 02 ☿ ☌ ♂
10 39 ☽ ⚼ ☉
14 36 ☽ ⊥ ♇
16 20 ☽ △ ♀
18 06 ☽ □ ♆

30 Thursday
01 13 ☽ ⊻ ♇
03 49 ☽ ± ☉
04 49 ♀ ⊥ ♂
06 14 ☽ □ ♅
06 40 ☽ Q ♃
08 00 ☽ ⊼ ♄
08 01 ♂ ∠ ♄
15 02 ☽ ⊼ ☉
17 50 ☽ ⚼ ♀
18 26 ☽ ± ♄
22 28 ☿ ⊥ ♀

31 Friday
01 22 ☽ Q ♆
01 39 ☽ # ☿
03 12 ☽ ✱ ♃
08 43 ☽ ± ♀
09 13 ☽ ☍ ☿
09 50 ☽ ⚼ ♄
11 05 ☽ ☍ ♂
13 00 ☽ Q ♅
17 38 ☉ ± ♃
18 29 ☽ ⊼ ♀
21 32 ☽ ✱ ♆

All ephemeris data is given at 12.00 UT and the Moon's longitude is additionally given for 24.00 UT

JUNE 1996

Date	Sidereal time h m s	Sun ☉ ° ' "	Moon ☽ ° ' "	Moon ☽ 24.00 ° ' "	Mercury ☿ ° '	Venus ♀ ° '	Mars ♂ ° '	Jupiter ♃ ° '	Saturn ♄ ° '	Uranus ♅ ° '	Neptune ♆ ° '	Pluto ♇ ° '
01	04 41 00	11 ♊ 15 48	06 ♐ 14 29	13 ♐ 35 30	20 ♉ 29	25 ♊ 19	21 ♉ 59	16 ♑ 28	05 ♈ 35	04 ♒ 21	27 ♑ 28	01 ♐ 28
02	04 44 56	12 13 16	21 ♐ 00 08	28 ♐ 27 17	20 52	24 R 50	22 43	16 R 23	05 39	04 R 20	27 R 27	01 R 27
03	04 48 53	13 10 42	05 ♑ 55 50	13 ♑ 24 38	21 19	24 19	23 26	16 18	05 43	04 19	27 26	01 25
04	04 52 49	14 08 08	20 ♑ 52 32	28 ♑ 18 30	21 51	23 46	24 10	16 13	05 48	04 18	27 25	01 23
05	04 56 46	15 05 33	05 ♒ 41 35	13 ♒ 00 59	22 27	23 12	24 53	16 07	05 52	04 17	27 24	01 22
06	05 00 42	16 02 57	20 ♒ 16 05	27 ♒ 26 26	23 07	22 37	25 36	16 02	05 56	04 15	27 23	01 20
07	05 04 39	17 00 21	04 ♓ 31 44	11 ♓ 31 51	23 51	22 01	26 20	15 56	06 00	04 14	27 22	01 19
08	05 08 35	17 57 44	18 ♓ 26 47	25 ♓ 16 38	24 38	21 24	27 03	15 50	06 04	04 13	27 21	01 17
09	05 12 32	18 55 06	02 ♈ 01 35	08 ♈ 41 53	25 30	20 47	27 46	15 44	06 07	04 11	27 20	01 15
10	05 16 29	19 52 28	15 ♈ 17 49	21 ♈ 49 42	26 25	20 10	28 29	15 38	06 11	04 10	27 18	01 14
11	05 20 25	20 49 50	28 ♈ 17 49	04 ♉ 42 30	27 24	19 32	29 12	15 32	06 15	04 08	27 17	01 12
12	05 24 22	21 47 11	11 ♉ 04 00	17 ♉ 22 34	28 26	18 55	29 ♉ 55	15 25	06 18	04 07	27 16	01 11
13	05 28 18	22 44 31	23 ♉ 38 26	29 ♉ 51 48	29 ♉ 32	18 18	00 ♊ 38	15 19	06 22	04 05	27 15	01 09
14	05 32 15	23 41 51	06 ♊ 02 50	12 ♊ 11 40	00 ♊ 42	17 41	01 21	15 12	06 25	04 04	27 13	01 08
15	05 36 11	24 39 11	18 ♊ 18 26	24 ♊ 23 15	01 54	17 06	02 04	15 06	06 29	04 02	27 12	01 06
16	05 40 08	25 36 30	00 ♋ 26 15	06 ♋ 27 32	03 10	16 32	02 47	14 59	06 32	04 00	27 11	01 05
17	05 44 04	26 33 48	12 ♋ 27 16	18 ♋ 25 36	04 29	15 59	03 29	14 52	06 35	03 58	27 10	01 03
18	05 48 01	27 31 06	24 ♋ 22 45	00 ♌ 18 56	05 52	15 28	04 12	14 45	06 38	03 57	27 08	01 02
19	05 51 58	28 28 23	06 ♌ 14 26	12 ♌ 09 34	07 17	14 58	04 54	14 38	06 41	03 55	27 07	01 01
20	05 55 54	29 ♊ 25 39	18 ♌ 04 43	24 ♌ 00 17	08 46	14 30	05 37	14 31	06 44	03 53	27 05	00 59
21	05 59 51	00 ♋ 22 55	29 ♌ 56 45	05 ♍ 54 37	10 18	14 04	06 19	14 24	06 46	03 51	27 04	00 58
22	06 03 47	01 20 10	11 ♍ 54 27	17 ♍ 56 48	11 53	13 40	07 02	14 16	06 49	03 49	27 03	00 56
23	06 07 44	02 17 24	24 ♍ 02 20	00 ♎ 11 39	13 32	13 18	07 44	14 09	06 52	03 48	27 01	00 55
24	06 11 40	03 14 38	06 ♎ 25 24	12 ♎ 44 13	15 13	12 58	08 26	14 02	06 54	03 46	27 00	00 54
25	06 15 37	04 11 51	19 ♎ 08 42	25 ♎ 39 26	16 57	12 41	09 09	13 54	06 56	03 44	26 58	00 52
26	06 19 33	05 09 04	02 ♏ 16 52	09 ♏ 01 25	18 44	12 26	09 51	13 47	06 59	03 42	26 57	00 51
27	06 23 30	06 06 16	15 ♏ 53 19	22 ♏ 52 40	20 34	12 14	10 33	13 39	07 01	03 40	26 55	00 50
28	06 27 27	07 03 28	29 ♏ 59 23	07 ♐ 13 20	22 27	12 03	11 15	13 32	07 03	03 38	26 54	00 49
29	06 31 23	08 00 39	14 ♐ 33 29	21 ♐ 59 34	24 22	11 56	11 57	13 24	07 05	03 35	26 52	00 47
30	06 35 20	08 ♋ 57 50	29 ♐ 30 27	07 ♑ 04 58	26 ♊ 20	11 ♊ 50	12 ♊ 39	13 ♑ 16	07 ♈ 07	03 ♒ 33	26 ♑ 51	00 ♐ 46

Date	Moon True ☊ ° '	Moon Mean ☊ ° '	Moon ☽ Latitude ° '
01	15 ♎ 44	14 ♎ 22	03 N 56
02	15 R 36	14 18	04 37
03	15 28	14 15	05 01
04	15 19	14 12	05 03
05	15 13	14 09	04 46
06	15 08	14 06	04 10
07	15 05	14 03	03 19
08	15 D 05	13 59	02 17
09	15 05	13 56	01 N 09
10	15 R 06	13 53	00 S 01
11	15 05	13 50	01 10
12	15 02	13 47	02 14
13	14 57	13 44	03 10
14	14 48	13 40	03 56
15	14 38	13 37	04 30
16	14 26	13 34	04 52
17	14 13	13 31	05 01
18	14 00	13 28	04 56
19	13 49	13 24	04 39
20	13 41	13 21	04 09
21	13 35	13 18	03 29
22	13 31	13 15	02 39
23	13 30	13 12	01 41
24	13 D 30	13 09	00 S 37
25	13 R 30	13 05	00 N 30
26	13 29	13 02	01 38
27	13 26	12 59	02 42
28	13 20	12 56	03 39
29	13 12	12 53	04 24
30	13 ♎ 03	12 ♎ 50	04 N 52

DECLINATIONS

Date	Sun ☉ ° '	Moon ☽ ° '	Mercury ☿ ° '	Venus ♀ ° '	Mars ♂ ° '	Jupiter ♃ ° '	Saturn ♄ ° '	Uranus ♅ ° '	Neptune ♆ ° '	Pluto ♇ ° '
01	22 N 08	17 S 28	14 N 09	24 N 56	18 N 03	22 S 25	00 N 12	19 S 45	20 S 11	07 S 16
02	22 15	18 31	14 13	24 43	18 15	22 25	00 14	19 45	20 11	07 16
03	22 23	18 18	14 20	24 28	18 27	22 26	00 15	19 45	20 11	07 16
04	22 30	16 49	14 28	24 13	18 38	22 27	00 17	19 46	20 12	07 16
05	22 36	14 14	14 38	23 58	18 49	22 28	00 18	19 46	20 12	07 16
06	22 42	10 47	14 50	23 42	19 00	22 29	00 19	19 46	20 12	07 16
07	22 48	06 46	15 04	23 26	19 11	22 29	00 21	19 47	20 12	07 16
08	22 54	02 S 28	15 19	23 10	19 22	22 30	00 22	19 47	20 13	07 15
09	22 58	01 N 52	15 36	22 53	19 32	22 31	00 23	19 47	20 13	07 15
10	23 03	06 00	15 54	22 37	19 43	22 32	00 25	19 48	20 13	07 15
11	23 07	09 47	16 13	22 20	19 53	22 33	00 26	19 48	20 13	07 15
12	23 11	13 02	16 33	22 03	20 02	22 34	00 27	19 48	20 13	07 15
13	23 14	15 37	16 54	21 46	20 12	22 34	00 28	19 49	20 14	07 15
14	23 17	17 26	17 16	21 29	20 22	22 35	00 29	19 49	20 14	07 15
15	23 20	18 26	17 39	21 12	20 31	22 36	00 30	19 50	20 14	07 15
16	23 22	18 34	18 02	20 56	20 40	22 37	00 31	19 50	20 14	07 15
17	23 24	17 52	18 26	20 40	20 49	22 38	00 32	19 51	20 15	07 15
18	23 25	16 23	18 50	20 24	20 58	22 39	00 33	19 51	20 15	07 15
19	23 26	14 12	19 14	20 09	21 06	22 40	00 34	19 51	20 15	07 15
20	23 26	11 27	19 39	19 55	21 14	22 41	00 35	19 52	20 16	07 15
21	23 26	08 14	20 03	19 41	21 22	22 41	00 36	19 52	20 16	07 15
22	23 26	04 39	20 27	19 27	21 30	22 42	00 37	19 53	20 16	07 15
23	23 25	00 N 49	20 51	19 15	21 38	22 43	00 37	19 53	20 16	07 15
24	23 24	03 S 07	21 14	19 03	21 45	22 44	00 38	19 54	20 17	07 15
25	23 22	07 02	21 37	18 52	21 52	22 45	00 39	19 54	20 17	07 15
26	23 20	10 44	21 58	18 42	21 59	22 46	00 40	19 55	20 17	07 15
27	23 18	14 00	22 19	18 32	22 06	22 47	00 40	19 55	20 17	07 15
28	23 15	16 34	22 38	18 24	22 13	22 48	00 41	19 56	20 18	07 16
29	23 12	18 10	22 55	18 16	22 19	22 49	00 41	19 56	20 18	07 16
30	23 N 08	18 S 34	23 N 11	18 N 09	22 N 25	22 S 50	00 N 42	19 S 57	20 S 18	07 S 16

ZODIAC SIGN ENTRIES

Date	h	m	Planets
01	01	43	☽ ♐
03	02	29	☽ ♑
05	02	44	☽ ♒
07	04	19	☽ ♓
09	08	23	☽ ♈
11	15	11	☽ ♉
12	14	42	♂ ♊
13	21	45	☿ ♊
14	00	16	☽ ♊
16	11	08	☽ ♋
18	23	22	☽ ♌
21	02	24	☉ ♋
21	12	07	☽ ♍
23	23	37	☽ ♎
26	07	53	☽ ♏
28	12	01	☽ ♐
30	12	47	☽ ♑

LATITUDES

Date	Mercury ☿ ° '	Venus ♀ ° '	Mars ♂ ° '	Jupiter ♃ ° '	Saturn ♄ ° '	Uranus ♅ ° '	Neptune ♆ ° '	Pluto ♇ ° '
01	03 S 52	01 N 35	00 S 13	00 N 01	02 S 12	00 S 36	00 N 29	13 N 27
04	03 53	00 56	00 11	00 00	02 12	00 36	00 29	13 26
07	03 47	00 N 15	00 09	00 00	02 13	00 36	00 29	13 26
10	03 33	00 S 28	00 07	00 S 01	02 14	00 36	00 29	13 25
13	03 13	01 10	00 05	00 01	02 14	00 36	00 29	13 24
16	02 48	01 50	00 03	00 01	02 15	00 36	00 29	13 23
19	02 19	02 27	00 S 01	00 02	02 16	00 36	00 29	13 23
22	01 46	03 00	00 N 01	00 02	02 17	00 36	00 29	13 22
25	01 11	03 29	00 03	00 02	02 18	00 36	00 29	13 21
28	00 36	03 52	00 05	00 03	02 18	00 36	00 29	13 20
31	00 S 01	04 S 11	00 N 07	00 S 04	02 S 19	00 S 36	00 N 29	13 N 19

DATA

Julian Date	2450236
Delta T	+62 seconds
Ayanamsa	23° 48' 29"
Synetic vernal point	05° ♓ 18' 31"
True obliquity of ecliptic	23° 26' 14"

MOON'S PHASES, APSIDES AND POSITIONS ☽

Date	h	m	Phase	Longitude ° '	Eclipse Indicator
01	20	47	○	11 ♐ 37	
08	11	05	☾	17 ♓ 56	
16	01	36	●	25 ♊ 12	
24	05	23	☽	02 ♎ 59	

Day	h	m		
03	16	24	Perigee	
19	06	31	Apogee	
02	19	57	Max dec	18° S 35'
09	01	36	ON	
16	03	42	Max dec	18° N 37'
23	17	01	OS	
30	07	23	Max dec	18° S 35'

LONGITUDES

Date	Chiron ⚷ ° '	Ceres ⚳ ° '	Pallas ⚴ ° '	Juno ⚵ ° '	Vesta ⚶ ° '	Black Moon Lilith ⚸ ° '
01	07 ♎ 55	08 ♐ 18	19 ♎ 48	28 ♓ 09	13 ♏ 25	27 ♋ 35
11	07 ♎ 57	06 ♐ 57	19 ♎ 35	01 ♈ 45	12 ♏ 14	28 ♋ 42
21	08 ♎ 10	04 ♐ 20	19 ♎ 07	05 ♈ 08	11 ♏ 52	29 ♋ 49
31	08 ♎ 35	02 ♐ 59	21 ♎ 17	08 ♈ 16	12 ♏ 18	00 ♌ 56

ASPECTARIAN

h m	Aspects
01 Saturday	
04 10	☽ ☌ ♇
04 11	☽ ∠ ♃
08 55	☽ ✶ ♅
10 54	☽ △ ♄
11 27	♃ ∠ ♇
18 53	☽ ⊥ ♃
20 22	☿ ∠ ♄
20 47	☽ ☍ ☉
22 10	☽ ∠ ♆
02 Sunday	
00 11	☽ # ♂
04 35	☽ ⊻ ♃
09 19	☽ ∠ ♅
11 46	☽ ⊼ ☿
12 44	☽ ⊥ ♆
14 54	☽ ⊼ ♂
17 44	☉ BQ ♆
17 58	☽ ☍ ♀
21 44	☽ ± ☿
22 23	☽ ⊻ ♆
23 48	☽ ⊥ ♅
03 Monday	
01 03	☽ ± ♂
04 46	☽ ⊻ ♇
08 22	☽ # ♂
09 25	☽ ⊻ ♅
11 40	☽ □ ♄
12 39	☽ BQ ☿
14 23	☽ ⊥ ♇
16 14	☽ BQ ♂
04 Tuesday	
00 25	☽ ⊼ ☉
01 07	☉ # ♃
04 32	☽ ☌ ♃
04 40	♀ ⊻ ♂
04 48	☽ ∠ ♇
10 44	☽ ± ☉
13 38	☽ △ ☿
16 30	☽ ⊼ ♀
16 43	☽ Q ♄
17 34	☽ △ ♂
22 33	☽ ☌ ♆
05 Wednesday	
01 51	☽ ± ♀
02 16	☽ BQ ☉
04 58	☽ ✶ ♇
08 55	☽ # ☿
09 42	☽ ☌ ♅
12 17	☽ ✶ ♄
15 57	☽ BQ ♀
06 Thursday	
00 33	☽ Q ♇
02 45	☿ ⊻ ♀
04 31	☽ △ ☉
05 01	☽ ⊻ ♃
11 29	☉ ⊼ ♃
13 06	☽ ∠ ♄
14 55	☽ ⊥ ♃
15 46	☽ △ ♀
16 59	☽ □ ☿
21 24	☽ □ ♂
23 53	☽ ⊻ ♆
07 Friday	
04 16	☽ ⊥ ♄
05 56	☽ ∠ ♃
06 33	☽ □ ♇
09 09	☽ ∥ ♇
10 01	☽ ⊥ ♆
11 30	☽ ⊻ ♅
14 31	☽ ⊻ ♄
21 45	☽ ⊥ ♅
08 Saturday	
01 17	☽ Q ☿
01 25	☽ ∠ ♆
05 46	☽ Q ♂
07 29	☽ ✶ ♃
11 05	☽ □ ☉
13 20	☽ ∠ ♅
14 28	♀ ± ♆
14 38	☉ Q ♄
16 58	☽ □ ♀
21 40	♂ △ ♆
23 31	☽ # ♄
23 36	☽ ✶ ☿
09 Sunday	
03 38	☽ ✶ ♆
03 44	☽ ∥ ♄
03 59	☽ ✶ ♂
04 25	☽ Q ♃
06 19	☉ ∥ ♀
10 38	☽ △ ♇
15 52	☽ ✶ ♅
18 35	☉ BQ ♅
19 23	☽ ☌ ♄
21 28	☽ Q ☉
23 37	☽ Q ♀
10 Monday	
01 07	☽ Q ♆
04 24	☽ ∠ ☿
08 30	☽ ∠ ♂
12 37	☽ □ ♃
13 35	☽ Q ♅
13 42	☽ BQ ♇
16 19	☉ ☌ ♀
18 22	♀ # ♃
19 38	☽ # ♇
20 31	☽ ✶ ♀
21 04	☽ ✶ ☉
22 09	☽ ⊥ ☿
11 Tuesday	
00 45	♂ # ♅
01 59	☽ ⊥ ♂
06 16	☽ ± ♇
09 20	☿ △ ♆
10 07	☽ □ ♆
10 11	☽ ⊻ ☿
13 47	☽ ⊻ ♂
17 25	☽ ⊼ ♇
22 54	☽ □ ♅
23 08	☽ ∠ ♀
23 13	☉ ± ♆
12 Wednesday	
02 58	☽ ⊻ ♄
03 15	☽ ∠ ☉
03 49	♀ BQ ♅
14 22	☽ ⊥ ♄
15 20	☽ ⊥ ♀
20 12	☽ △ ♃
21 42	☽ ⊥ ☉
13 Thursday	
02 14	☽ ⊻ ♀
02 41	♂ BQ ♃
07 37	☽ ∠ ♄
09 29	♀ Q ♄
10 08	☽ ⊻ ☉
16 01	♂ # ♆
18 56	☽ △ ♆
14 Friday	
00 31	☽ ☌ ☿
00 46	☽ BQ ♃
02 19	☽ ☌ ♂
02 28	☽ ☍ ♇
02 53	☿ BQ ♃
04 54	♂ ☍ ♇
08 09	☽ △ ♅
08 13	☽ ∥ ☿
12 44	☽ ✶ ♄
18 06	☽ ± ♃
20 38	☿ ☍ ♇
15 Saturday	
00 02	☽ BQ ♆
05 45	☽ ⊼ ♃
09 44	☽ ☌ ♀
12 20	☽ Q ♄
13 25	☽ BQ ♅
17 42	☽ ± ♆
19 12	☿ ☌ ♂
16 Sunday	
01 36	☽ ☌ ☉
05 33	☽ ⊼ ♆
07 11	☽ ± ♅
13 17	☽ ⊼ ♇
16 57	☽ ⊻ ♂
18 06	☽ ⊻ ☿
19 05	☽ ⊼ ♅
17 Monday	
00 12	☽ □ ♄
01 13	☽ ± ♇
01 44	☽ ∥ ☿
02 58	☿ △ ♅
03 13	♀ ∥ ♂
05 41	☽ ⊥ ♂
07 34	☽ ⊥ ☿
16 48	☽ ☍ ♃
18 47	☽ ⊻ ♀
19 13	☽ BQ ♇
18 Tuesday	
00 53	☽ ∠ ♂
02 38	☉ ⊼ ♆
03 49	♂ △ ♅
03 58	☽ ∠ ☿
06 21	☽ ⊥ ♀
17 34	☽ ☍ ♆
18 54	☽ ⊻ ☉
22 25	☉ ± ♅
23 47	☽ ∠ ♀
19 Wednesday	
01 26	☽ △ ♇
01 29	☿ ✶ ♄
02 23	♀ # ♆
07 18	☽ ☍ ♅
08 06	☽ ⊥ ☉
09 08	☽ ✶ ♂
12 54	☽ △ ♄
14 26	☽ ✶ ☿
20 Thursday	
03 57	☽ ∠ ☉
04 51	☽ ⊼ ♃
05 00	☽ ✶ ♀
08 12	☿ ± ♃
10 41	♀ ⊼ ♃
11 00	☽ Q ♂
16 31	♀ # ♅
16 53	☽ ± ♃
18 15	☽ Q ☿
19 25	☽ BQ ♄
21 50	☿ ∥ ♀
21 Friday	
01 03	☿ # ♅
04 26	☽ Q ♀
06 12	☽ ⊼ ♆
10 54	☽ BQ ♃
12 57	☽ ✶ ☉
13 40	☽ ± ♄
14 03	☽ □ ♇
18 17	☽ ± ♆
18 45	☽ # ♇
19 51	☽ ⊼ ♅
22 Saturday	
00 34	☿ # ♆
01 38	☽ □ ♂
01 47	☽ ⊼ ♄
02 16	☉ ⊼ ♇
04 16	♂ ✶ ♄
07 51	☽ ± ♅
11 58	☽ □ ☿
12 16	☽ BQ ♆
14 16	☿ BQ ♆
15 06	☽ Q ☉
15 24	☽ □ ♀
16 40	☽ △ ♃
23 Sunday	
01 42	☽ BQ ♅
01 56	☽ Q ♇
09 19	☿ ☌ ♀
13 11	☽ ∥ ♄
17 49	☽ △ ♆
20 23	☿ ⊼ ♃
20 52	☽ # ♄
24 Monday	
00 02	♂ ± ♃
01 23	☽ ✶ ♇
05 23	☽ □ ☉
06 54	☽ △ ♅
12 55	☽ ☍ ♄
16 05	☽ △ ♂
25 Tuesday	
00 10	☽ △ ♀
00 33	☉ ⊼ ♅
02 18	☽ □ ♃
05 54	☽ ∠ ♇
07 16	☽ △ ☿
13 23	☽ ∥ ♇
21 45	☽ BQ ♂
22 33	☽ ⊥ ♇
26 Wednesday	
02 22	☽ □ ♆
03 24	☽ BQ ♀
09 26	☽ ⊻ ♇
11 06	☽ Q ♃
11 28	☿ BQ ♅
14 02	☿ ∥ ♂
14 32	☽ □ ♅
14 58	☽ ± ♂
15 01	☽ BQ ☿
15 18	☿ Q ♄
17 32	☽ △ ☉
19 18	☽ ± ♀
20 24	☽ ⊼ ♄
27 Thursday	
02 11	☽ ⊼ ♂
05 43	☽ ⊼ ♀
06 59	☽ ± ♄
08 09	☽ ✶ ♃
09 21	☽ ± ☿
10 20	☽ Q ♆
16 31	☿ ± ♆
21 18	☽ ⊼ ☿
21 38	☽ BQ ☉
21 54	☽ Q ♅
22 33	☽ BQ ♄
28 Friday	
05 54	☉ ± ♇
06 49	☽ ✶ ♆
09 34	☽ ∠ ♃
11 48	☉ □ ♄
13 22	☽ ☌ ♇
13 55	☽ ± ☉
18 03	☽ ✶ ♅
23 45	☽ △ ♄
29 Saturday	
00 24	☽ ⊥ ♃
00 33	☽ ⊼ ☉
02 45	☿ # ♃
07 33	☽ ☍ ♂
07 38	☽ ∠ ♆
07 45	☽ ☍ ♀
09 26	♂ BQ ♆
10 08	☽ ⊻ ♃
11 20	♀ ☌ ♂
14 10	☽ # ♀
18 31	☽ ∠ ♅
22 11	☽ ⊥ ♆
30 Sunday	
06 11	☽ ☍ ☿
07 08	♀ BQ ♆
07 46	☽ ⊻ ♆
08 18	☉ ∥ ☿
08 54	☽ ⊥ ♅
14 00	☽ ⊻ ♇
18 08	☿ ⊼ ♆
18 25	☽ ⊻ ♅
23 29	☽ ⊥ ♇

All ephemeris data is given at 12.00 UT and the Moon's longitude is additionally given for 24.00 UT

JULY 1996

LONGITUDES

Date	Sidereal time h m s	Sun ☉ ° ' "	Moon ☽ ° ' "	Moon ☽ 24.00 ° ' "	Mercury ☿ ° '	Venus ♀ ° '	Mars ♂ ° '	Jupiter ♃ ° '	Saturn ♄ ° '	Uranus ♅ ° '	Neptune ♆ ° '	Pluto ♇ ° '
01	06 39 16	09 ♋ 55 01	14 ♑ 41 49	22 ♑ 19 35	28 ♊ 20	11 ♊ 47	13 ♊ 21	13 ♑ 09	07 ♈ 07	03 ♒ 31	26 ♑ 49	00 ♐ 45
02	06 43 13	10 52 12	29 ♑ 56 54	07 ♒ 32 21	00 ♋ 21	11 D 47	14 03	13 R 01	07 10	03 R 29	26 R 48	00 R 44
03	06 47 09	11 49 23	15 ♒ 04 43	22 ♒ 32 54	02 27	11 48	14 44	12 53	07 12	03 27	26 46	00 43
04	06 51 06	12 46 34	29 ♒ 55 59	07 ♓ 13 18	04 33	11 52	15 26	12 46	07 14	03 25	26 45	00 42
05	06 55 02	13 43 45	14 ♓ 24 23	21 ♓ 28 58	06 40	11 58	16 08	12 38	07 15	03 23	26 43	00 40
06	06 58 59	14 40 56	28 ♓ 27 00	05 ♈ 18 33	08 48	12 07	16 49	12 30	07 16	03 20	26 42	00 39
07	07 02 56	15 38 08	12 ♈ 03 50	18 ♈ 43 12	10 57	12 17	17 31	12 23	07 17	03 18	26 40	00 38
08	07 06 52	16 35 20	25 ♈ 17 01	01 ♉ 45 43	13 07	12 30	18 12	12 15	07 19	03 16	26 38	00 37
09	07 10 49	17 32 33	08 ♉ 09 47	14 ♉ 29 38	15 17	12 44	18 54	12 07	07 20	03 14	26 37	00 36
10	07 14 45	18 29 46	20 ♉ 45 46	26 ♉ 58 36	17 27	13 01	19 35	12 00	07 20	03 11	26 35	00 35
11	07 18 42	19 26 59	03 ♊ 08 31	09 ♊ 15 54	19 36	13 19	20 17	11 52	07 21	03 09	26 34	00 34
12	07 22 38	20 24 13	15 ♊ 21 04	21 ♊ 24 20	21 45	13 40	20 58	11 45	07 22	03 07	26 32	00 34
13	07 26 35	21 21 28	27 ♊ 25 55	03 ♋ 26 05	23 53	14 02	21 39	11 37	07 23	03 04	26 30	00 33
14	07 30 31	22 18 42	09 ♋ 25 01	15 ♋ 22 54	26 00	14 25	22 20	11 30	07 23	03 02	26 29	00 32
15	07 34 28	23 15 58	21 ♋ 19 54	27 ♋ 16 12	28 ♋ 06	14 51	23 01	11 22	07 23	03 00	26 27	00 31
16	07 38 25	24 13 13	03 ♌ 11 58	09 ♌ 07 23	00 ♌ 11	15 18	23 42	11 15	07 24	02 57	26 26	00 30
17	07 42 21	25 10 29	15 ♌ 02 40	20 ♌ 58 02	02 14	15 46	24 23	11 07	07 24	02 55	26 24	00 29
18	07 46 18	26 07 45	26 ♌ 53 48	02 ♍ 50 14	04 16	16 16	25 04	11 00	07 24	02 53	26 22	00 29
19	07 50 14	27 05 01	08 ♍ 47 42	14 ♍ 46 36	06 16	16 47	25 45	10 53	07 R 24	02 50	26 21	00 28
20	07 54 11	28 02 18	20 ♍ 47 22	26 ♍ 50 29	08 14	17 20	26 26	10 46	07 24	02 48	26 19	00 27
21	07 58 07	28 59 35	02 ♎ 56 28	09 ♎ 05 52	10 11	17 54	27 07	10 39	07 24	02 45	26 17	00 27
22	08 02 04	29 ♋ 56 52	15 ♎ 19 15	21 ♎ 37 13	12 06	18 29	27 47	10 32	07 23	02 43	26 16	00 26
23	08 06 00	00 ♌ 54 10	28 ♎ 00 21	04 ♏ 29 12	14 00	19 06	28 28	10 25	07 23	02 41	26 14	00 25
24	08 09 57	01 51 28	11 ♏ 04 17	17 ♏ 46 04	15 51	19 43	29 08	10 19	07 22	02 38	26 13	00 25
25	08 13 54	02 48 46	24 ♏ 34 53	01 ♐ 30 59	17 41	20 22	29 ♊ 49	10 12	07 22	02 36	26 11	00 24
26	08 17 50	03 46 05	08 ♐ 34 26	15 ♐ 45 05	19 30	21 02	00 ♋ 29	10 05	07 21	02 33	26 09	00 24
27	08 21 47	04 43 24	23 ♐ 02 38	00 ♑ 26 30	21 16	21 43	01 10	09 59	07 20	02 31	26 08	00 23
28	08 25 43	05 40 44	07 ♑ 55 55	15 ♑ 29 51	23 01	22 25	01 50	09 53	07 19	02 29	26 06	00 23
29	08 29 40	06 38 04	23 ♑ 07 07	00 ♒ 46 23	24 44	23 07	02 30	09 47	07 18	02 26	26 05	00 22
30	08 33 36	07 35 25	08 ♒ 26 12	16 ♒ 05 08	26 25	23 51	03 11	09 41	07 17	02 24	26 03	00 22
31	08 37 33	08 ♌ 32 47	23 ♒ 41 49	01 ♓ 14 56	28 ♌ 05	24 ♊ 36	03 ♋ 51	09 ♑ 35	07 ♈ 16	02 ♒ 22	26 ♑ 01	00 ♐ 22

Date	Moon True ☊ ° '	Moon Mean ☊ ° '	Moon ☽ Latitude ° '
01	12 ♎ 52	12 ♎ 46	05 N 00
02	12 R 41	12 43	04 47
03	12 33	12 40	04 14
04	12 26	12 37	03 24
05	12 22	12 34	02 22
06	12 20	12 30	01 13
07	12 D 20	12 27	00 N 01
08	12 R 20	12 24	01 S 08
09	12 19	12 21	02 12
10	12 16	12 18	03 08
11	12 10	12 15	03 54
12	12 01	12 11	04 29
13	11 50	12 08	04 51
14	11 37	12 05	05 00
15	11 24	12 02	04 56
16	11 11	11 59	04 39
17	10 59	11 56	04 10
18	10 50	11 52	03 30
19	10 43	11 49	02 40
20	10 39	11 46	01 43
21	10 37	11 43	00 S 41
22	10 D 37	11 40	00 N 25
23	10 37	11 36	01 31
24	10 R 37	11 33	02 35
25	10 35	11 30	03 32
26	10 31	11 27	04 18
27	10 25	11 24	04 50
28	10 17	11 21	05 03
29	10 08	11 17	04 56
30	09 59	11 14	04 27
31	09 ♎ 51	11 ♎ 11	03 N 40

DECLINATIONS

Date	Sun ☉	Moon ☽	Mercury ☿	Venus ♀	Mars ♂	Jupiter ♃	Saturn ♄	Uranus ♅	Neptune ♆	Pluto ♇
01	23 N 04	17 S 39	23 N 25	18 N 03	22 N 31	22 S 51	00 N 42	19 S 57	20 S 19	07 S 16
02	23 00	15 29	23 37	17 57	22 37	22 52	00 43	19 58	20 19	07 16
03	22 55	12 16	23 46	17 53	22 43	22 53	00 43	19 58	20 19	07 16
04	22 49	08 19	23 53	17 49	22 48	22 53	00 43	19 59	20 20	07 16
05	22 44	03 S 57	23 58	17 46	22 53	22 54	00 44	19 59	20 20	07 17
06	22 38	00 N 30	23 59	17 43	22 58	22 55	00 44	20 00	20 20	07 17
07	22 31	04 47	23 59	17 41	23 03	22 56	00 44	20 00	20 20	07 17
08	22 24	08 44	23 55	17 40	23 07	22 57	00 44	20 01	20 21	07 17
09	22 17	12 09	23 48	17 40	23 11	22 58	00 44	20 01	20 21	07 17
10	22 10	14 55	23 39	17 40	23 15	22 59	00 45	20 02	20 21	07 18
11	22 02	16 57	23 27	17 40	23 19	22 59	00 45	20 02	20 22	07 18
12	21 53	18 11	23 13	17 41	23 23	23 00	00 45	20 03	20 22	07 18
13	21 45	18 34	22 56	17 43	23 26	23 01	00 45	20 04	20 22	07 18
14	21 35	18 07	22 36	17 45	23 29	23 02	00 45	20 04	20 23	07 19
15	21 26	16 52	22 14	17 47	23 32	23 02	00 44	20 05	20 23	07 19
16	21 16	14 55	21 51	17 50	23 35	23 03	00 44	20 05	20 23	07 19
17	21 06	12 20	21 25	17 53	23 38	23 04	00 44	20 06	20 24	07 20
18	20 55	09 16	20 57	17 56	23 40	23 05	00 44	20 06	20 24	07 20
19	20 44	05 48	20 28	18 00	23 42	23 05	00 44	20 07	20 24	07 20
20	20 33	02 N 04	19 57	18 04	23 44	23 06	00 43	20 08	20 25	07 21
21	20 22	01 S 48	19 24	18 08	23 46	23 07	00 43	20 08	20 25	07 21
22	20 10	05 39	18 51	18 12	23 47	23 08	00 43	20 09	20 25	07 21
23	19 57	09 21	18 16	18 17	23 48	23 08	00 42	20 09	20 26	07 21
24	19 45	12 42	17 40	18 21	23 49	23 09	00 42	20 10	20 26	07 22
25	19 32	15 29	17 03	18 26	23 50	23 09	00 41	20 10	20 26	07 22
26	19 18	17 29	16 26	18 30	23 51	23 10	00 41	20 11	20 26	07 22
27	19 05	18 26	15 47	18 35	23 51	23 11	00 40	20 11	20 27	07 23
28	18 51	18 09	15 09	18 40	23 52	23 11	00 40	20 12	20 27	07 23
29	18 37	16 36	14 29	18 45	23 52	23 12	00 39	20 13	20 27	07 24
30	18 22	13 51	13 49	18 50	23 52	23 12	00 38	20 13	20 28	07 24
31	18 N 07	10 S 10	13 N 09	18 N 54	23 N 51	23 S 13	00 N 38	20 S 14	20 S 28	07 S 24

ZODIAC SIGN ENTRIES

Date	h	m	Planets
02	07	37	☿ ♋
02	12	05	☽ ♒
04	12	07	☽ ♓
06	14	42	☽ ♈
08	20	43	☽ ♉
11	05	52	☽ ♊
13	17	08	☽ ♋
16	05	31	☽ ♌
16	09	56	☿ ♌
18	18	16	☽ ♍
21	06	14	☽ ♎
22	13	19	☉ ♌
23	15	43	☽ ♏
25	18	32	♂ ♋
25	21	24	☽ ♐
27	23	17	☽ ♑
29	22	47	☽ ♒
31	22	00	☽ ♓

LATITUDES

Date	Mercury ☿	Venus ♀	Mars ♂	Jupiter ♃	Saturn ♄	Uranus ♅	Neptune ♆	Pluto ♇
01	00 S 01	04 S 11	00 N 07	00 S 04	02 S 19	00 S 36	00 N 29	13 N 19
04	00 N 32	04 26	00 09	00 04	02 20	00 36	00 29	13 17
07	01 00	04 37	00 12	00 04	02 21	00 36	00 29	13 16
10	01 22	04 44	00 14	00 05	02 22	00 36	00 29	13 15
13	01 38	04 48	00 16	00 05	02 23	00 37	00 29	13 14
16	01 46	04 50	00 18	00 06	02 23	00 37	00 29	13 12
19	01 49	04 48	00 20	00 06	02 24	00 37	00 29	13 11
22	01 45	04 45	00 22	00 06	02 25	00 37	00 29	13 10
25	01 36	04 40	00 24	00 07	02 26	00 37	00 29	13 08
28	01 22	04 34	00 26	00 07	02 27	00 37	00 29	13 07
31	01 N 05	04 S 25	00 N 28	00 S 08	02 S 27	00 S 37	00 N 29	13 N 05

DATA

Julian Date	2450266
Delta T	+62 seconds
Ayanamsa	23° 48' 34"
Synetic vernal point	05° ♓ 18' 26"
True obliquity of ecliptic	23° 26' 13"

MOON'S PHASES, APSIDES AND POSITIONS ☽

Date	h	m	Phase	Longitude ° '	Eclipse Indicator
01	03	58	○	09 ♑ 36	
07	18	55	☾	15 ♈ 55	
15	16	15	●	23 ♋ 26	
23	17	49	☽	01 ♏ 08	
30	10	35	○	07 ♒ 32	

Day	h	m		
01	22	18	Perigee	
16	13	47	Apogee	
30	07	41	Perigee	
06	09	18	0N	
13	11	01	Max dec	18° N 34'
21	00	54	0S	
27	18	51	Max dec	18° S 29'

LONGITUDES

Date	Chiron ⚷ ° '	Ceres ⚳ ° '	Pallas ⚴ ° '	Juno ⚵ ° '	Vesta ⚶ ° '	Black Moon Lilith ⚸ ° '
01	08 ♎ 35	02 ♐ 59	21 ♎ 17	08 ♈ 16	12 ♏ 18	00 ♌ 56
11	09 ♎ 09	02 ♐ 13	23 ♎ 01	11 ♈ 05	13 ♏ 30	02 ♌ 03
21	09 ♎ 54	02 ♐ 04	25 ♎ 13	13 ♈ 32	15 ♏ 23	03 ♌ 11
31	10 ♎ 47	02 ♐ 30	27 ♎ 48	15 ♈ 30	17 ♏ 50	04 ♌ 18

ASPECTARIAN

h m Aspects

01 Monday
00 05 ☽ □ ♄
02 27 ☿ ± ♅
03 58 ☽ ☍ ☉
04 30 ☽ # ♀
06 09 ♂ ⊼ ♃
07 26 ☽ ⊼ ♀
09 35 ☽ ☌ ♃
09 46 ☽ ⊼ ♂
13 39 ☽ ∠ ♇
16 51 ☽ ± ♀
19 40 ☽ ± ♂

02 Tuesday
04 28 ☽ Q ♄
06 51 ♀ St D
07 00 ☽ ⚼ ♀
07 03 ☽ ☌ ♆
10 30 ☽ ⚼ ♂
12 47 ☽ ⊼ ☿
13 14 ☽ ⚹ ♇
16 06 ☿ ⊼ ♇
17 34 ☽ ☌ ♅
23 22 ♀ ⚼ ♆
23 26 ☽ ⚹ ♄
23 45 ☽ ± ☿

03 Wednesday
06 27 ☽ ⊼ ☉
06 46 ☽ △ ♀
08 14 ☽ Q ♇
08 32 ☽ ⊻ ♃
11 26 ☽ △ ♂
11 30 ☉ ⊻ ♀
16 24 ☽ ⚼ ☿
16 42 ☽ ± ☉
18 03 ☽ ⊥ ♃
20 34 ☉ # ♃
23 21 ☿ ⊼ ♅
23 27 ☽ ∠ ♄

04 Thursday
06 49 ☽ ⊻ ♆
08 14 ☽ ⚼ ☉
08 29 ☽ ∠ ♃
11 41 ☉ ☍ ♃
13 14 ☽ □ ♇
14 07 ☽ ⊥ ♄
15 24 ☉ ∥ ♂
16 36 ☽ ⊥ ♆
17 42 ☽ ⊻ ♅
17 51 ☽ ∥ ♇
20 51 ☽ △ ☿

05 Friday
00 02 ☽ ⊻ ♄
03 36 ☽ ⊥ ♅
07 30 ☽ ∠ ♆
07 53 ☽ □ ♀
09 03 ☽ ⚹ ♃
10 47 ☽ △ ☉
12 08 ☿ ± ♇
15 03 ☽ □ ♂
18 40 ☿ □ ♄
18 42 ☽ ∠ ♅
19 30 ♂ # ♃

06 Saturday
05 15 ☽ Q ♃
05 21 ☽ # ♄
08 58 ☽ ⚹ ♆
13 17 ☽ ∥ ♄
14 55 ☽ Q ♀
15 50 ☽ △ ♇
20 31 ☽ ⚹ ♅
23 45 ☽ Q ♂

07 Sunday
03 29 ☽ ☌ ♄
05 57 ☽ Q ♆
09 38 ☽ □ ☿
12 05 ☉ ⚼ ♇
12 24 ☽ ⚹ ♀
12 33 ☽ □ ♃
17 48 ☽ Q ♅
18 25 ☽ ⚼ ♇
18 40 ♀ ⊼ ♃
18 55 ☽ □ ☉
22 21 ☽ ⚹ ♂

08 Monday
02 53 ☽ # ♇
02 56 ☿ ☍ ♃
04 22 ☿ ⊻ ♀
10 47 ☽ ± ♇
13 53 ♂ ⚼ ♅
14 30 ☽ □ ♆
16 09 ☽ ∠ ♀
21 52 ☽ ⊼ ♇

09 Tuesday
00 58 ☽ Q ☿
02 46 ☽ □ ♅
03 32 ☽ ∠ ♂
06 40 ☽ Q ☉
09 16 ☽ ⊥ ♀
10 25 ☽ ⊻ ♄
15 36 ☿ ⚼ ♇
19 25 ☽ △ ♃
20 51 ☽ ⊻ ♀
21 29 ☽ ⊥ ♂
21 47 ☽ ⊥ ♄

10 Wednesday
03 12 ♂ Q ♄
04 18 ☽ ⚹ ☿
07 17 ☽ ⚹ ☉
09 37 ☽ ⊻ ♂
15 02 ☽ ∠ ♄
23 13 ☽ △ ♆
23 55 ☽ ⚼ ♃

11 Thursday
07 00 ☽ ☍ ♇
07 10 ☉ ⊥ ♀
08 22 ☿ ⊥ ♀
09 01 ☉ ☌ ☿
12 01 ☽ △ ♅
14 46 ☽ ∠ ☉
15 27 ☽ ∠ ☿
17 17 ☽ ± ♃
20 15 ☽ ⚹ ♄
21 31 ♂ ± ♆
23 07 ☿ ⊻ ♂
23 11 ☿ ∥ ♂

12 Friday
00 10 ☽ ∥ ♀
04 29 ☽ ⚼ ♆
04 57 ☽ ⊼ ♃
08 34 ☽ ☌ ♀
09 58 ☽ ⊥ ☉
12 57 ☽ ⊥ ☿
17 27 ☽ ⚼ ♅
19 57 ☽ Q ♄
22 15 ☽ ± ♆
22 52 ☽ ⊻ ☉
23 48 ☽ ☌ ♂

13 Saturday
03 25 ☽ ⊻ ☿
05 16 ☿ # ♃
10 09 ☽ ⊼ ♆
11 17 ☽ ± ♅
18 12 ☽ ⊼ ♇
23 14 ☽ ⊼ ♅

14 Sunday
06 13 ☽ ± ♇
07 55 ☽ □ ♄
14 18 ☉ ⊻ ♂
16 08 ☽ ☍ ♃
17 23 ☿ ☍ ♆
20 38 ☽ ∥ ♀
22 26 ☽ ⊻ ♀

15 Monday
00 17 ☽ ⚼ ♇
10 59 ☽ ⊥ ♀
15 37 ☽ ⊻ ♂
16 15 ☽ ☌ ☉
22 19 ☽ ☍ ♆

16 Tuesday
03 50 ☿ ⊥ ♂
04 30 ☽ ⊥ ♂
04 35 ☽ ☌ ☿
05 53 ☽ ∠ ♀
06 33 ☽ △ ♇
11 31 ☽ ☍ ♅
13 43 ☿ ∠ ♀
15 45 ☿ △ ♇
20 30 ☽ △ ♄
23 50 ☽ ∠ ♂

17 Wednesday
04 08 ☽ ⊼ ♃
13 32 ☽ ⚹ ♀
16 10 ☽ ± ♃
19 55 ☿ ☍ ♅

18 Thursday
02 54 ☽ ⚼ ♄
08 05 ☽ ⚹ ♂
10 13 ☽ ⚼ ♃
10 19 ☽ ⊻ ☉
10 56 ☽ ⊼ ♆
14 13 ☉ ∥ ☿
14 53 ☽ Q ♀
17 55 ☉ ☍ ♆
19 14 ☽ □ ♇
20 28 ♄ St R
21 06 ☽ ± ♄
23 02 ☽ ± ♆
23 30 ☽ ⊥ ☉

19 Friday
00 02 ☽ ⊼ ♅
01 37 ☽ # ♇
05 53 ☽ ⊻ ☿
09 12 ☽ ⊼ ♄
09 47 ☽ Q ♂
12 05 ☽ ± ♅
14 34 ☿ # ♆
16 09 ☽ △ ♃
17 07 ☽ ⚼ ♆
19 11 ☽ ∠ ☉
20 21 ☽ ⊥ ☿

20 Saturday
01 45 ☿ △ ♄
03 45 ☿ # ♅
04 46 ☽ □ ♀
06 03 ☽ ⚼ ♅
07 21 ☽ Q ♇
08 06 ♂ ⊼ ♆
17 48 ☽ ∠ ☿
20 26 ☽ ∥ ♄
22 56 ☽ △ ♆
23 51 ☽ □ ♂

21 Sunday
00 12 ♂ ± ♅
03 35 ☽ ⚹ ☉
05 18 ☉ # ♆
05 22 ☽ # ♄
06 32 ♀ ⚼ ♅
07 06 ☽ ⚹ ♇
11 39 ☽ △ ♅
17 26 ☿ ⊼ ♃
20 41 ☽ ☍ ♄

22 Monday
02 52 ☽ □ ♃
04 42 ☽ ⚹ ☿
04 58 ☽ Q ☉
12 13 ☽ ∠ ♇
13 40 ☉ # ♅
18 21 ☽ △ ♀
22 52 ☽ ∥ ♇

23 Tuesday
00 03 ☉ △ ♇
01 27 ☿ ∠ ♂
05 18 ☽ ⊥ ♇
07 36 ☽ Q ☿
08 42 ☽ □ ♆
11 31 ☿ ∥ ♀
12 46 ☽ Q ♃
12 54 ☽ △ ♂
16 30 ☽ ⊻ ♇
17 49 ☽ □ ☉
20 39 ☽ □ ♅
23 02 ♀ Q ♄
23 50 ☽ ⚼ ♀

24 Wednesday
05 17 ☽ ⊼ ♄
10 38 ☽ ⚹ ♃
16 09 ☽ ± ♄
17 01 ☽ ± ♀
17 33 ☿ ± ♃
17 38 ☽ Q ♆
17 49 ☽ ⚼ ♂
21 58 ☽ □ ☿

25 Thursday
04 14 ☽ ⊼ ♀
05 02 ☽ Q ♅
05 32 ♀ ± ♆
06 49 ☉ ☍ ♅
08 07 ☽ ⚼ ♄
10 36 ☽ ± ♂
13 04 ☽ ∠ ♃
14 47 ☽ ⚹ ♆
21 32 ☽ ⊼ ♂
22 05 ☽ ☌ ♇

26 Friday
01 16 ☽ # ☿
01 49 ☽ ⚹ ♅
03 15 ☽ △ ☉
04 28 ☽ ⊥ ♃
08 40 ♂ ⊼ ♇
09 56 ☽ △ ♄
14 32 ☽ ⊻ ♃
16 20 ☽ ∠ ♆

27 Saturday
02 57 ☽ ∠ ♅
06 11 ☽ ⚼ ☉
07 14 ☽ ⊥ ♆
08 42 ☽ △ ☿
09 43 ☽ ☍ ♀
17 01 ☽ ⊻ ♆
17 38 ☽ ⊥ ♅
21 52 ☽ ± ☉
22 02 ☿ ⚹ ♀
23 54 ☽ ⊻ ♇

28 Sunday
01 47 ☽ ☍ ♂
02 32 ☿ ⚼ ♄
03 18 ☽ ⊻ ♅
08 10 ☽ ⊼ ☉
09 32 ☽ ⊥ ♇
11 02 ☽ □ ♄
12 09 ☽ ⚼ ☿
15 05 ☽ ☌ ♃
23 49 ☽ ∠ ♇

29 Monday
01 50 ☉ ∥ ♀
04 14 ☽ ± ☿
09 41 ♂ ⊼ ♅
12 01 ☽ ⊼ ♀
12 37 ☿ ⚼ ♃
14 51 ☽ ⊼ ☿
15 26 ☽ Q ♄
16 38 ☽ ☌ ♆
21 53 ☽ ± ♀
23 22 ☽ ⚹ ♇

30 Tuesday
02 34 ☽ ☌ ♅
03 23 ☽ ⊼ ♂
04 31 ☉ △ ♄
06 47 ☿ ⊼ ♆
10 12 ☽ ⚹ ♄
10 35 ☽ ☍ ☉
12 14 ☽ # ☿
12 41 ☽ ⚼ ♀
13 13 ☽ ± ♂
13 56 ☽ ⊻ ♃
18 10 ☽ Q ♇
23 17 ☽ ⊥ ♃

31 Wednesday
03 59 ☽ ⚼ ♂
09 44 ☽ ∠ ♄
13 23 ☽ ∠ ♃
13 30 ☽ △ ♀
15 41 ☽ ⊻ ♆
19 48 ☽ ☍ ☿
22 35 ☽ □ ♇

All ephemeris data is given at 12.00 UT and the Moon's longitude is additionally given for 24.00 UT

AUGUST 1996

LONGITUDES

Date	Sidereal time h m s	Sun ☉ ° ' "	Moon ☽ ° ' "	Moon ☽ 24.00 ° ' "	Mercury ☿ ° '	Venus ♀ ° '	Mars ♂ ° '	Jupiter ♃ ° '	Saturn ♄ ° '	Uranus ♅ ° '	Neptune ♆ ° '	Pluto ♇ ° '
01	08 41 29	09 ♌ 30 09	08 ♓ 43 24	16 ♓ 06 17	29 ♌ 43	25 ♊ 22	04 ♋ 31	09 ♑ 29	07 ♈ 15	02 ♒ 19	26 ♑ 00	00 ♐ 21
02	08 45 26	10 27 33	23 ♓ 22 54	00 ♈ 32 46	01 ♍ 19	26 08	05 11	09 R 23	07 R 13	02 R 17	25 R 58	00 R 21
03	08 49 23	11 24 58	07 ♈ 35 37	14 ♈ 31 24	02 53	26 55	05 51	09 18	07 12	02 14	25 57	00 21
04	08 53 19	12 22 24	21 ♈ 20 12	28 ♈ 02 15	04 26	27 43	06 31	09 12	07 10	02 12	25 55	00 21
05	08 57 16	13 19 51	04 ♉ 37 55	11 ♉ 07 36	05 57	28 32	07 11	09 07	07 08	02 10	25 54	00 20
06	09 01 12	14 17 20	17 ♉ 31 49	23 ♉ 51 03	07 27	29 ♊ 22	07 50	09 02	07 07	02 07	25 52	00 20
07	09 05 09	15 14 49	00 ♊ 05 50	06 ♊ 16 44	08 55	00 ♋ 12	08 30	08 57	07 05	02 05	25 51	00 20
08	09 09 05	16 12 21	12 ♊ 24 14	18 ♊ 28 51	10 21	01 03	09 10	08 52	07 03	02 03	25 49	00 20
09	09 13 02	17 09 53	24 ♊ 31 03	00 ♋ 31 16	11 45	01 55	09 49	08 48	07 01	02 00	25 48	00 20
10	09 16 58	18 07 27	06 ♋ 29 55	12 ♋ 27 20	13 07	02 47	10 29	08 43	06 58	01 58	25 46	00 20
11	09 20 55	19 05 02	18 ♋ 23 51	24 ♋ 19 47	14 28	03 40	11 09	08 39	06 56	01 56	25 45	00 D 20
12	09 24 52	20 02 39	00 ♌ 15 24	06 ♌ 10 55	15 47	04 33	11 48	08 35	06 54	01 54	25 43	00 20
13	09 28 48	21 00 16	12 ♌ 06 35	18 ♌ 02 36	17 04	05 27	12 27	08 31	06 51	01 51	25 42	00 20
14	09 32 45	21 57 55	23 ♌ 59 09	29 ♌ 56 28	18 19	06 22	13 07	08 27	06 49	01 49	25 40	00 20
15	09 36 41	22 55 35	05 ♍ 54 46	11 ♍ 54 16	19 31	07 17	13 46	08 23	06 46	01 47	25 39	00 20
16	09 40 38	23 53 17	17 ♍ 55 12	23 ♍ 57 52	20 42	08 12	14 25	08 20	06 44	01 45	25 38	00 21
17	09 44 34	24 50 59	00 ♎ 02 33	06 ♎ 09 35	21 51	09 08	15 04	08 17	06 41	01 43	25 36	00 21
18	09 48 31	25 48 43	12 ♎ 19 20	18 ♎ 32 11	22 57	10 05	15 43	08 14	06 38	01 41	25 35	00 21
19	09 52 27	26 46 27	24 ♎ 48 33	01 ♏ 08 52	24 01	11 02	16 22	08 11	06 35	01 38	25 33	00 21
20	09 56 24	27 44 13	07 ♏ 33 33	14 ♏ 03 03	25 03	11 59	17 01	08 08	06 32	01 36	25 32	00 22
21	10 00 21	28 42 00	20 ♏ 37 47	27 ♏ 18 06	26 02	12 57	17 40	08 05	06 29	01 34	25 31	00 22
22	10 04 17	29 ♌ 39 49	04 ♐ 04 21	10 ♐ 56 44	26 58	13 56	18 19	08 03	06 26	01 32	25 30	00 22
23	10 08 14	00 ♍ 37 38	17 ♐ 55 22	25 ♐ 00 16	27 51	14 54	18 58	08 01	06 22	01 30	25 28	00 23
24	10 12 10	01 35 29	02 ♑ 11 14	09 ♑ 27 57	28 41	15 54	19 37	07 59	06 19	01 28	25 27	00 23
25	10 16 07	02 33 21	16 ♑ 49 52	24 ♑ 16 17	29 ♍ 28	16 53	20 15	07 57	06 16	01 26	25 26	00 24
26	10 20 03	03 31 14	01 ♒ 46 19	09 ♒ 18 55	00 ♎ 12	17 53	20 54	07 55	06 12	01 24	25 25	00 24
27	10 24 00	04 29 08	16 ♒ 52 56	24 ♒ 27 09	00 52	18 53	21 32	07 54	06 09	01 23	25 23	00 25
28	10 27 56	05 27 04	02 ♓ 00 19	09 ♓ 31 14	01 28	19 54	22 11	07 53	06 05	01 21	25 22	00 25
29	10 31 53	06 25 01	16 ♓ 58 47	24 ♓ 21 57	02 00	20 55	22 49	07 52	06 01	01 19	25 21	00 26
30	10 35 50	07 23 00	01 ♈ 39 54	08 ♈ 51 58	02 27	21 56	23 28	07 51	05 57	01 17	25 20	00 26
31	10 39 46	08 ♍ 21 01	15 ♈ 57 40	22 ♈ 56 43	02 ♎ 50	22 ♋ 58	24 ♋ 06	07 ♑ 50	05 ♈ 54	01 ♒ 15	25 ♑ 19	00 ♐ 27

ASPECTARIAN

01 Thursday
- 00 01 ☽ ⊥ ♄
- 01 13 ☽ ⊥ ♆
- 01 44 ☽ ⊻ ♅
- 03 37 ☽ ∥ ♇
- 04 54 ☽ △ ♂
- 09 37 ☽ ⊻ ♄
- 11 21 ☽ ⊥ ♅
- 11 31 ☉ ⊼ ♃
- 13 13 ☽ ⚹ ♃
- 13 21 ☽ ⊼ ☉
- 15 40 ☽ ∠ ♆
- 21 34 ☿ □ ♇
- 23 47 ☽ ± ☉

02 Friday
- 01 57 ☽ ∠ ♅
- 07 06 ♀ ⊼ ♆
- 08 43 ☽ Q ♃
- 10 35 ☿ ± ♄
- 15 42 ☽ bQ ☉
- 15 45 ☽ # ♄
- 16 12 ♀ ± ♅
- 16 18 ☽ ⚹ ♆
- 16 51 ☽ □ ♀
- 21 46 ☿ ± ♆
- 22 02 ☽ ∥ ♄
- 23 40 ☽ △ ♇

03 Saturday
- 02 16 ☿ ⊼ ♅
- 02 53 ☽ ⚹ ♅
- 02 58 ☽ ⊼ ☿
- 08 51 ☽ □ ♂
- 11 19 ☽ ☌ ♄
- 12 36 ☽ Q ♆
- 14 31 ☽ ± ☿
- 14 55 ☽ □ ♃
- 19 05 ☽ △ ☉
- 23 28 ☽ Q ♅

04 Sunday
- 01 26 ☽ bQ ♇
- 01 29 ☽ Q ♀
- 05 57 ♂ ± ♇
- 08 12 ☽ bQ ☿
- 12 27 ☽ # ♇
- 17 22 ☽ ± ♇
- 17 57 ☽ Q ♂
- 20 10 ☽ □ ♆
- 23 24 ☉ ⊥ ♂

05 Monday
- 00 11 ☽ ⚹ ♀
- 04 10 ☽ ⊼ ♇
- 04 41 ☽ ∥ ☿
- 07 30 ☽ □ ♅
- 10 44 ♂ □ ♄
- 14 45 ☽ △ ☿
- 16 36 ☽ ⊻ ♄
- 16 56 ☽ ⚹ ♂
- 20 13 ☽ △ ♃

06 Tuesday
- 03 43 ☽ ⊥ ♄
- 05 25 ☽ □ ☉
- 05 38 ☽ ∠ ♀
- 06 37 ☿ ⊼ ♄
- 20 40 ☽ ∠ ♄
- 22 38 ☽ ∠ ♂
- 22 43 ☿ ± ♅
- 23 38 ☿ ⚹ ♂
- 23 52 ☽ ⊥ ♀

07 Wednesday
- 00 16 ☽ bQ ♃
- 02 46 ☉ ∠ ♀
- 03 50 ☽ △ ♆
- 05 12 ☉ ± ♃
- 11 27 ☽ ∥ ☉
- 12 13 ☽ ⊻ ♀
- 12 28 ☽ ☍ ♇
- 12 39 ☿ △ ♃
- 15 49 ♀ ⊼ ♇
- 15 50 ☽ △ ♅
- 16 55 ☽ ⊥ ♂
- 17 29 ☽ ± ♃
- 18 37 ☽ Q ☉

08 Thursday
- 01 31 ☽ ⚹ ♄
- 02 33 ♂ ☍ ♃
- 05 07 ☽ ⊼ ♃
- 05 17 ☽ ⊻ ♂
- 07 25 ☽ □ ☿
- 08 53 ☽ bQ ♆
- 19 54 ☿ bQ ♆
- 20 08 ☽ ⚹ ☉
- 21 08 ☽ bQ ♅
- 21 34 ☿ # ♇

09 Friday
- 01 05 ☽ Q ♄
- 02 37 ☽ ± ♆
- 14 32 ☽ ⊼ ♆
- 14 33 ♀ ⊼ ♅
- 14 58 ☽ ± ♅
- 23 37 ☽ ⊼ ♇
- 23 49 ☽ Q ☿

10 Saturday
- 02 56 ☽ ⊼ ♅
- 03 57 ☽ ☌ ♀
- 04 38 ☽ ∠ ☉
- 11 40 ☽ ± ♇
- 12 33 ♇ St D
- 12 57 ☽ □ ♄
- 16 27 ☽ ☍ ♃
- 20 30 ☽ ☌ ♂

11 Sunday
- 00 20 ☽ ⊥ ☉
- 03 03 ☽ ⚹ ☿
- 05 49 ☽ bQ ♇
- 13 31 ☽ ⊻ ☉

12 Monday
- 02 50 ☽ ☍ ♆
- 12 10 ☽ △ ♇
- 13 11 ☽ ∠ ☿
- 15 18 ☽ ☍ ♅
- 20 04 ☽ ∥ ☉
- 21 25 ☽ ⊻ ♀

13 Tuesday
- 01 24 ☽ △ ♄
- 04 46 ☽ ⊼ ♃
- 08 14 ☿ bQ ♅
- 09 37 ☽ ⊥ ☿
- 10 34 ☽ ⊥ ♀
- 12 45 ☽ ⊻ ♂
- 16 50 ☽ ± ♃
- 23 12 ☽ ⊻ ☿

14 Wednesday
- 01 35 ☽ ⊥ ♂
- 06 16 ☽ ∠ ♀
- 07 34 ☽ ☌ ☉
- 07 38 ☽ bQ ♄
- 08 24 ☉ bQ ♄
- 10 56 ☽ bQ ♃
- 11 23 ♀ ± ♇
- 12 33 ☿ Q ♇
- 15 24 ☽ ⊼ ♆
- 15 56 ☿ Q ♀
- 20 48 ☽ ∠ ♂
- 23 20 ♀ □ ♄

15 Thursday
- 00 48 ☽ □ ♇
- 01 42 ☽ ± ♄
- 03 27 ☽ ± ♆
- 03 44 ☽ ⊼ ♅
- 06 40 ☽ # ♇
- 13 43 ☽ ⊼ ♄
- 14 58 ☽ ⚹ ♀
- 15 44 ☽ ± ♅
- 16 57 ☽ △ ♃
- 21 28 ☽ bQ ♆
- 22 56 ☉ bQ ♃

16 Friday
- 04 26 ♂ # ♃
- 04 37 ☽ ⚹ ♂
- 09 40 ☽ bQ ♅
- 12 51 ☽ Q ♇
- 15 06 ♀ ☍ ♃
- 15 57 ☽ ∥ ☿
- 16 55 ☽ Q ♀
- 18 07 ☽ ☌ ☿

17 Saturday
- 00 52 ☽ ⊻ ☉
- 03 15 ☽ △ ♆
- 05 23 ☽ ∥ ♄
- 05 49 ☽ Q ♂
- 09 33 ☽ # ♄
- 12 36 ☽ ⚹ ♇
- 13 43 ☽ ⊥ ☉
- 15 16 ☽ △ ♅
- 18 42 ☽ # ☿
- 22 11 ♂ bQ ♇

18 Sunday
- 00 58 ☽ ☍ ♄
- 04 04 ☽ □ ♃
- 06 21 ☉ ⊼ ♆
- 07 17 ☽ □ ♀
- 08 49 ☽ ∠ ☉
- 17 52 ☽ ∠ ♇
- 18 57 ☽ □ ♂

19 Monday
- 07 21 ☽ ∥ ♇
- 10 22 ☽ ⊻ ☿
- 11 08 ☽ ⊥ ♇
- 13 25 ☽ □ ♆
- 14 36 ☽ Q ♃
- 16 03 ☽ ⚹ ☉
- 22 31 ☽ ⊻ ♇
- 22 45 ☽ ⊥ ☿

20 Tuesday
- 00 54 ☽ □ ♅
- 10 06 ☽ ⊼ ♄
- 10 56 ☿ ∥ ♄
- 13 04 ☽ ⚹ ♃
- 16 11 ☽ # ☉
- 16 22 ☽ Q ☉
- 17 00 ☽ ∠ ☿
- 20 52 ☽ △ ♀
- 21 10 ☽ ± ♄
- 23 02 ☽ Q ♆
- 23 35 ☿ △ ♆

21 Wednesday
- 06 21 ☽ △ ♂
- 09 05 ☿ # ♄
- 10 05 ☽ Q ♅
- 13 32 ☽ bQ ♄
- 16 26 ☽ ∠ ♃
- 20 47 ☽ ⚹ ♆
- 22 28 ☽ ⚹ ☿

22 Thursday
- 02 11 ☽ bQ ♀
- 03 36 ☽ □ ☉
- 05 28 ☽ ☌ ♇
- 07 32 ☽ ⚹ ♅
- 08 27 ☽ ⊥ ♃
- 10 36 ☽ bQ ♂
- 16 07 ☽ △ ♄
- 18 57 ☽ ⊻ ♃
- 19 16 ☽ ± ♀
- 21 10 ☽ Q ☿
- 23 12 ☽ ∠ ♆

23 Friday
- 03 05 ☽ ± ♂
- 05 49 ☉ □ ♇
- 05 59 ☉ ± ♄
- 06 27 ☽ ⊼ ♀
- 09 35 ☽ ∠ ♅
- 13 52 ☽ ⊼ ♂
- 14 38 ☽ ⊥ ♆
- 23 40 ♀ bQ ♇

24 Saturday
- 00 46 ☽ ⊻ ♆
- 00 49 ☽ ⊥ ♅
- 05 49 ☽ □ ☿
- 08 35 ☉ ± ♆
- 09 01 ☽ ⊻ ♇
- 09 06 ☉ ⊼ ♅
- 10 49 ☽ ⊻ ♅
- 10 56 ☽ △ ☉
- 18 48 ☽ □ ♄
- 18 57 ☽ ⊥ ♇
- 21 33 ☽ ☌ ♃

25 Sunday
- 09 40 ☽ ∠ ♇
- 12 05 ☽ ☍ ♀
- 13 15 ☽ bQ ☉
- 17 47 ☽ ☍ ♂
- 23 56 ☽ Q ♄

26 Monday
- 01 51 ☽ ☌ ♆
- 04 44 ☽ ± ☉
- 09 22 ☽ △ ☿
- 09 49 ☽ ⚹ ♇
- 11 25 ☽ ☌ ♅
- 14 59 ☽ ⊼ ☉
- 19 02 ☽ ⚹ ♄
- 19 23 ☿ ⚹ ♇
- 21 46 ☽ ⊻ ♃

27 Tuesday
- 04 55 ☽ Q ♇
- 07 17 ☽ ⊥ ♃
- 10 19 ☽ bQ ☿
- 10 24 ☿ Q ♀
- 15 24 ☽ ⊼ ♀
- 18 43 ☽ ∠ ♄
- 19 42 ☽ ⊼ ♂
- 21 31 ☽ ∠ ♃

28 Wednesday
- 01 11 ☽ ± ☿
- 01 28 ☽ ⊻ ♆
- 01 36 ☽ ± ♀
- 01 48 ☽ # ☉
- 05 39 ☽ ± ♂
- 07 20 ☿ △ ♅
- 08 57 ☽ ⊥ ♄
- 09 29 ☽ □ ♇
- 10 57 ☽ ⊻ ♅
- 11 00 ☽ ⊥ ♆
- 11 06 ☽ ⊼ ☿
- 13 15 ☽ ∥ ♇
- 16 57 ☽ bQ ♀
- 17 52 ☽ ☍ ☉
- 18 28 ☽ ⊻ ♄
- 20 30 ☽ ⊥ ♅
- 20 37 ☽ bQ ♂
- 21 22 ☽ ⚹ ♃

29 Thursday
- 01 21 ☽ ∠ ♆
- 02 44 ☉ ⊼ ♄
- 09 44 ☽ ∥ ☿
- 10 56 ☽ ∠ ♅
- 16 40 ☽ Q ♃
- 18 51 ☽ △ ♀
- 21 54 ☽ △ ♂

30 Friday
- 01 36 ☽ ⚹ ♆
- 05 35 ☽ # ♄
- 05 43 ☽ ∥ ♄
- 09 36 ☉ ± ♅
- 09 59 ☽ △ ♇
- 11 22 ☽ ⚹ ♅
- 13 20 ☽ ☍ ☿
- 19 06 ☽ ☌ ♄
- 21 25 ☽ Q ♆
- 22 12 ☽ ⊼ ☉
- 22 17 ☽ □ ♃
- 23 22 ☉ △ ♃

31 Saturday
- 05 00 ☽ # ☿
- 07 25 ☽ Q ♅
- 09 03 ☽ ± ☉
- 11 08 ☽ bQ ♇

Moon Nodes and Latitude

Date	Moon True ☊ ° '	Moon Mean ☊ ° '	Moon ☽ Latitude ° '
01	09 ♎ 45	11 ♎ 08	02 N 38
02	09 R 42	11 05	01 26
03	09 41	11 01	00 N 11
04	09 D 42	10 58	01 S 02
05	09 42	10 55	02 10
06	09 R 43	10 52	03 08
07	09 41	10 49	03 57
08	09 38	10 46	04 33
09	09 33	10 42	04 56
10	09 25	10 39	05 06
11	09 16	10 36	05 02
12	09 07	10 33	04 46
13	08 58	10 30	04 17
14	08 50	10 27	03 37
15	08 43	10 23	02 47
16	08 39	10 20	01 50
17	08 37	10 17	00 S 46
18	08 D 37	10 14	00 N 20
19	08 38	10 11	01 27
20	08 39	10 07	02 31
21	08 40	10 04	03 28
22	08 R 40	10 01	04 17
23	08 39	09 58	04 51
24	08 36	09 55	05 09
25	08 32	09 52	05 08
26	08 27	09 48	04 46
27	08 22	09 45	04 04
28	08 18	09 42	03 05
29	08 15	09 39	01 54
30	08 14	09 36	00 36
31	08 ♎ 14	09 ♎ 33	00 S 42

DECLINATIONS

Date	Sun ☉	Moon ☽	Mercury ☿	Venus ♀	Mars ♂	Jupiter ♃	Saturn ♄	Uranus ♅	Neptune ♆	Pluto ♇
01	17 N 52	05 S 52	12 N 29	18 N 59	23 N 51	23 S 14	00 N 37	20 S 14	20 S 28	07 S 25
02	17 37	01 S 19	11 48	19 04	23 50	23 14	00 36	20 15	20 29	07 25
03	17 21	03 N 11	11 08	19 08	23 49	23 15	00 35	20 15	20 29	07 26
04	17 05	07 22	10 27	19 12	23 48	23 15	00 34	20 16	20 29	07 26
05	16 49	11 02	09 46	19 16	23 46	23 16	00 33	20 16	20 30	07 27
06	16 33	14 03	09 05	19 20	23 45	23 16	00 32	20 17	20 30	07 27
07	16 16	16 19	08 25	19 24	23 43	23 16	00 31	20 17	20 30	07 28
08	15 59	17 46	07 44	19 27	23 41	23 17	00 30	20 18	20 30	07 28
09	15 41	18 24	07 04	19 31	23 39	23 17	00 29	20 18	20 31	07 29
10	15 24	18 11	06 24	19 34	23 37	23 18	00 28	20 19	20 31	07 29
11	15 06	17 11	05 45	19 37	23 34	23 18	00 27	20 19	20 31	07 30
12	14 48	15 26	05 06	19 39	23 31	23 19	00 26	20 20	20 32	07 30
13	14 30	13 02	04 27	19 41	23 28	23 19	00 25	20 20	20 32	07 31
14	14 11	10 06	03 49	19 43	23 25	23 19	00 24	20 21	20 32	07 31
15	13 52	06 45	03 11	19 45	23 22	23 20	00 22	20 21	20 32	07 32
16	13 33	03 N 06	02 34	19 46	23 19	23 20	00 21	20 22	20 33	07 32
17	13 14	00 S 43	01 58	19 47	23 15	23 20	00 20	20 22	20 33	07 33
18	12 55	04 34	01 23	19 47	23 11	23 21	00 18	20 23	20 33	07 33
19	12 35	08 16	00 48	19 47	23 07	23 21	00 17	20 23	20 34	07 34
20	12 15	11 39	00 N 14	19 47	23 03	23 21	00 16	20 24	20 34	07 34
21	11 56	14 33	00 S 18	19 46	22 59	23 21	00 14	20 24	20 34	07 35
22	11 35	16 45	00 50	19 45	22 54	23 22	00 13	20 25	20 34	07 36
23	11 15	18 03	01 20	19 44	22 50	23 22	00 11	20 25	20 35	07 36
24	10 54	18 16	01 49	19 42	22 45	23 22	00 10	20 26	20 35	07 37
25	10 34	17 17	02 17	19 39	22 40	23 22	00 08	20 26	20 35	07 37
26	10 13	15 07	02 43	19 36	22 35	23 22	00 07	20 26	20 35	07 38
27	09 52	11 54	03 08	19 33	22 30	23 23	00 05	20 27	20 35	07 39
28	09 31	07 53	03 31	19 30	22 24	23 23	00 04	20 27	20 36	07 39
29	09 09	03 S 24	03 52	19 25	22 18	23 23	00 N 02	20 28	20 36	07 40
30	08 48	01 N 13	04 11	19 21	22 13	23 23	00 00	20 28	20 36	07 40
31	08 N 26	05 N 38	04 S 27	19 N 16	22 N 07	23 S 23	00 S 01	20 S 28	20 S 36	07 S 41

ZODIAC SIGN ENTRIES

Date	h	m	Planets
01	16	17	☿ ♍
02	23	05	☽ ♈
05	03	33	☽ ♉
07	06	15	♀ ♋
07	11	49	☽ ♊
09	22	57	☽ ♋
12	11	29	☽ ♌
15	00	07	☽ ♍
17	11	55	☽ ♎
19	21	50	☽ ♏
22	04	48	☽ ♐
22	20	23	☉ ♍
24	08	22	☽ ♑
26	05	17	☿ ♎
26	09	10	☽ ♒
28	08	49	☽ ♓
30	09	15	☽ ♈

LATITUDES

Date	Mercury ☿	Venus ♀	Mars ♂	Jupiter ♃	Saturn ♄	Uranus ♅	Neptune ♆	Pluto ♇
01	00 N 58	04 S 23	00 N 29	00 S 08	02 S 28	00 S 37	00 N 29	13 N 05
04	00 36	04 13	00 31	00 08	02 29	00 37	00 29	13 03
07	00 N 12	04 02	00 33	00 08	02 29	00 37	00 29	13 02
10	00 S 15	03 51	00 35	00 09	02 30	00 37	00 29	13 00
13	00 43	03 38	00 37	00 09	02 31	00 37	00 29	12 59
16	01 13	03 26	00 40	00 09	02 31	00 37	00 29	12 57
19	01 43	03 12	00 42	00 10	02 32	00 37	00 29	12 56
22	02 13	02 58	00 44	00 10	02 33	00 37	00 29	12 54
25	02 43	02 44	00 46	00 10	02 33	00 37	00 29	12 53
28	03 12	02 30	00 48	00 11	02 34	00 37	00 28	12 51
31	03 S 38	02 S 15	00 N 50	00 S 11	02 S 34	00 S 36	00 N 28	12 N 50

DATA

Julian Date	2450297
Delta T	+62 seconds
Ayanamsa	23° 48' 38"
Synetic vernal point	05° ♓ 18' 21"
True obliquity of ecliptic	23° 26' 14"

MOON'S PHASES, APSIDES AND POSITIONS ☽

Date	h	m	Phase	Longitude ° '	Eclipse Indicator
06	05	25	☾	14 ♉ 02	
14	07	34	●	21 ♌ 47	
22	03	36	☽	29 ♏ 20	
28	17	52	○	05 ♓ 41	

Day	h	m		
12	16	45	Apogee	
27	16	59	Perigee	
02	18	54	0N	
09	17	53	Max dec	18° N 25'
17	07	29	0S	
24	04	25	Max dec	18° S 19'
30	05	39	0N	

LONGITUDES

Date	Chiron ⚷ ° '	Ceres ⚳ ° '	Pallas ⚴ ° '	Juno ⚵ ° '	Vesta ⚶ ° '	Black Moon Lilith ⚸ ° '
01	10 ♎ 53	02 ♐ 35	28 ♎ 05	15 ♈ 40	18 ♏ 06	04 ♌ 24
11	11 ♎ 55	03 ♐ 38	01 ♏ 02	17 ♈ 02	21 ♏ 05	05 ♌ 32
21	13 ♎ 04	05 ♐ 10	04 ♏ 15	17 ♈ 43	24 ♏ 29	06 ♌ 39
31	14 ♎ 18	07 ♐ 07	07 ♏ 42	17 ♈ 39	28 ♏ 13	07 ♌ 46

All ephemeris data is given at 12.00 UT and the Moon's longitude is additionally given for 24.00 UT

SEPTEMBER 1996

Date	Sidereal time h m s	Sun ☉ ° ' "	Moon ☽ ° ' "	Moon ☽ 24.00 ° ' "	Mercury ☿ ° '	Venus ♀ ° '	Mars ♂ ° '	Jupiter ♃ ° '	Saturn ♄ ° '	Uranus ♅ ° '	Neptune ♆ ° '	Pluto ♇ ° '
01	10 43 43	09 ♍ 19 03	29 ♈ 48 59	06 ♉ 34 29	03 ♎ 08	24 ♋ 00	24 ♋ 44	07 ♑ 50	05 ♈ 50	01 ♒ 14	25 ♑ 18	00 ♐ 28
02	10 47 39	10 17 08	13 ♉ 13 25	19 ♉ 46 02	03 21	25 02	25 22	07 R 49	05 R 46	01 R 12	25 R 17	00 29
03	10 51 36	11 15 14	26 ♉ 12 43	02 ♊ 33 55	03 28	26 05	26 00	07 49	05 42	01 10	25 16	00 30
04	10 55 32	12 13 22	08 ♊ 50 08	15 ♊ 01 53	03 R 29	27 08	26 38	07 D 49	05 38	01 09	25 15	00 30
05	10 59 29	13 11 33	21 ♊ 09 44	27 ♊ 14 14	03 25	28 11	27 16	07 50	05 34	01 07	25 14	00 31
06	11 03 25	14 09 45	03 ♋ 15 56	09 ♋ 15 23	03 14	29 ♋ 15	27 54	07 50	05 30	01 05	25 13	00 32
07	11 07 22	15 08 00	15 ♋ 13 06	21 ♋ 09 35	02 56	00 ♌ 18	28 32	07 51	05 25	01 04	25 12	00 33
08	11 11 19	16 06 16	27 ♋ 05 19	03 ♌ 00 42	02 33	01 22	29 10	07 51	05 21	01 02	25 11	00 34
09	11 15 15	17 04 34	08 ♌ 56 10	14 ♌ 52 05	02 02	02 27	29 ♋ 47	07 53	05 17	01 01	25 10	00 35
10	11 19 12	18 02 55	20 ♌ 48 46	26 ♌ 46 32	01 26	03 31	00 ♌ 25	07 54	05 12	01 00	25 10	00 36
11	11 23 08	19 01 17	02 ♍ 45 39	08 ♍ 46 23	00 ♎ 43	04 36	01 03	07 55	05 08	00 58	25 09	00 37
12	11 27 05	19 59 41	14 ♍ 48 56	20 ♍ 53 31	29 ♍ 55	05 41	01 40	07 57	05 04	00 57	25 08	00 38
13	11 31 01	20 58 07	27 ♍ 00 19	03 ♎ 09 31	29 01	06 47	02 18	07 59	04 59	00 56	25 07	00 39
14	11 34 58	21 56 34	09 ♎ 21 18	15 ♎ 35 51	28 04	07 52	02 55	08 00	04 55	00 54	25 07	00 40
15	11 38 54	22 55 04	21 ♎ 53 18	28 ♎ 13 52	27 03	08 58	03 32	08 03	04 50	00 53	25 06	00 42
16	11 42 51	23 53 35	04 ♏ 37 43	11 ♏ 05 03	26 01	10 04	04 09	08 05	04 46	00 52	25 05	00 43
17	11 46 48	24 52 08	17 ♏ 36 03	24 ♏ 10 53	24 58	11 10	04 46	08 08	04 41	00 51	25 05	00 44
18	11 50 44	25 50 43	00 ♐ 49 45	07 ♐ 32 49	23 56	12 17	05 24	08 10	04 37	00 50	25 04	00 45
19	11 54 41	26 49 20	14 ♐ 20 10	21 ♐ 11 55	22 56	13 23	06 01	08 13	04 32	00 49	25 03	00 47
20	11 58 37	27 47 58	28 ♐ 08 06	05 ♑ 08 40	22 00	14 30	06 37	08 16	04 27	00 48	25 03	00 48
21	12 02 34	28 46 38	12 ♑ 13 29	19 ♑ 22 21	21 10	15 37	07 14	08 20	04 23	00 47	25 02	00 49
22	12 06 30	29 ♍ 45 19	26 ♑ 34 57	03 ♒ 50 50	20 27	16 44	07 51	08 23	04 18	00 46	25 02	00 51
23	12 10 27	00 ♎ 44 02	11 ♒ 09 28	18 ♒ 30 12	19 51	17 52	08 28	08 27	04 13	00 45	25 01	00 52
24	12 14 23	01 42 47	25 ♒ 52 18	03 ♓ 14 58	19 25	18 59	09 04	08 30	04 09	00 45	25 01	00 53
25	12 18 20	02 41 33	10 ♓ 37 19	17 ♓ 58 29	19 07	20 07	09 41	08 34	04 04	00 44	25 01	00 55
26	12 22 17	03 40 22	25 ♓ 17 36	02 ♈ 33 48	19 00	21 15	10 17	08 39	03 59	00 43	25 00	00 56
27	12 26 13	04 39 12	09 ♈ 46 20	16 ♈ 54 32	19 D 03	22 23	10 54	08 43	03 55	00 42	25 00	00 58
28	12 30 10	05 38 04	23 ♈ 57 48	00 ♉ 55 41	19 16	23 31	11 30	08 47	03 50	00 42	25 00	00 59
29	12 34 06	06 36 59	07 ♉ 47 53	14 ♉ 34 11	19 39	24 40	12 06	08 52	03 45	00 41	24 59	01 01
30	12 38 03	07 ♎ 35 56	21 ♉ 14 33	27 ♉ 49 02	20 ♍ 12	25 ♌ 49	12 ♌ 43	08 ♑ 57	03 ♈ 41	00 ♒ 41	24 ♑ 59	01 ♐ 03

Date	Moon True ☊ ° '	Moon Mean ☊ ° '	Moon ☽ Latitude ° '
01	08 ♎ 15	09 ♎ 29	01 S 56
02	08 D 17	09 26	03 00
03	08 18	09 23	03 54
04	08 19	09 20	04 34
05	08 R 19	09 17	05 00
06	08 17	09 13	05 13
07	08 15	09 10	05 12
08	08 12	09 07	04 58
09	08 08	09 04	04 30
10	08 05	09 01	03 51
11	08 02	08 58	03 02
12	08 00	08 54	02 04
13	07 59	08 51	01 S 00
14	07 D 58	08 48	00 N 08
15	07 59	08 45	01 16
16	08 00	08 42	02 22
17	08 01	08 39	03 22
18	08 02	08 35	04 13
19	08 03	08 32	04 50
20	08 R 03	08 29	05 12
21	08 03	08 26	05 16
22	08 02	08 23	05 01
23	08 01	08 19	04 26
24	08 00	08 16	03 33
25	08 00	08 13	02 26
26	07 59	08 10	01 N 10
27	07 D 59	08 07	00 S 10
28	07 59	08 04	01 28
29	08 00	08 00	02 38
30	08 ♎ 00	07 ♎ 57	03 S 38

DECLINATIONS

Date	Sun ☉ ° '	Moon ☽ ° '	Mercury ☿ ° '	Venus ♀ ° '	Mars ♂ ° '	Jupiter ♃ ° '	Saturn ♄ ° '	Uranus ♅ ° '	Neptune ♆ ° '	Pluto ♇ ° '
01	08 N 05	09 N 36	04 S 41	19 N 10	22 N 01	23 S 23	00 S 03	20 S 29	20 S 37	07 S 42
02	07 43	12 56	04 53	19 04	21 55	23 24	00 05	20 29	20 37	07 42
03	07 21	15 31	05 01	18 57	21 48	23 24	00 06	20 29	20 37	07 43
04	06 58	17 16	05 07	18 50	21 42	23 24	00 08	20 30	20 37	07 44
05	06 36	18 09	05 10	18 43	21 35	23 24	00 10	20 30	20 37	07 44
06	06 14	18 11	05 09	18 35	21 28	23 24	00 12	20 30	20 38	07 45
07	05 51	17 24	05 04	18 26	21 21	23 24	00 13	20 31	20 38	07 46
08	05 29	15 52	04 56	18 17	21 14	23 24	00 15	20 31	20 38	07 46
09	05 06	13 40	04 43	18 08	21 07	23 24	00 17	20 31	20 38	07 47
10	04 43	10 54	04 27	17 58	21 00	23 24	00 19	20 32	20 38	07 48
11	04 21	07 39	04 07	17 47	20 52	23 24	00 21	20 32	20 39	07 48
12	03 58	04 04	03 43	17 36	20 45	23 24	00 23	20 32	20 39	07 49
13	03 35	00 N 16	03 15	17 25	20 37	23 24	00 24	20 32	20 39	07 50
14	03 12	03 S 35	02 43	17 13	20 29	23 24	00 26	20 33	20 39	07 50
15	02 49	07 21	02 09	17 00	20 21	23 24	00 28	20 33	20 39	07 51
16	02 26	10 50	01 32	16 47	20 13	23 24	00 30	20 33	20 39	07 52
17	02 02	13 51	00 54	16 34	20 05	23 24	00 32	20 33	20 39	07 52
18	01 39	16 12	00 S 14	16 20	19 56	23 24	00 34	20 34	20 40	07 53
19	01 16	17 43	00 N 26	16 06	19 48	23 24	00 36	20 34	20 40	07 54
20	00 53	18 13	01 05	15 51	19 40	23 23	00 38	20 34	20 40	07 54
21	00 29	17 38	01 43	15 36	19 31	23 23	00 40	20 34	20 40	07 55
22	00 N 06	15 55	02 18	15 20	19 22	23 23	00 41	20 34	20 40	07 56
23	00 S 18	13 10	02 50	15 04	19 13	23 23	00 43	20 34	20 40	07 56
24	00 41	09 33	03 19	14 47	19 04	23 23	00 45	20 35	20 40	07 57
25	01 04	05 20	03 44	14 30	18 55	23 23	00 47	20 35	20 40	07 58
26	01 28	00 S 48	04 04	14 12	18 46	23 22	00 49	20 35	20 40	07 59
27	01 51	03 N 43	04 19	13 54	18 37	23 22	00 51	20 35	20 40	07 59
28	02 14	07 56	04 30	13 36	18 28	23 22	00 53	20 35	20 41	08 00
29	02 38	11 37	04 35	13 17	18 18	23 22	00 55	20 35	20 41	08 01
30	03 S 01	14 N 34	04 N 36	12 N 58	18 N 09	23 S 22	00 S 56	20 S 35	20 S 41	08 S 01

ZODIAC SIGN ENTRIES

Date	h	m	Planets
01	12	19	☽ ♉
03	19	08	☽ ♊
06	05	29	☽ ♋
07	05	07	♀ ♌
08	17	54	☽ ♌
09	20	02	♂ ♌
11	06	28	☽ ♍
12	09	32	☿ ♍
13	17	51	☽ ♎
16	03	20	☽ ♏
18	10	31	☽ ♐
20	15	12	☽ ♑
22	17	39	☽ ♒
22	18	00	☉ ♎
24	18	43	☽ ♓
26	19	46	☽ ♈
28	22	24	☽ ♉

LATITUDES

Date	Mercury ☿ ° '	Venus ♀ ° '	Mars ♂ ° '	Jupiter ♃ ° '	Saturn ♄ ° '	Uranus ♅ ° '	Neptune ♆ ° '	Pluto ♇ ° '
01	03 S 45	02 S 10	00 N 51	00 S 11	02 S 35	00 S 36	00 N 28	12 N 49
04	04 04	01 56	00 53	00 11	02 35	00 36	00 28	12 48
07	04 15	01 41	00 55	00 12	02 35	00 36	00 28	12 46
10	04 14	01 27	00 57	00 12	02 36	00 36	00 28	12 45
13	03 58	01 12	01 00	00 12	02 36	00 36	00 28	12 43
16	03 24	00 58	01 02	00 12	02 36	00 36	00 28	12 42
19	02 35	00 44	01 04	00 13	02 37	00 36	00 28	12 40
22	01 37	00 31	01 06	00 13	02 37	00 36	00 28	12 39
25	00 S 37	00 18	01 08	00 13	02 37	00 36	00 28	12 38
28	00 N 16	00 S 05	01 11	00 13	02 37	00 36	00 28	12 37
31	00 N 59	00 N 07	01 N 13	00 S 14	02 S 37	00 S 36	00 N 28	12 N 35

DATA

Julian Date	2450328
Delta T	+62 seconds
Ayanamsa	23° 48' 41"
Synetic vernal point	05° ♓ 18' 18"
True obliquity of ecliptic	23° 26' 14"

MOON'S PHASES, APSIDES AND POSITIONS ☽

Date	h	m	Phase	Longitude ° '	Eclipse Indicator
04	19	06	☾	12 ♊ 31	
12	23	07	●	20 ♍ 27	
20	11	23	☽	27 ♐ 46	
27	02	51	○	04 ♈ 17	total

Day	h	m		
09	01	57	Apogee	
24	21	49	Perigee	
06	00	56	Max dec	18° N 16'
13	13	40	0S	
20	11	19	Max dec	18° S 13'
26	16	12	0N	

LONGITUDES

Date	Chiron ⚷ ° '	Ceres ⚳ ° '	Pallas ⚴ ° '	Juno ⚵ ° '	Vesta ⚶ ° '	Black Moon Lilith ⚸ ° '
01	14 ♎ 26	07 ♐ 20	08 ♏ 03	17 ♈ 36	28 ♏ 36	07 ♌ 52
11	15 ♎ 46	09 ♐ 41	11 ♏ 41	16 ♈ 41	02 ♐ 39	08 ♌ 59
21	17 ♎ 09	12 ♐ 21	15 ♏ 29	15 ♈ 02	06 ♐ 55	10 ♌ 06
31	18 ♎ 35	15 ♐ 17	19 ♏ 24	12 ♈ 52	11 ♐ 23	11 ♌ 13

All ephemeris data is given at 12.00 UT and the Moon's longitude is additionally given for 24.00 UT

ASPECTARIAN

01 Sunday
00 01 ☽ # ♇
01 00 ☽ □ ♀
01 39 ☽ ⚼ ☉
02 38 ☽ ± ♇
02 40 ☽ □ ♂
03 12 ☽ ∥ ☉
04 06 ☽ □ ♆
13 09 ☽ ⊼ ♇
14 29 ☽ □ ♅
17 58 ☽ ⊼ ☿
22 37 ☽ ⊻ ♄

02 Monday
02 14 ☽ △ ♃
04 52 ☽ ± ☿
06 15 ☽ △ ☉
08 48 ♂ ☍ ♆
09 22 ☽ ⊥ ♄
11 38 ☽ Q ♀
11 55 ☉ ⚼ ♆
12 17 ☽ Q ♂
12 25 ☉ # ♇
17 34 ♀ ☍ ♆
18 01 ☉ ∠ ♂
21 29 ☽ ⚼ ☿

03 Tuesday
01 46 ☽ ∠ ♄
05 40 ☽ ⚼ ♃
07 32 ♀ ☌ ♂
10 14 ☽ △ ♆
11 35 ☽ ⚹ ♂
11 44 ☽ ⚹ ♀
14 36 ♃ St D
20 05 ☽ ☍ ♇
21 20 ☽ △ ♅
22 35 ☽ ± ♃

04 Wednesday
01 45 ☽ △ ☿
05 47 ☿ St R
05 53 ☽ ⚹ ♄
10 03 ☽ ⊼ ♃
14 43 ☽ ⚼ ♆
17 42 ☽ ∠ ♂
18 57 ☽ ∠ ♀
19 06 ☽ □ ☉

05 Thursday
02 08 ☽ ⚼ ♅
04 59 ☽ Q ♄
08 13 ☽ ± ♆
12 13 ☽ ⊥ ♂
14 12 ☽ ⊥ ♀
14 31 ☉ ∠ ♀
19 47 ☽ ± ♅
20 01 ☽ ⊼ ♆

06 Friday
00 44 ☽ ⊻ ♂
03 13 ☽ ⊻ ♀
06 33 ☽ ⊼ ♇
07 40 ☽ ⊼ ♅
09 36 ☽ Q ☉
11 56 ☽ □ ☿
16 26 ☽ □ ♄
18 33 ☽ ± ♇
21 09 ☽ ☍ ♃

07 Saturday
11 49 ☽ ⚹ ☉
12 40 ☽ ⚼ ♇
17 38 ♀ △ ♇
23 13 ☽ Q ☿

08 Sunday
04 42 ♀ ☍ ♅
08 09 ☽ ☍ ♆
10 27 ☉ ⚼ ♅
16 26 ☽ ☌ ♂
19 04 ☽ △ ♇
19 59 ☽ ☍ ♅
20 52 ☽ ∠ ☉
21 33 ☽ ☌ ♀
22 38 ☽ ⚹ ☿

09 Monday
04 38 ☽ △ ♄
06 00 ☿ ⚹ ♀
09 51 ☽ ⊼ ♃
16 43 ☽ ⊥ ☉
22 01 ☽ ± ♃

10 Tuesday
03 37 ☽ ∠ ☿
05 56 ☽ ⊻ ☉
10 47 ☽ ⚼ ♄
16 12 ☽ ⚼ ♃
19 15 ♂ △ ♇
20 44 ☽ ⊼ ♆
20 48 ☽ ⊥ ☿

11 Wednesday
01 53 ☉ Q ♇
03 37 ☿ △ ♅
04 47 ☽ ± ♄
06 20 ☿ ⚹ ♂
07 42 ☽ □ ♇
08 09 ☽ ⊻ ☿
08 22 ☽ ⊻ ♂
08 25 ☽ ⊼ ♅
08 46 ☽ ± ♆
09 18 ♂ ☍ ♅
10 57 ☽ # ♇
15 04 ☿ ⚹ ♇
16 03 ☽ ⊻ ♀
16 43 ☽ ⊼ ♄
20 24 ☽ ± ♅
21 01 ☽ ⊥ ♂
22 19 ☽ △ ♃
23 03 ♀ △ ♄

12 Thursday
02 43 ☽ ⚼ ♆
05 11 ☽ ⊥ ♀
12 45 ☽ ∥ ☉
14 14 ☽ ⚼ ♅
14 36 ☽ # ☿
15 52 ☽ ∠ ♂
19 34 ☽ Q ♇
23 07 ☽ ☌ ☉

13 Friday
00 44 ☽ ∠ ♀
05 58 ♂ # ♆
08 19 ☽ △ ♆
11 09 ☽ # ♄
15 40 ☽ ☌ ☿
16 15 ☽ ∥ ♄
19 08 ☽ ⚹ ♇
19 39 ☽ △ ♅
22 52 ☽ ⚹ ♂

14 Saturday
01 14 ♂ # ♅
03 27 ☽ ☍ ♄
07 16 ☽ ∥ ☿
08 51 ☽ ⚹ ♀
09 23 ☽ □ ♃
09 46 ☽ # ☉
15 09 ♀ ⊼ ♃
23 15 ☽ Q ♂

15 Sunday
00 10 ☽ ∠ ♇
10 05 ☽ Q ♀
14 07 ☽ ⊻ ☉
15 19 ☽ ∥ ♇
17 20 ☽ ⊥ ♇
18 05 ☽ □ ♆
19 54 ☽ Q ♃
21 03 ☽ ⊻ ☿

16 Monday
02 24 ☽ ⊥ ☉
04 39 ☽ ⊻ ♇
04 58 ☽ □ ♅
07 29 ☽ ⊥ ☿
11 04 ☽ □ ♂
12 15 ☽ ⊼ ♄
18 27 ☽ ⚹ ♃
20 35 ☽ ∠ ☉
22 33 ☿ ∠ ♀
22 59 ☽ ∠ ☿
23 03 ☽ □ ♀
23 20 ☽ ± ♄

17 Tuesday
03 41 ☽ Q ♆
08 56 ♂ △ ♄
09 20 ☿ △ ♆
13 05 ☉ ☌ ☿
14 17 ☽ Q ♅
15 48 ☽ ⚼ ♄
17 02 ☉ △ ♆
22 07 ☽ ∠ ♃

18 Wednesday
00 26 ☽ ⚹ ☿
00 38 ☿ ∥ ♄
01 37 ☽ ⚹ ♆
02 18 ☽ ⚹ ☉
11 52 ☽ ☌ ♇
12 00 ☽ ⚹ ♅
13 30 ☽ # ♀
14 25 ☽ ⊥ ♃
18 44 ☽ △ ♄
20 29 ☽ Q ☿
20 33 ☽ △ ♂

19 Thursday
01 09 ☽ ⊻ ♃
01 30 ☽ Q ☉
04 27 ☽ ∠ ♆
10 11 ☽ △ ♀
14 36 ☽ ∠ ♅
18 25 ☿ # ♄
20 16 ☽ ⊥ ♆

20 Friday
00 13 ☽ ⚼ ♂
02 03 ☽ □ ☿
06 15 ☽ ⊥ ♅
06 41 ☽ ⊻ ♆
06 50 ♀ ± ♃
07 14 ☉ ∥ ☿
11 23 ☽ □ ☉
12 44 ♅ ⚹ ♇
14 33 ☽ ⚼ ♀
16 28 ☽ ± ♂
16 34 ☽ ⊻ ♅
16 35 ☽ ⊻ ♇
18 07 ☿ ∠ ♂
22 46 ☽ □ ♄

21 Saturday
02 10 ☉ # ♄
02 50 ☽ ⊥ ♇
03 11 ☽ ⊼ ♂
05 23 ☽ ☌ ♃
06 22 ☿ ⊥ ♀
07 13 ☽ ± ♀
18 04 ☽ ∠ ♇
18 12 ☽ ⊼ ♀

22 Sunday
02 15 ☽ △ ☿
04 55 ☽ Q ♄
09 26 ☽ ☌ ♆
17 38 ☽ △ ☉
18 35 ☽ # ♀
18 55 ☽ ☌ ♅
19 04 ☽ ⚹ ♇

23 Monday
00 41 ☽ ⚹ ♄
02 02 ☽ ⚼ ☿
07 23 ☽ ☍ ♂
07 32 ☽ ⊻ ♃
11 09 ♂ ⊼ ♃
12 30 ☉ △ ♅
14 48 ☽ Q ♇
15 19 ☉ ⚹ ♇
16 16 ☽ ± ☿
17 24 ☽ ⊥ ♃
20 01 ☽ ⚼ ☉
23 52 ☽ ☍ ♀

24 Tuesday
01 06 ☽ ∠ ♄
01 45 ☽ ⊼ ☿
08 08 ☽ ∠ ♃
10 37 ☽ ⊻ ♆
11 43 ☽ ± ☉
15 09 ♀ ⚼ ♄
15 41 ☽ ⊥ ♄
16 51 ☉ ∥ ♄
18 52 ☿ ⊻ ♀
19 55 ☽ ⊻ ♅
20 11 ☽ □ ♇
20 22 ☽ ⊥ ♆
21 22 ☽ ∥ ♇
22 11 ☽ ⊼ ☉

25 Wednesday
01 23 ☽ ⊻ ♄
05 40 ☽ ⊥ ♅
08 39 ☽ ⚹ ♃
10 24 ☽ ⊼ ♂
11 00 ☽ ∠ ♆
19 56 ☽ # ☿
20 19 ☽ ∠ ♅
20 37 ☽ ± ♂

26 Thursday
01 44 ☽ ☍ ☿
04 20 ☽ Q ♃
04 48 ☽ ⊼ ♀
08 49 ☽ ∥ ☉
11 32 ☽ ⚹ ♆
11 55 ☽ ∥ ♄
12 00 ☽ ⚼ ♂
15 30 ☽ ± ♀
17 08 ☿ St D
19 11 ☉ ☍ ♄
20 33 ☽ # ♄
20 56 ☽ ⚹ ♅
21 20 ☽ △ ♇

27 Friday
01 01 ☽ # ☉
02 17 ☽ ☌ ♄
02 51 ☽ ☍ ☉
07 22 ☽ Q ♆
07 40 ☽ ⚼ ♀
10 13 ☽ □ ♃
13 58 ☽ △ ♂
15 28 ☽ ∥ ☿
16 55 ☽ Q ♅
22 25 ☽ ⚼ ♇

28 Saturday
03 50 ☽ ⊼ ☿
11 11 ☽ △ ♀
12 22 ☽ # ♇
13 46 ☽ ± ♇
13 46 ☽ □ ♆
14 17 ☽ ± ☿
17 59 ♀ ⚼ ♃
23 36 ☽ □ ♅

29 Sunday
00 08 ☽ ⊼ ♇
04 57 ☽ ⊻ ♄
06 18 ☽ ⚼ ☿
09 46 ☽ ⊼ ☉
13 54 ☽ △ ♃
15 26 ☽ ⊥ ♄
18 48 ♀ ⊼ ♆
19 58 ☽ □ ♂
21 11 ☽ ± ☉
23 33 ☽ ∥ ♀

30 Monday
07 24 ☽ ∠ ♄
10 01 ☽ △ ☿
14 40 ☽ ⚼ ☉
16 57 ☽ ⚼ ♃
18 49 ☽ △ ♆
21 07 ☽ □ ♀

OCTOBER 1996

LONGITUDES

Date	Sidereal time h m s	Sun ☉ ° ′ ″	Moon ☽ ° ′ ″	Moon ☽ 24.00 ° ′ ″	Mercury ☿ ° ′	Venus ♀ ° ′	Mars ♂ ° ′	Jupiter ♃ ° ′	Saturn ♄ ° ′	Uranus ♅ ° ′	Neptune ♆ ° ′	Pluto ♇ ° ′
01	12 41 59	08 ♎ 34 55	04 ♊ 17 47	10 ♊ 41 05	20 ♍ 53	26 ♌ 57	13 ♌ 19	09 ♑ 02	03 ♈ 36	00 ♒ 40	24 ♑ 59	01 ♐ 04
02	12 45 56	09 33 56	16 ♊ 59 17	23 ♊ 12 48	21 43	28 06	13 55	09 07	03 R 31	00 R 40	24 R 59	01 06
03	12 49 52	10 32 59	29 ♊ 22 06	05 ♋ 27 43	22 41	29 ♌ 16	14 31	09 12	03 27	00 40	24 59	01 07
04	12 53 49	11 32 05	11 ♋ 30 12	17 ♋ 30 07	23 46	00 ♍ 25	15 06	09 18	03 22	00 39	24 59	01 09
05	12 57 46	12 31 14	23 ♋ 28 03	29 ♋ 24 35	24 58	01 34	15 42	09 24	03 18	00 39	24 59	01 11
06	13 01 42	13 30 24	05 ♌ 20 18	11 ♌ 15 45	26 15	02 44	16 18	09 29	03 13	00 39	24 59	01 13
07	13 05 39	14 29 37	17 ♌ 11 30	23 ♌ 08 04	27 37	03 54	16 54	09 35	03 08	00 39	24 D 59	01 14
08	13 09 35	15 28 52	29 ♌ 05 55	05 ♍ 05 30	29 ♍ 03	05 04	17 29	09 42	03 04	00 39	24 59	01 16
09	13 13 32	16 28 09	11 ♍ 07 16	17 ♍ 11 33	00 ♎ 33	06 14	18 05	09 48	02 59	00 39	24 59	01 18
10	13 17 28	17 27 28	23 ♍ 18 40	29 ♍ 28 55	02 06	07 24	18 40	09 54	02 55	00 D 39	24 59	01 20
11	13 21 25	18 26 50	05 ♎ 42 29	11 ♎ 59 34	03 42	08 34	19 15	10 01	02 50	00 39	24 59	01 22
12	13 25 21	19 26 13	18 ♎ 20 16	24 ♎ 44 40	05 19	09 45	19 50	10 08	02 46	00 39	24 59	01 24
13	13 29 18	20 25 39	01 ♏ 12 46	07 ♏ 44 32	06 59	10 55	20 25	10 15	02 42	00 39	24 59	01 25
14	13 33 15	21 25 07	14 ♏ 19 56	20 ♏ 58 50	08 39	12 06	21 00	10 22	02 37	00 39	25 00	01 27
15	13 37 11	22 24 36	27 ♏ 41 08	04 ♐ 26 40	10 20	13 17	21 35	10 29	02 33	00 39	25 00	01 29
16	13 41 08	23 24 08	11 ♐ 15 16	18 ♐ 06 45	12 03	14 28	22 10	10 36	02 29	00 40	25 00	01 31
17	13 45 04	24 23 41	25 ♐ 00 56	01 ♑ 57 38	13 45	15 39	22 45	10 44	02 25	00 40	25 01	01 33
18	13 49 01	25 23 16	08 ♑ 56 37	15 ♑ 57 43	15 28	16 50	23 19	10 52	02 21	00 40	25 01	01 35
19	13 52 57	26 22 53	23 ♑ 00 41	00 ♒ 05 20	17 11	18 01	23 54	10 59	02 17	00 41	25 01	01 37
20	13 56 54	27 22 32	07 ♒ 11 24	14 ♒ 18 39	18 54	19 13	24 28	11 07	02 13	00 41	25 02	01 39
21	14 00 50	28 22 12	21 ♒ 26 46	28 ♒ 35 27	20 36	20 24	25 03	11 16	02 09	00 42	25 02	01 41
22	14 04 47	29 ♎ 21 54	05 ♓ 44 21	12 ♓ 53 05	22 19	21 36	25 37	11 24	02 05	00 43	25 03	01 43
23	14 08 44	00 ♏ 21 37	20 ♓ 01 14	27 ♓ 08 20	24 01	22 48	26 11	11 32	02 01	00 43	25 03	01 45
24	14 12 40	01 21 23	04 ♈ 13 55	11 ♈ 17 29	25 43	23 59	26 45	11 41	01 57	00 44	25 04	01 48
25	14 16 37	02 21 10	18 ♈ 18 33	25 ♈ 16 39	27 24	25 11	27 19	11 49	01 53	00 45	25 05	01 50
26	14 20 33	03 20 59	02 ♉ 11 18	09 ♉ 02 06	29 ♎ 05	26 23	27 53	11 58	01 50	00 45	25 05	01 52
27	14 24 30	04 20 50	15 ♉ 48 42	22 ♉ 30 48	00 ♏ 46	27 35	28 26	12 07	01 46	00 46	25 06	01 54
28	14 28 26	05 20 43	29 ♉ 08 13	05 ♊ 40 49	02 26	28 ♍ 48	29 00	12 16	01 43	00 47	25 07	01 56
29	14 32 23	06 20 38	12 ♊ 08 35	18 ♊ 31 34	04 06	00 ♎ 00	29 ♌ 33	12 25	01 39	00 48	25 07	01 58
30	14 36 19	07 20 36	24 ♊ 49 56	01 ♋ 03 54	05 45	01 12	00 ♍ 07	12 34	01 36	00 49	25 08	02 01
31	14 40 16	08 ♏ 20 35	07 ♋ 13 49	13 ♋ 20 04	07 ♏ 24	02 ♎ 25	00 ♍ 40	12 ♑ 44	01 ♈ 33	00 ♒ 50	25 ♑ 09	02 ♐ 03

Date	Moon True ☊ ° ′	Moon Mean ☊ ° ′	Moon ☽ Latitude ° ′
01	08 ♎ 00	07 ♎ 54	04 S 24
02	08 R 00	07 51	04 56
03	07 59	07 48	05 14
04	07 D 59	07 45	05 17
05	08 00	07 41	05 06
06	08 00	07 38	04 42
07	08 00	07 35	04 06
08	08 01	07 32	03 20
09	08 02	07 29	02 24
10	08 03	07 25	01 21
11	08 03	07 22	00 S 13
12	08 R 03	07 19	00 N 57
13	08 02	07 16	02 05
14	08 01	07 13	03 08
15	07 59	07 10	04 01
16	07 57	07 06	04 42
17	07 55	07 03	05 08
18	07 53	07 00	05 16
19	07 53	06 57	05 05
20	07 D 53	06 54	04 35
21	07 54	06 50	03 49
22	07 55	06 47	02 48
23	07 57	06 44	01 37
24	07 57	06 41	00 N 21
25	07 R 57	06 38	00 S 57
26	07 55	06 35	02 09
27	07 52	06 31	03 13
28	07 48	06 28	04 05
29	07 44	06 25	04 42
30	07 39	06 22	05 05
31	07 ♎ 35	06 ♎ 19	05 S 13

DECLINATIONS

Date	Sun ☉ ° ′	Moon ☽ ° ′	Mercury ☿ ° ′	Venus ♀ ° ′	Mars ♂ ° ′	Jupiter ♃ ° ′	Saturn ♄ ° ′	Uranus ♅ ° ′	Neptune ♆ ° ′	Pluto ♇ ° ′
01	03 S 24	16 N 40	04 N 31	12 N 38	17 N 59	23 S 21	00 S 58	20 S 35	20 S 41	08 S 02
02	03 47	17 53	04 22	12 18	17 49	23 21	01 00	20 35	20 41	08 03
03	04 11	18 12	04 09	11 58	17 40	23 21	01 02	20 35	20 41	08 04
04	04 34	17 41	03 51	11 37	17 30	23 20	01 04	20 35	20 41	08 04
05	04 57	16 22	03 30	11 16	17 20	23 20	01 06	20 36	20 41	08 05
06	05 20	14 22	03 05	10 55	17 10	23 20	01 07	20 36	20 41	08 06
07	05 43	11 46	02 37	10 33	17 00	23 19	01 09	20 36	20 41	08 06
08	06 06	08 40	02 06	10 10	16 50	23 19	01 11	20 36	20 41	08 07
09	06 28	05 10	01 33	09 48	16 40	23 19	01 13	20 36	20 41	08 08
10	06 51	01 N 25	00 57	09 25	16 30	23 18	01 14	20 36	20 41	08 08
11	07 14	02 S 28	00 N 20	09 02	16 19	23 18	01 16	20 35	20 41	08 09
12	07 36	06 19	00 S 18	08 38	16 09	23 17	01 18	20 35	20 41	08 10
13	07 59	09 57	00 58	08 15	15 58	23 17	01 20	20 35	20 41	08 11
14	08 21	13 09	01 39	07 51	15 48	23 16	01 21	20 35	20 41	08 11
15	08 43	15 43	02 21	07 26	15 38	23 16	01 23	20 35	20 41	08 12
16	09 05	17 28	03 04	07 02	15 27	23 15	01 24	20 35	20 41	08 13
17	09 27	18 13	03 46	06 37	15 16	23 15	01 26	20 35	20 41	08 13
18	09 49	17 53	04 30	06 12	15 06	23 14	01 28	20 35	20 41	08 14
19	10 11	16 28	05 13	05 46	14 55	23 14	01 29	20 35	20 41	08 15
20	10 32	14 02	05 56	05 21	14 44	23 13	01 31	20 35	20 41	08 15
21	10 54	10 44	06 39	04 55	14 34	23 12	01 32	20 35	20 41	08 16
22	11 15	06 48	07 22	04 29	14 23	23 12	01 33	20 34	20 41	08 17
23	11 36	02 S 28	08 05	04 03	14 12	23 11	01 35	20 34	20 41	08 17
24	11 57	02 N 00	08 48	03 37	14 01	23 11	01 36	20 34	20 40	08 18
25	12 17	06 18	09 30	03 10	13 50	23 10	01 38	20 34	20 40	08 19
26	12 38	10 12	10 11	02 44	13 39	23 09	01 39	20 34	20 40	08 19
27	12 58	13 29	10 53	02 17	13 28	23 08	01 40	20 33	20 40	08 20
28	13 18	15 59	11 33	01 50	13 17	23 08	01 42	20 33	20 40	08 21
29	13 38	17 35	12 13	01 23	13 06	23 07	01 43	20 33	20 40	08 21
30	13 58	18 15	12 52	00 56	12 55	23 06	01 44	20 33	20 40	08 22
31	14 S 17	18 N 02	13 S 31	00 N 28	12 N 44	23 S 05	01 S 45	20 S 32	20 S 40	08 S 23

ZODIAC SIGN ENTRIES

Date	h	m	Planets
01	04	01	☽ ♊
03	13	14	☽ ♋
04	03	22	♀ ♍
06	01	12	☽ ♌
08	13	49	☽ ♍
09	03	13	☿ ♎
11	01	00	☽ ♎
13	09	46	☽ ♏
15	16	07	☽ ♐
17	20	37	☽ ♑
19	23	51	☽ ♒
22	02	22	☽ ♓
23	03	19	☉ ♏
24	04	50	☽ ♈
26	08	11	☽ ♉
27	01	01	☿ ♏
28	13	34	☽ ♊
29	12	02	♀ ♎
30	07	13	♂ ♍
30	21	56	☽ ♋

LATITUDES

Date	Mercury ☿ ° ′	Venus ♀ ° ′	Mars ♂ ° ′	Jupiter ♃ ° ′	Saturn ♄ ° ′	Uranus ♅ ° ′	Neptune ♆ ° ′	Pluto ♇ ° ′
01	00 N 59	00 N 07	01 N 13	00 S 14	02 S 37	00 S 36	00 N 28	12 N 35
04	01 30	00 19	01 15	00 14	02 37	00 36	00 28	12 34
07	01 49	00 30	01 17	00 14	02 37	00 36	00 27	12 33
10	01 57	00 40	01 20	00 14	02 37	00 36	00 27	12 32
13	01 57	00 50	01 22	00 14	02 37	00 36	00 27	12 31
16	01 51	00 59	01 24	00 15	02 36	00 36	00 27	12 30
19	01 40	01 08	01 27	00 15	02 36	00 35	00 27	12 29
22	01 25	01 15	01 29	00 15	02 36	00 35	00 27	12 28
25	01 08	01 22	01 32	00 15	02 36	00 35	00 27	12 27
28	00 49	01 28	01 34	00 15	02 35	00 35	00 27	12 26
31	00 N 29	01 N 34	01 N 36	00 S 16	02 S 35	00 S 35	00 N 27	12 N 25

DATA

Julian Date	2450358
Delta T	+62 seconds
Ayanamsa	23° 48′ 44″
Synetic vernal point	05° ♓ 18′ 15″
True obliquity of ecliptic	23° 26′ 14″

MOON'S PHASES, APSIDES AND POSITIONS ☽

Date	h	m	Phase	Longitude ° ′	Eclipse Indicator
04	12	04	☾	11 ♋ 32	
12	14	14	●	19 ♎ 32	Partial
19	18	09	☽	26 ♑ 38	
26	14	11	○	03 ♉ 26	

Day	h	m		
06	17	57	Apogee	
22	08	50	Perigee	
03	08	52	Max dec	18° N 13′
10	20	48	0S	
17	16	48	Max dec	18° S 14′
24	01	15	0N	
30	17	54	Max dec	18° N 17′

LONGITUDES

Date	Chiron ⚷ ° ′	Ceres ⚳ ° ′	Pallas ⚴ ° ′	Juno ⚵ ° ′	Vesta ⚶ ° ′	Black Moon Lilith ⚸ ° ′
01	18 ♎ 35	15 ♐ 17	19 ♏ 24	12 ♈ 52	11 ♐ 23	11 ♌ 13
11	20 ♎ 02	18 ♐ 26	23 ♏ 25	10 ♈ 28	16 ♐ 01	12 ♌ 20
21	21 ♎ 29	21 ♐ 47	27 ♏ 31	08 ♈ 16	20 ♐ 47	13 ♌ 27
31	22 ♎ 55	25 ♐ 16	01 ♐ 40	06 ♈ 35	25 ♐ 40	14 ♌ 34

All ephemeris data is given at 12.00 UT and the Moon's longitude is additionally given for 24.00 UT

ASPECTARIAN

h m	Aspects
01 Tuesday	
05 16	☽ △ ♅
05 59	☽ ☍ ♇
06 11	☽ Q ♂
09 38	☽ ± ♃
10 43	☽ ✶ ♄
20 42	☽ △ ☉
20 56	☽ ⚻ ♃
22 40	☽ ⚼ ♆
02 Wednesday	
00 01	☉ □ ♃
00 34	♀ ± ♄
05 50	☽ ✶ ♂
09 13	☽ Q ♄
09 28	☽ ⚼ ♅
10 09	☽ Q ♀
10 32	☽ ∥ ♂
15 50	☽ ± ♆
21 51	☽ □ ☿
03 Thursday	
02 49	☽ ± ♅
03 26	☽ ⚻ ♆
10 45	☉ ⋕ ☿
11 46	☽ ✶ ♀
12 17	☽ ∠ ♂
14 32	☽ ⚻ ♅
15 27	☽ ⚻ ♇
19 58	☽ □ ♄
04 Friday	
03 20	☽ ± ♇
06 59	☽ ⊥ ♂
07 35	☽ ☍ ♃
12 04	☽ □ ☉
12 36	☽ Q ☿
16 56	♀ ⚻ ♅
17 04	☽ ∥ ♂
19 35	☽ ⚺ ♂
20 39	☽ ∠ ♀
21 09	♂ ± ♃
21 19	☽ ⚼ ♇
23 38	♀ ± ♆
05 Saturday	
03 39	♀ □ ♇
12 15	☿ △ ♆
15 03	☽ ☍ ♆
15 22	☽ ✶ ☿
16 43	☽ ⊥ ♀
06 Sunday	
02 30	☽ ☍ ♅
03 32	☽ Q ☉
03 37	☽ △ ♇
06 09	☽ ⚺ ♀
07 44	☽ △ ♄
15 59	♆ St D
20 29	☽ ⚻ ♃
21 19	♀ ⚻ ♄
07 Monday	
01 31	☽ ∠ ☿
06 03	☽ ✶ ☉
08 44	☽ ± ♃
11 22	☽ ☌ ♂
13 54	☽ ⚼ ♄
22 09	☽ ⊥ ☿
23 13	☽ ∥ ♀
08 Tuesday	
03 04	☽ ⚼ ♃
03 43	☽ ⚻ ♆
07 56	☽ ± ♄
11 54	☽ ⚺ ☿
15 01	☽ ∠ ☉
15 06	☽ ⚻ ♅
15 46	☽ ± ♆
15 55	☽ ⋕ ♇
16 22	☽ □ ♇
19 54	☽ ⚻ ♄
09 Wednesday	
01 14	☽ ☌ ♀
03 06	☽ ± ♅
04 10	☽ ⋕ ☉
07 45	☉ ∠ ♇
08 54	♂ ⚼ ♄
09 21	☽ △ ♃
09 44	☽ ⚼ ♆
10 35	☽ ⊥ ☉
13 22	☿ △ ♅
20 27	♀ ± ♅
20 57	☽ ⚼ ♅
23 31	☽ ⚺ ☉
23 51	☿ ✶ ♇
10 Thursday	
00 55	♅ St D
01 05	☿ ⋕ ♄
02 26	☽ ⚺ ♂
04 11	☽ Q ♇
13 06	☽ ⋕ ♄
14 46	☽ ⊥ ♂
15 15	☽ △ ♆
15 25	☽ ∥ ☿
23 44	☿ ☍ ♄
11 Friday	
00 42	☽ ⋕ ☿
02 15	☽ △ ♅
03 37	☽ ✶ ♇
04 35	☽ ∥ ♄
06 31	☽ ☍ ♄
07 34	☽ ☌ ☿
09 04	☽ ∠ ♂
18 03	☽ ⚺ ♀
20 19	☽ □ ♃
12 Saturday	
00 55	☿ ∠ ♂
06 37	☽ ⊥ ♀
08 19	☽ ∠ ♇
14 14	☽ ☌ ☉
14 57	☽ ✶ ♂
16 54	♀ ⚼ ♆
20 40	♀ △ ♃
21 15	☽ ∥ ☉
13 Sunday	
00 00	☽ ∥ ♇
00 27	☽ □ ♆
01 13	☽ ∠ ♀
01 14	☽ ⊥ ♇
01 38	☽ ⋕ ♀
06 28	☽ Q ♃
10 57	☽ □ ♅
11 44	☉ ✶ ♂
12 23	☽ ⚺ ♇
14 20	☽ Q ♂
14 43	☽ ⚻ ♄
16 00	♀ ⋕ ♇
20 14	☉ ⋕ ♀
14 Monday	
00 09	☽ ⚺ ☿
00 57	☿ ∥ ♄
01 04	☉ ∥ ♇
01 40	☽ ± ♄
04 43	☽ ✶ ♃
07 33	☽ ✶ ♀
09 34	☽ Q ♆
12 40	☽ ⊥ ☿
17 55	☽ ⚼ ♄
19 48	☽ Q ♅
15 Tuesday	
00 36	☽ □ ♂
01 49	☽ ⚺ ☉
07 12	☽ ✶ ♆
07 13	☽ ∠ ☿
07 18	☽ Q ♀
08 02	☽ ∠ ♃
11 01	☽ ⋕ ♂
13 24	☽ ⊥ ☉
14 12	☿ □ ♃
17 17	☽ ✶ ♅
18 47	☽ ☌ ♇
20 36	☽ △ ♄
16 Wednesday	
00 11	☽ ⊥ ♃
06 35	☽ ∠ ☉
09 48	☽ ∠ ♆
10 51	☽ ⚺ ♃
13 35	☽ ✶ ☿
18 10	☽ □ ♀
19 43	☽ ∠ ♅
17 Thursday	
01 34	☽ ⊥ ♆
07 53	☽ △ ♂
10 50	☽ ✶ ☉
11 24	☽ ⊥ ♅
11 59	☽ ⚺ ♆
12 21	♀ ⚼ ♅
13 27	☽ Q ☿
21 46	☽ ⚺ ♅
23 20	☽ ⚺ ♇
18 Friday	
00 43	☽ □ ♄
02 57	☉ □ ♆
09 08	☽ Q ☉
09 40	☽ ⊥ ♇
10 53	☽ ⚼ ♂
15 19	☽ ☌ ♃
16 59	☉ ⊥ ♇
19 Saturday	
00 42	☽ □ ☿
01 06	☽ ∠ ♇
02 44	☽ △ ♀
02 56	☽ ± ♂
04 01	☿ ∠ ♇
07 22	☽ Q ♄
13 34	☽ ⚻ ♂
15 25	☽ ☌ ♆
18 09	☽ □ ☉
23 43	☿ ⋕ ♀
20 Sunday	
01 01	☽ ☌ ♅
02 38	☽ ✶ ♇
03 37	☽ ✶ ♄
05 26	☽ ⋕ ♂
06 31	☽ ⚼ ♀
18 42	☽ ⚺ ♃
21 10	♀ Q ♇
22 55	☽ Q ♇
23 05	☽ ± ♀
21 Monday	
02 39	☿ ⚺ ♀
04 48	☽ ∠ ♄
04 54	☽ ⊥ ♃
10 05	☽ ⚻ ♀
10 24	☽ △ ☿
11 02	☽ ∥ ☉
11 48	♂ ⚻ ♆
18 02	☽ ⚺ ♆
18 18	☽ ☍ ♂
19 51	☽ ⊥ ♄
20 10	☽ ∠ ♃
22 Tuesday	
00 30	☽ △ ☉
03 21	☽ ∥ ♇
03 33	☽ ⚺ ♅
04 07	☽ ⊥ ♆
05 15	☽ □ ♇
05 53	☽ ⚺ ♄
09 09	☽ ∥ ☿
12 52	☉ Q ♃
13 38	☽ ⊥ ♅
15 00	☽ ⚼ ☿
19 14	☽ ∠ ♆
21 36	☽ ✶ ♃
23 Wednesday	
02 24	☽ ⋕ ♀
03 34	☽ ⚼ ☉
04 46	☽ ∠ ♅
05 38	♂ ± ♄
08 10	☽ ± ☿
16 43	☽ ∥ ♄
17 06	☽ ☍ ♀
17 59	☽ Q ♃
18 57	☿ ∥ ♇
19 39	☽ ⚻ ☿
19 52	☽ ± ☉
20 29	☽ ✶ ♆
20 45	☉ □ ♅
22 49	☽ ⚻ ♂
24 Thursday	
02 45	☿ □ ♆
06 04	☽ ✶ ♅
06 46	☽ ⚻ ☉
07 52	☽ △ ♇
08 00	♂ ⚼ ♃
08 09	☽ ☌ ♄
09 23	☽ ± ♂
09 52	☽ ⋕ ♄
13 07	☿ ⊥ ♇
16 49	☽ Q ♆
20 01	☽ ∥ ♀
22 55	☉ ⚺ ♇
25 Friday	
00 47	☽ □ ♃
01 19	☽ ⚼ ♂
01 29	☉ ⚻ ♄
02 28	☽ Q ♅
09 27	☽ ⚼ ♇
09 44	♀ △ ♆
10 00	☿ ✶ ♂
23 40	☽ □ ♆
23 59	☽ ⋕ ♇
26 Saturday	
00 58	☽ ⚻ ♀
00 59	☽ ± ♇
03 03	♄ △ ♇
04 11	☽ △ ♂
05 52	☽ ☍ ☿
09 30	☽ □ ♅
11 23	☽ ⚺ ♄
11 26	☽ ⚻ ♇
11 52	☽ ⋕ ☿
12 23	☽ ± ♀
14 11	☽ ☍ ☉
21 50	☽ ⊥ ♄
27 Sunday	
01 45	☿ Q ♃
05 22	☽ △ ♃
05 43	☽ ⚼ ♀
07 17	☽ ⋕ ☉
11 52	☽ ∥ ♂
12 05	☿ □ ♅
13 42	☽ ∠ ♄
28 Monday	
01 55	☿ ⚻ ♄
04 40	☿ ⚺ ♇
04 41	☽ △ ♆
08 33	☽ ⚼ ♃
11 19	☽ △ ♀
11 23	☉ ⋕ ♂
11 44	☽ □ ♂
15 01	☽ △ ♅
16 41	☽ ✶ ♄
17 08	☽ ☍ ♇
18 54	☽ ⚻ ☿
19 02	♀ ⋕ ♄
19 34	♀ ⚺ ♂
29 Tuesday	
00 20	☽ ⚻ ☉
01 14	☽ ± ♃
07 37	☽ ± ☿
08 14	☽ ⚼ ♆
12 24	☽ ± ☉
12 31	☽ ⚻ ♃
14 49	☽ Q ♄
18 52	☽ ⚼ ♅
22 38	☽ Q ♂
30 Wednesday	
01 09	☽ ± ♆
03 03	☽ ⚼ ☿
04 15	♀ △ ♅
06 50	☽ ⚼ ☉
11 59	☽ ± ♅
12 35	☽ ⚻ ♆
13 26	☿ ⋕ ♂
17 48	☉ ± ♄
19 28	♀ ☍ ♄
22 38	☽ ✶ ♂
23 33	☽ ⚻ ♅
31 Thursday	
00 58	☽ □ ♄
01 36	☽ □ ♀
01 52	☽ ⚻ ♇
03 57	☉ ⊥ ♀
04 29	♀ ✶ ♇
12 22	☽ △ ☿
13 36	☽ ± ♇
14 08	☿ ± ♄
14 22	☽ △ ☉
19 47	♂ ⚻ ♅
22 57	☽ ☍ ♃

NOVEMBER 1996

LONGITUDES

Date	Sidereal time h m s	Sun ☉ ° ' "	Moon ☽ ° ' "	Moon ☽ 24.00 ° ' "	Mercury ☿ ° '	Venus ♀ ° '	Mars ♂ ° '	Jupiter ♃ ° '	Saturn ♄ ° '	Uranus ♅ ° '	Neptune ♆ ° '	Pluto ⯓ ° '
01	14 44 13	09 ♏ 20 36	19 ♋ 23 05	25 ♋ 23 23	09 ♏ 02	03 ♎ 37	01 ♍ 13	12 ♑ 53	01 ♈ 29	00 ♒ 51	25 ♑ 10	02 ♐ 05
02	14 48 09	10 20 40	01 ♌ 21 31	07 ♌ 18 04	10 40	04 50	01 46	13 03	01 R 26	00 53	25 11	02 07
03	14 52 06	11 20 46	13 ♌ 13 39	19 ♌ 08 54	12 17	06 03	02 19	13 13	01 23	00 54	25 12	02 09
04	14 56 02	12 20 53	25 ♌ 04 26	01 ♍ 00 54	13 54	07 16	02 52	13 22	01 20	00 55	25 13	02 12
05	14 59 59	13 21 03	06 ♍ 58 55	12 ♍ 59 06	15 30	08 29	03 24	13 32	01 17	00 56	25 13	02 14
06	15 03 55	14 21 15	19 ♍ 02 00	25 ♍ 08 11	17 06	09 42	03 57	13 42	01 15	00 58	25 14	02 16
07	15 07 52	15 21 29	01 ♎ 18 08	07 ♎ 32 16	18 42	10 55	04 29	13 53	01 12	00 59	25 16	02 19
08	15 11 48	16 21 45	13 ♎ 50 56	20 ♎ 14 25	20 17	12 08	05 02	14 03	01 09	01 01	25 17	02 21
09	15 15 45	17 22 02	26 ♎ 42 54	03 ♏ 16 27	21 52	13 21	05 34	14 13	01 07	01 02	25 18	02 23
10	15 19 42	18 22 22	09 ♏ 55 03	16 ♏ 38 35	23 27	14 34	06 06	14 24	01 04	01 04	25 19	02 25
11	15 23 38	19 22 43	23 ♏ 26 46	00 ♐ 19 18	25 01	15 48	06 38	14 34	01 02	01 05	25 20	02 28
12	15 27 35	20 23 06	07 ♐ 15 44	14 ♐ 15 34	26 35	17 01	07 09	14 45	01 00	01 07	25 21	02 30
13	15 31 31	21 23 31	21 ♐ 18 14	28 ♐ 23 09	28 09	18 14	07 41	14 56	00 58	01 09	25 22	02 32
14	15 35 28	22 23 57	05 ♑ 29 41	12 ♑ 37 17	29 ♏ 42	19 28	08 12	15 07	00 56	01 10	25 24	02 35
15	15 39 24	23 24 25	19 ♑ 45 23	26 ♑ 53 28	01 ♐ 15	20 42	08 44	15 18	00 54	01 12	25 25	02 37
16	15 43 21	24 24 54	04 ♒ 01 07	11 ♒ 07 57	02 48	21 55	09 15	15 29	00 52	01 14	25 26	02 40
17	15 47 17	25 25 24	18 ♒ 13 40	25 ♒ 18 03	04 21	23 09	09 46	15 40	00 50	01 16	25 28	02 42
18	15 51 14	26 25 56	02 ♓ 20 55	09 ♓ 22 09	05 53	24 23	10 16	15 52	00 49	01 18	25 29	02 44
19	15 55 11	27 26 28	16 ♓ 21 39	23 ♓ 19 21	07 25	25 36	10 47	16 03	00 47	01 20	25 30	02 47
20	15 59 07	28 27 02	00 ♈ 15 11	07 ♈ 09 04	08 57	26 50	11 18	16 15	00 46	01 22	25 32	02 49
21	16 03 04	29 ♏ 27 37	14 ♈ 00 55	20 ♈ 50 38	10 29	28 04	11 48	16 26	00 44	01 24	25 33	02 51
22	16 07 00	00 ♐ 28 13	27 ♈ 38 03	04 ♉ 23 02	12 00	29 ♎ 18	12 18	16 38	00 43	01 26	25 35	02 54
23	16 10 57	01 28 51	11 ♉ 05 23	17 ♉ 44 55	13 31	00 ♏ 32	12 48	16 49	00 42	01 28	25 36	02 56
24	16 14 53	02 29 30	24 ♉ 21 24	00 ♊ 54 39	15 02	01 46	13 18	17 01	00 41	01 30	25 38	02 58
25	16 18 50	03 30 10	07 ♊ 24 30	13 ♊ 50 49	16 33	03 00	13 48	17 13	00 40	01 32	25 39	03 01
26	16 22 46	04 30 52	20 ♊ 13 28	26 ♊ 32 26	18 04	04 14	14 17	17 25	00 39	01 35	25 41	03 03
27	16 26 43	05 31 35	02 ♋ 47 44	08 ♋ 59 26	19 34	05 29	14 46	17 37	00 38	01 37	25 42	03 06
28	16 30 40	06 32 19	15 ♋ 07 42	21 ♋ 12 47	21 04	06 43	15 16	17 49	00 38	01 39	25 44	03 08
29	16 34 36	07 33 05	27 ♋ 14 57	03 ♌ 14 36	22 33	07 57	15 45	18 02	00 37	01 42	25 46	03 10
30	16 38 33	08 ♐ 33 53	09 ♌ 12 10	15 ♌ 08 08	24 ♐ 02	09 ♏ 11	16 ♍ 13	18 ♑ 14	00 ♈ 37	01 ♒ 44	25 ♑ 47	03 ♐ 13

Date	Moon True ☊ ° '	Moon Mean ☊ ° '	Moon ☽ Latitude ° '
01	07 ♎ 31	06 ♎ 16	05 S 06
02	07 R 30	06 12	04 46
03	07 D 29	06 09	04 14
04	07 30	06 06	03 31
05	07 31	06 03	02 39
06	07 33	06 00	01 39
07	07 34	05 56	00 S 34
08	07 R 34	05 53	00 N 34
09	07 33	05 50	01 42
10	07 29	05 47	02 46
11	07 23	05 44	03 43
12	07 16	05 41	04 27
13	07 09	05 37	04 57
14	07 02	05 34	05 08
15	06 56	05 31	05 01
16	06 52	05 28	04 35
17	06 50	05 25	03 52
18	06 D 50	05 22	02 55
19	06 51	05 18	01 48
20	06 52	05 15	00 N 36
21	06 R 52	05 12	00 S 38
22	06 50	05 09	01 49
23	06 45	05 06	02 53
24	06 38	05 02	03 46
25	06 29	04 59	04 26
26	06 18	04 56	04 52
27	06 07	04 53	05 03
28	05 56	04 50	05 00
29	05 47	04 47	04 43
30	05 ♎ 40	04 ♎ 43	04 S 14

DECLINATIONS

Date	Sun ☉ ° '	Moon ☽ ° '	Mercury ☿ ° '	Venus ♀ ° '	Mars ♂ ° '	Jupiter ♃ ° '	Saturn ♄ ° '	Uranus ♅ ° '	Neptune ♆ ° '	Pluto ⯓ ° '
01	14 S 36	16 N 59	14 S 09	00 N 01	12 N 33	23 S 04	01 S 46	20 S 32	20 S 40	08 S 23
02	14 55	15 12	14 46	00 S 26	12 22	23 03	01 47	20 32	20 39	08 24
03	15 14	12 47	15 23	00 54	12 11	23 03	01 48	20 32	20 39	08 25
04	15 32	09 51	15 58	01 21	12 00	23 02	01 49	20 31	20 39	08 25
05	15 51	06 25	16 33	01 49	11 49	23 01	01 50	20 31	20 39	08 26
06	16 09	02 N 49	17 07	02 17	11 38	23 00	01 51	20 31	20 39	08 26
07	16 26	01 S 02	17 40	02 44	11 27	22 59	01 52	20 30	20 39	08 27
08	16 44	04 56	18 13	03 12	11 16	22 58	01 53	20 30	20 39	08 28
09	17 01	08 43	18 44	03 40	11 05	22 57	01 54	20 30	20 38	08 28
10	17 18	12 09	19 14	04 07	10 54	22 56	01 55	20 29	20 38	08 29
11	17 34	15 02	19 44	04 35	10 43	22 55	01 56	20 29	20 38	08 29
12	17 51	17 07	20 12	05 02	10 32	22 54	01 56	20 29	20 38	08 30
13	18 06	18 13	20 40	05 30	10 21	22 52	01 57	20 28	20 38	08 31
14	18 22	18 12	21 06	05 57	10 10	22 51	01 58	20 28	20 37	08 31
15	18 37	17 01	21 32	06 24	09 59	22 50	01 58	20 27	20 37	08 32
16	18 52	14 48	21 56	06 52	09 48	22 49	01 59	20 27	20 37	08 32
17	19 07	11 41	22 19	07 19	09 38	22 48	01 59	20 26	20 37	08 33
18	19 21	07 55	22 42	07 46	09 27	22 46	02 00	20 26	20 37	08 33
19	19 35	03 S 43	23 03	08 13	09 16	22 45	02 00	20 26	20 36	08 34
20	19 49	00 N 39	23 23	08 39	09 05	22 44	02 00	20 25	20 36	08 34
21	20 02	04 56	23 41	09 06	08 54	22 42	02 01	20 25	20 36	08 35
22	20 15	08 56	23 59	09 32	08 44	22 41	02 01	20 24	20 36	08 35
23	20 27	12 25	24 15	09 58	08 33	22 40	02 01	20 24	20 35	08 36
24	20 39	15 12	24 30	10 24	08 23	22 38	02 01	20 23	20 35	08 37
25	20 51	17 10	24 44	10 50	08 12	22 37	02 02	20 23	20 35	08 37
26	21 02	18 13	24 56	11 16	08 01	22 35	02 02	20 22	20 35	08 38
27	21 13	18 21	25 08	11 41	07 51	22 34	02 02	20 22	20 34	08 38
28	21 24	17 36	25 17	12 06	07 41	22 32	02 02	20 21	20 34	08 38
29	21 34	16 04	25 26	12 31	07 30	22 31	02 02	20 21	20 34	08 39
30	21 S 44	13 N 52	25 S 33	12 S 55	07 N 20	22 S 29	02 S 02	20 S 20	20 S 33	08 S 39

ZODIAC SIGN ENTRIES

Date	h	m	Planets
02	09	16	☽ ♌
04	21	57	☽ ♍
07	09	29	☽ ♎
09	18	02	☽ ♏
11	23	26	☽ ♐
14	02	44	☽ ♑
14	16	36	☿ ♐
16	05	14	☽ ♒
18	08	00	☽ ♓
20	11	34	☽ ♈
22	00	49	☉ ♐
22	16	12	☽ ♉
23	01	34	♀ ♏
24	22	20	☽ ♊
27	06	37	☽ ♋
29	17	30	☽ ♌

LATITUDES

Date	Mercury ☿ ° '	Venus ♀ ° '	Mars ♂ ° '	Jupiter ♃ ° '	Saturn ♄ ° '	Uranus ♅ ° '	Neptune ♆ ° '	Pluto ⯓ ° '
01	00 N 22	01 N 35	01 N 37	00 S 16	02 S 35	00 S 35	00 N 27	12 N 25
04	00 N 02	01 39	01 40	00 16	02 34	00 35	00 27	12 25
07	00 S 18	01 43	01 42	00 16	02 34	00 35	00 27	12 24
10	00 38	01 46	01 45	00 16	02 33	00 35	00 27	12 23
13	00 57	01 47	01 48	00 16	02 32	00 35	00 27	12 23
16	01 14	01 48	01 50	00 17	02 32	00 35	00 26	12 22
19	01 31	01 49	01 53	00 17	02 31	00 35	00 26	12 22
22	01 46	01 48	01 56	00 17	02 31	00 35	00 26	12 22
25	01 59	01 47	01 59	00 17	02 30	00 35	00 26	12 22
28	02 10	01 45	02 02	00 17	02 29	00 35	00 26	12 21
31	02 S 17	01 N 43	02 N 05	00 S 18	02 S 29	00 S 34	00 N 26	12 N 21

DATA

Julian Date	2450389
Delta T	+62 seconds
Ayanamsa	23° 48' 47"
Synetic vernal point	05° ♓ 18' 12"
True obliquity of ecliptic	23° 26' 14"

MOON'S PHASES, APSIDES AND POSITIONS ☽

Date	h	m	Phase	Longitude ° '	Eclipse Indicator
03	07	50	☾	11 ♌ 10	
11	04	16	●	19 ♏ 03	
18	01	09	☽	25 ♒ 59	
25	04	10	○	03 ♊ 10	

Day	h	m		
03	13	38	Apogee	
16	04	36	Perigee	
07	05	36	OS	
13	23	29	Max dec	18° S 21'
20	08	28	ON	
27	03	29	Max dec	18° N 24'

LONGITUDES

Date	Chiron ⚷ ° '	Ceres ⚳ ° '	Pallas ⚴ ° '	Juno ⚵ ° '	Vesta ⚶ ° '	Black Moon Lilith ⚸ ° '
01	23 ♎ 04	25 ♐ 38	02 ♐ 05	06 ♈ 28	26 ♐ 09	14 ♌ 41
11	24 ♎ 28	29 ♐ 16	06 ♐ 17	05 ♈ 40	01 ♑ 09	15 ♌ 48
21	25 ♎ 48	03 ♑ 01	10 ♐ 31	05 ♈ 47	06 ♑ 12	16 ♌ 55
31	27 ♎ 04	06 ♑ 51	14 ♐ 44	06 ♈ 49	11 ♑ 19	18 ♌ 02

All ephemeris data is given at 12.00 UT and the Moon's longitude is additionally given for 24.00 UT

ASPECTARIAN

01 Friday
05 24 ☽ ∠ ♂
07 25 ☽ ♇ ⯓
09 35 ♂ ± ♆
16 58 ☽ Q ♀
22 52 ♂ ⊼ ♄
23 34 ☽ ☍ ♆
23 55 ☉ ☌ ☿

02 Saturday
00 13 ☽ ⊥ ♂
11 02 ☽ ☍ ♅
12 10 ☽ △ ♄
12 52 ☽ ⊻ ♂
13 32 ☽ △ ⯓
14 42 ☽ # ☉
15 42 ☽ # ☿
19 49 ☽ ✶ ♀
22 08 ☿ ⊥ ♀

03 Sunday
00 10 ☉ ∥ ☿
04 35 ♂ □ ⯓
07 50 ☽ □ ☉
09 47 ☽ □ ☿
11 58 ☽ ⊼ ♃
17 33 ☽ ∥ ♂
18 23 ☽ ♇ ♄

04 Monday
00 18 ☽ ± ♃
01 39 ☿ Q ♆
03 20 ☿ ✶ ♃
05 39 ☽ ∠ ♀
12 16 ☽ ⊼ ♆
12 32 ☽ ± ♄
18 46 ☽ ♇ ♃
22 27 ☽ # ⯓
23 38 ☽ Q ☉
23 50 ☽ ⊼ ♅

05 Tuesday
00 24 ☽ ± ♆
00 36 ☽ ⊼ ♄
01 55 ☽ ⊥ ♀
02 25 ☽ □ ⯓
03 56 ☽ Q ☿
04 28 ☽ ☌ ♂
08 56 ☉ Q ♆
09 47 ☿ Q ♂
11 55 ☽ ± ♅
13 13 ♀ ∥ ♄
15 20 ☽ ⊻ ♀
17 25 ☉ ✶ ♃
18 30 ☽ ♇ ♆
23 26 ☿ ♇ ♄

06 Wednesday
01 17 ☽ △ ♃
01 53 ☽ ✶ ☉
05 54 ☽ ♇ ♅
07 36 ☽ ✶ ☿
14 27 ☽ Q ⯓
15 02 ☽ # ♀
18 01 ☽ # ♄
21 06 ♀ ⊥ ♂

07 Thursday
00 13 ☽ △ ♆
10 01 ☽ ∠ ☉
11 23 ☽ △ ♅
11 48 ☽ ☍ ♄
13 57 ☽ ✶ ⯓
16 22 ☿ Q ♅
17 08 ☽ ∥ ♄
17 19 ☽ ∠ ☿
18 26 ☽ ⊻ ♂
23 50 ☽ ∥ ♀

08 Friday
04 49 ☽ ⊥ ☉
06 25 ☽ ⊥ ♂
07 16 ☉ ♇ ♄
08 24 ☽ ☌ ♀
12 23 ☽ □ ♃
12 57 ☽ ⊥ ☿
17 09 ☽ ⊻ ☉
18 36 ☽ ∠ ⯓

09 Saturday
00 06 ☽ ∠ ♂
01 47 ☽ ⊻ ☿
09 23 ☽ □ ♆
10 25 ☽ ∥ ⯓
11 24 ☽ ⊥ ⯓
19 56 ☽ □ ♅
20 02 ☽ ⊼ ♄
21 53 ☉ Q ♂
22 13 ☽ Q ♃
22 25 ☽ ⊻ ⯓

10 Sunday
03 22 ☽ # ♂
04 50 ☽ ✶ ♂
06 54 ☽ ± ♄
08 03 ♀ □ ♃
16 50 ♄ ✶ ♅
18 05 ☽ Q ♆
20 07 ☽ ✶ ♃
21 09 ☽ ⊻ ♀
22 58 ☽ ♇ ♄

11 Monday
03 10 ☽ Q ♂
04 16 ☽ ☌ ☉
04 19 ☽ Q ♅
04 52 ☉ Q ♅
08 49 ☽ ⊥ ♀
15 07 ☽ ☌ ☿
15 19 ☽ ✶ ♆
16 52 ☿ ✶ ♆
22 51 ☽ ∠ ♃

12 Tuesday
01 12 ☽ △ ♄
01 21 ☽ ✶ ♅
02 04 ☽ ∠ ♀
03 46 ☽ ☌ ⯓
10 06 ♂ ± ♅
11 48 ☽ □ ♂
14 36 ☽ ⊥ ♃
17 19 ☽ ∠ ♆
21 51 ♀ ∠ ⯓

13 Wednesday
01 01 ☽ ⊻ ♃
01 52 ☿ ∥ ♅
03 12 ☽ ∠ ♅
05 49 ☽ ∥ ☉
06 18 ☽ ✶ ♀
08 43 ☽ ⊥ ♆
10 04 ☿ ∥ ♆
12 10 ☽ ⊻ ☉
18 32 ☽ ⊥ ♅
18 55 ☽ ⊻ ♆
23 06 ☽ ⊥ ☉

14 Thursday
01 01 ☽ ⊻ ☿
04 19 ☽ □ ♄
04 34 ☽ Q ♀
04 42 ☽ ⊻ ♅
06 09 ☽ ∥ ☉
07 04 ☽ ⊻ ⯓
12 24 ☽ ⊥ ☿
15 27 ☽ ∠ ☉
16 44 ☽ △ ♂
17 13 ☽ ⊥ ⯓
19 15 ☿ ∠ ♃

15 Friday
04 24 ☽ ☌ ♃
05 24 ☽ ∠ ☿
06 34 ☿ △ ♄
08 24 ☽ ∠ ⯓
10 33 ☽ Q ♄
11 10 ☿ ✶ ♅
13 43 ☽ □ ♀
18 36 ☽ ✶ ☉
18 56 ☽ ♇ ♂
21 32 ☽ ☌ ♆

16 Saturday
06 42 ☽ ✶ ♄
07 18 ☽ ☌ ♅
09 42 ☽ ✶ ☿
09 42 ☽ ✶ ⯓
09 44 ☿ ☌ ⯓
10 39 ☽ ± ♂
16 21 ☽ Q ☉
21 09 ☽ ⊼ ♂

17 Sunday
06 01 ☽ Q ⯓
07 37 ☽ ⊻ ♃
07 58 ☽ ∠ ♄
08 25 ☽ Q ☿
12 51 ☉ ✶ ♆
17 55 ☽ ⊥ ♃
21 08 ☽ △ ♀
23 11 ☽ ⊥ ♄

18 Monday
00 17 ☽ ⊻ ♆
01 09 ☽ □ ☉
02 12 ☽ # ♂
08 08 ☽ ∥ ⯓
09 23 ☽ ⊻ ♄
09 26 ☽ ∠ ♃
10 12 ☽ ⊻ ♅
10 31 ☽ ⊥ ♆
12 40 ☽ □ ⯓
12 49 ☽ ∥ ♀
17 01 ☿ ∥ ♃
18 47 ☽ □ ☿
20 28 ☽ ⊥ ♅
22 09 ♂ ♇ ♆

19 Tuesday
01 10 ☽ ♇ ♀
01 56 ☽ ∠ ♆
02 04 ☽ ☍ ♂
09 57 ♀ □ ♆
11 28 ☽ ✶ ♃
11 57 ☽ ∠ ♅
17 55 ♀ ∠ ♂
18 08 ☽ ± ♀
21 29 ☽ ∥ ♄

20 Wednesday
03 48 ☽ ✶ ♆
05 30 ☽ ⊼ ♀
07 38 ♀ ∥ ⯓
08 28 ☽ Q ♃
08 38 ☽ △ ☉
11 34 ♀ ⊥ ⯓
12 53 ☽ ☌ ♄
13 56 ☽ ✶ ♅
16 28 ☽ △ ⯓
19 31 ☽ # ♄

21 Thursday
00 41 ☽ Q ♆
04 42 ♀ # ♂
05 02 ☽ △ ☿
07 58 ☽ ⊼ ♂
10 55 ☽ Q ♅
11 11 ☿ ⊥ ♃
12 51 ☽ ♇ ☉
13 09 ☿ ∠ ♆
16 18 ☽ □ ♃
18 46 ☽ ♇ ⯓
18 54 ☽ ± ♂

22 Friday
05 57 ☽ ± ☉
08 21 ☽ □ ♆
09 51 ☽ # ⯓
10 41 ☽ ± ⯓
10 44 ☽ ♇ ☿
10 46 ☽ ∥ ♂
11 23 ☽ ♇ ♂
15 15 ☽ ☍ ♀
16 24 ☽ # ♀
17 26 ☽ ⊼ ☉
17 28 ☽ ⊻ ♄
17 45 ☉ △ ♄
18 45 ☽ □ ♅
19 01 ☿ □ ♂
21 23 ☽ ⊼ ⯓

23 Saturday
04 09 ☽ ⊥ ♄
04 48 ☽ ± ☿
05 09 ☉ ∥ ♅
05 44 ♂ # ⯓
11 38 ☉ ✶ ♅
15 07 ♀ ⊼ ♄
15 12 ☽ △ ♂
16 56 ☽ ⊼ ☿
20 17 ☽ ∠ ♄
22 08 ☉ ∠ ♃
22 29 ☽ △ ♃

24 Sunday
03 26 ☉ ∥ ♆
06 38 ♀ □ ♅
14 19 ☽ △ ♆
23 34 ☽ ✶ ♄
23 56 ☉ ☌ ⯓

25 Monday
01 07 ☽ △ ♅
02 16 ☽ ♇ ♃
03 00 ☽ ⊼ ♀
03 51 ☽ ☍ ⯓
04 10 ☽ ☍ ☉
11 47 ☿ ∠ ♅
12 11 ♀ ⊻ ⯓
15 17 ☽ ± ♀
18 03 ☽ ♇ ♆
19 12 ☽ ± ♃
21 47 ☽ Q ♄

26 Tuesday
00 14 ☿ ⊻ ♃
00 23 ☽ □ ♂
05 06 ☽ ♇ ♅
06 37 ☽ ⊼ ♃
07 22 ☽ ☍ ☿
09 57 ☽ ♇ ♀
10 58 ☽ ± ♆
22 11 ☽ ± ♅
22 23 ☽ ⊼ ♆

27 Wednesday
07 51 ☽ □ ♄
09 43 ☽ ⊼ ♅
11 57 ☽ Q ♂
12 35 ☽ ⊼ ⯓
14 18 ☿ ⊥ ♆
15 19 ♀ Q ♃
17 13 ☉ ⊻ ♀
17 45 ☽ ⊼ ☉
17 45 ☽ △ ♀

28 Thursday
00 14 ☽ ± ⯓
06 28 ☽ ± ☉
10 24 ♀ ± ♄
12 16 ☽ ✶ ♂
17 23 ☽ ☍ ♃
17 56 ☽ ♇ ⯓

29 Friday
01 21 ☽ ⊼ ☿
01 48 ☽ ♇ ☉
09 01 ☽ ☍ ♆
14 58 ☽ ± ☿
18 45 ☽ △ ♄
19 16 ☽ ∠ ♂
20 55 ☽ ☍ ♅
23 54 ☽ △ ⯓

30 Saturday
09 19 ♃ ∠ ⯓
10 36 ☽ △ ☉
11 37 ☽ ♇ ☿
11 58 ☽ □ ♀
14 09 ☽ ⊥ ♂
19 34 ☽ # ♀

DECEMBER 1996

LONGITUDES

Date	Sidereal time h m s	Sun ☉ ° ' "	Moon ☽ ° ' "	Moon ☽ 24.00 ° ' "	Mercury ☿ ° '	Venus ♀ ° '	Mars ♂ ° '	Jupiter ♃ ° '	Saturn ♄ ° '	Uranus ♅ ° '	Neptune ♆ ° '	Pluto ♇ ° '
01	16 42 29	09 ♐ 34 41	21 ♌ 03 03	26 ♌ 57 30	25 ♐ 31	10 ♏ 26	16 ♍ 42	18 ♑ 26	00 ♈ 37	01 ♒ 46	25 ♑ 49	03 ♐ 15
02	16 46 26	10 35 32	02 ♍ 52 08	08 ♍ 47 36	26 59	11 40	17 10	18 39	00 R 37	01 49	25 51	03 17
03	16 50 22	11 36 23	14 ♍ 44 34	20 ♍ 43 44	28 27	12 55	17 38	18 51	00 37	01 51	25 52	03 20
04	16 54 19	12 37 16	26 ♍ 45 48	02 ♎ 51 25	29 ♐ 54	14 09	18 06	19 04	00 D 37	01 54	25 54	03 22
05	16 58 15	13 38 10	09 ♎ 01 15	15 ♎ 15 53	01 ♑ 20	15 24	18 34	19 16	00 37	01 57	25 56	03 25
06	17 02 12	14 39 06	21 ♎ 35 53	28 ♎ 01 41	02 45	16 38	19 02	19 29	00 37	01 59	25 58	03 27
07	17 06 09	15 40 02	04 ♏ 33 39	11 ♏ 12 03	04 09	17 53	19 29	19 42	00 37	02 02	26 00	03 29
08	17 10 05	16 41 00	17 ♏ 56 56	24 ♏ 48 17	05 31	19 07	19 56	19 55	00 38	02 05	26 01	03 32
09	17 14 02	17 42 00	01 ♐ 45 50	08 ♐ 49 12	06 52	20 22	20 23	20 08	00 38	02 07	26 03	03 34
10	17 17 58	18 43 00	15 ♐ 57 49	23 ♐ 10 55	08 12	21 37	20 49	20 20	00 39	02 10	26 05	03 36
11	17 21 55	19 44 01	00 ♑ 27 41	07 ♑ 47 08	09 29	22 51	21 16	20 33	00 40	02 13	26 07	03 39
12	17 25 51	20 45 03	15 ♑ 08 17	22 ♑ 30 05	10 44	24 06	21 42	20 47	00 41	02 16	26 09	03 41
13	17 29 48	21 46 05	29 ♑ 51 34	07 ♒ 11 50	11 56	25 21	22 08	21 00	00 42	02 19	26 11	03 43
14	17 33 44	22 47 08	14 ♒ 30 05	21 ♒ 45 39	13 05	26 36	22 33	21 13	00 43	02 21	26 13	03 45
15	17 37 41	23 48 12	28 ♒ 58 01	06 ♓ 06 50	14 11	27 50	22 59	21 26	00 44	02 24	26 15	03 48
16	17 41 38	24 49 16	13 ♓ 11 50	20 ♓ 12 55	15 11	29 ♏ 05	23 24	21 39	00 46	02 27	26 17	03 50
17	17 45 34	25 50 20	27 ♓ 10 04	04 ♈ 03 23	16 07	00 ♐ 20	23 49	21 53	00 47	02 30	26 19	03 52
18	17 49 31	26 51 24	10 ♈ 52 58	17 ♈ 39 00	16 57	01 35	24 13	22 06	00 49	02 33	26 21	03 54
19	17 53 27	27 52 29	24 ♈ 21 39	01 ♉ 01 06	17 41	02 50	24 37	22 19	00 50	02 36	26 23	03 57
20	17 57 24	28 53 34	07 ♉ 37 32	14 ♉ 11 04	18 17	04 05	25 01	22 33	00 52	02 39	26 25	03 59
21	18 01 20	29 ♐ 54 40	20 ♉ 41 49	27 ♉ 09 53	18 45	05 20	25 25	22 46	00 54	02 42	26 27	04 01
22	18 05 17	00 ♑ 55 45	03 ♊ 35 17	09 ♊ 58 03	19 04	06 34	25 48	23 00	00 56	02 46	26 29	04 03
23	18 09 13	01 56 51	16 ♊ 18 11	22 ♊ 35 41	19 13	07 49	26 11	23 13	00 58	02 49	26 31	04 05
24	18 13 10	02 57 58	28 ♊ 50 30	05 ♋ 02 41	19 R 11	09 04	26 34	23 27	01 00	02 52	26 33	04 08
25	18 17 07	03 59 05	11 ♋ 12 12	17 ♋ 19 08	18 58	10 19	26 57	23 41	01 03	02 55	26 36	04 10
26	18 21 03	05 00 12	23 ♋ 23 32	29 ♋ 25 34	18 32	11 34	27 19	23 54	01 05	02 58	26 38	04 12
27	18 25 00	06 01 19	05 ♌ 25 25	11 ♌ 23 18	17 55	12 49	27 41	24 08	01 08	03 01	26 40	04 14
28	18 28 56	07 02 27	17 ♌ 19 33	23 ♌ 14 29	17 07	14 04	28 02	24 22	01 10	03 05	26 42	04 16
29	18 32 53	08 03 35	29 ♌ 08 34	05 ♍ 02 15	16 08	15 19	28 23	24 36	01 13	03 08	26 44	04 18
30	18 36 49	09 04 44	10 ♍ 56 03	16 ♍ 50 35	15 01	16 34	28 44	24 49	01 16	03 11	26 46	04 20
31	18 40 46	10 ♑ 05 53	22 ♍ 46 26	28 ♍ 44 17	13 ♑ 46	17 ♐ 49	29 ♍ 04	25 ♑ 03	01 ♈ 19	03 ♒ 15	26 ♑ 49	04 ♐ 22

Date	Moon True ☊ ° '	Moon Mean ☊ ° '	Moon ☽ Latitude ° '
01	05 ♎ 36	04 ♎ 40	03 S 34
02	05 R 34	04 37	02 45
03	05 D 34	04 34	01 49
04	05 34	04 31	00 S 47
05	05 R 34	04 28	00 N 18
06	05 33	04 24	01 24
07	05 30	04 21	02 28
08	05 24	04 18	03 25
09	05 15	04 15	04 12
10	05 04	04 12	04 45
11	04 52	04 08	05 00
12	04 40	04 05	04 56
13	04 31	04 02	04 33
14	04 23	03 59	03 51
15	04 19	03 56	02 55
16	04 17	03 53	01 49
17	04 D 17	03 49	00 N 38
18	04 R 17	03 46	00 S 35
19	04 15	03 43	01 44
20	04 12	03 40	02 46
21	04 06	03 37	03 39
22	03 56	03 34	04 19
23	03 44	03 30	04 46
24	03 30	03 27	04 59
25	03 15	03 24	04 57
26	03 01	03 21	04 41
27	02 49	03 18	04 13
28	02 39	03 14	03 34
29	02 32	03 11	02 46
30	02 28	03 08	01 51
31	02 ♎ 27	03 ♎ 05	00 S 51

DECLINATIONS

Date	Sun ☉ ° '	Moon ☽ ° '	Mercury ☿ ° '	Venus ♀ ° '	Mars ♂ ° '	Jupiter ♃ ° '	Saturn ♄ ° '	Uranus ♅ ° '	Neptune ♆ ° '	Pluto ♇ ° '
01	21 S 53	11 N 06	25 S 39	13 S 19	07 N 10	22 S 28	02 S 02	20 S 19	20 S 33	08 S 40
02	22 02	07 53	25 43	13 43	07 00	22 26	02 02	20 19	20 33	08 40
03	22 10	04 20	25 46	14 07	06 50	22 24	02 01	20 18	20 33	08 41
04	22 18	00 N 34	25 48	14 30	06 40	22 23	02 01	20 18	20 32	08 41
05	22 26	03 S 18	25 48	14 53	06 30	22 21	02 01	20 17	20 32	08 42
06	22 33	07 07	25 46	15 15	06 20	22 19	02 01	20 16	20 32	08 42
07	22 40	10 43	25 43	15 38	06 10	22 17	02 00	20 16	20 31	08 42
08	22 46	13 54	25 39	15 59	06 00	22 16	02 00	20 15	20 31	08 43
09	22 52	16 24	25 33	16 21	05 51	22 14	01 59	20 15	20 31	08 43
10	22 57	17 59	25 26	16 42	05 41	22 12	01 59	20 14	20 30	08 44
11	23 02	18 26	25 18	17 02	05 32	22 10	01 58	20 13	20 30	08 44
12	23 07	17 40	25 08	17 22	05 22	22 08	01 58	20 13	20 30	08 44
13	23 11	15 44	24 57	17 42	05 13	22 06	01 57	20 12	20 29	08 45
14	23 14	12 48	24 45	18 01	05 04	22 04	01 56	20 11	20 29	08 45
15	23 18	09 06	24 31	18 20	04 55	22 02	01 56	20 11	20 29	08 45
16	23 20	04 55	24 17	18 38	04 46	22 00	01 55	20 10	20 28	08 46
17	23 22	00 S 33	24 02	18 56	04 37	21 58	01 54	20 09	20 28	08 46
18	23 24	03 N 46	23 46	19 13	04 29	21 56	01 53	20 09	20 27	08 46
19	23 25	07 50	23 29	19 30	04 20	21 54	01 52	20 08	20 27	08 47
20	23 26	11 26	23 12	19 46	04 12	21 52	01 51	20 07	20 27	08 47
21	23 26	14 24	22 55	20 01	04 03	21 50	01 50	20 06	20 26	08 47
22	23 26	16 37	22 38	20 16	03 55	21 47	01 49	20 06	20 26	08 47
23	23 25	17 59	22 21	20 31	03 47	21 45	01 48	20 05	20 26	08 48
24	23 24	18 27	22 04	20 45	03 39	21 43	01 47	20 04	20 25	08 48
25	23 23	18 02	21 48	20 58	03 31	21 41	01 46	20 04	20 25	08 48
26	23 21	16 47	21 33	21 11	03 23	21 38	01 45	20 03	20 24	08 48
27	23 18	14 49	21 18	21 23	03 16	21 36	01 44	20 02	20 24	08 49
28	23 15	12 14	21 05	21 34	03 08	21 34	01 43	20 01	20 24	08 49
29	23 12	09 10	20 53	21 45	03 01	21 31	01 41	20 01	20 23	08 49
30	23 08	05 45	20 42	21 55	02 54	21 29	01 40	20 00	20 23	08 49
31	23 S 03	02 N 05	20 S 32	22 S 05	02 N 47	21 S 27	01 S 39	19 S 59	20 S 22	08 S 49

ZODIAC SIGN ENTRIES

Date	h m	Planets
02	06 11	☽ ♍
04	13 48	☿ ♑
04	18 23	☽ ♎
07	03 39	☽ ♏
09	08 58	☽ ♐
11	11 15	☽ ♑
13	12 14	☽ ♒
15	13 44	☽ ♓
17	05 34	♀ ♐
17	16 55	☽ ♈
19	22 09	☽ ♉
21	14 06	☉ ♑
22	05 17	☽ ♊
24	14 14	☽ ♋
27	01 09	☽ ♌
29	13 45	☽ ♍

LATITUDES

Date	Mercury ☿ ° '	Venus ♀ ° '	Mars ♂ ° '	Jupiter ♃ ° '	Saturn ♄ ° '	Uranus ♅ ° '	Neptune ♆ ° '	Pluto ♇ ° '
01	02 S 17	01 N 43	02 N 05	00 S 18	02 S 29	00 S 34	00 N 26	12 N 21
04	02 21	01 40	02 08	00 18	02 28	00 34	00 26	12 21
07	02 21	01 36	02 11	00 18	02 27	00 34	00 26	12 21
10	02 16	01 31	02 14	00 18	02 26	00 34	00 26	12 21
13	02 03	01 26	02 17	00 18	02 26	00 34	00 26	12 22
16	01 43	01 21	02 20	00 19	02 25	00 34	00 26	12 22
19	01 14	01 15	02 24	00 19	02 24	00 34	00 26	12 22
22	00 S 33	01 09	02 27	00 19	02 24	00 34	00 26	12 22
25	00 N 18	01 02	02 31	00 19	02 23	00 34	00 26	12 22
28	01 16	00 55	02 34	00 20	02 23	00 34	00 26	12 23
31	02 N 12	00 N 48	02 N 38	00 S 20	02 S 22	00 S 34	00 N 26	12 N 23

DATA

Julian Date	2450419
Delta T	+62 seconds
Ayanamsa	23° 48' 51"
Synetic vernal point	05° ♓ 18' 08"
True obliquity of ecliptic	23° 26' 13"

MOON'S PHASES, APSIDES AND POSITIONS ☽

Date	h m	Phase	Longitude ° '	Eclipse Indicator
03	05 06	☾	11 ♍ 19	
10	16 56	●	18 ♐ 56	
17	09 31	☽	25 ♓ 44	
24	20 41	○	03 ♋ 20	

Day	h m		
01	10 27	Apogee	
13	04 10	Perigee	
29	05 16	Apogee	
04	15 34	0S	
11	09 07	Max dec	18° S 26'
17	15 01	0N	
24	12 29	Max dec	18° N 27'

LONGITUDES

Date	Chiron ⚷ ° '	Ceres ⚳ ° '	Pallas ⚴ ° '	Juno ⚵ ° '	Vesta ⚶ ° '	Black Moon Lilith ⚸ ° '
01	27 ♎ 04	06 ♑ 51	14 ♐ 44	06 ♈ 49	11 ♑ 19	18 ♌ 02
11	28 ♎ 13	10 ♑ 45	18 ♐ 58	08 ♈ 40	16 ♑ 29	19 ♌ 09
21	29 ♎ 16	14 ♑ 41	23 ♐ 05	10 ♈ 01	21 ♑ 41	20 ♌ 15
31	00 ♏ 10	18 ♑ 40	27 ♐ 18	11 ♈ 30	26 ♑ 53	21 ♌ 22

ASPECTARIAN

h m Aspects

01 Sunday
00 58 ☽ ⚼ ♄
02 48 ☽ ⊻ ♂
03 18 ☿ ∠ ♀
06 36 ☽ ⊼ ♃
16 15 ♂ ⚼ ♅
16 19 ☿ ⊥ ♅
16 58 ☿ ⊻ ♆
19 00 ☽ ± ♃
19 14 ☽ ± ♄
21 42 ☽ ⊼ ♆
22 22 ☽ △ ☿

02 Monday
04 45 ☽ Q ♀
06 22 ☽ # ♇
07 25 ☽ ⊼ ♄
08 05 ♃ Q ♄
09 51 ☽ ⊼ ♅
09 55 ☽ ± ♆
12 52 ☽ □ ♇
13 36 ☽ ⚼ ♃
18 09 ☉ ∠ ♆
18 30 ☽ ∥ ♂
22 03 ☽ ± ♅

03 Tuesday
04 11 ☽ ⚼ ♆
05 06 ☽ □ ☉
07 53 ☽ ⚹ ♀
12 40 ♄ St D
16 16 ☽ ⚼ ♅
18 03 ☽ ☌ ♂
20 24 ☽ △ ♃

04 Wednesday
01 15 ☽ Q ♇
02 53 ☽ # ♄
07 03 ♀ Q ♆
10 17 ☽ △ ♆
17 15 ☽ ∠ ♀
19 00 ☽ □ ☿
19 35 ☽ ☍ ♄
20 18 ☽ Q ☉
22 10 ☽ △ ♅
22 25 ☉ ∥ ♃

05 Thursday
00 00 ☿ □ ♄
01 03 ☽ ⚹ ♇
01 09 ☉ ⊥ ♃
04 05 ☽ ∥ ♄
12 48 ☽ ⊥ ♀
16 13 ♀ ⚼ ♄
21 40 ☽ ⚹ ☉
22 45 ☿ ⊻ ♅

06 Friday
01 35 ☽ ⊻ ♀
06 02 ☽ ∠ ♇
06 58 ☽ ⊻ ♂
07 12 ☽ # ♂
07 57 ☽ □ ♃
10 12 ☽ Q ☿
18 39 ☽ ⊥ ♂
20 11 ☽ □ ♆
22 18 ☽ ∥ ♇
22 58 ☽ ⊥ ♇

07 Saturday
00 23 ☿ ⊻ ♇
04 16 ☽ ∠ ☉
04 47 ☽ ⊼ ♄
07 22 ☽ □ ♅
10 02 ☽ ⊻ ♇
11 09 ☽ ⚹ ☿
11 51 ☽ ∠ ♂
15 45 ☽ ± ♄
17 47 ☽ Q ♃
22 00 ☽ ⊥ ☉

08 Sunday
05 02 ☽ Q ♆
07 54 ☽ ⚼ ♄
09 35 ☽ ⊻ ☉
09 43 ♂ △ ♃
14 17 ☽ ☌ ♀
15 30 ☽ ⚹ ♃
15 37 ☽ ⚹ ♂
15 45 ☽ Q ♅
17 02 ☽ ∠ ☿
21 43 ☉ ∠ ♅

09 Monday
02 09 ☽ ⚹ ♆
06 20 ♀ ⚹ ♃
07 04 ♀ Q ♅
10 04 ☽ △ ♄
10 19 ☽ ⊥ ☿
11 18 ☽ ∥ ♀
12 22 ♀ ⚹ ♂
12 37 ☽ ⚹ ♅
13 05 ☽ Q ♂
15 05 ☽ ☌ ♇
17 49 ☽ ∠ ♃
21 37 ☽ ⊻ ☿

10 Tuesday
03 48 ☽ ∠ ♆
09 15 ☽ ⊥ ♃
14 01 ☽ ∠ ♅
16 56 ☽ ☌ ☉
18 53 ☽ ⊥ ♆
19 24 ☽ ⊻ ♃
20 21 ☽ □ ♂
22 17 ☽ ⊻ ♀

11 Wednesday
04 50 ☽ ⊻ ♆
04 59 ☽ ⊥ ♅
09 07 ☽ ⊥ ♀
12 20 ☽ □ ♄
14 53 ☽ ⊻ ♅
15 00 ☿ ⊥ ♇
17 14 ☽ ⊻ ♇
21 22 ☉ ⊥ ♆

12 Thursday
01 14 ☽ ∠ ♀
03 04 ☽ ⊥ ♇
04 10 ☽ ☌ ☿
10 55 ♂ Q ♇
12 45 ☉ ⊻ ♃
16 00 ☽ ∥ ♀
17 47 ☽ Q ♄
17 47 ☽ ∠ ♇
21 20 ☽ ☌ ♃
21 50 ☽ ⊻ ☉
23 01 ☽ △ ♂

13 Friday
03 58 ☽ ⚹ ♀
05 59 ☽ ☌ ♆
08 20 ☽ ⊥ ☉
13 22 ☽ ⚹ ♄
16 01 ☽ ☌ ♅
18 19 ☽ ⚹ ♇

14 Saturday
00 08 ☽ ∠ ☉
00 15 ☽ ⚼ ♂
01 23 ☽ Q ♀
02 42 ☉ □ ♂
04 30 ♀ ⚹ ♆
09 29 ☽ ⊻ ☿
14 01 ☽ ∠ ♄
14 04 ☽ Q ♇
15 29 ☽ ± ♂
20 12 ☽ ⊥ ☿
23 16 ☽ ⊻ ♃

15 Sunday
01 43 ☽ ⊼ ♂
02 44 ☽ ⚹ ☉
04 56 ☽ ⊥ ♄
07 27 ☽ ⊻ ♆
09 24 ☽ ⊥ ♃
09 56 ☽ □ ♀
12 23 ☽ ∠ ☿
14 02 ☽ ∥ ♇
14 58 ☽ ⊻ ♄
17 31 ☽ ⊥ ♆
17 47 ☽ ⊻ ♅
20 07 ☽ □ ♇

16 Monday
00 22 ☽ Q ☉
00 44 ☽ ∠ ♃
03 56 ☽ ⊥ ♅
08 44 ☽ ∠ ♆
12 53 ☽ # ♂
15 39 ☽ ⚹ ☿
19 18 ☽ ∠ ♅

17 Tuesday
02 43 ☽ ⚹ ♃
04 36 ☽ ∥ ♄
06 01 ☽ ☍ ♂
09 31 ☽ □ ☉
10 31 ☽ ⚹ ♆
13 46 ☽ Q ☿
18 03 ☽ △ ♀
18 18 ☽ ☌ ♄
20 53 ♀ △ ♄
21 19 ☽ ⚹ ♅
23 37 ☉ ⊻ ♆
23 42 ☽ △ ♇
23 53 ☽ Q ♃

18 Wednesday
01 28 ☽ # ♄
04 31 ☉ ⊥ ♅
07 31 ☽ Q ♆
15 54 ☽ ∥ ♂
18 31 ☽ Q ♅
23 08 ☽ ⚼ ♀
23 24 ☽ □ ☿

19 Thursday
02 17 ☽ ⚼ ♇
05 49 ☿ ∠ ♀
07 31 ♀ ⚹ ♅
08 17 ☽ □ ♃
12 29 ☽ ⊼ ♂
15 39 ☽ □ ♆
16 54 ☽ ± ♀
17 25 ☉ ∥ ☿
18 00 ☽ # ♇
18 28 ☽ ± ♇
18 51 ☽ △ ☉
23 38 ☽ ± ♂
23 42 ☽ ⊻ ♄

20 Friday
02 56 ☽ □ ♅
04 52 ☽ ⊼ ♀
05 21 ☽ ⊼ ♇
10 06 ♀ ☌ ♇
10 37 ☽ ⊥ ♄
16 31 ☽ ⚼ ♂

21 Saturday
00 26 ☽ ⚼ ☉
03 08 ☽ ∠ ♄
08 18 ☽ △ ☿
15 55 ☽ △ ♃
18 28 ☽ ± ☉
19 33 ♀ ∥ ♅
21 01 ☽ △ ♂
22 42 ☽ △ ♆
22 50 ☿ Q ♄

22 Sunday
06 36 ☽ ⊼ ☉
07 01 ☽ ⚹ ♄
10 12 ☿ ∠ ♇
10 26 ☽ △ ♅
12 08 ☉ □ ♄
12 53 ☽ ☍ ♇
12 55 ☽ ⚼ ☿
18 13 ☽ ☍ ♀
20 26 ☽ ⚼ ♃

23 Monday
02 55 ☽ ⚼ ♆
03 13 ♀ ∥ ♆
05 40 ☽ Q ♄
06 06 ☽ ± ☿
13 47 ☽ ± ♃
14 53 ☽ ⚼ ♅
17 34 ☽ ⊼ ☿
19 45 ☿ St R
20 04 ☽ ± ♆
21 25 ♀ ∠ ♃
22 10 ♀ Q ♂

24 Tuesday
01 27 ☽ ⊼ ♃
07 29 ☽ □ ♂
07 35 ☽ ⊼ ♆
08 11 ☽ ± ♅
09 28 ☉ ⊻ ♅
11 10 ♂ △ ♆
16 11 ☽ □ ♄
19 10 ☿ ∠ ♇
19 48 ☽ ⊼ ♅
20 41 ☽ ☍ ☉
22 15 ☽ ⊼ ♇

25 Wednesday
05 53 ☿ Q ♄
09 57 ☽ ± ♇
10 05 ☽ ⊼ ♀
16 19 ☉ ⊻ ♇
19 33 ☽ Q ♂
23 11 ☽ ± ♀

26 Thursday
00 55 ☿ ∥ ♃
02 46 ☽ ☍ ☿
03 41 ☽ ⚼ ♇
13 02 ☽ ☍ ♃
13 10 ♀ ∠ ♆
18 27 ☽ ☍ ♆
19 02 ☽ ⚼ ♀
20 02 ☽ ⚹ ♂

27 Friday
00 47 ☿ ⊥ ♀
03 22 ☽ △ ♄
07 10 ☽ ☍ ♅
07 39 ☿ ∥ ♀
09 36 ☽ △ ♇
13 19 ☽ ⊼ ☉

28 Saturday
02 31 ☽ ± ☉
03 03 ☽ ∠ ♂
04 38 ☽ △ ♀
09 39 ☽ ⚼ ♄
11 04 ♀ ∥ ♃
11 37 ☽ ⊼ ☿
21 50 ☽ ⊥ ♂
22 28 ☽ ⚼ ☉
22 53 ☽ ± ☿

29 Sunday
02 34 ☽ ⊼ ♃
03 59 ☽ ± ♄
07 05 ☽ ⊼ ♆
10 25 ☽ ⊻ ♂
14 34 ☽ # ♇
15 01 ☽ ± ♃
15 43 ☽ ⚼ ☿
16 14 ☽ ⊼ ♄
19 20 ☽ ± ♆
20 10 ☽ ⊼ ♅
20 24 ☿ ⊻ ♀
22 32 ☽ □ ♇

30 Monday
07 52 ☽ △ ☉
08 26 ☽ ± ♅
09 42 ☽ ⚼ ♃
13 43 ☽ ⚼ ♆
19 31 ☽ △ ☿

31 Tuesday
00 48 ☽ □ ♀
02 48 ☽ ⚼ ♅
07 22 ☽ ∥ ♂
11 11 ☽ Q ♇
14 51 ☽ # ♄
16 41 ☽ △ ♃
18 37 ☉ ⊥ ♇
20 09 ☽ △ ♆
20 27 ♀ ∠ ♅

All ephemeris data is given at 12.00 UT and the Moon's longitude is additionally given for 24.00 UT

JANUARY 1997

LONGITUDES

Date	Sidereal time h m s	Sun ☉ ° ' "	Moon ☽ ° ' "	Moon ☽ 24.00 ° ' "	Mercury ☿ ° '	Venus ♀ ° '	Mars ♂ ° '	Jupiter ♃ ° '	Saturn ♄ ° '	Uranus ♅ ° '	Neptune ♆ ° '	Pluto ♇ ° '
01	18 44 42	11 ♑ 07 02	04 ♎ 44 47	10 ♎ 48 39	12 ♑ 26	19 ♐ 04	29 ♍ 24	25 ♑ 17	01 ♈ 22	03 ♒ 18	26 ♑ 51	04 ♐ 24
02	18 48 39	12 08 12	16 ♎ 56 35	23 ♎ 09 15	11 R 05	20 19	29 ♍ 44	25 31	01 25	03 21	26 53	04 26
03	18 52 36	13 09 22	29 ♎ 27 21	05 ♏ 51 27	09 44	21 34	00 ♎ 03	25 45	01 28	03 25	26 55	04 28
04	18 56 32	14 10 32	12 ♏ 22 06	18 ♏ 59 44	08 27	22 50	00 22	25 59	01 31	03 28	26 57	04 30
05	19 00 29	15 11 42	25 ♏ 44 39	02 ♐ 37 01	07 15	24 05	00 40	26 13	01 35	03 31	27 00	04 32
06	19 04 25	16 12 53	09 ♐ 36 46	16 ♐ 43 41	06 10	25 20	00 58	26 27	01 38	03 35	27 02	04 34
07	19 08 22	17 14 04	23 ♐ 57 18	01 ♑ 16 58	05 14	26 35	01 16	26 41	01 42	03 38	27 04	04 36
08	19 12 18	18 15 15	08 ♑ 41 46	16 ♑ 10 39	04 27	27 50	01 33	26 55	01 45	03 41	27 06	04 38
09	19 16 15	19 16 25	23 ♑ 42 27	01 ♒ 15 50	03 50	29 ♐ 05	01 50	27 09	01 49	03 45	27 09	04 40
10	19 20 11	20 17 36	08 ♒ 49 31	16 ♒ 22 13	03 23	00 ♑ 20	02 06	27 23	01 53	03 48	27 11	04 41
11	19 24 08	21 18 45	23 ♒ 52 45	01 ♓ 20 04	03 06	01 35	02 22	27 37	01 57	03 52	27 13	04 43
12	19 28 05	22 19 55	08 ♓ 43 19	16 ♓ 01 48	02 59	02 51	02 37	27 51	02 01	03 55	27 16	04 45
13	19 32 01	23 21 03	23 ♓ 15 02	00 ♈ 22 42	03 D 00	04 06	02 52	28 05	02 05	03 59	27 18	04 47
14	19 35 58	24 22 11	07 ♈ 24 42	14 ♈ 21 01	03 09	05 21	03 06	28 19	02 09	04 02	27 20	04 48
15	19 39 54	25 23 19	21 ♈ 11 47	27 ♈ 57 13	03 26	06 36	03 20	28 33	02 14	04 06	27 22	04 50
16	19 43 51	26 24 25	04 ♉ 37 37	11 ♉ 13 17	03 50	07 51	03 33	28 48	02 18	04 09	27 25	04 52
17	19 47 47	27 25 31	17 ♉ 44 35	24 ♉ 11 52	04 20	09 06	03 46	29 02	02 22	04 13	27 27	04 53
18	19 51 44	28 26 36	00 ♊ 35 27	06 ♊ 55 40	04 56	10 21	03 58	29 16	02 27	04 16	27 29	04 55
19	19 55 40	29 ♑ 27 40	13 ♊ 12 49	19 ♊ 27 09	05 37	11 36	04 10	29 30	02 32	04 20	27 31	04 57
20	19 59 37	00 ♒ 28 43	25 ♊ 38 53	01 ♋ 48 15	06 22	12 52	04 21	29 44	02 36	04 23	27 34	04 58
21	20 03 34	01 29 46	07 ♋ 55 23	14 ♋ 00 29	07 12	14 07	04 32	29 ♑ 58	02 41	04 27	27 36	05 00
22	20 07 30	02 30 48	20 ♋ 03 40	26 ♋ 05 05	08 05	15 22	04 42	00 ♒ 12	02 46	04 30	27 38	05 01
23	20 11 27	03 31 49	02 ♌ 04 51	08 ♌ 03 08	09 02	16 37	04 51	00 26	02 51	04 34	27 41	05 03
24	20 15 23	04 32 49	14 ♌ 00 06	19 ♌ 55 56	10 02	17 52	05 00	00 40	02 56	04 37	27 43	05 04
25	20 19 20	05 33 48	25 ♌ 50 51	01 ♍ 45 07	11 05	19 07	05 08	00 55	03 01	04 41	27 45	05 06
26	20 23 16	06 34 47	07 ♍ 39 02	13 ♍ 32 56	12 10	20 22	05 16	01 09	03 06	04 44	27 47	05 07
27	20 27 13	07 35 45	19 ♍ 27 14	25 ♍ 22 20	13 18	21 38	05 23	01 23	03 12	04 48	27 50	05 08
28	20 31 09	08 36 42	01 ♎ 18 44	07 ♎ 16 57	14 27	22 53	05 29	01 37	03 17	04 51	27 52	05 10
29	20 35 06	09 37 38	13 ♎ 17 32	19 ♎ 21 05	15 39	24 08	05 35	01 51	03 22	04 55	27 54	05 11
30	20 39 03	10 38 34	25 ♎ 28 13	01 ♏ 39 32	16 52	25 23	05 40	02 05	03 28	04 58	27 56	05 12
31	20 42 59	11 ♒ 39 29	07 ♏ 55 41	14 ♏ 17 16	18 ♑ 07	26 ♑ 38	05 ♎ 44	02 ♒ 19	03 ♈ 33	05 ♒ 02	27 ♑ 59	05 ♐ 14

Date	Moon True ☊ ° '	Moon Mean ☊ ° '	Moon ☽ Latitude ° '
01	02 ♎ 26	03 ♎ 02	00 N 12
02	02 R 26	02 59	01 16
03	02 25	02 55	02 18
04	02 22	02 52	03 15
05	02 17	02 49	04 03
06	02 09	02 46	04 39
07	01 58	02 43	04 59
08	01 47	02 39	04 59
09	01 36	02 36	04 40
10	01 26	02 33	04 00
11	01 18	02 30	03 04
12	01 14	02 27	01 57
13	01 12	02 24	00 N 42
14	01 D 12	02 20	00 S 33
15	01 12	02 17	01 44
16	01 R 12	02 14	02 48
17	01 10	02 11	03 42
18	01 06	02 08	04 22
19	00 58	02 05	04 50
20	00 49	02 01	05 03
21	00 38	01 58	05 01
22	00 26	01 55	04 47
23	00 14	01 52	04 19
24	00 04	01 49	03 40
25	29 ♍ 57	01 45	02 52
26	29 51	01 42	01 56
27	29 49	01 39	00 S 55
28	29 D 48	01 36	00 N 08
29	29 49	01 33	01 12
30	29 50	01 30	02 14
31	29 ♍ 51	01 ♎ 26	03 N 11

DECLINATIONS

Date	Sun ☉ ° '	Moon ☽ ° '	Mercury ☿ ° '	Venus ♀ ° '	Mars ♂ ° '	Jupiter ♃ ° '	Saturn ♄ ° '	Uranus ♅ ° '	Neptune ♆ ° '	Pluto ♇ ° '
01	22 S 58	01 S 42	20 S 24	22 S 14	02 N 40	21 S 24	01 S 37	19 S 58	20 S 22	08 S 50
02	22 53	05 29	20 16	22 22	02 34	21 22	01 36	19 57	20 22	08 50
03	22 47	09 08	20 11	22 30	02 27	21 19	01 34	19 57	20 21	08 50
04	22 41	12 27	20 06	22 37	02 21	21 17	01 33	19 56	20 21	08 50
05	22 34	15 15	20 04	22 43	02 15	21 14	01 31	19 55	20 20	08 50
06	22 27	17 17	20 02	22 48	02 09	21 12	01 30	19 54	20 20	08 50
07	22 20	18 20	20 03	22 53	02 03	21 09	01 28	19 53	20 20	08 50
08	22 12	18 10	20 04	22 57	01 57	21 06	01 26	19 53	20 19	08 51
09	22 03	16 46	20 07	23 01	01 52	21 04	01 25	19 52	20 19	08 51
10	21 54	14 11	20 12	23 04	01 47	21 01	01 23	19 51	20 18	08 51
11	21 45	10 39	20 17	23 06	01 42	20 58	01 21	19 50	20 18	08 51
12	21 35	06 30	20 24	23 07	01 37	20 56	01 19	19 49	20 17	08 51
13	21 25	02 S 02	20 31	23 08	01 32	20 53	01 18	19 49	20 17	08 51
14	21 15	02 N 26	20 39	23 07	01 28	20 50	01 16	19 48	20 17	08 51
15	21 04	06 39	20 47	23 07	01 23	20 48	01 14	19 47	20 16	08 51
16	20 52	10 25	20 56	23 05	01 19	20 45	01 12	19 46	20 16	08 51
17	20 40	13 34	21 04	23 03	01 15	20 42	01 10	19 45	20 15	08 51
18	20 28	16 00	21 13	23 00	01 12	20 39	01 08	19 44	20 15	08 51
19	20 16	17 35	21 21	22 56	01 08	20 36	01 06	19 44	20 14	08 51
20	20 03	18 19	21 29	22 52	01 05	20 33	01 04	19 43	20 14	08 51
21	19 49	18 11	21 37	22 47	01 02	20 30	01 02	19 42	20 13	08 51
22	19 36	17 13	21 44	22 41	01 00	20 28	01 00	19 41	20 13	08 51
23	19 22	15 29	21 50	22 35	00 57	20 25	00 58	19 40	20 13	08 51
24	19 07	13 06	21 56	22 28	00 55	20 22	00 56	19 39	20 12	08 51
25	18 53	10 12	22 00	22 20	00 53	20 19	00 53	19 39	20 12	08 51
26	18 38	06 54	22 04	22 12	00 51	20 16	00 51	19 38	20 11	08 51
27	18 22	03 N 19	22 08	22 02	00 50	20 13	00 49	19 37	20 11	08 51
28	18 06	00 S 24	22 10	21 53	00 48	20 10	00 47	19 36	20 10	08 51
29	17 50	04 08	22 11	21 42	00 47	20 07	00 44	19 35	20 10	08 51
30	17 34	07 46	22 11	21 31	00 47	20 04	00 42	19 34	20 10	08 51
31	17 S 17	11 S 08	22 S 10	21 S 19	00 N 46	20 S 01	00 S 40	19 S 34	20 S 09	08 S 51

ZODIAC SIGN ENTRIES

Date	h	m	Planets
01	02	32	☽ ♎
03	08	10	♂ ♎
03	13	02	☽ ♏
05	19	27	☽ ♐
07	21	55	☽ ♑
09	22	00	☽ ♒
10	05	32	♀ ♑
11	21	51	☽ ♓
13	23	22	☽ ♈
16	03	40	☽ ♉
18	10	53	☽ ♊
20	00	43	☉ ♒
20	20	29	☽ ♋
21	15	13	♃ ♒
23	07	50	☽ ♌
25	20	26	☽ ♍
28	09	21	☽ ♎
30	20	48	☽ ♏

LATITUDES

Date	Mercury ☿ ° '	Venus ♀ ° '	Mars ♂ ° '	Jupiter ♃ ° '	Saturn ♄ ° '	Uranus ♅ ° '	Neptune ♆ ° '	Pluto ♇ ° '
01	02 N 28	00 N 46	02 N 39	00 S 20	02 S 21	00 S 34	00 N 26	12 N 24
04	03 04	00 38	02 43	00 20	02 21	00 34	00 26	12 24
07	03 18	00 30	02 47	00 20	02 20	00 34	00 26	12 25
10	03 12	00 23	02 51	00 21	02 19	00 34	00 25	12 25
13	02 53	00 15	02 55	00 21	02 19	00 34	00 25	12 26
16	02 27	00 N 07	02 59	00 21	02 18	00 34	00 25	12 27
19	01 58	00 S 01	03 03	00 21	02 18	00 34	00 25	12 27
22	01 28	00 08	03 07	00 22	02 17	00 34	00 25	12 28
25	00 58	00 16	03 11	00 22	02 17	00 34	00 25	12 29
28	00 30	00 23	03 15	00 22	02 16	00 34	00 25	12 30
31	00 N 03	00 S 30	03 N 19	00 S 23	02 S 16	00 S 34	00 N 25	12 N 31

DATA

Julian Date	2450450
Delta T	+62 seconds
Ayanamsa	23° 48' 56"
Synetic vernal point	05° ♓ 18' 03"
True obliquity of ecliptic	23° 26' 13"

MOON'S PHASES, APSIDES AND POSITIONS ☽

Date	h	m	Phase	Longitude ° '	Eclipse Indicator
02	01	45	☾	11 ♎ 42	
09	04	26	●	18 ♑ 57	
15	20	02	☽	25 ♈ 44	
23	15	11	○	03 ♌ 40	
31	19	40	☾	11 ♏ 59	

Day	h	m		
10	08	44	Perigee	
25	16	29	Apogee	
01	01	17	0S	
07	21	04	Max dec	18° S 25'
13	22	50	0N	
20	20	06	Max dec	18° N 22'
28	09	28	0S	

LONGITUDES

Date	Chiron ⚷ ° '	Ceres ⚳ ° '	Pallas ⚴ ° '	Juno ⚵ ° '	Vesta ⚶ ° '	Black Moon Lilith ⚸ ° '
01	00 ♏ 15	19 ♑ 04	27 ♐ 43	14 ♈ 51	27 ♑ 25	21 ♌ 29
11	00 ♏ 59	23 ♑ 03	01 ♑ 48	18 ♈ 39	02 ♒ 38	22 ♌ 36
21	01 ♏ 32	27 ♑ 02	05 ♑ 47	22 ♈ 52	07 ♒ 50	23 ♌ 42
31	01 ♏ 54	01 ♒ 00	09 ♑ 41	27 ♈ 26	13 ♒ 02	24 ♌ 49

All ephemeris data is given at 12.00 UT and the Moon's longitude is additionally given for 24.00 UT

ASPECTARIAN

h m	Aspects
01 Wednesday	
01 02	☽ ☌ ♂
05 13	☽ ☍ ♄
09 06	☽ △ ♅
11 19	☽ ⚹ ♇
11 31	☽ ∥ ♄
17 00	☿ ∥ ♆
17 01	♀ ⊥ ♃
17 09	☽ Q ♀
17 58	☽ ⋕ ♂
02 Thursday	
01 22	☉ ☌ ☿
01 41	☽ □ ☿
01 45	☽ □ ☉
16 51	☽ ∠ ♇
19 17	☽ ⚹ ♀
23 05	♀ ⊥ ♆
23 10	☿ ⊥ ♇
03 Friday	
04 50	☽ □ ♃
07 11	☽ □ ♆
09 03	☽ Q ☿
10 00	☽ ∥ ♇
10 08	☽ ⊥ ♇
13 09	☽ ⊻ ♂
15 29	☽ Q ☉
15 48	☽ ⊼ ♄
19 28	☽ □ ♅
21 26	☽ ⊻ ♇
04 Saturday	
00 40	☽ ⊥ ♂
02 45	☽ ∠ ♀
03 03	☽ ± ♄
05 26	☽ ⚹ ☿
15 00	☽ Q ♃
15 34	☽ ⚹ ☉
16 44	☽ Q ♆
17 35	☽ ∠ ♂
19 35	☽ ⚼ ♄
20 06	☉ ∥ ♀
20 56	☽ ⊥ ♀
05 Sunday	
04 29	☽ Q ♅
06 17	☽ ∠ ☿
08 45	☽ ⊻ ♀
12 50	☽ ⚹ ♃
14 12	☽ ⚹ ♆
20 25	☽ ∠ ☉
20 49	☽ ⚹ ♂
20 54	☽ ⊥ ☿
22 14	☽ △ ♄
06 Monday	
01 37	☽ ⚹ ♅
03 21	☽ ☌ ♇
06 31	☽ ⊻ ☿
13 06	☽ ⊥ ☉
15 10	☽ ∠ ♃
16 07	☽ ∠ ♆
17 49	☽ Q ♂
07 Tuesday	
00 00	☽ ⊻ ☉
03 09	☽ ∠ ♅
06 30	☽ ⊥ ♃
07 13	☽ ⊥ ♆
14 20	♀ ⊻ ♃
16 33	☽ ⊻ ♃
16 43	☽ ☌ ♀
17 08	☽ ⊻ ♆
18 04	☽ ⊥ ♅
21 39	♀ ⊻ ♆
08 Wednesday	
00 13	☽ □ ♂
00 43	☽ □ ♄
03 53	☽ ⊻ ♅
05 25	☽ ⊻ ♇
05 27	☽ ☌ ☿
06 12	☿ ⊻ ♇
09 09	♀ ⊥ ♅
15 07	☽ ⊥ ♇
09 Thursday	
04 26	☽ ☌ ☉
05 33	☽ ∠ ♇
05 47	☽ Q ♄
10 53	♂ ☍ ♄
11 39	♃ ☌ ♆
15 38	☿ ⊻ ♅
17 29	☽ ☌ ♆
17 33	☽ ☌ ♃
21 19	☽ ⊻ ♀
21 22	☉ ∠ ♇
10 Friday	
00 56	☽ ⚹ ♄
01 08	☽ △ ♂
01 42	☉ Q ♄
03 35	☽ ⊻ ☿
04 00	☽ ☌ ♅
05 25	☽ ⚹ ♇
07 42	☽ ⊥ ♀
12 52	☽ ⊥ ☿
23 18	☽ ∠ ♀
11 Saturday	
00 32	☽ Q ♇
00 52	☽ ∠ ♄
01 24	☽ ⚼ ♂
02 55	☽ ∠ ☿
07 35	☽ ⊻ ☉
13 58	☿ ∥ ♆
15 20	☽ ⊥ ♄
16 03	☽ ± ♂
17 23	☽ ⊻ ♆
17 55	☽ ⊥ ☉
18 06	☽ ⊻ ♃
19 16	♀ □ ♄
22 46	☽ ∥ ♇
12 Sunday	
01 03	☽ ⊻ ♄
01 33	☽ ⚹ ♀
01 54	☽ ⊼ ♂
02 43	☽ ⚹ ☿
03 05	☽ ⊥ ♆
03 57	☽ ⊥ ♃
04 09	☽ ⊻ ♅
05 31	☽ □ ♇
06 36	♀ □ ♂
09 34	☽ ∠ ☉
13 58	☽ ⊥ ♅
14 31	☿ ☌ ♀
17 48	☽ ∠ ♆
18 52	☽ ∠ ♃
20 42	☿ St D
22 15	☽ Q ☿
22 59	☽ Q ♀
13 Monday	
04 51	☽ ∠ ♅
09 41	♀ ⊻ ♅
12 11	☽ ⚹ ☉
14 42	☽ ⋕ ♂
15 58	☽ ∥ ♄
18 49	☽ ⚹ ♆
20 16	☽ ⚹ ♃
14 Tuesday	
01 25	♀ ⊻ ♇
02 58	☽ ☌ ♄
04 30	☽ ☍ ♂
04 37	☽ □ ☿
05 40	☽ ⋕ ♄
06 12	☽ ⚹ ♅
06 46	☽ ∥ ♂
07 32	☽ △ ♇
08 07	☽ □ ♀
10 04	☽ Q ☉
15 19	☽ Q ♆
17 06	☽ Q ♃
15 Wednesday	
03 00	☽ Q ♅
09 36	☽ ⚼ ♇
13 03	☿ ∥ ♃
20 02	☽ □ ☉
23 00	☽ □ ♆
16 Thursday	
01 19	☽ □ ♃
01 35	☽ ⋕ ♇
01 36	☽ ± ♇
07 47	☽ ⊻ ♄
07 59	☉ ∥ ☿
10 02	☽ ⊼ ♂
10 31	☽ △ ☿
11 08	☽ □ ♅
12 26	☽ ⊼ ♇
18 28	☽ △ ♀
18 42	☽ ⊥ ♄
21 06	☽ ± ♂
17 Friday	
05 59	☿ ⊻ ♅
08 01	☉ ∥ ♃
11 19	☽ ∠ ♄
12 34	☉ ☌ ♆
13 56	☽ ⚼ ♂
15 05	☽ ⚼ ☿
18 Saturday	
01 06	☽ ⚼ ♀
06 08	☽ △ ♆
07 36	☽ △ ☉
08 42	☽ ± ☿
09 27	☽ △ ♃
11 36	☿ ⊻ ♇
15 32	☽ ⚹ ♄
18 29	☽ △ ♂
18 59	☽ △ ♅
19 54	☽ ± ♀
20 12	☽ ☍ ♇
20 39	☽ ⊼ ☿
23 00	♀ ⊥ ♇
19 Sunday	
08 35	☽ ⊼ ♀
10 41	☽ ⚼ ♆
13 07	☉ ☌ ♃
14 31	☽ ⚼ ♃
14 32	☽ Q ♄
14 36	☽ ⚼ ☉
14 39	☉ ∥ ♆
23 49	☽ ⚼ ♅
20 Monday	
04 03	☽ ± ♆
08 13	☽ ± ♃
09 32	☽ ± ☉
15 44	☽ ⊼ ♆
17 21	☽ ± ♅
18 56	♂ △ ♅
20 07	☽ ⊼ ♃
22 16	☽ ⊼ ☉
21 Tuesday	
01 39	☽ □ ♄
05 08	☽ ⊼ ♅
05 14	☽ □ ♂
06 14	☽ ⊼ ♇
10 28	☽ ☍ ☿
18 04	☽ ± ♇
22 Wednesday	
01 37	☽ ☍ ♀
02 08	☉ ∥ ♅
04 24	♂ ⋕ ♄
11 55	☽ ⚼ ♇
17 19	☽ Q ♂
18 31	☉ ⚹ ♄
23 Thursday	
03 09	☽ ☍ ♆
08 39	☽ ☍ ♃
13 33	☽ △ ♄
15 11	☽ ☍ ☉
17 00	☽ ☍ ♅
17 38	☽ ⚹ ♂
17 58	☽ △ ♇
24 Friday	
03 15	☽ ⊼ ☿
13 53	☉ ☌ ♅
16 30	☽ ± ☿
20 01	☽ ⚼ ♄
20 45	☽ ⊼ ♀
25 Saturday	
00 17	☽ ∠ ♂
00 25	☉ △ ♂
00 39	☉ ⚹ ♇
02 31	♂ ⚹ ♇
10 21	☽ ± ♀
12 19	☿ ⊥ ♇
12 31	☽ ⚼ ☿
14 24	☽ ± ♄
15 53	☽ ⊼ ♆
18 46	☽ ⊥ ♂
22 07	☽ ⋕ ♇
22 30	☽ ⊼ ♃
26 Sunday	
02 41	☽ ⊼ ♄
04 07	☽ ± ♆
06 03	☽ ⊼ ♅
06 49	☽ ⚼ ♀
06 50	☽ □ ♇
06 59	♀ ∠ ♇
07 06	☽ ⊻ ♂
08 58	♂ ⋕ ♄
09 37	☽ ⊼ ☉
10 57	☽ ± ♃
18 19	☽ ± ♅
22 09	☽ △ ☿
22 29	☽ ⚼ ♆
22 58	☽ ± ☉
27 Monday	
01 50	☿ ∥ ♀
03 02	♀ Q ♄
05 38	☽ ⚼ ♃
12 42	☽ ⚼ ♅
16 56	☽ △ ♀
18 58	☽ ⚼ ☉
19 30	☽ Q ♇
28 Tuesday	
04 15	☽ ∥ ♂
04 24	☽ ⋕ ♄
05 01	☽ △ ♆
05 09	♃ ∥ ♆
12 37	☽ △ ♃
14 24	☽ ∥ ♄
14 37	☽ ⋕ ♂
16 00	☽ ☍ ♄
19 10	☽ △ ♅
19 46	☽ ⚹ ♇
20 28	☽ ☌ ♂
29 Wednesday	
04 01	☽ △ ☉
17 12	☽ □ ☿
30 Thursday	
01 40	☽ ∠ ♇
11 49	☽ □ ♀
16 49	☽ □ ♆
19 16	☽ ⊥ ♇
19 27	☽ ∥ ♇
31 Friday	
01 04	☽ □ ♃
03 35	☽ ⊼ ♄
06 27	☽ □ ♅
06 50	☽ ⊻ ♇
07 48	☽ ⊻ ♂
08 11	☽ Q ☿
15 06	☽ ± ♄
19 15	☽ ⊥ ♂
19 40	☽ □ ☉

FEBRUARY 1997

LONGITUDES

Date	Sidereal time h m s	Sun ☉ ° ' "	Moon ☽ ° ' "	Moon ☽ 24.00 ° ' "	Mercury ☿ ° '	Venus ♀ ° '	Mars ♂ ° '	Jupiter ♃ ° '	Saturn ♄ ° '	Uranus ♅ ° '	Neptune ♆ ° '	Pluto ♇ ° '
01	20 46 56	12 ♒ 40 23	20 ♏ 44 51	27 ♏ 18 57	19 ♑ 24	27 ♑ 53	05 ♎ 48	02 ♒ 33	03 ♈ 39	05 ♒ 05	28 ♑ 01	05 ♐ 15
02	20 50 52	13 41 17	04 ♐ 00 00	10 ♐ 48 18	20 42	29 ♑ 08	05 51	02 47	03 44	05 09	28 03	05 16
03	20 54 49	14 42 10	17 ♐ 44 02	24 ♐ 47 11	22 01	00 ♒ 24	05 53	03 01	03 50	05 12	28 05	05 17
04	20 58 45	15 43 02	01 ♑ 57 36	09 ♑ 14 50	23 22	01 39	05 55	03 15	03 56	05 16	28 07	05 18
05	21 02 42	16 43 53	16 ♑ 38 17	24 ♑ 07 07	24 44	02 54	05 55	03 29	04 02	05 19	28 10	05 19
06	21 06 38	17 44 43	01 ♒ 40 17	09 ♒ 16 36	26 07	04 09	05 R 55	03 43	04 08	05 23	28 12	05 20
07	21 10 35	18 45 32	16 ♒ 54 29	24 ♒ 33 21	27 31	05 24	05 55	03 57	04 14	05 26	28 14	05 21
08	21 14 32	19 46 19	02 ♓ 11 04	09 ♓ 46 35	28 ♑ 56	06 39	05 53	04 11	04 20	05 30	28 16	05 22
09	21 18 28	20 47 05	17 ♓ 18 43	24 ♓ 46 28	00 ♒ 22	07 54	05 51	04 25	04 26	05 33	28 18	05 23
10	21 22 25	21 47 50	02 ♈ 08 58	09 ♈ 25 36	01 49	09 09	05 48	04 39	04 32	05 37	28 20	05 24
11	21 26 21	22 48 33	16 ♈ 35 54	23 ♈ 39 39	03 17	10 25	05 44	04 52	04 38	05 40	28 23	05 25
12	21 30 18	23 49 14	00 ♉ 36 45	07 ♉ 27 17	04 46	11 40	05 39	05 06	04 45	05 43	28 25	05 26
13	21 34 14	24 49 54	14 ♉ 11 25	20 ♉ 49 27	06 16	12 55	05 34	05 20	04 51	05 47	28 27	05 27
14	21 38 11	25 50 32	27 ♉ 21 45	03 ♊ 48 43	07 47	14 10	05 28	05 34	04 57	05 50	28 29	05 28
15	21 42 07	26 51 09	10 ♊ 10 45	16 ♊ 28 26	09 18	15 25	05 21	05 47	05 04	05 54	28 31	05 28
16	21 46 04	27 51 44	22 ♊ 42 05	28 ♊ 52 13	10 51	16 40	05 13	06 01	05 10	05 57	28 33	05 29
17	21 50 01	28 52 17	04 ♋ 59 15	11 ♋ 03 35	12 25	17 55	05 05	06 15	05 17	06 00	28 35	05 30
18	21 53 57	29 ♒ 52 48	17 ♋ 05 37	23 ♋ 05 42	13 59	19 10	04 56	06 28	05 23	06 03	28 37	05 30
19	21 57 54	00 ♓ 53 17	29 ♋ 04 10	05 ♌ 01 18	15 34	20 25	04 46	06 42	05 30	06 07	28 39	05 31
20	22 01 50	01 53 45	10 ♌ 57 25	16 ♌ 52 46	17 11	21 40	04 35	06 55	05 36	06 10	28 41	05 32
21	22 05 47	02 54 11	22 ♌ 47 35	28 ♌ 42 08	18 48	22 55	04 23	07 09	05 43	06 13	28 43	05 32
22	22 09 43	03 54 35	04 ♍ 36 38	10 ♍ 31 20	20 26	24 10	04 11	07 22	05 50	06 17	28 45	05 33
23	22 13 40	04 54 58	16 ♍ 26 29	22 ♍ 22 21	22 05	25 25	03 58	07 35	05 57	06 20	28 47	05 33
24	22 17 36	05 55 19	28 ♍ 19 12	04 ♎ 17 19	23 45	26 40	03 44	07 49	06 04	06 23	28 49	05 34
25	22 21 33	06 55 38	10 ♎ 17 04	16 ♎ 18 46	25 26	27 55	03 30	08 02	06 10	06 26	28 51	05 34
26	22 25 30	07 55 56	22 ♎ 22 47	28 ♎ 29 33	27 08	29 ♒ 10	03 14	08 15	06 17	06 29	28 53	05 34
27	22 29 26	08 56 12	04 ♏ 39 29	10 ♏ 53 00	28 ♒ 51	00 ♓ 25	02 58	08 28	06 24	06 32	28 54	05 35
28	22 33 23	09 ♓ 56 27	17 ♏ 10 34	23 ♏ 32 38	00 ♓ 35	01 ♓ 40	02 ♎ 42	08 ♒ 42	06 ♈ 31	06 ♒ 35	28 ♑ 56	05 ♐ 35

MOON

Date	Moon True ☊ ° '	Moon Mean ☊ ° '	Moon ☽ Latitude ° '
01	29 ♍ 51	01 ♎ 23	04 N 01
02	29 R 49	01 20	04 39
03	29 45	01 17	05 03
04	29 39	01 14	05 09
05	29 32	01 11	04 56
06	29 25	01 07	04 22
07	29 19	01 04	03 30
08	29 15	01 01	02 22
09	29 13	00 58	01 N 04
10	29 D 12	00 55	00 S 16
11	29 13	00 51	01 34
12	29 15	00 48	02 43
13	29 16	00 45	03 41
14	29 R 16	00 42	04 26
15	29 15	00 39	04 56
16	29 13	00 36	05 11
17	29 09	00 32	05 11
18	29 04	00 29	04 58
19	28 59	00 26	04 31
20	28 54	00 23	03 53
21	28 50	00 20	03 05
22	28 47	00 17	02 09
23	28 45	00 13	01 07
24	28 44	00 10	00 S 02
25	28 D 44	00 07	01 N 03
26	28 46	00 04	02 07
27	28 47	00 ♎ 01	03 06
28	28 ♍ 49	29 ♍ 57	03 N 57

DECLINATIONS

Date	Sun ☉ ° '	Moon ☽ ° '	Mercury ☿ ° '	Venus ♀ ° '	Mars ♂ ° '	Jupiter ♃ ° '	Saturn ♄ ° '	Uranus ♅ ° '	Neptune ♆ ° '	Pluto ♇ ° '
01	17 S 00	14 S 04	22 S 08	21 S 07	00 N 46	19 S 57	00 S 37	19 S 33	20 S 09	08 S 51
02	16 43	16 23	22 05	20 54	00 46	19 54	00 35	19 32	20 08	08 50
03	16 25	17 51	22 00	20 40	00 47	19 51	00 33	19 31	20 08	08 50
04	16 07	18 16	21 55	20 26	00 47	19 48	00 30	19 30	20 07	08 50
05	15 49	17 30	21 48	20 11	00 48	19 45	00 28	19 29	20 07	08 50
06	15 31	15 31	21 40	19 55	00 49	19 42	00 25	19 29	20 07	08 50
07	15 12	12 26	21 31	19 39	00 51	19 39	00 23	19 28	20 06	08 50
08	14 53	08 29	21 20	19 23	00 52	19 35	00 20	19 27	20 06	08 50
09	14 34	04 S 01	21 08	19 05	00 54	19 32	00 18	19 26	20 05	08 49
10	14 14	00 N 37	20 55	18 48	00 57	19 29	00 15	19 25	20 05	08 49
11	13 55	05 05	20 41	18 29	00 59	19 26	00 12	19 24	20 04	08 49
12	13 35	09 08	20 25	18 10	01 02	19 23	00 10	19 24	20 04	08 49
13	13 15	12 34	20 08	17 51	01 05	19 19	00 07	19 23	20 04	08 49
14	12 54	15 15	19 50	17 31	01 09	19 16	00 05	19 22	20 03	08 48
15	12 34	17 06	19 31	17 11	01 13	19 13	00 S 02	19 21	20 03	08 48
16	12 13	18 04	19 10	16 50	01 16	19 10	00 N 01	19 20	20 02	08 48
17	11 52	18 10	18 48	16 29	01 21	19 06	00 03	19 20	20 02	08 48
18	11 31	17 25	18 24	16 07	01 25	19 03	00 06	19 19	20 01	08 47
19	11 09	15 55	17 59	15 45	01 30	19 00	00 09	19 18	20 01	08 47
20	10 48	13 45	17 33	15 22	01 35	18 56	00 12	19 17	20 01	08 47
21	10 26	11 00	17 06	14 59	01 40	18 53	00 14	19 16	20 00	08 47
22	10 04	07 49	16 37	14 35	01 46	18 50	00 17	19 16	20 00	08 46
23	09 42	04 19	16 07	14 11	01 52	18 47	00 20	19 15	20 00	08 46
24	09 20	00 N 38	15 35	13 47	01 58	18 43	00 23	19 14	19 59	08 46
25	08 58	03 S 06	15 02	13 22	02 04	18 40	00 25	19 13	19 59	08 46
26	08 36	06 45	14 28	12 57	02 11	18 37	00 28	19 12	19 58	08 45
27	08 13	10 09	13 53	12 32	02 17	18 33	00 31	19 12	19 58	08 45
28	07 S 50	13 S 10	13 S 16	12 S 06	02 N 24	18 S 30	00 N 34	19 S 11	19 S 58	08 S 45

ZODIAC SIGN ENTRIES

Date	h m	Planets
02	04 51	☽ ♐
03	04 28	♀ ♒
04	08 44	☽ ♑
06	09 21	☽ ♒
08	08 34	☽ ♓
09	05 53	☿ ♒
10	08 29	☽ ♈
12	10 56	☽ ♉
14	16 53	☽ ♊
17	02 13	☽ ♋
18	14 51	☉ ♓
19	13 52	☽ ♌
22	02 38	☽ ♍
24	15 23	☽ ♎
27	02 57	☽ ♏
27	04 01	♀ ♓
28	03 54	☿ ♓

LATITUDES

Date	Mercury ☿ ° '	Venus ♀ ° '	Mars ♂ ° '	Jupiter ♃ ° '	Saturn ♄ ° '	Uranus ♅ ° '	Neptune ♆ ° '	Pluto ♇ ° '
01	00 S 06	00 S 33	03 N 21	00 S 23	02 S 15	00 S 34	00 N 25	12 N 31
04	00 30	00 39	03 25	00 23	02 15	00 34	00 25	12 32
07	00 52	00 46	03 28	00 23	02 15	00 34	00 25	12 33
10	01 12	00 52	03 32	00 24	02 14	00 34	00 25	12 34
13	01 29	00 57	03 36	00 24	02 14	00 34	00 25	12 35
16	01 43	01 02	03 39	00 24	02 13	00 34	00 25	12 36
19	01 55	01 07	03 42	00 25	02 13	00 34	00 25	12 37
22	02 03	01 11	03 44	00 25	02 13	00 34	00 25	12 38
25	02 07	01 15	03 46	00 25	02 12	00 35	00 25	12 39
28	02 08	01 18	03 47	00 26	02 12	00 35	00 25	12 40
31	02 S 06	01 S 21	03 N 48	00 S 26	02 S 12	00 S 35	00 N 25	12 N 41

DATA

Julian Date	2450481
Delta T	+62 seconds
Ayanamsa	23° 49' 01"
Synetic vernal point	05° ♓ 17' 59"
True obliquity of ecliptic	23° 26' 13"

MOON'S PHASES, APSIDES AND POSITIONS ☽

Date	h m	Phase	Longitude ° '	Eclipse Indicator
07	15 06	●	18 ♒ 53	
14	08 58	☽	25 ♉ 43	
22	10 27	○	03 ♍ 51	

Day	h m		
07	20 33	Perigee	
21	16 52	Apogee	
04	08 52	Max dec	18° S 17'
10	08 50	ON	
17	02 40	Max dec	18° N 13'
24	16 04	OS	

LONGITUDES

Date	Chiron ⚷ ° '	Ceres ⚳ ° '	Pallas ⚴ ° '	Juno ⚵ ° '	Vesta ⚶ ° '	Black Moon Lilith ⚸ ° '
01	01 ♏ 56	01 ♒ 24	10 ♑ 04	27 ♈ 55	13 ♒ 33	24 ♌ 56
11	02 ♏ 05	05 ♒ 19	13 ♑ 49	02 ♉ 47	18 ♒ 43	26 ♌ 03
21	02 ♏ 02	09 ♒ 12	17 ♑ 25	07 ♉ 53	23 ♒ 51	27 ♌ 09
31	01 ♏ 48	13 ♒ 00	20 ♑ 49	13 ♉ 09	28 ♒ 55	28 ♌ 16

ASPECTARIAN

h m Aspects

01 Saturday
02 01 ☽ Q ♀
03 12 ☽ Q ♆
08 05 ☽ ⚼ ♄
09 14 ☽ ⚹ ☿
11 38 ☽ Q ♃
12 06 ☽ ∠ ♂
14 27 ♀ ☌ ♆
16 19 ☽ Q ♅

02 Sunday
01 18 ☽ ⚹ ♆
02 24 ☽ ⚹ ♀
03 54 ☿ ∠ ♇
07 32 ☽ Q ☉
09 48 ☽ ⚹ ♃
11 32 ☽ △ ♄
14 03 ☽ ⚹ ♅
14 15 ☽ ☌ ♇
15 17 ☽ ⚹ ♂
15 21 ☽ ∠ ☿
15 45 ☽ ∥ ☉

03 Monday
03 57 ☽ ∠ ♆
06 22 ☽ ⚹ ☉
07 34 ☽ ∠ ♀
08 22 ☿ Q ♄
08 45 ☽ ⊥ ☿
12 15 ☽ Q ♂
12 30 ☽ ∠ ♃
16 15 ☽ ∠ ♅
19 27 ☽ ⊥ ♆
20 05 ☽ ⊻ ☿

04 Tuesday
00 26 ☽ ⊥ ♀
04 01 ☽ ⊥ ♃
05 35 ☽ ⊻ ♆
07 30 ☽ ⊥ ♅
09 47 ☽ ∠ ☉
11 26 ☽ ⊻ ♀
14 11 ☽ ⊻ ♃
15 17 ☽ □ ♄
17 29 ☽ ⊻ ♅
17 32 ☽ ⊻ ♇
18 32 ☽ □ ♂

05 Wednesday
01 43 ☽ ⊥ ☉
03 22 ☽ ⊥ ♇
11 21 ♅ ⚹ ♇
12 10 ☽ ⊻ ☉
17 56 ☽ ∠ ♇
18 18 ♀ ∥ ♆
20 43 ☽ Q ♄

06 Thursday
00 37 ♂ St R
01 48 ♀ ☌ ♃
02 14 ☉ Q ♇
02 17 ☽ ☌ ☿
06 29 ☽ ☌ ♆
11 34 ♀ ⚹ ♄
12 07 ☽ ∥ ☉
15 17 ☽ ☌ ♃
15 55 ☽ ⚹ ♄
16 16 ☽ ☌ ♀
17 48 ☽ ⚹ ♇
17 53 ☽ ☌ ♅
18 43 ☽ △ ♂

07 Friday
11 07 ♀ ⚹ ♇
12 42 ☽ Q ♇
12 44 ♀ ☌ ♅
13 12 ♀ ∥ ♃
15 06 ☽ ☌ ☉
15 40 ☽ ∠ ♄
18 16 ☽ ⚼ ♂
21 33 ♀ △ ♂

08 Saturday
00 22 ☉ ∠ ♄
00 31 ☿ ☌ ♆
05 35 ♀ ∥ ♅
05 49 ☽ ⊻ ♆
05 53 ☽ ⊥ ♄
06 21 ☽ ⊻ ☿
08 23 ☽ ± ♂
10 06 ☽ ∥ ♇
15 12 ☽ ⊻ ♃
15 18 ☽ ⊥ ♆
15 24 ☽ ⊻ ♄
16 47 ☽ ⊥ ☿
17 02 ☽ □ ♇
17 15 ☽ ⊻ ♅
17 50 ☽ ⊼ ♂
19 41 ☽ ⊻ ♀

09 Sunday
00 50 ☽ ⊥ ♃
02 47 ☽ ⊥ ♅
05 35 ☽ ∠ ♆
06 04 ☽ ⊥ ♀
08 34 ☽ ∠ ☿
13 24 ☉ ⚼ ♂
15 25 ☽ ∠ ♃
15 33 ♃ ⚹ ♄
17 13 ☽ ∠ ♅
17 58 ☽ ⊻ ☉
21 49 ☽ ∠ ♀

10 Monday
04 00 ☽ # ♂
04 23 ☽ ⊥ ☉
05 46 ☽ ⚹ ♆
07 29 ☽ ∥ ♄
10 09 ☽ # ♄
11 24 ☽ ⚹ ☿
13 46 ☽ ∥ ♂
15 56 ☽ ☌ ♄
16 10 ☽ ⚹ ♃
17 21 ☽ △ ♇
17 43 ☽ ⚹ ♅
17 58 ☽ ☍ ♂
20 13 ☽ ∠ ☉

11 Tuesday
00 39 ☽ ⚹ ♀
01 33 ☽ Q ♆
09 32 ☽ Q ☿
12 28 ☽ Q ♃
13 49 ☽ Q ♅
18 29 ☽ ⚼ ♇
22 50 ☽ Q ♀
23 22 ☽ ⚹ ☉

12 Wednesday
00 07 ♂ △ ♅
01 56 ♃ ∥ ♅
08 10 ☽ □ ♆
09 57 ☽ ± ♇
09 57 ☽ # ♇
11 33 ☿ ⚹ ♄
18 22 ☿ ☌ ♃
19 17 ☽ ⊻ ♄
19 59 ☽ □ ♃
20 09 ☽ □ ☿
20 27 ☽ ⊼ ♇
20 46 ☽ ⊼ ♂
20 59 ☽ □ ♅
21 51 ☽ Q ☉
22 48 ☿ ⚹ ♇

13 Thursday
01 29 ☿ △ ♂
03 56 ☿ ☌ ♅
05 58 ☽ ⊥ ♄
07 20 ☽ ± ♂
09 28 ☽ □ ♀
16 50 ☽ # ☉
18 45 ☿ ∥ ♆
22 18 ☽ ∠ ♄
23 27 ☽ ⚼ ♂

14 Friday
00 47 ♃ ⚹ ♇
05 07 ♂ △ ♃
08 58 ☽ □ ☉
12 54 ♂ ⚹ ♇
14 04 ☽ △ ♆

15 Saturday
02 16 ☽ ⚹ ♄
02 58 ☽ △ ♂
03 06 ☽ ☍ ♇
03 34 ☽ △ ♃
03 52 ☽ △ ♅
10 07 ☽ △ ☿
13 11 ☽ # ♀
18 22 ☽ ⚼ ♆
23 05 ☽ △ ♀
23 36 ☿ ∥ ♅

16 Sunday
01 14 ☽ Q ♄
02 22 ♃ ☌ ♅
08 36 ☽ ⚼ ♅
08 41 ☽ ⚼ ♃
11 42 ☽ ± ♆
12 18 ☿ ∥ ♃
17 14 ♂ ☍ ♄
19 00 ☽ ⚼ ☿
22 56 ☽ △ ☉
23 24 ☽ ⊼ ♆

17 Monday
02 10 ☽ ± ♅
02 31 ☽ ± ♃
03 52 ♀ Q ♇
04 54 ☉ ⊻ ♆
07 28 ☽ ⚼ ♀
12 11 ☽ □ ♂
12 35 ☽ □ ♄
13 00 ☽ ⊼ ♇
14 01 ☽ ⊼ ♅
14 31 ☽ ⊼ ♃
15 13 ☽ ± ☿
16 20 ☉ ± ♂
22 50 ☉ ⊥ ♄

18 Tuesday
00 53 ☽ ± ♇
03 17 ☽ ± ♀
04 53 ☽ ⊼ ☿
07 11 ☽ ⚼ ☉
16 37 ☽ ⊼ ♀
18 49 ☽ ⚼ ♇
23 30 ☽ Q ♂

19 Wednesday
00 53 ♀ ⚼ ♂
02 49 ☽ ± ☉
11 09 ☽ ☍ ♆
13 40 ♀ ∠ ♄
14 46 ☽ # ♀
16 00 ☽ ⊼ ☉
16 56 ♄ △ ♇
23 18 ☽ ⚹ ♂

20 Thursday
01 01 ☽ △ ♇
01 05 ☽ △ ♄
02 16 ☽ ☍ ♅
03 41 ☽ ☍ ♃
17 13 ☿ Q ♇

21 Friday
02 36 ☽ ☍ ☿
05 12 ☽ ∠ ♂
07 45 ☽ ⚼ ♄
12 17 ☽ ☍ ♀
17 04 ☽ # ☉
19 43 ☿ ⚼ ♂
23 10 ☽ ⊥ ♂

22 Saturday
00 04 ☽ ⊼ ♆
02 12 ☽ ± ♄
05 03 ☽ # ♇
10 27 ☽ ☍ ☉
11 09 ☽ ⊻ ♂
12 17 ☽ ± ♆
13 54 ☽ □ ♇
14 30 ☽ ⊼ ♄
15 24 ☽ ⊼ ♅
17 21 ☉ ⊼ ♂
17 42 ☽ ⊼ ♃
18 15 ☿ ∠ ♄

23 Sunday
03 38 ☽ ± ♅
06 07 ☽ ± ♃
06 36 ☽ ⚼ ♆
08 40 ☉ ⊥ ♆
21 56 ☽ ⚼ ♅

24 Monday
00 41 ☽ ⚼ ♃
01 17 ☽ ⊼ ☿
02 24 ☽ Q ♇
03 18 ☉ □ ♇
03 38 ☽ ∥ ♂
08 17 ☽ ⊼ ♀
13 00 ☽ △ ♆
13 38 ☽ ∥ ♄
15 21 ☽ ± ☿
15 41 ☉ ⊻ ♄
18 34 ☽ # ♄
21 46 ☽ ± ♀
22 40 ☽ ☌ ♂
23 35 ☉ ⊻ ♅

25 Tuesday
02 33 ☽ ⚹ ♇
03 42 ☽ ☍ ♄
04 16 ☽ △ ♅
04 40 ☽ ⊼ ☉
05 10 ☽ # ♂
05 11 ♀ ± ♂
07 25 ☽ △ ♃
12 21 ☽ ⚼ ☿
17 45 ☽ ± ☉
17 51 ☽ ⚼ ♀

26 Wednesday
01 29 ☉ ∥ ♇
06 16 ♀ ⊻ ♆
08 26 ☽ ∠ ♇
13 11 ☽ ⚼ ☉
13 15 ☿ ± ♂
21 53 ☉ ⊻ ♃
22 52 ☽ △ ☿
23 29 ☽ ∥ ☉

27 Thursday
00 47 ☽ □ ♆
01 52 ☽ ∥ ♇
02 07 ☽ ⊥ ♇
02 49 ☽ △ ♀
08 48 ☽ ⊻ ♂
11 45 ♀ ⊥ ♄
12 46 ☿ ⊻ ♆
13 47 ☽ ⊻ ♇
15 25 ☽ ⊼ ♄
15 39 ☽ □ ♅
19 30 ☽ □ ♃
20 09 ☽ ⊥ ♂
20 59 ☽ △ ☉

28 Friday
03 03 ☽ ± ♄
04 12 ☽ ∥ ♀
11 01 ☿ ⊥ ♄
11 33 ☽ Q ♆
12 41 ☽ ∥ ☿
12 58 ☽ ∠ ♂
20 17 ☽ ⚼ ♄

All ephemeris data is given at 12.00 UT and the Moon's longitude is additionally given for 24.00 UT

MARCH 1997

LONGITUDES

Date	Sidereal time h m s	Sun ☉ ° ' "	Moon ☽ ° ' "	Moon ☽ 24.00 ° ' "	Mercury ☿ ° '	Venus ♀ ° '	Mars ♂ ° '	Jupiter ♃ ° '	Saturn ♄ ° '	Uranus ♅ ° '	Neptune ♆ ° '	Pluto ♇ ° '
01	22 37 19	10 ♓ 56 41	29 ♏ 59 39	06 ♐ 32 02	02 ♓ 20	02 ♓ 55	02 ♎ 24	08 ♒ 55	06 ♈ 38	06 ♒ 38	28 ♑ 58	05 ♐ 35
02	22 41 16	11 56 53	13 ♐ 10 10	19 ♐ 54 21	04 06	04 10	02 R 07	09 08	06 45	06 42	29 00	05 36
03	22 45 12	12 57 03	26 ♐ 44 48	03 ♑ 41 41	05 54	05 25	01 48	09 21	06 52	06 45	29 02	05 36
04	22 49 09	13 57 12	10 ♑ 44 58	17 ♑ 54 31	07 42	06 40	01 29	09 34	06 59	06 48	29 03	05 36
05	22 53 05	14 57 19	25 ♑ 10 00	02 ♒ 30 57	09 31	07 55	01 09	09 46	07 07	06 50	29 05	05 36
06	22 57 02	15 57 25	09 ♒ 56 41	17 ♒ 26 23	11 22	09 09	00 49	09 59	07 14	06 53	29 07	05 36
07	23 00 59	16 57 29	24 ♒ 59 04	02 ♓ 33 37	13 13	10 24	00 28	10 12	07 21	06 56	29 09	05 36
08	23 04 55	17 57 31	10 ♓ 08 52	17 ♓ 43 37	15 05	11 39	00 ♎ 07	10 25	07 28	06 59	29 10	05 36
09	23 08 52	18 57 32	25 ♓ 16 38	02 ♈ 46 49	16 59	12 54	29 ♍ 45	10 37	07 36	07 02	29 12	05 R 36
10	23 12 48	19 57 30	10 ♈ 13 06	17 ♈ 34 36	18 54	14 09	29 23	10 50	07 43	07 05	29 13	05 36
11	23 16 45	20 57 26	24 ♈ 50 32	02 ♉ 00 22	20 49	15 24	29 01	11 02	07 50	07 08	29 15	05 36
12	23 20 41	21 57 21	09 ♉ 03 42	16 ♉ 00 17	22 46	16 39	28 38	11 15	07 57	07 10	29 17	05 36
13	23 24 38	22 57 13	22 ♉ 50 05	29 ♉ 33 09	24 43	17 53	28 15	11 27	08 05	07 13	29 18	05 36
14	23 28 34	23 57 03	06 ♊ 09 42	12 ♊ 40 02	26 41	19 08	27 52	11 39	08 12	07 16	29 20	05 36
15	23 32 31	24 56 51	19 ♊ 04 34	25 ♊ 23 43	28 ♓ 40	20 23	27 29	11 51	08 20	07 18	29 21	05 35
16	23 36 28	25 56 36	01 ♋ 37 59	07 ♋ 47 55	00 ♈ 39	21 38	27 06	12 04	08 27	07 21	29 23	05 35
17	23 40 24	26 56 19	13 ♋ 54 03	19 ♋ 56 55	02 38	22 52	26 42	12 16	08 34	07 24	29 24	05 35
18	23 44 21	27 56 00	25 ♋ 57 04	01 ♌ 55 02	04 38	24 07	26 19	12 28	08 42	07 26	29 25	05 34
19	23 48 17	28 55 39	07 ♌ 51 18	13 ♌ 46 23	06 38	25 22	25 55	12 39	08 49	07 29	29 27	05 34
20	23 52 14	29 ♓ 55 15	19 ♌ 40 43	25 ♌ 34 44	08 37	26 36	25 32	12 51	08 57	07 31	29 28	05 34
21	23 56 10	00 ♈ 54 50	01 ♍ 28 51	07 ♍ 23 24	10 36	27 51	25 08	13 03	09 04	07 34	29 30	05 33
22	00 00 07	01 54 22	13 ♍ 18 44	19 ♍ 15 11	12 33	29 ♓ 06	24 45	13 15	09 12	07 36	29 31	05 33
23	00 04 03	02 53 51	25 ♍ 13 00	01 ♎ 12 27	14 30	00 ♈ 20	24 22	13 26	09 19	07 38	29 32	05 32
24	00 08 00	03 53 19	07 ♎ 13 48	13 ♎ 17 15	16 25	01 35	23 59	13 38	09 27	07 41	29 33	05 32
25	00 11 57	04 52 45	19 ♎ 23 02	25 ♎ 31 20	18 17	02 50	23 37	13 49	09 34	07 43	29 35	05 31
26	00 15 53	05 52 09	01 ♏ 42 21	07 ♏ 56 17	20 08	04 04	23 14	14 00	09 42	07 45	29 36	05 31
27	00 19 50	06 51 31	14 ♏ 13 20	20 ♏ 33 41	21 55	05 19	22 52	14 11	09 49	07 47	29 37	05 30
28	00 23 46	07 50 51	26 ♏ 57 33	03 ♐ 25 07	23 39	06 33	22 31	14 23	09 57	07 50	29 38	05 29
29	00 27 43	08 50 09	09 ♐ 56 35	16 ♐ 32 08	25 19	07 48	22 10	14 34	10 04	07 52	29 39	05 29
30	00 31 39	09 49 26	23 ♐ 11 57	29 ♐ 56 11	26 55	09 02	21 49	14 44	10 12	07 54	29 40	05 28
31	00 35 36	10 ♈ 48 41	06 ♑ 44 57	13 ♑ 38 20	28 ♈ 27	10 ♈ 17	21 ♍ 29	14 ♒ 55	10 ♈ 19	07 ♒ 56	29 ♑ 41	05 ♐ 27

Date	Moon True ☊ ° '	Moon Mean ☊ ° '	Moon ☽ Latitude ° '
01	28 ♍ 50	29 ♍ 54	04 N 38
02	28 R 50	29 51	05 05
03	28 50	29 48	05 17
04	28 49	29 45	05 10
05	28 47	29 42	04 44
06	28 45	29 38	03 59
07	28 44	29 35	02 56
08	28 43	29 32	01 41
09	28 42	29 29	00 N 19
10	28 D 43	29 26	01 S 04
11	28 43	29 23	02 20
12	28 44	29 19	03 26
13	28 44	29 16	04 18
14	28 45	29 13	04 54
15	28 45	29 10	05 13
16	28 R 45	29 07	05 18
17	28 45	29 03	05 07
18	28 45	29 00	04 43
19	28 D 45	28 57	04 07
20	28 45	28 54	03 21
21	28 45	28 51	02 26
22	28 45	28 48	01 25
23	28 45	28 44	00 S 20
24	28 R 45	28 41	00 N 47
25	28 45	28 38	01 52
26	28 44	28 35	02 53
27	28 43	28 32	03 47
28	28 42	28 28	04 30
29	28 41	28 25	05 00
30	28 40	28 22	05 16
31	28 ♍ 40	28 ♍ 19	05 N 14

DECLINATIONS

Date	Sun ☉ ° '	Moon ☽ ° '	Mercury ☿ ° '	Venus ♀ ° '	Mars ♂ ° '	Jupiter ♃ ° '	Saturn ♄ ° '	Uranus ♅ ° '	Neptune ♆ ° '	Pluto ♇ ° '
01	07 S 28	15 S 37	12 S 38	11 S 40	02 N 32	18 S 27	00 N 37	19 S 10	19 S 57	08 S 44
02	07 05	17 20	11 58	11 14	02 39	18 23	00 40	19 09	19 57	08 44
03	06 42	18 07	11 18	10 47	02 46	18 20	00 42	19 09	19 56	08 44
04	06 19	17 51	10 36	10 20	02 54	18 17	00 45	19 08	19 56	08 44
05	05 56	16 27	09 52	09 53	03 02	18 13	00 48	19 07	19 56	08 43
06	05 32	13 55	09 08	09 25	03 10	18 10	00 51	19 07	19 55	08 43
07	05 09	10 25	08 22	08 58	03 18	18 07	00 54	19 06	19 55	08 43
08	04 46	06 12	07 35	08 30	03 26	18 04	00 57	19 05	19 55	08 42
09	04 22	01 S 35	06 47	08 01	03 34	18 00	01 00	19 04	19 54	08 42
10	03 59	03 N 04	05 57	07 33	03 43	17 57	01 03	19 04	19 54	08 41
11	03 35	07 27	05 07	07 04	03 51	17 54	01 06	19 03	19 54	08 41
12	03 11	11 15	04 15	06 35	04 00	17 50	01 09	19 02	19 53	08 41
13	02 48	14 19	03 23	06 06	04 08	17 47	01 11	19 02	19 53	08 40
14	02 24	16 31	02 29	05 37	04 17	17 44	01 14	19 01	19 53	08 40
15	02 00	17 47	01 35	05 08	04 25	17 41	01 17	19 00	19 52	08 40
16	01 37	18 08	00 S 40	04 39	04 33	17 37	01 20	19 00	19 52	08 39
17	01 13	17 37	00 N 16	04 09	04 42	17 34	01 23	18 59	19 52	08 39
18	00 49	16 19	01 12	03 39	04 50	17 31	01 26	18 58	19 52	08 39
19	00 26	14 19	02 08	03 09	04 58	17 28	01 29	18 58	19 51	08 38
20	00 S 02	11 44	03 05	02 40	05 06	17 25	01 32	18 57	19 51	08 38
21	00 N 22	08 40	04 02	02 10	05 14	17 21	01 35	18 57	19 51	08 37
22	00 45	05 15	04 58	01 40	05 22	17 18	01 38	18 56	19 50	08 37
23	01 09	01 N 36	05 54	01 10	05 30	17 15	01 41	18 56	19 50	08 37
24	01 33	02 S 09	06 49	00 39	05 37	17 12	01 44	18 55	19 50	08 36
25	01 56	05 51	07 43	00 S 09	05 45	17 09	01 47	18 54	19 50	08 36
26	02 20	09 21	08 36	00 N 21	05 52	17 06	01 50	18 54	19 49	08 35
27	02 43	12 30	09 28	00 51	05 59	17 03	01 53	18 53	19 49	08 35
28	03 07	15 06	10 18	01 21	06 05	17 00	01 56	18 53	19 49	08 35
29	03 30	16 59	11 06	01 51	06 12	16 57	01 59	18 52	19 49	08 34
30	03 53	18 00	11 51	02 22	06 18	16 53	02 01	18 52	19 48	08 34
31	04 N 17	18 S 02	12 N 35	02 N 52	06 N 24	16 S 50	02 N 04	18 S 51	19 S 48	08 S 33

ZODIAC SIGN ENTRIES

Date	h m	Planets
01	12 01	☽ ♐
03	17 39	☽ ♑
05	19 54	☽ ♒
07	19 57	☽ ♓
08	19 49	♂ ♍
09	19 33	☽ ♈
11	20 37	☽ ♉
14	00 48	☽ ♊
16	04 13	☿ ♈
16	08 51	☽ ♋
18	20 08	☽ ♌
20	13 55	☉ ♈
21	08 59	☽ ♍
23	05 26	♀ ♈
23	21 35	☽ ♎
26	08 42	☽ ♏
28	17 40	☽ ♐
31	00 07	☽ ♑

LATITUDES

Date	Mercury ☿ ° '	Venus ♀ ° '	Mars ♂ ° '	Jupiter ♃ ° '	Saturn ♄ ° '	Uranus ♅ ° '	Neptune ♆ ° '	Pluto ♇ ° '
01	02 S 08	01 S 19	03 N 48	00 S 26	02 S 12	00 S 35	00 N 25	12 N 41
04	02 04	01 22	03 48	00 26	02 12	00 35	00 25	12 42
07	01 55	01 24	03 48	00 27	02 12	00 35	00 25	12 43
10	01 42	01 25	03 47	00 27	02 12	00 35	00 25	12 44
13	01 23	01 26	03 45	00 28	02 11	00 35	00 25	12 45
16	01 00	01 26	03 42	00 28	02 11	00 35	00 25	12 46
19	00 S 32	01 26	03 39	00 28	02 11	00 35	00 25	12 47
22	00 00	01 25	03 35	00 29	02 11	00 35	00 25	12 48
25	00 N 35	01 24	03 30	00 29	02 11	00 35	00 25	12 49
28	01 12	01 22	03 24	00 30	02 11	00 35	00 25	12 50
31	01 N 47	01 S 19	03 N 17	00 S 30	02 S 11	00 S 35	00 N 25	12 N 51

DATA

Julian Date	2450509
Delta T	+62 seconds
Ayanamsa	23° 49' 04"
Synetic vernal point	05° ♓ 17' 56"
True obliquity of ecliptic	23° 26' 14"

MOON'S PHASES, APSIDES AND POSITIONS ☽

Date	h m	Phase	Longitude ° '	Eclipse Indicator
02	09 38	☾	11 ♐ 51	
09	01 15	●	18 ♓ 31	Total
16	00 06	☽	25 ♊ 27	
24	04 45	○	03 ♎ 35	partial
31	19 38	☾	11 ♑ 08	

Day	h m		
08	08 54	Perigee	
20	23 31	Apogee	
03	18 11	Max dec	18° S 10'
09	20 06	ON	
16	09 33	Max dec	18° N 08'
23	22 16	OS	
31	00 46	Max dec	18° S 09'

LONGITUDES

Date	Chiron ⚷ ° '	Ceres ⚳ ° '	Pallas ⚴ ° '	Juno ⚵ ° '	Vesta ⚶ ° '	Black Moon Lilith ⚸ ° '
01	01 ♏ 52	12 ♒ 15	20 ♑ 09	12 ♉ 05	27 ♒ 54	28 ♌ 03
11	01 ♏ 29	15 ♒ 59	23 ♑ 24	17 ♉ 28	02 ♓ 56	29 ♌ 09
21	00 ♏ 57	19 ♒ 36	26 ♑ 24	22 ♉ 57	07 ♓ 54	00 ♍ 16
31	00 ♏ 18	23 ♒ 06	29 ♑ 07	28 ♉ 30	12 ♓ 46	01 ♍ 22

All ephemeris data is given at 12.00 UT and the Moon's longitude is additionally given for 24.00 UT

ASPECTARIAN

h m	Aspects
01 Saturday	
02 01	☽ Q ♅
04 07	♀ ⊼ ♂
06 11	☽ Q ♃
10 06	☽ ⚹ ♆
12 48	☿ ⊼ ♂
13 23	♄ ⚹ ♅
16 21	☽ ⚹ ♂
17 00	☽ □ ☿
17 57	☽ □ ♀
22 17	☽ ☌ ♇
02 Sunday	
00 15	☽ ⚹ ♅
00 18	☽ △ ♄
04 36	☽ ⚹ ♃
09 38	☽ □ ☉
13 29	☽ ∠ ♆
13 39	☽ Q ♂
14 32	☿ ☌ ♀
03 Monday	
00 12	☿ ⊥ ♆
03 12	☽ ∠ ♅
04 26	♀ ⊥ ♆
05 29	☽ ⊥ ♆
05 35	☽ Q ♀
06 16	☽ Q ☿
06 45	☉ ⊥ ♅
07 44	☽ ∠ ♃
08 00	☿ □ ♇
15 32	♀ □ ♇
15 58	☽ ⊻ ♆
18 57	☽ ⊥ ♅
19 51	☽ Q ☉
20 33	☽ □ ♂
23 35	☽ ⊥ ♃
23 39	☿ ⊻ ♅
04 Tuesday	
01 59	☿ ⊻ ♄
03 15	☽ ⊻ ♇
04 23	☽ ⚹ ♀
05 16	☽ ⊻ ♅
05 35	☽ □ ♄
06 04	☽ ⚹ ☿
09 58	☽ ⊻ ♃
13 26	☽ ⊥ ♇
14 33	☉ ∠ ♆
14 39	♀ ⊻ ♅
17 48	☽ ⚹ ☉
19 03	♀ ⊻ ♄
05 Wednesday	
04 28	☽ ∠ ♇
07 56	☽ ∠ ♀
10 47	☽ ∠ ☿
11 08	☿ ∥ ♀
11 54	☽ Q ♄
15 47	☿ ⊻ ♃
18 26	☽ ☌ ♆
20 24	☽ ∠ ☉
21 34	☽ △ ♂
06 Thursday	
00 02	☽ ⊥ ♀
03 33	☽ ⊥ ☿
05 00	☽ ⚹ ♇
07 04	☽ ☌ ♅
07 36	☽ ⚹ ♄
10 37	☽ ⊻ ♀
12 01	☽ ⊥ ☉
12 04	☽ ☌ ♃
12 56	☉ ⊥ ♃
14 36	☽ ⊻ ☿
21 12	☽ ⚼ ♂
22 19	☽ ⊻ ☉
07 Friday	
00 16	☽ Q ♇
01 10	☿ ∥ ♇
07 15	♀ ⊻ ♃
07 47	☽ ∠ ♄
08 20	☿ ⊥ ♅
11 12	☽ ± ♂
18 36	☽ ⊻ ♆
20 30	☽ ⊼ ♂
21 46	☽ ∥ ♀
22 08	☽ ∥ ♇
22 10	☽ ⊥ ♄
08 Saturday	
00 03	☿ ∠ ♆
01 07	♀ ∥ ♇
02 48	☽ ∥ ☿
04 07	☽ ⊥ ♆
04 49	☽ □ ♇
06 59	☽ ⊻ ♅
07 44	☽ ⊻ ♄
12 25	☽ ⊻ ♃
12 55	♇ St R
14 36	☽ ☌ ♀
16 30	☽ ⊥ ♅
18 23	☽ ∠ ♆
20 20	☽ ∥ ☉
20 56	☽ ☌ ☿
22 03	☽ ⊥ ♃
09 Sunday	
01 15	☽ ☌ ☉
02 05	☽ # ♂
06 49	☽ ∠ ♅
06 51	☿ ⊥ ♃
12 33	☽ ∠ ♃
14 40	♀ ⊥ ♅
14 59	☽ # ♄
18 16	☽ ⚹ ♆
18 59	☽ ☍ ♂
10 Monday	
01 21	☽ ∥ ♄
04 32	☽ △ ♇
06 54	☽ ⚹ ♅
07 55	☽ ☌ ♄
13 00	☽ ⚹ ♃
13 30	♀ ∠ ♆
13 38	☽ Q ♆
15 30	☽ ∥ ♂
16 24	☽ # ☉
18 59	☽ ⊻ ♀
21 59	♂ △ ♆
23 54	☉ # ♂
11 Tuesday	
01 01	☽ # ☿
02 31	☽ Q ♅
04 19	☽ ⊻ ☿
04 58	☽ ⚼ ♇
05 05	☽ ⊻ ☉
05 45	☽ ⊥ ♀
08 58	☽ Q ♃
10 04	☽ # ♀
15 34	☉ ☌ ☿
15 47	☽ ⊥ ☉
15 48	☽ ⊥ ☿
18 48	☽ ⊼ ♂
19 22	☽ # ♇
19 23	☽ □ ♆
19 57	☽ ± ♇
22 11	☽ ∠ ♀
12 Wednesday	
04 36	☿ ∠ ♅
04 39	☽ ± ♂
06 05	☽ ⊼ ♇
08 07	☽ ∠ ☉
08 46	☽ □ ♅
09 25	☽ ∠ ☿
10 06	☽ ⊻ ♄
15 49	☽ □ ♃
17 29	☉ ∠ ♅
18 07	☿ # ♂
19 41	☽ ⚼ ♂
20 31	☽ ⊥ ♄
13 Thursday	
01 54	♀ ⊥ ♃
02 25	☽ ⚹ ♀
12 14	☽ ⚹ ☉
12 26	☽ ∠ ♄
15 54	☽ ⚹ ☿
21 24	☽ △ ♂
23 34	☽ △ ♆
14 Friday	
01 55	☽ Q ♀
10 58	☽ ☍ ♇
11 35	☽ Q ☉
11 39	☿ ∠ ♃
14 02	☽ △ ♅
15 47	☽ ⚹ ♄
15 53	☉ ∥ ☿
17 27	☽ Q ☿
22 17	☽ △ ♃
15 Saturday	
00 06	☿ ☍ ♂
03 07	☽ ⚼ ♆
09 17	☽ # ♃
14 23	☽ Q ♄
14 44	☽ □ ♀
18 08	☽ ⚼ ♅
19 14	☿ # ♄
20 07	☽ ± ♆
20 30	☿ ⚹ ♆
16 Sunday	
00 06	☽ □ ☉
03 03	☽ ⚼ ♃
03 31	☽ □ ♂
07 38	☽ ⊼ ♆
09 44	☽ □ ☿
11 27	☽ ± ♅
13 23	♂ ⚼ ♃
15 17	♀ # ♂
19 40	☽ ⊼ ♇
20 45	☽ ± ♃
23 10	☽ ⊼ ♅
17 Monday	
01 24	☽ □ ♄
02 27	♀ ∠ ♅
02 50	☉ # ♄
07 25	☽ ± ♇
07 55	☉ ☍ ♂
08 43	☽ ⊼ ♃
13 26	☽ # ♃
13 32	☽ Q ♂
21 42	☉ ∠ ♃
18 Tuesday	
01 15	☽ ⚼ ♇
05 11	☉ # ☿
07 54	☽ △ ♀
12 42	☽ ⚹ ♂
16 20	☽ △ ☉
18 21	☿ ∥ ♄
19 00	☽ ☍ ♆
23 18	☿ △ ♇
19 Wednesday	
07 23	☽ △ ♇
09 01	☽ △ ☿
11 14	☽ ☍ ♅
13 59	☽ △ ♄
17 41	☽ ⚼ ♀
18 00	☽ ∠ ♂
20 08	♀ ☍ ♂
21 54	☽ ☍ ♃
22 30	☿ ⚹ ♅
20 Thursday	
00 51	☉ ⚹ ♆
01 27	☽ ⚼ ☉
04 57	☿ # ♀
11 42	☽ ⊥ ♂
14 07	☽ ± ♀
16 15	☿ ☌ ♄
20 46	☽ ⚼ ♄
21 25	☽ ± ☉
21 37	☽ ⚼ ☿
23 31	☽ ⊻ ♂
21 Friday	
03 45	☽ ⊼ ♀
07 57	☽ ⊼ ♆
10 45	☽ ⊼ ☉
12 22	☽ # ♇
15 16	☽ ± ♄
16 30	♀ ∠ ♃
19 35	☽ ± ☿
20 10	☽ ± ♆
20 16	☽ □ ♇
23 05	☿ Q ♆
22 Saturday	
00 23	☽ ⊼ ♅
03 34	☽ ⊼ ♄
10 10	☽ ⊼ ☿
11 15	☽ ∥ ♂
11 51	☽ ⊼ ♃
12 35	☽ ± ♅
13 13	♀ # ♄
13 32	☽ ∥ ☿
14 26	☽ ⚼ ♆
20 12	♀ ⚹ ♆
21 23	☿ ⚹ ♃
23 Sunday	
00 02	☿ ∥ ♂
00 11	☽ ± ♃
06 48	☽ ⚼ ♅
08 38	☽ Q ♇
10 21	☽ ☌ ♂
11 29	☽ ∥ ♄
12 11	☉ # ♀
14 36	☽ ∥ ☉
15 17	☽ # ♀
18 34	☽ ⚼ ♃
20 40	☽ △ ♆
23 27	☽ ☍ ♀
24 Monday	
03 35	☽ ∥ ♀
04 45	☽ ☍ ☉
07 41	☽ # ☉
08 37	☽ ⚹ ♇
09 17	☽ # ♄
12 54	☽ △ ♅
16 26	☽ ☍ ♄
25 Tuesday	
00 52	☽ △ ♃
00 54	☉ ∥ ♄
09 28	☽ ☍ ☿
11 14	☽ # ♂
14 14	☽ ∠ ♇
17 45	♂ Q ♇
20 01	☽ ⊻ ♂
26 Wednesday	
03 25	☉ △ ♇
04 48	☽ # ☿
06 34	☽ ∥ ♇
06 58	☿ Q ♅
07 22	☽ ⊥ ♂
07 45	☽ ⊥ ♇
07 55	☽ □ ♆
11 44	☿ # ♇
17 04	☿ ⚼ ♇
17 04	☽ ⊼ ♀
19 20	☽ ⊻ ♇
20 43	☽ ⊼ ☉
23 41	☽ □ ♅
27 Thursday	
00 13	☽ ∠ ♂
03 31	☽ ⊼ ♄
05 51	☽ ± ♀
09 11	☽ ± ☉
11 56	☽ □ ♃
15 04	☽ ± ♄
15 36	♀ △ ♇
16 51	♂ ⚼ ♅
18 27	☽ Q ♆
22 51	☿ ⊼ ♂
28 Friday	
00 47	☽ ⚼ ♀
03 39	☽ ⚼ ☉
03 54	☽ ⚹ ♂
04 51	☽ ⊼ ☿
08 12	☽ ⚼ ♄
09 53	☽ Q ♅
11 30	☉ ⚹ ♅
16 59	☽ ⚹ ♆
17 46	☽ ± ☿
22 13	☽ Q ♃
29 Saturday	
01 39	☽ Q ♂
03 49	☽ ☌ ♇
07 39	☽ △ ♀
08 11	☽ ⚹ ♅
09 48	☽ △ ☉
11 20	☽ ∥ ♃
12 14	☽ △ ♄
12 48	☽ ⚼ ☿
13 17	♀ ⚹ ♅
18 18	♀ ∥ ♄
20 32	☽ ⚹ ♃
20 36	☽ ∠ ♆
30 Sunday	
08 50	☿ Q ♃
09 35	☽ □ ♂
11 28	☽ ∠ ♅
12 51	☽ ⊥ ♆
19 31	☽ △ ☿
22 20	☉ ☌ ♄
23 20	☿ ± ♂
23 33	☽ ⊻ ♆
23 49	☽ ∠ ♃
31 Monday	
03 30	☽ ⊥ ♅
09 44	☽ ⊻ ♇
12 48	♀ ☌ ♄
14 04	☽ ⊻ ♅
15 51	☽ ⊥ ♃
18 18	☽ □ ♄
18 47	☽ □ ♀
19 38	☽ □ ☉
20 12	☽ ⊥ ♇

APRIL 1997

LONGITUDES

Date	Sidereal time h m s	Sun ⊙ ° ' "	Moon ☽ ° ' "	Moon ☽ 24.00 ° ' "
01	00 39 32	11 ♈ 47 54	20 ♑ 36 20	27 ♑ 38 54
02	00 43 29	12 47 05	04 ♒ 45 55	11 ♒ 57 07
03	00 47 26	13 46 15	19 ♒ 12 11	26 ♒ 30 40
04	00 51 22	14 45 23	03 ♓ 51 59	11 ♓ 15 27
05	00 55 19	15 44 28	18 ♓ 40 18	26 ♓ 05 41
06	00 59 15	16 43 32	03 ♈ 30 40	10 ♈ 54 19
07	01 03 12	17 42 34	18 ♈ 15 43	25 ♈ 33 56
08	01 07 08	18 41 34	02 ♉ 48 09	09 ♉ 57 40
09	01 11 05	19 40 32	17 ♉ 01 50	24 ♉ 00 12
10	01 15 01	20 39 28	00 ♊ 52 26	07 ♊ 38 22
11	01 18 58	21 38 22	14 ♊ 17 57	20 ♊ 51 18
12	01 22 55	22 37 13	27 ♊ 18 37	03 ♋ 40 13
13	01 26 51	23 36 02	09 ♋ 56 32	16 ♋ 08 01
14	01 30 48	24 34 49	22 ♋ 15 13	28 ♋ 18 43
15	01 34 44	25 33 34	04 ♌ 19 07	10 ♌ 17 01
16	01 38 41	26 32 16	16 ♌ 13 02	22 ♌ 07 48
17	01 42 37	27 30 56	28 ♌ 01 54	03 ♍ 55 55
18	01 46 34	28 29 34	09 ♍ 50 25	15 ♍ 45 53
19	01 50 30	29 ♈ 28 10	21 ♍ 42 49	27 ♍ 41 40
20	01 54 27	00 ♉ 26 43	03 ♎ 42 48	09 ♎ 46 34
21	01 58 24	01 25 15	15 ♎ 53 16	22 ♎ 03 07
22	02 02 20	02 23 44	28 ♎ 16 19	04 ♏ 32 59
23	02 06 17	03 22 12	10 ♏ 53 13	17 ♏ 17 01
24	02 10 13	04 20 38	23 ♏ 44 25	00 ♐ 15 19
25	02 14 10	05 19 02	06 ♐ 49 41	13 ♐ 27 24
26	02 18 06	06 17 25	20 ♐ 08 22	26 ♐ 52 27
27	02 22 03	07 15 46	03 ♑ 39 32	10 ♑ 29 29
28	02 25 59	08 14 05	17 ♑ 22 13	24 ♑ 17 36
29	02 29 56	09 12 23	01 ♒ 15 31	08 ♒ 15 52
30	02 33 53	10 ♉ 10 39	15 ♒ 18 33	22 ♒ 23 23

Date	Mercury ☿ ° '	Venus ♀ ° '	Mars ♂ ° '	Jupiter ♃ ° '	Saturn ♄ ° '	Uranus ♅ ° '	Neptune ♆ ° '	Pluto ♇ ° '
01	29 ♈ 54	11 ♈ 31	21 ♍ 09	15 ♒ 06	10 ♈ 27	07 ♒ 58	29 ♑ 42	05 ♐ 27
02	01 ♉ 15	12 46	20 R 50	15 17	10 34	08 00	29 43	05 R 26
03	02 32	14 00	20 31	15 27	10 42	08 02	29 44	05 25
04	03 42	15 15	20 13	15 38	10 49	08 04	29 45	05 24
05	04 47	16 29	19 56	15 48	10 57	08 06	29 46	05 23
06	05 45	17 44	19 39	15 58	11 04	08 07	29 47	05 22
07	06 37	18 58	19 24	16 08	11 12	08 09	29 48	05 21
08	07 23	20 13	19 08	16 18	11 19	08 11	29 48	05 20
09	08 02	21 27	18 54	16 28	11 27	08 12	29 49	05 19
10	08 34	22 41	18 40	16 38	11 34	08 14	29 50	05 18
11	09 00	23 56	18 27	16 47	11 42	08 16	29 51	05 17
12	09 19	25 10	18 15	16 57	11 49	08 17	29 51	05 16
13	09 32	26 24	18 03	17 06	11 57	08 19	29 52	05 15
14	09 39	27 39	17 52	17 16	12 04	08 20	29 52	05 14
15	09 R 39	28 ♈ 53	17 42	17 25	12 11	08 21	29 53	05 13
16	09 32	00 ♉ 07	17 33	17 34	12 19	08 23	29 54	05 12
17	09 21	01 21	17 25	17 43	12 26	08 24	29 54	05 11
18	09 03	02 36	17 17	17 52	12 34	08 25	29 54	05 09
19	08 41	03 50	17 11	18 00	12 41	08 26	29 55	05 08
20	08 14	05 04	17 05	18 09	12 48	08 28	29 55	05 07
21	07 43	06 18	16 59	18 17	12 55	08 29	29 56	05 06
22	07 09	07 32	16 55	18 25	13 03	08 30	29 56	05 04
23	06 32	08 46	16 51	18 34	13 10	08 31	29 56	05 03
24	05 53	10 00	16 48	18 42	13 17	08 32	29 57	05 02
25	05 13	11 14	16 46	18 49	13 24	08 33	29 57	05 00
26	04 32	12 28	16 45	18 57	13 32	08 33	29 57	04 59
27	03 52	13 42	16 44	19 05	13 39	08 34	29 57	04 58
28	03 13	14 56	16 D 44	19 12	13 46	08 35	29 57	04 56
29	02 35	16 11	16 45	19 19	13 53	08 36	29 57	04 55
30	02 ♉ 00	17 ♉ 24	16 ♍ 47	19 ♒ 27	14 ♈ 00	08 ♒ 36	29 ♑ 57	04 ♐ 53

Date	Moon True ☊ ° '	Moon Mean ☊ ° '	Moon ☽ Latitude ° '
01	28 ♍ 40	28 ♍ 16	04 N 54
02	28 D 40	28 13	04 16
03	28 41	28 09	03 22
04	28 43	28 06	02 13
05	28 43	28 03	00 N 55
06	28 R 44	28 00	00 S 26
07	28 43	27 57	01 46
08	28 41	27 54	02 57
09	28 38	27 50	03 56
10	28 35	27 47	04 39
11	28 32	27 44	05 05
12	28 29	27 41	05 15
13	28 27	27 38	05 09
14	28 27	27 34	04 48
15	28 D 27	27 31	04 16
16	28 28	27 28	03 32
17	28 30	27 25	02 40
18	28 32	27 22	01 41
19	28 33	27 19	00 S 37
20	28 R 33	27 15	00 N 28
21	28 31	27 12	01 34
22	28 28	27 09	02 35
23	28 24	27 06	03 31
24	28 18	27 03	04 16
25	28 12	27 00	04 49
26	28 06	26 56	05 08
27	28 01	26 53	05 09
28	27 58	26 50	04 53
29	27 56	26 47	04 20
30	27 ♍ 56	26 ♍ 44	03 N 31

DECLINATIONS

Date	Sun ⊙ ° '	Moon ☽ ° '	Mercury ☿ ° '	Venus ♀ ° '	Mars ♂ ° '	Jupiter ♃ ° '	Saturn ♄ ° '	Uranus ♅ ° '	Neptune ♆ ° '	Pluto ♇ ° '
01	04 N 40	17 S 00	13 N 16	03 N 22	06 N 30	16 S 47	02 N 07	18 S 51	19 S 48	08 S 33
02	05 03	14 55	13 55	03 52	06 35	16 45	02 10	18 50	19 48	08 33
03	05 26	11 52	14 30	04 22	06 41	16 42	02 13	18 50	19 48	08 32
04	05 49	08 01	15 03	04 51	06 45	16 39	02 16	18 49	19 47	08 32
05	06 12	03 S 38	15 33	05 21	06 50	16 36	02 19	18 49	19 47	08 31
06	06 34	00 N 59	16 00	05 51	06 54	16 33	02 22	18 48	19 47	08 31
07	06 57	05 32	16 23	06 20	06 58	16 30	02 25	18 48	19 47	08 31
08	07 19	09 40	16 44	06 49	07 02	16 27	02 28	18 48	19 47	08 30
09	07 42	13 09	17 01	07 19	07 05	16 24	02 31	18 47	19 47	08 30
10	08 04	15 47	17 14	07 48	07 08	16 22	02 33	18 47	19 46	08 29
11	08 26	17 27	17 25	08 16	07 11	16 19	02 36	18 46	19 46	08 29
12	08 48	18 10	17 32	08 45	07 14	16 16	02 39	18 46	19 46	08 29
13	09 10	17 56	17 35	09 14	07 16	16 14	02 42	18 46	19 46	08 28
14	09 31	16 51	17 35	09 42	07 17	16 11	02 45	18 45	19 46	08 28
15	09 53	15 02	17 32	10 10	07 19	16 08	02 48	18 45	19 46	08 27
16	10 14	12 36	17 26	10 38	07 20	16 06	02 50	18 45	19 46	08 27
17	10 35	09 39	17 16	11 05	07 21	16 03	02 53	18 44	19 46	08 27
18	10 56	06 19	17 03	11 32	07 21	16 01	02 56	18 44	19 45	08 26
19	11 17	02 N 43	16 47	11 59	07 22	15 58	02 59	18 44	19 45	08 26
20	11 38	01 S 03	16 29	12 26	07 22	15 56	03 02	18 44	19 45	08 25
21	11 58	04 49	16 08	12 52	07 21	15 53	03 04	18 43	19 45	08 25
22	12 18	08 26	15 44	13 18	07 21	15 51	03 07	18 43	19 45	08 25
23	12 38	11 45	15 19	13 44	07 20	15 49	03 10	18 43	19 45	08 24
24	12 58	14 34	14 52	14 10	07 18	15 46	03 13	18 43	19 45	08 24
25	13 18	16 41	14 24	14 35	07 17	15 44	03 15	18 43	19 45	08 24
26	13 37	17 57	13 55	14 59	07 15	15 42	03 18	18 42	19 45	08 23
27	13 56	18 14	13 25	15 24	07 13	15 40	03 21	18 42	19 45	08 23
28	14 15	17 28	12 56	15 48	07 10	15 38	03 23	18 42	19 45	08 22
29	14 34	15 39	12 27	16 11	07 08	15 35	03 26	18 42	19 45	08 22
30	14 N 52	12 S 53	11 N 58	16 N 35	07 N 05	15 S 33	03 N 29	18 S 42	19 S 45	08 S 22

ZODIAC SIGN ENTRIES

Date	h	m	Planets
01	13	45	☿ ♉
02	03	59	☽ ♒
04	05	42	☽ ♓
06	06	19	☽ ♈
08	07	20	☽ ♉
10	10	28	☽ ♊
12	17	03	☽ ♋
15	03	22	☽ ♌
16	09	43	♀ ♉
17	16	00	☽ ♍
20	01	03	⊙ ♉
20	04	36	☽ ♎
22	15	19	☽ ♏
24	23	32	☽ ♐
27	05	32	☽ ♑
29	09	50	☽ ♒

LATITUDES

Date	Mercury ☿ ° '	Venus ♀ ° '	Mars ♂ ° '	Jupiter ♃ ° '	Saturn ♄ ° '	Uranus ♅ ° '	Neptune ♆ ° '	Pluto ♇ ° '
01	01 N 58	01 S 18	03 N 15	00 S 31	02 S 11	00 S 35	00 N 25	12 N 51
04	02 27	01 15	03 08	00 31	02 11	00 36	00 25	12 52
07	02 49	01 11	03 01	00 32	02 11	00 36	00 25	12 53
10	03 03	01 07	02 53	00 32	02 12	00 36	00 25	12 53
13	03 05	01 02	02 45	00 33	02 12	00 36	00 25	12 54
16	02 55	00 57	02 38	00 33	02 12	00 36	00 25	12 55
19	02 32	00 51	02 30	00 34	02 12	00 36	00 25	12 55
22	01 57	00 45	02 22	00 34	02 12	00 36	00 25	12 56
25	01 12	00 39	02 14	00 35	02 13	00 36	00 25	12 56
28	00 N 22	00 33	02 06	00 36	02 13	00 36	00 25	12 57
31	00 S 29	00 S 26	01 N 59	00 S 36	02 S 13	00 S 37	00 N 25	12 N 57

DATA

Julian Date	2450540
Delta T	+62 seconds
Ayanamsa	23° 49' 06"
Synetic vernal point	05° ♓ 17' 53"
True obliquity of ecliptic	23° 26' 14"

MOON'S PHASES, APSIDES AND POSITIONS ☽

Date	h	m	Phase	Longitude ° '	Eclipse Indicator
07	11	02	●	17 ♈ 40	
14	17	00	☽	24 ♋ 47	
22	20	33	○	02 ♏ 45	
30	02	37	☾	09 ♒ 48	

Day	h	m		
05	16	40	Perigee	
17	15	22	Apogee	
06	06	52	ON	
12	17	53	Max dec	18° N 11'
20	05	23	OS	
27	06	29	Max dec	18° S 16'

LONGITUDES

Date	Chiron ⚷ ° '	Ceres ⚳ ° '	Pallas ⚴ ° '	Juno ⚵ ° '	Vesta ⚶ ° '	Black Moon Lilith ⚸ ° '
01	00 ♏ 14	23 ♒ 27	29 ♑ 22	29 ♉ 04	13 ♓ 15	01 ♍ 29
11	29 ♎ 29	26 ♒ 47	01 ♒ 43	04 ♊ 40	18 ♓ 02	02 ♍ 36
21	28 ♎ 43	29 ♒ 57	03 ♒ 40	10 ♊ 18	22 ♓ 42	03 ♍ 42
31	27 ♎ 57	02 ♓ 54	05 ♒ 10	15 ♊ 56	27 ♓ 15	04 ♍ 49

All ephemeris data is given at 12.00 UT and the Moon's longitude is additionally given for 24.00 UT

ASPECTARIAN

h m	Aspects
01 Tuesday	
02 24	☽ ⊻ ♃
04 21	☿ ± ♇
08 39	☿ □ ♆
09 40	⊙ Q ♆
11 43	☽ ∠ ♇
12 55	☽ △ ♂
14 27	♂ ± ♃
15 16	☽ ∥ ♃
15 31	♀ Q ♆
02 Wednesday	
01 28	☽ Q ♄
03 30	☽ ☌ ♆
04 38	☽ Q ♀
04 48	☽ Q ⊙
05 30	☽ □ ☿
13 07	☽ ✶ ♇
13 45	⊙ ☌ ♀
13 45	☽ ⚼ ♂
17 25	☽ ☌ ♅
19 13	☽ # ☿
21 47	☽ ✶ ♄
03 Thursday	
02 22	☽ ✶ ⊙
02 36	☽ ✶ ♀
04 26	☽ ± ♂
05 44	☽ ☌ ♃
09 03	☽ Q ♇
14 08	☽ ⊼ ♂
14 23	☽ Q ☿
22 46	☽ ∠ ♄
04 Friday	
04 49	☽ ∠ ⊙
05 17	☽ ⊻ ♆
05 34	☽ ∠ ♀
09 01	☽ ∥ ♇
11 42	☽ ✶ ☿
13 34	☽ ⊥ ♄
14 30	☽ □ ♇
15 04	☽ ⊥ ♆
18 50	☽ ⊻ ♅
19 00	☽ # ♂
20 29	♀ ✶ ♃
20 31	☽ ⊥ ⊙
21 33	☽ ⊥ ♀
23 20	☽ # ⊙
23 23	☽ ⊻ ♄
05 Saturday	
03 45	☽ # ♀
04 35	☽ ⊥ ♅
05 41	☽ ∠ ♆
06 55	☽ ⊻ ⊙
07 18	☽ ⊻ ♃
08 09	☽ ⊻ ♀
13 40	⊙ ✶ ♃
13 55	☽ ∠ ☿
14 00	☽ ☍ ♂
14 58	☿ ⚼ ♂
17 07	☽ ⊥ ♃
18 48	☽ # ♄
19 10	☽ ∠ ♅
06 Sunday	
02 34	☿ ⊼ ♇
05 30	☽ ⊥ ☿
05 57	☽ ✶ ♆
07 50	☽ ∠ ♃
15 01	☽ △ ♇
15 52	☽ ⊻ ☿
19 13	☽ ∥ ♄
19 30	☽ ✶ ♅
07 Monday	
00 23	☽ ☌ ♄
01 26	☽ Q ♆
08 29	☽ ✶ ♃
11 02	☽ ☌ ⊙
13 16	☽ ☌ ♀
13 45	⊙ ∥ ♂
13 49	☽ ⊼ ♂
15 06	☽ Q ♅
15 26	☽ ⚼ ♇
17 03	☽ ∥ ♀
18 43	☿ # ♃
18 45	♀ ⊼ ♂
20 10	☽ ∥ ♂
20 41	☽ ∥ ⊙
23 31	☽ ± ♂
08 Tuesday	
04 26	☽ Q ♃
04 57	☽ # ♇
06 15	☽ ± ♇
07 01	☽ □ ♆
11 24	♀ Q ♅
14 11	☽ ⚼ ♂
14 29	♀ ⚼ ♇
16 14	☽ ⊼ ♇
20 02	☽ ☌ ☿
20 41	⊙ ⊼ ♂
21 01	☽ □ ♅
23 44	♀ ∥ ♂
09 Wednesday	
02 26	☽ ⊻ ♄
11 02	☽ □ ♃
12 43	☽ ⊥ ♄
15 08	☽ △ ♂
16 52	☽ ⊻ ⊙
19 47	☿ □ ♅
20 20	☽ ⊻ ♀
10 Thursday	
01 22	⊙ Q ♅
03 33	⊙ ⚼ ♇
04 03	☽ ⊥ ⊙
04 24	☽ ∠ ♄
07 47	☽ ⊥ ♀
10 10	☽ △ ♆
18 45	☽ # ♃
19 50	☽ ☍ ♇
21 08	☽ ∠ ⊙
11 Friday	
01 05	☽ △ ♅
01 19	☽ ∠ ♀
02 09	☽ ⊻ ☿
07 15	☽ ✶ ♄
10 58	☽ ∥ ☿
12 59	☽ ⚼ ♆
13 19	☽ ⊥ ☿
15 10	⊙ # ♇
16 36	☽ △ ♃
19 27	☽ □ ♂
20 37	♀ ± ♂
22 21	♀ # ♇
12 Saturday	
02 33	☽ ✶ ⊙
04 29	☽ ⚼ ♅
05 26	☽ Q ♄
05 33	☽ ± ♆
06 18	☽ ∠ ☿
07 34	☽ ✶ ♀
16 47	☽ ⊼ ♆
19 06	♄ Q ♆
20 51	☽ ⚼ ♃
21 23	☽ ± ♅
22 24	⊙ ∥ ♀
13 Sunday	
02 58	☽ Q ⊙
03 02	☽ ⊼ ♇
04 39	☽ Q ♂
08 43	☽ Q ♀
08 51	☽ ⊼ ♅
11 13	☽ ✶ ☿
14 16	☽ ± ♃
14 31	☽ ± ♇
15 54	☽ □ ♄
21 18	⊙ ± ♂
21 39	☽ ∥ ☿
14 Monday	
02 04	☽ ⊼ ♃
03 31	☽ ✶ ♂
08 02	☽ ⚼ ♇
08 23	♃ Q ♇
10 47	☽ Q ☿
17 00	☽ □ ⊙
22 15	☽ # ♃
23 53	☽ □ ♀
15 Tuesday	
00 02	☿ St R
03 07	☽ ☍ ♆
08 49	☽ ∠ ♂
13 48	☽ △ ♇
18 23	♀ ± ♇
20 08	☽ ☍ ♅
22 39	☽ □ ☿
23 46	♀ Q ♃
16 Wednesday	
02 41	☽ ⊥ ♂
04 01	☽ △ ♄
07 35	♀ □ ♆
11 13	♂ ⊼ ♃
14 41	☽ ⊻ ♂
14 46	☽ ☍ ♃
17 Thursday	
02 17	☽ ∥ ♀
05 31	☽ ∥ ⊙
10 47	☽ ⚼ ♄
10 51	☽ △ ⊙
15 48	☽ ⊼ ♆
19 33	☽ △ ♀
21 01	☽ # ♇
18 Friday	
02 30	☽ □ ♇
04 01	☽ ± ♆
04 47	☽ ∥ ♂
05 16	☽ ± ♄
06 38	♀ ⚼ ♂
09 07	☽ ⊼ ♅
10 27	☽ △ ☿
17 34	☽ ⊼ ♄
20 04	☽ ⚼ ⊙
21 18	☽ ± ♅
22 16	☽ ⚼ ♆
19 Saturday	
02 56	☽ ☌ ♂
03 57	⊙ ± ♇
04 26	☽ ⊼ ♃
05 31	☽ ⚼ ♀
10 17	☽ ∥ ♄
14 51	☽ Q ♇
15 29	☽ ⚼ ♅
15 49	☽ ⚼ ☿
15 51	☽ ± ⊙
16 39	☽ ± ♃
23 02	⊙ □ ♆
20 Sunday	
00 49	☿ □ ♅
01 41	☽ ± ♀
03 22	⊙ Q ♃
04 27	☽ △ ♆
04 55	☽ ⊼ ⊙
09 10	☽ ± ☿
10 51	☽ ⚼ ♃
12 57	♀ ⊼ ♇
14 47	☽ ✶ ♇
14 59	☽ ⊼ ♀
20 36	☽ ⊼ ☿
21 25	☽ △ ♅
21 Monday	
00 44	☽ # ♄
06 08	☽ ☍ ♄
14 08	☽ ⊻ ♂
16 44	☽ △ ♃
20 11	☽ ∠ ♇
22 Tuesday	
00 58	⊙ ⚼ ♂
01 44	☽ ⊥ ♂
04 36	☽ # ♂
04 41	☿ # ♃
06 55	☿ ☌ ♀
11 49	☽ ∥ ♇
13 32	☽ ⊥ ♇
15 11	☽ □ ♆
18 56	☽ ∠ ♂
20 33	☽ ☍ ⊙
23 Wednesday	
00 58	☽ ⊻ ♇
04 09	☽ ☍ ☿
06 56	♀ □ ♅
07 31	☽ □ ♅
07 34	☽ ☍ ♀
16 20	☽ ⊼ ♄
19 57	☽ # ⊙
23 09	☽ ✶ ♂
24 Thursday	
01 14	☽ Q ♆
02 32	☽ □ ♃
03 39	☽ ± ♄
07 29	☽ # ♀
14 28	☽ # ☿
17 09	☽ Q ♅
20 28	☽ ⚼ ♄
21 19	☽ Q ♂
23 26	☽ ✶ ♆
25 Friday	
00 19	☽ ∥ ♃
04 28	⊙ ⊼ ♇
07 01	☿ ∥ ♀
08 41	☽ ☌ ♇
09 02	☽ ⊼ ⊙
09 13	☽ ⊼ ☿
10 32	⊙ ☌ ☿
11 59	☽ Q ♃
15 07	☽ ✶ ♅
19 34	☽ ± ☿
19 48	☿ ⊼ ♇
20 47	☽ ± ⊙
20 49	☽ ⊼ ♀
26 Saturday	
00 01	☽ △ ♄
02 41	☽ ∠ ♆
05 56	☽ □ ♂
08 43	☽ ± ♀
09 51	☽ ✶ ♃
10 59	☽ ⚼ ☿
14 13	☽ ⚼ ⊙
18 07	☽ ∠ ♅
18 48	☽ ⊥ ♆
20 41	⊙ ∥ ☿
27 Sunday	
02 22	☽ ⚼ ♀
05 27	☽ ⊻ ♆
10 05	☽ ⊥ ♅
10 38	♀ ⊻ ♄
12 21	☽ △ ☿
12 45	☽ ∠ ♃
14 17	☽ ⊻ ♇
18 50	☽ △ ⊙
19 09	♂ St D
20 39	☽ ⊻ ♅
28 Monday	
00 48	☽ ⊥ ♇
02 32	♀ # ♃
04 40	☽ ⊥ ♃
05 40	☽ □ ♄
07 21	☽ △ ♀
10 54	☽ △ ♂
15 13	☽ ⊻ ♃
16 27	☽ ∠ ♇
20 44	⊙ □ ♅
29 Tuesday	
07 01	☽ # ♀
09 46	☽ ☌ ♆
12 36	☽ ∥ ♃
12 51	☽ ⚼ ♂
13 05	☽ Q ♄
14 11	☽ □ ☿
18 15	☽ ✶ ♇
21 19	☽ # ⊙
23 32	♀ △ ♂
30 Wednesday	
00 35	☽ ☌ ♅
02 37	☽ □ ⊙
04 17	☽ ± ♂
09 45	☽ ✶ ♄
14 31	☽ ⊼ ♂
14 41	☽ Q ♇
15 54	☽ □ ♀
19 04	☽ ☌ ♃
19 39	☽ Q ☿
19 41	☽ # ☿
20 30	☿ ⚼ ♂

MAY 1997

LONGITUDES

Date	Sidereal time h m s	Sun ☉ ° ' "	Moon ☽ ° ' "	Moon ☽ 24.00 ° ' "	Mercury ☿ ° '	Venus ♀ ° '	Mars ♂ ° '	Jupiter ♃ ° '	Saturn ♄ ° '	Uranus ♅ ° '	Neptune ♆ ° '	Pluto ♇ ° '
01	02 37 49	11 ♉ 08 54	29 ♒ 30 14	06 ♓ 38 52	01 ♉ 27	18 ♉ 38	16 ♍ 49	19 ♒ 34	14 ♈ 07	08 ♒ 37	29 ♑ 57	04 ♐ 52
02	02 41 46	12 07 07	13 ♓ 49 00	21 ♓ 00 21	00 R 58	19 52	16 52	19 40	14 14	08 38	29 R 57	04 R 50
03	02 45 42	13 05 19	28 ♓ 12 29	05 ♈ 24 57	00 32	21 06	16 56	19 47	14 21	08 38	29 57	04 49
04	02 49 39	14 03 29	12 ♈ 37 13	19 ♈ 48 42	00 ♉ 10	22 20	17 01	19 54	14 28	08 39	29 57	04 47
05	02 53 35	15 01 38	26 ♈ 58 47	04 ♉ 06 48	29 ♈ 53	23 34	17 06	20 00	14 35	08 39	29 57	04 46
06	02 57 32	15 59 45	11 ♉ 12 05	18 ♉ 14 01	29 40	24 48	17 12	20 06	14 41	08 39	29 57	04 44
07	03 01 28	16 57 51	25 ♉ 12 01	02 ♊ 05 34	29 32	26 02	17 18	20 12	14 48	08 40	29 57	04 43
08	03 05 25	17 55 55	08 ♊ 51 23	15 ♊ 37 41	29 27	27 16	17 25	20 18	14 55	08 40	29 57	04 41
09	03 09 22	18 53 57	22 ♊ 15 44	28 ♊ 48 18	29 D 30	28 30	17 33	20 24	15 02	08 40	29 57	04 40
10	03 13 18	19 51 58	05 ♋ 15 24	11 ♋ 37 13	29 36	29 ♉ 44	17 42	20 30	15 08	08 40	29 56	04 38
11	03 17 15	20 49 57	17 ♋ 53 58	24 ♋ 06 01	29 ♈ 46	00 ♊ 57	17 51	20 35	15 15	08 40	29 56	04 36
12	03 21 11	21 47 54	00 ♌ 13 49	06 ♌ 17 51	00 ♉ 01	02 11	18 00	20 40	15 22	08 40	29 56	04 35
13	03 25 08	22 45 49	12 ♌ 18 41	18 ♌ 16 55	00 21	03 25	18 10	20 46	15 28	08 R 40	29 55	04 33
14	03 29 04	23 43 42	24 ♌ 13 11	00 ♍ 08 09	00 44	04 39	18 21	20 51	15 35	08 40	29 55	04 32
15	03 33 01	24 41 33	06 ♍ 02 29	11 ♍ 56 51	01 13	05 52	18 33	20 55	15 41	08 40	29 55	04 30
16	03 36 57	25 39 23	17 ♍ 51 54	23 ♍ 48 17	01 45	07 06	18 45	21 00	15 47	08 40	29 54	04 28
17	03 40 54	26 37 11	29 ♍ 46 36	05 ♎ 47 26	02 21	08 20	18 58	21 04	15 54	08 40	29 54	04 27
18	03 44 51	27 34 58	11 ♎ 51 18	17 ♎ 58 40	03 01	09 34	19 11	21 09	16 00	08 40	29 53	04 25
19	03 48 47	28 32 43	24 ♎ 09 56	00 ♏ 25 26	03 45	10 47	19 24	21 13	16 06	08 39	29 53	04 24
20	03 52 44	29 ♉ 30 26	06 ♏ 45 25	13 ♏ 10 02	04 32	12 01	19 39	21 17	16 12	08 39	29 52	04 22
21	03 56 40	00 ♊ 28 08	19 ♏ 39 20	26 ♏ 13 16	05 23	13 14	19 53	21 20	16 19	08 39	29 51	04 20
22	04 00 37	01 25 48	02 ♐ 51 44	09 ♐ 34 29	06 18	14 28	20 09	21 24	16 25	08 38	29 51	04 19
23	04 04 33	02 23 28	16 ♐ 21 13	23 ♐ 11 35	07 15	15 42	20 24	21 27	16 31	08 38	29 50	04 17
24	04 08 30	03 21 06	00 ♑ 05 10	07 ♑ 01 30	08 16	16 55	20 40	21 30	16 37	08 37	29 49	04 15
25	04 12 26	04 18 43	14 ♑ 00 08	21 ♑ 00 37	09 20	18 09	20 57	21 33	16 43	08 37	29 49	04 14
26	04 16 23	05 16 19	28 ♑ 02 31	05 ♒ 05 28	10 27	19 22	21 14	21 36	16 48	08 36	29 48	04 12
27	04 20 20	06 13 54	12 ♒ 09 06	19 ♒ 13 08	11 37	20 36	21 32	21 39	16 54	08 35	29 47	04 10
28	04 24 16	07 11 28	26 ♒ 17 21	03 ♓ 21 32	12 50	21 49	21 50	21 41	17 00	08 35	29 46	04 09
29	04 28 13	08 09 01	10 ♓ 25 33	17 ♓ 29 16	14 05	23 03	22 08	21 44	17 06	08 34	29 46	04 07
30	04 32 09	09 06 33	24 ♓ 32 33	01 ♈ 35 17	15 24	24 16	22 27	21 46	17 11	08 33	29 45	04 05
31	04 36 06	10 ♊ 04 05	08 ♈ 37 20	15 ♈ 38 31	16 ♉ 45	25 ♊ 30	22 ♍ 46	21 ♒ 48	17 ♈ 17	08 ♒ 32	29 ♑ 44	04 ♐ 04

Date	Moon True ☊ ° '	Moon Mean ☊ ° '	Moon ☽ Latitude ° '
01	27 ♍ 57	26 ♍ 40	02 N 28
02	27 D 58	26 37	01 N 16
03	27 R 58	26 34	00 S 01
04	27 57	26 31	01 18
05	27 55	26 28	02 30
06	27 49	26 25	03 32
07	27 42	26 21	04 20
08	27 34	26 18	04 51
09	27 26	26 15	05 06
10	27 18	26 12	05 04
11	27 11	26 09	04 47
12	27 07	26 06	04 18
13	27 04	26 03	03 37
14	27 D 03	25 59	02 47
15	27 04	25 56	01 51
16	27 05	25 53	00 S 49
17	27 R 05	25 50	00 N 14
18	27 04	25 46	01 18
19	27 01	25 43	02 20
20	26 55	25 40	03 15
21	26 47	25 37	04 02
22	26 37	25 34	04 38
23	26 27	25 31	04 58
24	26 16	25 27	05 02
25	26 07	25 24	04 49
26	26 00	25 21	04 17
27	25 56	25 18	03 30
28	25 54	25 15	02 30
29	25 D 53	25 12	01 21
30	25 53	25 08	00 07
31	25 ♍ 53	25 ♍ 05	01 S 07

DECLINATIONS

Date	Sun ☉ ° '	Moon ☽ ° '	Mercury ☿ ° '	Venus ♀ ° '	Mars ♂ ° '	Jupiter ♃ ° '	Saturn ♄ ° '	Uranus ♅ ° '	Neptune ♆ ° '	Pluto ♇ ° '
01	15 N 10	09 S 20	11 N 31	16 N 57	07 N 01	15 S 31	03 N 31	18 S 42	19 S 45	08 S 21
02	15 28	05 12	11 05	17 20	06 58	15 29	03 34	18 41	19 45	08 21
03	15 46	00 S 44	10 41	17 41	06 54	15 28	03 36	18 41	19 45	08 21
04	16 03	03 N 47	10 19	18 03	06 50	15 26	03 39	18 41	19 45	08 20
05	16 21	08 03	09 59	18 24	06 46	15 24	03 41	18 41	19 45	08 20
06	16 37	11 50	09 42	18 44	06 42	15 22	03 44	18 41	19 45	08 20
07	16 54	14 52	09 27	19 04	06 37	15 20	03 46	18 41	19 45	08 19
08	17 10	16 59	09 14	19 24	06 32	15 19	03 48	18 41	19 45	08 19
09	17 26	18 07	09 04	19 43	06 27	15 17	03 51	18 41	19 45	08 19
10	17 42	18 16	08 56	20 01	06 21	15 16	03 54	18 41	19 45	08 18
11	17 58	17 29	08 51	20 19	06 16	15 14	03 56	18 41	19 45	08 18
12	18 13	15 54	08 48	20 36	06 10	15 13	03 59	18 41	19 45	08 18
13	18 28	13 38	08 48	20 53	06 04	15 11	04 01	18 41	19 45	08 17
14	18 42	10 49	08 50	21 09	05 57	15 10	04 03	18 41	19 45	08 17
15	18 56	07 35	08 54	21 25	05 51	15 08	04 06	18 41	19 45	08 17
16	19 10	04 02	09 00	21 40	05 44	15 07	04 08	18 41	19 45	08 17
17	19 24	00 N 19	09 09	21 54	05 37	15 06	04 10	18 41	19 46	08 16
18	19 37	03 S 29	09 20	22 08	05 30	15 05	04 13	18 41	19 46	08 16
19	19 50	07 12	09 32	22 21	05 23	15 04	04 15	18 42	19 46	08 16
20	20 03	10 41	09 47	22 33	05 15	15 03	04 17	18 42	19 46	08 16
21	20 15	13 45	10 03	22 45	05 08	15 02	04 19	18 42	19 46	08 15
22	20 27	16 11	10 21	22 57	05 00	15 01	04 21	18 42	19 46	08 15
23	20 38	17 47	10 41	23 07	04 52	15 00	04 23	18 42	19 46	08 15
24	20 49	18 24	11 02	23 17	04 44	14 59	04 26	18 42	19 46	08 15
25	21 00	17 55	11 24	23 26	04 36	14 58	04 28	18 42	19 47	08 15
26	21 11	16 21	11 48	23 35	04 27	14 58	04 30	18 43	19 47	08 14
27	21 21	13 47	12 13	23 43	04 18	14 57	04 32	18 43	19 47	08 14
28	21 31	10 24	12 39	23 50	04 10	14 57	04 34	18 43	19 47	08 14
29	21 40	06 24	13 06	23 56	04 01	14 56	04 36	18 43	19 47	08 14
30	21 49	02 03	13 34	24 02	03 51	14 56	04 38	18 44	19 47	08 14
31	21 N 57	02 N 23	14 N 03	24 N 07	03 N 42	14 S 55	04 N 40	18 S 44	19 S 48	08 S 14

ZODIAC SIGN ENTRIES

Date	h	m	Planets
01	12	50	☽ ♓
03	14	59	☽ ♈
05	01	48	☿ ♈
05	17	04	☽ ♉
07	20	21	☽ ♊
10	02	13	☽ ♋
10	17	20	♀ ♊
12	10	25	☿ ♉
12	11	33	☽ ♌
14	23	43	☽ ♍
17	12	27	☽ ♎
19	23	11	☽ ♏
21	00	18	☉ ♊
22	06	51	☽ ♐
24	11	51	☽ ♑
26	15	20	☽ ♒
28	18	18	☽ ♓
30	21	18	☽ ♈

LATITUDES

Date	Mercury ☿ ° '	Venus ♀ ° '	Mars ♂ ° '	Jupiter ♃ ° '	Saturn ♄ ° '	Uranus ♅ ° '	Neptune ♆ ° '	Pluto ♇ ° '
01	00 S 29	00 S 26	01 N 59	00 S 36	02 S 13	00 S 37	00 N 25	12 N 57
04	01 18	00 19	01 51	00 37	02 14	00 37	00 25	12 57
07	02 00	00 12	01 44	00 37	02 14	00 37	00 25	12 57
10	02 34	00 S 05	01 37	00 38	02 14	00 37	00 25	12 58
13	02 59	00 N 03	01 30	00 39	02 15	00 37	00 25	12 58
16	03 16	00 10	01 24	00 40	02 15	00 37	00 25	12 58
19	03 26	00 17	01 18	00 40	02 16	00 37	00 25	12 58
22	03 27	00 25	01 11	00 41	02 16	00 37	00 25	12 57
25	03 23	00 32	01 06	00 42	02 17	00 37	00 25	12 57
28	03 12	00 39	01 00	00 42	02 17	00 37	00 25	12 57
31	02 S 55	00 N 46	00 N 54	00 S 43	02 S 18	00 S 38	00 N 25	12 N 57

DATA

Julian Date	2450570
Delta T	+62 seconds
Ayanamsa	23° 49' 09"
Synetic vernal point	05° ♓ 17' 50"
True obliquity of ecliptic	23° 26' 14"

LONGITUDES

Date	Chiron ⚷ ° '	Ceres ⚳ ° '	Pallas ⚴ ° '	Juno ⚵ ° '	Vesta ⚶ ° '	Black Moon Lilith ⚸ ° '
01	27 ♎ 57	02 ♓ 54	05 ♒ 10	15 ♊ 56	27 ♓ 15	04 ♍ 49
11	27 ♎ 14	05 ♓ 36	06 ♒ 08	21 ♊ 34	01 ♈ 40	05 ♍ 55
21	26 ♎ 38	08 ♓ 00	06 ♒ 32	27 ♊ 56	05 ♈ 56	07 ♍ 02
31	26 ♎ 09	10 ♓ 05	06 ♒ 18	02 ♋ 43	10 ♈ 00	08 ♍ 08

MOON'S PHASES, APSIDES AND POSITIONS ☽

Date	h	m	Phase	Longitude ° '	Eclipse Indicator
06	20	47	●	16 ♉ 21	
14	10	55	☽	23 ♌ 41	
22	09	13	○	01 ♐ 19	
29	07	51	☾	07 ♓ 59	

Day	h	m		
03	11	04	Perigee	
15	10	09	Apogee	
29	06	57	Perigee	
03	15	52	0N	
10	03	34	Max dec	18° N 20'
17	13	58	0S	
24	13	34	Max dec	18° S 24'
30	23	05	0N	

ASPECTARIAN

01 Thursday
07 52 ☿ Q ♃
11 20 ☽ ∠ ♄
11 21 ☽ Q ☉
12 46 ☽ ⊻ ♆
15 10 ☽ ✱ ☿
17 55 ☽ ∥ ♇
20 59 ☽ □ ♇
22 51 ☽ ⊥ ♆
23 17 ♆ St R

02 Friday
01 07 ☽ Q ♀
01 54 ☽ # ♂
02 35 ☽ ⊥ ♄
03 19 ☽ ⊻ ♅
07 43 ♀ □ ♃
08 57 ☽ ✱ ☉
12 42 ☽ ⊻ ♄
13 21 ☽ ⊥ ♅
13 28 ☉ # ♃
13 54 ☽ ∠ ♆
15 28 ☽ ∠ ☿
17 08 ☽ ☍ ♂
19 40 ♀ ⊥ ♄
20 49 ☽ # ♄
21 51 ☽ ⊻ ♃
23 04 ☽ ✱ ♀

03 Saturday
04 23 ☽ ∠ ♅
06 03 ☽ ⊥ ☿
07 56 ☽ ⊥ ♃
11 47 ☽ ∠ ☉
14 55 ☽ ✱ ♆
15 46 ☽ ⊻ ☿
22 59 ☽ △ ♇
23 02 ☽ ∠ ♃

04 Sunday
02 23 ☽ ∠ ♀
03 51 ☽ ⊥ ☉
05 22 ☽ ✱ ♅
10 54 ☽ Q ♆
11 17 ☽ ∥ ♄
14 34 ☽ ⊻ ☉
15 06 ☽ ☌ ♄
18 47 ☽ ⊥ ♀
19 22 ☽ ⊼ ♂
23 20 ☉ ⊻ ♄
23 56 ☽ Ꝗ ♇

05 Monday
00 14 ☽ ✱ ♃
01 24 ☽ Q ♅
04 39 ☽ ∥ ♂
05 27 ☽ ± ♂
05 43 ☿ □ ♆
05 45 ☽ ⊻ ♀
13 38 ☽ # ♇
14 59 ☽ ± ♇
16 48 ☽ ☌ ☿
17 00 ☽ □ ♆
20 30 ☽ Q ♃
20 39 ☽ Ꝗ ♂
22 55 ☽ ∥ ☿

06 Tuesday
01 04 ☽ ⊼ ♇
07 41 ☽ □ ♅
08 23 ♀ # ♅
17 59 ☽ ⊻ ♄
20 47 ☽ ☌ ☉
22 18 ☽ △ ♂

07 Wednesday
03 20 ☽ □ ♃
04 21 ☽ ⊥ ♄
13 35 ☽ ☌ ♀
16 31 ☽ # ♃
19 29 ☽ ⊻ ☿
20 04 ☽ ∠ ♄
20 15 ☽ △ ♆
21 30 ☉ △ ♂

08 Thursday
04 34 ☽ ☍ ♇
05 57 ☽ ⊥ ☿
11 35 ☽ △ ♅
15 26 ☽ ∥ ☉
18 05 ☿ St D
21 56 ☽ ∠ ☿
22 46 ☽ Ꝗ ♆
22 49 ☽ ✱ ♄

09 Friday
03 22 ☽ □ ♂
05 25 ☽ ⊻ ☉
08 36 ☽ △ ♃
14 34 ☽ Ꝗ ♅
15 04 ☽ ± ♆
15 05 ♀ # ♆
17 12 ☽ ⊥ ☉
20 48 ☽ Q ♄

10 Saturday
00 37 ☽ ⊻ ♀
01 22 ☽ ✱ ☿
02 06 ☽ ⊼ ♆
07 10 ☽ ± ♅
09 06 ☿ ⊻ ♀
10 50 ☽ ⊼ ♇
11 13 ☽ ∠ ☉
12 27 ☽ Ꝗ ♃
12 49 ☽ Q ♂
12 58 ☽ ⊥ ♀
16 07 ♀ △ ♆
18 25 ☽ ⊼ ♅
20 50 ♀ ∠ ♄
22 07 ☽ ± ♇

11 Sunday
00 06 ☽ Q ☿
03 24 ☽ ∥ ☉
05 16 ☉ □ ♃
05 36 ☽ ± ♃
06 52 ☽ □ ♄
07 52 ☽ ∠ ♀
11 53 ☽ ✱ ♂
15 17 ☽ Ꝗ ♇
17 13 ☽ ⊼ ♃
18 08 ☽ ✱ ☉
23 40 ☉ ⊥ ♄

12 Monday
04 15 ☿ □ ♆
11 24 ☽ ☍ ♆
11 34 ☽ □ ☿
16 17 ☽ ✱ ♀
17 33 ☽ ∠ ♂
19 39 ☽ Q ☉
20 09 ☽ # ♃
20 35 ☽ △ ♇

13 Tuesday
04 06 ♅ St R
04 44 ☽ ☍ ♅
11 44 ☽ ⊥ ♂
18 24 ☽ △ ♄
18 57 ☽ Q ♀
23 58 ☽ ⊻ ♂

14 Wednesday
05 07 ☽ ☍ ♃
09 46 ♀ ☍ ♇
10 15 ☉ # ♅
10 55 ☽ □ ☉
23 33 ☽ ⊼ ♆

15 Thursday
01 01 ☽ Ꝗ ♄
01 46 ☽ △ ☿
02 47 ☽ ∥ ☿
06 59 ☽ # ♇
08 53 ☽ □ ♇
11 37 ☽ □ ♀
11 44 ☽ ± ♆
17 21 ☽ ⊼ ♅
19 28 ☽ ± ♄

16 Friday
00 20 ☽ ∥ ♂
05 31 ☽ ± ♅
06 00 ☽ Ꝗ ♆
07 45 ☽ ⊼ ♄
09 37 ☽ Ꝗ ☿
11 23 ☽ ∥ ♄
13 49 ☽ ☌ ♂
18 23 ☽ ⊼ ♃
21 18 ☽ Q ♇
23 43 ☽ Ꝗ ♅

17 Saturday
04 45 ☽ ± ☿
05 07 ☽ △ ☉
06 32 ☽ ± ♃
12 14 ☽ △ ♆
12 48 ☿ ⊥ ♀
17 26 ☽ ⊼ ☿
18 31 ♀ △ ♅
21 18 ☽ ✱ ♇

18 Sunday
00 38 ☽ Ꝗ ♃
05 42 ☽ △ ♅
06 58 ☽ △ ♀
13 33 ☽ Ꝗ ☉
16 39 ☽ # ♄
16 41 ☿ Q ♃
20 12 ☽ ☍ ♄

19 Monday
00 30 ☽ # ♂
02 36 ☽ ⊻ ♂
02 47 ☽ ∠ ♇
03 50 ☉ # ♆
06 15 ☽ △ ♃
08 36 ☽ ± ☉
14 26 ☽ ⊥ ♂
15 28 ☽ Ꝗ ♀
19 05 ☽ ∥ ♇
20 06 ☽ ⊥ ♇
21 07 ☽ ⊼ ☉
22 57 ☽ □ ♆

20 Tuesday
04 59 ☽ # ☿
07 01 ☿ ⊼ ♇
07 30 ☽ ⊻ ♇
07 32 ☽ ☍ ☿
07 56 ☽ ∠ ♂
10 27 ☽ ± ♀
15 34 ☽ □ ♅
16 15 ☿ Ꝗ ♂
20 52 ☉ △ ♆
22 53 ☽ ⊼ ♀

21 Wednesday
05 47 ☽ ⊼ ♄
08 41 ☽ Q ♆
12 26 ☽ ✱ ♂
15 06 ☽ □ ♃
16 54 ☽ ± ♄
23 38 ☽ ∥ ♃

22 Thursday
00 46 ☽ Q ♅
06 34 ☽ ✱ ♆
09 13 ☽ ☍ ☉
09 22 ☽ Ꝗ ♄
10 41 ☽ Q ♂
11 27 ☉ ∠ ♄
14 36 ☽ ☌ ♇
18 37 ☽ ⊼ ☿
19 19 ♀ Ꝗ ♆
22 20 ☽ ✱ ♅
23 44 ☽ Q ♃

23 Friday
06 06 ☽ ± ☿
09 19 ☽ ∠ ♆
10 43 ☽ ☍ ♀
12 17 ☽ △ ♄
19 16 ☽ □ ♂
21 00 ☽ ✱ ♃
23 10 ☽ Ꝗ ☿

24 Saturday
00 45 ☽ ∠ ♅
01 07 ☽ ⊥ ♆
05 23 ♀ ✱ ♄
11 33 ☽ ⊻ ♆
16 23 ☽ ⊥ ♅
18 05 ☽ ⊼ ☉
19 12 ☽ ⊻ ♇
20 00 ☿ □ ♅
23 09 ☽ ∠ ♃

25 Sunday
02 44 ☽ ⊻ ♅
03 18 ☽ △ ☿
05 12 ☽ ± ☉
05 32 ☽ ⊥ ♇
09 59 ☉ ☍ ♇
14 41 ☽ ⊥ ♃
16 40 ☽ □ ♄
19 47 ☽ ⊼ ♀
20 56 ☽ ∠ ♇
21 46 ☽ Ꝗ ☉

26 Monday
00 09 ☽ △ ♂
00 59 ☽ ⊻ ♃
05 46 ♂ ∥ ♄
07 01 ☽ ± ♀
14 59 ☽ ☌ ♆
22 28 ☽ ✱ ♇
23 36 ☽ Q ♄
23 48 ☽ Ꝗ ♀

27 Tuesday
01 12 ☽ △ ☉
02 02 ☽ ∥ ♃
02 15 ☽ Ꝗ ♂
05 57 ☽ ☌ ♅
11 00 ☽ □ ☿
17 52 ☽ ± ♂
18 49 ☽ Q ♇
20 07 ☽ ✱ ♄
22 26 ☽ # ☿
23 16 ♂ ⊼ ♃

28 Wednesday
03 42 ☽ △ ♀
04 10 ☽ ☌ ♃
04 16 ☽ ⊼ ♂
09 20 ♀ △ ♃
12 10 ♀ □ ♂
17 54 ☽ ⊻ ♆
20 27 ☽ Q ☿
21 45 ☽ ∠ ♄

29 Thursday
01 19 ☽ □ ♇
01 21 ☽ ∥ ♇
04 05 ☽ ⊥ ♆
07 51 ☽ □ ☉
08 51 ☽ ⊻ ♅
10 44 ♂ Q ♇
13 08 ☽ ⊥ ♄
18 51 ☽ ✱ ☿
19 02 ☽ ⊥ ♅
19 21 ☽ ∠ ♆
22 03 ♀ Ꝗ ♅
22 04 ☽ # ♄
22 16 ☉ △ ♅
23 24 ☽ ⊻ ♄

30 Friday
01 47 ♀ ± ♆
01 52 ☽ # ♂
07 16 ☽ ⊻ ♃
08 22 ☽ ☍ ♂
10 19 ☽ ∠ ♅
11 30 ☽ □ ♀
16 41 ☽ Q ☉
17 30 ☽ ⊥ ♃
20 51 ☽ ✱ ♆
23 01 ☽ ∠ ☿

31 Saturday
04 14 ☽ △ ♇
08 53 ☽ ∠ ♃
11 51 ☽ ✱ ♅
14 39 ☽ ✱ ☉
16 02 ☽ ⊥ ☿
17 18 ☽ Q ♆
18 55 ☽ ∥ ♂
21 08 ☽ Q ♀
21 47 ☿ ⊻ ♄

All ephemeris data is given at 12.00 UT and the Moon's longitude is additionally given for 24.00 UT

JUNE 1997

LONGITUDES

Date	Sidereal time h m s	Sun ☉ ° ' "	Moon ☽ ° ' "	Moon ☽ 24.00 ° ' "	Mercury ☿ ° '	Venus ♀ ° '	Mars ♂ ° '	Jupiter ♃ ° '	Saturn ♄ ° '	Uranus ♅ ° '	Neptune ♆ ° '	Pluto ♇ ° '
01	04 40 02	11 ♊ 01 36	22 ♈ 38 39	29 ♈ 37 27	18 ♉ 09	26 ♊ 43	23 ♍ 06	21 ♒ 49	17 ♈ 22	08 ♒ 31	29 ♑ 43	04 ♐ 02
02	04 43 59	11 59 06	06 ♉ 34 39	13 ♉ 29 53	19 36	27 57	23 26	21 51	17 28	08 R 30	29 R 42	04 R 01
03	04 47 55	12 56 35	20 ♉ 22 49	27 ♉ 13 03	21 05	29 ♊ 10	23 47	21 52	17 33	08 29	29 41	03 59
04	04 51 52	13 54 04	04 ♊ 00 13	10 ♊ 43 56	22 37	00 ♋ 24	24 08	21 54	17 38	08 28	29 40	03 57
05	04 55 49	14 51 31	17 ♊ 23 53	23 ♊ 59 47	24 11	01 37	24 29	21 54	17 44	08 27	29 39	03 56
06	04 59 45	15 48 58	00 ♋ 31 26	06 ♋ 58 40	25 48	02 50	24 51	21 55	17 49	08 26	29 38	03 54
07	05 03 42	16 46 24	13 ♋ 21 27	19 ♋ 39 50	27 28	04 04	25 13	21 56	17 54	08 25	29 37	03 53
08	05 07 38	17 43 48	25 ♋ 53 56	02 ♌ 04 00	29 ♉ 10	05 17	25 35	21 56	17 59	08 24	29 36	03 51
09	05 11 35	18 41 12	08 ♌ 10 19	14 ♌ 13 16	00 ♊ 55	06 30	25 58	21 56	18 04	08 23	29 35	03 49
10	05 15 31	19 38 35	20 ♌ 13 20	26 ♌ 11 00	02 43	07 44	26 21	21 R 56	18 09	08 21	29 34	03 48
11	05 19 28	20 35 56	02 ♍ 06 53	08 ♍ 01 33	04 33	08 57	26 45	21 56	18 13	08 20	29 32	03 46
12	05 23 24	21 33 17	13 ♍ 55 41	19 ♍ 49 56	06 25	10 10	27 08	21 56	18 18	08 19	29 31	03 45
13	05 27 21	22 30 37	25 ♍ 45 00	01 ♎ 41 34	08 20	11 24	27 32	21 55	18 23	08 17	29 30	03 43
14	05 31 18	23 27 55	07 ♎ 40 19	13 ♎ 41 55	10 17	12 37	27 57	21 55	18 27	08 16	29 29	03 42
15	05 35 14	24 25 13	19 ♎ 46 59	25 ♎ 56 08	12 16	13 50	28 22	21 54	18 32	08 14	29 28	03 40
16	05 39 11	25 22 30	02 ♏ 09 52	08 ♏ 28 40	14 18	15 03	28 47	21 52	18 36	08 13	29 26	03 39
17	05 43 07	26 19 46	14 ♏ 52 53	21 ♏ 22 46	16 21	16 16	29 12	21 51	18 40	08 11	29 25	03 37
18	05 47 04	27 17 02	27 ♏ 58 28	04 ♐ 40 00	18 26	17 29	29 ♍ 38	21 50	18 45	08 10	29 24	03 36
19	05 51 00	28 14 17	11 ♐ 27 15	18 ♐ 19 55	20 33	18 43	00 ♎ 04	21 48	18 49	08 08	29 22	03 34
20	05 54 57	29 ♊ 11 31	25 ♐ 17 37	02 ♑ 19 49	22 42	19 56	00 30	21 46	18 53	08 06	29 21	03 33
21	05 58 53	00 ♋ 08 45	09 ♑ 25 53	16 ♑ 35 07	24 51	21 09	00 57	21 44	18 57	08 05	29 20	03 31
22	06 02 50	01 05 58	23 ♑ 46 44	00 ♒ 59 59	27 02	22 22	01 23	21 42	19 01	08 03	29 18	03 30
23	06 06 46	02 03 11	08 ♒ 14 05	15 ♒ 28 20	29 ♊ 13	23 35	01 50	21 39	19 05	08 01	29 17	03 28
24	06 10 43	03 00 24	22 ♒ 42 04	29 ♒ 54 44	01 ♋ 24	24 48	02 18	21 36	19 09	07 59	29 16	03 27
25	06 14 40	03 57 37	07 ♓ 05 52	14 ♓ 15 06	03 35	26 01	02 45	21 34	19 12	07 58	29 14	03 26
26	06 18 36	04 54 50	21 ♓ 22 09	28 ♓ 26 52	05 46	27 14	03 13	21 31	19 16	07 56	29 13	03 24
27	06 22 33	05 52 03	05 ♈ 29 05	12 ♈ 28 47	07 57	28 27	03 41	21 27	19 19	07 54	29 11	03 23
28	06 26 29	06 49 16	19 ♈ 25 53	26 ♈ 20 25	10 07	29 ♋ 40	04 10	21 24	19 23	07 52	29 10	03 22
29	06 30 26	07 46 29	03 ♉ 12 22	10 ♉ 01 43	12 16	00 ♌ 53	04 38	21 21	19 26	07 50	29 08	03 20
30	06 34 22	08 ♋ 43 42	16 ♉ 48 25	23 ♉ 32 27	14 ♋ 24	02 ♌ 06	05 ♎ 07	21 ♒ 17	19 ♈ 29	07 ♒ 48	29 ♑ 07	03 ♐ 19

Date	Moon True ☊ ° '	Moon Mean ☊ ° '	Moon ☽ Latitude ° '
01	25 ♍ 51	25 ♍ 02	02 S 17
02	25 R 46	24 59	03 18
03	25 38	24 56	04 07
04	25 28	24 52	04 40
05	25 16	24 49	04 58
06	25 04	24 46	05 00
07	24 52	24 43	04 46
08	24 42	24 40	04 18
09	24 34	24 37	03 39
10	24 29	24 33	02 51
11	24 26	24 30	01 55
12	24 D 25	24 27	00 S 55
13	24 R 25	24 24	00 N 07
14	24 25	24 21	01 10
15	24 23	24 17	02 10
16	24 20	24 14	03 06
17	24 13	24 11	03 54
18	24 05	24 08	04 31
19	23 54	24 05	04 54
20	23 43	24 02	05 01
21	23 31	23 58	04 50
22	23 21	23 55	04 20
23	23 13	23 52	03 34
24	23 08	23 49	02 34
25	23 06	23 46	01 24
26	23 05	23 43	00 N 09
27	23 D 05	23 39	01 S 05
28	23 R 05	23 36	02 15
29	23 02	23 33	03 16
30	22 ♍ 58	23 ♍ 30	04 S 05

DECLINATIONS

Date	Sun ☉ ° '	Moon ☽ ° '	Mercury ☿ ° '	Venus ♀ ° '	Mars ♂ ° '	Jupiter ♃ ° '	Saturn ♄ ° '	Uranus ♅ ° '	Neptune ♆ ° '	Pluto ♇ ° '
01	22 N 06	06 N 41	14 N 32	24 N 12	03 N 33	14 S 55	04 N 42	18 S 44	19 S 48	08 S 13
02	22 13	10 36	15 03	24 15	03 23	14 55	04 44	18 44	19 48	08 13
03	22 21	13 53	15 33	24 18	03 13	14 54	04 45	18 45	19 48	08 13
04	22 28	16 21	16 05	24 21	03 04	14 54	04 47	18 45	19 48	08 13
05	22 35	17 53	16 36	24 22	02 54	14 54	04 49	18 45	19 49	08 13
06	22 41	18 27	17 08	24 23	02 43	14 54	04 51	18 46	19 49	08 13
07	22 47	18 02	17 40	24 23	02 33	14 54	04 52	18 46	19 49	08 13
08	22 52	16 44	18 12	24 23	02 23	14 54	04 54	18 46	19 49	08 13
09	22 57	14 42	18 44	24 21	02 12	14 55	04 56	18 47	19 49	08 13
10	23 02	12 03	19 16	24 19	02 02	14 55	04 57	18 47	19 50	08 12
11	23 06	08 55	19 47	24 16	01 51	14 55	04 59	18 47	19 50	08 12
12	23 10	05 28	20 17	24 13	01 40	14 55	05 01	18 48	19 50	08 12
13	23 14	01 N 48	20 47	24 09	01 29	14 56	05 02	18 48	19 50	08 12
14	23 17	01 S 59	21 16	24 04	01 18	14 56	05 04	18 49	19 50	08 12
15	23 19	05 44	21 43	23 58	01 07	14 57	05 05	18 49	19 51	08 12
16	23 21	09 19	22 09	23 52	00 55	14 58	05 07	18 49	19 51	08 12
17	23 23	12 34	22 34	23 45	00 44	14 58	05 08	18 50	19 51	08 12
18	23 25	15 18	22 57	23 37	00 32	14 59	05 09	18 50	19 52	08 12
19	23 26	17 18	23 17	23 29	00 21	15 00	05 11	18 51	19 52	08 12
20	23 26	18 21	23 36	23 20	00 N 09	15 01	05 12	18 51	19 52	08 12
21	23 26	18 17	23 52	23 10	00 S 03	15 01	05 13	18 52	19 52	08 12
22	23 26	17 04	24 06	23 00	00 15	15 02	05 15	18 52	19 53	08 12
23	23 25	14 46	24 17	22 49	00 27	15 03	05 16	18 53	19 53	08 12
24	23 24	11 32	24 26	22 37	00 39	15 05	05 17	18 53	19 53	08 12
25	23 23	07 37	24 32	22 25	00 51	15 06	05 18	18 54	19 54	08 12
26	23 21	03 S 17	24 35	22 12	01 03	15 07	05 19	18 54	19 54	08 12
27	23 18	01 N 11	24 35	21 58	01 16	15 08	05 20	18 55	19 54	08 13
28	23 16	05 31	24 32	21 44	01 28	15 09	05 22	18 55	19 55	08 13
29	23 13	09 31	24 26	21 29	01 41	15 11	05 23	18 56	19 55	08 13
30	23 N 09	12 N 57	24 N 18	21 N 13	01 S 53	15 S 12	05 N 24	18 S 56	19 S 55	08 S 13

ZODIAC SIGN ENTRIES

Date	h m	Planets
02	00 39	☽ ♉
04	04 18	♀ ♋
04	04 55	☽ ♊
06	11 02	☽ ♋
08	19 58	☽ ♌
08	23 25	☿ ♊
11	07 43	☽ ♍
13	20 35	☽ ♎
16	07 51	☽ ♏
18	15 39	☽ ♐
19	08 30	♂ ♎
20	20 02	☽ ♑
21	08 20	☉ ♋
22	22 20	☽ ♒
23	20 41	☿ ♋
25	00 09	☽ ♓
27	02 38	☽ ♈
28	18 38	♀ ♌
29	06 23	☽ ♉

LATITUDES

Date	Mercury ☿ ° '	Venus ♀ ° '	Mars ♂ ° '	Jupiter ♃ ° '	Saturn ♄ ° '	Uranus ♅ ° '	Neptune ♆ ° '	Pluto ♇ ° '
01	02 S 48	00 N 48	00 N 53	00 S 43	02 S 18	00 S 38	00 N 25	12 N 56
04	02 25	00 55	00 48	00 44	02 19	00 38	00 25	12 56
07	01 58	01 01	00 42	00 45	02 19	00 38	00 25	12 56
10	01 28	01 07	00 38	00 46	02 20	00 38	00 25	12 55
13	00 55	01 12	00 33	00 46	02 21	00 38	00 25	12 54
16	00 S 22	01 17	00 28	00 47	02 21	00 38	00 25	12 54
19	00 N 11	01 22	00 24	00 48	02 22	00 38	00 25	12 53
22	00 42	01 26	00 20	00 49	02 23	00 38	00 25	12 52
25	01 08	01 30	00 16	00 49	02 23	00 38	00 25	12 51
28	01 29	01 33	00 12	00 50	02 24	00 38	00 25	12 50
31	01 N 43	01 N 35	00 N 08	00 S 51	02 S 25	00 S 38	00 N 25	12 N 49

DATA

Julian Date	2450601
Delta T	+62 seconds
Ayanamsa	23° 49' 13"
Synetic vernal point	05° ♓ 17' 46"
True obliquity of ecliptic	23° 26' 13"

MOON'S PHASES, APSIDES AND POSITIONS ☽

Date	h m	Phase	Longitude ° '	Eclipse Indicator
05	07 04	●	14 ♊ 40	
13	04 51	☽	22 ♍ 14	
20	19 09	○	29 ♐ 29	
27	12 42	☾	05 ♈ 54	

Day	h m		
12	05 06	Apogee	
24	05 10	Perigee	
06	13 27	Max dec	18° N 27'
13	23 28	0S	
20	22 52	Max dec	18° S 28'
27	05 39	0N	

LONGITUDES

Date	Chiron ⚷ ° '	Ceres ⚳ ° '	Pallas ⚴ ° '	Juno ⚵ ° '	Vesta ⚶ ° '	Black Moon Lilith ⚸ ° '
01	26 ♎ 07	10 ♓ 16	06 ♒ 14	03 ♋ 16	10 ♈ 24	08 ♍ 15
11	25 ♎ 49	11 ♓ 55	05 ♒ 16	08 ♋ 46	14 ♈ 15	09 ♍ 22
21	25 ♎ 41	13 ♓ 07	03 ♒ 38	14 ♋ 13	17 ♈ 51	10 ♍ 28
31	25 ♎ 44	13 ♓ 49	01 ♒ 28	19 ♋ 36	21 ♈ 10	11 ♍ 35

ASPECTARIAN

01 Sunday
00 35 ☽ ∥ ♄
02 54 ☽ ☌ ♄
03 26 ☽ ⊻ ☿
05 50 ☽ bQ ♇
08 22 ☽ Q ♅
10 35 ☽ ✶ ♃
12 48 ☽ ⊼ ♂
18 14 ☽ ∠ ☉
19 41 ☽ ✶ ♀
21 04 ☽ # ♇
21 15 ☽ ± ♇
23 22 ☽ ± ♂

02 Monday
00 09 ☽ □ ♆
05 51 ☿ # ♃
07 17 ☽ Q ♃
07 34 ☽ ⊼ ♇
10 54 ☽ ⊥ ☉
15 18 ☽ bQ ♂
15 20 ☽ □ ♅
16 45 ♂ bQ ♅
22 04 ☽ ⊻ ☉

03 Tuesday
00 07 ☽ ∠ ♀
07 01 ☽ ⊻ ♄
13 23 ☽ ☌ ☿
14 37 ☽ □ ♃
17 22 ☽ ⊥ ♀
17 35 ☽ ⊥ ♄
18 07 ☽ △ ♂
20 03 ♀ Q ♄
20 59 ☽ # ♃
21 57 ♀ ⊼ ♆

04 Wednesday
00 39 ☿ □ ♃
04 20 ☽ △ ♆
04 58 ☽ ⊻ ♀
07 51 ☽ ∥ ☿
09 34 ☽ ∠ ♄
11 55 ☽ ☍ ♇
19 57 ☽ △ ♅

05 Thursday
04 37 ☿ ⊥ ♄
06 52 ☉ bQ ♆
07 03 ☽ bQ ♆
07 04 ☽ ☌ ☉
12 36 ☽ ✶ ♄
17 42 ☿ △ ♂
20 12 ☽ △ ♃
23 00 ☽ bQ ♅
23 21 ☽ ± ♆

06 Friday
01 15 ☽ □ ♂
02 05 ☽ ⊻ ☿
04 15 ♀ ± ♅
10 21 ☽ ⊼ ♆
10 40 ☽ Q ♄
14 43 ☽ ⊥ ☿
15 32 ☽ ± ♅
16 44 ☽ ☌ ♀
18 15 ☽ ⊼ ♇
23 54 ☽ bQ ♃

07 Saturday
02 43 ☽ ⊼ ♅
05 27 ☽ ± ♇
08 25 ♀ ⊼ ♇
10 04 ☽ ∠ ☿
11 43 ☽ Q ♂
16 53 ☽ ± ♃
17 39 ☽ ∥ ☿
19 01 ☽ ⊻ ☉
20 41 ☽ □ ♄
22 28 ☽ bQ ♇

08 Sunday
04 21 ☽ ⊼ ♃
07 28 ☽ ⊥ ☉
11 23 ☽ ✶ ♂
17 13 ☿ ⊥ ♀
17 49 ☿ △ ♆
18 51 ☉ ✶ ♄
19 10 ☽ ☍ ♆
19 24 ☽ ✶ ☿

09 Monday
02 25 ☽ ∠ ☉
03 27 ☽ △ ♇
08 21 ☽ ⊻ ♀
09 47 ☽ # ♃
12 24 ☽ ☍ ♅
13 51 ☿ # ♅
17 43 ☽ ∠ ♂
20 32 ♀ bQ ♃
21 33 ☽ ⊥ ♀
23 02 ☽ Q ☿

10 Tuesday
00 23 ♃ St R
07 48 ☽ △ ♄
10 44 ☽ ✶ ☉
12 16 ☽ ⊥ ♂
15 27 ☽ ☍ ♃
17 37 ☽ ∠ ♀
17 58 ☿ ∠ ♄

11 Wednesday
00 07 ♀ ⊼ ♅
00 45 ☽ ⊻ ♂
02 05 ☿ ☍ ♇
06 48 ☽ ⊼ ♆
13 04 ☽ Q ☉
14 16 ☽ bQ ♄
14 23 ☿ # ♆
15 21 ☽ □ ♇
17 09 ☽ # ♇
17 51 ☽ □ ☿
18 56 ☽ ± ♆

12 Thursday
00 36 ☽ ⊼ ♅
03 29 ☽ ✶ ♀
03 48 ♀ ± ♇
08 40 ☽ ± ♄
12 47 ☽ ± ♅
13 12 ☽ bQ ♆
15 02 ☽ ∥ ♄
20 57 ☽ ⊼ ♄
21 22 ☉ △ ♃

13 Friday
03 51 ☽ Q ♇
04 15 ☽ ⊼ ♃
04 51 ☽ □ ☉
06 40 ☽ Q ♀
07 01 ☽ bQ ♅
11 30 ☿ △ ♅
14 07 ☽ ∥ ♂
15 45 ☽ ☌ ♂
16 23 ☽ ± ♃
19 34 ☽ △ ♆

14 Saturday
04 02 ☽ ✶ ♇
07 04 ☉ bQ ♅
07 54 ☽ # ♂
09 42 ♂ ± ♃
10 28 ☽ bQ ♃
12 23 ☉ ± ♆
13 11 ☽ △ ♅
18 14 ☽ △ ☿
22 57 ☽ □ ♀

15 Sunday
07 50 ☽ # ♄
09 31 ☽ ☍ ♄
09 49 ☽ ∠ ♇
16 07 ☽ △ ♃
21 49 ☽ △ ☉

16 Monday
03 19 ☽ ⊥ ♇
04 23 ☽ ∥ ♇
05 16 ☽ ⊻ ♂
05 25 ☽ bQ ☿
06 47 ☽ □ ♆
13 41 ☿ bQ ♆
14 49 ☽ ⊻ ♇
17 10 ☽ ⊥ ♂
22 00 ♄ bQ ♇
23 29 ☽ □ ♅

17 Tuesday
01 54 ☽ ± ☿
03 53 ♀ ± ♃
04 50 ☽ bQ ☉
09 43 ☿ ⊻ ♀
10 42 ☽ ∠ ♂
14 51 ☽ △ ♀
15 15 ☽ ⊼ ☿
16 42 ☽ Q ♆
19 04 ☽ ⊼ ♄
22 52 ☽ ± ☉
23 32 ♂ △ ♆

18 Wednesday
00 50 ☽ □ ♃
06 07 ☽ ± ♄
08 44 ☽ Q ♅
08 48 ☽ ∥ ♃
10 39 ☽ ⊼ ☉
14 34 ☽ ✶ ♆
15 05 ☽ ✶ ♂
15 35 ☿ ✶ ♄
16 15 ♀ Q ♂
20 56 ☽ bQ ♀
22 04 ☽ ☌ ♇
22 25 ☽ bQ ♄

19 Thursday
06 10 ☽ ✶ ♅
09 06 ☽ Q ♃
09 17 ♀ bQ ♇
13 06 ☽ Q ♂
14 12 ♀ □ ♄
14 25 ☽ ± ♀
17 06 ☽ ∠ ♆
20 25 ☉ ∥ ♀
21 44 ☿ ∥ ♀
22 25 ☉ ∥ ☿

20 Friday
00 54 ☽ △ ♄
01 45 ☿ △ ♃
01 52 ☽ ⊼ ♀
05 57 ☽ ✶ ♃
06 43 ☽ ☍ ☿
08 15 ☽ ∠ ♅
08 40 ☽ ⊥ ♆
15 57 ☉ ⊼ ♆
16 32 ☿ bQ ♅
18 56 ☽ ⊻ ♆
19 09 ☽ ☍ ☉
19 16 ☿ ± ♆
21 11 ☽ □ ♂
23 36 ☽ ⊥ ♅

21 Saturday
02 02 ☽ ⊻ ♇
07 28 ☽ ∠ ♃
09 44 ☽ ⊻ ♅
12 09 ☽ ⊥ ♇
22 33 ☽ ⊥ ♃
23 11 ♀ ⊼ ♃

22 Sunday
03 13 ☽ ∠ ♇
04 01 ☽ □ ♄
08 32 ☽ ⊻ ♃
09 26 ☽ ☍ ♀
09 44 ☉ Q ♄
18 22 ☽ ⊼ ☿
21 11 ☽ ☌ ♆

23 Monday
01 01 ☽ ⊼ ☉
01 03 ☽ △ ♂
01 49 ☉ □ ♂
04 07 ☽ ✶ ♇
06 05 ☽ ± ☿
09 27 ☽ ∥ ♃
10 05 ☽ Q ♄
11 12 ☉ ± ♅
11 39 ☽ ☌ ♅
11 41 ☽ ± ☉
12 49 ☿ ⊼ ♆
23 40 ☽ bQ ☿
23 59 ☽ Q ♇

24 Tuesday
02 44 ☽ bQ ♂
03 39 ☽ bQ ☉
06 04 ☽ ✶ ♄
09 07 ☿ Q ♄
10 11 ☽ ☌ ♃
15 49 ☽ ⊼ ♀
18 10 ☽ ± ♂
18 24 ☿ ± ♅
22 54 ☽ ⊻ ♆
22 56 ☉ ⊼ ♇

25 Wednesday
00 26 ☿ □ ♂
02 43 ☽ ± ♀
04 30 ☽ ⊼ ♂
05 05 ☽ △ ☿
05 53 ☽ □ ♇
06 23 ☽ △ ☉
07 08 ☽ ∠ ♄
08 32 ☽ ∥ ♇
08 54 ☽ ⊥ ♆
10 16 ☿ ⊼ ♇
13 26 ☽ ⊻ ♅
19 11 ☽ bQ ♀
19 14 ☉ ☌ ☿
22 17 ☽ ⊥ ♄
23 29 ☽ ⊥ ♅
23 57 ☽ ∠ ♆

26 Thursday
00 55 ☽ # ♄
08 26 ☽ ⊻ ♄
12 14 ☽ ⊻ ♃
14 38 ☽ ∠ ♅
19 54 ☿ bQ ♃
21 12 ♂ ✶ ♇
22 22 ☽ ⊥ ♃
22 52 ☽ △ ♀
23 27 ☽ ∥ ♂

27 Friday
01 17 ☽ ✶ ♆
08 25 ☽ △ ♇
08 49 ☽ ☍ ♂
11 23 ☿ ⊼ ♅
12 29 ☽ # ♂
12 42 ☽ □ ☉
13 40 ☽ ∠ ♃
16 07 ☽ ✶ ♅
17 00 ☽ □ ☿
21 46 ☽ Q ♆

28 Saturday
02 03 ☉ bQ ♃
02 21 ♀ ☍ ♆
03 40 ☿ ± ♇
10 09 ☽ bQ ♇
11 04 ☽ ∥ ♄
11 54 ☽ ☌ ♄
12 45 ☽ Q ♅
15 24 ☽ ✶ ♃
22 03 ☽ Q ☉

29 Sunday
01 46 ☽ ± ♇
03 52 ☽ # ♇
04 54 ☽ □ ♆
05 55 ☽ Q ☿
07 32 ☽ □ ♀
12 14 ☽ ⊼ ♇
12 14 ☽ Q ♃
13 26 ☉ ⊼ ♅
14 36 ☽ ⊼ ♂
20 07 ☽ □ ♅
20 38 ☽ ✶ ☉

30 Monday
01 33 ☽ ± ♂
06 57 ☽ ✶ ☿
16 47 ☽ ⊻ ♄
18 06 ☽ bQ ♂
18 26 ☽ Q ♀
19 56 ☽ □ ♃
21 39 ☿ ± ♃

All ephemeris data is given at 12.00 UT and the Moon's longitude is additionally given for 24.00 UT

JULY 1997

LONGITUDES

Date	Sidereal time h m s	Sun ☉ ° ' "	Moon ☽ ° ' "	Moon ☽ 24.00 ° ' "	Mercury ☿ ° '	Venus ♀ ° '	Mars ♂ ° '	Jupiter ♃ ° '	Saturn ♄ ° '	Uranus ♅ ° '	Neptune ♆ ° '	Pluto ♇ ° '
01	06 38 19	09 ♋ 40 55	00 ♊ 13 44	06 ♊ 52 10	16 ♋ 31	03 ♌ 19	05 ♎ 36	21 ♒ 13	19 ♈ 32	07 ♒ 46	29 ♑ 05	03 ♐ 18
02	06 42 15	10 38 09	13 ♊ 27 38	20 ♊ 00 01	18 36	04 31	06 05	21 R 09	19 35	07 R 44	29 R 04	03 R 17
03	06 46 12	11 35 22	26 ♊ 29 12	02 ♋ 55 05	20 40	05 44	06 35	21 05	19 38	07 42	29 02	03 16
04	06 50 09	12 32 35	09 ♋ 17 35	15 ♋ 36 39	22 42	06 57	07 05	21 00	19 41	07 40	29 01	03 14
05	06 54 05	13 29 49	21 ♋ 52 15	28 ♋ 04 28	24 42	08 10	07 35	20 56	19 44	07 38	28 59	03 13
06	06 58 02	14 27 02	04 ♌ 13 22	10 ♌ 19 08	26 40	09 23	08 05	20 51	19 47	07 36	28 58	03 12
07	07 01 58	15 24 16	16 ♌ 21 59	22 ♌ 22 12	28 ♋ 36	10 36	08 35	20 46	19 49	07 33	28 56	03 11
08	07 05 55	16 21 29	28 ♌ 20 08	04 ♍ 16 13	00 ♌ 31	11 48	09 06	20 41	19 52	07 31	28 55	03 10
09	07 09 51	17 18 42	10 ♍ 10 52	16 ♍ 04 39	02 24	13 01	09 37	20 36	19 54	07 29	28 53	03 09
10	07 13 48	18 15 56	21 ♍ 58 05	27 ♍ 51 48	04 14	14 14	10 08	20 30	19 56	07 27	28 52	03 08
11	07 17 44	19 13 09	03 ♎ 46 24	09 ♎ 42 34	06 03	15 26	10 39	20 25	19 59	07 25	28 50	03 07
12	07 21 41	20 10 22	15 ♎ 40 57	21 ♎ 42 14	07 50	16 39	11 11	20 19	20 01	07 22	28 48	03 06
13	07 25 38	21 07 35	27 ♎ 47 05	03 ♏ 56 08	09 35	17 52	11 42	20 13	20 03	07 20	28 47	03 05
14	07 29 34	22 04 48	10 ♏ 10 01	16 ♏ 29 16	11 18	19 04	12 14	20 08	20 05	07 18	28 45	03 04
15	07 33 31	23 02 01	22 ♏ 54 24	29 ♏ 25 46	12 59	20 17	12 46	20 01	20 06	07 16	28 43	03 03
16	07 37 27	23 59 14	06 ♐ 03 41	12 ♐ 48 18	14 38	21 29	13 18	19 55	20 08	07 13	28 42	03 02
17	07 41 24	24 56 28	19 ♐ 39 36	26 ♐ 37 25	16 15	22 42	13 51	19 49	20 10	07 11	28 40	03 01
18	07 45 20	25 53 42	03 ♑ 41 26	10 ♑ 51 08	17 50	23 54	14 23	19 43	20 11	07 09	28 39	03 00
19	07 49 17	26 50 56	18 ♑ 05 51	25 ♑ 24 47	19 23	25 07	14 56	19 36	20 13	07 06	28 37	03 00
20	07 53 13	27 48 10	02 ♒ 47 00	10 ♒ 11 31	20 54	26 19	15 29	19 29	20 14	07 04	28 35	02 59
21	07 57 10	28 45 25	17 ♒ 37 17	25 ♒ 03 19	22 23	27 32	16 02	19 23	20 15	07 02	28 34	02 58
22	08 01 07	29 ♋ 42 41	02 ♓ 28 37	09 ♓ 52 19	23 51	28 44	16 36	19 16	20 16	06 59	28 32	02 57
23	08 05 03	00 ♌ 39 57	17 ♓ 13 38	24 ♓ 31 54	25 16	29 ♌ 56	17 09	19 09	20 17	06 57	28 31	02 57
24	08 09 00	01 37 14	01 ♈ 46 37	08 ♈ 57 24	26 39	01 ♍ 08	17 43	19 02	20 18	06 55	28 29	02 56
25	08 12 56	02 34 32	16 ♈ 03 57	23 ♈ 06 09	28 00	02 21	18 17	18 55	20 19	06 52	28 27	02 55
26	08 16 53	03 31 51	00 ♉ 03 56	06 ♉ 57 18	29 ♌ 19	03 33	18 50	18 47	20 20	06 50	28 26	02 55
27	08 20 49	04 29 11	13 ♉ 46 21	20 ♉ 31 10	00 ♍ 36	04 45	19 25	18 40	20 20	06 47	28 24	02 54
28	08 24 46	05 26 32	27 ♉ 11 54	03 ♊ 48 43	01 51	05 57	19 59	18 33	20 21	06 45	28 22	02 54
29	08 28 42	06 23 55	10 ♊ 21 45	16 ♊ 51 11	03 03	07 09	20 33	18 25	20 21	06 43	28 21	02 53
30	08 32 39	07 21 18	23 ♊ 17 10	29 ♊ 39 48	04 13	08 21	21 08	18 18	20 21	06 40	28 19	02 53
31	08 36 36	08 ♌ 18 42	05 ♋ 59 15	12 ♋ 15 37	05 ♍ 21	09 ♍ 34	21 ♎ 43	18 ♒ 10	20 ♈ 22	06 ♒ 38	28 ♑ 18	02 ♐ 52

Date	Moon True ☊ ° '	Moon Mean ☊ ° '	Moon ☽ Latitude ° '
01	22 ♍ 51	23 ♍ 27	04 S 39
02	22 R 41	23 23	04 58
03	22 30	23 20	05 01
04	22 18	23 17	04 49
05	22 06	23 14	04 23
06	21 57	23 11	03 45
07	21 49	23 08	02 57
08	21 44	23 04	02 01
09	21 41	23 01	01 S 01
10	21 D 41	22 58	00 N 02
11	21 41	22 55	01 04
12	21 42	22 52	02 05
13	21 R 42	22 49	03 01
14	21 40	22 45	03 50
15	21 36	22 42	04 29
16	21 30	22 39	04 55
17	21 23	22 36	05 06
18	21 14	22 33	04 59
19	21 06	22 29	04 33
20	20 59	22 26	03 49
21	20 53	22 23	02 49
22	20 49	22 20	01 37
23	20 48	22 17	00 N 19
24	20 D 48	22 14	00 S 59
25	20 49	22 10	02 12
26	20 50	22 07	03 16
27	20 R 50	22 04	04 07
28	20 48	22 01	04 44
29	20 44	21 58	05 04
30	20 38	21 55	05 09
31	20 ♍ 31	21 ♍ 51	04 S 58

DECLINATIONS

Date	Sun ☉ ° '	Moon ☽ ° '	Mercury ☿ ° '	Venus ♀ ° '	Mars ♂ ° '	Jupiter ♃ ° '	Saturn ♄ ° '	Uranus ♅ ° '	Neptune ♆ ° '	Pluto ♇ ° '
01	23 N 05	15 N 39	24 N 07	20 N 57	02 S 06	15 S 14	05 N 24	18 S 57	19 S 56	08 S 13
02	23 01	17 29	23 54	20 41	02 19	15 15	05 25	18 57	19 56	08 13
03	22 56	18 22	23 39	20 23	02 32	15 17	05 26	18 58	19 56	08 13
04	22 51	18 18	23 21	20 06	02 44	15 19	05 27	18 58	19 57	08 13
05	22 45	17 20	23 01	19 47	02 57	15 20	05 28	18 59	19 57	08 14
06	22 39	15 34	22 39	19 28	03 10	15 22	05 29	18 59	19 57	08 14
07	22 33	13 07	22 16	19 09	03 23	15 24	05 29	19 00	19 58	08 14
08	22 26	10 09	21 50	18 49	03 37	15 26	05 30	19 01	19 58	08 14
09	22 19	06 48	21 24	18 29	03 50	15 27	05 31	19 01	19 58	08 14
10	22 11	03 N 13	20 56	18 08	04 03	15 29	05 31	19 02	19 59	08 14
11	22 04	00 S 31	20 26	17 46	04 16	15 31	05 32	19 02	19 59	08 15
12	21 55	04 15	19 56	17 25	04 30	15 33	05 32	19 03	19 59	08 15
13	21 47	07 52	19 24	17 02	04 43	15 35	05 33	19 04	20 00	08 15
14	21 38	11 14	18 52	16 40	04 57	15 37	05 33	19 04	20 00	08 15
15	21 28	14 10	18 18	16 16	05 10	15 40	05 34	19 05	20 00	08 16
16	21 19	16 28	17 44	15 53	05 24	15 42	05 34	19 05	20 01	08 16
17	21 08	17 57	17 10	15 29	05 37	15 44	05 35	19 06	20 01	08 16
18	20 58	18 24	16 34	15 04	05 51	15 46	05 35	19 07	20 01	08 16
19	20 47	17 42	15 59	14 40	06 04	15 48	05 35	19 07	20 02	08 17
20	20 36	15 49	15 23	14 14	06 18	15 51	05 35	19 08	20 02	08 17
21	20 24	12 52	14 47	13 49	06 32	15 53	05 36	19 08	20 02	08 17
22	20 13	09 05	14 11	13 23	06 45	15 55	05 36	19 09	20 03	08 18
23	20 00	04 45	13 34	12 57	06 59	15 58	05 36	19 10	20 03	08 18
24	19 48	00 S 12	12 58	12 30	07 13	16 00	05 36	19 10	20 03	08 18
25	19 35	04 N 17	12 21	12 03	07 27	16 02	05 36	19 11	20 04	08 19
26	19 22	08 26	11 45	11 36	07 41	16 05	05 36	19 11	20 04	08 19
27	19 08	12 02	11 09	11 09	07 54	16 07	05 36	19 12	20 04	08 19
28	18 54	14 56	10 33	10 41	08 08	16 10	05 36	19 13	20 05	08 20
29	18 40	16 59	09 57	10 13	08 22	16 12	05 36	19 13	20 05	08 20
30	18 26	18 08	09 22	09 45	08 36	16 15	05 36	19 14	20 05	08 20
31	18 N 11	18 N 20	08 N 47	09 N 16	08 S 50	16 S 17	05 N 35	19 S 14	20 S 06	08 S 21

ZODIAC SIGN ENTRIES

Date	h	m	Planets
01	11	35	☽ ♊
03	18	33	☽ ♋
06	03	45	☽ ♌
08	05	28	☿ ♌
08	15	22	☽ ♍
11	04	21	☽ ♎
13	16	20	☽ ♏
16	01	02	☽ ♐
18	05	45	☽ ♑
20	07	29	☽ ♒
22	07	59	☽ ♓
22	19	15	☉ ♌
23	13	16	♀ ♍
24	09	03	☽ ♈
26	11	53	☽ ♉
27	00	42	☿ ♍
28	17	04	☽ ♊
31	00	38	☽ ♋

LATITUDES

Date	Mercury ☿ ° '	Venus ♀ ° '	Mars ♂ ° '	Jupiter ♃ ° '	Saturn ♄ ° '	Uranus ♅ ° '	Neptune ♆ ° '	Pluto ♇ ° '
01	01 N 43	01 N 35	00 N 08	00 S 51	02 S 25	00 S 38	00 N 25	12 N 49
04	01 51	01 37	00 04	00 52	02 26	00 39	00 25	12 48
07	01 52	01 38	00 N 01	00 52	02 26	00 39	00 25	12 47
10	01 47	01 39	00 S 03	00 53	02 27	00 39	00 25	12 46
13	01 36	01 38	00 06	00 54	02 28	00 39	00 25	12 45
16	01 21	01 38	00 09	00 54	02 29	00 39	00 25	12 43
19	01 02	01 36	00 12	00 55	02 30	00 39	00 25	12 42
22	00 38	01 34	00 15	00 55	02 31	00 39	00 25	12 41
25	00 N 12	01 31	00 18	00 56	02 31	00 39	00 25	12 39
28	00 S 17	01 27	00 21	00 57	02 32	00 39	00 25	12 38
31	00 S 49	01 N 23	00 S 24	00 S 57	02 S 33	00 S 39	00 N 25	12 N 37

DATA

Julian Date	2450631
Delta T	+62 seconds
Ayanamsa	23° 49' 18"
Synetic vernal point	05° ♓ 17' 41"
True obliquity of ecliptic	23° 26' 13"

MOON'S PHASES, APSIDES AND POSITIONS ☽

Date	h	m	Phase	Longitude ° '	Eclipse Indicator
04	18	40	●	12 ♋ 48	
12	21	44	☽	20 ♎ 34	
20	03	20	○	27 ♑ 28	
26	18	28	☾	03 ♉ 47	

Day	h	m		
09	22	54	Apogee	
21	23	08	Perigee	
03	22	17	Max dec	18° N 27'
11	08	42	0S	
18	09	42	Max dec	18° S 25'
24	13	02	0N	
31	05	29	Max dec	18° N 22'

LONGITUDES

Date	Chiron ⚷ ° '	Ceres ⚳ ° '	Pallas ⚴ ° '	Juno ⚵ ° '	Vesta ⚶ ° '	Black Moon Lilith ⚸ ° '
01	25 ♎ 44	13 ♓ 49	01 ♒ 28	19 ♋ 36	21 ♈ 10	11 ♍ 35
11	25 ♎ 58	13 ♓ 57	28 ♑ 54	24 ♋ 55	24 ♈ 09	12 ♍ 41
21	26 ♎ 23	13 ♓ 29	26 ♑ 12	00 ♌ 09	26 ♈ 45	13 ♍ 48
31	26 ♎ 58	12 ♓ 27	23 ♑ 35	05 ♌ 18	28 ♈ 53	14 ♍ 54

All ephemeris data is given at 12.00 UT and the Moon's longitude is additionally given for 24.00 UT

ASPECTARIAN

h m Aspects

01 Tuesday
01 17 ☽ ∠ ☉
02 35 ☉ ± ♇
03 33 ☽ ⊥ ♄
07 44 ☽ # ♃
09 57 ☽ △ ♆
11 48 ♀ △ ♇
14 45 ☽ ∠ ☿
17 32 ☽ ☍ ♇
18 07 ☽ ⚹ ♀
18 43 ☽ ⊥ ☉
19 49 ☽ ∠ ♄
22 04 ☽ △ ♂

02 Wednesday
01 36 ☽ △ ♅
06 27 ☽ ⊻ ☉
08 19 ☿ ⚼ ♇
10 08 ☽ ⊥ ☿
13 06 ☽ ⚼ ♆
14 29 ♂ ⚼ ♃
23 12 ☽ ⊻ ☿
23 17 ☽ ⚹ ♄
23 46 ☿ □ ♄

03 Thursday
00 16 ☽ ∠ ♀
02 02 ☽ △ ♃
05 00 ☽ ⚼ ♅
05 37 ☽ ± ♆
16 43 ☿ ⊼ ♃
16 45 ☽ ⊼ ♆
18 41 ☽ ⊥ ♀
21 39 ☽ Q ♄
21 41 ☽ ± ♅

04 Friday
00 37 ☽ ⊼ ♇
05 17 ♄ Q ♅
05 50 ☽ ⚼ ♃
07 07 ☽ ⊻ ♀
07 39 ☽ □ ♂
08 56 ☽ ⊼ ♅
11 54 ☽ ± ♇
16 11 ♀ ⚹ ♂
18 40 ☽ ☌ ☉
22 47 ☽ ± ♃
23 47 ♀ # ♆

05 Saturday
01 40 ♀ ☍ ♅
05 00 ☽ ⚼ ♇
07 53 ☽ □ ♄
10 12 ☽ ⊼ ♃
14 17 ♂ △ ♅
18 29 ☽ ☌ ☿
19 28 ☽ Q ♂

06 Sunday
01 45 ☽ ☍ ♆
02 22 ☿ Q ♂
10 00 ☽ △ ♇
12 08 ☉ ∥ ☿
14 08 ☽ # ♃
18 36 ☽ ☍ ♅
19 55 ☽ ⚹ ♂
21 14 ☉ ± ♃
23 16 ☽ ☌ ♀

07 Monday
09 55 ☽ ⊻ ☉
16 04 ☿ ☍ ♆
18 55 ☽ △ ♄
20 44 ☽ ☍ ♃
22 39 ♀ # ♅
22 56 ☽ ⊥ ☉

08 Tuesday
03 05 ☽ ∠ ♂
13 10 ☽ ⊼ ♆
17 14 ☽ ⊻ ☿
18 38 ☽ ∠ ☉
21 45 ☽ □ ♇
22 04 ☽ ⊥ ♂

09 Wednesday
01 15 ☽ ⚼ ♄
01 16 ☽ ± ♆
02 01 ☽ # ♇
06 32 ☽ ⊼ ♅
07 41 ☽ ⊥ ☿
10 48 ☽ ⊻ ♂
18 26 ☽ ⊻ ♀
18 42 ☽ ± ♅
19 31 ☽ ⚼ ♆
19 36 ☽ ± ♄
20 46 ☽ ∥ ♄
21 39 ☿ △ ♇

10 Thursday
03 48 ☽ ⚹ ☉
05 25 ☽ ∠ ☿
06 48 ☽ # ♂
07 51 ☽ ⊼ ♄
08 03 ☽ ⊥ ♀
08 39 ☉ ⚼ ♇
09 03 ☽ ⊼ ♃
10 18 ☽ Q ♇
12 59 ☽ ⚼ ♅
21 10 ☽ ± ♃

11 Friday
02 00 ☽ △ ♆
04 28 ☽ ∠ ♀
06 22 ☽ Q ☉
10 40 ☽ ⚹ ♇
15 18 ☽ ⚼ ♃
17 26 ☽ ⚹ ☿
19 20 ☽ △ ♅

12 Saturday
02 32 ☽ ☌ ♂
05 55 ☿ ☍ ♅
07 48 ☉ □ ♄
09 16 ☿ # ♆
13 42 ☽ ∥ ♂
14 09 ☽ ⚹ ♀
15 23 ☉ ⊼ ♃
16 49 ☽ ∠ ♇
20 27 ☽ # ♄
20 40 ☽ ☍ ♄
21 11 ☽ △ ♃
21 41 ☽ Q ☿
21 44 ☽ □ ☉

13 Sunday
10 37 ☽ ⊥ ♇
13 57 ☽ □ ♆
14 38 ☽ ∥ ♇
16 31 ☽ Q ♀
22 20 ☽ ⊻ ♇

14 Monday
02 56 ☿ # ♅
06 31 ☽ □ ♅
14 29 ☽ □ ☿
16 07 ☽ ⊻ ♂
20 57 ♃ ⚹ ♄

15 Tuesday
00 28 ☽ Q ♆
03 57 ☽ ⊥ ♂
06 36 ☽ □ ♀
06 41 ☽ □ ♃
06 47 ☽ ⊼ ♄
07 19 ♀ ☍ ♃
07 39 ☿ ⚹ ♂
08 28 ♀ △ ♄
12 15 ☽ △ ☉
16 20 ☽ Q ♅
17 56 ☽ ± ♄
21 21 ☽ ∠ ♂
22 41 ☽ ⚹ ♆

16 Wednesday
02 55 ☽ ∥ ♃
06 09 ☽ # ♀
06 33 ☽ ☌ ♇
10 20 ☽ ⚼ ♄
14 04 ☽ ⚹ ♅
15 18 ☽ Q ♃
17 38 ☽ ⚼ ☉
22 09 ♀ # ♃

17 Thursday
01 25 ☽ ⚹ ♂
01 26 ☽ # ☿
01 33 ☽ ∠ ♆
05 15 ☽ △ ☿
07 14 ♂ # ♄
10 39 ☽ ± ☉
12 16 ☽ ⚹ ♃
12 52 ☽ △ ♄
16 21 ☽ ∠ ♅
17 12 ☽ ⊥ ♆
17 45 ☽ △ ♀
21 47 ☽ ⊼ ☉
23 06 ☽ Q ♂

18 Friday
03 28 ☽ ⊻ ♆
07 42 ☽ ⊥ ♅
10 22 ☽ ⚼ ☿
10 51 ☽ ⊻ ♇
13 42 ☽ ∠ ♃
17 47 ☽ ⊻ ♅
20 55 ☽ ⊥ ♇
21 33 ☽ ⚼ ♀

19 Saturday
03 16 ☽ ± ☿
04 37 ☽ ⊥ ♃
06 35 ☽ □ ♂
11 50 ☽ ∠ ♇
13 49 ☽ ± ♀
14 22 ☽ ⊼ ☿
14 27 ☽ ⊻ ♃
15 12 ☿ ☍ ♃
15 29 ☽ □ ♄
18 39 ☿ # ♃

20 Sunday
00 32 ☽ ⊼ ♀
01 13 ☿ △ ♄
03 20 ☽ ☍ ☉
05 12 ☽ ☌ ♆
11 45 ☽ ∥ ♃
12 19 ☽ ⚹ ♇
17 19 ☽ # ☿
18 56 ☽ ☌ ♅
20 51 ☽ Q ♄

21 Monday
04 01 ☽ # ♀
07 15 ☉ ☍ ♆
07 43 ☽ Q ♇
09 21 ☽ △ ♂
14 49 ☽ ☌ ♃
16 15 ☽ ⚹ ♄
20 32 ☽ ☍ ☿

22 Tuesday
05 24 ☽ ☍ ♀
05 38 ☽ ⊻ ♆
07 13 ☽ ⊼ ☉
08 12 ♀ ⊼ ♆
10 31 ☽ ⚼ ♂
12 47 ☽ □ ♇
15 20 ☽ ⊥ ♆
16 32 ☽ ∥ ♇
16 32 ☽ ∠ ♄
17 36 ☽ ± ☉
19 18 ☽ ⊻ ♅

23 Wednesday
00 28 ☽ ∥ ♂
01 41 ☽ ± ♂
05 02 ☽ ⊥ ♅
05 56 ☽ ∠ ♆
07 11 ☉ # ♆
07 11 ☽ ⊥ ♄
07 27 ☽ # ♄
09 16 ☽ ⚼ ☉
11 52 ☽ ⊼ ♂
15 07 ☽ ⊻ ♃
17 01 ☽ ⊻ ♄
19 43 ☽ ∠ ♅

24 Thursday
00 55 ☽ ⊥ ♃
02 36 ☽ ⊼ ☿
06 32 ☽ ⚹ ♆
10 51 ☽ ⊼ ♀
11 43 ☽ △ ☉
13 36 ☽ ± ☿
13 56 ☽ △ ♇
15 43 ☽ ∠ ♃
20 32 ☽ ⚹ ♅
21 22 ♂ ∠ ♇
21 47 ☽ ± ♀

25 Friday
02 32 ☽ Q ♆
06 16 ☽ ⚼ ☿
14 22 ☽ ⚼ ♀
15 09 ☽ ⚼ ♇
15 55 ☽ ☍ ♂
16 45 ☽ Q ♅
16 48 ☽ ⚹ ♃
19 14 ☽ ☌ ♄
19 20 ☽ ∥ ♄
20 02 ☿ ⊼ ♆
20 41 ☉ △ ♇
23 29 ♀ □ ♇

26 Saturday
06 34 ☽ ± ♇
07 08 ☽ # ♂
09 10 ☽ □ ♆
10 16 ♂ △ ♃
10 20 ☉ ⊻ ♀
10 34 ☽ △ ☿
11 16 ☽ # ♇
13 15 ☽ Q ♃
16 57 ☽ ⊼ ♇
18 28 ☽ □ ☉
18 38 ☽ △ ♀
23 04 ♀ ∠ ♂
23 45 ☽ □ ♅

27 Sunday
05 09 ♀ ± ♆
05 39 ☉ # ♅
06 23 ☽ ∥ ♀
06 35 ☽ ∥ ☿
12 13 ☿ ∥ ♀
20 37 ☽ □ ♃
22 28 ☽ ⊼ ♂
23 41 ☽ ⊻ ♄
23 47 ♀ ⚼ ♄

28 Monday
04 43 ☽ Q ☉
09 42 ☽ ± ♂
10 28 ☽ ⊥ ♄
14 07 ☽ △ ♆
21 16 ☽ □ ☿
22 20 ☽ ☍ ♇

29 Tuesday
01 26 ☽ # ♃
02 47 ☽ ⚼ ♂
02 48 ☽ ∠ ♄
03 22 ♀ ⊼ ♅
03 25 ♂ ☍ ♄
04 09 ☽ ⚹ ☉
05 19 ☽ △ ♅
05 31 ☽ □ ♀
08 12 ♂ ∥ ♇
08 48 ☿ □ ♇
17 30 ☽ ⚼ ♆
19 29 ☉ ☍ ♅

30 Wednesday
02 47 ☽ △ ♃
06 31 ☽ ⚹ ♄
07 47 ☽ △ ♂
08 59 ☽ ⚼ ♅
09 48 ☽ Q ☿
10 07 ☽ ∠ ☉
10 12 ☽ ± ♆
14 09 ☿ ± ♆
18 22 ☽ Q ♀
21 27 ☽ ⊼ ♆
22 08 ☽ ∥ ☉

31 Thursday
01 51 ☽ ± ♅
04 27 ☽ ⊥ ☉
05 06 ☽ Q ♄
06 05 ☽ ⊼ ♇
06 42 ☽ ⚼ ♃
10 39 ☽ ⚹ ☿
10 51 ☿ # ♂
12 23 ☿ ⚼ ♄
13 13 ☽ ⊼ ♅
16 48 ☽ ⊻ ☉
17 30 ☽ ± ♇
19 33 ☽ ⚹ ♀
23 43 ☽ ± ♃

AUGUST 1997

LONGITUDES

Date	Sidereal time h m s	Sun ☉ ° ' "	Moon ☽ ° ' "	Moon ☽ 24.00 ° ' "	Mercury ☿ ° '	Venus ♀ ° '	Mars ♂ ° '	Jupiter ♃ ° '	Saturn ♄ ° '	Uranus ♅ ° '	Neptune ♆ ° '	Pluto ♇ ° '
01	08 40 32	09 ♌ 16 07	18 ♋ 29 02	24 ♋ 39 36	06 ♍ 26	10 ♍ 46	22 ♎ 18	18 ♒ 03	20 ♈ 22	06 ♒ 35	28 ♑ 16	02 ♐ 52
02	08 44 29	10 13 33	00 ♌ 47 29	06 ♌ 52 47	07 28	11 58	22 53	17 R 55	20 R 22	06 R 33	28 R 14	02 R 52
03	08 48 25	11 11 00	12 ♌ 55 41	18 ♌ 56 22	08 28	13 09	23 28	17 47	20 21	06 31	28 13	02 51
04	08 52 22	12 08 28	24 ♌ 55 02	00 ♍ 51 56	09 25	14 21	24 03	17 40	20 21	06 28	28 11	02 51
05	08 56 18	13 05 57	06 ♍ 47 21	12 ♍ 41 36	10 19	15 33	24 39	17 32	20 21	06 26	28 10	02 51
06	09 00 15	14 03 26	18 ♍ 35 03	24 ♍ 28 05	11 10	16 45	25 15	17 24	20 20	06 23	28 08	02 50
07	09 04 11	15 00 57	00 ♎ 21 10	06 ♎ 14 46	11 57	17 57	25 50	17 16	20 20	06 21	28 07	02 50
08	09 08 08	15 58 28	12 ♎ 09 23	18 ♎ 05 34	12 42	19 09	26 26	17 08	20 19	06 19	28 05	02 50
09	09 12 05	16 56 00	24 ♎ 03 55	00 ♏ 04 59	13 22	20 20	27 02	17 01	20 18	06 16	28 04	02 50
10	09 16 01	17 53 33	06 ♏ 09 24	12 ♏ 17 46	13 59	21 32	27 39	16 53	20 18	06 14	28 02	02 50
11	09 19 58	18 51 07	18 ♏ 30 41	24 ♏ 48 44	14 32	22 44	28 15	16 45	20 17	06 12	28 01	02 50
12	09 23 54	19 48 42	01 ♐ 12 26	07 ♐ 42 16	15 01	23 55	28 52	16 37	20 16	06 09	27 59	02 50
13	09 27 51	20 46 18	14 ♐ 18 38	21 ♐ 01 49	15 25	25 07	29 ♎ 28	16 29	20 14	06 07	27 58	02 D 50
14	09 31 47	21 43 55	27 ♐ 52 01	04 ♑ 49 14	15 45	26 18	00 ♏ 05	16 21	20 13	06 05	27 56	02 50
15	09 35 44	22 41 32	11 ♑ 53 20	19 ♑ 04 00	15 59	27 30	00 42	16 14	20 12	06 03	27 55	02 50
16	09 39 40	23 39 11	26 ♑ 20 45	03 ♒ 42 54	16 09	28 41	01 19	16 06	20 10	06 00	27 53	02 50
17	09 43 37	24 36 51	11 ♒ 09 36	18 ♒ 39 52	16 14	29 ♍ 53	01 56	15 58	20 09	05 58	27 52	02 50
18	09 47 34	25 34 32	26 ♒ 12 37	03 ♓ 46 40	16 R 13	01 ♎ 04	02 34	15 51	20 07	05 56	27 50	02 50
19	09 51 30	26 32 15	11 ♓ 20 53	18 ♓ 54 04	16 06	02 15	03 11	15 43	20 05	05 54	27 49	02 50
20	09 55 27	27 29 58	26 ♓ 25 10	03 ♈ 53 10	15 54	03 26	03 48	15 36	20 04	05 51	27 48	02 51
21	09 59 23	28 27 44	11 ♈ 17 14	18 ♈ 36 40	15 36	04 37	04 26	15 28	20 02	05 49	27 46	02 51
22	10 03 20	29 ♌ 25 31	25 ♈ 50 55	02 ♉ 59 35	15 12	05 48	05 04	15 21	20 00	05 47	27 45	02 51
23	10 07 16	00 ♍ 23 19	10 ♉ 02 27	16 ♉ 59 25	14 42	07 00	05 42	15 13	19 57	05 45	27 44	02 51
24	10 11 13	01 21 10	23 ♉ 50 28	00 ♊ 35 45	14 07	08 10	06 20	15 06	19 55	05 43	27 42	02 52
25	10 15 09	02 19 02	07 ♊ 15 27	13 ♊ 49 48	13 27	09 21	06 58	14 59	19 53	05 41	27 41	02 52
26	10 19 06	03 16 56	20 ♊ 19 09	26 ♊ 43 47	12 43	10 32	07 36	14 52	19 51	05 39	27 40	02 53
27	10 23 03	04 14 52	03 ♋ 04 06	09 ♋ 20 25	11 54	11 43	08 15	14 45	19 48	05 37	27 39	02 53
28	10 26 59	05 12 50	15 ♋ 33 07	21 ♋ 42 32	11 02	12 54	08 53	14 38	19 45	05 35	27 38	02 53
29	10 30 56	06 10 49	27 ♋ 49 00	03 ♌ 52 52	10 07	14 05	09 32	14 31	19 43	05 33	27 36	02 54
30	10 34 52	07 08 50	09 ♌ 54 26	15 ♌ 54 00	09 11	15 15	10 11	14 25	19 40	05 31	27 35	02 55
31	10 38 49	08 ♍ 06 53	21 ♌ 51 50	27 ♌ 48 14	08 ♍ 15	16 ♎ 26	10 ♏ 50	14 ♒ 18	19 ♈ 37	05 ♒ 29	27 ♑ 34	02 ♐ 55

Date	Moon True ☊ ° '	Moon Mean ☊ ° '	Moon ☽ Latitude ° '
01	20 ♍ 24	21 ♍ 48	04 S 33
02	20 R 17	21 45	03 56
03	20 11	21 42	03 08
04	20 06	21 39	02 13
05	20 04	21 35	01 12
06	20 03	21 32	00 S 08
07	20 D 03	21 29	00 N 56
08	20 05	21 26	01 58
09	20 06	21 23	02 56
10	20 08	21 20	03 46
11	20 08	21 16	04 28
12	20 R 08	21 13	04 57
13	20 06	21 10	05 12
14	20 03	21 07	05 11
15	19 59	21 04	04 52
16	19 55	21 01	04 13
17	19 52	20 57	03 17
18	19 50	20 54	02 07
19	19 49	20 51	00 N 47
20	19 D 48	20 48	00 S 36
21	19 49	20 45	01 56
22	19 51	20 41	03 06
23	19 52	20 38	04 03
24	19 53	20 35	04 44
25	19 R 53	20 32	05 09
26	19 52	20 29	05 16
27	19 51	20 26	05 08
28	19 49	20 22	04 45
29	19 47	20 19	04 10
30	19 45	20 16	03 23
31	19 ♍ 43	20 ♍ 13	02 S 28

DECLINATIONS

Date	Sun ☉ ° '	Moon ☽ ° '	Mercury ☿ ° '	Venus ♀ ° '	Mars ♂ ° '	Jupiter ♃ ° '	Saturn ♄ ° '	Uranus ♅ ° '	Neptune ♆ ° '	Pluto ♇ ° '
01	17 N 56	17 N 39	08 N 13	08 N 47	09 S 04	16 S 20	05 N 35	19 S 15	20 S 06	08 S 21
02	17 41	16 08	07 40	08 18	09 18	16 22	05 35	19 16	20 06	08 22
03	17 25	13 55	07 07	07 49	09 31	16 25	05 35	19 16	20 07	08 22
04	17 09	11 08	06 36	07 20	09 45	16 27	05 34	19 17	20 07	08 22
05	16 53	07 54	06 05	06 50	09 59	16 30	05 34	19 17	20 07	08 23
06	16 37	04 24	05 35	06 20	10 13	16 32	05 33	19 18	20 08	08 23
07	16 20	00 N 43	05 06	05 51	10 27	16 35	05 33	19 19	20 08	08 24
08	16 03	02 S 59	04 38	05 20	10 41	16 37	05 32	19 19	20 08	08 24
09	15 46	06 37	04 12	04 50	10 55	16 40	05 32	19 20	20 09	08 25
10	15 28	10 01	03 47	04 20	11 08	16 42	05 31	19 20	20 09	08 25
11	15 10	13 03	03 24	03 50	11 22	16 45	05 31	19 21	20 09	08 25
12	14 52	15 33	03 03	03 19	11 36	16 47	05 30	19 22	20 09	08 26
13	14 34	17 20	02 43	02 48	11 50	16 50	05 29	19 22	20 10	08 26
14	14 16	18 14	02 26	02 18	12 03	16 52	05 29	19 23	20 10	08 27
15	13 57	18 04	02 11	01 47	12 17	16 54	05 28	19 23	20 10	08 27
16	13 38	16 44	01 58	01 16	12 31	16 57	05 27	19 24	20 11	08 28
17	13 19	14 16	01 48	00 45	12 44	16 59	05 26	19 24	20 11	08 28
18	13 00	10 48	01 40	00 N 14	12 58	17 02	05 25	19 25	20 11	08 29
19	12 40	06 35	01 35	00 S 17	13 12	17 04	05 25	19 25	20 12	08 29
20	12 20	01 S 59	01 34	00 48	13 25	17 06	05 24	19 26	20 12	08 30
21	12 00	02 N 41	01 35	01 19	13 38	17 09	05 23	19 27	20 12	08 31
22	11 40	07 06	01 40	01 50	13 52	17 11	05 22	19 27	20 12	08 31
23	11 20	10 58	01 48	02 21	14 05	17 13	05 21	19 28	20 13	08 32
24	11 00	14 08	01 59	02 51	14 19	17 15	05 20	19 28	20 13	08 32
25	10 39	16 26	02 14	03 22	14 32	17 17	05 19	19 29	20 13	08 33
26	10 18	17 49	02 32	03 53	14 45	17 19	05 17	19 29	20 14	08 33
27	09 57	18 16	02 54	04 24	14 58	17 22	05 16	19 30	20 14	08 34
28	09 36	17 48	03 18	04 55	15 11	17 24	05 15	19 30	20 14	08 34
29	09 15	16 31	03 45	05 25	15 24	17 26	05 14	19 30	20 14	08 35
30	08 53	14 30	04 14	05 56	15 37	17 28	05 13	19 31	20 15	08 36
31	08 N 32	11 N 53	04 N 45	06 S 26	15 S 50	17 S 30	05 N 11	19 S 31	20 S 15	08 S 36

ZODIAC SIGN ENTRIES

Date	h	m	Planets
02	10	27	☽ ♌
04	22	15	☽ ♍
07	11	17	☽ ♎
09	23	50	☽ ♏
12	09	45	☽ ♐
14	08	42	♂ ♏
14	15	42	☽ ♑
16	17	58	☽ ♒
17	14	31	♀ ♎
18	18	01	☽ ♓
20	17	45	☽ ♈
22	18	57	☽ ♉
23	02	19	☉ ♍
24	22	56	☽ ♊
27	06	10	☽ ♋
29	16	19	☽ ♌

LATITUDES

Date	Mercury ☿ ° '	Venus ♀ ° '	Mars ♂ ° '	Jupiter ♃ ° '	Saturn ♄ ° '	Uranus ♅ ° '	Neptune ♆ ° '	Pluto ♇ ° '
01	01 S 00	01 N 21	00 S 25	00 S 57	02 S 33	00 S 39	00 N 25	12 N 36
04	01 34	01 16	00 27	00 58	02 34	00 39	00 25	12 35
07	02 08	01 10	00 30	00 58	02 35	00 39	00 25	12 33
10	02 43	01 04	00 33	00 58	02 36	00 39	00 25	12 32
13	03 17	00 57	00 35	00 59	02 37	00 39	00 25	12 30
16	03 48	00 49	00 37	00 59	02 37	00 39	00 25	12 29
19	04 13	00 40	00 40	00 59	02 38	00 39	00 25	12 27
22	04 31	00 31	00 42	00 59	02 39	00 39	00 25	12 26
25	04 37	00 22	00 44	01 00	02 40	00 39	00 25	12 24
28	04 28	00 12	00 46	01 00	02 40	00 39	00 24	12 23
31	04 S 01	00 N 02	00 S 48	01 S 00	02 S 41	00 S 39	00 N 24	12 N 21

DATA

Julian Date	2450662
Delta T	+62 seconds
Ayanamsa	23° 49' 23"
Synetic vernal point	05° ♓ 17' 37"
True obliquity of ecliptic	23° 26' 13"

MOON'S PHASES, APSIDES AND POSITIONS ☽

Date	h	m	Phase	Longitude ° '	Eclipse Indicator
03	08	14	●	11 ♌ 02	
11	12	42	☽	18 ♏ 53	
18	10	55	○	25 ♒ 32	
25	02	24	☾	01 ♊ 56	

Day	h	m		
06	13	41	Apogee	
19	05	09	Perigee	
07	16	38	0S	
14	20	25	Max dec	18° S 18'
20	22	07	0N	
27	11	33	Max dec	18° N 16'

LONGITUDES

Date	Chiron ⚷ ° '	Ceres ⚳ ° '	Pallas ⚴ ° '	Juno ⚵ ° '	Vesta ⚶ ° '	Black Moon Lilith ⚸ ° '
01	27 ♎ 02	12 ♓ 19	23 ♑ 20	05 ♌ 48	29 ♈ 04	15 ♍ 01
11	27 ♎ 47	10 ♓ 42	21 ♑ 06	10 ♌ 51	00 ♉ 36	16 ♍ 07
21	28 ♎ 41	08 ♓ 41	19 ♑ 24	15 ♌ 48	01 ♉ 31	17 ♍ 14
31	29 ♎ 43	06 ♓ 29	18 ♑ 18	20 ♌ 39	01 ♉ 44	18 ♍ 20

ASPECTARIAN

01 Friday
02 04 ☽ ∥ ☉
02 45 ♀ # ♂
06 37 ☿ # ♇
10 48 ☽ ⅊ ♇
11 10 ☽ ⊼ ♃
15 32 ☿ ⊼ ♅
15 38 ☽ □ ♄
16 56 ♄ St R
18 14 ☽ ∠ ☿
19 46 ☽ □ ♂

02 Saturday
03 41 ☽ ∠ ♀
07 01 ☽ ☍ ♆
09 02 ☽ # ♃
09 22 ♀ # ♇
13 27 ☽ ⊥ ☿
16 04 ☽ △ ♇
23 18 ☽ ⊥ ♀
23 19 ☽ ☍ ♅
23 26 ♀ ± ♅

03 Sunday
02 21 ☽ ⊻ ☿
08 14 ☽ ☌ ☉
08 57 ☽ Q ♂
12 03 ☿ ∠ ♂
12 31 ☽ ⊻ ♀
13 06 ♀ ⅊ ♆
21 36 ☽ ☍ ♃

04 Monday
02 50 ☽ △ ♄
06 23 ☉ Q ♂
10 11 ☽ ✶ ♂
11 57 ♀ ± ♄
18 35 ☽ ⊼ ♆
21 51 ☽ # ♂

05 Tuesday
04 01 ☽ □ ♇
06 41 ☽ ± ♆
08 38 ☽ # ♇
09 05 ☽ ⅊ ♄
11 16 ☽ ⊼ ♅
18 07 ☽ ∠ ♂
19 44 ☽ ☌ ☿
20 44 ☽ ∥ ♀
23 26 ☽ ± ♅

06 Wednesday
00 56 ☽ ⅊ ♆
01 58 ☽ ⊻ ☉
02 47 ☽ ∥ ☿
03 22 ☽ ± ♄
04 12 ☽ ∥ ♄
07 51 ☽ ☌ ♀
09 37 ☽ ⊼ ♃
13 05 ☿ ∥ ♄
13 25 ☽ ⊥ ♂
15 16 ☽ ⊥ ☉
15 35 ☽ ⊼ ♄
16 36 ☽ Q ♇
17 23 ☉ # ♃
17 42 ☽ ⅊ ♅
21 43 ☽ ± ♃
23 43 ♀ ⊼ ♃

07 Thursday
02 18 ☽ ⊻ ♂
07 26 ☽ △ ♆
11 15 ☽ ∠ ☉
15 52 ☽ ⅊ ♃
17 04 ☽ ✶ ♇
23 58 ☿ ± ♅

08 Friday
00 10 ☽ △ ♅
02 17 ♀ ∥ ♄
13 09 ☽ ⊻ ☿
20 24 ☽ ✶ ☉
21 38 ☽ # ☿
21 58 ☽ △ ♃
23 29 ☽ ∠ ♇

09 Saturday
01 07 ☿ ⅊ ♆
01 34 ☽ # ♀
02 01 ☽ ⊥ ☿
03 42 ☽ ⊻ ♀
03 43 ♂ ⊥ ♇
04 28 ☽ ☍ ♄
04 45 ☽ # ♄
11 22 ♀ ⊼ ♄
13 39 ☉ ☍ ♃
17 03 ☽ ⊥ ♀
17 32 ☽ ⊥ ♇
18 16 ☽ ☌ ♂
19 58 ☽ □ ♆
21 04 ☽ ∠ ☿
21 52 ♀ Q ♇
22 33 ☽ Q ☉

10 Sunday
00 29 ☽ ∥ ♇
05 27 ☽ ⊻ ♇
06 08 ♀ ⅊ ♅
12 09 ☽ □ ♅
12 49 ☽ ∠ ♀
16 29 ♀ ⊥ ♂
21 15 ☽ ∥ ♂

11 Monday
00 48 ☿ ± ♄
02 49 ♂ □ ♆
04 00 ☽ ✶ ☿
07 12 ☽ Q ♆
08 39 ☽ □ ♃
12 21 ♀ ± ♃
12 42 ☽ □ ☉
15 23 ☽ ⊼ ♄
20 53 ☽ ✶ ♀
22 48 ☽ Q ♅

12 Tuesday
02 45 ☽ ± ♄
03 51 ☽ Q ☿
05 42 ☽ # ☉
05 59 ☽ ✶ ♆
07 24 ☽ ⊻ ♂
15 01 ☽ ☌ ♇
18 16 ☽ Q ♃
19 06 ☽ ⊥ ♂
19 30 ☽ ⅊ ♄
21 08 ☽ ✶ ♅
21 36 ☽ Q ♀
23 00 ☉ △ ♄

13 Wednesday
03 41 ☽ ∥ ♃
08 33 ♇ St D
09 34 ☽ ∠ ♆
12 18 ☽ ∠ ♂
14 03 ☽ □ ☿
15 52 ☽ ✶ ♃
21 23 ☿ ∥ ♀
22 35 ☽ △ ♄

14 Thursday
00 07 ☽ ∠ ♅
00 25 ☽ △ ☉
01 37 ☽ ⊥ ♆
09 01 ☽ □ ♀
12 07 ☽ ⊻ ♆
15 50 ☽ ⊥ ♅
16 02 ☽ ✶ ♂
18 00 ☽ ∠ ♃
20 35 ☽ ⊻ ♇

15 Friday
02 07 ☽ ⊻ ♅
04 23 ☽ ⅊ ☉
06 50 ☽ ⊥ ♇
09 13 ☽ ⊥ ♃
13 26 ☽ Q ♂
18 58 ☽ △ ☿
19 13 ☽ ⊻ ♃
20 13 ♀ △ ♆
20 38 ☽ ± ☉
21 57 ☽ ∠ ♇

16 Saturday
01 51 ☽ □ ♄
03 49 ☿ ∠ ♂
07 11 ☿ ⊼ ♃
07 16 ☽ ⊼ ☉
09 18 ☽ ∥ ♃
14 31 ☽ ☌ ♆
16 10 ☽ △ ♀
19 54 ☽ ⅊ ☿
20 28 ☽ □ ♂
22 34 ☽ ✶ ♇

17 Sunday
03 40 ☽ ☌ ♅
07 10 ☽ Q ♄
10 30 ☽ ± ☿
17 53 ☽ Q ♇
18 28 ☽ ⅊ ♀
19 38 ☽ ☌ ♃
19 48 ☿ St R
19 59 ☽ # ☉
20 07 ☽ ⊼ ☿
22 36 ☽ ∥ ♂

18 Monday
02 20 ☽ ✶ ♄
08 01 ♀ ⅊ ♃
10 01 ☽ ± ♀
10 55 ☽ ☍ ☉
13 11 ☉ # ♂
14 35 ☽ ⊻ ♆
20 21 ☽ ⊼ ♀
22 30 ☽ △ ♂
22 30 ☽ □ ♇
22 42 ♂ ⊻ ♇

19 Tuesday
00 05 ☽ ⊥ ♆
01 34 ☽ ∥ ♇
02 06 ☽ ∠ ♄
03 22 ☽ ⊻ ♅
12 52 ☽ ⊥ ♅
14 20 ☽ ∠ ♆
16 21 ☽ ⊥ ♄
18 18 ☽ # ♄
18 53 ☽ ⊻ ♃
19 28 ☽ ☍ ☿
23 20 ☽ ⅊ ♂
23 55 ♀ ✶ ♇

20 Wednesday
01 52 ☽ ⊻ ♄
03 08 ☽ ∠ ♅
04 21 ☽ ⊥ ♃
13 51 ☽ ⊼ ☉
14 10 ☽ # ☿
14 12 ☽ ✶ ♆
14 19 ☽ ± ♂
17 27 ☽ ∥ ♀
18 39 ☽ ∠ ♃
18 40 ☉ ⊥ ♀
19 13 ☉ ⊼ ♆
22 19 ☽ △ ♇

21 Thursday
00 10 ☽ ± ☉
00 15 ☽ ☍ ♀
00 24 ☽ ⊼ ♂
03 09 ☽ ✶ ♅
03 56 ♀ ⊻ ♂
03 58 ☽ # ♀
06 12 ☽ ∥ ☿
09 32 ☽ Q ♆
15 48 ☽ ⅊ ☉
18 47 ☽ ✶ ♃
18 52 ☽ ⊼ ☿
22 40 ☽ Q ♅
22 45 ☽ ⅊ ♇

22 Friday
00 03 ☿ ⊼ ♃
02 18 ☽ ☌ ♄
02 20 ☽ ∥ ♄
02 32 ☿ # ♀
04 30 ☽ ± ☿
11 34 ♀ △ ♅
13 40 ☽ ± ♇
14 29 ☽ Q ♃
15 11 ☽ □ ♆
18 25 ☽ △ ☉
19 04 ☽ ⅊ ☿
20 24 ☽ # ♇
23 46 ☽ ⊼ ♇

23 Saturday
04 14 ☽ ☍ ♂
04 42 ☽ □ ♅
06 19 ☽ ⊼ ♀
13 53 ♂ □ ♅
14 14 ☽ ∥ ☉
17 33 ☽ ± ♀
19 44 ☽ △ ☿
20 51 ☽ □ ♃

24 Sunday
05 08 ☽ ⊻ ♄
10 43 ☽ ⅊ ♀
13 41 ☽ # ♂
15 40 ☽ ⊥ ♄
18 50 ☽ △ ♆
20 08 ☿ ± ♄

25 Monday
02 24 ☽ □ ☉
04 04 ☽ ☍ ♇
07 43 ☽ ∠ ♄
09 09 ☽ △ ♅
11 27 ☽ ⊼ ♂
16 12 ☽ △ ♀
21 53 ☽ ⅊ ♆
22 43 ☽ □ ☿
22 57 ☽ ± ♂

26 Tuesday
01 02 ☽ # ♃
01 49 ☉ □ ♇
02 00 ☽ △ ♃
10 10 ♃ Q ♇
11 07 ☽ ✶ ♄
12 37 ☽ ⅊ ♅
13 27 ☿ ⅊ ♆
13 56 ☽ Q ☉
14 30 ☽ ± ♆
16 29 ☽ ⅊ ♂
21 20 ☉ ± ♆

27 Wednesday
01 44 ☽ ⊼ ♆
05 28 ☽ ± ♅
05 45 ☽ ⅊ ♃
06 21 ☽ Q ☿
09 36 ☽ Q ♄
11 39 ☽ ⊼ ♇
14 08 ☿ ⊻ ♀
14 26 ☽ ✶ ☉
16 51 ☽ ⊼ ♅
20 22 ☿ ± ♅
22 26 ☽ △ ♂
22 46 ☽ ± ♃
23 08 ☽ ± ♇

28 Thursday
01 09 ☉ ⅊ ♄
03 50 ☽ ✶ ☿
06 19 ☽ □ ♀
10 14 ☽ ⊼ ♃
16 33 ☽ ⅊ ♇
20 09 ☽ □ ♄
20 48 ☉ ⊼ ♅
21 11 ☽ # ♃
21 51 ☽ ∠ ☉

29 Friday
03 25 ♀ # ♄
07 04 ☽ ∠ ☿
11 35 ☽ ☍ ♆
17 04 ☽ ⊥ ☉
20 16 ♀ △ ♃
20 59 ☿ ✶ ♂
21 20 ☽ Q ♀
22 04 ☽ △ ♇
23 35 ☽ ⊥ ☿

30 Saturday
00 52 ☽ # ♂
03 16 ☽ ☍ ♅
06 01 ☽ ⊻ ☉
10 40 ☽ ⊻ ☿
11 15 ☿ ⊥ ♀
12 34 ☽ □ ♂
20 56 ☽ ☍ ♃
23 53 ☽ ✶ ♀

31 Sunday
06 56 ☉ # ♇
07 30 ☽ △ ♄
13 43 ☉ ☌ ☿
23 30 ☽ ⊼ ♆

All ephemeris data is given at 12.00 UT and the Moon's longitude is additionally given for 24.00 UT

SEPTEMBER 1997

LONGITUDES

Date	Sidereal time h m s	Sun ☉ ° ' "	Moon ☽ ° ' "	Moon ☽ 24.00 ° ' "	Mercury ☿ ° '	Venus ♀ ° '	Mars ♂ ° '	Jupiter ♃ ° '	Saturn ♄ ° '	Uranus ♅ ° '	Neptune ♆ ° '	Pluto ⯓ ° '
01	10 42 45	09 ♍ 04 58	03 ♍ 43 28	09 ♍ 37 48	07 ♍ 20	17 ♎ 36	11 ♏ 29	14 ♒ 12	19 ♈ 34	05 ♒ 27	27 ♑ 33	02 ♐ 56
02	10 46 42	10 03 04	15 ♍ 31 31	21 ♍ 24 54	06 R 26	18 47	12 08	14 R 05	19 R 31	05 R 25	27 32	02 56
03	10 50 38	11 01 12	27 ♍ 18 13	03 ♎ 11 47	05 36	19 57	12 47	13 59	19 28	05 23	27 31	02 57
04	10 54 35	11 59 21	09 ♎ 05 57	15 ♎ 01 02	04 51	21 08	13 26	13 53	19 25	05 22	27 30	02 58
05	10 58 32	12 57 32	20 ♎ 57 24	26 ♎ 55 27	04 11	22 18	14 06	13 47	19 22	05 20	27 29	02 59
06	11 02 28	13 55 45	02 ♏ 55 36	08 ♏ 58 18	03 37	23 28	14 45	13 41	19 18	05 18	27 28	02 59
07	11 06 25	14 53 59	15 ♏ 03 59	21 ♏ 13 07	03 11	24 38	15 25	13 36	19 15	05 17	27 27	03 00
08	11 10 21	15 52 15	27 ♏ 26 13	03 ♐ 43 44	02 53	25 49	16 05	13 30	19 11	05 15	27 26	03 01
09	11 14 18	16 50 32	10 ♐ 06 08	16 ♐ 33 54	02 44	26 59	16 45	13 25	19 08	05 13	27 25	03 02
10	11 18 14	17 48 51	23 ♐ 07 25	29 ♐ 47 02	02 D 43	28 09	17 25	13 19	19 04	05 12	27 24	03 03
11	11 22 11	18 47 11	06 ♑ 33 03	13 ♑ 25 38	02 51	29 ♎ 18	18 05	13 14	19 01	05 10	27 23	03 04
12	11 26 07	19 45 33	20 ♑ 24 52	27 ♑ 30 40	03 09	00 ♏ 28	18 45	13 10	18 57	05 09	27 22	03 05
13	11 30 04	20 43 56	04 ♒ 42 49	12 ♒ 00 55	03 36	01 38	19 25	13 05	18 53	05 07	27 21	03 06
14	11 34 01	21 42 21	19 ♒ 24 26	26 ♒ 52 36	04 11	02 48	20 06	13 00	18 49	05 06	27 21	03 07
15	11 37 57	22 40 48	04 ♓ 24 33	11 ♓ 59 14	04 55	03 57	20 46	12 56	18 45	05 05	27 20	03 08
16	11 41 54	23 39 16	19 ♓ 35 32	27 ♓ 12 14	05 47	05 07	21 27	12 52	18 41	05 03	27 19	03 09
17	11 45 50	24 37 46	04 ♈ 48 07	12 ♈ 21 59	06 47	06 16	22 08	12 48	18 37	05 02	27 18	03 10
18	11 49 47	25 36 19	19 ♈ 52 42	27 ♈ 19 16	07 54	07 25	22 48	12 44	18 33	05 01	27 18	03 11
19	11 53 43	26 34 53	04 ♉ 40 47	11 ♉ 56 34	09 07	08 34	23 29	12 40	18 29	04 59	27 17	03 12
20	11 57 40	27 33 29	19 ♉ 06 03	26 ♉ 08 55	10 26	09 44	24 10	12 36	18 25	04 58	27 16	03 14
21	12 01 36	28 32 08	03 ♊ 04 57	09 ♊ 54 09	11 51	10 53	24 51	12 33	18 20	04 57	27 16	03 15
22	12 05 33	29 ♍ 30 49	16 ♊ 36 36	23 ♊ 12 33	13 20	12 02	25 33	12 30	18 16	04 56	27 15	03 16
23	12 09 30	00 ♎ 29 32	29 ♊ 42 20	06 ♋ 06 21	14 53	13 10	26 14	12 27	18 12	04 55	27 15	03 18
24	12 13 26	01 28 18	12 ♋ 25 04	18 ♋ 38 59	16 30	14 19	26 55	12 24	18 07	04 54	27 14	03 19
25	12 17 23	02 27 05	24 ♋ 48 36	00 ♌ 54 29	18 09	15 28	27 37	12 22	18 03	04 53	27 14	03 20
26	12 21 19	03 25 55	06 ♌ 57 08	12 ♌ 57 05	19 51	16 36	28 18	12 19	17 58	04 52	27 13	03 22
27	12 25 16	04 24 47	18 ♌ 54 50	24 ♌ 50 51	21 34	17 45	29 00	12 17	17 54	04 51	27 13	03 23
28	12 29 12	05 23 42	00 ♍ 45 35	06 ♍ 39 27	23 20	18 53	29 ♏ 42	12 15	17 49	04 51	27 13	03 25
29	12 33 09	06 22 38	12 ♍ 32 51	18 ♍ 26 09	25 06	20 01	00 ♐ 24	12 13	17 45	04 50	27 12	03 26
30	12 37 05	07 ♎ 21 36	24 ♍ 19 40	00 ♎ 13 43	26 ♍ 53	21 ♏ 09	01 ♐ 06	12 ♒ 11	17 ♈ 40	04 ♒ 49	27 ♑ 12	03 ♐ 28

Date	Moon True ☊ ° '	Moon Mean ☊ ° '	Moon ☽ Latitude ° '
01	19 ♍ 42	20 ♍ 10	01 S 28
02	19 R 42	20 06	00 S 23
03	19 D 42	20 03	00 N 42
04	19 42	20 00	01 46
05	19 43	19 57	02 45
06	19 44	19 54	03 38
07	19 44	19 51	04 22
08	19 44	19 47	04 54
09	19 45	19 44	05 14
10	19 R 45	19 41	05 18
11	19 45	19 38	05 05
12	19 D 45	19 35	04 34
13	19 45	19 32	03 45
14	19 45	19 28	02 41
15	19 45	19 25	01 24
16	19 45	19 22	00 N 01
17	19 R 45	19 19	01 S 23
18	19 45	19 16	02 40
19	19 44	19 12	03 40
20	19 43	19 09	04 34
21	19 42	19 06	05 04
22	19 42	19 03	05 07
23	19 41	19 00	05 13
24	19 D 41	18 57	04 53
25	19 42	18 53	04 20
26	19 43	18 50	03 36
27	19 45	18 47	02 43
28	19 46	18 44	01 44
29	19 47	18 41	00 40
30	19 ♍ 47	18 ♍ 38	00 N 25

DECLINATIONS

Date	Sun ☉ ° '	Moon ☽ ° '	Mercury ☿ ° '	Venus ♀ ° '	Mars ♂ ° '	Jupiter ♃ ° '	Saturn ♄ ° '	Uranus ♅ ° '	Neptune ♆ ° '	Pluto ⯓ ° '
01	08 N 10	08 N 47	05 N 17	06 S 56	16 S 03	17 S 31	05 N 10	19 S 32	20 S 15	08 S 37
02	07 48	05 21	05 49	07 27	16 15	17 33	05 09	19 32	20 15	08 37
03	07 26	01 N 43	06 22	07 57	16 28	17 35	05 07	19 33	20 16	08 38
04	07 04	01 S 59	06 54	08 27	16 40	17 37	05 06	19 33	20 16	08 38
05	06 42	05 37	07 25	08 56	16 53	17 39	05 05	19 34	20 16	08 39
06	06 19	09 04	07 55	09 26	17 05	17 40	05 03	19 34	20 16	08 40
07	05 57	12 11	08 22	09 55	17 17	17 42	05 02	19 34	20 16	08 41
08	05 34	14 48	08 46	10 24	17 30	17 43	05 00	19 35	20 17	08 41
09	05 12	16 47	09 08	10 53	17 42	17 45	04 59	19 35	20 17	08 42
10	04 49	17 58	09 26	11 22	17 53	17 46	04 57	19 35	20 17	08 42
11	04 26	18 12	09 40	11 50	18 05	17 48	04 56	19 36	20 17	08 43
12	04 03	17 22	09 51	12 19	18 17	17 49	04 54	19 36	20 18	08 44
13	03 40	15 26	09 57	12 47	18 29	17 51	04 52	19 36	20 18	08 44
14	03 17	12 27	10 00	13 15	18 40	17 52	04 51	19 37	20 18	08 45
15	02 54	08 35	09 58	13 42	18 51	17 53	04 49	19 37	20 18	08 46
16	02 31	04 S 06	09 52	14 09	19 03	17 54	04 48	19 37	20 18	08 46
17	02 08	00 N 38	09 42	14 36	19 14	17 55	04 46	19 38	20 18	08 47
18	01 45	05 18	09 29	15 03	19 25	17 56	04 44	19 38	20 18	08 48
19	01 22	09 33	09 11	15 29	19 36	17 57	04 43	19 38	20 19	08 48
20	00 58	13 06	08 50	15 55	19 46	17 58	04 41	19 39	20 19	08 49
21	00 35	15 47	08 26	16 20	19 57	17 59	04 39	19 39	20 19	08 50
22	00 N 12	17 30	07 58	16 46	20 07	18 00	04 37	19 39	20 19	08 50
23	00 S 12	18 13	07 28	17 10	20 18	18 01	04 36	19 39	20 19	08 51
24	00 35	18 00	06 55	17 35	20 28	18 02	04 34	19 39	20 19	08 52
25	00 58	16 54	06 19	17 59	20 38	18 02	04 32	19 40	20 19	08 52
26	01 22	15 03	05 42	18 23	20 48	18 03	04 30	19 40	20 19	08 53
27	01 45	12 34	05 03	18 46	20 57	18 04	04 28	19 40	20 19	08 54
28	02 09	09 35	04 22	19 09	21 07	18 04	04 27	19 40	20 20	08 54
29	02 32	06 14	03 40	19 31	21 16	18 05	04 25	19 40	20 20	08 55
30	02 S 55	02 N 38	02 N 57	19 S 53	21 S 25	18 S 05	04 N 23	19 S 41	20 S 20	08 S 56

ZODIAC SIGN ENTRIES

Date	h	m	Planets	
01	04	27	☽	♍
03	17	30	☽	♎
06	06	10	☽	♏
08	16	54	☽	♐
11	00	23	☽	♑
12	02	17	♀	♏
13	04	10	☽	♒
15	04	59	☽	♓
17	04	25	☽	♈
19	04	21	☽	♉
21	06	38	☽	♊
22	23	56	☉	♎
23	12	33	☽	♋
25	22	12	☽	♌
28	10	27	☽	♍
28	22	22	♂	♐
30	23	32	☽	♎

LATITUDES

Date	Mercury ☿ ° '	Venus ♀ ° '	Mars ♂ ° '	Jupiter ♃ ° '	Saturn ♄ ° '	Uranus ♅ ° '	Neptune ♆ ° '	Pluto ⯓ ° '
01	03 S 49	00 S 02	00 S 49	01 S 00	02 S 41	00 S 39	00 N 24	12 N 21
04	03 02	00 13	00 50	01 00	02 42	00 39	00 24	12 19
07	02 07	00 24	00 52	01 00	02 42	00 39	00 24	12 18
10	01 09	00 36	00 54	01 00	02 43	00 39	00 24	12 16
13	00 S 15	00 47	00 55	01 00	02 43	00 39	00 24	12 15
16	00 N 31	00 59	00 57	01 00	02 44	00 38	00 24	12 14
19	01 07	01 11	00 59	01 00	02 44	00 38	00 24	12 12
22	01 32	01 23	01 00	00 59	02 45	00 38	00 24	12 11
25	01 47	01 35	01 02	00 59	02 45	00 38	00 24	12 10
28	01 53	01 47	01 03	00 59	02 45	00 38	00 24	12 08
31	01 N 51	01 S 58	01 S 04	00 S 59	02 S 45	00 S 38	00 N 24	12 N 07

DATA

Julian Date	2450693
Delta T	+62 seconds
Ayanamsa	23° 49' 26"
Synetic vernal point	05° ♓ 17' 33"
True obliquity of ecliptic	23° 26' 14"

MOON'S PHASES, APSIDES AND POSITIONS ☽

Date	h	m	Phase	Longitude ° '	Eclipse Indicator
01	23	52	●	09 ♍ 34	Partial
10	01	31	☽	17 ♐ 23	
16	18	50	○	23 ♓ 56	total
23	13	35	☾	00 ♋ 33	

Day	h	m		
02	21	32	Apogee	
16	15	28	Perigee	
29	23	46	Apogee	
03	23	08	0S	
11	05	29	Max dec	18° S 15'
17	08	46	0N	
23	17	57	Max dec	18° N 15'

LONGITUDES

Date	Chiron ⚷ ° '	Ceres ⚳ ° '	Pallas ⚴ ° '	Juno ⚵ ° '	Vesta ⚶ ° '	Black Moon Lilith ⚸ ° '
01	29 ♎ 49	06 ♓ 16	18 ♑ 14	21 ♌ 08	01 ♉ 43	18 ♍ 27
11	00 ♏ 58	04 ♓ 07	17 ♑ 50	25 ♌ 51	01 ♉ 07	19 ♍ 33
21	02 ♏ 12	02 ♓ 15	18 ♑ 02	00 ♍ 26	29 ♈ 46	20 ♍ 39
31	03 ♏ 30	00 ♓ 50	18 ♑ 46	04 ♍ 53	27 ♈ 46	21 ♍ 46

All ephemeris data is given at 12.00 UT and the Moon's longitude is additionally given for 24.00 UT

ASPECTARIAN

h m Aspects

01 Monday
02 53 ☽ Q ♂
07 23 ☿ ∥ ♄
09 29 ☽ ∠ ♀
10 23 ☽ □ ⯓
11 39 ☽ ± ♆
13 13 ☽ # ⯓
13 43 ☽ ♇ ♄
15 30 ☽ ⊼ ♅
17 01 ☽ ∥ ☉
18 38 ♀ ∠ ⯓
18 48 ☽ ☌ ☿
23 26 ☽ # ♀
23 52 ☽ ☌ ☉

02 Tuesday
03 40 ☽ ± ♅
04 41 ☽ ⚹ ♂
05 48 ☽ ⊥ ♀
05 55 ☽ ♇ ♆
07 56 ☽ ± ♄
09 06 ☽ ⊼ ♃
09 14 ☽ ∥ ☿
13 23 ☽ ∥ ♄
19 22 ☽ ⊻ ♀
20 06 ☽ ⊼ ♄
21 13 ☽ ± ♃
21 50 ☉ # ♀
21 57 ☽ ♇ ♅
23 03 ☽ Q ⯓

03 Wednesday
02 26 ♀ ☍ ♄
12 25 ☽ △ ♆
13 02 ☽ ∠ ♂
15 24 ☽ ♇ ♃
18 43 ☿ ⊼ ♅
19 55 ☿ ∠ ♀
20 56 ☉ ± ♅
23 31 ☽ ⚹ ⯓

04 Thursday
03 51 ☽ ⊻ ☿
04 25 ☽ △ ♅
08 26 ☽ ⊥ ♂
15 21 ☽ ⊥ ☿
16 17 ☉ ∥ ☿
18 23 ☽ ⊻ ☉
21 19 ☽ ⊻ ♂
21 37 ☽ △ ♃
21 58 ♀ ∥ ⯓

05 Friday
00 17 ☉ ♇ ♆
02 05 ♂ □ ♃
04 24 ☿ ♇ ♄
05 59 ☽ ∠ ⯓
07 37 ☽ ⊥ ☉
08 21 ☽ # ♄
08 35 ☽ ∠ ☿
08 48 ☽ ☍ ♄
15 00 ☽ ☌ ♀
18 34 ☽ # ☉
21 24 ☉ ± ♄

06 Saturday
00 07 ☽ ⊥ ⯓
01 05 ☽ □ ♆
02 23 ☽ # ☿
03 19 ☽ ∠ ☉
06 33 ☉ ⊼ ♃
09 05 ☽ ∥ ⯓
12 08 ☽ ⊻ ⯓
13 20 ☽ ⚹ ☿
15 07 ☽ ∥ ♀
16 43 ☽ □ ♅
20 20 ☿ ± ♆

07 Sunday
06 40 ☿ Q ♂
09 08 ☽ □ ♃
11 39 ☽ ⚹ ☉
12 14 ☽ Q ☿
12 44 ☽ ☌ ♂
12 44 ☽ Q ♆
12 57 ♂ Q ♆
20 08 ☽ ⊼ ♄

08 Monday
00 33 ☿ □ ⯓
03 57 ☽ Q ♅
06 24 ☿ # ⯓
07 42 ☽ ± ♄
08 33 ☽ ⊻ ♀
11 59 ☽ ⚹ ♆
12 54 ☽ Q ☉
19 43 ☽ Q ♃
21 12 ☽ ⊥ ♀
22 14 ☽ □ ☿
22 40 ☽ ☌ ⯓

09 Tuesday
00 49 ☽ ♇ ♄
02 51 ☽ ⚹ ♅
04 27 ☉ ⚹ ♂
13 09 ♀ ⊥ ⯓
15 51 ☽ ∠ ♀
16 18 ☽ ∠ ♆
18 07 ☽ ⚹ ♃
19 58 ♂ ∥ ♃
20 51 ♀ □ ♆

10 Wednesday
01 00 ☽ ⊻ ♂
01 31 ☽ □ ☉
01 42 ☿ St D
02 42 ☉ ∥ ♄
04 38 ☽ △ ♄
06 13 ☽ ∥ ♃
06 41 ☽ ∠ ♅
08 44 ☽ ∥ ♂
08 52 ☽ ⊥ ♆
12 33 ☽ ⊥ ♂
19 43 ☽ ⊻ ♆
21 20 ☽ ∠ ♃
21 56 ☽ ⚹ ♀
22 56 ☽ ⊥ ♅

11 Thursday
05 22 ☽ △ ☿
05 33 ☽ ∠ ♂
05 50 ☽ ⊻ ⯓
09 34 ☽ ⊻ ♅
13 12 ☽ ⊥ ♃
16 25 ☽ ⊥ ⯓
16 44 ☽ ∥ ♂
17 09 ☉ ⊼ ♄
21 05 ☽ Q ♀
22 20 ☉ ± ♃
23 36 ☽ ⊻ ♃

12 Friday
03 00 ☽ ∥ ♃
07 04 ☿ □ ⯓
08 00 ☽ ∠ ⯓
08 02 ☽ ♇ ☿
09 01 ☽ ⚹ ♂
09 30 ☽ □ ♄
10 48 ☽ △ ☉
18 22 ♂ ⊼ ♄
21 18 ☉ ♇ ♅
23 43 ☽ ± ☿
23 45 ☽ ☌ ♆

13 Saturday
00 34 ☿ ± ♆
06 16 ☽ Q ♂
06 26 ☽ □ ♀
09 19 ☽ ⚹ ⯓
10 05 ☽ ⊼ ☿
12 40 ☽ ☌ ♅
13 48 ☽ ♇ ☉
15 34 ☽ Q ♄
21 08 ☉ Q ⯓
23 16 ☿ ♇ ♄

14 Sunday
01 40 ☽ ☌ ♃
05 02 ☽ Q ⯓
05 35 ☽ ± ☉
07 02 ☽ ∥ ♀
11 03 ☽ ⚹ ♄
13 10 ☽ □ ♂
15 58 ☽ ⊼ ☉
18 44 ♀ ⊻ ⯓

15 Monday
00 44 ☽ ⊻ ♆
03 49 ☽ # ☿
09 58 ☽ □ ⯓
10 17 ☽ ⊥ ♆
10 58 ☽ ∠ ♄
10 59 ☽ ∥ ⯓
11 13 ☽ △ ♀
12 51 ☽ ☍ ☿
13 03 ☽ ⊻ ♅
16 32 ☿ ⊼ ♅
22 33 ☽ ⊥ ♅

16 Tuesday
00 32 ☽ ∠ ♆
01 09 ☽ ⊥ ♄
01 26 ☽ ⊻ ♃
08 26 ☽ # ♄
10 35 ☽ ⊻ ♄
10 50 ♀ □ ♅
10 51 ☽ ⊥ ♃
12 44 ☽ ∠ ♅
12 53 ☽ ♇ ♀
15 04 ☽ △ ♂
18 50 ☽ ☍ ☉
20 48 ☽ # ☉

17 Wednesday
00 10 ☽ ⚹ ♆
00 59 ☽ ∠ ♃
04 15 ☽ ± ♀
09 25 ☽ △ ⯓
12 22 ☽ ⚹ ♅
14 31 ☽ ⊼ ♀
15 22 ☽ ⊼ ☿
15 51 ☽ ♇ ♂
18 59 ☽ ∥ ☉
19 08 ☽ Q ♆

18 Thursday
00 37 ☽ ⚹ ♃
01 39 ☽ ± ☿
06 51 ☽ ± ♂
07 25 ☽ Q ♅
09 01 ☽ ∥ ♄
09 17 ☽ ♇ ⯓
09 53 ☽ ☌ ♄
16 56 ☽ ⊼ ♂
17 16 ☽ ♇ ☿
18 57 ♂ Q ♅
19 46 ☽ Q ♃
21 52 ☽ ⊼ ☉
23 48 ☽ ± ⯓
23 57 ☽ □ ♆

19 Friday
07 32 ☽ # ⯓
08 19 ☽ ± ☉
09 35 ☽ ⊼ ⯓
09 58 ☽ ∥ ☿
12 31 ☽ □ ♅
18 07 ♂ ∥ ♅
18 58 ☽ ☍ ♀
20 02 ☽ △ ☿

20 Saturday
00 14 ☽ ♇ ☉
01 09 ☽ □ ♃
05 06 ☉ △ ♆
10 50 ☽ ⊻ ♄
13 09 ☿ # ⯓
13 09 ☉ ♇ ♃
19 30 ♂ ± ♄
20 58 ☽ ⊥ ♄
21 03 ☽ ☍ ♂
21 08 ☿ ± ♅

21 Sunday
01 56 ☽ △ ♆
03 31 ☽ △ ☉
12 17 ☽ ☍ ⯓
12 27 ☽ ∠ ♄
15 16 ☽ △ ♅
18 51 ☿ ♇ ♆
19 43 ☿ ± ♄
20 08 ☽ # ♀
23 10 ☿ ⊼ ♃

22 Monday
03 01 ☽ ⊼ ♀
04 12 ☽ ♇ ♆
04 39 ☽ △ ♃
05 22 ☽ □ ☿
10 42 ♄ ♇ ⯓
14 48 ☽ ± ♀
14 59 ☽ ⚹ ♄
18 01 ☽ ♇ ♅
18 06 ☿ Q ♂
20 25 ☽ ± ♆
21 29 ♀ □ ♃

23 Tuesday
00 52 ☽ # ♃
05 12 ☽ ⊼ ♂
07 26 ☽ ⊼ ♆
07 50 ☽ ♇ ♃
08 53 ☽ ♇ ♀
10 32 ☽ ± ♅
12 54 ☽ Q ♄
13 35 ☽ □ ☉
15 27 ♂ ∥ ♆
16 59 ☽ ± ♂
18 43 ☽ ⊼ ⯓
18 47 ☽ Q ☿
21 45 ☽ ⊼ ♅

24 Wednesday
00 36 ☽ ± ♃
06 04 ☽ ± ⯓
10 42 ☽ # ♃
11 00 ☽ ♇ ♂
11 58 ☽ ⊼ ♃
16 01 ☽ △ ♀
20 03 ☽ # ♀
21 01 ☽ ⚹ ☿
22 51 ♂ ⚹ ♆
22 55 ☽ □ ♄
23 22 ☽ ♇ ⯓

25 Thursday
02 46 ☽ Q ☉
07 10 ♀ Q ♆
10 35 ☿ ⊼ ♄
14 56 ☿ ± ♃
15 33 ♀ ∥ ♃
16 45 ☽ ☍ ♆
17 50 ☽ △ ♂

26 Friday
04 23 ☽ ⚹ ☉
04 51 ☽ △ ⯓
07 07 ☽ ∠ ☿
07 52 ☽ ☍ ♅
10 13 ☉ ⚹ ⯓
12 20 ☿ ♇ ♅
22 42 ☽ ☍ ♃

27 Saturday
04 07 ☽ ⊥ ☿
09 21 ☿ Q ⯓
09 24 ☽ □ ♀
09 58 ☽ △ ♄
13 06 ☽ ∠ ☉
14 57 ♀ ⊼ ♄
18 18 ☽ ⊻ ☿
22 40 ☉ △ ♅

28 Sunday
04 48 ☽ ⊼ ♆
08 59 ☽ ⊥ ☉
09 23 ☿ ∥ ♄
09 43 ☽ □ ♂
16 10 ☽ ♇ ♄
16 59 ☽ ± ♆
17 04 ☽ # ⯓
17 24 ☽ □ ⯓
20 18 ☽ ⊼ ♅
22 17 ☽ ⊻ ☉

29 Monday
01 48 ☽ Q ♀
06 07 ♂ Q ♃
08 30 ☽ ± ♅
10 22 ☽ ± ♄
11 18 ☽ ♇ ♆
11 20 ☽ ⊼ ♃
22 00 ♀ ∥ ♅
22 31 ☽ ⊼ ♄
23 32 ☽ ± ♃

30 Tuesday
00 24 ☽ ∥ ♄
00 40 ☽ Q ♂
02 49 ☽ ♇ ♅
04 51 ☽ ⚹ ♀
06 09 ☽ Q ⯓
09 24 ☽ ∥ ☿
10 20 ☽ # ☉
12 47 ☉ # ☿
16 02 ☿ ♇ ♃
16 13 ☿ △ ♆
17 49 ☽ ♇ ♃
17 51 ☽ △ ♆
18 08 ☽ ☌ ☿

OCTOBER 1997

LONGITUDES

Date	Sidereal time h m s	Sun ☉ ° ' "	Moon ☽ ° ' "	Moon ☽ 24.00 ° ' "	Mercury ☿ ° '	Venus ♀ ° '	Mars ♂ ° '	Jupiter ♃ ° '	Saturn ♄ ° '	Uranus ♅ ° '	Neptune ♆ ° '	Pluto ♇ ° '
01	12 41 02	08 ♎ 20 37	06 ♎ 08 35	12 ♎ 04 33	28 ♍ 41	22 ♏ 17	01 ♐ 48	12 ♒ 10	17 ♈ 35	04 ♒ 48	27 ♑ 12	03 ♐ 29
02	12 44 59	09 19 40	18 ♎ 01 50	24 ♎ 00 42	00 ♎ 29	23 25	02 30	12 R 09	17 R 31	04 R 48	27 R 12	03 31
03	12 48 55	10 18 44	00 ♏ 01 22	06 ♏ 04 05	02 17	24 33	03 12	12 08	17 26	04 47	27 11	03 32
04	12 52 52	11 17 51	12 ♏ 09 05	18 ♏ 16 36	04 04	25 41	03 55	12 07	17 21	04 47	27 11	03 34
05	12 56 48	12 16 59	24 ♏ 26 53	00 ♐ 40 12	05 52	26 48	04 37	12 06	17 17	04 46	27 11	03 35
06	13 00 45	13 16 10	06 ♐ 56 49	13 ♐ 17 02	07 39	27 56	05 20	12 06	17 12	04 46	27 11	03 37
07	13 04 41	14 15 22	19 ♐ 41 09	26 ♐ 09 27	09 26	29 ♏ 03	06 02	12 06	17 07	04 45	27 11	03 39
08	13 08 38	15 14 36	02 ♑ 42 14	09 ♑ 19 47	11 13	00 ♐ 10	06 45	12 D 06	17 02	04 45	27 11	03 41
09	13 12 34	16 13 52	16 ♑ 02 22	22 ♑ 50 12	12 58	01 17	07 28	12 06	16 58	04 45	27 D 11	03 42
10	13 16 31	17 13 09	29 ♑ 43 26	06 ♒ 42 11	14 43	02 24	08 10	12 06	16 53	04 45	27 11	03 44
11	13 20 28	18 12 28	13 ♒ 46 26	20 ♒ 56 05	16 28	03 30	08 53	12 07	16 48	04 44	27 11	03 46
12	13 24 24	19 11 49	28 ♒ 10 53	05 ♓ 30 27	18 12	04 37	09 36	12 07	16 43	04 44	27 11	03 48
13	13 28 21	20 11 12	12 ♓ 54 15	20 ♓ 21 35	19 55	05 43	10 20	12 08	16 39	04 44	27 11	03 49
14	13 32 17	21 10 37	27 ♓ 51 37	05 ♈ 23 23	21 37	06 49	11 03	12 10	16 34	04 D 44	27 11	03 51
15	13 36 14	22 10 03	12 ♈ 55 48	20 ♈ 27 44	23 19	07 55	11 46	12 11	16 29	04 44	27 11	03 53
16	13 40 10	23 09 31	27 ♈ 58 02	05 ♉ 25 33	25 00	09 01	12 29	12 12	16 25	04 44	27 12	03 55
17	13 44 07	24 09 02	12 ♉ 49 13	20 ♉ 08 05	26 41	10 07	13 13	12 14	16 20	04 44	27 12	03 57
18	13 48 03	25 08 34	27 ♉ 20 44	04 ♊ 28 19	28 20	11 12	13 56	12 16	16 15	04 45	27 12	03 59
19	13 52 00	26 08 09	11 ♊ 28 37	18 ♊ 21 55	29 ♎ 59	12 17	14 40	12 18	16 11	04 45	27 13	04 01
20	13 55 57	27 07 46	25 ♊ 08 09	01 ♋ 47 23	01 ♏ 38	13 22	15 23	12 21	16 06	04 45	27 13	04 03
21	13 59 53	28 07 26	08 ♋ 19 50	14 ♋ 45 50	03 16	14 27	16 07	12 23	16 01	04 45	27 13	04 05
22	14 03 50	29 ♎ 07 07	21 ♋ 05 51	27 ♋ 20 23	04 53	15 31	16 51	12 26	15 57	04 46	27 14	04 07
23	14 07 46	00 ♏ 06 51	03 ♌ 30 02	09 ♌ 35 25	06 30	16 36	17 35	12 29	15 52	04 46	27 14	04 09
24	14 11 43	01 06 37	15 ♌ 37 10	21 ♌ 35 55	08 06	17 40	18 19	12 32	15 48	04 47	27 15	04 11
25	14 15 39	02 06 26	27 ♌ 32 21	03 ♍ 27 04	09 41	18 44	19 03	12 35	15 43	04 47	27 15	04 13
26	14 19 36	03 06 16	09 ♍ 20 40	15 ♍ 13 46	11 16	19 47	19 47	12 39	15 39	04 48	27 16	04 15
27	14 23 32	04 06 09	21 ♍ 06 51	27 ♍ 00 28	12 51	20 51	20 31	12 42	15 35	04 49	27 16	04 17
28	14 27 29	05 06 03	02 ♎ 55 02	08 ♎ 50 59	14 25	21 54	21 15	12 46	15 30	04 49	27 17	04 19
29	14 31 26	06 06 00	14 ♎ 48 38	20 ♎ 48 20	15 58	22 57	22 00	12 50	15 26	04 50	27 18	04 21
30	14 35 22	07 05 59	26 ♎ 50 18	02 ♏ 54 46	17 31	23 59	22 44	12 54	15 22	04 51	27 18	04 24
31	14 39 19	08 ♏ 06 00	09 ♏ 01 52	15 ♏ 11 45	19 ♏ 04	25 ♐ 01	23 ♐ 29	12 ♒ 59	15 ♈ 18	04 ♒ 52	27 ♑ 19	04 ♐ 26

Date	Moon True ☊ ° '	Moon Mean ☊ ° '	Moon ☽ Latitude ° '
01	19 ♍ 46	18 ♍ 34	01 N 29
02	19 R 44	18 31	02 30
03	19 41	18 28	03 24
04	19 38	18 25	04 10
05	19 34	18 22	04 45
06	19 31	18 18	05 07
07	19 28	18 15	05 15
08	19 26	18 12	05 06
09	19 26	18 09	04 41
10	19 D 26	18 06	04 00
11	19 27	18 03	03 04
12	19 29	17 59	01 54
13	19 30	17 56	00 N 36
14	19 R 30	17 53	00 S 46
15	19 28	17 50	02 05
16	19 25	17 47	03 15
17	19 20	17 44	04 11
18	19 15	17 40	04 49
19	19 09	17 37	05 09
20	19 04	17 34	05 10
21	19 01	17 31	04 54
22	18 59	17 28	04 24
23	18 D 59	17 24	03 43
24	18 59	17 21	02 52
25	19 01	17 18	01 54
26	19 02	17 15	00 S 52
27	19 R 03	17 12	00 N 11
28	19 02	17 09	01 14
29	18 58	17 05	02 15
30	18 53	17 02	03 10
31	18 ♍ 45	16 ♍ 59	03 N 56

DECLINATIONS

Date	Sun ☉ ° '	Moon ☽ ° '	Mercury ☿ ° '	Venus ♀ ° '	Mars ♂ ° '	Jupiter ♃ ° '	Saturn ♄ ° '	Uranus ♅ ° '	Neptune ♆ ° '	Pluto ♇ ° '
01	03 S 19	01 S 04	02 N 14	20 S 15	21 S 34	18 S 05	04 N 21	19 S 41	20 S 20	08 S 56
02	03 42	04 46	01 29	20 36	21 43	18 06	04 19	19 41	20 20	08 57
03	04 05	08 17	00 N 44	20 56	21 52	18 06	04 18	19 41	20 20	08 58
04	04 28	11 30	00 S 02	21 17	22 00	18 06	04 16	19 41	20 20	08 58
05	04 51	14 16	00 48	21 36	22 08	18 06	04 14	19 41	20 20	08 59
06	05 14	16 25	01 34	21 55	22 17	18 06	04 12	19 41	20 20	09 00
07	05 37	17 49	02 20	22 14	22 24	18 06	04 10	19 41	20 20	09 00
08	06 00	18 18	03 05	22 32	22 32	18 06	04 08	19 41	20 20	09 01
09	06 23	17 49	03 51	22 50	22 40	18 06	04 07	19 41	20 20	09 02
10	06 46	16 17	04 37	23 07	22 47	18 06	04 05	19 41	20 20	09 02
11	07 08	13 45	05 22	23 23	22 54	18 05	04 03	19 41	20 20	09 03
12	07 31	10 19	06 07	23 39	23 01	18 05	04 01	19 41	20 20	09 04
13	07 53	06 09	06 51	23 54	23 08	18 05	03 59	19 41	20 20	09 04
14	08 16	01 S 33	07 35	24 09	23 14	18 04	03 58	19 41	20 20	09 05
15	08 38	03 N 11	08 19	24 23	23 21	18 04	03 56	19 41	20 20	09 06
16	09 00	07 43	09 02	24 37	23 27	18 03	03 54	19 41	20 20	09 06
17	09 22	11 42	09 45	24 50	23 32	18 03	03 52	19 41	20 20	09 07
18	09 44	14 52	10 26	25 03	23 38	18 02	03 51	19 41	20 20	09 08
19	10 05	17 03	11 08	25 14	23 43	18 02	03 49	19 41	20 20	09 08
20	10 27	18 11	11 48	25 26	23 49	18 01	03 47	19 41	20 20	09 09
21	10 48	18 17	12 28	25 36	23 54	18 00	03 45	19 41	20 20	09 10
22	11 10	17 26	13 08	25 46	23 58	17 59	03 44	19 41	20 20	09 10
23	11 31	15 46	13 46	25 56	24 03	17 58	03 42	19 41	20 20	09 11
24	11 52	13 25	14 24	26 04	24 07	17 57	03 40	19 40	20 20	09 12
25	12 12	10 32	15 01	26 12	24 11	17 56	03 39	19 40	20 20	09 12
26	12 33	07 15	15 37	26 20	24 15	17 55	03 37	19 40	20 20	09 13
27	12 53	03 N 42	16 12	26 27	24 18	17 54	03 36	19 40	20 20	09 13
28	13 13	00 S 01	16 47	26 33	24 22	17 53	03 34	19 40	20 19	09 14
29	13 33	03 46	17 21	26 39	24 25	17 52	03 33	19 40	20 19	09 15
30	13 53	07 24	17 54	26 43	24 27	17 50	03 31	19 39	20 19	09 15
31	14 S 12	10 S 46	18 S 25	26 S 48	24 S 30	17 S 49	03 N 29	19 S 39	20 S 19	09 S 16

ZODIAC SIGN ENTRIES

Date	h	m	Planets
02	05	38	☿ ♎
03	11	57	☽ ♏
05	22	43	☽ ♐
08	07	04	☽ ♑
08	08	25	♀ ♐
10	12	29	☽ ♒
12	14	59	☽ ♓
14	15	25	☽ ♈
16	15	16	☽ ♉
18	16	26	☽ ♊
19	12	08	☿ ♏
20	20	45	☽ ♋
23	05	10	☽ ♌
23	09	15	☉ ♏
25	16	59	☽ ♍
28	06	05	☽ ♎
30	18	15	☽ ♏

LATITUDES

Date	Mercury ☿ ° '	Venus ♀ ° '	Mars ♂ ° '	Jupiter ♃ ° '	Saturn ♄ ° '	Uranus ♅ ° '	Neptune ♆ ° '	Pluto ♇ ° '
01	01 N 51	01 S 58	01 S 04	00 S 59	02 S 45	00 S 38	00 N 24	12 N 07
04	01 44	02 10	01 06	00 59	02 46	00 38	00 24	12 06
07	01 32	02 21	01 07	00 58	02 46	00 38	00 24	12 05
10	01 17	02 31	01 08	00 58	02 46	00 38	00 24	12 04
13	01 00	02 42	01 09	00 58	02 46	00 38	00 23	12 03
16	00 41	02 51	01 10	00 58	02 46	00 38	00 23	12 01
19	00 22	03 00	01 11	00 57	02 46	00 38	00 23	12 01
22	00 N 01	03 08	01 11	00 57	02 45	00 38	00 23	12 00
25	00 S 19	03 16	01 12	00 57	02 45	00 37	00 23	11 59
28	00 39	03 22	01 13	00 57	02 45	00 37	00 23	11 58
31	00 S 59	03 S 27	01 S 13	00 S 56	02 S 44	00 S 37	00 N 23	11 N 57

DATA

Julian Date	2450723
Delta T	+62 seconds
Ayanamsa	23° 49' 28"
Synetic vernal point	05° ♓ 17' 31"
True obliquity of ecliptic	23° 26' 14"

MOON'S PHASES, APSIDES AND POSITIONS ☽

Date	h	m	Phase	Longitude ° '	Eclipse Indicator
01	16	52	●	08 ♎ 33	
09	12	22	☽	16 ♑ 15	
16	03	46	○	22 ♈ 49	
23	04	48	☾	29 ♋ 49	
31	10	01	●	08 ♏ 01	

Day	h	m		
15	02	08	Perigee	
27	09	14	Apogee	
01	05	05	0S	
08	12	22	Max dec	18° S 18'
14	19	50	0N	
21	02	12	Max dec	18° N 22'
28	11	52	0S	

LONGITUDES

Date	Chiron ⚷ ° '	Ceres ⚳ ° '	Pallas ⚴ ° '	Juno ⚵ ° '	Vesta ⚶ ° '	Black Moon Lilith ⚸ ° '
01	03 ♏ 30	00 ♓ 50	18 ♑ 46	04 ♍ 53	27 ♈ 46	21 ♍ 46
11	04 ♏ 51	00 ♓ 20	20 ♑ 00	09 ♍ 11	25 ♈ 19	22 ♍ 52
21	06 ♏ 15	29 ♒ 46	21 ♑ 38	13 ♍ 18	22 ♈ 44	23 ♍ 59
31	07 ♏ 39	00 ♓ 08	23 ♑ 38	17 ♍ 13	20 ♈ 20	25 ♍ 05

ASPECTARIAN

h m	Aspects
01 Wednesday	
02 38	☽ ⚹ ♂
06 36	☽ ⚹ ♇
09 18	☽ △ ♅
14 34	☽ ∠ ♀
16 52	☽ ☌ ☉
17 42	♀ ∥ ♆
18 12	☽ # ☿
22 50	♀ Q ♅
02 Thursday	
00 10	☽ △ ♃
04 11	☽ ∥ ☉
09 09	☽ # ♄
10 39	☽ ⊥ ♀
10 52	☽ ∠ ♂
10 58	☽ ☍ ♄
12 21	♂ ⅊ ♄
12 58	☽ ∠ ♇
13 47	♀ ± ♄
23 57	☽ ⊻ ♀
03 Friday	
06 02	☽ ⊥ ♂
06 21	☽ □ ♆
07 02	☽ ⊥ ♇
16 48	☽ ∥ ♇
17 16	☽ ⊻ ☿
18 43	☽ ⊻ ♂
19 00	☽ ⊻ ♇
21 27	☽ □ ♅
23 46	♂ ☌ ♇
04 Saturday	
00 07	☉ # ♄
05 04	☿ ⚹ ♇
07 12	☽ ⊥ ☿
08 22	☿ ⚹ ♂
10 10	☽ ⊻ ☉
11 56	☽ □ ♃
17 57	☽ Q ♆
21 21	☿ △ ♅
22 08	☽ ⊼ ♄
22 58	☽ ⊥ ☉
05 Sunday	
03 52	☽ ∠ ☿
07 41	☉ △ ♃
08 45	☽ Q ♅
09 45	☽ ± ♄
17 01	☽ ☌ ♀
17 08	♂ ⚹ ♅
17 17	☽ ⚹ ♆
17 57	☽ ∠ ☉
20 05	♀ ⚹ ♆
22 54	☽ Q ♃
06 Monday	
02 59	☽ ⅊ ♄
05 38	☽ ☌ ♇
07 50	☽ ⚹ ♅
08 44	☽ ☌ ♂
13 34	☽ ⚹ ☿
21 45	☽ ⚹ ♃
21 55	☽ ∠ ♆
07 Tuesday	
00 59	☽ ⚹ ☉
07 14	☽ △ ♄
12 08	☽ ∠ ♅
14 47	☽ ⊥ ♆
15 47	☽ Q ☿
20 49	☽ ∥ ♃
08 Wednesday	
01 11	☽ Q ☉
01 43	☽ ∠ ♃
01 53	☽ ⊻ ♆
04 36	♃ St D
04 46	☽ ⊥ ♅
06 56	☽ ⊻ ♀
10 24	♀ Q ♃
12 11	♀ ∥ ♂
13 46	☽ ⊻ ♇
15 43	☽ ⊻ ♅
18 09	☽ ⊥ ♃
18 52	☽ ⊥ ♀
19 45	☽ ⊻ ♂
09 Thursday	
00 01	☿ △ ♃
00 39	☽ ⊥ ♇
01 31	♆ St D
02 16	☉ ∠ ♀
03 47	☽ ∥ ♃
04 58	☽ ⊻ ♃
05 43	☽ □ ☿
07 09	☽ ⊥ ♂
12 22	☽ □ ☉
12 28	☽ ∠ ♀
13 38	☽ □ ♄
16 44	☽ ∠ ♇
19 49	☿ # ♄
23 58	☽ ∠ ♂
10 Friday	
01 40	♀ ⅊ ♄
04 26	☉ ☍ ♄
07 35	☽ ☌ ♆
17 01	☽ ⚹ ♀
18 56	☽ ⚹ ♇
20 39	☽ ☌ ♅
20 50	☽ Q ♄
11 Saturday	
03 17	☽ ⚹ ♂
09 11	☽ ☌ ♃
15 10	☽ Q ♀
15 21	☽ Q ♇
16 28	☿ ☍ ♄
17 04	☽ ⚹ ♄
17 09	☽ △ ☿
17 42	♀ ☌ ♇
20 00	☽ △ ☉
12 Sunday	
00 33	☽ Q ♂
01 54	☉ ∠ ♇
07 49	♄ Q ♅
10 21	☽ ⊻ ♆
14 40	♀ ⚹ ♅
17 47	☽ ∠ ♄
19 36	☽ ∥ ♇
20 12	☽ ⊥ ♆
20 28	☿ ∠ ♇
21 13	☽ □ ♇
21 19	☽ ⅊ ☿
22 34	☽ ⅊ ☉
22 45	☽ ⊻ ♅
23 24	☽ □ ♀
13 Monday	
03 11	☽ ∥ ☉
07 37	☽ □ ♂
08 22	☽ ⊥ ♄
08 30	☽ ⊥ ♅
08 44	☽ ∥ ☿
10 46	☽ ⊻ ♃
10 50	☽ ∠ ♆
13 51	☽ ± ☿
14 13	☽ ± ☉
18 00	☽ ⊻ ♄
20 27	☽ ⊥ ♃
21 02	☉ ☌ ☿
23 00	☽ ∠ ♅
23 34	☽ # ♄
14 Tuesday	
00 33	☽ ⊼ ☉
00 45	☽ ⊼ ☿
10 49	♅ St D
10 53	☽ ∠ ♃
10 56	☽ ⚹ ♆
19 59	☿ ∠ ♀
21 35	☽ △ ♇
22 58	☽ ⚹ ♅
15 Wednesday	
03 24	☽ △ ♀
06 03	☽ Q ♆
10 03	☽ △ ♂
10 48	☽ ⚹ ♃
15 47	☽ ∥ ♄
17 38	☽ ☌ ♄
18 04	☽ Q ♅
21 30	☽ ⅊ ♇
16 Thursday	
02 12	♂ ∠ ♆
02 16	♂ ⚹ ♃
03 46	☽ ☍ ☉
05 11	☽ ⅊ ♀
05 58	☽ Q ♃
06 39	☽ ☍ ☿
09 32	☉ ∥ ☿
10 46	☽ □ ♆
11 12	☽ ⅊ ♂
11 55	☽ ± ♇
14 24	☿ ∥ ♇
19 09	☉ ∥ ♇
19 58	☽ # ♇
20 02	☽ # ☉
20 46	☽ ± ♀
21 07	☽ # ☿
21 35	☽ ⊼ ♇
22 53	☽ □ ♅
17 Friday	
02 25	☽ ± ♂
07 14	☽ ⊼ ♀
11 03	☽ □ ♃
12 40	☽ ⊼ ♂
17 43	☽ ⊻ ♄
19 33	☿ □ ♆
18 Saturday	
03 33	☽ ⊥ ♄
06 44	☿ ⊥ ♇
08 02	☽ ⊼ ☉
11 45	☽ △ ♆
13 52	☽ ⊼ ☿
18 31	☽ ∠ ♄
18 50	☽ ± ☉
23 12	☽ ☍ ♇
19 Sunday	
00 28	☽ △ ♅
01 20	☽ ± ☿
03 29	☿ ∠ ♂
10 19	♀ ∠ ♆
11 22	☽ ⅊ ☉
12 24	♀ ⚹ ♃
13 16	☽ ⅊ ♆
13 26	☽ △ ♃
13 31	☽ ☍ ♀
17 50	☽ ☍ ♂
18 55	☽ ⅊ ☿
20 07	☽ ⚹ ♄
20 Monday	
02 26	☽ ⅊ ♅
05 01	☽ ± ♆
06 19	☽ # ♃
14 07	☉ □ ♆
15 44	☽ ⊼ ♆
15 52	☽ △ ☉
15 58	☽ ⅊ ♃
17 17	☽ Q ♄
18 30	☽ ± ♅
21 Tuesday	
01 22	☽ △ ☿
04 10	☽ ⊼ ♇
05 25	☽ ⊼ ♅
08 24	☽ ± ♃
09 09	♂ △ ♄
10 55	☉ ⊥ ♇
15 15	☽ ± ♇
19 34	☽ ⊼ ♃
23 16	☽ # ♃
22 Wednesday	
00 21	☿ ⊻ ♇
00 27	☽ ⊼ ♀
02 17	☽ □ ♄
03 26	☽ ⊼ ♂
08 13	☽ ⅊ ♇
10 14	☿ □ ♅
12 54	☽ ± ♀
15 34	☽ ± ♂
20 51	♀ △ ♄
23 48	☽ ☍ ♆
23 Thursday	
04 48	☽ □ ☉
07 55	☽ ⅊ ♀
10 05	☽ ⅊ ♂
13 16	☽ △ ♇
14 30	☽ ☍ ♅
18 46	☽ □ ☿
24 Friday	
04 36	☽ # ☿
05 49	☽ ☍ ♃
12 21	☽ △ ♄
16 30	☽ △ ♀
17 45	☽ △ ♂
19 38	☽ Q ☉
25 Saturday	
00 01	☽ # ☉
11 25	☽ ⊼ ♆
12 21	☽ Q ☿
18 25	☽ ⅊ ♄
22 02	☽ # ♇
22 07	☽ ⚹ ☉
23 37	☽ ± ♆
26 Sunday	
01 36	☽ □ ♇
02 44	☽ ⊼ ♅
11 34	♀ ☌ ♂
12 12	♀ ∠ ♅
12 29	♂ ∠ ♅
12 37	☽ ± ♄
14 58	☽ ± ♅
16 32	☽ ⚹ ☿
17 58	☽ ⅊ ♆
18 46	☽ ⊼ ♃
27 Monday	
00 47	☽ ⊼ ♄
07 04	☽ ± ♃
07 31	☽ ∠ ☉
09 20	☽ ⅊ ♅
09 46	☿ □ ♃
10 42	☽ □ ♂
11 24	☽ □ ♀
12 40	☽ ∥ ♄
14 24	☽ Q ♇
16 35	☉ ⊻ ♇
21 53	♀ ⊥ ♆
28 Tuesday	
00 33	☽ △ ♆
01 29	☽ ⅊ ♃
03 32	☽ ⊥ ☉
03 48	☽ ∠ ☿
05 11	☉ □ ♅
12 54	♂ ⊥ ♆
14 51	☽ ⚹ ♇
15 52	☽ △ ♅
16 50	☽ ⊻ ☉
29 Wednesday	
00 47	☽ ⊥ ☿
01 34	☿ Q ♆
01 40	☽ Q ♂
03 29	☽ Q ♀
04 06	☿ ⊼ ♄
08 00	☽ △ ♃
10 35	☽ # ♄
12 51	☿ ⊥ ♂
13 15	☽ ☍ ♄
14 40	☽ ⊻ ☿
21 08	☽ ∠ ♇
30 Thursday	
03 19	☽ ⚹ ♂
05 48	☽ ⚹ ♀
09 44	☿ ∥ ♃
12 56	☽ □ ♆
15 05	☽ ⊥ ♇
31 Friday	
00 59	☽ ∥ ♇
02 57	☽ ⊻ ♇
03 49	☽ □ ♅
10 01	☽ ☌ ☉
10 18	☿ ⊥ ♀
10 51	☽ ∠ ♂
14 07	☽ ∠ ♀
19 45	☽ □ ♃

All ephemeris data is given at 12.00 UT and the Moon's longitude is additionally given for 24.00 UT

NOVEMBER 1997

LONGITUDES

Date	Sidereal time h m s	Sun ☉ ° ' "	Moon ☽ ° ' "	Moon ☽ 24.00 ° ' "	Mercury ☿ ° '	Venus ♀ ° '	Mars ♂ ° '	Jupiter ♃ ° '	Saturn ♄ ° '	Uranus ♅ ° '	Neptune ♆ ° '	Pluto ♇ ° '
01	14 43 15	09 ♏ 06 03	21 ♏ 24 29	27 ♏ 40 09	20 ♏ 36	26 ♐ 03	24 ♐ 13	13 ♒ 03	15 ♈ 13	04 ♒ 53	27 ♑ 20	04 ♐ 28
02	14 47 12	10 06 07	03 ♐ 58 45	10 ♐ 20 21	22 07	27 05	24 58	13 08	15 R 09	04 53	27 21	04 30
03	14 51 08	11 06 13	16 ♐ 44 57	23 ♐ 12 35	23 38	28 06	25 43	13 13	15 05	04 54	27 22	04 32
04	14 55 05	12 06 21	29 ♐ 43 17	06 ♑ 17 06	25 09	29 ♐ 07	26 27	13 18	15 01	04 55	27 22	04 35
05	14 59 01	13 06 31	12 ♑ 54 06	19 ♑ 34 22	26 40	00 ♑ 08	27 12	13 23	14 57	04 57	27 23	04 37
06	15 02 58	14 06 42	26 ♑ 18 01	03 ♒ 05 08	28 09	01 08	27 57	13 29	14 54	04 58	27 24	04 39
07	15 06 55	15 06 55	09 ♒ 55 50	16 ♒ 50 14	29 ♏ 39	02 08	28 42	13 35	14 50	04 59	27 25	04 41
08	15 10 51	16 07 09	23 ♒ 48 22	00 ♓ 50 17	01 ♐ 08	03 08	29 ♐ 27	13 40	14 46	05 00	27 26	04 43
09	15 14 48	17 07 24	07 ♓ 55 57	15 ♓ 05 13	02 36	04 07	00 ♑ 12	13 46	14 43	05 01	27 27	04 46
10	15 18 44	18 07 41	22 ♓ 17 52	29 ♓ 33 34	04 04	05 05	00 57	13 53	14 39	05 03	27 28	04 48
11	15 22 41	19 08 00	06 ♈ 51 51	14 ♈ 12 07	05 32	06 03	01 42	13 59	14 36	05 04	27 29	04 50
12	15 26 37	20 08 19	21 ♈ 33 39	28 ♈ 55 37	06 59	07 01	02 28	14 05	14 32	05 06	27 30	04 53
13	15 30 34	21 08 41	06 ♉ 17 08	13 ♉ 37 13	08 25	07 58	03 13	14 12	14 29	05 07	27 32	04 55
14	15 34 30	22 09 04	20 ♉ 54 54	28 ♉ 09 15	09 51	08 55	03 58	14 19	14 26	05 09	27 33	04 57
15	15 38 27	23 09 28	05 ♊ 19 22	12 ♊ 24 32	11 16	09 51	04 44	14 26	14 22	05 10	27 34	05 00
16	15 42 24	24 09 55	19 ♊ 24 05	26 ♊ 17 33	12 40	10 47	05 29	14 33	14 19	05 12	27 35	05 02
17	15 46 20	25 10 23	03 ♋ 04 37	09 ♋ 45 10	14 04	11 42	06 15	14 40	14 16	05 14	27 37	05 04
18	15 50 17	26 10 53	16 ♋ 19 13	22 ♋ 46 56	15 27	12 36	07 00	14 48	14 13	05 15	27 38	05 07
19	15 54 13	27 11 24	29 ♋ 08 36	05 ♌ 24 39	16 48	13 30	07 46	14 55	14 11	05 17	27 39	05 09
20	15 58 10	28 11 57	11 ♌ 35 35	17 ♌ 41 59	18 09	14 23	08 32	15 03	14 08	05 19	27 40	05 11
21	16 02 06	29 ♏ 12 32	23 ♌ 44 29	29 ♌ 43 45	19 28	15 16	09 18	15 11	14 05	05 21	27 42	05 14
22	16 06 03	00 ♐ 13 09	05 ♍ 40 30	11 ♍ 35 24	20 46	16 08	10 03	15 19	14 03	05 23	27 43	05 16
23	16 09 59	01 13 48	17 ♍ 29 10	23 ♍ 22 30	22 02	16 59	10 49	15 27	14 00	05 24	27 45	05 18
24	16 13 56	02 14 28	29 ♍ 16 04	05 ♎ 10 29	23 16	17 49	11 35	15 36	13 58	05 26	27 46	05 21
25	16 17 53	03 15 09	11 ♎ 06 21	17 ♎ 04 12	24 28	18 39	12 21	15 44	13 56	05 28	27 48	05 23
26	16 21 49	04 15 52	23 ♎ 04 31	29 ♎ 07 45	25 38	19 28	13 07	15 53	13 54	05 31	27 49	05 25
27	16 25 46	05 16 37	05 ♏ 14 13	11 ♏ 24 12	26 45	20 16	13 53	16 01	13 51	05 33	27 51	05 28
28	16 29 42	06 17 23	17 ♏ 37 55	23 ♏ 55 27	27 48	21 03	14 39	16 10	13 49	05 35	27 52	05 30
29	16 33 39	07 18 11	00 ♐ 16 52	06 ♐ 42 08	28 48	21 50	15 26	16 19	13 48	05 37	27 54	05 33
30	16 37 35	08 ♐ 19 00	13 ♐ 11 07	19 ♐ 43 41	29 ♐ 44	22 ♑ 35	16 ♑ 12	16 ♒ 28	13 ♈ 46	05 ♒ 39	27 ♑ 55	05 ♐ 35

Date	Moon True ☊ ° '	Moon Mean ☊ ° '	Moon ☽ Latitude ° '
01	18 ♍ 36	16 ♍ 56	04 N 33
02	18 R 26	16 53	04 57
03	18 17	16 50	05 06
04	18 08	16 46	05 00
05	18 02	16 43	04 38
06	17 58	16 40	04 01
07	17 57	16 37	03 09
08	17 D 57	16 34	02 06
09	17 57	16 30	00 N 53
10	17 R 58	16 27	00 S 23
11	17 56	16 24	01 39
12	17 52	16 21	02 49
13	17 46	16 18	03 48
14	17 37	16 15	04 31
15	17 26	16 11	04 56
16	17 15	16 08	05 03
17	17 05	16 05	04 52
18	16 57	16 02	04 25
19	16 51	15 59	03 45
20	16 47	15 55	02 56
21	16 46	15 52	01 59
22	16 D 46	15 49	00 S 59
23	16 R 46	15 46	00 N 04
24	16 45	15 43	01 06
25	16 43	15 40	02 05
26	16 38	15 36	03 00
27	16 29	15 33	03 47
28	16 19	15 30	04 24
29	16 06	15 27	04 49
30	15 ♍ 52	15 ♍ 24	05 N 00

DECLINATIONS

Date	Sun ☉ ° '	Moon ☽ ° '	Mercury ☿ ° '	Venus ♀ ° '	Mars ♂ ° '	Jupiter ♃ ° '	Saturn ♄ ° '	Uranus ♅ ° '	Neptune ♆ ° '	Pluto ♇ ° '
01	14 S 32	13 S 43	18 S 57	26 S 51	24 S 32	17 S 48	03 N 28	19 S 39	20 S 19	09 S 17
02	14 51	16 05	19 27	26 55	24 34	17 46	03 27	19 39	20 19	09 17
03	15 09	17 42	19 56	26 57	24 36	17 45	03 25	19 38	20 19	09 18
04	15 28	18 26	20 24	26 59	24 37	17 43	03 24	19 38	20 19	09 18
05	15 46	18 12	20 51	27 00	24 39	17 41	03 22	19 38	20 19	09 19
06	16 04	16 57	21 17	27 00	24 40	17 40	03 21	19 37	20 18	09 20
07	16 22	14 43	21 42	27 00	24 40	17 38	03 20	19 37	20 18	09 20
08	16 40	11 36	22 06	27 00	24 41	17 36	03 18	19 37	20 18	09 21
09	16 57	07 46	22 29	26 58	24 41	17 35	03 17	19 37	20 18	09 21
10	17 14	03 S 25	22 51	26 57	24 41	17 33	03 16	19 36	20 18	09 22
11	17 30	01 N 12	23 12	26 54	24 40	17 31	03 15	19 36	20 17	09 22
12	17 47	05 47	23 31	26 51	24 40	17 29	03 14	19 35	20 17	09 23
13	18 03	10 01	23 50	26 48	24 39	17 27	03 13	19 35	20 17	09 24
14	18 18	13 37	24 07	26 43	24 38	17 25	03 12	19 35	20 17	09 24
15	18 34	16 19	24 23	26 39	24 36	17 23	03 10	19 34	20 17	09 25
16	18 49	17 59	24 38	26 34	24 35	17 20	03 09	19 34	20 16	09 25
17	19 03	18 33	24 51	26 28	24 33	17 18	03 08	19 33	20 16	09 26
18	19 18	18 04	25 03	26 21	24 30	17 16	03 08	19 33	20 16	09 26
19	19 32	16 39	25 14	26 15	24 28	17 14	03 07	19 32	20 16	09 27
20	19 45	14 29	25 23	26 07	24 25	17 11	03 06	19 32	20 15	09 27
21	19 59	11 44	25 31	26 00	24 22	17 09	03 05	19 32	20 15	09 28
22	20 12	08 31	25 38	25 51	24 19	17 06	03 04	19 31	20 15	09 28
23	20 24	05 00	25 43	25 43	24 15	17 04	03 03	19 31	20 15	09 29
24	20 36	01 N 18	25 47	25 33	24 11	17 01	03 03	19 30	20 14	09 29
25	20 48	02 S 28	25 49	25 24	24 07	16 59	03 02	19 30	20 14	09 30
26	21 00	06 11	25 50	25 14	24 03	16 56	03 01	19 29	20 14	09 30
27	21 11	09 42	25 49	25 03	23 58	16 53	03 01	19 28	20 13	09 31
28	21 21	12 52	25 47	24 53	23 53	16 51	03 00	19 28	20 13	09 31
29	21 32	15 30	25 44	24 41	23 48	16 48	03 00	19 27	20 13	09 32
30	21 S 41	17 S 25	25 S 39	24 S 30	23 S 42	16 S 45	02 N 59	19 S 27	20 S 13	09 S 32

ZODIAC SIGN ENTRIES

Date	h	m	Planets	
02	04	27	☽	♐
04	12	31	☽	♑
05	08	50	♀	♑
06	18	33	☽	♒
07	17	42	☿	♐
08	22	35	☽	♓
09	05	33	♂	♑
11	00	44	☽	♈
13	01	45	☽	♉
15	03	05	☽	♊
17	06	32	☽	♋
19	13	38	☽	♌
22	00	33	☽	♍
22	06	48	☉	♐
24	13	29	☽	♎
27	01	43	☽	♏
29	11	28	☽	♐
30	19	11	☿	♑

LATITUDES

Date	Mercury ☿ ° '	Venus ♀ ° '	Mars ♂ ° '	Jupiter ♃ ° '	Saturn ♄ ° '	Uranus ♅ ° '	Neptune ♆ ° '	Pluto ♇ ° '
01	01 S 05	03 S 29	01 S 14	00 S 56	02 S 44	00 S 37	00 N 23	11 N 57
04	01 23	03 33	01 14	00 56	02 44	00 37	00 23	11 56
07	01 40	03 35	01 14	00 56	02 44	00 37	00 23	11 55
10	01 56	03 37	01 15	00 55	02 43	00 37	00 23	11 55
13	02 09	03 36	01 15	00 55	02 43	00 37	00 23	11 54
16	02 20	03 34	01 15	00 55	02 42	00 37	00 23	11 54
19	02 27	03 31	01 15	00 55	02 41	00 37	00 23	11 53
22	02 31	03 25	01 16	00 54	02 41	00 37	00 23	11 53
25	02 30	03 17	01 16	00 54	02 40	00 37	00 22	11 53
28	02 22	03 07	01 15	00 54	02 39	00 37	00 22	11 52
31	02 S 07	02 S 55	01 S 15	00 S 54	02 S 39	00 S 37	00 N 22	11 N 52

DATA

Julian Date	2450754
Delta T	+62 seconds
Ayanamsa	23° 49' 31"
Synetic vernal point	05° ♓ 17' 28"
True obliquity of ecliptic	23° 26' 14"

MOON'S PHASES, APSIDES AND POSITIONS ☽

Date	h	m	Phase	Longitude ° '	Eclipse Indicator
07	21	43	☽	15 ♒ 31	
14	14	12	○	22 ♉ 15	
21	23	58	☾	29 ♌ 43	
30	02	14	●	07 ♐ 54	

Day	h	m		
12	08	01	Perigee	
24	02	31	Apogee	
04	18	17	Max dec	18° S 28'
11	05	47	0N	
17	12	44	Max dec	18° N 33'
24	20	17	0S	

LONGITUDES

Date	Chiron ⚷ ° '	Ceres ⚳ ° '	Pallas ⚴ ° '	Juno ⚵ ° '	Vesta ⚶ ° '	Black Moon Lilith ⚸ ° '
01	07 ♏ 48	00 ♓ 12	23 ♑ 51	17 ♍ 36	20 ♈ 06	25 ♍ 18
11	09 ♏ 12	01 ♓ 12	26 ♑ 15	21 ♍ 16	18 ♈ 13	26 ♍ 12
21	10 ♏ 35	02 ♓ 41	28 ♑ 44	24 ♍ 40	17 ♈ 00	27 ♍ 25
31	11 ♏ 55	04 ♓ 38	01 ♒ 30	27 ♍ 46	16 ♈ 32	28 ♍ 31

All ephemeris data is given at 12.00 UT and the Moon's longitude is additionally given for 24.00 UT

ASPECTARIAN

h m Aspects

01 Saturday
00 07 ☽ ⊼ ♄
00 15 ☽ Q ♆
05 28 ☽ ⊥ ♂
09 10 ☽ ⊥ ♀
10 13 ☽ ☌ ☿
11 39 ☽ ± ♄
14 49 ☽ Q ♅
17 45 ☽ ⊻ ♂
20 36 ☽ ∥ ☉
21 27 ☿ ± ♄
21 43 ☽ ⊻ ♀
23 12 ☉ ∠ ♂
23 22 ☽ ✶ ♆

02 Sunday
04 47 ☽ ⚼ ♄
06 34 ☽ Q ♃
13 00 ☽ ☌ ♇
13 44 ☽ ✶ ♅
18 13 ♀ ⊻ ♆
21 42 ☿ ∥ ♅

03 Monday
00 16 ☿ Q ♅
00 32 ☽ ⊻ ☉
03 47 ☽ ∠ ♆
05 21 ☽ ✶ ♃
08 55 ☽ △ ♄
12 43 ☽ ⊥ ☉
12 51 ☽ ∥ ♃
14 51 ♀ ∠ ♃
17 53 ☽ ∠ ♅
20 35 ☽ ⊥ ♆

04 Tuesday
02 29 ☽ ⊻ ☿
05 38 ☽ ☌ ♂
06 47 ☽ ∠ ☉
07 14 ♀ ⊥ ♅
07 20 ☿ ∥ ♆
07 41 ☽ ⊻ ♆
09 23 ☽ ∠ ♃
10 32 ☽ ⊥ ♅
10 48 ☽ ☌ ♀
14 58 ☽ ⊥ ☿
20 54 ☽ ⊻ ♇
21 32 ☽ ⊻ ♅

05 Wednesday
01 57 ☽ ⊥ ♃
07 51 ☽ ⊥ ♇
09 28 ☽ ∠ ☿
12 24 ☽ ✶ ☉
12 53 ☽ ⊻ ♃
15 42 ☽ □ ♄
18 06 ♂ ⊻ ♆
19 27 ☉ □ ♃
23 49 ☿ ✶ ♆

06 Thursday
00 06 ☽ ∠ ♇
00 24 ☽ ∥ ♃
05 23 ☿ ⊻ ♂
11 38 ☽ Q ☉
13 58 ☽ ☌ ♆
15 06 ☽ ⊻ ♂
15 42 ☽ ✶ ☿
21 15 ☽ ⊻ ♀
21 23 ☽ ∥ ☉
23 37 ☽ Q ♄

07 Friday
02 19 ☽ ⊥ ♂
02 48 ☽ ✶ ♇
03 19 ☽ ☌ ♅
05 37 ☉ ⊼ ♄
07 29 ♂ ∠ ♃
08 37 ☽ ⊥ ♀
14 52 ☿ ⚼ ♄
15 21 ☽ Q ☿
18 24 ☽ ☌ ♃
18 56 ☽ ∠ ♂
19 25 ☉ Q ♆
20 29 ☽ ✶ ♄
21 16 ♂ ⊥ ♅
21 43 ☽ □ ☉
23 46 ☽ Q ♇

08 Saturday
01 29 ☽ ∠ ♀
18 13 ☽ ⊻ ♆
21 29 ☿ Q ♃
22 08 ☽ ∠ ♄
22 11 ☽ ✶ ♂

09 Sunday
01 57 ☽ □ ☿
02 33 ☽ ∥ ♇
04 26 ☽ ⊥ ♆
05 04 ☽ ✶ ♀
06 38 ☽ □ ♇
07 05 ☽ ⊻ ♅
13 18 ☽ ⊥ ♄
17 12 ☽ ⊥ ♅
19 34 ☽ Q ♂
19 36 ☽ ∠ ♆
21 53 ☽ ⊻ ♃
23 19 ☽ ⊻ ♄

10 Monday
02 42 ☽ Q ♀
04 33 ☽ △ ☉
04 40 ♀ ⊻ ♇
07 57 ☽ ⊥ ♃
08 15 ☽ ∠ ♅
10 59 ♀ ⊻ ♅
12 46 ☽ ⋕ ♄
20 34 ☽ ✶ ♆
22 57 ☽ ∠ ♃

11 Tuesday
00 18 ☿ ☌ ♇
03 04 ☽ □ ♂
04 17 ☿ ✶ ♅
07 11 ☽ ⚼ ☉
08 40 ☽ △ ♇
09 03 ☽ ✶ ♅
09 34 ☽ △ ☿
10 35 ☽ □ ♀
12 36 ☉ ∥ ♃
16 18 ☽ Q ♆
22 32 ☽ ∥ ♄
23 00 ☽ ± ☉
23 44 ☽ ✶ ♃

12 Wednesday
00 35 ☽ ☌ ♄
04 42 ☽ Q ♅
09 15 ☽ ⚼ ♇
09 31 ☽ ⊼ ☉
12 45 ☽ ⚼ ☿
13 53 ☿ ⊻ ♀
19 26 ☽ Q ♃
20 59 ☉ ± ♄
21 42 ☽ □ ♆
23 57 ☽ ± ♇

13 Thursday
05 01 ☽ ± ☿
06 43 ☽ △ ♂
08 13 ☽ ⋕ ♇
09 45 ☽ ⊼ ♇
10 05 ☽ □ ♅
14 57 ☽ △ ♀
15 52 ☽ ⊼ ☿
18 38 ♀ ⊥ ♃

14 Friday
01 03 ☽ □ ♃
01 22 ☽ ⊻ ♄
08 37 ☽ ⚼ ♂
11 12 ☽ ⊥ ♄
14 12 ☽ ☍ ☉
17 18 ☽ ⚼ ♀
23 00 ☽ △ ♆

15 Saturday
00 21 ☽ ± ♂
02 04 ☽ ∠ ♄
03 53 ♃ ✶ ♄
09 21 ☽ ± ♀
10 57 ☽ ⊼ ♂
11 27 ☽ ☍ ♇
11 45 ☽ △ ♅
12 18 ☉ Q ♅
20 11 ☽ ⊼ ♀
20 45 ♂ ⊻ ♇
23 10 ☽ ☍ ☿

16 Sunday
00 17 ☽ ⚼ ♆
01 01 ☽ ⋕ ♃
02 26 ♂ ⊻ ♅
03 18 ☽ ✶ ♄
03 35 ☽ △ ♃
07 42 ☿ ∥ ♂
10 31 ☿ ∠ ♆
13 23 ☽ ⚼ ♅
15 48 ☽ ± ♆
18 53 ♀ ⊥ ♇
20 56 ☽ ⊼ ☉

17 Monday
00 01 ☽ Q ♄
02 18 ☽ ⊼ ♆
05 09 ☽ ± ♅
05 54 ☽ ⚼ ♃
08 20 ☽ ± ☉
15 29 ☿ △ ♄
15 34 ☽ ⊼ ♇
15 51 ☽ ⊼ ♅
18 01 ☽ ☍ ♂
22 08 ☽ ± ♃
23 35 ☿ ✶ ♃

18 Tuesday
01 49 ☽ ⚼ ☉
02 26 ☽ ± ♇
04 41 ☽ ☍ ♀
08 10 ☽ □ ♄
09 10 ☽ ⊼ ♃
10 12 ☽ ⊼ ☿
19 02 ☽ ⚼ ♇
22 37 ☽ ± ☿

19 Wednesday
03 37 ☽ ⋕ ♃
07 59 ☽ △ ☉
09 10 ☽ ☍ ♆
13 12 ☉ ∥ ♅
17 41 ☽ ⚼ ☿
23 14 ☉ ✶ ♆
23 32 ☽ △ ♇
23 47 ☽ ☍ ♅

20 Thursday
05 22 ♀ □ ♄
05 39 ☽ ⊼ ♂
16 57 ☽ △ ♄
17 54 ☽ ⊼ ♀
18 08 ☽ ± ♂
18 51 ☽ ☍ ♃

21 Friday
02 27 ☽ △ ☿
06 41 ☽ ± ♀
07 48 ♂ ⊥ ♃
09 15 ☉ ⚼ ♄
09 19 ♀ ⊻ ♃
13 11 ☽ ⚼ ♂
19 56 ☽ ⊼ ♆
22 40 ☽ ⚼ ♄
23 58 ☽ □ ☉

22 Saturday
02 06 ☽ ⚼ ♀
04 39 ☿ ∠ ♅
05 10 ☽ ⋕ ♇
08 02 ☽ ± ♆
11 10 ☽ □ ♇
11 24 ☽ ⊼ ♅
16 47 ☽ ± ♄
17 50 ☉ ∥ ♆
21 30 ☽ △ ♂
23 36 ☽ ± ♅

23 Sunday
02 20 ☽ ⚼ ♆
04 56 ☽ ⊼ ♄
06 28 ☿ ⊥ ♆
07 49 ☽ ⊼ ♃
10 54 ☽ △ ♀
11 32 ☿ ∥ ♀
15 53 ☽ Q ☉
17 58 ☽ ⚼ ♅
20 11 ☽ ± ♃
22 21 ☽ □ ☿
23 54 ☽ Q ♇

24 Monday
00 45 ☽ ∥ ♄
04 01 ♂ ⊥ ♇
08 57 ☽ △ ♆
14 44 ☽ ⚼ ♃
18 37 ☽ ✶ ☉

25 Tuesday
00 23 ☽ ✶ ♇
00 35 ☽ △ ♅
14 41 ☽ □ ♂
15 03 ☽ Q ☿
15 36 ☽ ⋕ ♄
17 40 ☽ ☍ ♄
21 26 ☽ △ ♃

26 Wednesday
01 16 ☉ Q ♃
03 42 ☽ ∠ ☉
04 17 ☽ □ ♀
06 42 ☽ ∠ ♇
17 36 ☽ ✶ ☿
21 26 ☽ □ ♆
23 14 ☽ ⊥ ☉

27 Thursday
00 37 ☽ ⊥ ♇
05 00 ☽ Q ♂
10 41 ☽ ∥ ♇
11 01 ☉ ∠ ♀
11 05 ♂ □ ♄
12 05 ☽ ⊻ ☉
12 27 ☽ ⊻ ♇
12 36 ☽ □ ♅
16 35 ☉ ☌ ♇
18 13 ♀ ∠ ♇
18 19 ☽ Q ♀
18 34 ☉ ✶ ♅

28 Friday
01 51 ☽ ∠ ☿
04 42 ☽ ⊼ ♄
05 54 ☽ ✶ ♂
08 37 ☽ Q ♆
09 10 ☽ □ ♃
13 34 ☿ ⊻ ♆
16 11 ☽ ± ♄
18 58 ☽ ✶ ♀
20 40 ☽ ⊥ ☿
23 23 ☽ Q ♅

29 Saturday
07 30 ☽ ✶ ♆
09 00 ☽ ⊻ ☿
09 13 ☽ ⚼ ♄
12 17 ☽ ∠ ♂
19 39 ☽ Q ♃
21 52 ☽ ☌ ♇
22 01 ☽ ✶ ♅

30 Sunday
01 00 ☽ ∠ ♀
02 14 ☽ ☌ ☉
02 45 ☽ ∥ ♃
06 08 ☽ ⊥ ♂
09 43 ☿ ⊥ ♅
11 31 ☽ ∠ ♆
13 04 ☽ △ ♄
17 53 ☽ ⊻ ♂
18 07 ☽ ✶ ♃
18 37 ☽ ⊥ ♀
22 47 ♂ ⊻ ♃

DECEMBER 1997

Date	Sidereal time h m s	Sun ☉	Moon ☽	Moon ☽ 24.00	Mercury ☿	Venus ♀	Mars ♂	Jupiter ♃	Saturn ♄	Uranus ♅	Neptune ♆	Pluto ♇
01	16 41 32	09 ♐ 19 50	26 ♐ 19 37	02 ♑ 58 41	00 ♑ 35	23 ♑ 19	16 ♑ 58	16 ♒ 38	13 ♈ 44	05 ♒ 42	27 ♑ 57	05 ♐ 37
02	16 45 28	10 20 41	09 ♑ 40 38	16 ♑ 25 13	01 21	24 03	17 44	16 47	13 R 43	05 44	27 59	05 40
03	16 49 25	11 21 34	23 ♑ 12 11	00 ♒ 01 20	02 01	24 45	18 31	16 57	13 41	05 46	28 00	05 42
04	16 53 22	12 22 27	06 ♒ 52 28	13 ♒ 45 28	02 33	25 26	19 17	17 06	13 40	05 49	28 02	05 44
05	16 57 18	13 23 21	20 ♒ 40 13	27 ♒ 36 39	02 58	26 06	20 04	17 16	13 39	05 51	28 04	05 47
06	17 01 15	14 24 15	04 ♓ 34 44	11 ♓ 34 26	03 15	26 45	20 50	17 26	13 37	05 53	28 06	05 49
07	17 05 11	15 25 11	18 ♓ 35 45	25 ♓ 38 37	03 22	27 22	21 37	17 36	13 36	05 56	28 07	05 51
08	17 09 08	16 26 07	02 ♈ 42 57	09 ♈ 48 39	03 R 19	27 59	22 23	17 46	13 35	05 59	28 09	05 54
09	17 13 04	17 27 03	16 ♈ 55 30	24 ♈ 03 13	03 05	28 34	23 10	17 57	13 35	06 01	28 11	05 56
10	17 17 01	18 28 01	01 ♉ 11 28	08 ♉ 19 46	02 40	29 07	23 56	18 07	13 34	06 04	28 13	05 58
11	17 20 57	19 28 59	15 ♉ 27 36	22 ♉ 34 22	02 03	29 ♑ 39	24 43	18 17	13 33	06 06	28 15	06 01
12	17 24 54	20 29 57	29 ♉ 39 26	06 ♊ 42 08	01 15	00 ♒ 09	25 30	18 28	13 33	06 09	28 17	06 03
13	17 28 51	21 30 57	13 ♊ 41 49	20 ♊ 37 52	00 ♑ 16	00 38	26 16	18 39	13 33	06 12	28 18	06 05
14	17 32 47	22 31 57	27 ♊ 29 42	04 ♋ 16 51	29 ♐ 09	01 05	27 03	18 50	13 32	06 14	28 20	06 08
15	17 36 44	23 32 58	10 ♋ 58 59	17 ♋ 35 49	27 53	01 30	27 50	19 01	13 32	06 17	28 22	06 10
16	17 40 40	24 34 00	24 ♋ 07 15	00 ♌ 33 17	26 33	01 54	28 37	19 12	13 D 32	06 20	28 24	06 12
17	17 44 37	25 35 02	06 ♌ 54 03	13 ♌ 09 47	25 10	02 15	29 ♑ 24	19 23	13 32	06 23	28 26	06 14
18	17 48 33	26 36 05	19 ♌ 20 50	25 ♌ 27 37	23 48	02 35	00 ♒ 11	19 34	13 32	06 26	28 28	06 17
19	17 52 30	27 37 09	01 ♍ 30 40	07 ♍ 30 33	22 28	02 53	00 57	19 45	13 33	06 29	28 30	06 19
20	17 56 26	28 38 14	13 ♍ 27 53	19 ♍ 23 20	21 15	03 08	01 44	19 57	13 33	06 32	28 32	06 21
21	18 00 23	29 ♐ 39 20	25 ♍ 17 35	01 ♎ 11 20	20 08	03 22	02 31	20 08	13 33	06 35	28 34	06 23
22	18 04 20	00 ♑ 40 26	07 ♎ 05 18	13 ♎ 00 10	19 11	03 33	03 18	20 20	13 34	06 38	28 36	06 26
23	18 08 16	01 41 34	18 ♎ 56 37	24 ♎ 55 18	18 25	03 43	04 05	20 32	13 35	06 41	28 38	06 28
24	18 12 13	02 42 41	00 ♏ 56 49	07 ♏ 01 45	17 48	03 49	04 52	20 43	13 36	06 44	28 41	06 30
25	18 16 09	03 43 50	13 ♏ 10 35	19 ♏ 23 45	17 23	03 54	05 39	20 55	13 37	06 47	28 43	06 32
26	18 20 06	04 44 59	25 ♏ 41 34	02 ♐ 04 17	17 08	03 56	06 27	21 07	13 38	06 50	28 45	06 34
27	18 24 02	05 46 09	08 ♐ 32 02	15 ♐ 04 51	17 D 04	03 R 56	07 14	21 19	13 39	06 53	28 47	06 36
28	18 27 59	06 47 19	21 ♐ 42 38	28 ♐ 25 11	17 08	03 53	08 01	21 32	13 40	06 56	28 49	06 39
29	18 31 55	07 48 29	05 ♑ 12 13	12 ♑ 03 21	17 22	03 48	08 48	21 44	13 41	06 59	28 51	06 41
30	18 35 52	08 49 40	18 ♑ 58 06	25 ♑ 55 58	17 43	03 40	09 35	21 56	13 43	07 02	28 53	06 43
31	18 39 49	09 ♑ 50 50	02 ♒ 56 25	09 ♒ 58 54	18 ♐ 12	03 ♒ 30	10 ♒ 22	22 ♒ 09	13 ♈ 45	07 ♒ 06	28 ♑ 55	06 ♐ 45

DECLINATIONS

Date	Moon True ☊	Moon Mean ☊	Moon ☽ Latitude	Sun ☉	Moon ☽	Mercury ☿	Venus ♀	Mars ♂	Jupiter ♃	Saturn ♄	Uranus ♅	Neptune ♆	Pluto ♇
01	15 ♍ 39	15 ♍ 21	04 N 55	21 S 51	18 S 28	25 S 33	24 S 18	23 S 36	16 S 42	02 N 59	19 S 26	20 S 12	09 S 33
02	15 R 27	15 17	04 34	22 00	18 31	25 26	24 06	23 30	16 39	02 59	19 26	20 12	09 33
03	15 17	15 14	03 58	22 08	17 32	25 17	23 53	23 24	16 36	02 58	19 25	20 12	09 33
04	15 11	15 11	03 07	22 17	15 32	25 07	23 41	23 17	16 33	02 58	19 24	20 11	09 34
05	15 08	15 08	02 05	22 24	12 37	24 55	23 28	23 11	16 30	02 58	19 24	20 11	09 34
06	15 D 06	15 05	00 N 55	22 32	08 58	24 43	23 15	23 03	16 27	02 57	19 23	20 11	09 35
07	15 R 06	15 01	00 S 18	22 38	04 48	24 29	23 01	22 56	16 24	02 57	19 23	20 10	09 35
08	15 06	14 58	01 32	22 45	00 S 19	24 14	22 47	22 48	16 21	02 57	19 22	20 10	09 35
09	15 04	14 55	02 40	22 51	04 N 11	23 58	22 34	22 41	16 18	02 57	19 21	20 10	09 36
10	14 59	14 52	03 38	22 56	08 29	23 41	22 20	22 32	16 15	02 57	19 21	20 09	09 36
11	14 51	14 49	04 22	23 01	12 17	23 23	22 06	22 24	16 11	02 57	19 20	20 09	09 37
12	14 41	14 46	04 50	23 06	15 21	23 04	21 51	22 15	16 08	02 57	19 19	20 09	09 37
13	14 28	14 42	05 00	23 10	17 28	22 44	21 37	22 06	16 04	02 57	19 19	20 08	09 37
14	14 15	14 39	04 53	23 14	18 32	22 24	21 23	21 57	16 01	02 57	19 18	20 08	09 38
15	14 03	14 36	04 28	23 17	18 31	22 03	21 08	21 48	15 58	02 58	19 17	20 07	09 38
16	13 53	14 33	03 50	23 20	17 30	21 43	20 54	21 38	15 54	02 58	19 16	20 07	09 38
17	13 45	14 30	03 02	23 22	15 37	21 22	20 40	21 28	15 51	02 58	19 16	20 07	09 39
18	13 40	14 27	02 05	23 24	13 02	21 03	20 25	21 18	15 47	02 58	19 15	20 06	09 39
19	13 38	14 23	01 04	23 25	09 56	20 45	20 11	21 08	15 43	02 59	19 14	20 06	09 39
20	13 37	14 20	00 S 01	23 26	06 29	20 29	19 57	20 57	15 40	02 59	19 13	20 06	09 40
21	13 D 37	14 17	01 N 02	23 26	02 N 49	20 16	19 42	20 46	15 36	03 00	19 13	20 05	09 40
22	13 R 37	14 14	02 01	23 26	00 S 57	20 05	19 28	20 35	15 32	03 00	19 12	20 05	09 40
23	13 35	14 11	02 56	23 26	04 42	19 57	19 14	20 24	15 29	03 01	19 11	20 04	09 40
24	13 32	14 07	03 44	23 25	08 18	19 51	19 00	20 13	15 25	03 01	19 10	20 04	09 41
25	13 25	14 04	04 22	23 23	11 37	19 49	18 47	20 01	15 21	03 02	19 10	20 04	09 41
26	13 17	14 01	04 49	23 21	14 30	19 49	18 33	19 49	15 17	03 02	19 09	20 03	09 41
27	13 06	13 58	05 02	23 19	16 45	19 51	18 20	19 37	15 13	03 03	19 08	20 03	09 41
28	12 54	13 55	04 59	23 16	18 12	19 56	18 07	19 24	15 09	03 04	19 07	20 02	09 41
29	12 42	13 52	04 40	23 12	18 40	20 02	17 55	19 12	15 05	03 05	19 06	20 02	09 42
30	12 31	13 48	04 04	23 09	18 04	20 10	17 42	18 59	15 01	03 05	19 06	20 01	09 42
31	12 ♍ 23	13 ♍ 45	03 N 14	23 S 04	16 S 21	20 S 19	17 S 30	18 S 46	14 S 57	03 N 06	19 S 05	20 S 01	09 S 42

ZODIAC SIGN ENTRIES

Date	h	m	Planets
01	18	38	☽ ♑
03	23	58	☽ ♒
06	04	07	☽ ♓
08	07	24	☽ ♈
10	10	00	☽ ♉
12	04	39	♀ ♒
12	12	35	☽ ♊
13	18	06	☿ ♐
14	16	25	☽ ♋
16	22	58	☽ ♌
18	06	37	♂ ♒
19	09	00	☽ ♍
21	20	07	☉ ♑
21	21	35	☽ ♎
24	10	07	☽ ♏
26	20	07	☽ ♐
29	02	48	☽ ♑
31	06	58	☽ ♒

LATITUDES

Date	Mercury ☿	Venus ♀	Mars ♂	Jupiter ♃	Saturn ♄	Uranus ♅	Neptune ♆	Pluto ♇
01	02 S 07	02 S 55	01 S 15	00 S 54	02 S 39	00 S 37	00 N 22	11 N 52
04	01 42	02 40	01 15	00 54	02 38	00 37	00 22	11 52
07	01 05	02 23	01 15	00 53	02 37	00 36	00 22	11 52
10	00 S 16	02 02	01 15	00 53	02 36	00 36	00 22	11 52
13	00 N 42	01 39	01 14	00 53	02 36	00 36	00 22	11 52
16	01 41	01 12	01 14	00 53	02 35	00 36	00 22	11 52
19	02 28	00 41	01 13	00 53	02 34	00 36	00 22	11 52
22	02 56	00 S 07	01 12	00 53	02 33	00 36	00 22	11 53
25	03 03	00 N 31	01 12	00 52	02 33	00 36	00 22	11 53
28	02 54	01 12	01 11	00 52	02 32	00 36	00 22	11 53
31	02 N 36	01 N 55	01 S 10	00 S 52	02 S 31	00 S 36	00 N 22	11 N 54

DATA

Julian Date	2450784
Delta T	+62 seconds
Ayanamsa	23° 49' 35"
Synetic vernal point	05° ♓ 17' 24"
True obliquity of ecliptic	23° 26' 13"

MOON'S PHASES, APSIDES AND POSITIONS ☽

Date	h	m	Phase	Longitude	Eclipse Indicator
07	06	09	☽	15 ♓ 10	
14	02	37	○	22 ♊ 08	
21	21	43	☾	00 ♎ 04	
29	16	57	●	08 ♑ 01	

Day	h	m		
09	17	00	Perigee	
21	23	24	Apogee	
02	01	20	Max dec	18° S 38'
08	13	43	0N	
14	23	38	Max dec	18° N 40'
22	05	57	0S	
29	10	40	Max dec	18° S 40'

LONGITUDES

Date	Chiron ⚷	Ceres ⚳	Pallas ⚴	Juno ⚵	Vesta ⚶	Black Moon Lilith ⚸
01	11 ♏ 55	04 ♓ 38	01 ♒ 30	27 ♍ 46	16 ♈ 32	28 ♍ 31
11	13 ♏ 12	06 ♓ 57	04 ♒ 25	00 ♎ 30	16 ♈ 48	29 ♍ 38
21	14 ♏ 23	09 ♓ 36	07 ♒ 29	02 ♎ 50	17 ♈ 46	00 ♎ 44
31	15 ♏ 28	12 ♓ 31	10 ♒ 39	04 ♎ 40	19 ♈ 19	01 ♎ 51

All ephemeris data is given at 12.00 UT and the Moon's longitude is additionally given for 24.00 UT

ASPECTARIAN

01 Monday
01 44 ☽ ∠ ♅
04 02 ☽ ⊥ ♆
06 14 ☽ ⊻ ♀
14 57 ☽ ⊻ ♆
18 06 ☽ ⊥ ♅
20 11 ☽ ☌ ☿
21 41 ☽ ∠ ♃

02 Tuesday
04 48 ☽ ⊻ ♇
04 55 ☽ ⊻ ♅
13 17 ☽ ⊻ ☉
14 00 ☽ ⊥ ♃
15 33 ☽ ⊥ ♇
19 10 ☽ □ ♄

03 Wednesday
00 48 ☽ ⊻ ♃
00 50 ☽ ⊥ ☉
03 12 ☽ ☌ ♂
07 34 ☽ ∠ ♇
08 31 ☿ ∠ ♃
14 52 ☽ ☌ ♀
18 01 ☽ ∠ ☉
20 29 ☽ ☌ ♆

04 Thursday
00 52 ☽ ∥ ♃
02 54 ☽ Q ♄
04 09 ☽ ⊻ ☿
10 01 ☽ ✱ ♇
10 08 ☽ ☌ ♅
15 03 ☽ ⊥ ☿
22 21 ☽ ✱ ☉
23 49 ☽ ✱ ♄

05 Friday
04 05 ☉ ∠ ♆
06 02 ☽ ☌ ♃
06 58 ☽ Q ♇
07 12 ☽ ∠ ☿
10 53 ☽ ⊻ ♂
17 53 ☉ △ ♄
20 48 ☽ Q ☉
21 52 ☽ ⊻ ♀
21 52 ☽ ⊥ ♂

06 Saturday
00 49 ☽ ⊻ ♆
01 46 ☽ ∠ ♄
08 17 ☽ ∥ ♇
08 42 ☽ ⊥ ♀
09 41 ☽ ✱ ☿
11 10 ☽ ⊥ ♆
11 26 ♂ ∠ ♇
14 08 ☽ □ ♇
14 16 ☽ ⊻ ♅
14 17 ☽ ∠ ♂
17 13 ☽ ⊥ ♄

07 Sunday
00 35 ☽ ⊥ ♅
00 53 ☽ ∠ ♀
02 38 ☽ ∠ ♆
03 29 ☽ ⊻ ♄
06 09 ☽ □ ☉
06 28 ☽ Q ☿
10 17 ☽ ⊻ ♃
16 00 ☽ ∠ ♅
16 55 ☿ St R
17 26 ☽ ✱ ♂
20 38 ☽ ⊥ ♃
21 59 ☽ # ♄

08 Monday
03 37 ☽ ✱ ♀
04 15 ☽ ✱ ♆
07 02 ☉ ⊥ ♂
08 12 ♀ ∥ ♂
12 06 ☽ ∠ ♃
13 00 ☽ □ ☿
14 59 ☽ Q ♂
15 16 ☉ ∥ ♀
17 24 ☽ △ ♇
17 32 ☽ ✱ ♅
18 18 ☉ ∥ ♂
19 30 ♀ ☌ ♆

09 Tuesday
00 36 ☽ Q ♆
00 49 ☽ Q ♀
05 22 ☽ ∥ ♄
06 22 ☽ ☌ ♄
10 25 ♃ Q ♇
12 57 ☽ △ ☉
13 44 ☽ ✱ ♃
13 51 ☽ Q ♅
18 22 ☿ ∠ ♃
18 46 ☽ bQ ♇
23 06 ☽ □ ♂

10 Wednesday
02 00 ☉ ✱ ♃
06 59 ☽ □ ♆
08 22 ☽ □ ♀
09 57 ☽ ± ♇
10 10 ☽ Q ♃
14 23 ☽ △ ☿
16 07 ☽ bQ ☉
18 43 ☽ # ♇
20 04 ☽ ⊼ ♇
20 13 ☽ □ ♅

11 Thursday
08 25 ☽ ± ☉
08 48 ☽ ⊻ ♄
14 33 ☽ bQ ☿
16 50 ☽ □ ♃
18 54 ☽ ⊥ ♄
19 18 ☽ ⊼ ☉

12 Friday
04 32 ☽ △ ♂
04 57 ☽ ± ☿
09 39 ☽ △ ♆
10 07 ☽ ∠ ♄
10 13 ☉ ∥ ☿
12 52 ☽ △ ♀
14 33 ☽ ⊼ ☿
19 28 ☽ # ♃
22 55 ☽ ☍ ♇
23 06 ☽ △ ♅

13 Saturday
04 05 ☉ ∠ ♅
06 19 ☿ ⊻ ♀
07 35 ☽ bQ ♂
11 20 ☽ bQ ♆
11 44 ☽ ✱ ♄
13 41 ☿ ⊥ ♅
15 27 ☽ bQ ♀
20 40 ☽ △ ♃

14 Sunday
00 04 ☽ ± ♂
01 02 ☽ bQ ♅
02 37 ☽ ☍ ☉
02 57 ☽ ± ♆
07 18 ☉ ⊥ ♆
07 38 ☽ ± ♀
08 34 ☽ Q ♄
11 11 ☽ ⊼ ♂
13 29 ☽ ⊼ ♆
14 40 ☽ ☍ ☿
16 51 ☽ ± ♅
18 32 ☽ ⊼ ♀
23 21 ☽ bQ ♃

15 Monday
03 15 ☿ ⊻ ♆
03 20 ☽ ⊼ ♇
03 33 ☽ ⊼ ♅
12 40 ☿ ⊻ ♂
13 53 ♀ Q ♄
14 08 ☽ ± ♇
15 43 ☽ ± ♃
16 37 ☽ □ ♄

16 Tuesday
02 47 ☽ ⊼ ♃
05 18 ♂ ☌ ♆
06 36 ☽ bQ ♇
10 29 ♄ St D
12 54 ☽ ⊼ ☉
16 05 ☽ ⊼ ☿
20 00 ☽ ☍ ♆
20 54 ☽ ☍ ♂
21 03 ☿ ⊥ ♀
22 06 ☿ ∥ ♂

17 Wednesday
01 04 ☽ ± ☉
02 13 ☽ ± ☿
02 57 ☽ ☍ ♀
07 53 ☉ ☌ ☿
09 29 ☽ # ♃
10 44 ☽ △ ♇
11 01 ☽ ☍ ♅
17 38 ☽ bQ ☿
19 40 ☽ bQ ☉

18 Thursday
00 43 ☽ △ ♄
07 46 ☿ ⊥ ♂
11 24 ☉ ⊥ ♀
12 26 ☽ ☍ ♃
19 51 ☽ △ ☿

19 Friday
03 34 ☽ △ ☉
06 00 ☽ ⊼ ♆
06 06 ☽ bQ ♄
10 49 ☽ ⊼ ♂
11 27 ☿ ⊥ ♆
14 05 ☽ # ♇
14 48 ☽ ⊼ ♀
18 00 ☽ ± ♆
20 21 ♀ ∥ ♆
21 38 ☽ □ ♇
21 58 ☽ ⊼ ♅
23 39 ☽ ± ♂

20 Saturday
00 04 ☽ ± ♄
03 06 ☽ ± ♀
06 07 ♂ Q ♄
06 30 ☿ ∠ ♅
09 35 ☉ ⊻ ♆
10 06 ☽ ± ♅
12 09 ☽ bQ ♆
12 10 ☽ ⊼ ♄
19 06 ☽ bQ ♂
21 40 ☽ bQ ♀

21 Sunday
01 21 ☽ ⊼ ♃
02 23 ☽ □ ☿
04 25 ☽ bQ ♅
10 09 ☽ Q ♇
10 51 ☽ ∥ ♄
12 01 ☿ ✱ ♃
13 45 ☽ ± ♃
18 42 ☽ △ ♆
21 43 ☽ □ ☉

22 Monday
03 46 ☽ △ ♂
04 43 ☽ △ ♀
08 22 ☽ bQ ♃
10 39 ☽ ✱ ♇
10 48 ☉ ⊥ ♅
11 03 ☽ △ ♅
12 02 ☿ ∥ ♆
12 11 ☽ Q ☿
21 43 ♀ ☌ ♂

23 Tuesday
01 05 ☽ # ♄
01 09 ☽ ☍ ♄
01 14 ☿ ∠ ♂
03 39 ☿ ∠ ♀
10 59 ☽ ✱ ☿
13 39 ☽ Q ☉
15 15 ☽ △ ♃
17 05 ☽ ∠ ♇
17 36 ♀ ∥ ♅

24 Wednesday
07 29 ☽ □ ♆
11 07 ☽ ⊥ ♇
15 32 ☽ ∠ ☿
15 49 ☽ ✱ ☉
17 44 ☽ □ ♀
20 18 ☽ □ ♂
21 38 ☽ ∥ ♇
23 00 ☽ ⊻ ♇
23 27 ☽ □ ♅

25 Thursday
06 21 ♂ ∥ ♆
08 37 ☽ ⊥ ☿
12 50 ☽ ⊼ ♄
16 11 ☉ ⊻ ♀
18 51 ☽ Q ♆
19 56 ☽ ⊻ ☿
23 40 ☽ ∠ ☉

26 Friday
00 26 ☽ ± ♄
03 10 ☽ □ ♃
04 51 ☽ Q ♀
09 28 ☽ Q ♂
10 21 ☽ Q ♅
12 32 ☿ ∥ ♂
16 07 ♂ ✱ ♇
17 33 ☽ bQ ♄
17 47 ☽ ✱ ♆
18 16 ☽ ⊥ ☉
19 20 ☽ ∥ ♃
21 21 ♀ St R

27 Saturday
00 39 ♂ ☌ ♅
03 29 ☽ ✱ ♀
06 27 ☽ ⊻ ☉
08 26 ☽ ☌ ♇
08 56 ☽ ✱ ♅
09 26 ☽ ✱ ♂
11 42 ☿ St D
13 29 ☽ Q ♃
21 24 ☽ △ ♄
21 39 ☽ ∠ ♆

28 Sunday
03 40 ☽ ☌ ☿
04 19 ☉ ∠ ♃
06 56 ☽ ∠ ♀
08 26 ☉ ⊻ ♇
10 27 ☽ ∥ ♀
11 40 ☽ ✱ ♃
12 24 ☽ ∠ ♅
14 00 ☽ ⊥ ♆
14 29 ☽ ∠ ♂
15 36 ☉ ⊻ ♅
22 59 ☽ ⊥ ♀

29 Monday
00 44 ☽ ⊻ ♆
04 32 ☽ ⊥ ♅
07 30 ☽ ⊥ ♂
09 33 ☽ ⊻ ♀
11 22 ☿ ∥ ♆
14 36 ☽ ⊻ ♇
14 44 ☽ ∠ ♃
15 09 ☽ ⊻ ♅
16 57 ☽ ☌ ☉
18 42 ☽ ⊻ ♂
22 44 ♂ ∥ ♅

30 Tuesday
01 07 ☽ ⊥ ♇
02 52 ☽ □ ♄
06 40 ☽ ⊥ ♃
09 46 ☽ ⊻ ☿
16 45 ☽ ∠ ♇
17 12 ☽ ⊻ ♃
19 34 ☽ ∥ ♀
20 28 ☽ ⊥ ☿

31 Wednesday
05 07 ☽ ☌ ♆
09 57 ☽ Q ♄
12 27 ☽ ∠ ☿
12 57 ☽ ☌ ♀
18 31 ☽ ✱ ♇
19 07 ☽ ☌ ♅
21 40 ☿ ∠ ♀

JANUARY 1998

Date	Sidereal time (h m s)	LONGITUDES Sun ☉ (° ' ")	Moon ☽ (° ' ")	Moon ☽ 24.00 (° ' ")	Mercury ☿ (° ')	Venus ♀ (° ')	Mars ♂ (° ')	Jupiter ♃ (° ')	Saturn ♄ (° ')	Uranus ♅ (° ')	Neptune ♆ (° ')	Pluto ♇ (° ')
01	18 43 45	10 ♑ 52 01	17 ♒ 02 53	24 ♒ 07 52	18 ♐ 47	03 ♒ 17	11 ♒ 10	22 ♒ 21	13 ♈ 46	07 ♒ 09	28 ♑ 58	06 ♐ 47
02	18 47 42	11 53 11	01 ♓ 13 26	08 ♓ 19 11	19 28	03 R 02	11 57	22 34	13 48	07 12	29 00	06 49
03	18 51 38	12 54 21	15 ♓ 24 48	22 ♓ 30 01	20 14	02 44	12 44	22 46	13 50	07 15	29 02	06 51
04	18 55 35	13 55 31	29 ♓ 34 39	06 ♈ 38 31	21 05	02 24	13 31	22 59	13 52	07 19	29 04	06 53
05	18 59 31	14 56 41	13 ♈ 41 31	20 ♈ 43 32	21 59	02 02	14 19	23 12	13 54	07 22	29 07	06 55
06	19 03 28	15 57 50	27 ♈ 44 28	04 ♉ 44 11	22 58	01 38	15 06	23 25	13 56	07 25	29 09	06 57
07	19 07 24	16 58 59	11 ♉ 42 34	18 ♉ 39 27	24 00	01 11	15 53	23 38	13 59	07 29	29 11	06 59
08	19 11 21	18 00 07	25 ♉ 34 38	02 ♊ 27 53	25 04	00 42	16 40	23 51	14 01	07 32	29 13	07 01
09	19 15 18	19 01 15	09 ♊ 18 56	16 ♊ 07 32	26 12	00 ♒ 12	17 28	24 04	14 04	07 35	29 15	07 03
10	19 19 14	20 02 23	22 ♊ 53 21	29 ♊ 36 06	27 21	29 ♑ 40	18 15	24 17	14 06	07 39	29 18	07 04
11	19 23 11	21 03 30	06 ♋ 15 32	12 ♋ 51 21	28 33	29 06	19 02	24 30	14 09	07 42	29 20	07 06
12	19 27 07	22 04 37	19 ♋ 23 23	25 ♋ 51 27	29 ♐ 47	28 32	19 50	24 43	14 12	07 45	29 22	07 08
13	19 31 04	23 05 43	02 ♌ 15 29	08 ♌ 35 26	01 ♑ 02	27 56	20 37	24 56	14 15	07 49	29 24	07 10
14	19 35 00	24 06 49	14 ♌ 51 22	21 ♌ 03 25	02 19	27 20	21 24	25 10	14 18	07 52	29 27	07 12
15	19 38 57	25 07 55	27 ♌ 11 48	03 ♍ 16 46	03 37	26 43	22 12	25 23	14 21	07 56	29 29	07 13
16	19 42 53	26 09 00	09 ♍ 18 42	15 ♍ 17 59	04 56	26 06	22 59	25 37	14 24	07 59	29 31	07 15
17	19 46 50	27 10 06	21 ♍ 15 06	27 ♍ 10 34	06 17	25 29	23 46	25 50	14 28	08 03	29 34	07 17
18	19 50 47	28 11 10	03 ♎ 04 57	08 ♎ 58 52	07 39	24 53	24 34	26 04	14 31	08 06	29 36	07 19
19	19 54 43	29 ♑ 12 15	14 ♎ 52 55	20 ♎ 47 45	09 02	24 17	25 21	26 17	14 35	08 10	29 38	07 20
20	19 58 40	00 ♒ 13 19	26 ♎ 44 03	02 ♏ 42 28	10 26	23 41	26 08	26 31	14 38	08 13	29 40	07 22
21	20 02 36	01 14 23	08 ♏ 43 38	14 ♏ 48 13	11 50	23 07	26 56	26 45	14 42	08 17	29 43	07 23
22	20 06 33	02 15 26	20 ♏ 56 47	27 ♏ 09 54	13 16	22 34	27 43	26 58	14 46	08 20	29 45	07 25
23	20 10 29	03 16 29	03 ♐ 28 04	09 ♐ 51 41	14 42	22 03	28 30	27 12	14 50	08 24	29 47	07 27
24	20 14 26	04 17 31	16 ♐ 21 04	22 ♐ 56 26	16 10	21 33	29 ♒ 18	27 26	14 54	08 27	29 50	07 28
25	20 18 22	05 18 33	29 ♐ 37 54	06 ♑ 25 24	17 38	21 06	00 ♓ 05	27 40	14 58	08 31	29 52	07 30
26	20 22 19	06 19 35	13 ♑ 18 45	20 ♑ 17 39	19 06	20 40	00 52	27 54	15 02	08 34	29 54	07 31
27	20 26 16	07 20 35	27 ♑ 21 38	04 ♒ 30 08	20 36	20 16	01 40	28 08	15 06	08 38	29 56	07 33
28	20 30 12	08 21 35	11 ♒ 42 28	18 ♒ 57 51	22 06	19 54	02 27	28 22	15 11	08 41	29 ♑ 59	07 34
29	20 34 09	09 22 34	26 ♒ 15 30	03 ♓ 34 34	23 37	19 35	03 14	28 36	15 15	08 45	00 ♒ 01	07 35
30	20 38 05	10 23 32	10 ♓ 54 14	18 ♓ 13 42	25 08	19 18	04 02	28 50	15 19	08 48	00 03	07 37
31	20 42 02	11 ♒ 24 28	25 ♓ 32 15	02 ♈ 49 14	26 ♑ 41	19 ♑ 04	04 ♓ 49	29 ♒ 04	15 ♈ 24	08 ♒ 52	00 ♒ 05	07 ♐ 38

Date	Moon True ☊ (° ')	Moon Mean ☊ (° ')	Moon ☽ Latitude (° ')
01	12 ♍ 18	13 ♍ 42	02 N 10
02	12 R 15	13 39	00 N 59
03	12 D 15	13 36	00 S 17
04	12 15	13 33	01 32
05	12 16	13 29	02 40
06	12 R 15	13 26	03 39
07	12 13	13 23	04 24
08	12 08	13 20	04 54
09	12 00	13 17	05 06
10	11 51	13 13	05 01
11	11 42	13 10	04 39
12	11 33	13 07	04 03
13	11 25	13 04	03 15
14	11 20	13 01	02 18
15	11 16	12 58	01 15
16	11 15	12 54	00 S 11
17	11 D 16	12 51	00 N 54
18	11 17	12 48	01 56
19	11 19	12 45	02 53
20	11 20	12 42	03 42
21	11 R 19	12 39	04 23
22	11 18	12 35	04 52
23	11 14	12 32	05 09
24	11 09	12 29	05 10
25	11 03	12 26	04 55
26	10 57	12 23	04 24
27	10 51	12 19	03 35
28	10 47	12 16	02 32
29	10 45	12 13	01 N 19
30	10 D 44	12 10	00 S 01
31	10 ♍ 45	12 ♍ 07	01 S 20

DECLINATIONS

Sun ☉ (° ')	Moon ☽ (° ')	Mercury ☿ (° ')	Venus ♀ (° ')	Mars ♂ (° ')	Jupiter ♃ (° ')	Saturn ♄ (° ')	Uranus ♅ (° ')	Neptune ♆ (° ')	Pluto ♇ (° ')
23 S 00	13 S 39	20 S 30	17 S 18	18 S 33	14 S 53	03 N 07	19 S 04	20 S 01	09 S 42
22 54	10 07	20 41	17 07	18 19	14 49	03 08	19 03	20 00	09 42
22 49	06 01	20 53	16 56	18 06	14 45	03 09	19 02	20 00	09 43
22 43	01 S 34	21 05	16 45	17 52	14 40	03 10	19 01	19 59	09 43
22 36	02 N 56	21 17	16 35	17 38	14 36	03 11	19 01	19 59	09 43
22 29	07 16	21 29	16 25	17 24	14 32	03 12	19 00	19 58	09 43
22 22	11 09	21 41	16 16	17 09	14 28	03 14	18 59	19 58	09 43
22 14	14 24	21 53	16 07	16 55	14 23	03 15	18 58	19 57	09 43
22 05	16 48	22 04	15 58	16 40	14 19	03 16	18 57	19 57	09 43
21 57	18 15	22 14	15 50	16 25	14 15	03 17	18 56	19 57	09 43
21 47	18 39	22 25	15 42	16 10	14 10	03 19	18 55	19 56	09 44
21 38	18 02	22 34	15 35	15 55	14 06	03 20	18 55	19 56	09 44
21 28	16 30	22 42	15 29	15 40	14 01	03 21	18 54	19 55	09 44
21 17	14 11	22 50	15 23	15 24	13 57	03 23	18 53	19 55	09 44
21 06	11 16	22 57	15 17	15 08	13 52	03 24	18 52	19 54	09 44
20 55	07 55	23 03	15 12	14 53	13 48	03 26	18 51	19 54	09 44
20 43	04 18	23 08	15 07	14 37	13 43	03 27	18 50	19 53	09 44
20 31	00 N 33	23 12	15 03	14 21	13 39	03 29	18 49	19 53	09 44
20 19	03 S 13	23 14	15 00	14 04	13 34	03 30	18 48	19 52	09 44
20 06	06 51	23 16	14 57	13 48	13 29	03 32	18 47	19 52	09 44
19 53	10 15	23 16	14 54	13 31	13 25	03 34	18 46	19 51	09 44
19 39	13 17	23 16	14 52	13 15	13 20	03 35	18 46	19 51	09 44
19 25	15 48	23 14	14 50	12 58	13 15	03 37	18 45	19 51	09 44
19 11	17 36	23 11	14 49	12 41	13 11	03 39	18 44	19 50	09 44
18 56	18 31	23 06	14 49	12 24	13 06	03 40	18 43	19 50	09 44
18 41	18 24	23 01	14 48	12 07	13 01	03 42	18 42	19 49	09 44
18 26	17 10	22 54	14 48	11 49	12 56	03 44	18 41	19 49	09 44
18 10	14 50	22 45	14 49	11 32	12 51	03 46	18 40	19 48	09 44
17 54	11 32	22 36	14 50	11 15	12 46	03 48	18 39	19 48	09 44
17 38	07 29	22 25	14 51	10 57	12 42	03 50	18 38	19 47	09 44
17 S 21	03 S 00	22 S 13	14 S 53	10 S 39	12 S 37	03 N 52	18 S 37	19 S 47	09 S 44

ZODIAC SIGN ENTRIES

Date	h	m	Planets
02	09	56	☽ ♓
04	12	43	☽ ♈
06	15	52	☽ ♉
08	19	42	☽ ♊
09	21	03	♀ ♑
11	00	43	☽ ♋
12	16	20	☿ ♑
13	07	45	☽ ♌
15	17	31	☽ ♍
18	05	44	☽ ♎
20	06	46	☉ ♒
20	18	34	☽ ♏
23	05	25	☽ ♐
25	09	26	♂ ♓
25	12	39	☽ ♑
27	16	27	☽ ♒
29	02	52	♆ ♒
29	18	08	☽ ♓
31	19	21	☽ ♈

LATITUDES

Date	Mercury ☿ (° ')	Venus ♀ (° ')	Mars ♂ (° ')	Jupiter ♃ (° ')	Saturn ♄ (° ')	Uranus ♅ (° ')	Neptune ♆ (° ')	Pluto ♇ (° ')
01	02 N 29	02 N 10	01 S 10	00 S 52	02 S 30	00 S 36	00 N 22	11 N 54
04	02 04	02 57	01 09	00 52	02 30	00 36	00 22	11 54
07	01 37	03 43	01 08	00 52	02 29	00 36	00 22	11 55
10	01 10	04 29	01 07	00 52	02 28	00 36	00 22	11 55
13	00 44	05 12	01 06	00 52	02 27	00 36	00 22	11 56
16	00 N 18	05 50	01 05	00 52	02 27	00 36	00 22	11 56
19	00 S 06	06 21	01 04	00 52	02 26	00 36	00 22	11 57
22	00 29	06 46	01 03	00 52	02 25	00 36	00 22	11 58
25	00 50	07 03	01 01	00 52	02 24	00 36	00 22	11 59
28	01 09	07 13	01 00	00 52	02 24	00 36	00 21	11 59
31	01 S 25	07 N 16	00 S 59	00 S 52	02 S 23	00 S 36	00 N 21	12 N 00

DATA

Julian Date	2450815
Delta T	+63 seconds
Ayanamsa	23° 49' 41"
Synetic vernal point	05° ♓ 17' 19"
True obliquity of ecliptic	23° 26' 13"

MOON'S PHASES, APSIDES AND POSITIONS ☽

Date	h	m	Phase	Longitude (° ')	Eclipse Indicator
05	14	18	☽	15 ♈ 03	
12	17	24	○	22 ♋ 18	
20	19	40	☾	00 ♏ 33	
28	06	01	●	08 ♒ 06	

Day	h	m		
03	08	26	Perigee	
18	20	34	Apogee	
30	14	03	Perigee	
04	20	19	ON	
11	09	18	Max dec	18° N 39'
18	15	28	OS	
25	21	33	Max dec	18° S 36'

LONGITUDES

Date	Chiron ⚷ (° ')	Ceres ⚳ (° ')	Pallas ⚴ (° ')	Juno ⚵ (° ')	Vesta ⚶ (° ')	Black Moon Lilith ⚸ (° ')
01	15 ♏ 34	12 ♓ 50	10 ♒ 59	04 ♎ 50	19 ♈ 30	01 ♎ 58
11	16 ♏ 31	16 ♓ 00	14 ♒ 14	06 ♎ 04	21 ♈ 38	03 ♎ 04
21	17 ♏ 19	19 ♓ 22	17 ♒ 32	06 ♎ 42	24 ♈ 11	04 ♎ 11
31	17 ♏ 57	22 ♓ 54	20 ♒ 53	06 ♎ 39	27 ♈ 06	05 ♎ 17

All ephemeris data is given at 12.00 UT and the Moon's longitude is additionally given for 24.00 UT

ASPECTARIAN

h m	Aspects
01 Thursday	
00 41	☽ ⊻ ☉
01 25	☽ ☌ ♂
01 49	☽ ∥ ♃
06 26	☽ ⚹ ♄
11 40	☽ ⊥ ☉
14 57	☽ Q ♇
15 05	☽ ⚹ ☿
21 07	☽ ☌ ♃
02 Friday	
04 06	☽ ∠ ☉
07 54	☽ ∠ ♄
08 14	☽ ⊻ ♆
12 26	☽ Q ☿
14 34	☽ ∥ ♇
15 00	☽ ⊻ ♀
18 15	☉ ⊻ ♂
18 24	☽ ⊥ ♆
21 29	☽ □ ♇
22 09	☽ ⊻ ♅
23 09	☽ ⊥ ♄
03 Saturday	
00 57	☽ ⊥ ♀
07 12	☽ ⊻ ♂
07 26	☽ ⚹ ☉
08 20	☽ ⊥ ♅
09 19	☽ ⊻ ♄
09 40	☽ ∠ ♆
10 36	☉ ⊥ ♇
15 51	☽ ∠ ♀
17 57	☽ ⊥ ♂
20 40	☽ □ ☿
23 38	☽ ∠ ♅
04 Sunday	
00 39	☽ ⊻ ♃
03 29	☽ ⋕ ♄
05 19	☽ Q ☉
10 06	☽ ∠ ♂
10 33	☉ □ ♄
10 59	☽ ⊥ ♃
11 08	☽ ⚹ ♆
16 41	☽ ⚹ ♀
22 58	♂ ⚹ ♄
05 Monday	
00 26	☽ △ ♇
01 11	☽ ⚹ ♅
02 30	☽ ∠ ♃
07 35	☽ Q ♆
12 21	☽ ☌ ♄
12 34	☽ Q ♀
13 07	☽ ⚹ ♂
13 21	☽ ∥ ♄
14 18	☽ □ ☉
19 22	♀ Q ♄
21 43	☽ Q ♅
21 57	☿ ∠ ♅
06 Tuesday	
02 04	☽ ⚼ ♇
03 12	☽ △ ☿
04 29	☽ ⚹ ♃
10 50	☽ Q ♂
14 25	☽ □ ♆
16 27	☿ ⊥ ♆
17 31	☽ ± ♇
18 28	☽ □ ♀
07 Wednesday	
01 20	☿ ⚹ ♃
01 22	☽ Q ♃
02 46	☽ ⋕ ♇
03 50	☽ ⊼ ♇
04 41	☽ □ ♅
06 57	☽ ⚼ ☿
15 55	☽ ⊻ ♄
19 38	☽ □ ♂
21 50	☽ △ ☉
23 45	☽ ± ☿
08 Thursday	
02 20	☽ ⊥ ♄
06 26	☿ ⊥ ♀
07 15	☉ ⊥ ♃
08 56	☽ □ ♃
11 03	☽ ⊼ ☿
11 58	☽ ⋕ ♃
18 00	☽ ∠ ♄
18 21	☽ △ ♆
20 37	☽ △ ♀
09 Friday	
01 59	☽ ⚼ ☉
03 07	☽ ⋕ ♀
08 00	☽ ☍ ♇
08 57	☽ △ ♅
10 33	☽ ⋕ ♂
13 52	☉ ∥ ☿
19 03	☽ ± ☉
20 23	☽ ⚹ ♄
20 43	☽ ⚼ ♆
21 59	☽ ⚼ ♀
10 Saturday	
03 15	☽ △ ♂
06 31	☽ ⊼ ☉
11 34	☽ ⚼ ♅
12 44	☽ ± ♆
13 20	☽ ± ♀
14 31	☽ △ ♃
17 45	☽ Q ♄
20 45	☽ ☍ ☿
23 29	☽ ⊼ ♆
23 38	☽ ⊼ ♀
11 Sunday	
03 01	♀ ☌ ♆
03 44	☽ ± ♅
07 44	☽ ⚼ ♂
13 32	☽ ⊼ ♇
14 05	♂ Q ♇
14 37	☽ ⊼ ♅
17 59	☽ ⚼ ♃
19 30	☿ ⊻ ♀
12 Monday	
00 29	☽ ± ♇
01 07	☽ ± ♂
02 25	☽ □ ♄
03 54	☿ ⊻ ♆
10 44	☽ ± ♃
12 52	☽ ⊼ ♂
13 26	☉ ∠ ♇
17 05	☽ ⚼ ♇
17 24	☽ ☍ ☉
22 03	☽ ⊼ ♃
13 Tuesday	
04 15	☽ ☍ ♀
06 38	☽ ☍ ♆
09 26	☽ ⊼ ☿
21 19	☽ △ ♇
22 03	☽ ± ☿
22 34	☽ ☍ ♅
22 42	☽ ⋕ ♂
23 56	☽ ⋕ ♀
14 Wednesday	
03 26	☿ ⊥ ♅
10 55	☽ △ ♄
14 08	☽ ⋕ ♃
15 30	♀ ∥ ♂
17 17	☽ ⚼ ☿
15 Thursday	
01 33	☽ ☍ ♂
07 35	☽ ⊼ ☉
08 23	☽ ☍ ♃
11 06	☽ ⊼ ♀
16 15	☽ ⚼ ♄
16 31	☽ ⊼ ♆
19 41	☉ ⊻ ♃
20 28	☽ ± ☉
22 22	☽ ± ♀
23 17	☽ ⋕ ♇
16 Friday	
02 13	☽ △ ☿
04 26	☽ ± ♆
07 53	☽ □ ♇
09 21	☽ ⊼ ♅
10 11	☽ ± ♄
11 18	☉ ☌ ♀
15 24	☽ ⚼ ♀
16 01	☽ ⚼ ☉
21 24	☽ ± ♅
22 15	☽ ⊼ ♄
22 28	☽ ⚼ ♆
17 Saturday	
02 02	♀ ⊻ ♃
15 39	☽ ⚼ ♅
17 25	☽ ∥ ♄
17 28	☽ ⊼ ♂
20 09	☽ △ ♀
20 11	☽ Q ♇
21 28	☽ ⊼ ♃
18 Sunday	
01 07	☽ △ ☉
04 54	☽ △ ♆
05 57	☿ ⊻ ♇
06 30	☽ ± ♂
09 53	☽ ± ♃
17 26	♀ ⊻ ♂
20 15	☿ ⊻ ♅
20 37	☽ ⚹ ♇
22 16	☽ △ ♅
22 31	☽ □ ☿
19 Monday	
02 08	☽ ⚼ ♂
04 33	☽ ⚼ ♃
11 23	☽ ☍ ♄
13 55	☽ ⋕ ♄
22 34	☉ ☌ ♆
20 Tuesday	
03 09	☽ ∠ ♇
06 09	☽ □ ♀
10 43	☽ △ ♂
11 33	☽ △ ♃
15 52	☽ Q ☿
17 56	☽ □ ♆
19 40	☽ □ ☉
21 20	☽ ⊥ ♇
21 Wednesday	
04 08	♂ ☌ ♃
08 09	☽ ∥ ♇
09 20	☽ ⊻ ♇
10 06	☿ ∠ ♃
11 06	☽ □ ♅
14 33	☉ ∥ ♆
15 25	☿ ∠ ♂
16 32	☽ Q ♀
18 59	☽ ⚹ ☿
23 51	☽ ⊼ ♄
22 Thursday	
01 13	♂ ∥ ♃
05 45	☽ Q ♆
10 32	☽ Q ☉
11 38	☽ ∥ ♂
11 38	☽ ± ♄
12 22	☽ ∥ ♃
14 36	☿ ⊥ ♇
15 02	☽ ⚹ ♀
18 43	♀ ∠ ♇
22 27	☽ Q ♅
23 51	☽ □ ♃
23 Friday	
00 44	☉ Q ♄
01 56	☽ □ ♂
02 15	☽ ∥ ♀
03 37	♀ ⊥ ♂
03 56	☽ ∠ ☿
05 00	☽ ⚹ ♆
05 03	☽ ⚼ ♄
11 36	☽ ⚹ ☉
14 08	☿ □ ♄
18 29	☽ ∠ ♀
19 30	☽ ☌ ♇
21 18	☽ ⚹ ♅
23 06	☽ ⊥ ☿
24 Saturday	
09 11	☽ ∠ ♆
09 19	☽ △ ♄
10 17	☽ Q ♃
10 36	☽ ⊥ ♀
11 36	☽ ⊻ ☿
13 50	☽ Q ♂
16 08	♀ ⊥ ♃
17 50	☽ ∠ ☉
21 10	☽ ⊻ ♀
25 Sunday	
00 59	☽ ∠ ♅
01 38	☽ ⊥ ♆
04 56	♂ ⊻ ♆
07 57	♂ ∠ ♄
08 26	☽ ⚹ ♃
11 23	☽ ⊥ ☉
12 25	☽ ⊻ ♆
12 51	☽ ⚹ ♂
17 08	☽ ⊥ ♅
22 51	☽ ⊻ ☉
26 Monday	
01 54	☽ ⊻ ♇
03 43	☽ ⊻ ♅
11 07	☉ ∥ ♅
11 16	☽ ∠ ♃
12 21	☽ ⊥ ♇
14 59	☽ □ ♄
16 41	☽ ∠ ♂
23 09	☽ ☌ ☿
27 Tuesday	
00 16	☽ ☌ ♀
02 59	☽ ⊥ ♃
03 49	☽ ∠ ♇
07 47	☿ ☌ ♀
08 57	☽ ⊥ ♂
13 19	☽ ⊻ ♃
16 21	☽ ☌ ♆
16 49	☉ ⚹ ♇
19 40	☽ ⊻ ♂
21 42	☽ Q ♄
28 Wednesday	
05 06	☽ ⚹ ♇
06 01	☽ ☌ ☉
06 57	☽ ☌ ♅
12 09	☽ ∥ ♀
17 00	☿ ⊥ ♃
17 46	☽ ⚹ ♄
19 33	☿ ∠ ♇
20 08	☉ ☌ ♅
29 Thursday	
01 01	☽ Q ♇
01 15	☽ ⊻ ♀
03 28	☽ ∥ ♃
07 09	☽ ⊻ ☿
10 55	☽ ⊥ ♀
14 01	☽ ∥ ♂
15 54	☽ ☌ ♃
18 09	☽ ⊥ ☿
18 11	☽ ⊻ ♆
18 35	☽ ∠ ♄
23 09	☽ ∥ ♇
30 Friday	
00 06	☽ ☌ ♂
01 23	☽ ∠ ♀
04 02	☽ ⊥ ♆
06 36	☽ □ ♇
08 33	☽ ⊻ ♅
09 24	☽ ⊥ ♄
10 36	☽ ∠ ☿
11 06	☽ ⊻ ☉
18 21	♀ ∠ ♂
18 25	☽ ⊥ ♅
18 49	☽ ∠ ♆
19 17	☽ ⊻ ♄
21 40	☽ ⊥ ☉
31 Saturday	
01 32	☽ ⚹ ♀
07 32	☽ ⋕ ♄
09 14	☽ ∠ ♅
13 32	☽ ∠ ☉
14 06	☽ ⚹ ☿
17 54	☽ ⊻ ♃
19 31	☽ ⚹ ♆
20 58	☽ Q ♀

FEBRUARY 1998

LONGITUDES

Date	Sidereal time h m s	Sun ☉ ° ' "	Moon ☽ ° ' "	Moon ☽ 24.00 ° ' "	Mercury ☿ ° '	Venus ♀ ° '	Mars ♂ ° '	Jupiter ♃ ° '	Saturn ♄ ° '	Uranus ♅ ° '	Neptune ♆ ° '	Pluto ♇ ° '
01	20 45 58	12 ♒ 25 23	10 ♈ 04 06	17 ♈ 16 21	28 ♑ 14	18 ♑ 52	05 ♓ 36	29 ♒ 18	15 ♈ 29	08 ♒ 55	00 ♒ 08	07 ♐ 39
02	20 49 55	13 26 17	24 ♈ 25 39	01 ♉ 31 43	29 ♑ 47	18 R 42	06 23	29 32	15 34	08 59	00 10	07 41
03	20 53 51	14 27 10	08 ♉ 34 20	15 ♉ 33 23	01 ♒ 22	18 35	07 11	29 ♒ 46	15 38	09 02	00 12	07 42
04	20 57 48	15 28 01	22 ♉ 28 47	29 ♉ 20 31	02 57	18 30	07 58	00 ♓ 01	15 43	09 06	00 14	07 43
05	21 01 45	16 28 51	06 ♊ 08 36	12 ♊ 53 03	04 33	18 28	08 45	00 15	15 48	09 09	00 16	07 44
06	21 05 41	17 29 39	19 ♊ 33 55	26 ♊ 11 14	06 09	18 D 28	09 32	00 29	15 53	09 13	00 19	07 45
07	21 09 38	18 30 26	02 ♋ 45 06	09 ♋ 15 33	07 47	18 31	10 20	00 43	15 59	09 16	00 21	07 46
08	21 13 34	19 31 12	15 ♋ 42 39	22 ♋ 06 27	09 25	18 36	11 07	00 58	16 04	09 20	00 23	07 47
09	21 17 31	20 31 56	28 ♋ 27 02	04 ♌ 44 29	11 04	18 43	11 54	01 12	16 09	09 23	00 25	07 48
10	21 21 27	21 32 39	10 ♌ 58 52	17 ♌ 10 18	12 43	18 53	12 41	01 26	16 14	09 26	00 27	07 49
11	21 25 24	22 33 20	23 ♌ 18 54	29 ♌ 24 50	14 24	19 05	13 28	01 41	16 20	09 30	00 30	07 50
12	21 29 20	23 34 00	05 ♍ 28 17	11 ♍ 29 26	16 05	19 19	14 15	01 55	16 25	09 33	00 32	07 51
13	21 33 17	24 34 38	17 ♍ 28 33	23 ♍ 25 56	17 47	19 35	15 02	02 09	16 31	09 37	00 34	07 52
14	21 37 14	25 35 15	29 ♍ 21 54	05 ♎ 16 48	19 31	19 53	15 49	02 24	16 37	09 40	00 36	07 53
15	21 41 10	26 35 51	11 ♎ 11 03	17 ♎ 05 04	21 14	20 13	16 36	02 38	16 42	09 44	00 38	07 54
16	21 45 07	27 36 26	22 ♎ 59 22	28 ♎ 54 25	22 59	20 35	17 23	02 53	16 48	09 47	00 40	07 55
17	21 49 03	28 36 59	04 ♏ 50 45	10 ♏ 48 58	24 45	20 59	18 10	03 07	16 54	09 50	00 42	07 56
18	21 53 00	29 ♒ 37 31	16 ♏ 49 35	22 ♏ 53 14	26 32	21 24	18 57	03 22	17 00	09 54	00 44	07 56
19	21 56 56	00 ♓ 38 02	29 ♏ 00 29	05 ♐ 11 54	28 ♒ 19	21 51	19 44	03 36	17 06	09 57	00 46	07 57
20	22 00 53	01 38 32	11 ♐ 28 03	17 ♐ 49 26	00 ♓ 07	22 20	20 31	03 50	17 12	10 00	00 48	07 58
21	22 04 49	02 39 00	24 ♐ 16 33	00 ♑ 49 46	01 57	22 50	21 18	04 05	17 18	10 04	00 50	07 58
22	22 08 46	03 39 27	07 ♑ 29 24	14 ♑ 15 39	03 47	23 22	22 05	04 19	17 24	10 07	00 52	07 59
23	22 12 43	04 39 53	21 ♑ 08 35	28 ♑ 08 09	05 38	23 56	22 52	04 34	17 30	10 10	00 54	08 00
24	22 16 39	05 40 17	05 ♒ 14 07	12 ♒ 26 05	07 30	24 31	23 38	04 48	17 37	10 14	00 56	08 00
25	22 20 36	06 40 40	19 ♒ 43 30	27 ♒ 05 40	09 22	25 07	24 25	05 03	17 43	10 17	00 58	08 01
26	22 24 32	07 41 01	04 ♓ 31 44	12 ♓ 00 42	11 15	25 44	25 12	05 17	17 49	10 20	01 00	08 01
27	22 28 29	08 41 20	19 ♓ 31 33	27 ♓ 03 10	13 09	26 23	25 59	05 32	17 56	10 23	01 02	08 01
28	22 32 25	09 ♓ 41 37	04 ♈ 34 27	12 ♈ 04 19	15 ♓ 04	27 ♑ 02	26 ♓ 45	05 ♓ 46	18 ♈ 02	10 ♒ 26	01 ♒ 04	08 ♐ 02

DECLINATIONS

Date	Moon True ☊ ° '	Moon Mean ☊ ° '	Moon ☽ Latitude ° '	Sun ☉ ° '	Moon ☽ ° '	Mercury ☿ ° '	Venus ♀ ° '	Mars ♂ ° '	Jupiter ♃ ° '	Saturn ♄ ° '	Uranus ♅ ° '	Neptune ♆ ° '	Pluto ♇ ° '
01	10 ♍ 46	12 ♍ 04	02 S 34	17 S 04	01 N 37	21 S 59	14 S 54	10 S 21	12 S 32	03 N 54	18 S 36	19 S 46	09 S 44
02	10 D 48	12 00	03 37	16 47	06 06	21 44	14 57	10 04	12 27	03 56	18 35	19 46	09 43
03	10 49	11 57	04 26	16 30	10 09	21 28	14 59	09 46	12 22	03 58	18 35	19 45	09 43
04	10 R 49	11 54	04 59	16 12	13 34	21 11	15 02	09 28	12 17	04 00	18 34	19 45	09 43
05	10 47	11 51	05 14	15 54	16 11	20 52	15 04	09 09	12 12	04 02	18 33	19 44	09 43
06	10 45	11 48	05 11	15 35	17 52	20 31	15 07	08 51	12 07	04 04	18 32	19 44	09 43
07	10 42	11 44	04 52	15 17	18 33	20 09	15 10	08 33	12 02	04 06	18 31	19 44	09 43
08	10 38	11 41	04 18	14 58	18 14	19 46	15 13	08 15	11 57	04 09	18 30	19 43	09 43
09	10 35	11 38	03 32	14 39	17 01	19 22	15 17	07 56	11 52	04 11	18 29	19 43	09 43
10	10 32	11 35	02 36	14 19	14 58	18 56	15 20	07 38	11 47	04 13	18 28	19 42	09 42
11	10 31	11 32	01 34	14 00	12 16	18 28	15 23	07 19	11 42	04 15	18 27	19 42	09 42
12	10 30	11 29	00 S 28	13 40	09 04	17 59	15 26	07 00	11 36	04 18	18 26	19 41	09 42
13	10 D 30	11 25	00 N 39	13 20	05 32	17 29	15 30	06 42	11 31	04 20	18 25	19 41	09 42
14	10 31	11 22	01 43	12 59	01 N 49	16 57	15 33	06 23	11 26	04 22	18 25	19 40	09 42
15	10 32	11 19	02 42	12 39	01 S 56	16 24	15 36	06 04	11 21	04 25	18 24	19 40	09 42
16	10 33	11 16	03 35	12 18	05 37	15 50	15 39	05 46	11 16	04 27	18 23	19 39	09 41
17	10 34	11 13	04 18	11 57	09 05	15 14	15 41	05 27	11 11	04 29	18 22	19 39	09 41
18	10 35	11 10	04 51	11 36	12 13	14 37	15 44	05 08	11 06	04 32	18 21	19 39	09 41
19	10 35	11 06	05 11	11 15	14 52	13 58	15 47	04 49	11 00	04 34	18 20	19 38	09 41
20	10 R 35	11 03	05 18	10 53	16 55	13 18	15 49	04 30	10 55	04 37	18 19	19 38	09 40
21	10 35	11 00	05 09	10 32	18 10	12 36	15 51	04 11	10 50	04 39	18 18	19 37	09 40
22	10 34	10 57	04 43	10 10	18 31	11 54	15 52	03 52	10 45	04 42	18 17	19 37	09 40
23	10 33	10 54	04 02	09 48	17 48	11 10	15 54	03 33	10 39	04 44	18 17	19 36	09 40
24	10 33	10 50	03 04	09 26	15 59	10 24	15 55	03 14	10 34	04 47	18 16	19 36	09 40
25	10 32	10 47	01 53	09 04	13 06	09 37	15 56	02 55	10 29	04 49	18 15	19 36	09 39
26	10 32	10 44	00 N 33	08 41	09 20	08 49	15 56	02 36	10 24	04 52	18 14	19 35	09 39
27	10 D 32	10 41	00 S 50	08 19	04 55	08 00	15 57	02 17	10 18	04 54	18 13	19 35	09 39
28	10 ♍ 32	10 ♍ 38	02 S 10	07 S 56	00 S 10	07 S 10	15 S 57	01 S 58	10 S 13	04 N 57	18 S 12	19 S 34	09 S 38

ZODIAC SIGN ENTRIES

Date	h	m	Planets
02	15	15	☿ ♒
02	21	25	☽ ♉
04	10	52	♃ ♓
05	01	09	☽ ♊
07	06	57	☽ ♋
09	14	57	☽ ♌
12	01	09	☽ ♍
14	13	17	☽ ♎
17	02	13	☽ ♏
18	20	55	☉ ♓
19	13	56	☽ ♐
20	10	22	☿ ♓
21	22	30	☽ ♑
24	03	10	☽ ♒
26	04	42	☽ ♓
28	04	42	☽ ♈

LATITUDES

Date	Mercury ☿ ° '	Venus ♀ ° '	Mars ♂ ° '	Jupiter ♃ ° '	Saturn ♄ ° '	Uranus ♅ ° '	Neptune ♆ ° '	Pluto ♇ ° '
01	01 S 30	07 N 16	00 S 58	00 S 52	02 S 23	00 S 36	00 N 21	12 N 01
04	01 43	07 12	00 57	00 52	02 22	00 36	00 21	12 02
07	01 54	07 03	00 55	00 52	02 22	00 36	00 21	12 02
10	02 01	06 50	00 54	00 52	02 21	00 36	00 21	12 03
13	02 05	06 35	00 52	00 52	02 20	00 36	00 21	12 04
16	02 06	06 17	00 51	00 53	02 20	00 36	00 21	12 05
19	02 02	05 58	00 49	00 53	02 19	00 36	00 21	12 06
22	01 54	05 37	00 47	00 53	02 19	00 36	00 21	12 07
25	01 41	05 16	00 46	00 53	02 18	00 36	00 21	12 08
28	01 24	04 54	00 44	00 53	02 18	00 36	00 21	12 09
31	01 S 00	04 N 31	00 S 42	00 S 53	02 S 17	00 S 36	00 N 21	12 N 10

DATA

Julian Date	2450846
Delta T	+63 seconds
Ayanamsa	23° 49' 45"
Synetic vernal point	05° ♓ 17' 14"
True obliquity of ecliptic	23° 26' 14"

MOON'S PHASES, APSIDES AND POSITIONS ☽

Date	h	m	Phase	Longitude ° '	Eclipse Indicator
03	22	53	☽	14 ♉ 55	
11	10	23	○	22 ♌ 29	
19	15	27	☾	00 ♐ 47	
26	17	26	●	07 ♓ 55	Total

Day	h	m		
15	14	34	Apogee	
27	19	47	Perigee	
01	03	34	0N	
07	16	29	Max dec	18° N 34'
14	23	37	0S	
22	08	00	Max dec	18° S 31'
28	12	51	0N	

LONGITUDES

Date	Chiron ⚷ ° '	Ceres ⚳ ° '	Pallas ⚴ ° '	Juno ⚵ ° '	Vesta ⚶ ° '	Black Moon Lilith ⚸ ° '
01	18 ♏ 00	23 ♓ 15	21 ♒ 13	06 ♎ 37	27 ♈ 25	05 ♎ 24
11	18 ♏ 27	26 ♓ 55	24 ♒ 34	05 ♎ 48	00 ♉ 39	06 ♎ 30
21	18 ♏ 43	00 ♈ 41	27 ♒ 55	04 ♎ 21	04 ♉ 07	07 ♎ 37
31	18 ♏ 47	04 ♈ 32	01 ♓ 15	02 ♎ 20	07 ♉ 47	08 ♎ 44

ASPECTARIAN

h m Aspects

01 Sunday
03 58 ☽ ⊥ ♃
04 10 ☽ ⊻ ♂
07 59 ☽ △ ♇
10 05 ☽ ✶ ♅
12 18 ☽ Q ☿
14 42 ☽ ⊥ ♂
15 26 ☽ Q ♆
16 13 ☽ ✶ ☉
19 10 ☽ ∠ ♃
21 03 ☽ ☌ ♄

02 Monday
00 06 ☽ ∥ ♄
02 29 ☽ □ ♀
04 44 ♂ ⊥ ♆
06 11 ☽ Q ♅
06 36 ☽ ∠ ♂
07 29 ☿ ⊻ ♃
09 03 ☽ bQ ♇
13 50 ☽ Q ☉
17 53 ☿ ☌ ♆
20 46 ☽ ✶ ♃
21 43 ☽ □ ♆
22 11 ☽ □ ☿

03 Tuesday
00 16 ☽ ± ♇
06 26 ☿ ⊥ ♂
09 19 ☽ # ♇
09 29 ☽ ✶ ♂
09 43 ☽ # ♂
10 30 ☽ ⊼ ♇
12 48 ☽ □ ♅
14 58 ♂ ∥ ♇
17 35 ☽ Q ♃
22 53 ☽ □ ☉

04 Wednesday
00 13 ☽ ⊻ ♄
02 34 ☽ # ♃
04 12 ♂ □ ♇
05 08 ☽ △ ♀
07 22 ☽ Q ♂
10 40 ☽ ⊥ ♄
18 34 ☉ ✶ ♄

05 Thursday
00 18 ☿ Q ♄
00 36 ☽ # ♀
01 25 ☽ □ ♃
01 37 ☽ △ ♆
02 31 ☽ ∠ ♄
07 17 ☽ bQ ♀
08 47 ☽ △ ☿
09 18 ☽ # ☉
14 50 ☽ ☍ ♇
15 10 ♃ ⊻ ♆
16 55 ☽ □ ♂
17 22 ☽ △ ♅
21 26 ♀ St D
23 15 ☽ ± ♀

06 Friday
01 09 ♂ ⊻ ♅
04 20 ☽ bQ ♆
05 21 ☽ ✶ ♄
07 58 ☽ △ ☉
10 02 ☽ ⊼ ♀
15 16 ☽ bQ ☿
20 27 ☽ bQ ♅
20 37 ☽ ± ♆

07 Saturday
00 01 ♂ ⊥ ♄
03 12 ☽ Q ♄
07 35 ☽ ⊼ ♆
08 13 ☽ △ ♃
09 14 ☽ # ♅
09 58 ☽ ± ☿
11 56 ☿ ✶ ♇
12 18 ☉ ⊻ ♀
12 57 ☽ ± ♅
13 30 ☽ bQ ☉
19 04 ☉ ∥ ♀
21 16 ☽ ⊼ ♇
22 35 ☽ ⊼ ☿

08 Sunday
00 04 ☽ ⊼ ♅
00 30 ☽ # ♅
02 53 ☽ △ ♂
07 34 ☽ ± ☉
08 25 ☽ ± ♇
10 41 ☿ ☌ ♅
12 29 ☽ bQ ♃
12 40 ☽ □ ♄
15 07 ☿ ∥ ♆
17 27 ☽ ☍ ♀
18 32 ☉ Q ♇
19 45 ☽ ⊼ ☉

09 Monday
01 18 ☽ bQ ♇
04 06 ♃ ∠ ♄
05 43 ☽ ± ♃
08 51 ☽ bQ ♂
15 46 ☽ ☍ ♆
17 20 ☽ ⊼ ♃

10 Tuesday
03 10 ☽ ± ♂
05 55 ☽ △ ♇
08 24 ☽ # ♀
09 01 ☽ ☍ ♅
10 53 ☿ ⊻ ♂
15 31 ☽ ⊼ ♂
15 54 ☽ ☍ ☿
19 12 ☽ # ☉
22 16 ☽ △ ♄

11 Wednesday
03 35 ☽ ⊼ ♀
10 23 ☽ ☍ ☉
12 48 ☿ ∥ ♅
15 32 ☽ ± ♀
16 44 ☽ # ♃

12 Thursday
02 10 ☽ ⊼ ♆
03 55 ☽ bQ ♄
04 49 ☽ ☍ ♃
07 31 ☽ # ♇
09 39 ☽ bQ ♀
14 07 ☽ ± ♆
16 45 ☽ □ ♇
17 01 ☿ ✶ ♄
20 10 ☽ ⊼ ♅
21 57 ☽ ± ♄

13 Friday
03 36 ☽ # ♂
06 45 ☽ ☍ ♂
08 09 ☽ bQ ♆
08 14 ☽ ± ♅
10 03 ☽ ⊼ ♄
12 44 ☽ ⊼ ☿
16 20 ☽ △ ♀
19 48 ☽ ∥ ♄

14 Saturday
02 27 ☽ bQ ♅
02 52 ☽ ± ☿
03 39 ☽ ⊼ ☉
04 54 ♂ ∠ ♆
04 57 ☽ Q ♇
07 02 ♂ ⊥ ♅
14 31 ☽ △ ♆
16 56 ☽ ± ☉
17 17 ☿ Q ♇
18 17 ☽ ⊼ ♃
18 22 ☿ ⊻ ♀
22 17 ☉ ⊥ ♀

15 Sunday
00 13 ☽ bQ ☿
05 19 ☽ ✶ ♇
06 43 ☽ ± ♃
09 01 ☽ △ ♅
12 55 ☽ bQ ☉
15 34 ♂ ⊻ ♄
23 19 ☽ ☍ ♄
23 48 ☽ ⊼ ♂

16 Monday
01 24 ☽ bQ ♃
04 12 ☽ # ♄
06 57 ☽ □ ♀
11 51 ☽ ∠ ♇
12 00 ☽ △ ☿
12 52 ☽ ± ♂
12 54 ☽ ∥ ♂
19 02 ☿ ∥ ♀
22 14 ☽ △ ☉

17 Tuesday
03 37 ☽ □ ♆
06 06 ☽ ⊥ ♇
08 23 ☽ bQ ♂
08 27 ☽ △ ♃
16 25 ☽ ∥ ♇
18 12 ☽ ⊻ ♇
20 36 ☽ Q ♀
22 05 ☽ □ ♅

18 Wednesday
03 18 ☽ ∥ ♃
07 36 ☽ ∥ ☉
12 21 ☽ ⊼ ♄
15 48 ☽ Q ♆
16 31 ☽ △ ♂
21 25 ☽ ✶ ♀

19 Thursday
00 19 ☽ ± ♄
03 40 ☿ ⊥ ♀
05 03 ☽ ∥ ☿
09 56 ☽ Q ♅
10 25 ☽ □ ☿
15 24 ☉ ⊻ ♆
15 27 ☽ ✶ ♆
15 27 ☽ □ ☉
18 03 ☽ bQ ♄
21 06 ☽ □ ♃
21 51 ☽ ∥ ♀

20 Friday
03 47 ☽ ∠ ♀
04 27 ♂ # ♄
05 18 ☽ ☌ ♇
09 12 ☽ ✶ ♅
09 26 ☉ ∥ ♃
20 14 ☽ ∠ ♆
21 11 ☿ ⊻ ♆
21 35 ☽ ⊥ ♀
22 55 ☽ △ ♄

21 Saturday
02 38 ☽ Q ☿
02 45 ☉ ∠ ♄
04 43 ☽ Q ☉
06 08 ☽ □ ♂
07 52 ☽ Q ♃
09 14 ☽ ⊻ ♀
13 03 ☽ ⊥ ♆
13 27 ☽ ∠ ♅
16 11 ☽ ∥ ♅
16 57 ☿ ∠ ♄
18 08 ♀ ∠ ♇

22 Sunday
00 03 ☽ ⊻ ♆
04 16 ☽ ✶ ☿
04 33 ☽ ✶ ☉
05 55 ☽ ⊥ ♅
06 13 ☽ ✶ ♃
08 29 ☉ ☌ ☿
12 53 ☽ ⊻ ♇
16 42 ☽ ⊻ ♅
16 54 ☽ Q ♂
20 08 ☿ ☌ ♃
23 31 ☽ ⊥ ♇

23 Monday
00 12 ☽ ∥ ♅
05 38 ☽ □ ♄
08 51 ☉ ☌ ♃
09 13 ☽ ∠ ♃
09 14 ☽ ∠ ☉
10 58 ☽ ∠ ☿
15 08 ☽ ✶ ♂
15 12 ☽ ∠ ♇
17 00 ☽ ☌ ♀
21 00 ☉ ∥ ♇

24 Tuesday
00 57 ☽ ⊥ ♃
01 53 ☽ ⊥ ☉
04 44 ☽ ⊥ ☿
04 44 ☽ ☌ ♆
04 45 ☿ ⊥ ♆
06 00 ☿ ∥ ♃
11 16 ☽ ⊻ ♃
12 38 ☽ Q ♄
12 38 ☽ ∥ ♀
12 47 ☽ ⊻ ☉
16 21 ☽ ⊻ ☿
16 38 ☽ ✶ ♇
18 01 ☽ ∠ ♂
18 33 ☿ □ ♇
20 22 ☽ ☌ ♅

25 Wednesday
08 41 ☽ ✶ ♄
09 44 ☽ ⊥ ♂
11 01 ☿ ∥ ♇
12 28 ☽ Q ♇
19 14 ☉ ⊥ ♆
20 05 ☽ ⊻ ♂
21 10 ☽ ⊻ ♀
23 57 ☿ ⊻ ♅

26 Thursday
02 03 ☿ ∠ ♀
05 34 ☽ ∥ ♃
06 18 ☽ ⊻ ♆
07 18 ☽ ⊥ ♀
09 14 ☽ ∠ ♄
10 08 ☽ ∥ ♇
13 14 ☽ ☌ ♃
15 36 ☽ ∥ ☿
15 59 ☽ ⊥ ♆
16 03 ☽ ∥ ☉
16 30 ♂ ∠ ♅
17 26 ☽ ☌ ☉
17 36 ☽ □ ♇
19 33 ☉ ∥ ☿
19 37 ☿ ⊥ ♄
20 00 ☉ □ ♇
21 21 ☽ ⊻ ♅
22 24 ☽ ∠ ♀
23 47 ☽ ⊥ ♄

27 Friday
00 21 ☽ ☌ ☿
06 25 ☽ ∠ ♆
06 58 ☽ ⊥ ♅
09 26 ☽ ⊻ ♄
12 01 ☽ # ♄
21 23 ☽ ∠ ♅
22 51 ☽ ☌ ♂
23 25 ☽ ✶ ♀

28 Saturday
02 22 ☽ ∥ ♂
06 23 ☽ ✶ ♆
13 56 ☽ ⊻ ♃
17 32 ☽ △ ♇
19 29 ☽ Q ♀
20 47 ☽ ⊻ ☉
21 25 ☽ ✶ ♅
22 03 ☽ # ♂
23 42 ☽ ⊥ ♃

All ephemeris data is given at 12.00 UT and the Moon's longitude is additionally given for 24.00 UT

MARCH 1998

LONGITUDES

Date	Sidereal time h m s	Sun ☉ ° ' "	Moon ☽ ° ' "	Moon ☽ 24.00 ° ' "	Mercury ☿ ° '	Venus ♀ ° '	Mars ♂ ° '	Jupiter ♃ ° '	Saturn ♄ ° '	Uranus ♅ ° '	Neptune ♆ ° '	Pluto ♇ ° '
01	22 36 22	10 ♓ 41 53	19 ♈ 31 46	26 ♈ 55 55	16 ♓ 58	27 ♑ 43	27 ♓ 32	06 ♓ 01	18 ♈ 09	10 ♒ 30	01 ♒ 06	08 ♐ 02
02	22 40 18	11 42 07	04 ♉ 16 00	11 ♉ 31 23	18 53	28 25	28 19	06 15	18 15	10 33	01 08	08 02
03	22 44 15	12 42 18	18 ♉ 41 37	25 ♉ 46 22	20 48	29 08	29 05	06 29	18 22	10 36	01 10	08 03
04	22 48 12	13 42 28	02 ♊ 45 26	09 ♊ 38 47	22 43	29 ♑ 52	29 ♓ 52	06 44	18 29	10 39	01 11	08 03
05	22 52 08	14 42 35	16 ♊ 26 28	23 ♊ 08 38	24 38	00 ♒ 37	00 ♈ 38	06 58	18 35	10 42	01 13	08 03
06	22 56 05	15 42 40	29 ♊ 45 29	06 ♋ 17 19	26 32	01 23	01 25	07 13	18 42	10 45	01 15	08 03
07	23 00 01	16 42 44	12 ♋ 44 26	19 ♋ 07 11	28 ♓ 25	02 10	02 11	07 27	18 49	10 48	01 17	08 04
08	23 03 58	17 42 45	25 ♋ 25 54	01 ♌ 40 57	00 ♈ 16	02 57	02 58	07 41	18 56	10 51	01 18	08 04
09	23 07 54	18 42 44	07 ♌ 52 41	14 ♌ 01 25	02 06	03 46	03 44	07 56	19 03	10 54	01 20	08 04
10	23 11 51	19 42 40	20 ♌ 07 29	26 ♌ 11 12	03 54	04 35	04 30	08 10	19 10	10 57	01 22	08 04
11	23 15 47	20 42 35	02 ♍ 12 51	08 ♍ 12 42	05 39	05 25	05 16	08 24	19 17	11 00	01 23	08 R 04
12	23 19 44	21 42 28	14 ♍ 11 02	20 ♍ 08 06	07 20	06 16	06 03	08 39	19 24	11 03	01 25	08 04
13	23 23 41	22 42 18	26 ♍ 04 09	01 ♎ 59 25	08 59	07 07	06 49	08 53	19 31	11 06	01 27	08 04
14	23 27 37	23 42 07	07 ♎ 54 11	13 ♎ 48 42	10 33	07 59	07 35	09 07	19 38	11 08	01 28	08 04
15	23 31 34	24 41 54	19 ♎ 43 15	25 ♎ 38 06	12 02	08 52	08 21	09 21	19 45	11 11	01 30	08 03
16	23 35 30	25 41 39	01 ♏ 33 36	07 ♏ 30 03	13 26	09 45	09 07	09 35	19 52	11 14	01 31	08 03
17	23 39 27	26 41 22	13 ♏ 27 51	19 ♏ 27 21	14 45	10 39	09 53	09 50	19 59	11 17	01 33	08 03
18	23 43 23	27 41 04	25 ♏ 29 00	01 ♐ 33 14	15 58	11 34	10 39	10 04	20 06	11 19	01 34	08 03
19	23 47 20	28 40 43	07 ♐ 40 30	13 ♐ 51 16	17 04	12 29	11 25	10 18	20 14	11 22	01 36	08 03
20	23 51 16	29 ♓ 40 21	20 ♐ 06 03	26 ♐ 25 20	18 03	13 25	12 11	10 32	20 21	11 25	01 37	08 02
21	23 55 13	00 ♈ 39 58	02 ♑ 49 35	09 ♑ 19 17	18 56	14 21	12 57	10 46	20 28	11 27	01 38	08 02
22	23 59 10	01 39 32	15 ♑ 54 49	22 ♑ 36 34	19 40	15 18	13 43	11 00	20 36	11 30	01 40	08 02
23	00 03 06	02 39 05	29 ♑ 24 49	06 ♒ 19 44	20 18	16 15	14 29	11 14	20 43	11 32	01 41	08 01
24	00 07 03	03 38 36	13 ♒ 21 23	20 ♒ 29 41	20 47	17 13	15 15	11 28	20 50	11 35	01 43	08 01
25	00 10 59	04 38 05	27 ♒ 44 21	05 ♓ 04 58	21 09	18 11	16 00	11 42	20 58	11 37	01 44	08 00
26	00 14 56	05 37 32	12 ♓ 30 54	20 ♓ 01 20	21 23	19 09	16 46	11 56	21 05	11 40	01 45	08 00
27	00 18 52	06 36 58	27 ♓ 35 18	05 ♈ 11 40	21 29	20 08	17 32	12 09	21 13	11 42	01 46	07 59
28	00 22 49	07 36 21	12 ♈ 49 12	20 ♈ 26 35	21 R 28	21 07	18 17	12 23	21 20	11 44	01 47	07 59
29	00 26 45	08 35 43	28 ♈ 02 33	05 ♉ 35 49	21 19	22 07	19 03	12 37	21 27	11 47	01 49	07 58
30	00 30 42	09 35 02	13 ♉ 05 14	20 ♉ 29 49	21 04	23 07	19 48	12 51	21 35	11 49	01 50	07 58
31	00 34 39	10 ♈ 34 19	27 ♉ 48 43	05 ♊ 01 17	20 ♈ 42	24 ♒ 07	20 ♈ 34	13 ♓ 04	21 ♈ 42	11 ♒ 51	01 ♒ 51	07 ♐ 57

Date	Moon True ☊ ° '	Moon Mean ☊ ° '	Moon ☽ Latitude ° '
01	10 ♍ 32	10 ♍ 35	03 S 20
02	10 R 32	10 31	04 17
03	10 32	10 28	04 55
04	10 32	10 25	05 15
05	10 D 32	10 22	05 16
06	10 32	10 19	05 00
07	10 32	10 16	04 29
08	10 33	10 12	03 46
09	10 34	10 09	02 52
10	10 35	10 06	01 51
11	10 35	10 03	00 S 46
12	10 R 35	10 00	00 N 20
13	10 35	09 56	01 25
14	10 34	09 53	02 26
15	10 32	09 50	03 20
16	10 29	09 47	04 06
17	10 27	09 44	04 42
18	10 24	09 41	05 05
19	10 22	09 37	05 16
20	10 21	09 34	05 11
21	10 D 21	09 31	04 52
22	10 22	09 28	04 17
23	10 23	09 25	03 27
24	10 24	09 22	02 24
25	10 26	09 18	01 N 09
26	10 R 26	09 15	00 S 11
27	10 25	09 12	01 33
28	10 23	09 09	02 49
29	10 20	09 06	03 52
30	10 15	09 02	04 39
31	10 ♍ 11	08 ♍ 59	05 S 06

DECLINATIONS

Date	Sun ☉ ° '	Moon ☽ ° '	Mercury ☿ ° '	Venus ♀ ° '	Mars ♂ ° '	Jupiter ♃ ° '	Saturn ♄ ° '	Uranus ♅ ° '	Neptune ♆ ° '	Pluto ♇ ° '
01	07 S 33	04 N 33	06 S 19	15 S 56	01 S 39	10 S 08	05 N 00	18 S 11	19 S 34	09 S 38
02	07 10	08 55	05 27	15 55	01 20	10 03	05 02	18 11	19 34	09 38
03	06 47	12 39	04 34	15 54	01 00	09 57	05 05	18 10	19 33	09 38
04	06 24	15 33	03 40	15 52	00 41	09 52	05 07	18 09	19 33	09 37
05	06 01	17 30	02 46	15 50	00 22	09 47	05 10	18 08	19 32	09 37
06	05 38	18 26	01 52	15 48	00 S 03	09 42	05 13	18 07	19 32	09 37
07	05 15	18 22	00 57	15 45	00 N 16	09 36	05 15	18 07	19 32	09 36
08	04 51	17 21	00 S 03	15 42	00 35	09 31	05 18	18 06	19 31	09 36
09	04 28	15 32	00 N 52	15 38	00 54	09 26	05 21	18 05	19 31	09 36
10	04 04	13 01	01 46	15 34	01 13	09 21	05 24	18 04	19 31	09 36
11	03 41	09 58	02 39	15 29	01 31	09 15	05 26	18 03	19 30	09 35
12	03 17	06 32	03 31	15 24	01 50	09 10	05 29	18 03	19 30	09 35
13	02 54	02 N 52	04 21	15 18	02 09	09 05	05 32	18 02	19 29	09 35
14	02 30	00 S 54	05 10	15 12	02 28	08 59	05 35	18 01	19 29	09 34
15	02 06	04 37	05 57	15 06	02 47	08 54	05 37	18 00	19 29	09 34
16	01 43	08 10	06 42	14 59	03 05	08 49	05 40	18 00	19 28	09 34
17	01 19	11 24	07 25	14 52	03 24	08 44	05 43	17 59	19 28	09 33
18	00 55	14 11	08 04	14 44	03 43	08 38	05 46	17 58	19 28	09 33
19	00 32	16 24	08 41	14 35	04 01	08 33	05 48	17 58	19 27	09 32
20	00 S 08	17 54	09 15	14 27	04 20	08 28	05 51	17 57	19 27	09 32
21	00 N 16	18 32	09 45	14 17	04 38	08 23	05 54	17 56	19 27	09 32
22	00 40	18 14	10 11	14 07	04 57	08 17	05 57	17 55	19 27	09 31
23	01 03	16 53	10 34	13 57	05 15	08 12	06 00	17 55	19 26	09 31
24	01 27	14 31	10 53	13 46	05 33	08 07	06 03	17 54	19 26	09 31
25	01 51	11 10	11 08	13 35	05 51	08 02	06 05	17 53	19 26	09 30
26	02 14	07 02	11 19	13 24	06 10	07 57	06 08	17 53	19 25	09 30
27	02 38	02 S 23	11 25	13 12	06 28	07 51	06 11	17 52	19 25	09 30
28	03 01	02 N 28	11 28	12 59	06 46	07 46	06 14	17 52	19 25	09 29
29	03 24	07 09	11 26	12 46	07 03	07 41	06 17	17 51	19 25	09 29
30	03 48	11 20	11 20	12 33	07 21	07 36	06 19	17 50	19 24	09 29
31	04 N 11	14 N 42	11 N 11	12 S 19	07 N 39	07 S 31	06 N 22	17 S 50	19 S 24	09 S 28

ZODIAC SIGN ENTRIES

Date	h	m	Planets
02	05	00	☽ ♉
04	07	15	☽ ♊
04	16	14	♀ ♒
04	16	18	♂ ♈
06	12	27	☽ ♋
08	08	28	☿ ♈
08	20	46	☽ ♌
11	07	35	☽ ♍
13	19	58	☽ ♎
16	08	51	☽ ♏
18	20	56	☽ ♐
20	19	55	☉ ♈
21	06	43	☽ ♑
23	13	02	☽ ♒
25	15	43	☽ ♓
27	15	49	☽ ♈
29	15	06	☽ ♉
31	15	37	☽ ♊

LATITUDES

Date	Mercury ☿ ° '	Venus ♀ ° '	Mars ♂ ° '	Jupiter ♃ ° '	Saturn ♄ ° '	Uranus ♅ ° '	Neptune ♆ ° '	Pluto ♇ ° '
01	01 S 16	04 N 46	00 S 43	00 S 53	02 S 18	00 S 36	00 N 21	12 N 09
04	00 51	04 24	00 42	00 54	02 17	00 36	00 21	12 10
07	00 S 21	04 02	00 40	00 54	02 17	00 37	00 21	12 11
10	00 N 14	03 39	00 38	00 54	02 17	00 37	00 21	12 12
13	00 52	03 17	00 36	00 54	02 16	00 37	00 21	12 13
16	01 31	02 56	00 34	00 55	02 16	00 37	00 21	12 14
19	02 09	02 34	00 32	00 55	02 16	00 37	00 21	12 15
22	02 42	02 13	00 30	00 55	02 16	00 37	00 21	12 16
25	03 07	01 53	00 29	00 55	02 15	00 37	00 21	12 17
28	03 20	01 33	00 27	00 56	02 15	00 37	00 21	12 18
31	03 N 21	01 N 14	00 S 25	00 S 56	02 S 15	00 S 37	00 N 21	12 N 19

DATA

Julian Date	2450874
Delta T	+63 seconds
Ayanamsa	23° 49' 48"
Synetic vernal point	05° ♓ 17' 11"
True obliquity of ecliptic	23° 26' 14"

MOON'S PHASES, APSIDES AND POSITIONS ☽

Date	h	m	Phase	Longitude ° '	Eclipse Indicator
05	08	41	☽	14 ♊ 34	
13	04	34	○	22 ♍ 24	
21	07	38	☾	00 ♑ 29	
28	03	14	●	07 ♈ 15	

Day	h	m		
15	00	22	Apogee	
28	06	58	Perigee	
06	22	10	Max dec	18° N 31'
14	06	15	OS	
21	16	33	Max dec	18° S 33'
27	23	47	ON	

LONGITUDES

Date	Chiron ⚷ ° '	Ceres ⚳ ° '	Pallas ⚴ ° '	Juno ⚵ ° '	Vesta ⚶ ° '	Black Moon Lilith ⚸ ° '
01	18 ♏ 47	03 ♈ 45	00 ♓ 35	02 ♎ 47	07 ♉ 02	08 ♎ 30
11	18 ♏ 42	07 ♈ 39	03 ♓ 53	00 ♎ 28	10 ♉ 50	09 ♎ 37
21	18 ♏ 26	11 ♈ 35	07 ♓ 07	28 ♍ 01	14 ♉ 46	10 ♎ 44
31	18 ♏ 01	15 ♈ 33	10 ♓ 16	25 ♍ 40	18 ♉ 48	11 ♎ 50

All ephemeris data is given at 12.00 UT and the Moon's longitude is additionally given for 24.00 UT

ASPECTARIAN

01 Sunday
00 51 ☿ ∠ ♆
01 37 ☽ Q ♆
05 49 ☿ ⊥ ♅
06 49 ☉ ⊻ ♅
07 06 ☽ ⊥ ☉
07 16 ☽ ⊻ ☿
09 45 ☽ ☌ ♄
14 22 ☽ ‖ ♄
14 26 ☽ ∠ ♃
16 49 ☽ Q ♅
17 40 ☽ bQ ♇
18 24 ☽ ⊥ ☿
19 53 ☽ # ☿
22 43 ☽ ∠ ☉

02 Monday
01 42 ☽ ⊻ ♂
01 57 ☽ □ ♀
02 55 ☽ # ☉
03 36 ☿ ⊻ ♄
06 50 ☽ □ ♆
08 21 ☽ ± ♇
11 17 ☽ ∠ ☿
12 04 ☽ ⊥ ♂
15 19 ☽ ✶ ♃
16 17 ☽ # ♇
18 14 ☽ ⊼ ♇
18 40 ☽ # ♃
22 25 ☽ □ ♅
22 45 ☿ # ♄

03 Tuesday
01 13 ☽ ✶ ☉
02 55 ☉ ⊥ ♄
03 50 ☽ ∠ ♂
11 27 ☽ ⊻ ♄
11 39 ☽ Q ♃
16 07 ☽ ✶ ☿
21 41 ☽ ⊥ ♄
22 58 ☽ Q ☉

04 Wednesday
06 43 ☽ ✶ ♂
06 44 ☽ △ ♀
09 17 ☽ △ ♆
13 16 ☽ ∠ ♄
15 09 ☽ # ♀
15 57 ☽ Q ☿
17 47 ♀ ✶ ♂
19 01 ☽ □ ♃
21 12 ☽ ☍ ♇

05 Thursday
01 49 ☽ △ ♅
04 51 ☽ Q ♂
08 41 ☽ □ ☉
10 27 ☽ bQ ♀
11 36 ☽ bQ ♆
15 52 ☽ ✶ ♄

06 Friday
00 55 ☽ # ♅
01 50 ☿ ∠ ♅
03 33 ☽ ± ♀
03 47 ☽ ± ♆
04 11 ♀ ⊥ ♃
04 41 ☽ bQ ♅
05 09 ☽ □ ☿
06 45 ♂ ✶ ♆
07 39 ♀ ☌ ♆
13 45 ☽ Q ♄
14 44 ☽ ⊼ ♆
15 09 ☽ ⊼ ♀
15 13 ☽ □ ♂
16 21 ♃ ⊥ ♆
21 12 ☽ ± ♅

07 Saturday
01 15 ☉ ∠ ♆
01 58 ☽ △ ♃
03 17 ☽ ⊼ ♇
08 22 ☽ ⊼ ♅
11 18 ☉ # ♄
11 25 ♃ ‖ ♇
14 13 ☉ ⊥ ♅
14 28 ☽ ± ♇
20 05 ☽ △ ☉
20 31 ☽ # ♅
23 32 ☽ □ ♄

08 Sunday
01 32 ☿ # ♂
06 40 ☽ bQ ♃
07 29 ☽ bQ ♇
14 21 ♀ ✶ ♂
22 53 ☽ △ ☿
23 18 ☽ ☍ ♆

09 Monday
00 15 ☽ ± ♃
01 44 ☿ ✶ ♆
03 13 ☽ bQ ☉
03 25 ☽ △ ♂
03 28 ☽ ☍ ♀
10 48 ☽ # ♀
12 06 ☽ ⊼ ♃
12 22 ☽ △ ♇
13 18 ☿ ‖ ♂
17 55 ☽ ☍ ♅
18 39 ☉ ∠ ♀
21 05 ☉ ⊻ ♄
22 16 ☽ ± ☉

10 Tuesday
01 36 ♃ □ ♇
09 10 ☽ bQ ☿
10 05 ☽ △ ♄
10 41 ☽ bQ ♂
11 07 ☽ ⊼ ☉

11 Wednesday
02 48 ☿ ☌ ♂
04 55 ♇ St R
05 44 ☽ ± ♂
05 53 ☿ ✶ ♀
06 01 ☽ ± ☿
10 21 ☽ ⊼ ♆
14 46 ☽ # ♇
16 10 ☽ bQ ♄
17 18 ☽ # ♃
18 32 ☽ ⊼ ♂
18 53 ☽ ⊼ ♀
20 00 ☽ ⊼ ☿
22 23 ☽ ± ♆
23 42 ☽ □ ♇

12 Thursday
00 38 ☽ ☍ ♃
05 40 ☽ ⊼ ♅
07 44 ☉ # ☿
07 50 ☽ ± ♀
10 24 ☽ ± ♄
16 30 ☽ bQ ♆
17 47 ☽ ± ♅
18 52 ☽ ‖ ♄
22 28 ☿ △ ♇
22 37 ☽ ⊼ ♄

13 Friday
03 23 ☽ bQ ♀
04 08 ☽ ‖ ☿
04 34 ☽ ☍ ☉
10 17 ☿ ⊻ ♃
11 45 ☽ # ☉
11 59 ☽ Q ♇
12 03 ☽ bQ ♅
16 11 ☽ ‖ ♂
22 55 ☽ △ ♆

14 Saturday
00 37 ♀ Q ♄
11 19 ☽ ☍ ♂
12 11 ☽ △ ♀
12 19 ☽ ✶ ♇
13 10 ☉ # ♂
14 00 ♀ ✶ ♇
14 31 ☽ ⊼ ♃
18 10 ☽ ☍ ☿
18 36 ☽ △ ♅
21 15 ☽ ‖ ☉
21 42 ☿ ✶ ♅
22 56 ☽ # ♂

15 Sunday
01 00 ☿ ‖ ♄
02 48 ♂ △ ♇
02 57 ☽ ± ♃
12 03 ☽ ☍ ♄
18 43 ☽ # ♄
18 46 ☽ ∠ ♇
21 36 ☽ bQ ♃
23 02 ☽ ⊼ ☉
23 16 ☽ # ☿

16 Monday
05 56 ♀ ⊻ ♃
11 55 ☽ □ ♆
12 18 ☽ ± ☉
13 00 ☽ ⊥ ♇
13 27 ☿ Q ♆
16 31 ☽ ‖ ♃
22 01 ☽ ‖ ♇

17 Tuesday
01 07 ☽ ⊻ ♇
01 33 ☉ ∠ ♅
04 19 ☽ ⊼ ♂
04 32 ☽ △ ♃
05 54 ☽ □ ♀
07 35 ☽ □ ♅
08 07 ☽ bQ ☉
09 09 ♂ ⊻ ♃
14 53 ☽ ⊼ ☿
17 12 ☽ ± ♂

18 Wednesday
00 12 ☽ Q ♆
01 12 ☽ ⊼ ♄
04 15 ☽ ± ☿
05 13 ♀ ☌ ♅
12 22 ☽ bQ ♂
13 15 ☽ ± ♄
14 36 ☿ ⊥ ♃
16 45 ☽ △ ☉
17 02 ☽ ‖ ♀
19 37 ☽ Q ♅
20 44 ☽ Q ♀
23 57 ☽ bQ ☿

19 Thursday
00 03 ☽ ✶ ♆
07 10 ☽ bQ ♄
07 18 ☿ # ♃
10 05 ♂ ✶ ♅
12 43 ☽ ☌ ♇
17 12 ☽ □ ♃
19 12 ☽ ✶ ♅
19 46 ☽ △ ♂
22 06 ☽ ✶ ♀

20 Friday
05 19 ☽ ∠ ♆
07 46 ☽ △ ☿
12 29 ☽ △ ♄
13 14 ☽ ‖ ♅
22 30 ☽ ⊥ ♆

21 Saturday
00 01 ☽ ∠ ♅
01 26 ☿ # ♇
04 16 ☽ Q ♃
05 00 ☽ ∠ ♀
07 38 ☽ □ ☉
09 47 ☽ ⊻ ♆
16 53 ☽ ⊥ ♅
21 38 ☽ ⊻ ♇
23 01 ☽ ⊥ ♀

22 Sunday
02 55 ☽ ✶ ♃
03 57 ☽ ⊻ ♅
07 47 ☽ □ ♂
08 35 ☽ ⊥ ♇
10 17 ♂ Q ♆
10 48 ☽ ⊻ ♀
12 08 ☉ ✶ ♆
19 07 ☽ □ ☿
19 16 ☽ Q ☉
19 34 ☽ ‖ ♅
20 29 ☽ □ ♄

23 Monday
00 44 ☽ ∠ ♇
06 19 ☽ ∠ ♃
15 58 ☽ ☌ ♆
17 40 ☽ Q ♂
18 05 ☽ ✶ ☉
22 17 ☽ ⊥ ♃

24 Tuesday
02 54 ☽ ✶ ♇
03 58 ☽ Q ☿
04 14 ☽ Q ♄
08 44 ☽ ⊻ ♃
08 58 ☽ ☌ ♅
15 22 ☽ ✶ ♂
16 04 ☿ ☌ ♄
18 16 ☽ ‖ ♀
18 58 ☽ ☌ ♀
21 34 ☽ ∠ ☉
23 11 ☽ Q ♇

25 Wednesday
00 41 ☽ ✶ ♄
00 50 ☽ ✶ ☿
02 17 ♃ ⊻ ♅
12 14 ☽ # ☿
13 35 ☽ ⊥ ☉
17 39 ☽ ∠ ♂
18 33 ☽ ⊻ ♆
22 10 ☽ ‖ ♇

26 Thursday
00 05 ☽ ⊻ ☉
01 32 ☽ ∠ ♄
01 59 ☽ ∠ ☿
04 19 ☽ ⊥ ♆
04 44 ☽ □ ♇
06 57 ☽ ‖ ♃
09 02 ☽ ⊥ ♂
09 51 ♂ ‖ ♄
10 37 ☽ ⊻ ♅
11 02 ☽ ☌ ♃
16 09 ☽ ⊥ ♄
16 25 ☽ # ♂
16 39 ☽ ⊥ ☿
16 47 ☽ # ♄
18 48 ☽ ∠ ♆
19 10 ☽ ⊻ ♂
20 15 ☽ ⊥ ♅
23 21 ☽ ⊻ ♀

27 Friday
01 48 ☽ ⊻ ♄
02 17 ☽ ⊻ ☿
08 29 ♀ Q ♇
09 32 ☽ ⊥ ♀
10 35 ☽ ∠ ♅
10 52 ☽ # ☉
18 37 ☽ ✶ ♆
19 42 ☿ St R

28 Saturday
00 44 ☽ ∠ ♀
03 14 ☽ ☌ ☉
04 23 ☽ △ ♇
10 18 ☽ ✶ ♅
11 19 ☽ ⊻ ♃
13 32 ☽ Q ♆
14 58 ☽ ‖ ☉
16 30 ♂ ⊥ ♃
17 49 ♀ ✶ ♄
19 31 ☿ ✶ ♀
20 54 ☽ ⊥ ♃
20 58 ☉ △ ♇
21 03 ☽ ☌ ♂

29 Sunday
01 11 ☿ ☌ ♄
01 30 ☽ ☌ ☿
01 31 ☽ ☌ ♄
01 59 ☽ ✶ ♀
03 59 ☽ bQ ♇
05 14 ☽ Q ♅
07 17 ☽ ‖ ♄
11 19 ☽ ∠ ♃
11 26 ☽ ‖ ♂
14 48 ☽ # ♃
17 59 ☽ □ ♆
18 13 ☽ ± ♇
22 20 ☽ Q ♀

30 Monday
00 52 ☽ # ♇
03 47 ☽ ⊼ ♇
05 59 ☽ ⊻ ☉
09 57 ☽ □ ♅
11 36 ☽ ✶ ♃
12 04 ☽ ‖ ☿
16 19 ☽ ⊥ ☉
19 27 ☽ # ♀
23 28 ☽ ⊻ ♂

31 Tuesday
00 38 ☽ ⊻ ☿
01 53 ☽ ⊻ ♄
03 33 ♂ # ♃
05 29 ☽ □ ♀
07 25 ☽ Q ♃
08 02 ☽ ∠ ☉
09 50 ☽ ⊥ ♂
10 12 ☽ ⊥ ☿
11 50 ☽ ⊥ ♄
14 37 ☿ ☌ ♂
18 42 ☽ △ ♆

APRIL 1998

LONGITUDES

Date	Sidereal time h m s	Sun ☉ ° ' "	Moon ☽ ° ' "	Moon ☽ 24.00 ° ' "	Mercury ☿ ° '	Venus ♀ ° '	Mars ♂ ° '	Jupiter ♃ ° '	Saturn ♄ ° '	Uranus ♅ ° '	Neptune ♆ ° '	Pluto ♇ ° '
01	00 38 35	11 ♈ 33 34	12 ♊ 07 07	19 ♊ 05 56	20 ♈ 13	25 ♒ 08	21 ♈ 19	13 ♓ 18	21 ♈ 50	11 ♒ 53	01 ♒ 52	07 ♐ 56
02	00 42 32	12 32 46	25 ♊ 57 42	02 ♋ 42 31	19 R 40	26 09	22 05	13 31	21 57	11 55	01 53	07 R 56
03	00 46 28	13 31 56	09 ♋ 20 37	15 ♋ 52 20	19 02	27 11	22 50	13 45	22 05	11 57	01 54	07 55
04	00 50 25	14 31 04	22 ♋ 18 07	28 ♋ 38 27	18 20	28 12	23 35	13 58	22 13	11 59	01 55	07 54
05	00 54 21	15 30 10	04 ♌ 53 52	11 ♌ 04 55	17 36	29 ♒ 14	24 20	14 12	22 20	12 02	01 56	07 53
06	00 58 18	16 29 13	17 ♌ 12 09	23 ♌ 16 06	16 49	00 ♓ 17	25 06	14 25	22 28	12 03	01 57	07 52
07	01 02 14	17 28 13	29 ♌ 17 19	05 ♍ 16 18	16 02	01 19	25 51	14 39	22 35	12 05	01 58	07 52
08	01 06 11	18 27 12	11 ♍ 13 31	17 ♍ 09 24	15 15	02 22	26 36	14 52	22 43	12 07	01 59	07 51
09	01 10 08	19 26 08	23 ♍ 04 22	28 ♍ 58 47	14 28	03 25	27 21	15 05	22 50	12 09	02 00	07 50
10	01 14 04	20 25 03	04 ♎ 52 58	10 ♎ 47 14	13 43	04 28	28 06	15 18	22 58	12 11	02 00	07 49
11	01 18 01	21 23 55	16 ♎ 41 51	22 ♎ 37 03	13 00	05 32	28 51	15 31	23 06	12 13	02 01	07 48
12	01 21 57	22 22 45	28 ♎ 33 05	04 ♏ 30 08	12 21	06 36	29 ♈ 36	15 44	23 13	12 14	02 02	07 47
13	01 25 54	23 21 33	10 ♏ 28 25	16 ♏ 28 09	11 45	07 40	00 ♉ 20	15 57	23 21	12 16	02 03	07 46
14	01 29 50	24 20 19	22 ♏ 29 31	28 ♏ 32 45	11 14	08 44	01 05	16 10	23 28	12 18	02 03	07 45
15	01 33 47	25 19 04	04 ♐ 38 05	10 ♐ 45 47	10 47	09 48	01 50	16 23	23 36	12 19	02 04	07 44
16	01 37 43	26 17 47	16 ♐ 56 06	23 ♐ 09 22	10 25	10 53	02 35	16 36	23 44	12 21	02 04	07 43
17	01 41 40	27 16 28	29 ♐ 25 54	05 ♑ 46 03	10 08	11 58	03 19	16 48	23 51	12 22	02 05	07 41
18	01 45 37	28 15 07	12 ♑ 10 12	18 ♑ 38 44	09 56	13 03	04 04	17 01	23 59	12 24	02 06	07 40
19	01 49 33	29 ♈ 13 44	25 ♑ 12 01	01 ♒ 50 26	09 49	14 08	04 48	17 13	24 06	12 25	02 06	07 39
20	01 53 30	00 ♉ 12 20	08 ♒ 34 19	15 ♒ 23 57	09 D 48	15 14	05 33	17 26	24 14	12 26	02 07	07 38
21	01 57 26	01 10 55	22 ♒ 19 34	29 ♒ 21 17	09 51	16 20	06 17	17 38	24 22	12 28	02 07	07 37
22	02 01 23	02 09 27	06 ♓ 29 04	13 ♓ 42 48	10 00	17 25	07 02	17 51	24 29	12 29	02 07	07 36
23	02 05 19	03 07 58	21 ♓ 02 08	28 ♓ 26 33	10 13	18 31	07 46	18 03	24 37	12 30	02 08	07 34
24	02 09 16	04 06 27	05 ♈ 55 23	13 ♈ 27 42	10 31	19 38	08 30	18 15	24 44	12 31	02 08	07 33
25	02 13 12	05 04 55	21 ♈ 02 27	28 ♈ 38 27	10 54	20 44	09 15	18 27	24 52	12 33	02 08	07 32
26	02 17 09	06 03 21	06 ♉ 14 23	13 ♉ 48 55	11 21	21 50	09 59	18 39	24 59	12 34	02 09	07 30
27	02 21 06	07 01 45	21 ♉ 20 46	28 ♉ 48 43	11 52	22 57	10 43	18 51	25 07	12 35	02 09	07 29
28	02 25 02	08 00 07	06 ♊ 11 39	13 ♊ 28 42	12 27	24 04	11 27	19 03	25 14	12 36	02 09	07 28
29	02 28 59	08 58 27	20 ♊ 39 08	27 ♊ 42 28	13 06	25 11	12 11	19 15	25 22	12 36	02 09	07 26
30	02 32 55	09 ♉ 56 45	04 ♋ 38 27	11 ♋ 27 00	13 ♈ 49	26 ♓ 18	12 ♉ 55	19 ♓ 26	25 ♈ 29	12 ♒ 37	02 ♒ 10	07 ♐ 25

Date	Moon True ☊ ° '	Moon Mean ☊ ° '	Moon ☽ Latitude ° '
01	10 ♍ 07	08 ♍ 56	05 S 13
02	10 R 05	08 53	05 01
03	10 04	08 50	04 33
04	10 D 04	08 47	03 52
05	10 05	08 43	03 01
06	10 07	08 40	02 02
07	10 08	08 37	00 S 59
08	10 R 09	08 34	00 N 06
09	10 08	08 31	01 10
10	10 05	08 28	02 10
11	10 00	08 24	03 05
12	09 54	08 21	03 52
13	09 46	08 18	04 29
14	09 37	08 15	04 55
15	09 29	08 12	05 07
16	09 22	08 08	05 05
17	09 17	08 05	04 49
18	09 13	08 02	04 19
19	09 12	07 59	03 35
20	09 D 12	07 56	02 37
21	09 13	07 53	01 30
22	09 14	07 49	00 N 15
23	09 R 13	07 46	01 S 03
24	09 11	07 43	02 18
25	09 06	07 40	03 25
26	08 58	07 37	04 17
27	08 50	07 33	04 51
28	08 41	07 30	05 05
29	08 32	07 27	04 58
30	08 ♍ 26	07 ♍ 24	04 S 34

DECLINATIONS

Date	Sun ☉ ° '	Moon ☽ ° '	Mercury ☿ ° '	Venus ♀ ° '	Mars ♂ ° '	Jupiter ♃ ° '	Saturn ♄ ° '	Uranus ♅ ° '	Neptune ♆ ° '	Pluto ♇ ° '
01	04 N 34	17 N 04	10 N 57	12 S 04	07 N 57	07 S 26	06 N 25	17 S 49	19 S 24	09 S 28
02	04 57	18 21	10 40	11 50	08 14	07 21	06 28	17 49	19 24	09 27
03	05 20	18 34	10 19	11 34	08 32	07 16	06 31	17 48	19 23	09 27
04	05 43	17 46	09 56	11 19	08 49	07 11	06 34	17 48	19 23	09 27
05	06 06	16 07	09 29	11 03	09 06	07 06	06 36	17 47	19 23	09 26
06	06 29	13 44	09 01	10 46	09 23	07 00	06 39	17 46	19 23	09 26
07	06 51	10 48	08 31	10 30	09 40	06 55	06 42	17 46	19 23	09 26
08	07 14	07 27	08 00	10 13	09 57	06 50	06 45	17 45	19 22	09 25
09	07 36	03 49	07 29	09 55	10 14	06 46	06 48	17 45	19 22	09 25
10	07 59	00 N 03	06 57	09 37	10 31	06 41	06 51	17 45	19 22	09 24
11	08 21	03 S 42	06 26	09 19	10 47	06 36	06 53	17 44	19 22	09 24
12	08 43	07 20	05 56	09 00	11 04	06 31	06 56	17 44	19 22	09 24
13	09 04	10 42	05 26	08 41	11 20	06 26	06 59	17 43	19 21	09 23
14	09 26	13 38	04 59	08 22	11 36	06 21	07 02	17 43	19 21	09 23
15	09 48	16 02	04 33	08 02	11 52	06 16	07 05	17 42	19 21	09 23
16	10 09	17 44	04 10	07 42	12 08	06 11	07 07	17 42	19 21	09 22
17	10 30	18 37	03 48	07 22	12 24	06 07	07 10	17 42	19 21	09 22
18	10 51	18 35	03 30	07 01	12 40	06 02	07 13	17 41	19 21	09 21
19	11 12	17 35	03 13	06 40	12 56	05 57	07 16	17 41	19 21	09 21
20	11 33	15 35	03 00	06 19	13 11	05 52	07 18	17 40	19 21	09 21
21	11 53	12 39	02 49	05 57	13 26	05 48	07 21	17 40	19 20	09 20
22	12 13	08 54	02 40	05 36	13 42	05 43	07 24	17 40	19 20	09 20
23	12 33	04 S 31	02 34	05 14	13 57	05 38	07 27	17 40	19 20	09 20
24	12 53	00 N 14	02 31	04 51	14 12	05 34	07 29	17 39	19 20	09 19
25	13 13	05 03	02 30	04 29	14 26	05 29	07 32	17 39	19 20	09 19
26	13 32	09 33	02 32	04 06	14 41	05 25	07 35	17 39	19 20	09 19
27	13 52	13 25	02 36	03 43	14 55	05 20	07 37	17 38	19 20	09 18
28	14 10	16 20	02 42	03 20	15 10	05 16	07 40	17 38	19 20	09 18
29	14 29	18 09	02 50	02 57	15 24	05 11	07 43	17 38	19 20	09 18
30	14 N 48	18 N 48	03 N 01	02 S 33	15 N 38	05 S 07	07 N 45	17 S 38	19 S 20	09 S 17

ZODIAC SIGN ENTRIES

Date	h	m	Planets
02	19	09	☽ ♋
05	02	36	☽ ♌
06	05	38	♀ ♓
07	13	25	☽ ♍
10	02	04	☽ ♎
12	14	55	☽ ♏
13	01	05	♂ ♉
15	02	52	☽ ♐
17	13	05	☽ ♑
19	20	41	☽ ♒
20	06	57	☉ ♉
22	01	06	☽ ♓
24	02	30	☽ ♈
26	02	09	☽ ♉
28	01	55	☽ ♊
30	03	57	☽ ♋

LATITUDES

Date	Mercury ☿ ° '	Venus ♀ ° '	Mars ♂ ° '	Jupiter ♃ ° '	Saturn ♄ ° '	Uranus ♅ ° '	Neptune ♆ ° '	Pluto ♇ ° '
01	03 N 17	01 N 08	00 S 24	00 S 56	02 S 15	00 S 37	00 N 21	12 N 19
04	02 58	00 50	00 22	00 57	02 15	00 37	00 21	12 20
07	02 24	00 33	00 20	00 57	02 15	00 37	00 21	12 21
10	01 40	00 16	00 18	00 58	02 15	00 38	00 21	12 21
13	00 52	00 N 01	00 16	00 58	02 15	00 38	00 21	12 22
16	00 N 02	00 S 14	00 14	00 58	02 15	00 38	00 21	12 23
19	00 S 44	00 28	00 12	00 59	02 15	00 38	00 21	12 23
22	01 24	00 41	00 10	00 59	02 15	00 38	00 21	12 24
25	01 58	00 53	00 08	01 00	02 15	00 38	00 21	12 24
28	02 25	01 04	00 07	01 00	02 15	00 38	00 21	12 25
31	02 S 45	01 S 14	00 S 05	01 S 01	02 S 15	00 S 38	00 N 21	12 N 25

DATA

Julian Date	2450905
Delta T	+63 seconds
Ayanamsa	23° 49' 51"
Synetic vernal point	05° ♓ 17' 09"
True obliquity of ecliptic	23° 26' 14"

MOON'S PHASES, APSIDES AND POSITIONS ☽

Date	h	m	Phase	Longitude ° '	Eclipse Indicator
03	20	18	☽	13 ♋ 52	
11	22	23	○	21 ♎ 49	
19	19	53	☾	29 ♑ 33	
26	11	41	●	06 ♉ 03	

Day	h	m		
11	01	31	Apogee	
25	17	45	Perigee	
03	04	42	Max dec	18° N 37'
10	12	22	0S	
17	23	17	Max dec	18° S 43'
24	10	50	0N	
30	13	33	Max dec	18° N 48'

LONGITUDES

Date	Chiron ⚷ ° '	Ceres ⚳ ° '	Pallas ⚴ ° '	Juno ⚵ ° '	Vesta ⚶ ° '	Black Moon Lilith ° '
01	17 ♏ 58	15 ♈ 57	10 ♓ 35	25 ♍ 27	19 ♉ 13	11 ♎ 57
11	17 ♏ 24	19 ♈ 56	13 ♓ 38	23 ♍ 29	23 ♉ 21	13 ♎ 04
21	16 ♏ 43	23 ♈ 54	16 ♓ 34	22 ♍ 03	27 ♉ 33	14 ♎ 10
31	15 ♏ 59	27 ♈ 52	19 ♓ 22	21 ♍ 12	01 ♊ 48	15 ♎ 17

ASPECTARIAN

h m Aspects

01 Wednesday
00 44 ☽ ∠ ☿
01 38 ☽ ∠ ♂
02 58 ☽ ∠ ♄
04 55 ☽ ☍ ♇
10 59 ☽ ⚹ ☉
11 36 ☽ △ ♅
14 03 ☽ □ ♃
20 09 ☽ ⚼ ♆
20 15 ☉ ⚹ ♅
23 23 ☽ # ♅

02 Thursday
01 26 ☽ ⚹ ☿
04 47 ☽ ⚹ ♂
04 55 ☽ ⚹ ♄
07 29 ♂ ☌ ♄
09 19 ☽ Q ☉
11 52 ☽ ± ♆
12 22 ☽ △ ♀
13 42 ☽ ⚼ ♅
16 09 ☿ ⊥ ♃
21 42 ☽ Q ☿
22 32 ☽ ⊼ ♆

03 Friday
02 23 ☽ Q ♄
03 20 ☽ Q ♂
05 50 ☽ ± ♅
09 24 ☽ ⊼ ♇
14 32 ♂ ⚼ ♇
16 48 ☽ ⊼ ♅
17 38 ☽ ⚼ ♀
18 52 ☉ ⊻ ♃
20 13 ☽ △ ♃
20 18 ☽ □ ☉
20 22 ☽ ± ♇
21 08 ☉ Q ♆

04 Saturday
04 59 ☽ □ ☿
11 35 ☽ # ♅
11 48 ☽ ± ♀
11 50 ☽ □ ♄
13 07 ☽ ⚼ ♇
14 34 ☽ □ ♂

05 Sunday
00 10 ☽ ⊼ ♀
00 52 ☽ ⚼ ♃
01 30 ♂ Q ♅
06 17 ☽ ☍ ♆
14 46 ☿ # ♇
17 47 ☽ △ ♇
18 30 ☽ ± ♃

06 Monday
00 26 ☿ ∥ ♂
01 53 ☽ ☍ ♅
06 26 ☽ ⊼ ♃
10 28 ☽ △ ☉
11 18 ☽ △ ☿
15 39 ♂ # ♇
16 35 ☉ ☌ ☿
22 31 ☽ △ ♄

07 Tuesday
00 39 ☉ ∥ ♄
04 40 ☽ △ ♂
08 17 ☿ ∠ ♀
14 31 ☽ # ♀
15 17 ☽ ⚼ ☿
15 30 ☉ # ♃
16 27 ☽ ☍ ♀
17 22 ☽ ⊼ ♆
18 57 ☽ ⚼ ☉
19 42 ☽ ∥ ♂
22 09 ☽ # ♇

08 Wednesday
03 02 ♀ ⊻ ♆
04 51 ☽ ⚼ ♄
05 11 ☽ □ ♇
05 26 ☽ ± ♆
07 28 ☽ ∥ ☿
08 15 ☽ ± ☿
12 48 ☽ ⚼ ♂
13 19 ☽ ∥ ☉
13 49 ☽ ⊼ ♅
14 42 ☽ ± ☉
16 12 ☽ # ♃
16 40 ☽ ∥ ♄
19 30 ☽ ☍ ♃
19 37 ☽ ⊼ ☿
21 05 ☿ ⊻ ♃
22 42 ♀ # ♂
23 13 ☽ ± ♄
23 39 ☽ ⚼ ♆

09 Thursday
01 59 ☽ ± ♅
03 57 ☽ ⊼ ☉
05 04 ♃ # ♄
08 16 ☽ ± ♂
08 41 ☉ ∥ ☿
09 49 ♄ ⚼ ♇
11 31 ☽ ⊼ ♄
17 35 ☽ Q ♇
20 18 ☽ ⚼ ♅
21 16 ☽ ⊼ ♂

10 Friday
02 45 ☿ Q ♆
06 09 ☽ △ ♆
11 05 ☽ ⊼ ♀
16 39 ☿ ∥ ♄
17 57 ☽ ⚹ ♇

11 Saturday
00 29 ☽ ± ♀
02 52 ☽ △ ♅
03 07 ☿ # ♃
04 55 ♀ ∥ ♇
04 55 ☽ ☍ ☿
09 34 ☽ ⊼ ♃
15 47 ☉ ⊥ ♃
20 32 ☽ ⚼ ♀
21 57 ☽ ± ♃
22 23 ☽ ☍ ☉

12 Sunday
00 21 ☽ ∠ ♇
01 06 ☽ ☍ ♄
03 40 ☽ # ☿
06 32 ☽ ∥ ♃
08 34 ☿ ⊥ ♀
09 14 ☽ # ♄
14 15 ☽ ☍ ♂
16 07 ☿ ⚹ ♅
16 29 ☽ ⚼ ♃
18 30 ☽ ⊥ ♇
19 02 ☽ □ ♆
21 39 ☉ ⚼ ♇
22 16 ☉ # ♀
22 36 ☽ ∥ ♀
22 40 ☽ # ☉

13 Monday
02 23 ☽ ∥ ♇
05 48 ☽ △ ♀
06 34 ☽ ⊻ ♇
11 41 ☉ ☌ ♄
14 14 ♀ □ ♇
14 28 ☽ ⊼ ☿
15 36 ☽ □ ♅
17 23 ☽ # ♂
20 38 ♀ ⊥ ♆
23 10 ☽ △ ♃

14 Tuesday
01 57 ☽ ± ☿
05 27 ♀ ∠ ♄
07 09 ☽ Q ♆
08 33 ☉ # ♇
10 54 ☉ Q ♅
13 58 ☽ ⊼ ♄
15 36 ♂ ∠ ♃
15 59 ☽ ⊼ ☉
19 08 ☽ ⚼ ☿

15 Wednesday
01 59 ☽ ± ♄
03 29 ☽ Q ♅
04 54 ☽ ± ☉
06 08 ☽ ⊼ ♂
06 56 ☽ ⚹ ♆
08 44 ♂ ± ♇
18 03 ☽ ☌ ♇
18 41 ☽ ± ♂
19 35 ♂ □ ♆
19 51 ☽ ⚼ ♄
23 07 ☽ □ ♀
23 40 ☽ △ ☿

16 Thursday
00 06 ☽ ⚼ ☉
03 04 ☽ ⚹ ♅
04 03 ☿ ⊻ ♀
11 19 ☽ □ ♃
11 28 ☽ ∥ ♅
12 16 ☽ ∠ ♆
13 19 ☽ ⚼ ♂

17 Friday
01 14 ☽ △ ♄
05 36 ☽ ⊥ ♆
07 32 ☽ △ ☉
08 04 ☽ ∠ ♅
13 07 ☽ Q ♀
17 02 ☽ ⊻ ♆
19 50 ☽ △ ♂
21 09 ♀ ⊻ ♅
22 21 ☽ Q ♃
23 53 ♀ # ♄

18 Saturday
01 10 ☽ ⊥ ♅
03 36 ☽ ⊻ ♇
07 52 ☽ □ ☿
12 25 ☽ ⊻ ♅
13 48 ☽ ⚹ ♀
14 47 ☽ ⊥ ♇
21 08 ☽ ⚹ ♃
21 27 ♃ ∠ ♆

19 Sunday
07 22 ☽ ∠ ♇
09 59 ☽ □ ♄
10 16 ☽ ∥ ♅
16 44 ☽ Q ☿
19 47 ☽ ∠ ♀
19 53 ☽ □ ☉

20 Monday
00 29 ☽ ☌ ♆
00 53 ☽ ∠ ♃
06 19 ☽ □ ♂
06 23 ☉ ∠ ♀
07 31 ☿ St D
10 20 ☽ ⚹ ♇
13 16 ☽ ⊥ ♀
14 10 ☽ ⚹ ☿
17 07 ☽ ⊥ ♃
18 31 ☽ Q ♄
18 50 ☽ ☌ ♅

21 Tuesday
00 43 ☽ ⊻ ♀
03 46 ☽ ⊻ ♃
06 10 ☽ Q ☉
06 38 ☽ # ♂
07 20 ☽ Q ♇
15 31 ☽ ⚹ ♄
15 33 ☽ Q ♂
16 22 ☽ ∠ ☿
16 52 ☽ # ☉
22 22 ☉ ± ♇

22 Wednesday
01 52 ♀ ∥ ♃
04 12 ☽ ⚹ ☉
04 41 ☽ ⊻ ♆
05 23 ♀ ∠ ♆
07 46 ☽ ⊥ ☿
09 27 ☽ ∥ ♇
11 10 ☉ □ ♆
11 36 ♄ Q ♅
12 57 ☽ ⚹ ♂
13 51 ☽ □ ♇
14 44 ☽ ⊥ ♆
17 03 ☽ ∠ ♄
17 55 ☽ ⊻ ☿
20 29 ☽ # ♄
21 59 ☽ ⊻ ♅
23 13 ♀ ☌ ♃

23 Thursday
05 37 ☽ ∠ ♆
05 49 ♂ ⊼ ♇
06 01 ☽ ∥ ♃
06 55 ☽ ∠ ☉
07 03 ☽ ☌ ♃
07 34 ☽ ☌ ♀
07 52 ☽ ⊥ ♅
07 58 ☽ ∥ ♀
08 01 ☽ ⊥ ♄
09 21 ☉ ∠ ♃
11 33 ♀ ⊥ ♅
14 08 ♀ ⊥ ♄
14 58 ☽ ∠ ♂
17 52 ☽ ⊻ ♄
22 07 ☽ # ☿
22 30 ☽ ∠ ♅
22 35 ☽ ⊥ ☉

24 Friday
05 56 ☽ ⚹ ♆
06 15 ☽ ⊥ ♂
08 54 ☽ ⊻ ☉
14 36 ☽ △ ♇
16 20 ☽ ⊻ ♂
19 30 ☽ ☌ ☿
22 31 ☽ ⚹ ♅
23 15 ☽ ∥ ☿

25 Saturday
01 04 ☽ Q ♆
07 51 ☽ ⊻ ♃
09 22 ☽ # ♀
11 29 ☽ ⊻ ♀
14 13 ☽ # ♃
14 21 ☽ ⚼ ♇
17 28 ☽ ⊥ ♃
17 32 ☽ Q ♅
18 05 ☽ ☌ ♄
21 42 ☽ ⊥ ♀
23 51 ♃ ⊥ ♅

26 Sunday
01 05 ☽ ∥ ♄
04 32 ☽ ± ♇
05 32 ☽ □ ♆
07 51 ☽ ∠ ♃
10 39 ☽ # ♇
11 41 ☽ ☌ ☉
13 02 ☽ ∠ ♀
14 00 ☽ ⊼ ♇
18 13 ☽ ☌ ♂
20 21 ☽ ⊻ ☿
22 01 ☽ □ ♅

27 Monday
06 14 ☽ ⊥ ☿
07 58 ☽ ⚹ ♃
14 47 ☽ ⚹ ♀
15 33 ☽ ∥ ☉
18 05 ☽ ⊻ ♄
21 13 ☽ ∠ ☿
22 58 ☉ ⊼ ♇

28 Tuesday
00 21 ☽ ∥ ♂
03 30 ☽ Q ♃
03 52 ☽ ⊥ ♄
05 25 ☽ △ ♆
11 46 ☽ Q ♀
14 04 ☽ ☍ ♇
15 10 ☽ ⊻ ☉
17 27 ☿ ⚹ ♅
18 42 ☽ ∠ ♄
21 06 ☽ ⊻ ♂
22 33 ☽ △ ♅
22 46 ☽ ⚹ ☿

29 Wednesday
01 48 ☽ ⊥ ☉
03 25 ☽ # ♅
06 08 ☽ ⚼ ♆
07 37 ☽ ⊥ ♂
09 36 ☽ □ ♃
16 27 ♀ ⊻ ♄
16 39 ☿ # ♀
18 02 ☽ ∠ ☉
19 57 ☽ Q ☿
20 04 ☽ ⚹ ♄
20 20 ☽ □ ♀
21 21 ☽ ± ♆
23 43 ☽ ∠ ♂
23 50 ☽ ⚼ ♅

30 Thursday
02 12 ♂ □ ♅
07 41 ☽ ⊼ ♆
15 28 ☽ ± ♅
16 51 ☽ ⊼ ♇
17 02 ☽ Q ♄
22 03 ☽ ⚹ ☉
22 41 ☿ Q ♆

All ephemeris data is given at 12.00 UT and the Moon's longitude is additionally given for 24.00 UT

MAY 1998

Date	Sidereal time h m s	LONGITUDES Sun ☉ ° ' "	Moon ☽ ° ' "	Moon ☽ 24.00 ° ' "	Mercury ☿ ° '	Venus ♀ ° '	Mars ♂ ° '	Jupiter ♃ ° '	Saturn ♄ ° '	Uranus ♅ ° '	Neptune ♆ ° '	Pluto ♇ ° '
01	02 36 52	10 ♉ 55 02	18 ♋ 08 14	24 ♋ 42 25	14 ♈ 36	27 ♓ 25	13 ♉ 39	19 ♓ 38	25 ♈ 37	12 ♒ 38	02 ♒ 10	07 ♐ 23
02	02 40 48	11 53 16	01 ♌ 09 57	07 ♌ 31 19	15 26	28 32	14 23	19 49	25 44	12 39	02 10	07 R 22
03	02 44 45	12 51 28	13 ♌ 47 07	19 ♌ 57 56	16 19	29 ♓ 40	15 07	20 01	25 52	12 40	02 10	07 21
04	02 48 41	13 49 38	26 ♌ 04 26	02 ♍ 07 17	17 15	00 ♈ 47	15 50	20 12	25 59	12 40	02 R 10	07 19
05	02 52 38	14 47 46	08 ♍ 07 07	14 ♍ 04 35	18 15	01 55	16 34	20 23	26 06	12 41	02 10	07 18
06	02 56 35	15 45 52	20 ♍ 00 18	25 ♍ 54 51	19 17	03 02	17 18	20 35	26 14	12 42	02 10	07 16
07	03 00 31	16 43 56	01 ♎ 48 46	07 ♎ 42 32	20 22	04 10	18 01	20 46	26 21	12 42	02 10	07 15
08	03 04 28	17 41 58	13 ♎ 36 36	19 ♎ 31 23	21 30	05 18	18 45	20 56	26 28	12 43	02 10	07 13
09	03 08 24	18 39 59	25 ♎ 27 14	01 ♏ 24 26	22 40	06 26	19 28	21 07	26 35	12 43	02 09	07 12
10	03 12 21	19 37 58	07 ♏ 23 14	13 ♏ 23 51	23 53	07 34	20 12	21 18	26 43	12 43	02 09	07 10
11	03 16 17	20 35 55	19 ♏ 26 27	25 ♏ 31 11	25 09	08 43	20 55	21 29	26 50	12 44	02 09	07 09
12	03 20 14	21 33 51	01 ♐ 38 09	07 ♐ 47 26	26 27	09 51	21 39	21 39	26 57	12 44	02 09	07 07
13	03 24 10	22 31 45	13 ♐ 59 09	20 ♐ 13 22	27 47	11 00	22 22	21 50	27 04	12 44	02 08	07 05
14	03 28 07	23 29 38	26 ♐ 30 11	02 ♑ 49 41	29 ♈ 10	12 08	23 05	22 00	27 11	12 44	02 08	07 04
15	03 32 04	24 27 29	09 ♑ 12 02	15 ♑ 37 22	00 ♉ 35	13 17	23 48	22 10	27 18	12 44	02 08	07 02
16	03 36 00	25 25 20	22 ♑ 05 52	28 ♑ 37 45	02 02	14 26	24 31	22 20	27 25	12 45	02 07	07 01
17	03 39 57	26 23 09	05 ♒ 13 14	11 ♒ 52 34	03 32	15 34	25 15	22 30	27 32	12 45	02 07	06 59
18	03 43 53	27 20 56	18 ♒ 36 00	25 ♒ 23 46	05 04	16 43	25 58	22 40	27 39	12 R 45	02 07	06 58
19	03 47 50	28 18 43	02 ♓ 16 05	09 ♓ 13 06	06 38	17 52	26 41	22 50	27 46	12 45	02 06	06 56
20	03 51 46	29 ♉ 16 28	16 ♓ 14 54	23 ♓ 21 29	08 14	19 01	27 24	23 00	27 53	12 45	02 06	06 54
21	03 55 43	00 ♊ 14 13	00 ♈ 32 43	07 ♈ 48 19	09 53	20 11	28 06	23 09	28 00	12 44	02 05	06 53
22	03 59 39	01 11 56	15 ♈ 07 52	22 ♈ 30 46	11 34	21 20	28 49	23 18	28 07	12 44	02 05	06 51
23	04 03 36	02 09 39	29 ♈ 56 16	07 ♉ 23 26	13 16	22 29	29 ♉ 32	23 28	28 14	12 44	02 04	06 49
24	04 07 33	03 07 20	14 ♉ 51 17	22 ♉ 18 39	15 02	23 39	00 ♊ 15	23 37	28 20	12 43	02 03	06 48
25	04 11 29	04 05 00	29 ♉ 44 26	07 ♊ 07 29	16 49	24 48	00 57	23 46	28 27	12 43	02 03	06 46
26	04 15 26	05 02 39	14 ♊ 26 44	21 ♊ 41 13	18 38	25 57	01 40	23 55	28 34	12 43	02 02	06 45
27	04 19 22	06 00 17	28 ♊ 50 09	05 ♋ 52 55	20 30	27 07	02 23	24 04	28 40	12 42	02 01	06 43
28	04 23 19	06 57 53	12 ♋ 49 50	19 ♋ 38 19	22 24	28 17	03 05	24 12	28 47	12 42	02 01	06 41
29	04 27 15	07 55 28	26 ♋ 20 40	02 ♌ 56 12	24 20	29 ♈ 26	03 48	24 21	28 53	12 41	02 00	06 40
30	04 31 12	08 53 02	09 ♌ 25 11	15 ♌ 48 00	26 18	00 ♉ 36	04 30	24 29	29 00	12 41	01 59	06 38
31	04 35 08	09 ♊ 50 34	22 ♌ 05 08	28 ♌ 17 08	28 ♉ 18	01 ♉ 46	05 ♊ 13	24 ♓ 37	29 ♈ 06	12 ♒ 40	01 ♒ 58	06 ♐ 36

Date	Moon True ☊ ° '	Moon Mean ☊ ° '	Moon ☽ Latitude ° '
01	08 ♍ 21	07 ♍ 21	03 S 55
02	08 R 19	07 18	03 05
03	08 D 18	07 14	02 07
04	08 19	07 11	01 05
05	08 19	07 08	00 S 01
06	08 R 18	07 05	01 N 02
07	08 16	07 02	02 02
08	08 11	06 59	02 56
09	08 03	06 55	03 43
10	07 53	06 52	04 20
11	07 41	06 49	04 46
12	07 28	06 46	05 00
13	07 15	06 43	04 59
14	07 04	06 39	04 44
15	06 55	06 36	04 15
16	06 49	06 33	03 33
17	06 45	06 30	02 39
18	06 44	06 27	01 34
19	06 D 43	06 24	00 N 24
20	06 R 43	06 20	00 S 50
21	06 42	06 17	02 02
22	06 38	06 14	03 08
23	06 32	06 11	04 02
24	06 23	06 08	04 40
25	06 12	06 05	04 59
26	06 01	06 01	04 57
27	05 50	05 58	04 37
28	05 41	05 55	04 00
29	05 34	05 52	03 11
30	05 30	05 49	02 13
31	05 ♍ 29	05 ♍ 45	01 S 10

DECLINATIONS

Date	Sun ☉ ° '	Moon ☽ ° '	Mercury ☿ ° '	Venus ♀ ° '	Mars ♂ ° '	Jupiter ♃ ° '	Saturn ♄ ° '	Uranus ♅ ° '	Neptune ♆ ° '	Pluto ♇ ° '
01	15 N 06	18 N 20	03 N 13	02 S 10	15 N 52	05 S 02	07 N 48	17 S 38	19 S 20	09 S 17
02	15 24	16 53	03 28	01 46	16 05	04 58	07 51	17 37	19 20	09 17
03	15 42	14 39	03 44	01 22	16 19	04 54	07 53	17 37	19 20	09 16
04	15 59	11 48	04 02	00 58	16 32	04 49	07 56	17 37	19 20	09 16
05	16 16	08 30	04 22	00 33	16 45	04 45	07 59	17 37	19 20	09 16
06	16 33	04 54	04 44	00 S 09	16 58	04 41	08 01	17 37	19 20	09 15
07	16 50	01 N 08	05 07	00 N 16	17 11	04 37	08 04	17 37	19 20	09 15
08	17 07	02 S 40	05 31	00 40	17 24	04 33	08 06	17 37	19 20	09 15
09	17 23	06 23	05 57	01 05	17 36	04 29	08 09	17 37	19 20	09 15
10	17 38	09 52	06 24	01 30	17 49	04 25	08 11	17 36	19 20	09 14
11	17 54	13 00	06 52	01 55	18 01	04 21	08 14	17 36	19 20	09 14
12	18 09	15 36	07 22	02 20	18 13	04 17	08 16	17 36	19 20	09 14
13	18 24	17 32	07 53	02 45	18 25	04 13	08 19	17 36	19 20	09 13
14	18 39	18 39	08 25	03 10	18 36	04 09	08 21	17 36	19 20	09 13
15	18 53	18 52	08 58	03 35	18 48	04 05	08 24	17 36	19 20	09 13
16	19 07	18 07	09 31	04 00	18 59	04 01	08 26	17 36	19 20	09 12
17	19 21	16 24	10 06	04 25	19 10	03 57	08 28	17 36	19 20	09 12
18	19 34	13 45	10 42	04 50	19 21	03 54	08 31	17 36	19 21	09 12
19	19 47	10 18	11 18	05 15	19 31	03 50	08 33	17 36	19 21	09 12
20	20 00	06 12	11 55	05 40	19 42	03 46	08 36	17 37	19 21	09 11
21	20 12	01 S 39	12 32	06 05	19 52	03 43	08 38	17 37	19 21	09 11
22	20 24	03 N 04	13 10	06 30	20 02	03 39	08 40	17 37	19 21	09 11
23	20 36	07 40	13 48	06 55	20 12	03 36	08 43	17 37	19 21	09 11
24	20 47	11 50	14 27	07 19	20 22	03 33	08 45	17 37	19 21	09 11
25	20 58	15 14	15 06	07 44	20 31	03 29	08 47	17 37	19 21	09 11
26	21 08	17 45	15 45	08 08	20 41	03 26	08 49	17 37	19 22	09 10
27	21 18	18 49	16 24	08 33	20 50	03 23	08 51	17 37	19 22	09 10
28	21 28	18 50	17 02	08 58	20 59	03 19	08 54	17 38	19 22	09 10
29	21 38	17 45	17 41	09 22	21 07	03 16	08 56	17 38	19 22	09 10
30	21 47	15 45	18 19	09 46	21 16	03 13	08 58	17 38	19 22	09 10
31	21 N 55	13 N 02	18 N 56	10 N 10	21 N 24	03 S 10	09 N 00	17 S 38	19 S 22	09 S 10

ZODIAC SIGN ENTRIES

Date	h	m	Planets
02	09	49	☽ ♌
03	19	16	♀ ♈
04	19	47	☽ ♍
07	08	19	☽ ♎
09	21	10	☽ ♏
12	08	48	☽ ♐
14	18	39	☽ ♑
15	02	10	☿ ♉
17	02	30	☽ ♒
19	08	03	☽ ♓
21	06	05	☉ ♊
21	11	06	☽ ♈
23	12	06	☽ ♉
24	03	42	♂ ♊
25	12	25	☽ ♊
27	13	58	☽ ♋
29	18	38	☽ ♌
29	23	32	♀ ♉

LATITUDES

Date	Mercury ☿ ° '	Venus ♀ ° '	Mars ♂ ° '	Jupiter ♃ ° '	Saturn ♄ ° '	Uranus ♅ ° '	Neptune ♆ ° '	Pluto ♇ ° '
01	02 S 45	01 S 14	00 S 05	01 S 01	02 S 15	00 S 38	00 N 21	12 N 25
04	02 58	01 23	00 03	01 02	02 15	00 38	00 21	12 25
07	03 04	01 31	00 S 01	01 02	02 15	00 39	00 21	12 26
10	03 05	01 38	00 N 01	01 03	02 16	00 39	00 21	12 26
13	03 00	01 45	00 03	01 03	02 16	00 39	00 21	12 26
16	02 50	01 50	00 05	01 04	02 16	00 39	00 21	12 26
19	02 34	01 54	00 07	01 05	02 17	00 39	00 21	12 26
22	02 14	01 58	00 09	01 05	02 17	00 39	00 21	12 26
25	01 50	02 00	00 11	01 06	02 17	00 39	00 21	12 26
28	01 23	02 02	00 13	01 07	02 18	00 39	00 21	12 25
31	00 S 52	02 S 03	00 N 14	01 S 07	02 S 18	00 S 39	00 N 21	12 N 25

DATA

Julian Date	2450935
Delta T	+63 seconds
Ayanamsa	23° 49' 54"
Synetic vernal point	05° ♓ 17' 06"
True obliquity of ecliptic	23° 26' 14"

MOON'S PHASES, APSIDES AND POSITIONS ☽

Date	h	m	Phase	Longitude ° '	Eclipse Indicator
03	10	04	☽	12 ♌ 47	
11	14	29	○	20 ♏ 42	
19	04	35	☾	28 ♒ 01	
25	19	32	●	04 ♊ 23	

Day	h	m		
08	08	47	Apogee	
23	23	52	Perigee	
07	19	11	0S	
15	05	32	Max dec	18° S 55'
21	20	27	0N	
28	00	15	Max dec	18° N 58'

LONGITUDES

Date	Chiron ⚷ ° '	Ceres ⚳ ° '	Pallas ⚴ ° '	Juno ⚵ ° '	Vesta ⚶ ° '	Black Moon Lilith ⚸ ° '
01	15 ♏ 59	27 ♈ 52	19 ♓ 22	21 ♍ 12	01 ♊ 48	15 ♎ 17
11	15 ♏ 13	01 ♉ 49	21 ♓ 59	20 ♍ 58	06 ♊ 05	16 ♎ 24
21	14 ♏ 29	05 ♉ 43	24 ♓ 24	21 ♍ 19	10 ♊ 24	17 ♎ 30
31	13 ♏ 49	09 ♉ 34	26 ♓ 35	22 ♍ 11	14 ♊ 43	18 ♎ 37

All ephemeris data is given at 12.00 UT and the Moon's longitude is additionally given for 24.00 UT

ASPECTARIAN

h m Aspects

01 Friday
02 06 ☽ ⊼ ♅
03 27 ☽ ✶ ♂
03 29 ☽ ± ♇
04 35 ♃ ⊥ ♄
05 13 ☽ □ ☿
14 45 ☽ △ ♃
16 48 ♀ ∠ ♅
19 44 ☽ bQ ♇
21 24 ☽ Q ☉

02 Saturday
01 31 ☽ # ♅
01 48 ☽ □ ♄
02 33 ☽ Q ♂
06 37 ☽ △ ♀
13 52 ☽ ☍ ♆
18 59 ☽ bQ ♃
20 36 ☽ ∥ ♂
23 41 ☽ △ ♇

03 Sunday
02 54 ☽ ∥ ☉
07 05 ☉ □ ♅
09 50 ☽ ☍ ♅
10 04 ☽ □ ☉
12 27 ☽ ± ♃
13 51 ☽ bQ ♀
14 43 ☽ □ ♂
17 17 ☽ △ ☿

04 Monday
00 17 ☽ ⊼ ♃
09 12 ☽ ± ♀
10 39 ♆ St R
11 49 ☽ △ ♄
15 18 ♀ ∠ ♂
22 18 ☽ ⊼ ♀

05 Tuesday
00 05 ☽ ⊼ ♆
01 21 ☽ bQ ☿
06 44 ☽ # ♇
10 21 ☽ □ ♇
12 05 ☽ ± ♆
15 36 ☽ ∥ ♄
17 20 ♀ ✶ ♆
18 04 ☽ bQ ♄
21 05 ☽ ± ☿
21 12 ☽ ⊼ ♅

06 Wednesday
02 39 ☽ △ ☉
06 09 ☽ △ ♂
06 14 ☽ bQ ♆
09 20 ☽ ± ♅
09 38 ☿ # ♃
10 23 ☽ ⊼ ☿
12 27 ☽ ± ♄
13 04 ☽ ∥ ☿
13 11 ☽ ☍ ♃
13 29 ☽ # ♃
22 40 ☽ Q ♇

07 Thursday
00 46 ☽ ⊼ ♄
03 38 ☽ bQ ♅
11 49 ☽ bQ ☉
12 42 ☽ △ ♆
14 37 ☽ bQ ♂
17 00 ☽ ∥ ♀
17 19 ☽ ☍ ♀
21 54 ☽ # ♀
22 05 ☿ ⊻ ♃
23 02 ☽ ✶ ♇

08 Friday
07 46 ☽ ± ☉
10 08 ☽ ± ♂
10 10 ☽ △ ♅
21 02 ☽ ⊼ ☉
23 07 ☽ ⊼ ♂
23 48 ☽ ∥ ♃

09 Saturday
02 33 ☿ bQ ♇
03 06 ☽ ⊼ ♃
05 25 ☽ ∠ ♇
05 45 ☽ ☍ ☿
08 45 ☽ # ☿
12 16 ♂ # ♅
14 19 ☽ ☍ ♄
15 25 ☽ ± ♃
23 33 ☽ ⊥ ♇

10 Sunday
00 02 ☽ # ♄
01 30 ☽ □ ♆
03 38 ♀ △ ♇
07 29 ☽ ∥ ♇
08 58 ☉ # ♅
09 48 ☽ bQ ♃
11 34 ☽ ⊻ ♇
12 25 ☽ ⊼ ♀
22 40 ☽ □ ♅

11 Monday
01 38 ☽ ± ♀
04 02 ☿ Q ♅
13 24 ☽ Q ♆
14 29 ☽ ☍ ☉
15 07 ☽ ☍ ♂
16 05 ☽ △ ♃
21 19 ☽ bQ ♀

12 Tuesday
00 36 ☽ ⊼ ☿
02 43 ☽ ⊼ ♄
10 14 ☽ Q ♅
12 28 ♂ ✶ ♃
13 00 ☽ ✶ ♆
13 47 ☽ ± ☿
14 36 ☽ ± ♄
14 43 ☉ ✶ ♃
19 46 ☉ ☌ ♂
21 57 ☿ ☌ ♄
22 40 ☽ ☌ ♇

13 Wednesday
05 37 ☽ △ ♀
08 16 ☽ bQ ♄
09 24 ☽ bQ ☿
09 35 ☽ ✶ ♅
12 47 ☿ ⊥ ♃
13 14 ☽ ∥ ♅
15 50 ☉ ∥ ♂
18 05 ☽ ∠ ♆

14 Thursday
03 17 ☽ □ ♃
05 05 ☽ ⊼ ♂
05 47 ☽ ⊼ ☉
09 10 ☿ ∥ ♄
09 42 ☽ # ♂
11 18 ☽ ⊥ ♆
11 29 ☽ # ☉
13 19 ☽ △ ♄
14 21 ☽ ∠ ♅
17 12 ☽ ± ♂
17 42 ☽ △ ☿
18 09 ☽ ± ☉
22 41 ☽ ⊻ ♆

15 Friday
00 42 ♀ ✶ ♅
07 23 ☽ ⊥ ♅
07 57 ☽ ⊻ ♇
11 13 ☽ bQ ♂
11 31 ☽ # ☉
12 31 ☽ bQ ☉
13 50 ☽ Q ♃
15 40 ☽ # ♂
18 38 ☽ ⊻ ♅
19 10 ☽ ⊥ ♇
19 24 ☿ ± ♇
20 23 ☽ □ ♀
22 50 ☿ # ♇

16 Saturday
05 44 ♀ Q ♆
11 51 ☽ ∠ ♇
12 27 ☽ ✶ ♃
13 13 ♀ # ♃
13 21 ☿ □ ♆
16 44 ☽ △ ♂
18 37 ☽ △ ☉
20 44 ☽ ∥ ♅
21 53 ☽ □ ♄

17 Sunday
06 22 ☽ ☌ ♆
08 33 ☽ □ ☿
08 44 ☽ Q ♀
11 39 ☉ # ♆
15 03 ♅ St R
15 11 ☽ ✶ ♇
16 11 ☽ ∠ ♃

18 Monday
01 33 ☽ ☌ ♅
06 43 ☽ Q ♄
08 21 ☽ ✶ ♀
08 31 ☽ ⊥ ♃
11 30 ♂ # ♆
12 38 ☽ Q ♇
19 17 ☽ ⊻ ♃
20 41 ☉ ⊻ ♄
20 55 ☽ Q ☿

19 Tuesday
01 42 ☽ □ ♂
04 06 ☽ ✶ ♄
04 35 ☽ □ ☉
06 29 ☽ # ☿
11 43 ☽ ⊻ ♆
13 09 ☽ ∠ ♀
16 26 ☿ ⊼ ♇
18 47 ☽ ∥ ♇
20 03 ☽ □ ♇
20 32 ☽ ✶ ☿
22 05 ☽ ⊥ ♆
22 31 ☽ # ♄

20 Wednesday
06 01 ☽ ⊥ ♀
06 02 ☽ ⊻ ♅
06 14 ☽ ∠ ♄
07 57 ☿ ∠ ♃
10 28 ☽ Q ♂
13 26 ☽ ∠ ♆
13 52 ☽ Q ☉
14 40 ☽ # ♀
16 13 ☽ ⊥ ♅
17 07 ☽ ⊻ ♀
21 36 ☽ ⊥ ♄
23 31 ☽ ☌ ♃

21 Thursday
01 11 ☽ ∥ ♃
01 19 ☽ ∠ ☿
07 20 ☽ ∠ ♅
07 44 ☽ ✶ ♂
07 44 ☽ ⊻ ♄
07 44 ♂ ⊻ ♄
11 27 ☽ ✶ ☉
14 33 ☽ ✶ ♆
18 15 ☽ ⊥ ☿
22 27 ☽ △ ♇

22 Friday
05 24 ☽ ⊻ ☿
08 05 ☽ ✶ ♅
09 45 ☽ ∠ ♂
10 17 ☽ Q ♆
13 52 ☽ ∠ ☉
14 58 ☽ # ♃
22 35 ♀ bQ ♇
22 54 ☽ bQ ♇
22 56 ☽ ☌ ♀

23 Saturday
01 08 ☽ ⊥ ♂
01 26 ☽ ⊻ ♃
03 35 ☽ Q ♅
04 27 ☿ □ ♅
05 29 ☽ ⊥ ☉
07 31 ☽ ∥ ♀
09 13 ☽ ☌ ♄
09 41 ☉ △ ♆
11 14 ☽ ⊥ ♃
11 19 ☽ ⊻ ♂
13 26 ☽ ± ♇
15 26 ☽ □ ♆
15 50 ☽ ⊻ ☉
17 45 ☽ ∥ ♄
20 21 ☽ # ♇
23 04 ☽ ⊼ ♇

24 Sunday
01 52 ☽ ∠ ♃
08 35 ☽ □ ♅
11 22 ♀ ⊻ ♃
12 19 ☽ ☌ ☿

25 Monday
02 15 ☽ ✶ ♃
03 20 ☽ ⊻ ♀
09 54 ☽ ⊻ ♄
10 19 ♀ Q ♅
10 39 ☽ ∥ ☿
13 52 ☽ ⊥ ♀
14 04 ☽ ☌ ♂
15 44 ☽ △ ♆
19 32 ☽ ☌ ☉
19 42 ☽ ⊥ ♄
20 29 ♀ ⊥ ♂
21 53 ☽ Q ♃
22 22 ☉ ⊥ ♄
23 24 ☽ ☍ ♇

26 Tuesday
05 47 ☽ ∠ ♀
09 09 ☽ △ ♅
10 32 ☽ ∠ ♄
12 11 ☽ # ♅
16 16 ☽ bQ ♆
19 57 ☽ ⊻ ☿

27 Wednesday
00 11 ♂ △ ♆
03 53 ☽ □ ♃
07 16 ☽ ± ♆
07 28 ☽ ⊥ ☿
08 51 ☽ ✶ ♀
10 05 ☽ bQ ♅
11 43 ☽ ✶ ♄
13 38 ☉ Q ♃
17 24 ☽ ⊼ ♆
18 20 ☽ ⊻ ♂

28 Thursday
01 07 ☽ ⊻ ☉
01 07 ☽ ∠ ☿
01 24 ☽ ± ♅
01 24 ☽ ⊼ ♇
05 10 ☽ ⊥ ♂
05 15 ☉ ☍ ♇
07 11 ☽ Q ♀
07 44 ♀ ∥ ♄
08 26 ☽ Q ♄
11 46 ☽ ± ♇
11 47 ☽ ⊼ ♅
12 17 ☽ ⊥ ☉
21 46 ☽ ∠ ♂
23 25 ♀ ☌ ♄

29 Friday
00 14 ♀ # ♇
03 37 ☽ bQ ♇
05 23 ☽ ∠ ☉
07 45 ☽ ✶ ☿
08 22 ☽ △ ♃
10 01 ☿ # ♅
12 13 ☿ ✶ ♃
12 45 ☽ ∥ ☿
13 49 ☽ # ♅
16 39 ☽ □ ♄
18 09 ☽ □ ♀
22 16 ☽ ☍ ♆

30 Saturday
02 21 ☽ ✶ ♂
06 50 ☽ △ ♇
09 13 ♀ ⊥ ♃
09 32 ☽ Q ☿
10 55 ☽ ✶ ☉
12 07 ☽ bQ ♃
12 34 ♀ ± ♇
18 06 ☽ ☍ ♅

31 Sunday
02 07 ☽ Q ♂
05 18 ☽ ± ♃
07 46 ♂ ⊥ ♄
11 30 ☽ Q ☉
16 08 ♀ □ ♆
16 57 ☽ ⊼ ♃
22 05 ☿ ⊻ ♄

JUNE 1998

LONGITUDES

Date	Sidereal time h m s	Sun ☉ ° ' "	Moon ☽ ° ' "	Moon ☽ 24.00 ° ' "	Mercury ☿ ° '	Venus ♀ ° '	Mars ♂ ° '	Jupiter ♃ ° '	Saturn ♄ ° '	Uranus ♅ ° '	Neptune ♆ ° '	Pluto ♇ ° '
01	04 39 05	10 ♊ 48 05	04 ♍ 24 38	10 ♍ 28 16	00 ♊ 20	02 ♉ 56	05 ♊ 55	24 ♓ 45	29 ♈ 13	12 ♒ 39	01 ♒ 57	06 ♐ 35
02	04 43 02	11 45 35	16 ♍ 28 44	22 ♍ 26 42	02 24	04 06	06 37	24 53	29 19	12 R 38	01 R 57	06 R 33
03	04 46 58	12 43 03	28 ♍ 22 51	04 ♎ 17 50	04 29	05 16	07 19	25 01	29 25	12 38	01 56	06 31
04	04 50 55	13 40 30	10 ♎ 12 17	16 ♎ 06 47	06 36	06 26	08 02	25 09	29 31	12 37	01 55	06 30
05	04 54 51	14 37 56	22 ♎ 01 55	27 ♎ 58 10	08 45	07 36	08 44	25 16	29 37	12 36	01 54	06 28
06	04 58 48	15 35 21	03 ♏ 56 00	09 ♏ 55 47	10 55	08 46	09 26	25 24	29 44	12 35	01 53	06 27
07	05 02 44	16 32 45	15 ♏ 57 53	22 ♏ 02 35	13 05	09 57	10 08	25 31	29 50	12 34	01 52	06 25
08	05 06 41	17 30 07	28 ♏ 10 04	04 ♐ 20 30	15 17	11 07	10 50	25 38	29 ♈ 56	12 33	01 51	06 23
09	05 10 37	18 27 29	10 ♐ 34 00	16 ♐ 50 36	17 28	12 17	11 32	25 45	00 ♉ 01	12 32	01 50	06 22
10	05 14 34	19 24 50	23 ♐ 10 18	29 ♐ 33 05	19 41	13 28	12 13	25 52	00 07	12 31	01 49	06 20
11	05 18 31	20 22 10	05 ♑ 58 54	12 ♑ 27 41	21 53	14 38	12 55	25 59	00 13	12 30	01 48	06 19
12	05 22 27	21 19 30	18 ♑ 59 21	25 ♑ 33 52	24 04	15 49	13 37	26 05	00 19	12 29	01 47	06 17
13	05 26 24	22 16 49	02 ♒ 11 09	08 ♒ 51 13	26 15	16 59	14 19	26 11	00 25	12 27	01 45	06 15
14	05 30 20	23 14 08	15 ♒ 34 01	22 ♒ 19 36	28 ♊ 25	18 10	15 00	26 18	00 30	12 26	01 44	06 14
15	05 34 17	24 11 26	29 ♒ 08 00	05 ♓ 59 15	00 ♋ 35	19 20	15 42	26 24	00 36	12 25	01 43	06 12
16	05 38 13	25 08 43	12 ♓ 53 26	19 ♓ 50 34	02 42	20 31	16 24	26 30	00 41	12 23	01 42	06 11
17	05 42 10	26 06 01	26 ♓ 50 41	03 ♈ 53 44	04 49	21 42	17 05	26 35	00 47	12 22	01 41	06 09
18	05 46 06	27 03 18	10 ♈ 59 37	18 ♈ 08 11	06 53	22 53	17 47	26 41	00 52	12 21	01 39	06 08
19	05 50 03	28 00 34	25 ♈ 19 08	02 ♉ 32 06	08 56	24 04	18 28	26 46	00 57	12 19	01 38	06 06
20	05 54 00	28 57 51	09 ♉ 46 35	17 ♉ 02 01	10 57	25 14	19 09	26 51	01 02	12 18	01 37	06 05
21	05 57 56	29 ♊ 55 07	24 ♉ 17 42	01 ♊ 32 53	12 56	26 25	19 51	26 56	01 08	12 16	01 35	06 03
22	06 01 53	00 ♋ 52 24	08 ♊ 46 44	15 ♊ 58 28	14 53	27 36	20 32	27 01	01 13	12 15	01 34	06 02
23	06 05 49	01 49 40	23 ♊ 07 16	00 ♋ 12 22	16 48	28 48	21 13	27 06	01 18	12 13	01 33	06 00
24	06 09 46	02 46 55	07 ♋ 13 07	14 ♋ 08 58	18 41	29 ♉ 59	21 54	27 10	01 23	12 11	01 31	05 59
25	06 13 42	03 44 11	20 ♋ 59 27	27 ♋ 44 17	20 32	01 ♊ 10	22 36	27 15	01 27	12 10	01 30	05 58
26	06 17 39	04 41 26	04 ♌ 23 19	10 ♌ 56 32	22 20	02 21	23 17	27 19	01 32	12 08	01 29	05 56
27	06 21 35	05 38 40	17 ♌ 24 03	23 ♌ 46 06	24 06	03 32	23 58	27 23	01 37	12 06	01 27	05 55
28	06 25 32	06 35 54	00 ♍ 03 00	06 ♍ 15 11	25 50	04 44	24 39	27 27	01 42	12 04	01 26	05 53
29	06 29 29	07 33 07	12 ♍ 23 08	18 ♍ 27 26	27 32	05 55	25 20	27 30	01 46	12 03	01 24	05 52
30	06 33 25	08 ♋ 30 20	24 ♍ 28 40	00 ♎ 27 27	29 ♋ 12	07 ♊ 06	26 ♊ 01	27 ♓ 34	01 ♉ 51	12 ♒ 01	01 ♒ 23	05 ♐ 51

Date	Moon True ☊ ° '	Moon Mean ☊ ° '	Moon ☽ Latitude ° '
01	05 ♍ 28	05 ♍ 42	00 S 06
02	05 R 28	05 39	00 N 58
03	05 27	05 36	01 58
04	05 25	05 33	02 53
05	05 20	05 30	03 40
06	05 13	05 26	04 18
07	05 03	05 23	04 45
08	04 51	05 20	04 59
09	04 38	05 17	04 59
10	04 26	05 14	04 45
11	04 15	05 11	04 17
12	04 06	05 07	03 34
13	04 00	05 04	02 40
14	03 56	05 01	01 36
15	03 55	04 58	00 N 25
16	03 D 55	04 55	00 S 48
17	03 R 55	04 51	01 59
18	03 55	04 48	03 04
19	03 52	04 45	03 58
20	03 47	04 42	04 38
21	03 40	04 39	05 00
22	03 31	04 36	05 02
23	03 21	04 32	04 46
24	03 12	04 29	04 12
25	03 04	04 26	03 25
26	02 58	04 23	02 26
27	02 55	04 20	01 22
28	02 54	04 17	00 S 15
29	02 D 54	04 13	00 N 51
30	02 ♍ 55	04 ♍ 10	01 N 53

DECLINATIONS

Date	Sun ☉ ° '	Moon ☽ ° '	Mercury ☿ ° '	Venus ♀ ° '	Mars ♂ ° '	Jupiter ♃ ° '	Saturn ♄ ° '	Uranus ♅ ° '	Neptune ♆ ° '	Pluto ♇ ° '
01	22 N 04	09 N 48	19 N 33	10 N 34	21 N 32	03 S 07	09 N 02	17 S 38	19 S 23	09 S 09
02	22 12	06 14	20 08	10 57	21 40	03 04	09 04	17 39	19 23	09 09
03	22 19	02 N 27	20 43	11 21	21 48	03 01	09 06	17 39	19 23	09 09
04	22 26	01 S 23	21 16	11 44	21 55	02 58	09 08	17 39	19 23	09 09
05	22 33	05 11	21 47	12 07	22 03	02 56	09 10	17 39	19 23	09 09
06	22 39	08 47	22 17	12 30	22 10	02 53	09 12	17 40	19 24	09 09
07	22 45	12 04	22 45	12 52	22 16	02 50	09 14	17 40	19 24	09 09
08	22 51	14 53	23 10	13 15	22 23	02 48	09 16	17 40	19 24	09 09
09	22 56	17 05	23 34	13 37	22 30	02 45	09 18	17 41	19 24	09 09
10	23 01	18 31	23 54	13 58	22 36	02 43	09 20	17 41	19 25	09 08
11	23 05	19 02	24 12	14 20	22 42	02 40	09 22	17 41	19 25	09 08
12	23 09	18 33	24 28	14 41	22 47	02 38	09 24	17 42	19 25	09 08
13	23 13	17 04	24 40	15 02	22 53	02 36	09 26	17 42	19 25	09 08
14	23 16	14 38	24 50	15 23	22 58	02 34	09 27	17 43	19 26	09 08
15	23 19	11 23	24 57	15 43	23 04	02 32	09 29	17 43	19 26	09 08
16	23 21	07 27	25 02	16 03	23 08	02 29	09 31	17 43	19 26	09 08
17	23 23	03 S 04	25 03	16 23	23 13	02 27	09 33	17 44	19 26	09 08
18	23 24	01 N 32	25 02	16 42	23 18	02 25	09 34	17 44	19 27	09 08
19	23 25	06 06	24 58	17 01	23 22	02 24	09 36	17 45	19 27	09 08
20	23 26	10 21	24 52	17 19	23 26	02 22	09 37	17 45	19 27	09 08
21	23 26	14 00	24 43	17 37	23 30	02 20	09 39	17 46	19 28	09 08
22	23 26	16 47	24 32	17 55	23 33	02 18	09 41	17 46	19 28	09 08
23	23 25	18 30	24 19	18 12	23 37	02 17	09 42	17 46	19 28	09 08
24	23 24	19 02	24 04	18 29	23 40	02 15	09 44	17 47	19 28	09 08
25	23 23	18 26	23 47	18 45	23 43	02 14	09 45	17 47	19 29	09 08
26	23 21	16 47	23 28	19 01	23 46	02 12	09 47	17 48	19 29	09 08
27	23 19	14 19	23 08	19 17	23 48	02 11	09 48	17 48	19 29	09 08
28	23 16	11 13	22 47	19 32	23 51	02 10	09 49	17 49	19 30	09 08
29	23 13	07 42	22 24	19 46	23 53	02 09	09 51	17 50	19 30	09 09
30	23 N 10	03 N 56	21 N 59	20 N 00	23 N 55	02 S 08	09 N 52	17 S 50	19 S 30	09 S 09

ZODIAC SIGN ENTRIES

Date	h	m	Planets
01	03	21	☽ ♍
01	08	07	☿ ♊
03	15	17	☽ ♎
06	04	06	☽ ♏
08	15	34	☽ ♐
09	06	07	♄ ♉
11	00	50	☽ ♑
13	08	03	☽ ♒
15	05	33	☿ ♋
15	13	31	☽ ♓
17	17	23	☽ ♈
19	19	47	☽ ♉
21	14	03	☉ ♋
21	21	26	☽ ♊
23	23	39	☽ ♋
24	12	27	♀ ♊
26	04	04	☽ ♌
28	11	54	☽ ♍
30	23	05	☽ ♎
30	23	52	☿ ♌

LATITUDES

Date	Mercury ☿ ° '	Venus ♀ ° '	Mars ♂ ° '	Jupiter ♃ ° '	Saturn ♄ ° '	Uranus ♅ ° '	Neptune ♆ ° '	Pluto ♇ ° '
01	00 S 41	02 S 03	00 N 15	01 S 08	02 S 18	00 S 39	00 N 21	12 N 25
04	00 S 09	02 03	00 17	01 08	02 19	00 40	00 21	12 25
07	00 N 23	02 02	00 19	01 09	02 19	00 40	00 21	12 24
10	00 52	02 00	00 20	01 10	02 20	00 40	00 21	12 24
13	01 17	01 57	00 22	01 11	02 20	00 40	00 21	12 23
16	01 37	01 54	00 24	01 12	02 21	00 40	00 21	12 23
19	01 50	01 50	00 26	01 13	02 21	00 40	00 21	12 22
22	01 56	01 45	00 28	01 13	02 22	00 40	00 21	12 21
25	01 56	01 40	00 29	01 14	02 22	00 40	00 21	12 20
28	01 50	01 35	00 31	01 15	02 23	00 40	00 21	12 19
31	01 N 38	01 S 28	00 N 33	01 S 16	02 S 24	00 S 40	00 N 21	12 N 18

DATA

Julian Date	2450966
Delta T	+63 seconds
Ayanamsa	23° 49' 58"
Synetic vernal point	05° ♓ 17' 01"
True obliquity of ecliptic	23° 26' 13"

MOON'S PHASES, APSIDES AND POSITIONS ☽

Date	h	m	Phase	Longitude ° '	Eclipse Indicator
02	01	45	☽	11 ♍ 21	
10	04	18	○	19 ♐ 06	
17	10	38	☾	26 ♓ 03	
24	03	50	●	02 ♋ 27	

Day	h	m		
04	23	39	Apogee	
20	17	11	Perigee	
04	03	19	0S	
11	12	41	Max dec	19° S 02'
18	04	04	0N	
24	11	02	Max dec	19° N 02'

LONGITUDES

Date	Chiron ⚷ ° '	Ceres ⚳ ° '	Pallas ⚴ ° '	Juno ⚵ ° '	Vesta ⚶ ° '	Black Moon Lilith ⚸ ° '
01	13 ♏ 45	09 ♉ 57	26 ♓ 47	22 ♍ 18	15 ♊ 09	18 ♎ 44
11	13 ♏ 12	13 ♉ 43	28 ♓ 39	23 ♍ 40	19 ♊ 29	19 ♎ 51
21	12 ♏ 47	17 ♉ 24	00 ♈ 11	25 ♍ 25	23 ♊ 49	20 ♎ 58
31	12 ♏ 31	20 ♉ 59	01 ♈ 20	27 ♍ 31	28 ♊ 08	22 ♎ 05

ASPECTARIAN

h m	Aspects
01 Monday	
01 43	☽ △ ♄
02 24	☽ □ ☿
05 11	☿ # ♆
07 11	☽ ⊼ ♆
07 13	☽ ‖ ♀
08 47	☽ △ ♀
15 09	☽ □ ♂
16 16	☽ □ ♇
16 29	☽ # ♇
17 16	☽ ‖ ♄
19 00	☽ ± ♆
02 Tuesday	
01 45	☽ □ ☉
04 20	☽ ⊼ ♅
06 49	☿ △ ♆
07 38	☽ ꝑ ♄
09 46	♂ ☍ ♇
12 56	☽ ꝑ ♆
16 20	☽ ± ♅
17 50	☽ ꝑ ♀
23 24	♂ Q ♃
03 Wednesday	
01 53	☽ ± ♄
04 13	☽ Q ♇
05 08	☽ ☍ ♃
08 23	☽ # ♃
09 47	☉ △ ♅
10 29	☽ ꝑ ♅
13 59	☽ ± ♀
14 07	☽ ⊼ ♄
19 11	☽ △ ♆
23 08	☿ ⊥ ♄
04 Thursday	
03 05	☽ △ ☿
03 30	☽ ⊼ ♀
04 29	☽ ⚹ ♇
07 18	☽ △ ♂
07 43	☿ ⊻ ♀
10 47	☿ ☍ ♇
13 16	♀ ⊼ ♇
16 53	☽ △ ♅
18 30	☿ Q ♃
19 40	☽ △ ☉
19 43	♄ # ♇
21 51	☽ ‖ ♃
05 Friday	
10 52	☽ ∠ ♇
11 40	☿ ☌ ♂
11 46	☉ ∠ ♄
15 39	☽ ꝑ ♂
16 14	☽ ꝑ ☿
18 38	☽ ⊼ ♃
06 Saturday	
03 28	☽ ☍ ♄
03 54	☿ ‖ ♂
04 42	☽ ꝑ ☉
05 00	☽ ⊥ ♇
06 51	☽ ± ♃
07 53	☽ □ ♆
10 55	☽ ± ♂
14 23	☽ ± ☿
14 33	☽ ‖ ♇
14 59	☽ # ♄
17 01	☽ ⊻ ♇
22 44	☽ ☍ ♀
23 41	☽ ⊼ ♂
07 Sunday	
00 18	☽ ± ☉
01 04	☽ ꝑ ♃
05 02	☽ ⊼ ☿
05 16	☽ □ ♅
06 20	☿ △ ♅
12 45	☉ ‖ ☿
13 15	☽ ⊼ ☉
19 19	☽ # ♀
19 42	☽ Q ♆
19 51	☉ ꝑ ♆
21 28	♀ ⊻ ♂
08 Monday	
01 09	♀ ∠ ♃
07 00	☽ △ ♃
07 59	☿ ∠ ♄
15 27	☽ ⊼ ♄
16 38	☽ Q ♅
19 09	☽ ⚹ ♆
09 Tuesday	
03 11	☽ ± ♄
03 55	☽ ☌ ♇
05 01	☿ ꝑ ♆
13 57	☽ ☍ ♂
15 38	☽ ⊼ ♀
15 46	☽ ⚹ ♅
16 59	♀ □ ♅
20 20	☽ ‖ ♅
20 36	☽ ꝑ ♄
23 57	☽ ∠ ♆
10 Wednesday	
03 59	☽ ☍ ☿
04 15	☽ ± ♀
04 18	☽ ☍ ☉
06 56	☿ ⊥ ♀
06 57	☉ ☌ ☿
07 00	☉ ⊥ ♀
16 58	☽ ⊥ ♆
17 08	☽ □ ♃
20 10	☽ ∠ ♅
21 46	♂ △ ♅
22 57	☽ ꝑ ♀
11 Thursday	
01 10	☽ △ ♄
04 12	☽ ⊻ ♆
12 36	☽ ⊻ ♇
12 57	☽ ⊥ ♅
23 42	☽ ⊥ ♇
12 Friday	
00 03	☽ ⊻ ♅
01 34	☽ ⊼ ♂
02 55	☽ Q ♃
05 35	☽ △ ♀
05 54	♄ ± ♇
13 13	☽ ± ♂
16 11	☽ ∠ ♇
16 36	☽ ⊼ ☉
23 08	☽ ⊼ ☿
13 Saturday	
01 03	☽ ⚹ ♃
03 39	☽ ‖ ♅
04 23	☽ ± ☉
06 31	☽ ꝑ ♂
06 33	☿ ± ♆
08 46	☽ □ ♄
11 13	☽ ☌ ♆
11 16	☿ □ ♃
12 09	☽ ± ☿
19 19	☽ ⚹ ♇
21 53	☽ ꝑ ☉
14 Sunday	
01 08	☿ ꝑ ♅
04 19	☽ ∠ ♃
06 19	☽ # ♀
06 25	☽ ☌ ♅
07 27	☽ ꝑ ☿
10 57	☽ △ ♂
16 44	☽ Q ♇
17 03	☽ □ ♀
17 15	☽ Q ♄
20 28	☽ ⊥ ♃
15 Monday	
02 38	☽ △ ☉
07 09	☽ ⊻ ♃
07 47	♂ ∠ ♄
12 13	☿ ⚹ ♄
14 35	☽ ⚹ ♄
15 00	☽ △ ☿
16 32	☽ ⊻ ♆
23 55	☽ # ♄
16 Tuesday	
00 21	☽ □ ♇
00 42	☿ ⊼ ♆
02 07	☽ ‖ ♇
03 00	☽ ⊥ ♆
03 42	☽ Q ♀
11 08	☽ ⊻ ♅
16 52	☽ ∠ ♄
18 22	☽ □ ♂
18 34	☽ ∠ ♆
21 29	☽ ⊥ ♅
22 14	♂ ꝑ ♆
17 Wednesday	
01 35	☉ ± ♆
02 22	☽ ⚹ ♀
08 26	☽ ⊥ ♄
10 38	☽ □ ☉
11 33	☽ ☌ ♃
12 54	☽ ∠ ♅
15 16	☽ ‖ ♃
18 45	☽ ⊻ ♄
20 13	☽ ⚹ ♆
18 Thursday	
01 34	☉ □ ♃
02 44	☽ Q ♂
03 17	☿ ⊼ ♇
03 48	☽ △ ♇
03 53	☽ □ ☿
05 45	☿ ± ♅
06 16	☽ ∠ ♀
14 16	☽ ⚹ ♅
16 28	☽ Q ♆
16 37	☽ # ♃
19 06	☉ ꝑ ♅
19 19	☽ Q ☉
22 47	☽ ⊥ ♀
23 58	☽ ⚹ ♂
19 Friday	
04 59	☽ ꝑ ♇
09 43	☽ ⊻ ♀
10 20	☽ Q ♅
14 26	☽ ⊻ ♃
15 08	☽ Q ☿
15 26	☿ ∠ ♀
16 48	☽ ⚹ ☉
19 57	☽ ± ♇
21 26	☽ ☌ ♄
22 29	☽ □ ♆
20 Saturday	
00 28	☽ ⊥ ♃
02 13	☽ ∠ ♂
04 53	☽ # ♇
05 53	☽ ⊼ ♇
07 41	☽ ‖ ♄
11 55	☉ ‖ ♂
14 16	☽ ⚹ ☿
15 28	☽ ∠ ♃
16 10	☽ □ ♅
17 52	☽ ⊥ ♂
19 25	☽ ∠ ☉
21 Sunday	
01 24	☿ ± ♇
03 58	☿ ⊼ ♅
04 17	☽ ⊻ ♂
11 20	☽ ⊥ ☉
14 24	☿ Q ♄
15 50	☽ ☌ ♀
16 24	☽ ⚹ ♃
18 58	☽ ∠ ☿
21 57	☽ ⊻ ☉
23 14	♀ ⚹ ♃
23 22	☽ ⊻ ♄
23 31	♀ # ♅
22 Monday	
00 03	☽ △ ♆
07 27	☽ ☍ ♇
09 23	☽ ⊥ ♄
12 13	☽ ⊥ ☿
12 24	☽ Q ♃
17 45	☽ △ ♅
21 18	☉ ⚹ ♄
23 46	☽ ⊻ ☿
23 49	☽ # ♅
23 Tuesday	
00 28	☽ ∠ ♄
00 59	☽ ꝑ ♆
05 03	☽ ‖ ♀
05 06	☉ ⊼ ♆
08 38	☽ ☌ ♂
16 05	☽ ± ♆
18 46	☽ □ ♃
18 55	☽ ꝑ ♅
22 28	☽ ⊻ ♀
24 Wednesday	
01 56	☽ ⚹ ♄
02 16	☽ ⊼ ♆
03 50	☽ ☌ ☉
09 40	☽ ⊥ ♀
09 53	☽ ⊼ ♇
10 14	☽ ± ♅
15 34	♃ ∠ ♅
20 13	☽ ± ♇
20 34	☽ ⊼ ♅
22 43	☽ Q ♄
25 Thursday	
02 43	☽ ∠ ♀
05 55	☽ ‖ ♀
11 03	☽ ☌ ☿
11 57	☽ ꝑ ♇
14 59	☽ ⊻ ♂
16 41	☿ ‖ ♂
17 38	☿ ꝑ ♇
18 22	♀ ⊻ ♄
18 43	♀ △ ♆
22 24	♄ □ ♆
23 00	☽ # ♅
23 11	☽ △ ♃
26 Friday	
02 16	☽ ⊥ ♂
06 44	☽ ☍ ♆
06 48	☽ □ ♄
07 56	☽ ⚹ ♀
12 35	☽ ⊻ ☉
14 49	☽ △ ♇
19 30	☽ ∠ ♂
21 52	☉ ‖ ☿
27 Saturday	
00 27	☽ ⊥ ☉
02 10	☽ ☍ ♅
02 37	☽ ꝑ ♃
08 10	☽ Q ♀
08 51	☿ ⊻ ♂
18 35	☽ ∠ ☉
18 38	☉ ⊼ ♇
19 31	☽ ± ♃
23 12	☉ ± ♅
28 Sunday	
01 05	☽ ⚹ ♂
02 40	☽ ⊻ ☿
06 59	☽ ⊼ ♃
08 26	♀ # ♆
14 39	☽ ⊼ ♆
15 11	☽ △ ♄
16 00	☽ ⊥ ☿
21 42	☽ ‖ ♄
21 59	☽ □ ♀
23 17	☽ □ ♇
29 Monday	
01 31	☽ Q ♂
01 44	☽ ⚹ ☉
02 16	☽ ± ♆
02 23	☽ # ♇
11 07	♀ ☍ ♇
11 20	☽ ⊼ ♅
11 35	☿ △ ♃
12 20	☽ ∠ ☿
14 39	♂ ± ♆
19 55	☽ ꝑ ♆
20 42	☽ ꝑ ♄
23 09	☽ ± ♅
30 Tuesday	
03 24	☽ Q ☉
10 45	☽ Q ♇
14 45	☽ ± ♄
15 15	☽ □ ♂
17 04	☽ ꝑ ♅
18 13	☽ ☍ ♃
22 57	☽ ⚹ ☿
23 15	☽ # ♃

All ephemeris data is given at 12.00 UT and the Moon's longitude is additionally given for 24.00 UT

JULY 1998

LONGITUDES

Date	Sidereal time h m s	Sun ☉ ° ' "	Moon ☽ ° ' "	Moon ☽ 24.00 ° ' "	Mercury ☿ ° '	Venus ♀ ° '	Mars ♂ ° '	Jupiter ♃ ° '	Saturn ♄ ° '	Uranus ♅ ° '	Neptune ♆ ° '	Pluto ♇ ° '
01	06 37 22	09 ♋ 27 33	06 ♎ 24 26	12 ♎ 20 16	00 ♌ 49	08 ♊ 18	26 ♊ 42	27 ♓ 37	01 ♉ 55	11 ♒ 59	01 ♒ 22	05 ♐ 50
02	06 41 18	10 24 45	18 ♎ 15 36	24 ♎ 11 04	02 24	09 29	27 22	27 40	01 59	11 R 57	01 R 20	05 R 48
03	06 45 15	11 21 57	00 ♏ 07 18	06 ♏ 04 51	03 57	10 40	28 03	27 43	02 03	11 55	01 19	05 47
04	06 49 11	12 19 09	12 ♏ 04 17	18 ♏ 06 06	05 28	11 52	28 44	27 46	02 08	11 53	01 17	05 46
05	06 53 08	13 16 20	24 ♏ 10 45	00 ♐ 18 36	06 56	13 04	29 ♊ 24	27 48	02 12	11 51	01 16	05 45
06	06 57 04	14 13 31	06 ♐ 30 00	12 ♐ 45 11	08 22	14 15	00 ♋ 05	27 51	02 16	11 49	01 14	05 43
07	07 01 01	15 10 42	19 ♐ 04 20	25 ♐ 27 33	09 45	15 27	00 46	27 53	02 20	11 47	01 12	05 42
08	07 04 58	16 07 53	01 ♑ 54 52	08 ♑ 26 13	11 07	16 39	01 26	27 55	02 23	11 45	01 11	05 41
09	07 08 54	17 05 05	15 ♑ 01 32	21 ♑ 40 37	12 26	17 50	02 07	27 56	02 27	11 43	01 09	05 40
10	07 12 51	18 02 16	28 ♑ 23 17	05 ♒ 09 16	13 42	19 02	02 47	27 58	02 31	11 41	01 08	05 39
11	07 16 47	18 59 27	11 ♒ 58 19	18 ♒ 50 08	14 56	20 14	03 27	27 59	02 34	11 39	01 06	05 38
12	07 20 44	19 56 39	25 ♒ 44 26	02 ♓ 40 57	16 07	21 26	04 08	28 01	02 38	11 37	01 05	05 37
13	07 24 40	20 53 51	09 ♓ 39 26	16 ♓ 39 38	17 16	22 38	04 48	28 02	02 41	11 34	01 03	05 36
14	07 28 37	21 51 04	23 ♓ 41 18	00 ♈ 44 16	18 22	23 50	05 28	28 02	02 44	11 32	01 01	05 35
15	07 32 33	22 48 17	07 ♈ 48 18	14 ♈ 53 12	19 26	25 02	06 09	28 03	02 48	11 30	01 00	05 34
16	07 36 30	23 45 31	21 ♈ 58 46	29 ♈ 04 47	20 26	26 14	06 49	28 03	02 51	11 28	00 58	05 33
17	07 40 27	24 42 45	06 ♉ 11 01	13 ♉ 17 10	21 23	27 26	07 29	28 04	02 54	11 25	00 57	05 32
18	07 44 23	25 40 00	20 ♉ 22 56	27 ♉ 27 57	22 18	28 38	08 09	28 R 04	02 57	11 23	00 55	05 31
19	07 48 20	26 37 16	04 ♊ 31 53	11 ♊ 34 17	23 09	29 ♊ 50	08 49	28 03	02 59	11 21	00 53	05 30
20	07 52 16	27 34 33	18 ♊ 34 43	25 ♊ 32 47	23 56	01 ♋ 02	09 29	28 03	03 02	11 19	00 52	05 29
21	07 56 13	28 31 50	02 ♋ 28 01	09 ♋ 20 02	24 41	02 15	10 09	28 02	03 05	11 16	00 50	05 28
22	08 00 09	29 ♋ 29 09	16 ♋ 08 27	22 ♋ 52 56	25 21	03 27	10 49	28 02	03 07	11 14	00 49	05 28
23	08 04 06	00 ♌ 26 27	29 ♋ 33 14	06 ♌ 09 09	25 58	04 39	11 29	28 01	03 10	11 12	00 47	05 27
24	08 08 02	01 23 46	12 ♌ 40 35	19 ♌ 07 29	26 31	05 52	12 09	28 00	03 12	11 09	00 45	05 26
25	08 11 59	02 21 06	25 ♌ 29 55	01 ♍ 48 00	26 59	07 04	12 49	27 58	03 14	11 07	00 44	05 25
26	08 15 56	03 18 26	08 ♍ 01 57	14 ♍ 12 03	27 24	08 17	13 28	27 57	03 16	11 05	00 42	05 25
27	08 19 52	04 15 47	20 ♍ 18 39	26 ♍ 22 08	27 44	09 29	14 08	27 55	03 18	11 02	00 40	05 24
28	08 23 49	05 13 08	02 ♎ 23 00	08 ♎ 21 43	27 59	10 42	14 48	27 53	03 20	11 00	00 39	05 23
29	08 27 45	06 10 30	14 ♎ 18 50	20 ♎ 14 55	28 10	11 54	15 27	27 51	03 22	10 57	00 37	05 23
30	08 31 42	07 07 52	26 ♎ 10 34	02 ♏ 06 21	28 15	13 07	16 07	27 48	03 24	10 55	00 36	05 22
31	08 35 38	08 ♌ 05 15	08 ♏ 02 53	14 ♏ 00 47	28 ♌ 16	14 ♋ 20	16 ♋ 46	27 ♓ 46	03 ♉ 26	10 ♒ 53	00 ♒ 34	05 ♐ 22

Date	Moon True ☊ ° '	Moon Mean ☊ ° '	Moon ☽ Latitude ° '
01	02 ♍ 56	04 ♍ 07	02 N 50
02	02 R 56	04 04	03 39
03	02 54	04 01	04 19
04	02 50	03 57	04 48
05	02 44	03 54	05 04
06	02 36	03 51	05 07
07	02 28	03 48	04 55
08	02 20	03 45	04 28
09	02 13	03 42	03 46
10	02 07	03 38	02 52
11	02 04	03 35	01 47
12	02 02	03 32	00 N 34
13	02 D 02	03 29	00 S 41
14	02 03	03 26	01 55
15	02 05	03 22	03 02
16	02 05	03 19	03 58
17	02 R 05	03 16	04 40
18	02 03	03 13	05 05
19	01 59	03 10	05 11
20	01 55	03 07	04 58
21	01 49	03 03	04 28
22	01 44	03 00	03 43
23	01 40	02 57	02 46
24	01 38	02 54	01 42
25	01 36	02 51	00 S 33
26	01 D 36	02 48	00 N 35
27	01 37	02 44	01 41
28	01 39	02 41	02 41
29	01 41	02 38	03 34
30	01 42	02 35	04 17
31	01 ♍ 42	02 ♍ 32	04 N 49

DECLINATIONS

Date	Sun ☉ ° '	Moon ☽ ° '	Mercury ☿ ° '	Venus ♀ ° '	Mars ♂ ° '	Jupiter ♃ ° '	Saturn ♄ ° '	Uranus ♅ ° '	Neptune ♆ ° '	Pluto ♇ ° '
01	23 N 06	00 N 04	21 N 34	20 N 14	23 N 56	02 S 07	09 N 53	17 S 51	19 S 31	09 S 09
02	23 02	03 S 47	21 07	20 27	23 58	02 06	09 55	17 51	19 31	09 09
03	22 57	07 28	20 40	20 39	23 59	02 05	09 56	17 52	19 31	09 09
04	22 52	10 53	20 12	20 51	24 00	02 04	09 57	17 52	19 32	09 09
05	22 47	13 54	19 43	21 03	24 01	02 03	09 58	17 53	19 32	09 09
06	22 41	16 21	19 14	21 14	24 02	02 03	09 59	17 53	19 32	09 09
07	22 34	18 05	18 44	21 24	24 02	02 02	10 00	17 54	19 33	09 10
08	22 28	18 58	18 14	21 34	24 02	02 02	10 02	17 55	19 33	09 10
09	22 21	18 50	17 44	21 43	24 02	02 01	10 03	17 55	19 33	09 10
10	22 13	17 41	17 13	21 52	24 02	02 01	10 04	17 56	19 34	09 10
11	22 06	15 30	16 42	22 00	24 02	02 01	10 05	17 56	19 34	09 10
12	21 57	12 24	16 11	22 07	24 01	02 00	10 06	17 57	19 34	09 11
13	21 49	08 35	15 41	22 14	24 00	02 00	10 06	17 58	19 35	09 11
14	21 40	04 S 16	15 10	22 20	23 59	02 00	10 07	17 58	19 35	09 11
15	21 31	00 N 18	14 40	22 26	23 58	02 00	10 08	17 59	19 35	09 11
16	21 21	04 52	14 10	22 31	23 57	02 00	10 09	18 00	19 36	09 11
17	21 11	09 10	13 40	22 35	23 55	02 01	10 10	18 00	19 36	09 12
18	21 00	12 56	13 11	22 39	23 53	02 01	10 11	18 01	19 37	09 12
19	20 50	15 57	12 43	22 42	23 51	02 01	10 11	18 02	19 37	09 12
20	20 39	18 00	12 15	22 44	23 49	02 02	10 12	18 02	19 37	09 12
21	20 27	18 57	11 49	22 46	23 46	02 02	10 13	18 03	19 38	09 13
22	20 15	18 46	11 23	22 47	23 44	02 03	10 13	18 03	19 38	09 13
23	20 03	17 32	10 58	22 48	23 41	02 03	10 14	18 04	19 38	09 13
24	19 51	15 22	10 35	22 48	23 38	02 04	10 14	18 05	19 39	09 14
25	19 38	12 30	10 12	22 47	23 35	02 05	10 15	18 05	19 39	09 14
26	19 25	09 06	09 52	22 45	23 31	02 06	10 15	18 06	19 39	09 14
27	19 11	05 23	09 33	22 43	23 28	02 07	10 16	18 07	19 40	09 15
28	18 58	01 N 31	09 15	22 41	23 24	02 08	10 16	18 07	19 40	09 15
29	18 44	02 S 22	09 00	22 37	23 20	02 09	10 17	18 08	19 41	09 15
30	18 29	06 07	08 46	22 33	23 16	02 10	10 17	18 09	19 41	09 16
31	18 N 15	09 S 38	08 N 35	22 N 28	23 N 12	02 S 11	10 N 17	18 S 09	19 S 41	09 S 16

ZODIAC SIGN ENTRIES

Date	h	m	Planets
03	11	45	☽ ♏
05	23	24	☽ ♐
06	09	00	♂ ♋
08	08	27	☽ ♑
10	14	52	☽ ♒
12	19	22	☽ ♓
14	22	45	☽ ♈
17	01	33	☽ ♉
19	04	18	☽ ♊
19	15	17	♀ ♋
21	07	43	☽ ♋
23	00	55	☉ ♌
23	12	48	☽ ♌
25	20	34	☽ ♍
28	07	14	☽ ♎
30	19	44	☽ ♏

LATITUDES

Date	Mercury ☿ ° '	Venus ♀ ° '	Mars ♂ ° '	Jupiter ♃ ° '	Saturn ♄ ° '	Uranus ♅ ° '	Neptune ♆ ° '	Pluto ♇ ° '
01	01 N 38	01 S 28	00 N 33	01 S 16	02 S 24	00 S 40	00 N 21	12 N 18
04	01 20	01 22	00 34	01 17	02 24	00 40	00 21	12 17
07	00 58	01 15	00 36	01 18	02 25	00 41	00 21	12 16
10	00 32	01 08	00 38	01 19	02 26	00 41	00 21	12 15
13	00 N 01	01 00	00 39	01 20	02 26	00 41	00 21	12 14
16	00 S 32	00 52	00 41	01 21	02 27	00 41	00 21	12 13
19	01 09	00 44	00 43	01 21	02 28	00 41	00 21	12 12
22	01 48	00 36	00 44	01 22	02 29	00 41	00 21	12 11
25	02 27	00 28	00 46	01 23	02 29	00 41	00 21	12 09
28	03 06	00 20	00 47	01 24	02 30	00 41	00 21	12 08
31	03 S 44	00 S 12	00 N 49	01 S 25	02 S 31	00 S 41	00 N 21	12 N 07

DATA

Julian Date	2450996
Delta T	+63 seconds
Ayanamsa	23° 50' 03"
Synetic vernal point	05° ♓ 16' 57"
True obliquity of ecliptic	23° 26' 13"

MOON'S PHASES, APSIDES AND POSITIONS ☽

Date	h	m	Phase	Longitude ° '	Eclipse Indicator
01	18	43	☽	09 ♎ 44	
09	16	01	○	17 ♑ 15	
16	15	13	☾	23 ♈ 53	
23	13	44	●	00 ♌ 31	
31	12	05	☽	08 ♏ 05	

Day	h	m		
02	17	28	Apogee	
16	13	51	Perigee	
30	12	09	Apogee	
01	12	22	0S	
08	21	15	Max dec	19° S 02'
15	10	24	0N	
21	20	11	Max dec	19° N 01'
28	21	21	0S	

LONGITUDES

Date	Chiron ⚷ ° '	Ceres ⚳ ° '	Pallas ⚴ ° '	Juno ⚵ ° '	Vesta ⚶ ° '	Black Moon Lilith ⚸ ° '
01	12 ♏ 31	20 ♉ 59	01 ♈ 20	27 ♍ 31	28 ♊ 08	22 ♎ 05
11	12 ♏ 25	24 ♉ 26	02 ♈ 03	29 ♍ 52	02 ♋ 26	23 ♎ 11
21	12 ♏ 30	27 ♉ 44	02 ♈ 15	02 ♎ 28	06 ♋ 42	24 ♎ 18
31	12 ♏ 45	00 ♊ 50	01 ♈ 53	05 ♎ 15	10 ♋ 55	25 ♎ 25

All ephemeris data is given at 12.00 UT and the Moon's longitude is additionally given for 24.00 UT

ASPECTARIAN

h m	Aspects
01 Wednesday	
01 50	☽ △ ♆
02 53	☽ ⊼ ♄
03 53	♀ ⊥ ♄
10 50	☽ ✶ ♇
16 14	☽ △ ♀
18 43	☽ □ ☉
20 02	☿ ☍ ♆
21 46	♂ ꝑ ♅
23 15	☽ △ ♅
02 Thursday	
01 27	☽ ∥ ♃
02 59	☽ Q ☿
05 22	☿ □ ♄
15 53	♀ Q ♃
17 09	☽ ∠ ♇
23 17	♂ □ ♃
03 Friday	
02 01	☽ ꝑ ♀
07 07	☽ ⊼ ♃
07 34	☽ △ ♂
11 19	☽ ⊥ ♇
12 27	☿ ∥ ♀
14 23	☽ □ ♆
14 55	☿ ⊥ ♂
15 56	☽ ☍ ♄
19 16	☽ ± ♃
20 50	☽ □ ☿
22 12	☽ ± ♀
22 18	☉ ± ♇
23 23	☽ ⊻ ♇
23 31	☽ ∥ ♇
04 Saturday	
01 26	☉ ⊼ ♅
05 08	☽ # ♄
11 33	☽ ⊼ ♀
11 38	☽ □ ♅
12 21	♀ △ ♅
12 32	☽ △ ☉
13 23	☽ ꝑ ♃
15 30	☽ ꝑ ♂
16 51	☿ △ ♇
05 Sunday	
02 19	☽ Q ♆
10 24	☽ ± ♂
19 08	☽ △ ♃
20 42	☽ ꝑ ☉
21 17	☿ # ♆
22 50	☽ ⊼ ♂
23 05	☽ Q ♅
06 Monday	
01 49	☽ ✶ ♆
03 45	☽ ⊼ ♄
09 14	☉ ⊻ ♀
10 30	☽ ☌ ♇
12 57	☉ Q ♄
15 24	☽ ± ♄
15 36	☽ ± ☉
16 03	☽ △ ☿
22 11	☽ ✶ ♅
07 Tuesday	
04 01	☽ ⊼ ☉
04 25	☽ ☍ ♀
06 35	☽ ∠ ♆
08 41	☽ ꝑ ♄
08 41	☽ ∥ ♅
21 24	☽ # ☿
23 30	☽ ⊥ ♆
23 58	☽ ꝑ ☿
08 Wednesday	
02 26	☽ ∠ ♅
02 57	♀ ꝑ ♆
03 17	♂ ⊼ ♆
04 33	☽ □ ♃
10 39	☽ ⊻ ♆
11 04	☽ ☍ ♂
12 53	☽ △ ♄
18 34	☽ ± ☿
18 56	☽ ⊻ ♇
19 03	☽ ⊥ ♅
23 14	☿ ☍ ♅
09 Thursday	
03 04	☿ # ♅
03 48	♀ ∠ ♄
05 54	☽ ⊥ ♇
06 00	☽ ⊻ ♅
06 46	☽ ⊼ ☿
13 40	☽ Q ♃
16 01	☽ ☍ ☉
17 35	☽ ⊼ ♀
21 45	☿ ꝑ ♃
22 10	☽ ∠ ♇
10 Friday	
01 17	♂ ✶ ♄
05 26	☽ ± ♀
08 13	☽ ∥ ♅
11 15	☽ ✶ ♃
16 52	☽ ☌ ♆
19 21	☽ □ ♄
20 12	☽ # ☿
20 13	☽ ⊼ ♂
22 59	☽ ꝑ ♀
11 Saturday	
00 51	☽ ✶ ♇
07 21	☽ ± ♂
11 26	☽ ☌ ♅
13 47	☽ ∠ ♃
17 41	☽ ☍ ☿
20 57	☉ ∥ ♀
21 53	☽ Q ♇
23 55	☽ ꝑ ♂
12 Sunday	
01 11	☽ ⊼ ☉
03 05	☽ Q ♄
03 48	☽ △ ♀
05 31	☽ ⊥ ♃
12 23	☽ ± ☉
15 56	☽ ⊻ ♃
21 13	☽ ⊻ ♆
23 57	☽ ✶ ♄
13 Monday	
02 55	☽ # ♄
03 14	☽ △ ♂
04 31	☉ ꝑ ♇
05 02	☽ □ ♇
05 04	☽ ꝑ ☉
07 32	☽ ⊥ ♆
08 30	☽ ∥ ♇
15 17	☽ ⊻ ♅
22 56	☽ ∠ ♆
14 Tuesday	
01 31	☽ ⊥ ♅
01 48	☽ ∠ ♄
02 10	☽ ⊼ ☿
08 38	☽ △ ☉
12 16	☽ □ ♀
13 16	☽ ± ☿
14 07	♂ ± ♅
15 39	♂ ⊼ ♇
16 50	☽ ∠ ♅
17 13	☽ ⊥ ♄
19 25	☽ ☌ ♃
23 57	☽ ∥ ♃
15 Wednesday	
00 28	☽ ✶ ♆
03 28	☽ ⊻ ♄
05 49	☽ ꝑ ☿
08 12	☽ △ ♇
09 03	☽ □ ♂
11 25	♀ ± ♆
18 15	☽ ✶ ♅
20 47	☽ Q ♆
20 50	☽ # ♃
21 40	☽ Q ♀
16 Thursday	
09 11	☽ △ ☿
09 35	☽ ꝑ ♇
14 30	☽ Q ♅
15 13	☽ □ ☉
16 32	♀ ꝑ ♅
17 02	☽ Q ♂
19 51	☽ ✶ ♀
22 16	☽ ⊻ ♃
17 Friday	
00 46	☽ ± ♇
03 10	☽ □ ♆
06 25	☽ ☌ ♄
08 25	☽ ⊥ ♃
10 54	☽ ⊼ ♇
12 09	☽ # ♇
14 18	☽ ✶ ♂
17 59	☽ ∥ ♄
20 50	☽ □ ♅
23 32	☽ ∠ ♀
23 37	☽ ∠ ♃
23 50	☽ Q ☉
18 Saturday	
00 34	♀ □ ♃
01 47	♃ St R
05 42	☿ ± ♃
13 32	☽ ∥ ☿
15 27	☽ □ ☿
16 10	☽ ⊥ ♀
16 55	☽ ∠ ♂
21 36	☽ ✶ ☉
19 Sunday	
01 00	☽ ✶ ♃
03 17	☽ ⊻ ♀
05 49	☽ △ ♆
08 57	☽ ⊥ ♂
09 22	☽ ⊻ ♄
13 39	☽ ☍ ♇
19 37	☽ ⊥ ♄
19 40	☽ ⊻ ♂
21 25	☽ Q ♃
23 35	☽ △ ♅
23 57	☽ Q ☿
20 Monday	
00 58	☽ ∠ ☉
07 21	☽ ꝑ ♆
08 33	♀ ⊼ ♆
11 04	☽ ∠ ♄
12 42	☽ # ♅
17 32	☽ ⊥ ☉
21 45	☽ ✶ ☿
22 48	☽ ± ♆
23 49	☉ △ ♃
21 Tuesday	
01 17	☽ ꝑ ♅
04 19	☽ □ ♃
04 39	☽ ⊻ ☉
09 10	☽ ⊼ ♆
11 34	☽ ☌ ♀
13 04	☽ ✶ ♄
16 53	☽ ± ♅
17 14	☽ ⊼ ♇
22 Wednesday	
01 16	☽ ∠ ☿
02 07	☽ ☌ ♂
03 22	☽ ⊼ ♅
03 45	☽ ± ♇
05 13	♀ ✶ ♄
10 11	☽ Q ♄
17 59	☽ ⊥ ☿
19 40	☽ ꝑ ♇
23 Thursday	
02 13	♂ ⊼ ♅
03 49	☽ # ♅
05 14	☽ ⊻ ☿
09 13	☽ △ ♃
10 48	♂ ± ♇
13 44	☽ ☌ ☉
14 13	☽ ☍ ♆
18 34	☽ □ ♄
20 19	☉ ☍ ♆
22 12	☽ ⊻ ♀
22 23	♀ ± ♅
22 42	☽ △ ♇
24 Friday	
03 36	♀ ⊼ ♇
09 12	☽ ☍ ♅
10 20	☽ ⊥ ♀
10 58	☽ ⊻ ♂
12 35	☽ ꝑ ♃
22 43	☽ ⊥ ♂
25 Saturday	
04 51	☽ ∠ ♀
05 21	☽ ± ♃
09 23	☿ ∥ ♄
10 02	☉ # ♆
14 56	☽ ☌ ☿
16 38	☽ ∠ ♂
16 41	☽ ⊼ ♃
21 56	☽ ⊼ ♆
26 Sunday	
02 08	☽ ⊻ ☉
02 48	☽ △ ♄
04 10	☽ ∥ ♄
06 18	☽ ∥ ☿
06 57	☽ □ ♇
09 26	☽ ± ♆
11 05	☽ # ♇
11 05	☉ □ ♄
12 32	☽ ✶ ♀
14 41	☽ ⊥ ☉
17 53	☽ ⊼ ♅
23 11	☽ ✶ ♂
27 Monday	
02 54	☽ ꝑ ♆
05 35	☽ ± ♅
08 02	☽ ꝑ ♄
09 46	☽ ∠ ☉
14 35	☽ Q ♀
18 06	☽ Q ♇
23 18	☽ ꝑ ♅
28 Tuesday	
00 12	☽ Q ♂
01 54	☽ ± ♄
02 40	☿ ⊼ ♃
03 02	☽ ☍ ♃
03 03	☽ ⊻ ☿
08 14	☽ # ♃
08 32	☽ △ ♆
12 28	☿ # ♇
13 55	☽ ⊼ ♄
15 16	☽ ⊥ ☿
16 17	☉ △ ♇
17 48	♀ ⊼ ♅
18 01	☽ ✶ ♇
18 11	☽ ✶ ☉
29 Wednesday	
01 42	♀ ± ♇
05 15	☽ △ ♅
06 35	☽ □ ♀
08 46	♂ Q ♄
09 38	☽ ∠ ☿
10 40	☽ ∥ ♃
14 26	☽ □ ♂
20 29	☽ Q ☉
30 Thursday	
00 16	☽ ∠ ♇
14 48	☿ ∠ ♀
15 17	☽ ⊼ ♃
16 13	☽ ✶ ☿
18 28	☽ ⊥ ♇
20 55	☽ □ ♆
31 Friday	
02 27	☿ St R
02 39	☽ ☍ ♄
03 23	☽ ± ♃
04 56	☽ # ☿
06 35	☽ ⊻ ♇
09 23	☽ ∥ ♇
12 05	☽ □ ☉
16 27	☽ Q ☿
16 44	☽ # ♄
17 41	☽ □ ♅
20 03	☉ # ♅
21 27	☽ ꝑ ♃

AUGUST 1998

LONGITUDES

Date	Sidereal time h m s	Sun ☉ ° ' "	Moon ☽ ° ' "	Moon ☽ 24.00 ° ' "	Mercury ☿ ° '	Venus ♀ ° '	Mars ♂ ° '	Jupiter ♃ ° '	Saturn ♄ ° '	Uranus ♅ ° '	Neptune ♆ ° '	Pluto ♇ ° '
01	08 39 35	09 ♌ 02 39	20 ♏ 00 38	26 ♏ 02 59	28 ♌ 11	15 ♋ 32	17 ♋ 26	27 ♓ 43	03 ♉ 27	10 ♒ 50	00 ♒ 32	05 ♐ 21
02	08 43 31	10 00 03	02 ♐ 08 24	08 ♐ 17 22	28 R 01	16 45	18 05	27 R 40	03 28	10 R 48	00 R 31	05 R 21
03	08 47 28	10 57 27	14 ♐ 30 20	20 ♐ 47 42	27 46	17 58	18 45	27 37	03 30	10 46	00 29	05 20
04	08 51 25	11 54 53	27 ♐ 09 47	03 ♑ 36 51	27 26	19 11	19 24	27 34	03 31	10 43	00 28	05 20
05	08 55 21	12 52 19	10 ♑ 09 02	16 ♑ 46 24	27 01	20 24	20 03	27 30	03 32	10 41	00 26	05 20
06	08 59 18	13 49 46	23 ♑ 28 56	00 ♒ 16 30	26 31	21 37	20 43	27 27	03 33	10 38	00 24	05 19
07	09 03 14	14 47 14	07 ♒ 08 51	14 ♒ 05 40	25 56	22 50	21 22	27 23	03 34	10 36	00 23	05 19
08	09 07 11	15 44 43	21 ♒ 06 32	28 ♒ 10 59	25 18	24 03	22 01	27 19	03 35	10 34	00 21	05 19
09	09 11 07	16 42 13	05 ♓ 18 28	12 ♓ 28 24	24 36	25 16	22 40	27 15	03 36	10 31	00 20	05 19
10	09 15 04	17 39 44	19 ♓ 40 11	26 ♓ 53 11	23 51	26 29	23 19	27 10	03 36	10 29	00 18	05 18
11	09 19 00	18 37 16	04 ♈ 06 49	11 ♈ 20 30	23 03	27 42	23 58	27 06	03 37	10 26	00 17	05 18
12	09 22 57	19 34 50	18 ♈ 33 41	25 ♈ 45 53	22 14	28 ♋ 55	24 37	27 01	03 37	10 24	00 15	05 18
13	09 26 54	20 32 25	02 ♉ 56 39	10 ♉ 05 36	21 25	00 ♌ 08	25 16	26 56	03 37	10 22	00 14	05 18
14	09 30 50	21 30 02	17 ♉ 12 23	24 ♉ 16 43	20 36	01 21	25 55	26 51	03 38	10 19	00 12	05 18
15	09 34 47	22 27 40	01 ♊ 18 23	08 ♊ 17 09	19 48	02 35	26 34	26 46	03 38	10 17	00 11	05 18
16	09 38 43	23 25 20	15 ♊ 12 54	22 ♊ 05 29	19 03	03 48	27 13	26 41	03 R 38	10 15	00 09	05 D 18
17	09 42 40	24 23 01	28 ♊ 54 47	05 ♋ 40 45	18 20	05 01	27 52	26 35	03 38	10 12	00 08	05 18
18	09 46 36	25 20 44	12 ♋ 23 17	19 ♋ 02 22	17 42	06 15	28 31	26 30	03 37	10 10	00 06	05 18
19	09 50 33	26 18 29	25 ♋ 37 57	02 ♌ 10 02	17 09	07 28	29 10	26 24	03 37	10 08	00 05	05 18
20	09 54 29	27 16 15	08 ♌ 38 36	15 ♌ 03 40	16 41	08 42	29 ♋ 48	26 18	03 37	10 06	00 03	05 18
21	09 58 26	28 14 02	21 ♌ 25 18	27 ♌ 43 33	16 20	09 55	00 ♌ 27	26 12	03 36	10 03	00 02	05 18
22	10 02 23	29 ♌ 11 51	03 ♍ 58 30	10 ♍ 10 19	16 06	11 09	01 06	26 06	03 35	10 01	00 ♒ 01	05 18
23	10 06 19	00 ♍ 09 41	16 ♍ 19 08	22 ♍ 25 09	15 58	12 23	01 44	25 59	03 35	09 59	29 ♑ 59	05 19
24	10 10 16	01 07 32	28 ♍ 28 37	04 ♎ 29 48	15 D 59	13 36	02 23	25 53	03 34	09 57	29 58	05 19
25	10 14 12	02 05 25	10 ♎ 29 02	16 ♎ 26 39	16 07	14 50	03 01	25 46	03 33	09 55	29 57	05 19
26	10 18 09	03 03 20	22 ♎ 23 04	28 ♎ 18 41	16 23	16 04	03 40	25 40	03 32	09 52	29 55	05 20
27	10 22 05	04 01 15	04 ♏ 13 59	10 ♏ 09 27	16 48	17 18	04 18	25 33	03 30	09 50	29 54	05 20
28	10 26 02	04 59 12	16 ♏ 05 37	22 ♏ 03 01	17 20	18 31	04 57	25 26	03 29	09 48	29 53	05 20
29	10 29 58	05 57 10	28 ♏ 02 12	04 ♐ 03 44	18 00	19 45	05 35	25 19	03 28	09 46	29 52	05 21
30	10 33 55	06 55 10	10 ♐ 08 13	16 ♐ 16 10	18 48	20 59	06 13	25 12	03 26	09 44	29 50	05 21
31	10 37 52	07 ♍ 53 11	22 ♐ 28 10	28 ♐ 44 42	19 ♌ 43	22 ♌ 13	06 ♌ 52	25 ♓ 04	03 ♉ 25	09 ♒ 42	29 ♑ 49	05 ♐ 22

Date	Moon True ☊ ° '	Moon Mean ☊ ° '	Moon ☽ Latitude ° '
01	01 ♍ 42	02 ♍ 28	05 N 08
02	01 R 40	02 25	05 15
03	01 38	02 22	05 07
04	01 35	02 19	04 44
05	01 33	02 16	04 07
06	01 30	02 13	03 15
07	01 28	02 09	02 10
08	01 27	02 06	00 N 57
09	01 D 27	02 03	00 S 21
10	01 28	02 00	01 39
11	01 29	01 57	02 51
12	01 30	01 54	03 52
13	01 30	01 50	04 38
14	01 31	01 47	05 07
15	01 R 31	01 44	05 17
16	01 30	01 41	05 08
17	01 29	01 38	04 41
18	01 29	01 34	04 00
19	01 28	01 31	03 06
20	01 28	01 28	02 03
21	01 27	01 25	00 S 55
22	01 D 27	01 22	00 N 14
23	01 27	01 19	01 22
24	01 27	01 15	02 24
25	01 R 27	01 12	03 20
26	01 27	01 09	04 07
27	01 27	01 06	04 42
28	01 27	01 03	05 06
29	01 27	01 00	05 17
30	01 D 26	00 56	05 14
31	01 ♍ 27	00 ♍ 53	04 N 56

DECLINATIONS

Sun ☉ ° '	Moon ☽ ° '	Mercury ☿ ° '	Venus ♀ ° '	Mars ♂ ° '	Jupiter ♃ ° '	Saturn ♄ ° '	Uranus ♅ ° '	Neptune ♆ ° '	Pluto ♇ ° '
18 N 00	12 S 47	08 N 26	22 N 23	23 N 07	02 S 13	10 N 18	18 S 10	19 S 42	09 S 16
17 44	15 26	08 19	22 17	23 02	02 14	10 18	18 11	19 42	09 17
17 29	17 27	08 14	22 10	22 57	02 16	10 18	18 11	19 42	09 17
17 13	18 40	08 13	22 03	22 52	02 17	10 18	18 12	19 43	09 17
16 57	18 57	08 14	21 55	22 47	02 19	10 18	18 13	19 43	09 18
16 40	18 12	08 17	21 47	22 42	02 20	10 18	18 13	19 43	09 18
16 24	16 23	08 24	21 37	22 36	02 22	10 18	18 14	19 44	09 19
16 07	13 34	08 32	21 28	22 30	02 24	10 18	18 15	19 44	09 19
15 50	09 54	08 44	21 17	22 24	02 26	10 18	18 15	19 44	09 20
15 32	05 37	08 58	21 06	22 18	02 28	10 18	18 16	19 45	09 20
15 15	00 S 59	09 14	20 54	22 12	02 30	10 18	18 17	19 45	09 20
14 57	03 N 42	09 33	20 42	22 06	02 32	10 18	18 17	19 45	09 21
14 39	08 08	09 53	20 29	21 59	02 34	10 18	18 18	19 46	09 21
14 20	12 03	10 15	20 16	21 52	02 36	10 18	18 19	19 46	09 22
14 01	15 15	10 38	20 02	21 46	02 39	10 18	18 19	19 46	09 22
13 43	17 31	11 01	19 47	21 38	02 41	10 17	18 20	19 47	09 23
13 24	18 45	11 25	19 32	21 31	02 43	10 17	18 20	19 47	09 23
13 04	18 53	11 49	19 16	21 24	02 46	10 17	18 21	19 47	09 24
12 45	17 58	12 13	19 00	21 16	02 48	10 16	18 22	19 48	09 24
12 25	16 07	12 36	18 43	21 09	02 51	10 16	18 22	19 48	09 25
12 05	13 29	12 59	18 26	21 01	02 54	10 16	18 23	19 48	09 25
11 45	10 16	13 19	18 08	20 53	02 56	10 15	18 23	19 49	09 26
11 25	06 39	13 38	17 49	20 45	02 59	10 15	18 24	19 49	09 26
11 04	02 N 49	13 56	17 30	20 37	03 02	10 14	18 25	19 49	09 27
10 44	01 S 05	14 10	17 11	20 28	03 04	10 14	18 25	19 50	09 27
10 23	04 54	14 23	16 51	20 20	03 07	10 13	18 26	19 50	09 28
10 02	08 30	14 33	16 30	20 11	03 10	10 12	18 26	19 50	09 28
09 41	11 46	14 40	16 10	20 02	03 13	10 12	18 27	19 50	09 29
09 20	14 34	14 44	15 48	19 53	03 16	10 11	18 27	19 51	09 29
08 58	16 48	14 45	15 26	19 44	03 19	10 10	18 28	19 51	09 30
08 N 37	18 S 18	14 N 43	15 N 04	19 N 35	03 S 22	10 N 09	18 S 28	19 S 51	09 S 30

ZODIAC SIGN ENTRIES

Date	h m	Planets
02	07 48	☽ ♐
04	17 18	☽ ♑
06	23 31	☽ ♒
09	03 04	☽ ♓
11	05 10	☽ ♈
13	07 04	☽ ♉
13	09 19	♀ ♌
15	09 46	☽ ♊
17	13 55	☽ ♋
19	20 00	☽ ♌
20	19 16	♂ ♌
22	04 21	☽ ♍
23	00 13	♆ ♑
23	07 59	☉ ♍
24	15 02	☽ ♎
27	03 25	☽ ♏
29	15 55	☽ ♐

LATITUDES

Date	Mercury ☿ ° '	Venus ♀ ° '	Mars ♂ ° '	Jupiter ♃ ° '	Saturn ♄ ° '	Uranus ♅ ° '	Neptune ♆ ° '	Pluto ♇ ° '
01	03 S 55	00 S 09	00 N 49	01 S 25	02 S 31	00 S 41	00 N 21	12 N 06
04	04 25	00 S 01	00 51	01 26	02 32	00 41	00 21	12 05
07	04 45	00 N 07	00 52	01 27	02 33	00 41	00 21	12 03
10	04 52	00 15	00 54	01 28	02 34	00 41	00 21	12 02
13	04 44	00 23	00 55	01 29	02 34	00 41	00 21	12 01
16	04 18	00 30	00 57	01 29	02 35	00 41	00 21	11 59
19	03 38	00 37	00 58	01 30	02 36	00 41	00 21	11 58
22	02 49	00 44	01 00	01 31	02 37	00 41	00 21	11 56
25	01 55	00 50	01 01	01 31	02 38	00 41	00 20	11 55
28	01 01	00 56	01 02	01 32	02 38	00 41	00 20	11 53
31	00 S 12	01 N 01	01 N 04	01 S 32	02 S 39	00 S 41	00 N 20	11 N 52

DATA

Julian Date	2451027
Delta T	+63 seconds
Ayanamsa	23° 50' 07"
Synetic vernal point	05° ♓ 16' 52"
True obliquity of ecliptic	23° 26' 14"

MOON'S PHASES, APSIDES AND POSITIONS ☽

Date	h m	Phase	Longitude ° '	Eclipse Indicator
08	02 10	○	15 ♒ 21	
14	19 48	☾	21 ♉ 49	
22	02 03	●	28 ♌ 48	Annular
30	05 07	☽	06 ♐ 39	

Day	h m		
11	11 56	Perigee	
27	06 28	Apogee	
05	06 47	Max dec	18° S 59'
11	17 01	0N	
18	03 02	Max dec	18° N 57'
25	05 19	0S	

LONGITUDES

Date	Chiron ⚷ ° '	Ceres ⚳ ° '	Pallas ⚴ ° '	Juno ⚵ ° '	Vesta ⚶ ° '	Black Moon Lilith ⚸ ° '
01	12 ♏ 47	01 ♊ 08	01 ♈ 49	05 ♎ 33	11 ♋ 20	25 ♎ 32
11	13 ♏ 13	04 ♊ 01	00 ♈ 49	08 ♎ 31	15 ♋ 30	26 ♎ 39
21	13 ♏ 49	06 ♊ 37	29 ♓ 13	11 ♎ 36	19 ♋ 36	27 ♎ 46
31	14 ♏ 34	08 ♊ 53	27 ♓ 07	14 ♎ 49	23 ♋ 37	28 ♎ 53

ASPECTARIAN

h m Aspects

01 Saturday
02 03 ☽ △ ♀
06 33 ☽ △ ♂
09 04 ☽ Q ♆
10 13 ♀ Q ♄

02 Sunday
03 14 ☽ △ ♃
04 01 ☽ □ ☿
05 27 ☽ Q ♅
08 49 ☽ ✶ ♆
11 09 ☽ bQ ♀
13 58 ☽ bQ ♂
14 37 ☽ ⊼ ♄
18 16 ☽ ☌ ♇

03 Monday
02 19 ☽ ± ♄
04 36 ☽ △ ☉
04 48 ☽ ✶ ♅
06 35 ☽ ± ♀
07 12 ☉ ☍ ♅
08 26 ☽ ± ♂
12 21 ☽ # ☉
13 52 ☽ ∠ ♆
19 19 ☽ ⊼ ♀
19 39 ☽ bQ ♄
20 33 ☽ ⊼ ♂

04 Tuesday
00 30 ☽ ∥ ♅
01 41 ☿ ⊼ ♃
06 56 ☽ ⊥ ♆
09 18 ☽ ∠ ♅
11 30 ☽ bQ ☉
12 30 ☽ △ ☿
12 45 ☽ □ ♃
18 08 ☽ ⊻ ♆
21 30 ♀ ☌ ♂
23 50 ☽ △ ♄

05 Wednesday
02 00 ☽ ⊥ ♅
03 00 ♃ ⊥ ♄
03 10 ☽ ⊻ ♇
03 23 ☉ bQ ♃
05 32 ☽ ± ☉
10 42 ♀ bQ ♇
12 58 ☽ ⊻ ♅
14 09 ☽ ⊥ ♇
15 17 ☽ bQ ☿
17 20 ☽ ⊼ ☉
20 49 ☿ ⊥ ♀
21 40 ☽ Q ♃
21 54 ♂ bQ ♇

06 Thursday
06 22 ☽ ∠ ♇
06 48 ☽ ☍ ♂
06 55 ☽ ± ☿
08 03 ☿ ⊥ ♂
08 20 ☽ ☍ ♀
11 33 ☽ ∥ ♅
17 10 ☽ ⊼ ☿
18 59 ☽ ✶ ♃

07 Friday
00 12 ☽ ☌ ♆
05 46 ☽ □ ♄
08 49 ☽ ✶ ♇
11 49 ☽ # ☉
17 58 ☽ ☌ ♅
21 01 ☽ ∠ ♃

08 Saturday
02 10 ☽ ☍ ☉
05 31 ☽ Q ♇
12 21 ☽ ⊥ ♃
12 48 ☽ Q ♄
13 37 ☽ ⊼ ♂
17 27 ☽ ⊼ ♀
18 47 ☽ ☍ ☿
22 29 ☽ ⊻ ♃

09 Sunday
00 17 ☽ ± ♂
03 38 ☽ ⊻ ♆
03 47 ☿ ⊻ ♀
04 33 ☽ ± ♀
09 07 ☽ ✶ ♄
09 31 ☽ # ♄
12 00 ☽ □ ♇
13 43 ☽ ⊥ ♆
15 21 ☽ ∥ ♇
16 09 ☽ bQ ♂
18 27 ☽ # ☿
20 43 ☽ ⊻ ♅
21 04 ☽ bQ ♀

10 Monday
04 44 ☽ ∠ ♆
06 42 ☽ ⊥ ♅
08 25 ☽ ⊼ ☉
10 13 ☽ ∠ ♄
18 22 ☽ △ ♂
18 35 ☽ ⊼ ☿
19 07 ☽ ± ☉
20 45 ☿ ⊻ ♂
21 38 ☽ ∠ ♅

11 Tuesday
00 22 ☽ △ ♀
00 25 ☽ ☌ ♃
00 56 ♀ △ ♃
01 12 ☽ ⊥ ♄
04 02 ☽ ± ☿
04 18 ☽ ∥ ♃
05 39 ☽ ✶ ♆
11 07 ☽ bQ ☉
11 10 ☽ ⊻ ♄
13 58 ☽ △ ♇
18 12 ☽ bQ ☿
20 21 ☿ # ♇
22 29 ☽ ✶ ♅

12 Wednesday
01 32 ☽ Q ♆
05 57 ☽ # ♃
13 49 ☽ △ ☉
14 54 ☽ bQ ♇
17 48 ☽ △ ☿
18 23 ☽ Q ♅
22 34 ☽ □ ♂

13 Thursday
02 01 ☽ ⊻ ♃
05 54 ☽ ± ♇
06 52 ☽ □ ♀
07 28 ☽ □ ♆
12 00 ☽ ⊥ ♃
13 08 ☽ ☌ ♄
13 45 ♀ ☍ ♆
15 57 ☽ ⊼ ♇
19 07 ☽ # ♇
21 13 ☉ ± ♃
23 16 ☽ ∥ ☿
23 47 ☉ ☌ ☿

14 Friday
00 25 ☽ □ ♅
00 47 ☽ ∥ ♄
03 01 ☽ ∠ ♃
03 28 ☿ ± ♃
06 11 ☽ Q ♂
15 22 ☿ ∥ ♄
15 59 ☽ Q ♀
17 26 ☽ □ ☿
19 48 ☽ □ ☉

15 Saturday
02 57 ☽ ∥ ☉
03 31 ☽ ✶ ♂
04 18 ☽ ✶ ♃
10 04 ☽ △ ♆
14 23 ☽ ✶ ♀
15 59 ☽ ⊻ ♄
18 30 ♂ △ ♃
18 51 ☽ ☍ ♇
19 09 ♄ St R
22 35 ☽ Q ☿

16 Sunday
00 45 ☽ Q ♃
02 19 ☽ ⊥ ♄
03 25 ☽ △ ♅
04 56 ☽ Q ☉
06 10 ♇ St D
06 33 ☽ ∠ ♂
08 36 ♀ □ ♄
11 54 ☽ bQ ♆
12 31 ♀ # ♆
17 57 ☽ ∠ ♄
18 20 ☽ ✶ ☿
18 51 ☽ ∠ ♀
23 00 ☽ ⊥ ♂

17 Monday
01 24 ☽ # ♅
03 25 ☽ ✶ ☉
03 35 ☽ ± ♆
05 29 ☽ bQ ♅
07 56 ☽ □ ♃
10 04 ☽ ⊻ ♂
12 13 ☽ ⊥ ♀
14 09 ☽ ⊼ ♆
17 20 ♀ △ ♇
19 28 ☽ ∠ ☿
20 21 ☽ ✶ ♄
21 21 ☽ ± ♅
23 19 ☽ ⊼ ♇
23 55 ☽ ⊻ ♀

18 Tuesday
08 02 ☽ ⊼ ♅
08 03 ☽ ∠ ☉
10 03 ☽ ± ♇
10 49 ☽ ⊥ ☿
17 49 ☽ Q ♄
21 11 ☽ ⊻ ☿

19 Wednesday
01 32 ☽ ⊥ ☉
02 17 ☽ bQ ♇
04 35 ☽ # ♅
13 20 ☽ ⊻ ☉
13 24 ☽ △ ♃
14 05 ☉ ⊼ ♃
18 48 ☽ ☌ ♂
20 09 ☽ ☍ ♆

20 Thursday
02 40 ☽ □ ♄
05 36 ☉ ∥ ☿
05 48 ☽ △ ♇
12 07 ☽ ☌ ♀
14 42 ☽ ☍ ♅
16 55 ☽ bQ ♃
21 05 ♂ ☍ ♆

21 Friday
02 38 ☽ ☌ ☿
09 42 ☽ ± ♃
14 30 ♀ ☍ ♅
15 36 ♀ # ♅
15 39 ☽ ∥ ☿
21 01 ☽ ⊼ ♃

22 Saturday
00 09 ☽ ∥ ☉
02 03 ☽ ☌ ☉
04 23 ☽ ⊼ ♆
06 09 ☽ ⊻ ♂
11 01 ♀ bQ ♃
11 15 ☽ △ ♄
12 06 ☽ ∥ ♄
14 34 ☽ □ ♇
15 56 ☽ ± ♆
17 45 ☽ # ♇
18 21 ☽ ⊥ ♂
23 40 ☽ ⊼ ♅

23 Sunday
03 27 ☽ ⊻ ♀
07 48 ☉ ⊼ ♆
09 24 ☽ bQ ♆
11 20 ☽ ⊻ ☿
11 21 ☽ ± ♅
12 52 ☽ ∠ ♂
16 26 ☽ bQ ♄
16 30 ☽ ⊥ ♀
22 35 ☿ St D
23 06 ☽ ⊥ ☿

24 Monday
01 46 ☽ Q ♇
05 01 ☽ bQ ♅
06 54 ☽ ☍ ♃
10 11 ☽ ± ♄
10 42 ☽ # ♃
12 17 ☽ ∠ ♀
14 57 ☽ △ ♆
17 01 ☽ ∠ ☿
17 44 ☽ ⊻ ☉
20 13 ☽ ✶ ♂
22 07 ☽ ⊼ ♄

25 Tuesday
01 39 ☽ ✶ ♇
06 46 ☽ ⊥ ☉
10 51 ☽ △ ♅
21 39 ☽ Q ♂
21 46 ☽ ✶ ♀
23 35 ☽ ✶ ☿

26 Wednesday
00 35 ☽ ∥ ♃
02 29 ☽ ∠ ☉
07 05 ♂ □ ♄
07 50 ☽ ∠ ♇
18 34 ☽ ⊼ ♃
21 05 ☿ ☌ ♀
23 31 ☉ △ ♄
23 53 ☉ ∥ ♄

27 Thursday
00 33 ☽ Q ☿
00 50 ☽ Q ♀
02 04 ☽ ⊥ ♇
03 14 ☽ □ ♆
06 36 ☽ ± ♃
10 32 ☽ ☍ ♄
11 32 ☽ ✶ ☉
12 09 ☽ □ ♂
14 14 ☽ ⊻ ♇
18 53 ☽ ∥ ♇
21 53 ☽ # ☉
23 19 ☽ □ ♅

28 Friday
00 09 ☽ # ♄
00 40 ☽ bQ ♃
08 49 ☉ ⊻ ♂
13 58 ☽ Q ☉
14 38 ☽ □ ☿
15 36 ☽ Q ♆
17 28 ☽ □ ♀
20 47 ☉ □ ♇

29 Saturday
01 19 ☉ # ♇
02 58 ♂ △ ♇
04 11 ♀ ± ♃
06 36 ☽ △ ♃
09 44 ☉ ± ♆
11 28 ☽ Q ♅
13 33 ☽ # ☿
15 38 ☽ ✶ ♆
18 45 ♂ # ♆
22 41 ☽ # ♀
22 47 ☽ ⊼ ♄

30 Sunday
02 33 ☽ ☌ ♇
03 51 ☽ △ ♂
05 07 ☽ □ ☉
10 38 ☽ ± ♄
11 13 ☽ ✶ ♅
21 12 ☽ ∠ ♆
21 36 ☿ ± ♃

31 Monday
04 11 ☽ bQ ♄
06 14 ☽ △ ☿
10 46 ☽ bQ ♂
11 28 ☽ △ ♀
14 35 ☽ ⊥ ♆
16 16 ☽ ∠ ♅
16 16 ☽ ∥ ♅
16 57 ☽ □ ♃

All ephemeris data is given at 12.00 UT and the Moon's longitude is additionally given for 24.00 UT

SEPTEMBER 1998

LONGITUDES

Date	Sidereal time h m s	Sun ☉ ° ' "	Moon ☽ ° ' "	Moon ☽ 24.00 ° ' "	Mercury ☿ ° '	Venus ♀ ° '	Mars ♂ ° '	Jupiter ♃ ° '	Saturn ♄ ° '	Uranus ♅ ° '	Neptune ♆ ° '	Pluto ♇ ° '
01	10 41 48	08 ♍ 51 13	05 ♑ 06 17	11 ♑ 33 18	20 ♌ 45	23 ♌ 27	07 ♌ 30	24 ♓ 57	03 ♉ 23	09 ♒ 40	29 ♑ 48	05 ♐ 22
02	10 45 45	09 49 17	18 ♑ 06 07	24 ♑ 45 00	21 54	24 41	08 08	24 R 50	03 R 21	09 R 38	29 R 47	05 23
03	10 49 41	10 47 22	01 ♒ 30 05	08 ♒ 21 27	23 09	25 55	08 46	24 42	03 19	09 36	29 46	05 23
04	10 53 38	11 45 28	15 ♒ 18 58	22 ♒ 22 25	24 30	27 09	09 24	24 35	03 17	09 34	29 45	05 24
05	10 57 34	12 43 36	29 ♒ 31 25	06 ♓ 45 27	25 56	28 23	10 02	24 27	03 15	09 32	29 44	05 25
06	11 01 31	13 41 46	14 ♓ 03 49	21 ♓ 25 46	27 27	29 ♌ 37	10 40	24 19	03 13	09 30	29 42	05 25
07	11 05 27	14 39 57	28 ♓ 50 23	06 ♈ 16 43	29 ♌ 03	00 ♍ 51	11 18	24 12	03 11	09 29	29 41	05 26
08	11 09 24	15 38 10	13 ♈ 43 45	21 ♈ 10 28	00 ♍ 42	02 05	11 56	24 04	03 09	09 27	29 40	05 27
09	11 13 21	16 36 26	28 ♈ 35 56	05 ♉ 59 13	02 24	03 20	12 34	23 56	03 06	09 25	29 39	05 28
10	11 17 17	17 34 43	13 ♉ 19 33	20 ♉ 36 13	04 10	04 34	13 12	23 48	03 04	09 23	29 38	05 28
11	11 21 14	18 33 02	27 ♉ 48 41	04 ♊ 56 32	05 57	05 48	13 50	23 40	03 01	09 22	29 37	05 29
12	11 25 10	19 31 24	11 ♊ 59 30	18 ♊ 57 24	07 46	07 02	14 28	23 32	02 58	09 20	29 37	05 30
13	11 29 07	20 29 47	25 ♊ 50 13	02 ♋ 37 59	09 37	08 17	15 05	23 24	02 56	09 18	29 36	05 31
14	11 33 03	21 28 13	09 ♋ 20 51	15 ♋ 58 58	11 29	09 31	15 43	23 16	02 53	09 17	29 35	05 32
15	11 37 00	22 26 41	22 ♋ 32 37	29 ♋ 02 01	13 21	10 46	16 21	23 08	02 50	09 15	29 34	05 33
16	11 40 56	23 25 11	05 ♌ 27 28	11 ♌ 49 14	15 14	12 00	16 59	23 00	02 47	09 14	29 33	05 34
17	11 44 53	24 23 43	18 ♌ 07 36	24 ♌ 22 49	17 07	13 15	17 36	22 52	02 43	09 12	29 32	05 35
18	11 48 50	25 22 17	00 ♍ 35 10	06 ♍ 44 52	19 00	14 29	18 14	22 44	02 40	09 11	29 32	05 36
19	11 52 46	26 20 53	12 ♍ 52 09	18 ♍ 57 14	20 53	15 44	18 51	22 36	02 37	09 09	29 31	05 37
20	11 56 43	27 19 31	25 ♍ 00 19	01 ♎ 01 36	22 45	16 58	19 29	22 28	02 34	09 08	29 30	05 38
21	12 00 39	28 18 11	07 ♎ 01 18	12 ♎ 59 36	24 37	18 13	20 06	22 20	02 30	09 07	29 30	05 39
22	12 04 36	29 ♍ 16 53	18 ♎ 56 45	24 ♎ 52 59	26 28	19 27	20 44	22 13	02 27	09 06	29 29	05 41
23	12 08 32	00 ♎ 15 37	00 ♏ 48 33	06 ♏ 43 44	28 ♍ 18	20 42	21 21	22 05	02 23	09 04	29 28	05 42
24	12 12 29	01 14 23	12 ♏ 38 52	18 ♏ 34 16	00 ♎ 08	21 57	21 58	21 57	02 19	09 03	29 28	05 43
25	12 16 25	02 13 10	24 ♏ 30 21	00 ♐ 27 31	01 57	23 11	22 35	21 49	02 15	09 02	29 27	05 44
26	12 20 22	03 11 59	06 ♐ 26 13	12 ♐ 26 56	03 45	24 26	23 13	21 41	02 12	09 01	29 27	05 46
27	12 24 19	04 10 51	18 ♐ 30 11	24 ♐ 36 30	05 32	25 41	23 50	21 34	02 08	09 00	29 26	05 47
28	12 28 15	05 09 43	00 ♑ 46 26	07 ♑ 00 33	07 18	26 55	24 27	21 26	02 04	08 59	29 26	05 48
29	12 32 12	06 08 38	13 ♑ 19 23	19 ♑ 43 30	09 04	28 10	25 04	21 19	02 00	08 58	29 25	05 50
30	12 36 08	07 ♎ 07 34	26 ♑ 13 22	02 ♒ 49 28	10 ♎ 48	29 ♍ 25	25 ♌ 41	21 ♓ 11	01 ♉ 56	08 ♒ 57	29 ♑ 25	05 ♐ 51

Date	Moon True ☊ ° '	Moon Mean ☊ ° '	Moon ☽ Latitude ° '
01	01 ♍ 27	00 ♍ 50	04 N 24
02	01 D 28	00 47	03 38
03	01 29	00 44	02 39
04	01 29	00 40	01 28
05	01 30	00 37	00 N 11
06	01 R 30	00 34	01 S 09
07	01 29	00 31	02 26
08	01 27	00 28	03 33
09	01 25	00 25	04 26
10	01 24	00 21	05 00
11	01 22	00 18	05 15
12	01 21	00 15	05 10
13	01 D 21	00 12	04 48
14	01 22	00 09	04 09
15	01 23	00 05	03 18
16	01 25	00 ♍ 02	02 18
17	01 26	29 ♌ 59	01 13
18	01 27	29 56	00 S 05
19	01 R 26	29 53	01 N 02
20	01 24	29 50	02 06
21	01 21	29 46	03 03
22	01 17	29 43	03 51
23	01 12	29 40	04 30
24	01 06	29 37	04 56
25	01 01	29 34	05 10
26	00 57	29 31	05 11
27	00 55	29 27	04 58
28	00 53	29 24	04 31
29	00 D 53	29 21	03 51
30	00 ♍ 55	29 ♌ 18	02 N 58

DECLINATIONS

Date	Sun ☉ ° '	Moon ☽ ° '	Mercury ☿ ° '	Venus ♀ ° '	Mars ♂ ° '	Jupiter ♃ ° '	Saturn ♄ ° '	Uranus ♅ ° '	Neptune ♆ ° '	Pluto ♇ ° '
01	08 N 15	18 S 56	14 N 38	14 N 42	19 N 26	03 S 25	10 N 09	18 S 29	19 S 52	09 S 31
02	07 53	18 37	14 29	14 19	19 16	03 28	10 08	18 29	19 52	09 32
03	07 31	17 15	14 17	13 55	19 07	03 31	10 07	18 30	19 52	09 32
04	07 09	14 50	14 01	13 31	18 57	03 34	10 06	18 30	19 52	09 33
05	06 47	11 28	13 43	13 07	18 47	03 37	10 05	18 31	19 53	09 33
06	06 25	07 20	13 21	12 43	18 37	03 41	10 04	18 31	19 53	09 34
07	06 02	02 S 41	12 57	12 18	18 27	03 44	10 03	18 32	19 53	09 35
08	05 40	02 N 09	12 29	11 52	18 17	03 47	10 02	18 32	19 53	09 35
09	05 17	06 50	11 59	11 27	18 07	03 50	10 01	18 33	19 54	09 36
10	04 54	11 04	11 26	11 01	17 57	03 53	10 00	18 33	19 54	09 36
11	04 32	14 33	10 51	10 35	17 46	03 57	09 59	18 34	19 54	09 37
12	04 09	17 06	10 14	10 08	17 36	04 00	09 58	18 34	19 54	09 38
13	03 46	18 35	09 35	09 41	17 25	04 03	09 57	18 35	19 54	09 38
14	03 23	18 58	08 55	09 14	17 14	04 06	09 56	18 35	19 55	09 39
15	03 00	18 17	08 13	08 47	17 03	04 09	09 55	18 35	19 55	09 39
16	02 37	16 40	07 30	08 19	16 52	04 13	09 53	18 36	19 55	09 40
17	02 14	14 15	06 46	07 51	16 41	04 16	09 52	18 36	19 55	09 41
18	01 50	11 11	06 00	07 23	16 30	04 19	09 51	18 36	19 55	09 41
19	01 27	07 41	05 15	06 55	16 19	04 22	09 50	18 37	19 55	09 42
20	01 04	03 55	04 28	06 26	16 08	04 25	09 48	18 37	19 56	09 42
21	00 40	00 N 01	03 42	05 58	15 56	04 28	09 47	18 37	19 56	09 43
22	00 N 17	03 S 51	02 55	05 29	15 45	04 32	09 46	18 38	19 56	09 44
23	00 S 06	07 32	02 07	05 00	15 33	04 35	09 44	18 38	19 56	09 44
24	00 30	10 56	01 20	04 31	15 21	04 38	09 43	18 38	19 56	09 45
25	00 53	13 53	00 N 33	04 01	15 10	04 41	09 41	18 39	19 56	09 46
26	01 16	16 17	00 S 15	03 32	14 58	04 44	09 40	18 39	19 56	09 46
27	01 40	18 00	01 02	03 02	14 46	04 47	09 38	18 39	19 57	09 47
28	02 03	18 55	01 49	02 33	14 34	04 50	09 37	18 39	19 57	09 47
29	02 26	18 57	02 35	02 03	14 22	04 53	09 35	18 40	19 57	09 48
30	02 S 50	18 S 00	03 S 22	01 N 33	14 N 10	04 S 55	09 N 34	18 S 40	19 S 57	09 S 49

ZODIAC SIGN ENTRIES

Date	h	m	Planets
01	02	23	☽ ♑
03	09	21	☽ ♒
05	12	48	☽ ♓
06	19	24	♀ ♍
07	13	52	☽ ♈
08	01	58	☿ ♍
09	14	16	☽ ♉
11	15	40	☽ ♊
13	19	20	☽ ♋
16	01	48	☽ ♌
18	10	52	☽ ♍
20	21	57	☽ ♎
23	05	37	☉ ♎
23	10	22	☽ ♏
24	10	13	☿ ♎
25	23	05	☽ ♐
28	10	30	☽ ♑
30	18	53	☽ ♒
30	23	13	♀ ♎

LATITUDES

Date	Mercury ☿ ° '	Venus ♀ ° '	Mars ♂ ° '	Jupiter ♃ ° '	Saturn ♄ ° '	Uranus ♅ ° '	Neptune ♆ ° '	Pluto ♇ ° '
01	00 N 03	01 N 03	01 N 04	01 S 32	02 S 39	00 S 41	00 N 20	11 N 51
04	00 43	01 08	01 06	01 33	02 40	00 41	00 20	11 50
07	01 13	01 12	01 07	01 33	02 41	00 41	00 20	11 48
10	01 34	01 16	01 08	01 33	02 41	00 41	00 20	11 47
13	01 45	01 19	01 10	01 34	02 42	00 41	00 20	11 46
16	01 49	01 22	01 11	01 34	02 42	00 40	00 20	11 44
19	01 46	01 24	01 12	01 34	02 43	00 40	00 20	11 43
22	01 38	01 25	01 14	01 34	02 44	00 40	00 20	11 42
25	01 26	01 26	01 15	01 34	02 44	00 40	00 20	11 40
28	01 11	01 26	01 16	01 34	02 44	00 40	00 20	11 39
31	00 N 54	01 N 26	01 N 17	01 S 33	02 S 45	00 S 40	00 N 20	11 N 38

DATA

Julian Date	2451058
Delta T	+63 seconds
Ayanamsa	23° 50' 11"
Synetic vernal point	05° ♓ 16' 49"
True obliquity of ecliptic	23° 26' 14"

MOON'S PHASES, APSIDES AND POSITIONS ☽

Date	h	m	Phase	Longitude ° '	Eclipse Indicator
06	11	21	○	13 ♓ 40	
13	01	58	☾	20 ♊ 05	
20	17	01	●	27 ♍ 32	
28	21	11	☽	05 ♑ 32	

Day	h	m		
08	06	08	Perigee	
23	22	03	Apogee	
01	16	14	Max dec	18° S 57'
08	01	22	ON	
14	08	28	Max dec	18° N 59'
21	12	05	OS	
29	00	39	Max dec	19° S 03'

LONGITUDES

Date	Chiron ⚷ ° '	Ceres ⚳ ° '	Pallas ⚴ ° '	Juno ⚵ ° '	Vesta ⚶ ° '	Black Moon Lilith ⚸ ° '
01	14 ♏ 39	09 ♊ 06	26 ♓ 53	15 ♎ 08	24 ♋ 01	29 ♎ 00
11	15 ♏ 33	10 ♊ 57	24 ♓ 23	18 ♎ 26	27 ♋ 55	00 ♏ 07
21	16 ♏ 34	12 ♊ 21	21 ♓ 43	21 ♎ 48	01 ♌ 41	01 ♏ 14
31	17 ♏ 40	13 ♊ 14	19 ♓ 09	25 ♎ 13	05 ♌ 18	02 ♏ 21

All ephemeris data is given at 12.00 UT and the Moon's longitude is additionally given for 24.00 UT

ASPECTARIAN

h m	Aspects
01 Tuesday	
02 01	☽ ⊻ ♆
04 51	☽ ± ♂
08 47	☽ △ ♄
09 19	☽ ⊥ ♅
12 30	☽ ⊻ ♇
13 19	☽ ⚼ ☿
16 42	☽ ⊼ ♂
18 22	☿ ∥ ♀
18 54	☽ ⚼ ♀
19 34	☽ △ ☉
20 29	☽ ⊻ ♅
23 40	☽ ⊥ ♇
02 Wednesday	
02 26	☽ Q ♃
07 31	☉ ⊼ ♅
07 34	☽ ± ☿
13 10	☽ ± ♀
14 36	♀ ⊼ ♃
15 08	☽ ∥ ♅
16 08	☽ ∠ ♇
19 34	☽ ⊼ ☿
03 Thursday	
00 02	☽ ∗ ♃
01 04	☽ ⚼ ☉
01 05	☽ ⊼ ♀
08 56	☽ ☌ ♆
15 12	☽ □ ♄
18 13	☽ ± ☉
18 50	☽ ∗ ♇
04 Friday	
01 20	☽ ☍ ♂
02 08	☽ ☌ ♅
02 12	☽ ∠ ♃
05 26	☽ ⊼ ☉
13 17	☿ ⊼ ♃
13 56	♃ ∠ ♅
15 34	☽ Q ♇
17 29	♂ ⚼ ♃
17 31	☽ ⊥ ♃
18 00	♂ ☍ ♅
19 01	☽ ⋕ ☿
22 09	☽ Q ♄
23 30	☽ ⋕ ♀
05 Saturday	
03 35	☽ ⊻ ♃
05 19	☽ ☍ ☿
09 55	☽ ☍ ♀
12 20	☽ ⊻ ♆
18 12	☽ ∗ ♄
20 29	☽ ⋕ ♄
21 47	☽ □ ♇
22 17	☽ ⊥ ♆
23 32	☽ ∥ ♇
06 Sunday	
04 33	☽ ⊻ ♅
06 11	☽ ⊼ ♂
11 21	☽ ☍ ☉
13 03	☽ ∠ ♆
13 42	♀ ⊼ ♆
14 21	☽ ⊥ ♅
16 27	☽ ± ♂
17 22	☽ ⋕ ☉
18 46	☽ ∠ ♄
07 Monday	
01 48	♂ ⋕ ♅
04 33	☽ ☌ ♃
04 57	☽ ∠ ♅
06 49	☽ ∥ ♃
07 43	☽ ⚼ ♂
09 20	☽ ⊥ ♄
12 22	☽ ⊼ ☿
12 35	☉ ⚼ ♆
13 22	☽ ∗ ♆
15 33	☽ ⊼ ♀
19 00	☽ ⊻ ♄
21 22	☿ ⊼ ♆
22 39	☽ △ ♇
23 15	☽ ± ☿
08 Tuesday	
02 06	☽ ± ♀
05 07	☽ ∗ ♅
07 28	☉ ± ♅
08 42	☽ Q ♆
08 59	☽ △ ♂
15 17	☽ ⊼ ☉
15 35	☽ ⚼ ☿
17 54	☽ ⚼ ♀
20 18	☽ ⋕ ♃
22 50	☽ ⚼ ♇
09 Wednesday	
00 25	☽ Q ♅
01 38	☽ ± ☉
04 30	☽ ∥ ☉
04 31	☽ ⊻ ♃
07 48	♀ △ ♄
13 24	☽ ± ♇
13 43	☽ □ ♆
14 09	☽ ⊥ ♃
17 13	☽ ⚼ ☉
19 00	☽ △ ☿
19 17	☽ ☌ ♄
20 22	☽ △ ♀
21 23	☿ △ ♄
23 09	☽ ⊼ ♇
10 Thursday	
03 19	☽ ⋕ ♇
04 39	☽ ∠ ♃
05 34	☽ □ ♅
05 41	☽ ∥ ♄
11 45	☽ ∥ ♀
11 47	☽ □ ♂
14 03	☽ ∥ ☿
19 30	☽ △ ☉
23 25	☉ ⚼ ♄
11 Friday	
05 09	☽ ∗ ♃
05 40	☿ ☌ ♀
05 46	☿ □ ♇
05 48	♀ □ ♇
07 41	☿ ± ♆
08 36	♀ ± ♆
15 02	☽ △ ♆
19 04	☽ Q ♂
20 43	☽ ⊻ ♄
12 Saturday	
00 56	☽ ☍ ♇
01 07	☽ Q ♃
02 45	☽ □ ♀
03 44	☽ □ ☿
06 52	☽ ⊥ ♄
07 28	☽ △ ♅
16 26	☽ ∗ ♂
16 29	☽ ⚼ ♆
17 45	☽ ∥ ♂
20 21	☉ ⋕ ♃
21 17	♀ ∥ ♄
22 16	☽ ∠ ♄
22 24	☿ ∥ ♄
13 Sunday	
00 48	☿ ∥ ♀
01 58	☽ □ ☉
07 47	☽ □ ♃
08 01	☿ ⊼ ♅
08 05	☽ ± ♆
09 19	☽ ⚼ ♅
10 21	☿ ⋕ ♇
11 51	☽ ⋕ ♅
12 51	☽ Q ♀
14 38	♀ ⋕ ♇
15 38	☽ Q ☿
18 37	☽ ⊼ ♆
19 52	☽ ∠ ♂
14 Monday	
00 29	☽ ∗ ♄
01 10	☽ ± ♅
05 09	☽ ⊼ ♇
07 27	♀ ⊼ ♅
11 53	☽ ⊼ ♅
12 14	☽ Q ☉
12 21	☽ ∗ ♀
12 42	☽ ⊥ ♂
15 57	☽ ± ♇
16 28	☽ ∗ ☿
21 57	☽ Q ♄
15 Tuesday	
00 06	☽ ⊻ ♂
05 08	☽ ⋕ ♅
05 19	☉ ⊥ ♂
08 20	☽ ⚼ ♇
11 48	☽ ∗ ☉
13 05	☽ △ ♃
18 34	☽ ∠ ♀
16 Wednesday	
00 34	☽ ∠ ☿
00 59	☽ ☍ ♆
03 02	☉ ☍ ♃
03 21	☿ ⚼ ♆
07 00	☽ □ ♄
09 16	☽ ∥ ♂
11 54	☿ ± ♅
12 12	☽ △ ♇
12 57	♂ ± ♃
13 08	☽ ⊥ ♀
15 40	☉ Q ♇
16 45	☽ ⚼ ♃
18 02	☽ ∠ ☉
19 05	☽ ☍ ♅
20 21	☽ ⊥ ☿
17 Thursday	
01 41	☽ ⊻ ♀
07 26	☉ ⚼ ♅
09 38	☽ ± ♃
09 44	☽ ⊻ ☿
10 57	☽ ☌ ♂
12 33	☽ ⊥ ☉
19 29	☿ ⚼ ♄
21 00	☽ ⊼ ♃
21 12	☿ ⊻ ♂
18 Friday	
01 03	☽ ⊻ ☉
09 57	☽ ⊼ ♆
12 51	♀ ⚼ ♆
16 02	☽ △ ♄
21 34	☽ ∥ ♄
21 37	☽ ± ♆
21 47	☽ □ ♇
22 35	☽ ⋕ ♇
19 Saturday	
01 13	♀ ± ♅
04 44	☽ ⊼ ♅
15 14	☽ ⚼ ♆
16 30	☽ ± ♅
17 46	☽ ∥ ♀
18 13	☉ ± ♄
18 16	☽ ☌ ♀
21 19	☽ ⚼ ♄
20 Sunday	
00 27	☽ ⊻ ♂
06 42	☽ ☌ ☿
07 02	☽ ☍ ♃
07 37	☽ ∥ ☿
08 39	☿ ☍ ♃
08 51	☽ ⋕ ♃
09 17	☽ Q ♇
10 16	☽ ⚼ ♅
13 00	☽ ⊥ ♂
13 28	☿ ⋕ ♃
15 05	☽ ± ♄
17 01	☽ ☌ ☉
20 57	☽ △ ♆
22 53	♀ ⚼ ♄
23 31	☿ Q ♇
21 Monday	
02 59	☽ ⊼ ♄
05 36	☿ ⚼ ♅
07 29	☽ ∥ ☉
07 57	☽ ∠ ♂
09 16	☽ ∗ ♇
15 51	☽ ⋕ ☉
16 11	☽ △ ♅
22 Tuesday	
07 06	☽ ⋕ ☿
11 41	☿ ± ♄
13 09	☽ ⊻ ♀
15 30	☽ ∠ ♇
15 48	☽ ∗ ♂
16 21	☽ ∥ ♃
16 54	☉ △ ♆
17 04	☿ ⊥ ♂
18 31	☽ ⊼ ♃
21 13	☽ ⋕ ♀
23 Wednesday	
02 42	☽ ⊥ ♀
06 00	☽ ⊻ ☿
06 32	☽ ± ♃
09 18	☽ □ ♆
09 45	☽ ⊥ ♇
10 17	♄ ⋕ ♇
10 47	☽ ⊻ ☉
15 10	☽ ☍ ♄
17 26	☽ Q ♂
20 23	☽ ⊥ ☿
21 56	☽ ⊻ ♇
23 05	☽ ∠ ♀
24 Thursday	
00 03	☽ ⊥ ☉
00 34	☽ ⚼ ♃
03 10	☽ ⋕ ♄
03 12	☿ △ ♆
03 20	☽ ∥ ♇
04 43	☽ □ ♅
06 42	♀ ⋕ ♃
11 19	♂ ⊼ ♃
12 05	♀ ☍ ♃
13 01	♀ ⊻ ♂
17 57	☽ ∠ ☿
19 56	☽ ∠ ☉
21 45	☽ Q ♆
25 Friday	
05 06	☉ ⋕ ☿
06 38	☽ △ ♃
07 55	☽ □ ♂
09 02	☽ ∗ ♀
12 53	☉ ⊼ ♄
15 57	☿ ⊼ ♄
17 06	☽ Q ♅
19 50	☉ ☌ ☿
21 58	☽ ∗ ♆
22 50	♀ Q ♇
23 09	☽ ⋕ ♂
26 Saturday	
03 32	☽ ⊼ ♄
04 03	♀ ⚼ ♅
04 56	☽ ∗ ☉
05 40	☽ ∗ ☿
10 39	☽ ☌ ♇
12 00	☽ Q ♀
15 30	☽ ± ♄
17 09	☽ ∗ ♅
27 Sunday	
03 58	☽ ∠ ♆
07 00	☽ Q ☉
09 18	☽ ⚼ ♄
09 46	☽ Q ☿
15 24	☿ ∗ ♇
17 58	☽ □ ♃
20 15	♀ ± ♄
21 42	☽ ⊥ ♆
22 47	☽ ∠ ♅
23 02	☽ △ ♂
28 Monday	
02 24	☽ ∥ ♅
03 41	☽ □ ♀
09 24	☽ ⊻ ♆
14 29	☽ △ ♄
16 15	☽ ⊥ ♅
21 11	☽ □ ☉
21 43	☽ ⊻ ♇
29 Tuesday	
01 22	☉ ⋕ ♀
01 46	☿ ⋕ ♀
02 37	☽ □ ☿
02 43	☉ ∥ ☿
03 45	☽ ⊻ ♅
04 09	☉ ∗ ♇
04 28	☽ Q ♃
05 31	☽ ⚼ ♂
09 10	☽ ⊥ ♇
10 39	☿ △ ♅
22 12	☽ ∥ ♅
23 19	☽ ± ♂
30 Wednesday	
02 05	☽ ∠ ♇
02 49	☽ ∗ ♃
09 28	☿ ∠ ♂
10 58	☽ ⊼ ♂
12 01	♀ △ ♆
17 50	☽ ☌ ♆
18 26	☽ △ ♀
22 20	☽ □ ♄

OCTOBER 1998

Date	Sidereal time h m s	Sun ☉	Moon ☽	Moon ☽ 24.00	Mercury ☿	Venus ♀	Mars ♂	Jupiter ♃	Saturn ♄	Uranus ♅	Neptune ♆	Pluto ♇
01	12 40 05	08 ≏ 06 32	09 ♒ 32 09	16 ♒ 21 42	12 ≏ 32	00 ≏ 40	26 ♌ 18	21 ♓ 04	01 ♉ 52	08 ♒ 56	29 ♑ 25	05 ♐ 53
02	12 44 01	09 05 32	23 ♒ 18 16	00 ♓ 21 52	14 15	01 55	26 55	20 R 57	01 R 47	08 R 55	29 R 24	05 54
03	12 47 58	10 04 34	07 ♓ 32 18	14 ♓ 49 12	15 56	03 09	27 32	20 50	01 43	08 54	29 24	05 56
04	12 51 54	11 03 37	22 ♓ 12 02	29 ♓ 40 00	17 37	04 24	28 09	20 43	01 39	08 54	29 24	05 57
05	12 55 51	12 02 42	07 ♈ 12 09	14 ♈ 47 21	19 17	05 39	28 46	20 36	01 35	08 53	29 24	05 59
06	12 59 48	13 01 50	22 ♈ 24 20	00 ♉ 01 47	20 57	06 54	29 23	20 29	01 30	08 52	29 23	06 00
07	13 03 44	14 00 59	07 ♉ 38 21	15 ♉ 12 43	22 35	08 09	29 ♌ 59	20 22	01 26	08 52	29 23	06 02
08	13 07 41	15 00 11	22 ♉ 43 41	00 ♊ 10 11	24 13	09 24	00 ♍ 36	20 16	01 21	08 51	29 23	06 04
09	13 11 37	15 59 24	07 ♊ 31 21	14 ♊ 46 29	25 50	10 39	01 13	20 09	01 17	08 51	29 23	06 05
10	13 15 34	16 58 41	21 ♊ 55 09	28 ♊ 57 05	27 26	11 54	01 49	20 03	01 12	08 50	29 23	06 07
11	13 19 30	17 57 59	05 ♋ 52 13	12 ♋ 40 37	29 ≏ 02	13 09	02 26	19 57	01 08	08 50	29 23	06 09
12	13 23 27	18 57 20	19 ♋ 22 32	25 ♋ 58 17	00 ♏ 36	14 24	03 02	19 51	01 03	08 50	29 D 23	06 10
13	13 27 23	19 56 43	02 ♌ 28 17	08 ♌ 53 00	02 11	15 39	03 39	19 45	00 58	08 50	29 23	06 12
14	13 31 20	20 56 09	15 ♌ 12 55	21 ♌ 28 34	03 44	16 54	04 15	19 39	00 54	08 49	29 23	06 14
15	13 35 17	21 55 37	27 ♌ 40 27	03 ♍ 49 03	05 17	18 09	04 52	19 33	00 49	08 49	29 23	06 16
16	13 39 13	22 55 07	09 ♍ 54 49	15 ♍ 58 12	06 49	19 24	05 28	19 28	00 44	08 49	29 23	06 18
17	13 43 10	23 54 39	21 ♍ 59 35	27 ♍ 59 20	08 20	20 39	06 04	19 23	00 40	08 49	29 24	06 19
18	13 47 06	24 54 13	03 ≏ 57 46	09 ≏ 55 09	09 51	21 54	06 41	19 18	00 35	08 49	29 24	06 21
19	13 51 03	25 53 49	15 ≏ 51 46	21 ≏ 47 49	11 21	23 09	07 17	19 13	00 30	08 D 49	29 24	06 23
20	13 54 59	26 53 28	27 ≏ 43 31	03 ♏ 39 02	12 50	24 25	07 53	19 08	00 25	08 49	29 24	06 25
21	13 58 56	27 53 08	09 ♏ 34 35	15 ♏ 30 19	14 19	25 40	08 29	19 03	00 21	08 49	29 25	06 27
22	14 02 52	28 52 51	21 ♏ 26 27	27 ♏ 23 12	15 47	26 55	09 05	18 59	00 16	08 49	29 25	06 29
23	14 06 49	29 ≏ 52 35	03 ♐ 20 46	09 ♐ 19 27	17 14	28 10	09 41	18 54	00 11	08 49	29 25	06 31
24	14 10 46	00 ♏ 52 22	15 ♐ 19 31	21 ♐ 21 19	18 41	29 ≏ 25	10 17	18 50	00 06	08 49	29 26	06 33
25	14 14 42	01 52 10	27 ♐ 25 14	03 ♑ 31 41	20 07	00 ♏ 40	10 53	18 46	00 ♉ 01	08 50	29 26	06 35
26	14 18 39	02 51 59	09 ♑ 41 06	15 ♑ 54 00	21 33	01 56	11 29	18 43	29 ♈ 57	08 50	29 27	06 37
27	14 22 35	03 51 52	22 ♑ 10 51	28 ♑ 32 14	22 57	03 11	12 04	18 39	29 52	08 50	29 27	06 39
28	14 26 32	04 51 44	04 ♒ 58 39	11 ♒ 30 37	24 21	04 26	12 40	18 36	29 47	08 51	29 28	06 41
29	14 30 28	05 51 39	18 ♒ 08 39	24 ♒ 53 09	25 44	05 41	13 16	18 33	29 42	08 51	29 28	06 43
30	14 34 25	06 51 35	01 ♓ 45 29	08 ♓ 42 52	27 06	06 56	13 51	18 30	29 37	08 52	29 29	06 45
31	14 38 21	07 ♏ 51 33	15 ♓ 48 23	23 ♓ 00 56	28 ♏ 27	08 ♏ 12	14 ♍ 27	18 ♓ 27	29 ♈ 33	08 ♒ 53	29 ♑ 30	06 ♐ 47

Date	Moon True ☊	Moon Mean ☊	Moon ☽ Latitude
01	00 ♍ 56	29 ♌ 15	01 N 54
02	00 D 57	29 11	00 N 42
03	00 R 57	29 08	00 S 36
04	00 56	29 05	01 53
05	00 52	29 02	03 04
06	00 47	28 59	04 03
07	00 40	28 56	04 44
08	00 34	28 52	05 06
09	00 28	28 49	05 06
10	00 24	28 46	04 47
11	00 21	28 43	04 12
12	00 D 20	28 40	03 23
13	00 21	28 37	02 25
14	00 22	28 33	01 21
15	00 23	28 30	00 S 15
16	00 R 22	28 27	00 N 51
17	00 20	28 24	01 53
18	00 15	28 21	02 50
19	00 07	28 17	03 39
20	29 ♌ 58	28 14	04 18
21	29 46	28 11	04 45
22	29 34	28 08	05 00
23	29 23	28 05	05 03
24	29 13	28 02	04 52
25	29 05	27 58	04 28
26	29 00	27 55	03 51
27	28 58	27 52	03 03
28	28 D 57	27 49	02 04
29	28 57	27 46	00 N 57
30	28 R 57	27 43	00 S 15
31	28 ♌ 56	27 ♌ 39	01 S 28

DECLINATIONS

Date	Sun ☉	Moon ☽	Mercury ☿	Venus ♀	Mars ♂	Jupiter ♃	Saturn ♄	Uranus ♅	Neptune ♆	Pluto ♇
01	03 S 13	16 S 02	04 S 08	01 N 03	13 N 58	04 S 58	09 N 32	18 S 40	19 S 57	09 S 49
02	03 36	13 06	04 53	00 33	13 45	05 01	09 31	18 40	19 57	09 50
03	03 59	09 18	05 38	00 N 03	13 33	05 04	09 29	18 40	19 57	09 51
04	04 23	04 S 49	06 23	00 S 27	13 21	05 07	09 28	18 41	19 57	09 51
05	04 46	00 N 03	07 07	00 57	13 08	05 09	09 26	18 41	19 57	09 52
06	05 09	04 58	07 51	01 27	12 56	05 12	09 25	18 41	19 57	09 53
07	05 32	09 34	08 34	01 57	12 43	05 14	09 23	18 41	19 57	09 53
08	05 55	13 31	09 16	02 27	12 31	05 17	09 21	18 41	19 57	09 54
09	06 17	16 32	09 58	02 57	12 18	05 19	09 20	18 41	19 58	09 55
10	06 40	18 25	10 40	03 27	12 05	05 22	09 18	18 41	19 58	09 55
11	07 03	19 07	11 20	03 57	11 53	05 24	09 17	18 41	19 58	09 56
12	07 25	18 41	12 00	04 27	11 40	05 26	09 15	18 41	19 58	09 56
13	07 48	17 15	12 39	04 57	11 27	05 28	09 13	18 42	19 58	09 57
14	08 10	14 59	13 18	05 26	11 14	05 30	09 12	18 42	19 58	09 58
15	08 32	12 03	13 56	05 56	11 01	05 33	09 10	18 42	19 58	09 58
16	08 55	08 38	14 33	06 25	10 48	05 35	09 08	18 42	19 58	09 59
17	09 17	04 55	15 09	06 55	10 35	05 36	09 07	18 42	19 58	10 00
18	09 39	01 N 02	15 44	07 24	10 22	05 38	09 05	18 42	19 58	10 00
19	10 00	02 S 53	16 19	07 53	10 09	05 40	09 03	18 42	19 58	10 01
20	10 22	06 39	16 53	08 22	09 56	05 42	09 02	18 42	19 57	10 01
21	10 43	10 13	17 26	08 50	09 43	05 43	09 00	18 41	19 57	10 02
22	11 05	13 17	17 58	09 19	09 30	05 45	08 58	18 41	19 57	10 03
23	11 26	15 52	18 29	09 47	09 16	05 47	08 57	18 41	19 57	10 03
24	11 47	17 48	18 59	10 15	09 03	05 48	08 55	18 41	19 57	10 04
25	12 07	18 57	19 28	10 43	08 50	05 49	08 53	18 41	19 57	10 05
26	12 28	19 15	19 57	11 11	08 37	05 51	08 52	18 41	19 57	10 05
27	12 48	18 36	20 24	11 38	08 23	05 52	08 50	18 41	19 57	10 06
28	13 08	17 01	20 50	12 05	08 10	05 53	08 48	18 41	19 57	10 06
29	13 28	14 29	21 15	12 32	07 57	05 54	08 47	18 41	19 57	10 07
30	13 48	11 05	21 40	12 58	07 43	05 55	08 45	18 40	19 57	10 08
31	14 S 08	06 S 57	22 S 03	13 S 24	07 N 30	05 S 56	08 N 44	18 S 40	19 S 57	10 S 08

ZODIAC SIGN ENTRIES

Date	h	m	Planets
02	23	23	☽ ♓
05	00	32	☽ ♈
06	23	57	☽ ♉
07	12	28	♂ ♍
08	23	43	☽ ♊
11	01	48	☽ ♋
12	02	44	☿ ♏
13	07	25	☽ ♌
15	16	32	☽ ♍
18	04	02	☽ ≏
20	16	36	☽ ♏
23	05	16	☽ ♐
23	14	59	☉ ♏
24	23	06	♀ ♏
25	17	05	☽ ♑
25	18	41	♄ ♈
28	02	44	☽ ♒
30	08	58	☽ ♓

LATITUDES

Date	Mercury ☿	Venus ♀	Mars ♂	Jupiter ♃	Saturn ♄	Uranus ♅	Neptune ♆	Pluto ♇
01	00 N 54	01 N 26	01 N 17	01 S 33	02 S 45	00 S 40	00 N 20	11 N 38
04	00 35	01 25	01 19	01 33	02 45	00 40	00 20	11 37
07	00 N 14	01 24	01 20	01 33	02 45	00 40	00 20	11 35
10	00 S 06	01 22	01 21	01 33	02 46	00 40	00 20	11 34
13	00 27	01 19	01 22	01 32	02 46	00 40	00 19	11 33
16	00 48	01 16	01 24	01 32	02 46	00 40	00 19	11 32
19	01 08	01 12	01 25	01 31	02 46	00 40	00 19	11 31
22	01 28	01 08	01 26	01 31	02 46	00 40	00 19	11 30
25	01 46	01 03	01 27	01 30	02 46	00 39	00 19	11 29
28	02 03	00 58	01 29	01 30	02 46	00 39	00 19	11 28
31	02 S 17	00 N 53	01 N 30	01 S 29	02 S 46	00 S 39	00 N 19	11 N 28

DATA

Julian Date	2451088
Delta T	+63 seconds
Ayanamsa	23° 50' 13"
Synetic vernal point	05° ♓ 16' 46"
True obliquity of ecliptic	23° 26' 15"

MOON'S PHASES, APSIDES AND POSITIONS ☽

Date	h	m	Phase	Longitude	Eclipse Indicator
05	20	12	○	12 ♈ 23	
12	11	11	☾	18 ♋ 55	
20	10	09	●	26 ≏ 49	
28	11	46	☽	04 ♒ 51	

Day	h	m		
06	13	08	Perigee	
21	05	31	Apogee	
05	11	47	ON	
11	14	39	Max dec	19° N 07'
18	18	17	OS	
26	07	44	Max dec	19° S 16'

LONGITUDES

Date	Chiron ⚷	Ceres ⚳	Pallas ⚴	Juno ⚵	Vesta ⚶	Black Moon Lilith ⚸
01	17 ♏ 40	13 ♊ 14	19 ♓ 09	25 ≏ 13	05 ♌ 18	02 ♏ 21
11	18 ♏ 52	13 ♊ 31	16 ♓ 54	28 ≏ 40	08 ♌ 43	03 ♏ 28
21	20 ♏ 07	13 ♊ 09	15 ♓ 09	02 ♏ 08	11 ♌ 54	04 ♏ 35
31	21 ♏ 25	12 ♊ 08	14 ♓ 00	05 ♏ 36	14 ♌ 48	05 ♏ 42

ASPECTARIAN

01 Thursday
05 28 ☽ ⚹ ♇
05 53 ☽ ∠ ♃
09 16 ☽ △ ☉
10 56 ☽ ☌ ♅
18 03 ☽ △ ☿
21 39 ☽ ⊥ ♃
23 51 ☽ bQ ♀

02 Friday
02 40 ☽ Q ♇
05 59 ☽ Q ♄
06 52 ☽ # ♂
07 52 ☉ △ ♅
07 59 ☽ ⊻ ♃
09 49 ♀ ⊼ ♄
13 27 ☽ bQ ☉
16 24 ☿ ∥ ♃
16 53 ☽ ± ♀
18 27 ☽ ☍ ♂
22 23 ☽ ⊻ ♆
23 29 ☽ bQ ☿

03 Saturday
02 20 ☽ ⚹ ♄
04 00 ☽ ⊼ ♀
05 48 ☽ ± ☉
08 27 ☽ ⊥ ♆
08 48 ☽ ∥ ♇
09 19 ☽ □ ♇
10 52 ☽ # ♄
14 16 ☽ ⊻ ♅
16 30 ☽ ± ☿
16 30 ☽ ⊼ ☉
23 19 ☽ ∠ ♆

04 Sunday
00 08 ☽ ⊥ ♅
02 16 ♀ ⊥ ♂
03 02 ☽ ∠ ♄
03 37 ☽ ⊼ ☿
05 03 ☽ ∥ ☿
09 37 ☽ ☌ ♃
10 33 ☽ ∥ ♃
14 05 ☽ ∥ ☉
14 44 ☽ ∠ ♅
17 32 ☽ ⊥ ♄
21 59 ☽ ⊼ ♂
23 34 ☽ ⚹ ♆

05 Monday
03 06 ☽ ⊻ ♄
07 37 ☽ ∥ ♀
07 58 ☽ ± ♂
09 19 ☽ ☍ ♀
10 03 ☽ △ ♇
14 40 ☽ ⚹ ♅
16 53 ☽ # ♀
18 24 ♀ ⚹ ♇
18 38 ☽ Q ♆
20 12 ☽ ☍ ☉
22 49 ☽ bQ ♂

06 Tuesday
05 41 ☿ ⊼ ♃
09 00 ☽ ⊻ ♃
09 25 ☽ ☍ ☿
09 35 ☽ Q ♅
09 48 ☽ bQ ♇
12 37 ♂ ⊼ ♆
12 53 ☿ ∠ ♇
12 59 ☽ # ☉
13 10 ☽ # ♃
15 33 ☉ ∥ ♃
18 22 ☽ ⊥ ♃
23 00 ☽ □ ♆
23 26 ☽ △ ♂
23 59 ☽ ± ♇

07 Wednesday
02 15 ☽ ☌ ♄
05 29 ☽ # ☿
08 27 ☽ ∠ ♃
09 28 ☽ ⊼ ♇
10 58 ☽ ∥ ♄
12 53 ☽ ⊼ ♀
13 46 ☽ # ♇
13 56 ☽ □ ♅
22 48 ☽ ⊼ ☉
23 15 ☽ ± ♀

08 Thursday
01 36 ♀ △ ♅
05 43 ☽ ∥ ♂
08 05 ☽ ⚹ ♃
09 02 ☽ ± ☉
14 41 ☽ ⊼ ☿
14 45 ☿ # ♄
14 56 ☽ bQ ♀
22 44 ☽ △ ♆

09 Friday
00 34 ☽ bQ ☉
01 15 ☽ □ ♂
01 34 ☽ ± ☿
01 51 ☽ ⊻ ♄
03 17 ☽ Q ♃
09 39 ☽ ☍ ♇
09 48 ☿ ∥ ♇
11 36 ☽ ⊥ ♄
14 11 ☽ △ ♅
14 26 ♂ △ ♄
16 29 ☿ ± ♃
17 38 ☽ △ ♀
18 08 ☽ bQ ☿
23 21 ☽ bQ ♆

10 Saturday
02 02 ☉ ∠ ♂
02 26 ☽ ∠ ♄
03 04 ☽ △ ☉
08 18 ☽ Q ♂
08 52 ☽ □ ♃
14 29 ☽ ± ♆
15 15 ☽ bQ ♅
17 47 ☽ # ♅
22 37 ☽ △ ☿

11 Sunday
00 45 ☽ ⊼ ♆
03 48 ☽ ⚹ ♄
05 44 ☽ ⚹ ♂
06 43 ☽ ± ♅
12 29 ☽ ⊼ ♇
14 03 ♆ St D
17 12 ☽ ⊼ ♅
17 23 ☿ □ ♆
23 05 ☽ ± ♇

12 Monday
00 44 ☽ Q ♄
02 09 ☽ □ ♀
02 45 ☿ # ♂
05 15 ☿ ⊥ ♇
09 29 ☽ ∠ ♂
11 11 ☽ □ ☉
12 00 ☽ # ♅
12 50 ☽ △ ♃
15 16 ☽ bQ ♇
18 27 ☿ ☍ ♄

13 Tuesday
02 39 ☽ ⊥ ♂
06 17 ☽ ☍ ♆
07 36 ☉ ⊼ ♃
09 14 ☽ □ ♄
11 22 ☽ □ ☿
14 18 ☽ ⊻ ♂
14 26 ☽ Q ♀
16 12 ☽ bQ ♃
18 59 ☽ △ ♇
23 06 ☽ Q ☉
23 53 ☽ ☍ ♅

14 Wednesday
09 03 ☽ ± ♃
15 34 ♀ ∥ ♃
15 34 ☽ ⚹ ♀
19 25 ☉ ∠ ♇
20 25 ☽ ⊼ ♃
23 50 ☽ # ☿
23 54 ☽ ⚹ ☉

15 Thursday
01 20 ☿ ⚹ ♂
01 26 ☿ bQ ♃
02 16 ☽ Q ☿
15 20 ☽ ⊼ ♆
18 05 ☽ △ ♄
20 06 ☽ ∥ ♂
23 55 ☽ ∠ ♀

16 Friday
02 47 ☽ ☌ ♂
02 54 ☽ # ♇
03 05 ☽ ± ♆
03 43 ☿ ⊻ ♇
04 51 ☽ □ ♇
05 00 ☽ ⚹ ☿
07 43 ☽ ∠ ☉
08 38 ☽ ∥ ♄
08 58 ♂ ± ♆
09 50 ☽ ⊼ ♅
10 21 ☽ # ☉
13 08 ♀ ⊼ ♃
19 42 ☽ ⊥ ♀
20 52 ☽ bQ ♆
21 42 ☽ ± ♅
23 28 ☽ bQ ♄

17 Saturday
00 47 ☽ # ♀
01 50 ☉ # ♄
03 08 ☽ ⊥ ☉
06 49 ☽ ☍ ♃
07 40 ☽ # ♃
09 01 ☽ ⊻ ♀
15 04 ☽ ∠ ☿
15 38 ☽ bQ ♅
16 11 ☽ ⊻ ☉
16 40 ☽ Q ♇
17 18 ☽ ± ♄
19 33 ☿ □ ♅
22 33 ♂ □ ♇

18 Sunday
01 10 ♀ ∠ ♇
02 49 ☽ △ ♆
03 30 ♀ ∠ ♂
05 15 ☽ ⊼ ♄
11 44 ☽ ⊥ ☿
16 50 ☽ ⚹ ♇
17 45 ☽ ⊻ ♂
20 39 ☉ ± ♃
21 26 ♅ St D
21 46 ☽ △ ♅

19 Monday
01 34 ☽ ⊻ ☿
06 30 ☽ ⊥ ♂
12 42 ☉ ∥ ♇
18 04 ☉ # ♂
18 43 ☽ ⊼ ♃
23 12 ☽ ∠ ♇

20 Tuesday
01 40 ☽ ∠ ♂
02 17 ♂ # ♇
04 30 ☽ ☌ ♀
05 45 ☽ ∥ ♃
06 47 ☽ ± ♃
10 09 ☽ ☌ ☉
12 49 ♄ ± ♇
15 24 ☽ □ ♆
17 25 ☽ ☍ ♄
17 28 ☽ ⊥ ♇

21 Wednesday
00 53 ☽ bQ ♃
00 59 ♀ ± ♃
01 12 ☽ ∥ ♀
03 50 ☽ # ♄
05 39 ☽ ⊻ ♇
08 58 ☽ # ♂
09 40 ☽ ⚹ ♂
10 27 ☽ □ ♅
11 03 ☽ ∥ ♇
16 31 ☽ ∥ ☉
19 34 ♀ # ♄
22 57 ☽ ☌ ☿

22 Thursday
01 14 ♂ ⊼ ♅
03 52 ☽ Q ♆
07 03 ☽ △ ♃
11 14 ☽ Q ♂
18 11 ♀ # ♂
22 51 ☽ Q ♅

23 Friday
00 21 ☽ ⊻ ♀
01 01 ☉ □ ♆
04 06 ☽ ⚹ ♆
04 23 ☽ ⊻ ☉
05 41 ☽ ⊼ ♄
13 51 ☽ ⊥ ♀
15 02 ☿ Q ♆
17 33 ☽ ⊥ ☉
17 40 ☽ ± ♄
18 23 ☽ ☌ ♇
18 49 ☉ ☍ ♄
21 48 ☿ ∥ ♅
22 59 ☽ ⚹ ♅

24 Saturday
01 23 ☽ □ ♂
02 07 ♀ ∥ ♇
03 56 ☉ ⊥ ♇
09 59 ☽ ∠ ♀
10 13 ☽ ∠ ♆
11 33 ☽ bQ ♄
12 12 ♀ □ ♆
13 11 ☽ ∠ ☉
14 26 ☿ △ ♃
18 58 ☽ □ ♃
19 36 ☽ ⊻ ☿

25 Sunday
00 16 ♀ ☍ ♄
04 07 ☽ ⊥ ♆
04 39 ☽ ∥ ♅
04 54 ☽ ∠ ♅
04 58 ♂ ∥ ♄
09 06 ☽ ⊥ ☿
10 12 ♀ ⊥ ♇
15 59 ☽ ⊻ ♆
17 05 ☽ △ ♄
19 08 ☽ ⚹ ♀
21 32 ☽ ⚹ ☉
22 38 ☽ ⊥ ♅

26 Monday
05 06 ☽ ∠ ☿
06 01 ☽ ⊻ ♇
06 15 ☽ Q ♃
10 21 ☽ ⊻ ♅
12 26 ☿ ∥ ♆
15 39 ☽ △ ♂
17 41 ☽ ⊥ ♇
21 07 ☽ Q ♀
22 53 ☽ Q ☉

27 Tuesday
05 18 ☽ ⚹ ♃
07 12 ☉ bQ ♃
10 23 ☽ ∥ ♅
10 59 ☽ ∠ ♇
13 39 ☽ ⚹ ☿
20 40 ♀ bQ ♃
21 42 ☽ bQ ♂

28 Wednesday
01 44 ☽ ☌ ♆
02 24 ☽ □ ♄
09 27 ☽ ∠ ♃
10 53 ☽ □ ♀
11 46 ☽ □ ☉
14 50 ☽ Q ☿
15 10 ☽ ⚹ ♇
15 16 ☽ ± ♂
19 08 ☽ ☌ ♅
21 45 ☿ Q ♂

29 Thursday
01 55 ☽ ⊥ ♃
02 46 ☽ ⊼ ♂
11 13 ☽ Q ♄
12 43 ☽ ⊻ ♃
13 02 ☽ Q ♇
19 02 ☽ ∥ ☉

30 Friday
00 51 ☽ ∥ ♀
02 59 ☽ □ ☿
04 22 ☉ ☌ ♀
07 58 ☿ Q ♅
08 04 ☽ ⊻ ♆
08 19 ♀ ⊻ ♇
08 20 ☽ ⚹ ♄
09 22 ☉ ⊻ ♇
17 55 ☽ ∥ ♇
18 28 ☽ ⊥ ♆
20 40 ☽ □ ♇
21 30 ☽ △ ☉
21 51 ☽ △ ♀

31 Saturday
00 16 ☽ ⊻ ♅
02 05 ☽ # ♄
08 51 ☽ # ♂
09 37 ☽ ☍ ♂
09 48 ☽ ∠ ♆
09 54 ☽ ∠ ♄
10 26 ☽ ⊥ ♅
13 57 ♂ bQ ♆
15 31 ♂ bQ ♄
16 25 ☽ ☌ ♃
17 27 ☽ ∥ ♃

All ephemeris data is given at 12.00 UT and the Moon's longitude is additionally given for 24.00 UT

NOVEMBER 1998

LONGITUDES

Date	Sidereal time h m s	Sun ☉ ° ' "	Moon ☽ ° ' "	Moon ☽ 24.00 ° ' "	Mercury ☿ ° '	Venus ♀ ° '	Mars ♂ ° '	Jupiter ♃ ° '	Saturn ♄ ° '	Uranus ♅ ° '	Neptune ♆ ° '	Pluto ♇ ° '
01	14 42 18	08 ♏ 51 33	00 ♈ 20 13	07 ♈ 45 42	29 ♏ 47	09 ♏ 27	15 ♍ 02	18 ♓ 25	29 ♈ 28	08 ♒ 53	29 ♑ 30	06 ♐ 49
02	14 46 15	09 51 34	15 ♈ 16 38	22 ♈ 52 01	01 ♐ 05	10 42	15 38	18 R 22	29 R 23	08 54	29 31	06 52
03	14 50 11	10 51 37	00 ♉ 30 39	08 ♉ 11 11	02 23	11 57	16 13	18 20	29 19	08 55	29 32	06 54
04	14 54 08	11 51 42	15 ♉ 52 07	23 ♉ 31 59	03 39	13 13	16 48	18 18	29 14	08 56	29 33	06 56
05	14 58 04	12 51 48	01 ♊ 09 17	08 ♊ 42 41	04 54	14 28	17 24	18 16	29 09	08 57	29 33	06 58
06	15 02 01	13 51 57	16 ♊ 10 58	23 ♊ 33 09	06 07	15 43	17 59	18 15	29 05	08 57	29 34	07 01
07	15 05 57	14 52 08	00 ♋ 48 30	07 ♋ 56 31	07 18	16 58	18 34	18 13	29 00	08 58	29 35	07 03
08	15 09 54	15 52 20	14 ♋ 56 55	21 ♋ 49 40	08 28	18 14	19 09	18 12	28 56	08 59	29 36	07 05
09	15 13 50	16 52 35	28 ♋ 34 56	05 ♌ 12 59	09 35	19 29	19 44	18 11	28 51	09 01	29 37	07 07
10	15 17 47	17 52 51	11 ♌ 44 16	18 ♌ 09 18	10 39	20 44	20 19	18 11	28 47	09 02	29 38	07 09
11	15 21 44	18 53 10	24 ♌ 28 40	00 ♍ 42 59	11 41	22 00	20 54	18 10	28 43	09 03	29 39	07 11
12	15 25 40	19 53 31	06 ♍ 52 54	12 ♍ 59 02	12 39	23 15	21 29	18 10	28 38	09 04	29 40	07 14
13	15 29 37	20 53 53	19 ♍ 02 01	25 ♍ 02 26	13 34	24 30	22 04	18 10	28 34	09 05	29 41	07 16
14	15 33 33	21 54 17	01 ♎ 00 51	06 ♎ 57 46	14 24	25 45	22 38	18 D 10	28 30	09 07	29 42	07 18
15	15 37 30	22 54 44	12 ♎ 53 40	18 ♎ 48 56	15 10	27 01	23 13	18 10	28 26	09 08	29 43	07 21
16	15 41 26	23 55 11	24 ♎ 43 58	00 ♏ 39 03	15 51	28 16	23 47	18 11	28 22	09 09	29 44	07 23
17	15 45 23	24 55 41	06 ♏ 34 29	12 ♏ 30 29	16 26	29 ♏ 31	24 22	18 11	28 18	09 11	29 46	07 25
18	15 49 19	25 56 13	18 ♏ 27 14	24 ♏ 24 55	16 54	00 ♐ 47	24 56	18 12	28 14	09 12	29 47	07 27
19	15 53 16	26 56 46	00 ♐ 23 40	06 ♐ 23 36	17 16	02 02	25 31	18 13	28 10	09 14	29 48	07 30
20	15 57 13	27 57 20	12 ♐ 24 52	18 ♐ 27 36	17 29	03 17	26 05	18 15	28 06	09 16	29 49	07 32
21	16 01 09	28 57 56	24 ♐ 31 57	00 ♑ 38 04	17 R 33	04 33	26 39	18 16	28 02	09 17	29 51	07 34
22	16 05 06	29 ♏ 58 34	06 ♑ 46 10	12 ♑ 56 30	17 28	05 48	27 13	18 18	27 58	09 19	29 52	07 37
23	16 09 02	00 ♐ 59 12	19 ♑ 09 20	25 ♑ 25 00	17 13	07 03	27 47	18 20	27 55	09 21	29 53	07 39
24	16 12 59	01 59 52	01 ♒ 43 49	08 ♒ 06 13	16 48	08 19	28 21	18 22	27 51	09 22	29 55	07 41
25	16 16 55	03 00 33	14 ♒ 32 35	21 ♒ 03 23	16 11	09 34	28 55	18 24	27 48	09 24	29 56	07 44
26	16 20 52	04 01 15	27 ♒ 39 00	04 ♓ 19 54	15 24	10 49	29 ♍ 29	18 27	27 44	09 26	29 58	07 46
27	16 24 48	05 01 58	11 ♓ 06 25	17 ♓ 58 52	14 27	12 05	00 ♎ 03	18 30	27 41	09 28	29 ♑ 59	07 48
28	16 28 45	06 02 42	24 ♓ 57 27	02 ♈ 02 16	13 21	13 20	00 36	18 33	27 38	09 30	00 ♒ 01	07 51
29	16 32 42	07 03 27	09 ♈ 13 14	16 ♈ 30 06	12 07	14 36	01 10	18 36	27 35	09 32	00 02	07 53
30	16 36 38	08 ♐ 04 13	23 ♈ 52 24	01 ♉ 19 28	10 ♐ 48	15 ♐ 51	01 ♎ 43	18 ♓ 39	27 ♈ 32	09 ♒ 34	00 ♒ 04	07 ♐ 56

DECLINATIONS

Date	Moon True ☊ ° '	Moon Mean ☊ ° '	Moon ☽ Latitude ° '	Sun ☉ ° '	Moon ☽ ° '	Mercury ☿ ° '	Venus ♀ ° '	Mars ♂ ° '	Jupiter ♃ ° '	Saturn ♄ ° '	Uranus ♅ ° '	Neptune ♆ ° '	Pluto ♇ ° '
01	28 ♌ 53	27 ♌ 36	02 S 38	14 S 27	02 S 17	22 S 24	13 S 50	07 N 17	05 S 57	08 N 42	18 S 40	19 S 56	10 S 09
02	28 R 47	27 33	03 40	14 46	02 N 38	22 45	14 15	07 03	05 57	08 40	18 40	19 56	10 09
03	28 38	27 30	04 26	15 05	07 29	23 05	14 40	06 50	05 58	08 39	18 40	19 56	10 10
04	28 28	27 27	04 54	15 24	11 53	23 15	15 05	06 37	05 58	08 37	18 39	19 56	10 10
05	28 17	27 23	05 01	15 42	15 29	23 40	15 30	06 23	05 59	08 36	18 39	19 56	10 11
06	28 07	27 20	04 47	16 00	17 58	23 56	15 53	06 10	05 59	08 34	18 39	19 56	10 12
07	27 59	27 17	04 14	16 18	19 12	24 10	16 17	05 56	06 00	08 33	18 38	19 56	10 12
08	27 53	27 14	03 26	16 35	19 11	24 23	16 40	05 43	06 00	08 31	18 38	19 55	10 13
09	27 50	27 11	02 28	16 53	18 01	24 34	17 03	05 30	06 00	08 30	18 38	19 55	10 13
10	27 49	27 08	01 24	17 10	15 55	24 44	17 25	05 16	06 00	08 28	18 38	19 55	10 14
11	27 D 49	27 04	00 S 18	17 26	13 05	24 53	17 47	05 03	06 00	08 27	18 37	19 55	10 14
12	27 R 48	27 01	00 N 48	17 43	09 44	25 00	18 08	04 50	06 00	08 26	18 37	19 55	10 15
13	27 47	26 58	01 50	17 59	06 01	25 05	18 29	04 36	06 00	08 24	18 36	19 54	10 15
14	27 43	26 55	02 46	18 14	02 N 08	25 09	18 49	04 23	05 59	08 23	18 36	19 54	10 16
15	27 36	26 52	03 34	18 30	01 S 48	25 11	19 09	04 10	05 59	08 22	18 36	19 54	10 17
16	27 27	26 49	04 13	18 45	05 39	25 11	19 28	03 56	05 59	08 20	18 35	19 54	10 17
17	27 14	26 45	04 40	19 00	09 18	25 09	19 47	03 43	05 58	08 19	18 35	19 54	10 18
18	27 00	26 42	04 56	19 14	12 34	25 05	20 05	03 30	05 58	08 18	18 34	19 53	10 18
19	26 45	26 39	04 59	19 28	15 22	24 59	20 23	03 16	05 57	08 17	18 34	19 53	10 19
20	26 30	26 36	04 48	19 42	17 31	24 51	20 40	03 03	05 56	08 15	18 34	19 53	10 19
21	26 17	26 33	04 24	19 56	18 55	24 41	20 56	02 50	05 55	08 14	18 33	19 53	10 20
22	26 06	26 29	03 48	20 09	19 25	24 28	21 12	02 37	05 54	08 13	18 33	19 52	10 20
23	25 59	26 26	03 01	20 21	19 05	24 13	21 27	02 24	05 53	08 12	18 32	19 52	10 21
24	25 55	26 23	02 04	20 34	17 46	23 56	21 42	02 11	05 52	08 11	18 32	19 52	10 21
25	25 53	26 20	01 N 00	20 45	15 31	23 35	21 56	01 57	05 51	08 10	18 31	19 52	10 21
26	25 D 52	26 17	00 S 09	20 57	12 26	23 12	22 09	01 44	05 50	08 09	18 31	19 51	10 22
27	25 R 52	26 14	01 20	21 08	08 38	22 47	22 22	01 31	05 49	08 08	18 30	19 51	10 22
28	25 51	26 10	02 27	21 19	04 S 15	22 19	22 34	01 18	05 47	08 07	18 30	19 51	10 23
29	25 48	26 07	03 28	21 29	00 N 28	21 49	22 45	01 05	05 46	08 06	18 29	19 50	10 23
30	25 ♌ 42	26 ♌ 04	04 S 16	21 S 39	05 N 17	21 S 18	22 S 56	00 N 52	05 S 44	08 N 05	18 S 28	19 S 50	10 S 24

ZODIAC SIGN ENTRIES

Date	h	m	Planets
01	11	27	☽ ♈
01	16	02	☿ ♐
03	11	12	☽ ♉
05	10	11	☽ ♊
07	10	39	☽ ♋
09	14	33	☽ ♌
11	22	37	☽ ♍
14	09	58	☽ ♎
16	22	41	☽ ♏
17	21	06	♀ ♐
19	11	13	☽ ♐
21	22	45	☽ ♑
22	12	34	☉ ♐
24	08	43	☽ ♒
26	16	14	☽ ♓
27	10	10	♂ ♎
28	01	19	♆ ♒
28	20	34	☽ ♈
30	21	53	☽ ♉

LATITUDES

Date	Mercury ☿ ° '	Venus ♀ ° '	Mars ♂ ° '	Jupiter ♃ ° '	Saturn ♄ ° '	Uranus ♅ ° '	Neptune ♆ ° '	Pluto ♇ ° '
01	02 S 22	00 N 51	01 N 30	01 S 29	02 S 46	00 S 39	00 N 19	11 N 27
04	02 33	00 45	01 31	01 28	02 45	00 39	00 19	11 27
07	02 40	00 39	01 33	01 27	02 45	00 39	00 19	11 26
10	02 43	00 32	01 34	01 26	02 45	00 39	00 19	11 25
13	02 41	00 26	01 35	01 25	02 44	00 39	00 19	11 25
16	02 30	00 19	01 36	01 25	02 44	00 39	00 19	11 24
19	02 10	00 12	01 37	01 24	02 43	00 39	00 19	11 24
22	01 38	00 N 04	01 39	01 24	02 43	00 39	00 19	11 23
25	00 S 52	00 S 03	01 40	01 23	02 42	00 39	00 19	11 23
28	00 N 05	00 10	01 41	01 22	02 42	00 39	00 19	11 23
31	01 N 06	00 S 17	01 N 42	01 S 21	02 S 41	00 S 38	00 N 18	11 N 22

DATA

Julian Date	2451119
Delta T	+63 seconds
Ayanamsa	23° 50' 16"
Synetic vernal point	05° ♓ 16' 43"
True obliquity of ecliptic	23° 26' 14"

MOON'S PHASES, APSIDES AND POSITIONS ☽

Date	h	m	Phase	Longitude ° '	Eclipse Indicator
04	05	18	○	11 ♉ 35	
11	00	28	☾	18 ♌ 24	
19	04	27	●	26 ♏ 38	
27	00	23	☽	04 ♓ 33	

Day	h	m		
04	00	45	Perigee	
17	06	41	Apogee	
01	23	12	0N	
07	23	31	Max dec	19° N 21'
15	00	59	0S	
22	14	16	Max dec	19° S 28'
29	09	40	0N	

LONGITUDES

Date	Chiron ⚷ ° '	Ceres ⚳ ° '	Pallas ⚴ ° '	Juno ⚵ ° '	Vesta ⚶ ° '	Black Moon Lilith ⚸ ° '
01	21 ♏ 33	12 ♊ 00	13 ♓ 55	05 ♏ 56	15 ♌ 04	05 ♏ 49
11	22 ♏ 53	10 ♊ 20	13 ♓ 29	09 ♏ 23	17 ♌ 35	06 ♏ 56
21	24 ♏ 13	08 ♊ 13	13 ♓ 42	12 ♏ 48	19 ♌ 39	08 ♏ 03
31	25 ♏ 33	05 ♊ 53	14 ♓ 30	16 ♏ 09	21 ♌ 13	09 ♏ 10

All ephemeris data is given at 12.00 UT and the Moon's longitude is additionally given for 24.00 UT

ASPECTARIAN

h m Aspects

01 Sunday
00 37 ☽ ⚼ ☉
00 48 ☽ ⊥ ♄
01 12 ♀ □ ♅
01 26 ☽ ∠ ♅
01 27 ☽ ⚼ ♀
01 36 ♄ □ ♆
05 47 ♂ ± ♅
06 42 ☿ ⊼ ♄
07 03 ☿ ✱ ♆
10 36 ☽ ⊻ ♄
10 39 ☽ ✱ ♆
11 00 ☽ △ ☿
12 44 ☉ □ ♅
16 23 ☽ ± ☉
17 31 ☽ ± ♀
22 31 ☽ △ ♇

02 Monday
01 49 ☽ ✱ ♅
02 45 ☽ ⊼ ☉
04 03 ☽ ⊼ ♀
06 01 ☽ Q ♆
12 35 ☽ ⊼ ♂
13 25 ☽ ⚼ ☿
16 53 ☽ ⊻ ♃
20 55 ☽ Q ♅
22 26 ☽ ⚼ ♇
22 27 ☽ ± ♂

03 Tuesday
02 20 ☽ ⊥ ♃
04 17 ☽ # ♃
04 56 ☽ ± ☿
08 49 ☽ ∥ ♂
10 08 ☽ ☌ ♄
10 28 ☽ □ ♆
12 36 ☽ ± ♇
13 09 ☽ ⚼ ♂
15 12 ☽ ⊼ ☿
16 24 ☽ ∠ ♃
17 59 ☽ ∥ ♄
22 00 ☽ ⊼ ♇

04 Wednesday
01 09 ☽ □ ♅
02 12 ☽ # ♇
05 18 ☽ ☍ ☉
07 29 ☽ ☍ ♀
13 32 ☽ △ ♂
15 48 ☽ ✱ ♃

05 Thursday
08 52 ☽ ⊻ ♄
09 29 ☽ △ ♆
10 36 ☽ Q ♃
12 08 ☽ # ♀
13 57 ☽ # ☉
16 43 ☿ ± ♄
18 19 ☽ ⊥ ♄
18 28 ☽ ☍ ☿
21 15 ☽ ☍ ♇

06 Friday
00 23 ☽ △ ♅
08 00 ☽ ⊼ ☉
08 38 ☽ ∠ ♄
09 24 ☽ ⚼ ♆
11 11 ☽ ⊼ ♀
15 02 ☽ □ ♂
15 20 ☽ □ ♃
18 25 ☽ ± ☉
21 50 ☽ ± ♀
22 05 ☽ # ♅
22 25 ♂ ☍ ♃

07 Saturday
00 03 ☽ ± ♆
00 41 ☽ ⚼ ♅
06 23 ☿ ☌ ♇
06 28 ♂ # ♃
09 01 ☽ ✱ ♄
09 58 ☽ ⊼ ♆
10 19 ☽ ⚼ ☉
14 08 ☽ ⚼ ♀
15 27 ☉ ∥ ♀
15 38 ☽ ± ♅
22 05 ☽ Q ♂
22 30 ☽ ⊼ ♇
23 52 ♀ Q ♆
23 54 ☽ ⊼ ☿

08 Sunday
01 46 ☽ ⊼ ♅
05 08 ☽ Q ♄
08 46 ☽ ± ♇
11 05 ☽ ± ☿
11 34 ♀ △ ♃
13 43 ☽ △ ☉
17 39 ☽ △ ♃
18 16 ☽ △ ♀
19 38 ☽ ✱ ♂
23 26 ☿ ✱ ♅

09 Monday
00 29 ☽ ⚼ ♇
02 02 ☽ # ♅
04 14 ☽ ⚼ ☿
12 30 ☽ □ ♄
13 52 ☽ ☍ ♆
20 18 ☽ ⚼ ♃
21 03 ♀ ✱ ♂
22 29 ☽ # ♀
23 38 ☽ ∠ ♂

10 Tuesday
00 35 ☽ # ☉
03 31 ☽ △ ♇
05 59 ☉ Q ♆
06 59 ☽ ☍ ♅
09 49 ☽ △ ☿
12 49 ☽ ± ♃
17 02 ☽ ⊥ ♂
19 00 ☉ △ ♃

11 Wednesday
00 02 ☽ ⊼ ♃
00 28 ☽ □ ☉
04 52 ☽ ⊻ ♂
06 44 ☽ □ ♀
20 05 ☽ △ ♄
21 57 ☽ ⊼ ♆

12 Thursday
08 27 ☽ # ♇
09 37 ☽ ± ♆
12 41 ☽ □ ♇
14 09 ☽ Q ☉
16 18 ☽ ⊼ ♅
20 40 ☽ ∥ ♄
21 34 ☽ Q ♀

13 Friday
00 17 ☽ □ ☿
01 13 ☽ ⚼ ♄
03 21 ☽ ⚼ ♆
04 09 ☽ ± ♅
10 16 ☽ ☍ ♃
12 10 ☽ # ♃
12 13 ☿ ⚼ ♄
13 04 ♃ St D
16 03 ☽ ✱ ☉
18 20 ☽ ☌ ♂
19 01 ☽ ± ♄
20 49 ♀ ∥ ♅
21 23 ☽ ∥ ♂
22 07 ☽ ⚼ ♅

14 Saturday
00 12 ☽ ✱ ♀
00 30 ☽ Q ♇
06 51 ♂ ± ♄
06 58 ☽ ⊼ ♄
09 21 ☽ △ ♆
15 00 ☽ Q ☿
21 20 ☿ ∠ ♆

15 Sunday
00 44 ☽ ✱ ♇
00 59 ☽ ∠ ☉
04 22 ☽ △ ♅
10 00 ☽ ∠ ♀
14 21 ♀ Q ♅
16 54 ☽ ✱ ☿
20 52 ☉ ∥ ♅
20 54 ☽ ⊥ ☉
22 42 ☽ ⊼ ♃

16 Monday
01 45 ☽ # ♂
04 51 ☉ ✱ ♂
06 25 ☽ ⊥ ♀
07 13 ☽ ∠ ♇
10 00 ☽ ⊻ ♂
10 12 ☽ ⊻ ☉
10 52 ☽ ± ♃
13 42 ♀ ⊼ ♄
14 03 ☽ ∥ ♃
19 19 ☽ ☍ ♄
20 01 ☽ ⊻ ♀
22 10 ☽ □ ♆
22 47 ☽ ⊥ ♂

17 Tuesday
01 04 ☽ ∠ ☿
01 31 ☽ ⊥ ♇
03 57 ♂ ⚼ ♅
05 08 ☽ ⚼ ♃
05 24 ☽ # ♄
13 43 ☽ ⊻ ♇
16 37 ♀ ✱ ♆
17 17 ☽ □ ♅
17 56 ☽ ∠ ♂
19 01 ☽ ∥ ♇
20 09 ☽ ⊥ ☿
20 44 ♀ ∥ ♆

18 Wednesday
08 46 ☽ ⊻ ☿
10 39 ☽ Q ♆
11 30 ☽ △ ♃

19 Thursday
01 43 ☽ ✱ ♂
04 27 ☽ ☌ ☉
05 39 ☽ Q ♅
07 33 ☽ ⊼ ♄
10 49 ☽ ✱ ♆
11 19 ♂ Q ♇
15 40 ☽ ☌ ♀
19 00 ☉ Q ♅
19 30 ☽ ± ♄

20 Friday
02 15 ☽ ☌ ♇
02 56 ☽ Q ♂
05 42 ☽ ✱ ♅
13 21 ☽ ⚼ ♄
15 13 ☉ ⊼ ♄
16 48 ☽ ∠ ♆
22 10 ☽ ☌ ☿
23 36 ☽ □ ♃

21 Saturday
02 43 ♀ ± ♄
04 08 ☽ ∥ ♅
06 47 ☉ ∥ ♆
10 39 ☽ ⊥ ♆
11 31 ☽ ∠ ♅
11 45 ☿ St R
16 23 ☽ □ ♂
18 52 ☽ △ ♄
21 31 ☽ ⊻ ☉
22 28 ☽ ⊻ ♆

22 Sunday
05 14 ☽ ⊥ ♅
09 22 ☉ ✱ ♆
09 54 ☽ ⊻ ♀
10 19 ☽ ⊥ ☉
11 05 ☽ Q ♃
13 39 ☽ ⊻ ♇
16 58 ☽ ⊻ ♅
22 54 ☽ ⊥ ♀

23 Monday
01 21 ☽ ⊥ ♇
05 21 ☽ ∠ ☉
08 22 ☽ ⊻ ☿
10 25 ☽ ✱ ♃
16 51 ♂ ⊼ ♄
18 12 ☽ ∠ ♀
18 44 ☽ ∠ ♇
19 35 ☽ ⊥ ☿
23 43 ♀ ☌ ♇

24 Tuesday
00 09 ☽ ∥ ♅
04 41 ☽ □ ♄
05 18 ☽ △ ♂
08 33 ☽ ☌ ♆
12 07 ☽ ∠ ☿
12 33 ☽ ✱ ☉
15 06 ☽ ∠ ♃
23 16 ☽ ✱ ♇

25 Wednesday
01 44 ☽ ✱ ♀
02 25 ☽ ☌ ♅
08 01 ☽ ⊥ ♃
08 46 ♀ ✱ ♅
10 47 ☽ ⚼ ♂
12 56 ☽ Q ☉
14 19 ☽ Q ♄
14 53 ☽ ✱ ☿
19 10 ☽ ⊻ ♃
21 36 ☽ Q ♇

26 Thursday
02 18 ☽ Q ♀
04 06 ☽ ± ♂
05 43 ☉ ± ♄
11 35 ☽ Q ☿
12 10 ☽ ✱ ♄
15 27 ☽ ⊼ ♂
16 11 ☽ ⊻ ♆
23 23 ☿ ∠ ♆

27 Friday
00 23 ☽ □ ☉
01 31 ☽ ∥ ♇
02 56 ☽ ⊥ ♆
06 10 ☽ □ ♇
09 06 ☽ ⊻ ♅
09 28 ♂ △ ♆
10 42 ♀ Q ♂
13 53 ☽ □ ♀
14 46 ☽ ∠ ♄
14 53 ☽ # ♄
17 27 ☽ □ ☿
18 48 ☽ ∠ ♆
19 39 ☽ ⊥ ♅
23 06 ♀ ⚼ ♄

28 Saturday
00 56 ☽ ☌ ♃
03 11 ☿ ∥ ♀
03 50 ☽ ∥ ♃
06 19 ☽ ⊥ ♄
11 13 ☽ ∠ ♅
12 06 ☿ ☌ ♀
16 32 ☽ ⊻ ♄
20 36 ☽ ✱ ♆
21 58 ☽ ☍ ♂
22 08 ☿ Q ♂

29 Sunday
02 47 ☿ ⚼ ♄
03 51 ☽ # ♂
08 08 ☽ △ ☉
09 47 ☽ △ ♇
12 31 ☽ ✱ ♅
14 56 ☽ ∥ ♂
16 24 ☽ △ ☿
16 40 ☽ Q ♆
20 41 ♀ ∠ ♆
21 42 ☽ △ ♀
23 54 ☉ ∥ ☿

30 Monday
03 29 ☽ ⊻ ♃
08 15 ☽ Q ♅
08 26 ☉ ☌ ♇
10 28 ☽ ⚼ ♇
10 36 ☽ ⚼ ☉
13 16 ☽ ⊥ ♃
14 15 ☽ # ♃
14 51 ☽ ⚼ ☿
17 53 ☽ ☌ ♄
22 00 ☽ □ ♆

DECEMBER 1998

Date	Sidereal time h m s	Sun ☉ ° ' "	Moon ☽ ° ' "	Moon ☽ 24.00 ° ' "	Mercury ☿ ° '	Venus ♀ ° '	Mars ♂ ° '	Jupiter ♃ ° '	Saturn ♄ ° '	Uranus ♅ ° '	Neptune ♆ ° '	Pluto ♇ ° '
01	16 40 35	09 ♐ 05 00	08 ♉ 50 26	16 ♉ 24 12	09 ♐ 25	17 ♐ 06	02 ♎ 17	18 ♓ 43	27 ♈ 29	09 ♒ 36	00 ♒ 05	07 ♐ 58
02	16 44 31	10 05 48	23 ♉ 59 34	01 ♊ 35 11	08 R 03	18 21	02 50	18 47	27 R 26	09 38	00 07	08 00
03	16 48 28	11 06 37	09 ♊ 09 41	16 ♊ 41 43	06 43	19 37	03 23	18 50	27 23	09 40	00 09	08 03
04	16 52 24	12 07 27	24 ♊ 10 00	01 ♋ 33 25	05 28	20 52	03 56	18 55	27 20	09 43	00 10	08 05
05	16 56 21	13 08 19	08 ♋ 51 01	16 ♋ 02 03	04 21	22 07	04 29	18 59	27 18	09 45	00 12	08 07
06	17 00 17	14 09 11	23 ♋ 06 03	00 ♌ 02 41	03 23	23 23	05 02	19 03	27 15	09 47	00 14	08 10
07	17 04 14	15 10 05	06 ♌ 51 55	13 ♌ 33 50	02 36	24 38	05 35	19 08	27 13	09 50	00 15	08 12
08	17 08 11	16 11 00	20 ♌ 08 42	26 ♌ 36 55	02 00	25 53	06 07	19 13	27 11	09 52	00 17	08 14
09	17 12 07	17 11 56	02 ♍ 58 59	09 ♍ 15 29	01 35	27 09	06 40	19 18	27 08	09 54	00 19	08 17
10	17 16 04	18 12 54	15 ♍ 27 02	21 ♍ 34 17	01 21	28 24	07 13	19 23	27 06	09 57	00 21	08 19
11	17 20 00	19 13 52	27 ♍ 37 55	03 ♎ 38 34	01 D 18	29 ♐ 39	07 45	19 29	27 04	09 59	00 22	08 21
12	17 23 57	20 14 52	09 ♎ 36 54	15 ♎ 33 33	01 26	00 ♑ 55	08 17	19 34	27 02	10 02	00 24	08 24
13	17 27 53	21 15 52	21 ♎ 29 05	27 ♎ 24 04	01 42	02 10	08 50	19 40	27 00	10 04	00 26	08 26
14	17 31 50	22 16 54	03 ♏ 18 59	09 ♏ 14 18	02 07	03 25	09 22	19 46	26 59	10 07	00 28	08 28
15	17 35 46	23 17 57	15 ♏ 10 24	21 ♏ 07 39	02 40	04 41	09 54	19 52	26 57	10 10	00 30	08 31
16	17 39 43	24 19 01	27 ♏ 06 20	03 ♐ 06 42	03 19	05 56	10 26	19 58	26 56	10 12	00 32	08 33
17	17 43 40	25 20 05	09 ♐ 08 56	15 ♐ 13 13	04 05	07 11	10 57	20 05	26 54	10 15	00 34	08 35
18	17 47 36	26 21 10	21 ♐ 19 38	27 ♐ 28 18	04 56	08 27	11 29	20 11	26 53	10 18	00 36	08 37
19	17 51 33	27 22 16	03 ♑ 39 16	09 ♑ 52 36	05 52	09 42	12 00	20 18	26 52	10 20	00 38	08 40
20	17 55 29	28 23 23	16 ♑ 08 22	22 ♑ 26 38	06 51	10 57	12 32	20 25	26 51	10 23	00 40	08 42
21	17 59 26	29 ♐ 24 30	28 ♑ 47 28	05 ♒ 10 57	07 55	12 13	13 03	20 32	26 50	10 26	00 42	08 44
22	18 03 22	00 ♑ 25 37	11 ♒ 37 14	18 ♒ 06 28	09 01	13 28	13 34	20 40	26 49	10 29	00 44	08 46
23	18 07 19	01 26 45	24 ♒ 38 49	01 ♓ 14 29	10 11	14 43	14 05	20 47	26 48	10 32	00 46	08 49
24	18 11 15	02 27 52	07 ♓ 53 41	14 ♓ 36 38	11 23	15 59	14 36	20 55	26 48	10 35	00 48	08 51
25	18 15 12	03 29 00	21 ♓ 23 32	28 ♓ 14 34	12 37	17 14	15 07	21 02	26 47	10 38	00 50	08 53
26	18 19 09	04 30 08	05 ♈ 09 52	12 ♈ 09 30	13 53	18 29	15 37	21 10	26 47	10 41	00 52	08 55
27	18 23 05	05 31 16	19 ♈ 13 27	26 ♈ 21 34	15 10	19 44	16 08	21 18	26 46	10 44	00 54	08 57
28	18 27 02	06 32 23	03 ♉ 33 37	10 ♉ 49 12	16 30	21 00	16 38	21 27	26 46	10 47	00 56	08 59
29	18 30 58	07 33 31	18 ♉ 07 46	25 ♉ 28 40	17 50	22 15	17 08	21 35	26 46	10 50	00 58	09 02
30	18 34 55	08 34 39	02 ♊ 51 05	10 ♊ 14 06	19 12	23 30	17 38	21 43	26 D 46	10 53	01 00	09 04
31	18 38 51	09 ♑ 35 47	17 ♊ 36 46	24 ♊ 58 05	20 ♐ 34	24 ♑ 45	18 ♎ 08	21 ♓ 52	26 ♈ 46	10 ♒ 56	01 ♒ 02	09 ♐ 06

Date	Moon True ☊ ° '	Moon Mean ☊ ° '	Moon ☽ Latitude ° '
01	25 ♌ 34	26 ♌ 01	04 S 48
02	25 R 23	25 58	05 01
03	25 12	25 54	04 52
04	25 01	25 51	04 23
05	24 52	25 48	03 38
06	24 46	25 45	02 39
07	24 42	25 42	01 33
08	24 41	25 39	00 S 24
09	24 D 41	25 35	00 N 44
10	24 41	25 32	01 48
11	24 R 41	25 29	02 46
12	24 39	25 26	03 35
13	24 34	25 23	04 15
14	24 27	25 20	04 43
15	24 18	25 16	05 00
16	24 07	25 13	05 03
17	23 55	25 10	04 53
18	23 43	25 07	04 30
19	23 33	25 04	03 53
20	23 24	25 00	03 05
21	23 19	24 57	02 07
22	23 16	24 54	01 N 02
23	23 D 15	24 51	00 S 08
24	23 16	24 48	01 18
25	23 17	24 45	02 26
26	23 R 18	24 41	03 26
27	23 17	24 38	04 16
28	23 14	24 35	04 50
29	23 09	24 32	05 07
30	23 02	24 29	05 04
31	22 ♌ 55	24 ♌ 26	04 S 40

DECLINATIONS

Date	Sun ☉	Moon ☽	Mercury ☿	Venus ♀	Mars ♂	Jupiter ♃	Saturn ♄	Uranus ♅	Neptune ♆	Pluto ♇
01	21 S 49	09 N 53	20 S 47	23 S 06	00 N 39	05 S 43	08 N 04	18 S 28	19 S 50	10 S 24
02	21 58	13 54	20 15	23 15	00 27	05 41	08 03	18 27	19 49	10 25
03	22 06	17 00	19 45	23 24	00 14	05 39	08 03	18 27	19 49	10 25
04	22 15	18 55	19 17	23 32	00 N 01	05 37	08 02	18 26	19 49	10 25
05	22 22	19 31	18 52	23 39	00 S 12	05 35	08 01	18 25	19 48	10 26
06	22 30	18 51	18 30	23 45	00 24	05 33	08 01	18 25	19 48	10 26
07	22 37	17 03	18 12	23 51	00 37	05 31	08 00	18 24	19 48	10 27
08	22 43	14 23	17 58	23 56	00 50	05 29	07 59	18 23	19 47	10 27
09	22 49	11 06	17 49	24 00	01 02	05 27	07 59	18 23	19 47	10 27
10	22 55	07 24	17 43	24 04	01 15	05 25	07 58	18 22	19 47	10 28
11	23 00	03 N 29	17 42	24 07	01 27	05 22	07 58	18 21	19 46	10 28
12	23 05	00 S 30	17 43	24 09	01 39	05 20	07 57	18 21	19 46	10 28
13	23 09	04 26	17 49	24 10	01 52	05 18	07 57	18 20	19 46	10 29
14	23 13	08 10	17 56	24 11	02 04	05 15	07 57	18 19	19 45	10 29
15	23 16	11 36	18 07	24 10	02 16	05 12	07 56	18 19	19 45	10 29
16	23 19	14 35	18 19	24 09	02 28	05 10	07 56	18 18	19 44	10 30
17	23 21	17 00	18 33	24 08	02 40	05 07	07 56	18 17	19 44	10 30
18	23 23	18 40	18 48	24 05	02 52	05 04	07 56	18 16	19 44	10 30
19	23 25	19 30	19 04	24 02	03 04	05 01	07 55	18 16	19 43	10 31
20	23 26	19 24	19 21	23 58	03 16	04 58	07 55	18 15	19 43	10 31
21	23 26	18 19	19 39	23 53	03 28	04 55	07 55	18 14	19 42	10 31
22	23 26	16 18	19 57	23 48	03 40	04 52	07 55	18 13	19 42	10 31
23	23 26	13 25	20 15	23 42	03 51	04 49	07 55	18 12	19 41	10 32
24	23 25	09 49	20 33	23 35	04 03	04 46	07 55	18 12	19 41	10 32
25	23 23	05 39	20 50	23 27	04 14	04 43	07 55	18 11	19 41	10 32
26	23 22	01 S 06	21 08	23 19	04 26	04 39	07 55	18 10	19 40	10 32
27	23 19	03 N 35	21 25	23 10	04 37	04 36	07 56	18 09	19 40	10 32
28	23 17	08 09	21 41	23 00	04 48	04 33	07 56	18 08	19 39	10 33
29	23 13	12 18	21 57	22 50	04 59	04 29	07 56	18 07	19 39	10 33
30	23 10	15 45	22 12	22 39	05 10	04 26	07 56	18 07	19 38	10 33
31	23 S 05	18 N 12	22 S 27	22 S 27	05 S 21	04 S 22	07 N 57	18 S 06	19 S 38	10 S 33

ZODIAC SIGN ENTRIES

Date	h	m	Planets
02	21	30	☽ ♊
04	21	28	☽ ♋
06	23	55	☽ ♌
09	06	21	☽ ♍
11	16	43	☽ ♎
11	18	33	♀ ♑
14	05	16	☽ ♏
16	17	47	☽ ♐
19	04	55	☽ ♑
21	14	17	☽ ♒
22	01	56	☉ ♑
23	21	45	☽ ♓
26	03	03	☽ ♈
28	06	05	☽ ♉
30	07	22	☽ ♊

LATITUDES

Date	Mercury ☿	Venus ♀	Mars ♂	Jupiter ♃	Saturn ♄	Uranus ♅	Neptune ♆	Pluto ♇
01	01 N 06	00 S 17	01 N 42	01 S 21	02 S 41	00 S 38	00 N 18	11 N 22
04	01 58	00 24	01 43	01 21	02 40	00 38	00 18	11 22
07	02 32	00 31	01 44	01 20	02 39	00 38	00 18	11 22
10	02 46	00 38	01 46	01 19	02 39	00 38	00 18	11 22
13	02 45	00 45	01 47	01 18	02 38	00 38	00 18	11 22
16	02 33	00 51	01 48	01 18	02 37	00 38	00 18	11 22
19	02 15	00 57	01 49	01 17	02 36	00 38	00 18	11 22
22	01 53	01 03	01 50	01 16	02 35	00 38	00 18	11 22
25	01 29	01 08	01 51	01 16	02 35	00 38	00 18	11 22
28	01 04	01 13	01 53	01 15	02 34	00 38	00 18	11 23
31	00 N 40	01 S 18	01 N 54	01 S 14	02 S 33	00 S 38	00 N 18	11 N 23

DATA

Julian Date	2451149
Delta T	+63 seconds
Ayanamsa	23° 50' 20"
Synetic vernal point	05° ♓ 16' 39"
True obliquity of ecliptic	23° 26' 14"

MOON'S PHASES, APSIDES AND POSITIONS ☽

Date	h	m	Phase	Longitude ° '	Eclipse Indicator
03	15	19	○	11 ♊ 15	
10	17	54	☾	18 ♍ 28	
18	22	42	●	26 ♐ 48	
26	10	46	☽	04 ♈ 27	

Day	h	m		
02	12	23	Perigee	
14	17	13	Apogee	
30	17	54	Perigee	
05	10	57	Max dec	19° N 31'
12	08	56	0S	
19	21	25	Max dec	19° S 34'
26	17	41	0N	

LONGITUDES

Date	Chiron ⚷ ° '	Ceres ⚳ ° '	Pallas ⚴ ° '	Juno ⚵ ° '	Vesta ⚶ ° '	Black Moon Lilith ⚸ ° '
01	25 ♏ 33	03 ♊ 53	14 ♓ 30	16 ♏ 09	21 ♌ 13	09 ♏ 10
11	26 ♏ 50	03 ♊ 37	15 ♓ 50	19 ♏ 26	22 ♌ 06	10 ♏ 17
21	28 ♏ 05	01 ♊ 40	17 ♓ 38	22 ♏ 38	22 ♌ 26	11 ♏ 25
31	29 ♏ 15	00 ♊ 15	19 ♓ 51	25 ♏ 41	21 ♌ 57	12 ♏ 32

All ephemeris data is given at 12.00 UT and the Moon's longitude is additionally given for 24.00 UT

ASPECTARIAN

h m	Aspects
01 Tuesday	
00 16	☽ bQ ♀
01 00	☽ ± ♇
01 07	☽ ⊼ ♂
02 09	☽ ± ☉
02 18	☽ ∥ ♄
03 47	☽ ∠ ♃
04 05	☽ ± ☿
08 55	☿ ✶ ♅
10 36	☽ ⊼ ♇
11 04	☽ ± ♂
12 25	☽ ⊼ ☉
12 51	☽ ⊼ ☿
13 13	☽ □ ♅
14 52	☽ ⋕ ♇
15 23	☉ ☌ ☿
15 55	☽ ± ♀
02 Wednesday	
00 45	☉ ✶ ♅
01 53	☽ bQ ♂
02 17	☽ ⊼ ♀
03 43	☽ ✶ ♃
12 43	☿ ☌ ♇
17 25	☽ ⊻ ♄
20 24	♀ □ ♃
21 41	☽ △ ♆
22 46	☽ Q ♃
03 Thursday	
02 30	☽ △ ♂
02 52	☽ ⊥ ♄
08 26	☽ ☍ ☿
08 46	☿ ∥ ♆
10 13	☽ ☍ ♇
12 49	☽ △ ♅
15 19	☽ ☍ ☉
17 07	☽ ∠ ♄
21 32	☽ bQ ♆
04 Friday	
03 30	☽ □ ♃
04 09	☽ ⋕ ♅
06 12	☽ ☍ ♀
12 00	☽ ± ♆
12 53	☽ bQ ♅
16 54	☉ bQ ♄
17 07	☽ ✶ ♄
17 35	☽ ⋕ ☿
21 45	☽ ⊼ ♆
05 Saturday	
03 34	☽ ± ♅
04 31	☽ □ ♂
05 05	☽ ⊼ ☿
10 01	☿ ✶ ♂
10 47	☽ ⊼ ♇
12 44	☽ Q ♄
13 30	☽ ⊼ ♅
14 20	☽ ± ☿
19 41	☽ ⊼ ☉
20 49	☽ ± ♇
06 Sunday	
04 27	☽ bQ ☿
05 04	☽ △ ♃
06 35	☽ ± ☉
11 53	☽ Q ♂
12 06	☽ bQ ♇
12 32	☽ ⊼ ♀
15 51	☿ ± ♄
18 31	☿ ∥ ♅
19 08	☽ □ ♄
19 15	☽ ⋕ ♅
19 25	☽ ⋕ ☿
23 17	☽ bQ ☉
23 56	☽ ± ♀
07 Monday	
00 21	☽ ☍ ♆
04 33	♀ ⊥ ♆
04 51	☽ △ ☿
07 08	☽ bQ ♃
09 38	☽ ✶ ♂
14 07	☉ ∠ ♆
14 23	☽ △ ♇
15 46	♀ ∠ ♅
17 18	☽ ☍ ♅
17 27	☽ bQ ♀
23 18	☽ ± ♃
08 Tuesday	
04 09	☽ △ ☉
10 17	☽ ⊼ ♃
13 53	☽ ∠ ♂
23 48	☽ △ ♀
09 Wednesday	
01 01	☽ △ ♄
06 56	☽ ⊼ ♆
07 25	☽ ⊥ ♂
09 25	☽ □ ☿
11 52	♀ △ ♄
16 18	☽ ⋕ ♇
18 22	☽ ± ♆
19 21	☽ ⊻ ♂
22 09	☽ □ ♇
10 Thursday	
01 18	☽ ⊼ ♅
05 31	☽ bQ ♄
08 23	☽ ∥ ♄
11 47	☽ bQ ♆
12 58	☽ ± ♅
17 54	☽ □ ☉
19 34	☽ Q ☿
19 46	☽ ☍ ♃
23 03	☽ ± ♄
11 Friday	
00 23	☽ ⋕ ♃
06 30	☿ St D
06 44	☽ bQ ♅
09 27	☽ Q ♇
10 53	☽ ⊼ ♄
16 30	☽ □ ♀
17 29	☽ △ ♆
18 23	☉ □ ♃
19 22	☽ ✶ ☿
23 36	☽ ⋕ ♂
12 Saturday	
02 02	♀ ⊻ ♆
08 59	☽ Q ☉
09 12	☽ ☌ ♂
09 32	☽ ✶ ♇
12 50	☽ △ ♅
14 05	☉ Q ♂
17 01	♂ ✶ ♇
19 20	☽ ∥ ♂
13 Sunday	
00 11	☿ ⊻ ♀
02 03	☽ ∠ ☿
08 17	☽ ⊼ ♃
09 01	☽ Q ♀
11 31	☽ ✶ ☉
15 58	☽ ∠ ♇
17 19	☽ ∥ ♃
20 33	☽ ± ♃
20 50	☽ ⊥ ☿
23 11	☽ ☍ ♄
14 Monday	
06 12	☽ □ ♆
09 29	☽ ⊻ ☿
10 17	☽ ⊥ ♇
10 28	☽ ⋕ ♄
12 15	☽ ✶ ♀
14 58	☽ bQ ♃
20 48	☽ ∠ ☉
22 29	☽ ⊻ ♇
15 Tuesday	
00 50	☽ ⊻ ♂
01 43	♀ ⊥ ♅
01 50	☽ □ ♅
03 53	☽ ∥ ♇
13 31	☽ ⊥ ♂
16 41	☽ ⊥ ☉
18 43	☽ Q ♆
21 33	☽ △ ♃
22 09	☽ ∠ ♀
22 31	☿ ± ♄
16 Wednesday	
01 05	♂ △ ♅
05 53	☽ ⊻ ☉
08 29	☽ ∠ ♂
10 05	☿ ∥ ♅
11 39	☽ ⊼ ♄
14 12	☽ Q ♅
17 11	☉ ⊥ ♆
18 19	☽ ⊥ ♀
18 52	☽ ✶ ♆
23 37	☽ ± ♄
17 Thursday	
01 14	☽ ☌ ☿
07 40	☽ ⊻ ♀
09 53	☉ ∠ ♅
10 53	☽ ☌ ♇
14 11	☽ ✶ ♅
15 44	☽ ✶ ♂
17 27	☽ bQ ♄
18 Friday	
00 42	☽ ∠ ♆
05 11	☽ ∥ ♅
06 38	♀ Q ♃
09 45	☽ □ ♃
15 18	☽ ∥ ☿
15 30	♀ ⊻ ♇
16 24	☽ Q ♂
18 24	☽ ⊥ ♆
19 47	☽ ∠ ♅
22 42	☽ ☌ ☉
22 50	☽ △ ♄
19 Saturday	
00 15	☉ △ ♄
06 07	☽ ⊻ ♆
13 20	☽ ⊥ ♅
16 37	☽ ⊻ ☿
21 03	☽ Q ♃
21 41	☽ ⊻ ♇
20 Sunday	
00 42	♀ ⊻ ♅
00 56	☽ ⊻ ♅
00 58	☽ ☌ ♀
04 48	☽ □ ♂
05 10	☽ ⊥ ☿
09 14	☽ ⊥ ♇
13 05	☽ ∥ ☿
20 14	☽ ✶ ♃
23 52	☽ ∠ ☿
21 Monday	
02 25	☽ ∠ ♇
08 18	☽ □ ♄
13 16	☽ ⊻ ☉
13 22	☽ ∥ ♅
15 35	☽ ☌ ♆
16 28	☿ ∥ ♆
22 Tuesday	
00 47	☽ ∠ ♃
01 30	☽ ⊥ ☉
06 30	☿ ☌ ♇
06 41	☽ ✶ ♇
06 42	☽ ✶ ☿
09 53	☽ ☌ ♅
15 21	♀ □ ♂
15 46	☽ △ ♂
15 47	☽ ⊻ ♀
17 41	☽ ⊥ ♃
17 55	☽ Q ♄
19 19	☉ ⊻ ♆
19 39	☽ ∠ ☉
23 Wednesday	
04 03	☽ ⊥ ♀
04 51	☽ ⊻ ♃
04 57	☽ Q ♇
07 02	☽ Q ☿
13 44	♀ ⊥ ♇
15 27	♃ ⊥ ♄
15 56	☽ ✶ ♄
19 24	☿ ✶ ♅
20 25	☽ bQ ♂
22 13	☽ ∠ ♀
23 09	☽ ⊻ ♆
24 Thursday	
01 24	☽ ✶ ☉
07 33	☽ ∥ ♇
10 01	☽ ⊥ ♆
13 19	☽ ± ♂
13 43	☽ □ ♇
16 50	☽ ⊻ ♅
18 52	☽ □ ☿
18 58	☽ ∠ ♄
20 04	☿ bQ ♄
23 14	☽ ⋕ ♄
25 Friday	
00 27	☽ ⊼ ♂
00 42	☽ Q ☉
02 08	☽ ∠ ♆
03 33	☽ ⊥ ♅
03 54	☽ ✶ ♀
10 56	☽ ⊥ ♄
11 22	☽ ☌ ♃
17 07	☽ ∥ ♃
19 17	☽ ∥ ♂
19 28	☽ ∠ ♅
21 27	☽ ⊻ ♄
26 Saturday	
02 50	☉ ∥ ♀
03 06	☽ Q ♀
04 32	☽ ✶ ♆
10 46	☽ □ ☉
16 20	☉ ⊥ ♅
18 28	☽ △ ♇
21 30	☽ ✶ ♅
27 Sunday	
01 14	☽ Q ♆
04 26	☽ △ ☿
06 34	☽ ☍ ♂
10 18	♂ ∥ ♃
12 57	☽ □ ♀
15 33	☽ ⊻ ♃
17 12	☽ ⋕ ♃
17 35	☽ ⋕ ♂
17 56	☽ Q ♅
19 59	☽ bQ ♇
28 Monday	
00 41	☽ ☌ ♄
01 37	☿ ∠ ♆
01 43	☽ ⊥ ♃
07 38	☽ □ ♆
08 13	☽ bQ ☿
10 50	☽ ∥ ♄
11 03	☽ ± ♇
16 06	☿ ✶ ♂
16 49	☽ ∠ ♃
17 19	☽ △ ☉
21 00	☽ ⊼ ♇
21 38	♀ ✶ ♃
23 58	☽ □ ♅
29 Tuesday	
00 37	☽ ± ☿
01 27	☽ ⋕ ♇
10 19	☽ ⊼ ♂
11 28	☽ ⊼ ☿
15 45	♄ St D
17 42	☽ ✶ ♃
19 22	☽ △ ♀
19 47	☽ bQ ☉
20 28	☽ ± ♂
30 Wednesday	
02 06	☽ ⊻ ♄
08 59	☽ △ ♆
11 31	☽ ± ☉
11 38	☽ bQ ♂
11 52	☽ ⊥ ♄
13 26	☽ Q ♃
22 00	☽ ⊼ ☉
22 02	☽ bQ ♀
22 07	☽ ☍ ♇
23 01	♀ ∠ ♇
23 49	☉ ⊻ ♇
31 Thursday	
01 06	☽ △ ♅
02 30	☽ ∠ ♄
09 26	☽ bQ ♆
10 36	☽ ⋕ ♅
12 20	☿ ∥ ♀
12 53	☽ △ ♂
14 02	☽ ± ♀
17 20	☽ ☍ ☿
19 00	☽ □ ♃
19 27	☉ Q ♃

JANUARY 1999

LONGITUDES

Date	Sidereal time h m s	Sun ☉ ° ' "	Moon ☽ ° ' "	Moon ☽ 24.00 ° ' "	Mercury ☿ ° '	Venus ♀ ° '	Mars ♂ ° '	Jupiter ♃ ° '	Saturn ♄ ° '	Uranus ♅ ° '	Neptune ♆ ° '	Pluto ♇ ° '
01	18 42 48	10 ♑ 36 55	02 ♋ 17 01	09 ♋ 32 40	21 ♐ 58	26 ♑ 01	18 ♎ 38	22 ♓ 01	26 ♈ 47	10 ♒ 59	01 ♒ 05	09 ♐ 08
02	18 46 44	11 38 02	16 ♋ 44 09	23 ♋ 50 44	23 23	27 16	19 07	22 10	26 47	11 02	01 07	09 10
03	18 50 41	12 39 10	00 ♌ 51 51	07 ♌ 47 03	24 48	28 31	19 37	22 19	26 47	11 05	01 09	09 12
04	18 54 38	13 40 18	14 ♌ 36 04	21 ♌ 18 47	26 14	29 ♑ 46	20 06	22 28	26 48	11 08	01 11	09 14
05	18 58 34	14 41 26	27 ♌ 55 14	04 ♍ 25 36	27 41	01 ♒ 01	20 35	22 37	26 49	11 12	01 13	09 16
06	19 02 31	15 42 35	10 ♍ 50 08	17 ♍ 09 15	29 ♐ 08	02 16	21 04	22 47	26 49	11 15	01 16	09 18
07	19 06 27	16 43 43	23 ♍ 23 24	29 ♍ 33 07	00 ♑ 37	03 32	21 33	22 56	26 50	11 18	01 18	09 20
08	19 10 24	17 44 51	05 ♎ 38 57	11 ♎ 41 32	02 05	04 47	22 01	23 06	26 51	11 21	01 20	09 22
09	19 14 20	18 46 00	17 ♎ 41 29	23 ♎ 39 24	03 34	06 02	22 30	23 16	26 53	11 25	01 22	09 24
10	19 18 17	19 47 08	29 ♎ 35 57	05 ♏ 31 44	05 04	07 17	22 58	23 26	26 54	11 28	01 24	09 26
11	19 22 13	20 48 17	11 ♏ 27 20	17 ♏ 23 20	06 34	08 32	23 26	23 36	26 55	11 31	01 27	09 28
12	19 26 10	21 49 26	23 ♏ 20 15	29 ♏ 18 35	08 05	09 47	23 54	23 46	26 57	11 35	01 29	09 30
13	19 30 07	22 50 34	05 ♐ 18 46	11 ♐ 21 13	09 36	11 02	24 21	23 57	26 58	11 38	01 31	09 32
14	19 34 03	23 51 43	17 ♐ 26 16	23 ♐ 34 12	11 08	12 17	24 49	24 07	27 00	11 41	01 33	09 33
15	19 38 00	24 52 51	29 ♐ 45 15	05 ♑ 59 35	12 40	13 33	25 16	24 18	27 02	11 45	01 36	09 35
16	19 41 56	25 53 59	12 ♑ 17 19	18 ♑ 38 31	14 13	14 48	25 43	24 28	27 04	11 48	01 38	09 37
17	19 45 53	26 55 06	25 ♑ 03 12	01 ♒ 31 20	15 46	16 03	26 10	24 39	27 06	11 52	01 40	09 39
18	19 49 49	27 56 13	08 ♒ 02 52	14 ♒ 37 43	17 20	17 18	26 36	24 50	27 08	11 55	01 43	09 40
19	19 53 46	28 57 19	21 ♒ 15 46	27 ♒ 56 54	18 54	18 33	27 03	25 01	27 10	11 58	01 45	09 42
20	19 57 42	29 ♑ 58 25	04 ♓ 40 58	11 ♓ 27 53	20 29	19 48	27 29	25 12	27 12	12 02	01 47	09 44
21	20 01 39	00 ♒ 59 30	18 ♓ 17 28	25 ♓ 09 38	22 05	21 03	27 54	25 23	27 15	12 05	01 49	09 45
22	20 05 36	02 00 34	02 ♈ 04 15	09 ♈ 01 12	23 41	22 18	28 20	25 35	27 17	12 09	01 52	09 47
23	20 09 32	03 01 36	16 ♈ 00 20	23 ♈ 01 30	25 17	23 32	28 45	25 46	27 20	12 12	01 54	09 49
24	20 13 29	04 02 38	00 ♉ 04 34	07 ♉ 09 19	26 54	24 47	29 11	25 58	27 23	12 16	01 56	09 50
25	20 17 25	05 03 39	14 ♉ 15 31	21 ♉ 22 53	28 ♑ 32	26 02	29 ♎ 35	26 09	27 26	12 19	01 58	09 52
26	20 21 22	06 04 39	28 ♉ 31 06	05 ♊ 39 48	00 ♒ 10	27 17	00 ♏ 00	26 21	27 29	12 23	02 01	09 53
27	20 25 18	07 05 37	12 ♊ 48 33	19 ♊ 56 52	01 49	28 32	00 24	26 33	27 32	12 26	02 03	09 55
28	20 29 15	08 06 35	27 ♊ 04 16	04 ♋ 10 12	03 29	29 ♒ 47	00 48	26 45	27 35	12 30	02 05	09 56
29	20 33 11	09 07 31	11 ♋ 14 10	18 ♋ 15 37	05 09	01 ♓ 01	01 12	26 57	27 38	12 33	02 08	09 58
30	20 37 08	10 08 27	25 ♋ 14 02	02 ♌ 08 58	06 50	02 16	01 36	27 09	27 41	12 37	02 10	09 59
31	20 41 05	11 ♒ 09 21	09 ♌ 00 00	15 ♌ 46 48	08 ♒ 31	03 ♓ 31	01 ♏ 59	27 ♓ 21	27 ♈ 45	12 ♒ 40	02 ♒ 12	10 ♐ 01

Date	Moon True ☊ ° '	Moon Mean ☊ ° '	Moon ☽ Latitude ° '
01	22 ♌ 48	24 ♌ 22	03 S 59
02	22 R 43	24 19	03 02
03	22 39	24 16	01 55
04	22 37	24 13	00 S 43
05	22 D 36	24 10	00 N 29
06	22 38	24 06	01 38
07	22 39	24 03	02 40
08	22 41	24 00	03 33
09	22 42	23 57	04 16
10	22 R 41	23 54	04 47
11	22 39	23 51	05 06
12	22 35	23 47	05 12
13	22 31	23 44	05 05
14	22 26	23 41	04 43
15	22 21	23 38	04 09
16	22 16	23 35	03 22
17	22 13	23 32	02 23
18	22 11	23 28	01 17
19	22 10	23 25	00 N 05
20	22 D 10	23 22	01 S 08
21	22 11	23 19	02 19
22	22 13	23 16	03 23
23	22 14	23 12	04 15
24	22 15	23 09	04 52
25	22 R 15	23 06	05 13
26	22 14	23 03	05 14
27	22 12	23 00	04 56
28	22 10	22 57	04 19
29	22 08	22 53	03 27
30	22 07	22 50	02 23
31	22 ♌ 06	22 ♌ 47	01 S 12

DECLINATIONS

Date	Sun ☉ ° '	Moon ☽ ° '	Mercury ☿ ° '	Venus ♀ ° '	Mars ♂ ° '	Jupiter ♃ ° '	Saturn ♄ ° '	Uranus ♅ ° '	Neptune ♆ ° '	Pluto ♇ ° '
01	23 S 01	19 N 26	22 S 40	22 S 14	05 S 32	04 S 18	07 N 57	18 S 05	19 S 37	10 S 34
02	22 56	19 23	22 53	22 01	05 43	04 15	07 57	18 04	19 37	10 34
03	22 50	18 05	23 04	21 48	05 54	04 11	07 58	18 03	19 37	10 34
04	22 44	15 46	23 15	21 33	06 05	04 07	07 58	18 02	19 36	10 34
05	22 38	12 39	23 25	21 18	06 15	04 03	07 59	18 01	19 36	10 34
06	22 31	09 01	23 33	21 02	06 26	03 59	07 59	18 00	19 35	10 34
07	22 23	05 04	23 41	20 46	06 36	03 55	08 00	17 59	19 35	10 35
08	22 16	01 N 01	23 47	20 29	06 46	03 51	08 01	17 59	19 34	10 35
09	22 07	03 S 00	23 52	20 11	06 56	03 47	08 01	17 58	19 34	10 35
10	21 59	06 51	23 56	19 53	07 06	03 43	08 02	17 57	19 33	10 35
11	21 50	10 24	23 58	19 34	07 16	03 39	08 03	17 56	19 33	10 35
12	21 40	13 34	24 00	19 15	07 26	03 35	08 04	17 55	19 32	10 35
13	21 30	16 11	24 00	18 55	07 36	03 30	08 04	17 54	19 32	10 35
14	21 20	18 09	23 59	18 35	07 46	03 26	08 05	17 53	19 31	10 35
15	21 09	19 18	23 56	18 14	07 55	03 22	08 06	17 52	19 31	10 35
16	20 58	19 31	23 53	17 53	08 05	03 17	08 07	17 51	19 30	10 35
17	20 46	18 46	23 47	17 31	08 14	03 13	08 08	17 50	19 30	10 35
18	20 34	17 01	23 41	17 08	08 23	03 08	08 09	17 49	19 29	10 35
19	20 22	14 20	23 33	16 45	08 32	03 04	08 10	17 48	19 29	10 35
20	20 09	10 51	23 24	16 22	08 41	02 59	08 11	17 47	19 28	10 36
21	19 56	06 46	23 13	15 58	08 50	02 55	08 12	17 46	19 28	10 36
22	19 43	02 S 16	23 01	15 34	08 59	02 50	08 14	17 45	19 27	10 36
23	19 29	02 N 22	22 48	15 09	09 08	02 45	08 15	17 44	19 27	10 36
24	19 15	06 56	22 33	14 44	09 16	02 41	08 16	17 43	19 26	10 36
25	19 00	11 08	22 16	14 19	09 24	02 36	08 17	17 42	19 26	10 36
26	18 45	14 43	21 59	13 53	09 33	02 31	08 19	17 41	19 25	10 36
27	18 30	17 27	21 39	13 27	09 41	02 26	08 20	17 40	19 25	10 35
28	18 14	19 05	21 18	13 00	09 49	02 22	08 21	17 39	19 24	10 35
29	17 58	19 31	20 56	12 33	09 57	02 17	08 23	17 38	19 24	10 35
30	17 42	18 44	20 33	12 06	10 05	02 12	08 24	17 38	19 23	10 35
31	17 S 26	16 N 51	20 S 07	11 S 39	10 S 12	02 S 07	08 N 26	17 S 37	19 S 23	10 S 35

ZODIAC SIGN ENTRIES

Date	h	m	Planets
01	08	15	☽ ♋
03	10	31	☽ ♌
04	16	25	♀ ♒
05	15	49	☽ ♍
07	02	04	☿ ♑
08	00	53	☽ ♎
10	12	49	☽ ♏
13	01	23	☽ ♐
15	12	29	☽ ♑
17	21	11	☽ ♒
20	03	40	☽ ♓
20	12	37	☉ ♒
22	08	25	☽ ♈
24	11	52	☽ ♉
26	09	32	☿ ♒
26	11	59	♂ ♏
26	14	29	☽ ♊
28	16	17	♀ ♓
28	16	57	☽ ♋
30	20	16	☽ ♌

LATITUDES

Date	Mercury ☿ ° '	Venus ♀ ° '	Mars ♂ ° '	Jupiter ♃ ° '	Saturn ♄ ° '	Uranus ♅ ° '	Neptune ♆ ° '	Pluto ♇ ° '
01	00 N 32	01 S 19	01 N 54	01 S 14	02 S 33	00 S 38	00 N 18	11 N 23
04	00 N 08	01 23	01 55	01 14	02 32	00 38	00 18	11 23
07	00 S 14	01 26	01 56	01 13	02 31	00 38	00 18	11 24
10	00 35	01 29	01 58	01 13	02 30	00 38	00 18	11 24
13	00 55	01 31	01 59	01 12	02 29	00 38	00 18	11 25
16	01 12	01 33	02 00	01 11	02 28	00 38	00 18	11 25
19	01 28	01 34	02 01	01 11	02 27	00 38	00 18	11 26
22	01 41	01 34	02 02	01 10	02 27	00 38	00 18	11 27
25	01 52	01 34	02 03	01 10	02 26	00 38	00 18	11 27
28	01 59	01 33	02 04	01 10	02 25	00 38	00 18	11 28
31	02 S 04	01 S 32	02 N 05	01 S 09	02 S 24	00 S 38	00 N 18	11 N 29

DATA

Julian Date	2451180
Delta T	+64 seconds
Ayanamsa	23° 50' 26"
Synetic vernal point	05° ♓ 16' 34"
True obliquity of ecliptic	23° 26' 14"

MOON'S PHASES, APSIDES AND POSITIONS ☽

Date	h	m	Phase	Longitude ° '	Eclipse Indicator
02	02	50	○	11 ♋ 15	
09	14	22	☾	18 ♎ 52	
17	15	46	●	27 ♑ 05	
24	19	15	☽	04 ♉ 21	
31	16	07	○	11 ♌ 20	

Day	h	m		
11	11	45	Apogee	
26	21	31	Perigee	
01	22	47	Max dec	19° N 34'
08	18	02	OS	
16	05	47	Max dec	19° S 33'
22	23	46	ON	
29	08	28	Max dec	19° N 32'

LONGITUDES

Date	Chiron ⚷ ° '	Ceres ⚳ ° '	Pallas ⚴ ° '	Juno ⚵ ° '	Vesta ⚶ ° '	Black Moon Lilith ⚸ ° '
01	29 ♏ 21	00 ♊ 08	20 ♓ 05	25 ♏ 59	21 ♌ 51	12 ♏ 39
11	00 ♐ 25	29 ♉ 26	22 ♓ 42	28 ♏ 53	20 ♌ 32	13 ♏ 46
21	01 ♐ 22	29 ♉ 25	25 ♓ 37	01 ♐ 36	18 ♌ 32	14 ♏ 53
31	02 ♐ 11	00 ♊ 04	28 ♓ 47	04 ♐ 06	16 ♌ 03	16 ♏ 00

All ephemeris data is given at 12.00 UT and the Moon's longitude is additionally given for 24.00 UT

ASPECTARIAN

h m	Aspects
01 Friday	
00 09	☽ ± ♆
00 45	☽ ⊼ ♀
01 37	☽ bQ ♅
02 57	☽ ✶ ♄
10 01	☽ ⊼ ♆
12 51	☿ □ ♃
16 28	☽ ± ♅
21 08	☉ ⊻ ♅
22 44	☽ Q ♄
23 21	☽ ⊼ ♇
02 Saturday	
02 27	☽ ⊼ ♅
02 44	♀ □ ♄
02 50	☽ ☍ ☉
09 22	☽ ± ♇
16 06	☉ ∥ ☿
16 09	☽ □ ♂
21 15	☽ △ ♃
03 Sunday	
00 28	☽ ⊼ ☿
00 35	☽ bQ ♇
05 01	☽ □ ♄
07 34	☽ ☍ ♀
11 53	☽ ± ☿
12 28	☽ # ♅
12 30	☽ ☍ ♆
17 58	☿ ⊥ ♆
23 18	☽ bQ ♃
04 Monday	
00 08	☽ Q ♂
02 31	☽ △ ♇
05 21	☽ bQ ☿
05 52	☽ ☍ ♅
10 13	☽ ⊼ ☉
10 20	☿ ∠ ♅
15 21	☽ ± ♃
21 25	☿ △ ♄
21 48	☽ ± ☉
22 11	☽ ✶ ♂
05 Tuesday	
02 15	☽ ⊼ ♃
09 58	☽ △ ♄
11 31	☽ △ ☿
15 31	☽ bQ ☉
15 57	♀ ☌ ♆
18 05	☽ ⊼ ♆
18 18	☽ ⊼ ♀
06 Wednesday	
02 00	☽ # ♇
02 04	☉ ⊥ ♇
02 42	☽ ∠ ♂
05 16	☽ ± ♆
06 40	☽ ± ♀
09 06	☽ □ ♇
12 47	☽ ⊼ ♅
13 52	☽ bQ ♄
18 19	☽ ∥ ♄
20 20	☽ ⊥ ♂
22 03	☽ △ ☉
22 19	☽ bQ ♆
07 Thursday	
00 14	☽ ± ♅
01 35	☽ bQ ♀
03 12	☽ # ♂
07 04	☽ ± ♄
08 17	☽ ⊻ ♂
11 07	☽ ☍ ♃
17 41	☽ bQ ♅
18 43	☽ ⊼ ♄
18 57	☽ # ♃
19 41	☽ Q ♇
23 28	☿ ⊻ ♆
08 Friday	
03 28	☽ △ ♆
04 00	☽ □ ☿
10 05	☽ △ ♀
19 23	☽ ✶ ♇
23 23	☽ △ ♅
09 Saturday	
14 22	☽ □ ☉
16 45	☽ ∥ ♃
20 55	☽ Q ☿
22 03	☽ ☌ ♂
23 22	☽ ⊼ ♃
10 Sunday	
01 32	☽ ∠ ♇
06 32	☽ ☍ ♄
09 35	☿ Q ♂
11 39	☽ ± ♃
13 47	☽ ∥ ♂
15 40	☽ □ ♆
18 38	☿ ⊥ ♅
19 46	☽ ⊥ ♇
19 48	☽ # ♄
11 Monday	
00 40	☽ ✶ ☿
05 23	☽ □ ♀
06 08	☽ Q ☉
06 08	☽ bQ ♃
07 57	☽ ⊻ ♇
12 08	☽ □ ♅
13 16	☽ ∥ ♇
13 25	♀ ∠ ♃
14 10	♀ ∥ ♆
12 Tuesday	
01 51	♂ ⊼ ♃
04 12	☽ Q ♆
06 14	♀ ✶ ♇
08 40	☽ ✶ ☉
11 25	☽ ∠ ☿
12 53	☽ △ ♃
13 10	☽ ⊻ ♂
19 16	☽ ⊼ ♄
22 00	☽ Q ♀
13 Wednesday	
00 36	☽ Q ♅
01 42	☽ ⊥ ♂
04 24	☽ ✶ ♆
07 19	☽ ± ♄
08 06	☽ ⊥ ☿
10 44	☿ ⊻ ♇
17 30	☽ ∠ ☉
20 21	☽ ∠ ♂
20 24	☽ ☌ ♇
21 36	♂ ∠ ♇
21 46	☽ ⊻ ☿
23 56	♀ ☌ ♅
14 Thursday	
00 37	☽ ✶ ♅
00 41	☽ ✶ ♀
01 15	☽ bQ ♄
08 15	☽ ∥ ♅
10 16	☽ ∠ ♆
12 55	☽ ⊥ ☉
17 49	☽ ∥ ♀
19 15	☉ ✶ ♃
21 02	☿ ⊻ ♅
15 Friday	
01 15	☽ □ ♃
01 42	☽ ⊻ ☉
02 58	☽ ✶ ♂
03 55	☽ ⊥ ♆
04 50	☉ ∠ ♇
05 21	☿ Q ♃
06 09	☽ ∠ ♅
06 43	☽ △ ♄
09 24	☽ ∠ ♀
15 34	☽ ⊻ ♆
22 21	☽ ∥ ♆
23 35	☽ ⊥ ♅
16 Saturday	
02 58	☽ Q ♂
04 12	☉ □ ♂
04 37	☽ ⊥ ♀
06 54	☽ ⊻ ♇
11 04	☽ ⊻ ♅
12 21	☽ Q ♃
13 36	☽ ∥ ♆
13 47	♀ ∥ ♅
16 10	☽ ☌ ☿
17 16	☽ ⊻ ♀
17 18	♀ Q ♄
18 19	☽ ⊥ ♇
19 14	♂ # ♄
17 Sunday	
10 01	☿ ⊥ ♇
11 14	☽ ∠ ♇
11 14	☽ ✶ ♃
14 08	☽ □ ♂
15 46	☽ ☌ ☉
15 49	☽ □ ♄
16 19	☉ □ ♄
18 Monday	
00 19	☽ ☌ ♆
02 29	☽ ∥ ♅
08 56	☿ ⊻ ♀
10 23	☽ ∥ ♀
14 59	☽ ✶ ♇
15 19	☽ ∠ ♃
19 06	☽ ☌ ♅
19 Tuesday	
00 57	☽ Q ♄
06 35	☽ ☌ ♀
07 11	☽ ⊻ ☿
07 53	☽ ⊥ ♃
12 48	☽ Q ♇
18 51	☽ ⊻ ♃
19 26	☽ ⊥ ☿
19 37	♂ ☍ ♄
22 38	☽ ✶ ♄
22 39	♀ ⊥ ♃
22 44	☽ △ ♂
20 Wednesday	
02 56	☽ ⊻ ☉
06 50	☽ ⊻ ♆
13 37	☽ ∠ ☿
13 38	☽ ∥ ♇
14 29	☽ ⊥ ☉
17 31	☽ ⊥ ♆
20 57	☽ □ ♇
21 Thursday	
00 37	☽ ∥ ♂
01 03	☽ ⊻ ♅
01 21	☽ ∠ ♄
02 14	☽ bQ ♂
03 53	☽ # ♄
07 38	☽ ∠ ☉
09 25	☽ ∠ ♆
11 39	☽ ⊥ ♅
17 11	☽ ⊥ ♄
17 18	☽ ⊻ ♀
18 32	☽ ± ♂
19 29	☽ ✶ ☿
22 Friday	
00 34	☽ ☌ ♃
02 03	♀ Q ♇
03 25	☽ ∠ ♅
03 41	☽ ⊻ ♄
04 48	☽ ⊥ ♀
05 19	☽ ⊼ ♂
08 22	☉ ☌ ♆
09 01	☽ ∥ ♃
11 38	☽ ✶ ♆
11 53	☽ ✶ ☉
19 03	☽ Q ☿
21 55	☽ ∠ ♀
23 Saturday	
01 21	☽ △ ♇
04 52	☿ ∠ ♇
05 27	☽ ✶ ♅
08 23	☽ Q ♆
10 11	☽ Q ☉
13 58	☽ # ♃
15 24	☉ ∥ ♆
20 09	☿ ✶ ♃
24 Sunday	
02 04	☽ Q ♅
02 08	☽ ✶ ♀
03 04	☽ bQ ♇
04 54	☽ ⊻ ♃
05 54	☽ □ ☿
07 24	☽ ☌ ♄
10 25	☽ ☍ ♂
15 10	☽ □ ♆
15 14	☽ ⊥ ♃
18 24	☽ ± ♇
19 15	☽ □ ☉
19 16	☿ □ ♄
19 24	☽ ∥ ♄
25 Monday	
00 29	☽ Q ♀
01 27	☽ # ♂
04 34	☽ ⊼ ♇
06 41	☽ ∠ ♃
08 43	☽ □ ♅
08 44	☽ # ♇
14 39	♀ ⊻ ♃
26 Tuesday	
06 36	☽ # ♀
08 18	☽ ✶ ♃
08 43	☿ □ ♂
09 44	☽ □ ♀
10 15	☽ ⊻ ♄
14 34	☽ ⊼ ♂
15 08	☽ △ ☿
15 52	♀ ✶ ♄
17 53	☽ △ ♆
20 22	☽ ⊥ ♄
27 Wednesday	
00 56	☽ ± ♂
01 40	☽ △ ☉
04 44	☽ Q ♃
07 08	☽ ☍ ♇
11 22	☽ △ ♅
11 32	☽ ∠ ♄
14 33	☽ # ♅
15 27	☿ ☌ ♆
16 29	☽ bQ ♂
19 09	☽ bQ ♆
19 37	☽ bQ ☿
23 29	☽ # ☉
28 Thursday	
04 49	☽ bQ ☉
10 20	☽ ± ♆
11 26	☽ □ ♃
12 43	☽ bQ ♅
12 47	☽ ± ☿
12 52	☽ ✶ ♄
17 01	☽ △ ♀
18 29	☽ △ ♂
20 30	☽ ⊼ ♆
21 10	☽ ± ☉
21 27	☽ # ♆
29 Friday	
00 16	☽ ⊼ ☿
04 00	☽ ± ♅
08 08	☽ ⊼ ☉
09 16	☽ Q ♄
09 50	☽ ⊼ ♇
14 15	☽ ⊼ ♅
17 04	♀ △ ♂
19 50	☽ # ♆
20 05	☽ ± ♇
20 58	☽ bQ ♀
30 Saturday	
08 18	☉ ✶ ♇
09 54	♀ ⊻ ♆
11 34	☽ bQ ♇
13 58	☽ ± ♀
15 21	☽ △ ♃
16 16	☽ □ ♄
19 09	☉ ∥ ♅
23 21	☽ □ ♂
31 Sunday	
00 04	☽ ☍ ♆
01 26	☽ ⊼ ♀
03 43	☽ # ♅
04 58	☽ # ☉
11 03	☽ ☍ ☿
13 47	☽ △ ♇
16 07	☽ ☍ ☉
18 00	☽ bQ ♃
18 30	☽ ☍ ♅

FEBRUARY 1999

LONGITUDES

Date	Sidereal time	Sun ☉	Moon ☽	Moon ☽ 24.00	Mercury ☿	Venus ♀	Mars ♂	Jupiter ♃	Saturn ♄	Uranus ♅	Neptune ♆	Pluto ♇
	h m s	° ′ ″	° ′ ″	° ′ ″	° ′	° ′	° ′	° ′	° ′	° ′	° ′	° ′
01	20 45 01	12 ♒ 10 14	22 ♌ 29 06	29 ♌ 06 43	10 ♒ 14	04 ♓ 46	02 ♏ 22	27 ♓ 33	27 ♈ 48	12 ♒ 44	02 ♒ 14	10 ♐ 02
02	20 48 58	13 11 06	05 ♍ 39 34	12 ♍ 07 38	11 57	06 00	02 45	27 45	27 52	12 47	02 17	10 03
03	20 52 54	14 11 58	18 ♍ 31 01	24 ♍ 49 53	13 41	07 15	03 07	27 58	27 56	12 51	02 19	10 05
04	20 56 51	15 12 48	01 ♎ 04 29	07 ♎ 15 07	15 25	08 29	03 29	28 10	28 00	12 54	02 21	10 06
05	21 00 47	16 13 37	13 ♎ 22 12	19 ♎ 26 09	17 10	09 44	03 51	28 23	28 04	12 58	02 23	10 07
06	21 04 44	17 14 26	25 ♎ 27 27	01 ♏ 26 37	18 56	10 59	04 12	28 35	28 08	13 01	02 26	10 08
07	21 08 40	18 15 13	07 ♏ 24 13	13 ♏ 20 49	20 43	12 13	04 33	28 48	28 12	13 05	02 28	10 10
08	21 12 37	19 16 00	19 ♏ 17 00	25 ♏ 13 22	22 30	13 28	04 54	29 01	28 16	13 08	02 30	10 11
09	21 16 33	20 16 45	01 ♐ 10 31	07 ♐ 09 02	24 18	14 42	05 14	29 14	28 20	13 12	02 32	10 12
10	21 20 30	21 17 30	13 ♐ 09 28	19 ♐ 12 23	26 06	15 56	05 34	29 27	28 24	13 15	02 34	10 13
11	21 24 27	22 18 13	25 ♐ 18 16	01 ♑ 27 36	27 55	17 11	05 54	29 40	28 29	13 19	02 37	10 14
12	21 28 23	23 18 56	07 ♑ 40 47	13 ♑ 58 11	29 ♒ 44	18 25	06 13	29 ♓ 53	28 33	13 22	02 39	10 15
13	21 32 20	24 19 37	20 ♑ 20 04	26 ♑ 46 40	01 ♓ 34	19 40	06 32	00 ♈ 06	28 38	13 25	02 41	10 16
14	21 36 16	25 20 17	03 ♒ 18 07	09 ♒ 54 25	03 24	20 54	06 50	00 19	28 43	13 29	02 43	10 17
15	21 40 13	26 20 56	16 ♒ 35 33	23 ♒ 21 22	05 14	22 08	07 08	00 32	28 48	13 32	02 45	10 18
16	21 44 09	27 21 33	00 ♓ 11 38	07 ♓ 06 03	07 04	23 22	07 26	00 46	28 52	13 36	02 47	10 19
17	21 48 06	28 22 09	14 ♓ 04 11	21 ♓ 05 37	08 54	24 36	07 43	00 59	28 57	13 39	02 49	10 20
18	21 52 02	29 ♒ 22 43	28 ♓ 09 51	05 ♈ 16 19	10 43	25 51	08 00	01 12	29 02	13 43	02 51	10 20
19	21 55 59	00 ♓ 23 15	12 ♈ 24 29	19 ♈ 33 47	12 31	27 05	08 16	01 26	29 08	13 46	02 54	10 21
20	21 59 56	01 23 46	26 ♈ 43 40	03 ♉ 53 39	14 19	28 19	08 32	01 39	29 13	13 49	02 56	10 22
21	22 03 52	02 24 15	11 ♉ 03 14	18 ♉ 12 00	16 05	29 ♓ 33	08 47	01 53	29 18	13 53	02 58	10 23
22	22 07 49	03 24 42	25 ♉ 19 34	02 ♊ 25 38	17 49	00 ♈ 47	09 02	02 07	29 23	13 56	03 00	10 23
23	22 11 45	04 25 07	09 ♊ 29 53	16 ♊ 32 07	19 30	02 01	09 17	02 20	29 29	13 59	03 02	10 24
24	22 15 42	05 25 30	23 ♊ 32 08	00 ♋ 29 46	21 09	03 15	09 31	02 34	29 34	14 03	03 04	10 25
25	22 19 38	06 25 51	07 ♋ 24 53	14 ♋ 17 21	22 44	04 28	09 44	02 48	29 40	14 06	03 06	10 25
26	22 23 35	07 26 11	21 ♋ 07 06	27 ♋ 53 59	24 16	05 42	09 57	03 02	29 45	14 09	03 08	10 26
27	22 27 31	08 26 28	04 ♌ 37 56	11 ♌ 18 51	25 43	06 56	10 09	03 15	29 51	14 13	03 10	10 26
28	22 31 28	09 ♓ 26 43	17 ♌ 56 39	24 ♌ 31 15	27 ♓ 05	08 ♈ 10	10 ♏ 21	03 ♈ 29	29 ♈ 57	14 ♒ 16	03 ♒ 12	10 ♐ 27

Date	Moon True ☊	Moon Mean ☊	Moon ☽ Latitude
	° ′	° ′	° ′
01	22 ♌ 06	22 ♌ 44	00 N 02
02	22 D 06	22 41	01 14
03	22 07	22 38	02 21
04	22 07	22 34	03 20
05	22 08	22 31	04 08
06	22 09	22 28	04 44
07	22 09	22 25	05 07
08	22 09	22 22	05 17
09	22 R 09	22 18	05 13
10	22 09	22 15	04 56
11	22 D 09	22 12	04 26
12	22 09	22 09	03 43
13	22 09	22 06	02 48
14	22 09	22 03	01 43
15	22 10	21 59	00 N 31
16	22 R 10	21 56	00 S 44
17	22 09	21 53	01 59
18	22 09	21 50	03 07
19	22 08	21 47	04 04
20	22 07	21 43	04 46
21	22 06	21 40	05 11
22	22 05	21 37	05 16
23	22 D 05	21 34	05 02
24	22 06	21 31	04 30
25	22 07	21 28	03 43
26	22 08	21 24	02 43
27	22 09	21 21	01 35
28	22 ♌ 10	21 ♌ 18	00 S 23

DECLINATIONS

Date	Sun ☉	Moon ☽	Mercury ☿	Venus ♀	Mars ♂	Jupiter ♃	Saturn ♄	Uranus ♅	Neptune ♆	Pluto ♇
	° ′	° ′	° ′	° ′	° ′	° ′	° ′	° ′	° ′	° ′
01	17 S 09	14 N 03	19 S 41	11 S 11	10 S 20	02 S 02	08 N 27	17 S 36	19 S 22	10 S 35
02	16 52	10 35	19 12	10 43	10 27	01 57	08 29	17 35	19 22	10 35
03	16 34	06 42	18 43	10 14	10 35	01 52	08 30	17 34	19 21	10 35
04	16 16	02 N 38	18 12	09 45	10 42	01 47	08 32	17 33	19 21	10 35
05	15 58	01 S 28	17 39	09 16	10 49	01 42	08 34	17 32	19 20	10 35
06	15 40	05 26	17 05	08 47	10 56	01 36	08 35	17 31	19 20	10 35
07	15 21	09 08	16 29	08 18	11 02	01 31	08 37	17 30	19 19	10 35
08	15 03	12 28	15 52	07 48	11 09	01 26	08 39	17 29	19 19	10 35
09	14 43	15 17	15 14	07 18	11 16	01 21	08 40	17 28	19 18	10 34
10	14 24	17 29	14 34	06 48	11 22	01 16	08 42	17 27	19 18	10 34
11	14 05	18 56	13 52	06 18	11 28	01 10	08 44	17 26	19 17	10 34
12	13 45	19 31	13 10	05 48	11 34	01 05	08 46	17 25	19 17	10 34
13	13 25	19 08	12 26	05 17	11 40	01 00	08 48	17 24	19 16	10 34
14	13 04	17 45	11 41	04 47	11 46	00 54	08 50	17 23	19 16	10 34
15	12 44	15 22	10 54	04 16	11 51	00 49	08 51	17 22	19 16	10 34
16	12 23	12 06	10 07	03 45	11 57	00 44	08 53	17 21	19 15	10 33
17	12 02	08 05	09 19	03 14	12 02	00 38	08 55	17 20	19 15	10 33
18	11 41	03 S 35	08 30	02 43	12 07	00 33	08 57	17 19	19 14	10 33
19	11 20	01 N 10	07 40	02 12	12 12	00 27	08 59	17 18	19 14	10 33
20	10 59	05 51	06 50	01 40	12 17	00 22	09 01	17 17	19 13	10 33
21	10 37	10 13	05 59	01 09	12 22	00 16	09 03	17 16	19 13	10 32
22	10 15	13 59	05 08	00 38	12 27	00 11	09 05	17 15	19 12	10 32
23	09 53	16 54	04 18	00 S 06	12 31	00 S 05	09 08	17 14	19 12	10 32
24	09 31	18 47	03 27	00 N 25	12 35	00 00	09 10	17 13	19 11	10 32
25	09 09	19 31	02 38	00 56	12 39	00 N 06	09 12	17 12	19 11	10 32
26	08 47	19 05	01 49	01 28	12 43	00 11	09 14	17 11	19 10	10 31
27	08 24	17 34	01 02	01 59	12 47	00 17	09 16	17 10	19 10	10 31
28	08 S 02	15 N 05	00 S 17	02 N 30	12 S 51	00 N 22	09 N 18	17 S 09	19 S 10	10 S 31

ZODIAC SIGN ENTRIES

Date	h m	Planets
02	01 37	☽ ♍
04	09 56	☽ ♎
06	21 06	☽ ♏
09	09 38	☽ ♐
11	21 10	☽ ♑
12	15 28	☿ ♓
13	01 22	♃ ♈
14	05 57	☽ ♒
16	11 40	☽ ♓
18	15 06	☽ ♈
19	02 47	☉ ♓
20	17 29	☽ ♉
21	20 49	♀ ♈
22	19 54	☽ ♊
24	23 09	☽ ♋
27	03 44	☽ ♌

LATITUDES

Date	Mercury ☿	Venus ♀	Mars ♂	Jupiter ♃	Saturn ♄	Uranus ♅	Neptune ♆	Pluto ♇
	° ′	° ′	° ′	° ′	° ′	° ′	° ′	° ′
01	02 S 05	01 S 31	02 N 05	01 S 09	02 S 24	00 S 38	00 N 18	11 N 29
04	02 04	01 29	02 06	01 09	02 23	00 38	00 18	11 30
07	02 00	01 26	02 07	01 08	02 22	00 38	00 18	11 30
10	01 51	01 22	02 08	01 08	02 21	00 38	00 18	11 31
13	01 37	01 18	02 09	01 08	02 21	00 38	00 18	11 32
16	01 18	01 13	02 09	01 07	02 20	00 38	00 17	11 33
19	00 52	01 07	02 10	01 07	02 19	00 38	00 17	11 34
22	00 S 21	01 01	02 11	01 07	02 19	00 38	00 17	11 35
25	00 N 16	00 55	02 11	01 07	02 18	00 38	00 17	11 36
28	00 57	00 48	02 11	01 06	02 17	00 38	00 17	11 37
31	01 N 40	00 S 40	02 N 11	01 S 06	02 S 17	00 S 38	00 N 17	11 N 38

DATA

Julian Date	2451211
Delta T	+64 seconds
Ayanamsa	23° 50′ 31″
Synetic vernal point	05° ♓ 16′ 29″
True obliquity of ecliptic	23° 26′ 14″

LONGITUDES

Date	Chiron ⚷	Ceres ⚳	Pallas ⚴	Juno ⚵	Vesta ⚶	Black Moon Lilith ⚸
	° ′	° ′	° ′	° ′	° ′	° ′
01	02 ♐ 16	00 ♊ 10	29 ♓ 07	04 ♐ 20	15 ♌ 47	16 ♏ 07
11	02 ♐ 55	01 ♊ 29	02 ♈ 33	06 ♐ 32	13 ♌ 09	17 ♏ 14
21	03 ♐ 25	03 ♊ 19	06 ♈ 11	08 ♐ 25	10 ♌ 43	18 ♏ 21
31	03 ♐ 44	05 ♊ 37	10 ♈ 00	09 ♐ 56	08 ♌ 47	19 ♏ 29

MOON'S PHASES, APSIDES AND POSITIONS ☽

Date	h m	Phase	Longitude ° ′	Eclipse Indicator
08	11 58	☾	19 ♏ 16	
16	06 39	●	27 ♒ 08	Annular
23	02 43	☽	04 ♊ 02	

Day	h m		
08	08 51	Apogee	
20	14 28	Perigee	
05	03 19	0S	
12	14 53	Max dec	19° S 31′
19	06 10	0N	
25	15 04	Max dec	19° N 32′

All ephemeris data is given at 12.00 UT and the Moon's longitude is additionally given for 24.00 UT

ASPECTARIAN

01 Monday
03 11 ♂ □ ♆
08 05 ☽ Q ♂
09 12 ☿ ⚹ ♇
10 18 ☽ ± ♃
21 18 ☽ ⊼ ♃
21 40 ☽ △ ♄
23 16 ☉ ∠ ♃

02 Tuesday
01 59 ☉ ☌ ♅
03 56 ☿ ∥ ♆
05 46 ☽ ⊼ ♆
06 29 ☽ ⚹ ♂
11 07 ☽ # ♀
12 01 ☽ # ♇
12 42 ☽ ☍ ♀
12 50 ☽ # ♂
16 51 ☽ ± ♆
16 52 ♃ ∠ ♅
18 17 ♀ ∥ ♇
20 09 ☽ □ ♇
22 15 ♀ ∥ ♂

03 Wednesday
00 04 ☿ ☌ ♅
00 46 ☿ ∠ ♃
01 11 ☽ ∥ ♄
01 17 ☽ ⊼ ♅
01 27 ☽ ⚻ ♄
01 28 ☽ ⊼ ☿
03 11 ☽ ⊼ ☉
06 40 ♃ ⊻ ♄
09 43 ☽ ⚻ ♆
11 13 ☽ ∠ ♂
12 37 ☽ ± ♅
13 35 ♂ ∥ ♇
14 32 ☽ ± ☿
15 28 ☽ ± ☉
18 30 ☽ ± ♄

04 Thursday
04 52 ☽ ⊥ ♂
05 20 ☉ ☌ ☿
05 51 ☽ ⚻ ♅
06 02 ☽ ⊼ ♄
06 15 ☽ Q ♇
06 18 ☽ ☍ ♃
09 13 ♀ ⊥ ♆
10 11 ☽ ⚻ ☉
10 31 ☽ ⚻ ☿
14 29 ☽ △ ♆
16 48 ☽ ⊻ ♂
17 03 ☽ # ♃
20 14 ☿ Q ♄

05 Friday
04 03 ☽ ⊼ ♀
05 36 ☽ ⚹ ♇
07 45 ☉ Q ♄
11 11 ☽ △ ♅
13 16 ☽ ∥ ♃
17 12 ☽ ± ♀
17 22 ☿ ∥ ♅
18 09 ☽ △ ☉
19 34 ♀ □ ♇
20 47 ☽ △ ☿

06 Saturday
07 35 ♂ ⊥ ♇
11 22 ☽ ∠ ♇
13 09 ☽ ⚻ ♀
17 22 ☽ ☍ ♄
18 23 ☽ ⊼ ♃
21 17 ♀ # ♄

07 Sunday
02 01 ☽ □ ♆
05 27 ☽ ⊥ ♇
06 05 ☽ ☌ ♂
06 40 ☽ ± ♃
07 01 ☽ ∥ ♀
08 26 ☽ # ♄
17 34 ☽ ⊻ ♇
22 00 ☽ ∥ ♇
22 51 ☽ △ ♀
23 31 ☽ □ ♅

08 Monday
01 09 ☽ ⚻ ♃
01 48 ☽ ∥ ♂
05 25 ♀ ⊻ ♅
07 42 ☿ Q ♇
07 59 ♀ ∠ ♄
11 58 ☽ □ ☉
14 28 ☽ Q ♆
19 39 ☽ □ ☿
19 54 ☿ ⊥ ♃

09 Tuesday
06 15 ☽ ⊼ ♄
07 26 ☽ ∥ ☉
08 00 ☽ △ ♃
10 43 ♂ ± ♃
11 36 ☽ ∥ ☿
12 02 ☽ Q ♅
14 45 ☽ ⚹ ♆
18 23 ☽ ± ♄
20 24 ☽ ⊻ ♂

10 Wednesday
03 34 ☽ Q ☉
06 07 ☽ ☌ ♇
08 45 ☽ ⊥ ♂
11 35 ☽ ∥ ♅
12 11 ☽ ⚹ ♅
12 30 ☽ ⚻ ♄
14 12 ☽ Q ☿
18 10 ☽ □ ♀
20 48 ☽ ∠ ♆
22 38 ☉ ∥ ☿

11 Thursday
03 06 ☽ ∠ ♂
05 35 ☽ ⚹ ☉
10 16 ☉ Q ♇
14 34 ☽ ⊥ ♆
17 54 ☽ ∠ ♅
17 59 ☽ ⚹ ☿
18 15 ☽ △ ♄
19 49 ☿ ⚹ ♄
20 32 ♀ ∠ ♆
20 39 ☽ □ ♃
22 12 ☽ ∥ ♆

12 Friday
02 16 ☽ ⊻ ♆
09 07 ☽ ⚹ ♂
09 19 ☽ Q ♀
11 24 ☽ ⊥ ♅
13 20 ☽ ∠ ☉
14 07 ☿ ⊻ ♃
16 56 ☽ ⊻ ♇
22 54 ☽ ⊻ ♅

13 Saturday
03 43 ☽ ∠ ☿
04 20 ☽ ⊥ ♇
05 02 ☉ ⊥ ♃
07 14 ♀ ⊥ ♅
07 43 ☽ ∥ ♆
07 43 ☽ Q ♃
07 54 ☽ ⊥ ☉
08 32 ☽ Q ♂
10 36 ☽ ⚹ ♀
20 05 ☽ ⊻ ☉
21 13 ☽ ∠ ♇
23 22 ☽ ⊥ ☿

14 Sunday
02 54 ☿ ⊻ ♆
03 31 ☽ □ ♄
06 26 ☽ ⚹ ♃
09 35 ☿ ∥ ♂
10 56 ☽ ☌ ♆
12 12 ☽ ⊻ ☿
16 31 ☽ ∥ ♅
17 13 ☽ ∠ ♀
18 36 ☽ □ ♂

15 Monday
00 41 ☽ ⚹ ♇
06 31 ☽ ☌ ♅
10 05 ☽ ∠ ♃
11 06 ☽ ⊥ ♀
12 04 ♀ ⚻ ♂
12 22 ☽ Q ♄
22 08 ☽ Q ♇
22 41 ☿ ∥ ♇
22 50 ☽ ⊻ ♀

16 Tuesday
01 40 ♀ ⊥ ♄
02 19 ☽ ⊥ ♃
06 39 ☽ ☌ ☉
09 41 ☽ ⚹ ♄
09 52 ☽ ∥ ☉
12 56 ☽ ∥ ♂
13 00 ☽ ⊻ ♃
16 32 ☽ ⊻ ♆
17 41 ☿ △ ♂
21 41 ☽ ∥ ♇

17 Wednesday
00 50 ☽ △ ♂
01 45 ☽ ☌ ☿
02 57 ☽ ⊥ ♆
03 21 ☽ ∥ ☿
05 34 ☽ □ ♇
07 20 ☽ # ♄
11 03 ☿ ⊥ ♆
11 17 ☽ ⊻ ♅
11 48 ☽ ∠ ♄
12 08 ☉ ∥ ♂
18 26 ☽ ∠ ♆
21 35 ☽ ⊥ ♅
23 02 ☿ # ♄

18 Thursday
03 04 ☽ ⚻ ♂
03 13 ☉ ⚹ ♄
03 16 ☽ ⊥ ♄
07 00 ☿ □ ♇
07 42 ☽ ☌ ♀
12 56 ☽ ∠ ♅
13 29 ☽ ⊻ ♄
14 13 ☽ ⊻ ☉
17 00 ☽ ∥ ♀
17 13 ☽ ☌ ♃
18 36 ☽ ± ♂
19 57 ☽ ⚹ ♆

19 Friday
01 06 ☽ ⊥ ☉
03 43 ☽ ∥ ♃
04 55 ☽ ⊼ ♂
08 32 ☽ # ♃
08 33 ☽ △ ♇
12 13 ☽ ⊻ ☿
14 17 ☽ ⚹ ♅
16 11 ☽ Q ♆
16 42 ☽ # ♀
17 23 ☽ ∠ ☉
23 43 ☽ ⊥ ☿

20 Saturday
05 13 ☿ ⊻ ♅
09 43 ☽ ⚻ ♇
10 29 ☽ Q ♅
10 36 ☿ ∠ ♄
14 54 ☽ ⊻ ♀
16 11 ☽ ☌ ♄
16 20 ☽ # ☿
16 56 ☽ ∠ ☿
19 58 ☉ ⊻ ♃
20 23 ☽ ⊻ ♃
20 24 ☽ ⚹ ☉
22 24 ♀ ∠ ♅
22 24 ☽ □ ♆

21 Sunday
00 48 ☽ ± ♇
01 54 ☽ ⊥ ♀
05 18 ☽ ∥ ♄
06 36 ☽ ⊥ ♃
06 49 ♀ ⊻ ♄
08 08 ☽ ☍ ♂
10 52 ☽ ⊼ ♇
13 55 ☽ # ♇
14 11 ☽ # ☉
16 46 ☽ □ ♅
17 11 ☉ ∥ ♇
18 03 ☽ Q ☉
18 25 ☽ ∠ ♀
21 36 ☽ ⚹ ☿
21 57 ☽ ∠ ♃

22 Monday
01 26 ☽ # ♂
01 44 ☉ ⊻ ♆
14 40 ☿ ∠ ♆
18 54 ☽ ⊻ ♄
20 36 ☽ Q ☿
22 05 ☽ ⚹ ♀
23 39 ☽ ⚹ ♃

23 Tuesday
01 00 ☽ △ ♆
02 43 ☽ □ ☉
05 08 ☽ ⊥ ♄
11 37 ☽ ⊼ ♂
12 56 ♀ ∥ ♃
13 32 ☽ ☍ ♇
15 27 ☽ # ♅
19 17 ☿ ⊥ ♅
19 41 ☽ △ ♅
19 45 ♀ # ♃
19 48 ♀ ☌ ♃
20 23 ☽ Q ♃
20 26 ☽ Q ♀
20 32 ☽ ∠ ♄
22 01 ☽ ± ♂

24 Wednesday
02 35 ☽ ⚻ ♆
07 22 ☽ □ ☿
08 23 ♀ ⚹ ♆
13 42 ☽ ⚻ ♂
18 05 ☽ ± ♆
18 24 ♀ ± ♂
20 44 ☽ # ♆
21 32 ☽ ⚻ ♅
22 28 ☽ ⚹ ♄

25 Thursday
03 51 ☽ □ ♃
04 29 ☽ ⊼ ♆
06 23 ☽ □ ♀
09 18 ☉ # ♄
10 09 ☽ △ ☉
13 12 ☽ ± ♅
16 06 ☽ △ ♂
17 14 ☽ ⊼ ♇
19 27 ☽ Q ♄
23 43 ☽ ⊼ ♅

26 Friday
03 17 ☿ ⊥ ♄
03 45 ☽ ± ♇
09 51 ☽ # ♆
14 31 ☽ ⚻ ☉
18 14 ☽ △ ☿
18 39 ☿ # ♀
19 37 ☽ ⚻ ♇

27 Saturday
00 24 ♃ ⚹ ♆
01 02 ☿ ⚻ ♂
03 24 ☽ □ ♄
07 46 ☽ ± ☉
09 22 ☽ ☍ ♆
09 30 ☽ △ ♃
16 26 ☽ # ♅
16 32 ☽ △ ♀
19 23 ☽ ⊼ ☉
22 04 ☽ □ ♂
22 26 ☽ △ ♇

28 Sunday
00 12 ☽ ⚻ ☿
05 18 ☽ ☍ ♅
05 46 ☉ ⊥ ♆
09 26 ☿ # ♃
13 00 ☽ ⚻ ♃
18 20 ☽ ± ☿
22 29 ☽ ⚻ ♀
23 36 ♂ ⊻ ♇

MARCH 1999

LONGITUDES

Date	Sidereal time h m s	Sun ☉ ° ' "	Moon ☽ ° ' "	Moon ☽ 24.00 ° ' "	Mercury ☿ ° '	Venus ♀ ° '	Mars ♂ ° '	Jupiter ♃ ° '	Saturn ♄ ° '	Uranus ♅ ° '	Neptune ♆ ° '	Pluto ♇ ° '
01	22 35 25	10 ♓ 26 57	01 ♍ 02 34	07 ♍ 30 34	28 ♓ 21	09 ♈ 23	10 ♏ 33	03 ♈ 43	00 ♉ 03	14 ♒ 19	03 ♒ 14	10 ♐ 27
02	22 39 21	11 27 09	13 ♍ 55 13	20 ♍ 16 30	29 ♓ 31	10 37	10 44	03 57	00 08	14 22	03 15	10 28
03	22 43 18	12 27 18	26 ♍ 34 28	02 ♎ 49 10	00 ♈ 34	11 50	10 54	04 11	00 14	14 26	03 17	10 28
04	22 47 14	13 27 27	09 ♎ 00 42	15 ♎ 09 16	01 29	13 04	11 04	04 25	00 20	14 29	03 19	10 28
05	22 51 11	14 27 33	21 ♎ 15 02	27 ♎ 18 16	02 16	14 17	11 13	04 39	00 26	14 32	03 21	10 29
06	22 55 07	15 27 38	03 ♏ 19 17	09 ♏ 18 27	02 55	15 30	11 21	04 53	00 33	14 35	03 23	10 29
07	22 59 04	16 27 41	15 ♏ 16 09	21 ♏ 12 50	03 26	16 44	11 29	05 08	00 39	14 38	03 25	10 29
08	23 03 00	17 27 42	27 ♏ 09 00	03 ♐ 05 10	03 47	17 57	11 36	05 22	00 45	14 41	03 26	10 29
09	23 06 57	18 27 42	09 ♐ 01 53	14 ♐ 59 45	04 00	19 10	11 43	05 36	00 51	14 44	03 28	10 29
10	23 10 54	19 27 41	20 ♐ 59 20	27 ♐ 01 15	04 R 03	20 23	11 49	05 50	00 58	14 47	03 30	10 30
11	23 14 50	20 27 37	03 ♑ 06 07	09 ♑ 14 31	03 58	21 36	11 54	06 05	01 04	14 50	03 32	10 30
12	23 18 47	21 27 33	15 ♑ 27 01	21 ♑ 44 10	03 44	22 49	11 59	06 19	01 11	14 53	03 33	10 30
13	23 22 43	22 27 26	28 ♑ 06 27	04 ♒ 34 17	03 21	24 02	12 03	06 33	01 17	14 56	03 35	10 30
14	23 26 40	23 27 18	11 ♒ 08 01	17 ♒ 47 53	02 52	25 15	12 06	06 47	01 24	14 59	03 37	10 R 30
15	23 30 36	24 27 08	24 ♒ 33 59	01 ♓ 26 19	02 15	26 28	12 09	07 02	01 30	15 02	03 38	10 30
16	23 34 33	25 26 56	08 ♓ 24 43	15 ♓ 28 51	01 33	27 41	12 11	07 16	01 37	15 05	03 40	10 30
17	23 38 29	26 26 42	22 ♓ 38 14	29 ♓ 52 15	00 ♈ 45	28 ♈ 53	12 12	07 31	01 44	15 08	03 41	10 30
18	23 42 26	27 26 26	07 ♈ 10 08	14 ♈ 30 59	29 ♓ 54	00 ♉ 06	12 12	07 45	01 50	15 11	03 43	10 29
19	23 46 23	28 26 08	21 ♈ 53 51	29 ♈ 17 45	29 01	01 19	12 R 12	07 59	01 57	15 13	03 44	10 29
20	23 50 19	29 ♓ 25 48	06 ♉ 41 39	14 ♉ 04 36	28 06	02 31	12 11	08 14	02 04	15 16	03 46	10 29
21	23 54 16	00 ♈ 25 26	21 ♉ 25 41	28 ♉ 44 10	27 10	03 44	12 09	08 28	02 11	15 19	03 47	10 29
22	23 58 12	01 25 01	05 ♊ 59 21	13 ♊ 10 44	26 16	04 56	12 06	08 43	02 18	15 22	03 49	10 29
23	00 02 09	02 24 35	20 ♊ 17 56	27 ♊ 20 43	25 24	06 08	12 03	08 57	02 25	15 24	03 50	10 28
24	00 06 05	03 24 06	04 ♋ 18 57	11 ♋ 12 39	24 34	07 20	11 59	09 12	02 32	15 27	03 52	10 28
25	00 10 02	04 23 34	18 ♋ 01 52	24 ♋ 46 45	23 49	08 33	11 54	09 26	02 39	15 29	03 53	10 28
26	00 13 58	05 23 01	01 ♌ 27 30	08 ♌ 04 19	23 08	09 45	11 49	09 41	02 46	15 32	03 54	10 27
27	00 17 55	06 22 25	14 ♌ 37 27	21 ♌ 07 07	22 32	10 57	11 42	09 55	02 53	15 35	03 56	10 27
28	00 21 52	07 21 46	27 ♌ 33 33	03 ♍ 56 58	22 01	12 09	11 35	10 10	03 00	15 37	03 57	10 26
29	00 25 48	08 21 06	10 ♍ 17 31	16 ♍ 35 23	21 36	13 20	11 27	10 24	03 07	15 39	03 58	10 26
30	00 29 45	09 20 23	22 ♍ 50 43	29 ♍ 03 37	21 16	14 32	11 19	10 39	03 15	15 42	03 59	10 25
31	00 33 41	10 ♈ 19 38	05 ♎ 14 13	11 ♎ 22 36	21 ♓ 02	15 ♉ 44	11 ♏ 09	10 ♈ 53	03 ♉ 22	15 ♒ 44	04 ♒ 00	10 ♐ 25

MOON NODES, MOON LATITUDE AND DECLINATIONS

Date	Moon True ☊ ° '	Moon Mean ☊ ° '	Moon ☽ Latitude ° '	Sun ☉ ° '	Moon ☽ ° '	Mercury ☿ ° '	Venus ♀ ° '	Mars ♂ ° '	Jupiter ♃ ° '	Saturn ♄ ° '	Uranus ♅ ° '	Neptune ♆ ° '	Pluto ♇ ° '
01	22 ♌ 09	21 ♌ 15	00 N 49	07 S 39	11 N 52	00 N 26	03 N 01	12 S 54	00 N 28	09 N 21	17 S 08	19 S 09	10 S 31
02	22 R 08	21 12	01 57	07 16	08 08	01 07	03 33	12 58	00 34	09 23	17 08	19 09	10 30
03	22 06	21 09	02 58	06 53	04 N 05	01 45	04 04	13 01	00 39	09 25	17 07	19 08	10 30
04	22 03	21 05	03 50	06 30	00 S 03	02 20	04 35	13 04	00 45	09 27	17 06	19 08	10 30
05	21 59	21 02	04 30	06 07	04 07	02 52	05 06	13 07	00 50	09 30	17 05	19 07	10 30
06	21 55	20 59	04 58	05 44	07 57	03 20	05 36	13 09	00 56	09 32	17 04	19 07	10 29
07	21 52	20 56	05 12	05 21	11 26	03 44	06 07	13 12	01 02	09 34	17 03	19 07	10 29
08	21 49	20 53	05 12	04 57	14 27	04 03	06 38	13 14	01 07	09 37	17 02	19 06	10 29
09	21 47	20 49	05 00	04 34	16 52	04 19	07 08	13 16	01 13	09 39	17 01	19 06	10 28
10	21 D 47	20 46	04 34	04 10	18 34	04 29	07 38	13 18	01 19	09 41	17 00	19 05	10 28
11	21 48	20 43	03 56	03 47	19 28	04 35	08 08	13 20	01 24	09 44	17 00	19 05	10 28
12	21 49	20 40	03 06	03 23	19 27	04 36	08 38	13 21	01 30	09 46	16 59	19 05	10 28
13	21 51	20 37	02 06	03 00	18 28	04 32	09 08	13 23	01 36	09 48	16 58	19 04	10 27
14	21 52	20 34	00 N 58	02 36	16 30	04 24	09 38	13 24	01 42	09 51	16 57	19 04	10 27
15	21 R 53	20 30	00 S 15	02 12	13 34	04 11	10 07	13 25	01 47	09 53	16 56	19 03	10 27
16	21 51	20 27	01 29	01 49	09 48	03 54	10 36	13 26	01 53	09 56	16 55	19 03	10 26
17	21 49	20 24	02 40	01 25	05 22	03 34	11 05	13 27	01 59	09 58	16 55	19 03	10 26
18	21 44	20 21	03 42	01 01	00 S 33	03 10	11 33	13 27	02 04	10 01	16 54	19 02	10 26
19	21 39	20 18	04 30	00 37	04 N 21	02 44	12 02	13 27	02 10	10 03	16 53	19 02	10 25
20	21 33	20 15	05 00	00 S 14	09 01	02 15	12 30	13 27	02 16	10 06	16 52	19 02	10 25
21	21 28	20 11	05 10	00 N 10	13 07	01 44	12 58	13 27	02 22	10 08	16 51	19 01	10 25
22	21 24	20 08	05 00	00 34	16 22	01 13	13 25	13 27	02 27	10 11	16 51	19 01	10 24
23	21 22	20 05	04 32	00 57	18 34	00 41	13 52	13 27	02 33	10 13	16 50	19 01	10 24
24	21 D 21	20 02	03 47	01 21	19 35	00 N 10	14 19	13 26	02 39	10 16	16 49	19 01	10 24
25	21 21	19 59	02 51	01 45	19 24	00 S 21	14 45	13 25	02 44	10 18	16 48	19 00	10 23
26	21 23	19 55	01 46	02 08	18 07	00 51	15 11	13 24	02 50	10 21	16 48	19 00	10 23
27	21 24	19 52	00 S 37	02 32	15 52	01 19	15 37	13 23	02 56	10 23	16 47	18 59	10 23
28	21 R 24	19 49	00 N 33	02 55	12 50	01 46	16 03	13 21	03 01	10 26	16 46	18 59	10 22
29	21 22	19 46	01 40	03 19	09 15	02 10	16 28	13 20	03 07	10 28	16 45	18 59	10 22
30	21 18	19 43	02 41	03 42	05 18	02 32	16 52	13 18	03 13	10 31	16 45	18 59	10 22
31	21 ♌ 12	19 ♌ 40	03 N 34	04 N 05	01 N 11	02 S 52	17 N 16	13 S 16	03 N 19	10 N 33	16 S 44	18 S 58	10 S 21

ZODIAC SIGN ENTRIES

Date	h	m	Planets
01	01	26	♄ ♉
01	10	04	☽ ♍
02	22	50	☿ ♈
03	18	34	☽ ♎
06	05	22	☽ ♏
08	17	46	☽ ♐
11	05	54	☽ ♑
13	15	36	☽ ♒
15	21	30	☽ ♓
18	00	13	☽ ♈
18	09	23	☿ ♓
18	09	59	♀ ♉
20	01	09	☽ ♉
21	01	46	☉ ♈
22	02	05	☽ ♊
24	04	33	☽ ♋
26	09	22	☽ ♌
28	16	34	☽ ♍
31	01	49	☽ ♎

LATITUDES

Date	Mercury ☿ ° '	Venus ♀ ° '	Mars ♂ ° '	Jupiter ♃ ° '	Saturn ♄ ° '	Uranus ♅ ° '	Neptune ♆ ° '	Pluto ♇ ° '
01	01 N 12	00 S 45	02 N 11	01 S 06	02 S 17	00 S 38	00 N 17	11 N 37
04	01 55	00 38	02 11	01 06	02 17	00 38	00 17	11 38
07	02 35	00 30	02 11	01 06	02 16	00 38	00 17	11 39
10	03 08	00 21	02 11	01 06	02 16	00 38	00 17	11 40
13	03 29	00 12	02 10	01 06	02 15	00 38	00 17	11 41
16	03 35	00 S 03	02 10	01 05	02 15	00 38	00 17	11 42
19	03 24	00 N 06	02 08	01 05	02 14	00 38	00 17	11 42
22	02 57	00 16	02 07	01 05	02 14	00 39	00 17	11 43
25	02 17	00 26	02 05	01 05	02 13	00 39	00 17	11 44
28	01 32	00 36	02 03	01 05	02 13	00 39	00 17	11 45
31	00 N 45	00 N 46	02 N 00	01 S 05	02 S 13	00 S 39	00 N 17	11 N 46

DATA

Julian Date	2451239
Delta T	+64 seconds
Ayanamsa	23° 50' 34"
Synetic vernal point	05° ♓ 16' 26"
True obliquity of ecliptic	23° 26' 15"

MOON'S PHASES, APSIDES AND POSITIONS ☽

Date	h	m	Phase	Longitude ° '	Eclipse Indicator
02	06	58	○	11 ♍ 15	
10	08	40	☾	19 ♐ 19	
17	18	48	●	26 ♓ 44	
24	10	18	☽	03 ♋ 20	
31	22	49	○	10 ♎ 46	

Day	h	m		
08	05	03	Apogee	
20	00	09	Perigee	
04	11	44	0S	
11	23	46	Max dec	19° S 35'
18	14	41	0N	
24	20	19	Max dec	19° N 39'
31	18	53	0S	

LONGITUDES

Date	Chiron ⚷ ° '	Ceres ⚳ ° '	Pallas ⚴ ° '	Juno ⚵ ° '	Vesta ⚶ ° '	Black Moon Lilith ⚸ ° '
01	03 ♐ 41	05 ♊ 07	09 ♈ 13	09 ♐ 40	09 ♌ 07	19 ♏ 15
11	03 ♐ 47	07 ♊ 43	13 ♈ 09	10 ♐ 47	07 ♌ 44	20 ♏ 22
21	03 ♐ 52	10 ♊ 40	17 ♈ 14	11 ♐ 36	07 ♌ 08	21 ♏ 30
31	03 ♐ 41	13 ♊ 53	21 ♈ 26	11 ♐ 50	07 ♌ 18	22 ♏ 37

All ephemeris data is given at 12.00 UT and the Moon's longitude is additionally given for 24.00 UT

ASPECTARIAN

h m Aspects

01 Monday
04 51 ☽ # ♂
05 46 ☽ ± ♃
06 31 ☽ ⊼ ☿
07 20 ☽ Q ♂
10 08 ☽ △ ♄
12 05 ☉ □ ♇
13 02 ☿ ∥ ♃
14 50 ☉ △ ♂
16 03 ☽ ⊼ ♆
16 47 ☽ ± ♀
17 03 ☽ ⊼ ♃
21 01 ☽ # ♇

02 Tuesday
03 14 ☽ ± ♆
04 18 ☽ ∥ ♄
05 08 ☽ ⊼ ♀
05 30 ☽ □ ♇
05 55 ☽ ✱ ♂
06 58 ☽ ☍ ☉
08 51 ☿ ∠ ♅
09 00 ♀ △ ♇
12 51 ☽ ⊼ ♅
14 19 ☽ bQ ♄
14 36 ♀ ⊼ ♂
17 46 ☽ # ☉
20 12 ☽ bQ ♆

03 Wednesday
00 14 ☽ ± ♅
03 34 ☿ ⊻ ♄
07 30 ☽ ± ♄
10 41 ☽ ∠ ♂
12 09 ☽ ∥ ♀
15 38 ☽ Q ♇
17 29 ☽ bQ ♅
19 05 ☽ ⊼ ♄
20 17 ☽ ☍ ☿
23 50 ☽ ∥ ☿

04 Thursday
00 56 ☽ △ ♆
02 55 ☽ ☍ ♃
04 14 ☽ ⊥ ♂
07 30 ☽ ∥ ♃
14 51 ☽ ✱ ♇
16 02 ☽ ⊻ ♂
16 10 ☽ # ♃
20 47 ☽ ☍ ♀
21 27 ☽ ⊼ ☉
22 43 ☽ △ ♅

05 Friday
03 29 ☽ # ☿
10 18 ☽ ± ☉
13 50 ☉ ⊻ ♅
17 04 ♀ ✱ ♅
18 53 ☽ # ♀
20 22 ☽ ∠ ♇
23 11 ☽ ∥ ☉

06 Saturday
05 46 ☽ bQ ☉
06 24 ☽ ☍ ♄
06 56 ☉ ⊻ ♀
09 26 ♀ Q ♆
11 10 ☽ ⊼ ☿
12 07 ☽ □ ♆
14 12 ☉ ∠ ♄
14 19 ☽ ⊥ ♇
15 12 ☽ ⊼ ♃
15 17 ☉ # ♀
22 38 ☽ # ♄
23 46 ☽ ± ☿

07 Sunday
02 22 ☽ ⊻ ♇
03 29 ☽ ± ♃
04 18 ☽ ☌ ♂
05 08 ☽ ∥ ♇
10 43 ☽ □ ♅
10 47 ☿ ✱ ♆
14 38 ☽ △ ☉
15 17 ☽ ⊼ ♀
18 36 ☽ bQ ☿
22 00 ☽ bQ ♃

08 Monday
00 26 ☽ Q ♆
01 39 ☽ ∥ ♂
04 47 ☽ ± ♀
19 21 ☽ ⊼ ♄
23 15 ☽ Q ♅

09 Tuesday
00 45 ☽ ✱ ♆
01 04 ☽ bQ ♀
01 42 ☽ △ ☿
04 56 ☽ △ ♃
07 34 ☽ ± ♄
12 10 ☉ ∠ ♆
13 50 ☽ ∥ ♅
14 57 ☽ ☌ ♇
17 27 ☽ ⊻ ♂
22 23 ☉ # ☿
23 32 ☽ ✱ ♅

10 Wednesday
01 51 ☽ bQ ♄
05 36 ☽ ⊥ ♂
07 00 ☽ ∠ ♆
08 07 ♂ ± ♃
08 40 ☽ □ ☉
09 11 ☿ St R
10 40 ☽ △ ♀
22 54 ☽ ∥ ♆
23 41 ☽ ∠ ♂

11 Thursday
00 58 ☽ ⊥ ♆
05 33 ☽ ∠ ♅
07 58 ☽ △ ♄
12 50 ☽ ⊻ ♆
13 40 ☽ □ ☿
17 57 ☽ □ ♃
21 34 ☉ ⊥ ♅
23 16 ☽ ⊥ ♅
23 24 ☽ Q ☉

12 Friday
02 26 ☽ ⊻ ♇
05 16 ☽ ✱ ♂
10 55 ☽ ⊻ ♅
14 00 ☽ ⊥ ♇
23 30 ☿ ✱ ♆
23 40 ☽ Q ☿

13 Saturday
00 24 ☽ ∥ ♆
00 27 ☽ ✱ ☉
03 33 ☽ □ ♀
04 20 ☽ Q ♂
05 12 ☽ Q ♃
07 06 ☽ ∠ ♇
17 58 ☽ □ ♄
21 26 ☽ ✱ ☿
21 36 ♇ St R
22 12 ☽ ☌ ♆

14 Sunday
03 56 ☽ ✱ ♃
06 43 ☽ ∠ ☉
07 23 ☽ ∥ ♅
10 51 ☽ ✱ ♇
13 46 ☽ □ ♂
16 13 ☽ Q ♀
16 49 ♀ bQ ♇
18 59 ☽ ☌ ♅
23 37 ☽ ∠ ☿
23 47 ♀ ∥ ♄

15 Monday
00 18 ☽ ⊥ ☉
02 58 ☽ Q ♄
07 27 ☽ ∠ ♃
08 21 ☽ Q ♇
11 47 ☽ ⊻ ☉
13 02 ☽ ∥ ♂
14 49 ☽ ⊥ ☿
15 40 ☽ ✱ ♀
23 30 ☽ ⊥ ♃
23 43 ♀ Q ♅

16 Tuesday
00 13 ☽ ✱ ♄
00 47 ☽ ⊻ ☿
03 50 ☽ ⊻ ♆
04 06 ♀ # ♇
07 45 ☽ # ♀
08 11 ☽ ∥ ♇
08 22 ☉ # ♃
10 01 ☽ ⊻ ♃
10 03 ☿ ⊻ ♄
11 13 ☽ # ♄
14 08 ☽ ⊥ ♆
15 33 ☽ □ ♇
16 31 ☉ ⊥ ♄
18 25 ☽ △ ♂
19 57 ☽ ∠ ♀
23 22 ☽ ⊻ ♅

17 Wednesday
02 01 ☽ ∠ ♄
05 23 ☽ ∠ ♆
09 29 ☽ ⊥ ♅
12 28 ☽ ⊥ ♀
17 11 ☽ ⊥ ♄
18 48 ☽ ☌ ☉
19 35 ☽ bQ ♂
21 56 ☽ # ☿
23 20 ☽ ⊻ ♀

18 Thursday
00 28 ☽ ∠ ♅
00 44 ☽ ☌ ☿
03 11 ☽ ⊻ ♄
04 41 ☽ # ♃
04 52 ☿ ∠ ♅
06 16 ☉ bQ ♂
06 19 ☽ ✱ ♆
09 31 ☽ ∥ ☉
09 44 ☿ ⊻ ♀
10 25 ☽ ± ♂
12 58 ☽ ☌ ♃
13 41 ♂ St R
17 26 ☽ △ ♇
19 04 ☽ # ☉
20 14 ☽ ⊼ ♂

19 Friday
01 02 ☽ ∥ ♃
01 07 ☽ ✱ ♅
01 58 ☽ Q ♆
04 39 ☽ ∥ ☿
17 49 ☽ bQ ♇
19 14 ☉ ☌ ☿
20 40 ☽ Q ♅
22 52 ☽ ⊻ ☿
23 22 ☽ ⊻ ☉

20 Saturday
02 05 ♀ ☌ ♄
04 26 ☽ ☌ ♄
04 38 ☽ ☌ ♀
07 14 ☽ □ ♆
08 02 ☽ ⊥ ☿
08 25 ☽ ± ♇
09 48 ☽ ⊥ ☉
11 16 ☿ ∥ ♃
14 32 ☽ ⊻ ♃
17 57 ☽ ∥ ♄
18 09 ☽ ⊼ ♇
19 43 ☽ # ♇
20 54 ☽ ☍ ♂
21 47 ☽ ∠ ☿

21 Sunday
00 27 ☽ ⊥ ♃
01 29 ☽ ∠ ☉
01 59 ☽ □ ♅
05 44 ☿ ⊥ ♀
09 15 ☉ ∠ ♅
10 48 ☽ ∥ ♀
12 36 ☿ bQ ♂
13 15 ♀ □ ♆
14 10 ☽ # ♂
15 24 ☽ ∠ ♃
20 52 ☽ ✱ ☿

22 Monday
02 56 ♀ ± ♇
03 52 ☽ ✱ ☉
05 50 ☽ ⊻ ♄
08 23 ☽ △ ♆
10 05 ☽ ⊻ ♀
11 16 ☿ ⊥ ♄
13 50 ♀ # ♂
15 34 ☽ Q ☿
15 52 ☽ ⊥ ♄
16 13 ☽ # ♅
16 37 ☽ ✱ ♃
19 28 ☽ ☍ ♇
21 00 ☽ ⊥ ♀
22 10 ☽ ⊼ ♂

23 Tuesday
01 19 ☽ Q ☉
03 43 ☽ △ ♅
05 01 ☉ ∥ ☿
07 05 ☽ ∠ ♄
08 13 ☽ ± ♂
09 31 ☽ bQ ♆
12 07 ☉ ⊻ ♄
13 08 ☽ Q ♃
13 33 ☽ ∠ ♀
19 27 ☽ # ♆
20 11 ☽ □ ☿
23 27 ☽ bQ ♂

24 Wednesday
00 52 ☽ ± ♆
05 18 ☽ bQ ♅
08 53 ☽ ✱ ♄
10 18 ☽ □ ☉
11 12 ☽ ⊼ ♆
17 45 ☽ ✱ ♀
20 38 ☽ □ ♃
20 57 ☽ ± ♅
22 41 ☽ ⊼ ♇
23 20 ☉ ✱ ♆

25 Thursday
01 17 ☽ △ ♂
05 59 ☽ Q ♄
07 30 ☽ ⊼ ♅
09 13 ☽ ± ♇
15 24 ☿ ∠ ♀
16 53 ☽ Q ♀
21 46 ☽ △ ☿
21 52 ☽ # ♆

26 Friday
01 13 ☽ bQ ♇
10 26 ♀ ⊻ ♃
14 23 ☽ □ ♄
16 26 ☽ ☍ ♆
19 41 ☽ △ ☉
21 23 ☉ ± ♂
23 33 ☽ bQ ☿

27 Saturday
02 04 ♀ ⊼ ♇
03 13 ☽ # ♅
03 13 ☽ △ ♃
04 20 ☽ △ ♇
04 34 ☽ □ ♀
06 41 ☽ □ ♂
09 06 ♄ # ♇
13 45 ☽ ☍ ♅
13 51 ☽ ∥ ♀
15 22 ☽ ± ☿

28 Sunday
01 30 ☽ bQ ☉
01 53 ♀ ☍ ♂
02 02 ☽ ⊼ ☿
07 27 ☽ bQ ♃
08 13 ☽ # ♂
15 46 ☽ Q ♂
19 44 ☽ ± ☉
20 29 ☉ ∥ ♃
22 19 ☽ △ ♄

29 Monday
00 01 ☽ ⊼ ♆
00 39 ☽ ± ♃
04 17 ☽ ∥ ♄
04 51 ☽ # ♇
08 01 ☽ ⊼ ☉
08 15 ☿ ⊥ ♅
11 23 ☽ ± ♆
12 13 ☽ ⊼ ♃
12 16 ☽ □ ♇
14 00 ♃ △ ♇
14 11 ☽ ✱ ♂
18 24 ☽ △ ♀
22 15 ☽ ⊼ ♅

30 Tuesday
03 05 ☽ bQ ♄
04 35 ☽ bQ ♆
04 56 ♀ # ♅
09 02 ☽ ☍ ☿
09 47 ☽ ± ♅
18 36 ☽ ∠ ♂
20 34 ☽ ± ♄
20 36 ☽ ∥ ☉
22 45 ☽ Q ♇
23 58 ☽ ∥ ♃

31 Wednesday
02 18 ☽ bQ ♀
02 59 ☽ # ☿
03 13 ☽ bQ ♅
08 19 ☽ ⊼ ♄
09 36 ☽ △ ♆
11 50 ☽ ⊥ ♂
12 13 ♀ □ ♅
13 59 ☉ △ ♇
21 43 ☽ ± ♀
22 06 ☽ ✱ ♇
22 49 ☽ ☍ ☉
23 16 ☽ ☍ ♃
23 24 ☽ ⊻ ♂

APRIL 1999

Date	Sidereal time h m s	LONGITUDES Sun ☉ ° ' "	Moon ☽ ° ' "	Moon ☽ 24.00 ° ' "	Mercury ☿ ° '	Venus ♀ ° '	Mars ♂ ° '	Jupiter ♃ ° '	Saturn ♄ ° '	Uranus ♅ ° '	Neptune ♆ ° '	Pluto ♇ ° '
01	00 37 38	11 ♈ 18 51	17 ♎ 28 54	23 ♎ 33 12	20 ♓ 55	16 ♉ 55	10 ♏ 59	11 ♈ 08	03 ♉ 29	15 ♒ 47	04 ♒ 02	10 ♐ 24
02	00 41 34	12 18 02	29 ♎ 35 39	05 ♏ 36 24	20 D 52	18 07	10 R 48	11 23	03 37	15 49	04 03	10 R 23
03	00 45 31	13 17 11	11 ♏ 35 38	17 ♏ 33 35	20 56	19 18	10 36	11 37	03 44	15 51	04 04	10 23
04	00 49 27	14 16 19	23 ♏ 30 29	29 ♏ 26 40	21 04	20 29	10 24	11 52	03 51	15 53	04 05	10 22
05	00 53 24	15 15 24	05 ♐ 22 28	11 ♐ 18 18	21 18	21 40	10 11	12 06	03 59	15 56	04 06	10 21
06	00 57 21	16 14 28	17 ♐ 14 36	23 ♐ 11 51	21 37	22 52	09 57	12 21	04 06	15 58	04 07	10 21
07	01 01 17	17 13 29	29 ♐ 10 35	05 ♑ 11 23	22 01	24 03	09 42	12 35	04 13	16 00	04 08	10 20
08	01 05 14	18 12 29	11 ♑ 14 49	17 ♑ 21 32	22 29	25 14	09 27	12 50	04 21	16 02	04 09	10 19
09	01 09 10	19 11 28	23 ♑ 32 08	29 ♑ 47 16	23 02	26 24	09 11	13 04	04 28	16 04	04 10	10 18
10	01 13 07	20 10 24	06 ♒ 07 31	12 ♒ 33 29	23 38	27 35	08 55	13 19	04 36	16 06	04 11	10 17
11	01 17 03	21 09 19	19 ♒ 05 40	25 ♒ 44 31	24 19	28 46	08 38	13 33	04 43	16 08	04 11	10 16
12	01 21 00	22 08 12	02 ♓ 30 21	09 ♓ 23 23	25 03	29 ♉ 56	08 20	13 47	04 51	16 10	04 12	10 15
13	01 24 56	23 07 03	16 ♓ 23 37	23 ♓ 30 55	25 51	01 ♊ 07	08 01	14 02	04 59	16 12	04 13	10 15
14	01 28 53	24 05 52	00 ♈ 44 55	08 ♈ 05 03	26 42	02 17	07 42	14 16	05 06	16 13	04 14	10 14
15	01 32 50	25 04 40	15 ♈ 30 31	23 ♈ 00 20	27 36	03 27	07 23	14 31	05 14	16 15	04 15	10 13
16	01 36 46	26 03 25	00 ♉ 33 19	08 ♉ 08 13	28 34	04 37	07 03	14 45	05 21	16 17	04 15	10 12
17	01 40 43	27 02 09	15 ♉ 43 40	23 ♉ 18 19	29 ♓ 34	05 47	06 43	14 59	05 29	16 19	04 16	10 10
18	01 44 39	28 00 51	00 ♊ 50 55	08 ♊ 20 15	00 ♈ 37	06 57	06 22	15 14	05 37	16 20	04 16	10 09
19	01 48 36	28 59 30	15 ♊ 45 21	23 ♊ 05 22	01 42	08 07	06 01	15 28	05 44	16 22	04 17	10 08
20	01 52 32	29 ♈ 58 08	00 ♋ 19 44	07 ♋ 28 02	02 50	09 17	05 39	15 42	05 52	16 23	04 18	10 07
21	01 56 29	00 ♉ 56 43	14 ♋ 30 04	21 ♋ 25 48	04 01	10 26	05 18	15 57	05 59	16 25	04 18	10 06
22	02 00 25	01 55 16	28 ♋ 15 23	04 ♌ 59 01	05 14	11 36	04 56	16 11	06 07	16 26	04 19	10 05
23	02 04 22	02 53 47	11 ♌ 37 03	18 ♌ 09 52	06 29	12 45	04 34	16 25	06 15	16 28	04 19	10 04
24	02 08 19	03 52 15	24 ♌ 37 53	01 ♍ 01 34	07 46	13 54	04 11	16 39	06 22	16 29	04 20	10 02
25	02 12 15	04 50 41	07 ♍ 21 21	13 ♍ 37 39	09 05	15 03	03 49	16 53	06 30	16 30	04 20	10 01
26	02 16 12	05 49 06	19 ♍ 50 52	26 ♍ 01 23	10 27	16 12	03 26	17 08	06 38	16 32	04 20	10 00
27	02 20 08	06 47 28	02 ♎ 09 31	08 ♎ 15 34	11 51	17 21	03 04	17 22	06 45	16 33	04 21	09 59
28	02 24 05	07 45 48	14 ♎ 19 48	20 ♎ 22 26	13 16	18 30	02 41	17 36	06 53	16 34	04 21	09 57
29	02 28 01	08 44 06	26 ♎ 23 39	02 ♏ 23 38	14 44	19 38	02 19	17 50	07 01	16 35	04 21	09 56
30	02 31 58	09 ♉ 42 22	08 ♏ 22 33	14 ♏ 20 32	16 ♈ 13	20 ♊ 46	01 ♏ 57	18 ♈ 04	07 ♉ 08	16 ♒ 36	04 ♒ 21	09 ♐ 55

Date	Moon True ☊ ° '	Moon Mean ☊ ° '	Moon ☽ Latitude ° '
01	21 ♌ 03	19 ♌ 36	04 N 15
02	20 R 54	19 33	04 45
03	20 44	19 30	05 02
04	20 34	19 27	05 05
05	20 26	19 24	04 55
06	20 19	19 21	04 33
07	20 15	19 17	03 58
08	20 13	19 14	03 13
09	20 D 13	19 11	02 18
10	20 14	19 08	01 15
11	20 14	19 05	00 N 06
12	20 R 14	19 01	01 S 05
13	20 11	18 58	02 15
14	20 06	18 55	03 18
15	19 58	18 52	04 10
16	19 49	18 49	04 46
17	19 39	18 46	05 02
18	19 29	18 42	04 57
19	19 21	18 39	04 31
20	19 16	18 36	03 49
21	19 13	18 33	02 53
22	19 12	18 30	01 49
23	19 D 12	18 27	00 S 40
24	19 R 12	18 23	00 N 29
25	19 11	18 20	01 35
26	19 07	18 17	02 35
27	19 01	18 14	03 26
28	18 52	18 11	04 08
29	18 40	18 07	04 38
30	18 ♌ 27	18 ♌ 04	04 N 55

DECLINATIONS

Date	Sun ☉ ° '	Moon ☽ ° '	Mercury ☿ ° '	Venus ♀ ° '	Mars ♂ ° '	Jupiter ♃ ° '	Saturn ♄ ° '	Uranus ♅ ° '	Neptune ♆ ° '	Pluto ♇ ° '
01	04 N 29	02 S 56	03 S 09	17 N 40	13 S 14	03 N 24	10 N 36	16 S 43	18 S 58	10 S 21
02	04 52	06 53	03 23	18 03	13 11	03 30	10 38	16 43	18 58	10 21
03	05 15	10 31	03 35	18 26	13 09	03 36	10 41	16 42	18 58	10 20
04	05 38	13 43	03 45	18 49	13 06	03 41	10 44	16 41	18 57	10 20
05	06 00	16 21	03 52	19 11	13 03	03 47	10 46	16 41	18 57	10 20
06	06 23	18 18	03 56	19 32	13 00	03 53	10 49	16 40	18 57	10 19
07	06 46	19 28	03 58	19 53	12 57	03 58	10 51	16 40	18 57	10 19
08	07 08	19 46	03 58	20 14	12 53	04 04	10 54	16 39	18 56	10 19
09	07 31	19 07	03 55	20 34	12 49	04 10	10 56	16 39	18 56	10 18
10	07 53	17 32	03 50	20 53	12 46	04 15	10 59	16 38	18 56	10 18
11	08 15	15 00	03 43	21 12	12 42	04 21	11 02	16 37	18 56	10 18
12	08 37	11 36	03 34	21 31	12 37	04 26	11 04	16 37	18 56	10 17
13	08 59	07 26	03 23	21 48	12 33	04 32	11 07	16 36	18 55	10 17
14	09 21	02 S 44	03 10	22 06	12 29	04 37	11 09	16 36	18 55	10 16
15	09 42	02 N 15	02 55	22 22	12 24	04 43	11 12	16 35	18 55	10 16
16	10 04	07 12	02 38	22 39	12 19	04 49	11 14	16 35	18 55	10 16
17	10 25	11 44	02 20	22 54	12 15	04 54	11 17	16 34	18 55	10 15
18	10 46	15 29	02 00	23 09	12 10	05 00	11 20	16 34	18 55	10 15
19	11 07	18 11	01 38	23 24	12 05	05 05	11 22	16 33	18 54	10 15
20	11 28	19 37	01 15	23 37	12 00	05 10	11 25	16 33	18 54	10 15
21	11 48	19 47	00 50	23 51	11 54	05 16	11 27	16 33	18 54	10 14
22	12 08	18 44	00 S 24	24 03	11 49	05 21	11 30	16 32	18 54	10 14
23	12 29	16 39	00 N 03	24 15	11 44	05 27	11 32	16 32	18 54	10 14
24	12 48	13 46	00 32	24 26	11 38	05 32	11 35	16 31	18 54	10 13
25	13 08	10 16	01 02	24 37	11 33	05 38	11 37	16 31	18 54	10 13
26	13 28	06 23	01 34	24 47	11 28	05 43	11 40	16 31	18 54	10 13
27	13 47	02 N 18	02 06	24 56	11 22	05 48	11 42	16 30	18 54	10 12
28	14 06	01 S 50	02 40	25 05	11 17	05 54	11 45	16 30	18 53	10 12
29	14 25	05 52	03 15	25 13	11 12	05 59	11 47	16 30	18 53	10 12
30	14 N 43	09 S 38	03 N 51	25 N 20	11 S 06	06 N 04	11 N 50	16 S 30	18 S 53	10 S 11

ZODIAC SIGN ENTRIES

Date	h	m	Planets
02	12	48	☽ ♏
05	01	07	☽ ♐
07	13	39	☽ ♑
10	00	24	☽ ♒
12	07	35	☽ ♓
12	13	17	♀ ♊
14	10	46	☽ ♈
16	11	07	☽ ♉
17	22	09	☿ ♈
18	10	39	☽ ♊
20	11	27	☽ ♋
20	12	46	☉ ♉
22	15	06	☽ ♌
24	22	04	☽ ♍
27	07	46	☽ ♎
29	19	12	☽ ♏

LATITUDES

Date	Mercury ☿ ° '	Venus ♀ ° '	Mars ♂ ° '	Jupiter ♃ ° '	Saturn ♄ ° '	Uranus ♅ ° '	Neptune ♆ ° '	Pluto ♇ ° '
01	00 N 30	00 N 49	01 N 59	01 S 05	02 S 13	00 S 39	00 N 17	11 N 46
04	00 S 14	00 59	01 56	01 05	02 12	00 39	00 17	11 47
07	00 52	01 08	01 52	01 05	02 12	00 39	00 17	11 48
10	01 26	01 18	01 48	01 05	02 12	00 39	00 17	11 48
13	01 53	01 27	01 43	01 05	02 12	00 39	00 17	11 49
16	02 15	01 37	01 38	01 05	02 11	00 39	00 17	11 50
19	02 31	01 45	01 32	01 05	02 11	00 39	00 17	11 50
22	02 42	01 54	01 26	01 06	02 11	00 40	00 17	11 51
25	02 47	02 02	01 19	01 06	02 11	00 40	00 17	11 51
28	02 47	02 09	01 11	01 06	02 11	00 40	00 17	11 52
31	02 S 42	02 N 16	01 N 04	01 S 06	02 S 11	00 S 40	00 N 17	11 N 52

DATA

Julian Date	2451270
Delta T	+64 seconds
Ayanamsa	23° 50' 36"
Synetic vernal point	05° ♓ 16' 23"
True obliquity of ecliptic	23° 26' 15"

MOON'S PHASES, APSIDES AND POSITIONS ☽

Date	h	m	Phase	Longitude ° '	Eclipse Indicator
09	02	51	☾	18 ♑ 49	
16	04	22	●	25 ♈ 45	
22	19	02	☽	02 ♌ 12	
30	14	55	○	09 ♏ 49	

Day	h	m		
04	21	20	Apogee	
17	05	19	Perigee	
08	07	47	Max dec	19° S 46'
15	01	13	0N	
21	02	55	Max dec	19° N 52'
28	01	18	0S	

LONGITUDES

Date	Chiron ⚷ ° '	Ceres ⚳ ° '	Pallas ⚴ ° '	Juno ⚵ ° '	Vesta ⚶ ° '	Black Moon Lilith ° '
01	03 ♐ 40	14 ♊ 13	21 ♈ 52	11 ♐ 49	07 ♌ 22	22 ♏ 44
11	03 ♐ 19	17 ♊ 41	26 ♈ 12	11 ♐ 27	08 ♌ 20	23 ♏ 51
21	02 ♐ 49	21 ♊ 22	00 ♉ 39	10 ♐ 32	09 ♌ 57	24 ♏ 58
31	02 ♐ 12	25 ♊ 12	05 ♉ 12	09 ♐ 06	12 ♌ 08	26 ♏ 05

ASPECTARIAN

h m	Aspects
01 Thursday	
03 21	♂ ⊼ ♃
05 10	☉ ⊼ ♂
06 10	☉ ☌ ♃
08 38	☽ △ ♅
10 46	☽ ⊼ ♀
13 23	☽ ∥ ☿
14 54	☽ # ♃
17 24	♀ ⊥ ♃
18 44	☽ ⊼ ☿
22 13	☽ # ☉
02 Friday	
03 39	☽ ∠ ♇
06 35	☽ ± ☿
09 20	☿ St D
20 05	☽ ☍ ♄
20 53	☽ □ ♆
21 33	☽ ⊥ ♇
03 Saturday	
00 34	☽ bQ ☿
09 34	☽ ⊻ ♇
10 03	☽ ☌ ♂
10 22	☉ ⊥ ♀
10 44	☽ ∥ ♇
12 03	☽ ⊼ ♃
13 08	☽ # ♄
13 17	☿ # ♃
15 42	☽ ⊼ ☉
20 35	☽ □ ♅
04 Sunday	
00 22	☽ ± ♃
04 53	☽ ± ☉
05 14	☽ ☍ ♀
07 01	☽ △ ☿
07 05	☽ ∥ ♂
09 07	☽ Q ♆
15 57	♂ ⊻ ♇
18 55	☽ bQ ♃
21 07	♀ # ♆
05 Monday	
00 42	☽ bQ ☉
02 29	☿ ⚹ ♀
09 04	☽ Q ♅
09 09	☽ ⊼ ♄
09 25	☽ ⚹ ♆
15 29	☽ ∥ ♅
21 25	☽ ± ♄
21 33	☽ ⊻ ♂
22 04	☽ ☌ ♇
06 Tuesday	
01 54	☽ △ ♃
04 56	☉ ⚹ ♅
08 52	☉ Q ♆
09 24	☽ ⚹ ♅
09 26	☽ ⊥ ♂
09 48	☽ △ ☉
15 05	♄ □ ♆
15 47	☽ ∠ ♆
15 47	☽ bQ ♄
21 07	☽ □ ☿
23 10	☽ ∥ ♆
07 Wednesday	
00 34	☽ ⊼ ♀
03 13	☽ ∠ ♂
09 54	☽ ⊥ ♆
10 45	☿ ⊥ ♅
11 30	☿ # ♃
13 55	☽ ± ♀
15 39	☽ ∠ ♅
21 54	☽ ⊻ ♆
22 11	☽ △ ♄
08 Thursday	
06 14	♄ ± ♇
08 32	☽ ⚹ ♂
09 36	☽ ⊥ ♅
09 46	☽ bQ ♀
10 10	☽ ⊻ ♇
10 26	☽ Q ☿
15 10	☽ □ ♃
21 26	☽ ⊻ ♅
21 57	☽ ⊥ ♇
09 Friday	
02 51	☽ □ ☉
07 33	☽ Q ♂
10 59	☽ ⚹ ☿
15 24	☽ ∠ ♇
15 49	☽ ∥ ♆
18 06	☽ △ ♀
10 Saturday	
02 43	☽ Q ♃
08 19	☽ ☌ ♆
09 06	☽ □ ♄
16 10	☽ Q ☉
16 58	☽ ∠ ☿
17 07	☽ □ ♂
18 57	☿ bQ ♂
19 47	☽ ⚹ ♇
21 37	☽ ∥ ♅
11 Sunday	
01 38	☽ ⚹ ♃
06 33	♀ ∠ ♃
06 34	☽ ☌ ♅
10 30	☽ ⊥ ☿
16 02	☽ ⚹ ☉
17 45	☽ Q ♇
18 38	☽ Q ♄
21 59	☽ ⊻ ☿
12 Monday	
05 10	☽ ∥ ♂
05 19	☽ ∠ ♃
07 02	☽ □ ♀
14 59	☽ ⊻ ♆
15 13	☽ # ♄
16 09	☽ ⚹ ♄
19 58	☽ ∥ ♇
20 43	☽ ∠ ☉
21 24	☽ ⊥ ♃
21 57	☽ △ ♂
13 Tuesday	
01 25	☽ ⊥ ♆
01 29	☽ □ ♇
04 10	☽ # ☉
07 54	☽ ⊻ ♃
11 40	☽ ⊻ ♅
13 19	☽ ⊥ ☉
16 47	☽ ∠ ♆
17 01	☽ Q ♀
18 07	☽ ∠ ♄
21 48	☽ ⊥ ♅
22 56	☽ bQ ♂
14 Wednesday	
00 10	☽ ⊻ ☉
02 47	☽ # ♃
04 53	☽ ☌ ☿
09 16	☽ ⊥ ♄
09 46	☽ ∥ ☿
12 47	☽ ∠ ♅
13 33	☽ ± ♂
14 45	☽ ⚹ ♀
17 43	☽ ⚹ ♆
19 12	☽ ⊻ ♄
23 09	☽ ⊼ ♂
15 Thursday	
03 27	☽ △ ♇
10 22	☽ ☌ ♃
13 11	☽ Q ♆
13 12	☽ ⚹ ♅
15 00	☽ # ☿
15 10	☉ bQ ♇
17 08	☽ ∠ ♀
16 Friday	
00 02	☽ ∥ ♃
03 30	☽ bQ ♇
04 20	♀ △ ♆
04 22	☽ ☌ ☉
08 23	☽ Q ♅
08 37	☽ ⊻ ☿
08 41	☽ ⊥ ♀
17 45	☽ ± ♇
17 52	☽ □ ♆
18 47	☽ ⊥ ☿
18 59	☽ ⊻ ♀
19 40	☽ ☌ ♄
22 04	☽ ☍ ♂
17 Saturday	
01 36	☉ # ♇
03 14	☽ ⊼ ♇
03 52	☽ # ♇
04 04	☽ ∥ ☉
04 54	♀ ⊻ ♄
09 27	☽ ∥ ♄
09 52	☿ ⊥ ♄
10 01	☽ ∠ ☿
10 49	☽ ⊻ ♃
12 55	☽ □ ♅
14 55	☽ # ♂
20 28	☽ ⊥ ♃
18 Sunday	
02 40	♀ ⊼ ♂
07 10	☽ ⊻ ☉
07 51	☿ ± ♂
11 00	☽ ∠ ♃
11 35	☽ ⚹ ☿
17 25	☽ ⊥ ☉
17 29	☽ △ ♆
19 41	☽ ⊻ ♄
20 09	☉ Q ♅
20 25	☽ # ♅
20 38	☽ ⊼ ♂
22 36	☽ ☌ ♀
19 Monday	
02 55	☽ ☍ ♇
04 28	☿ ∠ ♅
05 26	☽ ⊥ ♄
06 04	☽ ± ♂
08 24	☽ Q ☿
08 56	☽ ∠ ☉
11 31	☽ ⚹ ♃
12 59	☽ △ ♅
17 46	☽ bQ ♆
20 12	☽ ∠ ♄
20 23	☽ bQ ♂
21 30	☽ # ♆
20 Tuesday	
01 45	♂ ☍ ♄
07 34	☽ Q ♃
08 07	☉ ∥ ♄
08 36	☽ ± ♆
11 21	☽ ⚹ ☉
13 46	☽ bQ ♅
16 34	☽ □ ☿
18 39	☽ ⊼ ♆
20 43	☽ △ ♂
21 22	☽ ⚹ ♄
21 Wednesday	
04 25	☽ ⊻ ♀
04 29	☽ ⊼ ♇
05 00	☽ ± ♅
05 07	♀ ☍ ♇
09 08	☽ Q ☉
14 32	☽ □ ♃
14 45	☽ ± ♇
15 18	☽ ⊼ ♅
15 38	☽ ⊥ ♀
17 51	☉ # ♂
17 52	☿ ⚹ ♆
18 05	☽ Q ♄
22 Thursday	
01 29	♀ ± ♂
06 24	☽ bQ ♇
07 31	☿ ⊼ ♂
08 48	☽ ∠ ♀
09 24	☽ # ♆
19 02	☽ □ ☉
22 48	☽ ☍ ♆
23 35	☽ □ ♂
23 Friday	
00 13	♀ ⊥ ♄
01 36	♃ Q ♆
01 43	☽ △ ☿
02 11	☽ □ ♄
07 08	☿ ⊻ ♄
09 10	☽ △ ♇
13 11	☽ # ♅
14 16	☽ ⚹ ♀
16 55	♃ ⚹ ♅
20 53	☽ ☍ ♅
20 57	☽ △ ♃
24 Saturday	
03 12	♂ □ ♆
07 34	☽ Q ♂
08 08	☽ bQ ☿
14 37	☽ Q ♀
16 06	☉ ± ♇
17 37	☉ ☍ ♂
18 19	☽ ∥ ☉
21 55	♂ ⊥ ♇
22 58	♂ # ♄
23 17	☉ □ ♆
25 Sunday	
01 26	☽ bQ ♃
02 57	☽ ± ☿
03 12	☽ ∥ ♄
03 22	☽ # ♂
05 28	☽ ⚹ ♂
06 15	☽ ⊼ ♆
06 50	☽ △ ☉
10 22	☽ △ ♄
12 22	☽ # ♇
15 42	☽ ⊼ ☿
17 05	☽ □ ♇
17 41	☽ ± ♆
18 53	☽ ± ♃
26 Monday	
04 14	☿ △ ♇
04 15	☽ □ ♀
05 34	☽ ⊼ ♅
06 38	☽ ⊼ ♃
09 21	☽ ∠ ♂
11 01	☽ bQ ♆
14 02	☽ bQ ☉
15 29	☽ bQ ♄
15 56	☽ ∥ ♃
17 12	☽ ± ♅
18 55	♀ △ ♅
27 Tuesday	
02 19	☽ ⊥ ♂
03 50	☽ Q ♇
09 05	☽ ± ☉
09 14	☽ ± ♄
10 48	☽ bQ ♅
11 04	☉ ☌ ♄
12 19	♀ ⚹ ♃
12 58	☽ ∥ ☿
13 43	☽ ⊻ ♂
16 18	☽ △ ♆
21 08	☽ ⊼ ♄
21 54	☽ ⊼ ☉
23 17	♀ bQ ♂
28 Wednesday	
03 22	☽ ⚹ ♇
09 37	☽ ☍ ☿
16 27	☽ △ ♅
17 39	☽ # ☿
18 36	☽ ☍ ♃
21 07	☽ △ ♀
29 Thursday	
06 04	♀ bQ ♆
09 06	☽ ∠ ♇
12 44	☽ # ♃
23 29	☽ ☌ ♂
30 Friday	
03 04	☽ ⊥ ♇
03 56	☽ □ ♆
06 14	☽ bQ ♀
09 30	☽ ☍ ♄
14 08	☿ Q ♆
14 55	☽ ☍ ☉
15 05	☽ ⊻ ♇
15 45	☽ ∥ ♇
16 59	☉ ⊼ ♇
18 07	☿ ⚹ ♅
21 50	☽ ∥ ♂

All ephemeris data is given at 12.00 UT and the Moon's longitude is additionally given for 24.00 UT

LONGITUDES

Date	Sidereal time h m s	Sun ☉ ° ' "	Moon ☽ ° ' "	Moon ☽ 24.00 ° ' "	Mercury ☿ ° '	Venus ♀ ° '	Mars ♂ ° '	Jupiter ♃ ° '	Saturn ♄ ° '	Uranus ♅ ° '	Neptune ♆ ° '	Pluto ♇ ° '
01	02 35 54	10 ♉ 40 37	20 ♏ 17 45	26 ♏ 14 21	17 ♈ 45	21 ♊ 55	01 ♏ 34	18 ♈ 18	07 ♉ 16	16 ♒ 37	04 ♒ 22	09 ♐ 53
02	02 39 51	11 38 50	02 ♐ 10 32	08 ♐ 06 30	19 18	23 03	01 R 13	18 32	07 24	16 38	04 22	09 R 52
03	02 43 48	12 37 01	14 ♐ 02 29	19 ♐ 58 46	20 54	24 10	00 51	18 45	07 31	16 39	04 22	09 51
04	02 47 44	13 35 11	25 ♐ 55 41	01 ♑ 53 35	22 31	25 18	00 29	18 59	07 39	16 40	04 22	09 49
05	02 51 41	14 33 19	07 ♑ 52 52	13 ♑ 54 02	24 10	26 26	00 ♏ 08	19 13	07 47	16 41	04 22	09 48
06	02 55 37	15 31 26	19 ♑ 57 32	26 ♑ 03 57	25 51	27 33	29 ♎ 48	19 27	07 54	16 42	04 22	09 46
07	02 59 34	16 29 31	02 ♒ 13 50	08 ♒ 27 46	27 34	28 40	29 27	19 41	08 02	16 42	04 R 22	09 45
08	03 03 30	17 27 35	14 ♒ 44 51	21 ♒ 10 14	29 ♈ 19	29 ♊ 48	29 07	19 54	08 10	16 43	04 22	09 43
09	03 07 27	18 25 38	27 ♒ 39 55	04 ♓ 15 59	01 ♉ 05	00 ♋ 54	28 48	20 08	08 17	16 44	04 22	09 42
10	03 11 23	19 23 39	10 ♓ 58 50	17 ♓ 48 51	02 54	02 01	28 29	20 21	08 25	16 44	04 22	09 40
11	03 15 20	20 21 39	24 ♓ 46 14	01 ♈ 51 01	04 44	03 08	28 11	20 35	08 32	16 45	04 22	09 39
12	03 19 17	21 19 37	09 ♈ 03 04	16 ♈ 21 59	06 37	04 14	27 53	20 48	08 40	16 45	04 22	09 37
13	03 23 13	22 17 34	23 ♈ 47 10	01 ♉ 17 46	08 31	05 20	27 36	21 02	08 47	16 46	04 21	09 36
14	03 27 10	23 15 30	08 ♉ 52 42	16 ♉ 30 42	10 27	06 26	27 19	21 15	08 55	16 46	04 21	09 34
15	03 31 06	24 13 24	24 ♉ 10 24	01 ♊ 50 18	12 26	07 32	27 03	21 29	09 03	16 47	04 21	09 33
16	03 35 03	25 11 17	09 ♊ 28 57	17 ♊ 04 56	14 25	08 38	26 48	21 42	09 10	16 47	04 21	09 31
17	03 38 59	26 09 09	24 ♊ 37 00	02 ♋ 04 03	16 27	09 43	26 33	21 55	09 17	16 47	04 20	09 30
18	03 42 56	27 06 59	09 ♋ 25 13	16 ♋ 39 51	18 30	10 48	26 20	22 08	09 25	16 47	04 20	09 28
19	03 46 52	28 04 47	23 ♋ 47 35	00 ♌ 48 12	20 35	11 53	26 07	22 21	09 32	16 47	04 20	09 27
20	03 50 49	29 ♉ 02 33	07 ♌ 41 42	14 ♌ 28 17	22 42	12 58	25 54	22 34	09 40	16 48	04 19	09 25
21	03 54 46	00 ♊ 00 18	21 ♌ 08 13	27 ♌ 41 54	24 50	14 02	25 43	22 47	09 47	16 48	04 19	09 23
22	03 58 42	00 58 01	04 ♍ 09 50	10 ♍ 32 31	26 59	15 07	25 32	23 00	09 55	16 R 48	04 18	09 22
23	04 02 39	01 55 43	16 ♍ 50 30	23 ♍ 04 18	29 ♉ 09	16 11	25 22	23 13	10 02	16 48	04 18	09 20
24	04 06 35	02 53 23	29 ♍ 14 29	05 ♎ 21 33	01 ♊ 20	17 14	25 13	23 26	10 09	16 47	04 17	09 18
25	04 10 32	03 51 01	11 ♎ 25 59	17 ♎ 28 13	03 31	18 18	25 05	23 39	10 16	16 47	04 17	09 17
26	04 14 28	04 48 38	23 ♎ 28 41	29 ♎ 27 44	05 43	19 21	24 58	23 51	10 24	16 47	04 16	09 15
27	04 18 25	05 46 14	05 ♏ 25 41	11 ♏ 22 50	07 55	20 24	24 51	24 04	10 31	16 47	04 15	09 14
28	04 22 21	06 43 48	17 ♏ 19 27	23 ♏ 15 44	10 07	21 27	24 45	24 16	10 38	16 47	04 15	09 12
29	04 26 18	07 41 21	29 ♏ 11 54	05 ♐ 08 09	12 18	22 29	24 40	24 29	10 45	16 46	04 14	09 10
30	04 30 15	08 38 53	11 ♐ 04 38	17 ♐ 01 33	14 28	23 31	24 36	24 41	10 52	16 46	04 13	09 09
31	04 34 11	09 ♊ 36 24	22 ♐ 59 05	28 ♐ 57 27	16 ♊ 38	24 ♋ 33	24 ♎ 32	24 ♈ 53	10 ♉ 59	16 ♒ 45	04 ♒ 13	09 ♐ 07

Date	Moon True ☊ ° '	Moon Mean ☊ ° '	Moon ☽ Latitude ° '
01	18 ♌ 13	18 ♌ 01	04 N 59
02	18 R 00	17 58	04 51
03	17 48	17 55	04 29
04	17 38	17 52	03 56
05	17 31	17 48	03 13
06	17 27	17 45	02 20
07	17 25	17 42	01 19
08	17 D 25	17 39	00 N 14
09	17 R 25	17 36	00 S 54
10	17 24	17 32	02 01
11	17 21	17 29	03 04
12	17 15	17 26	03 57
13	17 07	17 23	04 36
14	16 57	17 20	04 58
15	16 47	17 17	04 58
16	16 36	17 13	04 37
17	16 27	17 10	03 57
18	16 21	17 07	03 02
19	16 17	17 04	01 56
20	16 16	17 01	00 S 45
21	16 D 15	16 58	00 N 26
22	16 R 15	16 54	01 33
23	16 15	16 51	02 35
24	16 12	16 48	03 27
25	16 07	16 45	04 09
26	15 59	16 42	04 39
27	15 48	16 38	04 57
28	15 36	16 35	05 01
29	15 24	16 32	04 53
30	15 12	16 29	04 32
31	15 ♌ 01	16 ♌ 26	03 N 59

DECLINATIONS

Date	Sun ☉ ° '	Moon ☽ ° '	Mercury ☿ ° '	Venus ♀ ° '	Mars ♂ ° '	Jupiter ♃ ° '	Saturn ♄ ° '	Uranus ♅ ° '	Neptune ♆ ° '	Pluto ♇ ° '
01	15 N 02	13 S 00	04 N 28	25 N 27	11 S 01	06 N 09	11 N 52	16 S 29	18 S 53	10 S 11
02	15 20	15 51	05 05	25 33	10 56	06 15	11 55	16 29	18 53	10 11
03	15 37	18 01	05 44	25 38	10 51	06 20	11 57	16 29	18 53	10 11
04	15 55	19 26	06 24	25 43	10 46	06 25	12 00	16 29	18 53	10 10
05	16 12	20 00	07 04	25 47	10 41	06 30	12 02	16 28	18 53	10 10
06	16 29	19 39	07 46	25 51	10 36	06 35	12 05	16 28	18 53	10 10
07	16 46	18 22	08 27	25 53	10 31	06 40	12 07	16 28	18 53	10 09
08	17 02	16 11	09 10	25 55	10 27	06 45	12 10	16 28	18 53	10 09
09	17 19	13 08	09 53	25 57	10 22	06 51	12 12	16 28	18 53	10 09
10	17 35	09 19	10 37	25 57	10 18	06 56	12 14	16 27	18 53	10 09
11	17 50	04 53	11 21	25 57	10 14	07 01	12 17	16 27	18 53	10 08
12	18 05	00 S 03	12 05	25 57	10 10	07 06	12 19	16 27	18 53	10 08
13	18 20	04 N 57	12 50	25 55	10 07	07 10	12 22	16 27	18 53	10 08
14	18 35	09 45	13 35	25 53	10 03	07 15	12 24	16 27	18 53	10 07
15	18 50	14 00	14 20	25 51	10 00	07 20	12 26	16 27	18 53	10 07
16	19 04	17 18	15 04	25 48	09 57	07 25	12 29	16 27	18 54	10 07
17	19 17	19 23	15 49	25 44	09 54	07 30	12 31	16 27	18 54	10 07
18	19 31	20 05	16 33	25 39	09 52	07 35	12 33	16 27	18 54	10 07
19	19 44	19 26	17 16	25 34	09 49	07 40	12 36	16 27	18 54	10 06
20	19 57	17 37	17 59	25 28	09 47	07 44	12 38	16 27	18 54	10 06
21	20 09	14 52	18 41	25 22	09 45	07 49	12 40	16 27	18 54	10 06
22	20 21	11 26	19 22	25 15	09 44	07 54	12 42	16 27	18 54	10 06
23	20 33	07 34	20 01	25 08	09 43	07 58	12 45	16 27	18 54	10 06
24	20 44	03 N 28	20 39	25 00	09 42	08 03	12 47	16 27	18 54	10 05
25	20 55	00 S 42	21 15	24 51	09 41	08 08	12 49	16 27	18 55	10 05
26	21 06	04 48	21 49	24 42	09 40	08 12	12 51	16 27	18 55	10 05
27	21 16	08 40	22 21	24 32	09 40	08 17	12 53	16 27	18 55	10 05
28	21 26	12 11	22 51	24 21	09 40	08 21	12 56	16 27	18 55	10 05
29	21 35	15 12	23 18	24 10	09 41	08 26	12 58	16 28	18 55	10 05
30	21 45	17 37	23 43	23 59	09 41	08 30	13 00	16 28	18 55	10 04
31	21 N 53	19 S 17	24 N 05	23 N 47	09 S 42	08 N 34	13 N 02	16 S 28	18 S 56	10 S 04

ZODIAC SIGN ENTRIES

Date	h	m	Planets
02	07	36	☽ ♐
04	20	12	☽ ♑
05	21	32	♂ ♎
07	07	40	☽ ♒
08	16	29	♀ ♋
08	21	22	☿ ♉
09	16	16	☽ ♓
11	20	53	☽ ♈
13	21	56	☽ ♉
15	21	07	☽ ♊
17	20	39	☽ ♋
19	22	37	☽ ♌
21	11	52	☉ ♊
22	04	15	☽ ♍
23	21	22	☿ ♊
24	13	29	☽ ♎
27	01	05	☽ ♏
29	13	37	☽ ♐

LATITUDES

Date	Mercury ☿ ° '	Venus ♀ ° '	Mars ♂ ° '	Jupiter ♃ ° '	Saturn ♄ ° '	Uranus ♅ ° '	Neptune ♆ ° '	Pluto ♇ ° '
01	02 S 42	02 N 16	01 N 04	01 S 06	02 S 11	00 S 40	00 N 17	11 N 52
04	02 33	02 22	00 56	01 06	02 11	00 40	00 17	11 52
07	02 18	02 27	00 48	01 06	02 11	00 40	00 17	11 52
10	01 59	02 32	00 40	01 07	02 11	00 40	00 17	11 53
13	01 36	02 36	00 32	01 07	02 11	00 40	00 17	11 53
16	01 09	02 39	00 24	01 07	02 11	00 40	00 17	11 53
19	00 39	02 40	00 17	01 07	02 11	00 41	00 17	11 53
22	00 S 07	02 41	00 09	01 08	02 11	00 41	00 17	11 53
25	00 N 24	02 41	00 N 02	01 08	02 12	00 41	00 17	11 53
28	00 54	02 40	00 S 05	01 08	02 12	00 41	00 17	11 53
31	01 N 20	02 N 37	00 S 12	01 S 09	02 S 12	00 S 41	00 N 17	11 N 52

DATA

Julian Date	2451300
Delta T	+64 seconds
Ayanamsa	23° 50' 39"
Synetic vernal point	05° ♓ 16' 21"
True obliquity of ecliptic	23° 26' 15"

MOON'S PHASES, APSIDES AND POSITIONS ☽

Date	h	m	Phase	Longitude ° '	Eclipse Indicator
08	17	29	☾	17 ♒ 41	
15	12	05	●	24 ♉ 14	
22	05	34	☽	00 ♍ 43	
30	06	40	○	08 ♐ 26	

Day	h	m		
02	05	58	Apogee	
15	15	01	Perigee	
29	08	00	Apogee	
05	14	57	Max dec	20° S 00'
12	12	12	0N	
18	12	13	Max dec	20° N 05'
25	07	56	0S	

LONGITUDES

Date	Chiron ⚷ ° '	Ceres ⚳ ° '	Pallas ⚴ ° '	Juno ⚵ ° '	Vesta ⚶ ° '	Black Moon Lilith ⚸ ° '
01	02 ♐ 12	25 ♊ 12	05 ♉ 12	09 ♐ 06	12 ♌ 08	26 ♏ 05
11	01 ♐ 31	29 ♊ 10	09 ♉ 51	07 ♐ 14	14 ♌ 46	27 ♏ 12
21	00 ♐ 48	03 ♋ 14	14 ♉ 35	05 ♐ 05	17 ♌ 49	28 ♏ 20
31	00 ♐ 04	07 ♋ 25	19 ♉ 24	02 ♐ 51	21 ♌ 12	29 ♏ 27

All ephemeris data is given at 12.00 UT and the Moon's longitude is additionally given for 24.00 UT

ASPECTARIAN

h m Aspects

01 Saturday
02 13 ☽ ± ♀
03 28 ☽ # ♄
04 35 ☽ □ ♅
06 06 ☽ ⚻ ☿
07 53 ☽ ⚻ ♃
15 36 ☽ ⚻ ♀
16 10 ☽ Q ♆
20 00 ☽ ± ☿
20 14 ☽ ± ♃
20 34 ♀ ∠ ♄
21 59 ☿ ☌ ♃

02 Sunday
06 37 ☽ # ☉
10 06 ☽ ⊻ ♂
14 47 ☽ bQ ♃
16 25 ☽ ✱ ♆
16 58 ☽ bQ ☿
16 59 ☽ Q ♅
18 17 ☽ ∥ ♅
21 53 ☽ ⊥ ♂
22 41 ☽ ⚻ ♄

03 Monday
03 32 ☽ ☌ ♇
08 52 ☽ ⚻ ☉
10 57 ☽ ± ♄
15 33 ☽ ∠ ♂
17 17 ☽ ✱ ♅
21 43 ☽ △ ♃
22 04 ☽ ± ☉
22 46 ☽ ∠ ♆

04 Tuesday
00 56 ☽ ∥ ♆
04 01 ☽ △ ☿
05 19 ☽ bQ ♄
10 37 ☽ ☍ ♀
12 45 ☿ ∥ ♃
16 55 ☽ ⊥ ♆
17 50 ☽ bQ ☉
20 55 ☽ ✱ ♂
23 34 ☽ ∠ ♅

05 Wednesday
04 58 ☽ ⊻ ♆
11 48 ☽ △ ♄
15 49 ☽ ⊻ ♇
17 36 ☽ ⊥ ♅
20 15 ☽ Q ♂
20 57 ☿ bQ ♇

06 Thursday
02 28 ☽ △ ☉
03 44 ☽ ⊥ ♇
05 32 ☽ ⊻ ♅
10 21 ☉ # ♅
10 58 ☽ □ ♃
21 27 ☽ ∠ ♇

07 Friday
00 49 ♆ St R
01 27 ☽ □ ☿
04 15 ☽ ∥ ♆
04 24 ☽ ⚻ ♀
05 27 ♀ ± ♆
06 45 ☽ □ ♂
16 08 ☽ ☌ ♆
17 11 ☽ ± ♀
17 24 ☉ □ ♅
22 41 ☽ Q ♃
23 18 ☽ □ ♄

08 Saturday
00 53 ♀ △ ♂
02 26 ☽ ✱ ♇
03 53 ☿ Q ♅
04 32 ☽ # ☉
09 22 ☽ ∥ ♅
09 51 ☿ ☍ ♂
12 02 ☽ bQ ♀
15 40 ☽ ☌ ♅
17 29 ☽ □ ☉
17 33 ☽ Q ☿
21 49 ☽ ✱ ♃

09 Sunday
01 00 ☽ Q ♇
05 35 ☿ ✱ ♀
09 27 ☽ Q ♄
14 01 ☽ △ ♂
18 12 ☽ # ♄
18 28 ☽ △ ♀
19 14 ☽ ✱ ☿
20 37 ☿ # ♇

10 Monday
00 11 ☽ ⊻ ♆
01 48 ☽ ∠ ♃
02 43 ☿ # ♂
05 07 ☽ Q ☉
05 30 ☽ # ☿
05 54 ♀ bQ ♅
06 04 ☽ ∥ ♂
07 08 ☽ ∥ ♇
07 24 ☽ ✱ ♄
09 41 ☽ □ ♇
10 55 ☽ ⊥ ♆
16 19 ☽ bQ ♂
18 03 ☽ ⊥ ♃
21 11 ♀ Q ♃
22 02 ☿ ± ♇
22 08 ☽ ⊻ ♅

11 Tuesday
01 03 ☽ # ♃
02 00 ☽ ∠ ☿
02 42 ☽ ∠ ♆
03 51 ☽ ✱ ☉
04 41 ☽ ⊻ ♃
07 07 ☿ □ ♆
07 39 ☽ ± ♂
08 32 ☽ ⊥ ♅
09 52 ☽ ∠ ♄
17 41 ☽ ⚻ ♂
19 12 ☉ ⊻ ♃
19 47 ☽ ⊥ ☿
23 50 ☽ ∠ ♅

12 Wednesday
01 16 ☽ ⊥ ♄
03 20 ☽ □ ♀
04 12 ☽ ✱ ♆
07 10 ☽ ∠ ☉
07 21 ☽ ⊻ ☿
11 21 ☽ ⊻ ♄
12 57 ☽ △ ♇
14 44 ♀ ⚻ ♆
19 55 ☿ ∥ ♄
23 02 ☽ ⊥ ☉
23 59 ☽ Q ♆

13 Thursday
00 38 ☽ ✱ ♅
03 55 ♂ ∥ ♇
07 30 ☽ ☌ ♃
09 26 ☽ ⊻ ☉
11 13 ☽ Q ♀
13 18 ☽ bQ ♇
15 37 ☿ ☌ ♄
17 59 ☽ ☍ ♂
19 58 ☽ Q ♅
23 05 ☽ ∥ ♃

14 Friday
01 14 ☿ ⚻ ♇
03 38 ☽ ± ♇
04 51 ☽ □ ♆
07 51 ☽ ✱ ♀
12 04 ☽ ☌ ♄
13 06 ☽ ⚻ ♇
13 34 ☽ # ♂
13 58 ☽ # ♇
14 51 ☽ ☌ ☿

15 Saturday
00 25 ☽ □ ♅
02 35 ☽ ∥ ♄
07 43 ☽ ⊻ ♃
09 14 ☽ ∠ ♀
12 05 ☽ ☌ ☉
14 37 ☽ ∥ ☿
16 26 ☽ ⚻ ♂
17 15 ☽ ⊥ ♃
18 43 ☉ # ♆

16 Sunday
00 25 ☽ ⊥ ♀
01 40 ☽ ± ♂
03 56 ☽ △ ♆
04 58 ☽ # ♅
07 33 ☽ ∠ ♃
10 33 ☽ ⊻ ♀
11 30 ☽ ⊻ ♄
12 04 ☽ ☍ ♇
15 35 ☽ bQ ♂
20 59 ☽ ⊻ ☿
21 02 ☽ ⊥ ♄
23 32 ☽ △ ♅

17 Monday
01 24 ♀ ✱ ♄
03 35 ☽ bQ ♆
04 45 ☽ # ♆
07 12 ♀ ⚻ ♇
07 37 ☽ ✱ ♃
08 00 ☽ ⊥ ☿
10 13 ☽ ∥ ☉
11 29 ☽ ∠ ♄
14 38 ☽ ⊻ ☉
15 04 ☽ △ ♂
15 55 ☿ □ ♅
17 58 ☽ ± ♆
20 06 ☉ ⚻ ♂
23 33 ☽ bQ ♅

18 Tuesday
00 46 ☽ ∠ ☿
00 59 ☽ ⊥ ☉
03 14 ☽ Q ♃
03 41 ☽ ⚻ ♆
08 42 ☿ # ♅
11 40 ♀ ± ♅
12 00 ☽ ✱ ♄
12 05 ☽ ⚻ ♇
14 15 ☽ ± ♅
14 28 ☽ ☌ ♀
16 46 ☽ ∠ ☉
20 24 ♄ ⚻ ♇
21 59 ☽ ± ♇

19 Wednesday
00 13 ☽ ⚻ ♅
05 39 ☽ ✱ ☿
06 51 ☽ ∥ ☉
08 09 ☽ Q ♄
09 32 ☽ □ ♃
13 06 ☽ bQ ♇
15 53 ☽ □ ♂
19 51 ☽ ✱ ☉
20 46 ☽ # ♆
20 53 ☉ ⊥ ♃

20 Thursday
05 48 ☽ Q ☿
06 06 ☽ ☍ ♆
09 01 ☽ ∥ ☿
10 24 ☿ ⊻ ♃
15 01 ☽ △ ♇
15 30 ☽ □ ♄
18 21 ☽ Q ☉
22 07 ☽ ⊻ ♀
22 50 ☽ Q ♂
23 03 ☽ # ♅

21 Friday
04 09 ☽ ☍ ♅
09 50 ☽ ⊥ ♀
15 03 ☽ △ ♃
19 37 ☿ # ♆
20 03 ☽ □ ☿
20 14 ☽ ✱ ♂
21 07 ☿ ⚻ ♂
22 25 ♅ St R

22 Saturday
03 36 ☽ ∥ ♄
03 47 ☽ ∠ ♀
05 34 ☽ □ ☉
12 16 ☽ ⚻ ♆
17 30 ♀ ± ♇
19 20 ☽ bQ ♃
20 32 ☽ # ♇
21 45 ☽ □ ♇
22 53 ☽ # ♂
22 54 ☽ △ ♄
23 33 ☽ ± ♆
23 50 ☽ ∠ ♂

23 Sunday
00 06 ☉ ± ♂
09 36 ☽ ∥ ♃
10 37 ☽ ✱ ♀
11 54 ☽ ⚻ ♅
12 44 ☽ ± ♃
12 50 ☿ ⊥ ♃
16 42 ☽ bQ ♆
16 48 ☽ ⊥ ♂
23 27 ☽ ± ♅

24 Monday
00 30 ☽ ⚻ ♃
01 51 ♀ ⚻ ♅
03 57 ☽ bQ ♄
04 16 ☽ ⊻ ♂
08 14 ☽ Q ♇
10 52 ☿ ± ♂
12 00 ☽ Q ♀
16 48 ☉ ∥ ☿
16 59 ☽ △ ☿
16 59 ☽ bQ ♅
19 45 ☽ △ ☉
21 43 ☽ ± ♄
21 53 ☽ △ ♆

25 Tuesday
07 16 ☿ ∠ ♀
07 45 ☽ ✱ ♇
09 41 ☽ ⚻ ♄
18 22 ☉ ☌ ☿
20 12 ☿ △ ♆
22 33 ☉ △ ♆
22 38 ☽ △ ♅

26 Wednesday
02 58 ☽ □ ♀
04 02 ☽ bQ ☉
05 15 ☽ bQ ☿
12 46 ☽ ☍ ♃
13 33 ☽ ∠ ♇
14 56 ☽ ☌ ♂
23 38 ☽ ± ☉

27 Thursday
03 20 ☽ ± ☿
07 35 ☽ ⊥ ♇
09 29 ☽ # ♃
09 39 ☽ □ ♆
12 45 ☽ ⚻ ☉
18 09 ☽ ⚻ ☿
18 35 ☽ ∥ ♂
19 38 ☽ ⊻ ♇
21 20 ☽ ∥ ♇
22 21 ☽ ☍ ♄

28 Friday
01 51 ☿ ∠ ♃
02 09 ☿ ☍ ♇
04 39 ♃ bQ ♇
08 13 ☿ bQ ♂
10 54 ☽ □ ♅
17 36 ☽ # ♄
18 04 ☿ ⊻ ♄
21 08 ☽ △ ♀
21 56 ☽ Q ♆

29 Saturday
02 17 ☽ ⚻ ♃
02 54 ☽ ⊻ ♂
14 38 ☽ ± ♃
14 57 ☽ ⊥ ♂
19 02 ♀ Q ♄
22 10 ☽ ✱ ♆
23 15 ☽ Q ♅
23 40 ☽ ∥ ♅

30 Sunday
04 18 ♂ ☍ ♃
06 20 ☽ bQ ♀
06 40 ☽ ☍ ☉
08 06 ☽ ☌ ♇
09 02 ☽ ∠ ♂
09 08 ☽ bQ ♃
11 35 ☽ ⚻ ♄
20 22 ☽ ☍ ☿
22 58 ☿ ∥ ♀
23 28 ☽ ✱ ♅
23 48 ☽ ± ♄

31 Monday
00 06 ☉ ☍ ♇
02 13 ☽ ± ♀
02 14 ♀ bQ ♇
04 25 ☽ ∠ ♆
05 49 ☽ ∥ ♆
10 26 ☉ bQ ♂
11 52 ♀ □ ♂
13 28 ☿ △ ♅
15 07 ☽ ✱ ♂
15 26 ☽ ⚻ ♀
15 54 ☽ △ ♃
16 19 ☿ ⊥ ♄
18 06 ☽ bQ ♄
20 59 ☉ ∠ ♃
22 00 ♀ □ ♃
22 29 ☽ ⊥ ♆

JUNE 1999

LONGITUDES

Date	Sidereal time h m s	Sun ☉ ° ' "	Moon ☽ ° ' "	Moon ☽ 24.00 ° ' "	Mercury ☿ ° '	Venus ♀ ° '	Mars ♂ ° '	Jupiter ♃ ° '	Saturn ♄ ° '	Uranus ♅ ° '	Neptune ♆ ° '	Pluto ♇ ° '
01	04 38 08	10 ♊ 33 54	04 ♑ 56 51	10 ♑ 57 34	18 ♊ 46	25 ♋ 34	24 ♎ 30	25 ♈ 06	11 ♉ 06	16 ♒ 45	04 ♒ 12	09 ♐ 05
02	04 42 04	11 31 23	16 ♑ 59 51	23 ♑ 04 04	20 52	26 35	24 R 28	25 18	11 13	16 R 44	04 R 11	09 R 04
03	04 46 01	12 28 51	29 ♑ 10 33	05 ♒ 19 41	22 57	27 36	24 27	25 30	11 20	16 44	04 10	09 02
04	04 49 57	13 26 18	11 ♒ 31 56	17 ♒ 47 45	25 00	28 36	24 D 27	25 42	11 27	16 43	04 09	09 01
05	04 53 54	14 23 44	24 ♒ 07 36	00 ♓ 31 59	27 01	29 ♋ 37	24 27	25 54	11 34	16 42	04 08	08 59
06	04 57 50	15 21 10	07 ♓ 01 22	13 ♓ 36 15	29 ♊ 00	00 ♌ 36	24 29	26 05	11 41	16 42	04 08	08 57
07	05 01 47	16 18 35	20 ♓ 17 02	27 ♓ 04 04	00 ♋ 57	01 36	24 31	26 17	11 48	16 41	04 07	08 56
08	05 05 44	17 16 00	03 ♈ 57 37	10 ♈ 57 48	02 51	02 34	24 34	26 29	11 55	16 40	04 06	08 54
09	05 09 40	18 13 24	18 ♈ 04 36	25 ♈ 17 51	04 43	03 33	24 38	26 40	12 01	16 39	04 05	08 52
10	05 13 37	19 10 47	02 ♉ 37 07	10 ♉ 01 49	06 32	04 31	24 42	26 52	12 08	16 38	04 04	08 51
11	05 17 33	20 08 10	17 ♉ 31 08	25 ♉ 04 03	08 19	05 29	24 47	27 03	12 15	16 37	04 03	08 49
12	05 21 30	21 05 32	02 ♊ 39 26	10 ♊ 15 58	10 04	06 26	24 53	27 14	12 21	16 37	04 02	08 48
13	05 25 26	22 02 54	17 ♊ 52 20	25 ♊ 27 11	11 46	07 23	25 00	27 26	12 28	16 35	04 01	08 46
14	05 29 23	23 00 15	02 ♋ 59 14	10 ♋ 27 20	13 25	08 19	25 07	27 37	12 34	16 34	03 59	08 45
15	05 33 19	23 57 36	17 ♋ 50 27	25 ♋ 07 47	15 02	09 15	25 15	27 48	12 41	16 33	03 58	08 43
16	05 37 16	24 54 56	02 ♌ 18 43	09 ♌ 22 52	16 36	10 10	25 24	27 59	12 47	16 32	03 57	08 41
17	05 41 13	25 52 14	16 ♌ 20 01	23 ♌ 10 10	18 08	11 05	25 33	28 09	12 53	16 31	03 56	08 40
18	05 45 09	26 49 32	29 ♌ 53 25	06 ♍ 30 03	19 37	12 00	25 43	28 20	13 00	16 30	03 55	08 38
19	05 49 06	27 46 49	13 ♍ 00 27	19 ♍ 25 04	21 03	12 53	25 54	28 31	13 06	16 28	03 54	08 37
20	05 53 02	28 44 06	25 ♍ 44 23	01 ♎ 58 59	22 27	13 47	26 05	28 41	13 12	16 27	03 52	08 35
21	05 56 59	29 ♊ 41 21	08 ♎ 09 24	14 ♎ 16 12	23 48	14 39	26 17	28 52	13 18	16 26	03 51	08 34
22	06 00 55	00 ♋ 38 36	20 ♎ 19 59	26 ♎ 21 15	25 06	15 31	26 30	29 02	13 24	16 24	03 50	08 32
23	06 04 52	01 35 50	02 ♏ 20 32	08 ♏ 18 20	26 21	16 22	26 43	29 12	13 30	16 23	03 48	08 31
24	06 08 48	02 33 03	14 ♏ 15 05	20 ♏ 11 13	27 34	17 13	26 57	29 22	13 36	16 21	03 47	08 29
25	06 12 45	03 30 16	26 ♏ 07 07	02 ♐ 03 07	28 44	18 03	27 12	29 32	13 42	16 20	03 46	08 28
26	06 16 42	04 27 29	07 ♐ 59 32	13 ♐ 56 39	29 ♋ 50	18 52	27 27	29 42	13 48	16 18	03 44	08 26
27	06 20 38	05 24 41	19 ♐ 54 42	25 ♐ 53 54	00 ♌ 54	19 40	27 43	29 ♈ 51	13 54	16 17	03 43	08 25
28	06 24 35	06 21 53	01 ♑ 54 30	07 ♑ 56 39	01 54	20 28	27 59	00 ♉ 01	13 59	16 15	03 42	08 24
29	06 28 31	07 19 05	14 ♑ 00 33	20 ♑ 06 23	02 51	21 15	28 15	00 10	14 05	16 13	03 40	08 22
30	06 32 28	08 ♋ 16 16	26 ♑ 14 21	02 ♒ 24 39	03 ♌ 45	22 ♌ 01	28 ♎ 33	00 ♉ 20	14 ♉ 10	16 ♒ 12	03 ♒ 39	08 ♐ 21

DECLINATIONS

Date	Moon True ☊ ° '	Moon Mean ☊ ° '	Moon ☽ Latitude ° '	Sun ☉ ° '	Moon ☽ ° '	Mercury ☿ ° '	Venus ♀ ° '	Mars ♂ ° '	Jupiter ♃ ° '	Saturn ♄ ° '	Uranus ♅ ° '	Neptune ♆ ° '	Pluto ♇ ° '
01	14 ♌ 53	16 ♌ 23	03 N 15	22 N 02	20 S 06	24 N 25	23 N 35	09 S 43	08 N 39	13 N 04	16 S 28	18 S 56	10 S 04
02	14 R 47	16 19	02 22	22 10	20 00	24 41	23 22	09 44	08 43	13 06	16 28	18 56	10 04
03	14 43	16 16	01 22	22 17	18 59	24 55	23 08	09 46	08 47	13 08	16 28	18 56	10 04
04	14 42	16 13	00 N 17	22 25	17 03	25 07	22 54	09 48	08 52	13 10	16 29	18 56	10 04
05	14 D 42	16 10	00 S 50	22 32	14 16	25 15	22 40	09 50	08 56	13 12	16 29	18 56	10 04
06	14 43	16 07	01 57	22 38	10 44	25 21	22 25	09 53	09 00	13 14	16 29	18 57	10 04
07	14 R 43	16 04	02 58	22 44	06 35	25 24	22 10	09 55	09 04	13 16	16 29	18 57	10 03
08	14 42	16 00	03 52	22 50	01 S 59	25 25	21 55	09 58	09 08	13 18	16 30	18 57	10 03
09	14 38	15 57	04 34	22 55	02 N 52	25 24	21 39	10 01	09 12	13 20	16 30	18 57	10 03
10	14 33	15 54	04 59	23 00	07 41	25 20	21 22	10 05	09 16	13 22	16 30	18 58	10 03
11	14 26	15 51	05 05	23 04	12 10	25 14	21 05	10 08	09 20	13 24	16 31	18 58	10 03
12	14 18	15 48	04 50	23 08	15 56	25 07	20 48	10 12	09 24	13 26	16 31	18 58	10 03
13	14 11	15 44	04 15	23 12	18 39	24 57	20 31	10 16	09 28	13 28	16 31	18 58	10 03
14	14 04	15 41	03 22	23 15	20 02	24 45	20 13	10 21	09 32	13 29	16 32	18 59	10 03
15	14 00	15 38	02 16	23 18	20 00	24 32	19 55	10 25	09 35	13 31	16 32	18 59	10 03
16	13 57	15 35	01 S 02	23 20	18 38	24 18	19 36	10 30	09 39	13 33	16 32	18 59	10 03
17	13 D 57	15 32	00 N 13	23 22	16 09	24 02	19 18	10 35	09 43	13 35	16 33	18 59	10 03
18	13 57	15 29	01 25	23 24	12 50	23 45	18 59	10 40	09 47	13 37	16 33	19 00	10 03
19	13 59	15 25	02 30	23 25	08 59	23 26	18 40	10 46	09 50	13 38	16 34	19 00	10 03
20	13 59	15 22	03 26	23 26	04 51	23 07	18 20	10 51	09 54	13 40	16 34	19 00	10 03
21	13 R 59	15 19	04 11	23 26	00 N 37	22 46	18 00	10 57	09 57	13 42	16 35	19 01	10 03
22	13 57	15 16	04 43	23 26	03 S 34	22 25	17 40	11 03	10 01	13 43	16 35	19 01	10 03
23	13 53	15 13	05 03	23 26	07 32	22 03	17 20	11 09	10 04	13 45	16 35	19 01	10 03
24	13 48	15 10	05 09	23 25	11 11	21 41	17 00	11 16	10 07	13 47	16 36	19 02	10 03
25	13 41	15 06	05 02	23 23	14 23	21 18	16 40	11 22	10 11	13 48	16 36	19 02	10 03
26	13 34	15 03	04 42	23 22	17 00	20 54	16 19	11 29	10 14	13 50	16 37	19 02	10 03
27	13 27	15 00	04 09	23 20	18 55	20 31	15 58	11 36	10 17	13 51	16 37	19 03	10 03
28	13 21	14 57	03 26	23 17	20 00	20 07	15 37	11 43	10 21	13 53	16 38	19 03	10 03
29	13 16	14 54	02 32	23 14	20 11	19 43	15 17	11 50	10 24	13 55	16 38	19 03	10 03
30	13 ♌ 13	14 ♌ 50	01 N 31	23 N 11	19 S 25	19 N 19	14 N 56	11 S 58	10 N 27	13 N 56	16 S 39	19 S 04	10 S 03

ZODIAC SIGN ENTRIES

Date	h	m	Planets
01	02	05	☽ ♑
03	13	37	☽ ♒
05	21	25	♀ ♌
05	23	00	☽ ♓
07	00	18	☿ ♋
08	05	08	☽ ♈
10	07	44	☽ ♉
12	07	48	☽ ♊
14	07	14	☽ ♋
16	08	07	☽ ♌
18	12	12	☽ ♍
20	20	10	☽ ♎
21	19	49	☉ ♋
23	07	18	☽ ♏
25	19	51	☽ ♐
26	15	39	☿ ♌
28	08	12	☽ ♑
28	09	29	♃ ♉
30	19	19	☽ ♒

LATITUDES

Date	Mercury ☿ ° '	Venus ♀ ° '	Mars ♂ ° '	Jupiter ♃ ° '	Saturn ♄ ° '	Uranus ♅ ° '	Neptune ♆ ° '	Pluto ♇ ° '
01	01 N 27	02 N 36	00 S 14	01 S 09	02 S 12	00 S 41	00 N 17	11 N 52
04	01 46	02 31	00 21	01 09	02 12	00 41	00 17	11 52
07	01 58	02 26	00 27	01 09	02 13	00 41	00 17	11 52
10	02 04	02 18	00 33	01 10	02 13	00 41	00 17	11 51
13	02 02	02 10	00 38	01 10	02 13	00 41	00 17	11 51
16	01 54	01 59	00 44	01 11	02 14	00 42	00 17	11 50
19	01 40	01 47	00 49	01 11	02 14	00 42	00 17	11 50
22	01 20	01 34	00 53	01 12	02 14	00 42	00 17	11 49
25	00 54	01 18	00 58	01 12	02 15	00 42	00 17	11 48
28	00 N 23	01 01	01 02	01 13	02 15	00 42	00 17	11 47
31	00 S 12	00 N 41	01 S 06	01 S 13	02 S 16	00 S 42	00 N 17	11 N 46

DATA

Julian Date	2451331
Delta T	+64 seconds
Ayanamsa	23° 50' 43"
Synetic vernal point	05° ♓ 16' 16"
True obliquity of ecliptic	23° 26' 15"

MOON'S PHASES, APSIDES AND POSITIONS ☽

Date	h	m	Phase	Longitude ° '	Eclipse Indicator
07	04	20	☾	16 ♓ 00	
13	19	03	●	22 ♊ 20	
20	18	13	☽	28 ♍ 59	
28	21	37	○	06 ♑ 45	

Day	h	m		
13	00	30	Perigee	
25	15	23	Apogee	
01	21	43	Max dec	20° S 10'
08	21	53	0N	
14	23	22	Max dec	20° N 12'
21	15	28	0S	
29	04	38	Max dec	20° S 13'

LONGITUDES

Date	Chiron ⚷ ° '	Ceres ⚳ ° '	Pallas ⚴ ° '	Juno ⚵ ° '	Vesta ⚶ ° '	Black Moon Lilith ⚸ ° '
01	00 ♐ 00	07 ♋ 50	19 ♉ 54	02 ♐ 37	21 ♌ 34	29 ♏ 34
11	29 ♏ 19	12 ♋ 06	24 ♉ 48	00 ♐ 30	25 ♌ 16	00 ♐ 41
21	28 ♏ 42	16 ♋ 26	29 ♉ 48	28 ♏ 41	29 ♌ 12	01 ♐ 48
31	28 ♏ 13	20 ♋ 48	04 ♊ 52	27 ♏ 18	03 ♍ 20	02 ♐ 55

All ephemeris data is given at 12.00 UT and the Moon's longitude is additionally given for 24.00 UT

ASPECTARIAN

h m Aspects

01 Tuesday
05 36 ☽ ∠ ♅
09 52 ☉ ∠ ♀
10 30 ☽ ⊻ ♆
15 05 ☽ Q ♂
16 55 ☿ bQ ♆
20 15 ☽ ⊻ ♇
23 34 ☽ ⊥ ♅

02 Wednesday
00 11 ☽ ⊼ ☉
00 25 ☽ △ ♄
03 27 ☉ ⊻ ♄
05 47 ☿ ⊥ ♀
08 10 ☽ ⊥ ♇
11 29 ☽ ⊻ ♅
13 08 ☽ ± ☉
21 15 ☽ ⊼ ☿

03 Thursday
01 56 ☽ ∠ ♇
02 44 ☽ □ ♂
04 40 ☽ □ ♃
08 24 ☽ bQ ☉
08 38 ☽ ☍ ♀
11 28 ☽ ± ☿
12 49 ☽ ∥ ♆
21 44 ☽ ☌ ♆

04 Friday
05 29 ☿ △ ♂
06 11 ♂ St D
07 09 ☽ ✶ ♇
08 28 ☽ bQ ☿
11 51 ☽ □ ♄
15 58 ☽ △ ☉
16 13 ☽ Q ♃
17 33 ☽ ∥ ♅
21 07 ☿ ✶ ♃
21 56 ☽ ☌ ♅

05 Saturday
06 04 ☽ Q ♇
06 19 ☿ ∠ ♄
12 38 ☽ △ ♂
15 23 ☽ ✶ ♃
18 26 ☽ △ ☿
19 41 ☽ # ♄
21 48 ☉ ∥ ♀
22 18 ☽ Q ♄
23 08 ☽ ⊼ ♀

06 Sunday
01 27 ☿ ± ♆
06 40 ☽ ⊻ ♆
11 10 ☽ ± ♀
15 32 ☽ □ ♇
16 08 ☽ ∥ ♇
16 31 ☽ bQ ♂
17 11 ☽ ∥ ♂
17 40 ☽ ⊥ ♆
19 33 ☽ ∠ ♃
20 35 ☽ ✶ ♄
22 16 ☽ # ♃

07 Monday
04 20 ☽ □ ☉
04 52 ☽ bQ ♀
05 34 ☽ ⊻ ♅
08 50 ☽ ± ♂
09 54 ☽ ∠ ♆
12 00 ☽ ⊥ ♃
16 16 ☽ ⊥ ♅
19 32 ☽ ⊼ ♂
21 13 ☿ bQ ♅
21 15 ☉ △ ♅
22 47 ☽ ⊻ ♃
23 37 ☽ ∠ ♄

08 Tuesday
04 49 ☿ ⊻ ♀
08 02 ☽ ∠ ♅
09 25 ☽ △ ♀
09 46 ☽ □ ☿
12 14 ☽ ✶ ♆
14 25 ☽ Q ☉
15 23 ☽ ⊥ ♄
20 28 ☽ △ ♇

09 Wednesday
01 43 ☽ ⊻ ♄
03 51 ☿ ⊼ ♆
06 17 ☉ ⊥ ♄
08 39 ☽ Q ♆
09 37 ☽ ✶ ♅
12 16 ☽ ✶ ☉
20 51 ☽ Q ☿
21 38 ☽ bQ ♇
22 57 ☽ ☍ ♂

10 Thursday
00 50 ♀ ☍ ♆
02 15 ♂ ∥ ♇
02 27 ☽ ☌ ♃
05 30 ☽ Q ♅
09 05 ☉ bQ ♆
12 22 ☽ ± ♇
14 21 ☽ □ ♆
14 43 ☽ ∠ ☉
15 18 ☽ □ ♀
19 14 ☽ ✶ ☿
20 16 ☽ ∥ ♃
22 05 ☽ ⊼ ♇

11 Friday
00 18 ☽ # ♇
00 36 ☽ # ♂
03 30 ☽ ☌ ♄
06 13 ☽ ⊥ ☉
10 34 ☽ □ ♅
16 27 ☽ ⊻ ☉
18 44 ☿ ⊼ ♇
19 18 ☽ ∥ ♄
22 07 ☽ Q ♀
22 26 ☽ ∠ ☿
23 14 ☿ Q ♃
23 37 ☽ ⊼ ♂

12 Saturday
03 20 ☽ ⊻ ♃
09 11 ☽ ± ♂
12 56 ☽ ⊥ ♃
14 09 ☽ △ ♆
14 30 ☽ ⊥ ☿
16 20 ☽ # ♅
18 21 ☽ ✶ ♀
19 35 ☿ ± ♅
21 40 ☽ ☍ ♇
23 29 ☽ bQ ♂

13 Sunday
01 09 ☽ ⊻ ☿
03 18 ☽ ∠ ♃
03 24 ☽ ⊻ ♄
09 59 ☽ △ ♅
12 56 ☽ ⊥ ♄
13 48 ☽ bQ ♆
15 56 ☽ # ♆
19 03 ☽ ☌ ☉
19 36 ☽ ∠ ♀
22 48 ☿ ✶ ♄
23 22 ☽ △ ♂

14 Monday
03 18 ☽ ∠ ♄
03 19 ☽ ✶ ♃
04 02 ☽ ± ♆
09 45 ☽ bQ ♅
10 52 ☽ ⊥ ♀
13 36 ☽ ⊼ ♆
17 09 ☽ ∥ ♀
21 08 ☽ ⊻ ♀
21 13 ☽ ⊼ ♇
22 32 ♀ △ ♇
22 46 ☽ Q ♃

15 Tuesday
00 11 ☽ ± ♅
03 32 ☽ ✶ ♄
06 52 ☽ ☌ ☿
06 55 ☽ ± ♇
07 20 ☿ ± ♇
09 54 ☽ ⊼ ♅
16 01 ☽ ∥ ♀
21 39 ☽ bQ ♇
22 46 ☽ ⊻ ☉
23 20 ☽ Q ♄

16 Wednesday
00 19 ☽ □ ♂
04 39 ☽ □ ♃
07 26 ☽ # ♆
09 29 ☽ ⊥ ☉
10 58 ☿ ⊼ ♅
14 46 ☽ ☍ ♆
22 48 ☽ △ ♇

17 Thursday
01 52 ☽ ∠ ☉
02 18 ☽ ☌ ♀
02 22 ☉ △ ♂
05 59 ☽ □ ♄
07 07 ☽ Q ♂
08 40 ☽ # ♅
12 19 ☽ ☍ ♅
15 31 ☽ ⊻ ☿

18 Friday
03 24 ☽ ⊥ ☿
04 26 ☽ ✶ ♂
06 05 ☽ ✶ ☉
06 49 ☽ ∥ ♄
09 10 ☽ △ ♃
10 44 ♀ # ♆
19 16 ☽ ⊼ ♆
21 37 ☽ ∠ ☿

19 Saturday
01 31 ☽ # ♂
03 53 ☽ □ ♇
05 33 ☽ Q ☉
05 38 ☽ # ♇
06 14 ☽ ± ♆
07 00 ☽ ∥ ♃
07 23 ♃ Q ♅
08 02 ☽ ∠ ♂
11 46 ☽ ⊻ ♀
12 10 ☽ △ ♄
12 57 ☽ bQ ♃
13 16 ☉ ∥ ☿
14 45 ☉ ± ♆
18 22 ♀ □ ♄
18 27 ☽ ⊼ ♅
20 57 ☉ ∠ ♄
23 00 ☽ bQ ♆
23 50 ☽ ⊥ ♀

20 Sunday
01 06 ☽ ⊥ ♂
04 58 ☽ ✶ ☿
05 45 ☽ ± ♅
06 06 ☽ ± ♃
10 29 ☉ ✶ ♃
12 41 ☽ ⊻ ♂
13 37 ☽ Q ♇
16 45 ☽ bQ ♄
17 43 ☽ ⊼ ♃
18 13 ☽ □ ☉
18 15 ☽ ∠ ♀
22 57 ☽ bQ ♅
23 05 ♀ Q ♂

21 Monday
00 51 ☉ ∠ ♀
03 38 ☽ △ ♆
06 50 ☽ Q ☿
07 52 ☿ bQ ♇
10 19 ☽ ± ♄
12 48 ☽ ✶ ♇
22 11 ☽ ⊼ ♄

22 Tuesday
01 44 ☽ ✶ ♀
04 14 ☽ △ ♅
18 13 ☿ Q ♄
18 22 ☽ ∠ ♇
22 37 ☽ □ ☿

23 Wednesday
00 31 ☽ ☌ ♂
03 26 ☽ Q ♀
03 45 ♃ # ♇
05 36 ☽ ☍ ♃
06 41 ☉ bQ ♅
10 22 ☽ △ ☉
12 15 ♀ ☍ ♅
12 21 ☽ ⊥ ♇
14 56 ☽ □ ♆
20 52 ☿ □ ♂

24 Thursday
00 24 ☽ ⊻ ♇
04 13 ☽ ∥ ♇
04 36 ☽ # ♃
10 41 ☽ ☍ ♄
12 31 ☽ ∥ ♂
16 15 ☽ □ ♅
18 27 ☽ □ ♀
19 15 ☽ bQ ☉

25 Friday
03 12 ☽ Q ♆
07 17 ☽ # ♄
14 14 ☽ ⊻ ♂
15 03 ☽ ± ☉
15 38 ♀ # ♅
17 50 ☽ △ ☿
18 22 ☉ ⊼ ♆
19 00 ☽ ⊼ ♃

26 Saturday
02 37 ☽ ⊥ ♂
03 26 ☽ ✶ ♆
04 14 ☽ ⊼ ☉
04 34 ☽ Q ♅
05 57 ☽ # ♀
07 18 ☽ ± ♃
08 01 ☽ ∥ ♅
08 22 ☿ □ ♃
12 54 ☽ ☌ ♇
21 11 ☽ ∠ ♂
23 48 ☽ ⊼ ♄

27 Sunday
01 42 ☽ bQ ♃
03 09 ☽ bQ ☿
04 43 ☽ ✶ ♅
09 36 ☽ ∠ ♆
11 29 ☽ △ ♀
11 58 ☽ ± ♄
14 10 ☽ ∥ ♆
22 55 ☽ ± ☿

28 Monday
03 37 ☽ ⊥ ♆
03 59 ☽ ✶ ♂
06 08 ☽ bQ ♄
08 11 ☽ △ ♃
10 42 ☽ ∠ ♅
11 59 ☽ ⊼ ☿
14 46 ☽ # ☿
15 33 ☽ ⊻ ♆
19 35 ☽ bQ ♀
21 37 ☽ ☍ ☉

29 Tuesday
00 52 ☽ ⊻ ♇
04 24 ☽ Q ♂
04 32 ☽ ⊥ ♅
12 09 ☽ △ ♄
12 43 ☽ ⊥ ♇
14 37 ☽ ± ♀
16 21 ☽ ⊻ ♅

30 Wednesday
03 12 ☽ ⊼ ♀
06 22 ☽ ∠ ♇
09 12 ☿ ☍ ♆
13 55 ☉ ⊼ ♇
14 38 ☽ # ☿
16 36 ☽ □ ♂
18 07 ☽ ∥ ♆
20 04 ☽ □ ♃

JULY 1999

LONGITUDES

Date	Sidereal time h m s	Sun ☉ ° ' "	Moon ☽ ° ' "	Moon ☽ 24.00 ° ' "	Mercury ☿ ° '	Venus ♀ ° '	Mars ♂ ° '	Jupiter ♃ ° '	Saturn ♄ ° '	Uranus ♅ ° '	Neptune ♆ ° '	Pluto ♇ ° '
01	06 36 24	09 ♋ 13 27	08 ♒ 37 28	14 ♒ 53 04	04 ♌ 36	22 ♌ 46	28 ♎ 50	00 ♉ 29	14 ♉ 16	16 ♒ 10	03 ♒ 37	08 ♐ 20
02	06 40 21	10 10 39	21 ♒ 11 40	27 ♒ 33 33	05 22	23 30	29 09	00 38	14 21	16 R 08	03 R 36	08 R 18
03	06 44 17	11 07 50	03 ♓ 58 58	10 ♓ 28 13	06 06	24 14	29 28	00 47	14 27	16 06	03 35	08 17
04	06 48 14	12 05 01	17 ♓ 01 35	23 ♓ 39 20	06 45	24 56	29 ♎ 47	00 56	14 32	16 05	03 33	08 16
05	06 52 11	13 02 13	00 ♈ 21 43	07 ♈ 08 56	07 21	25 37	00 ♏ 07	01 05	14 37	16 03	03 32	08 14
06	06 56 07	13 59 25	14 ♈ 01 09	20 ♈ 58 26	07 52	26 18	00 27	01 13	14 42	16 01	03 30	08 13
07	07 00 04	14 56 37	28 ♈ 00 45	05 ♉ 08 00	08 19	26 57	00 48	01 21	14 47	15 59	03 29	08 12
08	07 04 00	15 53 50	12 ♉ 19 54	19 ♉ 36 03	08 42	27 35	01 09	01 30	14 52	15 57	03 27	08 11
09	07 07 57	16 51 03	26 ♉ 55 55	04 ♊ 18 49	09 01	28 12	01 30	01 38	14 57	15 55	03 26	08 10
10	07 11 53	17 48 17	11 ♊ 43 56	19 ♊ 10 22	09 15	28 48	01 53	01 46	15 02	15 53	03 24	08 08
11	07 15 50	18 45 31	26 ♊ 37 08	04 ♋ 03 12	09 24	29 22	02 15	01 54	15 07	15 51	03 22	08 07
12	07 19 46	19 42 45	11 ♋ 27 33	18 ♋ 49 12	09 29	29 ♌ 56	02 38	02 02	15 12	15 49	03 21	08 06
13	07 23 43	20 40 00	26 ♋ 07 14	03 ♌ 20 52	09 R 29	00 ♍ 27	03 02	02 10	15 16	15 47	03 19	08 05
14	07 27 40	21 37 14	10 ♌ 29 25	17 ♌ 32 22	09 24	00 58	03 25	02 17	15 21	15 45	03 18	08 04
15	07 31 36	22 34 29	24 ♌ 29 22	01 ♍ 20 11	09 14	01 27	03 50	02 24	15 25	15 43	03 16	08 03
16	07 35 33	23 31 44	08 ♍ 04 45	14 ♍ 43 07	09 00	01 54	04 14	02 32	15 30	15 41	03 15	08 02
17	07 39 29	24 28 59	21 ♍ 15 30	27 ♍ 42 08	08 41	02 20	04 39	02 39	15 34	15 38	03 13	08 01
18	07 43 26	25 26 15	04 ♎ 03 26	10 ♎ 19 47	08 18	02 45	05 05	02 46	15 38	15 36	03 11	08 00
19	07 47 22	26 23 30	16 ♎ 31 43	22 ♎ 39 44	07 50	03 07	05 30	02 52	15 42	15 34	03 10	07 59
20	07 51 19	27 20 46	28 ♎ 44 24	04 ♏ 46 15	07 19	03 28	05 57	02 59	15 46	15 32	03 08	07 58
21	07 55 15	28 18 01	10 ♏ 45 52	16 ♏ 43 49	06 44	03 47	06 23	03 05	15 50	15 29	03 06	07 57
22	07 59 12	29 ♋ 15 17	22 ♏ 40 38	28 ♏ 36 51	06 07	04 05	06 50	03 11	15 54	15 27	03 05	07 56
23	08 03 09	00 ♌ 12 34	04 ♐ 32 59	10 ♐ 29 29	05 27	04 20	07 17	03 18	15 58	15 25	03 03	07 56
24	08 07 05	01 09 51	16 ♐ 26 48	22 ♐ 25 21	04 46	04 33	07 45	03 24	16 02	15 23	03 02	07 55
25	08 11 02	02 07 08	28 ♐ 25 31	04 ♑ 27 36	04 03	04 44	08 13	03 29	16 05	15 20	03 00	07 54
26	08 14 58	03 04 26	10 ♑ 31 55	16 ♑ 38 43	03 20	04 53	08 41	03 35	16 09	15 18	02 58	07 53
27	08 18 55	04 01 44	22 ♑ 48 14	29 ♑ 00 38	02 38	05 00	09 09	03 40	16 12	15 16	02 57	07 52
28	08 22 51	04 59 03	05 ♒ 16 05	11 ♒ 34 41	01 57	05 05	09 38	03 46	16 15	15 13	02 55	07 52
29	08 26 48	05 56 23	17 ♒ 56 33	24 ♒ 21 44	01 18	05 07	10 07	03 51	16 19	15 11	02 54	07 51
30	08 30 44	06 53 43	00 ♓ 50 18	07 ♓ 22 17	00 41	05 R 08	10 37	03 56	16 22	15 09	02 52	07 50
31	08 34 41	07 ♌ 51 04	13 ♓ 57 42	20 ♓ 36 32	00 ♌ 09	05 ♍ 05	11 ♏ 06	04 ♉ 00	16 ♉ 25	15 ♒ 06	02 ♒ 50	07 ♐ 50

Date	Moon True ☊ ° '	Moon Mean ☊ ° '	Moon ☽ Latitude ° '
01	13 ♌ 12	14 ♌ 47	00 N 25
02	13 D 12	14 44	00 S 44
03	13 13	14 41	01 51
04	13 15	14 38	02 55
05	13 16	14 35	03 50
06	13 17	14 31	04 34
07	13 R 16	14 28	05 02
08	13 14	14 25	05 13
09	13 11	14 22	05 05
10	13 08	14 19	04 36
11	13 05	14 16	03 48
12	13 02	14 12	02 45
13	13 00	14 09	01 32
14	13 00	14 06	00 S 14
15	13 D 00	14 03	01 N 03
16	13 01	14 00	02 14
17	13 02	13 56	03 16
18	13 04	13 53	04 06
19	13 04	13 50	04 43
20	13 R 05	13 47	05 06
21	13 04	13 44	05 16
22	13 03	13 41	05 11
23	13 02	13 37	04 54
24	13 00	13 34	04 24
25	12 58	13 31	03 42
26	12 57	13 28	02 50
27	12 56	13 25	01 49
28	12 56	13 21	00 N 42
29	12 D 56	13 18	00 S 28
30	12 56	13 15	01 38
31	12 ♌ 56	13 ♌ 12	02 S 44

DECLINATIONS

Date	Sun ☉ ° '	Moon ☽ ° '	Mercury ☿ ° '	Venus ♀ ° '	Mars ♂ ° '	Jupiter ♃ ° '	Saturn ♄ ° '	Uranus ♅ ° '	Neptune ♆ ° '	Pluto ♇ ° '
01	23 N 07	17 S 42	18 N 55	14 N 35	12 S 05	10 N 30	13 N 57	16 S 40	19 S 04	10 S 03
02	23 03	15 07	18 31	14 13	12 13	10 33	13 59	16 40	19 04	10 04
03	22 58	11 46	18 08	13 52	12 21	10 36	14 00	16 41	19 05	10 04
04	22 53	07 48	17 45	13 31	12 29	10 39	14 02	16 41	19 05	10 04
05	22 48	03 S 22	17 22	13 10	12 37	10 42	14 03	16 42	19 05	10 04
06	22 42	01 N 19	17 01	12 49	12 45	10 44	14 04	16 42	19 06	10 04
07	22 36	06 03	16 40	12 28	12 53	10 47	14 06	16 43	19 06	10 04
08	22 29	10 33	16 20	12 07	13 02	10 50	14 07	16 44	19 06	10 04
09	22 23	14 32	16 00	11 46	13 10	10 52	14 08	16 44	19 07	10 05
10	22 15	17 38	15 42	11 26	13 19	10 55	14 09	16 45	19 07	10 05
11	22 07	19 36	15 26	11 05	13 28	10 58	14 11	16 45	19 07	10 05
12	21 59	20 12	15 10	10 45	13 37	11 00	14 12	16 46	19 08	10 05
13	21 51	19 25	14 56	10 25	13 46	11 02	14 13	16 47	19 08	10 05
14	21 42	17 23	14 44	10 05	13 55	11 05	14 14	16 47	19 09	10 05
15	21 33	14 21	14 33	09 45	14 04	11 07	14 15	16 48	19 09	10 06
16	21 23	10 37	14 24	09 25	14 13	11 09	14 16	16 49	19 09	10 06
17	21 13	06 28	14 16	09 06	14 22	11 12	14 17	16 49	19 10	10 06
18	21 03	02 N 09	14 11	08 47	14 32	11 14	14 18	16 50	19 10	10 06
19	20 52	02 S 09	14 07	08 29	14 41	11 16	14 19	16 51	19 10	10 07
20	20 41	06 15	14 05	08 11	14 51	11 18	14 20	16 51	19 11	10 07
21	20 30	10 03	14 06	07 53	15 00	11 20	14 21	16 52	19 11	10 07
22	20 18	13 25	14 08	07 36	15 10	11 22	14 22	16 53	19 12	10 07
23	20 06	16 14	14 12	07 19	15 20	11 24	14 23	16 53	19 12	10 08
24	19 54	18 22	14 18	07 03	15 30	11 25	14 24	16 54	19 12	10 08
25	19 41	19 44	14 25	06 47	15 39	11 27	14 25	16 55	19 13	10 08
26	19 28	20 12	14 34	06 32	15 49	11 29	14 26	16 55	19 13	10 08
27	19 15	19 43	14 45	06 18	15 59	11 31	14 27	16 56	19 14	10 09
28	19 01	18 16	14 56	06 04	16 09	11 32	14 27	16 57	19 14	10 09
29	18 47	15 54	15 09	05 51	16 19	11 34	14 28	16 58	19 14	10 09
30	18 33	12 42	15 23	05 38	16 29	11 35	14 29	16 58	19 15	10 10
31	18 N 18	08 S 50	15 N 38	05 N 27	16 S 39	11 N 36	14 N 29	16 S 59	19 S 15	10 S 10

ZODIAC SIGN ENTRIES

Date	h	m	Planets
03	04	34	☽ ♓
05	03	59	♂ ♏
05	11	21	☽ ♈
07	15	22	☽ ♉
09	17	00	☽ ♊
11	17	27	☽ ♋
12	15	16	♀ ♍
13	18	26	☽ ♌
15	21	39	☽ ♍
18	04	19	☽ ♎
20	14	30	☽ ♏
23	02	48	☽ ♐
23	06	44	☉ ♌
25	15	08	☽ ♑
28	01	54	☽ ♒
30	10	27	☽ ♓
31	18	44	☿ ♋

LATITUDES

Date	Mercury ☿ ° '	Venus ♀ ° '	Mars ♂ ° '	Jupiter ♃ ° '	Saturn ♄ ° '	Uranus ♅ ° '	Neptune ♆ ° '	Pluto ♇ ° '
01	00 S 12	00 N 41	01 S 06	01 S 13	02 S 16	00 S 42	00 N 17	11 N 46
04	00 52	00 20	01 09	01 14	02 16	00 42	00 17	11 46
07	01 35	00 S 04	01 13	01 14	02 17	00 42	00 17	11 45
10	02 19	00 30	01 16	01 15	02 17	00 42	00 17	11 44
13	03 03	00 58	01 19	01 16	02 18	00 42	00 17	11 42
16	03 45	01 28	01 22	01 16	02 18	00 42	00 17	11 41
19	04 20	02 01	01 25	01 17	02 19	00 42	00 17	11 40
22	04 46	02 35	01 27	01 17	02 20	00 42	00 17	11 39
25	04 58	03 12	01 30	01 18	02 20	00 43	00 17	11 38
28	04 54	03 51	01 32	01 19	02 21	00 43	00 17	11 37
31	04 S 35	04 S 30	01 S 34	01 S 19	02 S 21	00 S 43	00 N 17	11 N 35

DATA

Julian Date	2451361
Delta T	+64 seconds
Ayanamsa	23° 50' 48"
Synetic vernal point	05° ♓ 16' 11"
True obliquity of ecliptic	23° 26' 15"

MOON'S PHASES, APSIDES AND POSITIONS ☽

Date	h	m	Phase	Longitude ° '	Eclipse Indicator
06	11	57	☾	13 ♈ 59	
13	02	24	●	20 ♋ 17	
20	09	00	☽	27 ♎ 14	
28	11	25	○	04 ♒ 58	partial

Day	h	m		
11	06	00	Perigee	
23	05	41	Apogee	
06	05	19	0N	
12	10	22	Max dec	20° N 13'
18	23	56	0S	
26	12	01	Max dec	20° S 12'

LONGITUDES

Date	Chiron ⚷ ° '	Ceres ⚳ ° '	Pallas ⚴ ° '	Juno ⚵ ° '	Vesta ⚶ ° '	Black Moon Lilith ⚸ ° '
01	28 ♏ 13	20 ♋ 48	04 ♊ 52	27 ♏ 18	03 ♍ 20	02 ♐ 55
11	27 ♏ 51	25 ♋ 13	09 ♊ 59	26 ♏ 26	07 ♍ 40	04 ♐ 02
21	27 ♏ 38	29 ♋ 40	15 ♊ 10	26 ♏ 07	12 ♍ 08	05 ♐ 10
31	27 ♏ 36	04 ♌ 08	20 ♊ 24	26 ♏ 21	16 ♍ 45	06 ♐ 17

All ephemeris data is given at 12.00 UT and the Moon's longitude is additionally given for 24.00 UT

ASPECTARIAN

h m	Aspects
01 Thursday	
02 22	☽ ☌ ♆
02 53	☿ # ♆
03 40	☽ ☍ ☿
11 26	☽ ⚹ ♇
13 15	☽ ⊼ ☉
22 37	☽ ∥ ♅
22 54	☽ □ ♄
02 Friday	
01 41	☽ ± ☉
02 25	☽ ☌ ♅
07 05	☽ Q ♃
10 19	☽ Q ♇
11 01	☉ ± ♅
16 38	☽ ☍ ♀
19 49	☽ # ♀
20 08	☽ ⚼ ☉
20 42	☽ # ♄
03 Saturday	
03 21	☽ △ ♂
03 30	♀ ∥ ♄
05 58	☽ ⚹ ♃
08 22	☽ ∥ ♂
09 07	☽ Q ♄
11 15	☽ ⊻ ♆
16 08	☽ ⊼ ☿
19 23	☽ # ♃
19 57	☽ □ ♇
22 20	☽ ⊥ ♆
22 45	☽ ∥ ♇
04 Sunday	
02 15	☽ △ ☉
03 47	☽ ± ☿
07 25	☽ ⚹ ♄
07 48	☽ ⚼ ♂
09 59	☽ ∠ ♃
10 16	☽ ⊻ ♅
14 46	☽ ∠ ♆
20 59	☽ ⚼ ☿
21 08	☽ ⊥ ♅
05 Monday	
00 32	☽ ± ♂
02 27	☽ ⊥ ♃
03 05	☽ ⊼ ♀
10 40	☽ ∠ ♄
11 33	☽ ⊼ ♂
13 13	☽ ∠ ♅
13 13	☉ Q ♃
13 17	☽ ⊻ ♃
14 21	☽ ± ♀
17 36	☽ ⚹ ♆
06 Tuesday	
00 51	☽ △ ☿
01 54	☽ △ ♇
02 41	☽ ⊥ ♄
07 01	☽ ⚼ ♀
11 57	☽ □ ☉
13 12	☽ ⊻ ♄
14 34	☽ Q ♆
15 25	♀ # ♂
15 27	☽ ⚹ ♅
17 39	☉ ± ♇
07 Wednesday	
03 49	☽ ⚼ ♇
05 28	☿ △ ♇
07 48	☉ ⚹ ♄
08 16	☿ # ♅
10 06	☽ △ ♀
11 57	☽ Q ♅
16 49	☽ ☍ ♂
17 43	☽ ☌ ♃
19 04	☽ ± ♇
20 55	☽ Q ☉
21 12	☽ □ ♆
08 Thursday	
05 06	☽ ⊼ ♇
05 49	☽ □ ☿
09 18	☽ # ♇
13 17	☉ ⊼ ♅
13 33	☽ ∥ ♃
16 14	☽ ☌ ♄
17 58	☽ □ ♅
18 19	☽ ⚹ ☉
20 16	☽ ∥ ♀
09 Friday	
02 59	☽ # ♂
09 25	☽ ∥ ♄
12 08	☽ Q ☿
14 09	☽ □ ♀
19 38	☽ ⊼ ♂
19 44	☽ ⊻ ♃
20 33	☽ ∠ ☉
21 33	☽ ∥ ☿
22 33	☽ △ ♆
10 Saturday	
01 23	♂ ☍ ♃
04 13	☽ # ♅
05 33	☽ ⊥ ♃
05 37	☽ ± ♂
06 12	☽ ☍ ♇
07 56	☽ ⚹ ☿
12 07	☽ ⊥ ☉
17 22	☽ ⊻ ♄
18 41	☽ △ ♅
20 12	☽ ∠ ♃
20 30	☽ Q ♀
20 31	☽ ⚼ ♂
22 28	☽ ⊻ ☉
22 44	☽ ⚼ ♆
11 Sunday	
03 05	☽ ⊥ ♄
04 04	♂ ⊥ ♇
04 28	☽ # ♆
08 24	☽ ∠ ☿
13 13	☽ ± ♆
16 37	☽ ⚹ ♀
17 40	☽ ∠ ♄
18 49	☽ ⚼ ♅
19 58	♀ ∥ ♃
20 36	☽ ⚹ ♃
21 20	☽ △ ♂
22 53	☽ ⊼ ♆
23 01	☽ ⊥ ☿
12 Monday	
06 34	☽ ⊼ ♇
08 47	☽ ⊻ ☿
09 20	☽ ± ♅
16 13	☽ Q ♃
16 18	☽ ± ♇
17 51	☽ ∠ ♀
18 07	☽ ⚹ ♄
19 05	☽ ⊼ ♅
23 23	♃ ± ♇
23 33	☿ St R
13 Tuesday	
02 24	☽ ☌ ☉
07 00	☽ ⚼ ♇
09 09	☽ ⊥ ♀
13 55	☽ Q ♄
16 24	☽ # ♆
19 27	☽ ⊻ ♀
22 06	☽ □ ♃
23 47	☽ □ ♂
23 56	☽ ☍ ♆
14 Wednesday	
04 53	♂ □ ♆
07 55	☽ △ ♇
10 10	☽ ☌ ☿
11 02	♀ # ♇
17 18	☽ # ♅
20 18	☽ □ ♄
20 55	☽ ☍ ♅
15 Thursday	
07 14	☽ Q ♂
08 26	☽ ⊻ ☉
10 33	☽ ∥ ☿
12 39	☽ ∥ ♄
13 53	☽ # ♂
19 40	☽ ⊥ ☉
23 43	☉ ⚼ ♇
16 Friday	
00 38	☽ ☌ ♀
02 01	☽ △ ♃
03 23	☽ ⊼ ♆
04 55	☽ ⚹ ♂
08 43	☽ ∥ ♃
11 55	☽ □ ♇
12 52	☽ ∠ ☉
13 37	☽ ⊻ ☿
14 05	☽ ± ♆
15 04	☽ # ♇
19 38	☽ ∥ ♀
21 31	♂ # ♄
17 Saturday	
00 14	☽ ⊥ ☿
01 29	☽ △ ♄
01 42	☽ ⊼ ♅
02 42	☿ # ♂
05 17	☽ ⚼ ♃
06 24	☽ ⚼ ♆
08 04	☿ ∥ ♄
08 56	☽ ∠ ♂
12 42	☽ ± ♅
16 22	☽ ∠ ☿
18 28	☽ ⚹ ☉
20 50	☽ Q ♇
22 06	☽ ± ♃
18 Sunday	
02 15	☽ ⊥ ♂
03 54	♄ □ ♅
05 29	☽ ⚼ ♅
05 29	☽ ⚼ ♄
09 26	☽ ⊻ ♀
09 31	☽ ⊼ ♃
10 21	☽ △ ♆
13 06	♀ △ ♃
14 00	☽ ⊻ ♂
18 59	☽ Q ☉
19 31	☽ ⚹ ♇
19 49	☽ ⚹ ☿
21 14	☽ ⊥ ♀
22 44	☽ ± ♄
19 Monday	
04 29	☿ △ ♇
10 08	☽ △ ♅
10 24	☽ ⊼ ♄
14 26	♀ ⊼ ♆
15 12	☽ ∠ ♀
18 13	☽ Q ☿
20 Tuesday	
00 37	☽ ∠ ♇
09 00	☽ □ ☉
16 48	☉ ⊥ ♀
18 24	☽ ⊥ ♇
20 30	☽ ☍ ♃
20 43	☽ □ ♆
21 40	☽ ⚹ ♀
23 02	☽ # ♀
23 35	☉ Q ♄
21 Wednesday	
02 53	☽ ☌ ♂
04 19	☽ □ ☿
06 22	☽ ⊻ ♇
12 27	☽ ∥ ♇
15 52	♃ □ ♆
20 07	☿ □ ♂
20 49	☽ # ♃
21 28	☽ □ ♅
22 16	☽ ☍ ♄
22 22	☽ Q ♀
22 Thursday	
08 47	☽ Q ♆
17 44	☽ # ☿
19 38	☽ # ♄
23 Friday	
02 27	☽ △ ☉
03 13	☽ ∥ ♂
08 59	☽ ⚹ ♆
09 26	☽ ⊼ ♃
09 43	☽ Q ♅
11 33	☽ □ ♀
13 44	☽ △ ☿
17 45	☽ ⊻ ♂
18 38	☽ ∥ ♅
18 49	☽ ☌ ♇
21 40	☽ ± ♃
24 Saturday	
06 20	☽ ⊥ ♂
09 51	☽ ⚹ ♅
11 09	☽ ⊼ ♄
11 23	☽ ⚼ ☉
15 10	☽ ∠ ♆
15 57	☽ ⚼ ♃
17 35	☿ ⊻ ♀
18 17	☽ ⚼ ☿
20 25	♂ ⊻ ♇
23 16	☽ ± ♄
25 Sunday	
01 00	☽ ∥ ♆
01 09	☽ ∠ ♂
07 00	☽ ± ☉
09 10	☽ ⊥ ♆
11 10	☽ # ☉
11 18	☽ ± ☿
11 18	☿ ∥ ♄
15 48	☽ ∠ ♅
17 20	☽ ⚼ ♄
19 59	☽ ⊼ ☉
21 05	☽ ⊻ ♆
22 09	☽ △ ♃
22 34	☽ ⊼ ☿
26 Monday	
00 43	☽ △ ♀
04 44	☿ □ ♃
06 47	☽ ⊻ ♇
08 12	☽ ⚹ ♂
09 32	☉ ☍ ♆
09 35	☽ ⊥ ♅
15 48	☉ ☌ ☿
18 35	☽ ⊥ ♇
21 20	☽ ⊻ ♅
23 05	☽ △ ♄
27 Tuesday	
00 49	☿ ☍ ♆
02 06	☉ □ ♃
06 31	☽ ⚼ ♀
08 40	☽ Q ♂
12 08	☽ ∠ ♇
14 11	☉ # ♆
21 59	☽ ∥ ♆
23 32	☽ # ☉
28 Wednesday	
00 04	☽ ± ♀
05 58	☽ ☍ ☿
07 31	☽ ☌ ♆
09 06	☽ □ ♃
11 25	☽ ☍ ☉
11 39	☽ ⊼ ♀
14 38	☉ ⊻ ♀
16 57	☽ ⚹ ♇
20 39	☽ □ ♂
29 Thursday	
02 20	☽ ∥ ♅
06 50	☽ ☌ ♅
08 32	☽ ∥ ♂
08 56	☽ □ ♄
15 35	☽ Q ♇
17 40	☽ # ☿
19 21	☽ Q ♃
23 24	☽ # ♄
30 Friday	
01 41	♀ St R
11 44	☽ ⊼ ☿
15 44	☽ ⊻ ♆
17 43	☽ ⚹ ♃
18 31	☽ Q ♄
19 15	☽ # ♃
19 52	☽ ☍ ♀
22 18	☽ ± ☿
31 Saturday	
00 00	☽ ⊼ ☉
00 51	☽ □ ♇
02 42	☽ ⊥ ♆
04 06	☽ ∥ ♇
06 37	☽ △ ♂
11 30	☉ △ ♇
11 47	☽ ± ☉
14 04	☽ ⚼ ☿
14 04	☽ ⊻ ♅
16 28	☽ ⚹ ♄
19 00	☽ ∠ ♆
21 10	☽ ∠ ♃

AUGUST 1999

Date	Sidereal time h m s	Sun ☉ ° ' "	Moon ☽ ° ' "	Moon ☽ 24.00 ° ' "	Mercury ☿ ° '	Venus ♀ ° '	Mars ♂ ° '	Jupiter ♃ ° '	Saturn ♄ ° '	Uranus ♅ ° '	Neptune ♆ ° '	Pluto ♇ ° '
01	08 38 38	08 ♌ 48 26	27 ♓ 18 48	04 ♈ 04 27	29 ♋ 40	05 ♍ 01	11 ♏ 37	04 ♉ 05	16 ♉ 28	15 ♒ 04	02 ♒ 49	07 ♐ 49
02	08 42 34	09 45 50	10 ♈ 53 28	17 ♈ 45 47	29 R 16	04 R 54	12 07	04 09	16 31	15 R 02	02 R 47	07 R 49
03	08 46 31	10 43 14	24 ♈ 41 19	01 ♉ 39 57	28 57	04 45	12 37	04 14	16 34	14 59	02 45	07 48
04	08 50 27	11 40 40	08 ♉ 41 32	15 ♉ 45 52	28 44	04 33	13 08	04 18	16 37	14 57	02 44	07 48
05	08 54 24	12 38 07	22 ♉ 52 43	00 ♊ 01 48	28 36	04 19	13 39	04 22	16 39	14 54	02 42	07 47
06	08 58 20	13 35 35	07 ♊ 12 43	14 ♊ 25 06	28 D 35	04 03	14 11	04 25	16 42	14 52	02 41	07 47
07	09 02 17	14 33 05	21 ♊ 38 26	28 ♊ 52 14	28 41	03 44	14 43	04 29	16 44	14 50	02 39	07 46
08	09 06 13	15 30 36	06 ♋ 05 54	13 ♋ 18 51	28 53	03 23	15 15	04 32	16 46	14 47	02 37	07 46
09	09 10 10	16 28 08	20 ♋ 30 26	27 ♋ 40 03	29 12	03 00	15 47	04 35	16 49	14 45	02 36	07 46
10	09 14 07	17 25 41	04 ♌ 47 05	11 ♌ 50 57	29 ♋ 38	02 35	16 19	04 38	16 51	14 42	02 34	07 45
11	09 18 03	18 23 16	18 ♌ 51 07	25 ♌ 47 08	00 ♌ 11	02 08	16 52	04 41	16 53	14 40	02 33	07 45
12	09 22 00	19 20 51	02 ♍ 38 38	09 ♍ 25 17	00 51	01 39	17 25	04 44	16 55	14 38	02 31	07 45
13	09 25 56	20 18 28	16 ♍ 06 55	22 ♍ 43 25	01 37	01 09	17 58	04 46	16 56	14 35	02 30	07 45
14	09 29 53	21 16 05	29 ♍ 14 47	05 ♎ 41 07	02 30	00 37	18 32	04 48	16 58	14 33	02 28	07 45
15	09 33 49	22 13 44	12 ♎ 02 36	18 ♎ 19 28	03 29	00 ♍ 03	19 05	04 50	17 00	14 31	02 27	07 44
16	09 37 46	23 11 24	24 ♎ 32 05	00 ♏ 40 50	04 35	29 ♌ 28	19 39	04 52	17 01	14 28	02 25	07 44
17	09 41 42	24 09 05	06 ♏ 46 10	12 ♏ 48 35	05 46	28 53	20 14	04 53	17 02	14 26	02 24	07 44
18	09 45 39	25 06 46	18 ♏ 48 38	24 ♏ 46 52	07 04	28 16	20 48	04 55	17 04	14 23	02 22	07 44
19	09 49 36	26 04 29	00 ♐ 43 50	06 ♐ 40 09	08 26	27 39	21 23	04 56	17 05	14 21	02 21	07 D 44
20	09 53 32	27 02 13	12 ♐ 36 24	18 ♐ 33 08	09 54	27 02	21 57	04 57	17 06	14 19	02 19	07 44
21	09 57 29	27 59 59	24 ♐ 30 57	00 ♑ 30 22	11 27	26 25	22 33	04 58	17 07	14 17	02 18	07 44
22	10 01 25	28 57 45	06 ♑ 31 56	12 ♑ 36 06	13 04	25 48	23 08	04 59	17 08	14 14	02 16	07 44
23	10 05 22	29 ♌ 55 32	18 ♑ 43 18	24 ♑ 53 58	14 45	25 11	23 43	04 59	17 09	14 12	02 15	07 45
24	10 09 18	00 ♍ 53 21	01 ♒ 08 24	07 ♒ 26 52	16 30	24 35	24 19	04 59	17 09	14 10	02 14	07 45
25	10 13 15	01 51 11	13 ♒ 49 36	20 ♒ 16 43	18 17	24 00	24 55	04 R 59	17 10	14 07	02 12	07 45
26	10 17 11	02 49 03	26 ♒ 48 16	03 ♓ 24 14	20 08	23 26	25 31	04 59	17 10	14 05	02 11	07 45
27	10 21 08	03 46 56	10 ♓ 04 31	16 ♓ 48 56	22 00	22 53	26 07	04 59	17 11	14 03	02 10	07 45
28	10 25 05	04 44 50	23 ♓ 37 17	00 ♈ 29 13	23 54	22 22	26 43	04 58	17 11	14 01	02 08	07 46
29	10 29 01	05 42 46	07 ♈ 24 25	14 ♈ 22 28	25 50	21 52	27 20	04 57	17 11	13 59	02 07	07 46
30	10 32 58	06 40 44	21 ♈ 22 58	28 ♈ 25 30	27 47	21 25	27 57	04 56	17 R 11	13 56	02 06	07 46
31	10 36 54	07 ♍ 38 44	05 ♉ 29 38	12 ♉ 34 56	29 ♌ 44	20 ♌ 59	28 ♏ 34	04 ♉ 55	17 ♉ 11	13 ♒ 54	02 ♒ 04	07 ♐ 47

Date	Moon True ☊ ° '	Moon Mean ☊ ° '	Moon ☽ Latitude ° '
01	12 ♌ 57	13 ♌ 09	03 S 42
02	12 D 57	13 06	04 29
03	12 57	13 02	05 01
04	12 57	12 59	05 16
05	12 R 57	12 56	05 12
06	12 D 57	12 53	04 49
07	12 58	12 50	04 08
08	12 58	12 47	03 11
09	12 58	12 43	02 01
10	12 58	12 40	00 S 45
11	12 R 58	12 37	00 N 33
12	12 58	12 34	01 47
13	12 57	12 31	02 54
14	12 56	12 27	03 49
15	12 55	12 24	04 32
16	12 54	12 21	05 00
17	12 53	12 18	05 15
18	12 52	12 15	05 15
19	12 D 52	12 12	05 01
20	12 53	12 08	04 35
21	12 54	12 05	03 57
22	12 55	12 02	03 08
23	12 57	11 59	02 10
24	12 58	11 56	01 N 05
25	12 R 58	11 53	00 S 05
26	12 58	11 49	01 16
27	12 56	11 46	02 24
28	12 54	11 43	03 25
29	12 50	11 40	04 16
30	12 47	11 37	04 52
31	12 ♌ 44	11 ♌ 33	05 S 11

DECLINATIONS

Date	Sun ☉ ° '	Moon ☽ ° '	Mercury ☿ ° '	Venus ♀ ° '	Mars ♂ ° '	Jupiter ♃ ° '	Saturn ♄ ° '	Uranus ♅ ° '	Neptune ♆ ° '	Pluto ♇ ° '
01	18 N 03	04 S 28	15 N 53	05 N 16	16 S 49	11 N 38	14 N 30	17 S 00	19 S 15	10 S 10
02	17 48	00 N 11	16 08	05 06	16 59	11 39	14 30	17 00	19 16	10 11
03	17 33	04 53	16 24	04 57	17 09	11 40	14 31	17 01	19 16	10 11
04	17 17	09 24	16 39	04 49	17 19	11 41	14 32	17 02	19 17	10 11
05	17 01	13 27	16 55	04 41	17 29	11 42	14 32	17 02	19 17	10 12
06	16 45	16 45	17 09	04 35	17 39	11 43	14 33	17 03	19 17	10 12
07	16 28	19 03	17 23	04 30	17 49	11 44	14 33	17 04	19 18	10 13
08	16 11	20 07	17 37	04 25	17 59	11 45	14 33	17 05	19 18	10 13
09	15 54	19 52	17 49	04 22	18 09	11 46	14 34	17 05	19 18	10 13
10	15 37	18 20	18 00	04 20	18 18	11 47	14 34	17 06	19 19	10 14
11	15 19	15 41	18 09	04 19	18 28	11 48	14 35	17 07	19 19	10 14
12	15 01	12 12	18 17	04 18	18 38	11 48	14 35	17 07	19 20	10 15
13	14 43	08 08	18 23	04 19	18 48	11 49	14 35	17 08	19 20	10 15
14	14 25	03 N 48	18 28	04 21	18 58	11 49	14 35	17 09	19 20	10 15
15	14 06	00 S 35	18 30	04 24	19 08	11 50	14 36	17 10	19 21	10 16
16	13 47	04 51	18 29	04 28	19 17	11 50	14 36	17 10	19 21	10 16
17	13 28	08 49	18 27	04 32	19 27	11 50	14 36	17 11	19 21	10 17
18	13 09	12 22	18 21	04 38	19 37	11 51	14 36	17 12	19 22	10 17
19	12 50	15 23	18 14	04 45	19 46	11 51	14 36	17 12	19 22	10 18
20	12 30	17 46	18 03	04 52	19 56	11 51	14 36	17 13	19 22	10 18
21	12 10	19 23	17 49	05 00	20 05	11 51	14 36	17 14	19 23	10 19
22	11 50	20 09	17 33	05 09	20 15	11 51	14 36	17 14	19 23	10 19
23	11 30	19 59	17 14	05 18	20 24	11 51	14 36	17 15	19 23	10 20
24	11 09	18 51	16 52	05 28	20 33	11 51	14 36	17 16	19 24	10 20
25	10 49	16 45	16 27	05 39	20 42	11 51	14 36	17 16	19 24	10 21
26	10 28	13 46	16 00	05 49	20 51	11 50	14 36	17 17	19 24	10 21
27	10 07	10 01	15 30	06 01	21 00	11 50	14 36	17 17	19 25	10 22
28	09 46	05 40	14 58	06 12	21 09	11 50	14 36	17 18	19 25	10 22
29	09 25	00 S 59	14 24	06 23	21 18	11 49	14 36	17 19	19 25	10 23
30	09 04	03 N 49	13 48	06 35	21 26	11 49	14 35	17 19	19 26	10 23
31	08 N 42	08 N 27	13 N 10	06 N 47	21 S 35	11 N 48	14 N 35	17 S 20	19 S 26	10 S 24

ZODIAC SIGN ENTRIES

Date	h	m	Planets
01	16	47	☽ ♈
03	21	09	☽ ♉
05	23	57	☽ ♊
08	01	52	☽ ♋
10	03	55	☽ ♌
11	04	25	☿ ♌
12	07	21	☽ ♍
14	13	24	☽ ♎
15	14	12	♀ ♌
16	22	40	☽ ♏
19	10	31	☽ ♐
21	22	59	☽ ♑
23	13	51	☉ ♍
24	09	49	☽ ♒
26	17	50	☽ ♓
28	23	09	☽ ♈
31	02	41	☽ ♉
31	15	15	☿ ♍

LATITUDES

Date	Mercury ☿ ° '	Venus ♀ ° '	Mars ♂ ° '	Jupiter ♃ ° '	Saturn ♄ ° '	Uranus ♅ ° '	Neptune ♆ ° '	Pluto ♇ ° '
01	04 S 26	04 S 44	01 S 34	01 S 20	02 S 22	00 S 43	00 N 17	11 N 35
04	03 50	05 24	01 36	01 20	02 22	00 43	00 17	11 34
07	03 06	06 03	01 38	01 21	02 23	00 43	00 17	11 32
10	02 17	06 39	01 40	01 22	02 24	00 43	00 17	11 31
13	01 27	07 12	01 41	01 22	02 24	00 43	00 17	11 30
16	00 S 39	07 40	01 42	01 23	02 25	00 43	00 17	11 28
19	00 N 05	08 01	01 44	01 24	02 26	00 43	00 17	11 27
22	00 41	08 15	01 45	01 25	02 26	00 43	00 16	11 25
25	01 10	08 20	01 46	01 25	02 27	00 43	00 16	11 24
28	01 30	08 18	01 47	01 26	02 28	00 43	00 16	11 23
31	01 N 42	08 S 08	01 S 47	01 S 27	02 S 29	00 S 43	00 N 16	11 N 21

DATA

Julian Date	2451392
Delta T	+64 seconds
Ayanamsa	23° 50' 53"
Synetic vernal point	05° ♓ 16' 06"
True obliquity of ecliptic	23° 26' 15"

MOON'S PHASES, APSIDES AND POSITIONS ☽

Date	h	m	Phase	Longitude ° '	Eclipse Indicator
04	17	27	☾	11 ♉ 54	
11	11	09	●	18 ♌ 21	Total
19	01	47	☽	25 ♏ 40	
26	23	48	○	03 ♓ 17	

Day	h	m		
07	23	24	Perigee	
19	23	25	Apogee	
02	11	04	0N	
08	19	27	Max dec	20° N 11'
15	08	46	0S	
22	19	55	Max dec	20° S 12'
29	16	55	0N	

LONGITUDES

Date	Chiron ⚷ ° '	Ceres ⚳ ° '	Pallas ⚴ ° '	Juno ⚵ ° '	Vesta ⚶ ° '	Black Moon Lilith ⚸ ° '
01	27 ♏ 36	04 ♌ 35	20 ♊ 55	26 ♏ 24	17 ♍ 13	06 ♐ 23
11	27 ♏ 44	09 ♌ 04	26 ♊ 11	27 ♏ 10	21 ♍ 57	07 ♐ 30
21	28 ♏ 03	13 ♌ 33	01 ♋ 26	28 ♏ 24	26 ♍ 48	08 ♐ 38
31	28 ♏ 30	18 ♌ 01	06 ♋ 41	00 ♐ 01	01 ♎ 44	09 ♐ 45

All ephemeris data is given at 12.00 UT and the Moon's longitude is additionally given for 24.00 UT

ASPECTARIAN

h m	Aspects
01 Sunday	
00 51	☽ ⊥ ♅
05 15	☽ ⚼ ☉
07 33	☽ ⋕ ♀
10 42	☽ ⚼ ♂
13 23	☽ ⊥ ♃
16 03	☽ △ ☿
16 53	☽ ∠ ♅
19 25	☽ ∠ ♄
21 45	☽ ✶ ♆
02 Monday	
00 05	☽ ⊻ ♃
01 33	☽ ⊼ ♀
03 17	☽ ± ♂
06 36	☽ △ ♇
09 52	☽ △ ☉
11 21	☽ ⊥ ♄
12 01	☽ ± ♀
14 13	☽ ⊼ ♂
16 15	♂ ∥ ♅
18 48	☽ Q ♆
19 13	☽ ✶ ♅
21 52	☽ ⊻ ♄
03 Tuesday	
03 33	☽ ⚼ ♀
08 45	☽ ⚼ ♇
12 18	☽ ∥ ♀
15 57	☽ Q ♅
19 12	☽ □ ☿
04 Wednesday	
00 14	☽ ± ♇
01 51	☽ □ ♆
04 28	☽ ☌ ♃
05 02	☽ △ ♀
10 21	☉ ⋕ ♂
10 28	☽ ⊼ ♇
16 26	☽ ⋕ ♇
17 27	☽ □ ☉
19 50	☽ ☍ ♂
22 35	☽ □ ♅
05 Thursday	
01 11	☽ ∥ ♃
01 28	☽ ☌ ♄
01 29	☽ Q ☿
03 21	☿ Q ♄
08 51	♀ △ ♃
09 42	☉ ⋕ ♅
16 49	☉ ∥ ☿
19 14	☽ ∥ ♄
21 35	☽ ✶ ☿
06 Friday	
01 21	☿ ⋕ ♅
01 57	☽ Q ☉
03 26	☿ St D
04 27	☽ △ ♆
06 49	☽ □ ♀
07 19	☽ ⊻ ♃
11 54	☽ ∥ ☉
12 57	☽ ☍ ♇
14 34	☽ ⋕ ♅
15 48	☽ ∥ ☿
17 22	☽ ⊥ ♃
20 34	☽ ⋕ ♂
22 40	☽ ∠ ☿
23 23	☽ ✶ ☉
07 Saturday	
00 03	☽ ⊼ ♂
00 43	☽ △ ♅
03 50	☽ ⊻ ♄
05 23	☽ ⚼ ♆
08 24	☽ ∠ ♃
10 24	☽ ± ♂
12 09	☽ Q ♀
13 45	☽ ⊥ ☿
13 49	☽ ⊥ ♄
15 38	☽ ⋕ ♆
16 55	♂ □ ♅
18 38	☉ ☍ ♅
20 18	☽ ± ♆
20 56	☉ □ ♂
21 11	☿ Q ♄
23 50	☽ ⊻ ☿
08 Sunday	
01 33	☽ ⚼ ♅
01 54	☽ ⚼ ♂
02 04	☽ ∠ ☉
04 48	☽ ∠ ♄
06 14	☽ ⊼ ♆
07 37	☽ ✶ ♀
09 24	☽ ✶ ♃
14 46	☽ ⊼ ♇
15 53	♀ Q ♂
16 27	☽ ± ♅
18 04	☽ ⊥ ☉
09 Monday	
00 45	☽ ± ♇
02 25	☽ ⊼ ♅
03 48	☽ △ ♂
04 47	☽ ⊻ ☉
05 26	☽ Q ♃
05 49	☽ ✶ ♄
07 56	☽ ∠ ♀
15 46	☽ ⚼ ♇
20 51	☉ □ ♄
22 35	☽ ⊥ ♀
23 08	☽ ⋕ ♆
10 Tuesday	
01 57	☽ Q ♄
03 01	☽ ☌ ☿
08 16	☽ ☍ ♆
08 24	☽ ⊻ ♀
11 45	☽ □ ♃
12 17	☽ ⋕ ♂
12 58	♀ ⊼ ♆
15 25	☽ ∥ ☿
17 02	☽ △ ♇
11 Wednesday	
00 16	☽ ⋕ ♅
04 50	☽ ☍ ♅
08 27	☽ □ ♂
08 36	☽ □ ♄
11 09	☽ ☌ ☉
12 29	♂ ☍ ♄
15 06	☽ ∥ ☉
20 11	☽ ∥ ♄
12 Thursday	
08 40	☽ ⊻ ☿
10 19	☽ ☌ ♀
11 47	☽ ⊼ ♆
14 25	☽ ∥ ♃
15 41	☽ △ ♃
17 06	☽ Q ♂
19 51	☽ ⊥ ☿
21 01	☽ □ ♇
22 23	☽ ± ♆
23 50	☽ ⋕ ♇
13 Friday	
03 31	☿ ⊻ ♀
09 15	☽ ⊼ ♅
12 58	☽ ∠ ☿
13 29	☽ △ ♄
14 29	☽ ⚼ ♆
15 30	☽ ✶ ♂
18 38	☽ ⚼ ♃
20 05	☽ ± ♅
20 11	☽ ⊻ ☉
22 09	☉ ∥ ♄
14 Saturday	
05 32	☽ Q ♇
08 03	☽ ⊥ ☉
09 02	☽ ∥ ♀
11 10	☽ ± ♃
11 20	☿ ☍ ♆
12 33	☽ ⚼ ♅
14 26	☽ ⊻ ♀
17 04	☽ ⚼ ♄
17 58	☽ △ ♆
18 31	☽ ✶ ☿
20 20	☽ ∠ ♂
22 22	☽ ⊼ ♃
15 Sunday	
01 10	☽ ⊥ ♀
02 10	☽ ∠ ☉
03 52	☽ ✶ ♇
10 00	☽ ± ♄
14 05	☽ ⊥ ♂
16 41	☽ △ ♅
17 29	☽ ∠ ♀
19 10	☽ Q ☿
21 28	☽ ⊼ ♄
16 Monday	
02 07	☽ ⊻ ♂
08 31	☽ ∠ ♇
09 10	☽ ✶ ☉
09 44	☽ ⋕ ♀
18 07	☿ □ ♃
21 12	☽ ✶ ♀
21 18	♂ ∥ ♆
17 Tuesday	
02 05	☽ ⊥ ♇
03 23	☽ □ ♆
08 17	☽ ☍ ♃
09 48	☽ □ ☿
10 40	☽ Q ☉
13 55	☽ ⊻ ♇
18 44	♂ Q ♆
19 46	☽ Q ♀
21 32	☽ ∥ ♇
18 Wednesday	
03 11	☽ □ ♅
08 15	☽ ⋕ ♃
08 29	☽ ☍ ♄
15 07	☽ Q ♆
16 12	☽ ☌ ♂
17 15	☽ ⋕ ☉
19 Thursday	
00 00	☿ △ ♇
01 44	♇ St D
01 47	☽ □ ☉
05 17	☽ ⋕ ♄
06 06	☽ □ ♀
15 15	☽ ✶ ♆
15 16	☽ Q ♅
20 30	☽ ⊼ ♃
20 Friday	
02 10	☽ ☌ ♇
05 45	☽ △ ☿
05 45	☽ ∥ ♅
08 39	☽ ± ♃
11 58	☉ ☌ ♀
15 10	☽ ⋕ ☿
15 26	☽ ✶ ♅
21 05	☽ ⊼ ♄
21 30	☽ ∠ ♆
21 Saturday	
02 50	☽ ⚼ ♃
07 50	☽ ⊻ ♂
09 11	☽ ± ♄
11 57	☽ ∥ ♆
15 34	☽ ⊥ ♆
15 37	☽ △ ♀
16 29	☽ ⚼ ☿
19 36	☽ △ ☉
20 29	☽ ⊥ ♂
21 18	♂ ∥ ♆
21 31	☽ ∠ ♅
22 Sunday	
03 14	☽ ⚼ ♄
03 33	☽ ⊻ ♆
08 54	☽ △ ♃
10 47	☉ ∥ ♃
13 14	☽ ± ☿
14 24	☽ ⊻ ♇
15 20	☽ ∠ ♂
15 22	☽ ⊥ ♅
20 02	☽ ⚼ ♀
23 Monday	
02 15	☽ ⊥ ♇
02 58	☽ ⊼ ☿
03 10	☽ ⊻ ♅
03 56	☽ ⚼ ☉
04 22	☿ ☍ ♅
08 55	☽ △ ♄
11 00	☿ ⋕ ♅
12 51	☽ ± ♀
19 50	☽ ∠ ♇
22 12	☽ ✶ ♂
22 58	☽ ± ☉
23 58	☽ ⊼ ♀
24 Tuesday	
02 55	☽ ∥ ♆
11 29	☽ ⊼ ☉
14 04	☽ ☌ ♆
17 23	♀ □ ♂
19 20	☽ □ ♃
20 55	☿ □ ♄
22 20	☽ Q ♂
25 Wednesday	
00 34	☽ ✶ ♇
02 39	♃ St R
06 58	☽ ∥ ♅
12 33	☽ ☌ ♅
15 13	☽ ⋕ ☿
18 14	☽ □ ♄
20 33	☉ ⊼ ♆
21 41	☽ ☍ ☿
23 01	☽ Q ♇
26 Thursday	
05 00	☽ Q ♃
05 53	☽ ⋕ ♄
06 04	☽ ☍ ♀
09 31	☽ □ ♂
19 57	☉ ⋕ ♇
21 46	☽ ⊻ ♆
23 48	☽ ☍ ☉
27 Friday	
00 51	☽ ⋕ ♃
02 51	☽ ✶ ♃
03 12	☽ Q ♄
07 51	☽ □ ♇
08 34	☽ ⊥ ♆
09 56	☽ ∥ ♇
11 17	☽ ⋕ ☉
19 04	☽ ⊻ ♅
20 42	☿ ☌ ♀
28 Saturday	
00 35	☽ ∠ ♆
00 38	☽ ✶ ♄
05 35	☽ ∠ ♃
05 40	☽ ⊥ ♅
09 20	☽ ⋕ ♀
09 52	☽ ⊼ ♀
12 35	☽ ⊼ ☿
17 25	☉ △ ♃
17 41	☽ △ ♂
20 00	☽ ± ♀
21 21	☽ ⊥ ♃
21 24	☽ ∠ ♅
29 Sunday	
00 46	☽ ± ☿
02 51	☽ ✶ ♆
02 57	☽ ∠ ♄
04 03	☿ ∥ ♄
07 46	☽ ⊻ ♃
08 51	☽ ⊼ ☉
11 06	☽ ⚼ ♀
12 37	☽ △ ♇
18 31	☽ ⊥ ♄
18 52	☽ ⚼ ☿
19 59	☽ ± ☉
20 53	☽ ⚼ ♂
23 17	☽ ✶ ♅
23 32	☽ Q ♆
30 Monday	
01 23	♄ St R
04 49	☽ ⊻ ♄
12 03	☽ △ ♀
12 33	☽ ⚼ ☉
13 00	☽ ± ♂
14 23	☽ ⚼ ♇
15 02	☿ □ ♂
19 45	☽ Q ♅
23 42	☽ ⊼ ♂
31 Tuesday	
00 39	☽ △ ☿
02 45	☽ ∥ ♀
05 42	☽ ± ♇
06 12	☽ □ ♆
11 02	☽ ☌ ♃
13 13	☽ ∥ ☉
15 22	☉ □ ♇
15 53	☽ ⊼ ♇
15 55	☽ △ ☉
22 25	☉ ± ♆
22 42	☽ ⋕ ♇

SEPTEMBER 1999

LONGITUDES

Date	Sidereal time h m s	Sun ☉ ° ' "	Moon ☽ ° ' "	Moon ☽ 24.00 ° ' "	Mercury ☿ ° '	Venus ♀ ° '	Mars ♂ ° '	Jupiter ♃ ° '	Saturn ♄ ° '	Uranus ♅ ° '	Neptune ♆ ° '	Pluto ♇ ° '
01	10 40 51	08 ♍ 36 45	19 ♉ 41 02	26 ♉ 47 32	01 ♍ 42	20 ♌ 35	29 ♏ 11	04 ♉ 54	17 ♉ 11	13 ♒ 52	02 ♒ 03	07 ♐ 47
02	10 44 47	09 34 49	03 ♊ 54 08	11 ♊ 00 31	03 39	20 R 14	29 ♏ 48	04 R 52	17 R 10	13 R 50	02 R 02	07 48
03	10 48 44	10 32 55	18 ♊ 06 23	25 ♊ 11 31	05 37	19 55	00 ♐ 26	04 50	17 10	13 48	02 01	07 48
04	10 52 40	11 31 02	02 ♋ 15 40	09 ♋ 18 37	07 34	19 38	01 04	04 48	17 09	13 46	02 00	07 49
05	10 56 37	12 29 12	16 ♋ 20 08	23 ♋ 20 01	09 31	19 24	01 41	04 46	17 09	13 44	01 58	07 49
06	11 00 34	13 27 24	00 ♌ 18 02	07 ♌ 13 56	11 27	19 12	02 19	04 44	17 08	13 42	01 57	07 50
07	11 04 30	14 25 37	14 ♌ 07 30	20 ♌ 58 28	13 23	19 02	02 58	04 41	17 07	13 40	01 56	07 51
08	11 08 27	15 23 52	27 ♌ 46 35	04 ♍ 31 36	15 17	18 55	03 36	04 39	17 06	13 38	01 55	07 51
09	11 12 23	16 22 10	11 ♍ 13 16	17 ♍ 51 24	17 11	18 50	04 14	04 36	17 05	13 36	01 54	07 52
10	11 16 20	17 20 28	24 ♍ 25 47	00 ♎ 56 18	19 03	18 48	04 53	04 32	17 04	13 34	01 53	07 53
11	11 20 16	18 18 49	07 ♎ 22 51	13 ♎ 45 25	20 55	18 D 48	05 32	04 29	17 03	13 33	01 52	07 53
12	11 24 13	19 17 12	20 ♎ 04 03	26 ♎ 18 50	22 45	18 50	06 11	04 26	17 01	13 31	01 51	07 54
13	11 28 09	20 15 36	02 ♏ 29 58	08 ♏ 37 40	24 34	18 54	06 50	04 22	17 00	13 29	01 50	07 55
14	11 32 06	21 14 02	14 ♏ 42 15	20 ♏ 44 07	26 23	19 01	07 30	04 18	16 58	13 27	01 49	07 56
15	11 36 03	22 12 29	26 ♏ 43 40	02 ♐ 41 24	28 10	19 10	08 09	04 14	16 57	13 26	01 48	07 57
16	11 39 59	23 10 58	08 ♐ 37 50	14 ♐ 33 32	29 ♍ 56	19 21	08 49	04 10	16 55	13 24	01 47	07 58
17	11 43 56	24 09 29	20 ♐ 29 06	26 ♐ 25 09	01 ♎ 41	19 35	09 28	04 05	16 53	13 22	01 47	07 59
18	11 47 52	25 08 02	02 ♑ 22 19	08 ♑ 21 15	03 25	19 50	10 08	04 01	16 51	13 21	01 46	08 00
19	11 51 49	26 06 36	14 ♑ 22 34	20 ♑ 26 53	05 08	20 07	10 48	03 56	16 49	13 19	01 45	08 01
20	11 55 45	27 05 12	26 ♑ 34 48	02 ♒ 46 53	06 50	20 26	11 28	03 51	16 47	13 18	01 44	08 02
21	11 59 42	28 03 49	09 ♒ 03 37	15 ♒ 25 27	08 31	20 47	12 09	03 46	16 44	13 16	01 44	08 03
22	12 03 38	29 ♍ 02 29	21 ♒ 52 43	28 ♒ 25 41	10 11	21 10	12 49	03 40	16 42	13 15	01 43	08 04
23	12 07 35	00 ♎ 01 10	05 ♓ 04 30	11 ♓ 49 09	11 50	21 34	13 30	03 35	16 40	13 13	01 42	08 05
24	12 11 32	00 59 52	18 ♓ 39 32	25 ♓ 35 23	13 28	22 01	14 11	03 29	16 37	13 12	01 42	08 06
25	12 15 28	01 58 37	02 ♈ 36 18	09 ♈ 41 45	15 05	22 28	14 51	03 24	16 34	13 10	01 41	08 07
26	12 19 25	02 57 24	16 ♈ 51 03	24 ♈ 03 30	16 41	22 58	15 32	03 18	16 32	13 09	01 40	08 08
27	12 23 21	03 56 13	01 ♉ 18 15	08 ♉ 34 28	18 16	23 29	16 13	03 12	16 29	13 08	01 40	08 10
28	12 27 18	04 55 04	15 ♉ 51 19	23 ♉ 07 58	19 50	24 01	16 55	03 05	16 26	13 07	01 39	08 11
29	12 31 14	05 53 57	00 ♊ 23 43	07 ♊ 37 53	21 23	24 35	17 36	02 59	16 23	13 06	01 39	08 12
30	12 35 11	06 ♎ 52 52	14 ♊ 49 56	21 ♊ 59 26	22 ♎ 56	25 ♌ 10	18 ♐ 17	02 ♉ 53	16 ♉ 20	13 ♒ 04	01 ♒ 38	08 ♐ 14

Date	Moon True ☊ ° '	Moon Mean ☊ ° '	Moon ☽ Latitude ° '
01	12 ♌ 42	11 ♌ 30	05 S 11
02	12 R 41	11 27	04 53
03	12 D 41	11 24	03 16
04	12 42	11 21	03 24
05	12 43	11 18	02 20
06	12 45	11 14	01 S 08
07	12 R 45	11 11	00 N 08
08	12 44	11 08	01 21
09	12 42	11 05	02 29
10	12 38	11 02	03 28
11	12 32	10 59	04 14
12	12 26	10 55	04 47
13	12 19	10 52	05 06
14	12 13	10 49	05 10
15	12 09	10 46	05 00
16	12 06	10 43	04 38
17	12 04	10 39	04 03
18	12 D 05	10 36	03 18
19	12 06	10 33	02 25
20	12 07	10 30	01 23
21	12 08	10 27	00 N 17
22	12 R 08	10 24	00 S 52
23	12 05	10 20	02 01
24	12 01	10 17	03 04
25	11 55	10 14	03 57
26	11 47	10 11	04 38
27	11 38	10 08	05 01
28	11 31	10 05	05 05
29	11 24	10 01	04 49
30	11 ♌ 20	09 ♌ 58	04 S 16

DECLINATIONS

Date	Sun ☉ ° '	Moon ☽ ° '	Mercury ☿ ° '	Venus ♀ ° '	Mars ♂ ° '	Jupiter ♃ ° '	Saturn ♄ ° '	Uranus ♅ ° '	Neptune ♆ ° '	Pluto ♇ ° '
01	08 N 20	12 N 39	12 N 30	06 N 58	21 S 43	11 N 47	14 N 35	17 S 21	19 S 26	10 S 24
02	07 59	16 08	11 49	07 10	21 52	11 47	14 35	17 21	19 26	10 25
03	07 37	18 39	11 06	07 21	22 00	11 46	14 34	17 22	19 27	10 25
04	07 15	20 01	10 23	07 33	22 08	11 45	14 34	17 22	19 27	10 26
05	06 52	20 08	09 38	07 44	22 16	11 44	14 33	17 23	19 27	10 26
06	06 30	18 59	08 53	07 54	22 24	11 43	14 33	17 23	19 28	10 27
07	06 08	16 43	08 07	08 05	22 31	11 42	14 33	17 24	19 28	10 27
08	05 45	13 31	07 21	08 15	22 39	11 41	14 32	17 24	19 28	10 28
09	05 23	09 39	06 34	08 24	22 46	11 40	14 32	17 25	19 28	10 29
10	05 00	05 23	05 46	08 33	22 54	11 38	14 31	17 26	19 29	10 29
11	04 37	00 N 58	04 59	08 42	23 01	11 37	14 31	17 26	19 29	10 30
12	04 14	03 S 25	04 11	08 50	23 08	11 36	14 30	17 27	19 29	10 30
13	03 52	07 33	03 24	08 58	23 15	11 34	14 29	17 27	19 29	10 31
14	03 29	11 19	02 36	09 05	23 21	11 33	14 29	17 28	19 30	10 31
15	03 05	14 33	01 49	09 12	23 28	11 31	14 28	17 28	19 30	10 32
16	02 42	17 10	01 02	09 18	23 34	11 30	14 27	17 28	19 30	10 32
17	02 19	19 03	00 N 15	09 23	23 40	11 28	14 27	17 29	19 30	10 33
18	01 56	20 07	00 S 32	09 28	23 46	11 26	14 26	17 29	19 30	10 33
19	01 33	20 16	01 19	09 32	23 52	11 25	14 25	17 30	19 31	10 34
20	01 10	19 29	02 05	09 36	23 58	11 23	14 24	17 30	19 31	10 35
21	00 46	17 43	02 51	09 39	24 03	11 21	14 24	17 31	19 31	10 35
22	00 N 23	15 02	03 36	09 41	24 09	11 19	14 23	17 31	19 31	10 36
23	00 00	11 31	04 21	09 43	24 14	11 17	14 22	17 31	19 31	10 37
24	00 S 24	07 18	05 06	09 45	24 19	11 15	14 21	17 32	19 31	10 37
25	00 47	02 S 36	05 50	09 45	24 23	11 13	14 20	17 32	19 32	10 38
26	01 11	02 N 21	06 33	09 45	24 28	11 11	14 19	17 32	19 32	10 38
27	01 34	07 13	07 16	09 45	24 32	11 09	14 18	17 33	19 32	10 39
28	01 57	11 43	07 59	09 43	24 36	11 06	14 17	17 33	19 32	10 40
29	02 21	15 31	08 41	09 42	24 40	11 04	14 16	17 33	19 32	10 40
30	02 S 44	18 N 21	09 S 22	09 N 39	24 S 44	11 N 02	14 N 15	17 S 34	19 S 32	10 S 41

ZODIAC SIGN ENTRIES

Date	h	m	Planets
02	05	25	☽ ♊
02	19	29	♂ ♐
04	08	09	☽ ♋
06	11	29	☽ ♌
08	15	57	☽ ♍
10	22	16	☽ ♎
13	07	08	☽ ♏
15	18	35	☽ ♐
16	12	53	☿ ♎
18	07	13	☽ ♑
20	18	38	☽ ♒
23	02	51	☽ ♓
23	11	32	☉ ♎
25	07	34	☽ ♈
27	09	51	☽ ♉
29	11	21	☽ ♊

LATITUDES

Date	Mercury ☿ ° '	Venus ♀ ° '	Mars ♂ ° '	Jupiter ♃ ° '	Saturn ♄ ° '	Uranus ♅ ° '	Neptune ♆ ° '	Pluto ♇ ° '
01	01 N 45	08 S 03	01 S 47	01 S 27	02 S 29	00 S 43	00 N 16	11 N 21
04	01 47	07 46	01 48	01 28	02 29	00 43	00 16	11 19
07	01 43	07 24	01 49	01 28	02 30	00 42	00 16	11 18
10	01 34	06 58	01 49	01 29	02 31	00 42	00 16	11 16
13	01 21	06 30	01 49	01 29	02 31	00 42	00 16	11 15
16	01 05	06 01	01 49	01 30	02 32	00 42	00 16	11 14
19	00 47	05 31	01 50	01 30	02 33	00 42	00 16	11 12
22	00 28	05 01	01 50	01 31	02 33	00 42	00 16	11 11
25	00 N 07	04 30	01 49	01 31	02 34	00 42	00 16	11 10
28	00 S 15	04 01	01 49	01 32	02 34	00 42	00 16	11 09
31	00 S 36	03 S 32	01 S 49	01 S 32	02 S 35	00 S 42	00 N 16	11 N 07

DATA

Julian Date	2451423
Delta T	+64 seconds
Ayanamsa	23° 50' 56"
Synetic vernal point	05° ♓ 16' 03"
True obliquity of ecliptic	23° 26' 16"

MOON'S PHASES, APSIDES AND POSITIONS ☽

Date	h	m	Phase	Longitude ° '	Eclipse Indicator
02	22	17	☾	10 ♊ 00	
09	22	02	●	16 ♍ 47	
17	20	06	☽	24 ♐ 29	
25	10	51	○	01 ♈ 56	

Day	h	m		
02	18	05	Perigee	
16	18	41	Apogee	
28	16	52	Perigee	
05	01	59	Max dec	20° N 14'
11	17	13	0S	
19	04	06	Max dec	20° S 19'
26	00	40	0N	

LONGITUDES

Date	Chiron ⚷ ° '	Ceres ⚳ ° '	Pallas ⚴ ° '	Juno ⚵ ° '	Vesta ⚶ ° '	Black Moon Lilith ⚸ ° '
01	28 ♏ 34	18 ♌ 28	07 ♋ 12	00 ♐ 12	02 ♎ 14	09 ♐ 51
11	29 ♏ 11	22 ♌ 54	12 ♋ 24	02 ♐ 13	07 ♎ 15	10 ♐ 58
21	29 ♏ 57	27 ♌ 18	17 ♋ 29	04 ♐ 32	12 ♎ 23	12 ♐ 05
31	00 ♐ 50	01 ♍ 39	22 ♋ 26	07 ♐ 06	17 ♎ 29	13 ♐ 12

All ephemeris data is given at 12.00 UT and the Moon's longitude is additionally given for 24.00 UT

ASPECTARIAN

h m	Aspects
01 Wednesday	
02 12	☽ □ ♅
06 47	☽ ∥ ♃
07 46	☽ ☌ ♄
11 12	☽ ∥ ☿
13 29	☽ □ ♀
16 20	☿ ⊼ ♆
02 Thursday	
00 32	☽ ∥ ♄
04 46	☽ ☍ ♂
08 51	☽ △ ♆
11 31	☽ □ ☿
13 16	☿ ∥ ♃
13 38	☽ ⊻ ♃
18 35	☽ ☍ ♇
19 09	☽ Q ♀
22 17	☽ □ ☉
22 22	☽ # ♅
23 45	☽ ⊥ ♃
03 Friday	
02 37	☿ △ ♃
04 44	☽ △ ♅
10 09	☽ Bq ♆
10 24	☽ ⊻ ♄
14 56	☽ ∠ ♃
15 00	☽ ⚹ ♀
20 33	☽ ⊥ ♄
22 50	☽ Q ☿
22 57	☉ ∥ ♀
23 13	☽ # ♆
04 Saturday	
01 22	☽ ± ♆
06 05	☽ Bq ♅
07 00	☽ Q ☉
09 52	☽ ⊼ ♂
10 23	☿ # ♇
11 33	☽ ⊼ ♆
11 49	☽ ∠ ♄
14 58	☿ □ ♇
15 58	☽ ∠ ♀
16 19	☽ ⚹ ♃
17 08	☿ ± ♆
20 33	☽ ± ♂
21 21	☽ ± ♅
21 27	☽ ⊼ ♇
22 29	☽ ⚹ ☿
05 Sunday	
04 56	☽ ⚹ ☉
07 03	☽ ⊥ ♀
07 34	☽ ⊼ ♅
07 42	☽ ± ♇
12 38	☽ Bq ♂
12 45	☽ Q ♃
13 23	☽ ⚹ ♄
13 36	♂ Q ♅
17 09	☽ ⊻ ♀
22 31	♂ ⚹ ♆
23 08	☽ Bq ♇
06 Monday	
04 18	☽ ∠ ☿
04 43	☽ # ♆
08 35	☽ ∠ ☉
09 59	☽ Q ♄
14 51	☽ ☍ ♆
15 40	☽ △ ♂
17 50	☉ ⊼ ♅
19 38	☽ □ ♃
22 21	☽ ⊥ ☿
07 Tuesday	
01 03	☽ △ ♇
01 19	☽ ⊥ ☉
05 47	☽ # ♅
10 29	☽ ⊻ ☿
11 12	☽ ☍ ♅
12 34	☽ ⊻ ☉
12 58	☿ ∥ ♀
15 36	☿ ⊼ ♅
17 14	☽ □ ♄
20 30	☽ ☌ ♀
08 Wednesday	
04 55	☽ ∥ ♄
14 58	☉ ☌ ☿
19 21	☽ ⊼ ♆
22 52	☽ □ ♂
23 52	☽ ∥ ♃
09 Thursday	
00 10	☽ △ ♃
05 58	☽ □ ♇
06 03	☽ ± ♆
07 10	☽ # ♇
08 33	☿ Bq ♆
10 51	☿ △ ♄
16 17	☽ ⊼ ♅
18 57	☽ ∥ ♀
22 02	☽ ☌ ☉
22 15	☽ Bq ♆
22 35	☽ △ ♄
10 Friday	
00 09	♂ ⊼ ♃
00 33	☽ ☌ ☿
00 56	☉ Bq ♆
01 44	☽ ⊻ ♀
03 06	☽ Bq ♃
03 09	☽ ± ♅
05 19	☉ △ ♄
08 43	☿ ⊻ ♀
09 02	☽ Q ♂
09 25	☽ ∥ ☿
12 40	☽ ⊥ ♀
14 19	☽ ∥ ☉
14 40	☽ Q ♇
18 07	☿ Bq ♃
18 36	☿ ± ♅
19 32	☽ ± ♃
19 36	☽ Bq ♅
11 Saturday	
00 24	♀ St D
01 44	☽ △ ♆
02 04	☽ Bq ♄
05 18	☽ ∠ ♀
06 37	☽ ⊼ ♃
08 22	☽ ⚹ ♂
12 57	☽ ⚹ ♇
18 52	☽ ± ♄
23 34	☽ △ ♅
12 Sunday	
00 10	☉ ⊻ ♀
06 12	☽ ⊼ ♄
09 07	☉ ∥ ☿
09 38	☽ ⚹ ♀
10 23	☽ ⊻ ☉
14 15	☽ ∠ ♂
15 15	☉ Bq ♃
15 42	☽ # ☿
16 17	☽ # ☉
17 24	☉ ± ♅
17 26	☽ ∠ ♇
18 01	☽ ⊻ ☿
22 52	☽ ⊥ ☉
13 Monday	
07 36	☽ ⊥ ☿
08 35	☽ ⊥ ♂
08 53	☽ Q ♀
10 43	☽ □ ♆
10 52	☽ ⊥ ♇
15 37	☽ ☍ ♃
16 39	☿ ⊥ ♀
17 27	☿ Q ♂
17 51	☽ ∠ ☉
20 58	☽ ⊻ ♂
20 58	☽ # ♀
22 37	☽ ⊻ ♇
14 Tuesday	
04 17	☽ ∠ ☿
05 58	☿ Q ♇
06 42	☽ ∥ ♇
09 32	☽ □ ♅
13 37	☽ # ♃
16 29	☽ ☍ ♄
20 41	☽ □ ♀
22 10	☽ Q ♆
15 Wednesday	
02 09	☽ ⚹ ☉
04 22	♂ ☌ ♇
11 18	☽ # ♄
12 51	☿ ± ♃
15 24	☽ ⚹ ☿
15 27	☿ Bq ♅
21 26	☽ Q ♅
22 12	☽ ⚹ ♆
16 Thursday	
03 01	☽ ⊼ ♃
04 25	☽ Q ☉
10 39	☽ ☌ ♇
12 23	☽ ☌ ♂
15 05	☽ ± ♃
15 20	☽ ∥ ♅
19 51	☽ Q ☿
21 38	☽ ⚹ ♅
17 Friday	
04 30	☽ ∠ ♆
04 43	☽ ⊼ ♄
09 11	☽ Bq ♃
10 07	☽ △ ♀
13 15	☿ △ ♆
14 39	☿ Bq ♄
16 50	☽ ± ♄
20 06	☽ ∥ ♆
20 06	☽ □ ☉
22 41	☽ ⊥ ♆
18 Saturday	
03 54	☽ ∠ ♅
07 51	♂ ± ♃
10 47	☽ ⊻ ♆
10 57	☽ Bq ♄
14 28	☽ □ ☿
15 16	☽ △ ♃
17 03	☽ Bq ♀
19 52	☿ ⊼ ♃
21 57	☽ ⊥ ♅
23 18	☽ ⊻ ♇
19 Sunday	
04 29	☽ ⊻ ♂
09 29	☉ Q ♇
09 54	☽ ⊻ ♅
11 16	☽ ⊥ ♇
11 28	☽ ± ♀
11 41	☿ ∠ ♀
12 14	☉ ⊥ ♀
16 48	☉ # ☿
16 49	☽ △ ♄
17 06	☽ ⊥ ♂
23 39	☽ ⊼ ♀
20 Monday	
05 03	☽ ∠ ♇
11 18	☽ ∥ ♆
11 47	☽ ∠ ♂
13 04	☽ △ ☉
21 59	☽ ☌ ♆
21 Tuesday	
01 57	☽ □ ♃
05 10	☉ ± ♃
05 11	☿ ⚹ ♇
10 04	☽ ⚹ ♇
10 48	☽ △ ☿
14 14	☽ ∥ ♅
16 53	☉ Bq ♅
18 10	☽ ⚹ ♂
19 56	☽ ☌ ♅
20 12	☽ Bq ☉
22 Wednesday	
02 25	☽ □ ♄
08 38	☽ Q ♇
10 38	☽ ☍ ♀
11 37	☽ Q ♃
14 19	☽ ± ☉
16 59	☽ # ♄
17 42	☽ Q ♂
18 57	☽ Bq ☿
19 22	☿ ± ♄
23 Thursday	
02 10	☽ ⊼ ☉
02 30	♂ ⚹ ♅
05 56	☽ ⊻ ♆
09 20	☽ ⚹ ♃
11 15	☽ Q ♄
13 27	☽ # ♃
13 32	☽ ± ☿
16 42	☽ ⊥ ♆
17 23	☽ □ ♇
17 28	☽ ∥ ♇
22 35	☽ # ♀
24 Friday	
01 39	☽ ⊼ ☿
02 27	☽ ⊻ ♅
03 44	☽ □ ♂
08 10	☿ △ ♅
08 27	☽ ⚹ ♄
08 34	☽ ∠ ♆
11 42	☽ ∠ ♃
12 56	☽ ⊥ ♅
18 01	☽ ⊼ ♀
21 55	☽ ∥ ☿
25 Saturday	
02 31	☉ Bq ♄
03 09	☽ ⊥ ♃
04 27	☽ ∠ ♅
04 42	☽ ± ♀
04 51	☉ △ ♆
06 18	☿ ⚹ ♂
10 15	☽ ∠ ♄
10 26	☽ ⚹ ♆
10 51	☽ ☍ ☉
13 20	☽ ⊻ ♃
20 12	☽ ∥ ☉
20 33	☽ Bq ♀
21 22	☽ △ ♇
26 Sunday	
01 26	☽ ⊥ ♄
05 49	☽ ⚹ ♅
05 53	☽ # ☉
06 41	☽ Q ♆
09 42	☽ △ ♂
09 48	☿ ⊼ ♄
11 28	☽ ⊻ ♄
11 40	☽ ☍ ☿
19 30	☉ ⊼ ♃
22 30	☽ Bq ♇
22 33	☽ △ ♀
27 Monday	
01 48	☽ Q ♅
11 51	☽ Bq ♂
12 20	☽ # ☿
12 36	☽ □ ♆
13 25	☽ ± ♇
15 06	☽ ☌ ♃
16 40	☽ ⊼ ☉
20 25	♂ ⊼ ♄
23 20	☽ ⊼ ♇
28 Tuesday	
01 04	☽ ∥ ♀
03 13	♂ ∠ ♆
03 16	☽ ± ☉
03 27	☽ ± ♂
06 05	☽ # ♇
07 29	☽ □ ♅
08 37	☽ ∥ ♃
11 17	☉ Q ♂
12 57	☽ ☌ ♄
13 49	☽ ⊼ ♂
19 11	☽ Bq ☉
19 21	☽ ⊼ ☿
29 Wednesday	
02 00	☽ □ ♀
03 34	☽ ∥ ♄
06 26	☽ ± ☿
14 04	☽ △ ♆
16 15	☽ ⊻ ♃
21 47	☽ △ ☉
23 07	☽ Bq ☿
23 38	♂ Bq ♃
30 Thursday	
00 58	☽ ☍ ♇
02 09	☽ ⊥ ♃
04 21	☽ # ♅
09 04	☽ △ ♅
09 06	☽ Q ♀
14 30	☽ ⊻ ♄
15 01	☽ Bq ♆
16 45	☿ ∠ ♇
17 03	☽ ∠ ♃
18 05	☽ ☍ ♂
21 24	☿ # ♀

OCTOBER 1999

LONGITUDES

Date	Sidereal time h m s	Sun ☉ ° ' "	Moon ☽ ° ' "	Moon ☽ 24.00 ° ' "	Mercury ☿ ° '	Venus ♀ ° '	Mars ♂ ° '	Jupiter ♃ ° '	Saturn ♄ ° '	Uranus ♅ ° '	Neptune ♆ ° '	Pluto ♇ ° '
01	12 39 07	07 ♎ 51 50	29 ♊ 06 04	06 ♋ 09 37	24 ♎ 27	25 ♌ 47	18 ♐ 59	02 ♉ 46	16 ♉ 17	13 ♒ 03	01 ♒ 38	08 ♐ 15
02	12 43 04	08 50 51	13 ♋ 09 58	20 ♋ 07 04	25 58	26 25	19 40	02 R 39	16 R 13	13 R 02	01 R 37	08 16
03	12 47 01	09 49 53	27 ♋ 00 56	03 ♌ 51 38	27 28	27 04	20 22	02 32	16 10	13 01	01 37	08 18
04	12 50 57	10 48 58	10 ♌ 39 14	17 ♌ 23 50	28 ♎ 57	27 44	21 04	02 25	16 07	13 00	01 37	08 19
05	12 54 54	11 48 05	24 ♌ 05 31	00 ♍ 44 20	00 ♏ 25	28 25	21 46	02 18	16 03	13 00	01 36	08 21
06	12 58 50	12 47 15	07 ♍ 20 20	13 ♍ 53 33	01 52	29 08	22 28	02 11	16 00	12 59	01 36	08 22
07	13 02 47	13 46 26	20 ♍ 23 59	26 ♍ 51 37	03 18	29 ♌ 51	23 11	02 04	15 56	12 58	01 36	08 24
08	13 06 43	14 45 40	03 ♎ 16 24	09 ♎ 38 18	04 44	00 ♍ 36	23 53	01 56	15 52	12 57	01 36	08 25
09	13 10 40	15 44 56	15 ♎ 57 17	22 ♎ 13 21	06 08	01 21	24 35	01 49	15 48	12 56	01 36	08 27
10	13 14 36	16 44 14	28 ♎ 26 28	04 ♏ 36 41	07 32	02 07	25 18	01 41	15 45	12 56	01 35	08 29
11	13 18 33	17 43 34	10 ♏ 44 06	16 ♏ 48 50	08 54	02 55	26 01	01 34	15 41	12 55	01 35	08 30
12	13 22 30	18 42 55	22 ♏ 51 03	28 ♏ 51 02	10 16	03 43	26 43	01 26	15 37	12 55	01 35	08 32
13	13 26 26	19 42 19	04 ♐ 49 02	10 ♐ 45 28	11 36	04 32	27 26	01 18	15 33	12 54	01 35	08 34
14	13 30 23	20 41 45	16 ♐ 40 43	22 ♐ 35 15	12 55	05 21	28 09	01 10	15 28	12 54	01 D 35	08 35
15	13 34 19	21 41 12	28 ♐ 29 38	04 ♑ 24 24	14 13	06 12	28 52	01 02	15 24	12 53	01 35	08 37
16	13 38 16	22 40 42	10 ♑ 20 11	16 ♑ 17 37	15 30	07 03	29 ♐ 36	00 54	15 20	12 53	01 35	08 39
17	13 42 12	23 40 13	22 ♑ 17 23	28 ♑ 20 09	16 46	07 55	00 ♑ 19	00 46	15 16	12 53	01 35	08 41
18	13 46 09	24 39 46	04 ♒ 26 36	10 ♒ 37 24	18 00	08 48	01 02	00 38	15 11	12 52	01 36	08 42
19	13 50 05	25 39 20	16 ♒ 53 13	23 ♒ 14 38	19 12	09 41	01 46	00 30	15 07	12 52	01 36	08 44
20	13 54 02	26 38 56	29 ♒ 42 12	06 ♓ 16 19	20 22	10 35	02 29	00 22	15 03	12 52	01 36	08 46
21	13 57 59	27 38 34	12 ♓ 57 20	19 ♓ 45 26	21 31	11 30	03 13	00 14	14 58	12 52	01 36	08 48
22	14 01 55	28 38 14	26 ♓ 40 37	03 ♈ 42 44	22 38	12 25	03 56	00 ♉ 06	14 54	12 52	01 36	08 50
23	14 05 52	29 ♎ 37 56	10 ♈ 51 24	18 ♈ 06 03	23 42	13 21	04 40	29 ♈ 58	14 49	12 D 52	01 37	08 52
24	14 09 48	00 ♏ 37 39	25 ♈ 25 55	02 ♉ 50 03	24 44	14 18	05 24	29 50	14 44	12 52	01 37	08 54
25	14 13 45	01 37 25	10 ♉ 17 22	17 ♉ 46 41	25 44	15 15	06 08	29 42	14 40	12 52	01 37	08 55
26	14 17 41	02 37 12	25 ♉ 16 46	02 ♊ 46 26	26 40	16 12	06 52	29 34	14 35	12 52	01 38	08 57
27	14 21 38	03 37 02	10 ♊ 14 32	17 ♊ 40 02	27 33	17 10	07 36	29 25	14 30	12 52	01 38	08 59
28	14 25 34	04 36 54	25 ♊ 02 04	02 ♋ 19 57	28 22	18 09	08 20	29 17	14 26	12 52	01 39	09 01
29	14 29 31	05 36 48	09 ♋ 33 09	16 ♋ 41 20	29 07	19 08	09 05	29 09	14 21	12 53	01 39	09 03
30	14 33 28	06 36 44	23 ♋ 44 21	00 ♌ 42 08	29 ♏ 47	20 07	09 49	29 01	14 16	12 53	01 40	09 06
31	14 37 24	07 ♏ 36 42	07 ♌ 34 49	14 ♌ 22 32	00 ♐ 23	21 ♍ 07	10 ♑ 33	28 ♈ 53	14 ♉ 11	12 ♒ 53	01 ♒ 40	09 ♐ 08

Moon / DECLINATIONS

Date	Moon True ☊ ° '	Moon Mean ☊ ° '	Moon ☽ Latitude ° '	Sun ☉ ° '	Moon ☽ ° '	Mercury ☿ ° '	Venus ♀ ° '	Mars ♂ ° '	Jupiter ♃ ° '	Saturn ♄ ° '	Uranus ♅ ° '	Neptune ♆ ° '	Pluto ♇ ° '
01	11 ♌ 18	09 ♌ 55	03 S 26	03 S 07	20 00	10 S 03	09 N 36	24 S 47	11 N 00	14 N 14	17 S 34	19 S 32	10 S 41
02	11 D 18	09 52	02 25	03 30	20 23	10 43	09 32	24 51	10 57	14 13	17 34	19 32	10 42
03	11 18	09 49	01 16	03 54	19 31	11 22	09 28	24 54	10 55	14 12	17 34	19 33	10 43
04	11 19	09 45	00 S 04	04 17	17 30	12 00	09 23	24 57	10 52	14 11	17 35	19 33	10 43
05	11 R 18	09 42	01 N 08	04 40	14 33	12 38	09 18	24 59	10 50	14 10	17 35	19 33	10 44
06	11 15	09 39	02 14	05 03	10 53	13 15	09 12	25 02	10 47	14 08	17 35	19 33	10 44
07	11 10	09 36	03 12	05 26	06 45	13 52	09 05	25 04	10 45	14 07	17 35	19 33	10 45
08	11 01	09 33	04 00	05 49	02 N 22	14 27	08 58	25 06	10 42	14 06	17 36	19 33	10 46
09	10 51	09 30	04 35	06 12	02 S 03	15 02	08 50	25 07	10 39	14 05	17 36	19 33	10 46
10	10 39	09 26	04 55	06 35	06 19	15 36	08 41	25 09	10 37	14 04	17 36	19 33	10 47
11	10 26	09 23	05 02	06 57	10 15	16 09	08 32	25 10	10 34	14 03	17 36	19 33	10 47
12	10 14	09 20	04 55	07 20	13 44	16 41	08 23	25 11	10 31	14 01	17 36	19 33	10 48
13	10 04	09 17	04 35	07 42	16 35	17 12	08 13	25 12	10 29	14 00	17 36	19 33	10 49
14	09 57	09 14	04 03	08 05	18 45	17 43	08 02	25 12	10 26	13 59	17 36	19 33	10 49
15	09 52	09 10	03 21	08 27	20 05	18 12	07 51	25 12	10 23	13 57	17 36	19 33	10 50
16	09 49	09 07	02 30	08 49	20 33	18 40	07 40	25 12	10 21	13 56	17 37	19 33	10 50
17	09 49	09 04	01 32	09 11	20 05	19 07	07 28	25 12	10 18	13 55	17 37	19 33	10 51
18	09 D 49	09 01	00 N 28	09 33	18 41	19 34	07 15	25 12	10 15	13 54	17 37	19 33	10 52
19	09 R 49	08 58	00 S 37	09 55	16 22	19 59	07 02	25 11	10 12	13 52	17 37	19 33	10 52
20	09 47	08 55	01 43	10 17	13 11	20 22	06 48	25 10	10 09	13 51	17 37	19 33	10 53
21	09 43	08 51	02 46	10 38	09 15	20 45	06 34	25 09	10 07	13 49	17 37	19 33	10 53
22	09 37	08 48	03 41	10 59	04 S 42	21 06	06 20	25 07	10 04	13 48	17 37	19 33	10 54
23	09 28	08 45	04 24	11 21	00 N 15	21 26	06 05	25 05	10 01	13 47	17 37	19 33	10 55
24	09 17	08 42	04 52	11 41	05 18	21 44	05 49	25 03	09 58	13 45	17 37	19 33	10 55
25	09 06	08 39	05 00	12 02	10 09	22 01	05 33	25 01	09 55	13 44	17 37	19 33	10 56
26	08 55	08 36	04 48	12 23	14 25	22 16	05 17	24 59	09 53	13 43	17 37	19 33	10 56
27	08 45	08 32	04 17	12 43	17 45	22 30	05 00	24 56	09 50	13 41	17 36	19 33	10 57
28	08 38	08 29	03 28	13 04	19 53	22 42	04 43	24 53	09 47	13 40	17 36	19 33	10 58
29	08 35	08 26	02 27	13 24	20 39	22 52	04 26	24 49	09 44	13 38	17 36	19 33	10 58
30	08 33	08 23	01 17	13 43	20 05	23 00	04 08	24 46	09 42	13 37	17 36	19 32	10 59
31	08 ♌ 33	08 ♌ 20	00 S 05	14 S 03	18 N 18	23 S 06	03 N 49	24 S 42	09 N 39	13 N 36	17 S 36	19 S 32	10 S 59

ZODIAC SIGN ENTRIES

Date	h	m	Planets
01	13	31	☽ ♋
03	17	13	☽ ♌
05	05	12	☿ ♏
05	22	40	☽ ♍
07	16	51	♀ ♍
08	05	52	☽ ♎
10	15	01	☽ ♏
13	02	18	☽ ♐
15	15	04	☽ ♑
17	01	35	♂ ♑
18	03	17	☽ ♒
20	12	33	☽ ♓
22	17	42	☽ ♈
23	05	48	♃ ♈
23	20	52	☉ ♏
24	19	25	☽ ♉
26	19	33	☽ ♊
28	20	09	☽ ♋
30	20	08	☿ ♐
30	22	47	☽ ♌

LATITUDES

Date	Mercury ☿ ° '	Venus ♀ ° '	Mars ♂ ° '	Jupiter ♃ ° '	Saturn ♄ ° '	Uranus ♅ ° '	Neptune ♆ ° '	Pluto ♇ ° '
01	00 S 36	03 S 32	01 S 49	01 S 32	02 S 35	00 S 42	00 N 16	11 N 07
04	00 58	03 04	01 49	01 32	02 35	00 42	00 16	11 06
07	01 20	02 36	01 48	01 32	02 36	00 42	00 16	11 05
10	01 40	02 10	01 48	01 32	02 36	00 42	00 16	11 04
13	02 00	01 45	01 47	01 32	02 36	00 42	00 16	11 03
16	02 17	01 21	01 46	01 33	02 37	00 42	00 15	11 02
19	02 33	00 59	01 45	01 33	02 37	00 41	00 15	11 01
22	02 46	00 37	01 44	01 32	02 37	00 41	00 15	11 00
25	02 55	00 S 17	01 43	01 32	02 37	00 41	00 15	10 59
28	02 59	00 N 02	01 42	01 32	02 37	00 41	00 15	10 58
31	02 S 56	00 N 20	01 S 41	01 S 32	02 S 37	00 S 41	00 N 15	10 N 57

DATA

Julian Date	2451453
Delta T	+64 seconds
Ayanamsa	23° 50' 59"
Synetic vernal point	05° ♓ 16' 00"
True obliquity of ecliptic	23° 26' 16"

MOON'S PHASES, APSIDES AND POSITIONS ☽

Date	h	m	Phase	Longitude ° '	Eclipse Indicator
02	04	02	☾	08 ♋ 31	
09	11	34	●	15 ♎ 44	
17	15	00	☽	23 ♑ 48	
24	21	02	○	01 ♉ 00	
31	12	04	☾	07 ♌ 37	

Day	h	m		
14	13	53	Apogee	
26	13	06	Perigee	
02	07	12	Max dec	20° N 25'
09	00	49	0S	
16	12	09	Max dec	20° S 33'
23	10	51	0N	
29	13	35	Max dec	20° N 39'

LONGITUDES

Date	Chiron ⚷ ° '	Ceres ⚳ ° '	Pallas ⚴ ° '	Juno ⚵ ° '	Vesta ⚶ ° '	Black Moon Lilith ⚸ ° '
01	00 ♐ 50	01 ♍ 39	22 ♋ 26	07 ♐ 06	17 ♎ 29	13 ♐ 12
11	01 ♐ 49	05 ♍ 56	27 ♋ 10	09 ♐ 54	22 ♎ 42	14 ♐ 19
21	02 ♐ 54	10 ♍ 07	01 ♌ 37	12 ♐ 54	27 ♎ 56	15 ♐ 26
31	04 ♐ 03	14 ♍ 11	05 ♌ 41	16 ♐ 03	03 ♏ 13	16 ♐ 33

All ephemeris data is given at 12.00 UT and the Moon's longitude is additionally given for 24.00 UT

ASPECTARIAN

h m	Aspects
01 Friday	
00 31	☽ ⊥ ♄
03 13	☽ △ ☿
03 13	☽ # ♆
06 08	☽ ⚹ ♀
06 08	☽ ± ♆
10 14	☽ ⚼ ♅
15 41	☽ ∠ ♄
16 17	☽ ⊼ ♆
18 10	☽ ⚹ ♃
21 40	☉ ⚹ ♇
02 Saturday	
01 31	☽ ± ♅
03 36	☽ ⊼ ♇
04 02	☽ □ ☉
08 50	☽ ∠ ♀
11 42	☿ ∥ ♇
11 47	☽ ⊼ ♅
13 55	☽ ± ♇
14 32	☽ Q ♃
17 15	☽ ⚹ ♄
20 22	☿ # ♃
23 50	☽ ⊼ ♂
03 Sunday	
00 30	☿ ⚹ ♀
01 07	☽ ⊥ ♀
05 31	☽ ⚼ ♇
10 49	☽ ± ♂
11 32	☽ # ♆
12 05	☽ ⊻ ♀
12 53	☽ □ ☿
13 32	☽ Q ☉
14 00	☽ Q ♄
19 44	☉ ± ♄
20 03	☽ ☍ ♆
21 36	☽ □ ♃
04 Monday	
03 27	☽ ⚼ ♂
07 52	☽ △ ♇
11 19	☽ # ♅
12 19	☽ ⚹ ☉
16 10	☽ ☍ ♅
21 40	☽ □ ♄
05 Tuesday	
00 34	☽ Q ☿
05 06	♃ ± ♇
07 36	☽ △ ♂
14 49	☽ ∥ ♄
17 16	☽ ∠ ☉
20 14	☽ ☌ ♀
20 54	♂ ± ♄
23 09	☽ # ☿
06 Wednesday	
00 49	☽ ⚹ ☿
01 34	☽ ⊼ ♆
02 42	☽ △ ♃
07 38	☿ □ ♆
10 55	☽ ⊥ ☉
12 29	☽ ± ♆
12 38	☽ ∥ ♃
12 54	☽ # ♇
13 53	☽ □ ♇
16 36	☉ △ ♅
16 53	☿ ☍ ♃
20 32	☿ ⊥ ♇
22 19	☽ ⊼ ♅
22 22	☽ ∥ ♀
22 47	☽ ⊻ ☉
07 Thursday	
03 47	☽ △ ♄
04 59	☽ ⚼ ♆
05 54	☽ ⚼ ♃
07 39	☽ ∠ ☿
09 08	♃ # ♇
09 21	☽ ± ♅
17 27	☽ □ ♂
18 43	☽ # ☉
22 06	☿ # ♄
22 25	☽ ± ♃
23 10	☽ Q ♇
08 Friday	
02 03	☽ ⚼ ♅
02 26	☽ ⊥ ☿
06 40	☽ ⊻ ♀
07 31	☽ ⚼ ♄
08 51	☽ △ ♆
09 31	☽ ⊼ ♃
15 05	☽ ⊻ ☿
18 38	☽ ⊥ ♀
21 43	☽ ⚹ ♇
09 Saturday	
00 23	☽ ± ♄
05 13	☽ Q ♂
06 16	☽ △ ♅
11 34	☽ ☌ ☉
11 43	☽ ⊼ ♄
12 48	☽ ∠ ♀
13 19	☉ ⊼ ♄
19 36	♀ ⊼ ♆
10 Sunday	
00 30	♀ △ ♃
02 23	☽ ∠ ♇
05 33	☽ ⚹ ♂
13 41	☽ ∥ ☉
18 07	☽ □ ♆
18 14	☽ ☍ ♃
19 38	☽ ⚹ ♀
19 51	☽ ⊥ ♇
21 50	♂ ⊥ ♆
11 Monday	
01 39	☽ # ♀
04 51	☿ ⊻ ♇
06 47	♃ □ ♆
07 36	☽ ⊻ ♇
07 57	☽ ☌ ☿
12 35	☽ ∠ ♂
14 00	☽ # ♃
15 29	☽ ∥ ♇
16 18	☽ □ ♅
20 49	☽ Q ♀
21 42	☽ ☍ ♄
12 Tuesday	
03 02	☽ ⊻ ☉
05 30	☽ Q ♆
07 30	☽ ⊥ ♂
11 32	☉ ∠ ♀
14 14	☽ # ♄
16 03	☽ ⊥ ☉
20 14	☽ ⊻ ♂
13 Wednesday	
04 07	☽ Q ♅
05 00	☽ ⊼ ♃
05 30	☽ ⚹ ♆
11 22	☽ □ ♀
11 45	☽ ∠ ☉
16 58	☽ ± ♃
19 34	☽ ☌ ♇
19 48	☽ ∥ ☿
22 16	☽ ∥ ♅
14 Thursday	
01 34	♆ St D
03 23	♂ ∠ ♅
03 26	☽ ⊻ ☿
04 20	☽ ⚹ ♅
06 53	☿ ∥ ♅
09 34	☽ ⊼ ♄
10 14	☉ # ♀
10 59	☽ ⚼ ♃
11 30	☿ □ ♅
11 49	☽ ∠ ♆
17 07	☽ ⊥ ☿
20 54	☽ ⚹ ☉
21 27	☿ ∠ ♂
21 41	☽ ± ♄
15 Friday	
00 37	☽ ∥ ♆
06 06	☽ ⊥ ♆
10 46	☽ ∠ ♅
12 49	☽ ☌ ♂
13 40	☽ ∠ ☿
15 51	☽ ⚼ ♄
17 07	☽ △ ♃
18 17	☽ ⊻ ♆
23 30	☽ Q ☉
16 Saturday	
04 51	☽ △ ♀
05 01	☽ ⊥ ♅
08 34	☽ ⊻ ♇
08 58	☿ ☍ ♄
17 08	☽ ⊻ ♅
20 43	☽ ⊥ ♇
22 01	☽ △ ♄
23 38	☽ ⚹ ☿
17 Sunday	
02 54	♀ ± ♆
10 27	♂ ⚼ ♄
12 08	☉ ∠ ♇
13 21	☽ ⚼ ♀
14 46	☽ ∠ ♇
15 00	☽ □ ☉
23 09	☽ ∥ ♆
18 Monday	
00 56	♂ △ ♃
02 03	☽ ∥ ☿
02 17	☽ Q ☿
04 38	☽ □ ♃
04 54	☽ ⊻ ♂
06 25	☽ ☌ ♆
08 32	☽ ± ♀
09 23	♀ □ ♇
11 38	☿ ∥ ♆
17 22	☽ ⊥ ♂
20 19	☽ ⚹ ♇
21 07	☽ ⊼ ♀
19 Tuesday	
00 18	☽ ∥ ♅
04 19	☽ ☌ ♅
06 33	♂ ⊻ ♆
08 39	☽ □ ♄
11 45	☽ ∠ ♂
15 02	☽ Q ♃
16 50	☽ □ ☿
19 18	☽ Q ♇
20 04	☿ Q ♆
20 Wednesday	
04 56	☉ # ♃
05 53	☽ △ ☉
07 28	☽ # ♄
13 13	☽ ⚹ ♃
15 29	☽ ⊻ ♆
17 24	☽ ⚹ ♂
18 06	☽ Q ♄
21 Thursday	
02 24	☽ ⊥ ♆
02 31	☽ ∥ ♇
04 32	☽ □ ♇
04 40	☽ ∥ ☉
07 01	☽ # ♃
09 13	☽ ☍ ♀
11 24	☽ ⚼ ☉
11 50	☽ ⊻ ♅
15 33	☽ ⚹ ♄
16 00	☽ ∠ ♃
16 14	☽ Q ♂
18 28	☽ ∠ ♆
22 26	☽ ⊥ ♅
22 Friday	
03 15	☽ # ♀
04 24	☽ △ ☿
04 29	☽ ± ☉
05 49	☉ ∥ ♇
07 36	☽ ⊥ ♃
14 02	☽ ∠ ♅
15 37	☽ ⊼ ☉
17 29	☽ ∠ ♄
17 48	☽ ⊻ ♃
20 26	☽ ⚹ ♆
23 26	♀ ⊼ ♅
23 Saturday	
01 03	☽ □ ♂
06 14	♅ St D
08 07	☽ ⚼ ☿
08 37	☽ ⊥ ♄
08 39	☽ △ ♇
15 20	☽ ⚹ ♅
16 26	☽ ⊼ ♀
16 35	☽ Q ♆
18 33	☽ ⊻ ♄
19 04	☉ ☍ ♃
24 Sunday	
00 14	☽ ± ☿
03 01	☽ ± ♀
09 29	☽ ⚼ ♇
10 47	☽ ⊼ ☿
11 04	☽ Q ♅
14 21	☽ ∥ ♀
18 42	☽ ⚼ ♀
19 05	☽ ☌ ♃
21 02	☽ ☍ ☉
22 02	☽ □ ♆
22 30	♀ △ ♄
23 55	♀ ⚼ ♃
25 Monday	
00 07	☽ ± ♇
04 58	☽ △ ♂
09 48	☽ ⊼ ♇
10 51	☽ ∥ ♃
12 01	☉ □ ♆
16 07	☽ # ♇
16 08	☽ □ ♅
18 59	☽ ☌ ♄
20 29	☽ △ ♀
23 00	☽ # ☉
26 Tuesday	
06 16	☽ ⚼ ♂
07 45	☽ ∥ ♄
11 59	♂ ⊥ ♅
14 21	☽ ☍ ☿
18 47	☽ ⊻ ♃
20 24	☉ ⊥ ♇
21 24	☽ ± ♂
22 10	☽ △ ♆
22 46	♀ ⚼ ♆
27 Wednesday	
00 36	☽ ⊼ ☉
04 19	☽ ⊥ ♃
07 32	☽ ⊼ ♂
09 59	☽ ☍ ♇
10 48	☽ # ♅
10 55	☽ ± ☉
16 14	☽ △ ♅
18 41	☽ ∠ ♃
18 51	☽ ⊻ ♄
22 20	☽ ⚼ ♆
23 59	☽ □ ♀
28 Thursday	
02 31	☽ ⚼ ☉
04 31	☽ ⊥ ♄
07 02	☽ # ♆
13 00	☽ ± ♆
16 39	☽ ⚼ ♅
17 46	☽ ⊼ ☿
18 55	☽ ⚹ ♃
19 10	☽ ∠ ♄
22 52	☽ ⊼ ♆
29 Friday	
04 13	☽ ± ☿
04 57	☽ △ ☉
05 53	♀ ± ♅
07 32	☽ ± ♅
07 40	☽ Q ♀
11 10	☽ ☍ ♂
11 10	☽ ⊼ ♇
11 23	♂ ⊻ ♇
13 08	☿ ⊼ ♃
14 40	☽ Q ♃
17 35	☽ ⊼ ♅
20 00	☽ ⚹ ♄
20 03	☽ ⚼ ☿
21 16	☽ ± ♇
30 Saturday	
04 50	☉ # ♄
05 21	☽ ⚹ ♀
12 36	☽ ⚼ ♇
16 19	☽ Q ♄
21 00	☽ □ ♃
21 01	☽ # ♆
22 54	☽ △ ☿
31 Sunday	
01 41	☽ ☍ ♆
09 14	☽ ∠ ♀
12 04	☽ □ ☉
14 43	☽ △ ♇
17 32	☽ ⊼ ♂
18 45	☽ # ♅
21 22	☽ ☍ ♅
23 36	☽ □ ♄

NOVEMBER 1999

LONGITUDES

Date	Sidereal time h m s	Sun ☉ ° ' "	Moon ☽ ° ' "	Moon ☽ 24.00 ° ' "	Mercury ☿ ° '	Venus ♀ ° '	Mars ♂ ° '	Jupiter ♃ ° '	Saturn ♄ ° '	Uranus ♅ ° '	Neptune ♆ ° '	Pluto ♇ ° '
01	14 41 21	08 ♏ 36 43	21 ♌ 05 34	27 ♌ 44 11	00 ♐ 53	22 ♍ 08	11 ♑ 18	28 ♈ 45	14 ♉ 07	12 ♒ 54	01 ♒ 41	09 ♐ 10
02	14 45 17	09 36 46	04 ♍ 18 43	10 ♍ 49 27	01 16	23 09	12 02	28 R 38	14 R 02	12 54	01 42	09 12
03	14 49 14	10 36 51	17 ♍ 16 41	23 ♍ 40 42	01 33	24 10	12 47	28 30	13 57	12 55	01 42	09 14
04	14 53 10	11 36 58	00 ♎ 01 45	06 ♎ 20 02	01 42	25 12	13 32	28 22	13 52	12 56	01 43	09 16
05	14 57 07	12 37 06	12 ♎ 35 44	18 ♎ 49 00	01 R 43	26 14	14 16	28 14	13 47	12 56	01 44	09 18
06	15 01 03	13 37 17	24 ♎ 59 56	01 ♏ 08 40	01 35	27 16	15 01	28 07	13 42	12 57	01 44	09 20
07	15 05 00	14 37 30	07 ♏ 15 15	13 ♏ 19 48	01 18	28 19	15 46	27 59	13 38	12 58	01 45	09 22
08	15 08 57	15 37 45	19 ♏ 22 24	25 ♏ 23 09	00 51	29 ♍ 22	16 31	27 52	13 33	12 58	01 46	09 25
09	15 12 53	16 38 01	01 ♐ 22 11	07 ♐ 19 40	00 ♐ 15	00 ♎ 26	17 16	27 45	13 28	12 59	01 47	09 27
10	15 16 50	17 38 19	13 ♐ 15 48	19 ♐ 10 51	29 ♏ 28	01 30	18 01	27 37	13 23	13 00	01 48	09 29
11	15 20 46	18 38 39	25 ♐ 05 07	00 ♑ 58 55	28 32	02 34	18 46	27 30	13 18	13 01	01 49	09 31
12	15 24 43	19 39 00	06 ♑ 52 41	12 ♑ 46 53	27 28	03 38	19 32	27 23	13 13	13 02	01 50	09 33
13	15 28 39	20 39 23	18 ♑ 41 59	24 ♑ 38 34	26 16	04 43	20 17	27 17	13 09	13 03	01 51	09 36
14	15 32 36	21 39 47	00 ♒ 37 13	06 ♒ 38 34	25 00	05 48	21 02	27 10	13 04	13 04	01 52	09 38
15	15 36 32	22 40 13	12 ♒ 43 17	18 ♒ 52 01	23 40	06 53	21 47	27 03	12 59	13 06	01 53	09 40
16	15 40 29	23 40 40	25 ♒ 05 28	01 ♓ 24 16	22 19	07 59	22 33	26 57	12 54	13 07	01 54	09 42
17	15 44 26	24 41 08	07 ♓ 49 04	14 ♓ 20 25	21 00	09 05	23 18	26 50	12 50	13 08	01 55	09 45
18	15 48 22	25 41 38	20 ♓ 58 48	27 ♓ 44 36	19 46	10 11	24 04	26 44	12 45	13 09	01 56	09 47
19	15 52 19	26 42 08	04 ♈ 38 02	11 ♈ 39 09	18 40	11 17	24 49	26 38	12 40	13 11	01 57	09 49
20	15 56 15	27 42 40	18 ♈ 47 47	26 ♈ 03 34	17 42	12 24	25 35	26 32	12 36	13 12	01 59	09 51
21	16 00 12	28 43 14	03 ♉ 25 53	10 ♉ 53 53	16 54	13 31	26 21	26 27	12 31	13 13	02 00	09 54
22	16 04 08	29 ♏ 43 48	18 ♉ 26 31	26 ♉ 02 33	16 18	14 38	27 06	26 21	12 27	13 15	02 01	09 56
23	16 08 05	00 ♐ 44 25	03 ♊ 40 37	11 ♊ 19 18	15 53	15 45	27 52	26 16	12 22	13 17	02 02	09 58
24	16 12 01	01 45 02	18 ♊ 57 11	26 ♊ 32 55	15 40	16 53	28 38	26 10	12 18	13 18	02 04	10 01
25	16 15 58	02 45 41	04 ♋ 05 17	11 ♋ 33 15	15 D 38	18 00	29 ♑ 24	26 05	12 13	13 20	02 05	10 03
26	16 19 55	03 46 22	18 ♋ 55 58	26 ♋ 12 49	15 47	19 08	00 ♒ 10	26 00	12 09	13 21	02 06	10 05
27	16 23 51	04 47 04	03 ♌ 23 23	10 ♌ 27 26	16 06	20 17	00 56	25 56	12 05	13 23	02 08	10 08
28	16 27 48	05 47 48	17 ♌ 24 58	24 ♌ 16 04	16 33	21 25	01 42	25 51	12 00	13 25	02 09	10 10
29	16 31 44	06 48 33	01 ♍ 00 59	07 ♍ 40 01	17 09	22 34	02 27	25 47	11 56	13 27	02 11	10 12
30	16 35 41	07 ♐ 49 20	14 ♍ 13 33	20 ♍ 42 01	17 ♏ 52	23 ♎ 42	03 ♒ 13	25 ♈ 43	11 ♉ 52	13 ♒ 29	02 ♒ 12	10 ♐ 15

Date	Moon True ☊ ° '	Moon Mean ☊ ° '	Moon ☽ Latitude ° '
01	08 ♌ 32	08 ♌ 16	01 N 06
02	08 R 31	08 13	02 12
03	08 27	08 10	03 09
04	08 20	08 07	03 56
05	08 11	08 04	04 31
06	07 58	08 01	04 52
07	07 44	07 57	05 00
08	07 29	07 54	04 53
09	07 16	07 51	04 34
10	07 03	07 48	04 03
11	06 54	07 45	03 21
12	06 47	07 42	02 31
13	06 44	07 38	01 34
14	06 42	07 35	00 N 32
15	06 D 42	07 32	00 S 32
16	06 R 42	07 29	01 36
17	06 42	07 26	02 37
18	06 39	07 22	03 32
19	06 34	07 19	04 18
20	06 26	07 16	04 49
21	06 17	07 13	05 02
22	06 06	07 10	04 56
23	05 56	07 07	04 29
24	05 48	07 03	03 42
25	05 42	07 00	02 40
26	05 38	06 57	01 28
27	05 37	06 54	00 S 12
28	05 D 37	06 51	01 N 03
29	05 38	06 48	02 12
30	05 ♌ 38	06 ♌ 44	03 N 11

DECLINATIONS

Date	Sun ☉ ° '	Moon ☽ ° '	Mercury ☿ ° '	Venus ♀ ° '	Mars ♂ ° '	Jupiter ♃ ° '	Saturn ♄ ° '	Uranus ♅ ° '	Neptune ♆ ° '	Pluto ♇ ° '
01	14 S 22	15 N 30	23 S 10	03 N 31	24 S 38	09 N 36	13 N 34	17 S 36	19 S 32	11 S 00
02	14 41	11 58	23 11	03 12	24 34	09 34	13 33	17 36	19 32	11 00
03	15 00	07 56	23 09	02 52	24 29	09 31	13 31	17 35	19 32	11 01
04	15 19	03 N 36	23 05	02 33	24 24	09 28	13 30	17 35	19 32	11 01
05	15 37	00 S 49	22 58	02 13	24 19	09 26	13 29	17 35	19 32	11 02
06	15 56	05 08	22 47	01 52	24 14	09 23	13 27	17 35	19 32	11 03
07	16 13	09 12	22 33	01 32	24 08	09 21	13 26	17 35	19 31	11 03
08	16 31	12 52	22 15	01 11	24 02	09 18	13 24	17 34	19 31	11 04
09	16 48	15 57	21 54	00 50	23 56	09 16	13 23	17 34	19 31	11 04
10	17 05	18 22	21 29	00 29	23 50	09 13	13 22	17 34	19 31	11 05
11	17 22	20 00	21 00	00 N 07	23 43	09 11	13 20	17 33	19 31	11 05
12	17 39	20 45	20 27	00 S 15	23 36	09 09	13 19	17 33	19 31	11 06
13	17 55	20 35	19 52	00 37	23 29	09 07	13 18	17 33	19 30	11 06
14	18 11	19 29	19 13	00 59	23 21	09 04	13 16	17 32	19 30	11 07
15	18 26	17 30	18 34	01 22	23 14	09 02	13 15	17 32	19 30	11 07
16	18 41	14 40	17 54	01 44	23 06	09 00	13 14	17 32	19 30	11 08
17	18 56	11 04	17 14	02 07	22 58	08 58	13 12	17 31	19 29	11 08
18	19 11	06 50	16 37	02 30	22 49	08 56	13 11	17 31	19 29	11 09
19	19 25	02 S 06	16 02	02 53	22 41	08 54	13 10	17 30	19 29	11 09
20	19 39	02 N 54	15 31	03 16	22 32	08 52	13 08	17 30	19 29	11 10
21	19 52	07 55	15 05	03 39	22 23	08 50	13 07	17 30	19 28	11 10
22	20 05	12 34	14 43	04 03	22 13	08 48	13 06	17 29	19 28	11 11
23	20 18	16 29	14 27	04 26	22 04	08 47	13 05	17 29	19 28	11 11
24	20 31	19 18	14 17	04 50	21 54	08 45	13 04	17 28	19 28	11 12
25	20 43	20 43	14 11	05 14	21 44	08 43	13 02	17 28	19 27	11 12
26	20 54	20 39	14 10	05 37	21 33	08 42	13 01	17 27	19 27	11 12
27	21 05	19 12	14 13	06 01	21 23	08 40	13 00	17 27	19 27	11 13
28	21 16	16 37	14 21	06 25	21 12	08 39	12 59	17 26	19 26	11 13
29	21 27	13 10	14 31	06 48	21 01	08 38	12 58	17 26	19 26	11 14
30	21 S 37	09 N 09	14 S 45	07 S 12	20 S 50	08 N 36	12 N 57	17 S 25	19 S 26	11 S 14

ZODIAC SIGN ENTRIES

Date	h	m	Planets
02	04	07	☽ ♍
04	11	57	☽ ♎
06	21	46	☽ ♏
09	02	19	♀ ♎
09	09	15	☽ ♐
09	20	13	☿ ♏
11	22	00	☽ ♑
14	10	46	☽ ♒
16	21	21	☽ ♓
19	03	57	☽ ♈
21	06	26	☽ ♉
22	18	25	☉ ♐
23	06	13	☽ ♊
25	05	29	☽ ♋
26	06	56	♂ ♒
27	06	19	☽ ♌
29	10	11	☽ ♍

LATITUDES

Date	Mercury ☿ ° '	Venus ♀ ° '	Mars ♂ ° '	Jupiter ♃ ° '	Saturn ♄ ° '	Uranus ♅ ° '	Neptune ♆ ° '	Pluto ♇ ° '
01	02 S 53	00 N 25	01 S 41	01 S 32	02 S 37	00 S 41	00 N 15	10 N 57
04	02 38	00 41	01 40	01 31	02 37	00 41	00 15	10 56
07	02 11	00 56	01 38	01 31	02 37	00 41	00 15	10 55
10	01 29	01 10	01 37	01 30	02 37	00 41	00 15	10 55
13	00 S 33	01 22	01 35	01 30	02 37	00 41	00 15	10 54
16	00 N 28	01 33	01 34	01 29	02 36	00 41	00 15	10 53
19	01 24	01 43	01 32	01 28	02 36	00 40	00 15	10 53
22	02 04	01 52	01 31	01 28	02 36	00 40	00 15	10 52
25	02 26	01 59	01 29	01 27	02 35	00 40	00 15	10 52
28	02 33	02 05	01 27	01 26	02 35	00 40	00 15	10 52
31	02 N 28	02 N 11	01 S 25	01 S 25	02 S 34	00 S 40	00 N 15	10 N 51

DATA

Julian Date	2451484
Delta T	+64 seconds
Ayanamsa	23° 51' 02"
Synetic vernal point	05° ♓ 15' 57"
True obliquity of ecliptic	23° 26' 16"

MOON'S PHASES, APSIDES AND POSITIONS ☽

Date	h	m	Phase	Longitude ° '	Eclipse Indicator
08	03	53	●	15 ♏ 17	
16	09	03	☽	23 ♒ 33	
23	07	04	○	00 ♊ 32	
29	23	19	☾	07 ♍ 17	

Day	h	m		
11	05	39	Apogee	
23	22	01	Perigee	
05	07	33	0S	
12	19	45	Max dec	20° S 48'
19	22	09	0N	
25	22	56	Max dec	20° N 52'

LONGITUDES

Date	Chiron ⚷ ° '	Ceres ⚳ ° '	Pallas ⚴ ° '	Juno ⚵ ° '	Vesta ⚶ ° '	Black Moon Lilith ⚸ ° '
01	04 ♐ 10	14 ♍ 35	06 ♌ 04	16 ♐ 23	03 ♏ 44	16 ♐ 40
11	05 ♐ 22	18 ♍ 29	09 ♌ 34	19 ♐ 42	09 ♏ 02	17 ♐ 47
21	06 ♐ 36	22 ♍ 13	12 ♌ 21	23 ♐ 07	14 ♏ 21	18 ♐ 54
31	07 ♐ 51	25 ♍ 41	14 ♌ 29	26 ♐ 39	14 ♏ 39	20 ♐ 01

All ephemeris data is given at 12.00 UT and the Moon's longitude is additionally given for 24.00 UT

ASPECTARIAN

h m Aspects

01 Monday
02 24 ☽ ⊥ ♀
04 48 ☽ ± ♂
13 17 ☿ Q ♅
14 01 ☽ ⊻ ♀
19 29 ☽ # ☉
21 57 ☽ ⚼ ♂
22 46 ☽ Q ☉

02 Tuesday
01 11 ♀ ± ♃
01 38 ☉ ⊻ ♇
01 43 ☽ △ ♃
01 45 ☽ ∥ ♄
06 17 ☽ □ ☿
07 12 ☽ ⊼ ♆
17 56 ☽ # ♇
18 13 ☽ ± ♆
21 01 ☽ □ ♇
22 34 ☽ ✶ ☉

03 Wednesday
02 44 ☽ ∥ ♃
03 07 ☽ △ ♂
03 52 ☽ ⊼ ♅
05 02 ☽ ⚼ ♃
05 50 ☽ △ ♄
10 56 ☽ ⚼ ♆
15 04 ☽ ± ♅
16 19 ☽ Q ☿
16 23 ♂ ⊻ ♅
21 41 ☽ ± ♃

04 Thursday
02 03 ☽ ☌ ♀
04 59 ☽ ∠ ☉
06 45 ☽ Q ♇
08 01 ☽ ⚼ ♅
08 53 ☽ ⊼ ♃
09 49 ☽ ⚼ ♄
15 11 ☽ ✶ ☿
15 12 ☽ △ ♆
18 12 ☽ ∥ ♀
19 50 ☿ ✶ ♆
21 54 ♂ △ ♄
23 33 ☽ ⊥ ☉

05 Friday
02 50 ☽ ± ♄
02 58 ☿ St R
05 39 ☽ ✶ ♇
05 58 ☿ ✶ ♆
12 03 ☽ ⊻ ☉
12 39 ☽ △ ♅
14 17 ☽ ⊼ ♄
15 26 ☽ □ ♂
19 06 ☽ # ♀
19 43 ☉ □ ♅
19 53 ☽ ∠ ☿

06 Saturday
10 42 ☽ ∠ ♇
13 08 ☽ ⊥ ☿
13 36 ♀ Q ♇
13 53 ☉ ☍ ♄
16 50 ☽ ⊻ ♀
18 01 ☽ ☍ ♃
22 41 ♂ ⊥ ♇

07 Sunday
00 37 ☽ ⊻ ☿
01 11 ☽ □ ♆
03 46 ♀ ⚼ ♅
04 21 ☽ ⊥ ♇
04 42 ☽ Q ♂
05 18 ♀ ⊼ ♃
05 41 ☽ ⊥ ♀
12 51 ☽ # ♃
16 11 ☽ ⊻ ♇
18 33 ♀ ⚼ ♄
23 10 ☿ ∠ ♂
23 17 ☽ □ ♅
23 45 ☽ ∥ ♇

08 Monday
00 30 ☽ ☍ ♄
01 07 ☽ ∠ ♀
03 53 ☽ ☌ ☉
05 57 ☽ ✶ ♂
06 38 ☿ Q ♅
12 47 ☽ Q ♆
15 53 ☽ # ♄

09 Tuesday
04 48 ☽ ⊼ ♃
09 31 ☿ ✶ ♀
09 52 ☽ ☌ ☿
09 55 ☽ ✶ ♀
11 14 ☽ Q ♅
12 50 ☽ ✶ ♆
13 56 ☽ ∠ ♂
16 44 ☽ ± ♃
20 37 ☽ ∥ ☉

10 Wednesday
03 06 ☽ ∥ ♅
04 20 ☽ ☌ ♇
09 19 ☽ ⊥ ♂
10 43 ☽ ⚼ ♃
11 28 ☽ ✶ ♅
12 14 ☽ ⊼ ♄
12 30 ☽ Q ♀
18 59 ♀ △ ♆
19 10 ☽ ∠ ♆
21 42 ☽ ⊻ ☉
22 18 ☽ ⊻ ♂

11 Thursday
00 20 ☽ ± ♄
03 28 ☽ ∥ ♆
11 01 ☽ ⊥ ☉
13 29 ☽ ⊥ ♆
16 53 ☽ △ ♃
17 59 ☽ ∠ ♅
18 28 ☽ ⊻ ☿
18 30 ☽ ⚼ ♄

12 Friday
00 09 ☉ ✶ ♂
01 43 ☽ ⊻ ♆
03 55 ☉ ∥ ♅
04 44 ☽ □ ♀
04 58 ☽ ∥ ☿
05 40 ☽ ⊥ ☿
07 03 ☽ ∠ ☉
12 19 ☽ ⊥ ♅
13 49 ☿ ⊼ ♃
16 21 ☉ Q ♆
17 28 ☽ ⊻ ♇
22 21 ☽ ∠ ☿

13 Saturday
00 32 ☽ ⊻ ♅
00 49 ☽ △ ♄
05 41 ☽ ⊥ ♇
15 25 ☽ ☌ ♂
16 19 ☽ ✶ ☉
23 56 ☽ ∠ ♇

14 Sunday
01 36 ☿ ∥ ♆
01 49 ☽ ✶ ☿
05 08 ☽ □ ♃
09 36 ♄ □ ♅
11 50 ☽ ∥ ♆
14 29 ☽ ☌ ♆
18 20 ☽ ∥ ☿
18 38 ☽ Q ☉
23 21 ☽ △ ♀
23 26 ☽ Q ☿

15 Monday
03 05 ☽ ∥ ☉
05 58 ☽ ✶ ♇
11 40 ☽ ∥ ♅
12 31 ☽ □ ♄
12 44 ☽ ☌ ♅
13 59 ♀ ± ♄
15 17 ☉ ∥ ☿
16 32 ☽ Q ♃
22 04 ☉ ☌ ☿

16 Tuesday
05 26 ☿ ∠ ♀
05 28 ☽ Q ♇
06 48 ☽ ⊻ ♂
07 11 ☽ □ ☿
07 34 ☽ ⚼ ♀
09 03 ☽ □ ☉
09 19 ☿ ✶ ♂
15 31 ☽ ✶ ♃
19 01 ☽ ⊥ ♂
22 15 ☽ # ♄
22 59 ☽ Q ♄

17 Wednesday
00 57 ☽ ⊻ ♆
01 25 ☿ ∥ ♅
02 19 ☽ ± ♀
11 36 ☽ ∥ ♇
12 11 ☽ ⊥ ♆
12 58 ☽ ∠ ♂
14 33 ☽ ⊼ ♀
15 35 ☽ □ ♇
19 22 ☽ ∠ ♃
21 11 ☽ ✶ ♄
21 49 ☽ ⊻ ♅

18 Thursday
00 25 ☽ # ♃
03 01 ♀ ✶ ♇
04 42 ☽ ∠ ♆
08 43 ☽ ⊥ ♅
08 46 ☿ Q ♆
10 01 ☽ △ ☿
11 34 ☽ ⊥ ♃
17 50 ☽ ✶ ♂
21 04 ☽ △ ☉
22 09 ☽ ⊻ ♃
23 57 ☽ ∠ ♄

19 Friday
00 45 ☽ ∠ ♅
07 21 ☽ ✶ ♆
08 26 ☽ ∥ ♀
10 26 ☽ ⚼ ☿
10 34 ☉ ⊼ ♃
11 50 ♂ ∠ ♇
15 29 ☽ ⊥ ♄
15 59 ☽ Q ♂
18 35 ☉ ∥ ♆
20 55 ☽ △ ♇

20 Saturday
00 21 ☽ ☍ ♀
00 48 ☽ ± ☿
01 01 ☽ ⚼ ☉
01 39 ☽ ⊻ ♄
02 36 ☽ ✶ ♅
03 39 ☿ ⊥ ♀
03 55 ☽ Q ♆
10 16 ☽ ⊼ ☿
13 52 ☽ # ♀
15 59 ♀ ⊼ ♄
17 12 ☽ ± ☉
22 03 ☽ ⚼ ♇
22 36 ☽ Q ♅
23 50 ☽ □ ♂

21 Sunday
00 42 ☽ ☌ ♃
03 47 ☽ ⊼ ☉
05 41 ♀ △ ♅
09 40 ☽ □ ♆
12 45 ☽ ± ♇
14 43 ♂ □ ♃
16 32 ☽ ∥ ♃
22 25 ☽ ⊼ ♇

22 Monday
02 31 ☽ ☌ ♄
03 44 ☽ □ ♅
04 29 ☽ # ♇
05 28 ☽ ⊼ ♀
08 42 ☽ ☍ ☿
14 56 ☽ ∥ ♄
15 44 ☽ ± ♀
23 37 ☽ # ☿

23 Tuesday
00 25 ☽ ⊻ ♃
02 24 ☽ △ ♂
07 03 ☽ ⚼ ♀
07 04 ☽ ☍ ☉
09 22 ☉ ∠ ♀
09 25 ☽ △ ♆
09 47 ☽ ⊥ ♃
12 14 ☿ Q ♂
14 09 ☿ ⊻ ♀
19 18 ☽ # ♅
21 54 ☽ ☍ ♇
23 50 ☽ ∠ ♃

24 Wednesday
01 03 ☉ Q ♅
01 35 ☽ ⊻ ♄
03 06 ☽ △ ♅
03 11 ☽ ⚼ ♂
06 52 ☽ ⊼ ☿
08 28 ☽ △ ♀
09 01 ☽ ⚼ ♆
10 58 ☽ ⊥ ♄
13 55 ☽ # ♆
16 15 ☽ ± ☿
18 07 ☽ ± ♂
19 33 ☉ ✶ ♆
21 14 ☉ ± ♃
23 15 ☽ ± ♆
23 20 ☽ ✶ ♃

25 Thursday
01 07 ☽ ∠ ♄
02 48 ☽ ⚼ ♅
03 56 ☿ St D
04 07 ☽ ⊼ ♂
06 28 ☽ ⚼ ☿
08 48 ☽ ⊼ ♆
09 44 ☽ ⊼ ☉
11 55 ☽ # ☉
17 12 ☽ ± ♅
18 23 ☽ Q ♃
20 02 ☽ ± ☉
21 36 ☽ ⊼ ♇

26 Friday
01 01 ☽ ✶ ♄
02 54 ☽ ⊼ ♅
03 38 ☽ # ☉
06 47 ☽ △ ☿
07 21 ☽ ± ♇
11 43 ☽ ⚼ ☉
12 22 ☽ □ ♀
20 32 ☽ Q ♄
22 10 ☽ ⚼ ♇
23 36 ☽ □ ♃

27 Saturday
07 38 ☽ ☍ ♂
09 04 ☽ # ♆
09 53 ☽ ☍ ♆
14 32 ☽ △ ☉
21 00 ☽ Q ♀
23 28 ☽ △ ♇

28 Sunday
02 42 ☽ □ ♄
05 04 ☽ ☍ ♅
05 15 ☽ # ♅
07 13 ☉ ∥ ♂
10 27 ☽ □ ☿
19 37 ☽ ✶ ♀

29 Monday
02 43 ☽ △ ♃
02 58 ♂ ☌ ♆
03 34 ☽ # ☿
13 16 ☽ ∥ ♄
14 05 ☽ ⊼ ♆
14 45 ☽ ⊼ ♂
19 51 ☽ Q ☿
23 19 ☽ □ ☉
23 52 ☽ # ♇

30 Tuesday
00 56 ☽ ∠ ♀
00 57 ☽ ± ♆
02 16 ☽ ± ♂
04 41 ☽ □ ♇
05 35 ☽ ⚼ ♃
07 42 ☽ △ ♄
10 37 ☽ ⊼ ♅
15 05 ☽ ∥ ♃
17 31 ☽ ⚼ ♆
19 04 ☽ ⊥ ♀
19 11 ☽ ✶ ☿
19 52 ☽ ⚼ ♂
21 45 ☽ ± ♅
22 01 ☽ # ♀
22 06 ☽ ± ♃

DECEMBER 1999

LONGITUDES

Date	Sidereal time h m s	Sun ☉ ° ' "	Moon ☽ ° ' "	Moon ☽ 24.00 ° ' "	Mercury ☿ ° '	Venus ♀ ° '	Mars ♂ ° '	Jupiter ♃ ° '	Saturn ♄ ° '	Uranus ♅ ° '	Neptune ♆ ° '	Pluto ♇ ° '
01	16 39 37	08 ♐ 50 09	27 ♍ 05 52	03 ♎ 25 32	18 ♏ 42	24 ♎ 51	04 ♒ 00	25 ♈ 39	11 ♉ 48	13 ♒ 30	02 ♒ 14	10 ♐ 17
02	16 43 34	09 50 58	09 ♎ 41 27	15 ♎ 54 02	19 37	26 01	04 46	25 R 35	11 R 44	13 32	02 15	10 19
03	16 47 30	10 51 50	22 ♎ 03 40	28 ♎ 10 42	20 37	27 10	05 32	25 31	11 40	13 34	02 17	10 22
04	16 51 27	11 52 42	04 ♏ 15 28	10 ♏ 18 14	21 42	28 19	06 18	25 28	11 36	13 36	02 18	10 24
05	16 55 24	12 53 36	16 ♏ 19 16	22 ♏ 18 48	22 50	29 ♎ 29	07 04	25 24	11 33	13 39	02 20	10 26
06	16 59 20	13 54 31	28 ♏ 17 02	04 ♐ 14 09	24 02	00 ♏ 39	07 50	25 21	11 29	13 41	02 22	10 29
07	17 03 17	14 55 27	10 ♐ 10 20	16 ♐ 05 47	25 16	01 49	08 36	25 19	11 25	13 43	02 23	10 31
08	17 07 13	15 56 24	22 ♐ 00 40	27 ♐ 55 12	26 33	02 59	09 23	25 16	11 22	13 45	02 25	10 33
09	17 11 10	16 57 22	03 ♑ 49 37	09 ♑ 44 09	27 51	04 09	10 09	25 14	11 18	13 47	02 27	10 36
10	17 15 06	17 58 22	15 ♑ 39 05	21 ♑ 34 46	29 ♏ 12	05 19	10 55	25 11	11 15	13 50	02 28	10 38
11	17 19 03	18 59 21	27 ♑ 31 31	03 ♒ 29 46	00 ♐ 34	06 30	11 42	25 09	11 11	13 52	02 30	10 40
12	17 22 59	20 00 22	09 ♒ 29 57	15 ♒ 32 33	01 57	07 40	12 28	25 07	11 08	13 54	02 32	10 43
13	17 26 56	21 01 23	21 ♒ 38 03	27 ♒ 47 01	03 22	08 51	13 14	25 06	11 05	13 57	02 34	10 45
14	17 30 53	22 02 24	04 ♓ 00 01	10 ♓ 17 35	04 47	10 02	14 01	25 05	11 02	13 59	02 36	10 47
15	17 34 49	23 03 27	16 ♓ 40 19	23 ♓ 08 45	06 14	11 13	14 47	25 03	10 59	14 01	02 37	10 50
16	17 38 46	24 04 29	29 ♓ 43 21	06 ♈ 24 34	07 41	12 24	15 34	25 02	10 56	14 04	02 39	10 52
17	17 42 42	25 05 32	13 ♈ 12 45	20 ♈ 08 04	09 08	13 35	16 20	25 02	10 53	14 07	02 41	10 54
18	17 46 39	26 06 35	27 ♈ 10 36	04 ♉ 20 15	10 37	14 46	17 07	25 01	10 51	14 09	02 43	10 56
19	17 50 35	27 07 39	11 ♉ 36 39	18 ♉ 59 19	12 06	15 58	17 53	25 01	10 48	14 12	02 45	10 59
20	17 54 32	28 08 43	26 ♉ 27 29	04 ♊ 00 12	13 35	17 09	18 40	25 01	10 46	14 14	02 47	11 01
21	17 58 28	29 ♐ 09 47	11 ♊ 36 20	19 ♊ 14 38	15 05	18 21	19 26	25 D 01	10 43	14 17	02 49	11 03
22	18 02 25	00 ♑ 10 52	26 ♊ 53 46	04 ♋ 32 19	16 35	19 32	20 13	25 01	10 41	14 20	02 51	11 06
23	18 06 22	01 11 57	12 ♋ 08 59	19 ♋ 42 30	18 05	20 44	20 59	25 01	10 39	14 23	02 53	11 08
24	18 10 18	02 13 03	27 ♋ 11 46	04 ♌ 35 50	19 36	21 56	21 46	25 02	10 37	14 25	02 55	11 10
25	18 14 15	03 14 10	11 ♌ 53 57	19 ♌ 05 36	21 07	23 08	22 32	25 03	10 35	14 28	02 57	11 12
26	18 18 11	04 15 16	26 ♌ 10 26	03 ♍ 08 18	22 39	24 20	23 19	25 04	10 33	14 31	02 59	11 14
27	18 22 08	05 16 24	09 ♍ 59 12	16 ♍ 43 18	24 10	25 32	24 05	25 06	10 31	14 34	03 01	11 17
28	18 26 04	06 17 32	23 ♍ 20 51	29 ♍ 52 15	25 42	26 44	24 52	25 07	10 29	14 37	03 03	11 19
29	18 30 01	07 18 40	06 ♎ 17 54	12 ♎ 38 18	27 15	27 57	25 38	25 09	10 28	14 40	03 05	11 21
30	18 33 57	08 19 49	18 ♎ 53 56	25 ♎ 05 21	28 ♐ 47	29 ♏ 09	26 25	25 11	10 26	14 43	03 07	11 23
31	18 37 54	09 ♑ 20 58	01 ♏ 13 04	07 ♏ 17 36	00 ♑ 20	00 ♐ 21	27 ♒ 11	25 ♈ 13	10 ♉ 25	14 ♒ 46	03 ♒ 09	11 ♐ 25

MOON NODES AND LATITUDE; DECLINATIONS

Date	Moon True ☊ ° '	Moon Mean ☊ ° '	Moon ☽ Latitude ° '	Sun ☉ ° '	Moon ☽ ° '	Mercury ☿ ° '	Venus ♀ ° '	Mars ♂ ° '	Jupiter ♃ ° '	Saturn ♄ ° '	Uranus ♅ ° '	Neptune ♆ ° '	Pluto ♇ ° '
01	05 ♌ 36	06 ♌ 41	04 N 00	21 S 46	04 N 49	15 S 01	07 S 36	20 S 38	08 N 35	12 N 56	17 S 24	19 S 26	11 S 14
02	05 R 32	06 38	04 35	21 56	00 N 23	15 19	08 00	20 27	08 34	12 55	17 24	19 25	11 15
03	05 25	06 35	04 57	22 04	03 S 59	15 40	08 23	20 15	08 33	12 54	17 23	19 25	11 15
04	05 17	06 32	05 06	22 13	08 08	16 01	08 47	20 03	08 32	12 53	17 23	19 24	11 16
05	05 07	06 28	05 00	22 21	11 56	16 24	09 10	19 50	08 31	12 52	17 22	19 24	11 16
06	04 56	06 25	04 41	22 28	15 12	16 47	09 34	19 38	08 30	12 51	17 21	19 24	11 16
07	04 46	06 22	04 10	22 35	17 51	17 11	09 57	19 25	08 30	12 50	17 21	19 23	11 17
08	04 37	06 19	03 28	22 42	19 44	17 35	10 20	19 12	08 29	12 49	17 20	19 23	11 17
09	04 30	06 16	02 37	22 48	20 46	18 00	10 43	18 59	08 28	12 48	17 19	19 23	11 17
10	04 26	06 13	01 40	22 54	20 52	18 24	11 06	18 46	08 28	12 47	17 19	19 22	11 18
11	04 24	06 09	00 N 37	22 59	20 03	18 49	11 29	18 33	08 27	12 46	17 18	19 22	11 18
12	04 D 24	06 06	00 S 28	23 04	18 19	19 13	11 52	18 19	08 27	12 46	17 17	19 22	11 19
13	04 25	06 03	01 32	23 08	15 45	19 36	12 14	18 05	08 27	12 45	17 17	19 21	11 19
14	04 26	06 00	02 34	23 12	12 26	20 00	12 36	17 51	08 27	12 44	17 16	19 21	11 19
15	04 27	05 57	03 30	23 15	08 29	20 22	12 58	17 37	08 26	12 44	17 15	19 20	11 19
16	04 R 27	05 54	04 16	23 18	04 S 02	20 44	13 20	17 22	08 26	12 43	17 14	19 20	11 20
17	04 26	05 50	04 50	23 21	00 N 45	21 06	13 42	17 08	08 26	12 42	17 14	19 20	11 20
18	04 23	05 47	05 09	23 23	05 40	21 26	14 03	16 53	08 26	12 42	17 13	19 19	11 20
19	04 18	05 44	05 09	23 24	10 25	21 46	14 24	16 38	08 27	12 41	17 12	19 19	11 21
20	04 12	05 41	04 48	23 25	14 41	22 04	14 45	16 23	08 27	12 41	17 11	19 18	11 21
21	04 07	05 38	04 08	23 26	18 05	22 22	15 05	16 08	08 27	12 40	17 11	19 18	11 21
22	04 02	05 34	03 09	23 26	20 15	22 39	15 25	15 53	08 28	12 40	17 10	19 17	11 21
23	03 59	05 31	01 57	23 26	20 57	22 55	15 45	15 37	08 28	12 39	17 09	19 17	11 22
24	03 57	05 28	00 S 37	23 25	20 07	23 09	16 05	15 21	08 29	12 39	17 08	19 17	11 22
25	03 D 57	05 25	00 N 43	23 24	17 55	23 23	16 24	15 06	08 29	12 38	17 07	19 16	11 22
26	03 59	05 22	01 59	23 22	14 39	23 35	16 43	14 50	08 30	12 38	17 06	19 16	11 22
27	04 00	05 19	03 05	23 20	10 41	23 47	17 01	14 33	08 31	12 38	17 06	19 15	11 22
28	04 02	05 15	03 59	23 17	06 18	23 57	17 19	14 17	08 32	12 38	17 05	19 15	11 23
29	04 02	05 12	04 38	23 14	01 N 46	24 06	17 37	14 01	08 32	12 37	17 04	19 14	11 23
30	04 R 02	05 09	05 04	23 11	02 S 43	24 14	17 54	13 44	08 33	12 37	17 03	19 14	11 23
31	04 ♌ 00	05 ♌ 06	05 N 14	23 S 06	06 S 59	24 S 20	18 S 11	13 S 28	08 N 35	12 N 37	17 S 02	19 S 13	11 S 23

ZODIAC SIGN ENTRIES

Date	h	m	Planets
01	17	29	☽ ♎
04	03	35	☽ ♏
05	22	41	♀ ♏
06	15	27	☽ ♐
09	04	14	☽ ♑
11	02	09	☿ ♐
11	16	59	☽ ♒
14	04	18	☽ ♓
16	12	30	☽ ♈
18	16	45	☽ ♉
20	17	39	☽ ♊
22	07	44	☉ ♑
22	16	52	☽ ♋
24	16	32	☽ ♌
26	18	34	☽ ♍
29	00	14	☽ ♎
31	04	54	♀ ♐
31	06	48	☿ ♑
31	09	36	☽ ♏

LATITUDES

Date	Mercury ☿ ° '	Venus ♀ ° '	Mars ♂ ° '	Jupiter ♃ ° '	Saturn ♄ ° '	Uranus ♅ ° '	Neptune ♆ ° '	Pluto ♇ ° '
01	02 N 28	02 N 11	01 S 25	01 S 25	02 S 34	00 S 40	00 N 15	10 N 51
04	02 15	02 15	01 24	01 24	02 34	00 40	00 15	10 51
07	01 57	02 17	01 22	01 23	02 33	00 40	00 15	10 51
10	01 37	02 19	01 20	01 23	02 32	00 40	00 14	10 51
13	01 14	02 20	01 18	01 22	02 32	00 40	00 14	10 51
16	00 52	02 20	01 16	01 21	02 31	00 40	00 14	10 51
19	00 29	02 19	01 14	01 20	02 30	00 40	00 14	10 51
22	00 N 07	02 17	01 11	01 19	02 29	00 40	00 14	10 51
25	00 S 15	02 14	01 09	01 18	02 29	00 40	00 14	10 51
28	00 35	02 10	01 07	01 17	02 28	00 40	00 14	10 51
31	00 S 54	02 N 06	01 S 05	01 S 16	02 S 27	00 S 40	00 N 14	10 N 51

DATA

Julian Date	2451514
Delta T	+64 seconds
Ayanamsa	23° 51' 06"
Synetic vernal point	05° ♓ 15' 53"
True obliquity of ecliptic	23° 26' 16"

MOON'S PHASES, APSIDES AND POSITIONS ☽

Date	h	m	Phase	Longitude ° '	Eclipse Indicator
07	22	32	●	15 ♐ 22	
16	00	50	◐	23 ♓ 36	
22	17	31	○	00 ♋ 25	
29	14	04	☾	07 ♎ 24	

Day	h	m		
08	11	19	Apogee	
22	11	01	Perigee	
02	14	05	0S	
10	02	51	Max dec	20° S 56'
17	08	16	0N	
23	10	44	Max dec	20° N 57'
29	21	21	0S	

LONGITUDES

Date	Chiron ⚷ ° '	Ceres ⚳ ° '	Pallas ⚴ ° '	Juno ⚵ ° '	Vesta ⚶ ° '	Black Moon Lilith ☾ ° '
01	07 ♐ 51	25 ♍ 41	14 ♌ 29	26 ♐ 39	19 ♏ 39	20 ♐ 01
11	09 ♐ 06	28 ♍ 53	15 ♌ 33	00 ♑ 15	24 ♏ 57	21 ♐ 08
21	10 ♐ 20	01 ♎ 46	15 ♌ 29	03 ♑ 55	00 ♐ 13	22 ♐ 14
31	11 ♐ 30	04 ♎ 14	14 ♌ 14	07 ♑ 37	05 ♐ 27	23 ♐ 21

All ephemeris data is given at 12.00 UT and the Moon's longitude is additionally given for 24.00 UT

ASPECTARIAN

h m	Aspects
01 Wednesday	
07 22	☽ ⊻ ♀
09 16	☽ ⊼ ♃
11 27	☽ bQ ♄
11 28	☽ Q ☉
14 15	☽ Q ♇
14 40	☽ bQ ♅
21 13	♀ ∠ ♇
21 44	☽ △ ♆
02 Thursday	
01 29	☽ ∠ ☿
01 56	☽ △ ♂
03 30	♀ ☍ ♃
04 27	☽ ± ♄
12 20	☽ ⚹ ☉
13 13	☽ ⚹ ♇
15 55	☽ ⊼ ♄
19 27	☽ △ ♅
20 14	☽ ⊥ ☿
21 44	♃ Q ♅
23 39	☉ ☌ ♇
03 Friday	
03 45	☿ Q ♆
04 16	☉ bQ ♃
08 56	☽ ⊻ ☿
18 29	☽ ∠ ♇
18 44	☽ ☍ ♃
20 07	☽ ∠ ☉
21 37	♀ # ♃
23 03	☽ ☌ ♀
04 Saturday	
05 56	☉ ⊼ ♄
08 08	☽ □ ♆
12 17	☽ ⊥ ♇
14 23	☽ # ♃
15 30	☽ ⊥ ☉
16 19	☽ □ ♂
16 19	☽ ∥ ♀
05 Sunday	
00 14	☽ ⊻ ♇
02 31	☽ ☍ ♄
03 30	♃ bQ ♇
04 32	☽ ⊻ ☉
06 38	☽ □ ♅
07 36	☽ ∥ ♇
18 25	☽ # ♄
20 03	☽ Q ♆
06 Monday	
02 29	☽ ☌ ☿
06 08	☽ ⊼ ♃
06 20	☉ ⚹ ♅
06 44	☽ Q ♂
17 16	☽ ⊻ ♀
18 10	☽ ± ♃
18 51	☽ Q ♅
20 14	☽ ⚹ ♆
07 Tuesday	
04 01	☽ ∥ ☿
06 42	☽ ⊥ ♀
06 55	☽ ∥ ♅
08 37	☽ ⚹ ♂
12 17	☽ bQ ♃
12 42	☽ ☌ ♇
12 49	☿ ⊼ ♃
14 31	☽ ⊼ ♄
15 36	♂ ∥ ♆
19 12	☽ ⚹ ♅
21 28	☿ ∥ ♅
22 32	☽ ☌ ☉
08 Wednesday	
00 11	♀ □ ♆
02 37	☽ ± ♄
02 39	☽ ∠ ♆
02 55	☽ ∠ ♀
05 01	☽ ∥ ♂
06 41	☽ ∥ ♆
17 09	☽ ∠ ♂
18 35	☽ △ ♃
20 47	☽ bQ ♄
20 58	☽ ⊥ ♆
22 21	☽ ⊻ ☿
09 Thursday	
01 44	☽ ∠ ♅
09 11	☽ ⊻ ♆
12 04	☽ ⊥ ☿
12 42	☽ ⊥ ♂
12 44	☽ ⚹ ♀
19 46	☉ ± ♄
20 04	☽ ⊥ ♅
21 28	♀ ⊥ ♇
23 54	☉ ∠ ♆
10 Friday	
00 37	☿ Q ♂
01 44	☽ ⊻ ♂
01 47	☽ ⊻ ♇
02 33	♂ ⚹ ♇
03 07	☽ △ ♄
08 17	☽ ⊻ ♅
08 40	☽ ∠ ☿
14 00	☽ ⊥ ♇
15 45	☽ Q ♀
17 09	☽ ⊻ ☉
21 23	♂ □ ♄
11 Saturday	
00 18	♀ ∥ ♇
01 46	☿ ∥ ♂
06 25	☽ ⊥ ☉
07 14	☽ □ ♃
08 15	☽ ∠ ♇
18 55	☽ ⚹ ☿
22 02	☿ ± ♃
22 02	☽ ☌ ♆
23 01	☽ ∥ ♆
12 Sunday	
02 12	☽ ∠ ☉
03 04	☽ ∥ ☿
07 58	☽ □ ♀
11 07	☿ Q ♅
12 03	☽ ∥ ♂
14 25	☽ ⚹ ♇
15 15	☽ □ ♄
18 19	☽ ☌ ♂
19 12	☽ Q ♃
20 41	☿ ∥ ♆
20 47	☽ ☌ ♅
22 01	☽ Q ☿
22 08	☿ ⚹ ♆
22 34	☽ ∥ ♅
13 Monday	
07 42	♂ Q ♃
10 42	☽ ⚹ ☉
14 12	☽ Q ♇
18 46	☽ ⚹ ♃
14 Tuesday	
02 28	☽ Q ♄
09 17	☽ ⊻ ♆
09 58	☽ # ♄
10 57	☽ ∥ ♀
10 59	♂ ☌ ♅
12 05	☽ Q ☉
13 42	☽ □ ☿
19 06	☽ ∥ ♇
20 17	♀ # ♄
20 48	☽ ⊥ ♆
23 34	☽ ∠ ♃
15 Wednesday	
00 41	☽ △ ♀
00 59	☽ □ ♇
01 21	☽ ⚹ ♄
03 54	♀ ⊻ ♇
07 01	☽ ⊻ ♅
07 33	♀ ☍ ♄
08 15	☽ ⊻ ♂
12 12	☽ # ♃
13 47	☽ ∠ ♆
16 26	☽ ⊥ ♃
18 15	☽ ⊥ ♅
20 08	☽ ⊥ ♂
16 Thursday	
00 50	☽ □ ☉
03 29	☽ ⊻ ♃
05 09	☽ ∠ ♄
07 22	☽ bQ ♀
10 48	☽ ∠ ♅
13 37	☽ ∠ ♂
17 18	☽ ⚹ ♆
21 21	☽ ⊥ ♄
17 Friday	
01 08	☽ ± ♀
01 47	♂ ∥ ♅
03 59	☽ △ ☿
07 56	☽ △ ♇
07 57	☽ ⊻ ♄
08 29	♄ ⊼ ♇
10 28	☉ △ ♃
12 43	☽ ⊼ ♀
13 34	☽ ⚹ ♅
14 35	☽ Q ♆
17 46	☽ ⚹ ♂
23 00	♀ □ ♅
18 Saturday	
02 22	☿ bQ ♃
06 03	☉ bQ ♄
08 21	☽ ☌ ♃
09 03	☽ bQ ☿
09 54	☽ bQ ♇
10 03	☽ △ ☉
10 16	☽ Q ♅
15 27	☽ Q ♂
15 41	☿ ⊼ ♄
17 28	☿ ☌ ♇
21 20	☽ □ ♆
19 Sunday	
01 02	☽ ± ♇
01 51	☽ ∥ ♃
01 53	☽ ± ☿
02 50	☉ ⊥ ♆
10 41	☽ ☌ ♄
10 58	☽ ⊼ ♇
12 53	☽ ⊼ ☿
12 55	☽ bQ ☉
16 14	☽ □ ♅
16 56	☽ # ♇
19 43	☽ ☍ ♀
22 47	☽ □ ♂
20 Monday	
00 15	☽ ∥ ♄
04 35	☽ ± ☉
09 41	☽ ⊻ ♃
12 24	☽ # ♀
14 49	♃ St D
14 53	☽ ⊼ ☉
19 15	☽ ⊥ ♃
22 05	☽ △ ♆
22 19	☽ # ♂
22 53	☿ ⚹ ♅
21 Tuesday	
04 48	☽ # ♅
09 29	☽ ∠ ♃
10 37	☽ ⊻ ♄
11 08	☽ ☍ ♇
14 59	☉ ∠ ♅
16 13	☽ △ ♅
18 03	☽ ☍ ☿
20 01	☽ ⊥ ♄
21 47	☽ bQ ♆
23 26	☽ # ♆
23 29	☽ ⊼ ♀
22 Wednesday	
00 57	☽ △ ♂
09 03	☽ ⚹ ♃
09 42	☽ ± ♀
10 06	☽ ∠ ♄
11 56	☽ ± ♆
13 37	☿ ± ♄
15 50	☽ bQ ♅
17 31	☽ ☍ ☉
21 22	☽ ⊼ ♆
23 Thursday	
01 01	☽ bQ ♀
01 45	☽ bQ ♂
03 54	☽ Q ♃
06 01	☽ ± ♅
06 29	♀ ∥ ♂
08 40	☿ ∠ ♆
09 38	☽ ⚹ ♄
10 23	☽ ⊼ ♇
15 01	♀ Q ♆
15 32	☽ ⊼ ♅
16 44	☽ ± ♂
19 55	☽ ± ♇
22 28	☽ ⊼ ☿
24 Friday	
02 07	♀ □ ♂
02 48	☽ ⊼ ♂
02 49	☽ △ ♀
04 39	☽ Q ♄
08 31	☽ □ ♃
09 09	☽ ± ☿
10 20	☽ bQ ♇
20 44	☽ ⊼ ☉
21 17	☽ ☍ ♆
22 52	☽ # ♆
25 Saturday	
01 23	☽ bQ ☿
05 01	☉ ⊻ ♆
07 16	☽ ± ☉
09 49	☽ □ ♄
10 51	☽ △ ♇
13 27	☉ ∥ ☿
16 17	☽ ☍ ♅
18 33	☽ # ♅
22 57	☽ # ♀
23 22	☽ bQ ☉
26 Sunday	
05 16	☽ △ ☿
06 50	☽ ☍ ♂
08 34	☽ □ ♀
09 43	♂ Q ♇
10 07	☽ △ ♃
10 48	☽ # ♂
23 46	☽ ⊼ ♆
27 Monday	
00 37	☽ ∥ ♄
02 58	♀ ⊼ ♃
03 03	☽ △ ☉
08 00	☽ # ♇
09 16	☿ ⚹ ♂
10 17	☽ ± ♆
12 11	☽ bQ ♃
12 56	☽ △ ♄
14 17	☽ □ ♇
17 53	♀ ∥ ♅
18 55	☽ Q ♀
20 10	☽ ⊼ ♅
28 Tuesday	
00 00	☽ ∥ ♃
02 22	☽ bQ ♆
02 39	☿ △ ♃
04 18	☽ ± ♃
07 00	☽ ± ♅
08 42	☿ bQ ♄
14 57	☽ ⊼ ♂
15 15	☽ ⊼ ♃
15 55	☽ bQ ♄
16 53	☽ □ ☿
18 51	☽ ⚹ ♀
20 13	♂ ⚹ ♃
23 00	☽ Q ♇
23 34	☽ bQ ♅
29 Wednesday	
02 43	☽ ± ♂
05 58	☽ △ ♆
08 34	☽ ± ♄
09 30	☿ ⊥ ♆
14 04	☽ □ ☉
19 51	☽ ⊼ ♄
20 44	☽ bQ ♂
21 34	☽ ⚹ ♇
30 Thursday	
01 55	☽ ∠ ♀
03 55	☽ △ ♅
07 22	☽ Q ☿
21 07	☽ ⊥ ♀
21 23	☉ ⊥ ♅
31 Friday	
00 12	☽ ☍ ♃
02 34	☽ ∠ ♇
02 47	☿ ∠ ♅
03 34	☽ △ ♂
03 44	☽ Q ☉
10 01	☽ ⚹ ☿
10 07	☽ ⊻ ♀
13 31	☿ ⊻ ♀
15 50	☽ □ ♆
20 19	☽ ⊥ ♇
21 30	☽ # ♃

JANUARY 2000

LONGITUDES

Date	Sidereal time h m s	Sun ☉ ° ' "	Moon ☽ ° ' "	Moon ☽ 24.00 ° ' "	Mercury ☿ ° '	Venus ♀ ° '	Mars ♂ ° '	Jupiter ♃ ° '	Saturn ♄ ° '	Uranus ♅ ° '	Neptune ♆ ° '	Pluto ♇ ° '
01	18 41 51	10 ♑ 22 08	13 ♏ 19 26	19 ♏ 19 02	01 ♑ 53	01 ♐ 34	27 ♒ 58	25 ♈ 15	10 ♉ 24	14 ♒ 49	03 ♒ 12	11 ♐ 27
02	18 45 47	11 23 18	25 ♏ 16 51	01 ♐ 13 19	03 27	02 47	28 44	25 18	10 R 23	14 52	03 14	11 29
03	18 49 44	12 24 29	07 ♐ 08 47	13 ♐ 03 37	05 01	03 59	29 ♒ 31	25 20	10 22	14 55	03 16	11 31
04	18 53 40	13 25 39	18 ♐ 58 09	24 ♐ 52 40	06 35	05 12	00 ♓ 17	25 23	10 21	14 58	03 18	11 34
05	18 57 37	14 26 50	00 ♑ 47 26	06 ♑ 42 43	08 10	06 25	01 04	25 27	10 20	15 01	03 20	11 36
06	19 01 33	15 28 01	12 ♑ 38 45	18 ♑ 35 46	09 44	07 37	01 50	25 30	10 19	15 04	03 22	11 38
07	19 05 30	16 29 11	24 ♑ 34 00	00 ♒ 33 40	11 20	08 50	02 37	25 33	10 19	15 07	03 25	11 40
08	19 09 26	17 30 22	06 ♒ 35 01	12 ♒ 38 18	12 55	10 03	03 24	25 37	10 18	15 10	03 27	11 42
09	19 13 23	18 31 32	18 ♒ 43 46	24 ♒ 51 43	14 31	11 16	04 10	25 41	10 18	15 14	03 29	11 44
10	19 17 20	19 32 42	01 ♓ 02 26	07 ♓ 16 15	16 08	12 29	04 57	25 45	10 17	15 17	03 31	11 46
11	19 21 16	20 33 51	13 ♓ 33 29	19 ♓ 54 29	17 45	13 42	05 43	25 50	10 17	15 20	03 33	11 48
12	19 25 13	21 35 00	26 ♓ 19 36	02 ♈ 49 11	19 22	14 55	06 30	25 54	10 D 17	15 23	03 36	11 49
13	19 29 09	22 36 08	09 ♈ 23 34	16 ♈ 03 03	21 00	16 09	07 16	25 59	10 17	15 26	03 38	11 51
14	19 33 06	23 37 15	22 ♈ 47 53	29 ♈ 38 17	22 38	17 22	08 03	26 04	10 18	15 30	03 40	11 53
15	19 37 02	24 38 22	06 ♉ 34 20	13 ♉ 36 03	24 17	18 35	08 49	26 09	10 18	15 33	03 42	11 55
16	19 40 59	25 39 28	20 ♉ 43 21	27 ♉ 55 57	25 57	19 48	09 35	26 14	10 18	15 36	03 45	11 57
17	19 44 55	26 40 34	05 ♊ 13 29	12 ♊ 35 25	27 36	21 02	10 22	26 19	10 19	15 40	03 47	11 59
18	19 48 52	27 41 38	20 ♊ 01 01	27 ♊ 29 28	29 ♑ 17	22 15	11 08	26 25	10 20	15 43	03 49	12 01
19	19 52 49	28 42 42	04 ♋ 59 49	12 ♋ 31 02	00 ♒ 57	23 28	11 55	26 31	10 20	15 46	03 52	12 02
20	19 56 45	29 ♑ 43 46	20 ♋ 02 01	27 ♋ 31 38	02 39	24 42	12 41	26 37	10 21	15 50	03 54	12 04
21	20 00 42	00 ♒ 44 48	04 ♌ 58 51	12 ♌ 22 37	04 21	25 55	13 27	26 43	10 22	15 53	03 56	12 06
22	20 04 38	01 45 50	19 ♌ 42 03	26 ♌ 56 22	06 03	27 09	14 14	26 49	10 23	15 57	03 58	12 07
23	20 08 35	02 46 51	04 ♍ 04 56	11 ♍ 07 18	07 46	28 22	15 00	26 56	10 25	16 00	04 01	12 09
24	20 12 31	03 47 52	18 ♍ 03 09	24 ♍ 52 22	09 29	29 ♐ 36	15 46	27 02	10 26	16 03	04 03	12 11
25	20 16 28	04 48 52	01 ♎ 34 56	08 ♎ 11 00	11 13	00 ♑ 49	16 32	27 09	10 27	16 07	04 05	12 12
26	20 20 24	05 49 52	14 ♎ 40 49	21 ♎ 04 45	12 57	02 03	17 19	27 16	10 29	16 10	04 07	12 14
27	20 24 21	06 50 50	27 ♎ 23 13	03 ♏ 36 43	14 41	03 17	18 05	27 23	10 30	16 14	04 10	12 16
28	20 28 18	07 51 49	09 ♏ 45 46	15 ♏ 50 58	16 26	04 30	18 51	27 30	10 32	16 17	04 12	12 17
29	20 32 14	08 52 47	21 ♏ 52 53	27 ♏ 52 05	18 11	05 44	19 37	27 38	10 34	16 21	04 14	12 19
30	20 36 11	09 53 44	03 ♐ 49 11	09 ♐ 44 44	19 56	06 58	20 23	27 45	10 36	16 24	04 17	12 20
31	20 40 07	10 ♒ 54 40	15 ♐ 39 18	21 ♐ 33 24	21 ♒ 41	08 ♑ 12	21 ♓ 09	27 ♈ 53	10 ♉ 38	16 ♒ 28	04 ♒ 19	12 ♐ 21

DECLINATIONS

Date	Moon True ☊ ° '	Moon Mean ☊ ° '	Moon ☽ Latitude ° '	Sun ☉ ° '	Moon ☽ ° '	Mercury ☿ ° '	Venus ♀ ° '	Mars ♂ ° '	Jupiter ♃ ° '	Saturn ♄ ° '	Uranus ♅ ° '	Neptune ♆ ° '	Pluto ♇ ° '
01	03 ♌ 57	05 ♌ 03	05 N 10	23 S 02	10 S 54	24 S 25	18 S 27	13 S 11	08 N 36	12 N 37	17 S 01	19 S 13	11 S 24
02	03 R 54	04 59	04 53	22 57	14 20	24 29	18 43	12 54	08 37	12 37	17 00	19 12	11 24
03	03 50	04 56	04 23	22 52	17 10	24 32	18 58	12 37	08 38	12 37	16 59	19 12	11 24
04	03 46	04 53	03 42	22 46	19 17	24 33	19 13	12 20	08 40	12 37	16 58	19 11	11 24
05	03 43	04 50	02 52	22 39	20 34	24 33	19 28	12 03	08 41	12 37	16 58	19 11	11 24
06	03 41	04 47	01 54	22 32	20 57	24 31	19 42	11 45	08 42	12 37	16 57	19 10	11 24
07	03 39	04 44	00 N 50	22 25	20 23	24 28	19 55	11 28	08 44	12 37	16 56	19 10	11 25
08	03 D 39	04 40	00 S 16	22 18	18 53	24 24	20 08	11 10	08 46	12 37	16 55	19 09	11 25
09	03 40	04 37	01 23	22 09	16 31	24 18	20 21	10 53	08 47	12 37	16 54	19 09	11 25
10	03 41	04 34	02 26	22 01	13 23	24 11	20 32	10 35	08 49	12 37	16 53	19 08	11 25
11	03 42	04 31	03 24	21 52	09 36	24 02	20 44	10 17	08 51	12 38	16 52	19 08	11 25
12	03 43	04 28	04 12	21 42	05 19	23 52	20 54	09 59	08 53	12 38	16 51	19 07	11 25
13	03 44	04 25	04 49	21 33	00 S 43	23 40	21 05	09 41	08 55	12 38	16 50	19 07	11 25
14	03 44	04 21	05 12	21 22	04 N 02	23 27	21 14	09 23	08 57	12 39	16 49	19 06	11 25
15	03 R 44	04 18	05 17	21 12	08 43	23 12	21 23	09 05	08 59	12 39	16 48	19 06	11 25
16	03 44	04 15	05 04	21 01	13 03	22 56	21 32	08 47	09 01	12 39	16 47	19 05	11 25
17	03 43	04 12	04 31	20 49	16 44	22 38	21 40	08 29	09 03	12 40	16 46	19 05	11 26
18	03 42	04 09	03 40	20 37	19 25	22 19	21 47	08 11	09 06	12 40	16 45	19 04	11 26
19	03 42	04 05	02 32	20 25	20 48	21 58	21 53	07 52	09 08	12 41	16 44	19 04	11 26
20	03 41	04 02	01 S 15	20 12	20 42	21 36	21 59	07 34	09 10	12 41	16 43	19 03	11 26
21	03 D 41	03 59	00 N 07	19 59	19 08	21 12	22 05	07 15	09 13	12 42	16 42	19 03	11 26
22	03 41	03 56	01 28	19 46	16 18	20 47	22 09	06 57	09 15	12 43	16 41	19 02	11 26
23	03 41	03 53	02 41	19 32	12 31	20 20	22 13	06 38	09 18	12 43	16 40	19 02	11 26
24	03 R 41	03 50	03 43	19 18	08 08	19 51	22 17	06 19	09 21	12 44	16 39	19 01	11 26
25	03 41	03 46	04 29	19 04	03 N 29	19 21	22 20	06 01	09 23	12 45	16 38	19 01	11 26
26	03 41	03 43	05 01	18 49	01 S 10	18 50	22 22	05 42	09 26	12 45	16 37	19 00	11 26
27	03 41	03 40	05 16	18 34	05 37	18 17	22 23	05 23	09 29	12 46	16 36	19 00	11 26
28	03 D 41	03 37	05 16	18 18	09 44	17 43	22 24	05 04	09 32	12 47	16 35	18 59	11 26
29	03 41	03 34	05 02	18 02	13 22	17 07	22 24	04 46	09 35	12 48	16 34	18 58	11 26
30	03 41	03 31	04 35	17 46	16 24	16 29	22 24	04 27	09 38	12 49	16 33	18 58	11 26
31	03 ♌ 42	03 ♌ 27	03 N 57	17 S 30	18 S 45	15 S 51	22 S 22	04 S 08	09 N 41	12 N 50	16 S 32	18 S 57	11 S 26

ZODIAC SIGN ENTRIES

Date	h	m	Planets
02	21	32	☽ ♐
04	03	01	♂ ♓
05	10	24	☽ ♑
07	22	53	☽ ♒
10	09	59	☽ ♓
12	18	48	☽ ♈
15	00	38	☽ ♉
17	03	25	☽ ♊
18	22	20	☿ ♒
19	04	01	☽ ♋
20	18	23	☉ ♒
21	03	58	☽ ♌
23	05	07	☽ ♍
24	19	52	♀ ♑
25	09	09	☽ ♎
27	17	01	☽ ♏
30	04	17	☽ ♐

LATITUDES

Date	Mercury ☿ ° '	Venus ♀ ° '	Mars ♂ ° '	Jupiter ♃ ° '	Saturn ♄ ° '	Uranus ♅ ° '	Neptune ♆ ° '	Pluto ♇ ° '
01	01 S 00	02 N 04	01 S 04	01 S 16	02 S 27	00 S 39	00 N 14	10 N 51
04	01 16	01 59	01 02	01 15	02 26	00 39	00 14	10 52
07	01 31	01 53	01 00	01 14	02 25	00 39	00 14	10 52
10	01 44	01 46	00 57	01 13	02 24	00 39	00 14	10 52
13	01 54	01 39	00 55	01 12	02 23	00 39	00 14	10 53
16	02 01	01 31	00 53	01 11	02 22	00 39	00 14	10 53
19	02 05	01 23	00 50	01 10	02 21	00 39	00 14	10 54
22	02 05	01 15	00 48	01 09	02 21	00 39	00 14	10 54
25	02 02	01 07	00 46	01 09	02 20	00 39	00 14	10 55
28	01 53	00 58	00 43	01 08	02 19	00 39	00 14	10 55
31	01 S 39	00 N 49	00 S 41	01 S 07	02 S 18	00 S 39	00 N 14	10 N 56

DATA

Julian Date	2451545
Delta T	+64 seconds
Ayanamsa	23° 51' 12"
Synetic vernal point	05° ♓ 15' 48"
True obliquity of ecliptic	23° 26' 16"

MOON'S PHASES, APSIDES AND POSITIONS ☽

Date	h	m	Phase	Longitude ° '	Eclipse Indicator
06	18	14	●	15 ♑ 44	
14	13	34	☽	23 ♈ 41	
21	04	40	○	00 ♌ 26	total
28	07	57	☾	07 ♏ 42	

Day	h	m		
04	12	38	Apogee	
19	22	54	Perigee	
06	09	41	Max dec	20° S 57'
13	15	38	0N	
19	22	30	Max dec	20° N 57'
26	05	57	0S	

LONGITUDES

Date	Chiron ⚷ ° '	Ceres ⚳ ° '	Pallas ⚴ ° '	Juno ⚵ ° '	Vesta ⚶ ° '	Black Moon Lilith ⚸ ° '
01	11 ♐ 37	04 ♎ 27	14 ♌ 03	08 ♑ 00	05 ♐ 58	23 ♐ 28
11	12 ♐ 44	06 ♎ 24	11 ♌ 35	11 ♑ 44	11 ♐ 09	24 ♐ 35
21	13 ♐ 45	07 ♎ 46	08 ♌ 28	15 ♑ 28	16 ♐ 14	25 ♐ 41
31	14 ♐ 40	08 ♎ 29	05 ♌ 01	19 ♑ 12	21 ♐ 14	26 ♐ 48

All ephemeris data is given at 12.00 UT and the Moon's longitude is additionally given for 24.00 UT

ASPECTARIAN

h m	Aspects
01 Saturday	
05 34	☽ ✶ ☉
05 35	♀ ± ♃
06 10	☽ ☍ ♄
08 16	☽ ⊻ ♇
12 37	☉ △ ♄
14 59	☽ □ ♅
15 14	☽ ∥ ♇
20 12	☽ ∠ ☿
23 29	☽ # ♄
02 Sunday	
01 03	♂ Q ♄
02 18	☽ ∥ ♂
03 49	☽ Q ♆
08 33	☿ ⊻ ♆
12 02	☽ ⊼ ♃
13 45	♀ Q ♅
14 27	☽ ∠ ☉
14 28	☉ ⊻ ♇
17 02	☽ ⊥ ☿
19 28	☽ □ ♂
21 16	♀ ✶ ♆
03 Monday	
00 12	☽ ± ♃
03 23	☽ Q ♅
04 07	☽ ✶ ♆
04 52	☽ ☌ ♀
07 01	☽ ⊻ ☿
10 15	☽ ∥ ♅
10 22	☽ ⊥ ☉
12 27	♂ # ♄
18 30	☽ ⚼ ♃
18 31	☽ ⊼ ♄
20 54	☽ ☌ ♇
23 41	☽ ⊻ ☉
04 Tuesday	
03 50	☽ ✶ ♅
06 40	☽ ± ♄
09 02	♀ ∥ ♆
10 31	☽ Q ♂
10 38	☽ ∠ ♆
10 39	☽ ∥ ♆
10 55	☽ ∥ ♀
05 Wednesday	
00 56	☽ ⚼ ♄
01 06	☽ △ ♃
04 58	☽ ⊥ ♆
10 25	☽ ∠ ♅
12 36	☽ ✶ ♂
17 11	☽ ⊻ ♆
06 Thursday	
00 41	☽ ⊻ ♀
01 26	☿ ⊥ ♅
02 04	☉ ⊻ ♅
04 44	☽ ⊥ ♅
05 13	☽ ☌ ☿
07 18	☽ △ ♄
09 56	☽ ⊻ ♇
14 12	☽ ⊥ ♀
16 54	☽ ⊻ ♅
18 14	☽ ☌ ☉
20 42	☿ △ ♄
21 03	☽ ∠ ♂
22 05	☽ ⊥ ♇
07 Friday	
10 22	☽ ∠ ♀
14 00	☽ □ ♃
16 13	☽ ∠ ♇
16 24	☽ ⊥ ♂
16 39	♂ ∥ ♇
17 07	☿ ⊻ ♇
19 56	☽ ∥ ♀
08 Saturday	
05 13	☽ ⊻ ♂
05 44	☽ ☌ ♆
08 34	☽ ∥ ♆
13 45	♂ ⊻ ♆
16 35	☉ ⊥ ♇
16 51	♀ ⊼ ♄
19 22	☽ □ ♄
19 39	☽ ✶ ♀
22 10	☽ ✶ ♇
23 48	♀ ⚼ ♃
09 Sunday	
02 01	☽ Q ♃
02 28	☽ ⊻ ☿
05 04	☽ ☌ ♅
08 39	☽ ∥ ♅
11 34	☽ ⊻ ☉
16 03	☽ ⊥ ☿
21 15	♀ ☌ ♇
21 49	☽ Q ♇
21 52	☽ Q ♀
22 50	☿ ⊻ ♅
10 Monday	
00 22	☽ ⊥ ☉
01 41	☽ ✶ ♃
06 40	☽ Q ♄
12 12	☽ ∠ ☿
16 48	☽ ⊻ ♆
17 06	☽ # ♄
19 22	☽ ∠ ☉
20 02	☽ ☌ ♂
11 Tuesday	
00 54	☽ ∥ ♇
04 21	☽ ⊥ ♆
05 47	☽ ✶ ♄
06 46	☽ ∠ ♃
07 31	☽ ∥ ♂
08 38	☽ □ ♇
12 19	☽ □ ♀
12 40	☿ ⊥ ♇
15 23	☽ ⊻ ♅
16 20	☽ # ♃
21 05	☽ ✶ ☿
21 29	☽ ∠ ♆
23 55	☽ ⊥ ♃
12 Wednesday	
02 23	☽ ✶ ☉
02 44	☽ ⊥ ♅
03 08	☉ ∠ ♂
04 59	♄ St D
10 04	☽ ∠ ♄
11 12	☽ ⊻ ♃
19 33	☽ ∠ ♅
21 32	♀ ✶ ♅
22 40	☽ Q ☿
13 Thursday	
01 28	☽ ✶ ♆
02 32	☽ Q ☉
02 42	☽ ⊥ ♄
07 54	☽ ⊻ ♂
13 38	☽ ⊻ ♄
14 54	♀ ± ♄
16 28	☽ △ ♇
19 26	☽ ⊥ ♂
22 57	☽ ✶ ♅
23 17	☽ Q ♆
14 Friday	
01 23	☽ △ ♀
11 41	☽ □ ☿
12 27	☽ ∠ ♂
13 34	☽ □ ☉
17 47	☽ ☌ ♃
19 13	☽ ⚼ ♇
20 18	☽ Q ♅
21 50	☉ ∥ ♀
23 04	☿ ∠ ♂
15 Saturday	
06 21	☽ ⚼ ♀
07 03	☽ □ ♆
10 52	☽ ± ♇
13 26	☽ ∥ ♃
13 52	☽ # ♂
14 30	♀ ∠ ♆
16 04	☽ ✶ ♂
18 23	☽ ☌ ♄
18 36	☉ ⊥ ♀
19 27	♂ # ♃
21 10	☽ ⊼ ♇
23 15	☽ ± ♀
16 Sunday	
01 19	☉ ☌ ☿
02 41	☽ # ♇
03 22	☽ □ ♅
04 27	☿ ⊥ ♀
09 42	☽ ∥ ♄
10 19	☽ ⊼ ♀
13 32	☽ Q ♂
16 26	☿ □ ♃
17 03	♂ ⊥ ♆
20 51	☽ △ ☉
21 14	☽ ⊻ ♃
21 50	☽ △ ☿
17 Monday	
02 48	☿ ∠ ♇
02 51	☉ □ ♃
07 13	☽ ⊥ ♃
09 38	☽ △ ♆
10 28	♂ ✶ ♄
12 18	☽ # ♅
19 21	☉ ∠ ♇
20 19	☽ ⊻ ♄
20 51	☽ □ ♂
22 00	☽ ∠ ♃
23 02	☽ ☍ ♇
23 18	☽ ⚼ ☉
18 Tuesday	
01 33	☽ ⚼ ☿
05 02	☽ △ ♅
06 03	☽ ⊥ ♄
08 16	☽ # ♆
10 04	☽ ⚼ ♆
14 54	☽ ± ☉
15 55	☽ ☍ ♀
17 54	☽ ± ☿
20 32	☽ ∠ ♄
21 52	♂ ∠ ♃
22 21	☽ ✶ ♃
19 Wednesday	
00 34	☽ ± ♆
01 13	☽ ⊼ ☉
03 51	☽ # ☉
04 44	☽ ⊼ ☿
05 14	☽ ⚼ ♅
10 11	☽ ⊼ ♆
16 08	♂ □ ♇
16 23	☿ ∥ ♀
17 39	☽ Q ♃
19 39	☽ ± ♅
20 32	☽ ✶ ♄
23 15	☽ ⊼ ♇
23 38	☽ △ ♂
20 Thursday	
05 16	☽ ⊼ ♅
08 51	☽ ± ♇
15 43	☽ Q ♄
20 08	☽ ⊼ ♀
22 36	☽ □ ♃
23 17	☽ ⚼ ♇
23 57	☽ # ☉
21 Friday	
00 55	☽ ⚼ ♂
01 03	♀ ⚼ ♄
04 40	☽ ☍ ☉
06 04	☿ ☌ ♆
06 37	☽ ± ♀
10 18	☽ ☍ ♆
10 50	☽ ☍ ☿
12 58	☽ # ♆
16 13	☽ ± ♂
20 45	☽ □ ♄
22 30	☽ ⚼ ♀
23 34	☽ △ ♇
22 Saturday	
02 31	☽ ⊼ ♂
04 58	♀ △ ♃
05 48	☽ ☍ ♅
09 13	☽ # ♅
23 53	☽ △ ♃
23 Sunday	
01 30	☽ △ ♀
04 42	♀ ⊥ ♆
09 38	☽ ⊼ ☉
10 47	☽ ∥ ♄
11 53	☽ ⊼ ♆
18 11	☽ # ♇
19 08	☽ ⊼ ☿
20 37	☽ ± ☉
22 07	☽ ± ♆
22 48	☽ △ ♄
24 Monday	
01 29	☽ ⚼ ♃
01 48	☽ □ ♇
05 39	☽ ∥ ♃
06 54	☽ ± ☿
07 48	☽ ☍ ♂
08 31	☽ ⊼ ♅
13 24	☽ ⚼ ☉
13 45	☽ ⚼ ♆
17 16	☽ ± ♃
18 08	☉ ☌ ♆
19 03	☽ ± ♅
19 07	☿ ⊥ ♂
21 42	♂ ⊻ ♅
22 08	☽ # ♂
25 Tuesday	
00 58	☽ ⚼ ☿
01 01	☽ ⚼ ♄
01 18	☿ □ ♄
03 59	☽ ⊼ ♃
09 31	☽ Q ♇
10 30	☽ □ ♀
11 09	☽ ⚼ ♅
16 33	☽ △ ♆
17 05	☉ ∥ ♆
17 12	☽ ± ♄
17 58	♀ ∠ ♅
18 20	☽ △ ☉
26 Wednesday	
01 56	☿ ✶ ♇
03 36	♃ ⚼ ♇
04 12	☽ ⊼ ♄
04 17	☿ ∥ ♆
07 27	☽ ✶ ♇
08 17	☽ △ ☿
13 39	☉ ∥ ☿
14 48	☽ △ ♅
17 14	☽ ⊼ ♂
23 08	☽ Q ♀
27 Thursday	
05 17	☽ ± ♂
10 46	☽ ∥ ♂
11 45	☽ ∠ ♇
11 59	☽ ☍ ♃
22 14	☿ Q ♃
23 42	☽ ⚼ ♂
28 Friday	
00 36	☽ ✶ ♀
01 06	☽ □ ♆
05 11	☽ ⊥ ♇
05 48	♀ ⊻ ♆
07 57	☽ □ ☉
09 55	☿ ☌ ♅
10 43	☽ # ♃
13 31	☽ ☍ ♄
16 58	☽ ⊻ ♇
22 45	☽ ∥ ♇
23 27	♂ ∠ ♆
29 Saturday	
00 56	☽ □ ♅
03 23	☽ □ ☿
07 11	☽ △ ♂
07 58	☽ # ♄
09 27	☽ ∠ ♀
12 43	☽ Q ♆
22 57	☽ Q ☉
23 38	☽ ⊼ ♃
30 Sunday	
05 34	☽ ⊥ ♀
09 55	☿ ∥ ♅
11 52	☽ ± ♃
12 36	☽ ∥ ☿
12 55	☽ ✶ ♆
13 11	☽ Q ♅
13 12	☽ ∥ ♅
19 06	☽ ⊻ ♀
21 45	☽ Q ☿
23 18	☿ ⊻ ♂
23 39	☽ ∥ ☉
31 Monday	
01 28	☽ ✶ ☉
01 46	☽ ⊼ ♄
05 15	☉ □ ♄
05 17	☽ ☌ ♇
06 18	☽ ⚼ ♃
13 39	☽ ✶ ♅
14 00	☽ ± ♄
14 40	☽ ∥ ♆
19 28	☽ ∠ ♆
23 58	☽ □ ♂

FEBRUARY 2000

LONGITUDES

Date	Sidereal time h m s	Sun ☉ ° ' "	Moon ☽ ° ' "	Moon ☽ 24.00 ° ' "	Mercury ☿ ° '	Venus ♀ ° '	Mars ♂ ° '	Jupiter ♃ ° '	Saturn ♄ ° '	Uranus ♅ ° '	Neptune ♆ ° '	Pluto ♇ ° '
01	20 44 04	11 ♒ 55 36	27 ♐ 27 31	03 ♑ 22 08	23 ♒ 25	09 ♑ 25	21 ♓ 56	28 ♈ 01	10 ♉ 40	16 ♒ 31	04 ♒ 21	12 ♐ 23
02	20 48 00	12 56 31	09 ♑ 17 41	15 ♑ 14 32	25 09	10 39	22 42	28 09	10 43	16 35	04 23	12 24
03	20 51 57	13 57 25	21 ♑ 13 02	27 ♑ 13 31	26 53	11 53	23 28	28 17	10 45	16 38	04 26	12 26
04	20 55 53	14 58 18	03 ♒ 16 15	09 ♒ 21 28	28 ♒ 35	13 07	24 14	28 25	10 48	16 42	04 28	12 27
05	20 59 50	15 59 09	15 ♒ 29 22	21 ♒ 40 07	00 ♓ 16	14 21	25 00	28 34	10 50	16 45	04 30	12 28
06	21 03 47	17 00 00	27 ♒ 53 51	04 ♓ 10 41	01 56	15 35	25 46	28 42	10 53	16 49	04 32	12 30
07	21 07 43	18 00 49	10 ♓ 30 42	16 ♓ 53 59	03 33	16 49	26 32	28 51	10 56	16 52	04 35	12 31
08	21 11 40	19 01 37	23 ♓ 20 36	29 ♓ 50 35	05 08	18 03	27 18	29 00	10 59	16 56	04 37	12 32
09	21 15 36	20 02 24	06 ♈ 23 59	13 ♈ 00 50	06 40	19 17	28 03	29 09	11 02	16 59	04 39	12 33
10	21 19 33	21 03 09	19 ♈ 41 10	26 ♈ 25 02	08 09	20 30	28 49	29 18	11 05	17 03	04 41	12 34
11	21 23 29	22 03 52	03 ♉ 12 26	10 ♉ 03 22	09 33	21 44	29 ♓ 35	29 27	11 08	17 06	04 43	12 36
12	21 27 26	23 04 34	16 ♉ 57 50	23 ♉ 55 46	10 52	22 58	00 ♈ 21	29 37	11 11	17 10	04 46	12 37
13	21 31 22	24 05 15	00 ♊ 57 07	08 ♊ 01 44	12 05	24 12	01 07	29 46	11 15	17 13	04 48	12 38
14	21 35 19	25 05 53	15 ♊ 09 25	22 ♊ 19 55	13 13	25 26	01 52	29 ♈ 56	11 18	17 16	04 50	12 39
15	21 39 16	26 06 30	29 ♊ 32 55	06 ♋ 47 59	14 13	26 40	02 38	00 ♉ 06	11 22	17 20	04 52	12 40
16	21 43 12	27 07 05	14 ♋ 04 36	21 ♋ 22 13	15 05	27 54	03 24	00 16	11 26	17 23	04 54	12 41
17	21 47 09	28 07 39	28 ♋ 40 11	05 ♌ 57 46	15 49	29 ♑ 08	04 09	00 26	11 29	17 27	04 56	12 42
18	21 51 05	29 ♒ 08 11	13 ♌ 14 14	20 ♌ 28 48	16 24	00 ♒ 22	04 55	00 36	11 33	17 30	04 59	12 43
19	21 55 02	00 ♓ 08 41	27 ♌ 40 44	04 ♍ 49 17	16 50	01 36	05 40	00 46	11 37	17 34	05 01	12 43
20	21 58 58	01 09 10	11 ♍ 53 47	18 ♍ 53 40	17 05	02 51	06 26	00 57	11 41	17 37	05 03	12 44
21	22 02 55	02 09 37	25 ♍ 48 27	02 ♎ 37 45	17 11	04 05	07 11	01 07	11 46	17 41	05 05	12 45
22	22 06 51	03 10 02	09 ♎ 21 20	15 ♎ 59 06	17 R 06	05 19	07 57	01 18	11 50	17 44	05 07	12 46
23	22 10 48	04 10 26	22 ♎ 31 04	28 ♎ 57 20	16 51	06 33	08 42	01 28	11 54	17 47	05 09	12 47
24	22 14 45	05 10 49	05 ♏ 18 10	11 ♏ 33 54	16 27	07 47	09 27	01 39	11 59	17 51	05 11	12 47
25	22 18 41	06 11 10	17 ♏ 44 57	23 ♏ 51 49	15 53	09 01	10 13	01 50	12 03	17 54	05 13	12 48
26	22 22 38	07 11 30	29 ♏ 55 02	05 ♐ 55 11	15 12	10 15	10 58	02 01	12 08	17 57	05 15	12 49
27	22 26 34	08 11 48	11 ♐ 52 54	17 ♐ 48 49	14 23	11 29	11 43	02 12	12 12	18 01	05 17	12 49
28	22 30 31	09 12 05	23 ♐ 43 34	29 ♐ 37 49	13 29	12 43	12 28	02 23	12 17	18 04	05 19	12 50
29	22 34 27	10 ♓ 12 21	05 ♑ 32 10	11 ♑ 27 16	12 ♓ 30	13 ♒ 57	13 ♈ 13	02 ♉ 35	12 ♉ 22	18 ♒ 07	05 ♒ 21	12 ♐ 50

DECLINATIONS

Date	Moon True ☊ ° '	Moon Mean ☊ ° '	Moon ☽ Latitude ° '	Sun ☉ ° '	Moon ☽ ° '	Mercury ☿ ° '	Venus ♀ ° '	Mars ♂ ° '	Jupiter ♃ ° '	Saturn ♄ ° '	Uranus ♅ ° '	Neptune ♆ ° '	Pluto ♇ ° '
01	03 ♌ 43	03 ♌ 24	03 N 08	17 S 13	20 S 17	15 S 11	22 S 21	03 S 49	09 N 44	12 N 51	16 S 31	18 S 57	11 S 26
02	03 D 44	03 21	02 12	16 56	20 55	14 30	22 18	03 30	09 47	12 52	16 29	18 56	11 25
03	03 45	03 18	01 09	16 38	20 38	13 48	22 15	03 11	09 50	12 53	16 28	18 56	11 25
04	03 45	03 15	00 N 03	16 21	19 23	13 05	22 11	02 52	09 53	12 54	16 27	18 55	11 25
05	03 R 45	03 11	01 S 05	16 03	17 13	12 22	22 06	02 33	09 56	12 55	16 26	18 55	11 25
06	03 44	03 08	02 10	15 44	14 14	11 37	22 01	02 14	10 00	12 56	16 25	18 54	11 25
07	03 42	03 05	03 10	15 26	10 33	10 52	21 55	01 55	10 03	12 57	16 24	18 54	11 25
08	03 40	03 02	04 01	15 07	06 20	10 08	21 49	01 36	10 06	12 58	16 23	18 53	11 25
09	03 37	02 59	04 41	14 48	01 S 46	09 23	21 42	01 17	10 10	12 59	16 22	18 53	11 25
10	03 35	02 56	05 07	14 29	02 N 58	08 38	21 34	00 58	10 13	13 01	16 21	18 52	11 25
11	03 33	02 52	05 16	14 09	07 38	07 55	21 26	00 39	10 17	13 02	16 20	18 52	11 25
12	03 33	02 49	05 07	13 49	11 59	07 12	21 17	00 21	10 20	13 03	16 19	18 51	11 25
13	03 D 33	02 46	04 41	13 30	15 46	06 31	21 07	00 S 02	10 24	13 05	16 18	18 51	11 24
14	03 33	02 43	03 56	13 09	18 42	05 51	20 57	00 N 17	10 27	13 06	16 17	18 50	11 24
15	03 35	02 40	02 57	12 49	20 29	05 14	20 46	00 36	10 31	13 07	16 16	18 50	11 24
16	03 36	02 37	01 46	12 28	20 57	04 39	20 34	00 55	10 35	13 09	16 15	18 49	11 24
17	03 37	02 33	00 S 27	12 07	19 59	04 08	20 22	01 14	10 38	13 10	16 14	18 49	11 24
18	03 R 37	02 30	00 N 53	11 46	17 41	03 40	20 09	01 33	10 42	13 11	16 13	18 48	11 24
19	03 35	02 27	02 09	11 25	14 17	03 15	19 56	01 51	10 46	13 13	16 12	18 48	11 23
20	03 32	02 24	03 15	11 04	10 06	02 55	19 42	02 10	10 50	13 14	16 11	18 47	11 23
21	03 28	02 21	04 08	10 42	05 28	02 40	19 27	02 29	10 54	13 16	16 10	18 47	11 23
22	03 23	02 17	04 46	10 21	00 N 41	02 29	19 12	02 47	10 57	13 18	16 09	18 46	11 23
23	03 19	02 14	05 07	09 59	04 S 00	02 23	18 57	03 06	11 01	13 19	16 08	18 46	11 23
24	03 14	02 11	05 13	09 37	08 22	02 22	18 41	03 24	11 05	13 21	16 07	18 45	11 23
25	03 11	02 08	05 03	09 15	12 16	02 26	18 24	03 43	11 09	13 22	16 05	18 45	11 22
26	03 09	02 05	04 40	08 52	15 35	02 35	18 07	04 01	11 13	13 24	16 04	18 44	11 22
27	03 D 09	02 02	04 04	08 30	18 10	02 48	17 49	04 20	11 17	13 26	16 03	18 44	11 22
28	03 10	01 58	03 19	08 07	19 58	03 05	17 30	04 38	11 21	13 27	16 02	18 43	11 22
29	03 ♌ 12	01 ♌ 55	02 N 25	07 S 44	20 S 54	03 S 26	17 S 12	04 N 56	11 N 25	13 N 29	16 S 01	18 S 43	11 S 21

ZODIAC SIGN ENTRIES

Date	h	m	Planets
01	17	10	☽ ♑
04	05	31	☽ ♒
05	08	09	☿ ♓
06	16	02	☽ ♓
09	00	17	☽ ♈
11	06	21	☽ ♉
12	01	04	♂ ♈
13	10	23	☽ ♊
14	21	39	♃ ♉
15	12	45	☽ ♋
17	14	11	☽ ♌
18	04	43	♀ ♒
19	08	33	☉ ♓
19	15	53	☽ ♍
21	19	21	☽ ♎
24	01	58	☽ ♏
26	12	10	☽ ♐
29	00	45	☽ ♑

LATITUDES

Date	Mercury ☿ ° '	Venus ♀ ° '	Mars ♂ ° '	Jupiter ♃ ° '	Saturn ♄ ° '	Uranus ♅ ° '	Neptune ♆ ° '	Pluto ♇ ° '
01	01 S 34	00 N 46	00 S 40	01 S 07	02 S 18	00 S 39	00 N 14	10 N 56
04	01 12	00 37	00 38	01 06	02 17	00 39	00 14	10 57
07	00 43	00 28	00 35	01 05	02 16	00 39	00 14	10 58
10	00 S 08	00 19	00 33	01 05	02 15	00 39	00 14	10 59
13	00 N 34	00 10	00 31	01 04	02 14	00 39	00 14	10 59
16	01 19	00 N 01	00 28	01 03	02 13	00 39	00 14	11 00
19	02 07	00 S 08	00 26	01 02	02 13	00 39	00 14	11 01
22	02 50	00 16	00 24	01 02	02 12	00 39	00 14	11 02
25	03 23	00 25	00 21	01 01	02 11	00 39	00 14	11 03
28	03 41	00 32	00 19	01 01	02 10	00 39	00 14	11 03
31	03 N 39	00 S 40	00 S 17	01 S 00	02 S 10	00 S 40	00 N 14	11 N 04

DATA

Julian Date	2451576
Delta T	+64 seconds
Ayanamsa	23° 51' 17"
Synetic vernal point	05° ♓ 15' 43"
True obliquity of ecliptic	23° 26' 16"

MOON'S PHASES, APSIDES AND POSITIONS ☽

Date	h	m	Phase	Longitude ° '	Eclipse Indicator
05	13	03	●	16 ♒ 02	Partial
12	23	21	☽	23 ♉ 33	
19	16	27	○	00 ♍ 20	
27	03	53	☾	07 ♐ 51	

Day	h	m		
01	01	29	Apogee	
17	02	38	Perigee	
28	20	49	Apogee	
02	16	35	Max dec	20° S 56'
09	20	59	0N	
16	07	44	Max dec	20° N 58'
22	15	25	0S	
29	23	54	Max dec	21° S 01'

LONGITUDES

Date	Chiron ⚷ ° '	Ceres ⚳ ° '	Pallas ⚴ ° '	Juno ⚵ ° '	Vesta ⚶ ° '	Black Moon Lilith ⚸ ° '
01	14 ♐ 45	08 ♎ 31	04 ♌ 41	19 ♑ 34	21 ♐ 44	26 ♐ 55
11	15 ♐ 33	08 ♎ 28	01 ♌ 45	23 ♑ 16	26 ♐ 36	28 ♐ 02
21	16 ♐ 12	07 ♎ 41	29 ♋ 41	26 ♑ 55	01 ♑ 20	29 ♐ 08
31	16 ♐ 42	06 ♎ 14	28 ♋ 39	00 ♒ 30	05 ♑ 53	00 ♑ 15

ASPECTARIAN

h m	Aspects
01 Tuesday	
02 22	☽ ∗ ☿
08 21	☽ BQ ♄
10 49	☽ ∠ ☉
13 08	☽ △ ♃
13 49	☽ ⊥ ♆
15 16	♂ ⊥ ♃
20 17	☽ ∠ ♅
23 01	☉ ∗ ♇
02 Wednesday	
01 31	☿ Q ♇
02 02	☽ ⊻ ♆
06 48	☽ ⊥ ☉
08 03	♂ ⊥ ♅
10 25	♀ ⊥ ♅
13 07	♀ △ ♄
14 02	☽ ∠ ☿
14 02	♀ Q ♂
14 36	☽ ⊥ ♅
14 52	☽ △ ♄
15 01	☽ Q ♂
15 04	☽ ☌ ♀
18 18	☽ ⊻ ♇
20 03	☽ ⊻ ☉
03 Thursday	
02 46	☽ ⊻ ♅
06 24	☽ ⊥ ♇
11 12	☽ ⊥ ☿
12 31	☿ ∠ ♀
16 48	☽ ∗ ♂
22 46	♀ ⊻ ♇
04 Friday	
00 26	☽ ∠ ♇
01 10	☽ ⊻ ☿
02 16	☽ □ ♃
02 17	☉ ∥ ♅
09 31	☿ ∗ ♃
14 22	☽ ☌ ♆
15 02	☿ Q ♄
18 03	☽ ∥ ♆
18 10	☿ # ♄
05 Saturday	
00 32	☽ ∠ ♂
02 52	☽ □ ♄
06 06	☽ ∗ ♇
09 32	☽ ⊻ ♀
13 03	☽ ☌ ☉
14 07	☽ Q ♃
14 28	☽ ☌ ♅
18 58	☽ ∥ ♅
19 16	☽ ⊥ ♂
22 29	☽ ⊥ ♀
23 23	☽ ∥ ☉
06 Sunday	
03 54	☉ Q ♃
05 27	☽ Q ♇
07 14	☉ ☌ ♅
07 38	☽ ⊻ ♂
13 34	☽ ∗ ♃
13 54	☽ Q ♄
15 59	♂ ∠ ♄
17 42	☽ ∠ ♀
18 29	☿ ∥ ♇
20 52	☽ ☌ ☿
20 55	☽ # ♄
07 Monday	
00 43	☽ ⊻ ♆
06 42	☽ ∥ ♇
09 35	☽ ∥ ☿
12 07	☽ ⊥ ♆
12 47	☽ ∗ ♄
13 10	♀ ⊻ ♅
14 58	☽ # ♃
15 47	☽ □ ♇
16 22	♃ Q ♅
18 22	☽ ∠ ♃
08 Tuesday	
00 00	☽ ⊻ ♅
01 05	☽ ∗ ♀
03 17	☽ ⊻ ☉
03 48	☿ ⊻ ♆
05 03	☽ ∠ ♆
11 13	☽ ⊥ ♅
11 21	☽ ⊥ ♃
12 38	☿ # ♃
15 23	☽ ⊥ ☉
16 54	☽ ∠ ♄
19 46	☽ ☌ ♂
21 43	♀ ⊥ ♇
22 34	☽ ⊻ ♃
09 Wednesday	
01 39	☽ Q ♀
03 54	☽ ∠ ♅
08 48	☽ ∗ ♆
09 19	☽ ∠ ☉
09 29	☽ ⊥ ♄
12 33	☽ ⊻ ☿
14 36	☽ ∥ ♂
20 26	☽ ⊻ ♄
23 11	☽ △ ♇
10 Thursday	
00 49	☽ ⊥ ☿
02 33	☽ # ♂
06 36	☽ Q ♆
07 14	☽ ∗ ♅
13 37	☽ □ ♀
14 39	☽ ∗ ☉
18 54	☽ ∠ ☿
11 Friday	
02 04	☽ BQ ♇
04 43	☽ Q ♅
05 14	☽ ⊻ ♂
05 19	☽ ☌ ♃
06 59	♂ ⊻ ♃
13 17	☽ # ☿
13 38	☽ Q ☉
14 40	☽ □ ♆
16 25	☽ ⊥ ♂
17 57	☽ ± ♇
12 Saturday	
00 18	☽ ∗ ☿
01 56	☽ ☌ ♄
02 29	☽ ∥ ♃
04 26	☽ ⊼ ♇
08 40	☽ # ♇
09 02	☽ ∠ ♂
10 00	☿ ⊥ ♆
12 20	☽ □ ♅
18 23	☽ ∥ ♄
18 28	☿ ∗ ♄
22 13	☽ # ☉
23 06	☉ ⊻ ♀
23 10	☽ Q ☿
23 21	☽ □ ☉
23 22	☽ △ ♀
13 Sunday	
09 58	☽ ⊻ ♃
12 17	☽ ∗ ♂
15 47	☽ # ♅
18 33	☽ △ ♆
20 17	☽ ⊥ ♃
23 24	☿ □ ♇
14 Monday	
01 04	☉ Q ♇
03 19	☽ BQ ♀
05 30	☽ ⊻ ♄
07 47	☽ ☍ ♇
08 28	☽ □ ☿
09 43	☽ Q ♂
11 37	☽ ∠ ♃
13 25	☽ # ♆
15 34	☽ △ ♅
15 37	☽ ⊥ ♄
15 47	☉ # ♄
19 51	☽ ± ♀
19 51	☽ BQ ♆
15 Tuesday	
01 45	♂ ∠ ♅
05 51	☽ △ ☉
06 41	☽ ∠ ♄
06 47	☽ ⊼ ♀
10 52	☽ ± ♆
12 55	☽ ∗ ♃
16 38	☽ BQ ♅
17 24	☽ □ ♂
17 33	☽ # ♀
20 50	☽ ⊼ ♆
16 Wednesday	
07 30	♀ ∠ ♇
07 33	☽ ± ♅
07 37	☽ ∗ ♄
08 32	☽ BQ ☉
08 59	☽ Q ♃
09 42	☽ ⊼ ♇
13 46	☽ △ ☿
17 28	☽ ⊼ ♅
18 46	☿ ∠ ♃
19 35	☽ ± ♇
17 Thursday	
00 27	☽ ± ☉
03 27	☽ Q ♄
04 33	☽ # ♀
10 24	☽ BQ ♇
11 03	☽ ⊼ ☉
12 51	☽ ☍ ♀
14 28	☉ ⊥ ♂
14 56	☽ □ ♃
15 42	☽ BQ ☿
21 31	☽ △ ♂
22 21	☽ ☍ ♆
18 Friday	
01 53	☽ # ♆
07 09	☽ ± ☿
09 12	☽ □ ♄
11 08	☽ △ ♇
14 07	♂ ∗ ♆
17 04	♀ □ ♃
17 25	☽ ⊼ ☿
19 05	☽ ☍ ♅
22 39	☉ Q ♄
23 19	☽ # ♅
23 40	☽ BQ ♂
19 Saturday	
10 16	♂ ⊥ ♄
14 00	☉ ∥ ♇
15 31	☽ ± ♂
16 27	☽ ☍ ☉
17 15	☽ △ ♃
17 49	☿ ∠ ♀
18 28	☽ ∥ ♄
19 13	☽ ⊼ ♀
20 Sunday	
00 21	☽ ⊼ ♆
02 12	☽ ⊼ ♂
04 59	☽ # ♇
05 57	☉ ∗ ♃
06 18	☽ # ☉
06 18	☽ ± ♀
08 07	☽ ∥ ♃
10 33	☽ ± ♆
11 39	☽ △ ♄
13 26	☽ □ ♇
19 00	☽ BQ ♃
20 59	☽ ☍ ☿
21 51	☽ ⊼ ♅
23 11	☽ BQ ♀
21 Monday	
01 23	☉ # ♃
02 01	☽ BQ ♆
08 16	☽ ± ♅
10 47	☽ ± ♃
12 46	☿ St R
13 40	☽ BQ ♄
20 41	☽ Q ♇
20 42	☿ # ♂
21 27	☽ ⊼ ♃
22 Tuesday	
00 04	☽ ⊼ ☉
00 08	☽ BQ ♅
02 03	☽ ∥ ♂
02 37	☽ # ☿
04 02	☽ △ ♀
04 24	☽ △ ♆
05 39	☽ ± ♄
08 06	♀ ☌ ♆
09 19	☽ ☍ ♂
11 38	☽ ± ☉
16 29	☽ ⊼ ♄
18 09	☽ ∗ ♇
23 Wednesday	
01 49	☽ ⊼ ☿
03 15	☽ △ ♅
03 43	☽ ∥ ☿
05 19	☽ BQ ☉
06 56	☽ # ♂
12 36	☽ ± ☿
21 48	☽ ∠ ♇
24 Thursday	
04 20	☿ ∠ ♃
04 57	☽ BQ ☿
04 59	☽ ☍ ♃
05 02	♀ ∥ ♆
11 45	☽ △ ☉
11 46	☽ □ ♆
12 06	☉ ⊻ ♆
14 50	☽ ⊥ ♇
17 15	☽ □ ♀
18 40	☽ ∥ ☉
20 27	☽ ⊼ ♂
25 Friday	
00 52	☽ ☍ ♄
02 22	☽ ⊻ ♇
04 38	☽ # ♃
06 10	☽ ∥ ♇
08 33	☽ △ ☿
08 48	☽ ± ♂
12 18	☽ □ ♅
19 32	☽ # ♄
22 45	☽ Q ♆
23 07	☿ ⊥ ♀
26 Saturday	
03 38	☽ BQ ♂
08 18	☽ Q ♀
16 06	☽ ∥ ♅
16 15	☽ ⊼ ♃
22 41	☽ ∗ ♆
27 Sunday	
00 08	☽ Q ♅
03 53	☽ □ ☉
04 28	☽ ± ♃
08 38	☽ ∥ ♀
11 06	☽ ∗ ♀
11 39	☽ △ ♂
12 39	☽ ⊼ ♄
13 54	☽ ☌ ♇
16 43	☽ □ ☿
18 17	☽ ∥ ♆
22 56	☽ BQ ♃
23 29	♀ ∗ ♂
28 Monday	
00 28	☽ ∗ ♅
00 53	☽ ± ♄
02 56	♀ □ ♄
05 04	☽ ∠ ♆
05 25	♂ ⊻ ♄
14 07	♀ ∗ ♇
14 53	♃ # ♇
19 17	☽ BQ ♄
19 43	☽ Q ☉
20 21	☿ ⊻ ♀
21 04	☽ ∠ ♀
23 24	☽ ⊥ ♆
23 37	♂ △ ♇
29 Tuesday	
02 11	☿ ⊻ ♂
02 35	☽ Q ☿
04 06	☿ □ ♇
05 54	☽ △ ♃
07 05	☽ ∠ ♅
11 37	☽ ⊻ ♆
15 06	☿ ∗ ♄
17 29	☽ ⊥ ♀
22 21	☽ ∗ ☉

All ephemeris data is given at 12.00 UT and the Moon's longitude is additionally given for 24.00 UT

MARCH 2000

LONGITUDES

Date	Sidereal time h m s	Sun ☉ ° ′ ″	Moon ☽ ° ′ ″	Moon ☽ 24.00 ° ′ ″	Mercury ☿ ° ′	Venus ♀ ° ′	Mars ♂ ° ′	Jupiter ♃ ° ′	Saturn ♄ ° ′	Uranus ♅ ° ′	Neptune ♆ ° ′	Pluto ♇ ° ′
01	22 38 24	11 ♓ 12 35	17 ♑ 23 41	23 ♑ 21 58	11 ♓ 29	15 ♒ 12	13 ♈ 58	02 ♉ 46	12 ♉ 27	18 ♒ 11	05 ♒ 23	12 ♐ 51
02	22 42 20	12 12 48	29 ♑ 22 39	05 ♒ 26 10	10 R 26	16 26	14 43	02 58	12 32	18 14	05 25	12 51
03	22 46 17	13 12 58	11 ♒ 32 57	17 ♒ 43 19	09 23	17 40	15 28	03 09	12 37	18 17	05 27	12 52
04	22 50 14	14 13 07	23 ♒ 57 33	00 ♓ 15 51	08 22	18 54	16 13	03 21	12 42	18 20	05 29	12 52
05	22 54 10	15 13 15	06 ♓ 38 21	13 ♓ 05 04	07 24	20 08	16 58	03 33	12 47	18 24	05 31	12 52
06	22 58 07	16 13 20	19 ♓ 35 59	26 ♓ 11 00	06 30	21 22	17 43	03 44	12 53	18 27	05 32	12 53
07	23 02 03	17 13 24	02 ♈ 49 55	09 ♈ 32 32	05 40	22 36	18 28	03 56	12 58	18 30	05 34	12 53
08	23 06 00	18 13 26	16 ♈ 18 33	23 ♈ 07 40	04 56	23 51	19 13	04 08	13 04	18 33	05 36	12 53
09	23 09 56	19 13 25	29 ♈ 59 33	06 ♉ 53 51	04 19	25 05	19 58	04 20	13 09	18 36	05 38	12 54
10	23 13 53	20 13 23	13 ♉ 50 17	20 ♉ 48 30	03 47	26 19	20 42	04 33	13 15	18 39	05 40	12 54
11	23 17 49	21 13 18	27 ♉ 48 14	04 ♊ 49 14	03 22	27 33	21 27	04 45	13 20	18 42	05 41	12 54
12	23 21 46	22 13 12	11 ♊ 51 17	18 ♊ 54 12	03 04	28 ♒ 47	22 12	04 57	13 26	18 46	05 43	12 54
13	23 25 43	23 13 03	25 ♊ 57 48	03 ♋ 01 57	02 52	00 ♓ 01	22 56	05 10	13 32	18 49	05 45	12 54
14	23 29 39	24 12 52	10 ♋ 06 30	17 ♋ 11 16	02 47	01 15	23 41	05 22	13 38	18 52	05 46	12 54
15	23 33 36	25 12 38	24 ♋ 16 05	01 ♌ 20 44	02 D 48	02 29	24 25	05 35	13 44	18 55	05 48	12 R 54
16	23 37 32	26 12 22	08 ♌ 24 57	15 ♌ 28 26	02 55	03 44	25 10	05 47	13 50	18 58	05 50	12 54
17	23 41 29	27 12 04	22 ♌ 30 50	29 ♌ 31 45	03 07	04 58	25 54	06 00	13 56	19 00	05 51	12 54
18	23 45 25	28 11 44	06 ♍ 30 46	13 ♍ 27 25	03 25	06 12	26 38	06 13	14 02	19 03	05 53	12 54
19	23 49 22	29 ♓ 11 22	20 ♍ 21 14	27 ♍ 11 47	03 48	07 26	27 23	06 26	14 08	19 06	05 54	12 54
20	23 53 18	00 ♈ 10 57	03 ♎ 58 39	10 ♎ 41 27	04 16	08 40	28 07	06 38	14 14	19 09	05 56	12 54
21	23 57 15	01 10 31	17 ♎ 19 53	23 ♎ 53 44	04 48	09 54	28 51	06 51	14 21	19 12	05 57	12 54
22	00 01 12	02 10 02	00 ♏ 22 53	06 ♏ 47 17	05 24	11 08	29 ♈ 35	07 04	14 27	19 15	05 59	12 53
23	00 05 08	03 09 32	13 ♏ 07 00	19 ♏ 22 11	06 05	12 22	00 ♉ 19	07 17	14 33	19 17	06 00	12 53
24	00 09 05	04 09 00	25 ♏ 33 07	01 ♐ 40 08	06 49	13 36	01 03	07 30	14 40	19 20	06 02	12 53
25	00 13 01	05 08 26	07 ♐ 43 38	13 ♐ 44 08	07 37	14 51	01 48	07 44	14 46	19 23	06 03	12 52
26	00 16 58	06 07 51	19 ♐ 42 10	25 ♐ 38 19	08 28	16 05	02 31	07 57	14 53	19 25	06 05	12 52
27	00 20 54	07 07 13	01 ♑ 33 15	07 ♑ 27 35	09 22	17 19	03 15	08 10	15 00	19 28	06 06	12 52
28	00 24 51	08 06 34	13 ♑ 22 02	19 ♑ 17 17	10 19	18 33	03 59	08 23	15 06	19 31	06 07	12 51
29	00 28 47	09 05 53	25 ♑ 13 59	01 ♒ 12 50	11 19	19 47	04 43	08 37	15 13	19 33	06 08	12 51
30	00 32 44	10 05 11	07 ♒ 14 28	13 ♒ 19 30	12 22	21 01	05 27	08 50	15 20	19 36	06 10	12 50
31	00 36 41	11 ♈ 04 26	19 ♒ 28 28	25 ♒ 41 54	13 ♓ 27	22 ♓ 15	06 ♉ 11	09 ♉ 04	15 ♉ 27	19 ♒ 38	06 ♒ 11	12 ♐ 50

Date	Moon True ☊ ° ′	Moon Mean ☊ ° ′	Moon ☽ Latitude ° ′
01	03 ♌ 14	01 ♌ 52	01 N 25
02	03 D 15	01 49	00 N 21
03	03 R 14	01 46	00 S 45
04	03 12	01 43	01 50
05	03 08	01 39	02 51
06	03 02	01 36	03 45
07	02 55	01 33	04 27
08	02 47	01 30	04 56
09	02 39	01 27	05 08
10	02 33	01 23	05 02
11	02 28	01 20	04 39
12	02 26	01 17	03 58
13	02 D 25	01 14	03 04
14	02 26	01 11	01 57
15	02 27	01 08	00 S 44
16	02 R 27	01 04	00 N 32
17	02 25	01 01	01 46
18	02 21	00 58	02 52
19	02 14	00 55	03 48
20	02 06	00 52	04 29
21	01 55	00 48	04 55
22	01 45	00 45	05 04
23	01 35	00 42	04 58
24	01 26	00 39	04 38
25	01 20	00 36	04 06
26	01 17	00 33	03 23
27	01 15	00 29	02 32
28	01 D 15	00 26	01 34
29	01 16	00 23	00 N 32
30	01 R 16	00 20	00 S 32
31	01 ♌ 14	00 ♌ 17	01 S 35

DECLINATIONS

Date	Sun ☉ ° ′	Moon ☽ ° ′	Mercury ☿ ° ′	Venus ♀ ° ′	Mars ♂ ° ′	Jupiter ♃ ° ′	Saturn ♄ ° ′	Uranus ♅ ° ′	Neptune ♆ ° ′	Pluto ♇ ° ′
01	07 S 22	20 S 54	03 S 50	16 S 52	05 N 15	11 N 29	13 N 31	16 S 00	18 S 42	11 S 21
02	06 59	19 56	04 16	16 33	05 33	11 33	13 32	15 59	18 42	11 21
03	06 36	18 02	04 44	16 13	05 51	11 38	13 34	15 58	18 41	11 21
04	06 13	15 16	05 13	15 52	06 09	11 42	13 36	15 57	18 41	11 21
05	05 49	11 43	05 42	15 31	06 27	11 46	13 38	15 57	18 40	11 20
06	05 26	07 34	06 12	15 09	06 44	11 50	13 40	15 56	18 40	11 20
07	05 03	02 S 57	06 40	14 47	07 02	11 54	13 41	15 55	18 40	11 20
08	04 39	01 N 52	07 08	14 25	07 20	11 58	13 43	15 54	18 39	11 20
09	04 16	06 40	07 34	14 02	07 38	12 03	13 45	15 53	18 39	11 19
10	03 52	11 11	07 58	13 39	07 55	12 07	13 47	15 52	18 38	11 19
11	03 29	15 08	08 20	13 15	08 12	12 11	13 49	15 51	18 38	11 19
12	03 05	18 16	08 40	12 51	08 30	12 15	13 51	15 50	18 37	11 19
13	02 42	20 19	08 58	12 27	08 47	12 20	13 53	15 49	18 37	11 18
14	02 18	21 06	09 13	12 03	09 04	12 24	13 55	15 48	18 37	11 18
15	01 54	20 32	09 27	11 38	09 21	12 28	13 57	15 47	18 36	11 18
16	01 30	18 40	09 37	11 13	09 38	12 33	13 59	15 46	18 36	11 17
17	01 07	15 40	09 46	10 47	09 55	12 37	14 00	15 45	18 35	11 17
18	00 43	11 47	09 52	10 21	10 12	12 41	14 02	15 44	18 35	11 17
19	00 S 19	07 18	09 56	09 55	10 28	12 46	14 04	15 44	18 35	11 17
20	00 N 04	02 N 32	09 58	09 29	10 45	12 50	14 06	15 43	18 34	11 16
21	00 28	02 S 16	09 58	09 02	11 01	12 55	14 09	15 42	18 34	11 16
22	00 52	06 51	09 55	08 35	11 18	12 59	14 11	15 41	18 34	11 16
23	01 15	11 02	09 51	08 08	11 34	13 03	14 13	15 40	18 33	11 15
24	01 39	14 39	09 45	07 41	11 50	13 08	14 15	15 39	18 33	11 15
25	02 03	17 33	09 36	07 13	12 06	13 12	14 17	15 39	18 33	11 15
26	02 26	19 40	09 26	06 45	12 21	13 16	14 19	15 38	18 32	11 14
27	02 50	20 54	09 15	06 17	12 37	13 21	14 21	15 37	18 32	11 14
28	03 13	21 12	09 01	05 49	12 53	13 25	14 23	15 36	18 32	11 14
29	03 36	20 34	08 46	05 21	13 08	13 30	14 25	15 35	18 31	11 14
30	04 00	18 58	08 29	04 53	13 23	13 34	14 27	15 35	18 31	11 13
31	04 N 23	16 S 29	08 S 10	04 S 24	13 N 39	13 N 39	14 N 29	15 S 34	18 S 31	11 S 13

ZODIAC SIGN ENTRIES

Date	h	m	Planets
02	13	14	☽ ♒
04	23	30	☽ ♓
07	06	54	☽ ♈
09	12	01	☽ ♉
11	15	46	☽ ♊
13	11	36	♀ ♓
13	18	51	☽ ♋
15	21	43	☽ ♌
18	00	48	☽ ♍
20	04	57	☽ ♎
20	07	35	☉ ♈
22	11	17	☽ ♏
23	01	25	♂ ♉
24	20	43	☽ ♐
27	08	51	☽ ♑
29	21	34	☽ ♒

LATITUDES

Date	Mercury ☿ ° ′	Venus ♀ ° ′	Mars ♂ ° ′	Jupiter ♃ ° ′	Saturn ♄ ° ′	Uranus ♅ ° ′	Neptune ♆ ° ′	Pluto ♇ ° ′
01	03 N 42	00 S 37	00 S 18	01 S 00	02 S 10	00 S 40	00 N 14	11 N 04
04	03 28	00 45	00 15	01 00	02 09	00 40	00 13	11 05
07	02 58	00 51	00 13	00 59	02 08	00 40	00 13	11 06
10	02 19	00 58	00 11	00 59	02 08	00 40	00 13	11 07
13	01 35	01 03	00 09	00 58	02 07	00 40	00 13	11 07
16	00 52	01 09	00 06	00 58	02 07	00 40	00 13	11 08
19	00 N 12	01 14	00 04	00 57	02 06	00 40	00 13	11 09
22	00 S 25	01 18	00 S 02	00 57	02 05	00 40	00 13	11 10
25	00 58	01 21	00 00	00 56	02 05	00 40	00 13	11 11
28	01 26	01 24	00 N 02	00 56	02 04	00 40	00 13	11 12
31	01 S 49	01 S 27	00 N 04	00 S 55	02 S 04	00 S 40	00 N 13	11 N 12

DATA

Julian Date	2451605
Delta T	+64 seconds
Ayanamsa	23° 51′ 20″
Synetic vernal point	05° ♓ 15′ 39″
True obliquity of ecliptic	23° 26′ 17″

MOON'S PHASES, APSIDES AND POSITIONS ☽

Date	h	m	Phase	Longitude ° ′	Eclipse Indicator
06	05	17	●	15 ♓ 57	
13	06	59	☽	23 ♊ 01	
20	04	44	○	29 ♍ 53	
28	00	21	☾	07 ♑ 38	

Day	h	m		
14	23	44	Perigee	
27	17	19	Apogee	
08	02	47	0N	
14	14	00	Max dec	21° N 06′
21	00	36	0S	
28	07	44	Max dec	21° S 13′

LONGITUDES

Date	Chiron ⚷ ° ′	Ceres ⚳ ° ′	Pallas ⚴ ° ′	Juno ⚵ ° ′	Vesta ⚶ ° ′	Black Moon Lilith ⚸ ° ′
01	16 ♐ 39	06 ♎ 25	28 ♋ 42	00 ♒ 09	05 ♑ 26	00 ♑ 08
11	17 ♐ 01	04 ♎ 29	28 ♋ 35	03 ♒ 40	09 ♑ 49	01 ♑ 15
21	17 ♐ 12	02 ♎ 14	29 ♋ 22	07 ♒ 03	13 ♑ 57	02 ♑ 22
31	17 ♐ 14	29 ♍ 58	00 ♌ 55	10 ♒ 19	17 ♑ 48	03 ♑ 28

All ephemeris data is given at 12.00 UT and the Moon's longitude is additionally given for 24.00 UT

ASPECTARIAN

h m	Aspects
01 Wednesday	
01 01	☽ ⚹ ☿
01 25	☽ ⊥ ♅
01 56	☽ △ ♄
02 49	☽ ⊻ ♇
04 38	☽ □ ♂
07 02	☽ ⊻ ♀
13 35	☽ ⊻ ♅
14 11	☿ ⊥ ♆
14 56	☽ ⊥ ♇
15 10	☉ ☌ ☿
16 16	☉ ⊥ ♆
02 Thursday	
04 46	☽ ∠ ☿
07 18	☽ ∠ ☉
08 58	☽ ∠ ♇
19 05	☽ Q ♂
19 13	☽ □ ♃
20 19	☉ ⚹ ♄
21 13	☽ ⊥ ☿
23 59	☽ ☌ ♆
03 Friday	
02 45	☽ ⊥ ☉
03 27	☉ □ ♇
04 57	☽ ∥ ♆
08 06	☽ ⊻ ☿
10 50	☿ ⊥ ♂
14 06	☽ □ ♄
14 34	☽ ⚹ ♇
15 32	☽ ⊻ ☉
20 08	☽ ⚹ ♂
04 Saturday	
00 39	♀ ☌ ♅
01 09	☽ ☌ ♅
01 12	☽ ☌ ♀
05 10	♀ ∥ ♅
06 36	☽ ∥ ♅
06 47	☽ ∥ ♀
06 55	☽ Q ♃
13 44	☽ Q ♇
14 16	☉ # ♂
23 48	☽ # ♄
05 Sunday	
00 55	☽ Q ♄
02 41	☽ ∠ ♂
06 06	☽ ⚹ ♃
09 53	☽ ⊻ ♆
11 45	☽ # ♃
13 19	☽ ☌ ☿
14 22	☽ ∥ ♇
15 12	☉ ∥ ☿
20 35	☽ ⊥ ♂
21 06	☽ ⊥ ♆
23 32	☽ ⚹ ♄
23 37	☽ □ ♇
06 Monday	
05 17	☽ ☌ ☉
05 57	♂ Q ♆
08 21	☽ ⊻ ♂
09 53	☽ ⊻ ♅
10 24	☽ ∠ ♃
12 14	♄ ⊼ ♇
13 44	☽ ∠ ♆
15 35	☽ ⊻ ♀
16 08	☽ # ♂
18 37	☽ ∥ ☿
20 32	♀ Q ♃
20 53	☽ ⊥ ♅
07 Tuesday	
00 20	☽ ∥ ☉
03 03	☽ ⊥ ♃
03 10	☽ ∠ ♄
03 36	☽ ⊥ ♀
13 06	♂ ⚹ ♅
13 12	☽ ∠ ♅
14 01	☽ ⊻ ♃
15 01	☿ ⊻ ♆
16 49	☽ ⊻ ☿
16 55	☽ ⚹ ♆
19 28	☽ ⊥ ♄
21 25	☽ ∠ ♀
08 Wednesday	
02 57	☽ ⊥ ☿
05 57	☽ △ ♇
06 13	☽ ⊻ ♄
14 17	☽ Q ♆
15 39	☽ ⊻ ☉
15 58	☽ ⚹ ♅
17 25	☽ ☌ ♂
18 06	☽ ∠ ☿
20 19	☉ ⊻ ♅
09 Thursday	
00 26	☿ ∠ ♂
00 50	☽ # ☉
02 34	☽ ⚹ ♀
03 01	☽ ⊥ ☉
08 20	☽ bQ ♇
08 23	♀ Q ♇
11 05	☿ ⚹ ♃
13 04	☽ Q ♅
15 31	☉ ∠ ♃
17 03	☽ # ☿
17 16	☽ ∥ ♂
19 13	☽ ⚹ ☿
19 41	☽ ☌ ♃
19 56	☽ ∠ ☉
21 49	☽ □ ♆
10 Friday	
00 00	☽ ± ♇
01 31	☽ Q ♀
01 35	☿ # ♂
04 23	♀ # ♄
10 22	☽ ⊼ ♇
10 58	☽ ☌ ♄
12 46	☽ # ♇
15 15	☽ Q ☿
17 24	☽ ∥ ♃
20 20	☽ □ ♅
22 48	☉ ∠ ♆
23 50	☽ ⚹ ☉
11 Saturday	
00 29	☽ ⊻ ♂
01 08	☽ # ♀
03 23	☽ ∥ ♄
11 22	☽ ⊥ ♂
11 31	☽ □ ♀
16 49	☽ # ♅
21 18	☽ □ ☿
21 59	☽ Q ☉
12 Sunday	
00 03	☽ ⊻ ♃
01 31	☽ △ ♆
03 37	☽ ∠ ♂
09 37	☉ ⊻ ♂
10 26	☽ ⊥ ♃
13 47	☽ ☍ ♇
14 43	☽ ⊻ ♄
15 19	☽ # ♆
23 48	☽ △ ♅
13 Monday	
01 00	☽ ⊥ ♄
01 59	☽ ∠ ♃
03 07	☽ bQ ♆
06 34	☽ ⚹ ♂
06 59	☽ □ ☉
16 24	☽ ∠ ♄
18 14	♀ # ♃
18 26	☽ ± ♆
19 33	☽ △ ♀
23 38	☽ △ ☿
14 Tuesday	
01 22	☽ bQ ♅
03 51	☽ ⚹ ♃
04 05	☽ Q ♂
04 38	☽ ⊼ ♆
16 41	☽ ± ♅
16 44	☽ ⊼ ♇
18 01	☽ ⚹ ♄
19 54	♀ Q ♄
20 40	☿ St D
23 25	☽ bQ ♀
15 Wednesday	
00 29	☽ Q ♃
01 00	☽ bQ ☿
02 53	☽ ⊼ ♅
02 54	☽ ± ♇
04 22	☉ ⊥ ♅
11 50	♇ St R
12 16	☽ □ ♂
13 43	☽ △ ☉
14 30	☽ Q ♄
16 08	☽ ± ♀
16 19	☽ ± ☿
18 10	☽ bQ ♇
18 23	☿ ☌ ♀
16 Thursday	
02 34	☽ ⊼ ☿
03 17	☽ ⊼ ♀
07 19	♀ ∥ ♇
07 28	☽ □ ♃
07 36	☽ ☍ ♆
09 17	☿ # ♂
12 45	☽ # ♆
17 06	☽ bQ ☉
17 14	♃ □ ♆
19 37	☽ △ ♇
21 16	☽ □ ♄
17 Friday	
06 00	☽ ☍ ♅
09 35	☽ ± ☉
11 25	☽ # ♅
18 07	☽ △ ♂
20 37	☽ ⊼ ☉
22 45	☽ ∥ ♄
18 Saturday	
05 43	♀ ⊻ ♆
06 33	☽ ☍ ☿
06 53	☽ ∥ ♃
10 55	☽ ⊼ ♆
11 24	☽ ☍ ♀
11 28	☽ △ ♃
12 20	♀ ⚹ ♃
14 50	☽ # ♇
17 17	♀ # ♂
20 17	☽ ∥ ♂
20 49	☽ # ♀
21 17	☽ ± ♆
21 21	☽ bQ ♂
22 23	☽ # ☿
23 02	☽ □ ♇
19 Sunday	
01 06	☽ △ ♄
09 49	☽ ⊼ ♅
10 33	☉ ∠ ♄
11 04	☿ ∥ ♀
12 58	☽ bQ ♆
13 54	☽ ± ♂
13 54	☽ bQ ♃
20 21	☽ ± ♅
20 Monday	
01 02	☽ ⊼ ♂
03 32	☽ bQ ♄
04 44	☽ ☍ ☉
04 49	♂ bQ ♇
05 59	☽ ± ♃
06 32	☽ Q ♇
12 19	☽ bQ ♅
12 31	☽ ⊼ ☿
15 29	☽ △ ♆
16 49	☽ ⊼ ♃
19 40	☽ ± ♄
21 13	☽ ⊼ ♀
23 19	☽ ∥ ☉
23 41	☽ ± ☿
21 Tuesday	
02 07	☉ ⊥ ♃
02 08	☽ # ☉
03 58	☽ ⚹ ♇
06 33	☽ ⊼ ♄
09 08	☽ ± ♀
15 25	☽ △ ♅
15 59	♃ ± ♇
16 42	☽ bQ ☿
22 Wednesday	
03 18	☽ bQ ♀
07 22	☽ ∠ ♇
09 14	♂ # ♇
10 26	☽ ☍ ♂
15 37	☽ ⊼ ☉
20 42	☽ ∥ ♀
21 54	☽ △ ☿
22 30	☽ □ ♆
23 Thursday	
00 11	☽ ⊥ ♇
00 45	☽ ☍ ♃
03 50	☽ ± ☉
04 43	♀ ⊥ ♆
05 03	☽ ∥ ☿
09 26	☿ ⊻ ♆
10 26	☽ △ ♀
11 33	☽ ⊻ ♇
13 23	☽ ∥ ♇
14 47	☽ ☍ ♄
15 32	☽ # ♂
21 54	♀ □ ♇
22 30	☽ bQ ☉
23 53	☽ □ ♅
24 Friday	
01 11	☽ # ♃
09 02	☽ Q ♆
09 06	☽ # ♄
16 42	☉ ∠ ♅
19 38	☽ ∥ ♅
21 40	♂ Q ♅
23 29	☽ ⊼ ♂
25 Saturday	
06 24	☽ △ ☉
08 40	☽ ⚹ ♆
10 31	♀ ⚹ ♄
11 18	☽ Q ♅
11 45	☽ □ ☿
12 00	☽ ⊼ ♃
12 08	☽ ± ♂
16 33	☿ ⚹ ♃
22 01	☽ ∥ ♆
22 16	☽ ☌ ♇
26 Sunday	
00 12	☽ ± ♃
02 13	☽ ⊼ ♄
03 52	☽ □ ♀
07 19	☽ bQ ♂
10 37	☉ ⚹ ♆
11 26	☽ ⚹ ♅
14 24	☽ ± ♄
14 47	☽ ∠ ♆
18 41	☽ bQ ♃
27 Monday	
02 47	☽ Q ☿
08 48	☽ bQ ♄
09 02	☽ ⊥ ♆
15 41	☽ △ ♂
17 57	☽ ∠ ♅
20 32	☽ Q ♀
21 15	☽ ⊻ ♆
28 Tuesday	
00 21	☽ □ ☉
01 42	☽ △ ♃
05 16	☽ ⚹ ☿
10 58	☽ ⊻ ♇
12 18	☽ ⊥ ♅
15 34	☽ △ ♄
20 49	☉ ⊻ ♃
23 07	☽ ⊥ ♇
23 43	☽ ⚹ ♀
29 Wednesday	
00 30	☽ ⊻ ♅
07 24	♀ ⊻ ♅
08 56	♀ ∠ ♂
14 24	☽ ∠ ☿
15 15	☉ ⊥ ♄
16 05	☽ Q ☉
17 15	☽ ∠ ♇
30 Thursday	
07 16	☿ ⊥ ♆
08 13	☽ □ ♂
09 18	☽ ∠ ♀
09 51	☽ ☌ ♆
10 06	☽ ⊥ ☿
14 51	♀ ∠ ♆
15 13	☽ □ ♃
17 09	☽ ∥ ♆
18 07	☽ ⚹ ☉
22 27	☿ □ ♇
23 02	☽ ⚹ ♇
23 06	☽ ⊻ ☿
31 Friday	
04 05	☽ □ ♄
05 02	☽ ⊥ ♀
11 51	♂ ∥ ♃
12 10	♂ □ ♆
12 19	☽ ☌ ♅
12 35	☉ # ♀
17 58	☽ ⊻ ♀
19 20	☽ ∥ ♅
21 39	☽ Q ♂
22 20	☽ Q ♇

APRIL 2000

LONGITUDES

Date	Sidereal time h m s	Sun ☉	Moon ☽	Moon ☽ 24.00	Mercury ☿	Venus ♀	Mars ♂	Jupiter ♃	Saturn ♄	Uranus ♅	Neptune ♆	Pluto ⯓
01	00 40 37	12 ♈ 03 40	02 ♓ 00 11	08 ♓ 23 40	14 ♓ 35	23 ♓ 29	06 ♉ 54	09 ♉ 17	15 ♉ 33	19 ♒ 41	06 ♒ 12	12 ♐ 49
02	00 44 34	13 02 51	14 ♓ 52 34	21 ♓ 27 00	15 45	24 43	07 38	09 31	15 40	19 43	06 13	12 R 49
03	00 48 30	14 02 01	28 ♓ 06 55	04 ♈ 52 11	16 57	25 57	08 22	09 44	15 47	19 46	06 14	12 48
04	00 52 27	15 01 09	11 ♈ 42 32	18 ♈ 37 34	18 11	27 11	09 05	09 58	15 54	19 48	06 16	12 47
05	00 56 23	16 00 15	25 ♈ 36 46	02 ♉ 39 33	19 28	28 26	09 49	10 12	16 01	19 50	06 17	12 47
06	01 00 20	16 59 19	09 ♉ 45 14	16 ♉ 53 08	20 46	29 ♓ 40	10 32	10 26	16 08	19 52	06 18	12 46
07	01 04 16	17 58 20	24 ♉ 02 33	01 ♊ 12 48	22 06	00 ♈ 54	11 16	10 39	16 15	19 55	06 19	12 45
08	01 08 13	18 57 20	08 ♊ 23 14	15 ♊ 33 20	23 28	02 08	11 59	10 53	16 23	19 57	06 20	12 45
09	01 12 10	19 56 17	22 ♊ 42 35	29 ♊ 50 37	24 52	03 22	12 42	11 07	16 30	19 59	06 21	12 44
10	01 16 06	20 55 12	06 ♋ 57 08	14 ♋ 01 55	26 17	04 36	13 25	11 21	16 37	20 01	06 22	12 43
11	01 20 03	21 54 05	21 ♋ 04 50	28 ♋ 05 48	27 45	05 50	14 09	11 35	16 44	20 03	06 22	12 42
12	01 23 59	22 52 55	05 ♌ 04 46	12 ♌ 01 43	29 ♓ 14	07 04	14 52	11 49	16 52	20 05	06 23	12 41
13	01 27 56	23 51 43	18 ♌ 56 38	25 ♌ 49 29	00 ♈ 47	08 18	15 35	12 03	16 59	20 07	06 24	12 40
14	01 31 52	24 50 29	02 ♍ 40 14	09 ♍ 28 48	02 17	09 32	16 18	12 17	17 06	20 09	06 25	12 39
15	01 35 49	25 49 12	16 ♍ 15 05	22 ♍ 58 57	03 51	10 46	17 01	12 31	17 14	20 11	06 26	12 39
16	01 39 45	26 47 54	29 ♍ 40 13	06 ♎ 18 44	05 27	12 00	17 44	12 45	17 21	20 13	06 26	12 38
17	01 43 42	27 46 33	12 ♎ 54 17	19 ♎ 26 40	07 04	13 14	18 27	12 59	17 28	20 15	06 27	12 37
18	01 47 39	28 45 10	25 ♎ 55 44	02 ♏ 21 19	08 43	14 27	19 10	13 13	17 36	20 17	06 27	12 36
19	01 51 35	29 ♈ 43 45	08 ♏ 43 19	15 ♏ 01 41	10 23	15 41	19 53	13 27	17 43	20 18	06 28	12 34
20	01 55 32	00 ♉ 42 18	21 ♏ 16 25	27 ♏ 27 36	12 06	16 55	20 35	13 41	17 51	20 20	06 29	12 33
21	01 59 28	01 40 50	03 ♐ 35 24	09 ♐ 40 01	13 50	18 09	21 18	13 55	17 58	20 22	06 30	12 32
22	02 03 25	02 39 20	15 ♐ 41 45	21 ♐ 40 57	15 35	19 23	22 01	14 09	18 06	20 23	06 30	12 31
23	02 07 21	03 37 48	27 ♐ 38 04	03 ♑ 33 35	17 23	20 37	22 43	14 23	18 13	20 25	06 31	12 30
24	02 11 18	04 36 15	09 ♑ 28 02	15 ♑ 22 00	19 11	21 51	23 26	14 38	18 21	20 26	06 31	12 29
25	02 15 14	05 34 40	21 ♑ 16 07	27 ♑ 11 02	21 02	23 05	24 09	14 52	18 29	20 28	06 32	12 28
26	02 19 11	06 33 03	03 ♒ 07 27	09 ♒ 06 02	22 54	24 19	24 51	15 06	18 36	20 29	06 32	12 26
27	02 23 08	07 31 25	15 ♒ 07 29	21 ♒ 12 29	24 48	25 33	25 33	15 20	18 44	20 31	06 32	12 25
28	02 27 04	08 29 45	27 ♒ 21 41	03 ♓ 35 42	26 44	26 47	26 15	15 35	18 51	20 32	06 33	12 24
29	02 31 01	09 28 03	09 ♓ 55 04	16 ♓ 20 17	28 ♈ 41	28 01	26 58	15 49	18 59	20 33	06 33	12 23
30	02 34 57	10 ♉ 26 20	22 ♓ 51 43	29 ♓ 29 36	00 ♉ 40	29 ♈ 14	27 ♉ 41	16 ♉ 03	19 ♉ 07	20 ♒ 34	06 ♒ 33	12 ♐ 21

DECLINATIONS

Date	Moon True ☊	Moon Mean ☊	Moon ☽ Latitude	Sun ☉	Moon ☽	Mercury ☿	Venus ♀	Mars ♂	Jupiter ♃	Saturn ♄	Uranus ♅	Neptune ♆	Pluto ⯓
01	01 ♌ 10	00 ♌ 14	02 S 36	04 N 46	13 S 11	07 S 50	03 S 55	13 N 54	13 N 43	14 N 31	15 S 33	18 S 30	11 S 13
02	01 R 04	00 10	03 30	05 09	09 11	07 29	03 27	14 08	13 47	14 34	15 32	18 30	11 12
03	00 55	00 07	04 14	05 32	04 S 38	07 06	02 58	14 23	13 52	14 36	15 32	18 30	11 12
04	00 44	00 04	04 45	05 55	00 N 15	06 42	02 29	14 38	13 56	14 38	15 31	18 30	11 12
05	00 32	00 ♌ 01	05 00	06 18	05 14	06 16	02 00	14 52	14 01	14 40	15 30	18 29	11 11
06	00 20	29 ♋ 58	04 57	06 40	10 02	05 49	01 31	15 07	14 05	14 42	15 30	18 29	11 11
07	00 10	29 54	04 36	07 03	14 19	05 20	01 01	15 21	14 09	14 44	15 29	18 29	11 10
08	00 03	29 51	03 57	07 25	17 48	04 51	00 32	15 35	14 14	14 46	15 28	18 29	11 10
09	29 ♋ 58	29 48	03 03	07 48	20 11	04 20	00 S 03	15 49	14 18	14 49	15 28	18 29	11 10
10	29 56	29 45	01 58	08 10	21 17	03 47	00 N 27	16 02	14 23	14 51	15 27	18 28	11 10
11	29 D 56	29 42	00 S 47	08 32	21 01	03 14	00 56	16 16	14 27	14 53	15 26	18 28	11 09
12	29 R 56	29 39	00 N 27	08 54	19 26	02 40	01 25	16 29	14 31	14 55	15 26	18 28	11 09
13	29 55	29 35	01 39	09 16	16 43	02 04	01 54	16 42	14 36	14 57	15 25	18 27	11 09
14	29 52	29 32	02 44	09 37	13 04	01 27	02 24	16 55	14 40	14 59	15 25	18 27	11 09
15	29 46	29 29	03 39	09 59	08 47	00 49	02 53	17 08	14 44	15 02	15 24	18 27	11 08
16	29 38	29 26	04 21	10 20	04 N 07	00 S 10	03 22	17 21	14 49	15 04	15 23	18 27	11 08
17	29 27	29 23	04 48	10 41	00 S 40	00 N 30	03 51	17 34	14 53	15 06	15 23	18 27	11 08
18	29 14	29 20	05 00	11 02	05 22	01 11	04 20	17 46	14 57	15 08	15 22	18 27	11 07
19	29 00	29 16	04 56	11 23	09 44	01 53	04 49	17 58	15 02	15 10	15 22	18 26	11 07
20	28 48	29 13	04 38	11 43	13 36	02 35	05 18	18 10	15 06	15 12	15 21	18 26	11 07
21	28 37	29 10	04 07	12 03	16 50	03 19	05 47	18 22	15 10	15 15	15 21	18 26	11 06
22	28 28	29 07	03 25	12 24	19 16	04 04	06 16	18 34	15 15	15 17	15 20	18 26	11 06
23	28 23	29 04	02 35	12 44	20 50	04 49	06 44	18 45	15 19	15 19	15 20	18 26	11 06
24	28 20	29 00	01 38	13 03	21 28	05 35	07 12	18 57	15 23	15 21	15 19	18 26	11 06
25	28 19	28 57	00 N 37	13 23	21 09	06 22	07 41	19 08	15 27	15 23	15 19	18 26	11 05
26	28 D 18	28 54	00 S 26	13 42	19 52	07 10	08 09	19 19	15 32	15 25	15 19	18 25	11 05
27	28 R 18	28 51	01 28	14 01	17 42	07 58	08 37	19 29	15 36	15 28	15 18	18 25	11 05
28	28 17	28 48	02 28	14 20	14 42	08 46	09 04	19 40	15 40	15 30	15 18	18 25	11 04
29	28 14	28 45	03 22	14 39	10 58	09 35	09 32	19 51	15 44	15 32	15 17	18 25	11 04
30	28 ♋ 08	28 ♋ 41	04 S 07	14 N 57	06 S 37	10 N 24	09 N 59	20 N 01	15 N 48	15 N 34	15 S 17	18 S 25	11 S 04

ZODIAC SIGN ENTRIES

Date	h	m	Planets
01	08	12	☽ ♓
03	15	22	☽ ♈
05	19	29	☽ ♉
06	18	37	♀ ♈
07	21	58	☽ ♊
10	00	16	☽ ♋
12	03	16	☽ ♌
13	00	17	☿ ♈
14	07	19	☽ ♍
16	12	36	☽ ♎
18	19	35	☽ ♏
19	18	40	☉ ♉
21	04	58	☽ ♐
23	16	47	☽ ♑
26	05	42	☽ ♒
28	17	06	☽ ♓
30	03	53	☿ ♉

LATITUDES

Date	Mercury ☿	Venus ♀	Mars ♂	Jupiter ♃	Saturn ♄	Uranus ♅	Neptune ♆	Pluto ⯓
01	01 S 55	01 S 27	00 N 05	00 S 55	02 S 04	00 S 40	00 N 13	11 N 13
04	02 12	01 29	00 07	00 55	02 03	00 40	00 13	11 13
07	02 24	01 30	00 09	00 54	02 03	00 40	00 13	11 14
10	02 31	01 30	00 11	00 54	02 02	00 41	00 13	11 14
13	02 34	01 30	00 13	00 54	02 02	00 41	00 13	11 15
16	02 32	01 29	00 14	00 54	02 02	00 41	00 13	11 16
19	02 26	01 28	00 16	00 53	02 01	00 41	00 13	11 16
22	02 15	01 26	00 18	00 53	02 01	00 41	00 13	11 17
25	01 59	01 24	00 20	00 53	02 01	00 41	00 13	11 17
28	01 39	01 20	00 22	00 52	02 01	00 41	00 13	11 18
31	01 S 15	01 S 17	00 N 23	00 S 52	02 S 00	00 S 41	00 N 13	11 N 18

DATA

Julian Date	2451636
Delta T	+64 seconds
Ayanamsa	23° 51' 23"
Synetic vernal point	05° ♓ 15' 37"
True obliquity of ecliptic	23° 26' 17"

MOON'S PHASES, APSIDES AND POSITIONS ☽

Date	h	m	Phase	Longitude	Eclipse Indicator
04	18	12	●	15 ♈ 16	
11	13	30	☽	21 ♋ 58	
18	17	42	○	28 ♎ 59	
26	19	30	☾	06 ♒ 51	

Day	h	m		
08	22	00	Perigee	
24	12	20	Apogee	
04	10	46	ON	
10	19	17	Max dec	21° N 21'
17	08	37	OS	
24	15	51	Max dec	21° S 29'

LONGITUDES

Date	Chiron ⚷	Ceres ⚳	Pallas ⚴	Juno ⚵	Vesta ⚶	Black Moon Lilith ⚸
01	17 ♐ 13	29 ♍ 45	01 ♌ 07	10 ♒ 39	18 ♑ 10	03 ♑ 35
11	17 ♐ 04	27 ♍ 48	03 ♌ 20	13 ♒ 44	21 ♑ 38	04 ♑ 58
21	16 ♐ 45	26 ♍ 21	06 ♌ 03	16 ♒ 37	24 ♑ 41	05 ♑ 48
31	16 ♐ 18	25 ♍ 34	09 ♌ 08	19 ♒ 15	27 ♑ 16	06 ♑ 54

ASPECTARIAN

h m	Aspects
01 Saturday	
01 48	☽ ∠ ☉
02 53	☽ Q ♃
03 01	☽ # ♄
07 36	☽ # ♂
08 33	☽ # ♃
09 16	♂ ± ⯓
14 58	☽ Q ♄
19 55	☽ ⊻ ♆
20 17	☽ ⊥ ☉
21 46	☽ ⚹ ♂
02 Sunday	
00 19	☽ ∥ ⯓
01 54	☽ ⚹ ♃
06 19	☉ △ ⯓
07 03	♀ ∠ ♃
07 06	☽ ⊥ ♆
08 12	☽ □ ⯓
08 21	☽ ⊻ ☉
10 19	☿ ⚹ ♄
13 29	☽ ⚹ ♄
13 46	☽ ☌ ☿
20 53	☽ ⊻ ♅
22 05	☽ ∥ ☿
23 36	☽ ∠ ♆
03 Monday	
02 58	☽ ∠ ♂
05 50	☽ ∠ ♃
07 44	☽ ☌ ♀
07 46	☽ # ☉
07 46	☽ ⊥ ♅
08 02	♀ ⊥ ♅
16 49	☽ ∠ ♄
19 59	☽ ⊥ ♂
21 15	☽ ∥ ♀
22 11	☽ ⊥ ♃
23 50	☽ ∠ ♅
04 Tuesday	
02 26	☽ ⚹ ♆
07 09	☽ ⊻ ♂
08 49	☽ ⊥ ♄
08 54	☽ ⊻ ♃
12 06	♂ ∥ ♄
13 53	☽ △ ⯓
18 12	☽ ☌ ☉
19 21	☽ ⊻ ♄
21 45	☽ # ♀
23 23	☽ Q ♆
05 Wednesday	
00 22	☽ ⊻ ☿
02 04	☽ ⚹ ♅
11 07	☉ # ☿
11 43	☽ ⊥ ☿
12 30	☉ ⊻ ♄
15 42	☽ ♇ ⯓
16 36	☽ # ☿
17 16	☽ ⊻ ♀
17 36	☽ ∥ ☉
19 13	☿ ⊻ ♅
22 38	☽ Q ♅
06 Thursday	
04 25	☽ ⊥ ♀
04 34	☽ ∠ ☿
06 09	☽ □ ♆
06 41	♂ ☌ ♃
06 58	☽ ± ⯓
13 09	☽ ☌ ♃
13 23	☽ ☌ ♂
17 04	☽ ⊼ ⯓
18 06	☽ # ⯓
21 02	☽ ∠ ♀
21 44	☿ ∠ ♆
22 50	☽ ☌ ♄
07 Friday	
01 04	☽ ⊻ ☉
05 04	☽ □ ♅
08 24	☽ ⚹ ☿
10 58	☽ ∥ ♃
11 52	☽ ⊥ ☉
14 35	☽ ∥ ♄
18 51	☽ ∥ ♂
19 17	☽ # ♅
19 51	♀ ∠ ♄
20 25	☉ Q ♆
08 Saturday	
00 33	☽ ⚹ ♀
01 21	♂ # ♅
04 02	☽ ∠ ☉
06 36	☽ Q ☿
08 33	☽ △ ♆
16 15	☽ ⊻ ♃
17 42	☽ # ♆
18 20	☽ ⊻ ♂
19 17	☽ ☍ ⯓
22 31	☽ Q ♀
09 Sunday	
01 29	☽ ⊻ ♄
02 28	☽ ⊥ ♃
04 55	☽ ⊥ ♂
07 00	☽ ⚹ ☉
07 25	☽ △ ♅
09 42	☽ ♇ ♆
11 38	☽ ⊥ ♄
12 54	♂ ⊼ ⯓
13 09	☉ ⚹ ♅
16 01	☽ □ ☿
17 49	☽ ∠ ♃
20 51	☽ ∠ ♂
10 Monday	
00 51	☽ ± ♆
02 55	☽ ∠ ♄
04 41	☽ Q ☉
07 22	☿ ⊥ ♅
07 38	☽ □ ♀
08 43	☽ ♇ ♅
11 00	☽ ⊼ ♆
13 07	☿ ∠ ♃
19 34	☽ ⚹ ♃
20 29	♀ ∠ ♅
21 45	☽ ⊼ ⯓
23 33	☽ ⚹ ♂
11 Tuesday	
00 00	☽ ± ♅
04 32	☽ ⚹ ♄
06 00	♀ ⊥ ♃
07 57	☽ ± ⯓
10 15	☽ ⊼ ♅
13 30	☽ □ ☉
16 20	☽ Q ♃
21 07	☽ Q ♂
22 45	♀ ⚹ ♆
23 19	☽ ♇ ⯓
12 Wednesday	
00 45	☽ △ ☿
01 13	☽ Q ♄
14 15	☽ ☍ ♆
15 45	☽ △ ♀
21 43	☽ # ♆
23 49	☽ □ ♃
13 Thursday	
01 08	☽ △ ⯓
05 45	☽ ♇ ☿
05 50	☽ □ ♂
07 16	☿ ∠ ♂
08 33	☽ □ ♄
12 01	☽ ∥ ♂
14 03	☽ ☍ ♅
15 26	☿ # ♀
20 20	☽ ♇ ♀
21 09	☽ # ♅
21 14	☽ △ ☉
23 23	☽ ± ☿
14 Friday	
00 05	☽ ∥ ♄
02 16	☽ ∥ ♃
09 01	☿ ∠ ♄
11 14	☽ ⊼ ☿
13 39	☽ ± ♀
18 36	☽ ⊼ ♆
23 09	☽ # ⯓
15 Saturday	
01 17	☽ ⊼ ♀
01 37	☽ ♇ ☉
05 13	☽ ± ♆
05 15	☽ △ ♃
05 36	☽ □ ⯓
06 04	☽ ∥ ☉
13 26	☽ △ ♂
13 45	☽ △ ♄
18 51	☽ ± ☉
19 01	☽ ⊼ ♅
20 27	♂ ☌ ♄
21 14	☽ ♇ ♆
22 04	♀ ⊥ ♄
23 56	♀ ⊥ ♂
16 Sunday	
00 47	♃ ⊼ ⯓
05 47	☽ ± ♅
06 26	☽ ⊼ ☉
08 28	☽ ♇ ♃
08 32	☿ ∠ ♅
13 43	☽ Q ⯓
15 25	☽ ∥ ♀
16 53	☽ ♇ ♄
17 50	☽ ♇ ♂
22 02	☽ ♇ ♅
23 52	☽ ☍ ☿
17 Monday	
00 10	♀ △ ⯓
00 15	☽ △ ♆
01 01	☽ ± ♃
02 55	☿ ⚹ ♆
06 02	♀ ⊻ ♃
06 51	☽ ∥ ☿
07 59	☉ ♇ ⯓
09 22	☽ ± ♄
10 30	☿ ⊥ ♃
10 58	☽ # ☿
11 07	☽ ± ♂
11 28	☽ ⚹ ⯓
12 08	☽ ⊼ ♃
12 39	☽ ☍ ♀
20 27	☽ ⊼ ♄
22 45	☽ ⊼ ♂
18 Tuesday	
01 30	☽ △ ♅
06 03	☽ # ♀
15 05	☽ ∠ ⯓
17 42	☽ ☍ ☉
18 14	☉ # ⯓
19 Wednesday	
07 45	☽ □ ♆
07 57	☽ ⊥ ⯓
15 39	☽ ⊼ ☿
19 18	☽ ⊻ ⯓
20 11	☽ ∥ ⯓
21 09	☽ ☍ ♃
22 42	☽ # ☉
20 Thursday	
02 43	☽ ⊼ ♀
02 59	♂ □ ♅
04 54	☽ ± ☿
05 20	☽ ☍ ♄
08 14	☿ ⊥ ♄
10 11	☽ □ ♅
10 36	☽ ☍ ♂
15 32	☽ ± ♀
18 13	☽ Q ♆
18 21	☿ △ ⯓
22 42	☽ # ♃
23 23	☽ # ♄
21 Friday	
00 19	☽ ∥ ♅
01 08	☽ ♇ ☿
07 56	☽ ⊼ ☉
08 02	♀ ⊻ ♄
11 03	☽ ♇ ♀
13 26	☿ ⊻ ♃
17 44	☽ ⚹ ♆
18 40	♀ Q ♆
20 06	♂ # ♆
20 46	☽ ± ☉
21 26	☽ Q ♅
22 Saturday	
02 48	☽ ∥ ♆
03 25	☽ # ♂
05 11	☉ Q ♅
05 41	☽ ☌ ⯓
08 52	☽ ⊼ ♃
11 45	☽ △ ☿
16 16	☽ ♇ ☉
16 51	☽ ⊼ ♄
20 14	☽ △ ♀
21 07	☽ ± ♃
21 25	☽ ⚹ ♅
21 33	☿ ⊥ ♂
23 39	☽ ∠ ♆
23 Sunday	
01 28	☽ ⊼ ♂
05 02	☽ ± ♄
07 54	♀ ⚹ ♅
13 22	♃ ∥ ♄
14 20	☽ ± ♂
15 37	☽ ♇ ♃
17 07	♃ # ♅
17 49	☽ ⊥ ♆
20 08	♄ # ♅
23 26	☽ ♇ ♄
24 Monday	
00 06	☿ ⊻ ♄
01 14	☽ △ ☉
03 08	☿ Q ♆
03 48	☽ ∠ ♅
06 00	☽ ⊻ ♆
09 46	☽ ♇ ♂
18 07	☽ ⊻ ⯓
22 08	☽ ⊥ ♅
22 43	☽ △ ♃
25 Tuesday	
04 29	☿ ⚹ ♅
06 16	☽ △ ♄
06 18	☽ ⊥ ⯓
10 21	☽ ⊻ ♅
11 26	☽ □ ☿
16 07	☽ □ ♀
18 12	☽ △ ♂
26 Wednesday	
00 32	☽ ∠ ⯓
09 21	☉ ± ⯓
11 35	☉ □ ♆
18 37	☽ # ♂
18 52	☽ ☌ ♆
19 30	☽ □ ☉
27 Thursday	
05 00	☽ ∥ ♆
06 32	☽ Q ☿
06 38	☽ ⚹ ⯓
08 31	☽ Q ♀
12 26	☽ □ ♃
12 33	♀ ⊻ ♂
19 12	☽ □ ♄
22 39	☽ ☌ ♅
28 Friday	
02 50	☿ ⊻ ♂
05 03	☽ # ♃
06 15	☽ Q ⯓
06 15	☽ # ♄
07 40	☽ ∥ ♅
09 45	☽ □ ♂
10 11	☽ Q ☉
10 33	☽ ⚹ ☿
10 45	☽ ⚹ ♀
13 27	☿ ☌ ♀
14 20	☽ # ☉
20 07	☿ ♇ ⯓
23 56	♀ ♇ ⯓
29 Saturday	
00 12	☽ Q ♃
05 38	☽ ⊻ ♆
06 24	☽ Q ♄
08 17	☿ ∥ ♀
11 05	☽ ⚹ ☉
11 22	☽ ∥ ⯓
16 37	☽ □ ⯓
16 57	☽ ⊥ ♆
18 25	☽ ∠ ♀
18 38	☽ # ☿
19 26	☽ # ♀
20 21	☽ ∠ ☿
22 00	☽ Q ♂
23 14	☽ ⚹ ♃
30 Sunday	
05 04	☽ ⚹ ♄
07 48	☽ ⊻ ♅
09 37	☽ ∠ ♆
12 46	☽ ⊥ ♀
15 53	☽ ⊥ ☿
17 03	☽ ∠ ☉
18 45	☽ ⊥ ♅
21 13	☽ ⚹ ♂

All ephemeris data is given at 12.00 UT and the Moon's longitude is additionally given for 24.00 UT

MAY 2000

LONGITUDES

Date	Sidereal time h m s	Sun ☉ ° ' "	Moon ☽ ° ' "	Moon ☽ 24.00 ° ' "	Mercury ☿ ° '	Venus ♀ ° '	Mars ♂ ° '	Jupiter ♃ ° '	Saturn ♄ ° '	Uranus ♅ ° '	Neptune ♆ ° '	Pluto ⯓ ° '
01	02 38 54	11 ♉ 24 36	06 ♈ 14 04	13 ♈ 05 04	02 ♉ 41	00 ♉ 28	28 ♉ 23	16 ♉ 17	19 ♉ 14	20 ♒ 36	06 ♒ 34	12 ♐ 20
02	02 42 50	12 22 49	20 ♈ 02 23	27 ♈ 05 39	04 43	01 42	29 05	16 32	19 22	20 37	06 34	12 R 19
03	02 46 47	13 21 02	04 ♉ 14 18	11 ♉ 27 38	06 47	02 56	29 ♉ 47	16 46	19 30	20 38	06 34	12 17
04	02 50 43	14 19 12	18 ♉ 44 49	26 ♉ 04 55	08 52	04 10	00 ♊ 29	17 00	19 38	20 39	06 34	12 16
05	02 54 40	15 17 21	03 ♊ 26 56	10 ♊ 49 52	10 58	05 24	01 11	17 14	19 45	20 40	06 34	12 15
06	02 58 37	16 15 29	18 ♊ 12 44	25 ♊ 34 38	13 06	06 38	01 53	17 29	19 53	20 41	06 34	12 13
07	03 02 33	17 13 34	02 ♋ 54 46	10 ♋ 12 26	15 15	07 51	02 35	17 43	20 01	20 42	06 34	12 12
08	03 06 30	18 11 38	17 ♋ 27 06	24 ♋ 38 19	17 24	09 05	03 17	17 57	20 08	20 43	06 34	12 10
09	03 10 26	19 09 39	01 ♌ 45 50	08 ♌ 49 27	19 34	10 19	03 59	18 12	20 16	20 43	06 R 34	12 09
10	03 14 23	20 07 39	15 ♌ 49 07	22 ♌ 44 50	21 45	11 33	04 41	18 26	20 24	20 44	06 34	12 07
11	03 18 19	21 05 37	29 ♌ 36 40	06 ♍ 24 43	23 56	12 47	05 23	18 40	20 32	20 45	06 34	12 06
12	03 22 16	22 03 32	13 ♍ 09 08	19 ♍ 50 01	26 06	14 00	06 05	18 54	20 39	20 45	06 34	12 04
13	03 26 12	23 01 26	26 ♍ 27 32	03 ♎ 01 47	28 ♉ 16	15 14	06 46	19 09	20 47	20 46	06 34	12 03
14	03 30 09	23 59 19	09 ♎ 32 53	16 ♎ 00 54	00 ♊ 26	16 28	07 28	19 23	20 55	20 47	06 34	12 01
15	03 34 06	24 57 09	22 ♎ 25 54	28 ♎ 47 57	02 35	17 42	08 10	19 37	21 02	20 47	06 34	12 00
16	03 38 02	25 54 58	05 ♏ 07 05	11 ♏ 23 19	04 42	18 56	08 51	19 51	21 10	20 48	06 33	11 58
17	03 41 59	26 52 46	17 ♏ 36 44	23 ♏ 47 21	06 48	20 09	09 33	20 06	21 18	20 48	06 33	11 57
18	03 45 55	27 50 32	29 ♏ 55 17	06 ♐ 00 36	08 52	21 23	10 14	20 20	21 26	20 48	06 33	11 55
19	03 49 52	28 48 16	12 ♐ 03 29	18 ♐ 04 06	10 54	22 37	10 56	20 34	21 33	20 49	06 32	11 54
20	03 53 48	29 ♉ 46 00	24 ♐ 02 41	29 ♐ 59 30	12 54	23 51	11 37	20 48	21 41	20 49	06 32	11 52
21	03 57 45	00 ♊ 43 42	05 ♑ 54 53	11 ♑ 49 13	14 52	25 04	12 18	21 02	21 49	20 49	06 32	11 51
22	04 01 41	01 41 23	17 ♑ 42 55	23 ♑ 36 27	16 47	26 18	13 00	21 16	21 56	20 49	06 31	11 49
23	04 05 38	02 39 03	29 ♑ 30 21	05 ♒ 25 09	18 40	27 32	13 41	21 31	22 04	20 49	06 31	11 47
24	04 09 35	03 36 42	11 ♒ 21 28	17 ♒ 19 53	20 30	28 45	14 22	21 45	22 12	20 49	06 30	11 46
25	04 13 31	04 34 19	23 ♒ 21 04	29 ♒ 25 40	22 17	29 ♉ 59	15 03	21 59	22 19	20 R 49	06 30	11 44
26	04 17 28	05 31 56	05 ♓ 34 18	11 ♓ 47 37	24 01	01 ♊ 13	15 45	22 13	22 27	20 49	06 29	11 43
27	04 21 24	06 29 32	18 ♓ 06 13	24 ♓ 30 38	25 42	02 27	16 26	22 27	22 34	20 49	06 29	11 41
28	04 25 21	07 27 07	01 ♈ 01 22	07 ♈ 38 47	27 21	03 40	17 07	22 41	22 42	20 49	06 28	11 39
29	04 29 17	08 24 41	14 ♈ 23 09	21 ♈ 14 36	28 ♊ 56	04 54	17 48	22 55	22 50	20 49	06 27	11 38
30	04 33 14	09 22 14	28 ♈ 13 05	05 ♉ 18 23	00 ♋ 29	06 08	18 29	23 09	22 57	20 49	06 27	11 36
31	04 37 10	10 ♊ 19 46	12 ♉ 30 06	19 ♉ 47 39	01 ♋ 58	07 ♊ 22	19 ♊ 10	23 ♉ 23	23 ♉ 05	20 ♒ 49	06 ♒ 26	11 ♐ 34

Date	Moon True ☊ ° '	Moon Mean ☊ ° '	Moon ☽ Latitude ° '
01	28 ♋ 00	28 ♋ 38	04 S 41
02	27 R 50	28 35	04 59
03	27 39	28 32	05 00
04	27 28	28 29	04 41
05	27 18	28 26	04 04
06	27 11	28 22	03 11
07	27 06	28 19	02 05
08	27 05	28 16	00 S 51
09	27 D 04	28 13	00 N 25
10	27 05	28 10	01 38
11	27 R 04	28 06	02 43
12	27 03	28 03	03 40
13	26 58	28 00	04 22
14	26 52	27 57	04 50
15	26 43	27 54	05 03
16	26 32	27 51	05 01
17	26 22	27 47	04 44
18	26 11	27 44	04 14
19	26 02	27 41	03 32
20	25 56	27 38	02 42
21	25 51	27 35	01 45
22	25 49	27 32	00 N 43
23	25 D 49	27 28	00 S 20
24	25 50	27 25	01 23
25	25 51	27 22	02 23
26	25 51	27 19	03 18
27	25 R 50	27 16	04 05
28	25 48	27 12	04 41
29	25 44	27 09	05 03
30	25 38	27 06	05 08
31	25 ♋ 31	27 ♋ 03	04 S 55

DECLINATIONS

Date	Sun ☉ ° '	Moon ☽ ° '	Mercury ☿ ° '	Venus ♀ ° '	Mars ♂ ° '	Jupiter ♃ ° '	Saturn ♄ ° '	Uranus ♅ ° '	Neptune ♆ ° '	Pluto ⯓ ° '
01	15 N 15	01 S 49	11 N 14	10 N 26	20 N 11	15 N 53	15 N 36	15 S 17	18 S 25	11 S 04
02	15 33	03 N 13	12 04	10 53	20 21	15 57	15 38	15 16	18 25	11 03
03	15 51	08 13	12 53	11 20	20 30	16 01	15 40	15 16	18 25	11 03
04	16 08	12 53	13 43	11 46	20 40	16 05	15 43	15 16	18 25	11 03
05	16 25	16 51	14 32	12 12	20 49	16 09	15 45	15 16	18 25	11 02
06	16 42	19 45	15 21	12 38	20 58	16 13	15 47	15 15	18 25	11 02
07	16 59	21 20	16 09	13 03	21 07	16 17	15 49	15 15	18 25	11 02
08	17 15	21 27	16 56	13 28	21 16	16 21	15 51	15 15	18 25	11 02
09	17 31	20 10	17 42	13 53	21 24	16 25	15 53	15 15	18 25	11 01
10	17 46	17 39	18 27	14 17	21 32	16 29	15 55	15 14	18 25	11 01
11	18 02	14 10	19 10	14 42	21 41	16 33	15 57	15 14	18 25	11 01
12	18 17	10 00	19 52	15 05	21 48	16 37	15 59	15 14	18 25	11 01
13	18 32	05 25	20 31	15 29	21 56	16 41	16 01	15 14	18 25	11 01
14	18 46	00 N 40	21 09	15 52	22 04	16 45	16 03	15 14	18 25	11 00
15	19 00	04 S 02	21 45	16 14	22 11	16 49	16 05	15 14	18 25	11 00
16	19 14	08 30	22 18	16 37	22 18	16 53	16 07	15 13	18 25	11 00
17	19 28	12 33	22 49	16 58	22 25	16 57	16 10	15 13	18 25	11 00
18	19 41	16 00	23 17	17 20	22 31	17 00	16 12	15 13	18 25	11 00
19	19 54	18 44	23 42	17 41	22 38	17 04	16 14	15 13	18 25	10 59
20	20 06	20 36	24 05	18 01	22 44	17 08	16 16	15 13	18 26	10 59
21	20 18	21 34	24 26	18 21	22 50	17 12	16 17	15 13	18 26	10 59
22	20 30	21 33	24 43	18 41	22 56	17 15	16 19	15 13	18 26	10 59
23	20 41	20 35	24 58	19 00	23 02	17 19	16 21	15 13	18 26	10 59
24	20 52	18 42	25 11	19 18	23 07	17 23	16 23	15 13	18 26	10 58
25	21 03	15 59	25 21	19 36	23 12	17 27	16 25	15 13	18 26	10 58
26	21 14	12 32	25 29	19 54	23 17	17 30	16 27	15 13	18 26	10 58
27	21 24	08 27	25 34	20 10	23 22	17 34	16 29	15 13	18 26	10 58
28	21 33	03 S 53	25 38	20 27	23 26	17 37	16 31	15 13	18 27	10 58
29	21 42	01 N 01	25 39	20 43	23 31	17 41	16 33	15 13	18 27	10 58
30	21 51	06 02	25 38	20 58	23 35	17 44	16 35	15 14	18 27	10 58
31	22 N 00	10 N 54	25 N 36	21 N 13	23 N 39	17 N 48	16 N 37	15 S 14	18 S 27	10 S 57

ZODIAC SIGN ENTRIES

Date	h	m	Planets
01	00	55	☽ ♈
01	02	49	♀ ♉
03	04	54	☽ ♉
03	19	18	♂ ♊
05	06	23	☽ ♊
07	07	14	☽ ♋
09	09	01	☽ ♌
11	12	41	☽ ♍
13	18	27	☽ ♎
14	07	10	☿ ♊
16	02	16	☽ ♏
18	12	09	☽ ♐
20	17	49	☉ ♊
21	00	01	☽ ♑
23	13	00	☽ ♒
25	12	15	♀ ♊
26	01	07	☽ ♓
28	10	08	☽ ♈
30	04	27	☿ ♋
30	15	02	☽ ♉

LATITUDES

Date	Mercury ☿ ° '	Venus ♀ ° '	Mars ♂ ° '	Jupiter ♃ ° '	Saturn ♄ ° '	Uranus ♅ ° '	Neptune ♆ ° '	Pluto ⯓ ° '
01	01 S 15	01 S 17	00 N 23	00 S 52	02 S 00	00 S 41	00 N 13	11 N 18
04	00 47	01 13	00 25	00 52	02 00	00 41	00 13	11 18
07	00 S 17	01 08	00 27	00 52	02 00	00 42	00 13	11 18
10	00 N 15	01 03	00 29	00 52	02 00	00 42	00 13	11 19
13	00 46	00 58	00 30	00 51	02 00	00 42	00 13	11 19
16	01 15	00 52	00 32	00 51	02 00	00 42	00 13	11 19
19	01 39	00 46	00 33	00 51	02 00	00 42	00 13	11 19
22	01 57	00 40	00 35	00 51	02 00	00 42	00 13	11 19
25	02 08	00 33	00 36	00 51	02 00	00 42	00 13	11 19
28	02 13	00 27	00 38	00 51	02 00	00 42	00 13	11 19
31	02 N 10	00 S 20	00 N 39	00 S 51	02 S 00	00 S 42	00 N 13	11 N 19

DATA

Julian Date	2451666
Delta T	+64 seconds
Ayanamsa	23° 51' 26"
Synetic vernal point	05° ♓ 15' 34"
True obliquity of ecliptic	23° 26' 17"

MOON'S PHASES, APSIDES AND POSITIONS ☽

Date	h	m	Phase	Longitude ° '	Eclipse Indicator
04	04	12	●	14 ♉ 00	
10	20	01	☽	20 ♌ 27	
18	07	34	○	27 ♏ 40	
26	11	55	☾	05 ♓ 32	

Day	h	m		
06	09	00	Perigee	
22	03	47	Apogee	
01	20	45	ON	
08	02	06	Max dec	21° N 35'
14	15	22	OS	
21	23	41	Max dec	21° S 40'
29	07	06	ON	

LONGITUDES

Date	Chiron ⚷ ° '	Ceres ⚳ ° '	Pallas ⚴ ° '	Juno ⚵ ° '	Vesta ⚶ ° '	Black Moon Lilith ⚸ ° '
01	16 ♐ 18	25 ♍ 34	09 ♌ 08	19 ♒ 15	27 ♑ 16	06 ♑ 54
11	15 ♐ 45	25 ♍ 28	12 ♌ 31	21 ♒ 36	29 ♑ 35	08 ♑ 01
21	15 ♐ 06	25 ♍ 42	16 ♌ 09	23 ♒ 36	00 ♒ 36	09 ♑ 08
31	14 ♐ 24	27 ♍ 13	19 ♌ 57	25 ♒ 12	01 ♒ 12	10 ♑ 14

All ephemeris data is given at 12.00 UT and the Moon's longitude is additionally given for 24.00 UT

ASPECTARIAN

h m Aspects

01 Monday
00 43 ☽ ⊻ ♀
03 03 ☽ ∠ ♃
04 36 ☽ ⊻ ☿
06 58 ☿ ⋕ ⯓
08 26 ☽ ∠ ♄
10 26 ☽ ⊥ ☉
10 52 ☽ ∠ ♅
10 54 ☿ Q ♅
12 34 ☽ ⚹ ♆
13 58 ☉ ⋕ ♅
19 15 ☽ ⊥ ♃
21 46 ☽ ⊻ ☉
22 41 ☽ △ ⯓

02 Tuesday
00 23 ☽ ⊥ ♄
01 11 ☽ ∠ ♂
05 51 ☽ ⊻ ♃
09 28 ☽ Q ♆
10 21 ☉ ⊼ ⯓
10 50 ☽ ⊻ ♄
12 59 ☽ ⚹ ♅
17 28 ☽ ⊥ ♂
20 03 ☉ ∥ ♄
21 13 ♀ ⋕ ⯓

03 Wednesday
00 21 ☽ ♇ ⯓
04 09 ☽ ⊻ ♂
05 59 ♀ Q ♅
06 22 ☿ ± ⯓
09 18 ☽ Q ♅
09 30 ☿ □ ♆
09 37 ☽ ☌ ♀
15 25 ☽ ± ⯓
15 53 ☽ □ ♆
16 57 ☽ ☌ ☿

04 Thursday
01 21 ☽ ⊼ ⯓
02 12 ☽ ⋕ ⯓
04 12 ☽ ☌ ☉
05 19 ☽ ∥ ♀
06 13 ☉ ∥ ♃
09 05 ☽ ☌ ♃
13 27 ☽ ☌ ♄
15 07 ☽ □ ♅
17 44 ☽ ∥ ☿

05 Friday
01 45 ☽ ⋕ ♅
04 42 ☽ ∥ ♄
07 17 ☽ ∥ ♃
08 08 ☽ ☌ ♂
08 54 ☽ ∥ ☉
15 27 ☽ ⊻ ♀
17 05 ☽ △ ♆
23 46 ☽ ⋕ ♆

06 Saturday
02 05 ☽ ⊥ ♀
02 12 ☿ ⊼ ⯓
02 16 ☽ ☍ ⯓
02 17 ☽ ⊻ ☿
04 14 ♀ ± ⯓
08 36 ☽ ⊻ ☉
09 22 ☿ ⋕ ♅
10 47 ☽ ⊻ ♃
10 57 ♀ □ ♆
13 41 ☽ ⊥ ☿
14 45 ☽ ⊻ ♄
16 01 ☽ △ ♅
17 28 ☽ ♇ ♆
18 04 ☽ ∠ ♀
19 03 ☽ ⊥ ☉
20 43 ☽ ⊥ ♃

07 Sunday
00 37 ☽ ⊥ ♄
01 40 ☿ ∥ ♄
06 18 ☽ ∥ ♂
06 53 ☽ ∠ ☿
08 10 ☽ ± ♆
10 48 ☽ ∠ ☉
11 27 ☽ ⊻ ♂
11 40 ☽ ∠ ♃
15 29 ☽ ∠ ♄
16 34 ☽ ♇ ♅
16 43 ☿ ∥ ♃
18 01 ☽ ⊼ ♆
20 52 ☽ ⚹ ♀
21 48 ☽ ⊥ ♂

08 Monday
03 16 ☽ ⊼ ⯓
04 08 ☉ ☌ ♃
07 26 ☽ ± ♅
11 54 ☽ ⚹ ☿
12 29 ♆ St R
12 51 ☽ ⚹ ♃
13 12 ☽ ± ⯓
13 19 ☽ ⚹ ☉
13 28 ☽ ∠ ♂
16 31 ☽ ⚹ ♄
17 15 ☽ ∥ ♂
17 26 ☽ ⊼ ♅
18 37 ☽ Q ♀
18 53 ☿ ☌ ♃

09 Tuesday
03 08 ☉ ∥ ☿
03 49 ☉ ☌ ☿
04 14 ☽ ♇ ⯓
09 18 ☽ Q ♃
10 54 ☽ Q ☉
11 37 ☽ Q ☿
12 52 ☽ Q ♄
15 58 ☽ ⚹ ♂
20 10 ☽ ☍ ♆
20 11 ☿ ☌ ♄

10 Wednesday
00 46 ☿ □ ♅
03 57 ☽ □ ♀
05 40 ☽ △ ⯓
05 45 ☽ ⋕ ♆
06 52 ☽ ∥ ☿
11 07 ☿ ⋕ ♆
11 08 ☽ ∥ ☉
13 34 ☽ Q ♂
16 35 ☽ □ ♃
19 45 ☉ ☌ ♄
20 00 ☽ □ ♄
20 01 ☽ □ ☉
20 31 ☽ ☍ ♅
20 33 ☽ ∥ ♃
23 02 ♀ ⊼ ⯓

11 Thursday
00 11 ☽ □ ☿
00 32 ☽ ∥ ♄
03 16 ☉ □ ♅
05 13 ☽ ⋕ ♅
09 02 ☽ ∥ ♀
22 44 ☽ □ ♂

12 Friday
00 17 ☽ ⊼ ♆
06 26 ☽ ⋕ ⯓
10 05 ☽ □ ⯓
10 57 ☽ ± ♆
13 41 ☽ △ ♀
20 43 ♀ ⋕ ♅
22 31 ☽ △ ♃

13 Saturday
01 10 ☉ ⋕ ♆
01 37 ☽ △ ♄
01 41 ☽ ⊼ ♅
03 08 ☽ ♇ ♆
04 53 ♂ △ ♆
05 17 ☽ △ ☉
08 34 ♄ □ ♅
12 34 ☽ ± ♅
15 57 ☽ △ ☿
18 32 ☽ Q ⯓
19 36 ☽ ♇ ♀

14 Sunday
02 18 ☽ ♇ ♃
05 02 ☽ ♇ ♅
05 14 ☽ ♇ ♄
06 30 ☽ △ ♆
07 57 ☽ △ ♂
10 53 ☽ ♇ ☉
13 53 ☽ ± ♀
16 35 ☽ ⚹ ⯓
19 14 ☽ ± ♃
22 03 ☽ ± ♄

15 Monday
01 06 ☽ ♇ ☿
01 33 ♀ ∥ ♄
02 12 ☽ ⊼ ♀
04 57 ☽ ± ☉
06 38 ☽ ⊼ ♃
08 55 ☽ △ ♅
09 22 ☽ ⊼ ♄
13 27 ☽ ♇ ♂
17 08 ☽ ⊼ ☉
20 35 ☽ ∠ ⯓
21 22 ☽ ± ☿

16 Tuesday
07 26 ☽ ± ♂
11 03 ☽ ⊼ ☿
12 03 ☿ ∥ ♂
13 38 ☽ ⊥ ⯓
14 45 ☽ □ ♆
19 33 ☽ ⊼ ♂

17 Wednesday
01 06 ☽ ⊻ ⯓
02 27 ☽ ∥ ⯓
09 10 ☿ △ ♆
09 36 ♀ ∥ ♃
10 30 ♀ ☌ ♃
16 54 ☽ ☍ ♃
17 28 ☽ ☍ ♀
18 11 ☽ □ ♅
19 14 ☽ ☍ ♄

18 Thursday
00 39 ♀ □ ♅
01 29 ☽ Q ♆
06 09 ☽ ∥ ♅
07 34 ☽ ☍ ☉
12 56 ♀ ☌ ♄
13 30 ☽ ⋕ ♄
20 16 ☽ ⋕ ♃

19 Friday
00 24 ☽ ⋕ ♀
01 03 ☽ ⚹ ♆
05 33 ☽ Q ♅
08 56 ☽ ∥ ♆
09 15 ☽ ☍ ☿
09 37 ☽ ☍ ♂
11 41 ☽ ☌ ⯓
12 26 ☿ ☌ ♂
23 41 ☿ ☍ ⯓

20 Saturday
03 13 ☽ ⋕ ☉
05 21 ☽ ⊼ ♃
05 30 ☽ ⚹ ♅
06 57 ☽ ∠ ♆
07 12 ☽ ⊼ ♄
11 33 ☽ ⊼ ♀
13 16 ♃ □ ♅
17 40 ☽ ± ♃
19 25 ☽ ± ♄
20 27 ♂ ☍ ⯓

21 Sunday
00 34 ☽ ⊼ ☉
01 03 ☽ ± ♀
01 05 ☽ ⊥ ♆
11 48 ☽ ∠ ♅
12 15 ☽ ♇ ♃
13 15 ☽ ⊻ ♆
13 48 ☽ ± ☉
13 50 ☽ ♇ ♄
17 41 ♀ ⋕ ♆
21 25 ☽ ♇ ♀

22 Monday
00 01 ☽ ⊻ ⯓
01 48 ☽ ⊼ ♂
06 06 ☽ ⊥ ♅
09 44 ☽ ♇ ☉
09 45 ☽ ⊼ ☿
12 12 ☽ ⊥ ⯓
14 46 ☽ ± ♂
18 20 ☽ ⊻ ♅
19 24 ☽ △ ♃
20 42 ☽ △ ♄

23 Tuesday
00 18 ☽ ± ☿
06 29 ☽ ∠ ⯓
07 31 ☽ △ ♀
10 13 ☽ ♇ ♂
10 19 ☽ ⋕ ☉
18 57 ☽ △ ☉
22 00 ☽ ♇ ☿

24 Wednesday
02 12 ☽ ☌ ♆
06 15 ☽ ⋕ ♀
12 49 ☽ ⚹ ⯓
14 41 ☽ ∥ ♆
16 23 ☿ △ ♅
18 26 ☽ △ ♂

25 Thursday
00 15 ☽ ⋕ ♃
01 25 ☿ ♇ ♆
06 59 ☽ ☌ ♅
07 17 ☿ ⊻ ♃
08 20 ♅ St R
08 36 ☽ ⋕ ♄
09 13 ☽ □ ♃
09 31 ☽ △ ☿
09 56 ☽ □ ♄
12 35 ☿ ⊻ ♄
12 46 ☽ Q ⯓
17 46 ☽ ∥ ♅

26 Friday
02 34 ☽ □ ♀
11 55 ☽ □ ☉
13 46 ☽ ⊻ ♆
21 08 ☽ Q ♃
21 31 ☽ Q ♄
21 37 ☽ ∥ ⯓
23 49 ☽ □ ⯓

27 Saturday
01 19 ☽ ⊥ ♆
08 39 ☽ □ ♂
11 37 ☉ △ ♆
16 53 ☽ Q ♀
17 07 ☽ ⊻ ♅
18 20 ☽ ∠ ♆
20 19 ☽ ⚹ ♃
20 29 ☽ ⚹ ♄

28 Sunday
00 56 ☽ Q ☉
04 17 ☽ ⊥ ♅
04 17 ☽ □ ☿
16 04 ♃ ☌ ♄
17 19 ☽ ⚹ ♀
19 50 ☽ Q ♂
20 43 ☽ ∠ ♅
21 52 ☽ ⚹ ♆

29 Monday
00 13 ☽ ∠ ♄
00 17 ☽ ∠ ♃
00 33 ☽ ⚹ ☉
07 08 ☽ △ ⯓
10 07 ☿ ⊥ ♄
11 34 ☿ ⊥ ♃
16 20 ☽ ⊥ ♄
16 32 ☽ ⊥ ♃
17 04 ☽ Q ☿
18 18 ☽ ⚹ ♂
19 08 ☽ Q ♆
22 37 ☽ ∠ ♀
23 15 ☽ ⚹ ♅

30 Tuesday
02 53 ☽ ⊻ ♄
03 09 ☽ ⊻ ♃
04 55 ☽ ∠ ☉
09 15 ☽ ♇ ⯓
11 28 ☿ ± ♆
15 34 ☽ ⊥ ♀
16 18 ☽ ⚹ ☿
18 03 ♀ △ ♆
19 48 ☽ Q ♅
21 22 ☽ ∠ ♂
21 22 ☽ ⊥ ☉

31 Wednesday
00 28 ☽ ± ⯓
01 54 ☽ □ ♆
02 38 ☽ ⊻ ♀
08 08 ☽ ⊻ ☉
10 28 ☽ ⊼ ⯓
12 19 ☽ ⋕ ⯓
13 09 ☽ ⊥ ♂
20 11 ☽ ∠ ☿
23 30 ☽ ⊻ ♂

JUNE 2000

LONGITUDES

Date	Sidereal time h m s	Sun ☉ ° ' "	Moon ☽ ° ' "	Moon ☽ 24.00 ° ' "	Mercury ☿ ° '	Venus ♀ ° '	Mars ♂ ° '	Jupiter ♃ ° '	Saturn ♄ ° '	Uranus ♅ ° '	Neptune ♆ ° '	Pluto ♇ ° '
01	04 41 07	11 ♊ 17 18	27 ♉ 10 14	04 ♊ 36 54	03 ♋ 24	08 ♊ 35	19 ♊ 51	23 ♉ 37	23 ♉ 12	20 ♒ 48	06 ♒ 25	11 ♐ 33
02	04 45 04	12 14 49	12 ♊ 06 38	19 ♊ 38 15	04 47	09 49	20 31	23 51	23 20	20 R 48	06 R 24	11 R 31
03	04 49 00	13 12 18	27 ♊ 06 35	04 ♋ 42 28	06 07	11 03	21 12	24 04	23 27	20 47	06 24	11 30
04	04 52 57	14 09 47	12 ♋ 12 49	19 ♋ 40 36	07 24	12 17	21 53	24 18	23 34	20 47	06 23	11 28
05	04 56 53	15 07 14	27 ♋ 04 58	04 ♌ 25 12	08 38	13 30	22 34	24 32	23 42	20 46	06 22	11 26
06	05 00 50	16 04 41	11 ♌ 40 44	18 ♌ 51 10	09 48	14 44	23 14	24 46	23 49	20 46	06 21	11 25
07	05 04 46	17 02 06	25 ♌ 56 15	02 ♍ 55 51	10 55	15 58	23 55	25 00	23 57	20 45	06 20	11 23
08	05 08 43	17 59 30	09 ♍ 49 59	16 ♍ 38 44	11 58	17 11	24 36	25 13	24 04	20 45	06 19	11 21
09	05 12 39	18 56 52	23 ♍ 22 16	00 ♎ 00 49	12 58	18 25	25 16	25 27	24 11	20 44	06 18	11 20
10	05 16 36	19 54 14	06 ♎ 34 37	13 ♎ 03 58	13 54	19 39	25 57	25 40	24 18	20 43	06 17	11 18
11	05 20 33	20 51 35	19 ♎ 29 09	25 ♎ 50 28	14 47	20 53	26 37	25 54	24 26	20 42	06 16	11 17
12	05 24 29	21 48 54	02 ♏ 08 13	08 ♏ 22 39	15 35	22 06	27 18	26 08	24 33	20 42	06 15	11 15
13	05 28 26	22 46 13	14 ♏ 34 04	20 ♏ 42 42	16 20	23 20	27 58	26 21	24 40	20 41	06 14	11 13
14	05 32 22	23 43 31	26 ♏ 47 47	02 ♐ 52 34	17 01	24 34	28 39	26 34	24 47	20 40	06 13	11 12
15	05 36 19	24 40 48	08 ♐ 54 15	14 ♐ 54 04	17 38	25 47	29 19	26 48	24 54	20 39	06 12	11 10
16	05 40 15	25 38 05	20 ♐ 52 15	26 ♐ 49 00	18 11	27 01	29 ♊ 59	27 01	25 01	20 38	06 11	11 09
17	05 44 12	26 35 21	02 ♑ 44 36	08 ♑ 39 16	18 40	28 15	00 ♋ 39	27 14	25 08	20 37	06 10	11 07
18	05 48 08	27 32 36	14 ♑ 33 19	20 ♑ 27 02	19 04	29 ♊ 29	01 20	27 28	25 15	20 36	06 09	11 06
19	05 52 05	28 29 51	26 ♑ 20 44	02 ♒ 14 48	19 24	00 ♋ 42	02 00	27 41	25 22	20 34	06 08	11 04
20	05 56 02	29 ♊ 27 06	08 ♒ 09 38	14 ♒ 05 38	19 39	01 56	02 40	27 54	25 29	20 33	06 06	11 03
21	05 59 58	00 ♋ 24 20	20 ♒ 03 15	26 ♒ 02 59	19 50	03 10	03 20	28 07	25 36	20 32	06 05	11 01
22	06 03 55	01 21 34	02 ♓ 05 19	08 ♓ 10 48	19 56	04 23	04 00	28 20	25 43	20 31	06 04	11 00
23	06 07 51	02 18 48	14 ♓ 19 57	20 ♓ 33 19	19 R 58	05 37	04 40	28 33	25 49	20 29	06 03	10 58
24	06 11 48	03 16 02	26 ♓ 51 26	03 ♈ 14 48	19 55	06 51	05 20	28 46	25 56	20 28	06 01	10 57
25	06 15 44	04 13 16	09 ♈ 43 54	16 ♈ 19 09	19 47	08 05	06 00	28 59	26 03	20 27	06 00	10 55
26	06 19 41	05 10 29	23 ♈ 00 53	29 ♈ 49 22	19 36	09 18	06 40	29 12	26 09	20 25	05 59	10 54
27	06 23 37	06 07 43	06 ♉ 44 41	13 ♉ 46 50	19 20	10 32	07 20	29 24	26 16	20 24	05 57	10 52
28	06 27 34	07 04 57	20 ♉ 55 37	28 ♉ 10 42	19 00	11 46	08 00	29 37	26 22	20 22	05 56	10 51
29	06 31 31	08 02 11	05 ♊ 31 31	12 ♊ 57 22	18 36	13 00	08 40	29 ♉ 50	26 29	20 21	05 55	10 49
30	06 35 27	08 ♋ 59 24	20 ♊ 27 21	28 ♊ 00 28	18 ♋ 09	14 ♋ 13	09 ♋ 20	00 ♊ 02	26 ♉ 35	20 ♒ 19	05 ♒ 53	10 ♐ 48

Date	Moon True ☊ ° '	Moon Mean ☊ ° '	Moon ☽ Latitude ° '
01	25 ♋ 24	27 ♋ 00	04 S 23
02	25 R 18	26 57	03 32
03	25 14	26 53	02 26
04	25 11	26 50	01 S 10
05	25 D 11	26 47	00 N 10
06	25 12	26 44	01 29
07	25 13	26 41	02 40
08	25 14	26 37	03 39
09	25 R 14	26 34	04 25
10	25 13	26 31	04 56
11	25 10	26 28	05 10
12	25 06	26 25	05 10
13	25 01	26 22	04 54
14	24 56	26 18	04 25
15	24 50	26 15	03 45
16	24 46	26 12	02 55
17	24 43	26 09	01 58
18	24 41	26 06	00 N 55
19	24 D 41	26 03	00 S 09
20	24 41	25 59	01 14
21	24 43	25 56	02 15
22	24 44	25 53	03 12
23	24 46	25 50	04 01
24	24 47	25 47	04 39
25	24 R 47	25 43	05 05
26	24 46	25 40	05 16
27	24 44	25 37	05 09
28	24 42	25 34	04 44
29	24 40	25 31	03 59
30	24 ♋ 38	25 ♋ 28	02 S 58

DECLINATIONS

Date	Sun ☉ ° '	Moon ☽ ° '	Mercury ☿ ° '	Venus ♀ ° '	Mars ♂ ° '	Jupiter ♃ ° '	Saturn ♄ ° '	Uranus ♅ ° '	Neptune ♆ ° '	Pluto ♇ ° '
01	22 N 08	15 N 15	25 N 31	21 N 27	23 N 42	17 N 51	16 N 39	15 S 14	18 S 27	10 S 57
02	22 16	18 44	25 25	21 40	23 46	17 55	16 40	15 14	18 28	10 57
03	22 23	20 58	25 18	21 53	23 49	17 58	16 42	15 14	18 28	10 57
04	22 30	21 43	25 09	22 06	23 52	18 02	16 44	15 14	18 28	10 57
05	22 36	20 55	24 59	22 17	23 55	18 05	16 46	15 15	18 28	10 57
06	22 43	18 42	24 47	22 28	23 58	18 08	16 48	15 15	18 28	10 57
07	22 48	15 23	24 35	22 39	24 00	18 12	16 49	15 15	18 29	10 57
08	22 54	11 16	24 21	22 48	24 02	18 15	16 51	15 15	18 29	10 57
09	22 59	06 41	24 07	22 57	24 04	18 18	16 53	15 15	18 29	10 57
10	23 03	01 N 55	23 52	23 06	24 06	18 22	16 55	15 16	18 29	10 56
11	23 07	02 S 50	23 36	23 14	24 08	18 25	16 56	15 16	18 30	10 56
12	23 11	07 22	23 20	23 21	24 09	18 28	16 58	15 16	18 30	10 56
13	23 14	11 31	23 03	23 27	24 10	18 31	17 00	15 17	18 30	10 56
14	23 17	15 08	22 46	23 33	24 11	18 34	17 01	15 17	18 30	10 56
15	23 20	18 05	22 28	23 38	24 12	18 37	17 03	15 17	18 31	10 56
16	23 22	20 13	22 10	23 42	24 13	18 40	17 05	15 18	18 31	10 56
17	23 24	21 27	21 53	23 46	24 13	18 43	17 06	15 18	18 31	10 56
18	23 25	21 43	21 35	23 49	24 13	18 46	17 08	15 18	18 31	10 56
19	23 26	21 02	21 17	23 51	24 13	18 49	17 09	15 19	18 32	10 56
20	23 26	19 25	21 00	23 53	24 13	18 52	17 11	15 19	18 32	10 56
21	23 26	16 56	20 43	23 53	24 12	18 55	17 12	15 20	18 32	10 56
22	23 26	13 43	20 26	23 54	24 11	18 58	17 14	15 20	18 33	10 56
23	23 25	09 52	20 10	23 53	24 10	19 01	17 15	15 21	18 33	10 56
24	23 24	05 31	19 54	23 52	24 09	19 04	17 17	15 21	18 33	10 56
25	23 22	00 S 49	19 39	23 50	24 08	19 07	17 18	15 21	18 34	10 56
26	23 20	04 N 03	19 25	23 47	24 06	19 09	17 20	15 22	18 34	10 56
27	23 18	08 54	19 11	23 44	24 05	19 12	17 21	15 22	18 34	10 57
28	23 15	13 25	18 59	23 39	24 03	19 15	17 23	15 23	18 35	10 57
29	23 12	17 18	18 47	23 35	24 01	19 18	17 24	15 23	18 35	10 57
30	23 N 08	20 N 08	18 N 37	23 N 29	23 N 58	19 N 20	17 N 26	15 S 24	18 S 35	10 S 57

ZODIAC SIGN ENTRIES

Date	h	m	Planets
01	16	34	☽ ♊
03	16	30	☽ ♋
05	16	45	☽ ♌
07	18	57	☽ ♍
09	23	59	☽ ♎
12	07	55	☽ ♏
14	18	18	☽ ♐
16	12	30	♂ ♋
17	06	26	☽ ♑
18	22	15	♀ ♋
19	19	26	☽ ♒
21	01	48	☉ ♋
22	07	52	☽ ♓
24	17	55	☽ ♈
27	00	19	☽ ♉
29	02	59	☽ ♊
30	07	35	♃ ♊

LATITUDES

Date	Mercury ☿ ° '	Venus ♀ ° '	Mars ♂ ° '	Jupiter ♃ ° '	Saturn ♄ ° '	Uranus ♅ ° '	Neptune ♆ ° '	Pluto ♇ ° '
01	02 N 08	00 S 17	00 N 40	00 S 51	02 S 00	00 S 42	00 N 13	11 N 19
04	01 55	00 10	00 41	00 51	02 00	00 43	00 13	11 18
07	01 36	00 S 03	00 42	00 50	02 00	00 43	00 13	11 18
10	01 09	00 N 04	00 44	00 50	02 00	00 43	00 13	11 18
13	00 N 37	00 11	00 45	00 50	02 00	00 43	00 13	11 17
16	00 S 02	00 18	00 46	00 50	02 00	00 43	00 13	11 17
19	00 45	00 25	00 48	00 50	02 01	00 43	00 13	11 16
22	01 33	00 32	00 49	00 50	02 01	00 43	00 13	11 16
25	02 21	00 38	00 50	00 50	02 01	00 43	00 13	11 15
28	03 09	00 45	00 51	00 50	02 01	00 43	00 13	11 14
31	03 S 51	00 N 51	00 N 52	00 S 50	02 S 01	00 S 44	00 N 13	11 N 14

DATA

Julian Date	2451697
Delta T	+64 seconds
Ayanamsa	23° 51' 30"
Synetic vernal point	05° ♓ 15' 29"
True obliquity of ecliptic	23° 26' 17"

LONGITUDES

Date	Chiron ⚷ ° '	Ceres ⚳ ° '	Pallas ⚴ ° '	Juno ⚵ ° '	Vesta ⚶ ° '	Black Moon Lilith ⚸ ° '
01	14 ♐ 20	27 ♍ 22	20 ♌ 20	25 ♒ 20	01 ♒ 12	10 ♑ 21
11	13 ♐ 38	29 ♍ 07	24 ♌ 18	26 ♒ 24	00 ♒ 55	11 ♑ 27
21	12 ♐ 57	01 ♎ 20	28 ♌ 23	26 ♒ 35	29 ♑ 34	12 ♑ 34
31	12 ♐ 21	03 ♎ 56	02 ♍ 33	26 ♒ 49	28 ♑ 02	13 ♑ 40

MOON'S PHASES, APSIDES AND POSITIONS ☽

Date	h	m	Phase	Longitude ° '	Eclipse Indicator
02	12	14	●	12 ♊ 15	
09	03	29	☽	18 ♍ 37	
16	22	27	○	26 ♐ 03	
25	01	00	☾	03 ♈ 47	

Day	h	m		
03	13	12	Perigee	
18	12	41	Apogee	
04	11	24	Max dec	21° N 43'
10	21	37	0S	
18	06	48	Max dec	21° S 45'
25	16	06	0N	

ASPECTARIAN

h m Aspects

01 Thursday
01 39 ☽ □ ♅
05 30 ☽ ☌ ♄
06 08 ☽ ☌ ♃
11 51 ☽ # ♅
12 25 ☽ ⊥ ☿
18 18 ☉ ☍ ♇
20 47 ☽ ∥ ♄
23 06 ☽ ⊻ ☿

02 Friday
02 53 ☽ △ ♆
05 29 ☽ ∥ ♃
08 01 ☽ ☌ ♀
09 45 ☽ # ♆
11 04 ☽ ☍ ♇
12 14 ☽ ☌ ☉
21 35 ♂ △ ♅

03 Saturday
01 50 ☽ △ ♅
02 03 ☽ ☌ ♂
02 48 ☽ ⚼ ♆
05 57 ☿ ⚼ ♅
06 01 ☽ ⊻ ♄
06 59 ☽ ⊻ ♃
15 39 ☽ ⊥ ♄
16 41 ☽ ⊥ ♃
16 59 ☿ ⊼ ♆
17 07 ☽ ± ♆
18 38 ♂ ⚼ ♆
20 31 ♀ ☍ ♇

04 Sunday
01 43 ☽ ⚼ ♅
02 41 ☽ ⊼ ♆
03 36 ☽ ☌ ☿
06 08 ☽ ∠ ♄
07 16 ☽ ∠ ♃
10 48 ☽ ⊼ ♇
12 07 ☽ ⊻ ♀
15 20 ☽ ⊻ ☉
16 07 ☽ ± ♅
20 25 ☽ ± ♇
22 37 ☽ ⊥ ♀

05 Monday
01 40 ☽ ⊥ ☉
01 47 ☽ ⊼ ♅
04 19 ☽ ⊻ ♂
06 27 ☽ ⚹ ♄
07 48 ☽ ⚹ ♃
10 57 ☽ ⚼ ♇
13 35 ☿ ∠ ♄
14 32 ☽ ⊥ ♂
14 32 ☽ ∠ ♀
17 18 ☽ ∠ ☉

06 Tuesday
02 13 ☽ Q ♄
03 12 ☽ ☍ ♆
03 44 ☽ Q ♃
06 01 ☽ ∠ ♂
08 36 ☽ ⊻ ☿
11 09 ☿ ∠ ♃
11 33 ☽ △ ♇
13 55 ☽ # ♆
16 30 ☽ ∥ ♃
17 34 ☽ ⚹ ♀
19 28 ☽ ⊥ ☿
19 52 ☽ ⚹ ☉

07 Wednesday
02 27 ☽ ∥ ♄
03 13 ☽ ☍ ♅
08 23 ☽ ⚹ ♂
08 35 ☽ □ ♄
10 22 ☽ □ ♃
11 57 ☽ ∠ ☿
12 48 ☽ # ♅
13 04 ♂ ⊻ ♄
15 47 ☽ Q ♀
17 41 ☽ Q ☉
22 23 ☿ ⊼ ♇

08 Thursday
05 53 ☽ ⊼ ♆
06 04 ☽ Q ♂
13 45 ☽ # ♇
14 40 ☽ □ ♇
16 03 ☽ ⚹ ☿
16 22 ☽ ± ♆

09 Friday
02 16 ☽ □ ♀
03 29 ☽ □ ☉
07 17 ☽ ⊼ ♅
08 18 ☽ ⚼ ♆
13 29 ☽ △ ♄
15 05 ☽ Q ☿
15 33 ☿ ∥ ♂
15 36 ☽ □ ♂
15 48 ☽ △ ♃
18 03 ☽ ± ♅
19 20 ☉ ∥ ♀
21 26 ♂ ⊻ ♃
22 44 ☽ Q ♇

10 Saturday
10 26 ☽ ⚼ ♅
11 29 ☽ △ ♆
17 05 ☽ ⚼ ♄
19 42 ☽ ⚼ ♃
20 43 ☽ ⚹ ♇

11 Sunday
02 33 ☽ □ ☿
08 12 ☉ △ ♅
08 42 ♀ △ ♅
10 00 ☽ ± ♄
10 02 ☿ ± ♅
10 31 ☉ ☌ ♀
12 48 ☽ ± ♃
14 18 ☽ △ ♅
14 48 ☽ △ ☉
14 54 ☽ △ ♀
19 40 ♀ ⚼ ♆
21 25 ☽ ⊼ ♄
22 14 ☉ ⚼ ♆

12 Monday
00 20 ☽ ⊼ ♃
00 48 ☽ ∠ ♇
02 15 ☽ △ ♂
10 52 ☿ ∥ ♀
17 58 ☽ ⊥ ♇
19 54 ☽ □ ♆
21 44 ☽ ⚼ ☉
22 10 ☉ ∥ ☿
22 35 ☽ ⚼ ♀

13 Tuesday
03 02 ♃ # ♆
05 31 ☽ ⊻ ♇
08 28 ☽ ∥ ♇
08 43 ☽ ⚼ ♂
15 40 ☽ △ ☿
16 39 ☽ ± ☉
17 59 ☽ ± ♀
23 55 ☽ □ ♅

14 Wednesday
03 19 ☽ ± ♂
05 24 ☽ ⊼ ☉
06 54 ☽ Q ♆
07 04 ☽ ⊼ ♀
07 58 ☽ ☍ ♄
11 31 ☽ ☍ ♃
13 04 ☽ ∥ ♅
15 49 ☽ ⊼ ♂
16 49 ♀ ⊻ ♄
18 18 ☿ ± ♇
22 53 ☽ ⚼ ☿

15 Thursday
02 47 ☽ # ♄
06 37 ☽ ⚹ ♆
11 29 ☽ Q ♅
16 11 ☽ ∥ ♆
16 31 ☽ ☌ ♇
17 25 ☽ # ♃
17 44 ☽ ± ☿
18 23 ☉ ⊻ ♄

16 Friday
06 21 ☽ ⊼ ☿
11 31 ☽ ⚹ ♅
12 01 ♀ ⊻ ♃
12 38 ☽ ∠ ♆
18 55 ♂ ± ♆
20 27 ☽ ⊼ ♄
22 27 ☽ ☍ ☉

17 Saturday
00 39 ☽ ⊼ ♃
01 50 ☽ ☍ ♀
06 47 ☽ ⊥ ♆
07 31 ☽ ☍ ♂
08 43 ☽ ± ♄
13 02 ☽ ± ♃
17 49 ☽ ∠ ♅
18 56 ☽ ⊻ ♆

18 Sunday
00 52 ☽ # ☿
03 10 ☽ ⚼ ♄
04 59 ☽ ⊻ ♇
07 40 ☽ ⚼ ♃
08 46 ♂ ⊥ ♄
09 19 ☉ ⊻ ♃
12 05 ☽ ⊥ ♅
17 09 ☽ ⊥ ♇
21 28 ☽ ☍ ☿

19 Monday
00 16 ☽ ⊻ ♅
00 55 ♀ ± ♆
03 19 ☽ # ☿
10 00 ☽ △ ♄
11 26 ☽ ∠ ♇
14 46 ☽ △ ♃
16 46 ☽ ⊼ ☉
21 54 ☽ ⊼ ♀

20 Tuesday
00 11 ☽ ⊼ ♂
02 19 ♀ ⊥ ♄
06 02 ☽ ± ☉
07 51 ☽ ☌ ♆
11 29 ☽ ± ♀
13 05 ☽ ± ♂
17 49 ☽ # ♃
17 49 ☽ ⚹ ♇
21 24 ☽ ∥ ♆

21 Wednesday
01 50 ☽ ⚼ ☉
04 09 ☉ ± ♆
07 46 ☽ ⚼ ♀
08 20 ☽ ⚼ ♂
09 44 ☽ # ♄
11 33 ☽ ⊼ ☿
12 58 ☽ ☌ ♅
17 56 ☽ Q ♇
19 27 ♀ ☌ ♂
23 12 ☽ □ ♄
23 41 ☽ ± ☿

22 Thursday
00 36 ☽ ∥ ♅
04 25 ☽ □ ♃
10 26 ☽ △ ☉
10 43 ♀ ⊥ ♃
16 00 ☽ △ ♂
17 03 ☽ △ ♀
17 39 ☽ ⚼ ☿
19 50 ☽ ⊻ ♆
22 01 ☉ ⊥ ♄

23 Friday
05 28 ☽ □ ♇
05 39 ☽ ∥ ♇
05 40 ♂ ⊥ ♃
07 34 ☽ ⊥ ♆
08 32 ☿ St R
09 33 ♀ ⚼ ♅
11 00 ☽ Q ♄
16 22 ☽ Q ♃
20 11 ♀ ⊼ ♆
22 50 ☽ △ ☿
23 51 ☽ ⊻ ♅

24 Saturday
00 55 ☽ ∠ ♆
10 14 ☽ ⚹ ♄
11 16 ☽ ⊥ ♅
15 40 ☽ ⚹ ♃
16 33 ♂ ⚼ ♅

25 Sunday
01 00 ☽ □ ☉
04 06 ☽ ∠ ♅
04 45 ☽ □ ♂
05 08 ☽ ⚹ ♆
08 38 ☽ □ ♀
11 56 ♂ ⊼ ♆
14 10 ☽ △ ♇
14 26 ☽ ∠ ♄
19 53 ☽ ∠ ♃

26 Monday
03 01 ☽ Q ♆
06 00 ☽ □ ☿
06 52 ☽ ⊥ ♄
07 23 ☽ ⚹ ♅
12 18 ☽ Q ☉
12 20 ☽ ⊥ ♃
12 39 ☉ ⊥ ♃
15 05 ☽ Q ♂
17 06 ☽ ⚼ ♇
17 36 ☽ ⊻ ♄
18 03 ☉ ⚼ ♅
20 20 ☽ Q ♀
23 04 ☽ ⊻ ♃

27 Tuesday
04 30 ☽ Q ♅
07 47 ☉ ⊼ ♆
08 47 ☽ ± ♇
10 32 ☿ ∥ ♃
10 39 ☽ □ ♆
10 52 ☽ ⚹ ☉
12 59 ☽ Q ☿
13 04 ☽ ⚹ ♂
18 28 ♀ ⊼ ♇
19 03 ☽ ⊼ ♇
19 06 ☽ ⚹ ♀
22 33 ☽ # ♇

28 Wednesday
03 41 ♀ ∠ ♄
08 51 ☽ ⚹ ☿
11 05 ☽ □ ♅
14 04 ☽ ∠ ☉
15 37 ☽ ∠ ♂
21 06 ☽ ☌ ♄
22 34 ☽ ∠ ♀
23 32 ☽ # ♅

29 Thursday
02 34 ☽ ☌ ♃
05 56 ☽ ⊥ ☉
07 07 ☽ ⊥ ♂
08 57 ☽ ∠ ☿
12 38 ☽ △ ♆
12 46 ☽ ∥ ♄
14 36 ☽ ⊥ ♀
16 21 ☽ ⊻ ☉
17 19 ☽ ⊻ ♂
20 33 ☽ ☍ ♇
21 41 ☽ # ♆
22 40 ☽ ∥ ☿
23 06 ☽ ⊥ ☿

30 Friday
01 08 ☽ ⊻ ♀
03 58 ☽ ∥ ♃
08 26 ☽ ⊻ ☿
11 47 ☽ △ ♅
12 41 ☽ ⚼ ♆
13 53 ♀ ± ♅
15 04 ☿ # ♆
21 49 ☽ ⊻ ♄

All ephemeris data is given at 12.00 UT and the Moon's longitude is additionally given for 24.00 UT

JULY 2000

LONGITUDES

Date	Sidereal time h m s	Sun ☉ ° ′ ″	Moon ☽ ° ′ ″	Moon ☽ 24.00 ° ′ ″	Mercury ☿ ° ′	Venus ♀ ° ′	Mars ♂ ° ′	Jupiter ♃ ° ′	Saturn ♄ ° ′	Uranus ♅ ° ′	Neptune ♆ ° ′	Pluto ♇ ° ′
01	06 39 24	09 ♋ 56 38	05 ♋ 35 35	13 ♋ 11 29	17 ♋ 39	15 ♋ 27	09 ♋ 59	00 ♊ 15	26 ♉ 42	20 ♒ 18	05 ♒ 52	10 ♐ 47
02	06 43 20	10 53 52	20 ♋ 46 59	28 ♋ 20 53	17 R 06	16 41	10 39	00 27	26 48	20 R 16	05 R 50	10 R 45
03	06 47 17	11 51 06	05 ♌ 52 05	13 ♌ 19 34	16 32	17 55	11 19	00 40	26 54	20 14	05 49	10 44
04	06 51 13	12 48 19	20 ♌ 42 29	28 ♌ 00 07	15 55	19 08	11 59	00 52	27 00	20 13	05 48	10 43
05	06 55 10	13 45 33	05 ♍ 11 56	12 ♍ 17 35	15 18	20 22	12 38	01 04	27 07	20 11	05 46	10 41
06	06 59 06	14 42 46	19 ♍ 16 50	26 ♍ 09 39	14 41	21 36	13 18	01 16	27 13	20 09	05 45	10 40
07	07 03 03	15 39 58	02 ♎ 56 05	09 ♎ 36 18	14 04	22 50	13 57	01 28	27 19	20 07	05 43	10 39
08	07 07 00	16 37 11	16 ♎ 10 36	22 ♎ 39 17	13 28	24 03	14 37	01 40	27 25	20 05	05 42	10 38
09	07 10 56	17 34 23	29 ♎ 02 45	05 ♏ 21 26	12 54	25 17	15 16	01 52	27 31	20 03	05 40	10 36
10	07 14 53	18 31 35	11 ♏ 35 45	17 ♏ 46 11	12 22	26 31	15 56	02 04	27 36	20 02	05 39	10 35
11	07 18 49	19 28 47	23 ♏ 53 10	29 ♏ 57 09	11 53	27 45	16 35	02 16	27 42	20 00	05 37	10 34
12	07 22 46	20 26 00	05 ♐ 58 33	11 ♐ 57 48	11 27	28 ♋ 58	17 15	02 27	27 48	19 58	05 35	10 33
13	07 26 42	21 23 12	17 ♐ 55 17	23 ♐ 51 23	11 05	00 ♌ 12	17 54	02 39	27 54	19 56	05 34	10 32
14	07 30 39	22 20 24	29 ♐ 46 26	05 ♑ 40 47	10 47	01 26	18 34	02 51	27 59	19 54	05 32	10 31
15	07 34 35	23 17 37	11 ♑ 34 45	17 ♑ 28 37	10 34	02 40	19 13	03 02	28 05	19 52	05 31	10 30
16	07 38 32	24 14 50	23 ♑ 22 43	29 ♑ 17 18	10 26	03 53	19 52	03 13	28 10	19 50	05 29	10 29
17	07 42 29	25 12 04	05 ♒ 12 39	11 ♒ 09 05	10 23	05 07	20 31	03 25	28 15	19 48	05 28	10 28
18	07 46 25	26 09 17	17 ♒ 06 51	23 ♒ 06 15	10 D 26	06 21	21 11	03 36	28 21	19 45	05 26	10 27
19	07 50 22	27 06 32	29 ♒ 07 37	05 ♓ 11 14	10 34	07 35	21 50	03 47	28 26	19 43	05 24	10 26
20	07 54 18	28 03 46	11 ♓ 17 26	17 ♓ 26 35	10 47	08 49	22 29	03 58	28 31	19 41	05 23	10 25
21	07 58 15	29 01 01	23 ♓ 39 01	29 ♓ 55 07	11 07	10 02	23 08	04 09	28 36	19 39	05 21	10 24
22	08 02 11	29 ♋ 58 18	06 ♈ 15 13	12 ♈ 39 43	11 32	11 16	23 47	04 19	28 41	19 37	05 20	10 23
23	08 06 08	00 ♌ 55 35	19 ♈ 08 57	25 ♈ 43 16	12 02	12 30	24 26	04 30	28 46	19 35	05 18	10 22
24	08 10 04	01 52 53	02 ♉ 22 57	09 ♉ 08 14	12 39	13 44	25 06	04 41	28 51	19 32	05 16	10 21
25	08 14 01	02 50 12	15 ♉ 59 19	22 ♉ 56 17	13 21	14 57	25 45	04 51	28 56	19 30	05 15	10 20
26	08 17 58	03 47 32	29 ♉ 59 08	07 ♊ 07 42	14 10	16 11	26 24	05 01	29 01	19 28	05 13	10 19
27	08 21 54	04 44 53	14 ♊ 21 45	21 ♊ 40 51	15 03	17 25	27 03	05 12	29 05	19 26	05 11	10 19
28	08 25 51	05 42 15	29 ♊ 04 27	06 ♋ 31 49	16 02	18 39	27 42	05 22	29 10	19 23	05 10	10 18
29	08 29 47	06 39 38	14 ♋ 02 06	21 ♋ 34 20	17 07	19 53	28 21	05 32	29 14	19 21	05 08	10 17
30	08 33 44	07 37 02	29 ♋ 07 28	06 ♌ 40 21	18 17	21 06	28 59	05 42	29 19	19 19	05 07	10 16
31	08 37 40	08 ♌ 34 26	14 ♌ 11 52	21 ♌ 40 55	19 ♋ 32	22 ♌ 20	29 ♋ 38	05 ♊ 52	29 ♉ 23	19 ♒ 16	05 ♒ 05	10 ♐ 16

Date	Moon True ☊ ° ′	Moon Mean ☊ ° ′	Moon ☽ Latitude ° ′
01	24 ♋ 37	25 ♋ 24	01 S 43
02	24 R 37	25 21	00 S 21
03	24 D 37	25 18	01 N 02
04	24 37	25 15	02 20
05	24 38	25 12	03 27
06	24 39	25 09	04 19
07	24 40	25 05	04 55
08	24 40	25 02	05 14
09	24 R 40	24 59	05 17
10	24 39	24 56	05 04
11	24 39	24 53	04 37
12	24 38	24 49	03 59
13	24 38	24 46	03 10
14	24 37	24 43	02 14
15	24 37	24 40	01 12
16	24 D 37	24 37	00 N 07
17	24 R 37	24 34	00 S 59
18	24 37	24 30	02 02
19	24 37	24 27	03 00
20	24 36	24 24	03 51
21	24 36	24 21	04 33
22	24 35	24 18	05 02
23	24 35	24 15	05 16
24	24 D 35	24 11	05 15
25	24 35	24 08	04 56
26	24 36	24 05	04 19
27	24 36	24 02	03 25
28	24 37	23 59	02 17
29	24 38	23 55	00 S 58
30	24 R 38	23 52	00 N 25
31	24 ♋ 37	23 ♋ 49	01 N 46

DECLINATIONS

Date	Sun ☉ ° ′	Moon ☽ ° ′	Mercury ☿ ° ′	Venus ♀ ° ′	Mars ♂ ° ′	Jupiter ♃ ° ′	Saturn ♄ ° ′	Uranus ♅ ° ′	Neptune ♆ ° ′	Pluto ♇ ° ′
01	23 N 04	21 N 36	18 N 27	23 N 23	23 N 56	19 N 23	17 N 27	15 S 25	18 S 36	10 S 57
02	22 59	21 29	18 19	23 16	23 53	19 25	17 28	15 25	18 36	10 57
03	22 55	19 48	18 12	23 08	23 50	19 28	17 30	15 26	18 36	10 57
04	22 49	16 48	18 07	23 00	23 47	19 30	17 31	15 26	18 37	10 57
05	22 44	12 48	18 02	22 51	23 44	19 33	17 32	15 27	18 37	10 57
06	22 38	08 13	17 59	22 42	23 40	19 35	17 33	15 27	18 38	10 57
07	22 31	03 N 21	17 58	22 31	23 37	19 38	17 35	15 28	18 38	10 58
08	22 24	01 S 32	17 57	22 20	23 33	19 40	17 36	15 29	18 38	10 58
09	22 17	06 12	17 58	22 09	23 29	19 43	17 37	15 29	18 39	10 58
10	22 09	10 29	18 01	21 56	23 24	19 45	17 38	15 30	18 39	10 58
11	22 01	14 16	18 04	21 44	23 20	19 47	17 40	15 30	18 39	10 58
12	21 53	17 23	18 09	21 30	23 15	19 49	17 41	15 31	18 40	10 58
13	21 44	19 44	18 14	21 16	23 10	19 52	17 42	15 32	18 40	10 59
14	21 35	21 12	18 21	21 01	23 05	19 54	17 43	15 32	18 41	10 59
15	21 26	21 44	18 29	20 46	23 00	19 56	17 44	15 33	18 41	10 59
16	21 16	21 18	18 38	20 30	22 55	19 58	17 45	15 34	18 41	10 59
17	21 06	19 55	18 47	20 13	22 49	20 00	17 46	15 34	18 42	10 59
18	20 55	17 38	18 57	19 56	22 44	20 02	17 47	15 35	18 42	11 00
19	20 44	14 35	19 08	19 39	22 38	20 04	17 48	15 36	18 43	11 00
20	20 33	10 54	19 19	19 20	22 32	20 06	17 49	15 36	18 43	11 00
21	20 21	06 41	19 30	19 02	22 26	20 08	17 50	15 37	18 43	11 00
22	20 09	02 S 08	19 41	18 42	22 19	20 10	17 51	15 38	18 44	11 01
23	19 57	02 N 37	19 52	18 23	22 13	20 12	17 52	15 39	18 44	11 01
24	19 44	07 22	20 03	18 02	22 06	20 14	17 53	15 39	18 45	11 01
25	19 31	11 54	20 14	17 42	21 59	20 16	17 54	15 40	18 45	11 01
26	19 18	15 55	20 24	17 21	21 52	20 18	17 55	15 41	18 45	11 02
27	19 05	19 07	20 33	16 59	21 45	20 20	17 56	15 42	18 46	11 02
28	18 51	21 09	20 41	16 36	21 37	20 21	17 56	15 42	18 46	11 02
29	18 36	21 44	20 48	16 14	21 30	20 23	17 57	15 43	18 47	11 03
30	18 22	20 44	20 54	15 50	21 22	20 25	17 58	15 44	18 47	11 03
31	18 N 07	18 N 16	20 N 58	15 N 27	21 N 14	20 N 26	17 N 59	15 S 44	18 S 47	11 S 03

ZODIAC SIGN ENTRIES

Date	h	m	Planets
01	03	09	☽ ♋
03	02	38	☽ ♌
05	03	19	☽ ♍
07	06	47	☽ ♎
09	13	48	☽ ♏
12	00	06	☽ ♐
13	08	02	♀ ♌
14	12	28	☽ ♑
17	01	27	☽ ♒
19	13	44	☽ ♓
22	00	09	☽ ♈
22	12	43	☉ ♌
24	07	44	☽ ♉
26	12	01	☽ ♊
28	13	30	☽ ♋
30	13	23	☽ ♌

LATITUDES

Date	Mercury ☿ ° ′	Venus ♀ ° ′	Mars ♂ ° ′	Jupiter ♃ ° ′	Saturn ♄ ° ′	Uranus ♅ ° ′	Neptune ♆ ° ′	Pluto ♇ ° ′
01	03 S 51	00 N 51	00 N 52	00 S 50	02 S 01	00 S 44	00 N 13	11 N 13
04	04 24	00 56	00 53	00 50	02 02	00 44	00 13	11 12
07	04 46	01 02	00 54	00 51	02 02	00 44	00 13	11 11
10	04 52	01 07	00 55	00 51	02 02	00 44	00 13	11 10
13	04 45	01 11	00 57	00 51	02 03	00 44	00 13	11 09
16	04 25	01 15	00 58	00 51	02 03	00 44	00 13	11 08
19	03 54	01 19	00 58	00 51	02 03	00 44	00 13	11 07
22	03 16	01 22	00 59	00 51	02 04	00 44	00 13	11 06
25	02 33	01 24	01 00	00 51	02 04	00 44	00 13	11 05
28	01 48	01 26	01 00	00 51	02 05	00 44	00 13	11 04
31	01 S 03	01 N 28	01 N 02	00 S 51	02 S 05	00 S 44	00 N 13	11 N 03

DATA

Julian Date	2451727
Delta T	+64 seconds
Ayanamsa	23° 51′ 35″
Synetic vernal point	05° ♓ 15′ 24″
True obliquity of ecliptic	23° 26′ 17″

MOON'S PHASES, APSIDES AND POSITIONS ☽

Date	h	m	Phase	Longitude ° ′	Eclipse Indicator
01	19	20	●	10 ♋ 14	Partial
08	12	53	☽	16 ♎ 39	
16	13	55	○	24 ♑ 19	total
24	11	02	☾	01 ♉ 51	
31	02	25	●	08 ♌ 12	Partial

Day	h	m		
01	22	10	Perigee	
15	15	18	Apogee	
30	07	37	Perigee	
01	22	17	Max dec	21° N 45′
08	04	25	0S	
15	13	09	Max dec	21° S 44′
22	22	50	0N	
29	08	57	Max dec	21° N 45′

LONGITUDES

Date	Chiron ⚷ ° ′	Ceres ⚳ ° ′	Pallas ⚴ ° ′	Juno ⚵ ° ′	Vesta ⚶ ° ′	Black Moon Lilith ⚸ ° ′
01	12 ♐ 21	03 ♎ 56	02 ♍ 33	26 ♒ 49	28 ♑ 02	13 ♑ 40
11	11 ♐ 49	06 ♎ 50	06 ♍ 48	26 ♒ 03	25 ♑ 46	14 ♑ 47
21	11 ♐ 25	10 ♎ 01	11 ♍ 07	24 ♒ 40	23 ♑ 21	15 ♑ 53
31	11 ♐ 10	13 ♎ 26	15 ♍ 28	22 ♒ 42	21 ♑ 07	17 ♑ 00

All ephemeris data is given at 12.00 UT and the Moon's longitude is additionally given for 24.00 UT

ASPECTARIAN

01 Saturday
02 57 ☽ ± ♆
03 26 ☽ ⊻ ♃
07 13 ♀ ∠ ♃
07 23 ☽ ⊥ ♄
11 32 ☽ ⚼ ♅
12 26 ☽ ⊼ ♆
13 03 ☽ ⊥ ♃
15 50 ☉ ☌ ♂
19 16 ☽ ☌ ♂
19 20 ☽ ☌ ☉
20 11 ☽ ⊼ ♇
21 42 ☽ ∠ ♄

02 Sunday
01 43 ☽ ± ♅
03 27 ☽ ∠ ♃
04 56 ☽ ☌ ♀
05 38 ☽ ± ♇
06 23 ☽ ☌ ☿
08 31 ☉ ⊼ ♇
11 11 ☽ ⊼ ♅
13 28 ♀ ± ♇
15 37 ♂ ⊼ ♇
17 40 ☿ ☌ ♀
19 52 ☽ ⚼ ♇
21 36 ☽ ✶ ♄

03 Monday
03 09 ☿ ± ♇
03 34 ☽ ✶ ♃
11 55 ☽ ☍ ♆
13 30 ☉ ∠ ♄
15 15 ☽ ∥ ♃
16 54 ☽ Q ♄
19 48 ☽ △ ♇
21 10 ☽ ⊻ ♂
22 17 ☽ ⊻ ☉
22 42 ☽ # ♆
23 05 ☽ Q ♃

04 Tuesday
02 21 ☽ ∥ ☿
04 31 ☽ ⊻ ☿
07 03 ☽ ∥ ♄
07 20 ☽ ⊥ ♂
08 41 ☽ ⊥ ☉
09 12 ☽ ⊻ ♀
11 11 ☽ ☍ ♅
13 22 ♂ ∠ ♄
13 44 ☿ ∠ ♃
13 55 ☽ ⊥ ☿
19 56 ☽ ⊥ ♀
20 44 ☽ # ♅
22 26 ☽ □ ♄
22 48 ☽ ∠ ♂

05 Wednesday
00 30 ☽ ∠ ☉
04 10 ☽ ∠ ☿
05 00 ☽ □ ♃
08 25 ♀ ⊼ ♅
12 19 ☽ ∠ ♀
12 57 ☽ ⊼ ♆
21 16 ☽ □ ♇
21 57 ☽ # ♇
22 17 ☉ ± ♅
23 05 ☽ ± ♆

06 Thursday
01 12 ☽ ✶ ♂
03 34 ☽ ✶ ☉
04 25 ☽ ✶ ☿
11 35 ☉ ☌ ☿
13 30 ☽ ⊼ ♅
14 32 ☽ ⚼ ♆
16 25 ☽ ✶ ♀
23 01 ☽ Q ♂
23 57 ☽ ± ♅

07 Friday
00 22 ☽ Q ☿
01 57 ☽ Q ☉
01 57 ☽ △ ♄
04 24 ☽ Q ♇
09 21 ☽ △ ♃
09 55 ☿ ± ♅
12 45 ☉ ∥ ♀
14 08 ☿ ☌ ♂
15 44 ☽ Q ♀
15 54 ☽ ⚼ ♅
16 59 ☽ △ ♆
17 40 ♂ ± ♅

08 Saturday
01 52 ☽ ✶ ♇
05 03 ☽ ⚼ ♄
07 15 ☽ □ ☿
08 59 ☽ □ ♂
12 12 ☉ ± ♇
12 53 ☽ □ ☉
12 55 ☽ ⚼ ♃
13 38 ☉ ∠ ♃
19 13 ☽ △ ♅
21 46 ☽ ± ♄

09 Sunday
04 10 ☽ □ ♀
05 32 ☽ ∠ ♇
05 55 ☽ ± ♃
09 05 ☽ ⊼ ♄
17 26 ☽ ⊼ ♃
18 11 ♀ ⚼ ♇
22 33 ☽ ⊥ ♇

10 Monday
00 34 ☽ □ ♆
02 41 ☿ ∠ ♄
10 03 ☽ ⊻ ♇
13 26 ☽ △ ☿
14 51 ☽ ∥ ♇
20 54 ☽ △ ♂

11 Tuesday
02 37 ☽ △ ☉
04 22 ☽ □ ♅
11 08 ♀ ✶ ♄
11 12 ♂ ± ♇
11 28 ☽ Q ♆
17 42 ☽ ⚼ ☿
19 36 ☽ ☍ ♄
20 29 ☽ △ ♀
20 58 ☽ ∥ ♅

12 Wednesday
00 31 ☉ ⊼ ♅
04 08 ☽ ⚼ ♂
04 52 ☽ ☍ ♃
10 49 ☽ ⚼ ☉
10 59 ☽ ± ☿
11 14 ☽ ✶ ♆
14 37 ☽ # ♄
15 58 ☽ Q ♅
19 10 ☽ # ☿
21 09 ☽ ☌ ♇
22 38 ☽ ⊼ ☿
22 52 ♂ ∠ ♃
23 11 ☽ ± ♂

13 Thursday
00 04 ☽ ∥ ♆
05 53 ☽ ⚼ ♀
06 27 ☽ ± ☉
11 58 ☽ ⊼ ♂
13 39 ☽ # ♃
16 03 ☽ ✶ ♅
17 20 ☽ ∠ ♆
19 37 ☽ ⊼ ☉

14 Friday
02 10 ☽ ± ♀
08 21 ☽ ⊼ ♄
08 36 ☽ # ♀
11 32 ☽ ⊥ ♆
15 46 ☽ ⊼ ♀
18 20 ☽ ⊼ ♃
20 37 ☽ ± ♄
21 13 ☽ # ☉
22 22 ☽ ∠ ♅
23 41 ☽ ⊻ ♆

15 Saturday
06 44 ☽ ± ♃
09 48 ☽ ⊻ ♇
09 59 ☽ ☍ ☿
15 04 ☽ ⚼ ♄
16 38 ☽ ⊥ ♅
20 33 ♀ ✶ ♃
21 59 ☽ ⊥ ♇

16 Sunday
01 21 ☽ ⚼ ♃
01 48 ☿ ⊼ ♇
04 27 ☽ ☍ ♂
04 48 ☽ ⊻ ♅
10 29 ♂ ⊼ ♅
13 08 ☽ # ☉
13 55 ☽ ☍ ☉
16 15 ☽ ∠ ♇
21 48 ☽ △ ♄
22 04 ☿ # ♆

17 Monday
06 55 ☽ # ♀
08 18 ☽ △ ♃
10 48 ☽ # ♃
11 48 ☽ ☍ ♀
12 30 ☽ ☌ ♆
13 20 ☿ St D
18 22 ☉ ⚼ ♇
18 29 ♀ ☍ ♆
22 28 ☽ ⊼ ☿
22 35 ☽ ✶ ♇

18 Tuesday
00 07 ☽ # ☿
01 52 ☽ ∥ ♆
04 35 ♀ ∥ ♃
10 37 ☽ ± ☿
10 41 ☽ # ♄
15 18 ☿ ⊼ ♇
17 17 ☽ ☌ ♅
20 37 ☽ ⊼ ♂
22 40 ☽ Q ♇

19 Wednesday
04 42 ☽ ∥ ♅
04 48 ☽ ⚼ ☿
07 39 ☽ ⊼ ☉
09 17 ☽ ± ♂
10 37 ☽ □ ♄
20 34 ☽ ± ☉
21 22 ☽ □ ♃

20 Thursday
00 24 ☽ ⊻ ♆
04 06 ☽ ⚼ ♂
06 35 ☽ ⊼ ♀
10 17 ☽ □ ♇
10 59 ☽ △ ☿
11 21 ☽ ∥ ♇
12 11 ☽ ⊥ ♆
13 27 ☿ ∥ ♀
15 46 ☽ ⚼ ☉
19 38 ☽ ± ♀
22 17 ☽ Q ♄

21 Friday
00 38 ☉ ✶ ♄
04 18 ☽ ⊻ ♅
05 39 ☽ ∠ ♆
09 03 ☽ Q ♃
10 58 ☽ △ ♂
14 58 ☽ ⚼ ♀
15 50 ☽ ⊥ ♅
18 51 ♀ △ ♇
21 34 ☽ ✶ ♄
23 08 ☽ △ ☉
23 52 ♀ Q ♄

22 Saturday
08 18 ☽ ✶ ♃
08 55 ☽ ∠ ♅
10 15 ☽ ✶ ♆
10 19 ♀ # ♆
10 22 ☉ ∥ ♃
19 44 ☽ △ ♇
20 18 ☿ ⊻ ♀
22 16 ☽ □ ☿
22 24 ☽ △ ♀

23 Sunday
02 00 ☽ ∠ ♄
08 36 ☽ Q ♆
12 39 ☽ ∠ ♃
12 47 ☽ ✶ ♅
16 51 ☉ ∥ ☿
18 40 ☽ ⊥ ♄
22 11 ☽ □ ♂
23 20 ☽ ⚼ ♇

24 Monday
05 15 ☽ ⊥ ♃
05 37 ☽ ⊻ ♄
08 44 ☽ Q ☿
10 30 ☽ Q ♅
11 02 ☽ □ ☉
15 31 ☽ ± ♇
16 09 ☽ ⊻ ♃
17 09 ☽ □ ♆
21 17 ♂ ⚼ ♇
22 26 ♀ ∥ ♄

25 Tuesday
02 07 ☽ ⊼ ♇
07 09 ☽ ✶ ☿
07 12 ☽ # ♇
07 53 ☽ Q ♂
10 02 ☽ □ ♀
16 17 ☿ ± ♅
18 04 ☽ □ ♅
18 13 ☿ ∥ ♃
21 00 ☽ Q ☉

26 Wednesday
05 37 ☽ ✶ ♂
07 20 ☿ ∠ ♄
10 20 ☽ ☌ ♄
10 25 ☽ # ♅
10 31 ☽ ∠ ☿
18 52 ☽ ✶ ☉
19 45 ☽ Q ♀
20 35 ☽ ☌ ♃
20 46 ☽ ∥ ♀
20 47 ☽ △ ♆

27 Thursday
02 04 ☽ ∥ ♄
02 36 ☽ ⊥ ☿
05 18 ☽ ☍ ♇
07 00 ♀ Q ♃
07 59 ☽ ∠ ♂
08 53 ☽ # ♆
11 22 ♃ △ ♆
11 38 ☽ ∥ ☉
13 13 ☽ ⊻ ☿
17 29 ☽ ✶ ♀
20 18 ☽ △ ♅
21 28 ☽ ∠ ☉
21 33 ☽ ⚼ ♆
22 49 ☉ ☍ ♆
23 28 ☽ ⊥ ♂

28 Friday
00 28 ☽ ∥ ♃
01 41 ☉ ✶ ♃
04 17 ☽ ∥ ☿
09 40 ☽ ⊻ ♂
12 09 ☽ ± ♆
12 09 ☽ ⊻ ♄
13 05 ☽ ⊥ ☉
17 51 ☿ ± ♇
19 14 ☉ # ♆
20 02 ☽ ∠ ♀
20 32 ☽ ⚼ ♅
21 46 ☽ ∥ ♂
21 47 ☽ ⊼ ♆
21 51 ☽ ⊥ ♄
22 15 ☽ ⊻ ♃
23 24 ☽ ⊻ ☉

29 Saturday
02 00 ♀ ☍ ♅
06 01 ☽ ⊼ ♇
07 58 ☽ ⊥ ♃
10 54 ☽ ± ♅
11 44 ☽ ⊥ ♀
12 20 ☽ ∠ ♄
15 35 ☽ ± ♇
17 19 ☽ ☌ ☿
20 26 ☽ ⊼ ♅
22 08 ☽ ⊻ ♀
22 28 ☽ ∠ ♃
23 55 ☽ ∥ ♂

30 Sunday
05 53 ☽ ⚼ ♇
09 47 ☽ ∥ ☿
11 47 ☽ ☌ ♂
12 18 ☽ ✶ ♄
16 05 ☽ ∥ ♃
18 44 ♀ # ♅
21 30 ☽ ☍ ♆
22 34 ☽ ✶ ♃

31 Monday
01 21 ♂ ✶ ♄
02 25 ☽ ☌ ☉
05 43 ☽ △ ♇
07 11 ☿ ⊼ ♅
07 29 ☽ Q ♄
07 50 ☽ # ♆
13 13 ☽ ∥ ☉
14 08 ☽ ∥ ♄
17 56 ☽ Q ♃
20 06 ☽ ☍ ♅
21 23 ☽ ⊻ ☿

AUGUST 2000

LONGITUDES

Date	Sidereal time h m s	Sun ☉ ° ' "	Moon ☽ ° ' "	Moon ☽ 24.00 ° ' "	Mercury ☿ ° '	Venus ♀ ° '	Mars ♂ ° '	Jupiter ♃ ° '	Saturn ♄ ° '	Uranus ♅ ° '	Neptune ♆ ° '	Pluto ♇ ° '
01	08 41 37	09 ♌ 31 52	29 ♌ 06 27	06 ♍ 27 32	20 ♋ 52	23 ♌ 34	00 ♌ 17	06 ♊ 02	29 ♉ 27	19 ♒ 14	05 ♒ 03	10 ♐ 15
02	08 45 33	10 29 18	13 ♍ 43 23	20 ♍ 53 21	22 18	24 48	00 56	06 11	29 31	19 R 12	05 R 02	10 R 15
03	08 49 30	11 26 45	27 ♍ 56 58	04 ♎ 53 56	23 47	26 02	01 35	06 21	29 35	19 09	05 00	10 14
04	08 53 27	12 24 12	11 ♎ 44 07	18 ♎ 27 31	25 22	27 15	02 14	06 30	29 39	19 07	04 58	10 13
05	08 57 23	13 21 40	25 ♎ 04 19	01 ♏ 34 45	27 00	28 29	02 52	06 39	29 43	19 04	04 57	10 13
06	09 01 20	14 19 09	07 ♏ 59 13	14 ♏ 18 10	28 ♋ 42	29 ♌ 43	03 31	06 48	29 47	19 02	04 55	10 12
07	09 05 16	15 16 39	20 ♏ 32 04	26 ♏ 41 28	00 ♌ 28	00 ♍ 57	04 10	06 57	29 51	19 00	04 54	10 12
08	09 09 13	16 14 10	02 ♐ 46 58	08 ♐ 49 06	02 17	02 11	04 49	07 06	29 54	18 57	04 52	10 12
09	09 13 09	17 11 41	14 ♐ 48 29	20 ♐ 45 39	04 09	03 24	05 27	07 15	29 ♉ 58	18 55	04 50	10 11
10	09 17 06	18 09 13	26 ♐ 41 11	02 ♑ 35 35	06 04	04 38	06 06	07 24	00 ♊ 01	18 53	04 49	10 11
11	09 21 02	19 06 46	08 ♑ 29 23	14 ♑ 23 01	08 01	05 52	06 45	07 32	00 05	18 50	04 47	10 11
12	09 24 59	20 04 21	20 ♑ 16 56	26 ♑ 11 33	09 59	07 06	07 23	07 41	00 08	18 48	04 46	10 10
13	09 28 56	21 01 56	02 ♒ 07 13	08 ♒ 04 15	11 59	08 19	08 02	07 49	00 11	18 45	04 44	10 10
14	09 32 52	21 59 32	14 ♒ 02 57	20 ♒ 03 36	14 00	09 33	08 40	07 57	00 14	18 43	04 43	10 10
15	09 36 49	22 57 10	26 ♒ 06 24	02 ♓ 11 34	16 01	10 47	09 19	08 05	00 17	18 41	04 41	10 10
16	09 40 45	23 54 48	08 ♓ 19 17	14 ♓ 29 42	18 03	12 00	09 57	08 13	00 20	18 38	04 40	10 09
17	09 44 42	24 52 28	20 ♓ 42 59	26 ♓ 59 15	20 05	13 14	10 36	08 20	00 23	18 36	04 38	10 09
18	09 48 38	25 50 10	03 ♈ 18 39	09 ♈ 41 18	22 07	14 28	11 14	08 28	00 25	18 33	04 37	10 09
19	09 52 35	26 47 53	16 ♈ 07 20	22 ♈ 36 54	24 09	15 42	11 53	08 35	00 28	18 31	04 35	10 09
20	09 56 31	27 45 37	29 ♈ 10 07	05 ♉ 47 09	26 10	16 55	12 31	08 43	00 30	18 29	04 34	10 09
21	10 00 28	28 43 23	12 ♉ 28 07	19 ♉ 13 09	28 ♌ 10	18 09	13 09	08 50	00 33	18 26	04 32	10 D 09
22	10 04 25	29 ♌ 41 11	26 ♉ 02 22	02 ♊ 55 50	00 ♍ 09	19 23	13 48	08 57	00 35	18 24	04 31	10 09
23	10 08 21	00 ♍ 39 01	09 ♊ 53 37	16 ♊ 55 41	02 07	20 36	14 26	09 04	00 37	18 22	04 29	10 09
24	10 12 18	01 36 53	24 ♊ 01 57	01 ♋ 12 16	04 05	21 50	15 04	09 10	00 39	18 19	04 28	10 09
25	10 16 14	02 34 46	08 ♋ 26 19	15 ♋ 43 45	06 01	23 04	15 43	09 17	00 41	18 17	04 27	10 09
26	10 20 11	03 32 41	23 ♋ 04 03	00 ♌ 26 35	07 56	24 17	16 21	09 23	00 43	18 15	04 25	10 10
27	10 24 07	04 30 38	07 ♌ 50 38	15 ♌ 15 20	09 49	25 31	16 59	09 29	00 45	18 13	04 24	10 10
28	10 28 04	05 28 36	22 ♌ 39 47	00 ♍ 03 01	11 42	26 45	17 38	09 35	00 47	18 10	04 22	10 10
29	10 32 00	06 26 36	07 ♍ 24 04	14 ♍ 42 00	13 33	27 59	18 16	09 41	00 48	18 08	04 21	10 10
30	10 35 57	07 24 38	21 ♍ 55 54	29 ♍ 05 01	15 23	29 ♍ 12	18 54	09 47	00 50	18 06	04 20	10 11
31	10 39 54	08 ♍ 22 41	06 ♎ 08 42	13 ♎ 06 25	17 ♍ 11	00 ♎ 26	19 ♌ 32	09 ♊ 52	00 ♊ 51	18 ♒ 04	04 ♒ 19	10 ♐ 11

Date	Moon True ☊ ° '	Moon Mean ☊ ° '	Moon ☽ Latitude ° '
01	24 ♋ 36	23 ♋ 46	03 N 00
02	24 R 34	23 43	04 00
03	24 32	23 40	04 43
04	24 30	23 36	05 08
05	24 29	23 33	05 16
06	24 28	23 30	05 07
07	24 D 28	23 27	04 44
08	24 28	23 24	04 08
09	24 30	23 21	03 22
10	24 32	23 17	02 28
11	24 33	23 14	01 27
12	24 34	23 11	00 N 24
13	24 R 34	23 08	00 S 41
14	24 32	23 05	01 45
15	24 29	23 01	02 45
16	24 25	22 58	03 37
17	24 21	22 55	04 20
18	24 15	22 52	04 52
19	24 11	22 49	05 09
20	24 07	22 46	05 11
21	24 04	22 42	04 56
22	24 03	22 39	04 25
23	24 D 03	22 36	03 38
24	24 04	22 33	02 37
25	24 06	22 30	01 24
26	24 06	22 26	00 S 06
27	24 R 05	22 23	01 N 14
28	24 03	22 20	02 29
29	23 58	22 17	03 33
30	23 52	22 14	04 23
31	23 ♋ 45	22 ♋ 11	04 N 54

DECLINATIONS

Date	Sun ☉ ° '	Moon ☽ ° '	Mercury ☿ ° '	Venus ♀ ° '	Mars ♂ ° '	Jupiter ♃ ° '	Saturn ♄ ° '	Uranus ♅ ° '	Neptune ♆ ° '	Pluto ♇ ° '
01	17 N 52	14 N 35	21 N 01	15 N 03	21 N 06	20 N 28	18 N 00	15 S 45	18 S 48	11 S 04
02	17 37	10 05	21 02	14 39	20 58	20 30	18 00	15 46	18 48	11 04
03	17 21	05 08	21 01	14 14	20 50	20 31	18 01	15 47	18 49	11 04
04	17 05	00 N 06	20 57	13 49	20 41	20 33	18 02	15 47	18 49	11 05
05	16 49	04 S 48	20 51	13 23	20 33	20 34	18 02	15 48	18 50	11 05
06	16 32	09 19	20 43	12 57	20 24	20 36	18 03	15 49	18 50	11 05
07	16 15	13 19	20 32	12 31	20 15	20 37	18 04	15 50	18 50	11 06
08	15 58	16 39	20 18	12 05	20 06	20 39	18 04	15 50	18 51	11 06
09	15 41	19 14	20 02	11 38	19 57	20 40	18 05	15 51	18 51	11 06
10	15 23	20 56	19 43	11 11	19 48	20 41	18 05	15 52	18 51	11 07
11	15 05	21 43	19 22	10 43	19 38	20 43	18 06	15 53	18 52	11 07
12	14 47	21 31	18 58	10 15	19 29	20 44	18 06	15 53	18 52	11 08
13	14 29	20 22	18 31	09 47	19 19	20 45	18 07	15 54	18 53	11 08
14	14 11	18 17	18 02	09 19	19 09	20 46	18 07	15 55	18 53	11 08
15	13 52	15 24	17 31	08 51	18 59	20 48	18 08	15 56	18 53	11 09
16	13 33	11 48	16 58	08 22	18 49	20 49	18 08	15 56	18 54	11 09
17	13 14	07 40	16 22	07 53	18 39	20 50	18 09	15 57	18 54	11 10
18	12 54	03 S 09	15 46	07 24	18 29	20 51	18 09	15 58	18 55	11 10
19	12 35	01 N 35	15 07	06 54	18 18	20 52	18 09	15 59	18 55	11 10
20	12 15	06 20	14 27	06 25	18 07	20 53	18 10	15 59	18 55	11 11
21	11 55	10 52	13 46	05 55	17 57	20 54	18 10	16 00	18 56	11 11
22	11 35	14 58	13 04	05 25	17 46	20 55	18 10	16 01	18 56	11 12
23	11 15	18 21	12 20	04 55	17 35	20 56	18 10	16 02	18 56	11 12
24	10 54	20 42	11 36	04 25	17 24	20 57	18 11	16 02	18 57	11 13
25	10 33	21 46	10 52	03 55	17 13	20 58	18 11	16 03	18 57	11 13
26	10 12	21 22	10 06	03 24	17 02	20 59	18 11	16 04	18 57	11 14
27	09 51	19 30	09 20	02 54	16 50	21 00	18 11	16 04	18 58	11 14
28	09 30	16 18	08 34	02 23	16 39	21 01	18 11	16 05	18 58	11 15
29	09 09	12 05	07 48	01 52	16 27	21 01	18 12	16 06	18 58	11 15
30	08 47	07 13	07 01	01 22	16 16	21 02	18 12	16 06	18 59	11 16
31	08 N 26	02 N 04	06 N 14	00 N 51	16 N 04	21 N 03	18 N 12	16 S 07	18 S 59	11 S 16

ZODIAC SIGN ENTRIES

Date	h	m	Planets
01	01	21	♂ ♌
01	13	27	☽ ♍
03	15	31	☽ ♎
05	21	04	☽ ♏
06	17	32	♀ ♍
07	05	42	☿ ♌
08	06	30	☽ ♐
10	02	26	♄ ♊
10	18	44	☽ ♑
13	07	43	☽ ♒
15	19	41	☽ ♓
18	05	44	☽ ♈
20	13	31	☽ ♉
22	10	11	☿ ♍
22	18	55	☽ ♊
22	19	49	☉ ♍
24	22	00	☽ ♋
26	23	17	☽ ♌
28	23	55	☽ ♍
31	01	33	☽ ♎
31	03	35	♀ ♎

LATITUDES

Date	Mercury ☿ ° '	Venus ♀ ° '	Mars ♂ ° '	Jupiter ♃ ° '	Saturn ♄ ° '	Uranus ♅ ° '	Neptune ♆ ° '	Pluto ♇ ° '
01	00 S 49	01 N 28	01 N 02	00 S 51	02 S 05	00 S 44	00 N 13	11 N 02
04	00 S 07	01 29	01 03	00 52	02 06	00 44	00 13	11 01
07	00 N 30	01 29	01 04	00 52	02 06	00 44	00 13	11 00
10	01 00	01 28	01 05	00 52	02 07	00 44	00 12	10 58
13	01 22	01 27	01 06	00 52	02 07	00 44	00 12	10 57
16	01 37	01 25	01 06	00 52	02 08	00 44	00 12	10 56
19	01 45	01 22	01 07	00 53	02 08	00 44	00 12	10 54
22	01 45	01 19	01 08	00 53	02 09	00 44	00 12	10 53
25	01 40	01 16	01 08	00 53	02 10	00 44	00 12	10 52
28	01 31	01 11	01 09	00 53	02 10	00 44	00 12	10 50
31	01 N 17	01 N 07	01 N 10	00 S 53	02 S 11	00 S 44	00 N 12	10 N 49

DATA

Julian Date	2451758
Delta T	+64 seconds
Ayanamsa	23° 51' 40"
Synetic vernal point	05° ♓ 15' 19"
True obliquity of ecliptic	23° 26' 17"

MOON'S PHASES, APSIDES AND POSITIONS ☽

Date	h	m	Phase	Longitude ° '	Eclipse Indicator
07	01	02	☽	14 ♏ 50	
15	05	13	○	22 ♒ 41	
22	18	51	☾	29 ♉ 58	
29	10	19	●	06 ♍ 23	

Day	h	m		
11	22	16	Apogee	
27	13	49	Perigee	
04	12	26	0S	
11	19	15	Max dec	21° S 45'
19	04	01	0N	
25	17	44	Max dec	21° N 49'
31	21	36	0S	

LONGITUDES

Date	Chiron ⚷ ° '	Ceres ⚳ ° '	Pallas ⚴ ° '	Juno ⚵ ° '	Vesta ⚶ ° '	Black Moon Lilith ⚸ ° '
01	11 ♐ 09	13 ♎ 47	15 ♍ 55	22 ♒ 29	20 ♑ 55	17 ♑ 06
11	11 ♐ 03	17 ♎ 23	20 ♍ 19	20 ♒ 08	19 ♑ 15	18 ♑ 13
21	11 ♐ 06	21 ♎ 09	24 ♍ 45	17 ♒ 41	18 ♑ 17	19 ♑ 19
31	11 ♐ 19	25 ♎ 03	29 ♍ 13	15 ♒ 25	18 ♑ 07	20 ♑ 26

All ephemeris data is given at 12.00 UT and the Moon's longitude is additionally given for 24.00 UT

ASPECTARIAN

h m Aspects

01 Tuesday
00 33 ☉ ∥ ♄
02 14 ☽ ☌ ♀
05 05 ☽ # ♅
08 01 ☽ ⊥ ☿
09 00 ☽ ∥ ♀
12 34 ☽ □ ♄
14 00 ☽ ⊻ ♂
14 59 ☿ ∠ ♃
21 41 ☽ ⊼ ♆
23 25 ☽ □ ♃

02 Wednesday
00 12 ☽ ∠ ☿
00 16 ☽ ⊥ ♂
01 18 ☿ ∥ ♂
05 54 ☉ △ ♇
06 14 ☽ □ ♇
06 16 ☽ ⊻ ☉
07 02 ☽ # ♇
07 32 ☽ ± ♆
15 52 ☽ ∠ ♂
16 56 ☽ ⊥ ☉
21 07 ☽ ⊼ ♅
22 32 ☽ bQ ♆

03 Thursday
04 03 ☽ ⚹ ☿
07 14 ☽ ± ♅
08 24 ☽ ⊻ ♀
09 14 ☽ ∠ ☉
12 29 ☽ Q ♇
14 50 ☽ △ ♄
15 54 ☉ Q ♄
18 33 ☽ ⚹ ♂
19 42 ☽ ⊥ ♀
22 40 ☽ bQ ♅

04 Friday
00 09 ☽ △ ♆
02 41 ☽ △ ♃
03 17 ☽ Q ☿
09 20 ☽ ⚹ ♇
10 00 ☿ bQ ♇
13 01 ☽ ∠ ♀
13 16 ☽ ⚹ ☉
16 39 ☽ Q ♂
17 13 ☽ bQ ♄

05 Saturday
01 08 ☽ △ ♅
05 42 ☽ bQ ♃
08 29 ♂ ∥ ♃
09 31 ☽ ± ♄
12 16 ☽ ∠ ♇
12 34 ☽ Q ☉
16 03 ☽ □ ☿
18 56 ☽ ⚹ ♀
20 36 ☽ ⊼ ♄
22 24 ☽ ± ♃

06 Sunday
03 10 ☽ □ ♂
04 54 ☽ ⊥ ♇
06 15 ☽ □ ♆
09 45 ☽ ⊼ ♃
13 25 ♀ □ ♄
16 12 ☽ ⊻ ♇
19 50 ☽ Q ♀
22 13 ☽ ∥ ♇

07 Monday
01 02 ☽ □ ☉
02 21 ☿ ∥ ♃
03 20 ☿ ⚹ ♄
07 27 ☽ # ♀
09 02 ☽ □ ♅
16 34 ☽ Q ♆

08 Tuesday
05 37 ☽ ∥ ♅
06 17 ☽ ☍ ♄
07 05 ☽ # ☉
07 34 ☿ ⊻ ♀
10 40 ☽ □ ♀
10 51 ☽ △ ☿
14 03 ♂ ☍ ♆
16 07 ☽ ⚹ ♆
16 15 ☽ △ ♂
20 15 ☽ Q ♅
20 41 ☽ ☍ ♃
22 19 ☉ # ♅

09 Wednesday
00 18 ☽ # ♄
02 44 ☽ ☌ ♇
07 57 ☽ ∥ ♆
17 13 ☽ △ ☉
20 02 ☽ # ♂
20 15 ☽ ⚹ ♅
20 27 ☽ # ☿
20 33 ☿ ☍ ♆
22 07 ☽ ∠ ♆
22 25 ☽ bQ ☿

10 Thursday
00 02 ☽ bQ ♂
01 43 ☿ ∥ ♂
07 31 ☽ # ♃
12 37 ☿ ☌ ♂
15 21 ♀ # ♇
15 27 ♀ ⊼ ♆
16 19 ☽ ⊥ ♆
18 48 ☽ ⊼ ♄
19 20 ☽ ± ♂
20 12 ☽ ± ☿

11 Friday
02 18 ☽ bQ ☉
02 34 ☽ ∠ ♅
04 29 ☽ ⊻ ♆
05 20 ☉ ☍ ♅
05 44 ☿ ⚹ ♃
06 02 ☽ △ ♀
07 04 ☽ ± ♄
08 14 ☽ ⊼ ♂
10 02 ☽ ⊼ ♃
10 50 ☽ ⊼ ☿
15 26 ☽ ⊻ ♇
20 49 ☽ ⊥ ♅
22 15 ☽ ± ☉
22 24 ☽ ± ♃

12 Saturday
00 25 ☉ Q ♃
01 29 ☽ bQ ♄
03 39 ☽ ⊥ ♇
08 59 ☽ ⊻ ♅
11 32 ☽ ⊼ ☉
14 14 ☿ △ ♇
16 06 ☽ bQ ♀
16 55 ☽ bQ ♃
16 56 ☿ # ♆
21 56 ☽ ∠ ♇
23 58 ♀ ⊻ ♂

13 Sunday
00 50 ♀ □ ♃
01 50 ♂ ⚹ ♃
05 50 ☽ # ♃
08 04 ☽ △ ♄
12 27 ☽ ± ♀
14 28 ☿ Q ♄
17 16 ☽ ☌ ♆
23 37 ☽ △ ♃

14 Monday
00 35 ☽ ☍ ♂
01 56 ☽ ⊼ ♀
02 27 ☽ # ♂
04 13 ☽ ⚹ ♇
06 03 ☽ ∥ ♆
07 54 ☿ ∥ ♄
11 53 ☽ ☍ ☿
13 35 ☽ # ♄
14 58 ☽ # ☿
21 17 ☽ ☌ ♅
23 56 ♀ □ ♇

15 Tuesday
04 11 ☽ Q ♇
05 13 ☽ ☍ ☉
08 01 ☽ ∥ ♅
10 12 ♀ ± ♆
20 17 ☽ □ ♄
23 47 ☽ # ☉

16 Wednesday
01 12 ♂ # ♆
04 51 ☽ ⊻ ♆
11 47 ☽ □ ♃
15 21 ☽ ⊼ ♂
15 35 ☽ □ ♇
15 58 ☽ ∥ ♇
16 33 ☽ ⊥ ♆
18 43 ☿ ☍ ♅
19 36 ♂ △ ♇
19 58 ☽ ☍ ♀

17 Thursday
03 38 ☽ ± ♂
07 29 ☽ Q ♄
07 56 ☽ ⊻ ♅
09 56 ☽ ∠ ♆
10 34 ☽ ⊼ ☿
10 40 ☽ # ♀
15 09 ☿ Q ♃
19 25 ☽ ⊥ ♅
20 38 ☽ ⊼ ☉
21 51 ☽ bQ ♂
22 52 ☽ Q ♃

18 Friday
00 16 ☽ ± ☿
04 16 ☿ # ♅
06 31 ☽ ⚹ ♄
08 59 ☽ ± ☉
12 28 ☽ ∠ ♅
14 27 ☽ ⚹ ♆
20 32 ☽ bQ ☿
21 48 ☽ ⚹ ♃

19 Saturday
00 52 ☽ △ ♇
03 18 ☽ bQ ☉
03 41 ☽ △ ♂
10 47 ☽ ∠ ♄
11 07 ☽ ⊼ ♀
12 51 ☽ Q ♆
16 26 ☽ ⚹ ♅
23 23 ☽ ± ♀

20 Sunday
01 55 ☽ ∠ ♃
03 27 ☽ ⊥ ♄
04 40 ☽ bQ ♇
05 31 ☽ △ ☿
07 22 ♂ ∥ ♄
09 14 ☽ △ ☉
11 40 ♂ Q ♄
12 24 ☽ ∥ ♀
14 23 ☽ Q ♅
14 27 ☽ ⊻ ♄
17 31 ☽ bQ ♀
18 30 ☽ ⊥ ♃
21 03 ☽ ± ♇
21 46 ☽ □ ♆
22 44 ♇ St D

21 Monday
05 25 ☽ ⊻ ♃
07 51 ☽ ⊼ ♇
13 17 ☽ □ ♂
13 46 ☽ # ♇
17 24 ☽ ∥ ☉
17 28 ♀ ⊼ ♅
22 35 ☽ □ ♅
23 07 ☽ △ ♀

22 Tuesday
01 05 ☉ ☌ ☿
02 08 ☽ ∥ ☿
14 34 ♀ bQ ♆
17 23 ☿ □ ♄
18 51 ☽ # ♅
18 51 ☽ □ ☉
19 57 ☽ ☌ ♄
20 22 ☽ □ ☿
22 31 ☽ Q ♂

23 Wednesday
02 43 ☽ △ ♆
04 59 ♀ ⊥ ♂
06 20 ☽ ∥ ♂
10 33 ☽ ☌ ♃
10 38 ☽ ∥ ♄
11 16 ☉ □ ♄
12 27 ☽ ☍ ♇
14 37 ☉ # ♇
17 07 ☽ # ♆
20 08 ☽ ⚹ ♂

24 Thursday
02 23 ☽ △ ♅
04 00 ☽ Q ☉
04 19 ☽ bQ ♆
07 57 ☽ □ ♀
08 11 ☽ Q ☿
15 45 ☽ ∥ ♃
16 46 ☿ ⊼ ♆
19 25 ☽ ± ♆
22 35 ☽ ∠ ♂
23 07 ☽ ⊻ ♄

25 Friday
00 36 ☿ # ♇
01 36 ☽ ⚹ ☉
03 29 ☽ bQ ♅
05 24 ☽ ⊼ ♆
07 22 ☽ ⚹ ☿
09 06 ☽ ⊥ ♄
13 24 ☽ ⊻ ♃
14 12 ☽ ⊥ ♂
14 50 ☽ ⊼ ♇
16 44 ☽ Q ♀
18 19 ☽ ± ♅
23 21 ☽ ⊥ ♃
23 58 ☽ ∠ ♄

26 Saturday
00 31 ☽ ⊻ ♂
00 42 ☽ ± ♇
04 06 ☽ ∠ ☉
04 09 ☽ ⊼ ♅
06 02 ☉ ∥ ☿
11 09 ♀ ± ♅
11 44 ☽ ∠ ☿
14 10 ☽ ∠ ♃
14 11 ☽ ⚹ ♀
15 25 ☽ bQ ♇
18 51 ☽ ∥ ♃
19 48 ☽ ⊥ ☉

27 Sunday
00 29 ☽ ⚹ ♄
04 32 ☽ ⊥ ☿
06 13 ☽ ⊻ ☉
06 25 ☽ ☍ ♆
07 31 ☿ □ ♃
09 14 ☉ ⊼ ♆
14 41 ☽ ⚹ ♃
15 40 ☽ ⊻ ☿
15 46 ☽ △ ♇
16 24 ☿ □ ♇
16 43 ☽ ∠ ♀
16 46 ☽ # ♆
19 16 ☿ ± ♆
19 58 ☽ Q ♄
22 57 ☽ ∥ ♄

28 Monday
03 28 ☽ ☌ ♂
04 44 ☽ ☍ ♅
08 37 ☽ ⊥ ♀
09 40 ☽ ∥ ♂
10 15 ☽ Q ♃
13 23 ☽ # ♅
19 14 ☽ ⊻ ♀

29 Tuesday
01 12 ☽ □ ♄
07 01 ☽ ⊼ ♆
07 22 ♂ ☍ ♅
10 19 ☽ ☌ ☉
15 46 ☽ □ ♃
15 52 ♀ Q ♇
16 18 ☽ # ♇
16 33 ☽ □ ♇
16 50 ☽ ± ♆
23 33 ☽ ☌ ☿

30 Wednesday
03 55 ☽ ∥ ☉
05 38 ☽ ⊼ ♅
06 43 ☽ ⊻ ♂
07 41 ☽ bQ ♆
13 07 ☽ ∥ ☿
15 36 ☽ ± ♅
17 12 ☽ ⊥ ♂
22 29 ☽ Q ♇

31 Thursday
01 21 ☽ ☌ ♀
02 58 ☽ △ ♄
05 39 ♂ # ♅
06 45 ☽ bQ ♅
08 52 ☽ △ ♆
09 07 ☽ ∠ ♂
16 07 ☽ ⊻ ☉
18 17 ☽ ∥ ♀
18 27 ☽ △ ♃
18 57 ☽ ⚹ ♇
20 21 ♀ △ ♄
23 24 ☿ ⊼ ♅

SEPTEMBER 2000

LONGITUDES

Date	Sidereal time h m s	Sun ☉ ° ' "	Moon ☽ ° ' "	Moon ☽ 24.00 ° ' "	Mercury ☿ ° '	Venus ♀ ° '	Mars ♂ ° '	Jupiter ♃ ° '	Saturn ♄ ° '	Uranus ♅ ° '	Neptune ♆ ° '	Pluto ♇ ° '
01	10 43 50	09 ♍ 20 46	19 ♎ 57 50	26 ♎ 42 45	18 ♍ 59	01 ♎ 39	20 ♌ 10	09 ♊ 58	00 ♊ 52	18 ♒ 01	04 ♒ 17	10 ♐ 11
02	10 47 47	10 18 52	03 ♏ 21 10	09 ♏ 53 10	20 45	02 53	20 49	10 03	00 53	17 R 59	04 R 16	10 12
03	10 51 43	11 16 59	16 ♏ 19 01	22 ♏ 39 04	22 30	04 07	21 27	10 08	00 54	17 57	04 15	10 12
04	10 55 40	12 15 08	28 ♏ 53 46	05 ♐ 03 39	24 13	05 20	22 05	10 13	00 55	17 55	04 14	10 13
05	10 59 36	13 13 19	11 ♐ 09 18	17 ♐ 11 19	25 56	06 34	22 43	10 17	00 56	17 53	04 12	10 13
06	11 03 33	14 11 31	23 ♐ 10 23	29 ♐ 07 08	27 37	07 47	23 21	10 22	00 57	17 51	04 11	10 14
07	11 07 29	15 09 44	05 ♑ 02 14	10 ♑ 56 20	29 ♍ 17	09 01	23 59	10 26	00 57	17 49	04 10	10 14
08	11 11 26	16 07 59	16 ♑ 50 04	22 ♑ 44 02	00 ♎ 56	10 15	24 37	10 30	00 58	17 47	04 09	10 15
09	11 15 23	17 06 16	28 ♑ 38 48	04 ♒ 34 54	02 34	11 28	25 15	10 34	00 58	17 45	04 08	10 15
10	11 19 19	18 04 34	10 ♒ 32 48	16 ♒ 32 56	04 10	12 42	25 53	10 38	00 59	17 43	04 07	10 16
11	11 23 16	19 02 54	22 ♒ 35 41	28 ♒ 41 21	05 46	13 55	26 31	10 42	00 59	17 41	04 06	10 17
12	11 27 12	20 01 15	04 ♓ 50 11	11 ♓ 02 21	07 20	15 09	27 09	10 45	00 R 59	17 39	04 05	10 18
13	11 31 09	20 59 38	17 ♓ 18 00	23 ♓ 37 10	08 53	16 22	27 47	10 48	00 59	17 37	04 04	10 18
14	11 35 05	21 58 03	29 ♓ 59 53	06 ♈ 26 05	10 26	17 36	28 25	10 52	00 58	17 35	04 03	10 19
15	11 39 02	22 56 30	12 ♈ 55 42	19 ♈ 28 36	11 57	18 49	29 03	10 54	00 58	17 33	04 02	10 20
16	11 42 58	23 54 59	26 ♈ 04 40	02 ♉ 43 44	13 27	20 02	29 ♌ 41	10 57	00 58	17 32	04 01	10 21
17	11 46 55	24 53 30	09 ♉ 25 40	16 ♉ 10 20	14 55	21 16	00 ♍ 18	11 00	00 57	17 30	04 00	10 22
18	11 50 51	25 52 03	22 ♉ 57 37	29 ♉ 47 25	16 23	22 29	00 56	11 02	00 57	17 28	03 59	10 23
19	11 54 48	26 50 39	06 ♊ 39 40	13 ♊ 34 18	17 50	23 43	01 34	11 04	00 56	17 26	03 58	10 24
20	11 58 45	27 49 16	20 ♊ 31 17	27 ♊ 30 35	19 15	24 56	02 12	11 06	00 55	17 25	03 58	10 25
21	12 02 41	28 47 56	04 ♋ 32 10	11 ♋ 35 57	20 40	26 09	02 50	11 08	00 54	17 23	03 57	10 26
22	12 06 38	29 ♍ 46 38	18 ♋ 41 50	25 ♋ 49 39	22 03	27 23	03 28	11 09	00 53	17 22	03 56	10 27
23	12 10 34	00 ♎ 45 23	02 ♌ 59 12	10 ♌ 10 09	23 25	28 36	04 05	11 10	00 52	17 20	03 55	10 28
24	12 14 31	01 44 10	17 ♌ 22 06	24 ♌ 34 35	24 46	29 ♎ 50	04 43	11 12	00 51	17 19	03 55	10 29
25	12 18 27	02 42 58	01 ♍ 47 00	08 ♍ 58 43	26 05	01 ♏ 03	05 21	11 12	00 50	17 17	03 54	10 30
26	12 22 24	03 41 49	16 ♍ 09 02	23 ♍ 17 13	27 23	02 16	05 59	11 13	00 48	17 16	03 53	10 31
27	12 26 20	04 40 42	00 ♎ 22 33	07 ♎ 24 20	28 40	03 29	06 36	11 14	00 47	17 14	03 53	10 32
28	12 30 17	05 39 37	14 ♎ 21 55	21 ♎ 14 45	29 ♎ 55	04 43	07 14	11 14	00 45	17 13	03 52	10 33
29	12 34 14	06 38 34	28 ♎ 02 25	04 ♏ 44 35	01 ♏ 09	05 56	07 52	11 14	00 43	17 12	03 52	10 35
30	12 38 10	07 ♎ 37 33	11 ♏ 21 06	17 ♏ 51 53	02 ♏ 21	07 ♏ 09	08 ♍ 29	11 ♊ 14	00 ♊ 41	17 ♒ 10	03 ♒ 51	10 ♐ 36

Date	Moon True ☊ ° '	Moon Mean ☊ ° '	Moon ☽ Latitude ° '
01	23 ♋ 38	22 ♋ 07	05 N 08
02	23 R 32	22 04	05 04
03	23 27	22 01	04 45
04	23 24	21 58	04 12
05	23 23	21 55	03 28
06	23 D 24	21 52	02 36
07	23 25	21 48	01 38
08	23 26	21 45	00 N 36
09	23 R 26	21 42	00 S 28
10	23 24	21 39	01 31
11	23 20	21 36	02 30
12	23 14	21 32	03 23
13	23 06	21 29	04 07
14	22 56	21 26	04 40
15	22 45	21 23	05 00
16	22 35	21 20	05 03
17	22 26	21 17	04 51
18	22 20	21 13	04 22
19	22 16	21 10	03 38
20	22 14	21 07	02 41
21	22 D 14	21 04	01 33
22	22 14	21 01	00 S 19
23	22 R 14	20 58	00 N 57
24	22 12	20 54	02 09
25	22 07	20 51	03 14
26	21 59	20 48	04 05
27	21 49	20 45	04 41
28	21 38	20 42	04 59
29	21 26	20 38	05 00
30	21 ♋ 15	20 ♋ 35	04 N 44

DECLINATIONS

Date	Sun ☉ ° '	Moon ☽ ° '	Mercury ☿ ° '	Venus ♀ ° '	Mars ♂ ° '	Jupiter ♃ ° '	Saturn ♄ ° '	Uranus ♅ ° '	Neptune ♆ ° '	Pluto ♇ ° '
01	08 N 04	03 S 03	05 N 28	00 N 20	15 N 52	21 N 04	18 N 12	16 S 08	18 S 59	11 S 17
02	07 42	07 52	04 41	00 S 11	15 40	21 04	18 12	16 08	19 00	11 17
03	07 20	12 10	03 54	00 42	15 28	21 05	18 12	16 09	19 00	11 18
04	06 58	15 49	03 08	01 13	15 16	21 06	18 12	16 10	19 00	11 18
05	06 36	18 41	02 21	01 44	15 03	21 06	18 12	16 10	19 01	11 19
06	06 13	20 40	01 35	02 15	14 51	21 07	18 12	16 11	19 01	11 19
07	05 51	21 43	00 49	02 45	14 39	21 07	18 12	16 12	19 01	11 20
08	05 28	21 47	00 N 03	03 16	14 26	21 08	18 12	16 12	19 01	11 20
09	05 06	20 53	00 S 42	03 47	14 14	21 08	18 12	16 13	19 02	11 21
10	04 43	19 03	01 27	04 18	14 01	21 09	18 12	16 13	19 02	11 21
11	04 20	16 21	02 12	04 48	13 48	21 09	18 11	16 14	19 02	11 22
12	03 57	12 53	02 56	05 19	13 35	21 10	18 11	16 14	19 02	11 22
13	03 34	08 49	03 39	05 49	13 22	21 10	18 11	16 15	19 03	11 23
14	03 11	04 S 17	04 23	06 20	13 09	21 10	18 11	16 16	19 03	11 23
15	02 48	00 N 30	05 05	06 50	12 56	21 11	18 11	16 16	19 03	11 24
16	02 25	05 21	05 48	07 20	12 43	21 11	18 10	16 17	19 03	11 24
17	02 02	10 02	06 29	07 50	12 30	21 11	18 10	16 17	19 04	11 25
18	01 39	14 17	07 10	08 20	12 17	21 11	18 10	16 18	19 04	11 26
19	01 15	17 50	07 51	08 49	12 03	21 12	18 09	16 18	19 04	11 26
20	00 52	20 26	08 30	09 19	11 50	21 12	18 09	16 19	19 04	11 27
21	00 29	21 49	09 09	09 48	11 36	21 12	18 09	16 19	19 04	11 27
22	00 N 05	21 49	09 48	10 17	11 23	21 12	18 08	16 20	19 05	11 28
23	00 S 18	20 25	10 25	10 46	11 09	21 12	18 08	16 20	19 05	11 28
24	00 41	17 41	11 02	11 14	10 56	21 12	18 08	16 20	19 05	11 29
25	01 05	13 51	11 38	11 42	10 42	21 12	18 07	16 21	19 05	11 29
26	01 28	09 14	12 14	12 10	10 28	21 12	18 07	16 21	19 05	11 30
27	01 52	04 N 09	12 48	12 38	10 14	21 12	18 06	16 22	19 06	11 31
28	02 15	01 S 04	13 22	13 06	10 01	21 12	18 06	16 22	19 06	11 31
29	02 38	06 07	13 54	13 33	09 47	21 12	18 05	16 22	19 06	11 32
30	03 S 02	10 S 44	14 S 26	14 S 00	09 N 33	21 N 12	18 N 05	16 S 23	19 S 06	11 S 32

ZODIAC SIGN ENTRIES

Date	h	m	Planets
02	05	55	☽ ♏
04	14	08	☽ ♐
07	01	47	☽ ♑
07	22	22	☿ ♎
09	14	44	☽ ♒
12	02	34	☽ ♓
14	12	00	☽ ♈
16	19	05	☽ ♉
17	00	19	♂ ♍
19	00	22	☽ ♊
21	04	16	☽ ♋
22	17	28	☉ ♎
23	07	00	☽ ♌
24	15	26	♀ ♏
25	09	02	☽ ♍
27	11	22	☽ ♎
28	13	28	☿ ♏
29	15	29	☽ ♏

LATITUDES

Date	Mercury ☿ ° '	Venus ♀ ° '	Mars ♂ ° '	Jupiter ♃ ° '	Saturn ♄ ° '	Uranus ♅ ° '	Neptune ♆ ° '	Pluto ♇ ° '
01	01 N 12	01 N 05	01 N 10	00 S 54	02 S 11	00 S 44	00 N 12	10 N 48
04	00 54	00 59	01 11	00 54	02 11	00 44	00 12	10 47
07	00 35	00 53	01 11	00 54	02 12	00 44	00 12	10 46
10	00 N 14	00 47	01 12	00 54	02 12	00 44	00 12	10 44
13	00 S 09	00 40	01 12	00 54	02 13	00 44	00 12	10 43
16	00 32	00 33	01 13	00 55	02 14	00 44	00 12	10 42
19	00 55	00 25	01 13	00 55	02 14	00 44	00 12	10 41
22	01 18	00 17	01 14	00 55	02 15	00 44	00 12	10 39
25	01 41	00 N 08	01 14	00 55	02 15	00 44	00 12	10 38
28	02 03	00 00	01 15	00 56	02 16	00 44	00 12	10 37
31	02 S 23	00 S 09	01 N 15	00 S 56	02 S 16	00 S 44	00 N 12	10 N 36

DATA

Julian Date	2451789
Delta T	+64 seconds
Ayanamsa	23° 51' 44"
Synetic vernal point	05° ♓ 15' 16"
True obliquity of ecliptic	23° 26' 18"

MOON'S PHASES, APSIDES AND POSITIONS ☽

Date	h	m	Phase	Longitude ° '	Eclipse Indicator
05	16	27	☽	13 ♐ 24	
13	19	37	○	21 ♓ 18	
21	01	28	☾	28 ♊ 22	
27	19	53	●	05 ♎ 00	

Day	h	m		
08	12	32	Apogee	
24	08	11	Perigee	
08	01	48	Max dec	21° S 53'
15	09	31	0N	
22	00	11	Max dec	22° N 00'
28	07	05	0S	

LONGITUDES

Date	Chiron ⚷ ° '	Ceres ⚳ ° '	Pallas ⚴ ° '	Juno ⚵ ° '	Vesta ⚶ ° '	Black Moon Lilith ⚸ ° '
01	11 ♐ 21	25 ♎ 26	29 ♍ 40	15 ♒ 13	18 ♑ 08	20 ♑ 32
11	11 ♐ 44	29 ♎ 27	04 ♎ 09	13 ♒ 28	18 ♑ 49	21 ♑ 39
21	12 ♐ 16	03 ♏ 32	08 ♎ 38	12 ♒ 22	20 ♑ 11	22 ♑ 45
31	12 ♐ 56	07 ♏ 42	13 ♎ 08	12 ♒ 00	22 ♑ 10	23 ♑ 52

All ephemeris data is given at 12.00 UT and the Moon's longitude is additionally given for 24.00 UT

ASPECTARIAN

h m Aspects

01 Friday
00 20 ☽ # ♀
03 17 ☽ ⊥ ☉
04 48 ☽ ⅊ ♄
08 36 ☽ △ ♅
10 00 ☽ ⊻ ☿
12 23 ☽ ⚹ ♂
16 07 ☿ ⅊ ♆
20 22 ☽ ∠ ☉
20 43 ☽ ± ♄
20 56 ☽ ⅊ ♃
21 17 ☽ ∠ ♇
22 09 ☽ # ☿
22 15 ☽ ⊥ ☿

02 Saturday
04 46 ☉ □ ♃
07 31 ☽ ⊼ ♄
09 02 ☉ □ ♇
10 51 ☉ ± ♆
10 58 ☽ Q ♂
11 04 ☽ ⊻ ♀
11 14 ☽ # ☉
13 17 ☽ ± ♃
13 19 ☿ ⊻ ♂
13 32 ☽ ⊥ ♇
13 40 ☽ □ ♆
13 57 ♀ ⅊ ♅
17 03 ☽ ∠ ☿
23 12 ☽ ⊥ ♀

03 Sunday
00 23 ☽ ⊼ ♃
00 35 ☽ ⊻ ♇
01 50 ☽ ⚹ ☉
06 50 ☽ ∥ ♇
14 36 ♀ △ ♆
15 04 ☽ □ ♅
17 50 ☽ ∠ ♀
22 13 ☽ □ ♂
23 13 ☽ Q ♆

04 Monday
01 34 ☽ ⚹ ☿
02 19 ☽ Q ☉
07 48 ☿ ± ♅
08 15 ☽ # ♂
11 13 ♃ ☍ ♇
14 35 ☽ ∥ ♅
15 56 ☽ ☍ ♄
17 45 ♂ Q ♃
22 21 ☽ ⚹ ♆

05 Tuesday
01 38 ☽ Q ♅
01 56 ☽ ⚹ ♀
04 36 ☽ Q ☿
07 27 ☽ # ♄
10 09 ☽ ☌ ♇
10 17 ☽ ☍ ♃
15 21 ☽ ∥ ♆
16 27 ☽ □ ☉
23 40 ☿ # ♀

06 Wednesday
01 21 ☽ ⚹ ♅
04 01 ☽ ∠ ♆
04 26 ☽ Q ♀
12 22 ☽ △ ♂
19 50 ☽ # ♃
20 48 ☿ Q ♇
22 06 ☽ ⊥ ♆
22 26 ☽ □ ☿

07 Thursday
03 43 ☽ ⊼ ♄
07 30 ☽ ∠ ♅
10 14 ☽ ⊻ ♆
10 38 ♀ ∠ ♂
15 54 ☽ ± ♄
20 29 ☽ ⅊ ♂
21 02 ☽ □ ♀
22 35 ☽ ⊻ ♇
23 03 ☽ ⊼ ♃

08 Friday
01 44 ☽ ⊥ ♅
04 28 ☿ ⊥ ♂
10 14 ☽ ⅊ ♄
10 27 ☽ △ ☉
10 48 ☽ ⊥ ♇
11 20 ☽ ± ♃
12 06 ♀ ⚹ ♇
12 27 ☿ △ ♄
13 55 ☽ ⊻ ♅
15 50 ☽ ± ♂
17 29 ♀ △ ♃

09 Saturday
04 43 ☽ ⊼ ♂
05 07 ☽ ∠ ♇
05 44 ☽ ⅊ ♃
07 22 ☽ # ♃
14 40 ☿ ⅊ ♅
16 42 ☽ △ ♄
19 37 ☽ ⅊ ☉
21 10 ☽ △ ☿
23 05 ☽ ☌ ♆

10 Sunday
03 19 ☉ ⊼ ♅
11 08 ☿ △ ♆
11 27 ☽ ⚹ ♇
12 10 ☽ ∥ ♆
12 11 ☽ △ ♃
15 20 ☽ ± ☉
16 47 ☽ △ ♀
20 26 ☽ # ♄
23 18 ☉ # ♀

11 Monday
02 17 ☽ ☌ ♅
04 21 ☽ ⊼ ☉
07 50 ☽ ⅊ ☿
11 23 ☽ Q ♇
12 52 ☽ ∥ ♅
13 10 ☉ ⅊ ♆
20 09 ☽ ☍ ♂

12 Tuesday
01 50 ☽ ⅊ ♀
04 11 ☽ ± ☿
04 29 ☽ □ ♄
07 13 ☽ # ♂
10 32 ☽ ⊻ ♆
11 34 ♄ St R
17 33 ☽ ⊼ ☿
21 15 ☽ ± ♀
21 17 ☽ ∥ ♇
22 08 ☽ ⊥ ♆
22 34 ☽ □ ♇
23 30 ☽ □ ♃

13 Wednesday
04 21 ♀ ⅊ ♄
10 02 ☽ ⊼ ♀
10 06 ☉ # ☿
12 36 ☽ ⊻ ♅
15 12 ☽ Q ♄
15 21 ☽ ∠ ♆
19 37 ☽ ☍ ☉
23 58 ☽ ⊥ ♅

14 Thursday
02 28 ☽ ∥ ♀
08 52 ☽ ⊼ ♂
09 51 ☽ Q ♃
10 19 ☿ ⚹ ♇
11 36 ☽ ∥ ☿
11 53 ♀ △ ♅
13 50 ☽ ⚹ ♄
16 50 ☽ ∠ ♅
18 06 ☽ # ☉
19 03 ☿ △ ♃
19 33 ☽ ⚹ ♆
20 40 ☽ ± ♂

15 Friday
07 13 ☽ △ ♇
08 16 ☽ ⚹ ♃
09 57 ☽ ☍ ☿
14 09 ☽ ⅊ ♂
17 35 ☽ ∠ ♄
17 42 ☽ Q ♆
20 28 ☽ ⚹ ♅
22 30 ☽ ∥ ☉
23 54 ☽ ☍ ♀

16 Saturday
02 41 ☉ ± ♅
07 46 ☽ ⊼ ☉
09 59 ☽ ⊥ ♄
10 40 ☽ ⅊ ♇
11 46 ☽ ∠ ♃
14 34 ☽ # ☿
18 13 ☽ Q ♅
18 50 ☽ △ ♂
19 29 ☽ ± ☉
20 49 ☽ ⊻ ♄
23 10 ☽ # ♀

17 Sunday
02 18 ☽ □ ♆
02 55 ☽ ± ♇
04 02 ☽ ⊥ ♃
12 54 ☽ ⅊ ☉
13 40 ☽ ⊼ ♇
14 48 ☽ ⊻ ♃
19 31 ☽ # ♇
22 58 ☽ ⊼ ☿
23 02 ☿ ∠ ♂

18 Monday
00 53 ☽ ∥ ♂
02 19 ☽ □ ♅
04 49 ☿ ⅊ ♄
10 52 ☽ ± ☿
11 05 ☽ ⊼ ♀
12 18 ♂ □ ♄
17 31 ☽ △ ☉
22 40 ☽ ± ♀

19 Tuesday
00 54 ☽ # ♅
02 01 ☽ ☌ ♄
02 41 ☽ □ ♂
04 32 ☽ ⅊ ☿
05 38 ☿ △ ♅
07 19 ☽ △ ♆
14 29 ☽ ∥ ♄
15 55 ☽ ⅊ ♀
18 30 ☽ ☍ ♇
19 40 ☽ ☌ ♃
22 09 ☽ # ♆

20 Wednesday
06 39 ☽ △ ♅
09 19 ☽ ⅊ ♆
09 34 ☽ △ ☿
11 25 ☽ Q ♂
11 44 ♀ ± ♄
20 19 ☽ △ ♀
21 26 ♀ ∠ ♇
22 33 ☽ ∥ ♃

21 Thursday
00 46 ☽ ± ♆
01 28 ☽ □ ☉
02 41 ☉ Q ♇
05 49 ☽ ⊻ ♄
08 21 ☽ ⅊ ♅
08 57 ☽ ⚹ ♂
11 00 ☽ ⊼ ♆
11 22 ♀ ⅊ ♃
16 02 ☽ ⊥ ♄
22 01 ☽ ⊼ ♇
23 13 ☽ ⊻ ♃
23 37 ☽ ± ♅

22 Friday
03 47 ♂ # ♇
07 16 ☽ ∠ ♄
08 11 ☽ ± ♇
09 23 ☽ ⊥ ♃
09 45 ☽ ⊼ ♅
10 20 ☽ Q ☉
11 35 ☽ ∠ ♂
18 15 ☽ □ ☿
23 22 ☽ ⅊ ♇

23 Saturday
00 34 ☽ ∠ ♃
01 27 ☽ ∥ ♃
03 25 ☽ ⊥ ♂
03 58 ☽ □ ♀
05 48 ♂ ⊼ ♆
08 00 ☽ ⚹ ☉
08 28 ☽ ⚹ ♄
13 34 ☽ ☍ ♆
13 56 ☽ ⊻ ♂
14 43 ☉ △ ♄

24 Sunday
00 30 ☽ △ ♇
01 02 ☽ # ♆
01 31 ♀ # ♂
01 42 ☽ ⚹ ♃
03 32 ☽ Q ☿
04 29 ☽ Q ♄
08 43 ☽ ∥ ♄
08 50 ☿ # ♂
10 52 ☽ ∠ ☉
11 54 ☽ ☍ ♅
12 50 ☽ Q ♀
13 34 ☿ ± ♄
21 07 ☽ # ♅
21 42 ☽ Q ♃

25 Monday
00 47 ♀ ∥ ♇
01 10 ☿ ∠ ♇
01 33 ☽ ⚹ ☿
01 42 ☉ ⅊ ♅
02 57 ☽ ⊥ ☉
05 53 ☿ ∥ ♇
07 45 ♀ ⊼ ♄
10 25 ☽ □ ♄
10 40 ☽ ⚹ ♀
13 40 ☽ ⊻ ☉
14 15 ☿ ⅊ ♃
15 31 ☽ ⊼ ♆
18 13 ☽ ☌ ♂
22 29 ☽ # ♀
22 32 ☽ # ☿

26 Tuesday
00 38 ☽ # ♇
00 54 ☿ ∥ ♀
01 32 ☽ ± ♆
02 34 ☽ □ ♇
03 44 ☽ □ ♃
05 04 ☽ ∠ ☿
05 33 ☽ ∥ ♂
13 52 ☽ ⊼ ♅
14 03 ☽ ∠ ♀
16 36 ☽ ⅊ ♆
16 40 ☉ △ ♆
21 40 ☽ ⊥ ☿
23 56 ☽ ± ♅

27 Wednesday
06 39 ☽ ⊥ ♀
08 49 ☽ ⊻ ☿
08 53 ☽ Q ♇
12 41 ☽ △ ♄
15 10 ☽ ⅊ ♅
17 49 ☽ ⊻ ♀
17 58 ☽ △ ♆
19 34 ♀ □ ♆
19 53 ☽ ☌ ☉
21 48 ☽ # ☉
23 07 ☽ ⊻ ♂

28 Thursday
05 25 ☽ ⚹ ♇
06 35 ☽ △ ♃
08 55 ♀ ⊥ ♇
09 57 ☽ ⊥ ♂
14 24 ☽ ⅊ ♄
16 57 ☽ △ ♅
17 58 ☽ ∥ ☉
22 15 ♀ ± ♃

29 Friday
02 24 ☽ ∠ ♂
03 39 ☿ ⊼ ♄
06 08 ☽ ± ♄
07 38 ☽ ∠ ♇
08 48 ☽ ⅊ ♃
12 56 ♃ St R
16 46 ☽ ⊼ ♄
18 06 ☽ ☌ ☿
22 24 ☽ □ ♆
23 43 ☽ ⊥ ♇

30 Saturday
00 53 ☽ ± ♃
03 35 ☽ ☌ ♀
04 40 ☽ ⊻ ☉
05 51 ☽ # ♂
06 31 ☽ ⚹ ♂
10 38 ☽ ⊻ ♇
11 47 ☽ ⊼ ♃
16 29 ☽ ∥ ♇
16 31 ☽ ⊥ ☉
22 42 ☽ □ ♅

OCTOBER 2000

Date	Sidereal time h m s	Sun ☉ ° ' "	Moon ☽ ° ' "	Moon ☽ 24.00 ° ' "	Mercury ☿ ° '	Venus ♀ ° '	Mars ♂ ° '	Jupiter ♃ ° '	Saturn ♄ ° '	Uranus ♅ ° '	Neptune ♆ ° '	Pluto ♇ ° '
01	12 42 07	08 ♎ 36 34	24 ♏ 17 03	00 ♐ 36 47	03 ♏ 32	08 ♏ 23	09 ♍ 07	11 ♊ 14	00 ♊ 39	17 ♒ 09	03 ♒ 51	10 ♐ 37
02	12 46 03	09 35 37	06 ♐ 51 25	13 ♐ 01 23	04 41	09 36	09 44	11 R 13	00 R 37	17 R 08	03 R 50	10 39
03	12 50 00	10 34 41	19 ♐ 07 09	25 ♐ 09 18	05 47	10 49	10 22	11 12	00 35	17 07	03 50	10 40
04	12 53 56	11 33 48	01 ♑ 08 27	07 ♑ 05 14	06 52	12 02	11 00	11 12	00 33	17 06	03 49	10 41
05	12 57 53	12 32 56	13 ♑ 00 22	18 ♑ 54 32	07 54	13 15	11 37	11 10	00 30	17 05	03 49	10 43
06	13 01 49	13 32 06	24 ♑ 48 25	00 ♒ 42 43	08 54	14 28	12 15	11 09	00 28	17 04	03 49	10 44
07	13 05 46	14 31 17	06 ♒ 38 06	12 ♒ 35 14	09 51	15 42	12 52	11 08	00 25	17 03	03 49	10 46
08	13 09 43	15 30 31	18 ♒ 34 41	24 ♒ 37 02	10 45	16 55	13 30	11 06	00 23	17 02	03 48	10 47
09	13 13 39	16 29 46	00 ♓ 42 47	06 ♓ 52 21	11 36	18 08	14 07	11 04	00 20	17 01	03 48	10 49
10	13 17 36	17 29 03	13 ♓ 06 04	19 ♓ 24 14	12 23	19 21	14 45	11 02	00 17	17 00	03 48	10 50
11	13 21 32	18 28 22	25 ♓ 46 59	02 ♈ 14 25	13 07	20 34	15 22	11 00	00 14	16 59	03 48	10 52
12	13 25 29	19 27 43	08 ♈ 46 29	15 ♈ 23 03	13 46	21 47	15 59	10 57	00 11	16 59	03 48	10 54
13	13 29 25	20 27 06	22 ♈ 03 54	28 ♈ 48 45	14 21	23 00	16 37	10 54	00 08	16 58	03 47	10 55
14	13 33 22	21 26 30	05 ♉ 37 12	12 ♉ 28 52	14 50	24 13	17 14	10 52	00 05	16 57	03 47	10 57
15	13 37 18	22 25 57	19 ♉ 23 16	26 ♉ 19 59	15 14	25 26	17 52	10 49	00 ♊ 02	16 57	03 47	10 59
16	13 41 15	23 25 27	03 ♊ 18 34	10 ♊ 18 36	15 32	26 39	18 29	10 45	29 ♉ 58	16 56	03 D 47	11 00
17	13 45 12	24 24 58	17 ♊ 19 42	24 ♊ 21 35	15 44	27 52	19 06	10 42	29 55	16 56	03 47	11 02
18	13 49 08	25 24 32	01 ♋ 23 59	08 ♋ 26 41	15 48	29 ♏ 04	19 44	10 38	29 51	16 55	03 48	11 04
19	13 53 05	26 24 08	15 ♋ 29 33	22 ♋ 32 26	15 R 45	00 ♐ 17	20 21	10 34	29 48	16 55	03 48	11 05
20	13 57 01	27 23 47	29 ♋ 35 16	06 ♌ 37 57	15 33	01 30	20 58	10 30	29 44	16 55	03 48	11 07
21	14 00 58	28 23 28	13 ♌ 40 24	20 ♌ 42 29	15 13	02 43	21 36	10 26	29 40	16 54	03 48	11 09
22	14 04 54	29 ♎ 23 11	27 ♌ 44 04	04 ♍ 44 56	14 44	03 56	22 13	10 22	29 37	16 54	03 48	11 11
23	14 08 51	00 ♏ 22 56	11 ♍ 44 50	18 ♍ 43 28	14 06	05 08	22 50	10 17	29 33	16 54	03 48	11 13
24	14 12 47	01 22 43	25 ♍ 40 29	02 ♎ 35 29	13 20	06 21	23 27	10 12	29 29	16 54	03 49	11 15
25	14 16 44	02 22 33	09 ♎ 28 03	16 ♎ 17 46	12 25	07 34	24 04	10 07	29 25	16 54	03 49	11 17
26	14 20 41	03 22 25	23 ♎ 04 13	29 ♎ 47 00	11 22	08 47	24 42	10 02	29 21	16 54	03 49	11 18
27	14 24 37	04 22 18	06 ♏ 25 48	13 ♏ 00 20	10 13	09 59	25 19	09 57	29 17	16 D 54	03 50	11 20
28	14 28 34	05 22 14	19 ♏ 30 25	25 ♏ 55 56	09 00	11 12	25 56	09 52	29 12	16 54	03 50	11 22
29	14 32 30	06 22 11	02 ♐ 16 53	08 ♐ 33 22	07 43	12 25	26 33	09 46	29 08	16 54	03 51	11 24
30	14 36 27	07 22 11	14 ♐ 45 33	20 ♐ 53 44	06 26	13 37	27 10	09 40	29 04	16 54	03 51	11 26
31	14 40 23	08 ♏ 22 12	26 ♐ 58 15	02 ♑ 59 32	05 ♏ 11	14 ♐ 50	27 ♍ 47	09 ♊ 34	28 ♉ 59	16 ♒ 54	03 ♒ 52	11 ♐ 28

Date	Moon True ☊ ° '	Moon Mean ☊ ° '	Moon ☽ Latitude ° '
01	21 ♋ 06	20 ♋ 32	04 N 13
02	20 R 59	20 29	03 31
03	20 55	20 26	02 40
04	20 54	20 23	01 43
05	20 D 53	20 19	00 N 42
06	20 R 53	20 16	00 S 21
07	20 53	20 13	01 23
08	20 50	20 10	02 21
09	20 45	20 07	03 14
10	20 37	20 04	03 59
11	20 27	20 00	04 33
12	20 15	19 57	04 54
13	20 02	19 54	05 00
14	19 49	19 51	04 49
15	19 38	19 48	04 21
16	19 30	19 44	03 38
17	19 24	19 41	02 41
18	19 21	19 38	01 33
19	19 D 21	19 35	00 S 20
20	19 R 21	19 32	00 N 54
21	19 20	19 29	02 05
22	19 18	19 25	03 08
23	19 13	19 22	04 00
24	19 05	19 19	04 37
25	18 54	19 16	04 57
26	18 42	19 13	05 00
27	18 29	19 10	04 47
28	18 18	19 06	04 18
29	18 08	19 03	03 37
30	18 00	19 00	02 47
31	17 ♋ 56	18 ♋ 57	01 N 49

DECLINATIONS

Date	Sun ☉ ° '	Moon ☽ ° '	Mercury ☿ ° '	Venus ♀ ° '	Mars ♂ ° '	Jupiter ♃ ° '	Saturn ♄ ° '	Uranus ♅ ° '	Neptune ♆ ° '	Pluto ♇ ° '
01	03 S 25	14 S 45	14 S 56	14 S 26	09 N 19	21 N 12	18 N 04	16 S 23	19 S 06	11 S 33
02	03 48	17 59	15 26	14 52	09 05	21 12	18 04	16 23	19 06	11 33
03	04 11	20 20	15 54	15 18	08 50	21 12	18 03	16 24	19 06	11 34
04	04 34	21 43	16 21	15 44	08 36	21 12	18 02	16 24	19 06	11 35
05	04 57	22 07	16 47	16 09	08 22	21 12	18 02	16 24	19 07	11 35
06	05 20	21 30	17 12	16 34	08 08	21 11	18 01	16 25	19 07	11 36
07	05 43	19 57	17 35	16 58	07 54	21 11	18 00	16 25	19 07	11 36
08	06 06	17 30	17 56	17 22	07 39	21 11	18 00	16 25	19 07	11 37
09	06 29	14 15	18 16	17 45	07 25	21 10	17 59	16 25	19 07	11 37
10	06 52	10 19	18 34	18 08	07 11	21 10	17 58	16 26	19 07	11 38
11	07 14	05 51	18 50	18 31	06 56	21 10	17 58	16 26	19 07	11 39
12	07 37	01 S 02	19 04	18 53	06 42	21 09	17 57	16 26	19 07	11 39
13	07 59	03 N 57	19 16	19 15	06 27	21 09	17 56	16 26	19 07	11 40
14	08 22	08 51	19 26	19 36	06 13	21 08	17 55	16 26	19 07	11 40
15	08 44	13 23	19 33	19 57	05 58	21 08	17 54	16 26	19 07	11 41
16	09 06	17 15	19 37	20 17	05 44	21 07	17 54	16 27	19 07	11 41
17	09 28	20 10	19 39	20 36	05 29	21 07	17 53	16 27	19 07	11 42
18	09 50	21 52	19 37	20 56	05 15	21 06	17 52	16 27	19 07	11 42
19	10 11	22 12	19 31	21 14	05 00	21 06	17 51	16 27	19 07	11 43
20	10 33	21 07	19 22	21 32	04 46	21 05	17 50	16 27	19 07	11 44
21	10 54	18 43	19 09	21 49	04 31	21 05	17 49	16 27	19 07	11 44
22	11 15	15 12	18 51	22 06	04 16	21 04	17 49	16 27	19 07	11 45
23	11 36	10 51	18 30	22 23	04 02	21 03	17 48	16 27	19 07	11 45
24	11 57	05 57	18 03	22 38	03 47	21 03	17 47	16 27	19 07	11 46
25	12 18	00 N 48	17 33	22 53	03 32	21 02	17 46	16 27	19 07	11 46
26	12 38	04 S 19	16 58	23 08	03 18	21 01	17 45	16 27	19 07	11 47
27	12 59	09 09	16 19	23 21	03 03	21 00	17 44	16 27	19 07	11 47
28	13 19	13 28	15 38	23 34	02 48	21 00	17 43	16 27	19 07	11 48
29	13 39	17 04	14 54	23 47	02 34	20 59	17 42	16 27	19 07	11 49
30	13 58	19 49	14 09	23 59	02 19	20 58	17 41	16 27	19 06	11 49
31	14 S 18	21 S 35	13 S 25	24 S 10	02 N 04	20 N 57	17 N 40	16 S 27	19 S 06	11 S 50

ZODIAC SIGN ENTRIES

Date	h	m	Planets
01	22	50	☽ ♐
04	09	42	☽ ♑
06	22	33	☽ ♒
09	10	36	☽ ♓
11	19	51	☽ ♈
14	02	06	☽ ♉
16	00	44	♄ ♉
16	06	19	☽ ♊
18	09	37	☽ ♋
19	06	18	♀ ♐
20	12	42	☽ ♌
22	15	52	☽ ♍
23	02	47	☉ ♏
24	19	30	☽ ♎
27	00	23	☽ ♏
29	07	40	☽ ♐
31	18	01	☽ ♑

LATITUDES

Date	Mercury ☿ ° '	Venus ♀ ° '	Mars ♂ ° '	Jupiter ♃ ° '	Saturn ♄ ° '	Uranus ♅ ° '	Neptune ♆ ° '	Pluto ♇ ° '
01	02 S 23	00 S 09	01 N 15	00 S 56	02 S 16	00 S 44	00 N 12	10 N 36
04	02 42	00 18	01 15	00 56	02 17	00 43	00 12	10 34
07	02 58	00 27	01 16	00 56	02 17	00 43	00 12	10 33
10	03 10	00 36	01 16	00 56	02 17	00 43	00 12	10 32
13	03 17	00 45	01 16	00 56	02 18	00 43	00 11	10 31
16	03 17	00 54	01 17	00 57	02 18	00 43	00 11	10 30
19	03 06	01 03	01 17	00 57	02 18	00 43	00 11	10 29
22	02 43	01 12	01 17	00 57	02 19	00 43	00 11	10 28
25	02 05	01 20	01 17	00 57	02 19	00 43	00 11	10 27
28	01 12	01 28	01 18	00 57	02 19	00 43	00 11	10 26
31	00 S 10	01 S 36	01 N 18	00 S 57	02 S 19	00 S 43	00 N 11	10 N 25

DATA

Julian Date	2451819
Delta T	+64 seconds
Ayanamsa	23° 51' 46"
Synetic vernal point	05° ♓ 15' 13"
True obliquity of ecliptic	23° 26' 18"

MOON'S PHASES, APSIDES AND POSITIONS ☽

Date	h	m	Phase	Longitude ° '	Eclipse Indicator
05	10	59	☽	12 ♑ 30	
13	08	53	○	20 ♈ 19	
20	07	59	☾	27 ♋ 14	
27	07	58	●	04 ♏ 12	

Day	h	m		
06	07	02	Apogee	
19	22	02	Perigee	
05	09	21	Max dec	22° S 07'
12	17	00	ON	
19	05	32	Max dec	22° N 15'
25	15	44	OS	

LONGITUDES

Date	Chiron ⚷ ° '	Ceres ⚳ ° '	Pallas ⚴ ° '	Juno ⚵ ° '	Vesta ⚶ ° '	Black Moon Lilith ⚸ ° '
01	12 ♐ 56	07 ♏ 42	13 ♎ 08	12 ♒ 00	22 ♑ 10	23 ♑ 52
11	13 ♐ 43	11 ♏ 55	17 ♎ 37	12 ♒ 23	24 ♑ 41	24 ♑ 58
21	14 ♐ 37	16 ♏ 10	22 ♎ 05	13 ♒ 29	27 ♑ 38	26 ♑ 04
31	15 ♐ 36	20 ♏ 26	26 ♎ 33	15 ♒ 13	00 ♒ 57	27 ♑ 11

All ephemeris data is given at 12.00 UT and the Moon's longitude is additionally given for 24.00 UT

ASPECTARIAN

h m	Aspects
01 Sunday	
05 45	☽ Q ♂
07 25	☽ Q ♆
09 42	☽ ‖ ♀
10 37	☽ ∠ ☉
13 28	☽ ‖ ☿
18 26	☿ □ ♆
23 23	☽ ‖ ♅
02 Monday	
00 03	☽ ☍ ♄
06 11	☽ ✶ ♆
07 22	☽ ⊻ ☿
08 41	☽ Q ♅
11 19	☿ ⊥ ♇
11 45	☉ ⊻ ♀
12 37	☽ # ♄
15 40	♂ ± ♆
17 46	☽ ✶ ☉
17 50	♀ ✶ ♂
17 54	☽ □ ♂
17 54	☽ ⊻ ♀
19 22	☽ ☌ ♇
20 10	☽ ⊥ ☿
20 28	☽ ☍ ♃
21 51	☉ ⊻ ♂
22 16	☽ ‖ ♆
23 32	☿ ± ♃
03 Tuesday	
06 57	☽ ⊥ ♀
08 03	☽ ✶ ♅
09 01	♀ ⊻ ♇
11 26	☽ ∠ ♆
14 14	☉ ✶ ♇
15 38	☽ ∠ ☿
19 28	☽ Q ☉
19 38	♀ ⊼ ♃
23 58	♂ □ ♇
04 Wednesday	
00 49	☽ # ♃
02 50	☽ ∠ ♀
03 08	☉ △ ♃
05 21	☽ ⊥ ♆
10 49	☽ ⊼ ♄
13 55	☽ ∠ ♅
14 36	☿ ‖ ♅
17 24	☽ ⊻ ♆
19 28	♂ □ ♃
22 52	☽ ± ♄
05 Thursday	
00 40	☽ ✶ ☿
07 20	☽ ⊻ ♇
08 06	☽ ⊥ ♅
08 17	☽ ⊼ ♃
09 02	☽ △ ♂
10 59	☽ □ ☉
12 34	☽ ✶ ♀
17 04	☽ ♇ ♄
19 33	☽ ⊥ ♇
20 16	☽ ⊻ ♅
20 27	☽ ± ♃
06 Friday	
03 05	♀ ‖ ♅
03 19	☽ Q ☿
13 54	☽ ∠ ♇
14 44	☽ ♇ ♃
15 47	☽ Q ♀
17 14	☽ ♇ ♂
18 18	☽ # ♃
23 28	☽ △ ♄
07 Saturday	
06 17	☽ ☌ ♆
12 30	☽ ± ♂
19 02	☽ □ ☿
20 21	☽ ✶ ♇
21 03	☽ △ ♃
21 08	☽ ‖ ♆
08 Sunday	
01 16	☽ ⊼ ♂
05 19	☽ △ ☉
07 40	☽ # ♄
08 18	☽ □ ♀
08 39	☽ ‖ ☿
08 55	☽ ☌ ♅
09 01	☉ ♇ ♄
12 56	☽ ‖ ♀
13 06	☿ ⊻ ♇
14 19	♀ □ ♅
16 07	☿ # ♄
20 24	☽ Q ♇
20 33	☽ ‖ ♅
21 23	☿ ⊼ ♃
09 Monday	
11 16	☽ □ ♄
13 40	☽ ♇ ☉
18 02	☽ ⊻ ♆
10 Tuesday	
00 28	☉ △ ♅
01 39	♀ # ♄
04 23	☽ ‖ ♇
05 39	☽ ⊥ ♆
07 39	☽ □ ♇
08 03	☽ □ ♃
08 38	☽ ± ☉
10 33	☽ △ ☿
15 18	☽ ☍ ♂
19 26	☽ ⊻ ♅
21 04	☽ ⊼ ☉
21 51	☽ Q ♄
22 51	☽ ∠ ♆
11 Wednesday	
00 12	☉ # ♂
01 09	☽ △ ♀
05 21	☽ ‖ ☉
06 05	☽ # ♂
06 46	☽ ⊥ ♅
16 35	☽ ♇ ☿
17 58	☽ Q ♃
20 16	☽ ✶ ♄
23 32	☽ ∠ ♅
12 Thursday	
02 52	☽ ✶ ♆
07 59	☽ ♇ ♀
10 04	☽ ± ☿
12 14	♀ Q ♆
15 52	☽ △ ♇
15 57	☽ ✶ ♃
16 51	☿ ‖ ♆
21 30	☽ ⊼ ☿
23 36	☽ ∠ ♄
13 Friday	
00 44	☽ Q ♆
01 44	☽ ⊼ ♂
02 00	☽ ± ♀
02 52	☽ ✶ ♅
03 14	♀ ‖ ♆
08 04	♃ ☍ ♇
08 53	☽ ☍ ☉
13 02	☽ ± ♂
13 50	☽ ⊼ ♀
15 25	☿ ‖ ♀
15 41	☽ ⊥ ♄
18 49	☽ ∠ ♃
18 53	☽ ♇ ♆
23 35	☽ ‖ ♂
14 Saturday	
00 16	☽ Q ♅
01 20	♂ ⊼ ♅
02 17	☽ ⊻ ♄
05 46	☽ ♇ ♂
08 47	☽ □ ♆
09 22	☽ # ☉
10 40	☽ ⊥ ♃
10 49	☽ ± ♇
21 08	☽ ⊻ ♃
21 21	☽ ⊼ ♇
15 Sunday	
02 39	☽ # ♇
04 36	☽ ☍ ♀
07 46	☽ □ ♅
09 14	☽ △ ♂
14 10	♆ St D
17 40	☽ ⊼ ☉
23 26	☽ ☍ ☿
16 Monday	
04 49	☽ ± ☉
06 17	☽ ☌ ♄
06 31	☽ # ♅
12 50	☽ △ ♆
16 35	☽ ‖ ♄
21 26	☽ ♇ ☉
23 51	♂ ♇ ♆
17 Tuesday	
00 34	☉ ± ♄
00 42	☽ ☌ ♃
01 13	☽ ☍ ♇
02 15	☽ # ♆
06 55	☽ # ☿
09 14	☽ ⊼ ☿
11 19	☽ △ ♅
14 30	☽ ♇ ♆
15 11	☽ □ ♂
17 36	☽ # ♀
19 34	☽ ± ☿
22 58	☽ ‖ ♃
18 Wednesday	
01 01	☽ △ ☉
05 51	☽ ± ♆
07 40	☽ ⊼ ♀
09 23	☽ ⊻ ♄
10 59	☽ ♇ ☿
12 53	☽ ♇ ♅
13 40	☿ St R
16 05	☽ ⊼ ♆
17 08	☉ ♇ ♃
18 51	☽ ± ♀
19 34	☽ ⊥ ♄
23 17	☽ Q ♂
19 Thursday	
01 30	♀ # ♃
02 46	♀ ☍ ♃
03 40	☽ ⊻ ♃
04 13	☽ ± ♅
04 16	☉ ∠ ♇
04 29	☽ ⊼ ♆
08 36	☉ ⊥ ♂
10 49	☽ ∠ ♄
11 37	☽ ♇ ♀
12 25	☽ △ ☿
13 50	☽ ⊥ ♃
14 25	☽ ⊼ ♅
14 44	☽ ± ♇
20 39	☽ ✶ ♂
20 Friday	
05 05	☽ ∠ ♃
06 05	☽ ♇ ♀
06 46	☽ # ♀
07 59	☽ □ ☉
12 15	☽ ✶ ♄
12 24	☽ ‖ ♃
15 34	☽ △ ♀
19 10	☽ ☍ ♆
23 23	☽ ∠ ♂
21 Saturday	
06 30	☽ ✶ ♃
07 42	☽ △ ♆
08 05	☽ # ♅
08 36	☽ Q ♄
08 42	☽ # ♆
14 33	☽ □ ☿
14 49	☿ ‖ ♆
15 25	☽ ⊥ ♂
16 59	☽ Q ☉
17 31	☽ ☍ ♅
18 49	☽ ‖ ♄
22 Sunday	
02 08	☽ ⊻ ♂
02 52	☽ Q ♃
04 12	☽ # ♅
09 31	♀ ✶ ♆
15 02	☽ ✶ ☉
15 12	☽ □ ♄
17 05	☉ ⊼ ♄
20 13	☽ Q ☿
22 23	☽ ⊼ ♆
23 33	☽ Q ♂
23 36	☽ □ ♀
23 Monday	
07 13	♀ Q ♅
07 19	☽ # ♇
08 22	☽ # ☉
08 40	☽ ± ♆
09 30	☽ □ ♃
11 05	☽ ⊼ ♇
14 30	♂ ± ♅
15 51	☽ ✶ ☿
18 44	☽ ∠ ☉
20 51	☽ ⊼ ♅
22 33	☉ ‖ ♇
24 Tuesday	
00 09	☽ ♇ ♆
07 12	☽ ± ♅
07 59	☽ ☌ ♂
09 30	☽ Q ♀
11 27	☽ ⊥ ☉
16 20	☽ ∠ ☿
18 12	☽ Q ♇
18 34	☽ △ ♄
22 40	☽ ⊻ ☉
22 41	☽ ‖ ♂
22 47	☽ ♇ ♅
25 Wednesday	
01 42	☿ # ♄
02 08	☽ △ ♆
07 01	☽ ⊥ ☿
08 21	☽ ✶ ♀
13 08	☽ △ ♃
15 11	☽ ✶ ♇
16 49	☽ ⊻ ☿
20 38	☽ ♇ ♄
26 Thursday	
01 03	☽ △ ♅
07 22	☽ # ♂
12 29	☽ ± ♄
13 21	☿ ⊻ ♆
13 23	☽ ∠ ♀
15 02	☽ ⊻ ♂
15 22	♅ St D
15 29	☽ ♇ ♃
17 47	☽ ∠ ♇
22 53	☉ □ ♆
23 09	☽ ⊼ ♄
27 Friday	
02 18	☽ ⊥ ♂
02 41	☉ ± ♃
07 08	☽ ⊥ ♀
07 17	☿ ‖ ♅
07 17	☽ □ ♆
07 32	☽ ± ♃
07 58	☽ ☌ ☉
10 01	☽ ⊥ ♇
10 48	☿ ∠ ♂
11 17	♀ ☍ ♃
14 20	☿ ⊻ ♀
17 54	☿ ⊼ ♃
18 20	☽ ☌ ☿
18 22	☽ ⊼ ♃
19 08	☽ ⊻ ♇
19 25	☽ ∠ ♂
20 58	☽ ⊻ ♆
28 Saturday	
02 20	☽ ‖ ♇
07 10	☽ □ ♅
11 02	☽ ‖ ☉
12 04	☉ ⊥ ♇
15 33	♀ ☌ ♇
16 20	☽ Q ♆
23 29	☽ ‖ ☿
29 Sunday	
00 37	☽ ✶ ♂
06 04	☽ ☍ ♄
07 30	☽ ‖ ♅
07 31	☉ ⊥ ♀
14 59	☽ ✶ ♆
16 53	☽ # ♄
16 59	☽ Q ♅
20 29	☽ ⊻ ☉
21 25	☽ ⊻ ☿
30 Monday	
00 35	☿ ⊥ ♀
00 37	☽ Q ♂
02 09	☉ ☌ ☿
02 13	☽ ☍ ♃
05 02	☽ ‖ ♆
05 33	☽ ☌ ♇
07 54	☽ ⊥ ☿
09 04	☽ ⊥ ☉
09 33	☽ ☌ ♀
16 07	☉ ‖ ☿
16 11	☽ ✶ ♅
20 00	☽ ∠ ♆
23 50	☽ ∠ ☿
31 Tuesday	
01 49	☽ # ♃
04 14	☽ ∠ ☉
06 23	☿ ⊥ ♇
13 43	☽ □ ♂
13 46	☽ ⊥ ♆
15 59	☽ ⊼ ♄
21 50	☽ ∠ ♅

NOVEMBER 2000

LONGITUDES

Date	Sidereal time h m s	Sun ☉ ° ' "	Moon ☽ ° ' "	Moon ☽ 24.00 ° ' "	Mercury ☿ ° '	Venus ♀ ° '	Mars ♂ ° '	Jupiter ♃ ° '	Saturn ♄ ° '	Uranus ♅ ° '	Neptune ♆ ° '	Pluto ♇ ° '
01	14 44 20	09 ♏ 22 15	08 ♑ 58 05	14 ♑ 54 29	03 ♏ 59	16 ♐ 02	28 ♍ 24	09 ♊ 28	28 ♉ 55	16 ♒ 55	03 ♒ 52	11 ♐ 30
02	14 48 16	10 22 19	20 ♑ 49 18	26 ♑ 43 11	02 R 55	17 55	29 01	09 R 22	28 R 50	16 55	03 53	11 32
03	14 52 13	11 22 25	02 ♒ 36 49	08 ♒ 30 53	01 58	18 27	29 ♍ 38	09 16	28 46	16 55	03 53	11 35
04	14 56 10	12 22 33	14 ♒ 26 04	20 ♒ 23 04	01 11	19 40	00 ♎ 15	09 09	28 41	16 56	03 54	11 37
05	15 00 06	13 22 42	26 ♒ 22 33	02 ♓ 25 10	00 35	20 52	00 52	09 03	28 37	16 56	03 55	11 39
06	15 04 03	14 22 53	08 ♓ 31 33	14 ♓ 42 16	00 ♏ 11	22 04	01 29	08 56	28 32	16 57	03 56	11 41
07	15 07 59	15 23 05	20 ♓ 57 47	27 ♓ 18 31	29 ♎ 58	23 17	02 06	08 49	28 27	16 57	03 56	11 43
08	15 11 56	16 23 18	03 ♈ 44 49	10 ♈ 16 51	29 ♎ 57	24 29	02 43	08 42	28 23	16 58	03 57	11 45
09	15 15 52	17 23 34	16 ♈ 54 43	23 ♈ 38 21	00 ♏ 07	25 41	03 20	08 35	28 18	16 59	03 58	11 47
10	15 19 49	18 23 50	00 ♉ 27 35	07 ♉ 22 06	00 D 27	26 53	03 57	08 28	28 13	16 59	03 59	11 49
11	15 23 45	19 24 09	14 ♉ 21 26	21 ♉ 25 02	00 57	28 05	04 34	08 20	28 08	17 00	04 00	11 52
12	15 27 42	20 24 29	28 ♉ 32 14	05 ♊ 42 20	01 35	29 ♐ 17	05 11	08 13	28 04	17 01	04 01	11 54
13	15 31 39	21 24 51	12 ♊ 54 35	20 ♊ 08 14	02 22	00 ♑ 29	05 47	08 05	27 59	17 02	04 01	11 56
14	15 35 35	22 25 15	27 ♊ 22 33	04 ♋ 36 52	03 15	01 41	06 24	07 58	27 54	17 03	04 02	11 58
15	15 39 32	23 25 40	11 ♋ 50 33	19 ♋ 03 05	04 14	02 53	07 01	07 50	27 49	17 04	04 03	12 00
16	15 43 28	24 26 08	26 ♋ 14 03	03 ♌ 23 06	05 18	04 05	07 38	07 42	27 44	17 05	04 04	12 03
17	15 47 25	25 26 37	10 ♌ 29 57	17 ♌ 34 25	06 27	05 17	08 14	07 35	27 39	17 06	04 06	12 05
18	15 51 21	26 27 08	24 ♌ 36 23	01 ♍ 35 45	07 40	06 29	08 51	07 27	27 34	17 07	04 07	12 07
19	15 55 18	27 27 41	08 ♍ 32 28	15 ♍ 26 30	08 56	07 40	09 28	07 19	27 30	17 08	04 08	12 09
20	15 59 14	28 28 16	22 ♍ 17 49	29 ♍ 06 22	10 15	08 52	10 04	07 11	27 25	17 09	04 09	12 12
21	16 03 11	29 ♏ 28 52	05 ♎ 52 08	12 ♎ 35 02	11 36	10 03	10 41	07 03	27 20	17 11	04 10	12 14
22	16 07 08	00 ♐ 29 30	19 ♎ 15 00	25 ♎ 51 56	13 00	11 15	11 18	06 55	27 15	17 12	04 11	12 16
23	16 11 04	01 30 10	02 ♏ 25 45	08 ♏ 56 22	14 25	12 26	11 54	06 47	27 10	17 13	04 13	12 18
24	16 15 01	02 30 51	15 ♏ 23 40	21 ♏ 47 37	15 52	13 38	12 31	06 38	27 05	17 15	04 14	12 21
25	16 18 57	03 31 34	28 ♏ 08 10	04 ♐ 25 17	17 20	14 49	13 07	06 30	27 00	17 16	04 15	12 23
26	16 22 54	04 32 19	10 ♐ 39 01	16 ♐ 49 27	18 49	16 01	13 44	06 22	26 55	17 18	04 17	12 25
27	16 26 50	05 33 04	22 ♐ 56 43	29 ♐ 01 00	20 18	17 12	14 20	06 14	26 51	17 19	04 18	12 28
28	16 30 47	06 33 51	05 ♑ 02 32	11 ♑ 01 38	21 49	18 23	14 56	06 06	26 46	17 21	04 19	12 30
29	16 34 43	07 34 39	16 ♑ 58 38	22 ♑ 53 58	23 20	19 34	15 33	05 58	26 41	17 23	04 21	12 32
30	16 38 40	08 ♐ 35 28	28 ♑ 48 05	04 ♒ 41 29	24 ♏ 51	20 ♑ 45	16 ♎ 09	05 ♊ 49	26 ♉ 36	17 ♒ 24	04 ♒ 22	12 ♐ 35

DECLINATIONS

Date	Moon True ☊ ° '	Moon Mean ☊ ° '	Moon ☽ Latitude ° '	Sun ☉ ° '	Moon ☽ ° '	Mercury ☿ ° '	Venus ♀ ° '	Mars ♂ ° '	Jupiter ♃ ° '	Saturn ♄ ° '	Uranus ♅ ° '	Neptune ♆ ° '	Pluto ♇ ° '
01	17 ♋ 54	18 ♋ 54	00 N 47	14 S 37	22 S 21	12 S 42	24 S 20	01 N 50	20 N 56	17 N 39	16 S 27	19 S 06	11 S 50
02	17 D 53	18 50	00 S 16	14 56	22 05	12 01	24 30	01 35	20 55	17 38	16 27	19 06	11 51
03	17 54	18 47	01 18	15 15	20 50	11 24	24 39	01 20	20 54	17 37	16 26	19 06	11 51
04	17 R 54	18 44	02 17	15 33	18 41	10 52	24 47	01 06	20 53	17 36	16 26	19 06	11 52
05	17 53	18 41	03 10	15 51	15 42	10 25	24 55	00 51	20 52	17 35	16 26	19 06	11 52
06	17 50	18 38	03 56	16 09	12 01	10 04	25 02	00 36	20 51	17 34	16 26	19 06	11 53
07	17 45	18 35	04 32	16 27	07 45	09 49	25 08	00 22	20 50	17 33	16 26	19 05	11 53
08	17 37	18 31	04 56	16 44	03 S 02	09 39	25 14	00 N 07	20 49	17 32	16 25	19 05	11 54
09	17 27	18 28	05 05	17 01	01 N 57	09 35	25 19	00 S 08	20 48	17 31	16 25	19 05	11 54
10	17 17	18 25	04 57	17 18	07 00	09 36	25 23	00 22	20 47	17 30	16 25	19 05	11 55
11	17 07	18 22	04 31	17 35	11 49	09 42	25 26	00 37	20 46	17 29	16 25	19 05	11 55
12	16 58	18 19	03 49	17 51	16 07	09 52	25 28	00 51	20 45	17 28	16 24	19 05	11 56
13	16 52	18 15	02 51	18 07	19 31	10 06	25 30	01 06	20 44	17 27	16 24	19 04	11 56
14	16 48	18 12	01 42	18 22	21 42	10 24	25 31	01 20	20 43	17 26	16 24	19 04	11 57
15	16 46	18 09	00 S 26	18 38	22 28	10 44	25 32	01 35	20 41	17 25	16 23	19 04	11 57
16	16 D 46	18 06	00 N 51	18 53	21 44	11 07	25 31	01 49	20 40	17 24	16 23	19 04	11 58
17	16 47	18 03	02 04	19 07	19 36	11 32	25 30	02 04	20 39	17 23	16 23	19 03	11 58
18	16 48	18 00	03 09	19 22	16 18	11 58	25 28	02 18	20 38	17 22	16 22	19 03	11 58
19	16 R 48	17 56	04 02	19 36	12 07	12 26	25 26	02 33	20 37	17 21	16 22	19 03	11 59
20	16 45	17 53	04 41	19 49	07 21	12 55	25 23	02 47	20 35	17 20	16 22	19 03	11 59
21	16 40	17 50	05 03	20 02	02 N 18	13 25	25 19	03 01	20 34	17 19	16 21	19 02	12 00
22	16 34	17 47	05 08	20 15	02 S 47	13 55	25 14	03 16	20 33	17 18	16 21	19 02	12 00
23	16 26	17 44	04 56	20 28	07 40	14 25	25 08	03 30	20 32	17 17	16 20	19 02	12 01
24	16 17	17 41	04 30	20 40	12 09	14 57	25 02	03 44	20 30	17 15	16 20	19 02	12 01
25	16 10	17 37	03 50	20 51	16 00	15 28	24 55	03 58	20 29	17 14	16 19	19 01	12 01
26	16 03	17 34	03 00	21 03	19 05	15 58	24 48	04 13	20 28	17 13	16 19	19 01	12 02
27	15 58	17 31	02 02	21 14	21 13	16 29	24 39	04 27	20 27	17 12	16 18	19 01	12 02
28	15 56	17 28	00 N 59	21 24	22 22	16 59	24 30	04 41	20 25	17 11	16 18	19 00	12 03
29	15 D 55	17 25	00 S 06	21 34	22 27	17 29	24 21	04 55	20 24	17 11	16 17	19 00	12 03
30	15 ♋ 56	17 ♋ 21	01 S 10	21 S 44	21 S 32	17 S 58	24 S 10	05 S 09	20 N 23	17 N 10	16 S 17	19 S 00	12 S 04

ZODIAC SIGN ENTRIES

Date	h	m	Planets
03	06	41	☽ ♒
04	02	00	♂ ♎
05	19	13	☽ ♓
07	07	28	☿ ♎
08	05	02	☽ ♈
08	21	42	☿ ♏
10	11	12	☽ ♉
12	14	27	☽ ♊
13	02	14	♀ ♑
14	16	21	☽ ♋
16	18	19	☽ ♌
18	21	15	☽ ♍
21	01	35	☽ ♎
22	00	19	☉ ♐
23	07	33	☽ ♏
25	15	33	☽ ♐
28	01	57	☽ ♑
30	14	26	☽ ♒

LATITUDES

Date	Mercury ☿ ° '	Venus ♀ ° '	Mars ♂ ° '	Jupiter ♃ ° '	Saturn ♄ ° '	Uranus ♅ ° '	Neptune ♆ ° '	Pluto ♇ ° '
01	00 N 10	01 S 38	01 N 18	00 S 57	02 S 19	00 S 43	00 N 11	10 N 25
04	01 05	01 46	01 18	00 57	02 19	00 42	00 11	10 24
07	01 46	01 52	01 18	00 57	02 19	00 42	00 11	10 24
10	02 10	01 59	01 18	00 56	02 19	00 42	00 11	10 23
13	02 20	02 04	01 18	00 56	02 19	00 42	00 11	10 22
16	02 18	02 09	01 18	00 56	02 19	00 42	00 11	10 22
19	02 09	02 13	01 18	00 56	02 19	00 42	00 11	10 21
22	01 55	02 17	01 19	00 56	02 19	00 42	00 11	10 21
25	01 37	02 20	01 18	00 55	02 18	00 42	00 11	10 20
28	01 17	02 21	01 18	00 55	02 18	00 42	00 11	10 20
31	00 N 56	02 S 22	01 N 18	00 S 54	02 S 18	00 S 42	00 N 11	10 N 19

DATA

Julian Date	2451850
Delta T	+64 seconds
Ayanamsa	23° 51' 50"
Synetic vernal point	05° ♓ 15' 10"
True obliquity of ecliptic	23° 26' 18"

MOON'S PHASES, APSIDES AND POSITIONS ☽

Date	h	m	Phase	Longitude ° '	Eclipse Indicator
04	07	27	☽	12 ♒ 11	
11	21	15	○	19 ♉ 47	
18	15	24	☾	26 ♌ 36	
25	23	11	●	04 ♐ 00	

Day	h	m		
03	03	31	Apogee	
14	23	12	Perigee	
30	23	42	Apogee	
01	17	45	Max dec	22° S 23'
09	02	42	ON	
15	12	10	Max dec	22° N 28'
21	22	49	OS	
29	02	11	Max dec	22° S 33'

LONGITUDES

Date	Chiron ⚷ ° '	Ceres ⚳ ° '	Pallas ⚴ ° '	Juno ⚵ ° '	Vesta ⚶ ° '	Black Moon Lilith ⚸ ° '
01	15 ♐ 42	20 ♏ 52	26 ♎ 59	15 ♒ 26	01 ♒ 18	27 ♑ 18
11	16 ♐ 45	25 ♏ 09	01 ♏ 24	17 ♒ 48	04 ♒ 56	28 ♑ 56
21	17 ♐ 52	29 ♏ 26	05 ♏ 46	20 ♒ 40	08 ♒ 50	29 ♑ 31
31	19 ♐ 01	03 ♐ 42	10 ♏ 05	23 ♒ 58	12 ♒ 56	00 ♒ 37

All ephemeris data is given at 12.00 UT and the Moon's longitude is additionally given for 24.00 UT

ASPECTARIAN

h m Aspects

01 Wednesday
01 45 ☽ ⊻ ♆
02 53 ☽ ✶ ☿
03 54 ☽ ± ♄
06 21 ☿ ⊥ ♂
12 53 ☽ ✶ ☉
13 00 ☽ ⊼ ♃
14 10 ☉ ⊼ ♃
14 29 ☿ □ ♆
15 55 ☽ ⊥ ♅
17 08 ☽ ⊻ ♇
21 56 ☽ bQ ♄

02 Thursday
00 21 ☿ ± ♃
00 59 ☽ Q ☿
01 02 ☽ ± ♃
03 55 ☽ ⊻ ♀
04 04 ☽ ⊻ ♅
05 19 ☽ ⊥ ♇
05 25 ♀ ✶ ♅
05 41 ♂ △ ♄
15 27 ☽ Q ☉
17 29 ☽ ⊥ ♀
18 26 ☿ ∥ ♇
19 09 ☽ bQ ♃
19 14 ☿ ∠ ♀
23 40 ☽ ∠ ♇

03 Friday
04 13 ☽ △ ♄
05 37 ☽ △ ♂
09 20 ♂ Q ♇
10 46 ☽ □ ☿
11 04 ☽ # ♃
13 54 ☽ ∠ ♀
14 36 ☽ ☌ ♆
17 00 ☉ ⊻ ♇
20 48 ♀ ∠ ♆

04 Saturday
01 24 ☽ △ ♃
06 16 ☽ ✶ ♇
07 27 ☽ □ ☉
08 02 ☽ ∥ ♆
13 45 ☽ bQ ♂
17 03 ☽ ☌ ♅
21 31 ☽ # ♄
23 44 ☽ ✶ ♀

05 Sunday
06 05 ☿ ⊻ ♂
06 32 ☽ Q ♇
06 40 ☽ ∥ ♅
08 50 ☽ ± ♂
11 02 ☽ ∥ ☉
16 26 ☽ □ ♄
20 04 ☽ △ ☿
21 25 ☽ ⊼ ♂

06 Monday
02 18 ☽ Q ♀
02 58 ☽ ⊻ ♆
12 47 ☽ □ ♃
12 52 ☽ ∥ ♇
14 44 ☽ ⊥ ♆
18 10 ☽ □ ♇

07 Tuesday
00 10 ☽ ∥ ☿
00 22 ☽ △ ☉
00 40 ☽ bQ ☿
03 26 ☽ Q ♄
04 20 ☽ ⊻ ♅
06 06 ♂ bQ ♅
08 08 ☽ ∠ ♆
10 18 ☉ ∥ ♅
15 47 ☽ ⊥ ♅
16 51 ☽ □ ♀
17 41 ☽ ± ☿
22 59 ☽ Q ♃

08 Wednesday
02 04 ☽ ✶ ♄
02 29 ☿ St D
04 56 ☽ ⊼ ☿
07 15 ☽ bQ ☉
08 42 ☽ ∠ ♅
10 00 ☽ ☍ ♂
12 23 ☽ ✶ ♆
21 02 ☽ ✶ ♃

09 Thursday
01 12 ☽ ± ☉
01 57 ☉ □ ♅
02 33 ☽ ∥ ♂
02 43 ☽ △ ♇
02 52 ☽ # ♂
05 32 ☽ ∠ ♄
10 18 ☽ Q ♆
12 07 ☽ ✶ ♅
12 56 ☽ ⊼ ☉
21 34 ☽ ⊥ ♄
23 48 ☽ ∠ ♃

10 Friday
05 07 ☽ △ ♀
05 36 ☽ bQ ♇
08 06 ☽ ⊻ ♄
09 25 ☽ Q ♅
12 00 ☽ ☍ ☿
13 13 ♂ △ ♆
15 28 ☽ ⊥ ♃
18 08 ☽ □ ♆
18 22 ☽ ⊼ ♂
21 21 ☽ ± ♇

11 Saturday
00 56 ☽ # ☿
01 46 ☽ ⊻ ♃
04 01 ☉ # ♄
05 12 ☽ ± ♂
07 43 ☽ ⊼ ♇
09 38 ☽ bQ ♀
10 06 ♀ ⊥ ♆
12 30 ☽ # ♇
13 01 ♀ ⊼ ♄
16 31 ☽ □ ♅
21 15 ☽ ☍ ☉
21 16 ☽ bQ ♂
21 47 ☉ ∠ ♂

12 Sunday
02 21 ☽ ± ♀
11 12 ☽ ☌ ♄
13 23 ☽ ⊼ ♀
13 50 ☽ # ♅
17 23 ☽ ⊼ ☿
20 40 ☽ ∥ ♄
21 11 ☽ △ ♆
23 37 ☽ △ ♂

13 Monday
00 18 ☽ # ☉
03 58 ☽ ± ☿
04 03 ☽ ☌ ♃
05 09 ☿ ± ♃
08 24 ☽ # ♆
10 22 ☽ ☍ ♇
18 51 ☽ △ ♅
19 51 ☽ bQ ☿
22 10 ☽ bQ ♆
23 23 ☽ ∥ ♃

14 Tuesday
02 47 ☉ Q ♆
03 10 ☽ ⊼ ☉
12 52 ☽ ⊻ ♄
13 06 ☽ ± ♆
13 52 ☽ ± ☉
19 18 ♀ ∠ ♅
19 45 ☽ bQ ♅
19 47 ☽ ☍ ♀
22 25 ☽ △ ☿
22 45 ☽ ⊥ ♄
23 04 ☽ ⊼ ♆

15 Wednesday
03 38 ☽ □ ♂
05 24 ☽ ⊻ ♃
05 54 ☽ bQ ☉
07 58 ☿ □ ♆
10 42 ☽ ± ♅
12 16 ☽ ⊼ ♇
13 37 ☽ ∠ ♄
15 17 ☽ ⊥ ♃
20 42 ☽ ⊼ ♅
22 17 ☽ ± ♇

16 Thursday
05 30 ♀ ± ♄
06 09 ☽ ∠ ♃
08 46 ☽ △ ☉
10 56 ☽ Q ♂
11 50 ♀ ⊻ ♆
13 22 ☽ bQ ♇
14 30 ☽ ✶ ♄
14 38 ♂ △ ♃

17 Friday
01 11 ☽ ☍ ♆
01 51 ☽ ∥ ♃
02 23 ☽ ⊼ ♀
04 15 ☿ ⊥ ♇
04 33 ☽ □ ☿
05 33 ☉ ∥ ♆
07 07 ☽ ✶ ♃
08 01 ☽ ✶ ♂
10 35 ☽ Q ♄
13 27 ☽ ± ♀
14 41 ☽ △ ♇
15 42 ☽ # ☉
16 32 ☽ # ♆
23 12 ☽ ☍ ♅

18 Saturday
03 16 ☽ Q ♃
04 58 ☽ ∥ ♄
06 09 ☽ bQ ♀
08 14 ☿ ⊼ ♃
10 39 ☽ ∠ ♂
11 30 ☽ # ♅
12 15 ☿ ∥ ♇
13 59 ☽ Q ☿
15 24 ☽ □ ☉
17 03 ☽ □ ♄

19 Sunday
02 49 ☽ ⊥ ♂
04 22 ☽ ⊼ ♆
05 33 ♀ ⊼ ♃
09 54 ☽ □ ♃
10 21 ☽ △ ♀
10 26 ☽ # ☿
12 41 ☽ # ♇
12 41 ☉ ☍ ♄
12 45 ☽ ✶ ☿
13 40 ☽ ⊻ ♂
14 46 ☽ ± ♆
18 18 ☽ □ ♇

20 Monday
00 59 ☽ Q ☉
02 59 ☽ ⊼ ♅
06 07 ☿ ⊻ ♂
06 28 ☽ bQ ♆
13 31 ☽ ± ♅
17 46 ☽ ∠ ☿
20 57 ☽ △ ♄
23 45 ☽ ✶ ☉

21 Tuesday
01 58 ☽ Q ♇
05 26 ☽ bQ ♅
08 46 ☽ # ♂
08 58 ☽ △ ♆
11 29 ☽ ⊥ ☿
14 05 ☽ △ ♃
20 12 ☽ □ ♀
21 00 ☽ ☌ ♂
23 08 ☿ ⊻ ♇
23 24 ☽ ✶ ♇
23 26 ☽ ⊻ ☿
23 28 ☽ bQ ♄

22 Wednesday
04 40 ☽ ∠ ☉
08 18 ☽ △ ♅
11 00 ♀ ⊥ ♅
13 43 ♀ □ ♂
14 24 ☽ ∥ ♂
15 36 ☽ ± ♄
16 46 ☽ bQ ♃
22 17 ☽ ⊥ ☉

23 Thursday
02 26 ☽ ⊼ ♄
02 36 ☽ ∠ ♇
06 49 ♀ bQ ♄
08 00 ☽ Q ♀
09 00 ☽ ± ♃
09 13 ♀ ⊻ ♇
10 09 ☽ ⊻ ☉
15 17 ☽ □ ♆
18 04 ♀ ± ♃
19 10 ☽ ⊥ ♇
19 17 ☉ # ♃
19 55 ☽ ⊼ ♃
21 14 ♂ bQ ♄

24 Friday
05 06 ♂ ✶ ♇
06 18 ☽ ⊻ ♇
06 22 ☽ ⊻ ♂
08 23 ☽ ✶ ♀
11 16 ☽ ∥ ♇
13 00 ☽ ☌ ☿
15 28 ☽ □ ♅
18 07 ☽ ⊥ ♂

25 Saturday
00 51 ☽ Q ♆
07 43 ☽ ∥ ☿
09 52 ☽ ☍ ♄
11 02 ☿ □ ♅
11 58 ☽ ∠ ♂
14 12 ☽ ∥ ♅
15 32 ☽ ∠ ♀
20 50 ☽ # ♄
23 11 ☽ ☌ ☉
23 42 ☽ ✶ ♆

26 Sunday
01 40 ☽ Q ♅
03 50 ☽ ☍ ♃
05 38 ☉ ✶ ♆
10 38 ☽ ⊥ ♀
11 27 ☽ ∥ ♆
15 27 ☽ ☌ ♇
18 17 ☽ ✶ ♂
23 31 ☽ ⊻ ♀

27 Monday
00 57 ☽ ✶ ♅
02 03 ☽ # ♃
03 48 ☿ ∥ ♅
04 50 ☽ ∠ ♆
06 06 ☽ ⊻ ☿
06 29 ☉ Q ♅
12 05 ☽ ∥ ☉
12 42 ☿ ⊥ ♂
14 40 ♀ ⊻ ♅
19 02 ☽ Q ♂
19 34 ☽ ⊥ ☿
19 39 ☽ ⊼ ♄
22 36 ☽ ⊥ ♆

28 Tuesday
02 12 ☉ ☍ ♃
06 37 ☽ ∠ ♅
07 29 ☽ ± ♄
10 34 ☽ ⊻ ♆
14 05 ☽ ⊼ ♃
14 30 ♀ ⊥ ♇
15 20 ☽ ⊻ ☉
16 04 ☽ ∠ ☿
20 10 ☿ Q ♆
21 41 ☿ # ♄

29 Wednesday
00 41 ☽ ⊥ ♅
01 24 ☽ bQ ♄
01 59 ☽ ± ♃
03 01 ☽ ⊻ ♇
04 30 ☽ ⊥ ☉
08 57 ☽ □ ♂
12 49 ☽ ⊻ ♅
15 10 ☽ ⊥ ♇
17 49 ☽ ☌ ♀
19 58 ☽ bQ ♃

30 Thursday
00 25 ☽ ∠ ☉
02 47 ☽ ✶ ☿
07 34 ☽ △ ♄
08 50 ☽ ∥ ☉
09 30 ☽ ∠ ♇
13 21 ♀ bQ ♃
23 22 ☽ ☌ ♆

DECEMBER 2000

LONGITUDES

Date	Sidereal time h m s	Sun ☉ ° ′ ″	Moon ☽ ° ′ ″	Moon ☽ 24.00 ° ′ ″	Mercury ☿ ° ′	Venus ♀ ° ′	Mars ♂ ° ′	Jupiter ♃ ° ′	Saturn ♄ ° ′	Uranus ♅ ° ′	Neptune ♆ ° ′	Pluto ♇ ° ′
01	16 42 37	09 ♐ 36 18	10 ♒ 34 43	16 ♒ 28 21	26 ♏ 23	21 ♑ 56	16 ♎ 46	05 ♊ 41	26 ♉ 32	17 ♒ 26	04 ♒ 24	12 ♐ 37
02	16 46 33	10 37 09	22 ♒ 23 01	28 ♒ 19 19	27 55	23 07	17 22	05 R 33	26 R 27	17 28	04 25	12 39
03	16 50 30	11 38 00	04 ♓ 17 55	10 ♓ 19 27	29 ♏ 28	24 17	17 58	05 25	26 22	17 30	04 27	12 42
04	16 54 26	12 38 53	16 ♓ 24 35	22 ♓ 33 54	01 ♐ 00	25 28	18 34	05 17	26 18	17 32	04 28	12 44
05	16 58 23	13 39 46	28 ♓ 48 02	05 ♈ 07 29	02 33	26 38	19 11	05 09	26 13	17 34	04 30	12 46
06	17 02 19	14 40 40	11 ♈ 32 45	18 ♈ 04 14	04 06	27 49	19 47	05 01	26 09	17 36	04 31	12 49
07	17 06 16	15 41 35	24 ♈ 42 11	01 ♉ 26 47	05 38	28 ♑ 59	20 23	04 53	26 04	17 38	04 33	12 51
08	17 10 12	16 42 30	08 ♉ 18 03	15 ♉ 15 51	07 12	00 ♒ 09	20 59	04 45	26 00	17 40	04 35	12 53
09	17 14 09	17 43 27	22 ♉ 19 54	29 ♉ 29 44	08 45	01 19	21 35	04 37	25 55	17 42	04 36	12 56
10	17 18 06	18 44 24	06 ♊ 44 45	14 ♊ 04 11	10 18	02 29	22 11	04 30	25 51	17 44	04 38	12 58
11	17 22 02	19 45 22	21 ♊ 27 10	28 ♊ 52 43	11 51	03 39	22 47	04 22	25 47	17 46	04 40	13 00
12	17 25 59	20 46 21	06 ♋ 19 51	13 ♋ 47 32	13 25	04 49	23 23	04 14	25 43	17 48	04 42	13 03
13	17 29 55	21 47 21	21 ♋ 14 46	28 ♋ 40 37	14 58	05 59	23 59	04 07	25 38	17 51	04 43	13 05
14	17 33 52	22 48 22	06 ♌ 04 13	13 ♌ 24 50	16 31	07 08	24 35	04 00	25 34	17 53	04 45	13 07
15	17 37 48	23 49 23	20 ♌ 41 51	27 ♌ 54 47	18 05	08 17	25 11	03 52	25 30	17 55	04 47	13 09
16	17 41 45	24 50 26	05 ♍ 03 16	12 ♍ 07 04	19 39	09 27	25 46	03 45	25 26	17 58	04 49	13 12
17	17 45 41	25 51 29	19 ♍ 06 03	26 ♍ 00 10	21 13	10 36	26 22	03 38	25 23	18 00	04 51	13 14
18	17 49 38	26 52 34	02 ♎ 49 28	09 ♎ 34 04	22 47	11 44	26 58	03 31	25 19	18 03	04 53	13 16
19	17 53 35	27 53 39	16 ♎ 14 06	22 ♎ 49 45	24 21	12 53	27 34	03 24	25 15	18 05	04 54	13 19
20	17 57 31	28 54 45	29 ♎ 21 14	05 ♏ 48 45	25 55	14 02	28 09	03 18	25 11	18 08	04 56	13 21
21	18 01 28	29 ♐ 55 52	12 ♏ 12 32	18 ♏ 32 48	27 30	15 10	28 45	03 11	25 08	18 10	04 58	13 23
22	18 05 24	00 ♑ 57 00	24 ♏ 49 45	01 ♐ 03 37	29 ♐ 04	16 19	29 21	03 05	25 04	18 13	05 00	13 25
23	18 09 21	01 58 08	07 ♐ 14 34	13 ♐ 22 50	00 ♑ 39	17 27	29 ♎ 56	02 58	25 01	18 15	05 02	13 28
24	18 13 17	02 59 17	19 ♐ 28 35	25 ♐ 32 01	02 15	18 35	00 ♏ 32	02 52	24 58	18 18	05 04	13 30
25	18 17 14	04 00 26	01 ♑ 33 19	07 ♑ 32 44	03 50	19 42	01 07	02 46	24 54	18 21	05 06	13 32
26	18 21 10	05 01 36	13 ♑ 30 27	19 ♑ 26 44	05 25	20 50	01 42	02 41	24 51	18 23	05 08	13 34
27	18 25 07	06 02 45	25 ♑ 21 50	01 ♒ 16 02	07 01	21 57	02 18	02 35	24 48	18 26	05 10	13 36
28	18 29 04	07 03 55	07 ♒ 09 40	13 ♒ 03 05	08 37	23 05	02 53	02 29	24 45	18 29	05 12	13 39
29	18 33 00	08 05 05	18 ♒ 56 39	24 ♒ 50 47	10 14	24 12	03 28	02 24	24 42	18 32	05 14	13 41
30	18 36 57	09 06 15	00 ♓ 45 56	06 ♓ 42 34	11 51	25 18	04 04	02 19	24 39	18 35	05 16	13 43
31	18 40 53	10 ♑ 07 25	12 ♓ 41 12	18 ♓ 42 21	13 ♑ 28	26 ♒ 25	04 ♏ 39	02 ♊ 14	24 ♉ 37	18 ♒ 38	05 ♒ 19	13 ♐ 45

Date	Moon True ☊ ° ′	Moon Mean ☊ ° ′	Moon ☽ Latitude ° ′
01	15 ♋ 58	17 ♋ 18	02 S 11
02	15 D 59	17 15	03 06
03	16 01	17 12	03 54
04	16 R 01	17 09	04 33
05	16 00	17 06	04 59
06	15 57	17 02	05 12
07	15 53	16 59	05 10
08	15 49	16 56	04 50
09	15 44	16 53	04 12
10	15 40	16 50	03 18
11	15 37	16 47	02 09
12	15 36	16 43	00 S 51
13	15 D 36	16 40	00 N 31
14	15 37	16 37	01 51
15	15 38	16 34	03 02
16	15 39	16 31	04 00
17	15 40	16 27	04 43
18	15 R 40	16 24	05 08
19	15 39	16 21	05 16
20	15 38	16 18	05 07
21	15 35	16 15	04 43
22	15 33	16 12	04 05
23	15 31	16 08	03 16
24	15 29	16 05	02 19
25	15 28	16 02	01 16
26	15 28	15 59	00 N 11
27	15 D 28	15 56	00 S 55
28	15 29	15 53	01 58
29	15 29	15 49	02 56
30	15 30	15 46	03 46
31	15 ♋ 31	15 ♋ 43	04 S 28

DECLINATIONS

Date	Sun ☉ ° ′	Moon ☽ ° ′	Mercury ☿ ° ′	Venus ♀ ° ′	Mars ♂ ° ′	Jupiter ♃ ° ′	Saturn ♄ ° ′	Uranus ♅ ° ′	Neptune ♆ ° ′	Pluto ♇ ° ′
01	21 S 53	19 S 41	18 S 27	23 S 59	05 S 23	20 N 22	17 N 09	16 S 16	18 S 59	12 S 04
02	22 02	16 59	18 56	23 47	05 37	20 20	17 08	16 16	18 59	12 04
03	22 11	13 34	19 22	23 35	05 51	20 19	17 07	16 15	18 59	12 05
04	22 19	09 33	19 48	23 22	06 04	20 18	17 06	16 14	18 58	12 05
05	22 26	05 03	20 14	23 08	06 18	20 16	17 05	16 14	18 58	12 05
06	22 34	00 S 14	20 38	22 54	06 32	20 15	17 04	16 13	18 58	12 06
07	22 40	04 N 46	21 02	22 39	06 45	20 14	17 03	16 12	18 57	12 06
08	22 46	09 42	21 25	22 24	06 59	20 13	17 02	16 12	18 57	12 06
09	22 52	14 17	21 47	22 08	07 12	20 11	17 01	16 11	18 56	12 07
10	22 58	18 11	22 08	21 51	07 26	20 10	17 01	16 10	18 56	12 07
11	23 03	21 01	22 27	21 34	07 39	20 09	17 00	16 10	18 56	12 07
12	23 07	22 26	22 46	21 16	07 53	20 08	16 59	16 09	18 55	12 08
13	23 11	22 16	23 04	20 58	08 06	20 07	16 58	16 08	18 55	12 08
14	23 15	20 32	23 20	20 39	08 19	20 05	16 57	16 08	18 54	12 08
15	23 18	17 28	23 35	20 20	08 32	20 04	16 57	16 07	18 54	12 09
16	23 20	13 23	23 49	20 00	08 45	20 03	16 56	16 06	18 54	12 09
17	23 22	08 39	24 02	19 39	08 58	20 02	16 55	16 05	18 53	12 09
18	23 24	03 N 36	24 14	19 19	09 11	20 01	16 54	16 05	18 53	12 09
19	23 25	01 S 31	24 24	18 57	09 24	20 00	16 54	16 04	18 52	12 10
20	23 26	06 27	24 33	18 36	09 37	19 59	16 53	16 03	18 52	12 10
21	23 26	11 00	24 41	18 13	09 50	19 58	16 52	16 02	18 51	12 10
22	23 26	15 00	24 47	17 51	10 02	19 57	16 52	16 01	18 51	12 10
23	23 25	18 17	24 52	17 28	10 15	19 56	16 51	16 01	18 50	12 11
24	23 24	20 42	24 56	17 04	10 28	19 55	16 51	16 00	18 50	12 11
25	23 22	22 09	24 58	16 40	10 40	19 54	16 50	15 59	18 49	12 11
26	23 21	22 34	24 59	16 16	10 52	19 53	16 50	15 58	18 49	12 11
27	23 18	21 58	24 59	15 52	11 05	19 52	16 49	15 57	18 48	12 11
28	23 15	20 23	24 57	15 27	11 17	19 51	16 49	15 56	18 48	12 12
29	23 12	17 55	24 53	15 01	11 29	19 50	16 48	15 55	18 48	12 12
30	23 08	14 43	24 48	14 36	11 41	19 49	16 48	15 54	18 47	12 12
31	23 S 03	10 S 55	24 S 42	14 S 10	11 S 53	19 N 49	16 N 47	15 S 53	18 S 47	12 S 12

ZODIAC SIGN ENTRIES

Date	h	m	Planets
03	03	23	☽ ♓
03	20	26	☿ ♐
05	14	17	☽ ♈
07	21	27	☽ ♉
08	08	48	♀ ♒
10	00	50	☽ ♊
12	01	48	☽ ♋
14	02	09	☽ ♌
16	03	30	☽ ♍
18	07	01	☽ ♎
20	13	12	☽ ♏
21	13	37	☉ ♑
22	21	57	☽ ♐
23	02	03	☿ ♑
23	14	37	♂ ♏
25	08	54	☽ ♑
27	21	25	☽ ♒
30	10	27	☽ ♓

LATITUDES

Date	Mercury ☿ ° ′	Venus ♀ ° ′	Mars ♂ ° ′	Jupiter ♃ ° ′	Saturn ♄ ° ′	Uranus ♅ ° ′	Neptune ♆ ° ′	Pluto ♇ ° ′
01	00 N 56	02 S 22	01 N 18	00 S 54	02 S 18	00 S 42	00 N 11	10 N 19
04	00 34	02 22	01 18	00 54	02 18	00 42	00 11	10 19
07	00 N 13	02 21	01 18	00 54	02 17	00 41	00 11	10 19
10	00 S 08	02 19	01 18	00 53	02 16	00 41	00 11	10 19
13	00 28	02 15	01 18	00 52	02 16	00 41	00 10	10 19
16	00 47	02 11	01 17	00 52	02 15	00 41	00 10	10 19
19	01 05	02 06	01 17	00 51	02 15	00 41	00 10	10 18
22	01 21	01 59	01 17	00 51	02 14	00 41	00 10	10 19
25	01 35	01 52	01 16	00 50	02 13	00 41	00 10	10 19
28	01 47	01 43	01 16	00 49	02 13	00 41	00 10	10 19
31	01 S 57	01 S 33	01 N 16	00 S 49	02 S 12	00 S 41	00 N 10	10 N 19

DATA

Julian Date	2451880
Delta T	+64 seconds
Ayanamsa	23° 51′ 54″
Synetic vernal point	05° ♓ 15′ 05″
True obliquity of ecliptic	23° 26′ 18″

MOON'S PHASES, APSIDES AND POSITIONS ☽

Date	h	m	Phase	Longitude ° ′	Eclipse Indicator
04	03	55	☽	12 ♓ 18	
11	09	03	○	19 ♊ 38	
18	00	41	☾	26 ♍ 24	
25	17	22	●	04 ♑ 14	Partial

Day	h	m		
12	22	29	Perigee	
28	15	14	Apogee	
06	13	07	0N	
12	21	31	Max dec	22° N 34′
19	04	50	0S	
26	09	39	Max dec	22° S 35′

LONGITUDES

Date	Chiron ⚷ ° ′	Ceres ⚳ ° ′	Pallas ⚴ ° ′	Juno ⚵ ° ′	Vesta ⚶ ° ′	Black Moon Lilith ⚸ ° ′
01	19 ♐ 01	03 ♐ 42	10 ♏ 05	23 ♒ 58	12 ♒ 56	00 ♒ 37
11	20 ♐ 11	07 ♐ 56	14 ♏ 18	27 ♒ 38	17 ♒ 12	01 ♒ 44
21	21 ♐ 20	12 ♐ 08	18 ♏ 25	01 ♓ 38	21 ♒ 37	02 ♒ 50
31	22 ♐ 29	16 ♐ 15	22 ♏ 25	05 ♓ 55	26 ♒ 08	03 ♒ 57

All ephemeris data is given at 12.00 UT and the Moon's longitude is additionally given for 24.00 UT

ASPECTARIAN

h m Aspects

01 Friday
02 09 ☽ △ ♃
04 23 ☽ # ♃
06 52 ☽ Q ☿
09 50 ☽ ✱ ☉
14 07 ☿ ☍ ♄
16 10 ☽ ✱ ♇
18 53 ☽ ∥ ♆
22 06 ☽ ∥ ☿

02 Saturday
01 16 ☽ △ ♂
02 00 ☽ ☌ ♅
10 52 ☽ # ♄
12 31 ☽ Q ☉
13 38 ☽ ⊻ ♀
15 50 ☿ ∥ ♆
16 18 ♂ △ ♅
16 37 ☽ Q ♇
17 32 ☽ ∥ ♅
20 10 ☽ □ ♄

03 Sunday
00 51 ☽ □ ☿
03 05 ☽ ⊥ ♀
09 12 ☽ bQ ♂
12 18 ☽ ⊻ ♆
14 12 ☽ □ ♃
21 16 ☽ ∥ ♇
23 01 ☽ ∠ ♀

04 Monday
00 16 ☽ ⊥ ♆
03 55 ☽ □ ☉
04 03 ☽ ± ♂
04 44 ☽ □ ♇
07 52 ☽ Q ♄
14 05 ☉ ☌ ♇
14 12 ☽ ⊻ ♅
16 27 ☽ ⊼ ♂
18 00 ☽ ∠ ♆

05 Tuesday
01 15 ☽ Q ♃
01 54 ☽ ⊥ ♅
03 54 ♀ △ ♄
05 52 ☽ ∥ ♂
07 05 ☽ ✱ ♄
07 26 ☽ ✱ ♀
14 28 ☿ # ♃
19 11 ☽ ∠ ♅
20 07 ☽ △ ☿
22 50 ☽ ✱ ♆
23 55 ☽ ✱ ♃

06 Wednesday
08 28 ☽ Q ♀
11 15 ☽ ∠ ♄
11 55 ♀ ∠ ♇
14 21 ☽ △ ♇
18 16 ☽ △ ☉
18 49 ☿ ✱ ♆
19 43 ♂ bQ ♃
21 12 ☽ Q ♆
23 10 ☽ ✱ ♅

07 Thursday
00 57 ♂ ± ♄
01 10 ☿ ☍ ♃
03 23 ☽ ∠ ♃
03 41 ☽ ⊥ ♄
03 42 ☽ bQ ☿
03 50 ☽ ☍ ♂
05 23 ☿ ∠ ♂
11 02 ☉ ∥ ♀
11 47 ☿ Q ♅
14 26 ☽ ⊻ ♄
17 38 ☽ bQ ♇
19 23 ☽ ⊥ ♃
20 22 ☽ □ ♀
20 48 ☽ Q ♅
21 57 ☽ ± ☿
22 04 ☽ # ♂
23 32 ☽ bQ ☉

08 Friday
05 30 ☽ □ ♆
05 52 ☽ ⊻ ♃
09 32 ☽ ± ♇
09 50 ☽ ⊼ ☿
16 30 ☽ ± ☉
19 57 ☽ ⊼ ♇

09 Saturday
00 17 ☽ # ♇
03 35 ☽ ⊼ ☉
04 08 ☽ □ ♅
10 41 ☽ ⊼ ♂
11 20 ☉ ✱ ♅
14 16 ♃ △ ♆
18 00 ☽ ☌ ♄
21 12 ☽ ± ♂
22 59 ☽ # ♅

10 Sunday
01 24 ☿ ∥ ♀
04 10 ☽ ∥ ♄
04 21 ☽ △ ♀
08 19 ☽ ☌ ♃
08 31 ☽ △ ♆
12 45 ☽ bQ ♂
17 25 ☽ # ♆
18 32 ☽ ☍ ☿
22 13 ☽ ☍ ♇

11 Monday
03 28 ☽ ∥ ♃
06 00 ☽ △ ♅
07 04 ☽ bQ ♀
09 03 ☽ ☍ ☉
09 06 ☽ bQ ♆
09 46 ☉ ∠ ♆
14 15 ☽ △ ♂
17 45 ☽ # ♀
18 58 ☽ ⊻ ♄
22 52 ☽ ± ♀
23 41 ☽ ± ♆

12 Tuesday
01 16 ♀ △ ♃
04 36 ☽ ⊥ ♄
06 12 ☿ ☌ ♇
06 19 ☽ bQ ♅
08 40 ☽ ⊻ ♃
09 21 ☽ ⊼ ♀
09 22 ☽ ⊼ ♆
09 24 ♀ ☌ ♆
18 14 ☽ ⊥ ♃
19 01 ☽ ∠ ♄
20 50 ☽ ± ♅
22 49 ☽ ⊼ ♇

13 Wednesday
00 43 ☽ ⊼ ☿
06 30 ☽ ⊼ ♅
08 30 ☽ ± ♇
08 36 ☽ ∠ ♃
11 30 ☽ ± ☿
12 56 ☽ ⊼ ☉
16 36 ☽ □ ♂
19 04 ☽ ✱ ♄
23 04 ☽ bQ ♇
23 20 ☽ ± ☉

14 Thursday
01 42 ☉ ∥ ☿
03 45 ☽ bQ ☿
08 39 ☽ ✱ ♃
09 51 ☽ ☍ ♆
10 44 ☽ # ♀
13 53 ☽ ☍ ♀
14 26 ☽ Q ♄
15 02 ☽ bQ ☉
16 14 ☽ ∥ ♃
23 05 ☽ Q ♂
23 33 ☽ △ ♇

15 Friday
01 54 ☽ # ♆
04 06 ☽ Q ♃
07 10 ☽ △ ☿
07 24 ☽ ☍ ♅
09 24 ☿ ✱ ♅
15 23 ☽ ∥ ♄
17 35 ☽ △ ☉
19 45 ☽ ✱ ♂
19 57 ☽ □ ♄
20 32 ☽ # ♅
23 54 ♂ ⊼ ♄

16 Saturday
07 52 ♀ # ♃
09 49 ☽ □ ♃
11 36 ☽ ⊼ ♆
14 35 ☿ ∠ ♆
18 30 ☽ # ♇
20 06 ☽ ⊼ ♀
21 48 ☽ ± ♆
22 08 ☽ ∠ ♂

17 Sunday
01 18 ☉ ⊼ ♄
01 53 ☽ □ ♇
07 17 ☽ ± ♀
10 06 ☽ ⊼ ♅
10 29 ☽ # ♂
13 17 ☽ bQ ♆
14 18 ☽ ⊥ ♂
16 07 ☽ □ ☿
20 32 ☽ ± ♅
22 51 ☽ △ ♄

18 Monday
00 19 ☽ bQ ♀
00 41 ☽ □ ☉
01 13 ☽ ⊻ ♂
09 15 ☽ Q ♇
12 23 ☽ bQ ♅
13 13 ☽ △ ♃
15 39 ☽ △ ♆
17 09 ☉ ✱ ♂

19 Tuesday
01 16 ☽ bQ ♄
04 03 ☽ Q ☿
05 23 ☽ △ ♀
06 42 ☽ ✱ ♇
06 44 ♂ ± ♃
11 20 ☽ Q ☉
13 16 ☉ ∠ ♀
15 22 ☽ △ ♅
15 54 ☽ bQ ♃
17 27 ☽ ± ♄
17 46 ♀ ∥ ♆
21 08 ♀ ✱ ♇

20 Wednesday
01 15 ☿ ⊼ ♄
04 22 ☽ ⊼ ♄
04 48 ☽ ✱ ☿
08 14 ☽ ± ♃
09 41 ☽ ☌ ♂
10 08 ☽ ∠ ♇
11 07 ☽ ✱ ☉
12 39 ☉ ⊥ ♆
19 15 ☽ ⊼ ♃
20 13 ♂ ∠ ♇
22 24 ☽ □ ♆

21 Thursday
02 55 ☽ ⊥ ♇
05 14 ☽ ∥ ♂
12 37 ☽ ∠ ☿
14 13 ☽ ⊻ ♇
17 35 ☽ ∠ ☉
18 09 ☽ □ ♀
18 37 ☽ ∥ ♇
23 19 ☽ □ ♅

22 Friday
08 09 ☽ ⊥ ☿
08 30 ☽ Q ♆
10 54 ☿ ⊥ ♆
12 15 ☽ ⊥ ☉
12 28 ☽ ☍ ♄
14 43 ☉ ± ♄
18 31 ☿ ✱ ♂
18 49 ☽ ∥ ♅
21 07 ☽ ⊻ ♂
21 21 ☽ ⊻ ☿

23 Saturday
00 49 ☽ # ♄
00 50 ☽ ⊻ ☉
03 46 ☽ ☍ ♃
06 06 ☽ ∥ ♀
07 42 ☽ ✱ ♆
08 09 ☽ Q ♀
09 20 ☽ ⊥ ♂
10 04 ☽ Q ♅
16 44 ☽ ∥ ♆
17 15 ☿ ± ♄

24 Sunday
00 11 ☽ ☌ ♇
03 05 ☽ # ♃
03 49 ☽ ∠ ♂
05 52 ♀ ☌ ♅
09 31 ☉ ⊼ ♃
09 40 ☽ ✱ ♅
10 03 ☽ ✱ ♀
13 11 ☽ ∠ ♆
19 41 ☉ ∠ ♅
20 58 ☿ ⊼ ♃
22 49 ☽ ⊼ ♄

25 Monday
02 01 ♀ # ♄
04 27 ☿ ∠ ♅
07 06 ☽ ⊥ ♆
10 42 ☽ ± ♄
11 05 ☽ ✱ ♂
14 25 ☽ ⊼ ♃
15 36 ☽ ∠ ♅
17 15 ☽ ☌ ☿
17 22 ☽ ☌ ☉
18 58 ☽ ∠ ♀
19 07 ☽ ⊻ ♆
19 23 ☉ ☌ ☿

26 Tuesday
02 21 ☽ ± ♃
04 40 ☽ bQ ♄
07 35 ☿ ⊻ ♆
09 44 ☽ ⊥ ♅
12 08 ☽ ⊻ ♇
12 25 ☽ Q ♂
14 42 ☉ ⊻ ♆
14 57 ☽ ⊥ ♀
20 21 ☽ bQ ♃
21 54 ☽ ⊻ ♅

27 Wednesday
00 17 ☽ ⊥ ♇
04 22 ☽ ⊻ ♀
06 26 ♀ ∥ ♅
08 42 ☿ ∠ ♀
10 52 ☽ △ ♄
18 36 ☽ ∠ ♇
22 00 ♂ ⊼ ♃

28 Thursday
02 34 ☽ △ ♃
02 50 ☽ □ ♂
08 00 ☽ ☌ ♆
10 05 ☿ ± ♃
11 47 ☽ ⊻ ☉
15 27 ☽ ⊻ ☿
17 56 ☽ # ♃

29 Friday
01 10 ☽ ⊥ ☉
01 15 ☽ ✱ ♇
04 21 ☽ ∥ ♆
04 21 ☿ bQ ♄
05 36 ☽ ⊥ ☿
11 09 ☽ ☌ ♅
18 50 ☉ ± ♃
21 03 ☽ # ♄
21 13 ☽ ∠ ☉
22 35 ♀ □ ♄
23 40 ☽ □ ♄
23 47 ☽ ☌ ♀

30 Saturday
01 44 ☽ Q ♇
02 48 ☽ ∠ ☿
03 40 ☽ ∥ ♅
13 00 ☽ ∥ ♀
15 07 ☽ □ ♃
19 00 ☽ △ ♂
21 08 ☽ ⊻ ♆
21 09 ♀ Q ♇
23 14 ☿ ⊥ ♅

31 Sunday
00 28 ☉ bQ ♄
04 14 ☽ ∥ ♇
06 23 ☽ ✱ ☉
06 29 ☽ ∥ ♂
09 14 ☽ ⊥ ♆
11 51 ☽ Q ♄
13 47 ☽ ✱ ☿
14 08 ☽ □ ♇
16 23 ☿ ⊻ ♇
23 53 ☽ ⊻ ♅

JANUARY 2001

LONGITUDES

Date	Sidereal time h m s	Sun ☉ ° ' "	Moon ☽ ° ' "	Moon ☽ 24.00 ° ' "	Mercury ☿ ° '	Venus ♀ ° '	Mars ♂ ° '	Jupiter ♃ ° '	Saturn ♄ ° '	Uranus ♅ ° '	Neptune ♆ ° '	Pluto ♇ ° '
01	18 44 50	11 ♑ 08 35	24 ♓ 46 33	00 ♈ 54 23	15 ♑ 05	27 ♒ 31	05 ♏ 14	02 ♊ 09	24 ♉ 34	18 ♒ 40	05 ♒ 21	13 ♐ 47
02	18 48 46	12 09 45	07 ♈ 06 24	13 ♈ 23 08	16 43	28 37	05 49	02 R 04	24 R 32	18 43	05 23	13 49
03	18 52 43	13 10 54	19 ♈ 45 07	26 ♈ 12 50	18 21	29 ♒ 43	06 24	02 00	24 29	18 46	05 25	13 51
04	18 56 39	14 12 03	02 ♉ 46 44	09 ♉ 27 09	19 59	00 ♓ 49	06 59	01 56	24 27	18 49	05 27	13 53
05	19 00 36	15 13 12	16 ♉ 14 21	23 ♉ 08 29	21 37	01 54	07 34	01 52	24 25	18 52	05 29	13 55
06	19 04 33	16 14 20	00 ♊ 09 32	07 ♊ 17 22	23 16	02 59	08 08	01 48	24 23	18 55	05 31	13 58
07	19 08 29	17 15 28	14 ♊ 31 38	21 ♊ 51 49	24 55	04 03	08 43	01 44	24 21	18 59	05 34	14 00
08	19 12 26	18 16 36	29 ♊ 17 15	06 ♋ 47 04	26 35	05 08	09 18	01 40	24 19	19 02	05 36	14 02
09	19 16 22	19 17 44	14 ♋ 20 15	21 ♋ 55 40	28 14	06 12	09 53	01 37	24 17	19 05	05 38	14 04
10	19 20 19	20 18 51	29 ♋ 32 08	07 ♌ 08 25	29 ♑ 54	07 16	10 27	01 34	24 15	19 08	05 40	14 06
11	19 24 15	21 19 58	14 ♌ 43 15	22 ♌ 15 32	01 ♒ 34	08 19	11 02	01 31	24 14	19 11	05 42	14 08
12	19 28 12	22 21 04	29 ♌ 44 09	07 ♍ 08 14	03 14	09 22	11 36	01 28	24 12	19 14	05 45	14 09
13	19 32 08	23 22 11	14 ♍ 27 01	21 ♍ 39 54	04 53	10 25	12 11	01 26	24 11	19 17	05 47	14 11
14	19 36 05	24 23 17	28 ♍ 46 31	05 ♎ 46 37	06 33	11 27	12 45	01 23	24 10	19 21	05 49	14 13
15	19 40 02	25 24 23	12 ♎ 40 08	19 ♎ 27 09	08 12	12 29	13 19	01 21	24 09	19 24	05 51	14 15
16	19 43 58	26 25 29	26 ♎ 07 51	02 ♏ 42 31	09 51	13 31	13 53	01 19	24 08	19 27	05 54	14 17
17	19 47 55	27 26 35	09 ♏ 11 30	15 ♏ 35 13	11 29	14 32	14 28	01 18	24 07	19 30	05 56	14 19
18	19 51 51	28 27 41	21 ♏ 54 07	28 ♏ 08 40	13 06	15 33	15 02	01 16	24 06	19 34	05 58	14 21
19	19 55 48	29 ♑ 28 46	04 ♐ 19 19	10 ♐ 26 33	14 42	16 33	15 36	01 15	24 05	19 37	06 00	14 22
20	19 59 44	00 ♒ 29 51	16 ♐ 30 48	22 ♐ 32 31	16 17	17 33	16 10	01 14	24 05	19 40	06 03	14 24
21	20 03 41	01 30 55	28 ♐ 32 05	04 ♑ 29 54	17 50	18 33	16 44	01 13	24 04	19 44	06 05	14 26
22	20 07 37	02 31 59	10 ♑ 26 18	16 ♑ 21 38	19 20	19 32	17 17	01 12	24 04	19 47	06 07	14 28
23	20 11 34	03 33 03	22 ♑ 16 11	28 ♑ 10 15	20 48	20 30	17 51	01 12	24 04	19 50	06 10	14 29
24	20 15 31	04 34 05	04 ♒ 04 05	09 ♒ 57 57	22 12	21 28	18 25	01 11	24 04	19 54	06 12	14 31
25	20 19 27	05 35 07	15 ♒ 52 06	21 ♒ 46 46	23 33	22 26	18 58	01 D 11	24 D 04	19 57	06 14	14 33
26	20 23 24	06 36 08	27 ♒ 42 13	03 ♓ 38 41	24 49	23 23	19 32	01 11	24 04	20 01	06 16	14 34
27	20 27 20	07 37 08	09 ♓ 36 27	15 ♓ 35 49	25 59	24 19	20 05	01 12	24 04	20 04	06 19	14 36
28	20 31 17	08 38 07	21 ♓ 37 05	27 ♓ 40 35	27 04	25 15	20 38	01 12	24 04	20 07	06 21	14 38
29	20 35 13	09 39 05	03 ♈ 46 40	09 ♈ 55 44	28 02	26 10	21 12	01 13	24 05	20 11	06 23	14 39
30	20 39 10	10 40 01	16 ♈ 08 10	22 ♈ 24 24	28 52	27 05	21 45	01 14	24 05	20 14	06 25	14 41
31	20 43 06	11 ♒ 40 56	28 ♈ 44 51	05 ♉ 09 57	29 ♒ 33	27 ♓ 59	22 ♏ 18	01 ♊ 15	24 ♉ 06	20 ♒ 18	06 ♒ 28	14 ♐ 42

Date	Moon True ☊ ° '	Moon Mean ☊ ° '	Moon ☽ Latitude ° '
01	15 ♋ 31	15 ♋ 40	04 S 58
02	15 D 31	15 37	05 15
03	15 R 31	15 33	05 17
04	15 31	15 30	05 04
05	15 31	15 27	04 34
06	15 D 31	15 24	03 47
07	15 31	15 21	02 44
08	15 32	15 18	01 29
09	15 32	15 14	00 S 07
10	15 R 32	15 11	01 N 17
11	15 31	15 08	02 35
12	15 31	15 05	03 42
13	15 30	15 02	04 32
14	15 29	14 59	05 04
15	15 28	14 55	05 17
16	15 28	14 52	05 12
17	15 D 28	14 49	04 51
18	15 29	14 46	04 16
19	15 30	14 43	03 29
20	15 32	14 39	02 34
21	15 33	14 36	01 33
22	15 33	14 33	00 N 28
23	15 R 33	14 30	00 S 37
24	15 32	14 27	01 41
25	15 30	14 24	02 40
26	15 27	14 20	03 32
27	15 23	14 17	04 16
28	15 19	14 14	04 48
29	15 15	14 11	05 08
30	15 12	14 08	05 14
31	15 ♋ 10	14 ♋ 04	05 S 06

DECLINATIONS

Date	Sun ☉ ° '	Moon ☽ ° '	Mercury ☿ ° '	Venus ♀ ° '	Mars ♂ ° '	Jupiter ♃ ° '	Saturn ♄ ° '	Uranus ♅ ° '	Neptune ♆ ° '	Pluto ♇ ° '
01	22 S 58	06 S 38	24 S 34	13 S 44	12 S 05	19 N 48	16 N 47	15 S 53	18 S 46	12 S 13
02	22 53	02 S 00	24 25	13 17	12 17	19 47	16 47	15 52	18 46	12 13
03	22 47	02 N 50	24 14	12 51	12 28	19 47	16 46	15 51	18 45	12 13
04	22 41	07 40	24 01	12 24	12 40	19 46	16 46	15 50	18 45	12 13
05	22 34	12 19	23 47	11 56	12 51	19 46	16 46	15 49	18 44	12 13
06	22 27	16 29	23 31	11 29	13 03	19 45	16 45	15 48	18 43	12 13
07	22 19	19 49	23 14	11 01	13 14	19 44	16 45	15 47	18 43	12 13
08	22 11	21 57	22 56	10 34	13 25	19 44	16 45	15 46	18 42	12 14
09	22 03	22 33	22 35	10 06	13 36	19 44	16 45	15 45	18 42	12 14
10	21 54	21 30	22 13	09 37	13 47	19 43	16 45	15 44	18 41	12 14
11	21 45	18 54	21 50	09 09	13 58	19 43	16 45	15 43	18 41	12 14
12	21 35	15 02	21 25	08 41	14 09	19 43	16 44	15 42	18 40	12 14
13	21 25	10 18	20 59	08 12	14 20	19 42	16 44	15 41	18 40	12 14
14	21 14	05 N 08	20 31	07 43	14 30	19 42	16 44	15 40	18 39	12 14
15	21 03	00 S 08	20 02	07 15	14 41	19 42	16 44	15 39	18 39	12 14
16	20 52	05 14	19 31	06 46	14 51	19 42	16 44	15 38	18 38	12 14
17	20 40	09 57	18 59	06 17	15 02	19 42	16 44	15 37	18 38	12 14
18	20 28	14 07	18 26	05 48	15 12	19 42	16 44	15 36	18 37	12 14
19	20 16	17 35	17 52	05 19	15 22	19 42	16 45	15 35	18 37	12 14
20	20 03	20 12	17 16	04 50	15 32	19 42	16 45	15 34	18 36	12 14
21	19 49	21 53	16 40	04 21	15 42	19 42	16 45	15 33	18 35	12 14
22	19 36	22 33	16 03	03 52	15 52	19 42	16 45	15 31	18 35	12 14
23	19 22	22 12	15 26	03 22	16 02	19 42	16 45	15 30	18 34	12 15
24	19 07	20 52	14 49	02 53	16 11	19 42	16 46	15 29	18 34	12 15
25	18 52	18 37	14 11	02 24	16 21	19 42	16 46	15 28	18 33	12 15
26	18 37	15 35	13 34	01 55	16 30	19 43	16 46	15 27	18 33	12 15
27	18 22	11 55	12 58	01 26	16 39	19 43	16 46	15 26	18 32	12 15
28	18 06	07 44	12 22	00 58	16 49	19 43	16 47	15 25	18 32	12 14
29	17 50	03 S 13	11 48	00 S 29	16 58	19 44	16 47	15 24	18 31	12 14
30	17 34	01 N 30	11 16	00 00	17 06	19 44	16 48	15 23	18 30	12 14
31	17 S 17	06 N 16	10 S 46	00 N 28	17 S 15	19 N 45	16 N 48	15 S 22	18 S 30	12 S 14

ZODIAC SIGN ENTRIES

Date	h	m	Planets
01	22	14	☽ ♈
03	18	14	♀ ♓
04	06	57	☽ ♉
06	11	44	☽ ♊
08	13	09	☽ ♋
10	12	44	☽ ♌
10	13	26	☿ ♒
12	12	26	☽ ♍
14	14	05	☽ ♎
16	19	02	☽ ♏
19	03	36	☽ ♐
20	00	16	☉ ♒
21	14	57	☽ ♑
24	03	43	☽ ♒
26	16	39	☽ ♓
29	04	35	☽ ♈
31	14	21	☽ ♉

LATITUDES

Date	Mercury ☿ ° '	Venus ♀ ° '	Mars ♂ ° '	Jupiter ♃ ° '	Saturn ♄ ° '	Uranus ♅ ° '	Neptune ♆ ° '	Pluto ♇ ° '
01	02 S 00	01 S 29	01 N 15	00 S 48	02 S 12	00 S 41	00 N 10	10 N 19
04	02 05	01 17	01 15	00 48	02 11	00 41	00 10	10 19
07	02 08	01 05	01 14	00 47	02 10	00 41	00 10	10 19
10	02 06	00 50	01 14	00 46	02 09	00 41	00 10	10 20
13	02 00	00 35	01 13	00 45	02 09	00 41	00 10	10 20
16	01 48	00 S 18	01 12	00 45	02 08	00 41	00 10	10 21
19	01 30	00 00	01 11	00 44	02 07	00 41	00 10	10 21
22	01 05	00 N 19	01 11	00 43	02 06	00 41	00 10	10 21
25	00 S 33	00 39	01 10	00 42	02 05	00 41	00 10	10 22
28	00 N 07	01 01	01 09	00 42	02 04	00 41	00 10	10 22
31	00 N 55	01 N 23	01 N 07	00 S 41	02 S 03	00 S 41	00 N 10	10 N 23

DATA

Julian Date	2451911
Delta T	+64 seconds
Ayanamsa	23° 52' 00"
Synetic vernal point	05° ♓ 15' 00"
True obliquity of ecliptic	23° 26' 18"

MOON'S PHASES, APSIDES AND POSITIONS ☽

Date	h	m	Phase	Longitude ° '	Eclipse Indicator
02	22	32	☽	12 ♈ 37	
09	20	24	○	19 ♋ 39	total
16	12	35	☾	26 ♎ 27	
24	13	07	●	04 ♒ 37	

Day	h	m		
10	09	07	Perigee	
24	19	21	Apogee	
02	22	01	0N	
09	08	52	Max dec	22° N 34'
15	11	24	0S	
22	15	47	Max dec	22° S 34'
30	04	23	0N	

LONGITUDES

Date	Chiron ⚷ ° '	Ceres ⚳ ° '	Pallas ⚴ ° '	Juno ⚵ ° '	Vesta ⚶ ° '	Black Moon Lilith ⚸ ° '
01	22 ♐ 35	16 ♐ 40	22 ♏ 48	06 ♓ 21	26 ♒ 35	04 ♒ 03
11	23 ♐ 41	20 ♐ 42	26 ♏ 38	10 ♓ 54	01 ♓ 12	05 ♒ 10
21	24 ♐ 43	24 ♐ 38	00 ♐ 16	15 ♓ 40	05 ♓ 52	06 ♒ 16
31	25 ♐ 40	28 ♐ 27	03 ♐ 40	20 ♓ 37	10 ♓ 36	07 ♒ 23

All ephemeris data is given at 12.00 UT and the Moon's longitude is additionally given for 24.00 UT

ASPECTARIAN

h m	Aspects
01 Monday	
02 34	☽ bQ ♂
02 56	☽ Q ♃
03 14	☽ ∠ ♆
08 29	☽ Q ☉
11 36	☽ ⚹ ♄
11 48	☽ ⊥ ♅
17 01	♂ □ ♆
17 13	☽ Q ☿
17 55	☽ ⊻ ♀
21 10	☽ ± ♂
02 Tuesday	
02 20	☽ ⚹ ♃
04 02	♂ ∥ ♇
05 27	☽ ∠ ♅
06 44	☽ ⊥ ♀
08 40	☽ ⚹ ♆
09 23	☽ ⊼ ♂
16 38	☽ ∠ ♄
17 06	☿ bQ ♃
22 32	☽ □ ☉
03 Wednesday	
00 52	☽ △ ♇
01 38	☽ ∠ ♀
01 53	☉ ⊥ ♅
06 52	☽ ∠ ♃
07 36	☽ Q ♆
08 58	☽ □ ☿
09 38	☽ ⊥ ♄
10 10	☽ ⚹ ♅
13 14	☿ Q ♂
18 30	☿ ⊻ ♅
20 47	☽ ⊻ ♄
23 32	☽ ⊥ ♃
04 Thursday	
01 56	♀ ∥ ♂
04 26	☉ ⊻ ♇
04 54	☽ bQ ♇
08 06	☽ ⚹ ♀
08 26	☽ Q ♅
10 28	☽ ⊻ ♃
10 39	☿ ⊥ ♇
16 51	☽ □ ♆
19 55	☽ ☍ ♂
21 14	☽ ± ♇
21 19	♀ ∥ ♇
05 Friday	
07 31	☽ Q ♀
07 55	☽ ⊼ ♇
10 04	☽ △ ☉
10 09	☽ # ♀
11 14	♀ □ ♃
11 27	☽ # ♇
15 01	☽ # ♂
16 37	☽ □ ♅
22 38	☽ △ ☿
06 Saturday	
02 09	☽ ☌ ♄
04 01	♂ ⊥ ♇
07 45	☽ # ♅
13 42	☽ ∥ ♄
13 58	☽ bQ ☉
14 45	☽ ☌ ♃
17 10	☽ □ ♀
21 04	☽ △ ♆
07 Sunday	
00 19	☉ bQ ♃
01 59	☽ ⊼ ♂
03 11	☽ # ♆
03 24	☽ bQ ☿
03 46	☿ △ ♄
06 12	☽ ± ☉
11 07	☽ ☍ ♇
11 18	☽ ∥ ♃
12 20	☽ ± ♂
16 49	☽ ⊼ ☉
19 19	☽ △ ♅
20 07	☽ ± ☿
21 54	☽ bQ ♆
08 Monday	
03 37	☽ bQ ♂
04 00	☽ ⊻ ♄
07 05	☽ ⊼ ☿
12 30	☽ ± ♆
13 39	☽ ⊥ ♄
15 49	☽ ⊻ ♃
16 05	☽ # ☉
19 37	☽ bQ ♅
22 05	☽ △ ♀
22 08	☽ ⊼ ♆
22 50	♀ ⊻ ♆
09 Tuesday	
01 22	☽ ⊥ ♃
04 00	☽ ∠ ♄
04 38	☽ △ ♂
06 37	☉ ⊻ ♅
10 00	☽ ± ♅
11 34	☽ ⊼ ♇
15 36	☽ ∠ ♃
19 32	☽ ⊼ ♅
20 24	☽ ☍ ☉
21 04	☽ ± ♇
23 40	☽ bQ ♀
10 Wednesday	
00 06	☿ ∠ ♇
03 41	☽ ⚹ ♄
05 40	☽ # ☉
06 37	☉ ⊥ ♇
11 18	☽ bQ ♇
12 39	☽ ☍ ☿
14 56	☽ ± ♀
15 12	☽ ⚹ ♃
21 42	☽ ☍ ♆
22 35	☽ Q ♄
11 Thursday	
01 06	☽ ⊼ ♀
05 42	☽ ∥ ♃
05 55	☽ □ ♂
10 06	☽ Q ♃
11 03	☽ △ ♇
11 21	☿ △ ♃
13 32	☽ # ♆
19 07	☽ ☍ ♅
20 30	☉ ∥ ☿
23 17	☽ ⊼ ☉
12 Friday	
02 08	☽ ∥ ♄
03 08	☽ □ ♄
08 14	☽ # ♅
09 36	☽ ± ☉
11 47	☽ Q ♂
14 48	☽ □ ♃
16 32	☽ # ♂
17 32	☿ ⊥ ♀
18 22	☽ ⊼ ☿
21 45	☽ ⊼ ♆
13 Saturday	
01 16	☽ bQ ☉
02 35	☽ # ♇
04 51	☽ ☍ ♀
05 23	☽ ± ☿
07 35	☽ ± ♆
08 06	☽ ⚹ ♂
11 34	☽ □ ♇
20 04	☽ ⊼ ♅
22 13	☽ bQ ☿
22 33	☽ bQ ♆
22 57	☽ # ♀
14 Sunday	
01 10	☿ ☌ ♆
04 00	☽ △ ☉
04 12	☽ △ ♄
06 10	☽ ± ♅
06 49	☉ △ ♄
10 11	☽ ∠ ♂
16 27	☽ △ ♃
17 54	☽ Q ♇
20 44	♀ ⊥ ♆
21 34	☽ bQ ♅
15 Monday	
00 06	☽ △ ♆
02 16	☽ ⊥ ♂
03 09	☽ △ ☿
04 09	♀ Q ♄
05 51	☽ bQ ♄
07 24	☉ Q ♂
11 39	☽ ⊼ ♀
13 12	☽ ⊻ ♂
14 47	☽ ⚹ ♇
18 29	☽ bQ ♃
21 40	☽ ± ♄
23 08	☽ ± ♀
23 57	☽ △ ♅
16 Tuesday	
03 40	☿ # ♃
08 23	☽ ⊼ ♄
10 33	☽ ± ♃
12 35	☽ □ ☉
16 42	☽ bQ ♀
17 45	☽ ∠ ♇
18 49	☽ ∥ ♀
21 26	☽ ⊼ ♃
17 Wednesday	
05 29	♂ ⊻ ♇
05 56	☽ □ ♆
06 37	♀ □ ♇
07 58	♀ △ ♂
10 22	☽ ⊥ ♇
16 55	☽ □ ☿
21 37	☽ ⊻ ♇
22 20	☽ ☌ ♂
22 53	☽ △ ♀
18 Thursday	
00 42	☽ ∥ ♇
00 45	☽ Q ☉
03 51	☿ ∥ ♆
07 31	☽ □ ♅
15 58	☽ Q ♆
16 12	☽ ☍ ♄
19 18	☽ ∥ ♂
21 33	☽ ∥ ♅
19 Friday	
01 44	☽ ⚹ ☉
05 41	☽ # ♄
06 01	☽ ☍ ♃
06 54	☿ ⚹ ♇
08 23	☽ Q ☿
09 27	☉ ∠ ♇
13 53	☽ ∥ ☿
15 18	☽ ⚹ ♆
18 29	☽ Q ♅
20 30	☽ ∥ ♆
20 Saturday	
06 40	☽ # ♃
07 49	☽ ☌ ♇
09 05	☿ □ ♂
09 48	☽ ∠ ☉
10 28	☽ ∥ ☉
11 16	☽ ⊻ ♂
11 28	☽ ⚹ ☿
14 15	☽ □ ♀
15 06	♂ ∥ ♅
18 18	☽ ⚹ ♅
21 02	☽ ∠ ♆
23 48	☽ ⊥ ♂
21 Sunday	
03 04	☽ ⊼ ♄
04 58	☉ △ ♃
05 23	☽ ⊥ ☉
08 56	☿ # ♄
15 05	☽ ± ♄
15 07	☽ ⊥ ♆
17 23	☽ ⊼ ♃
18 33	☽ ⊻ ☉
18 44	☽ ∠ ♂
21 54	☽ ∠ ☿
22 Monday	
00 31	☽ ∠ ♅
01 28	☉ # ♃
03 15	☽ ⊻ ♆
04 03	♀ Q ♃
05 28	☽ ± ♃
05 35	☽ Q ♀
09 13	☽ bQ ♄
17 55	☿ ∥ ♂
18 36	♀ ⊻ ♅
18 42	☽ ⊥ ☿
18 48	☽ ⊥ ♅
19 35	☿ ☌ ♅
20 10	☽ ⊻ ♇
21 26	☿ ⊻ ♀
23 40	☽ bQ ♃
23 Tuesday	
02 35	☽ ⚹ ♂
07 02	☽ ⊻ ♅
08 06	☽ ⚹ ♀
08 22	☽ ⊥ ♇
08 35	☽ ⊻ ☿
09 14	☿ ∥ ♅
15 39	☽ △ ♄
24 Wednesday	
02 43	☽ ∠ ♇
04 11	☽ Q ♂
04 49	♀ ∠ ♆
06 09	☽ △ ♃
13 07	☽ ☌ ☉
16 21	☽ ☌ ♆
17 20	☽ ∠ ♀
25 Thursday	
00 23	♄ St D
01 33	☽ # ♃
08 39	♃ St D
09 18	☽ ⚹ ♇
09 26	☽ ∥ ☉
12 36	☽ ∥ ♆
13 15	☽ ⊥ ♀
18 37	☽ □ ♂
20 20	☽ ☌ ♅
21 35	☿ □ ♄
26 Friday	
02 29	☽ ⊻ ♀
03 19	☽ # ♄
03 54	☉ ☌ ♆
04 37	☽ □ ♄
05 28	☽ ☌ ☿
05 38	☽ ∥ ♂
09 42	☽ Q ♇
12 57	☽ ∥ ♅
19 03	☽ □ ♃
19 36	☉ ∥ ♆
27 Saturday	
04 16	☽ ∥ ☿
05 21	☽ ⊻ ♆
05 25	♀ ⚹ ♄
07 38	☽ ⊻ ☉
09 58	☽ ∥ ♇
11 09	♂ □ ♅
16 56	☽ Q ♄
17 26	☽ ⊥ ♆
20 47	☽ ⊥ ☉
22 02	☽ □ ♇
28 Sunday	
01 38	☿ Q ♇
07 12	☽ Q ♃
07 27	♂ # ♄
09 01	☽ ⊻ ♅
09 58	☽ △ ♂
11 28	☽ ∠ ♆
16 22	☽ ∠ ☉
16 52	☽ ⚹ ♄
17 26	☿ ∥ ♇
19 48	☽ ☌ ♀
20 58	☽ ⊥ ♅
23 45	☽ ⊻ ☿
29 Monday	
06 58	☽ ⚹ ♃
12 17	♀ ⊥ ♅
12 32	☽ ⊥ ☿
14 46	☽ ∠ ♅
16 57	☽ bQ ♂
17 07	☽ ⚹ ♆
22 21	☽ ∠ ♄
30 Tuesday	
00 29	☽ ⚹ ☉
03 31	☽ ∥ ♀
05 07	☽ # ♀
07 20	☽ ∠ ☿
09 11	☽ △ ♇
11 13	☽ ± ♂
12 11	☽ ∠ ♃
15 45	☽ ⊥ ♄
16 24	☽ Q ♆
19 54	☽ ⚹ ♅
23 14	☽ ⊼ ♂
31 Wednesday	
01 35	☽ Q ☉
03 13	☽ ⊻ ♄
05 24	☽ ⊥ ♃
10 27	☽ ⊻ ♀
13 30	☉ ∥ ♂
13 36	☽ ⚹ ☿
13 48	☽ bQ ♇
16 43	☽ ⊻ ♃
18 41	☽ Q ♅
22 31	☽ ⊥ ♀

FEBRUARY 2001

LONGITUDES

Date	Sidereal time h m s	Sun ☉ ° ' "	Moon ☽ ° ' "	Moon ☽ 24.00 ° ' "	Mercury ☿ ° '	Venus ♀ ° '	Mars ♂ ° '	Jupiter ♃ ° '	Saturn ♄ ° '	Uranus ♅ ° '	Neptune ♆ ° '	Pluto ♇ ° '
01	20 47 03	12 ♒ 41 50	11 ♉ 40 09	18 ♉ 15 49	00 ♓ 06	28 ♓ 52	22 ♏ 51	01 ♊ 17	24 ♉ 07	20 ♒ 21	06 ♒ 30	14 ♐ 44
02	20 51 00	13 42 43	24 ♉ 57 21	01 ♊ 45 01	00 28	29 ♓ 44	23 23	01 18	24 08	20 25	06 32	14 45
03	20 54 56	14 43 34	08 ♊ 39 04	15 ♊ 39 36	00 40	00 ♈ 36	23 56	01 20	24 09	20 28	06 35	14 46
04	20 58 53	15 44 24	22 ♊ 46 36	29 ♊ 59 55	00 R 41	01 27	24 29	01 22	24 10	20 32	06 37	14 48
05	21 02 49	16 45 13	07 ♋ 19 13	14 ♋ 43 58	00 31	02 17	25 01	01 24	24 11	20 35	06 39	14 49
06	21 06 46	17 46 00	22 ♋ 13 28	29 ♋ 46 49	00 ♓ 09	03 06	25 34	01 26	24 12	20 39	06 41	14 50
07	21 10 42	18 46 46	07 ♌ 22 57	15 ♌ 00 41	29 ♒ 38	03 55	26 06	01 29	24 14	20 42	06 44	14 52
08	21 14 39	19 47 30	22 ♌ 38 44	00 ♍ 15 45	28 56	04 42	26 38	01 32	24 15	20 46	06 46	14 53
09	21 18 35	20 48 13	07 ♍ 50 25	15 ♍ 21 30	28 06	05 29	27 10	01 35	24 17	20 49	06 48	14 54
10	21 22 32	21 48 55	22 ♍ 47 53	00 ♎ 08 38	27 08	06 15	27 42	01 38	24 19	20 52	06 50	14 55
11	21 26 29	22 49 36	07 ♎ 22 57	14 ♎ 30 20	26 04	06 59	28 14	01 41	24 21	20 56	06 52	14 57
12	21 30 25	23 50 15	21 ♎ 30 25	28 ♎ 23 04	24 57	07 43	28 46	01 45	24 23	20 59	06 55	14 58
13	21 34 22	24 50 54	05 ♏ 08 21	11 ♏ 46 28	23 47	08 25	29 18	01 48	24 25	21 03	06 57	14 59
14	21 38 18	25 51 31	18 ♏ 17 46	24 ♏ 42 40	22 37	09 07	29 ♏ 49	01 52	24 27	21 06	06 59	15 00
15	21 42 15	26 52 07	01 ♐ 01 43	07 ♐ 15 29	21 29	09 47	00 ♐ 21	01 56	24 29	21 10	07 01	15 01
16	21 46 11	27 52 42	13 ♐ 24 35	19 ♐ 29 39	20 24	10 26	00 52	02 01	24 32	21 13	07 03	15 02
17	21 50 08	28 53 16	25 ♐ 31 18	01 ♑ 30 09	19 24	11 03	01 23	02 05	24 34	21 17	07 06	15 03
18	21 54 04	29 ♒ 53 49	07 ♑ 26 48	13 ♑ 21 48	18 29	11 40	01 54	02 10	24 37	21 20	07 08	15 04
19	21 58 01	00 ♓ 54 20	19 ♑ 15 42	25 ♑ 08 59	17 41	12 15	02 25	02 14	24 40	21 24	07 10	15 05
20	22 01 58	01 54 50	01 ♒ 02 04	06 ♒ 55 23	17 00	12 48	02 56	02 19	24 43	21 27	07 12	15 06
21	22 05 54	02 55 18	12 ♒ 49 17	18 ♒ 44 04	16 26	13 20	03 27	02 25	24 46	21 31	07 14	15 06
22	22 09 51	03 55 45	24 ♒ 40 01	00 ♓ 37 21	15 59	13 51	03 57	02 30	24 49	21 34	07 16	15 07
23	22 13 47	04 56 10	06 ♓ 36 17	12 ♓ 37 00	15 41	14 20	04 28	02 35	24 52	21 37	07 18	15 08
24	22 17 44	05 56 33	18 ♓ 39 38	24 ♓ 44 21	15 29	14 47	04 58	02 41	24 55	21 41	07 20	15 09
25	22 21 40	06 56 55	00 ♈ 51 16	07 ♈ 00 31	15 25	15 13	05 28	02 47	24 58	21 44	07 22	15 10
26	22 25 37	07 57 15	13 ♈ 12 16	19 ♈ 26 40	15 D 27	15 37	05 58	02 53	25 02	21 48	07 24	15 10
27	22 29 33	08 57 33	25 ♈ 43 54	02 ♉ 04 10	15 36	15 59	06 27	02 59	25 05	21 51	07 26	15 11
28	22 33 30	09 ♓ 57 49	08 ♉ 27 42	14 ♉ 54 46	15 ♒ 50	16 ♈ 19	06 ♐ 57	03 ♊ 05	25 ♉ 09	21 ♒ 54	07 ♒ 28	15 ♐ 11

DECLINATIONS

Date	Moon True ☊ ° '	Moon Mean ☊ ° '	Moon ☽ Latitude ° '	Sun ☉ ° '	Moon ☽ ° '	Mercury ☿ ° '	Venus ♀ ° '	Mars ♂ ° '	Jupiter ♃ ° '	Saturn ♄ ° '	Uranus ♅ ° '	Neptune ♆ ° '	Pluto ♇ ° '
01	15 ♋ 10	14 ♋ 01	04 S 41	17 S 00	10 N 52	10 S 19	00 N 56	17 S 24	19 N 45	16 N 49	15 S 21	18 S 29	12 S 14
02	15 D 10	13 58	04 02	16 43	15 05	09 56	01 25	17 33	19 46	16 49	15 20	18 29	12 14
03	15 11	13 55	03 07	16 25	18 40	09 35	01 53	17 41	19 46	16 50	15 18	18 28	12 14
04	15 13	13 52	02 00	16 07	21 15	09 19	02 20	17 49	19 47	16 50	15 17	18 28	12 14
05	15 14	13 49	00 S 43	15 49	22 31	09 07	02 48	17 58	19 48	16 51	15 16	18 27	12 14
06	15 R 14	13 45	00 N 38	15 30	22 14	09 00	03 15	18 06	19 48	16 51	15 15	18 27	12 14
07	15 13	13 42	01 58	15 12	20 20	08 57	03 43	18 14	19 49	16 52	15 14	18 26	12 14
08	15 09	13 39	03 10	14 53	16 58	08 58	04 09	18 22	19 50	16 53	15 13	18 25	12 14
09	15 04	13 36	04 08	14 34	12 28	09 05	04 36	18 30	19 51	16 53	15 12	18 25	12 14
10	14 59	13 33	04 48	14 14	07 16	09 15	05 02	18 37	19 51	16 54	15 11	18 24	12 14
11	14 53	13 30	05 09	13 54	01 N 48	09 29	05 28	18 45	19 52	16 55	15 10	18 24	12 14
12	14 48	13 26	05 09	13 34	03 S 36	09 46	05 54	18 52	19 53	16 56	15 08	18 23	12 13
13	14 45	13 23	04 52	13 14	08 39	10 06	06 19	19 00	19 54	16 56	15 07	18 23	12 13
14	14 43	13 20	04 20	12 54	13 07	10 28	06 44	19 07	19 55	16 57	15 06	18 22	12 13
15	14 D 42	13 17	03 35	12 33	16 51	10 51	07 09	19 14	19 56	16 58	15 05	18 22	12 13
16	14 43	13 14	02 42	12 13	19 44	11 15	07 33	19 21	19 57	16 59	15 04	18 21	12 13
17	14 45	13 10	01 43	11 52	21 39	11 39	07 56	19 28	19 59	17 00	15 03	18 21	12 13
18	14 46	13 07	00 N 40	11 30	22 34	12 03	08 20	19 35	20 00	17 01	15 02	18 20	12 13
19	14 R 46	13 04	00 S 24	11 09	22 27	12 26	08 42	19 42	20 01	17 02	15 01	18 19	12 12
20	14 45	13 01	01 27	10 48	21 20	12 48	09 04	19 48	20 02	17 02	15 00	18 19	12 12
21	14 41	12 58	02 26	10 26	19 17	13 09	09 26	19 55	20 03	17 03	14 59	18 18	12 12
22	14 35	12 55	03 18	10 04	16 25	13 28	09 47	20 01	20 05	17 04	14 57	18 18	12 12
23	14 26	12 51	04 02	09 42	12 50	13 45	10 08	20 07	20 06	17 05	14 56	18 17	12 12
24	14 16	12 48	04 36	09 20	08 43	14 01	10 28	20 13	20 07	17 06	14 55	18 17	12 12
25	14 06	12 45	04 58	08 58	04 S 13	14 15	10 47	20 19	20 08	17 08	14 54	18 16	12 11
26	13 56	12 42	05 06	08 35	00 N 31	14 27	11 05	20 25	20 10	17 09	14 53	18 16	12 11
27	13 47	12 39	04 59	08 13	05 18	14 37	11 23	20 31	20 11	17 10	14 52	18 15	12 11
28	13 ♋ 40	12 ♋ 36	04 S 38	07 S 50	09 N 56	14 S 46	11 N 40	20 S 37	20 N 13	17 N 11	14 S 51	18 S 15	12 S 11

ZODIAC SIGN ENTRIES

Date	h	m	Planets
01	07	13	☿ ♓
02	19	14	♀ ♈
02	20	56	☽ ♊
05	00	00	☽ ♋
06	19	57	☿ ♒
07	00	21	☽ ♌
08	23	35	☽ ♍
10	23	46	☽ ♎
13	02	51	☽ ♏
14	20	06	♂ ♐
15	10	02	☽ ♐
17	20	59	☽ ♑
18	14	27	☉ ♓
20	09	53	☽ ♒
22	22	45	☽ ♓
25	10	20	☽ ♈
27	20	06	☽ ♉

LATITUDES

Date	Mercury ☿ ° '	Venus ♀ ° '	Mars ♂ ° '	Jupiter ♃ ° '	Saturn ♄ ° '	Uranus ♅ ° '	Neptune ♆ ° '	Pluto ♇ ° '
01	01 N 11	01 N 31	01 N 07	00 S 41	02 S 03	00 S 41	00 N 10	10 N 23
04	02 03	01 55	01 06	00 40	02 02	00 41	00 10	10 24
07	02 50	02 21	01 05	00 39	02 01	00 41	00 10	10 25
10	03 25	02 47	01 03	00 39	02 01	00 41	00 10	10 25
13	03 42	03 15	01 02	00 38	02 00	00 41	00 10	10 26
16	03 38	03 43	01 00	00 37	01 59	00 41	00 10	10 27
19	03 15	04 12	00 58	00 36	01 58	00 41	00 10	10 27
22	02 42	04 42	00 56	00 36	01 57	00 41	00 10	10 28
25	02 03	05 12	00 54	00 35	01 56	00 41	00 10	10 29
28	01 23	05 42	00 52	00 34	01 56	00 41	00 10	10 30
31	00 N 45	06 N 11	00 N 50	00 S 34	01 S 55	00 S 41	00 N 10	10 N 30

DATA

Julian Date	2451942
Delta T	+64 seconds
Ayanamsa	23° 52' 04"
Synetic vernal point	05° ♓ 14' 55"
True obliquity of ecliptic	23° 26' 19"

MOON'S PHASES, APSIDES AND POSITIONS ☽

Date	h	m	Phase	Longitude ° '	Eclipse Indicator
01	14	02	☽	12 ♉ 47	
08	07	12	○	19 ♌ 35	
15	03	24	☾	26 ♏ 30	
23	08	21	●	04 ♓ 47	

Day	h	m		
07	22	22	Perigee	
20	21	53	Apogee	
05	19	52	Max dec	22° N 36'
11	19	52	0S	
18	21	21	Max dec	22° S 39'
26	09	24	0N	

LONGITUDES

Date	Chiron ⚷ ° '	Ceres ⚳ ° '	Pallas ⚴ ° '	Juno ⚵ ° '	Vesta ⚶ ° '	Black Moon Lilith ⚸ ° '
01	25 ♐ 46	28 ♐ 50	03 ♐ 59	21 ♓ 08	11 ♓ 04	07 ♒ 29
11	26 ♐ 37	02 ♑ 29	07 ♐ 05	26 ♓ 16	15 ♓ 49	08 ♒ 36
21	27 ♐ 21	05 ♑ 58	09 ♐ 50	01 ♈ 32	20 ♓ 35	09 ♒ 43
31	27 ♐ 57	09 ♑ 14	12 ♐ 11	06 ♈ 57	25 ♓ 22	10 ♒ 49

All ephemeris data is given at 12.00 UT and the Moon's longitude is additionally given for 24.00 UT

ASPECTARIAN

h m Aspects

01 Thursday
02 27 ☽ □ ♆
06 35 ☽ ± ♇
09 21 ☽ # ☿
12 48 ☽ Q ☿
14 02 ☽ □ ☉
16 18 ☽ ∠ ♀
17 36 ☽ ⊼ ♇
19 30 ☽ # ♇

02 Friday
03 17 ☉ # ♄
03 50 ☽ □ ♅
09 05 ☽ ☍ ♂
10 31 ☽ ☌ ♄
13 25 ☽ # ♅
21 02 ☽ ✶ ♀
21 28 ☽ # ☉
21 56 ☽ □ ☿
22 58 ☽ ∥ ♄
23 14 ☽ ☌ ♃

03 Saturday
04 33 ☽ # ♂
08 24 ☽ △ ♆
10 34 ☽ # ♆
13 07 ☉ ✶ ♇
13 59 ☿ ⊻ ♀
19 14 ☽ Q ♀
21 07 ☽ ∥ ♃
21 33 ♂ ☍ ♄
22 30 ☽ ☍ ♇
23 13 ☽ △ ☉

04 Sunday
01 58 ☿ St R
08 13 ☽ △ ♅
09 26 ♀ ✶ ♃
10 03 ☽ ꝑ ♆
14 19 ☽ ⊻ ♄
14 57 ☽ ⊼ ♂
18 22 ♂ Q ♆

05 Monday
00 17 ☽ ⊥ ♄
01 00 ☽ △ ☿
01 03 ☽ ± ♆
01 17 ☽ ± ♂
02 13 ☽ ꝑ ☉
02 17 ☽ ⊻ ♃
03 16 ☽ □ ♀
09 09 ☽ ꝑ ♅
10 54 ☽ ⊼ ♆
12 08 ☽ ⊥ ♃
15 02 ☽ ∠ ♄
16 33 ☽ ꝑ ♂
17 59 ☽ ± ☉
23 48 ☽ ± ♅

06 Tuesday
00 09 ☽ ⊼ ♇
00 59 ☽ ꝑ ☿
02 43 ☽ ∠ ♃
04 21 ☽ ⊼ ☉
09 28 ☽ ⊼ ♅
09 47 ☽ ± ♇
14 59 ☽ ± ☿
15 10 ☽ ✶ ♄
17 31 ☽ △ ♂

07 Wednesday
00 07 ☽ ꝑ ♇
00 12 ☽ ⊼ ☿
02 40 ☽ ✶ ♃
06 14 ☽ △ ♀
08 54 ☉ ∥ ♅
10 11 ☽ Q ♄
10 58 ☽ ☍ ♆
16 22 ☽ ∥ ♃
21 37 ☽ Q ♃
23 47 ☽ △ ♇

08 Thursday
02 40 ☽ # ♆
02 57 ☉ ∠ ♀
03 25 ☽ # ♂
07 08 ☽ ꝑ ♀
07 12 ☽ ☍ ☉
09 01 ☽ ☍ ♅
12 30 ☽ ∥ ♄
14 32 ☽ □ ♄
15 29 ☿ ⊥ ♀
18 31 ☽ □ ♂
21 25 ☽ ☍ ☿
21 58 ☽ # ♅
22 04 ☽ ± ♀
22 19 ♂ ∥ ♆

09 Friday
00 37 ☽ # ☉
02 03 ☽ □ ♃
08 04 ☽ ⊼ ♀
10 21 ☽ ⊼ ♆
12 19 ☉ ☌ ♅
13 09 ☽ # ♇
19 55 ☽ ± ♆
23 17 ☽ □ ♇
23 19 ♀ ∠ ♅

10 Saturday
00 08 ☽ Q ♂
02 58 ☿ □ ♂
03 26 ☽ # ☿
08 52 ☽ ⊼ ♅
10 17 ☽ ⊼ ☉
10 26 ☽ ꝑ ♆
14 28 ☽ △ ♄
16 48 ☿ Q ♇
18 36 ☽ ⊼ ☿
18 40 ☽ ± ♅
20 18 ☽ ✶ ♂
20 47 ☽ ± ☉
21 07 ☽ ∥ ♀

11 Sunday
02 30 ☽ △ ♃
03 48 ☽ ± ☿
04 36 ☽ Q ♇
08 08 ♀ ✶ ♆
09 34 ☽ ꝑ ♅
11 09 ☽ △ ♆
11 18 ☽ ☍ ♀
12 48 ☽ ꝑ ☉
15 18 ☽ ꝑ ♄
17 45 ☽ ꝑ ☿
22 14 ☽ ∠ ♂

12 Monday
00 46 ☽ ✶ ♇
03 47 ☽ ꝑ ♃
06 36 ☽ ± ♄
11 06 ☽ △ ♅
14 16 ☽ ⊥ ♂
16 22 ☽ △ ☉
17 00 ☽ ⊼ ♄
17 31 ☽ △ ☿
19 24 ☽ ± ♃
23 25 ☿ □ ♄
23 37 ☽ # ♀

13 Tuesday
00 17 ☉ ☌ ☿
01 11 ☽ ⊻ ♂
01 21 ☉ □ ♄
02 48 ☽ ∠ ♇
06 01 ☽ ⊼ ♃
15 15 ☽ □ ♆
16 43 ☿ ∠ ♀
18 14 ☽ ⊼ ♀
18 56 ☽ ⊥ ♇
20 02 ☽ ∥ ☿

14 Wednesday
05 47 ☽ ± ♀
05 54 ☽ ⊻ ♇
06 53 ☽ ∥ ♇
10 51 ☽ ∥ ☉
17 15 ☽ □ ♅
19 24 ☽ □ ☿
23 28 ☽ ꝑ ♀
23 33 ☽ ☍ ♄

15 Thursday
00 04 ☽ ∥ ♅
00 33 ☽ Q ♆
00 58 ♀ ∠ ♄
03 24 ☽ □ ☉
10 38 ☽ ☌ ♂
12 50 ☽ # ♄
13 45 ☽ ☍ ♃
15 32 ☉ Q ♇
18 38 ☿ ☌ ♅
23 34 ☽ ✶ ♆
23 37 ☽ ∥ ♆

16 Friday
03 00 ☽ Q ☿
03 47 ☽ Q ♅
05 51 ☽ △ ♀
08 12 ☽ ∥ ♂
11 39 ☉ ∥ ♇
14 21 ☽ # ♃
15 11 ☽ ☌ ♇
17 18 ☽ Q ☉

17 Saturday
00 43 ☽ ✶ ☿
03 30 ☽ ✶ ♅
05 08 ☽ ∠ ♆
10 06 ☽ ⊼ ♄
18 48 ☉ ∥ ☿
19 22 ☽ ✶ ☉
22 10 ☽ ± ♄
23 12 ☽ ⊥ ♆

18 Sunday
00 18 ☽ ⊻ ♂
01 15 ☽ ⊼ ♃
04 32 ☽ ∠ ☿
09 45 ☽ ∠ ♅
11 21 ☽ ⊻ ♆
12 58 ☽ ⊥ ♂
13 27 ☽ ± ♃
16 25 ☽ ꝑ ♄
21 00 ☽ □ ♀
21 32 ☽ ⊥ ☿
22 08 ☿ ∥ ♇

19 Monday
01 57 ♂ ☍ ♃
03 29 ☽ ⊻ ♇
04 05 ☽ ⊥ ♅
04 32 ☽ ∠ ☉
07 51 ☽ ꝑ ♃
08 05 ☽ ∠ ♂
08 58 ☽ ⊻ ☿
15 42 ☽ ⊥ ♇
16 22 ☽ ⊻ ♅
23 03 ☽ △ ♄

20 Tuesday
00 35 ☽ ⊥ ☉
10 05 ☽ ∠ ♇
11 31 ☽ Q ♀
13 58 ☽ ⊻ ☉
14 39 ☽ △ ♃
16 03 ☽ ✶ ♂
22 39 ☉ □ ♃

21 Wednesday
00 36 ☽ ☌ ♆
04 15 ☽ # ♃
06 01 ☽ ∥ ♂
13 06 ☽ ✶ ♀
16 39 ☽ ✶ ♇
17 34 ☽ Q ♂
19 02 ☽ ☌ ☿
20 58 ☽ ∥ ♆

22 Thursday
05 42 ☽ ☌ ♅
07 01 ☽ # ♄
12 18 ☽ □ ♄
13 01 ☿ Q ♂
13 11 ☉ □ ♂
16 57 ☽ Q ♇
20 48 ☽ ∠ ♀
21 28 ☉ # ♀
22 20 ☽ ∥ ♅

23 Friday
03 53 ☽ □ ♃
05 37 ♂ # ♃
06 37 ☽ ∥ ☿
07 31 ☽ □ ♂
08 21 ☽ ☌ ☉
13 24 ☽ ⊻ ♆
15 36 ☽ ⊥ ♀
15 56 ☽ ∥ ♇

24 Saturday
00 33 ☽ Q ♄
01 24 ☽ ⊥ ♆
02 53 ☽ # ♀
04 02 ☽ ⊻ ♀
05 02 ☽ □ ♇
05 47 ☽ ⊻ ☿
08 15 ☽ ∥ ☉
16 02 ☽ Q ♃
17 32 ☽ ⊥ ☿
18 00 ☽ ⊻ ♅
19 17 ☽ ∠ ♆

25 Sunday
00 25 ☽ ✶ ♄
05 52 ☽ ⊥ ♅
08 34 ♀ △ ♇
11 08 ☽ ∠ ☿
15 42 ☿ St D
15 48 ☽ ✶ ♃
21 22 ☽ △ ♂
22 27 ☉ ⊻ ♆
23 31 ☽ ∠ ♅
23 50 ☿ ✶ ♀

26 Monday
00 44 ☽ ✶ ♆
00 56 ☽ ⊻ ☉
05 50 ☽ ∠ ♄
13 34 ☽ ⊥ ☉
15 48 ☽ △ ♇
16 22 ☽ ✶ ☿
16 47 ☽ ☌ ♀
21 05 ☽ ∠ ♃
23 16 ☽ ⊥ ♄
23 58 ☽ Q ♆

27 Tuesday
03 31 ☽ ꝑ ♂
04 34 ☽ ✶ ♅
08 20 ☽ ∠ ☉
10 47 ☽ ⊻ ♄
14 24 ☽ ⊥ ♃
15 36 ☽ Q ☿
20 27 ☽ ꝑ ♇
21 19 ☽ ± ♂

28 Wednesday
01 48 ☽ # ☉
01 50 ☽ ⊻ ♃
03 25 ☽ Q ♅
09 03 ☽ ⊼ ♂
10 09 ☽ □ ♆
13 22 ☽ ± ♇
15 02 ☽ ✶ ☉
22 02 ☽ ∥ ♀

MARCH 2001

LONGITUDES

Date	Sidereal time h m s	Sun ☉ ° ' "	Moon ☽ ° ' "	Moon ☽ 24.00 ° ' "	Mercury ☿ ° '	Venus ♀ ° '	Mars ♂ ° '	Jupiter ♃ ° '	Saturn ♄ ° '	Uranus ♅ ° '	Neptune ♆ ° '	Pluto ♇ ° '
01	22 37 27	10 ♓ 58 03	21 ♉ 25 37	28 ♉ 00 34	16 ♒ 11	16 ♈ 37	07 ♐ 26	03 ♊ 12	25 ♉ 13	21 ♒ 58	07 ♒ 30	15 ♐ 12
02	22 41 23	11 58 15	04 ♊ 39 53	11 ♊ 23 52	16 36	16 53	07 56	03 19	25 17	22 01	07 32	15 13
03	22 45 20	12 58 25	18 ♊ 12 47	25 ♊ 06 49	17 07	17 07	08 25	03 25	25 21	22 04	07 34	15 13
04	22 49 16	13 58 33	02 ♋ 06 08	09 ♋ 10 47	17 42	17 19	08 53	03 32	25 25	22 08	07 36	15 14
05	22 53 13	14 58 39	16 ♋ 20 42	23 ♋ 35 40	18 21	17 29	09 22	03 39	25 29	22 11	07 38	15 14
06	22 57 09	15 58 42	00 ♌ 55 20	08 ♌ 19 10	19 04	17 36	09 51	03 47	25 33	22 14	07 40	15 14
07	23 01 06	16 58 44	15 ♌ 46 26	23 ♌ 16 17	19 51	17 41	10 19	03 54	25 37	22 17	07 42	15 15
08	23 05 02	17 58 43	00 ♍ 47 40	08 ♍ 19 26	20 42	17 43	10 47	04 02	25 42	22 21	07 44	15 15
09	23 08 59	18 58 41	15 ♍ 50 23	23 ♍ 19 16	21 35	17 R 44	11 15	04 09	25 46	22 24	07 46	15 16
10	23 12 56	19 58 36	00 ♎ 44 55	08 ♎ 06 13	22 32	17 41	11 43	04 17	25 51	22 27	07 47	15 16
11	23 16 52	20 58 30	15 ♎ 22 12	22 ♎ 32 05	23 31	17 36	12 10	04 25	25 55	22 30	07 49	15 16
12	23 20 49	21 58 21	29 ♎ 35 16	06 ♏ 31 23	24 33	17 29	12 38	04 33	26 00	22 33	07 51	15 16
13	23 24 45	22 58 11	13 ♏ 20 15	20 ♏ 01 52	25 38	17 19	13 05	04 42	26 05	22 36	07 53	15 16
14	23 28 42	23 58 00	26 ♏ 36 24	03 ♐ 04 11	26 45	17 07	13 32	04 50	26 10	22 40	07 54	15 17
15	23 32 38	24 57 47	09 ♐ 25 40	15 ♐ 41 23	27 54	16 52	13 58	04 59	26 15	22 43	07 56	15 17
16	23 36 35	25 57 32	21 ♐ 51 56	27 ♐ 57 58	29 ♒ 05	16 35	14 25	05 07	26 20	22 46	07 58	15 17
17	23 40 31	26 57 15	04 ♑ 00 10	09 ♑ 59 44	00 ♓ 18	16 15	14 51	05 16	26 25	22 49	07 59	15 17
18	23 44 28	27 56 57	15 ♑ 55 52	21 ♑ 50 44	01 33	15 53	15 17	05 25	26 30	22 52	08 01	15 R 17
19	23 48 25	28 56 37	27 ♑ 44 30	03 ♒ 37 46	02 50	15 29	15 43	05 34	26 35	22 55	08 03	15 17
20	23 52 21	29 ♓ 56 15	09 ♒ 31 09	15 ♒ 25 08	04 09	15 03	16 08	05 43	26 41	22 58	08 04	15 17
21	23 56 18	00 ♈ 55 51	21 ♒ 20 15	27 ♒ 16 54	05 29	14 34	16 33	05 52	26 46	23 01	08 06	15 17
22	00 00 14	01 55 25	03 ♓ 15 27	09 ♓ 16 13	06 51	14 04	16 58	06 02	26 51	23 04	08 07	15 16
23	00 04 11	02 54 58	15 ♓ 19 27	21 ♓ 25 20	08 15	13 33	17 23	06 11	26 57	23 06	08 09	15 16
24	00 08 07	03 54 28	27 ♓ 34 01	03 ♈ 45 34	09 40	12 59	17 47	06 21	27 03	23 09	08 10	15 16
25	00 12 04	04 53 57	10 ♈ 00 03	16 ♈ 17 27	11 07	12 25	18 11	06 31	27 08	23 12	08 11	15 16
26	00 16 00	05 53 23	22 ♈ 37 45	29 ♈ 00 56	12 35	11 49	18 35	06 41	27 14	23 15	08 13	15 15
27	00 19 57	06 52 47	05 ♉ 26 57	11 ♉ 55 47	14 05	11 12	18 59	06 51	27 20	23 18	08 15	15 15
28	00 23 54	07 52 10	18 ♉ 27 24	25 ♉ 01 50	15 36	10 35	19 22	07 01	27 26	23 20	08 16	15 15
29	00 27 50	08 51 30	01 ♊ 39 07	08 ♊ 19 17	17 09	09 57	19 45	07 11	27 32	23 23	08 17	15 14
30	00 31 47	09 50 47	15 ♊ 02 28	21 ♊ 48 44	18 43	09 20	20 07	07 21	27 38	23 26	08 18	15 14
31	00 35 43	10 ♈ 50 03	28 ♊ 38 15	05 ♋ 31 06	20 ♓ 18	08 ♈ 42	20 ♐ 29	07 ♊ 32	27 ♉ 44	23 ♒ 28	08 ♒ 20	15 ♐ 14

MOON NODES, LATITUDE AND DECLINATIONS

Date	Moon True ☊ ° '	Moon Mean ☊ ° '	Moon ☽ Latitude ° '	Sun ☉ ° '	Moon ☽ ° '	Mercury ☿ ° '	Venus ♀ ° '	Mars ♂ ° '	Jupiter ♃ ° '	Saturn ♄ ° '	Uranus ♅ ° '	Neptune ♆ ° '	Pluto ♇ ° '
01	13 ♋ 36	12 ♋ 32	04 S 02	07 S 27	14 N 13	14 S 52	11 N 56	20 S 42	20 N 14	17 N 12	14 S 50	18 S 14	12 S 11
02	13 R 34	12 29	03 13	07 04	17 55	14 57	12 12	20 48	20 16	17 13	14 49	18 14	12 10
03	13 D 34	12 26	02 11	06 41	20 44	15 00	12 26	20 53	20 17	17 14	14 48	18 13	12 10
04	13 35	12 23	01 S 01	06 18	22 24	15 01	12 40	20 58	20 19	17 16	14 47	18 13	12 10
05	13 R 35	12 20	00 N 15	05 55	22 41	15 00	12 52	21 03	20 20	17 17	14 46	18 12	12 10
06	13 34	12 16	01 31	05 32	21 26	14 58	13 04	21 09	20 22	17 18	14 45	18 12	12 10
07	13 30	12 13	02 43	05 09	18 42	14 54	13 14	21 13	20 23	17 19	14 44	18 11	12 09
08	13 24	12 10	03 44	04 45	14 41	14 48	13 23	21 18	20 25	17 20	14 43	18 11	12 09
09	13 15	12 07	04 29	04 22	09 43	14 41	13 32	21 23	20 27	17 22	14 41	18 11	12 09
10	13 05	12 04	04 56	03 58	04 N 14	14 32	13 38	21 28	20 28	17 23	14 40	18 10	12 09
11	12 55	12 01	05 03	03 35	01 S 24	14 22	13 44	21 32	20 30	17 24	14 39	18 10	12 08
12	12 45	11 57	04 50	03 11	06 48	14 10	13 48	21 37	20 32	17 26	14 38	18 09	12 08
13	12 37	11 54	04 21	02 47	11 41	13 56	13 51	21 41	20 33	17 27	14 37	18 09	12 08
14	12 31	11 51	03 38	02 24	15 51	13 41	13 53	21 45	20 35	17 28	14 36	18 08	12 08
15	12 28	11 48	02 46	02 00	19 08	13 25	13 53	21 50	20 37	17 30	14 35	18 08	12 07
16	12 27	11 45	01 47	01 36	21 24	13 07	13 52	21 54	20 38	17 31	14 34	18 07	12 07
17	12 D 27	11 42	00 N 45	01 13	22 38	12 48	13 49	21 58	20 40	17 33	14 33	18 07	12 07
18	12 R 27	11 38	00 S 18	00 49	22 48	12 27	13 44	22 02	20 42	17 34	14 33	18 07	12 07
19	12 26	11 35	01 20	00 25	21 55	12 06	13 39	22 06	20 44	17 35	14 32	18 06	12 06
20	12 23	11 32	02 18	00 S 01	20 05	11 42	13 31	22 09	20 46	17 37	14 31	18 06	12 06
21	12 18	11 29	03 10	00 N 22	17 24	11 18	13 22	22 13	20 47	17 38	14 30	18 05	12 06
22	12 09	11 26	03 55	00 46	13 57	10 52	13 12	22 17	20 49	17 40	14 29	18 05	12 06
23	11 58	11 22	04 29	01 10	09 55	10 25	13 00	22 20	20 51	17 41	14 28	18 05	12 05
24	11 45	11 19	04 51	01 33	05 25	09 56	12 47	22 24	20 53	17 43	14 27	18 04	12 05
25	11 31	11 16	05 00	01 57	00 S 38	09 26	12 32	22 27	20 55	17 44	14 26	18 04	12 05
26	11 17	11 13	04 54	02 20	04 N 15	08 55	12 17	22 31	20 56	17 46	14 25	18 03	12 05
27	11 05	11 10	04 34	02 44	09 02	08 23	11 59	22 34	20 58	17 47	14 24	18 03	12 04
28	10 56	11 07	03 59	03 07	13 29	07 50	11 41	22 37	21 00	17 49	14 23	18 03	12 04
29	10 49	11 03	03 11	03 31	17 23	07 15	11 22	22 40	21 02	17 50	14 23	18 02	12 04
30	10 45	11 00	02 11	03 54	20 25	06 40	11 02	22 43	21 04	17 52	14 22	18 02	12 03
31	10 ♋ 44	10 ♋ 57	01 S 04	04 N 17	22 N 22	06 S 03	10 N 41	22 S 46	21 N 06	17 N 53	14 S 21	18 S 02	12 S 03

ZODIAC SIGN ENTRIES

Date	h	m	Planets
02	03	36	☽ ♊
04	08	24	☽ ♋
06	10	30	☽ ♌
08	10	44	☽ ♍
10	10	47	☽ ♎
12	12	42	☽ ♏
14	18	17	☽ ♐
17	04	02	☽ ♑
17	06	05	☿ ♓
19	16	36	☽ ♒
20	13	31	☉ ♈
22	05	28	☽ ♓
24	16	44	☽ ♈
27	01	51	☽ ♉
29	09	01	☽ ♊
31	14	23	☽ ♋

LATITUDES

Date	Mercury ☿ ° '	Venus ♀ ° '	Mars ♂ ° '	Jupiter ♃ ° '	Saturn ♄ ° '	Uranus ♅ ° '	Neptune ♆ ° '	Pluto ♇ ° '
01	01 N 10	05 N 52	00 N 52	00 S 34	01 S 55	00 S 41	00 N 10	10 N 30
04	00 N 32	06 21	00 49	00 34	01 55	00 41	00 10	10 31
07	00 S 02	06 49	00 47	00 33	01 54	00 41	00 10	10 32
10	00 34	07 16	00 44	00 32	01 54	00 41	00 10	10 32
13	01 01	07 39	00 41	00 32	01 53	00 41	00 10	10 33
16	01 25	07 58	00 38	00 31	01 52	00 41	00 09	10 34
19	01 45	08 11	00 35	00 31	01 51	00 41	00 09	10 35
22	02 01	08 19	00 31	00 30	01 50	00 41	00 09	10 35
25	02 12	08 18	00 28	00 30	01 50	00 41	00 09	10 36
28	02 20	08 06	00 24	00 29	01 49	00 41	00 09	10 37
31	02 S 24	07 N 53	00 N 19	00 S 29	01 S 49	00 S 41	00 N 09	10 N 38

DATA

Julian Date	2451970
Delta T	+64 seconds
Ayanamsa	23° 52' 08"
Synetic vernal point	05° ♓ 14' 52"
True obliquity of ecliptic	23° 26' 19"

MOON'S PHASES, APSIDES AND POSITIONS ☽

Date	h	m	Phase	Longitude ° '	Eclipse Indicator
03	02	03	☽	12 ♊ 33	
09	17	23	○	19 ♍ 12	
16	20	45	☾	26 ♐ 19	
25	01	21	●	04 ♈ 28	

Day	h	m	
08	09	00	Perigee
20	11	33	Apogee

Day	h	m		
05	04	31	Max dec	22° N 45'
11	06	00	0S	
18	03	40	Max dec	22° S 51'
25	15	08	0N	

LONGITUDES

Date	Chiron ⚷ ° '	Ceres ⚳ ° '	Pallas ⚴ ° '	Juno ⚵ ° '	Vesta ⚶ ° '	Black Moon Lilith ⚸ ° '
01	27 ♐ 51	08 ♑ 36	11 ♐ 45	05 ♈ 51	24 ♓ 25	10 ♒ 36
11	28 ♐ 20	11 ♑ 41	13 ♐ 42	11 ♈ 21	29 ♓ 11	11 ♒ 43
21	28 ♐ 41	14 ♑ 29	15 ♐ 07	16 ♈ 56	03 ♈ 56	12 ♒ 49
31	28 ♐ 53	16 ♑ 58	15 ♐ 53	22 ♈ 37	08 ♈ 39	13 ♒ 56

ASPECTARIAN

h m	Aspects
01 Thursday	
00 13	☉ ⊥ ♀
00 14	☽ # ♇
00 32	☽ ⊼ ♇
02 04	☽ □ ☿
02 57	☽ ⊻ ♀
03 39	☿ ∥ ♅
12 59	☽ □ ♅
14 14	☽ ⊥ ♀
15 03	☽ Q ☉
15 32	♂ ✶ ♆
15 39	☽ # ♅
16 00	☽ # ☿
18 57	☽ ☌ ♄
02 Friday	
06 55	☽ ∠ ♀
07 04	☽ ∥ ♄
09 33	☽ ☌ ♃
10 00	♀ # ♇
14 20	☽ # ♆
17 09	☽ △ ♆
18 03	☽ ☍ ♂
03 Saturday	
02 03	☽ □ ☉
06 45	☽ ☍ ♇
07 28	☽ ∥ ♃
09 59	☽ △ ☿
10 03	☽ ✶ ♀
12 36	☿ ✶ ♀
13 40	☽ # ♂
18 45	☽ △ ♅
19 37	☽ ⚼ ♆
21 29	☉ Q ♄
04 Sunday	
00 27	☽ ⊻ ♄
02 47	☉ ⊥ ♆
07 10	☽ Q ♀
10 49	☽ ⊥ ♄
11 09	☽ ± ♆
13 04	☽ ⚼ ☿
14 28	☽ ⊻ ♃
20 34	☽ ⚼ ♅
21 22	☽ ⊼ ♆
23 55	☽ ⊼ ♂
05 Monday	
00 43	☽ ⊥ ♃
02 08	☽ ∠ ♄
05 00	☽ ± ☿
09 33	☽ △ ☉
10 09	☽ ⊼ ♇
10 19	☽ ± ♂
11 44	☽ ± ♅
13 54	☽ □ ♀
15 30	☽ ⊼ ☿
15 53	☽ ∠ ♃
18 12	☉ □ ♇
20 07	☽ ± ♇
21 42	☽ ⊼ ♅
06 Tuesday	
01 43	☽ ⚼ ♂
03 10	☽ ✶ ♄
10 53	☽ ⚼ ♇
12 06	☽ ⚼ ☉
15 13	☽ # ♂
16 41	☽ ✶ ♃
22 48	☽ Q ♄
22 51	☽ ∥ ♃
22 58	☽ ☍ ♆
07 Wednesday	
02 56	☽ △ ♂
03 44	☽ ± ☉
10 29	♂ Q ♅
11 09	☽ △ ♇
12 13	☽ Q ♃
14 04	☽ ⊼ ☉
15 04	☽ △ ♀
15 28	☽ # ♆
18 55	☽ ☍ ☿
20 59	☽ ∥ ♄
22 28	☽ ☍ ♅
08 Thursday	
03 50	☽ □ ♄
05 42	☉ ⊻ ♀
11 19	☽ # ☿
11 50	☽ # ♅
15 05	☽ ⚼ ♀
17 12	☽ □ ♃
18 24	☽ ∥ ♀
23 04	☽ ⊼ ♆
09 Friday	
00 41	☽ # ♇
01 06	♀ St R
04 26	☽ □ ♂
05 26	☽ ± ♀
08 40	☽ ± ♆
10 14	☿ ∥ ♅
11 04	☽ □ ♇
15 01	☽ ⊼ ♀
17 23	☽ ☍ ☉
21 50	☽ ⊼ ☿
22 33	☽ ⊼ ♅
23 07	☽ ⚼ ♆
10 Saturday	
04 01	☽ △ ♄
08 10	☽ ± ☿
08 16	☽ ± ♅
09 50	☿ ☌ ♅
10 16	☽ Q ♂
13 11	☽ # ☉
16 05	☽ Q ♇
17 49	☽ △ ♃
22 58	☽ ⚼ ♅
23 31	☽ △ ♆
23 51	☽ ⚼ ☿
11 Sunday	
04 36	☽ ⚼ ♄
06 32	☽ ✶ ♂
11 50	☽ ✶ ♇
15 42	☽ ☍ ♀
18 50	☽ ⚼ ♃
19 39	☽ ± ♄
20 51	☽ ∥ ☉
22 05	☽ ⊼ ☉
23 59	☽ △ ♅
12 Monday	
02 44	☽ △ ☿
05 50	☽ ⊼ ♄
08 32	☽ ∠ ♂
09 01	☽ ± ☉
10 13	☽ ± ♃
13 10	☽ ∠ ♇
14 50	☿ Q ♂
20 40	☽ ⊼ ♃
13 Tuesday	
00 36	☽ ⊥ ♂
01 47	☽ ⚼ ☉
02 21	☽ □ ♆
02 48	☉ ⊻ ♅
04 18	☉ Q ♃
04 49	☽ ⊥ ♇
09 44	☉ ∠ ♆
11 32	☽ ⊻ ♂
14 20	☽ ∥ ♇
15 27	☽ ⊻ ♇
19 01	☽ ⊼ ♀
19 14	☿ # ♀
22 30	☿ □ ♄
23 41	☽ ∥ ☿
23 59	☽ # ♀
14 Wednesday	
04 19	☽ ∥ ♅
04 45	☽ □ ♅
05 42	☽ ± ♀
06 46	☽ △ ☉
10 43	☽ Q ♆
11 10	☽ ☍ ♄
12 17	☽ □ ☿
22 02	☽ ⚼ ♀
23 05	☽ # ♄
23 09	☿ Q ♇
15 Thursday	
03 29	☽ ☍ ♃
03 56	☽ ∥ ♆
09 10	☽ ✶ ♆
14 27	☽ Q ♅
21 01	☽ ☌ ♂
23 12	☽ ☌ ♇
16 Friday	
01 42	☽ Q ☿
01 58	☽ △ ♀
02 31	☽ # ♃
13 46	☽ ✶ ♅
14 09	☽ ∠ ♆
19 46	☽ ∥ ♂
20 45	☽ □ ☉
20 49	☽ ⊼ ♄
21 42	☉ ✶ ♄
17 Saturday	
03 48	☽ ✶ ☿
07 59	☽ ⊥ ♆
08 48	☽ ± ♄
14 34	☽ ⊼ ♃
19 40	☽ ∠ ♅
20 00	☽ ⊻ ♆
18 Sunday	
02 11	☿ ∠ ♀
02 39	♇ St R
02 46	☽ ± ♃
02 59	☽ ⚼ ♄
10 38	☽ ⊻ ♂
10 41	☽ ⊻ ♇
11 49	♂ ☌ ♇
11 55	☽ □ ♀
12 02	☽ Q ☉
13 25	☽ ∠ ☿
13 54	☽ ⊥ ♅
21 13	☽ ⚼ ♃
22 51	☽ ⊥ ♇
23 16	☽ ⊥ ♂
19 Monday	
02 08	☽ ⊻ ♅
05 30	♀ △ ♂
09 00	☽ ∥ ♂
09 38	☽ △ ♄
09 56	☽ ⊥ ☿
11 01	☿ ∥ ♇
11 12	☉ ⊥ ♅
14 40	☽ ✶ ☉
17 10	☽ ∠ ♇
18 17	☽ ∠ ♂
23 17	☽ Q ♀
23 28	♀ △ ♇
23 41	☽ ⊻ ☿
20 Tuesday	
04 09	☽ △ ♃
04 38	☽ # ♃
09 03	☽ ☌ ♆
22 49	☽ ✶ ♀
23 43	☽ ✶ ♇
21 Wednesday	
00 02	☽ ∠ ☉
01 57	☽ ✶ ♂
06 25	☽ ∥ ♆
10 06	☽ # ♄
15 24	☽ ☌ ♅
19 41	☿ □ ♃
19 55	☽ ⊥ ☉
23 03	☽ □ ♄
23 59	☽ Q ♇
22 Thursday	
03 05	☽ Q ♂
03 57	☽ ∠ ♀
08 36	☽ ∥ ♅
09 05	☽ ⊻ ☉
16 58	☽ # ♀
17 37	☽ □ ♃
20 08	☽ ☌ ☿
21 13	☽ ⊥ ♀
21 44	☽ ⊻ ♆
23 27	☽ ∥ ♇
23 Friday	
03 14	☿ ⊥ ♀
08 38	☽ ⊻ ♀
08 52	☽ ∥ ☿
09 40	☽ ⊥ ♆
10 14	☿ ⊻ ♆
11 15	☽ Q ♄
11 54	☽ □ ♇
16 12	☽ □ ♂
24 Saturday	
03 21	☽ ⊻ ♅
03 25	☽ ∠ ♆
05 39	☽ Q ♃
10 58	☽ ✶ ♄
15 06	☽ ⊥ ♅
25 Sunday	
01 21	☽ ☌ ☉
05 13	☽ ✶ ♃
06 00	☽ # ☉
08 32	☽ ∠ ♅
08 32	☽ ✶ ♆
14 25	☽ ⊻ ☿
16 07	☽ ∠ ♄
16 24	☽ ☌ ♀
21 27	♀ ∠ ♄
22 03	☽ △ ♇
26 Monday	
01 48	☽ ∥ ☉
03 04	☿ ⊻ ♀
03 21	☽ ⊥ ☿
04 07	☽ △ ♂
07 27	☽ Q ♆
09 21	☽ ⊥ ♄
10 11	☽ ∠ ♃
13 10	☽ ✶ ♅
20 44	☽ ⊻ ♄
22 33	☽ ∠ ☿
27 Tuesday	
02 19	☽ ⚼ ♇
03 18	☽ ⊥ ♃
05 15	♀ # ♇
09 02	☽ # ☿
09 11	☽ ⚼ ♂
10 57	☉ ✶ ♃
11 43	☽ Q ♅
14 37	☽ ⊻ ♃
14 38	☿ ⊥ ♆
14 53	☽ ⊻ ☉
17 12	☽ □ ♆
19 03	☽ ± ♇
22 11	☽ ⊻ ♀
28 Wednesday	
02 21	☽ ± ♂
02 39	☽ ∥ ♀
02 53	☽ ⊥ ☉
04 04	☽ # ♇
06 04	☽ ✶ ☿
06 07	☽ ⊼ ♇
06 29	☿ □ ♇
08 43	☽ ⊥ ♀
09 08	☿ Q ♄
13 42	☽ ⊼ ♂
17 10	☽ # ♅
20 43	☽ ∠ ☉
20 57	☽ □ ♅
21 54	☉ ✶ ♆
23 55	☉ ∠ ♅
29 Thursday	
00 25	☽ ∠ ♀
04 29	☽ ☌ ♄
06 52	☽ Q ☿
15 13	☽ ∥ ♄
16 38	☽ # ♆
22 05	☽ ☌ ♃
23 58	☽ △ ♆
30 Friday	
02 00	☽ ✶ ☉
02 15	☽ ✶ ♀
04 17	☉ ☌ ♀
12 21	☽ ☍ ♇
18 32	☽ ∥ ♃
19 23	☽ □ ☿
21 16	☽ ☍ ♂
22 39	☽ Q ♀
31 Saturday	
01 00	☽ Q ☉
02 40	☽ ⚼ ♆
02 54	☽ △ ♅
10 24	☽ ⊻ ♄
15 36	☿ □ ♂
18 28	☽ ± ♆
20 07	♀ ∠ ♅
20 58	☽ ⊥ ♄
21 36	☽ # ♂

All ephemeris data is given at 12.00 UT and the Moon's longitude is additionally given for 24.00 UT

APRIL 2001

Date	Sidereal time h m s	Sun ☉ ° ' "	Moon ☽ ° ' "	Moon ☽ 24.00 ° ' "	Mercury ☿ ° '	Venus ♀ ° '	Mars ♂ ° '	Jupiter ♃ ° '	Saturn ♄ ° '	Uranus ♅ ° '	Neptune ♆ ° '	Pluto ♇ ° '
01	00 39 40	11 ♈ 49 16	12 ♋ 27 26	19 ♋ 27 16	21 ♓ 55	08 ♈ 05	20 ♐ 51	07 ♊ 42	27 ♉ 50	23 ♒ 31	08 ♒ 21	15 ♐ 13
02	00 43 36	12 48 27	26 ♋ 30 39	03 ♌ 37 30	23 33	07 R 28	21 13	07 53	27 56	23 33	08 23	15 R 13
03	00 47 33	13 47 35	10 ♌ 47 38	18 ♌ 00 48	25 13	06 52	21 34	08 03	28 03	23 36	08 24	15 12
04	00 51 29	14 46 42	25 ♌ 16 33	02 ♍ 34 23	26 54	06 17	21 55	08 14	28 09	23 38	08 25	15 12
05	00 55 26	15 45 45	09 ♍ 53 36	17 ♍ 13 26	28 ♓ 37	05 43	22 15	08 25	28 15	23 41	08 26	15 11
06	00 59 23	16 44 47	24 ♍ 33 00	01 ♎ 51 22	00 ♈ 21	05 11	22 35	08 36	28 22	23 43	08 27	15 11
07	01 03 19	17 43 46	09 ♎ 07 36	16 ♎ 20 45	02 06	04 41	22 55	08 47	28 28	23 46	08 28	15 10
08	01 07 16	18 42 43	23 ♎ 29 57	00 ♏ 34 27	03 53	04 12	23 14	08 58	28 35	23 48	08 29	15 09
09	01 11 12	19 41 39	07 ♏ 33 37	14 ♏ 26 56	05 42	03 46	23 33	09 10	28 41	23 50	08 30	15 08
10	01 15 09	20 40 32	21 ♏ 14 07	27 ♏ 54 59	07 32	03 21	23 52	09 21	28 48	23 53	08 31	15 08
11	01 19 05	21 39 24	04 ♐ 29 32	10 ♐ 57 55	09 23	02 59	24 10	09 32	28 55	23 55	08 32	15 07
12	01 23 02	22 38 14	17 ♐ 20 25	23 ♐ 37 23	11 16	02 39	24 27	09 44	29 01	23 57	08 33	15 06
13	01 26 58	23 37 02	29 ♐ 49 20	05 ♑ 56 48	13 10	02 22	24 45	09 55	29 08	23 59	08 34	15 05
14	01 30 55	24 35 48	12 ♑ 00 24	18 ♑ 00 47	15 06	02 07	25 01	10 07	29 15	24 01	08 35	15 05
15	01 34 52	25 34 33	23 ♑ 58 37	29 ♑ 54 36	17 04	01 54	25 18	10 19	29 22	24 03	08 36	15 04
16	01 38 48	26 33 16	05 ♒ 49 25	11 ♒ 43 43	19 03	01 44	25 34	10 31	29 29	24 05	08 37	15 03
17	01 42 45	27 31 57	17 ♒ 38 12	23 ♒ 33 27	21 03	01 36	25 49	10 42	29 36	24 07	08 38	15 02
18	01 46 41	28 30 36	29 ♒ 30 05	05 ♓ 28 38	23 05	01 31	26 04	10 54	29 43	24 09	08 38	15 01
19	01 50 38	29 ♈ 29 11	11 ♓ 29 35	17 ♓ 33 22	25 08	01 28	26 18	11 06	29 50	24 11	08 39	15 00
20	01 54 34	00 ♉ 27 50	23 ♓ 40 20	29 ♓ 50 46	27 12	01 D 27	26 32	11 19	29 ♉ 57	24 13	08 40	14 59
21	01 58 31	01 26 24	06 ♈ 04 54	12 ♈ 22 51	29 ♈ 17	01 29	26 45	11 31	00 ♊ 04	24 15	08 40	14 58
22	02 02 27	02 24 57	18 ♈ 44 41	25 ♈ 10 24	01 ♉ 24	01 34	26 58	11 43	00 11	24 17	08 41	14 57
23	02 06 24	03 23 28	01 ♉ 39 54	08 ♉ 13 04	03 31	01 40	27 10	11 55	00 19	24 19	08 42	14 56
24	02 10 21	04 21 56	14 ♉ 49 42	21 ♉ 29 36	05 39	01 49	27 22	12 08	00 26	24 20	08 42	14 55
25	02 14 17	05 20 23	28 ♉ 12 32	04 ♊ 58 15	07 47	02 00	27 33	12 20	00 33	24 22	08 43	14 53
26	02 18 14	06 18 49	11 ♊ 46 31	18 ♊ 37 07	09 56	02 13	27 43	12 33	00 40	24 24	08 43	14 52
27	02 22 10	07 17 12	25 ♊ 29 51	02 ♋ 24 34	12 05	02 27	27 53	12 45	00 48	24 25	08 44	14 51
28	02 26 07	08 15 33	09 ♋ 21 07	16 ♋ 19 23	14 13	02 44	28 02	12 58	00 55	24 27	08 44	14 50
29	02 30 03	09 13 52	23 ♋ 19 16	00 ♌ 20 41	16 21	03 03	28 11	13 10	01 02	24 28	08 44	14 49
30	02 34 00	10 ♉ 12 09	07 ♌ 23 32	14 ♌ 27 43	18 ♉ 28	03 ♈ 24	28 ♐ 19	13 ♊ 23	01 ♊ 10	24 ♒ 30	08 ♒ 45	14 ♐ 48

Date	Moon True ☊ ° '	Moon Mean ☊ ° '	Moon ☽ Latitude ° '	DECLINATIONS Sun ☉ ° '	Moon ☽ ° '	Mercury ☿ ° '	Venus ♀ ° '	Mars ♂ ° '	Jupiter ♃ ° '	Saturn ♄ ° '	Uranus ♅ ° '	Neptune ♆ ° '	Pluto ♇ ° '
01	10 ♋ 44	10 ♋ 54	00 N 09	04 N 40	23 N 00	05 S 25	10 N 19	22 S 49	21 N 08	17 N 55	14 S 20	18 S 01	12 S 03
02	10 R 44	10 51	01 22	05 04	22 12	04 45	09 57	22 52	21 10	17 57	14 19	18 01	12 03
03	10 42	10 48	02 32	05 27	19 57	04 05	09 35	22 55	21 11	17 58	14 18	18 01	12 02
04	10 38	10 44	03 32	05 49	16 25	03 24	09 12	22 58	21 13	18 00	14 18	18 01	12 02
05	10 31	10 41	04 19	06 12	11 51	02 42	08 49	23 01	21 15	18 01	14 17	18 00	12 02
06	10 22	10 38	04 49	06 35	06 35	01 58	08 27	23 04	21 17	18 03	14 16	18 00	12 02
07	10 10	10 35	05 00	06 57	00 N 59	01 14	08 04	23 07	21 19	18 04	14 15	18 00	12 01
08	09 58	10 32	04 52	07 20	04 S 36	00 S 28	07 42	23 09	21 21	18 06	14 15	17 59	12 01
09	09 47	10 28	04 26	07 42	09 50	00 N 18	07 20	23 12	21 23	18 08	14 14	17 59	12 01
10	09 38	10 25	03 46	08 04	14 26	01 05	06 58	23 15	21 25	18 09	14 13	17 59	12 00
11	09 31	10 22	02 54	08 26	18 11	01 53	06 37	23 17	21 26	18 11	14 12	17 59	12 00
12	09 27	10 19	01 54	08 48	20 56	02 43	06 17	23 20	21 28	18 12	14 12	17 58	12 00
13	09 25	10 16	00 N 51	09 10	22 36	03 32	05 58	23 23	21 30	18 14	14 11	17 58	12 00
14	09 D 25	10 13	00 S 14	09 32	23 07	04 23	05 39	23 25	21 32	18 16	14 10	17 58	11 59
15	09 25	10 09	01 17	09 53	22 34	05 14	05 21	23 28	21 34	18 17	14 10	17 58	11 59
16	09 R 25	10 06	02 15	10 15	21 00	06 06	05 05	23 30	21 36	18 19	14 09	17 58	11 59
17	09 23	10 03	03 08	10 36	18 32	06 58	04 49	23 33	21 37	18 21	14 08	17 57	11 58
18	09 19	10 00	03 53	10 57	15 17	07 51	04 34	23 36	21 39	18 22	14 08	17 57	11 58
19	09 12	09 57	04 28	11 17	11 23	08 44	04 20	23 38	21 41	18 24	14 07	17 57	11 58
20	09 03	09 53	04 52	11 38	06 59	09 37	04 08	23 41	21 43	18 25	14 07	17 57	11 58
21	08 52	09 50	05 02	11 58	02 S 12	10 31	03 56	23 43	21 45	18 27	14 06	17 57	11 57
22	08 40	09 47	04 58	12 19	02 N 45	11 24	03 45	23 46	21 47	18 29	14 05	17 56	11 57
23	08 29	09 44	04 38	12 39	07 42	12 17	03 36	23 49	21 48	18 30	14 05	17 56	11 57
24	08 19	09 41	04 04	12 59	12 24	13 10	03 27	23 51	21 50	18 32	14 04	17 56	11 57
25	08 11	09 38	03 16	13 18	16 35	14 02	03 20	23 54	21 52	18 34	14 04	17 56	11 56
26	08 05	09 34	02 15	13 37	19 58	14 53	03 14	23 56	21 54	18 35	14 03	17 56	11 56
27	08 02	09 31	01 S 06	13 57	22 15	15 43	03 08	23 59	21 55	18 37	14 03	17 56	11 56
28	08 D 02	09 28	00 N 07	14 16	23 14	16 32	03 04	24 02	21 57	18 39	14 02	17 56	11 55
29	08 02	09 25	01 21	14 34	22 45	17 20	03 01	24 05	21 59	18 40	14 02	17 56	11 55
30	08 ♋ 03	09 ♋ 22	02 N 30	14 N 53	20 N 51	18 N 06	02 N 58	24 S 08	22 N 00	18 N 42	14 S 01	17 S 55	11 S 55

ZODIAC SIGN ENTRIES

Date	h	m	Planets
02	17	54	☽ ♌
04	19	46	☽ ♍
06	07	14	☿ ♈
06	20	57	☽ ♎
08	23	01	☽ ♏
11	03	47	☽ ♐
13	12	21	☽ ♑
16	00	11	☽ ♒
18	13	00	☽ ♓
20	00	36	☉ ♉
20	21	59	♄ ♊
21	00	18	☽ ♈
21	20	08	☿ ♉
23	08	56	☽ ♉
25	15	11	☽ ♊
27	19	49	☽ ♋
29	23	25	☽ ♌

LATITUDES

Date	Mercury ☿ ° '	Venus ♀ ° '	Mars ♂ ° '	Jupiter ♃ ° '	Saturn ♄ ° '	Uranus ♅ ° '	Neptune ♆ ° '	Pluto ♇ ° '
01	02 S 24	07 N 46	00 N 18	00 S 28	01 S 48	00 S 41	00 N 09	10 N 38
04	02 22	07 19	00 13	00 28	01 48	00 42	00 09	10 39
07	02 15	06 46	00 08	00 27	01 47	00 42	00 09	10 39
10	02 04	06 09	00 N 03	00 27	01 47	00 42	00 09	10 40
13	01 48	05 28	00 S 02	00 26	01 46	00 42	00 09	10 41
16	01 28	04 47	00 08	00 26	01 46	00 42	00 09	10 41
19	01 04	04 06	00 15	00 26	01 45	00 42	00 09	10 42
22	00 36	03 25	00 22	00 25	01 45	00 42	00 09	10 42
25	00 S 05	02 46	00 29	00 25	01 45	00 42	00 09	10 43
28	00 N 27	02 09	00 37	00 24	01 44	00 42	00 09	10 43
31	00 N 59	01 N 35	00 S 45	00 S 24	01 S 44	00 S 42	00 N 09	10 N 43

DATA

Julian Date	2452001
Delta T	+64 seconds
Ayanamsa	23° 52' 11"
Synetic vernal point	05° ♓ 14' 48"
True obliquity of ecliptic	23° 26' 20"

MOON'S PHASES, APSIDES AND POSITIONS ☽

Date	h	m	Phase	Longitude ° '	Eclipse Indicator
01	10	49	☽	11 ♋ 46	
08	03	22	○	18 ♎ 22	
15	15	31	☾	25 ♑ 43	
23	15	26	●	03 ♉ 32	
30	17	08	☽	10 ♌ 25	

Day	h	m		
05	10	11	Perigee	
17	06	09	Apogee	
01	10	44	Max dec	23° N 01'
07	16	12	0S	
14	11	32	Max dec	23° S 08'
21	22	45	0N	
28	16	15	Max dec	23° N 15'

LONGITUDES

Date	Chiron ⚷ ° '	Ceres ⚳ ° '	Pallas ⚴ ° '	Juno ⚵ ° '	Vesta ⚶ ° '	Black Moon Lilith ⚸ ° '
01	28 ♐ 54	17 ♑ 12	15 ♐ 55	23 ♈ 12	09 ♈ 07	14 ♒ 03
11	28 ♐ 55	19 ♑ 17	15 ♐ 53	28 ♈ 57	13 ♈ 48	15 ♒ 09
21	28 ♐ 47	20 ♑ 56	15 ♐ 04	04 ♉ 46	18 ♈ 26	16 ♒ 16
31	28 ♐ 30	22 ♑ 06	13 ♐ 27	10 ♉ 38	23 ♈ 01	17 ♒ 23

All ephemeris data is given at 12.00 UT and the Moon's longitude is additionally given for 24.00 UT

ASPECTARIAN

h m	Aspects
01 Sunday	
01 30	♀ ∗ ♆
03 41	☽ ⊻ ♃
04 45	☽ □ ♀
04 54	☽ ⊼ ♆
05 10	☽ ⚼ ♅
10 49	☽ □ ☉
12 39	☽ ∠ ♄
14 10	☽ ⊥ ♃
16 45	☽ ⊼ ♇
20 43	☽ ± ♅
22 07	☽ # ♂
23 18	♀ ∗ ♃
02 Monday	
02 46	☽ ⊼ ♂
03 00	☽ ± ♇
05 05	☉ # ☿
05 45	☽ ∠ ♃
06 19	☽ △ ☿
06 58	☽ ⊼ ♅
09 21	☿ ∠ ♆
12 02	☿ ⊻ ♅
13 13	☽ ± ♂
14 26	☽ ∗ ♄
15 34	☉ ∠ ♄
18 15	☽ ⚼ ♇
03 Tuesday	
00 48	☽ ∥ ♃
04 45	☽ ⚼ ♂
05 42	☽ △ ♀
07 22	☽ ∗ ♃
07 59	☽ ☍ ♆
10 44	☽ Q ♄
10 55	☽ ⚼ ☿
17 22	☽ △ ☉
19 20	☽ △ ♇
04 Wednesday	
01 26	☿ Q ♃
02 11	☽ # ♆
02 21	☽ ∥ ♄
03 35	☽ Q ♃
03 50	☽ ± ☿
05 40	☽ ⚼ ♀
06 19	☽ △ ♂
09 18	☽ ☍ ♅
15 02	☽ ⊼ ☿
16 46	☽ □ ♄
19 56	☽ ± ♀
19 57	☽ ⚼ ☉
22 04	☉ △ ♇
22 58	♄ # ♆
23 48	☽ # ♅
05 Thursday	
05 25	☽ ⊼ ♀
06 40	☿ ∗ ♄
09 33	☽ □ ♃
09 37	☽ ⊼ ♆
11 10	☽ # ♇
11 46	☽ ± ☉
14 24	♃ △ ♆
19 27	☽ ± ♆
20 39	☽ □ ♇
22 18	☽ ⊼ ☉
06 Friday	
03 11	☽ ∥ ♀
03 11	☿ ⊥ ♅
08 42	☽ □ ♂
10 12	☽ ⚼ ♆
10 38	☽ ⊼ ♅
12 02	☽ ∥ ☉
18 18	☽ △ ♄
20 31	☽ ± ♅
22 49	☽ ☍ ☿
07 Saturday	
02 10	☽ Q ♇
04 54	☽ ☍ ♀
10 49	☽ # ☿
10 55	☽ △ ♆
11 24	☽ ⚼ ♅
11 26	☽ △ ♃
15 02	☽ Q ♂
19 16	☽ ⚼ ♄
20 19	☽ ∥ ☿
22 01	☽ ∗ ♇
08 Sunday	
03 22	☽ ☍ ☉
10 26	☽ ± ♄
11 33	☽ ∗ ♂
12 31	☽ △ ♅
12 49	☽ ⚼ ♃
15 25	☿ ☌ ♀
20 40	☽ ⊼ ♄
23 16	☽ ∠ ♇
23 43	☉ ∥ ♀
09 Monday	
01 00	☽ # ♀
01 12	☽ # ☉
04 19	☽ ± ♃
05 39	☽ ⊼ ♀
08 17	♂ ∠ ♆
08 18	☽ ⊼ ☿
13 38	☽ □ ♆
13 45	☽ ∠ ♂
14 44	☽ ⊥ ♇
14 49	☽ ⊼ ♃
15 43	☽ ± ♀
20 17	☽ ± ☿
22 52	☽ ∥ ♇
10 Tuesday	
01 12	☽ ⊻ ♇
05 52	☽ ⊥ ♂
07 02	☽ ⚼ ♀
08 13	☉ Q ♆
10 46	☽ ∥ ♅
10 55	☽ ⊼ ☉
13 27	♂ ∗ ♅
14 40	☽ ⚼ ☿
16 44	☽ □ ♅
16 48	☽ ⊻ ♂
21 30	☽ Q ♆
22 32	☽ ± ☉
11 Wednesday	
01 02	☿ ∗ ♆
01 43	☽ ☍ ♄
05 51	☿ ∠ ♅
09 19	☽ △ ♀
10 27	☽ ∥ ♆
11 56	☽ # ♄
14 13	☿ ∗ ♃
16 19	☽ ⚼ ☉
19 29	☽ ∗ ♆
21 29	☽ ☍ ♃
22 36	☽ △ ☿
12 Thursday	
01 49	☽ Q ♅
07 46	☽ ☌ ♇
18 15	☽ # ♃
22 43	☉ ⊥ ♄
22 58	☽ △ ☉
23 53	☽ ∠ ♆
13 Friday	
00 40	☽ ∗ ♅
01 56	☽ ☌ ♂
10 39	☽ ⊼ ♄
16 51	☽ □ ♀
17 23	☽ ⊥ ♆
21 24	☉ ∗ ♅
22 30	☽ ± ♄
14 Saturday	
00 46	☿ ∠ ♄
05 12	☽ ⊻ ♆
06 04	☽ ∠ ♅
08 11	☽ ⊼ ♃
11 38	☿ △ ♇
16 31	☽ ⚼ ♄
18 07	☽ ⊻ ♇
19 23	☽ □ ☿
20 20	☽ ± ♃
15 Sunday	
00 03	☽ ⊥ ♅
02 31	☉ △ ♂
03 55	☽ Q ♀
03 57	☉ ∠ ♃
06 08	☽ ⊥ ♇
12 10	☽ ⊻ ♅
14 31	☿ ∥ ♀
14 43	☽ ⊻ ♂
14 45	☽ ⚼ ♃
15 31	☽ □ ☉
23 00	☽ △ ♄
16 Monday	
00 17	☽ ∠ ♇
03 09	☽ ⊥ ♂
03 48	☽ ∗ ♀
04 38	☽ # ♃
14 59	☽ Q ☿
17 40	☽ ☌ ♆
21 41	☽ △ ♃
21 50	☽ ∠ ♂
17 Tuesday	
06 43	☽ ∗ ♇
06 56	☿ Q ♆
07 21	☽ Q ☉
09 55	☽ ∠ ♀
13 33	☽ # ♄
16 42	☽ ∥ ♆
20 21	☽ ∗ ☿
18 Wednesday	
01 11	☽ ☌ ♅
04 00	☽ ⊥ ♀
04 55	☽ ∗ ♂
07 00	☽ Q ♇
09 49	☽ ∗ ☉
12 26	☽ □ ♄
16 02	☽ ⊻ ♀
19 31	☽ ∥ ♅
19 57	☿ ⊥ ♄
19 Thursday	
00 53	☿ ∗ ♅
05 31	☽ Q ♂
06 20	☽ ⊻ ♆
08 38	☽ ∥ ♇
08 43	☽ ∠ ☿
11 13	☽ □ ♃
12 30	☽ # ☉
18 16	☽ ⊥ ♆
18 28	☽ ∠ ☉
18 56	☽ □ ♇
21 39	☉ ⊻ ♄
20 Friday	
00 15	☽ # ☿
00 21	☉ ⚼ ♇
00 38	☿ ∠ ♃
00 40	☽ Q ♄
03 21	☿ △ ♂
04 34	♀ St D
06 10	☽ ⊥ ☿
11 59	☽ ∠ ♆
13 04	☽ ⊻ ♅
13 41	☽ ⊥ ☉
17 41	☽ □ ♂
20 16	☽ ⊻ ☿
23 09	☽ Q ♃
21 Saturday	
00 19	☽ ∗ ♄
00 45	☽ ⊥ ♅
02 19	☽ ⊻ ☉
03 08	☽ # ♀
03 09	☽ ☌ ♀
10 39	☉ # ♇
13 18	☉ ⊻ ♀
16 57	☽ ∗ ♆
18 04	☽ ∠ ♅
19 40	☿ ⚼ ♇
21 28	☿ ⊻ ♄
22 31	☽ ∗ ♃
22 Sunday	
04 52	☽ △ ♇
05 15	☽ ∠ ♄
13 58	☿ ⊻ ♀
15 38	☽ Q ♆
16 41	☽ ∥ ♀
22 16	☽ ⊥ ♄
22 22	☽ ∗ ♅
23 Monday	
02 47	☿ # ♇
03 06	☽ ∠ ♃
03 34	☽ △ ♂
08 48	☽ ⚼ ♇
09 24	☉ ☌ ☿
09 29	☽ ⊻ ♄
12 00	☽ ⊻ ♀
15 26	☽ ☌ ☉
16 04	☽ ☌ ☿
19 56	☽ ⊥ ♃
20 32	☽ Q ♅
23 07	☽ ⊥ ♀
24 Tuesday	
00 53	☽ □ ♆
01 17	☽ ± ♇
03 42	☉ ∥ ☿
07 02	☽ ⊻ ♃
07 28	☽ ⚼ ♂
09 35	☽ # ♇
12 09	☽ ⊼ ♇
15 22	☽ ∥ ☉
15 38	☽ ∠ ♀
17 11	☽ ∥ ☿
17 58	☿ ⊥ ♃
19 51	☿ Q ♅
21 09	☽ # ♅
23 56	☽ ± ♂
25 Wednesday	
05 08	☽ □ ♅
10 48	☽ ⊼ ♂
12 54	☿ # ♅
14 32	☿ ⊥ ♀
16 12	☽ ☌ ♄
18 50	☽ ∗ ♀
20 49	☽ # ♆
22 23	☿ □ ♆
26 Thursday	
00 15	☿ ± ♇
01 19	☽ ∥ ♄
01 38	☽ ⊻ ☉
06 37	☽ △ ♆
08 09	☽ ⊻ ☿
13 01	☽ ⊥ ☉
13 22	☽ ☌ ♃
13 59	☉ Q ♅
16 21	☽ Q ♀
17 26	☽ ☍ ♇
19 12	☉ ⊥ ♃
20 39	☽ ⊥ ☿
27 Friday	
05 59	☽ ∠ ☉
07 27	☽ ∥ ♃
08 55	☽ ⚼ ♆
10 07	☽ △ ♅
15 15	☽ ∠ ☿
16 12	☽ ☍ ♂
19 38	☉ # ♅
20 23	☿ ⊻ ♃
21 17	☽ ⊻ ♄
21 45	☿ ⚼ ♂
28 Saturday	
00 20	☽ □ ♀
00 33	☽ ± ♆
07 46	☽ ⊥ ♄
09 58	☽ ∗ ☉
10 56	☽ ⊼ ♆
12 10	☽ ⚼ ♅
18 19	☽ ⊻ ♃
18 52	☿ ⊼ ♇
21 26	☽ ⊼ ♇
21 53	☽ ∗ ☿
23 24	☽ ∠ ♄
23 49	☉ □ ♆
29 Sunday	
01 53	☉ ± ♇
03 40	☽ ± ♅
04 47	☽ ⊥ ♃
05 24	☉ ⊥ ♀
07 43	☽ ± ♇
08 09	☽ Q ☉
13 58	☽ ⊼ ♅
20 23	☽ ⊼ ♂
20 25	☽ ∠ ♃
22 07	☽ Q ☿
23 05	☽ ⚼ ♇
23 45	☽ ∥ ♃
30 Monday	
01 18	☽ ∗ ♄
05 02	☽ △ ♀
06 32	☿ # ♆
06 43	☽ ± ♂
11 07	☿ ∠ ♀
14 18	☽ ☍ ♆
17 08	☽ □ ☉
21 53	☽ Q ♄
22 09	☽ ⚼ ♂
22 20	☽ ∗ ♃

MAY 2001

LONGITUDES

Date	Sidereal time h m s	Sun ☉ ° ' "	Moon ☽ ° ' "	Moon ☽ 24.00 ° ' "	Mercury ☿ ° '	Venus ♀ ° '	Mars ♂ ° '	Jupiter ♃ ° '	Saturn ♄ ° '	Uranus ♅ ° '	Neptune ♆ ° '	Pluto ♇ ° '
01	02 37 56	11 ♉ 10 23	21 ♌ 33 04	28 ♌ 39 23	20 ♉ 33	03 ♈ 46	28 ♐ 26	13 ♊ 36	01 ♊ 17	24 ♒ 31	08 ♒ 45	14 ♐ 46
02	02 41 53	12 08 36	05 ♍ 46 26	12 ♍ 53 52	22 38	04 10	28 33	13 49	01 25	24 32	08 45	14 R 45
03	02 45 50	13 06 47	20 ♍ 01 20	27 ♍ 08 21	24 40	04 36	28 39	14 02	01 32	24 34	08 46	14 44
04	02 49 46	14 04 55	04 ♎ 14 27	11 ♎ 19 03	26 40	05 04	28 44	14 14	01 40	24 35	08 46	14 42
05	02 53 43	15 03 02	18 ♎ 21 35	25 ♎ 21 30	28 ♉ 38	05 32	28 49	14 27	01 47	24 36	08 46	14 41
06	02 57 39	16 01 07	02 ♏ 18 15	09 ♏ 11 17	00 ♊ 34	06 03	28 53	14 40	01 55	24 37	08 46	14 40
07	03 01 36	16 59 10	16 ♏ 00 12	22 ♏ 44 35	02 27	06 35	28 57	14 53	02 03	24 38	08 46	14 38
08	03 05 32	17 57 12	29 ♏ 24 13	05 ♐ 58 53	04 17	07 08	28 59	15 07	02 10	24 39	08 47	14 37
09	03 09 29	18 55 12	12 ♐ 28 32	18 ♐ 53 13	06 03	07 42	29 01	15 20	02 18	24 40	08 47	14 36
10	03 13 25	19 53 11	25 ♐ 13 04	01 ♑ 28 19	07 47	08 18	29 02	15 33	02 25	24 41	08 47	14 34
11	03 17 22	20 51 08	07 ♑ 39 18	13 ♑ 45 29	09 27	08 55	29 03	15 46	02 33	24 42	08 R 47	14 33
12	03 21 19	21 49 04	19 ♑ 50 07	25 ♑ 50 56	11 04	09 34	29 R 03	15 59	02 41	24 43	08 47	14 31
13	03 25 15	22 46 58	01 ♒ 49 25	07 ♒ 46 11	12 37	10 13	29 02	16 13	02 48	24 44	08 47	14 30
14	03 29 12	23 44 52	13 ♒ 41 50	19 ♒ 37 01	14 07	10 53	29 00	16 26	02 56	24 45	08 46	14 28
15	03 33 08	24 42 44	25 ♒ 32 21	01 ♓ 28 30	15 33	11 35	28 57	16 39	03 04	24 45	08 46	14 27
16	03 37 05	25 40 34	07 ♓ 26 05	13 ♓ 25 42	16 55	12 17	28 54	16 53	03 12	24 46	08 46	14 25
17	03 41 01	26 38 24	19 ♓ 27 55	25 ♓ 33 16	18 13	13 01	28 50	17 06	03 19	24 47	08 46	14 24
18	03 44 58	27 36 12	01 ♈ 42 13	07 ♈ 55 12	19 28	13 45	28 45	17 20	03 27	24 47	08 46	14 22
19	03 48 54	28 33 59	14 ♈ 12 34	20 ♈ 34 34	20 39	14 31	28 40	17 33	03 35	24 48	08 45	14 21
20	03 52 51	29 ♉ 31 45	27 ♈ 01 23	03 ♉ 33 07	21 46	15 17	28 33	17 47	03 43	24 48	08 45	14 19
21	03 56 48	00 ♊ 29 30	10 ♉ 09 45	16 ♉ 51 10	22 49	16 04	28 26	18 00	03 50	24 49	08 45	14 18
22	04 00 44	01 27 14	23 ♉ 37 12	00 ♊ 27 33	23 48	16 51	28 18	18 14	03 58	24 49	08 44	14 16
23	04 04 41	02 24 57	07 ♊ 21 51	14 ♊ 19 42	24 43	17 40	28 10	18 27	04 06	24 49	08 44	14 15
24	04 08 37	03 22 38	21 ♊ 20 38	28 ♊ 24 09	25 34	18 29	28 01	18 41	04 14	24 50	08 44	14 13
25	04 12 34	04 20 18	05 ♋ 29 46	12 ♋ 36 58	26 20	19 19	27 50	18 55	04 21	24 50	08 43	14 11
26	04 16 30	05 17 56	19 ♋ 45 16	26 ♋ 53 26	27 02	20 10	27 40	19 08	04 29	24 50	08 43	14 10
27	04 20 27	06 15 33	04 ♌ 03 24	11 ♌ 12 25	27 40	21 01	27 28	19 22	04 37	24 50	08 42	14 08
28	04 24 23	07 13 09	18 ♌ 20 55	25 ♌ 28 37	28 13	21 53	27 16	19 36	04 45	24 50	08 42	14 07
29	04 28 20	08 10 43	02 ♍ 35 12	09 ♍ 40 27	28 42	22 45	27 03	19 49	04 53	24 50	08 41	14 05
30	04 32 17	09 08 16	16 ♍ 44 05	23 ♍ 45 50	29 06	23 38	26 50	20 03	05 00	24 R 50	08 41	14 03
31	04 36 13	10 ♊ 05 47	00 ♎ 45 44	07 ♎ 43 18	29 ♊ 26	24 ♈ 32	26 ♐ 36	20 ♊ 17	05 ♊ 08	24 ♒ 50	08 ♒ 40	14 ♐ 02

DECLINATIONS

Date	Moon True ☊ ° '	Moon Mean ☊ ° '	Moon ☽ Latitude ° '	Sun ☉ ° '	Moon ☽ ° '	Mercury ☿ ° '	Venus ♀ ° '	Mars ♂ ° '	Jupiter ♃ ° '	Saturn ♄ ° '	Uranus ♅ ° '	Neptune ♆ ° '	Pluto ♇ ° '
01	08 ♋ 03	09 ♋ 19	03 N 31	15 N 11	17 N 39	18 N 50	02 N 57	24 S 10	22 N 02	18 N 43	14 S 01	17 S 55	11 S 55
02	08 R 00	09 15	04 19	15 29	13 24	19 32	02 56	24 13	22 04	18 45	14 01	17 55	11 54
03	07 56	09 12	04 51	15 46	08 25	20 12	02 57	24 16	22 05	18 47	14 00	17 55	11 54
04	07 49	09 09	05 05	16 04	02 N 59	20 50	02 58	24 19	22 07	18 48	14 00	17 55	11 54
05	07 41	09 06	05 01	16 21	02 S 34	21 25	03 00	24 22	22 09	18 50	13 59	17 55	11 54
06	07 33	09 03	04 38	16 38	07 55	21 58	03 03	24 25	22 10	18 51	13 59	17 55	11 54
07	07 25	08 59	04 00	16 55	12 48	22 28	03 07	24 28	22 12	18 53	13 59	17 55	11 53
08	07 18	08 56	03 09	17 11	16 57	22 56	03 11	24 31	22 14	18 55	13 58	17 55	11 53
09	07 13	08 53	02 09	17 27	20 09	23 22	03 17	24 35	22 15	18 56	13 58	17 55	11 53
10	07 10	08 50	01 N 04	17 43	22 17	23 45	03 23	24 38	22 17	18 58	13 58	17 55	11 53
11	07 D 10	08 47	00 S 03	17 58	23 16	24 05	03 29	24 41	22 18	18 59	13 58	17 55	11 52
12	07 10	08 44	01 08	18 13	23 06	24 23	03 37	24 44	22 20	19 01	13 57	17 55	11 52
13	07 12	08 40	02 10	18 28	21 51	24 38	03 45	24 48	22 21	19 03	13 57	17 55	11 52
14	07 13	08 37	03 05	18 43	19 40	24 52	03 53	24 51	22 23	19 04	13 57	17 55	11 52
15	07 R 14	08 34	03 52	18 57	16 39	25 03	04 03	24 54	22 24	19 06	13 57	17 55	11 52
16	07 13	08 31	04 30	19 11	12 57	25 11	04 13	24 58	22 26	19 07	13 57	17 55	11 51
17	07 10	08 28	04 56	19 24	08 42	25 18	04 23	25 01	22 27	19 09	13 56	17 55	11 51
18	07 06	08 25	05 09	19 37	04 S 03	25 23	04 34	25 05	22 28	19 10	13 56	17 55	11 51
19	07 01	08 21	05 08	19 50	00 N 52	25 26	04 46	25 08	22 30	19 12	13 56	17 55	11 51
20	06 54	08 18	04 52	20 03	05 52	25 27	04 57	25 12	22 31	19 13	13 56	17 56	11 51
21	06 48	08 15	04 20	20 15	10 45	25 27	05 10	25 16	22 32	19 15	13 56	17 56	11 51
22	06 43	08 12	03 33	20 27	15 14	25 25	05 23	25 19	22 34	19 16	13 56	17 56	11 50
23	06 39	08 09	02 33	20 39	19 02	25 21	05 36	25 23	22 35	19 18	13 56	17 56	11 50
24	06 36	08 05	01 22	20 50	21 47	25 16	05 50	25 26	22 36	19 19	13 56	17 56	11 50
25	06 35	08 02	00 S 06	21 01	23 13	25 09	06 04	25 30	22 38	19 21	13 56	17 56	11 50
26	06 D 36	07 59	01 N 11	21 11	23 13	25 01	06 18	25 34	22 39	19 22	13 56	17 56	11 50
27	06 37	07 56	02 25	21 21	21 35	24 52	06 33	25 37	22 40	19 24	13 56	17 56	11 50
28	06 39	07 53	03 29	21 31	18 38	24 42	06 48	25 41	22 41	19 25	13 56	17 56	11 49
29	06 40	07 50	04 20	21 40	14 36	24 31	07 03	25 44	22 42	19 27	13 56	17 57	11 49
30	06 R 39	07 46	04 55	21 49	09 46	24 19	07 19	25 48	22 43	19 28	13 56	17 57	11 49
31	06 ♋ 38	07 ♋ 43	05 N 12	21 N 58	04 N 28	24 N 06	07 N 35	25 S 51	22 N 45	19 N 30	13 S 56	17 S 57	11 S 49

ZODIAC SIGN ENTRIES

Date	h m	Planets
02	02 16	☽ ♍
04	04 51	☽ ♎
06	04 53	☿ ♊
06	08 00	☽ ♏
08	13 05	☽ ♐
10	21 10	☽ ♑
13	08 20	☽ ♒
15	21 01	☽ ♓
18	08 41	☽ ♈
20	17 29	☽ ♉
20	23 44	☉ ♊
22	23 12	☽ ♊
25	02 42	☽ ♋
27	05 12	☽ ♌
29	07 38	☽ ♍
31	10 41	☽ ♎

LATITUDES

Date	Mercury ☿ ° '	Venus ♀ ° '	Mars ♂ ° '	Jupiter ♃ ° '	Saturn ♄ ° '	Uranus ♅ ° '	Neptune ♆ ° '	Pluto ♇ ° '
01	00 N 59	01 N 35	00 S 45	00 S 24	01 S 44	00 S 42	00 N 09	10 N 43
04	01 27	01 02	00 53	00 24	01 43	00 43	00 09	10 44
07	01 52	00 33	01 02	00 23	01 43	00 43	00 09	10 44
10	02 10	00 N 05	01 12	00 23	01 43	00 43	00 09	10 44
13	02 21	00 S 20	01 22	00 22	01 43	00 43	00 09	10 44
16	02 24	00 42	01 32	00 22	01 42	00 43	00 09	10 44
19	02 20	01 03	01 43	00 22	01 42	00 43	00 09	10 44
22	02 07	01 21	01 54	00 21	01 42	00 43	00 09	10 44
25	01 46	01 37	02 05	00 21	01 42	00 43	00 09	10 44
28	01 17	01 51	02 16	00 21	01 42	00 44	00 09	10 44
31	00 N 40	02 S 04	02 S 28	00 S 20	01 S 41	00 S 44	00 N 09	10 N 44

DATA

Julian Date	2452031
Delta T	+64 seconds
Ayanamsa	23° 52' 14"
Synetic vernal point	05° ♓ 14' 45"
True obliquity of ecliptic	23° 26' 19"

MOON'S PHASES, APSIDES AND POSITIONS ☽

Date	h	m	Phase	Longitude ° '	Eclipse Indicator
07	13	53	○	17 ♏ 04	
15	10	11	☾	24 ♒ 38	
23	02	46	●	02 ♊ 03	
29	22	09	☽	08 ♍ 35	

Day	h	m		
02	03	45	Perigee	
15	01	28	Apogee	
27	06	55	Perigee	
05	00	54	0S	
11	20	24	Max dec	23° S 20'
19	07	48	0N	
25	23	03	Max dec	23° N 23'

LONGITUDES

Date	Chiron ⚷ ° '	Ceres ⚳ ° '	Pallas ⚴ ° '	Juno ⚵ ° '	Vesta ⚶ ° '	Black Moon Lilith ⚸ ° '
01	28 ♐ 30	22 ♑ 36	13 ♐ 27	10 ♉ 38	23 ♈ 01	17 ♒ 23
11	28 ♐ 05	22 ♑ 43	11 ♐ 09	16 ♉ 32	27 ♈ 32	18 ♒ 29
21	27 ♐ 34	22 ♑ 45	08 ♐ 21	22 ♉ 27	01 ♉ 58	19 ♒ 36
31	26 ♐ 58	22 ♑ 10	05 ♐ 22	28 ♉ 24	06 ♉ 19	20 ♒ 43

All ephemeris data is given at 12.00 UT and the Moon's longitude is additionally given for 24.00 UT

ASPECTARIAN

h m Aspects

01 Tuesday
00 32 ☽ △ ⯓
04 54 ☽ ∥ ♄
05 28 ☽ ∥ ☿
07 10 ☽ ♇ ♀
08 15 ☿ ∥ ♄
10 02 ☽ □ ☿
10 14 ☽ # ♆
17 01 ☽ ☍ ♅
18 56 ☽ Q ♃
22 48 ☽ ± ♀
23 44 ☽ △ ♂

02 Wednesday
01 38 ☽ ∥ ☉
04 35 ☽ □ ♄
08 52 ☽ # ♅
09 13 ☽ ⊼ ♀
11 03 ☿ ± ♂
17 02 ☽ ⊼ ♆
19 32 ☽ # ⯓
23 31 ☽ △ ☉

03 Thursday
01 45 ☽ □ ♃
03 06 ☽ □ ⯓
03 08 ☽ ± ♆
10 42 ☿ □ ♅
18 18 ☽ ♇ ♆
19 39 ☽ ⊼ ♅
21 07 ☽ △ ☿

04 Friday
02 39 ☽ ♇ ☉
02 39 ☽ □ ♂
02 43 ☉ ♇ ♂
05 48 ☽ ± ♅
07 36 ☽ △ ♄
09 25 ☽ Q ⯓
12 05 ☽ ∥ ♀
13 26 ☽ ☍ ♀
17 04 ☉ ⊻ ♃
18 59 ☽ ± ☉
19 40 ☽ △ ♆
21 04 ☽ ♇ ♅

05 Saturday
02 39 ☽ ♇ ☿
03 07 ☉ ⊼ ⯓
05 14 ☽ △ ♃
05 44 ☽ ⚹ ⯓
05 56 ☽ ⊼ ☉
09 18 ☽ ♇ ♄
09 21 ☽ Q ♂
13 56 ☽ # ♀
14 17 ☿ ⊼ ♂
20 31 ☽ ± ☿
22 43 ☽ △ ♅

06 Sunday
00 52 ☽ ± ♄
06 03 ☽ ⚹ ♂
07 22 ☽ ♇ ♃
07 26 ☽ ∠ ⯓
08 30 ☽ ⊼ ☿
10 47 ♃ ☍ ⯓
11 19 ☽ ⊼ ♄
18 46 ☽ ⊼ ♀
21 59 ☿ ∥ ♃
23 04 ☽ ⊥ ⯓
23 16 ☽ □ ♆
23 17 ☽ ± ♃

07 Monday
05 42 ☽ ± ♀
06 26 ☿ ☌ ♄
07 18 ☽ ∥ ⯓
08 21 ☽ ∠ ♂
09 35 ☽ ⊻ ⯓
10 00 ☽ ⊼ ♃
13 53 ☽ ☍ ☉
18 22 ☽ ∥ ♅
22 20 ☽ ♇ ♀

08 Tuesday
00 24 ☽ ⊥ ♂
03 25 ☽ □ ♅
07 15 ☽ Q ♆
11 15 ☽ ⊻ ♂
13 39 ☽ # ☉
17 05 ☽ ☍ ♄
18 31 ☽ ∥ ♆
22 17 ☽ ☍ ☿

09 Wednesday
01 52 ☽ # ♄
02 46 ☽ △ ♀
05 09 ☽ ⚹ ♆
12 22 ☽ Q ♅
15 56 ☽ ☌ ⯓
17 25 ☽ ☍ ♃

10 Thursday
01 03 ☽ ⊼ ☉
09 15 ☽ ∠ ♆
10 59 ☽ ⚹ ♅
11 53 ☽ # ♃
13 23 ☽ ± ☉
19 20 ☽ ☌ ♂
23 47 ☿ ⚹ ♀

11 Friday
01 13 ♆ St R
01 59 ☽ ⊼ ♄
02 15 ☿ △ ♆
02 31 ☽ ⊥ ♆
06 24 ♀ ⚹ ♆
07 21 ☉ # ♆
08 11 ☽ ♇ ☉
13 46 ☽ ± ♄
14 11 ☽ ⊻ ♆
14 37 ☽ □ ♀
16 00 ☽ ∠ ♅
16 03 ☽ ⊼ ☿
16 08 ♂ St R

12 Saturday
01 30 ☽ ⊻ ⯓
04 14 ☽ ⊼ ♃
05 40 ☽ ± ☿
07 40 ☽ ♇ ♄
09 47 ☽ ⊥ ♅
13 22 ☽ ⊥ ⯓
16 17 ☽ △ ☉
16 22 ☽ ± ♃
17 56 ♀ ∠ ♅
21 45 ☽ ⊻ ♅

13 Sunday
02 18 ☽ ♇ ☿
04 19 ☽ Q ♀
04 39 ☽ # ♃
06 23 ☽ ⊻ ♂
07 20 ☽ ∠ ⯓
10 45 ☽ ♇ ♃
14 00 ☽ △ ♄
17 57 ☉ ± ♂
18 27 ☽ ⊥ ♂

14 Monday
02 02 ☽ ☌ ♆
05 58 ☽ ⚹ ♀
10 32 ☿ # ♂
12 37 ☽ ∠ ♂
12 57 ☽ △ ☿
13 34 ☽ ⚹ ⯓
17 12 ☽ # ♄
17 39 ☽ △ ♃
17 53 ☿ ☍ ⯓
19 38 ☽ # ☉

15 Tuesday
02 35 ☽ ∥ ♆
10 11 ☽ □ ☉
10 25 ☽ ☌ ♅
13 06 ☉ □ ♅
13 50 ☽ Q ⯓
14 15 ☽ ∠ ♀
18 53 ☽ ⚹ ♂

16 Wednesday
03 22 ☽ □ ♄
05 16 ☉ ∥ ♄
05 55 ☽ ∥ ♅
09 33 ☽ ⊥ ♀
11 15 ☿ ☌ ♃
14 41 ☽ ⊻ ♆
18 26 ☽ ∥ ⯓
18 55 ☽ Q ♂
22 21 ☽ ⊻ ♀

17 Thursday
01 35 ☽ Q ☉
01 57 ☽ □ ⯓
02 40 ☽ ⊥ ♆
07 14 ☽ □ ♃
09 15 ☽ □ ☿
15 43 ☽ Q ♄
20 29 ☽ ∠ ♆
22 29 ☽ ⊻ ♅

18 Friday
03 20 ☽ ⚹ ☉
06 18 ☽ □ ♂
09 29 ☽ # ♀
10 13 ☽ ⊥ ♅
15 25 ☽ ⚹ ♄
19 08 ☽ Q ♃

19 Saturday
00 19 ☽ Q ☿
01 37 ☽ ⚹ ♆
03 35 ☽ ∠ ♅
06 58 ♀ △ ⯓
10 41 ☽ ∠ ☉
12 16 ☽ △ ⯓
12 36 ☽ ☌ ♀
14 10 ☉ ⊼ ♂
18 26 ☽ ⚹ ♃
20 20 ☽ ∠ ♄

20 Sunday
00 20 ☽ Q ♆
01 19 ☽ ⚹ ☿
05 00 ☽ ⊥ ☉
07 24 ☽ ∥ ♀
07 53 ☽ ⚹ ♅
13 17 ☽ ⊥ ♄
14 49 ☽ △ ♂
16 14 ☽ ♇ ⯓
17 00 ☽ ⊻ ☉
22 46 ☽ ∠ ♃

21 Monday
00 25 ☽ ⊻ ♄
05 56 ☽ Q ♅
07 24 ☽ ∠ ☿
08 38 ☽ ± ⯓
09 27 ☽ □ ♆
15 22 ☽ ⊥ ♃
17 35 ☽ # ⯓
17 50 ☽ ♇ ♂
19 25 ☽ ⊼ ⯓
23 15 ☽ ⊻ ♀

22 Tuesday
00 54 ☽ ⊥ ☿
02 17 ☽ ⊻ ♃
04 40 ☽ # ♅
09 42 ☽ ± ♂
10 30 ☿ ♇ ♆
10 34 ☽ ⊥ ♀
12 21 ☽ ⊻ ☿
14 07 ☽ □ ♅
20 09 ☽ ⊼ ♂

23 Wednesday
02 46 ☽ ☌ ☉
03 20 ☽ ∠ ♀
04 26 ☽ # ♆
05 48 ☿ # ♂
06 17 ☽ ☌ ♄
13 59 ☽ ∥ ♄
14 22 ☽ △ ♆
14 56 ☿ △ ♅
23 50 ☽ ☍ ⯓

24 Thursday
01 43 ☽ ∥ ☉
06 49 ☽ ⚹ ♀
07 23 ☽ ☌ ♃
16 04 ☽ ♇ ♆
17 56 ☽ △ ♅
19 36 ☽ ☌ ☿
19 53 ♀ ⚹ ♃
22 51 ☽ ∥ ♃
23 12 ☽ ☍ ♂

25 Friday
04 30 ☽ Q ♀
07 19 ☽ ± ♆
08 12 ☉ ∠ ♀
09 54 ☽ ⊻ ☉
10 04 ☽ ⊻ ♄
12 33 ☉ ☌ ♄
13 19 ♀ ∠ ♄
17 26 ☽ ⊼ ♆
19 19 ☽ ♇ ♅
20 16 ☽ ⊥ ♄
20 45 ☽ ⊥ ☉

26 Saturday
02 37 ☽ ⊼ ⯓
10 27 ☽ ± ♅
10 57 ☽ ⊻ ♃
11 33 ☽ ∠ ♄
12 41 ☽ ± ⯓
12 44 ☽ □ ♀
12 59 ☽ ∠ ☉
20 32 ☽ ⊼ ♅
21 11 ☽ ⊥ ♃
22 29 ☽ ∥ ♃

27 Sunday
00 48 ☽ ⊻ ☿
01 06 ☽ ⊼ ♂
03 23 ♀ Q ♆
03 46 ☽ ♇ ⯓
06 03 ☿ ☍ ♂
11 02 ☽ ± ♂
11 19 ☽ ⊥ ☿
12 32 ☽ ∠ ♃
12 57 ☽ ⚹ ♄
14 12 ☽ ∥ ☉
15 58 ☽ ⚹ ☉
19 48 ☽ ☍ ♆

28 Monday
01 56 ☽ ♇ ♂
03 03 ☽ ∠ ☿
04 53 ☽ △ ⯓
06 33 ☽ ∥ ♄
09 17 ☽ Q ♄
13 34 ☽ Q ☉
14 08 ☽ ⚹ ♃
16 35 ☽ # ♆
18 20 ☽ △ ♀
22 55 ☽ ☍ ♅

29 Tuesday
02 48 ☽ △ ♂
05 13 ☽ ⚹ ☿
10 41 ☽ Q ♃
15 13 ♅ St R
15 31 ☽ # ♅
15 54 ☽ □ ♄
21 19 ☽ ♇ ♀
22 09 ☽ □ ☉
22 19 ☽ ⊼ ♆

30 Wednesday
00 34 ☉ △ ♆
02 08 ☽ # ⯓
02 10 ☽ Q ☿
07 27 ☽ □ ⯓
08 30 ☽ ± ♆
13 38 ☽ ± ♀
17 45 ☽ □ ♃
22 43 ☽ ∥ ♀
23 50 ☽ ♇ ♆

31 Thursday
00 35 ☽ ⊼ ♀
01 50 ☽ ⊼ ♅
04 58 ☽ □ ♂
09 40 ☽ □ ☿
12 08 ☽ ± ♅
14 11 ☽ Q ⯓
19 36 ☽ △ ♄
20 12 ♀ ⚹ ♅

JUNE 2001

Date	Sidereal time h m s	Sun ☉ ° ' "	Moon ☽ ° ' "	Moon ☽ 24.00 ° ' "	Mercury ☿ ° '	Venus ♀ ° '	Mars ♂ ° '	Jupiter ♃ ° '	Saturn ♄ ° '	Uranus ♅ ° '	Neptune ♆ ° '	Pluto ⯓ ° '
01	04 40 10	11 ♊ 03 17	14 ♎ 38 24	21 ♎ 30 52	29 ♊ 41	25 ♈ 26	26 ♐ 21	20 ♊ 31	05 ♊ 16	24 ♒ 50	08 ♒ 39	14 ♐ 00
02	04 44 06	12 00 46	28 ♎ 20 27	05 ♏ 06 58	29 51	26 20	26 R 06	20 45	05 24	24 R 50	08 R 39	13 R 59
03	04 48 03	12 58 13	11 ♏ 50 14	18 ♏ 30 05	29 57	27 16	25 50	20 58	05 31	24 50	08 38	13 57
04	04 51 59	13 55 39	25 ♏ 06 22	01 ♐ 38 59	29 R 58	28 11	25 34	21 12	05 39	24 49	08 37	13 55
05	04 55 56	14 53 05	08 ♐ 07 50	14 ♐ 32 55	29 54	29 ♈ 07	25 17	21 26	05 47	24 49	08 36	13 54
06	04 59 52	15 50 29	20 ♐ 54 13	27 ♐ 11 48	29 46	00 ♉ 04	25 00	21 40	05 54	24 49	08 36	13 52
07	05 03 49	16 47 53	03 ♑ 25 48	09 ♑ 36 23	29 34	01 01	24 43	21 54	06 02	24 48	08 35	13 50
08	05 07 46	17 45 16	15 ♑ 43 46	21 ♑ 48 14	29 18	01 58	24 25	22 07	06 10	24 48	08 34	13 49
09	05 11 42	18 42 38	27 ♑ 50 07	03 ♒ 49 46	28 58	02 56	24 06	22 21	06 17	24 47	08 33	13 47
10	05 15 39	19 39 59	09 ♒ 47 38	15 ♒ 44 09	28 35	03 54	23 48	22 35	06 25	24 47	08 32	13 46
11	05 19 35	20 37 20	21 ♒ 39 49	27 ♒ 35 09	28 09	04 53	23 29	22 49	06 33	24 46	08 31	13 44
12	05 23 32	21 34 40	03 ♓ 30 43	09 ♓ 27 04	27 40	05 52	23 10	23 03	06 40	24 46	08 30	13 42
13	05 27 28	22 32 00	15 ♓ 24 47	21 ♓ 24 28	27 09	06 51	22 50	23 17	06 48	24 45	08 29	13 41
14	05 31 25	23 29 20	27 ♓ 26 41	03 ♈ 32 00	26 37	07 51	22 31	23 30	06 56	24 44	08 28	13 39
15	05 35 21	24 26 39	09 ♈ 40 58	15 ♈ 54 07	26 03	08 51	22 11	23 44	07 03	24 43	08 27	13 38
16	05 39 18	25 23 57	22 ♈ 11 54	28 ♈ 34 45	25 29	09 51	21 52	23 58	07 11	24 43	08 26	13 36
17	05 43 15	26 21 16	05 ♉ 03 00	11 ♉ 36 54	24 56	10 52	21 32	24 12	07 18	24 42	08 25	13 34
18	05 47 11	27 18 34	18 ♉ 16 38	25 ♉ 02 14	24 22	11 53	21 13	24 26	07 26	24 41	08 24	13 33
19	05 51 08	28 15 51	01 ♊ 53 39	08 ♊ 50 41	23 50	12 54	20 53	24 40	07 33	24 40	08 23	13 31
20	05 55 04	29 ♊ 13 09	15 ♊ 53 00	23 ♊ 00 12	23 20	13 55	20 34	24 53	07 41	24 39	08 22	13 30
21	05 59 01	00 ♋ 10 26	00 ♋ 11 42	07 ♋ 26 50	22 52	14 57	20 15	25 07	07 48	24 38	08 20	13 28
22	06 02 57	01 07 43	14 ♋ 44 54	22 ♋ 05 04	22 27	15 59	19 56	25 21	07 55	24 37	08 19	13 27
23	06 06 54	02 04 59	29 ♋ 26 30	06 ♌ 48 22	22 06	17 02	19 38	25 35	08 03	24 36	08 18	13 25
24	06 10 50	03 02 15	14 ♌ 09 52	21 ♌ 30 12	21 47	18 04	19 20	25 49	08 10	24 35	08 17	13 24
25	06 14 47	03 59 30	28 ♌ 48 40	06 ♍ 04 39	21 33	19 07	19 02	26 02	08 17	24 33	08 16	13 22
26	06 18 44	04 56 44	13 ♍ 17 36	20 ♍ 27 05	21 23	20 10	18 44	26 16	08 25	24 32	08 14	13 21
27	06 22 40	05 53 58	27 ♍ 32 46	04 ♎ 34 24	21 17	21 13	18 28	26 30	08 32	24 31	08 13	13 19
28	06 26 37	06 51 11	11 ♎ 31 50	18 ♎ 25 00	21 D 16	22 17	18 11	26 44	08 39	24 29	08 12	13 18
29	06 30 33	07 48 24	25 ♎ 13 51	01 ♏ 58 27	21 20	23 21	17 55	26 57	08 46	24 28	08 10	13 16
30	06 34 30	08 ♋ 45 36	08 ♏ 38 53	15 ♏ 15 14	21 ♊ 28	24 ♉ 25	17 ♐ 40	27 ♊ 11	08 ♊ 53	24 ♒ 27	08 ♒ 09	13 ♐ 15

Date	Moon True ☊ ° '	Moon Mean ☊ ° '	Moon ☽ Latitude ° '
01	06 ♋ 36	07 ♋ 40	05 N 11
02	06 R 33	07 37	04 51
03	06 29	07 34	04 16
04	06 26	07 31	03 28
05	06 23	07 27	02 29
06	06 21	07 24	01 24
07	06 21	07 21	00 N 16
08	06 D 21	07 18	00 S 52
09	06 22	07 15	01 56
10	06 23	07 11	02 54
11	06 25	07 08	03 45
12	06 26	07 05	04 26
13	06 26	07 02	04 56
14	06 R 26	06 59	05 13
15	06 26	06 56	05 16
16	06 25	06 52	05 05
17	06 24	06 49	04 38
18	06 23	06 46	03 56
19	06 22	06 43	03 00
20	06 21	06 40	01 51
21	06 21	06 37	00 S 34
22	06 D 21	06 33	00 N 46
23	06 21	06 30	02 05
24	06 21	06 27	03 15
25	06 22	06 24	04 12
26	06 22	06 21	04 52
27	06 22	06 17	05 14
28	06 22	06 14	05 16
29	06 22	06 11	05 00
30	06 ♋ 22	06 ♋ 08	04 N 28

DECLINATIONS

Date	Sun ☉ ° '	Moon ☽ ° '	Mercury ☿ ° '	Venus ♀ ° '	Mars ♂ ° '	Jupiter ♃ ° '	Saturn ♄ ° '	Uranus ♅ ° '	Neptune ♆ ° '	Pluto ⯓ ° '
01	22 N 06	01 S 00	23 N 52	07 N 51	25 S 55	22 N 46	19 N 31	13 S 56	17 S 57	11 S 49
02	22 14	06 21	23 38	08 08	25 58	22 47	19 32	13 56	17 57	11 49
03	22 21	11 19	23 22	08 24	26 02	22 48	19 34	13 56	17 58	11 49
04	22 28	15 41	23 07	08 41	26 05	22 49	19 35	13 56	17 58	11 49
05	22 35	19 12	22 50	08 58	26 08	22 50	19 36	13 56	17 58	11 49
06	22 41	21 44	22 33	09 15	26 11	22 51	19 38	13 56	17 58	11 49
07	22 47	23 08	22 16	09 32	26 14	22 52	19 39	13 57	17 58	11 49
08	22 52	23 22	21 59	09 50	26 17	22 53	19 41	13 57	17 59	11 48
09	22 58	22 29	21 42	10 07	26 20	22 54	19 42	13 57	17 59	11 48
10	23 02	20 36	21 25	10 25	26 22	22 54	19 43	13 57	17 59	11 48
11	23 06	17 50	21 07	10 42	26 25	22 55	19 44	13 57	17 59	11 48
12	23 10	14 21	20 50	11 00	26 28	22 56	19 46	13 58	18 00	11 48
13	23 14	10 17	20 34	11 17	26 30	22 57	19 47	13 58	18 00	11 48
14	23 17	05 48	20 18	11 36	26 32	22 58	19 48	13 58	18 00	11 48
15	23 19	01 S 01	20 03	11 53	26 34	22 58	19 50	13 58	18 00	11 48
16	23 22	03 N 56	19 48	12 11	26 36	22 59	19 51	13 59	18 01	11 48
17	23 23	08 50	19 35	12 29	26 38	23 00	19 52	13 59	18 01	11 48
18	23 25	13 29	19 23	12 47	26 40	23 01	19 53	13 59	18 01	11 48
19	23 26	17 36	19 12	13 04	26 41	23 01	19 55	14 00	18 02	11 48
20	23 26	20 51	19 02	13 22	26 43	23 02	19 56	14 00	18 02	11 48
21	23 26	22 52	18 53	13 40	26 44	23 02	19 57	14 01	18 02	11 48
22	23 26	23 24	18 47	13 57	26 45	23 03	19 58	14 01	18 02	11 48
23	23 25	22 18	18 41	14 14	26 46	23 04	19 59	14 01	18 03	11 48
24	23 24	19 41	18 38	14 32	26 47	23 04	20 00	14 02	18 03	11 48
25	23 23	15 49	18 36	14 49	26 48	23 05	20 02	14 02	18 03	11 48
26	23 21	11 03	18 35	15 06	26 49	23 05	20 03	14 03	18 04	11 48
27	23 18	05 46	18 37	15 22	26 49	23 06	20 04	14 03	18 04	11 48
28	23 16	00 N 17	18 39	15 39	26 50	23 06	20 05	14 04	18 04	11 49
29	23 13	05 S 06	18 44	15 55	26 50	23 06	20 06	14 04	18 05	11 49
30	23 N 09	10 S 09	18 N 49	16 N 11	26 S 51	23 N 07	20 N 07	14 S 05	18 S 05	11 S 49

ZODIAC SIGN ENTRIES

Date	h	m	Planets
02	14	56	☽ ♏
04	20	58	☽ ♐
06	10	25	♀ ♉
07	05	23	☽ ♑
09	16	20	☽ ♒
12	04	53	☽ ♓
14	17	03	☽ ♈
17	02	39	☽ ♉
19	08	42	☽ ♊
21	07	38	☉ ♋
21	11	41	☽ ♋
23	12	55	☽ ♌
25	13	57	☽ ♍
27	16	11	☽ ♎
29	20	28	☽ ♏

LATITUDES

Date	Mercury ☿ ° '	Venus ♀ ° '	Mars ♂ ° '	Jupiter ♃ ° '	Saturn ♄ ° '	Uranus ♅ ° '	Neptune ♆ ° '	Pluto ⯓ ° '
01	00 N 26	02 S 08	02 S 32	00 S 20	01 S 41	00 S 44	00 N 09	10 N 44
04	00 S 20	02 18	02 43	00 20	01 41	00 44	00 09	10 44
07	01 10	02 26	02 54	00 20	01 41	00 44	00 09	10 43
10	02 01	02 33	03 05	00 19	01 41	00 44	00 09	10 43
13	02 51	02 39	03 15	00 19	01 41	00 44	00 09	10 43
16	03 33	02 43	03 25	00 19	01 41	00 44	00 09	10 42
19	04 06	02 46	03 34	00 19	01 41	00 44	00 09	10 42
22	04 27	02 47	03 43	00 18	01 41	00 44	00 09	10 41
25	04 35	02 48	03 50	00 18	01 41	00 45	00 09	10 40
28	04 30	02 47	03 56	00 18	01 41	00 45	00 09	10 40
31	04 S 15	02 S 45	04 S 01	00 S 18	01 S 41	00 S 45	00 N 09	10 N 39

DATA

Julian Date	2452062
Delta T	+64 seconds
Ayanamsa	23° 52' 19"
Synetic vernal point	05° ♓ 14' 40"
True obliquity of ecliptic	23° 26' 19"

MOON'S PHASES, APSIDES AND POSITIONS ☽

Date	h	m	Phase	Longitude ° '	Eclipse Indicator
06	01	39	○	15 ♐ 26	
14	03	28	☾	23 ♓ 09	
21	11	58	●	00 ♋ 10	Total
28	03	20	☽	06 ♎ 31	

Day	h	m		
11	19	42	Apogee	
23	17	12	Perigee	
01	07	38	0S	
08	04	59	Max dec	23° S 25'
15	16	56	0N	
22	07	49	Max dec	23° N 25'
28	13	16	0S	

LONGITUDES

Date	Chiron ⚷ ° '	Ceres ⚳ ° '	Pallas ⚴ ° '	Juno ⚵ ° '	Vesta ⚶ ° '	Black Moon Lilith ⚸ ° '
01	26 ♐ 54	22 ♑ 05	05 ♐ 04	28 ♉ 59	06 ♉ 45	20 ♒ 50
11	26 ♐ 14	20 ♑ 51	02 ♐ 17	04 ♊ 55	10 ♉ 59	21 ♒ 57
21	25 ♐ 34	19 ♑ 08	29 ♏ 57	10 ♊ 50	15 ♉ 06	23 ♒ 03
31	24 ♐ 54	17 ♑ 04	28 ♏ 18	16 ♊ 43	19 ♉ 04	24 ♒ 10

ASPECTARIAN

h m Aspects

01 Friday
01 37 ☽ △ ♆
03 39 ☽ ♇ ♅
05 19 ☽ △ ☉
10 54 ☽ ✱ ⯓
11 30 ☽ Q ♂
21 54 ☽ ♇ ♄
22 25 ☽ △ ♃

02 Saturday
05 49 ☽ △ ♅
07 03 ♀ △ ♂
08 07 ☽ ✱ ♂
08 13 ☽ ☍ ♀
09 29 ☽ ♇ ☉
13 07 ☽ ∠ ⯓
13 52 ☽ ± ♄
14 42 ☽ △ ☿
20 49 ☽ # ♀

03 Sunday
00 37 ☽ ⊼ ♄
01 20 ☽ ♇ ♃
02 38 ☽ ± ☉
05 04 ☽ ⊥ ⯓
06 16 ☽ □ ♆
10 15 ☽ ∠ ♂
14 11 ☽ ⊼ ☉
14 33 ☽ ∥ ⯓
15 47 ☽ ⊻ ⯓
17 36 ☽ ♇ ☿
17 44 ☽ ± ♃

04 Monday
01 54 ☽ ∥ ♅
02 08 ☽ ⊥ ♂
04 46 ☽ ⊼ ♃
05 22 ☿ St R
09 55 ☽ ± ☿
11 29 ☽ □ ♅
11 51 ☉ ☍ ⯓
12 49 ☽ ⊻ ♂
14 46 ☽ Q ♆
18 04 ☽ ⊼ ♀
20 52 ☽ ⊼ ☿

05 Tuesday
02 47 ☽ ∥ ♆
05 59 ☽ ± ♀
06 23 ♀ ♇ ⯓
07 35 ☽ ☍ ♄
12 34 ☿ ∥ ♃
12 53 ☽ ✱ ♆
15 15 ☽ # ♄
20 45 ☽ Q ♅
22 45 ☽ ☌ ⯓

06 Wednesday
00 05 ☽ ♇ ♀
01 39 ☽ ☍ ☉
03 58 ☉ ∥ ☿
05 41 ☿ ✱ ♀
07 26 ♀ ⊥ ♄
13 28 ☽ ☍ ♃
17 06 ☽ ∠ ♆
19 26 ☽ ✱ ♅
19 38 ☽ ☌ ♂
21 51 ☽ # ☿

07 Thursday
03 13 ☽ # ☉
03 54 ♂ ✱ ♅
04 41 ☽ ☍ ☿
05 25 ☽ # ♃
06 57 ☽ △ ♀
10 21 ☽ ⊥ ♆
17 06 ☽ ⊼ ♄
21 59 ☽ ⊻ ♆

08 Friday
00 23 ☽ ∠ ♅
04 55 ☽ ± ♄
08 15 ☽ ⊻ ⯓
12 58 ☉ ∥ ♃
16 20 ☽ ⊼ ☉
18 03 ☽ ⊥ ♅
20 02 ☽ ⊥ ⯓
22 51 ☽ ♇ ♄

09 Saturday
00 53 ☽ ⊼ ♃
03 19 ☽ # ☉
04 15 ☽ # ♃
04 45 ☽ ⊻ ♂
05 14 ☽ ± ☉
05 56 ☽ ⊻ ♅
13 03 ☽ ± ♃
13 54 ☽ ∠ ⯓
14 12 ☽ ⊼ ☿
16 25 ☽ ⊥ ♂
23 06 ☽ □ ♀

10 Sunday
00 47 ☽ ♇ ☉
01 34 ☽ # ☿
01 52 ☽ ± ☿
05 08 ☽ △ ♄
07 28 ☽ ♇ ♃
09 28 ☽ ☌ ♆
10 02 ☽ ∠ ♂
19 24 ☽ ♇ ☿
19 59 ☽ ✱ ⯓
20 20 ☽ # ♄

11 Monday
08 51 ♂ ∠ ♆
09 42 ☽ △ ☉
10 48 ☽ ∥ ♆
14 23 ☽ △ ♃
14 41 ☽ Q ♀
15 35 ☽ ✱ ♂
18 18 ☽ ☌ ♅
20 14 ☽ Q ⯓

12 Tuesday
00 39 ☽ △ ☿
14 25 ☽ ∥ ♅
15 15 ☽ Q ♂
17 01 ♂ ☍ ♃
17 11 ☽ ✱ ♀
18 28 ☽ □ ♄
22 04 ☽ ⊻ ♆

13 Wednesday
03 22 ☽ ∥ ⯓
06 40 ☽ # ♀
08 32 ☽ □ ⯓
09 35 ♀ Q ♅
10 09 ☽ ⊥ ♆
10 34 ♀ ⊻ ♄
17 46 ☉ ☍ ♂

14 Thursday
02 03 ☽ ∠ ♀
02 28 ☽ □ ♂
03 28 ☽ □ ☉
04 02 ☽ □ ♃
04 07 ☽ ∠ ♆
06 02 ♀ ♇ ♂
06 38 ☽ ⊻ ♅
06 57 ☽ Q ♄
07 30 ♀ ± ⯓
08 20 ♃ ♇ ♆
10 26 ☽ □ ☿
11 32 ☉ ♇ ♆
12 38 ☉ ☌ ♃
18 30 ☽ ⊥ ♅
21 27 ☽ ⊥ ♀

15 Friday
02 45 ♀ □ ♆
04 56 ♀ # ⯓
06 50 ☽ ✱ ♄
08 40 ♀ ∠ ♃
09 37 ☽ ✱ ♆
10 14 ☽ ⊻ ♀
12 05 ☽ ∠ ♅
16 03 ☽ Q ♃
17 47 ☽ Q ☉
18 57 ☉ △ ♅
19 37 ☽ △ ⯓
20 05 ☽ Q ☿

16 Saturday
08 06 ☿ ∥ ♄
08 40 ☽ Q ♆
11 23 ☽ △ ♂
11 58 ☽ ∠ ♄
13 26 ☉ ☌ ☿
15 25 ☽ ✱ ♃
16 44 ☽ ✱ ♅
17 57 ☽ ✱ ☿
18 32 ☽ ✱ ☉
21 44 ☿ ∠ ♀

17 Sunday
00 01 ☽ ♇ ⯓
05 00 ☽ ⊥ ♄
14 40 ☽ ♇ ♂
15 01 ☽ Q ♅
16 11 ☽ ⊻ ♄
16 38 ☽ ± ⯓
18 10 ☽ □ ♆
19 44 ☽ ∠ ♃
20 34 ☽ ∠ ☿
22 08 ☿ △ ♅
23 31 ☽ ☌ ♀

18 Monday
00 25 ☽ ∠ ☉
03 04 ☽ # ⯓
03 31 ☽ ⊼ ⯓
06 38 ☽ ± ♂
07 56 ☽ ∥ ♀
10 12 ☿ ☌ ♃
12 10 ☽ ⊥ ☿
12 17 ☽ ⊥ ♃
14 44 ☽ # ♅
17 07 ☽ ⊼ ♂
17 49 ☽ ⊥ ☉
22 25 ☽ ⊻ ☿
23 07 ☽ ⊻ ♃
23 22 ☽ □ ♅

19 Tuesday
05 12 ☽ ⊻ ☉
12 25 ♃ △ ♅
14 44 ☽ # ♆
21 52 ☽ ☌ ♄
22 16 ☽ ∥ ☿
23 11 ☽ △ ♆

20 Wednesday
02 15 ♀ ⊼ ⯓
04 16 ☽ ∥ ♄
07 57 ☽ ☍ ⯓
08 25 ☽ ⊻ ♀
10 49 ☿ ♇ ♆
19 21 ☽ ⊥ ♀
19 44 ☽ ☍ ♂
23 30 ♀ ± ♂

21 Thursday
00 09 ☽ ☌ ☿
00 35 ☽ ♇ ♆
02 44 ☽ △ ♅
03 25 ☽ ☌ ♃
11 34 ☽ ∠ ♀
11 58 ☽ ☌ ☉
15 27 ☽ ∥ ♃
15 33 ☽ ± ♆

22 Friday
00 41 ☽ ⊻ ♄
01 27 ☽ ⊼ ♆
03 35 ☽ ♇ ♅
09 52 ☽ ⊼ ⯓
10 38 ☽ ⊥ ♄
14 11 ☽ ✱ ♀
17 33 ♀ # ♅
18 19 ☽ ± ♅
19 41 ☽ ± ⯓
19 54 ☿ ⊥ ♀
20 19 ☽ ⊼ ♂
23 48 ☽ ∥ ♃

23 Saturday
00 18 ☽ ⊻ ☿
01 29 ☽ ∠ ♄
04 06 ☽ ⊼ ♅
05 36 ☽ ⊻ ♃
05 55 ☽ ± ♂
09 51 ☽ ⊥ ☿
10 20 ☽ ♇ ⯓
11 16 ☽ Q ♀
15 33 ☽ ⊥ ♃
16 36 ☽ ⊻ ☉
17 22 ☉ ± ♆
20 17 ☽ ♇ ♂

24 Sunday
00 12 ☽ ∠ ☿
02 08 ☽ ✱ ♄
02 25 ☽ ☍ ♆
03 03 ☉ ∠ ♀
03 03 ☽ ⊥ ☉
06 27 ☽ ∠ ♃
09 38 ☽ ∥ ♄
10 45 ☽ △ ⯓
18 46 ☽ ∠ ☉
18 52 ☽ □ ♀
19 21 ☽ ∥ ☿
20 16 ☽ △ ♂
21 54 ☽ Q ♄
22 57 ☽ # ♆

25 Monday
00 15 ☽ ✱ ☿
05 01 ☽ ☍ ♅
07 15 ♄ △ ♆
07 22 ☽ ✱ ♃
10 26 ♀ ⊼ ♂
17 05 ☽ ∥ ♀
19 43 ☽ Q ☿
21 09 ☽ ✱ ☉
21 24 ☽ # ♅

26 Tuesday
03 30 ☽ Q ♃
03 36 ☽ ⊼ ♆
03 48 ☽ □ ♄
08 26 ☽ # ⯓
12 05 ☽ □ ⯓
13 35 ☽ ± ♆
14 57 ♀ ⊥ ♃
18 33 ☽ Q ☉
20 57 ☽ □ ♂

27 Wednesday
00 27 ☽ △ ♀
01 28 ☽ □ ☿
04 41 ☽ ♇ ♆
06 52 ☽ ⊼ ♅
10 12 ☽ □ ♃
13 22 ☿ ⊻ ♀
17 03 ☽ ± ♅
18 25 ☽ Q ⯓

28 Thursday
02 57 ☽ Q ♂
03 20 ☽ □ ☉
04 03 ☽ ♇ ♀
05 49 ☿ St D
06 14 ☽ △ ♆
06 58 ☽ △ ♄
08 28 ☽ ♇ ♅
15 04 ☽ ✱ ⯓
20 57 ☽ ± ♀
23 22 ☽ ✱ ♂

29 Friday
05 04 ☽ △ ☿
08 23 ☽ ⊼ ♀
09 23 ☽ ♇ ♄
10 39 ☽ △ ♅
15 07 ☽ △ ♃
17 23 ☽ ∠ ⯓
21 00 ☉ ⊼ ♆

30 Saturday
01 26 ☽ ∠ ♂
01 32 ☽ ± ♄
08 01 ☽ ♇ ☿
09 29 ☽ ⊥ ⯓
11 06 ☽ □ ♆
12 13 ☽ △ ☉
12 26 ☽ ⊼ ♄
12 46 ♀ □ ♅
15 38 ☉ ⊻ ♄
17 22 ☽ ⊥ ♂
18 31 ☽ ♇ ♃
20 20 ☽ ⊻ ⯓
20 32 ☽ ∥ ⯓
23 44 ☉ ∥ ♃

All ephemeris data is given at 12.00 UT and the Moon's longitude is additionally given for 24.00 UT

JULY 2001

LONGITUDES

Date	Sidereal time h m s	Sun ☉ ° ' "	Moon ☽ ° ' "	Moon ☽ 24.00 ° ' "	Mercury ☿ ° '	Venus ♀ ° '	Mars ♂ ° '	Jupiter ♃ ° '	Saturn ♄ ° '	Uranus ♅ ° '	Neptune ♆ ° '	Pluto ♇ ° '
01	06 38 26	09 ♋ 42 48	21 ♏ 47 41	28 ♏ 16 21	21 ♊ 41	25 ♉ 29	17 ♐ 25	27 ♊ 25	09 ♊ 00	24 ♒ 25	08 ♒ 08	13 ♐ 14
02	06 42 23	10 40 00	04 ♐ 41 24	11 ♐ 03 01	22 00	26 33	17 R 11	27 38	09 07	24 R 24	08 R 06	13 R 12
03	06 46 19	11 37 11	17 ♐ 21 21	23 ♐ 36 36	22 23	27 38	16 58	27 52	09 14	24 22	08 05	13 11
04	06 50 16	12 34 23	29 ♐ 48 56	05 ♑ 58 31	22 51	28 42	16 45	28 06	09 21	24 21	08 03	13 09
05	06 54 13	13 31 34	12 ♑ 05 34	18 ♑ 10 16	23 24	29 ♉ 47	16 33	28 19	09 28	24 19	08 02	13 08
06	06 58 09	14 28 45	24 ♑ 12 49	00 ♒ 13 28	24 02	00 ♊ 52	16 22	28 33	09 35	24 17	08 01	13 07
07	07 02 06	15 25 57	06 ♒ 12 26	12 ♒ 10 02	24 45	01 58	16 11	28 46	09 42	24 16	07 59	13 06
08	07 06 02	16 23 08	18 ♒ 06 31	24 ♒ 02 14	25 32	03 03	16 01	29 00	09 49	24 14	07 58	13 04
09	07 09 59	17 20 20	29 ♒ 57 32	05 ♓ 52 47	26 24	04 09	15 52	29 13	09 55	24 12	07 56	13 03
10	07 13 55	18 17 32	11 ♓ 48 25	17 ♓ 44 51	27 21	05 15	15 44	29 27	10 02	24 11	07 55	13 02
11	07 17 52	19 14 44	23 ♓ 42 35	29 ♓ 42 05	28 23	06 21	15 37	29 40	10 09	24 09	07 53	13 00
12	07 21 48	20 11 57	05 ♈ 43 53	11 ♈ 48 31	29 ♊ 29	07 27	15 30	29 ♊ 53	10 15	24 07	07 52	12 59
13	07 25 45	21 09 10	17 ♈ 56 30	24 ♈ 08 24	00 ♋ 39	08 33	15 24	00 ♋ 07	10 22	24 05	07 50	12 58
14	07 29 42	22 06 24	00 ♉ 24 44	06 ♉ 46 00	01 54	09 39	15 19	00 20	10 28	24 03	07 49	12 57
15	07 33 38	23 03 39	13 ♉ 12 40	19 ♉ 45 10	03 14	10 46	15 15	00 33	10 35	24 01	07 47	12 56
16	07 37 35	24 00 54	26 ♉ 23 51	03 ♊ 08 57	04 37	11 53	15 12	00 46	10 41	23 59	07 45	12 55
17	07 41 31	24 58 10	10 ♊ 00 38	16 ♊ 58 55	06 05	13 00	15 09	01 00	10 48	23 57	07 44	12 54
18	07 45 28	25 55 26	24 ♊ 03 39	01 ♋ 14 35	07 37	14 07	15 07	01 13	10 54	23 55	07 42	12 53
19	07 49 24	26 52 43	08 ♋ 31 13	15 ♋ 52 57	09 13	15 14	15 07	01 26	11 00	23 53	07 41	12 51
20	07 53 21	27 50 01	23 ♋ 18 58	00 ♌ 48 20	10 53	16 21	15 D 07	01 39	11 06	23 51	07 39	12 50
21	07 57 17	28 47 19	08 ♌ 20 00	15 ♌ 52 49	12 36	17 29	15 08	01 52	11 12	23 49	07 37	12 49
22	08 01 14	29 ♋ 44 38	23 ♌ 25 38	00 ♍ 57 15	14 23	18 37	15 09	02 05	11 18	23 47	07 36	12 48
23	08 05 11	00 ♌ 41 57	08 ♍ 26 36	15 ♍ 52 39	16 13	19 44	15 12	02 18	11 24	23 45	07 34	12 48
24	08 09 07	01 39 16	23 ♍ 14 32	00 ♎ 31 32	18 06	20 52	15 15	02 31	11 30	23 43	07 33	12 47
25	08 13 04	02 36 35	07 ♎ 43 05	14 ♎ 48 50	20 01	22 00	15 20	02 44	11 36	23 41	07 31	12 46
26	08 17 00	03 33 55	21 ♎ 48 32	28 ♎ 42 09	21 59	23 08	15 25	02 56	11 42	23 38	07 29	12 45
27	08 20 57	04 31 16	05 ♏ 29 44	12 ♏ 11 27	23 59	24 16	15 31	03 09	11 48	23 36	07 28	12 44
28	08 24 53	05 28 37	18 ♏ 47 36	25 ♏ 18 28	26 01	25 25	15 37	03 22	11 54	23 34	07 26	12 43
29	08 28 50	06 25 58	01 ♐ 44 28	08 ♐ 05 58	28 ♋ 04	26 33	15 45	03 34	11 59	23 32	07 24	12 42
30	08 32 46	07 23 20	14 ♐ 23 24	20 ♐ 37 10	00 ♌ 09	27 42	15 53	03 47	12 05	23 29	07 23	12 42
31	08 36 43	08 ♌ 20 42	26 ♐ 47 40	02 ♑ 55 19	02 ♌ 14	28 ♊ 50	16 ♐ 02	03 ♋ 59	12 ♊ 10	23 ♒ 27	07 ♒ 21	12 ♐ 41

Date	Moon True ☊ ° '	Moon Mean ☊ ° '	Moon ☽ Latitude ° '
01	06 ♋ 22	06 ♋ 05	03 N 43
02	06 D 23	06 02	02 47
03	06 23	05 58	01 44
04	06 23	05 55	00 N 36
05	06 R 23	05 52	00 S 32
06	06 23	05 49	01 37
07	06 22	05 46	02 38
08	06 21	05 42	03 31
09	06 19	05 39	04 15
10	06 18	05 36	04 48
11	06 16	05 33	05 09
12	06 15	05 30	05 16
13	06 15	05 27	05 10
14	06 D 15	05 23	04 49
15	06 16	05 20	04 13
16	06 17	05 17	03 23
17	06 18	05 14	02 20
18	06 19	05 11	01 S 07
19	06 R 19	05 08	00 N 12
20	06 19	05 04	01 32
21	06 17	05 01	02 47
22	06 14	04 58	03 51
23	06 11	04 55	04 38
24	06 08	04 52	05 06
25	06 06	04 48	05 14
26	06 04	04 45	05 02
27	06 D 03	04 42	04 34
28	06 04	04 39	03 51
29	06 05	04 36	02 57
30	06 07	04 33	01 56
31	06 ♋ 08	04 ♋ 29	00 N 51

DECLINATIONS

Date	Sun ☉ ° '	Moon ☽ ° '	Mercury ☿ ° '	Venus ♀ ° '	Mars ♂ ° '	Jupiter ♃ ° '	Saturn ♄ ° '	Uranus ♅ ° '	Neptune ♆ ° '	Pluto ♇ ° '
01	23 N 05	14 S 37	18 N 56	16 N 27	26 S 51	23 N 07	20 N 08	14 S 05	18 S 06	11 S 49
02	23 01	18 20	19 04	16 43	26 51	23 08	20 09	14 06	18 06	11 49
03	22 56	21 07	19 14	16 59	26 51	23 08	20 10	14 06	18 06	11 49
04	22 51	22 50	19 24	17 14	26 51	23 08	20 11	14 07	18 07	11 49
05	22 45	23 25	19 35	17 29	26 51	23 08	20 12	14 07	18 07	11 49
06	22 39	22 52	19 48	17 43	26 51	23 09	20 13	14 08	18 07	11 49
07	22 33	21 16	20 00	17 58	26 51	23 09	20 14	14 08	18 08	11 49
08	22 26	18 45	20 14	18 12	26 51	23 09	20 15	14 09	18 08	11 50
09	22 19	15 28	20 27	18 25	26 51	23 09	20 16	14 10	18 09	11 50
10	22 11	11 34	20 41	18 39	26 51	23 09	20 17	14 10	18 09	11 50
11	22 03	07 13	20 55	18 52	26 51	23 10	20 18	14 11	18 09	11 50
12	21 55	02 S 34	21 08	19 05	26 51	23 10	20 19	14 11	18 10	11 50
13	21 47	02 N 16	21 21	19 17	26 51	23 10	20 19	14 12	18 10	11 50
14	21 37	07 06	21 34	19 29	26 51	23 10	20 20	14 13	18 11	11 51
15	21 28	11 47	21 46	19 40	26 51	23 10	20 21	14 13	18 11	11 51
16	21 18	16 03	21 57	19 51	26 50	23 10	20 22	14 14	18 11	11 51
17	21 08	19 38	22 07	20 02	26 50	23 10	20 23	14 15	18 12	11 51
18	20 58	22 11	22 16	20 12	26 50	23 10	20 24	14 15	18 12	11 51
19	20 47	23 22	22 24	20 22	26 50	23 10	20 25	14 16	18 13	11 52
20	20 36	22 56	22 29	20 32	26 50	23 10	20 25	14 17	18 13	11 52
21	20 24	20 52	22 33	20 41	26 50	23 10	20 26	14 18	18 14	11 52
22	20 12	17 20	22 35	20 49	26 50	23 09	20 27	14 18	18 14	11 52
23	20 00	12 42	22 34	20 57	26 50	23 09	20 28	14 19	18 14	11 53
24	19 47	07 22	22 31	21 05	26 50	23 09	20 28	14 20	18 15	11 53
25	19 35	01 N 45	22 26	21 12	26 51	23 09	20 29	14 20	18 16	11 53
26	19 21	03 S 49	22 18	21 18	26 51	23 09	20 30	14 21	18 16	11 53
27	19 08	09 03	22 08	21 24	26 51	23 09	20 31	14 22	18 16	11 54
28	18 54	13 43	21 54	21 30	26 51	23 08	20 31	14 23	18 16	11 54
29	18 40	17 37	21 38	21 35	26 51	23 08	20 32	14 23	18 17	11 54
30	18 25	20 36	21 21	21 39	26 52	23 08	20 33	14 24	18 17	11 54
31	18 N 11	22 S 33	20 N 59	21 N 43	26 S 52	23 N 08	20 N 33	14 S 25	18 S 18	11 S 55

ZODIAC SIGN ENTRIES

Date	h	m	Planets	
02	03	13	☽	♐
04	12	21	☽	♑
05	16	44	♀	♊
06	23	33	☽	♒
09	12	05	☽	♓
12	00	36	☽	♈
12	22	47	☿	♋
13	00	03	♃	♋
14	11	13	☽	♉
16	18	26	☽	♊
18	21	56	☽	♋
20	22	43	☽	♌
22	18	26	☉	♌
22	22	29	☽	♍
24	23	08	☽	♎
27	02	17	☽	♏
29	08	44	☽	♐
30	10	18	☿	♌
31	18	16	☽	♑

LATITUDES

Date	Mercury ☿ ° '	Venus ♀ ° '	Mars ♂ ° '	Jupiter ♃ ° '	Saturn ♄ ° '	Uranus ♅ ° '	Neptune ♆ ° '	Pluto ♇ ° '
01	04 S 15	02 S 45	04 S 01	00 S 18	01 S 41	00 S 45	00 N 09	10 N 39
04	03 51	02 42	04 06	00 17	01 41	00 45	00 09	10 38
07	03 20	02 39	04 10	00 17	01 41	00 45	00 09	10 37
10	02 44	02 34	04 12	00 17	01 41	00 45	00 09	10 37
13	02 05	02 29	04 14	00 17	01 42	00 45	00 09	10 36
16	01 24	02 23	04 15	00 16	01 42	00 45	00 09	10 35
19	00 44	02 16	04 15	00 16	01 42	00 45	00 09	10 34
22	00 S 05	02 08	04 15	00 16	01 42	00 45	00 09	10 33
25	00 N 30	02 00	04 14	00 16	01 42	00 45	00 09	10 31
28	00 59	01 52	04 13	00 15	01 43	00 45	00 09	10 30
31	01 N 21	01 S 43	04 S 11	00 S 15	01 S 43	00 S 45	00 N 08	10 N 29

DATA

Julian Date	2452092
Delta T	+64 seconds
Ayanamsa	23° 52' 24"
Synetic vernal point	05° ♓ 14' 35"
True obliquity of ecliptic	23° 26' 19"

MOON'S PHASES, APSIDES AND POSITIONS ☽

Date	h	m	Phase	Longitude ° '	Eclipse Indicator
05	15	04	○	13 ♑ 39	partial
13	18	45	☾	21 ♈ 25	
20	19	44	●	28 ♋ 08	
27	10	08	☽	04 ♏ 27	

Day	h	m		
09	11	12	Apogee	
21	20	39	Perigee	
05	12	13	Max dec	23° S 25'
13	00	48	0N	
19	17	53	Max dec	23° N 25'
25	19	26	0S	

LONGITUDES

Date	Chiron ⚷ ° '	Ceres ⚳ ° '	Pallas ⚴ ° '	Juno ⚵ ° '	Vesta ⚶ ° '	Black Moon Lilith ⚸ ° '
01	24 ♐ 54	17 ♑ 04	28 ♏ 18	16 ♊ 43	19 ♉ 04	24 ♒ 10
11	24 ♐ 18	14 ♑ 53	27 ♏ 25	22 ♊ 33	22 ♉ 53	25 ♒ 17
21	23 ♐ 46	12 ♑ 47	27 ♏ 16	28 ♊ 20	26 ♉ 31	26 ♒ 24
31	23 ♐ 20	10 ♑ 59	27 ♏ 49	04 ♋ 02	29 ♉ 56	27 ♒ 31

All ephemeris data is given at 12.00 UT and the Moon's longitude is additionally given for 24.00 UT

ASPECTARIAN

h m	Aspects
01 Sunday	
00 35	☽ ± ☿
04 06	☽ ⊻ ♂
04 47	☉ bQ ♅
08 54	☽ ∥ ♅
11 17	☽ ± ♃
11 48	☽ ⊼ ☿
16 50	☽ □ ♅
17 49	☽ bQ ☉
19 25	☽ ☍ ♀
20 00	☽ Q ♆
22 35	☽ ⊼ ♃
02 Monday	
00 00	☽ # ♀
10 16	☽ ∥ ♆
11 57	☽ ± ☉
17 52	☽ # ☿
18 25	☽ ✶ ♆
20 26	☽ ☍ ♄
03 Tuesday	
00 11	☽ ⊼ ☉
02 31	☽ Q ♅
02 42	☽ # ♄
04 03	☽ ☌ ♇
11 16	☽ ☌ ♂
18 47	♀ ⊻ ♃
21 59	☽ ☍ ☿
22 58	☽ ∠ ♆
04 Wednesday	
01 26	☽ ✶ ♅
08 36	☽ ☍ ♃
09 38	☽ ⊼ ♀
12 14	☽ # ☉
16 21	☽ ⊥ ♆
19 33	☽ # ♃
21 06	☿ bQ ♆
22 26	☽ ± ♀
05 Thursday	
02 23	☉ ⊼ ♇
04 03	☽ ⊻ ♆
06 33	☽ ∠ ♅
06 48	☽ ⊼ ♄
14 03	☽ ⊻ ♇
15 04	☽ ☍ ☉
17 50	☽ bQ ♀
18 43	☽ ± ♄
20 39	☽ ⊻ ♂
06 Friday	
00 16	☽ ⊥ ♅
01 53	☽ ⊥ ♇
04 50	☽ # ♃
08 22	☽ ⊥ ♂
11 37	☽ ⊼ ☿
12 09	☽ ⊻ ♅
12 45	☽ bQ ♄
16 42	☽ # ☉
19 46	☽ ∠ ♇
20 42	☿ △ ♅
20 48	☽ ⊼ ♃
07 Saturday	
00 19	☽ ± ☿
02 04	☽ ∠ ♂
02 37	☽ △ ♀
09 03	☽ ± ♃
15 34	☽ ☌ ♆
19 06	☽ △ ♄
19 36	☽ bQ ☿
19 36	☉ ⊥ ♄
22 46	☽ # ♄
23 59	☽ # ☿
08 Sunday	
01 50	☽ ✶ ♇
03 32	☽ bQ ♃
04 10	☉ ⊼ ♂
05 48	♀ # ♆
07 50	☽ ✶ ♂
08 13	☽ ⊼ ☉
14 55	☿ ∥ ♄
16 06	☽ # ♀
16 51	☽ ∥ ♆
21 25	☽ ± ☉
09 Monday	
00 22	☽ ☌ ♅
02 04	☽ Q ♇
04 12	☽ △ ☿
07 49	☽ Q ♂
10 28	☽ △ ♃
17 15	☽ bQ ☉
20 22	☽ ∥ ♅
21 21	☽ □ ♀
10 Tuesday	
04 08	☽ ⊻ ♆
08 23	☽ □ ♄
09 09	☉ ± ♅
10 27	☽ ∥ ♇
14 28	☽ □ ♇
16 15	☽ ⊥ ♆
19 51	☽ □ ♂
11 Wednesday	
02 14	☽ △ ☉
06 09	☉ ± ♇
10 21	☽ ∠ ♆
12 52	☽ ⊻ ♅
13 24	☽ Q ♀
20 59	☽ Q ♄
22 17	☽ □ ☿
12 Thursday	
00 09	☽ □ ♃
00 51	☽ ⊥ ♅
15 44	☽ ✶ ♀
16 12	☽ ✶ ♆
18 41	☽ ∠ ♅
20 49	♀ △ ♆
21 02	☽ ✶ ♄
22 29	☿ ☌ ♃
13 Friday	
02 18	☽ △ ♇
07 05	☽ △ ♂
12 20	☽ Q ♃
13 33	☽ Q ☿
15 40	☽ Q ♆
17 43	☉ ± ♂
18 45	☽ □ ☉
23 52	☽ ✶ ♅
23 55	☽ ∠ ♀
14 Saturday	
02 29	☽ ∠ ♄
07 19	☽ bQ ♇
10 11	☿ ± ♆
11 49	☽ bQ ♂
11 51	☽ ✶ ♃
15 10	☽ ✶ ☿
15 41	☉ ∥ ☿
18 44	☽ ⊥ ♀
19 45	☽ ⊥ ♄
22 38	☽ Q ♅
15 Sunday	
00 19	☽ ± ♇
01 55	☽ □ ♆
04 41	☽ ± ♂
07 03	☽ ⊻ ♀
07 05	☽ ⊻ ♄
07 32	♀ ☌ ♄
07 42	☽ Q ☉
11 29	☽ ⊼ ♇
12 20	☽ # ♇
15 44	☽ ⊼ ♂
16 24	☽ ∠ ♃
22 18	☽ ∠ ☿
16 Monday	
01 22	☽ # ♅
07 23	☽ ✶ ☉
07 41	☽ □ ♅
09 02	☽ ⊥ ♃
11 20	☉ ⊼ ♅
16 27	☽ ⊥ ☿
19 56	☽ ⊻ ♃
17 Tuesday	
01 37	☽ # ♆
04 20	☽ ⊻ ☿
08 02	☽ △ ♆
09 47	♀ ☍ ♇
11 55	☽ ∠ ☉
13 22	☽ ☌ ♄
15 19	☽ ∥ ♀
16 59	☽ ☍ ♇
17 37	☽ ☌ ♀
18 00	☽ ∥ ♄
20 51	☽ ☍ ♂
23 52	☽ ∥ ☉
18 Wednesday	
04 30	☽ ⊥ ☉
09 43	☽ bQ ♆
11 17	☉ ∠ ♄
11 46	☽ △ ♅
13 09	☽ ∥ ☿
13 16	☿ ⊼ ♆
15 21	☽ ⊻ ☉
19 Thursday	
00 08	☽ ☌ ♃
00 44	☽ ± ♆
04 32	☽ ∥ ♃
07 14	☿ bQ ♅
09 19	♀ ☍ ♂
10 37	☽ ⊼ ♆
12 36	☽ bQ ♅
13 17	☽ ☌ ☿
16 05	☽ ⊻ ♄
18 24	♀ ∥ ♄
19 05	☽ ⊼ ♇
22 45	☽ ⊼ ♂
22 45	♂ St D
23 51	☽ ⊻ ♀
20 Friday	
01 55	☽ ⊥ ♄
03 13	☽ ± ♅
04 48	☽ ± ♇
07 08	☽ ∥ ♃
08 27	☽ ± ♂
10 20	☽ ⊥ ♀
12 09	♃ ± ♆
12 10	☉ bQ ♇
12 52	☽ ⊼ ♅
15 26	☿ ⊻ ♄
16 31	☽ ∠ ♄
16 39	☉ ∥ ♀
18 56	☽ ∥ ☿
19 15	☽ bQ ♇
19 44	☽ ☌ ☉
22 54	☽ bQ ♂
21 Saturday	
01 33	☽ ⊻ ♃
01 55	☽ ∠ ♀
08 01	☉ ∥ ♄
10 52	☽ ☍ ♆
11 15	☽ ⊥ ♃
13 32	☽ ∥ ♀
15 03	☿ ⊼ ♇
15 31	☽ ∥ ♄
16 03	☽ ∥ ☉
16 36	☽ ✶ ♄
19 08	☽ △ ♇
19 41	☽ ⊻ ☿
22 49	☽ △ ♂
22 Sunday	
01 46	☽ ∠ ♃
03 43	☽ ✶ ♀
06 30	☽ ⊥ ☿
06 41	☽ # ♆
11 49	☽ Q ♄
12 34	☽ ☍ ♅
22 28	☿ ⊼ ♂
22 45	☽ ⊻ ☉
22 45	☉ bQ ♂
22 47	☽ ∠ ☿
23 Monday	
00 23	☽ Q ♀
02 00	☽ ✶ ♃
04 09	☽ # ♅
09 00	☽ ⊥ ☉
10 36	☽ ⊼ ♆
15 51	☽ # ♇
16 48	☽ □ ♄
19 00	☽ □ ♇
20 15	☽ ± ♆
21 35	☽ Q ♃
22 56	☽ □ ♂
24 Tuesday	
00 31	☽ ∠ ☉
02 22	☽ ✶ ☿
04 11	☿ ⊥ ♄
07 17	☿ ± ♅
07 48	☽ □ ♀
10 51	☽ bQ ♆
12 46	☽ ⊼ ♅
20 31	☿ ± ♇
22 38	☽ ± ♅
25 Wednesday	
00 24	☽ Q ♇
01 00	☽ Q ☿
02 51	☽ ✶ ☉
03 32	☽ □ ♃
04 37	☽ Q ♂
11 40	☽ △ ♆
13 36	☽ bQ ♅
15 47	☉ ⊻ ♃
18 36	☽ △ ♄
20 30	☽ ✶ ♇
22 39	♀ bQ ♆
26 Thursday	
00 30	☽ Q ☉
00 57	☽ ✶ ♂
04 40	☿ ± ♂
12 22	☽ □ ☿
14 30	☽ △ ♀
15 10	☽ △ ♅
20 34	☽ bQ ♄
22 18	♀ △ ♅
22 19	☽ ∠ ♇
27 Friday	
03 06	☽ ∠ ♂
07 29	☿ ⊼ ♅
07 46	☽ △ ♃
10 08	☽ □ ☉
12 32	☽ ± ♄
14 12	☽ ⊥ ♇
15 30	☽ □ ♆
19 14	☽ ⊥ ♂
19 23	☽ bQ ♀
19 42	☿ ⊻ ♀
23 22	☽ ⊼ ♄
28 Saturday	
00 58	☽ ⊻ ♇
02 14	☽ ∥ ♇
06 10	☽ ⊻ ♂
11 12	☽ bQ ♃
13 14	☽ ± ♀
15 47	☽ ∥ ♅
20 45	☽ □ ♅
22 42	☿ ∠ ♄
29 Sunday	
00 13	☽ Q ♆
01 22	☽ ⊼ ♀
03 50	☽ △ ☿
04 05	☽ ± ♃
07 44	☿ bQ ♇
15 30	☽ ⊼ ♃
16 15	☿ ∥ ♀
16 46	☽ ∥ ♆
19 04	☽ # ☉
21 34	☽ △ ☉
22 40	☽ ✶ ♆
30 Monday	
06 28	☽ Q ♅
07 33	☽ ☍ ♄
08 45	☽ ☌ ♇
11 26	☽ # ♄
11 48	☉ ☍ ♆
13 45	☽ bQ ☿
14 54	☽ ☌ ♂
18 33	☽ # ☿
21 07	☿ bQ ♂
23 38	☽ # ♀
31 Tuesday	
00 50	☉ # ♆
03 23	☽ ∠ ♆
04 43	☽ bQ ☉
05 31	☽ ✶ ♅
10 41	☽ ± ☿
16 24	☽ ☍ ♀
20 54	☽ ⊥ ♆
23 47	☽ ± ☉

AUGUST 2001

LONGITUDES

Date	Sidereal time h m s	Sun ☉ ° ' "	Moon ☽ ° ' "	Moon ☽ 24.00 ° ' "	Mercury ☿ ° '	Venus ♀ ° '	Mars ♂ ° '	Jupiter ♃ ° '	Saturn ♄ ° '	Uranus ♅ ° '	Neptune ♆ ° '	Pluto ♇ ° '
01	08 40 40	09 ♌ 18 06	09 ♑ 00 27	15 ♑ 03 25	04 ♌ 19	29 ♊ 59	16 ♐ 12	04 ♋ 12	12 ♊ 16	23 ♒ 25	07 ♒ 20	12 ♐ 40
02	08 44 36	10 15 29	21 ♑ 04 32	27 ♑ 04 06	06 25	01 ♋ 08	16 22	04 24	12 21	23 R 23	07 R 18	12 R 40
03	08 48 33	11 12 54	03 ♒ 02 22	08 ♒ 59 36	08 31	02 17	16 34	04 37	12 26	23 20	07 16	12 39
04	08 52 29	12 10 20	14 ♒ 56 03	20 ♒ 51 55	10 36	03 26	16 46	04 49	12 31	23 18	07 15	12 38
05	08 56 26	13 07 46	26 ♒ 47 26	02 ♓ 42 51	12 40	04 35	16 58	05 01	12 36	23 16	07 13	12 38
06	09 00 22	14 05 14	08 ♓ 38 23	14 ♓ 34 19	14 44	05 45	17 12	05 13	12 41	23 13	07 11	12 37
07	09 04 19	15 02 42	20 ♓ 30 53	26 ♓ 28 26	16 47	06 54	17 26	05 25	12 46	23 11	07 10	12 37
08	09 08 15	16 00 12	02 ♈ 27 15	08 ♈ 27 42	18 49	08 04	17 41	05 37	12 51	23 09	07 08	12 36
09	09 12 12	16 57 43	14 ♈ 30 11	20 ♈ 35 07	20 50	09 13	17 56	05 49	12 56	23 06	07 07	12 36
10	09 16 09	17 55 15	26 ♈ 42 56	02 ♉ 54 07	22 49	10 23	18 12	06 01	13 01	23 04	07 05	12 35
11	09 20 05	18 52 49	09 ♉ 09 10	15 ♉ 28 34	24 47	11 33	18 29	06 13	13 05	23 01	07 03	12 35
12	09 24 02	19 50 24	21 ♉ 52 50	28 ♉ 22 27	26 44	12 43	18 46	06 25	13 10	22 59	07 02	12 34
13	09 27 58	20 48 00	04 ♊ 57 53	11 ♊ 39 30	28 ♌ 39	13 53	19 04	06 36	13 15	22 57	07 00	12 34
14	09 31 55	21 45 38	18 ♊ 27 41	25 ♊ 22 37	00 ♍ 33	15 03	19 22	06 48	13 19	22 54	06 59	12 34
15	09 35 51	22 43 16	02 ♋ 24 25	09 ♋ 33 02	02 25	16 13	19 42	07 00	13 23	22 52	06 57	12 33
16	09 39 48	23 40 59	16 ♋ 48 15	24 ♋ 09 36	04 16	17 24	20 01	07 11	13 28	22 49	06 56	12 33
17	09 43 44	24 38 42	01 ♌ 36 28	09 ♌ 08 00	06 06	18 34	20 22	07 22	13 32	22 47	06 54	12 33
18	09 47 41	25 36 26	16 ♌ 43 09	24 ♌ 20 44	07 54	19 45	20 42	07 34	13 36	22 45	06 53	12 33
19	09 51 38	26 34 11	01 ♍ 59 24	09 ♍ 37 47	09 40	20 55	21 04	07 45	13 40	22 42	06 51	12 33
20	09 55 34	27 31 58	17 ♍ 14 30	24 ♍ 48 12	11 26	22 06	21 26	07 56	13 44	22 40	06 50	12 33
21	09 59 31	28 29 46	02 ♎ 17 42	09 ♎ 41 57	13 10	23 17	21 48	08 07	13 47	22 38	06 48	12 32
22	10 03 27	29 ♌ 27 35	17 ♎ 00 09	24 ♎ 11 41	14 52	24 28	22 11	08 18	13 51	22 35	06 47	12 32
23	10 07 24	00 ♍ 25 25	01 ♏ 16 08	08 ♏ 13 21	16 33	25 39	22 35	08 29	13 55	22 33	06 45	12 32
24	10 11 20	01 23 17	15 ♏ 03 20	21 ♏ 46 16	18 13	26 50	22 59	08 39	13 58	22 30	06 44	12 D 32
25	10 15 17	02 21 09	28 ♏ 22 28	04 ♐ 52 20	19 51	28 01	23 24	08 50	14 02	22 28	06 42	12 32
26	10 19 13	03 19 03	11 ♐ 16 22	17 ♐ 35 07	21 28	29 ♋ 12	23 49	09 01	14 05	22 26	06 41	12 33
27	10 23 10	04 16 59	23 ♐ 49 10	29 ♐ 59 05	23 04	00 ♌ 23	24 14	09 11	14 08	22 23	06 39	12 33
28	10 27 07	05 14 55	06 ♑ 05 27	12 ♑ 08 49	24 38	01 35	24 40	09 21	14 11	22 21	06 38	12 33
29	10 31 03	06 12 53	18 ♑ 09 45	24 ♑ 08 42	26 11	02 46	25 06	09 32	14 14	22 19	06 37	12 33
30	10 35 00	07 10 52	00 ♒ 06 10	06 ♒ 02 33	27 43	03 57	25 33	09 42	14 17	22 16	06 35	12 33
31	10 38 56	08 ♍ 08 53	11 ♒ 58 13	17 ♒ 53 31	29 ♍ 13	05 ♌ 09	26 ♐ 00	09 ♋ 52	14 ♊ 20	22 ♒ 14	06 ♒ 34	12 ♐ 33

Date	Moon True ☊ ° '	Moon Mean ☊ ° '	Moon ☽ Latitude ° '
01	06 ♋ 09	04 ♋ 26	00 S 16
02	06 R 07	04 23	01 21
03	06 05	04 20	02 21
04	06 00	04 17	03 15
05	05 54	04 14	04 01
06	05 48	04 10	04 36
07	05 41	04 07	04 59
08	05 34	04 04	05 09
09	05 28	04 01	05 06
10	05 24	03 58	04 48
11	05 22	03 54	04 17
12	05 D 22	03 51	03 33
13	05 22	03 48	02 37
14	05 24	03 45	01 30
15	05 24	03 42	00 S 16
16	05 R 24	03 39	01 N 01
17	05 21	03 35	02 16
18	05 17	03 32	03 23
19	05 10	03 29	04 17
20	05 03	03 26	04 52
21	04 55	03 23	05 06
22	04 48	03 20	04 59
23	04 43	03 16	04 34
24	04 39	03 13	03 54
25	04 38	03 10	03 02
26	04 D 38	03 07	02 02
27	04 39	03 04	00 N 58
28	04 R 39	03 00	00 S 08
29	04 38	02 57	01 12
30	04 35	02 54	02 11
31	04 ♋ 30	02 ♋ 51	03 S 05

DECLINATIONS

Date	Sun ☉ ° '	Moon ☽ ° '	Mercury ☿ ° '	Venus ♀ ° '	Mars ♂ ° '	Jupiter ♃ ° '	Saturn ♄ ° '	Uranus ♅ ° '	Neptune ♆ ° '	Pluto ♇ ° '
01	17 N 56	23 S 24	20 N 35	21 N 46	26 S 53	23 N 07	20 N 34	14 S 26	18 S 18	11 S 55
02	17 40	23 07	20 09	21 49	26 53	23 07	20 34	14 27	18 19	11 55
03	17 25	21 46	19 41	21 51	26 53	23 07	20 35	14 28	18 19	11 56
04	17 09	19 28	19 11	21 53	26 54	23 06	20 36	14 28	18 19	11 56
05	16 53	16 21	18 39	21 54	26 54	23 06	20 36	14 29	18 20	11 56
06	16 36	12 35	18 05	21 54	26 55	23 05	20 37	14 30	18 20	11 57
07	16 19	08 20	17 29	21 54	26 55	23 05	20 37	14 30	18 21	11 57
08	16 02	03 S 45	16 52	21 54	26 56	23 05	20 38	14 31	18 21	11 57
09	15 45	01 N 01	16 13	21 52	26 56	23 04	20 38	14 32	18 22	11 58
10	15 28	05 49	15 34	21 51	26 57	23 04	20 39	14 33	18 22	11 58
11	15 10	10 28	14 53	21 48	26 57	23 03	20 39	14 33	18 22	11 58
12	14 52	14 48	14 12	21 45	26 58	23 03	20 40	14 34	18 23	11 59
13	14 34	18 33	13 29	21 42	26 58	23 02	20 40	14 35	18 23	11 59
14	14 15	21 26	12 46	21 37	26 59	23 02	20 41	14 36	18 24	11 59
15	13 56	23 09	12 03	21 33	26 59	23 01	20 41	14 37	18 24	12 00
16	13 38	23 24	11 19	21 27	27 00	23 01	20 41	14 37	18 24	12 00
17	13 18	22 01	10 34	21 21	27 00	23 00	20 42	14 38	18 25	12 01
18	12 59	19 03	09 49	21 15	27 00	23 00	20 42	14 39	18 25	12 01
19	12 40	14 45	09 04	21 07	27 01	22 59	20 42	14 40	18 26	12 01
20	12 20	09 31	08 19	21 00	27 01	22 58	20 43	14 41	18 26	12 02
21	12 00	03 N 46	07 34	20 51	27 01	22 58	20 43	14 41	18 26	12 02
22	11 40	02 S 04	06 49	20 42	27 01	22 57	20 44	14 42	18 27	12 03
23	11 19	07 38	06 04	20 33	27 01	22 57	20 44	14 43	18 27	12 03
24	10 59	12 38	05 19	20 23	27 02	22 56	20 44	14 44	18 28	12 04
25	10 38	16 51	04 34	20 12	27 02	22 55	20 44	14 44	18 28	12 04
26	10 17	20 07	03 49	20 01	27 02	22 55	20 45	14 45	18 28	12 04
27	09 56	22 20	03 04	19 49	27 01	22 54	20 45	14 46	18 29	12 05
28	09 35	23 26	02 20	19 37	27 01	22 54	20 45	14 47	18 29	12 05
29	09 14	23 23	01 36	19 24	27 01	22 53	20 45	14 47	18 29	12 06
30	08 52	22 16	00 53	19 10	27 01	22 52	20 46	14 48	18 30	12 06
31	08 N 31	20 S 10	00 N 09	18 N 56	27 S 00	22 N 52	20 N 46	14 S 49	18 S 30	12 S 07

ZODIAC SIGN ENTRIES

Date	h	m	Planets	
01	12	18	♀	♋
03	05	53	☽	♒
05	18	30	☽	♓
08	07	05	☽	♈
10	18	23	☽	♉
13	02	59	☽	♊
14	05	04	☿	♍
15	07	55	☽	♋
17	09	25	☽	♌
19	08	53	☽	♍
21	08	19	☽	♎
23	01	27	☉	♍
23	09	50	☽	♏
25	14	59	☽	♐
27	04	12	♀	♌
28	00	02	☽	♑
30	11	48	☽	♒

LATITUDES

Date	Mercury ☿ ° '	Venus ♀ ° '	Mars ♂ ° '	Jupiter ♃ ° '	Saturn ♄ ° '	Uranus ♅ ° '	Neptune ♆ ° '	Pluto ♇ ° '
01	01 N 27	01 S 40	04 S 11	00 S 15	01 S 43	00 S 45	00 N 08	10 N 29
04	01 40	01 31	04 08	00 15	01 43	00 45	00 08	10 28
07	01 45	01 21	04 06	00 15	01 43	00 45	00 08	10 26
10	01 45	01 12	04 03	00 14	01 44	00 46	00 08	10 25
13	01 39	01 02	04 00	00 14	01 44	00 46	00 08	10 24
16	01 28	00 52	03 56	00 14	01 44	00 46	00 08	10 23
19	01 13	00 42	03 53	00 14	01 45	00 46	00 08	10 21
22	00 56	00 32	03 49	00 14	01 45	00 46	00 08	10 20
25	00 35	00 22	03 45	00 13	01 45	00 46	00 08	10 19
28	00 N 13	00 12	03 42	00 13	01 46	00 46	00 08	10 18
31	00 S 10	00 S 02	03 S 38	00 S 13	01 S 46	00 S 45	00 N 08	10 N 16

DATA

Julian Date	2452123
Delta T	+64 seconds
Ayanamsa	23° 52' 29"
Synetic vernal point	05° ♓ 14' 30"
True obliquity of ecliptic	23° 26' 20"

MOON'S PHASES, APSIDES AND POSITIONS ☽

Date	h	m	Phase	Longitude ° '	Eclipse Indicator
04	05	56	○	11 ♒ 56	
12	07	53	☾	19 ♉ 41	
19	02	55	●	26 ♌ 12	
25	19	55	☽	02 ♐ 40	

Day	h	m		
05	20	49	Apogee	
19	05	34	Perigee	
01	17	58	Max dec	23° S 26'
09	06	54	0N	
16	03	48	Max dec	23° N 29'
22	03	25	0S	
28	23	09	Max dec	23° S 33'

LONGITUDES

Date	Chiron ⚷ ° '	Ceres ⚳ ° '	Pallas ⚴ ° '	Juno ⚵ ° '	Vesta ⚶ ° '	Black Moon Lilith ° '
01	23 ♐ 18	10 ♑ 50	27 ♏ 54	04 ♋ 36	00 ♊ 16	27 ♒ 38
11	23 ♐ 01	09 ♑ 34	29 ♏ 07	10 ♋ 11	03 ♊ 14	28 ♒ 45
21	22 ♐ 52	08 ♑ 53	00 ♐ 50	15 ♋ 40	06 ♊ 14	29 ♒ 52
31	22 ♐ 52	08 ♑ 47	02 ♐ 59	21 ♋ 00	08 ♊ 42	00 ♓ 59

All ephemeris data is given at 12.00 UT and the Moon's longitude is additionally given for 24.00 UT

ASPECTARIAN

h m Aspects

01 Wednesday
00 18 ☽ # ♃
00 51 ☽ ⊼ ☿
02 21 ☽ ☍ ♃
08 41 ☽ ⊻ ♆
10 24 ☿ ⊻ ♃
10 50 ☽ ∠ ♅
12 38 ☽ ⊼ ☉
13 21 ☿ ∥ ♄
18 29 ☽ ⊼ ♄
19 15 ☽ ⊻ ♇
02 Thursday
02 29 ☽ ⊻ ♂
04 38 ☽ ⊥ ♅
06 31 ☽ ± ♄
07 11 ☽ ⊥ ♇
12 00 ☽ # ♃
14 38 ☽ ⊥ ♂
15 23 ♀ ± ♆
16 35 ☽ ⊻ ♅
16 43 ☉ ⊥ ♃
21 57 ☿ ☍ ♆
03 Friday
00 39 ☽ Ⴒ ♄
01 10 ☽ ∠ ♇
06 11 ☿ ⊥ ♀
08 59 ☽ ∠ ♂
10 19 ☽ ⊼ ♀
10 57 ☽ # ♀
15 13 ☽ ⊼ ♃
20 31 ☽ ☌ ♆
23 42 ☽ ± ♀
04 Saturday
01 23 ☽ ☍ ☿
01 28 ☽ # ♄
03 32 ☽ ± ♃
05 56 ☽ ☍ ☉
07 05 ☽ △ ♄
07 22 ☽ ⚹ ♇
14 47 ☿ ⊥ ♃
15 04 ☽ # ☿
15 46 ☽ ⚹ ♂
19 51 ☽ Ⴒ ♀
21 28 ☽ ∥ ♆
21 37 ☉ ⚹ ♄
22 03 ☽ Ⴒ ♃
23 33 ☉ △ ♇
05 Sunday
04 52 ☽ ☌ ♅
07 37 ☽ Q ♇
08 00 ☽ # ☉
11 11 ☿ ⚹ ♄
11 28 ☿ △ ♇
16 30 ☽ Q ♂
17 20 ♄ ☍ ♇
21 51 ☉ ☌ ☿
22 50 ♀ ☌ ♃
06 Monday
00 20 ☽ ∥ ♅
01 08 ☿ # ♆
04 58 ☽ △ ♃
05 30 ☽ △ ♀
09 04 ☽ ⊻ ♆
15 47 ☽ ∥ ♇
20 03 ☽ □ ♇
20 15 ☽ □ ♄
21 11 ☽ ⊥ ♆
23 59 ☽ ⊼ ☉
07 Tuesday
02 55 ☽ ⊼ ☿
05 39 ☽ □ ♂
13 10 ☽ ± ☉
15 19 ☽ ∠ ♆
17 18 ♀ ⊼ ♆
17 21 ☽ ⊻ ♅
17 31 ☽ ± ☿
20 38 ☿ △ ♂
08 Wednesday
05 23 ☽ ⊥ ♅
08 46 ☽ Q ♄
08 50 ☽ Ⴒ ☉
13 38 ♀ Ⴒ ♅
15 16 ☽ Ⴒ ☿
18 27 ☽ □ ♃
21 20 ☽ ⚹ ♆
23 20 ☽ ∠ ♅
09 Thursday
00 24 ☽ □ ♀
08 13 ☽ △ ♇
08 52 ☽ ⚹ ♄
11 58 ☿ ∠ ♃
17 17 ☽ △ ☉
18 56 ☽ △ ♂
21 05 ☽ Q ♆
10 Friday
02 54 ☽ △ ☿
04 53 ☽ ⚹ ♅
06 39 ☽ Q ♃
13 42 ☽ Ⴒ ♇
14 33 ☽ ∠ ♄
14 57 ☿ ☍ ♅
15 35 ☽ Q ♀
18 29 ☉ ∥ ☿
21 47 ☉ △ ♂
11 Saturday
00 52 ☽ Ⴒ ♂
04 07 ☽ Q ♅
06 17 ☽ ⚹ ♃
07 05 ☽ ± ♇
08 00 ☽ □ ♆
08 02 ☽ ⊥ ♄
15 57 ☿ Q ♄
17 02 ☽ ⚹ ♀
18 28 ☽ ± ♂
18 31 ☽ ⊼ ♇
19 32 ☽ ⊻ ♄
20 04 ☽ # ♇
23 07 ☿ # ♅
12 Sunday
06 03 ☽ ⊼ ♂
07 53 ☽ □ ☉
08 58 ☽ ∥ ☿
09 06 ♀ ⊼ ♇
10 40 ☽ # ♅
11 07 ☽ ∠ ♃
12 21 ☽ ∥ ☉
14 03 ☽ □ ♅
21 57 ♀ ⊻ ♄
22 32 ☽ □ ☿
23 51 ☽ ∠ ♀
13 Monday
03 58 ☽ ⊥ ♃
10 06 ☉ # ♅
10 50 ☽ # ♆
15 00 ☽ ⊻ ♃
15 40 ☽ △ ♆
17 45 ☽ ⊥ ♀
19 26 ☽ Q ☉
19 36 ☿ ∠ ♀
14 Tuesday
01 37 ☽ ☍ ♇
02 54 ☽ ☌ ♄
04 43 ☽ ∥ ♄
05 27 ☽ ⊻ ♀
12 10 ☽ Q ☿
13 14 ☉ ∠ ♃
13 38 ☽ ☍ ♂
13 51 ☽ ∥ ♀
18 07 ☽ Ⴒ ♆
18 11 ☽ ⚹ ☉
19 43 ☽ △ ♅
15 Wednesday
07 41 ♃ ⊼ ♆
09 16 ☽ ∥ ♃
09 32 ☽ ± ♆
12 02 ☽ ⚹ ☿
13 29 ☿ # ♇
15 25 ☉ ☍ ♅
19 39 ☽ ⊼ ♆
19 50 ☽ ☌ ♃
21 10 ☽ Ⴒ ♅
21 35 ☽ ∠ ☉
16 Thursday
00 43 ♀ ± ♅
04 59 ☽ ⊼ ♇
06 28 ☽ ⊻ ♄
12 02 ☽ ± ♅
13 03 ☽ ☌ ♀
13 33 ☽ ⊥ ☉
14 52 ☽ ± ♇
16 22 ☽ ⊥ ♄
16 37 ☽ ∠ ☿
17 23 ☽ ⊼ ♂
21 48 ☽ ⊼ ♅
22 13 ☽ ∥ ♃
17 Friday
00 00 ☽ ⊻ ☉
03 22 ☽ ± ♂
05 29 ☽ Ⴒ ♇
07 02 ☽ ∠ ♄
09 14 ☽ ⊥ ☿
11 37 ♀ ± ♇
18 08 ☽ Ⴒ ♂
18 53 ☽ ∥ ♀
20 09 ☽ ⊻ ☿
20 26 ☽ ☍ ♆
20 38 ☉ ⊥ ♀
21 19 ☽ ⊻ ♃
22 33 ☿ ⊼ ♆
18 Saturday
00 12 ☽ ∥ ♄
05 25 ☽ △ ♇
06 57 ☽ ⊥ ♃
06 57 ☿ ⚹ ♃
07 03 ☽ ⚹ ♄
08 47 ♀ ⊥ ♄
11 41 ☉ Q ♄
16 01 ☽ # ♆
17 10 ☽ ⊻ ♀
18 26 ☽ △ ♂
21 18 ☽ ∠ ♃
21 28 ☽ ☍ ♅
19 Sunday
02 01 ☽ Q ♄
02 55 ☽ ☌ ☉
03 23 ☽ ⊥ ♀
07 39 ♃ Ⴒ ♅
12 28 ☽ # ♅
16 15 ♀ ⊼ ♂
18 41 ☽ ∠ ♀
19 37 ☽ ⊼ ♆
21 09 ☽ ⚹ ♃
22 41 ☽ ∥ ☉
20 Monday
00 54 ☽ # ♇
01 39 ☽ ☌ ☿
04 35 ☽ □ ♇
05 03 ☽ ± ♆
06 26 ☽ □ ♄
16 19 ☽ Q ♃
17 51 ☽ ∥ ☿
18 48 ☽ □ ♂
19 15 ☽ Ⴒ ♆
20 21 ☽ ⚹ ♀
20 34 ☽ ⊼ ♅
23 08 ♀ ⊼ ♅
21 Tuesday
03 24 ☿ □ ♇
05 29 ☽ ⊻ ☉
06 07 ☽ ± ♅
07 04 ☿ ± ♆
09 06 ☉ # ♇
09 11 ☽ Q ♇
11 40 ♂ ∠ ♆
15 48 ☽ ⊥ ☉
17 14 ☽ Q ♀
19 16 ☽ △ ♆
20 36 ☽ Ⴒ ♅
21 10 ☿ □ ♄
21 32 ☽ □ ♃
22 Wednesday
00 30 ☽ Q ♂
04 39 ☽ ⚹ ♇
06 47 ☽ △ ♄
07 30 ☽ ∠ ☉
08 00 ☽ ⊻ ☿
09 16 ♀ ∥ ♄
19 17 ☽ ⊥ ☿
20 53 ☽ ⚹ ♂
21 16 ☽ △ ♅
23 Thursday
01 34 ☽ □ ♀
05 39 ☽ ∠ ♇
05 53 ☽ # ☿
07 58 ☽ Ⴒ ♄
09 59 ♂ ⚹ ♅
10 27 ☽ ⚹ ☉
12 33 ☽ ∠ ☿
16 07 ♇ St D
21 05 ☽ ⊥ ♇
21 26 ☽ □ ♆
23 13 ☽ ∠ ♂
23 31 ☽ ± ♄
24 Friday
00 36 ☽ △ ♃
04 16 ☽ # ☉
07 33 ☽ ⊻ ♇
08 50 ☽ Q ☉
09 06 ☽ ∥ ♇
10 04 ☽ ⊼ ♄
15 32 ☽ ⊥ ♂
18 24 ☽ ⚹ ☿
23 23 ☽ ∥ ♅
25 Saturday
01 17 ☽ □ ♅
02 28 ♀ Ⴒ ♇
02 38 ☽ ⊻ ♂
03 37 ☽ Ⴒ ♃
05 19 ☽ Q ♆
11 16 ☽ △ ♀
19 19 ☽ Q ☿
19 55 ☽ □ ☉
20 20 ☽ ± ♃
22 59 ☽ ∥ ♆
26 Sunday
03 23 ☽ ⚹ ♆
04 19 ☿ Q ♃
07 41 ☽ ⊼ ♃
09 34 ♀ ∠ ♄
10 25 ☽ Q ♅
11 10 ☽ # ♀
14 24 ☽ ☌ ♇
15 08 ☿ Ⴒ ♆
17 21 ☽ ☍ ♄
17 43 ☽ # ♄
18 07 ☽ Ⴒ ♀
27 Monday
02 05 ☿ ⊼ ♅
07 13 ♀ ± ♂
07 50 ☽ ∠ ♆
09 15 ☽ ⚹ ♅
10 19 ☽ □ ☿
12 50 ☽ ☌ ♂
13 13 ☽ ± ♀
21 25 ☽ # ♃
28 Tuesday
01 17 ☽ ⊥ ♆
02 10 ☽ ⊼ ♀
10 12 ☽ △ ☉
12 42 ☿ □ ♂
13 04 ☽ ⊻ ♆
14 29 ☽ ∠ ♅
18 33 ☽ ☍ ♃
29 Wednesday
00 48 ☽ ⊻ ♇
04 08 ☽ ⊼ ♄
08 19 ☽ ⊥ ♅
12 46 ☽ ⊥ ♇
16 11 ☽ ± ♄
18 39 ☽ Ⴒ ☉
20 18 ☽ ⊻ ♅
21 36 ☉ ⊼ ♆
30 Thursday
01 44 ☽ # ♃
02 28 ☽ ⊻ ♂
06 28 ☽ △ ☿
06 51 ☽ ∠ ♇
10 21 ☽ Ⴒ ♄
14 22 ☽ ± ☉
15 03 ☽ ⊥ ♂
20 39 ☽ ☍ ♀
20 43 ☿ ± ♅
31 Friday
01 05 ☽ ☌ ♆
03 34 ☽ ⊼ ☉
06 07 ☽ # ♄
07 40 ☽ ⊼ ♃
09 58 ☽ ∠ ♂
13 11 ☽ ⚹ ♇
16 49 ☽ △ ♄
17 13 ☽ Ⴒ ☿
20 00 ☽ ± ♃
23 39 ☽ # ♀

SEPTEMBER 2001

LONGITUDES

Date	Sidereal time h m s	Sun ☉ ° ' "	Moon ☽ ° ' "	Moon ☽ 24.00 ° ' "	Mercury ☿ ° '	Venus ♀ ° '	Mars ♂ ° '	Jupiter ♃ ° '	Saturn ♄ ° '	Uranus ♅ ° '	Neptune ♆ ° '	Pluto ♇ ° '
01	10 42 53	09 ♍ 06 55	23 ♒ 48 45	29 ♒ 44 09	00 ♎ 42	06 ♌ 21	26 ♐ 28	10 ♋ 02	14 ♊ 23	22 ♒ 12	06 ♒ 33	12 ♐ 34
02	10 46 49	10 04 59	05 ♓ 39 57	11 ♓ 36 21	02 10	07 32	26 56	10 12	14 26	22 R 10	06 R 31	12 34
03	10 50 46	11 03 04	17 ♓ 33 33	23 ♓ 31 42	03 36	08 44	27 25	10 21	14 28	22 07	06 30	12 34
04	10 54 42	12 01 11	29 ♓ 30 58	05 ♈ 31 32	05 01	09 56	27 54	10 31	14 31	22 05	06 29	12 35
05	10 58 39	12 59 19	11 ♈ 33 36	17 ♈ 37 21	06 25	11 08	28 23	10 40	14 33	22 03	06 28	12 35
06	11 02 36	13 57 30	23 ♈ 43 02	29 ♈ 50 53	07 47	12 20	28 52	10 50	14 35	22 01	06 26	12 36
07	11 06 32	14 55 42	06 ♉ 01 13	12 ♉ 14 22	09 07	13 32	29 22	10 59	14 37	21 59	06 25	12 36
08	11 10 29	15 53 57	18 ♉ 30 42	24 ♉ 50 36	10 27	14 45	29 ♐ 53	11 08	14 40	21 57	06 24	12 37
09	11 14 25	16 52 13	01 ♊ 14 30	07 ♊ 42 49	11 44	15 57	00 ♑ 23	11 17	14 41	21 54	06 23	12 37
10	11 18 22	17 50 32	14 ♊ 16 01	20 ♊ 54 30	13 00	17 09	00 54	11 26	14 43	21 52	06 22	12 38
11	11 22 18	18 48 53	27 ♊ 38 39	04 ♋ 28 49	14 15	18 22	01 26	11 35	14 45	21 50	06 21	12 38
12	11 26 15	19 47 16	11 ♋ 25 14	18 ♋ 28 01	15 28	19 34	01 57	11 44	14 47	21 48	06 20	12 39
13	11 30 11	20 45 41	25 ♋ 37 10	02 ♌ 52 30	16 39	20 47	02 29	11 52	14 48	21 46	06 18	12 40
14	11 34 08	21 44 08	10 ♌ 13 36	17 ♌ 39 55	17 48	22 00	03 02	12 01	14 50	21 44	06 17	12 40
15	11 38 05	22 42 37	25 ♌ 10 36	02 ♍ 44 39	18 55	23 12	03 34	12 09	14 51	21 42	06 16	12 41
16	11 42 01	23 41 08	10 ♍ 20 54	17 ♍ 58 00	20 00	24 25	04 07	12 17	14 52	21 40	06 15	12 42
17	11 45 58	24 39 41	25 ♍ 34 36	03 ♎ 09 16	21 02	25 38	04 40	12 25	14 53	21 39	06 14	12 43
18	11 49 54	25 38 16	10 ♎ 40 41	18 ♎ 07 38	22 03	26 51	05 14	12 33	14 54	21 37	06 14	12 43
19	11 53 51	26 36 53	25 ♎ 29 04	02 ♏ 44 10	23 00	28 04	05 48	12 41	14 55	21 35	06 13	12 44
20	11 57 47	27 35 32	09 ♏ 52 17	16 ♏ 53 05	23 55	29 ♌ 17	06 22	12 49	14 56	21 33	06 12	12 45
21	12 01 44	28 34 12	23 ♏ 46 23	00 ♐ 32 13	24 47	00 ♍ 30	06 56	12 56	14 57	21 31	06 11	12 46
22	12 05 40	29 ♍ 32 54	07 ♐ 10 49	13 ♐ 42 32	25 36	01 43	07 31	13 03	14 57	21 30	06 10	12 47
23	12 09 37	00 ♎ 31 38	20 ♐ 07 49	26 ♐ 27 15	26 22	02 56	08 06	13 11	14 57	21 28	06 09	12 48
24	12 13 34	01 30 24	02 ♑ 41 25	08 ♑ 50 59	27 03	04 10	08 41	13 18	14 58	21 26	06 09	12 49
25	12 17 30	02 29 11	14 ♑ 56 35	20 ♑ 58 55	27 41	05 23	09 16	13 25	14 58	21 25	06 08	12 50
26	12 21 27	03 28 00	26 ♑ 58 36	02 ♒ 56 15	28 14	06 36	09 52	13 31	14 58	21 23	06 07	12 51
27	12 25 23	04 26 50	08 ♒ 52 28	14 ♒ 47 47	28 42	07 50	10 28	13 38	14 R 58	21 21	06 06	12 52
28	12 29 20	05 25 43	20 ♒ 42 42	26 ♒ 37 40	29 06	09 03	11 04	13 44	14 58	21 20	06 06	12 53
29	12 33 16	06 24 37	02 ♓ 33 04	08 ♓ 29 16	29 24	10 17	11 40	13 51	14 58	21 18	06 05	12 55
30	12 37 13	07 ♎ 23 33	14 ♓ 26 34	20 ♓ 25 11	29 ♎ 35	11 ♍ 31	12 ♑ 17	13 ♋ 57	14 ♊ 58	21 ♒ 17	06 ♒ 05	12 ♐ 56

Date	Moon True ☊ ° '	Moon Mean ☊ ° '	Moon ☽ Latitude ° '
01	04 ♋ 21	02 ♋ 48	03 S 50
02	04 R 11	02 45	04 26
03	03 59	02 41	04 50
04	03 46	02 38	05 01
05	03 34	02 35	04 59
06	03 24	02 32	04 43
07	03 15	02 29	04 14
08	03 10	02 26	03 33
09	03 07	02 22	02 40
10	03 06	02 19	01 38
11	03 D 06	02 16	00 S 29
12	03 R 06	02 13	00 N 44
13	03 04	02 10	01 56
14	03 01	02 06	03 03
15	02 54	02 03	03 58
16	02 45	02 00	04 38
17	02 34	01 57	04 58
18	02 23	01 54	04 57
19	02 13	01 51	04 36
20	02 04	01 47	03 58
21	01 58	01 44	03 07
22	01 55	01 41	02 07
23	01 D 54	01 38	01 N 02
24	01 R 54	01 35	00 S 04
25	01 54	01 31	01 08
26	01 52	01 28	02 08
27	01 48	01 25	03 02
28	01 42	01 22	03 47
29	01 33	01 19	04 23
30	01 ♋ 21	01 ♋ 16	04 S 47

DECLINATIONS

Date	Sun ☉ ° '	Moon ☽ ° '	Mercury ☿ ° '	Venus ♀ ° '	Mars ♂ ° '	Jupiter ♃ ° '	Saturn ♄ ° '	Uranus ♅ ° '	Neptune ♆ ° '	Pluto ♇ ° '
01	08 N 09	17 S 12	00 S 34	18 N 42	27 S 00	22 N 51	20 N 46	14 S 50	18 S 30	12 S 07
02	07 47	13 33	01 16	18 27	26 59	22 50	20 46	14 50	18 31	12 08
03	07 25	09 21	01 58	18 11	26 58	22 50	20 46	14 51	18 31	12 08
04	07 03	04 47	02 39	17 55	26 57	22 49	20 47	14 52	18 31	12 09
05	06 41	00 S 01	03 20	17 38	26 56	22 48	20 47	14 53	18 32	12 09
06	06 19	04 N 49	04 00	17 21	26 55	22 47	20 47	14 53	18 32	12 10
07	05 56	09 31	04 40	17 04	26 54	22 47	20 47	14 54	18 32	12 10
08	05 34	13 55	05 19	16 46	26 53	22 46	20 47	14 55	18 33	12 11
09	05 11	17 48	05 57	16 27	26 51	22 45	20 47	14 55	18 33	12 11
10	04 48	20 53	06 35	16 08	26 50	22 45	20 47	14 56	18 33	12 12
11	04 26	22 56	07 11	15 48	26 48	22 44	20 47	14 57	18 34	12 12
12	04 03	23 41	07 47	15 28	26 46	22 43	20 47	14 57	18 34	12 13
13	03 40	22 55	08 22	15 08	26 44	22 43	20 47	14 58	18 34	12 13
14	03 17	20 37	08 56	14 47	26 42	22 42	20 47	14 58	18 35	12 14
15	02 54	16 52	09 29	14 26	26 40	22 41	20 47	14 59	18 35	12 14
16	02 30	11 59	10 02	14 04	26 37	22 41	20 47	15 00	18 35	12 15
17	02 07	06 19	10 32	13 42	26 35	22 40	20 47	15 00	18 35	12 15
18	01 44	00 N 20	11 02	13 19	26 32	22 39	20 47	15 01	18 35	12 16
19	01 21	05 S 34	11 30	12 57	26 29	22 39	20 47	15 01	18 36	12 16
20	00 57	11 00	11 58	12 33	26 26	22 38	20 47	15 02	18 36	12 17
21	00 34	15 42	12 23	12 10	26 23	22 37	20 47	15 03	18 36	12 17
22	00 N 11	19 26	12 47	11 46	26 19	22 37	20 47	15 03	18 36	12 18
23	00 S 13	22 03	13 09	11 21	26 16	22 36	20 47	15 04	18 37	12 18
24	00 36	23 29	13 30	10 57	26 12	22 36	20 47	15 04	18 37	12 19
25	00 59	23 44	13 48	10 32	26 08	22 35	20 47	15 05	18 37	12 19
26	01 23	22 52	14 05	10 06	26 04	22 34	20 47	15 05	18 37	12 20
27	01 46	20 58	14 18	09 41	25 59	22 34	20 47	15 06	18 37	12 20
28	02 09	18 11	14 30	09 15	25 55	22 33	20 47	15 06	18 37	12 21
29	02 33	14 39	14 39	08 49	25 50	22 33	20 46	15 07	18 38	12 21
30	02 S 56	10 S 32	14 S 44	08 N 23	25 S 45	22 N 32	20 N 46	15 S 07	18 S 38	12 S 22

ZODIAC SIGN ENTRIES

Date	h	m	Planets	
01	00	37	☿	♎
02	00	32	☽	♓
04	12	58	☽	♈
07	00	18	☽	♉
08	17	51	♂	♑
09	09	41	☽	♊
11	16	09	☽	♋
13	19	16	☽	♌
15	19	39	☽	♍
17	19	00	☽	♎
19	19	27	☽	♏
21	02	09	♀	♍
21	23	02	☽	♐
22	23	04	☉	♎
24	06	48	☽	♑
26	18	05	☽	♒
29	06	50	☽	♓

LATITUDES

Date	Mercury ☿ ° '	Venus ♀ ° '	Mars ♂ ° '	Jupiter ♃ ° '	Saturn ♄ ° '	Uranus ♅ ° '	Neptune ♆ ° '	Pluto ♇ ° '
01	00 S 18	00 N 01	03 S 36	00 S 13	01 S 46	00 S 45	00 N 08	10 N 16
04	00 43	00 10	03 32	00 13	01 47	00 45	00 08	10 14
07	01 09	00 19	03 28	00 12	01 47	00 45	00 08	10 13
10	01 34	00 28	03 24	00 12	01 47	00 45	00 08	10 12
13	01 59	00 36	03 19	00 12	01 48	00 45	00 08	10 11
16	02 23	00 44	03 15	00 12	01 48	00 45	00 08	10 09
19	02 46	00 51	03 10	00 11	01 48	00 45	00 08	10 08
22	03 06	00 58	03 06	00 11	01 49	00 45	00 08	10 07
25	03 23	01 04	03 01	00 11	01 49	00 45	00 08	10 05
28	03 35	01 10	02 57	00 11	01 50	00 45	00 08	10 04
31	03 S 39	01 N 15	02 S 52	00 S 10	01 S 50	00 S 45	00 N 08	10 N 03

DATA

Julian Date	2452154
Delta T	+64 seconds
Ayanamsa	23° 52' 33"
Synetic vernal point	05° ♓ 14' 26"
True obliquity of ecliptic	23° 26' 21"

MOON'S PHASES, APSIDES AND POSITIONS ☽

Date	h	m	Phase	Longitude ° '	Eclipse Indicator
02	21	43	○	10 ♓ 28	
10	19	00	☾	18 ♊ 08	
17	10	27	●	24 ♍ 36	
24	09	31	☽	01 ♑ 24	

Day	h	m		
01	23	11	Apogee	
16	15	40	Perigee	
29	05	25	Apogee	
05	12	04	0N	
12	12	10	Max dec	23° N 41'
18	13	20	0S	
25	05	13	Max dec	23° S 47'

LONGITUDES

Date	Chiron ⚷ ° '	Ceres ⚳ ° '	Pallas ⚴ ° '	Juno ⚵ ° '	Vesta ⚶ ° '	Black Moon Lilith ⚸ ° '
01	22 ♐ 53	08 ♑ 49	03 ♐ 13	21 ♋ 31	08 ♊ 11	01 ♓ 05
11	23 ♐ 03	09 ♑ 21	05 ♐ 46	26 ♋ 41	10 ♊ 55	02 ♓ 05
21	23 ♐ 22	10 ♑ 26	08 ♐ 37	01 ♌ 39	12 ♊ 23	03 ♓ 19
31	23 ♐ 49	11 ♑ 59	11 ♐ 42	06 ♌ 23	13 ♊ 16	04 ♓ 27

ASPECTARIAN

h m Aspects

01 Saturday
02 12 ☽ ∥ ♆
08 44 ☽ ☌ ♅
09 43 ☿ Q ♇
13 31 ☽ Q ♇
14 03 ☽ ± ☿
14 30 ☽ bQ ♃
15 54 ♀ ☍ ♆
17 36 ☽ ✶ ♂

02 Sunday
03 55 ☽ ⊼ ☿
03 59 ☽ ∥ ♅
05 54 ♀ # ♆
13 44 ☽ ⊻ ♆
15 16 ☉ ✶ ♃
16 13 ☽ ⊼ ♀
18 53 ☽ Q ♂
20 25 ☽ ∥ ♇
21 16 ☽ △ ♃
21 43 ☽ ☍ ☉

03 Monday
01 49 ☽ ⊥ ♆
01 57 ☽ □ ♇
05 41 ☽ ± ♀
05 45 ☽ □ ♄
19 55 ☽ ∠ ♆
21 09 ☽ ⊻ ♅
23 17 ☽ # ☉

04 Tuesday
01 48 ☽ bQ ♀
08 37 ☽ □ ♂
09 09 ☽ ⊥ ♅
18 00 ☽ Q ♄
21 28 ☽ ∥ ☿
23 10 ☉ ± ♆

05 Wednesday
00 26 ☽ ☍ ☿
01 20 ♀ ⊻ ♃
01 53 ☽ ✶ ♆
01 56 ☉ □ ♇
03 04 ☽ ∠ ♅
10 13 ☽ □ ♃
11 04 ☽ △ ♀
12 51 ☿ △ ♆
14 02 ☽ △ ♇
15 05 ☽ ⊼ ☉
17 57 ☽ ✶ ♄
22 54 ☿ bQ ♅

06 Thursday
01 38 ☽ Q ♆
03 58 ☽ ± ☉
07 18 ☽ # ☿
08 40 ☽ ✶ ♅
17 06 ♀ △ ♇
18 58 ☽ ∥ ☉
19 36 ☽ bQ ♇
22 08 ☽ Q ♃
22 31 ☽ △ ♂
23 09 ☽ bQ ☉
23 32 ☽ ∠ ♄

07 Friday
04 11 ☉ □ ♄
08 03 ☽ Q ♅
12 46 ☽ □ ♆
13 07 ☽ ± ♇
17 03 ☽ ⊥ ♄
18 43 ☽ ⊼ ☿
21 42 ☽ ✶ ♃

08 Saturday
00 42 ☽ ⊼ ♇
02 10 ☽ # ♇
04 02 ☽ □ ♀
04 37 ☽ ⊻ ♄
04 46 ☽ bQ ♂
06 36 ☽ △ ☉
07 36 ☽ ± ☿
10 14 ♀ ✶ ♄
16 33 ♀ bQ ♂
17 47 ☉ # ☿
17 47 ☽ # ♅
18 30 ☽ □ ♅
22 36 ☽ ± ♂

09 Sunday
02 30 ☿ □ ♃
02 37 ☽ ∠ ♃
02 37 ☽ bQ ☿
03 52 ☽ ∥ ♀
10 20 ☽ ⊼ ♂
11 42 ♂ ⊥ ♆
17 17 ☽ # ♆
17 33 ☽ Q ♀
19 37 ☽ ⊥ ♃
21 32 ☽ △ ♆

10 Monday
04 41 ☿ ✶ ♇
06 47 ☽ ⊻ ♃
09 01 ☽ ☍ ♇
09 28 ☽ △ ☿
11 05 ☽ ∥ ♄
12 50 ☽ ☌ ♄
17 46 ☽ ✶ ♀
18 24 ♀ ⊥ ♃
19 00 ☽ □ ☉

11 Tuesday
00 48 ☽ bQ ♆
01 42 ☽ △ ♅
08 50 ☽ ∥ ♃
16 45 ☽ ± ♆
18 56 ☽ ☍ ♂
22 06 ☿ △ ♄
23 01 ☽ ∠ ♀

12 Wednesday
03 13 ☽ ⊼ ♆
04 04 ☽ bQ ♅
05 16 ☽ Q ☉
12 32 ☽ ☌ ♃
14 06 ☽ ⊼ ♇
16 02 ☽ ⊥ ♀
17 45 ☽ ⊻ ♄
19 28 ☽ ± ♅
19 32 ☽ □ ☿

13 Thursday
00 19 ☽ ± ♇
03 09 ☽ ⊻ ♀
03 16 ☽ ✶ ☉
03 56 ☽ ⊥ ♄
05 35 ☽ ⊼ ♅
09 57 ☉ ⊻ ♀
15 05 ☽ ∥ ♃
15 24 ☽ bQ ♇
18 57 ☽ ∠ ♄
23 15 ♀ # ♅
23 48 ☽ ⊼ ♂

14 Friday
01 13 ☉ bQ ♆
04 11 ☽ Q ☿
05 36 ☽ ☍ ♆
05 55 ☽ ∠ ☉
07 06 ♀ ☍ ♅
09 58 ☽ ± ♂
10 35 ☽ ∥ ♄
12 05 ☉ ⊼ ♅
14 55 ☽ ⊻ ♃
15 58 ☽ △ ♇
19 27 ☽ ✶ ♄
21 31 ☽ ⊥ ☉

15 Saturday
00 41 ☽ ⊥ ♃
01 03 ☽ bQ ♂
01 12 ☽ ✶ ☿
02 05 ☽ # ♆
06 29 ☽ ☍ ♅
07 48 ☽ ⊻ ☉
08 35 ☽ ☌ ♀
14 40 ☽ Q ♄
15 10 ☽ ∠ ♃
21 48 ☽ # ♅

16 Sunday
01 33 ☽ ∥ ♀
01 48 ☽ △ ♂
02 55 ☽ ∠ ☿
05 33 ☽ ⊼ ♆
10 48 ☽ # ♇
15 00 ☽ ± ♆
15 05 ☽ ✶ ♃
15 42 ☽ □ ♇
18 10 ☽ ⊥ ☿
19 08 ☽ □ ♄
19 49 ☽ # ☿

17 Monday
04 19 ☽ ⊻ ☿
05 10 ☽ bQ ♆
05 11 ☉ Q ♃
05 48 ☽ ⊼ ♅
10 10 ☽ Q ♃
10 27 ☽ ☌ ☉
12 06 ☽ ⊻ ♀
15 16 ☽ ± ♅
20 08 ☽ Q ♇
22 25 ☽ ⊥ ♀

18 Tuesday
01 51 ☿ △ ♅
02 58 ☽ □ ♂
04 54 ☽ △ ♆
05 31 ☽ bQ ♅
06 02 ☽ ∥ ☉
13 07 ♀ Q ♄
14 02 ☽ ∠ ♀
15 02 ☽ □ ♃
15 17 ☽ ✶ ♇
18 48 ☽ △ ♄
19 47 ☽ # ☉

19 Wednesday
03 36 ♀ ∠ ♃
05 38 ☽ △ ♅
07 40 ☽ ☌ ☿
09 07 ☽ Q ♂
14 00 ☽ ⊻ ☉
15 43 ☽ ∠ ♇
16 38 ☽ ✶ ♀
19 19 ☽ bQ ♄
23 23 ♃ ⊼ ♇

20 Thursday
00 39 ☽ ⊥ ☉
05 12 ♂ ⊻ ♆
05 48 ☽ □ ♆
05 50 ☽ ✶ ♂
06 43 ☽ ⊥ ♇
10 24 ☽ ± ♄
11 01 ☉ ± ♅
14 37 ☽ Q ♀
16 54 ☽ ⊻ ♇
16 59 ☽ ∠ ☉
16 59 ☽ ∥ ☿
17 03 ☽ △ ♃
18 06 ☽ ∥ ♇
18 54 ☽ # ♀
19 33 ♂ ∠ ♅
20 39 ☽ ⊼ ♄

21 Friday
04 35 ♀ # ♇
05 22 ☿ # ♀
06 09 ☿ ∥ ♇
08 04 ☽ □ ♅
08 21 ☽ ∥ ♅
08 38 ☽ ∠ ♂
12 43 ☽ Q ♆
13 54 ☽ ⊻ ☿
19 26 ☽ bQ ♃
21 09 ☽ ✶ ☉

22 Saturday
01 08 ☽ □ ♀
01 16 ☽ ⊥ ☿
01 17 ☽ ⊥ ♂
01 57 ☿ Q ♂
06 05 ☽ ∥ ♆
10 10 ☽ ✶ ♆
11 46 ☽ ± ♃
12 38 ☽ ⊻ ♂
16 13 ☽ Q ♅
18 40 ☽ ∠ ☿
20 39 ☽ Q ☉
22 18 ☽ ☌ ♇
22 54 ☽ ⊼ ♃
23 09 ☽ # ♄

23 Sunday
02 19 ☽ ☍ ♄
13 56 ☽ ∠ ♆
14 31 ☽ ✶ ♅
18 47 ☉ Q ♇
19 11 ☽ # ♃

24 Monday
00 32 ☽ ✶ ☿
07 05 ☽ ⊥ ♆
09 31 ☽ □ ☉
15 10 ☽ △ ♀
18 42 ☽ ⊻ ♆
19 16 ☽ ∠ ♅

25 Tuesday
00 15 ☽ ☌ ♂
01 05 ☽ Q ☿
07 50 ☽ ⊻ ♇
08 57 ☽ ☍ ♃
12 03 ☽ ⊼ ♄
12 55 ☽ ⊥ ♅
18 34 ☿ ∠ ♇
19 44 ☽ ⊥ ♇
23 58 ☽ ± ♄

26 Wednesday
00 02 ☽ bQ ♀
00 50 ☽ ⊻ ♅
02 31 ♀ ⊼ ♆
13 46 ☽ ∠ ♇
14 38 ☽ □ ☿
16 39 ☽ # ♃
18 01 ☽ bQ ♄
20 08 ☽ ± ♀

27 Thursday
00 05 ♄ St R
02 15 ☽ △ ☉
06 24 ☽ ☌ ♆
09 39 ☽ ⊼ ♀
13 52 ☽ # ♄
15 23 ☽ ⊻ ♂
20 06 ☽ ✶ ♇
21 44 ☽ ⊼ ♃

28 Friday
00 21 ☽ △ ♄
04 12 ☽ ⊥ ♂
08 35 ☽ ∥ ♆
10 01 ☽ ± ♃
11 22 ☽ bQ ☉
13 15 ☽ ☌ ♅
14 54 ☿ Q ♂
20 29 ☽ Q ♇
23 26 ☽ ∠ ♂

29 Saturday
04 10 ☉ △ ♆
04 26 ☽ bQ ♃
05 28 ☽ △ ☿
07 16 ☽ ± ☉
09 07 ☽ ∥ ♅
09 33 ☉ bQ ♅
12 02 ☽ ∥ ☿
19 09 ☽ ⊻ ♆
20 30 ☽ ⊼ ☉

30 Sunday
01 42 ☽ ∥ ♇
05 25 ☽ ☍ ♀
07 14 ☽ ⊥ ♆
07 24 ☽ ✶ ♂
08 57 ☽ □ ♇
11 00 ☽ △ ♃
12 18 ☽ bQ ☿
13 02 ☽ □ ♄
23 02 ♀ ± ♆

All ephemeris data is given at 12.00 UT and the Moon's longitude is additionally given for 24.00 UT

OCTOBER 2001

LONGITUDES

Date	Sidereal time h m s	Sun ☉ ° ′ ″	Moon ☽ ° ′ ″	Moon ☽ 24.00 ° ′ ″	Mercury ☿ ° ′	Venus ♀ ° ′	Mars ♂ ° ′	Jupiter ♃ ° ′	Saturn ♄ ° ′	Uranus ♅ ° ′	Neptune ♆ ° ′	Pluto ♇ ° ′
01	12 41 09	08 ♎ 22 31	26 ♓ 25 20	02 ♈ 27 10	29 ♎ 41	12 ♍ 44	12 ♑ 53	14 ♋ 03	14 ♊ 57	21 ♒ 16	06 ♒ 04	12 ♐ 57
02	12 45 06	09 21 31	08 ♈ 30 51	14 ♈ 36 27	29 R 39	13 58	13 30	14 09	14 R 57	21 R 14	06 R 04	12 58
03	12 49 03	10 20 33	20 ♈ 44 04	26 ♈ 53 48	29 31	15 12	14 08	14 14	14 56	21 13	06 03	12 59
04	12 52 59	11 19 37	03 ♉ 05 43	09 ♉ 19 56	29 15	16 25	14 45	14 20	14 55	21 12	06 03	13 01
05	12 56 56	12 18 43	15 ♉ 36 33	21 ♉ 55 43	28 51	17 39	15 23	14 25	14 54	21 10	06 02	13 02
06	13 00 52	13 17 51	28 ♉ 17 37	04 ♊ 42 26	28 19	18 53	16 00	14 30	14 53	21 09	06 02	13 03
07	13 04 49	14 17 02	11 ♊ 10 25	17 ♊ 41 50	27 39	20 07	16 38	14 35	14 52	21 08	06 01	13 05
08	13 08 45	15 16 15	24 ♊ 16 58	00 ♋ 56 06	26 52	21 21	17 16	14 40	14 51	21 07	06 01	13 06
09	13 12 42	16 15 30	07 ♋ 39 32	14 ♋ 27 33	25 58	22 35	17 55	14 45	14 50	21 06	06 01	13 08
10	13 16 38	17 14 48	21 ♋ 20 21	28 ♋ 18 06	24 57	23 50	18 33	14 49	14 48	21 05	06 00	13 09
11	13 20 35	18 14 08	05 ♌ 20 51	12 ♌ 28 35	23 50	25 04	19 12	14 54	14 47	21 04	06 00	13 11
12	13 24 32	19 13 30	19 ♌ 41 05	26 ♌ 58 01	22 40	26 18	19 51	14 58	14 45	21 03	06 00	13 12
13	13 28 28	20 12 55	04 ♍ 18 50	11 ♍ 42 52	21 28	27 32	20 30	15 02	14 43	21 02	06 00	13 14
14	13 32 25	21 12 21	19 ♍ 09 15	26 ♍ 37 00	20 16	28 ♍ 47	21 09	15 06	14 42	21 01	06 00	13 15
15	13 36 21	22 11 50	04 ♎ 05 01	11 ♎ 32 07	19 05	00 ♎ 01	21 48	15 09	14 40	21 01	06 00	13 17
16	13 40 18	23 11 22	18 ♎ 57 11	26 ♎ 19 03	17 59	01 15	22 27	15 13	14 38	21 00	06 00	13 19
17	13 44 14	24 10 55	03 ♏ 36 44	10 ♏ 49 20	16 58	02 30	23 07	15 16	14 35	20 59	06 00	13 20
18	13 48 11	25 10 30	17 ♏ 56 07	24 ♏ 56 34	16 05	03 44	23 47	15 19	14 33	20 58	06 D 00	13 22
19	13 52 07	26 10 07	01 ♐ 50 21	08 ♐ 37 18	15 21	04 59	24 27	15 22	14 31	20 58	06 00	13 24
20	13 56 04	27 09 46	15 ♐ 17 27	21 ♐ 50 57	14 48	06 13	25 07	15 24	14 28	20 57	06 00	13 25
21	14 00 01	28 09 27	28 ♐ 18 09	04 ♑ 39 26	14 25	07 28	25 47	15 27	14 26	20 57	06 00	13 27
22	14 03 57	29 ♎ 09 10	10 ♑ 55 20	17 ♑ 06 25	14 14	08 42	26 27	15 29	14 23	20 56	06 00	13 29
23	14 07 54	00 ♏ 08 54	23 ♑ 13 19	29 ♑ 16 39	14 D 13	09 57	27 08	15 31	14 20	20 56	06 00	13 31
24	14 11 50	01 08 40	05 ♒ 17 06	11 ♒ 15 19	14 24	11 12	27 48	15 33	14 18	20 56	06 00	13 33
25	14 15 47	02 08 27	17 ♒ 11 57	23 ♒ 07 38	14 46	12 27	28 29	15 35	14 15	20 55	06 00	13 34
26	14 19 43	03 08 17	29 ♒ 02 57	04 ♓ 58 28	15 17	13 41	29 10	15 36	14 12	20 55	06 00	13 36
27	14 23 40	04 08 08	10 ♓ 54 42	16 ♓ 52 07	15 57	14 56	29 ♑ 51	15 38	14 09	20 55	06 01	13 38
28	14 27 36	05 08 00	22 ♓ 51 08	28 ♓ 52 07	16 45	16 11	00 ♒ 32	15 39	14 05	20 55	06 01	13 40
29	14 31 33	06 07 55	04 ♈ 55 22	11 ♈ 01 07	17 41	17 26	01 13	15 40	14 02	20 55	06 02	13 42
30	14 35 30	07 07 51	17 ♈ 09 35	23 ♈ 20 53	18 43	18 41	01 54	15 41	13 59	20 55	06 02	13 44
31	14 39 26	08 ♏ 07 49	29 ♈ 35 07	05 ♉ 52 19	19 ♎ 51	19 ♎ 55	02 ♒ 36	15 ♋ 41	13 ♊ 55	20 ♒ 55	06 ♒ 03	13 ♐ 46

Date	Moon True ☊ ° ′	Moon Mean ☊ ° ′	Moon ☽ Latitude ° ′
01	01 ♋ 08	01 ♋ 12	04 S 59
02	00 R 54	01 09	04 57
03	00 40	01 06	04 42
04	00 28	01 03	04 13
05	00 19	01 00	03 32
06	00 12	00 57	02 40
07	00 09	00 53	01 39
08	00 07	00 50	00 S 31
09	00 D 07	00 47	00 N 40
10	00 R 08	00 44	01 50
11	00 06	00 41	02 55
12	00 03	00 37	03 51
13	29 ♊ 58	00 34	04 33
14	29 49	00 31	04 57
15	29 39	00 28	05 01
16	29 29	00 25	04 45
17	29 19	00 22	04 10
18	29 11	00 18	03 20
19	29 06	00 15	02 20
20	29 03	00 12	01 13
21	29 02	00 09	00 N 04
22	29 D 02	00 06	01 S 03
23	29 03	00 ♋ 03	02 06
24	29 R 03	29 ♊ 59	03 01
25	29 02	29 56	03 48
26	28 58	29 53	04 25
27	28 52	29 50	04 51
28	28 44	29 47	05 04
29	28 35	29 43	05 04
30	28 25	29 40	04 50
31	28 ♊ 15	29 ♊ 37	04 S 22

DECLINATIONS

Date	Sun ☉ ° ′	Moon ☽ ° ′	Mercury ☿ ° ′	Venus ♀ ° ′	Mars ♂ ° ′	Jupiter ♃ ° ′	Saturn ♄ ° ′	Uranus ♅ ° ′	Neptune ♆ ° ′	Pluto ♇ ° ′
01	03 S 19	05 S 59	14 S 47	07 N 56	25 S 40	22 N 32	20 N 46	15 S 07	18 S 38	12 S 22
02	03 43	01 S 11	14 46	07 29	25 35	22 31	20 46	15 08	18 38	12 23
03	04 06	03 N 44	14 41	07 02	25 30	22 30	20 46	15 08	18 38	12 23
04	04 29	08 34	14 32	06 35	25 24	22 30	20 46	15 09	18 38	12 24
05	04 52	13 08	14 19	06 07	25 18	22 30	20 45	15 09	18 38	12 25
06	05 15	17 11	14 02	05 39	25 12	22 29	20 45	15 09	18 38	12 25
07	05 38	20 29	13 40	05 11	25 06	22 29	20 45	15 10	18 39	12 26
08	06 01	22 48	13 14	04 43	25 00	22 28	20 45	15 10	18 39	12 26
09	06 24	23 53	12 43	04 15	24 53	22 28	20 44	15 10	18 39	12 27
10	06 46	23 33	12 08	03 47	24 46	22 27	20 44	15 11	18 39	12 27
11	07 09	21 46	11 29	03 18	24 39	22 27	20 44	15 11	18 39	12 28
12	07 32	18 34	10 47	02 49	24 32	22 27	20 44	15 11	18 39	12 28
13	07 54	14 10	10 03	02 21	24 25	22 26	20 43	15 11	18 39	12 29
14	08 16	08 51	09 18	01 52	24 17	22 26	20 43	15 12	18 39	12 29
15	08 39	02 N 59	08 32	01 23	24 10	22 26	20 43	15 12	18 39	12 30
16	09 01	03 S 01	07 48	00 54	24 02	22 25	20 42	15 12	18 39	12 30
17	09 23	08 47	07 06	00 N 25	23 53	22 25	20 42	15 12	18 39	12 31
18	09 44	13 58	06 27	00 S 05	23 45	22 25	20 42	15 12	18 39	12 31
19	10 06	18 15	05 52	00 34	23 36	22 25	20 41	15 13	18 39	12 32
20	10 28	21 25	05 22	01 03	23 28	22 24	20 41	15 13	18 39	12 32
21	10 49	23 22	04 58	01 32	23 19	22 24	20 41	15 13	18 39	12 33
22	11 10	24 02	04 39	02 02	23 09	22 24	20 40	15 13	18 39	12 34
23	11 31	23 30	04 26	02 31	23 00	22 24	20 40	15 13	18 39	12 34
24	11 52	21 53	04 19	03 00	22 51	22 24	20 39	15 13	18 39	12 35
25	12 13	19 18	04 18	03 29	22 41	22 24	20 39	15 13	18 39	12 35
26	12 34	15 57	04 22	03 58	22 31	22 23	20 39	15 13	18 39	12 36
27	12 54	11 57	04 31	04 27	22 21	22 23	20 38	15 13	18 39	12 36
28	13 14	07 29	04 44	04 56	22 10	22 23	20 38	15 13	18 39	12 37
29	13 34	02 S 41	05 01	05 25	22 00	22 23	20 37	15 13	18 39	12 37
30	13 54	02 N 17	05 22	05 54	21 49	22 23	20 37	15 13	18 39	12 38
31	14 S 13	07 N 14	05 S 46	06 S 23	21 S 38	22 N 23	20 N 36	15 S 13	18 S 39	12 S 38

ZODIAC SIGN ENTRIES

Date	h	m	Planets
01	19	08	☽ ♈
04	06	01	☽ ♉
06	15	12	☽ ♊
08	22	19	☽ ♋
11	02	54	☽ ♌
13	04	58	☽ ♍
15	05	26	☽ ♎
15	11	42	♀ ♎
17	06	03	☽ ♏
19	08	47	☽ ♐
21	15	11	☽ ♑
23	08	26	☉ ♏
24	01	26	☽ ♒
26	13	56	☽ ♓
27	17	19	♂ ♒
29	02	15	☽ ♈
31	12	48	☽ ♉

LATITUDES

Date	Mercury ☿ ° ′	Venus ♀ ° ′	Mars ♂ ° ′	Jupiter ♃ ° ′	Saturn ♄ ° ′	Uranus ♅ ° ′	Neptune ♆ ° ′	Pluto ♇ ° ′
01	03 S 39	01 N 15	02 S 52	00 S 10	01 S 50	00 S 45	00 N 08	10 N 03
04	03 34	01 19	02 48	00 10	01 50	00 45	00 08	10 02
07	03 15	01 23	02 43	00 10	01 51	00 45	00 08	10 01
10	02 40	01 27	02 39	00 10	01 51	00 45	00 07	10 00
13	01 49	01 29	02 34	00 09	01 51	00 45	00 07	09 59
16	00 S 48	01 31	02 29	00 09	01 52	00 45	00 07	09 58
19	00 N 12	01 32	02 25	00 09	01 52	00 44	00 07	09 57
22	01 02	01 33	02 20	00 09	01 52	00 44	00 07	09 56
25	01 39	01 33	02 16	00 08	01 52	00 44	00 07	09 55
28	02 00	01 32	02 11	00 08	01 53	00 44	00 07	09 54
31	02 N 09	01 N 31	02 S 07	00 S 08	01 S 53	00 S 44	00 N 07	09 N 53

DATA

Julian Date	2452184
Delta T	+64 seconds
Ayanamsa	23° 52′ 36″
Synetic vernal point	05° ♓ 14′ 24″
True obliquity of ecliptic	23° 26′ 21″

MOON'S PHASES, APSIDES AND POSITIONS ☽

Date	h	m	Phase	Longitude ° ′	Eclipse Indicator
02	13	49	○	09 ♈ 26	
10	04	20	☾	16 ♋ 56	
16	19	23	●	23 ♎ 30	
24	02	58	☽	00 ♒ 46	

Day	h	m	
14	22	52	Perigee
26	20	09	Apogee

Day	h	m		
02	17	45	0N	
09	18	42	Max dec	23° N 56′
15	23	55	0S	
22	13	08	Max dec	24° S 02′
30	01	03	0N	

LONGITUDES

Date	Chiron ⚷ ° ′	Ceres ⚳ ° ′	Pallas ⚴ ° ′	Juno ⚵ ° ′	Vesta ⚶ ° ′	Black Moon Lilith ⚸ ° ′
01	23 ♐ 49	11 ♑ 59	11 ♐ 42	06 ♌ 23	13 ♊ 16	04 ♓ 27
11	24 ♐ 24	13 ♑ 57	14 ♐ 59	10 ♌ 52	13 ♊ 28	05 ♓ 34
21	25 ♐ 06	16 ♑ 17	18 ♐ 26	15 ♌ 02	12 ♊ 57	06 ♓ 41
31	25 ♐ 54	18 ♑ 56	22 ♐ 00	18 ♌ 51	11 ♊ 42	07 ♓ 48

ASPECTARIAN

h m	Aspects
01 Monday	
00 53	☽ # ♀
01 18	☽ ∠ ♆
01 42	☽ ⊻ ♅
06 30	☽ ± ☿
08 47	☽ Q ♂
13 40	☽ ⊥ ♅
14 20	♂ ⊻ ♇
16 14	♀ □ ♇
18 03	♀ △ ♂
18 30	☽ ⊼ ☿
19 22	☿ St R
02 Tuesday	
00 25	☽ ∥ ☉
00 59	☽ Q ♄
07 09	☽ ⚹ ♆
07 31	☽ ∠ ♅
13 49	☽ ☍ ☉
15 49	♀ ⚹ ♃
20 48	☽ △ ♇
22 22	☽ □ ♂
23 11	☽ □ ♃
23 56	☽ ⊼ ♀
03 Wednesday	
00 24	☿ ∠ ♀
00 39	☽ ⚹ ♄
06 45	☽ Q ♆
06 54	♀ □ ♄
12 56	☽ ⚹ ♅
13 00	☽ ± ♀
13 54	☽ # ☉
17 09	♂ ☍ ♃
04 Thursday	
02 09	☽ ⚼ ♇
02 50	☽ ∥ ♀
04 45	☽ ☍ ☿
05 52	☽ ∠ ♄
08 25	☽ ⚼ ♀
10 31	☽ Q ♃
12 11	☽ Q ♅
17 41	☽ □ ♆
18 20	♂ ⊼ ♄
19 34	☽ ± ♇
23 12	☽ ⊥ ♄
05 Friday	
04 32	♂ ⊥ ♅
05 10	☽ ⊼ ☉
07 05	☽ ⊼ ♇
08 03	☽ # ♇
09 43	☽ ⚹ ♃
10 39	☽ ⊻ ♄
11 32	☽ △ ♂
16 19	☽ △ ♀
17 35	☽ ± ☉
18 19	☽ # ☿
22 33	☽ □ ♅
23 32	☽ # ♅
06 Saturday	
06 02	☉ ⚹ ♇
12 00	☽ ⚼ ☉
12 03	☽ ⊼ ☿
14 18	☽ ∠ ♃
17 21	☽ ⚼ ♂
21 42	☿ ∠ ♇
21 51	☽ # ♆
22 45	☽ ± ☿
23 27	☉ # ♀
07 Sunday	
02 27	☽ △ ♆
07 11	☽ ⊥ ♃
10 57	☽ ± ♂
14 13	☽ ∥ ♄
14 36	☽ ⚼ ☿
15 32	☽ ☍ ♇
18 12	☽ △ ☉
18 20	☽ ⊻ ♃
18 48	☽ ☌ ♄
20 08	☉ □ ♃
22 34	☽ ⊼ ♂
08 Monday	
01 58	☉ △ ♄
05 28	♀ ⚼ ♆
06 04	☽ ⚼ ♆
06 08	☽ □ ♀
06 15	☽ △ ♅
06 21	☿ ⊥ ♀
07 25	♀ ⊼ ♅
07 41	☽ ∥ ♃
16 24	☽ △ ☿
22 21	☽ ± ♆
09 Tuesday	
09 05	☽ ⊼ ♆
09 14	☽ ⚼ ♅
17 43	☽ Q ♀
21 41	☽ ⊼ ♇
23 17	☿ ∥ ♇
10 Wednesday	
00 35	☽ ☌ ♃
00 37	☽ ⊻ ♄
01 06	☽ ± ♅
04 20	☽ □ ☉
06 55	☽ ☍ ♂
07 03	♃ ⊻ ♄
08 12	☽ ± ♇
11 04	☽ ⊥ ♄
11 33	☽ ⊼ ♅
16 43	☽ ⚹ ♀
17 47	☽ □ ☿
23 34	☿ ⊻ ♀
23 46	☽ ⚼ ♇
11 Thursday	
02 33	☽ ∠ ♄
04 51	☽ ∥ ♃
11 21	♂ ⊥ ♇
13 07	☽ ☍ ♆
13 37	☽ Q ☉
20 42	☽ ∠ ♀
20 56	☽ ∥ ♄
22 07	☽ Q ☿
12 Friday	
01 12	☽ △ ♇
03 49	☽ ⚹ ♄
04 07	☽ ⊻ ♃
11 11	☽ ⚹ ☉
11 28	☽ # ♆
12 16	☽ ⊼ ♂
13 07	☽ ⊥ ♀
14 08	☽ ⊥ ♃
14 15	☽ ☍ ♅
16 34	☽ ⚹ ☿
22 38	☽ ± ♂
23 37	☽ Q ♄
23 55	☽ ⊻ ♀
13 Saturday	
01 39	♀ Q ♃
02 23	♀ ± ♅
04 59	☽ ∠ ♃
06 52	☽ # ♅
13 18	♃ ± ♅
13 34	☽ ∠ ☉
14 00	☽ ⚼ ♂
14 44	☽ ⊼ ♆
15 14	☽ ∠ ☿
19 57	☽ # ♇
20 08	♂ ± ♄
20 45	☿ △ ♅
14 Sunday	
00 27	☽ ± ♆
00 38	☿ □ ♂
01 43	☉ ☌ ☿
02 28	☽ □ ♇
04 43	☽ ⊥ ☿
04 49	☽ □ ♄
05 11	☽ ⊥ ☉
05 26	☽ ⚹ ♃
07 35	♂ ⊻ ♅
07 36	☉ △ ♅
07 37	☉ □ ♂
09 46	☽ # ☿
13 39	☽ ⊻ ☿
14 16	☽ # ☉
14 58	☽ ⚼ ♆
15 00	☽ ⊼ ♅
15 21	☽ △ ♂
15 32	☽ ⊻ ☉
15 Monday	
00 38	☽ ± ♅
00 49	☽ Q ♃
04 52	☽ ☌ ♀
07 29	☽ Q ♇
09 48	☉ ∥ ☿
15 04	☽ △ ♆
15 06	☽ ⚼ ♅
18 59	☽ ∥ ♀
16 Tuesday	
02 51	☽ ⚹ ♇
04 07	☽ # ♀
05 00	☽ △ ♄
05 55	☽ □ ♃
10 32	☽ ☌ ☿
13 05	♀ Q ♇
15 19	☽ △ ♅
17 58	☽ □ ♂
19 23	☽ ☌ ☉
17 Wednesday	
03 18	☽ ∠ ♇
05 23	☽ ⚼ ♄
05 32	☽ ∥ ☿
09 59	☽ ⊻ ♀
14 45	☽ ∥ ☉
15 57	☽ □ ♆
18 12	☽ ⊥ ♇
20 15	☽ ± ♄
20 53	☽ ⊥ ♀
18 Thursday	
01 06	☽ Q ♂
01 50	♆ St D
04 15	☽ ⊻ ♇
04 56	☽ ∥ ♇
06 17	☽ ⊼ ♄
07 33	☽ △ ♃
09 02	☽ ⊻ ☿
13 30	☽ ∠ ♀
17 11	☽ □ ♅
18 25	☽ ∥ ♅
18 43	☽ ⊥ ☿
22 22	☽ Q ♆
22 30	☽ ⚹ ♂
19 Friday	
01 21	☽ ⊻ ☉
09 24	☽ ⚼ ♃
09 31	☽ ∠ ☿
11 48	☿ □ ♃
12 37	☽ ⊥ ☉
14 38	☽ ∥ ♆
18 05	☽ ⚹ ♀
19 20	☽ ⚹ ♆
20 Saturday	
00 36	☽ Q ♅
01 22	☽ ± ♃
02 11	☽ ∠ ♂
05 31	☽ # ♄
05 53	☽ ∠ ☉
06 53	♀ ⚼ ♅
07 36	♀ △ ♆
08 37	☽ ☌ ♇
10 31	☽ ☍ ♄
11 08	☽ ⚹ ☿
12 13	☽ ⊼ ♃
17 54	☽ Q ♀
19 21	☽ ⊥ ♂
22 12	☽ # ♃
22 21	☽ ⚹ ♅
22 25	☽ ∠ ♆
21 Sunday	
07 02	☽ ⊻ ♂
08 33	☽ Q ☿
10 46	☿ △ ♄
11 08	☽ ∥ ♂
11 42	☽ ⚹ ☉
15 11	☽ ⊥ ♆
19 19	☉ ∠ ♇
22 Monday	
02 27	☽ ∠ ♅
02 33	☽ ⊻ ♆
07 16	☽ □ ♀
12 29	☽ Q ☉
16 58	☽ ⊻ ♇
17 24	☉ ⚼ ♄
18 21	☽ □ ☿
18 41	☽ ⊼ ♄
19 46	☽ ⊥ ♅
20 52	☽ ☍ ♃
23 Tuesday	
00 23	☿ St D
04 41	☽ ⊥ ♇
06 21	☽ ± ♄
07 30	☽ ⊻ ♅
20 11	☽ ☌ ♂
22 17	☽ ∥ ♂
22 30	☽ ∠ ♇
24 Wednesday	
00 05	☽ ⚼ ♄
02 42	☿ △ ♄
02 58	☽ □ ☉
05 48	☽ # ♃
13 27	☽ ☌ ♆
25 Thursday	
00 26	☽ # ♄
01 16	☽ △ ♀
04 39	☽ ⚹ ♇
06 03	☽ △ ♄
06 53	☽ △ ☿
08 43	☽ ⊼ ♃
15 13	♂ ∠ ♇
17 05	☽ ∥ ♆
19 32	☽ ☌ ♅
20 53	☽ ± ♃
26 Friday	
05 00	☽ Q ♇
10 20	♀ ⚹ ♇
11 11	☽ ⚼ ♀
12 15	☽ ⊻ ♂
12 58	♂ ⚼ ♄
14 27	☉ ∥ ♇
14 37	☽ ⚼ ☿
15 10	☽ ⚼ ♃
16 37	☽ ∥ ♅
21 03	☽ △ ☉
21 24	♀ △ ♄
27 Saturday	
00 55	☿ □ ♃
01 09	☽ ⊥ ♂
02 06	☽ ⊻ ♆
05 25	♂ # ♃
07 01	☽ ∥ ☉
07 32	☽ ± ♀
08 20	☽ ∥ ♇
09 55	☽ ± ☿
14 14	☽ ⊥ ♆
16 17	☿ ∥ ♀
17 30	☽ □ ♇
18 29	☽ □ ♄
20 25	☽ ∠ ♂
21 03	☽ ⊼ ♀
21 31	☽ △ ♃
22 51	☽ ⊼ ☿
28 Sunday	
01 36	♀ □ ♃
06 04	☽ ⚼ ☉
08 07	☽ ⊻ ♅
08 20	☽ ∠ ♆
20 06	☽ ⊥ ♅
23 43	☽ ∥ ♀
29 Monday	
01 12	☽ ∥ ☿
01 39	☽ ± ☉
04 14	☽ ⚹ ♂
06 19	☽ Q ♄
09 31	☉ □ ♆
13 57	☽ ∠ ♅
14 11	☽ ⚹ ♆
14 36	☽ ⊼ ☉
30 Tuesday	
05 17	☽ Q ♂
05 18	☽ △ ♇
05 50	☽ ⚹ ♄
09 06	☽ □ ♃
13 42	☽ Q ♆
15 17	☽ ☍ ♀
15 19	☽ ☍ ☿
18 54	☿ ☌ ♀
19 17	☽ ⚹ ♅
22 57	♅ St D
31 Wednesday	
02 53	☉ ⊥ ♇
04 09	☽ # ☿
07 18	☉ ± ♄
07 21	☽ # ♀
10 25	☽ ⚼ ♇
10 44	☽ ∠ ♄
18 06	☽ □ ♂
18 21	☽ Q ♅
19 50	☽ Q ♃

All ephemeris data is given at 12.00 UT and the Moon's longitude is additionally given for 24.00 UT

NOVEMBER 2001

LONGITUDES

Date	Sidereal time h m s	Sun ☉ ° ' "	Moon ☽ ° ' "	Moon ☽ 24.00 ° ' "	Mercury ☿ ° '	Venus ♀ ° '	Mars ♂ ° '	Jupiter ♃ ° '	Saturn ♄ ° '	Uranus ♅ ° '	Neptune ♆ ° '	Pluto ♇ ° '
01	14 43 23	09 ♏ 07 49	12 ♉ 12 30	18 ♉ 35 40	21 ♎ 04	21 ♎ 10	03 ♒ 17	15 ♋ 41	13 ♊ 52	20 ♒ 55	06 ♒ 03	13 ♐ 48
02	14 47 19	10 07 50	25 ♉ 01 45	01 ♊ 30 44	22 21	22 25	03 59	15 41	13 R 48	20 D 55	06 04	13 50
03	14 51 16	11 07 54	08 ♊ 02 35	14 ♊ 37 14	23 42	23 40	04 40	15 R 41	13 45	20 55	06 04	13 52
04	14 55 12	12 08 00	21 ♊ 14 42	27 ♊ 54 57	25 05	24 55	05 22	15 41	13 41	20 55	06 05	13 54
05	14 59 09	13 08 08	04 ♋ 38 00	11 ♋ 23 54	26 32	26 10	06 04	15 41	13 37	20 55	06 05	13 56
06	15 03 05	14 08 18	18 ♋ 12 40	25 ♋ 04 22	28 00	27 25	06 46	15 40	13 33	20 56	06 06	13 58
07	15 07 02	15 08 30	01 ♌ 59 01	08 ♌ 56 37	29 ♎ 30	28 40	07 28	15 39	13 29	20 56	06 07	14 00
08	15 10 59	16 08 44	15 ♌ 57 09	23 ♌ 00 33	01 ♏ 02	29 ♎ 55	08 10	15 38	13 25	20 56	06 07	14 02
09	15 14 55	17 09 00	00 ♍ 06 38	07 ♍ 15 12	02 34	01 ♏ 11	08 52	15 37	13 21	20 57	06 08	14 04
10	15 18 52	18 09 18	14 ♍ 25 54	21 ♍ 38 21	04 08	02 26	09 35	15 35	13 17	20 57	06 09	14 06
11	15 22 48	19 09 38	28 ♍ 52 00	06 ♎ 06 17	05 42	03 41	10 17	15 34	13 13	20 58	06 10	14 08
12	15 26 45	20 09 59	13 ♎ 20 30	20 ♎ 33 55	07 17	04 56	10 59	15 32	13 08	20 59	06 10	14 11
13	15 30 41	21 10 23	27 ♎ 45 48	04 ♏ 55 23	08 52	06 11	11 42	15 30	13 04	20 59	06 11	14 13
14	15 34 38	22 10 49	12 ♏ 01 55	19 ♏ 04 45	10 28	07 26	12 24	15 27	13 00	21 00	06 12	14 15
15	15 38 34	23 11 16	26 ♏ 03 18	02 ♐ 57 03	12 04	08 42	13 07	15 25	12 55	21 01	06 13	14 17
16	15 42 31	24 11 45	09 ♐ 45 38	16 ♐ 28 50	13 40	09 57	13 50	15 22	12 51	21 02	06 14	14 19
17	15 46 28	25 12 16	23 ♐ 06 29	29 ♐ 38 36	15 15	11 12	14 32	15 19	12 46	21 02	06 15	14 21
18	15 50 24	26 12 48	06 ♑ 05 19	12 ♑ 26 50	16 51	12 27	15 15	15 16	12 42	21 03	06 16	14 24
19	15 54 21	27 13 21	18 ♑ 43 27	24 ♑ 55 36	18 27	13 43	15 58	15 13	12 37	21 04	06 17	14 26
20	15 58 17	28 13 55	01 ♒ 03 43	07 ♒ 08 19	20 03	14 58	16 41	15 10	12 32	21 05	06 18	14 28
21	16 02 14	29 ♏ 14 31	13 ♒ 09 59	19 ♒ 09 16	21 39	16 13	17 24	15 06	12 28	21 06	06 19	14 30
22	16 06 10	00 ♐ 15 08	25 ♒ 06 49	01 ♓ 03 13	23 14	17 29	18 07	15 02	12 23	21 08	06 21	14 33
23	16 10 07	01 15 46	06 ♓ 59 07	12 ♓ 55 06	24 49	18 44	18 50	14 58	12 18	21 09	06 22	14 35
24	16 14 03	02 16 25	18 ♓ 51 47	24 ♓ 49 44	26 25	19 59	19 34	14 54	12 13	21 10	06 23	14 37
25	16 18 00	03 17 06	00 ♈ 49 30	06 ♈ 51 34	28 00	21 15	20 17	14 50	12 08	21 11	06 24	14 39
26	16 21 57	04 17 47	12 ♈ 56 24	19 ♈ 04 24	29 ♏ 35	22 30	21 00	14 45	12 04	21 13	06 25	14 42
27	16 25 53	05 18 30	25 ♈ 15 54	01 ♉ 31 13	01 ♐ 10	23 46	21 43	14 40	11 59	21 14	06 27	14 44
28	16 29 50	06 19 13	07 ♉ 50 31	14 ♉ 13 58	02 44	25 01	22 27	14 36	11 54	21 15	06 28	14 46
29	16 33 46	07 19 58	20 ♉ 41 38	27 ♉ 13 30	04 19	26 16	23 10	14 30	11 49	21 17	06 29	14 48
30	16 37 43	08 ♐ 20 44	03 ♊ 49 31	10 ♊ 29 32	05 ♐ 53	27 ♏ 32	23 ♒ 54	14 ♋ 25	11 ♊ 44	21 ♒ 18	06 ♒ 31	14 ♐ 51

DECLINATIONS

Date	Moon True ☊ ° '	Moon Mean ☊ ° '	Moon ☽ Latitude ° '	Sun ☉ ° '	Moon ☽ ° '	Mercury ☿ ° '	Venus ♀ ° '	Mars ♂ ° '	Jupiter ♃ ° '	Saturn ♄ ° '	Uranus ♅ ° '	Neptune ♆ ° '	Pluto ♇ ° '
01	28 ♊ 07	29 ♊ 34	03 S 41	14 S 32	12 N 00	06 S 12	06 S 51	21 S 27	22 N 24	20 N 36	15 S 13	18 S 39	12 S 39
02	28 R 00	29 31	02 48	14 51	16 19	06 41	07 20	21 16	22 24	20 35	15 13	18 39	12 39
03	27 56	29 28	01 45	15 10	19 55	07 12	07 48	21 04	22 24	20 35	15 13	18 39	12 40
04	27 54	29 24	00 S 36	15 29	22 33	07 45	08 16	20 53	22 24	20 34	15 13	18 38	12 40
05	27 D 54	29 21	00 N 36	15 47	23 58	08 18	08 44	20 41	22 24	20 34	15 13	18 38	12 41
06	27 55	29 18	01 48	16 05	23 59	08 53	09 12	20 29	22 24	20 33	15 13	18 38	12 41
07	27 57	29 15	02 54	16 23	22 33	09 28	09 40	20 17	22 24	20 33	15 13	18 38	12 41
08	27 57	29 12	03 51	16 40	19 43	10 04	10 07	20 04	22 25	20 32	15 12	18 38	12 42
09	27 R 56	29 09	04 34	16 57	15 43	10 41	10 34	19 52	22 25	20 32	15 12	18 38	12 42
10	27 54	29 05	05 02	17 14	10 46	11 17	11 01	19 39	22 25	20 31	15 12	18 37	12 43
11	27 50	29 02	05 10	17 31	05 N 11	11 54	11 28	19 26	22 25	20 31	15 12	18 37	12 43
12	27 44	28 59	04 59	17 47	00 S 41	12 30	11 54	19 13	22 26	20 30	15 12	18 37	12 44
13	27 38	28 56	04 29	18 03	06 30	13 07	12 20	19 00	22 26	20 30	15 11	18 37	12 44
14	27 32	28 53	03 42	18 19	11 55	13 42	12 46	18 46	22 27	20 29	15 11	18 37	12 45
15	27 28	28 49	02 43	18 34	16 38	14 18	13 12	18 33	22 27	20 29	15 11	18 36	12 45
16	27 25	28 46	01 35	18 49	20 21	14 53	13 37	18 19	22 27	20 28	15 10	18 36	12 46
17	27 24	28 43	00 N 23	19 04	22 52	15 28	14 02	18 05	22 28	20 28	15 10	18 36	12 46
18	27 D 24	28 40	00 S 47	19 18	24 05	16 02	14 26	17 51	22 28	20 27	15 10	18 36	12 46
19	27 25	28 37	01 54	19 32	24 01	16 35	14 50	17 37	22 29	20 26	15 09	18 35	12 47
20	27 27	28 34	02 54	19 46	22 45	17 08	15 14	17 23	22 29	20 26	15 09	18 35	12 47
21	27 29	28 30	03 45	19 59	20 27	17 39	15 38	17 08	22 30	20 25	15 09	18 35	12 48
22	27 30	28 27	04 25	20 12	17 19	18 10	16 01	16 54	22 30	20 25	15 08	18 35	12 48
23	27 R 30	28 24	04 54	20 25	13 30	18 41	16 23	16 39	22 31	20 24	15 08	18 34	12 49
24	27 28	28 21	05 10	20 37	09 10	19 10	16 45	16 24	22 31	20 24	15 08	18 34	12 49
25	27 26	28 18	05 13	20 49	04 S 28	19 38	17 07	16 09	22 32	20 23	15 07	18 34	12 49
26	27 23	28 14	05 03	21 00	00 N 28	20 06	17 28	15 54	22 33	20 22	15 07	18 33	12 50
27	27 19	28 11	04 38	21 11	05 28	20 32	17 49	15 38	22 33	20 22	15 06	18 33	12 50
28	27 16	28 08	03 59	21 22	10 21	20 58	18 09	15 23	22 34	20 21	15 06	18 33	12 50
29	27 12	28 05	03 08	21 32	14 55	21 22	18 29	15 07	22 35	20 21	15 05	18 32	12 51
30	27 ♊ 10	28 ♊ 02	02 S 05	21 S 42	18 N 52	21 S 46	18 S 48	14 S 52	22 N 35	20 N 20	15 S 05	18 S 32	12 S 51

ZODIAC SIGN ENTRIES

Date	h	m	Planets
02	21	13	☽ ♊
05	03	44	☽ ♋
07	08	34	☽ ♌
07	19	53	☿ ♏
08	13	18	♀ ♏
09	11	49	☽ ♍
11	13	53	☽ ♎
13	15	44	☽ ♏
15	18	51	☽ ♐
18	00	40	☽ ♑
20	09	55	☽ ♒
22	06	00	☉ ♐
22	21	52	☽ ♓
25	10	21	☽ ♈
26	18	23	☿ ♐
27	21	06	☽ ♉
30	05	04	☽ ♊

LATITUDES

Date	Mercury ☿ ° '	Venus ♀ ° '	Mars ♂ ° '	Jupiter ♃ ° '	Saturn ♄ ° '	Uranus ♅ ° '	Neptune ♆ ° '	Pluto ♇ ° '
01	02 N 10	01 N 31	02 S 05	00 S 08	01 S 53	00 S 44	00 N 07	09 N 53
04	02 07	01 29	02 01	00 07	01 53	00 44	00 07	09 52
07	01 57	01 26	01 56	00 07	01 53	00 44	00 07	09 51
10	01 43	01 23	01 52	00 07	01 53	00 44	00 07	09 51
13	01 25	01 19	01 47	00 06	01 53	00 44	00 07	09 50
16	01 06	01 15	01 43	00 06	01 53	00 43	00 07	09 49
19	00 46	01 10	01 39	00 06	01 53	00 43	00 07	09 49
22	00 24	01 05	01 34	00 05	01 53	00 43	00 07	09 48
25	00 N 05	00 59	01 30	00 05	01 53	00 43	00 07	09 48
28	00 S 16	00 54	01 26	00 04	01 53	00 43	00 07	09 47
31	00 S 35	00 N 47	01 S 21	00 S 04	01 S 53	00 S 43	00 N 07	09 N 47

DATA

Julian Date	2452215
Delta T	+64 seconds
Ayanamsa	23° 52' 39"
Synetic vernal point	05° ♓ 14' 20"
True obliquity of ecliptic	23° 26' 21"

MOON'S PHASES, APSIDES AND POSITIONS ☽

Date	h	m	Phase	Longitude ° '	Eclipse Indicator
01	05	41	○	08 ♉ 52	
08	12	21	☾	16 ♌ 10	
15	06	40	●	22 ♏ 58	
22	23	21	☽	00 ♓ 44	
30	20	49	○	08 ♊ 43	

Day	h	m		
11	17	13	Perigee	
23	15	45	Apogee	
06	00	18	Max dec	24° N 09'
12	09	15	0S	
18	22	32	Max dec	24° S 13'
26	09	46	0N	

LONGITUDES

Date	Chiron ⚷ ° '	Ceres ⚳ ° '	Pallas ⚴ ° '	Juno ⚵ ° '	Vesta ⚶ ° '	Black Moon Lilith ⚸ ° '
01	25 ♐ 59	19 ♑ 13	22 ♐ 12	19 ♌ 12	11 ♊ 32	07 ♓ 55
11	26 ♐ 53	22 ♑ 08	26 ♐ 03	22 ♌ 33	09 ♊ 36	09 ♓ 02
21	27 ♐ 52	25 ♑ 17	29 ♐ 49	25 ♌ 23	07 ♊ 08	10 ♓ 09
31	28 ♐ 53	28 ♑ 37	03 ♑ 38	27 ♌ 37	04 ♊ 31	11 ♓ 16

All ephemeris data is given at 12.00 UT and the Moon's longitude is additionally given for 24.00 UT

ASPECTARIAN

h m Aspects

01 Thursday
00 20 ☽ □ ♆
03 38 ☽ ± ♇
03 49 ☽ ⊥ ♄
05 41 ☽ ☍ ☉
06 59 ♀ △ ♅
09 04 ☿ △ ♅
15 00 ☽ ⊼ ♇
15 06 ☽ ⊻ ♄
15 25 ☽ # ♇
18 33 ☽ ⚹ ♃

02 Friday
02 51 ☽ # ☉
04 20 ☽ □ ♅
05 36 ☽ # ♅
05 50 ♄ ☍ ♇
06 27 ☽ ⊼ ☿
06 37 ☽ ⊼ ♀
15 30 ♃ St R
18 51 ☽ ± ☿
18 57 ☽ ± ♀
22 29 ☽ ∠ ♃

03 Saturday
02 50 ☽ # ♆
05 28 ☽ △ ♂
07 28 ☿ ☌ ♀
08 23 ☽ △ ♆
13 16 ☽ bQ ♀
13 20 ☽ bQ ☿
15 01 ☽ ⊥ ♃
15 48 ☉ ∥ ♅
17 11 ☽ ∥ ♄
18 07 ☽ ⊼ ☉
20 42 ☽ # ♂
22 21 ☽ ☌ ♄
22 39 ☽ ☍ ♇

04 Sunday
01 56 ☽ ⊻ ♃
05 55 ☽ ± ☉
10 14 ☽ ∥ ♃
10 20 ☽ bQ ♂
11 25 ☽ △ ♅
11 42 ☽ bQ ♆
19 18 ☽ △ ♀
19 45 ☽ △ ☿
23 27 ☽ bQ ☉

05 Monday
03 24 ☽ ± ♂
03 53 ☽ ± ♆
12 43 ♂ ☌ ♆
14 18 ☽ bQ ♅
14 35 ☽ ⊼ ♆
14 41 ☽ ⊼ ♂
22 50 ☉ ⊼ ♄

06 Tuesday
02 12 ♂ # ♄
03 51 ☽ ⊻ ♄
04 16 ☽ △ ☉
04 31 ☽ ⊼ ♇
06 14 ☽ ± ♅
07 32 ☽ ☌ ♃
07 41 ☉ ⊻ ♇
14 20 ☽ ⊥ ♄
15 05 ☽ ± ♇
16 46 ☽ ⊼ ♅
20 31 ☿ bQ ♄

07 Wednesday
03 50 ☿ ∠ ♇
05 41 ☽ □ ♀
05 58 ☽ ∠ ♄
06 49 ☽ bQ ♇
07 10 ☽ □ ☿
08 37 ♀ bQ ♄
13 29 ☽ ∥ ♃
18 27 ♀ ∠ ♇
19 08 ☽ ☍ ♆
21 58 ☽ ☍ ♂

08 Thursday
00 01 ☉ △ ♃
06 03 ☽ ∥ ♄
07 42 ☽ ⚹ ♄
08 43 ☽ △ ♇
09 24 ☽ # ♂
11 27 ☽ ⊻ ♃
12 21 ☽ □ ☉
15 41 ☽ Q ♀
17 53 ☽ Q ☿
19 15 ☽ # ♆
19 15 ☿ ∥ ♀
20 30 ☽ ☍ ♅
21 39 ☽ ⊥ ♃

09 Friday
04 00 ☽ Q ♄
05 37 ☽ # ☉
12 51 ☽ ∠ ♃
13 58 ☽ ⚹ ♀
14 39 ☽ # ♅
16 39 ☽ ⚹ ☿
21 07 ☽ Q ☉
22 08 ☽ ⊼ ♆

10 Saturday
02 59 ☽ # ♇
03 28 ☽ ⊼ ♂
08 11 ☽ ± ♆
09 52 ☽ # ☿
10 05 ☽ □ ♄
10 56 ☽ # ♀
11 27 ☽ □ ♇
13 56 ☽ ⚹ ♃
14 00 ☽ ± ♂
17 28 ☽ ∠ ♀
18 40 ☽ ⚹ ☉
20 47 ☽ ∠ ☿
22 52 ☽ ⊼ ♅
23 12 ☽ bQ ♆

11 Sunday
05 45 ☽ bQ ♂
08 51 ☽ ± ♅
09 50 ☽ Q ♃
09 51 ☽ ⊥ ♀
13 34 ☽ ⊥ ☿
17 26 ☽ Q ♇
18 59 ☿ □ ♆
20 44 ☽ ⊻ ♀
21 26 ☽ ∠ ☉
23 47 ☽ bQ ♅

12 Monday
00 06 ☽ △ ♆
00 44 ☽ ⊻ ☿
07 54 ☽ △ ♂
09 52 ☿ ± ♄
11 40 ☽ △ ♄
13 23 ☽ ⚹ ♇
13 28 ☽ ⊥ ☉
15 37 ☽ □ ♃
21 03 ☿ ∥ ♇

13 Tuesday
00 11 ☽ ⊻ ☉
00 42 ☽ △ ♅
01 45 ☿ ⊥ ♇
07 32 ☉ □ ♅
12 02 ♀ □ ♆
12 30 ☽ bQ ♄
14 26 ☽ ∠ ♇

14 Wednesday
02 09 ☽ □ ♆
03 29 ☽ ☌ ♀
03 32 ☽ ± ♄
03 55 ♀ ± ♄
05 35 ☽ ⊥ ♇
09 01 ☽ ☌ ☿
10 37 ♀ ∥ ♇
12 40 ☽ □ ♂
13 37 ☽ ⊼ ♄
15 46 ☽ ⊻ ♇
15 55 ☽ ∥ ♇
16 24 ☽ ∥ ♀
17 48 ☽ △ ♃
21 49 ☽ ∥ ☿

15 Thursday
03 18 ☽ □ ♅
03 54 ♀ ⊥ ♇
04 06 ☽ ∥ ♅
05 35 ♂ ∥ ♆
06 00 ♂ △ ♄
06 40 ☽ ☌ ☉
08 50 ☽ Q ♆
10 50 ☉ ∥ ♂
15 23 ☉ ∥ ♆
19 33 ☽ bQ ♃
21 16 ☽ Q ♂
22 48 ☽ ∥ ♂
23 49 ☽ ∥ ♆

16 Friday
00 18 ☿ ⊼ ♄
00 27 ☽ ∥ ☉
05 46 ☽ ⚹ ♆
10 42 ☽ Q ♅
11 19 ☽ ± ♃
12 22 ☽ ⊻ ♀
12 54 ☽ # ♄
12 57 ☉ Q ♆
16 33 ☿ □ ♂
17 27 ☽ ☍ ♄
19 39 ☽ ⚹ ♂
19 53 ☽ ⊻ ☿
20 09 ☽ ☌ ♇
21 58 ☽ ⊼ ♃
22 09 ☿ ⊻ ♇
23 53 ☿ ∥ ♅

17 Saturday
00 11 ☽ ⊥ ♀
05 28 ♂ ⚹ ♇
07 07 ☽ # ♃
08 10 ☽ ⊥ ☿
08 14 ☽ ⚹ ♅
08 37 ☽ ∠ ♆
12 57 ☿ △ ♃
16 09 ☽ ⊻ ☉
18 16 ☽ ∠ ♀

18 Sunday
00 30 ☽ ∠ ♂
01 09 ☽ ⊥ ♆
02 59 ☽ ∠ ☿
04 09 ☽ ⊥ ☉
11 56 ☽ ∠ ♅
12 20 ☽ ⊻ ♆
12 32 ♂ ⊼ ♃
16 13 ♀ ⊼ ♄
18 19 ☽ ⊥ ♂
22 30 ☽ ∠ ☉

19 Monday
00 23 ☽ ⊼ ♄
01 21 ☽ ⚹ ♀
03 45 ☽ ⊻ ♇
04 59 ☽ ⊥ ♅
05 19 ☽ ☍ ♃
06 24 ☽ ⊻ ♂
11 24 ☽ ⚹ ☿
11 48 ☽ ± ♄
15 18 ☽ ⊥ ♇
16 32 ☽ ⊻ ♅

20 Tuesday
02 08 ♀ ⊻ ♇
03 04 ☽ Q ♀
05 08 ☽ bQ ♄
05 57 ☽ ⚹ ☉
07 00 ♀ ∥ ♅
08 52 ☽ ∠ ♇
14 14 ☽ Q ☿
15 21 ☽ # ♃
15 31 ♀ △ ♃
19 48 ☿ ∥ ♂
22 22 ☽ ☌ ♆

21 Wednesday
03 51 ☿ □ ♅
07 48 ☽ Q ☉
09 57 ♃ ± ♅
10 36 ☽ △ ♄
12 19 ☽ # ♄
14 41 ☽ ⚹ ♇
15 43 ☽ ∥ ☉
15 51 ☽ ⊼ ♃
18 50 ☽ □ ♀
21 02 ☽ ☌ ♂

22 Thursday
03 01 ☽ ∥ ♆
03 50 ☽ ± ♃
03 57 ☽ ☌ ♅
06 46 ☽ ∥ ☿
07 11 ☉ bQ ♃
07 38 ☽ □ ☿
14 54 ☽ Q ♇
15 03 ☽ ∥ ♂
19 50 ☽ ∥ ♀
21 53 ☽ bQ ♃
23 21 ☽ □ ☉

23 Friday
02 08 ☽ ∥ ♅
04 57 ☿ Q ♆
07 03 ☿ ∥ ♆
10 44 ☉ # ♄
10 44 ☽ ⊻ ♆
15 59 ☽ ∥ ♇
16 43 ♀ □ ♂
22 09 ♀ ∥ ♂
22 41 ☽ □ ♄
22 54 ☽ ⊥ ♆

24 Saturday
03 24 ☽ □ ♇
04 03 ☽ △ ♃
13 30 ☽ ⊻ ♂
14 32 ☽ △ ♀
16 39 ☽ ⊻ ♅
17 05 ☽ ∠ ♆

25 Sunday
02 20 ☽ ⊥ ♂
04 43 ☽ ⊥ ♅
05 29 ☽ △ ☿
10 39 ☽ Q ♄
10 50 ♀ □ ♅
17 21 ☽ △ ☉
21 26 ☽ ∠ ♂
22 41 ☽ ∠ ♅
23 07 ☽ ⚹ ♆

26 Monday
00 02 ☽ bQ ♀
04 30 ♂ ± ♃
10 17 ☽ ⚹ ♄
14 30 ☿ bQ ♃
15 27 ☽ △ ♇
15 32 ☽ □ ♃
15 42 ☽ bQ ☿
19 07 ♂ ☌ ♅
19 47 ☽ ± ♀
22 45 ☽ Q ♆

27 Tuesday
00 03 ♃ ⊼ ♇
01 33 ☽ bQ ☉
02 39 ☿ # ♄
04 11 ☽ ⚹ ♅
04 44 ☽ ⚹ ♂
08 46 ☽ ⊼ ♀
11 46 ☽ ± ☿
15 17 ☽ ∠ ♄
20 27 ☽ ± ☉
20 37 ☽ bQ ♇

28 Wednesday
00 56 ☽ ⊼ ☿
01 22 ♀ Q ♆
02 07 ☽ Q ♃
03 17 ☽ Q ♅
05 11 ☽ Q ♂
08 21 ☽ ⊥ ♄
08 52 ☽ ⊼ ☉
09 24 ☽ □ ♆
13 45 ☽ ± ♇
15 35 ☉ ⚹ ♆
19 35 ☽ ⊻ ♄

29 Thursday
00 35 ☽ ⚹ ♃
00 49 ☽ # ♇
01 02 ☽ ⊼ ♇
12 59 ☽ # ♅
13 05 ☽ □ ♅
13 07 ☽ # ♂
15 20 ♂ ∥ ♅
16 10 ♀ ∥ ♆
16 50 ☽ □ ♂
23 21 ☽ ☍ ♀

30 Friday
04 04 ☽ ∠ ♃
05 12 ☉ ∥ ☿
09 48 ☽ # ♆
11 33 ☽ # ♀
13 39 ☉ ± ♃
16 14 ☽ ☍ ☿
16 52 ☽ △ ♆
20 14 ☽ ⊥ ♃
20 49 ☽ ☍ ☉
21 39 ☿ ⚹ ♆
22 29 ☽ ∥ ♄

DECEMBER 2001

Date	Sidereal time h m s	Sun ☉ ° ' "	Moon ☽ ° ' "	Moon ☽ 24.00 ° ' "	Mercury ☿ ° '	Venus ♀ ° '	Mars ♂ ° '	Jupiter ♃ ° '	Saturn ♄ ° '	Uranus ♅ ° '	Neptune ♆ ° '	Pluto ⯓ ° '
01	16 41 39	09 ♐ 21 32	17 ♊ 13 22	24 ♊ 00 46	07 ♐ 28	28 ♏ 47	24 ♒ 37	14 ♋ 20	11 ♊ 39	21 ♒ 20	06 ♒ 32	14 ♐ 53
02	16 45 36	10 22 20	00 ♋ 51 27	07 ♋ 45 07	09 02	00 ♐ 03	25 20	14 R 14	11 R 34	21 21	06 34	14 55
03	16 49 32	11 23 10	14 ♋ 41 26	21 ♋ 40 03	10 36	01 18	26 04	14 09	11 29	21 23	06 35	14 58
04	16 53 29	12 24 01	28 ♋ 40 39	05 ♌ 42 53	12 11	02 33	26 48	14 03	11 24	21 25	06 37	15 00
05	16 57 26	13 24 53	12 ♌ 46 26	19 ♌ 50 59	13 45	03 49	27 31	13 57	11 19	21 26	06 38	15 02
06	17 01 22	14 25 47	26 ♌ 56 15	04 ♍ 01 56	15 19	05 04	28 15	13 51	11 14	21 28	06 40	15 05
07	17 05 19	15 26 42	11 ♍ 07 45	18 ♍ 13 26	16 53	06 20	28 58	13 44	11 10	21 30	06 41	15 07
08	17 09 15	16 27 38	25 ♍ 18 44	02 ♎ 23 22	18 28	07 35	29 ♒ 42	13 38	11 05	21 32	06 43	15 09
09	17 13 12	17 28 36	09 ♎ 27 03	16 ♎ 29 31	20 02	08 51	00 ♓ 26	13 31	11 00	21 34	06 44	15 12
10	17 17 08	18 29 34	23 ♎ 30 26	00 ♏ 29 32	21 36	10 06	01 09	13 25	10 55	21 36	06 46	15 14
11	17 21 05	19 30 34	07 ♏ 26 30	14 ♏ 21 01	23 10	11 22	01 53	13 18	10 50	21 38	06 48	15 16
12	17 25 01	20 31 35	21 ♏ 12 46	28 ♏ 01 30	24 45	12 37	02 37	13 11	10 45	21 40	06 49	15 18
13	17 28 58	21 32 37	04 ♐ 46 54	11 ♐ 28 46	26 19	13 53	03 21	13 04	10 40	21 42	06 51	15 21
14	17 32 55	22 33 40	18 ♐ 06 53	24 ♐ 41 06	27 54	15 08	04 05	12 57	10 36	21 44	06 53	15 23
15	17 36 51	23 34 43	01 ♑ 11 18	07 ♑ 37 28	29 ♐ 29	16 24	04 48	12 50	10 31	21 46	06 55	15 25
16	17 40 48	24 35 48	13 ♑ 59 35	20 ♑ 17 46	01 ♑ 04	17 39	05 32	12 42	10 26	21 48	06 57	15 28
17	17 44 44	25 36 53	26 ♑ 32 08	02 ♒ 42 55	02 39	18 55	06 16	12 35	10 22	21 51	06 58	15 30
18	17 48 41	26 37 58	08 ♒ 50 21	14 ♒ 54 48	04 14	20 10	07 00	12 27	10 17	21 53	07 00	15 32
19	17 52 37	27 39 04	20 ♒ 56 36	26 ♒ 56 13	05 49	21 26	07 44	12 20	10 12	21 55	07 02	15 35
20	17 56 34	28 40 10	02 ♓ 54 07	08 ♓ 50 47	07 24	22 41	08 28	12 12	10 08	21 58	07 04	15 37
21	18 00 30	29 ♐ 41 16	14 ♓ 46 46	20 ♓ 42 39	08 59	23 57	09 12	12 04	10 03	22 00	07 06	15 39
22	18 04 27	00 ♑ 42 22	26 ♓ 39 00	02 ♈ 36 24	10 35	25 12	09 56	11 56	09 59	22 03	07 08	15 41
23	18 08 24	01 43 29	08 ♈ 35 27	14 ♈ 36 46	12 10	26 28	10 40	11 48	09 54	22 05	07 10	15 44
24	18 12 20	02 44 36	20 ♈ 40 54	26 ♈ 48 25	13 45	27 43	11 23	11 41	09 50	22 08	07 12	15 46
25	18 16 17	03 45 43	02 ♉ 59 49	09 ♉ 15 36	15 21	28 ♐ 59	12 07	11 33	09 46	22 10	07 13	15 48
26	18 20 13	04 46 50	15 ♉ 36 10	22 ♉ 01 53	16 56	00 ♑ 14	12 51	11 25	09 42	22 13	07 15	15 50
27	18 24 10	05 47 57	28 ♉ 32 59	05 ♊ 09 40	18 31	01 30	13 35	11 16	09 37	22 15	07 17	15 52
28	18 28 06	06 49 05	11 ♊ 51 59	18 ♊ 39 54	20 06	02 45	14 19	11 08	09 33	22 18	07 19	15 55
29	18 32 03	07 50 12	25 ♊ 33 14	02 ♋ 31 41	21 41	04 01	15 03	11 00	09 29	22 21	07 22	15 57
30	18 35 59	08 51 20	09 ♋ 34 52	16 ♋ 42 15	23 15	05 16	15 47	10 52	09 25	22 23	07 24	15 59
31	18 39 56	09 ♑ 52 27	23 ♋ 53 12	01 ♌ 07 03	24 ♑ 49	06 ♑ 32	16 ♓ 31	10 ♋ 44	09 ♊ 21	22 ♒ 26	07 ♒ 26	16 ♐ 01

Date	Moon True ☊ ° '	Moon Mean ☊ ° '	Moon ☽ Latitude ° '
01	27 ♊ 09	27 ♊ 59	00 S 55
02	27 D 09	27 55	00 N 20
03	27 09	27 52	01 35
04	27 10	27 49	02 46
05	27 12	27 46	03 46
06	27 13	27 43	04 34
07	27 13	27 40	05 04
08	27 R 13	27 36	05 16
09	27 12	27 33	05 09
10	27 11	27 30	04 44
11	27 10	27 27	04 02
12	27 10	27 24	03 06
13	27 09	27 20	02 01
14	27 09	27 17	00 N 50
15	27 D 09	27 14	00 S 22
16	27 09	27 11	01 32
17	27 09	27 08	02 36
18	27 09	27 05	03 31
19	27 R 09	27 01	04 16
20	27 09	26 58	04 49
21	27 09	26 55	05 10
22	27 09	26 52	05 17
23	27 D 09	26 49	05 11
24	27 09	26 46	04 51
25	27 09	26 42	04 17
26	27 10	26 39	03 31
27	27 11	26 36	02 32
28	27 12	26 33	01 24
29	27 12	26 30	00 S 09
30	27 R 12	26 26	01 N 08
31	27 ♊ 11	26 ♊ 23	02 N 22

DECLINATIONS

Date	Sun ☉ ° '	Moon ☽ ° '	Mercury ☿ ° '	Venus ♀ ° '	Mars ♂ ° '	Jupiter ♃ ° '	Saturn ♄ ° '	Uranus ♅ ° '	Neptune ♆ ° '	Pluto ⯓ ° '
01	21 S 51	21 N 55	22 S 08	19 S 07	14 S 36	22 N 36	20 N 19	15 S 04	18 S 32	12 S 52
02	22 00	23 47	22 29	19 25	14 20	22 37	20 19	15 04	18 31	12 52
03	22 09	24 13	22 49	19 43	14 04	22 38	20 18	15 03	18 31	12 52
04	22 17	23 08	23 08	20 00	13 48	22 38	20 18	15 02	18 31	12 53
05	22 25	20 35	23 26	20 17	13 31	22 39	20 17	15 02	18 30	12 53
06	22 32	16 49	23 43	20 33	13 15	22 40	20 17	15 01	18 30	12 53
07	22 39	12 05	23 58	20 48	12 59	22 41	20 16	15 01	18 30	12 54
08	22 45	06 42	24 12	21 03	12 42	22 41	20 15	15 00	18 29	12 54
09	22 51	01 N 00	24 25	21 17	12 25	22 42	20 15	14 59	18 29	12 54
10	22 56	04 S 44	24 37	21 31	12 08	22 43	20 14	14 59	18 28	12 55
11	23 01	10 11	24 47	21 44	11 52	22 44	20 14	14 58	18 28	12 55
12	23 06	15 04	24 56	21 56	11 35	22 45	20 13	14 57	18 28	12 55
13	23 10	19 07	25 04	22 08	11 18	22 45	20 13	14 57	18 27	12 55
14	23 14	22 05	25 10	22 19	11 00	22 46	20 12	14 56	18 27	12 56
15	23 17	23 48	25 15	22 29	10 43	22 47	20 11	14 55	18 26	12 56
16	23 20	24 14	25 19	22 39	10 26	22 48	20 11	14 54	18 26	12 56
17	23 22	23 24	25 21	22 48	10 09	22 49	20 10	14 54	18 25	12 57
18	23 24	21 27	25 22	22 56	09 51	22 50	20 10	14 53	18 25	12 57
19	23 25	18 33	25 21	23 04	09 34	22 50	20 09	14 52	18 25	12 57
20	23 26	14 56	25 18	23 11	09 16	22 51	20 09	14 51	18 24	12 57
21	23 26	10 45	25 15	23 17	08 58	22 52	20 08	14 51	18 24	12 58
22	23 26	06 11	25 09	23 23	08 41	22 53	20 08	14 50	18 23	12 58
23	23 26	01 S 21	25 02	23 28	08 23	22 54	20 07	14 49	18 23	12 58
24	23 25	03 N 35	24 54	23 32	08 05	22 55	20 07	14 48	18 22	12 58
25	23 23	08 29	24 44	23 35	07 47	22 55	20 07	14 47	18 22	12 58
26	23 21	13 09	24 33	23 38	07 29	22 56	20 06	14 46	18 21	12 59
27	23 19	17 22	24 20	23 39	07 11	22 57	20 06	14 46	18 21	12 59
28	23 16	20 49	24 05	23 41	06 53	22 58	20 05	14 45	18 20	12 59
29	23 12	23 13	23 49	23 41	06 35	22 59	20 05	14 44	18 20	12 59
30	23 09	24 13	23 32	23 41	06 17	23 00	20 04	14 43	18 19	12 59
31	23 S 04	23 N 40	23 S 13	23 S 40	05 S 59	23 N 00	20 N 04	14 S 42	18 S 19	13 S 00

ZODIAC SIGN ENTRIES

Date	h	m	Planets
02	10	30	☽ ♋
02	11	11	♀ ♐
04	14	15	☽ ♌
06	17	11	☽ ♍
08	19	57	☽ ♎
08	21	52	♂ ♓
10	23	09	☽ ♏
13	03	30	☽ ♐
15	09	48	☽ ♑
15	19	55	☿ ♑
17	18	43	☽ ♒
20	06	09	☽ ♓
21	19	21	☉ ♑
22	18	45	☽ ♈
25	06	12	☽ ♉
26	07	25	♀ ♑
27	14	39	☽ ♊
29	19	40	☽ ♋
31	22	09	☽ ♌

LATITUDES

Date	Mercury ☿ ° '	Venus ♀ ° '	Mars ♂ ° '	Jupiter ♃ ° '	Saturn ♄ ° '	Uranus ♅ ° '	Neptune ♆ ° '	Pluto ⯓ ° '
01	00 S 35	00 N 47	01 S 21	00 S 04	01 S 53	00 S 43	00 N 07	09 N 47
04	00 54	00 41	01 17	00 04	01 52	00 43	00 07	09 47
07	01 11	00 34	01 13	00 03	01 52	00 43	00 07	09 46
10	01 27	00 27	01 09	00 03	01 52	00 43	00 07	09 46
13	01 41	00 20	01 05	00 02	01 51	00 43	00 07	09 46
16	01 53	00 13	01 01	00 02	01 51	00 43	00 07	09 46
19	02 02	00 N 06	00 57	00 02	01 51	00 42	00 06	09 46
22	02 09	00 S 02	00 54	00 01	01 50	00 42	00 06	09 46
25	02 12	00 09	00 50	00 S 01	01 50	00 42	00 06	09 46
28	02 11	00 16	00 46	00 00	01 49	00 42	00 06	09 46
31	02 S 05	00 S 23	00 S 43	00 00	01 S 48	00 S 42	00 N 06	09 N 46

DATA

Julian Date	2452245
Delta T	+64 seconds
Ayanamsa	23° 52' 44"
Synetic vernal point	05° ♓ 14' 16"
True obliquity of ecliptic	23° 26' 20"

MOON'S PHASES, APSIDES AND POSITIONS ☽

Date	h	m	Phase	Longitude ° '	Eclipse Indicator
07	19	52	☾	15 ♍ 47	
14	20	47	●	22 ♐ 56	Annular
22	20	56	☽	01 ♈ 05	
30	10	41	○	08 ♋ 48	

Day	h	m		
06	22	49	Perigee	
21	13	03	Apogee	
03	06	55	Max dec	24° N 15'
09	16	09	0S	
16	07	51	Max dec	24° S 15'
23	18	37	0N	
30	15	40	Max dec	24° N 15'

LONGITUDES

Date	Chiron ⚷ ° '	Ceres ⚳ ° '	Pallas ⚴ ° '	Juno ⚵ ° '	Vesta ⚶ ° '	Black Moon Lilith ⚸ ° '
01	28 ♐ 53	28 ♑ 37	03 ♑ 38	27 ♌ 37	04 ♊ 31	11 ♓ 16
11	29 ♐ 57	02 ♒ 07	07 ♑ 29	29 ♌ 11	02 ♊ 02	12 ♓ 23
21	01 ♑ 02	05 ♒ 44	11 ♑ 21	29 ♌ 58	29 ♉ 59	13 ♓ 30
31	02 ♑ 07	09 ♒ 27	15 ♑ 12	29 ♌ 55	28 ♉ 33	14 ♓ 38

All ephemeris data is given at 12.00 UT and the Moon's longitude is additionally given for 24.00 UT

ASPECTARIAN

h m	Aspects
01 Saturday	
02 08	☽ ☌ ♄
06 54	☽ ⊻ ♃
07 50	☽ ☍ ⯓
11 18	☉ Q ♅
11 19	☽ # ☉
14 27	☽ # ☿
19 04	☽ ∥ ♃
19 17	☽ △ ♅
19 39	☽ ♇ ♆
21 43	♀ ♇ ♃
02 Sunday	
00 31	☿ ± ♃
01 48	☽ △ ♂
10 26	☽ ⊼ ♀
11 29	☽ ± ♆
16 59	☿ Q ♅
21 09	☿ # ♃
21 36	☽ ♇ ♅
21 56	☽ ± ♀
21 57	☽ ⊼ ♆
03 Monday	
04 03	☽ ⊼ ☿
05 24	☽ ♇ ♂
05 51	☽ ⊼ ☉
06 30	☽ ⊻ ♄
11 04	☽ ☌ ♃
12 28	☽ ⊼ ⯓
13 12	☽ ± ♅
14 13	☉ ☍ ♄
15 03	☽ ♇ ♀
15 43	☽ ± ☿
16 47	☽ ⊥ ♄
17 00	☽ ± ☉
21 46	☽ ± ♂
22 49	☽ ± ⯓
23 32	☽ ⊼ ♅
04 Tuesday	
00 47	☿ ☍ ♄
08 08	☽ ∠ ♄
08 36	☽ ⊼ ♂
09 07	☽ ♇ ☿
09 39	☽ ♇ ☉
11 51	☽ # ☿
14 16	☽ ♇ ⯓
17 48	☽ ∥ ♃
19 15	♂ Q ⯓
19 16	☽ △ ♀
21 04	☽ # ☉
21 36	☉ ☌ ☿
05 Wednesday	
01 33	☽ ☍ ♆
09 33	☽ ⚹ ♄
12 19	♀ # ♄
13 10	☽ △ ☉
13 52	☽ △ ☿
13 59	☽ ⊻ ♃
14 07	☽ # ♀
14 16	☽ ∥ ♄
14 51	☿ ⊼ ♃
15 51	☽ △ ⯓
23 26	☉ ⊼ ♃
06 Thursday	
00 05	☽ ⊥ ♃
02 07	☽ # ♆
02 44	☽ ☍ ♅
05 47	☽ Q ♄
08 13	☿ ☌ ⯓
14 20	☽ ☍ ♂
15 12	☽ ∠ ♃
21 36	☽ # ♅
07 Friday	
03 06	☽ □ ♀
03 55	☉ ☌ ⯓
04 29	☽ ⊼ ♆
05 17	♂ ♇ ♃
07 29	☽ # ♂
08 08	☽ # ⯓
12 03	☽ □ ♄
14 25	☿ Q ♂
14 38	☽ ± ♆
16 23	☽ ⚹ ♃
18 46	☽ □ ⯓
18 59	♂ ∥ ⯓
18 59	♀ ⚹ ♆
19 52	☽ □ ☉
20 03	☉ # ♃
22 57	☽ □ ☿
08 Saturday	
05 35	☽ ⊼ ♅
05 54	☽ ♇ ♆
12 31	☽ Q ♀
12 32	☽ Q ♃
12 48	♀ ± ♃
15 46	☽ ± ♅
19 50	☽ ⊼ ♂
09 Sunday	
01 20	☽ Q ⯓
04 43	☽ Q ☉
06 35	☽ ± ♂
07 05	☽ ♇ ♅
07 23	☽ △ ♆
09 17	☽ Q ☿
10 52	☽ ⚹ ♀
14 37	☽ △ ♄
18 53	☽ □ ♃
21 49	☽ ⚹ ⯓
22 45	☽ ♇ ♂
10 Monday	
02 05	♀ Q ♅
02 45	☽ ⚹ ☉
08 19	☽ ⚹ ☿
08 43	☽ △ ♅
11 56	☿ ⚹ ♅
14 36	☿ ∠ ♆
15 01	☽ ∠ ♀
16 06	☽ ♇ ♄
23 35	☽ ∠ ⯓
11 Tuesday	
01 52	☽ △ ♂
02 32	♀ ☍ ♄
06 32	☽ ∠ ☉
07 31	☽ ± ♄
08 03	☽ ⊥ ♀
10 53	☽ □ ♆
13 26	☽ ∠ ☿
15 11	☽ ⊥ ⯓
17 51	☽ ⊼ ♄
19 28	☽ ∥ ♂
19 29	☽ ⊻ ♀
22 05	☽ △ ♃
23 22	☽ ⊥ ☉
12 Wednesday	
01 01	☽ ∥ ⯓
01 38	☽ ⊻ ⯓
07 07	☽ ⊥ ☿
10 42	☽ ⊻ ☉
11 25	☽ ∥ ♅
12 48	☽ □ ♅
18 22	☽ Q ♆
19 02	☽ ⊻ ☿
19 27	☉ Q ♂
21 49	♀ ⊼ ♃
13 Thursday	
00 11	☽ ♇ ♃
07 40	☽ ∥ ♆
09 18	☽ □ ♂
15 43	☽ ⚹ ♆
15 49	☉ ⚹ ♅
16 03	☽ ± ♃
19 32	☉ ∠ ♆
19 49	☽ # ♄
20 49	☽ Q ♅
22 29	☽ ☍ ♄
14 Friday	
02 44	☽ ⊼ ♃
06 02	☽ ☌ ♀
07 02	☽ ☌ ⯓
14 42	☽ ∥ ♀
16 52	♀ ☌ ⯓
18 37	☽ ⚹ ♅
18 53	☽ ∠ ♆
19 39	☽ Q ♂
19 42	☽ # ♃
20 47	☽ ☌ ☉
15 Saturday	
02 16	☽ ∥ ☉
08 24	☽ ☌ ☿
11 29	☽ ⊥ ♆
19 08	☽ ⚹ ♂
22 26	☽ ∠ ♅
22 41	☽ ⊻ ♆
16 Sunday	
05 19	☽ ⊼ ♄
09 35	☽ ☍ ♃
10 11	☿ ⊥ ♆
14 48	☽ ⊻ ⯓
15 27	☽ ⊥ ♅
16 37	☽ ± ♄
19 44	☽ ⊻ ♀
17 Monday	
01 14	☽ ∠ ♂
02 17	☽ ⊥ ⯓
02 56	☽ ⊻ ♅
08 31	☽ ⊥ ♀
09 44	☽ ♇ ♄
10 04	☽ ⊻ ☉
12 31	☽ ∥ ☉
13 52	♀ # ♃
19 41	☽ ⊥ ♂
19 42	☽ ∠ ⯓
20 09	☽ ∥ ♀
20 36	☽ # ♃
22 45	☽ ⊥ ☉
18 Tuesday	
01 37	☽ ⊻ ☿
03 59	☽ ∠ ♀
08 09	☽ ⊻ ♂
08 23	☽ ☌ ♆
12 06	♂ ⊻ ♆
14 49	☽ △ ♄
15 08	☽ ⊥ ☿
18 01	☽ ∠ ☉
19 03	☽ ⊼ ♃
23 32	☽ # ♄
19 Wednesday	
01 17	☽ ⚹ ⯓
06 50	☽ ± ♃
11 42	☽ ∠ ☿
13 04	☽ ∥ ♆
13 05	☽ ⚹ ♀
13 58	☽ ☌ ♅
21 41	♀ ⚹ ♅
23 47	♀ ∠ ♆
20 Thursday	
00 39	☽ ♇ ♃
01 19	☽ Q ⯓
02 41	☽ ⚹ ☉
05 14	☿ ∠ ♅
06 51	☿ ⊻ ♆
12 28	☽ ∥ ♅
16 02	☽ Q ♀
20 25	☽ ⊻ ♆
22 29	☽ ⚹ ☿
23 41	☽ ∥ ⯓
23 58	☽ ☌ ♂
21 Friday	
02 30	☽ □ ♄
05 09	☽ Q ☉
06 35	☽ △ ♃
08 35	☽ ⊥ ♆
13 46	☽ □ ⯓
17 47	☿ ⚹ ♂
22 13	☽ ∥ ♂
22 Saturday	
02 30	☽ Q ☿
02 40	☽ ⊻ ♅
02 50	☽ ∠ ♆
03 23	☿ ⊼ ♄
08 44	☽ □ ♀
13 35	♂ □ ♄
14 40	☽ Q ♄
14 49	☽ ⊥ ♅
20 56	☽ □ ☉
22 15	☉ ⊥ ♆
23 Sunday	
03 32	☉ ∥ ♀
06 59	☿ ☍ ♃
08 58	☽ ∠ ♅
09 08	☽ ⚹ ♆
14 37	☽ ⚹ ♄
16 24	☽ ⊻ ♂
18 21	☽ □ ♃
20 13	☽ □ ☿
24 Monday	
02 15	☽ △ ⯓
05 05	☽ ⊥ ♂
09 03	☽ Q ♆
14 51	☽ ⚹ ♅
19 54	♂ △ ♃
20 06	☽ ∠ ♄
23 54	☽ ∠ ♂
25 Tuesday	
03 21	☽ △ ♀
05 23	☽ Q ♃
07 45	☽ ♇ ⯓
08 45	☽ # ♂
13 28	☽ ⊥ ♄
13 36	☽ △ ☉
14 16	☽ Q ♅
17 59	☿ ± ♄
18 59	☿ ⊻ ⯓
20 08	☽ □ ♆
26 Wednesday	
00 44	☿ ⊥ ♅
00 53	☽ ⊻ ♄
01 04	☽ ± ⯓
04 10	☽ ⚹ ♃
06 30	☽ ⚹ ♂
11 05	☽ # ⯓
11 15	☽ ♇ ♀
12 26	☽ ⊼ ⯓
14 51	☽ △ ☿
20 29	☽ ♇ ☉
20 50	☽ # ♅
27 Thursday	
00 23	☽ □ ♅
01 50	♂ ⊥ ♆
05 48	☽ ± ♀
06 15	☽ Q ♂
07 52	☽ ∠ ♃
07 56	♀ ⊥ ♆
14 28	☽ ± ☉
16 02	♀ Q ♂
17 56	☽ ⊼ ♀
18 13	☽ # ♆
22 16	☽ ♇ ☿
28 Friday	
00 05	☽ ⊥ ♃
02 14	☽ ⊼ ☉
03 52	☽ △ ♆
06 19	☽ ∥ ♄
07 54	☽ ☌ ♄
10 43	☽ ⊻ ♃
16 29	☽ ± ☿
16 36	☽ □ ♂
19 11	☽ ☍ ⯓
23 52	☉ ∠ ♅
29 Saturday	
00 21	☉ ⊻ ♆
04 24	☽ ⊼ ☿
06 25	☽ △ ♅
06 26	☽ ♇ ♆
08 56	☽ ∥ ♃
11 56	☽ # ☉
16 09	☿ ⊥ ⯓
19 24	☽ # ♀
20 07	☽ # ☿
22 01	☽ ± ♆
22 25	☿ ⊻ ♅
23 23	☿ ∥ ♀
30 Sunday	
03 58	☽ ☍ ♀
08 16	☽ ♇ ♅
08 17	☽ ⊼ ♆
10 41	☽ ☍ ☉
11 44	☽ ⊻ ♄
14 10	☽ ☌ ♃
18 53	♂ □ ⯓
21 48	☽ ⊥ ♄
22 49	☽ ⊼ ⯓
23 01	☽ △ ♂
23 31	☽ ± ♅
31 Monday	
00 31	☉ ⊼ ♄
05 17	☿ ♇ ♄
08 53	☽ ± ⯓
09 34	☽ ⊼ ♅
12 05	☽ # ♀
12 47	☽ ∠ ♄
13 43	☽ ☍ ☿
20 21	☽ # ☿
20 58	☽ # ☉
21 21	☽ ∥ ♃
23 52	☽ ♇ ⯓

JANUARY 2002

LONGITUDES

Date	Sidereal time h m s	Sun ☉ ° ' "	Moon ☽ ° ' "	Moon ☽ 24.00 ° ' "	Mercury ☿ ° '	Venus ♀ ° '	Mars ♂ ° '	Jupiter ♃ ° '	Saturn ♄ ° '	Uranus ♅ ° '	Neptune ♆ ° '	Pluto ♇ ° '
01	18 43 53	10 ♑ 53 35	08 ♌ 23 00	15 ♌ 40 16	26 ♑ 21	07 ♑ 47	17 ♓ 15	10 ♋ 36	09 ♊ 18	22 ♒ 29	07 ♒ 28	16 ♐ 03
02	18 47 49	11 54 44	22 ♌ 58 05	00 ♍ 15 40	27 53	09 03	17 59	10 R 28	09 R 14	22 32	07 30	16 05
03	18 51 46	12 55 52	07 ♍ 32 17	14 ♍ 47 18	29 ♑ 24	10 18	18 43	10 20	09 10	22 35	07 32	16 08
04	18 55 42	13 57 00	22 ♍ 00 07	29 ♍ 10 18	00 ♒ 53	11 34	19 27	10 12	09 07	22 37	07 34	16 10
05	18 59 39	14 58 09	06 ♎ 17 27	13 ♎ 21 17	02 21	12 49	20 10	10 04	09 03	22 40	07 36	16 12
06	19 03 35	15 59 18	20 ♎ 21 37	27 ♎ 18 20	03 46	14 05	20 54	09 55	09 00	22 43	07 38	16 14
07	19 07 32	17 00 27	04 ♏ 11 23	11 ♏ 00 47	05 09	15 20	21 38	09 47	08 56	22 46	07 40	16 16
08	19 11 28	18 01 37	17 ♏ 46 34	24 ♏ 28 50	06 29	16 36	22 22	09 39	08 53	22 49	07 43	16 18
09	19 15 25	19 02 46	01 ♐ 07 39	07 ♐ 43 07	07 45	17 51	23 06	09 32	08 50	22 52	07 45	16 20
10	19 19 22	20 03 56	14 ♐ 15 21	20 ♐ 44 25	08 57	19 07	23 50	09 24	08 47	22 55	07 47	16 22
11	19 23 18	21 05 05	27 ♐ 10 25	03 ♑ 33 26	10 04	20 22	24 34	09 16	08 44	22 58	07 49	16 24
12	19 27 15	22 06 15	09 ♑ 53 32	16 ♑ 10 47	11 06	21 38	25 17	09 08	08 41	23 01	07 51	16 26
13	19 31 11	23 07 24	22 ♑ 25 16	28 ♑ 37 04	12 01	22 53	26 01	09 00	08 38	23 04	07 54	16 28
14	19 35 08	24 08 32	04 ♒ 46 16	10 ♒ 53 01	12 48	24 09	26 45	08 53	08 35	23 08	07 56	16 30
15	19 39 04	25 09 41	16 ♒ 57 27	22 ♒ 59 44	13 28	25 24	27 29	08 45	08 33	23 11	07 58	16 32
16	19 43 01	26 10 48	29 ♒ 00 06	04 ♓ 58 47	13 58	26 40	28 13	08 38	08 30	23 14	08 00	16 34
17	19 46 57	27 11 55	10 ♓ 56 07	16 ♓ 52 25	14 19	27 55	28 56	08 30	08 28	23 17	08 03	16 36
18	19 50 54	28 13 01	22 ♓ 48 05	28 ♓ 43 33	14 28	29 ♑ 11	29 ♓ 40	08 23	08 25	23 20	08 05	16 37
19	19 54 51	29 ♑ 14 06	04 ♈ 39 16	10 ♈ 35 47	14 R 27	00 ♒ 26	00 ♈ 24	08 16	08 23	23 24	08 07	16 39
20	19 58 47	00 ♒ 15 11	16 ♈ 33 36	22 ♈ 33 20	14 14	01 42	01 08	08 09	08 21	23 27	08 09	16 41
21	20 02 44	01 16 14	28 ♈ 35 32	04 ♉ 40 51	13 49	02 57	01 51	08 02	08 19	23 30	08 12	16 43
22	20 06 40	02 17 17	10 ♉ 49 53	17 ♉ 03 13	13 13	04 12	02 35	07 55	08 17	23 33	08 14	16 45
23	20 10 37	03 18 19	23 ♉ 21 28	29 ♉ 45 10	12 26	05 28	03 19	07 48	08 15	23 37	08 16	16 46
24	20 14 33	04 19 20	06 ♊ 14 48	12 ♊ 50 46	11 30	06 43	04 02	07 42	08 14	23 40	08 18	16 48
25	20 18 30	05 20 19	19 ♊ 33 24	26 ♊ 22 53	10 26	07 59	04 46	07 35	08 12	23 43	08 21	16 50
26	20 22 26	06 21 18	03 ♋ 19 16	10 ♋ 22 26	09 16	09 14	05 29	07 29	08 11	23 47	08 23	16 51
27	20 26 23	07 22 16	17 ♋ 32 03	24 ♋ 47 40	08 02	10 29	06 13	07 23	08 09	23 50	08 25	16 53
28	20 30 20	08 23 13	02 ♌ 08 34	09 ♌ 33 53	06 46	11 45	06 57	07 17	08 08	23 53	08 27	16 55
29	20 34 16	09 24 08	17 ♌ 02 37	24 ♌ 33 37	05 31	13 00	07 40	07 11	08 07	23 57	08 30	16 56
30	20 38 13	10 25 03	02 ♍ 05 40	09 ♍ 37 32	04 19	14 15	08 23	07 05	08 06	24 00	08 32	16 58
31	20 42 09	11 ♒ 25 57	17 ♍ 08 01	24 ♍ 36 00	03 ♒ 11	15 ♒ 31	09 ♈ 07	07 ♋ 00	08 ♊ 05	24 ♒ 04	08 ♒ 34	16 ♐ 59

Moon Nodes and Latitude; DECLINATIONS

Date	Moon True ☊ ° '	Moon Mean ☊ ° '	Moon ☽ Latitude ° '	Sun ☉ ° '	Moon ☽ ° '	Mercury ☿ ° '	Venus ♀ ° '	Mars ♂ ° '	Jupiter ♃ ° '	Saturn ♄ ° '	Uranus ♅ ° '	Neptune ♆ ° '	Pluto ♇ ° '
01	27 ♊ 09	26 ♊ 20	03 N 29	23 S 00	21 N 31	22 S 52	23 S 38	05 S 40	23 N 01	20 N 04	14 S 41	18 S 18	13 S 00
02	27 R 08	26 17	04 21	22 54	17 58	22 30	23 35	05 22	23 02	20 03	14 40	18 18	13 00
03	27 06	26 14	04 58	22 49	13 20	22 07	23 32	05 04	23 03	20 03	14 39	18 17	13 00
04	27 04	26 11	05 14	22 42	07 59	21 43	23 28	04 46	23 03	20 03	14 38	18 17	13 00
05	27 03	26 07	05 12	22 36	02 N 16	21 17	23 23	04 27	23 04	20 02	14 37	18 16	13 00
06	27 D 03	26 04	04 50	22 29	03 S 28	20 51	23 18	04 09	23 05	20 02	14 36	18 16	13 01
07	27 03	26 01	04 12	22 21	08 58	20 23	23 12	03 51	23 06	20 02	14 35	18 15	13 01
08	27 05	25 58	03 20	22 14	13 56	19 55	23 05	03 32	23 06	20 01	14 34	18 14	13 01
09	27 06	25 55	02 18	22 05	18 08	19 26	22 57	03 14	23 07	20 01	14 33	18 14	13 01
10	27 08	25 52	01 N 10	21 56	21 21	18 57	22 49	02 56	23 08	20 01	14 32	18 13	13 01
11	27 08	25 48	00 00	21 47	23 25	18 28	22 40	02 37	23 09	20 01	14 31	18 13	13 01
12	27 R 08	25 45	01 S 10	21 38	24 14	17 59	22 30	02 19	23 09	20 00	14 30	18 12	13 01
13	27 05	25 42	02 14	21 27	23 47	17 31	22 20	02 00	23 10	20 00	14 29	18 12	13 01
14	27 02	25 39	03 12	21 17	22 10	17 03	22 09	01 42	23 11	20 00	14 28	18 11	13 01
15	26 57	25 36	04 00	21 06	19 33	16 37	21 57	01 24	23 11	20 00	14 27	18 11	13 01
16	26 52	25 32	04 36	20 55	16 08	16 13	21 45	01 05	23 12	20 00	14 26	18 10	13 01
17	26 46	25 29	05 00	20 43	12 05	15 50	21 32	00 47	23 12	19 59	14 25	18 09	13 02
18	26 41	25 26	05 11	20 31	07 37	15 30	21 18	00 28	23 13	19 59	14 24	18 09	13 02
19	26 37	25 23	05 09	20 19	02 S 52	15 13	21 04	00 S 10	23 14	19 59	14 23	18 08	13 02
20	26 34	25 20	04 53	20 06	02 N 00	15 00	20 49	00 N 08	23 14	19 59	14 22	18 08	13 02
21	26 33	25 17	04 25	19 53	06 51	14 49	20 33	00 27	23 15	19 59	14 21	18 07	13 02
22	26 D 34	25 13	03 44	19 39	11 32	14 42	20 17	00 45	23 15	19 59	14 20	18 07	13 02
23	26 35	25 10	02 51	19 25	15 51	14 39	20 00	01 03	23 16	19 59	14 18	18 06	13 02
24	26 37	25 07	01 49	19 11	19 34	14 40	19 43	01 21	23 16	19 59	14 17	18 05	13 02
25	26 38	25 04	00 S 38	18 56	22 23	14 44	19 25	01 40	23 17	19 59	14 16	18 05	13 02
26	26 R 38	25 01	00 N 36	18 41	24 00	14 51	19 06	01 58	23 17	19 59	14 15	18 04	13 02
27	26 36	24 58	01 51	18 26	24 07	15 00	18 47	02 16	23 18	19 59	14 14	18 04	13 02
28	26 32	24 54	03 00	18 10	22 37	15 12	18 28	02 34	23 18	19 59	14 13	18 03	13 02
29	26 27	24 51	03 59	17 54	19 31	15 26	18 08	02 52	23 19	19 59	14 12	18 03	13 02
30	26 20	24 48	04 41	17 38	15 06	15 40	17 47	03 10	23 19	19 59	14 11	18 02	13 02
31	26 ♊ 13	24 ♊ 45	05 N 04	17 S 21	09 N 45	15 S 56	17 S 26	03 N 28	23 N 20	19 N 59	14 S 09	18 S 01	13 S 02

ZODIAC SIGN ENTRIES

Date	h	m	Planets	
02	23	34	☽	♍
03	21	38	☿	♒
05	01	24	☽	♎
07	04	41	☽	♏
09	09	57	☽	♐
11	17	18	☽	♑
14	02	41	☽	♒
16	14	00	☽	♓
18	22	53	♂	♈
19	02	35	☽	♈
19	03	42	♀	♒
20	06	02	☉	♒
21	14	47	☽	♉
24	00	28	☽	♊
26	06	17	☽	♋
28	08	31	☽	♌
30	08	40	☽	♍

LATITUDES

Date	Mercury ☿ ° '	Venus ♀ ° '	Mars ♂ ° '	Jupiter ♃ ° '	Saturn ♄ ° '	Uranus ♅ ° '	Neptune ♆ ° '	Pluto ♇ ° '
01	02 S 01	00 S 25	00 S 41	00 00	01 S 48	00 S 42	00 N 06	09 N 46
04	01 48	00 32	00 38	00 N 01	01 48	00 42	00 06	09 46
07	01 27	00 39	00 34	00 01	01 47	00 42	00 06	09 46
10	00 58	00 45	00 31	00 01	01 46	00 42	00 06	09 46
13	00 S 20	00 50	00 28	00 02	01 46	00 42	00 06	09 47
16	00 N 27	00 56	00 25	00 02	01 45	00 42	00 06	09 47
19	01 20	01 02	00 21	00 03	01 44	00 42	00 06	09 47
22	02 14	01 07	00 18	00 03	01 44	00 42	00 06	09 48
25	03 00	01 11	00 15	00 04	01 43	00 42	00 06	09 48
28	03 29	01 15	00 12	00 04	01 42	00 42	00 06	09 49
31	03 N 36	01 S 18	00 S 09	00 N 04	01 S 41	00 S 42	00 N 06	09 N 49

DATA

Julian Date	2452276
Delta T	+64 seconds
Ayanamsa	23° 52' 50"
Synetic vernal point	05° ♓ 14' 10"
True obliquity of ecliptic	23° 26' 21"

MOON'S PHASES, APSIDES AND POSITIONS ☽

Date	h	m	Phase	Longitude ° '	Eclipse Indicator
06	03	55	☾	15 ♎ 39	
13	13	29	●	23 ♑ 11	
21	17	47	☽	01 ♉ 31	
28	22	50	○	08 ♌ 51	

Day	h	m		
02	07	23	Perigee	
18	08	53	Apogee	
30	09	07	Perigee	
05	21	26	OS	
12	15	24	Max dec	24° S 14'
20	02	12	ON	
27	01	53	Max dec	24° N 16'

LONGITUDES

Date	Chiron ⚷ ° '	Ceres ⚳ ° '	Pallas ⚴ ° '	Juno ⚵ ° '	Vesta ⚶ ° '	Black Moon Lilith ⚸ ° '
01	02 ♑ 13	09 ♒ 49	15 ♑ 35	29 ♌ 52	28 ♉ 27	14 ♓ 44
11	03 ♑ 17	13 ♒ 37	19 ♑ 26	28 ♌ 54	27 ♉ 50	15 ♓ 52
21	04 ♑ 18	17 ♒ 29	23 ♑ 13	27 ♌ 09	27 ♉ 57	16 ♓ 59
31	05 ♑ 16	21 ♒ 23	26 ♑ 58	24 ♌ 50	28 ♉ 45	18 ♓ 06

All ephemeris data is given at 12.00 UT and the Moon's longitude is additionally given for 24.00 UT

ASPECTARIAN

h m	Aspects
01 Tuesday	
00 48	☉ ∥ ☿
01 20	☽ bQ ♂
01 56	☿ # ♃
05 05	☉ # ♃
05 33	♀ ⊻ ♆
05 53	♀ ∠ ♅
05 53	☉ ☍ ♃
10 29	☽ ☍ ♆
10 56	☽ ⊼ ♀
13 30	☽ ⚹ ♄
15 37	☽ ⊻ ♃
16 27	☽ ⊼ ☉
16 58	☽ ± ♂
21 45	☽ ± ♀
22 58	☽ ∥ ♄
02 Wednesday	
00 40	☽ △ ♇
01 24	☽ ⊥ ♃
03 04	☽ ± ☉
03 22	☽ ⊼ ♂
09 09	☽ Q ♄
10 07	☽ # ♆
11 16	☽ ☍ ♅
13 57	☽ bQ ♀
15 17	♀ ⊼ ♄
16 04	☽ ∠ ♃
18 58	☽ bQ ☉
21 02	☽ ⊼ ☿
03 Thursday	
05 37	☽ # ♅
08 04	☽ ± ☿
11 59	☽ ⊼ ♆
12 22	♀ ☍ ♃
13 36	☽ # ♇
14 41	☽ □ ♄
16 34	☽ ⚹ ♃
17 01	☽ △ ♀
21 36	☽ △ ☉
21 56	☽ ± ♆
04 Friday	
00 40	☽ bQ ☿
02 15	☽ □ ♇
07 30	☽ ☍ ♂
12 19	☽ Q ♃
12 57	☽ bQ ♆
13 02	☽ ⊼ ♅
16 34	☿ ∠ ♇
23 07	☽ ± ♅
05 Saturday	
02 24	☽ # ♂
04 35	☽ △ ☿
08 27	☽ Q ♇
13 48	☉ ± ♄
14 14	☽ △ ♆
14 21	☽ bQ ♅
16 39	☽ △ ♄
18 20	☽ □ ♃
06 Sunday	
00 11	☽ □ ♀
03 55	☽ □ ☉
04 54	☽ ⚹ ♇
12 59	☽ ⊼ ♂
14 44	☽ ∥ ♂
16 05	☽ △ ♅
17 54	☉ ⊻ ♇
18 14	☽ bQ ♄
23 56	☽ ± ♂
07 Monday	
04 38	♀ ± ♄
06 07	☉ ⊥ ♅
06 53	☽ ∠ ♇
09 49	☽ ± ♄
10 22	☽ Q ♀
13 33	☽ Q ☉
13 52	☽ □ ☿
16 32	☽ bQ ♂
18 08	☽ □ ♆
20 18	☽ ⊼ ♄
21 45	☽ △ ♃
22 42	☽ ⊥ ♇
08 Tuesday	
06 08	♀ ⊻ ♇
06 36	☿ # ♄
07 08	♀ # ♃
07 20	☽ ∥ ♇
09 22	☽ ⊻ ♇
09 41	☽ ⚹ ♀
12 29	☽ ⚹ ☉
15 22	☽ ∥ ♅
16 24	♀ ⊥ ♅
20 41	☽ △ ♂
21 03	☽ □ ♅
23 51	♂ ∠ ♆
09 Wednesday	
00 12	☽ bQ ♃
01 17	☽ Q ☿
02 15	☽ Q ♆
03 56	♂ ⊻ ♅
11 56	☿ ☌ ♆
12 41	☽ ∥ ♆
15 28	☽ ∠ ♀
16 19	☽ ± ♃
17 45	☽ ∠ ☉
19 44	☽ ∥ ☿
10 Thursday	
00 05	☽ ⚹ ♆
01 05	☽ # ♄
01 17	☽ ⚹ ☿
01 59	☽ ☍ ♄
03 09	☽ ⊼ ♃
05 24	☿ ∠ ♂
05 50	☽ Q ♅
08 36	☿ △ ♄
09 40	☽ ⊥ ♀
11 37	☽ ⊥ ☉
15 54	☽ ☌ ♇
17 18	☽ ∥ ☉
20 19	☿ ⊼ ♃
21 57	☽ ⊻ ♀
23 40	☽ ⊻ ☉
11 Friday	
02 21	☽ ∥ ♀
03 51	☽ ∠ ♆
04 07	☽ ⚹ ♅
06 49	☽ □ ♂
07 42	☽ ∠ ☿
07 49	☽ # ♃
20 45	☽ ⊥ ♆
12 Saturday	
00 49	☿ ∥ ♆
02 08	☽ ⊥ ☿
08 07	☽ ⊻ ♆
08 26	☽ ∠ ♅
09 42	☽ ⊼ ♄
10 34	☽ ☍ ♃
14 29	☽ ⊻ ☿
18 52	☽ Q ♂
19 59	☉ ⊥ ♇
21 06	☽ ± ♄
13 Sunday	
00 31	☽ ⊻ ♇
01 40	☽ ⊥ ♅
03 42	♀ ⊥ ♇
10 47	☉ ⊻ ♅
12 05	☽ ⊥ ♇
13 00	☽ ☌ ♀
13 16	☽ ⊻ ♅
13 29	☽ ☌ ☉
14 20	☽ bQ ♄
15 41	♀ ⊻ ♅
19 24	☽ ⚹ ♂
23 11	☽ # ♃
23 30	☉ bQ ♄
14 Monday	
01 43	♀ bQ ♄
05 35	☽ ∠ ♇
06 58	♂ Q ♄
11 32	☉ ☌ ♀
12 18	☽ ∥ ♀
18 13	☽ ☌ ♆
19 27	☽ △ ♄
19 58	☽ ⊼ ♃
21 51	☽ ∥ ☉
15 Tuesday	
02 35	☽ ∠ ♂
04 44	☽ ☌ ☿
07 41	☽ ± ♃
08 28	☽ # ♄
11 09	☽ ⚹ ♇
21 34	☽ ⊥ ♂
22 18	☽ ∥ ♆
16 Wednesday	
00 25	☽ ☌ ♅
01 22	☽ bQ ♃
05 50	☽ ⊻ ☉
06 46	☽ ⊻ ♀
10 19	☽ ⊻ ♂
11 07	☽ Q ♇
11 25	☽ ∥ ☿
18 58	☽ ⊥ ☉
20 12	☽ ⊥ ♀
22 30	☽ ∥ ♅
17 Thursday	
06 09	☽ ⊻ ♆
06 41	☽ ∥ ♇
07 02	☽ □ ♄
07 09	☽ △ ♃
14 47	☽ ∠ ☉
16 29	☽ ∠ ♀
18 18	☽ ⊥ ♆
18 57	☽ ⊻ ☿
23 28	☽ □ ♇
18 Friday	
00 15	♂ ⊥ ♅
00 33	♃ ⊻ ♄
02 48	☿ ± ♃
04 29	☿ ∠ ♂
07 15	☽ ⊥ ☿
12 34	☽ ∠ ♆
13 06	☽ ⊻ ♅
18 25	♃ bQ ♅
19 19	☽ Q ♄
20 49	☿ St R
19 Saturday	
00 00	☽ ⚹ ☉
01 18	☽ ⊥ ♅
01 32	☽ ∠ ☿
02 27	☽ ⚹ ♀
02 49	☽ ☌ ♂
10 20	♀ ⚹ ♂
19 01	☽ ⚹ ♆
19 14	☽ □ ♃
19 31	☽ ⚹ ♄
19 35	☽ ∠ ♅
20 Sunday	
02 08	☽ # ♂
02 16	☽ ∥ ♂
02 32	☽ Q ☉
05 33	☽ Q ♀
07 26	☽ ⚹ ☿
10 58	♃ ⊼ ♆
11 51	♀ ∠ ♇
12 15	☽ △ ♇
19 14	☽ Q ♆
20 09	☿ ± ♃
21 Monday	
00 11	☉ # ♄
01 33	☽ ∠ ♄
01 50	☽ ⚹ ♅
06 42	☽ Q ☿
06 58	☽ Q ♃
17 47	☽ □ ☉
18 11	☽ bQ ♇
18 51	☽ ⊻ ♂
19 20	☽ ⊥ ♄
21 35	☽ □ ♀
22 46	☉ ∠ ♇
22 Tuesday	
01 40	☽ Q ♅
06 23	☽ ⚹ ♃
06 56	☽ □ ♆
07 04	☽ ⊻ ♄
07 21	☽ ⊥ ♂
11 50	☽ ± ♇
16 22	☽ □ ☿
20 03	☽ # ♇
23 26	☽ ⊼ ♇
23 Wednesday	
01 49	☽ ∠ ♂
03 09	☽ # ♅
05 08	☽ # ☿
08 19	♄ △ ♆
10 58	☽ ∠ ♃
12 25	☉ ⚹ ♂
12 29	☽ □ ♅
13 45	♀ # ♄
24 Thursday	
01 55	☽ # ♆
03 41	☽ ⊥ ♃
07 42	☽ ⚹ ♂
08 10	☽ △ ☉
09 25	☽ # ☉
12 57	☽ △ ♀
12 59	☽ # ♀
14 38	☽ ⊻ ♃
15 04	☽ ∥ ♄
15 37	☽ ☌ ♄
15 47	☽ △ ♆
20 52	☽ △ ☿
25 Friday	
05 15	♀ ⊼ ♃
06 45	☽ Q ♂
07 08	☽ ☍ ♇
13 30	☽ bQ ☉
16 17	♀ △ ♄
18 39	☽ bQ ♀
18 42	☽ bQ ♆
19 16	♀ ☌ ♆
19 23	☽ △ ♅
21 32	☽ bQ ☿
22 48	☽ ∥ ♃
26 Saturday	
06 30	☽ ± ☉
10 23	☽ ± ♆
11 50	☽ ± ♀
11 54	☽ ± ☿
12 17	☿ ☌ ♀
15 55	☽ □ ♂
17 35	☽ ⊼ ☉
19 04	☽ ☌ ♃
20 16	☽ ⊻ ♄
20 39	☽ ⊼ ♆
21 19	☽ ⊼ ☿
21 20	☽ bQ ♅
23 03	☽ ⊼ ♀
27 Sunday	
04 39	☿ ☌ ♆
06 22	☽ ⊥ ♄
09 26	☿ △ ♄
10 55	☽ ⊼ ♇
12 17	☉ ⊼ ♃
12 30	☽ ± ♅
18 55	☉ ☌ ☿
20 52	☽ ± ♇
21 18	☽ ∠ ♄
22 28	☽ ⊼ ♅
28 Monday	
01 18	☿ ⊼ ♃
03 56	☽ ∥ ♃
06 12	☉ △ ♄
09 52	☿ ⚹ ♂
11 37	☽ bQ ♇
13 45	☉ ☌ ♆
18 54	☽ ☍ ☿
20 10	☽ △ ♂
20 16	☽ ⊻ ♃
21 41	☽ ⚹ ♄
21 58	♂ □ ♃
22 15	☽ ☍ ♆
22 47	☉ ∥ ♆
22 50	☽ ☍ ☉
29 Tuesday	
04 55	☽ ☍ ♀
05 52	☽ ⊥ ♃
08 58	☽ ∥ ♄
11 50	☽ △ ♇
15 20	♀ ± ♃
16 54	☽ Q ♄
18 15	♀ ∥ ♆
20 10	☽ ∠ ♃
20 45	☽ # ♆
20 58	☽ # ♀
21 26	☽ bQ ♂
22 10	☽ # ☉
23 04	☽ ☍ ♅
30 Wednesday	
02 35	♂ ⚹ ♄
09 19	☽ # ☿
12 30	☽ ± ♂
15 16	☽ ⊼ ☿
16 25	☽ # ♅
16 59	♂ ⚹ ♆
19 55	☽ ⚹ ♃
21 34	☽ □ ♄
21 41	☽ # ♇
22 17	☽ ⊼ ♆
22 32	☽ ⊼ ♂
31 Thursday	
00 09	☽ ± ☿
02 13	☽ ⊼ ☉
07 53	☽ ± ♆
09 10	☽ ⊼ ♀
09 58	♂ ∠ ♅
11 46	☽ □ ♇
12 31	☽ ± ☉
13 34	☽ bQ ☿
14 58	☽ Q ♃
19 40	☽ ± ♀
22 22	☽ bQ ♆
23 10	☽ ⊼ ♅

FEBRUARY 2002

LONGITUDES

Date	Sidereal time h m s	Sun ☉ ° ' "	Moon ☽ ° ' "	Moon ☽ 24.00 ° ' "	Mercury ☿ ° '	Venus ♀ ° '	Mars ♂ ° '	Jupiter ♃ ° '	Saturn ♄ ° '	Uranus ♅ ° '	Neptune ♆ ° '	Pluto ⯓ ° '
01	20 46 06	12 ♒ 26 50	02 ♎ 00 29	09 ♎ 20 40	02 ♒ 09	16 ♒ 46	09 ♈ 50	06 ♋ 55	08 ♊ 04	24 ♒ 07	08 ♒ 37	17 ♐ 01
02	20 50 02	13 27 42	16 ♎ 35 52	23 ♎ 45 39	01 R 14	18 01	10 34	06 R 49	08 R 04	24 10	08 39	17 02
03	20 53 59	14 28 34	00 ♏ 49 44	07 ♏ 47 59	00 ♒ 27	19 16	11 17	06 44	08 03	24 14	08 41	17 04
04	20 57 55	15 29 24	14 ♏ 40 27	21 ♏ 27 17	29 ♑ 49	20 32	12 00	06 39	08 03	24 17	08 43	17 05
05	21 01 52	16 30 14	28 ♏ 08 43	04 ♐ 45 04	29 19	21 47	12 44	06 35	08 02	24 21	08 46	17 07
06	21 05 49	17 31 03	11 ♐ 16 42	17 ♐ 43 59	28 57	23 02	13 27	06 30	08 02	24 24	08 48	17 08
07	21 09 45	18 31 51	24 ♐ 07 19	00 ♑ 27 05	28 44	24 17	14 10	06 26	08 02	24 28	08 50	17 09
08	21 13 42	19 32 38	06 ♑ 43 39	12 ♑ 57 20	28 38	25 33	14 54	06 21	08 D 02	24 31	08 52	17 11
09	21 17 38	20 33 24	19 ♑ 08 28	25 ♑ 17 17	28 D 40	26 48	15 37	06 17	08 02	24 35	08 55	17 12
10	21 21 35	21 34 09	01 ♒ 24 02	07 ♒ 28 56	28 49	28 03	16 20	06 14	08 02	24 38	08 57	17 13
11	21 25 31	22 34 53	13 ♒ 32 09	19 ♒ 33 51	29 04	29 ♒ 18	17 03	06 10	08 03	24 41	08 59	17 15
12	21 29 28	23 35 35	25 ♒ 34 11	01 ♓ 33 18	29 26	00 ♓ 34	17 46	06 06	08 03	24 45	09 01	17 16
13	21 33 24	24 36 16	07 ♓ 31 20	13 ♓ 28 27	29 ♑ 53	01 49	18 29	06 03	08 04	24 48	09 04	17 17
14	21 37 21	25 36 55	19 ♓ 24 50	25 ♓ 20 41	00 ♒ 26	03 04	19 12	06 00	08 04	24 52	09 06	17 18
15	21 41 18	26 37 33	01 ♈ 16 15	07 ♈ 11 48	01 03	04 19	19 55	05 57	08 05	24 55	09 08	17 19
16	21 45 14	27 38 09	13 ♈ 07 40	19 ♈ 04 13	01 44	05 34	20 39	05 54	08 06	24 59	09 10	17 20
17	21 49 11	28 38 43	25 ♈ 01 51	01 ♉ 01 04	02 30	06 49	21 21	05 52	08 07	25 02	09 12	17 21
18	21 53 07	29 ♒ 39 16	07 ♉ 02 20	13 ♉ 06 13	03 19	08 04	22 04	05 50	08 08	25 06	09 15	17 22
19	21 57 04	00 ♓ 39 47	19 ♉ 13 18	25 ♉ 24 09	04 11	09 19	22 47	05 48	08 09	25 09	09 17	17 23
20	22 01 00	01 40 16	01 ♊ 39 26	07 ♊ 59 44	05 07	10 34	23 30	05 46	08 11	25 13	09 19	17 24
21	22 04 57	02 40 43	14 ♊ 25 39	20 ♊ 57 46	06 05	11 49	24 13	05 44	08 12	25 16	09 21	17 25
22	22 08 53	03 41 09	27 ♊ 36 34	04 ♋ 22 28	07 07	13 04	24 56	05 42	08 14	25 20	09 23	17 26
23	22 12 50	04 41 33	11 ♋ 15 45	18 ♋ 16 33	08 10	14 19	25 39	05 41	08 15	25 23	09 25	17 27
24	22 16 47	05 41 54	25 ♋ 24 49	02 ♌ 40 18	09 17	15 34	26 21	05 40	08 17	25 26	09 27	17 28
25	22 20 43	06 42 14	10 ♌ 02 29	17 ♌ 30 39	10 25	16 49	27 04	05 39	08 19	25 30	09 29	17 29
26	22 24 40	07 42 32	25 ♌ 03 50	02 ♍ 40 51	11 35	18 04	27 47	05 38	08 21	25 33	09 32	17 29
27	22 28 36	08 42 48	10 ♍ 20 20	18 ♍ 00 51	12 47	19 19	28 29	05 38	08 23	25 37	09 34	17 30
28	22 32 33	09 ♓ 43 02	25 ♍ 40 53	03 ♎ 18 59	14 ♒ 01	20 ♓ 34	29 ♈ 12	05 ♋ 38	08 ♊ 25	25 ♒ 40	09 ♒ 36	17 ♐ 31

Moon nodes and latitude

Date	Moon True ☊ ° '	Moon Mean ☊ ° '	Moon ☽ Latitude ° '
01	26 ♊ 06	24 ♊ 42	05 N 06
02	26 R 02	24 38	04 49
03	25 59	24 35	04 13
04	25 58	24 32	03 24
05	25 D 58	24 29	02 24
06	25 59	24 26	01 18
07	26 00	24 23	00 N 10
08	25 R 59	24 19	00 S 57
09	25 57	24 16	02 01
10	25 51	24 13	02 58
11	25 43	24 10	03 46
12	25 33	24 07	04 23
13	25 21	24 03	04 49
14	25 09	24 00	05 02
15	24 57	23 57	05 01
16	24 47	23 54	04 48
17	24 39	23 51	04 22
18	24 34	23 48	03 44
19	24 31	23 44	02 56
20	24 D 31	23 41	01 58
21	24 31	23 38	00 S 53
22	24 R 31	23 35	00 N 16
23	24 30	23 32	01 28
24	24 27	23 29	02 36
25	24 21	23 25	03 36
26	24 12	23 22	04 23
27	24 02	23 19	04 52
28	23 ♊ 51	23 ♊ 16	05 N 01

DECLINATIONS

Date	Sun ☉ ° '	Moon ☽ ° '	Mercury ☿ ° '	Venus ♀ ° '	Mars ♂ ° '	Jupiter ♃ ° '	Saturn ♄ ° '	Uranus ♅ ° '	Neptune ♆ ° '	Pluto ⯓ ° '
01	17 S 04	03 N 53	16 S 12	17 S 04	03 N 46	23 N 20	19 N 59	14 S 08	18 S 01	13 S 02
02	16 47	02 S 05	16 28	16 42	04 04	23 20	20 00	14 07	18 00	13 02
03	16 29	07 48	16 44	16 20	04 22	23 21	20 00	14 06	18 00	13 02
04	16 12	13 00	17 00	15 57	04 40	23 21	20 00	14 05	17 59	13 02
05	15 53	17 24	17 14	15 33	04 58	23 21	20 00	14 04	17 58	13 01
06	15 35	20 50	17 28	15 09	05 15	23 22	20 00	14 03	17 58	13 01
07	15 16	23 08	17 41	14 45	05 33	23 22	20 01	14 01	17 57	13 01
08	14 57	24 13	17 53	14 20	05 50	23 22	20 01	14 00	17 57	13 01
09	14 38	24 04	18 04	13 55	06 08	23 23	20 01	13 59	17 56	13 01
10	14 19	22 44	18 14	13 30	06 25	23 23	20 01	13 58	17 56	13 01
11	13 59	20 22	18 23	13 04	06 43	23 23	20 02	13 57	17 55	13 01
12	13 39	17 07	18 30	12 38	07 00	23 24	20 02	13 56	17 54	13 01
13	13 19	13 13	18 36	12 11	07 17	23 24	20 02	13 55	17 54	13 01
14	12 59	08 49	18 41	11 44	07 34	23 24	20 03	13 53	17 53	13 01
15	12 38	04 S 06	18 45	11 17	07 51	23 24	20 03	13 52	17 53	13 01
16	12 18	00 N 46	18 47	10 50	08 08	23 25	20 04	13 51	17 52	13 00
17	11 57	05 37	18 48	10 22	08 25	23 25	20 04	13 50	17 52	13 00
18	11 36	10 20	18 48	09 54	08 42	23 25	20 04	13 49	17 51	13 00
19	11 14	14 42	18 46	09 25	08 59	23 25	20 05	13 48	17 50	13 00
20	10 53	18 34	18 43	08 57	09 16	23 26	20 05	13 46	17 50	13 00
21	10 31	21 39	18 39	08 28	09 32	23 26	20 06	13 45	17 49	13 00
22	10 09	23 41	18 33	07 59	09 49	23 26	20 06	13 44	17 49	13 00
23	09 47	24 25	18 26	07 30	10 05	23 26	20 07	13 43	17 48	12 59
24	09 25	23 37	18 18	07 01	10 21	23 26	20 07	13 42	17 48	12 59
25	09 03	21 12	18 08	06 31	10 37	23 26	20 08	13 41	17 47	12 59
26	08 41	17 18	17 57	06 01	10 53	23 26	20 08	13 39	17 47	12 59
27	08 18	12 12	17 45	05 31	11 09	23 27	20 09	13 38	17 46	12 59
28	07 S 56	06 N 19	17 S 31	05 S 01	11 N 25	23 N 27	20 N 10	13 S 37	17 S 46	12 S 59

ZODIAC SIGN ENTRIES

Date	h	m	Planets
01	08	44	☽ ♎
03	10	35	☽ ♏
04	04	18	☿ ♑
05	15	21	☽ ♐
07	23	08	☽ ♑
10	09	15	☽ ♒
12	01	18	♀ ♓
12	20	53	☽ ♓
13	17	20	☿ ♒
15	09	26	☽ ♈
17	21	58	☽ ♉
18	20	13	☉ ♓
20	08	50	☽ ♊
22	16	16	☽ ♋
24	19	36	☽ ♌
26	19	47	☽ ♍
28	18	47	☽ ♎

LATITUDES

Date	Mercury ☿ ° '	Venus ♀ ° '	Mars ♂ ° '	Jupiter ♃ ° '	Saturn ♄ ° '	Uranus ♅ ° '	Neptune ♆ ° '	Pluto ⯓ ° '
01	03 N 34	01 S 19	00 S 08	00 N 04	01 S 41	00 S 42	00 N 06	09 N 49
04	03 16	01 22	00 05	00 05	01 40	00 42	00 06	09 50
07	02 47	01 24	00 S 03	00 05	01 39	00 42	00 06	09 50
10	02 12	01 26	00 00	00 06	01 39	00 42	00 06	09 51
13	01 36	01 27	00 N 03	00 06	01 38	00 42	00 06	09 52
16	01 01	01 28	00 05	00 06	01 37	00 42	00 06	09 52
19	00 N 28	01 27	00 08	00 07	01 36	00 42	00 06	09 53
22	00 S 03	01 27	00 10	00 07	01 36	00 42	00 06	09 54
25	00 31	01 25	00 13	00 07	01 35	00 42	00 06	09 54
28	00 57	01 24	00 15	00 08	01 34	00 42	00 06	09 55
31	01 S 18	01 S 21	00 N 17	00 N 08	01 S 33	00 S 42	00 N 06	09 N 56

DATA

Julian Date	2452307
Delta T	+64 seconds
Ayanamsa	23° 52' 55"
Synetic vernal point	05° ♓ 14' 04"
True obliquity of ecliptic	23° 26' 21"

MOON'S PHASES, APSIDES AND POSITIONS ☽

Date	h	m	Phase	Longitude ° '	Eclipse Indicator
04	13	33	☾	15 ♏ 33	
12	07	41	●	23 ♒ 25	
20	12	02	☽	01 ♊ 40	
27	09	17	○	08 ♍ 36	

Day	h	m		
14	22	30	Apogee	
27	19	50	Perigee	
02	03	35	0S	
08	20	54	Max dec	24° S 18'
16	08	16	0N	
23	11	46	Max dec	24° N 25'

LONGITUDES

Date	Chiron ⚷ ° '	Ceres ⚳ ° '	Pallas ⚴ ° '	Juno ⚵ ° '	Vesta ⚶ ° '	Black Moon Lilith ⚸ ° '
01	05 ♑ 21	21 ♒ 47	27 ♑ 20	24 ♌ 34	28 ♉ 52	18 ♓ 13
11	06 ♑ 14	25 ♒ 42	00 ♒ 59	21 ♌ 59	00 ♊ 21	19 ♓ 20
21	07 ♑ 01	29 ♒ 38	04 ♒ 33	19 ♌ 30	02 ♊ 20	20 ♓ 27
31	07 ♑ 42	03 ♓ 33	07 ♒ 59	17 ♌ 27	04 ♊ 46	21 ♓ 34

ASPECTARIAN

01 Friday
04 03 ☽ ♇ ☉
08 55 ☽ ± ♅
11 34 ☽ ♇ ♀
12 13 ☽ △ ☿
12 27 ☽ ∥ ♂
13 07 ☉ ∥ ♀
15 09 ☿ ∠ ⯓
16 55 ♀ ⚹ ⯓
16 55 ☽ Q ⯓
19 57 ☽ □ ♃
21 54 ☽ △ ♄
22 03 ☉ ± ♃
22 49 ☽ △ ♆
23 40 ☽ ♇ ♅

02 Saturday
01 29 ☽ ☍ ♂
06 24 ☽ △ ☉
12 44 ☽ ⚹ ⯓
14 36 ☽ △ ♀
20 36 ☽ # ♂
20 40 ☿ ∥ ♀
22 49 ☽ ♇ ♄

03 Sunday
00 45 ☽ △ ♅
01 17 ☉ ∥ ☿
11 23 ☽ □ ☿
14 05 ☽ ± ♄
14 07 ☽ ∠ ⯓
22 06 ☽ △ ♃

04 Monday
00 26 ☽ ⊼ ♄
01 34 ☽ □ ♆
05 43 ☽ ⊥ ⯓
07 04 ☽ ⊼ ♂
08 21 ☿ Q ♂
12 10 ☽ ∥ ⯓
13 33 ☽ □ ☉
16 16 ☽ ⊻ ⯓
17 19 ☽ Q ☿
17 29 ☽ ∥ ♅
18 13 ☽ ± ♂
23 25 ☽ □ ♀

05 Tuesday
00 17 ☽ ♇ ♃
02 13 ☽ ∥ ♀
03 47 ☽ ∥ ☉
05 08 ☽ □ ♅
08 17 ♀ ♇ ♃
09 30 ☽ Q ♆
10 57 ☽ ∥ ☿
11 12 ☽ ♇ ♂
14 02 ☽ ⚹ ☿
15 31 ☽ ∥ ♆
16 22 ☽ ± ♃

06 Wednesday
00 31 ☽ Q ☉
02 45 ☉ ⚹ ⯓
03 15 ☽ ⊼ ♃
05 26 ☽ # ♄
06 01 ☽ ☍ ♄
07 25 ☽ ⚹ ♆
11 30 ☽ Q ♀
14 05 ☽ Q ♅
16 16 ☽ △ ♂
16 51 ☽ ∠ ☿
22 54 ☽ ☌ ⯓

07 Thursday
00 35 ☽ ⚹ ☉
09 24 ☽ ⊥ ☿
11 28 ☽ ∠ ♆
12 21 ☽ ⚹ ♀
12 38 ☽ ⚹ ♅
15 23 ♀ ☌ ♅
15 25 ☽ # ♃
20 38 ☽ ⊻ ☿

08 Friday
01 33 ♄ St D
04 36 ☽ ⊥ ♆
07 27 ☽ ∠ ☉
11 18 ☽ ☍ ♃
14 30 ☽ ⊼ ♄
16 08 ☽ ⊻ ♆
17 23 ☽ ∠ ♅
17 29 ☿ St D
18 34 ☿ ∥ ♆
20 10 ☽ ∠ ♀

09 Saturday
02 05 ☽ ± ♄
02 18 ☽ ⊥ ☉
04 43 ☽ □ ♂
08 01 ♀ ∥ ♅
08 13 ☽ ⊻ ⯓
10 54 ☽ ⊥ ♅
15 00 ☽ ⊻ ☉
15 36 ☽ ⊥ ♀
19 36 ☽ ♇ ♄
19 56 ☽ ⊥ ⯓
22 39 ☽ ⊻ ♅

10 Sunday
02 42 ☽ # ♃
04 20 ☉ ♇ ♃
04 40 ☽ ⊻ ♀
06 50 ☽ ☌ ☿
13 37 ☽ ∠ ⯓
18 09 ☽ Q ♂
21 28 ☽ ⊼ ♃

11 Monday
01 06 ☽ △ ♄
02 57 ☽ ☌ ♆
06 12 ☿ ⊻ ♀
09 17 ☽ ± ♃
10 47 ♀ Q ⯓
14 36 ♀ ∥ ⯓
14 46 ☽ # ♄
15 07 ☉ ∥ ♅
18 34 ♂ △ ⯓
19 23 ☽ ⚹ ⯓
19 26 ☽ ⚹ ♂

12 Tuesday
02 47 ☽ ∥ ☿
03 07 ☽ ♇ ♃
06 40 ☽ ∥ ♆
07 41 ☽ ☌ ☉
10 21 ☽ ☌ ♅
19 25 ☽ Q ⯓
20 02 ☽ ⊻ ☿
23 10 ☽ ☌ ♀

13 Wednesday
03 22 ☽ ∠ ♂
07 56 ☽ ∥ ♅
08 34 ☽ ⊥ ☿
09 03 ☽ △ ♃
11 18 ☽ ∥ ☉
13 05 ☽ □ ♄
13 07 ☽ ∥ ⯓
15 06 ☽ ⊻ ♆
17 06 ☉ ☌ ♅
18 28 ☽ ∥ ♀
22 39 ☽ ⊥ ♂

14 Thursday
03 15 ☽ ⊥ ♆
03 32 ☽ ∠ ☿
07 44 ☽ □ ⯓
09 55 ☉ ∥ ⯓
11 33 ☽ ⊻ ♂
18 03 ☽ # ♂
21 30 ☽ ∠ ♆
23 05 ☽ ⊻ ♅

15 Friday
01 29 ☽ Q ♄
01 43 ☽ ⊻ ☉
11 18 ☽ ⊥ ♅
11 31 ☽ ⚹ ☿
15 00 ☽ ⊥ ☉
18 54 ☽ ⊻ ♀
21 27 ☽ □ ♃

16 Saturday
01 49 ☽ ⚹ ♄
03 58 ☽ ⚹ ♆
05 36 ☽ ∠ ♅
08 28 ☽ ⊥ ♀
10 55 ☽ ∠ ☉
13 18 ☽ Q ☿
15 17 ♀ ∠ ♂
18 18 ♀ △ ♃
20 31 ☽ △ ⯓

17 Sunday
04 08 ☽ ☌ ♂
04 17 ☽ Q ♆
04 47 ☽ ∠ ♀
06 40 ♂ Q ♆
07 45 ☿ ∠ ⯓
08 09 ☽ ∠ ♄
09 40 ☽ Q ♃
12 01 ☽ ⚹ ♅
19 55 ☽ ⚹ ☉

18 Monday
02 13 ☽ ⊥ ♄
02 42 ☽ ♇ ⯓
03 02 ☽ ∥ ♂
04 01 ☽ □ ☿
05 13 ☉ Q ⯓
09 36 ☽ ⚹ ♃
09 57 ☽ # ♀
12 07 ☽ Q ♅
13 14 ♀ □ ♄
14 11 ☽ ⊻ ♄
14 17 ☽ ⚹ ♀
16 23 ☽ □ ♆
18 14 ☽ # ☉
20 36 ☽ ± ⯓
21 58 ☽ Q ☉

19 Tuesday
02 23 ☽ # ⯓
06 49 ☽ # ♅
08 25 ☽ ⊼ ⯓
11 09 ♀ ⊻ ♆
15 03 ☽ ∠ ♃
16 33 ☽ Q ♀
19 22 ☽ ⊻ ♂
23 34 ☽ □ ♅

20 Wednesday
00 41 ♂ ∠ ♄
02 07 ♀ # ♂
07 09 ☽ # ♆
07 38 ☽ ⊥ ♂
08 23 ☽ ⊥ ♃
12 02 ☽ □ ☉
13 03 ☽ # ☿
19 06 ☽ △ ☿
19 46 ☽ ⊻ ♃
20 18 ♂ Q ♃
23 03 ☽ ∥ ♄

21 Thursday
00 22 ☽ ☌ ♄
01 43 ☽ ∠ ♂
02 31 ☽ △ ♆
03 34 ☿ ⊼ ♃
06 39 ☽ □ ♀
17 32 ☽ ☍ ⯓

22 Friday
01 15 ☽ ♇ ☿
06 11 ☽ ♇ ♆
06 55 ☽ ⚹ ♂
07 53 ☽ △ ♅
07 56 ☽ ∥ ♃
16 31 ☿ ⊥ ♀
18 46 ☽ ± ☿
22 17 ☽ ± ♆
23 39 ☽ △ ☉

23 Saturday
01 03 ☉ # ♂
02 19 ☽ ☌ ♃
02 28 ♂ ⚹ ♅
05 23 ☽ Q ♂
06 12 ☽ ⊼ ☿
06 47 ☽ ⊻ ♄
08 48 ☽ ⊼ ♆
10 29 ☽ ♇ ♅
13 52 ☿ △ ♄
17 10 ☽ ⊥ ♄
17 47 ☽ △ ♀
22 37 ☽ ⊼ ⯓

24 Sunday
01 56 ☽ ± ♅
03 30 ☽ ♇ ☉
08 26 ☽ ∠ ♄
08 44 ☽ ± ⯓
09 42 ♀ ⊥ ♆
11 16 ☉ △ ♃
12 03 ☽ ⊼ ♅
13 39 ☽ □ ♂
14 28 ☽ ∥ ♃
15 59 ☿ ☌ ♆
19 38 ☽ ± ☉
21 21 ☽ ♇ ♀
23 40 ☽ ♇ ⯓

25 Monday
04 53 ☽ ⊻ ♃
06 11 ☽ ⊼ ☉
09 12 ☽ ⚹ ♄
11 06 ☽ ☍ ♆
12 39 ☽ ☍ ☿
13 23 ☽ ± ♀
14 36 ☽ ⊥ ♃
19 28 ☽ ∥ ♄
23 53 ☽ ⊼ ♀
23 57 ☽ △ ⯓

26 Tuesday
00 46 ♀ □ ⯓
04 31 ☽ Q ♄
04 59 ☽ ∠ ♃
08 20 ☽ # ☿
09 27 ☽ # ♆
12 47 ☽ ☍ ♅
13 05 ☿ ± ♃
16 30 ☽ △ ♂

27 Wednesday
03 54 ☉ □ ♄
04 38 ☽ ⚹ ♃
05 39 ☽ # ♅
08 36 ☽ # ⯓
08 56 ☽ □ ♄
09 17 ☽ ☍ ☉
09 46 ☿ ∥ ♆
10 47 ☽ ⊼ ♆
16 10 ☽ ⊼ ☿
16 15 ☽ ∥ ♂
17 10 ☽ ♇ ♂
20 11 ☽ ± ♆
23 13 ☽ □ ⯓
23 24 ☽ Q ♃

28 Thursday
02 22 ☽ ± ☿
03 17 ☽ ☍ ♀
05 13 ☽ # ☉
07 55 ☽ ± ♂
08 56 ☉ ⊻ ♆
10 17 ☽ ♇ ♆
11 59 ☽ ⊼ ♅
17 30 ☽ # ♀
17 43 ☽ ♇ ☿
17 47 ☽ ⊼ ♂
21 26 ☽ ± ♅

All ephemeris data is given at 12.00 UT and the Moon's longitude is additionally given for 24.00 UT

MARCH 2002

LONGITUDES

Date	Sidereal time h m s	Sun ☉ ° ' "	Moon ☽ ° ' "	Moon ☽ 24.00 ° ' "	Mercury ☿ ° '	Venus ♀ ° '	Mars ♂ ° '	Jupiter ♃ ° '	Saturn ♄ ° '	Uranus ♅ ° '	Neptune ♆ ° '	Pluto ♇ ° '
01	22 36 29	10 ♓ 43 15	10 ♎ 53 45	18 ♎ 24 01	15 ♒ 17	21 ♓ 49	29 ♈ 55	05 ♋ 37	08 ♊ 28	25 ♒ 44	09 ♒ 38	17 ♐ 32
02	22 40 26	11 43 26	25 ♎ 48 45	03 ♏ 07 12	16 35	23 04	00 ♉ 37	05 D 37	08 30	25 47	09 40	17 32
03	22 44 22	12 43 35	10 ♏ 18 50	17 ♏ 23 21	17 54	24 18	01 20	05 38	08 33	25 50	09 42	17 33
04	22 48 19	13 43 43	24 ♏ 20 41	01 ♐ 10 54	19 14	25 33	02 02	05 38	08 35	25 54	09 44	17 33
05	22 52 16	14 43 50	07 ♐ 54 18	14 ♐ 31 12	20 36	26 48	02 44	05 39	08 38	25 57	09 46	17 34
06	22 56 12	15 43 55	21 ♐ 02 05	27 ♐ 27 27	21 59	28 03	03 27	05 40	08 41	26 00	09 48	17 34
07	23 00 09	16 43 58	03 ♑ 47 52	10 ♑ 03 51	23 24	29 ♓ 17	04 09	05 41	08 44	26 04	09 49	17 35
08	23 04 05	17 44 00	16 ♑ 15 59	22 ♑ 24 47	24 50	00 ♈ 32	04 51	05 42	08 47	26 07	09 51	17 35
09	23 08 02	18 44 00	28 ♑ 30 44	04 ♒ 34 19	26 18	01 47	05 34	05 44	08 50	26 10	09 53	17 36
10	23 11 58	19 43 59	10 ♒ 35 56	16 ♒ 35 57	27 46	03 01	06 16	05 45	08 53	26 13	09 55	17 36
11	23 15 55	20 43 55	22 ♒ 34 42	28 ♒ 32 28	29 ♒ 16	04 16	06 58	05 47	08 57	26 17	09 57	17 36
12	23 19 51	21 43 50	04 ♓ 29 30	10 ♓ 26 00	00 ♓ 47	05 31	07 40	05 49	09 00	26 20	09 59	17 37
13	23 23 48	22 43 43	16 ♓ 22 10	22 ♓ 18 10	02 20	06 45	08 23	05 51	09 04	26 23	10 01	17 37
14	23 27 45	23 43 34	28 ♓ 14 08	04 ♈ 10 16	03 53	08 00	09 05	05 54	09 07	26 26	10 02	17 37
15	23 31 41	24 43 23	10 ♈ 06 42	16 ♈ 03 37	05 28	09 14	09 47	05 56	09 11	26 29	10 04	17 37
16	23 35 38	25 43 10	22 ♈ 01 13	27 ♈ 59 45	07 04	10 29	10 29	05 59	09 15	26 33	10 06	17 37
17	23 39 34	26 42 55	03 ♉ 59 29	10 ♉ 00 43	08 42	11 43	11 11	06 02	09 19	26 36	10 08	17 38
18	23 43 31	27 42 38	16 ♉ 03 50	22 ♉ 09 14	10 20	12 58	11 53	06 05	09 23	26 39	10 09	17 38
19	23 47 27	28 42 18	28 ♉ 17 21	04 ♊ 28 40	12 00	14 12	12 35	06 08	09 27	26 42	10 11	17 38
20	23 51 24	29 ♓ 41 57	10 ♊ 43 42	17 ♊ 03 00	13 41	15 26	13 16	06 12	09 31	26 45	10 13	17 38
21	23 55 20	00 ♈ 41 33	23 ♊ 27 06	29 ♊ 56 34	15 23	16 41	13 58	06 16	09 35	26 48	10 14	17 R 38
22	23 59 17	01 41 07	06 ♋ 31 55	13 ♋ 13 36	17 06	17 55	14 40	06 20	09 40	26 51	10 16	17 38
23	00 03 14	02 40 38	20 ♋ 02 00	26 ♋ 57 25	18 51	19 09	15 22	06 24	09 44	26 54	10 17	17 38
24	00 07 10	03 40 08	03 ♌ 59 59	11 ♌ 09 39	20 37	20 24	16 04	06 28	09 49	26 57	10 19	17 37
25	00 11 07	04 39 35	18 ♌ 26 11	25 ♌ 49 08	22 25	21 38	16 45	06 32	09 53	27 00	10 20	17 37
26	00 15 03	05 38 59	03 ♍ 17 46	10 ♍ 51 10	24 13	22 52	17 27	06 37	09 58	27 03	10 22	17 37
27	00 19 00	06 38 22	18 ♍ 28 11	26 ♍ 07 30	26 03	24 06	18 09	06 42	10 03	27 06	10 23	17 37
28	00 22 56	07 37 42	03 ♎ 47 42	11 ♎ 27 17	27 54	25 20	18 50	06 46	10 08	27 08	10 25	17 37
29	00 26 53	08 37 00	19 ♎ 04 47	26 ♎ 38 51	29 ♓ 47	26 34	19 32	06 52	10 13	27 11	10 26	17 36
30	00 30 49	09 36 16	04 ♏ 08 15	11 ♏ 31 59	01 ♈ 41	27 48	20 13	06 57	10 18	27 14	10 28	17 36
31	00 34 46	10 ♈ 35 30	18 ♏ 49 14	25 ♏ 59 28	03 ♈ 36	29 ♈ 02	20 ♉ 55	07 ♋ 02	10 ♊ 23	27 ♒ 17	10 ♒ 29	17 ♐ 36

Date	Moon True ☊ ° '	Moon Mean ☊ ° '	Moon ☽ Latitude ° '
01	23 ♊ 40	23 ♊ 13	04 N 48
02	23 R 32	23 09	04 15
03	23 26	23 06	03 27
04	23 23	23 03	02 27
05	23 22	23 00	01 21
06	23 D 22	22 57	00 N 12
07	23 R 21	22 54	00 S 55
08	23 20	22 50	01 58
09	23 16	22 47	02 54
10	23 09	22 44	03 42
11	22 59	22 41	04 19
12	22 46	22 38	04 44
13	22 32	22 35	04 57
14	22 17	22 31	04 57
15	22 02	22 28	04 45
16	21 50	22 25	04 19
17	21 39	22 22	03 42
18	21 32	22 19	02 55
19	21 28	22 15	01 59
20	21 26	22 12	00 S 56
21	21 D 26	22 09	00 N 11
22	21 R 26	22 06	01 19
23	21 25	22 03	02 25
24	21 22	22 00	03 25
25	21 17	21 56	04 14
26	21 09	21 53	04 47
27	20 59	21 50	05 01
28	20 48	21 47	04 54
29	20 38	21 44	04 25
30	20 30	21 41	03 39
31	20 ♊ 24	21 ♊ 37	02 N 39

DECLINATIONS

Date	Sun ☉ ° '	Moon ☽ ° '	Mercury ☿ ° '	Venus ♀ ° '	Mars ♂ ° '	Jupiter ♃ ° '	Saturn ♄ ° '	Uranus ♅ ° '	Neptune ♆ ° '	Pluto ♇ ° '
01	07 S 33	00 N 06	17 S 16	04 S 31	11 N 41	23 N 27	20 N 10	13 S 36	17 S 45	12 S 58
02	07 10	06 S 01	17 00	04 01	11 57	23 27	20 11	13 35	17 44	12 58
03	06 47	11 39	16 43	03 30	12 12	23 27	20 12	13 34	17 44	12 58
04	06 24	16 29	16 24	03 00	12 27	23 27	20 12	13 33	17 43	12 58
05	06 01	20 18	16 04	02 29	12 43	23 27	20 13	13 32	17 43	12 58
06	05 38	22 56	15 42	01 58	12 58	23 27	20 14	13 30	17 42	12 58
07	05 14	24 18	15 20	01 28	13 13	23 27	20 14	13 29	17 42	12 57
08	04 51	24 24	14 56	00 57	13 28	23 27	20 15	13 28	17 41	12 57
09	04 27	23 18	14 30	00 S 26	13 43	23 27	20 16	13 27	17 41	12 57
10	04 04	21 08	14 04	00 N 05	13 57	23 27	20 16	13 26	17 40	12 57
11	03 40	18 04	13 36	00 36	14 12	23 27	20 17	13 25	17 40	12 56
12	03 17	14 16	13 07	01 07	14 26	23 27	20 18	13 24	17 39	12 56
13	02 53	09 56	12 37	01 37	14 41	23 27	20 19	13 23	17 39	12 56
14	02 29	05 15	12 05	02 08	14 55	23 27	20 19	13 22	17 38	12 56
15	02 06	00 S 22	11 33	02 39	15 09	23 27	20 20	13 21	17 38	12 56
16	01 42	04 N 34	10 59	03 10	15 23	23 27	20 21	13 20	17 38	12 55
17	01 18	09 22	10 24	03 40	15 36	23 27	20 22	13 19	17 37	12 55
18	00 55	13 51	09 47	04 11	15 50	23 27	20 23	13 18	17 37	12 55
19	00 31	17 51	09 10	04 41	16 03	23 27	20 23	13 16	17 36	12 55
20	00 S 07	21 08	08 31	05 12	16 17	23 27	20 24	13 15	17 36	12 54
21	00 N 17	23 27	07 51	05 42	16 30	23 27	20 25	13 14	17 35	12 54
22	00 40	24 36	07 10	06 12	16 43	23 27	20 26	13 13	17 35	12 54
23	01 04	24 20	06 28	06 42	16 56	23 27	20 27	13 12	17 35	12 54
24	01 28	22 34	05 44	07 12	17 08	23 27	20 28	13 11	17 34	12 54
25	01 51	19 19	05 00	07 42	17 21	23 27	20 29	13 10	17 34	12 53
26	02 15	14 45	04 14	08 12	17 33	23 26	20 30	13 10	17 33	12 53
27	02 38	09 10	03 27	08 41	17 45	23 26	20 31	13 09	17 33	12 53
28	03 02	02 N 59	02 40	09 10	17 58	23 26	20 31	13 08	17 33	12 53
29	03 25	03 S 23	01 51	09 39	18 09	23 26	20 32	13 07	17 32	12 52
30	03 48	09 28	01 01	10 08	18 21	23 26	20 33	13 06	17 32	12 52
31	04 N 12	14 S 52	00 S 10	10 N 36	18 N 33	23 N 26	20 N 34	13 S 05	17 S 31	12 S 52

ZODIAC SIGN ENTRIES

Date	h	m	Planets
01	15	05	♂ ♉
02	18	51	☽ ♏
04	21	55	☽ ♐
07	04	48	☽ ♑
08	01	42	♀ ♈
09	14	56	☽ ♒
11	23	34	☿ ♓
12	02	56	☽ ♓
14	15	34	☽ ♈
17	04	01	☽ ♉
19	15	20	☽ ♊
20	19	16	☉ ♈
22	00	06	☽ ♋
24	05	13	☽ ♌
26	06	44	☽ ♍
28	06	04	☽ ♎
29	14	44	☿ ♈
30	05	21	☽ ♏

LATITUDES

Date	Mercury ☿ ° '	Venus ♀ ° '	Mars ♂ ° '	Jupiter ♃ ° '	Saturn ♄ ° '	Uranus ♅ ° '	Neptune ♆ ° '	Pluto ♇ ° '
01	01 S 04	01 S 23	00 N 15	00 N 08	01 S 34	00 S 42	00 N 06	09 N 55
04	01 25	01 20	00 18	00 08	01 33	00 42	00 06	09 56
07	01 42	01 17	00 20	00 08	01 32	00 42	00 06	09 57
10	01 56	01 13	00 22	00 09	01 32	00 42	00 06	09 57
13	02 07	01 09	00 24	00 09	01 31	00 42	00 06	09 58
16	02 13	01 04	00 26	00 09	01 30	00 42	00 06	09 59
19	02 16	00 59	00 28	00 09	01 30	00 42	00 06	10 00
22	02 15	00 53	00 30	00 10	01 29	00 42	00 05	10 00
25	02 09	00 47	00 32	00 10	01 28	00 42	00 05	10 01
28	02 00	00 41	00 33	00 10	01 28	00 42	00 05	10 02
31	01 S 45	00 S 34	00 N 35	00 N 10	01 S 27	00 S 42	00 N 05	10 N 02

DATA

Julian Date	2452335
Delta T	+64 seconds
Ayanamsa	23° 52' 58"
Synetic vernal point	05° ♓ 14' 01"
True obliquity of ecliptic	23° 26' 22"

MOON'S PHASES, APSIDES AND POSITIONS ☽

Date	h	m	Phase	Longitude ° '	Eclipse Indicator
06	01	25	☾	15 ♐ 17	
14	02	03	●	23 ♓ 19	
22	02	28	☽	01 ♋ 17	
28	18	25	○	07 ♎ 54	

Day	h	m		
14	01	26	Apogee	
28	07	48	Perigee	
01	12	23	0S	
08	01	52	Max dec	24° S 31'
15	13	45	0N	
22	19	51	Max dec	24° N 40'
28	23	14	0S	

LONGITUDES

Date	Chiron ⚷ ° '	Ceres ⚳ ° '	Pallas ⚴ ° '	Juno ⚵ ° '	Vesta ⚶ ° '	Black Moon Lilith ⚸ ° '
01	07 ♑ 35	02 ♓ 46	07 ♒ 18	17 ♌ 49	04 ♊ 15	21 ♓ 21
11	08 ♑ 10	06 ♓ 40	10 ♒ 38	16 ♌ 16	06 ♊ 59	22 ♓ 28
21	08 ♑ 37	10 ♓ 32	13 ♒ 47	15 ♌ 25	10 ♊ 02	23 ♓ 35
31	08 ♑ 57	14 ♓ 20	16 ♒ 46	15 ♌ 18	13 ♊ 21	24 ♓ 43

All ephemeris data is given at 12.00 UT and the Moon's longitude is additionally given for 24.00 UT

ASPECTARIAN

h m Aspects

01 Friday
03 29 ☽ Q ♇
03 39 ☽ □ ♃
08 08 ☽ △ ♄
09 59 ☽ △ ♆
11 42 ☽ ⊼ ☉
11 44 ☽ bQ ♅
15 14 ♃ St D
19 40 ☽ △ ☿
21 58 ☽ ± ☉
22 36 ☽ ✶ ♇

02 Saturday
04 37 ☽ ∥ ♀
07 07 ☽ ⊼ ♀
08 14 ☽ bQ ♄
11 57 ☽ △ ♅
13 36 ☽ bQ ☉
16 25 ☽ ∥ ☉
17 48 ☽ ± ♀
20 16 ☽ ☍ ♂
23 01 ☽ ± ♄
23 03 ☽ ∠ ♇

03 Sunday
04 10 ☽ △ ♃
05 40 ☿ ✶ ♇
09 01 ☽ ⊼ ♄
10 09 ☽ bQ ♀
10 57 ☽ □ ♆
14 05 ☽ ⊥ ♇
14 42 ☽ # ♂
16 23 ☽ △ ☉
18 09 ☽ ∥ ♇
19 40 ♀ ∠ ♆
20 58 ☽ ∥ ♅

04 Monday
00 17 ☽ ⊻ ♇
02 13 ☽ □ ☿
05 34 ☽ bQ ♃
11 34 ☽ ∥ ☿
14 19 ☽ △ ♀
14 43 ☽ □ ♅
17 56 ☽ Q ♆
18 52 ♀ ⊻ ♅
19 04 ☽ ∥ ♆
21 17 ☽ ± ♃

05 Tuesday
02 15 ☽ ⊼ ♂
05 57 ♂ bQ ♇
07 57 ☽ ⊼ ♃
08 10 ♂ ⊥ ♄
08 43 ♀ Q ♄
09 24 ♀ ⊥ ♂
11 23 ☽ # ♄
12 49 ☿ bQ ♃
13 19 ☽ ☍ ♄
13 24 ☽ Q ☿
13 35 ☽ ± ♂
15 21 ☽ ✶ ♆
16 56 ☿ Q ♂
23 00 ☽ Q ♅

06 Wednesday
01 25 ☽ □ ☉
05 36 ☽ ☌ ♇
06 56 ☽ bQ ♂
11 29 ♂ # ♇
13 29 ☉ ⊥ ♆
14 00 ☽ ✶ ☿
18 52 ☽ # ♃
19 01 ☽ ∠ ♆
19 59 ☿ ⊥ ♀
21 19 ☽ ✶ ♅

07 Thursday
02 31 ☽ □ ♀
12 03 ☽ ⊥ ♆
12 43 ☽ △ ♂
13 56 ☽ Q ☉
15 36 ☽ ☍ ♃
21 28 ☽ ⊼ ♄
21 57 ☽ ∠ ☿
23 34 ☽ ⊻ ♆

08 Friday
01 59 ☽ ∠ ♅
08 29 ☉ □ ♇
09 06 ☽ ± ♄
12 38 ♂ # ♅
14 34 ☽ ⊻ ♇
15 06 ☽ ✶ ☉
16 55 ☽ Q ♀
17 41 ☽ ⊥ ☿
19 32 ☽ ⊥ ♅

09 Saturday
02 19 ☽ ⊥ ♇
02 45 ☽ bQ ♄
07 02 ☽ ⊻ ☿
07 22 ☽ ⊻ ♅
09 39 ☽ # ♃
09 52 ☿ ☌ ♅
17 46 ♂ ✶ ♃
19 12 ☽ ✶ ♀
19 52 ♀ ⊥ ♅
20 05 ☽ ∠ ♇
23 16 ☽ ∠ ☉

10 Sunday
02 19 ☽ ⊼ ♃
02 50 ☽ □ ♂
08 34 ☽ △ ♄
10 38 ☽ ☌ ♆
14 19 ☽ ± ♃
15 50 ☿ # ♂
18 50 ☽ ⊥ ☉
19 22 ☽ # ♄

11 Monday
02 01 ☽ ✶ ♇
04 35 ☽ ∠ ♀
07 57 ☽ ⊻ ☉
08 23 ☽ bQ ♃
14 43 ☽ ∥ ♆
17 07 ☽ Q ♂
17 20 ☿ Q ♇
19 29 ☽ ☌ ♅
21 47 ☿ ∥ ♅

12 Tuesday
00 47 ☽ ⊥ ♀
02 09 ☽ Q ♇
03 26 ☽ ☌ ☿
11 04 ☽ # ♂
14 18 ☽ ⊻ ♀
14 41 ☽ △ ♃
17 04 ☽ ∥ ♅
18 05 ♀ □ ♃
18 50 ☽ ✶ ♂
19 31 ☽ ∥ ☿
19 40 ☽ ∥ ♇
20 49 ☿ ∥ ♇
21 09 ☽ □ ♄
23 07 ☽ ⊻ ♆

13 Wednesday
11 16 ☽ ⊥ ♆
12 21 ♂ Q ♅
14 31 ☽ □ ♇

14 Thursday
02 03 ☽ ☌ ☉
03 04 ☽ ∠ ♂
05 31 ☽ ∠ ♆
08 21 ☽ ⊻ ♅
09 44 ☽ Q ♄
13 40 ♂ ⊻ ♄
20 32 ☽ ⊥ ♅
21 24 ☉ # ♀
22 24 ☽ ⊥ ♂

15 Friday
01 11 ☽ ⊻ ☿
01 53 ☽ # ♀
02 47 ☽ ∥ ☉
03 32 ☽ □ ♃
10 02 ☽ ☌ ♀
10 07 ☽ ✶ ♄
10 53 ♀ ✶ ♄
11 17 ☽ ⊻ ♂
11 55 ☽ ✶ ♆
14 48 ☽ ∠ ♅
15 10 ☽ ⊥ ☿
16 27 ☉ ∠ ♂
19 13 ☿ △ ♃
20 34 ☉ ∠ ♆
22 23 ♂ □ ♆
23 03 ☽ # ☉

16 Saturday
03 09 ☽ △ ♇
04 20 ☽ ∥ ♀
04 26 ♀ ✶ ♆
11 57 ♀ ⊻ ♂
12 07 ☽ ∠ ☿
12 09 ☽ Q ♆
15 58 ☽ Q ♃
16 30 ☽ ∠ ♄
20 06 ☽ ⊻ ☉
21 08 ☽ ✶ ♅

17 Sunday
08 56 ☉ ⊻ ♅
09 13 ☽ ⊥ ☉
09 16 ☽ bQ ♇
09 28 ♀ ∠ ♅
10 38 ☽ ⊥ ♄
16 06 ☽ ✶ ♃
16 44 ☽ # ☿
21 13 ☽ Q ♅
21 29 ☿ □ ♄
22 40 ☽ ⊻ ♄
22 51 ☽ ✶ ☿

18 Monday
00 15 ☽ □ ♆
03 12 ☽ ☌ ♂
03 13 ☽ ± ♇
03 24 ♂ ± ♇
03 26 ☉ Q ♄
04 46 ☽ ∠ ☉
05 09 ☽ ⊻ ♀
06 49 ☽ # ♇
08 54 ☽ # ♅
09 20 ☿ ⊻ ♆
15 05 ☽ ⊼ ♇
18 22 ☽ ⊥ ♀
21 57 ☽ ∠ ♃

19 Tuesday
00 08 ☽ ∥ ♂
02 17 ☽ Q ☿
08 53 ☽ □ ♅
10 27 ☽ # ♆
12 53 ☽ ✶ ☉
13 58 ☽ ∠ ♀
15 37 ☽ ⊥ ♃

20 Wednesday
02 12 ☿ ✶ ♂
03 17 ☽ ⊻ ♃
06 08 ☽ ∥ ♄
09 40 ☽ ☌ ♄
11 00 ☽ △ ♆
14 01 ☽ Q ☉
14 57 ♇ St R
17 08 ☽ ⊻ ♂
18 29 ☽ □ ☿
21 56 ☽ ✶ ♀

21 Thursday
01 05 ☽ ☍ ♇
05 07 ☽ ⊥ ♂
11 56 ☽ ∥ ♃
15 19 ☽ bQ ♆
18 14 ☽ △ ♅
22 42 ☽ Q ♀
22 48 ☽ ∠ ♂

22 Friday
00 06 ☿ ⊥ ♆
02 28 ☽ □ ☉
06 22 ♀ △ ♇
07 53 ☽ ± ♆
11 38 ☽ ☌ ♃
17 40 ☽ ⊻ ♄
18 44 ☽ ⊼ ♆
19 11 ☿ □ ♇
21 35 ☽ bQ ♅

23 Saturday
03 21 ☽ ✶ ♂
04 24 ☽ ⊥ ♄
07 06 ☿ # ♀
07 47 ☽ ⊼ ♇
09 38 ☽ △ ☿
10 19 ☽ □ ♀
13 31 ☽ ± ♅
17 39 ☉ ⊥ ♅
18 15 ☽ ± ♇
20 13 ☽ ∠ ♄
23 57 ☽ ⊼ ♅

24 Sunday
01 22 ☽ Q ♂
01 52 ☿ ⊻ ♀
02 35 ☽ ∥ ♃
09 40 ☽ bQ ♇
11 24 ☽ △ ☉
15 07 ☽ bQ ☿
16 10 ☽ ⊻ ♃
21 48 ☽ ✶ ♄
22 37 ☽ ☍ ♆

25 Monday
02 14 ☽ ⊥ ♃
04 35 ☽ ∥ ♄
08 12 ☽ ± ☿
09 06 ☽ □ ♂
10 40 ☽ △ ♇
14 09 ☽ bQ ☉
17 05 ☽ ∠ ♃
17 40 ☽ Q ♄
17 41 ☽ △ ♀
19 23 ☽ ⊼ ☿
22 00 ☽ # ♆
22 38 ☽ ∥ ♂

26 Tuesday
01 57 ☽ ☍ ♅
02 03 ♀ Q ♆
05 45 ☽ ± ☉
12 07 ♂ # ♆
16 01 ☽ ⊼ ☉
17 19 ☽ ✶ ♃
17 49 ♂ ⊼ ♇
19 15 ☽ # ♅
19 55 ☽ bQ ♀
20 27 ☽ # ♇
22 39 ☽ □ ♄
23 15 ☽ ⊼ ♆

27 Wednesday
03 14 ☿ ∠ ♆
08 44 ☽ ± ♆
10 40 ☽ □ ♇
11 22 ☽ ± ♀
11 28 ☽ △ ♂
12 21 ☽ Q ♃
13 24 ☉ □ ♃
13 49 ☽ ∥ ♀
21 37 ☽ ⊼ ♀
22 52 ☽ bQ ♆

28 Thursday
00 14 ♀ Q ♃
01 31 ☽ ☍ ☿
01 33 ☽ ⊼ ♅
01 51 ☿ ⊻ ♅
04 37 ☉ # ☿
07 37 ♀ ∠ ♄
10 58 ☽ ± ♅
11 50 ☽ ∥ ☉
12 04 ☽ bQ ♂
13 23 ☽ # ☿
14 51 ☽ Q ♇
14 58 ☿ Q ♄
16 41 ☽ □ ♃
18 25 ☽ ☍ ☉
21 58 ☽ △ ♄
22 23 ☽ △ ♆

29 Friday
01 07 ☽ bQ ♅
02 51 ☽ ± ♂
06 51 ☽ ∥ ☿
09 41 ☽ ✶ ♇
12 10 ☽ # ☉
12 45 ☽ ⊼ ♂
21 46 ☽ bQ ♄

30 Saturday
00 25 ♀ ✶ ♅
00 54 ☽ △ ♅
00 57 ☽ ☍ ♀
07 29 ☽ ⊼ ☿
09 32 ☽ ∠ ♇
12 15 ☽ ± ♄
15 01 ☽ # ♀
16 34 ☽ △ ♃
18 35 ☽ ± ☿
21 29 ☽ ⊼ ☉
22 02 ☽ ⊼ ♄
22 16 ☽ □ ♆

31 Sunday
00 06 ☽ ⊥ ♇
02 38 ☽ ∥ ♇
03 38 ☽ ∥ ♅
06 22 ☉ ✶ ♄
07 52 ☿ ⊥ ♅
08 02 ☽ ± ☉
09 19 ☉ ✶ ♆
09 58 ☽ ⊻ ♇
11 35 ☽ bQ ☿
15 39 ☽ ☍ ♂
17 23 ☽ bQ ♃

APRIL 2002

LONGITUDES

Date	Sidereal time h m s	Sun ☉ ° ' "	Moon ☽ ° ' "	Moon ☽ 24.00 ° ' "	Mercury ☿ ° '	Venus ♀ ° '	Mars ♂ ° '	Jupiter ♃ ° '	Saturn ♄ ° '	Uranus ♅ ° '	Neptune ♆ ° '	Pluto ⯓ ° '
01	00 38 43	11 ♈ 34 43	03 ♐ 02 21	09 ♐ 57 46	05 ♈ 33	00 ♉ 16	21 ♉ 36	07 ♋ 08	10 ♊ 28	27 ♒ 20	10 ♒ 30	17 ♐ 35
02	00 42 39	12 33 53	16 ♐ 45 49	23 ♐ 26 44	07 31	01 30	22 17	07 13	10 33	27 22	10 32	17 R 35
03	00 46 36	13 33 02	00 ♑ 00 53	06 ♑ 28 44	09 30	02 44	22 59	07 19	10 39	27 25	10 33	17 35
04	00 50 32	14 32 10	12 ♑ 50 49	19 ♑ 07 43	11 30	03 58	23 40	07 25	10 44	27 28	10 34	17 34
05	00 54 29	15 31 15	25 ♑ 20 01	01 ♒ 28 21	13 32	05 12	24 21	07 32	10 49	27 30	10 35	17 34
06	00 58 25	16 30 19	07 ♒ 33 18	13 ♒ 35 27	15 34	06 26	25 03	07 38	10 55	27 33	10 37	17 33
07	01 02 22	17 29 21	19 ♒ 35 21	25 ♒ 33 30	17 38	07 40	25 44	07 44	11 01	27 35	10 38	17 32
08	01 06 18	18 28 21	01 ♓ 30 22	07 ♓ 26 24	19 42	08 53	26 25	07 51	11 06	27 38	10 39	17 32
09	01 10 15	19 27 19	13 ♓ 21 57	19 ♓ 17 23	21 47	10 07	27 06	07 58	11 12	27 40	10 40	17 31
10	01 14 12	20 26 15	25 ♓ 12 59	01 ♈ 09 01	23 52	11 21	27 47	08 05	11 18	27 43	10 41	17 31
11	01 18 08	21 25 10	07 ♈ 05 42	13 ♈ 03 14	25 58	12 35	28 28	08 12	11 24	27 45	10 42	17 30
12	01 22 05	22 24 02	19 ♈ 01 48	25 ♈ 01 33	28 ♈ 04	13 48	29 09	08 19	11 30	27 47	10 43	17 29
13	01 26 01	23 22 53	01 ♉ 02 40	07 ♉ 05 19	00 ♉ 10	15 02	29 ♉ 50	08 26	11 36	27 50	10 44	17 28
14	01 29 58	24 21 41	13 ♉ 09 38	19 ♉ 15 51	02 15	16 15	00 ♊ 31	08 34	11 42	27 52	10 45	17 28
15	01 33 54	25 20 28	25 ♉ 24 09	01 ♊ 34 47	04 19	17 29	01 12	08 41	11 48	27 54	10 46	17 27
16	01 37 51	26 19 12	07 ♊ 48 02	14 ♊ 04 11	06 22	18 42	01 53	08 49	11 54	27 56	10 47	17 26
17	01 41 47	27 17 55	20 ♊ 23 34	26 ♊ 46 33	08 23	19 56	02 34	08 57	12 00	27 59	10 48	17 25
18	01 45 44	28 16 35	03 ♋ 13 30	09 ♋ 44 47	10 23	21 09	03 15	09 05	12 07	28 01	10 49	17 24
19	01 49 41	29 ♈ 15 13	16 ♋ 20 48	23 ♋ 01 52	12 21	22 23	03 56	09 13	12 13	28 03	10 49	17 23
20	01 53 37	00 ♉ 13 49	29 ♋ 48 18	06 ♌ 40 21	14 16	23 36	04 36	09 22	12 20	28 05	10 50	17 23
21	01 57 34	01 12 22	13 ♌ 38 08	20 ♌ 41 43	16 08	24 49	05 17	09 30	12 26	28 07	10 51	17 22
22	02 01 30	02 10 54	27 ♌ 50 57	05 ♍ 05 36	17 57	26 02	05 58	09 38	12 33	28 09	10 52	17 21
23	02 05 27	03 09 23	12 ♍ 25 11	19 ♍ 49 06	19 43	27 16	06 39	09 47	12 39	28 11	10 52	17 20
24	02 09 23	04 07 49	27 ♍ 16 32	04 ♎ 46 32	21 25	28 29	07 19	09 56	12 46	28 13	10 53	17 19
25	02 13 20	05 06 14	12 ♎ 18 00	19 ♎ 49 47	23 04	29 ♉ 42	08 00	10 05	12 53	28 14	10 54	17 17
26	02 17 16	06 04 37	27 ♎ 20 39	04 ♏ 49 24	24 39	00 ♊ 55	08 40	10 14	12 59	28 16	10 54	17 16
27	02 21 13	07 02 58	12 ♏ 14 57	19 ♏ 36 14	26 09	02 08	09 21	10 23	13 06	28 18	10 55	17 15
28	02 25 10	08 01 17	26 ♏ 52 26	04 ♐ 02 49	27 35	03 21	10 01	10 32	13 13	28 20	10 55	17 14
29	02 29 06	08 59 35	11 ♐ 06 54	18 ♐ 04 21	28 ♉ 57	04 34	10 42	10 41	13 20	28 21	10 56	17 13
30	02 33 03	09 ♉ 57 51	24 ♐ 55 01	01 ♑ 38 56	00 ♊ 15	05 ♊ 47	11 ♊ 22	10 ♋ 51	13 ♊ 27	28 ♒ 23	10 ♒ 56	17 ♐ 12

DECLINATIONS

Date	Moon True ☊ ° '	Moon Mean ☊ ° '	Moon ☽ Latitude ° '	Sun ☉ ° '	Moon ☽ ° '	Mercury ☿ ° '	Venus ♀ ° '	Mars ♂ ° '	Jupiter ♃ ° '	Saturn ♄ ° '	Uranus ♅ ° '	Neptune ♆ ° '	Pluto ⯓ ° '
01	20 ♊ 20	21 ♊ 34	01 N 30	04 N 35	19 S 17	00 N 41	11 N 04	18 N 44	23 N 25	20 N 35	13 S 04	17 S 31	12 S 52
02	20 R 19	21 31	00 N 19	04 58	22 28	01 34	11 32	18 55	23 25	20 36	13 03	17 31	12 51
03	20 D 19	21 28	00 S 51	05 21	24 18	02 27	12 00	19 07	23 25	20 37	13 02	17 30	12 51
04	20 20	21 25	01 57	05 44	24 45	03 21	12 28	19 17	23 25	20 38	13 01	17 30	12 51
05	20 R 19	21 21	02 55	06 07	23 56	04 15	12 55	19 28	23 24	20 39	13 00	17 30	12 51
06	20 17	21 18	03 44	06 29	21 59	05 10	13 21	19 39	23 24	20 40	12 59	17 29	12 50
07	20 12	21 15	04 22	06 52	19 05	06 06	13 48	19 49	23 24	20 41	12 59	17 29	12 50
08	20 05	21 12	04 48	07 14	15 25	07 01	14 14	19 59	23 24	20 42	12 58	17 29	12 50
09	19 56	21 09	05 01	07 37	11 10	07 57	14 40	20 09	23 23	20 43	12 57	17 29	12 50
10	19 45	21 06	05 02	07 59	06 31	08 53	15 05	20 19	23 23	20 44	12 56	17 28	12 49
11	19 33	21 02	04 50	08 21	01 S 37	09 49	15 30	20 29	23 22	20 45	12 56	17 28	12 49
12	19 22	20 59	04 24	08 43	03 N 22	10 44	15 55	20 38	23 22	20 46	12 55	17 28	12 49
13	19 12	20 56	03 47	09 05	08 17	11 38	16 19	20 48	23 22	20 47	12 54	17 27	12 48
14	19 05	20 53	02 59	09 27	12 56	12 32	16 43	20 57	23 21	20 48	12 53	17 27	12 48
15	18 59	20 50	02 03	09 48	17 08	13 25	17 07	21 06	23 21	20 49	12 52	17 27	12 48
16	18 57	20 47	00 S 59	10 10	20 38	14 17	17 30	21 15	23 20	20 50	12 52	17 27	12 48
17	18 D 56	20 43	00 N 08	10 31	23 13	15 07	17 52	21 23	23 20	20 51	12 51	17 26	12 47
18	18 57	20 40	01 16	10 52	24 40	15 55	18 15	21 32	23 19	20 52	12 50	17 26	12 47
19	18 58	20 37	02 22	11 13	24 47	16 42	18 36	21 40	23 19	20 53	12 49	17 26	12 47
20	18 59	20 34	03 22	11 33	23 29	17 27	18 57	21 48	23 18	20 54	12 49	17 26	12 47
21	18 R 58	20 31	04 12	11 54	20 45	18 09	19 18	21 56	23 18	20 55	12 48	17 26	12 47
22	18 55	20 27	04 48	12 14	16 43	18 49	19 38	22 03	23 17	20 56	12 47	17 25	12 46
23	18 51	20 24	05 06	12 34	11 37	19 27	19 58	22 11	23 17	20 57	12 47	17 25	12 46
24	18 45	20 21	05 05	12 54	05 N 45	20 03	20 17	22 18	23 16	20 58	12 46	17 25	12 46
25	18 38	20 18	04 43	13 13	00 S 31	20 36	20 36	22 25	23 16	20 59	12 46	17 25	12 46
26	18 32	20 15	04 01	13 33	06 47	21 06	20 54	22 32	23 15	21 00	12 45	17 25	12 45
27	18 27	20 12	03 03	13 52	12 36	21 34	21 11	22 39	23 14	21 01	12 44	17 25	12 45
28	18 23	20 08	01 54	14 11	17 37	21 59	21 28	22 45	23 14	21 02	12 44	17 25	12 45
29	18 21	20 05	00 39	14 30	21 28	22 22	21 44	22 51	23 13	21 03	12 43	17 24	12 45
30	18 ♊ 21	20 ♊ 02	00 S 36	14 N 48	23 S 56	22 N 42	22 N 00	22 N 57	23 N 12	21 N 04	12 S 43	17 S 24	12 S 44

ZODIAC SIGN ENTRIES

Date	h	m	Planets
01	06	39	♀ ♉
01	06	48	☽ ♐
03	11	58	☽ ♑
05	21	07	☽ ♒
08	08	57	☽ ♓
10	21	41	☽ ♈
13	09	55	☽ ♉
13	10	10	☿ ♉
13	17	36	♂ ♊
15	20	56	☽ ♊
18	06	01	☽ ♋
20	06	20	☉ ♉
20	12	21	☽ ♌
22	15	35	☽ ♍
24	16	22	☽ ♎
25	17	57	♀ ♊
26	16	15	☽ ♏
28	17	13	☽ ♐
30	07	15	☿ ♊
30	21	03	☽ ♑

LATITUDES

Date	Mercury ☿ ° '	Venus ♀ ° '	Mars ♂ ° '	Jupiter ♃ ° '	Saturn ♄ ° '	Uranus ♅ ° '	Neptune ♆ ° '	Pluto ⯓ ° '
01	01 S 39	00 S 32	00 N 36	00 N 10	01 S 27	00 S 42	00 N 05	10 N 03
04	01 19	00 24	00 37	00 11	01 26	00 42	00 05	10 03
07	00 53	00 17	00 39	00 11	01 26	00 43	00 05	10 04
10	00 S 25	00 09	00 40	00 11	01 25	00 43	00 05	10 04
13	00 N 07	00 S 01	00 42	00 11	01 24	00 43	00 05	10 05
16	00 40	00 N 07	00 43	00 12	01 24	00 43	00 05	10 06
19	01 13	00 15	00 45	00 12	01 23	00 43	00 05	10 06
22	01 43	00 23	00 46	00 12	01 23	00 43	00 05	10 07
25	02 07	00 31	00 47	00 12	01 22	00 43	00 05	10 07
28	02 26	00 39	00 48	00 12	01 22	00 43	00 05	10 07
31	02 N 36	00 N 47	00 N 50	00 N 13	01 S 22	00 S 43	00 N 05	10 N 08

DATA

Julian Date	2452366
Delta T	+64 seconds
Ayanamsa	23° 53' 01"
Synetic vernal point	05° ♓ 13' 58"
True obliquity of ecliptic	23° 26' 22"

MOON'S PHASES, APSIDES AND POSITIONS ☽

Date	h	m	Phase	Longitude ° '	Eclipse Indicator
04	15	29	☾	14 ♑ 41	
12	19	21	●	22 ♈ 42	
20	12	48	☽	00 ♌ 16	
27	03	00	○	06 ♏ 41	

Day	h	m		
10	05	44	Apogee	
25	16	30	Perigee	
04	08	19	Max dec	24° S 46'
11	19	47	0N	
19	02	07	Max dec	24° N 55'
25	10	01	0S	

LONGITUDES

Date	Chiron ⚷ ° '	Ceres ⚳ ° '	Pallas ⚴ ° '	Juno ⚵ ° '	Vesta ⚶ ° '	Black Moon Lilith ⚸ ° '
01	08 ♑ 58	14 ♓ 42	17 ♒ 03	15 ♌ 19	13 ♊ 42	24 ♓ 49
11	09 ♑ 07	18 ♓ 26	19 ♒ 47	15 ♌ 56	17 ♊ 15	25 ♓ 57
21	09 ♑ 08	22 ♓ 04	22 ♒ 15	17 ♌ 09	20 ♊ 59	27 ♓ 04
31	09 ♑ 00	25 ♓ 36	24 ♒ 26	18 ♌ 51	24 ♊ 52	28 ♓ 11

ASPECTARIAN

h m Aspects

01 Monday
00 10 ☽ ♇ ☉
01 38 ☽ ∥ ♆
02 13 ☽ □ ♅
04 15 ☽ Q ♆
06 49 ☽ ⊼ ♀
08 27 ☽ # ♂
08 42 ☽ ± ♃
17 02 ☽ △ ☿
18 08 ☽ ± ♀
19 07 ☽ ⊼ ♃
20 42 ☽ # ♄

02 Tuesday
00 58 ☽ ☍ ♄
00 58 ☽ ⚹ ♆
02 37 ♄ △ ♆
03 59 ☽ △ ☉
07 03 ☉ ∠ ♅
07 52 ☿ ∠ ♂
08 19 ☿ □ ♃
09 15 ♂ ∠ ♃
09 31 ☽ Q ♅
11 30 ☽ ♇ ♀
13 28 ☽ ☌ ⯓
22 14 ☽ # ♃
22 27 ☽ ⊼ ♂

03 Wednesday
03 48 ☽ ∠ ♆
07 13 ☽ ⚹ ♅
08 50 ♀ ♇ ⯓
10 00 ☽ ± ♂
17 34 ☽ △ ♀
20 25 ☽ ⊥ ♆

04 Thursday
00 45 ☿ ⚹ ♆
01 41 ☽ ☍ ♃
02 23 ☿ ⚹ ♄
03 40 ☽ ♇ ♂
07 41 ☽ ⊻ ♆
07 58 ☽ ⊼ ♄
08 58 ☽ □ ☿
11 16 ☽ ∠ ♅
15 29 ☽ □ ☉
19 28 ☽ ± ♄
21 00 ☽ ⊻ ⯓
23 36 ☿ ∠ ♅

05 Friday
04 03 ♀ ⊥ ♄
04 33 ☽ ⊥ ♅
08 21 ♀ # ⯓
08 33 ☽ ⊥ ⯓
09 59 ☽ △ ♂
12 58 ☽ ♇ ♄
16 14 ☽ ⊻ ♅
16 55 ♀ # ♅
20 00 ☽ # ♃

06 Saturday
02 08 ☽ ∠ ⯓
02 32 ☽ Q ☿
05 26 ☽ Q ☉
09 32 ☽ □ ♀
12 09 ☽ ⊼ ♃
18 04 ☽ ☌ ♆
18 44 ☽ △ ♄
23 43 ☽ # ♄

07 Sunday
00 11 ☽ ± ♃
06 47 ☽ # ♂
07 15 ☽ ⚹ ☿
07 25 ☽ ⚹ ☉
07 54 ☽ ⚹ ⯓
08 55 ☉ ☌ ☿
11 00 ☿ △ ⯓
13 16 ☉ △ ⯓
13 38 ♀ ⚹ ♃
18 23 ☽ ♇ ♃
23 00 ☽ ∥ ♆

08 Monday
01 06 ☽ □ ♂
01 37 ☽ Q ♀
04 09 ☽ ☌ ♅
08 01 ☽ Q ⯓
16 20 ☽ ∠ ☉
18 19 ☽ # ♀
19 50 ☽ ∠ ☿
21 21 ☉ ∥ ☿

09 Tuesday
00 21 ☿ ⊥ ♂
00 57 ☽ △ ♃
02 15 ☽ ∥ ♅
02 53 ♀ Q ♅
02 58 ☽ ∥ ⯓
04 40 ☽ ⚹ ♀
06 32 ☽ ⊻ ♆
07 35 ☽ □ ♄
12 12 ☽ ⊥ ☉
15 44 ☽ Q ♂
17 57 ☽ ⊥ ☿
18 42 ☽ ⊥ ♆
20 25 ☽ □ ⯓
22 15 ☿ Q ♆
22 51 ♀ □ ♆

10 Wednesday
01 27 ☽ ⊻ ☉
02 00 ☽ # ☿
05 09 ☽ # ☉
08 42 ☽ ⊻ ☿
09 04 ♂ □ ♅
10 56 ♀ ⊻ ♄
12 57 ☽ ∠ ♆
14 33 ☽ ∠ ♀
15 08 ♀ ± ⯓
17 04 ☽ ⊻ ♅
17 31 ☽ ⚹ ♂
20 20 ☽ Q ♄

11 Thursday
05 14 ☽ ⊥ ♅
10 50 ☽ ⊥ ♀
14 14 ☽ □ ♃
14 44 ☿ Q ♃
17 08 ☿ ∠ ♄
19 17 ☽ ⚹ ♆
20 44 ☽ ⚹ ♄
23 26 ☽ ∠ ♅

12 Friday
00 18 ☽ ⊻ ♀
01 37 ☽ ∠ ♂
08 46 ☿ ⚹ ♅
08 54 ☽ △ ⯓
19 21 ☽ ☌ ☉
19 24 ☽ Q ♆
19 56 ☉ Q ♆
20 46 ☽ ⊥ ♂

13 Saturday
02 44 ☽ Q ♃
03 04 ☽ ∠ ♄
05 34 ☽ ⚹ ♅
06 34 ☿ ⊻ ♂
09 08 ♂ ∥ ♄
09 28 ☽ ⊻ ♂
09 52 ☽ ☌ ☿
14 50 ☽ ♇ ⯓
16 19 ☽ ∥ ☉
21 07 ☽ ⊥ ♄

14 Sunday
02 50 ☽ ⚹ ♃
05 29 ☽ Q ♅
07 14 ☽ □ ♆
08 39 ☽ ± ⯓
09 05 ☽ ⊻ ♄
09 20 ☽ ∥ ☿
11 17 ☽ # ⯓
11 43 ☽ # ♅
14 31 ☿ ♇ ⯓
18 46 ☽ ☌ ♀
19 10 ☿ # ⯓
20 27 ☽ ⊼ ⯓
21 15 ☿ # ♅

15 Monday
01 04 ☉ ⊥ ♂
08 38 ☽ ∠ ♃
11 21 ♀ ⊼ ⯓
11 52 ☽ ⊻ ☉
11 55 ☽ ∥ ♀
14 01 ☽ # ♆
16 53 ☽ □ ♅
23 56 ☽ ☌ ♂

16 Tuesday
00 32 ☽ ⊥ ☉
02 18 ☽ ⊥ ♃
06 18 ☿ ⊥ ♄
08 41 ♀ # ♆
08 42 ☽ ⊻ ☿
13 34 ☽ ∥ ♄
13 59 ☽ ⊻ ♃
17 10 ☽ ∥ ♂
17 44 ☽ △ ♆
19 19 ☽ ∠ ☉
19 56 ☽ ☌ ♄
22 26 ☽ ⊥ ☿

17 Wednesday
02 09 ☉ Q ♃
04 02 ☉ ∠ ♄
06 23 ☽ ☍ ⯓
11 02 ☽ ⊻ ♀
13 18 ☽ ∥ ♃
18 42 ☽ ∠ ☿
19 12 ☿ ⚹ ♃
22 11 ☽ ♇ ♆
23 31 ☽ ⊥ ♀

18 Thursday
02 03 ☽ ⚹ ☉
02 17 ☽ △ ♅
04 48 ♂ ⊥ ♃
05 15 ☉ ⚹ ♅
07 23 ☿ Q ♅
12 03 ☽ ⊻ ♂
14 56 ☽ ± ♆
17 13 ☿ □ ♆
17 58 ☽ ∠ ♀
22 54 ☽ ☌ ♃
23 42 ☽ ⊥ ♂

19 Friday
00 21 ☿ ± ⯓
01 58 ☽ ⊼ ♆
02 01 ☽ Q ☉
03 29 ☽ ⚹ ☿
04 27 ☽ ⊻ ♄
06 00 ☽ ♇ ♅
10 23 ☿ ⊻ ♄
13 53 ☽ ⊼ ⯓
15 24 ☽ ⊥ ♄
16 54 ☽ ∠ ♂
22 16 ☽ ± ♅
23 55 ☽ ⚹ ♀

20 Saturday
00 38 ☽ ± ⯓
04 43 ☽ Q ☿
07 36 ☽ ∠ ♄
08 57 ☽ ⊼ ♅
11 37 ☿ # ♆
12 48 ☽ □ ☉
14 00 ☽ ∥ ♃
16 30 ☽ ♇ ⯓
20 51 ☽ ⚹ ♂
23 07 ☽ Q ♀

21 Sunday
03 26 ☽ ∥ ♂
04 49 ☽ ⊻ ♃
04 53 ♀ ∠ ♃
07 13 ☽ ☍ ♆
09 56 ☽ ⚹ ♄
10 50 ☽ ∥ ♄
15 13 ☽ ⊥ ♃
16 54 ☽ □ ☿
18 21 ☽ △ ⯓
18 32 ☽ Q ♂
20 43 ☽ ∥ ♀

22 Monday
01 59 ☽ ∥ ☿
03 57 ☿ ⊼ ⯓
06 26 ☽ Q ♄
06 35 ☽ ∠ ♃
08 15 ☽ # ♆
08 42 ☽ □ ♀
12 30 ☽ ☍ ♅
15 54 ☉ ♇ ⯓
19 43 ☽ △ ☉

23 Tuesday
02 05 ☽ □ ♂
06 51 ☽ # ♅
06 55 ☽ # ⯓
07 39 ☽ ⚹ ♃
08 04 ☽ ∥ ☉
09 28 ☽ ⊼ ♆
12 23 ☽ □ ♄
19 14 ☽ ± ♆
19 58 ☽ □ ⯓
21 58 ☽ ♇ ☉

24 Wednesday
01 23 ☽ △ ☿
02 26 ☉ # ⯓
03 08 ☉ # ♅
03 19 ☽ Q ♃
06 32 ♀ □ ♅
09 46 ☽ ♇ ♆
13 28 ☽ ± ☉
13 30 ☽ ⊼ ♅
14 06 ☽ △ ♀
23 07 ☽ ± ♅
23 44 ☽ ⊼ ☉

25 Thursday
00 50 ☽ Q ⯓
04 26 ☽ ♇ ☿
04 49 ☽ △ ♂
08 26 ☽ □ ♃
09 45 ☽ △ ♆
12 17 ☿ ∥ ♀
12 56 ☽ △ ♄
13 30 ☽ ♇ ♅
16 09 ☽ ♇ ♀
17 10 ♅ ∥ ⯓
19 57 ☽ ⚹ ⯓
20 30 ☽ ± ☿

26 Friday
05 51 ☽ ♇ ♂
06 48 ☿ ∥ ♄
07 11 ☽ ⊼ ☿
07 47 ☽ ± ♀
13 02 ☽ ♇ ♄
13 29 ☽ △ ♅
18 14 ☽ ⊼ ♀
19 53 ☽ ∠ ⯓
20 45 ♀ ∥ ♄
20 57 ☽ ± ♂
22 13 ☿ ∠ ♃

27 Saturday
03 00 ☽ ☍ ☉
03 36 ☽ ± ♄
07 04 ☽ ⊼ ♂
08 56 ☽ △ ♃
09 50 ☽ □ ♆
10 23 ☽ ⊥ ⯓
12 35 ☽ ∥ ♅
12 38 ☽ ∥ ⯓
13 24 ☽ ⊼ ♄
13 29 ☉ ⊥ ♄
17 59 ☽ # ☉
20 09 ☽ ⊻ ⯓

28 Sunday
09 45 ☽ ♇ ♃
10 55 ☽ ∥ ♆
13 19 ☽ ☍ ☿
14 25 ☽ □ ♅
15 24 ☽ Q ♆
23 50 ☽ ☍ ♀

29 Monday
00 58 ☽ ± ♃
01 03 ☿ □ ♅
08 07 ☽ ⊼ ☉
08 57 ☽ # ♄
11 15 ☽ ☍ ♂
11 16 ☽ ⊼ ♃
11 41 ☽ ⚹ ♆
11 53 ♂ ⊻ ♃
14 21 ☽ # ♀
14 51 ♀ ⊥ ♃
15 50 ☽ ☍ ♄
19 10 ☽ ± ☉
20 24 ☽ # ☿
20 26 ♂ △ ♆
21 02 ☽ Q ♅
22 30 ☽ ☌ ⯓

30 Tuesday
00 16 ☽ # ♂
03 21 ☽ # ♃
12 05 ☽ ♇ ☉
13 48 ☽ ∠ ♆
18 10 ☽ ⚹ ♅
22 27 ☽ ⊼ ☿
22 37 ☉ Q ♅

All ephemeris data is given at 12.00 UT and the Moon's longitude is additionally given for 24.00 UT

MAY 2002

LONGITUDES

Date	Sidereal time h m s	Sun ☉ ° ' "	Moon ☽ ° ' "	Moon ☽ 24.00 ° ' "	Mercury ☿ ° '	Venus ♀ ° '	Mars ♂ ° '	Jupiter ♃ ° '	Saturn ♄ ° '	Uranus ♅ ° '	Neptune ♆ ° '	Pluto ♇ ° '
01	02 36 59	10 ♉ 56 06	08 ♑ 16 15	14 ♑ 47 15	01 ♊ 28	07 ♊ 00	12 ♊ 02	11 ♋ 00	13 ♊ 34	28 ♒ 24	10 ♒ 57	17 ♐ 11
02	02 40 56	11 54 18	21 ♑ 12 20	27 ♑ 31 56	02 36	08 12	12 43	11 10	13 41	28 26	10 57	17 R 09
03	02 44 52	12 52 30	03 ♒ 46 36	09 ♒ 56 52	03 40	09 25	13 23	11 20	13 48	28 28	10 57	17 08
04	02 48 49	13 50 40	16 ♒ 03 20	22 ♒ 06 35	04 39	10 38	14 03	11 30	13 55	28 29	10 58	17 07
05	02 52 45	14 48 48	28 ♒ 07 12	04 ♓ 05 46	05 33	11 51	14 43	11 40	14 02	28 30	10 58	17 06
06	02 56 42	15 46 55	10 ♓ 02 52	15 ♓ 59 00	06 22	13 03	15 24	11 50	14 09	28 32	10 58	17 04
07	03 00 38	16 45 01	21 ♓ 54 41	27 ♓ 50 24	07 07	14 16	16 04	12 00	14 16	28 33	10 58	17 03
08	03 04 35	17 43 05	03 ♈ 46 33	09 ♈ 43 33	07 46	15 28	16 44	12 10	14 24	28 34	10 58	17 02
09	03 08 32	18 41 08	15 ♈ 41 45	21 ♈ 41 26	08 20	16 41	17 24	12 20	14 31	28 35	10 59	17 00
10	03 12 28	19 39 09	27 ♈ 42 55	03 ♉ 46 23	08 49	17 53	18 04	12 31	14 38	28 37	10 59	16 59
11	03 16 25	20 37 09	09 ♉ 52 05	16 ♉ 00 10	09 14	19 06	18 44	12 41	14 46	28 38	10 59	16 58
12	03 20 21	21 35 07	22 ♉ 10 47	28 ♉ 24 05	09 32	20 18	19 24	12 52	14 53	28 39	10 59	16 56
13	03 24 18	22 33 04	04 ♊ 40 09	10 ♊ 59 07	09 46	21 31	20 04	13 03	15 00	28 40	10 59	16 55
14	03 28 14	23 30 59	17 ♊ 21 05	23 ♊ 46 10	09 55	22 43	20 44	13 14	15 08	28 41	10 R 59	16 53
15	03 32 11	24 28 53	00 ♋ 14 26	06 ♋ 46 03	09 59	23 55	21 24	13 24	15 15	28 42	10 59	16 52
16	03 36 07	25 26 45	13 ♋ 21 05	19 ♋ 59 40	09 R 58	25 07	22 04	13 35	15 23	28 43	10 59	16 50
17	03 40 04	26 24 36	26 ♋ 41 54	03 ♌ 27 53	09 52	26 20	22 44	13 46	15 30	28 43	10 59	16 49
18	03 44 01	27 22 24	10 ♌ 17 40	17 ♌ 11 19	09 42	27 32	23 24	13 57	15 38	28 44	10 58	16 47
19	03 47 57	28 20 11	24 ♌ 08 48	01 ♍ 10 04	09 27	28 44	24 04	14 09	15 45	28 45	10 58	16 46
20	03 51 54	29 ♉ 17 57	08 ♍ 14 57	15 ♍ 23 15	09 09	29 ♊ 56	24 43	14 20	15 53	28 46	10 58	16 44
21	03 55 50	00 ♊ 15 40	22 ♍ 34 39	29 ♍ 48 42	08 47	01 ♋ 08	25 23	14 31	16 01	28 46	10 58	16 43
22	03 59 47	01 13 22	07 ♎ 04 56	14 ♎ 22 43	08 21	02 19	26 03	14 43	16 08	28 47	10 58	16 41
23	04 03 43	02 11 02	21 ♎ 41 23	29 ♎ 00 11	07 53	03 31	26 43	14 54	16 16	28 47	10 57	16 40
24	04 07 40	03 08 41	06 ♏ 18 20	13 ♏ 35 02	07 23	04 43	27 22	15 06	16 23	28 48	10 57	16 38
25	04 11 36	04 06 19	20 ♏ 49 29	28 ♏ 00 57	06 50	05 55	28 02	15 17	16 31	28 48	10 57	16 37
26	04 15 33	05 03 55	05 ♐ 08 44	12 ♐ 12 14	06 17	07 06	28 41	15 29	16 39	28 49	10 56	16 35
27	04 19 30	06 01 30	19 ♐ 10 58	26 ♐ 04 31	05 43	08 18	29 ♊ 21	15 41	16 47	28 49	10 56	16 34
28	04 23 26	06 59 04	02 ♑ 52 39	09 ♑ 35 11	05 09	09 29	00 ♋ 00	15 53	16 54	28 49	10 55	16 32
29	04 27 23	07 56 37	16 ♑ 12 07	22 ♑ 43 31	04 36	10 41	00 40	16 04	17 02	28 50	10 55	16 31
30	04 31 19	08 54 09	29 ♑ 09 35	05 ♒ 30 33	04 04	11 52	01 19	16 16	17 10	28 50	10 54	16 29
31	04 35 16	09 ♊ 51 40	11 ♒ 46 47	17 ♒ 58 43	03 ♊ 33	13 ♋ 04	01 ♋ 59	16 ♋ 28	17 ♊ 17	28 ♒ 50	10 ♒ 54	16 ♐ 27

Date	Moon True ☊ ° '	Moon Mean ☊ ° '	Moon ☽ Latitude ° '
01	18 ♊ 23	19 ♊ 59	01 S 46
02	18 D 24	19 56	02 49
03	18 25	19 52	03 42
04	18 R 26	19 49	04 24
05	18 25	19 46	04 52
06	18 22	19 43	05 08
07	18 19	19 40	05 11
08	18 14	19 37	05 00
09	18 09	19 33	04 37
10	18 04	19 30	04 01
11	17 59	19 27	03 13
12	17 56	19 24	02 16
13	17 54	19 21	01 12
14	17 53	19 18	00 S 03
15	17 D 54	19 14	01 N 07
16	17 55	19 11	02 16
17	17 56	19 08	03 18
18	17 58	19 05	04 10
19	17 58	19 02	04 48
20	17 R 58	18 58	05 11
21	17 57	18 55	05 14
22	17 56	18 52	04 58
23	17 54	18 49	04 22
24	17 52	18 46	03 30
25	17 50	18 43	02 24
26	17 49	18 39	01 N 10
27	17 D 49	18 36	00 S 08
28	17 50	18 33	01 22
29	17 50	18 30	02 31
30	17 51	18 27	03 29
31	17 ♊ 52	18 ♊ 24	04 S 16

DECLINATIONS

Date	Sun ☉ ° '	Moon ☽ ° '	Mercury ☿ ° '	Venus ♀ ° '	Mars ♂ ° '	Jupiter ♃ ° '	Saturn ♄ ° '	Uranus ♅ ° '	Neptune ♆ ° '	Pluto ♇ ° '
01	15 N 06	24 S 57	23 N 00	22 N 15	23 N 03	23 N 12	21 N 05	12 S 42	17 S 24	12 S 44
02	15 24	24 33	23 16	22 30	23 09	23 11	21 06	12 42	17 24	12 44
03	15 42	22 54	23 29	22 44	23 14	23 10	21 07	12 41	17 24	12 44
04	16 00	20 13	23 40	22 57	23 20	23 09	21 08	12 41	17 24	12 44
05	16 17	16 42	23 48	23 10	23 25	23 09	21 09	12 40	17 24	12 43
06	16 34	12 33	23 55	23 22	23 30	23 08	21 10	12 40	17 24	12 43
07	16 51	07 58	23 59	23 33	23 34	23 07	21 11	12 40	17 24	12 43
08	17 07	03 S 06	24 01	23 44	23 39	23 06	21 12	12 39	17 24	12 43
09	17 23	01 N 55	24 01	23 54	23 43	23 05	21 13	12 39	17 24	12 43
10	17 39	06 55	23 59	24 03	23 47	23 04	21 14	12 38	17 24	12 42
11	17 54	11 43	23 55	24 12	23 51	23 03	21 14	12 38	17 24	12 42
12	18 10	16 07	23 49	24 20	23 54	23 02	21 15	12 38	17 24	12 42
13	18 25	19 54	23 41	24 27	23 58	23 01	21 16	12 37	17 24	12 42
14	18 39	22 47	23 31	24 33	24 01	23 00	21 17	12 37	17 24	12 42
15	18 53	24 34	23 20	24 39	24 04	22 59	21 18	12 37	17 24	12 41
16	19 07	25 01	23 07	24 44	24 07	22 58	21 19	12 37	17 24	12 41
17	19 21	24 03	22 52	24 49	24 10	22 57	21 20	12 36	17 24	12 41
18	19 34	21 40	22 36	24 53	24 12	22 56	21 21	12 36	17 24	12 41
19	19 47	18 00	22 18	24 56	24 14	22 55	21 22	12 36	17 24	12 41
20	20 00	13 16	22 00	24 58	24 16	22 54	21 23	12 36	17 24	12 41
21	20 12	07 45	21 40	25 00	24 18	22 53	21 24	12 35	17 24	12 41
22	20 24	01 N 45	21 19	25 01	24 20	22 52	21 25	12 35	17 24	12 40
23	20 36	04 S 24	20 57	25 01	24 21	22 51	21 25	12 35	17 24	12 40
24	20 47	10 19	20 35	25 00	24 22	22 49	21 26	12 35	17 24	12 40
25	20 58	15 39	20 12	24 59	24 23	22 48	21 27	12 35	17 24	12 40
26	21 09	20 01	19 49	24 57	24 24	22 47	21 28	12 35	17 24	12 40
27	21 19	23 07	19 27	24 54	24 24	22 46	21 29	12 35	17 25	12 40
28	21 29	24 47	19 05	24 51	24 25	22 44	21 30	12 35	17 25	12 40
29	21 38	24 57	18 43	24 47	24 25	22 43	21 31	12 35	17 25	12 39
30	21 47	23 44	18 22	24 42	24 25	22 42	21 31	12 35	17 25	12 39
31	21 N 56	21 S 21	18 N 03	24 N 37	24 N 25	22 N 40	21 N 32	12 S 35	17 S 25	12 S 39

ZODIAC SIGN ENTRIES

Date	h	m	Planets
03	04	43	☽ ♒
05	15	46	☽ ♓
08	04	22	☽ ♈
10	16	32	☽ ♉
13	03	04	☽ ♊
15	11	33	☽ ♋
17	17	52	☽ ♌
19	22	01	☽ ♍
20	13	27	♀ ♋
21	05	29	☉ ♊
22	00	19	☽ ♎
24	01	38	☽ ♏
26	03	20	☽ ♐
28	06	54	☽ ♑
28	11	43	♂ ♋
30	13	35	☽ ♒

LATITUDES

Date	Mercury ☿ ° '	Venus ♀ ° '	Mars ♂ ° '	Jupiter ♃ ° '	Saturn ♄ ° '	Uranus ♅ ° '	Neptune ♆ ° '	Pluto ♇ ° '
01	02 N 36	00 N 47	00 N 50	00 N 13	01 S 22	00 S 43	00 N 05	10 N 08
04	02 38	00 55	00 51	00 13	01 21	00 43	00 05	10 08
07	02 31	01 03	00 52	00 13	01 21	00 44	00 05	10 08
10	02 14	01 10	00 53	00 13	01 20	00 44	00 05	10 09
13	01 47	01 17	00 54	00 14	01 20	00 44	00 05	10 09
16	01 11	01 24	00 55	00 14	01 20	00 44	00 05	10 09
19	00 N 27	01 30	00 56	00 14	01 19	00 44	00 05	10 09
22	00 S 24	01 35	00 57	00 14	01 19	00 44	00 05	10 09
25	01 16	01 41	00 58	00 14	01 19	00 44	00 05	10 09
28	02 07	01 45	00 58	00 15	01 18	00 44	00 05	10 09
31	02 S 52	01 N 49	00 N 59	00 N 15	01 S 18	00 S 45	00 N 05	10 N 09

DATA

Julian Date	2452396
Delta T	+64 seconds
Ayanamsa	23° 53' 05"
Synetic vernal point	05° ♓ 13' 55"
True obliquity of ecliptic	23° 26' 22"

MOON'S PHASES, APSIDES AND POSITIONS ☽

Date	h	m	Phase	Longitude ° '	Eclipse Indicator
04	07	16	☾	13 ♒ 39	
12	10	45	●	21 ♉ 32	
19	19	42	☽	28 ♌ 39	
26	11	51	○	05 ♐ 04	

Day	h	m		
07	19	22	Apogee	
23	15	38	Perigee	
01	16	58	Max dec	24° S 59'
09	02	51	0N	
16	07	47	Max dec	25° N 02'
22	18	50	0S	
29	02	44	Max dec	25° S 03'

LONGITUDES

Date	Chiron ⚷ ° '	Ceres ⚳ ° '	Pallas ⚴ ° '	Juno ⚵ ° '	Vesta ⚶ ° '	Black Moon Lilith ⚸ ° '
01	09 ♑ 00	25 ♓ 36	24 ♒ 26	18 ♌ 51	24 ♊ 52	28 ♓ 11
11	08 ♑ 45	29 ♓ 00	26 ♒ 15	20 ♌ 59	28 ♊ 52	29 ♓ 18
21	08 ♑ 21	02 ♈ 14	27 ♒ 40	23 ♌ 27	02 ♋ 59	00 ♈ 26
31	07 ♑ 51	05 ♈ 17	28 ♒ 39	26 ♌ 12	07 ♋ 11	01 ♈ 33

ASPECTARIAN

h m	Aspects
01 Wednesday	
01 44	♃ ⊼ ♆
05 56	☽ ⊥ ♆
09 26	☽ ⊼ ♀
10 23	☽ ± ☿
12 10	☉ □ ♆
14 09	☉ ⚹ ♃
16 54	☽ ⊻ ♆
17 05	☽ ☍ ♃
17 17	☽ △ ☉
17 52	☉ ± ♇
18 36	☿ ∥ ♂
19 18	☽ ⊼ ♂
21 28	☽ ∠ ♅
21 34	☽ ± ♀
21 49	☽ ⊼ ♄
02 Thursday	
04 25	☽ ⊻ ♇
04 27	☿ ∥ ♃
04 37	☽ bQ ☿
07 03	☽ ± ♂
09 06	☽ ± ♄
14 19	☽ ⊥ ♅
15 41	☽ ⊥ ♇
16 11	☽ bQ ♀
19 53	♂ ∥ ♃
03 Friday	
01 02	☽ bQ ♂
01 45	☽ ⊻ ♅
02 20	☽ bQ ♄
05 54	☽ # ☿
08 22	☽ # ♂
08 50	☽ ∠ ♇
09 00	☽ # ♃
11 46	☽ △ ☿
13 42	☽ # ♀
04 Saturday	
00 10	☽ △ ♀
01 58	☽ ☌ ♆
02 54	☽ ⊼ ♃
04 44	☽ # ♄
05 58	♂ ☌ ♄
07 16	☽ □ ☉
07 44	☽ △ ♄
07 50	☽ △ ♂
14 00	☉ ⊻ ♄
14 05	☽ ⚹ ♇
14 53	☽ ± ♃
18 29	♀ △ ♆
05 Sunday	
04 52	☉ ⊻ ♂
07 34	☽ ∥ ♆
07 47	♀ ⊻ ♃
09 02	☽ bQ ♃
10 14	♀ ∥ ♃
12 46	☽ ☌ ♅
13 57	☽ Q ♇
14 23	☽ # ☉
15 45	☿ ⊥ ♃
22 15	☽ Q ☉
06 Monday	
04 04	☽ □ ☿
11 06	☽ ∥ ♇
11 23	☽ ∥ ♅
13 52	☽ ⊻ ♆
15 39	☽ △ ♃
18 46	☽ □ ♀
20 23	☽ □ ♄
23 27	☽ □ ♂
07 Tuesday	
00 37	☽ ⚹ ☉
02 00	☽ ⊥ ♆
02 11	☽ □ ♇
12 11	♀ ☌ ♄
16 45	♀ ∥ ♂
18 53	☽ Q ☿
19 16	☉ ⊼ ♇
20 13	☽ ∠ ♆
08 Wednesday	
01 28	☽ ⊻ ♅
09 11	☽ Q ♄
09 40	☽ ∠ ☉
11 19	☽ Q ♀
13 37	☽ ⊥ ♅
14 03	☽ Q ♂
20 29	☽ ⚹ ☿
22 14	♂ ☍ ♇
09 Thursday	
02 31	☽ ⚹ ♆
05 10	☽ □ ♃
05 25	☽ ⊥ ☉
07 46	☽ ∠ ♅
09 36	☽ ⚹ ♄
13 00	☉ # ♆
14 12	☽ ⚹ ♀
14 37	☽ △ ♇
15 37	☽ ⚹ ♂
18 17	♀ ☍ ♇
18 31	☽ ⊻ ☉
10 Friday	
02 34	☽ Q ♆
03 43	☿ ∥ ♀
03 57	☽ ∠ ☿
13 47	☽ ⚹ ♅
15 51	☽ ∠ ♄
17 38	☽ Q ♃
19 56	♀ ☌ ♂
20 27	☽ bQ ♇
22 29	☽ ⊥ ☿
23 14	☽ ∠ ♂
23 23	☽ ∠ ♀
11 Saturday	
09 48	☽ ⊥ ♄
10 42	☽ ⊻ ☿
13 30	☽ Q ♅
14 08	☽ ± ♇
14 11	☽ □ ♆
16 48	☽ # ♅
17 10	☽ # ♇
17 37	☽ ⚹ ♃
17 57	☽ ⊥ ♂
19 01	☽ ⊥ ♀
21 40	☽ ⊻ ♄
22 43	☿ ∥ ♂
12 Sunday	
01 50	☽ ⊼ ♇
06 19	☽ ⊻ ♂
07 58	☽ ⊻ ♀
10 45	☽ ☌ ☉
19 37	☽ # ♆
23 08	☽ ∠ ♃
13 Monday	
00 29	☽ □ ♅
01 16	☽ ∥ ☉
12 11	♆ St R
16 35	☽ ⊥ ♃
21 50	☽ ☌ ☿
22 28	☽ ∥ ♄
14 Tuesday	
00 00	☽ △ ♆
04 07	☽ ⊻ ♃
07 47	☽ ☌ ♄
11 08	☽ ☍ ♇
14 15	☽ ∥ ♃
18 41	☽ ☌ ♂
19 26	☽ ∥ ☿
23 04	☽ ☌ ♀
15 Wednesday	
00 28	☽ ⊻ ☉
03 06	☽ ∥ ♂
04 07	☽ bQ ♆
09 08	☽ △ ♅
12 29	☽ ⊥ ☉
14 14	☽ ∥ ♀
18 50	☿ St R
20 43	☽ ± ♆
16 Thursday	
05 52	☽ ⊻ ☿
06 18	☽ ∠ ☉
07 41	☽ ⊼ ♆
12 26	☽ ☌ ♃
12 39	☽ bQ ♅
15 43	☽ ⊻ ♄
16 43	☽ ⊥ ☿
18 18	☽ ⊼ ♇
22 22	☽ ∥ ♀
17 Friday	
02 37	☽ ⊥ ♄
03 17	☿ ∥ ♃
04 33	☽ ⊻ ♂
04 54	☽ ± ♅
05 03	♀ bQ ♆
05 05	☽ ± ♇
05 10	♃ bQ ♅
08 46	☽ ∠ ☿
10 24	☽ ∥ ♂
11 16	☽ ⊻ ♀
11 27	☽ ⚹ ☉
15 37	☽ ⊼ ♅
15 48	☽ ⊥ ♂
18 50	☽ ∠ ♄
20 30	☉ ⊻ ♀
21 04	☽ bQ ♇
22 57	☽ ⊥ ♀
18 Saturday	
00 48	☽ ∥ ♃
03 08	☽ ∥ ☿
08 30	☽ ∠ ♂
10 16	☽ Q ☉
10 58	☽ ⚹ ☿
13 11	☽ ☍ ♆
14 25	☽ ∥ ♄
16 16	☽ ∠ ♀
18 29	☽ ⊻ ♃
21 23	☽ ⚹ ♄
23 17	☽ △ ♇
19 Sunday	
01 50	☽ ∥ ☉
05 01	☽ ⊥ ♃
07 28	☽ Q ☿
11 51	☽ ⚹ ♂
12 26	♀ △ ♅
15 22	☽ # ♆
18 14	☽ Q ♄
19 42	☽ □ ☉
19 53	☽ ☍ ♅
20 34	☽ ⚹ ♀
20 40	☽ ∠ ♃
22 25	☉ □ ♅
23 00	♅ Q ♇
20 Monday	
09 18	☽ Q ♂
13 02	☉ ∠ ♃
13 29	☽ □ ☿
14 45	☽ # ♇
15 07	☽ # ♅
16 35	☽ ⊼ ♆
18 46	☽ Q ♀
22 22	☽ ⚹ ♃
21 Tuesday	
00 57	☽ □ ♄
02 14	☽ □ ♇
02 38	☽ ± ♆
16 53	☽ □ ♂
17 37	☽ bQ ♆
18 38	☽ Q ♃
22 14	☿ ⊥ ♃
22 17	☽ ⊼ ♅
22 Wednesday	
01 39	☽ △ ☉
03 27	☽ □ ♀
05 31	☿ ∥ ♄
08 04	☽ Q ♇
08 12	☽ ± ♅
08 50	♂ bQ ♆
14 02	☽ △ ☿
18 23	☽ △ ♆
23 02	☽ bQ ♅
23 Thursday	
00 43	☽ □ ♃
03 01	☽ △ ♄
03 46	☽ ⚹ ♇
04 05	☽ bQ ☉
13 54	☽ bQ ☿
19 54	☽ ± ☉
20 37	☽ △ ♂
23 39	☽ △ ♅
24 Friday	
03 00	☉ ∥ ☿
03 51	☽ bQ ♄
04 10	☽ ± ☿
04 21	☽ ∠ ♇
06 26	☽ ⊼ ☉
09 09	☽ △ ♀
13 42	☽ ⊼ ☿
16 38	♀ ± ♆
18 47	☽ ± ♄
19 08	☽ ⊥ ♇
19 39	☽ □ ♆
21 46	☽ ∥ ♅
22 08	☽ ∥ ♇
22 28	☽ bQ ♂
25 Saturday	
02 42	☽ △ ♃
04 47	☽ ⊼ ♄
05 01	☽ ⊻ ♇
12 09	☽ bQ ♀
14 06	☽ ± ♂
20 56	☽ ∥ ♆
26 Sunday	
00 36	☽ ⊼ ♂
00 44	☿ ⊻ ♀
01 20	☽ □ ♅
01 33	☽ Q ♆
02 44	♄ ☍ ♇
04 02	☽ bQ ♃
04 34	☽ ± ♀
10 52	☽ # ☿
11 51	☽ ☍ ☉
13 51	☽ ☍ ☿
15 38	☽ ⊼ ♀
16 30	♂ △ ♅
19 28	☽ ± ♃
19 58	☽ # ☉
21 50	☽ ⚹ ♆
21 58	☽ # ♄
27 Monday	
05 52	☽ ⊼ ♃
07 09	☉ ☌ ☿
07 29	☽ ☌ ♇
07 48	☽ ☍ ♄
07 55	☽ Q ♅
08 32	☽ # ♃
23 44	☽ ∠ ♆
28 Tuesday	
04 18	☽ # ♂
04 50	☽ ⚹ ♅
06 40	☽ ☍ ♂
13 52	☽ # ♀
15 16	☉ ∥ ♄
15 38	☽ ⊥ ♆
15 53	☽ ⊼ ☿
19 53	☽ ⊼ ☉
29 Wednesday	
00 59	☽ ☍ ♀
02 13	☽ ± ☿
02 24	☽ ⊻ ♆
07 34	☽ ± ☉
07 40	☽ ∠ ♅
10 49	☿ ⊥ ♀
11 46	☽ ☍ ♃
12 34	☽ ⊻ ♇
13 32	☽ ⊼ ♄
16 38	♀ ⊼ ♆
17 58	☽ bQ ☿
18 08	☽ # ♀
23 34	☽ ⊥ ♇
30 Thursday	
00 11	☽ ⊥ ♅
00 42	☽ ± ♄
01 24	☽ bQ ☉
01 41	☽ # ♂
11 23	☽ ⊻ ♅
16 18	☽ ⊼ ♂
16 22	☽ ∠ ♇
17 43	☽ bQ ♄
20 53	☽ △ ☿
23 58	☽ # ♃
31 Friday	
04 19	☽ ± ♂
07 18	☽ # ☉
08 00	☽ △ ☉
10 05	♃ ⊼ ♇
10 18	☽ ☌ ♆
10 27	☽ # ♄
14 44	☽ ⊼ ♀
21 01	☽ ⚹ ♇
21 13	☽ ⊼ ♃
22 38	☽ bQ ♂
22 47	☽ △ ♄

All ephemeris data is given at 12.00 UT and the Moon's longitude is additionally given for 24.00 UT

JUNE 2002

LONGITUDES

Date	Sidereal time h m s	Sun ☉ ° ' "	Moon ☽ ° ' "	Moon ☽ 24.00 ° ' "	Mercury ☿ ° '	Venus ♀ ° '	Mars ♂ ° '	Jupiter ♃ ° '	Saturn ♄ ° '	Uranus ♅ ° '	Neptune ♆ ° '	Pluto ♇ ° '
01	04 39 12	10 ♊ 49 11	24 ♒ 06 46	00 ♓ 11 29	03 ♊ 05	14 ♋ 15	02 ♋ 38	16 ♋ 40	17 ♊ 25	28 ♒ 50	10 ♒ 53	16 ♐ 26
02	04 43 09	11 46 40	06 ♓ 13 24	12 ♓ 13 03	02 R 39	15 26	03 18	16 53	17 33	28 50	10 R 52	16 R 24
03	04 47 05	12 44 09	18 ♓ 11 02	24 ♓ 07 54	02 17	16 37	03 57	17 05	17 41	28 R 50	10 52	16 23
04	04 51 02	13 41 37	00 ♈ 04 15	06 ♈ 00 36	01 58	17 48	04 37	17 17	17 49	28 50	10 51	16 21
05	04 54 59	14 39 04	11 ♈ 57 32	17 ♈ 55 32	01 43	18 59	05 16	17 29	17 56	28 50	10 50	16 19
06	04 58 55	15 36 30	23 ♈ 55 06	29 ♈ 56 41	01 31	20 10	05 55	17 42	18 04	28 50	10 50	16 18
07	05 02 52	16 33 56	06 ♉ 00 43	12 ♉ 07 32	01 24	21 21	06 34	17 54	18 12	28 50	10 49	16 16
08	05 06 48	17 31 22	18 ♉ 17 29	24 ♉ 30 50	01 21	22 32	07 14	18 06	18 20	28 49	10 48	16 14
09	05 10 45	18 28 46	00 ♊ 47 47	07 ♊ 08 30	01 D 23	23 43	07 53	18 19	18 28	28 49	10 47	16 13
10	05 14 41	19 26 10	13 ♊ 33 06	20 ♊ 01 38	01 29	24 54	08 32	18 31	18 35	28 49	10 47	16 11
11	05 18 38	20 23 33	26 ♊ 34 05	03 ♋ 10 24	01 40	26 04	09 11	18 44	18 43	28 48	10 46	16 10
12	05 22 34	21 20 56	09 ♋ 50 30	16 ♋ 34 13	01 55	27 15	09 51	18 57	18 51	28 48	10 45	16 08
13	05 26 31	22 18 17	23 ♋ 21 23	00 ♌ 11 47	02 15	28 25	10 30	19 09	18 59	28 47	10 44	16 06
14	05 30 28	23 15 38	07 ♌ 05 11	14 ♌ 01 19	02 39	29 ♋ 36	11 09	19 22	19 07	28 47	10 43	16 05
15	05 34 24	24 12 58	20 ♌ 59 56	28 ♌ 00 44	03 07	00 ♌ 46	11 48	19 35	19 14	28 46	10 42	16 03
16	05 38 21	25 10 17	05 ♍ 03 27	12 ♍ 07 46	03 40	01 56	12 27	19 48	19 22	28 46	10 41	16 02
17	05 42 17	26 07 35	19 ♍ 13 23	26 ♍ 20 01	04 17	03 07	13 06	20 00	19 30	28 45	10 40	16 00
18	05 46 14	27 04 52	03 ♎ 27 21	10 ♎ 35 03	04 59	04 17	13 45	20 13	19 38	28 44	10 39	15 59
19	05 50 10	28 02 08	17 ♎ 42 49	24 ♎ 50 17	05 44	05 27	14 24	20 26	19 45	28 44	10 38	15 57
20	05 54 07	28 59 24	01 ♏ 57 06	09 ♏ 02 55	06 34	06 37	15 03	20 39	19 53	28 43	10 37	15 55
21	05 58 03	29 ♊ 56 39	16 ♏ 07 21	23 ♏ 10 01	07 28	07 46	15 42	20 52	20 01	28 42	10 36	15 54
22	06 02 00	00 ♋ 53 53	00 ♐ 10 33	07 ♐ 08 34	08 25	08 56	16 21	21 05	20 09	28 41	10 34	15 52
23	06 05 57	01 51 07	14 ♐ 03 42	20 ♐ 55 36	09 27	10 06	17 00	21 18	20 16	28 40	10 33	15 51
24	06 09 53	02 48 20	27 ♐ 43 57	04 ♑ 28 30	10 32	11 15	17 39	21 31	20 24	28 39	10 32	15 49
25	06 13 50	03 45 33	11 ♑ 09 01	17 ♑ 45 20	11 41	12 25	18 18	21 44	20 32	28 38	10 31	15 48
26	06 17 46	04 42 45	24 ♑ 17 20	00 ♒ 45 01	12 54	13 34	18 57	21 57	20 39	28 37	10 30	15 46
27	06 21 43	05 39 58	07 ♒ 08 23	13 ♒ 27 34	14 10	14 43	19 35	22 10	20 47	28 36	10 28	15 45
28	06 25 39	06 37 10	19 ♒ 42 43	25 ♒ 54 07	15 31	15 53	20 14	22 23	20 55	28 35	10 27	15 43
29	06 29 36	07 34 22	02 ♓ 02 02	08 ♓ 06 51	16 54	17 02	20 53	22 37	21 02	28 34	10 26	15 42
30	06 33 32	08 ♋ 31 34	14 ♓ 09 00	20 ♓ 08 55	18 ♊ 21	18 ♌ 11	21 ♋ 32	22 ♋ 50	21 ♊ 10	28 ♒ 32	10 ♒ 25	15 ♐ 40

Date	Moon True ☊ ° '	Moon Mean ☊ ° '	Moon ☽ Latitude ° '
01	17 ♊ 52	18 ♊ 20	04 S 50
02	17 D 53	18 17	05 10
03	17 R 53	18 14	05 17
04	17 53	18 11	05 10
05	17 52	18 08	04 49
06	17 52	18 04	04 16
07	17 52	18 01	03 32
08	17 52	17 58	02 37
09	17 D 52	17 55	01 33
10	17 52	17 52	00 S 24
11	17 R 52	17 49	00 N 48
12	17 51	17 45	01 59
13	17 51	17 42	03 04
14	17 51	17 39	04 00
15	17 50	17 36	04 43
16	17 49	17 33	05 09
17	17 49	17 30	05 16
18	17 D 49	17 26	05 05
19	17 49	17 23	04 35
20	17 50	17 20	03 47
21	17 51	17 17	02 47
22	17 52	17 14	01 36
23	17 52	17 10	00 N 21
24	17 R 52	17 07	00 S 54
25	17 51	17 04	02 05
26	17 50	17 01	03 08
27	17 47	16 58	03 59
28	17 44	16 55	04 38
29	17 42	16 51	05 03
30	17 ♊ 39	16 ♊ 48	05 S 14

DECLINATIONS

Date	Sun ☉ ° '	Moon ☽ ° '	Mercury ☿ ° '	Venus ♀ ° '	Mars ♂ ° '	Jupiter ♃ ° '	Saturn ♄ ° '	Uranus ♅ ° '	Neptune ♆ ° '	Pluto ♇ ° '
01	22 N 04	18 S 02	17 N 45	24 N 31	24 N 24	22 N 39	21 N 33	12 S 35	17 S 25	12 S 39
02	22 12	14 02	17 28	24 24	24 24	22 37	21 34	12 35	17 26	12 39
03	22 19	09 32	17 13	24 16	24 23	22 36	21 35	12 35	17 26	12 39
04	22 27	04 S 42	17 00	24 08	24 22	22 34	21 35	12 35	17 26	12 39
05	22 33	00 N 17	16 49	23 59	24 21	22 33	21 36	12 35	17 26	12 39
06	22 40	05 18	16 40	23 50	24 19	22 31	21 37	12 35	17 26	12 39
07	22 46	10 12	16 33	23 40	24 17	22 30	21 38	12 35	17 27	12 39
08	22 51	14 46	16 28	23 29	24 16	22 28	21 39	12 35	17 27	12 39
09	22 56	18 48	16 25	23 17	24 14	22 27	21 39	12 35	17 27	12 39
10	23 01	22 02	16 24	23 05	24 11	22 25	21 40	12 35	17 27	12 39
11	23 05	24 12	16 25	22 52	24 09	22 23	21 41	12 36	17 27	12 39
12	23 09	25 03	16 28	22 39	24 06	22 22	21 41	12 36	17 28	12 39
13	23 13	24 27	16 33	22 25	24 04	22 20	21 42	12 36	17 28	12 39
14	23 16	22 22	16 40	22 11	24 01	22 18	21 43	12 36	17 28	12 38
15	23 19	18 57	16 48	21 56	23 57	22 16	21 44	12 36	17 28	12 38
16	23 21	14 26	16 58	21 40	23 54	22 15	21 44	12 37	17 29	12 38
17	23 23	09 07	17 10	21 24	23 50	22 13	21 45	12 37	17 29	12 38
18	23 24	03 N 17	17 23	21 07	23 47	22 11	21 46	12 37	17 29	12 38
19	23 25	02 S 43	17 37	20 50	23 43	22 09	21 46	12 37	17 30	12 38
20	23 26	08 35	17 53	20 32	23 39	22 07	21 47	12 38	17 30	12 38
21	23 26	14 00	18 09	20 13	23 34	22 05	21 47	12 38	17 30	12 39
22	23 26	18 37	18 26	19 54	23 30	22 03	21 48	12 38	17 31	12 39
23	23 26	22 08	18 45	19 35	23 25	22 02	21 49	12 39	17 31	12 39
24	23 25	24 19	19 04	19 15	23 20	22 00	21 49	12 39	17 31	12 39
25	23 23	25 03	19 23	18 55	23 15	21 58	21 50	12 40	17 32	12 39
26	23 21	24 20	19 43	18 34	23 10	21 56	21 50	12 40	17 32	12 39
27	23 19	22 20	20 03	18 13	23 05	21 54	21 51	12 40	17 32	12 39
28	23 16	19 18	20 23	17 51	22 59	21 51	21 52	12 41	17 33	12 39
29	23 13	15 28	20 43	17 29	22 53	21 49	21 52	12 41	17 33	12 39
30	23 N 10	11 S 04	21 N 03	17 N 06	22 N 48	21 N 47	21 N 53	12 S 42	17 S 33	12 S 39

ZODIAC SIGN ENTRIES

Date	h m	Planets
01	23 37	☽ ♓
04	11 51	☽ ♈
07	00 07	☽ ♉
09	10 29	☽ ♊
11	18 15	☽ ♋
13	23 39	☽ ♌
14	20 16	♀ ♌
16	03 23	☽ ♍
18	06 11	☽ ♎
20	08 42	☽ ♏
21	13 24	☉ ♋
22	11 42	☽ ♐
24	16 01	☽ ♑
26	22 36	☽ ♒
29	08 00	☽ ♓

LATITUDES

Date	Mercury ☿ ° '	Venus ♀ ° '	Mars ♂ ° '	Jupiter ♃ ° '	Saturn ♄ ° '	Uranus ♅ ° '	Neptune ♆ ° '	Pluto ♇ ° '
01	03 S 05	01 N 51	01 N 00	00 N 15	01 S 18	00 S 45	00 N 05	10 N 09
04	03 37	01 54	01 00	00 15	01 18	00 45	00 05	10 08
07	03 58	01 56	01 01	00 15	01 17	00 45	00 05	10 08
10	04 09	01 58	01 02	00 16	01 17	00 45	00 05	10 08
13	04 08	01 59	01 02	00 16	01 17	00 45	00 05	10 07
16	03 59	01 59	01 03	00 16	01 17	00 45	00 05	10 07
19	03 42	01 59	01 03	00 16	01 17	00 45	00 05	10 07
22	03 19	01 57	01 04	00 16	01 17	00 45	00 05	10 06
25	02 50	01 55	01 05	00 17	01 16	00 46	00 05	10 05
28	02 17	01 52	01 05	00 17	01 16	00 46	00 05	10 05
31	01 S 42	01 N 47	01 N 05	00 N 17	01 S 16	00 S 46	00 N 05	10 N 04

DATA

Julian Date	2452427
Delta T	+64 seconds
Ayanamsa	23° 53' 10"
Synetic vernal point	05° ♓ 13' 50"
True obliquity of ecliptic	23° 26' 22"

MOON'S PHASES, APSIDES AND POSITIONS ☽

Date	h m	Phase	Longitude ° '	Eclipse Indicator
03	00 05	☾	12 ♓ 16	
10	23 47	●	19 ♊ 54	Annular
18	00 29	☽	26 ♍ 37	
24	21 42	○	03 ♑ 11	

Day	h m		
04	13 01	Apogee	
19	07 34	Perigee	
05	10 38	0N	
12	14 14	Max dec	25° N 03'
19	01 09	0S	
25	11 51	Max dec	25° S 03'

LONGITUDES

Date	Chiron ⚷ ° '	Ceres ⚳ ° '	Pallas ⚴ ° '	Juno ⚵ ° '	Vesta ⚶ ° '	Black Moon Lilith ⚸ ° '
01	07 ♑ 48	05 ♈ 35	28 ♒ 43	26 ♌ 29	07 ♋ 36	01 ♈ 39
11	07 ♑ 13	08 ♈ 24	29 ♒ 07	29 ♌ 29	11 ♋ 53	02 ♈ 47
21	06 ♑ 35	10 ♈ 57	28 ♒ 58	02 ♍ 39	16 ♋ 13	03 ♈ 54
31	05 ♑ 56	13 ♈ 13	28 ♒ 12	05 ♍ 58	20 ♋ 36	05 ♈ 01

All ephemeris data is given at 12.00 UT and the Moon's longitude is additionally given for 24.00 UT

ASPECTARIAN

h m	Aspects
01 Saturday	
03 37	☽ ± ♀
03 37	♀ ♇ ♅
07 24	☉ ⊥ ♃
09 08	☽ ± ♃
13 37	☉ △ ♆
14 01	☽ ♯ ☿
15 56	☽ ∥ ♆
20 29	☽ Q ♇
21 19	☽ ☌ ♅
21 40	☿ ⊻ ♂
23 14	☽ ♇ ♀
02 Sunday	
03 12	☽ ♇ ♃
05 07	☽ □ ☿
05 50	☽ △ ♂
12 34	♀ ∥ ♂
16 12	☿ ♯ ♆
19 35	☽ ∥ ♇
19 59	☽ ∥ ♅
21 18	☽ ⊻ ♆
03 Monday	
00 05	☽ □ ☉
00 10	♅ St R
07 08	♀ ⊼ ♇
08 22	☽ □ ♇
08 30	☽ △ ♀
09 21	☽ ⊥ ♆
09 44	☽ △ ♃
10 58	☽ □ ♄
16 07	☽ Q ☿
20 57	☿ ∠ ♃
22 25	☿ ∠ ♀
23 12	♀ ☌ ♃
04 Tuesday	
03 29	☽ ∠ ♆
09 30	☽ ⊻ ♅
12 04	♀ ⊻ ♄
15 34	☽ Q ☉
15 44	☽ ⚹ ☿
20 48	♂ ± ♆
21 37	☽ ⊥ ♅
21 42	☽ □ ♂
23 43	☽ Q ♄
05 Wednesday	
09 45	☽ ⚹ ♆
10 28	☉ ∥ ♃
15 46	☽ ∠ ♅
17 54	☽ ⚹ ☉
20 46	☽ △ ♇
21 24	☽ ∠ ☿
23 19	☽ □ ♃
06 Thursday	
00 10	☽ ⚹ ♄
03 41	☽ □ ♀
09 50	☽ Q ♆
12 00	☽ Q ♂
15 09	☽ ⊥ ☿
21 47	☽ ⚹ ♅
07 Friday	
02 28	☽ ∠ ☉
02 39	☽ ♇ ♇
02 58	☽ ⊻ ☿
04 44	☉ ☍ ♇
06 23	☽ ∠ ♄
11 46	☽ Q ♃
13 10	☽ ⚹ ♂
19 17	☽ Q ♀
20 21	☽ ± ♇
21 26	☽ □ ♆
21 27	☽ Q ♅
21 42	☽ ⊥ ☉
08 Saturday	
00 16	☽ ♯ ♅
00 16	☽ ⊥ ♄
00 36	☽ ♯ ♇
06 09	♀ ± ♇
08 02	☽ ⊼ ♇
10 23	☽ ⊻ ☉
11 38	☽ ⚹ ♃
12 04	☽ ⊻ ♄
15 11	☿ St D
16 37	☿ ⊥ ♂
17 50	♀ ± ♅
20 02	☽ ∠ ♂
21 03	☽ ⚹ ♀
21 31	☽ ∥ ☿
09 Sunday	
03 31	☽ ♯ ♆
06 43	☉ ⊻ ♃
08 14	☽ □ ♅
11 24	☉ ☌ ♄
13 07	☽ ☌ ☿
14 11	☽ ⊥ ♂
16 51	☽ ∠ ♃
10 Monday	
02 07	☽ ⊻ ♂
04 28	☽ ∠ ♀
05 01	♀ ⊥ ♄
06 49	☽ △ ♆
08 52	☽ ∥ ♄
10 03	☽ ⊥ ♃
15 25	☽ ∥ ♃
16 53	☽ ☍ ♇
17 40	☉ ∥ ♀
21 14	☽ ∥ ♀
21 23	☽ ⊻ ♃
21 26	☽ ☌ ♄
21 41	☽ ∥ ☉
22 54	☽ ⊥ ♀
23 47	☽ ☌ ☉
11 Tuesday	
07 36	♃ ⊻ ♄
10 32	☽ ♇ ♆
11 00	☽ ⊻ ♀
11 18	☽ ∥ ♂
16 05	☽ △ ♅
21 26	☽ ⊻ ☿
12 Wednesday	
02 51	☽ ± ♆
08 28	☽ ⊥ ☿
12 00	☽ ☌ ♂
13 37	☽ ⊼ ♆
19 04	☽ ♇ ♅
23 08	☉ ⊥ ♀
23 12	☽ ⊼ ♇
13 Thursday	
00 55	☽ ∠ ☿
04 12	☽ ⊻ ♄
04 28	☽ ☌ ♃
09 48	☽ ± ♇
10 00	☽ ⊻ ☉
11 00	☽ ± ♅
14 53	☽ ⊥ ♄
18 11	☽ ∥ ♂
19 29	♀ ⊼ ♅
20 29	♂ ⊼ ♆
21 21	☽ ⊥ ☉
21 32	☽ ⊼ ♅
21 44	☽ ☌ ♀
22 03	♀ ∥ ♃
14 Friday	
01 34	☽ ♇ ♇
03 40	☽ ∥ ☉
04 02	☽ ⚹ ☿
06 46	☽ ∠ ♄
12 34	☽ ∥ ♃
13 46	☽ ∥ ♀
14 11	☽ ∠ ☉
17 21	☽ ∥ ♄
18 17	☽ ☍ ♆
19 23	☽ ⊻ ♂
15 Saturday	
01 31	☽ Q ☿
03 31	☽ △ ♇
06 14	☽ ⊥ ♂
08 57	☽ ⚹ ♄
09 32	☽ ⊻ ♃
17 44	♀ ♇ ♇
17 55	☽ ⚹ ☉
19 58	☽ ⊥ ♃
20 27	☽ ♯ ♆
22 25	☽ ∠ ♂
23 39	☽ ∥ ☿
16 Sunday	
01 17	☽ ☍ ♅
05 40	☽ Q ♄
05 44	♀ ∥ ♄
06 13	☽ ⊻ ♀
09 32	☽ □ ☿
11 33	☽ ∠ ♃
15 51	☽ Q ☉
17 20	☽ ⊥ ♀
20 27	☽ ♯ ♇
20 35	☽ ♯ ♅
21 32	☽ ⊼ ♆
17 Monday	
00 38	☉ ♇ ♆
01 09	☽ ⚹ ♂
06 34	☽ □ ♇
07 41	☽ ± ♆
09 57	☽ ∠ ♀
12 28	☽ □ ♄
13 21	☽ ⚹ ♃
22 24	☽ Q ♂
22 52	☽ ♇ ♆
18 Tuesday	
00 29	☽ □ ☉
04 04	☽ ⊼ ♅
09 53	☽ Q ♃
11 33	♂ ♇ ♅
12 52	☽ Q ♇
13 30	☽ ⚹ ♀
14 10	☽ ± ♅
14 42	☽ △ ☿
20 05	♀ ∠ ♄
22 51	☿ ∠ ♃
23 27	☿ ♯ ♆
19 Wednesday	
00 06	☽ △ ♆
05 18	☽ ♇ ♅
06 10	☽ □ ♂
09 02	☽ ⚹ ♇
11 30	☽ Q ♀
15 28	☽ △ ♄
16 39	☽ □ ♃
17 24	☽ ♇ ☿
20 Thursday	
05 08	☉ △ ♅
06 32	☽ △ ♅
06 39	☽ △ ☉
08 52	☿ ⚹ ♀
09 31	☽ ± ☿
10 16	☽ ∠ ♇
17 00	☽ ♇ ♄
20 19	☽ ⊼ ☿
20 35	☽ □ ♀
21 Friday	
01 27	☽ ⊥ ♇
02 38	☽ □ ♆
05 39	☽ ∥ ♅
05 41	☽ ∥ ♇
08 23	☽ ± ♄
09 51	☽ ♇ ☉
11 15	☽ △ ♂
11 37	☽ ⊻ ♇
18 41	☽ ⊼ ♄
18 59	♂ ⊼ ♇
20 12	☽ △ ♃
22 Saturday	
02 18	☽ ± ☉
05 45	☽ ∥ ♆
09 15	☽ Q ♆
09 27	☽ □ ♅
10 54	☽ ♯ ☿
13 20	☽ ⊼ ☉
14 07	☽ ♇ ♂
17 35	♅ ∥ ♇
19 15	☽ ♯ ♀
22 20	☽ ♇ ♃
23 Sunday	
03 20	☽ ☍ ☿
04 29	☽ △ ♀
05 55	☽ ⚹ ♆
06 25	☽ ± ♂
09 20	☽ ♯ ♄
09 36	☉ ∥ ♂
11 04	☽ ♯ ♃
14 11	☽ ± ♃
15 06	☽ ☌ ♇
16 32	☽ Q ♅
17 23	☽ ⊼ ♂
21 21	♀ ☍ ♆
22 57	☽ ☍ ♄
23 41	☽ ♯ ♂
24 Monday	
00 06	☽ ♯ ☉
00 52	☽ ⊼ ♃
08 07	☽ ∠ ♆
09 09	☽ ♇ ♀
11 59	☿ △ ♆
13 38	☽ ⚹ ♅
19 06	☿ ∥ ♀
21 42	☽ ☍ ☉
25 Tuesday	
00 05	☽ ⊥ ♆
02 40	☽ ± ♀
10 51	☽ ⊻ ♆
13 04	☽ ⊼ ☿
14 30	☽ ⊼ ♀
16 30	☽ ∠ ♅
18 01	♃ ± ♇
20 25	☽ ⊻ ♇
26 Wednesday	
01 04	☽ ± ☿
01 40	☽ ☍ ♂
05 15	☽ ⊼ ♄
06 38	☉ ± ♆
07 22	☽ ⊥ ♇
07 37	☽ ☍ ♃
08 55	☽ ⊥ ♅
13 42	☿ ⊥ ♂
16 25	☽ ± ♄
19 25	☽ ♇ ☿
20 01	☽ ⊻ ♅
27 Thursday	
00 01	☽ ∠ ♇
01 45	☽ ♯ ☉
04 17	☽ ♯ ♂
09 00	☽ ⊼ ☉
09 25	☽ ♇ ♄
16 06	☽ ♯ ♃
16 24	☽ ♯ ♄
18 18	☽ ☌ ♆
21 17	☽ ± ☉
28 Friday	
02 57	☽ △ ☿
03 53	☽ ☍ ♀
04 20	☽ ⚹ ♇
04 58	☽ ♯ ☿
08 50	♀ △ ♇
11 06	♃ ∥ ♄
13 04	☽ ⊼ ♂
14 20	☽ △ ♄
15 39	☿ ☍ ♇
16 00	☽ ♇ ☉
17 16	☽ ⊼ ♃
22 34	☽ ♯ ♀
23 28	☽ ∥ ♆
29 Saturday	
01 21	☽ ± ♂
03 31	☽ Q ♇
05 10	☽ ± ♃
05 12	☽ ☌ ♅
06 07	☿ ⊥ ♃
06 56	♃ ± ♅
07 45	♀ ♯ ♆
19 01	♂ ⊻ ♄
20 01	☽ ♇ ♂
22 16	☿ ⚹ ♀
23 12	☽ ♇ ♃
23 52	☽ △ ☉
30 Sunday	
03 22	☽ ∥ ♅
03 36	☽ ∥ ♇
04 34	☽ ⊻ ♆
15 02	☽ □ ♇
16 30	☽ ⊥ ♆
17 06	♂ ± ♇
20 54	☽ ⊼ ♀
21 37	☽ □ ☿

JULY 2002

Date	Sidereal time h m s	Sun ☉ ° ' "	Moon ☽ ° ' "	Moon ☽ 24.00 ° ' "	Mercury ☿ ° '	Venus ♀ ° '	Mars ♂ ° '	Jupiter ♃ ° '	Saturn ♄ ° '	Uranus ♅ ° '	Neptune ♆ ° '	Pluto ♇ ° '
01	06 37 29	09 ♋ 28 47	26 ♓ 07 09	02 ♈ 04 12	19 ♊ 52	19 ♌ 19	22 ♋ 11	23 ♋ 03	21 ♊ 17	28 ♒ 31	10 ♒ 23	15 ♐ 39
02	06 41 26	10 25 59	08 ♈ 00 39	13 ♈ 57 04	21 26	20 28	22 49	23 16	21 25	28 R 30	10 R 22	15 R 37
03	06 45 22	11 23 12	19 ♈ 54 04	25 ♈ 52 14	23 04	21 37	23 28	23 30	21 32	28 28	10 21	15 36
04	06 49 19	12 20 24	01 ♉ 52 08	07 ♉ 54 23	24 45	22 45	24 07	23 43	21 40	28 27	10 19	15 35
05	06 53 15	13 17 37	13 ♉ 59 31	20 ♉ 08 03	26 29	23 54	24 45	23 56	21 47	28 26	10 18	15 33
06	06 57 12	14 14 51	26 ♉ 20 28	02 ♊ 37 11	28 ♊ 16	25 02	25 24	24 09	21 55	28 24	10 16	15 32
07	07 01 08	15 12 04	08 ♊ 58 34	15 ♊ 24 54	00 ♋ 07	26 10	26 03	24 23	22 02	28 23	10 15	15 31
08	07 05 05	16 09 18	21 ♊ 56 23	28 ♊ 33 06	02 00	27 18	26 41	24 36	22 09	28 21	10 14	15 29
09	07 09 01	17 06 33	05 ♋ 15 03	12 ♋ 02 07	03 56	28 26	27 20	24 49	22 17	28 19	10 12	15 28
10	07 12 58	18 03 47	18 ♋ 54 05	25 ♋ 50 36	05 54	29 ♌ 34	27 59	25 03	22 24	28 18	10 11	15 27
11	07 16 55	19 01 01	02 ♌ 51 15	09 ♌ 55 29	07 55	00 ♍ 42	28 37	25 16	22 31	28 16	10 09	15 25
12	07 20 51	19 58 16	17 ♌ 02 44	24 ♌ 12 20	09 57	01 49	29 16	25 30	22 39	28 15	10 08	15 24
13	07 24 48	20 55 31	01 ♍ 23 37	08 ♍ 35 54	12 02	02 57	29 ♋ 55	25 43	22 46	28 13	10 06	15 23
14	07 28 44	21 52 45	15 ♍ 48 32	23 ♍ 00 55	14 07	04 04	00 ♌ 33	25 56	22 53	28 11	10 05	15 22
15	07 32 41	22 50 00	00 ♎ 12 29	07 ♎ 22 45	16 14	05 11	01 12	26 10	23 00	28 09	10 03	15 20
16	07 36 37	23 47 15	14 ♎ 31 18	21 ♎ 37 50	18 22	06 18	01 50	26 23	23 07	28 07	10 02	15 19
17	07 40 34	24 44 29	28 ♎ 42 03	05 ♏ 43 49	20 30	07 25	02 29	26 37	23 14	28 06	10 00	15 18
18	07 44 30	25 41 44	12 ♏ 42 57	19 ♏ 39 24	22 39	08 32	03 07	26 50	23 21	28 04	09 58	15 17
19	07 48 27	26 38 59	26 ♏ 33 05	03 ♐ 23 58	24 48	09 39	03 46	27 03	23 28	28 02	09 57	15 16
20	07 52 24	27 36 15	10 ♐ 12 03	16 ♐ 57 16	26 56	10 45	04 24	27 17	23 35	28 00	09 55	15 15
21	07 56 20	28 33 30	23 ♐ 39 36	00 ♑ 19 01	29 ♋ 03	11 51	05 03	27 30	23 42	27 58	09 54	15 14
22	08 00 17	29 ♋ 30 46	06 ♑ 55 28	13 ♑ 28 53	01 ♌ 10	12 57	05 41	27 44	23 49	27 56	09 52	15 13
23	08 04 13	00 ♌ 28 03	19 ♑ 59 13	26 ♑ 26 25	03 16	14 03	06 20	27 57	23 56	27 54	09 50	15 12
24	08 08 10	01 25 20	02 ♒ 50 25	09 ♒ 11 14	05 21	15 09	06 58	28 10	24 03	27 52	09 49	15 11
25	08 12 06	02 22 37	15 ♒ 28 50	21 ♒ 43 16	07 25	16 14	07 36	28 24	24 09	27 50	09 47	15 10
26	08 16 03	03 19 55	27 ♒ 54 37	04 ♓ 03 00	09 27	17 20	08 15	28 37	24 16	27 48	09 46	15 09
27	08 19 59	04 17 14	10 ♓ 08 37	16 ♓ 11 41	11 28	18 25	08 53	28 50	24 23	27 46	09 44	15 08
28	08 23 56	05 14 34	22 ♓ 12 30	28 ♓ 11 23	13 27	19 30	09 32	29 04	24 29	27 43	09 42	15 07
29	08 27 53	06 11 55	04 ♈ 08 46	10 ♈ 05 04	15 25	20 35	10 10	29 17	24 36	27 41	09 41	15 06
30	08 31 49	07 09 16	16 ♈ 00 47	21 ♈ 56 28	17 21	21 39	10 48	29 30	24 42	27 39	09 39	15 05
31	08 35 46	08 ♌ 06 39	27 ♈ 52 40	03 ♉ 50 01	19 ♌ 16	22 ♍ 43	11 ♌ 27	29 ♋ 44	24 ♊ 49	27 ♒ 37	09 ♒ 38	15 ♐ 05

Date	Moon True ☊ ° '	Moon Mean ☊ ° '	Moon ☽ Latitude ° '
01	17 ♊ 38	16 ♊ 45	05 S 11
02	17 R 37	16 42	04 55
03	17 D 37	16 39	04 26
04	17 38	16 36	03 46
05	17 40	16 32	02 55
06	17 42	16 29	01 55
07	17 43	16 26	00 S 48
08	17 R 43	16 23	00 N 23
09	17 42	16 20	01 35
10	17 39	16 16	02 42
11	17 36	16 13	03 42
12	17 31	16 10	04 29
13	17 26	16 07	04 59
14	17 22	16 04	05 11
15	17 19	16 01	05 03
16	17 17	15 57	04 37
17	17 D 17	15 54	03 54
18	17 17	15 51	02 57
19	17 19	15 48	01 50
20	17 20	15 45	00 N 39
21	17 R 20	15 41	00 S 34
22	17 18	15 38	01 44
23	17 15	15 35	02 47
24	17 09	15 32	03 41
25	17 02	15 29	04 22
26	16 54	15 26	04 51
27	16 45	15 22	05 05
28	16 38	15 19	05 06
29	16 31	15 16	04 53
30	16 27	15 13	04 27
31	16 ♊ 24	15 ♊ 10	03 S 51

DECLINATIONS

Date	Sun ☉ ° '	Moon ☽ ° '	Mercury ☿ ° '	Venus ♀ ° '	Mars ♂ ° '	Jupiter ♃ ° '	Saturn ♄ ° '	Uranus ♅ ° '	Neptune ♆ ° '	Pluto ♇ ° '
01	23 N 06	06 S 18	21 N 22	16 N 44	22 N 41	21 N 45	21 N 53	12 S 42	17 S 34	12 S 39
02	23 02	01 S 20	21 41	16 20	22 35	21 43	21 54	12 43	17 34	12 39
03	22 57	03 N 40	21 59	15 57	22 29	21 41	21 54	12 43	17 34	12 39
04	22 52	08 36	22 16	15 32	22 22	21 39	21 55	12 44	17 35	12 39
05	22 47	13 16	22 32	15 08	22 15	21 36	21 55	12 44	17 35	12 40
06	22 41	17 29	22 46	14 43	22 09	21 34	21 56	12 45	17 36	12 40
07	22 34	21 01	22 59	14 18	22 01	21 32	21 56	12 45	17 36	12 40
08	22 28	23 35	23 10	13 53	21 54	21 30	21 57	12 46	17 36	12 40
09	22 21	24 55	23 19	13 27	21 47	21 27	21 57	12 47	17 37	12 40
10	22 13	24 47	23 26	13 01	21 39	21 25	21 58	12 47	17 37	12 40
11	22 05	23 07	23 31	12 35	21 32	21 23	21 58	12 48	17 38	12 40
12	21 57	19 59	23 34	12 08	21 24	21 20	21 58	12 48	17 38	12 40
13	21 49	15 38	23 33	11 41	21 16	21 18	21 59	12 49	17 38	12 41
14	21 40	10 22	23 31	11 14	21 07	21 16	21 59	12 50	17 39	12 41
15	21 30	04 N 33	23 25	10 47	20 59	21 13	22 00	12 50	17 39	12 41
16	21 21	01 S 28	23 17	10 19	20 51	21 11	22 00	12 51	17 40	12 41
17	21 11	07 22	23 06	09 52	20 42	21 08	22 00	12 52	17 40	12 41
18	21 00	12 51	22 52	09 24	20 33	21 06	22 01	12 52	17 41	12 42
19	20 50	17 36	22 35	08 55	20 24	21 03	22 01	12 53	17 41	12 42
20	20 38	21 21	22 17	08 27	20 15	21 01	22 01	12 54	17 41	12 42
21	20 27	23 51	21 55	07 59	20 06	20 58	22 02	12 54	17 42	12 42
22	20 15	24 59	21 32	07 30	19 57	20 56	22 02	12 55	17 42	12 42
23	20 03	24 42	21 06	07 01	19 47	20 53	22 02	12 56	17 43	12 43
24	19 51	23 06	20 38	06 32	19 38	20 50	22 03	12 56	17 43	12 43
25	19 38	20 22	20 08	06 03	19 28	20 48	22 03	12 57	17 44	12 43
26	19 25	16 44	19 37	05 34	19 18	20 45	22 03	12 58	17 44	12 43
27	19 11	12 28	19 04	05 05	19 08	20 43	22 03	12 59	17 45	12 44
28	18 57	07 46	18 29	04 35	18 58	20 40	22 04	12 59	17 45	12 44
29	18 43	02 S 50	17 53	04 06	18 48	20 37	22 04	13 00	17 45	12 44
30	18 29	02 N 11	17 17	03 36	18 37	20 35	22 04	13 01	17 46	12 44
31	18 N 14	07 N 08	16 N 39	03 N 07	18 N 27	20 N 32	22 N 04	13 S 02	17 S 46	12 S 45

ZODIAC SIGN ENTRIES

Date	h	m	Planets
01	19	49	☽ ♈
04	08	16	☽ ♉
06	19	01	☽ ♊
07	10	35	☿ ♋
09	02	36	☽ ♋
10	21	09	♀ ♍
11	07	08	☽ ♌
13	09	41	☽ ♍
13	15	23	♂ ♌
15	11	39	☽ ♎
17	14	13	☽ ♏
19	18	02	☽ ♐
21	22	41	☿ ♌
21	23	26	☽ ♑
23	00	15	☉ ♌
24	06	40	☽ ♒
26	16	04	☽ ♓
29	03	39	☽ ♈
31	16	17	☽ ♉

LATITUDES

Date	Mercury ☿ ° '	Venus ♀ ° '	Mars ♂ ° '	Jupiter ♃ ° '	Saturn ♄ ° '	Uranus ♅ ° '	Neptune ♆ ° '	Pluto ♇ ° '
01	01 S 42	01 N 47	01 N 05	00 N 17	01 S 16	00 S 46	00 N 05	10 N 04
04	01 04	01 42	01 06	00 17	01 16	00 46	00 05	10 03
07	00 S 27	01 36	01 06	00 18	01 16	00 46	00 05	10 03
10	00 N 08	01 29	01 07	00 18	01 16	00 46	00 05	10 02
13	00 40	01 21	01 07	00 18	01 16	00 46	00 05	10 01
16	01 07	01 13	01 07	00 18	01 16	00 46	00 04	10 00
19	01 27	01 03	01 08	00 19	01 16	00 46	00 04	09 59
22	01 40	00 52	01 08	00 19	01 16	00 46	00 04	09 58
25	01 47	00 40	01 08	00 19	01 16	00 46	00 04	09 57
28	01 47	00 28	01 08	00 20	01 16	00 46	00 04	09 56
31	01 N 41	00 N 14	01 N 08	00 N 20	01 S 16	00 S 46	00 N 04	09 N 55

DATA

Julian Date	2452457
Delta T	+64 seconds
Ayanamsa	23° 53' 15"
Synetic vernal point	05° ♓ 13' 44"
True obliquity of ecliptic	23° 26' 22"

MOON'S PHASES, APSIDES AND POSITIONS ☽

Date	h	m	Phase	Longitude ° '	Eclipse Indicator
02	17	19	☾	10 ♈ 39	
10	10	26	●	18 ♋ 00	
17	04	47	☽	24 ♎ 27	
24	09	07	○	01 ♒ 18	

Day	h	m		
02	07	34	Apogee	
14	13	15	Perigee	
30	01	38	Apogee	
02	18	25	0N	
09	22	04	Max dec	25° N 03'
16	06	09	0S	
22	19	06	Max dec	25° S 03'
30	01	33	0N	

LONGITUDES

Date	Chiron ⚷ ° '	Ceres ⚳ ° '	Pallas ⚴ ° '	Juno ⚵ ° '	Vesta ⚶ ° '	Black Moon Lilith ⚸ ° '
01	05 ♑ 56	13 ♈ 13	28 ♒ 12	05 ♍ 58	20 ♋ 36	05 ♈ 01
11	05 ♑ 18	15 ♈ 06	26 ♒ 51	09 ♍ 25	25 ♋ 02	06 ♈ 08
21	04 ♑ 42	16 ♈ 35	24 ♒ 58	12 ♍ 57	29 ♋ 29	07 ♈ 15
31	04 ♑ 11	17 ♈ 36	22 ♒ 40	16 ♍ 33	03 ♌ 58	08 ♈ 22

All ephemeris data is given at 12.00 UT and the Moon's longitude is additionally given for 24.00 UT

ASPECTARIAN

h m	Aspects
01 Monday	
02 11	☽ □ ♄
03 37	☽ △ ♂
05 43	☽ △ ♃
10 14	☽ ± ♀
10 32	☽ ∠ ♆
16 50	☽ ⊻ ♅
02 Tuesday	
00 18	♂ ± ♅
04 55	☽ ⊥ ♅
06 19	☽ ⚼ ♀
10 20	☉ ⊼ ♆
11 34	☿ ☌ ♄
14 43	☿ ∥ ♃
14 52	☽ Q ♄
15 20	☽ Q ☿
16 45	☽ ⚹ ♆
17 19	☽ □ ☉
23 04	☽ ∠ ♅
03 Wednesday	
03 21	☽ △ ♇
05 45	☿ ∥ ♄
10 14	♀ ⚹ ♄
13 24	♂ ☌ ♃
15 20	☽ ⚹ ♄
15 49	☽ △ ♀
16 54	☽ Q ♆
19 06	☿ ⊻ ♃
19 22	☽ □ ♃
19 24	☽ ⚹ ☿
19 35	☽ □ ♂
21 27	☿ ⊻ ♂
04 Thursday	
05 11	☽ ⚹ ♅
08 41	☽ Q ☉
09 26	☽ ⚼ ♇
18 45	☿ ∥ ♂
19 53	☿ ⚼ ♆
21 38	☽ ∠ ♄
05 Friday	
03 17	☽ ± ♇
04 44	☽ □ ♆
05 00	☽ Q ♅
06 14	☽ ∠ ☿
07 53	☽ Q ♃
08 48	☽ # ♇
09 13	☽ # ♅
09 26	☽ Q ♂
10 31	☽ ⚹ ☉
13 01	♀ ⊻ ♃
15 04	☽ ⊼ ♇
15 15	☉ ⚼ ♅
15 33	☽ ⊥ ♄
21 23	☽ ∥ ♀
06 Saturday	
02 48	☽ ⊥ ☿
03 22	☽ ⊻ ♄
05 28	☉ ∥ ☿
07 43	☽ ⚹ ♃
09 14	☽ □ ♀
10 06	☽ ⚹ ♂
12 43	☽ # ♆
13 42	☿ △ ♅
15 57	☽ □ ♅
16 20	☽ ⊻ ☿
18 02	☽ ∠ ☉
07 Sunday	
05 55	♀ ⊻ ♂
12 27	☽ ⊥ ☉
12 46	☽ ∠ ♃
14 23	☽ △ ♆
16 05	☽ ∠ ♂
16 07	☽ ∥ ♃
19 35	☉ ⊼ ♇
19 35	☽ ∥ ♄
19 59	☽ ∥ ♂
22 38	☽ Q ♀
08 Monday	
00 09	☽ ☍ ♇
00 32	☽ ⊻ ☉
00 46	☽ ∥ ☉
04 33	♂ ∥ ♄
05 46	☽ ⊥ ♃
06 58	☽ ∥ ☿
09 36	☽ ⊥ ♂
12 24	☽ ☌ ♄
16 56	☽ ⊻ ♃
17 58	☽ ⚼ ♆
21 05	☽ ⊻ ♂
22 40	☽ ⚹ ♀
23 37	☽ △ ♅
09 Tuesday	
09 14	☽ ☌ ☿
09 38	♀ ☍ ♅
10 08	☽ ± ♆
15 20	☿ ± ♆
20 45	☽ ⊼ ♆
10 Wednesday	
02 14	☽ ⚼ ♅
03 46	☽ ∠ ♀
05 59	☽ ⊼ ♇
10 26	☽ ☌ ☉
16 24	☽ ± ♇
17 52	☽ ± ♅
18 07	☽ ⊻ ♄
20 48	☽ ⊥ ♀
22 48	☽ ☌ ♃
23 23	♂ ⊼ ♅
11 Thursday	
00 27	♀ # ♅
04 11	☽ ⊼ ♅
04 25	☽ ☌ ♂
04 32	☽ ⊥ ♄
06 55	♀ # ♇
07 28	♂ ⊥ ♄
07 51	☽ ⚼ ♇
07 53	☽ ∥ ☿
08 00	☽ ⊻ ♀
20 00	☽ ∠ ♄
21 32	☽ ∥ ☉
22 02	☽ ⊻ ☿
22 05	☽ ∥ ♄
12 Friday	
00 22	☽ ☍ ♆
02 03	☽ ∥ ♂
02 45	☽ ∥ ♃
03 10	♀ ⊥ ♃
08 54	☉ ∥ ♄
09 14	☽ △ ♇
09 51	☽ ⊥ ☿
14 00	☿ ⊼ ♆
17 16	☽ ⊻ ☉
21 28	☽ ⚹ ♄
13 Saturday	
01 40	☽ # ♆
02 17	♂ ∥ ♃
02 22	☽ ⊻ ♃
03 29	☽ ∠ ☿
04 01	☽ ⊥ ☉
06 42	☽ ☍ ♅
09 25	☽ ⊻ ♂
12 33	☽ ⊥ ♃
14 49	☽ ☌ ♀
17 40	☽ Q ♄
19 52	☽ ⊥ ♂
20 05	☽ ∠ ☉
23 14	☉ ± ♇
14 Sunday	
01 14	☽ # ♅
01 26	☿ ⚼ ♅
01 52	☽ # ♇
02 29	☽ ⊼ ♆
03 46	☽ ∠ ♃
05 05	♂ ⚼ ♇
07 55	☽ ∥ ♀
08 43	☽ ⚹ ☿
11 15	☽ □ ♇
11 33	☽ ∠ ♂
12 27	☽ ± ♆
19 26	☉ ± ♅
22 49	☽ ⚹ ☉
23 53	☽ □ ♄
15 Monday	
01 56	☿ ⊼ ♇
03 25	☽ ⚼ ♆
05 08	☽ ⚹ ♃
07 32	♀ Q ♄
08 09	☽ Q ☿
08 35	☽ ⊼ ♅
13 44	☽ ⚹ ♂
16 55	☉ ⊻ ♄
17 14	☽ Q ♇
18 35	☽ ± ♅
20 17	☽ Q ☉
21 02	☽ ⊻ ♀
16 Tuesday	
01 31	☽ Q ♃
04 27	☽ △ ♆
07 57	☽ ⊥ ♀
09 39	☽ ⚼ ♅
10 48	☽ Q ♂
13 21	☽ ⚹ ♇
19 38	☽ □ ☿
17 Wednesday	
00 26	☽ ∠ ♀
02 39	☽ △ ♄
04 47	☽ □ ☉
08 23	☽ □ ♃
10 58	☽ △ ♅
14 43	☽ ∠ ♇
18 45	☽ □ ♂
19 35	☉ ∥ ♃
20 50	☿ ± ♇
21 47	☽ # ♀
18 Thursday	
04 11	☽ ⚹ ♀
04 26	☽ ⚼ ♄
05 30	☿ ± ♅
06 06	☽ ⊥ ♇
07 17	☽ □ ♆
11 18	☽ ∥ ♇
12 07	☽ ∥ ♅
16 25	☽ ⊻ ♇
20 05	☽ ± ♄
20 23	☿ ⊻ ♄
19 Friday	
02 42	☽ Q ♀
06 35	☽ ⊼ ♄
08 22	☽ △ ☿
08 32	☿ ∠ ♀
12 11	☽ △ ☉
12 31	☽ ∥ ♆
12 54	☽ △ ♃
14 26	☽ Q ♆
14 35	☽ □ ♅
18 07	♀ ⊥ ♂
18 26	♀ ⊼ ♆
20 Saturday	
01 16	☽ △ ♂
01 19	☉ ☌ ♃
04 34	☽ # ♂
07 10	☽ # ☉
09 35	☽ # ♃
11 30	☽ ⚹ ♆
13 04	☽ □ ♀
15 38	☽ ⚼ ☿
15 45	☽ ⚼ ♃
16 24	☿ ☌ ♃
16 35	☽ ⚼ ☉
17 25	☽ # ♄
18 44	☽ # ☿
20 57	☽ ☌ ♇
21 34	☉ ⊼ ♅
22 16	☽ Q ♅
23 51	☿ ⊼ ♅
21 Sunday	
01 47	☉ ☌ ☿
05 05	☿ ∥ ♄
05 12	☽ ⚼ ♂
08 04	☽ ± ♃
09 52	☽ ± ☉
10 43	☽ ± ☿
12 05	☽ ☍ ♄
14 13	☽ ∠ ♆
19 02	☽ ⊼ ♃
19 44	☽ ⚹ ♅
19 46	☿ ⊥ ♄
21 30	☽ ⊼ ☉
22 11	☽ ± ♂
23 34	☽ ⊼ ☿
22 Monday	
01 11	☿ ⚼ ♇
05 43	♀ ∠ ♃
06 27	☽ ⊥ ♆
09 38	☽ ⊼ ♂
17 22	☽ ⊻ ♆
20 44	☉ ⊥ ♄
22 58	☽ ∠ ♅
23 Tuesday	
00 03	☽ △ ♀
03 10	☽ ⊻ ♇
05 17	☉ ⚼ ♇
07 18	♃ ⊼ ♅
14 14	☽ ⊥ ♇
15 32	☽ ⊥ ♅
19 23	☽ ⊼ ♄
24 Wednesday	
00 14	☿ ∥ ♃
02 41	☽ ⊻ ♅
03 05	☽ ☍ ♃
06 28	☽ ⚼ ♀
06 42	☽ ± ♄
07 00	☽ ∠ ♇
09 07	☽ ☍ ☉
12 41	♀ □ ♇
17 39	☽ ☍ ☿
20 12	☽ ☍ ♂
22 22	☽ # ♄
23 50	☽ ⚼ ♄
25 Thursday	
01 03	☽ ± ♀
01 10	☽ ☌ ♆
02 18	♀ ± ♆
08 41	☽ # ♃
11 24	☽ ⚹ ♇
13 36	☽ ⊼ ♀
14 01	☽ # ☿
15 20	☿ ☌ ♂
17 41	☽ # ☉
18 48	☽ # ♂
26 Friday	
04 52	☽ △ ♄
05 53	☽ ∥ ♆
09 45	☿ ∠ ♄
10 31	☽ Q ♇
11 47	☽ ☌ ♅
13 24	☽ ⊼ ♃
15 39	☿ ☍ ♆
23 29	☽ ⊼ ☉
27 Saturday	
01 21	☽ ± ♃
02 53	☉ ∥ ☿
07 34	☿ ∥ ♂
09 17	☽ ∥ ♅
09 23	☽ ⊼ ♂
10 37	☽ ∥ ♇
11 12	☽ ⊻ ♆
12 18	☽ ± ☉
15 07	☽ ⊼ ☿
19 27	☽ ⚼ ♃
21 53	☽ □ ♇
21 55	☽ ± ♂
23 04	☽ ⊥ ♆
28 Sunday	
05 24	☽ ± ☿
06 03	☽ ☍ ♀
07 44	☽ ⚼ ☉
09 25	☉ ∥ ♂
10 19	♂ ∠ ♄
13 13	☿ ⊥ ♀
16 37	☽ □ ♄
16 54	☽ ⚼ ♂
17 00	☽ ∠ ♆
18 31	♂ ☍ ♆
23 02	☽ ⊻ ♅
29 Monday	
02 01	☽ △ ♃
03 00	☽ ⚼ ☿
05 15	☽ # ♀
08 12	☿ △ ♇
11 05	☽ ⊥ ♅
16 30	☽ △ ☉
17 14	☿ # ♆
23 09	☽ ⚹ ♆
30 Tuesday	
00 51	☽ △ ♂
02 39	♄ ⚼ ♆
05 13	☽ ∠ ♅
05 15	☽ Q ♄
10 08	☽ △ ♇
15 14	☽ △ ☿
18 13	☽ ∥ ♀
23 23	☽ Q ♆
31 Wednesday	
00 33	☽ ⊼ ♀
05 45	☽ ⚹ ♄
11 28	☽ ⚹ ♅
13 52	☽ ± ♀
15 48	☽ □ ♃
16 26	☽ ⚼ ♇

AUGUST 2002

Date	Sidereal time h m s	Sun ☉ ° ' "	Moon ☽ ° ' "	Moon ☽ 24.00 ° ' "	Mercury ☿ ° '	Venus ♀ ° '	Mars ♂ ° '	Jupiter ♃ ° '	Saturn ♄ ° '	Uranus ♅ ° '	Neptune ♆ ° '	Pluto ♇ ° '
01	08 39 42	09 ♌ 04 03	09 ♉ 49 07	15 ♉ 50 36	21 ♌ 09	23 ♍ 47	12 ♌ 05	29 ♋ 57	24 ♊ 55	27 ♒ 35	09 ♒ 36	15 ♐ 04
02	08 43 39	10 01 28	21 ♉ 55 07	28 ♉ 03 19	23 00	24 51	12 43	00 ♌ 10	25 01	27 R 32	09 R 34	15 R 03
03	08 47 35	10 58 54	04 ♊ 15 46	10 ♊ 33 06	24 50	25 55	13 22	00 24	25 08	27 30	09 33	15 02
04	08 51 32	11 56 21	16 ♊ 55 47	23 ♊ 24 19	26 38	26 59	14 00	00 37	25 14	27 28	09 31	15 02
05	08 55 28	12 53 50	29 ♊ 59 01	06 ♋ 40 09	28 ♌ 24	28 02	14 38	00 50	25 20	27 26	09 29	15 01
06	08 59 25	13 51 20	13 ♋ 27 49	20 ♋ 21 59	00 ♍ 09	29 ♍ 05	15 17	01 03	25 26	27 23	09 28	15 00
07	09 03 22	14 48 51	27 ♋ 22 27	04 ♌ 28 49	01 53	00 ♎ 07	15 55	01 17	25 32	27 21	09 26	15 00
08	09 07 18	15 46 23	11 ♌ 40 32	18 ♌ 56 52	03 34	01 10	16 33	01 30	25 38	27 19	09 24	14 59
09	09 11 15	16 43 56	26 ♌ 16 59	03 ♍ 39 52	05 15	02 12	17 11	01 43	25 44	27 16	09 23	14 59
10	09 15 11	17 41 30	11 ♍ 04 30	18 ♍ 29 47	06 53	03 14	17 50	01 56	25 50	27 14	09 21	14 58
11	09 19 08	18 39 04	25 ♍ 54 40	03 ♎ 18 11	08 30	04 16	18 28	02 09	25 56	27 12	09 20	14 58
12	09 23 04	19 36 40	10 ♎ 39 25	17 ♎ 57 38	10 06	05 17	19 06	02 22	26 01	27 09	09 18	14 57
13	09 27 01	20 34 17	25 ♎ 12 12	02 ♏ 22 42	11 40	06 18	19 45	02 35	26 07	27 07	09 16	14 57
14	09 30 57	21 31 55	09 ♏ 28 48	16 ♏ 30 22	13 12	07 19	20 23	02 48	26 12	27 05	09 15	14 56
15	09 34 54	22 29 33	23 ♏ 27 20	00 ♐ 19 47	14 43	08 19	21 01	03 01	26 18	27 02	09 13	14 56
16	09 38 51	23 27 13	07 ♐ 07 50	13 ♐ 51 40	16 12	09 19	21 39	03 14	26 23	27 00	09 12	14 56
17	09 42 47	24 24 54	20 ♐ 31 32	27 ♐ 07 38	17 40	10 19	22 17	03 27	26 29	26 57	09 10	14 55
18	09 46 44	25 22 36	03 ♑ 40 15	10 ♑ 09 34	19 06	11 18	22 56	03 40	26 34	26 55	09 09	14 55
19	09 50 40	26 20 18	16 ♑ 35 48	22 ♑ 59 09	20 30	12 17	23 34	03 53	26 39	26 53	09 07	14 55
20	09 54 37	27 18 02	29 ♑ 19 44	05 ♒ 37 42	21 53	13 16	24 12	04 06	26 44	26 50	09 06	14 55
21	09 58 33	28 15 47	11 ♒ 53 09	18 ♒ 06 09	23 14	14 14	24 50	04 18	26 49	26 48	09 04	14 54
22	10 02 30	29 ♌ 13 34	24 ♒ 16 47	00 ♓ 25 07	24 33	15 12	25 29	04 31	26 54	26 45	09 02	14 54
23	10 06 26	00 ♍ 11 22	06 ♓ 31 15	12 ♓ 35 16	25 51	16 10	26 07	04 44	26 59	26 43	09 01	14 54
24	10 10 23	01 09 11	18 ♓ 37 18	24 ♓ 37 29	27 07	17 07	26 45	04 56	27 04	26 41	08 59	14 54
25	10 14 20	02 07 02	00 ♈ 36 01	06 ♈ 33 09	28 21	18 03	27 23	05 09	27 09	26 38	08 58	14 54
26	10 18 16	03 04 54	12 ♈ 29 09	18 ♈ 24 22	29 ♍ 33	18 59	28 01	05 22	27 13	26 36	08 57	14 D 54
27	10 22 13	04 02 48	24 ♈ 19 11	00 ♉ 14 02	00 ♎ 43	19 55	28 39	05 34	27 18	26 34	08 55	14 54
28	10 26 09	05 00 44	06 ♉ 09 25	12 ♉ 05 51	01 51	20 50	29 18	05 47	27 22	26 31	08 54	14 54
29	10 30 06	05 58 41	18 ♉ 03 55	24 ♉ 04 15	02 57	21 44	29 ♌ 56	05 59	27 27	26 29	08 52	14 54
30	10 34 02	06 56 41	00 ♊ 07 28	06 ♊ 14 13	04 00	22 38	00 ♍ 34	06 11	27 31	26 26	08 51	14 54
31	10 37 59	07 ♍ 54 42	12 ♊ 25 12	18 ♊ 41 02	05 ♎ 01	23 ♎ 32	01 ♍ 12	06 ♌ 24	27 ♊ 35	26 ♒ 24	08 ♒ 49	14 ♐ 54

Date	Moon True ☊ ° '	Moon Mean ☊ ° '	Moon ☽ Latitude ° '
01	16 ♊ 24	15 ♊ 07	03 S 04
02	16 D 24	15 03	02 08
03	16 25	15 00	01 S 05
04	16 R 26	14 57	00 N 03
05	16 25	14 54	01 12
06	16 22	14 51	02 19
07	16 17	14 47	03 21
08	16 09	14 44	04 11
09	16 01	14 41	04 46
10	15 51	14 38	05 02
11	15 43	14 35	04 59
12	15 36	14 32	04 35
13	15 31	14 28	03 54
14	15 29	14 25	02 59
15	15 D 28	14 22	01 54
16	15 28	14 19	00 N 44
17	15 R 29	14 16	00 S 27
18	15 27	14 13	01 35
19	15 24	14 09	02 37
20	15 17	14 06	03 30
21	15 08	14 03	04 12
22	14 57	14 00	04 41
23	14 44	13 57	04 57
24	14 31	13 53	05 00
25	14 18	13 50	04 49
26	14 08	13 47	04 25
27	14 00	13 44	03 50
28	13 54	13 41	03 05
29	13 51	13 38	02 12
30	13 50	13 34	01 12
31	13 ♊ 50	13 ♊ 31	00 S 07

DECLINATIONS

Date	Sun ☉ ° '	Moon ☽ ° '	Mercury ☿ ° '	Venus ♀ ° '	Mars ♂ ° '	Jupiter ♃ ° '	Saturn ♄ ° '	Uranus ♅ ° '	Neptune ♆ ° '	Pluto ♇ ° '
01	17 N 59	11 N 51	16 N 00	02 N 37	18 N 16	20 N 29	22 N 05	13 S 02	17 S 47	12 S 45
02	17 44	16 11	15 20	02 07	18 05	20 26	22 05	13 03	17 47	12 45
03	17 29	19 56	14 40	01 37	17 54	20 24	22 05	13 04	17 48	12 46
04	17 13	22 51	13 59	01 08	17 43	20 21	22 05	13 05	17 48	12 46
05	16 57	24 38	13 18	00 38	17 32	20 18	22 06	13 06	17 49	12 46
06	16 40	25 04	12 36	00 N 08	17 21	20 15	22 06	13 06	17 49	12 47
07	16 23	23 58	11 54	00 S 22	17 10	20 13	22 06	13 07	17 49	12 47
08	16 07	21 18	11 12	00 51	16 58	20 10	22 06	13 08	17 50	12 47
09	15 49	17 14	10 29	01 21	16 47	20 07	22 06	13 09	17 50	12 47
10	15 32	12 04	09 46	01 51	16 35	20 04	22 06	13 10	17 51	12 48
11	15 14	06 N 12	09 04	02 20	16 23	20 01	22 07	13 11	17 51	12 48
12	14 56	00 00	08 21	02 50	16 11	19 58	22 07	13 11	17 52	12 48
13	14 38	06 S 07	07 38	03 19	15 59	19 56	22 07	13 12	17 52	12 49
14	14 20	11 49	06 56	03 49	15 47	19 53	22 07	13 13	17 53	12 49
15	14 01	16 48	06 13	04 18	15 35	19 50	22 07	13 14	17 53	12 50
16	13 42	20 46	05 31	04 47	15 22	19 47	22 07	13 15	17 53	12 50
17	13 23	23 33	04 49	05 16	15 10	19 44	22 07	13 15	17 54	12 50
18	13 04	24 58	04 08	05 45	14 58	19 41	22 07	13 16	17 54	12 51
19	12 44	25 00	03 27	06 14	14 45	19 38	22 08	13 17	17 55	12 51
20	12 25	23 43	02 46	06 43	14 32	19 35	22 08	13 18	17 55	12 51
21	12 05	21 15	02 05	07 11	14 20	19 32	22 08	13 19	17 55	12 52
22	11 45	17 51	01 26	07 39	14 07	19 29	22 08	13 20	17 56	12 52
23	11 24	13 43	00 46	08 08	13 54	19 27	22 08	13 20	17 56	12 53
24	11 04	09 06	00 N 08	08 36	13 41	19 24	22 08	13 21	17 57	12 53
25	10 43	04 S 10	00 S 30	09 03	13 28	19 21	22 08	13 22	17 57	12 53
26	10 22	00 N 52	01 08	09 31	13 14	19 18	22 08	13 23	17 57	12 54
27	10 02	05 52	01 44	09 58	13 01	19 15	22 08	13 24	17 58	12 54
28	09 40	10 40	02 20	10 25	12 48	19 12	22 08	13 24	17 58	12 55
29	09 19	15 06	02 55	10 52	12 34	19 09	22 08	13 25	17 59	12 55
30	08 58	19 00	03 29	11 19	12 21	19 06	22 08	13 26	17 59	12 56
31	08 N 36	22 N 10	04 S 01	11 S 45	12 N 07	19 N 03	22 N 08	13 S 27	17 S 59	12 S 56

ZODIAC SIGN ENTRIES

Date	h m	Planets
01	17 20	♃ ♌
03	03 47	☽ ♊
05	12 02	☽ ♋
06	09 51	☿ ♍
07	09 09	♀ ♎
07	16 27	☽ ♌
09	18 03	☽ ♍
11	18 38	☽ ♎
13	20 01	☽ ♏
15	23 25	☽ ♐
18	05 15	☽ ♑
20	13 16	☽ ♒
22	23 11	☽ ♓
23	07 17	☉ ♍
25	10 48	☽ ♈
26	21 10	☿ ♎
27	23 32	☽ ♉
29	13 38	♂ ♍
30	11 45	☽ ♊

LATITUDES

Date	Mercury ☿ ° '	Venus ♀ ° '	Mars ♂ ° '	Jupiter ♃ ° '	Saturn ♄ ° '	Uranus ♅ ° '	Neptune ♆ ° '	Pluto ♇ ° '
01	01 N 38	00 N 10	01 N 08	00 N 20	01 S 16	00 S 46	00 N 04	09 N 55
04	01 26	00 S 05	01 09	00 20	01 16	00 47	00 04	09 54
07	01 10	00 20	01 09	00 21	01 16	00 47	00 04	09 52
10	00 51	00 36	01 09	00 21	01 16	00 47	00 04	09 51
13	00 29	00 53	01 09	00 21	01 16	00 47	00 04	09 50
16	00 N 05	01 11	01 09	00 21	01 16	00 47	00 04	09 49
19	00 S 21	01 30	01 09	00 22	01 16	00 47	00 04	09 48
22	00 48	01 49	01 09	00 22	01 16	00 47	00 04	09 46
25	01 16	02 08	01 09	00 23	01 17	00 47	00 04	09 45
28	01 44	02 29	01 09	00 23	01 17	00 47	00 04	09 44
31	02 S 13	02 S 49	01 N 09	00 N 23	01 S 17	00 S 47	00 N 04	09 N 43

DATA

Julian Date	2452488
Delta T	+64 seconds
Ayanamsa	23° 53' 20"
Synetic vernal point	05° ♓ 13' 40"
True obliquity of ecliptic	23° 26' 23"

MOON'S PHASES, APSIDES AND POSITIONS ☽

Date	h m	Phase	Longitude ° '	Eclipse Indicator
01	10 22	☾	09 ♉ 00	
08	19 15	●	16 ♌ 04	
15	10 12	☽	22 ♏ 25	
22	22 29	○	29 ♒ 39	
31	02 31	☾	07 ♊ 32	

Day	h m		
10	23 24	Perigee	
26	17 31	Apogee	
06	06 58	Max dec	25° N 06'
12	12 01	0S	
19	00 34	Max dec	25° S 10'
26	07 53	0N	

LONGITUDES

Date	Chiron ⚷ ° '	Ceres ⚳ ° '	Pallas ⚴ ° '	Juno ⚵ ° '	Vesta ⚶ ° '	Black Moon Lilith ⚸ ° '
01	04 ♑ 08	17 ♈ 40	22 ♒ 25	16 ♍ 55	04 ♌ 25	08 ♈ 29
11	03 ♑ 43	18 ♈ 05	19 ♒ 51	20 ♍ 35	08 ♌ 52	09 ♈ 36
21	03 ♑ 25	17 ♈ 55	17 ♒ 18	24 ♍ 17	13 ♌ 25	10 ♈ 43
31	03 ♑ 15	17 ♈ 08	14 ♒ 58	28 ♍ 01	17 ♌ 55	11 ♈ 50

All ephemeris data is given at 12.00 UT and the Moon's longitude is additionally given for 24.00 UT

ASPECTARIAN

h m Aspects

01 Thursday
09 45 ☽ ⊑ ♀
10 22 ☽ □ ☉
10 29 ☽ ± ♇
11 31 ☽ Q ♅
11 34 ☽ □ ♆
12 12 ☽ ∠ ♄
16 46 ☽ # ♇
16 46 ☽ □ ♂
18 21 ☽ # ♅
22 26 ☽ ⊼ ♇
23 30 ♃ ⊑ ♇
02 Friday
00 57 ☉ ☍ ♆
04 29 ☽ Q ♃
05 43 ♀ ⊑ ♆
06 15 ☽ ⊥ ♄
07 13 ☉ # ♆
07 44 ☽ ∥ ☿
11 58 ☉ ∠ ♄
14 30 ☽ □ ☿
16 10 ♀ □ ♄
18 08 ☽ ⊻ ♄
18 18 ☽ △ ♀
20 47 ☽ ∥ ☉
21 43 ☽ # ♆
22 58 ☽ □ ♅
23 04 ☽ ∥ ♂
03 Saturday
00 57 ☽ Q ☉
04 24 ☽ ⚹ ♃
06 06 ☽ Q ♂
15 17 ☽ ∥ ♃
16 11 ☿ ⚹ ♄
22 04 ☽ △ ♆
04 Sunday
01 52 ☽ ⚹ ☉
02 00 ♂ # ♆
03 11 ☉ ∠ ♀
04 56 ☽ ∥ ♄
06 13 ☽ ⚹ ♂
07 00 ☽ Q ☿
08 26 ☽ ☍ ♇
09 30 ☽ ∠ ♃
22 45 ♀ ⊼ ♅
22 59 ☿ ☍ ♅
23 20 ☿ ⊻ ♀
05 Monday
02 01 ☽ ⊑ ♆
02 28 ☽ ⊥ ♃
03 28 ☽ ☌ ♄
07 22 ☽ △ ♅
07 55 ☽ ∠ ☉
08 09 ☽ □ ♀
08 42 ☽ ⚹ ☿
11 21 ☽ ∠ ♂
13 34 ☽ ⊻ ♃
18 18 ☽ ± ♆
18 53 ☿ # ♅
06 Tuesday
01 21 ☽ ⊥ ☉
01 58 ♂ △ ♇
04 16 ☽ ⊥ ♂
04 58 ☽ ⊼ ♆
06 04 ☿ # ♇
10 07 ☽ ⊑ ♅
12 44 ☽ ⊻ ☉
14 42 ☽ ⊼ ♇
15 19 ☽ ⊻ ♂
15 23 ☽ ∠ ☿
18 49 ☽ Q ♀
07 Wednesday
01 06 ☽ ± ♇
01 43 ☽ ± ♅
02 20 ☿ ⊻ ♃
08 51 ☽ ⊻ ♄
09 06 ☽ ⊥ ☿
11 58 ☽ ⊼ ♅
16 27 ☽ ⊑ ♇
16 29 ☉ △ ♇
17 02 ☽ ⚹ ♀
18 43 ☽ ☌ ♃
19 06 ☽ ⊥ ♄
20 40 ☽ ⊻ ☿
08 Thursday
06 01 ☽ ∥ ♄
08 14 ☽ ☍ ♆
10 16 ☽ ∠ ♄
17 28 ☽ △ ♇
19 15 ☽ ☌ ☉
19 37 ☽ ∥ ♃
20 00 ☽ ∠ ♀
20 26 ☽ ☌ ♂
21 41 ♀ ⚹ ♃
09 Friday
08 51 ☽ # ♆
11 06 ☽ ⚹ ♄
11 21 ♀ ∠ ♂
11 51 ☽ ⊥ ♀
13 37 ☽ ☍ ♅
14 26 ☽ ∥ ♂
15 26 ♃ ⊥ ♄
19 27 ☽ ∥ ☉
20 58 ☽ ⊻ ♃
22 21 ☽ ⊻ ♀
10 Saturday
04 23 ☽ ☌ ☿
05 51 ♀ Q ♇
06 43 ☽ Q ♄
06 50 ☽ ⊥ ♃
07 16 ☽ # ♅
08 52 ☽ # ♇
09 13 ☽ ⊼ ♆
11 59 ♀ ± ♅
18 17 ☽ □ ♇
18 54 ☽ ± ♆
21 37 ☽ ∠ ♃
22 17 ☉ ☌ ♂
22 59 ☽ ∥ ☿
23 25 ☽ ⊻ ♂
23 26 ☽ ⊻ ☉
11 Sunday
02 50 ☿ Q ♄
05 56 ☿ ⊥ ♃
09 26 ☽ ⊑ ♆
09 33 ☽ ⊥ ♂
09 49 ☽ ⊥ ☉
12 01 ☽ □ ♄
14 04 ☽ ⊼ ♅
22 17 ☽ ⚹ ♃
23 26 ☽ Q ♇
23 48 ☽ ± ♅
12 Monday
00 11 ☿ ⊼ ♆
00 49 ☽ ∠ ♂
01 27 ☽ ∠ ☉
01 54 ☽ # ♀
02 34 ☽ ☌ ♀
09 47 ☽ △ ♆
10 58 ☽ ⊻ ☿
14 27 ☽ ⊑ ♅
18 11 ☽ Q ♃
19 03 ☽ ⚹ ♇
22 00 ☽ ⊥ ☿
23 57 ☽ ∥ ♀
13 Tuesday
02 32 ☽ ⚹ ♂
03 46 ☽ ⚹ ☉
13 32 ☽ △ ♄
14 43 ☽ ∠ ☿
15 11 ☽ △ ♅
17 31 ☽ # ☿
19 55 ☽ ∠ ♇
23 27 ☽ Q ♂
14 Wednesday
00 33 ☽ □ ♃
01 13 ☽ Q ☉
08 03 ☽ ⊻ ♀
11 05 ☽ ⊥ ♇
11 36 ☽ □ ♆
14 57 ☽ ⊑ ♄
16 32 ☽ ∥ ♇
18 22 ☽ ∥ ♅
19 02 ☽ ⊥ ♀
19 06 ☽ ⚹ ☿
21 18 ☽ ⊻ ♇
22 55 ☽ # ☉
15 Thursday
06 00 ☽ # ♂
06 30 ☽ ± ♄
07 34 ☽ □ ♂
10 12 ☽ □ ☉
11 45 ☽ ∠ ♀
15 31 ☿ □ ♇
16 59 ☽ ⊼ ♄
17 58 ☽ ∥ ♆
18 13 ☽ □ ♅
18 22 ☽ Q ☿
18 33 ☽ Q ♆
20 01 ☿ ± ♆
16 Friday
05 00 ☽ △ ♃
05 26 ☽ # ♃
09 01 ♀ △ ♆
15 40 ☽ ⚹ ♆
16 12 ☽ ⚹ ♀
22 18 ☽ # ♄
17 Saturday
01 54 ☽ ☌ ♇
02 00 ☽ Q ♅
02 55 ☿ # ♀
06 12 ☽ □ ☿
08 12 ☽ ⊑ ♃
15 22 ☽ △ ♂
15 31 ☽ Q ♀
18 36 ☽ ∠ ♆
19 37 ☽ △ ☉
21 05 ☉ # ♅
22 53 ☽ ☍ ♄
23 39 ☽ ⚹ ♅
18 Sunday
00 48 ☽ ± ♃
03 30 ☿ ∠ ♃
11 02 ☽ ⊥ ♆
11 28 ♅ Q ♇
12 00 ☽ ⊼ ♃
20 16 ☽ ⊑ ♂
22 06 ☽ ⊻ ♆
19 Monday
01 24 ☽ ⊑ ☉
02 15 ♀ ⊑ ♅
03 13 ☽ ∠ ♅
03 18 ☽ □ ♀
03 49 ☉ # ♇
08 51 ☽ ⊻ ♇
13 55 ☽ ± ♂
19 35 ☽ ± ☉
20 00 ☽ ⊥ ♅
20 06 ☽ ⊥ ♇
20 13 ☽ △ ☿
20 35 ☉ ⚹ ♄
20 Tuesday
00 54 ☉ ☍ ♅
01 47 ☽ ⊼ ♂
07 03 ☽ ⊼ ♄
07 17 ☽ ⊻ ♅
07 50 ☽ ⊼ ☉
13 06 ☽ ∠ ♇
18 32 ☽ ± ♄
21 14 ☽ ☍ ♃
21 Wednesday
04 09 ☽ ⊑ ☿
04 36 ☽ # ♄
06 36 ☽ ☌ ♆
07 18 ♄ △ ♅
11 53 ☽ ⊑ ♄
16 55 ☽ △ ♀
17 49 ☽ ⚹ ♇
23 34 ☽ ± ☿
22 Thursday
01 00 ☽ # ♃
02 45 ☿ ⊑ ♆
04 35 ♀ ⚹ ♇
11 28 ☽ ∥ ♆
12 36 ☽ ⊼ ☿
14 28 ☽ ☍ ♂
16 49 ☽ ☌ ♅
17 07 ☽ Q ♇
17 09 ☽ △ ♄
22 29 ☽ ☍ ☉
23 Friday
00 34 ☽ ⊑ ♀
08 25 ☽ ⊼ ♃
10 59 ☽ # ♂
14 03 ☽ ∥ ♅
16 31 ☽ ∥ ♇
16 55 ☽ ⊻ ♆
19 48 ☽ ± ♀
20 28 ☽ ± ♃
21 48 ☿ ⊻ ♂
24 Saturday
01 13 ☽ # ☉
03 52 ☿ ⊼ ♅
04 36 ☽ □ ♇
04 47 ☽ ⊥ ♆
06 32 ♀ Q ♃
08 44 ☽ ⊼ ♀
09 29 ♂ ☍ ♅
10 58 ☿ □ ♄
14 16 ☽ ∥ ♀
14 41 ☽ ⊑ ♃
22 43 ☽ ∠ ♆
25 Sunday
01 40 ♂ ⚹ ♄
04 04 ☽ ⊻ ♅
05 01 ☽ □ ♄
05 10 ☽ ⊼ ♂
06 58 ☽ ☍ ☿
15 19 ☽ ⊼ ☉
16 05 ☽ ⊥ ♅
17 56 ☽ ± ♂
21 20 ☽ △ ♃
21 29 ♂ # ♅
26 Monday
03 34 ☽ ∥ ☿
04 30 ☽ ± ☉
04 51 ☽ ⚹ ♆
10 13 ☽ ∠ ♅
11 02 ♇ St D
13 09 ☽ ⊑ ♂
13 25 ☽ # ☿
16 54 ☽ △ ♇
17 35 ☽ Q ♄
27 Tuesday
00 21 ☽ ⊑ ☉
02 18 ☽ ☍ ♀
05 07 ☽ Q ♆
13 35 ☉ # ♀
16 32 ☽ ⚹ ♅
18 05 ☽ ⚹ ♄
21 18 ☽ △ ♂
23 20 ☽ ⊑ ♇
23 50 ♂ # ♇
28 Wednesday
02 22 ☽ ⊼ ☿
07 18 ☽ ∥ ☉
09 29 ☽ △ ☉
10 40 ☽ # ♀
11 13 ☽ □ ♃
15 47 ☽ ± ☿
16 46 ☽ Q ♅
17 32 ☽ □ ♆
17 33 ☽ ± ♇
22 42 ☽ ∥ ♂
23 53 ☽ # ♇
29 Thursday
00 38 ☽ ∠ ♄
02 03 ☿ ± ♅
02 36 ☽ # ♅
05 39 ☽ ⊼ ♇
11 03 ☿ Q ♇
11 44 ☽ ⊑ ☿
12 11 ☉ ⊻ ♃
18 49 ☽ ⊥ ♄
19 57 ☽ ⊼ ♀
30 Friday
00 02 ☽ Q ♃
04 44 ☽ □ ♅
05 18 ☽ # ♆
06 49 ☽ ⊻ ♄
08 50 ☽ ± ♀
12 37 ☽ ∥ ♃
12 55 ☽ □ ♂
20 20 ☽ △ ☿
31 Saturday
00 07 ☽ ⚹ ♃
02 31 ☽ □ ☉
03 53 ☽ ⊑ ♀
05 03 ☽ △ ♆
11 48 ☽ ∥ ♄
16 47 ☽ ☍ ♇

SEPTEMBER 2002

LONGITUDES

Date	Sidereal time h m s	Sun ☉ ° ' "	Moon ☽ ° ' "	Moon ☽ 24.00 ° ' "	Mercury ☿ ° '	Venus ♀ ° '	Mars ♂ ° '	Jupiter ♃ ° '	Saturn ♄ ° '	Uranus ♅ ° '	Neptune ♆ ° '	Pluto ♇ ° '
01	10 41 55	08 ♍ 52 46	25 ♊ 02 22	01 ♋ 29 46	06 ♎ 00	24 ♎ 25	01 ♍ 50	06 ♌ 36	27 ♊ 40	26 ♒ 22	08 ♒ 48	14 ♐ 55
02	10 45 52	09 50 51	08 ♋ 03 45	14 ♋ 44 44	06 55	25 17	02 29	06 48	27 44	26 R 19	08 R 47	14 55
03	10 49 49	10 48 58	21 ♋ 33 00	28 ♋ 28 41	07 48	26 09	03 07	07 00	27 48	26 17	08 45	14 55
04	10 53 45	11 47 07	05 ♌ 31 43	12 ♌ 41 51	08 38	27 00	03 45	07 12	27 51	26 15	08 44	14 55
05	10 57 42	12 45 18	19 ♌ 58 36	27 ♌ 21 16	09 25	27 51	04 23	07 25	27 55	26 13	08 43	14 56
06	11 01 38	13 43 31	04 ♍ 48 55	12 ♍ 20 28	10 08	28 40	05 01	07 37	27 59	26 10	08 42	14 56
07	11 05 35	14 41 45	19 ♍ 54 40	27 ♍ 30 11	10 47	29 ♎ 30	05 39	07 48	28 03	26 08	08 40	14 56
08	11 09 31	15 40 01	05 ♎ 05 38	12 ♎ 39 42	11 23	00 ♏ 18	06 18	08 00	28 06	26 06	08 39	14 57
09	11 13 28	16 38 19	20 ♎ 11 10	27 ♎ 38 56	11 54	01 06	06 56	08 12	28 10	26 04	08 38	14 57
10	11 17 24	17 36 38	05 ♏ 02 08	12 ♏ 20 05	12 20	01 52	07 34	08 24	28 13	26 02	08 37	14 58
11	11 21 21	18 34 59	19 ♏ 30 25	26 ♏ 38 24	12 42	02 38	08 12	08 36	28 16	25 59	08 36	14 58
12	11 25 18	19 33 22	03 ♐ 38 25	10 ♐ 32 20	12 58	03 24	08 50	08 47	28 19	25 57	08 34	14 59
13	11 29 14	20 31 46	17 ♐ 20 20	24 ♐ 02 42	13 09	04 08	09 28	08 59	28 22	25 55	08 33	14 59
14	11 33 11	21 30 12	00 ♑ 39 46	07 ♑ 11 55	13 14	04 51	10 07	09 10	28 25	25 53	08 32	15 00
15	11 37 07	22 28 39	13 ♑ 39 33	20 ♑ 03 06	13 R 13	05 33	10 45	09 21	28 28	25 51	08 31	15 01
16	11 41 04	23 27 08	26 ♑ 22 56	02 ♒ 39 28	13 06	06 15	11 23	09 33	28 31	25 49	08 30	15 01
17	11 45 00	24 25 39	08 ♒ 53 01	15 ♒ 03 55	12 51	06 55	12 01	09 44	28 33	25 47	08 29	15 02
18	11 48 57	25 24 11	21 ♒ 12 26	27 ♒ 18 49	12 30	07 34	12 39	09 55	28 36	25 45	08 28	15 03
19	11 52 53	26 22 45	03 ♓ 23 18	09 ♓ 26 03	12 01	08 12	13 17	10 06	28 38	25 43	08 27	15 04
20	11 56 50	27 21 20	15 ♓ 27 14	21 ♓ 27 01	11 26	08 49	13 56	10 17	28 41	25 41	08 26	15 04
21	12 00 47	28 19 58	27 ♓ 25 32	03 ♈ 22 58	10 44	09 25	14 34	10 28	28 43	25 39	08 25	15 05
22	12 04 43	29 ♍ 18 37	09 ♈ 19 27	15 ♈ 15 11	09 56	09 59	15 12	10 39	28 45	25 37	08 24	15 06
23	12 08 40	00 ♎ 17 19	21 ♈ 10 23	27 ♈ 05 17	09 01	10 32	15 50	10 50	28 47	25 35	08 24	15 07
24	12 12 36	01 16 02	03 ♉ 00 10	08 ♉ 55 22	08 02	11 04	16 28	11 00	28 49	25 34	08 23	15 08
25	12 16 33	02 14 48	14 ♉ 51 14	20 ♉ 48 13	06 59	11 34	17 06	11 11	28 51	25 32	08 22	15 09
26	12 20 29	03 13 36	26 ♉ 46 45	02 ♊ 47 21	05 53	12 03	17 45	11 21	28 53	25 30	08 21	15 10
27	12 24 26	04 12 26	08 ♊ 50 33	14 ♊ 56 57	04 47	12 30	18 23	11 32	28 54	25 28	08 20	15 11
28	12 28 22	05 11 18	21 ♊ 07 07	27 ♊ 21 42	03 40	12 56	19 01	11 42	28 56	25 27	08 20	15 12
29	12 32 19	06 10 13	03 ♋ 41 17	10 ♋ 06 29	02 36	13 19	19 39	11 52	28 57	25 25	08 19	15 13
30	12 36 16	07 ♎ 09 10	16 ♋ 37 52	23 ♋ 15 53	01 ♎ 37	13 ♏ 42	20 ♍ 17	12 ♌ 02	28 ♊ 58	25 ♒ 23	08 ♒ 18	15 ♐ 14

Date	Moon True ☊ ° '	Moon Mean ☊ ° '	Moon ☽ Latitude ° '
01	13 ♊ 50	13 ♊ 28	00 N 59
02	13 R 48	13 25	02 04
03	13 44	13 22	03 05
04	13 38	13 19	03 57
05	13 29	13 15	04 36
06	13 19	13 12	04 57
07	13 07	13 09	04 58
08	12 57	13 06	04 38
09	12 48	13 03	03 59
10	12 42	12 59	03 04
11	12 38	12 56	01 59
12	12 37	12 53	00 N 47
13	12 D 37	12 50	00 S 25
14	12 R 37	12 47	01 34
15	12 35	12 44	02 36
16	12 32	12 40	03 29
17	12 25	12 37	04 11
18	12 16	12 34	04 41
19	12 04	12 31	04 57
20	11 51	12 28	05 00
21	11 38	12 25	04 49
22	11 25	12 21	04 26
23	11 14	12 18	03 52
24	11 05	12 15	03 07
25	11 00	12 12	02 14
26	10 57	12 09	01 15
27	10 56	12 05	00 S 11
28	10 D 56	12 02	00 N 54
29	10 56	11 59	01 58
30	10 ♊ 56	11 ♊ 56	02 N 58

DECLINATIONS

Date	Sun ☉ ° '	Moon ☽ ° '	Mercury ☿ ° '	Venus ♀ ° '	Mars ♂ ° '	Jupiter ♃ ° '	Saturn ♄ ° '	Uranus ♅ ° '	Neptune ♆ ° '	Pluto ♇ ° '
01	08 N 14	24 N 20	04 S 33	12 S 12	11 N 53	19 N 00	22 N 08	13 S 28	18 S 00	12 S 56
02	07 53	25 16	05 04	12 37	11 40	18 57	22 08	13 28	18 00	12 57
03	07 31	24 46	05 33	13 03	11 26	18 54	22 08	13 29	18 00	12 57
04	07 09	22 43	06 01	13 28	11 12	18 51	22 08	13 30	18 01	12 58
05	06 46	19 11	06 27	13 53	10 58	18 48	22 08	13 31	18 01	12 58
06	06 24	14 21	06 52	14 18	10 44	18 45	22 08	13 31	18 02	12 59
07	06 02	08 34	07 14	14 42	10 30	18 42	22 08	13 32	18 02	12 59
08	05 39	02 N 14	07 35	15 06	10 16	18 39	22 08	13 33	18 02	13 00
09	05 16	04 S 12	07 54	15 30	10 02	18 36	22 08	13 34	18 03	13 00
10	04 54	10 18	08 11	15 53	09 47	18 34	22 08	13 34	18 03	13 00
11	04 31	15 43	08 25	16 16	09 33	18 31	22 08	13 35	18 03	13 01
12	04 08	20 06	08 37	16 38	09 19	18 28	22 08	13 36	18 03	13 01
13	03 45	23 15	08 46	17 00	09 04	18 25	22 08	13 37	18 04	13 02
14	03 22	25 00	08 52	17 22	08 50	18 22	22 08	13 37	18 04	13 02
15	02 59	25 20	08 54	17 43	08 35	18 19	22 08	13 38	18 04	13 03
16	02 36	24 18	08 53	18 04	08 21	18 16	22 08	13 39	18 05	13 03
17	02 13	22 05	08 48	18 24	08 06	18 13	22 08	13 39	18 05	13 04
18	01 50	18 52	08 39	18 44	07 51	18 10	22 08	13 40	18 05	13 04
19	01 26	14 53	08 26	19 03	07 36	18 08	22 08	13 41	18 05	13 05
20	01 03	10 20	08 09	19 22	07 22	18 05	22 08	13 41	18 06	13 05
21	00 40	05 27	07 48	19 40	07 07	18 02	22 08	13 42	18 06	13 06
22	00 N 16	00 S 23	07 22	19 58	06 52	17 59	22 08	13 42	18 06	13 06
23	00 S 07	04 N 41	06 53	20 15	06 37	17 56	22 08	13 43	18 06	13 07
24	00 30	09 35	06 19	20 31	06 22	17 54	22 08	13 44	18 07	13 07
25	00 54	14 10	05 42	20 47	06 07	17 51	22 08	13 44	18 07	13 08
26	01 17	18 14	05 03	21 02	05 52	17 48	22 08	13 45	18 07	13 08
27	01 40	21 36	04 21	21 17	05 37	17 45	22 08	13 45	18 07	13 09
28	02 04	24 03	03 39	21 31	05 22	17 43	22 08	13 46	18 08	13 09
29	02 27	25 22	02 56	21 44	05 07	17 40	22 08	13 46	18 08	13 10
30	02 S 50	25 N 21	02 S 14	21 S 57	04 N 52	17 N 37	22 N 08	13 S 47	18 S 08	13 S 10

ZODIAC SIGN ENTRIES

Date	h	m	Planets
01	21	14	☽ ♋
04	02	36	☽ ♌
06	04	16	☽ ♍
08	03	05	♀ ♏
08	03	57	☽ ♎
10	03	48	☽ ♏
12	05	44	☽ ♐
14	10	47	☽ ♑
16	18	54	☽ ♒
19	05	18	☽ ♓
21	17	11	☽ ♈
23	04	55	☉ ♎
24	05	53	☽ ♉
26	18	26	☽ ♊
29	05	01	☽ ♋

LATITUDES

Date	Mercury ☿ ° '	Venus ♀ ° '	Mars ♂ ° '	Jupiter ♃ ° '	Saturn ♄ ° '	Uranus ♅ ° '	Neptune ♆ ° '	Pluto ♇ ° '
01	02 S 22	02 S 56	01 N 09	00 N 23	01 S 17	00 S 47	00 N 04	09 N 42
04	02 49	03 17	01 09	00 24	01 17	00 47	00 04	09 41
07	03 14	03 39	01 09	00 24	01 17	00 47	00 04	09 40
10	03 36	04 00	01 08	00 24	01 17	00 47	00 04	09 39
13	03 53	04 22	01 08	00 25	01 17	00 46	00 04	09 37
16	04 02	04 44	01 08	00 25	01 18	00 46	00 04	09 36
19	04 00	05 05	01 08	00 26	01 18	00 46	00 04	09 35
22	03 45	05 26	01 07	00 26	01 18	00 46	00 04	09 34
25	03 12	05 46	01 07	00 27	01 18	00 46	00 04	09 32
28	02 23	06 05	01 07	00 27	01 18	00 46	00 04	09 31
31	01 S 24	06 S 23	01 N 06	00 N 28	01 S 19	00 S 46	00 N 04	09 N 30

DATA

Julian Date	2452519
Delta T	+64 seconds
Ayanamsa	23° 53' 24"
Synetic vernal point	05° ♓ 13' 35"
True obliquity of ecliptic	23° 26' 23"

MOON'S PHASES, APSIDES AND POSITIONS ☽

Date	h	m	Phase	Longitude ° '	Eclipse Indicator
07	03	10	●	14 ♍ 20	
13	18	08	☽	20 ♐ 47	
21	13	59	○	28 ♓ 25	
29	17	03	☾	06 ♋ 23	

Day	h	m		
08	03	12	Perigee	
23	03	09	Apogee	
02	15	59	Max dec	25° N 17'
08	20	18	0S	
15	05	32	Max dec	25° S 23'
22	13	49	0N	
29	23	59	Max dec	25° N 32'

LONGITUDES

Date	Chiron ⚷ ° '	Ceres ⚳ ° '	Pallas ⚴ ° '	Juno ⚵ ° '	Vesta ⚶ ° '	Black Moon Lilith ⚸ ° '
01	03 ♑ 15	17 ♈ 01	14 ♒ 45	28 ♍ 23	18 ♌ 22	11 ♈ 57
11	03 14	15 ♈ 37	12 ♒ 53	02 ♎ 08	22 ♌ 51	13 ♈ 04
21	03 ♑ 22	13 ♈ 44	11 ♒ 33	05 ♎ 53	27 ♌ 19	14 ♈ 11
31	03 ♑ 38	11 ♈ 33	10 ♒ 49	09 ♎ 36	01 ♍ 44	15 ♈ 18

All ephemeris data is given at 12.00 UT and the Moon's longitude is additionally given for 24.00 UT

ASPECTARIAN

h m	Aspects
01 Sunday	
01 04	♀ # ♂
01 41	☽ Q ♂
05 25	☽ ∠ ♃
09 41	☽ bQ ♆
10 08	☉ ⊼ ♆
10 44	☽ △ ♀
14 28	☽ △ ♅
15 43	☽ Q ☉
16 55	☽ ☌ ♄
22 31	☽ ⊥ ♃
02 Monday	
01 17	☽ ⚹ ♂
02 23	☽ ± ♆
07 53	☿ ⚹ ♃
08 47	☉ Q ♄
09 41	☽ ⊻ ♃
09 47	☽ □ ☿
13 18	☽ ⊼ ♆
15 29	☽ ⚹ ☉
17 52	☽ bQ ♅
03 Tuesday	
00 18	☽ ⊼ ♇
05 40	☽ ∠ ♂
06 34	♀ ∥ ♇
09 48	☽ ± ♅
10 54	☽ ± ♇
15 40	♀ △ ♅
19 53	☽ Q ☿
19 58	☽ ∠ ☉
20 12	☽ ⊼ ♅
20 31	☽ □ ♀
22 07	☽ ⊥ ♂
22 52	☽ ⊻ ♄
04 Wednesday	
02 28	☽ bQ ♇
08 51	☽ ⊻ ♂
09 10	☽ ⊥ ♄
12 28	☽ ⊥ ☉
13 44	♀ ∥ ♅
14 52	☽ ☌ ♃
14 53	☿ △ ♆
16 41	☽ ∥ ♄
17 23	☽ ☍ ♆
17 32	☽ ⚹ ☿
23 14	☽ ⊻ ☉
05 Thursday	
00 19	☽ ∠ ♄
03 42	☽ △ ♇
04 48	☽ Q ♀
14 07	☽ ∥ ♃
14 26	♀ △ ♄
18 18	☽ # ♆
19 37	☽ ∠ ☿
21 51	☉ # ☿
22 07	☽ ☍ ♅
06 Friday	
00 58	☽ ⚹ ♄
01 33	☽ ⚹ ♀
03 21	☉ ∠ ♀
08 23	☉ ⊥ ♃
10 51	☽ ⊥ ☿
12 13	☽ # ♀
12 21	☽ ☌ ♂
15 37	☽ # ♅
16 31	☽ ⊻ ♃
17 59	☽ # ♇
18 11	☽ ⊼ ♆
20 17	☽ Q ♄
20 53	☽ ⊻ ☿
07 Saturday	
02 12	☽ ⊥ ♃
02 56	☽ ∠ ♀
03 10	☽ ☌ ☉
03 43	☽ ± ♆
03 59	☽ ∥ ♂
04 07	☽ □ ♇
11 26	☉ ± ♆
16 38	☽ ∠ ♃
16 51	☽ # ☿
17 56	☽ bQ ♆
17 59	☽ ⊥ ♀
18 05	☉ □ ♇
21 49	☽ ⊼ ♅
22 23	☽ ∥ ☉
08 Sunday	
00 54	☽ □ ♄
01 01	☿ bQ ♅
01 25	♀ ∠ ♇
04 00	☽ ⊻ ♀
07 16	☽ ± ♅
08 36	☽ Q ♇
13 59	☽ ⊻ ♂
16 40	☽ ⚹ ♃
17 38	☽ △ ♆
21 30	☽ bQ ♅
22 20	☽ ☌ ☿
23 55	☽ ⊥ ♂
09 Monday	
03 39	☽ ⚹ ♇
05 56	☽ ⊻ ☉
12 02	☽ Q ♃
14 55	☽ ∠ ♂
15 53	☽ # ☉
16 12	☽ ⊥ ☉
21 25	☽ △ ♅
10 Tuesday	
00 52	☽ △ ♄
03 03	☽ ∥ ☿
03 44	☽ ∠ ♇
06 34	☽ ☌ ♀
07 46	☽ ∠ ☉
09 58	☽ # ♂
16 20	☽ ⚹ ♂
17 35	☽ □ ♃
17 51	☽ □ ♆
18 26	☽ ⊥ ♇
23 32	☽ ∥ ♇
11 Wednesday	
00 20	☽ ⊻ ☿
01 30	☽ bQ ♄
02 04	☽ ∥ ♅
04 22	☽ ⊻ ♇
10 17	☽ ⚹ ☉
10 34	☽ ⊥ ☿
12 06	♃ ☍ ♆
13 10	☽ Q ♂
14 55	☽ ∥ ♀
16 37	☽ ± ♄
22 52	☽ □ ♅
23 54	☽ Q ♆
12 Thursday	
00 01	☽ ∥ ♆
02 06	☽ ∠ ☿
02 18	♂ ⊼ ♆
02 22	☽ # ♃
02 50	☽ ⊼ ♄
08 08	☽ Q ☉
09 07	♂ ⊻ ♃
11 33	☽ ⊻ ♀
20 33	☽ ⚹ ♆
21 04	☽ △ ♃
21 28	☽ □ ♂
22 34	☽ ⊥ ♀
13 Friday	
02 16	☽ # ♄
04 31	☽ ⚹ ☿
05 58	☽ Q ♅
07 50	☽ ☌ ♇
15 22	☽ ∠ ♀
18 08	☽ □ ☉
23 06	☽ ∠ ♆
14 Saturday	
00 03	☽ bQ ♃
02 07	☽ Q ☿
03 20	☽ ⚹ ♅
07 54	☽ ☍ ♄
09 35	☿ # ♂
15 25	☽ ⊥ ♆
16 39	☽ ± ♃
19 38	☿ St R
20 07	☽ ⚹ ♀
15 Sunday	
00 39	♂ Q ♄
02 27	☽ ⊻ ♆
03 53	☽ ⊼ ♃
06 18	☽ △ ♂
06 47	☽ ∠ ♅
11 11	☽ □ ☿
14 32	☽ ⊻ ♇
19 43	☽ Q ♀
23 35	☽ ⊥ ♅
16 Monday	
01 49	☽ ⊥ ♇
05 58	☽ △ ☉
10 55	☽ ⊻ ♅
12 00	☽ bQ ♂
13 12	☉ bQ ♆
13 13	♀ ∥ ♆
16 05	☽ ⊼ ♄
18 57	☽ ∠ ♇
17 Tuesday	
00 56	♀ # ♃
03 37	☽ ± ♄
06 10	☽ ± ♂
07 59	☽ □ ♀
11 14	☽ ☌ ♆
11 28	☽ # ♄
13 09	☽ bQ ☉
13 40	☽ ☍ ♃
18 24	☽ ⊼ ♂
19 30	☽ △ ☿
21 06	☽ bQ ♄
21 18	☉ ∠ ♃
23 57	☽ ⚹ ♇
18 Wednesday	
08 10	☽ ± ☉
08 22	☿ ⊻ ♂
12 48	☽ ∥ ♀
16 30	☽ # ♃
16 59	☽ ∥ ♆
20 14	☉ ⊼ ♅
20 54	☽ ☌ ♅
20 57	☽ ⊼ ☉
23 29	☽ Q ♇
23 56	☽ bQ ☿
19 Thursday	
02 35	☽ △ ♄
17 00	☽ ± ☿
18 35	☽ ∥ ♅
21 19	♀ □ ♆
21 47	☽ ∥ ♇
22 02	☽ ⊻ ♆
22 04	☽ △ ♀
20 Friday	
01 32	☽ ⊼ ♃
04 21	♃ # ♆
04 23	☽ ⊼ ☿
08 47	☽ ☍ ♂
09 58	☽ ⊥ ♆
11 14	☽ □ ♇
13 41	☽ ± ♃
22 22	♀ ⊥ ♇
23 41	☽ ∥ ☿
21 Saturday	
03 32	☽ # ♂
03 58	☽ ∠ ♆
05 38	☽ bQ ♀
06 46	♂ ± ♆
08 00	☽ bQ ♃
08 27	☽ ⊻ ♅
13 59	☽ ☍ ☉
14 36	☽ □ ♄
14 44	☿ bQ ♅
18 44	☿ ⚹ ♃
20 29	☽ ⊥ ♅
21 47	☉ □ ♄
22 Sunday	
00 41	☽ ± ♀
08 12	♂ □ ♇
10 09	☽ ⚹ ♆
11 00	☿ ⊻ ♀
12 34	☽ # ☉
13 08	☽ ☍ ☿
13 24	☽ ⊼ ♀
14 37	☽ ∠ ♅
14 43	☽ △ ♃
14 54	☽ ∥ ☉
23 42	☽ △ ♇
23 Monday	
00 34	☽ ⊼ ♂
03 05	☽ Q ♄
10 25	☽ Q ♆
13 25	☽ ± ♂
20 55	☽ ∥ ♂
20 56	☽ ⚹ ♅
21 34	☽ # ☿
24 Tuesday	
03 29	☽ ⚹ ♄
03 46	☿ △ ♆
06 10	☽ bQ ♇
07 53	♀ □ ♃
08 10	☽ ⊼ ☉
08 18	☿ # ♂
08 43	☽ bQ ♂
18 58	☉ ± ♅
21 13	☽ Q ♅
21 23	☽ ⊼ ☿
21 26	☽ ± ☉
22 53	☽ □ ♆
25 Wednesday	
00 26	☽ ± ♇
04 28	☽ □ ♃
05 04	☽ ☍ ♀
06 23	☽ # ♇
08 32	☽ ± ☿
09 41	☽ # ♅
09 58	☽ ∠ ♄
12 36	☽ ⊼ ♇
15 49	♂ ⊥ ♃
16 49	☽ △ ♂
17 16	☽ bQ ☉
26 Thursday	
01 10	☽ bQ ☿
04 09	☽ ⊥ ♄
09 20	☽ ∥ ♃
09 27	☽ □ ♅
09 36	☿ ⊥ ♀
10 26	☉ Q ♇
11 18	☽ # ♆
16 12	☽ ⊻ ♄
17 14	☽ Q ♃
27 Friday	
02 00	☽ △ ☉
04 38	☽ △ ☿
09 17	☽ # ♀
11 00	☽ △ ♆
16 31	☽ ∥ ♄
17 22	☽ ⚹ ♃
18 31	☉ ☌ ☿
19 28	☽ ⊼ ♀
28 Saturday	
00 28	☽ ☍ ♇
07 36	☽ ± ♀
07 42	☽ □ ♂
16 16	☽ bQ ♆
20 19	☽ △ ♅
22 22	☿ Q ♇
22 53	☽ ∠ ♃
29 Sunday	
01 31	☽ bQ ♀
03 01	☽ ☌ ♄
09 25	☽ ± ♆
10 07	☽ □ ☿
16 09	☽ ⊥ ♃
17 03	☽ □ ☉
19 49	☽ Q ♂
20 40	☽ ⊼ ♆
22 39	☉ ∥ ☿
30 Monday	
00 33	☽ bQ ♅
03 28	☽ ⊻ ♃
06 28	☽ △ ♀
09 27	☽ ⊼ ♇
17 00	☽ ± ♅
17 03	☽ Q ☿
17 43	☿ ± ♅
18 59	☽ ⚹ ♂
20 22	☽ ± ♇

OCTOBER 2002

LONGITUDES

Date	Sidereal time h m s	Sun ☉ ° ' "	Moon ☽ ° ' "	Moon ☽ 24.00 ° ' "	Mercury ☿ ° '	Venus ♀ ° '	Mars ♂ ° '	Jupiter ♃ ° '	Saturn ♄ ° '	Uranus ♅ ° '	Neptune ♆ ° '	Pluto ♇ ° '
01	12 40 12	08 ♎ 08 09	00 ♌ 00 59	06 ♌ 53 25	00 ♎ 42	14 ♏ 02	20 ♍ 56	12 ♌ 12	29 ♊ 01	25 ♒ 22	08 ♒ 18	15 ♐ 15
02	12 44 09	09 07 11	13 ♌ 53 19	21 ♌ 00 38	29 ♍ 55	14 21	21 34	12 22	29 01	25 R 20	08 R 17	15 16
03	12 48 05	10 06 14	28 ♌ 15 05	05 ♍ 36 12	29 R 17	14 37	22 12	12 32	29 02	25 19	08 16	15 18
04	12 52 02	11 05 20	13 ♍ 03 14	20 ♍ 35 15	28 47	14 52	22 50	12 42	29 02	25 17	08 16	15 19
05	12 55 58	12 04 28	28 ♍ 11 06	05 ♎ 49 29	28 28	15 05	23 29	12 51	29 03	25 16	08 15	15 20
06	12 59 55	13 03 39	13 ♎ 28 59	21 ♎ 08 11	28 20	15 16	24 07	13 01	29 04	25 15	08 15	15 21
07	13 03 51	14 02 51	28 ♎ 45 40	06 ♏ 20 08	28 D 22	15 24	24 45	13 10	29 04	25 13	08 15	15 23
08	13 07 48	15 02 05	13 ♏ 50 26	21 ♏ 15 34	28 34	15 31	25 23	13 19	29 05	25 12	08 14	15 24
09	13 11 45	16 01 21	28 ♏ 34 47	05 ♐ 47 34	28 57	15 35	26 02	13 29	29 05	25 11	08 14	15 25
10	13 15 41	17 00 39	12 ♐ 53 35	19 ♐ 52 41	29 ♍ 30	15 36	26 40	13 38	29 05	25 10	08 13	15 27
11	13 19 38	17 59 59	26 ♐ 44 56	03 ♑ 30 29	00 ♎ 11	15 R 36	27 18	13 47	29 05	25 08	08 13	15 28
12	13 23 34	18 59 21	10 ♑ 09 39	16 ♑ 42 49	01 01	15 33	27 56	13 55	29 R 05	25 07	08 13	15 30
13	13 27 31	19 58 44	23 ♑ 10 26	29 ♑ 32 58	01 59	15 28	28 35	14 04	29 05	25 06	08 13	15 31
14	13 31 27	20 58 09	05 ♒ 50 56	12 ♒ 04 49	03 04	15 20	29 13	14 13	29 05	25 05	08 12	15 33
15	13 35 24	21 57 36	18 ♒ 15 07	24 ♒ 22 19	04 15	15 10	29 ♍ 51	14 21	29 04	25 04	08 12	15 34
16	13 39 20	22 57 05	00 ♓ 26 52	06 ♓ 29 09	05 31	14 57	00 ♎ 29	14 29	29 04	25 03	08 12	15 36
17	13 43 17	23 56 35	12 ♓ 29 36	18 ♓ 28 31	06 52	14 42	01 08	14 37	29 03	25 02	08 12	15 37
18	13 47 14	24 56 07	24 ♓ 26 14	00 ♈ 23 02	08 16	14 25	01 46	14 46	29 02	25 02	08 12	15 39
19	13 51 10	25 55 41	06 ♈ 19 11	12 ♈ 14 54	09 44	14 05	02 24	14 53	29 02	25 01	08 12	15 41
20	13 55 07	26 55 17	18 ♈ 10 25	24 ♈ 05 57	11 15	13 43	03 02	15 01	29 01	25 00	08 12	15 42
21	13 59 03	27 54 55	00 ♉ 01 41	05 ♉ 57 50	12 48	13 19	03 41	15 09	29 00	24 59	08 D 12	15 44
22	14 03 00	28 54 35	11 ♉ 54 38	17 ♉ 52 19	14 23	12 53	04 19	15 16	28 58	24 59	08 12	15 46
23	14 06 56	29 ♎ 54 17	23 ♉ 51 08	29 ♉ 51 22	16 00	12 25	04 57	15 24	28 57	24 58	08 12	15 47
24	14 10 53	00 ♏ 54 01	05 ♊ 53 21	11 ♊ 57 24	17 38	11 55	05 36	15 31	28 56	24 57	08 12	15 49
25	14 14 49	01 53 48	18 ♊ 03 55	24 ♊ 13 17	19 16	11 24	06 14	15 38	28 54	24 57	08 12	15 51
26	14 18 46	02 53 36	00 ♋ 25 57	06 ♋ 42 22	20 55	10 51	06 52	15 45	28 53	24 56	08 12	15 53
27	14 22 43	03 53 27	13 ♋ 03 00	19 ♋ 28 19	22 35	10 17	07 31	15 52	28 51	24 56	08 13	15 54
28	14 26 39	04 53 20	25 ♋ 58 46	02 ♌ 34 46	24 15	09 42	08 09	15 59	28 49	24 56	08 13	15 56
29	14 30 36	05 53 15	09 ♌ 16 41	16 ♌ 04 49	25 55	09 06	08 47	16 05	28 47	24 55	08 13	15 58
30	14 34 32	06 53 12	22 ♌ 59 22	00 ♍ 00 24	27 35	08 30	09 26	16 11	28 45	24 55	08 13	16 00
31	14 38 29	07 ♏ 53 11	07 ♍ 07 50	14 ♍ 21 26	29 ♎ 15	07 ♏ 54	10 ♎ 04	16 ♌ 18	28 ♊ 43	24 ♒ 55	08 ♒ 14	16 ♐ 02

Date	Moon True ☊ ° '	Moon Mean ☊ ° '	Moon ☽ Latitude ° '
01	10 ♊ 54	11 ♊ 53	03 N 51
02	10 R 49	11 50	04 32
03	10 43	11 46	04 57
04	10 34	11 43	05 04
05	10 25	11 40	04 50
06	10 17	11 37	04 15
07	10 10	11 34	03 22
08	10 05	11 30	02 16
09	10 02	11 27	01 N 01
10	10 D 02	11 24	00 S 15
11	10 02	11 21	01 29
12	10 03	11 18	02 35
13	10 R 04	11 15	03 31
14	10 02	11 11	04 15
15	09 59	11 08	04 46
16	09 54	11 05	05 04
17	09 46	11 02	05 07
18	09 38	10 59	04 58
19	09 29	10 56	04 35
20	09 21	10 52	04 01
21	09 14	10 49	03 16
22	09 09	10 46	02 22
23	09 06	10 43	01 22
24	09 04	10 40	00 S 17
25	09 D 05	10 36	00 N 49
26	09 06	10 33	01 54
27	09 08	10 30	02 55
28	09 09	10 27	03 48
29	09 R 09	10 24	04 31
30	09 08	10 21	05 01
31	09 ♊ 06	10 ♊ 17	05 N 13

DECLINATIONS

Date	Sun ☉ ° '	Moon ☽ ° '	Mercury ☿ ° '	Venus ♀ ° '	Mars ♂ ° '	Jupiter ♃ ° '	Saturn ♄ ° '	Uranus ♅ ° '	Neptune ♆ ° '	Pluto ♇ ° '
01	03 S 14	23 N 54	01 S 34	22 S 08	04 N 37	17 N 35	22 N 08	13 S 47	18 S 08	13 S 11
02	03 37	21 00	00 57	22 19	04 22	17 32	22 08	13 48	18 08	13 11
03	04 00	16 44	00 S 23	22 29	04 06	17 29	22 08	13 48	18 08	13 12
04	04 23	11 20	00 N 07	22 39	03 51	17 27	22 07	13 49	18 08	13 12
05	04 46	05 N 09	00 32	22 47	03 36	17 24	22 07	13 49	18 09	13 13
06	05 09	01 S 24	00 52	22 55	03 21	17 22	22 07	13 50	18 09	13 13
07	05 32	07 53	01 06	23 01	03 05	17 19	22 07	13 50	18 09	13 14
08	05 55	13 50	01 16	23 07	02 50	17 17	22 07	13 51	18 09	13 14
09	06 18	18 51	01 19	23 11	02 35	17 14	22 07	13 51	18 09	13 15
10	06 41	22 36	01 18	23 15	02 20	17 12	22 07	13 51	18 09	13 15
11	07 04	24 53	01 11	23 17	02 04	17 09	22 07	13 52	18 09	13 15
12	07 26	25 37	01 00	23 18	01 49	17 07	22 07	13 52	18 09	13 16
13	07 49	24 55	00 44	23 18	01 33	17 05	22 07	13 52	18 10	13 17
14	08 11	22 56	00 25	23 17	01 18	17 02	22 07	13 53	18 10	13 17
15	08 33	19 53	00 N 02	23 14	01 03	17 00	22 07	13 53	18 10	13 18
16	08 55	16 03	00 S 25	23 11	00 47	16 58	22 07	13 53	18 10	13 18
17	09 17	11 36	00 54	23 05	00 32	16 56	22 07	13 54	18 10	13 19
18	09 39	06 46	01 26	22 59	00 17	16 53	22 07	13 54	18 10	13 19
19	10 01	01 S 42	01 59	22 51	00 N 01	16 51	22 07	13 54	18 10	13 20
20	10 23	03 N 25	02 35	22 41	00 S 14	16 49	22 07	13 54	18 10	13 20
21	10 44	08 26	03 12	22 30	00 29	16 47	22 07	13 54	18 10	13 20
22	11 05	13 09	03 50	22 18	00 45	16 45	22 06	13 55	18 10	13 21
23	11 26	17 25	04 30	22 04	01 00	16 43	22 06	13 55	18 10	13 21
24	11 47	21 00	05 09	21 49	01 15	16 41	22 06	13 55	18 10	13 22
25	12 08	23 43	05 50	21 33	01 31	16 39	22 06	13 55	18 10	13 22
26	12 29	25 20	06 31	21 16	01 46	16 37	22 06	13 55	18 10	13 23
27	12 49	25 42	07 12	20 57	02 01	16 35	22 06	13 55	18 10	13 23
28	13 09	24 41	07 53	20 37	02 17	16 34	22 06	13 55	18 10	13 24
29	13 29	22 18	08 34	20 16	02 32	16 32	22 06	13 56	18 10	13 24
30	13 49	18 35	09 15	19 54	02 47	16 30	22 06	13 56	18 10	13 25
31	14 S 08	13 N 44	09 S 56	19 S 31	03 S 03	16 N 28	22 N 06	13 S 56	18 S 09	13 S 25

ZODIAC SIGN ENTRIES

Date	h	m	Planets	
01	11	58	☽	♌
02	09	26	☿	♍
03	14	52	☽	♍
05	14	51	☽	♎
07	13	57	☽	♏
09	14	21	☽	♐
11	05	56	☿	♎
11	17	45	☽	♑
14	00	51	☽	♒
15	17	38	♂	♎
16	11	07	☽	♓
18	23	13	☽	♈
21	11	57	☽	♉
23	14	18	☉	♏
24	00	17	☽	♊
26	11	10	☽	♋
28	19	20	☽	♌
30	23	59	☽	♍
31	22	43	☿	♏

LATITUDES

Date	Mercury ☿ ° '	Venus ♀ ° '	Mars ♂ ° '	Jupiter ♃ ° '	Saturn ♄ ° '	Uranus ♅ ° '	Neptune ♆ ° '	Pluto ♇ ° '
01	01 S 24	06 S 23	01 N 06	00 N 28	01 S 19	00 S 46	00 N 04	09 N 30
04	00 S 54	06 39	01 06	00 28	01 19	00 46	00 03	09 29
07	00 N 30	06 53	01 06	00 28	01 19	00 46	00 03	09 28
10	01 12	07 03	01 05	00 29	01 19	00 46	00 03	09 27
13	01 40	07 09	01 05	00 29	01 19	00 46	00 03	09 26
16	01 56	07 11	01 04	00 30	01 19	00 46	00 03	09 25
19	02 02	07 07	01 04	00 31	01 20	00 46	00 03	09 24
22	01 59	06 56	01 03	00 31	01 20	00 45	00 03	09 23
25	01 51	06 38	01 03	00 32	01 20	00 45	00 03	09 22
28	01 38	06 13	01 02	00 32	01 20	00 45	00 03	09 21
31	01 N 22	05 S 42	01 N 02	00 N 33	01 S 20	00 S 45	00 N 03	09 N 20

DATA

Julian Date	2452549
Delta T	+64 seconds
Ayanamsa	23° 53' 27"
Synetic vernal point	05° ♓ 13' 32"
True obliquity of ecliptic	23° 26' 24"

MOON'S PHASES, APSIDES AND POSITIONS ☽

Date	h	m	Phase	Longitude ° '	Eclipse Indicator
06	11	18	●	13 ♎ 02	
13	05	33	☽	19 ♑ 43	
21	07	20	○	27 ♈ 43	
29	05	28	☾	05 ♌ 37	

Day	h	m		
06	13	10	Perigee	
20	04	25	Apogee	
06	06	55	0S	
12	11	56	Max dec	25° S 37'
19	19	58	0N	
27	06	31	Max dec	25° N 44'

LONGITUDES

Date	Chiron ⚷ ° '	Ceres ⚳ ° '	Pallas ⚴ ° '	Juno ⚵ ° '	Vesta ⚶ ° '	Black Moon Lilith ⚸ ° '
01	03 ♑ 38	11 ♈ 33	10 ♒ 49	09 ♎ 36	01 ♍ 44	15 ♈ 18
11	04 ♑ 02	09 ♈ 17	10 ♒ 39	13 ♎ 18	06 ♍ 07	16 ♈ 25
21	04 ♑ 33	07 ♈ 12	11 ♒ 03	16 ♎ 58	10 ♍ 25	17 ♈ 32
31	05 ♑ 12	05 ♈ 31	11 ♒ 56	20 ♎ 34	14 ♍ 38	18 ♈ 39

ASPECTARIAN

h m	Aspects
01 Tuesday	
03 46	☽ ⊼ ♅
04 35	☽ Q ☉
08 19	☉ ⊥ ♀
08 38	♀ ⚻ ♄
10 11	☽ ⊻ ♄
10 30	♀ # ♄
12 25	☽ ⚻ ♇
13 09	☽ ⚹ ☿
15 50	☉ △ ♆
20 43	☽ ⊥ ♄
22 50	☽ ∠ ♂
02 Wednesday	
02 25	☽ ☍ ♆
02 59	☽ # ♀
03 14	☽ ⚹ ☉
03 58	☽ ∥ ♄
09 23	☽ ☌ ♃
12 12	☽ ∠ ♄
12 48	☽ □ ♀
13 40	☽ ∠ ☿
14 21	☽ △ ♇
14 59	☽ ⊥ ♂
03 Thursday	
01 31	☽ ⊻ ♂
02 26	☿ ∠ ♀
04 06	☽ ⊥ ☿
04 47	☽ # ♆
06 26	☽ ∠ ☉
07 10	☽ ☍ ♅
08 08	☽ ∥ ♃
13 16	☽ ⚹ ♄
13 37	☽ ⊻ ☿
15 59	☉ # ♂
17 00	☉ ⚻ ♅
19 17	☽ Q ♀
22 15	☽ ⊥ ☉
23 08	☿ □ ♄
04 Friday	
01 31	☽ # ♅
04 11	☽ # ♇
04 19	☽ ⊼ ♆
08 38	☽ ⊻ ☉
08 46	☽ Q ♄
11 25	☽ ⊻ ♃
13 56	☽ ± ♆
14 57	☽ ⚹ ♀
15 37	☽ □ ♇
21 06	☽ ⊥ ♃
05 Saturday	
03 52	♂ ⚻ ♆
04 14	☽ ⚻ ♆
04 15	☽ ☌ ♂
07 24	☽ ⊼ ♅
11 29	☽ ∠ ♃
12 27	☽ ☌ ☿
13 21	☽ # ☉
13 22	☽ □ ♄
15 01	☽ ∠ ♀
16 50	☽ ± ♅
18 00	☽ ∥ ♂
20 06	☽ Q ♇
06 Sunday	
03 48	☽ △ ♆
04 08	☽ ∥ ☿
05 19	☽ ⊥ ♀
06 56	☽ ⚻ ♅
09 58	☽ # ☿
10 38	☉ ⚹ ♃
11 15	☽ ⚹ ♃
11 18	☽ ☌ ☉
14 49	☽ ⊻ ♀
14 56	☽ ⚹ ♇
18 50	☽ # ♂
19 26	☿ St D
07 Monday	
02 39	☽ ∥ ☉
05 24	☽ ⊻ ♂
06 08	♀ ⊻ ♇
06 17	☽ Q ♃
06 26	☽ △ ♅
11 22	☽ ⊻ ☿
12 29	☽ △ ♄
14 33	☽ ∠ ♇
15 17	☽ ⊥ ♂
20 57	☽ ⊥ ☿
08 Tuesday	
03 02	☽ □ ♆
04 52	☽ ⊥ ♇
05 10	♂ ⊼ ♅
06 13	☽ ∠ ♂
09 25	☽ ∥ ♇
11 10	☽ □ ♃
11 34	☽ ∠ ☿
12 02	☽ ∥ ♅
12 23	☽ ⚻ ♄
14 03	☽ ⊻ ☉
14 31	☽ ⊻ ♇
14 42	☽ ☌ ♀
21 05	☉ ⚹ ♇
09 Wednesday	
00 28	☽ ⊥ ☉
00 31	☉ ⊻ ♀
02 58	☽ ± ♄
03 44	☽ # ♃
06 25	☽ □ ♅
07 36	☽ ⚹ ♂
08 08	☽ Q ♆
08 18	☽ ∥ ♆
12 38	☽ ⚹ ☿
12 50	☽ ⊼ ♄
16 20	☽ ∠ ☉
18 28	☿ □ ♄
10 Thursday	
04 05	☽ ⚹ ♆
04 29	☽ Q ♂
08 20	☽ # ♄
09 31	☽ Q ☿
12 27	☽ Q ♅
13 16	☽ △ ♃
16 22	☽ ☌ ♇
16 39	☽ ⊻ ♀
17 30	☽ ∥ ♀
18 35	♀ St R
19 35	☽ ⚹ ☉
11 Friday	
03 00	☽ ⊥ ♀
05 49	☽ ∠ ♆
09 11	☽ ⚹ ♅
13 01	☽ □ ♂
13 01	♄ St R
15 37	☽ ⚻ ♃
16 08	☽ ☍ ♄
18 12	☽ Q ☉
18 28	☽ □ ☿
18 48	☽ ∠ ♀
21 41	☽ ⊥ ♆
23 44	☿ ∠ ♀
12 Saturday	
07 54	☽ ± ♃
08 28	☽ ⊻ ♆
11 56	☽ ∠ ♅
14 31	☿ ± ♅
18 57	☽ ⊼ ♃
21 46	☽ ⊻ ♇
21 48	☽ ⚹ ♀
13 Sunday	
01 05	♀ ⊻ ♇
04 26	☽ ⊥ ♅
05 33	☽ □ ☉
08 54	☽ ⊥ ♇
14 24	♃ ∠ ♄
15 37	☽ ⊻ ♅
19 59	☽ Q ♀
22 42	☽ △ ♂
23 07	☽ ⊼ ♄
14 Monday	
01 52	☽ ∠ ♇
06 10	☽ △ ☿
06 56	♂ □ ♄
08 29	☽ ∥ ♀
10 31	☽ ± ♄
11 51	♂ ∠ ♃
16 31	☽ ☌ ♆
19 12	☽ # ♄
22 12	☿ Q ♇
15 Tuesday	
03 52	☽ ⚻ ♄
04 19	☽ ☍ ♃
05 01	☽ ⚻ ♂
06 04	☽ □ ♀
06 46	☽ ⚹ ♇
14 09	☽ ⚻ ☿
19 54	☽ △ ☉
20 59	♀ ∠ ♂
23 19	☽ ∥ ♆
23 35	☽ ± ♂
16 Wednesday	
01 21	☽ ☌ ♅
06 21	☽ Q ♇
06 34	☽ # ♃
09 16	☽ △ ♄
09 56	☽ ± ☿
12 05	☽ ⊼ ♂
23 18	☽ ⊼ ☿
23 58	☽ ∥ ♅
17 Thursday	
00 23	☿ # ♂
03 06	☽ ∥ ♇
03 25	☽ ⊻ ♆
04 16	☽ ⚻ ☉
08 49	♂ ± ♅
15 25	☽ ⊥ ♆
16 19	☽ ⊼ ♃
16 20	☽ △ ♀
16 34	♀ □ ♃
18 17	☽ □ ♇
22 50	☽ ∥ ☉
23 55	☽ ± ☉
18 Friday	
04 31	☽ ± ♃
09 30	☽ ∠ ♆
10 46	☿ △ ♆
13 06	☽ ⊼ ☉
13 11	☽ ⊻ ♅
13 57	☿ ⊥ ♀
14 08	☉ △ ♅
21 17	☽ □ ♄
21 47	☽ ⚻ ♀
22 52	☽ ⚻ ♃
19 Saturday	
01 17	☽ ⊥ ♅
03 38	☽ ☍ ♂
10 47	☽ ∥ ☿
15 28	☽ ± ♀
15 48	☽ ⚹ ♆
16 10	♀ ⚻ ♄
16 21	☿ ⚻ ♅
19 28	☽ ∠ ♅
19 41	☽ ∥ ♂
19 55	☽ ☍ ☿
20 17	☽ # ♂
20 Sunday	
03 15	☽ ⊼ ♀
05 33	☽ △ ♃
06 59	☽ △ ♇
07 33	☽ # ☿
09 39	☽ Q ♄
13 52	♆ St D
14 45	☉ Q ♃
16 06	☽ Q ♆
21 Monday	
01 49	☽ ⚹ ♅
07 20	☽ ☍ ☉
09 55	☽ ⚹ ♄
13 25	☽ ⚻ ♇
14 04	♂ Q ♇
18 10	☿ ⊻ ♀
19 48	☽ ⊼ ♂
22 Tuesday	
00 25	☽ # ☉
02 03	☽ Q ♅
04 30	☽ □ ♆
07 39	☽ ± ♇
08 36	☽ ± ♂
13 03	☽ # ♇
13 31	☉ △ ♄
13 53	☽ ☍ ♀
16 03	☽ # ♅
16 09	☽ ∠ ♄
17 46	☽ ⊼ ☿
18 51	☽ □ ♃
19 46	☽ ⊼ ♇
23 Wednesday	
02 20	☿ ⚹ ♃
03 45	☽ ⚻ ♂
07 42	☽ ± ☿
07 52	☽ ∥ ♃
08 45	♀ # ♄
08 50	☿ ⚹ ♇
10 12	☽ ⊥ ♄
14 14	☽ □ ♅
16 37	☽ # ♆
22 11	☽ ⊻ ♄
24 Thursday	
01 11	☽ ⊼ ☉
04 30	☽ ⚻ ☿
07 15	☽ Q ♃
09 55	☉ ∠ ♇
11 23	☽ △ ♂
14 11	☽ ± ☉
16 35	☽ △ ♆
17 53	☽ # ♀
18 47	♀ ⊥ ♂
20 45	☽ ∥ ♄
23 26	☽ ⊼ ♀
25 Friday	
07 11	☽ ⚹ ♃
07 38	☽ ☍ ♇
09 30	☽ ⚻ ☉
10 45	☽ ± ♀
14 43	☽ △ ☿
22 01	☽ ⚻ ♆
26 Saturday	
01 24	☽ △ ♅
03 32	☽ ⚻ ♀
09 01	☽ ☌ ♄
12 37	☽ ∠ ♃
15 24	☽ ± ♆
17 08	☽ △ ☉
27 Sunday	
00 58	☽ □ ♂
02 51	☽ ⊼ ♆
05 56	☽ ⊥ ♃
06 08	☽ ⚻ ♅
07 01	☽ △ ♀
17 20	☽ ⊻ ♃
17 22	☽ ⊼ ♇
23 00	☽ ± ♅
28 Monday	
00 02	♃ △ ♇
02 50	♀ ⊥ ♇
04 33	☽ ± ♇
08 22	☽ □ ☿
10 04	☽ ⊼ ♅
12 19	☽ Q ♂
14 27	♂ △ ♆
17 11	☽ ⊻ ♄
21 03	☽ ⚻ ♇
21 43	☿ △ ♅
29 Tuesday	
04 00	☽ ⊥ ♄
05 28	☽ □ ☉
06 07	☉ ∥ ♇
10 07	☽ ☍ ♆
11 05	☽ ⚹ ♂
11 42	☽ □ ♀
13 28	☽ ∥ ♄
18 09	♀ ⊻ ♂
19 58	☽ ∠ ♄
21 21	☽ Q ☿
23 50	☽ △ ♇
30 Wednesday	
00 06	☽ ☌ ♃
03 33	☽ # ♀
14 19	☽ # ♆
14 35	☽ ∠ ♂
15 19	☽ ☍ ♅
15 31	☽ Q ☉
17 47	☽ Q ♀
20 18	☉ ∥ ♅
20 57	☽ ⚹ ☿
21 15	☿ Q ♃
21 51	☽ ⚹ ♄
22 53	♀ □ ♆
22 59	☽ ∥ ♃
31 Thursday	
04 28	☿ △ ♄
06 21	♂ ⚻ ♅
06 37	☽ ⊥ ♂
10 14	☽ # ☉
11 06	☽ # ♅
12 06	☉ ☌ ♀
13 13	☽ ⚹ ♀
13 21	☽ ⚹ ☉
13 22	☽ # ♇
13 50	☽ ⊼ ♆
17 07	☽ ⊻ ♂
17 58	☽ Q ♄
20 16	☉ □ ♆
23 48	☽ ± ♆

All ephemeris data is given at 12.00 UT and the Moon's longitude is additionally given for 24.00 UT

NOVEMBER 2002

LONGITUDES

Date	Sidereal time h m s	Sun ☉ ° ' "	Moon ☽ ° ' "	Moon ☽ 24.00 ° ' "	Mercury ☿ ° '	Venus ♀ ° '	Mars ♂ ° '	Jupiter ♃ ° '	Saturn ♄ ° '	Uranus ♅ ° '	Neptune ♆ ° '	Pluto ♇ ° '
01	14 42 25	08 ♏ 53 13	21 ♍ 40 47	29 ♍ 05 13	00 ♏ 55	07 ♏ 17	10 ♎ 42	16 ♌ 24	28 ♊ 41	24 ♒ 55	08 ♒ 14	16 ♐ 04
02	14 46 22	09 53 17	06 ♎ 33 58	14 ♎ 06 03	02 35	06 R 41	11 21	16 30	28 R 39	24 R 55	08 15	16 06
03	14 50 18	10 53 22	21 ♎ 40 21	29 ♎ 15 39	04 15	06 05	11 59	16 35	28 36	24 55	08 15	16 08
04	14 54 15	11 53 30	06 ♏ 50 42	14 ♏ 24 15	05 54	05 30	12 37	16 41	28 34	24 D 54	08 15	16 10
05	14 58 12	12 53 39	21 ♏ 55 05	29 ♏ 22 07	07 33	04 56	13 16	16 46	28 31	24 55	08 16	16 12
06	15 02 08	13 53 51	06 ♐ 44 23	14 ♐ 01 08	09 11	04 23	13 54	16 52	28 28	24 55	08 17	16 14
07	15 06 05	14 54 04	21 ♐ 11 44	28 ♐ 15 47	10 49	03 51	14 33	16 57	28 26	24 55	08 17	16 16
08	15 10 01	15 54 19	05 ♑ 13 03	12 ♑ 03 29	12 27	03 21	15 11	17 02	28 23	24 55	08 18	16 18
09	15 13 58	16 54 35	18 ♑ 47 09	25 ♑ 24 16	14 05	02 52	15 49	17 06	28 20	24 55	08 18	16 20
10	15 17 54	17 54 53	01 ♒ 55 10	08 ♒ 20 15	15 42	02 26	16 28	17 11	28 17	24 56	08 19	16 22
11	15 21 51	18 55 12	14 ♒ 39 59	20 ♒ 54 52	17 19	02 01	17 06	17 15	28 14	24 56	08 20	16 24
12	15 25 47	19 55 32	27 ♒ 05 27	03 ♓ 12 16	18 55	01 39	17 45	17 20	28 10	24 56	08 21	16 26
13	15 29 44	20 55 54	09 ♓ 15 53	15 ♓ 16 51	20 31	01 18	18 23	17 24	28 07	24 57	08 21	16 28
14	15 33 41	21 56 18	21 ♓ 15 42	27 ♓ 12 55	22 07	01 00	19 02	17 28	28 04	24 57	08 22	16 30
15	15 37 37	22 56 42	03 ♈ 08 23	09 ♈ 04 21	23 43	00 45	19 40	17 31	28 00	24 58	08 23	16 32
16	15 41 34	23 57 08	14 ♈ 59 27	20 ♈ 54 39	25 18	00 32	20 19	17 35	27 56	24 58	08 24	16 34
17	15 45 30	24 57 36	26 ♈ 50 18	02 ♉ 46 43	26 53	00 21	20 57	17 38	27 53	24 59	08 25	16 37
18	15 49 27	25 58 05	08 ♉ 44 12	14 ♉ 43 01	28 ♏ 28	00 13	21 35	17 41	27 49	25 00	08 26	16 39
19	15 53 23	26 58 35	20 ♉ 43 23	26 ♉ 45 32	00 ♐ 02	00 07	22 14	17 44	27 45	25 00	08 27	16 41
20	15 57 20	27 59 07	02 ♊ 49 40	08 ♊ 55 59	01 36	00 04	22 52	17 47	27 41	25 01	08 28	16 43
21	16 01 16	28 ♏ 59 41	15 ♊ 04 39	21 ♊ 15 52	03 10	00 D 03	23 31	17 50	27 37	25 02	08 29	16 45
22	16 05 13	00 ♐ 00 16	27 ♊ 29 50	03 ♋ 46 42	04 44	00 05	24 09	17 52	27 33	25 03	08 30	16 48
23	16 09 10	01 00 53	10 ♋ 06 42	16 ♋ 30 01	06 18	00 09	24 48	17 54	27 29	25 04	08 31	16 50
24	16 13 06	02 01 31	22 ♋ 56 51	29 ♋ 27 25	07 51	00 16	25 26	17 56	27 25	25 05	08 32	16 52
25	16 17 03	03 02 11	06 ♌ 01 55	12 ♌ 40 32	09 25	00 24	26 05	17 58	27 21	25 06	08 33	16 54
26	16 20 59	04 02 52	19 ♌ 23 27	26 ♌ 10 49	10 58	00 35	26 43	18 00	27 17	25 07	08 35	16 56
27	16 24 56	05 03 35	03 ♍ 02 42	09 ♍ 59 11	12 31	00 49	27 22	18 01	27 12	25 08	08 36	16 59
28	16 28 52	06 04 20	17 ♍ 00 11	24 ♍ 05 37	14 04	01 04	28 00	18 03	27 08	25 10	08 37	17 01
29	16 32 49	07 05 06	01 ♎ 15 15	08 ♎ 28 46	15 37	01 21	28 39	18 04	27 03	25 11	08 38	17 03
30	16 36 45	08 ♐ 05 54	15 ♎ 45 42	23 ♎ 05 30	17 ♐ 10	01 ♏ 41	29 ♎ 18	18 ♌ 05	26 ♊ 59	25 ♒ 12	08 ♒ 40	17 ♐ 05

Date	Moon True ☊ ° '	Moon Mean ☊ ° '	Moon ☽ Latitude ° '
01	09 ♊ 02	10 ♊ 14	05 N 05
02	08 R 58	10 11	04 38
03	08 54	10 08	03 50
04	08 50	10 05	02 46
05	08 48	10 02	01 31
06	08 47	09 58	00 N 11
07	08 D 48	09 55	01 S 08
08	08 49	09 52	02 21
09	08 50	09 49	03 23
10	08 52	09 46	04 12
11	08 52	09 42	04 48
12	08 R 52	09 39	05 09
13	08 51	09 36	05 16
14	08 49	09 33	05 08
15	08 47	09 30	04 48
16	08 45	09 27	04 15
17	08 42	09 23	03 31
18	08 41	09 20	02 38
19	08 39	09 17	01 38
20	08 39	09 14	00 S 32
21	08 D 39	09 11	00 N 36
22	08 39	09 08	01 43
23	08 40	09 04	02 46
24	08 41	09 01	03 42
25	08 41	08 58	04 28
26	08 42	08 55	05 00
27	08 42	08 52	05 16
28	08 R 42	08 48	05 14
29	08 42	08 45	04 53
30	08 ♊ 41	08 ♊ 42	04 N 14

DECLINATIONS

Date	Sun ☉ ° '	Moon ☽ ° '	Mercury ☿ ° '	Venus ♀ ° '	Mars ♂ ° '	Jupiter ♃ ° '	Saturn ♄ ° '	Uranus ♅ ° '	Neptune ♆ ° '	Pluto ♇ ° '
01	14 S 28	07 N 58	10 S 37	19 S 08	03 S 18	16 N 27	22 N 06	13 S 56	18 S 09	13 S 26
02	14 47	01 N 38	11 17	18 44	03 33	16 25	22 06	13 56	18 09	13 26
03	15 06	04 S 53	11 56	18 19	03 48	16 24	22 06	13 56	18 09	13 27
04	15 24	11 11	12 35	17 55	04 03	16 22	22 06	13 56	18 09	13 27
05	15 43	16 46	13 14	17 30	04 18	16 21	22 06	13 56	18 09	13 27
06	16 01	21 15	13 52	17 05	04 34	16 19	22 06	13 55	18 09	13 28
07	16 18	24 17	14 29	16 40	04 49	16 18	22 06	13 55	18 09	13 28
08	16 36	25 41	15 05	16 15	05 04	16 17	22 06	13 55	18 09	13 29
09	16 53	25 28	15 41	15 52	05 19	16 15	22 06	13 55	18 08	13 29
10	17 10	23 50	16 16	15 28	05 34	16 14	22 05	13 55	18 08	13 30
11	17 27	21 01	16 50	15 05	05 49	16 13	22 05	13 55	18 08	13 30
12	17 43	17 19	17 23	14 43	06 03	16 12	22 05	13 55	18 08	13 31
13	17 59	12 58	17 56	14 22	06 18	16 11	22 05	13 55	18 08	13 31
14	18 15	08 11	18 27	14 01	06 33	16 10	22 05	13 54	18 07	13 31
15	18 31	03 S 09	18 58	13 42	06 48	16 09	22 05	13 54	18 07	13 32
16	18 46	01 N 59	19 28	13 23	07 03	16 08	22 05	13 54	18 07	13 32
17	19 00	07 04	19 56	13 06	07 17	16 07	22 05	13 54	18 07	13 33
18	19 15	11 55	20 24	12 49	07 32	16 07	22 05	13 53	18 07	13 33
19	19 29	16 22	20 51	12 34	07 46	16 06	22 05	13 53	18 06	13 33
20	19 43	20 12	21 17	12 20	08 01	16 05	22 05	13 53	18 06	13 34
21	19 56	23 12	21 41	12 07	08 15	16 05	22 05	13 52	18 06	13 34
22	20 09	25 08	22 05	11 55	08 30	16 04	22 05	13 52	18 06	13 35
23	20 22	25 49	22 27	11 45	08 44	16 04	22 05	13 52	18 05	13 35
24	20 34	25 08	22 49	11 36	08 59	16 03	22 05	13 51	18 05	13 35
25	20 46	23 05	23 09	11 28	09 13	16 03	22 05	13 51	18 05	13 36
26	20 57	19 45	23 28	11 21	09 27	16 03	22 05	13 51	18 04	13 36
27	21 09	15 18	23 46	11 15	09 41	16 02	22 05	13 50	18 04	13 36
28	21 19	09 57	24 02	11 10	09 55	16 02	22 04	13 50	18 04	13 37
29	21 30	03 59	24 18	11 06	10 09	16 02	22 04	13 49	18 03	13 37
30	21 S 40	02 S 18	24 S 32	11 S 03	10 S 23	16 N 02	22 N 04	13 S 49	18 S 03	13 S 37

ZODIAC SIGN ENTRIES

Date	h	m	Planets
02	01	28	☽ ♎
04	01	10	☽ ♏
06	01	01	☽ ♐
08	02	59	☽ ♑
10	08	27	☽ ♒
12	17	42	☽ ♓
15	05	38	☽ ♈
17	18	23	☽ ♉
19	11	29	☿ ♐
20	06	25	☽ ♊
22	11	54	☉ ♐
22	16	48	☽ ♋
25	01	00	☽ ♌
27	06	42	☽ ♍
29	09	54	☽ ♎

LATITUDES

Date	Mercury ☿ ° '	Venus ♀ ° '	Mars ♂ ° '	Jupiter ♃ ° '	Saturn ♄ ° '	Uranus ♅ ° '	Neptune ♆ ° '	Pluto ♇ ° '
01	01 N 16	05 S 30	01 N 01	00 N 33	01 S 20	00 S 45	00 N 03	09 N 20
04	00 57	04 50	01 01	00 34	01 20	00 45	00 03	09 19
07	00 38	04 07	01 00	00 34	01 20	00 45	00 03	09 18
10	00 N 17	03 22	00 59	00 35	01 20	00 45	00 03	09 18
13	00 S 03	02 36	00 59	00 35	01 20	00 45	00 03	09 17
16	00 22	01 51	00 58	00 36	01 20	00 45	00 03	09 16
19	00 42	01 07	00 57	00 37	01 20	00 44	00 03	09 16
22	01 01	00 S 27	00 56	00 37	01 20	00 44	00 03	09 15
25	01 18	00 N 10	00 55	00 38	01 20	00 44	00 03	09 15
28	01 34	00 44	00 54	00 39	01 20	00 44	00 03	09 14
31	01 S 48	01 N 14	00 N 53	00 N 39	01 S 20	00 S 44	00 N 03	09 N 14

DATA

Julian Date	2452580
Delta T	+64 seconds
Ayanamsa	23° 53' 31"
Synetic vernal point	05° ♓ 13' 29"
True obliquity of ecliptic	23° 26' 24"

MOON'S PHASES, APSIDES AND POSITIONS ☽

Date	h	m	Phase	Longitude ° '	Eclipse Indicator
04	20	34	●	12 ♏ 15	
11	20	52	☽	19 ♒ 17	
20	01	34	○	27 ♉ 33	
27	15	46	☾	05 ♍ 13	

Day	h	m		
04	00	39	Perigee	
16	11	23	Apogee	
02	18	03	0S	
08	20	44	Max dec	25° S 47'
16	02	44	0N	
23	12	14	Max dec	25° N 49'
30	03	17	0S	

LONGITUDES

Date	Chiron ⚷ ° '	Ceres ⚳ ° '	Pallas ⚴ ° '	Juno ⚵ ° '	Vesta ⚶ ° '	Black Moon Lilith ⚸ ° '
01	05 ♑ 16	05 ♈ 22	12 ♒ 03	20 ♎ 56	15 ♍ 03	18 ♈ 46
11	06 ♑ 30	04 ♈ 17	13 ♒ 26	24 ♎ 27	19 ♍ 07	19 ♈ 53
21	06 ♑ 51	03 ♈ 49	15 ♒ 11	27 ♎ 53	23 ♍ 03	21 ♈ 00
31	07 ♑ 45	04 ♈ 00	17 ♒ 16	01 ♏ 12	26 ♍ 48	22 ♈ 07

ASPECTARIAN

h m	Aspects
01 Friday	
01 22	☽ ∠ ☿
02 29	☽ # ☿
02 47	☽ □ ♇
03 18	☽ ⊻ ♃
12 57	☽ ∠ ♀
13 10	☽ ⊥ ♃
14 04	☿ ∠ ♇
14 32	☽ ⚼ ♆
15 51	☽ ∠ ☉
17 15	☽ ⊼ ♅
17 56	☽ ⊥ ☿
23 19	☽ □ ♄
02 Saturday	
02 56	☽ ⊥ ♀
02 56	☽ ± ♅
03 49	☽ ∠ ♃
04 50	☽ ⊻ ☿
05 11	☽ # ♂
07 24	☽ ⊥ ☉
08 02	☽ Q ♇
12 10	☽ ⊻ ♀
14 41	☽ △ ♆
17 06	☉ ⊥ ♇
17 20	☽ ⚼ ♅
17 41	☽ ⊻ ☉
19 57	☽ ☌ ♂
03 Sunday	
03 12	☽ ✶ ♇
03 54	☽ ✶ ♃
07 50	☽ ∥ ♂
17 07	☽ △ ♅
21 58	♀ ∥ ♆
22 56	☽ △ ♄
23 00	☽ Q ♃
04 Monday	
02 59	☽ ∠ ♇
06 26	♅ St D
07 43	☿ ☌ ♀
09 57	☽ ☌ ♀
10 19	☽ ☌ ☿
14 14	☽ □ ♆
17 16	☽ ⊥ ♇
18 25	☽ ∥ ☿
20 34	☽ ☌ ☉
21 18	☽ ∥ ♇
21 35	☽ ⊻ ♂
22 38	☽ ⚼ ♄
23 18	☽ ∥ ♅
05 Tuesday	
02 50	☽ ⊻ ♇
03 43	☽ □ ♃
06 48	☽ ∥ ☉
07 34	☽ ⊥ ♂
10 01	☽ # ♃
12 58	☽ ± ♄
15 11	☽ ∥ ♀
16 48	☽ □ ♅
18 41	☽ ∥ ♆
19 00	☽ Q ♆
20 50	☿ ∥ ♇
22 36	☽ ⊼ ♄
22 37	☿ □ ♆
22 40	☽ ∠ ♂
06 Wednesday	
02 20	☉ ⚼ ♄
08 16	☽ ⊻ ♀
12 27	☉ ⊻ ♂
14 31	☽ ✶ ♆
14 33	☿ ∥ ♅
16 31	☽ ⊻ ☿
17 36	☽ # ♄
17 46	☽ ⊥ ♀
22 10	☽ Q ♅
07 Thursday	
00 21	☽ ✶ ♂
00 41	☽ ⊻ ☉
03 33	☿ ⊥ ♇
03 42	☽ ☌ ♇
03 44	☽ ⊥ ☿
04 50	☽ △ ♃
08 12	☽ ∠ ♀
11 28	☽ ⊥ ☉
11 29	☉ # ♃
15 32	☽ ∠ ♆
18 17	☽ ✶ ♅
20 52	☽ ∠ ☿
21 30	☽ Q ♂
08 Friday	
00 14	☽ ☍ ♄
00 27	☉ ∥ ♀
03 57	☽ ∠ ☉
06 26	☽ ⚼ ♃
06 56	☽ ⊥ ♆
08 52	☽ ✶ ♀
11 04	♀ # ♃
17 23	☽ ⊻ ♆
20 13	☽ ∠ ♅
21 36	☉ ⊻ ♇
22 14	☽ ± ♃
09 Saturday	
01 16	☿ ⚼ ♄
02 26	☽ ✶ ☿
05 14	☽ Q ♀
06 26	☽ □ ♂
07 35	☽ ⊻ ♇
08 22	☽ ✶ ☉
08 58	☽ ⊼ ♃
12 15	☽ ⊥ ♅
16 26	☿ ∥ ♀
17 06	☉ □ ♃
18 25	☽ ⊥ ♇
23 07	☽ ⊻ ♅
10 Sunday	
03 06	☽ Q ☿
05 18	☽ ⊼ ♄
07 54	♂ ✶ ♇
07 59	☽ Q ☉
10 58	☽ ∠ ♇
11 00	☿ # ♃
12 55	☽ □ ♀
16 22	☽ ± ♄
22 05	☿ ⊻ ♇
23 59	☽ ☌ ♆
11 Monday	
03 50	☽ # ♄
06 55	☿ ⊻ ♂
09 16	☽ ⚼ ♄
11 09	☿ □ ♃
15 19	☽ ✶ ♇
16 55	☽ △ ♂
16 59	☽ ☍ ♃
17 49	☽ □ ☿
18 22	♂ ✶ ♃
20 52	☽ □ ☉
12 Tuesday	
07 06	☽ ∥ ♆
07 48	☽ ☌ ♅
09 44	☽ ∥ ☉
11 37	☽ ∥ ☿
14 06	☽ △ ♄
14 38	☽ Q ♇
18 28	☽ # ♃
20 41	☽ △ ♀
23 43	☽ ⚼ ♂
13 Wednesday	
01 19	♀ ∠ ♇
03 58	☽ ∥ ♀
07 01	☽ ∥ ♅
09 07	☽ ∥ ♇
10 12	☽ ⊻ ♆
17 31	☉ ∥ ☿
18 34	☽ ± ♂
20 58	☿ ∥ ♆
22 10	☽ ⊥ ♆
14 Thursday	
00 24	☉ ∥ ♆
01 42	☽ ⚼ ♀
02 25	☽ □ ♇
04 19	☽ ⊼ ♃
04 39	☉ ☌ ☿
07 15	☽ ⊼ ♂
11 08	☿ ± ♄
13 29	☽ △ ☉
13 59	☽ △ ☿
14 43	☉ ± ♄
16 15	☽ ∠ ♆
16 27	☽ ± ♃
19 22	☽ ± ♀
19 26	☽ ⊻ ♅
19 30	☽ ∥ ♂
20 14	♀ ∥ ♅
15 Friday	
01 38	☽ □ ♄
07 14	☽ ⊼ ♀
07 34	☽ ⊥ ♅
10 43	☽ ⚼ ♃
22 37	☽ ⚼ ☉
22 37	☽ ✶ ♆
16 Saturday	
00 11	♀ ∥ ♇
01 01	☽ ⚼ ☿
01 49	☽ ∠ ♅
07 03	☿ □ ♅
13 55	☽ Q ♄
15 13	☽ △ ♇
17 16	☽ △ ♃
18 34	☽ ± ☉
22 05	☽ ± ☿
22 59	☽ Q ♆
23 24	☽ ☍ ♂
17 Sunday	
04 52	☿ Q ♆
07 51	☽ ⊼ ☉
08 15	☽ ✶ ♅
12 06	☽ ⊼ ☿
12 34	☉ □ ♅
13 09	☽ # ♂
13 46	☿ ⊥ ♂
14 06	☽ ✶ ♄
19 01	☽ ☍ ♀
21 40	☽ ⚼ ♇
18 Monday	
02 36	☿ ⊼ ♄
08 30	☽ Q ♅
11 23	☽ □ ♆
15 51	☽ ± ♇
16 26	☽ # ♀
20 09	☽ ∠ ♄
20 32	☽ # ♇
22 19	☽ # ♅
23 12	☉ Q ♆
19 Tuesday	
03 54	☽ ⊼ ♇
06 01	☽ □ ♃
10 30	☽ ∥ ♃
13 16	☿ ⊻ ♀
14 02	☽ ⊥ ♄
15 10	☽ ⊼ ♂
20 32	☽ □ ♅
22 20	☽ # ♆
20 Wednesday	
01 34	☽ ☍ ☉
01 54	☽ ⊻ ♄
03 45	☽ ± ♂
05 21	☉ ⊼ ♄
06 34	☽ ⊼ ♀
08 27	☽ # ☉
09 14	☽ ☍ ☿
17 51	☽ Q ♃
18 22	☽ ± ♀
20 59	☽ # ☿
22 28	☽ ⚼ ♂
23 06	☽ △ ♆
21 Thursday	
02 09	☽ ∥ ♄
07 13	♀ St D
11 57	☽ ⚼ ♀
15 16	☽ ☍ ♇
17 22	☽ ✶ ♃
22 Friday	
04 18	☽ ⚼ ♆
05 14	☽ △ ♂
07 18	☽ △ ♅
12 04	☿ # ♄
12 07	☽ ☌ ♄
13 58	☉ ⊻ ♀
16 58	☽ △ ♀
17 13	☽ ⊼ ☉
21 35	☽ ± ♆
21 50	☉ ⊥ ♂
22 18	☽ ∠ ♃
23 Saturday	
03 46	☽ ⊼ ☿
05 38	☽ ± ☉
08 59	☽ ⊼ ♆
09 37	☿ ⊥ ♀
11 55	☽ ⚼ ♅
15 23	☽ ⊥ ♃
16 41	☽ ± ☿
22 20	♂ △ ♅
24 Sunday	
00 02	☽ ⚼ ☉
00 39	☽ ⊼ ♇
02 40	☽ ⊻ ♃
04 49	☽ ± ♅
11 48	☽ ⚼ ☿
11 51	☽ ± ♇
15 57	☽ ⊼ ♅
16 51	☽ □ ♂
20 12	☽ ⊻ ♄
22 38	☿ ✶ ♆
25 Monday	
01 37	☽ □ ♀
04 27	☽ ⚼ ♇
06 06	☽ △ ☉
07 08	☽ ⊥ ♄
11 28	☽ # ☿
16 35	☽ ☍ ♆
18 56	☽ △ ☿
20 12	☽ ∥ ♄
23 21	☽ ∠ ♄
26 Tuesday	
03 15	☽ Q ♂
04 37	☽ # ☉
07 37	☽ △ ♇
09 31	☽ ☌ ♃
10 33	☽ Q ♀
21 41	☽ # ♆
22 09	☽ ☍ ♅
27 Wednesday	
01 36	☽ ✶ ♂
01 51	☽ ✶ ♄
06 33	♂ △ ♄
07 57	☿ ∠ ♂
08 02	☽ ✶ ♀
08 19	☽ ∥ ♃
15 46	☽ □ ☉
18 55	☽ # ♅
19 57	☽ # ♇
21 37	☽ ⊼ ♆
21 43	☿ Q ♅
22 36	☽ Q ♄
28 Thursday	
04 51	☽ ∠ ♂
06 23	☽ □ ☿
06 46	☽ # ♀
07 56	☽ ± ♆
10 22	☽ ∠ ♀
12 01	☽ □ ♇
12 08	☽ # ♂
13 46	☽ ⊻ ♃
20 53	☽ ⊥ ♂
23 13	☽ ⚼ ♆
23 56	☽ ⊥ ♃
29 Friday	
00 53	☽ Q ☉
01 49	☽ ⊼ ♅
01 55	☽ ⊥ ♀
05 01	☽ □ ♄
07 27	☽ ⊻ ♂
11 52	☽ ± ♅
12 11	☽ ⊻ ♀
15 01	☽ ∠ ♃
16 24	☽ Q ☿
18 20	☽ Q ♇
21 21	☉ ⊥ ♀
22 25	☽ ✶ ☉
30 Saturday	
00 17	☽ △ ♆
02 27	☿ ∠ ♀
02 50	☽ ⚼ ♅
10 51	☿ ☌ ♇
14 11	☽ ✶ ♇
14 34	☽ ✶ ☿
15 48	☽ ✶ ♃

All ephemeris data is given at 12.00 UT and the Moon's longitude is additionally given for 24.00 UT

DECEMBER 2002

LONGITUDES

Date	Sidereal time h m s	Sun ☉ ° ′ ″	Moon ☽ ° ′ ″	Moon ☽ 24.00 ° ′ ″	Mercury ☿ ° ′	Venus ♀ ° ′	Mars ♂ ° ′	Jupiter ♃ ° ′	Saturn ♄ ° ′	Uranus ♅ ° ′	Neptune ♆ ° ′	Pluto ♇ ° ′
01	16 40 42	09 ♐ 06 43	00 ♏ 27 31	07 ♏ 50 58	18 ♐ 43	02 ♏ 02	29 ♎ 56	18 ♌ 05	26 ♊ 54	25 ♒ 13	08 ♒ 41	17 ♐ 08
02	16 44 39	10 07 34	15 ♏ 15 01	22 ♏ 38 46	20 15	02 26	00 ♏ 35	18 06	26 R 50	25 15	08 42	17 10
03	16 48 35	11 08 26	00 ♐ 01 17	07 ♐ 21 41	21 48	02 51	01 13	18 06	26 45	25 16	08 44	17 12
04	16 52 32	12 09 19	14 ♐ 39 03	21 ♐ 52 36	23 20	03 18	01 52	18 06	26 40	25 18	08 45	17 15
05	16 56 28	13 10 14	29 ♐ 01 38	06 ♑ 05 31	24 52	03 46	02 30	18 R 06	26 36	25 19	08 47	17 17
06	17 00 25	14 11 09	13 ♑ 03 50	19 ♑ 56 14	26 24	04 16	03 09	18 06	26 31	25 21	08 48	17 19
07	17 04 21	15 12 06	26 ♑ 42 31	03 ♒ 22 40	27 56	04 48	03 48	18 05	26 26	25 23	08 50	17 22
08	17 08 18	16 13 03	09 ♒ 56 44	16 ♒ 24 55	29 ♐ 28	05 21	04 26	18 05	26 21	25 24	08 51	17 24
09	17 12 14	17 14 00	22 ♒ 47 31	29 ♒ 04 54	01 ♑ 00	05 56	05 05	18 04	26 16	25 26	08 53	17 26
10	17 16 11	18 14 59	05 ♓ 17 31	11 ♓ 25 54	02 31	06 32	05 43	18 03	26 12	25 28	08 54	17 28
11	17 20 08	19 15 58	17 ♓ 30 34	23 ♓ 32 06	04 02	07 09	06 22	18 01	26 07	25 30	08 56	17 31
12	17 24 04	20 16 57	29 ♓ 31 06	05 ♈ 28 11	05 32	07 48	07 00	18 00	26 02	25 31	08 58	17 33
13	17 28 01	21 17 57	11 ♈ 23 56	17 ♈ 18 56	07 02	08 28	07 39	17 58	25 57	25 33	08 59	17 35
14	17 31 57	22 18 58	23 ♈ 13 46	29 ♈ 08 59	08 32	09 09	08 18	17 56	25 52	25 35	09 01	17 38
15	17 35 54	23 19 59	05 ♉ 05 07	11 ♉ 02 37	10 00	09 51	08 56	17 54	25 47	25 37	09 03	17 40
16	17 39 50	24 21 01	17 ♉ 01 57	23 ♉ 03 31	11 28	10 34	09 35	17 52	25 42	25 39	09 04	17 42
17	17 43 47	25 22 03	29 ♉ 07 40	05 ♊ 14 43	12 55	11 19	10 14	17 50	25 37	25 41	09 06	17 44
18	17 47 43	26 23 06	11 ♊ 24 54	17 ♊ 38 26	14 20	12 04	10 52	17 47	25 32	25 43	09 08	17 47
19	17 51 40	27 24 10	23 ♊ 55 27	00 ♋ 16 02	15 44	12 51	11 31	17 44	25 27	25 46	09 10	17 49
20	17 55 37	28 25 14	06 ♋ 40 16	13 ♋ 08 06	17 06	13 38	12 09	17 41	25 22	25 48	09 12	17 51
21	17 59 33	29 ♐ 26 18	19 ♋ 39 32	26 ♋ 14 27	18 26	14 26	12 48	17 38	25 17	25 50	09 13	17 53
22	18 03 30	00 ♑ 27 23	02 ♌ 52 46	09 ♌ 34 21	19 44	15 15	13 27	17 35	25 12	25 52	09 15	17 56
23	18 07 26	01 28 29	16 ♌ 19 02	23 ♌ 06 40	20 59	16 05	14 05	17 31	25 07	25 55	09 17	17 58
24	18 11 23	02 29 36	29 ♌ 57 06	06 ♍ 50 09	22 10	16 56	14 44	17 27	25 03	25 57	09 19	18 00
25	18 15 19	03 30 43	13 ♍ 45 41	20 ♍ 43 32	23 18	17 48	15 23	17 23	24 58	25 59	09 21	18 02
26	18 19 16	04 31 50	27 ♍ 43 31	04 ♎ 45 30	24 21	18 40	16 01	17 19	24 53	26 02	09 23	18 05
27	18 23 12	05 32 58	11 ♎ 49 18	18 ♎ 54 44	25 18	19 33	16 40	17 15	24 48	26 04	09 25	18 07
28	18 27 09	06 34 07	26 ♎ 01 34	03 ♏ 09 34	26 10	20 27	17 19	17 10	24 43	26 07	09 27	18 09
29	18 31 06	07 35 17	10 ♏ 18 26	17 ♏ 27 49	26 55	21 21	17 57	17 06	24 38	26 09	09 29	18 11
30	18 35 02	08 36 27	24 ♏ 37 21	01 ♐ 46 34	27 32	22 17	18 36	17 01	24 34	26 12	09 31	18 14
31	18 38 59	09 ♑ 37 37	08 ♐ 55 01	16 ♐ 02 09	28 ♑ 00	23 ♏ 12	19 ♏ 15	16 ♌ 56	24 ♊ 29	26 ♒ 14	09 ♒ 33	18 ♐ 16

Date	Moon True ☊ ° ′	Moon Mean ☊ ° ′	Moon ☽ Latitude ° ′
01	08 ♊ 41	08 ♊ 39	03 N 17
02	08 D 42	08 36	02 07
03	08 42	08 33	00 N 48
04	08 R 42	08 29	00 S 33
05	08 41	08 26	01 51
06	08 41	08 23	02 59
07	08 40	08 20	03 56
08	08 39	08 17	04 38
09	08 38	08 14	05 05
10	08 38	08 10	05 17
11	08 37	08 07	05 13
12	08 D 38	08 04	04 56
13	08 38	08 01	04 27
14	08 39	07 58	03 46
15	08 41	07 54	02 56
16	08 42	07 51	01 57
17	08 43	07 48	00 S 53
18	08 R 43	07 45	00 N 15
19	08 43	07 42	01 23
20	08 41	07 39	02 28
21	08 38	07 35	03 27
22	08 35	07 32	04 16
23	08 31	07 29	04 51
24	08 28	07 26	05 11
25	08 26	07 23	05 13
26	08 25	07 19	04 57
27	08 D 25	07 16	04 22
28	08 26	07 13	03 32
29	08 27	07 10	02 28
30	08 28	07 07	01 15
31	08 ♊ 29	07 ♊ 04	00 S 02

DECLINATIONS

Date	Sun ☉ ° ′	Moon ☽ ° ′	Mercury ☿ ° ′	Venus ♀ ° ′	Mars ♂ ° ′	Jupiter ♃ ° ′	Saturn ♄ ° ′	Uranus ♅ ° ′	Neptune ♆ ° ′	Pluto ♇ ° ′
01	21 S 49	08 S 33	24 S 45	11 S 02	10 S 37	16 N 02	22 N 04	13 S 48	18 S 03	13 S 38
02	21 58	14 23	24 57	11 01	10 51	16 02	22 04	13 48	18 02	13 38
03	22 07	19 22	25 07	11 01	11 05	16 02	22 04	13 47	18 02	13 38
04	22 15	23 06	25 16	11 02	11 18	16 02	22 04	13 47	18 02	13 39
05	22 23	25 17	25 23	11 04	11 32	16 03	22 04	13 46	18 01	13 39
06	22 30	25 47	25 29	11 07	11 45	16 03	22 04	13 46	18 01	13 39
07	22 37	24 41	25 34	11 11	11 59	16 03	22 04	13 45	18 01	13 40
08	22 44	22 13	25 38	11 15	12 12	16 04	22 04	13 44	18 00	13 40
09	22 50	18 43	25 39	11 20	12 25	16 04	22 04	13 44	18 00	13 40
10	22 55	14 28	25 40	11 26	12 38	16 05	22 04	13 43	17 59	13 41
11	23 00	09 44	25 39	11 32	12 51	16 05	22 04	13 43	17 59	13 41
12	23 05	04 S 43	25 36	11 39	13 04	16 06	22 04	13 42	17 59	13 41
13	23 09	00 N 25	25 32	11 47	13 17	16 07	22 04	13 41	17 58	13 41
14	23 13	05 31	25 27	11 55	13 30	16 08	22 03	13 41	17 58	13 42
15	23 16	10 27	25 20	12 04	13 43	16 08	22 03	13 40	17 57	13 42
16	23 19	15 03	25 12	12 13	13 55	16 09	22 03	13 39	17 57	13 42
17	23 22	19 06	25 02	12 22	14 08	16 10	22 03	13 38	17 56	13 42
18	23 23	22 24	24 51	12 32	14 20	16 11	22 03	13 38	17 56	13 43
19	23 25	24 41	24 38	12 43	14 33	16 12	22 03	13 37	17 55	13 43
20	23 26	25 44	24 24	12 54	14 45	16 14	22 03	13 36	17 55	13 43
21	23 26	25 25	24 09	13 05	14 57	16 15	22 03	13 35	17 54	13 43
22	23 26	23 39	23 52	13 16	15 09	16 16	22 03	13 35	17 54	13 44
23	23 26	20 34	23 35	13 28	15 21	16 17	22 03	13 34	17 54	13 44
24	23 25	16 20	23 16	13 40	15 33	16 19	22 03	13 33	17 53	13 44
25	23 24	11 12	22 57	13 52	15 45	16 20	22 03	13 32	17 53	13 44
26	23 22	05 N 26	22 37	14 04	15 56	16 22	22 03	13 31	17 52	13 44
27	23 19	00 S 39	22 16	14 17	16 08	16 23	22 03	13 30	17 52	13 45
28	23 17	06 45	21 55	14 30	16 19	16 25	22 03	13 30	17 51	13 45
29	23 13	12 34	21 34	14 42	16 30	16 26	22 03	13 29	17 50	13 45
30	23 10	17 43	21 14	14 55	16 41	16 28	22 02	13 28	17 50	13 45
31	23 S 05	21 S 50	20 S 53	15 S 08	16 S 52	16 N 30	22 N 02	13 S 27	17 S 49	13 S 45

ZODIAC SIGN ENTRIES

Date	h	m	Planets
01	11	15	☽ ♏
01	14	26	♂ ♏
03	11	58	☽ ♐
05	13	39	☽ ♑
07	17	54	☽ ♒
08	20	21	☿ ♑
10	01	46	☽ ♓
12	12	58	☽ ♈
15	01	43	☽ ♉
17	13	43	☽ ♊
19	23	30	☽ ♋
22	01	14	☉ ♑
22	06	48	☽ ♌
24	12	05	☽ ♍
26	15	53	☽ ♎
28	18	41	☽ ♏
30	21	01	☽ ♐

LATITUDES

Date	Mercury ☿ ° ′	Venus ♀ ° ′	Mars ♂ ° ′	Jupiter ♃ ° ′	Saturn ♄ ° ′	Uranus ♅ ° ′	Neptune ♆ ° ′	Pluto ♇ ° ′
01	01 S 48	01 N 14	00 N 53	00 N 39	01 S 20	00 S 44	00 N 03	09 N 14
04	02 00	01 40	00 52	00 40	01 20	00 44	00 03	09 13
07	02 09	02 04	00 51	00 41	01 20	00 44	00 03	09 13
10	02 15	02 24	00 50	00 41	01 19	00 44	00 03	09 13
13	02 17	02 41	00 49	00 42	01 19	00 44	00 03	09 12
16	02 16	02 56	00 48	00 43	01 19	00 44	00 03	09 12
19	02 08	03 07	00 47	00 44	01 19	00 44	00 03	09 12
22	01 54	03 17	00 45	00 44	01 18	00 43	00 03	09 12
25	01 32	03 24	00 44	00 45	01 18	00 43	00 02	09 12
28	01 01	03 30	00 43	00 46	01 18	00 43	00 02	09 12
31	00 S 20	03 N 33	00 N 41	00 N 46	01 S 17	00 S 43	00 N 02	09 N 12

DATA

Julian Date	2452610
Delta T	+64 seconds
Ayanamsa	23° 53′ 35″
Synetic vernal point	05° ♓ 13′ 24″
True obliquity of ecliptic	23° 26′ 23″

MOON'S PHASES, APSIDES AND POSITIONS ☽

Date	h	m	Phase	Longitude ° ′	Eclipse Indicator
04	07	34	●	11 ♐ 58	Total
11	15	49	☽	19 ♓ 26	
19	19	10	○	27 ♊ 42	
27	00	31	☾	05 ♎ 04	

Day	h	m		
02	08	42	Perigee	
14	03	54	Apogee	
30	00	55	Perigee	
06	07	08	Max dec	25° S 49′
13	10	04	0N	
20	18	30	Max dec	25° N 47′
27	09	29	0S	

LONGITUDES

Date	Chiron ⚷ ° ′	Ceres ⚳ ° ′	Pallas ⚴ ° ′	Juno ⚵ ° ′	Vesta ⚶ ° ′	Black Moon Lilith ⚸ ° ′
01	07 ♑ 45	04 ♈ 00	17 ♒ 16	01 ♏ 12	26 ♍ 48	22 ♈ 07
11	08 ♑ 42	04 ♈ 46	19 ♒ 38	04 ♏ 22	00 ♎ 18	23 ♈ 13
21	09 ♑ 42	06 ♈ 06	22 ♒ 13	07 ♏ 23	03 ♎ 31	24 ♈ 20
31	10 ♑ 42	07 ♈ 54	25 ♒ 00	10 ♏ 11	06 ♎ 22	25 ♈ 27

All ephemeris data is given at 12.00 UT and the Moon's longitude is additionally given for 24.00 UT

ASPECTARIAN

h m	Aspects
01 Sunday	
00 54	☽ ∠ ☉
01 37	☉ ⚹ ♆
02 19	☿ △ ♃
03 28	☽ △ ♅
06 15	☽ △ ♄
11 07	☽ ☌ ♂
11 24	☽ Q ♃
14 38	☽ ☌ ♀
14 43	☽ ∠ ♇
16 38	☽ ⊥ ☉
17 52	♂ Q ♃
17 54	☽ ∠ ☿
18 25	♀ ∠ ♇
20 31	☽ ∥ ♂
21 50	☽ ∥ ♀
02 Monday	
01 22	☽ □ ♆
03 05	☽ ⊻ ☉
05 22	☽ ⊥ ♇
06 29	☽ ⚼ ♄
08 43	☽ ∥ ♇
09 26	☽ ∥ ♅
10 12	☽ ⊥ ☿
15 07	☽ ⊻ ♇
16 37	☽ □ ♃
19 24	☽ ⫲ ♃
21 00	☽ ± ♄
21 03	☽ ⊻ ☿
03 Tuesday	
04 15	☽ □ ♅
04 49	☉ ⫲ ♄
05 01	☽ ∥ ♆
06 02	♀ ∥ ♂
06 38	☽ Q ♆
06 42	☽ ⊼ ♄
14 03	☽ ⊻ ♂
16 45	☽ ⊻ ♀
04 Wednesday	
00 19	☽ ⊥ ♂
02 16	☽ ⚹ ♆
02 53	☽ ⊥ ♀
04 20	☽ ⫲ ♄
05 20	☽ ∥ ☉
07 34	☽ ☌ ☉
09 45	☽ Q ♅
12 24	♃ St R
15 50	☽ ∠ ♂
16 18	☽ ☌ ♇
17 43	☽ △ ♃
18 14	☽ ∠ ♀
18 39	☿ ∠ ♆
05 Thursday	
03 06	♂ ∠ ♇
03 10	☽ ∠ ♆
04 10	☽ ☌ ☿
05 45	☽ ⚹ ♅
07 55	☽ ☍ ♄
14 18	☽ ∥ ☿
15 40	☉ Q ♅
18 10	☽ ⚹ ♂
18 22	☽ ⊥ ♆
18 54	☽ ⚼ ♃
19 09	☿ ⚹ ♅
20 20	☽ ⚹ ♀
06 Friday	
04 38	☽ ⊻ ♆
07 18	☽ ∠ ♅
10 20	☽ ± ♃
13 36	☿ ☍ ♄
14 06	☽ ⊻ ☉
15 48	☽ Q ♂
17 48	☽ Q ♀
19 26	☽ ⊻ ♇
20 46	☽ ⊼ ♃
21 22	☽ ∥ ☿
22 59	☽ ⊥ ♅
07 Saturday	
01 26	☽ ⊥ ☉
06 01	☽ ⊥ ♇
09 37	☽ ⊻ ♅
11 31	☽ ⊼ ♄
14 29	☽ ⊻ ☿
18 47	☽ ∠ ☉
22 11	☽ ∠ ♇
22 14	☽ ± ♄
08 Sunday	
01 24	☽ □ ♂
02 44	☽ ⊥ ☿
03 13	☽ □ ♀
08 00	☽ ∥ ☉
09 59	☽ ☌ ♆
13 13	☽ ⫲ ♄
14 35	☽ ⚼ ♄
21 30	☽ ∠ ☿
09 Monday	
00 38	☽ ⚹ ☉
01 53	☽ ⚹ ♇
03 06	☽ ☍ ♃
16 21	☽ ∥ ♆
16 57	☉ ☌ ♇
17 02	☽ ☌ ♅
18 35	☽ △ ♄
10 Tuesday	
00 43	☽ Q ♇
01 23	☽ Q ☉
03 19	☽ ⫲ ♃
05 52	☽ ⚹ ☿
07 16	☉ △ ♃
12 53	☽ △ ♂
14 32	☽ △ ♀
15 58	☽ ∥ ♅
16 10	☽ ∥ ♇
18 18	☿ ⊥ ♆
19 04	☽ ⊻ ♆
20 17	☿ ⚼ ♃
21 06	☽ ∥ ♂
11 Wednesday	
03 18	☽ ∥ ♀
06 53	☽ ⊥ ♆
08 39	☽ Q ☿
12 00	☽ □ ♇
13 01	☽ ⊼ ♃
15 49	☽ □ ☉
20 06	☽ ⚼ ♂
21 45	☽ ⚼ ♀
12 Thursday	
00 49	☽ ∠ ♆
00 57	☽ ± ♃
03 57	☽ ⊻ ♅
05 02	☽ □ ♄
15 10	☽ ± ♂
16 03	☽ ⊥ ♅
16 51	☽ ± ♀
19 00	☽ ⚼ ♃
13 Friday	
01 54	☽ □ ☿
03 59	☽ ⊼ ♂
05 42	☽ ⊼ ♀
07 06	☽ ⚹ ♆
07 26	♅ ∥ ♇
10 17	☽ ∠ ♅
17 08	☽ Q ♄
14 Saturday	
00 36	☽ △ ♇
01 18	☽ △ ♃
05 25	☿ ⚹ ♂
07 18	♀ □ ♆
07 30	☽ Q ♆
09 58	☽ △ ☉
16 48	☽ ⚹ ♅
17 18	☽ ⚹ ♄
20 05	☿ ⊻ ♆
15 Sunday	
06 41	♂ ∥ ♅
07 06	☽ ⚼ ♇
07 09	☿ ⚹ ♀
10 19	♂ ∥ ♇
16 07	♂ □ ♆
17 08	☽ Q ♅
19 10	☽ ⚼ ☉
20 00	☽ □ ♆
20 13	☽ ☍ ♂
20 24	☽ ⫲ ♀
22 13	☽ ☍ ♀
22 20	☿ ∠ ♅
23 18	☽ △ ☿
23 24	☽ ∠ ♄
16 Monday	
01 17	☽ ± ♇
04 28	☽ ⫲ ♅
04 43	☽ ⫲ ♇
05 16	☉ ∠ ♆
05 36	☽ ⫲ ♂
13 20	☽ ⊼ ♇
13 40	☽ □ ♃
14 52	☽ ± ☉
15 47	♀ ⚼ ♄
17 17	☽ ⊥ ♄
18 13	☽ ∥ ♃
18 28	☿ ± ♃
21 35	♄ △ ♅
17 Tuesday	
02 53	☉ ∠ ♂
03 54	☽ ⊼ ☉
04 42	☽ ⫲ ♆
05 07	☽ ⊻ ♄
05 11	☽ □ ♅
09 17	☽ ⚼ ☿
17 28	☉ ☍ ♄
19 50	☉ ⚹ ♅
18 Wednesday	
00 58	♂ ⚼ ♄
01 05	☽ Q ♃
02 22	♀ ⊥ ♇
05 15	☽ ± ☿
07 34	☽ △ ♆
09 09	☽ ∥ ♄
10 53	☽ ⊼ ♂
13 20	♃ △ ♇
13 21	☽ ⊼ ♀
18 22	☽ ⊼ ☿
21 10	☽ ⫲ ☉
23 05	☽ ± ♂
19 Thursday	
00 14	☽ ⚹ ♃
00 18	☽ ☍ ♇
01 40	☽ ± ♀
11 24	☽ ⫲ ☿
12 27	☽ ⚼ ♆
14 53	☽ ☌ ♄
15 30	☽ △ ♅
17 10	☽ ⚼ ♂
19 10	☽ ☍ ☉
19 55	☽ ⚼ ♀
20 Friday	
00 00	♂ ⊥ ♇
04 34	☽ ∠ ♃
05 29	☽ ± ♆
16 42	☽ ⊼ ♆
19 42	☽ ⚼ ♅
21 17	☽ ⊥ ♃
21 58	☿ ⊼ ♃
22 43	☽ △ ♂
21 Saturday	
01 46	☽ △ ♀
01 47	☿ ⊻ ♇
08 18	☽ ⊻ ♃
08 45	☽ ⊼ ♇
09 31	☽ ☍ ☿
11 48	☉ ∠ ♀
12 19	☽ ± ♅
19 45	☽ ± ♇
22 12	☽ ⊻ ♄
23 18	☽ ⊼ ♅
22 Sunday	
07 16	☽ ⊼ ☉
09 00	☽ ⊥ ♄
09 33	☽ ⫲ ☿
12 05	☽ ⚼ ♇
14 06	☽ ⫲ ☉
14 41	☿ ⊥ ♅
18 57	☽ ± ☉
23 27	☽ ☍ ♆
23 Monday	
01 03	☽ ∠ ♄
01 42	☽ ∥ ♄
07 51	☽ □ ♂
11 34	☽ □ ♀
12 18	☽ ⚼ ☉
14 07	☽ ☌ ♃
14 56	☽ △ ♇
21 03	☽ ⊼ ☿
23 17	♀ ∥ ♅
24 Tuesday	
00 13	☉ ∥ ☿
03 27	☽ ⚹ ♄
03 50	☽ ⫲ ♆
04 58	☽ ☍ ♅
08 36	☽ ± ☿
11 06	☉ ⚼ ♃
12 06	☽ ∥ ♃
15 45	☽ ⫲ ♂
16 48	☽ △ ☉
17 06	☽ Q ♂
20 42	♀ ∥ ♇
21 16	☽ Q ♀
25 Wednesday	
00 17	☽ Q ♄
00 25	☽ ⫲ ♀
00 34	☽ ⫲ ♇
01 26	♀ □ ♃
01 28	☽ ⫲ ♅
01 43	☽ ⚼ ☿
04 21	☽ ⊼ ♆
08 02	☉ ⊥ ♆
14 45	☽ ± ♆
14 56	☽ ⚹ ♂
18 14	☽ ⊻ ♃
19 04	♀ ⊻ ♇
19 24	☽ □ ♇
19 25	☽ ⚹ ♀
26 Thursday	
04 29	☽ ⊥ ♃
05 31	☿ ⊥ ♇
05 45	☽ △ ☿
06 16	☽ ⚼ ♆
07 10	☽ □ ♄
09 05	☽ ⊼ ♅
17 18	♀ ± ♄
17 54	☽ ∠ ♂
19 22	☽ ± ♅
19 48	☽ ∠ ♃
22 49	☽ ∠ ♀
27 Friday	
00 07	☿ ⊼ ♄
00 31	☽ □ ☉
02 17	☽ Q ♇
07 54	☽ △ ♆
09 57	☽ ⊥ ♂
10 43	☽ ⚼ ♅
15 08	☽ ⊥ ♀
20 36	☽ ⊻ ♂
21 08	☽ ⚹ ♃
22 41	☽ ⚹ ♇
28 Saturday	
01 58	☽ ⊻ ♀
03 35	☿ ⫲ ♄
07 22	♂ □ ♃
09 21	☽ Q ☉
09 49	☽ △ ♄
10 19	☿ ⊻ ♅
12 09	☽ △ ♅
12 15	☽ □ ☿
17 16	☽ Q ♃
29 Sunday	
00 01	☽ ∠ ♇
02 02	♂ ⫲ ♃
07 05	☽ ⚹ ☉
10 37	☽ □ ♆
10 53	☽ ⚼ ♄
15 10	☽ ⊥ ♇
16 00	☽ ∥ ♅
17 13	☽ ∥ ♇
20 06	☽ Q ☿
21 11	♂ ⊻ ♇
21 56	☽ ∥ ♀
23 19	☽ □ ♃
30 Monday	
01 15	☽ ⊻ ♇
01 26	☽ ☌ ♂
01 54	☽ ± ♄
05 47	☽ ⫲ ♃
06 43	☽ ∥ ♂
07 48	☽ ☌ ♀
10 10	☽ ∠ ☉
10 43	♂ ± ♄
11 54	☽ ⊼ ♄
12 38	☽ ∥ ♆
14 39	☽ □ ♅
16 52	☽ Q ♆
17 04	☽ ⚹ ☿
22 30	♄ ⚼ ♆
31 Tuesday	
02 25	☽ ⊥ ☉
06 20	☽ ∥ ☿
10 03	☉ ⊻ ♆
13 04	☽ ⚹ ♆
13 17	☽ ⊻ ☉
13 30	☽ ⫲ ♄
19 04	☽ ∠ ☿
21 00	☽ Q ♅
21 13	☽ ∥ ☉

JANUARY 2003

LONGITUDES

Date	Sidereal time h	m	s	Sun ☉ °	'	"	Moon ☽ °	'	"	Moon ☽ 24.00 °	'	"	Mercury ☿ °	'	Venus ♀ °	'	Mars ♂ °	'	Jupiter ♃ °	'	Saturn ♄ °	'	Uranus ♅ °	'	Neptune ♆ °	'	Pluto ♇ °	'
01	18	42	55	10 ♑	38	48	23 ♐	07	25	00 ♑	10	16	28 ♑	19	24 ♏	08	19 ♏	53	16 ♌	51	24 ♊	24	26 ♒	17	09 ♒	35	18 ♐	18
02	18	46	52	11	39	59	07 ♑	10	09	14 ♑	06	33	28	27	25	05	20	32	16 R	45	24 R	20	26	20	09	37	18	20
03	18	50	48	12	41	10	20 ♑	58	58	27 ♑	47	01	28 R	25	26	03	21	11	16	40	24	15	26	22	09	39	18	22
04	18	54	45	13	42	21	04 ♒	30	22	11 ♒	08	46	28	10	27	01	21	49	16	34	24	11	26	25	09	41	18	24
05	18	58	41	14	43	32	17 ♒	42	05	24 ♒	10	18	27	44	27	59	22	28	16	29	24	06	26	28	09	43	18	26
06	19	02	38	15	44	42	00 ♓	33	28	06 ♓	51	46	27	06	28	58	23	07	16	23	24	02	26	31	09	45	18	29
07	19	06	35	16	45	52	13 ♓	05	29	19 ♓	14	57	26	17	29 ♏	57	23	46	16	17	23	57	26	34	09	47	18	31
08	19	10	31	17	47	02	25 ♓	20	35	01 ♈	22	55	25	18	00 ♐	57	24	24	16	10	23	53	26	36	09	50	18	33
09	19	14	28	18	48	12	07 ♈	22	28	13 ♈	19	50	24	10	01	57	25	03	16	04	23	49	26	39	09	52	18	35
10	19	18	24	19	49	20	19 ♈	15	37	25 ♈	10	30	22	56	02	58	25	42	15	58	23	45	26	42	09	54	18	37
11	19	22	21	20	50	29	01 ♉	05	07	07 ♉	00	07	21	38	03	59	26	20	15	51	23	40	26	45	09	56	18	39
12	19	26	17	21	51	37	12 ♉	56	11	18 ♉	53	55	20	18	05	00	26	59	15	44	23	36	26	48	09	58	18	41
13	19	30	14	22	52	44	24 ♉	53	58	00 ♊	56	52	18	59	06	02	27	38	15	38	23	32	26	51	10	00	18	43
14	19	34	10	23	53	51	07 ♊	03	10	13 ♊	13	21	17	43	07	04	28	16	15	31	23	29	26	54	10	03	18	45
15	19	38	07	24	54	57	19 ♊	27	48	25 ♊	46	51	16	33	08	07	28	55	15	24	23	25	26	57	10	05	18	47
16	19	42	04	25	56	02	02 ♋	10	45	08 ♋	39	37	15	30	09	10	29 ♏	34	15	17	23	21	27	00	10	07	18	49
17	19	46	00	26	57	07	15 ♋	13	30	21 ♋	52	19	14	35	10	13	00 ♐	12	15	09	23	17	27	03	10	09	18	51
18	19	49	57	27	58	12	28 ♋	35	54	05 ♌	23	58	13	49	11	16	00	51	15	02	23	14	27	07	10	12	18	52
19	19	53	53	28 ♑	59	16	12 ♌	16	08	19 ♌	11	57	13	13	12	20	01	30	14	55	23	10	27	10	10	14	18	54
20	19	57	50	00 ♒	00	19	26 ♌	10	55	03 ♍	12	27	12	46	13	24	02	08	14	47	23	07	27	13	10	16	18	56
21	20	01	46	01	01	22	10 ♍	15	59	17 ♍	20	59	12	28	14	29	02	47	14	40	23	03	27	16	10	18	18	58
22	20	05	43	02	02	24	24 ♍	26	52	01 ♎	33	11	12	19	15	33	03	26	14	32	23	00	27	19	10	21	19	00
23	20	09	39	03	03	26	08 ♎	39	28	15 ♎	45	23	12 D	19	16	38	04	04	14	24	22	57	27	22	10	23	19	02
24	20	13	36	04	04	27	22 ♎	50	36	29 ♎	54	56	12	26	17	44	04	43	14	17	22	54	27	26	10	25	19	03
25	20	17	33	05	05	28	06 ♏	58	11	14 ♏	00	15	12	41	18	49	05	22	14	09	22	51	27	29	10	27	19	05
26	20	21	29	06	06	28	21 ♏	01	02	28 ♏	00	27	13	02	19	55	06	00	14	01	22	48	27	32	10	30	19	07
27	20	25	26	07	07	28	04 ♐	58	26	11 ♐	54	55	13	30	21	01	06	39	13	53	22	45	27	35	10	32	19	09
28	20	29	22	08	08	28	18 ♐	49	48	25 ♐	42	56	14	03	22	07	07	18	13	45	22	42	27	39	10	34	19	10
29	20	33	19	09	09	27	02 ♑	34	10	09 ♑	23	17	14	41	23	13	07	56	13	37	22	40	27	42	10	36	19	12
30	20	37	15	10	10	25	16 ♑	10	02	22 ♑	54	12	15	24	24	20	08	35	13	29	22	37	27	45	10	39	19	14
31	20	41	12	11 ♒	11	22	29 ♑	35	29	06 ♒	13	38	16 ♑	12	25 ♐	27	09 ♐	14	13 ♌	21	22 ♊	35	27 ♒	49	10 ♒	41	19 ♐	15

Moon Nodes and Latitude

Date	Moon True ☊ °	'	Moon Mean ☊ °	'	Moon ☽ Latitude °	'
01	08 ♊	28	07 ♊	00	01 S	19
02	08 R	26	06	57	02	30
03	08	21	06	54	03	31
04	08	16	06	51	04	19
05	08	09	06	48	04	51
06	08	03	06	45	05	08
07	07	57	06	41	05	09
08	07	52	06	38	04	56
09	07	50	06	35	04	30
10	07	49	06	32	03	53
11	07 D	50	06	29	03	06
12	07	51	06	25	02	11
13	07	53	06	22	01	10
14	07	53	06	19	00 S	05
15	07 R	53	06	16	01 N	02
16	07	50	06	13	02	07
17	07	45	06	10	03	07
18	07	37	06	06	03	58
19	07	29	06	03	04	37
20	07	19	06	00	05	00
21	07	11	05	57	05	05
22	07	04	05	54	04	52
23	06	59	05	51	04	21
24	06	56	05	47	03	33
25	06 D	55	05	44	02	33
26	06	56	05	41	01	24
27	06	57	05	38	00 N	11
28	06 R	56	05	35	01 S	03
29	06	53	05	31	02	12
30	06	48	05	28	03	13
31	06 ♊	39	05 ♊	25	04 S	02

DECLINATIONS

Date	Sun ☉ °	'	Moon ☽ °	'	Mercury ☿ °	'	Venus ♀ °	'	Mars ♂ °	'	Jupiter ♃ °	'	Saturn ♄ °	'	Uranus ♅ °	'	Neptune ♆ °	'	Pluto ♇ °	'
01	23 S	01	24 S	35	20 S	34	15 S	21	17 S	03	16 N	31	22 N	02	13 S	26	17 S	49	13 S	46
02	22	56	25	45	20	15	15	34	17	14	16	33	22	02	13	25	17	48	13	46
03	22	50	25	17	19	58	15	47	17	25	16	35	22	02	13	24	17	48	13	46
04	22	44	23	19	19	43	16	00	17	35	16	37	22	02	13	23	17	47	13	46
05	22	38	20	09	19	29	16	13	17	46	16	39	22	02	13	22	17	47	13	46
06	22	31	16	04	19	18	16	26	17	56	16	41	22	02	13	21	17	46	13	46
07	22	23	11	24	19	08	16	39	18	06	16	43	22	02	13	20	17	46	13	46
08	22	15	06	23	19	01	16	52	18	16	16	45	22	02	13	19	17	45	13	46
09	22	07	01 S	13	18	56	17	04	18	26	16	47	22	02	13	18	17	44	13	47
10	21	59	03 N	56	18	53	17	17	18	36	16	49	22	02	13	17	17	44	13	47
11	21	49	08	56	18	52	17	29	18	45	16	51	22	02	13	16	17	43	13	47
12	21	40	13	38	18	52	17	41	18	55	16	53	22	02	13	15	17	43	13	47
13	21	30	17	52	18	54	17	53	19	04	16	56	22	02	13	14	17	42	13	47
14	21	20	21	25	18	58	18	05	19	13	16	58	22	02	13	13	17	42	13	47
15	21	09	24	03	19	02	18	16	19	22	17	00	22	02	13	12	17	41	13	47
16	20	58	25	33	19	08	18	28	19	31	17	02	22	02	13	11	17	40	13	47
17	20	46	25	40	19	15	18	38	19	40	17	05	22	02	13	10	17	40	13	47
18	20	34	24	20	19	22	18	49	19	49	17	07	22	02	13	09	17	39	13	47
19	20	22	21	33	19	30	19	00	19	57	17	09	22	02	13	08	17	39	13	47
20	20	09	17	29	19	38	19	10	20	06	17	12	22	02	13	07	17	38	13	47
21	19	56	12	25	19	47	19	19	20	14	17	14	22	02	13	05	17	37	13	47
22	19	42	06	40	19	56	19	29	20	22	17	16	22	02	13	04	17	37	13	47
23	19	28	00 N	34	20	05	19	38	20	30	17	19	22	02	13	03	17	36	13	47
24	19	14	05 S	35	20	13	19	46	20	37	17	21	22	02	13	02	17	36	13	47
25	19	00	11	26	20	22	19	55	20	45	17	24	22	02	13	01	17	35	13	47
26	18	45	16	39	20	30	20	02	20	52	17	26	22	02	13	00	17	34	13	47
27	18	29	20	57	20	38	20	10	21	00	17	28	22	02	12	59	17	34	13	47
28	18	14	24	01	20	45	20	17	21	07	17	31	22	02	12	58	17	33	13	47
29	17	58	25	37	20	52	20	24	21	14	17	33	22	02	12	56	17	33	13	47
30	17	42	25	39	20	58	20	30	21	21	17	36	22	02	12	55	17	32	13	47
31	17 S	25	24 S	11	21 S	03	20 S	35	21 S	27	17 N	38	22 N	02	12 S	54	17 S	31	13 S	47

ZODIAC SIGN ENTRIES

Date	h	m	Planets
01	23	42	☽ ♑
04	03	56	☽ ♒
06	10	57	☽ ♓
07	13	07	♀ ♐
08	21	15	☽ ♈
11	09	48	☽ ♉
13	22	08	☽ ♊
16	07	56	☽ ♋
17	04	22	♂ ♐
18	14	29	☽ ♌
20	11	53	☉ ♒
20	18	32	☽ ♍
22	21	23	☽ ♎
25	00	09	☽ ♏
27	03	26	☽ ♐
29	07	30	☽ ♑
31	12	44	☽ ♒

LATITUDES

Date	Mercury ☿ °	'	Venus ♀ °	'	Mars ♂ °	'	Jupiter ♃ °	'	Saturn ♄ °	'	Uranus ♅ °	'	Neptune ♆ °	'	Pluto ♇ °	'
01	00 S	04	03 N	34	00 N	41	00 N	46	01 S	17	00 S	43	00 N	02	09 N	12
04	00 N	50	03	35	00	39	00	47	01	17	00	43	00	02	09	12
07	01	47	03	34	00	38	00	48	01	16	00	43	00	02	09	12
10	02	39	03	32	00	36	00	48	01	16	00	43	00	02	09	13
13	03	13	03	29	00	35	00	49	01	15	00	43	00	02	09	13
16	03	26	03	24	00	33	00	49	01	15	00	43	00	02	09	13
19	03	18	03	19	00	31	00	50	01	14	00	43	00	02	09	13
22	02	57	03	12	00	29	00	50	01	13	00	43	00	02	09	13
25	02	29	03	04	00	27	00	51	01	13	00	43	00	02	09	14
28	01	57	02	56	00	25	00	51	01	12	00	43	00	02	09	14
31	01 N	25	02 N	46	00 N	23	00 N	51	01 S	12	00 S	43	00 N	02	09 N	15

DATA

Julian Date	2452641
Delta T	+65 seconds
Ayanamsa	23° 53' 41"
Synetic vernal point	05° ♓ 13' 18"
True obliquity of ecliptic	23° 26' 23"

MOON'S PHASES, APSIDES AND POSITIONS ☽

Date	h	m	Phase	Longitude °	'	Eclipse Indicator
02	20	23	●	12 ♑	01	
10	13	15	☽	19 ♈	53	
18	10	48	○	27 ♋	55	
25	08	33	☾	04 ♏	57	

Day	h	m	
11	00	42	Apogee
23	22	39	Perigee

Day	h	m		
02	16	59	Max dec	25° S 47'
09	17	38	0N	
17	02	12	Max dec	25° N 48'
23	14	11	0S	
30	00	30	Max dec	25° S 50'

LONGITUDES

Date	Chiron ⚷ °	'	Ceres ⚳ °	'	Pallas ⚴ °	'	Juno ⚵ °	'	Vesta ⚶ °	'	Black Moon Lilith ⚸ °	'
01	10 ♑	08	08 ♈	07	25 ♒	17	10 ♏	27	06 ♎	38	25 ♈	34
11	11 ♑	49	10 ♈	23	28 ♒	14	12 ♏	59	09 ♎	01	26 ♈	41
21	12 ♑	48	13 ♈	00	01 ♓	19	15 ♏	14	10 ♎	52	27 ♈	47
31	13 ♑	45	15 ♈	54	04 ♓	30	17 ♏	09	12 ♎	05	28 ♈	54

All ephemeris data is given at 12.00 UT and the Moon's longitude is additionally given for 24.00 UT

ASPECTARIAN

h m	Aspects
01 Wednesday	
01 26	☽ △ ♃
03 48	☽ ☌ ♇
06 15	☽ ⊻ ♂
10 36	☽ ⊥ ☿
13 51	☽ ⊻ ♀
14 10	☽ ☍ ♄
14 29	☽ ∠ ♆
16 19	☉ ± ♃
16 56	☽ ⊥ ♂
17 23	☽ ⚹ ♅
18 14	♀ ⊼ ♄
20 57	☽ ⊻ ☿
02 Thursday	
00 49	☽ ⊥ ♀
02 46	☽ ⚼ ♃
03 42	☉ ∠ ♅
05 53	☽ ⊥ ♆
09 03	☽ ∠ ♂
16 14	☽ ⊻ ♆
17 24	☽ ∠ ♀
18 09	☽ ± ♃
18 18	☿ St R
19 12	☽ ∠ ♅
20 23	☽ ☌ ☉
03 Friday	
04 30	☽ ⊼ ♃
07 25	☽ ⊻ ♇
10 56	☽ ⊥ ♅
12 22	☽ ⚹ ♂
17 43	☽ ⊼ ♄
17 58	☽ ⊥ ♇
20 38	♀ □ ♅
21 32	☽ ⊻ ♅
21 36	☽ ⚹ ♀
04 Saturday	
00 56	☽ ☌ ☿
04 18	☽ ± ♄
10 01	☽ ∠ ♇
10 43	☽ Q ♂
17 18	☽ ∥ ☉
20 23	☽ ⚼ ♄
20 46	☽ Q ♀
21 22	☽ ☌ ♆
22 41	☽ # ♄
05 Sunday	
05 17	♀ Q ♆
06 04	☽ ⊻ ☉
08 01	☿ ⚹ ♀
09 46	☽ ☍ ♃
13 22	☽ ⚹ ♇
14 14	♂ ∥ ♆
16 26	☽ ∥ ☿
18 04	☽ ⊥ ☉
21 18	☽ □ ♂
23 48	☽ △ ♄
06 Monday	
01 58	☽ ∥ ♂
02 28	☽ ∥ ♆
04 21	☽ ☌ ♅
05 50	☽ ⊻ ☿
08 40	☽ # ♃
08 44	☽ □ ♀
10 05	☽ ∥ ♀
11 51	☽ Q ♇
12 23	☽ ∠ ☉
16 33	☽ ⊥ ☿
07 Tuesday	
00 06	☽ ∥ ♇
01 34	☉ ⊼ ♃
02 17	☽ ∥ ♅
04 50	☿ ⊻ ♅
05 36	☽ ⊻ ♆
08 44	☽ ∠ ☿
17 16	☽ ⊥ ♆
18 08	☽ ⊼ ♃
18 36	♂ ⊼ ♄
19 47	☽ ⚹ ☉
20 34	♀ # ♃
22 35	☽ □ ♇
08 Wednesday	
05 48	☽ ± ♃
09 08	☽ □ ♄
10 02	☽ △ ♂
10 58	☽ ∠ ♆
11 55	☽ ⚹ ☿
14 31	☽ ⊻ ♅
21 38	☽ Q ☉
23 29	☽ ⚼ ♃
09 Thursday	
00 09	☽ △ ♀
00 18	☿ ⚹ ♂
02 30	☽ ⊥ ♅
03 49	☿ ⊥ ♇
06 32	☉ ⊻ ♇
09 48	☽ Q ☿
17 01	☽ ⚹ ♆
17 41	☽ ⚼ ♂
19 30	☿ ⊼ ♄
20 39	☽ ∠ ♅
20 53	☽ Q ♄
10 Friday	
02 39	☉ # ♄
05 23	☽ △ ♃
09 08	☽ ⚼ ♀
10 41	☽ △ ♇
12 56	☽ ± ♂
13 15	☽ □ ☉
17 22	☽ Q ♆
18 43	☽ □ ☿
21 02	☽ ⚹ ♄
11 Saturday	
01 48	☽ ⊼ ♂
03 10	☽ ⚹ ♅
05 06	☽ ± ♀
09 49	☉ ⊥ ♅
17 13	☽ ⚼ ♇
18 26	☽ ⊼ ♀
20 02	☉ ☌ ☿
12 Sunday	
03 12	☿ ⊥ ♅
03 18	☽ ∠ ♄
03 37	☽ Q ♅
04 19	☿ ∥ ♂
04 47	♂ □ ♅
05 59	☽ □ ♆
09 57	☽ # ♅
11 29	☽ ± ♇
12 46	☽ # ♇
14 56	♀ ∥ ♆
14 58	☿ ∠ ♀
17 36	☽ □ ♃
21 21	☽ ⊥ ♄
23 36	☽ ⊼ ♇
13 Monday	
01 20	☽ △ ☿
06 20	☽ ∥ ♃
07 36	☽ △ ☉
09 18	☽ ⊻ ♄
11 02	☽ # ♆
12 09	☽ # ♀
15 54	☽ □ ♅
16 53	☿ ⊻ ♇
17 44	☽ ☍ ♂
18 37	☽ # ☿
19 55	☽ # ♂
14 Tuesday	
02 39	☉ ⊼ ♄
03 01	♂ Q ♆
04 17	☽ ⚼ ☿
05 07	☽ Q ♃
11 22	☽ # ☉
12 02	☽ ☍ ♀
15 56	☽ ⚼ ☉
16 52	☽ ∥ ♄
17 51	☽ △ ♆
20 17	☽ ± ☿
15 Wednesday	
04 16	☽ ⚹ ♃
06 53	☽ ⊼ ☿
08 39	☉ ⊥ ♇
10 41	☽ ☍ ♇
10 52	☽ ± ☉
19 29	☽ ☌ ♄
22 42	☽ ⚼ ♆
23 16	☽ ⊼ ☉
16 Thursday	
02 16	☽ △ ♅
06 51	☽ ⊼ ♂
08 29	☽ ∠ ♃
15 37	☽ ± ♆
16 00	☿ ⊥ ♀
18 20	☿ ⊼ ♃
18 36	☽ ± ♂
17 Friday	
01 01	☽ ⊥ ♃
02 03	☽ ⊼ ♀
02 11	☿ ∠ ♂
02 43	☽ ⊼ ♆
06 12	☽ ⚼ ♅
10 35	♀ ⚹ ♆
10 54	☽ ☍ ☿
11 52	☽ ⊻ ♃
11 58	☽ ⚼ ♂
13 57	☽ ± ♀
14 36	☉ ⊻ ♅
18 34	☽ ⊼ ♇
22 35	☽ ± ♅
18 Saturday	
02 28	☽ ⊻ ♄
05 21	☽ ± ♇
07 31	☽ ⚼ ♀
09 21	☽ ⊼ ♅
10 48	☽ ☍ ☉
13 07	☽ ⊥ ♄
16 11	☽ △ ♂
21 20	☽ ⚼ ♇
19 Sunday	
04 53	☽ ∠ ♄
08 26	☽ ☍ ♆
08 33	☽ ∥ ♄
12 08	☽ △ ♀
13 35	☽ ⊼ ☿
16 01	☉ ± ♄
16 33	☽ ☌ ♃
20 13	☽ # ☉
21 55	☽ # ♂
23 31	☽ △ ♇
23 37	☽ ± ☿
20 Monday	
00 33	☽ # ☿
01 29	☿ ⊻ ♀
03 15	☽ # ♀
06 45	☽ ⚹ ♄
11 14	☽ # ♆
13 30	☽ ∥ ♃
13 46	☽ ☍ ♅
14 38	☽ ⚼ ☿
15 52	☉ ∥ ♂
19 03	☽ ⊼ ☉
22 40	☽ □ ♂
21 Tuesday	
03 11	☽ Q ♄
05 54	☽ # ♇
06 04	☽ ± ☉
09 02	☽ # ♅
12 04	☽ ⊼ ♆
15 35	♀ △ ♃
15 40	☽ △ ☿
19 23	☽ ⊻ ♃
19 44	☽ □ ♀
21 44	☉ ∥ ☿
22 15	☽ ± ♆
22 30	☽ ⚼ ☉
22 Wednesday	
02 46	☽ □ ♇
05 27	☽ ⊥ ♃
06 26	♀ Q ♅
06 39	☽ Q ♂
09 34	☽ □ ♄
11 16	☿ ∠ ♅
13 31	☽ ⚼ ♆
16 52	☽ ⊼ ♅
20 31	☽ ∠ ♃
23 Thursday	
01 09	☿ St D
01 49	☽ △ ☉
02 21	☉ ∥ ♀
03 02	☽ ± ♅
03 53	☽ ⚹ ♂
04 39	☽ Q ♀
09 14	☽ Q ♇
14 55	☽ △ ♆
18 12	☽ □ ☿
18 18	☽ ⚼ ♅
21 38	☽ ⚹ ♃
24 Friday	
02 37	☽ ⚹ ♀
05 34	☽ ⚹ ♇
06 27	☽ ∠ ♂
10 52	☿ ∠ ♅
11 34	☉ ∠ ♇
12 05	☽ △ ♄
17 46	☽ Q ♃
19 48	☽ △ ♅
22 26	☽ ⊥ ♂
25 Saturday	
01 05	☽ Q ☿
06 11	☽ ∠ ♀
07 05	☽ ∠ ♇
08 33	☽ □ ☉
09 08	☽ ⊻ ♂
13 29	☽ ⚼ ♄
17 57	☽ □ ♆
17 59	♀ ☌ ♇
18 55	☽ ∥ ♅
21 58	☽ ⚹ ☿
22 25	☽ ∥ ♇
22 27	☽ ⊥ ♇
22 49	☽ ⊥ ♀
26 Sunday	
00 08	☽ □ ♃
04 48	☽ ± ♄
05 26	☉ ⚹ ♂
08 44	☽ ⊻ ♇
09 57	☽ ⊻ ♀
15 02	☽ ⊼ ♄
15 58	☽ # ♃
16 39	☽ ∥ ♆
17 43	☽ Q ☉
22 19	☽ ∥ ☉
23 14	☽ □ ♅
27 Monday	
00 27	☽ ∠ ☿
00 52	☽ Q ♆
06 56	☽ ∥ ♀
09 53	☽ ∥ ☿
12 16	☽ ∥ ♂
15 02	☽ ☌ ♂
16 00	☽ ⚹ ☉
16 32	☽ ⊥ ☿
19 20	☽ # ♄
21 38	☽ ⚹ ♆
28 Tuesday	
01 51	☿ ⊼ ♃
02 05	☉ ⚼ ♄
03 16	☽ △ ♃
03 20	☽ ⊻ ☿
06 27	☽ Q ♅
12 36	☽ ☌ ♇
18 13	☽ ☌ ♀
18 43	☽ ☍ ♄
20 06	☽ ∠ ☉
23 47	☽ ∠ ♆
29 Wednesday	
00 13	♀ ☍ ♄
03 26	☽ ⚹ ♅
05 09	☽ ⚼ ♃
07 02	♃ # ♆
13 07	☽ ⊥ ☉
15 35	☽ ⊥ ♆
20 48	☽ ± ♃
21 55	☽ ⊻ ♂
30 Thursday	
00 32	☽ ⊻ ☉
02 12	☽ ⊻ ♆
05 56	☽ ∠ ♅
07 17	☽ ⊼ ♃
09 03	☽ ⊥ ♂
10 34	☽ ☌ ☿
17 27	☽ ⊻ ♇
19 38	☉ # ♃
22 00	☽ ⊥ ♅
23 27	☽ ⊼ ♄
23 34	☉ ☌ ♆
31 Friday	
01 53	☽ ∠ ♂
02 33	☉ ∥ ♆
03 53	☽ ⊻ ♀
04 11	☽ ⊥ ♇
08 47	☽ ⊻ ♅
10 11	☽ ± ♄
15 39	☽ ⊥ ♀
17 14	♀ ∠ ♆
20 26	☽ ∠ ♇

FEBRUARY 2003

LONGITUDES

Date	Sidereal time h m s	Sun ☉ ° ' "	Moon ☽ ° ' "	Moon ☽ 24.00 ° ' "	Mercury ☿ ° '	Venus ♀ ° '	Mars ♂ ° '	Jupiter ♃ ° '	Saturn ♄ ° '	Uranus ♅ ° '	Neptune ♆ ° '	Pluto ♇ ° '
01	20 45 08	12 ♒ 12 18	12 ♒ 48 23	19 ♒ 19 31	17 ♑ 02	26 ♐ 34	09 ♐ 52	13 ♌ 13	22 ♊ 32	27 ♒ 52	10 ♒ 43	19 ♐ 17
02	20 49 05	13 13 13	25 ♒ 46 52	02 ♓ 10 18	17 57	27 41	10 31	13 R 05	22 R 30	27 56	10 46	19 18
03	20 53 02	14 14 07	08 ♓ 29 47	14 ♓ 45 20	18 54	28 48	11 10	12 57	22 28	27 59	10 48	19 20
04	20 56 58	15 15 00	20 ♓ 57 04	27 ♓ 05 10	19 55	29 ♐ 56	11 48	12 49	22 26	28 02	10 50	19 21
05	21 00 55	16 15 51	03 ♈ 09 54	09 ♈ 11 37	20 58	01 ♑ 04	12 27	12 41	22 24	28 06	10 52	19 23
06	21 04 51	17 16 41	15 ♈ 10 43	21 ♈ 07 42	22 03	02 11	13 05	12 33	22 22	28 09	10 55	19 24
07	21 08 48	18 17 29	27 ♈ 03 05	02 ♉ 57 29	23 11	03 19	13 44	12 25	22 20	28 13	10 57	19 26
08	21 12 44	19 18 16	08 ♉ 51 31	14 ♉ 45 50	24 21	04 28	14 23	12 17	22 19	28 16	10 59	19 27
09	21 16 41	20 19 02	20 ♉ 41 07	26 ♉ 38 06	25 33	05 36	15 01	12 09	22 17	28 19	11 01	19 28
10	21 20 37	21 19 46	02 ♊ 37 26	08 ♊ 39 51	26 46	06 44	15 40	12 02	22 16	28 23	11 04	19 30
11	21 24 34	22 20 29	14 ♊ 46 00	20 ♊ 56 31	28 01	07 53	16 18	11 54	22 15	28 26	11 06	19 31
12	21 28 31	23 21 10	27 ♊ 11 58	03 ♋ 32 52	29 ♑ 18	09 02	16 57	11 46	22 14	28 30	11 08	19 32
13	21 32 27	24 21 49	09 ♋ 59 38	16 ♋ 32 34	00 ♒ 36	10 11	17 35	11 39	22 12	28 33	11 10	19 33
14	21 36 24	25 22 27	23 ♋ 11 51	29 ♋ 57 30	01 56	11 20	18 14	11 31	22 12	28 37	11 13	19 35
15	21 40 20	26 23 03	06 ♌ 49 24	13 ♌ 47 15	03 16	12 29	18 52	11 23	22 11	28 40	11 15	19 36
16	21 44 17	27 23 37	20 ♌ 50 35	27 ♌ 58 49	04 38	13 38	19 31	11 16	22 10	28 44	11 17	19 37
17	21 48 13	28 24 10	05 ♍ 11 10	12 ♍ 26 47	06 02	14 47	20 09	11 09	22 09	28 47	11 19	19 38
18	21 52 10	29 ♒ 24 42	19 ♍ 44 45	27 ♍ 04 06	07 26	15 57	20 48	11 01	22 09	28 51	11 21	19 39
19	21 56 06	00 ♓ 25 12	04 ♎ 23 53	11 ♎ 43 12	08 52	17 06	21 26	10 54	22 09	28 54	11 24	19 40
20	22 00 03	01 25 40	19 ♎ 01 15	26 ♎ 17 21	10 19	18 16	22 05	10 47	22 08	28 57	11 26	19 41
21	22 04 00	02 26 08	03 ♏ 30 54	10 ♏ 41 29	11 47	19 26	22 43	10 40	22 08	29 01	11 28	19 42
22	22 07 56	03 26 34	17 ♏ 48 49	24 ♏ 52 42	13 15	20 36	23 22	10 33	22 D 08	29 04	11 30	19 43
23	22 11 53	04 26 59	01 ♐ 53 02	08 ♐ 49 52	14 45	21 46	24 00	10 26	22 08	29 08	11 32	19 44
24	22 15 49	05 27 22	15 ♐ 43 14	22 ♐ 33 16	16 16	22 56	24 38	10 20	22 08	29 11	11 34	19 45
25	22 19 46	06 27 44	29 ♐ 20 04	06 ♑ 03 47	17 48	24 06	25 17	10 13	22 09	29 15	11 37	19 46
26	22 23 42	07 28 05	12 ♑ 44 32	19 ♑ 22 25	19 21	25 16	25 55	10 07	22 09	29 18	11 39	19 47
27	22 27 39	08 28 24	25 ♑ 57 30	02 ♒ 29 50	20 56	26 27	26 34	10 00	22 10	29 22	11 41	19 48
28	22 31 35	09 ♓ 28 42	08 ♒ 59 25	15 ♒ 26 15	22 ♒ 31	27 ♑ 37	27 ♐ 12	09 ♌ 54	22 ♊ 10	29 ♒ 25	11 ♒ 43	19 ♐ 48

Date	Moon True ☊ ° '	Moon Mean ☊ ° '	Moon ☽ Latitude ° '
01	06 ♊ 28	05 ♊ 22	04 S 37
02	06 R 16	05 19	04 57
03	06 04	05 16	05 02
04	05 52	05 12	04 52
05	05 42	05 09	04 29
06	05 35	05 06	03 54
07	05 30	05 03	03 09
08	05 28	05 00	02 16
09	05 D 27	04 57	01 18
10	05 28	04 53	00 S 15
11	05 R 27	04 50	00 N 49
12	05 26	04 47	01 53
13	05 21	04 44	02 52
14	05 14	04 41	03 44
15	05 05	04 37	04 25
16	04 53	04 34	04 51
17	04 41	04 31	05 00
18	04 29	04 28	04 50
19	04 19	04 25	04 20
20	04 12	04 22	03 34
21	04 07	04 18	02 34
22	04 05	04 15	01 25
23	04 D 05	04 12	00 N 12
24	04 R 05	04 09	01 S 01
25	04 03	04 06	02 09
26	04 00	04 03	03 09
27	03 54	03 59	03 58
28	03 ♊ 44	03 ♊ 56	04 S 33

DECLINATIONS

Date	Sun ☉ ° '	Moon ☽ ° '	Mercury ☿ ° '	Venus ♀ ° '	Mars ♂ ° '	Jupiter ♃ ° '	Saturn ♄ ° '	Uranus ♅ ° '	Neptune ♆ ° '	Pluto ♇ ° '
01	17 S 08	21 S 24	21 S 07	20 S 41	21 S 34	17 N 41	22 N 02	12 S 53	17 S 31	13 S 47
02	16 51	17 35	21 11	20 45	21 40	17 43	22 02	12 52	17 30	13 47
03	16 34	13 03	21 13	20 49	21 46	17 45	22 03	12 51	17 30	13 47
04	16 16	08 03	21 15	20 53	21 52	17 48	22 03	12 49	17 29	13 47
05	15 58	02 S 51	21 15	20 56	21 58	17 50	22 03	12 48	17 28	13 47
06	15 39	02 N 23	21 15	20 59	22 04	17 52	22 03	12 47	17 28	13 47
07	15 21	07 29	21 13	21 01	22 09	17 55	22 03	12 46	17 27	13 47
08	15 02	12 18	21 10	21 03	22 14	17 57	22 03	12 45	17 27	13 47
09	14 43	16 41	21 06	21 04	22 20	17 59	22 03	12 43	17 26	13 47
10	14 24	20 26	21 01	21 05	22 25	18 02	22 03	12 42	17 25	13 47
11	14 04	23 23	20 55	21 05	22 29	18 04	22 03	12 41	17 25	13 47
12	13 44	25 17	20 47	21 04	22 34	18 06	22 04	12 40	17 24	13 47
13	13 24	25 55	20 38	21 03	22 38	18 08	22 04	12 39	17 24	13 47
14	13 04	25 08	20 28	21 02	22 43	18 10	22 04	12 37	17 23	13 47
15	12 43	22 50	20 17	20 59	22 47	18 13	22 04	12 36	17 22	13 46
16	12 23	19 09	20 04	20 57	22 51	18 15	22 04	12 35	17 22	13 46
17	12 02	14 16	19 51	20 53	22 55	18 17	22 04	12 34	17 21	13 46
18	11 41	08 30	19 36	20 49	22 58	18 19	22 05	12 33	17 21	13 46
19	11 19	02 N 14	19 19	20 45	23 02	18 21	22 05	12 31	17 20	13 46
20	10 58	04 S 09	19 02	20 40	23 05	18 23	22 05	12 30	17 19	13 46
21	10 36	10 16	18 43	20 34	23 08	18 25	22 05	12 29	17 19	13 46
22	10 15	15 47	18 22	20 28	23 11	18 27	22 05	12 28	17 18	13 46
23	09 53	20 21	18 01	20 21	23 14	18 29	22 06	12 27	17 18	13 45
24	09 31	23 41	17 38	20 14	23 16	18 30	22 06	12 25	17 17	13 45
25	09 08	25 35	17 14	20 06	23 19	18 32	22 06	12 24	17 16	13 45
26	08 46	25 58	16 49	19 58	23 21	18 34	22 06	12 23	17 16	13 45
27	08 24	24 51	16 22	19 49	23 23	18 36	22 07	12 22	17 15	13 45
28	08 S 01	22 S 24	15 S 54	19 S 39	23 S 25	18 N 37	22 N 07	12 S 21	17 S 15	13 S 45

ZODIAC SIGN ENTRIES

Date	h	m	Planets
02	19	54	☽ ♓
04	13	27	♀ ♑
05	05	44	☽ ♈
07	17	59	☽ ♉
10	06	45	☽ ♊
12	17	19	☽ ♋
13	01	00	☿ ♒
15	00	04	☽ ♌
17	03	22	☽ ♍
19	02	00	☉ ♓
19	04	48	☽ ♎
21	06	09	☽ ♏
23	08	46	☽ ♐
25	13	11	☽ ♑
27	19	24	☽ ♒

LATITUDES

Date	Mercury ☿ ° '	Venus ♀ ° '	Mars ♂ ° '	Jupiter ♃ ° '	Saturn ♄ ° '	Uranus ♅ ° '	Neptune ♆ ° '	Pluto ♇ ° '
01	01 N 14	02 N 43	00 N 22	00 N 52	01 S 12	00 S 43	00 N 02	09 N 15
04	00 43	02 33	00 20	00 52	01 11	00 43	00 02	09 15
07	00 N 14	02 23	00 18	00 52	01 10	00 43	00 02	09 16
10	00 S 13	02 12	00 16	00 52	01 10	00 43	00 02	09 16
13	00 38	02 00	00 13	00 53	01 09	00 43	00 02	09 17
16	01 00	01 49	00 11	00 53	01 08	00 43	00 02	09 17
19	01 19	01 37	00 08	00 53	01 08	00 43	00 02	09 18
22	01 36	01 25	00 05	00 53	01 07	00 43	00 02	09 18
25	01 50	01 13	00 N 03	00 53	01 06	00 43	00 02	09 19
28	02 00	01 00	00 00	00 53	01 06	00 43	00 02	09 20
31	02 S 07	00 N 49	00 S 03	00 N 53	01 S 05	00 S 43	00 N 02	09 N 20

DATA

Julian Date	2452672
Delta T	+65 seconds
Ayanamsa	23° 53' 47"
Synetic vernal point	05° ♓ 13' 12"
True obliquity of ecliptic	23° 26' 24"

MOON'S PHASES, APSIDES AND POSITIONS ☽

Date	h	m	Phase	Longitude ° '	Eclipse Indicator
01	10	48	●	12 ♒ 09	
09	11	11	☽	20 ♉ 17	
16	23	51	○	27 ♌ 54	
23	16	46	☾	04 ♐ 39	

Day	h	m		
07	21	59	Apogee	
19	16	25	Perigee	
06	01	02	0N	
13	10	59	Max dec	25° N 55'
19	20	23	0S	
26	05	49	Max dec	26° S 01'

LONGITUDES

Date	Chiron ⚷ ° '	Ceres ⚳ ° '	Pallas ⚴ ° '	Juno ⚵ ° '	Vesta ⚶ ° '	Black Moon Lilith ⚸ ° '
01	13 ♑ 50	16 ♈ 12	04 ♓ 50	17 ♏ 20	12 ♎ 10	29 ♈ 01
11	14 ♑ 43	19 ♈ 23	08 ♓ 06	18 ♏ 49	12 ♎ 35	00 ♉ 07
21	15 ♑ 32	22 ♈ 46	11 ♓ 25	19 ♏ 52	12 ♎ 13	01 ♉ 14
31	16 ♑ 16	26 ♈ 19	14 ♓ 47	20 ♏ 25	11 ♎ 03	02 ♉ 21

ASPECTARIAN

h m Aspects

01 Saturday
02 25 ☽ ⊡ ♄
06 21 ☽ ⚹ ♂
07 14 ☽ # ♄
08 10 ☽ ☌ ♆
09 31 ☽ ∠ ♀
10 48 ☽ ☌ ☉
10 50 ☽ ∥ ♂
12 45 ☽ ☍ ♃
13 53 ☽ ∥ ☿
16 53 ☽ ∥ ♀
20 21 ☽ ⊻ ☿
23 56 ☽ ⚹ ♇

02 Sunday
05 36 ☽ Q ♂
05 54 ☽ △ ♄
08 19 ☽ ⊥ ☿
09 12 ☉ ☍ ♃
11 15 ☽ # ♃
12 26 ☽ ∥ ♆
15 54 ☽ ⚹ ♀
16 02 ☽ ☌ ♅
16 23 ☽ ∥ ☉
17 27 ♀ ⚹ ♅
19 41 ♀ ⊡ ♃
21 39 ♂ ⚹ ♆
22 23 ☽ Q ♇

03 Monday
02 33 ☽ ∠ ☿
08 15 ☽ ∥ ♇
13 00 ☽ ∥ ♅
16 25 ☽ ⊻ ♆
16 51 ☽ Q ♀
17 22 ☽ □ ♂
20 26 ☽ ⊼ ♃
22 27 ☿ ⊻ ♇
23 58 ☽ ⊻ ☉

04 Tuesday
03 59 ☽ ⊥ ♆
07 54 ☽ ± ♃
08 53 ☽ □ ♇
09 48 ☽ ⚹ ☿
12 38 ☽ ⊥ ☉
14 53 ☽ □ ♄
21 34 ☽ ∠ ♆

05 Wednesday
01 18 ☽ ⊡ ♃
01 56 ☽ ⊻ ♅
07 24 ☽ □ ♀
07 54 ☽ ∠ ☉
11 34 ☽ Q ☿
13 51 ☽ ⊥ ♅
19 25 ♂ △ ♃

06 Thursday
02 22 ☽ Q ♄
03 24 ☽ ⚹ ♆
06 47 ☽ △ ♃
07 34 ☽ △ ♂
07 55 ☽ ∠ ♅
08 44 ♂ # ♄
14 09 ☿ ⊥ ♅
16 37 ☽ ⚹ ☉
18 31 ☿ ⊼ ♄
20 32 ☽ △ ♇

07 Friday
02 28 ☽ ⚹ ♄
03 20 ☽ □ ☿
03 39 ☽ Q ♆
05 24 ☉ ∠ ♀
14 22 ☽ ⚹ ♅
15 37 ☽ ⊡ ♂
19 12 ☽ Q ☉

08 Saturday
02 06 ☽ △ ♀
03 01 ☽ ⊡ ♇
08 52 ☽ ∠ ♄
10 58 ☽ ± ♂
14 18 ☽ # ♅
14 53 ☽ Q ♅
15 31 ☉ ⚹ ♇
16 20 ☽ □ ♆
18 54 ☽ □ ♃
19 50 ☽ # ♇
21 21 ☽ ± ♇
23 28 ♀ ⊥ ♆
23 51 ☽ ⊼ ♂

09 Sunday
01 39 ☽ # ☉
03 07 ☽ ⊥ ♄
09 32 ☽ ⊼ ♇
10 33 ☿ ⊥ ♇
11 11 ☽ □ ☉
11 48 ☽ ⊡ ♀
15 14 ☽ ⊻ ♄
16 29 ☽ # ♆
19 57 ☽ ∥ ♃
20 46 ☿ ∥ ♀
22 33 ♀ ± ♃
22 55 ☽ △ ☿

10 Monday
03 28 ☽ □ ♅
06 52 ☽ Q ♃
07 50 ☽ ± ♀
16 03 ☽ # ☿
16 38 ☽ # ♀
21 03 ☽ ⊼ ♀

11 Tuesday
00 14 ☽ ∥ ♄
03 36 ☽ # ♂
04 47 ☽ △ ♆
06 26 ☽ ⚹ ♃
08 11 ☽ ⊡ ☿
09 45 ☉ △ ♄
15 10 ☽ ☍ ♂
16 12 ☿ ± ♄
17 33 ♂ Q ♅
20 18 ☿ ⊻ ♅
21 16 ☽ ☍ ♇

12 Wednesday
02 29 ☽ ☌ ♄
03 41 ☽ ± ☿
04 00 ☽ △ ☉
08 50 ☉ ∥ ♇
09 58 ☽ ⊡ ♆
11 11 ☽ ∠ ♃
14 29 ☽ △ ♅
16 26 ☽ ⊼ ☿

13 Thursday
03 01 ☽ ± ♆
04 00 ☽ ⊥ ♃
10 44 ☽ ⊡ ☉
12 22 ☽ ☍ ♀
14 11 ☽ ⊼ ♆
15 01 ☽ ⊻ ♃
18 35 ☽ ⊡ ♅

14 Friday
02 36 ☽ ⊼ ♂
04 34 ☽ ± ☉
05 29 ☽ ⊼ ♇
09 31 ♀ ⊻ ♆
10 12 ☽ ⊻ ♄
10 57 ☽ ± ♅
13 56 ☽ ± ♂
15 34 ♀ ⊼ ♃
16 12 ☽ ⊼ ☉
16 15 ☽ ± ♇
20 52 ☽ ⊥ ♄
21 40 ☽ ⊼ ♅

15 Saturday
05 08 ☽ ☍ ☿
06 37 ☽ ⊡ ♂
08 07 ☽ ⊡ ♇
12 27 ☽ # ♂
12 37 ☽ ∠ ♄
17 48 ☽ ∥ ♄
19 40 ☽ ☍ ♆
19 49 ☽ ☌ ♃
20 48 ☉ ∥ ♅
22 38 ☽ ⊼ ♀

16 Sunday
01 14 ☽ # ♀
06 24 ☽ # ☿
07 50 ☿ ∠ ♂
09 12 ♃ ☍ ♆
09 39 ☽ △ ♂
09 46 ☽ ± ♀
09 55 ☽ △ ♇
11 35 ☿ ∠ ♇
14 05 ♀ ∠ ♅
14 14 ☽ ⚹ ♄
16 00 ♂ ☌ ♇
16 49 ☽ ∥ ♃
21 24 ☽ # ♆
23 51 ☽ ☍ ☉

17 Monday
01 18 ☽ ☍ ♅
02 14 ☽ ⊡ ♀
10 18 ☽ Q ♄
13 33 ☽ ⊼ ☿
14 10 ☽ # ♇
19 25 ☽ # ♅
21 38 ☉ ☌ ♅
21 46 ☽ ⊻ ♃
22 10 ☽ ⊼ ♆
22 17 ☽ # ☉

18 Tuesday
00 31 ☽ ± ☿
05 13 ☽ △ ♀
07 08 ☿ ⊡ ♄
07 34 ☽ ⊥ ♃
08 04 ☽ ± ♆
11 51 ☽ □ ♇
13 48 ☽ □ ♂
15 56 ☽ □ ♄
16 53 ☽ ⊡ ☿
22 12 ☽ ∠ ♃
22 52 ☽ ⊡ ♆

19 Wednesday
02 58 ☽ ⊼ ♅
05 00 ☽ ⊼ ☉
12 49 ☽ ± ♅
15 33 ☽ ± ☉
17 22 ☽ Q ♇
20 07 ☽ △ ☿
20 38 ☽ Q ♂
22 34 ☽ ⚹ ♃
23 30 ☽ △ ♆

20 Thursday
03 38 ☽ ⊡ ♅
07 25 ☽ ⊡ ☉
10 39 ☽ □ ♀
13 06 ☽ ⚹ ♇
14 14 ♂ ☍ ♄
17 08 ☽ △ ♄
17 16 ☽ ⚹ ♂
18 09 ☽ Q ♃
18 19 ☉ Q ♇
19 12 ☿ ☍ ♃

21 Friday
04 29 ☽ △ ♅
06 50 ☿ ☌ ♆
10 04 ☽ △ ☉
13 17 ☽ ∥ ☉
13 59 ☽ ∠ ♇
17 47 ♀ ⊻ ♇
18 03 ☽ ⊡ ♄
19 07 ☽ Q ♀
19 21 ☽ ∠ ♂
21 11 ☽ ∥ ♅
23 52 ☽ □ ♃

22 Saturday
01 20 ☽ □ ♆
02 47 ☽ ∥ ♇
03 25 ☽ □ ☿
05 05 ☽ ⊥ ♇
07 30 ☿ # ♃
07 40 ♄ St D
09 10 ☽ ± ♄
11 12 ☽ ⊥ ♂
15 14 ☽ ⊻ ♇
17 08 ☽ ⚹ ♀
19 20 ☽ ⊼ ♄
19 22 ☽ ∥ ♆
21 52 ☽ ⊻ ♂
23 58 ☽ ∥ ☿

23 Sunday
01 22 ☽ # ♃
07 15 ☽ □ ♅
07 58 ☽ Q ♆
12 02 ☽ ∥ ♀
13 41 ☽ Q ☿
16 46 ☽ □ ☉
19 45 ♀ ⊼ ♄
21 11 ☽ ∠ ♀
23 21 ☽ # ♄

24 Monday
02 40 ☽ △ ♃
04 45 ☽ ⚹ ♆
08 20 ☽ ∥ ♂
13 05 ☽ ⚹ ☿
14 19 ☽ ⊥ ♀
14 35 ☽ Q ♅
17 36 ♀ ⊥ ♅
19 05 ☽ ☌ ♇
23 16 ☽ ☍ ♄

25 Tuesday
01 51 ☽ ⊻ ♀
02 41 ☽ Q ☉
04 28 ☽ ☌ ♂
04 46 ☽ ⊡ ♃
07 09 ☽ ∠ ♆
09 41 ☿ ∥ ♆
10 01 ♂ ⊡ ♃
11 50 ☽ ⚹ ♅
18 59 ☽ ∠ ☿
20 38 ☽ ± ♃
23 13 ☽ ⊥ ♆

26 Wednesday
01 45 ☽ ⚹ ☉
07 18 ☽ ⊼ ♃
10 01 ☽ ⊻ ♆
13 15 ☽ ⊥ ☿
14 50 ☽ ∠ ♅
18 34 ☿ ⚹ ♇
22 36 ♀ ⊥ ♇

27 Thursday
00 45 ☽ ⊻ ♇
01 35 ☽ ⊻ ☿
05 04 ☽ ⊼ ♄
07 05 ☽ ∠ ☉
07 14 ☽ ⊥ ♅
11 42 ☽ ⊥ ♇
12 58 ☽ ☌ ♀
13 10 ☽ ⊻ ♂
16 02 ☽ ± ♄
16 42 ♂ ∠ ♆
17 20 ♀ ⊻ ♂
17 44 ☉ Q ♂
18 16 ☽ ⊻ ♅

28 Friday
00 45 ☽ ⊥ ♂
00 57 ☽ ⊥ ☉
03 27 ☽ ∥ ♂
04 15 ☽ ∠ ♇
06 51 ☿ △ ♄
08 38 ☽ ⊡ ♄
12 59 ☽ ⊻ ☉
13 41 ☽ ☍ ♃
14 10 ☽ # ♄
17 04 ☽ ☌ ♆
18 16 ☽ ∠ ♂
21 16 ☉ ⊼ ♃
23 28 ♀ ± ♄

All ephemeris data is given at 12.00 UT and the Moon's longitude is additionally given for 24.00 UT

MARCH 2003

Date	Sidereal time h m s	Sun ☉ ° ' "	Moon ☽ ° ' "	Moon ☽ 24.00 ° ' "	Mercury ☿ ° '	Venus ♀ ° '	Mars ♂ ° '	Jupiter ♃ ° '	Saturn ♄ ° '	Uranus ♅ ° '	Neptune ♆ ° '	Pluto ♇ ° '
01	22 35 32	10 ♓ 28 58	21 ♒ 50 19	28 ♒ 11 32	24 ♒ 07	28 ♑ 47	27 ♐ 50	09 ♌ 48	22 ♊ 11	29 ♒ 28	11 ♒ 45	19 ♐ 49
02	22 39 29	11 29 12	04 ♓ 29 54	10 ♓ 45 22	25 44	29 ♑ 58	28 29	09 R 42	22 12	29 32	11 47	19 50
03	22 43 25	12 29 25	16 ♓ 57 55	23 ♓ 07 36	27 22	01 ♒ 09	29 07	09 37	22 13	29 35	11 49	19 51
04	22 47 22	13 29 35	29 ♓ 14 28	05 ♈ 18 38	29 ♒ 01	02 19	29 ♐ 45	09 31	22 14	29 39	11 51	19 51
05	22 51 18	14 29 44	11 ♈ 20 16	17 ♈ 19 37	00 ♓ 42	03 30	00 ♑ 23	09 26	22 15	29 42	11 53	19 52
06	22 55 15	15 29 51	23 ♈ 16 56	29 ♈ 12 36	02 23	04 41	01 02	09 20	22 16	29 45	11 55	19 52
07	22 59 11	16 29 56	05 ♉ 07 01	11 ♉ 00 39	04 05	05 52	01 40	09 15	22 18	29 49	11 57	19 53
08	23 03 08	17 29 59	16 ♉ 54 00	22 ♉ 47 40	05 49	07 03	02 18	09 10	22 19	29 52	11 59	19 53
09	23 07 04	18 29 59	28 ♉ 42 14	04 ♊ 38 21	07 34	08 14	02 56	09 05	22 21	29 55	12 01	19 54
10	23 11 01	19 29 58	10 ♊ 36 42	16 ♊ 37 57	09 19	09 25	03 34	09 01	22 23	29 ♒ 59	12 03	19 54
11	23 14 58	20 29 54	22 ♊ 42 49	28 ♊ 51 59	11 06	10 36	04 12	08 56	22 25	00 ♓ 02	12 05	19 55
12	23 18 54	21 29 48	05 ♋ 06 07	11 ♋ 25 49	12 54	11 47	04 50	08 52	22 27	00 05	12 07	19 55
13	23 22 51	22 29 41	17 ♋ 51 39	24 ♋ 24 06	14 44	12 59	05 29	08 48	22 29	00 09	12 08	19 56
14	23 26 47	23 29 30	01 ♌ 03 30	07 ♌ 50 05	16 34	14 10	06 07	08 44	22 31	00 12	12 10	19 56
15	23 30 44	24 29 18	14 ♌ 43 55	21 ♌ 44 53	18 26	15 21	06 45	08 40	22 33	00 15	12 12	19 56
16	23 34 40	25 29 03	28 ♌ 52 38	06 ♍ 06 39	20 18	16 33	07 23	08 37	22 36	00 18	12 14	19 56
17	23 38 37	26 28 46	13 ♍ 26 13	20 ♍ 50 25	22 12	17 44	08 00	08 33	22 38	00 22	12 16	19 57
18	23 42 33	27 28 28	28 ♍ 18 13	05 ♎ 48 26	24 07	18 56	08 38	08 30	22 41	00 25	12 17	19 57
19	23 46 30	28 28 07	13 ♎ 19 51	20 ♎ 51 16	26 03	20 07	09 16	08 27	22 43	00 28	12 19	19 57
20	23 50 27	29 ♓ 27 44	28 ♎ 21 32	05 ♏ 49 34	28 01	21 19	09 54	08 24	22 46	00 31	12 21	19 57
21	23 54 23	00 ♈ 27 19	13 ♏ 14 26	20 ♏ 35 24	29 ♓ 59	22 31	10 32	08 21	22 49	00 34	12 22	19 57
22	23 58 20	01 26 53	27 ♏ 51 52	05 ♐ 03 25	01 ♈ 58	23 42	11 10	08 19	22 52	00 37	12 24	19 57
23	00 02 16	02 26 25	12 ♐ 09 47	19 ♐ 10 52	03 58	24 54	11 48	08 17	22 55	00 40	12 26	19 R 57
24	00 06 13	03 25 55	26 ♐ 06 41	02 ♑ 57 19	05 59	26 06	12 25	08 14	22 59	00 43	12 27	19 57
25	00 10 09	04 25 24	09 ♑ 42 58	16 ♑ 23 53	08 00	27 18	13 03	08 13	23 02	00 46	12 29	19 57
26	00 14 06	05 24 50	23 ♑ 00 18	29 ♑ 32 32	10 02	28 29	13 41	08 11	23 05	00 49	12 30	19 57
27	00 18 02	06 24 15	06 ♒ 00 51	12 ♒ 25 32	12 04	29 ♒ 41	14 18	08 09	23 09	00 52	12 32	19 57
28	00 21 59	07 23 39	18 ♒ 46 51	25 ♒ 05 00	14 06	00 ♓ 53	14 56	08 08	23 12	00 55	12 34	19 57
29	00 25 56	08 23 00	01 ♓ 20 14	07 ♓ 32 44	16 08	02 05	15 33	08 07	23 16	00 58	12 35	19 56
30	00 29 52	09 22 19	13 ♓ 42 39	19 ♓ 50 11	18 09	03 17	16 11	08 06	23 20	01 01	12 36	19 56
31	00 33 49	10 ♈ 21 37	25 ♓ 55 26	01 ♈ 58 34	20 ♈ 10	04 ♓ 29	16 ♑ 48	08 ♌ 05	23 ♊ 24	01 ♓ 04	12 ♒ 38	19 ♐ 56

Date	Moon True ☊ ° '	Moon Mean ☊ ° '	Moon ☽ Latitude ° '
01	03 ♊ 32	03 ♊ 53	04 S 54
02	03 R 18	03 50	05 00
03	03 04	03 47	04 51
04	02 51	03 43	04 29
05	02 40	03 40	03 55
06	02 31	03 37	03 11
07	02 25	03 34	02 19
08	02 22	03 31	01 21
09	02 21	03 28	00 S 19
10	02 D 21	03 24	00 N 44
11	02 R 21	03 21	01 46
12	02 20	03 18	02 45
13	02 17	03 15	03 37
14	02 12	03 12	04 20
15	02 04	03 08	04 49
16	01 55	03 05	05 03
17	01 45	03 02	04 57
18	01 35	02 59	04 31
19	01 26	02 56	03 46
20	01 20	02 53	02 46
21	01 16	02 49	01 35
22	01 15	02 46	00 N 18
23	01 D 15	02 43	00 S 58
24	01 16	02 40	02 09
25	01 R 16	02 37	03 11
26	01 14	02 34	04 01
27	01 11	02 30	04 37
28	01 04	02 27	04 59
29	00 56	02 24	05 06
30	00 47	02 21	04 58
31	00 ♊ 37	02 ♊ 18	04 S 37

DECLINATIONS

Date	Sun ☉ ° '	Moon ☽ ° '	Mercury ☿ ° '	Venus ♀ ° '	Mars ♂ ° '	Jupiter ♃ ° '	Saturn ♄ ° '	Uranus ♅ ° '	Neptune ♆ ° '	Pluto ♇ ° '
01	07 S 38	18 S 51	15 S 25	19 S 29	23 S 27	18 N 39	22 N 07	12 S 19	17 S 14	13 S 45
02	07 15	14 31	14 54	19 18	23 28	18 41	22 07	12 18	17 14	13 44
03	06 52	09 37	14 23	19 07	23 30	18 42	22 08	12 17	17 13	13 44
04	06 29	04 S 25	13 49	18 55	23 31	18 44	22 08	12 16	17 13	13 44
05	06 06	00 N 52	13 15	18 43	23 32	18 45	22 08	12 15	17 12	13 44
06	05 43	06 05	12 39	18 30	23 33	18 47	22 08	12 13	17 11	13 44
07	05 20	11 03	12 02	18 17	23 33	18 48	22 09	12 12	17 11	13 44
08	04 56	15 35	11 24	18 03	23 34	18 49	22 09	12 11	17 10	13 43
09	04 33	19 33	10 45	17 48	23 34	18 50	22 09	12 10	17 10	13 43
10	04 09	22 46	10 04	17 33	23 34	18 52	22 10	12 09	17 09	13 43
11	03 46	25 00	09 22	17 18	23 34	18 53	22 10	12 08	17 09	13 43
12	03 22	26 05	08 39	17 02	23 34	18 54	22 10	12 07	17 08	13 43
13	02 59	25 50	07 54	16 46	23 34	18 55	22 11	12 05	17 08	13 42
14	02 35	24 09	07 09	16 29	23 33	18 56	22 11	12 04	17 07	13 42
15	02 11	21 01	06 22	16 11	23 33	18 57	22 11	12 03	17 07	13 42
16	01 48	16 35	05 34	15 54	23 32	18 58	22 12	12 02	17 06	13 42
17	01 24	11 05	04 45	15 35	23 31	18 59	22 12	12 01	17 06	13 42
18	01 00	04 N 49	03 55	15 17	23 30	19 00	22 12	12 00	17 05	13 41
19	00 37	01 S 47	03 04	14 57	23 28	19 00	22 13	11 59	17 05	13 41
20	00 S 13	08 18	02 11	14 38	23 27	19 01	22 13	11 58	17 04	13 41
21	00 N 11	14 19	01 18	14 18	23 25	19 02	22 13	11 56	17 04	13 41
22	00 35	19 23	00 S 24	13 58	23 23	19 02	22 14	11 55	17 03	13 41
23	00 58	23 13	00 N 30	13 37	23 21	19 03	22 14	11 54	17 03	13 40
24	01 22	25 32	01 26	13 16	23 19	19 03	22 14	11 53	17 03	13 40
25	01 45	26 15	02 22	12 54	23 17	19 04	22 15	11 52	17 02	13 40
26	02 09	25 26	03 18	12 32	23 14	19 04	22 15	11 51	17 02	13 40
27	02 33	23 15	04 15	12 10	23 12	19 05	22 16	11 50	17 01	13 40
28	02 56	19 56	05 12	11 47	23 09	19 05	22 16	11 49	17 01	13 39
29	03 19	15 45	06 09	11 24	23 06	19 05	22 16	11 48	17 00	13 39
30	03 43	10 59	07 05	11 01	23 03	19 05	22 17	11 47	17 00	13 39
31	04 N 06	05 S 51	08 N 01	10 S 38	23 S 00	19 N 06	22 N 17	11 S 46	17 S 00	13 S 39

ZODIAC SIGN ENTRIES

Date	h	m	Planets
02	03	26	☽ ♓
02	12	40	♀ ♒
04	13	30	☽ ♈
04	21	17	♂ ♑
05	02	04	☿ ♓
07	01	36	☽ ♉
09	14	38	☽ ♊
10	20	53	♅ ♓
12	02	12	☽ ♋
14	10	06	☽ ♌
16	13	52	☽ ♍
18	14	43	☽ ♎
20	14	38	☽ ♏
21	01	00	☉ ♈
21	12	16	☿ ♈
22	15	33	☽ ♐
24	18	48	☽ ♑
27	00	51	☽ ♒
27	18	14	♀ ♓
29	09	26	☽ ♓
31	20	04	☽ ♈

LATITUDES

Date	Mercury ☿ ° '	Venus ♀ ° '	Mars ♂ ° '	Jupiter ♃ ° '	Saturn ♄ ° '	Uranus ♅ ° '	Neptune ♆ ° '	Pluto ♇ ° '
01	02 S 03	00 N 56	00 S 01	00 N 53	01 S 06	00 S 43	00 N 02	09 N 20
04	02 09	00 45	00 04	00 53	01 05	00 43	00 02	09 21
07	02 11	00 33	00 08	00 53	01 04	00 43	00 02	09 21
10	02 09	00 21	00 11	00 53	01 04	00 43	00 02	09 22
13	02 03	00 N 10	00 14	00 53	01 03	00 43	00 02	09 23
16	01 53	00 S 01	00 18	00 53	01 02	00 43	00 02	09 23
19	01 37	00 12	00 21	00 53	01 02	00 43	00 02	09 24
22	01 18	00 22	00 25	00 53	01 01	00 43	00 02	09 24
25	00 53	00 31	00 29	00 53	01 01	00 43	00 02	09 25
28	00 S 24	00 40	00 33	00 53	01 00	00 43	00 01	09 26
31	00 N 09	00 S 49	00 S 37	00 N 52	01 S 00	00 S 43	00 N 01	09 N 26

DATA

Julian Date	2452700
Delta T	+65 seconds
Ayanamsa	23° 53' 50"
Synetic vernal point	05° ♓ 13' 09"
True obliquity of ecliptic	23° 26' 25"

MOON'S PHASES, APSIDES AND POSITIONS ☽

Date	h	m	Phase	Longitude ° '	Eclipse Indicator
03	02	35	●	12 ♓ 06	
11	07	15	☽	20 ♊ 18	
18	10	35	○	27 ♍ 25	
25	01	51	☾	04 ♑ 00	

Day	h	m		
07	16	37	Apogee	
19	19	09	Perigee	
05	08	02	ON	
12	19	44	Max dec	26° N 10'
19	05	33	OS	
25	10	57	Max dec	26° S 16'

LONGITUDES

Date	Chiron ⚷ ° '	Ceres ⚳ ° '	Pallas ⚴ ° '	Juno ⚵ ° '	Vesta ⚶ ° '	Black Moon Lilith ⚸ ° '
01	16 ♑ 07	25 ♈ 36	14 ♓ 07	20 ♏ 21	11 ♎ 20	02 ♉ 07
11	16 ♑ 46	29 ♈ 16	17 ♓ 29	20 ♏ 27	09 ♎ 35	03 ♉ 14
21	17 ♑ 18	03 ♉ 02	20 ♓ 53	20 ♏ 00	07 ♎ 16	04 ♉ 21
31	17 ♑ 43	06 ♉ 55	24 ♓ 15	18 ♏ 59	04 ♎ 43	05 ♉ 27

All ephemeris data is given at 12.00 UT and the Moon's longitude is additionally given for 24.00 UT

ASPECTARIAN

h m	Aspects
01 Saturday	
08 00	☽ ∥ ♀
08 12	☽ ✶ ♇
12 39	☽ △ ♄
13 14	☽ # ♃
16 54	☽ ☌ ☿
21 25	☽ ∥ ♆
23 56	☽ ✶ ♂
02 Sunday	
02 29	☽ ⊻ ♀
02 30	☽ ☌ ♅
02 39	♀ ⊻ ♅
06 55	☽ Q ♇
09 41	☽ ∥ ☿
15 06	☽ ⊥ ♀
15 55	☽ ∥ ♇
19 19	☉ ⊻ ♆
21 54	☽ ⊼ ♃
23 07	☽ ∥ ♅
03 Monday	
00 05	☽ Q ♂
02 01	☽ ⊻ ♆
02 35	☽ ☌ ☉
09 24	☽ ± ♃
10 15	☽ ∠ ♀
13 39	☽ ⊥ ♆
17 36	☽ □ ♇
22 14	☽ □ ♄
04 Tuesday	
01 47	☽ ∥ ☉
02 48	☽ ⚼ ♃
07 17	☽ ∠ ♆
07 30	♂ ✶ ♅
11 30	☽ ⊻ ☿
12 48	☽ ⊻ ♅
13 04	☽ □ ♂
15 47	☿ ∥ ♇
18 44	☽ ✶ ♀
21 16	☿ ☌ ♅
05 Wednesday	
00 43	☽ ⊥ ♅
01 15	☽ ⊥ ☿
05 01	☿ ✶ ♂
08 05	♀ # ♃
08 13	☽ △ ♃
09 50	☽ Q ♄
13 06	☽ ✶ ♆
18 46	☽ ∠ ♅
18 54	☽ ⊻ ☉
21 15	☽ Q ♀
22 09	☽ ∠ ☿
06 Thursday	
04 44	☿ Q ♇
05 07	☽ △ ♇
08 04	☽ ⊥ ☉
08 30	☉ ± ♃
09 58	☽ ✶ ♄
10 24	☽ # ☉
13 17	☽ Q ♆
15 52	♀ ∠ ♇
07 Friday	
01 10	☽ ✶ ♅
03 58	☽ ∠ ☉
04 35	☽ △ ♂
05 26	☿ ∥ ♅
09 34	☽ ✶ ☿
11 31	☽ ⚼ ♇
13 42	☽ □ ♀
16 27	☽ ∠ ♄
16 28	☽ # ☿
17 53	☽ # ♅
20 22	☽ □ ♃
08 Saturday	
01 42	☽ Q ♅
01 50	☽ # ♇
01 57	☽ □ ♆
05 52	☽ ± ♇
10 49	☽ ⊥ ♄
12 52	☽ ⚼ ♂
13 20	☽ ✶ ☉
14 11	☽ Q ☿
17 42	♀ ⚼ ♄
18 06	☽ ⊼ ♇
21 05	☽ # ♆
23 04	☽ ⊻ ♄
23 57	☉ ⊥ ♆
09 Sunday	
01 32	☽ # ♀
07 19	☽ ∥ ♃
08 13	☽ ± ♂
08 45	☽ Q ♃
14 29	☽ □ ♅
15 58	☽ Q ☉
17 11	♂ ± ♃
21 03	☽ ⊼ ♂
10 Monday	
04 18	♀ ☍ ♃
06 56	☽ ∥ ♄
07 57	☿ ⊼ ♃
08 49	☽ ✶ ♃
08 58	☽ □ ☿
09 21	☽ △ ♀
14 53	☽ △ ♆
15 50	☿ ⊻ ♀
18 41	♀ ⊥ ♂
19 28	☽ # ♂
21 51	☉ □ ♇
11 Tuesday	
06 29	☽ ☍ ♇
07 15	☽ □ ☉
11 24	☽ ☌ ♄
14 23	☽ ∠ ♃
18 16	☽ ⚼ ♀
20 33	☽ ⚼ ♆
12 Wednesday	
01 11	☿ ⊻ ♆
02 18	♀ ∥ ♆
02 19	☽ △ ♅
07 45	☽ ⊥ ♃
11 29	☽ ☍ ♂
13 27	☽ ± ♀
13 56	☽ ± ♆
18 35	♀ ☌ ♆
19 07	☽ ⊻ ♃
13 Thursday	
01 18	☽ ⊼ ♆
01 59	☽ ⊼ ♀
05 13	☽ △ ☿
06 56	☽ ⚼ ♅
11 38	☉ □ ♄
12 53	☿ ± ♃
15 49	☽ ⊼ ♇
20 31	☽ ⊻ ♄
21 13	☽ △ ☉
23 35	☽ ± ♅
14 Friday	
02 46	☽ ± ♇
07 26	☽ ⊥ ♄
10 27	☽ ⊼ ♅
13 03	☽ ⚼ ☿
14 24	♂ ⊥ ♆
17 28	☉ ⚼ ♃
17 32	☽ # ♂
18 53	☽ ⚼ ♇
21 24	☽ ⊼ ♂
23 28	☽ ∠ ♄
15 Saturday	
01 31	☽ ☌ ♃
02 11	☽ ⚼ ☉
04 15	☽ ∥ ♄
07 24	☽ ± ☿
07 37	☽ ☍ ♆
08 24	☽ ± ♂
09 02	☿ ⊥ ♆
13 11	☽ ☍ ♀
18 06	☿ Q ♂
18 57	☽ ± ☉
19 19	☽ ⊼ ☿
20 55	☽ △ ♇
16 Sunday	
00 00	☽ ∥ ♃
00 33	☽ ⚼ ♂
01 24	☽ ✶ ♄
05 53	☽ ⊼ ☉
07 19	☿ □ ♇
09 30	☽ # ♆
14 24	☽ ☍ ♅
15 28	☽ # ♀
21 32	☽ Q ♄
17 Monday	
01 05	☽ # ♇
02 43	☽ △ ♂
04 03	☽ ⊻ ♃
08 09	☽ # ♅
10 05	☽ ⊼ ♆
13 49	☽ ⊥ ♃
17 34	☿ □ ♄
19 36	☽ ⊼ ♀
19 51	☽ ± ♆
22 33	☽ □ ♇
18 Tuesday	
02 56	☽ □ ♄
04 17	☽ ☍ ☿
04 19	☽ ∠ ♃
04 28	☿ ⚼ ♃
06 07	☽ ± ♀
07 05	♂ ⊼ ♃
07 23	☉ ∠ ♆
10 22	☽ ⚼ ♆
10 35	☽ ☍ ☉
15 23	☽ ⊼ ♅
15 50	☽ # ☿
21 47	☽ ⚼ ♀
19 Wednesday	
01 01	☽ ± ♅
02 48	☽ # ☉
03 25	☽ Q ♇
04 14	☽ ✶ ♃
05 15	☽ □ ♂
08 00	☽ ∥ ☉
08 29	♀ ✶ ♇
10 23	☽ △ ♆
14 07	☿ ⊥ ♀
15 25	☽ ⚼ ♅
16 05	☽ ∥ ☿
22 33	☽ ✶ ♇
23 19	☽ Q ♃
23 46	☽ △ ♀
20 Thursday	
03 02	☽ △ ♄
03 45	☿ ∠ ♆
11 14	☽ Q ♂
11 21	☽ ⊼ ☿
13 54	☽ ⊼ ☉
15 28	☽ △ ♅
22 27	☽ ± ☿
22 35	☽ ∠ ♇
21 Friday	
00 14	☽ ± ☉
02 13	☽ ∥ ♅
03 11	☽ ⚼ ♄
04 06	☽ □ ♃
07 25	☽ ✶ ♂
09 20	☽ ∥ ♇
10 35	☽ □ ♆
11 58	☽ ∥ ♀
13 09	☽ ⊥ ♇
14 55	☉ ⊻ ♅
15 16	☽ ⚼ ☿
15 52	☽ ⚼ ☉
17 51	☽ ± ♄
18 31	♀ △ ♄
19 21	☿ ⊻ ♅
22 57	☽ ⊻ ♇
23 34	☉ ☌ ☿
22 Saturday	
00 21	☽ ∥ ♆
03 43	☽ ⊼ ♄
04 30	☽ □ ♀
08 50	☉ # ☿
09 03	☽ ∠ ♂
10 08	☽ # ♃
16 13	☽ Q ♆
16 36	☽ □ ♅
18 24	☽ △ ☉
19 55	☽ △ ☿
23 Sunday	
00 44	☽ ⊥ ♂
04 55	☽ # ♄
05 13	♇ St R
05 26	☽ △ ♃
07 51	♀ ∥ ♇
11 20	☽ ⊻ ♂
12 27	☽ ✶ ♆
13 07	☽ ∥ ♂
13 22	☽ Q ♀
23 10	☽ Q ♅
24 Monday	
01 20	☽ ☌ ♇
06 32	☽ ☍ ♄
07 01	☽ ⚼ ♃
08 59	☉ ∥ ☿
11 58	☽ ✶ ♀
13 20	♂ ⊻ ♆
14 21	☽ ∠ ♆
20 06	☽ ✶ ♅
21 06	☿ ⊥ ♅
22 43	☽ ± ♃
25 Tuesday	
01 51	☽ □ ☉
06 14	☽ ⊥ ♆
08 24	☽ □ ☿
09 19	☽ ⊼ ♃
14 27	☿ △ ♃
16 58	☽ ⊻ ♆
17 04	☽ ∠ ♀
18 16	☽ ☌ ♂
22 55	☽ ∠ ♅
26 Wednesday	
06 26	☽ ⊻ ♇
10 58	☽ ⊥ ♀
12 09	☽ ⊼ ♄
12 48	☽ Q ☉
15 20	☽ ⊥ ♅
17 23	☽ ⊥ ♇
19 53	♀ ∠ ♂
22 55	☽ Q ☿
23 05	☽ ⊻ ♀
23 13	☽ ± ♄
27 Thursday	
00 51	☿ Q ♄
02 25	☽ ⊻ ♅
10 01	☽ ∠ ♇
12 23	☽ ∥ ♂
12 47	☽ ✶ ☉
14 14	♃ ∠ ♄
15 59	☽ ☍ ♃
16 00	☽ ⚼ ♄
17 36	☿ ✶ ♆
19 55	☽ # ♄
23 59	☉ ⊥ ♅
28 Friday	
00 14	☽ ☌ ♆
01 28	☽ ✶ ☿
04 21	☽ ⊻ ♂
10 02	♀ ∥ ♅
12 44	♀ ☌ ♅
14 12	☽ ✶ ♇
16 18	☽ ⊥ ♂
17 14	☽ # ♃
19 27	☽ ∠ ☉
20 27	☽ △ ♄
29 Saturday	
02 12	☿ □ ♂
05 12	☽ ∥ ♆
05 32	☉ △ ♃
09 04	♀ Q ♇
10 04	☿ ∠ ♅
10 25	☽ ∠ ♂
11 18	☽ ☌ ♅
11 32	☽ ∠ ☿
13 10	☽ Q ♇
13 36	☽ ☌ ♀
14 11	☽ ⊥ ☉
22 55	☽ ∥ ♇
30 Sunday	
01 05	☽ ⊼ ♃
02 49	☽ ⊻ ☉
05 14	♂ ∠ ♅
08 09	☽ ∥ ♅
08 22	☽ ⊥ ☿
09 51	☽ ⊻ ♆
11 52	☽ ∥ ♀
12 45	☽ ± ♃
15 54	☿ ∠ ♀
17 06	☽ ✶ ♂
21 36	☽ ⊥ ♆
21 37	☉ ⊥ ♀
22 25	☽ ⊻ ☿
31 Monday	
00 12	☽ □ ♇
03 33	☽ # ☿
06 24	☽ ⚼ ♃
06 59	☽ □ ♄
09 13	☿ △ ♇
15 23	☽ ∠ ♆
18 01	☽ Q ♂
19 27	☽ # ☉
22 14	☽ ⊻ ♅

APRIL 2003

Date	Sidereal time h m s	Sun ☉ ° ' "	Moon ☽ ° ' "	Moon ☽ 24.00 ° ' "	Mercury ☿ ° '	Venus ♀ ° '	Mars ♂ ° '	Jupiter ♃ ° '	Saturn ♄ ° '	Uranus ♅ ° '	Neptune ♆ ° '	Pluto ♇ ° '
01	00 37 45	11 ♈ 20 52	07 ♈ 59 45	13 ♈ 59 07	22 ♈ 09	05 ♓ 41	17 ♑ 26	08 ♌ 04	23 ♊ 28	01 ♓ 07	12 ♒ 39	19 ♐ 56
02	00 41 42	12 20 05	19 ♈ 56 51	25 ♈ 53 09	24 07	06 53	18 03	08 R 04	23 32	01 10	12 41	19 R 55
03	00 45 38	13 19 17	01 ♉ 48 17	07 ♉ 42 29	26 03	08 05	18 40	08 04	23 36	01 13	12 42	19 55
04	00 49 35	14 18 26	13 ♉ 36 04	19 ♉ 29 23	27 57	09 18	19 18	08 D 04	23 40	01 15	12 43	19 55
05	00 53 31	15 17 33	25 ♉ 22 49	01 ♊ 16 49	29 ♈ 48	10 30	19 55	08 04	23 44	01 18	12 45	19 54
06	00 57 28	16 16 38	07 ♊ 11 52	13 ♊ 08 27	01 ♉ 36	11 42	20 32	08 04	23 49	01 21	12 46	19 54
07	01 01 25	17 15 41	19 ♊ 07 07	25 ♊ 08 29	03 21	12 54	21 09	08 05	23 53	01 23	12 47	19 53
08	01 05 21	18 14 41	01 ♋ 13 06	07 ♋ 21 38	05 02	14 06	21 46	08 05	23 58	01 26	12 48	19 53
09	01 09 18	19 13 39	13 ♋ 34 40	19 ♋ 52 48	06 38	15 19	22 23	08 06	24 03	01 29	12 50	19 52
10	01 13 14	20 12 35	26 ♋ 16 37	02 ♌ 46 37	08 11	16 31	23 00	08 06	24 07	01 31	12 51	19 52
11	01 17 11	21 11 29	09 ♌ 23 17	16 ♌ 06 56	09 39	17 43	23 37	08 09	24 12	01 34	12 52	19 51
12	01 21 07	22 10 20	22 ♌ 57 49	29 ♌ 55 59	11 01	18 55	24 14	08 10	24 17	01 36	12 53	19 50
13	01 25 04	23 09 09	07 ♍ 01 22	14 ♍ 13 39	12 19	20 08	24 51	08 12	24 22	01 39	12 54	19 50
14	01 29 00	24 07 55	21 ♍ 32 21	28 ♍ 56 47	13 32	21 20	25 27	08 14	24 27	01 41	12 55	19 49
15	01 32 57	25 06 40	06 ♎ 26 03	13 ♎ 59 07	14 39	22 32	26 04	08 16	24 32	01 43	12 56	19 48
16	01 36 54	26 05 22	21 ♎ 34 47	29 ♎ 11 47	15 40	23 45	26 40	08 18	24 38	01 46	12 57	19 48
17	01 40 50	27 04 03	06 ♏ 48 50	14 ♏ 24 39	16 36	24 57	27 17	08 20	24 43	01 48	12 58	19 47
18	01 44 47	28 02 41	21 ♏ 58 05	29 ♏ 28 01	17 26	26 10	27 53	08 23	24 48	01 50	12 59	19 46
19	01 48 43	29 01 18	06 ♐ 53 33	14 ♐ 13 57	18 10	27 22	28 30	08 26	24 54	01 53	13 00	19 45
20	01 52 40	29 ♈ 59 53	21 ♐ 28 38	28 ♐ 37 14	18 49	28 35	29 06	08 29	24 59	01 55	13 01	19 44
21	01 56 36	00 ♉ 58 27	05 ♑ 39 31	12 ♑ 35 26	19 21	29 ♓ 47	29 ♑ 42	08 32	25 05	01 57	13 01	19 43
22	02 00 33	01 56 59	19 ♑ 25 03	26 ♑ 08 32	19 47	01 ♈ 00	00 ♒ 18	08 35	25 10	01 59	13 02	19 42
23	02 04 29	02 55 29	02 ♒ 46 09	09 ♒ 18 14	20 07	02 12	00 54	08 39	25 16	02 01	13 03	19 42
24	02 08 26	03 53 57	15 ♒ 45 08	22 ♒ 07 16	20 22	03 25	01 30	08 42	25 22	02 03	13 04	19 41
25	02 12 23	04 52 24	28 ♒ 25 02	04 ♓ 38 51	20 30	04 37	02 06	08 46	25 28	02 05	13 04	19 40
26	02 16 19	05 50 50	10 ♓ 46 43	16 ♓ 56 18	20 R 33	05 50	02 42	08 50	25 34	02 07	13 05	19 39
27	02 20 16	06 49 13	23 ♓ 00 41	29 ♓ 02 41	20 30	07 02	03 18	08 54	25 40	02 09	13 06	19 38
28	02 24 12	07 47 35	05 ♈ 02 38	11 ♈ 00 50	20 22	08 15	03 53	08 58	25 46	02 11	13 06	19 36
29	02 28 09	08 45 56	16 ♈ 57 36	22 ♈ 53 13	20 09	09 27	04 29	09 03	25 52	02 13	13 07	19 35
30	02 32 05	09 ♉ 44 14	28 ♈ 47 57	04 ♉ 42 04	19 ♉ 51	10 ♈ 40	05 ♒ 04	09 ♌ 07	25 ♊ 58	02 ♓ 15	13 ♒ 07	19 ♐ 34

Date	Moon True ☊ ° '	Moon Mean ☊ ° '	Moon ☽ Latitude ° '	DECLINATIONS Sun ☉ ° '	Moon ☽ ° '	Mercury ☿ ° '	Venus ♀ ° '	Mars ♂ ° '	Jupiter ♃ ° '	Saturn ♄ ° '	Uranus ♅ ° '	Neptune ♆ ° '	Pluto ♇ ° '
01	00 ♊ 27	02 ♊ 14	04 S 04	04 N 29	00 S 33	08 N 56	10 S 14	22 S 57	19 N 06	22 N 17	11 S 45	16 S 59	13 S 38
02	00 R 20	02 11	03 20	04 52	04 N 43	09 51	09 49	22 53	19 06	22 18	11 44	16 59	13 38
03	00 14	02 08	02 27	05 16	09 48	10 44	09 25	22 50	19 06	22 18	11 43	16 59	13 38
04	00 10	02 05	01 28	05 38	14 31	11 36	09 00	22 46	19 06	22 19	11 42	16 58	13 38
05	00 08	02 02	00 S 26	06 01	18 42	12 27	08 35	22 42	19 06	22 19	11 41	16 58	13 38
06	00 D 08	01 59	00 N 38	06 24	22 08	13 15	08 10	22 38	19 05	22 19	11 40	16 57	13 37
07	00 09	01 55	01 41	06 47	24 41	14 02	07 45	22 34	19 05	22 20	11 39	16 57	13 37
08	00 11	01 52	02 41	07 09	26 07	14 46	07 19	22 29	19 05	22 20	11 38	16 57	13 37
09	00 12	01 49	03 34	07 32	26 18	15 28	06 53	22 25	19 05	22 21	11 38	16 56	13 37
10	00 R 12	01 46	04 19	07 54	25 08	16 08	06 27	22 20	19 04	22 21	11 37	16 56	13 36
11	00 10	01 43	04 51	08 16	22 35	16 45	06 00	22 16	19 04	22 21	11 36	16 56	13 36
12	00 07	01 40	05 09	08 38	18 43	17 19	05 34	22 11	19 03	22 22	11 35	16 56	13 36
13	00 03	01 36	05 09	09 00	13 43	17 51	05 07	22 06	19 03	22 22	11 34	16 55	13 36
14	29 ♉ 58	01 33	04 50	09 22	07 47	18 20	04 40	22 01	19 02	22 22	11 33	16 55	13 35
15	29 52	01 30	04 11	09 43	01 N 17	18 46	04 13	21 56	19 02	22 23	11 32	16 55	13 35
16	29 48	01 27	03 13	10 05	05 S 25	19 09	03 46	21 50	19 01	22 23	11 32	16 54	13 35
17	29 45	01 24	02 02	10 26	11 52	19 30	03 19	21 45	19 00	22 24	11 31	16 54	13 35
18	29 43	01 20	00 N 42	10 47	17 35	19 47	02 52	21 40	19 00	22 24	11 30	16 54	13 35
19	29 D 43	01 17	00 S 39	11 08	22 06	20 02	02 24	21 34	18 59	22 24	11 29	16 54	13 34
20	29 44	01 14	01 57	11 28	25 06	20 14	01 57	21 28	18 58	22 25	11 28	16 53	13 34
21	29 46	01 11	03 05	11 49	26 23	20 22	01 29	21 22	18 57	22 25	11 28	16 53	13 34
22	29 47	01 08	04 00	12 09	25 59	20 28	01 01	21 16	18 56	22 25	11 27	16 53	13 34
23	29 47	01 05	04 40	12 29	24 00	20 31	00 33	21 10	18 55	22 26	11 26	16 53	13 34
24	29 R 47	01 01	05 05	12 49	20 58	20 31	00 S 06	21 04	18 54	22 26	11 26	16 53	13 33
25	29 45	00 58	05 14	13 09	16 56	20 28	00 N 22	20 58	18 53	22 27	11 25	16 52	13 33
26	29 42	00 55	05 08	13 28	12 16	20 23	00 50	20 52	18 52	22 27	11 24	16 52	13 33
27	29 39	00 52	04 49	13 48	07 12	20 15	01 18	20 45	18 51	22 27	11 24	16 52	13 33
28	29 35	00 49	04 16	14 07	01 S 55	20 04	01 46	20 39	18 50	22 28	11 23	16 52	13 33
29	29 32	00 46	03 33	14 25	03 N 23	19 50	02 14	20 32	18 49	22 28	11 22	16 52	13 32
30	29 ♉ 29	00 ♊ 42	02 S 41	14 N 44	08 N 32	19 N 34	02 N 42	20 S 26	18 N 47	22 N 28	11 S 22	16 S 52	13 S 32

ZODIAC SIGN ENTRIES

Date	h	m	Planets
03	08	20	☽ ♉
05	14	37	☿ ♉
05	21	24	☽ ♊
08	09	36	☽ ♋
10	18	54	☽ ♌
13	00	07	☽ ♍
15	01	42	☽ ♎
17	01	16	☽ ♏
19	00	51	☽ ♐
20	12	03	☉ ♉
21	02	20	☽ ♑
21	16	18	♀ ♈
21	23	48	♂ ♒
23	06	58	☽ ♒
25	15	02	☽ ♓
28	01	54	☽ ♈
30	14	26	☽ ♉

LATITUDES

Date	Mercury ☿ ° '	Venus ♀ ° '	Mars ♂ ° '	Jupiter ♃ ° '	Saturn ♄ ° '	Uranus ♅ ° '	Neptune ♆ ° '	Pluto ♇ ° '
01	00 N 20	00 S 52	00 S 39	00 N 52	00 S 59	00 S 43	00 N 01	09 N 27
04	00 55	00 59	00 43	00 52	00 59	00 43	00 01	09 27
07	01 29	01 07	00 48	00 52	00 58	00 43	00 01	09 28
10	02 00	01 13	00 52	00 52	00 58	00 43	00 01	09 28
13	02 26	01 19	00 57	00 52	00 57	00 44	00 01	09 29
16	02 45	01 24	01 02	00 51	00 57	00 44	00 01	09 29
19	02 54	01 29	01 08	00 51	00 56	00 44	00 01	09 30
22	02 54	01 33	01 13	00 51	00 56	00 44	00 01	09 30
25	02 42	01 36	01 19	00 51	00 55	00 44	00 01	09 31
28	02 18	01 38	01 24	00 51	00 55	00 44	00 01	09 31
31	01 N 43	01 S 40	01 S 30	00 N 50	00 S 54	00 S 44	00 N 01	09 N 31

DATA

Julian Date	2452731
Delta T	+65 seconds
Ayanamsa	23° 53' 53"
Synetic vernal point	05° ♓ 13' 06"
True obliquity of ecliptic	23° 26' 25"

MOON'S PHASES, APSIDES AND POSITIONS ☽

Date	h	m	Phase	Longitude ° '	Eclipse Indicator
01	19	19	●	11 ♈ 39	
09	23	40	☽	19 ♋ 42	
16	19	36	○	26 ♎ 24	
23	12	18	☾	02 ♒ 56	

Day	h	m		
04	04	40	Apogee	
17	05	02	Perigee	
01	14	31	0N	
09	03	23	Max dec	26° N 23'
15	16	36	0S	
21	18	01	Max dec	26° S 27'
28	20	40	0N	

LONGITUDES

Date	Chiron ⚷ ° '	Ceres ⚳ ° '	Pallas ⚴ ° '	Juno ⚵ ° '	Vesta ⚶ ° '	Black Moon Lilith ⚸ ° '
01	17 ♑ 45	07 ♉ 19	24 ♓ 36	18 ♏ 51	04 ♎ 28	05 ♉ 34
11	18 ♑ 00	11 ♉ 16	27 ♓ 57	17 ♏ 15	02 ♎ 03	06 ♉ 41
21	18 ♑ 08	15 ♉ 17	01 ♈ 15	15 ♏ 17	00 ♎ 08	07 ♉ 47
31	18 ♑ 07	19 ♉ 20	04 ♈ 31	13 ♏ 04	28 ♍ 55	08 ♉ 54

All ephemeris data is given at 12.00 UT and the Moon's longitude is additionally given for 24.00 UT

ASPECTARIAN

h m Aspects

01 Tuesday
06 53 ☽ ⊻ ♀
10 14 ☽ ⊥ ♅
12 09 ☽ △ ♃
14 58 ☉ Q ♄
18 59 ☽ Q ♄
19 19 ☽ ☌ ☉
20 13 ☽ ⊥ ♀
21 21 ☽ ⚹ ♆

02 Wednesday
04 21 ☽ ∠ ♅
04 29 ☿ ⚹ ♄
07 58 ☽ □ ♂
11 32 ☿ # ♀
11 57 ☽ △ ♇
12 47 ☽ ∥ ☉
16 22 ☽ ∠ ♀
18 58 ☿ Q ♆
19 16 ☽ ⚹ ♄
20 32 ☉ ⚹ ♆
21 34 ☽ Q ♆
22 05 ☽ ☌ ☿

03 Thursday
10 16 ☽ # ♀
10 47 ☽ ⚹ ♅
11 24 ♀ ⊼ ♃
17 36 ☽ ∥ ☿
18 19 ☽ ⚼ ♇
21 27 ☽ # ♅

04 Friday
00 43 ☽ □ ♃
01 54 ☽ ∠ ♄
02 14 ☽ ⚹ ♀
03 03 ♃ St D
07 18 ☽ # ♇
10 12 ☽ □ ♆
11 18 ☽ Q ♅
12 38 ☽ ± ♇
13 34 ☽ ⊻ ☉
14 45 ☿ # ♅
20 20 ☽ ⊥ ♄

05 Saturday
00 15 ☽ △ ♂
00 51 ☽ ⊼ ♇
01 38 ☽ # ♆
02 55 ☽ ⊥ ☉
05 27 ☽ Q ♀
08 38 ☽ ⊻ ♄
11 29 ♂ ⊻ ♇
13 24 ☽ Q ♃
14 31 ☽ ∥ ♃
22 38 ☽ ⊻ ☿
22 54 ☽ ∠ ☉

06 Sunday
00 05 ☽ □ ♅
08 27 ☽ ⚼ ♂
08 27 ☿ ⚹ ♅
12 58 ☽ ⊥ ☿
13 27 ☽ ∥ ♄
13 44 ☉ ∠ ♅
13 46 ☽ ⚹ ♃
15 54 ☽ # ♂
22 07 ☽ □ ♀
23 13 ☿ # ♇
23 16 ☽ △ ♆

07 Monday
03 37 ☽ ± ♂
07 57 ☽ ⚹ ☉
09 38 ♀ ⊻ ♆
10 12 ☽ ∠ ☿
13 32 ☽ ☍ ♇
16 17 ☽ ⊼ ♂
19 55 ☽ ∠ ♃
21 34 ☽ ☌ ♄

08 Tuesday
05 16 ☽ ⚼ ♆
09 52 ☿ ⚼ ♇
09 55 ☽ Q ☉
11 42 ♀ ± ♃
12 25 ☽ △ ♅
13 43 ☽ ⊥ ♃
16 45 ☉ # ♀
20 36 ☽ ⚹ ☿
22 56 ☽ ± ♆

09 Wednesday
01 26 ☽ ⊻ ♃
10 33 ☽ ⊼ ♆
15 40 ☽ △ ♀
17 34 ☽ ⚼ ♅
23 00 ☽ Q ☿
23 40 ☽ □ ☉
23 58 ☽ ⊼ ♇

10 Thursday
03 32 ☉ △ ♇
05 34 ☽ ☍ ♂
07 57 ☽ ⊻ ♄
09 34 ♂ # ♄
10 35 ☽ ± ♅
11 08 ☿ □ ♃
11 13 ☽ ± ♇
19 10 ☽ ⊥ ♄
21 44 ☽ ⊼ ♅
22 40 ☽ ⚼ ♀

11 Friday
03 47 ☽ ⚼ ♇
04 14 ☿ ∠ ♄
09 46 ☽ ☌ ♃
11 40 ☽ ∠ ♄
12 31 ☽ □ ☿
13 38 ☽ ∥ ♄
14 22 ☽ # ♂
16 36 ☽ ± ♀
18 14 ☽ ☍ ♆
19 18 ☿ # ♆

12 Saturday
04 16 ☽ ⊼ ♀
06 33 ☽ △ ♇
10 10 ☽ ∥ ♃
10 31 ☽ △ ☉
11 08 ♀ ⊥ ♆
14 18 ☽ ⊼ ♂
14 18 ☽ ⚹ ♄
14 30 ♂ ⊼ ♄
18 29 ☽ ∥ ☿
21 12 ☽ # ♆

13 Sunday
01 05 ☽ ± ♂
02 53 ☽ ☍ ♅
06 03 ♀ □ ♇
10 54 ☽ Q ♄
12 30 ☽ # ♇
13 59 ☽ ⊻ ♃
14 02 ☽ ⚼ ☉
16 56 ☽ ⚼ ♂
21 05 ☽ # ♅
21 40 ☽ △ ☿
21 49 ☽ ⊼ ♆
23 27 ☿ □ ♆
23 59 ☽ ⊥ ♃

14 Monday
06 02 ☽ ± ☉
06 15 ☽ ∥ ☉
07 43 ☽ ± ♆
09 11 ☽ □ ♇
11 38 ☽ ☍ ♀
14 46 ☽ ∠ ♃
15 22 ☿ Q ♅
16 31 ☽ ⊼ ☉
16 46 ☽ □ ♄
17 56 ☿ ± ♇
18 38 ☽ △ ♂
20 38 ☉ ⚹ ♄
21 43 ♂ ⊥ ♅
22 21 ☽ ⚼ ♆

15 Tuesday
00 16 ☽ ⚼ ☿
00 32 ☽ # ♀
01 58 ♂ ⊥ ♇
04 27 ☽ ⊼ ♅
07 34 ☉ Q ♆
14 04 ☽ ± ♅
14 11 ☽ Q ♇
14 55 ☽ ⚹ ♃
15 48 ☽ ± ☿
22 21 ☽ △ ♆

16 Wednesday
02 01 ☽ ⊼ ☿
02 51 ♀ ⚼ ♃
03 20 ☿ ∥ ♃
04 23 ☽ ⚼ ♅
06 27 ☽ ∥ ♀
09 11 ☽ ⚹ ♇
09 59 ☽ Q ♃
15 43 ☽ ⊼ ♀
16 50 ☽ △ ♄
19 36 ☽ ☍ ☉
20 22 ☽ □ ♂

17 Thursday
00 58 ♅ Q ♇
01 58 ☽ ± ♀
04 05 ☽ △ ♅
06 08 ☽ # ☉
06 52 ♀ □ ♄
08 48 ☽ ∠ ♇
10 38 ☽ ∥ ♅
14 25 ☽ □ ♃
16 36 ☽ ⚼ ♄
17 23 ☽ ⚼ ♀
18 48 ☽ ∥ ♇
21 43 ☽ □ ♆
22 59 ☽ ⊥ ♇

18 Friday
01 55 ☉ □ ♂
01 56 ☽ Q ♂
04 23 ☽ ☍ ☿
06 56 ☽ ± ♄
08 30 ☽ ⊻ ♇
08 54 ☽ ∥ ♆
16 33 ☽ ⊼ ♄
18 46 ☽ # ♃
19 17 ☽ △ ♀
21 52 ☽ ⚹ ♂
22 24 ☽ ⊼ ☉
23 27 ☽ # ☿

19 Saturday
02 27 ☽ Q ♆
03 52 ☽ □ ♅
08 45 ☽ ± ☉
08 45 ☽ ∥ ♂
13 55 ☽ # ♄
14 31 ☽ △ ♃
21 59 ☽ ⚹ ♆
23 15 ☽ ∠ ♂

20 Sunday
00 29 ☽ ⚼ ☉
00 36 ♀ ∠ ♆
07 22 ☽ ⊼ ☿
09 06 ☽ ☌ ♇
09 24 ☽ Q ♅
12 11 ☉ # ♅
14 50 ☽ ⊥ ♂
15 21 ☽ ⚼ ♃
17 49 ☽ ± ☿
17 55 ☽ ☍ ♄
20 55 ☿ ⊥ ♄
22 59 ☽ ∠ ♆

21 Monday
01 02 ☽ □ ♀
01 23 ☽ ⊻ ♂
03 24 ☽ △ ☉
05 38 ☽ ⚹ ♅
06 37 ☽ ± ♃
08 50 ♀ ⚹ ♂
09 40 ☽ ⚼ ☿
14 21 ☽ ⊥ ♆
16 58 ☽ ⊼ ♃

22 Tuesday
00 46 ☽ ⊻ ♆
07 28 ☿ ⊼ ♇
07 42 ☽ ∠ ♅
11 10 ☽ Q ♀
12 31 ☽ ⊻ ♇
12 40 ☽ △ ☿
12 57 ☉ ⚹ ♅
22 20 ☽ ⊼ ♄
23 12 ☽ ⊥ ♇
23 45 ☽ ⊥ ♅

23 Wednesday
08 21 ♀ ⊻ ♅
08 27 ☽ ☌ ♂
09 15 ☽ ± ♄
10 38 ☽ ⊻ ♅
10 52 ☽ ⚹ ♀
12 18 ☽ □ ☉
15 31 ☽ ∠ ♇
22 50 ☽ ☍ ♃

24 Thursday
01 40 ☽ # ♄
01 53 ☽ ⚼ ♄
05 15 ♂ ± ♄
06 58 ☽ ☌ ♆
11 15 ☽ ∥ ♂
14 54 ☽ # ☿
17 31 ☽ ∠ ♀
19 22 ☽ ⚹ ♇
20 48 ☽ □ ☿

25 Friday
00 33 ☽ Q ☉
00 52 ☽ # ♃
06 19 ☽ △ ♄
06 49 ☉ ⚼ ♇
11 21 ♂ ⊻ ♅
12 18 ☽ ∥ ♆
12 26 ☽ ⊥ ♀
18 13 ☽ Q ♇
19 05 ☽ ☌ ♅
19 27 ☽ ⊻ ♂

26 Saturday
01 14 ☽ ⊻ ♀
01 30 ☽ ⚹ ☉
05 37 ☽ ∥ ♇
06 24 ☽ # ☉
06 30 ☿ ∠ ♀
07 35 ☽ Q ☿
07 40 ☽ ⊥ ♂
08 06 ☽ ⊼ ♃
11 59 ☿ St R
13 58 ☉ ⊻ ♀
16 11 ☽ ∥ ♅
16 26 ☽ ⊻ ♆
17 45 ☉ # ♇
19 54 ☽ ± ♃

27 Sunday
02 12 ☽ ∠ ♂
04 14 ☽ ⊥ ♆
05 19 ☽ □ ♇
07 05 ☽ ⚹ ☿
09 26 ☽ ∠ ☉
13 46 ☽ ⚼ ♃
17 18 ☽ □ ♄
22 07 ☽ ∠ ♆

28 Monday
04 55 ☽ ⊥ ☉
06 16 ☽ ⊻ ♅
09 34 ☽ ⚹ ♂
10 45 ♀ ⊥ ♅
12 37 ☽ # ♀
12 39 ☽ ∠ ☿
18 01 ☽ ⊻ ☉
18 20 ☽ ⊥ ♅
19 09 ☽ ☌ ♀
19 56 ☽ △ ♃

29 Tuesday
03 19 ♀ △ ♃
04 13 ☽ ⚹ ♆
05 42 ☽ Q ♄
06 18 ☽ ∥ ♀
06 27 ☽ ⊥ ☿
10 59 ☽ Q ♂
12 31 ☽ ∠ ♅
16 15 ♂ ∠ ♇
17 19 ☽ △ ♇
18 19 ☽ ⊻ ☿
19 32 ☉ □ ♃

30 Wednesday
04 32 ☽ Q ♆
05 43 ☿ ⊥ ♄
06 12 ☽ ⚹ ♄
19 02 ☽ ⚹ ♅
23 43 ☽ ⚼ ♇

MAY 2003

LONGITUDES

Date	Sidereal time h m s	Sun ☉ ° ' "	Moon ☽ ° ' "	Moon ☽ 24.00 ° ' "	Mercury ☿ ° '	Venus ♀ ° '	Mars ♂ ° '	Jupiter ♃ ° '	Saturn ♄ ° '	Uranus ♅ ° '	Neptune ♆ ° '	Pluto ♇ ° '
01	02 36 02	10 ♉ 42 31	10 ♉ 35 51	16 ♉ 29 33	19 ♉ 29	11 ♈ 53	05 ♒ 40	09 ♌ 12	26 ♊ 04	02 ♓ 17	13 ♒ 08	19 ♐ 33
02	02 39 58	11 40 46	22 ♉ 23 27	28 ♉ 17 48	19 R 03	13 05	06 15	09 17	26 11	02 18	13 08	19 R 32
03	02 43 55	12 39 00	04 ♊ 12 57	10 ♊ 09 10	18 33	14 18	06 50	09 22	26 17	02 20	13 09	19 31
04	02 47 52	13 37 11	16 ♊ 06 50	22 ♊ 06 17	18 01	15 31	07 25	09 28	26 23	02 22	13 09	19 30
05	02 51 48	14 35 21	28 ♊ 07 54	04 ♋ 12 07	17 26	16 43	08 00	09 33	26 30	02 23	13 09	19 28
06	02 55 45	15 33 29	10 ♋ 19 20	16 ♋ 30 01	16 50	17 56	08 34	09 38	26 36	02 25	13 10	19 27
07	02 59 41	16 31 35	22 ♋ 44 38	29 ♋ 03 36	16 13	19 09	09 09	09 44	26 43	02 26	13 10	19 26
08	03 03 38	17 29 39	05 ♌ 27 24	11 ♌ 56 28	15 36	20 21	09 43	09 50	26 49	02 28	13 10	19 25
09	03 07 34	18 27 41	18 ♌ 31 10	25 ♌ 11 52	14 59	21 34	10 18	09 56	26 56	02 29	13 10	19 23
10	03 11 31	19 25 42	01 ♍ 58 48	08 ♍ 52 11	14 23	22 47	10 52	10 02	27 03	02 30	13 11	19 22
11	03 15 27	20 23 40	15 ♍ 52 01	22 ♍ 58 16	13 49	23 59	11 26	10 09	27 10	02 32	13 11	19 21
12	03 19 24	21 21 36	00 ♎ 10 41	07 ♎ 28 51	13 17	25 12	12 00	10 15	27 16	02 33	13 11	19 19
13	03 23 21	22 19 31	14 ♎ 52 14	22 ♎ 20 03	12 47	26 25	12 34	10 22	27 23	02 34	13 11	19 18
14	03 27 17	23 17 24	29 ♎ 51 27	07 ♏ 25 23	12 21	27 38	13 08	10 28	27 30	02 35	13 11	19 17
15	03 31 14	24 15 15	15 ♏ 00 44	22 ♏ 36 20	11 59	28 ♈ 50	13 41	10 35	27 37	02 37	13 11	19 15
16	03 35 10	25 13 05	00 ♐ 10 58	07 ♐ 43 29	11 40	00 ♉ 03	14 14	10 42	27 44	02 38	13 R 11	19 14
17	03 39 07	26 10 53	15 ♐ 12 47	22 ♐ 37 54	11 25	01 16	14 48	10 49	27 51	02 39	13 11	19 12
18	03 43 03	27 08 40	29 ♐ 57 58	07 ♑ 12 20	11 15	02 29	15 21	10 57	27 58	02 40	13 11	19 11
19	03 47 00	28 06 26	14 ♑ 20 28	21 ♑ 22 02	11 09	03 41	15 54	11 04	28 05	02 41	13 11	19 09
20	03 50 56	29 ♉ 04 11	28 ♑ 16 52	05 ♒ 04 57	11 D 07	04 54	16 26	11 11	28 12	02 41	13 11	19 08
21	03 54 53	00 ♊ 01 55	11 ♒ 46 22	18 ♒ 21 22	11 11	06 07	16 59	11 19	28 20	02 42	13 11	19 06
22	03 58 50	00 59 37	24 ♒ 50 17	01 ♓ 13 29	11 18	07 20	17 31	11 27	28 27	02 43	13 10	19 05
23	04 02 46	01 57 18	07 ♓ 31 27	13 ♓ 44 40	11 30	08 33	18 04	11 35	28 34	02 44	13 10	19 03
24	04 06 43	02 54 59	19 ♓ 53 40	25 ♓ 58 57	11 47	09 46	18 36	11 43	28 41	02 45	13 10	19 02
25	04 10 39	03 52 38	02 ♈ 01 04	08 ♈ 00 33	12 08	10 59	19 07	11 51	28 49	02 45	13 10	19 00
26	04 14 36	04 50 16	13 ♈ 57 54	19 ♈ 53 37	12 34	12 11	19 39	11 59	28 56	02 46	13 09	18 59
27	04 18 32	05 47 54	25 ♈ 48 09	01 ♉ 41 57	13 03	13 24	20 10	12 08	29 03	02 46	13 09	18 57
28	04 22 29	06 45 30	07 ♉ 35 26	13 ♉ 28 59	13 37	14 37	20 42	12 16	29 11	02 47	13 09	18 56
29	04 26 25	07 43 05	19 ♉ 22 57	25 ♉ 17 39	14 15	15 50	21 13	12 25	29 18	02 47	13 08	18 54
30	04 30 22	08 40 39	01 ♊ 13 52	07 ♊ 10 33	14 56	17 03	21 43	12 33	29 26	02 48	13 08	18 53
31	04 34 19	09 ♊ 38 13	13 ♊ 09 16	19 ♊ 09 52	15 ♉ 42	18 ♉ 16	22 ♒ 14	12 ♌ 42	29 ♊ 33	02 ♓ 48	13 ♒ 07	18 ♐ 51

DECLINATIONS

Date	Moon True ☊ ° '	Moon Mean ☊ ° '	Moon ☽ Latitude ° '	Sun ☉ ° '	Moon ☽ ° '	Mercury ☿ ° '	Venus ♀ ° '	Mars ♂ ° '	Jupiter ♃ ° '	Saturn ♄ ° '	Uranus ♅ ° '	Neptune ♆ ° '	Pluto ♇ ° '
01	29 ♉ 27	00 ♊ 39	01 S 42	15 N 02	13 N 23	19 N 15	03 N 10	20 S 19	18 N 46	22 N 29	11 S 21	16 S 52	13 S 32
02	29 R 26	00 36	00 S 39	15 20	17 45	18 55	03 38	20 12	18 45	22 29	11 20	16 51	13 32
03	29 D 26	00 33	00 N 26	15 38	21 25	18 33	04 06	20 05	18 43	22 29	11 20	16 51	13 32
04	29 26	00 30	01 31	15 56	24 14	18 09	04 33	19 58	18 42	22 30	11 19	16 51	13 31
05	29 28	00 26	02 32	16 13	25 58	17 43	05 01	19 51	18 40	22 30	11 19	16 51	13 31
06	29 29	00 23	03 28	16 30	26 29	17 17	05 29	19 44	18 39	22 30	11 18	16 51	13 31
07	29 30	00 20	04 14	16 47	25 42	16 50	05 56	19 37	18 37	22 31	11 18	16 51	13 31
08	29 31	00 17	04 50	17 03	23 35	16 22	06 24	19 30	18 35	22 31	11 17	16 51	13 31
09	29 R 31	00 14	05 11	17 19	20 12	15 55	06 51	19 23	18 34	22 31	11 17	16 51	13 30
10	29 31	00 11	05 17	17 35	15 42	15 28	07 18	19 16	18 32	22 31	11 16	16 51	13 30
11	29 30	00 07	05 04	17 51	10 15	15 01	07 45	19 08	18 30	22 32	11 16	16 51	13 30
12	29 29	00 04	04 33	18 06	04 N 06	14 36	08 12	19 01	18 28	22 32	11 16	16 51	13 30
13	29 28	00 ♊ 01	03 43	18 21	02 S 26	14 12	08 39	18 54	18 27	22 32	11 15	16 51	13 30
14	29 28	29 ♉ 58	02 37	18 36	08 58	13 49	09 05	18 46	18 25	22 33	11 15	16 51	13 30
15	29 27	29 55	01 N 19	18 50	15 04	13 29	09 32	18 39	18 23	22 33	11 14	16 51	13 30
16	29 D 27	29 52	00 S 04	19 04	20 00	13 09	09 58	18 31	18 21	22 33	11 14	16 51	13 29
17	29 27	29 48	01 26	19 18	24 03	12 53	10 24	18 24	18 19	22 33	11 14	16 51	13 29
18	29 28	29 45	02 42	19 31	26 08	12 39	10 49	18 16	18 17	22 34	11 13	16 51	13 29
19	29 28	29 42	03 45	19 44	26 23	12 27	11 15	18 09	18 15	22 34	11 13	16 51	13 29
20	29 28	29 39	04 32	19 57	24 57	12 17	11 40	18 01	18 12	22 34	11 12	16 51	13 29
21	29 R 28	29 36	05 03	20 09	22 06	12 10	12 05	17 54	18 10	22 34	11 12	16 51	13 29
22	29 28	29 32	05 17	20 22	18 13	12 05	12 30	17 46	18 08	22 34	11 12	16 51	13 29
23	29 D 28	29 29	05 15	20 33	13 36	12 03	12 54	17 39	18 06	22 35	11 12	16 51	13 29
24	29 28	29 26	04 58	20 45	08 34	12 03	13 18	17 31	18 04	22 35	11 12	16 51	13 28
25	29 28	29 23	04 28	20 56	03 S 18	12 05	13 42	17 24	18 01	22 35	11 12	16 51	13 28
26	29 29	29 20	03 47	21 06	02 N 01	12 09	14 06	17 16	17 59	22 35	11 11	16 51	13 28
27	29 29	29 17	02 57	21 16	07 14	12 15	14 29	17 09	17 57	22 35	11 11	16 51	13 28
28	29 30	29 13	01 59	21 26	12 10	12 24	14 52	17 01	17 54	22 36	11 11	16 52	13 28
29	29 31	29 10	00 S 56	21 36	16 41	12 34	15 14	16 54	17 52	22 36	11 11	16 52	13 28
30	29 R 31	29 07	00 N 10	21 45	20 34	12 47	15 37	16 46	17 49	22 36	11 11	16 52	13 28
31	29 ♉ 30	29 ♉ 04	01 N 15	21 N 54	23 N 37	13 N 01	15 N 58	16 S 39	17 N 47	22 N 36	11 S 11	16 S 52	13 S 28

ZODIAC SIGN ENTRIES

Date	h m	Planets
03	03 27	☽ ♊
05	15 42	☽ ♋
08	01 46	☽ ♌
10	08 31	☽ ♍
12	11 42	☽ ♎
14	12 14	☽ ♏
16	10 58	♀ ♉
16	11 43	☽ ♐
18	12 03	☽ ♑
20	15 01	☽ ♒
21	11 12	☉ ♊
22	21 41	☽ ♓
25	07 59	☽ ♈
27	20 32	☽ ♉
30	09 32	☽ ♊

LATITUDES

Date	Mercury ☿ ° '	Venus ♀ ° '	Mars ♂ ° '	Jupiter ♃ ° '	Saturn ♄ ° '	Uranus ♅ ° '	Neptune ♆ ° '	Pluto ♇ ° '
01	01 N 43	01 S 40	01 S 30	00 N 50	00 S 54	00 S 44	00 N 01	09 N 31
04	00 59	01 41	01 36	00 50	00 54	00 44	00 01	09 32
07	00 N 09	01 41	01 43	00 50	00 53	00 44	00 01	09 32
10	00 S 44	01 41	01 49	00 50	00 53	00 44	00 01	09 32
13	01 33	01 40	01 56	00 50	00 53	00 45	00 01	09 33
16	02 17	01 38	02 03	00 50	00 53	00 45	00 01	09 33
19	02 52	01 36	02 10	00 49	00 52	00 45	00 01	09 33
22	03 18	01 33	02 18	00 49	00 51	00 45	00 01	09 33
25	03 34	01 29	02 26	00 49	00 51	00 45	00 01	09 33
28	03 42	01 25	02 34	00 49	00 51	00 45	00 01	09 33
31	03 S 41	01 S 21	02 S 42	00 N 49	00 S 51	00 S 45	00 N 01	09 N 32

DATA

Julian Date	2452761
Delta T	+65 seconds
Ayanamsa	23° 53' 57"
Synetic vernal point	05° ♓ 13' 03"
True obliquity of ecliptic	23° 26' 25"

MOON'S PHASES, APSIDES AND POSITIONS ☽

Date	h m	Phase	Longitude ° '	Eclipse Indicator
01	12 15	●	10 ♉ 43	
09	11 53	☽	18 ♌ 27	
16	03 36	○	24 ♏ 53	total
23	00 31	☾	01 ♓ 30	
31	04 20	●	09 ♊ 20	Annular

Day	h m		
01	07 55	Apogee	
15	15 44	Perigee	
28	13 18	Apogee	
06	09 45	Max dec	26° N 30'
13	03 10	0S	
19	03 20	Max dec	26° S 30'
26	02 51	0N	

LONGITUDES

Date	Chiron ⚷ ° '	Ceres ⚳ ° '	Pallas ⚴ ° '	Juno ⚵ ° '	Vesta ⚶ ° '	Black Moon Lilith ⚸ ° '
01	18 ♑ 07	19 ♉ 20	04 ♈ 31	13 ♏ 04	28 ♍ 55	08 ♉ 54
11	17 ♑ 59	23 ♉ 25	07 ♈ 41	10 ♏ 48	28 ♍ 32	10 ♉ 00
21	17 ♑ 42	27 ♉ 31	10 ♈ 46	08 ♏ 43	28 ♍ 57	11 ♉ 07
31	17 ♑ 19	01 ♊ 38	13 ♈ 44	06 ♏ 58	00 ♎ 08	12 ♉ 13

ASPECTARIAN

h m Aspects

01 Thursday
01 25 ☽ □ ♂
01 43 ☽ ⋕ ♅
07 38 ☿ ⚻ ♇
09 09 ☽ □ ♃
12 15 ☽ ☌ ☉
12 46 ☽ ⋕ ♇
12 58 ☽ ∠ ♄
14 54 ☽ ⊻ ♀
17 09 ☽ □ ♆
18 00 ☽ ± ♇
19 30 ☽ Q ♅
21 23 ☽ ∥ ☉
22 01 ☉ ∠ ♄

02 Friday
04 31 ☽ ⊥ ♀
05 27 ☽ ☌ ☿
06 12 ☽ ⚻ ♇
06 51 ☽ ⋕ ♆
07 27 ☽ ⊥ ♄
11 24 ☿ ⊥ ♀
12 57 ♀ ⚹ ♆
18 01 ☽ ∥ ♃
18 31 ☽ ∥ ☿
19 46 ☽ ⊻ ♄
22 01 ☽ Q ♃

03 Saturday
00 14 ☿ ∥ ♃
00 54 ☽ ∠ ♀
02 58 ☽ ⋕ ♂
08 11 ☽ □ ♅
11 36 ♀ Q ♄
17 34 ☽ △ ♂
20 13 ☽ ∥ ♄
22 30 ☽ ⚹ ♃

04 Sunday
00 18 ☉ □ ♆
06 02 ☽ △ ♆
06 33 ☽ ⊻ ☉
08 56 ☉ ± ♇
10 39 ☽ ⚹ ♀
15 39 ☽ ⊻ ☿
18 46 ☽ ☍ ♇
19 39 ☽ ⊥ ☉

05 Monday
01 15 ☽ ⊏ ♂
03 06 ☽ ⊥ ☿
04 49 ☽ ∠ ♃
06 50 ☉ Q ♅
08 43 ☽ ☌ ♄
12 03 ☽ ⊏ ♆
13 18 ☽ Q ♀
15 08 ☽ ∠ ☉
20 02 ☽ ± ♂
20 08 ☽ ∠ ☿
20 26 ☽ △ ♅
21 31 ☿ ⊻ ♀
22 48 ☽ ⊥ ♃

06 Tuesday
01 27 ♀ ∠ ♅
05 49 ☽ ± ♆
08 25 ☽ ⚻ ♂
10 40 ☽ ⊻ ♃
17 32 ☽ ⚻ ♆
23 02 ☽ ⚹ ☉

07 Wednesday
00 03 ☽ ⚹ ☿
01 47 ☽ ⊏ ♅
04 21 ☽ □ ♀
05 40 ☽ ⚻ ♇
07 21 ☉ ☌ ☿
10 54 ☿ ⋕ ♆
13 37 ☉ ∥ ☿
17 07 ☽ ± ♇
17 36 ♀ △ ♇
18 06 ☉ ⋕ ♆
19 03 ☽ ± ♅
19 37 ☽ ⊻ ♄
21 55 ☽ Q ☿
23 54 ☽ Q ☉

08 Thursday
06 23 ☽ ⚻ ♅
07 02 ☽ ⊥ ♄
10 03 ☽ ⊏ ♇
17 36 ♂ ☍ ♃
20 11 ☽ ☌ ♃
20 17 ☽ ☍ ♂
20 36 ☽ ∥ ♄
23 53 ☽ ∠ ♄

09 Friday
02 16 ☽ ☍ ♆
05 51 ☽ □ ☿
11 53 ☽ □ ☉
13 34 ☽ △ ♇
17 00 ☽ ⋕ ♂
18 03 ☽ △ ♀
21 29 ☽ ∥ ♃

10 Saturday
03 06 ☽ ∥ ☉
03 13 ☽ ⚹ ♄
06 22 ☽ ⋕ ♆
07 07 ☿ Q ♅
10 30 ☉ ⚻ ♇
12 56 ☽ ☍ ♅
13 14 ☽ ∥ ☿
15 17 ♀ Q ♂
22 06 ☽ ⋕ ♇
23 05 ☽ ⊏ ♀

11 Sunday
00 24 ☽ Q ♄
02 07 ☽ ⊻ ♃
04 06 ☽ ⚻ ♂
07 25 ☽ ⚻ ♆
07 46 ☽ ⋕ ♅
08 38 ☽ △ ☿
12 28 ☽ ⊥ ♃
14 46 ☽ ± ♂
15 57 ☽ ± ♀
17 37 ☽ ± ♆
17 53 ☽ □ ♇
20 14 ☽ △ ☉
21 20 ☽ ∥ ♀

12 Monday
02 58 ☽ ⚻ ♀
03 45 ☽ ∠ ♃
06 31 ☽ ⊏ ♂
07 09 ☽ □ ♄
08 42 ☽ ⊏ ♆
08 58 ☽ ⊏ ☿
09 32 ☉ ⊥ ♄
09 54 ☿ ± ♇
11 36 ♀ Q ♆
15 55 ☽ ⚻ ♅
16 30 ☿ □ ♆
22 53 ☽ ⊏ ☉
23 17 ☽ ± ☿
23 43 ☽ Q ♇

13 Tuesday
01 46 ☽ ± ♅
02 30 ♂ ⊏ ♄
04 38 ☽ ⚹ ♃
08 07 ☽ △ ♂
08 44 ☽ ⚻ ☿
09 16 ☽ △ ♆
14 31 ☽ ± ☉
16 22 ☽ ⊏ ♅
17 19 ☿ □ ♂
19 07 ☽ ⚹ ♇
19 49 ☉ ∥ ♃

14 Wednesday
00 08 ☽ Q ♃
00 49 ☽ ⚻ ☉
05 21 ☿ ∠ ♄
08 08 ☽ ☍ ♀
08 13 ☽ △ ♄
09 15 ♀ ⚹ ♄
12 29 ☽ ⋕ ♀
14 35 ♂ ☌ ♆
16 21 ☽ △ ♅
19 00 ☽ ∠ ♇
20 39 ☽ ∥ ♅
23 28 ☉ ⋕ ♂

15 Thursday
04 57 ☽ □ ♃
05 30 ☽ ∥ ♇
05 47 ☽ ⋕ ☿
07 19 ☽ ☍ ☿
08 11 ☽ ⊏ ♄
09 07 ☽ □ ♆
09 13 ☽ ⊥ ♇
09 49 ☽ □ ♂
10 46 ☿ ⋕ ♇
18 41 ☽ ⊻ ♇
19 38 ☽ ∥ ♆
22 31 ☽ ± ♄

16 Friday
00 44 ♆ St R
02 34 ☽ ⋕ ♃
03 33 ☽ ∥ ♂
03 36 ☽ ☍ ☉
05 41 ☽ ⋕ ☉
08 05 ☽ ⚻ ♄
11 46 ☽ ⚻ ♀
13 36 ☽ Q ♆
15 24 ☽ Q ♂
15 53 ☽ □ ♅
22 09 ☽ ± ♀

17 Saturday
01 17 ☽ ⋕ ♄
04 54 ☽ △ ♃
06 00 ☽ ⚻ ☿
08 44 ☽ ⚹ ♆
11 18 ☽ ⚹ ♂
13 51 ☽ ⊏ ♀
15 31 ☽ ± ☿
18 26 ☽ ☌ ♇
20 47 ☽ Q ♅

18 Sunday
05 21 ☽ ⊏ ♃
05 57 ☽ ⊏ ☿
07 02 ☽ ⚻ ☉
08 41 ☽ ☍ ♄
09 04 ☽ ∠ ♆
10 29 ♂ ⋕ ♃
12 39 ☽ ∠ ♂
15 39 ♀ ⚹ ♅
16 27 ☽ ⚹ ♅
16 31 ☽ △ ♀
17 37 ☽ ± ☉
20 18 ☽ ± ♃
23 58 ☽ ⊥ ♆

19 Monday
04 12 ☽ ⊥ ♂
06 25 ☽ ⚻ ♃
06 38 ☽ △ ☿
09 46 ☽ ⊏ ☉
10 02 ☽ ⊻ ♆
10 16 ♀ ⋕ ♅
11 25 ☉ ⊻ ♄
14 45 ☽ ⊻ ♂
17 41 ☽ ∠ ♅
20 11 ☽ ⊻ ♇
21 02 ♀ ⊏ ♇
23 47 ☿ □ ♃

20 Tuesday
06 31 ☽ ⊥ ♇
07 06 ♄ ⊏ ♆
07 32 ☿ St D
09 13 ☽ ⊥ ♅
11 52 ☽ ⚻ ♄
13 29 ☽ △ ☉
15 30 ☉ Q ♃
19 46 ☽ ⊻ ♅
22 18 ☽ ∠ ♇
22 32 ☽ ± ♄

21 Wednesday
00 50 ☽ □ ♀
08 38 ☽ ⋕ ♄
10 55 ☽ □ ☿
11 10 ☽ ☍ ♃
14 33 ☽ ☌ ♆
14 50 ☽ ⊏ ♄
15 50 ☿ ∥ ♀
21 53 ☽ ☌ ♂

22 Thursday
00 02 ☽ ⋕ ☉
01 21 ☽ ⚹ ♇
12 26 ☽ ⋕ ♃
13 01 ☽ Q ♀
14 31 ☽ ∥ ♂
18 49 ☽ △ ♄
19 25 ☽ ∥ ♆
20 30 ☽ Q ☿
23 43 ☽ Q ♇

23 Friday
00 31 ☽ □ ☉
02 51 ☽ ☌ ♅
12 39 ☽ ∥ ♇
14 10 ☽ ⚹ ♀
15 11 ☽ ⋕ ♀
19 38 ☽ ⋕ ☿
19 50 ☽ ⚹ ☿
19 54 ☽ ⚻ ♃
22 53 ☽ ⊻ ♆
23 40 ☽ ∥ ♅

24 Saturday
01 25 ☿ □ ♃
07 37 ☉ □ ♅
07 41 ☽ ± ♃
09 20 ☽ ⊻ ♂
10 19 ☽ □ ♇
10 34 ☽ ⊥ ♆
14 11 ☽ Q ☉
21 41 ☽ ⊥ ♂
21 55 ♀ ⋕ ♇
22 39 ☽ ∠ ♀

25 Sunday
01 36 ☽ ⊏ ♃
01 59 ☽ ∠ ☿
04 20 ☽ ∠ ♆
05 33 ☽ □ ♄
06 58 ♂ ⚹ ♇
13 28 ☽ ⊻ ♅
16 02 ☽ ⚹ ☉
16 24 ☽ ∠ ♂
18 35 ☽ ⊥ ♀
20 32 ☽ ⊥ ☿

26 Monday
01 30 ☽ ⊥ ♅
07 29 ♀ □ ♃
07 58 ☽ △ ♃
08 01 ☽ ⊻ ♀
09 03 ☽ ⊻ ☿
10 22 ☽ ⚹ ♆
18 04 ☽ Q ♄
19 41 ☽ ∠ ♅
22 08 ☽ △ ♇

27 Tuesday
00 02 ☽ ⚹ ♂
00 04 ☿ ☌ ♀
00 56 ☽ ∠ ☉
03 20 ♀ ± ♇
07 01 ♀ □ ♆
07 42 ☿ ± ♇
10 41 ☽ Q ♆
16 17 ☿ □ ♆
18 41 ☽ ⚹ ♄
20 51 ☽ ⊥ ☉

28 Wednesday
01 34 ☽ Q ♂
02 12 ☽ ⚹ ♅
02 18 ♀ ∠ ♄
04 34 ☽ ⊏ ♇
07 04 ☽ ⋕ ♅
10 09 ☽ ⊻ ☉
13 12 ☽ ∥ ☿
15 13 ♀ Q ♅
18 38 ☽ ⋕ ♇
21 39 ☽ □ ♃
22 51 ☽ ± ♇
23 18 ☽ □ ♆

29 Thursday
00 57 ☽ ☌ ☿
01 33 ☽ ∠ ♄
02 39 ☽ Q ♅
03 18 ☽ ∥ ♀
03 57 ☽ ☌ ♀
11 02 ☽ ⚻ ♇
13 03 ☽ ⋕ ♆
13 13 ☽ ⋕ ♂
14 32 ☿ ∠ ♄
15 53 ☽ □ ♂
18 46 ♂ ∥ ♆
18 48 ☽ ∥ ♃
20 03 ☽ ⊥ ♄

30 Friday
07 10 ☿ Q ♅
08 20 ☽ ⊻ ♄
10 38 ☽ Q ♃
15 11 ☽ □ ♅
20 56 ☽ ∥ ☉

31 Saturday
03 08 ☽ ∥ ♄
04 20 ☽ ☌ ☉
11 05 ☽ ⚹ ♃
11 56 ☽ △ ♆
17 27 ☽ ⊻ ☿
23 20 ♀ ⚻ ♇
23 21 ☽ ☍ ♇
23 21 ☽ ⊻ ♀

All ephemeris data is given at 12.00 UT and the Moon's longitude is additionally given for 24.00 UT

JUNE 2003

Date	Sidereal time h m s	LONGITUDES Sun ☉ ° ' "	Moon ☽ ° ' "	Moon ☽ 24.00 ° ' "	Mercury ☿ ° '	Venus ♀ ° '	Mars ♂ ° '	Jupiter ♃ ° '	Saturn ♄ ° '	Uranus ♅ ° '	Neptune ♆ ° '	Pluto ♇ ° '
01	04 38 15	10 ♊ 35 45	25 ♊ 12 33	01 ♋ 17 34	16 ♉ 31	19 ♉ 29	22 ♒ 44	12 ♌ 51	29 ♊ 41	02 ♓ 48	13 ♒ 07	18 ♐ 50
02	04 42 12	11 33 16	07 ♋ 25 09	13 ♋ 35 31	17 24	20 42	23 14	13 00	29 48	02 49	13 R 06	18 R 48
03	04 46 08	12 30 46	19 ♋ 48 54	26 ♋ 05 31	18 20	21 55	23 44	13 09	29 ♊ 56	02 49	13 06	18 46
04	04 50 05	13 28 15	02 ♌ 25 36	08 ♌ 49 25	19 19	23 08	24 13	13 19	00 ♋ 03	02 49	13 05	18 45
05	04 54 01	14 25 42	15 ♌ 17 09	21 ♌ 49 04	20 22	24 21	24 43	13 28	00 11	02 49	13 04	18 43
06	04 57 58	15 23 09	28 ♌ 25 22	05 ♍ 06 14	21 28	25 34	25 11	13 37	00 19	02 49	13 04	18 42
07	05 01 54	16 20 34	11 ♍ 51 52	18 ♍ 42 23	22 38	26 47	25 40	13 47	00 26	02 R 49	13 03	18 40
08	05 05 51	17 17 58	25 ♍ 37 51	02 ♎ 38 15	23 50	28 00	26 09	13 57	00 34	02 49	13 02	18 38
09	05 09 48	18 15 20	09 ♎ 43 32	16 ♎ 53 30	25 06	29 ♉ 13	26 37	14 06	00 42	02 49	13 02	18 37
10	05 13 44	19 12 42	24 ♎ 07 52	01 ♏ 26 13	26 25	00 ♊ 26	27 04	14 16	00 49	02 49	13 01	18 35
11	05 17 41	20 10 03	08 ♏ 48 01	16 ♏ 12 36	27 46	01 39	27 32	14 26	00 57	02 49	13 00	18 34
12	05 21 37	21 07 23	23 ♏ 39 12	01 ♐ 06 57	29 ♉ 11	02 52	27 59	14 36	01 05	02 49	12 59	18 32
13	05 25 34	22 04 41	08 ♐ 34 55	16 ♐ 02 05	00 ♊ 38	04 05	28 26	14 46	01 12	02 48	12 58	18 30
14	05 29 30	23 02 00	23 ♐ 27 27	00 ♑ 50 02	02 09	05 18	28 53	14 57	01 20	02 48	12 58	18 29
15	05 33 27	23 59 17	08 ♑ 07 57	15 ♑ 23 19	03 42	06 31	29 19	15 07	01 28	02 48	12 57	18 27
16	05 37 23	24 56 34	22 ♑ 32 28	29 ♑ 35 50	05 19	07 44	29 ♒ 45	15 17	01 36	02 47	12 56	18 26
17	05 41 20	25 53 51	06 ♒ 32 59	13 ♒ 23 41	06 58	08 57	00 ♓ 10	15 28	01 43	02 47	12 55	18 24
18	05 45 17	26 51 07	20 ♒ 07 48	26 ♒ 45 25	08 40	10 10	00 35	15 38	01 51	02 46	12 54	18 22
19	05 49 13	27 48 23	03 ♓ 16 40	09 ♓ 41 52	10 25	11 24	01 00	15 49	01 59	02 46	12 53	18 21
20	05 53 10	28 45 38	16 ♓ 01 24	22 ♓ 15 43	12 12	12 37	01 24	16 00	02 07	02 45	12 52	18 19
21	05 57 06	29 ♊ 42 53	28 ♓ 25 21	04 ♈ 30 52	14 03	13 50	01 48	16 10	02 15	02 44	12 51	18 18
22	06 01 03	00 ♋ 40 08	10 ♈ 32 52	16 ♈ 31 58	15 56	15 03	02 12	16 21	02 22	02 44	12 50	18 16
23	06 04 59	01 37 23	22 ♈ 28 47	28 ♈ 23 56	17 51	16 16	02 35	16 32	02 30	02 43	12 49	18 15
24	06 08 56	02 34 38	04 ♉ 18 01	10 ♉ 11 36	19 49	17 30	02 58	16 43	02 38	02 42	12 47	18 13
25	06 12 52	03 31 52	16 ♉ 05 16	21 ♉ 59 32	21 49	18 43	03 20	16 54	02 46	02 41	12 46	18 12
26	06 16 49	04 29 07	27 ♉ 54 51	03 ♊ 51 41	23 52	19 56	03 42	17 05	02 54	02 40	12 45	18 10
27	06 20 46	05 26 21	09 ♊ 50 26	15 ♊ 51 27	25 56	21 09	04 03	17 17	03 01	02 40	12 44	18 09
28	06 24 42	06 23 36	21 ♊ 55 01	28 ♊ 01 23	28 ♊ 02	22 23	04 24	17 28	03 09	02 39	12 43	18 07
29	06 28 39	07 20 50	04 ♋ 10 47	10 ♋ 23 21	00 ♋ 09	23 36	04 44	17 39	03 17	02 38	12 41	18 06
30	06 32 35	08 ♋ 18 04	16 ♋ 39 13	22 ♋ 58 26	02 ♋ 18	24 ♊ 49	05 ♓ 04	17 ♌ 51	03 ♋ 25	02 ♓ 37	12 ♒ 40	18 ♐ 04

Date	Moon True ☊ ° '	Moon Mean ☊ ° '	Moon ☽ Latitude ° '	DECLINATIONS Sun ☉ ° '	Moon ☽ ° '	Mercury ☿ ° '	Venus ♀ ° '	Mars ♂ ° '	Jupiter ♃ ° '	Saturn ♄ ° '	Uranus ♅ ° '	Neptune ♆ ° '	Pluto ♇ ° '
01	29 ♉ 30	29 ♉ 01	02 N 18	22 N 02	25 N 39	13 N 16	16 N 20	16 S 32	17 N 44	22 N 36	11 S 11	16 S 52	13 S 28
02	29 R 28	28 57	03 15	22 10	26 29	13 33	16 41	16 24	17 41	22 36	11 11	16 52	13 28
03	29 26	28 54	04 04	22 18	26 00	13 52	17 01	16 17	17 39	22 36	11 11	16 52	13 27
04	29 25	28 51	04 42	22 25	24 12	14 12	17 22	16 10	17 36	22 36	11 11	16 53	13 27
05	29 23	28 48	05 07	22 32	21 08	14 33	17 41	16 03	17 33	22 37	11 11	16 53	13 27
06	29 21	28 45	05 16	22 38	16 57	14 56	18 01	15 56	17 31	22 37	11 11	16 53	13 27
07	29 21	28 42	05 09	22 44	11 52	15 19	18 19	15 49	17 28	22 37	11 11	16 53	13 27
08	29 D 21	28 38	04 44	22 50	06 N 04	15 43	18 38	15 42	17 25	22 37	11 11	16 53	13 27
09	29 22	28 35	04 01	22 55	00 S 09	16 09	18 56	15 35	17 22	22 37	11 11	16 54	13 27
10	29 23	28 32	03 03	23 00	06 32	16 34	19 13	15 28	17 19	22 37	11 11	16 54	13 27
11	29 24	28 29	01 51	23 05	12 40	17 01	19 30	15 22	17 16	22 37	11 11	16 54	13 27
12	29 25	28 26	00 N 32	23 09	18 10	17 28	19 46	15 15	17 13	22 37	11 11	16 54	13 27
13	29 R 25	28 23	00 S 50	23 12	22 34	17 55	20 02	15 09	17 10	22 37	11 11	16 55	13 27
14	29 24	28 19	02 09	23 15	25 25	18 23	20 18	15 02	17 07	22 37	11 11	16 55	13 27
15	29 21	28 16	03 17	23 18	26 28	18 50	20 32	14 56	17 04	22 37	11 12	16 55	13 27
16	29 18	28 13	04 12	23 21	25 42	19 18	20 47	14 50	17 01	22 37	11 12	16 55	13 27
17	29 14	28 10	04 50	23 23	23 18	19 45	21 00	14 44	16 58	22 37	11 12	16 56	13 27
18	29 10	28 07	05 10	23 24	19 40	20 13	21 13	14 38	16 55	22 37	11 12	16 56	13 27
19	29 07	28 03	05 13	23 25	15 10	20 39	21 26	14 32	16 52	22 37	11 12	16 56	13 27
20	29 05	28 00	05 00	23 26	10 07	21 05	21 38	14 27	16 48	22 37	11 13	16 57	13 27
21	29 05	27 57	04 34	23 26	04 S 49	21 30	21 49	14 21	16 45	22 37	11 13	16 57	13 27
22	29 D 05	27 54	03 55	23 26	00 N 34	21 54	22 00	14 16	16 42	22 37	11 13	16 57	13 27
23	29 06	27 51	03 07	23 26	05 51	22 17	22 10	14 10	16 39	22 37	11 13	16 58	13 27
24	29 08	27 48	02 12	23 25	10 53	22 38	22 20	14 05	16 35	22 37	11 14	16 58	13 27
25	29 10	27 44	01 11	23 24	15 31	22 58	22 29	14 00	16 32	22 37	11 14	16 58	13 27
26	29 10	27 41	00 S 07	23 22	19 35	23 16	22 37	13 56	16 29	22 37	11 14	16 59	13 27
27	29 R 10	27 38	00 N 58	23 20	22 53	23 32	22 45	13 51	16 25	22 36	11 15	16 59	13 28
28	29 08	27 35	02 01	23 17	25 12	23 45	22 51	13 47	16 22	22 36	11 15	16 59	13 28
29	29 04	27 32	02 59	23 14	26 21	23 57	22 58	13 42	16 18	22 36	11 16	17 00	13 28
30	28 ♉ 58	27 ♉ 29	03 N 49	23 N 11	26 N 12	24 N 05	23 N 03	13 S 38	16 N 15	22 N 36	11 S 16	17 S 00	13 S 28

ZODIAC SIGN ENTRIES

Date	h	m	Planets
01	21	27	☽ ♋
04	01	28	♄ ♋
04	07	25	☽ ♌
06	14	51	☽ ♍
08	19	30	☽ ♎
10	03	32	♀ ♊
10	21	39	☽ ♏
12	22	12	☽ ♐
13	01	34	☿ ♊
14	22	38	☽ ♑
17	00	41	☽ ♒
17	02	25	♂ ♓
19	05	57	☽ ♓
21	15	06	☽ ♈
21	19	10	☉ ♋
24	03	15	☽ ♉
26	16	13	☽ ♊
29	03	52	☽ ♋
29	10	17	☿ ♋

LATITUDES

Date	Mercury ☿ ° '	Venus ♀ ° '	Mars ♂ ° '	Jupiter ♃ ° '	Saturn ♄ ° '	Uranus ♅ ° '	Neptune ♆ ° '	Pluto ♇ ° '
01	03 S 39	01 S 19	02 S 45	00 N 49	00 S 50	00 S 45	00 N 01	09 N 32
04	03 29	01 14	02 53	00 49	00 50	00 45	00 01	09 32
07	03 13	01 09	03 02	00 48	00 50	00 46	00 01	09 32
10	02 51	01 03	03 11	00 48	00 49	00 46	00 01	09 32
13	02 25	00 57	03 20	00 48	00 49	00 46	00 01	09 31
16	01 55	00 50	03 30	00 48	00 49	00 46	00 01	09 31
19	01 22	00 43	03 39	00 48	00 49	00 46	00 01	09 31
22	00 48	00 36	03 49	00 48	00 48	00 46	00 01	09 30
25	00 S 13	00 29	04 00	00 48	00 48	00 46	00 N 01	09 30
28	00 N 20	00 22	04 10	00 48	00 48	00 46	00 00	09 29
31	00 N 50	00 S 14	04 S 21	00 N 48	00 S 48	00 S 46	00 00	09 N 28

DATA

Julian Date	2452792
Delta T	+65 seconds
Ayanamsa	23° 54' 02"
Synetic vernal point	05° ♓ 12' 57"
True obliquity of ecliptic	23° 26' 24"

LONGITUDES

Date	Chiron ⚷ ° '	Ceres ⚳ ° '	Pallas ⚴ ° '	Juno ⚵ ° '	Vesta ⚶ ° '	Black Moon Lilith ⚸ ° '
01	17 ♑ 17	02 ♊ 14	14 ♈ 52	06 ♏ 49	00 ♎ 17	12 ♉ 20
11	16 ♑ 47	08 ♊ 08	16 ♈ 50	05 ♏ 35	02 ♎ 12	13 ♉ 26
21	16 ♑ 13	10 ♊ 13	19 ♈ 27	04 ♏ 53	04 ♎ 40	14 ♉ 33
31	15 ♑ 37	14 ♊ 16	21 ♈ 52	04 ♏ 43	07 ♎ 37	15 ♉ 39

MOON'S PHASES, APSIDES AND POSITIONS ☽

Date	h	m	Phase	Longitude ° '	Eclipse Indicator
07	20	28	☽	16 ♍ 41	
14	11	16	○	23 ♐ 00	
21	14	45	☾	29 ♓ 49	
29	18	39	●	07 ♋ 37	

Day	h	m		
12	23	23	Perigee	
25	02	32	Apogee	
02	15	23	Max dec	26° N 29'
09	11	25	0S	
15	13	34	Max dec	26° S 29'
22	09	28	0N	
29	21	13	Max dec	26° N 27'

ASPECTARIAN

h m	Aspects
01 Sunday	
06 16	☽ ⊥ ☿
06 53	☽ △ ♂
12 36	☽ ⊥ ♀
17 17	☽ ∠ ♃
17 44	☽ bQ ♆
20 55	☽ ☌ ♄
22 05	♀ # ♂
02 Monday	
01 23	☽ ∠ ☿
02 59	☽ △ ♅
04 03	☿ # ♇
08 16	☽ ∠ ♀
11 11	☽ ⊥ ♃
11 23	☽ ± ♆
13 39	☽ bQ ♂
20 44	☽ ⊻ ☉
23 00	☽ ⊻ ♃
23 03	☽ ⊼ ♆
03 Tuesday	
01 25	♀ # ♆
02 55	♃ ☍ ♆
07 50	☽ ± ♂
08 09	☽ bQ ♅
08 54	☽ ⚹ ☿
09 17	☽ ⊥ ☉
10 00	☽ ⊼ ♇
16 27	☽ ⚹ ♀
19 48	☽ ⊼ ♂
21 28	☽ ± ♇
22 37	☿ ⊼ ♇
04 Wednesday	
01 23	☽ ± ♅
02 27	☉ △ ♆
03 54	☽ ∠ ☉
07 14	☉ ⚹ ♃
07 29	☽ ⊻ ♄
09 44	☽ Q ☿
12 44	☽ ⊼ ♅
14 29	☽ bQ ♇
17 37	☽ Q ♀
18 53	☽ ⊥ ♄
05 Thursday	
01 34	☽ ∥ ♄
02 29	☽ ∥ ☉
03 27	♀ ∥ ♃
07 55	☽ ☍ ♆
08 24	♀ ⊥ ♄
08 36	☽ ☌ ♃
10 17	☽ ⚹ ☉
11 48	☽ ∠ ♄
18 19	☽ △ ♇
22 12	☽ □ ☿
23 53	♀ □ ♂
06 Friday	
05 24	☉ ∥ ♄
05 56	☽ ☍ ♂
06 18	☽ □ ♀
06 50	☽ ∥ ♀
09 06	☽ ∥ ♃
09 59	☽ Q ☉
12 23	☽ # ♆
15 26	☽ ⚹ ♄
17 18	☽ # ♂
19 55	☽ ☍ ♅
21 19	☽ ∥ ☿
07 Saturday	
04 54	☽ # ♇
06 56	♅ St R
13 01	☽ Q ♄
14 06	☽ ⊼ ♆
14 59	☽ # ♅
15 25	☽ ⊻ ♃
20 28	☽ □ ☉
23 54	☽ □ ♇
08 Sunday	
00 36	☽ ± ♆
02 02	☽ ⊥ ♃
08 36	☽ △ ☿
10 58	☿ # ♂
12 55	☽ ⊼ ♂
16 08	☽ bQ ♆
16 27	☽ △ ♀
17 46	☽ ∠ ♃
20 33	☽ □ ♄
23 32	☽ ± ♂
09 Monday	
00 19	☽ ⊼ ♅
03 31	☿ ⊥ ♄
06 45	☽ Q ♇
10 28	☽ ± ♅
12 41	☽ bQ ☿
15 17	☽ bQ ♂
17 32	☽ △ ♆
19 26	☽ ⚹ ♃
20 14	☽ bQ ♀
20 44	☉ ☍ ♇
10 Tuesday	
01 32	☽ bQ ♅
02 50	☽ ⚹ ♇
03 17	☽ △ ☉
05 13	☽ ± ☿
12 32	☽ ± ♀
15 34	☽ Q ♃
16 08	☽ ⊼ ☿
17 00	☽ △ ♂
20 38	♀ ⊻ ♄
23 05	☽ △ ♄
23 17	☽ ⊼ ♀
11 Wednesday	
02 15	☽ △ ♅
03 29	☽ ∠ ♇
05 41	☽ bQ ☉
05 48	☿ □ ♂
05 50	☿ # ♆
06 00	☽ ∥ ♅
15 11	☽ ∥ ♇
18 05	☽ ⊥ ♇
18 48	☽ □ ♆
21 14	☽ □ ♃
21 18	☽ ± ☉
23 03	☽ ∥ ♂
23 41	☽ bQ ♄
12 Thursday	
00 24	☿ ∥ ♃
03 46	☽ ⊻ ♇
06 04	♀ Q ♃
06 05	☽ ∥ ♆
07 35	☽ # ♃
07 39	☽ ⊼ ☉
08 20	☽ # ☿
10 56	♀ □ ♅
14 19	☽ ± ♄
19 11	☽ □ ♂
20 26	☽ # ♀
21 51	☽ ☍ ☿
23 47	☽ Q ♆
13 Friday	
00 03	☽ ⊼ ♄
02 43	☽ □ ♅
04 07	☽ ☍ ♀
12 21	☽ # ♄
16 24	☽ # ☉
19 04	☽ ⚹ ♆
21 57	☿ ⊻ ♄
22 05	☽ △ ♃
14 Saturday	
01 02	☽ Q ♂
03 58	☽ ☌ ♇
07 42	☽ Q ♅
11 16	☽ ☍ ☉
19 18	☽ ∠ ♆
21 05	☽ ⚹ ♂
22 04	☿ □ ♅
22 40	☽ bQ ♃
15 Sunday	
00 56	☽ ☍ ♄
01 51	☿ Q ♃
03 12	☽ ⚹ ♅
03 49	☽ ⊼ ☿
09 04	☽ ⊼ ♀
10 01	☽ ⊥ ♆
13 37	☽ ± ♃
14 53	☽ ± ☿
19 53	☽ ± ♀
19 55	☽ ⊻ ♆
22 31	☽ ∠ ♂
23 41	☽ ⊼ ♃
16 Monday	
04 01	☽ ∠ ♅
05 06	☽ ⊻ ♇
07 46	☽ bQ ☿
12 22	☽ bQ ♀
14 06	☽ ⊥ ♂
15 11	☽ ⊥ ♇
16 22	☽ ⊼ ☉
19 12	☽ ⊥ ♅
17 Tuesday	
00 38	☽ ⊻ ♂
03 22	☽ ± ☉
03 34	☽ ⊼ ♄
05 28	☽ ⊻ ♅
06 33	☽ ∠ ♇
11 28	☽ # ☉
12 49	☽ △ ☿
14 04	☽ ± ♄
16 36	☽ △ ♀
17 09	☽ # ♄
20 10	☽ bQ ☉
23 08	☽ ☌ ♆
18 Wednesday	
00 28	♂ Q ♇
03 04	☽ # ♀
03 52	☽ ☍ ♃
04 28	♃ # ♆
06 05	☽ bQ ♄
08 52	☽ ⚹ ♇
09 08	☽ # ☿
19 Thursday	
01 08	☽ △ ☉
02 58	☽ ∥ ♆
03 15	☽ # ♃
06 36	☽ Q ♇
07 39	☽ ☌ ♂
09 35	☽ △ ♄
11 03	☽ ☌ ♅
13 49	☉ bQ ♆
15 09	☽ ∥ ♂
20 21	☽ ∥ ♇
20 Friday	
03 32	☽ □ ☿
04 49	☽ □ ♀
05 59	☽ ⊻ ♆
06 58	☽ ∥ ♅
11 57	☽ ⊼ ♃
16 24	☽ □ ♇
16 52	♀ △ ♆
17 26	☽ ⊥ ♆
20 32	☿ △ ♆
23 39	☽ ± ♃
21 Saturday	
03 53	☿ ☌ ♀
10 52	☽ ∠ ♆
14 45	☽ □ ☉
17 29	☽ bQ ♃
18 52	☽ ⊻ ♂
19 27	☽ Q ♀
19 36	☽ □ ♄
20 25	☽ Q ☿
20 29	☽ ⊻ ♅
22 Sunday	
07 10	☽ ⊥ ♂
08 22	☽ ⊥ ♅
16 33	☽ ⚹ ♆
17 55	☿ ⚹ ♃
22 03	☽ ⚹ ♀
23 01	☿ ∥ ♀
23 49	☽ △ ♃
23 Monday	
00 51	☽ ⚹ ☿
01 47	☽ ∠ ♂
02 24	☽ ∠ ♅
03 28	☽ △ ♇
04 21	♂ △ ♄
05 43	☽ Q ☉
07 58	☽ Q ♄
09 18	☉ ∠ ♃
16 42	☽ Q ♆
16 46	☿ ☍ ♇
18 08	♀ ⚹ ♃
20 02	♂ ☌ ♅
24 Tuesday	
07 54	☽ ∠ ♀
08 11	☽ ⚹ ☉
08 34	☽ ⚹ ♄
08 45	☽ ⚹ ♅
09 11	☽ ⚹ ♂
09 48	☽ bQ ♇
10 09	☿ ∥ ♄
13 16	☽ ∠ ☿
13 39	☉ ☌ ♄
13 43	☽ # ♅
15 08	☉ △ ♅
23 34	♄ △ ♅
25 Wednesday	
01 01	☽ # ♇
02 00	♀ ☍ ♇
03 54	☉ △ ♂
04 01	☽ # ♂
04 05	☽ ± ♇
04 20	☽ ⊥ ♀
05 16	☽ □ ♆
09 09	☽ Q ♅
10 25	☽ Q ♂
11 21	☽ ⊥ ☿
13 41	☽ □ ♃
15 27	☽ ∠ ♄
16 16	☽ ⊼ ♇
17 24	☽ ∠ ☉
17 34	☽ ∥ ♃
17 57	☽ ⊻ ♀
20 08	☽ # ♆
26 Thursday	
02 04	☽ ⊻ ☿
09 55	☽ ⊥ ♄
11 02	♀ ∥ ♄
13 15	☽ ⊥ ☉
19 40	☉ ∥ ☿
21 36	☽ □ ♅
22 10	☽ ⊻ ♄
27 Friday	
00 02	☽ □ ♂
02 24	☽ ⊻ ☉
02 42	☽ Q ♃
09 44	☽ ∥ ♄
10 47	☽ ∥ ♀
15 49	☽ ∥ ☉
17 46	☽ △ ♆
18 17	☽ ∥ ☿
28 Saturday	
03 03	☽ ⚹ ♃
04 30	☽ ☍ ♇
08 25	☿ bQ ♆
13 01	☽ ☌ ♀
19 49	☉ ± ♆
23 22	☽ bQ ♆
29 Sunday	
02 31	☽ ☌ ☿
08 59	☽ △ ♅
08 59	☽ ∠ ♃
10 15	☽ ☌ ♄
13 07	☽ △ ♂
16 51	☽ ± ♆
18 39	☽ ☌ ☉
30 Monday	
02 39	☽ ⊥ ♃
04 24	☽ ⊼ ♆
13 49	☽ bQ ♅
14 18	☽ ⊻ ♃
14 41	☽ ⊼ ♇
15 27	☿ △ ♅
18 40	☽ bQ ♂
18 42	☿ ∠ ♃

All ephemeris data is given at 12.00 UT and the Moon's longitude is additionally given for 24.00 UT

JULY 2003

LONGITUDES

Date	Sidereal time h m s	Sun ☉ ° ' "	Moon ☽ ° ' "	Moon ☽ 24.00 ° ' "	Mercury ☿ ° '	Venus ♀ ° '	Mars ♂ ° '	Jupiter ♃ ° '	Saturn ♄ ° '	Uranus ♅ ° '	Neptune ♆ ° '	Pluto ♇ ° '
01	06 36 32	09 ♋ 15 18	29 ♋ 21 02	05 ♌ 47 03	04 ♋ 27	26 ♊ 03	05 ♓ 23	18 ♌ 02	03 ♋ 33	02 ♓ 35	12 ♒ 39	18 ♐ 03
02	06 40 28	10 12 32	12 ♌ 16 27	18 ♌ 49 12	06 38	27 16	05 42	18 14	03 40	02 R 34	12 R 38	18 R 01
03	06 44 25	11 09 45	25 ♌ 25 16	02 ♍ 04 37	08 48	28 29	06 00	18 25	03 48	02 33	12 36	18 00
04	06 48 21	12 06 58	08 ♍ 47 11	15 ♍ 32 57	10 59	29 ♊ 43	06 18	18 37	03 56	02 32	12 35	17 58
05	06 52 18	13 04 11	22 ♍ 21 52	29 ♍ 13 54	13 09	00 ♋ 56	06 35	18 49	04 04	02 31	12 34	17 57
06	06 56 15	14 01 23	06 ♎ 09 01	13 ♎ 07 11	15 19	02 10	06 52	19 01	04 11	02 29	12 32	17 55
07	07 00 11	14 58 35	20 ♎ 08 19	27 ♎ 12 19	17 28	03 23	07 08	19 12	04 19	02 28	12 31	17 54
08	07 04 08	15 55 47	04 ♏ 19 04	11 ♏ 28 21	19 37	04 36	07 23	19 24	04 27	02 27	12 29	17 53
09	07 08 04	16 52 59	18 ♏ 39 56	25 ♏ 53 26	21 44	05 50	07 38	19 36	04 35	02 25	12 28	17 51
10	07 12 01	17 50 10	03 ♐ 08 28	10 ♐ 24 30	23 50	07 03	07 52	19 48	04 42	02 24	12 27	17 50
11	07 15 57	18 47 22	17 ♐ 40 57	24 ♐ 57 07	25 55	08 17	08 06	20 00	04 50	02 22	12 25	17 49
12	07 19 54	19 44 34	02 ♑ 12 19	09 ♑ 25 46	27 58	09 30	08 18	20 12	04 58	02 21	12 24	17 47
13	07 23 50	20 41 46	16 ♑ 36 41	23 ♑ 44 21	29 ♋ 59	10 44	08 31	20 25	05 05	02 19	12 22	17 46
14	07 27 47	21 38 58	00 ♒ 48 04	07 ♒ 47 13	01 ♌ 59	11 58	08 42	20 37	05 13	02 18	12 21	17 45
15	07 31 44	22 36 10	14 ♒ 41 17	21 ♒ 29 52	03 57	13 11	08 53	20 49	05 20	02 16	12 19	17 44
16	07 35 40	23 33 23	28 ♒ 12 43	04 ♓ 49 42	05 53	14 25	09 03	21 01	05 28	02 14	12 18	17 42
17	07 39 37	24 30 36	11 ♓ 20 49	17 ♓ 46 13	07 48	15 38	09 13	21 14	05 36	02 13	12 16	17 41
18	07 43 33	25 27 50	24 ♓ 06 08	00 ♈ 20 55	09 41	16 52	09 22	21 26	05 43	02 11	12 15	17 40
19	07 47 30	26 25 04	06 ♈ 30 01	12 ♈ 36 57	11 32	18 06	09 30	21 38	05 51	02 09	12 13	17 39
20	07 51 26	27 22 19	18 ♈ 39 16	24 ♈ 38 37	13 21	19 19	09 37	21 51	05 58	02 07	12 12	17 38
21	07 55 23	28 19 35	00 ♉ 35 36	06 ♉ 30 56	15 08	20 33	09 44	22 03	06 05	02 05	12 10	17 37
22	07 59 19	29 ♋ 16 52	12 ♉ 25 15	18 ♉ 19 14	16 53	21 47	09 49	22 16	06 13	02 03	12 08	17 36
23	08 03 16	00 ♌ 14 10	24 ♉ 13 32	00 ♊ 08 48	18 37	23 00	09 54	22 28	06 20	02 02	12 07	17 34
24	08 07 13	01 11 28	06 ♊ 05 37	12 ♊ 04 34	20 19	24 14	09 59	22 41	06 28	02 00	12 05	17 33
25	08 11 09	02 08 47	18 ♊ 06 10	24 ♊ 10 52	21 59	25 28	10 02	22 54	06 35	01 58	12 04	17 32
26	08 15 06	03 06 07	00 ♋ 19 04	06 ♋ 31 06	23 37	26 42	10 05	23 06	06 42	01 56	12 02	17 31
27	08 19 02	04 03 28	12 ♋ 47 11	19 ♋ 07 31	25 14	27 56	10 07	23 19	06 49	01 54	12 00	17 30
28	08 22 59	05 00 50	25 ♋ 32 10	02 ♌ 01 06	26 48	29 ♋ 09	10 08	23 32	06 57	01 52	11 59	17 29
29	08 26 55	05 58 13	08 ♌ 34 16	15 ♌ 11 29	28 21	00 ♌ 23	10 R 08	23 44	07 04	01 50	11 57	17 29
30	08 30 52	06 55 36	21 ♌ 52 31	28 ♌ 37 04	29 ♌ 52	01 37	10 07	23 57	07 11	01 48	11 56	17 28
31	08 34 48	07 ♌ 53 00	05 ♍ 24 51	12 ♍ 15 29	01 ♍ 21	02 ♌ 51	10 ♓ 06	24 ♌ 10	07 ♋ 18	01 ♓ 45	11 ♒ 54	17 ♐ 27

Date	Moon True ☊ ° '	Moon Mean ☊ ° '	Moon ☽ Latitude ° '
01	28 ♉ 51	27 ♉ 25	04 N 30
02	28 R 44	27 22	04 57
03	28 38	27 19	05 09
04	28 33	27 16	05 04
05	28 29	27 13	04 43
06	28 27	27 09	04 05
07	28 D 27	27 06	03 12
08	28 28	27 03	02 07
09	28 29	27 00	00 N 53
10	28 R 29	26 57	00 S 25
11	28 28	26 54	01 42
12	28 24	26 50	02 51
13	28 18	26 47	03 49
14	28 10	26 44	04 32
15	28 02	26 41	04 58
16	27 53	26 38	05 06
17	27 45	26 35	04 57
18	27 39	26 31	04 34
19	27 35	26 28	03 59
20	27 33	26 25	03 13
21	27 D 32	26 22	02 19
22	27 33	26 19	01 20
23	27 33	26 15	00 S 18
24	27 R 33	26 12	00 N 45
25	27 31	26 09	01 47
26	27 27	26 06	02 45
27	27 20	26 03	03 36
28	27 11	26 00	04 18
29	27 00	25 56	04 47
30	26 48	25 53	05 01
31	26 ♉ 38	25 ♉ 50	04 N 59

DECLINATIONS

Date	Sun ☉ ° '	Moon ☽ ° '	Mercury ☿ ° '	Venus ♀ ° '	Mars ♂ ° '	Jupiter ♃ ° '	Saturn ♄ ° '	Uranus ♅ ° '	Neptune ♆ ° '	Pluto ♇ ° '
01	23 N 07	24 N 41	24 N 11	23 N 08	13 S 34	16 N 11	22 N 36	11 S 16	17 S 00	13 S 28
02	23 03	21 52	24 15	23 13	13 31	16 08	22 36	11 17	17 01	13 28
03	22 58	17 53	24 15	23 16	13 27	16 04	22 36	11 17	17 01	13 28
04	22 53	12 58	24 13	23 19	13 24	16 00	22 36	11 18	17 01	13 28
05	22 48	07 22	24 08	23 22	13 21	15 57	22 35	11 18	17 02	13 28
06	22 42	01 N 18	24 00	23 23	13 18	15 53	22 35	11 19	17 02	13 28
07	22 36	04 S 54	23 50	23 24	13 16	15 49	22 35	11 19	17 03	13 28
08	22 29	10 58	23 37	23 24	13 13	15 46	22 35	11 20	17 03	13 29
09	22 22	16 32	23 21	23 24	13 11	15 42	22 35	11 20	17 03	13 29
10	22 15	21 12	23 03	23 23	13 09	15 38	22 34	11 21	17 04	13 29
11	22 07	24 33	22 43	23 21	13 07	15 34	22 34	11 21	17 04	13 29
12	21 59	26 17	22 21	23 18	13 06	15 30	22 34	11 22	17 05	13 29
13	21 51	26 12	21 56	23 15	13 05	15 27	22 34	11 23	17 05	13 29
14	21 42	24 24	21 30	23 11	13 04	15 23	22 34	11 23	17 06	13 30
15	21 33	21 10	21 03	23 07	13 03	15 19	22 33	11 24	17 06	13 30
16	21 23	16 52	20 33	23 01	13 02	15 15	22 33	11 25	17 06	13 30
17	21 13	11 53	20 02	22 55	13 02	15 11	22 33	11 25	17 07	13 30
18	21 03	06 32	19 30	22 49	13 02	15 07	22 33	11 26	17 07	13 30
19	20 52	01 S 04	18 57	22 42	13 02	15 03	22 32	11 26	17 08	13 30
20	20 41	04 N 20	18 23	22 34	13 03	14 59	22 32	11 27	17 08	13 30
21	20 30	09 30	17 48	22 25	13 04	14 55	22 32	11 28	17 09	13 31
22	20 18	14 18	17 12	22 16	13 05	14 51	22 31	11 29	17 09	13 31
23	20 06	18 32	16 36	22 06	13 06	14 47	22 31	11 29	17 10	13 31
24	19 54	22 04	15 58	21 55	13 08	14 43	22 31	11 30	17 10	13 31
25	19 41	24 41	15 21	21 44	13 10	14 39	22 31	11 31	17 10	13 32
26	19 28	26 12	14 42	21 32	13 12	14 34	22 30	11 31	17 11	13 32
27	19 15	26 25	14 04	21 20	13 14	14 30	22 30	11 32	17 11	13 32
28	19 01	25 15	13 25	21 06	13 17	14 26	22 30	11 33	17 12	13 32
29	18 47	22 44	12 46	20 53	13 19	14 22	22 29	11 34	17 12	13 33
30	18 33	18 57	12 07	20 38	13 22	14 18	22 29	11 34	17 13	13 33
31	18 N 18	14 N 09	11 N 28	20 N 23	13 S 26	14 N 13	22 N 29	11 S 35	17 S 13	13 S 33

ZODIAC SIGN ENTRIES

Date	h	m	Planets
01	13	13	☽ ♌
03	20	16	☽ ♍
04	17	39	♀ ♋
06	01	20	☽ ♎
08	04	43	☽ ♏
10	06	48	☽ ♐
12	08	21	☽ ♑
13	12	10	☿ ♌
14	10	38	☽ ♒
16	15	14	☽ ♓
18	23	20	☽ ♈
21	10	48	☽ ♉
23	06	04	☉ ♌
23	23	42	☽ ♊
26	11	23	☽ ♋
28	20	17	☽ ♌
29	04	25	♀ ♌
30	14	05	☿ ♍
31	02	27	☽ ♍

LATITUDES

Date	Mercury ☿ ° '	Venus ♀ ° '	Mars ♂ ° '	Jupiter ♃ ° '	Saturn ♄ ° '	Uranus ♅ ° '	Neptune ♆ ° '	Pluto ♇ ° '
01	00 N 50	00 S 14	04 S 21	00 N 48	00 S 48	00 S 46	00 00	09 N 28
04	01 14	00 S 07	04 32	00 48	00 47	00 47	00 00	09 28
07	01 33	00 00	04 42	00 48	00 47	00 47	00 00	09 27
10	01 45	00 N 08	04 53	00 48	00 47	00 47	00 00	09 26
13	01 50	00 15	05 04	00 48	00 47	00 47	00 00	09 26
16	01 49	00 22	05 15	00 48	00 47	00 47	00 00	09 25
19	01 42	00 29	05 26	00 48	00 46	00 47	00 00	09 24
22	01 30	00 35	05 37	00 48	00 46	00 47	00 00	09 23
25	01 14	00 42	05 47	00 48	00 46	00 47	00 00	09 22
28	00 54	00 48	05 57	00 48	00 46	00 47	00 00	09 21
31	00 N 30	00 N 53	06 S 07	00 N 48	00 S 46	00 S 47	00 00	09 N 20

DATA

Julian Date	2452822
Delta T	+65 seconds
Ayanamsa	23° 54' 08"
Synetic vernal point	05° ♓ 12' 52"
True obliquity of ecliptic	23° 26' 24"

MOON'S PHASES, APSIDES AND POSITIONS ☽

Date	h	m	Phase	Longitude ° '	Eclipse Indicator
07	02	32	☽	14 ♎ 36	
13	19	21	○	20 ♑ 59	
21	07	01	☾	28 ♈ 08	
29	06	53	●	05 ♌ 46	

Day	h	m		
10	22	07	Perigee	
22	19	40	Apogee	
06	17	04	0S	
12	22	56	Max dec	26° S 28'
19	16	41	0N	
27	03	57	Max dec	26° N 30'

LONGITUDES

Date	Chiron ⚷ ° '	Ceres ⚳ ° '	Pallas ⚴ ° '	Juno ⚵ ° '	Vesta ⚶ ° '	Black Moon Lilith ⚸ ° '
01	15 ♑ 37	14 ♊ 16	21 ♈ 52	04 ♏ 43	07 ♎ 37	15 ♉ 39
11	15 ♑ 00	18 ♊ 17	24 ♈ 00	05 ♏ 05	10 ♎ 59	16 ♉ 46
21	14 ♑ 23	22 ♊ 14	25 ♈ 50	05 ♏ 54	14 ♎ 41	17 ♉ 52
31	13 ♑ 49	26 ♊ 07	27 ♈ 16	07 ♏ 10	18 ♎ 41	18 ♉ 59

All ephemeris data is given at 12.00 UT and the Moon's longitude is additionally given for 24.00 UT

ASPECTARIAN

h m	Aspects
01 Tuesday	
01 14	☿ ☌ ♄
02 02	☽ ± ♇
05 08	☽ ⊻ ♀
06 50	☽ ± ♅
08 09	☉ ∥ ♀
12 05	☽ ± ♂
12 47	♃ △ ♇
16 54	☽ ∥ ☿
17 34	☽ ⊥ ♀
18 03	☽ ⊼ ♅
18 53	☽ ⚼ ♇
19 55	☽ ⊻ ♄
23 27	☽ ⊻ ☿
23 33	☽ ⊼ ♂
02 Wednesday	
00 06	☿ △ ♂
02 00	☽ ∥ ♀
02 53	☽ ∥ ☉
06 37	☽ ∥ ♄
07 09	☽ ⊥ ♄
07 53	☽ ⊻ ☉
11 59	☽ ∠ ♀
12 00	☿ ± ♆
12 39	☽ ☍ ♆
12 47	☽ ⊥ ☿
18 57	♀ ⚼ ♆
19 48	☽ ⊥ ☉
22 31	☽ △ ♇
23 05	☽ ☌ ♃
23 51	☽ ∠ ♄
03 Thursday	
07 22	♂ ∥ ♇
08 29	☽ ∠ ☿
13 27	☽ ∠ ☉
16 33	☽ # ♆
18 06	☽ ⚹ ♀
21 29	☽ ∥ ♃
04 Friday	
00 50	☽ ☍ ♅
03 15	☽ ⚹ ♄
07 28	☽ ☍ ♂
09 44	☽ # ♇
10 02	☽ # ♂
16 39	☽ ⚹ ☿
17 43	☽ Q ♀
18 22	☽ ⚹ ☉
18 44	☽ ⊼ ♆
19 25	☽ # ♅
23 28	☉ ⊼ ♆
05 Saturday	
00 48	☽ Q ♄
03 52	☉ ⊥ ♃
04 15	☽ □ ♇
05 19	☽ ± ♆
05 30	☿ ⊼ ♆
05 40	☽ ⊻ ♃
07 51	☿ ⊥ ♃
10 20	☉ ☌ ☿
16 21	☽ ⊥ ♃
17 05	☽ Q ☉
17 48	☽ Q ☿
21 04	☽ ⚼ ♆
06 Sunday	
04 25	☽ □ ♀
05 40	☽ ⊼ ♅
08 15	☽ ∠ ♃
08 35	☽ □ ♄
11 37	☽ Q ♇
13 15	☽ ⊼ ♂
16 02	☽ ± ♅
18 21	♀ △ ♅
22 59	☽ △ ♆
23 47	☽ ± ♂
07 Monday	
02 32	☽ □ ☉
06 38	☽ □ ☿
07 27	☽ ⚼ ♅
08 11	☽ ⚹ ♇
10 23	☽ ⚹ ♃
11 55	☿ ⚼ ♅
15 20	☉ ∥ ♄
15 27	☽ ⚼ ♂
16 44	☿ ⊼ ♇
08 Tuesday	
07 01	☽ Q ♃
07 15	♀ ∠ ♃
08 30	♀ ☌ ♄
08 51	☽ △ ♅
09 24	☿ ⊻ ♃
09 35	☽ ∠ ♇
12 13	☽ △ ♄
12 32	☽ △ ♀
13 29	☽ ∥ ♅
17 15	☽ △ ♂
21 18	☽ ∥ ♂
22 27	☽ ∥ ♇
09 Wednesday	
00 40	☽ ⊥ ♇
01 41	☽ □ ♆
02 42	♃ ∠ ♄
08 01	☿ ∥ ♀
08 14	☽ # ♃
08 49	☽ △ ☉
10 39	☽ ⊻ ♇
13 32	☽ ⚼ ♄
13 35	☽ □ ♃
14 27	☽ ∥ ♆
15 56	☽ ⚼ ♀
17 59	☽ △ ☿
23 31	☿ ⚼ ♂
10 Thursday	
00 12	♀ ± ♆
01 11	☉ ⚼ ♅
04 36	☽ ± ♄
07 33	☽ Q ♆
08 14	☽ ± ♀
10 46	☽ □ ♅
11 28	☽ ⚼ ☉
11 57	☉ ⊼ ♇
12 00	☿ ± ♇
14 36	☽ ⊼ ♄
18 17	☽ # ☉
19 04	☽ ⊼ ♀
19 56	☽ □ ♂
20 37	☽ # ♄
22 54	☽ # ☿
22 58	☽ ⚼ ☿
11 Friday	
02 10	☽ # ♀
03 20	☽ ⚹ ♆
03 22	☽ ± ☉
07 28	♀ △ ♂
12 13	☽ ☌ ♇
13 57	☽ ⊼ ☉
15 53	☽ △ ♃
16 17	☽ ± ☿
16 26	☽ Q ♅
17 17	☿ ± ♅
21 48	☿ ∥ ♄
12 Saturday	
02 06	☽ Q ♂
03 49	☽ ⊼ ☿
04 03	☽ ∠ ♆
12 14	☽ ⚹ ♅
16 37	☽ ☍ ♄
17 03	☽ ⚼ ♃
18 56	☽ ⊥ ♆
22 17	☽ ⚹ ♂
13 Sunday	
01 15	☽ ☍ ♀
02 49	☉ ⊻ ♃
04 55	☽ ⊻ ♆
08 15	☽ ± ♃
13 11	☽ ∠ ♅
13 56	☽ ⊻ ♇
18 28	☽ ⊼ ♃
19 21	☽ ☍ ☉
20 05	☉ ∥ ☿
23 47	☽ ∠ ♂
14 Monday	
00 02	☽ ⊥ ♇
04 20	☽ ⊥ ♅
14 21	☽ ☍ ☿
14 33	☽ ⊻ ♅
15 18	☽ ⊥ ♂
15 20	☽ ∠ ♇
15 42	☿ ⊼ ♅
19 25	♀ ⊼ ♆
19 38	☽ ⊼ ♄
21 12	☿ ⚼ ♇
21 39	☿ ± ♂
22 24	☽ # ♀
15 Tuesday	
01 46	☽ ⊻ ♂
02 47	☽ # ♄
06 06	☽ ± ♄
07 52	☽ ☌ ♆
09 07	☽ ⊼ ♀
09 30	☽ # ☉
12 54	☽ # ☿
17 20	☽ ⚹ ♇
20 41	☽ ± ♀
22 02	☽ ⚼ ♄
22 57	☽ ☍ ♃
16 Wednesday	
03 02	☽ ⊼ ☉
06 21	☿ ⊻ ♄
10 47	☽ ∥ ♆
14 23	☽ ⚼ ♀
14 37	☽ ± ☉
14 42	☽ Q ♇
15 43	☉ ± ♇
19 16	☽ ☌ ♅
20 13	☽ # ♃
17 Thursday	
01 18	☽ △ ♄
02 19	♀ ⊥ ♃
03 08	☉ ⚼ ♂
04 20	☽ ⊼ ☿
04 30	☽ ∥ ♇
06 40	☽ ∥ ♂
08 00	☽ ☌ ♂
08 20	☽ ⚼ ☉
13 43	☽ ⊻ ♆
14 09	☽ ∥ ♅
17 21	☽ ± ☿
20 51	☽ △ ♀
23 50	☽ □ ♇
18 Friday	
00 55	☽ ⊥ ♆
06 50	☽ ⊼ ♃
07 37	☿ ⊼ ♂
13 17	☽ ⚼ ☿
14 49	☽ △ ☉
18 00	♀ ⚼ ♅
18 00	☽ ∠ ♆
18 29	☽ ± ♃
19 Saturday	
03 26	♀ ⊼ ♇
03 31	☽ ⊻ ♅
05 28	☉ ± ♅
10 40	☽ □ ♄
12 15	☽ ⚼ ♃
15 12	☽ ⊥ ♅
16 27	☿ ⊥ ♄
17 54	☽ ⊻ ♂
20 58	☿ ☍ ♆
23 11	☽ ⚹ ♆
23 35	☽ △ ☿
20 Sunday	
05 54	☽ ⊥ ♂
08 57	☽ ∠ ♅
09 58	☽ △ ♇
13 29	☽ □ ♀
16 30	♀ ∥ ♄
18 30	☽ △ ♃
22 45	☽ Q ♄
23 04	☽ Q ♆
21 Monday	
00 04	☽ ∠ ♂
07 01	☽ □ ☉
15 01	☽ ⚹ ♅
16 05	☽ ⚼ ♇
21 34	☽ # ♅
23 15	☽ ⚹ ♄
22 Tuesday	
00 04	♄ ± ♆
05 40	☽ # ♂
06 00	☽ Q ♀
06 41	☽ ⚹ ♂
07 57	☽ # ♇
10 19	☽ ± ♇
11 26	☽ □ ♆
13 57	☿ # ♆
14 55	☽ ∥ ♃
15 19	☽ Q ♅
21 38	☿ △ ♇
22 30	☽ ⊼ ♇
22 39	☽ □ ☿
22 45	☽ Q ☉
23 25	♀ ⊻ ♃
23 Wednesday	
02 00	☽ ∥ ☿
03 47	☽ # ♆
06 04	☽ ∠ ♄
07 16	☽ Q ♂
08 23	☽ □ ♃
09 14	☽ ⚹ ♀
21 24	☽ ∥ ☉
22 55	♀ ± ♇
24 Thursday	
00 31	☽ ⊥ ♄
01 15	☽ ⚹ ☉
03 45	☽ □ ♅
10 53	☽ ∥ ♀
12 45	☽ ⊻ ♄
15 29	☽ ∥ ♄
17 11	☽ Q ☿
19 02	☽ ∠ ♀
19 51	☽ □ ♂
21 23	☽ Q ♃
25 Friday	
00 00	☽ △ ♆
03 14	♀ ⚼ ♂
05 46	☿ ∠ ♄
07 31	☉ ⊼ ♅
09 56	☽ ∠ ☉
10 53	☽ ☍ ♇
15 00	☽ ⊥ ♀
20 52	☽ ⚹ ☿
21 25	♀ ± ♅
21 38	☽ ⚹ ♃
21 43	☉ ⚼ ♇
26 Saturday	
03 17	☿ ☌ ♃
04 09	☽ ⊻ ♀
05 12	☽ ⊥ ☉
05 36	☽ ⚼ ♆
15 07	☽ △ ♅
17 34	☿ ∥ ♃
17 51	☽ ⊻ ☉
23 03	☽ ± ♆
27 Sunday	
00 29	☽ ☌ ♄
03 18	☽ ∠ ♃
06 24	☽ ∠ ☿
06 53	☽ △ ♂
10 31	☽ ⊼ ♆
13 23	☉ ± ♂
19 46	☽ ⚼ ♅
20 44	☽ ⊥ ♃
20 56	☽ ⊼ ♇
28 Monday	
01 55	☽ ⊥ ☿
07 30	☿ # ♇
08 11	☽ ⊻ ♃
08 12	☽ ± ♇
11 15	☽ ⚼ ♂
12 36	☽ ± ♅
14 41	☽ ⊻ ☿
16 56	☿ # ♂
19 26	☽ ☌ ♀
23 41	☽ ⊼ ♅
29 Tuesday	
00 51	☽ ⚼ ♇
03 53	☽ ± ♂
06 53	☽ ☌ ☉
07 36	♂ St R
09 14	☽ ⊻ ♄
13 48	☽ ∥ ♄
14 51	☽ ⊼ ♂
18 08	☽ ☍ ♆
20 14	☽ ⊥ ♄
30 Wednesday	
01 32	☽ ∥ ♀
04 06	☽ △ ♇
12 33	☽ ∠ ♄
14 24	☽ ∥ ☉
15 16	♀ ⊼ ♅
15 46	☽ ☌ ♃
19 23	☉ ⊻ ♄
21 14	☽ # ♆
31 Thursday	
03 58	☽ ☌ ☿
04 11	♀ ⚼ ♇
05 34	☽ ☍ ♅
07 02	☽ ⊻ ♀
07 37	☿ # ♅
11 39	☽ ∥ ♃
14 42	☽ # ♇
15 14	☽ # ♂
15 21	☽ ⚹ ♄
16 40	☽ ⊻ ☉
18 24	☿ ☍ ♅
18 38	☽ ⊥ ♀
20 13	☽ ☍ ♂
23 21	☽ # ♅
23 21	☽ ⊼ ♆

AUGUST 2003

Date	Sidereal time h m s	Sun ☉ ° ' "	Moon ☽ ° ' "	Moon ☽ 24.00 ° ' "	Mercury ☿ ° '	Venus ♀ ° '	Mars ♂ ° '	Jupiter ♃ ° '	Saturn ♄ ° '	Uranus ♅ ° '	Neptune ♆ ° '	Pluto ♇ ° '
01	08 38 45	08 ♌ 50 24	19 ♍ 08 38	26 ♍ 03 57	02 ♍ 49	04 ♌ 05	10 ♓ 04	24 ♌ 23	07 ♋ 25	01 ♓ 43	11 ♒ 52	17 ♐ 26
02	08 42 42	09 47 49	03 ♎ 01 08	09 ♎ 59 52	04 14	05 19	10 R 01	24 36	07 32	01 R 41	11 R 51	17 R 25
03	08 46 38	10 45 15	16 ♎ 59 55	24 ♎ 01 04	05 38	06 33	09 57	24 49	07 39	01 39	11 49	17 24
04	08 50 35	11 42 41	01 ♏ 03 09	08 ♏ 06 03	07 00	07 47	09 53	25 01	07 46	01 37	11 47	17 24
05	08 54 31	12 40 08	15 ♏ 09 36	22 ♏ 13 43	08 20	09 01	09 48	25 14	07 53	01 35	11 46	17 23
06	08 58 28	13 37 36	29 ♏ 18 16	06 ♐ 23 06	09 37	10 15	09 42	25 27	08 00	01 32	11 44	17 22
07	09 02 24	14 35 05	13 ♐ 28 03	20 ♐ 32 53	10 53	11 29	09 35	25 40	08 07	01 30	11 42	17 21
08	09 06 21	15 32 34	27 ♐ 37 18	04 ♑ 40 58	12 07	12 43	09 27	25 53	08 14	01 28	11 41	17 21
09	09 10 17	16 30 04	11 ♑ 43 29	18 ♑ 44 25	13 18	13 57	09 19	26 06	08 20	01 26	11 39	17 20
10	09 14 14	17 27 35	25 ♑ 43 17	02 ♒ 39 35	14 28	15 11	09 11	26 19	08 27	01 23	11 38	17 20
11	09 18 11	18 25 07	09 ♒ 32 49	16 ♒ 22 30	15 35	16 25	09 01	26 32	08 34	01 21	11 36	17 19
12	09 22 07	19 22 40	23 ♒ 08 14	29 ♒ 49 39	16 39	17 39	08 51	26 45	08 40	01 19	11 34	17 18
13	09 26 04	20 20 14	06 ♓ 26 27	12 ♓ 58 28	17 41	18 53	08 40	26 58	08 47	01 16	11 33	17 18
14	09 30 00	21 17 50	19 ♓ 25 36	25 ♓ 47 53	18 40	20 08	08 29	27 11	08 53	01 14	11 31	17 18
15	09 33 57	22 15 26	02 ♈ 05 26	08 ♈ 18 29	19 37	21 22	08 17	27 24	09 00	01 12	11 30	17 17
16	09 37 53	23 13 05	14 ♈ 27 19	20 ♈ 32 22	20 31	22 36	08 04	27 37	09 06	01 09	11 28	17 17
17	09 41 50	24 10 44	26 ♈ 34 06	02 ♉ 33 02	21 21	23 50	07 51	27 51	09 12	01 07	11 26	17 16
18	09 45 46	25 08 25	08 ♉ 29 46	14 ♉ 24 57	22 09	25 04	07 38	28 04	09 19	01 05	11 25	17 16
19	09 49 43	26 06 08	20 ♉ 19 12	26 ♉ 13 14	22 53	26 18	07 24	28 17	09 25	01 02	11 23	17 16
20	09 53 40	27 03 52	02 ♊ 07 43	08 ♊ 03 22	23 33	27 33	07 10	28 30	09 31	01 00	11 22	17 15
21	09 57 36	28 01 39	14 ♊ 00 50	20 ♊ 00 48	24 10	28 ♌ 47	06 55	28 43	09 37	00 57	11 20	17 15
22	10 01 33	28 59 26	26 ♊ 03 53	02 ♋ 10 40	24 42	00 ♍ 01	06 40	28 56	09 43	00 55	11 19	17 15
23	10 05 29	29 ♌ 57 16	08 ♋ 21 39	14 ♋ 37 17	25 11	01 16	06 25	29 09	09 49	00 53	11 17	17 15
24	10 09 26	00 ♍ 55 07	20 ♋ 57 57	27 ♋ 23 53	25 34	02 30	06 09	29 22	09 55	00 50	11 16	17 14
25	10 13 22	01 53 00	03 ♌ 55 15	10 ♌ 32 04	25 53	03 44	05 53	29 35	10 01	00 48	11 14	17 14
26	10 17 19	02 50 54	17 ♌ 14 15	24 ♌ 01 35	26 07	04 59	05 37	29 ♌ 48	10 06	00 45	11 13	17 14
27	10 21 15	03 48 50	00 ♍ 53 42	07 ♍ 50 11	26 16	06 13	05 21	00 ♍ 01	10 12	00 43	11 11	17 14
28	10 25 12	04 46 47	14 ♍ 50 28	21 ♍ 53 57	26 19	07 27	05 05	00 14	10 18	00 41	11 10	17 14
29	10 29 09	05 44 46	28 ♍ 59 59	06 ♎ 07 55	26 R 17	08 42	04 49	00 28	10 23	00 38	11 08	17 D 14
30	10 33 05	06 42 46	13 ♎ 17 06	20 ♎ 26 54	26 08	09 56	04 33	00 41	10 29	00 36	11 07	17 14
31	10 37 02	07 ♍ 40 48	27 ♎ 36 48	04 ♏ 46 17	25 ♍ 54	11 ♍ 10	04 ♓ 17	00 ♍ 54	10 ♋ 34	00 ♓ 33	11 ♒ 05	17 ♐ 14

Date	Moon ☽ True ☊ ° '	Moon ☽ Mean ☊ ° '	Moon ☽ Latitude ° '
01	26 ♉ 28	25 ♉ 47	04 N 39
02	26 R 22	25 44	04 03
03	26 17	25 41	03 12
04	26 15	25 37	02 09
05	26 D 15	25 34	00 N 58
06	26 R 15	25 31	00 S 16
07	26 14	25 28	01 30
08	26 12	25 25	02 38
09	26 06	25 21	03 36
10	25 58	25 18	04 20
11	25 47	25 15	04 49
12	25 35	25 12	05 00
13	25 23	25 09	04 55
14	25 12	25 06	04 35
15	25 02	25 02	04 01
16	24 56	24 59	03 16
17	24 51	24 56	02 24
18	24 49	24 53	01 25
19	24 D 49	24 50	00 S 24
20	24 R 49	24 46	00 N 39
21	24 48	24 43	01 40
22	24 46	24 40	02 37
23	24 41	24 37	03 29
24	24 34	24 34	04 11
25	24 24	24 31	04 42
26	24 13	24 27	04 59
27	24 01	24 24	04 59
28	23 49	24 21	04 41
29	23 40	24 18	04 06
30	23 32	24 15	03 16
31	23 ♉ 28	24 ♉ 12	02 N 12

DECLINATIONS

Date	Sun ☉ ° '	Moon ☽ ° '	Mercury ☿ ° '	Venus ♀ ° '	Mars ♂ ° '	Jupiter ♃ ° '	Saturn ♄ ° '	Uranus ♅ ° '	Neptune ♆ ° '	Pluto ♇ ° '
01	18 N 03	08 N 34	10 N 49	20 N 08	13 S 29	14 N 09	22 N 28	11 S 36	17 S 14	13 S 33
02	17 48	02 N 31	10 10	19 52	13 33	14 05	22 28	11 37	17 14	13 34
03	17 32	03 S 43	09 31	19 35	13 37	14 01	22 28	11 38	17 15	13 34
04	17 16	09 49	08 52	19 18	13 41	13 56	22 27	11 38	17 15	13 34
05	17 00	15 27	08 14	19 00	13 46	13 52	22 27	11 39	17 16	13 35
06	16 44	20 16	07 35	18 42	13 50	13 48	22 26	11 40	17 16	13 35
07	16 27	23 54	06 58	18 23	13 55	13 43	22 26	11 41	17 16	13 35
08	16 11	26 03	06 20	18 04	14 00	13 39	22 26	11 42	17 17	13 35
09	15 53	26 30	05 44	17 44	14 05	13 35	22 25	11 42	17 17	13 36
10	15 36	25 16	05 08	17 23	14 10	13 30	22 25	11 43	17 18	13 36
11	15 18	22 30	04 32	17 03	14 15	13 26	22 25	11 44	17 18	13 36
12	15 01	18 32	03 57	16 41	14 21	13 22	22 24	11 45	17 19	13 37
13	14 42	13 43	03 24	16 19	14 26	13 17	22 24	11 46	17 19	13 37
14	14 24	08 24	02 51	15 57	14 32	13 13	22 23	11 47	17 20	13 37
15	14 06	02 S 51	02 19	15 35	14 37	13 08	22 23	11 47	17 20	13 38
16	13 47	02 N 41	01 48	15 11	14 43	13 04	22 22	11 48	17 21	13 38
17	13 28	08 01	01 18	14 48	14 49	12 59	22 22	11 49	17 21	13 38
18	13 09	12 59	00 49	14 24	14 54	12 55	22 22	11 50	17 22	13 39
19	12 49	17 27	00 N 22	14 00	15 00	12 50	22 21	11 51	17 22	13 39
20	12 29	21 13	00 S 03	13 35	15 06	12 46	22 21	11 52	17 22	13 39
21	12 10	24 08	00 27	13 10	15 11	12 41	22 21	11 53	17 23	13 40
22	11 50	26 00	00 49	12 44	15 17	12 37	22 20	11 53	17 23	13 40
23	11 29	26 39	01 09	12 19	15 22	12 32	22 20	11 54	17 24	13 41
24	11 09	25 57	01 27	11 52	15 28	12 28	22 19	11 55	17 24	13 41
25	10 48	23 50	01 43	11 26	15 33	12 23	22 19	11 56	17 25	13 41
26	10 28	20 24	01 57	10 59	15 38	12 19	22 19	11 57	17 25	13 42
27	10 07	15 48	02 07	10 32	15 44	12 14	22 18	11 58	17 25	13 42
28	09 46	10 17	02 15	10 05	15 48	12 10	22 18	11 58	17 26	13 42
29	09 24	04 N 10	02 20	09 37	15 53	12 05	22 17	11 59	17 26	13 43
30	09 03	02 S 14	02 22	09 09	15 57	12 01	22 17	12 00	17 27	13 43
31	08 N 41	08 S 34	02 S 20	08 N 41	16 S 01	11 N 56	22 N 17	12 S 01	17 S 27	13 S 44

ZODIAC SIGN ENTRIES

Date	h m	Planets
02	06 48	☽ ♎
04	10 12	☽ ♏
06	13 11	☽ ♐
08	16 02	☽ ♑
10	19 23	☽ ♒
13	00 19	☽ ♓
15	08 00	☽ ♈
17	18 52	☽ ♉
20	07 41	☽ ♊
22	11 36	♀ ♍
22	19 44	☽ ♋
23	13 08	☉ ♍
25	04 48	☽ ♌
27	09 26	♃ ♍
27	10 27	☽ ♍
29	13 41	☽ ♎
31	16 00	☽ ♏

LATITUDES

Date	Mercury ☿ ° '	Venus ♀ ° '	Mars ♂ ° '	Jupiter ♃ ° '	Saturn ♄ ° '	Uranus ♅ ° '	Neptune ♆ ° '	Pluto ♇ ° '
01	00 N 22	00 N 55	06 S 09	00 N 48	00 S 46	00 S 47	00 00	09 N 20
04	00 S 05	01 00	06 18	00 48	00 46	00 47	00 00	09 19
07	00 34	01 05	06 25	00 48	00 45	00 47	00 00	09 18
10	01 04	01 10	06 31	00 48	00 45	00 47	00 00	09 16
13	01 36	01 13	06 36	00 49	00 45	00 47	00 00	09 15
16	02 08	01 17	06 40	00 49	00 45	00 47	00 00	09 14
19	02 40	01 19	06 42	00 49	00 45	00 48	00 00	09 13
22	03 11	01 22	06 42	00 49	00 45	00 48	00 00	09 12
25	03 39	01 23	06 41	00 49	00 45	00 48	00 00	09 11
28	04 03	01 24	06 37	00 50	00 45	00 48	00 00	09 10
31	04 S 20	01 N 25	06 S 32	00 N 50	00 S 45	00 S 48	00 00	09 N 08

DATA

Julian Date	2452853
Delta T	+65 seconds
Ayanamsa	23° 54' 13"
Synetic vernal point	05° ♓ 12' 46"
True obliquity of ecliptic	23° 26' 25"

MOON'S PHASES, APSIDES AND POSITIONS ☽

Date	h m	Phase	Longitude ° '	Eclipse Indicator
05	07 28	☽	12 ♏ 29	
12	04 48	○	19 ♒ 05	
20	00 48	☾	26 ♉ 37	
27	17 26	●	04 ♍ 02	

Day	h m		
06	14 07	Perigee	
19	14 21	Apogee	
31	18 42	Perigee	
02	21 42	0S	
09	06 19	Max dec	26° S 33'
16	00 18	0N	
23	11 42	Max dec	26° N 39'
30	03 38	0S	

LONGITUDES

Date	Chiron ⚷ ° '	Ceres ⚳ ° '	Pallas ⚴ ° '	Juno ⚵ ° '	Vesta ⚶ ° '	Black Moon Lilith ⚸ ° '
01	13 ♑ 46	26 ♊ 30	27 ♈ 23	07 ♏ 19	19 ♎ 06	19 ♉ 06
11	13 ♑ 16	00 ♋ 17	28 ♈ 19	08 ♏ 59	23 ♎ 21	20 ♉ 12
21	12 ♑ 52	03 ♋ 57	28 ♈ 43	10 ♏ 59	27 ♎ 49	21 ♉ 18
31	12 ♑ 35	07 ♋ 28	28 ♈ 28	13 ♏ 15	02 ♏ 27	22 ♉ 25

ASPECTARIAN

h m Aspects

01 Friday
01 28 ☽ ∥ ☿
03 57 ☽ ⊥ ☉
09 02 ☽ □ ♇
09 48 ☽ ± ♆
11 40 ♀ ± ♂
11 53 ☽ ∠ ♀
12 29 ☽ Q ♄
20 45 ☽ ∠ ☉
21 14 ☽ ⊻ ♃
02 Saturday
01 22 ☽ ꝑ ♆
07 45 ☽ ⊥ ♃
09 43 ☽ ⊼ ♅
14 20 ☽ ⊻ ☿
16 08 ☽ Q ♇
16 12 ♂ ∥ ♇
16 20 ☽ ⚹ ♀
17 13 ☉ ⊼ ♂
19 50 ☽ □ ♄
20 01 ☽ ± ♅
23 29 ☽ ∠ ♃
23 59 ☽ ⊼ ♂
03 Sunday
00 31 ☽ ⚹ ☉
01 48 ☽ ⊥ ☿
03 08 ☽ △ ♆
10 13 ☽ ± ♂
11 24 ☽ ꝑ ♅
12 42 ☽ ⚹ ♇
14 55 ☽ Q ♀
18 54 ☽ ∠ ☿
22 34 ☽ Q ☉
04 Monday
01 32 ☽ ꝑ ♂
01 33 ☽ ⚹ ♃
08 32 ☽ # ☿
11 45 ♀ ⊻ ♄
12 57 ☽ △ ♅
13 54 ☉ ☍ ♆
14 01 ☉ # ♆
14 17 ☽ ∠ ♇
19 30 ☽ ∥ ♅
22 20 ☽ Q ♃
23 11 ☽ ⚹ ☿
23 32 ☽ △ ♄
05 Tuesday
00 33 ☽ □ ♀
02 56 ☽ △ ♂
03 12 ☿ ⚹ ♄
03 41 ☽ ∥ ♇
04 24 ☽ ∥ ♂
05 03 ☽ # ♃
05 35 ☽ ⊥ ♇
06 14 ☽ □ ♆
07 28 ☽ □ ☉
15 46 ☽ ⊻ ♇
18 52 ☽ # ☉
20 28 ☽ ∥ ♆
21 40 ☽ Q ☿
06 Wednesday
01 13 ☽ ꝑ ♄
02 02 ♀ ⊼ ♂
04 07 ☽ # ♀
05 22 ☽ □ ♃
05 29 ♂ # ♃
12 44 ☽ Q ♆
13 12 ☿ ☍ ♂
15 47 ☽ □ ♅
16 36 ☽ ± ♄
22 37 ☉ ⊥ ♄
07 Thursday
01 14 ☽ # ♄
02 51 ☽ ⊼ ♄
05 28 ☽ □ ♂
07 12 ☽ □ ☿
08 19 ☽ △ ♀
09 01 ☽ ⚹ ♆
14 02 ☽ △ ☉
16 18 ♀ ☍ ♆
18 35 ☽ ☌ ♇
22 12 ☽ Q ♅
08 Friday
03 37 ☿ ⊼ ♆
09 01 ☽ △ ♃
10 24 ☽ ∠ ♆
11 43 ☽ Q ♂
12 11 ☽ ꝑ ♀
17 19 ☽ ꝑ ☉
18 31 ☽ ⚹ ♅
09 Saturday
01 40 ☽ ⊥ ♆
04 57 ☽ ± ♀
06 11 ☽ ☍ ♄
06 32 ♃ # ♇
07 57 ☽ ⚹ ♂
09 46 ☽ ± ☉
10 55 ☽ ꝑ ♃
11 53 ☽ ⊻ ♆
14 57 ☽ △ ☿
16 10 ☽ ⊼ ♀
20 01 ☽ ∠ ♅
20 18 ♀ ⊥ ♄
20 46 ☽ ⊼ ☉
21 35 ☽ ⊻ ♇
10 Sunday
02 34 ☽ ± ♃
07 53 ☽ ⊥ ♇
08 42 ☉ △ ♇
09 22 ☽ ∠ ♂
11 26 ☽ ⊥ ♅
13 03 ☽ ⊼ ♃
18 15 ♀ # ♆
19 02 ☽ ꝑ ☿
21 46 ☽ ⊻ ♅
23 25 ☽ ∠ ♇
11 Monday
00 45 ☽ ⊥ ♂
10 16 ☽ ⊼ ♄
11 05 ☽ ⊻ ♂
12 03 ☽ ± ☿
12 36 ☽ # ♄
15 35 ☽ ☌ ♆
20 52 ☽ ± ♄
23 31 ☽ ⊼ ☿
12 Tuesday
01 17 ☽ ☍ ♀
01 39 ☽ ⚹ ♇
04 48 ☽ ☍ ☉
05 19 ♀ △ ♇
12 58 ☽ ꝑ ♄
18 23 ☽ ∥ ♆
18 35 ☽ ☍ ♃
22 22 ☽ # ♀
23 03 ☽ Q ♇
13 Wednesday
02 38 ☽ ☌ ♅
02 55 ♂ △ ♄
03 01 ☿ □ ♇
06 59 ☽ # ☉
08 39 ☽ ∥ ♂
08 49 ☿ ± ♆
12 27 ☽ ∥ ♇
14 02 ☽ # ♃
16 01 ☽ ☌ ♂
16 19 ☽ △ ♄
20 58 ☽ ∥ ♅
21 21 ☽ ⊻ ♆
14 Thursday
04 26 ☉ # ♂
08 01 ☽ □ ♇
08 27 ☽ ⊥ ♆
10 29 ☽ ☍ ☿
13 27 ☽ ⊼ ♀
15 48 ☽ ⊼ ☉
15 Friday
01 20 ☽ ∠ ♆
01 59 ☽ ± ♀
02 54 ☽ ⊼ ♃
04 04 ☽ ± ☉
10 17 ☽ ⊻ ♅
14 34 ☽ # ☿
14 34 ☽ ± ♃
21 08 ☽ ꝑ ♀
21 48 ☽ ⊥ ♅
22 48 ☽ ꝑ ☉
23 45 ☽ ⊻ ♂
16 Saturday
01 27 ☽ □ ♄
06 10 ☽ ⚹ ♆
08 21 ☽ ꝑ ♃
08 27 ☽ ∥ ☿
11 16 ☽ ⊥ ♂
15 20 ☽ ∠ ♅
17 33 ☽ △ ♇
22 49 ☉ # ♇
17 Sunday
00 51 ☽ ⊼ ☿
04 44 ☽ ∠ ♂
05 46 ☽ Q ♆
05 55 ☽ △ ♀
06 49 ☽ △ ☉
07 00 ☿ Q ♄
11 20 ♀ # ♂
12 44 ☉ ∠ ♄
13 17 ☽ Q ♄
13 41 ☽ ± ☿
14 36 ☽ △ ♃
19 52 ♀ ∠ ♄
21 05 ☽ ⚹ ♅
23 26 ☽ ꝑ ♇
18 Monday
06 12 ☽ # ♅
09 05 ☽ ꝑ ☿
10 17 ☽ ⚹ ♂
11 38 ☽ ∥ ♃
12 44 ☽ ∥ ☉
13 40 ☽ ⚹ ♄
15 21 ☽ # ♇
17 36 ☽ ± ♇
17 54 ☽ □ ♆
18 05 ☉ ☌ ♀
18 40 ☽ ∥ ♀
21 15 ☽ Q ♅
22 09 ☽ # ♂
19 Tuesday
05 47 ☽ ⊼ ♇
09 54 ☉ ∥ ♃
10 10 ☽ Q ♂
11 32 ☽ # ♆
17 32 ☽ △ ☿
20 24 ☽ ∠ ♄
20 Wednesday
00 48 ☽ □ ☉
01 36 ☽ □ ♀
04 29 ☽ □ ♃
07 37 ♀ # ♇
09 43 ☽ □ ♅
14 50 ☽ ⊥ ♄
20 24 ☽ ∥ ♄
21 59 ☽ □ ♂
21 Thursday
03 04 ☽ ⊻ ♄
06 38 ☽ △ ♆
10 23 ♀ ☌ ♃
16 23 ☽ Q ☉
17 31 ☽ Q ♃
18 11 ☽ Q ♀
18 29 ☽ ☍ ♇
22 Friday
07 37 ☉ # ♅
09 12 ☽ □ ☿
10 08 ☉ ☌ ♃
12 29 ☽ ꝑ ♆
17 45 ☽ ⚹ ♃
18 15 ☽ ⚹ ☉
20 26 ♀ ∥ ♃
20 39 ☽ ⚹ ♀
21 30 ☽ △ ♅
23 Saturday
04 49 ♀ ☍ ♅
06 04 ☽ ± ♆
08 18 ☽ △ ♂
14 50 ☽ ☌ ♄
17 37 ☽ ⊼ ♆
21 34 ☽ Q ☿
23 18 ☽ ∠ ♃
24 Sunday
01 41 ☽ ∠ ☉
02 20 ☽ ꝑ ♅
04 45 ☽ ∠ ♀
04 59 ☽ ⊼ ♇
09 38 ♀ # ♅
10 02 ☉ ☍ ♅
12 20 ☽ ꝑ ♂
16 16 ☽ ± ♇
16 35 ☽ ⊥ ♃
19 13 ☽ ± ♅
20 00 ☽ ⊥ ☉
20 51 ☽ ⚹ ☿
23 25 ☽ ⊥ ♀
25 Monday
03 55 ☽ ⊻ ♃
04 45 ☽ ± ♂
06 17 ☽ ⊼ ♅
07 58 ☽ ⊻ ☉
08 55 ☽ ꝑ ♇
11 38 ☽ ⊻ ♀
15 31 ☽ ⊼ ♂
23 08 ☽ ⊻ ♄
23 47 ☽ ∥ ♄
26 Tuesday
00 53 ☽ ∠ ☿
01 14 ☽ ☍ ♆
09 58 ☽ ⊥ ♄
12 00 ☽ △ ♇
17 11 ☽ ⊥ ☿
22 17 ♀ ☍ ♂
22 27 ☿ ꝑ ♆
27 Wednesday
02 00 ☽ ∠ ♄
03 52 ☽ ⊻ ☿
04 10 ☽ # ♆
10 28 ☽ ☌ ♃
11 41 ☽ ☍ ♅
12 24 ☽ # ♂
17 26 ☽ ☌ ☉
19 35 ☽ ☍ ♂
21 33 ☽ # ♇
22 06 ☽ ☌ ♀
28 Thursday
04 07 ☽ ∥ ♃
04 10 ☽ ⚹ ♄
05 02 ☽ # ♅
05 43 ☽ ⊼ ♆
12 56 ☽ ∥ ♀
13 41 ☿ St R
14 18 ☽ ∥ ☉
15 57 ☽ ± ♆
16 05 ☽ □ ♇
17 59 ☉ ☍ ♂
29 Friday
00 45 ☽ Q ♄
03 34 ♇ St D
07 10 ☽ ꝑ ♆
07 26 ☽ ☌ ☿
14 30 ☽ ⊻ ♃
14 45 ☽ ⊼ ♅
18 50 ☽ # ☿
21 37 ☽ ⊼ ♂
22 30 ☽ Q ♇
30 Saturday
00 11 ☽ ⊻ ☉
00 45 ☽ ⊥ ♃
00 49 ☽ ± ♅
04 37 ♃ ☍ ♅
05 51 ☽ ⊻ ♀
07 16 ☽ □ ♄
07 30 ☽ ± ♂
08 22 ☽ △ ♆
10 58 ☽ ⊥ ☉
12 29 ☽ ∥ ☿
13 52 ♃ # ♅
15 52 ☽ ꝑ ♅
16 04 ☿ ꝑ ♆
16 04 ☽ ∠ ♃
16 52 ☽ ⊥ ♀
18 37 ☽ ⚹ ♇
22 19 ☽ ꝑ ♂
23 23 ♀ ⚹ ♄
31 Sunday
03 08 ☽ ∠ ☉
09 11 ☽ ⊻ ☿
09 22 ☽ ∠ ♀
10 21 ♀ ⊼ ♆
10 59 ☉ ∥ ♀
12 26 ☽ # ♀
12 27 ☽ # ☉
16 55 ☽ △ ♅
17 35 ☽ ⚹ ♃
19 02 ☽ ⊥ ☿
19 45 ☽ ∠ ♇
22 59 ☽ △ ♂

All ephemeris data is given at 12.00 UT and the Moon's longitude is additionally given for 24.00 UT

SEPTEMBER 2003

Date	Sidereal time h m s	LONGITUDES Sun ☉ ° ' "	Moon ☽ ° ' "	Moon ☽ 24.00 ° ' "	Mercury ☿ ° '	Venus ♀ ° '	Mars ♂ ° '	Jupiter ♃ ° '	Saturn ♄ ° '	Uranus ♅ ° '	Neptune ♆ ° '	Pluto ♇ ° '
01	10 40 58	08 ♍ 38 51	11 ♏ 54 59	19 ♏ 02 33	25 ♍ 33	12 ♍ 25	04 ♓ 02	01 ♍ 07	10 ♋ 40	00 ♓ 31	11 ♒ 04	17 ♐ 14
02	10 44 55	09 36 55	26 ♏ 08 46	03 ♐ 13 25	25 R 06	13 39	03 R 46	01 20	10 45	00 R 29	11 R 02	17 14
03	10 48 51	10 35 01	10 ♐ 16 25	17 ♐ 17 39	24 33	14 54	03 31	01 33	10 50	00 26	11 01	17 15
04	10 52 48	11 33 09	24 ♐ 17 03	01 ♑ 14 33	23 55	16 08	03 16	01 46	10 55	00 24	11 00	17 15
05	10 56 44	12 31 17	08 ♑ 10 05	15 ♑ 03 33	23 10	17 23	03 01	01 59	11 00	00 22	10 58	17 15
06	11 00 41	13 29 27	21 ♑ 54 50	28 ♑ 43 49	22 21	18 37	02 47	02 12	11 05	00 19	10 57	17 15
07	11 04 38	14 27 39	05 ♒ 30 19	12 ♒ 14 07	21 28	19 52	02 33	02 25	11 10	00 17	10 56	17 16
08	11 08 34	15 25 52	18 ♒ 55 03	25 ♒ 32 53	20 31	21 06	02 20	02 38	11 15	00 15	10 55	17 16
09	11 12 31	16 24 07	02 ♓ 07 26	08 ♓ 38 29	19 32	22 21	02 07	02 50	11 20	00 13	10 53	17 16
10	11 16 27	17 22 23	15 ♓ 05 55	21 ♓ 29 36	18 31	23 35	01 55	03 03	11 24	00 10	10 52	17 17
11	11 20 24	18 20 41	27 ♓ 49 31	04 ♈ 05 38	17 31	24 50	01 43	03 16	11 29	00 08	10 51	17 17
12	11 24 20	19 19 01	10 ♈ 18 03	16 ♈ 26 55	16 33	26 04	01 32	03 29	11 33	00 06	10 50	17 18
13	11 28 17	20 17 23	22 ♈ 32 27	28 ♈ 34 55	15 37	27 19	01 21	03 42	11 38	00 04	10 48	17 18
14	11 32 13	21 15 47	04 ♉ 34 42	10 ♉ 32 11	14 46	28 33	01 11	03 55	11 42	00 ♓ 01	10 47	17 19
15	11 36 10	22 14 13	16 ♉ 27 52	22 ♉ 22 15	14 01	29 ♍ 48	01 02	04 07	11 46	29 ♒ 59	10 46	17 19
16	11 40 07	23 12 41	28 ♉ 15 56	04 ♊ 09 30	13 22	01 ♎ 02	00 53	04 20	11 50	29 57	10 45	17 20
17	11 44 03	24 11 12	10 ♊ 03 36	15 ♊ 58 53	12 51	02 17	00 45	04 33	11 54	29 55	10 44	17 20
18	11 48 00	25 09 44	21 ♊ 56 02	27 ♊ 55 43	12 29	03 31	00 38	04 45	11 58	29 53	10 43	17 21
19	11 51 56	26 08 19	03 ♋ 58 36	10 ♋ 05 20	12 16	04 46	00 32	04 58	12 02	29 51	10 42	17 22
20	11 55 53	27 06 56	16 ♋ 16 31	22 ♋ 32 43	12 D 12	06 01	00 26	05 11	12 06	29 49	10 41	17 22
21	11 59 49	28 05 35	28 ♋ 54 26	05 ♌ 22 04	12 18	07 15	00 21	05 23	12 10	29 47	10 40	17 23
22	12 03 46	29 ♍ 04 16	11 ♌ 55 54	18 ♌ 36 08	12 34	08 30	00 17	05 36	12 13	29 45	10 39	17 24
23	12 07 42	00 ♎ 02 59	25 ♌ 22 47	02 ♍ 15 45	13 00	09 44	00 13	05 48	12 17	29 43	10 38	17 25
24	12 11 39	01 01 45	09 ♍ 14 46	16 ♍ 19 23	13 34	10 59	00 10	06 01	12 20	29 41	10 37	17 25
25	12 15 36	02 00 32	23 ♍ 29 03	00 ♎ 43 01	14 18	12 14	00 08	06 13	12 24	29 39	10 36	17 26
26	12 19 32	02 59 22	08 ♎ 00 30	15 ♎ 20 35	15 10	13 28	00 07	06 25	12 27	29 37	10 35	17 27
27	12 23 29	03 58 13	22 ♎ 42 21	00 ♏ 04 52	16 09	14 43	00 D 07	06 38	12 30	29 35	10 35	17 28
28	12 27 25	04 57 06	07 ♏ 27 13	14 ♏ 48 34	17 16	15 57	00 08	06 50	12 33	29 33	10 34	17 29
29	12 31 22	05 56 02	22 ♏ 08 12	29 ♏ 25 28	18 29	17 12	00 09	07 02	12 36	29 32	10 33	17 30
30	12 35 18	06 ♎ 54 59	06 ♐ 39 51	13 ♐ 50 57	19 ♍ 48	18 ♎ 27	00 ♓ 11	07 ♍ 14	12 ♋ 39	29 ♒ 30	10 ♒ 32	17 ♐ 31

Date	Moon True ☊ ° '	Moon Mean ☊ ° '	Moon ☽ Latitude ° '	DECLINATIONS Sun ☉ ° '	Moon ☽ ° '	Mercury ☿ ° '	Venus ♀ ° '	Mars ♂ ° '	Jupiter ♃ ° '	Saturn ♄ ° '	Uranus ♅ ° '	Neptune ♆ ° '	Pluto ♇ ° '
01	23 ♉ 25	24 ♉ 08	01 N 01	08 N 20	14 S 27	02 S 15	08 N 13	16 S 05	11 N 51	22 N 16	12 S 02	17 S 27	13 S 44
02	23 D 25	24 05	00 S 14	07 58	19 31	02 07	07 44	16 09	11 47	22 16	12 03	17 28	13 44
03	23 25	24 02	01 28	07 36	23 27	01 54	07 15	16 12	11 42	22 15	12 03	17 28	13 45
04	23 R 25	23 59	02 36	07 14	25 55	01 38	06 46	16 15	11 38	22 15	12 04	17 29	13 45
05	23 22	23 56	03 34	06 52	26 45	01 18	06 17	16 18	11 33	22 15	12 05	17 29	13 46
06	23 17	23 52	04 19	06 29	25 55	00 54	05 48	16 20	11 29	22 14	12 06	17 29	13 46
07	23 10	23 49	04 49	06 07	23 33	00 S 27	05 18	16 23	11 24	22 14	12 07	17 30	13 47
08	23 00	23 46	05 02	05 45	19 56	00 N 03	04 48	16 25	11 19	22 13	12 08	17 30	13 47
09	22 49	23 43	04 59	05 22	15 22	00 35	04 19	16 26	11 15	22 13	12 08	17 30	13 47
10	22 38	23 40	04 40	04 59	10 10	01 10	03 49	16 27	11 10	22 13	12 09	17 31	13 48
11	22 27	23 37	04 08	04 37	04 S 39	01 46	03 19	16 28	11 06	22 12	12 10	17 31	13 48
12	22 19	23 33	03 24	04 14	00 N 57	02 24	02 48	16 29	11 01	22 12	12 11	17 31	13 49
13	22 12	23 30	02 31	03 51	06 26	03 01	02 18	16 29	10 57	22 12	12 11	17 32	13 49
14	22 08	23 27	01 32	03 28	11 36	03 37	01 48	16 29	10 52	22 11	12 12	17 32	13 50
15	22 06	23 24	00 S 30	03 05	16 17	04 12	01 17	16 28	10 48	22 11	12 13	17 32	13 50
16	22 D 06	23 21	00 N 33	02 42	20 19	04 45	00 47	16 28	10 43	22 10	12 14	17 33	13 50
17	22 07	23 18	01 35	02 19	23 31	05 15	00 N 16	16 27	10 38	22 10	12 14	17 33	13 51
18	22 08	23 14	02 33	01 55	25 45	05 42	00 S 14	16 25	10 34	22 10	12 15	17 33	13 51
19	22 R 08	23 11	03 25	01 32	26 48	06 05	00 45	16 23	10 29	22 09	12 16	17 34	13 52
20	22 06	23 08	04 09	01 09	26 34	06 24	01 16	16 21	10 25	22 09	12 17	17 34	13 52
21	22 02	23 05	04 43	00 46	24 59	06 38	01 46	16 19	10 20	22 09	12 17	17 34	13 53
22	21 55	23 02	05 02	00 N 22	22 03	06 48	02 17	16 16	10 16	22 09	12 18	17 34	13 53
23	21 48	22 58	05 06	00 S 01	17 52	06 54	02 47	16 13	10 11	22 08	12 19	17 35	13 54
24	21 39	22 55	04 53	00 25	12 37	06 55	03 18	16 10	10 07	22 08	12 19	17 35	13 54
25	21 31	22 52	04 21	00 48	06 35	06 51	03 48	16 06	10 02	22 08	12 20	17 35	13 55
26	21 24	22 49	03 32	01 11	00 N 04	06 42	04 19	16 02	09 58	22 07	12 21	17 35	13 55
27	21 19	22 46	02 28	01 35	06 S 32	06 30	04 49	15 58	09 53	22 07	12 21	17 36	13 55
28	21 16	22 43	01 N 14	01 58	12 50	06 13	05 19	15 53	09 49	22 07	12 22	17 36	13 56
29	21 D 16	22 39	00 S 05	02 21	18 23	05 53	05 49	15 48	09 45	22 07	12 22	17 36	13 56
30	21 ♉ 16	22 ♉ 36	01 S 23	02 S 45	22 S 47	05 N 29	06 S 19	15 S 43	09 N 40	22 N 06	12 S 23	17 S 36	13 S 57

ZODIAC SIGN ENTRIES

Date	h m	Planets
02	18 32	☽ ♐
04	21 51	☽ ♑
07	02 15	☽ ♒
09	08 07	☽ ♓
11	16 09	☽ ♈
14	02 50	☽ ♉
15	03 47	♅ ♒
15	15 58	♀ ♎
16	15 32	☽ ♊
19	04 07	☽ ♋
21	14 03	☽ ♌
23	10 47	☉ ♎
23	20 04	☽ ♍
25	22 49	☽ ♎
27	23 52	☽ ♏
30	00 57	☽ ♐

LATITUDES

Date	Mercury ☿ ° '	Venus ♀ ° '	Mars ♂ ° '	Jupiter ♃ ° '	Saturn ♄ ° '	Uranus ♅ ° '	Neptune ♆ ° '	Pluto ♇ ° '
01	04 S 23	01 N 25	06 S 30	00 N 50	00 S 45	00 S 48	00 00	09 N 08
04	04 25	01 25	06 23	00 50	00 45	00 47	00 00	09 07
07	04 11	01 24	06 14	00 50	00 45	00 47	00 00	09 06
10	03 40	01 22	06 04	00 51	00 45	00 47	00 00	09 04
13	02 53	01 20	05 53	00 51	00 45	00 47	00 00	09 03
16	01 56	01 18	05 41	00 51	00 44	00 47	00 00	09 02
19	00 57	01 15	05 28	00 52	00 44	00 47	00 00	09 01
22	00 S 02	01 11	05 14	00 52	00 44	00 47	00 00	09 00
25	00 N 43	01 07	05 00	00 52	00 44	00 47	00 S 01	08 59
28	01 18	01 02	04 46	00 53	00 44	00 47	00 01	08 57
31	01 N 40	00 N 57	04 S 32	00 N 53	00 S 44	00 S 47	00 S 01	08 N 56

DATA

Julian Date	2452884
Delta T	+65 seconds
Ayanamsa	23° 54' 17"
Synetic vernal point	05° ♓ 12' 43"
True obliquity of ecliptic	23° 26' 26"

MOON'S PHASES, APSIDES AND POSITIONS ☽

Date	h m	Phase	Longitude ° '	Eclipse Indicator
03	12 34	☽	10 ♐ 36	
10	16 36	○	17 ♓ 34	
18	19 03	☾	25 ♊ 27	
26	03 09	●	02 ♎ 38	

Day	h m		
16	09 16	Apogee	
28	05 53	Perigee	
05	11 52	Max dec	26° S 45'
12	07 54	ON	
19	19 55	Max dec	26° N 53'
26	12 15	0S	

LONGITUDES

Date	Chiron ⚷ ° '	Ceres ⚳ ° '	Pallas ⚴ ° '	Juno ⚵ ° '	Vesta ⚶ ° '	Black Moon Lilith ⚸ ° '
01	12 ♑ 33	07 ♋ 49	28 ♈ 24	13 ♏ 30	02 ♏ 05	22 ♉ 32
11	12 ♑ 24	11 ♋ 09	27 ♈ 24	16 ♏ 02	07 ♏ 43	23 ♉ 38
21	12 ♑ 23	14 ♋ 36	25 ♈ 42	18 ♏ 46	12 ♏ 38	24 ♉ 44
31	12 ♑ 29	17 ♋ 08	23 ♈ 20	21 ♏ 40	17 ♏ 40	25 ♉ 51

All ephemeris data is given at 12.00 UT and the Moon's longitude is additionally given for 24.00 UT

ASPECTARIAN

h m Aspects

01 Monday
01 15 ☽ # ♃
01 48 ☽ ∥ ♅
06 06 ☽ ⚹ ☉
08 56 ☽ ∥ ♇
09 46 ☽ ∠ ☿
09 52 ☽ △ ♄
10 34 ☽ □ ♆
10 52 ☽ ⊥ ♇
12 55 ☽ ⚹ ♀
14 02 ☽ Q ♃
19 20 ☽ ∥ ♂
20 58 ☽ ⊻ ♇

02 Tuesday
01 40 ☽ ∥ ♆
03 47 ☽ Q ☉
10 18 ☽ ⚹ ☿
11 05 ☽ Q ♀
11 19 ☽ bQ ♄
16 54 ☽ Q ♆
19 19 ☽ □ ♅
20 55 ☽ □ ♃

03 Wednesday
00 42 ☽ □ ♂
02 41 ☽ ± ♄
03 49 ☽ # ♄
05 55 ☽ Q ☿
12 34 ☽ □ ☉
12 58 ☽ ⊼ ♄
13 16 ☽ ⚹ ♆
18 48 ☉ ⚹ ♄
20 40 ☽ □ ♀
22 31 ☉ ⊼ ♆
23 55 ☽ ☌ ♇

04 Thursday
01 55 ☽ Q ♅
06 54 ☽ Q ♂
11 23 ☽ □ ☿
14 56 ☽ ∠ ♆
22 31 ☽ ⚹ ♅

05 Friday
01 06 ☽ △ ♃
03 14 ☽ ⚹ ♂
04 18 ♀ ± ♆
05 09 ♄ ⊼ ♆
06 28 ☽ ⊥ ♆
09 31 ♀ □ ♇
16 37 ☿ Q ♄
16 52 ☽ ⊻ ♆
16 58 ☽ ☍ ♄
20 09 ☽ △ ☉

06 Saturday
00 30 ☽ ∠ ♅
03 36 ☽ bQ ♃
03 50 ☽ ⊻ ♇
04 53 ☽ ∠ ♂
05 39 ☽ △ ♀
12 43 ☽ △ ☿
14 21 ☽ ⊥ ♇
16 13 ☽ ⊥ ♅
19 39 ☽ ± ♃
20 26 ☽ ⊥ ♂

07 Sunday
00 28 ☽ bQ ☉
02 46 ☽ ⊻ ♅
06 14 ☽ ∠ ♇
06 25 ☽ ⊼ ♃
06 51 ☽ ⊻ ♂
10 44 ☽ bQ ♀
13 35 ☽ bQ ☿
17 40 ☽ ± ☉
20 00 ♂ ☍ ♃
21 39 ☽ ☌ ♆
21 41 ☽ # ♄
22 09 ☽ ⊼ ♄

08 Monday
04 26 ☽ ± ♀
04 37 ☽ ± ☿
05 14 ☽ ⊼ ☉
05 35 ☿ ☌ ♀
08 58 ☽ ± ♄
09 01 ☽ ⚹ ♇
14 41 ☽ ⊼ ☿
16 21 ☽ ⊼ ♀

09 Tuesday
01 15 ☽ ∥ ♆
01 21 ☽ bQ ♄
06 44 ☽ ∥ ♂
06 47 ☽ Q ♇
08 30 ☽ ☌ ♅
12 00 ☽ ☌ ♂
13 20 ☽ ☍ ♃
19 30 ☽ ∥ ♇
23 45 ☉ ± ♆

10 Wednesday
03 07 ☽ ∥ ♅
04 08 ☽ ⊻ ♆
05 05 ☽ △ ♄
07 28 ☽ # ♃
08 17 ♀ Q ♄
09 37 ☉ □ ♇
15 18 ☽ ⊥ ♆
16 05 ☽ □ ♇
16 36 ☽ ☍ ☉
17 56 ☽ ☍ ☿

11 Thursday
01 57 ☉ ☌ ☿
05 41 ☽ ☍ ♀
08 15 ☽ ∠ ♆
12 12 ☽ # ☉
16 24 ☽ ⊻ ♅
17 42 ☿ □ ♇
18 19 ☽ # ♀
19 20 ☽ ⊻ ♂
22 36 ☽ ⊼ ♃
23 05 ☽ # ☿

12 Friday
03 53 ☽ ⊥ ♅
04 51 ☿ ± ♆
06 43 ☽ ⊥ ♂
07 24 ♀ bQ ♆
10 23 ☽ ± ♃
13 01 ☽ ⚹ ♆
14 27 ☽ □ ♄
19 00 ☽ ∥ ☿
19 20 ☽ ∥ ♀
20 50 ☿ ∥ ♀
21 19 ☽ ∠ ♅
23 19 ☽ ⊼ ☿
23 59 ☽ ∠ ♂

13 Saturday
01 18 ☽ ∥ ☉
01 40 ☽ △ ♇
04 17 ☽ bQ ♃
07 10 ☽ ⊼ ☉
10 18 ☽ ± ☿
12 32 ☽ Q ♆
20 05 ☽ ± ☉
22 33 ☽ ⊼ ♀

14 Sunday
02 10 ☽ Q ♄
02 54 ☽ ⚹ ♅
02 59 ☽ bQ ☿
05 18 ☽ ⚹ ♂
07 27 ☽ bQ ♇
08 17 ☉ ∥ ☿
08 33 ☽ ∥ ♃
10 38 ☽ △ ♃
11 57 ☽ ± ♀
14 58 ☽ # ♅
15 41 ☽ bQ ☉
23 05 ☽ # ♇

15 Monday
00 29 ☽ □ ♆
01 34 ☽ ± ♇
02 26 ☽ ⚹ ♄
02 58 ☽ Q ♅
05 08 ☽ Q ♂
07 19 ☽ △ ☿
08 13 ☽ bQ ♀
13 04 ☽ # ♂
13 44 ☽ ⊼ ♇
15 37 ♀ ⊼ ♅
19 04 ☽ # ♆

16 Tuesday
00 47 ☽ △ ☉
09 05 ☽ ∠ ♄
09 25 ♀ ⊼ ♂
15 25 ☽ □ ♅
17 17 ☽ □ ♂
18 19 ☽ △ ♀

17 Wednesday
00 35 ☽ □ ♃
01 03 ☽ ∥ ♄
03 31 ☽ ⊥ ♄
04 36 ☉ Q ♄
13 22 ☽ △ ♆
15 46 ☽ ⊻ ♄
17 28 ☽ □ ☿

18 Thursday
02 45 ☽ ☍ ♇
13 41 ☽ Q ♃
19 03 ☽ □ ☉
19 34 ☽ bQ ♆

19 Friday
01 21 ☉ bQ ♆
03 51 ☽ △ ♅
04 45 ☽ Q ☿
05 14 ☽ △ ♂
13 25 ☽ ± ♆
13 44 ☽ □ ♀
13 59 ☽ ⚹ ♃
16 41 ♀ ⊻ ♃
23 35 ♀ Q ♇

20 Saturday
01 10 ☽ ⊼ ♆
03 53 ☽ ☌ ♄
04 08 ☽ ⚹ ☿
08 19 ♀ ± ♅
08 52 ☿ St D
09 01 ☉ # ♀
09 11 ☽ bQ ♅
09 34 ☽ Q ☉
10 23 ☽ bQ ♂
14 07 ☽ ⊼ ♇
19 35 ♀ ± ♂
19 37 ☽ ∠ ♃

21 Sunday
01 35 ☽ ± ♇
02 22 ☽ ± ♅
03 28 ☽ ± ♂
04 23 ☽ Q ♀
08 58 ☽ ∠ ☿
10 21 ☽ ⚹ ☉
12 55 ☽ ⊥ ♃
13 38 ☽ ⊼ ♅
14 40 ☽ ⊼ ♂
18 29 ☽ bQ ♇

22 Monday
00 14 ☽ ⊻ ♃
01 59 ☽ ⊥ ☿
05 05 ☽ ⚹ ♀
09 40 ☽ ☍ ♆
11 22 ☽ ∥ ♄
12 32 ☽ ⊻ ♄
13 12 ☽ ⊻ ☿
16 11 ☽ ∠ ☉
21 51 ☽ △ ♇
23 22 ☽ ⊥ ♄

23 Tuesday
04 00 ☉ ⊼ ♅
09 29 ☽ ⊥ ☉
10 46 ☽ ∠ ♀
13 26 ☽ # ♆
15 21 ☽ ∠ ♄
15 54 ☉ ⊼ ♂
19 33 ☽ ☍ ♅
20 07 ☽ # ♂
20 25 ☽ ☍ ♂
20 47 ☽ ⊻ ☉

24 Wednesday
03 59 ☽ ⊥ ♀
05 03 ♀ △ ♆
06 23 ☽ ☌ ♃
06 31 ☽ # ♇
13 15 ☽ # ♅
14 20 ☽ ⊼ ♆
15 15 ☽ ⊻ ♀
17 17 ☽ ⚹ ♄
19 44 ☽ ☌ ☿
22 24 ☽ ∥ ♃

25 Thursday
00 29 ☽ ± ♆
01 52 ☽ □ ♇
10 59 ☽ ∥ ☿
11 48 ♀ ⊥ ♃
13 31 ☽ Q ♄
15 21 ♀ □ ♄
15 31 ☽ bQ ♆
21 37 ☽ # ♀
22 13 ☽ ⊼ ♅
23 02 ☽ ⊼ ♂

26 Friday
03 09 ☽ ☌ ☉
07 48 ☽ Q ♇
08 05 ☽ ± ♅
08 10 ☽ # ☉
08 54 ☽ ± ♂
09 22 ☽ ⊻ ♃
16 14 ☽ △ ♆
16 50 ☽ ∥ ☉
19 17 ☽ □ ♄
19 20 ☽ ⊥ ♃
21 46 ☽ ☌ ♀
22 48 ☽ bQ ♅
23 38 ☽ bQ ♂

27 Saturday
00 32 ☽ ⊻ ☿
03 28 ☽ ⚹ ♇
05 10 ☽ ∥ ♀
07 52 ♂ St D
09 36 ♀ bQ ♅
10 13 ☽ ∠ ♃
11 02 ☽ ⊥ ☿
11 51 ☽ # ☿
19 51 ♀ bQ ♂
21 13 ☿ ± ♆
23 10 ☽ △ ♅

28 Sunday
00 04 ☽ △ ♂
00 23 ☽ # ♃
02 51 ☽ ∠ ☿
03 54 ☽ ∠ ♇
07 38 ☽ ⊻ ☉
10 08 ☽ ∥ ♅
10 59 ☽ ⚹ ♃
16 28 ☽ ∥ ♇
16 31 ☿ □ ♇
17 04 ☽ □ ♆
18 07 ☽ ⊥ ☉
18 35 ☽ ⊥ ♇
20 20 ☽ △ ♄

29 Monday
00 30 ☽ ∥ ♂
01 16 ☉ Q ♇
02 20 ☉ ± ♅
03 10 ☽ ⊻ ♀
04 24 ☽ ⊻ ♇
05 27 ☽ ⚹ ☿
06 51 ☽ Q ♃
08 21 ☽ ∥ ♆
09 53 ☽ ∠ ☉
13 37 ☿ # ♀
13 55 ☽ ⊥ ♀
17 28 ☉ ± ♂
17 53 ♀ ⚹ ♇
21 01 ☽ bQ ♄
22 33 ☽ Q ♆

30 Tuesday
00 09 ☽ □ ♅
00 30 ♅ Q ♇
01 14 ☽ □ ♂
03 06 ☽ Q ☿
06 09 ☽ ∠ ♀
07 47 ☽ # ♄
11 58 ☽ ± ♄
12 27 ☽ ⚹ ☉
12 58 ☽ □ ♃
18 27 ☽ ⚹ ♆
21 57 ☉ ⊻ ♃
22 01 ☽ ⊼ ♄

OCTOBER 2003

Date	Sidereal time h m s	Sun ☉ ° ' "	Moon ☽ ° ' "	Moon ☽ 24.00 ° ' "	Mercury ☿ ° '	Venus ♀ ° '	Mars ♂ ° '	Jupiter ♃ ° '	Saturn ♄ ° '	Uranus ♅ ° '	Neptune ♆ ° '	Pluto ♇ ° '
01	12 39 15	07 ♎ 53 58	20 ♐ 58 29	28 ♐ 02 14	21 ♍ 11	19 ♎ 41	00 ♓ 14	07 ♍ 27	12 ♋ 41	29 ♒ 28	10 ♒ 32	17 ♐ 32
02	12 43 11	08 52 58	05 ♑ 02 06	11 ♑ 58 03	22 39	20 56	00 18	07 39	12 44	29 R 26	10 R 31	17 33
03	12 47 08	09 52 01	18 ♑ 50 05	25 ♑ 38 17	24 11	22 11	00 23	07 51	12 46	29 25	10 30	17 34
04	12 51 05	10 51 05	02 ♒ 22 42	09 ♒ 03 26	25 46	23 25	00 28	08 03	12 49	29 23	10 30	17 36
05	12 55 01	11 50 11	15 ♒ 40 34	22 ♒ 14 13	27 24	24 40	00 34	08 15	12 51	29 22	10 29	17 37
06	12 58 58	12 49 18	28 ♒ 44 27	05 ♓ 11 21	29 ♍ 03	25 55	00 41	08 26	12 53	29 20	10 28	17 38
07	13 02 54	13 48 27	11 ♓ 35 00	17 ♓ 55 28	00 ♎ 45	27 09	00 49	08 38	12 55	29 19	10 28	17 39
08	13 06 51	14 47 39	24 ♓ 12 51	00 ♈ 27 11	02 27	28 24	00 57	08 50	12 57	29 17	10 27	17 40
09	13 10 47	15 46 52	06 ♈ 38 36	12 ♈ 47 11	04 11	29 ♎ 38	01 06	09 02	12 59	29 16	10 27	17 42
10	13 14 44	16 46 07	18 ♈ 53 05	24 ♈ 56 27	05 55	00 ♏ 53	01 16	09 13	13 01	29 14	10 27	17 43
11	13 18 40	17 45 24	00 ♉ 57 29	06 ♉ 56 24	07 40	02 08	01 26	09 25	13 03	29 13	10 26	17 44
12	13 22 37	18 44 43	12 ♉ 53 29	18 ♉ 49 02	09 25	03 22	01 38	09 36	13 04	29 12	10 26	17 46
13	13 26 34	19 44 05	24 ♉ 43 24	00 ♊ 36 59	11 10	04 37	01 50	09 48	13 06	29 10	10 25	17 47
14	13 30 30	20 43 29	06 ♊ 30 13	12 ♊ 23 34	12 55	05 52	02 02	09 59	13 07	29 09	10 25	17 49
15	13 34 27	21 42 55	18 ♊ 17 32	24 ♊ 12 41	14 40	07 06	02 15	10 10	13 08	29 08	10 25	17 50
16	13 38 23	22 42 23	00 ♋ 09 35	06 ♋ 08 48	16 24	08 21	02 29	10 21	13 09	29 07	10 24	17 51
17	13 42 20	23 41 53	12 ♋ 10 58	18 ♋ 16 41	18 08	09 36	02 44	10 33	13 10	29 06	10 24	17 53
18	13 46 16	24 41 26	24 ♋ 26 34	00 ♌ 41 11	19 52	10 50	02 59	10 44	13 11	29 05	10 24	17 54
19	13 50 13	25 41 01	07 ♌ 01 07	13 ♌ 26 51	21 35	12 05	03 15	10 55	13 12	29 04	10 24	17 56
20	13 54 09	26 40 39	19 ♌ 58 50	26 ♌ 37 25	23 18	13 19	03 31	11 06	13 13	29 03	10 24	17 58
21	13 58 06	27 40 18	03 ♍ 22 49	10 ♍ 15 08	25 00	14 34	03 48	11 16	13 13	29 02	10 24	17 59
22	14 02 03	28 40 00	17 ♍ 14 20	24 ♍ 20 12	26 41	15 49	04 05	11 27	13 14	29 01	10 24	18 01
23	14 05 59	29 ♎ 39 44	01 ♎ 32 21	08 ♎ 50 11	28 ♎ 22	17 03	04 23	11 38	13 14	29 00	10 D 24	18 02
24	14 09 56	00 ♏ 39 30	16 ♎ 13 00	23 ♎ 39 54	00 ♏ 03	18 18	04 42	11 48	13 14	28 59	10 24	18 04
25	14 13 52	01 39 18	01 ♏ 09 52	08 ♏ 41 50	01 43	19 33	05 01	11 59	13 14	28 59	10 24	18 06
26	14 17 49	02 39 08	16 ♏ 14 38	23 ♏ 47 08	03 22	20 47	05 21	12 09	13 R 14	28 58	10 24	18 08
27	14 21 45	03 39 01	01 ♐ 18 15	08 ♐ 46 56	05 01	22 02	05 41	12 20	13 14	28 57	10 24	18 09
28	14 25 42	04 38 55	16 ♐ 12 19	23 ♐ 33 37	06 39	23 17	06 02	12 30	13 14	28 57	10 24	18 11
29	14 29 38	05 38 50	00 ♑ 50 11	08 ♑ 01 35	08 17	24 31	06 23	12 40	13 14	28 56	10 25	18 13
30	14 33 35	06 38 47	15 ♑ 07 29	22 ♑ 07 42	09 54	25 46	06 45	12 50	13 13	28 56	10 25	18 15
31	14 37 32	07 ♏ 38 47	29 ♑ 02 11	05 ♒ 50 59	11 ♏ 31	27 ♏ 01	07 ♓ 07	13 ♍ 00	13 ♋ 13	28 ♒ 55	10 ♒ 25	18 ♐ 17

Date	Moon True ☊ ° '	Moon Mean ☊ ° '	Moon ☽ Latitude ° '	DECLINATIONS Sun ☉ ° '	Moon ☽ ° '	Mercury ☿ ° '	Venus ♀ ° '	Mars ♂ ° '	Jupiter ♃ ° '	Saturn ♄ ° '	Uranus ♅ ° '	Neptune ♆ ° '	Pluto ♇ ° '
01	21 ♉ 18	22 ♉ 33	02 S 34	03 S 08	25 S 42	05 N 02	06 S 49	15 S 37	09 N 36	22 N 06	12 S 24	17 S 37	13 S 57
02	21 D 18	22 30	03 35	03 31	26 56	04 32	07 19	15 31	09 31	22 06	12 24	17 37	13 58
03	21 R 18	22 27	04 22	03 55	26 27	03 59	07 48	15 25	09 27	22 06	12 25	17 37	13 58
04	21 16	22 24	04 54	04 18	24 24	03 24	08 18	15 19	09 23	22 05	12 25	17 37	13 59
05	21 12	22 20	05 09	04 41	21 03	02 47	08 47	15 12	09 18	22 05	12 26	17 37	13 59
06	21 07	22 17	05 08	05 04	16 43	02 08	09 16	15 05	09 14	22 05	12 26	17 37	14 00
07	21 01	22 14	04 51	05 27	11 42	01 28	09 45	14 58	09 10	22 05	12 27	17 38	14 00
08	20 54	22 11	04 20	05 50	06 17	00 47	10 13	14 51	09 06	22 05	12 27	17 38	14 01
09	20 48	22 08	03 37	06 13	00 S 41	00 N 05	10 42	14 43	09 01	22 04	12 28	17 38	14 01
10	20 43	22 04	02 45	06 35	04 N 51	00 S 38	11 10	14 36	08 57	22 04	12 28	17 38	14 01
11	20 40	22 01	01 46	06 58	10 09	01 22	11 38	14 28	08 53	22 04	12 29	17 38	14 02
12	20 38	21 58	00 S 42	07 21	15 02	02 06	12 05	14 19	08 49	22 04	12 29	17 38	14 02
13	20 D 38	21 55	00 N 22	07 43	19 19	02 50	12 33	14 11	08 44	22 04	12 30	17 38	14 03
14	20 39	21 52	01 26	08 06	22 49	03 35	13 00	14 02	08 40	22 04	12 30	17 38	14 03
15	20 40	21 49	02 26	08 28	25 21	04 20	13 26	13 53	08 36	22 04	12 30	17 39	14 04
16	20 42	21 45	03 20	08 50	26 47	05 04	13 53	13 44	08 32	22 04	12 31	17 39	14 04
17	20 44	21 42	04 07	09 12	26 59	05 49	14 19	13 35	08 28	22 03	12 31	17 39	14 05
18	20 44	21 39	04 43	09 34	25 52	06 33	14 45	13 25	08 24	22 03	12 31	17 39	14 05
19	20 R 44	21 36	05 06	09 56	23 27	07 17	15 10	13 16	08 20	22 03	12 32	17 39	14 06
20	20 42	21 33	05 15	10 17	19 48	08 00	15 35	13 06	08 16	22 03	12 32	17 39	14 06
21	20 39	21 30	05 07	10 39	15 02	08 43	15 59	12 56	08 12	22 03	12 32	17 39	14 07
22	20 36	21 26	04 42	11 00	09 22	09 26	16 24	12 45	08 08	22 03	12 33	17 39	14 07
23	20 33	21 23	03 58	11 21	03 N 02	10 08	16 47	12 35	08 04	22 03	12 33	17 39	14 07
24	20 31	21 20	02 58	11 42	03 S 38	10 49	17 11	12 25	08 00	22 03	12 33	17 39	14 08
25	20 29	21 17	01 45	12 03	10 15	11 30	17 34	12 14	07 56	22 03	12 33	17 39	14 08
26	20 28	21 14	00 N 23	12 24	16 19	12 10	17 56	12 03	07 53	22 03	12 33	17 39	14 09
27	20 D 28	21 10	01 S 00	12 44	21 24	12 50	18 18	11 52	07 49	22 03	12 34	17 39	14 09
28	20 29	21 07	02 18	13 04	25 01	13 29	18 40	11 41	07 45	22 03	12 34	17 39	14 10
29	20 30	21 04	03 26	13 24	26 52	14 07	19 00	11 29	07 41	22 03	12 34	17 39	14 10
30	20 31	21 01	04 19	13 44	26 52	14 44	19 21	11 18	07 38	22 03	12 34	17 39	14 10
31	20 ♉ 32	20 ♉ 58	04 S 55	14 S 04	25 S 10	15 S 21	19 S 41	11 S 06	07 N 34	22 N 03	12 S 34	17 S 39	14 S 11

ZODIAC SIGN ENTRIES

Date	h m	Planets
02	03 21	☽ ♑
04	07 45	☽ ♒
06	14 20	☽ ♓
07	01 28	☿ ♎
08	23 08	☽ ♈
09	18 56	♀ ♏
11	10 05	☽ ♉
13	22 45	☽ ♊
16	11 41	☽ ♋
18	22 41	☽ ♌
21	06 01	☽ ♍
23	09 27	☽ ♎
23	20 08	☉ ♏
24	11 20	☿ ♏
25	10 08	☽ ♏
27	09 55	☽ ♐
29	10 37	☽ ♑
31	13 41	☽ ♒

LATITUDES

Date	Mercury ☿ ° '	Venus ♀ ° '	Mars ♂ ° '	Jupiter ♃ ° '	Saturn ♄ ° '	Uranus ♅ ° '	Neptune ♆ ° '	Pluto ♇ ° '
01	01 N 40	00 N 57	04 S 32	00 N 53	00 S 44	00 S 47	00 S 01	08 N 56
04	01 52	00 52	04 17	00 53	00 44	00 47	00 01	08 55
07	01 56	00 46	04 03	00 54	00 44	00 47	00 01	08 54
10	01 52	00 39	03 49	00 54	00 44	00 47	00 01	08 53
13	01 43	00 33	03 36	00 55	00 44	00 47	00 01	08 52
16	01 30	00 26	03 22	00 55	00 44	00 47	00 01	08 51
19	01 14	00 19	03 10	00 56	00 44	00 47	00 01	08 50
22	00 56	00 11	02 57	00 56	00 44	00 46	00 01	08 49
25	00 36	00 N 04	02 45	00 57	00 44	00 46	00 01	08 48
28	00 N 16	00 S 04	02 34	00 57	00 44	00 46	00 01	08 47
31	00 S 04	00 S 12	02 S 22	00 N 58	00 S 44	00 S 46	00 S 01	08 N 46

DATA

Julian Date	2452914
Delta T	+65 seconds
Ayanamsa	23° 54' 20"
Synetic vernal point	05° ♓ 12' 39"
True obliquity of ecliptic	23° 26' 26"

MOON'S PHASES, APSIDES AND POSITIONS ☽

Date	h m	Phase	Longitude ° '	Eclipse Indicator
02	19 09	☽	09 ♑ 11	
10	07 27	○	16 ♈ 35	
18	12 31	☾	24 ♋ 43	
25	12 50	●	01 ♏ 41	

Day	h m		
14	02 14	Apogee	
26	11 24	Perigee	
02	17 05	Max dec	26° S 58'
09	14 57	0N	
17	03 40	Max dec	27° N 03'
23	22 59	0S	
29	23 59	Max dec	27° S 06'

LONGITUDES

Date	Chiron ⚷ ° '	Ceres ⚳ ° '	Pallas ⚴ ° '	Juno ⚵ ° '	Vesta ⚶ ° '	Black Moon Lilith ⚸ ° '
01	12 ♑ 29	17 ♋ 08	23 ♈ 20	21 ♏ 40	17 ♏ 40	25 ♉ 51
11	12 ♑ 44	19 ♋ 40	20 ♈ 31	24 ♏ 13	22 ♏ 47	26 ♉ 57
21	13 ♑ 06	21 ♋ 51	17 ♈ 29	27 ♏ 54	27 ♏ 58	28 ♉ 04
31	13 ♑ 36	23 ♋ 33	14 ♈ 35	01 ♐ 10	03 ♐ 14	29 ♉ 10

ASPECTARIAN (h m — Aspects)

01 Wednesday
- 06 06 ☽ Q ♅
- 06 11 ☽ ☌ ♇
- 07 22 ☽ Q ♂
- 09 37 ☽ ⚹ ♀
- 10 03 ☽ Q ☉
- 12 24 ☽ □ ☿
- 19 43 ☽ ∠ ♆

02 Thursday
- 02 25 ☽ ⚹ ♅
- 03 50 ☽ ⚹ ♂
- 08 02 ☽ Q ♀
- 11 06 ☽ ⊥ ♆
- 16 34 ☽ △ ♃
- 19 09 ☽ □ ☉
- 21 28 ☽ ⊻ ♆

03 Friday
- 01 22 ☽ ☍ ♄
- 04 16 ☽ ∠ ♅
- 05 54 ☽ ∠ ♂
- 09 47 ☽ ⊻ ♇
- 13 53 ☉ # ☿
- 18 28 ☽ □ ♀
- 19 10 ☽ ꝑ ♃
- 20 03 ☽ ⊥ ♅
- 20 21 ☽ ⊥ ♇
- 21 12 ☿ Q ♄
- 21 50 ☽ ⊥ ♂
- 22 40 ☽ △ ☿

04 Saturday
- 03 19 ♀ ∠ ♃
- 03 22 ☉ △ ♆
- 06 40 ☽ ⊻ ♅
- 07 53 ☿ ꝑ ♆
- 08 34 ☽ ⊻ ♂
- 11 24 ☽ ± ♃
- 12 23 ☽ ∠ ♇
- 22 20 ☽ ⊼ ♃

05 Sunday
- 02 35 ☽ ☌ ♆
- 04 28 ☽ △ ☉
- 05 12 ☽ ꝑ ☿
- 05 22 ☽ # ♄
- 06 51 ☽ ⊼ ♄
- 15 32 ☽ ⚹ ♇
- 17 49 ☽ ± ♄
- 23 58 ☽ ± ☿

06 Monday
- 06 13 ☽ △ ♀
- 07 19 ☽ ∥ ♆
- 10 09 ☽ ꝑ ☉
- 10 25 ☽ ꝑ ♄
- 10 34 ♀ # ♃
- 12 40 ☽ ⊼ ☿
- 13 06 ☽ ☌ ♅
- 13 39 ☽ Q ♇
- 13 40 ☉ □ ♄
- 15 38 ☽ ☌ ♂
- 15 55 ☿ ⊼ ♅
- 20 15 ☽ ∥ ♂

07 Tuesday
- 01 19 ☽ ∥ ♇
- 04 19 ☽ ± ☉
- 06 22 ☽ ☍ ♃
- 08 34 ☽ ∥ ♅
- 09 54 ☽ ⊻ ♆
- 13 01 ☿ ⊼ ♂
- 13 11 ☽ ꝑ ♀
- 14 32 ☽ △ ♄
- 16 33 ☽ ⊼ ☉
- 20 04 ☽ ∥ ♀
- 21 13 ☽ ⊥ ♆
- 23 30 ☽ □ ♇
- 23 32 ☽ # ♃
- 23 54 ☉ ꝑ ♅

08 Wednesday
- 08 08 ☽ ± ♀
- 13 08 ☉ ⊥ ♃
- 13 48 ☽ ∥ ☉
- 14 23 ☽ ∠ ♆
- 20 56 ☽ ⊼ ♀
- 21 44 ☽ ⊻ ♅

09 Thursday
- 01 07 ☽ ⊻ ♂
- 04 48 ♀ △ ♅
- 06 26 ☽ ☍ ☿
- 09 19 ☽ ⊥ ♅
- 12 54 ☽ ⊥ ♂
- 14 55 ☽ ∥ ☿
- 14 59 ☽ # ☿
- 16 43 ☽ ⊼ ♃
- 19 25 ☽ ⚹ ♆
- 21 18 ☉ ꝑ ♂

10 Friday
- 00 25 ☽ □ ♄
- 02 43 ☿ ± ♅
- 02 52 ☽ ∠ ♅
- 04 40 ☽ ± ♃
- 06 46 ☽ ∠ ♂
- 07 27 ☽ ☍ ☉
- 09 10 ☿ Q ♇
- 09 41 ☽ △ ♇
- 19 02 ☽ Q ♆
- 20 17 ☽ # ☉
- 20 32 ♀ △ ♂
- 22 44 ☽ ꝑ ♃

11 Saturday
- 06 09 ☽ ∥ ♃
- 08 31 ☽ ⚹ ♅
- 08 35 ☿ ± ♂
- 11 33 ☉ ⚹ ♇
- 12 10 ☽ Q ♄
- 12 59 ☽ ⚹ ♂
- 14 37 ☽ ☍ ♀
- 15 34 ☽ ꝑ ♇
- 19 42 ☽ # ♀
- 23 07 ☽ # ♅

12 Sunday
- 00 00 ♀ ∠ ♇
- 03 46 ☽ ⊼ ☿
- 05 15 ☽ △ ♃
- 06 52 ☽ # ♇
- 07 02 ☽ □ ♆
- 08 25 ☽ # ♂
- 08 35 ☽ Q ♅
- 09 43 ☽ ± ♇
- 12 22 ☽ ⚹ ♄
- 13 31 ☽ Q ♂
- 14 55 ☿ ⊻ ♃
- 17 59 ☽ ± ☿
- 21 53 ☽ ⊼ ♇

13 Monday
- 00 56 ☽ ⊼ ☉
- 01 53 ☿ △ ♆
- 02 07 ☽ # ♆
- 09 07 ♀ ∥ ♅
- 14 15 ☽ ± ☉
- 15 27 ☽ ꝑ ☿
- 18 52 ☽ ∠ ♄
- 21 02 ☽ □ ♅

14 Tuesday
- 02 44 ☽ □ ♂
- 06 19 ☽ ∥ ♄
- 08 50 ♂ ∥ ♇
- 10 16 ☽ ꝑ ☉
- 10 32 ☽ ⊼ ♀
- 13 15 ☽ ⊥ ♄
- 14 49 ☿ □ ♄
- 19 12 ☽ □ ♃
- 19 59 ☽ △ ♆

15 Wednesday
- 00 12 ☽ ± ♀
- 01 30 ☽ ⊻ ♄
- 03 20 ☽ △ ☿
- 04 50 ☿ ꝑ ♅
- 11 04 ☽ ☍ ♇
- 19 35 ☽ △ ☉
- 19 37 ☉ # ♃
- 20 38 ☽ ꝑ ♀

16 Thursday
- 02 26 ☽ ꝑ ♆
- 06 03 ♀ ∥ ♂
- 08 19 ☽ Q ♃
- 09 54 ☽ △ ♅
- 11 20 ☿ ⊥ ♃
- 16 46 ☽ △ ♂
- 18 48 ♃ ⊼ ♆
- 20 32 ☽ ± ♆
- 22 37 ♀ ∥ ♇

17 Friday
- 05 27 ☿ ꝑ ♂
- 06 16 ☽ △ ♀
- 08 27 ☿ ⚹ ♇
- 08 29 ☽ ⊼ ♆
- 08 42 ☽ ⚹ ♃
- 13 58 ☽ ☌ ♄
- 15 46 ☽ ꝑ ♅
- 23 09 ☽ ꝑ ♂
- 23 15 ☽ ⊼ ♇

18 Saturday
- 01 39 ☽ □ ☿
- 03 42 ♀ □ ♆
- 09 22 ☽ ± ♅
- 09 33 ♀ ⚹ ♃
- 10 58 ☽ ± ♇
- 12 31 ☽ □ ☉
- 14 31 ☽ ∠ ♃
- 17 00 ☽ ± ♂
- 20 55 ☽ ⊼ ♅

19 Sunday
- 04 16 ☽ ꝑ ♇
- 04 43 ☽ ⊼ ♂
- 07 58 ☽ ⊥ ♃
- 09 07 ♀ ⊥ ♇
- 17 33 ☽ Q ☿
- 18 20 ☽ ☍ ♆
- 18 43 ☉ ∠ ♃
- 19 24 ☽ ⊻ ♃
- 22 07 ☽ ∥ ♄
- 22 28 ☽ □ ♀
- 23 33 ☽ ⊻ ♄

20 Monday
- 01 28 ☽ Q ☉
- 08 18 ☽ △ ♇
- 09 48 ♀ △ ♄
- 10 36 ☽ ⊥ ♄
- 18 54 ☽ ⚹ ☿
- 20 06 ☿ # ♃
- 23 29 ☽ # ♆

21 Tuesday
- 01 04 ☽ ⚹ ☉
- 02 51 ☽ ∠ ♄
- 04 18 ☽ ☍ ♅
- 07 54 ☽ # ♀
- 10 26 ☽ Q ♀
- 12 45 ☽ ☍ ♂
- 16 11 ☽ # ♇
- 21 37 ☽ # ♂
- 22 59 ☽ # ♅

22 Wednesday
- 00 15 ☽ ⊼ ♆
- 01 10 ☽ ∠ ☿
- 01 57 ☽ ☌ ♃
- 05 08 ☽ ⚹ ♄
- 05 25 ☽ ∠ ☉
- 05 47 ☽ # ☉
- 08 15 ☿ ∠ ♃
- 09 20 ☽ ⚹ ♀
- 10 34 ☽ ± ♆
- 11 47 ☽ # ☿
- 13 19 ☽ □ ♇
- 16 53 ☽ ∥ ♃
- 18 39 ☽ ⊥ ☿
- 20 18 ☉ △ ♅
- 21 53 ☽ ⊥ ☉

23 Thursday
- 01 30 ☽ Q ♄
- 01 47 ☽ ꝑ ♆
- 01 54 ♆ St D
- 06 03 ☽ ⊻ ☿
- 07 48 ☽ ⊼ ♅
- 08 40 ☽ ⊻ ☉
- 12 56 ☽ ∠ ♀
- 16 49 ☽ ⊼ ♂
- 17 05 ♂ ∥ ♅
- 17 42 ☽ ± ♅
- 19 26 ☽ Q ♇
- 20 57 ☿ △ ♅

24 Friday
- 02 33 ☽ △ ♆
- 02 51 ☽ ± ♂
- 04 45 ☽ ⊻ ♃
- 05 04 ☽ ⊥ ♀
- 07 10 ☽ □ ♄
- 07 25 ♀ ⊻ ♇
- 08 24 ☽ ꝑ ♅
- 14 36 ☽ ⊥ ♃
- 15 00 ☽ ⚹ ♇
- 15 41 ☽ ⊻ ♀
- 17 45 ☽ ꝑ ♂

25 Saturday
- 03 35 ☽ # ♃
- 05 14 ☽ ∠ ♃
- 08 31 ☽ △ ♅
- 09 58 ☉ ☌ ☿
- 12 50 ☽ ☌ ☉
- 12 59 ☽ ☌ ☿
- 15 05 ☽ ∠ ♇
- 17 18 ☽ ∥ ☿
- 17 31 ♀ ∥ ♆
- 18 17 ☽ △ ♂
- 19 14 ☽ ∥ ☉
- 19 19 ☽ ∥ ♂
- 20 12 ☉ ∥ ♂
- 20 47 ☽ ∥ ♅
- 23 42 ♄ St R

26 Sunday
- 02 43 ☽ □ ♆
- 03 02 ☽ ∥ ♇
- 05 25 ☽ ⚹ ♃
- 05 26 ☽ ⊥ ♇
- 07 13 ☽ △ ♄
- 08 26 ☿ ∠ ♇
- 08 32 ☿ ∥ ♂
- 15 00 ☽ ⊻ ♇
- 17 44 ☽ ∥ ♆
- 19 33 ☽ ∥ ♀
- 19 53 ☽ ☌ ♀
- 23 39 ☉ ∥ ♅
- 23 44 ☉ ∠ ♇

27 Monday
- 00 44 ☽ Q ♃
- 02 07 ☿ ∥ ♅
- 04 44 ☉ ∥ ☿
- 07 06 ☽ ꝑ ♄
- 07 22 ☽ Q ♆
- 08 15 ☽ □ ♅
- 15 41 ☽ # ♄
- 16 01 ☽ ⊻ ☉
- 18 40 ☽ ⊻ ☿
- 19 11 ☽ □ ♂
- 21 31 ☽ ± ♄

28 Tuesday
- 00 29 ☿ △ ♂
- 02 22 ☽ ⊥ ☉
- 02 37 ☽ ⚹ ♆
- 05 32 ☽ ⊥ ☿
- 05 55 ☽ □ ♃
- 07 11 ☽ ⊼ ♄
- 13 12 ☽ Q ♅
- 15 13 ☽ ☌ ♇
- 18 01 ☽ ∠ ☉
- 21 59 ☽ ∠ ☿

29 Wednesday
- 00 37 ☽ ⊻ ♀
- 01 05 ☽ Q ♂
- 03 02 ☽ ∠ ♆
- 08 51 ☽ ⚹ ♅
- 11 26 ☽ ⊥ ♀
- 14 04 ☿ ∥ ♇
- 15 12 ♀ Q ♃
- 17 57 ☽ ⊥ ♆
- 20 37 ☽ ⚹ ☉
- 21 29 ☽ ⚹ ♂

30 Thursday
- 02 01 ☽ ⚹ ☿
- 03 54 ☽ ∠ ♀
- 04 01 ☽ ⊻ ♆
- 08 04 ☽ △ ♃
- 08 46 ☽ ☍ ♄
- 09 58 ☽ ∠ ♅
- 15 55 ☉ △ ♂
- 17 20 ☽ ⊻ ♇
- 18 29 ☽ Q ☉
- 19 41 ☿ □ ♆
- 23 39 ☽ ∠ ♂

31 Friday
- 01 08 ☽ Q ☿
- 01 23 ☽ ⊥ ♅
- 03 42 ☽ ⊥ ♇
- 08 07 ☽ ⚹ ♀
- 10 10 ☽ ꝑ ♃
- 11 48 ☽ ⊻ ♅
- 15 45 ☽ ⊥ ♂
- 19 28 ☽ ∠ ♇
- 21 03 ☉ ∥ ♇
- 23 38 ☿ ⊥ ♇

All ephemeris data is given at 12.00 UT and the Moon's longitude is additionally given for 24.00 UT

NOVEMBER 2003

Date	Sidereal time h m s	LONGITUDES Sun ☉ ° ' "	Moon ☽ ° ' "	Moon ☽ 24.00 ° ' "	Mercury ☿ ° '	Venus ♀ ° '	Mars ♂ ° '	Jupiter ♃ ° '	Saturn ♄ ° '	Uranus ♅ ° '	Neptune ♆ ° '	Pluto ♇ ° '
01	14 41 28	08 ♏ 38 47	12 ♒ 34 15	19 ♒ 12 10	13 ♏ 07	28 ♏ 15	07 ♓ 30	13 ♍ 10	13 ♋ 12	28 ♒ 55	10 ♒ 26	18 ♐ 18
02	14 45 25	09 38 49	25 ♒ 45 02	02 ♓ 13 08	14 43	29 ♏ 30	07 53	13 19	13 R 11	28 R 55	10 26	18 20
03	14 49 21	10 38 53	08 ♓ 36 48	14 ♓ 56 21	16 18	00 ♐ 44	08 17	13 29	13 10	28 54	10 26	18 22
04	14 53 18	11 38 58	21 ♓ 12 09	27 ♓ 24 31	17 53	01 59	08 41	13 39	13 09	28 54	10 27	18 24
05	14 57 14	12 39 05	03 ♈ 33 47	09 ♈ 40 16	19 27	03 14	09 05	13 48	13 08	28 54	10 27	18 26
06	15 01 11	13 39 13	15 ♈ 44 16	21 ♈ 46 03	21 01	04 28	09 30	13 57	13 07	28 54	10 28	18 28
07	15 05 07	14 39 23	27 ♈ 45 55	03 ♉ 44 08	22 35	05 43	09 55	14 07	13 06	28 54	10 28	18 30
08	15 09 04	15 39 35	09 ♉ 40 56	15 ♉ 36 35	24 08	06 57	10 21	14 16	13 04	28 54	10 29	18 32
09	15 13 01	16 39 49	21 ♉ 31 21	27 ♉ 25 29	25 41	08 12	10 47	14 25	13 03	28 D 54	10 29	18 34
10	15 16 57	17 40 04	03 ♊ 19 16	09 ♊ 12 59	27 14	09 27	11 13	14 34	13 01	28 54	10 30	18 36
11	15 20 54	18 40 21	15 ♊ 06 56	21 ♊ 01 27	28 ♏ 46	10 41	11 40	14 42	12 59	28 54	10 30	18 38
12	15 24 50	19 40 40	26 ♊ 56 53	02 ♋ 53 36	00 ♐ 18	11 56	12 07	14 51	12 57	28 54	10 31	18 40
13	15 28 47	20 41 01	08 ♋ 52 40	14 ♋ 52 31	01 50	13 10	12 34	14 59	12 55	28 55	10 32	18 42
14	15 32 43	21 41 24	20 ♋ 55 36	27 ♋ 01 43	03 21	14 25	13 01	15 08	12 53	28 55	10 33	18 44
15	15 36 40	22 41 49	03 ♌ 11 21	09 ♌ 24 59	04 52	15 39	13 29	15 16	12 51	28 55	10 33	18 46
16	15 40 36	23 42 15	15 ♌ 43 08	22 ♌ 06 16	06 22	16 54	13 58	15 24	12 49	28 55	10 34	18 48
17	15 44 33	24 42 44	28 ♌ 34 51	05 ♍ 09 18	07 53	18 09	14 26	15 32	12 46	28 56	10 35	18 50
18	15 48 30	25 43 14	11 ♍ 49 59	18 ♍ 37 09	09 23	19 23	14 55	15 40	12 44	28 56	10 36	18 52
19	15 52 26	26 43 46	25 ♍ 31 00	02 ♎ 31 36	10 53	20 38	15 24	15 48	12 41	28 57	10 37	18 55
20	15 56 23	27 44 20	09 ♎ 38 50	16 ♎ 52 29	12 22	21 52	15 54	15 56	12 39	28 57	10 38	18 57
21	16 00 19	28 44 55	24 ♎ 12 05	01 ♏ 37 04	13 51	23 07	16 24	16 03	12 36	28 58	10 39	18 59
22	16 04 16	29 ♏ 45 32	09 ♏ 06 38	16 ♏ 39 51	15 19	24 21	16 54	16 10	12 33	28 59	10 40	19 01
23	16 08 12	00 ♐ 46 11	24 ♏ 15 37	01 ♐ 52 45	16 47	25 36	17 24	16 18	12 30	28 59	10 41	19 03
24	16 12 09	01 46 52	09 ♐ 29 59	17 ♐ 06 05	18 15	26 50	17 54	16 25	12 27	29 00	10 42	19 05
25	16 16 05	02 47 34	24 ♐ 39 48	02 ♑ 10 00	19 42	28 05	18 25	16 32	12 24	29 01	10 43	19 08
26	16 20 02	03 48 17	09 ♑ 35 41	16 ♑ 55 59	21 09	29 ♐ 19	18 56	16 39	12 21	29 02	10 44	19 10
27	16 23 59	04 49 01	24 ♑ 10 14	01 ♒ 17 56	22 34	00 ♑ 34	19 27	16 45	12 17	29 03	10 45	19 12
28	16 27 55	05 49 46	08 ♒ 18 49	15 ♒ 12 44	23 59	01 48	19 59	16 52	12 14	29 04	10 46	19 14
29	16 31 52	06 50 33	21 ♒ 59 44	28 ♒ 39 59	25 23	03 03	20 31	16 58	12 10	29 05	10 48	19 17
30	16 35 48	07 ♐ 51 20	05 ♓ 13 47	11 ♓ 41 31	26 ♐ 46	04 ♑ 17	21 ♓ 03	17 ♍ 04	12 ♋ 07	29 ♒ 06	10 ♒ 49	19 ♐ 19

Date	Moon True ☊ ° '	Moon Mean ☊ ° '	Moon ☽ Latitude ° '
01	20 ♉ 32	20 ♉ 55	05 S 14
02	20 R 32	20 51	05 16
03	20 31	20 48	05 02
04	20 30	20 45	04 33
05	20 28	20 42	03 52
06	20 27	20 39	03 01
07	20 27	20 35	02 03
08	20 26	20 32	00 S 59
09	20 D 26	20 29	00 N 06
10	20 26	20 26	01 11
11	20 26	20 23	02 13
12	20 27	20 20	03 09
13	20 R 27	20 16	03 58
14	20 27	20 13	04 37
15	20 26	20 10	05 04
16	20 26	20 07	05 17
17	20 D 26	20 04	05 15
18	20 26	20 01	04 56
19	20 27	19 57	04 20
20	20 27	19 54	03 28
21	20 28	19 51	02 21
22	20 29	19 48	01 N 03
23	20 R 29	19 45	00 S 21
24	20 28	19 41	01 44
25	20 27	19 38	02 59
26	20 26	19 35	04 00
27	20 24	19 32	04 44
28	20 22	19 29	05 10
29	20 21	19 26	05 17
30	20 ♉ 20	19 ♉ 22	05 S 06

DECLINATIONS

Date	Sun ☉ ° '	Moon ☽ ° '	Mercury ☿ ° '	Venus ♀ ° '	Mars ♂ ° '	Jupiter ♃ ° '	Saturn ♄ ° '	Uranus ♅ ° '	Neptune ♆ ° '	Pluto ♇ ° '
01	14 S 23	22 S 03	15 S 57	20 00	10 S 54	07 N 30	22 N 04	12 S 34	17 S 39	14 S 11
02	14 42	17 53	16 32	20 19	10 42	07 27	22 04	12 34	17 38	14 12
03	15 01	13 00	17 06	20 38	10 30	07 23	22 04	12 34	17 38	14 12
04	15 20	07 40	17 39	20 55	10 18	07 20	22 04	12 35	17 38	14 13
05	15 38	02 S 08	18 11	21 12	10 06	07 16	22 04	12 35	17 38	14 13
06	15 56	03 N 24	18 43	21 29	09 53	07 13	22 04	12 35	17 38	14 13
07	16 14	08 46	19 13	21 45	09 40	07 10	22 05	12 35	17 38	14 14
08	16 32	13 46	19 43	22 00	09 28	07 06	22 05	12 35	17 38	14 14
09	16 49	18 15	20 12	22 15	09 15	07 03	22 05	12 34	17 38	14 15
10	17 06	21 59	20 39	22 29	09 02	07 00	22 05	12 34	17 37	14 15
11	17 23	24 49	21 06	22 43	08 49	06 57	22 05	12 34	17 37	14 15
12	17 39	26 34	21 32	22 55	08 36	06 53	22 05	12 34	17 37	14 16
13	17 55	27 06	21 56	23 08	08 22	06 50	22 05	12 34	17 37	14 16
14	18 11	26 22	22 20	23 19	08 09	06 47	22 06	12 34	17 37	14 17
15	18 27	24 22	22 42	23 30	07 56	06 44	22 06	12 34	17 37	14 17
16	18 42	21 10	23 03	23 40	07 42	06 41	22 06	12 34	17 36	14 18
17	18 57	16 53	23 23	23 49	07 28	06 38	22 06	12 33	17 36	14 18
18	19 11	11 41	23 42	23 58	07 14	06 36	22 07	12 33	17 36	14 18
19	19 26	05 N 46	24 00	24 06	07 01	06 33	22 07	12 33	17 36	14 19
20	19 39	00 S 38	24 16	24 13	06 47	06 30	22 07	12 33	17 35	14 19
21	19 53	07 12	24 32	24 20	06 33	06 27	22 08	12 33	17 35	14 19
22	20 06	13 32	24 46	24 25	06 19	06 25	22 08	12 32	17 35	14 20
23	20 19	19 11	24 58	24 30	06 04	06 22	22 08	12 32	17 35	14 20
24	20 31	23 35	25 10	24 35	05 50	06 19	22 09	12 32	17 34	14 20
25	20 43	26 18	25 20	24 38	05 36	06 17	22 09	12 31	17 34	14 21
26	20 55	27 05	25 28	24 41	05 21	06 15	22 09	12 31	17 34	14 21
27	21 06	25 56	25 36	24 43	05 07	06 12	22 10	12 31	17 33	14 21
28	21 17	23 10	25 42	24 45	04 52	06 10	22 10	12 30	17 33	14 22
29	21 27	19 10	25 46	24 45	04 37	06 08	22 10	12 30	17 33	14 22
30	21 S 37	14 S 20	25 S 49	24 S 45	04 S 23	06 N 05	22 N 11	12 S 29	17 S 32	14 S 22

ZODIAC SIGN ENTRIES

Date	h	m	Planets	
02	19	52	☽	♓
02	21	42	♀	♐
05	05	02	☽	♈
07	16	29	☽	♉
10	05	14	☽	♊
12	07	19	☿	♐
12	18	10	☽	♋
15	05	48	☽	♌
17	14	36	☽	♍
19	19	42	☽	♎
21	21	24	☽	♏
22	17	43	☉	♐
23	21	03	☽	♐
25	20	31	☽	♑
27	01	07	♀	♑
27	21	48	☽	♒
30	02	25	☽	♓

LATITUDES

Date	Mercury ☿ ° '	Venus ♀ ° '	Mars ♂ ° '	Jupiter ♃ ° '	Saturn ♄ ° '	Uranus ♅ ° '	Neptune ♆ ° '	Pluto ♇ ° '
01	00 S 11	00 S 14	02 S 19	00 N 58	00 S 44	00 S 46	00 S 01	08 N 46
04	00 31	00 22	02 08	00 58	00 44	00 46	00 01	08 45
07	00 50	00 30	01 58	00 59	00 44	00 46	00 01	08 45
10	01 09	00 38	01 48	01 00	00 44	00 46	00 01	08 44
13	01 26	00 45	01 39	01 00	00 43	00 46	00 01	08 43
16	01 42	00 52	01 30	01 01	00 43	00 46	00 01	08 42
19	01 56	01 00	01 22	01 02	00 43	00 45	00 01	08 42
22	02 08	01 06	01 14	01 02	00 43	00 45	00 01	08 41
25	02 18	01 13	01 06	01 03	00 43	00 45	00 01	08 41
28	02 24	01 19	00 59	01 04	00 43	00 45	00 01	08 40
31	02 S 25	01 S 25	00 S 52	01 N 04	00 S 43	00 S 45	00 S 01	08 N 40

DATA

Julian Date	2452945
Delta T	+65 seconds
Ayanamsa	23° 54' 24"
Synetic vernal point	05° ♓ 12' 35"
True obliquity of ecliptic	23° 26' 26"

MOON'S PHASES, APSIDES AND POSITIONS ☽

Date	h	m	Phase	Longitude ° '	Eclipse Indicator
01	04	25	☽	08 ♒ 20	
09	01	13	○	16 ♉ 13	total
17	04	15	☾	24 ♌ 23	
23	22	59	●	01 ♐ 14	Total
30	17	16	☽	08 ♓ 05	

Day	h	m		
10	11	47	Apogee	
23	23	10	Perigee	
05	21	12	ON	
13	10	17	Max dec	27° N 06'
20	09	40	OS	
26	09	25	Max dec	27° S 06'

LONGITUDES

Date	Chiron ⚷ ° '	Ceres ⚳ ° '	Pallas ⚴ ° '	Juno ⚵ ° '	Vesta ⚶ ° '	Black Moon Lilith ⚸ ° '
01	13 ♑ 39	23 ♋ 42	14 ♈ 19	01 ♐ 30	03 ♐ 46	29 ♉ 17
11	14 ♑ 16	24 ♋ 50	11 ♈ 56	04 ♐ 52	09 ♐ 04	00 ♊ 23
21	14 ♑ 58	25 ♋ 20	10 ♈ 16	08 ♐ 17	14 ♐ 25	01 ♊ 30
31	15 ♑ 45	25 ♋ 10	09 ♈ 25	11 ♐ 44	19 ♐ 48	02 ♊ 36

All ephemeris data is given at 12.00 UT and the Moon's longitude is additionally given for 24.00 UT

ASPECTARIAN

01 Saturday
02 13 ☽ ± ♃
02 40 ☽ ⊻ ♂
04 25 ☽ □ ☉
07 25 ☽ Q ♀
08 09 ☽ ☌ ♆
10 59 ♀ ⚼ ♄
11 57 ☽ # ♄
12 48 ☿ ✶ ♃
13 05 ☽ ⊼ ♃
13 07 ☽ □ ☿
13 08 ☽ ⊼ ♄
13 17 ☿ △ ♄
15 19 ♀ Q ♆
17 12 ♃ ✶ ♄
22 24 ☽ ✶ ♇
23 32 ☽ ∥ ♀
23 59 ☽ ± ♄

02 Sunday
00 43 ♀ □ ♅
13 17 ☽ ∥ ♆
16 30 ☽ ⚼ ♄
17 51 ☽ ☌ ♅
18 13 ☽ ∥ ☿
19 40 ☽ □ ♀
20 31 ☽ Q ♇

03 Monday
02 58 ☽ ∥ ☉
06 20 ☽ ∥ ♇
06 54 ☉ □ ♆
11 21 ☽ ☌ ♂
13 57 ☽ ∥ ♅
15 27 ☽ ⊻ ♆
16 11 ☽ △ ☉
20 37 ☽ △ ♄
21 21 ☽ ☍ ♃
23 52 ☽ ∥ ♂

04 Tuesday
02 52 ☽ ⊥ ♆
04 42 ☽ △ ☿
06 36 ☽ □ ♇
11 26 ☿ ∥ ♆
13 29 ☽ # ♃
20 05 ☿ ⊻ ♇
20 12 ☽ ∠ ♆
23 27 ☽ ⚼ ☉

05 Wednesday
02 54 ☽ ⊻ ♅
06 35 ☉ ⊥ ♇
11 16 ☽ △ ♀
14 00 ☽ ⚼ ☿
14 37 ☽ ⊥ ♅
18 36 ☽ ± ☉
23 13 ☽ ⊻ ♂
23 24 ☉ △ ♄

06 Thursday
01 33 ☽ ✶ ♆
06 49 ☽ □ ♄
07 30 ☽ ⊼ ☉
08 21 ☽ ∠ ♅
08 25 ☽ ⊼ ♃
10 22 ☽ ± ☿
11 30 ☽ ⊥ ♂
17 26 ☽ △ ♇
20 17 ☽ ⚼ ♀
20 30 ☽ ± ♃
20 33 ☉ ✶ ♃

07 Friday
00 05 ☽ ⊼ ☿
01 23 ☽ Q ♆
04 45 ☽ ∥ ♃
06 05 ☽ ∠ ♂
14 16 ☽ ✶ ♅
14 44 ☽ ⚼ ♃
16 02 ☽ # ♂
16 22 ☽ ± ♀
18 40 ☽ Q ♄
23 33 ☽ ⚼ ♇

08 Saturday
05 51 ☽ ⊼ ♀
06 03 ☽ # ♅
12 42 ♅ St D
13 23 ☽ ✶ ♂
13 36 ☽ □ ♆
14 08 ♀ ± ♄
14 21 ☽ # ♇
14 27 ☽ Q ♅
17 47 ☽ ± ♇
18 38 ♀ # ♄
18 50 ☽ ✶ ♄
19 36 ♂ ⊻ ♆
21 23 ☽ △ ♃

09 Sunday
01 13 ☽ ☍ ☉
03 24 ☽ # ☉
05 58 ☽ ⊼ ♇
08 29 ☽ # ♆
14 39 ☽ Q ♂
21 44 ☽ ☍ ☿

10 Monday
00 28 ☿ Q ♃
01 14 ☽ ∠ ♄
01 34 ☽ # ☿
03 00 ☽ □ ♅
12 43 ☽ ∥ ♄
16 01 ☽ # ♀
19 30 ☽ ⊥ ♄

11 Tuesday
00 04 ☿ ⚼ ♄
01 56 ☽ ☍ ♀
02 37 ☽ △ ♆
04 42 ☽ □ ♂
07 41 ☽ ⊻ ♄
07 55 ☿ Q ♆
08 30 ♀ ✶ ♆
10 58 ☉ ⊻ ♇
11 09 ☽ □ ♃
14 05 ☿ □ ♅
19 10 ☽ ☍ ♇
19 54 ☽ ⊼ ☉

12 Wednesday
08 49 ☉ ∥ ♆
09 06 ☽ ⚼ ♆
09 12 ☽ ± ☉
15 57 ☽ △ ♅
17 27 ♀ □ ♂
19 46 ☽ ⊼ ☿

13 Thursday
00 03 ☽ Q ♃
03 17 ☽ ± ♆
05 02 ☽ ⚼ ☉
07 20 ♀ ⊼ ♄
09 36 ☽ ± ☿
15 20 ☽ ⊼ ♆
19 41 ☽ △ ♂
20 05 ☽ ☌ ♄
21 35 ☿ # ♄
21 36 ☽ ⊼ ♀
22 04 ☽ ⚼ ♅

14 Friday
00 22 ☽ ✶ ♃
05 27 ♂ △ ♄
06 10 ☽ ⚼ ☿
07 39 ☽ ⊼ ♇
10 52 ☽ ± ♀
13 39 ☽ △ ☉
15 55 ☽ ± ♅
19 31 ☽ ± ♇

15 Saturday
02 30 ☽ ⚼ ♂
03 34 ♀ □ ♃
03 41 ☽ ⊼ ♅
06 16 ☽ ∠ ♃
06 32 ☽ ⚼ ♀
13 08 ☽ ⚼ ♇
15 41 ☽ △ ☿
19 01 ☽ # ♀
20 38 ☽ ± ♂
23 51 ☽ ⊥ ♃

16 Sunday
00 12 ☽ # ☿
02 12 ☽ ☍ ♆
05 46 ☽ ∥ ♄
06 30 ☽ ⊻ ♄
08 32 ☽ ⊼ ♂
11 24 ☽ ⊻ ♃
12 26 ♀ Q ♅
14 28 ☽ △ ♀
17 49 ☽ ⊥ ♄
17 50 ☽ △ ♇
18 50 ☿ ± ♄

17 Monday
01 41 ☽ # ☉
04 15 ☽ □ ☉
08 19 ☽ # ♆
10 31 ☽ ∠ ♄
12 38 ☽ ☍ ♅

18 Tuesday
00 24 ☽ # ♇
01 51 ♀ ☌ ♇
07 04 ☽ □ ☿
08 13 ☽ # ♅
09 48 ☽ ⊼ ♆
13 36 ☽ ✶ ♄
15 38 ☽ Q ☉
17 41 ☽ ☍ ♂
18 53 ☽ ☌ ♃
20 27 ☽ ± ♆

19 Wednesday
00 29 ☽ □ ♇
02 40 ☽ □ ♀
06 57 ☽ # ♂
07 43 ☿ ✶ ♆
08 56 ☽ ∥ ♃
10 35 ☽ Q ♄
12 10 ☽ ⚼ ♆
14 15 ☽ ✶ ☉
17 54 ☽ ⊼ ♅
18 28 ☽ Q ☿

20 Thursday
03 42 ☿ ∥ ♀
04 07 ☽ ± ♅
07 27 ☽ Q ♇
09 53 ☉ ⚼ ♄
12 24 ☽ Q ♀
13 38 ☽ △ ♆
13 59 ♂ ☍ ♃
16 24 ☿ ⊼ ♄
16 59 ☽ □ ♄
17 03 ☽ ✶ ☿
17 07 ☉ Q ♃
17 32 ☽ ∠ ☉
19 11 ☽ ⚼ ♅
22 32 ☽ ⊻ ♃
22 45 ☽ ⊼ ♂

21 Friday
03 27 ☽ ✶ ♇
08 28 ☽ ⊥ ♃
08 57 ☽ ± ♂
09 17 ☽ # ♃
09 27 ☽ ⊥ ☉
09 29 ☉ Q ♆
09 41 ☽ ∥ ♂
10 04 ☽ ✶ ♀
17 14 ☉ □ ♅
19 44 ☽ △ ♅
19 55 ☽ ⊻ ☉
20 22 ☽ ∠ ☿
23 11 ☽ ∠ ♃
23 19 ♂ # ♃

22 Saturday
00 02 ☽ ⚼ ♂
00 33 ♂ ⊥ ♆
03 50 ☽ ∠ ♇
08 04 ☽ ∥ ♅
12 22 ☽ ⊥ ☿
12 25 ☽ ∠ ♀
14 28 ☽ □ ♆
15 08 ☽ ∥ ♇
17 28 ☽ △ ♄
18 14 ☽ ⊥ ♇
22 56 ☽ ⊻ ☿
23 19 ☽ ✶ ♃

23 Sunday
00 47 ☽ △ ♂
03 09 ☿ □ ♃
03 46 ☽ ⊻ ♇
03 59 ☽ ⊥ ♀
04 45 ☽ ∥ ♆
13 37 ♀ ∠ ♆
14 18 ☽ ⊻ ♀
15 18 ☿ Q ♅
17 05 ☽ ⚼ ♄
17 45 ☽ ∥ ☉
18 24 ☽ Q ♃
18 58 ☽ Q ♆
19 28 ☽ □ ♅
22 59 ☽ ☌ ☉

24 Monday
03 11 ☽ # ♄
03 13 ☿ □ ♂
07 13 ☽ ± ♄
13 53 ☽ ✶ ♆
16 38 ☽ ⊼ ♄
19 14 ☽ ∥ ♀
23 00 ☽ □ ♃
23 51 ☽ Q ♅

25 Tuesday
00 39 ☽ ∥ ☿
01 44 ☽ □ ♂
02 13 ☿ ☌ ♇
03 11 ☽ ☌ ♇
03 17 ☽ ☌ ☿
13 41 ☽ ∠ ♆
17 56 ☽ ☌ ♀
18 57 ☽ ✶ ♅

26 Wednesday
01 57 ☽ ⊻ ☉
04 07 ☽ ⊥ ♆
06 22 ♀ ✶ ♅
07 32 ☽ Q ♂
12 22 ☽ ⊥ ☉
13 51 ☽ ⊻ ♆
16 28 ☽ ☍ ♄
19 15 ☽ ∠ ♅
23 24 ♂ □ ♇
23 37 ☽ △ ♃

27 Thursday
03 43 ☽ ⊻ ♇
03 52 ☽ ✶ ♂
04 13 ☽ ∠ ☉
09 03 ☽ ⊻ ☿
10 08 ☽ ⊥ ♅
11 45 ♃ ± ♆
13 44 ☽ ⊥ ♇
15 41 ☽ ∥ ☿
20 12 ☽ ⊻ ♅
20 12 ☽ ⊥ ☿
23 47 ☽ ⊻ ♀

28 Friday
00 00 ☽ ∥ ♀
00 52 ☽ ⚼ ♃
04 59 ☽ ∠ ♇
06 03 ☽ ∠ ♂
07 24 ☽ ✶ ☉
11 02 ☽ ⊥ ♀
13 18 ☽ ∠ ☿
16 16 ☽ ☌ ♆
16 27 ☽ ± ♃
18 37 ☽ # ♄
18 46 ☽ ⊼ ♄
21 01 ☉ ± ♄
22 14 ☽ ⊥ ♂
23 34 ☽ ∥ ☉

29 Saturday
03 01 ☽ ⊼ ♃
04 17 ☽ ∠ ♀
05 15 ☽ ± ♄
05 57 ☽ Q ☉
07 09 ☽ ✶ ♇
09 15 ☽ ⊻ ♂
18 47 ☽ ✶ ☿
19 06 ☿ ∠ ♆
20 25 ☽ ∥ ♆
21 15 ☽ ⚼ ♄

30 Sunday
00 46 ☽ ☌ ♅
04 48 ☽ Q ♇
10 05 ☽ ✶ ♀
11 50 ☽ ∥ ♇
17 16 ☽ □ ☉
19 20 ☽ Q ☿
20 31 ☽ ∥ ♅
22 22 ♀ ⊥ ♆
22 22 ☽ ⊻ ♆

DECEMBER 2003

LONGITUDES

Date	Sidereal time h m s	Sun ☉ ° ' "	Moon ☽ ° ' "	Moon ☽ 24.00 ° ' "	Mercury ☿ ° '	Venus ♀ ° '	Mars ♂ ° '	Jupiter ♃ ° '	Saturn ♄ ° '	Uranus ♅ ° '	Neptune ♆ ° '	Pluto ♇ ° '
01	16 39 45	08 ♐ 52 08	18 ♓ 03 38	24 ♓ 20 37	28 ♐ 08	05 ♑ 32	21 ♓ 35	17 ♍ 10	12 ♋ 03	29 ♒ 07	10 ♒ 50	19 ♐ 21
02	16 43 41	09 52 56	00 ♈ 32 59	06 ♈ 41 19	29 ♐ 28	06 46	22 07	17 16	11 R 59	29 08	10 51	19 23
03	16 47 38	10 53 46	12 ♈ 46 07	18 ♈ 47 56	00 ♑ 47	08 00	22 40	17 22	11 56	29 10	10 53	19 26
04	16 51 34	11 54 37	24 ♈ 47 17	00 ♉ 44 38	02 04	09 15	23 12	17 28	11 52	29 11	10 54	19 28
05	16 55 31	12 55 29	06 ♉ 40 27	12 ♉ 35 11	03 19	10 29	23 45	17 33	11 48	29 12	10 55	19 30
06	16 59 28	13 56 21	18 ♉ 29 13	24 ♉ 22 54	04 31	11 44	24 18	17 38	11 44	29 14	10 57	19 32
07	17 03 24	14 57 15	00 ♊ 16 35	06 ♊ 10 33	05 41	12 58	24 52	17 43	11 40	29 15	10 58	19 35
08	17 07 21	15 58 09	12 ♊ 05 06	18 ♊ 00 28	06 47	14 12	25 25	17 48	11 36	29 17	11 00	19 37
09	17 11 17	16 59 05	23 ♊ 56 53	29 ♊ 54 36	07 50	15 27	25 59	17 53	11 31	29 18	11 01	19 39
10	17 15 14	18 00 01	05 ♋ 53 49	11 ♋ 54 45	08 48	16 41	26 32	17 58	11 27	29 20	11 03	19 41
11	17 19 10	19 00 59	17 ♋ 57 36	24 ♋ 02 37	09 41	17 55	27 06	18 02	11 23	29 21	11 04	19 44
12	17 23 07	20 01 57	00 ♌ 10 02	06 ♌ 20 07	10 29	19 09	27 40	18 06	11 18	29 23	11 06	19 46
13	17 27 03	21 02 57	12 ♌ 33 06	18 ♌ 49 19	11 10	20 24	28 15	18 10	11 14	29 25	11 08	19 48
14	17 31 00	22 03 57	25 ♌ 09 04	01 ♍ 32 41	11 44	21 38	28 49	18 14	11 09	29 26	11 09	19 51
15	17 34 56	23 04 58	08 ♍ 00 29	14 ♍ 32 48	12 10	22 52	29 23	18 18	11 05	29 28	11 11	19 53
16	17 38 53	24 06 01	21 ♍ 09 59	27 ♍ 52 20	12 27	24 06	29 ♓ 58	18 22	11 00	29 30	11 12	19 55
17	17 42 50	25 07 04	04 ♎ 40 06	11 ♎ 33 31	12 34	25 20	00 ♈ 33	18 25	10 56	29 32	11 14	19 57
18	17 46 46	26 08 08	18 ♎ 32 41	25 ♎ 37 38	12 R 30	26 35	01 08	18 28	10 51	29 34	11 16	20 00
19	17 50 43	27 09 14	02 ♏ 48 16	10 ♏ 04 21	12 15	27 49	01 43	18 31	10 46	29 36	11 18	20 02
20	17 54 39	28 10 20	17 ♏ 25 27	24 ♏ 51 01	11 49	29 ♑ 03	02 18	18 34	10 41	29 38	11 19	20 04
21	17 58 36	29 ♐ 11 27	02 ♐ 20 17	09 ♐ 52 21	11 10	00 ♒ 17	02 53	18 37	10 37	29 40	11 21	20 06
22	18 02 32	00 ♑ 12 35	17 ♐ 26 07	25 ♐ 00 28	10 21	01 31	03 28	18 39	10 32	29 42	11 23	20 09
23	18 06 29	01 13 43	02 ♑ 34 07	10 ♑ 05 50	09 21	02 45	04 04	18 42	10 27	29 44	11 25	20 11
24	18 10 25	02 14 52	17 ♑ 34 25	24 ♑ 58 45	08 12	03 59	04 40	18 44	10 22	29 46	11 27	20 13
25	18 14 22	03 16 01	02 ♒ 17 50	09 ♒ 30 52	06 56	05 13	05 15	18 46	10 17	29 49	11 29	20 15
26	18 18 19	04 17 10	16 ♒ 37 15	23 ♒ 36 33	05 36	06 27	05 51	18 47	10 12	29 51	11 30	20 18
27	18 22 15	05 18 19	00 ♓ 28 34	07 ♓ 13 17	04 14	07 41	06 27	18 49	10 07	29 53	11 32	20 20
28	18 26 12	06 19 28	13 ♓ 50 50	20 ♓ 21 31	02 52	08 55	07 03	18 50	10 02	29 55	11 34	20 22
29	18 30 08	07 20 37	26 ♓ 45 44	03 ♈ 04 01	01 34	10 09	07 39	18 51	09 58	29 ♒ 58	11 36	20 24
30	18 34 05	08 21 47	09 ♈ 16 56	15 ♈ 25 06	00 ♑ 22	11 22	08 15	18 52	09 53	00 ♓ 00	11 38	20 27
31	18 38 01	09 ♑ 22 55	21 ♈ 29 10	27 ♈ 29 49	29 ♐ 18	12 ♒ 36	08 ♈ 52	18 ♍ 53	09 ♋ 48	00 ♓ 03	11 ♒ 40	20 ♐ 29

Date	Moon True ☊ ° '	Moon Mean ☊ ° '	Moon ☽ Latitude ° '
01	20 ♉ 21	19 ♉ 19	04 S 40
02	20 D 22	19 16	04 02
03	20 23	19 13	03 13
04	20 25	19 10	02 17
05	20 26	19 07	01 15
06	20 27	19 03	00 S 11
07	20 R 26	19 00	00 N 54
08	20 25	18 57	01 56
09	20 22	18 54	02 54
10	20 17	18 51	03 44
11	20 12	18 47	04 25
12	20 07	18 44	04 54
13	20 02	18 41	05 10
14	19 58	18 38	05 11
15	19 55	18 35	04 57
16	19 54	18 32	04 27
17	19 D 54	18 28	03 42
18	19 55	18 25	02 43
19	19 57	18 22	01 32
20	19 58	18 19	00 N 14
21	19 R 57	18 16	01 S 07
22	19 55	18 13	02 24
23	19 50	18 09	03 31
24	19 44	18 06	04 22
25	19 37	18 03	04 55
26	19 30	18 00	05 09
27	19 23	17 57	05 03
28	19 19	17 53	04 41
29	19 16	17 50	04 06
30	19 D 16	17 47	03 19
31	19 ♉ 16	17 ♉ 44	02 S 24

DECLINATIONS

Date	Sun ☉ ° '	Moon ☽ ° '	Mercury ☿ ° '	Venus ♀ ° '	Mars ♂ ° '	Jupiter ♃ ° '	Saturn ♄ ° '	Uranus ♅ ° '	Neptune ♆ ° '	Pluto ♇ ° '
01	21 S 47	09 S 01	25 S 51	24 S 44	04 S 08	06 N 03	22 N 11	12 S 29	17 S 32	14 S 23
02	21 56	03 S 29	25 51	24 42	03 53	06 01	22 11	12 29	17 32	14 23
03	22 05	02 N 04	25 50	24 40	03 38	05 59	22 12	12 28	17 31	14 23
04	22 13	07 28	25 48	24 37	03 23	05 57	22 12	12 28	17 31	14 23
05	22 21	12 33	25 44	24 33	03 08	05 55	22 13	12 27	17 31	14 24
06	22 28	17 09	25 39	24 28	02 53	05 54	22 13	12 27	17 30	14 24
07	22 35	21 05	25 32	24 23	02 38	05 52	22 14	12 26	17 30	14 24
08	22 42	24 10	25 24	24 16	02 23	05 50	22 14	12 25	17 30	14 25
09	22 48	26 12	25 15	24 09	02 08	05 49	22 14	12 25	17 29	14 25
10	22 54	27 02	25 04	24 02	01 52	05 47	22 15	12 24	17 29	14 25
11	22 59	26 37	24 53	23 53	01 37	05 45	22 15	12 24	17 28	14 25
12	23 04	24 54	24 40	23 44	01 22	05 44	22 16	12 23	17 28	14 26
13	23 08	21 59	24 27	23 35	01 06	05 43	22 16	12 22	17 27	14 26
14	23 12	18 01	24 13	23 24	00 51	05 41	22 17	12 22	17 27	14 26
15	23 16	13 09	23 58	23 13	00 36	05 40	22 17	12 21	17 27	14 26
16	23 19	07 36	23 43	23 01	00 20	05 39	22 18	12 21	17 26	14 27
17	23 21	01 N 33	23 27	22 48	00 S 05	05 38	22 18	12 20	17 26	14 27
18	23 23	04 S 45	23 11	22 35	00 N 11	05 37	22 18	12 19	17 25	14 27
19	23 25	11 00	22 55	22 21	00 26	05 36	22 19	12 18	17 25	14 27
20	23 26	16 49	22 38	22 07	00 42	05 35	22 19	12 18	17 24	14 28
21	23 26	21 43	22 22	21 52	00 57	05 34	22 20	12 17	17 24	14 28
22	23 26	25 14	22 07	21 36	01 13	05 34	22 20	12 16	17 23	14 28
23	23 26	26 55	21 51	21 19	01 29	05 33	22 21	12 15	17 23	14 28
24	23 25	26 37	21 36	21 02	01 44	05 32	22 21	12 15	17 22	14 28
25	23 24	24 26	21 22	20 45	02 00	05 32	22 22	12 14	17 22	14 29
26	23 22	20 45	21 08	20 26	02 15	05 31	22 22	12 13	17 21	14 29
27	23 20	16 02	20 56	20 08	02 31	05 31	22 23	12 12	17 21	14 29
28	23 17	10 41	20 44	19 48	02 47	05 31	22 23	12 11	17 20	14 29
29	23 14	05 S 03	20 34	19 28	03 02	05 31	22 24	12 10	17 20	14 29
30	23 11	00 N 38	20 26	19 08	03 18	05 31	22 24	12 09	17 19	14 29
31	23 S 07	06 N 09	20 S 20	18 S 47	03 N 34	05 N 31	22 N 25	12 S 09	17 S 19	14 S 30

ZODIAC SIGN ENTRIES

Date	h	m	Planets
02	10	56	☽ ♈
02	21	34	☿ ♑
04	22	30	☽ ♉
07	11	26	☽ ♊
10	00	11	☽ ♋
12	11	40	☽ ♌
14	21	07	☽ ♍
16	13	24	♂ ♈
17	03	46	☽ ♎
19	07	20	☽ ♏
21	06	32	♀ ♒
21	08	16	☽ ♐
22	07	04	☉ ♑
23	07	55	☽ ♑
25	08	13	☽ ♒
27	11	10	☽ ♓
29	18	08	☽ ♈
30	09	14	♅ ♓
30	19	52	☿ ♐

LATITUDES

Date	Mercury ☿ ° '	Venus ♀ ° '	Mars ♂ ° '	Jupiter ♃ ° '	Saturn ♄ ° '	Uranus ♅ ° '	Neptune ♆ ° '	Pluto ♇ ° '
01	02 S 25	01 S 25	00 S 52	01 N 04	00 S 43	00 S 45	00 S 01	08 N 40
04	02 22	01 30	00 45	01 05	00 43	00 45	00 01	08 39
07	02 13	01 34	00 39	01 06	00 42	00 45	00 01	08 39
10	01 56	01 39	00 32	01 07	00 42	00 45	00 01	08 39
13	01 29	01 42	00 27	01 07	00 42	00 45	00 01	08 38
16	00 51	01 45	00 21	01 08	00 42	00 44	00 01	08 38
19	00 S 02	01 48	00 16	01 09	00 41	00 44	00 01	08 38
22	00 N 56	01 49	00 11	01 10	00 41	00 44	00 01	08 38
25	01 54	01 50	00 06	01 11	00 41	00 44	00 01	08 38
28	02 40	01 50	00 S 01	01 12	00 41	00 44	00 02	08 38
31	03 N 06	01 S 50	00 N 03	01 N 12	00 S 40	00 S 44	00 S 02	08 N 38

DATA

Julian Date	2452975
Delta T	+65 seconds
Ayanamsa	23° 54' 29"
Synetic vernal point	05° ♓ 12' 30"
True obliquity of ecliptic	23° 26' 25"

MOON'S PHASES, APSIDES AND POSITIONS ☽

Date	h	m	Phase	Longitude ° '	Eclipse Indicator
08	20	37	○	16 ♊ 20	
16	17	42	☾	24 ♍ 21	
23	09	43	●	01 ♑ 08	
30	10	03	☽	08 ♈ 17	

Day	h	m		
07	11	50	Apogee	
22	11	43	Perigee	
03	03	00	0N	
10	15	59	Max dec	27° N 03'
17	17	56	0S	
23	20	16	Max dec	27° S 03'
30	09	19	0N	

LONGITUDES

Date	Chiron ⚷ ° '	Ceres ⚳ ° '	Pallas ⚴ ° '	Juno ⚵ ° '	Vesta ⚶ ° '	Black Moon Lilith ⚸ ° '
01	15 ♑ 45	25 ♋ 10	09 ♈ 25	11 ♐ 44	19 ♐ 48	02 ♊ 36
11	16 ♑ 36	24 ♋ 18	09 ♈ 25	15 ♐ 13	25 ♐ 11	03 ♊ 43
21	17 ♑ 30	22 ♋ 47	10 ♈ 13	18 ♐ 43	00 ♑ 35	04 ♊ 49
31	18 ♑ 26	20 ♋ 44	11 ♈ 45	22 ♐ 11	05 ♑ 59	05 ♊ 56

All ephemeris data is given at 12.00 UT and the Moon's longitude is additionally given for 24.00 UT

ASPECTARIAN

h m Aspects

01 Monday
00 44 ☽ △ ♄
09 40 ☽ ⊥ ♆
10 18 ☽ ☍ ♃
10 53 ☽ Q ♀
14 27 ☽ □ ♇
19 00 ☽ ☌ ♂

02 Tuesday
01 00 ☽ # ♃
02 54 ☽ ∠ ♆
05 53 ☿ ✶ ♅
09 15 ☽ ⊻ ♅
09 39 ☽ □ ☿
10 11 ☽ ∥ ♂
20 59 ☽ ⊥ ♅

03 Wednesday
01 32 ☽ □ ♀
07 57 ☽ △ ☉
08 15 ☽ ✶ ♆
10 21 ☽ □ ♄
11 34 ☉ ✶ ♆
14 46 ☽ ∠ ♅
18 33 ☽ # ♂
21 13 ☽ ⊼ ♃

04 Thursday
01 18 ☽ △ ♇
05 11 ☽ ∥ ♃
08 12 ☽ Q ♆
08 40 ☽ ⊻ ♂
09 19 ☽ ± ♃
09 32 ☉ # ♄
10 56 ☉ ⊼ ♄
16 40 ☽ ꝑ ☉
20 52 ☽ ✶ ♅
21 19 ☽ ⊥ ♂
22 10 ☽ Q ♄

05 Friday
03 35 ☽ ꝑ ♃
04 25 ☽ △ ☿
07 35 ☽ ꝑ ♇
11 29 ☽ # ♅
12 33 ☽ ± ☉
16 25 ☽ ∠ ♂
20 38 ♀ ⊻ ♆
20 38 ☽ □ ♆
20 38 ☽ △ ♀
21 13 ☽ Q ♅
21 15 ☽ # ♇
22 20 ☽ ✶ ♄

06 Saturday
01 53 ☽ ⊼ ☉
01 54 ☽ ± ♇
10 15 ☽ △ ♃
12 04 ♀ ☍ ♄
13 58 ☽ # ♆
14 09 ☽ ⊼ ♇
14 21 ☽ ꝑ ☿
20 51 ☿ ⊥ ♆

07 Sunday
00 26 ☽ ✶ ♂
04 41 ☽ ∠ ♄
06 45 ☽ ꝑ ♀
08 16 ♀ Q ♂
09 55 ☽ □ ♅
10 40 ☽ ± ☿
20 05 ☽ ∥ ♄
22 53 ☽ ⊥ ♄
23 13 ☽ # ☉

08 Monday
00 09 ☽ ⊼ ☿
02 03 ☽ Q ♂
03 12 ☽ ± ♀
09 47 ☽ △ ♆
11 00 ☽ ⊻ ♄
12 59 ☽ # ♀
13 25 ♀ ∠ ♅
16 48 ☽ ⊼ ♀
20 37 ☽ ☍ ☉
23 40 ☽ □ ♃

09 Tuesday
00 04 ☽ # ☿
03 18 ☽ ☍ ♇
14 00 ♂ ∠ ♆
16 11 ☽ ꝑ ♆
16 17 ☽ □ ♂
19 41 ☉ Q ♅
22 48 ☽ △ ♅

10 Wednesday
10 18 ☽ ± ♆
10 58 ☉ □ ♃
12 08 ☽ Q ♃
18 17 ☽ ☍ ☿
22 18 ☽ ⊼ ♆
23 01 ☽ ☌ ♄

11 Thursday
04 50 ☽ ꝑ ♅
11 55 ☽ ☍ ♀
12 09 ☽ ✶ ♃
14 17 ☽ ⊼ ☉
14 22 ♀ △ ♃
15 30 ☽ ⊼ ♇
22 40 ☽ ± ♅

12 Friday
03 10 ☽ ± ☉
03 21 ☽ ± ♇
05 28 ☉ ☌ ♇
06 53 ☽ △ ♂
10 28 ☽ ⊼ ♅
14 29 ☽ # ☿
17 46 ☽ ∠ ♃
20 59 ☽ ꝑ ♇
22 19 ☽ ꝑ ☉
23 21 ☽ # ♀

13 Saturday
00 12 ♀ ⊻ ♇
03 46 ☽ # ☉
09 11 ☽ ⊼ ☿
09 15 ☽ ☍ ♆
09 28 ☽ ⊻ ♄
10 02 ☽ ∥ ♄
10 22 ☿ ⊻ ♆
11 16 ☽ ⊥ ♃
13 23 ☽ ꝑ ♂
14 13 ☿ ☍ ♄
20 55 ☽ ⊥ ♄
21 17 ☽ ± ☿
22 49 ☽ ⊻ ♃

14 Sunday
01 55 ☽ △ ♇
04 37 ☽ ⊼ ♀
05 39 ☽ △ ☉
07 22 ☽ ± ♂
12 48 ♄ ⊼ ♆
13 53 ☽ ∠ ♄
15 02 ☽ # ♆
15 06 ☽ ꝑ ☿
17 10 ☽ ± ♀
19 13 ☽ ⊼ ♂
20 05 ☽ ☍ ♅

15 Monday
06 02 ☽ # ♇
07 40 ☉ ∥ ♀
11 43 ☽ ꝑ ♀
15 33 ♂ ⊻ ♅
15 39 ☽ # ♅
17 37 ☽ ✶ ♄
17 51 ☽ ⊼ ♆
19 50 ☽ △ ☿
23 59 ♀ ⊥ ♅

16 Tuesday
04 50 ☽ ± ♆
06 55 ☽ ☌ ♃
09 45 ☽ □ ♇
11 35 ☉ ⊻ ♀
15 17 ☽ Q ♄
17 42 ☽ □ ☉
17 49 ☽ △ ♀
19 55 ☽ ∥ ♃
21 03 ☽ ꝑ ♆

17 Wednesday
02 55 ☽ ⊼ ♅
04 25 ☽ ☍ ♂
13 31 ☽ ± ♅
16 00 ☿ St R
17 46 ☽ Q ♇
17 52 ☽ # ♂
17 59 ☽ ∥ ♂
19 35 ☉ ∥ ☿
22 51 ☽ □ ♄
23 28 ☽ △ ♆

18 Thursday
00 21 ♀ ⊥ ♇
01 43 ☽ □ ☿
03 51 ☽ Q ☉
05 10 ☽ ꝑ ♅
11 52 ☽ ⊻ ♃
14 29 ☽ ✶ ♇
15 08 ☉ ∠ ♆
15 15 ☽ # ♃
22 05 ☽ ⊥ ♃

19 Friday
01 51 ☽ ✶ ☉
02 53 ☽ □ ♀
06 39 ☽ △ ♅
07 51 ☽ Q ☿
10 06 ☽ ⊼ ♂
13 12 ☽ ∠ ♃
15 42 ☽ ∠ ♇
15 59 ♀ # ♄
17 10 ☽ ∥ ♅
20 27 ☽ ± ♂

20 Saturday
01 04 ☽ △ ♄
01 56 ☽ ∥ ♇
02 02 ☽ □ ♆
03 09 ☽ ✶ ☿
04 34 ☽ ∠ ☉
06 32 ☽ ⊥ ♇
11 20 ☽ Q ♀
11 47 ☽ ꝑ ♂
13 52 ☽ ✶ ♃
14 38 ☽ ∥ ♆
16 18 ☽ ⊻ ♇
20 15 ☽ ⊥ ☉
23 41 ♀ ⊻ ♅

21 Sunday
01 17 ☽ ꝑ ♄
02 34 ☽ ∠ ☿
06 07 ☿ ⊻ ♆
06 36 ☽ ⊻ ☉
07 13 ☽ Q ♆
07 43 ☽ □ ♅
08 25 ☽ ✶ ♀
09 14 ☽ Q ♃
12 44 ☽ ∥ ♀
12 54 ☽ △ ♂
15 30 ☽ # ♄
15 30 ☽ ∥ ☿
15 36 ☽ ± ♄
15 40 ☿ # ♄
16 18 ☽ ⊥ ☿
22 23 ☽ ∥ ☉
23 34 ☉ ✶ ♅

22 Monday
01 06 ☽ ⊼ ♄
01 22 ☽ ⊻ ☿
02 23 ☽ ✶ ♆
06 36 ☿ ☍ ♄
10 25 ☽ ∠ ♀
12 25 ☽ Q ♅
13 56 ☽ □ ♃
16 18 ☽ ☌ ♇

23 Tuesday
01 56 ☽ ⊥ ♀
02 13 ☽ ∠ ♆
07 29 ☽ ✶ ♅
09 43 ☽ ☌ ☉
12 19 ☽ ⊻ ♀
14 29 ☽ □ ♂
16 32 ☽ ⊥ ♆
22 04 ☽ ☌ ☿

24 Wednesday
00 30 ☽ ☍ ♄
02 08 ☽ ⊻ ♆
06 54 ♀ ꝑ ♃
07 29 ☽ ∠ ♅
13 52 ☽ △ ♃
16 17 ☽ ⊻ ♇
20 34 ☽ Q ♂
22 04 ☽ ⊥ ♅

25 Thursday
02 04 ☽ ⊥ ♇
07 54 ☽ ⊻ ♅
12 50 ♀ ∠ ♇
13 29 ♀ ✶ ♂
13 43 ☽ ⊻ ☉
14 25 ☽ ꝑ ♃
16 55 ☽ ∠ ♇
17 06 ☽ ✶ ♂
17 17 ☽ ☌ ♀
19 03 ☽ ⊻ ☿
19 44 ☽ ∥ ☉

26 Friday
00 28 ☽ ⊥ ☉
01 13 ☽ ⊼ ♄
02 28 ☽ # ♄
03 20 ☽ ☌ ♆
04 07 ☿ ⊻ ♀
04 14 ☽ ⊥ ☿
05 29 ☽ ± ♃
08 53 ☿ □ ♂
09 43 ☽ ∥ ☿
11 18 ☽ ± ♄
11 48 ♂ ⊥ ♅
13 33 ☿ ⊥ ♆
13 53 ☽ ∥ ♀
15 42 ☽ ⊼ ♃
16 55 ☽ ∠ ☉
18 12 ☽ ∠ ☿
18 18 ☽ ✶ ♇
19 34 ☽ ∠ ♂

27 Saturday
01 11 ☉ ☌ ☿
02 41 ☽ ꝑ ♄
05 41 ☽ ∥ ♆
10 57 ☽ ☌ ♅
11 57 ☽ ⊥ ♂
15 17 ☽ Q ♇
17 42 ☉ ⊥ ♆
18 02 ☽ ✶ ☿
19 09 ☽ ∥ ♇
21 16 ☽ ✶ ☉
23 07 ☽ ⊻ ♂

28 Sunday
02 08 ☽ ⊻ ♀
05 07 ☽ △ ♄
05 24 ☽ ∥ ♅
07 51 ☽ ⊻ ♆
11 37 ☿ ⊥ ♀
13 42 ☽ Q ☿
14 09 ☽ ⊥ ♀
18 51 ☽ ⊥ ♆
20 56 ☽ Q ☉
21 11 ☽ ☍ ♃

29 Monday
00 03 ☽ □ ♇
08 37 ☽ ∠ ♀
08 38 ♀ ⊼ ♄
10 02 ☽ # ♃
11 42 ☽ ∠ ♆
18 05 ☽ ⊻ ♅
20 05 ☽ # ♂
20 19 ☽ □ ☿

30 Tuesday
05 38 ☽ ⊥ ♅
05 56 ☉ □ ♂
09 55 ☽ ☌ ♂
10 03 ☽ □ ☉
13 09 ☽ □ ♄
16 31 ☽ ✶ ♀
16 36 ☽ ✶ ♆
17 19 ♀ ☌ ♆
19 30 ☿ ✶ ♅
23 13 ☽ ∠ ♅

31 Wednesday
00 03 ☽ ∥ ♂
06 50 ☽ ⊼ ♃
09 11 ☽ ∥ ♃
10 00 ☽ △ ♇
16 22 ☽ Q ♆
17 34 ♀ ± ♃
18 47 ☽ ± ♃
18 55 ☽ Q ♀
20 57 ☉ ☍ ♄

JANUARY 2004

LONGITUDES

Date	Sidereal time h m s	Sun ☉ ° ' "	Moon ☽ ° ' "	Moon ☽ 24.00 ° ' "	Mercury ☿ ° '	Venus ♀ ° '	Mars ♂ ° '	Jupiter ♃ ° '	Saturn ♄ ° '	Uranus ♅ ° '	Neptune ♆ ° '	Pluto ♇ ° '
01	18 41 58	10 ♑ 24 04	03 ♉ 27 42	09 ♉ 23 29	28 ♐ 22	13 ♒ 50	09 ♈ 28	18 ♍ 54	09 ♋ 43	00 ♓ 05	11 ♒ 42	20 ♐ 31
02	18 45 54	11 25 13	15 ♉ 17 46	21 ♉ 11 10	27 R 37	15 04	10 05	18 54	09 R 38	00 08	11 44	20 33
03	18 49 51	12 26 22	27 ♉ 04 15	02 ♊ 57 30	27 01	16 17	10 41	18 54	09 33	00 10	11 46	20 35
04	18 53 48	13 27 30	08 ♊ 51 24	14 ♊ 46 22	26 36	17 31	11 18	18 R 54	09 28	00 13	11 48	20 37
05	18 57 44	14 28 39	20 ♊ 42 47	26 ♊ 40 56	26 21	18 44	11 55	18 54	09 23	00 16	11 50	20 39
06	19 01 41	15 29 47	02 ♋ 41 05	08 ♋ 43 28	26 16	19 58	12 31	18 54	09 18	00 18	11 52	20 42
07	19 05 37	16 30 55	14 ♋ 48 14	20 ♋ 55 31	26 D 20	21 11	13 08	18 53	09 13	00 21	11 55	20 44
08	19 09 34	17 32 03	27 ♋ 05 24	03 ♌ 17 58	26 32	22 25	13 45	18 53	09 08	00 24	11 57	20 46
09	19 13 30	18 33 11	09 ♌ 33 14	15 ♌ 51 16	26 52	23 38	14 22	18 51	09 03	00 26	11 59	20 48
10	19 17 27	19 34 18	22 ♌ 12 04	28 ♌ 35 43	27 18	24 51	14 59	18 50	08 58	00 29	12 01	20 50
11	19 21 23	20 35 26	05 ♍ 02 14	11 ♍ 31 42	27 51	26 05	15 36	18 49	08 54	00 32	12 03	20 52
12	19 25 20	21 36 33	18 ♍ 04 14	24 ♍ 39 57	28 30	27 18	16 13	18 47	08 49	00 35	12 05	20 54
13	19 29 17	22 37 41	01 ♎ 18 59	08 ♎ 01 31	29 ♐ 14	28 31	16 51	18 46	08 44	00 38	12 07	20 56
14	19 33 13	23 38 48	14 ♎ 47 42	21 ♎ 37 44	00 ♑ 02	29 ♒ 44	17 28	18 44	08 39	00 41	12 10	20 58
15	19 37 10	24 39 55	28 ♎ 31 44	05 ♏ 29 51	00 54	00 ♓ 57	18 05	18 41	08 35	00 44	12 12	21 00
16	19 41 06	25 41 02	12 ♏ 32 07	19 ♏ 38 30	01 50	02 10	18 43	18 39	08 30	00 46	12 14	21 02
17	19 45 03	26 42 09	26 ♏ 48 54	04 ♐ 03 02	02 50	03 23	19 20	18 37	08 26	00 49	12 16	21 04
18	19 48 59	27 43 16	11 ♐ 20 32	18 ♐ 40 52	03 52	04 36	19 57	18 34	08 21	00 52	12 18	21 06
19	19 52 56	28 44 22	26 ♐ 03 21	03 ♑ 27 10	04 57	05 49	20 35	18 31	08 17	00 56	12 21	21 08
20	19 56 52	29 ♑ 45 28	10 ♑ 51 24	18 ♑ 15 01	06 05	07 01	21 13	18 28	08 12	00 59	12 23	21 10
21	20 00 49	00 ♒ 46 34	25 ♑ 36 58	02 ♒ 56 12	07 14	08 14	21 50	18 25	08 08	01 02	12 25	21 12
22	20 04 46	01 47 39	10 ♒ 11 44	17 ♒ 22 39	08 26	09 27	22 28	18 21	08 03	01 05	12 27	21 13
23	20 08 42	02 48 43	24 ♒ 28 10	01 ♓ 27 41	09 39	10 39	23 06	18 18	07 59	01 08	12 30	21 15
24	20 12 39	03 49 46	08 ♓ 20 46	15 ♓ 07 09	10 54	11 51	23 43	18 14	07 55	01 11	12 32	21 17
25	20 16 35	04 50 49	21 ♓ 46 47	28 ♓ 19 44	12 11	13 04	24 21	18 10	07 51	01 14	12 34	21 19
26	20 20 32	05 51 50	04 ♈ 46 15	11 ♈ 06 42	13 29	14 16	24 59	18 06	07 47	01 17	12 36	21 21
27	20 24 28	06 52 50	17 ♈ 21 33	23 ♈ 31 22	14 48	15 28	25 37	18 02	07 43	01 21	12 39	21 22
28	20 28 25	07 53 49	29 ♈ 36 47	05 ♉ 38 26	16 09	16 40	26 15	17 57	07 39	01 24	12 41	21 24
29	20 32 21	08 54 47	11 ♉ 37 03	17 ♉ 33 19	17 30	17 52	26 53	17 52	07 35	01 27	12 43	21 26
30	20 36 18	09 55 44	23 ♉ 27 58	29 ♉ 21 40	18 53	19 04	27 31	17 48	07 31	01 30	12 45	21 27
31	20 40 15	10 ♒ 56 40	05 ♊ 15 06	11 ♊ 08 55	20 ♑ 17	20 ♓ 16	28 ♈ 09	17 ♍ 43	07 ♋ 27	01 ♓ 34	12 ♒ 48	21 ♐ 29

Moon Nodes, Latitude and DECLINATIONS

Date	Moon True ☊ ° '	Moon Mean ☊ ° '	Moon ☽ Latitude ° '	Sun ☉ ° '	Moon ☽ ° '	Mercury ☿ ° '	Venus ♀ ° '	Mars ♂ ° '	Jupiter ♃ ° '	Saturn ♄ ° '	Uranus ♅ ° '	Neptune ♆ ° '	Pluto ♇ ° '
01	19 ♉ 18	17 ♉ 41	01 S 25	23 S 02	11 N 21	20 S 16	18 S 25	03 N 49	05 N 31	22 N 25	12 S 08	17 S 18	14 S 30
02	19 D 19	17 38	00 S 22	22 57	16 05	20 14	18 03	04 05	05 31	22 25	12 07	17 18	14 30
03	19 R 18	17 34	00 N 42	22 52	20 11	20 13	17 41	04 20	05 31	22 26	12 06	17 17	14 30
04	19 16	17 31	01 43	22 46	23 29	20 15	17 18	04 36	05 31	22 26	12 05	17 16	14 30
05	19 11	17 28	02 40	22 39	25 47	20 18	16 54	04 51	05 31	22 27	12 04	17 16	14 30
06	19 04	17 25	03 31	22 32	26 56	20 23	16 30	05 07	05 32	22 27	12 03	17 15	14 30
07	18 55	17 22	04 12	22 25	26 48	20 30	16 06	05 23	05 32	22 28	12 02	17 15	14 30
08	18 43	17 19	04 43	22 17	25 22	20 37	15 41	05 38	05 33	22 28	12 01	17 14	14 31
09	18 32	17 15	05 00	22 09	22 40	20 45	15 16	05 54	05 33	22 29	12 00	17 14	14 31
10	18 20	17 12	05 03	22 01	18 53	20 54	14 51	06 09	05 34	22 29	11 59	17 13	14 31
11	18 11	17 09	04 51	21 52	14 10	21 03	14 25	06 24	05 35	22 30	11 58	17 12	14 31
12	18 04	17 06	04 23	21 42	08 45	21 13	13 58	06 40	05 36	22 30	11 57	17 12	14 31
13	17 59	17 03	03 42	21 32	02 N 52	21 23	13 32	06 55	05 37	22 30	11 56	17 11	14 31
14	17 57	16 59	02 47	21 22	03 S 16	21 32	13 05	07 10	05 38	22 31	11 55	17 11	14 31
15	17 D 56	16 56	01 41	21 12	09 22	21 41	12 37	07 26	05 39	22 31	11 54	17 10	14 31
16	17 57	16 53	00 N 29	21 00	15 09	21 51	12 10	07 41	05 40	22 32	11 53	17 09	14 31
17	17 R 57	16 50	00 S 47	20 49	20 12	21 59	11 42	07 56	05 42	22 32	11 52	17 09	14 31
18	17 55	16 47	02 01	20 37	24 08	22 07	11 13	08 11	05 43	22 33	11 51	17 08	14 31
19	17 50	16 44	03 08	20 25	26 30	22 14	10 45	08 27	05 44	22 33	11 50	17 08	14 31
20	17 42	16 40	04 02	20 12	27 01	22 21	10 16	08 42	05 45	22 33	11 48	17 07	14 31
21	17 32	16 37	04 40	19 59	25 36	22 27	09 47	08 57	05 47	22 34	11 47	17 06	14 31
22	17 21	16 34	04 59	19 46	22 29	22 32	09 18	09 12	05 48	22 34	11 46	17 06	14 31
23	17 09	16 31	04 59	19 32	18 03	22 36	08 48	09 27	05 50	22 35	11 45	17 05	14 31
24	16 58	16 28	04 41	19 18	12 47	22 39	08 19	09 41	05 52	22 35	11 44	17 04	14 31
25	16 49	16 24	04 08	19 03	07 03	22 41	07 48	09 56	05 54	22 35	11 43	17 04	14 32
26	16 43	16 21	03 22	18 48	01 S 12	22 42	07 18	10 11	05 55	22 36	11 42	17 03	14 32
27	16 39	16 18	02 28	18 33	04 N 32	22 42	06 48	10 26	05 57	22 36	11 40	17 03	14 32
28	16 38	16 15	01 29	18 18	09 57	22 41	06 18	10 40	05 59	22 36	11 39	17 02	14 32
29	16 D 37	16 12	00 S 27	18 02	14 54	22 38	05 47	10 55	06 01	22 37	11 38	17 01	14 32
30	16 R 37	16 09	00 N 36	17 46	19 13	22 35	05 16	11 09	06 03	22 37	11 37	17 01	14 31
31	16 ♉ 36	16 ♉ 05	01 N 37	17 S 29	22 N 46	22 S 30	04 S 45	11 N 24	06 N 06	22 N 38	11 S 36	17 S 00	14 S 31

ZODIAC SIGN ENTRIES

Date	h	m	Planets
01	05	02	☽ ♉
03	17	58	☽ ♊
06	06	38	☽ ♋
08	17	38	☽ ♌
11	02	37	☽ ♍
13	09	38	☽ ♎
14	11	02	☿ ♑
14	17	16	♀ ♓
15	14	33	☽ ♏
17	17	18	☽ ♐
19	18	24	☽ ♑
20	17	42	☉ ♒
21	19	11	☽ ♒
23	21	29	☽ ♓
26	03	06	☽ ♈
28	12	46	☽ ♉
31	01	18	☽ ♊

LATITUDES

Date	Mercury ☿ ° '	Venus ♀ ° '	Mars ♂ ° '	Jupiter ♃ ° '	Saturn ♄ ° '	Uranus ♅ ° '	Neptune ♆ ° '	Pluto ♇ ° '
01	03 N 10	01 S 49	00 N 04	01 N 13	00 S 40	00 S 44	00 S 02	08 N 38
04	03 09	01 48	00 08	01 13	00 40	00 44	00 02	08 38
07	02 54	01 45	00 12	01 14	00 40	00 44	00 02	08 38
10	02 31	01 42	00 16	01 15	00 39	00 44	00 02	08 38
13	02 04	01 38	00 20	01 16	00 39	00 44	00 02	08 38
16	01 35	01 34	00 23	01 17	00 39	00 44	00 02	08 38
19	01 06	01 28	00 26	01 18	00 38	00 44	00 02	08 38
22	00 39	01 22	00 29	01 18	00 38	00 44	00 02	08 39
25	00 N 12	01 15	00 32	01 19	00 37	00 44	00 02	08 39
28	00 S 13	01 07	00 35	01 20	00 37	00 43	00 02	08 39
31	00 S 36	00 S 59	00 N 37	01 N 21	00 S 36	00 S 43	00 S 02	08 N 40

DATA

Julian Date	2453006
Delta T	+65 seconds
Ayanamsa	23° 54' 34"
Synetic vernal point	05° ♓ 12' 25"
True obliquity of ecliptic	23° 26' 25"

MOON'S PHASES, APSIDES AND POSITIONS ☽

Date	h	m	Phase	Longitude ° '	Eclipse Indicator
07	15	40	○	16 ♋ 40	
15	04	46	☾	24 ♎ 21	
21	21	05	●	01 ♒ 10	
29	06	03	☽	08 ♉ 40	

Day	h	m		
03	20	11	Apogee	
19	19	16	Perigee	
31	13	56	Apogee	
06	21	39	Max dec	27° N 02'
13	23	17	0S	
20	06	16	Max dec	27° S 04'
26	16	58	0N	

LONGITUDES

Date	Chiron ⚷ ° '	Ceres ⚳ ° '	Pallas ⚴ ° '	Juno ⚵ ° '	Vesta ⚶ ° '	Black Moon Lilith ° '
01	18 ♑ 31	20 ♋ 31	11 ♈ 56	22 ♐ 32	06 ♑ 31	06 ♊ 02
11	19 ♑ 28	18 ♋ 11	14 ♈ 10	25 ♐ 58	11 ♑ 53	07 ♊ 09
21	20 ♑ 24	15 ♋ 52	16 ♈ 58	29 ♐ 21	17 ♑ 14	08 ♊ 16
31	21 ♑ 19	13 ♋ 45	20 ♈ 14	02 ♑ 40	22 ♑ 31	09 ♊ 22

All ephemeris data is given at 12.00 UT and the Moon's longitude is additionally given for 24.00 UT

ASPECTARIAN

h m Aspects

01 Thursday
00 30 ☽ Q ♄
02 27 ☽ △ ☿
05 11 ☽ ✱ ♅
06 42 ☿ ∠ ♀
12 52 ☽ ⚼ ♃
15 47 ☽ # ♅
16 10 ☽ ⚼ ♇
20 22 ♂ □ ♄

02 Friday
00 34 ☽ ✱ ♄
00 49 ☽ ⚺ ♂
03 22 ☽ △ ☉
03 40 ☽ # ♇
04 44 ☽ □ ♆
05 32 ☽ Q ♅
06 50 ☽ ⚼ ☿
10 29 ☽ ± ♇
11 28 ☽ □ ♀
13 41 ☽ ⊥ ♂
18 41 ☽ # ♆
19 21 ☽ △ ♃
19 43 ☉ ⚺ ♆
22 08 ☽ # ♀
22 26 ♀ ± ♄
22 44 ☽ ⚻ ♇

03 Saturday
00 12 ☽ ± ☿
06 53 ☽ ∠ ♄
09 01 ☽ ∠ ♂
11 54 ☽ ⚻ ☿
12 17 ☽ # ☿
12 49 ☽ ⚼ ☉
18 21 ☽ □ ♅
23 59 ♃ St R

04 Sunday
00 12 ☿ ∠ ♆
01 06 ☽ ⊥ ♄
03 41 ☽ ∥ ♄
06 21 ☽ # ☉
08 53 ☽ ± ☉
13 13 ☽ ⚺ ♄
13 26 ♀ ∥ ♆
17 14 ☽ ✱ ♂
18 00 ☽ △ ♆
22 13 ☽ ⚻ ☉

05 Monday
07 34 ☽ △ ♀
08 21 ☽ □ ♃
09 05 ♂ ✱ ♆
11 53 ☽ ☍ ♇
15 09 ♀ ⚻ ♃
18 47 ☽ Q ♂
23 14 ☽ ☍ ☿

06 Tuesday
00 21 ☽ ⚼ ♆
07 14 ☽ △ ♅
07 15 ☉ ∠ ♅
13 45 ☿ St D
17 03 ☽ ⚼ ♀
18 22 ☽ ± ♆
20 22 ☽ Q ♃

07 Wednesday
01 03 ☽ ☌ ♄
02 44 ♀ ✱ ♇
03 59 ☉ # ♄
06 17 ☽ ⚻ ♆
08 33 ☽ □ ♂
12 50 ☽ ± ♀
13 04 ☽ ⚼ ♅
15 40 ☽ ☍ ☉
20 00 ☽ ✱ ♃
23 39 ☽ ⚻ ♇

08 Thursday
01 54 ☽ ⚻ ♀
03 36 ♂ ∥ ♃
06 45 ☽ ± ♅
10 54 ☽ ⚻ ☿
11 22 ☽ ± ♇
18 25 ☽ ⚻ ♅
22 47 ☽ ± ☿

09 Friday
01 05 ☽ ∠ ♃
04 47 ☽ ⚼ ♇
11 03 ☽ ⚺ ♄
13 26 ☽ ∥ ♄
15 52 ☽ # ☉
16 33 ☽ ⚼ ☿
16 39 ☽ ☍ ♆
18 17 ☽ ⊥ ♃
19 00 ☉ △ ♃
19 34 ☿ ∠ ♆
19 46 ♀ ⚼ ♄
21 39 ☽ △ ♂
22 25 ☽ ⊥ ♄

10 Saturday
00 29 ☽ # ☿
05 39 ☽ ⚺ ♃
06 36 ☽ ⚻ ☉
09 25 ☽ △ ♇
15 19 ☽ ∠ ♄
17 31 ☽ ☍ ♀
18 53 ☽ ± ☉
20 59 ☽ # ♆
22 00 ☽ △ ☿

11 Sunday
03 20 ☽ ⚼ ♂
03 35 ☽ ☍ ♅
06 24 ♀ ∥ ♇
09 01 ♂ ∠ ♅
10 22 ☽ # ♇
10 45 ☽ # ♀
13 07 ☽ ⚼ ☉
18 44 ☉ ⚺ ♇
19 06 ☽ ✱ ♄
20 52 ☽ ± ♂
22 07 ☽ # ♅

12 Monday
01 00 ☽ ⚻ ♆
08 27 ☽ ⚻ ♂
12 02 ☽ ± ♆
13 18 ☽ ☌ ♃
16 59 ☽ Q ♄
17 10 ☽ □ ♇
18 59 ☽ △ ☉
20 20 ☽ ∥ ♂

13 Tuesday
01 01 ☽ ∥ ♃
04 26 ☽ ⚼ ♆
06 27 ☽ ⚻ ♀
08 01 ☽ □ ☿
10 46 ☽ ⚻ ♅
18 18 ☽ ± ♀
21 33 ☽ ± ♅

14 Wednesday
00 05 ☉ ∥ ☿
01 11 ☽ □ ♄
01 35 ☉ ⊥ ♀
01 39 ☽ Q ♇
07 20 ☽ △ ♆
11 53 ☽ ⚼ ♀
13 34 ☽ ⚼ ♅
16 56 ☽ ☍ ♂
18 05 ☽ Q ☿
18 54 ☽ ⚺ ♃
21 15 ☽ # ♃
22 52 ☽ ✱ ♇

15 Thursday
03 57 ☽ # ♂
04 46 ☽ □ ☉
05 21 ☽ ⊥ ♃
06 53 ☿ ✱ ♅
07 23 ♀ ☌ ♅
08 42 ☿ ✱ ♀
13 30 ☉ ⊥ ♅
15 49 ☽ △ ♅
16 24 ☽ ✱ ☿
16 35 ☽ △ ♀
20 53 ☽ ∠ ♃
22 13 ☽ ∥ ♅

16 Friday
00 17 ☽ ∥ ♀
00 53 ☽ ∠ ♇
05 11 ☽ △ ♄
09 17 ☽ ∥ ♇
09 56 ♂ ⚻ ♃
11 29 ☽ □ ♆
14 06 ☽ Q ☉
16 14 ☽ ⊥ ♇
19 49 ☽ ∠ ☿
21 02 ☽ ∥ ♆
22 18 ☽ ✱ ♃
22 55 ☽ ⚻ ♂

17 Saturday
02 22 ☽ ⚺ ♇
03 09 ♀ ∥ ♅
05 35 ♀ Q ♇
06 22 ☽ ⚼ ♄
09 25 ☽ ± ♂
11 48 ☽ ✱ ☉
12 02 ☽ ⊥ ☿
15 07 ☽ ∥ ☉
17 45 ☽ Q ♆
18 17 ☽ Q ♃
18 41 ☽ □ ♅
20 50 ☉ ⊥ ♇
21 16 ☽ ± ♄
22 12 ☽ ∥ ☿
22 45 ☽ ⚺ ☿
23 53 ☽ □ ♀

18 Sunday
01 02 ☽ ⚼ ♂
01 13 ☽ # ♄
07 07 ☽ ⚻ ♄
13 35 ☽ ✱ ♆
14 26 ☽ ∠ ☉
23 46 ☽ □ ♃

19 Monday
00 22 ☽ Q ♅
02 42 ☽ △ ♂
02 48 ♀ ∠ ♂
03 58 ☽ ☌ ♇
06 13 ☽ ⊥ ☉
08 01 ☽ Q ♀
14 06 ☽ ∠ ♆
16 41 ☽ ⚺ ☉
19 56 ☽ ✱ ♅

20 Tuesday
03 36 ☽ ☌ ☿
04 44 ☽ ⊥ ♆
05 14 ☽ ✱ ♀
07 43 ☽ ☍ ♄
10 02 ♂ △ ♇
14 29 ☽ ⚺ ♆
18 35 ☿ ⊥ ♆
20 20 ☽ ∠ ♅

21 Wednesday
00 19 ☽ △ ♃
04 46 ☽ ⚺ ♇
05 34 ☽ □ ♂
07 46 ☽ ∠ ♀
10 04 ♀ △ ♄
10 48 ♃ ± ♆
11 02 ☽ ⊥ ♅
14 35 ☽ ⊥ ♇
18 15 ☉ ⚺ ♅
20 54 ☽ ⚺ ♅
21 05 ☽ ☌ ☉
23 49 ☽ ⊥ ♀

22 Thursday
00 44 ☽ ⚼ ♃
04 59 ☿ ☍ ♄
05 24 ☽ ∠ ♇
08 28 ☽ ⚻ ♄
08 48 ☽ ⚺ ☿
10 38 ☽ ⚺ ♀
11 27 ☽ # ♄
11 42 ☽ ∥ ☿
12 28 ☽ Q ♂
15 19 ♀ # ♂
15 35 ☽ ± ♃
15 46 ☽ ☌ ♆
18 24 ☽ ± ♄
19 43 ☽ ⊥ ☿

23 Friday
01 35 ☽ ⚻ ♃
02 10 ☿ # ♄
04 13 ☽ ∥ ☉
06 32 ☽ ✱ ♇
09 29 ☽ ⚼ ♄
09 33 ☽ ✱ ♂
12 20 ☽ ∠ ☿
16 40 ☽ ∥ ♆
22 46 ☉ ⚼ ♃
23 28 ☽ ☌ ♅

24 Saturday
03 08 ☽ Q ♇
03 29 ☽ ⚺ ☉
04 22 ☽ ∥ ♇
11 15 ☽ △ ♄
12 42 ☽ ∠ ♂
14 49 ☽ ⊥ ☉
16 30 ☽ ∥ ♅
16 58 ☽ ✱ ☿
18 48 ☽ ☌ ♀
19 25 ☽ ⚺ ♆

25 Sunday
00 33 ☽ # ♂
01 49 ♀ ⚺ ♆
05 30 ☽ ⊥ ♂
05 30 ☽ ☍ ♃
05 34 ♂ ± ♃
06 10 ☽ ⊥ ♆
08 12 ☽ ∠ ☉
08 36 ☽ ∥ ♀
11 09 ☽ □ ♇
16 44 ☽ # ♃
16 51 ☽ Q ☿
16 56 ☽ ⚺ ♂
19 27 ☿ ⚺ ♆
20 40 ♂ Q ♆
22 38 ☽ ∠ ♆

26 Monday
05 28 ☽ ⚺ ♅
14 14 ☽ ✱ ☉
16 46 ☽ ⊥ ♅
17 38 ☽ □ ♄
23 39 ☉ ∠ ♇

27 Tuesday
02 54 ☽ ✱ ♆
06 29 ☽ □ ☿
07 58 ☽ ⚺ ♀
10 02 ☽ ∠ ♅
13 17 ☽ ⚻ ♃
15 13 ☽ Q ☉
15 15 ♂ Q ♄
18 12 ☽ ∥ ♃
19 49 ☽ △ ♇
20 51 ☽ ⊥ ♀
21 01 ☽ # ♀

28 Wednesday
00 55 ☽ ± ♃
02 15 ☽ Q ♆
04 12 ☽ Q ♄
04 59 ☽ ☌ ♂
06 25 ☉ ⚻ ♄
15 31 ☽ ∥ ♂
15 33 ☽ ✱ ♅
16 33 ☽ ∠ ♀
16 44 ☿ ∠ ♅
18 36 ☽ ⚼ ♃
19 57 ☽ # ♅

29 Thursday
01 25 ♀ # ♃
01 33 ☽ ⚼ ♇
03 56 ☽ ✱ ♄
06 03 ☽ □ ☉
10 05 ☽ # ♇
12 01 ♀ ☍ ♃
14 14 ☽ □ ♆
15 43 ☽ Q ♅
18 09 ☿ △ ♃
19 43 ☽ ± ♇
22 50 ☿ # ♄
23 16 ☽ # ♆

30 Friday
00 34 ☽ △ ♃
01 27 ☽ △ ☿
02 04 ☽ ✱ ♀
03 57 ☽ # ☉
05 29 ♀ ⊥ ♆
07 54 ☽ ⚻ ♇
10 05 ☽ ∠ ♄
20 42 ☽ ⚺ ♂

31 Saturday
04 18 ☽ ⊥ ♄
04 27 ☽ □ ♅
05 14 ☽ Q ♀
09 37 ☽ ⊥ ♂
10 02 ☽ # ☿
10 54 ☽ ∥ ♄
11 08 ☿ ✱ ♀
12 04 ☽ ⚼ ☿
16 28 ☽ ⚺ ♄

FEBRUARY 2004

Date	Sidereal time h m s	Sun ☉ LONGITUDES ° ' "	Moon ☽ ° ' "	Moon ☽ 24.00 ° ' "	Mercury ☿ ° '	Venus ♀ ° '	Mars ♂ ° '	Jupiter ♃ ° '	Saturn ♄ ° '	Uranus ♅ ° '	Neptune ♆ ° '	Pluto ♇ ° '
01	20 44 11	11 ♒ 57 34	17 ♊ 03 43	23 ♊ 00 04	21 ♑ 41	21 ♓ 28	28 ♈ 47	17 ♍ 38	07 ♋ 24	01 ♓ 37	12 ♒ 50	21 ♐ 31
02	20 48 08	12 58 27	28 ♊ 58 27	04 ♋ 59 19	23 07	22 40	29 ♈ 25	17 R 32	07 R 20	01 40	12 52	21 32
03	20 52 04	13 59 19	11 ♋ 03 02	17 ♋ 09 55	24 34	23 51	00 ♉ 03	17 27	07 17	01 44	12 55	21 34
04	20 56 01	15 00 10	23 ♋ 20 11	29 ♋ 34 00	26 01	25 03	00 41	17 21	07 13	01 47	12 57	21 35
05	20 59 57	16 00 59	05 ♌ 51 26	12 ♌ 12 30	27 30	26 14	01 19	17 16	07 10	01 50	12 59	21 37
06	21 03 54	17 01 47	18 ♌ 37 09	25 ♌ 05 16	28 ♑ 59	27 25	01 57	17 10	07 07	01 54	13 01	21 38
07	21 07 50	18 02 34	01 ♍ 36 42	08 ♍ 11 17	00 ♒ 29	28 36	02 36	17 04	07 04	01 57	13 04	21 40
08	21 11 47	19 03 20	14 ♍ 49 49	21 ♍ 29 05	02 00	29 ♓ 47	03 14	16 58	07 01	02 00	13 06	21 41
09	21 15 44	20 04 04	28 ♍ 11 54	04 ♎ 57 06	03 31	00 ♈ 58	03 52	16 52	06 58	02 04	13 08	21 43
10	21 19 40	21 04 48	11 ♎ 44 33	18 ♎ 34 09	05 04	02 09	04 30	16 45	06 55	02 07	13 10	21 44
11	21 23 37	22 05 30	25 ♎ 25 49	02 ♏ 19 32	06 37	03 19	05 09	16 39	06 52	02 11	13 13	21 46
12	21 27 33	23 06 12	09 ♏ 15 17	16 ♏ 13 03	08 12	04 30	05 47	16 32	06 49	02 14	13 15	21 47
13	21 31 30	24 06 52	23 ♏ 12 52	00 ♐ 14 41	09 47	05 40	06 25	16 25	06 47	02 17	13 17	21 48
14	21 35 26	25 07 31	07 ♐ 18 28	14 ♐ 24 07	11 22	06 51	07 03	16 19	06 44	02 21	13 20	21 49
15	21 39 23	26 08 10	21 ♐ 31 26	28 ♐ 40 11	12 59	08 01	07 42	16 12	06 42	02 24	13 22	21 51
16	21 43 19	27 08 47	05 ♑ 49 59	13 ♑ 00 23	14 37	09 11	08 20	16 05	06 40	02 28	13 24	21 52
17	21 47 16	28 09 23	20 ♑ 10 52	27 ♑ 20 48	16 15	10 20	08 58	15 58	06 38	02 31	13 26	21 53
18	21 51 13	29 ♒ 09 58	04 ♒ 29 29	11 ♒ 36 14	17 55	11 30	09 37	15 50	06 36	02 35	13 28	21 54
19	21 55 09	00 ♓ 10 31	18 ♒ 40 20	25 ♒ 41 03	19 35	12 40	10 15	15 43	06 34	02 38	13 31	21 55
20	21 59 06	01 11 02	02 ♓ 37 47	09 ♓ 29 59	21 16	13 49	10 53	15 36	06 32	02 42	13 33	21 57
21	22 03 02	02 11 32	16 ♓ 17 11	22 ♓ 59 04	22 58	14 58	11 32	15 28	06 30	02 45	13 35	21 58
22	22 06 59	03 12 01	29 ♓ 35 26	06 ♈ 06 14	24 41	16 08	12 10	15 21	06 28	02 49	13 37	21 59
23	22 10 55	04 12 27	12 ♈ 31 31	18 ♈ 51 29	26 25	17 17	12 49	15 13	06 27	02 52	13 39	22 00
24	22 14 52	05 12 52	25 ♈ 06 26	01 ♉ 16 45	28 10	18 25	13 27	15 06	06 25	02 55	13 42	22 01
25	22 18 48	06 13 15	07 ♉ 22 56	13 ♉ 25 30	29 ♒ 56	19 34	14 06	14 58	06 24	02 59	13 44	22 02
26	22 22 45	07 13 36	19 ♉ 25 04	25 ♉ 22 15	01 ♓ 43	20 42	14 44	14 50	06 23	03 02	13 46	22 03
27	22 26 42	08 13 55	01 ♊ 17 45	07 ♊ 12 15	03 30	21 51	15 22	14 42	06 22	03 06	13 48	22 03
28	22 30 38	09 14 12	13 ♊ 06 24	19 ♊ 00 56	05 19	22 59	16 01	14 35	06 21	03 09	13 50	22 04
29	22 34 35	10 ♓ 14 27	24 ♊ 56 29	00 ♋ 53 43	07 ♓ 09	24 ♈ 07	16 ♉ 39	14 ♍ 27	06 ♋ 20	03 ♓ 13	13 ♒ 52	22 ♐ 05

Date	Moon True ☊ ° '	Moon Mean ☊ ° '	Moon ☽ Latitude ° '
01	16 ♉ 33	16 ♉ 02	02 N 34
02	16 R 27	15 59	03 24
03	16 18	15 56	04 06
04	16 07	15 53	04 37
05	15 53	15 50	04 55
06	15 39	15 46	04 59
07	15 26	15 43	04 48
08	15 14	15 40	04 21
09	15 05	15 37	03 40
10	14 59	15 34	02 46
11	14 55	15 30	01 41
12	14 54	15 27	00 N 30
13	14 D 54	15 24	00 S 44
14	14 R 54	15 21	01 56
15	14 52	15 18	03 01
16	14 47	15 15	03 55
17	14 40	15 11	04 34
18	14 30	15 08	04 57
19	14 18	15 05	05 00
20	14 06	15 02	04 46
21	13 55	14 59	04 15
22	13 45	14 56	03 31
23	13 39	14 52	02 37
24	13 34	14 49	01 37
25	13 33	14 46	00 S 33
26	13 D 33	14 43	00 N 31
27	13 33	14 40	01 33
28	13 R 33	14 36	02 31
29	13 ♉ 32	14 ♉ 33	03 N 22

DECLINATIONS

Date	Sun ☉ ° '	Moon ☽ ° '	Mercury ☿ ° '	Venus ♀ ° '	Mars ♂ ° '	Jupiter ♃ ° '	Saturn ♄ ° '	Uranus ♅ ° '	Neptune ♆ ° '	Pluto ♇ ° '
01	17 S 12	25 N 22	22 S 24	04 S 14	11 N 38	06 N 08	22 N 38	11 S 35	16 S 59	14 S 31
02	16 55	26 50	22 17	03 43	11 52	06 10	22 38	11 33	16 59	14 31
03	16 38	27 04	22 09	03 12	12 07	06 12	22 39	11 32	16 58	14 31
04	16 20	25 58	21 59	02 41	12 21	06 15	22 39	11 31	16 58	14 31
05	16 02	23 34	21 48	02 09	12 35	06 17	22 39	11 30	16 57	14 31
06	15 44	19 59	21 36	01 38	12 49	06 20	22 40	11 29	16 56	14 31
07	15 25	15 23	21 22	01 07	13 03	06 22	22 40	11 27	16 56	14 31
08	15 07	10 00	21 07	00 35	13 17	06 25	22 40	11 26	16 55	14 31
09	14 48	04 N 05	20 51	00 S 04	13 30	06 27	22 40	11 25	16 54	14 31
10	14 28	02 S 06	20 33	00 N 28	13 44	06 30	22 41	11 24	16 54	14 31
11	14 09	08 16	20 15	00 59	13 57	06 33	22 41	11 22	16 53	14 31
12	13 49	14 06	19 54	01 31	14 11	06 35	22 41	11 21	16 52	14 31
13	13 29	19 17	19 33	02 02	14 24	06 38	22 42	11 20	16 52	14 31
14	13 09	23 26	19 10	02 34	14 37	06 41	22 42	11 19	16 51	14 31
15	12 48	26 11	18 46	03 05	14 51	06 44	22 42	11 18	16 51	14 31
16	12 28	27 13	18 20	03 36	15 04	06 47	22 42	11 16	16 50	14 31
17	12 07	26 27	17 53	04 07	15 17	06 50	22 43	11 15	16 49	14 30
18	11 46	23 56	17 25	04 38	15 29	06 53	22 43	11 14	16 49	14 30
19	11 25	19 59	16 55	05 09	15 42	06 56	22 43	11 13	16 48	14 30
20	11 03	14 59	16 24	05 40	15 55	06 59	22 43	11 11	16 47	14 30
21	10 42	09 20	15 51	06 11	16 07	07 02	22 44	11 10	16 47	14 30
22	10 20	03 S 23	15 18	06 42	16 20	07 05	22 44	11 09	16 46	14 30
23	09 58	02 N 32	14 42	07 12	16 32	07 08	22 44	11 08	16 46	14 30
24	09 36	08 13	14 06	07 43	16 44	07 11	22 44	11 06	16 45	14 30
25	09 14	13 28	13 28	08 13	16 56	07 14	22 45	11 05	16 44	14 30
26	08 51	18 05	12 49	08 43	17 08	07 17	22 45	11 04	16 44	14 29
27	08 29	21 57	12 08	09 13	17 20	07 20	22 45	11 03	16 43	14 29
28	08 06	24 52	11 26	09 42	17 32	07 23	22 45	11 01	16 43	14 29
29	07 S 44	26 N 43	10 S 43	10 N 12	17 N 43	07 N 26	22 N 45	11 S 00	16 S 42	14 S 29

ZODIAC SIGN ENTRIES

Date	h	m	Planets
02	14	03	☽ ♋
03	10	04	♂ ♉
05	00	50	☽ ♌
07	04	20	☿ ♒
07	09	03	☽ ♍
08	16	20	♀ ♈
09	15	12	☽ ♎
11	19	58	☽ ♏
13	23	35	☽ ♐
16	02	14	☽ ♑
18	04	27	☽ ♒
19	07	50	☉ ♓
20	07	27	☽ ♓
22	12	45	☽ ♈
24	21	30	☽ ♉
25	12	58	☿ ♓
27	09	22	☽ ♊
29	22	12	☽ ♋

LATITUDES

Date	Mercury ☿ ° '	Venus ♀ ° '	Mars ♂ ° '	Jupiter ♃ ° '	Saturn ♄ ° '	Uranus ♅ ° '	Neptune ♆ ° '	Pluto ♇ ° '
01	00 S 43	00 S 56	00 N 38	01 N 21	00 S 36	00 S 43	00 S 02	08 N 40
04	01 03	00 46	00 41	01 21	00 36	00 43	00 02	08 40
07	01 21	00 36	00 43	01 22	00 35	00 43	00 02	08 40
10	01 36	00 25	00 45	01 23	00 35	00 43	00 02	08 41
13	01 49	00 14	00 47	01 23	00 34	00 43	00 02	08 41
16	01 58	00 S 02	00 49	01 24	00 34	00 43	00 02	08 42
19	02 04	00 N 10	00 51	01 24	00 33	00 43	00 02	08 42
22	02 07	00 23	00 53	01 25	00 33	00 43	00 02	08 43
25	02 06	00 36	00 54	01 25	00 33	00 43	00 02	08 43
28	02 01	00 50	00 56	01 25	00 32	00 43	00 02	08 44
31	01 S 51	01 N 04	00 N 58	01 N 26	00 S 32	00 S 43	00 S 02	08 N 45

DATA

Julian Date	2453037
Delta T	+65 seconds
Ayanamsa	23° 54' 40"
Synetic vernal point	05° ♓ 12' 19"
True obliquity of ecliptic	23° 26' 26"

MOON'S PHASES, APSIDES AND POSITIONS ☽

Date	h	m	Phase	Longitude ° '	Eclipse Indicator
06	08	47	○	16 ♌ 54	
13	13	40	☾	24 ♏ 11	
20	09	18	●	01 ♓ 04	
28	03	24	☽	08 ♊ 53	

Day	h	m	
16	07	30	Perigee
28	10	43	Apogee

Day	h	m		
03	04	12	Max dec	27° N 08'
10	03	53	0S	
16	13	48	Max dec	27° S 14'
23	01	39	0N	

LONGITUDES

Date	Chiron ⚷ ° '	Ceres ⚳ ° '	Pallas ⚴ ° '	Juno ⚵ ° '	Vesta ⚶ ° '	Black Moon Lilith ⚸ ° '
01	21 ♑ 25	13 ♋ 43	20 ♈ 35	02 ♑ 59	23 ♑ 03	09 ♊ 29
11	22 ♑ 17	12 ♋ 22	24 ♈ 18	06 ♑ 11	28 ♑ 17	10 ♊ 35
21	23 ♑ 06	11 ♋ 41	27 ♈ 23	09 ♑ 15	03 ♒ 28	11 ♊ 42
31	23 ♑ 51	11 ♋ 44	02 ♉ 46	12 ♑ 09	08 ♒ 33	12 ♊ 49

ASPECTARIAN

h m Aspects

01 Sunday
00 40 ☽ △ ☉
03 24 ☽ △ ♆
04 45 ☉ ± ♃
04 58 ☽ ∠ ♂
06 25 ♂ # ♅
08 51 ☽ ± ☿
08 56 ☿ ⊻ ♇
12 57 ♀ □ ♇
13 08 ☽ □ ♃
21 01 ☽ ☍ ♇
21 54 ☽ □ ♀
22 38 ☽ ⊼ ☿

02 Monday
06 52 ☉ ∥ ♆
09 29 ☉ ☌ ♆
09 47 ☽ bQ ♆
09 49 ☽ bQ ☉
12 56 ☽ ⚹ ♂
17 25 ☽ △ ♅
20 05 ☉ ± ♄

03 Tuesday
01 00 ☽ Q ♃
03 47 ☽ ± ♆
04 35 ☽ ☌ ♄
05 24 ☽ ± ☉
14 05 ☽ Q ♂
15 40 ☽ ⊼ ♆
18 18 ☽ ⊼ ☉
20 35 ♀ ⊥ ♂
23 12 ☽ bQ ♅

04 Wednesday
00 28 ☽ ⚹ ♃
07 57 ☿ ⊥ ♅
08 37 ☽ ⊼ ♇
15 39 ☽ △ ♀
16 45 ☽ ± ♅
17 53 ☽ ☍ ☿
20 13 ☽ ± ♇

05 Thursday
02 54 ☽ □ ♂
04 19 ☽ ⊼ ♅
05 12 ☽ ∠ ♃
13 27 ☽ bQ ♇
14 02 ☿ ⊥ ♇
14 28 ☽ ⊻ ♄
18 54 ☽ ∥ ♄
22 08 ☽ ⊥ ♃
23 12 ☽ bQ ♀

06 Friday
01 30 ☽ ☍ ♆
01 31 ☽ # ☿
01 45 ☽ ⊥ ♄
08 47 ☽ ☍ ☉
09 18 ☽ ⊻ ♃
09 22 ♂ ⚹ ♅
14 54 ☉ ⊼ ♃
17 38 ☽ △ ♇
17 44 ☽ ± ♀
18 28 ☽ ∠ ♄
18 44 ♂ bQ ♃

07 Saturday
00 40 ♀ ∠ ♆
04 26 ☽ # ♆
05 56 ☽ ⊼ ♀
09 39 ☽ ⊼ ☿
11 47 ☽ # ☉
11 51 ♄ ± ♆
12 37 ☽ ☍ ♅
13 53 ☽ △ ♂
16 03 ☽ # ♇
21 55 ☽ ⚹ ♄
22 03 ☽ ± ☿
22 20 ☽ ∥ ♂

08 Sunday
05 50 ☽ # ♅
08 54 ☽ ⊼ ♆
11 31 ☿ bQ ♃
12 11 ☿ ⊻ ♅
15 51 ☽ ☌ ♃
16 26 ☽ bQ ☿
18 28 ☽ bQ ♂
19 32 ☽ Q ♄
19 44 ☽ ± ♆
20 16 ☽ ⊼ ☉

09 Monday
00 23 ☽ □ ♇
02 37 ☽ ∥ ♃
07 53 ☽ ± ☉
11 23 ☽ ± ♂
11 53 ☽ bQ ♆
17 24 ☽ ☍ ♀
18 54 ☽ ⊼ ♅
21 08 ☿ □ ♂
22 35 ☽ ⊼ ♂
22 41 ☽ △ ☿

10 Tuesday
01 11 ☽ bQ ☉
02 52 ☽ ∥ ♀
03 30 ☽ □ ♄
05 06 ☽ # ♀
05 35 ☽ ± ♅
08 27 ☽ Q ♇
08 40 ☉ ∥ ♇
11 26 ♀ ⊻ ♅
14 32 ☽ △ ♆
20 45 ☽ ⊻ ♃
21 30 ☽ bQ ♅

11 Wednesday
03 56 ☉ ⚹ ♇
05 10 ☽ # ♃
05 35 ☽ ⚹ ♇
05 42 ☽ △ ☉
06 55 ☉ bQ ♄
07 10 ☽ ⊥ ♃
14 08 ☿ ∠ ♇
15 40 ☿ ⊼ ♄
20 17 ☉ # ♂
22 44 ☽ ∠ ♃
23 47 ☽ △ ♅

12 Thursday
00 28 ☽ ∥ ♅
03 00 ☽ ⊼ ♀
05 42 ☽ ☍ ♂
07 43 ☽ ∠ ♇
07 49 ☽ △ ♄
09 56 ☽ □ ☿
10 49 ☽ ∥ ☉
12 20 ☽ # ♂
13 46 ☽ ∥ ♇
14 21 ☽ ± ♀
18 55 ☽ □ ♆
23 16 ☽ ⊥ ♇

13 Friday
00 19 ☽ ∥ ♆
00 27 ☽ ⚹ ♃
07 15 ☽ bQ ♀
09 33 ☽ bQ ♄
09 35 ☽ ⊻ ♇
13 14 ☽ ∥ ☿
13 40 ☽ □ ☉
20 47 ☽ Q ☿
20 49 ☽ Q ♃
21 06 ☿ ± ♃
23 57 ♂ # ♇

14 Saturday
00 50 ♂ ⚹ ♄
00 52 ☽ ± ♄
01 49 ☽ Q ♆
03 01 ♂ bQ ♇
03 33 ☽ □ ♅
07 10 ☽ # ♄
07 31 ☉ Q ♂
09 58 ♀ □ ♄
11 02 ☽ ⊼ ♄
11 08 ☽ △ ♀
11 33 ☽ ⊼ ♂
19 46 ☽ ⚹ ☿
21 39 ♀ ⊻ ♂
22 11 ☽ ± ♂
22 13 ☽ ⚹ ♆
22 36 ☽ Q ☉

15 Sunday
03 06 ☽ □ ♃
07 52 ☿ ± ♄
10 07 ☽ Q ♅
12 33 ☽ ☌ ♇
14 04 ☽ bQ ♂
17 43 ☿ ☌ ♆
20 20 ☽ ⚹ ☉
20 32 ♀ ⊥ ♅
23 31 ☽ ∠ ♆

16 Monday
00 14 ☽ ∠ ☿
06 20 ☽ ⚹ ♅
13 23 ☽ ☍ ♄
14 38 ☽ ⊥ ♆
16 23 ☽ △ ♂
17 15 ☽ ⊥ ☿
18 05 ☽ □ ♀
23 22 ☽ ∠ ☉

17 Tuesday
00 41 ☽ ⊻ ♆
04 35 ☽ ⊻ ☿
05 00 ☽ △ ♃
07 32 ☽ ∠ ♅
08 00 ☿ ⊼ ♃
14 51 ☽ ⊻ ♇
15 33 ☽ ⊥ ☉
22 40 ☽ ⊥ ♅

18 Wednesday
00 55 ☽ ⊥ ♇
02 22 ☽ ⊻ ☉
02 53 ☽ Q ♀
05 55 ☽ bQ ♃
08 46 ☽ ⊻ ♅
15 32 ☽ ⊼ ♄
16 04 ☽ ∠ ♇
20 16 ☽ # ♄
20 56 ☽ ± ♃
21 02 ☽ □ ♂

19 Thursday
00 53 ☽ ⚹ ♀
01 39 ☽ ± ♄
03 12 ☽ ☌ ♆
07 01 ☽ ⊼ ♃
13 45 ☽ ☌ ☿
16 55 ☽ bQ ♄
17 29 ☿ ∥ ♆
17 34 ☽ ⚹ ♇

20 Friday
02 27 ☉ ∥ ♅
03 44 ☽ ∥ ♆
04 48 ☽ ∠ ♀
04 51 ☽ ∥ ☿
05 13 ☽ Q ♂
06 09 ♀ ⚹ ♆
07 57 ☽ # ♂
09 18 ☽ ☌ ☉
12 07 ☽ ☌ ♅
14 06 ☽ ∥ ♇
14 17 ☽ Q ♇
15 41 ☿ bQ ♄
18 47 ☽ △ ♄
21 42 ☿ ⚹ ♇
21 53 ☽ ⊥ ♀

21 Saturday
03 10 ☽ ⚹ ♂
03 37 ☿ # ♂
04 21 ☽ ∥ ♅
05 58 ☽ ∥ ☉
07 11 ☽ ⊻ ♆
09 27 ☽ ⊻ ♀
10 34 ☽ ☍ ♃
17 54 ☽ ⊥ ♆
21 17 ☽ # ♃
21 19 ♀ ⊼ ♃
22 10 ☽ □ ♇
23 45 ☽ # ♀

22 Sunday
01 44 ☽ ⊻ ☿
02 07 ☉ ☌ ♅
07 22 ☽ ∠ ♂
10 13 ☽ ∠ ♆
14 18 ☽ ⊥ ☿
17 56 ☽ ⊻ ♅
19 11 ☽ ⊻ ☉

23 Monday
00 40 ☽ □ ♄
00 45 ☽ ⊥ ♂
05 07 ☽ ⊥ ♅
06 51 ☉ Q ♇
07 17 ☽ ⊥ ☉
07 57 ♀ ∥ ♃
09 35 ☽ ∠ ☿
12 34 ☽ ⊻ ♂
14 08 ☽ ⚹ ♆
17 02 ☽ ⊼ ♃
20 23 ☿ ∥ ♇
21 53 ☽ ☌ ♀
22 09 ☽ ∠ ♅

24 Tuesday
01 01 ♀ ∠ ♅
01 46 ☽ ∠ ☉
04 21 ☽ ± ♃
06 02 ☽ △ ♇
07 26 ☽ ∥ ♃
09 33 ☽ ∥ ♀
10 41 ☽ Q ♄
13 08 ☽ Q ♆
13 44 ♂ # ♆
17 42 ☽ # ☉
18 55 ☽ ⚹ ☿
21 32 ♂ □ ♆
21 35 ☽ bQ ♃

25 Wednesday
00 52 ☽ # ♅
03 18 ☽ ⚹ ♅
09 30 ☽ ⚹ ☉
10 04 ☽ ⚹ ♄
11 18 ☽ bQ ♇
12 01 ☽ # ☿
16 15 ☉ △ ♄
17 04 ☽ # ♇
22 35 ☽ Q ☿

26 Thursday
00 39 ☽ □ ♆
02 05 ☽ ☌ ♂
02 55 ☽ △ ♃
03 11 ☽ Q ♅
04 35 ☽ # ♆
05 13 ☽ ± ♇
06 30 ☽ ∥ ♂
11 35 ☽ Q ☉
14 29 ♀ ± ♃
14 52 ☽ ⊻ ♀
15 13 ♂ △ ♃
15 57 ☽ ∠ ♄
16 01 ☉ # ♀
17 17 ☽ ⊼ ♇

27 Friday
00 34 ♂ Q ♅
04 16 ☽ ⊥ ♀
06 21 ☿ ☌ ♅
09 16 ☿ Q ♂
10 07 ☽ ⊥ ♄
15 40 ☽ □ ♅
16 34 ♀ △ ♇
17 18 ☽ □ ☿
17 53 ☽ ∥ ♄
19 21 ☿ Q ♇
22 17 ☽ ⊻ ♄

28 Saturday
00 28 ☽ ∠ ♀
03 24 ☽ □ ☉
13 29 ☽ △ ♆
14 07 ♂ ± ♇
14 57 ☽ □ ♃
18 15 ☽ ⊻ ♂

29 Sunday
01 23 ☿ △ ♄
02 14 ☿ ∥ ♅
06 13 ☽ ☍ ♇
07 07 ☽ ⊥ ♂
10 08 ☽ ⚹ ♀
16 44 ♀ Q ♄
19 57 ☽ bQ ♆
22 15 ☿ # ♀

All ephemeris data is given at 12.00 UT and the Moon's longitude is additionally given for 24.00 UT

MARCH 2004

Date	Sidereal time h	m	s	Sun ☉ °		'	"	Moon ☽ °		'	"	Moon ☽ 24.00 °		'	"	Mercury ☿ °	'	Venus ♀ °	'	Mars ♂ °	'	Jupiter ♃ °	'	Saturn ♄ °	'	Uranus ♅ °	'	Neptune ♆ °	'	Pluto ♇ °	'
01	22	38	31	11	♓	14	40	06	♋	53	14	12	♋	55	36	09 ♓	00	25 ♈	14	17 ♉	18	14 ♍	19	06 ♋	19	03 ♓	16	13 ♒	54	22 ♐	06
02	22	42	28	12		14	51	19	♋	01	20	25	♋	10	52	10	52	26	22	17	56	14 R	11	06 R	19	03	20	13	56	22	07
03	22	46	24	13		15	00	01	♌	24	34	07	♌	42	44	12	45	27	29	18	35	14	03	06	18	03	23	13	58	22	07
04	22	50	21	14		15	07	14	♌	05	32	20	♌	33	05	14	38	28	36	19	13	13	56	06	18	03	26	14	00	22	08
05	22	54	17	15		15	12	27	♌	05	23	03	♍	42	19	16	33	29 ♈	43	19	52	13	48	06	17	03	30	14	02	22	09
06	22	58	14	16		15	15	10	♍	23	41	17	♍	09	14	18	28	00 ♉	49	20	30	13	40	06	17	03	33	14	04	22	09
07	23	02	11	17		15	16	23	♍	58	37	00	♎	51	24	20	24	01	56	21	09	13	32	06	17	03	37	14	06	22	10
08	23	06	07	18		15	15	07	♎	47	11	14	♎	45	29	22	21	03	02	21	47	13	24	06 D	17	03	40	14	08	22	10
09	23	10	04	19		15	13	21	♎	45	51	28	♎	47	52	24	18	04	08	22	26	13	16	06	17	03	43	14	10	22	11
10	23	14	00	20		15	08	05	♏	51	06	12	♏	55	11	26	16	05	13	23	04	13	09	06	18	03	47	14	12	22	11
11	23	17	57	21		15	02	19	♏	59	49	27	♏	04	43	28 ♓	13	06	19	23	42	13	01	06	18	03	50	14	14	22	12
12	23	21	53	22		14	55	04	♐	09	38	11	♐	14	24	00 ♈	11	07	24	24	21	12	53	06	18	03	53	14	16	22	12
13	23	25	50	23		14	46	18	♐	18	49	25	♐	22	45	02	09	08	29	24	59	12	46	06	19	03	57	14	18	22	13
14	23	29	46	24		14	35	02	♑	26	01	09	♑	28	28	04	06	09	33	25	38	12	38	06	20	04	00	14	20	22	13
15	23	33	43	25		14	23	16	♑	29	54	23	♑	30	06	06	02	10	38	26	16	12	31	06	21	04	03	14	22	22	13
16	23	37	40	26		14	08	00	♒	28	51	07	♒	25	52	07	57	11	42	26	55	12	23	06	22	04	07	14	24	22	14
17	23	41	36	27		13	53	14	♒	20	50	21	♒	13	27	09	50	12	45	27	33	12	16	06	23	04	10	14	25	22	14
18	23	45	33	28		13	35	28	♒	03	23	04	♓	50	17	11	41	13	49	28	12	12	08	06	24	04	13	14	27	22	14
19	23	49	29	29	♓	13	15	11	♓	33	53	18	♓	13	51	13	30	14	52	28	50	12	01	06	25	04	16	14	29	22	15
20	23	53	26	00	♈	12	54	24	♓	49	59	01	♈	22	05	15	16	15	55	29 ♉	29	11	54	06	26	04	19	14	31	22	15
21	23	57	22	01		12	30	07	♈	50	02	14	♈	13	46	16	59	16	57	00 ♊	07	11	47	06	28	04	23	14	32	22	15
22	00	01	19	02		12	04	20	♈	33	20	26	♈	48	49	18	38	17	59	00	45	11	40	06	29	04	26	14	34	22	15
23	00	05	15	03		11	37	03	♉	00	25	09	♉	08	21	20	13	19	01	01	24	11	33	06	31	04	29	14	36	22	15
24	00	09	12	04		11	07	15	♉	12	58	21	♉	14	39	21	43	20	02	02	02	11	26	06	33	04	32	14	37	22	15
25	00	13	09	05		10	35	27	♉	13	49	03	♊	11	00	23	09	21	03	02	41	11	19	06	35	04	35	14	39	22 R	15
26	00	17	05	06		10	00	09	♊	06	42	15	♊	01	30	24	28	22	04	03	19	11	13	06	37	04	38	14	41	22	15
27	00	21	02	07		09	24	20	♊	56	00	26	♊	50	49	25	42	23	04	03	58	11	06	06	39	04	41	14	42	22	15
28	00	24	58	08		08	45	02	♋	46	36	08	♋	43	58	26	50	24	04	04	36	11	00	06	41	04	44	14	44	22	15
29	00	28	55	09		08	04	14	♋	43	33	20	♋	45	58	27	51	25	03	05	14	10	54	06	44	04	47	14	45	22	14
30	00	32	51	10		07	21	26	♋	51	49	03	♌	01	38	28	46	26	02	05	53	10	48	06	46	04	50	14	47	22	14
31	00	36	48	11	♈	06	35	09	♌	15	56	15	♌	35	08	29 ♈	34	27 ♉	01	06 ♊	31	10 ♍	42	06 ♋	49	04 ♓	53	14 ♒	48	22 ♐	14

Date	Moon True ☊ °		'	Moon Mean ☊ °		'	Moon ☽ Latitude °		'
01	13	♉	29	14	♉	36	04	N	05
02	13	R	23	14		27	04		38
03	13		15	14		24	04		58
04	13		05	14		21	05		04
05	12		55	14		17	04		55
06	12		44	14		14	04		30
07	12		35	14		11	03		49
08	12		29	14		08	02		54
09	12		24	14		05	01		48
10	12		23	14		02	00	N	35
11	12	D	23	13		58	00	S	41
12	12		24	13		55	01		55
13	12		24	13		52	03		01
14	12	R	24	13		49	03		57
15	12		22	13		46	04		38
16	12		18	13		42	05		02
17	12		12	13		39	05		08
18	12		05	13		36	04		56
19	11		58	13		33	04		28
20	11		50	13		30	03		46
21	11		45	13		27	02		53
22	11		41	13		23	01		52
23	11		38	13		20	00	S	47
24	11	D	38	13		17	00	N	20
25	11		39	13		14	01		24
26	11		41	13		11	02		24
27	11		43	13		08	03		18
28	11		44	13		04	04		04
29	11	R	44	13		01	04		39
30	11		42	12		58	05		03
31	11	♉	39	12	♉	55	05	N	13

DECLINATIONS

Date	Sun ☉ °	'	Moon ☽ °	'	Mercury ☿ °	'	Venus ♀ °	'	Mars ♂ °	'	Jupiter ♃ °	'	Saturn ♄ °	'	Uranus ♅ °	'	Neptune ♆ °	'	Pluto ♇ °	'
01	07 S	21	27 N	20	09 S	58	10 N	41	17 N	55	07 N	29	22 N	46	10 S	59	16 S	41	14 S	29
02	06	58	26	40	09	13	11	10	18	06	07	32	22	46	10	58	16	41	14	29
03	06	35	24	41	08	26	11	38	18	17	07	35	22	46	10	56	16	40	14	29
04	06	12	21	27	07	37	12	07	18	28	07	38	22	46	10	55	16	40	14	28
05	05	49	17	06	06	48	12	35	18	39	07	42	22	46	10	54	16	39	14	28
06	05	25	11	50	05	57	13	03	18	50	07	45	22	47	10	53	16	39	14	28
07	05	02	05 N	54	05	06	13	30	19	01	07	48	22	47	10	52	16	38	14	28
08	04	39	00 S	25	04	13	13	58	19	11	07	51	22	47	10	50	16	37	14	28
09	04	15	06	48	03	20	14	25	19	21	07	54	22	47	10	49	16	37	14	28
10	03	52	12	55	02	26	14	51	19	32	07	57	22	47	10	48	16	36	14	27
11	03	28	18	24	01	31	15	18	19	42	08	00	22	47	10	47	16	36	14	27
12	03	05	22	51	00 S	35	15	44	19	52	08	03	22	47	10	45	16	35	14	27
13	02	41	25	56	00 N	20	16	09	20	01	08	06	22	47	10	44	16	35	14	27
14	02	17	27	22	01	16	16	35	20	11	08	09	22	48	10	43	16	34	14	27
15	01	54	27	01	02	12	17	00	20	21	08	11	22	48	10	42	16	34	14	26
16	01	30	24	57	03	08	17	24	20	30	08	14	22	48	10	41	16	33	14	26
17	01	06	21	26	04	04	17	48	20	39	08	17	22	48	10	40	16	33	14	26
18	00	42	16	47	04	58	18	12	20	48	08	20	22	48	10	38	16	32	14	26
19	00 S	19	11	22	05	52	18	36	20	57	08	23	22	48	10	37	16	32	14	26
20	00 N	05	05 S	31	06	45	18	59	21	06	08	25	22	48	10	36	16	31	14	26
21	00	29	00 N	27	07	36	19	21	21	14	08	28	22	48	10	35	16	31	14	25
22	00	53	06	18	08	25	19	43	21	23	08	31	22	48	10	34	16	30	14	25
23	01	16	11	47	09	13	20	05	21	31	08	33	22	48	10	33	16	30	14	25
24	01	40	16	43	09	58	20	26	21	39	08	36	22	48	10	32	16	29	14	25
25	02	03	20	54	10	41	20	47	21	47	08	38	22	49	10	31	16	29	14	25
26	02	27	24	12	11	21	21	08	21	55	08	41	22	49	10	29	16	28	14	24
27	02	50	26	26	11	58	21	28	22	02	08	43	22	49	10	28	16	28	14	24
28	03	14	27	28	12	33	21	47	22	10	08	45	22	49	10	27	16	27	14	24
29	03	37	27	15	13	04	22	06	22	17	08	48	22	49	10	26	16	27	14	24
30	04	01	25	44	13	32	22	24	22	24	08	50	22	49	10	25	16	27	14	24
31	04 N	24	22 N	57	13 N	57	22 N	42	22 N	31	08 N	52	22 N	49	10 S	24	16 S	26	14 S	23

ZODIAC SIGN ENTRIES

Date	h	m	Planets
03	09	18	☽ ♌
05	17	18	☽ ♍
05	18	12	♀ ♉
07	22	31	☽ ♎
10	02	03	☽ ♏
12	04	57	☽ ♐
12	09	44	☿ ♈
14	07	51	☽ ♑
16	11	10	☽ ♒
18	15	26	☽ ♓
20	06	49	☉ ♈
20	21	29	☽ ♈
21	07	39	♂ ♊
23	06	10	☽ ♉
25	17	35	☽ ♊
28	06	23	☽ ♋
30	18	07	☽ ♌

LATITUDES

Date	Mercury ☿ °	'	Venus ♀ °	'	Mars ♂ °	'	Jupiter ♃ °	'	Saturn ♄ °	'	Uranus ♅ °	'	Neptune ♆ °	'	Pluto ♇ °	'
01	01 S	55	00 N	59	00 N	57	01 N	25	00 S	32	00 S	43	00 S	02	08 N	44
04	01	42	01	13	00	59	01	26	00	31	00	43	00	02	08	45
07	01	24	01	27	01	00	01	26	00	31	00	43	00	02	08	46
10	01	01	01	41	01	01	01	26	00	30	00	43	00	02	08	46
13	00	33	01	56	01	02	01	26	00	30	00	43	00	02	08	47
16	00 S	01	02	10	01	03	01	26	00	29	00	43	00	02	08	47
19	00 N	35	02	24	01	05	01	26	00	29	00	44	00	02	08	48
22	01	12	02	38	01	06	01	26	00	29	00	44	00	02	08	48
25	01	49	02	52	01	06	01	26	00	28	00	44	00	02	08	49
28	02	22	03	05	01	07	01	25	00	28	00	44	00	02	08	50
31	02 N	48	03 N	18	01 N	08	01 N	25	00 S	27	00 S	44	00 S	03	08 N	50

DATA

Julian Date	2453066
Delta T	+65 seconds
Ayanamsa	23° 54' 44"
Synetic vernal point	05° ♓ 12' 15"
True obliquity of ecliptic	23° 26' 26"

MOON'S PHASES, APSIDES AND POSITIONS ☽

Date	h	m	Phase	Longitude °		'	Eclipse Indicator
06	23	14	○	16	♍	43	
13	21	01	☾	23	♐	37	
20	22	41	●	00	♈	39	
28	23	48	☽	08	♋	38	

Day	h	m		
12	04	03	Perigee	
27	07	03	Apogee	
01	11	48	Max dec	27° N 20'
08	10	26	0S	
14	19	16	Max dec	27° S 27'
21	10	09	0N	
28	19	52	Max dec	27° N 32'

LONGITUDES

Date	Chiron ⚷ °		'	Ceres ⚳ °		'	Pallas ⚴ °		'	Juno ⚵ °		'	Vesta ⚶ °		'	Black Moon Lilith ☾ °		'
01	23	♑	47	11	♋	42	02	♉	19	11	♑	52	08	♒	03	12	♊	42
11	24	♑	57	12	♋	22	06	♉	57	14	♑	37	13	♒	03	13	♊	49
21	25	♑	01	13	♋	41	11	♉	50	17	♑	07	17	♒	57	14	♊	55
31	25	♑	29	15	♋	32	16	♉	56	19	♑	22	22	♒	43	16	♊	02

All ephemeris data is given at 12.00 UT and the Moon's longitude is additionally given for 24.00 UT

ASPECTARIAN

h m	Aspects
01 Monday	
02 18	☽ ∠ ♂
02 57	☽ Q ♃
04 20	☉ # ♃
04 44	☽ △ ♅
10 52	☽ ☌ ♄
12 46	☽ Q ♀
14 02	☽ ± ♆
16 58	☽ △ ☿
21 27	☽ △ ☉
02 Tuesday	
01 58	☽ ⊼ ♆
02 27	♀ # ♅
02 36	☽ ⚹ ♃
02 41	♀ Q ♆
09 46	☽ ⚹ ♂
10 38	☽ bQ ♅
18 02	☽ ⊼ ♇
03 Wednesday	
03 42	☽ bQ ☿
03 42	☽ □ ♀
03 50	☿ ∠ ♀
04 13	☽ ± ♅
05 24	☽ bQ ☉
05 41	☽ ± ♇
07 32	☽ ∠ ♃
10 19	☽ Q ♂
15 47	☽ ⊼ ♅
21 19	☽ ⊻ ♄
22 54	☽ bQ ♇
23 57	☽ ± ☿
04 Thursday	
00 05	☽ ± ☉
00 12	♃ ⊼ ♆
00 31	☽ ⊥ ♃
01 43	☉ ☌ ☿
03 15	☽ ∥ ♄
03 36	☿ ☍ ♃
03 54	☿ ⊻ ♆
05 05	☉ ☍ ♃
05 55	☉ ⊻ ♆
08 38	☽ ⊥ ♄
11 27	☿ # ♃
11 42	☽ ⊻ ♃
11 50	☽ ☍ ♆
12 19	☽ ⊼ ☉
13 12	☽ ⊼ ☿
18 18	♀ bQ ♃
22 02	☽ □ ♂
05 Friday	
01 22	☽ ∠ ♄
02 56	☽ △ ♇
04 22	☽ ∥ ♂
14 12	☽ # ♆
17 13	☽ △ ♀
23 40	☽ ☍ ♅
06 Saturday	
00 27	☽ # ♇
04 39	☽ ⚹ ♄
07 11	☽ ∥ ♀
16 02	☽ # ♅
17 46	☽ ☌ ♃
18 34	☽ ⊼ ♆
22 30	☽ bQ ♀
23 14	☽ ☍ ☉
07 Sunday	
02 00	☽ Q ♄
04 38	☽ ∥ ♃
04 41	☽ ☍ ☿
05 11	☽ ± ♆
06 47	☽ △ ♂
08 18	☿ ⊥ ♆
08 49	☽ □ ♇
15 06	☉ ∥ ☿
15 35	☽ # ☉
15 37	☽ # ☿
15 43	☽ ± ♀
16 52	♄ St D
17 19	♂ ∠ ♄
20 59	☽ bQ ♆
08 Monday	
01 41	☿ ⚹ ♂
03 04	☽ ⊼ ♀
04 51	☽ ⊼ ♅
09 25	☽ □ ♄
09 52	☿ □ ♇
10 11	☽ bQ ♂
15 15	☽ ± ♅
16 07	☽ Q ♇
21 35	☽ ⊻ ♃
22 58	☽ △ ♆
09 Tuesday	
00 30	☽ ∥ ☿
02 25	☽ ± ♂
02 37	♀ ⚹ ♅
02 46	♂ ⊼ ♇
02 55	☽ ∥ ☉
06 46	☽ bQ ♅
07 23	☽ ⊼ ☉
07 47	☽ ⊥ ♃
12 43	☽ ⚹ ♇
13 11	☽ ⊼ ♂
14 36	♀ # ♇
16 12	☽ # ♃
17 02	☽ ⊼ ☿
18 25	☽ ± ☉
23 00	☽ ∠ ♃
10 Wednesday	
03 29	☽ ∥ ♅
04 54	☽ ± ☿
08 28	☽ △ ♅
10 49	☉ ⊥ ♆
10 50	☽ ☍ ♀
10 54	☽ bQ ☉
12 45	☽ △ ♄
14 17	☽ ∠ ♇
18 23	☽ ∥ ♇
19 51	☉ ∠ ♀
20 46	☽ # ♀
22 40	☽ bQ ☿
11 Thursday	
00 16	☽ ⚹ ♃
02 13	☽ □ ♆
03 42	☽ ∥ ♆
05 33	☽ ⊥ ♇
11 44	♀ ⚹ ♄
14 13	☽ bQ ♄
14 17	☽ △ ☉
15 44	☽ ⊻ ♇
18 35	☽ ☍ ♂
18 36	☽ # ♂
20 26	☽ Q ♃
12 Friday	
00 37	☿ ∠ ♆
04 11	☽ △ ☿
05 28	☽ ± ♄
07 44	♀ bQ ♇
08 47	☽ Q ♆
10 57	☉ □ ♇
11 32	☽ □ ♅
11 35	☽ # ♄
15 38	☽ ⊼ ♄
17 56	☽ ⊼ ♀
13 Saturday	
02 40	☽ □ ♃
04 57	☽ ± ♀
05 10	☽ ⚹ ♆
18 11	☽ Q ♅
18 37	☽ ☌ ♇
21 01	☽ □ ☉
21 10	☿ ⊥ ♀
21 30	☽ bQ ♀
23 53	☽ ⊼ ♂
14 Sunday	
06 43	☽ ∠ ♆
10 34	☽ ± ♂
10 49	☿ ⊻ ♅
11 32	♀ # ♆
14 41	☽ ⚹ ♅
15 17	☽ □ ☿
18 39	☽ ☍ ♄
22 04	☽ ⊥ ♆
15 Monday	
01 08	☽ △ ♀
02 38	☽ bQ ♂
05 14	☽ △ ♃
06 00	☽ Q ☉
06 18	☉ # ☿
08 20	☽ ⊻ ♆
15 57	☿ □ ♄
16 24	☽ ∠ ♅
21 49	☽ ⊻ ♇
16 Tuesday	
02 57	☽ Q ☿
04 08	☽ ⚹ ☉
05 34	☽ △ ♂
06 43	☽ bQ ♃
07 54	☽ ⊥ ♅
08 07	☽ ⊥ ♇
18 17	☽ ⊻ ♅
22 06	☽ ± ♃
22 09	☽ ⊼ ♄
23 39	☽ ∠ ♇
17 Wednesday	
02 00	♀ △ ♃
02 56	☽ ⚹ ☿
03 41	☽ # ♄
08 02	☽ ∠ ☉
08 24	☽ ⊼ ♃
08 34	☽ ± ♄
09 00	☽ □ ♀
12 08	☽ ☌ ♆
16 16	☽ # ♂
16 21	☿ ⊥ ♅
18 Thursday	
00 17	☽ bQ ♄
00 57	☽ ⊥ ☉
01 46	☽ ⚹ ♇
05 39	☽ # ♀
09 13	☽ ∠ ☿
09 48	☉ ⚹ ♂
12 15	☽ □ ♂
12 19	☽ ⊻ ☉
13 09	☽ ∥ ♆
17 31	☿ ⊼ ♃
19 11	☽ Q ♀
21 15	♅ Q ♇
22 44	☽ ∥ ♇
22 56	☽ Q ♇
22 56	☽ ☌ ♅
19 Friday	
02 47	☽ △ ♄
03 02	♀ □ ♆
03 38	☽ ⊥ ☿
12 48	☽ ☍ ♃
15 07	☽ ∥ ♅
16 01	☽ ⊻ ☿
17 15	☽ ⊻ ♆
18 26	☽ ⚹ ♀
18 29	☉ ∠ ♆
18 53	☿ ∠ ♂
21 57	☽ Q ♂
20 Saturday	
00 18	☽ # ♃
01 22	☿ ⚹ ♆
04 07	☽ ⊥ ♆
07 17	☽ □ ♇
07 40	☽ # ☿
19 40	♀ ± ♇
20 36	☽ ∠ ♆
20 57	☽ ⚹ ♂
22 03	♀ Q ♅
22 41	☽ ☌ ☉
21 Sunday	
00 08	☽ ∠ ♀
05 32	☽ ⊻ ♅
08 27	☽ # ☉
09 26	☽ □ ♄
10 35	☿ ⊻ ♀
12 06	☽ ∥ ☉
16 47	☽ ⊥ ♅
18 21	☽ ⊥ ♀
19 19	☽ ⊼ ♃
22 39	☿ ± ♃
22 Monday	
00 37	☽ ⚹ ♆
01 35	♂ ⊥ ♄
02 24	☽ ∠ ♂
06 33	☽ ± ♃
06 40	☽ ⊻ ♀
07 49	☽ ☌ ☿
09 51	☽ ∠ ♅
14 52	☿ ∥ ♃
15 14	☽ △ ♇
19 33	☽ Q ♄
20 29	☽ ⊥ ♂
21 31	☽ ∥ ♃
22 35	☽ ∥ ☿
23 33	☽ Q ♆
23 36	☽ bQ ♃
23 Tuesday	
00 16	☿ ∠ ♅
06 25	☽ # ♅
08 42	☽ ⊻ ♂
12 24	☽ ⊻ ☉
14 53	☽ ⚹ ♅
18 52	☽ ⚹ ♄
20 17	☽ bQ ♇
24 Wednesday	
00 26	☽ # ♇
01 11	☽ ⊥ ☉
04 35	☽ △ ♃
10 49	☽ □ ♆
10 50	☽ # ♆
14 03	☽ ± ♇
14 38	☽ Q ♅
15 09	♇ St R
20 36	☽ ∠ ☉
20 41	☿ △ ♇
20 54	☉ ⊻ ♅
22 29	☽ ☌ ♀
25 Thursday	
00 38	☽ ∠ ♄
02 00	☽ ⊼ ♇
02 43	☽ ⊻ ☿
06 15	☿ # ♅
11 10	☽ ∥ ♀
16 21	☽ ⊥ ☿
17 59	☽ ∥ ♂
18 46	☽ ⊥ ♄
23 36	☽ ☌ ♂
26 Friday	
00 54	♀ ∠ ♄
01 00	☽ ∥ ♄
02 54	☽ □ ♅
05 29	☽ ⚹ ☉
06 56	☽ ⊻ ♄
12 49	☽ ∠ ☿
14 47	☿ Q ♄
16 14	☽ □ ♃
16 19	♀ ⊼ ♇
23 15	☉ □ ♄
23 19	☽ △ ♆
27 Saturday	
08 04	☽ Q ☉
14 40	☽ ☍ ♇
16 44	☽ ⊻ ♀
19 39	☿ bQ ♃
22 44	☽ ⚹ ☿
28 Sunday	
04 26	☽ Q ♃
05 49	☽ bQ ♆
06 01	☽ ⊥ ♀
09 38	☿ Q ♆
15 53	☽ ⊻ ♂
15 59	☽ △ ♅
17 40	♂ □ ♅
19 55	☽ ☌ ♄
23 48	☽ □ ☉
29 Monday	
00 01	☽ ± ♆
01 22	☽ Q ☿
01 49	☽ ∠ ♀
04 25	☽ ⚹ ♃
04 38	☽ ⊥ ♂
12 03	☽ ⊼ ♆
22 06	☽ bQ ♅
23 34	☽ ∠ ♂
30 Tuesday	
02 55	☽ ⊼ ♇
09 56	☽ ∠ ♃
10 14	☽ ⚹ ♀
11 28	♀ ∥ ♂
14 41	☽ ± ♇
15 52	☽ ± ♅
16 00	☽ □ ☿
31 Wednesday	
02 53	☉ ⊼ ♃
03 18	☽ ⊥ ♃
03 34	☽ ⊼ ♅
06 18	☉ ⊥ ♅
06 27	☽ ⚹ ♂
07 17	☽ ⊻ ♄
08 07	☽ bQ ♇
11 28	☽ Q ♀
13 01	☽ ∥ ♄
13 37	☽ ∥ ♀
14 43	☽ ⊻ ♃
14 57	☽ ∥ ♂
15 49	☽ △ ☉
18 46	☽ ⊥ ♄
20 36	♀ ∥ ♄
22 33	☽ ☍ ♆
23 42	♂ ⊻ ♄

APRIL 2004

LONGITUDES

Date	Sidereal time h m s	Sun ☉ ° ' "	Moon ☽ ° ' "	Moon ☽ 24.00 ° ' "	Mercury ☿ ° '	Venus ♀ ° '	Mars ♂ ° '	Jupiter ♃ ° '	Saturn ♄ ° '	Uranus ♅ ° '	Neptune ♆ ° '	Pluto ♇ ° '
01	00 40 44	12 ♈ 05 47	21 ♌ 59 36	28 ♌ 29 37	00 ♉ 16	27 ♉ 59	07 ♊ 09	10 ♍ 36	06 ♋ 51	04 ♓ 56	14 ♒ 50	22 ♐ 14
02	00 44 41	13 04 56	05 ♍ 05 19	11 ♍ 46 48	00 50	28 56	07 48	10 R 30	06 54	04 59	14 51	22 R 14
03	00 48 38	14 04 04	18 ♍ 33 57	25 ♍ 26 37	01 17	29 ♉ 53	08 26	10 25	06 57	05 02	14 52	22 13
04	00 52 34	15 03 09	02 ♎ 24 28	09 ♎ 27 05	01 36	00 ♊ 49	09 05	10 19	07 00	05 05	14 54	22 13
05	00 56 31	16 02 12	16 ♎ 33 56	23 ♎ 44 24	01 49	01 45	09 43	10 14	07 03	05 08	14 55	22 13
06	01 00 27	17 01 13	00 ♏ 57 47	08 ♏ 13 22	01 55	02 41	10 21	10 09	07 06	05 10	14 56	22 12
07	01 04 24	18 00 12	15 ♏ 30 24	22 ♏ 48 09	01 R 54	03 35	11 00	10 04	07 09	05 13	14 58	22 12
08	01 08 20	18 59 09	00 ♐ 05 52	07 ♐ 22 55	01 47	04 30	11 38	09 59	07 13	05 16	14 59	22 11
09	01 12 17	19 58 05	14 ♐ 38 42	21 ♐ 52 39	01 33	05 23	12 16	09 55	07 16	05 19	15 00	22 11
10	01 16 13	20 56 59	29 ♐ 04 20	06 ♑ 13 23	01 13	06 16	12 54	09 50	07 20	05 21	15 01	22 11
11	01 20 10	21 55 51	13 ♑ 19 29	20 ♑ 22 25	00 48	07 08	13 33	09 46	07 23	05 24	15 03	22 10
12	01 24 07	22 54 41	27 ♑ 21 59	04 ♒ 18 06	00 ♉ 18	08 00	14 11	09 42	07 27	05 26	15 04	22 09
13	01 28 03	23 53 30	11 ♒ 10 40	17 ♒ 59 39	29 ♈ 45	08 51	14 49	09 38	07 31	05 29	15 05	22 08
14	01 32 00	24 52 17	24 ♒ 45 02	01 ♓ 26 49	29 07	09 41	15 28	09 34	07 35	05 32	15 06	22 08
15	01 35 56	25 51 02	08 ♓ 05 02	14 ♓ 39 41	28 27	10 30	16 06	09 30	07 39	05 34	15 07	22 07
16	01 39 53	26 49 45	21 ♓ 10 50	27 ♓ 38 31	27 45	11 19	16 44	09 27	07 43	05 37	15 08	22 06
17	01 43 49	27 48 27	04 ♈ 02 47	10 ♈ 23 42	27 02	12 06	17 22	09 23	07 47	05 39	15 09	22 06
18	01 47 46	28 47 06	16 ♈ 41 22	22 ♈ 55 52	26 19	12 53	18 00	09 20	07 51	05 41	15 10	22 05
19	01 51 42	29 ♈ 45 44	29 ♈ 07 20	05 ♉ 15 55	25 36	13 39	18 39	09 17	07 55	05 44	15 11	22 04
20	01 55 39	00 ♉ 44 20	11 ♉ 21 47	17 ♉ 25 08	24 54	14 24	19 17	09 15	08 00	05 46	15 12	22 03
21	01 59 36	01 42 54	23 ♉ 26 12	29 ♉ 25 17	24 15	15 08	19 55	09 12	08 04	05 48	15 13	22 03
22	02 03 32	02 41 26	05 ♊ 22 41	11 ♊ 18 45	23 38	15 52	20 33	09 10	08 09	05 51	15 13	22 02
23	02 07 29	03 39 56	17 ♊ 13 52	23 ♊ 08 27	23 04	16 34	21 12	09 07	08 14	05 53	15 14	22 01
24	02 11 25	04 38 24	29 ♊ 02 58	04 ♋ 57 53	22 34	17 15	21 50	09 05	08 18	05 55	15 15	22 00
25	02 15 22	05 36 50	10 ♋ 53 44	16 ♋ 51 02	22 08	17 55	22 28	09 03	08 23	05 57	15 16	21 59
26	02 19 18	06 35 14	22 ♋ 50 21	28 ♋ 52 15	21 47	18 34	23 06	09 02	08 28	05 59	15 16	21 58
27	02 23 15	07 33 36	04 ♌ 57 17	11 ♌ 06 03	21 30	19 11	23 44	09 00	08 33	06 01	15 17	21 57
28	02 27 11	08 31 55	17 ♌ 19 05	23 ♌ 36 54	21 17	19 48	24 22	08 59	08 38	06 03	15 18	21 56
29	02 31 08	09 30 13	29 ♌ 59 59	06 ♍ 28 45	21 10	20 23	25 00	08 58	08 43	06 05	15 18	21 55
30	02 35 05	10 ♉ 28 28	13 ♍ 03 34	19 ♍ 44 41	21 ♈ 07	20 ♊ 56	25 ♊ 39	08 ♍ 57	08 ♋ 49	06 ♓ 07	15 ♒ 19	21 ♐ 54

Date	Moon True ☊ ° '	Moon Mean ☊ ° '	Moon ☽ Latitude ° '
01	11 ♉ 36	12 ♉ 52	05 N 08
02	11 R 31	12 48	04 47
03	11 27	12 45	04 10
04	11 23	12 42	03 18
05	11 20	12 39	02 12
06	11 19	12 36	00 N 57
07	11 D 18	12 33	00 S 23
08	11 19	12 29	01 42
09	11 20	12 26	02 54
10	11 22	12 23	03 54
11	11 23	12 20	04 39
12	11 R 23	12 17	05 06
13	11 22	12 13	05 16
14	11 21	12 10	05 07
15	11 19	12 07	04 42
16	11 17	12 04	04 03
17	11 15	12 01	03 11
18	11 13	11 58	02 12
19	11 12	11 54	01 S 06
20	11 D 12	11 51	00 N 01
21	11 12	11 48	01 07
22	11 13	11 45	02 10
23	11 14	11 42	03 07
24	11 14	11 39	03 55
25	11 15	11 35	04 34
26	11 15	11 32	05 01
27	11 16	11 29	05 16
28	11 R 15	11 26	05 16
29	11 15	11 23	05 01
30	11 ♉ 15	11 ♉ 19	04 N 30

DECLINATIONS

Date	Sun ☉ ° '	Moon ☽ ° '	Mercury ☿ ° '	Venus ♀ ° '	Mars ♂ ° '	Jupiter ♃ ° '	Saturn ♄ ° '	Uranus ♅ ° '	Neptune ♆ ° '	Pluto ♇ ° '
01	04 N 47	19 N 02	14 N 18	23 N 00	22 N 38	08 N 54	22 N 49	10 S 23	16 S 26	14 S 23
02	05 10	14 06	14 36	23 17	22 44	08 56	22 49	10 22	16 25	14 23
03	05 33	08 21	14 50	23 33	22 51	08 58	22 49	10 21	16 25	14 23
04	05 56	02 N 04	15 00	23 49	22 57	09 00	22 49	10 20	16 25	14 23
05	06 19	04 S 29	15 07	24 05	23 03	09 02	22 49	10 19	16 24	14 23
06	06 41	10 55	15 09	24 20	23 09	09 04	22 49	10 18	16 24	14 22
07	07 04	16 51	15 08	24 34	23 15	09 06	22 49	10 17	16 23	14 22
08	07 26	21 50	15 04	24 48	23 20	09 07	22 49	10 16	16 23	14 22
09	07 48	25 26	14 56	25 02	23 26	09 09	22 49	10 15	16 23	14 22
10	08 11	27 20	14 44	25 15	23 31	09 11	22 49	10 14	16 22	14 22
11	08 33	27 24	14 29	25 27	23 36	09 12	22 48	10 13	16 22	14 21
12	08 55	25 41	14 10	25 39	23 41	09 14	22 48	10 12	16 22	14 21
13	09 16	22 28	13 49	25 50	23 45	09 15	22 48	10 11	16 21	14 21
14	09 38	18 05	13 26	26 01	23 50	09 16	22 48	10 10	16 21	14 21
15	09 59	12 54	13 00	26 11	23 54	09 17	22 48	10 09	16 21	14 21
16	10 21	07 13	12 32	26 21	23 58	09 19	22 48	10 08	16 21	14 20
17	10 42	01 S 19	12 03	26 30	24 02	09 20	22 48	10 08	16 20	14 20
18	11 03	04 N 32	11 33	26 39	24 06	09 21	22 48	10 07	16 20	14 20
19	11 23	10 08	11 02	26 47	24 09	09 22	22 48	10 06	16 20	14 20
20	11 44	15 15	10 32	26 55	24 13	09 23	22 48	10 05	16 19	14 20
21	12 04	19 43	10 02	27 02	24 16	09 23	22 47	10 04	16 19	14 19
22	12 24	23 20	09 32	27 08	24 19	09 24	22 47	10 03	16 19	14 19
23	12 44	25 56	09 04	27 14	24 21	09 25	22 47	10 03	16 19	14 19
24	13 04	27 22	08 37	27 20	24 24	09 25	22 47	10 02	16 19	14 19
25	13 24	27 33	08 12	27 25	24 26	09 26	22 47	10 01	16 18	14 19
26	13 43	26 27	07 49	27 30	24 29	09 27	22 47	10 00	16 18	14 19
27	14 02	24 08	07 29	27 34	24 31	09 27	22 46	10 00	16 18	14 18
28	14 21	20 39	07 11	27 37	24 33	09 27	22 46	09 59	16 18	14 18
29	14 39	16 09	06 55	27 41	24 34	09 27	22 46	09 58	16 18	14 18
30	14 N 58	10 N 49	06 N 41	27 N 43	24 N 36	09 N 28	22 N 46	09 S 57	16 S 17	14 S 18

ZODIAC SIGN ENTRIES

Date	h	m	Planets
01	02	27	☿ ♉
02	02	45	☽ ♍
03	14	57	♀ ♊
04	07	52	☽ ♎
06	10	24	☽ ♏
08	11	50	☽ ♐
10	13	33	☽ ♑
12	16	33	☽ ♒
13	01	23	☿ ♈
14	21	24	☽ ♓
17	04	24	☽ ♈
19	13	43	☽ ♉
19	17	50	☉ ♉
22	01	10	☽ ♊
24	13	56	☽ ♋
27	02	14	☽ ♌
29	12	00	☽ ♍

LATITUDES

Date	Mercury ☿ ° '	Venus ♀ ° '	Mars ♂ ° '	Jupiter ♃ ° '	Saturn ♄ ° '	Uranus ♅ ° '	Neptune ♆ ° '	Pluto ♇ ° '
01	02 N 55	03 N 23	01 N 08	01 N 25	00 S 27	00 S 44	00 S 03	08 N 50
04	03 10	03 35	01 09	01 25	00 27	00 44	00 03	08 51
07	03 12	03 46	01 10	01 24	00 26	00 44	00 03	08 51
10	03 01	03 57	01 11	01 24	00 26	00 44	00 03	08 52
13	02 37	04 07	01 11	01 23	00 25	00 44	00 03	08 52
16	02 00	04 15	01 12	01 23	00 25	00 44	00 03	08 53
19	01 14	04 23	01 12	01 23	00 25	00 44	00 03	08 53
22	00 N 23	04 29	01 13	01 22	00 24	00 44	00 03	08 54
25	00 S 27	04 33	01 13	01 22	00 24	00 44	00 03	08 54
28	01 13	04 35	01 13	01 21	00 23	00 45	00 03	08 54
31	01 S 53	04 N 36	01 N 14	01 N 20	00 S 23	00 S 45	00 S 03	08 N 55

DATA

Julian Date	2453097
Delta T	+65 seconds
Ayanamsa	23° 54' 47"
Synetic vernal point	05° ♓ 12' 12"
True obliquity of ecliptic	23° 26' 27"

MOON'S PHASES, APSIDES AND POSITIONS ☽

Date	h	m	Phase	Longitude ° '	Eclipse Indicator
05	11	03	○	16 ♎ 00	
12	03	46	☾	22 ♑ 35	
19	13	21	●	29 ♈ 49	Partial
27	17	32	☽	07 ♌ 47	

Day	h	m		
08	02	35	Perigee	
24	00	30	Apogee	
04	19	37	0S	
11	00	47	Max dec	27° S 36'
17	17	22	0N	
25	03	29	Max dec	27° N 38'

LONGITUDES

Date	Chiron ⚷ ° '	Ceres ⚳ ° '	Pallas ⚴ ° '	Juno ⚵ ° '	Vesta ⚶ ° '	Black Moon Lilith ⚸ ° '
01	25 ♑ 31	15 ♋ 45	17 ♉ 27	19 ♑ 34	23 ♒ 11	16 ♊ 09
11	25 ♑ 51	18 ♋ 08	22 ♉ 45	21 ♑ 28	27 ♒ 48	17 ♊ 15
21	26 ♑ 03	20 ♋ 54	28 ♉ 13	22 ♑ 59	02 ♓ 14	18 ♊ 22
31	26 ♑ 08	24 ♋ 01	03 ♊ 49	24 ♑ 04	06 ♓ 29	19 ♊ 29

ASPECTARIAN

h m	Aspects
01 Thursday	
06 26	☽ Q ♂
11 44	☽ ∠ ♄
12 26	☽ △ ♇
18 38	☿ # ♇
22 12	☽ ⚼ ☉
23 56	☽ □ ♀
02 Friday	
01 12	☽ # ♆
03 57	☽ △ ☿
09 51	☽ ∥ ☿
10 41	☽ # ♇
11 49	☽ ☍ ♅
15 17	☽ ⚹ ♄
15 53	☽ ± ☉
17 07	☽ □ ♂
21 40	☽ ☌ ♃
03 Saturday	
03 26	☽ ⊼ ☉
03 58	☽ # ♅
04 04	♂ ∥ ♄
05 29	☽ ⊼ ♆
07 51	☽ ⚼ ☿
09 34	☽ ∥ ♃
12 40	☽ Q ♄
16 03	☽ ± ♆
18 24	☽ □ ♇
22 17	☽ ∥ ☉
04 Sunday	
00 01	☽ ± ☿
07 41	☽ ⚼ ♆
08 06	☉ ⚹ ♆
09 05	☽ △ ♀
10 36	☽ ⊼ ☿
16 35	☽ ⊼ ♅
16 39	♀ ⊥ ♄
19 52	☽ □ ♄
23 54	☽ △ ♂
05 Monday	
01 17	☽ Q ♇
01 24	☽ ⊻ ♃
02 48	☽ ± ♅
09 14	☽ △ ♆
11 03	☽ ☍ ☉
11 27	☽ ⊥ ♃
12 21	☽ ⚼ ♀
14 01	☿ ⊻ ♀
16 28	☉ ± ♃
17 59	☽ ⚼ ♅
19 07	☽ # ☉
21 26	☽ ⚹ ♇
06 Tuesday	
02 15	☽ ⚼ ♂
02 24	☽ ∠ ♃
04 24	☽ ± ♀
04 55	☽ # ♃
05 16	♂ □ ♃
09 36	☽ ∥ ♅
13 35	☽ ☍ ☿
15 02	☽ ⊼ ♀
17 52	☽ ± ♂
18 59	☽ △ ♅
20 28	☿ St R
22 11	☽ △ ♄
22 19	☽ ∠ ♇
07 Wednesday	
01 35	☽ ∥ ♇
03 06	☽ ⚹ ♃
04 13	☽ ⊼ ♂
04 47	☽ # ☿
10 00	☽ ∥ ♆
11 06	☽ □ ♆
13 08	☽ ⊥ ♇
16 24	☽ ⊼ ☉
22 44	☽ Q ♃
22 58	☽ ⚼ ♄
23 00	☽ ⊻ ♇
08 Thursday	
02 59	☽ ± ☉
13 50	☽ ± ♄
14 44	☽ ⊼ ☿
16 45	☽ Q ♆
17 38	☽ # ♄
18 52	☽ ⚼ ☉
19 43	☽ ☍ ♀
20 32	☽ □ ♅
21 04	☽ # ♂
23 46	☽ ⊼ ♄
09 Friday	
00 28	☽ ± ☿
04 13	☽ □ ♃
07 53	☽ ☍ ♂
08 26	☽ # ♀
09 51	♀ □ ♅
12 36	☽ ⚹ ♆
15 05	☽ ⚼ ☿
20 45	☉ ∠ ♅
21 28	☽ △ ☉
10 Saturday	
00 30	☽ ☌ ♇
02 26	☽ Q ♅
13 36	☽ ∠ ♆
15 30	☽ △ ☿
22 34	☽ ⚹ ♅
11 Sunday	
00 52	☽ ⊼ ♀
01 55	☽ ☍ ♄
04 45	☽ ⊥ ♆
05 55	☿ ⊥ ♀
06 00	☽ △ ♃
11 40	☽ ± ♀
12 23	☽ ⊼ ♂
14 55	☽ ⊻ ♆
17 35	☉ △ ♇
19 27	♀ ⊻ ♄
22 19	☿ # ♇
23 05	☽ ± ♂
12 Monday	
00 05	☽ ∠ ♅
03 03	☽ ⊻ ♇
03 46	☽ □ ☉
04 00	☽ ⚼ ♀
07 26	☽ ⚼ ♃
12 24	☽ # ♀
13 21	☽ ⊥ ♇
15 17	☽ ⚼ ♂
15 35	☽ ⊥ ♅
16 53	☽ □ ☿
22 54	☽ ± ♃
13 Tuesday	
02 02	☽ ⊻ ♅
03 28	☉ ∠ ♀
03 46	☽ # ♂
04 57	☽ ∠ ♇
05 33	☽ ⊼ ♄
07 39	☽ △ ♀
09 18	☽ ⊼ ♃
09 55	☽ # ♄
10 23	☉ ∥ ♃
10 28	☿ ∠ ♂
13 21	☽ Q ☉
16 07	☽ ± ♄
18 43	☽ △ ♂
18 52	☽ ☌ ♆
22 02	♂ △ ♆
23 04	☽ Q ☿
14 Wednesday	
04 56	☉ ⚼ ♃
07 20	☽ ⚹ ♇
08 06	☽ ⚼ ♄
08 57	♀ □ ♃
12 14	☽ ⚹ ☉
19 27	☽ ⚹ ☿
20 23	☽ ∥ ♆
15 Thursday	
04 49	☽ Q ♇
05 35	☽ ∥ ♇
06 32	☉ Q ♄
07 25	☽ ☌ ♅
11 12	☽ △ ♄
11 31	☽ # ☿
14 34	☽ ☍ ♃
16 41	☽ □ ♀
17 26	☽ ∠ ☉
21 18	☽ ∠ ☿
22 46	☉ # ♅
23 43	☽ # ☉
23 47	☽ ∥ ♅
16 Friday	
00 51	☽ ⊻ ♆
03 19	☽ # ♃
03 23	☽ □ ♂
11 18	☽ ⊥ ☉
11 55	☽ ⊥ ♆
13 00	☽ ⊥ ☿
13 43	☽ □ ♇
19 33	☉ Q ♆
23 21	☽ ⊻ ☉
23 34	☽ ⊻ ☿
17 Saturday	
01 05	☉ ☌ ☿
04 07	☽ Q ♀
04 40	☽ ∠ ♆
08 18	☿ Q ♆
10 53	☿ ∠ ♀
14 38	☽ Q ♂
15 02	☽ ⊻ ♅
19 05	☽ □ ♄
22 03	☽ ⊼ ♃
18 Sunday	
02 26	☽ ⊥ ♅
04 16	☽ ⚹ ♀
09 05	☽ ⚹ ♆
09 26	☽ ± ♃
14 40	☽ ⚹ ♂
19 42	☽ ∠ ♅
22 21	☽ △ ♇
19 Monday	
02 02	☿ Q ♄
02 11	☉ ∥ ☿
02 40	☽ ⚼ ♃
05 32	☽ ☌ ☿
05 45	☽ Q ♄
08 13	☽ Q ♆
08 36	☽ ∥ ♃
11 02	☽ ∠ ♀
11 52	☽ # ♅
13 21	☽ ☌ ☉
15 44	☽ ∥ ☿
18 04	☽ ∥ ☉
21 19	☽ ∠ ♂
20 Tuesday	
00 57	☽ ⚹ ♅
03 31	☽ ⚼ ♇
05 20	☽ ⚹ ♄
05 47	☽ ⊥ ♀
07 27	☽ # ♇
07 50	☽ △ ♃
16 00	☽ ⊥ ♂
17 24	☽ # ♆
18 25	☽ ⊻ ♀
19 36	☽ □ ♆
21 17	☽ ± ♇
21 Wednesday	
00 44	☽ Q ♅
04 35	☽ ⊻ ♂
09 13	☽ ⊼ ♇
09 50	☿ # ♅
11 16	☽ ∠ ♄
13 33	☽ ⊻ ☿
14 00	☿ ⚼ ♃
14 17	♀ △ ♆
22 Thursday	
00 59	☽ ⊥ ☿
05 27	☽ ⊥ ♄
06 06	☽ ⊻ ☉
07 56	☽ ∥ ♄
12 57	☽ □ ♅
17 38	☽ ⊻ ♄
18 16	☽ ∠ ☿
18 30	☿ ∥ ♃
19 18	☽ ⊥ ☉
19 37	☽ □ ♃
20 04	☽ ∥ ♂
23 Friday	
07 57	☽ △ ♆
10 34	☽ ☌ ♀
15 10	☽ ∠ ☉
20 30	☽ ☌ ♂
21 42	☽ ☍ ♇
23 22	☽ ⚹ ☿
24 Saturday	
08 02	☽ Q ♃
11 10	☽ ∥ ♀
14 26	☽ ⚼ ♆
18 13	♂ ☍ ♇
22 47	☽ Q ☿
25 Sunday	
00 22	☽ ⚹ ☉
01 58	☽ △ ♅
04 30	☿ ⚹ ♂
06 54	☽ ☌ ♄
08 18	☽ ⚹ ♃
08 42	☽ ± ♆
16 37	☽ ∥ ♀
20 40	☉ ⚹ ♅
20 49	☽ ⊼ ♆
22 13	☿ △ ♇
26 Monday	
02 44	☽ Q ☉
02 57	☽ ⊻ ♀
08 17	☽ ⚼ ♅
09 56	☽ □ ☿
10 15	☽ ⊼ ♇
12 33	☽ ⊻ ♂
14 22	☽ ∠ ♃
15 37	☽ ⊥ ♀
21 11	☉ ⚼ ♇
22 11	☽ ± ♇
27 Tuesday	
01 09	☽ ⊥ ♂
02 15	☽ ± ♅
08 10	☽ ⊥ ♃
08 47	☽ ∥ ♂
10 25	☽ ∠ ♀
14 06	☽ ⊼ ♅
15 54	☽ ⚼ ♇
17 32	☽ □ ☉
19 05	☽ ⊻ ♄
19 48	☽ ∠ ♂
19 54	☽ ⊻ ♃
22 17	☽ ∥ ♄
28 Wednesday	
06 49	☽ ⊥ ♄
08 06	☽ ☍ ♆
08 36	☉ # ♇
14 52	☉ ⚹ ♄
16 58	☽ ⚹ ♀
19 29	☽ △ ☿
20 48	☽ △ ♇
22 54	☉ △ ♃
29 Thursday	
00 08	☽ ∠ ♄
02 08	☽ ⚹ ♂
11 20	☽ # ♆
16 38	☽ Q ♀
18 42	☽ ∥ ☉
20 44	☽ # ♇
23 19	☽ ☍ ♅
23 21	☽ ⚼ ☿
30 Friday	
01 38	☽ Q ♂
04 13	☽ ⚹ ♄
04 32	☽ ☌ ♃
06 56	☽ △ ☉
09 20	☿ ∠ ♅
13 05	☿ St D
15 33	☽ # ♅
15 43	☽ ± ☿
16 04	☽ ⊼ ♆
17 35	☽ ∥ ♃
20 02	☿ ⚹ ♀

All ephemeris data is given at 12.00 UT and the Moon's longitude is additionally given for 24.00 UT

MAY 2004

LONGITUDES

Date	Sidereal time h m s	Sun ☉ ° ' "	Moon ☽ ° ' "	Moon ☽ 24.00 ° ' "	Mercury ☿ ° '	Venus ♀ ° '	Mars ♂ ° '	Jupiter ♃ ° '	Saturn ♄ ° '	Uranus ♅ ° '	Neptune ♆ ° '	Pluto ♇ ° '
01	02 39 01	11 ♉ 26 41	26 ♍ 32 15	03 ♎ 26 19	21 ♈ 09	21 ♊ 28	26 ♊ 17	08 ♍ 56	08 ♋ 54	06 ♓ 09	15 ♒ 19	21 ♐ 53
02	02 42 58	12 24 53	10 ♎ 26 44	17 ♎ 33 16	21 D 16	21 59	26 55	08 R 56	08 59	06 11	15 20	21 R 52
03	02 46 54	13 23 02	24 ♎ 45 30	02 ♏ 02 51	21 28	22 29	27 33	08 55	09 05	06 13	15 20	21 50
04	02 50 51	14 21 10	09 ♏ 24 37	16 ♏ 49 58	21 44	22 56	28 11	08 55	09 10	06 15	15 21	21 49
05	02 54 47	15 19 16	24 ♏ 17 56	01 ♐ 47 30	22 05	23 22	28 49	08 D 55	09 16	06 16	15 21	21 48
06	02 58 44	16 17 20	09 ♐ 17 37	16 ♐ 47 14	22 31	23 47	29 ♊ 27	08 55	09 22	06 18	15 22	21 47
07	03 02 40	17 15 23	24 ♐ 15 18	01 ♑ 40 54	23 00	24 10	00 ♋ 05	08 55	09 27	06 20	15 22	21 46
08	03 06 37	18 13 25	09 ♑ 03 12	16 ♑ 21 27	23 34	24 30	00 43	08 56	09 33	06 21	15 22	21 44
09	03 10 34	19 11 25	23 ♑ 35 06	00 ♒ 43 42	24 11	24 49	01 21	08 57	09 39	06 23	15 22	21 43
10	03 14 30	20 09 24	07 ♒ 46 57	14 ♒ 44 43	24 53	25 07	01 59	08 58	09 45	06 24	15 23	21 42
11	03 18 27	21 07 21	21 ♒ 36 56	28 ♒ 23 40	25 38	25 22	02 37	08 59	09 51	06 26	15 23	21 41
12	03 22 23	22 05 17	05 ♓ 05 04	11 ♓ 41 21	26 26	25 35	03 15	09 00	09 57	06 27	15 23	21 39
13	03 26 20	23 03 12	18 ♓ 12 46	24 ♓ 39 39	27 18	25 46	03 53	09 01	10 03	06 29	15 23	21 38
14	03 30 16	24 01 06	01 ♈ 02 17	07 ♈ 21 01	28 13	25 55	04 31	09 03	10 09	06 30	15 23	21 37
15	03 34 13	24 58 58	13 ♈ 36 10	19 ♈ 48 05	29 ♈ 12	26 01	05 09	09 05	10 15	06 31	15 24	21 35
16	03 38 09	25 56 49	25 ♈ 57 04	02 ♉ 03 25	00 ♉ 13	26 06	05 47	09 07	10 22	06 32	15 24	21 34
17	03 42 06	26 54 39	08 ♉ 07 25	14 ♉ 09 21	01 18	26 08	06 25	09 09	10 28	06 34	15 24	21 33
18	03 46 03	27 52 27	20 ♉ 09 29	26 ♉ 08 04	02 25	26 R 08	07 03	09 11	10 34	06 35	15 R 24	21 31
19	03 49 59	28 50 14	02 ♊ 05 20	08 ♊ 01 34	03 35	26 05	07 41	09 14	10 41	06 36	15 24	21 30
20	03 53 56	29 ♉ 48 00	13 ♊ 56 59	19 ♊ 51 52	04 48	26 01	08 19	09 16	10 47	06 37	15 23	21 29
21	03 57 52	00 ♊ 45 44	25 ♊ 46 29	01 ♋ 41 08	06 04	25 53	08 57	09 19	10 54	06 38	15 23	21 27
22	04 01 49	01 43 27	07 ♋ 36 07	13 ♋ 31 46	07 22	25 43	09 35	09 22	11 00	06 39	15 23	21 25
23	04 05 45	02 41 09	19 ♋ 28 27	25 ♋ 26 34	08 43	25 31	10 13	09 25	11 07	06 40	15 23	21 24
24	04 09 42	03 38 49	01 ♌ 26 31	07 ♌ 28 46	10 06	25 17	10 51	09 29	11 14	06 41	15 23	21 22
25	04 13 38	04 36 27	13 ♌ 33 45	19 ♌ 41 59	11 32	25 00	11 29	09 32	11 21	06 41	15 23	21 21
26	04 17 35	05 34 04	25 ♌ 53 56	02 ♍ 10 08	13 00	24 40	12 07	09 36	11 27	06 42	15 22	21 19
27	04 21 32	06 31 40	08 ♍ 31 04	14 ♍ 57 14	14 31	24 19	12 45	09 40	11 34	06 43	15 22	21 18
28	04 25 28	07 29 14	21 ♍ 29 04	28 ♍ 06 59	16 04	23 55	13 23	09 44	11 41	06 44	15 22	21 16
29	04 29 25	08 26 47	04 ♎ 51 18	11 ♎ 42 15	17 40	23 29	14 00	09 48	11 48	06 44	15 21	21 15
30	04 33 21	09 24 18	18 ♎ 40 00	25 ♎ 44 29	19 18	23 01	14 38	09 52	11 55	06 45	15 21	21 13
31	04 37 18	10 ♊ 21 48	02 ♏ 55 35	10 ♏ 12 55	20 ♉ 58	22 ♊ 31	15 ♋ 16	09 ♍ 57	12 ♋ 02	06 ♓ 45	15 ♒ 20	21 ♐ 12

Date	Moon True ☊ ° '	Moon Mean ☊ ° '	Moon ☽ Latitude ° '
01	11 ♉ 15	11 ♉ 16	03 N 44
02	11 D 15	11 13	02 43
03	11 15	11 10	01 30
04	11 15	11 07	00 N 10
05	11 R 15	11 04	01 S 12
06	11 15	11 00	02 29
07	11 15	10 57	03 37
08	11 14	10 54	04 29
09	11 13	10 51	05 02
10	11 13	10 48	05 16
11	11 D 12	10 45	05 12
12	11 13	10 41	04 50
13	11 13	10 38	04 13
14	11 14	10 35	03 25
15	11 16	10 32	02 27
16	11 17	10 29	01 24
17	11 17	10 25	00 S 17
18	11 R 17	10 22	00 N 49
19	11 16	10 19	01 52
20	11 14	10 16	02 51
21	11 11	10 13	03 41
22	11 08	10 10	04 22
23	11 04	10 06	04 52
24	11 01	10 03	05 10
25	10 59	10 00	05 14
26	10 57	09 57	05 04
27	10 D 57	09 54	04 39
28	10 58	09 51	03 59
29	10 59	09 47	03 06
30	11 01	09 44	02 00
31	11 ♉ 02	09 ♉ 41	00 N 44

DECLINATIONS

Date	Sun ☉ ° '	Moon ☽ ° '	Mercury ☿ ° '	Venus ♀ ° '	Mars ♂ ° '	Jupiter ♃ ° '	Saturn ♄ ° '	Uranus ♅ ° '	Neptune ♆ ° '	Pluto ♇ ° '
01	15 N 16	04 N 48	06 N 31	27 N 45	24 N 37	09 N 28	22 N 46	09 S 57	16 S 17	14 S 18
02	15 34	01 S 38	06 23	27 47	24 38	09 28	22 45	09 56	16 17	14 18
03	15 51	08 11	06 17	27 48	24 39	09 28	22 45	09 56	16 17	14 17
04	16 09	14 28	06 14	27 49	24 40	09 28	22 45	09 55	16 17	14 17
05	16 26	20 00	06 13	27 49	24 40	09 28	22 45	09 54	16 17	14 17
06	16 43	24 18	06 15	27 49	24 40	09 27	22 44	09 54	16 17	14 17
07	16 59	26 55	06 19	27 48	24 41	09 27	22 44	09 53	16 17	14 17
08	17 15	27 36	06 26	27 46	24 41	09 27	22 44	09 53	16 17	14 17
09	17 31	26 20	06 34	27 45	24 40	09 26	22 43	09 52	16 16	14 16
10	17 47	23 25	06 45	27 43	24 40	09 26	22 43	09 52	16 16	14 16
11	18 02	19 13	06 58	27 40	24 39	09 25	22 43	09 51	16 16	14 16
12	18 17	14 08	07 12	27 37	24 38	09 24	22 42	09 51	16 16	14 16
13	18 32	08 33	07 29	27 33	24 37	09 24	22 42	09 50	16 16	14 16
14	18 47	02 S 43	07 47	27 29	24 36	09 23	22 42	09 50	16 16	14 16
15	19 01	03 N 06	08 06	27 24	24 35	09 22	22 41	09 49	16 16	14 16
16	19 15	08 43	08 28	27 19	24 33	09 21	22 41	09 49	16 16	14 16
17	19 28	13 56	08 51	27 13	24 32	09 20	22 41	09 48	16 16	14 15
18	19 41	18 34	09 15	27 07	24 30	09 19	22 40	09 48	16 16	14 15
19	19 54	22 25	09 40	27 00	24 28	09 18	22 40	09 48	16 16	14 15
20	20 07	25 18	10 07	26 52	24 25	09 17	22 39	09 47	16 16	14 15
21	20 19	27 04	10 35	26 44	24 23	09 16	22 39	09 47	16 16	14 15
22	20 30	27 35	11 04	26 36	24 20	09 15	22 39	09 47	16 16	14 15
23	20 42	26 51	11 34	26 27	24 17	09 13	22 38	09 46	16 17	14 15
24	20 53	24 52	12 05	26 17	24 14	09 12	22 38	09 46	16 17	14 15
25	21 04	21 46	12 36	26 07	24 11	09 11	22 37	09 46	16 17	14 15
26	21 14	17 39	13 09	25 56	24 08	09 09	22 37	09 45	16 17	14 15
27	21 24	12 41	13 42	25 44	24 04	09 07	22 36	09 45	16 17	14 14
28	21 34	07 02	14 16	25 32	24 01	09 05	22 36	09 45	16 17	14 14
29	21 43	00 N 55	14 50	25 19	23 57	09 03	22 35	09 45	16 17	14 14
30	21 52	05 S 28	15 24	25 05	23 53	09 02	22 35	09 45	16 17	14 14
31	22 N 00	11 S 48	15 N 59	24 N 51	23 N 48	09 N 00	22 N 34	09 S 45	16 S 17	14 S 14

ZODIAC SIGN ENTRIES

Date	h	m	Planets
01	18	03	☽ ♎
03	20	39	☽ ♏
05	21	08	☽ ♐
07	08	46	♂ ♋
07	21	17	☽ ♑
09	22	46	☽ ♒
12	02	52	☽ ♓
14	10	02	☽ ♈
16	06	54	☿ ♉
16	19	57	☽ ♉
19	07	47	☽ ♊
20	16	59	☉ ♊
21	20	35	☽ ♋
24	09	07	☽ ♌
26	19	52	☽ ♍
29	03	22	☽ ♎
31	07	08	☽ ♏

LATITUDES

Date	Mercury ☿ ° '	Venus ♀ ° '	Mars ♂ ° '	Jupiter ♃ ° '	Saturn ♄ ° '	Uranus ♅ ° '	Neptune ♆ ° '	Pluto ♇ ° '
01	01 S 53	04 N 36	01 N 14	01 N 20	00 S 23	00 S 45	00 S 03	08 N 55
04	02 25	04 34	01 14	01 20	00 23	00 45	00 03	08 55
07	02 49	04 29	01 14	01 19	00 22	00 45	00 03	08 55
10	03 06	04 22	01 14	01 19	00 22	00 45	00 03	08 55
13	03 15	04 11	01 15	01 18	00 22	00 45	00 03	08 56
16	03 18	03 56	01 15	01 18	00 21	00 45	00 03	08 56
19	03 14	03 37	01 15	01 17	00 21	00 45	00 03	08 56
22	03 04	03 13	01 15	01 16	00 21	00 45	00 03	08 56
25	02 49	02 46	01 15	01 16	00 20	00 46	00 03	08 56
28	02 29	02 14	01 15	01 15	00 20	00 46	00 03	08 56
31	02 S 05	01 N 37	01 N 15	01 N 15	00 S 20	00 S 46	00 S 03	08 N 56

DATA

Julian Date	2453127
Delta T	+65 seconds
Ayanamsa	23° 54' 51"
Synetic vernal point	05° ♓ 12' 08"
True obliquity of ecliptic	23° 26' 27"

MOON'S PHASES, APSIDES AND POSITIONS ☽

Date	h	m	Phase	Longitude ° '	Eclipse Indicator
04	20	33	○	14 ♏ 42	total
11	11	04	☾	21 ♒ 05	
19	04	52	●	28 ♉ 33	
27	07	57	☽	06 ♍ 22	

Day	h	m		
06	04	40	Perigee	
21	12	11	Apogee	
02	05	59	OS	
08	08	09	Max dec	27° S 37'
14	23	09	ON	
22	10	01	Max dec	27° N 36'
29	15	27	OS	

LONGITUDES

Date	Chiron ⚷ ° '	Ceres ⚳ ° '	Pallas ⚴ ° '	Juno ⚵ ° '	Vesta ⚶ ° '	Black Moon Lilith ⚸ ° '
01	26 ♑ 08	22 ♋ 01	03 ♊ 03	24 ♑ 04	06 ♓ 29	19 ♊ 29
11	26 ♑ 05	27 ♋ 24	09 ♊ 32	24 ♑ 39	10 ♓ 30	20 ♊ 35
21	25 ♑ 54	01 ♌ 01	15 ♊ 21	24 ♑ 42	14 ♓ 06	21 ♊ 42
31	25 ♑ 36	04 ♌ 50	21 ♊ 15	24 ♑ 09	17 ♓ 43	22 ♊ 49

All ephemeris data is given at 12.00 UT and the Moon's longitude is additionally given for 24.00 UT

ASPECTARIAN

h m	Aspects
01 Saturday	
00 05	☉ ∠ ♂
01 59	☽ Q ♄
02 28	☽ ⚻ ☿
02 43	☽ □ ♀
02 48	☽ ± ♆
03 48	☽ □ ♇
05 11	☽ ∥ ☿
11 31	☽ □ ♂
11 37	☿ ∠ ♅
11 50	☽ bQ ☉
18 37	☽ bQ ♆
20 31	♃ ✶ ♄
02 Sunday	
04 36	☽ ± ☉
04 42	☽ ⚻ ♅
06 07	♀ ☍ ♇
09 25	☽ ⚺ ♃
09 30	☽ □ ♄
11 00	☽ Q ♇
12 26	♂ Q ♃
14 57	☽ ± ♅
15 35	☽ ⚻ ☉
19 35	☽ ⊥ ♃
20 16	☽ △ ♆
03 Monday	
05 04	☽ # ☿
06 06	☽ bQ ♅
06 27	☽ ☍ ☿
07 10	☽ ✶ ♇
08 05	☽ △ ♀
10 37	☽ ∠ ♃
16 44	☽ # ♃
16 49	☽ △ ♂
18 28	☽ ∥ ♅
04 Tuesday	
06 50	☽ △ ♅
07 48	☽ ∠ ♇
09 32	☽ bQ ♀
11 12	☽ ✶ ♃
11 17	☽ ∥ ♇
11 37	☽ △ ♄
17 51	☿ △ ♇
18 23	☽ bQ ♂
19 13	☽ # ☉
19 25	☽ ∥ ♆
20 33	☽ ☍ ☉
21 37	☽ □ ♆
22 21	☽ ⊥ ♇
23 20	☉ # ♆
05 Wednesday	
00 33	☽ ± ♀
03 09	♃ St D
06 34	☽ Q ♃
08 00	☽ ⚺ ♇
08 22	☽ ⚻ ☿
09 31	☽ ± ♂
10 28	☽ ⚻ ♀
11 57	☽ bQ ♄
12 49	☉ □ ♆
18 14	☽ ± ☿
19 34	☽ ⚻ ♂
23 41	☉ ± ♇
06 Thursday	
02 15	☽ # ♄
02 27	☽ ± ♄
02 30	☽ Q ♆
07 12	☽ □ ♅
09 03	☽ bQ ☿
11 24	☽ □ ♃
12 06	☽ ⚻ ♄
12 18	♄ ± ♆
14 34	☽ # ♂
21 43	☽ ✶ ♆
23 59	☽ ⚻ ☉
07 Friday	
08 00	☽ ☌ ♇
09 54	☽ △ ☿
10 17	☽ ± ☉
11 50	☽ ☍ ♀
12 07	☽ Q ♅
21 50	☽ ☍ ♂
21 52	☽ ∠ ♆
22 42	♂ bQ ♆
08 Saturday	
01 50	☽ bQ ☉
07 35	☽ ✶ ♅
11 48	☽ △ ♃
12 31	☽ ⊥ ♆
12 49	☽ ☍ ♄
15 19	☉ Q ♅
22 23	☽ ⚺ ♆
22 36	☉ ⊥ ♀
09 Sunday	
02 48	☿ bQ ♃
04 10	☽ △ ☉
08 19	☽ ∠ ♅
08 54	☽ ⚺ ♇
12 36	☽ bQ ♃
13 03	☽ □ ☿
14 07	☽ ⚻ ♀
18 55	☽ ⊥ ♇
23 26	☽ ⊥ ♅
10 Monday	
00 25	☽ ± ♀
01 40	☽ ⚻ ♂
03 08	☽ # ♂
03 46	☽ ± ♃
09 38	☽ ⚺ ♅
10 09	☽ ∠ ♇
12 22	☽ ± ♂
14 01	☽ ⚻ ♃
15 24	☽ ⚻ ♄
16 04	☽ bQ ♀
16 29	☽ # ♄
21 15	☽ Q ☿
23 47	☿ ✶ ♀
11 Tuesday	
01 06	☽ ☌ ♆
01 50	☽ ± ♄
04 40	☽ bQ ♂
11 04	☽ □ ☉
12 07	☽ ✶ ♇
17 35	☽ # ☉
17 45	☽ bQ ♄
18 44	☽ △ ♀
19 31	☽ ✶ ☿
12 Wednesday	
01 29	☉ ⚻ ♇
02 18	☽ ∥ ♆
08 32	☽ △ ♂
09 26	☽ Q ♇
11 26	☽ ∥ ♇
14 29	☽ ☌ ♅
19 06	☽ ☍ ♃
20 53	☽ △ ♄
21 48	☽ Q ☉
13 Thursday	
00 20	☽ ∠ ☿
06 36	☽ ∥ ♅
06 47	☽ ⚺ ♆
08 26	☽ # ♃
14 20	☿ Q ♆
16 14	☽ # ☿
17 54	☽ ⊥ ♆
18 10	☽ ⊥ ☿
18 20	☽ □ ♇
21 44	☽ ✶ ☉
14 Friday	
02 14	☽ □ ♀
06 16	☽ ⚺ ☿
09 58	☿ Q ♄
10 47	☽ ∠ ♆
18 57	☽ □ ♂
22 23	☽ ⚺ ♅
15 Saturday	
03 17	☽ ⚻ ♃
04 27	☽ ∠ ☉
05 30	☽ □ ♄
09 55	☽ ⊥ ♅
12 49	☽ Q ♀
14 51	☽ ± ♃
15 27	☽ ✶ ♆
19 37	☉ ∠ ♄
23 17	☽ ⊥ ☉
16 Sunday	
03 22	☽ ∠ ♅
03 27	☽ △ ♇
07 32	☽ Q ♂
08 23	☽ bQ ♃
10 47	☽ ∥ ☿
11 59	☽ ⚺ ☉
12 17	☽ ✶ ♀
14 47	☽ ∥ ♃
14 50	☽ Q ♆
15 59	☉ ⚺ ♀
16 46	☽ Q ♄
16 50	☽ # ♅
21 11	☽ ☌ ☿
17 Monday	
08 26	☽ ✶ ♂
08 52	☽ bQ ♇
08 54	☽ ✶ ♅
12 13	♆ St R
13 33	☽ # ♇
14 02	☽ △ ♃
16 42	☽ ✶ ♄
17 28	♂ △ ♅
17 59	☽ ∠ ♀
22 28	♀ St R
23 38	☽ # ♆
18 Tuesday	
02 28	☽ □ ♆
02 44	☽ ± ♇
08 50	☽ Q ♅
11 57	☽ ⊥ ♀
14 43	☽ ⚻ ♇
16 01	☽ ∠ ♂
16 01	☿ ∥ ♃
18 49	☽ ∥ ☉
22 58	☽ ∠ ♄
23 58	☽ ⚺ ♀
19 Wednesday	
04 52	☽ ☌ ☉
11 08	☽ ⊥ ♂
13 46	☽ ∥ ♄
15 22	☽ ⚺ ☿
17 17	☽ ⊥ ♄
18 32	☿ # ♅
21 07	☽ □ ♅
23 57	☽ ⚺ ♂
20 Thursday	
02 29	☽ □ ♃
03 50	☽ ∥ ♂
04 53	☽ ⊥ ☿
05 32	☽ ⚺ ♄
14 55	☽ △ ♆
21 Friday	
01 17	☽ ∠ ☿
03 14	☽ ☍ ♇
06 40	☽ ∥ ♀
12 13	☽ ☌ ♀
15 09	☽ Q ♃
19 02	☿ bQ ♇
21 22	☽ bQ ♆
22 43	☿ ✶ ♅
23 01	☽ ⚺ ☉
22 Saturday	
03 05	♂ ✶ ♃
04 35	♂ ± ♆
10 04	☽ △ ♅
11 28	☽ ✶ ☿
12 16	☽ ⊥ ☉
15 36	☽ ✶ ♃
15 37	☽ ± ♆
16 14	☽ ☌ ♂
18 58	☽ ☌ ♄
23 Sunday	
03 45	☽ ⚻ ♆
08 05	☽ ∠ ☉
14 49	☽ Q ☿
15 52	☽ ⚻ ♇
16 25	☽ bQ ♅
18 58	☽ ∥ ♀
22 00	☽ ∠ ♃
23 55	☽ ⚺ ♀
24 Monday	
00 51	☿ △ ♃
03 53	☽ ± ♇
10 28	☽ ± ♅
11 41	☽ ⊥ ♀
14 33	☿ ∠ ♀
16 04	☽ ⊥ ♃
16 46	☽ ✶ ☉
17 44	☽ ∥ ♂
21 47	☽ bQ ♇
22 25	☽ ⚻ ♅
25 Tuesday	
04 02	☽ ⚺ ♃
05 08	☽ ∠ ♀
05 39	♂ ☌ ♄
06 07	☽ ∥ ♄
07 28	☽ □ ☿
07 35	☽ ⚺ ♄
07 41	☽ ⚺ ♂
08 39	☿ ✶ ♄
10 33	☿ ✶ ♂
15 33	☽ ☍ ♆
16 18	☽ ∥ ☉
18 28	☽ Q ☉
19 28	☽ ⊥ ♄
20 05	☽ ⊥ ♂
26 Wednesday	
03 10	☽ △ ♇
08 49	☉ ⊥ ♄
09 42	☽ ✶ ♀
13 05	☽ ∠ ♄
14 27	☽ ∠ ♂
18 59	☽ # ♆
27 Thursday	
04 50	☽ # ♇
07 48	☽ ∥ ☿
07 57	☽ □ ☉
07 58	☽ Q ♀
08 36	☽ ☍ ♅
14 09	☽ ☌ ♃
16 43	☉ □ ♅
17 46	☽ ✶ ♄
20 18	☽ ✶ ♂
28 Friday	
00 00	☿ ± ♇
00 42	☽ △ ☿
00 46	☽ ⚻ ♆
00 46	☽ # ♅
01 14	☿ □ ♆
03 32	☽ ∥ ♃
03 47	☉ ⊥ ♂
11 05	☿ # ♇
11 37	☽ □ ♇
11 47	☽ ± ♆
16 02	☽ Q ♄
16 17	☽ □ ♀
19 25	☽ Q ♂
29 Saturday	
04 01	☽ bQ ♆
07 35	☽ bQ ☿
09 57	☿ ⊥ ♀
15 19	☽ ⚻ ♅
18 47	☽ △ ☉
19 42	☽ Q ♇
20 43	☽ ⚺ ♃
30 Sunday	
00 16	☽ □ ♄
01 31	☽ ± ☿
01 48	☽ ± ♅
03 58	☿ Q ♅
04 45	☽ □ ♂
06 18	☽ △ ♆
07 10	☽ ⊥ ♃
13 13	☽ ⚻ ☿
16 21	☽ ✶ ♇
17 15	☽ bQ ♅
19 09	☽ △ ♀
22 27	☽ bQ ☉
22 35	☽ ∠ ♃
31 Monday	
00 34	☉ □ ♃
01 19	☽ # ♃
04 06	☽ ∥ ♅
14 33	☽ ± ☉
14 42	♂ ⚻ ♆
15 10	☿ ⚻ ♇
17 23	☽ ∠ ♇
18 20	☽ △ ♅
19 19	☽ bQ ♀
21 40	☽ ∥ ♇
23 37	☽ ✶ ♃

JUNE 2004

LONGITUDES

Date	Sidereal time h m s	Sun ☉ ° ' "	Moon ☽ ° ' "	Moon ☽ 24.00 ° ' "	Mercury ☿ ° '	Venus ♀ ° '	Mars ♂ ° '	Jupiter ♃ ° '	Saturn ♄ ° '	Uranus ♅ ° '	Neptune ♆ ° '	Pluto ♇ ° '
01	04 41 14	11 ♊ 19 16	17 ♏ 35 57	25 ♏ 03 59	22 ♉ 41	22 ♊ 00	15 ♋ 54	10 ♍ 01	12 ♋ 09	06 ♓ 46	15 ♒ 20	21 ♐ 10
02	04 45 11	12 16 44	02 ♐ 36 06	10 ♐ 11 13	24 26	21 R 27	16 32	10 06	12 16	06 46	15 R 20	21 R 09
03	04 49 07	13 14 11	17 ♐ 48 10	25 ♐ 25 40	26 14	20 52	17 10	10 11	12 23	06 46	15 19	21 07
04	04 53 04	14 11 36	03 ♑ 02 25	10 ♑ 37 07	28 04	20 17	17 48	10 16	12 30	06 47	15 18	21 05
05	04 57 01	15 09 01	18 ♑ 08 34	25 ♑ 35 41	29 ♉ 56	19 40	18 25	10 21	12 38	06 47	15 18	21 04
06	05 00 57	16 06 25	02 ♒ 57 35	10 ♒ 13 31	01 ♊ 51	19 03	19 03	10 27	12 45	06 47	15 17	21 02
07	05 04 54	17 03 49	17 ♒ 22 57	24 ♒ 25 35	03 48	18 26	19 41	10 32	12 52	06 48	15 17	21 01
08	05 08 50	18 01 12	01 ♓ 21 16	08 ♓ 10 03	05 47	17 48	20 19	10 38	12 59	06 48	15 16	20 59
09	05 12 47	18 58 34	14 ♓ 52 04	21 ♓ 27 39	07 48	17 11	20 57	10 44	13 07	06 48	15 15	20 58
10	05 16 43	19 55 55	27 ♓ 57 11	04 ♈ 21 06	09 51	16 33	21 35	10 50	13 14	06 48	15 15	20 56
11	05 20 40	20 53 17	10 ♈ 39 55	16 ♈ 54 09	11 56	15 56	22 12	10 56	13 21	06 R 48	15 14	20 54
12	05 24 36	21 50 37	23 ♈ 04 21	29 ♈ 11 01	14 03	15 20	22 50	11 02	13 29	06 48	15 13	20 53
13	05 28 33	22 47 57	05 ♉ 14 41	11 ♉ 15 50	16 11	14 45	23 28	11 08	13 36	06 48	15 12	20 51
14	05 32 30	23 45 17	17 ♉ 14 56	23 ♉ 12 24	18 20	14 12	24 06	11 15	13 44	06 47	15 11	20 50
15	05 36 26	24 42 37	29 ♉ 08 37	05 ♊ 03 57	20 31	13 39	24 44	11 22	13 51	06 47	15 10	20 48
16	05 40 23	25 39 56	10 ♊ 58 44	16 ♊ 53 14	22 42	13 08	25 21	11 29	13 59	06 47	15 10	20 46
17	05 44 19	26 37 14	22 ♊ 47 45	28 ♊ 42 30	24 54	12 39	25 59	11 35	14 06	06 47	15 09	20 45
18	05 48 16	27 34 32	04 ♋ 37 43	10 ♋ 33 37	27 05	12 12	26 37	11 43	14 14	06 46	15 08	20 43
19	05 52 12	28 31 49	16 ♋ 30 24	22 ♋ 28 19	29 ♊ 17	11 46	27 15	11 50	14 22	06 46	15 07	20 42
20	05 56 09	29 ♊ 29 06	28 ♋ 27 33	04 ♌ 28 22	01 ♋ 29	11 23	27 53	11 57	14 29	06 45	15 06	20 40
21	06 00 05	00 ♋ 26 22	10 ♌ 31 01	16 ♌ 35 47	03 39	11 02	28 30	12 05	14 37	06 45	15 05	20 39
22	06 04 02	01 23 38	22 ♌ 42 59	28 ♌ 52 58	05 49	10 43	29 08	12 12	14 44	06 44	15 04	20 37
23	06 07 59	02 20 53	05 ♍ 06 05	11 ♍ 22 45	07 58	10 27	29 ♋ 46	12 20	14 52	06 44	15 03	20 35
24	06 11 55	03 18 07	17 ♍ 43 22	24 ♍ 08 22	10 06	10 13	00 ♌ 24	12 28	15 00	06 43	15 02	20 34
25	06 15 52	04 15 21	00 ♎ 38 12	07 ♎ 13 15	12 12	10 01	01 02	12 36	15 08	06 42	15 00	20 32
26	06 19 48	05 12 34	13 ♎ 53 57	20 ♎ 40 38	14 17	09 52	01 39	12 44	15 15	06 42	14 59	20 31
27	06 23 45	06 09 46	27 ♎ 33 33	04 ♏ 32 54	16 20	09 45	02 17	12 52	15 23	06 41	14 58	20 29
28	06 27 41	07 06 59	11 ♏ 38 44	18 ♏ 50 55	18 21	09 40	02 55	13 00	15 31	06 40	14 57	20 28
29	06 31 38	08 04 10	26 ♏ 09 11	03 ♐ 33 05	20 20	09 38	03 33	13 09	15 38	06 39	14 56	20 26
30	06 35 34	09 ♋ 01 22	11 ♐ 01 54	18 ♐ 34 45	22 ♋ 18	09 ♊ 38	04 ♌ 11	13 ♍ 17	15 ♋ 46	06 ♓ 38	14 ♒ 54	20 ♐ 25

DECLINATIONS

Date	Moon True ☊ ° '	Moon Mean ☊ ° '	Moon ☽ Latitude ° '	Sun ☉ ° '	Moon ☽ ° '	Mercury ☿ ° '	Venus ♀ ° '	Mars ♂ ° '	Jupiter ♃ ° '	Saturn ♄ ° '	Uranus ♅ ° '	Neptune ♆ ° '	Pluto ♇ ° '
01	11 ♉ 02	09 ♉ 38	00 S 36	22 N 08	17 S 39	16 N 34	24 N 36	23 N 44	08 N 58	22 N 34	09 S 44	16 S 18	14 S 14
02	11 R 00	09 35	01 55	22 16	22 34	17 09	24 21	23 39	08 56	22 33	09 44	16 18	14 14
03	10 57	09 31	03 08	22 23	26 00	17 44	24 05	23 34	08 54	22 33	09 44	16 18	14 14
04	10 53	09 28	04 06	22 30	27 31	18 19	23 49	23 30	08 52	22 32	09 44	16 18	14 14
05	10 49	09 25	04 48	22 37	26 58	18 53	23 32	23 24	08 50	22 31	09 44	16 18	14 14
06	10 44	09 22	05 09	22 43	24 30	19 27	23 15	23 19	08 47	22 31	09 44	16 18	14 14
07	10 40	09 19	05 09	22 49	20 32	20 00	22 58	23 14	08 45	22 30	09 44	16 19	14 14
08	10 37	09 16	04 51	22 54	15 32	20 32	22 40	23 08	08 43	22 30	09 44	16 19	14 14
09	10 36	09 12	04 18	22 59	09 55	21 04	22 23	23 02	08 40	22 29	09 44	16 19	14 14
10	10 D 36	09 09	03 31	23 04	04 S 03	21 34	22 05	22 56	08 38	22 28	09 44	16 19	14 14
11	10 37	09 06	02 36	23 08	01 N 50	22 02	21 48	22 50	08 35	22 28	09 44	16 19	14 14
12	10 39	09 03	01 34	23 11	07 31	22 29	21 30	22 44	08 33	22 27	09 44	16 20	14 14
13	10 40	09 00	00 S 29	23 15	12 49	22 54	21 13	22 37	08 30	22 27	09 44	16 20	14 14
14	10 R 40	08 57	00 N 36	23 18	17 33	23 16	20 56	22 31	08 28	22 26	09 44	16 20	14 14
15	10 38	08 53	01 38	23 20	21 34	23 37	20 39	22 24	08 25	22 25	09 44	16 20	14 14
16	10 34	08 50	02 36	23 22	24 40	23 55	20 23	22 17	08 22	22 25	09 44	16 21	14 14
17	10 28	08 47	03 28	23 24	26 42	24 11	20 07	22 10	08 19	22 24	09 45	16 21	14 14
18	10 21	08 44	04 10	23 25	27 31	24 24	19 52	22 03	08 17	22 23	09 45	16 21	14 14
19	10 12	08 41	04 41	23 26	27 04	24 34	19 38	21 55	08 14	22 22	09 45	16 22	14 14
20	10 03	08 37	05 00	23 26	25 22	24 41	19 24	21 48	08 11	22 22	09 45	16 22	14 14
21	09 54	08 34	05 06	23 26	22 31	24 46	19 11	21 40	08 08	22 21	09 45	16 22	14 14
22	09 47	08 31	04 59	23 26	18 39	24 48	18 59	21 32	08 05	22 20	09 46	16 23	14 14
23	09 41	08 28	04 37	23 25	13 56	24 46	18 48	21 24	08 02	22 19	09 46	16 23	14 14
24	09 38	08 25	04 02	23 24	08 34	24 43	18 37	21 16	07 58	22 19	09 46	16 23	14 14
25	09 37	08 22	03 13	23 22	02 N 42	24 36	18 27	21 08	07 55	22 18	09 46	16 24	14 14
26	09 D 37	08 18	02 13	23 20	03 S 26	24 27	18 18	20 59	07 52	22 17	09 47	16 24	14 14
27	09 38	08 15	01 N 04	23 18	09 36	24 16	18 10	20 50	07 49	22 16	09 47	16 24	14 15
28	09 R 38	08 12	00 S 11	23 15	15 30	24 02	18 03	20 42	07 46	22 16	09 47	16 25	14 15
29	09 37	08 09	01 27	23 12	20 42	23 46	17 56	20 33	07 42	22 15	09 48	16 25	14 15
30	09 ♉ 34	08 ♉ 06	02 S 40	23 N 08	24 S 44	23 N 28	17 N 51	20 N 24	07 N 39	22 N 14	09 S 48	16 S 25	14 S 15

ZODIAC SIGN ENTRIES

Date	h	m	Planets
02	07	52	☽ ♐
04	07	12	☽ ♑
05	12	47	☿ ♊
06	07	10	☽ ♒
08	09	38	☽ ♓
10	15	49	☽ ♈
13	01	37	☽ ♉
15	13	44	☽ ♊
18	02	37	☽ ♋
19	19	49	☿ ♋
20	15	05	☽ ♌
21	00	57	☉ ♋
23	02	10	☽ ♍
23	20	50	♂ ♌
25	10	50	☽ ♎
27	16	13	☽ ♏
29	18	15	☽ ♐

LATITUDES

Date	Mercury ☿ ° '	Venus ♀ ° '	Mars ♂ ° '	Jupiter ♃ ° '	Saturn ♄ ° '	Uranus ♅ ° '	Neptune ♆ ° '	Pluto ♇ ° '
01	01 S 56	01 N 24	01 N 15	01 N 15	00 S 19	00 S 46	00 S 03	08 N 56
04	01 27	00 44	01 15	01 14	00 19	00 46	00 03	08 56
07	00 55	00 N 02	01 15	01 14	00 19	00 46	00 03	08 55
10	00 S 22	00 S 41	01 15	01 13	00 19	00 46	00 03	08 55
13	00 N 10	01 22	01 14	01 13	00 18	00 46	00 03	08 55
16	00 41	02 00	01 14	01 12	00 18	00 46	00 04	08 54
19	01 08	02 35	01 14	01 12	00 18	00 47	00 04	08 54
22	01 29	03 06	01 14	01 11	00 17	00 47	00 04	08 54
25	01 44	03 32	01 13	01 11	00 17	00 47	00 04	08 53
28	01 52	03 54	01 13	01 10	00 17	00 47	00 04	08 53
31	01 N 54	04 S 11	01 N 13	01 N 10	00 S 17	00 S 47	00 S 04	08 N 52

DATA

Julian Date	2453158
Delta T	+65 seconds
Ayanamsa	23° 54' 56"
Synetic vernal point	05° ♓ 12' 03"
True obliquity of ecliptic	23° 26' 26"

MOON'S PHASES, APSIDES AND POSITIONS ☽

Day	h	m	Phase	Longitude ° '	Eclipse Indicator
03	04	20	○	12 ♐ 56	
09	20	02	☾	19 ♓ 18	
17	20	27	●	26 ♊ 57	
25	19	08	☽	04 ♎ 32	

Day	h	m		
03	13	18	Perigee	
17	16	17	Apogee	
04	17	32	Max dec	27° S 34'
11	04	28	0N	
18	15	32	Max dec	27° N 32'
25	22	39	0S	

LONGITUDES

Date	Chiron ⚷ ° '	Ceres ⚳ ° '	Pallas ⚴ ° '	Juno ⚵ ° '	Vesta ⚶ ° '	Black Moon Lilith ⚸ ° '
01	25 ♑ 34	05 ♌ 13	21 ♊ 50	24 ♑ 04	18 ♓ 02	22 ♊ 56
11	25 ♑ 09	09 ♌ 02	27 ♊ 47	22 ♑ 54	21 ♓ 05	24 ♊ 03
21	24 ♑ 40	13 ♌ 20	03 ♋ 46	21 ♑ 11	23 ♓ 42	25 ♊ 09
31	24 ♑ 07	17 ♌ 34	09 ♋ 45	19 ♑ 05	25 ♓ 49	26 ♊ 16

ASPECTARIAN

h m	Aspects
01 Tuesday	
00 26	☿ # ♆
01 06	☽ ⊼ ☉
03 05	☽ △ ♄
04 43	☿ ⊻ ♀
06 08	☽ ∥ ♆
06 47	☽ # ☿
08 04	☽ ⊥ ♇
08 20	☽ □ ♆
09 08	☽ △ ♂
09 30	☽ ± ♀
17 44	☽ ⊻ ♇
18 50	☽ ⊼ ♀
19 09	☽ Q ♃
21 16	☽ ☍ ☿
02 Wednesday	
03 27	☽ bQ ♄
10 14	☽ bQ ♂
10 17	☽ # ☉
11 42	☉ ⊻ ♄
11 55	☽ # ♄
13 09	☽ Q ♆
17 51	☽ ± ♄
18 19	☽ # ♂
18 36	☽ □ ♅
22 14	☽ # ♀
23 56	☽ □ ♃
03 Thursday	
01 05	☽ ± ♂
01 24	♀ ☍ ♇
03 24	☽ ⊼ ♄
04 20	☽ ☍ ☉
08 05	☽ ⚹ ♆
10 57	☽ ⊼ ♂
16 39	☽ ☍ ♀
17 12	☽ ☌ ♇
22 59	☽ Q ♅
04 Friday	
03 04	☽ ⊼ ☿
04 15	☿ ∠ ♄
07 41	☽ ∠ ♆
13 51	☽ ± ☿
17 54	☉ ∥ ♄
17 55	☽ ⚹ ♅
21 55	☽ ⊥ ♆
23 30	☽ △ ♃
05 Saturday	
03 07	☽ ☍ ♄
06 08	☽ bQ ☿
06 53	☽ ⊼ ☉
07 27	☽ ⊻ ♆
12 28	☽ ☍ ♂
14 22	☽ ⊼ ♀
15 40	☉ △ ♆
16 41	☽ ⊻ ♇
17 10	☽ ± ☉
17 51	☽ ∠ ♅
23 39	☽ ± ♀
23 41	☽ bQ ♃
06 Sunday	
02 21	☽ ⊥ ♇
03 58	♀ ∥ ♂
08 27	☽ ⊥ ♅
08 45	☽ bQ ☉
09 54	☽ △ ☿
12 03	♀ ⊻ ♂
13 44	☽ bQ ♀
14 27	☽ ± ♃
17 03	☽ ∠ ♇
18 18	☽ ⊻ ♅
20 13	☽ # ♂
21 07	☽ # ♀
23 28	☽ # ☉
07 Monday	
00 27	☽ ⊼ ♃
00 59	☽ # ♄
04 21	☽ ⊼ ♄
08 27	☽ ☌ ♆
11 25	☽ △ ☉
13 42	☽ △ ♀
14 27	☽ # ☿
14 32	☽ ± ♄
16 05	☽ ⊼ ♂
18 09	☽ ⚹ ♇
21 39	☉ ∥ ♀
08 Tuesday	
02 50	☽ ± ♂
03 47	☿ ∠ ♂
06 06	☽ bQ ♄
08 28	☽ ∥ ♆
08 43	☉ ☌ ♀
14 51	☽ Q ♇
17 43	☽ ∥ ♇
19 17	☽ bQ ♂
21 08	☽ □ ☿
21 34	☽ ☌ ♅
09 Wednesday	
00 05	☿ □ ♅
03 08	♀ ∥ ♄
03 20	☿ ⊥ ♄
04 31	☽ ☍ ♃
08 48	☽ △ ♄
12 29	♂ ⊼ ♇
12 42	☽ ⊻ ♆
12 47	☽ ∥ ♅
15 59	☽ □ ♀
17 11	☽ # ♃
19 20	☉ ∥ ♂
20 02	☽ □ ☉
23 03	☽ □ ♇
23 37	☽ ⊥ ♆
23 37	☽ △ ♂
10 Thursday	
11 47	☽ Q ☿
15 49	♅ St R
16 16	☽ ∠ ♆
20 23	♂ bQ ♅
23 49	☽ Q ♀
23 51	☿ □ ♃
11 Friday	
04 23	☿ ∥ ♀
04 38	☽ ⊻ ♅
06 51	♀ ⊥ ♂
08 20	☽ Q ☉
12 26	☉ ☍ ♇
12 31	☽ ⊼ ♃
14 56	☽ ⚹ ☿
16 05	☽ ⊥ ♅
17 13	☽ □ ♄
20 46	☽ ⚹ ♆
21 40	☽ ⚹ ♀
12 Saturday	
00 09	☽ ± ♃
05 12	☿ ⊻ ♄
07 44	☽ △ ♇
09 24	☽ ⚹ ☉
09 30	☽ ∠ ♅
10 38	☿ ∥ ♄
11 31	☽ □ ♂
16 31	☽ ∥ ♃
17 08	♀ △ ♆
17 51	☽ bQ ♃
20 07	☽ Q ♆
21 49	☽ # ♅
23 24	☿ ☌ ♀
23 25	☿ ∥ ♂
13 Sunday	
01 05	☿ △ ♆
01 37	☽ ∠ ♀
02 13	☽ ∠ ☿
04 43	☽ Q ♄
13 12	☽ bQ ♇
15 05	☽ ⚹ ♅
17 31	☽ ∠ ☉
18 40	☽ ⊥ ♀
18 53	☽ # ♇
23 52	☽ △ ♃
23 59	☽ ⊥ ☿
14 Monday	
01 06	☽ Q ♂
04 52	☽ ⚹ ♄
05 31	☽ # ♆
06 08	☽ ⊻ ♀
07 09	☽ ± ♇
07 52	☽ □ ♆
08 12	☿ ⊥ ♂
13 06	☽ ⊥ ☉
13 33	☉ ∥ ☿
14 41	☽ ⊻ ☿
15 06	☽ Q ♅
19 11	☽ ⊼ ♇
15 Tuesday	
02 15	☽ ⊻ ☉
02 34	☽ ⚹ ♂
04 32	♀ ⊻ ♄
06 28	☽ ∥ ♀
07 08	♂ ∥ ♄
11 25	☽ ∠ ♄
13 15	☉ ⊻ ♂
15 07	☿ ☍ ♇
17 32	☽ ∥ ♂
17 52	☽ ∥ ♄
16 Wednesday	
00 51	☽ ∥ ☉
03 29	☽ □ ♅
04 39	☽ ∥ ☿
05 51	☽ ⊥ ♄
10 40	☽ ∠ ♂
13 01	☽ □ ♃
16 12	☽ ☌ ♀
18 10	☽ ⊻ ♄
20 29	☽ △ ♆
17 Thursday	
05 59	☽ ⊥ ♂
07 51	☽ ☍ ♇
17 14	☽ ☌ ☿
18 51	☽ ⊻ ♂
20 27	☽ ☌ ☉
18 Friday	
01 56	☽ Q ♃
02 54	☽ bQ ♆
04 47	☿ ⊻ ♂
15 45	♂ ± ♇
16 17	♂ ∠ ♃
16 20	☽ △ ♅
21 06	☽ ± ♆
21 24	☉ ☌ ☿
19 Saturday	
01 05	♀ ∠ ♂
02 28	☽ ⚹ ♃
02 46	☽ ⊻ ♀
07 37	☽ ☌ ♄
09 12	☽ ⊼ ♆
09 29	♀ □ ♃
14 28	☽ ⊥ ♀
18 18	☿ Q ♃
20 25	☽ ⊼ ♇
20 59	☿ bQ ♆
22 34	☽ bQ ♅
20 Sunday	
07 59	☽ ∠ ♀
08 25	☽ ± ♇
08 57	☽ ∠ ♃
10 46	☽ ☌ ♂
14 14	☽ ⊻ ☉
16 35	☽ ± ♅
18 25	☽ ∥ ☿
19 22	☽ ⊻ ☿
21 Monday	
01 28	☉ Q ♃
02 21	☽ bQ ♇
03 06	☉ bQ ♆
03 06	☽ ⊥ ♃
03 13	☽ ⊥ ☉
04 32	☽ ⊼ ♅
05 07	☽ ∥ ☉
09 55	☽ ⊥ ☿
13 00	☽ ⚹ ♀
13 09	☽ ∥ ♄
15 07	☽ ⊻ ♃
18 00	☽ ∥ ♂
20 11	☽ ⊻ ♄
21 00	☽ ☍ ♆
22 33	☽ ∠ ☉
22 Tuesday	
07 30	☽ ∠ ☿
07 54	☽ △ ♇
08 06	☽ ⊥ ♄
10 00	☽ ∥ ♀
12 01	☽ Q ♀
22 11	☿ △ ♅
23 Wednesday	
00 01	☽ # ♆
01 10	☽ ⊻ ♂
01 48	☽ ∠ ♄
06 16	☽ ⚹ ☉
10 33	☽ # ♇
13 21	☽ ⊥ ♂
15 07	☽ ☍ ♅
18 38	☽ ⚹ ☿
22 02	☽ □ ♀
23 58	☿ ± ♆
24 Thursday	
01 57	☽ ☌ ♃
06 48	☽ ⚹ ♄
06 49	☽ # ♅
06 55	☽ ⊼ ♆
07 04	☽ Q ☉
07 23	☽ ∠ ♂
13 11	☿ ⊻ ♀
14 30	☽ ∥ ♃
16 40	♄ ⊼ ♆
17 19	☽ □ ♇
18 11	☽ ± ♆
21 48	☽ Q ☿
25 Friday	
00 03	♂ ± ♅
05 28	☽ Q ♄
10 51	☽ bQ ♆
12 45	☽ ⚹ ♂
16 48	☿ ⚹ ♃
19 08	☽ □ ☉
23 04	☽ ⊼ ♅
26 Saturday	
02 21	☽ Q ♇
04 50	☽ △ ♀
09 51	☽ ± ♅
09 53	☽ ⊻ ♃
11 33	☽ Q ♂
12 48	☽ □ ☿
13 56	☽ △ ♆
14 26	☽ □ ♄
20 11	☿ ⊼ ♆
20 39	☽ ⊥ ♃
23 41	☽ ⚹ ♇
27 Sunday	
00 07	☿ ☌ ♄
01 47	☽ bQ ♅
05 03	☽ # ♃
05 27	☿ ⊥ ♀
07 09	☽ bQ ♀
12 32	☽ ∠ ♃
12 42	☽ ∥ ♅
20 31	☽ □ ♂
22 34	☽ ± ♀
28 Monday	
00 55	☉ △ ♅
01 34	☽ ∠ ♇
03 37	☽ △ ♅
03 49	☽ △ ☉
06 43	☽ ∥ ♇
08 41	☽ ⊼ ♀
14 18	☽ ⚹ ♃
15 56	☽ ∥ ♆
16 42	☽ ⊥ ♇
17 31	☽ □ ♆
18 31	☽ △ ♄
23 00	☽ # ♀
29 Tuesday	
00 57	☽ △ ☿
02 38	☽ ⊻ ♇
06 36	☽ bQ ☉
10 20	☽ Q ♃
10 22	♀ ⊥ ♄
11 13	☽ # ♂
13 12	☿ ⊼ ♇
19 22	☽ bQ ♄
20 13	☽ # ♄
22 16	☽ ± ☉
22 59	☽ Q ♆
23 15	♀ St D
30 Wednesday	
00 31	☽ △ ♂
01 35	☽ # ☉
04 00	☿ bQ ♅
04 08	☽ # ☿
04 59	☽ □ ♅
05 07	☽ bQ ☿
08 34	☽ ⊼ ☉
09 11	☉ ± ♆
09 46	☽ ☍ ♀
09 58	☽ ± ♄
15 38	☽ □ ♃
18 10	☽ ⚹ ♆
19 37	☽ ⊼ ♄
21 36	☽ ± ☿

All ephemeris data is given at 12.00 UT and the Moon's longitude is additionally given for 24.00 UT

LONGITUDES

Date	Sidereal time h m s	Sun ☉ ° ' "	Moon ☽ ° ' "	Moon ☽ 24.00 ° ' "	Mercury ☿ ° '	Venus ♀ ° '	Mars ♂ ° '	Jupiter ♃ ° '	Saturn ♄ ° '	Uranus ♅ ° '	Neptune ♆ ° '	Pluto ⯓ ° '
01	06 39 31	09 ♋ 58 33	26 ♐ 10 35	03 ♑ 48 11	24 ♋ 13	09 ♊ 40	04 ♌ 48	13 ♍ 26	15 ♋ 54	06 ♓ 37	14 ♒ 53	20 ♐ 23
02	06 43 28	10 55 44	11 ♑ 26 12	19 ♑ 03 15	26 06	09 D 45	05 26	13 34	16 02	06 R 36	14 R 52	20 R 22
03	06 47 24	11 52 55	26 ♑ 37 59	04 ♒ 09 06	27 58	09 52	06 04	13 43	16 09	06 35	14 51	20 20
04	06 51 21	12 50 06	11 ♒ 35 26	18 ♒ 56 02	29 ♋ 47	10 01	06 42	13 52	16 17	06 34	14 49	20 19
05	06 55 17	13 47 17	26 ♒ 10 07	03 ♓ 17 08	01 ♌ 34	10 12	07 20	14 01	16 25	06 33	14 48	20 18
06	06 59 14	14 44 28	10 ♓ 16 46	17 ♓ 08 54	03 20	10 25	07 57	14 10	16 33	06 32	14 47	20 16
07	07 03 10	15 41 39	23 ♓ 53 37	00 ♈ 31 10	05 03	10 40	08 35	14 20	16 41	06 31	14 45	20 15
08	07 07 07	16 38 51	07 ♈ 01 54	13 ♈ 26 19	06 44	10 57	09 13	14 29	16 48	06 30	14 44	20 13
09	07 11 03	17 36 03	19 ♈ 44 57	25 ♈ 58 24	08 23	11 16	09 51	14 38	16 56	06 28	14 43	20 12
10	07 15 00	18 33 16	02 ♉ 07 20	08 ♉ 12 22	10 00	11 37	10 28	14 48	17 04	06 27	14 41	20 11
11	07 18 57	19 30 29	14 ♉ 14 08	20 ♉ 13 16	11 35	12 00	11 06	14 58	17 12	06 26	14 40	20 09
12	07 22 53	20 27 43	26 ♉ 10 22	02 ♊ 06 00	13 08	12 24	11 44	15 07	17 20	06 24	14 38	20 08
13	07 26 50	21 24 57	08 ♊ 00 40	13 ♊ 54 50	14 39	12 50	12 22	15 17	17 27	06 23	14 37	20 07
14	07 30 46	22 22 11	19 ♊ 48 58	25 ♊ 43 25	16 07	13 17	13 00	15 27	17 35	06 21	14 35	20 05
15	07 34 43	23 19 27	01 ♋ 38 31	07 ♋ 34 35	17 34	13 46	13 38	15 37	17 43	06 20	14 34	20 04
16	07 38 39	24 16 42	13 ♋ 31 50	19 ♋ 30 29	18 58	14 17	14 15	15 47	17 51	06 18	14 32	20 03
17	07 42 36	25 13 58	25 ♋ 30 42	01 ♌ 32 39	20 21	14 48	14 53	15 57	17 58	06 17	14 31	20 02
18	07 46 32	26 11 15	07 ♌ 36 27	13 ♌ 42 15	21 41	15 22	15 31	16 08	18 06	06 15	14 29	20 00
19	07 50 29	27 08 31	19 ♌ 50 08	26 ♌ 00 16	22 59	15 56	16 09	16 18	18 14	06 13	14 28	19 59
20	07 54 26	28 05 48	02 ♍ 12 47	08 ♍ 27 51	24 14	16 32	16 47	16 28	18 22	06 12	14 26	19 58
21	07 58 22	29 ♋ 03 06	14 ♍ 45 40	21 ♍ 06 27	25 27	17 09	17 25	16 39	18 29	06 10	14 25	19 57
22	08 02 19	00 ♌ 00 24	27 ♍ 30 27	03 ♎ 57 58	26 38	17 47	18 02	16 50	18 37	06 08	14 23	19 56
23	08 06 15	00 57 42	10 ♎ 29 17	17 ♎ 04 43	27 46	18 26	18 40	17 00	18 45	06 06	14 22	19 55
24	08 10 12	01 55 00	23 ♎ 44 37	00 ♏ 29 16	28 52	19 06	19 18	17 11	18 52	06 05	14 20	19 54
25	08 14 08	02 52 19	07 ♏ 18 57	14 ♏ 13 53	29 ♌ 55	19 47	19 56	17 22	19 00	06 03	14 18	19 53
26	08 18 05	03 49 38	21 ♏ 14 12	28 ♏ 19 56	00 ♍ 55	20 29	20 34	17 33	19 08	06 01	14 17	19 51
27	08 22 01	04 46 58	05 ♐ 30 59	12 ♐ 47 05	01 52	21 13	21 12	17 44	19 15	05 59	14 15	19 50
28	08 25 58	05 44 18	20 ♐ 07 50	27 ♐ 32 35	02 47	21 57	21 50	17 55	19 23	05 57	14 14	19 49
29	08 29 55	06 41 39	05 ♑ 00 33	12 ♑ 30 47	03 38	22 42	22 28	18 06	19 30	05 55	14 12	19 49
30	08 33 51	07 39 00	20 ♑ 02 09	27 ♑ 33 30	04 26	23 28	23 05	18 17	19 38	05 53	14 10	19 48
31	08 37 48	08 ♌ 36 22	05 ♒ 03 33	12 ♒ 31 05	05 ♍ 10	24 ♊ 14	23 ♌ 43	18 ♍ 28	19 ♋ 46	05 ♓ 51	14 ♒ 09	19 ♐ 47

Date	Moon True ☊ ° '	Moon Mean ☊ ° '	Moon ☽ Latitude ° '
01	09 ♉ 28	08 ♉ 02	03 S 42
02	09 R 20	07 59	04 29
03	09 12	07 56	04 56
04	09 02	07 53	05 03
05	08 54	07 50	04 50
06	08 48	07 47	04 19
07	08 44	07 43	03 35
08	08 42	07 40	02 40
09	08 D 41	07 37	01 39
10	08 42	07 34	00 S 35
11	08 R 42	07 31	00 N 29
12	08 40	07 28	01 31
13	08 37	07 24	02 29
14	08 31	07 21	03 20
15	08 22	07 18	04 02
16	08 11	07 15	04 33
17	07 58	07 12	04 53
18	07 44	07 08	05 00
19	07 32	07 05	04 53
20	07 21	07 02	04 33
21	07 12	06 59	03 59
22	07 06	06 56	03 13
23	07 03	06 53	02 15
24	07 D 02	06 49	01 N 10
25	07 R 02	06 46	00 S 02
26	07 02	06 43	01 14
27	07 00	06 40	02 24
28	06 56	06 37	03 26
29	06 49	06 34	04 16
30	06 40	06 30	04 48
31	06 ♉ 29	06 ♉ 27	05 S 00

DECLINATIONS

Date	Sun ☉ ° '	Moon ☽ ° '	Mercury ☿ ° '	Venus ♀ ° '	Mars ♂ ° '	Jupiter ♃ ° '	Saturn ♄ ° '	Uranus ♅ ° '	Neptune ♆ ° '	Pluto ⯓ ° '
01	23 N 04	27 S 05	23 N 08	17 N 46	20 N 15	07 N 35	22 N 13	09 S 49	16 S 26	14 S 15
02	22 59	27 25	22 47	17 42	20 05	07 32	22 12	09 49	16 26	14 15
03	22 55	25 41	22 24	17 39	19 56	07 28	22 11	09 49	16 27	14 15
04	22 49	22 10	21 59	17 36	19 46	07 25	22 11	09 50	16 27	14 15
05	22 44	17 20	21 33	17 34	19 37	07 21	22 10	09 50	16 27	14 15
06	22 38	11 43	21 05	17 33	19 27	07 17	22 09	09 51	16 28	14 15
07	22 31	05 S 43	20 37	17 32	19 17	07 14	22 08	09 51	16 28	14 16
08	22 24	00 N 20	20 07	17 32	19 07	07 10	22 07	09 52	16 29	14 16
09	22 17	06 12	19 37	17 33	18 56	07 06	22 06	09 52	16 29	14 16
10	22 09	11 40	19 06	17 34	18 46	07 02	22 05	09 53	16 29	14 16
11	22 01	16 35	18 34	17 36	18 35	06 59	22 04	09 53	16 30	14 16
12	21 53	20 47	18 01	17 38	18 25	06 55	22 04	09 54	16 30	14 16
13	21 44	24 06	17 28	17 40	18 14	06 51	22 03	09 54	16 31	14 16
14	21 35	26 22	16 55	17 43	18 03	06 47	22 02	09 55	16 31	14 17
15	21 26	27 28	16 21	17 46	17 52	06 43	22 01	09 55	16 32	14 17
16	21 16	27 17	15 47	17 50	17 41	06 39	22 00	09 56	16 32	14 17
17	21 05	25 51	15 13	17 53	17 30	06 35	21 59	09 57	16 32	14 17
18	20 55	23 12	14 38	17 58	17 19	06 31	21 58	09 57	16 33	14 17
19	20 44	19 30	14 04	18 02	17 07	06 27	21 57	09 58	16 33	14 18
20	20 33	14 56	13 30	18 06	16 56	06 22	21 56	09 59	16 34	14 18
21	20 21	09 40	12 56	18 11	16 44	06 18	21 55	09 59	16 34	14 18
22	20 09	03 N 56	12 22	18 16	16 32	06 14	21 54	10 00	16 35	14 18
23	19 57	02 S 05	11 49	18 21	16 20	06 10	21 53	10 01	16 35	14 18
24	19 44	08 08	11 16	18 26	16 08	06 06	21 52	10 01	16 36	14 19
25	19 31	13 59	10 43	18 31	15 56	06 01	21 51	10 02	16 36	14 19
26	19 18	19 16	10 11	18 36	15 44	05 57	21 50	10 03	16 37	14 19
27	19 04	23 36	09 40	18 41	15 32	05 53	21 49	10 04	16 37	14 19
28	18 50	26 30	09 09	18 46	15 19	05 48	21 48	10 04	16 38	14 20
29	18 36	27 36	08 40	18 51	15 07	05 44	21 47	10 05	16 38	14 20
30	18 21	26 41	08 11	18 56	14 54	05 39	21 46	10 06	16 39	14 20
31	18 N 07	23 S 51	07 N 44	19 N 01	14 N 42	05 N 35	21 N 45	10 S 07	16 S 39	14 S 20

ZODIAC SIGN ENTRIES

Date	h	m	Planets
01	18	01	☽ ♑
03	17	22	☽ ♒
04	14	52	☿ ♌
05	18	26	☽ ♓
07	23	03	☽ ♈
10	07	51	☽ ♉
12	19	45	☽ ♊
15	08	40	☽ ♋
17	20	56	☽ ♌
20	07	48	☽ ♍
22	11	50	☉ ♌
22	16	39	☽ ♎
24	23	08	☽ ♏
25	13	58	☿ ♍
27	02	48	☽ ♐
29	03	57	☽ ♑
31	03	54	☽ ♒

LATITUDES

Date	Mercury ☿ ° '	Venus ♀ ° '	Mars ♂ ° '	Jupiter ♃ ° '	Saturn ♄ ° '	Uranus ♅ ° '	Neptune ♆ ° '	Pluto ⯓ ° '
01	01 N 54	04 S 11	01 N 13	01 N 10	00 S 17	00 S 47	00 S 04	08 N 52
04	01 50	04 24	01 12	01 09	00 16	00 47	00 04	08 51
07	01 39	04 33	01 12	01 09	00 16	00 47	00 04	08 51
10	01 24	04 39	01 12	01 09	00 16	00 47	00 04	08 50
13	01 04	04 42	01 11	01 08	00 16	00 47	00 04	08 49
16	00 41	04 43	01 11	01 08	00 15	00 47	00 04	08 48
19	00 N 13	04 42	01 11	01 08	00 15	00 48	00 04	08 48
22	00 S 17	04 38	01 10	01 07	00 15	00 48	00 04	08 47
25	00 50	04 33	01 10	01 07	00 15	00 48	00 04	08 46
28	01 25	04 26	01 09	01 07	00 14	00 48	00 04	08 45
31	02 S 02	04 S 18	01 N 09	01 N 06	00 S 14	00 S 48	00 S 04	08 N 44

DATA

Julian Date	2453188
Delta T	+65 seconds
Ayanamsa	23° 55' 02"
Synetic vernal point	05° ♓ 11' 58"
True obliquity of ecliptic	23° 26' 26"

MOON'S PHASES, APSIDES AND POSITIONS ☽

Date	h	m	Phase	Longitude ° '	Eclipse Indicator
02	11	09	○	10 ♑ 54	
09	07	34	☾	17 ♈ 25	
17	11	24	●	25 ♋ 13	
25	03	37	☽	02 ♏ 32	
31	18	05	○	08 ♒ 51	

Day	h	m		
01	23	03	Perigee	
14	21	21	Apogee	
30	06	27	Perigee	
02	03	45	Max dec	27° S 32'
08	10	39	0N	
15	20	47	Max dec	27° N 33'
23	03	47	0S	
29	13	12	Max dec	27° S 36'

LONGITUDES

Date	Chiron ⚷ ° '	Ceres ⚳ ° '	Pallas ⚴ ° '	Juno ⚵ ° '	Vesta ⚶ ° '	Black Moon Lilith ⚸ ° '
01	24 ♑ 07	17 ♌ 34	09 ♋ 45	19 ♑ 05	25 ♓ 49	26 ♊ 16
11	23 ♑ 31	21 ♌ 54	15 ♋ 43	16 ♑ 45	27 ♓ 21	27 ♊ 36
21	22 ♑ 55	26 ♌ 18	21 ♋ 40	14 ♑ 26	28 ♓ 13	28 ♊ 30
31	22 ♑ 20	00 ♍ 46	27 ♋ 34	12 ♑ 21	28 ♓ 20	29 ♊ 37

All ephemeris data is given at 12.00 UT and the Moon's longitude is additionally given for 24.00 UT

ASPECTARIAN

h m Aspects

01 Thursday
01 30 ☽ ♇ ♂
02 53 ☽ ☌ ⯓
03 52 ☉ ⊻ ♀
08 28 ☽ ⚻ ☿
09 33 ☽ Q ♅
16 19 ☽ ± ♂
17 50 ☽ ∠ ♆
17 55 ☿ ∠ ♀
18 33 ☉ ∥ ☿

02 Friday
02 10 ☽ ⚻ ♂
04 25 ☽ ⚹ ♅
07 58 ☽ ⊥ ♆
09 20 ☽ ⚻ ♀
09 21 ♂ ♇ ⯓
11 09 ☽ ☍ ☉
15 15 ☿ ± ⯓
15 24 ☽ △ ♃
17 23 ☽ ⊻ ♆
18 50 ☽ ± ♀
19 17 ☽ ☍ ♄

03 Saturday
02 03 ☽ ⊻ ⯓
04 01 ☽ ∠ ♅
09 10 ☽ ♇ ♀
11 32 ☽ ⊥ ⯓
14 24 ☽ ☍ ☿
15 21 ☽ ♇ ♃
18 18 ☽ ⊥ ♅
22 49 ☿ ∠ ♃

04 Sunday
00 22 ☿ ∥ ♄
01 53 ☽ ∠ ⯓
03 44 ☽ ☍ ♂
03 54 ☽ ⊻ ♅
05 55 ☽ ± ♃
07 27 ♂ ⚻ ♅
08 06 ☽ # ☉
09 25 ☽ △ ♀
11 54 ☽ # ♄
13 07 ☽ # ☿
14 10 ☽ ⚻ ☉
15 45 ☽ ⚻ ♃
17 15 ☽ ☌ ♆
19 43 ☽ ⚻ ♄
22 25 ☿ ± ♅

05 Monday
00 40 ☽ ± ☉
00 58 ☽ # ♂
02 15 ☽ ⚹ ⯓
05 42 ☽ ± ♄
10 57 ☽ # ♀
15 57 ☽ ∥ ♆
16 42 ☽ ♇ ☉
18 59 ☉ ⚹ ♃
20 55 ☽ ♇ ♄
22 18 ☽ Q ⯓
22 23 ☽ ⚻ ☿

06 Tuesday
01 27 ☽ ∥ ⯓
05 33 ☽ ☌ ♅
07 48 ☽ ⚻ ♂
10 07 ☽ ± ☿
12 15 ☽ □ ♀
12 56 ☉ ⚻ ♆
18 42 ☽ ± ♂
18 51 ☽ ☍ ♃
19 33 ☽ ∥ ♅
19 49 ☽ ⊻ ♆
20 21 ☽ △ ☉
23 01 ♂ ⊥ ♃
23 03 ☽ △ ♄

07 Wednesday
04 08 ☽ ♇ ☿
05 30 ☽ □ ⯓
05 56 ☽ # ♃
06 24 ☽ ⊥ ♆
11 25 ☽ ♇ ♂
13 02 ♀ ⊥ ♄
14 46 ☿ ♇ ⯓
20 49 ☽ Q ♀
22 35 ☽ ∠ ♆

08 Thursday
08 37 ☿ ⚻ ♅
11 00 ☽ ⊻ ♅
11 22 ☽ △ ☿
16 17 ☽ △ ♂
16 38 ☉ ☌ ♄
19 30 ☽ ⚹ ♀
22 12 ☽ ⊥ ♅
23 29 ☉ ⊥ ♀

09 Friday
02 09 ☽ ⚻ ♃
02 25 ☽ ⚹ ♆
06 34 ☽ □ ♄
07 34 ☽ □ ☉
12 52 ☽ △ ⯓
13 44 ☽ ± ♃
15 18 ☽ ∠ ♅
15 49 ☽ ∥ ♃
16 12 ☿ ⊥ ♃
21 10 ♃ ⚻ ♆

10 Saturday
00 56 ☽ ∠ ♀
01 24 ☽ Q ♆
03 55 ☽ # ♅
07 23 ☽ ♇ ♃
17 51 ☽ Q ♄
18 00 ☽ ♇ ⯓
19 06 ☽ ⊥ ♀
20 31 ☽ ⚹ ♅
21 29 ☽ Q ☉
23 50 ☿ ☌ ♂

11 Sunday
00 19 ☽ # ⯓
01 28 ☉ ∥ ♄
05 25 ☽ □ ♂
05 55 ☽ □ ☿
07 23 ☽ ⊻ ♀
10 02 ☿ ∥ ♂
11 34 ☽ # ♆
11 50 ☽ ± ⯓
12 51 ☽ □ ♆
13 28 ☽ △ ♃
17 26 ☽ ∥ ♀
17 59 ☽ ⚹ ♄
20 23 ☽ Q ♅
20 30 ☿ ⚹ ♀
21 34 ☽ ∥ ☿
22 30 ☽ ∥ ♂
23 29 ☽ ⚹ ☉
23 51 ☽ ⚻ ⯓

12 Monday
03 53 ☉ ⚻ ⯓
18 59 ☽ ∥ ☉
19 37 ☽ Q ♂
20 27 ☽ ∥ ♄
23 31 ☽ Q ☿

13 Tuesday
00 36 ☽ ∠ ♄
04 02 ☿ ∥ ♀
08 28 ☽ ∠ ☉
08 42 ☽ □ ♅
11 07 ☉ ♇ ♅
11 31 ☿ ☍ ♆
19 05 ☽ ⊥ ♄
21 21 ☽ ⚹ ♂
22 11 ☽ ☌ ♀
23 38 ☿ ⊻ ♃

14 Wednesday
01 24 ☽ △ ♆
03 00 ☽ □ ♃
03 26 ☽ ⚹ ☿
04 23 ☽ ⊥ ☉
07 25 ☽ ⊻ ♄
12 33 ☽ ☍ ⯓
17 39 ☽ ⊻ ☉

15 Thursday
04 32 ☿ # ♆
05 33 ☽ ∠ ♂
07 48 ☽ ♇ ♆
14 08 ☽ ∠ ☿
14 44 ☿ ⊻ ♄
16 03 ☽ Q ♃
21 28 ☽ △ ♅
22 16 ♀ ∥ ♂

16 Friday
00 47 ☽ ⊥ ♂
01 58 ☽ ± ♆
10 44 ☽ ⊥ ☿
13 32 ☽ ⊻ ♂
13 34 ☽ ⊻ ♀
14 02 ☽ ⚻ ♆
16 36 ☽ ⚹ ♃
16 45 ♀ ⚹ ♂
20 46 ☽ ☌ ♄
22 26 ♂ ☍ ♆
23 30 ♀ △ ♆

17 Saturday
00 21 ☽ ⊻ ☿
01 03 ☽ ⚻ ⯓
02 10 ☽ ⊥ ♀
03 34 ☽ ♇ ♅
06 26 ☿ △ ⯓
11 24 ☽ ☌ ☉
13 01 ☽ ± ⯓
20 57 ☽ ∠ ♀
21 28 ☽ ± ♅
22 59 ☽ ∠ ♃

18 Sunday
06 52 ☽ ♇ ⯓
07 33 ☉ ± ⯓
09 20 ☽ ⚻ ♅
17 02 ☽ ⊥ ♃
20 47 ☽ ∥ ♄

19 Monday
01 31 ☽ ☍ ♆
02 36 ☿ # ⯓
04 00 ☽ ⚹ ♀
04 19 ☽ ∥ ☉
04 24 ☽ ☌ ♂
04 59 ☽ ⊻ ♃
08 50 ☽ ⊻ ♄
12 18 ☽ △ ⯓
18 50 ☽ ☌ ☿
20 00 ♂ ⊻ ♃
20 04 ☽ ∥ ♀
20 39 ☽ ⊥ ♄

20 Tuesday
01 35 ☽ ∥ ♂
03 23 ☽ ⊻ ☉
03 53 ☽ # ♆
04 32 ☽ Q ♀
08 57 ♀ □ ♃
14 14 ☽ ∠ ♄
14 40 ☿ ⊥ ♄
15 02 ☽ # ⯓
15 55 ☽ ⊥ ☉
19 38 ☽ ☍ ♅
19 39 ☽ ∥ ☿

21 Wednesday
10 33 ☽ ∠ ☉
10 38 ☽ # ♅
11 20 ☽ ⚻ ♆
15 38 ☽ ☌ ♃
16 45 ☽ □ ♀
17 17 ☽ ⊻ ♂
19 08 ☽ ⚹ ♄
21 48 ☽ □ ⯓
22 40 ☽ ± ♆

22 Thursday
02 27 ☽ ∥ ♃
05 10 ☽ ⊥ ♂
07 00 ♂ # ♆
10 13 ☽ ⊻ ☿
15 09 ☉ ± ♅
15 30 ☽ ♇ ♆
17 02 ☽ ⚹ ☉
17 51 ☽ Q ♄
22 28 ☽ ⊥ ☿
22 49 ☽ ∠ ♂

23 Friday
03 58 ☽ ⚻ ♅
07 17 ☽ Q ⯓
14 57 ☽ ± ♅
15 32 ♂ ⊻ ♄
16 34 ☽ ∠ ☿
16 52 ☽ Q ☉
19 03 ☽ △ ♆

24 Saturday
00 02 ☽ ⊻ ♃
02 05 ♀ ⊻ ♄
03 10 ☽ □ ♄
03 12 ☽ △ ♀
03 37 ☽ ⚹ ♂
03 57 ☽ # ♃
05 06 ☽ ⚹ ⯓
07 13 ☽ ♇ ♅
10 59 ☽ ⊥ ♃
19 36 ☽ ∥ ♅
20 14 ☉ ∠ ♃
21 54 ☽ ⚹ ☿
23 34 ☽ # ☿

25 Sunday
02 06 ☽ Q ♂
03 12 ☽ ∠ ♃
03 37 ☽ □ ☉
07 20 ☽ ♇ ♀
07 44 ☽ ∠ ⯓
09 47 ☽ △ ♅
09 52 ♂ △ ⯓
13 26 ☽ ∥ ⯓
15 00 ♀ ☍ ⯓
20 12 ☽ # ♂
20 38 ☽ Q ☿
23 22 ☽ ⊥ ⯓
23 31 ☽ ∥ ♆
23 50 ☽ ± ♀

26 Monday
00 07 ☽ □ ♆
05 37 ☽ ⚹ ♃
08 22 ☽ △ ♄
08 43 ☽ # ♀
09 39 ☽ ⊻ ⯓
10 40 ☽ ⚻ ♀
10 48 ☽ □ ♂
12 09 ☽ # ☉
18 14 ☿ # ♅

27 Tuesday
01 24 ☽ # ♄
02 13 ☽ Q ♃
05 31 ☽ □ ☿
06 35 ☽ Q ♆
08 30 ♀ ⚹ ♂
09 53 ☽ ♇ ♄
10 42 ☽ △ ☉
12 46 ☽ □ ♅
13 27 ☉ ♇ ⯓

28 Wednesday
00 53 ☽ ± ♄
02 23 ☽ ⚹ ♆
08 20 ☽ □ ♃
10 46 ☽ ⚻ ♄
11 30 ☽ ☌ ⯓
13 03 ☽ ♇ ☉
14 53 ☽ △ ♂
15 06 ☽ ☍ ♀
16 56 ☉ ∥ ♀
17 07 ☉ ⚻ ♅
18 11 ☽ Q ♅

29 Thursday
02 41 ☽ ∠ ♆
04 36 ☽ ± ☉
09 40 ☽ △ ☿
13 27 ☽ ⚹ ♅
14 53 ☽ ⚻ ☉
16 06 ☽ ♇ ♂
17 06 ☽ ⊥ ♆

30 Friday
02 40 ☽ ⊻ ♆
07 06 ☽ ± ♂
09 10 ☽ △ ♃
10 59 ☽ ♇ ☿
11 21 ☽ ☍ ♄
11 37 ☽ ⊻ ⯓
13 21 ☽ ∠ ♅
17 05 ☽ ⚻ ♂
17 45 ☽ ⚻ ♀
19 46 ☿ ∠ ♄
21 10 ☽ ⊥ ⯓

31 Saturday
02 06 ☽ ± ☿
03 41 ☽ ⊥ ♅
03 51 ☽ ± ♀
09 25 ☽ ♇ ♃
11 33 ☽ ∠ ⯓
12 11 ☽ ⚻ ☿
13 16 ☽ ⊻ ♅
15 20 ♄ ⚻ ⯓
18 05 ☽ ☍ ☉
19 05 ☽ ♇ ♀

AUGUST 2004

LONGITUDES

Date	Sidereal time h m s	Sun ☉ ° ' "	Moon ☽ ° ' "	Moon ☽ 24.00 ° ' "	Mercury ☿ ° '	Venus ♀ ° '	Mars ♂ ° '	Jupiter ♃ ° '	Saturn ♄ ° '	Uranus ♅ ° '	Neptune ♆ ° '	Pluto ♇ ° '
01	08 41 44	09 ♌ 33 45	19 ♒ 54 57	27 ♒ 14 05	05 ♍ 51	25 ♊ 02	24 ♌ 21	18 ♍ 39	19 ♋ 53	05 ♓ 49	14 ♒ 07	19 ♐ 46
02	08 45 41	10 31 08	04 ♓ 27 37	11 ♓ 34 52	06 28	25 50	24 59	18 51	20 01	05 R 47	14 R 06	19 R 45
03	08 49 37	11 28 33	18 ♓ 35 20	25 ♓ 28 45	07 01	26 39	25 37	19 02	20 08	05 45	14 04	19 44
04	08 53 34	12 25 59	02 ♈ 15 00	08 ♈ 54 11	07 30	27 29	26 15	19 14	20 15	05 43	14 02	19 43
05	08 57 30	13 23 25	15 ♈ 26 34	21 ♈ 52 30	07 55	28 19	26 53	19 25	20 23	05 40	14 01	19 43
06	09 01 27	14 20 54	28 ♈ 12 28	04 ♉ 27 02	08 15	29 ♊ 10	27 31	19 37	20 30	05 38	13 59	19 42
07	09 05 24	15 18 23	10 ♉ 36 49	16 ♉ 42 28	08 30	00 ♋ 02	28 09	19 49	20 38	05 36	13 57	19 41
08	09 09 20	16 15 54	22 ♉ 44 38	28 ♉ 44 01	08 40	00 54	28 47	20 00	20 45	05 34	13 56	19 40
09	09 13 17	17 13 26	04 ♊ 41 15	10 ♊ 37 00	08 46	01 47	29 ♌ 25	20 12	20 52	05 32	13 54	19 40
10	09 17 13	18 11 00	16 ♊ 31 53	22 ♊ 26 28	08 R 46	02 41	00 ♍ 03	20 24	20 59	05 29	13 53	19 39
11	09 21 10	19 08 35	28 ♊ 21 17	04 ♋ 16 49	08 41	03 35	00 41	20 36	21 07	05 27	13 51	19 38
12	09 25 06	20 06 11	10 ♋ 13 31	16 ♋ 11 45	08 30	04 30	01 19	20 48	21 14	05 25	13 49	19 38
13	09 29 03	21 03 49	22 ♋ 11 50	28 ♋ 14 04	08 14	05 25	01 57	21 00	21 21	05 23	13 48	19 37
14	09 32 59	22 01 28	04 ♌ 18 38	10 ♌ 25 42	07 52	06 21	02 35	21 12	21 28	05 20	13 46	19 37
15	09 36 56	22 59 08	16 ♌ 35 23	22 ♌ 47 47	07 25	07 17	03 13	21 24	21 35	05 18	13 44	19 36
16	09 40 53	23 56 50	29 ♌ 02 54	05 ♍ 20 48	06 53	08 14	03 51	21 36	21 42	05 16	13 43	19 36
17	09 44 49	24 54 32	11 ♍ 41 27	18 ♍ 04 52	06 16	09 11	04 29	21 48	21 49	05 13	13 41	19 35
18	09 48 46	25 52 16	24 ♍ 31 04	01 ♎ 00 03	05 34	10 09	05 07	22 00	21 56	05 11	13 40	19 35
19	09 52 42	26 50 02	07 ♎ 31 53	14 ♎ 06 36	04 48	11 07	05 45	22 13	22 03	05 09	13 38	19 35
20	09 56 39	27 47 48	20 ♎ 44 19	27 ♎ 25 07	04 00	12 05	06 23	22 25	22 10	05 06	13 36	19 34
21	10 00 35	28 45 36	04 ♏ 09 08	10 ♏ 56 31	03 08	13 04	07 02	22 37	22 17	05 04	13 35	19 34
22	10 04 32	29 ♌ 43 24	17 ♏ 47 23	24 ♏ 41 50	02 16	14 03	07 40	22 50	22 23	05 01	13 33	19 34
23	10 08 28	00 ♍ 41 14	01 ♐ 39 57	08 ♐ 41 45	01 22	15 03	08 18	23 02	22 30	04 59	13 32	19 33
24	10 12 25	01 39 05	15 ♐ 47 10	22 ♐ 56 03	00 ♍ 29	16 03	08 56	23 14	22 37	04 57	13 30	19 33
25	10 16 22	02 36 58	00 ♑ 08 06	07 ♑ 22 57	29 ♌ 38	17 03	09 34	23 27	22 43	04 54	13 29	19 33
26	10 20 18	03 34 51	14 ♑ 40 05	21 ♑ 58 50	28 49	18 04	10 12	23 39	22 50	04 52	13 27	19 33
27	10 24 15	04 32 46	29 ♑ 18 28	06 ♒ 38 07	28 04	19 05	10 51	23 52	22 56	04 50	13 26	19 33
28	10 28 11	05 30 43	13 ♒ 56 53	21 ♒ 13 51	27 24	20 06	11 29	24 05	23 03	04 47	13 24	19 33
29	10 32 08	06 28 40	28 ♒ 28 06	05 ♓ 38 47	26 50	21 08	12 07	24 17	23 09	04 45	13 23	19 33
30	10 36 04	07 26 39	12 ♓ 45 08	19 ♓ 46 29	26 22	22 10	12 45	24 30	23 16	04 42	13 21	19 33
31	10 40 01	08 ♍ 24 40	26 ♓ 42 20	03 ♈ 32 19	26 ♌ 02	23 ♋ 12	13 ♍ 24	24 ♍ 42	23 ♋ 22	04 ♓ 40	13 ♒ 20	19 ♐ 33

Date	Moon True ☊ ° '	Moon Mean ☊ ° '	Moon ☽ Latitude ° '
01	06 ♉ 18	06 ♉ 24	04 S 52
02	06 R 08	06 21	04 25
03	06 00	06 18	03 42
04	05 54	06 14	02 48
05	05 51	06 11	01 46
06	05 50	06 08	00 S 40
07	05 D 49	06 05	00 N 25
08	05 R 49	06 02	01 28
09	05 48	05 59	02 27
10	05 45	05 55	03 18
11	05 39	05 52	04 00
12	05 31	05 49	04 33
13	05 21	05 46	04 53
14	05 09	05 43	05 01
15	04 56	05 40	04 55
16	04 44	05 36	04 35
17	04 33	05 33	04 01
18	04 25	05 30	03 14
19	04 19	05 27	02 17
20	04 16	05 24	01 11
21	04 16	05 20	00 N 01
22	04 D 16	05 17	01 S 11
23	04 R 16	05 14	02 21
24	04 16	05 11	03 22
25	04 13	05 08	04 13
26	04 08	05 05	04 47
27	04 00	05 01	05 03
28	03 52	04 58	05 00
29	03 43	04 55	04 37
30	03 34	04 52	03 57
31	03 ♉ 28	04 ♉ 49	03 S 03

DECLINATIONS

Date	Sun ☉ ° '	Moon ☽ ° '	Mercury ☿ ° '	Venus ♀ ° '	Mars ♂ ° '	Jupiter ♃ ° '	Saturn ♄ ° '	Uranus ♅ ° '	Neptune ♆ ° '	Pluto ♇ ° '
01	17 N 52	19 S 27	07 N 17	19 N 06	14 N 29	05 N 30	21 N 44	10 S 07	16 S 39	14 S 21
02	17 36	13 59	06 52	19 11	14 16	05 26	21 43	10 08	16 40	14 21
03	17 20	07 55	06 29	19 15	14 03	05 21	21 42	10 09	16 40	14 21
04	17 04	01 S 40	06 07	19 20	13 50	05 17	21 41	10 10	16 41	14 21
05	16 48	04 N 27	05 47	19 24	13 37	05 12	21 40	10 10	16 41	14 22
06	16 32	10 13	05 28	19 28	13 23	05 07	21 39	10 11	16 42	14 22
07	16 15	15 25	05 12	19 32	13 10	05 03	21 38	10 12	16 42	14 22
08	15 58	19 53	04 57	19 35	12 57	04 58	21 37	10 13	16 43	14 22
09	15 40	23 29	04 46	19 39	12 43	04 53	21 36	10 14	16 43	14 23
10	15 23	26 02	04 36	19 42	12 30	04 49	21 35	10 15	16 44	14 23
11	15 05	27 26	04 29	19 44	12 16	04 44	21 34	10 15	16 44	14 23
12	14 47	27 35	04 25	19 47	12 02	04 39	21 33	10 16	16 45	14 24
13	14 29	26 26	04 24	19 49	11 49	04 34	21 32	10 17	16 45	14 24
14	14 10	24 03	04 26	19 51	11 35	04 30	21 31	10 18	16 46	14 24
15	13 51	20 33	04 30	19 52	11 21	04 25	21 30	10 19	16 46	14 25
16	13 32	16 06	04 38	19 53	11 07	04 20	21 29	10 20	16 47	14 25
17	13 13	10 53	04 49	19 54	10 53	04 15	21 28	10 21	16 47	14 25
18	12 54	05 N 09	05 03	19 54	10 38	04 10	21 27	10 21	16 48	14 26
19	12 34	00 S 53	05 20	19 54	10 24	04 05	21 26	10 22	16 48	14 26
20	12 14	07 00	05 40	19 54	10 10	04 00	21 25	10 23	16 49	14 26
21	11 54	12 54	06 02	19 53	09 56	03 56	21 24	10 24	16 49	14 27
22	11 34	18 17	06 27	19 51	09 41	03 51	21 23	10 25	16 49	14 27
23	11 14	22 47	06 53	19 50	09 27	03 46	21 22	10 26	16 50	14 27
24	10 53	26 02	07 21	19 48	09 12	03 41	21 21	10 27	16 50	14 28
25	10 33	27 39	07 50	19 45	08 58	03 36	21 20	10 28	16 51	14 28
26	10 12	27 23	08 19	19 42	08 43	03 31	21 19	10 28	16 51	14 28
27	09 51	25 14	08 48	19 38	08 28	03 26	21 18	10 29	16 52	14 29
28	09 29	21 25	09 17	19 34	08 13	03 21	21 17	10 30	16 52	14 29
29	09 08	16 20	09 45	19 30	07 59	03 16	21 16	10 31	16 53	14 29
30	08 47	10 25	10 11	19 25	07 44	03 11	21 15	10 32	16 53	14 30
31	08 N 25	04 S 07	10 N 35	19 N 20	07 N 29	03 N 06	21 N 14	10 S 33	16 S 53	14 S 30

ZODIAC SIGN ENTRIES

Date	h m	Planets
02	04 34	☽ ♓
04	07 59	☽ ♈
06	15 26	☽ ♉
07	11 02	♀ ♋
09	02 33	☽ ♊
10	10 14	♂ ♍
11	15 20	☽ ♋
14	03 30	☽ ♌
16	13 49	☽ ♍
18	22 09	☽ ♎
21	04 37	☽ ♏
22	18 53	☉ ♍
23	09 08	☽ ♐
25	01 33	☿ ♌
25	11 47	☽ ♑
27	13 08	☽ ♒
29	14 33	☽ ♓
31	17 46	☽ ♈

LATITUDES

Date	Mercury ☿ ° '	Venus ♀ ° '	Mars ♂ ° '	Jupiter ♃ ° '	Saturn ♄ ° '	Uranus ♅ ° '	Neptune ♆ ° '	Pluto ♇ ° '
01	02 S 14	04 S 15	01 N 08	01 N 06	00 S 14	00 S 48	00 S 04	08 N 44
04	02 51	04 05	01 08	01 06	00 14	00 48	00 04	08 43
07	03 26	03 55	01 07	01 06	00 14	00 48	00 04	08 42
10	03 58	03 43	01 07	01 06	00 13	00 48	00 04	08 41
13	04 24	03 31	01 06	01 06	00 13	00 48	00 04	08 40
16	04 41	03 18	01 06	01 05	00 13	00 48	00 04	08 39
19	04 44	03 05	01 05	01 05	00 13	00 48	00 04	08 37
22	04 31	02 51	01 04	01 05	00 12	00 48	00 04	08 36
25	04 01	02 37	01 03	01 05	00 12	00 48	00 04	08 35
28	03 17	02 23	01 03	01 05	00 12	00 48	00 04	08 34
31	02 S 24	02 S 09	01 N 02	01 N 05	00 S 12	00 S 48	00 S 04	08 N 33

DATA

Julian Date	2453219
Delta T	+65 seconds
Ayanamsa	23° 55' 07"
Synetic vernal point	05° ♓ 11' 52"
True obliquity of ecliptic	23° 26' 27"

MOON'S PHASES, APSIDES AND POSITIONS ☽

Date	h m	Phase	Longitude ° '	Eclipse Indicator
07	22 01	☾	15 ♉ 42	
16	01 24	●	23 ♌ 31	
23	10 12	☽	00 ♐ 37	
30	02 22	○	07 ♓ 03	

Day	h m		
11	09 42	Apogee	
27	05 44	Perigee	
04	18 27	0N	
12	02 39	Max dec	27° N 40'
19	08 30	0S	
25	20 46	Max dec	27° S 47'

LONGITUDES

Date	Chiron ⚷ ° '	Ceres ⚳ ° '	Pallas ⚴ ° '	Juno ⚵ ° '	Vesta ⚶ ° '	Black Moon Lilith ⚸ ° '
01	22 ♑ 17	01 ♍ 13	28 ♋ 09	12 ♑ 10	28 ♓ 19	29 ♊ 44
11	21 ♑ 45	05 ♍ 45	03 ♌ 59	10 ♑ 35	27 ♓ 35	00 ♋ 51
21	21 ♑ 18	10 ♍ 18	09 ♌ 43	09 ♑ 33	26 ♓ 06	01 ♋ 58
31	20 ♑ 56	14 ♍ 54	15 ♌ 21	09 ♑ 10	24 ♓ 00	03 ♋ 05

All ephemeris data is given at 12.00 UT and the Moon's longitude is additionally given for 24.00 UT

ASPECTARIAN

h m	Aspects
01 Sunday	
00 04	☽ ± ♃
00 24	☽ # ♄
02 36	☽ ☌ ♆
09 55	☽ ⊼ ♃
10 50	☿ ☍ ♅
11 45	☽ ⚹ ♇
11 57	☽ ⊼ ♄
13 38	☽ # ♀
19 35	☽ ☍ ♂
19 46	☽ # ☉
20 51	☽ △ ♀
21 52	☽ ± ♄
02 Monday	
00 41	☽ ∥ ♆
03 05	♂ # ♇
07 29	☽ Q ♇
10 30	☽ ∥ ♇
10 47	☽ # ♂
12 56	☽ ⚼ ♄
14 12	☽ ☌ ♅
15 30	☽ ☍ ☿
22 56	☽ ⊼ ☉
03 Tuesday	
03 21	☽ ∥ ♅
04 15	☽ ⊻ ♆
09 56	☽ ± ☉
12 47	☽ ☍ ♃
13 59	☽ □ ♇
14 33	☽ ⊥ ♆
14 42	☽ △ ♄
17 53	☽ # ☿
21 59	☽ # ♃
04 Wednesday	
00 50	☽ ⊼ ♂
02 48	☽ ⚼ ☉
02 58	☽ □ ♀
06 18	☽ ∠ ♆
12 00	☽ ± ♂
12 20	♂ ⊥ ♄
18 12	☽ ⊻ ♅
21 16	☉ ∠ ♀
21 46	☽ ⊼ ☿
05 Thursday	
05 05	☽ ⊥ ♅
05 06	☽ ⚼ ♂
07 55	☽ △ ☉
09 05	☽ ± ☿
09 22	☽ ⚹ ♆
12 58	☉ ⊥ ♃
13 44	☽ Q ♀
14 58	☽ ∥ ♃
17 04	☽ ∥ ☿
19 31	☽ ⊼ ♃
19 56	☽ △ ♇
21 17	☽ □ ♄
21 32	☉ # ♆
21 43	☽ ∠ ♅
06 Friday	
02 20	☽ ⚼ ☿
03 07	☉ ☍ ♆
06 53	♀ ⚼ ♆
06 59	☽ ± ♃
07 46	☽ Q ♆
10 36	☽ △ ♂
11 54	☽ # ♅
13 59	☽ ⚹ ♀
21 31	♃ □ ♇
07 Saturday	
00 28	☽ ⚼ ♇
00 31	☽ ⚼ ♃
01 44	☽ ∥ ♂
02 16	☽ ⚹ ♅
06 56	☽ # ♇
07 47	☽ △ ☿
08 04	☽ Q ♄
08 26	♄ ⚼ ♅
15 57	☽ ∥ ☉
18 02	☽ ± ♇
18 33	☽ # ♆
18 33	☽ □ ♆
21 22	☽ ∠ ♀
22 01	☽ □ ☉
08 Sunday	
01 44	☽ Q ♅
03 57	♃ ± ♆
05 53	☽ ⊼ ♇
06 27	☽ △ ♃
07 59	☽ ⚹ ♄
10 09	☿ ∥ ♃
10 16	☽ ∥ ♀
16 40	☽ ⊥ ♀
22 45	☽ ∥ ♄
09 Monday	
00 46	☽ □ ♂
05 41	☽ ⊻ ♀
13 11	☽ Q ☉
13 42	☽ □ ♅
14 25	☽ ∠ ♄
20 16	☽ □ ☿
10 Tuesday	
00 32	☿ St R
06 37	☽ △ ♆
08 50	☽ ⊥ ♄
15 15	☽ Q ♂
15 39	☽ ⚹ ☉
18 20	☽ ☍ ♇
19 59	☽ □ ♃
21 09	☽ ⊻ ♄
11 Wednesday	
08 38	☽ Q ☿
13 00	☽ ⚼ ♆
16 59	☽ ⚹ ♂
23 29	☽ ☌ ♀
12 Thursday	
00 19	☉ △ ♇
00 45	☽ ∠ ☉
02 19	☽ △ ♅
07 10	☽ ± ♆
08 35	☽ ⚹ ☿
09 04	☽ Q ♃
19 13	☽ ⊼ ♆
20 29	☽ ⊥ ☉
13 Friday	
00 55	☽ ∠ ♂
06 52	☽ ⊼ ♇
08 23	☽ ⚼ ♅
09 33	☽ ⊻ ☉
09 34	☽ ⚹ ♃
09 49	☉ ⊻ ♃
10 17	☽ ☌ ♄
10 57	♀ △ ♅
14 00	☽ ∠ ☿
18 00	☉ # ♇
18 49	☽ ± ♇
19 53	☽ ⊥ ♂
20 07	☉ ⊻ ♄
14 Saturday	
02 13	☽ ± ♅
07 20	☽ ⊥ ☿
08 24	☽ ⊻ ♂
12 36	☽ ⚼ ♇
14 01	☽ ⊼ ♅
15 46	☽ ∠ ♃
16 20	☽ ⊻ ♀
18 45	☽ ⊻ ☿
23 06	☿ ∥ ♃
15 Sunday	
05 03	☽ ⊥ ♀
06 07	☽ ∥ ♄
06 28	☽ ☍ ♆
09 39	☽ ⊥ ♃
14 10	☿ ⚹ ♀
15 57	☽ ∥ ♀
17 50	☽ △ ♇
21 27	☽ ⊻ ♃
21 45	☽ ⊻ ♄
23 16	♀ ± ♆
23 55	☽ ∠ ♀
16 Monday	
01 24	☽ ☌ ☉
08 35	☽ # ♆
09 24	☽ ⊥ ♄
17 59	☿ ∠ ♄
20 04	☽ # ♇
21 39	☽ ☌ ♂
23 48	☽ ☍ ♅
17 Tuesday	
00 56	☽ ∥ ☉
02 14	☽ ☌ ☿
02 42	☽ ∠ ♄
06 53	☽ ⚹ ♀
12 03	☽ ∥ ♂
14 22	☽ # ♅
15 45	☽ ⊼ ♆
16 33	♃ ⚹ ♄
18 Wednesday	
02 49	☽ □ ♇
02 58	☽ ± ♆
07 09	☽ ⚹ ♄
07 13	☽ Q ♀
07 15	☽ ☌ ♃
07 40	♀ Q ♃
12 22	☽ ∥ ☿
14 14	♂ ☍ ♅
14 43	☽ ⊻ ☉
16 01	☽ ∥ ♃
19 40	☽ ⚼ ♆
19 47	☿ ☌ ♂
19 Thursday	
01 00	☿ ☍ ♅
02 41	☽ ⊥ ☉
05 33	☽ Q ♄
07 17	☽ ⊻ ☿
07 38	☽ ⊼ ♅
08 35	☽ ⊻ ♂
12 05	☽ Q ♇
15 00	♂ # ♅
17 39	☽ ⊥ ☿
18 35	☽ ± ♅
19 03	☽ □ ♀
20 06	☽ ⊥ ♂
20 29	☽ ∠ ☉
23 07	☽ △ ♆
20 Friday	
00 23	☽ # ♃
06 26	☽ # ☿
09 02	☽ ∠ ☿
09 54	☽ ⚹ ♇
10 52	☽ ⚼ ♅
13 14	☽ ∠ ♂
14 35	☽ □ ♄
15 04	☽ ⊻ ♃
22 20	☉ ⊥ ♄
21 Saturday	
00 13	☽ # ♂
01 39	☽ ∥ ♅
01 39	☽ ⚹ ☉
02 00	☽ ⊥ ♃
08 05	☽ # ☉
10 19	☽ ⚹ ☿
12 44	☽ ∠ ♇
13 37	☽ △ ♅
17 20	☽ ⚹ ♂
18 14	☽ ∠ ♃
18 37	☽ ∥ ♇
23 28	♂ ∠ ♄
22 Sunday	
00 18	♀ ⊼ ♆
00 34	☽ Q ☉
04 36	☽ ⊥ ♇
04 36	☽ □ ♆
04 57	☽ △ ♀
05 11	☽ ∥ ♆
06 12	☽ Q ☿
15 05	☽ ⊻ ♇
15 25	☽ Q ♂
19 45	☽ # ♀
20 04	☽ △ ♄
20 54	☽ ⚹ ♃
23 Monday	
03 46	☽ # ♄
09 00	☽ ⚼ ♀
10 12	☽ □ ☉
11 31	☽ □ ☿
11 46	☽ Q ♆
17 40	☽ □ ♅
17 51	☽ Q ♃
20 51	☉ ☌ ☿
22 03	☽ ⚼ ♄
23 51	☽ □ ♂
24 Tuesday	
01 33	☽ ± ♀
04 53	☿ ∠ ♀
08 09	☽ ⚹ ♆
12 28	☽ ⊼ ♀
13 24	☽ ± ♄
18 20	☽ ☌ ♇
23 33	☽ ⊼ ♄
23 59	☽ Q ♅
25 Wednesday	
00 42	☽ □ ♃
09 15	☽ ∠ ♆
11 13	☽ △ ☿
16 25	☽ △ ☉
17 29	☉ # ♅
19 53	☽ ⚹ ♅
26 Thursday	
00 08	☽ ⊥ ♆
04 19	☽ △ ♂
10 00	☽ ⊻ ♆
10 41	☽ ⚼ ☿
11 45	☿ ⊥ ♄
17 59	☽ ☍ ♀
18 53	☽ ⚼ ☉
20 01	☽ ⊻ ♇
20 31	☽ ∠ ♅
27 Friday	
00 43	☽ ± ☿
00 53	☿ ∥ ♂
01 30	☽ ☍ ♄
02 58	☽ △ ♃
05 51	☽ ⊥ ♇
06 04	☽ ⚼ ♂
10 04	☽ ⊼ ☿
10 40	☽ ± ☉
11 13	☽ ⊥ ♅
18 41	☉ ☍ ♅
20 35	☽ ∠ ♇
21 01	☽ ⊻ ♅
21 11	☽ ⊼ ☉
21 29	☽ ± ♂
22 58	♀ ⊼ ♇
28 Saturday	
03 53	☽ ⚼ ♃
04 53	♀ ⚼ ♅
07 46	☽ ⊼ ♂
11 06	☽ ☌ ♆
12 45	☽ # ♄
17 57	☉ ∥ ☿
18 54	☽ ± ♃
21 13	☽ ⚹ ♇
21 26	☽ # ♀
22 54	☽ ⊼ ♀
29 Sunday	
03 07	☽ ⊼ ♄
04 57	☽ ⊼ ♃
07 27	☿ ⊥ ♀
09 23	☽ ☍ ☿
09 36	☽ ± ♀
09 37	☽ ∥ ♆
13 09	☽ ± ♄
17 08	☽ Q ♇
19 43	☽ ∥ ♇
22 28	☽ ☌ ♅
30 Monday	
01 49	☽ ⚼ ♀
02 22	☽ ☍ ☉
04 20	☽ ⚼ ♄
11 33	☽ ∥ ♅
12 00	☽ ☍ ♂
12 50	☽ # ☿
13 01	☽ ⊻ ♆
18 42	☽ # ☉
19 39	♇ St D
22 45	☽ # ♂
23 16	☽ ⊥ ♆
23 36	☽ □ ♇
31 Tuesday	
05 25	☽ △ ♀
06 09	☽ △ ♄
08 28	☽ ☍ ♃
09 18	☿ # ♅
09 47	♂ ⊼ ♆
10 42	☉ ∠ ♄
10 50	☽ ⊼ ☿
14 50	☽ ∠ ♆
15 53	☽ # ♃
16 08	♀ ☌ ♄
21 10	☽ ± ☿

SEPTEMBER 2004

Date	Sidereal time h m s	Sun ☉ ° ' "	Moon ☽ ° ' "	Moon ☽ 24.00 ° ' "	Mercury ☿ ° '	Venus ♀ ° '	Mars ♂ ° '	Jupiter ♃ ° '	Saturn ♄ ° '	Uranus ♅ ° '	Neptune ♆ ° '	Pluto ♇ ° '
01	10 43 57	09 ♍ 22 42	10 ♈ 16 14	16 ♈ 54 02	25 ♌ 49	24 ♋ 15	14 ♍ 02	24 ♍ 55	23 ♋ 28	04 ♓ 38	13 ♒ 18	19 ♐ 33
02	10 47 54	10 20 46	23 ♈ 25 47	29 ♈ 51 42	25 R 44	25 18	14 40	25 08	23 34	04 R 35	13 R 17	19 D 33
03	10 51 51	11 18 52	06 ♉ 12 07	12 ♉ 27 26	25 D 48	26 21	15 18	25 21	23 40	04 33	13 16	19 33
04	10 55 47	12 17 01	18 ♉ 38 09	24 ♉ 44 49	26 01	27 25	15 57	25 33	23 46	04 30	13 14	19 33
05	10 59 44	13 15 11	00 ♊ 48 02	06 ♊ 48 24	26 22	28 28	16 35	25 46	23 52	04 28	13 13	19 33
06	11 03 40	14 13 23	12 ♊ 46 35	18 ♊ 43 13	26 52	29 ♋ 33	17 13	25 59	23 58	04 26	13 11	19 33
07	11 07 37	15 11 37	24 ♊ 38 57	00 ♋ 34 24	27 30	00 ♌ 37	17 52	26 12	24 04	04 23	13 10	19 34
08	11 11 33	16 09 53	06 ♋ 30 11	12 ♋ 26 52	28 16	01 41	18 30	26 25	24 10	04 21	13 09	19 34
09	11 15 30	17 08 11	18 ♋ 24 59	24 ♋ 25 02	29 ♌ 10	02 46	19 09	26 37	24 16	04 19	13 08	19 34
10	11 19 26	18 06 31	00 ♌ 27 27	06 ♌ 32 38	00 ♍ 12	03 51	19 47	26 50	24 21	04 16	13 06	19 34
11	11 23 23	19 04 54	12 ♌ 40 54	18 ♌ 52 30	01 20	04 57	20 25	27 03	24 27	04 14	13 05	19 35
12	11 27 20	20 03 18	25 ♌ 07 38	01 ♍ 26 25	02 35	06 02	21 04	27 16	24 33	04 12	13 04	19 35
13	11 31 16	21 01 44	07 ♍ 48 56	14 ♍ 15 10	03 56	07 08	21 42	27 29	24 38	04 10	13 03	19 36
14	11 35 13	22 00 12	20 ♍ 45 05	27 ♍ 18 34	05 22	08 14	22 21	27 42	24 43	04 07	13 01	19 36
15	11 39 09	22 58 41	03 ♎ 55 30	10 ♎ 35 41	06 52	09 20	22 59	27 55	24 49	04 05	13 00	19 37
16	11 43 06	23 57 13	17 ♎ 18 58	24 ♎ 05 08	08 27	10 27	23 38	28 08	24 54	04 03	12 59	19 37
17	11 47 02	24 55 47	00 ♏ 54 00	07 ♏ 45 23	10 05	11 34	24 17	28 21	24 59	04 01	12 58	19 38
18	11 50 59	25 54 22	14 ♏ 39 07	21 ♏ 35 00	11 46	12 40	24 55	28 34	25 04	03 59	12 57	19 38
19	11 54 55	26 52 59	28 ♏ 32 56	05 ♐ 32 43	13 30	13 47	25 34	28 47	25 09	03 56	12 56	19 39
20	11 58 52	27 51 37	12 ♐ 34 15	19 ♐ 37 21	15 15	14 55	26 12	29 00	25 14	03 54	12 55	19 40
21	12 02 49	28 50 18	26 ♐ 41 50	03 ♑ 47 32	17 03	16 02	26 51	29 13	25 19	03 52	12 54	19 40
22	12 06 45	29 ♍ 49 00	10 ♑ 54 11	18 ♑ 01 31	18 51	17 10	27 30	29 26	25 24	03 50	12 53	19 41
23	12 10 42	00 ♎ 47 44	25 ♑ 09 11	02 ♒ 16 49	20 40	18 17	28 08	29 39	25 29	03 48	12 52	19 42
24	12 14 38	01 46 29	09 ♒ 23 59	16 ♒ 30 13	22 30	19 25	28 47	29 ♍ 52	25 33	03 46	12 51	19 43
25	12 18 35	02 45 16	23 ♒ 35 01	00 ♓ 37 53	24 20	20 33	29 ♍ 26	00 ♎ 05	25 38	03 44	12 50	19 43
26	12 22 31	03 44 05	07 ♓ 38 16	14 ♓ 35 41	26 10	21 42	00 ♎ 04	00 18	25 42	03 42	12 49	19 44
27	12 26 28	04 42 55	21 ♓ 29 41	28 ♓ 19 50	28 00	22 50	00 43	00 31	25 46	03 40	12 48	19 45
28	12 30 24	05 41 48	05 ♈ 05 48	11 ♈ 47 18	29 ♍ 50	23 59	01 22	00 44	25 51	03 38	12 47	19 46
29	12 34 21	06 40 42	18 ♈ 24 09	24 ♈ 56 16	01 ♎ 39	25 08	02 01	00 56	25 55	03 36	12 47	19 47
30	12 38 18	07 ♎ 39 39	01 ♉ 23 38	07 ♉ 46 20	03 ♎ 28	26 ♌ 17	02 ♎ 40	01 ♎ 09	25 ♋ 59	03 ♓ 34	12 ♒ 46	19 ♐ 48

Date	Moon True ☊ ° '	Moon Mean ☊ ° '	Moon ☽ Latitude ° '
01	03 ♉ 23	04 ♉ 46	02 S 01
02	03 R 21	04 42	00 S 53
03	03 D 20	04 39	00 N 15
04	03 21	04 36	01 21
05	03 22	04 33	02 22
06	03 23	04 30	03 16
07	03 R 22	04 26	04 01
08	03 20	04 23	04 35
09	03 16	04 20	04 58
10	03 10	04 17	05 08
11	03 03	04 14	05 04
12	02 55	04 11	04 46
13	02 47	04 07	04 13
14	02 41	04 04	03 27
15	02 36	04 01	02 29
16	02 33	03 58	01 22
17	02 32	03 55	00 N 09
18	02 D 32	03 51	01 S 06
19	02 34	03 48	02 17
20	02 35	03 45	03 21
21	02 36	03 42	04 13
22	02 R 35	03 39	04 50
23	02 33	03 36	05 10
24	02 30	03 32	05 10
25	02 26	03 29	04 51
26	02 22	03 26	04 15
27	02 18	03 23	03 25
28	02 14	03 20	02 23
29	02 12	03 17	01 15
30	02 ♉ 12	03 ♉ 13	00 S 04

DECLINATIONS

Date	Sun ☉ ° '	Moon ☽ ° '	Mercury ☿ ° '	Venus ♀ ° '	Mars ♂ ° '	Jupiter ♃ ° '	Saturn ♄ ° '	Uranus ♅ ° '	Neptune ♆ ° '	Pluto ♇ ° '
01	08 N 03	02 N 13	10 N 57	19 N 14	07 N 14	03 N 01	21 N 13	10 S 34	16 S 54	14 S 31
02	07 41	08 17	11 17	19 07	06 59	02 56	21 12	10 35	16 54	14 31
03	07 19	13 50	11 33	19 00	06 44	02 50	21 11	10 35	16 55	14 31
04	06 57	18 41	11 46	18 53	06 29	02 45	21 10	10 36	16 55	14 32
05	06 35	22 38	11 56	18 45	06 13	02 40	21 09	10 37	16 55	14 32
06	06 13	25 34	12 03	18 37	05 58	02 35	21 08	10 38	16 56	14 33
07	05 50	27 21	12 05	18 28	05 43	02 30	21 07	10 39	16 56	14 33
08	05 28	27 52	12 04	18 19	05 28	02 25	21 06	10 40	16 57	14 33
09	05 05	27 06	12 00	18 09	05 12	02 20	21 05	10 41	16 57	14 34
10	04 42	25 04	11 51	17 58	04 57	02 15	21 04	10 41	16 57	14 34
11	04 19	21 51	11 39	17 48	04 42	02 10	21 03	10 42	16 58	14 35
12	03 56	17 38	11 23	17 36	04 26	02 05	21 02	10 43	16 58	14 35
13	03 33	12 33	11 04	17 24	04 11	01 59	21 01	10 44	16 58	14 35
14	03 10	06 50	10 41	17 12	03 55	01 54	21 00	10 45	16 59	14 36
15	02 47	00 N 43	10 15	16 59	03 40	01 49	21 00	10 45	16 59	14 36
16	02 24	05 S 32	09 47	16 46	03 24	01 44	20 59	10 46	16 59	14 37
17	02 01	11 39	09 15	16 32	03 09	01 39	20 58	10 47	17 00	14 37
18	01 38	17 17	08 42	16 18	02 53	01 34	20 57	10 48	17 00	14 37
19	01 14	22 04	08 06	16 03	02 38	01 29	20 56	10 49	17 00	14 38
20	00 51	25 38	07 28	15 48	02 22	01 23	20 55	10 49	17 01	14 38
21	00 28	27 37	06 48	15 33	02 06	01 18	20 55	10 50	17 01	14 39
22	00 N 04	27 49	06 07	15 16	01 51	01 13	20 54	10 51	17 01	14 39
23	00 S 19	26 11	05 24	15 00	01 35	01 08	20 53	10 52	17 02	14 39
24	00 42	22 53	04 40	14 43	01 19	01 03	20 52	10 52	17 02	14 40
25	01 06	18 14	03 56	14 26	01 04	00 58	20 51	10 53	17 02	14 40
26	01 29	12 39	03 11	14 08	00 48	00 53	20 51	10 54	17 02	14 41
27	01 52	06 30	02 25	13 49	00 32	00 47	20 50	10 54	17 03	14 41
28	02 16	00 S 10	01 39	13 31	00 17	00 42	20 49	10 55	17 03	14 42
29	02 39	06 N 03	00 52	13 12	00 01	00 37	20 49	10 56	17 03	14 42
30	03 S 02	11 N 53	00 N 06	12 N 52	00 S 15	00 N 32	20 N 48	10 S 56	17 S 03	14 S 42

ZODIAC SIGN ENTRIES

Date	h m	Planets
03	00 16	☽ ♉
05	10 24	☽ ♊
06	22 16	♀ ♌
07	22 50	☽ ♋
10	07 38	☿ ♍
10	11 06	☽ ♌
12	21 16	☽ ♍
15	04 54	☽ ♎
17	10 25	☽ ♏
19	14 30	☽ ♐
21	17 35	☽ ♑
22	16 30	☉ ♎
23	20 10	☽ ♒
25	03 23	♃ ♎
25	22 55	☽ ♓
26	09 15	♂ ♎
28	02 57	☽ ♈
28	14 13	☿ ♎
30	09 24	☽ ♉

LATITUDES

Date	Mercury ☿ ° '	Venus ♀ ° '	Mars ♂ ° '	Jupiter ♃ ° '	Saturn ♄ ° '	Uranus ♅ ° '	Neptune ♆ ° '	Pluto ♇ ° '
01	02 S 05	02 S 04	01 N 02	01 N 05	00 S 12	00 S 48	00 S 04	08 N 33
04	01 09	01 50	01 01	01 05	00 11	00 48	00 04	08 32
07	00 S 16	01 35	01 00	01 05	00 11	00 48	00 04	08 30
10	00 N 29	01 21	00 59	01 05	00 11	00 48	00 05	08 29
13	01 04	01 07	00 58	01 05	00 11	00 48	00 05	08 28
16	01 29	00 53	00 57	01 05	00 10	00 48	00 05	08 27
19	01 44	00 40	00 56	01 05	00 10	00 48	00 05	08 26
22	01 51	00 26	00 56	01 05	00 10	00 48	00 05	08 25
25	01 50	00 14	00 55	01 05	00 10	00 48	00 05	08 24
28	01 43	00 S 01	00 54	01 05	00 10	00 48	00 05	08 23
31	01 N 32	00 N 11	00 N 53	01 N 05	00 S 09	00 S 48	00 S 05	08 N 21

DATA

Julian Date	2453250
Delta T	+65 seconds
Ayanamsa	23° 55' 11"
Synetic vernal point	05° ♓ 11' 48"
True obliquity of ecliptic	23° 26' 27"

MOON'S PHASES, APSIDES AND POSITIONS ☽

Date	h m	Phase	Longitude ° '	Eclipse Indicator
06	15 11	☾	14 ♊ 21	
14	14 29	●	22 ♍ 06	
21	15 54	☽	29 ♐ 00	
28	13 09	○	05 ♈ 45	

Day	h m	
08	02 44	Apogee
22	21 01	Perigee

Day	h m		
01	03 31	0N	
08	09 41	Max dec	27° N 52'
15	14 46	0S	
22	02 33	Max dec	27° S 57'
28	12 38	0N	

LONGITUDES

Date	Chiron ⚷ ° '	Ceres ⚳ ° '	Pallas ⚴ ° '	Juno ⚵ ° '	Vesta ⚶ ° '	Black Moon Lilith ⚸ ° '
01	20 ♑ 54	15 ♍ 21	15 ♌ 55	09 ♑ 10	23 ♓ 46	03 ♋ 11
11	20 ♑ 39	19 ♍ 57	21 ♌ 26	09 ♑ 28	21 ♓ 17	04 ♋ 18
21	20 ♑ 31	24 ♍ 34	26 ♌ 49	10 ♑ 21	18 ♓ 47	05 ♋ 26
31	20 ♑ 31	29 ♍ 09	02 ♍ 02	11 ♑ 48	16 ♓ 35	06 ♋ 33

All ephemeris data is given at 12.00 UT and the Moon's longitude is additionally given for 24.00 UT

ASPECTARIAN

h m	Aspects
01 Wednesday	
01 57	☽ ⊻ ♅
10 16	☽ ⊼ ☉
12 38	☽ ⊥ ♅
12 58	☽ Ꝓ ☿
15 01	☽ ∥ ♃
17 28	☽ ⚹ ♆
19 07	☽ ⊼ ♂
21 58	☽ ± ☉
02 Thursday	
04 50	☽ △ ♇
04 56	☽ ∠ ♅
06 38	☽ ± ♂
06 55	☽ ∥ ♂
07 15	♀ ⚹ ♃
09 43	☽ ∥ ☉
12 16	☽ □ ♄
13 09	☿ St D
15 13	☽ ⊼ ♃
15 26	☽ Q ♆
15 46	☽ □ ♀
15 51	☽ Ꝓ ☉
16 17	☽ △ ☿
17 41	♀ ± ♇
21 38	☽ # ♅
22 15	☿ ⊻ ♀
03 Friday	
00 15	☽ Ꝓ ♂
01 19	☽ ∥ ☿
01 55	☉ ∠ ♀
02 38	☽ ± ♃
08 51	☽ Ꝓ ♇
08 52	☽ ⚹ ♅
15 12	☽ # ♇
20 04	☽ Ꝓ ♃
22 34	☽ Q ♄
22 37	☽ △ ☉
04 Saturday	
01 32	☽ □ ♆
02 07	☽ ± ♇
02 48	☽ # ♆
05 08	☽ Q ♀
06 28	☽ △ ♂
07 52	☽ Q ♅
13 06	☽ ∥ ♀
13 47	☽ ⊼ ♇
22 10	☽ ⚹ ♄
05 Sunday	
01 50	☽ △ ♃
02 17	☽ ∥ ♄
02 54	☽ □ ☿
06 56	☽ ⚹ ♀
11 03	☉ ⊼ ♆
11 53	♀ ± ♅
19 18	☽ □ ♅
06 Monday	
04 17	☽ ∠ ♄
12 50	☽ △ ♆
15 11	☽ □ ☉
15 55	☽ ∠ ♀
16 25	☽ Q ☿
21 29	☽ □ ♂
22 35	☽ ⊥ ♄
07 Tuesday	
01 42	☽ ☍ ♇
10 49	☽ ⊻ ♄
11 55	☽ ⊥ ♀
15 11	☽ □ ♃
18 08	☽ ⚹ ☿
19 07	☽ Ꝓ ♆
08 Wednesday	
01 17	☽ ⊻ ♀
06 51	☽ Q ☉
07 40	☽ △ ♅
11 11	☉ ∥ ♂
12 00	☽ Q ♂
13 18	☽ ± ♆
09 Thursday	
01 23	☽ ⊼ ♆
02 44	☽ ∠ ☿
04 15	☽ Q ♃
09 12	☽ ⚹ ☉
11 23	♂ ± ♆
13 32	☽ ⚹ ♂
13 47	☽ Ꝓ ♅
14 19	☽ ⊼ ♇
22 22	☽ ⊥ ☿
23 47	☽ ☌ ♄
10 Friday	
02 18	☽ ± ♇
04 06	♂ □ ♇
04 42	☽ ⚹ ♃
07 41	☽ ± ♅
11 26	☽ ⊻ ☿
15 54	☿ ⊥ ♄
17 42	☽ ∠ ☉
19 22	☽ ☌ ♀
19 31	☽ ⊼ ♅
20 08	☽ Ꝓ ♇
20 53	♀ ⊼ ♅
21 01	☽ ∠ ♂
11 Saturday	
03 54	♀ Ꝓ ♇
10 45	☽ ∠ ♃
12 04	☉ ± ♆
12 47	☽ ☍ ♆
12 51	☽ ⊥ ☉
15 34	☽ ⊥ ♂
17 02	☽ ∥ ♄
12 Sunday	
00 23	☉ □ ♇
01 22	☽ △ ♇
01 27	☽ ⊻ ☉
03 48	☽ ⊻ ♂
04 29	☽ ⊥ ♃
10 53	☽ ⊻ ♄
12 06	☽ ∥ ♀
13 23	♀ ∠ ♂
15 19	☽ # ♆
16 09	☽ ⊻ ♃
22 23	☽ ⊥ ♄
13 Monday	
02 47	☽ # ♇
03 49	☽ ☌ ☿
05 09	☽ ☍ ♅
10 36	☽ ⊻ ♀
15 26	☽ ∠ ♄
15 54	☿ ☍ ♅
18 54	☽ ∥ ☿
19 52	☽ # ♅
21 44	☽ ⊼ ♆
22 51	☽ ⊥ ♀
14 Tuesday	
08 36	☿ # ♅
08 49	☽ ± ♆
09 53	☽ □ ♇
14 29	☽ ☌ ☉
15 05	☽ ☌ ♂
16 59	☽ ∠ ♀
19 20	☽ ⚹ ♄
15 Wednesday	
00 04	☽ ∥ ♂
00 55	☽ ☌ ♃
01 17	☽ Ꝓ ♆
03 28	☽ ∥ ☉
07 42	☽ ∥ ♃
12 17	☽ ⊼ ♅
12 24	♀ # ♆
12 55	☉ ☌ ♂
17 15	☽ Q ♄
18 01	☽ ⊻ ☿
18 39	☽ Q ♇
20 59	♃ Ꝓ ♆
21 36	☽ # ♃
22 38	☽ ⚹ ♀
23 03	☽ ± ♅
16 Thursday	
00 40	☽ # ☉
04 08	☽ # ♂
04 17	☽ △ ♆
06 12	☽ ⊥ ☿
15 04	☽ Ꝓ ♅
16 06	☽ ⚹ ♇
21 55	☽ Q ♀
23 45	☽ ⊻ ♂
17 Friday	
00 21	☽ ∠ ☿
00 41	☽ ⊻ ☉
01 31	☽ □ ♄
03 13	☽ # ☿
07 27	☽ ⊻ ♃
08 31	☽ ∥ ♅
10 31	☿ ∠ ♄
10 51	☽ ⊥ ♂
12 03	☽ ⊥ ☉
13 32	☉ ⚹ ♄
17 26	☽ △ ♅
18 08	☽ ⊥ ♃
18 32	☽ ∠ ♇
18 Saturday	
00 19	☽ ∥ ♇
03 22	☽ ∠ ♂
05 00	☽ ∠ ☉
06 17	☽ ⚹ ☿
07 48	☽ # ♀
08 16	☽ □ ♀
09 03	☽ □ ♆
10 05	☽ ∠ ♃
10 14	☽ ⊥ ♇
10 45	☽ ∥ ♆
17 17	☉ ∥ ♃
17 50	♀ ☍ ♆
18 32	♂ ⚹ ♄
20 39	☽ ⊻ ♇
19 Sunday	
04 15	☿ ⊼ ♆
05 51	☽ # ♄
06 00	☽ Q ☿
06 07	☽ △ ♄
06 37	☽ ⚹ ♂
08 55	☽ ⚹ ☉
11 45	♀ ∠ ♃
12 24	☽ ⚹ ♃
16 05	☽ Q ♆
21 14	☽ □ ♅
23 05	☿ ⊻ ♀
20 Monday	
04 12	☽ Q ♂
07 02	☽ Q ☉
08 00	☽ Ꝓ ♄
09 16	☽ Q ♃
12 35	☽ ⚹ ♆
13 16	☉ Ꝓ ♆
16 20	☽ △ ♀
17 14	☽ □ ☿
23 25	☽ ± ♄
21 Tuesday	
00 04	☽ ☌ ♇
03 50	☽ Q ♅
09 39	☽ ⊼ ♄
12 16	☽ □ ♂
14 02	☽ ∠ ♆
15 54	☽ □ ☉
16 19	☽ □ ♃
19 58	☽ Ꝓ ♀
23 47	☉ ☌ ♃
22 Wednesday	
00 06	☽ ⚹ ♅
05 13	☽ ⊥ ♆
12 22	☿ ± ♆
12 28	☽ ± ♀
15 20	☽ ⊻ ♆
23 04	☿ □ ♇
23 27	☽ ⊼ ♀
23 Thursday	
01 20	☽ ∠ ♅
02 00	♂ Ꝓ ♆
02 48	☽ ⊻ ♇
03 21	☽ △ ☿
12 33	☽ ☍ ♄
12 55	☽ ⊥ ♇
16 27	☽ ⊥ ♅
17 16	☽ △ ♂
19 41	☽ △ ♃
22 12	☽ △ ☉
24 Friday	
02 32	☽ ⊻ ♅
04 05	☽ ∠ ♇
08 19	☽ Ꝓ ☿
16 06	♀ # ♇
17 49	☽ ☌ ♆
18 07	♀ △ ♇
19 45	☽ Ꝓ ♂
21 22	☽ Ꝓ ♃
23 09	☽ # ♄
25 Saturday	
01 23	☽ Ꝓ ☉
01 47	☽ ± ☿
05 12	☉ # ♃
05 27	☽ ⚹ ♇
06 25	☽ ☍ ♀
10 48	☉ # ♂
11 43	☽ ± ♂
12 51	☽ ± ♃
13 28	☽ ⊼ ☿
15 30	☽ ⊼ ♄
17 27	☽ ∥ ♆
17 48	☽ ± ☉
22 26	☽ ⊼ ♂
23 14	☽ ⊼ ♃
26 Sunday	
01 47	☽ ± ♄
01 53	☽ Q ♇
01 54	♂ ∥ ♃
03 39	☽ ∥ ♇
04 48	☽ ⊼ ☉
05 16	☽ ☌ ♅
05 37	☽ # ♀
05 38	☿ ⚹ ♄
11 10	☉ ⊼ ♅
17 18	☽ Ꝓ ♄
19 00	☽ ∥ ♅
20 55	☽ ⊻ ♆
27 Monday	
00 17	♂ ☌ ♃
02 10	♄ ± ♇
07 19	☽ ⊥ ♆
08 57	☽ □ ♇
09 24	☿ Ꝓ ♆
14 34	☽ ⊼ ♀
19 32	☽ △ ♄
23 04	☽ ∠ ♆
23 13	☉ # ☿
28 Tuesday	
01 12	☽ ☍ ☿
02 05	☽ ± ♀
04 06	☽ ☍ ♃
04 32	☽ ∥ ☉
05 02	☽ ☍ ♂
05 38	☽ # ☿
09 24	☽ ⊻ ♅
09 56	☽ # ♃
11 34	☽ # ♂
13 09	☽ ☍ ☉
13 37	☽ ∥ ♂
15 16	☽ ∥ ♃
17 19	☿ ⊥ ♀
18 08	☽ ∥ ☿
19 36	☽ Ꝓ ♀
20 06	☽ ⊥ ♅
21 52	☽ # ☉
29 Wednesday	
01 21	☿ ☌ ♃
01 48	☽ ⚹ ♆
07 11	♀ ⊥ ♃
12 22	☽ ∠ ♅
14 31	☽ △ ♇
19 21	☿ ☌ ♂
20 44	☿ ∥ ♃
23 41	☽ Q ♆
30 Thursday	
01 33	☽ △ ♀
01 53	☽ □ ♄
05 28	♀ ⊻ ♄
07 55	☽ # ♅
08 29	☿ # ♂
11 33	☽ ⊼ ♃
13 22	☿ ⊼ ♅
14 30	☽ ⊼ ♂
15 24	☉ Q ♇
16 01	☽ ∥ ♀
16 04	☽ ⚹ ♅
16 32	☽ ⊼ ☿
18 24	☽ Ꝓ ♇
20 28	☉ Q ♄
23 01	☽ ± ♃

OCTOBER 2004

Date	Sidereal time h m s	Sun ☉ ° ' "	Moon ☽ ° ' "	Moon ☽ 24.00 ° ' "	Mercury ☿ ° '	Venus ♀ ° '	Mars ♂ ° '	Jupiter ♃ ° '	Saturn ♄ ° '	Uranus ♅ ° '	Neptune ♆ ° '	Pluto ♇ ° '
01	12 42 14	08 ♎ 38 37	14 ♉ 04 32	20 ♉ 18 29	05 ♎ 16	27 ♌ 26	03 ♎ 18	01 ♎ 22	26 ♋ 03	03 ♓ 33	12 ♒ 45	19 ♐ 49
02	12 46 11	09 37 38	26 ♉ 28 31	02 ♊ 35 00	07 04	28 35	03 57	01 35	26 07	03 R 31	12 R 44	19 50
03	12 50 07	10 36 42	08 ♊ 38 23	14 ♊ 39 09	08 51	29 ♌ 45	04 36	01 48	26 11	03 29	12 44	19 51
04	12 54 04	11 35 47	20 ♊ 37 50	26 ♊ 34 59	10 37	00 ♍ 54	05 15	02 01	26 14	03 27	12 43	19 52
05	12 58 00	12 34 55	02 ♋ 31 12	08 ♋ 27 04	12 23	02 04	05 54	02 14	26 18	03 26	12 42	19 53
06	13 01 57	13 34 05	14 ♋ 23 11	20 ♋ 20 10	14 07	03 14	06 33	02 27	26 22	03 24	12 42	19 54
07	13 05 53	14 33 18	26 ♋ 18 35	02 ♌ 19 02	15 51	04 24	07 12	02 40	26 25	03 22	12 41	19 55
08	13 09 50	15 32 32	08 ♌ 22 03	14 ♌ 28 10	17 34	05 34	07 51	02 52	26 28	03 21	12 41	19 57
09	13 13 47	16 31 49	20 ♌ 37 49	26 ♌ 51 27	19 17	06 44	08 30	03 05	26 32	03 19	12 40	19 58
10	13 17 43	17 31 08	03 ♍ 09 25	09 ♍ 31 58	20 58	07 55	09 09	03 18	26 35	03 18	12 40	19 59
11	13 21 40	18 30 30	15 ♍ 59 20	22 ♍ 31 38	22 39	09 05	09 48	03 31	26 38	03 16	12 39	20 00
12	13 25 36	19 29 53	29 ♍ 08 53	05 ♎ 51 02	24 19	10 16	10 27	03 44	26 41	03 15	12 39	20 02
13	13 29 33	20 29 19	12 ♎ 37 54	19 ♎ 29 15	25 59	11 27	11 06	03 56	26 44	03 13	12 38	20 03
14	13 33 29	21 28 47	26 ♎ 24 46	03 ♏ 24 02	27 38	12 38	11 45	04 09	26 47	03 12	12 38	20 04
15	13 37 26	22 28 17	10 ♏ 26 35	17 ♏ 31 54	29 ♎ 16	13 49	12 25	04 22	26 49	03 11	12 38	20 06
16	13 41 22	23 27 49	24 ♏ 39 29	01 ♐ 48 44	00 ♏ 53	15 00	13 04	04 34	26 52	03 09	12 37	20 07
17	13 45 19	24 27 22	08 ♐ 59 08	16 ♐ 10 07	02 30	16 11	13 43	04 47	26 54	03 08	12 37	20 09
18	13 49 16	25 26 58	23 ♐ 21 11	00 ♑ 31 51	04 06	17 23	14 22	04 59	26 57	03 07	12 37	20 10
19	13 53 12	26 26 35	07 ♑ 41 41	14 ♑ 50 18	05 41	18 34	15 01	05 12	26 59	03 06	12 37	20 12
20	13 57 09	27 26 14	21 ♑ 57 22	29 ♑ 02 33	07 16	19 46	15 41	05 25	27 01	03 05	12 37	20 13
21	14 01 05	28 25 55	06 ♒ 05 38	13 ♒ 06 23	08 50	20 57	16 20	05 37	27 03	03 04	12 37	20 15
22	14 05 02	29 ♎ 25 37	20 ♒ 04 36	27 ♒ 00 07	10 23	22 09	16 59	05 49	27 05	03 03	12 36	20 16
23	14 08 58	00 ♏ 25 21	03 ♓ 52 49	10 ♓ 42 33	11 57	23 21	17 39	06 02	27 07	03 02	12 36	20 18
24	14 12 55	01 25 07	17 ♓ 29 13	24 ♓ 12 43	13 29	24 33	18 18	06 14	27 08	03 01	12 D 36	20 20
25	14 16 51	02 24 54	00 ♈ 52 57	07 ♈ 29 51	15 01	25 45	18 57	06 26	27 10	03 00	12 36	20 21
26	14 20 48	03 24 44	14 ♈ 03 21	20 ♈ 33 25	16 32	26 57	19 37	06 39	27 11	02 59	12 36	20 23
27	14 24 45	04 24 35	27 ♈ 00 01	03 ♉ 23 10	18 03	28 09	20 16	06 51	27 13	02 58	12 37	20 25
28	14 28 41	05 24 28	09 ♉ 42 53	15 ♉ 59 15	19 33	29 ♍ 22	20 56	07 03	27 14	02 57	12 37	20 26
29	14 32 38	06 24 23	22 ♉ 12 23	28 ♉ 22 24	21 03	00 ♎ 34	21 35	07 15	27 15	02 57	12 37	20 28
30	14 36 34	07 24 20	04 ♊ 29 31	10 ♊ 33 56	22 32	01 47	22 15	07 27	27 16	02 56	12 37	20 30
31	14 40 31	08 ♏ 24 19	16 ♊ 35 58	22 ♊ 35 54	24 ♏ 01	02 ♎ 59	22 ♎ 54	07 ♎ 39	27 ♋ 17	02 ♓ 56	12 ♒ 37	20 ♐ 32

Date	Moon True ☊ ° '	Moon Mean ☊ ° '	Moon ☽ Latitude ° '
01	02 ♉ 12	03 ♉ 10	01 N 05
02	02 D 13	03 07	02 10
03	02 15	03 04	03 07
04	02 17	03 01	03 56
05	02 18	02 57	04 34
06	02 R 18	02 54	05 00
07	02 18	02 51	05 14
08	02 16	02 48	05 14
09	02 15	02 45	04 59
10	02 12	02 42	04 31
11	02 10	02 38	03 48
12	02 09	02 35	02 52
13	02 07	02 32	01 46
14	02 07	02 29	00 N 32
15	02 D 07	02 26	00 S 46
16	02 07	02 23	02 02
17	02 08	02 19	03 10
18	02 09	02 16	04 07
19	02 09	02 13	04 49
20	02 09	02 10	05 12
21	02 R 09	02 07	05 16
22	02 09	02 03	05 02
23	02 09	02 00	04 30
24	02 09	01 57	03 43
25	02 D 09	01 54	02 45
26	02 09	01 51	01 39
27	02 09	01 48	00 S 29
28	02 R 09	01 44	00 N 42
29	02 09	01 41	01 49
30	02 08	01 38	02 50
31	02 ♉ 08	01 ♉ 35	03 N 42

DECLINATIONS

Date	Sun ☉ ° '	Moon ☽ ° '	Mercury ☿ ° '	Venus ♀ ° '	Mars ♂ ° '	Jupiter ♃ ° '	Saturn ♄ ° '	Uranus ♅ ° '	Neptune ♆ ° '	Pluto ♇ ° '
01	03 S 26	17 N 06	00 S 41	12 N 32	00 S 31	00 N 27	20 N 47	10 S 57	17 S 04	14 S 43
02	03 49	21 28	01 28	12 12	00 46	00 22	20 47	10 58	17 04	14 43
03	04 12	24 50	02 14	11 51	01 02	00 17	20 46	10 58	17 04	14 44
04	04 35	27 02	03 01	11 30	01 18	00 12	20 45	10 59	17 04	14 44
05	04 58	27 59	03 47	11 09	01 33	00 07	20 45	10 59	17 04	14 45
06	05 21	27 38	04 33	10 47	01 49	00 N 02	20 44	11 00	17 05	14 45
07	05 44	26 01	05 18	10 25	02 05	00 S 04	20 44	11 01	17 05	14 45
08	06 07	23 13	06 03	10 02	02 21	00 09	20 43	11 01	17 05	14 46
09	06 30	19 21	06 48	09 39	02 36	00 14	20 42	11 02	17 05	14 46
10	06 53	14 33	07 32	09 16	02 52	00 19	20 42	11 02	17 05	14 47
11	07 15	09 02	08 16	08 53	03 08	00 24	20 41	11 03	17 05	14 47
12	07 38	02 N 58	08 59	08 29	03 23	00 29	20 41	11 03	17 05	14 48
13	08 00	03 S 22	09 41	08 05	03 39	00 34	20 40	11 04	17 06	14 48
14	08 23	09 42	10 23	07 41	03 55	00 39	20 40	11 04	17 06	14 48
15	08 45	15 41	11 04	07 16	04 10	00 44	20 40	11 05	17 06	14 49
16	09 07	20 54	11 45	06 51	04 26	00 49	20 39	11 05	17 06	14 49
17	09 29	24 56	12 25	06 26	04 41	00 53	20 39	11 05	17 06	14 50
18	09 51	27 23	13 04	06 01	04 57	00 58	20 38	11 06	17 06	14 50
19	10 12	28 01	13 42	05 35	05 12	01 03	20 38	11 06	17 06	14 51
20	10 34	26 47	14 20	05 10	05 28	01 08	20 38	11 07	17 06	14 51
21	10 55	23 51	14 57	04 44	05 43	01 13	20 37	11 07	17 06	14 51
22	11 16	19 33	15 33	04 17	05 59	01 18	20 37	11 07	17 06	14 52
23	11 37	14 16	16 08	03 51	06 14	01 23	20 37	11 08	17 06	14 52
24	11 58	08 22	16 43	03 25	06 30	01 28	20 37	11 08	17 06	14 53
25	12 19	02 S 10	17 17	02 58	06 45	01 32	20 36	11 08	17 06	14 53
26	12 39	04 N 02	17 49	02 31	07 00	01 37	20 36	11 08	17 06	14 53
27	12 59	09 58	18 21	02 04	07 15	01 42	20 36	11 09	17 06	14 54
28	13 20	15 23	18 52	01 37	07 31	01 47	20 36	11 09	17 06	14 54
29	13 39	20 05	19 22	01 10	07 46	01 51	20 36	11 09	17 06	14 55
30	13 59	23 49	19 52	00 43	08 01	01 56	20 36	11 09	17 06	14 55
31	14 S 18	26 N 27	20 S 20	00 N 15	08 S 16	02 S 01	20 N 35	11 S 10	17 S 06	14 S 55

ZODIAC SIGN ENTRIES

Date	h	m	Planets
02	18	55	☽ ♊
03	17	20	♀ ♍
05	06	54	☽ ♋
07	19	23	☽ ♌
10	06	00	☽ ♍
12	13	32	☽ ♎
14	18	10	☽ ♏
15	22	57	☿ ♏
16	20	58	☽ ♐
18	23	07	☽ ♑
21	01	38	☽ ♒
23	01	49	☉ ♏
23	05	13	☽ ♓
25	10	24	☽ ♈
27	17	37	☽ ♉
29	00	39	♀ ♎
30	03	11	☽ ♊

LATITUDES

Date	Mercury ☿ ° '	Venus ♀ ° '	Mars ♂ ° '	Jupiter ♃ ° '	Saturn ♄ ° '	Uranus ♅ ° '	Neptune ♆ ° '	Pluto ♇ ° '
01	01 N 32	00 N 11	00 N 53	01 N 05	00 S 09	00 S 48	00 S 05	08 N 21
04	01 18	00 22	00 52	01 05	00 09	00 48	00 05	08 20
07	01 01	00 33	00 51	01 05	00 09	00 48	00 05	08 19
10	00 42	00 43	00 50	01 06	00 09	00 48	00 05	08 18
13	00 22	00 53	00 49	01 06	00 08	00 47	00 05	08 17
16	00 N 02	01 01	00 47	01 06	00 08	00 47	00 05	08 16
19	00 S 18	01 10	00 46	01 06	00 08	00 47	00 05	08 15
22	00 39	01 17	00 45	01 06	00 08	00 47	00 05	08 15
25	00 59	01 23	00 44	01 07	00 07	00 47	00 05	08 14
28	01 18	01 29	00 43	01 07	00 07	00 47	00 05	08 13
31	01 S 36	01 N 34	00 N 42	01 N 07	00 S 07	00 S 47	00 S 05	08 N 12

DATA

Julian Date	2453280
Delta T	+65 seconds
Ayanamsa	23° 55' 14"
Synetic vernal point	05° ♓ 11' 45"
True obliquity of ecliptic	23° 26' 27"

MOON'S PHASES, APSIDES AND POSITIONS ☽

Date	h	m	Phase	Longitude ° '	Eclipse Indicator
06	10	12	☾	13 ♋ 30	
14	02	48	●	21 ♎ 06	Partial
20	21	59	☽	27 ♑ 51	
28	03	07	○	05 ♉ 02	total

Day	h	m		
05	22	08	Apogee	
17	23	47	Perigee	
05	17	37	Max dec	28° N 01'
12	23	20	0S	
19	07	59	Max dec	28° S 03'
25	20	21	0N	

LONGITUDES

Date	Chiron ⚷ ° '	Ceres ⚳ ° '	Pallas ⚴ ° '	Juno ⚵ ° '	Vesta ⚶ ° '	Black Moon Lilith ⚸ ° '
01	20 ♑ 31	29 ♍ 09	02 ♍ 02	11 ♑ 48	16 ♓ 35	06 ♋ 33
11	20 ♑ 39	03 ♎ 44	07 ♍ 00	13 ♑ 43	14 ♓ 58	07 ♋ 40
21	20 ♑ 54	08 ♎ 17	12 ♍ 06	15 ♑ 39	14 ♓ 04	08 ♋ 47
31	21 ♑ 16	12 ♎ 46	16 ♍ 40	18 ♑ 47	13 ♓ 56	09 ♋ 54

ASPECTARIAN

h m	Aspects
01 Friday	
00 35	☽ # ♇
00 47	☽ ⊼ ☉
02 25	☽ ± ♂
03 48	☿ ∥ ♂
05 27	☿ # ♃
05 45	☽ ± ☿
06 06	♀ ⊥ ♂
07 54	♂ # ♃
09 28	☽ □ ♆
11 30	☽ ± ♇
11 49	☽ # ♆
11 57	☽ Q ♄
13 11	☽ ± ☉
14 48	☽ Q ♅
16 29	☽ ₽ ♃
20 21	♂ ⊼ ♅
20 35	☽ ₽ ♂
23 04	☽ ⊼ ♇
02 Saturday	
01 57	☽ ₽ ☿
07 50	☽ ∥ ♄
08 05	☽ ₽ ☉
09 16	☉ ± ♅
11 18	☽ ✶ ♄
16 34	☽ □ ♀
22 13	☽ △ ♃
22 24	☿ Q ♇
03 Sunday	
01 48	☽ □ ♅
02 39	☿ Q ♄
03 32	☽ △ ♂
12 29	☽ △ ☿
16 17	☽ △ ☉
17 05	☽ ∠ ♄
20 08	☽ △ ♆
20 27	☿ ± ♅
04 Monday	
08 09	☽ Q ♀
10 28	☽ ☍ ♇
11 13	☽ ⊥ ♄
23 22	☽ ⊻ ♄
05 Tuesday	
02 17	☽ ₽ ♆
10 59	☽ ✶ ♀
11 24	☽ □ ♃
13 50	☽ △ ♅
14 59	☉ △ ♆
16 14	♀ ⊻ ♃
16 30	☿ △ ♆
17 09	♀ ⊥ ♄
18 29	☉ ☌ ☿
19 14	☽ □ ♂
20 28	☽ ± ♆
21 46	♀ # ♅
06 Wednesday	
08 35	☽ ⊼ ♆
10 12	☽ □ ☉
11 22	☽ □ ☿
15 23	♀ ☍ ♅
20 05	☽ ₽ ♅
20 36	☽ ∠ ♀
23 09	☽ ⊼ ♇
07 Thursday	
00 27	☽ Q ♃
09 39	☽ Q ♂
11 14	☽ ± ♇
12 13	☽ ☌ ♄
14 07	☽ ± ♅
16 38	☽ ⊥ ♀
08 Friday	
00 55	☽ ✶ ♃
01 35	☽ Q ☉
02 04	☽ ⊼ ♅
05 13	☽ ₽ ♇
05 33	☽ Q ☿
05 52	☽ ⊻ ♀
10 55	☽ ✶ ♂
15 42	♂ Q ♇
16 13	☉ ∥ ☿
20 29	☽ ☍ ♆
22 40	☿ ₽ ♅
09 Saturday	
03 20	☽ ✶ ☉
04 12	☽ ∥ ♄
06 59	☽ ∠ ♃
08 58	☽ ✶ ☿
10 42	☽ △ ♇
13 15	♂ Q ♄
17 51	☽ ∠ ♂
21 49	☿ ✶ ♇
23 25	☽ ⊻ ♄
23 50	☽ # ♆
10 Sunday	
00 39	☽ ⊥ ♃
10 41	☽ ∠ ☉
10 54	☽ ⊥ ♄
10 57	☽ # ♇
11 07	♃ ⊼ ♅
11 59	☽ ⊥ ♂
12 15	☽ ☍ ♅
12 17	☽ ⊻ ♃
17 07	♂ ± ♅
18 08	☽ ∠ ☿
21 52	☽ ☌ ♀
23 53	☽ ⊻ ♂
11 Monday	
03 34	☽ # ♅
03 53	☽ ∠ ♄
05 01	☽ ⊥ ☉
05 49	☽ ⊼ ♆
06 19	☉ ₽ ♅
12 40	☽ ∥ ♀
13 25	☽ ⊥ ☿
14 48	☽ # ☿
16 54	☽ ± ♆
17 01	☽ ⊻ ☉
18 47	☽ # ☉
19 24	☽ □ ♇
12 Tuesday	
01 19	☿ # ♀
02 01	☽ ⊻ ☿
07 32	☽ ✶ ♄
09 18	☽ ₽ ♆
10 28	☽ # ♂
19 20	☽ ⊼ ♅
20 21	☽ ☌ ♃
20 26	♀ ⊻ ♂
21 24	☽ # ♃
13 Wednesday	
01 10	☉ ✶ ♇
01 19	☽ ∥ ♃
03 54	☽ Q ♇
05 05	☽ Q ♄
05 59	☽ ± ♅
09 10	☽ ☌ ♂
09 43	☽ ⊻ ♀
12 01	☽ △ ♆
13 06	☽ ∥ ♂
14 27	☉ # ♀
17 59	♀ ∠ ♄
21 14	☽ ⊥ ♀
21 47	☽ ₽ ♅
23 14	☿ □ ♄
14 Thursday	
01 00	☽ ✶ ♇
02 48	☽ ☌ ☉
04 43	☽ # ♀
06 35	☽ ∥ ☉
12 07	♀ ⊼ ♆
12 10	☿ ∠ ♀
12 38	☽ □ ♄
14 17	☽ ∠ ♀
14 22	☽ ☌ ☿
14 58	☽ ∥ ☿
17 19	☽ ∥ ♅
23 38	☽ △ ♅
15 Friday	
01 29	☽ ⊻ ♃
02 53	☽ ∠ ♇
08 21	☽ ∥ ♇
11 52	☽ ⊥ ♃
12 13	☿ ∥ ♅
15 30	☽ ⊻ ♂
15 42	☽ □ ♆
18 06	☽ ∥ ♆
18 12	☽ ⊥ ♇
18 14	☽ ✶ ♀
20 02	♂ △ ♆
16 Saturday	
02 08	☽ ⊥ ♂
03 19	☽ ∠ ♃
04 21	☽ ⊻ ♇
09 51	☽ ⊻ ☉
10 44	☽ # ♄
15 43	☽ △ ♄
16 17	☽ Q ♀
17 59	☽ ∠ ♂
20 40	☽ ⊥ ☉
22 00	☽ Q ♆
23 46	☽ ⊻ ☿
17 Sunday	
02 14	☽ □ ♅
04 52	☽ ✶ ♃
11 04	☽ ⊥ ☿
12 51	☽ ∠ ☉
16 53	☽ ₽ ♄
18 04	☽ ✶ ♆
20 17	☽ ✶ ♂
21 30	☿ △ ♅
18 Monday	
01 07	☽ □ ♀
01 13	☽ Q ♃
04 00	☽ ∠ ☿
06 40	☽ ☌ ♇
07 58	☽ ± ♄
08 16	☽ Q ♅
15 46	☽ ✶ ☉
17 17	☽ Q ♂
18 01	☽ ⊼ ♄
19 07	☽ ∠ ♆
19 Tuesday	
03 37	☿ ⊻ ♃
04 18	☽ ✶ ♅
04 31	☿ ∠ ♇
07 45	☽ □ ♃
08 12	☽ ✶ ☿
10 11	☽ ⊥ ♆
12 56	♀ ± ♆
13 21	☽ Q ☉
20 15	☽ ⊻ ♆
20 Wednesday	
00 54	☽ □ ♂
01 21	♀ # ♂
01 25	☉ □ ♄
05 28	☽ ∠ ♅
06 53	☽ Q ☿
07 57	☽ △ ♀
09 04	☽ ⊻ ♇
19 14	☽ ⊥ ♇
20 35	☽ ☍ ♄
20 39	☽ ⊥ ♅
21 28	♀ □ ♇
21 59	☽ □ ☉
21 Thursday	
06 50	☽ ⊻ ♅
08 15	☿ ∥ ♇
10 33	☽ ∠ ♇
11 10	☽ △ ♃
11 44	☽ ₽ ♀
17 16	☽ □ ☿
17 31	♄ ± ♅
23 09	☽ ☌ ♆
22 Friday	
01 42	☉ ∥ ♅
04 36	☽ ± ♀
06 25	☽ △ ♂
06 36	☽ # ♄
12 20	☽ ✶ ♇
13 19	☽ ₽ ♃
15 56	☽ ⊼ ♀
23 36	☽ ∥ ♆
23 Saturday	
00 10	☽ ⊼ ♄
04 41	☽ ∥ ☿
05 10	☽ ± ♃
05 29	☽ △ ☉
09 14	☽ Q ♇
09 27	☽ ∥ ♇
09 44	☽ ₽ ♂
10 30	☽ ☌ ♅
10 39	☽ ± ♄
13 35	☿ ⊥ ♃
15 50	☽ ⊼ ♃
22 20	☿ □ ♆
22 23	☽ ∥ ☉
24 Sunday	
01 00	☽ ∥ ♅
01 39	♂ ₽ ♅
02 21	☽ ± ♂
02 30	☽ ₽ ♄
03 21	☽ ⊻ ♆
04 00	☽ △ ☿
09 57	☽ ₽ ☉
11 55	♆ St D
13 31	☽ ⊼ ♂
13 59	☽ ⊥ ♆
17 04	☽ □ ♇
19 02	☽ ∥ ♂
25 Monday	
01 27	☿ ⊥ ♇
01 51	☽ ☍ ♀
03 18	☽ ± ☉
04 29	☿ ∥ ♆
05 17	☽ △ ♄
06 06	☽ ∠ ♆
08 42	☽ # ♀
10 14	☽ ₽ ☿
14 24	☽ ∥ ♃
15 00	☽ ⊼ ☉
15 49	☽ ⊻ ♅
22 14	☽ ☍ ♃
26 Tuesday	
01 49	☉ △ ♅
02 29	☽ # ♃
02 43	☽ ⊥ ♅
04 42	☽ ± ☿
06 30	☽ ∥ ♀
09 20	☽ ✶ ♆
16 49	♀ ✶ ♄
17 10	☽ ⊼ ☿
19 14	☽ ∠ ♅
22 48	☽ ☍ ♂
23 42	☽ △ ♇
27 Wednesday	
00 22	☽ # ♂
01 04	♀ ₽ ♆
07 32	☽ Q ♆
12 24	☽ □ ♄
14 24	☽ ⊼ ♀
17 01	☽ # ♅
17 17	♂ ✶ ♇
23 12	☽ ✶ ♅
28 Thursday	
01 55	☽ # ☉
02 52	☽ ± ♀
03 07	☽ ☍ ☉
03 52	☽ ₽ ♇
04 56	♀ # ♃
06 51	☽ ⊼ ♃
09 44	☽ # ♇
12 47	☉ ∠ ♇
17 04	☉ ⊥ ♀
17 32	☽ □ ♆
18 29	☽ ± ♃
20 18	☽ # ♆
21 03	☽ ± ♇
21 50	☽ ₽ ♀
22 01	☽ Q ♅
22 34	☽ Q ♄
29 Friday	
02 24	☿ ⊻ ♇
07 37	☽ # ☿
08 38	☽ ⊼ ♇
09 28	☽ ☍ ☿
10 44	☽ ⊼ ♂
12 06	☽ ₽ ♃
14 58	☽ ∥ ♄
21 50	☽ ✶ ♄
23 04	☽ ± ♂
30 Saturday	
03 27	☿ ⊻ ♂
06 05	☽ △ ♀
08 57	☽ □ ♅
10 24	☿ ∠ ♃
13 31	☉ ⊻ ♃
17 45	☽ ₽ ♂
17 57	☽ △ ♃
18 16	☽ ⊼ ☉
31 Sunday	
03 24	☽ ∠ ♄
04 05	☽ △ ♆
07 14	☽ ± ☉
10 44	♀ ⊼ ♅
19 52	☽ ☍ ♇
21 23	☽ ⊥ ♄

All ephemeris data is given at 12.00 UT and the Moon's longitude is additionally given for 24.00 UT

NOVEMBER 2004

LONGITUDES

Date	Sidereal time h m s	Sun ☉ ° ' "	Moon ☽ ° ' "	Moon ☽ 24.00 ° ' "	Mercury ☿ ° '	Venus ♀ ° '	Mars ♂ ° '	Jupiter ♃ ° '	Saturn ♄ ° '	Uranus ♅ ° '	Neptune ♆ ° '	Pluto ♇ ° '
01	14 44 27	09 ♏ 24 20	28 ♊ 34 07	04 ♋ 31 02	25 ♏ 29	04 ♎ 12	23 ♎ 34	07 ♎ 51	27 ♋ 18	02 ♓ 55	12 ♒ 37	20 ♐ 33
02	14 48 24	10 24 23	10 ♋ 27 05	16 ♋ 22 44	26 57	05 25	24 14	08 03	27 19	02 R 55	12 38	20 35
03	14 52 20	11 24 28	22 ♋ 18 31	28 ♋ 14 58	28 24	06 38	24 53	08 15	27 19	02 54	12 38	20 37
04	14 56 17	12 24 35	04 ♌ 12 39	10 ♌ 12 08	29 ♏ 50	07 51	25 33	08 27	27 20	02 54	12 38	20 39
05	15 00 14	13 24 45	16 ♌ 14 01	22 ♌ 18 52	01 ♐ 16	09 04	26 13	08 39	27 20	02 53	12 39	20 41
06	15 04 10	14 24 56	28 ♌ 27 16	04 ♍ 39 46	02 41	10 17	26 52	08 50	27 20	02 53	12 39	20 43
07	15 08 07	15 25 09	10 ♍ 56 53	17 ♍ 19 07	04 06	11 30	27 32	09 02	27 21	02 53	12 40	20 45
08	15 12 03	16 25 25	23 ♍ 46 51	00 ♎ 20 26	05 29	12 43	28 12	09 14	27 R 21	02 53	12 40	20 47
09	15 16 00	17 25 42	07 ♎ 00 06	13 ♎ 45 59	06 52	13 57	28 52	09 25	27 21	02 52	12 41	20 49
10	15 19 56	18 26 02	20 ♎ 38 05	27 ♎ 36 16	08 14	15 10	29 ♎ 31	09 37	27 20	02 52	12 41	20 51
11	15 23 53	19 26 23	04 ♏ 40 14	11 ♏ 49 34	09 35	16 24	00 ♏ 11	09 48	27 20	02 52	12 42	20 53
12	15 27 49	20 26 46	19 ♏ 03 41	26 ♏ 21 51	10 55	17 37	00 51	09 59	27 20	02 D 52	12 43	20 55
13	15 31 46	21 27 11	03 ♐ 43 15	11 ♐ 06 56	12 13	18 51	01 31	10 11	27 19	02 52	12 43	20 57
14	15 35 43	22 27 37	18 ♐ 31 57	25 ♐ 57 17	13 30	20 04	02 11	10 22	27 18	02 53	12 44	20 59
15	15 39 39	23 28 05	03 ♑ 21 58	10 ♑ 45 03	14 46	21 18	02 51	10 33	27 18	02 53	12 45	21 01
16	15 43 36	24 28 34	18 ♑ 05 43	25 ♑ 23 14	16 00	22 32	03 31	10 44	27 17	02 53	12 45	21 03
17	15 47 32	25 29 05	02 ♒ 37 00	09 ♒ 46 33	17 12	23 46	04 11	10 55	27 16	02 53	12 46	21 05
18	15 51 29	26 29 37	16 ♒ 51 34	23 ♒ 51 51	18 22	24 59	04 51	11 06	27 15	02 53	12 47	21 07
19	15 55 25	27 30 09	00 ♓ 47 19	07 ♓ 37 59	19 29	26 13	05 31	11 16	27 14	02 54	12 48	21 09
20	15 59 22	28 30 44	14 ♓ 23 57	21 ♓ 05 24	20 33	27 27	06 11	11 27	27 12	02 54	12 49	21 11
21	16 03 18	29 ♏ 31 19	27 ♓ 42 31	04 ♈ 15 32	21 34	28 41	06 51	11 38	27 11	02 55	12 50	21 13
22	16 07 15	00 ♐ 31 55	10 ♈ 44 43	17 ♈ 10 19	22 31	29 ♎ 55	07 32	11 48	27 09	02 55	12 51	21 15
23	16 11 12	01 32 33	23 ♈ 32 35	29 ♈ 51 45	23 24	01 ♏ 09	08 12	11 59	27 08	02 56	12 52	21 18
24	16 15 08	02 33 12	06 ♉ 08 01	12 ♉ 21 37	24 13	02 24	08 52	12 09	27 06	02 56	12 53	21 20
25	16 19 05	03 33 52	18 ♉ 32 44	24 ♉ 41 31	24 56	03 38	09 32	12 19	27 04	02 57	12 54	21 22
26	16 23 01	04 34 33	00 ♊ 48 08	06 ♊ 52 45	25 33	04 52	10 12	12 30	27 02	02 58	12 55	21 24
27	16 26 58	05 35 16	12 ♊ 55 31	18 ♊ 56 36	26 03	06 06	10 53	12 40	27 00	02 59	12 56	21 26
28	16 30 54	06 36 00	24 ♊ 56 10	00 ♋ 54 25	26 25	07 20	11 33	12 50	26 58	03 00	12 57	21 28
29	16 34 51	07 36 46	06 ♋ 51 34	12 ♋ 47 52	26 40	08 35	12 13	13 00	26 56	03 00	12 58	21 31
30	16 38 47	08 ♐ 37 33	18 ♋ 43 37	24 ♋ 39 07	26 ♐ 45	09 ♏ 49	12 ♏ 54	13 ♎ 09	26 ♋ 53	03 ♓ 01	12 ♒ 59	21 ♐ 33

Date	Moon True ☊ ° '	Moon Mean ☊ ° '	Moon ☽ Latitude ° '
01	02 ♉ 07	01 ♉ 32	04 N 24
02	02 R 06	01 29	04 54
03	02 05	01 25	05 12
04	02 04	01 22	05 16
05	02 D 04	01 19	05 07
06	02 04	01 16	04 44
07	02 05	01 13	04 07
08	02 06	01 09	03 16
09	02 08	01 06	02 15
10	02 09	01 03	01 N 03
11	02 R 09	01 00	00 S 14
12	02 08	00 57	01 32
13	02 07	00 54	02 45
14	02 04	00 50	03 48
15	02 01	00 47	04 36
16	01 58	00 44	05 05
17	01 56	00 41	05 14
18	01 54	00 38	05 04
19	01 D 54	00 35	04 35
20	01 55	00 31	03 52
21	01 56	00 28	02 57
22	01 58	00 25	01 54
23	01 59	00 22	00 S 46
24	02 R 00	00 19	00 N 23
25	01 58	00 15	01 29
26	01 55	00 12	02 31
27	01 51	00 09	03 25
28	01 45	00 06	04 09
29	01 38	00 03	04 42
30	01 ♉ 31	00 ♉ 00	05 N 02

DECLINATIONS

Date	Sun ☉ ° '	Moon ☽ ° '	Mercury ☿ ° '	Venus ♀ ° '	Mars ♂ ° '	Jupiter ♃ ° '	Saturn ♄ ° '	Uranus ♅ ° '	Neptune ♆ ° '	Pluto ♇ ° '
01	14 S 38	27 N 50	20 S 47	00 S 12	08 S 31	02 S 05	20 N 35	11 S 10	17 S 06	14 S 56
02	14 57	27 55	21 13	00 40	08 46	02 10	20 35	11 10	17 06	14 56
03	15 15	26 43	21 38	01 07	09 01	02 14	20 35	11 10	17 06	14 57
04	15 34	24 19	22 02	01 35	09 16	02 19	20 35	11 10	17 06	14 57
05	15 52	20 51	22 25	02 03	09 30	02 23	20 35	11 10	17 06	14 57
06	16 10	16 26	22 47	02 30	09 45	02 28	20 35	11 10	17 06	14 58
07	16 28	11 16	23 07	02 58	10 00	02 32	20 35	11 10	17 05	14 58
08	16 45	05 N 29	23 27	03 26	10 14	02 37	20 36	11 10	17 05	14 58
09	17 02	00 S 43	23 45	03 53	10 29	02 41	20 36	11 10	17 05	14 59
10	17 19	07 05	24 02	04 21	10 43	02 46	20 36	11 10	17 05	14 59
11	17 35	13 18	24 18	04 49	10 58	02 50	20 36	11 10	17 05	15 00
12	17 52	18 58	24 32	05 16	11 12	02 54	20 36	11 10	17 05	15 00
13	18 08	23 36	24 46	05 44	11 27	02 59	20 36	11 10	17 04	15 00
14	18 23	26 44	24 57	06 11	11 41	03 03	20 36	11 10	17 04	15 01
15	18 38	28 00	25 08	06 39	11 55	03 07	20 37	11 10	17 04	15 01
16	18 53	27 15	25 17	07 06	12 09	03 11	20 37	11 10	17 04	15 01
17	19 08	24 40	25 25	07 33	12 23	03 15	20 37	11 10	17 04	15 02
18	19 22	20 36	25 31	08 00	12 37	03 20	20 38	11 10	17 03	15 02
19	19 36	15 29	25 36	08 27	12 51	03 24	20 38	11 10	17 03	15 02
20	19 50	09 42	25 39	08 53	13 04	03 28	20 38	11 09	17 03	15 03
21	20 03	03 S 37	25 41	09 20	13 18	03 32	20 39	11 09	17 03	15 03
22	20 16	02 N 30	25 41	09 46	13 31	03 36	20 39	11 09	17 02	15 03
23	20 28	08 26	25 40	10 12	13 45	03 40	20 39	11 09	17 02	15 04
24	20 40	13 55	25 37	10 38	13 58	03 43	20 40	11 08	17 02	15 04
25	20 52	18 47	25 33	11 04	14 12	03 47	20 40	11 08	17 02	15 04
26	21 03	22 47	25 27	11 29	14 25	03 51	20 41	11 08	17 01	15 05
27	21 14	25 44	25 20	11 54	14 38	03 55	20 41	11 07	17 01	15 05
28	21 25	27 29	25 11	12 19	14 51	03 59	20 42	11 07	17 01	15 05
29	21 35	27 57	25 00	12 44	15 04	04 02	20 42	11 07	17 00	15 05
30	21 S 45	27 N 07	24 S 47	13 S 08	15 S 16	04 S 06	20 N 43	11 S 06	17 S 00	15 S 06

ZODIAC SIGN ENTRIES

Date	h	m	Planets
01	14	53	☽ ♋
04	03	32	☽ ♌
04	14	40	☿ ♐
06	15	00	☽ ♍
08	23	23	☽ ♎
11	04	05	☽ ♏
11	05	11	♂ ♏
13	05	56	☽ ♐
15	06	33	☽ ♑
17	07	39	☽ ♒
19	10	38	☽ ♓
21	16	11	☽ ♈
21	23	22	☉ ♐
22	13	31	♀ ♏
24	00	16	☽ ♉
26	10	25	☽ ♊
28	22	10	☽ ♋

LATITUDES

Date	Mercury ☿ ° '	Venus ♀ ° '	Mars ♂ ° '	Jupiter ♃ ° '	Saturn ♄ ° '	Uranus ♅ ° '	Neptune ♆ ° '	Pluto ♇ ° '
01	01 S 42	01 N 36	00 N 41	01 N 07	00 S 07	00 S 47	00 S 05	08 N 12
04	01 58	01 40	00 40	01 08	00 07	00 47	00 05	08 11
07	02 12	01 43	00 39	01 08	00 06	00 47	00 05	08 10
10	02 23	01 45	00 37	01 08	00 06	00 46	00 05	08 09
13	02 31	01 47	00 36	01 09	00 06	00 46	00 05	08 09
16	02 36	01 48	00 34	01 09	00 05	00 46	00 05	08 08
19	02 35	01 48	00 33	01 10	00 05	00 46	00 05	08 07
22	02 28	01 47	00 32	01 10	00 05	00 46	00 05	08 07
25	02 13	01 46	00 30	01 10	00 05	00 46	00 05	08 06
28	01 47	01 44	00 29	01 11	00 04	00 46	00 05	08 06
31	01 S 09	01 N 41	00 N 27	01 N 11	00 S 04	00 S 46	00 S 05	08 N 05

DATA

Julian Date	2453311
Delta T	+65 seconds
Ayanamsa	23° 55' 18"
Synetic vernal point	05° ♓ 11' 41"
True obliquity of ecliptic	23° 26' 27"

MOON'S PHASES, APSIDES AND POSITIONS ☽

Date	h	m	Phase	Longitude ° '	Eclipse Indicator
05	05	53	☾	13 ♌ 09	
12	14	27	●	20 ♏ 33	
19	05	50	☽	27 ♒ 15	
26	20	07	○	04 ♊ 55	

Day	h	m		
02	18	03	Apogee	
14	13	48	Perigee	
30	11	12	Apogee	
02	01	33	Max dec	28° N 03'
09	09	16	0S	
15	15	03	Max dec	28° S 01'
22	02	08	0N	
29	08	30	Max dec	27° N 58'

LONGITUDES

Date	Chiron ⚷ ° '	Ceres ⚳ ° '	Pallas ⚴ ° '	Juno ⚵ ° '	Vesta ⚶ ° '	Black Moon Lilith ⚸ ° '
01	21 ♑ 19	13 ♎ 13	17 ♍ 07	19 ♑ 04	13 ♓ 57	10 ♋ 01
11	21 ♑ 48	17 ♎ 38	21 ♍ 33	22 ♑ 09	14 ♓ 38	11 ♋ 08
21	22 ♑ 24	21 ♎ 58	25 ♍ 42	25 ♑ 30	15 ♓ 58	12 ♋ 15
31	23 ♑ 05	26 ♎ 12	29 ♍ 32	29 ♑ 06	17 ♓ 53	13 ♋ 22

All ephemeris data is given at 12.00 UT and the Moon's longitude is additionally given for 24.00 UT

ASPECTARIAN

h m	Aspects
01 Monday	
01 21	☽ △ ♂
01 52	☿ # ♄
02 52	☽ ⚼ ☉
04 56	☽ ⚻ ☿
09 27	☽ ⚺ ♄
10 06	☽ ⚼ ♆
18 43	☽ ± ☿
20 46	☽ △ ♅
02 Tuesday	
00 39	☽ □ ♀
04 16	☽ ± ♆
07 04	☽ □ ♃
11 28	☉ ∥ ♇
11 54	☽ △ ☉
15 27	☽ ⚼ ☿
16 25	☽ ⚻ ♆
18 01	☿ △ ♄
03 Wednesday	
03 05	☽ ⚼ ♅
08 34	☽ ⚻ ♇
17 14	☽ Q ♀
17 31	☽ □ ♂
20 06	☽ Q ♃
20 44	☽ ± ♇
21 16	☽ ± ♅
22 08	☽ ☌ ♄
04 Thursday	
02 00	☽ △ ☿
09 21	☽ ⚻ ♅
14 54	☽ ⚼ ♇
17 34	☉ □ ♆
20 07	☽ ⚹ ♀
20 38	☽ ⚹ ♃
05 Friday	
01 28	☿ Q ♆
02 11	♀ ☌ ♃
02 57	☽ # ☿
04 15	♀ Q ♇
04 52	☽ ☍ ♆
05 53	☽ □ ☉
07 45	☽ Q ♂
08 34	♀ ± ♅
13 33	☽ ∥ ♄
17 10	♃ Q ♇
17 23	♀ Q ♄
20 49	☽ △ ♇
06 Saturday	
02 51	☽ ∠ ♃
05 08	☽ ∠ ♀
08 43	☽ # ♆
08 45	☽ ⚹ ♂
09 28	♀ ∥ ♃
09 50	☽ ⚺ ♄
13 17	☽ # ☉
15 18	☿ □ ♅
17 17	♃ ± ♅
17 50	☿ ⊥ ♂
19 11	☽ # ♇
19 20	☉ ⊥ ♇
20 21	☽ Q ☉
20 34	☽ ☍ ♅
20 38	☽ ⊥ ♃
21 15	☽ □ ☿
21 27	☽ ⊥ ♄
07 Sunday	
00 29	☽ ⊥ ♀
00 36	☉ ⊥ ♃
05 03	♂ □ ♄
08 18	☽ ⚺ ♃
12 23	☽ # ♅
13 10	☽ ⚺ ♀
14 38	☽ ∠ ♄
15 10	☽ ∠ ♂
15 15	☽ ⚻ ♆
17 13	☽ # ♂
21 10	☽ ⚹ ☉
08 Monday	
02 31	☽ ± ♆
06 26	☽ □ ♇
06 53	♄ St R
08 55	☽ ⊥ ♂
10 58	♀ △ ♆
11 24	☽ Q ☿
18 32	☽ ⚹ ♄
19 09	☽ ⚼ ♆
19 30	☽ # ♀
20 32	☽ ⚺ ♂
23 06	☽ # ♃
09 Tuesday	
02 24	♃ Q ♄
03 06	☽ ∠ ☉
04 35	☽ ⚻ ♅
11 44	☽ ⚹ ☿
15 14	☽ Q ♇
15 20	☽ ± ♅
16 10	☽ Q ♄
16 14	☉ ∥ ♆
16 22	☽ ☌ ♃
19 32	☽ ∥ ♃
20 30	☽ ⊥ ☉
22 05	☽ △ ♆
10 Wednesday	
00 54	☽ ∥ ♀
01 32	☽ ☌ ♀
07 12	☽ ⚼ ♅
07 52	☽ ⚺ ☉
12 22	☽ ⚹ ♇
16 59	☽ ∠ ☿
23 33	☽ □ ♄
11 Thursday	
02 28	☽ ∥ ♂
03 38	☽ ∥ ♅
04 02	☽ ☌ ♂
08 58	☽ △ ♅
09 51	♅ Q ♇
09 58	☽ ⊥ ☿
14 02	☽ ∠ ♇
16 31	☿ ⚹ ♃
18 52	☽ ∥ ♇
19 13	♅ St D
20 44	☽ ⚺ ♃
21 06	☽ ⚺ ☿
12 Friday	
01 28	☽ □ ♆
03 38	☽ ∥ ♆
05 07	☽ ⊥ ♇
06 48	☽ ∥ ☉
06 51	☽ ⊥ ♃
08 43	♂ ∥ ♅
09 24	☽ ⚺ ♀
14 27	☽ ☌ ☉
15 03	☽ ⚺ ♇
16 59	♀ ⚼ ♅
19 47	☽ # ♄
20 11	☽ ⊥ ♀
21 52	☽ ∠ ♃
23 24	☉ ⚺ ♇
13 Saturday	
01 34	☽ △ ♄
07 07	☽ Q ♆
08 15	☽ ⚺ ♂
10 37	☽ □ ♅
12 13	☽ ∠ ♀
13 45	☿ ⚼ ♄
18 28	☽ ⊥ ♂
19 56	☽ ∥ ☿
21 19	☿ ⚹ ♆
22 37	☽ ⚹ ♃
14 Sunday	
01 56	☽ ⚼ ♄
02 37	☽ ⚹ ♆
03 06	☽ ☌ ☿
09 43	☽ ∠ ♂
14 43	☽ ⚹ ♀
15 47	☽ Q ♅
15 58	☽ ☌ ♇
16 29	☽ ± ♄
18 16	☽ Q ♃
18 49	☽ ⚺ ☉
15 Monday	
02 11	☽ ⚻ ♄
02 53	☽ ∠ ♆
05 13	☽ ⊥ ☉
06 11	♀ ⚹ ♇
11 08	☽ ⚹ ♂
11 13	☽ ⚹ ♅
11 53	☽ Q ♀
12 58	♂ △ ♅
17 29	☽ ⊥ ♆
20 54	☽ ∠ ☉
23 49	☽ □ ♃
16 Tuesday	
03 16	☽ ⚺ ♆
07 35	☽ Q ♂
08 15	☽ ⚺ ☿
11 39	☽ ∠ ♅
16 51	☽ ⚺ ♇
19 00	☽ ⊥ ☿
19 57	☽ □ ♀
23 17	☽ ⚹ ☉
17 Wednesday	
02 29	☽ ⊥ ♅
02 47	☽ ⊥ ♇
03 07	☽ ☍ ♄
06 40	☽ ∥ ☿
11 14	☽ ∠ ☿
12 27	☽ ⚺ ♅
14 45	☽ □ ♂
17 48	☽ ∠ ♇
20 46	☽ Q ☉
18 Thursday	
00 26	☉ ∠ ♃
02 06	☽ △ ♃
05 04	☽ ☌ ♆
11 53	☽ # ♄
14 47	☽ ⚹ ☿
17 53	☽ ∥ ☉
19 18	☽ ⚹ ♇
19 Friday	
03 18	☽ △ ♀
04 03	☽ ⚼ ♃
05 01	☽ ∥ ♆
05 33	☉ △ ♄
05 49	☽ ⚻ ♄
05 50	☽ □ ☉
13 18	☽ Q ☿
13 55	☽ ∥ ♇
15 41	☽ ☌ ♅
16 08	☽ Q ♇
16 15	☽ ± ♄
19 57	☽ ± ♃
20 42	☽ △ ♂
22 47	☽ ∥ ♂
20 Saturday	
06 09	☽ ∥ ♅
06 41	☽ ⚻ ♃
07 12	♀ □ ♄
08 06	☽ ⚼ ♄
08 11	☽ ⚼ ♀
09 10	☽ ⚺ ♆
11 51	♂ ∠ ♇
15 03	☽ ∥ ♀
19 41	♄ ± ♇
19 55	☽ ⊥ ♆
20 19	☿ Q ♅
23 57	☽ □ ☿
21 Sunday	
00 12	☽ □ ♇
00 49	☽ ⚼ ♂
01 57	☽ ± ♀
02 57	☿ ± ♄
03 24	☿ ☌ ♇
11 02	☽ △ ♄
12 13	☽ ∠ ♆
12 21	☽ ∥ ♃
13 58	☽ ⚻ ♀
15 35	☽ △ ☉
18 04	☽ ± ♂
21 32	☽ ⚺ ♅
22 Monday	
05 42	☽ ⚻ ♂
08 37	☽ ⊥ ♅
12 32	☿ ∠ ♂
14 00	☽ ☍ ♃
15 55	☽ ⚹ ♆
16 22	☽ # ♃
19 32	☉ Q ♆
21 41	☽ ⚼ ☉
23 Tuesday	
01 25	☽ ∠ ♅
07 44	☽ △ ♇
11 43	☽ △ ☿
14 30	☽ Q ♆
16 07	☽ ± ☉
18 47	☽ □ ♄
20 09	☽ # ♀
23 34	☽ # ♅
24 Wednesday	
04 03	☽ ☍ ♀
04 32	☽ ⚻ ☉
05 53	☽ ⚹ ♅
09 32	☿ Q ♃
10 55	☉ # ♄
12 14	☽ # ♂
12 22	☽ ⚼ ♇
17 21	☽ # ♇
17 33	☽ ☍ ♂
18 18	☽ ⚼ ☿
21 19	☉ □ ♅
22 46	♀ △ ♅
23 46	☽ ⚻ ♃
25 Thursday	
01 01	☽ □ ♆
02 55	☽ # ♆
05 01	☽ Q ♅
05 12	☉ ⚺ ♀
05 16	☽ Q ♄
05 48	☽ ± ♇
11 34	☽ ± ♃
12 47	☽ ± ☿
15 58	♀ ∥ ♅
17 31	☽ ⚻ ♇
22 41	☽ ∥ ♄
26 Friday	
00 25	☽ # ☉
01 09	☽ ⚻ ☿
04 37	☽ ⚹ ♄
05 24	☽ ⚼ ♃
16 16	☽ □ ♅
20 07	☽ ☍ ☉
20 56	☽ ⚻ ♀
27 Saturday	
07 42	☽ ⚻ ♂
08 17	☽ # ☿
10 10	☽ ∠ ♄
10 11	☽ ± ♀
11 28	☽ △ ♃
12 01	☽ △ ♆
18 40	♀ ∠ ♇
20 21	☽ ± ♂
23 07	♄ ± ♅
28 Sunday	
02 57	☿ ∠ ♂
04 04	☽ ⊥ ♄
05 02	☽ ☍ ♇
06 12	☽ ⚼ ♀
15 04	☽ ☍ ☿
15 26	☽ ⚼ ♂
16 04	☽ ⚺ ♄
18 04	☽ ⚼ ♆
29 Monday	
04 13	☽ △ ♅
08 26	♃ △ ♆
12 13	☽ ± ♆
13 40	☽ ⚻ ☉
15 29	♂ ∥ ♇
15 53	☽ △ ♀
23 30	☽ △ ♂
30 Tuesday	
00 22	☽ ⚻ ♆
00 34	☽ □ ♃
02 55	☽ ± ☉
10 34	☽ ⚼ ♅
12 16	☿ St R
15 25	♂ □ ♆
17 44	☽ ⚻ ♇
22 51	☽ ⚼ ☉

DECEMBER 2004

Date	Sidereal time h m s	Sun ☉ ° ' "	Moon ☽ ° ' "	Moon ☽ 24.00 ° ' "	Mercury ☿ ° '	Venus ♀ ° '	Mars ♂ ° '	Jupiter ♃ ° '	Saturn ♄ ° '	Uranus ♅ ° '	Neptune ♆ ° '	Pluto ♇ ° '
01	16 42 44	09 ♐ 38 21	00 ♌ 34 45	06 ♌ 30 55	26 ♐ 40	11 ♏ 04	13 ♏ 34	13 ♎ 19	26 ♋ 51	03 ♓ 02	13 ♒ 01	21 ♐ 35
02	16 46 41	10 39 10	12 ♌ 28 03	18 ♌ 26 38	26 R 24	12 18	14 15	13 29	26 R 48	03 03	13 02	21 37
03	16 50 37	11 40 01	24 ♌ 27 11	00 ♍ 30 16	25 57	13 33	14 55	13 38	26 46	03 04	13 03	21 40
04	16 54 34	12 40 54	06 ♍ 36 27	12 ♍ 46 19	25 20	14 47	15 36	13 47	26 43	03 06	13 05	21 42
05	16 58 30	13 41 47	19 ♍ 00 28	25 ♍ 19 30	24 31	16 02	16 16	13 57	26 40	03 07	13 06	21 44
06	17 02 27	14 42 42	01 ♎ 43 58	08 ♎ 14 24	23 32	17 16	16 57	14 06	26 37	03 08	13 07	21 46
07	17 06 23	15 43 38	14 ♎ 51 16	21 ♎ 34 57	22 23	18 31	17 37	14 15	26 34	03 09	13 09	21 49
08	17 10 20	16 44 35	28 ♎ 25 41	05 ♏ 23 37	21 08	19 46	18 18	14 24	26 31	03 11	13 10	21 51
09	17 14 16	17 45 34	12 ♏ 28 42	19 ♏ 40 41	19 48	21 00	18 59	14 33	26 28	03 12	13 12	21 53
10	17 18 13	18 46 34	26 ♏ 59 08	04 ♐ 23 23	18 25	22 15	19 39	14 41	26 24	03 13	13 13	21 55
11	17 22 10	19 47 34	11 ♐ 52 35	19 ♐ 25 38	17 02	23 30	20 20	14 50	26 21	03 15	13 15	21 58
12	17 26 06	20 48 36	27 ♐ 01 21	04 ♑ 38 23	15 43	24 45	21 01	14 59	26 17	03 16	13 16	22 00
13	17 30 03	21 49 39	12 ♑ 15 21	19 ♑ 50 55	14 30	25 59	21 42	15 07	26 14	03 18	13 18	22 02
14	17 33 59	22 50 42	27 ♑ 23 46	04 ♒ 52 44	13 24	27 14	22 22	15 15	26 10	03 19	13 19	22 04
15	17 37 56	23 51 46	12 ♒ 16 52	19 ♒ 35 23	12 28	28 29	23 03	15 23	26 07	03 21	13 21	22 07
16	17 41 52	24 52 50	26 ♒ 47 42	03 ♓ 53 28	11 42	29 ♏ 44	23 44	15 31	26 03	03 23	13 22	22 09
17	17 45 49	25 53 54	10 ♓ 52 33	17 ♓ 44 58	11 07	00 ♐ 59	24 25	15 39	25 59	03 25	13 24	22 11
18	17 49 45	26 54 59	24 ♓ 30 53	01 ♈ 10 37	10 43	02 14	25 06	15 47	25 55	03 26	13 26	22 13
19	17 53 42	27 56 04	07 ♈ 44 31	14 ♈ 13 03	10 30	03 29	25 47	15 54	25 51	03 28	13 28	22 16
20	17 57 39	28 57 09	20 ♈ 36 42	26 ♈ 55 56	10 D 27	04 43	26 28	16 02	25 47	03 30	13 29	22 18
21	18 01 35	29 ♐ 58 14	03 ♉ 11 17	09 ♉ 23 12	10 34	05 58	27 09	16 09	25 43	03 32	13 31	22 20
22	18 05 32	00 ♑ 59 20	15 ♉ 32 10	21 ♉ 38 34	10 50	07 13	27 50	16 16	25 38	03 34	13 33	22 22
23	18 09 28	02 00 26	27 ♉ 42 49	03 ♊ 45 14	11 14	08 28	28 31	16 23	25 34	03 36	13 35	22 25
24	18 13 25	03 01 32	09 ♊ 46 09	15 ♊ 45 47	11 45	09 43	29 12	16 30	25 30	03 38	13 36	22 27
25	18 17 21	04 02 39	21 ♊ 44 24	27 ♊ 42 12	12 23	10 58	29 ♏ 53	16 37	25 25	03 40	13 38	22 29
26	18 21 18	05 03 46	03 ♋ 39 20	09 ♋ 35 59	13 06	12 13	00 ♐ 34	16 44	25 21	03 42	13 40	22 31
27	18 25 14	06 04 53	15 ♋ 32 19	21 ♋ 28 28	13 55	13 28	01 15	16 50	25 17	03 44	13 42	22 34
28	18 29 11	07 06 00	27 ♋ 24 37	03 ♌ 20 58	14 48	14 43	01 57	16 56	25 12	03 47	13 44	22 36
29	18 33 08	08 07 08	09 ♌ 17 43	15 ♌ 15 07	15 46	15 58	02 38	17 03	25 07	03 49	13 46	22 38
30	18 37 04	09 08 16	21 ♌ 13 28	27 ♌ 13 05	16 47	17 13	03 19	17 09	25 03	03 51	13 48	22 40
31	18 41 01	10 ♑ 09 24	03 ♍ 14 21	09 ♍ 17 40	17 ♐ 51	18 ♐ 28	04 ♐ 00	17 ♎ 14	24 ♋ 58	03 ♓ 54	13 ♒ 50	22 ♐ 42

Date	Moon True ☊ ° '	Moon Mean ☊ ° '	Moon ☽ Latitude ° '
01	01 ♉ 24	29 ♈ 56	05 N 10
02	01 R 19	29 53	05 04
03	01 15	29 50	04 45
04	01 14	29 47	04 13
05	01 D 13	29 44	03 29
06	01 14	29 40	02 33
07	01 16	29 37	01 28
08	01 17	29 34	00 N 15
09	01 R 16	29 31	01 S 00
10	01 14	29 28	02 14
11	01 09	29 25	03 21
12	01 02	29 21	04 14
13	00 54	29 18	04 50
14	00 46	29 15	05 05
15	00 38	29 12	05 00
16	00 33	29 09	04 35
17	00 30	29 06	03 54
18	00 29	29 02	03 00
19	00 D 29	28 59	01 59
20	00 30	28 56	00 S 53
21	00 R 30	28 53	00 N 14
22	00 29	28 50	01 19
23	00 25	28 46	02 20
24	00 19	28 43	03 13
25	00 09	28 40	03 57
26	29 ♈ 58	28 37	04 31
27	29 45	28 34	04 52
28	29 31	28 31	05 01
29	29 18	28 27	04 57
30	29 07	28 24	04 40
31	28 ♈ 59	28 ♈ 21	04 N 10

DECLINATIONS

Date	Sun ☉ ° '	Moon ☽ ° '	Mercury ☿ ° '	Venus ♀ ° '	Mars ♂ ° '	Jupiter ♃ ° '	Saturn ♄ ° '	Uranus ♅ ° '	Neptune ♆ ° '	Pluto ♇ ° '
01	21 S 54	25 N 04	24 S 33	13 S 33	15 S 29	04 S 10	20 N 43	11 S 06	17 S 00	15 S 06
02	22 03	21 55	24 17	13 56	15 41	04 13	20 44	11 06	16 59	15 06
03	22 11	17 51	24 00	14 20	15 54	04 17	20 45	11 05	16 59	15 07
04	22 19	13 00	23 40	14 43	16 06	04 20	20 45	11 05	16 59	15 07
05	22 27	07 33	23 19	15 06	16 18	04 24	20 46	11 04	16 58	15 07
06	22 34	01 N 39	22 57	15 28	16 30	04 27	20 47	11 04	16 58	15 07
07	22 41	04 S 30	22 33	15 50	16 42	04 30	20 47	11 03	16 57	15 08
08	22 47	10 41	22 08	16 12	16 54	04 34	20 48	11 03	16 57	15 08
09	22 53	16 32	21 43	16 33	17 06	04 37	20 49	11 02	16 57	15 08
10	22 58	21 40	21 17	16 54	17 18	04 40	20 49	11 02	16 56	15 08
11	23 03	25 32	20 53	17 14	17 29	04 43	20 50	11 01	16 56	15 09
12	23 07	27 39	20 30	17 34	17 40	04 46	20 51	11 01	16 55	15 09
13	23 11	27 41	20 09	17 53	17 51	04 49	20 52	11 00	16 55	15 09
14	23 15	25 40	19 50	18 12	18 03	04 52	20 52	11 00	16 54	15 09
15	23 18	21 54	19 35	18 31	18 13	04 55	20 53	10 59	16 54	15 09
16	23 21	16 53	19 23	18 49	18 24	04 58	20 54	10 58	16 54	15 10
17	23 23	11 05	19 14	19 06	18 35	05 01	20 55	10 58	16 53	15 10
18	23 24	04 S 56	19 09	19 23	18 45	05 04	20 56	10 57	16 53	15 10
19	23 25	01 N 15	19 07	19 39	18 56	05 06	20 57	10 56	16 52	15 10
20	23 26	07 14	19 08	19 55	19 06	05 09	20 57	10 55	16 52	15 10
21	23 26	12 48	19 12	20 11	19 16	05 12	20 58	10 55	16 51	15 11
22	23 26	17 46	19 18	20 25	19 26	05 14	20 59	10 54	16 51	15 11
23	23 26	21 55	19 26	20 40	19 35	05 17	21 00	10 53	16 50	15 11
24	23 24	25 06	19 36	20 53	19 45	05 19	21 01	10 52	16 50	15 11
25	23 23	27 08	19 48	21 06	19 54	05 22	21 02	10 52	16 49	15 11
26	23 21	27 54	20 00	21 19	20 04	05 24	21 03	10 51	16 49	15 11
27	23 18	27 22	20 13	21 30	20 13	05 26	21 04	10 50	16 48	15 12
28	23 15	25 36	20 27	21 41	20 22	05 28	21 05	10 49	16 47	15 12
29	23 12	22 42	20 41	21 52	20 30	05 31	21 06	10 48	16 47	15 12
30	23 08	18 50	20 56	22 02	20 39	05 33	21 06	10 47	16 46	15 12
31	23 S 03	14 N 12	21 S 10	22 S 11	20 S 47	05 S 35	21 N 07	10 S 47	16 S 46	15 S 12

ZODIAC SIGN ENTRIES

Date	h m	Planets
01	10 50	☽ ♌
03	23 00	☽ ♍
06	08 46	☽ ♎
08	14 44	☽ ♏
10	16 54	☽ ♐
12	16 42	☽ ♑
14	16 10	☽ ♒
16	17 10	♀ ♐
16	17 24	☽ ♓
18	21 52	☽ ♈
21	05 52	☽ ♉
21	12 42	☉ ♑
23	16 32	☽ ♊
25	16 04	♂ ♐
26	04 38	☽ ♋
28	17 14	☽ ♌
31	05 33	☽ ♍

LATITUDES

Date	Mercury ☿ ° '	Venus ♀ ° '	Mars ♂ ° '	Jupiter ♃ ° '	Saturn ♄ ° '	Uranus ♅ ° '	Neptune ♆ ° '	Pluto ♇ ° '
01	01 S 09	01 N 41	00 N 27	01 N 11	00 S 04	00 S 46	00 S 05	08 N 05
04	00 S 19	01 38	00 26	01 12	00 04	00 46	00 05	08 05
07	00 N 40	01 34	00 24	01 13	00 03	00 45	00 05	08 05
10	01 39	01 29	00 22	01 13	00 03	00 45	00 05	08 04
13	02 25	01 24	00 21	01 14	00 03	00 45	00 05	08 04
16	02 50	01 19	00 19	01 14	00 02	00 45	00 05	08 04
19	02 56	01 13	00 17	01 15	00 02	00 45	00 05	08 03
22	02 48	01 07	00 15	01 16	00 02	00 45	00 05	08 03
25	02 30	01 00	00 14	01 16	00 02	00 45	00 05	08 03
28	02 08	00 53	00 12	01 17	00 01	00 45	00 05	08 03
31	01 N 43	00 N 46	00 N 10	01 N 18	00 S 01	00 S 45	00 S 05	08 N 03

DATA

Julian Date	2453341
Delta T	+65 seconds
Ayanamsa	23° 55' 24"
Synetic vernal point	05° ♓ 11' 36"
True obliquity of ecliptic	23° 26' 27"

MOON'S PHASES, APSIDES AND POSITIONS ☽

Date	h m	Phase	Longitude ° '	Eclipse Indicator
05	00 53	☾	13 ♍ 14	
12	01 29	●	20 ♐ 22	
18	16 40	☽	27 ♓ 07	
26	15 06	○	05 ♋ 12	

Day	h m		
12	21 21	Perigee	
27	18 57	Apogee	
06	18 30	0S	
13	00 31	Max dec	27° S 56'
19	07 06	0N	
26	14 13	Max dec	27° N 54'

LONGITUDES

Date	Chiron ⚷ ° '	Ceres ⚳ ° '	Pallas ⚴ ° '	Juno ⚵ ° '	Vesta ⚶ ° '	Black Moon Lilith ⚸ ° '
01	23 ♑ 05	26 ♎ 12	29 ♍ 32	29 ♑ 06	17 ♓ 53	13 ♋ 22
11	23 ♑ 50	00 ♏ 17	03 ♎ 01	02 ♒ 55	20 ♓ 17	14 ♋ 29
21	24 ♑ 39	04 ♏ 13	06 ♎ 05	06 ♒ 55	23 ♓ 06	15 ♋ 36
31	25 ♑ 30	07 ♏ 58	08 ♎ 40	11 ♒ 04	26 ♓ 15	16 ♋ 44

All ephemeris data is given at 12.00 UT and the Moon's longitude is additionally given for 24.00 UT

ASPECTARIAN

h m Aspects

01 Wednesday
00 05 ♂ ⊻ ♃
04 10 ☽ ⊼ ☿
04 28 ☽ ☌ ♄
04 49 ☽ ± ♅
05 55 ☽ ± ♇
13 31 ☽ Q ♃
16 09 ☽ ± ☿
16 57 ☽ # ☿
16 59 ☽ ⊼ ♅
21 57 ☿ ∠ ♀

02 Thursday
00 11 ☽ bQ ♇
08 00 ☽ △ ☉
09 55 ☽ bQ ☿
11 12 ☽ # ☉
11 38 ☽ □ ♀
13 08 ☽ ☍ ♆
14 03 ☽ ⚹ ♃
15 47 ☽ □ ♂
19 31 ☽ ∥ ♄

03 Friday
02 22 ♀ □ ♆
06 25 ☽ △ ♇
14 01 ♀ ⊻ ♃
14 06 ☉ bQ ♄
14 52 ☽ △ ☿
16 33 ☽ # ♆
16 34 ☽ ⊻ ♄
20 25 ☽ ∠ ♃
21 39 ☽ # ♂
22 32 ☿ Q ♃

04 Saturday
01 57 ☽ # ♇
03 39 ☽ Q ♀
04 23 ☽ ⊥ ♄
04 28 ☽ # ♀
05 05 ☽ ☍ ♅
05 45 ☽ Q ♂
14 21 ☽ ⊥ ♃
15 48 ♂ ⊥ ♇
20 44 ☽ # ♅
21 32 ☉ ⚹ ♆
21 55 ☽ ∠ ♄

05 Sunday
00 37 ☽ ⊼ ♆
00 53 ☽ □ ☉
02 09 ☽ ⊻ ♃
05 39 ☽ ⚹ ♀
06 07 ♀ ⊥ ♇
06 27 ☽ ⚹ ♂
12 10 ☽ ± ♆
13 33 ♀ ∥ ♇
17 13 ☽ □ ♇
18 57 ☉ ⚹ ♃
21 45 ☽ □ ☿
22 13 ♀ ☌ ♂

06 Monday
00 53 ☽ # ♃
02 28 ☽ ⚹ ♄
05 15 ☽ bQ ♆
12 25 ☽ ∠ ♂
13 07 ☽ ∠ ♀
13 58 ☽ Q ☉
14 36 ☽ ⊼ ♅
14 39 ☿ ⊥ ♀
19 54 ☿ ⊥ ♂

07 Tuesday
00 38 ☽ Q ♄
01 39 ☽ ± ♅
02 50 ☽ Q ♇
04 35 ☽ Q ☿
05 51 ☽ ⊥ ♂
06 03 ☉ ∥ ☿
07 21 ☽ ⊥ ♀
08 55 ☽ △ ♆
10 54 ☽ ☌ ♃
12 00 ☽ ∥ ♃
13 42 ☽ ⚹ ☉
17 14 ☽ ⊻ ♂
17 55 ☽ bQ ♅
19 13 ☽ ⊻ ♀
23 03 ☿ ☌ ♇

08 Wednesday
00 19 ☽ ⚹ ☿
00 26 ☽ ⚹ ♇
08 41 ☽ □ ♄
11 15 ☿ Q ♅
13 28 ☽ ∥ ♅
17 28 ♂ ∥ ♆
18 11 ☽ ∠ ☉
20 13 ☽ △ ♅
23 42 ☿ ± ♄

09 Thursday
00 07 ☽ ∠ ☿
00 49 ☿ ⊻ ♀
01 58 ♀ ⊥ ♃
02 31 ☽ ∠ ♇
06 03 ☽ ∥ ♇
10 42 ☽ ⊥ ☉
12 03 ☽ ∥ ♀
13 12 ☽ □ ♆
13 45 ☽ ∥ ♆
14 01 ☽ ⊥ ☿
14 32 ☽ ∥ ♂
15 30 ☽ ⊻ ♃
17 43 ☽ ⊥ ♇
21 29 ☽ ⊻ ☉
21 31 ☿ ⊻ ♂
23 08 ☽ ⊻ ☿
23 22 ☽ ☌ ♂

10 Friday
01 34 ☽ ⊥ ♃
03 31 ☽ ☌ ♀
03 41 ☽ ⊻ ♇
05 27 ♀ ⊻ ♇
07 43 ☽ # ♄
08 21 ☉ ☌ ☿
10 15 ☽ ∥ ☿
11 03 ☽ △ ♄
15 02 ♀ ∥ ♆
16 27 ☽ ∠ ♃
18 53 ☽ Q ♆
19 18 ☽ ∥ ☉
22 08 ☽ □ ♅

11 Saturday
11 10 ☽ bQ ♄
14 11 ☽ ⚹ ♆
14 44 ☿ # ♄
16 45 ☽ ⚹ ♃
19 32 ☽ ☌ ☿

12 Sunday
00 25 ☉ ± ♄
01 24 ☽ ± ♄
01 29 ☽ ☌ ☉
02 04 ☽ ⊻ ♂
02 54 ☽ Q ♅
04 03 ☽ ☌ ♇
08 05 ☽ ⊻ ♀
10 19 ♂ ⊥ ♃
10 51 ☽ ⊼ ♄
11 56 ☽ Q ♃
11 59 ☽ ⊥ ♂
13 58 ☽ ∠ ♆
18 23 ☽ ⊥ ♀
21 52 ☽ ⚹ ♅
23 11 ☉ Q ♅

13 Monday
00 46 ☿ ⚹ ♃
02 29 ☉ ⊻ ♂
02 50 ☽ ∠ ♂
04 10 ☽ ⊥ ♆
06 50 ♀ ∥ ♂
09 50 ☽ ∠ ♀
13 38 ☽ ⊻ ♆
15 17 ☽ ⊻ ☿
16 26 ♀ △ ♄
16 33 ☽ □ ♃
17 05 ☉ ☌ ♇
21 34 ☽ ∠ ♅

14 Tuesday
00 07 ☽ ⊥ ☿
00 45 ♂ ⊻ ♇
03 30 ☽ ⊻ ♇
03 38 ☽ ⚹ ♂
04 14 ☽ ⊻ ☉
10 03 ☽ ☍ ♄
11 43 ☽ ⚹ ♀
11 53 ☽ ⊥ ♅
13 05 ☽ ⊥ ♇
13 30 ☽ ∠ ☿
13 45 ☿ ⚹ ♆
14 29 ☽ ⊥ ☉
21 31 ☽ ⊻ ♅
23 43 ☽ Q ♂

15 Wednesday
03 35 ☽ ∠ ♇
04 16 ☽ ∥ ☉
06 02 ☽ ∠ ☉
08 48 ☽ Q ♀
12 16 ☽ ⚹ ☿
13 45 ☽ ☌ ♆
17 08 ☽ △ ♃
17 18 ☽ # ♄

16 Thursday
00 13 ☽ ∥ ☿
03 48 ☽ ∥ ♀
04 13 ☽ ⚹ ♇
05 26 ☽ ∥ ♂
06 38 ☽ □ ♂
07 03 ☽ Q ☿
08 33 ☽ ⚹ ☉
10 45 ☽ ⊼ ♄
11 58 ☽ ∥ ♆
17 25 ☽ □ ♀
18 20 ☽ bQ ♃
19 23 ☽ ∥ ♇
20 49 ☽ ± ♄
23 09 ☽ ☌ ♅

17 Friday
00 28 ☽ Q ♇
04 58 ♀ ∠ ♃
06 27 ☽ Q ☉
09 52 ☽ ± ♃
12 11 ☽ bQ ♄
12 24 ☽ □ ☿
12 32 ☽ ∥ ♅
13 50 ☉ ⊼ ♄
16 24 ☽ ⊻ ♆
20 11 ☿ bQ ♄
20 20 ♀ Q ♆
20 24 ☽ ⊼ ♃
20 33 ☿ ∥ ♀

18 Saturday
02 57 ☽ ⊥ ♆
07 54 ☽ □ ♇
11 32 ☽ ∥ ♃
13 06 ☽ △ ♂
14 30 ☽ △ ♄
16 40 ☽ □ ☉
19 03 ☽ ∠ ♆

19 Sunday
03 22 ☽ △ ♀
04 09 ☽ ⊻ ♅
11 15 ☉ Q ♃
11 53 ♀ □ ♅
14 10 ♂ △ ♄
15 12 ☽ ⊥ ♅
17 03 ☽ △ ☿
17 55 ☽ bQ ♂
22 37 ☽ ⚹ ♆

20 Monday
00 45 ☉ ∠ ♆
03 18 ☽ ☍ ♃
03 25 ☽ # ♃
06 30 ☿ St D
08 01 ☽ ∠ ♅
10 09 ☽ bQ ♀
11 42 ☽ ± ♂
15 12 ☽ △ ♇
20 39 ☿ ∥ ♂
21 14 ☽ bQ ☿
21 16 ☽ Q ♆
21 45 ☽ □ ♄
23 45 ☽ ⊼ ♂

21 Tuesday
03 36 ☽ # ♅
05 08 ☽ ± ♀
05 16 ☽ △ ☉
12 40 ☽ ⚹ ♅
14 43 ☽ ± ☿
17 59 ☽ ⊼ ♀
20 03 ☽ bQ ♇
23 05 ☽ # ♇
23 20 ☿ bQ ♄

22 Wednesday
02 35 ☽ ⊼ ☿
07 18 ☽ # ♆
08 06 ☽ □ ♆
08 19 ☽ Q ♄
12 04 ☽ Q ♅
12 58 ☽ bQ ☉
13 27 ☽ ⊼ ♃
13 39 ☽ ± ♇
20 35 ☽ # ☿
21 23 ☽ # ♂

23 Thursday
01 22 ☽ ± ♃
01 29 ☽ ⊼ ♇
03 41 ☽ # ♀
06 12 ☽ ∥ ♄
07 47 ☽ ⚹ ♄
08 19 ☽ ± ☉
13 41 ☽ ☍ ♂
19 22 ☽ bQ ♃
21 19 ☽ ⊼ ☉
22 20 ☽ # ☉
23 44 ☽ □ ♅

24 Friday
11 54 ☽ ☍ ♀
13 27 ☽ ∠ ♄
16 10 ☽ ☍ ☿
19 42 ☽ △ ♆

25 Saturday
01 37 ☽ △ ♃
02 06 ♀ bQ ♄
02 50 ☉ ⚹ ♅
03 15 ♀ # ♄
07 23 ☽ ⊥ ♄
13 30 ☽ ☍ ♇
19 22 ☽ ⊻ ♄

26 Sunday
01 55 ☽ bQ ♆
05 24 ☽ ⊼ ♂
12 06 ☽ △ ♅
15 06 ☽ ☍ ☉
18 14 ☽ ± ♂
20 07 ☽ ± ♆

27 Monday
05 35 ☿ ⚹ ♆
07 20 ☽ ⊼ ♀
08 17 ☽ ⊼ ♆
08 28 ☽ ⊼ ☿
09 00 ☿ ∥ ♂
13 32 ☽ bQ ♂
14 39 ☽ □ ♃
16 32 ♀ ⚹ ♆
18 30 ☽ bQ ♅
20 53 ☽ ± ♀
21 33 ☽ ± ☿

28 Tuesday
02 14 ☽ ⊼ ♇
04 20 ♂ Q ♆
07 34 ☽ ☌ ♄
11 53 ♂ ∠ ♃
12 45 ☽ ± ♅
14 24 ☽ ± ♇
17 13 ☽ bQ ♀
17 15 ☽ bQ ☿
18 23 ☿ ☌ ♀
21 43 ☽ △ ♂

29 Wednesday
00 54 ☽ ⊼ ♅
03 21 ☽ Q ♃
03 24 ☉ ⊥ ♆
08 21 ☽ # ☉
08 38 ☽ bQ ♇
09 24 ☽ ⊼ ☉
17 24 ☽ # ♀
21 02 ☽ ☍ ♆
22 35 ☽ ∥ ♄
22 38 ☽ ± ☉

30 Thursday
00 22 ☽ # ☿
01 45 ☽ # ♂
02 14 ☽ △ ☿
03 01 ☽ △ ♀
03 44 ☽ ⚹ ♃
10 19 ♀ ⚹ ♃
14 54 ☽ △ ♇
18 23 ☽ bQ ☉
19 37 ☽ ⊻ ♄
21 07 ☿ ⚹ ♃
23 08 ☽ # ♆

31 Friday
01 01 ☉ ⊥ ♂
06 58 ☿ # ♄
07 05 ☽ # ♇
07 31 ☽ ⊥ ♄
07 49 ♂ □ ♅
10 00 ☽ ∠ ♃
13 18 ☽ ☍ ♅
13 37 ☽ □ ♂
20 56 ♀ ± ♄

JANUARY 2005

LONGITUDES

Date	Sidereal time h m s	Sun ☉ ° ' "	Moon ☽ ° ' "	Moon ☽ 24.00 ° ' "	Mercury ☿ ° '	Venus ♀ ° '	Mars ♂ ° '	Jupiter ♃ ° '	Saturn ♄ ° '	Uranus ♅ ° '	Neptune ♆ ° '	Pluto ♇ ° '
01	18 44 57	11 ♑ 10 33	15 ♍ 23 31	21 ♍ 32 23	18 ♐ 58	19 ♐ 44	04 ♐ 42	17 ♎ 20	24 ♋ 53	03 ♓ 56	13 ♒ 52	22 ♐ 45
02	18 48 54	12 11 42	27 ♍ 44 50	04 ♎ 01 23	20 07	20 59	05 23	17 26	24 R 49	03 58	13 54	22 47
03	18 52 50	13 12 51	10 ♎ 22 39	16 ♎ 49 11	21 19	22 14	06 04	17 31	24 44	04 01	13 56	22 49
04	18 56 47	14 14 00	23 ♎ 21 33	00 ♏ 00 15	22 33	23 29	06 46	17 36	24 39	04 03	13 58	22 51
05	19 00 43	15 15 10	06 ♏ 45 43	13 ♏ 38 19	23 48	24 44	07 27	17 41	24 34	04 06	14 00	22 53
06	19 04 40	16 16 20	20 ♏ 38 13	27 ♏ 45 28	25 05	25 59	08 09	17 46	24 29	04 08	14 02	22 55
07	19 08 37	17 17 30	04 ♐ 59 55	12 ♐ 21 11	26 24	27 14	08 50	17 51	24 25	04 11	14 04	22 57
08	19 12 33	18 18 40	19 ♐ 48 36	27 ♐ 21 20	27 44	28 29	09 32	17 56	24 20	04 13	14 06	22 59
09	19 16 30	19 19 51	04 ♑ 58 16	12 ♑ 38 05	29 ♐ 05	29 ♐ 45	10 13	18 00	24 15	04 16	14 08	23 02
10	19 20 26	20 21 01	20 ♑ 19 21	28 ♑ 00 32	00 ♑ 27	01 ♑ 00	10 55	18 04	24 10	04 19	14 10	23 04
11	19 24 23	21 22 11	05 ♒ 40 06	13 ♒ 16 37	01 50	02 15	11 37	18 08	24 05	04 22	14 12	23 06
12	19 28 19	22 23 20	20 ♒ 48 45	28 ♒ 15 22	03 14	03 30	12 18	18 12	24 00	04 24	14 15	23 08
13	19 32 16	23 24 29	05 ♓ 35 36	12 ♓ 48 47	04 39	04 45	13 00	18 16	23 55	04 27	14 17	23 10
14	19 36 12	24 25 37	19 ♓ 54 33	26 ♓ 52 43	06 05	06 00	13 42	18 19	23 50	04 30	14 19	23 12
15	19 40 09	25 26 44	03 ♈ 43 21	10 ♈ 26 39	07 31	07 16	14 23	18 23	23 45	04 33	14 21	23 14
16	19 44 06	26 27 51	17 ♈ 02 59	23 ♈ 32 50	08 58	08 31	15 05	18 26	23 40	04 36	14 23	23 16
17	19 48 02	27 28 57	29 ♈ 56 44	06 ♉ 15 17	10 26	09 46	15 47	18 29	23 35	04 38	14 25	23 18
18	19 51 59	28 30 02	12 ♉ 29 05	18 ♉ 38 46	11 54	11 01	16 29	18 32	23 30	04 41	14 28	23 20
19	19 55 55	29 ♑ 31 06	24 ♉ 44 57	00 ♊ 48 13	13 24	12 16	17 10	18 34	23 25	04 44	14 30	23 22
20	19 59 52	00 ♒ 32 09	06 ♊ 49 07	12 ♊ 48 09	14 53	13 31	17 52	18 37	23 21	04 47	14 32	23 23
21	20 03 48	01 33 12	18 ♊ 45 48	24 ♊ 41 52	16 24	14 47	18 34	18 39	23 16	04 50	14 34	23 25
22	20 07 45	02 34 14	00 ♋ 38 32	06 ♋ 34 19	17 55	16 02	19 16	18 41	23 11	04 53	14 37	23 27
23	20 11 41	03 35 15	12 ♋ 30 04	18 ♋ 26 03	19 26	17 17	19 58	18 43	23 06	04 56	14 39	23 29
24	20 15 38	04 36 15	24 ♋ 22 26	00 ♌ 19 25	20 59	18 32	20 40	18 45	23 01	04 59	14 41	23 31
25	20 19 35	05 37 14	06 ♌ 17 08	12 ♌ 15 43	22 32	19 47	21 22	18 46	22 56	05 03	14 43	23 33
26	20 23 31	06 38 12	18 ♌ 15 19	24 ♌ 16 04	24 05	21 02	22 04	18 48	22 52	05 06	14 46	23 34
27	20 27 28	07 39 10	00 ♍ 18 08	06 ♍ 21 41	25 39	22 17	22 46	18 49	22 47	05 09	14 48	23 36
28	20 31 24	08 40 06	12 ♍ 28 57	18 ♍ 34 10	27 14	23 33	23 28	18 50	22 42	05 12	14 50	23 38
29	20 35 21	09 41 02	24 ♍ 43 37	00 ♎ 55 39	28 ♑ 50	24 48	24 10	18 51	22 38	05 15	14 52	23 40
30	20 39 17	10 41 57	07 ♎ 10 36	13 ♎ 28 55	00 ♒ 26	26 03	24 53	18 51	22 33	05 18	14 55	23 41
31	20 43 14	11 ♒ 42 52	19 ♎ 50 59	26 ♎ 17 18	02 ♒ 02	27 ♑ 18	25 ♐ 35	18 ♎ 52	22 ♋ 29	05 ♓ 22	14 ♒ 57	23 ♐ 43

Date	Moon True ☊ ° '	Moon Mean ☊ ° '	Moon ☽ Latitude ° '
01	28 ♈ 53	28 ♈ 18	03 N 28
02	28 R 50	28 15	02 37
03	28 49	28 12	01 36
04	28 D 49	28 08	00 N 29
05	28 R 49	28 05	00 S 42
06	28 48	28 02	01 53
07	28 44	27 59	02 59
08	28 37	27 56	03 55
09	28 27	27 52	04 36
10	28 16	27 49	04 57
11	28 04	27 46	04 57
12	27 53	27 43	04 37
13	27 45	27 40	03 58
14	27 39	27 37	03 05
15	27 35	27 33	02 02
16	27 D 34	27 30	00 S 55
17	27 R 34	27 27	00 N 13
18	27 34	27 24	01 18
19	27 32	27 21	02 18
20	27 28	27 18	03 11
21	27 20	27 14	03 55
22	27 10	27 11	04 28
23	26 57	27 08	04 50
24	26 43	27 05	04 59
25	26 28	27 02	04 55
26	26 14	26 58	04 38
27	26 01	26 55	04 08
28	25 51	26 52	03 27
29	25 45	26 49	02 36
30	25 41	26 46	01 36
31	25 ♈ 39	26 ♈ 43	00 N 31

DECLINATIONS

Date	Sun ☉	Moon ☽	Mercury ☿	Venus ♀	Mars ♂	Jupiter ♃	Saturn ♄	Uranus ♅	Neptune ♆	Pluto ♇
01	22 S 58	08 N 58	21 S 25	22 S 20	20 S 56	05 S 37	21 N 08	10 S 46	16 S 45	15 S 12
02	22 53	03 N 17	21 39	22 28	21 04	05 38	21 09	10 45	16 45	15 12
03	22 47	02 S 38	21 52	22 35	21 12	05 40	21 10	10 44	16 44	15 13
04	22 41	08 38	22 05	22 41	21 19	05 42	21 11	10 43	16 44	15 13
05	22 34	14 26	22 18	22 47	21 27	05 44	21 12	10 42	16 43	15 13
06	22 27	19 43	22 29	22 52	21 34	05 45	21 13	10 41	16 42	15 13
07	22 19	24 04	22 40	22 57	21 41	05 47	21 14	10 40	16 42	15 13
08	22 11	26 57	22 51	23 01	21 48	05 48	21 15	10 39	16 41	15 13
09	22 03	27 56	23 00	23 04	21 55	05 50	21 16	10 38	16 41	15 13
10	21 54	26 48	23 08	23 06	22 02	05 51	21 17	10 37	16 40	15 13
11	21 45	23 39	23 16	23 08	22 08	05 53	21 18	10 36	16 39	15 13
12	21 35	18 55	23 22	23 09	22 14	05 54	21 19	10 35	16 39	15 13
13	21 25	13 09	23 27	23 09	22 20	05 55	21 20	10 34	16 38	15 13
14	21 14	06 50	23 31	23 08	22 26	05 56	21 21	10 33	16 38	15 13
15	21 03	00 S 23	23 34	23 07	22 32	05 57	21 22	10 32	16 37	15 14
16	20 52	05 N 51	23 36	23 05	22 37	05 58	21 23	10 31	16 36	15 14
17	20 40	11 39	23 37	23 03	22 43	05 59	21 24	10 30	16 36	15 14
18	20 28	16 49	23 36	22 59	22 48	06 00	21 25	10 29	16 35	15 14
19	20 15	21 11	23 34	22 55	22 52	06 01	21 26	10 28	16 34	15 14
20	20 02	24 35	23 31	22 51	22 57	06 01	21 27	10 26	16 34	15 14
21	19 49	26 52	23 26	22 45	23 01	06 02	21 28	10 25	16 33	15 14
22	19 35	27 55	23 21	22 39	23 06	06 02	21 28	10 24	16 33	15 14
23	19 21	27 40	23 14	22 32	23 10	06 03	21 29	10 23	16 32	15 14
24	19 07	26 09	23 05	22 25	23 14	06 03	21 30	10 22	16 31	15 14
25	18 52	23 27	22 56	22 17	23 17	06 04	21 31	10 21	16 31	15 14
26	18 37	19 46	22 45	22 08	23 20	06 04	21 32	10 20	16 30	15 14
27	18 21	15 14	22 32	21 58	23 24	06 04	21 33	10 18	16 29	15 14
28	18 06	10 05	22 18	21 48	23 27	06 04	21 34	10 17	16 29	15 14
29	17 50	04 N 29	22 03	21 37	23 29	06 04	21 35	10 16	16 28	15 14
30	17 33	01 S 23	21 47	21 26	23 32	06 04	21 36	10 15	16 27	15 14
31	17 S 16	07 S 17	21 S 29	21 S 14	23 S 34	06 S 04	21 N 36	10 S 14	16 S 27	15 S 14

ZODIAC SIGN ENTRIES

Date	h	m	Planets
02	16	19	☽ ♎
05	00	00	☽ ♏
07	03	44	☽ ♐
09	04	11	☽ ♑
09	16	56	♀ ♑
10	04	09	☿ ♑
11	03	07	☽ ♒
13	02	50	☽ ♓
15	05	27	☽ ♈
17	12	06	☽ ♉
19	22	24	☽ ♊
19	23	22	☉ ♒
22	10	42	☽ ♋
24	23	21	☽ ♌
27	11	24	☽ ♍
29	22	13	☽ ♎
30	05	37	☿ ♒

LATITUDES

Date	Mercury ☿	Venus ♀	Mars ♂	Jupiter ♃	Saturn ♄	Uranus ♅	Neptune ♆	Pluto ♇
01	01 N 35	00 N 43	00 N 09	01 N 18	00 S 01	00 S 45	00 S 05	08 N 03
04	01 09	00 36	00 07	01 19	00 00	00 45	00 06	08 03
07	00 43	00 28	00 05	01 19	00 00	00 44	00 06	08 03
10	00 N 18	00 20	00 03	01 20	00 00	00 44	00 06	08 03
13	00 S 06	00 12	00 N 01	01 21	00 N 01	00 44	00 06	08 03
16	00 28	00 N 05	00 S 01	01 22	00 01	00 44	00 06	08 03
19	00 48	00 S 03	00 03	01 22	00 01	00 44	00 06	08 03
22	01 07	00 11	00 05	01 23	00 01	00 44	00 06	08 04
25	01 23	00 18	00 08	01 24	00 02	00 44	00 06	08 04
28	01 37	00 25	00 10	01 25	00 02	00 44	00 06	08 04
31	01 S 49	00 S 32	00 S 12	01 N 25	00 N 02	00 S 44	00 S 06	08 N 04

DATA

Julian Date	2453372
Delta T	+66 seconds
Ayanamsa	23° 55' 30"
Synetic vernal point	05° ♓ 11' 30"
True obliquity of ecliptic	23° 26' 27"

MOON'S PHASES, APSIDES AND POSITIONS ☽

Date	h	m	Phase	Longitude ° '	Eclipse Indicator
03	17	46	☾	13 ♎ 28	
10	12	03	●	20 ♑ 21	
17	06	57	☽	27 ♈ 16	
25	10	32	○	05 ♌ 34	

Day	h	m		
10	10	00	Perigee	
23	18	40	Apogee	
03	01	24	0S	
09	11	16	Max dec	27° S 56'
15	13	28	0N	
22	19	24	Max dec	27° N 58'
30	06	25	0S	

LONGITUDES

Date	Chiron ⚷ ° '	Ceres ⚳ ° '	Pallas ⚴ ° '	Juno ⚵ ° '	Vesta ⚶ ° '	Black Moon Lilith ⚸ ° '
01	25 ♑ 35	08 ♏ 19	08 ♎ 54	11 ♒ 30	26 ♓ 35	16 ♋ 50
11	26 ♑ 28	11 ♏ 49	10 ♎ 51	15 ♒ 49	00 ♈ 02	17 ♋ 58
21	27 ♑ 22	15 ♏ 01	12 ♎ 07	20 ♒ 15	03 ♈ 43	19 ♋ 05
31	28 ♑ 14	17 ♏ 53	12 ♎ 36	24 ♒ 48	07 ♈ 35	20 ♋ 12

All ephemeris data is given at 12.00 UT and the Moon's longitude is additionally given for 24.00 UT

ASPECTARIAN

01 Saturday
01 15 ☽ ∠ ♄
02 57 ☽ △ ☉
03 58 ☽ ⊥ ♃
03 59 ☽ ⋕ ♅
09 00 ☽ ⊼ ♆
10 32 ☿ ± ♄
15 50 ☽ ⊻ ♃
19 42 ☽ □ ☿
20 46 ☽ ± ♆
21 26 ☽ □ ♀

02 Sunday
02 17 ☽ ⋕ ♃
02 23 ☽ □ ♇
03 05 ☽ Q ♂
06 23 ☽ ✶ ♄
14 13 ☽ ꝑ ♆
23 56 ☽ ⊼ ♅

03 Monday
03 25 ☽ ✶ ♂
05 11 ☽ Q ♄
07 24 ♂ ⋕ ♄
07 42 ♀ Q ♅
09 48 ☽ Q ☿
11 19 ☽ ± ♅
11 41 ☽ Q ♀
12 49 ☽ Q ♇
17 46 ☽ □ ☉
18 39 ☽ △ ♆
23 33 ♀ ☌ ♇

04 Tuesday
00 12 ☽ ∥ ♃
01 23 ☽ ☌ ♃
02 05 ☿ Q ♅
04 06 ☽ ꝑ ♅
05 26 ☉ ⊻ ♆
08 56 ☽ ∠ ♂
10 22 ☽ ✶ ☿
11 04 ☽ ✶ ♇
11 08 ☉ ∥ ♀
12 15 ☽ ✶ ♀
14 20 ☽ □ ♄
17 58 ☿ ☌ ♇
20 27 ☽ ∥ ♅

05 Wednesday
02 05 ☽ ⊥ ♂
05 17 ☽ Q ☉
07 16 ☽ △ ♅
09 06 ♀ ⊼ ♄
13 17 ☽ ⊻ ♂
13 59 ☽ ∠ ♇
15 21 ☽ ∥ ♇
15 57 ☽ ∠ ☿
17 44 ☽ ∠ ♀
21 57 ☽ ∥ ♆

06 Thursday
00 39 ☽ □ ♆
01 30 ☿ ⊼ ♄
03 57 ☽ ✶ ☉
05 38 ☽ ⊥ ♇
07 05 ☽ ⊻ ♃
08 50 ☉ ∥ ☿
09 06 ☽ ⊥ ☿
10 47 ☽ ⊥ ♀
15 53 ☽ ⊻ ♇
17 20 ☽ ⊥ ♃
18 29 ☽ △ ♄
19 37 ☽ ⋕ ♄
20 17 ☽ ⊻ ☿
21 43 ☽ ∥ ♂
21 54 ☽ ⊻ ♀

07 Friday
01 53 ☽ ∥ ☉
03 11 ☽ ∥ ☿
05 02 ☽ ∥ ♀
07 09 ☽ Q ♆
07 12 ☽ ∠ ☉
08 26 ☽ ∠ ♃
10 39 ☽ □ ♅
18 36 ☽ ☌ ♂
19 11 ☽ ꝑ ♄
23 02 ☽ ⊥ ☉

08 Saturday
02 12 ☉ □ ♃
02 48 ☽ ✶ ♆
05 46 ♂ ꝑ ♄
08 58 ☽ ✶ ♃
09 25 ☽ ⊻ ☉
09 38 ☽ ± ♄
15 52 ☽ Q ♅
17 05 ☽ ☌ ♇
19 10 ☽ ⊼ ♄

09 Sunday
00 03 ♀ ∠ ♆
01 49 ☽ ☌ ☿
02 48 ☽ ∠ ♆
03 02 ☽ ☌ ♀
04 08 ☽ Q ♃
10 29 ☉ ∠ ♅
10 54 ☽ ✶ ♅
13 00 ☿ ∠ ♆
16 59 ☽ ⊥ ♆
17 12 ♀ Q ♃
20 37 ☽ ⊻ ♂
23 56 ☉ ∥ ♂

10 Monday
02 23 ☽ ⊻ ♆
02 57 ☿ ∥ ♀
05 00 ☿ Q ♃
06 26 ☽ ⊥ ♂
08 28 ☽ □ ♃
10 25 ☽ ∠ ♅
12 03 ☽ ☌ ☉
16 17 ☽ ⊻ ♇
17 58 ☽ ☍ ♄
21 09 ☽ ∠ ♂

11 Tuesday
00 31 ☽ ⊥ ♅
01 40 ☽ ⊥ ♇
05 23 ☽ ⊻ ☿
06 09 ☽ ⊻ ♀
09 56 ☽ ⊻ ♅
14 16 ☽ ∥ ☿
15 03 ☽ ∥ ♀
15 45 ☽ ⊥ ☿
15 49 ☽ ∠ ♇
16 25 ☽ ⊥ ♀
20 18 ☽ ∥ ♂
21 49 ☽ ✶ ♂
22 54 ☽ ∥ ☉

12 Wednesday
00 44 ☽ ⋕ ♄
01 30 ☽ ☌ ♆
07 27 ☽ ∠ ☿
07 49 ☽ △ ♃
07 58 ☽ ∠ ♀
14 43 ☽ ⊻ ☉
15 44 ☽ ✶ ♇
17 05 ☽ ⊼ ♄
17 53 ☽ Q ♂
21 52 ☽ ∥ ♆

13 Thursday
01 07 ☽ ⊥ ☉
02 45 ☽ ± ♄
03 43 ☽ ∥ ♇
05 59 ♀ ✶ ♅
06 01 ☉ ⊻ ♇
08 09 ☽ ꝑ ♃
08 33 ☿ ✶ ♅
10 07 ☽ ☌ ♅
10 17 ☽ ✶ ☿
10 29 ☽ ✶ ♀
11 17 ☽ Q ♇
17 00 ☽ ∠ ☉
17 28 ☽ ꝑ ♄
21 56 ☉ ⋕ ♄
21 58 ☽ ∥ ♅
23 06 ☉ ☍ ♄
23 08 ☽ ± ♃

14 Friday
00 56 ☽ □ ♂
02 30 ☽ ⊻ ♆
02 35 ☿ ☌ ♀
08 27 ☽ Q ♀
08 32 ☽ Q ☿
09 17 ☽ ⊼ ♃
12 42 ☽ ⊥ ♆
15 19 ☽ ∥ ♃
17 39 ☽ □ ♇
18 42 ☽ △ ♄
20 22 ☽ ✶ ☉

15 Saturday
04 17 ☽ ∠ ♆
10 39 ♂ ✶ ♆
13 28 ☽ ⊻ ♅
18 56 ☽ □ ♀
19 10 ☽ Q ☉
19 34 ☽ □ ☿

16 Sunday
00 14 ☽ ⊥ ♅
02 10 ☿ ⊥ ♆
07 08 ☽ ✶ ♆
08 12 ☽ △ ♂
09 33 ♀ ⊥ ♆
12 29 ☽ ⋕ ♃
14 33 ☽ ☍ ♃
16 41 ☽ ∠ ♅
23 30 ☽ △ ♇

17 Monday
00 09 ☽ □ ♄
05 21 ☽ Q ♆
06 57 ☽ □ ☉
07 03 ☽ ⋕ ♅
13 40 ☽ ꝑ ♂
20 57 ☽ ✶ ♅

18 Tuesday
03 57 ☽ ꝑ ♇
04 14 ☽ ⋕ ♇
07 53 ☽ ± ♂
08 51 ☽ △ ♀
10 07 ☽ Q ♄
10 44 ☽ △ ☿
10 49 ☽ ⋕ ♆
15 51 ☽ □ ♆
16 42 ☉ ⊥ ♅
20 13 ☽ Q ♅
20 14 ☽ ⊼ ♂
21 27 ☽ ± ♇
23 49 ☽ ⊼ ♃

19 Wednesday
06 41 ☽ ⋕ ☉
08 08 ☉ ⊥ ♇
09 15 ☽ ⊼ ♇
09 24 ☽ ✶ ♄
11 39 ☽ ± ♃
13 29 ☽ ∥ ♄
17 34 ☽ ꝑ ♀
19 40 ♂ ± ♄
19 53 ♀ ∥ ♂
20 14 ☽ ꝑ ☿
22 19 ☽ △ ☉
23 05 ☽ ⋕ ♀
23 14 ☽ ⋕ ♂

20 Thursday
01 37 ♄ ⊼ ♇
03 43 ☽ ⋕ ☿
05 34 ☽ ꝑ ♃
06 11 ☿ ⊻ ♆
07 56 ☽ □ ♅
13 34 ☽ ± ♀
15 02 ☽ ∠ ♄
16 45 ☽ ± ☿

21 Friday
03 01 ☽ ⊼ ♀
03 32 ☽ △ ♆
06 32 ☽ ⊼ ☿
07 08 ☽ ꝑ ☉
07 59 ♀ ⊻ ♆
09 00 ☽ ⊥ ♄
11 35 ☽ ☍ ♂
11 46 ☽ △ ♃
14 54 ♂ ✶ ♃
21 01 ☽ ⊻ ♄
21 26 ☽ ☍ ♇

22 Saturday
03 00 ☽ ± ☉
09 54 ☽ ꝑ ♆
16 16 ☽ ⊼ ☉
20 38 ☽ △ ♅

23 Sunday
00 24 ☿ □ ♃
04 11 ☽ ± ♆
16 21 ☽ ⊼ ♆
20 04 ☿ ∠ ♅
20 13 ☿ ∥ ♂
22 49 ☽ ☍ ♀

24 Monday
00 36 ☽ □ ♃
03 07 ☽ ꝑ ♅
03 08 ☿ ⊻ ♂
04 03 ☽ ⊼ ♂
04 07 ☽ ☍ ☿
09 17 ☽ ☌ ♄
10 16 ☽ ⊼ ♇
16 09 ♀ □ ♃
16 55 ☽ ± ♂
21 21 ☽ ± ♅
21 38 ☉ ⊻ ♅
22 24 ☽ ± ♇

25 Tuesday
09 29 ☽ ⊼ ♅
10 32 ☽ ☍ ☉
12 11 ☽ ꝑ ♂
12 59 ☽ Q ♃
13 16 ☽ ⋕ ♂
16 04 ☽ ⋕ ☿
16 33 ☽ ꝑ ♇
17 08 ♀ ∠ ♅
18 04 ☿ ☍ ♄
20 41 ☽ ⋕ ♀

26 Wednesday
01 15 ☽ ∥ ♄
04 00 ☿ ⊻ ♇
04 59 ☽ ☍ ♆
13 05 ☽ ✶ ♃
18 12 ☽ ⊼ ♀
18 51 ☽ ⋕ ☉
20 05 ☽ △ ♂
21 08 ☽ ⊻ ♄
22 39 ☽ △ ♇

27 Thursday
01 23 ☽ ⊼ ☿
05 42 ☽ ⋕ ♆
07 33 ☽ ± ♀
09 00 ☽ ⊥ ♄
12 01 ☽ ⋕ ♇
12 24 ♂ ⊼ ♄
15 05 ☽ ± ☿
18 58 ☽ ∠ ♃
20 52 ♀ ☍ ♄
20 57 ☉ ∠ ♂
21 39 ☽ ☍ ♅

28 Friday
01 58 ♂ Q ♅
02 43 ☽ ∠ ♄
03 26 ☽ ꝑ ♀
03 53 ☽ ⊼ ☉
08 51 ♀ ⊻ ♂
11 03 ☽ ⋕ ♅
11 09 ☉ ∠ ♇
11 31 ☽ ꝑ ☿
12 45 ☽ ⊥ ♃
13 45 ♀ ⊻ ♇
16 42 ☽ ⊼ ♆
16 45 ☽ ± ☉
17 47 ♂ ☌ ♇

29 Saturday
00 31 ☽ ⊻ ♃
04 29 ☽ ± ♆
05 18 ☽ ⋕ ♃
07 57 ☽ ✶ ♄
09 56 ☽ □ ♇
10 52 ☽ □ ♂
11 55 ☽ ꝑ ☉
12 09 ☽ △ ♀
17 09 ♀ ⋕ ♄
18 39 ☿ ⊥ ♅
21 07 ☽ △ ☿
22 00 ☽ ꝑ ♆

30 Sunday
00 47 ☿ ⊥ ♇
07 00 ☽ Q ♄
08 24 ☽ ⊼ ♅
19 18 ☽ △ ☉
19 54 ☽ ± ♅
20 37 ☽ Q ♇
23 29 ☽ Q ♂
23 52 ☿ ⊥ ♂

31 Monday
02 12 ☿ ⋕ ♄
02 45 ☽ △ ♆
07 02 ☽ ∥ ♃
10 08 ☽ ☌ ♃
12 58 ☽ ꝑ ♅
16 53 ☽ □ ♄
19 15 ☽ ✶ ♇
23 18 ☽ ✶ ♂

FEBRUARY 2005

LONGITUDES

Date	Sidereal time h m s	Sun ☉ ° ' "	Moon ☽ ° ' "	Moon ☽ 24.00 ° ' "	Mercury ☿ ° '	Venus ♀ ° '	Mars ♂ ° '	Jupiter ♃ ° '	Saturn ♄ ° '	Uranus ♅ ° '	Neptune ♆ ° '	Pluto ♇ ° '
01	20 47 10	12 ♒ 43 45	02 ♏ 48 19	09 ♏ 24 28	03 ♒ 40	28 ♑ 33	26 ♐ 17	18 ♎ 52	22 ♋ 24	05 ♓ 25	14 ♒ 59	23 ♐ 45
02	20 51 07	13 44 38	16 ♏ 06 12	22 ♏ 53 52	05 18	29 ♑ 48	26 59	18 R 52	22 R 20	05 28	15 01	23 46
03	20 55 04	14 45 30	29 ♏ 47 45	06 ♐ 48 01	06 57	01 ♒ 27	27 41	18 52	22 15	05 31	15 04	23 48
04	20 59 00	15 46 22	13 ♐ 54 43	21 ♐ 07 41	08 37	02 19	28 24	18 51	22 11	05 35	15 06	23 50
05	21 02 57	16 47 12	28 ♐ 24 36	05 ♑ 50 53	10 17	03 34	29 06	18 51	22 07	05 38	15 08	23 51
06	21 06 53	17 48 02	13 ♑ 19 47	20 ♑ 52 19	11 58	04 49	29 ♐ 48	18 50	22 02	05 41	15 11	23 53
07	21 10 50	18 48 50	28 ♑ 27 20	06 ♒ 03 33	13 40	06 04	00 ♑ 31	18 49	21 58	05 45	15 13	23 54
08	21 14 46	19 49 38	13 ♒ 39 35	21 ♒ 14 04	15 23	07 19	01 13	18 48	21 54	05 48	15 15	23 56
09	21 18 43	20 50 24	28 ♒ 45 39	06 ♓ 13 06	17 06	08 34	01 56	18 47	21 50	05 51	15 17	23 57
10	21 22 39	21 51 09	13 ♓ 35 23	20 ♓ 51 37	18 51	09 50	02 38	18 45	21 46	05 55	15 20	23 58
11	21 26 36	22 51 52	28 ♓ 01 10	05 ♈ 03 37	20 36	11 05	03 21	18 43	21 42	05 58	15 22	24 00
12	21 30 33	23 52 33	11 ♈ 58 44	18 ♈ 46 32	22 22	12 20	04 03	18 41	21 39	06 01	15 24	24 01
13	21 34 29	24 53 13	25 ♈ 27 11	02 ♉ 00 58	24 09	13 35	04 46	18 39	21 35	06 05	15 27	24 03
14	21 38 26	25 53 52	08 ♉ 28 20	14 ♉ 49 47	25 56	14 50	05 28	18 37	21 31	06 08	15 29	24 04
15	21 42 22	26 54 29	21 ♉ 05 53	27 ♉ 17 15	27 45	16 05	06 11	18 35	21 28	06 12	15 31	24 05
16	21 46 19	27 55 04	03 ♊ 24 32	09 ♊ 28 21	29 ♒ 34	17 20	06 53	18 32	21 24	06 15	15 33	24 06
17	21 50 15	28 55 37	15 ♊ 29 21	21 ♊ 28 08	01 ♓ 24	18 35	07 36	18 29	21 21	06 19	15 35	24 08
18	21 54 12	29 ♒ 56 08	27 ♊ 25 17	03 ♋ 21 21	03 14	19 50	08 19	18 26	21 17	06 22	15 38	24 09
19	21 58 08	00 ♓ 56 38	09 ♋ 16 51	15 ♋ 12 13	05 05	21 05	09 01	18 23	21 14	06 25	15 40	24 10
20	22 02 05	01 57 06	21 ♋ 07 54	27 ♋ 04 14	06 57	22 20	09 44	18 20	21 11	06 29	15 42	24 11
21	22 06 02	02 57 32	03 ♌ 01 33	09 ♌ 00 07	08 49	23 35	10 27	18 16	21 08	06 32	15 44	24 12
22	22 09 58	03 57 57	15 ♌ 00 11	21 ♌ 01 56	10 42	24 50	11 10	18 13	21 05	06 36	15 47	24 13
23	22 13 55	04 58 19	27 ♌ 05 31	03 ♍ 11 05	12 35	26 05	11 52	18 09	21 02	06 39	15 49	24 14
24	22 17 51	05 58 40	09 ♍ 18 45	15 ♍ 28 36	14 28	27 20	12 35	18 05	20 59	06 43	15 51	24 15
25	22 21 48	06 58 59	21 ♍ 40 45	27 ♍ 55 19	16 21	28 35	13 18	18 00	20 57	06 46	15 53	24 16
26	22 25 44	07 59 17	04 ♎ 12 23	10 ♎ 32 07	18 13	29 ♒ 50	14 01	17 56	20 54	06 50	15 55	24 17
27	22 29 41	08 59 33	16 ♎ 54 39	23 ♎ 20 10	20 05	01 ♓ 05	14 44	17 52	20 52	06 53	15 57	24 18
28	22 33 37	09 ♓ 59 47	29 ♎ 48 52	06 ♏ 20 58	21 ♓ 56	02 ♓ 20	15 ♑ 27	17 ♎ 47	20 ♋ 49	06 ♓ 56	15 ♒ 59	24 ♐ 19

DECLINATIONS

Date	Moon True ☊ ° '	Moon Mean ☊ ° '	Moon ☽ Latitude ° '	Sun ☉	Moon ☽	Mercury ☿	Venus ♀	Mars ♂	Jupiter ♃	Saturn ♄	Uranus ♅	Neptune ♆	Pluto ♇
01	25 ♈ 39	26 ♈ 39	00 S 38	16 S 59	13 S 02	21 S 09	21 S 01	23 S 36	06 S 04	21 N 37	10 S 12	16 S 26	15 S 14
02	25 R 39	26 36	01 46	16 42	18 21	20 48	20 47	23 38	06 04	21 38	10 11	16 25	15 14
03	25 39	26 33	02 51	16 24	22 53	20 26	20 33	23 40	06 04	21 39	10 10	16 25	15 14
04	25 36	26 30	03 46	16 07	26 13	20 02	20 19	23 41	06 03	21 40	10 09	16 24	15 14
05	25 30	26 27	04 30	15 48	27 55	19 37	20 04	23 42	06 03	21 41	10 08	16 23	15 14
06	25 22	26 23	04 56	15 30	27 41	19 10	19 48	23 43	06 02	21 41	10 06	16 23	15 14
07	25 12	26 20	05 02	15 11	25 24	18 42	19 31	23 44	06 02	21 42	10 05	16 22	15 13
08	25 02	26 17	04 47	14 52	21 18	18 13	19 14	23 45	06 01	21 43	10 04	16 21	15 13
09	24 52	26 14	04 12	14 33	15 50	17 42	18 57	23 45	06 00	21 44	10 03	16 21	15 13
10	24 44	26 11	03 20	14 13	09 32	17 10	18 39	23 45	05 59	21 44	10 01	16 20	15 13
11	24 39	26 08	02 17	13 54	02 S 53	16 36	18 20	23 45	05 59	21 45	10 00	16 19	15 13
12	24 36	26 04	01 S 07	13 34	03 N 43	16 00	18 01	23 44	05 58	21 46	09 59	16 19	15 13
13	24 D 35	26 01	00 N 05	13 14	09 55	15 24	17 41	23 44	05 57	21 46	09 58	16 18	15 13
14	24 36	25 58	01 14	12 53	15 29	14 46	17 21	23 43	05 56	21 47	09 56	16 17	15 13
15	24 36	25 55	02 17	12 33	20 14	14 06	17 00	23 42	05 54	21 48	09 55	16 17	15 13
16	24 R 36	25 52	03 12	12 12	23 59	13 25	16 39	23 41	05 53	21 48	09 54	16 16	15 13
17	24 35	25 49	03 58	11 51	26 35	12 43	16 18	23 39	05 52	21 49	09 53	16 15	15 13
18	24 31	25 45	04 32	11 30	27 57	11 59	15 56	23 38	05 51	21 50	09 51	16 15	15 13
19	24 24	25 42	04 55	11 08	28 01	11 14	15 33	23 36	05 49	21 50	09 50	16 14	15 13
20	24 16	25 39	05 05	10 47	26 48	10 28	15 10	23 33	05 48	21 51	09 49	16 14	15 12
21	24 07	25 36	05 02	10 25	24 22	09 41	14 47	23 31	05 46	21 51	09 47	16 13	15 12
22	23 56	25 33	04 45	10 03	20 53	08 52	14 23	23 29	05 45	21 52	09 46	16 12	15 12
23	23 47	25 29	04 16	09 41	16 29	08 02	13 59	23 26	05 43	21 52	09 45	16 12	15 12
24	23 38	25 26	03 35	09 19	11 23	07 12	13 34	23 23	05 41	21 53	09 44	16 11	15 12
25	23 32	25 23	02 43	08 57	05 N 47	06 20	13 09	23 19	05 39	21 54	09 42	16 10	15 12
26	23 28	25 20	01 42	08 34	00 S 07	05 28	12 44	23 16	05 37	21 54	09 41	16 10	15 12
27	23 26	25 17	00 35	08 12	06 06	04 36	12 19	23 12	05 36	21 54	09 40	16 09	15 12
28	23 ♈ 26	25 ♈ 14	00 S 35	07 S 49	11 S 57	03 S 43	11 S 53	23 S 08	05 S 34	21 N 55	09 S 39	16 S 08	15 S 12

ZODIAC SIGN ENTRIES

Date	h	m	Planets
01	06	51	☽ ♏
02	15	42	♀ ♒
03	12	21	☽ ♐
05	14	32	☽ ♑
06	18	32	♂ ♑
07	14	26	☽ ♒
09	13	59	☽ ♓
11	15	21	☽ ♈
13	20	18	☽ ♉
16	05	18	☽ ♊
16	17	46	☿ ♓
18	13	32	☉ ♓
18	17	13	☽ ♋
21	05	54	☽ ♌
23	17	44	☽ ♍
26	03	59	☽ ♎
26	15	07	♀ ♓
28	12	21	☽ ♏

LATITUDES

Date	Mercury ☿ ° '	Venus ♀ ° '	Mars ♂ ° '	Jupiter ♃ ° '	Saturn ♄ ° '	Uranus ♅ ° '	Neptune ♆ ° '	Pluto ♇ ° '
01	01 S 52	00 S 35	00 S 13	01 N 26	00 N 03	00 S 44	00 S 06	08 N 04
04	02 00	00 41	00 15	01 26	00 03	00 44	00 06	08 05
07	02 04	00 47	00 18	01 27	00 03	00 44	00 06	08 05
10	02 05	00 53	00 20	01 28	00 03	00 44	00 06	08 05
13	02 02	00 59	00 23	01 29	00 04	00 44	00 06	08 06
16	01 55	01 04	00 25	01 29	00 04	00 44	00 06	08 06
19	01 43	01 08	00 28	01 30	00 04	00 44	00 06	08 07
22	01 25	01 13	00 30	01 31	00 05	00 44	00 06	08 07
25	01 02	01 16	00 33	01 31	00 05	00 44	00 06	08 08
28	00 33	01 19	00 36	01 32	00 05	00 44	00 06	08 08
31	00 00	01 S 22	00 S 38	01 N 33	00 N 06	00 S 44	00 S 06	08 N 09

DATA

Julian Date	2453403
Delta T	+66 seconds
Ayanamsa	23° 55' 35"
Synetic vernal point	05° ♓ 11' 24"
True obliquity of ecliptic	23° 26' 27"

MOON'S PHASES, APSIDES AND POSITIONS ☽

Date	h	m	Phase	Longitude ° '	Eclipse Indicator
02	07	27	☾	13 ♏ 33	
08	22	28	●	20 ♒ 16	
16	00	16	☽	27 ♉ 25	
24	04	54	○	05 ♍ 41	

Day	h	m		
07	22	04	Perigee	
20	04	53	Apogee	
05	21	07	Max dec	28° S 04'
11	22	23	ON	
19	01	13	Max dec	28° N 09'
26	11	33	OS	

LONGITUDES

Date	Chiron ⚷ ° '	Ceres ⚳ ° '	Pallas ⚴ ° '	Juno ⚵ ° '	Vesta ⚶ ° '	Black Moon Lilith ° '
01	28 ♑ 20	18 ♏ 09	12 ♎ 37	25 ♒ 16	07 ♈ 59	20 ♋ 19
11	29 ♑ 11	20 ♏ 36	12 ♎ 10	29 ♒ 55	12 ♈ 01	21 ♋ 26
21	29 ♑ 59	22 ♏ 35	10 ♎ 49	04 ♓ 39	16 ♈ 09	22 ♋ 33
31	00 ♒ 44	24 ♏ 02	08 ♎ 39	09 ♓ 28	20 ♈ 24	23 ♋ 41

All ephemeris data is given at 12.00 UT and the Moon's longitude is additionally given for 24.00 UT

ASPECTARIAN

h m	Aspects
01 Tuesday	
00 05	☽ ∥ ♅
03 21	☽ □ ♀
13 48	☽ □ ☿
16 47	☽ △ ♅
21 35	☽ ∥ ♇
22 50	☽ ∠ ♇
02 Wednesday	
02 22	♃ St R
02 59	☽ ∥ ♆
04 14	☽ ∠ ♂
04 39	☽ ∥ ☉
05 13	♀ ⊥ ♅
07 27	☽ □ ☉
10 04	☽ □ ♆
11 20	♀ ⊥ ♇
14 27	☿ ∥ ♀
14 31	☿ ⊻ ♅
14 58	☽ ⊥ ♇
15 20	☽ Q ♀
16 54	☽ ⊻ ♃
21 07	☽ ⊥ ♂
22 56	☽ △ ♄
23 25	☽ ∥ ☿
23 41	☽ ∥ ♀
03 Thursday	
01 34	☽ ⊻ ♇
02 26	☽ Q ☿
03 26	☽ ⊥ ♃
04 53	☽ # ♄
08 10	☽ ⊻ ♂
11 43	☉ ∥ ♆
14 24	☽ ⚹ ♀
16 50	☽ ∥ ♂
17 30	☽ Q ☉
17 38	☽ Q ♆
18 59	☽ ∠ ♃
19 29	☉ ☌ ♆
21 52	☽ □ ♅
04 Friday	
00 42	☽ P ♄
01 53	☽ ⚹ ☿
14 00	☽ ⚹ ♆
15 09	☿ ∠ ♇
15 21	☽ ⚹ ☉
15 46	☽ ± ♄
18 13	☽ ∠ ♀
20 14	☽ ⚹ ♃
05 Saturday	
01 40	☽ ⚻ ♄
04 06	☽ Q ♅
04 28	☽ ☌ ♇
06 10	☽ ∠ ☿
10 26	☽ ⊥ ♀
13 08	☽ ☌ ♂
14 46	☽ ∠ ♆
15 54	☽ Q ♃
17 50	☽ ∠ ☉
21 05	☽ ⊻ ♀
22 41	☽ ⊥ ☿
23 42	☽ ⚹ ♅
06 Sunday	
05 20	☽ ⊥ ♆
09 23	☽ ⊥ ☉
09 33	☽ ⊻ ☿
14 57	☽ ⊻ ♆
19 38	☽ ⊻ ☉
20 45	☽ □ ♃
23 45	☽ ∠ ♅
07 Monday	
01 15	♂ ∠ ♆
01 47	☽ ☍ ♄
04 47	☽ ⊻ ♇
05 28	♀ ⊻ ♅
09 05	☉ ∥ ♇
12 03	☉ △ ♃
14 02	☽ ⊥ ♅
14 17	☽ ⊥ ♇
15 25	☽ ⊻ ♂
22 00	♂ Q ♃
22 53	☽ ∥ ♂
23 33	☽ ⊻ ♅
08 Tuesday	
01 06	☽ ☌ ♀
01 20	☽ ⊥ ♂
04 31	☽ ∠ ♇
07 35	♀ ⊥ ♂
09 56	☽ # ♄
10 10	☿ ☌ ♆
14 32	☽ ☌ ♆
15 04	☽ ☌ ☿
16 15	☽ ∠ ♂
20 07	☽ △ ♃
22 11	☽ ∥ ♀
22 28	☽ ☌ ☉
09 Wednesday	
01 00	☽ ⚻ ♄
03 32	☽ ∥ ☿
04 19	☽ ⚹ ♇
07 49	☿ ∠ ♂
09 56	☽ ∥ ♆
10 32	☽ ± ♄
14 26	☽ ∥ ♇
17 20	☽ ⚹ ♂
17 22	☽ ∥ ☉
19 23	♀ ∠ ♇
20 03	☽ P ♃
23 27	☽ ☌ ♅
23 35	☽ Q ♇
10 Thursday	
00 57	☽ P ♄
05 17	☽ ⊻ ♀
10 11	☽ ∥ ♅
10 11	☉ ⚻ ♄
10 38	☽ ± ♃
10 43	☿ △ ♃
13 48	☽ Q ♂
14 52	☽ ⊻ ♆
16 01	☽ ⊥ ♀
20 29	☽ ⚻ ♃
21 51	☽ ⊻ ☿
11 Friday	
00 49	☽ ⊥ ♆
00 51	☽ ∥ ♃
01 27	☽ △ ♄
02 41	☽ ⊻ ☉
05 13	☽ □ ♇
08 25	☽ ∠ ♀
09 16	☽ ⊥ ☿
13 32	☽ ⊥ ☉
15 59	☽ ∠ ♆
21 32	☽ □ ♂
23 26	☿ ∥ ♆
12 Saturday	
01 37	☽ ⊻ ♅
02 34	☿ ⚻ ♄
02 48	☽ ∠ ☿
06 10	☽ ∠ ☉
09 43	♅ Q ♇
12 05	☽ ⊥ ♅
12 41	☽ ⚹ ♀
15 31	☉ ⚹ ♇
18 02	☽ ⚹ ♆
20 27	☽ # ♃
23 49	☽ ☍ ♃
13 Sunday	
04 05	☽ ∠ ♅
05 03	☽ □ ♄
09 16	☽ ⚹ ☿
09 27	☽ △ ♇
10 38	☿ ⚹ ♇
10 53	☽ ⚹ ☉
12 11	☽ # ♅
12 15	☽ Q ♀
15 38	☽ Q ♆
18 48	☿ ∥ ♇
14 Monday	
01 04	☽ # ☉
06 04	☽ △ ♂
07 37	☽ ⚹ ♅
09 01	☽ # ☿
10 45	☽ # ♇
10 50	☉ ☌ ☿
10 50	☽ Q ☉
10 50	☽ Q ☿
13 07	☽ P ♇
13 57	☽ Q ♄
15 45	☽ # ♆
20 18	☽ # ♀
15 Tuesday	
00 47	♀ ☌ ♆
01 17	☽ □ ♆
01 20	☽ □ ♀
06 12	☽ ± ♇
06 24	☽ Q ♅
07 10	☽ ⚻ ♃
08 23	☿ ± ♄
12 10	☽ P ♂
12 31	♂ ⚹ ♅
12 42	☽ ⚹ ♄
17 47	☽ ⚻ ♇
18 42	☽ ± ♃
21 13	☽ ∥ ♄
16 Wednesday	
00 16	☽ □ ☉
00 25	☉ ± ♄
03 07	☽ □ ☿
06 45	☽ ± ♂
09 48	☽ # ♂
12 15	☽ P ♃
17 38	☽ □ ♅
17 53	☽ ∠ ♄
19 18	☽ ⚻ ♂
17 Thursday	
10 10	♀ △ ♃
11 43	☽ ⊥ ♄
12 12	☽ △ ♆
14 25	♀ ∥ ♆
17 59	☽ △ ♃
18 56	☽ △ ♀
19 59	♄ P ♅
23 42	☽ ⊻ ♄
18 Friday	
05 23	☽ ☍ ♇
14 33	☿ P ♃
17 33	☽ △ ☉
18 30	☽ P ♆
19 Saturday	
01 56	☽ △ ☿
04 46	☽ P ♀
06 11	☽ △ ♅
11 27	☽ ☍ ♂
12 47	☽ ± ♆
14 45	♀ ⚻ ♄
17 41	☉ ∥ ☿
20 Sunday	
00 59	☽ ⚻ ♆
01 09	☽ ± ♀
02 02	☿ Q ♇
02 23	☿ P ♄
02 45	☽ P ☉
05 44	☿ ☌ ♅
06 21	☽ □ ♃
09 35	♀ ∥ ♇
10 52	♂ ⊥ ♆
12 06	☽ ☌ ♄
12 43	☽ P ♅
13 58	☽ P ☿
14 44	☽ ⚻ ♀
18 11	☽ ⚻ ♇
22 39	☽ ± ☉
21 Monday	
06 19	☽ ± ♇
06 58	☽ ± ♅
08 25	☿ ∥ ♅
11 31	☽ ± ☿
11 51	☽ ⚻ ☉
18 29	☽ Q ♃
18 38	☽ # ♂
19 00	☉ P ♃
19 05	☽ ⚻ ♅
22 Tuesday	
00 00	♀ ⚹ ♇
00 25	☽ P ♇
01 48	☽ ⚻ ☿
03 50	☽ ⚻ ♂
05 52	☽ ∥ ♄
13 33	☽ ☍ ♆
16 34	☽ ± ♂
18 21	☽ ⚹ ♃
21 24	☿ ⚹ ♂
23 Wednesday	
00 03	☽ ⊻ ♄
06 21	☽ △ ♇
06 36	☿ ± ♃
07 50	☉ ∥ ♅
09 47	☽ ☍ ♀
11 32	☽ P ♂
11 53	☽ ⊥ ♄
13 28	☽ # ♆
18 20	☽ # ♇
23 51	☽ ∠ ♃
24 Thursday	
01 13	☽ # ♀
04 54	☽ ☍ ☉
05 32	☽ ∠ ♄
05 33	♀ ± ♄
06 53	☽ ☍ ♅
12 18	☉ P ♄
17 22	☽ ⊥ ♃
18 46	☉ Q ♇
18 46	☽ △ ♂
19 21	☽ # ♅
21 44	☽ # ☉
23 02	♀ ∠ ♂
23 50	☽ ☍ ☿
25 Friday	
00 46	☽ ⚻ ♆
04 57	☽ ⊻ ♃
06 00	☿ ⊻ ♆
06 33	☉ ☌ ♅
09 18	☽ # ☿
10 35	☽ ⚹ ♄
12 24	☽ ± ♆
12 33	☽ # ♃
17 00	☽ □ ♇
26 Saturday	
02 45	☽ ⚻ ♀
05 43	☽ P ♆
07 40	☿ ∥ ♃
08 28	☿ ⚻ ♃
09 32	☽ Q ♄
15 26	☽ ± ♀
17 00	☽ ⚻ ♅
19 48	☽ ⚻ ☉
27 Sunday	
03 20	☽ Q ♇
04 24	☽ ± ♅
06 43	☽ ∥ ☿
07 39	☽ □ ♂
08 05	☽ ± ☉
09 57	☽ ∥ ♃
10 12	☽ △ ♆
10 17	☽ P ♀
13 46	☽ ☌ ♃
18 56	☽ ⚻ ☿
19 22	☽ □ ♄
19 58	☽ ∥ ☉
21 20	☽ P ♅
21 51	☿ △ ♄
28 Monday	
01 49	☽ ⚹ ♇
02 20	☽ P ☉
02 26	☽ ∥ ♅
07 57	☽ ± ☿
11 44	☽ ∥ ♀
12 47	☿ ⊥ ♆
17 08	☽ △ ♀
19 04	☽ Q ♂
20 02	♀ P ♃

MARCH 2005

LONGITUDES

Date	Sidereal time h m s	Sun ☉ ° ′ ″	Moon ☽ ° ′ ″	Moon ☽ 24.00 ° ′ ″	Mercury ☿ ° ′	Venus ♀ ° ′	Mars ♂ ° ′	Jupiter ♃ ° ′	Saturn ♄ ° ′	Uranus ♅ ° ′	Neptune ♆ ° ′	Pluto ♇ ° ′
01	22 37 34	11 ♓ 00 00	12 ♏ 56 41	19 ♏ 36 15	23 ♓ 46	03 ♓ 35	16 ♑ 09	17 ♎ 42	20 ♋ 47	07 ♓ 00	16 ♒ 02	24 ♐ 20
02	22 41 31	12 00 12	26 ♏ 19 54	03 ♐ 07 49	25 34	04 50	16 52	17 R 37	20 R 45	07 03	16 04	24 21
03	22 45 27	13 00 22	10 ♐ 00 11	16 ♐ 57 05	27 20	06 05	17 35	17 32	20 43	07 07	16 06	24 22
04	22 49 24	14 00 30	23 ♐ 58 31	01 ♑ 04 25	29 ♓ 03	07 20	18 18	17 27	20 41	07 10	16 08	24 22
05	22 53 20	15 00 37	08 ♑ 14 35	15 ♑ 28 41	00 ♈ 43	08 35	19 01	17 21	20 39	07 14	16 10	24 23
06	22 57 17	16 00 43	22 ♑ 46 14	00 ♒ 06 37	02 20	09 50	19 45	17 16	20 37	07 17	16 12	24 24
07	23 01 13	17 00 47	07 ♒ 29 05	14 ♒ 52 48	03 52	11 05	20 28	17 10	20 35	07 20	16 14	24 25
08	23 05 10	18 00 49	22 ♒ 16 48	29 ♒ 40 06	05 20	12 20	21 11	17 04	20 34	07 24	16 16	24 25
09	23 09 06	19 00 49	07 ♓ 01 42	14 ♓ 20 37	06 43	13 34	21 54	16 58	20 32	07 27	16 18	24 26
10	23 13 03	20 00 48	21 ♓ 35 57	28 ♓ 46 53	08 00	14 49	22 37	16 52	20 31	07 31	16 20	24 26
11	23 17 00	21 00 44	05 ♈ 52 45	12 ♈ 53 02	09 11	16 04	23 20	16 46	20 30	07 34	16 22	24 27
12	23 20 56	22 00 39	19 ♈ 47 20	26 ♈ 35 28	10 15	17 19	24 03	16 39	20 29	07 37	16 24	24 27
13	23 24 53	23 00 31	03 ♉ 17 21	09 ♉ 53 03	11 11	18 34	24 46	16 33	20 28	07 41	16 26	24 28
14	23 28 49	24 00 21	16 ♉ 22 47	22 ♉ 46 51	12 00	19 48	25 30	16 26	20 27	07 44	16 28	24 28
15	23 32 46	25 00 09	29 ♉ 05 37	05 ♊ 19 35	12 42	21 03	26 13	16 20	20 26	07 47	16 30	24 29
16	23 36 42	25 59 55	11 ♊ 29 15	17 ♊ 35 10	13 15	22 18	26 56	16 13	20 25	07 51	16 32	24 29
17	23 40 39	26 59 39	23 ♊ 37 57	29 ♊ 38 10	13 40	23 33	27 39	16 06	20 25	07 54	16 33	24 29
18	23 44 35	27 59 21	05 ♋ 36 27	11 ♋ 33 22	13 57	24 48	28 23	15 59	20 24	07 57	16 35	24 30
19	23 48 32	28 59 00	17 ♋ 29 31	23 ♋ 25 28	14 05	26 02	29 06	15 52	20 24	08 01	16 37	24 30
20	23 52 29	29 ♓ 58 37	29 ♋ 21 45	05 ♌ 18 52	14 R 05	27 17	29 ♑ 49	15 45	20 24	08 04	16 39	24 30
21	23 56 25	00 ♈ 58 12	11 ♌ 17 16	17 ♌ 17 24	13 57	28 32	00 ♒ 32	15 38	20 24	08 07	16 41	24 30
22	00 00 22	01 57 44	23 ♌ 19 37	29 ♌ 24 15	13 42	29 ♓ 46	01 16	15 30	20 D 24	08 10	16 42	24 31
23	00 04 18	02 57 15	05 ♍ 31 35	11 ♍ 41 52	13 19	01 ♈ 01	01 59	15 23	20 24	08 14	16 44	24 31
24	00 08 15	03 56 43	17 ♍ 55 16	24 ♍ 11 55	12 50	02 16	02 42	15 16	20 24	08 17	16 46	24 31
25	00 12 11	04 56 08	00 ♎ 31 55	06 ♎ 55 20	12 14	03 30	03 26	15 08	20 24	08 20	16 47	24 31
26	00 16 08	05 55 32	13 ♎ 22 09	19 ♎ 52 23	11 34	04 45	04 09	15 01	20 25	08 23	16 49	24 31
27	00 20 04	06 54 54	26 ♎ 25 59	03 ♏ 02 54	10 50	05 59	04 53	14 53	20 25	08 26	16 51	24 R 31
28	00 24 01	07 54 14	09 ♏ 43 03	16 ♏ 26 20	10 02	07 14	05 36	14 45	20 26	08 29	16 52	24 31
29	00 27 58	08 53 32	23 ♏ 12 42	00 ♐ 02 01	09 13	08 28	06 20	14 38	20 27	08 33	16 54	24 31
30	00 31 54	09 52 49	06 ♐ 54 11	13 ♐ 49 06	08 22	09 43	07 03	14 30	20 28	08 36	16 55	24 31
31	00 35 51	10 ♈ 52 03	20 ♐ 46 36	27 ♐ 46 35	07 ♈ 31	10 ♈ 57	07 ♒ 46	14 ♎ 23	20 ♋ 29	08 ♓ 39	16 ♒ 57	24 ♐ 31

Date	Moon True ☊ ° ′	Moon Mean ☊ ° ′	Moon ☽ Latitude ° ′
01	23 ♈ 27	25 ♈ 10	01 S 44
02	23 D 28	25 07	02 49
03	23 29	25 04	03 46
04	23 R 29	25 01	04 30
05	23 27	24 58	05 00
06	23 23	24 55	05 10
07	23 18	24 51	05 01
08	23 12	24 48	04 32
09	23 07	24 45	03 44
10	23 03	24 42	02 43
11	23 00	24 39	01 32
12	22 59	24 35	00 S 17
13	23 D 00	24 32	00 N 56
14	23 01	24 29	02 05
15	23 03	24 26	03 05
16	23 04	24 23	03 55
17	23 05	24 20	04 33
18	23 R 05	24 16	04 59
19	23 03	24 13	05 12
20	23 01	24 10	05 12
21	22 58	24 07	04 58
22	22 54	24 04	04 31
23	22 51	24 01	03 51
24	22 48	23 57	03 00
25	22 46	23 54	01 59
26	22 45	23 51	00 N 52
27	22 D 45	23 48	00 S 20
28	22 45	23 45	01 32
29	22 46	23 41	02 40
30	22 47	23 38	03 40
31	22 ♈ 48	23 ♈ 35	04 S 28

DECLINATIONS

Date	Sun ☉ ° ′	Moon ☽ ° ′	Mercury ☿ ° ′	Venus ♀ ° ′	Mars ♂ ° ′	Jupiter ♃ ° ′	Saturn ♄ ° ′	Uranus ♅ ° ′	Neptune ♆ ° ′	Pluto ♇ ° ′
01	07 S 26	17 S 22	02 S 49	11 S 26	23 S 04	05 S 32	21 N 55	09 S 37	16 S 08	15 S 11
02	07 04	22 04	01 56	11 00	23 00	05 30	21 56	09 36	16 07	15 11
03	06 41	25 40	01 03	10 33	22 55	05 27	21 56	09 35	16 07	15 11
04	06 18	27 48	00 S 11	10 06	22 50	05 25	21 57	09 33	16 06	15 11
05	05 54	28 10	00 N 41	09 38	22 45	05 23	21 57	09 32	16 05	15 11
06	05 31	26 37	01 31	09 11	22 40	05 21	21 57	09 31	16 05	15 11
07	05 08	23 15	02 20	08 43	22 35	05 18	21 58	09 30	16 04	15 11
08	04 44	18 22	03 07	08 15	22 29	05 16	21 58	09 28	16 04	15 10
09	04 21	12 24	03 53	07 46	22 23	05 14	21 58	09 27	16 03	15 10
10	03 57	05 S 49	04 36	07 18	22 17	05 11	21 59	09 26	16 02	15 10
11	03 34	00 N 56	05 16	06 49	22 11	05 09	21 59	09 25	16 02	15 10
12	03 10	07 28	05 53	06 20	22 04	05 06	21 59	09 23	16 01	15 10
13	02 47	13 29	06 27	05 51	21 57	05 03	21 59	09 22	16 01	15 10
14	02 23	18 44	06 58	05 22	21 50	05 01	22 00	09 21	16 00	15 10
15	01 59	22 58	07 25	04 52	21 43	04 58	22 00	09 20	16 00	15 09
16	01 35	26 02	07 47	04 23	21 36	04 55	22 00	09 18	15 59	15 09
17	01 12	27 50	08 06	03 53	21 28	04 53	22 00	09 17	15 59	15 09
18	00 48	28 19	08 21	03 23	21 21	04 50	22 00	09 16	15 58	15 09
19	00 24	27 27	08 31	02 53	21 13	04 47	22 00	09 15	15 57	15 09
20	00 S 01	25 22	08 37	02 23	21 05	04 44	22 01	09 14	15 57	15 09
21	00 N 23	22 10	08 38	01 53	20 56	04 41	22 01	09 12	15 56	15 08
22	00 47	18 00	08 35	01 23	20 48	04 39	22 01	09 11	15 56	15 08
23	01 10	13 04	08 27	00 53	20 39	04 36	22 01	09 10	15 55	15 08
24	01 34	07 32	08 16	00 S 23	20 30	04 33	22 01	09 09	15 55	15 08
25	01 58	01 N 37	08 00	00 N 07	20 21	04 30	22 01	09 08	15 54	15 08
26	02 21	04 S 29	07 41	00 38	20 12	04 27	22 01	09 06	15 54	15 08
27	02 45	10 31	07 18	01 08	20 02	04 24	22 01	09 05	15 53	15 07
28	03 08	16 11	06 52	01 38	19 53	04 21	22 01	09 04	15 53	15 07
29	03 32	21 10	06 25	02 08	19 43	04 18	22 01	09 03	15 53	15 07
30	03 55	25 05	05 55	02 38	19 33	04 15	22 01	09 02	15 52	15 07
31	04 N 18	27 S 35	05 N 24	03 N 08	19 S 23	04 S 12	22 N 01	09 S 01	15 S 52	15 S 07

ZODIAC SIGN ENTRIES

Date	h	m	Planets
02	18	29	☽ ♐
04	22	12	☽ ♑
05	01	34	☿ ♈
06	23	49	☽ ♒
09	00	32	☽ ♓
11	02	03	☽ ♈
13	06	05	☽ ♉
15	13	44	☽ ♊
18	00	44	☽ ♋
20	12	33	☉ ♈
20	13	17	☽ ♌
20	18	02	♂ ♒
22	16	25	♀ ♈
23	01	10	☽ ♍
25	11	00	☽ ♎
27	18	29	☽ ♏
29	23	56	☽ ♐

LATITUDES

Date	Mercury ☿ ° ′	Venus ♀ ° ′	Mars ♂ ° ′	Jupiter ♃ ° ′	Saturn ♄ ° ′	Uranus ♅ ° ′	Neptune ♆ ° ′	Pluto ♇ ° ′
01	00 S 23	01 S 20	00 S 37	01 N 32	00 N 05	00 S 44	00 S 06	08 N 08
04	00 N 13	01 23	00 39	01 33	00 06	00 44	00 06	08 09
07	00 52	01 24	00 42	01 33	00 06	00 44	00 06	08 09
10	01 33	01 26	00 45	01 34	00 06	00 44	00 06	08 10
13	02 12	01 26	00 48	01 34	00 07	00 44	00 06	08 10
16	02 47	01 26	00 51	01 35	00 07	00 44	00 06	08 11
19	03 13	01 26	00 53	01 35	00 07	00 44	00 06	08 11
22	03 27	01 25	00 56	01 35	00 07	00 44	00 06	08 12
25	03 26	01 23	00 59	01 35	00 08	00 44	00 06	08 12
28	03 09	01 21	01 02	01 35	00 08	00 44	00 06	08 13
31	02 N 38	01 S 18	01 S 05	01 N 36	00 N 08	00 S 44	00 S 06	08 N 13

DATA

Julian Date	2453431
Delta T	+66 seconds
Ayanamsa	23° 55′ 39″
Synetic vernal point	05° ♓ 11′ 21″
True obliquity of ecliptic	23° 26′ 28″

MOON'S PHASES, APSIDES AND POSITIONS ☽

Date	h	m	Phase	Longitude ° ′	Eclipse Indicator
03	17	36	☾	13 ♐ 14	
10	09	10	●	19 ♓ 54	
17	19	19	☽	27 ♊ 18	
25	20	58	○	05 ♎ 18	

Day	h	m		
08	03	30	Perigee	
19	22	51	Apogee	
05	04	38	Max dec	28° S 15′
11	08	42	ON	
18	08	24	Max dec	28° N 19′
25	18	23	OS	

LONGITUDES

Date	Chiron ⚷ ° ′	Ceres ⚳ ° ′	Pallas ⚴ ° ′	Juno ⚵ ° ′	Vesta ⚶ ° ′	Black Moon Lilith ⚸ ° ′
01	00 ♒ 35	23 ♏ 47	09 ♎ 08	08 ♓ 30	19 ♈ 32	23 ♋ 27
11	01 ♒ 16	24 ♏ 46	06 ♎ 26	13 ♓ 21	23 ♈ 50	24 ♋ 34
21	01 ♒ 53	25 ♏ 06	03 ♎ 19	18 ♓ 16	28 ♈ 11	25 ♋ 42
31	02 ♒ 23	24 ♏ 45	00 ♎ 11	23 ♓ 14	02 ♉ 34	26 ♋ 49

All ephemeris data is given at 12.00 UT and the Moon's longitude is additionally given for 24.00 UT

ASPECTARIAN

01 Tuesday
01 08 ☽ △ ♅
02 03 ☽ ∥ ♇
03 11 ☽ bQ ☿
05 26 ☽ ∠ ♇
06 17 ☽ ∥ ♆
07 22 ♂ ⚺ ♆
08 11 ☽ △ ☉
17 35 ☽ □ ♆
18 08 ☽ ⚹ ♂
19 39 ☿ □ ♇
20 32 ☽ ⚺ ♃
21 44 ☽ ⊥ ♇

02 Wednesday
02 04 ☽ △ ♄
03 29 ☉ ± ♃
07 12 ☽ ⊥ ♃
08 28 ☽ ⚺ ♇
10 25 ☽ △ ☿
11 13 ☽ ⋕ ♄
17 21 ☽ ∥ ♂
22 20 ☽ ∠ ♂
23 02 ☽ ∠ ♃

03 Thursday
01 40 ☽ Q ♆
04 29 ☽ □ ♀
04 33 ☽ bQ ♄
05 04 ♀ bQ ♄
06 57 ☽ □ ♅
10 15 ♂ □ ♃
14 54 ☽ ⊥ ♂
17 23 ♀ Q ♇
17 36 ☽ □ ☉
20 08 ☽ ± ♄
22 33 ☽ ⚹ ♆

04 Friday
00 55 ☽ ⚹ ♃
01 48 ☽ ⚺ ♂
06 24 ☽ ⚻ ♄
08 44 ♀ ☌ ♅
12 41 ☽ ☌ ♇
14 02 ☽ Q ♅
14 32 ☽ Q ♀
21 12 ☽ Q ♃
21 45 ☽ □ ☿

05 Saturday
00 08 ☽ ∠ ♆
02 36 ☽ Q ☉
10 18 ☽ ⚹ ♅
12 37 ☽ ⚹ ♀
15 12 ☽ ⊥ ♆
17 37 ♀ ∥ ♅
18 41 ☿ ∠ ♆
19 58 ☿ Q ♂

06 Sunday
00 04 ☽ ⚹ ☉
01 10 ☽ ⚺ ♆
03 00 ☽ □ ♃
06 46 ☽ ☌ ♂
07 31 ☽ Q ☿
08 29 ☽ ☍ ♄
11 12 ☽ ∠ ♅
14 40 ☽ ⚺ ♇
15 41 ☽ ∠ ♀
16 40 ☉ ⚺ ♆
23 53 ☉ ∥ ♃

07 Monday
00 29 ☽ ⊥ ♇
01 58 ☽ ⊥ ♅
02 27 ☽ ∠ ☉
05 28 ☽ ⚹ ☿
07 44 ☽ ⊥ ♀
11 46 ☽ ⚺ ♅
13 34 ♀ ± ♃
15 08 ☽ ∠ ♇
15 19 ☉ ⚻ ♃
15 49 ☽ ∥ ♂
16 15 ♂ ☍ ♄
18 09 ☽ ⊥ ☉
18 22 ☽ ⚺ ♀
18 59 ☽ ⋕ ♄

08 Tuesday
02 14 ☽ ☌ ♆
03 36 ☽ △ ♃
04 35 ☽ ⚺ ☉
08 31 ☽ ∠ ☿
09 13 ☽ ⚻ ♄
10 07 ☽ ⚺ ♂
15 28 ☽ ⚹ ♇
18 57 ☽ ± ♄
20 21 ☽ ⊥ ♂
21 42 ☽ ∥ ♆

09 Wednesday
00 40 ☽ ⊥ ☿
01 15 ☽ ∥ ♇
03 48 ☽ bQ ♃
09 34 ☽ bQ ♄
11 01 ☽ Q ♇
11 26 ☽ ⚺ ☿
11 46 ☽ ∠ ♂
12 42 ☽ ☌ ♅
18 24 ☽ ± ♃
22 04 ☉ ⋕ ☿
22 58 ☽ ∥ ♅
23 44 ☽ ☌ ♀

10 Thursday
02 12 ☿ ⚺ ♅
03 16 ☽ ⚺ ♆
04 13 ☽ ⚻ ♃
06 20 ☽ ∥ ♀
08 13 ♂ ∠ ♅
09 10 ☽ ☌ ☉
10 13 ☽ △ ♄
13 13 ☽ ⊥ ♆
13 47 ☽ ⚹ ♂
14 18 ☽ ∥ ♃
15 59 ☽ ⋕ ☿
16 44 ☽ □ ♇
19 03 ☽ ∥ ☉
23 54 ☉ △ ♄

11 Friday
04 20 ☽ ∠ ♆
07 48 ☿ ⋕ ♃
11 01 ☽ Q ♂
14 53 ☽ ⚺ ♅
17 54 ♀ ⚺ ♆
18 07 ☽ ☌ ☿
20 59 ☽ ⋕ ☉

12 Saturday
00 18 ♀ ⚻ ♃
01 14 ☽ ⊥ ♅
03 12 ☽ ⋕ ♃
05 26 ☽ ∥ ☿
06 04 ☽ ⚹ ♆
06 34 ☽ ☍ ♃
07 15 ☽ ⚺ ♀
08 01 ☽ ⋕ ♀
13 13 ☽ □ ♄
16 12 ☽ ⚺ ☉
17 00 ☽ ∠ ♅
18 49 ☽ ⊥ ♀
19 21 ☽ ⋕ ♅
19 55 ☽ □ ♂
20 13 ☽ △ ♇
21 39 ☉ ⊥ ♆
22 02 ☿ ⋕ ♀

13 Sunday
01 34 ♂ ⚺ ♇
03 15 ☽ Q ♆
03 42 ☽ ⊥ ☉
04 58 ♂ ⋕ ♄
12 33 ☽ ∠ ♀
19 14 ☽ ⋕ ♇
20 00 ☽ ⚹ ♅
21 16 ☽ ∠ ☉
21 23 ☽ Q ♄
23 02 ☽ ⋕ ♆
23 14 ☽ bQ ♇

14 Monday
03 23 ☽ ⚺ ☿
07 46 ♃ △ ♆
12 06 ☽ ⚻ ♃
12 09 ☽ □ ♆
15 13 ☽ ⊥ ☿
15 54 ☽ ± ♇
18 18 ☽ Q ♅
19 06 ☽ ⚹ ♀
19 36 ☽ ⚹ ♄
23 15 ☽ ± ♃
23 16 ☉ □ ♇

15 Tuesday
00 11 ♀ △ ♄
03 12 ☽ ⚻ ♇
03 32 ☽ ⚹ ☉
04 30 ☽ ⋕ ♂
05 56 ☽ ∥ ♄
06 10 ☽ △ ♂
06 39 ♀ ∥ ♃
09 11 ☽ ∠ ☿
16 15 ☽ bQ ♃
20 27 ☽ Q ♀

16 Wednesday
00 12 ☽ ∠ ♄
04 35 ☽ Q ☉
04 52 ☽ □ ♅
12 56 ☽ bQ ♂
15 36 ☽ ⚹ ☿
16 27 ♀ ⊥ ♆
17 45 ☽ ⊥ ♄
21 12 ☽ △ ♃
21 56 ☽ △ ♆

17 Thursday
05 36 ☽ ⚺ ♄
07 49 ☽ ± ♂
11 49 ☽ □ ♀
13 43 ☽ ☍ ♇
16 11 ☽ Q ☿
19 19 ☽ □ ☉
20 33 ☽ ⚻ ♂

18 Friday
03 54 ☽ bQ ♆
06 14 ♀ □ ♇
13 59 ☿ ⊥ ♅
16 45 ☽ △ ♅
22 04 ☽ ± ♆

19 Saturday
05 03 ☽ □ ☿
08 45 ☽ □ ♃
10 14 ☽ ⚻ ♆
17 53 ☽ ☌ ♄
22 02 ☉ ⚹ ♂
23 13 ☽ bQ ♅

20 Sunday
00 15 ☿ St R
02 11 ☽ ⚻ ♇
07 18 ☽ △ ♀
12 59 ☽ ☍ ♂
13 21 ☽ △ ☉
14 18 ☽ ± ♇
15 27 ☿ ⊥ ♅
17 29 ☽ ± ♅
20 45 ☽ Q ♃

21 Monday
05 37 ☽ ⚻ ♅
08 26 ☽ bQ ♇
10 51 ♂ ⊥ ♇
12 58 ☽ ∥ ♄
17 00 ☽ bQ ♀
17 15 ☽ △ ☿
19 54 ☽ ⋕ ♂
20 36 ☽ ⚹ ♃
22 13 ☽ bQ ☉
22 48 ☽ ☍ ♆

22 Tuesday
02 54 ♄ St D
05 37 ☉ ∠ ♆
06 11 ☽ ⚺ ♄
12 59 ☽ ± ♀
14 20 ☽ △ ♇
17 41 ☽ ± ☉
18 04 ☽ ⊥ ♄
21 42 ☿ Q ♂
22 19 ☽ bQ ☿
22 31 ☽ ⋕ ♆

23 Wednesday
02 02 ☽ ∠ ♃
02 10 ☽ ⚻ ♀
02 21 ☽ ⋕ ♇
04 12 ☉ ⋕ ♀
04 38 ☽ ⚻ ♂
06 32 ☽ ⚻ ☉
11 45 ☽ ∠ ♄
15 22 ☽ ± ☿
17 05 ☽ ± ♂
17 17 ☽ ☍ ♅
19 26 ☽ ⊥ ♃
20 42 ♂ ⊥ ♅

24 Thursday
02 12 ♀ ∠ ♆
02 35 ☽ ⚻ ☿
05 13 ☽ ⋕ ♅
06 56 ☽ ⚺ ♃
08 51 ☽ ∥ ☿
09 46 ☽ ⚻ ♆
11 34 ☽ bQ ♂
16 45 ☽ ⚹ ♄
21 17 ☽ ± ♆

25 Friday
00 23 ☽ ⋕ ♃
00 36 ☽ □ ♇
08 41 ♀ ⚹ ♂
10 42 ☽ ∥ ☉
14 23 ☽ bQ ♆
15 32 ☽ Q ♄
17 27 ☽ ∥ ♀
17 47 ☽ △ ♂
18 12 ☽ ☍ ♀
19 29 ☽ ⋕ ♀
20 58 ☽ ☍ ☉

26 Saturday
02 42 ☽ ⚻ ♅
03 03 ☽ ⋕ ☉
08 50 ☽ ☍ ☿
10 25 ☽ Q ♇
11 51 ☽ ∥ ♃
13 53 ☽ ± ♅
15 01 ☽ ☌ ♃
18 23 ☽ △ ♆
23 53 ☽ ⋕ ☿

27 Sunday
01 00 ☽ □ ♄
02 28 ♇ St R
06 15 ☽ ∥ ♅
06 31 ☽ bQ ♅
08 30 ☽ ⚹ ♇

28 Monday
04 11 ☽ □ ♂
07 05 ☽ ⚻ ♀
07 18 ☽ ∥ ♇
08 29 ☽ ⚻ ☉
09 47 ☽ △ ♅
10 39 ☽ ∥ ♆
11 38 ☽ ∠ ♇
12 33 ☽ ⚻ ☿
18 55 ☽ ± ♀
20 05 ☽ ± ☉
20 55 ☽ ⚺ ♃
22 38 ☽ ± ☿

29 Tuesday
00 48 ☽ □ ♆
03 02 ☉ ⚺ ♅
03 41 ☽ ⊥ ♇
04 48 ☽ ∥ ♂
07 06 ☽ △ ♄
07 29 ☽ ⊥ ♃
12 31 ☽ bQ ♀
13 18 ☽ bQ ☉
13 23 ♀ ⚺ ♅
13 40 ☽ bQ ☿
14 04 ☽ Q ♂
14 18 ☽ ⚺ ♇
16 11 ☉ ☌ ☿
16 38 ☽ ⋕ ♄
20 29 ☿ ☌ ♀
23 11 ☽ ∠ ♃
23 52 ♃ ± ♅

30 Wednesday
05 55 ☿ ⚺ ♅
08 33 ☽ Q ♆
09 29 ☽ bQ ♄
12 16 ☽ ⚹ ♂
14 24 ☽ △ ☿
14 57 ☽ □ ♅
17 22 ☽ △ ♀
17 34 ☽ △ ☉

31 Thursday
01 04 ☽ ⚹ ♃
01 07 ☽ ± ♄
03 30 ☉ ☌ ♀
05 24 ☽ ⚹ ♆
06 23 ☉ ⋕ ♃
08 06 ☿ ⚹ ♂
11 29 ☽ ⚻ ♄
15 37 ☽ ∠ ♂
18 24 ☽ ☌ ♇
21 31 ☽ Q ♃
22 06 ☽ Q ♅

APRIL 2005

LONGITUDES

Date	Sidereal time h m s	Sun ☉ ° ' "	Moon ☽ ° ' "	Moon ☽ 24.00 ° ' "	Mercury ☿ ° '	Venus ♀ ° '	Mars ♂ ° '	Jupiter ♃ ° '	Saturn ♄ ° '	Uranus ♅ ° '	Neptune ♆ ° '	Pluto ♇ ° '
01	00 39 47	11 ♈ 51 16	04 ♑ 48 50	11 ♑ 53 10	06 ♈ 41	12 ♈ 12	08 ♒ 30	14 ♎ 15	20 ♋ 30	08 ♓ 42	16 ♒ 59	24 ♐ 30
02	00 43 44	12 50 28	18 ♑ 59 22	26 ♑ 07 06	05 R 54	13 26	09 13	14 R 07	20 31	08 45	17 00	24 R 30
03	00 47 40	13 49 37	03 ♒ 16 05	10 ♒ 25 56	05 08	14 41	09 57	13 59	20 32	08 48	17 01	24 30
04	00 51 37	14 48 45	17 ♒ 36 13	24 ♒ 46 27	04 27	15 55	10 40	13 52	20 34	08 51	17 03	24 30
05	00 55 33	15 47 50	01 ♓ 56 09	09 ♓ 04 47	03 49	17 10	11 24	13 44	20 35	08 54	17 04	24 29
06	00 59 30	16 46 54	16 ♓ 11 48	23 ♓ 16 39	03 16	18 24	12 08	13 36	20 37	08 56	17 06	24 29
07	01 03 27	17 45 56	00 ♈ 18 48	07 ♈ 17 45	02 47	19 39	12 51	13 28	20 38	08 59	17 07	24 29
08	01 07 23	18 44 56	14 ♈ 13 02	21 ♈ 04 17	02 24	20 53	13 35	13 21	20 40	09 02	17 08	24 28
09	01 11 20	19 43 55	27 ♈ 51 10	04 ♉ 33 26	02 06	22 07	14 18	13 13	20 42	09 05	17 10	24 28
10	01 15 16	20 42 51	11 ♉ 10 57	17 ♉ 43 37	01 54	23 22	15 02	13 06	20 44	09 08	17 11	24 28
11	01 19 13	21 41 45	24 ♉ 11 29	00 ♊ 34 39	01 47	24 36	15 45	12 58	20 46	09 11	17 12	24 27
12	01 23 09	22 40 37	06 ♊ 53 19	13 ♊ 07 43	01 D 45	25 50	16 29	12 50	20 49	09 13	17 13	24 27
13	01 27 06	23 39 26	19 ♊ 18 13	25 ♊ 25 11	01 48	27 05	17 12	12 43	20 51	09 16	17 15	24 26
14	01 31 02	24 38 14	01 ♋ 29 05	07 ♋ 30 24	01 57	28 19	17 56	12 35	20 54	09 19	17 16	24 25
15	01 34 59	25 36 59	13 ♋ 29 39	19 ♋ 27 23	02 11	29 ♈ 33	18 40	12 28	20 56	09 21	17 17	24 25
16	01 38 56	26 35 42	25 ♋ 24 10	01 ♌ 20 36	02 29	00 ♉ 48	19 23	12 21	20 59	09 24	17 18	24 24
17	01 42 52	27 34 23	07 ♌ 17 15	13 ♌ 14 41	02 52	02 02	20 07	12 13	21 02	09 27	17 19	24 24
18	01 46 49	28 33 02	19 ♌ 13 30	25 ♌ 14 13	03 20	03 16	20 50	12 06	21 04	09 29	17 20	24 23
19	01 50 45	29 ♈ 31 38	01 ♍ 17 23	07 ♍ 23 29	03 52	04 30	21 34	11 59	21 07	09 32	17 21	24 22
20	01 54 42	00 ♉ 30 12	13 ♍ 32 57	19 ♍ 46 11	04 28	05 44	22 17	11 52	21 10	09 34	17 22	24 21
21	01 58 38	01 28 44	26 ♍ 03 31	02 ♎ 25 14	05 07	06 58	23 01	11 45	21 14	09 36	17 23	24 21
22	02 02 35	02 27 14	08 ♎ 51 32	15 ♎ 22 32	05 51	08 13	23 44	11 38	21 17	09 39	17 24	24 20
23	02 06 31	03 25 42	21 ♎ 58 16	28 ♎ 38 42	06 38	09 27	24 28	11 32	21 20	09 41	17 25	24 19
24	02 10 28	04 24 08	05 ♏ 23 42	12 ♏ 13 02	07 28	10 41	25 12	11 25	21 24	09 44	17 26	24 18
25	02 14 25	05 22 33	19 ♏ 06 25	26 ♏ 03 29	08 22	11 55	25 55	11 18	21 27	09 46	17 26	24 17
26	02 18 21	06 20 55	03 ♐ 03 49	10 ♐ 06 55	09 19	13 09	26 39	11 12	21 31	09 48	17 27	24 16
27	02 22 18	07 19 16	17 ♐ 12 18	24 ♐ 19 27	10 18	14 23	27 22	11 06	21 35	09 50	17 28	24 16
28	02 26 14	08 17 35	01 ♑ 27 49	08 ♑ 36 54	11 21	15 37	28 06	10 59	21 38	09 53	17 29	24 15
29	02 30 11	09 15 53	15 ♑ 46 13	22 ♑ 55 19	12 26	16 51	28 49	10 53	21 42	09 55	17 29	24 14
30	02 34 07	10 ♉ 14 09	00 ♒ 03 47	07 ♒ 11 15	13 ♈ 34	18 ♉ 05	29 ♒ 33	10 ♎ 47	21 ♋ 46	09 ♓ 57	17 ♒ 30	24 ♐ 13

Date	Moon True ☊ ° '	Moon Mean ☊ ° '	Moon ☽ Latitude ° '
01	22 ♈ 49	23 ♈ 32	05 S 01
02	22 R 49	23 29	05 16
03	22 48	23 26	05 11
04	22 48	23 22	04 47
05	22 47	23 19	04 06
06	22 46	23 16	03 09
07	22 46	23 13	02 01
08	22 45	23 10	00 S 47
09	22 D 45	23 07	00 N 28
10	22 45	23 03	01 40
11	22 45	23 00	02 46
12	22 46	22 57	03 41
13	22 R 45	22 54	04 25
14	22 45	22 51	04 55
15	22 45	22 47	05 13
16	22 45	22 44	05 16
17	22 D 45	22 41	05 07
18	22 45	22 38	04 44
19	22 46	22 35	04 08
20	22 47	22 32	03 21
21	22 48	22 28	02 23
22	22 48	22 25	01 16
23	22 49	22 22	00 N 05
24	22 R 48	22 19	01 S 09
25	22 47	22 16	02 20
26	22 46	22 12	03 24
27	22 44	22 09	04 17
28	22 42	22 06	04 54
29	22 41	22 03	05 13
30	22 ♈ 40	22 ♈ 00	05 S 13

DECLINATIONS

Date	Sun ☉ ° '	Moon ☽ ° '	Mercury ☿ ° '	Venus ♀ ° '	Mars ♂ ° '	Jupiter ♃ ° '	Saturn ♄ ° '	Uranus ♅ ° '	Neptune ♆ ° '	Pluto ♇ ° '
01	04 N 41	28 S 22	04 N 52	03 N 38	19 S 12	04 S 09	22 N 01	09 S 00	15 S 51	15 S 07
02	05 04	27 18	04 20	04 08	19 02	04 06	22 01	08 58	15 51	15 06
03	05 27	24 28	03 48	04 38	18 51	04 03	22 01	08 57	15 50	15 06
04	05 50	20 07	03 18	05 08	18 40	04 00	22 00	08 56	15 50	15 06
05	06 13	14 36	02 48	05 38	18 29	03 57	22 00	08 55	15 49	15 06
06	06 36	08 21	02 20	06 07	18 18	03 54	22 00	08 54	15 49	15 06
07	06 58	01 S 44	01 54	06 37	18 07	03 51	22 00	08 53	15 49	15 05
08	07 21	04 N 53	01 30	07 06	17 55	03 48	22 00	08 52	15 48	15 05
09	07 43	11 09	01 08	07 35	17 44	03 45	21 59	08 51	15 48	15 05
10	08 05	16 46	00 49	08 04	17 32	03 42	21 59	08 50	15 48	15 05
11	08 27	21 30	00 32	08 33	17 20	03 39	21 59	08 49	15 47	15 05
12	08 49	25 05	00 18	09 01	17 08	03 36	21 59	08 48	15 47	15 05
13	09 11	27 24	00 N 06	09 30	16 56	03 34	21 58	08 47	15 46	15 04
14	09 33	28 21	00 S 03	09 58	16 43	03 31	21 58	08 46	15 46	15 04
15	09 54	27 56	00 09	10 26	16 31	03 28	21 58	08 45	15 46	15 04
16	10 16	26 14	00 13	10 53	16 18	03 25	21 58	08 44	15 45	15 04
17	10 37	23 23	00 14	11 21	16 05	03 22	21 57	08 43	15 45	15 04
18	10 58	19 33	00 13	11 48	15 52	03 20	21 57	08 42	15 45	15 04
19	11 18	14 53	00 09	12 15	15 39	03 17	21 56	08 41	15 45	15 04
20	11 39	09 33	00 S 04	12 41	15 26	03 14	21 56	08 40	15 44	15 03
21	11 59	03 N 45	00 N 04	13 07	15 13	03 12	21 56	08 39	15 44	15 03
22	12 20	02 S 20	00 14	13 33	14 59	03 09	21 55	08 39	15 44	15 03
23	12 40	08 29	00 26	13 59	14 46	03 07	21 55	08 38	15 44	15 03
24	12 59	14 24	00 40	14 24	14 32	03 04	21 54	08 37	15 43	15 03
25	13 19	19 45	00 56	14 49	14 18	03 02	21 54	08 36	15 43	15 03
26	13 38	24 07	01 14	15 14	14 05	02 59	21 53	08 35	15 43	15 02
27	13 57	27 05	01 33	15 38	13 51	02 57	21 53	08 34	15 43	15 02
28	14 16	28 20	01 54	16 02	13 37	02 55	21 52	08 34	15 42	15 02
29	14 35	27 41	02 17	16 25	13 22	02 53	21 52	08 33	15 42	15 02
30	14 N 53	25 S 14	02 N 41	16 N 48	13 S 08	02 S 50	21 N 51	08 S 32	15 S 42	15 S 02

ZODIAC SIGN ENTRIES

Date	h	m	Planets
01	03	48	☽ ♑
03	06	31	☽ ♒
05	08	45	☽ ♓
07	11	28	☽ ♈
09	15	50	☽ ♉
11	22	55	☽ ♊
14	09	03	☽ ♋
15	20	37	♀ ♉
16	21	17	☽ ♌
19	09	27	☽ ♍
19	23	37	☉ ♉
21	19	27	☽ ♎
24	02	25	☽ ♏
26	06	46	☽ ♐
28	09	33	☽ ♑
30	11	54	☽ ♒

LATITUDES

Date	Mercury ☿ ° '	Venus ♀ ° '	Mars ♂ ° '	Jupiter ♃ ° '	Saturn ♄ ° '	Uranus ♅ ° '	Neptune ♆ ° '	Pluto ♇ ° '
01	02 N 24	01 S 17	01 S 06	01 N 36	00 N 08	00 S 44	00 S 07	08 N 14
04	01 40	01 14	01 09	01 36	00 08	00 44	00 07	08 14
07	00 51	01 10	01 12	01 36	00 09	00 44	00 07	08 14
10	00 N 04	01 05	01 16	01 35	00 09	00 44	00 07	08 15
13	00 S 40	01 00	01 19	01 35	00 09	00 44	00 07	08 15
16	01 18	00 55	01 22	01 35	00 09	00 45	00 07	08 16
19	01 51	00 50	01 25	01 35	00 10	00 45	00 07	08 16
22	02 16	00 44	01 28	01 34	00 10	00 45	00 07	08 17
25	02 36	00 37	01 31	01 34	00 10	00 45	00 07	08 17
28	02 49	00 31	01 34	01 34	00 10	00 45	00 07	08 17
31	02 S 56	00 S 24	01 S 37	01 N 33	00 N 11	00 S 45	00 S 07	08 N 18

DATA

Julian Date	2453462
Delta T	+66 seconds
Ayanamsa	23° 55' 42"
Synetic vernal point	05° ♓ 11' 17"
True obliquity of ecliptic	23° 26' 28"

MOON'S PHASES, APSIDES AND POSITIONS ☽

Date	h	m	Phase	Longitude ° '	Eclipse Indicator
02	00	50	☾	12 ♑ 23	
08	20	32	●	19 ♈ 06	Ann-Total
16	14	37	☽	26 ♋ 42	
24	10	06	○	04 ♏ 20	

Day	h	m		
04	11	00	Perigee	
16	18	40	Apogee	
29	10	23	Perigee	
01	10	16	Max dec	28° S 22'
07	18	13	0N	
14	16	36	Max dec	28° N 23'
22	02	51	0S	
28	15	52	Max dec	28° S 21'

LONGITUDES

Date	Chiron ⚷ ° '	Ceres ⚳ ° '	Pallas ⚴ ° '	Juno ⚵ ° '	Vesta ⚶ ° '	Black Moon Lilith ⚸ ° '
01	02 ♒ 26	24 ♏ 40	29 ♍ 53	23 ♓ 44	03 ♉ 01	26 ♋ 56
11	02 ♒ 49	23 ♏ 35	27 ♍ 10	28 ♓ 44	07 ♉ 26	28 ♋ 03
21	03 ♒ 06	21 ♏ 56	25 ♍ 06	03 ♈ 46	11 ♉ 51	29 ♋ 10
31	03 ♒ 15	19 ♏ 52	23 ♍ 50	08 ♈ 50	16 ♉ 17	00 ♌ 17

ASPECTARIAN

01 Friday
07 10 ☽ ∠ ♆
07 51 ☽ ⊥ ♂
15 01 ☽ □ ☿
16 41 ☉ ∥ ☿
18 36 ☽ ⊻ ♂
18 37 ☽ ✶ ♅
18 58 ♂ ⊻ ♅
22 29 ☽ ⊥ ♆

02 Saturday
00 50 ☽ □ ☉
01 44 ☽ □ ♀
03 51 ☽ □ ♃
08 38 ☽ ⊻ ♆
10 17 ♀ # ♃
14 34 ☽ ☍ ♄
16 33 ☿ ∥ ♀
19 50 ☽ Q ☿
20 02 ☽ ∠ ♅
21 13 ♂ ∠ ♇
21 17 ☽ ⊻ ♇
23 43 ☿ # ♃
23 52 ♀ ☍ ♃

03 Sunday
07 22 ☽ ⊥ ♇
09 24 ☽ Q ☉
10 55 ☽ Q ♀
11 12 ☽ ⊥ ♅
14 16 ♀ ⊥ ♅
14 59 ☽ ✶ ☿
15 30 ☉ ☍ ♃
21 17 ☽ ⊻ ♅
22 26 ☽ ∠ ♇
23 47 ☽ ☌ ♂

04 Monday
02 26 ☽ # ♄
05 48 ☽ △ ♃
06 59 ☽ ✶ ☉
08 55 ☽ ✶ ♀
11 04 ☽ ☌ ♆
12 49 ☉ ⊥ ♅
14 56 ☽ ∠ ☿
16 57 ☽ ⊼ ♄
18 57 ☽ ∥ ♂
23 32 ☽ ✶ ♇

05 Tuesday
03 01 ☽ ± ♄
05 22 ☽ ⊥ ☿
06 41 ☽ bQ ♃
07 01 ☽ ∥ ♆
09 57 ☽ ∠ ☉
10 00 ☽ ∥ ♇
10 11 ♀ ✶ ♆
12 25 ☽ ∠ ♀
15 02 ☽ ⊻ ☿
18 08 ☽ bQ ♄
19 39 ☽ Q ♇
21 39 ☽ ± ♃
23 44 ☽ ☌ ♅

06 Wednesday
02 11 ☽ ⊥ ☉
04 46 ☽ ⊻ ♂
04 59 ☽ ⊥ ♀
07 40 ☽ ⊼ ♃
09 56 ☽ ∥ ♅
13 04 ☽ ⊻ ☉
13 31 ☽ ⊻ ♆
15 26 ☽ ⊥ ♂
16 06 ☽ ⊻ ♀
18 04 ☽ # ☉
19 29 ☽ △ ♄
19 35 ☽ # ♀
19 49 ☉ ✶ ♆
23 43 ☽ ⊥ ♆

07 Thursday
02 03 ☽ □ ♇
04 18 ☽ ∥ ♃
07 34 ☽ ∠ ♂
11 21 ☽ # ☿
15 06 ☽ ∠ ♆
16 07 ☽ ☌ ☿

08 Friday
00 19 ☽ ∥ ☿
02 59 ☽ ⊻ ♅
05 31 ♂ △ ♃
07 45 ♀ □ ♄
08 03 ☽ # ♃
10 30 ☽ ☍ ♃
10 49 ☽ ✶ ♂
13 26 ☽ ⊥ ♅
17 07 ☽ ✶ ♆
20 32 ☽ ☌ ☉
20 58 ☽ ∥ ♀
21 47 ☽ ∥ ☉
23 19 ☽ □ ♄

09 Saturday
00 50 ☽ ☌ ♀
02 59 ☽ # ♅
05 17 ☽ ∠ ♅
06 00 ☽ △ ♇
07 01 ☿ ∠ ♆
09 05 ☽ Q ♂
14 20 ☽ Q ♆
19 27 ☽ ⊻ ☿

10 Sunday
04 25 ☽ # ♇
06 06 ☽ ⊥ ☿
07 33 ☽ Q ♄
07 33 ☽ # ♆
08 15 ☽ ✶ ♅
08 52 ☽ bQ ♇
12 36 ☉ □ ♄
15 24 ☽ # ♂
15 27 ☽ ⊼ ♃
17 03 ☉ ∥ ♀
19 27 ☽ □ ♂
22 21 ☽ ∠ ☿
23 01 ☽ □ ♆

11 Monday
01 21 ☽ ± ♇
02 23 ☽ ± ♃
03 25 ♀ ∠ ♅
05 37 ☽ ✶ ♄
06 22 ☽ Q ♅
06 58 ☽ ⊻ ☉
09 05 ♀ △ ♇
12 29 ☽ ⊼ ♇
12 51 ☽ ⊻ ♀
14 51 ☽ ∥ ♄
19 00 ☽ bQ ♃
19 07 ☽ ⊥ ☉

12 Tuesday
01 10 ♀ # ♅
01 21 ☽ ⊥ ♀
02 13 ☽ ✶ ☿
07 46 ☿ St D
09 56 ☽ ∠ ♄
10 31 ☉ # ♅
13 38 ☽ ∠ ☉
16 29 ☽ □ ♅
20 25 ☽ ∠ ♀
21 12 ☿ ∠ ♂
23 20 ☽ △ ♃

13 Wednesday
01 14 ☽ Q ☿
03 18 ☽ ⊥ ♄
07 39 ☽ △ ♂
07 58 ☽ △ ♆
13 11 ♂ ☌ ♆
15 02 ☽ ⊻ ♄
21 17 ☽ ✶ ☉
22 03 ☽ ☍ ♇

14 Thursday
03 39 ☉ ∠ ♅
05 01 ☽ ✶ ♀
06 51 ☉ △ ♇
12 56 ☽ □ ☿
13 33 ☽ bQ ♆
15 04 ☽ bQ ♂
23 10 ☽ Q ☉

15 Friday
03 40 ☽ △ ♅
06 35 ♀ Q ♆
07 33 ☽ ± ♆
07 40 ☽ Q ♀
09 58 ☽ □ ♃
10 13 ☽ ± ♂
19 38 ☽ ⊼ ♆
21 10 ☿ ∠ ♆
23 04 ☽ ⊼ ♂
23 14 ♃ Q ♇

16 Saturday
03 02 ☽ ☌ ♄
09 58 ☽ bQ ♅
09 59 ☽ ⊼ ♇
14 37 ☽ □ ☉
21 53 ☽ Q ♃
22 06 ☽ ± ♇

17 Sunday
00 09 ☽ □ ♀
02 46 ☽ △ ☿
04 13 ☽ ± ♅
15 49 ♀ Q ♂
16 15 ☽ bQ ♇
16 22 ☽ ⊼ ♅
21 40 ☽ ∥ ♄
21 51 ☽ ✶ ♃

18 Monday
08 12 ☽ ☍ ♆
10 08 ☽ bQ ☿
14 06 ☿ ⊻ ♀
15 26 ☽ ☍ ♂
15 43 ☽ ⊻ ♄
20 23 ♂ ⊼ ♄
22 17 ☽ △ ♇

19 Tuesday
02 11 ♂ ∥ ♆
03 34 ☽ ∠ ♃
03 43 ☽ ⊥ ♄
04 53 ☽ ± ☿
07 34 ☉ Q ♆
07 49 ☽ # ♆
08 04 ☽ # ♂
08 13 ☽ △ ☉
11 08 ☽ # ♇
17 19 ☽ ⊼ ☿
19 03 ☽ △ ♀
21 09 ☽ ⊥ ♃
21 33 ☽ ∠ ♄
23 15 ☽ ∥ ♀

20 Wednesday
03 24 ☽ ∥ ☉
04 14 ☽ ☍ ♅
08 46 ☽ ⊻ ♃
15 46 ☽ # ♅
16 06 ☽ bQ ☉
19 23 ☽ ⊼ ♆

21 Thursday
02 45 ☽ ✶ ♄
03 22 ☽ bQ ♀
05 51 ☽ ⊼ ♂
06 54 ☽ ± ♆
08 45 ☽ □ ♇
10 48 ☽ ± ☉
14 14 ☽ # ♃
17 56 ☽ ± ♂
22 17 ☽ ± ♀
23 05 ☽ ⊼ ☉
23 56 ☽ bQ ♆

22 Friday
01 34 ☽ Q ♄
02 13 ☽ ∥ ☿
03 32 ☽ # ☿
05 33 ♂ ∥ ♇
06 04 ☽ ☍ ☿
10 40 ☽ ⊼ ♀
11 46 ☽ bQ ♂
13 28 ☽ ⊼ ♅
15 09 ☽ ∥ ♃
17 06 ☽ ☌ ♃
18 24 ☽ Q ♇

23 Saturday
00 32 ☽ ± ♅
03 43 ☽ △ ♆
07 10 ♂ ✶ ♇
09 34 ♀ bQ ♇
09 49 ♀ Q ♄
10 51 ☽ □ ♄
12 33 ☽ ∥ ♅
16 14 ☽ ✶ ♇
16 46 ☽ △ ♂
16 52 ♀ ✶ ♅
16 55 ☽ bQ ♅

24 Sunday
05 44 ☽ # ☉
10 06 ☽ ☍ ☉
12 00 ☽ # ♀
12 32 ☽ ∥ ♂
14 43 ☽ ∥ ♇
15 55 ☽ ⊼ ☿
16 56 ♀ # ♂
17 37 ☽ ∥ ♆
18 53 ☽ ∠ ♇
19 39 ☽ △ ♅
22 14 ☽ ☍ ♀
22 31 ☽ ⊻ ♃

25 Monday
01 08 ♀ ⊼ ♃
03 10 ☽ ± ☿
08 54 ☽ ⊥ ♃
09 06 ☽ □ ♆
10 35 ☽ ⊥ ♇
16 05 ☽ △ ♄
19 53 ☽ bQ ☿
20 57 ☽ ⊻ ♇
22 59 ☽ # ♄
23 09 ♂ bQ ♃

26 Tuesday
00 20 ☽ ∠ ♃
00 24 ☽ □ ♂
00 59 ♀ # ♇
16 05 ☽ Q ♆
17 55 ☽ bQ ♄
18 01 ☽ ⊼ ☉
23 26 ☽ △ ☿
23 30 ☽ □ ♅

27 Wednesday
00 29 ☿ ⊻ ♅
01 44 ☽ ✶ ♃
04 57 ☽ ± ☉
06 46 ☽ ⊼ ♀
07 06 ☉ # ♂
08 44 ☽ Q ♂
09 14 ☽ ± ♄
12 26 ☽ ✶ ♆
16 42 ♀ # ♆
17 52 ☽ ± ♀
19 24 ☽ ⊼ ♄
19 29 ♂ ± ♄
21 16 ☽ bQ ☉
21 51 ☽ Q ♃
23 53 ☽ ☌ ♇

28 Thursday
04 37 ☿ ☍ ♃
05 57 ☽ Q ♅
06 02 ☽ ✶ ♂
10 27 ☽ bQ ♀
13 42 ☽ ∠ ♆

29 Friday
00 18 ☽ △ ☉
02 09 ☽ ✶ ♅
03 52 ☽ □ ♃
04 49 ☽ ⊥ ♆
05 56 ☽ □ ☿
08 34 ☽ ∠ ♂
11 06 ☉ bQ ♇
12 42 ♀ ± ♃
13 59 ☽ △ ♀
14 53 ☽ ⊻ ♆
22 00 ☽ ☍ ♄
23 42 ☉ Q ♄

30 Saturday
00 28 ☽ ⊥ ♂
00 33 ♀ □ ♆
02 11 ☽ ⊻ ♇
03 23 ☽ ∠ ♅
04 35 ☉ ✶ ♅
11 05 ☽ ⊻ ♂
12 15 ☽ ⊥ ♇
14 28 ♀ ± ♇
14 45 ☽ Q ☿
18 33 ☽ ⊥ ♅
20 28 ☿ # ♃
23 06 ☉ # ♇

All ephemeris data is given at 12.00 UT and the Moon's longitude is additionally given for 24.00 UT

MAY 2005

LONGITUDES

Date	Sidereal time h m s	Sun ☉ ° ' "	Moon ☽ ° ' "	Moon ☽ 24.00 ° ' "	Mercury ☿ ° '	Venus ♀ ° '
01	02 38 04	11 ♉ 12 24	14 ♒ 17 25	21 ♒ 22 00	14 ♈ 45	19 ♉ 19
02	02 42 00	12 10 37	28 ♒ 24 45	05 ♓ 25 29	15 58	20 33
03	02 45 57	13 08 49	12 ♓ 23 59	19 ♓ 20 08	17 13	21 47
04	02 49 54	14 06 59	26 ♓ 13 46	03 ♈ 04 45	18 31	23 01
05	02 53 50	15 05 08	09 ♈ 52 55	16 ♈ 38 10	19 51	24 15
06	02 57 47	16 03 15	23 ♈ 20 22	29 ♈ 59 23	21 13	25 28
07	03 01 43	17 01 21	06 ♉ 35 07	13 ♉ 07 27	22 37	26 42
08	03 05 40	17 59 25	19 ♉ 36 20	26 ♉ 01 41	24 04	27 56
09	03 09 36	18 57 28	02 ♊ 23 32	08 ♊ 41 53	25 32	29 ♉ 10
10	03 13 33	19 55 29	14 ♊ 56 48	21 ♊ 08 27	27 03	00 ♊ 24
11	03 17 29	20 53 28	27 ♊ 16 58	03 ♋ 22 37	28 ♈ 36	01 38
12	03 21 26	21 51 26	09 ♋ 25 40	15 ♋ 26 28	00 ♉ 11	02 51
13	03 25 23	22 49 22	21 ♋ 25 24	27 ♋ 22 54	01 48	04 05
14	03 29 19	23 47 16	03 ♌ 19 28	09 ♌ 15 36	03 27	05 19
15	03 33 16	24 45 08	15 ♌ 11 51	21 ♌ 08 48	05 08	06 33
16	03 37 12	25 42 58	27 ♌ 07 02	03 ♍ 07 10	06 52	07 46
17	03 41 09	26 40 47	09 ♍ 09 49	15 ♍ 15 35	08 37	09 00
18	03 45 05	27 38 34	21 ♍ 25 02	27 ♍ 38 44	10 24	10 14
19	03 49 02	28 36 20	03 ♎ 57 11	10 ♎ 20 51	12 13	11 27
20	03 52 58	29 ♉ 34 03	16 ♎ 50 06	23 ♎ 25 13	14 05	12 41
21	03 56 55	00 ♊ 31 46	00 ♏ 06 24	06 ♏ 53 40	15 58	13 55
22	04 00 52	01 29 26	13 ♏ 46 57	20 ♏ 46 01	17 54	15 08
23	04 04 48	02 27 06	27 ♏ 50 29	04 ♐ 59 49	19 51	16 22
24	04 08 45	03 24 44	12 ♐ 13 22	19 ♐ 30 20	21 51	17 35
25	04 12 41	04 22 21	26 ♐ 49 51	04 ♑ 10 58	23 52	18 49
26	04 16 38	05 19 57	11 ♑ 32 46	18 ♑ 54 18	25 55	20 02
27	04 20 34	06 17 32	26 ♑ 14 41	03 ♒ 33 08	28 ♉ 00	21 16
28	04 24 31	07 15 06	10 ♒ 48 59	18 ♒ 01 39	00 ♊ 07	22 29
29	04 28 27	08 12 39	25 ♒ 10 45	02 ♓ 15 57	02 15	23 43
30	04 32 24	09 10 11	09 ♓ 17 05	16 ♓ 14 04	04 24	24 56
31	04 36 21	10 ♊ 07 43	23 ♓ 06 56	29 ♓ 55 45	06 ♊ 34	26 ♊ 10

Date	Mars ♂ ° '	Jupiter ♃ ° '	Saturn ♄ ° '	Uranus ♅ ° '	Neptune ♆ ° '	Pluto ♇ ° '
01	00 ♓ 16	10 ♎ 41	21 ♋ 50	09 ♓ 59	17 ♒ 31	24 ♐ 12
02	01 00	10 R 36	21 55	10 01	17 31	24 R 11
03	01 43	10 30	21 59	10 03	17 32	24 10
04	02 27	10 25	22 03	10 05	17 32	24 08
05	03 10	10 19	22 08	10 07	17 33	24 07
06	03 54	10 14	22 12	10 09	17 33	24 06
07	04 37	10 09	22 17	10 10	17 34	24 05
08	05 21	10 04	22 21	10 12	17 34	24 04
09	06 04	10 00	22 26	10 14	17 34	24 03
10	06 47	09 55	22 31	10 16	17 35	24 01
11	07 31	09 51	22 36	10 17	17 35	24 00
12	08 14	09 46	22 41	10 19	17 35	23 59
13	08 57	09 42	22 46	10 20	17 36	23 58
14	09 40	09 38	22 51	10 22	17 36	23 56
15	10 24	09 35	22 56	10 23	17 36	23 55
16	11 07	09 31	23 02	10 25	17 36	23 54
17	11 50	09 28	23 07	10 26	17 36	23 52
18	12 33	09 24	23 12	10 28	17 36	23 51
19	13 16	09 21	23 18	10 29	17 36	23 50
20	13 59	09 18	23 24	10 30	17 R 36	23 48
21	14 42	09 16	23 29	10 31	17 36	23 47
22	15 25	09 13	23 35	10 32	17 36	23 45
23	16 08	09 11	23 41	10 34	17 36	23 44
24	16 51	09 08	23 46	10 35	17 36	23 42
25	17 34	09 06	23 52	10 36	17 36	23 41
26	18 17	09 05	23 58	10 37	17 36	23 40
27	19 00	09 03	24 04	10 38	17 35	23 38
28	19 42	09 01	24 10	10 38	17 35	23 37
29	20 25	09 00	24 17	10 39	17 35	23 35
30	21 08	08 59	24 23	10 40	17 34	23 34
31	21 ♓ 50	08 ♎ 58	24 ♋ 29	10 ♓ 41	17 ♒ 34	23 ♐ 32

Date	Moon True ☊ ° '	Moon Mean ☊ ° '	Moon ☽ Latitude ° '
01	22 ♈ 40	21 ♈ 57	04 S 53
02	22 D 40	21 53	04 16
03	22 41	21 50	03 24
04	22 43	21 47	02 21
05	22 44	21 44	01 S 10
06	22 R 44	21 41	00 N 03
07	22 44	21 38	01 15
08	22 42	21 34	02 22
09	22 39	21 31	03 20
10	22 34	21 28	04 08
11	22 30	21 25	04 43
12	22 25	21 22	05 04
13	22 21	21 18	05 12
14	22 18	21 15	05 06
15	22 16	21 12	04 48
16	22 D 15	21 09	04 16
17	22 16	21 06	03 34
18	22 17	21 03	02 40
19	22 19	20 59	01 38
20	22 20	20 56	00 N 30
21	22 R 20	20 53	00 S 42
22	22 18	20 50	01 54
23	22 15	20 47	03 00
24	22 09	20 44	03 57
25	22 03	20 40	04 39
26	21 57	20 37	05 03
27	21 51	20 34	05 07
28	21 47	20 31	04 52
29	21 44	20 28	04 18
30	21 D 43	20 24	03 29
31	21 ♈ 44	20 ♈ 21	02 S 28

DECLINATIONS

Date	Sun ☉ ° '	Moon ☽ ° '	Mercury ☿ ° '	Venus ♀ ° '	Mars ♂ ° '	Jupiter ♃ ° '	Saturn ♄ ° '	Uranus ♅ ° '	Neptune ♆ ° '	Pluto ♇ ° '
01	15 N 12	21 S 13	03 N 06	17 N 10	12 S 54	02 S 48	21 N 51	08 S 31	15 S 42	15 S 02
02	15 29	16 02	03 33	17 33	12 39	02 46	21 50	08 31	15 42	15 02
03	15 47	10 03	04 02	17 54	12 25	02 44	21 49	08 30	15 42	15 01
04	16 05	03 S 39	04 31	18 15	12 10	02 42	21 49	08 29	15 41	15 01
05	16 22	02 N 50	05 02	18 36	11 55	02 40	21 48	08 28	15 41	15 01
06	16 39	09 07	05 34	18 56	11 40	02 38	21 47	08 28	15 41	15 01
07	16 55	14 54	06 07	19 16	11 25	02 37	21 47	08 27	15 41	15 01
08	17 12	19 55	06 41	19 35	11 10	02 35	21 46	08 26	15 41	15 01
09	17 28	23 55	07 16	19 54	10 55	02 33	21 45	08 26	15 41	15 01
10	17 43	26 42	07 52	20 12	10 40	02 32	21 45	08 25	15 41	15 01
11	17 59	28 07	08 29	20 29	10 25	02 30	21 44	08 25	15 41	15 01
12	18 14	28 10	09 07	20 46	10 10	02 29	21 43	08 24	15 41	15 00
13	18 29	26 52	09 45	21 03	09 54	02 27	21 42	08 23	15 41	15 00
14	18 43	24 23	10 24	21 19	09 39	02 26	21 42	08 23	15 40	15 00
15	18 57	20 51	11 04	21 34	09 23	02 25	21 41	08 22	15 40	15 00
16	19 11	16 29	11 44	21 49	09 08	02 23	21 40	08 22	15 40	15 00
17	19 25	11 26	12 25	22 03	08 52	02 22	21 39	08 21	15 40	15 00
18	19 38	05 N 52	13 06	22 16	08 37	02 21	21 38	08 21	15 40	15 00
19	19 51	00 S 04	13 47	22 29	08 21	02 20	21 37	08 20	15 40	15 00
20	20 04	06 09	14 28	22 41	08 06	02 19	21 37	08 20	15 40	15 00
21	20 16	12 10	15 10	22 53	07 50	02 18	21 36	08 20	15 40	15 00
22	20 28	17 47	15 51	23 04	07 34	02 18	21 35	08 19	15 40	15 00
23	20 39	22 37	16 33	23 14	07 18	02 17	21 34	08 19	15 40	14 59
24	20 50	26 11	17 14	23 23	07 02	02 16	21 33	08 18	15 41	14 59
25	21 01	28 03	17 54	23 32	06 47	02 16	21 32	08 18	15 41	14 59
26	21 12	27 59	18 34	23 40	06 31	02 15	21 31	08 18	15 41	14 59
27	21 22	25 56	19 13	23 48	06 15	02 15	21 30	08 17	15 41	14 59
28	21 31	22 12	19 50	23 55	05 59	02 14	21 29	08 17	15 41	14 59
29	21 41	17 10	20 27	24 01	05 43	02 14	21 28	08 17	15 41	14 59
30	21 50	11 18	21 02	24 06	05 27	02 14	21 27	08 17	15 41	14 59
31	21 N 58	05 S 00	21 N 36	24 N 11	05 S 11	02 S 14	21 N 26	08 S 16	15 S 41	14 S 59

ZODIAC SIGN ENTRIES

Date	h m	Planets
01	02 58	♂ ♓
02	14 43	☽ ♓
04	18 36	☽ ♈
07	00 01	☽ ♉
09	07 29	☽ ♊
10	04 14	♀ ♊
11	17 20	☽ ♋
12	09 13	☿ ♉
14	05 17	☽ ♌
16	17 46	☽ ♍
19	04 30	☽ ♎
20	22 47	☉ ♊
21	11 49	☽ ♏
23	15 38	☽ ♐
25	17 11	☽ ♑
27	18 10	☽ ♒
28	10 44	☿ ♊
29	20 09	☽ ♓

LATITUDES

Date	Mercury ☿ ° '	Venus ♀ ° '	Mars ♂ ° '	Jupiter ♃ ° '	Saturn ♄ ° '	Uranus ♅ ° '	Neptune ♆ ° '	Pluto ♇ ° '
01	02 S 56	00 S 24	01 S 37	01 N 33	00 N 11	00 S 45	00 S 07	08 N 18
04	02 57	00 17	01 40	01 33	00 11	00 45	00 07	08 18
07	02 53	00 10	01 43	01 32	00 11	00 45	00 07	08 18
10	02 44	00 S 02	01 47	01 31	00 11	00 45	00 07	08 18
13	02 30	00 N 05	01 50	01 31	00 11	00 45	00 07	08 18
16	02 12	00 12	01 53	01 30	00 12	00 46	00 07	08 18
19	01 49	00 20	01 56	01 29	00 12	00 46	00 07	08 19
22	01 22	00 27	01 59	01 29	00 12	00 46	00 07	08 19
25	00 52	00 34	02 01	01 28	00 12	00 46	00 07	08 19
28	00 S 21	00 41	02 04	01 27	00 13	00 46	00 07	08 19
31	00 N 11	00 N 48	02 S 07	01 N 26	00 N 13	00 S 46	00 S 07	08 N 19

DATA

Julian Date	2453492
Delta T	+66 seconds
Ayanamsa	23° 55' 46"
Synetic vernal point	05° ♓ 11' 13"
True obliquity of ecliptic	23° 26' 27"

MOON'S PHASES, APSIDES AND POSITIONS ☽

Date	h m	Phase	Longitude ° '	Eclipse Indicator
01	06 24	☾	10 ♒ 59	
08	08 45	●	17 ♉ 52	
16	08 57	☽	25 ♌ 36	
23	20 18	○	02 ♐ 47	
30	11 47	☾	09 ♓ 10	

Day	h m		
14	13 43	Apogee	
26	10 52	Perigee	
05	01 28	0N	
12	00 39	Max dec	28° N 19'
19	11 44	0S	
25	23 02	Max dec	28° S 16'

LONGITUDES

Date	Chiron ⚷ ° '	Ceres ⚳ ° '	Pallas ⚴ ° '	Juno ⚵ ° '	Vesta ⚶ ° '	Black Moon Lilith ⚸ ° '
01	03 ♒ 15	19 ♏ 52	23 ♍ 50	08 ♈ 50	16 ♉ 17	00 ♌ 17
11	03 ♒ 17	17 ♏ 37	23 ♍ 23	13 ♈ 55	20 ♉ 42	01 ♌ 24
21	03 ♒ 12	15 ♏ 28	23 ♍ 43	19 ♈ 00	25 ♉ 06	02 ♌ 32
31	02 ♒ 59	13 ♏ 39	24 ♍ 44	24 ♈ 06	29 ♉ 29	03 ♌ 39

All ephemeris data is given at 12.00 UT and the Moon's longitude is additionally given for 24.00 UT

ASPECTARIAN

h m Aspects

01 Sunday
00 25 ☉ ⊼ ♃
03 24 ☽ ∠ ♇
04 42 ☽ ⊻ ♅
05 57 ☽ △ ♃
06 24 ☽ □ ☉
08 42 ☽ # ♄
12 50 ☽ ⚹ ☿
17 27 ☽ ☌ ♆
21 20 ☽ □ ♀

02 Monday
00 52 ☽ ⊼ ♄
04 47 ☽ ⚹ ♇
05 50 ☽ # ♀
07 14 ☽ ᖷ ♃
11 08 ☽ ± ♄
13 07 ☿ ⊥ ♅
13 24 ☽ ∥ ♆
13 50 ☿ ∠ ♂
14 09 ☽ # ☉
15 14 ☽ Q ☉
16 12 ☽ ∥ ♇
16 40 ☽ ☌ ♂
16 46 ☽ ∠ ☿
22 30 ☽ ± ♃

03 Tuesday
01 16 ☽ Q ♇
02 25 ☽ ∥ ♂
02 37 ☽ ᖷ ♄
04 22 ☉ # ♆
07 03 ☽ Q ♀
07 56 ☽ ☌ ♅
08 45 ☽ ⊼ ♃
09 45 ☽ ⊥ ☿
13 23 ☽ ⚹ ☉
16 11 ♀ ⚹ ♄
17 22 ♀ Q ♅
17 55 ☿ ⚹ ♆
17 56 ☽ ∥ ♅
20 52 ☽ ⊻ ♆
21 10 ☽ ⊻ ☿

04 Wednesday
04 41 ☽ △ ♄
05 50 ☽ ⚹ ♀
07 18 ☽ ⊥ ♆
08 22 ☽ □ ♇
09 01 ☽ # ☿
15 31 ☽ ∥ ♃
17 26 ☽ ∠ ☉
23 03 ☽ ∠ ♆
23 30 ☽ ⊻ ♂

05 Thursday
09 40 ♀ ⊼ ♇
10 29 ☽ ⊥ ☉
10 40 ☽ ⊥ ♂
10 45 ☽ ∠ ♀
11 23 ☽ # ♃
12 25 ☽ ⊻ ♅
12 47 ☽ ☍ ♃
20 29 ☉ Q ♂
20 59 ☽ ∥ ☿
21 57 ☽ ⊻ ☉
23 05 ☽ ⊥ ♅

06 Friday
01 38 ☽ ⚹ ♆
03 35 ☽ ∠ ♂
04 22 ☽ ⊥ ♀
07 40 ♀ ᖷ ♃
07 45 ☽ ☌ ☿
09 25 ☽ # ♅
09 57 ☽ □ ♄
13 22 ☽ △ ♇
15 15 ☽ ∠ ♅
16 09 ☉ ± ♃
16 14 ☽ ⊻ ♀
21 51 ☽ # ♂
22 07 ♂ ± ♃
23 13 ☽ Q ♆

07 Saturday
05 55 ☿ □ ♄
07 37 ♃ ⊼ ♅
08 12 ☽ ⚹ ♂
12 30 ☽ # ♇
15 29 ☽ # ♆
16 34 ☽ ᖷ ♇
18 30 ☽ ⊼ ♃
18 35 ☽ ⚹ ♅
18 48 ☽ Q ♄
21 41 ☽ ∥ ☉

08 Sunday
01 26 ☉ □ ♆
05 29 ☽ ± ♃
07 33 ☽ Q ♂
08 13 ☽ □ ♆
08 45 ☽ ☌ ☉
09 08 ☽ ± ♇
10 08 ☽ ∥ ♀
12 01 ☿ △ ♇
13 47 ☉ ± ♇
16 51 ☽ Q ♅
17 10 ☽ ⚹ ♄
20 18 ☽ ⊼ ♇
21 24 ☽ ⊻ ☿
22 09 ☽ ᖷ ♃
22 16 ☽ ∥ ♄

09 Monday
05 15 ☽ ☌ ♀
06 56 ☿ ∠ ♅
10 10 ☽ ⊥ ☿
11 17 ♂ Q ♇
19 24 ☽ □ ♂
21 39 ☽ ∠ ♄

10 Tuesday
02 23 ☽ △ ♃
02 58 ☽ □ ♅
05 39 ☽ ∠ ☿
15 03 ☽ ⊥ ♄
17 05 ☽ △ ♆
22 27 ☽ ⊻ ☉

11 Wednesday
02 47 ☽ ⊻ ♄
05 35 ☽ ☍ ♇
09 16 ☿ # ♅
11 10 ☽ ⊥ ☉
14 58 ☽ ⚹ ☿
15 20 ♂ ᖷ ♄
21 30 ☽ ⊻ ♀
22 26 ☽ ᖷ ♆

12 Thursday
02 59 ☿ Q ♆
06 27 ☽ ∠ ☉
09 28 ☽ △ ♂
10 44 ☽ ⊥ ♀
12 41 ☽ □ ♃
13 46 ☽ △ ♅
16 18 ☽ ± ♆
18 20 ☽ Q ☿
23 42 ☉ Q ♅

13 Friday
04 18 ☽ ⊼ ♆
06 46 ☽ ∠ ♀
10 29 ☉ ⚹ ♄
14 43 ☽ ☌ ♄
15 04 ☽ ⚹ ☉
16 08 ☿ # ♂
17 06 ☽ ⊼ ♇
17 25 ☽ ᖷ ♂
19 54 ☽ ᖷ ♅

14 Saturday
00 35 ☽ Q ♃
05 10 ☽ ± ♇
10 55 ♂ ⊼ ♃
12 18 ☽ □ ☿
12 45 ☽ ± ♂
14 07 ☽ ± ♅
15 41 ☉ ⊼ ♇
16 29 ☽ ⚹ ♀
17 25 ☽ Q ☉
23 20 ☽ ᖷ ♇

15 Sunday
00 42 ☽ ⚹ ♃
01 40 ☽ ⊼ ♂
02 16 ☽ ⊼ ♅
06 53 ☽ ∥ ♄
07 53 ☉ ᖷ ♃
07 53 ☽ ∥ ♀
11 53 ♂ ☌ ♅
16 51 ☽ ☍ ♆
19 32 ☽ Q ♀
22 23 ☽ ∥ ☉
22 32 ♀ ∥ ♄

16 Monday
03 44 ☽ ⊻ ♄
05 33 ☽ △ ♇
06 48 ☽ ∠ ♃
08 57 ☽ □ ☉
15 51 ☽ ⊥ ♄
16 03 ☽ # ♆
17 22 ♀ ∠ ♄
19 21 ☽ # ♇

17 Tuesday
00 44 ☽ ⊥ ♃
08 04 ☽ ∥ ☿
09 55 ☽ ∠ ♄
10 43 ☽ △ ☿
11 39 ☽ □ ♀
12 35 ☽ ⊻ ♃
14 31 ☽ ☍ ♅
15 28 ☿ ᖷ ♇
17 36 ☽ ☍ ♂
20 34 ♀ △ ♃
23 04 ☿ ⊼ ♃
23 48 ☽ # ♂

18 Wednesday
01 30 ☽ # ♅
04 35 ☽ ⊼ ♆
04 47 ☿ ⊻ ♀
12 47 ☿ ⚹ ♅
15 29 ☽ ⚹ ♄
16 14 ☽ ± ♆
16 36 ♀ □ ♅
16 42 ☽ □ ♇
21 00 ☽ ᖷ ☿
23 14 ☿ Q ♄

19 Thursday
01 00 ☽ △ ☉
02 22 ☽ # ♃
09 27 ☽ ᖷ ♆
13 08 ♂ ∥ ♅
14 34 ☽ Q ♄
17 00 ☽ ± ☿
20 58 ☽ ∥ ♃
22 06 ☽ ☌ ♃
23 38 ♆ St R

20 Friday
00 16 ☽ ⊼ ♅
02 43 ☽ Q ♇
03 32 ☽ △ ♀
06 05 ☽ ⊼ ☿
06 27 ☽ ⊼ ♂
07 30 ☽ ᖷ ☉
10 02 ☿ ⚹ ♂
11 23 ☽ ± ♅
13 25 ☽ △ ♆
18 06 ☽ ± ♂
19 20 ☽ ∥ ♂
20 36 ☽ ∥ ♅

21 Saturday
00 02 ☽ □ ♄
00 40 ☽ ⚹ ♇
01 13 ☽ ± ☉
03 12 ☿ ± ♃
03 47 ☽ ᖷ ♅
06 03 ☿ # ♇
09 39 ☽ ᖷ ♀
11 15 ☽ ᖷ ♂
12 49 ☽ ⊼ ☉
23 46 ☽ ∥ ♇

22 Sunday
02 15 ☽ # ☿
02 41 ☽ ∥ ♆
03 08 ☽ ± ♀
03 17 ☽ ∠ ♇
04 05 ☽ ⊻ ♃
05 40 ☿ # ♆
06 22 ☽ △ ♅
08 20 ☿ □ ♆
10 16 ☿ ± ♇
14 28 ☽ ⊥ ♃
14 34 ☽ ⊼ ♀
14 59 ☽ △ ♂
18 35 ☽ □ ♆
18 50 ☽ ⊥ ♇
20 14 ☽ ☍ ☿

23 Monday
01 12 ☽ # ☉
01 21 ♀ □ ♂
04 54 ☽ △ ♄
05 04 ☽ ⊻ ♇
05 49 ☽ ∠ ♃
06 21 ☽ # ♄
15 43 ☽ # ♀
20 18 ☽ ☍ ☉
22 47 ♄ ⊼ ♇

24 Tuesday
01 00 ☽ Q ♆
06 15 ☽ ᖷ ♄
06 54 ☽ ⚹ ♃
09 16 ☽ □ ♅
12 11 ♀ △ ♆
15 58 ♀ ⊥ ♄
20 02 ☽ □ ♂
20 47 ☿ Q ♅
20 52 ☽ ⚹ ♆
21 13 ☽ ± ♄
21 40 ☽ ☍ ♀

25 Wednesday
02 39 ☽ Q ♃
06 23 ☽ ⊼ ☿
06 52 ☽ ☌ ♇
07 08 ☽ ⊼ ♄
09 50 ☿ ⊼ ♇
12 02 ☿ ⚹ ♄
13 00 ♂ ⊻ ♆
14 44 ☿ ᖷ ♃
14 53 ☽ Q ♅
17 46 ☽ ± ☿
21 25 ☽ ∠ ♆

26 Thursday
01 10 ☽ ⊼ ☉
02 59 ☽ Q ♂
07 59 ☽ □ ♃
10 28 ☽ ⚹ ♅
10 49 ☽ ᖷ ☿
11 38 ☽ ± ☉
12 05 ☽ ⊥ ♆
21 51 ☽ ⊻ ♆
23 32 ☽ ⚹ ♂

27 Friday
03 07 ☽ ⊼ ♀
03 20 ☽ ᖷ ☉
07 44 ☽ ⊻ ♇
08 25 ☽ ☍ ♄
10 59 ☽ ∠ ♅
13 49 ☽ ± ♀
15 22 ☽ △ ☿
17 33 ☽ ⊥ ♇

28 Saturday
01 23 ☽ ∠ ♂
01 47 ☽ ⊥ ♅
02 24 ☽ # ♀
05 41 ☽ △ ☉
05 59 ☽ ᖷ ♀
07 00 ☉ ∥ ♄
08 21 ☽ ∠ ♇
09 02 ☽ △ ♃
11 43 ☽ ⊻ ♅
15 25 ☽ # ☉
15 44 ☽ # ♄
17 03 ☽ ⊥ ♂
22 34 ☽ # ☿
23 15 ☽ ☌ ♆

29 Sunday
03 35 ☽ ⊻ ♂
09 18 ☽ △ ♀
09 19 ☽ ⚹ ♇
09 31 ♀ ☍ ♇
10 01 ☽ ᖷ ♃
10 28 ☽ ⊼ ♄
14 54 ☿ Q ♂
18 21 ☽ ∥ ♆
20 41 ☽ ± ♄
21 16 ☽ ∥ ♇

30 Monday
00 03 ♀ ⊻ ♄
01 14 ☽ ± ♃
02 07 ☽ □ ☿
05 38 ☽ Q ♇
07 23 ☉ △ ♃
11 29 ☽ ⊼ ♃
11 47 ☽ □ ☉
12 10 ☽ ᖷ ♄
14 23 ☽ ☌ ♅
17 54 ☉ ∠ ♄
23 40 ☽ ∥ ♅

31 Tuesday
02 19 ☽ ⊻ ♆
05 14 ☿ ∥ ♄
09 38 ☽ ☌ ♂
11 17 ☽ ∥ ♂
12 44 ☽ □ ♇
12 48 ☽ ⊥ ♆
14 25 ☽ △ ♄
15 02 ☽ Q ☿
17 53 ☽ □ ♀
21 29 ☽ Q ☉
22 20 ☽ ∥ ♃

JUNE 2005

LONGITUDES

Date	Sidereal time h m s	Sun ☉ ° ' "	Moon ☽ ° ' "	Moon ☽ 24.00 ° ' "	Mercury ☿ ° '	Venus ♀ ° '	Mars ♂ ° '	Jupiter ♃ ° '	Saturn ♄ ° '	Uranus ♅ ° '	Neptune ♆ ° '	Pluto ♇ ° '
01	04 40 17	11 ♊ 05 13	06 ♈ 40 39	13 ♈ 21 47	08 ♊ 45	27 ♊ 23	22 ♓ 33	08 ♎ 57	24 ♋ 35	10 ♓ 41	17 ♒ 34	23 ♐ 31
02	04 44 14	12 02 43	19 ♈ 59 21	26 ♈ 33 29	10 57	28 37	23 15	08 R 57	24 42	10 42	17 R 33	23 R 29
03	04 48 10	13 00 12	03 ♉ 04 23	09 ♉ 32 12	13 09	29 ♊ 50	23 58	08 56	24 48	10 43	17 33	23 27
04	04 52 07	13 57 41	15 ♉ 57 04	22 ♉ 19 06	15 21	01 ♋ 03	24 40	08 56	24 55	10 43	17 32	23 26
05	04 56 03	14 55 08	28 ♉ 38 23	04 ♊ 55 01	17 33	02 17	25 22	08 D 56	25 01	10 44	17 32	23 24
06	05 00 00	15 52 35	11 ♊ 09 04	17 ♊ 20 37	19 44	03 30	26 05	08 56	25 08	10 44	17 31	23 23
07	05 03 56	16 50 01	23 ♊ 29 43	29 ♊ 36 29	21 55	04 43	26 47	08 56	25 14	10 45	17 31	23 21
08	05 07 53	17 47 26	05 ♋ 41 02	11 ♋ 43 31	24 05	05 57	27 29	08 57	25 21	10 45	17 30	23 20
09	05 11 50	18 44 50	17 ♋ 44 05	23 ♋ 42 59	26 13	07 10	28 11	08 57	25 28	10 45	17 30	23 18
10	05 15 46	19 42 13	29 ♋ 40 28	05 ♌ 36 51	28 ♊ 20	08 23	28 53	08 58	25 34	10 45	17 29	23 16
11	05 19 43	20 39 36	11 ♌ 32 30	17 ♌ 27 50	00 ♋ 26	09 36	29 ♓ 35	08 59	25 41	10 46	17 28	23 15
12	05 23 39	21 36 57	23 ♌ 23 18	29 ♌ 19 25	02 29	10 50	00 ♈ 17	09 01	25 48	10 46	17 28	23 13
13	05 27 36	22 34 17	05 ♍ 16 43	11 ♍ 15 47	04 31	12 03	00 58	09 02	25 55	10 46	17 27	23 12
14	05 31 32	23 31 36	17 ♍ 17 14	23 ♍ 21 41	06 31	13 16	01 40	09 03	26 02	10 46	17 26	23 10
15	05 35 29	24 28 55	29 ♍ 29 48	05 ♎ 42 12	08 29	14 29	02 21	09 05	26 09	10 R 46	17 25	23 09
16	05 39 25	25 26 12	11 ♎ 59 30	18 ♎ 22 18	10 24	15 42	03 03	09 07	26 16	10 46	17 24	23 07
17	05 43 22	26 23 29	24 ♎ 51 09	01 ♏ 26 29	12 17	16 56	03 44	09 09	26 23	10 46	17 24	23 05
18	05 47 19	27 20 45	08 ♏ 08 40	14 ♏ 57 55	14 08	18 09	04 26	09 11	26 30	10 46	17 23	23 04
19	05 51 15	28 18 00	21 ♏ 54 19	28 ♏ 57 46	15 57	19 22	05 07	09 14	26 37	10 45	17 22	23 02
20	05 55 12	29 ♊ 15 15	06 ♐ 07 58	13 ♐ 24 23	17 43	20 35	05 48	09 16	26 44	10 45	17 21	23 01
21	05 59 08	00 ♋ 12 29	20 ♐ 46 19	28 ♐ 12 52	19 27	21 48	06 29	09 19	26 52	10 45	17 20	22 59
22	06 03 05	01 09 42	05 ♑ 42 56	13 ♑ 15 19	21 09	23 01	07 10	09 22	26 59	10 45	17 19	22 58
23	06 07 01	02 06 55	20 ♑ 48 44	28 ♑ 21 55	22 48	24 14	07 51	09 25	27 06	10 44	17 18	22 56
24	06 10 58	03 04 08	05 ♒ 53 35	13 ♒ 22 37	24 25	25 27	08 31	09 29	27 13	10 44	17 17	22 54
25	06 14 54	04 01 21	20 ♒ 48 00	28 ♒ 08 54	26 00	26 40	09 12	09 32	27 21	10 43	17 16	22 53
26	06 18 51	04 58 33	05 ♓ 24 42	12 ♓ 34 57	27 32	27 53	09 52	09 36	27 28	10 43	17 15	22 51
27	06 22 48	05 55 46	19 ♓ 39 24	26 ♓ 37 59	29 ♋ 01	29 ♋ 06	10 33	09 39	27 35	10 42	17 14	22 50
28	06 26 44	06 52 58	03 ♈ 30 44	10 ♈ 17 50	00 ♌ 29	00 ♌ 19	11 13	09 43	27 43	10 42	17 12	22 48
29	06 30 41	07 50 11	16 ♈ 59 33	23 ♈ 36 14	01 54	01 31	11 53	09 47	27 50	10 41	17 11	22 47
30	06 34 37	08 ♋ 47 24	00 ♉ 08 14	06 ♉ 35 55	03 ♌ 16	02 ♌ 44	12 ♈ 33	09 ♎ 52	27 ♋ 58	10 ♓ 40	17 ♒ 10	22 ♐ 45

Date	Moon True ☊ ° '	Moon Mean ☊ ° '	Moon ☽ Latitude ° '
01	21 ♈ 45	20 ♈ 18	01 S 20
02	21 D 46	20 15	00 S 10
03	21 R 45	20 12	01 N 00
04	21 43	20 09	02 06
05	21 38	20 05	03 04
06	21 30	20 02	03 53
07	21 21	19 59	04 29
08	21 11	19 56	04 53
09	21 00	19 53	05 04
10	20 50	19 50	05 00
11	20 42	19 46	04 44
12	20 36	19 43	04 16
13	20 32	19 40	03 37
14	20 31	19 37	02 47
15	20 D 31	19 34	01 50
16	20 31	19 30	00 N 45
17	20 R 31	19 27	00 S 23
18	20 30	19 24	01 32
19	20 27	19 21	02 39
20	20 21	19 18	03 37
21	20 13	19 15	04 23
22	20 03	19 11	04 52
23	19 53	19 08	05 01
24	19 44	19 05	04 50
25	19 36	19 02	04 18
26	19 31	18 59	03 31
27	19 28	18 56	02 31
28	19 28	18 52	01 23
29	19 D 28	18 49	00 13
30	19 ♈ 27	18 ♈ 46	00 N 56

DECLINATIONS

Date	Sun ☉ ° '	Moon ☽ ° '	Mercury ☿ ° '	Venus ♀ ° '	Mars ♂ ° '	Jupiter ♃ ° '	Saturn ♄ ° '	Uranus ♅ ° '	Neptune ♆ ° '	Pluto ♇ ° '
01	22 N 06	01 N 25	22 N 07	24 N 15	04 S 55	02 S 14	21 N 25	08 S 16	15 S 41	14 S 59
02	22 14	07 40	22 37	24 18	04 39	02 14	21 24	08 16	15 41	14 59
03	22 22	13 29	23 04	24 21	04 23	02 14	21 23	08 16	15 42	14 59
04	22 29	18 38	23 29	24 23	04 07	02 14	21 22	08 16	15 42	14 59
05	22 35	22 51	23 52	24 24	03 51	02 14	21 21	08 15	15 42	14 59
06	22 42	25 57	24 12	24 24	03 35	02 14	21 20	08 15	15 42	14 59
07	22 47	27 46	24 29	24 24	03 19	02 15	21 18	08 15	15 42	14 59
08	22 53	28 12	24 43	24 23	03 03	02 15	21 17	08 15	15 42	14 59
09	22 58	27 17	24 55	24 21	02 47	02 16	21 16	08 15	15 43	14 59
10	23 02	25 07	25 03	24 19	02 32	02 16	21 15	08 15	15 43	14 59
11	23 07	21 52	25 09	24 16	02 16	02 17	21 14	08 15	15 43	14 59
12	23 11	17 45	25 13	24 12	02 00	02 18	21 13	08 15	15 43	14 59
13	23 14	12 56	25 13	24 07	01 44	02 18	21 11	08 15	15 44	14 59
14	23 17	07 35	25 11	24 02	01 28	02 19	21 10	08 15	15 44	14 59
15	23 20	01 N 53	25 07	23 56	01 13	02 20	21 09	08 15	15 44	14 59
16	23 22	04 S 03	25 00	23 49	00 57	02 21	21 08	08 15	15 44	14 59
17	23 23	09 59	24 52	23 42	00 41	02 22	21 06	08 15	15 45	14 59
18	23 25	15 41	24 41	23 34	00 26	02 23	21 05	08 15	15 45	14 59
19	23 26	20 48	24 28	23 25	00 S 10	02 25	21 04	08 15	15 45	14 59
20	23 26	24 54	24 13	23 16	00 N 06	02 26	21 03	08 15	15 46	14 59
21	23 26	27 30	23 57	23 06	00 21	02 27	21 01	08 16	15 46	14 59
22	23 26	28 11	23 39	22 55	00 37	02 29	21 00	08 16	15 46	14 59
23	23 25	26 48	23 19	22 43	00 52	02 30	20 59	08 16	15 46	14 59
24	23 24	23 29	22 59	22 31	01 07	02 32	20 57	08 16	15 47	14 59
25	23 23	18 38	22 37	22 19	01 23	02 33	20 56	08 16	15 47	14 59
26	23 21	12 47	22 14	22 05	01 38	02 35	20 55	08 17	15 47	15 00
27	23 18	06 S 24	21 50	21 51	01 53	02 37	20 53	08 17	15 48	15 00
28	23 16	00 N 07	21 25	21 37	02 08	02 38	20 52	08 17	15 48	15 00
29	23 13	06 28	21 00	21 22	02 23	02 40	20 50	08 17	15 49	15 00
30	23 N 09	12 N 24	20 N 33	21 N 06	02 N 38	02 S 42	20 N 49	08 S 18	15 S 49	15 S 00

ZODIAC SIGN ENTRIES

Date	h m	Planets
01	00 08	☽ ♈
03	06 20	☽ ♉
03	15 18	♀ ♋
05	14 36	☽ ♊
08	00 46	☽ ♋
10	12 39	☽ ♌
11	07 03	☿ ♋
12	02 30	♂ ♈
13	01 22	☽ ♍
15	12 59	☽ ♎
17	21 24	☽ ♏
20	01 45	☽ ♐
21	06 46	☉ ♋
22	02 52	☽ ♑
24	02 36	☽ ♒
26	03 03	☽ ♓
28	04 01	☿ ♌
28	05 51	☽ ♈
28	05 53	♀ ♌
30	11 45	☽ ♉

LATITUDES

Date	Mercury ☿ ° '	Venus ♀ ° '	Mars ♂ ° '	Jupiter ♃ ° '	Saturn ♄ ° '	Uranus ♅ ° '	Neptune ♆ ° '	Pluto ♇ ° '
01	00 N 22	00 N 50	02 S 08	01 N 26	00 N 13	00 S 46	00 S 07	08 N 18
04	00 51	00 57	02 11	01 25	00 13	00 46	00 07	08 18
07	01 17	01 03	02 13	01 25	00 13	00 46	00 08	08 18
10	01 37	01 09	02 16	01 24	00 14	00 46	00 08	08 18
13	01 51	01 14	02 19	01 23	00 14	00 47	00 08	08 18
16	01 59	01 19	02 21	01 22	00 14	00 47	00 08	08 17
19	01 59	01 23	02 24	01 21	00 14	00 47	00 08	08 17
22	01 53	01 27	02 26	01 21	00 14	00 47	00 08	08 17
25	01 42	01 31	02 28	01 20	00 15	00 47	00 08	08 16
28	01 24	01 34	02 30	01 19	00 15	00 47	00 08	08 16
31	01 N 01	01 N 36	02 S 32	01 N 18	00 N 15	00 S 47	00 S 08	08 N 15

DATA

Julian Date	2453523
Delta T	+66 seconds
Ayanamsa	23° 55' 51"
Synetic vernal point	05° ♓ 11' 08"
True obliquity of ecliptic	23° 26' 27"

MOON'S PHASES, APSIDES AND POSITIONS ☽

Date	h m	Phase	Longitude ° '	Eclipse Indicator
06	21 55	●	16 ♊ 16	
15	01 22	☽	24 ♍ 04	
22	04 14	○	00 ♑ 51	
28	18 23	☾	07 ♈ 08	

Day	h m		
11	06 16	Apogee	
23	11 54	Perigee	
01	06 40	0N	
08	07 33	Max dec	28° N 13'
15	19 40	0S	
22	08 01	Max dec	28° S 13'
28	11 33	0N	

LONGITUDES

Date	Chiron ⚷ ° '	Ceres ⚳ ° '	Pallas ⚴ ° '	Juno ⚵ ° '	Vesta ⚶ ° '	Black Moon Lilith ⚸ ° '
01	02 ♒ 58	13 ♏ 30	24 ♍ 52	24 ♈ 37	29 ♉ 55	03 ♌ 46
11	02 ♒ 38	12 ♏ 15	26 ♍ 31	29 ♈ 42	04 ♊ 15	04 ♌ 53
21	02 ♒ 13	11 ♏ 37	28 ♍ 39	04 ♉ 45	08 ♊ 33	06 ♌ 00
31	01 ♒ 44	11 ♏ 37	01 ♎ 12	09 ♉ 46	12 ♊ 47	07 ♌ 07

All ephemeris data is given at 12.00 UT and the Moon's longitude is additionally given for 24.00 UT

ASPECTARIAN

h m	Aspects
01 Wednesday	
01 58	☉ □ ♅
04 40	☽ ∠ ♆
11 04	☉ ∥ ☿
14 11	☿ △ ♃
15 03	☽ # ♃
16 04	☽ ☍ ♃
16 26	☽ ✶ ☿
19 12	☽ ⊻ ♅
20 31	☽ ✶ ☉
21 37	☿ ∠ ♄
02 Thursday	
00 45	☽ # ♂
05 14	☽ Q ♀
06 02	☽ ⊥ ♅
07 35	☽ ✶ ♆
09 18	☿ □ ♅
14 22	☽ # ♅
18 18	☽ ⊻ ♂
18 21	☽ △ ♇
19 27	♂ □ ♇
20 39	☽ □ ♄
22 05	♂ ⊥ ♆
22 26	☽ ∠ ♅
03 Friday	
01 05	☽ ∠ ☿
01 55	☽ ∠ ☉
05 24	☽ ✶ ♀
05 30	☽ Q ♆
05 56	☽ ⊥ ♂
09 12	☉ ☌ ☿
18 38	☽ # ♇
19 52	☽ ⊥ ☉
21 07	☽ ⊥ ☿
21 52	☽ # ♆
21 58	☽ bQ ♇
22 53	☽ ⊼ ♃
23 34	☽ ∠ ♂
04 Saturday	
02 12	☽ ✶ ♅
06 15	☽ Q ♄
07 58	☽ ⊻ ☉
10 05	☽ ± ♃
10 38	☽ ⊻ ☿
12 13	☽ ∠ ♀
14 47	☽ ± ♇
14 59	☽ □ ♆
21 42	♂ △ ♄
05 Sunday	
00 46	☽ Q ♅
02 05	☽ ⊼ ♇
02 47	☽ ∥ ♄
03 03	☽ bQ ♃
05 03	☽ ✶ ♄
05 25	☽ ✶ ♂
07 01	☽ ⊥ ♀
07 22	♃ St D
10 15	☽ ∥ ☉
11 48	☿ △ ♆
16 57	♀ bQ ♆
19 38	☽ ∥ ☿
19 42	☽ ⊻ ♀
22 50	☽ ∥ ♀
06 Monday	
04 56	☿ ⊥ ♄
05 43	☽ Q ♂
07 43	☽ △ ♃
10 00	☽ ∠ ♄
11 12	☽ □ ♅
21 55	☽ ☌ ☉
07 Tuesday	
00 20	☽ △ ♆
03 37	☽ ⊥ ♄
05 29	☿ ∥ ♀
08 15	☽ ☌ ☿
11 43	☽ ☍ ♇
15 27	☽ ⊻ ♄
18 50	☽ □ ♂
08 Wednesday	
03 43	☿ ☍ ♇
04 52	☉ △ ♆
05 43	☽ bQ ♆
12 34	☽ ☌ ♀
18 29	☽ □ ♃
22 03	☽ △ ♅
23 33	☽ ± ♆
09 Thursday	
02 59	☿ ⊻ ♄
11 31	☽ ⊼ ♆
14 12	☽ ⊻ ☉
23 08	☽ ⊼ ♇
10 Friday	
03 18	☽ ⊥ ☉
03 40	☽ ☌ ♄
04 06	☽ bQ ♅
06 33	☽ Q ♃
08 20	☉ ⊥ ♄
08 44	☽ ⊻ ☿
10 18	☽ △ ♂
11 12	☽ ± ♇
12 29	☽ ∥ ☿
18 45	☽ ∥ ♀
21 18	☿ □ ♂
22 16	☽ ± ♅
23 03	☽ ∠ ☉
23 27	☽ ⊥ ☿
23 38	♀ □ ♃
11 Saturday	
03 48	☽ ∥ ☉
05 21	☽ bQ ♇
06 49	☽ ✶ ♃
07 38	☽ ⊻ ♀
10 06	♂ ∥ ♃
10 25	☽ ⊼ ♅
16 06	☽ ∥ ♄
18 32	☽ bQ ♂
21 11	☽ ⊥ ♀
21 33	☽ ∠ ☿
12 Sunday	
00 00	☽ ☍ ♆
03 24	♄ bQ ♅
08 06	☽ ✶ ☉
10 41	♀ △ ♅
11 38	☿ bQ ♆
11 40	☽ △ ♇
13 15	☽ ∠ ♃
13 54	☽ ± ♂
16 56	☽ ⊻ ♄
17 30	☽ ∠ ♀
22 29	☽ # ♆
13 Monday	
00 16	♀ ± ♆
02 10	☽ # ♇
02 47	☽ ⊼ ♂
05 10	☽ ⊥ ♄
07 28	☽ ⊥ ♃
10 10	☽ ✶ ☿
10 27	☽ Q ☉
19 33	☽ ⊻ ♃
23 00	☽ ☍ ♅
23 25	☽ ∠ ♄
14 Tuesday	
03 06	☽ ✶ ♀
03 15	☉ ☍ ♇
09 08	☽ # ♅
12 17	☽ ⊼ ♆
14 54	☽ Q ☿
22 36	♅ St R
23 36	☽ □ ♇
15 Wednesday	
00 08	☽ ± ♆
01 22	☽ □ ☉
05 24	☽ ✶ ♄
05 29	☽ Q ♀
10 06	☽ # ♃
14 09	♂ ∠ ♆
14 52	☽ # ♂
17 40	☽ bQ ♆
17 53	☽ ☍ ♂
19 42	☿ □ ♃
16 Thursday	
00 02	☽ ∥ ♂
04 51	☽ Q ♄
05 10	☽ ∥ ♃
06 31	☽ ☌ ♃
08 26	☽ □ ☿
09 40	☽ ⊼ ♅
10 21	☽ Q ♇
16 36	☿ △ ♅
19 45	☽ □ ♀
20 59	☽ ± ♅
22 11	☽ △ ♆
17 Friday	
00 39	☿ ± ♆
04 58	☽ ∥ ♅
08 46	☽ ✶ ♇
11 47	☉ ⊻ ♄
13 40	☽ bQ ♅
14 50	☽ □ ♄
15 03	☽ △ ☉
21 04	♀ ⊼ ♆
18 Saturday	
05 01	☽ ⊼ ♂
08 59	☽ ∥ ♇
11 51	☽ ∠ ♇
12 19	☽ ∥ ♆
13 52	☽ ⊻ ♃
16 15	☽ ± ♂
16 37	☽ △ ♅
19 58	☽ bQ ☉
19 Sunday	
00 10	☽ △ ☿
00 26	☽ ⊥ ♃
03 37	☽ ⊥ ♇
04 11	☽ □ ♆
07 12	☽ △ ♀
08 46	☽ bQ ♂
10 24	☉ ∥ ♀
12 44	☽ ± ☉
13 23	☽ # ♄
13 56	☽ ⊻ ♇
15 59	☽ ∠ ♃
20 06	☽ △ ♄
23 40	☽ ⊼ ☉
20 Monday	
01 56	☽ # ♀
02 34	☽ # ☉
05 31	☽ bQ ☿
06 58	☿ ⊼ ♆
07 45	☽ # ☿
10 42	☽ Q ♆
11 00	☽ bQ ♀
11 25	☽ △ ♂
17 13	☽ ✶ ♃
19 38	☽ □ ♅
21 21	☽ bQ ♄
22 29	☽ ± ☿
21 Tuesday	
03 11	☽ ± ♀
06 25	☽ ✶ ♆
09 35	☽ ⊼ ☿
12 09	☽ ± ♄
12 53	☽ Q ♃
13 49	☽ ⊼ ♀
15 34	☽ ☌ ♇
21 54	☽ ⊼ ♄
22 Wednesday	
00 51	☽ Q ♅
04 14	☽ ☍ ☉
06 35	☽ ∠ ♆
10 55	♀ ⊼ ♇
14 25	☽ □ ♂
17 50	☽ □ ♃
20 00	☽ ✶ ♅
20 54	☽ ⊥ ♆
23 Thursday	
04 23	☉ ∥ ☿
06 26	☽ ⊻ ♆
13 55	☿ ⊼ ♇
15 22	☽ ⊻ ♇
15 32	☽ ☍ ☿
16 31	☉ bQ ♆
17 54	☽ ☍ ♀
19 49	☽ ∠ ♅
20 22	☽ Q ♂
22 04	☽ ☍ ♄
24 Friday	
00 53	☽ ⊥ ♇
07 11	☽ ⊼ ☉
10 08	☽ ⊥ ♅
12 25	☽ # ☉
15 04	☽ # ☿
15 13	☽ ∠ ♇
16 24	☽ ✶ ♂
17 26	☽ ± ☉
17 30	♀ bQ ♅
17 33	☽ # ♀
17 45	☽ △ ♃
19 44	☽ ⊻ ♅
25 Saturday	
01 21	☽ # ♄
06 17	☽ ☌ ♆
07 50	☿ bQ ♅
08 55	☽ bQ ☉
15 23	☽ ✶ ♇
17 48	☽ ∠ ♂
18 06	☽ bQ ♃
21 28	☽ ⊼ ☿
22 26	☽ ⊼ ♀
22 46	☽ ⊼ ♄
26 Sunday	
00 06	☽ ∥ ♆
01 03	♂ ☍ ♃
02 59	♀ ☌ ♄
03 19	☽ ∥ ♇
06 03	♀ Q ♃
08 30	☽ ± ☿
08 45	☽ ± ♄
08 58	☽ ± ♃
09 13	☽ ± ♀
09 19	☽ ⊥ ♂
10 58	☿ ☌ ♄
11 05	☽ Q ♇
11 13	☽ △ ☉
13 04	☿ Q ♃
19 00	☽ ⊼ ♃
19 49	☽ ⊻ ♂
20 51	☽ ☌ ♅
23 55	☽ bQ ♄
27 Monday	
01 19	☽ bQ ☿
01 40	☽ bQ ♀
05 04	☽ ∥ ♅
06 53	♀ ± ♇
07 52	☽ ⊻ ♆
08 41	☿ ∥ ♀
08 55	☿ ± ♇
17 26	☽ □ ♇
17 29	♂ ⊻ ♅
18 07	☽ ⊥ ♆
18 39	☿ ☌ ♀
28 Tuesday	
01 47	☽ △ ♄
01 53	☽ ∥ ♃
04 00	☽ # ♂
05 51	☽ △ ♀
06 04	☽ △ ☿
09 43	☽ ∠ ♆
14 28	♃ Q ♄
18 23	☽ □ ☉
19 47	☽ ∥ ♂
21 25	☽ # ♃
23 02	☽ ☍ ♃
29 Wednesday	
00 42	☽ ⊻ ♅
02 21	☽ ☌ ♂
11 26	☽ ⊥ ♅
12 21	☽ ✶ ♆
19 08	☽ # ♅
20 55	☿ ∥ ♄
22 29	☽ △ ♇
30 Thursday	
03 47	☽ ∠ ♅
05 21	☽ Q ☉
07 57	☽ □ ♄
10 13	☽ Q ♆
17 19	☽ □ ♀
18 28	☽ □ ☿
19 34	♂ # ♃
23 26	☽ # ♇

JULY 2005

LONGITUDES

Date	Sidereal time h m s	Sun ☉ ° ' "	Moon ☽ ° ' "	Moon ☽ 24.00 ° ' "	Mercury ☿ ° '	Venus ♀ ° '	Mars ♂ ° '	Jupiter ♃ ° '	Saturn ♄ ° '	Uranus ♅ ° '	Neptune ♆ ° '	Pluto ♇ ° '
01	06 38 34	09 ♋ 44 37	12 ♉ 59 41	19 ♉ 19 55	04 ♌ 36	03 ♌ 57	13 ♈ 13	09 ♎ 56	28 ♋ 05	10 ♓ 39	17 ♒ 09	22 ♐ 44
02	06 42 30	10 41 50	25 ♉ 36 57	01 ♊ 51 06	05 53	05 10	13 53	10 01	28 13	10 R 39	17 R 08	22 R 42
03	06 46 27	11 39 03	08 ♊ 02 40	14 ♊ 11 53	07 08	06 23	14 33	10 05	28 20	10 38	17 07	22 41
04	06 50 23	12 36 17	20 ♊ 18 58	26 ♊ 24 08	08 20	07 36	15 12	10 10	28 28	10 37	17 05	22 39
05	06 54 20	13 33 30	02 ♋ 27 31	08 ♋ 29 18	09 29	08 48	15 52	10 15	28 35	10 36	17 04	22 38
06	06 58 17	14 30 44	14 ♋ 29 36	20 ♋ 28 34	10 36	10 01	16 31	10 20	28 43	10 35	17 03	22 36
07	07 02 13	15 27 57	26 ♋ 26 21	02 ♌ 23 08	11 39	11 14	17 10	10 26	28 51	10 34	17 01	22 35
08	07 06 10	16 25 11	08 ♌ 19 06	14 ♌ 14 28	12 40	12 26	17 49	10 31	28 58	10 33	17 00	22 34
09	07 10 06	17 22 25	20 ♌ 09 31	26 ♌ 04 32	13 37	13 39	18 28	10 37	29 06	10 32	16 59	22 32
10	07 14 03	18 19 38	01 ♍ 59 53	07 ♍ 55 57	14 32	14 52	19 06	10 42	29 14	10 31	16 57	22 31
11	07 17 59	19 16 52	13 ♍ 53 12	19 ♍ 52 06	15 23	16 04	19 45	10 48	29 21	10 29	16 56	22 29
12	07 21 56	20 14 05	25 ♍ 53 11	01 ♎ 57 03	16 11	17 17	20 23	10 54	29 29	10 28	16 55	22 28
13	07 25 52	21 11 19	08 ♎ 04 16	14 ♎ 15 27	16 55	18 29	21 01	11 01	29 37	10 27	16 53	22 27
14	07 29 49	22 08 32	20 ♎ 31 16	26 ♎ 52 18	17 35	19 42	21 39	11 07	29 44	10 26	16 52	22 25
15	07 33 46	23 05 46	03 ♏ 19 11	09 ♏ 52 26	18 12	20 54	22 17	11 13	29 ♋ 52	10 24	16 50	22 24
16	07 37 42	24 03 00	16 ♏ 32 31	23 ♏ 19 50	18 44	22 07	22 54	11 20	00 ♌ 00	10 23	16 49	22 23
17	07 41 39	25 00 13	00 ♐ 14 37	07 ♐ 16 55	19 12	23 19	23 32	11 27	00 08	10 21	16 47	22 22
18	07 45 35	25 57 28	14 ♐ 26 37	21 ♐ 43 22	19 37	24 32	24 09	11 34	00 15	10 20	16 46	22 20
19	07 49 32	26 54 42	29 ♐ 06 36	06 ♑ 35 30	19 56	25 44	24 46	11 41	00 23	10 18	16 44	22 19
20	07 53 28	27 51 57	14 ♑ 09 02	21 ♑ 45 59	20 11	26 56	25 23	11 48	00 31	10 17	16 43	22 18
21	07 57 25	28 49 12	29 ♑ 24 58	07 ♒ 04 34	20 22	28 09	25 59	11 55	00 39	10 15	16 41	22 17
22	08 01 21	29 ♋ 46 27	14 ♒ 43 19	22 ♒ 19 50	20 27	29 ♌ 21	26 36	12 02	00 46	10 14	16 40	22 16
23	08 05 18	00 ♌ 43 43	29 ♒ 52 51	07 ♓ 21 17	20 R 28	00 ♍ 33	27 12	12 10	00 54	10 12	16 38	22 14
24	08 09 15	01 41 00	14 ♓ 44 12	22 ♓ 00 58	20 23	01 45	27 48	12 18	01 02	10 10	16 37	22 13
25	08 13 11	02 38 18	29 ♓ 11 08	06 ♈ 14 29	20 14	02 57	28 24	12 25	01 10	10 09	16 35	22 12
26	08 17 08	03 35 36	13 ♈ 10 56	20 ♈ 00 38	20 00	04 09	28 59	12 33	01 17	10 07	16 33	22 11
27	08 21 04	04 32 56	26 ♈ 43 50	03 ♉ 20 52	19 41	05 22	29 ♈ 35	12 41	01 25	10 05	16 32	22 10
28	08 25 01	05 30 16	09 ♉ 52 10	16 ♉ 18 12	19 17	06 34	00 ♉ 10	12 49	01 33	10 03	16 30	22 09
29	08 28 57	06 27 38	22 ♉ 39 28	28 ♉ 56 28	18 48	07 46	00 45	12 58	01 41	10 01	16 29	22 08
30	08 32 54	07 25 01	05 ♊ 09 41	11 ♊ 19 35	18 16	08 58	01 19	13 06	01 48	10 00	16 27	22 07
31	08 36 50	08 ♌ 22 25	17 ♊ 26 38	23 ♊ 31 13	17 ♌ 39	10 ♍ 10	01 ♉ 54	13 ♎ 14	01 ♌ 56	09 ♓ 58	16 ♒ 26	22 ♐ 06

Date	Moon True ☊ ° '	Moon Mean ☊ ° '	Moon ☽ Latitude ° '
01	19 ♈ 26	18 ♈ 43	02 N 01
02	19 R 22	18 40	02 58
03	19 15	18 36	03 46
04	19 05	18 33	04 23
05	18 53	18 30	04 47
06	18 39	18 27	04 59
07	18 25	18 24	04 57
08	18 12	18 21	04 42
09	18 01	18 17	04 14
10	17 52	18 14	03 36
11	17 46	18 11	02 48
12	17 43	18 08	01 53
13	17 42	18 05	00 N 51
14	17 D 42	18 01	00 S 15
15	17 R 42	17 58	01 22
16	17 40	17 55	02 26
17	17 37	17 52	03 25
18	17 31	17 49	04 13
19	17 23	17 46	04 46
20	17 13	17 42	05 00
21	17 02	17 39	04 54
22	16 53	17 36	04 26
23	16 45	17 33	03 40
24	16 39	17 30	02 40
25	16 36	17 27	01 31
26	16 35	17 23	00 S 18
27	16 D 35	17 20	00 N 54
28	16 R 35	17 17	02 00
29	16 34	17 14	02 59
30	16 31	17 11	03 48
31	16 ♈ 25	17 ♈ 07	04 N 25

DECLINATIONS

Date	Sun ☉ ° '	Moon ☽ ° '	Mercury ☿ ° '	Venus ♀ ° '	Mars ♂ ° '	Jupiter ♃ ° '	Saturn ♄ ° '	Uranus ♅ ° '	Neptune ♆ ° '	Pluto ♇ ° '
01	23 N 05	17 N 40	20 N 07	20 N 49	02 N 53	02 S 44	20 N 48	08 S 18	15 S 49	15 S 00
02	23 01	22 03	19 39	20 32	03 08	02 46	20 46	08 18	15 50	15 00
03	22 56	25 23	19 12	20 15	03 22	02 48	20 45	08 19	15 50	15 00
04	22 51	27 28	18 44	19 57	03 37	02 50	20 43	08 19	15 50	15 00
05	22 45	28 12	18 16	19 38	03 51	02 52	20 42	08 19	15 51	15 00
06	22 39	27 36	17 48	19 19	04 06	02 55	20 40	08 20	15 51	15 01
07	22 33	25 43	17 20	18 59	04 20	02 57	20 39	08 20	15 51	15 01
08	22 26	22 43	16 52	18 39	04 34	02 59	20 37	08 20	15 52	15 01
09	22 19	18 47	16 24	18 19	04 49	03 02	20 36	08 21	15 52	15 01
10	22 11	14 08	15 57	17 57	05 03	03 04	20 34	08 22	15 53	15 01
11	22 03	08 56	15 30	17 36	05 17	03 07	20 33	08 22	15 53	15 01
12	21 55	03 N 21	15 04	17 14	05 30	03 09	20 31	08 22	15 54	15 01
13	21 46	02 S 26	14 38	16 51	05 44	03 12	20 30	08 23	15 54	15 02
14	21 37	08 15	14 13	16 28	05 58	03 15	20 28	08 24	15 55	15 02
15	21 28	13 54	13 49	16 05	06 11	03 17	20 27	08 24	15 55	15 02
16	21 18	19 07	13 26	15 41	06 25	03 20	20 25	08 25	15 56	15 02
17	21 08	23 32	13 04	15 17	06 38	03 23	20 24	08 25	15 56	15 02
18	20 57	26 43	12 43	14 52	06 51	03 26	20 22	08 26	15 56	15 03
19	20 47	28 12	12 24	14 27	07 04	03 29	20 20	08 27	15 57	15 03
20	20 35	27 40	12 06	14 02	07 17	03 32	20 19	08 27	15 57	15 03
21	20 24	25 04	11 50	13 36	07 30	03 35	20 17	08 28	15 58	15 03
22	20 12	20 40	11 35	13 10	07 43	03 38	20 16	08 28	15 58	15 03
23	20 00	14 57	11 22	12 43	07 55	03 41	20 14	08 29	15 59	15 03
24	19 47	08 28	11 12	12 17	08 08	03 44	20 12	08 30	15 59	15 04
25	19 34	01 S 43	11 03	11 50	08 20	03 47	20 11	08 30	16 00	15 04
26	19 21	04 N 56	10 56	11 22	08 32	03 51	20 09	08 31	16 00	15 04
27	19 08	11 08	10 52	10 55	08 44	03 54	20 08	08 32	16 01	15 04
28	18 54	16 40	10 50	10 27	08 56	03 57	20 06	08 32	16 01	15 05
29	18 40	21 19	10 51	09 59	09 08	04 01	20 04	08 33	16 02	15 05
30	18 25	24 54	10 53	09 30	09 20	04 04	20 03	08 34	16 02	15 05
31	18 N 10	27 N 15	10 N 59	09 N 02	09 N 31	04 S 08	20 N 01	08 S 35	16 S 03	15 S 05

ZODIAC SIGN ENTRIES

Date	h	m	Planets
02	20	26	☽ ♊
05	07	07	☽ ♋
07	19	11	☽ ♌
10	07	57	☽ ♍
12	20	09	☽ ♎
15	05	51	☽ ♏
16	12	30	♄ ♌
17	11	35	☽ ♐
19	13	26	☽ ♑
21	12	55	☽ ♒
22	17	41	☉ ♌
23	01	01	♀ ♍
23	12	11	☽ ♓
25	13	23	☽ ♈
27	17	54	☽ ♉
28	05	12	♂ ♉
30	02	02	☽ ♊

LATITUDES

Date	Mercury ☿ ° '	Venus ♀ ° '	Mars ♂ ° '	Jupiter ♃ ° '	Saturn ♄ ° '	Uranus ♅ ° '	Neptune ♆ ° '	Pluto ♇ ° '
01	01 N 01	01 N 36	02 S 32	01 N 18	00 N 15	00 S 47	00 S 08	08 N 15
04	01 01	01 38	02 34	01 18	00 16	00 47	00 08	08 15
07	00 N 48	01 38	02 36	01 17	00 16	00 48	00 08	08 14
10	00 S 33	01 39	02 38	01 16	00 16	00 48	00 08	08 13
13	01 11	01 39	02 40	01 16	00 16	00 48	00 08	08 13
16	01 52	01 38	02 41	01 15	00 16	00 48	00 08	08 12
19	02 34	01 36	02 43	01 14	00 17	00 48	00 08	08 11
22	03 15	01 34	02 44	01 14	00 17	00 48	00 08	08 10
25	03 53	01 30	02 45	01 13	00 17	00 48	00 08	08 10
28	04 25	01 27	02 46	01 12	00 17	00 48	00 08	08 09
31	04 S 47	01 N 22	02 S 47	01 N 12	00 N 18	00 S 48	00 S 08	08 N 08

DATA

Julian Date	2453553
Delta T	+66 seconds
Ayanamsa	23° 55' 57"
Synetic vernal point	05° ♓ 11' 02"
True obliquity of ecliptic	23° 26' 27"

MOON'S PHASES, APSIDES AND POSITIONS ☽

Date	h	m	Phase	Longitude ° '	Eclipse Indicator
06	12	02	●	14 ♋ 31	
14	15	20	☽	22 ♎ 16	
21	11	00	○	28 ♑ 47	
28	03	19	☾	05 ♉ 10	

Day	h	m		
08	17	50	Apogee	
21	19	51	Perigee	
05	13	07	Max dec	28° N 13'
13	02	00	0S	
19	17	54	Max dec	28° S 16'
25	18	06	0N	

LONGITUDES

Date	Chiron ⚷ ° '	Ceres ⚳ ° '	Pallas ⚴ ° '	Juno ⚵ ° '	Vesta ⚶ ° '	Black Moon Lilith ⚸ ° '
01	01 ♒ 44	11 ♏ 37	01 ♎ 12	09 ♉ 46	12 ♊ 47	07 ♌ 07
11	01 ♒ 11	12 ♏ 13	04 ♎ 05	14 ♉ 44	16 ♊ 57	08 ♌ 14
21	00 ♒ 37	13 ♏ 22	07 ♎ 15	19 ♉ 37	21 ♊ 02	09 ♌ 21
31	00 ♒ 02	15 ♏ 01	10 ♎ 40	24 ♉ 24	25 ♊ 01	10 ♌ 29

All ephemeris data is given at 12.00 UT and the Moon's longitude is additionally given for 24.00 UT

ASPECTARIAN

h m	Aspects
01 Friday	
02 08	☽ ⚼ ♇
03 11	☽ # ♆
05 24	☽ ✶ ☉
06 12	☽ ⊼ ♃
07 36	☽ ✶ ♅
12 27	☽ ⊻ ♂
13 04	☿ ± ♅
14 52	♀ ∥ ♄
17 10	☉ □ ♃
17 35	☽ ± ♃
17 54	☽ Q ♄
19 03	☽ ± ♇
19 51	☽ □ ♆
23 32	☽ ∥ ☿
02 Saturday	
00 26	☽ ⊥ ♂
01 47	♀ ± ♅
03 40	☽ ∥ ♀
04 26	☽ ∥ ♄
06 19	☽ Q ♅
06 26	☽ ⊼ ♇
06 48	☽ Q ♀
08 19	☽ Q ☿
10 40	☉ △ ♅
10 50	☽ ⚼ ♃
12 10	☽ ∠ ☉
17 02	☽ ✶ ♄
18 00	☽ ∥ ☉
18 38	☽ ∠ ♂
22 39	☉ ± ♆
03 Sunday	
06 58	☽ ⊥ ☉
08 25	☽ ✶ ♀
10 02	☽ ✶ ☿
16 00	☽ △ ♃
17 02	☽ □ ♅
19 37	☽ ⊻ ☉
22 26	☽ ∠ ♄
22 38	☿ ⚼ ♇
04 Monday	
01 24	☽ ✶ ♂
05 40	☽ △ ♆
13 14	♀ ⚼ ♇
16 16	☽ ⊥ ♄
16 36	☽ ☍ ♇
16 59	☽ ∠ ♀
18 34	☽ ∠ ☿
05 Tuesday	
02 22	☽ Q ♂
04 15	☽ ⊻ ♄
11 13	☽ ⚼ ♆
12 46	☽ ⊥ ♀
14 15	☽ ⊥ ☿
18 24	♄ ± ♇
06 Wednesday	
02 03	☽ ⊻ ♀
03 26	☽ ⊻ ☿
03 38	☽ □ ♃
04 11	☽ △ ♅
05 07	☽ ± ♆
05 52	☿ ✶ ♃
11 43	☿ ⊼ ♅
12 02	☽ ☌ ☉
14 31	♂ ⊥ ♅
16 17	☽ □ ♂
17 06	☽ ⊼ ♆
18 52	♀ ✶ ♃
23 03	♀ ⊼ ♅
07 Thursday	
04 15	☽ ⊼ ♇
06 59	♂ ✶ ♆
10 15	☽ ⚼ ♅
16 02	☽ Q ♃
16 19	☽ ± ♇
16 54	☽ ☌ ♄
08 Friday	
04 23	☽ ± ♅
10 28	☽ ⚼ ♇
13 58	☽ ∥ ☉
16 29	☽ ✶ ♃
16 30	☽ ⊼ ♅
18 21	♃ ⊼ ♅
20 37	♃ Q ♇
21 18	☽ ☌ ♀
21 36	☽ ☌ ☿
09 Saturday	
01 28	☽ ∥ ♄
02 17	☉ ⊼ ♆
05 34	☽ ☍ ♆
05 51	☽ ⊻ ☉
08 21	☽ △ ♂
09 43	☿ ☌ ♀
14 50	☽ ∥ ♀
16 49	☽ △ ♇
19 06	☽ ⊥ ☉
23 09	☽ ∠ ♃
10 Sunday	
02 01	☽ ∥ ☿
03 20	☽ # ♆
06 20	☽ ⊻ ♄
07 38	☽ # ♇
14 56	☽ ∠ ☉
15 33	☿ # ♆
16 30	☽ ⚼ ♂
17 32	☽ ⊥ ♃
18 36	☽ ⊥ ♄
11 Monday	
05 10	☽ ☍ ♅
05 45	☽ ⊻ ♃
11 42	☽ ± ♂
12 57	☽ ∠ ♄
14 28	☽ # ♅
15 14	☽ ⊻ ☿
16 53	☽ ⊻ ♀
18 06	☽ ⊼ ♆
23 46	☽ ✶ ☉
12 Tuesday	
00 25	☽ ⊼ ♂
03 15	☽ ∥ ♂
04 06	☽ ⊥ ☿
04 46	♀ ☍ ♆
05 12	☽ □ ♇
06 05	☽ ± ♆
06 14	☽ ⊥ ♀
12 51	☽ # ♃
14 05	☿ # ♇
19 12	☽ ✶ ♄
23 00	☉ □ ♂
23 10	☽ ∠ ☿
23 54	☽ ⚼ ♆
13 Wednesday	
01 38	☽ Q ☉
02 03	☽ ∠ ♀
11 12	☿ ☍ ♆
15 12	☽ ∥ ♃
16 37	☽ Q ♇
16 37	☽ ⊼ ♅
17 46	☽ ☌ ♃
18 57	☽ Q ♄
14 Thursday	
02 10	☽ # ♂
04 11	☽ ± ♅
05 01	☽ △ ♆
06 05	☽ ✶ ☿
10 16	☽ ✶ ♀
12 37	☽ ∥ ♅
14 15	☽ ☍ ♂
15 20	☽ □ ☉
15 36	☽ ✶ ♇
18 55	☉ ⊼ ♇
21 16	☽ ⚼ ♅
15 Friday	
05 32	☽ □ ♄
05 56	☽ Q ☿
11 09	☽ Q ♀
11 40	☽ # ☿
16 37	♂ △ ♇
17 00	☽ ∥ ♇
19 29	☽ ∠ ♇
20 59	☽ ∥ ♆
21 01	☽ # ♀
21 29	♀ # ♆
16 Saturday	
00 56	☽ △ ♅
02 34	☽ ⊻ ♃
11 43	☽ ⊥ ♇
12 29	☽ □ ♆
13 25	☽ ⊥ ♃
16 03	☽ □ ☿
17 13	♀ △ ♇
18 31	☽ # ♄
22 19	☽ ⊻ ♇
22 47	☽ # ☉
22 49	☽ □ ♀
23 47	☽ ⊼ ♂
17 Sunday	
02 15	☽ △ ☉
05 23	☽ ∠ ♃
10 42	☽ ± ♂
11 48	☽ △ ♄
19 46	☽ Q ♆
20 27	♀ △ ♂
20 40	☉ ⚼ ♅
18 Monday	
01 50	♀ # ♇
02 45	☽ ⚼ ♂
05 09	☽ □ ♅
05 46	☽ ⚼ ☉
07 09	☽ ✶ ♃
13 22	☽ ⚼ ♄
15 50	☽ ✶ ♆
20 45	☽ △ ☿
21 45	☽ ± ☉
19 Tuesday	
00 59	☽ ☌ ♇
03 07	☽ Q ♃
04 16	☽ ± ♄
04 39	☽ △ ♂
06 03	☽ △ ♀
08 12	☽ ⊼ ☉
10 42	☽ Q ♅
14 04	☽ ⊼ ♄
16 13	☽ ∠ ♆
21 32	☽ ⚼ ☿
20 Wednesday	
05 53	☽ ✶ ♅
06 34	☽ ⊥ ♆
08 12	☽ ⚼ ♀
08 15	☽ □ ♃
08 22	♂ ∠ ♅
08 52	♀ ∠ ♃
12 04	☽ ± ☿
16 02	☽ ⊻ ♆
21 39	☽ ⊼ ☿
22 39	☉ ± ♇
23 37	☽ ± ♀
21 Thursday	
00 49	☽ ⊻ ♇
05 29	☽ ∠ ♅
06 24	☽ □ ♂
09 50	☽ ⊼ ♀
10 13	☽ ⊥ ♇
11 00	☽ ☍ ☉
13 56	☽ ☍ ♄
19 34	☽ ⊥ ♅
22 Friday	
00 18	☽ ∠ ♇
03 33	☉ ∥ ♄
04 57	☽ ⊻ ♅
07 45	☽ △ ♃
11 48	☽ Q ♂
13 52	☽ # ♄
14 13	☽ # ☉
15 03	☽ ☌ ♆
19 42	☉ Q ♃
21 03	☽ ☍ ☿
23 52	☽ ✶ ♇
23 Saturday	
03 00	☿ St R
07 33	☽ ✶ ♂
07 38	☽ ⚼ ♃
07 58	☽ ∥ ♆
11 35	☽ ∥ ♇
13 10	☽ ☍ ♀
13 27	☽ ⊼ ☉
13 39	☽ ⊼ ♄
17 01	☉ ☌ ♄
18 58	☽ Q ♇
19 50	♀ ⊻ ♄
21 06	☽ # ♀
22 10	☽ ± ♃
23 22	☽ ± ♄
23 45	☽ ± ☉
24 Sunday	
01 52	☽ # ☿
04 35	☽ ☌ ♅
05 12	☉ ⊻ ♀
07 58	☽ ⊼ ♃
08 42	☽ ∠ ♂
11 55	☽ ∥ ♅
13 11	☽ # ♂
14 08	☽ ⚼ ♄
15 04	☽ ⊻ ♆
15 25	☽ ⚼ ☉
21 13	☽ ⊼ ☿
25 Monday	
00 09	☽ ⊥ ♂
00 19	☽ □ ♇
00 58	☽ ⊥ ♆
04 42	☽ ∥ ♃
07 06	☽ ± ☿
10 37	☽ ⊻ ♂
15 22	☽ △ ♄
16 03	☽ ∠ ♆
18 16	☽ △ ☉
18 59	☽ ⊼ ♀
19 14	♂ Q ♆
22 07	☽ ⚼ ☿
26 Tuesday	
06 15	☽ ± ♀
06 41	☽ ⊻ ♅
07 58	☽ # ♃
09 18	♂ # ♅
10 54	☽ ☍ ♃
17 07	☽ ⊥ ♅
17 54	☽ ✶ ♆
23 31	☽ ⚼ ♀
23 43	☽ △ ☿
27 Wednesday	
00 43	☉ ± ♅
01 38	☽ # ♅
02 09	☽ ∥ ♂
03 50	☽ △ ♇
09 03	☽ ∠ ♅
10 55	☽ ∥ ☿
11 08	☽ ∥ ♀
14 22	☿ ∥ ♀
15 15	☽ Q ♆
17 23	☽ ☌ ♂
20 34	☽ □ ♄
28 Thursday	
03 19	☽ □ ☉
04 42	☽ # ♇
05 16	☽ △ ♀
06 59	☽ ⚼ ♇
08 58	☽ # ♆
12 21	☽ ✶ ♅
17 33	☽ ⊼ ♃
17 54	♀ ⊥ ♃
22 20	☽ ∥ ☉
23 42	☽ ± ♇
29 Friday	
00 21	☽ □ ♆
04 55	☽ ± ♃
04 59	☽ □ ☿
05 04	☽ ∥ ♄
06 18	☽ Q ♄
10 06	♀ ⊥ ♄
10 48	☽ Q ♅
11 00	☽ ⊼ ♇
15 43	☽ Q ☉
22 14	☽ ⚼ ♃
30 Saturday	
04 13	☽ ⊻ ♂
04 33	☉ ⚼ ♇
05 27	☽ ✶ ♄
14 02	☽ Q ☿
16 24	☽ ⊥ ♂
16 45	☽ ✶ ☉
18 08	♀ ∥ ♂
20 11	☽ □ ♀
21 22	☽ □ ♅
31 Sunday	
03 39	☽ △ ♃
08 08	♀ ☍ ♅
10 00	☽ △ ♆
10 52	☽ ∠ ♂
10 59	☽ ∠ ♄
12 23	☽ ✶ ☿
13 59	♂ □ ♄
21 10	☽ ☍ ♇

AUGUST 2005

LONGITUDES

Date	Sidereal time h m s	Sun ☉ ° ' "	Moon ☽ ° ' "	Moon ☽ 24.00 ° ' "	Mercury ☿ ° '	Venus ♀ ° '	Mars ♂ ° '	Jupiter ♃ ° '	Saturn ♄ ° '	Uranus ♅ ° '	Neptune ♆ ° '	Pluto ♇ ° '
01	08 40 47	09 ♌ 19 50	29 ♊ 33 43	05 ♋ 34 29	16 ♌ 59	11 ♍ 21	02 ♉ 28	13 ♎ 23	02 ♌ 04	09 ♓ 56	16 ♒ 24	22 ♐ 05
02	08 44 43	10 17 15	11 ♋ 33 47	17 ♋ 31 55	16 R 17	12 33	03 02	13 32	02 11	09 R 54	16 R 22	22 R 04
03	08 48 40	11 14 42	23 ♋ 29 06	29 ♋ 25 33	15 32	13 45	03 35	13 40	02 19	09 52	16 21	22 03
04	08 52 37	12 12 10	05 ♌ 21 29	11 ♌ 17 05	14 46	14 57	04 09	13 49	02 27	09 50	16 19	22 02
05	08 56 33	13 09 39	17 ♌ 12 31	23 ♌ 08 01	14 00	16 09	04 42	13 58	02 35	09 48	16 17	22 01
06	09 00 30	14 07 09	29 ♌ 03 45	04 ♍ 59 59	13 14	17 20	05 14	14 07	02 42	09 46	16 16	22 01
07	09 04 26	15 04 40	10 ♍ 56 57	16 ♍ 54 57	12 29	18 32	05 47	14 17	02 50	09 44	16 14	22 00
08	09 08 23	16 02 11	22 ♍ 54 18	28 ♍ 55 22	11 46	19 44	06 19	14 26	02 57	09 41	16 12	21 59
09	09 12 19	16 59 44	04 ♎ 58 33	11 ♎ 04 18	11 06	20 55	06 51	14 35	03 05	09 39	16 11	21 58
10	09 16 16	17 57 17	17 ♎ 13 05	23 ♎ 25 24	10 29	22 07	07 22	14 45	03 13	09 37	16 09	21 58
11	09 20 12	18 54 51	29 ♎ 41 46	06 ♏ 02 44	09 57	23 18	07 53	14 54	03 20	09 35	16 08	21 57
12	09 24 09	19 52 27	12 ♏ 28 49	19 ♏ 00 32	09 31	24 30	08 24	15 04	03 28	09 33	16 06	21 56
13	09 28 06	20 50 03	25 ♏ 38 20	02 ♐ 22 38	09 10	25 41	08 55	15 14	03 35	09 31	16 04	21 56
14	09 32 02	21 47 40	09 ♐ 13 42	16 ♐ 11 43	08 55	26 53	09 25	15 24	03 43	09 28	16 03	21 55
15	09 35 59	22 45 18	23 ♐ 16 43	00 ♑ 28 32	08 47	28 04	09 54	15 34	03 50	09 26	16 01	21 54
16	09 39 55	23 42 57	07 ♑ 46 48	15 ♑ 10 56	08 D 46	29 ♍ 15	10 24	15 44	03 58	09 24	15 59	21 54
17	09 43 52	24 40 38	22 ♑ 40 10	00 ♒ 13 30	08 52	00 ♎ 26	10 53	15 54	04 05	09 22	15 58	21 53
18	09 47 48	25 38 19	07 ♒ 49 46	15 ♒ 27 40	09 05	01 38	11 21	16 04	04 13	09 19	15 56	21 53
19	09 51 45	26 36 01	23 ♒ 05 52	00 ♓ 42 57	09 26	02 49	11 50	16 15	04 20	09 17	15 55	21 53
20	09 55 41	27 33 45	08 ♓ 17 38	15 ♓ 48 42	09 55	04 00	12 18	16 25	04 27	09 15	15 53	21 52
21	09 59 38	28 31 30	23 ♓ 15 03	00 ♈ 35 51	10 31	05 11	12 45	16 36	04 35	09 12	15 51	21 52
22	10 03 35	29 ♌ 29 17	07 ♈ 50 25	14 ♈ 58 17	11 14	06 22	13 12	16 46	04 42	09 10	15 50	21 51
23	10 07 31	00 ♍ 27 05	21 ♈ 59 12	28 ♈ 53 05	12 04	07 33	13 39	16 57	04 49	09 08	15 48	21 51
24	10 11 28	01 24 55	05 ♉ 40 02	12 ♉ 20 16	13 02	08 43	14 05	17 08	04 57	09 05	15 47	21 51
25	10 15 24	02 22 47	18 ♉ 54 06	25 ♉ 21 59	14 06	09 54	14 31	17 18	05 04	09 03	15 45	21 50
26	10 19 21	03 20 40	01 ♊ 44 22	08 ♊ 01 46	15 17	11 05	14 56	17 29	05 11	09 00	15 44	21 50
27	10 23 17	04 18 36	14 ♊ 14 44	20 ♊ 23 49	16 34	12 15	15 21	17 40	05 18	08 58	15 42	21 50
28	10 27 14	05 16 33	26 ♊ 29 33	02 ♋ 32 28	17 56	13 26	15 45	17 51	05 25	08 56	15 41	21 50
29	10 31 10	06 14 32	08 ♋ 33 02	14 ♋ 31 46	19 24	14 37	16 09	18 02	05 32	08 53	15 39	21 50
30	10 35 07	07 12 33	20 ♋ 29 06	26 ♋ 25 25	20 56	15 47	16 33	18 14	05 39	08 51	15 38	21 50
31	10 39 04	08 ♍ 10 36	02 ♌ 21 07	08 ♌ 16 32	22 ♌ 33	16 ♎ 58	16 ♉ 56	18 ♎ 25	05 ♌ 46	08 ♓ 48	15 ♒ 36	21 ♐ 50

Moon Nodes and Latitude

Date	Moon True ☊ ° '	Moon Mean ☊ ° '	Moon ☽ Latitude ° '
01	16 ♈ 17	17 ♈ 04	04 N 50
02	16 R 07	17 01	05 01
03	15 55	16 58	05 00
04	15 43	16 55	04 45
05	15 31	16 52	04 18
06	15 22	16 48	03 40
07	15 14	16 45	02 52
08	15 09	16 42	01 56
09	15 07	16 39	00 N 54
10	15 D 06	16 36	00 S 11
11	15 07	16 33	01 17
12	15 08	16 29	02 21
13	15 R 08	16 26	03 20
14	15 06	16 23	04 09
15	15 03	16 20	04 45
16	14 58	16 17	05 04
17	14 51	16 13	05 04
18	14 44	16 10	04 43
19	14 37	16 07	04 01
20	14 31	16 04	03 02
21	14 27	16 01	01 52
22	14 26	15 58	00 S 36
23	14 D 26	15 54	00 N 41
24	14 27	15 51	01 52
25	14 28	15 48	02 56
26	14 29	15 45	03 48
27	14 R 28	15 42	04 28
28	14 26	15 39	04 55
29	14 22	15 35	05 09
30	14 17	15 32	05 08
31	14 ♈ 11	15 ♈ 29	04 N 55

DECLINATIONS

Date	Sun ☉ ° '	Moon ☽ ° '	Mercury ☿ ° '	Venus ♀ ° '	Mars ♂ ° '	Jupiter ♃ ° '	Saturn ♄ ° '	Uranus ♅ ° '	Neptune ♆ ° '	Pluto ♇ ° '
01	17 N 55	28 N 16	11 N 06	08 N 33	09 N 43	04 S 11	19 N 59	08 S 35	16 S 03	15 S 06
02	17 40	27 56	11 16	08 04	09 54	04 15	19 58	08 36	16 04	15 06
03	17 24	26 19	11 27	07 34	10 05	04 18	19 56	08 37	16 04	15 06
04	17 08	23 32	11 41	07 05	10 16	04 22	19 54	08 38	16 05	15 06
05	16 52	19 46	11 57	06 35	10 27	04 25	19 53	08 39	16 05	15 07
06	16 36	15 14	12 14	06 05	10 38	04 29	19 51	08 39	16 06	15 07
07	16 19	10 07	12 32	05 35	10 48	04 33	19 49	08 40	16 06	15 07
08	16 02	04 N 35	12 51	05 05	10 59	04 37	19 48	08 41	16 07	15 07
09	15 45	01 S 09	13 11	04 35	11 09	04 40	19 46	08 42	16 07	15 08
10	15 27	06 56	13 31	04 05	11 19	04 44	19 44	08 43	16 08	15 08
11	15 09	12 34	13 51	03 34	11 29	04 48	19 43	08 43	16 08	15 08
12	14 51	17 50	14 12	03 04	11 39	04 52	19 41	08 44	16 09	15 09
13	14 33	22 24	14 31	02 33	11 48	04 56	19 39	08 45	16 09	15 09
14	14 15	25 56	14 50	02 02	11 58	05 00	19 38	08 46	16 10	15 09
15	13 56	28 01	15 08	01 31	12 07	05 04	19 36	08 47	16 10	15 09
16	13 37	28 17	15 25	01 00	12 16	05 08	19 34	08 48	16 11	15 10
17	13 18	26 32	15 40	00 N 30	12 25	05 12	19 33	08 49	16 11	15 10
18	12 58	22 51	15 53	00 S 01	12 34	05 16	19 31	08 49	16 12	15 10
19	12 39	17 36	16 05	00 32	12 43	05 20	19 29	08 50	16 12	15 11
20	12 19	11 17	16 14	01 03	12 51	05 24	19 28	08 51	16 13	15 11
21	11 59	04 S 23	16 21	01 34	13 00	05 28	19 26	08 52	16 13	15 11
22	11 39	02 N 34	16 26	02 05	13 08	05 33	19 24	08 53	16 14	15 12
23	11 19	09 12	16 28	02 36	13 16	05 37	19 23	08 54	16 14	15 12
24	10 58	15 11	16 27	03 07	13 24	05 41	19 21	08 55	16 15	15 12
25	10 38	20 16	16 23	03 38	13 32	05 45	19 20	08 56	16 15	15 13
26	10 17	24 14	16 17	04 09	13 40	05 50	19 18	08 57	16 16	15 13
27	09 56	26 57	16 07	04 39	13 47	05 54	19 16	08 58	16 16	15 13
28	09 35	28 19	15 54	05 10	13 54	05 58	19 15	08 58	16 16	15 14
29	09 13	28 18	15 39	05 41	14 02	06 03	19 13	08 59	16 17	15 14
30	08 52	26 57	15 20	06 11	14 09	06 07	19 11	09 00	16 17	15 14
31	08 N 30	24 N 25	14 N 58	06 S 41	14 N 15	06 S 11	19 N 10	09 S 01	16 S 18	15 S 15

ZODIAC SIGN ENTRIES

Date	h	m	Planets
01	12	52	☽ ♋
04	01	10	☽ ♌
06	13	54	☽ ♍
09	02	08	☽ ♎
11	12	35	☽ ♏
13	19	47	☽ ♐
15	23	13	☽ ♑
17	03	05	♀ ♎
17	23	39	☽ ♒
19	22	52	☽ ♓
21	23	01	☽ ♈
23	00	45	☉ ♍
24	01	58	☽ ♉
26	08	43	☽ ♊
28	18	57	☽ ♋
31	07	14	☽ ♌

LATITUDES

Date	Mercury ☿ ° '	Venus ♀ ° '	Mars ♂ ° '	Jupiter ♃ ° '	Saturn ♄ ° '	Uranus ♅ ° '	Neptune ♆ ° '	Pluto ♇ ° '
01	04 S 52	01 N 20	02 S 47	01 N 12	00 N 18	00 S 48	00 S 08	08 N 08
04	04 56	01 15	02 48	01 11	00 18	00 48	00 08	08 07
07	04 43	01 09	02 48	01 10	00 18	00 48	00 08	08 06
10	04 15	01 02	02 48	01 10	00 19	00 48	00 08	08 05
13	03 34	00 55	02 49	01 09	00 19	00 48	00 08	08 04
16	02 45	00 47	02 48	01 09	00 19	00 48	00 08	08 03
19	01 53	00 38	02 48	01 08	00 19	00 49	00 08	08 02
22	01 01	00 29	02 48	01 08	00 20	00 49	00 08	08 01
25	00 S 13	00 19	02 47	01 08	00 20	00 49	00 09	08 00
28	00 N 28	00 N 09	02 46	01 07	00 20	00 49	00 09	07 59
31	01 N 01	00 S 02	02 S 45	01 N 07	00 N 21	00 S 49	00 S 09	07 N 58

DATA

Julian Date	2453584
Delta T	+66 seconds
Ayanamsa	23° 56' 03"
Synetic vernal point	05° ♓ 10' 57"
True obliquity of ecliptic	23° 26' 27"

MOON'S PHASES, APSIDES AND POSITIONS ☽

Date	h	m	Phase	Longitude ° '	Eclipse Indicator
05	03	05	●	12 ♌ 48	
13	02	39	☽	20 ♏ 28	
19	17	53	○	26 ♒ 50	
26	15	18	☾	03 ♊ 29	

Day	h	m		
04	22	05	Apogee	
19	05	42	Perigee	
01	18	07	Max dec	28° N 19'
09	07	14	0S	
16	03	15	Max dec	28° S 25'
22	03	05	0N	
28	23	41	Max dec	28° N 29'

LONGITUDES

Date	Chiron ⚷ ° '	Ceres ⚳ ° '	Pallas ⚴ ° '	Juno ⚵ ° '	Vesta ⚶ ° '	Black Moon Lilith ⚸ ° '
01	29 ♑ 59	15 ♏ 12	11 ♎ 01	24 ♉ 52	25 ♊ 24	10 ♌ 35
11	29 ♑ 26	17 ♏ 19	14 ♎ 38	29 ♉ 29	29 ♊ 16	11 ♌ 42
21	28 ♑ 56	19 ♏ 46	18 ♎ 25	03 ♊ 54	02 ♋ 59	12 ♌ 50
31	28 ♑ 30	22 ♏ 32	22 ♎ 21	08 ♊ 04	06 ♋ 32	13 ♌ 57

All ephemeris data is given at 12.00 UT and the Moon's longitude is additionally given for 24.00 UT

ASPECTARIAN

h m	Aspects
01 Monday	
00 43	☽ ∠ ☉
04 58	☽ ⊥ ♄
09 47	♀ # ♅
11 33	☽ Q ♀
15 39	☽ bQ ♆
16 34	☽ ∠ ☿
17 02	☽ ⊻ ♄
18 04	☽ ⚹ ♂
20 10	☽ ⊥ ☉
02 Tuesday	
02 31	☉ ⊼ ♅
08 40	☽ △ ♅
08 52	☿ ☍ ♆
09 13	☽ ⊻ ☉
09 34	☽ ⊥ ☿
09 37	☽ ± ♆
14 13	☽ ⚹ ♀
16 00	☽ □ ♃
19 19	☽ Q ♂
20 56	☽ ⊻ ☿
21 39	☽ ⊼ ♆
03 Wednesday	
09 07	☽ ⊼ ♇
10 10	♀ ⊻ ♃
14 46	☽ bQ ♅
21 13	☽ ± ♇
23 50	☽ ∠ ♀
04 Thursday	
04 45	☽ Q ♃
06 03	☽ ☌ ♄
08 55	☽ ± ♅
09 25	☽ □ ♂
09 50	☿ ⊻ ♀
15 24	☽ bQ ♇
20 05	☽ ⊥ ♀
21 02	☽ ⊼ ♅
05 Friday	
03 05	☽ ☌ ☉
05 21	☽ ⚹ ♃
05 54	☽ ☌ ☿
09 36	☽ ⊻ ♀
10 09	☽ ☍ ♆
11 23	☽ ‖ ♄
12 40	☿ ⚹ ♃
14 50	♀ ⊼ ♆
21 45	☽ △ ♇
23 36	☉ ☌ ☿
06 Saturday	
04 44	☽ ‖ ☉
07 42	☽ # ♆
12 07	☽ ∠ ♃
12 08	☉ ⚹ ♃
12 35	☽ # ♇
19 26	☽ ⊻ ♄
20 08	♀ ∠ ♄
07 Sunday	
01 05	☽ △ ♂
01 34	☽ ‖ ☿
06 33	☽ ⊥ ♃
07 41	☽ ⊥ ♄
08 58	☽ ‖ ♂
09 33	☽ ☍ ♅
11 17	☿ ⊥ ♀
14 54	☽ ⊻ ☿
18 23	☽ # ♅
18 47	☽ ⊻ ♃
21 02	☽ ⊻ ☉
22 37	☽ ⊼ ♆
08 Monday	
01 59	☽ ∠ ♄
02 16	☽ ⊥ ☿
04 56	☽ ☌ ♀
05 20	☉ # ♆
08 40	☽ bQ ♂
09 40	☽ ‖ ♀
10 07	☽ ⊥ ☉
10 10	☽ □ ♇
10 37	☽ ± ♆
11 55	☽ # ♃
16 11	☉ ☍ ♆
19 17	☽ ∠ ☿
09 Tuesday	
03 27	☽ ± ♂
04 30	☽ bQ ♆
05 36	☽ ∠ ☉
08 12	♀ # ♃
08 13	☽ ⚹ ♄
15 51	☽ ⊼ ♂
17 42	♂ bQ ♇
21 12	☽ ⊼ ♅
21 50	☽ Q ♇
23 28	☽ ⚹ ☿
10 Wednesday	
01 04	☽ # ♀
02 45	☽ ‖ ♃
07 08	☽ ☌ ♃
08 03	☽ Q ♄
08 54	☽ ± ♅
08 54	♀ □ ♇
09 56	☽ △ ♆
12 45	♀ ± ♆
13 33	☽ ⚹ ☉
19 28	☽ ‖ ♅
21 02	♀ bQ ♂
21 10	☽ ⚹ ♇
21 46	☽ Q ☿
22 29	☽ ⊻ ♀
11 Thursday	
02 15	☽ bQ ♅
07 07	☽ # ♂
11 11	☽ ⊥ ♀
13 25	☉ # ♇
14 30	☽ Q ☉
18 02	☽ # ☿
18 58	☽ □ ♄
22 53	☽ # ☉
23 26	☽ ‖ ♇
12 Friday	
01 41	☽ ∠ ♇
04 02	☽ ‖ ♆
04 06	☽ ☍ ♂
05 54	☽ ∠ ♀
06 34	☽ △ ♅
06 39	☽ □ ☿
09 42	☿ ⊼ ♅
12 11	☿ ∠ ♀
16 50	☽ ⊻ ♃
18 22	☽ ⊥ ♇
18 40	☽ □ ♆
21 10	☽ # ♄
13 Saturday	
00 08	♄ ± ♅
02 39	☽ □ ☉
03 57	☽ ⊥ ♃
05 19	☽ ⊻ ♇
12 06	☽ ⚹ ♀
13 09	☉ ‖ ☿
19 41	☿ □ ♂
20 18	☽ ∠ ♃
14 Sunday	
02 16	☽ △ ♄
02 57	☽ Q ♆
11 20	☽ Q ♀
11 28	☽ △ ☿
12 20	☽ ⊼ ♂
12 25	☽ □ ♅
14 45	♂ ⚹ ♅
15 03	☉ △ ♇
20 43	♃ ± ♅
22 46	☽ ⚹ ♃
23 03	☽ ± ♂
23 43	☽ ⚹ ♆
15 Monday	
04 26	☽ bQ ♄
09 42	☽ ☌ ♇
11 03	☽ △ ☉
12 50	☽ bQ ☿
13 51	☿ # ♇
14 50	☽ bQ ♂
18 56	☽ Q ♅
19 15	☽ Q ♃
19 42	☽ ± ♄
20 43	☽ □ ♀
16 Tuesday	
00 52	☽ ∠ ♆
03 46	☽ ± ☿
03 50	☿ St D
05 42	☽ ⊼ ♄
13 36	☽ ⊼ ☿
13 38	☽ bQ ☉
14 38	☽ ⚹ ♅
15 36	☽ ⊥ ♆
16 24	☽ △ ♂
17 Wednesday	
01 02	☽ □ ♃
01 17	☽ ⊻ ♆
05 11	☽ ± ☉
10 46	☽ ⊻ ♇
14 41	☽ ∠ ♅
15 25	☽ ⊼ ☉
19 36	♃ △ ♆
20 18	☽ ⊥ ♇
22 21	♀ bQ ♆
18 Thursday	
01 23	☽ △ ♀
04 55	☽ ⊥ ♅
06 15	☽ ☍ ♄
10 31	☽ ∠ ♇
13 15	☉ ⊥ ♀
14 01	☽ ☍ ☿
14 20	☽ ⊻ ♅
17 44	☽ □ ♂
19 Friday	
00 44	☽ ☌ ♆
01 06	☽ △ ♃
03 00	☽ bQ ♀
03 22	☿ ⊼ ♅
04 01	☽ # ♄
08 38	☉ ‖ ♂
10 05	☽ ⚹ ♇
17 37	☽ ‖ ♆
17 53	☽ ☍ ☉
17 56	☽ # ☿
18 20	☽ ± ♀
21 35	☽ ‖ ♇
22 56	☽ Q ♂
20 Saturday	
00 59	☽ bQ ♃
04 36	☽ ⊼ ♀
04 59	☽ Q ♇
05 52	☽ ⊼ ♄
06 23	☽ # ♂
07 21	☿ # ♆
08 02	☽ # ☉
13 30	☽ ☌ ♅
14 40	☽ ⊼ ☿
15 25	☽ ± ♃
15 28	☽ ± ♄
18 34	☽ ⚹ ♂
20 33	☽ ‖ ♅
22 27	♀ ⚹ ♄
21 Sunday	
00 06	☽ ⊻ ♆
00 38	☽ ± ☿
01 08	☽ ⊼ ♃
06 01	☽ bQ ♄
06 07	♃ Q ♄
08 19	☽ ‖ ♃
09 45	☽ ⊥ ♆
09 45	☽ □ ♇
15 51	☽ bQ ☿
19 34	☽ ∠ ♂
21 01	☽ ‖ ♀
21 12	☽ ⊼ ☉
22 Monday	
00 24	☽ ∠ ♆
06 44	☽ △ ♄
07 48	☽ ± ☉
09 19	☽ ☍ ♀
10 11	☽ # ♀
10 54	☽ ⊥ ♂
14 13	☽ ⊻ ♅
18 01	☽ △ ☿
21 18	☽ ⊻ ♂
22 40	☽ # ♃
23 Tuesday	
00 00	☽ bQ ☉
00 18	☽ ⊥ ♅
01 26	☽ ⚹ ♆
03 15	☽ ☍ ♃
10 52	☽ # ♅
11 46	☽ △ ♇
15 26	♀ ± ♂
15 41	☽ ∠ ♅
19 43	☽ ‖ ☉
22 05	☽ Q ♆
24 Wednesday	
03 53	☽ △ ☉
04 22	☽ ‖ ♂
10 42	☽ □ ♄
12 07	☽ # ♇
14 06	☽ bQ ♇
16 40	☽ # ♆
17 33	☽ ‖ ☿
18 00	☽ ⊼ ♀
18 07	☽ ⚹ ♅
19 10	♀ ⊼ ♅
25 Thursday	
02 24	☽ □ ☿
03 41	☽ ☌ ♂
05 57	☽ ± ♀
06 14	☽ □ ♆
06 23	☽ ± ♇
07 14	☽ ‖ ♄
09 01	☽ ⊼ ♃
09 45	☉ ∠ ♃
10 46	♀ Q ♇
15 57	☽ Q ♅
17 26	☽ ⊼ ♇
19 47	☽ Q ♄
20 16	☽ ± ♃
26 Friday	
00 15	☽ bQ ♀
01 30	☿ □ ♂
13 26	☽ bQ ♃
15 14	☿ # ♆
15 15	☽ Q ☿
15 18	☽ □ ☉
18 37	☽ ⚹ ♄
20 25	☿ ☍ ♆
27 Saturday	
01 50	☽ □ ♅
07 45	☽ △ ♀
14 13	☽ ⊻ ♂
14 49	☽ △ ♆
17 03	☽ ⚹ ☿
18 46	☽ △ ♃
23 56	☽ ∠ ♄
28 Sunday	
02 21	☽ ⊥ ♂
02 49	☽ ☍ ♇
05 07	☽ Q ☉
07 30	♂ □ ♆
10 28	☿ ⚹ ♃
16 06	☉ ⊻ ♄
16 24	♂ ± ♇
17 51	☽ ⊥ ♄
20 16	☽ bQ ♆
20 44	☽ ∠ ♂
29 Monday	
02 31	☽ ∠ ☿
05 55	☽ ⊻ ♄
06 59	☽ ⚹ ☉
12 40	☽ △ ♅
14 12	☽ ± ♆
17 30	♀ ± ♅
23 08	☽ ⊥ ☿
30 Tuesday	
01 29	☽ □ ♀
02 14	☽ ⊼ ♆
03 05	☉ # ♅
03 47	☽ ⚹ ♂
07 22	☽ □ ♃
08 19	♀ ‖ ♃
08 50	♀ △ ♆
13 03	☽ ⊻ ☿
14 43	☽ ⊼ ♇
15 47	☽ ∠ ☉
18 08	☿ # ♇
18 46	☽ bQ ♅
31 Wednesday	
01 25	☿ △ ♇
02 50	☽ ± ♇
04 50	☽ Q ♂
11 00	♀ ⊼ ♂
11 37	☽ ⊥ ☉
12 55	☽ ± ♅
17 52	☽ Q ♀
19 00	☽ ☌ ♄
20 22	☽ Q ♃
21 04	☽ bQ ♇

SEPTEMBER 2005

LONGITUDES

Date	Sidereal time h m s	Sun ☉ ° ' "	Moon ☽ ° ' "	Moon ☽ 24.00 ° ' "	Mercury ☿ ° '	Venus ♀ ° '	Mars ♂ ° '	Jupiter ♃ ° '	Saturn ♄ ° '	Uranus ♅ ° '	Neptune ♆ ° '	Pluto ♇ ° '
01	10 43 00	09 ♍ 08 40	14 ♌ 11 57	20 ♌ 07 41	24 ♌ 13	18 ♎ 08	17 ♉ 18	18 ♎ 36	05 ♌ 53	08 ♓ 46	15 ♒ 35	21 ♐ 49
02	10 46 57	10 06 47	26 ♌ 03 58	02 ♍ 01 03	25 57	19 18	17 40	18 48	06 00	08 R 44	15 R 33	21 D 49
03	10 50 53	11 04 55	07 ♍ 59 08	13 ♍ 58 26	27 44	20 28	18 01	18 59	06 07	08 41	15 32	21 49
04	10 54 50	12 03 04	19 ♍ 59 11	26 ♍ 01 34	29 ♌ 33	21 39	18 22	19 10	06 14	08 39	15 30	21 50
05	10 58 46	13 01 15	02 ♎ 05 49	08 ♎ 12 10	01 ♍ 24	22 49	18 42	19 22	06 21	08 37	15 29	21 50
06	11 02 43	13 59 28	14 ♎ 20 52	20 ♎ 32 12	03 17	23 59	19 01	19 34	06 27	08 34	15 28	21 50
07	11 06 39	14 57 43	26 ♎ 46 25	03 ♏ 03 52	05 10	25 09	19 20	19 45	06 34	08 32	15 26	21 50
08	11 10 36	15 55 59	09 ♏ 24 51	15 ♏ 49 43	07 05	26 19	19 38	19 57	06 41	08 29	15 25	21 50
09	11 14 33	16 54 16	22 ♏ 18 47	28 ♏ 52 24	08 59	27 28	19 56	20 09	06 47	08 27	15 23	21 50
10	11 18 29	17 52 36	05 ♐ 30 52	12 ♐ 14 27	10 55	28 38	20 13	20 21	06 54	08 25	15 22	21 51
11	11 22 26	18 50 56	19 ♐ 03 21	25 ♐ 57 44	12 50	29 ♎ 48	20 29	20 33	07 00	08 22	15 21	21 51
12	11 26 22	19 49 19	02 ♑ 57 39	10 ♑ 03 00	14 44	00 ♏ 57	20 45	20 45	07 07	08 20	15 20	21 51
13	11 30 19	20 47 43	17 ♑ 13 35	24 ♑ 29 04	16 39	02 07	21 00	20 57	07 13	08 18	15 18	21 51
14	11 34 15	21 46 08	01 ♒ 48 57	09 ♒ 12 34	18 33	03 16	21 15	21 09	07 19	08 15	15 17	21 52
15	11 38 12	22 44 35	16 ♒ 39 07	24 ♒ 07 41	20 26	04 25	21 28	21 21	07 25	08 13	15 16	21 52
16	11 42 08	23 43 04	01 ♓ 37 16	09 ♓ 06 46	22 18	05 35	21 41	21 33	07 32	08 11	15 15	21 53
17	11 46 05	24 41 34	16 ♓ 35 06	24 ♓ 01 13	24 10	06 44	21 53	21 45	07 38	08 08	15 13	21 53
18	11 50 02	25 40 06	01 ♈ 24 06	08 ♈ 42 51	26 00	07 53	22 05	21 58	07 44	08 06	15 12	21 54
19	11 53 58	26 38 41	15 ♈ 56 42	23 ♈ 05 01	27 50	09 02	22 16	22 10	07 50	08 04	15 11	21 54
20	11 57 55	27 37 17	00 ♉ 07 21	07 ♉ 03 23	29 ♍ 39	10 11	22 26	22 22	07 56	08 02	15 10	21 55
21	12 01 51	28 35 55	13 ♉ 52 58	20 ♉ 36 06	01 ♎ 27	11 19	22 35	22 35	08 02	08 00	15 09	21 55
22	12 05 48	29 ♍ 34 36	27 ♉ 12 55	03 ♊ 43 38	03 14	12 28	22 44	22 47	08 07	07 57	15 08	21 56
23	12 09 44	00 ♎ 33 19	10 ♊ 08 36	16 ♊ 28 13	05 00	13 36	22 51	23 00	08 13	07 55	15 07	21 57
24	12 13 41	01 32 04	22 ♊ 42 58	28 ♊ 53 21	06 45	14 45	22 58	23 12	08 19	07 53	15 06	21 57
25	12 17 37	02 30 51	04 ♋ 59 56	11 ♋ 03 15	08 29	15 53	23 04	23 25	08 24	07 51	15 05	21 58
26	12 21 34	03 29 41	17 ♋ 03 53	23 ♋ 02 24	10 12	17 01	23 09	23 37	08 30	07 49	15 04	21 59
27	12 25 31	04 28 33	28 ♋ 59 21	04 ♌ 55 16	11 54	18 10	23 14	23 50	08 36	07 47	15 03	22 00
28	12 29 27	05 27 27	10 ♌ 50 41	16 ♌ 46 03	13 35	19 18	23 17	24 02	08 41	07 45	15 02	22 00
29	12 33 24	06 26 23	22 ♌ 41 51	28 ♌ 38 30	15 15	20 25	23 20	24 15	08 46	07 43	15 01	22 01
30	12 37 20	07 ♎ 25 21	04 ♍ 36 22	10 ♍ 35 49	16 ♎ 55	21 ♏ 33	23 ♉ 21	24 ♎ 28	08 ♌ 52	07 ♓ 41	15 ♒ 00	22 ♐ 02

DECLINATIONS

Date	Moon True ☊ ° '	Moon Mean ☊ ° '	Moon ☽ Latitude ° '	Sun ☉ ° '	Moon ☽ ° '	Mercury ☿ ° '	Venus ♀ ° '	Mars ♂ ° '	Jupiter ♃ ° '	Saturn ♄ ° '	Uranus ♅ ° '	Neptune ♆ ° '	Pluto ♇ ° '
01	14 ♈ 04	15 ♈ 26	04 N 29	08 N 09	20 N 51	14 N 33	07 S 12	14 N 22	06 S 16	19 N 08	09 S 02	16 S 18	15 S 15
02	13 R 58	15 23	03 51	07 47	16 27	14 06	07 42	14 29	06 20	19 07	09 03	16 19	15 15
03	13 53	15 19	03 03	07 25	11 24	13 36	08 12	14 35	06 25	19 05	09 04	16 19	15 16
04	13 49	15 16	02 06	07 03	05 54	13 03	08 41	14 41	06 29	19 03	09 05	16 20	15 16
05	13 47	15 13	01 N 03	06 40	00 N 08	12 29	09 11	14 47	06 34	19 02	09 06	16 20	15 16
06	13 D 46	15 10	00 S 03	06 18	05 S 42	11 52	09 41	14 53	06 38	19 00	09 06	16 20	15 17
07	13 47	15 07	01 11	05 55	11 25	11 13	10 10	14 59	06 43	18 59	09 07	16 21	15 17
08	13 48	15 04	02 16	05 33	16 47	10 33	10 39	15 04	06 47	18 57	09 08	16 21	15 18
09	13 50	15 00	03 16	05 10	21 30	09 51	11 08	15 10	06 52	18 56	09 09	16 22	15 18
10	13 51	14 57	04 07	04 48	25 16	09 08	11 36	15 15	06 56	18 54	09 10	16 22	15 18
11	13 R 52	14 54	04 46	04 25	27 44	08 24	12 05	15 20	07 01	18 53	09 11	16 22	15 19
12	13 51	14 51	05 09	04 02	28 33	07 39	12 33	15 25	07 06	18 51	09 12	16 23	15 19
13	13 49	14 48	05 14	03 39	27 31	06 54	13 01	15 30	07 10	18 49	09 13	16 23	15 19
14	13 47	14 45	04 59	03 16	24 37	06 08	13 28	15 34	07 15	18 48	09 13	16 24	15 20
15	13 44	14 41	04 24	02 53	20 03	05 21	13 56	15 39	07 19	18 47	09 14	16 24	15 20
16	13 42	14 38	03 31	02 30	14 11	04 34	14 23	15 43	07 24	18 45	09 15	16 24	15 20
17	13 40	14 35	02 24	02 07	07 30	03 46	14 49	15 47	07 29	18 44	09 16	16 25	15 21
18	13 38	14 32	01 S 07	01 43	00 S 28	02 59	15 16	15 51	07 33	18 42	09 17	16 25	15 21
19	13 D 38	14 29	00 N 13	01 20	06 N 28	02 11	15 42	15 55	07 38	18 41	09 18	16 25	15 22
20	13 39	14 25	01 30	00 57	12 55	01 24	16 08	15 59	07 43	18 39	09 18	16 26	15 22
21	13 40	14 22	02 40	00 33	18 33	00 N 36	16 33	16 02	07 47	18 38	09 19	16 26	15 22
22	13 41	14 19	03 38	00 N 10	23 05	00 S 11	16 58	16 06	07 52	18 37	09 20	16 26	15 23
23	13 42	14 16	04 24	00 S 13	26 19	00 58	17 23	16 09	07 57	18 35	09 21	16 27	15 23
24	13 42	14 13	04 56	00 37	28 10	01 45	17 47	16 12	08 01	18 34	09 22	16 27	15 24
25	13 R 43	14 10	05 13	01 00	28 33	02 32	18 11	16 15	08 06	18 32	09 22	16 27	15 24
26	13 42	14 06	05 16	01 23	27 34	03 18	18 35	16 17	08 11	18 31	09 23	16 28	15 24
27	13 42	14 03	05 05	01 47	25 20	04 04	18 58	16 20	08 16	18 30	09 24	16 28	15 25
28	13 40	14 00	04 42	02 10	22 02	04 49	19 20	16 22	08 20	18 28	09 25	16 28	15 25
29	13 39	13 57	04 06	02 33	17 49	05 34	19 43	16 25	08 25	18 27	09 25	16 28	15 26
30	13 ♈ 38	13 ♈ 54	03 N 20	02 S 57	12 N 55	06 S 18	20 S 04	16 N 27	08 S 30	18 N 26	09 S 26	16 S 29	15 S 26

ZODIAC SIGN ENTRIES

Date	h	m	Planets
02	19	56	☽ ♍
04	17	52	☿ ♍
05	07	52	☽ ♎
07	18	10	☽ ♏
10	02	03	☽ ♐
11	16	14	♀ ♏
12	06	57	☽ ♑
14	09	02	☽ ♒
16	09	24	☽ ♓
18	09	43	☽ ♈
20	11	47	☽ ♉
20	16	40	☿ ♎
22	17	07	☽ ♊
22	22	23	☉ ♎
25	02	10	☽ ♋
27	14	03	☽ ♌
30	02	44	☽ ♍

LATITUDES

Date	Mercury ☿ ° '	Venus ♀ ° '	Mars ♂ ° '	Jupiter ♃ ° '	Saturn ♄ ° '	Uranus ♅ ° '	Neptune ♆ ° '	Pluto ♇ ° '
01	01 N 10	00 S 05	02 S 44	01 N 07	00 N 21	00 S 49	00 S 09	07 N 57
04	01 31	00 16	02 43	01 06	00 21	00 49	00 09	07 56
07	01 44	00 28	02 41	01 06	00 21	00 49	00 09	07 55
10	01 48	00 39	02 39	01 05	00 22	00 49	00 09	07 54
13	01 46	00 51	02 36	01 05	00 22	00 49	00 09	07 53
16	01 38	01 03	02 34	01 05	00 22	00 48	00 09	07 52
19	01 27	01 15	02 30	01 05	00 23	00 48	00 09	07 51
22	01 12	01 28	02 26	01 04	00 23	00 48	00 09	07 50
25	00 54	01 40	02 22	01 04	00 23	00 48	00 09	07 49
28	00 35	01 52	02 18	01 04	00 24	00 48	00 09	07 48
31	00 N 15	02 S 03	02 S 13	01 N 04	00 N 24	00 S 48	00 S 09	07 N 47

DATA

Julian Date	2453615
Delta T	+66 seconds
Ayanamsa	23° 56' 07"
Synetic vernal point	05° ♓ 10' 52"
True obliquity of ecliptic	23° 26' 28"

MOON'S PHASES, APSIDES AND POSITIONS ☽

Date	h	m	Phase	Longitude ° '	Eclipse Indicator
03	18	45	●	11 ♍ 21	
11	11	37	☽	18 ♐ 50	
18	02	01	○	25 ♓ 16	
25	06	41	☾	02 ♋ 18	

Day	h	m		
01	02	48	Apogee	
16	13	58	Perigee	
28	15	28	Apogee	
05	12	34	0S	
12	10	53	Max dec	28° S 33'
18	13	35	0N	
25	06	42	Max dec	28° N 36'

LONGITUDES

Date	Chiron ° '	Ceres ⚳ ° '	Pallas ⚴ ° '	Juno ⚵ ° '	Vesta ⚶ ° '	Black Moon Lilith ⚸ ° '
01	28 ♑ 28	22 ♏ 50	22 ♎ 45	08 ♊ 28	06 ♋ 53	14 ♌ 03
11	28 ♑ 08	25 ♏ 52	26 ♎ 48	12 ♊ 17	10 ♋ 12	15 ♌ 10
21	27 ♑ 55	29 ♏ 07	00 ♏ 56	15 ♊ 41	13 ♋ 17	16 ♌ 17
31	27 ♑ 49	02 ♐ 34	05 ♏ 10	18 ♊ 33	16 ♋ 05	17 ♌ 24

All ephemeris data is given at 12.00 UT and the Moon's longitude is additionally given for 24.00 UT

ASPECTARIAN

h m	Aspects
01 Thursday	
00 51	☽ ⚺ ☉
01 02	☽ ⚻ ♅
03 03	☉ ☍ ♅
06 28	♀ Q ♄
14 47	☽ ☍ ♆
18 28	☽ □ ♂
20 07	☿ ∥ ♂
20 50	☽ ⚹ ♀
21 03	☽ ⚹ ♃
21 55	☽ ∥ ♄
23 30	♀ ☌ ♃
02 Friday	
03 26	☽ △ ♇
10 52	♇ St D
11 44	☽ ☌ ☿
12 42	☽ ⫲ ♆
14 17	☉ ⫲ ♀
17 55	☽ ⫲ ♇
21 30	☽ ∥ ♂
03 Saturday	
00 45	☽ ∥ ☿
03 49	☽ ∠ ♃
06 24	☽ ∠ ♀
08 13	☽ ⚺ ♄
13 24	☽ ☍ ♅
18 45	☽ ☌ ☉
20 22	☽ ⊥ ♄
22 11	☽ ⊥ ♃
22 22	☽ ⫲ ♅
04 Sunday	
01 01	☽ ⫲ ♀
02 24	☽ ⊥ ♀
03 05	☽ ⚻ ♆
06 48	☽ ∥ ☉
08 40	☽ △ ♂
09 33	☽ ⫲ ♃
10 21	☽ ⚺ ♃
14 30	☽ ∠ ♄
15 01	☽ ± ♆
15 39	☽ ⚺ ♀
15 40	☽ □ ♇
15 46	♀ ⚹ ♇
17 01	☉ ⊥ ♄
05 Monday	
07 26	♀ ∥ ♅
08 49	☽ ⚼ ♆
10 23	☽ ⚺ ☿
15 14	☽ ⚼ ♂
17 54	☉ ⫲ ♃
20 26	☽ ⚹ ♄
22 43	☉ ⊥ ♃
06 Tuesday	
00 18	☽ ⊥ ☿
00 45	☽ ⚻ ♅
03 11	☽ Q ♇
03 52	♀ ⚼ ♅
09 20	☽ ± ♂
11 15	☽ ⚺ ☉
12 26	☽ ± ♅
14 09	☽ △ ♆
14 18	☽ ⫲ ☉
15 54	☽ ∥ ♃
20 03	☽ Q ♄
21 00	☽ ∠ ☿
21 18	☽ ⚻ ♂
22 17	☽ ☌ ♃
23 52	☽ ⊥ ☉
07 Wednesday	
02 12	☽ ∥ ♅
02 30	☽ ⚹ ♇
05 47	☽ ⚼ ♅
06 07	☽ ∥ ♀
06 11	☿ ∠ ♃
08 33	☽ ☌ ♀
11 14	☽ ⫲ ☿
18 36	☽ ∠ ☉
23 26	☉ ⚻ ♆
08 Thursday	
03 57	☽ ⫲ ♂
05 06	☽ ∥ ♇
06 40	☿ ⚺ ♄
06 48	☽ □ ♄
06 49	☽ ⚹ ☿
07 08	☽ ∠ ♇
10 00	☽ ∥ ♆
10 01	☿ ⫲ ♀
10 16	☽ △ ♅
22 30	☽ ⫲ ♄
23 12	☽ □ ♆
09 Friday	
00 01	☽ ⊥ ♇
01 11	☽ ⚹ ☉
05 22	☿ ☍ ♅
07 31	☽ ☍ ♂
07 57	☽ ⚺ ♃
09 09	☽ Q ☿
11 07	☽ ⚺ ♇
19 09	☽ ⊥ ♃
22 22	☽ ⚺ ♀
23 53	♄ ⚼ ♇
10 Saturday	
01 01	☽ Q ☉
08 09	☽ Q ♆
10 16	☽ ⊥ ♀
11 09	☿ ⫲ ♅
11 42	☽ ∠ ♃
14 30	☽ △ ♄
17 10	☽ □ ♅
23 14	☽ □ ☿
11 Sunday	
02 52	♂ Q ♅
03 49	☽ ∠ ♀
05 30	☽ ⚹ ♆
05 42	♂ ⫲ ♇
11 37	☽ □ ☉
14 20	☿ ⊥ ♄
14 33	☽ ⚻ ♂
14 38	☽ ⚹ ♃
16 52	☽ ☌ ♇
17 11	☽ ⚼ ♄
12 Monday	
00 40	☽ Q ♅
01 10	☽ ± ♂
07 30	☽ ∠ ♆
08 16	☽ ⚹ ♀
08 49	☽ ± ♄
09 30	♂ ⚻ ♃
11 38	☽ Q ♃
12 05	☿ ⊥ ♃
16 50	☽ ⚼ ♂
19 06	☽ ⚻ ♄
19 17	☿ ⚻ ♆
21 05	☽ ⚹ ♅
22 46	☽ ⊥ ♆
13 Tuesday	
04 15	☿ ⫲ ♃
06 21	☽ Q ♀
08 48	☽ ⚺ ♆
10 53	☽ △ ☿
16 41	☉ ⚺ ♃
18 15	☽ □ ♃
18 21	☽ △ ☉
18 22	☽ △ ♂
18 49	☉ △ ♂
19 40	☽ ⚺ ♇
22 01	☽ ∠ ♅
14 Wednesday	
00 17	☉ ± ♆
03 07	☿ ∠ ♀
05 33	☽ ⊥ ♇
12 43	☽ ⊥ ♅
14 21	☉ □ ♇
14 34	☽ □ ♀
15 13	☽ ⚼ ☿
20 12	☽ ∠ ♇
20 37	☽ ⚼ ☉
21 00	☽ ☍ ♄
22 26	☽ ⚺ ♅
15 Thursday	
03 12	☉ ∠ ♄
07 54	☽ ± ☿
09 46	☽ ☌ ♆
12 09	☽ ± ☉
17 38	☽ ⫲ ♄
18 56	☽ ⚻ ☿
19 39	☽ △ ♃
19 51	☽ □ ♂
20 23	☽ ⚹ ♇
22 28	☽ ⚻ ☉
22 34	☿ ± ♆
16 Friday	
01 12	☿ ⚺ ♃
03 05	☿ △ ♂
03 25	☽ ∥ ♆
06 11	☽ ⫲ ♂
06 32	☿ □ ♇
07 35	☽ ∥ ♇
11 19	☽ ∥ ♀
15 04	☿ ∠ ♄
15 37	☽ Q ♇
18 52	☽ △ ♀
20 00	☽ ⚼ ♃
21 31	☽ ⚻ ♄
22 29	☽ ☌ ♅
17 Saturday	
01 06	☽ Q ♂
05 52	☽ ∥ ♅
07 13	☽ ± ♄
09 49	☽ ⚺ ♆
10 39	☽ ± ♃
11 20	♂ ⚻ ♇
12 05	☽ ∥ ♃
15 16	♀ ∠ ♇
19 28	☽ ⊥ ♆
20 27	☽ ⚻ ♃
20 33	☽ □ ♇
20 40	☽ ⚹ ♂
21 00	☽ ⚼ ♀
21 49	☽ ⚼ ♄
18 Sunday	
01 58	☽ ☍ ☿
02 01	☽ ☍ ☉
02 23	☽ ⫲ ☿
02 38	☉ ☌ ☿
03 57	♃ ⚹ ♇
07 30	☽ ⫲ ☉
08 34	♀ □ ♄
10 03	☽ ∠ ♆
12 51	☽ ± ♀
16 31	♀ △ ♅
17 06	♀ ∥ ♇
19 05	☽ ∥ ☉
21 26	☽ ∠ ♂
22 27	☽ △ ♄
22 35	☽ ∥ ☿
22 58	☽ ⚺ ♅
23 32	☽ ⚻ ♀
19 Monday	
08 53	☽ ⊥ ♅
10 44	☽ ⚹ ♆
12 32	☽ ⊥ ♂
16 12	☽ ⫲ ♃
22 01	☽ △ ♇
22 14	☽ ⫲ ♅
22 36	☽ ☍ ♃
22 45	☽ ⚺ ♂
23 56	☽ ∠ ♅
20 Tuesday	
02 07	♀ ⫲ ♂
06 57	☽ Q ♆
07 24	☽ ⚻ ☉
11 04	☽ ⚻ ☿
18 29	☽ ± ☉
18 49	☿ ⚼ ♆
21 58	☽ ⫲ ♇
22 59	☽ ± ☿
23 45	☽ ⚼ ♇
21 Wednesday	
00 40	☽ ∥ ♂
01 37	☽ □ ♄
01 40	☽ ⚹ ♅
02 16	☽ ⫲ ♀
02 29	☽ ⫲ ♆
05 10	♀ ∥ ♆
05 49	♄ ⚻ ♅
07 04	☽ ☍ ♀
11 27	☽ ⚼ ☉
12 25	☽ ∥ ♄
14 15	☽ □ ♆
14 41	☉ ∥ ☿
15 06	♂ ⚻ ♃
15 38	☽ ± ♇
17 15	☽ ⚼ ☿
22 52	☽ Q ♅
22 Thursday	
02 24	☽ ⚻ ♇
03 45	☽ ☌ ♂
03 49	☽ ⚻ ♃
10 00	☽ Q ♄
11 38	☉ ⫲ ☿
14 55	☽ ± ♃
16 41	☽ △ ☉
23 Friday	
00 50	☽ △ ☿
01 23	☉ ⚼ ♆
04 55	♃ ⚼ ♅
07 50	☽ □ ♅
07 53	☽ ⚼ ♃
08 21	☽ ⚹ ♄
19 12	☽ ⚻ ♀
21 24	☽ △ ♆
24 Saturday	
07 49	☽ ± ♀
10 32	☽ ☍ ♇
12 30	☽ ⚺ ♂
12 57	☽ △ ♃
13 10	☽ ∠ ♄
19 15	♀ □ ♆
25 Sunday	
00 16	☽ ⊥ ♂
02 21	☽ ⚼ ♆
03 05	☽ ⚼ ♀
03 27	☿ ⚻ ♅
05 59	☿ ⚼ ♂
06 41	☽ □ ☉
06 52	☽ ⊥ ♄
10 59	☿ ⚹ ♄
13 43	♀ ⊥ ♇
17 37	☽ △ ♅
18 07	☽ ∠ ♂
18 48	☽ ⚺ ♄
20 02	☽ □ ☿
20 04	☽ ± ♆
26 Monday	
08 00	☽ ⚻ ♆
08 38	♀ ⫲ ♄
08 57	☿ Q ♇
11 55	☽ △ ♀
21 41	☽ Q ☉
21 53	☽ ⚻ ♇
23 31	☽ ⚼ ♅
27 Tuesday	
00 19	☽ ⚹ ♂
01 24	☽ □ ♃
09 59	☽ ± ♇
14 08	☽ Q ☿
17 38	☽ ± ♅
23 16	☿ ⊥ ♀
28 Wednesday	
00 06	☽ ⚹ ☉
00 41	☽ Q ♂
04 13	☽ ⚼ ♇
05 44	☽ ⚻ ♅
07 35	☽ ☌ ♄
14 16	☿ ± ♅
14 28	☽ Q ♃
18 28	☽ ⚹ ☿
20 29	☽ ☍ ♆
29 Thursday	
02 36	☽ ⫲ ♀
06 55	☽ □ ♀
08 38	☿ △ ♆
08 39	☽ ∥ ♄
09 14	☽ ∠ ☉
10 38	☽ △ ♇
13 17	☽ □ ♂
15 12	☽ ⚹ ♃
18 56	☽ ⫲ ♆
19 11	☽ ∥ ♂
30 Friday	
00 06	☽ ⫲ ♇
05 02	☽ ⊥ ☉
05 43	☽ ∠ ☿
18 04	☉ ⚻ ♅
18 09	☽ ☍ ♅
18 09	☽ ⚺ ☉
18 34	☿ ± ♂
20 35	☽ ⚺ ♄
21 55	☽ ∠ ♃
22 25	♀ ⚺ ♇
22 56	☽ Q ♀

OCTOBER 2005

Date	Sidereal time h m s	Sun ☉ ° ′ ″	Moon ☽ ° ′ ″	Moon ☽ 24.00 ° ′ ″	Mercury ☿ ° ′	Venus ♀ ° ′	Mars ♂ ° ′	Jupiter ♃ ° ′	Saturn ♄ ° ′	Uranus ♅ ° ′	Neptune ♆ ° ′	Pluto ♇ ° ′
01	12 41 17	08 ♎ 24 22	16 ♍ 37 10	22 ♍ 40 40	18 ♎ 33	22 ♏ 41	23 ♉ 22	24 ♎ 41	08 ♌ 57	07 ♓ 39	15 ♒ 00	22 ♐ 03
02	12 45 13	09 23 25	28 ♍ 46 34	04 ♎ 55 05	20 11	23 48	23 R 22	24 53	09 02	07 R 37	14 R 59	22 04
03	12 49 10	10 22 29	11 ♎ 06 23	17 ♎ 20 37	21 48	24 56	23 21	25 06	09 07	07 35	14 58	22 05
04	12 53 06	11 21 36	23 ♎ 37 54	29 ♎ 58 21	23 24	26 03	23 19	25 19	09 12	07 33	14 57	22 06
05	12 57 03	12 20 45	06 ♏ 22 03	12 ♏ 49 03	24 59	27 10	23 17	25 32	09 17	07 31	14 57	22 07
06	13 01 00	13 19 56	19 ♏ 19 27	25 ♏ 53 16	26 33	28 17	23 13	25 45	09 21	07 29	14 56	22 08
07	13 04 56	14 19 08	02 ♐ 30 34	09 ♐ 11 23	28 07	29 ♏ 24	23 08	25 58	09 26	07 28	14 55	22 09
08	13 08 53	15 18 23	15 ♐ 55 44	22 ♐ 43 37	29 ♎ 40	00 ♐ 31	23 03	26 11	09 31	07 26	14 55	22 10
09	13 12 49	16 17 39	29 ♐ 35 03	06 ♑ 29 57	01 ♏ 12	01 37	22 56	26 23	09 35	07 24	14 54	22 11
10	13 16 46	17 16 57	13 ♑ 28 17	20 ♑ 29 54	02 43	02 43	22 49	26 36	09 40	07 23	14 53	22 13
11	13 20 42	18 16 17	27 ♑ 34 38	04 ♒ 42 17	04 14	03 50	22 41	26 49	09 44	07 21	14 53	22 14
12	13 24 39	19 15 39	11 ♒ 52 31	19 ♒ 04 58	05 44	04 56	22 32	27 02	09 48	07 19	14 52	22 15
13	13 28 35	20 15 02	26 ♒ 19 13	03 ♓ 34 44	07 13	06 01	22 22	27 15	09 53	07 18	14 52	22 16
14	13 32 32	21 14 27	10 ♓ 50 56	18 ♓ 07 13	08 41	07 07	22 12	27 28	09 57	07 16	14 52	22 18
15	13 36 29	22 13 54	25 ♓ 22 52	02 ♈ 37 12	10 09	08 12	22 00	27 41	10 01	07 15	14 51	22 19
16	13 40 25	23 13 22	09 ♈ 49 32	16 ♈ 59 09	11 35	09 18	21 48	27 54	10 05	07 13	14 51	22 20
17	13 44 22	24 12 53	24 ♈ 05 25	01 ♉ 07 46	13 01	10 23	21 35	28 07	10 08	07 12	14 51	22 22
18	13 48 18	25 12 26	08 ♉ 05 39	14 ♉ 58 41	14 27	11 28	21 21	28 20	10 12	07 10	14 50	22 23
19	13 52 15	26 12 00	21 ♉ 46 33	28 ♉ 29 01	15 51	12 32	21 07	28 34	10 16	07 09	14 50	22 24
20	13 56 11	27 11 37	05 ♊ 06 02	11 ♊ 37 34	17 14	13 37	20 51	28 47	10 19	07 08	14 50	22 26
21	14 00 08	28 11 16	18 ♊ 03 47	24 ♊ 24 52	18 37	14 41	20 35	29 00	10 23	07 07	14 50	22 27
22	14 04 04	29 ♎ 10 58	00 ♋ 41 07	06 ♋ 52 57	19 59	15 45	20 19	29 13	10 26	07 05	14 49	22 29
23	14 08 01	00 ♏ 10 41	13 ♋ 00 46	19 ♋ 05 05	21 19	16 48	20 02	29 26	10 30	07 04	14 49	22 30
24	14 11 58	01 10 27	25 ♋ 06 27	01 ♌ 05 25	22 39	17 52	19 44	29 39	10 33	07 03	14 49	22 32
25	14 15 54	02 10 15	07 ♌ 02 36	12 ♌ 58 35	23 57	18 55	19 25	29 ♎ 52	10 36	07 02	14 49	22 34
26	14 19 51	03 10 06	18 ♌ 53 59	24 ♌ 49 24	25 15	19 58	19 06	00 ♏ 05	10 39	07 01	14 49	22 35
27	14 23 47	04 09 58	00 ♍ 45 27	06 ♍ 42 40	26 30	21 00	18 47	00 18	10 42	07 00	14 D 49	22 37
28	14 27 44	05 09 53	12 ♍ 41 37	18 ♍ 42 49	27 45	22 03	18 27	00 31	10 44	06 59	14 49	22 38
29	14 31 40	06 09 49	24 ♍ 46 43	00 ♎ 53 44	28 ♏ 58	23 05	18 07	00 44	10 47	06 58	14 49	22 40
30	14 35 37	07 09 48	07 ♎ 04 14	13 ♎ 18 32	00 ♐ 09	24 06	17 47	00 57	10 50	06 57	14 49	22 42
31	14 39 33	08 ♏ 09 49	19 ♎ 36 51	25 ♎ 59 22	01 ♐ 18	25 ♐ 08	17 ♉ 26	01 ♏ 10	10 ♌ 52	06 ♓ 57	14 ♒ 49	22 ♐ 44

Date	Moon True ☊ ° ′	Moon Mean ☊ ° ′	Moon ☽ Latitude ° ′
01	13 ♈ 38	13 ♈ 50	02 N 24
02	13 R 37	13 47	01 21
03	13 37	13 44	00 N 14
04	13 D 37	13 41	00 S 55
05	13 37	13 38	02 03
06	13 37	13 35	03 05
07	13 R 37	13 31	03 59
08	13 37	13 28	04 41
09	13 37	13 25	05 08
10	13 37	13 22	05 17
11	13 D 37	13 19	05 08
12	13 37	13 16	04 39
13	13 38	13 12	03 53
14	13 38	13 09	02 52
15	13 39	13 06	01 40
16	13 39	13 03	00 S 21
17	13 R 39	13 00	00 N 57
18	13 39	12 56	02 11
19	13 37	12 53	03 16
20	13 35	12 50	04 08
21	13 33	12 47	04 45
22	13 32	12 44	05 08
23	13 30	12 41	05 16
24	13 29	12 37	05 09
25	13 D 29	12 34	04 50
26	13 30	12 31	04 18
27	13 31	12 28	03 35
28	13 33	12 25	02 42
29	13 34	12 22	01 42
30	13 35	12 18	00 36
31	13 ♈ 35	12 ♈ 15	00 S 33

DECLINATIONS

Date	Sun ☉ ° ′	Moon ☽ ° ′	Mercury ☿ ° ′	Venus ♀ ° ′	Mars ♂ ° ′	Jupiter ♃ ° ′	Saturn ♄ ° ′	Uranus ♅ ° ′	Neptune ♆ ° ′	Pluto ♇ ° ′
01	03 S 20	07 N 30	07 S 02	20 S 26	16 N 29	08 S 34	18 N 25	09 S 27	16 S 29	15 S 26
02	03 43	01 N 44	07 46	20 47	16 30	08 39	18 23	09 28	16 29	15 27
03	04 06	04 S 11	08 28	21 07	16 32	08 44	18 22	09 28	16 29	15 27
04	04 30	10 02	09 11	21 27	16 33	08 49	18 21	09 29	16 30	15 28
05	04 53	15 35	09 52	21 46	16 35	08 53	18 20	09 30	16 30	15 28
06	05 16	20 32	10 33	22 05	16 36	08 58	18 19	09 30	16 30	15 28
07	05 39	24 34	11 13	22 24	16 37	09 03	18 18	09 31	16 30	15 29
08	06 02	27 21	11 53	22 42	16 38	09 07	18 17	09 32	16 30	15 29
09	06 25	28 34	12 32	22 59	16 38	09 12	18 15	09 32	16 30	15 30
10	06 47	28 01	13 10	23 16	16 39	09 17	18 14	09 33	16 31	15 30
11	07 10	25 40	13 48	23 32	16 39	09 22	18 13	09 33	16 31	15 30
12	07 32	21 42	14 24	23 48	16 39	09 26	18 12	09 34	16 31	15 31
13	07 55	16 24	15 00	24 03	16 39	09 31	18 11	09 34	16 31	15 31
14	08 17	10 09	15 35	24 18	16 39	09 36	18 10	09 35	16 31	15 31
15	08 39	03 S 22	16 10	24 32	16 38	09 40	18 09	09 36	16 31	15 32
16	09 01	03 N 34	16 43	24 45	16 38	09 45	18 09	09 36	16 31	15 32
17	09 23	10 14	17 16	24 58	16 37	09 50	18 08	09 37	16 32	15 33
18	09 45	16 17	17 48	25 10	16 37	09 54	18 07	09 37	16 32	15 33
19	10 07	21 21	18 18	25 22	16 36	09 59	18 06	09 37	16 32	15 33
20	10 28	25 12	18 48	25 33	16 34	10 04	18 05	09 38	16 32	15 34
21	10 50	27 38	19 17	25 43	16 33	10 08	18 04	09 38	16 32	15 34
22	11 11	28 34	19 45	25 53	16 32	10 13	18 03	09 39	16 32	15 35
23	11 32	28 02	20 12	26 02	16 30	10 18	18 03	09 39	16 32	15 35
24	11 53	26 11	20 37	26 11	16 28	10 22	18 02	09 40	16 32	15 35
25	12 14	23 11	21 02	26 19	16 27	10 27	18 01	09 40	16 32	15 36
26	12 34	19 14	21 26	26 26	16 25	10 32	18 01	09 40	16 32	15 36
27	12 55	14 33	21 48	26 33	16 22	10 36	18 00	09 41	16 32	15 36
28	13 15	09 17	22 09	26 39	16 20	10 41	17 59	09 41	16 32	15 37
29	13 35	03 N 38	22 29	26 44	16 18	10 45	17 59	09 41	16 32	15 37
30	13 54	02 S 15	22 48	26 49	16 15	10 50	17 58	09 41	16 32	15 37
31	14 S 14	08 S 11	23 S 05	26 S 53	16 N 13	10 S 54	17 N 58	09 S 42	16 S 32	15 S 38

ZODIAC SIGN ENTRIES

Date	h	m	Planets
02	14	24	☽ ♎
05	00	03	☽ ♏
07	07	28	☽ ♐
08	01	00	♀ ♐
08	17	15	☿ ♏
09	12	43	☽ ♑
11	16	05	☽ ♒
13	18	05	☽ ♓
15	19	39	☽ ♈
17	22	04	☽ ♉
20	02	44	☽ ♊
22	10	41	☽ ♋
23	07	42	☉ ♏
24	21	48	☽ ♌
26	02	52	♃ ♏
27	10	28	☽ ♍
29	22	15	☽ ♎
30	09	02	☿ ♐

LATITUDES

Date	Mercury ☿ ° ′	Venus ♀ ° ′	Mars ♂ ° ′	Jupiter ♃ ° ′	Saturn ♄ ° ′	Uranus ♅ ° ′	Neptune ♆ ° ′	Pluto ♇ ° ′
01	00 N 15	02 S 03	02 S 13	01 N 04	00 N 24	00 S 48	00 S 09	07 N 47
04	00 S 06	02 15	02 07	01 03	00 25	00 48	00 09	07 46
07	00 27	02 26	02 01	01 03	00 25	00 48	00 09	07 45
10	00 48	02 37	01 54	01 03	00 25	00 48	00 09	07 44
13	01 09	02 47	01 46	01 03	00 26	00 48	00 09	07 43
16	01 29	02 56	01 38	01 03	00 26	00 48	00 09	07 42
19	01 48	03 05	01 30	01 03	00 27	00 48	00 09	07 41
22	02 05	03 13	01 21	01 03	00 27	00 48	00 09	07 40
25	02 21	03 21	01 11	01 03	00 27	00 48	00 09	07 39
28	02 34	03 27	01 01	01 03	00 28	00 47	00 09	07 38
31	02 S 43	03 S 32	00 S 51	01 N 03	00 N 28	00 S 47	00 S 09	07 N 37

DATA

Julian Date	2453645
Delta T	+66 seconds
Ayanamsa	23° 56' 10"
Synetic vernal point	05° ♓ 10' 49"
True obliquity of ecliptic	23° 26' 28"

MOON'S PHASES, APSIDES AND POSITIONS ☽

Date	h	m	Phase	Longitude ° ′	Eclipse Indicator
03	10	28	●	10 ♎ 19	Annular
10	19	01	☽	17 ♑ 34	
17	12	14	○	24 ♈ 13	partial
25	01	17	☾	01 ♌ 44	

Day	h	m		
14	14	05	Perigee	
26	09	36	Apogee	
02	19	03	0S	
09	16	42	Max dec	28° S 36'
15	23	37	0N	
22	15	04	Max dec	28° N 35'
30	02	52	0S	

LONGITUDES

Date	Chiron ⚷ ° ′	Ceres ⚳ ° ′	Pallas ⚴ ° ′	Juno ⚵ ° ′	Vesta ⚶ ° ′	Black Moon Lilith ⚸ ° ′
01	27 ♑ 49	02 ♐ 34	05 ♏ 10	18 ♊ 33	16 ♋ 05	17 ♌ 24
11	27 ♑ 50	06 ♐ 09	09 ♏ 27	20 ♊ 48	18 ♋ 31	18 ♌ 31
21	27 ♑ 58	09 ♐ 51	13 ♏ 46	22 ♊ 16	20 ♋ 31	19 ♌ 38
31	28 ♑ 13	13 ♐ 40	18 ♏ 08	22 ♊ 52	22 ♋ 02	20 ♌ 45

All ephemeris data is given at 12.00 UT and the Moon's longitude is additionally given for 24.00 UT

ASPECTARIAN

01 Saturday
02 38 ☽ ⊥ ☿
03 36 ☽ ⋕ ♅
07 26 ☽ ⋕ ♃
08 39 ☽ ⊥ ♄
08 46 ☽ ⊼ ♆
11 08 ☉ ⊡ ♂
13 44 ☽ ⋕ ☿
14 27 ♂ ⋕ ♆
16 09 ☽ ⊥ ♃
16 26 ☽ ⊻ ☿
20 40 ☽ ± ♆
22 04 ♂ St R
22 47 ☽ □ ♇

02 Sunday
01 14 ☽ ⚹ ♀
01 22 ☽ △ ♂
02 23 ☉ ⚹ ♄
02 36 ☽ ∠ ♄
02 45 ♀ ☍ ♂
04 14 ☽ ⊻ ♃
04 20 ☽ ⋕ ☉
14 21 ☽ ⊡ ♆

03 Monday
01 16 ☿ Q ♄
04 46 ☉ Q ♇
05 12 ☽ ⊼ ♅
06 41 ☽ ⊡ ♂
08 07 ☽ ⚹ ♄
09 30 ☽ ∠ ♀
10 01 ☽ Q ♇
10 28 ☽ ☌ ☉
11 41 ☽ ∥ ☉
16 20 ☿ ⚹ ♇
16 37 ♀ ⊻ ♃
16 46 ☽ ± ♅
19 26 ☽ △ ♆
21 51 ☿ ∥ ♃
23 33 ☿ ⊡ ♅

04 Tuesday
00 00 ☽ ± ♂
04 30 ☽ ⊥ ♀
06 51 ☽ ∥ ♃
07 20 ☽ Q ♄
07 57 ☽ ∥ ☿
09 05 ☽ ⚹ ♇
09 42 ☽ ∥ ♅
09 57 ☽ ⊡ ♅
10 55 ☿ ⊼ ♂
11 25 ☽ ⊼ ♂
11 29 ☽ ☌ ☿
15 15 ☽ ☌ ♃
17 02 ☽ ⊻ ♀
22 43 ☿ ∥ ♅

05 Wednesday
11 30 ☽ ∥ ♇
13 24 ☽ ∠ ♇
14 09 ☽ △ ♅
16 12 ☽ ∥ ♆
16 36 ☽ ⋕ ♂
17 28 ☽ □ ♄
21 39 ☿ ☌ ♃

06 Thursday
00 03 ☽ ⊻ ☉
00 51 ☽ ⋕ ♄
03 55 ☽ □ ♆
06 07 ☽ ⊥ ♇
12 01 ☽ ⊥ ☉
15 44 ☉ ± ♅
17 09 ☽ ⊻ ♇
19 05 ☽ ☍ ♂
21 03 ☉ ∠ ♀
21 13 ☽ ∥ ♀
23 56 ☽ ⊻ ♃

07 Friday
02 59 ☽ ⊻ ☿
05 46 ☽ ∠ ☉
05 51 ☽ ☌ ♀
11 00 ☽ ⊥ ♃
12 44 ☽ Q ♆
15 17 ☽ ⊥ ☿
20 53 ☽ □ ♅

08 Saturday
00 31 ☽ △ ♄
02 28 ☉ △ ♆
03 25 ☽ ∠ ♃
09 28 ☽ ∠ ☿
10 12 ☽ ⚹ ♆
10 49 ☽ ⚹ ☉
23 02 ☽ ☌ ♇

09 Sunday
00 28 ☽ ⊼ ♂
03 13 ☽ ⊡ ♄
04 43 ☽ Q ♅
06 20 ☽ ⚹ ♃
09 35 ☽ Q ☉
10 53 ☽ ± ♂
12 33 ☽ ∠ ♆
15 10 ☽ ⚹ ☿
15 51 ☽ ⊻ ♀
19 00 ☽ ± ♄

10 Monday
01 32 ☽ ⚹ ♅
02 03 ☉ ± ♂
02 22 ☽ ⊡ ♂
03 08 ☽ ⊥ ♀
03 30 ☽ Q ♃
04 08 ☽ ⊥ ♆
05 25 ☽ ⊼ ♄
08 51 ♀ ⊥ ♃
12 10 ☿ ⊻ ♀
14 24 ☽ Q ☿
14 26 ☽ ⊻ ♆
15 37 ♀ Q ♆
19 01 ☽ □ ☉
19 53 ☽ ∠ ♀

11 Tuesday
02 56 ☽ ⊻ ♇
03 10 ☽ ∠ ♅
03 48 ☽ △ ♂
10 42 ☽ □ ♃
13 06 ☽ ⊥ ♇
18 21 ☽ ⊥ ♅
23 24 ☽ ⚹ ♀

12 Wednesday
00 31 ☽ □ ☿
01 09 ☽ ∥ ♀
04 16 ☽ ∠ ♇
04 24 ☽ ⊻ ♅
08 32 ☽ ☍ ♄
17 00 ☽ ☌ ♆
21 07 ☽ Q ♀

13 Thursday
01 12 ☽ △ ☉
04 25 ☽ ⋕ ♄
05 17 ☽ ⚹ ♇
05 32 ☽ □ ♂
10 58 ☽ ⋕ ♂
11 30 ☽ ∥ ♆
13 00 ☿ ∠ ♇
13 20 ☿ △ ♅
13 34 ☽ △ ♃
15 34 ☽ ∥ ♇
17 07 ☽ ∥ ☿

14 Friday
00 22 ♂ ⊼ ♇
01 10 ☽ Q ♇
03 50 ☽ ⊡ ☉
05 20 ☽ □ ♀
06 06 ☽ ☌ ♅
07 41 ♃ ∥ ♅
08 02 ☽ △ ☿
09 12 ☿ ∥ ♇
10 30 ☽ ⊼ ♄
10 56 ☽ Q ♂
13 59 ☽ ∥ ♃
14 03 ☽ ∥ ♅
14 43 ☽ ⊡ ♃
15 17 ♀ □ ♅
18 22 ☽ ∥ ☉
18 37 ☽ ⊻ ♆
19 47 ☽ ± ☉
20 27 ☽ ± ♄

15 Saturday
04 31 ☽ ⊥ ♆
05 48 ☽ ± ♃
06 18 ☉ Q ♄
06 25 ☽ ⊼ ☉
06 29 ☽ ⚹ ♂
06 55 ☽ □ ♇
07 25 ☉ ⊼ ♂
09 44 ☿ □ ♄
11 23 ☽ ⊡ ♄
11 34 ☽ ⊡ ☿
12 19 ☉ ⊡ ♅
14 05 ☉ ⚹ ♇
15 53 ☽ ⊼ ♃
19 24 ☽ ∠ ♆

16 Sunday
03 28 ☿ ∥ ♆
04 09 ☽ ± ☿
07 02 ☽ ∠ ♂
07 40 ☽ ⊻ ♅
08 16 ☿ ⋕ ♂
11 03 ☽ △ ♀
12 25 ☽ △ ♄
15 17 ☽ ⊼ ☿
17 40 ☽ ⊥ ♅
20 24 ☽ ⚹ ♆
21 52 ☽ ⊥ ♂

17 Monday
06 24 ♀ △ ♄
07 49 ☽ ⊻ ♂
08 40 ☽ ⋕ ☉
08 48 ☽ ∠ ♅
09 04 ☽ △ ♇
09 40 ☽ ⋕ ♅
10 28 ☽ ⋕ ♃
12 14 ☽ ☍ ☉
14 22 ☽ ⊡ ♀
16 40 ☽ Q ♆
18 58 ☽ ☍ ♃

18 Tuesday
02 43 ☉ ∥ ♅
07 04 ☽ ± ♀
08 56 ☽ ⋕ ♇
10 25 ☽ ⚹ ♅
10 46 ☽ ⊡ ♇
13 05 ☽ ⋕ ♆
13 26 ☽ ∥ ♂
15 41 ☽ □ ♄
18 21 ☽ ⊼ ♀
18 41 ☿ □ ♆
19 22 ☽ ⋕ ☿
20 04 ☽ ∥ ♄
23 45 ☽ □ ♆

19 Wednesday
00 20 ☽ ☍ ☿
00 58 ☉ ∥ ♃
02 29 ☿ ⋕ ♄
02 30 ☽ ± ♇
07 21 ☽ Q ♅
10 50 ☽ ☌ ♂
13 08 ☽ ⊼ ♇
20 32 ☽ ⊼ ☉
21 47 ☿ ⊥ ♇
23 39 ☽ Q ♄

20 Thursday
00 20 ☽ ⊼ ♃
08 15 ☽ ± ☉
11 24 ☽ ± ♃
14 48 ☽ ⋕ ♀
15 43 ☽ □ ♅
16 43 ♀ ∠ ♃
21 38 ☽ ⚹ ♄

21 Friday
02 08 ☽ ⊡ ☉
04 16 ☽ ⊡ ♃
05 06 ☽ ☍ ♀
05 57 ☽ △ ♆
13 10 ☽ ⊼ ☿
15 19 ♀ ⚹ ♆
16 39 ☽ ⊻ ♂
20 18 ☽ ☍ ♇

22 Saturday
01 53 ☽ ± ☿
01 54 ☽ ∠ ♄
03 48 ☽ ⊥ ♂
08 17 ♂ ⋕ ♆
08 52 ☽ △ ☉
09 07 ☽ △ ♃
10 20 ☽ ⊡ ♆
12 54 ☉ ☌ ♃
16 55 ☿ ☍ ♂
19 17 ☽ ⊥ ♄
20 45 ☽ ∠ ♂
21 19 ☽ ⊡ ☿

23 Sunday
00 23 ☽ △ ♅
03 47 ☽ ± ♆
07 02 ☽ ⊻ ♄
15 34 ☽ ⊼ ♆
20 12 ☽ ⊼ ♀

24 Monday
01 32 ☽ ⚹ ♂
05 55 ☽ ⊡ ♅
06 29 ☽ △ ☿
06 51 ☽ ⊼ ♇
09 16 ☽ ± ♀
09 51 ☿ ⊻ ♇
12 01 ☽ ⋕ ♀
18 53 ☽ ± ♇
21 16 ☽ □ ♃
23 54 ☽ ± ♅

25 Tuesday
00 57 ☽ Q ♂
01 17 ☽ □ ☉
05 04 ☽ ⊡ ♀
11 59 ☽ ⊼ ♅
13 03 ☽ ⊡ ♇
19 13 ☽ ☌ ♄
20 59 ♀ ⊼ ♂

26 Wednesday
00 25 ☽ ⋕ ☿
03 44 ☽ ☍ ♆
10 19 ☽ Q ♃
12 25 ☽ □ ♂
14 22 ☽ △ ♀
17 01 ☽ Q ☉
18 38 ☽ ∥ ♄
19 08 ♂ Q ♅
19 29 ☽ △ ♇
23 24 ♆ St D

27 Thursday
02 14 ☽ ⋕ ♆
02 23 ☽ □ ☿
02 58 ☽ ∥ ♂
06 52 ☽ ⋕ ♇
11 04 ☽ ⚹ ♃
19 15 ☽ ⋕ ☉
19 30 ☽ ⚹ ☉

28 Friday
00 34 ☽ ☍ ♅
05 57 ☽ ⋕ ♃
08 04 ☽ ⊻ ♄
10 17 ☽ ⋕ ♅
16 15 ☽ ⊼ ♆
17 45 ☽ ∠ ♃
18 47 ☽ Q ☿
20 06 ☽ ⊥ ♄
23 11 ☽ △ ♂

29 Saturday
02 15 ♀ ☌ ♇
04 11 ☽ ± ♆
04 13 ☽ ∠ ☉
07 50 ☽ □ ♇
08 20 ☽ □ ♀
11 55 ☽ ⊥ ♃
13 59 ☽ ∠ ♄
21 06 ☽ ⚹ ☿
21 54 ☽ ⊡ ♆
23 30 ☽ ⊥ ☉
23 54 ☽ ⊻ ♃

30 Sunday
03 54 ☽ ⊡ ♂
06 15 ♀ ± ♂
07 07 ☉ △ ♅
11 47 ☽ ⊼ ♅
12 12 ☽ ⊻ ☉
19 00 ☽ Q ♇
19 16 ☽ ⚹ ♄
20 49 ☽ ± ♂
22 33 ☽ Q ♀
23 19 ☽ ± ♅

31 Monday
01 11 ☉ ∠ ♇
02 53 ☽ △ ♆
05 05 ☽ ∠ ☿
07 43 ♀ Q ♅
07 58 ☽ ⊼ ♂
08 43 ☿ ⊻ ♃
16 24 ☽ ⊡ ♅
17 53 ☽ ⚹ ♇
18 10 ☽ Q ♄
18 13 ☽ ∥ ♅
23 17 ☽ ⚹ ♀
23 23 ☽ ∥ ♃
23 44 ☽ ⊥ ☿

NOVEMBER 2005

LONGITUDES

Date	Sidereal time h	m	s	Sun ☉ °	'	"	Moon ☽ °	'	"	Moon ☽ 24.00 °	'	"	Mercury ☿ °	'	Venus ♀ °	'	Mars ♂ °	'	Jupiter ♃ °	'	Saturn ♄ °	'	Uranus ♅ °	'	Neptune ♆ °	'	Pluto ♇ °	'
01	14	43	30	09 ♏	09	51	02 ♏	26	08	08 ♏	57	12	02 ♐	25	26 ♐	09	17 ♉	05	01 ♏	23	10 ♌	54	06 ♓	56	14 ♒	50	22 ♐	45
02	14	47	27	10	09	56	15 ♏	32	30	22 ♏	11	53	03	29	27	09	16 R	44	01	36	10	57	06 R	55	14	50	22	47
03	14	51	23	11	10	03	28 ♏	55	09	05 ♐	42	04	04	31	28	10	16	22	01	49	10	59	06	55	14	50	22	49
04	14	55	20	12	10	11	12 ♐	32	19	19 ♐	25	34	05	30	29 ♐	10	16	01	02	02	11	01	06	54	14	50	22	51
05	14	59	16	13	10	21	26 ♐	21	27	03 ♑	19	37	06	26	00 ♑	09	15	40	02	15	11	03	06	53	14	51	22	52
06	15	03	13	14	10	33	10 ♑	19	40	17 ♑	21	17	07	18	01	09	15	18	02	28	11	05	06	53	14	51	22	54
07	15	07	09	15	10	46	24 ♑	24	07	01 ♒	27	51	08	06	02	07	14	57	02	41	11	06	06	52	14	51	22	56
08	15	11	06	16	11	01	08 ♒	32	14	15 ♒	36	59	08	49	03	06	14	36	02	54	11	08	06	52	14	52	22	58
09	15	15	02	17	11	17	22 ♒	41	54	29 ♒	46	45	09	27	04	04	14	15	03	07	11	09	06	52	14	52	23	00
10	15	18	59	18	11	35	06 ♓	51	20	13 ♓	55	28	09	59	05	01	13	54	03	20	11	11	06	51	14	53	23	02
11	15	22	56	19	11	53	20 ♓	58	54	28 ♓	01	26	10	25	05	58	13	34	03	33	11	12	06	51	14	53	23	04
12	15	26	52	20	12	14	05 ♈	02	49	12 ♈	02	44	10	43	06	54	13	14	03	46	11	13	06	51	14	54	23	06
13	15	30	49	21	12	35	19 ♈	00	56	25 ♈	57	02	10	54	07	50	12	55	03	58	11	14	06	51	14	54	23	08
14	15	34	45	22	12	59	02 ♉	50	44	09 ♉	41	40	10 R	56	08	46	12	35	04	11	11	15	06	51	14	55	23	10
15	15	38	42	23	13	23	16 ♉	29	29	23 ♉	13	52	10	49	09	40	12	17	04	24	11	16	06	51	14	56	23	12
16	15	42	38	24	13	50	29 ♉	54	31	06 ♊	31	11	10	32	10	35	11	59	04	37	11	17	06 D	51	14	56	23	14
17	15	46	35	25	14	18	13 ♊	03	41	19 ♊	31	53	10	05	11	28	11	41	04	49	11	17	06	51	14	57	23	16
18	15	50	31	26	14	47	25 ♊	55	45	02 ♋	15	19	09	27	12	21	11	24	05	02	11	18	06	51	14	58	23	18
19	15	54	28	27	15	19	08 ♋	30	42	14 ♋	42	05	08	39	13	13	11	07	05	14	11	18	06	51	14	58	23	20
20	15	58	25	28	15	52	20 ♋	49	46	26 ♋	54	06	07	41	14	05	10	52	05	27	11	18	06	51	14	59	23	22
21	16	02	21	29 ♏	16	27	02 ♌	55	30	08 ♌	54	27	06	35	14	55	10	36	05	40	11	19	06	51	15	00	23	24
22	16	06	18	00 ♐	17	03	14 ♌	51	30	20 ♌	47	11	05	21	15	45	10	22	05	52	11 R	19	06	52	15	01	23	26
23	16	10	14	01	17	41	26 ♌	42	10	02 ♍	37	04	04	02	16	35	10	08	06	05	11	19	06	52	15	02	23	28
24	16	14	11	02	18	21	08 ♍	32	32	14 ♍	29	14	02	40	17	23	09	55	06	17	11	18	06	52	15	03	23	30
25	16	18	07	03	19	02	20 ♍	27	51	26 ♍	29	01	01	19	18	11	09	43	06	29	11	18	06	53	15	04	23	32
26	16	22	04	04	19	45	02 ♎	33	23	08 ♎	41	31	00 ♐	00	18	57	09	31	06	42	11	18	06	53	15	05	23	35
27	16	26	00	05	20	29	14 ♎	53	58	21 ♎	11	13	28 ♏	46	19	43	09	20	06	54	11	17	06	54	15	06	23	37
28	16	29	57	06	21	15	27 ♎	33	39	04 ♏	01	36	27	40	20	28	09	10	07	06	11	17	06	55	15	07	23	39
29	16	33	54	07	22	03	10 ♏	35	14	17 ♏	14	38	26	43	21	12	09	01	07	18	11	16	06	55	15	08	23	41
30	16	37	50	08 ♐	22	52	23 ♏	59	44	00 ♐	50	21	25 ♏	57	21 ♑	55	08 ♉	53	07 ♏	31	11 ♌	15	06 ♓	56	15 ♒	09	23 ♐	43

Moon True ☊ / Moon Mean ☊ / Moon ☽ Latitude

Date	Moon True ☊ °	'	Moon Mean ☊ °	'	Moon ☽ Latitude °	'
01	13 ♈	34	12 ♈	12	01 S	42
02	13 R	32	12	09	02	46
03	13	28	12	06	03	43
04	13	23	12	02	04	28
05	13	19	11	59	04	59
06	13	15	11	56	05	12
07	13	12	11	53	05	06
08	13	10	11	50	04	42
09	13 D	10	11	47	04	01
10	13	11	11	43	03	05
11	13	12	11	40	01	58
12	13	14	11	37	00 S	44
13	13 R	14	11	34	00 N	31
14	13	12	11	31	01	45
15	13	08	11	28	02	51
16	13	03	11	24	03	46
17	12	56	11	21	04	28
18	12	48	11	18	04	55
19	12	40	11	15	05	07
20	12	33	11	12	05	05
21	12	27	11	08	04	49
22	12	24	11	05	04	20
23	12	22	11	02	03	41
24	12 D	22	10	59	02	52
25	12	23	10	56	01	55
26	12	24	10	53	00 N	53
27	12 R	25	10	49	00 S	13
28	12	24	10	46	01	20
29	12	20	10	43	02	25
30	12 ♈	14	10 ♈	40	03 S	24

DECLINATIONS

Date	Sun ☉ °	'	Moon ☽ °	'	Mercury ☿ °	'	Venus ♀ °	'	Mars ♂ °	'	Jupiter ♃ °	'	Saturn ♄ °	'	Uranus ♅ °	'	Neptune ♆ °	'	Pluto ♇ °	'
01	14 S	33	13 S	55	23 S	21	26 S	57	16 N	10	10 S	59	17 N	57	09 S	42	16 S	32	15 S	38
02	14	52	19	09	23	36	27	00	16	08	11	03	17	57	09	42	16	32	15	39
03	15	11	23	33	23	49	27	02	16	05	11	08	17	56	09	42	16	32	15	39
04	15	29	26	44	24	00	27	04	16	02	11	12	17	56	09	43	16	32	15	39
05	15	48	28	22	24	10	27	05	15	59	11	17	17	55	09	43	16	32	15	40
06	16	06	28	13	24	18	27	05	15	56	11	21	17	55	09	43	16	32	15	40
07	16	23	26	16	24	25	27	05	15	54	11	26	17	55	09	43	16	31	15	40
08	16	41	22	40	24	29	27	04	15	51	11	30	17	54	09	43	16	31	15	41
09	16	58	17	45	24	32	27	03	15	48	11	34	17	54	09	43	16	31	15	41
10	17	15	11	52	24	32	27	01	15	45	11	39	17	54	09	43	16	31	15	41
11	17	32	05 S	23	24	30	26	59	15	42	11	43	17	54	09	43	16	31	15	42
12	17	48	01 N	20	24	26	26	56	15	39	11	47	17	54	09	43	16	31	15	42
13	18	04	07	56	24	20	26	52	15	37	11	52	17	53	09	43	16	31	15	42
14	18	19	14	06	24	11	26	48	15	34	11	56	17	53	09	43	16	30	15	43
15	18	35	19	29	23	59	26	43	15	31	12	00	17	53	09	43	16	30	15	43
16	18	50	23	48	23	44	26	38	15	29	12	04	17	53	09	43	16	30	15	43
17	19	05	26	48	23	26	26	32	15	27	12	09	17	53	09	43	16	30	15	43
18	19	19	28	18	23	04	26	26	15	24	12	13	17	53	09	43	16	30	15	44
19	19	33	28	17	22	40	26	19	15	22	12	17	17	53	09	43	16	29	15	44
20	19	46	26	51	22	12	26	12	15	20	12	21	17	53	09	43	16	29	15	44
21	20	00	24	11	21	42	26	05	15	18	12	25	17	53	09	43	16	29	15	45
22	20	13	20	31	21	09	25	56	15	17	12	29	17	54	09	43	16	29	15	45
23	20	25	16	04	20	34	25	48	15	15	12	34	17	54	09	43	16	28	15	45
24	20	37	11	02	19	59	25	39	15	13	12	38	17	54	09	42	16	28	15	46
25	20	49	05 N	33	19	23	25	30	15	12	12	42	17	54	09	42	16	28	15	46
26	21	01	00 S	13	18	49	25	20	15	11	12	46	17	54	09	42	16	28	15	46
27	21	12	06	05	18	17	25	10	15	10	12	50	17	55	09	42	16	27	15	46
28	21	22	11	51	17	47	24	59	15	09	12	54	17	55	09	42	16	27	15	47
29	21	32	17	18	17	22	24	48	15	09	12	57	17	55	09	41	16	27	15	47
30	21 S	42	22 S	03	17 S	01	24 S	37	15 N	09	13 S	01	17 N	56	09 S	41	16 S	26	15 S	47

ZODIAC SIGN ENTRIES

Date	h	m	Planets
01	07	29	☽ ♏
03	13	55	☽ ♐
05	08	10	♀ ♑
05	18	17	☽ ♑
07	21	31	☽ ♒
10	00	22	☽ ♓
12	03	22	☽ ♈
14	07	02	☽ ♉
16	12	10	☽ ♊
18	19	42	☽ ♋
21	06	10	☽ ♌
22	05	15	☉ ♐
23	18	41	☽ ♍
26	06	58	☽ ♎
26	11	53	☿ ♏
28	16	33	☽ ♏
30	22	32	☽ ♐

LATITUDES

Date	Mercury ☿ °	'	Venus ♀ °	'	Mars ♂ °	'	Jupiter ♃ °	'	Saturn ♄ °	'	Uranus ♅ °	'	Neptune ♆ °	'	Pluto ♇ °	'
01	02 S	46	03 S	34	00 S	48	01 N	02	00 N	28	00 S	47	00 S	09	07 N	37
04	02	50	03	37	00	37	01	02	00	29	00	47	00	09	07	36
07	02	47	03	40	00	27	01	02	00	29	00	47	00	09	07	36
10	02	37	03	41	00	17	01	03	00	30	00	47	00	09	07	35
13	02	16	03	40	00 S	07	01	03	00	30	00	47	00	09	07	34
16	01	43	03	38	00 N	03	01	03	00	31	00	47	00	09	07	33
19	00 S	56	03	33	00	13	01	03	00	31	00	47	00	09	07	33
22	00 N	03	03	27	00	22	01	03	00	32	00	47	00	09	07	32
25	01	03	03	19	00	30	01	03	00	32	00	46	00	09	07	32
28	01	54	03	08	00	38	01	03	00	33	00	46	00	09	07	31
31	02 N	26	02 S	55	00 N	46	01 N	03	00 N	33	00 S	46	00 S	09	07 N	31

DATA

Julian Date	2453676
Delta T	+66 seconds
Ayanamsa	23° 56' 14"
Synetic vernal point	05° ♓ 10' 46"
True obliquity of ecliptic	23° 26' 28"

MOON'S PHASES, APSIDES AND POSITIONS ☽

Date	h	m	Phase	Longitude °	'	Eclipse Indicator
02	01	25	●	09 ♏	43	
09	01	57	☽	16 ♒	46	
16	00	58	○	23 ♉	46	
23	22	11	☾	01 ♍	43	

Day	h	m		
10	00	26	Perigee	
23	06	15	Apogee	
05	22	04	Max dec	28° S 32'
12	07	16	0N	
18	23	42	Max dec	28° N 28'
26	11	08	0S	

LONGITUDES

Date	Chiron ⚷ °	'	Ceres ⚳ °	'	Pallas ⚴ °	'	Juno ⚵ °	'	Vesta ⚶ °	'	Black Moon Lilith ⚸ °	'
01	28 ♑	15	14 ♐	03	18 ♏	34	22 ♊	53	22 ♋	09	20 ♌	52
11	28 ♑	38	17 ♐	57	22 ♏	57	22 ♊	27	23 ♋	00	21 ♌	59
21	29 ♑	08	21 ♐	55	27 ♏	20	21 ♊	08	23 ♋	11	23 ♌	06
31	29 ♑	42	25 ♐	56	01 ♐	42	19 ♊	09	22 ♋	37	24 ♌	13

ASPECTARIAN

h m Aspects

01 Tuesday
06 07 ♀ ⚼ ♄
10 02 ☽ ☌ ♃
11 57 ☽ ⚺ ☿
14 58 ☽ ∥ ☉
19 36 ☽ ∥ ♇
20 17 ☽ △ ♅
21 07 ☿ Q ♆
21 49 ☽ ∠ ♇
21 56 ☽ # ♂
23 39 ☽ ∥ ♆

02 Wednesday
01 25 ☽ ☌ ☉
03 37 ☽ □ ♄
05 21 ☽ ∠ ♀
06 13 ☽ # ♄
08 30 ♂ ± ♇
10 42 ☽ □ ♆
14 05 ☽ ☍ ♂
14 15 ☽ ⊥ ♇
22 58 ☽ ⊥ ♀

03 Thursday
01 05 ☽ ⚺ ♇
07 21 ☉ □ ♄
10 33 ☽ ⚺ ♀
13 46 ☽ ∥ ☿
17 14 ☽ ⚺ ♃
18 56 ☽ Q ♆
22 42 ☽ ☌ ☿

04 Friday
02 07 ☽ □ ♅
03 59 ☽ ⊥ ♃
09 20 ☽ △ ♄
11 18 ☽ ⚺ ☉
15 22 ☽ ∥ ♀
16 01 ☽ ⚹ ♆
17 55 ☽ ⚻ ♂
19 58 ☽ ∠ ♃
22 35 ☽ ⊥ ☉

05 Saturday
01 09 ☉ ∥ ♇
04 05 ☽ ± ♂
04 20 ♀ ∠ ♆
05 58 ☽ ☌ ♇
09 28 ☽ Q ♅
11 28 ☽ ⚼ ♄
15 22 ☽ ∠ ☉
18 01 ☽ ∠ ♆
19 03 ☽ ☌ ♀
19 14 ☽ ⚼ ♂
20 58 ♀ ⚼ ♂
22 19 ☽ ⚹ ♃

06 Sunday
00 20 ☿ □ ♅
01 19 ☉ # ♂
02 59 ☽ ± ♄
06 06 ☽ ⚹ ♅
06 29 ☽ ⚺ ☿
09 28 ☽ ⊥ ♆
13 17 ☽ ⚻ ♄
17 24 ☽ ⊥ ☿
19 05 ☽ ⚹ ☉
19 11 ☽ Q ♃
19 44 ☽ ⚺ ♆
20 18 ☽ △ ♂

07 Monday
04 11 ☉ □ ♆
04 13 ☽ ∥ ♀
07 42 ☽ ∠ ♅
07 57 ☉ ☍ ♂
09 30 ☽ ⚺ ♇
09 39 ☽ ∠ ☿
17 05 ☽ Q ☉
18 26 ♂ □ ♆
19 43 ☽ ⊥ ♇
22 58 ☉ ∥ ♆
23 00 ☽ ⊥ ♅

08 Tuesday
01 17 ☽ ∥ ☿
02 06 ☽ ⚺ ♀
02 17 ☽ □ ♃
05 46 ♀ ⚹ ♃
09 10 ☽ ⚺ ♅
11 02 ☽ ∠ ♇
12 30 ☽ ⚹ ☿
13 01 ☽ ⊥ ♀
16 24 ☽ ☍ ♄
16 38 ☿ ⊥ ♃
22 02 ☽ □ ♂
22 44 ☽ ☌ ♆

09 Wednesday
01 57 ☽ □ ☉
05 23 ☽ ∠ ♀
07 19 ☉ ⊥ ♇
09 47 ☽ Q ☿
11 18 ☽ # ♄
12 31 ☽ ⚹ ♇
15 12 ☽ ∥ ☉
17 17 ☽ ∥ ♆
20 23 ☽ # ♂
20 47 ☽ ∥ ♇

10 Thursday
03 48 ☽ Q ♂
05 56 ☽ △ ♃
08 40 ☽ ⚹ ♀
08 54 ☽ Q ♇
12 00 ☽ ☌ ♅
12 49 ☽ ∥ ♃
16 08 ♀ ± ♄
17 29 ☽ □ ☿
19 21 ☽ ⚻ ♄
20 06 ☽ ∥ ♅
23 41 ☽ ⚹ ♂

11 Friday
01 38 ☽ ⚺ ♆
05 34 ☽ ± ♄
06 30 ☽ Q ♀
07 47 ☽ ⚼ ♃
08 44 ☽ △ ☉
11 50 ☽ ⊥ ♆
15 33 ☽ □ ♇
16 13 ♂ # ♇
20 54 ☽ ⚼ ♄
23 21 ☽ ± ♃

12 Saturday
00 38 ☽ ∠ ♂
03 11 ☽ ∠ ♆
09 46 ☽ ⚻ ♃
10 28 ♀ ⚹ ♅
12 17 ☽ ⚼ ☉
15 05 ☽ ⚺ ♅
15 25 ☽ □ ♀
15 40 ☽ ⊥ ♂
20 32 ☉ # ♄
21 52 ☽ △ ☿
22 36 ☽ △ ♄

13 Sunday
01 23 ☽ ⊥ ♅
01 43 ☽ ⚺ ♂
04 55 ☽ ⚹ ♆
04 56 ☽ ± ☉
16 05 ☽ ⚻ ☉
16 53 ☽ ∠ ♅
18 45 ☽ # ♅
19 07 ☽ △ ♇
23 58 ☽ ⚼ ☿

14 Monday
01 40 ☽ Q ♆
03 12 ☽ # ♃
05 41 ☿ St R
14 23 ☽ ☍ ♃
15 38 ☽ ± ☿
16 04 ♀ ⊥ ♆
18 07 ☽ ∥ ♂
18 46 ☽ # ♇
19 00 ☽ ⚹ ♅
21 20 ☽ ⚼ ♇
22 13 ☽ # ♆
23 06 ☽ △ ♀

15 Tuesday
02 05 ☽ ⚻ ☿
02 45 ☽ □ ♄
04 25 ☽ ∥ ♄
04 43 ☽ ☌ ♂
07 24 ☽ # ☉
09 13 ☽ □ ♆
11 17 ☉ ⚺ ♇
13 15 ☽ ± ♇
16 10 ☽ Q ♅
23 58 ☽ ⚻ ♇

16 Wednesday
00 08 ♅ St D
00 58 ☽ ☍ ☉
03 38 ☽ ⚼ ♀
08 41 ☿ ⊥ ♃
10 52 ☽ Q ♄
11 08 ☿ ⚺ ♀
11 31 ☽ # ☿
20 39 ☽ ⚻ ♃
21 04 ☽ ± ♀

17 Thursday
00 35 ☽ □ ♅
06 43 ☽ ☍ ☿
07 06 ♀ ⚻ ♄
07 48 ☽ ± ♃
08 44 ☽ ⚹ ♄
08 51 ☽ ⚻ ♀
09 30 ☽ # ♀
09 31 ☽ ⚺ ♂
15 29 ☽ △ ♆
16 23 ♀ △ ♂
20 22 ☽ ⊥ ♂

18 Friday
00 45 ☽ ⚼ ♃
07 02 ☽ ☍ ♇
12 39 ☽ ⚻ ☉
12 42 ☽ ∠ ♄
12 52 ☽ ∠ ♂
19 38 ☽ ⚼ ♆
20 28 ♂ □ ♄

19 Saturday
01 02 ☽ ± ☉
05 36 ☽ △ ♃
05 50 ☽ ⊥ ♄
08 48 ☽ △ ♅
12 15 ☽ ⚻ ☿
12 54 ☽ ± ♆
16 56 ☽ ⚹ ♂
17 24 ☽ ⚺ ♄
19 53 ☽ ⚼ ☉
21 48 ☽ ☍ ♀
23 04 ☽ ± ☿

20 Sunday
00 33 ☽ ⚻ ♆
07 01 ☿ ⊥ ♀
14 01 ☽ ⚼ ♅
15 22 ☽ ⚼ ☿
15 55 ☽ Q ♂
17 01 ☽ ⚻ ♇
19 14 ☽ # ♀

21 Monday
04 03 ☽ △ ☉
04 57 ☽ ± ♇
06 18 ☿ □ ♅
07 52 ☽ ± ♅
14 17 ♀ ⚺ ♆
17 34 ☽ □ ♃
18 40 ☽ △ ☿
19 53 ☽ ⚻ ♅
23 01 ☽ ⚼ ♇

22 Tuesday
03 07 ☽ □ ♂
03 33 ☿ ⚺ ♃
04 50 ☽ ☌ ♄
07 40 ☽ # ☿
09 01 ♄ St R
12 19 ☽ ☍ ♆
13 44 ☽ # ☉
13 57 ☽ ⚻ ♀

23 Wednesday
02 37 ☽ ∥ ♄
03 01 ☽ ± ♀
05 25 ☽ △ ♇
06 35 ☽ Q ♃
09 59 ☽ # ♆
13 36 ☽ # ♇
16 07 ☽ ∥ ♂
16 37 ☉ ∥ ☿
22 11 ☽ □ ☉
22 37 ☽ ⚼ ♀

24 Thursday
01 21 ☽ □ ☿
04 44 ☽ # ♃
05 32 ☿ Q ♆
07 21 ☽ ⚹ ♃
08 37 ☽ ☍ ♅
14 44 ☽ △ ♂
15 12 ☿ ∠ ♀
15 43 ☉ ☌ ☿
17 35 ☽ ⚺ ♄
17 55 ☽ # ♅
20 53 ☉ ∠ ♀

25 Friday
01 09 ☽ ⚻ ♆
05 40 ☽ ⊥ ♄
05 53 ☉ Q ♆
07 06 ☽ △ ♀
09 56 ☽ Q ☿
13 12 ☽ ± ♆
13 52 ☽ Q ☉
14 05 ☽ ∠ ♃
18 10 ☽ □ ♇
20 20 ☽ ⚼ ♂
23 38 ☽ ∠ ♄

26 Saturday
00 59 ♀ Q ♃
07 07 ☽ ⚼ ♆
07 26 ☽ ⚹ ☿
08 16 ☽ ⊥ ♃
13 51 ☽ ± ♂
15 48 ☽ ⚹ ☉
20 15 ☽ ⚺ ♃
20 30 ☽ ⚻ ♅

27 Sunday
01 24 ☽ ⚻ ♂
05 02 ☽ ⚹ ♄
05 39 ☽ Q ♇
08 09 ☽ ± ♅
10 00 ☽ ∠ ☿
11 56 ♃ △ ♅
12 23 ☽ △ ♆
21 48 ☽ □ ♀
23 18 ☽ ∠ ☉

28 Monday
01 22 ☽ ⚼ ♅
01 45 ☽ ⊥ ☿
02 55 ☽ ∥ ♅
03 57 ☽ Q ♄
04 38 ☽ ⚹ ♇
05 36 ☿ # ♄
12 10 ☽ ⚺ ☿
16 28 ☽ ∥ ♃
17 39 ☽ ⊥ ☉

29 Tuesday
01 17 ☉ □ ♅
02 16 ☽ # ♂
05 05 ☽ ∥ ♇
05 19 ☽ △ ♅
05 38 ☽ ⚺ ☉
05 56 ☽ ☌ ♃
08 05 ☽ ∥ ♆
08 32 ☽ ∠ ♇
09 10 ☽ ☍ ♂
09 19 ☽ Q ♀
10 12 ☉ ⚺ ♃
12 19 ☽ ∥ ☿
13 14 ☽ □ ♄
14 57 ☽ # ♄
20 14 ☽ □ ♆

30 Wednesday
00 49 ☽ ⊥ ♇
08 06 ☽ ⚹ ♀
09 59 ☽ ∥ ☉
11 31 ☽ ⚺ ♇
12 47 ♀ ∠ ♅
15 16 ☽ ☌ ☿
22 23 ☉ ⚻ ♂

All ephemeris data is given at 12.00 UT and the Moon's longitude is additionally given for 24.00 UT

DECEMBER 2005

LONGITUDES

Date	Sidereal time h m s	Sun ☉ ° ' "	Moon ☽ ° ' "	Moon ☽ 24.00 ° ' "	Mercury ☿ ° '	Venus ♀ ° '	Mars ♂ ° '	Jupiter ♃ ° '	Saturn ♄ ° '	Uranus ♅ ° '	Neptune ♆ ° '	Pluto ♇ ° '
01	16 41 47	09 ♐ 23 42	07 ♐ 46 07	14 ♐ 46 35	25 ♏ 22	22 ♑ 36	08 ♉ 45	07 ♏ 43	11 ♌ 14	06 ♓ 57	15 ♒ 10	23 ♐ 45
02	16 45 43	10 24 34	21 ♐ 51 10	28 ♐ 59 10	24 R 58	23 17	08 R 38	07 55	11 R 13	06 58	15 12	23 48
03	16 49 40	11 25 26	06 ♑ 09 50	13 ♑ 22 22	24 46	23 56	08 32	08 07	11 12	06 58	15 13	23 50
04	16 53 36	12 26 20	20 ♑ 35 58	27 ♑ 49 53	24 D 45	24 34	08 27	08 19	11 10	06 59	15 14	23 52
05	16 57 33	13 27 14	05 ♒ 03 24	12 ♒ 15 52	24 54	25 11	08 23	08 30	11 09	07 00	15 15	23 54
06	17 01 29	14 28 09	19 ♒ 26 47	26 ♒ 35 43	25 13	25 47	08 20	08 42	11 07	07 01	15 17	23 56
07	17 05 26	15 29 05	03 ♓ 42 20	10 ♓ 46 26	25 40	26 21	08 17	08 54	11 06	07 02	15 18	23 59
08	17 09 23	16 30 01	17 ♓ 47 53	24 ♓ 46 37	26 15	26 54	08 15	09 06	11 04	07 03	15 19	24 01
09	17 13 19	17 30 58	01 ♈ 42 40	08 ♈ 36 02	26 57	27 25	08 14	09 17	11 02	07 05	15 21	24 03
10	17 17 16	18 31 56	15 ♈ 26 47	22 ♈ 14 58	27 45	27 55	08 D 14	09 29	11 00	07 06	15 22	24 05
11	17 21 12	19 32 54	29 ♈ 00 38	05 ♉ 43 49	28 38	28 23	08 15	09 40	10 58	07 07	15 24	24 08
12	17 25 09	20 33 53	12 ♉ 24 29	19 ♉ 02 37	29 ♏ 36	28 49	08 16	09 52	10 56	07 08	15 25	24 10
13	17 29 05	21 34 53	25 ♉ 38 08	02 ♊ 10 58	00 ♐ 39	29 13	08 19	10 03	10 54	07 10	15 27	24 12
14	17 33 02	22 35 53	08 ♊ 40 59	15 ♊ 08 06	01 44	29 36	08 22	10 14	10 51	07 11	15 28	24 14
15	17 36 58	23 36 54	21 ♊ 32 10	27 ♊ 53 07	02 53	29 ♑ 57	08 25	10 26	10 49	07 13	15 30	24 17
16	17 40 55	24 37 56	04 ♋ 10 53	10 ♋ 25 26	04 05	00 ♒ 16	08 30	10 37	10 46	07 14	15 31	24 19
17	17 44 52	25 38 58	16 ♋ 36 48	22 ♋ 45 03	05 19	00 32	08 35	10 48	10 44	07 16	15 33	24 21
18	17 48 48	26 40 01	28 ♋ 50 20	04 ♌ 52 52	06 35	00 47	08 41	10 59	10 41	07 17	15 34	24 23
19	17 52 45	27 41 05	10 ♌ 52 54	16 ♌ 50 48	07 53	01 00	08 47	11 10	10 38	07 19	15 36	24 26
20	17 56 41	28 42 10	22 ♌ 46 58	28 ♌ 41 52	09 13	01 10	08 55	11 20	10 35	07 21	15 38	24 28
21	18 00 38	29 ♐ 43 15	04 ♍ 36 01	10 ♍ 30 00	10 34	01 18	09 03	11 31	10 32	07 22	15 39	24 30
22	18 04 34	00 ♑ 44 21	16 ♍ 24 25	22 ♍ 19 56	11 57	01 24	09 11	11 42	10 29	07 24	15 41	24 32
23	18 08 31	01 45 27	28 ♍ 17 14	04 ♎ 17 01	13 20	01 27	09 21	11 52	10 26	07 26	15 43	24 35
24	18 12 27	02 46 34	10 ♎ 19 58	16 ♎ 26 48	14 44	01 R 28	09 31	12 03	10 23	07 28	15 45	24 37
25	18 16 24	03 47 42	22 ♎ 38 11	28 ♎ 54 46	16 09	01 27	09 41	12 13	10 19	07 30	15 46	24 39
26	18 20 21	04 48 51	05 ♏ 17 06	11 ♏ 45 42	17 35	01 23	09 52	12 23	10 16	07 32	15 48	24 41
27	18 24 17	05 50 00	18 ♏ 20 58	25 ♏ 03 08	19 02	01 16	10 04	12 33	10 12	07 34	15 50	24 43
28	18 28 14	06 51 10	01 ♐ 52 20	08 ♐ 48 29	20 29	01 07	10 17	12 44	10 09	07 36	15 52	24 46
29	18 32 10	07 52 20	15 ♐ 51 40	23 ♐ 00 26	21 57	00 56	10 30	12 54	10 05	07 38	15 54	24 48
30	18 36 07	08 53 31	00 ♑ 15 07	07 ♑ 34 33	23 26	00 42	10 43	13 03	10 01	07 40	15 56	24 50
31	18 40 03	09 ♑ 54 41	14 ♑ 57 46	22 ♑ 23 39	24 ♐ 55	00 ♒ 26	10 ♉ 57	13 ♏ 13	09 ♌ 57	07 ♓ 42	15 ♒ 58	24 ♐ 52

Moon nodes, latitude and DECLINATIONS

Date	Moon True ☊ ° '	Moon Mean ☊ ° '	Moon ☽ Latitude ° '	Sun ☉ ° '	Moon ☽ ° '	Mercury ☿ ° '	Venus ♀ ° '	Mars ♂ ° '	Jupiter ♃ ° '	Saturn ♄ ° '	Uranus ♅ ° '	Neptune ♆ ° '	Pluto ♇ ° '
01	12 ♈ 06	10 ♈ 37	04 S 12	21 S 52	25 S 45	16 S 44	24 S 25	15 N 08	13 S 05	17 N 56	09 S 41	16 S 26	15 S 47
02	11 R 56	10 34	04 46	22 01	27 56	16 32	24 14	15 08	13 09	17 57	09 40	16 26	15 48
03	11 46	10 30	05 02	22 09	28 20	16 25	24 02	15 09	13 13	17 57	09 40	16 25	15 48
04	11 36	10 27	05 00	22 17	26 48	16 22	23 49	15 09	13 17	17 58	09 40	16 25	15 48
05	11 29	10 24	04 39	22 25	23 30	16 24	23 37	15 10	13 20	17 58	09 39	16 25	15 48
06	11 24	10 21	04 00	22 32	18 47	16 28	23 24	15 11	13 24	17 59	09 39	16 25	15 49
07	11 21	10 18	03 06	22 39	13 02	16 36	23 11	15 12	13 28	17 59	09 38	16 24	15 49
08	11 D 21	10 14	02 02	22 45	06 42	16 47	22 58	15 13	13 31	18 00	09 38	16 23	15 49
09	11 21	10 11	00 S 51	22 51	00 S 06	17 00	22 44	15 15	13 35	18 01	09 37	16 23	15 49
10	11 R 21	10 08	00 N 22	22 57	06 N 25	17 15	22 31	15 16	13 39	18 01	09 37	16 23	15 49
11	11 20	10 05	01 32	23 02	12 34	17 32	22 17	15 18	13 42	18 02	09 36	16 22	15 50
12	11 16	10 02	02 37	23 06	18 03	17 50	22 04	15 21	13 46	18 03	09 36	16 22	15 50
13	11 09	09 59	03 32	23 10	22 36	18 09	21 50	15 23	13 49	18 03	09 35	16 21	15 50
14	11 00	09 55	04 15	23 14	25 57	18 29	21 36	15 25	13 53	18 04	09 35	16 21	15 50
15	10 48	09 52	04 44	23 17	27 54	18 50	21 22	15 28	13 56	18 05	09 34	16 20	15 51
16	10 34	09 49	04 59	23 20	28 21	19 10	21 09	15 31	13 59	18 06	09 34	16 20	15 51
17	10 21	09 46	04 59	23 22	27 21	19 31	20 55	15 34	14 03	18 07	09 33	16 19	15 51
18	10 08	09 43	04 45	23 24	25 02	19 51	20 41	15 37	14 06	18 08	09 32	16 19	15 51
19	09 58	09 39	04 19	23 25	21 39	20 12	20 27	15 41	14 09	18 09	09 32	16 19	15 51
20	09 50	09 36	03 41	23 26	17 24	20 32	20 14	15 44	14 13	18 10	09 31	16 18	15 51
21	09 44	09 33	02 54	23 26	12 32	20 52	20 00	15 48	14 16	18 11	09 30	16 18	15 52
22	09 42	09 30	02 00	23 26	07 12	21 11	19 47	15 52	14 19	18 12	09 30	16 17	15 52
23	09 D 41	09 27	01 N 00	23 26	01 N 36	21 29	19 34	15 56	14 22	18 13	09 29	16 17	15 52
24	09 R 41	09 24	00 S 03	23 25	04 S 09	21 47	19 21	16 01	14 25	18 14	09 28	16 16	15 52
25	09 41	09 20	01 08	23 24	09 52	22 04	19 08	16 05	14 28	18 15	09 27	16 15	15 52
26	09 39	09 17	02 11	23 21	15 20	22 20	18 55	16 10	14 31	18 16	09 27	16 15	15 52
27	09 34	09 14	03 09	23 19	20 19	22 36	18 42	16 14	14 34	18 17	09 26	16 14	15 52
28	09 27	09 11	03 59	23 16	24 26	22 50	18 30	16 19	14 37	18 18	09 25	16 14	15 53
29	09 17	09 08	04 36	23 12	27 15	23 03	18 18	16 24	14 40	18 19	09 24	16 13	15 53
30	09 05	09 05	04 56	23 09	28 23	23 15	18 06	16 29	14 43	18 20	09 24	16 13	15 53
31	08 ♈ 53	09 ♈ 01	04 S 58	23 S 04	27 S 32	23 S 27	17 S 55	16 N 35	14 S 46	18 N 21	09 S 23	16 S 12	15 S 53

ZODIAC SIGN ENTRIES

Date	h	m	Planets
03	01	42	☽ ♑
05	03	36	☽ ♒
07	05	44	☽ ♓
09	09	02	☽ ♈
11	13	46	☽ ♉
12	21	19	☿ ♐
13	19	59	☽ ♊
15	15	57	♀ ♒
16	04	01	☽ ♋
18	14	18	☽ ♌
21	02	39	☽ ♍
21	18	35	☉ ♑
23	15	26	☽ ♎
26	02	04	☽ ♏
28	08	44	☽ ♐
30	11	35	☽ ♑

LATITUDES

Date	Mercury ☿ ° '	Venus ♀ ° '	Mars ♂ ° '	Jupiter ♃ ° '	Saturn ♄ ° '	Uranus ♅ ° '	Neptune ♆ ° '	Pluto ♇ ° '
01	02 N 26	02 S 55	00 N 46	01 N 03	00 N 33	00 S 46	00 S 09	07 N 31
04	02 40	02 39	00 53	01 03	00 33	00 46	00 09	07 30
07	02 39	02 20	00 59	01 04	00 34	00 46	00 09	07 30
10	02 28	01 58	01 05	01 04	00 34	00 46	00 09	07 29
13	02 11	01 33	01 10	01 04	00 35	00 46	00 09	07 29
16	01 50	01 04	01 15	01 04	00 35	00 46	00 09	07 29
19	01 27	00 S 32	01 19	01 04	00 36	00 45	00 09	07 29
22	01 03	00 N 04	01 23	01 05	00 36	00 45	00 09	07 28
25	00 39	00 44	01 26	01 05	00 37	00 45	00 09	07 28
28	00 N 16	01 27	01 30	01 05	00 37	00 45	00 09	07 28
31	00 S 06	02 N 12	01 N 32	01 N 06	00 N 38	00 S 45	00 S 09	07 N 28

DATA

Julian Date	2453706
Delta T	+66 seconds
Ayanamsa	23° 56' 19"
Synetic vernal point	05° ♓ 10' 40"
True obliquity of ecliptic	23° 26' 27"

MOON'S PHASES, APSIDES AND POSITIONS ☽

Date	h	m	Phase	Longitude ° '	Eclipse Indicator
01	15	01	●	09 ♐ 31	
08	09	36	☽	16 ♓ 24	
15	16	16	○	23 ♊ 48	
23	19	36	☾	02 ♎ 05	
31	03	12	●	09 ♑ 32	

Day	h	m		
05	04	26	Perigee	
21	02	43	Apogee	
03	04	54	Max dec	28° S 25'
09	12	22	ON	
16	07	18	Max dec	28° N 23'
23	18	43	OS	
30	13	59	Max dec	28° S 23'

LONGITUDES

Date	Chiron ⚷ ° '	Ceres ⚳ ° '	Pallas ⚴ ° '	Juno ⚵ ° '	Vesta ⚶ ° '	Black Moon Lilith ⚸ ° '
01	29 ♑ 42	25 ♐ 56	01 ♐ 42	19 ♊ 09	22 ♋ 37	24 ♌ 13
11	00 ♒ 22	29 ♐ 58	06 ♐ 16	16 ♊ 51	21 ♋ 19	25 ♌ 20
21	01 ♒ 06	04 ♑ 02	10 ♐ 20	14 ♊ 43	19 ♋ 21	26 ♌ 26
31	01 ♒ 53	08 ♑ 05	14 ♐ 33	13 ♊ 06	16 ♋ 55	27 ♌ 33

All ephemeris data is given at 12.00 UT and the Moon's longitude is additionally given for 24.00 UT

ASPECTARIAN

h m Aspects

01 Thursday
02 53 ☽ ∥ ♀
04 03 ☽ Q ♆
10 35 ☽ □ ♅
10 58 ♂ ⯓ ♇
11 42 ☽ ∠ ♀
11 54 ☽ ⊻ ♃
13 40 ☽ ⚻ ♂
15 01 ☽ ☌ ☉
17 56 ☽ △ ♄
22 20 ☽ ⊥ ♃
23 51 ☽ ± ♂

02 Friday
00 41 ☽ ⚹ ♆
03 53 ☽ ⊥ ♀
13 49 ☽ ∠ ♃
14 32 ☽ ⊻ ♀
14 59 ☽ ⯓ ♂
15 17 ☽ ☌ ♇
17 09 ☽ ⊻ ☿
17 14 ☽ Q ♅
19 20 ☽ ⯓ ♄

03 Saturday
02 02 ☽ ∠ ♆
03 05 ☽ ⊥ ☿
06 41 ☉ △ ♄
07 44 ♀ ⊻ ♇
10 23 ☽ ± ♄
11 42 ☿ ∥ ♆
13 21 ☽ ⚹ ♅
15 17 ☽ ⚹ ♃
15 56 ☽ △ ♂
17 05 ☽ ⊥ ♆
17 58 ☽ ∠ ☿
20 22 ☽ ⚻ ♄
21 25 ☽ ⊻ ☉

04 Sunday
02 25 ☿ St D
03 05 ☽ ⊻ ♆
08 09 ☽ ⊥ ☉
11 31 ☽ Q ♃
14 18 ☽ ∠ ♅
17 26 ☽ ⊻ ♇
18 54 ☽ ☌ ♀
18 56 ☽ ⚹ ☿
20 06 ☿ ⚹ ♀

05 Monday
00 12 ☽ ∠ ☉
00 44 ♂ ☍ ♃
03 25 ☽ ⊥ ♇
05 16 ☽ ⊥ ♅
11 22 ☽ ∥ ♀
15 08 ☽ Q ☿
15 15 ☽ ⊻ ♅
17 31 ☽ □ ♂
17 49 ☽ □ ♃
17 59 ☽ ∥ ☉
18 23 ☿ ∥ ♆
18 25 ☽ ∠ ♇
22 07 ☽ ☍ ♄

06 Tuesday
03 03 ☽ ⚹ ☉
05 01 ☽ ☌ ♆
08 47 ☉ ± ♂
15 35 ☽ # ♄
18 53 ☉ ⊥ ♃
19 33 ☽ ⚹ ♇
21 52 ☽ ∥ ☿
21 58 ☽ □ ☿
22 22 ☽ ∥ ♆
23 05 ☽ ⊻ ♀
23 31 ☽ Q ♂

07 Wednesday
00 42 ☽ Q ☉
00 50 ☽ ∥ ♇
03 23 ☽ # ♂
07 30 ☉ ⚹ ♆
09 37 ☽ ⊥ ♀
10 22 ☽ ∥ ♃
15 52 ☽ Q ♇
17 40 ☽ ☌ ♅
19 45 ☽ ⚹ ♂
20 56 ☽ △ ♃
23 48 ♃ ∠ ♇

08 Thursday
00 32 ☽ ⚻ ♄
01 05 ☽ ∥ ♅
01 31 ☽ ∠ ♀
07 45 ☽ ⊻ ♆
09 36 ☽ □ ☉
10 45 ☽ ± ♄
18 04 ☽ ⊥ ♆
21 22 ☽ ∠ ♂
22 43 ☽ □ ♇
22 59 ☽ ⯓ ♃

09 Friday
02 12 ☽ ⯓ ♄
03 16 ☉ ∥ ♀
03 17 ☽ △ ☿
04 17 ☽ ⚹ ♀
09 37 ☽ ∠ ♆
12 55 ☽ ⊥ ♂
14 47 ☽ ± ♃
21 21 ☽ ⊻ ♅
23 22 ☽ ⊻ ♂

10 Saturday
01 23 ☽ ⚻ ♃
01 56 ☽ Q ♀
04 04 ♂ St D
04 13 ☽ △ ♄
06 57 ☽ ⯓ ☿
07 52 ☽ ⊥ ♅
11 52 ☽ ⚹ ♆
17 52 ☽ △ ☉
22 00 ☿ ⚹ ♀
23 45 ☽ ∠ ♅
23 52 ☽ ± ☿

11 Sunday
00 13 ☽ # ♅
03 18 ☽ △ ♇
09 07 ☽ Q ♆
10 50 ☽ □ ♀
11 17 ☽ ⚻ ☿
16 46 ☽ # ♃
22 42 ☽ ⯓ ☉
23 38 ☽ ∥ ♂

12 Monday
01 52 ☽ # ♇
02 31 ☽ ⚹ ♅
04 14 ☽ # ♆
04 33 ☽ ☌ ♂
06 09 ☽ ⯓ ♇
07 21 ☽ ☍ ♃
09 21 ☽ □ ♄
10 56 ☽ # ☿
11 59 ☽ ∥ ♄
16 13 ☽ ± ☉
17 27 ☽ □ ♆
22 26 ☽ ± ♇

13 Tuesday
00 12 ☽ Q ♅
04 00 ☽ ⚻ ☉
04 22 ☿ # ♄
07 45 ☽ # ♀
09 22 ☽ ⚻ ♇
15 36 ☽ # ☉
17 57 ☽ Q ♄
18 46 ☽ △ ♀
22 00 ☽ ☍ ☿

14 Wednesday
09 14 ☽ □ ♅
11 24 ☽ ⊻ ♂
14 56 ☽ ⚻ ♃
16 01 ☽ ⚹ ♄
22 36 ☽ ⊥ ♂
23 19 ☽ ⯓ ♀

15 Thursday
00 39 ☽ △ ♆
02 16 ☽ ± ♃
07 05 ☉ ⯓ ♂
15 34 ☽ ∠ ♂
16 16 ☽ ☍ ☉
16 40 ☽ ± ♀
17 11 ☽ ☍ ♇
19 27 ☽ ⯓ ♃
20 03 ☽ ∠ ♄
23 26 ☉ ⊥ ♀

16 Friday
00 31 ☿ Q ♆
04 12 ☉ ☌ ♇
04 20 ☽ ⚻ ♀
05 00 ☽ ⯓ ♆
11 48 ☽ ⚻ ☿
13 08 ☽ ⊥ ♄
17 07 ♀ ⊥ ♇
17 52 ☽ △ ♅
20 20 ☽ ⚹ ♂
22 17 ☽ ± ♆

17 Saturday
00 33 ☽ △ ♃
00 35 ☽ ± ☿
00 38 ☽ ⊻ ♄
02 38 ☉ Q ♅
05 16 ♃ □ ♄
09 55 ☽ ⚻ ♆
13 50 ☉ ⯓ ♄
16 10 ☉ ∠ ♃
19 49 ☽ Q ♂
20 04 ☽ ⯓ ☿
23 04 ☽ ⯓ ♅

18 Sunday
03 12 ☽ ⚻ ♇
07 19 ☽ ⚻ ☉
15 05 ☽ ± ♇
15 55 ☽ ☍ ♀
16 52 ☽ ± ♅
20 18 ☽ ± ☉

19 Monday
00 22 ☽ # ☉
01 14 ☿ □ ♅
04 51 ☽ ⚻ ♅
05 17 ☽ △ ☿
07 46 ☽ □ ♂
09 04 ☽ ⯓ ♇
11 31 ☽ ☌ ♄
12 34 ☽ □ ♃
15 57 ☽ ⯓ ☉
19 35 ☽ # ♀
20 01 ☽ # ☿
21 31 ☽ ☍ ♆
22 58 ☿ ∥ ♀

20 Tuesday
05 56 ☿ ⚻ ♂
08 00 ☽ ∥ ♄
15 25 ☽ △ ♇
17 41 ☽ # ♆
19 54 ☽ # ♇
20 23 ☽ ∥ ♂

21 Wednesday
01 09 ☽ △ ☉
01 30 ☽ Q ♃
03 50 ☽ # ♃
05 13 ☽ ⚻ ♀
11 27 ☿ △ ♄
17 32 ☽ ± ♀
17 39 ☽ ☍ ♅
21 09 ☽ △ ♂

22 Thursday
00 02 ☽ ⊻ ♄
01 44 ☽ □ ☿
01 51 ☽ # ♅
02 17 ☽ ⚹ ♃
07 04 ☿ ⊻ ♃
09 27 ♂ # ♇
10 32 ☽ ⚻ ♆
10 43 ☉ ∠ ♆
11 58 ☽ ⯓ ♀
12 10 ☽ ⊥ ♄
17 09 ♀ ⊥ ♅
22 43 ☽ ± ♆

23 Friday
03 57 ☽ ⯓ ♂
04 27 ☉ ⊻ ♀
04 30 ☽ □ ♇
06 17 ☽ ∠ ♄
09 07 ☽ ∠ ♃
16 53 ☽ ⯓ ♆
18 22 ☽ △ ♀
18 55 ☽ Q ☿
19 36 ☽ □ ☉
22 16 ☽ ± ♂

24 Saturday
03 23 ☽ ⊥ ♃
06 19 ☽ ⚻ ♅
09 36 ♀ St R
10 21 ☽ ⚻ ♂
12 05 ☽ ⚹ ♄
13 20 ♀ ⊥ ♅
15 25 ☽ ⊻ ♃
16 30 ☽ Q ♇
18 11 ☽ ± ♅
21 47 ☽ ⚹ ☿
22 39 ☽ △ ♆

25 Sunday
02 56 ☿ ± ♂
05 25 ☿ ⚹ ♆
05 46 ♂ ⯓ ♇
10 14 ☽ Q ☉
10 18 ☽ ∥ ♅
11 24 ☽ Q ♄
11 44 ☽ ⯓ ♅
15 53 ☽ ⚹ ♇
16 37 ☿ ∠ ♀
23 44 ☉ ± ♄

26 Monday
04 42 ☽ □ ♀
06 18 ☽ ∠ ☿
08 16 ☽ ∥ ♃
11 03 ☽ ⚹ ☉
14 25 ☽ ∥ ♇
15 48 ☽ # ♂
16 09 ☽ ∥ ♆
16 11 ☽ △ ♅
20 12 ☽ ∠ ♇
20 39 ☽ ☍ ♂
21 12 ☽ □ ♄

27 Tuesday
01 08 ☽ ⊥ ☿
01 19 ☽ ☌ ♃
01 44 ☽ # ♄
03 02 ☿ ⊥ ♃
04 11 ☽ ∥ ♀
07 26 ☽ □ ♆
12 41 ☽ ⊥ ♇
12 41 ♂ # ♆
13 23 ☽ ⊻ ☿
13 39 ☽ Q ♀
16 50 ☽ ∠ ☉
23 27 ☽ ⊻ ♇

28 Wednesday
00 12 ♂ □ ♄
01 16 ☽ ∥ ☿
04 36 ☽ ∥ ☉
10 05 ☽ ⊥ ☉
10 42 ☽ ⚹ ♀
15 29 ☽ Q ♆
21 19 ☽ ⊻ ☉
21 57 ☽ □ ♅

29 Thursday
02 14 ☽ △ ♄
02 44 ☽ ⚻ ♂
06 07 ☉ ⚹ ♅
06 55 ☽ ⊻ ♃
10 11 ♀ # ♄
12 04 ☽ ⚹ ♆
12 08 ☽ ∠ ♀
13 06 ☽ ± ♂
17 11 ☽ ⊥ ♃
23 25 ☽ ☌ ☿

30 Friday
00 27 ♀ ⊥ ♇
01 37 ☉ ∥ ☿
02 58 ☽ ⊥ ♀
03 01 ☽ ☌ ♇
03 23 ☽ ⯓ ♄
04 23 ☽ ⯓ ♂
04 24 ☽ Q ♅
08 20 ☽ ∠ ♃
12 44 ☽ ⊻ ♀
13 07 ☽ ∠ ♆
18 10 ☽ ± ♄

31 Saturday
00 11 ☽ ⚹ ♅
03 12 ☽ ☌ ☉
03 52 ☽ ⊥ ♆
03 55 ☽ ⚻ ♄
05 24 ☽ △ ♂
05 26 ☿ ⊥ ♀
09 09 ☽ ⚹ ♃
11 18 ☿ ☌ ♇
12 39 ☿ ⯓ ♄
12 58 ☉ ⚻ ♄
13 13 ☉ ⊥ ♆
13 37 ☽ ⊻ ♆

JANUARY 2006

Date	Sidereal time h m s	Sun ☉ ° ' "	Moon ☽ ° ' "	Moon ☽ 24.00 ° ' "	Mercury ☿ ° '	Venus ♀ ° '	Mars ♂ ° '	Jupiter ♃ ° '	Saturn ♄ ° '	Uranus ♅ ° '	Neptune ♆ ° '	Pluto ♇ ° '
01	18 44 00	10 ♑ 55 52	29 ♑ 51 03	07 ♒ 18 48	26 ♐ 24	00 ♒ 07	11 ♉ 12	13 ♏ 23	09 ♌ 53	07 ♓ 44	16 ♒ 00	24 ♐ 54
02	18 47 56	11 57 03	14 ♒ 45 47	22 ♒ 10 56	27 54	29 ♑ 46	11 27	13 32	09 R 49	07 47	16 02	24 57
03	18 51 53	12 58 13	29 ♒ 33 23	06 ♓ 52 21	29 ♐ 24	29 R 23	11 43	13 42	09 45	07 49	16 04	24 59
04	18 55 50	13 59 23	14 ♓ 07 17	21 ♓ 17 46	00 ♑ 55	28 57	11 59	13 51	09 41	07 51	16 06	25 01
05	18 59 46	15 00 33	28 ♓ 23 34	05 ♈ 24 34	02 26	28 30	12 16	14 01	09 37	07 54	16 08	25 03
06	19 03 43	16 01 43	12 ♈ 20 47	19 ♈ 12 20	03 58	28 00	12 33	14 10	09 33	07 56	16 10	25 05
07	19 07 39	17 02 52	25 ♈ 59 25	02 ♉ 42 16	05 30	27 29	12 51	14 19	09 28	07 58	16 12	25 07
08	19 11 36	18 04 00	09 ♉ 21 08	15 ♉ 56 16	07 02	26 57	13 09	14 28	09 24	08 01	16 14	25 09
09	19 15 32	19 05 08	22 ♉ 27 57	28 ♉ 56 24	08 35	26 23	13 28	14 36	09 20	08 03	16 16	25 12
10	19 19 29	20 06 16	05 ♊ 21 49	11 ♊ 44 24	10 08	25 48	13 47	14 45	09 15	08 06	16 18	25 14
11	19 23 25	21 07 24	18 ♊ 04 17	24 ♊ 21 34	11 42	25 12	14 06	14 54	09 11	08 08	16 20	25 16
12	19 27 22	22 08 30	00 ♋ 36 22	06 ♋ 48 44	13 16	24 35	14 26	15 02	09 06	08 11	16 22	25 18
13	19 31 19	23 09 37	12 ♋ 58 45	19 ♋ 06 26	14 50	23 59	14 46	15 10	09 02	08 14	16 24	25 20
14	19 35 15	24 10 43	25 ♋ 11 54	01 ♌ 15 12	16 25	23 22	15 07	15 19	08 57	08 16	16 26	25 22
15	19 39 12	25 11 48	07 ♌ 16 28	13 ♌ 15 49	18 01	22 45	15 28	15 27	08 52	08 19	16 28	25 24
16	19 43 08	26 12 53	19 ♌ 13 28	25 ♌ 09 39	19 37	22 08	15 49	15 35	08 47	08 22	16 31	25 26
17	19 47 05	27 13 58	01 ♍ 04 37	06 ♍ 58 44	21 13	21 33	16 11	15 42	08 43	08 25	16 33	25 28
18	19 51 01	28 15 02	12 ♍ 52 23	18 ♍ 46 00	22 50	20 58	16 33	15 50	08 38	08 27	16 35	25 30
19	19 54 58	29 ♑ 16 06	24 ♍ 40 05	00 ♎ 35 10	24 28	20 24	16 56	15 58	08 33	08 30	16 37	25 32
20	19 58 54	00 ♒ 17 10	06 ♎ 31 50	12 ♎ 30 44	26 06	19 52	17 19	16 05	08 28	08 33	16 39	25 34
21	20 02 51	01 18 13	18 ♎ 32 28	24 ♎ 37 45	27 45	19 22	17 42	16 12	08 23	08 36	16 42	25 36
22	20 06 48	02 19 15	00 ♏ 47 14	07 ♏ 01 35	29 ♑ 24	18 53	18 05	16 19	08 19	08 39	16 44	25 38
23	20 10 44	03 20 18	13 ♏ 21 28	19 ♏ 47 29	01 ♒ 04	18 26	18 29	16 26	08 14	08 42	16 46	25 39
24	20 14 41	04 21 20	26 ♏ 20 08	02 ♐ 59 53	02 44	18 01	18 53	16 33	08 09	08 45	16 48	25 41
25	20 18 37	05 22 21	09 ♐ 47 02	16 ♐ 41 43	04 25	17 38	19 18	16 40	08 04	08 48	16 50	25 43
26	20 22 34	06 23 22	23 ♐ 43 55	00 ♑ 53 25	06 07	17 18	19 42	16 46	07 59	08 51	16 53	25 45
27	20 26 30	07 24 22	08 ♑ 09 45	15 ♑ 32 14	07 49	17 00	20 07	16 53	07 54	08 54	16 55	25 47
28	20 30 27	08 25 22	22 ♑ 59 58	00 ♒ 31 54	09 32	16 44	20 33	16 59	07 49	08 57	16 57	25 49
29	20 34 23	09 26 21	08 ♒ 06 47	15 ♒ 43 17	11 16	16 31	20 58	17 05	07 44	09 00	17 00	25 50
30	20 38 20	10 27 18	23 ♒ 20 03	00 ♓ 55 44	13 00	16 20	21 24	17 11	07 39	09 03	17 02	25 52
31	20 42 17	11 ♒ 28 15	08 ♓ 29 07	15 ♓ 59 04	14 ♒ 45	16 ♑ 11	21 ♉ 50	17 ♏ 17	07 ♌ 34	09 ♓ 06	17 ♒ 04	25 ♐ 54

Date	Moon True ☊ ° '	Moon Mean ☊ ° '	Moon ☽ Latitude ° '
01	08 ♈ 41	08 ♈ 58	04 S 40
02	08 R 31	08 55	04 03
03	08 24	08 52	03 09
04	08 20	08 49	02 04
05	08 19	08 45	00 S 52
06	08 D 19	08 42	00 N 21
07	08 R 19	08 39	01 32
08	08 17	08 36	02 36
09	08 13	08 33	03 31
10	08 06	08 30	04 13
11	07 56	08 26	04 43
12	07 44	08 23	04 58
13	07 31	08 20	04 59
14	07 17	08 17	04 46
15	07 04	08 14	04 20
16	06 52	08 11	03 43
17	06 44	08 07	02 57
18	06 38	08 04	02 03
19	06 36	08 01	01 N 03
20	06 35	07 58	00 00
21	06 D 35	07 55	01 S 03
22	06 36	07 51	02 05
23	06 R 35	07 48	03 03
24	06 33	07 45	03 53
25	06 28	07 42	04 33
26	06 21	07 39	04 57
27	06 12	07 36	05 04
28	06 03	07 32	04 52
29	05 54	07 29	04 18
30	05 46	07 26	03 27
31	05 ♈ 41	07 ♈ 23	02 S 20

DECLINATIONS

Date	Sun ☉ ° '	Moon ☽ ° '	Mercury ☿ ° '	Venus ♀ ° '	Mars ♂ ° '	Jupiter ♃ ° '	Saturn ♄ ° '	Uranus ♅ ° '	Neptune ♆ ° '	Pluto ♇ ° '
01	22 S 59	24 S 45	23 S 37	17 S 44	16 N 40	14 S 49	18 N 23	09 S 22	16 S 12	15 S 53
02	22 54	20 16	23 46	17 33	16 45	14 51	18 24	09 21	16 11	15 53
03	22 49	14 35	23 54	17 22	16 51	14 54	18 25	09 20	16 10	15 53
04	22 42	08 09	24 00	17 12	16 57	14 57	18 26	09 19	16 10	15 53
05	22 36	01 S 26	24 06	17 02	17 03	14 59	18 28	09 18	16 09	15 53
06	22 29	05 N 12	24 10	16 53	17 09	15 02	18 29	09 17	16 09	15 54
07	22 21	11 28	24 13	16 44	17 15	15 05	18 30	09 16	16 08	15 54
08	22 13	17 05	24 14	16 35	17 21	15 07	18 31	09 15	16 07	15 54
09	22 05	21 47	24 15	16 27	17 27	15 09	18 33	09 15	16 07	15 54
10	21 56	25 21	24 14	16 20	17 33	15 12	18 34	09 14	16 06	15 54
11	21 47	27 36	24 11	16 12	17 40	15 14	18 35	09 13	16 06	15 54
12	21 37	28 25	24 08	16 05	17 46	15 17	18 37	09 12	16 05	15 54
13	21 27	27 46	24 03	15 59	17 53	15 19	18 38	09 11	16 04	15 54
14	21 17	25 47	23 56	15 53	17 59	15 21	18 39	09 10	16 04	15 54
15	21 06	22 39	23 48	15 48	18 06	15 23	18 41	09 08	16 03	15 54
16	20 55	18 36	23 39	15 43	18 13	15 26	18 42	09 07	16 02	15 54
17	20 43	13 51	23 28	15 38	18 19	15 28	18 43	09 06	16 02	15 54
18	20 31	08 37	23 16	15 34	18 26	15 30	18 45	09 05	16 01	15 54
19	20 18	03 N 05	23 02	15 31	18 33	15 32	18 46	09 04	16 01	15 54
20	20 05	02 S 35	22 47	15 28	18 40	15 34	18 48	09 03	16 00	15 54
21	19 52	08 15	22 31	15 25	18 47	15 36	18 49	09 02	15 59	15 54
22	19 39	13 42	22 13	15 23	18 54	15 38	18 50	09 01	15 59	15 54
23	19 25	18 46	21 53	15 22	19 01	15 39	18 52	09 00	15 58	15 54
24	19 10	23 07	21 32	15 21	19 08	15 41	18 53	08 59	15 57	15 54
25	18 56	26 25	21 09	15 20	19 15	15 43	18 54	08 58	15 57	15 54
26	18 41	28 15	20 45	15 20	19 22	15 45	18 56	08 56	15 56	15 54
27	18 25	28 15	20 20	15 20	19 29	15 46	18 57	08 55	15 55	15 54
28	18 10	26 16	19 52	15 20	19 36	15 48	18 59	08 54	15 55	15 54
29	17 54	22 24	19 24	15 21	19 43	15 49	19 00	08 53	15 54	15 54
30	17 37	16 59	18 54	15 22	19 50	15 51	19 01	08 52	15 53	15 54
31	17 S 21	10 S 33	18 S 22	15 S 23	19 N 57	15 S 52	19 N 03	08 S 50	15 S 53	15 S 54

ZODIAC SIGN ENTRIES

Date	h	m	Planets
01	12	14	☽ ♒
01	20	18	♀ ♑
03	12	43	☽ ♓
03	21	26	☿ ♑
05	14	44	☽ ♈
07	19	09	☽ ♉
10	01	58	☽ ♊
12	10	50	☽ ♋
14	21	31	☽ ♌
17	09	49	☽ ♍
19	22	49	☽ ♎
20	05	15	☉ ♒
22	10	28	☽ ♏
22	20	41	☿ ♒
24	18	38	☽ ♐
26	22	31	☽ ♑
28	23	09	☽ ♒
30	22	32	☽ ♓

LATITUDES

Date	Mercury ☿ ° '	Venus ♀ ° '	Mars ♂ ° '	Jupiter ♃ ° '	Saturn ♄ ° '	Uranus ♅ ° '	Neptune ♆ ° '	Pluto ♇ ° '
01	00 S 13	02 N 27	01 N 33	01 N 06	00 N 38	00 S 45	00 S 09	07 N 28
04	00 34	03 14	01 36	01 06	00 38	00 45	00 09	07 28
07	00 53	04 00	01 38	01 07	00 39	00 45	00 09	07 28
10	01 11	04 44	01 39	01 07	00 39	00 45	00 09	07 28
13	01 26	05 25	01 41	01 07	00 39	00 45	00 10	07 28
16	01 40	05 59	01 42	01 08	00 40	00 45	00 10	07 28
19	01 50	06 27	01 44	01 08	00 40	00 45	00 10	07 28
22	01 59	06 47	01 45	01 09	00 40	00 44	00 10	07 28
25	02 04	07 00	01 45	01 09	00 41	00 44	00 10	07 28
28	02 05	07 07	01 46	01 09	00 41	00 44	00 10	07 28
31	02 S 03	07 N 07	01 N 47	01 N 10	00 N 41	00 S 44	00 S 10	07 N 29

DATA

Julian Date	2453737
Delta T	+66 seconds
Ayanamsa	23° 56' 25"
Synetic vernal point	05° ♓ 10' 34"
True obliquity of ecliptic	23° 26' 27"

LONGITUDES

Date	Chiron ⚷ ° '	Ceres ⚳ ° '	Pallas ⚴ ° '	Juno ⚵ ° '	Vesta ⚶ ° '	Black Moon Lilith ⚸ ° '
01	01 ♒ 58	08 ♑ 29	14 ♐ 58	12 ♊ 59	16 ♋ 39	27 ♌ 40
11	02 ♒ 47	12 ♑ 31	19 ♐ 06	12 ♊ 16	14 ♋ 01	28 ♌ 47
21	03 ♒ 37	16 ♑ 31	23 ♐ 07	12 ♊ 27	11 ♋ 32	29 ♌ 53
31	04 ♒ 27	20 ♑ 28	27 ♐ 01	13 ♊ 29	09 ♋ 31	01 ♍ 00

MOON'S PHASES, APSIDES AND POSITIONS ☽

Date	h	m	Phase	Longitude ° '	Eclipse Indicator
06	18	56	☽	16 ♈ 19	
14	09	48	○	24 ♋ 05	
22	15	14	☾	02 ♏ 27	
29	14	15	●	09 ♒ 32	

Day	h	m		
01	22	43	Perigee	
17	18	55	Apogee	
30	07	45	Perigee	
05	17	08	0N	
12	13	15	Max dec	28° N 25'
20	01	05	0S	
27	00	12	Max dec	28° S 30'

ASPECTARIAN

01 Sunday
00 32 ☽ ∠ ♅
01 03 ☿ Q ♅
04 02 ☽ ⊻ ♇
04 44 ☽ Q ♃
05 50 ☽ ⊻ ☿
07 53 ♄ ⚼ ♇
08 04 ☿ ⚼ ♂
12 25 ☽ ☌ ♀
13 42 ☽ ⊥ ♇
15 03 ☽ ⊥ ♅
16 34 ☽ ⊥ ☿
18 37 ☽ ‖ ☿
20 21 ☉ △ ♂
22 35 ☽ ‖ ☉

02 Monday
00 43 ☽ ⊻ ♅
04 04 ☽ ☍ ♄
04 13 ☽ ∠ ♇
06 34 ☽ □ ♂
07 08 ☽ ⊻ ☉
08 40 ☽ ∠ ☿
10 01 ☽ □ ♃
14 03 ☽ ☌ ♆
17 32 ☽ ⊥ ☉
20 21 ☽ # ♄
23 26 ☿ ∠ ♃

03 Tuesday
00 23 ☽ ‖ ♀
03 02 ☽ # ♂
04 31 ☽ ⚹ ♇
05 39 ☽ ‖ ♆
06 49 ☽ ‖ ♇
09 13 ☽ ∠ ☉
10 45 ☽ ‖ ♃
11 39 ☿ ⊻ ♀
11 43 ☽ ⊻ ♀
11 44 ☽ ⚹ ☿
12 16 ☽ Q ♂
21 17 ☽ ⊥ ♀

04 Wednesday
00 12 ☽ Q ♇
01 35 ☽ ☌ ♅
04 41 ☽ ⊼ ♄
07 46 ☽ ‖ ♅
08 16 ☉ ⚹ ♃
08 23 ☽ ⚹ ♂
09 46 ☽ Q ☿
11 33 ☽ △ ♃
11 44 ☽ ∠ ♀
11 46 ☽ ⚹ ☉
14 36 ☽ ± ♄
14 49 ☿ ∠ ♆
15 18 ☽ ⊻ ♆

05 Thursday
01 22 ☽ ⊥ ♆
05 38 ☽ ⚼ ♄
06 19 ☽ □ ♇
09 28 ☽ Q ☉
10 03 ☽ ∠ ♂
11 37 ♀ # ♂
12 10 ☽ ⚹ ♀
13 04 ☽ ⚼ ♃
16 40 ☽ ∠ ♆
19 45 ☽ □ ☿

06 Friday
01 45 ☽ ⊥ ♂
04 20 ☽ ⊻ ♅
04 40 ☽ ± ♃
05 44 ☿ ± ♄
07 10 ☽ △ ♄
08 05 ☽ Q ♀
12 22 ☽ ⊻ ♂
14 46 ☽ ⊥ ♅
15 12 ☉ ⊻ ♆
15 12 ☽ ⊼ ♃
18 40 ☽ ⚹ ♆
18 56 ☽ □ ☉

07 Saturday
03 22 ☽ # ♅
06 38 ☽ ∠ ♅
10 27 ☽ △ ♇
14 34 ☽ □ ♀
15 56 ☽ Q ♆

08 Sunday
03 09 ☽ # ♃
06 39 ☽ # ♇
07 15 ☽ △ ☿
07 41 ☽ # ♆
09 34 ☽ ⚹ ♅
09 50 ☽ # ♀
12 05 ☽ □ ♄
13 17 ☽ ‖ ♂
13 28 ☽ ⚼ ♇
18 53 ☽ ‖ ♄
19 04 ☽ ☌ ♂
21 24 ☽ ☍ ♃

09 Monday
00 34 ☽ □ ♆
03 37 ☿ ⚹ ♅
04 27 ♀ Q ♃
05 15 ☽ △ ☉
05 57 ☽ ± ♇
07 33 ☽ Q ♅
13 41 ☽ # ☉
14 21 ☽ ⚼ ☿
17 03 ☽ ⊼ ♇
18 56 ☽ △ ♀
20 57 ☽ Q ♄
23 00 ☿ ⊼ ♄

10 Tuesday
03 33 ☽ # ☿
09 23 ☽ ± ☿
11 28 ☽ ⚼ ☉
14 33 ☿ ⊥ ♆
17 09 ☽ □ ♅
19 16 ☽ ⚹ ♄
21 46 ☽ ⚼ ♀
22 14 ☽ ⊼ ☿

11 Wednesday
04 16 ☽ ⊻ ♂
05 54 ☽ ⊼ ♃
05 55 ☽ ± ☉
08 41 ☽ △ ♆
09 38 ♀ ⊻ ♇
14 03 ☽ ± ♀
15 58 ☽ ⊥ ♂
17 26 ☽ ± ♃
18 20 ☽ ⊼ ☉
23 35 ☽ ∠ ♄

12 Thursday
00 59 ☽ ⊼ ♀
01 46 ☽ ☍ ♇
09 41 ☽ ∠ ♂
10 53 ☽ ⚼ ♃
13 28 ☽ ⚼ ♆
13 37 ♀ ‖ ♆
16 47 ☽ ⊥ ♄

13 Friday
02 43 ☽ △ ♅
04 21 ☽ ⊻ ♄
06 58 ☽ ± ♆
10 40 ☿ △ ♂
13 42 ☉ ∠ ♅
15 36 ☽ ⚹ ♂
16 10 ☽ ☍ ☿
16 20 ☽ △ ♃
17 32 ☿ ⚹ ♃
18 43 ☽ ⊼ ♆
23 59 ☉ ☌ ♀

14 Saturday
08 08 ♀ ‖ ♇
08 11 ☽ ⚼ ♅
08 33 ☽ ☍ ♀
09 48 ☽ ☍ ☉
12 14 ☿ ⊻ ♆
12 20 ☽ ⊼ ♇
15 08 ♀ ∠ ♅
15 54 ☽ Q ♂

15 Sunday
00 15 ☽ ± ♇
02 05 ☽ ± ♅
03 41 ☽ # ☿
09 28 ♂ ☍ ♃
14 06 ☽ ⊼ ♅
15 10 ☽ ☌ ♄
16 54 ☉ ⊻ ♇
18 16 ☽ ⚼ ♇
22 14 ☽ # ☉

16 Monday
04 34 ☽ □ ♃
04 56 ☽ □ ♂
06 31 ☽ ☍ ♆
11 25 ☽ ‖ ♄
12 55 ☽ ⊼ ☿
14 00 ☽ ‖ ♂
17 36 ☽ ⊼ ♀

17 Tuesday
00 35 ☽ △ ♇
01 17 ☽ # ♆
01 57 ☽ # ♇
02 57 ☽ ± ☿
03 08 ☽ # ♀
03 28 ☽ ⊼ ☉
04 12 ☽ # ♃
05 10 ☽ ± ♀
15 31 ☿ ☌ ♀
16 48 ☽ ± ☉
17 24 ☽ Q ♃
22 36 ☽ ⚼ ♀

18 Wednesday
00 07 ☽ ⚼ ☿
00 45 ☉ Q ♃
02 58 ☽ ☍ ♅
03 25 ☽ ⊻ ♄
09 53 ☽ # ♅
12 50 ☽ ⚼ ☉
13 53 ♂ □ ♆
15 34 ☽ ⊥ ♄
18 06 ☽ ⚹ ♃
19 35 ☽ ⊼ ♆
19 45 ☽ △ ♂
21 24 ☿ ∠ ♅

19 Thursday
03 43 ☽ △ ♀
07 49 ☽ ± ♆
08 16 ♀ ‖ ♃
09 45 ☽ ∠ ♄
11 31 ☽ △ ☿
13 45 ☽ □ ♇
21 02 ♄ ⊼ ♅
22 13 ☽ △ ☉

20 Friday
00 53 ☽ ∠ ♃
02 08 ☽ ⚼ ♆
03 12 ☽ ⚼ ♂
03 58 ☿ ⊻ ♇
15 53 ☽ ⚹ ♄
16 05 ☽ ⊼ ♅
19 13 ☽ ⊥ ♃
21 55 ☽ ± ♂

21 Saturday
02 08 ☽ Q ♇
04 08 ☽ ± ♅
07 19 ☽ ⊻ ♃
08 19 ☽ △ ♆
10 16 ☽ ⊼ ♂
13 33 ☽ □ ♀
15 23 ☽ ‖ ♅
15 38 ☽ Q ♄
19 04 ☉ ⊥ ♇
19 11 ☿ Q ♃
21 32 ♂ ‖ ♄
22 01 ☽ ⚼ ♅

22 Sunday
01 56 ☽ ⚹ ♇
08 53 ☽ □ ☿
15 14 ☽ □ ☉
19 40 ☽ ‖ ♀
20 07 ☉ ⊥ ♅
20 53 ☽ ‖ ♃
22 08 ☽ ‖ ♇
22 27 ☽ ‖ ♆
23 18 ☽ Q ♀

23 Monday
02 21 ☽ □ ♄
03 09 ☽ △ ♅
06 53 ☽ ∠ ♇
10 21 ♀ △ ♂
12 31 ☽ # ♄
13 18 ☽ # ♂
15 10 ☽ ‖ ☉
17 49 ☽ ☌ ♃
18 24 ☽ □ ♆
20 42 ☿ ⊥ ♇
21 10 ☽ ⚹ ♀
21 53 ☽ ☍ ♂
23 47 ☽ ⊥ ♇

24 Tuesday
00 14 ☽ Q ☿
03 26 ☽ ‖ ☿
04 07 ☽ Q ☉
10 49 ☽ ⊻ ♇
12 09 ☿ ⊥ ♅
15 07 ☉ # ♂
23 41 ☽ ∠ ♀

25 Wednesday
01 11 ☽ ⚹ ☿
03 15 ☽ Q ♆
03 36 ☽ ⚹ ☉
09 00 ☽ △ ♄
10 16 ☽ □ ♅
13 49 ☉ # ♄
15 09 ☽ ⊥ ♀

26 Thursday
00 02 ☽ ⊻ ♃
00 17 ☽ ⚹ ♆
01 17 ☽ ⊻ ♀
04 57 ☽ ⊼ ♂
06 57 ☽ ∠ ☿
07 43 ☽ ∠ ☉
10 22 ☽ ⊥ ♃
10 44 ☽ ⚼ ♄
14 47 ♂ ± ♇
15 25 ☽ ☌ ♇
15 26 ☽ ± ♂
17 16 ☽ Q ♅
21 34 ☉ ☌ ☿

27 Friday
00 01 ☽ ⊥ ☉
00 09 ☽ ⊥ ☿
01 34 ☽ ∠ ♃
01 41 ☽ ∠ ♆
01 44 ☽ ± ♄
06 51 ☽ ⚼ ♂
10 40 ☽ ⊻ ☉
11 22 ☽ ⊻ ☿
11 34 ☽ ⊼ ♄
13 03 ☿ ☍ ♄
13 12 ☽ ⚹ ♅
16 31 ☽ ⊥ ♆
17 45 ♀ ⊻ ♆
19 07 ♀ ⚹ ♃
22 48 ☉ ☍ ♄

28 Saturday
01 24 ♃ □ ♆
02 06 ☽ ☌ ♀
02 16 ☽ ⊻ ♆
02 16 ☽ ⚹ ♃
03 32 ☿ ⊻ ♅
07 57 ☽ △ ♂
13 31 ☽ ∠ ♅
16 30 ☽ ⊻ ♇
21 02 ♆ ‖ ♇
21 36 ☽ Q ♃
23 29 ☿ # ♂

29 Sunday
01 05 ☉ ⊻ ♅
02 03 ☽ ⊥ ♇
03 53 ☽ ⊥ ♅
06 00 ☿ ∠ ♇
11 25 ☽ ☍ ♄
13 24 ☽ ⊻ ♅
14 04 ♂ Q ♅
14 15 ☽ ☌ ☉
16 19 ☽ ∠ ♇
17 37 ☽ ☌ ☿

30 Monday
00 23 ☽ # ♂
01 05 ☽ ⊻ ♀
02 02 ☽ ☌ ♆
02 14 ☽ □ ♃
03 23 ☽ ‖ ☿
03 39 ☽ # ♄
06 15 ☿ # ♄
08 51 ☽ □ ♂
09 20 ☽ ‖ ☉
10 26 ☽ ⊥ ♀
16 00 ☽ ⚹ ♇
16 17 ☽ ‖ ♇
16 21 ☽ ‖ ♆
16 28 ☽ ‖ ♃
18 20 ☽ ‖ ♀
22 02 ☉ ∠ ♇

31 Tuesday
00 31 ☽ ∠ ♀
10 33 ☽ ⊼ ♄
11 04 ☽ Q ♇
12 41 ♃ ‖ ♆
12 59 ☽ ☌ ♅
14 13 ☽ Q ♂
17 07 ☽ ⊻ ☉
18 03 ☽ ‖ ♅
20 05 ☽ ± ♄
23 21 ☽ ⊻ ☿

All ephemeris data is given at 12.00 UT and the Moon's longitude is additionally given for 24.00 UT

FEBRUARY 2006

LONGITUDES

Date	Sidereal time h m s	Sun ☉ ° ' "	Moon ☽ ° ' "	Moon ☽ 24.00 ° ' "	Mercury ☿ ° '	Venus ♀ ° '	Mars ♂ ° '	Jupiter ♃ ° '	Saturn ♄ ° '	Uranus ♅ ° '	Neptune ♆ ° '	Pluto ♇ ° '
01	20 46 13	12 ♒ 29 10	23 ♓ 24 38	00 ♈ 45 06	16 ♒ 30	16 ♑ 06	22 ♉ 16	17 ♏ 22	07 ♌ 29	09 ♓ 09	17 ♒ 06	25 ♐ 55
02	20 50 10	13 30 04	07 ♈ 59 55	15 ♈ 08 43	18 16	16 R 02	22 43	17 28	07 R 25	09 13	17 09	25 57
03	20 54 06	14 30 57	22 ♈ 11 20	29 ♈ 07 45	20 03	16 D 01	23 10	17 33	07 20	09 16	17 11	25 59
04	20 58 03	15 31 49	05 ♉ 58 06	12 ♉ 42 34	21 50	16 03	23 37	17 38	07 15	09 19	17 13	26 00
05	21 01 59	16 32 39	19 ♉ 21 29	25 ♉ 55 11	23 37	16 07	24 04	17 43	07 10	09 22	17 15	26 02
06	21 05 56	17 33 27	02 ♊ 24 02	08 ♊ 48 28	25 24	16 13	24 31	17 48	07 05	09 25	17 18	26 04
07	21 09 52	18 34 14	15 ♊ 08 50	21 ♊ 25 34	27 12	16 21	24 59	17 53	07 00	09 29	17 20	26 05
08	21 13 49	19 35 00	27 ♊ 39 00	03 ♋ 49 29	29 ♒ 00	16 32	25 27	17 57	06 56	09 32	17 22	26 07
09	21 17 46	20 35 44	09 ♋ 57 20	16 ♋ 02 50	00 ♓ 48	16 45	25 55	18 02	06 51	09 35	17 25	26 08
10	21 21 42	21 36 27	22 ♋ 06 15	28 ♋ 07 49	02 35	17 00	26 24	18 06	06 46	09 39	17 27	26 10
11	21 25 39	22 37 08	04 ♌ 07 44	10 ♌ 06 14	04 22	17 17	26 52	18 10	06 42	09 42	17 29	26 11
12	21 29 35	23 37 48	16 ♌ 03 30	21 ♌ 59 43	06 08	17 36	27 21	18 13	06 37	09 45	17 31	26 13
13	21 33 32	24 38 26	27 ♌ 55 06	03 ♍ 49 51	07 54	17 57	27 50	18 17	06 32	09 49	17 34	26 14
14	21 37 28	25 39 03	09 ♍ 44 12	15 ♍ 38 24	09 37	18 19	28 19	18 21	06 28	09 52	17 36	26 15
15	21 41 25	26 39 38	21 ♍ 32 45	27 ♍ 27 34	11 19	18 44	28 48	18 24	06 24	09 55	17 38	26 17
16	21 45 21	27 40 12	03 ♎ 23 11	09 ♎ 20 00	12 59	19 10	29 17	18 27	06 19	09 59	17 40	26 18
17	21 49 18	28 40 45	15 ♎ 18 27	21 ♎ 18 58	14 36	19 38	29 ♉ 47	18 30	06 15	10 02	17 43	26 19
18	21 53 15	29 ♒ 41 17	27 ♎ 22 04	03 ♏ 28 17	16 09	20 08	00 ♊ 16	18 33	06 10	10 06	17 45	26 20
19	21 57 11	00 ♓ 41 47	09 ♏ 38 08	15 ♏ 52 11	17 39	20 39	00 46	18 35	06 06	10 09	17 47	26 22
20	22 01 08	01 42 16	22 ♏ 10 59	28 ♏ 35 05	19 05	21 12	01 16	18 38	06 02	10 12	17 49	26 23
21	22 05 04	02 42 44	05 ♐ 04 59	11 ♐ 41 08	20 25	21 46	01 46	18 40	05 58	10 16	17 52	26 24
22	22 09 01	03 43 10	18 ♐ 23 56	25 ♐ 13 39	21 39	22 21	02 17	18 42	05 54	10 19	17 54	26 25
23	22 12 57	04 43 35	02 ♑ 10 25	09 ♑ 14 15	22 47	22 58	02 47	18 44	05 50	10 23	17 56	26 26
24	22 16 54	05 43 59	16 ♑ 24 58	23 ♑ 42 10	23 48	23 36	03 18	18 45	05 46	10 26	17 58	26 27
25	22 20 50	06 44 21	01 ♒ 05 19	08 ♒ 33 36	24 41	24 15	03 49	18 47	05 42	10 30	18 00	26 28
26	22 24 47	07 44 42	16 ♒ 06 04	23 ♒ 41 37	25 26	24 55	04 19	18 48	05 38	10 33	18 03	26 30
27	22 28 44	08 45 01	01 ♓ 18 59	08 ♓ 56 54	26 02	25 37	04 50	18 49	05 34	10 36	18 05	26 30
28	22 32 40	09 ♓ 45 19	16 ♓ 34 02	24 ♓ 09 08	26 ♓ 29	26 ♑ 19	05 ♊ 22	18 ♏ 50	05 ♌ 31	10 ♓ 40	18 ♒ 07	26 ♐ 31

Date	Moon True ☊ ° '	Moon Mean ☊ ° '	Moon ☽ Latitude ° '
01	05 ♈ 38	07 ♈ 20	01 S 05
02	05 D 37	07 17	00 N 13
03	05 38	07 13	01 28
04	05 39	07 10	02 36
05	05 R 39	07 07	03 33
06	05 38	07 04	04 18
07	05 35	07 01	04 49
08	05 29	06 57	05 05
09	05 22	06 54	05 07
10	05 14	06 51	04 55
11	05 05	06 48	04 30
12	04 57	06 45	03 53
13	04 50	06 42	03 06
14	04 45	06 38	02 12
15	04 43	06 35	01 11
16	04 42	06 32	00 N 07
17	04 D 42	06 29	00 S 58
18	04 44	06 26	02 01
19	04 45	06 23	02 59
20	04 47	06 19	03 51
21	04 R 47	06 16	04 32
22	04 46	06 13	05 01
23	04 44	06 10	05 13
24	04 40	06 07	05 07
25	04 36	06 03	04 41
26	04 32	06 00	03 55
27	04 29	05 57	02 52
28	04 ♈ 27	05 ♈ 54	01 S 37

DECLINATIONS

Date	Sun ☉ ° '	Moon ☽ ° '	Mercury ☿ ° '	Venus ♀ ° '	Mars ♂ ° '	Jupiter ♃ ° '	Saturn ♄ ° '	Uranus ♅ ° '	Neptune ♆ ° '	Pluto ♇ ° '
01	17 S 04	03 S 37	17 S 49	15 S 25	20 N 04	15 S 54	19 N 04	08 S 49	15 S 52	15 S 54
02	16 46	03 N 22	17 14	15 27	20 11	15 55	19 05	08 48	15 51	15 54
03	16 29	10 00	16 38	15 29	20 18	15 57	19 07	08 47	15 50	15 54
04	16 11	15 58	16 00	15 32	20 25	15 58	19 08	08 46	15 50	15 54
05	15 53	20 59	15 21	15 34	20 32	15 59	19 09	08 44	15 49	15 54
06	15 34	24 51	14 41	15 37	20 38	16 00	19 11	08 43	15 48	15 54
07	15 16	27 24	13 59	15 40	20 45	16 01	19 12	08 42	15 48	15 54
08	14 57	28 30	13 17	15 43	20 52	16 02	19 13	08 41	15 47	15 54
09	14 38	28 10	12 33	15 46	20 59	16 04	19 15	08 39	15 46	15 54
10	14 18	26 29	11 47	15 49	21 06	16 04	19 16	08 38	15 46	15 54
11	13 59	23 36	11 01	15 51	21 12	16 05	19 17	08 37	15 45	15 54
12	13 39	19 44	10 14	15 54	21 19	16 06	19 18	08 36	15 44	15 54
13	13 19	15 07	09 27	15 57	21 26	16 07	19 19	08 34	15 44	15 54
14	12 58	09 57	08 39	16 00	21 32	16 08	19 21	08 33	15 43	15 54
15	12 38	04 N 26	07 50	16 03	21 39	16 09	19 22	08 32	15 42	15 53
16	12 17	01 S 14	07 02	16 05	21 45	16 09	19 23	08 31	15 42	15 53
17	11 56	06 55	06 14	16 08	21 52	16 10	19 24	08 29	15 41	15 53
18	11 35	12 25	05 26	16 10	21 58	16 11	19 25	08 28	15 40	15 53
19	11 14	17 32	04 39	16 12	22 04	16 11	19 26	08 27	15 40	15 53
20	10 52	22 02	03 53	16 14	22 11	16 12	19 28	08 25	15 39	15 53
21	10 31	25 37	03 08	16 15	22 17	16 12	19 29	08 24	15 38	15 53
22	10 09	27 56	02 26	16 16	22 23	16 12	19 30	08 23	15 38	15 53
23	09 47	28 38	01 45	16 17	22 29	16 13	19 31	08 21	15 37	15 53
24	09 25	27 30	01 08	16 18	22 35	16 13	19 32	08 20	15 36	15 53
25	09 02	24 29	00 33	16 18	22 41	16 13	19 33	08 19	15 36	15 53
26	08 40	19 45	00 S 01	16 19	22 46	16 13	19 34	08 18	15 35	15 52
27	08 17	13 41	00 N 26	16 18	22 52	16 14	19 35	08 16	15 34	15 52
28	07 S 55	06 S 47	00 N 50	16 S 18	22 N 58	16 S 14	19 N 36	08 S 15	15 S 34	15 S 52

ZODIAC SIGN ENTRIES

Date	h	m	Planets
01	22	46	☽ ♈
04	01	31	☽ ♉
06	07	32	☽ ♊
08	16	33	☽ ♋
09	01	22	☿ ♓
11	03	44	☽ ♌
13	16	13	☽ ♍
16	05	09	☽ ♎
17	22	44	♂ ♊
18	17	11	☽ ♏
18	19	26	☉ ♓
21	02	38	☽ ♐
23	08	16	☽ ♑
25	10	14	☽ ♒
27	09	56	☽ ♓

LATITUDES

Date	Mercury ☿ ° '	Venus ♀ ° '	Mars ♂ ° '	Jupiter ♃ ° '	Saturn ♄ ° '	Uranus ♅ ° '	Neptune ♆ ° '	Pluto ♇ ° '
01	02 S 01	07 N 06	01 N 47	01 N 10	00 N 41	00 S 44	00 S 10	07 N 29
04	01 52	07 00	01 47	01 11	00 42	00 44	00 10	07 29
07	01 39	06 50	01 48	01 11	00 42	00 44	00 10	07 29
10	01 20	06 36	01 48	01 12	00 42	00 44	00 10	07 30
13	00 54	06 20	01 48	01 12	00 42	00 44	00 10	07 30
16	00 S 22	06 02	01 48	01 13	00 43	00 44	00 10	07 30
19	00 N 15	05 43	01 48	01 13	00 43	00 44	00 10	07 31
22	00 58	05 23	01 48	01 14	00 43	00 44	00 10	07 31
25	01 42	05 02	01 48	01 14	00 43	00 44	00 10	07 31
28	02 26	04 41	01 47	01 15	00 43	00 44	00 10	07 32
31	03 N 03	04 N 19	01 N 47	01 N 15	00 N 44	00 S 44	00 S 10	07 N 32

DATA

Julian Date	2453768
Delta T	+66 seconds
Ayanamsa	23° 56' 31"
Synetic vernal point	05° ♓ 10' 28"
True obliquity of ecliptic	23° 26' 28"

MOON'S PHASES, APSIDES AND POSITIONS ☽

Date	h	m	Phase	Longitude ° '	Eclipse Indicator
05	06	29	☽	16 ♉ 19	
13	04	44	○	24 ♌ 20	
21	07	17	☾	02 ♐ 31	
28	00	31	●	09 ♓ 16	

Day	h	m	
14	00	29	Apogee
27	20	19	Perigee

Day	h	m		
02	00	21	0N	
08	18	19	Max dec	28° N 33'
16	06	48	0S	
23	09	36	Max dec	28° S 39'

LONGITUDES

Date	Chiron ⚷ ° '	Ceres ⚳ ° '	Pallas ⚴ ° '	Juno ⚵ ° '	Vesta ⚶ ° '	Black Moon Lilith ⚸ ° '
01	04 ♒ 32	20 ♑ 52	27 ♐ 23	13 ♊ 38	09 ♋ 21	01 ♍ 07
11	05 ♒ 22	24 ♑ 44	01 ♑ 05	15 ♊ 30	08 ♋ 04	02 ♍ 14
21	06 ♒ 09	28 ♑ 32	04 ♑ 36	17 ♊ 59	07 ♋ 32	03 ♍ 20
31	06 ♒ 54	02 ♒ 13	07 ♑ 52	20 ♊ 58	07 ♋ 45	04 ♍ 27

ASPECTARIAN

h m	Aspects
01 Wednesday	
00 15	☽ ⚹ ♀
01 47	☽ ⊻ ♆
02 10	☽ △ ♃
03 27	☽ ⊥ ☉
06 39	☿ ⊻ ♀
10 06	☽ ⚹ ♂
10 20	☽ ⊥ ☿
10 31	☽ ꝑ ♄
11 30	☽ ⊥ ♆
16 06	☽ □ ♇
17 39	♃ ∥ ♇
19 08	☽ ∠ ☉
19 36	☽ Q ♀
20 22	☿ ☌ ♆
02 Thursday	
00 29	☿ □ ♃
02 16	☽ ∠ ♆
02 46	☽ ꝑ ♃
03 04	☽ ∠ ☿
11 01	☽ △ ♄
11 31	☽ ∠ ♂
14 02	☽ ⊻ ♅
17 50	☽ ± ♃
21 56	☽ ⚹ ☉
03 Friday	
00 09	☽ ⊥ ♅
01 29	☽ □ ♀
03 08	☽ ⊥ ♂
03 26	☽ ⚹ ♆
04 02	☽ ⊼ ♃
07 27	☽ ⋕ ♅
07 48	☽ ⚹ ☿
09 19	♀ St D
13 44	☽ ⊻ ♂
15 35	☽ ∠ ♅
18 33	☽ △ ♇
20 03	☽ Q ☉
23 24	☉ ∥ ☿
04 Saturday	
00 07	☽ Q ♆
07 39	☽ Q ☿
10 07	☽ ⋕ ♀
11 26	☽ ⋕ ♆
11 44	☽ ⋕ ♇
12 00	☽ ⋕ ♃
12 10	☽ ⋕ ☿
12 54	☽ ⋕ ☉
13 32	☿ ∥ ♃
14 15	☽ □ ♄
15 03	☿ ⊥ ♀
15 56	☿ ∥ ♇
17 58	☽ ⚹ ♅
18 43	☿ ∥ ♆
20 58	☽ ꝑ ♇
05 Sunday	
00 55	☉ ⊻ ♀
02 36	☽ ∥ ♄
04 10	☉ ∥ ♃
04 41	☿ ∥ ♀
06 05	☽ △ ♀
06 29	☽ □ ☉
08 11	☽ □ ♆
09 01	☽ ☍ ♃
09 29	☽ ∥ ♂
10 15	☉ ∥ ♇
13 14	☽ ± ♇
15 41	☽ Q ♅
16 56	☉ ∥ ♆
20 09	☿ □ ♂
20 54	☽ ☌ ♂
21 00	☽ □ ☿
22 33	☽ Q ♄
06 Monday	
00 14	☽ ⊼ ♇
05 33	☉ ☌ ♆
09 05	☉ ∥ ♀
09 46	☽ ꝑ ♀
18 16	☉ □ ♃
20 43	☽ ⚹ ♄
20 52	☿ ⚹ ♇
07 Tuesday	
01 13	☽ □ ♅
02 48	☽ ± ♀
12 56	♂ Q ♄
14 20	☽ ⊼ ♀
16 11	☽ △ ♆
17 14	☽ ⊼ ♃
19 06	☽ △ ☉
08 Wednesday	
01 02	☽ ∠ ♄
04 49	☽ ± ♃
07 35	☽ ⊻ ♂
09 01	☽ ☍ ♇
15 04	☽ △ ☿
18 19	☽ ⊥ ♄
19 40	☽ ⊥ ♂
21 12	☽ ꝑ ♆
22 22	☽ ꝑ ♃
09 Thursday	
02 41	☽ ꝑ ☉
05 57	☽ ⊻ ♄
11 17	☽ △ ♅
13 58	☽ ∠ ♂
14 52	☽ ± ♆
17 27	♀ ∥ ♆
21 58	☽ ± ☉
23 37	♂ ⊼ ♇
10 Friday	
01 30	☽ ꝑ ☿
01 39	☽ ☍ ♀
02 43	☿ ∠ ♀
02 44	☽ ⊼ ♆
04 00	☽ △ ♃
10 55	☽ ⊼ ☉
17 04	☽ ꝑ ♅
20 05	☽ ⊼ ♇
20 53	☽ ⚹ ♂
22 29	☽ ± ☿
11 Saturday	
08 06	☽ ± ♇
11 08	☽ ± ♅
12 34	☽ ⊼ ☿
17 07	☽ ☌ ♄
21 54	☽ Q ♂
23 14	☽ ⊼ ♅
12 Sunday	
02 12	☽ ꝑ ♇
03 00	☽ ∥ ♂
05 57	♀ ∥ ♇
06 05	♀ ⊻ ♆
10 45	☉ ⊥ ♀
14 21	☽ ∥ ♄
14 58	☽ ☍ ♆
15 12	☽ ⊼ ♀
16 24	☽ □ ♃
18 15	☿ ⊼ ♄
13 Monday	
03 42	☽ ± ♀
04 44	☽ ☍ ☉
07 02	☽ ⋕ ♃
07 53	☽ ⋕ ♀
08 09	☽ ⋕ ♇
08 34	☽ △ ♇
08 58	☽ ⋕ ♆
11 48	☽ □ ♂
16 45	☿ Q ♇
21 13	☽ ⋕ ☉
22 32	☽ ꝑ ♀
14 Tuesday	
05 04	☽ Q ♃
05 24	☽ ⊻ ♄
11 44	☽ ☍ ☿
12 16	☽ ☍ ♅
13 25	♀ ⚹ ♃
14 54	☿ ∥ ♅
15 33	☿ ☌ ♅
17 31	☽ ⊥ ♄
18 14	☽ ⋕ ♅
18 47	☽ ⋕ ☿
15 Wednesday	
02 41	☉ ⚹ ♇
04 02	☽ ⊼ ♆
05 34	☽ ⚹ ♃
06 04	☽ △ ♀
11 41	☽ ∠ ♄
16 15	☽ ± ♆
21 37	☽ □ ♇
23 21	☽ ⊼ ☉
16 Thursday	
02 46	☿ ± ♄
03 21	☽ △ ♂
10 33	☽ ꝑ ♆
12 08	☽ ∠ ♃
12 38	☽ ± ☉
17 53	☽ ⚹ ♄
17 Friday	
01 22	☽ ⊼ ♅
06 21	☽ ⊥ ♃
08 26	☽ ꝑ ☉
09 26	☽ ∥ ☿
10 01	☽ Q ♇
10 21	☽ ⊼ ☿
10 54	☽ ꝑ ♂
13 28	☽ ± ♅
16 50	☽ △ ♆
17 51	☽ Q ♄
18 24	☽ ⊻ ♃
18 43	☽ ∥ ♅
21 01	☽ □ ♀
18 Saturday	
00 09	☽ ± ☿
05 37	☽ ± ♂
07 29	☽ ꝑ ♅
08 31	☽ ∥ ☉
09 58	☽ ⚹ ♇
16 59	☽ △ ☉
17 58	☽ ⊼ ♂
20 31	☽ ꝑ ☿
23 18	♀ ∥ ♃
19 Sunday	
02 57	☽ ∥ ♆
04 00	☽ ∥ ♇
05 11	☽ □ ♄
05 26	☽ ∥ ♃
05 27	☽ ∥ ♀
10 00	☽ Q ♀
13 00	☽ △ ♅
14 14	☿ ⊻ ♆
15 21	☽ ∠ ♇
15 33	☉ □ ♂
21 44	☽ ⋕ ♄
20 Monday	
03 42	☽ □ ♆
04 03	☿ △ ♃
05 14	☽ ☌ ♃
05 23	☽ △ ☿
08 35	☽ ⊥ ♇
10 03	☽ ⚹ ♀
12 52	☽ ⋕ ♂
17 26	☿ Q ♂
19 54	☽ ⊻ ♇
21 Tuesday	
05 40	☽ ☍ ♂
07 17	☽ □ ☉
13 26	☽ Q ♆
13 36	☽ △ ♄
15 13	☽ ∠ ♀
21 28	☽ □ ♅
21 55	☿ ꝑ ♄
22 Wednesday	
08 11	☽ ⊥ ♀
11 06	☽ ⚹ ♆
12 32	☽ ⊻ ♃
16 23	☽ ꝑ ♄
18 17	☽ □ ☿
18 20	☽ Q ☉
19 18	☽ ⊻ ♀
23 06	☽ ⊥ ♃
23 Thursday	
02 06	☽ ☌ ♇
05 26	☽ Q ♅
08 00	☽ ± ♄
13 05	☽ ⊼ ♂
13 18	☽ ∠ ♆
14 40	☽ ∠ ♃
16 42	☽ ⚹ ☉
18 12	☽ ⊼ ♄
22 16	☿ ⚹ ♀
23 40	☽ ± ♂
24 Friday	
01 58	☽ ⚹ ♅
03 44	☽ Q ☿
04 34	☽ ⊥ ♆
12 41	☉ ⊼ ♄
14 35	☽ ⊻ ♆
15 14	☽ ꝑ ♂
15 53	☽ ⚹ ♃
16 38	☿ ⊥ ♆
19 39	☽ ∠ ☉
25 Saturday	
00 22	☽ ☌ ♀
00 58	☽ ⚹ ☿
02 53	☽ ∠ ♅
04 31	☽ ⊻ ♇
11 24	☽ ⊥ ☉
11 30	☽ Q ♃
14 14	☽ ⊥ ♇
16 33	☽ △ ♂
17 30	☽ ⊥ ♅
19 23	☽ ☍ ♄
21 44	☽ ⊻ ☉
21 54	☽ ⋕ ♂
26 Sunday	
02 33	☽ ∠ ☿
03 08	☽ ⊻ ♅
04 40	☽ ∠ ♇
12 48	☽ ⋕ ♄
15 05	☽ ☌ ♆
16 17	☽ □ ♃
17 31	☽ ⊥ ☿
27 Monday	
02 09	☽ ∥ ♀
02 27	☽ ∥ ♃
02 36	☽ ⊻ ♀
03 22	☽ ⊻ ☿
03 49	☽ ∥ ♇
04 26	☽ ⚹ ♇
04 58	☽ ∥ ♆
06 07	☉ Q ♇
11 42	♀ ∠ ♅
12 29	☽ ⊥ ♀
13 19	☉ ∥ ♅
17 44	☽ □ ♂
18 40	☽ ⊼ ♄
23 19	☽ Q ♇
28 Tuesday	
00 31	☽ ☌ ☉
02 40	☽ ☌ ♅
03 20	☽ ∠ ♀
04 04	☽ ± ♄
07 03	☽ ∥ ♅
08 00	☽ ∥ ☉
14 27	☽ ⊻ ♆
14 44	☿ □ ♇
15 35	☽ △ ♃
18 13	☽ ꝑ ♄
18 20	♂ ⚹ ♄
18 50	♀ ⊻ ♇
22 07	☿ ⚹ ♀
23 08	☽ Q ♂
23 58	☽ ⊥ ♆

All ephemeris data is given at 12.00 UT and the Moon's longitude is additionally given for 24.00 UT

MARCH 2006

LONGITUDES

Date	Sidereal time h m s	Sun ☉ ° ' "	Moon ☽ ° ' "	Moon ☽ 24.00 ° ' "	Mercury ☿ ° '	Venus ♀ ° '	Mars ♂ ° '	Jupiter ♃ ° '	Saturn ♄ ° '	Uranus ♅ ° '	Neptune ♆ ° '	Pluto ♇ ° '
01	22 36 37	10 ♓ 45 34	01 ♈ 41 01	09 ♈ 08 38	26 ♓ 47	27 ♑ 03	05 ♊ 53	18 ♏ 51	05 ♌ 27	10 ♓ 43	18 ♒ 09	26 ♐ 32
02	22 40 33	11 45 48	16 ♈ 31 07	23 ♈ 47 47	26 55	27 48	06 24	18 51	05 R 24	10 47	18 11	26 33
03	22 44 30	12 46 00	00 ♉ 58 08	08 ♉ 01 52	26 R 53	28 33	06 56	18 52	05 20	10 50	18 13	26 34
04	22 48 26	13 46 09	14 ♉ 58 48	21 ♉ 48 59	26 43	29 ♑ 19	07 27	18 52	05 17	10 54	18 15	26 35
05	22 52 23	14 46 17	28 ♉ 32 34	05 ♊ 09 47	26 23	00 ♒ 07	07 59	18 R 52	05 14	10 57	18 18	26 36
06	22 56 19	15 46 23	11 ♊ 40 59	18 ♊ 06 34	25 55	00 55	08 31	18 51	05 11	11 01	18 20	26 37
07	23 00 16	16 46 26	24 ♊ 27 00	00 ♋ 42 46	25 19	01 44	09 03	18 51	05 08	11 04	18 22	26 37
08	23 04 13	17 46 27	06 ♋ 54 21	13 ♋ 02 15	24 36	02 33	09 35	18 50	05 05	11 07	18 24	26 38
09	23 08 09	18 46 27	19 ♋ 06 58	25 ♋ 08 57	23 48	03 24	10 07	18 50	05 02	11 11	18 26	26 39
10	23 12 06	19 46 24	01 ♌ 08 41	07 ♌ 06 34	22 55	04 15	10 39	18 49	04 59	11 14	18 28	26 39
11	23 16 02	20 46 19	13 ♌ 03 02	18 ♌ 58 26	21 59	05 07	11 12	18 47	04 56	11 18	18 30	26 40
12	23 19 59	21 46 11	24 ♌ 53 09	00 ♍ 47 28	21 00	05 59	11 44	18 46	04 54	11 21	18 32	26 41
13	23 23 55	22 46 02	06 ♍ 41 44	12 ♍ 36 12	20 02	06 52	12 17	18 45	04 51	11 24	18 34	26 41
14	23 27 52	23 45 51	18 ♍ 31 10	24 ♍ 26 52	19 04	07 46	12 49	18 43	04 49	11 28	18 36	26 42
15	23 31 48	24 45 37	00 ♎ 23 35	06 ♎ 21 32	18 08	08 40	13 22	18 41	04 47	11 31	18 38	26 42
16	23 35 45	25 45 22	12 ♎ 21 00	18 ♎ 22 14	17 15	09 35	13 55	18 39	04 44	11 35	18 40	26 42
17	23 39 42	26 45 05	24 ♎ 25 31	00 ♏ 31 07	16 26	10 30	14 28	18 36	04 42	11 38	18 42	26 43
18	23 43 38	27 44 46	06 ♏ 39 20	12 ♏ 50 29	15 41	11 26	15 01	18 34	04 40	11 41	18 43	26 43
19	23 47 35	28 44 25	19 ♏ 04 54	25 ♏ 22 54	15 02	12 23	15 34	18 31	04 38	11 45	18 45	26 44
20	23 51 31	29 ♓ 44 02	01 ♐ 44 51	08 ♐ 11 05	14 28	13 20	16 07	18 29	04 37	11 48	18 47	26 44
21	23 55 28	00 ♈ 43 38	14 ♐ 41 56	21 ♐ 17 42	14 01	14 17	16 40	18 26	04 35	11 51	18 49	26 44
22	23 59 24	01 43 12	27 ♐ 58 41	04 ♑ 45 05	13 39	15 15	17 13	18 22	04 33	11 55	18 51	26 44
23	00 03 21	02 42 45	11 ♑ 37 05	18 ♑ 34 44	13 24	16 14	17 47	18 19	04 32	11 58	18 53	26 45
24	00 07 17	03 42 15	25 ♑ 38 01	02 ♒ 46 47	13 14	17 12	18 20	18 16	04 30	12 01	18 54	26 45
25	00 11 14	04 41 44	10 ♒ 00 46	17 ♒ 19 31	13 11	18 11	18 54	18 12	04 29	12 04	18 56	26 45
26	00 15 11	05 41 11	24 ♒ 42 27	02 ♓ 08 53	13 D 13	19 11	19 27	18 08	04 28	12 08	18 58	26 45
27	00 19 07	06 40 36	09 ♓ 37 56	17 ♓ 08 40	13 21	20 11	20 01	18 04	04 27	12 11	18 59	26 45
28	00 23 04	07 40 00	24 ♓ 40 01	02 ♈ 10 55	13 35	21 11	20 34	18 00	04 26	12 14	19 01	26 45
29	00 27 00	08 39 21	09 ♈ 40 17	17 ♈ 07 02	13 53	22 12	21 08	17 55	04 25	12 17	19 03	26 45
30	00 30 57	09 38 40	24 ♈ 30 11	01 ♉ 48 52	14 17	23 13	21 42	17 51	04 25	12 20	19 04	26 R 45
31	00 34 53	10 ♈ 37 57	09 ♉ 02 20	16 ♉ 09 58	14 ♓ 45	24 ♒ 14	22 ♊ 16	17 ♏ 46	04 ♌ 24	12 ♓ 23	19 ♒ 06	26 ♐ 45

Date	Moon True ☊ ° '	Moon Mean ☊ ° '	Moon ☽ Latitude ° '
01	04 ♈ 26	05 ♈ 51	00 S 15
02	04 D 26	05 48	01 N 06
03	04 27	05 44	02 21
04	04 29	05 41	03 25
05	04 30	05 38	04 16
06	04 31	05 35	04 51
07	04 R 31	05 32	05 11
08	04 30	05 28	05 16
09	04 28	05 25	05 06
10	04 26	05 22	04 42
11	04 24	05 19	04 07
12	04 22	05 16	03 21
13	04 20	05 13	02 27
14	04 19	05 09	01 26
15	04 19	05 06	00 N 22
16	04 D 19	05 03	00 S 44
17	04 19	05 00	01 49
18	04 20	04 57	02 50
19	04 20	04 54	03 44
20	04 21	04 50	04 28
21	04 21	04 47	04 59
22	04 21	04 44	05 16
23	04 R 21	04 41	05 15
24	04 21	04 38	04 56
25	04 D 21	04 34	04 18
26	04 21	04 31	03 23
27	04 22	04 28	02 13
28	04 22	04 25	00 S 54
29	04 R 22	04 22	00 N 29
30	04 21	04 19	01 50
31	04 ♈ 21	04 ♈ 15	03 N 01

DECLINATIONS

Date	Sun ☉ ° '	Moon ☽ ° '	Mercury ☿ ° '	Venus ♀ ° '	Mars ♂ ° '	Jupiter ♃ ° '	Saturn ♄ ° '	Uranus ♅ ° '	Neptune ♆ ° '	Pluto ♇ ° '
01	07 S 32	00 N 26	01 N 09	16 S 17	23 N 03	16 S 14	19 N 37	08 S 14	15 S 33	15 S 52
02	07 09	07 31	01 24	16 15	23 08	16 14	19 38	08 12	15 32	15 52
03	06 46	14 01	01 34	16 13	23 14	16 14	19 38	08 11	15 32	15 52
04	06 23	19 36	01 39	16 11	23 19	16 13	19 39	08 10	15 31	15 52
05	06 00	24 00	01 39	16 09	23 24	16 13	19 40	08 08	15 30	15 52
06	05 37	27 00	01 35	16 06	23 29	16 13	19 41	08 07	15 30	15 51
07	05 13	28 30	01 25	16 02	23 34	16 13	19 42	08 06	15 29	15 51
08	04 50	28 31	01 11	15 58	23 39	16 12	19 42	08 04	15 29	15 51
09	04 26	27 07	00 54	15 54	23 43	16 12	19 43	08 03	15 28	15 51
10	04 03	24 30	00 32	15 49	23 48	16 12	19 44	08 02	15 27	15 51
11	03 39	20 51	00 N 08	15 44	23 52	16 11	19 45	08 01	15 27	15 51
12	03 16	16 23	00 S 20	15 38	23 57	16 11	19 45	07 59	15 26	15 51
13	02 52	11 20	00 49	15 32	24 01	16 10	19 46	07 58	15 26	15 51
14	02 29	05 52	01 19	15 26	24 05	16 09	19 46	07 57	15 25	15 50
15	02 05	00 N 11	01 50	15 19	24 09	16 09	19 47	07 55	15 24	15 50
16	01 41	05 S 34	02 21	15 11	24 13	16 08	19 48	07 54	15 24	15 50
17	01 18	11 10	02 52	15 03	24 17	16 07	19 48	07 53	15 23	15 50
18	00 54	16 25	03 21	14 55	24 20	16 07	19 49	07 52	15 23	15 50
19	00 30	21 05	03 50	14 46	24 24	16 06	19 49	07 50	15 22	15 50
20	00 S 06	24 53	04 16	14 37	24 27	16 05	19 50	07 49	15 21	15 50
21	00 N 17	27 31	04 41	14 27	24 30	16 04	19 50	07 48	15 21	15 49
22	00 41	28 41	05 03	14 16	24 33	16 03	19 50	07 47	15 20	15 49
23	01 05	28 10	05 23	14 06	24 36	16 02	19 51	07 45	15 20	15 49
24	01 28	25 52	05 40	13 54	24 39	16 01	19 51	07 44	15 19	15 49
25	01 52	21 53	05 56	13 43	24 42	15 59	19 52	07 43	15 19	15 49
26	02 16	16 28	06 08	13 30	24 44	15 58	19 52	07 42	15 18	15 49
27	02 39	10 01	06 18	13 18	24 47	15 57	19 52	07 40	15 18	15 49
28	03 03	02 S 56	06 26	13 05	24 49	15 56	19 52	07 39	15 17	15 48
29	03 26	04 N 17	06 31	12 51	24 51	15 54	19 53	07 38	15 17	15 48
30	03 49	11 12	06 34	12 37	24 53	15 53	19 53	07 37	15 16	15 48
31	04 N 13	17 N 22	06 S 34	12 S 23	24 N 55	15 S 52	19 N 53	07 S 36	15 S 16	15 S 48

ZODIAC SIGN ENTRIES

Date	h	m	Planets
01	09	19	☽ ♈
03	10	22	☽ ♉
05	08	39	♀ ♒
05	14	38	☽ ♊
07	22	38	☽ ♋
10	09	42	☽ ♌
12	22	24	☽ ♍
15	11	12	☽ ♎
17	22	59	☽ ♏
20	08	43	☽ ♐
20	18	26	☉ ♈
22	15	36	☽ ♑
24	19	21	☽ ♒
26	20	33	☽ ♓
28	20	31	☽ ♈
30	21	01	☽ ♉

LATITUDES

Date	Mercury ☿ ° '	Venus ♀ ° '	Mars ♂ ° '	Jupiter ♃ ° '	Saturn ♄ ° '	Uranus ♅ ° '	Neptune ♆ ° '	Pluto ♇ ° '
01	02 N 39	04 N 33	01 N 47	01 N 15	00 N 43	00 S 44	00 S 10	07 N 32
04	03 13	04 12	01 47	01 15	00 44	00 44	00 10	07 32
07	03 34	03 50	01 47	01 16	00 44	00 44	00 10	07 33
10	03 39	03 28	01 46	01 16	00 44	00 44	00 10	07 33
13	03 25	03 07	01 46	01 17	00 44	00 44	00 10	07 34
16	02 55	02 46	01 45	01 17	00 44	00 44	00 10	07 34
19	02 15	02 25	01 45	01 18	00 44	00 44	00 10	07 34
22	01 30	02 05	01 44	01 18	00 44	00 44	00 10	07 35
25	00 44	01 45	01 44	01 19	00 44	00 44	00 10	07 35
28	00 N 02	01 26	01 43	01 19	00 44	00 44	00 10	07 36
31	00 S 37	01 N 08	01 N 42	01 N 19	00 N 44	00 S 44	00 S 10	07 N 36

DATA

Julian Date	2453796
Delta T	+66 seconds
Ayanamsa	23° 56' 35"
Synetic vernal point	05° ♓ 10' 25"
True obliquity of ecliptic	23° 26' 28"

MOON'S PHASES, APSIDES AND POSITIONS ☽

Date	h	m	Phase	Longitude ° '	Eclipse Indicator
06	20	16	☽	16 ♊ 07	
14	23	35	○	24 ♍ 15	
22	19	10	☾	02 ♑ 01	
29	10	15	●	08 ♈ 35	Total

Day	h	m		
13	01	33	Apogee	
28	07	05	Perigee	
01	10	33	ON	
08	00	06	Max dec	28° N 42'
15	12	44	OS	
22	16	53	Max dec	28° S 43'
28	21	44	ON	

LONGITUDES

Date	Chiron ⚷ ° '	Ceres ⚳ ° '	Pallas ⚴ ° '	Juno ⚵ ° '	Vesta ⚶ ° '	Black Moon Lilith ⚸ ° '
01	06 ♒ 45	01 ♒ 29	07 ♑ 14	20 ♊ 20	07 ♋ 39	04 ♍ 14
11	07 ♒ 27	05 ♒ 04	10 ♑ 16	23 ♊ 39	08 ♋ 26	05 ♍ 20
21	08 ♒ 05	08 ♒ 30	13 ♑ 00	27 ♊ 18	09 ♋ 51	06 ♍ 27
31	08 ♒ 37	11 ♒ 46	15 ♑ 21	01 ♋ 12	11 ♋ 48	07 ♍ 33

All ephemeris data is given at 12.00 UT and the Moon's longitude is additionally given for 24.00 UT

ASPECTARIAN

h m	Aspects
01 Wednesday	
03 47	☽ □ ♇
04 03	☽ ☌ ☿
04 14	☽ ⚹ ♀
06 56	☽ # ☿
11 02	☉ ☌ ♅
14 21	☽ ∠ ♆
14 28	☽ ∥ ☿
15 28	☽ ⚼ ♃
18 02	☽ △ ♄
18 59	☽ ⚹ ♂
02 Thursday	
00 28	☽ Q ♀
02 37	☽ ⊻ ♅
03 41	☽ ⊻ ☉
03 42	☉ ± ♄
06 01	☽ ± ♃
10 48	☽ # ☉
12 26	☽ ⊥ ♅
14 11	☽ ⊥ ☉
14 26	☽ # ♅
14 45	☽ ⚹ ♆
15 50	☽ ⊼ ♃
20 20	☽ ∠ ♂
20 29	☿ St R
03 Friday	
03 22	☽ ∠ ♅
04 37	☽ △ ♇
05 12	☽ ⊻ ☿
06 13	☽ ∠ ☉
07 42	☽ □ ♀
09 12	♀ ∥ ♃
10 44	☽ Q ♆
11 56	☽ ⊥ ♂
15 13	☽ ⊥ ☿
18 04	☽ # ♆
19 22	☽ □ ♄
19 27	☽ # ♇
20 54	☽ # ♀
20 58	☽ # ♃
22 31	☽ ⊻ ♂
04 Saturday	
04 53	☽ ⚹ ♅
06 06	☽ ⚼ ♇
06 26	☽ ∠ ☿
09 44	☽ ⚹ ☉
12 15	☽ ∥ ♄
17 45	☽ □ ♆
18 03	♃ St R
18 47	☽ ☍ ♃
21 50	☽ ± ♇
22 21	☿ □ ♇
05 Sunday	
01 58	☽ Q ♅
02 33	☽ Q ♄
08 12	☽ ∥ ♂
08 14	☽ ⚹ ☿
08 30	☽ ⊼ ♇
08 34	☽ Q ☉
15 00	☽ △ ♀
21 57	☿ Q ♂
06 Monday	
00 04	☽ ⚹ ♄
05 18	☽ Q ☿
05 54	☽ ☌ ♂
10 22	♀ Q ♃
10 45	☽ □ ♅
20 16	☽ □ ☉
20 25	☽ ⚼ ♀
07 Tuesday	
00 27	☽ △ ♆
01 24	☽ ⊼ ♃
03 50	☽ ∠ ♄
05 46	☉ ∠ ♀
12 46	☽ ± ♃
13 34	☽ □ ☿
14 36	☽ ± ♀
16 09	☽ ☍ ♇
20 55	☽ ⊥ ♄
08 Wednesday	
01 50	♄ ± ♅
02 57	☽ ⊼ ♀
05 10	☽ ⚼ ♆
06 03	☽ ⚼ ♃
08 28	☽ ⊻ ♄
14 18	♀ ⊥ ♇
17 28	☽ ⊻ ♂
18 10	☿ ⊥ ♆
20 17	☽ △ ♅
22 46	☽ ± ♆
09 Thursday	
03 27	☉ ⊻ ♆
05 48	☽ ⊥ ♂
10 38	☽ ⊼ ♆
11 16	☽ △ ☉
11 26	☽ △ ♃
13 14	☉ △ ♃
20 41	☽ △ ☿
10 Friday	
00 30	☽ ∠ ♂
02 08	☽ ⚼ ♅
03 00	☽ ⊼ ♇
03 50	♀ ∥ ♇
15 02	☽ ± ♇
16 50	☉ ⚼ ♄
16 59	☽ ∥ ♂
18 43	☽ ☍ ♀
19 41	☽ ☌ ♄
19 58	☽ ⚼ ☉
20 16	☽ ± ♅
11 Saturday	
00 38	☽ ⚼ ☿
07 31	♀ ☍ ♄
08 04	☽ ⚹ ♂
08 26	☽ ⊼ ♅
09 12	☽ ⚼ ♇
15 48	☽ ± ☉
16 49	♂ □ ♅
17 27	♀ ⊥ ♅
17 29	☽ ± ☿
18 18	☽ ∥ ♄
23 04	☽ ☍ ♆
23 36	☽ □ ♃
12 Sunday	
02 44	☉ ☌ ☿
04 44	☽ ⊼ ☿
05 05	☽ ⊼ ☉
09 33	☽ Q ♂
12 19	☿ ∠ ♀
13 03	☽ # ♃
14 43	☽ # ♇
15 38	☽ △ ♇
15 47	☽ # ♀
16 44	☽ # ♆
13 Monday	
08 16	☽ ⊻ ♄
12 06	☽ Q ♃
12 23	☽ ⊼ ♀
16 30	☿ ⚼ ♄
20 25	☽ ⊥ ♄
21 37	☽ ☍ ♅
23 53	☽ □ ♂
14 Tuesday	
01 34	☽ ± ♀
03 00	☽ # ♅
12 09	☽ ⊼ ♆
12 23	☽ ⚹ ♃
13 01	☽ ☍ ☿
14 37	☽ ∠ ♄
15 28	♀ ∥ ♆
21 10	☿ △ ♃
21 18	☽ ⚼ ♀
23 28	☿ ⊻ ♆
23 35	☽ ☍ ☉
15 Wednesday	
00 20	☽ ± ♆
03 26	☽ # ☉
04 33	☽ □ ♇
05 38	☽ # ☿
08 43	☉ ⊥ ♆
18 29	☉ ∥ ☿
18 32	☽ ⚼ ♆
18 36	☽ ∠ ♃
20 48	☽ ⚹ ♄
20 49	☽ ∥ ☉
21 13	☽ ∥ ☿
16 Thursday	
06 00	☽ △ ♀
07 05	♃ □ ♆
10 27	☽ ⊼ ♅
12 35	☽ ⊥ ♃
15 16	☽ △ ♂
16 43	☽ Q ♇
20 44	☽ Q ♄
21 07	☽ ⊼ ☿
21 52	☽ ∥ ♅
22 28	☽ ± ♅
17 Friday	
00 31	☽ ⊻ ♃
00 37	☽ △ ♆
08 17	☽ ± ☿
10 51	☿ ⊥ ♀
11 07	☉ □ ♇
16 23	☽ ⚼ ♅
16 31	☽ ⚹ ♇
17 00	☽ ⊼ ☉
22 24	☽ ⚼ ♂
18 Saturday	
00 58	☽ ⚼ ☿
05 07	☽ ∥ ♀
05 49	☽ ± ☉
07 05	☽ ∥ ♆
08 08	☽ □ ♄
09 13	☽ ∥ ♇
10 32	☽ ∥ ♃
16 48	☽ ± ♂
18 45	♀ ⊻ ♅
19 15	♀ ∠ ♇
21 49	☽ △ ♅
21 50	☽ ∠ ♇
22 03	☽ □ ♀
19 Sunday	
00 51	☽ ⚼ ☉
01 09	☿ □ ♂
04 35	☽ △ ☿
04 56	☽ ⊼ ♂
05 07	☽ # ♄
10 56	☽ ☌ ♃
11 22	☽ □ ♆
15 09	☽ ⊥ ♇
20 Monday	
02 33	☽ ⊻ ♇
07 54	☽ △ ☉
08 52	☽ # ♂
11 09	☽ Q ♀
17 20	☽ △ ♄
21 25	☽ Q ♆
21 Tuesday	
06 45	☽ □ ♅
07 09	☿ ⊻ ♀
10 46	☽ □ ☿
11 11	☽ ⚹ ♀
15 45	☽ ☍ ♂
18 46	☽ ⊻ ♃
19 31	☽ ⚹ ♆
20 53	☽ ⚼ ♄
22 Wednesday	
05 34	☽ ⊥ ♃
09 47	☽ ☌ ♇
13 02	☽ ± ♄
15 27	☽ Q ♅
16 22	☽ ∠ ♀
18 23	☽ Q ☿
19 10	☽ □ ☉
21 32	☽ ∠ ♃
22 26	☽ ∠ ♆
23 38	☽ ⊼ ♄
23 Thursday	
09 24	☽ ⊥ ♀
12 36	☽ ⚹ ♅
14 11	☽ ⊥ ♆
15 02	☽ ⚹ ☿
20 34	☽ ⊻ ♀
23 04	☽ ⊼ ♂
23 30	☽ ⚹ ♃
24 Friday	
00 32	☽ ⊻ ♆
01 50	☉ ⚼ ♃
04 50	☽ Q ☉
09 05	♂ ⊼ ♃
09 43	☽ ± ♂
13 53	☽ ⊻ ♇
14 21	☽ ∠ ♅
16 22	☽ ∠ ☿
16 59	☉ ∠ ♆
19 45	☽ Q ♃
20 18	☽ # ♂
23 57	☽ ⊥ ♇
25 Saturday	
01 27	☽ ⚼ ♂
02 33	☽ ⚹ ☉
02 52	☽ ☍ ♄
05 27	☽ ⊥ ♅
07 03	☉ △ ♄
07 19	☽ ⊥ ☿
12 09	♀ □ ♃
13 43	☿ St D
13 52	♂ △ ♆
14 52	☽ ∠ ♇
15 24	☽ ⊻ ♅
17 13	☽ ⊻ ☿
21 39	☽ # ♄
26 Sunday	
01 22	☽ □ ♃
02 23	☽ ☌ ♀
02 39	☽ ☌ ♆
03 08	☽ △ ♂
05 00	☽ ∠ ☉
06 30	♀ ☌ ♆
12 42	♂ ∠ ♄
14 00	☽ ∥ ♃
14 38	☽ ∥ ♇
15 18	☽ ⚹ ♇
16 36	☽ ∥ ♆
20 37	☽ ⊥ ☉
23 47	☽ ∥ ♀
27 Monday	
02 41	♀ △ ♂
03 42	☽ ⊼ ♄
06 56	☽ ⊻ ☉
10 36	☽ Q ♇
13 18	☽ ± ♄
16 05	☽ ☌ ♅
18 02	☽ ☌ ☿
20 07	☽ ∥ ♅
28 Tuesday	
00 30	☽ ∥ ☿
01 25	☽ △ ♃
02 58	☽ ⊻ ♆
03 40	☽ ⚼ ♄
05 13	☽ □ ♂
06 03	☽ ⊻ ♀
11 41	☽ # ☉
12 34	☽ ⊥ ♆
15 20	☽ □ ♇
16 19	☽ ⊥ ♀
29 Wednesday	
01 14	☽ ⚼ ♃
02 58	☽ ∠ ♆
03 35	☽ △ ♄
07 45	☽ ∠ ♀
08 59	☽ ∥ ☉
10 15	☽ ☌ ☉
11 06	☽ Q ♂
12 39	♇ St R
15 36	☽ ± ♃
16 13	☽ ⊻ ♅
18 57	☽ ⊻ ☿
19 38	☽ # ☿
23 22	☽ # ♅
30 Thursday	
01 14	☽ ⊼ ♃
01 57	☽ ⊥ ♅
03 09	☽ ⚹ ♆
04 56	☽ ⊥ ☿
07 15	☽ ⚹ ♂
09 44	☽ ⚹ ♀
15 05	☉ Q ♂
15 41	☽ △ ♇
16 39	☽ ∠ ♅
17 04	☽ # ♀
20 04	☽ ∠ ☿
22 48	☽ Q ♆
31 Friday	
03 24	☽ # ♆
04 17	☽ □ ♄
05 32	☽ # ♇
05 49	☽ # ♃
06 58	☽ Q ♀
08 55	☽ ∠ ♂
14 52	☽ ⊻ ☉
16 33	☽ ⚼ ♇
17 39	☽ ⚹ ♅
21 58	☽ ⚹ ☿
23 11	☽ ∥ ♄

APRIL 2006

LONGITUDES

Date	Sidereal time h m s	Sun ☉ ° ' "	Moon ☽ ° ' "	Moon ☽ 24.00 ° ' "	Mercury ☿ ° '	Venus ♀ ° '	Mars ♂ ° '	Jupiter ♃ ° '	Saturn ♄ ° '	Uranus ♅ ° '	Neptune ♆ ° '	Pluto ⯓ ° '
01	00 38 50	11 ♈ 37 12	23 ♉ 11 20	00 ♊ 06 11	15 ♓ 17	25 ♒ 16	22 ♊ 50	17 ♏ 41	04 ♌ 23	12 ♓ 27	19 ♒ 08	26 ♐ 45
02	00 42 46	12 36 25	06 ♊ 54 23	13 ♊ 35 57	15 54	26 18	23 24	17 R 36	04 R 23	12 30	19 09	26 R 45
03	00 46 43	13 35 36	20 ♊ 11 03	26 ♊ 39 58	16 34	27 20	23 58	17 31	04 23	12 33	19 11	26 45
04	00 50 40	14 34 44	03 ♋ 03 02	09 ♋ 20 44	17 19	28 22	24 32	17 26	04 23	12 36	19 12	26 45
05	00 54 36	15 33 50	15 ♋ 33 32	21 ♋ 41 58	18 06	29 ♒ 25	25 06	17 21	04 23	12 39	19 14	26 45
06	00 58 33	16 32 54	27 ♋ 46 38	03 ♌ 48 05	18 57	00 ♓ 28	25 40	17 15	04 D 23	12 42	19 15	26 44
07	01 02 29	17 31 55	09 ♌ 46 54	15 ♌ 43 40	19 52	01 31	26 15	17 09	04 23	12 45	19 16	26 44
08	01 06 26	18 30 54	21 ♌ 38 57	27 ♌ 33 17	20 49	02 35	26 49	17 04	04 23	12 48	19 18	26 44
09	01 10 22	19 29 51	03 ♍ 27 12	09 ♍ 21 09	21 49	03 38	27 23	16 58	04 23	12 51	19 19	26 43
10	01 14 19	20 28 45	15 ♍ 15 37	21 ♍ 11 01	22 52	04 42	27 58	16 52	04 24	12 54	19 20	26 43
11	01 18 15	21 27 38	27 ♍ 07 43	03 ♎ 06 05	23 57	05 47	28 32	16 45	04 24	12 57	19 22	26 43
12	01 22 12	22 26 28	09 ♎ 06 24	15 ♎ 08 57	25 05	06 51	29 07	16 39	04 25	12 59	19 23	26 42
13	01 26 09	23 25 16	21 ♎ 13 59	27 ♎ 21 41	26 15	07 56	29 ♊ 41	16 33	04 26	13 02	19 24	26 42
14	01 30 05	24 24 02	03 ♏ 32 13	09 ♏ 45 45	27 27	09 01	00 ♋ 16	16 26	04 27	13 05	19 26	26 41
15	01 34 02	25 22 46	16 ♏ 02 24	22 ♏ 22 15	28 42	10 06	00 51	16 20	04 28	13 08	19 27	26 41
16	01 37 58	26 21 29	28 ♏ 45 25	05 ♐ 11 58	29 ♓ 59	11 11	01 25	16 13	04 29	13 10	19 28	26 40
17	01 41 55	27 20 10	11 ♐ 41 58	18 ♐ 15 28	01 ♈ 18	12 16	02 00	16 06	04 30	13 13	19 29	26 40
18	01 45 51	28 18 49	24 ♐ 52 32	01 ♑ 33 13	02 39	13 22	02 35	15 59	04 32	13 16	19 30	26 39
19	01 49 48	29 ♈ 17 26	08 ♑ 17 34	15 ♑ 05 37	04 02	14 28	03 10	15 52	04 33	13 19	19 31	26 39
20	01 53 44	00 ♉ 16 02	21 ♑ 57 22	28 ♑ 52 50	05 26	15 34	03 44	15 45	04 35	13 21	19 32	26 38
21	01 57 41	01 14 36	05 ♒ 51 58	12 ♒ 54 41	06 53	16 40	04 19	15 38	04 36	13 24	19 33	26 37
22	02 01 38	02 13 08	20 ♒ 00 50	27 ♒ 10 12	08 22	17 46	04 54	15 31	04 38	13 26	19 34	26 36
23	02 05 34	03 11 39	04 ♓ 22 32	11 ♓ 37 27	09 52	18 52	05 29	15 23	04 40	13 29	19 35	26 36
24	02 09 31	04 10 08	18 ♓ 54 28	26 ♓ 13 04	11 24	19 59	06 04	15 16	04 42	13 31	19 36	26 35
25	02 13 27	05 08 35	03 ♈ 32 35	10 ♈ 52 20	12 58	21 06	06 39	15 09	04 44	13 34	19 37	26 34
26	02 17 24	06 07 01	18 ♈ 11 32	25 ♈ 29 23	14 34	22 13	07 14	15 01	04 46	13 36	19 38	26 33
27	02 21 20	07 05 25	02 ♉ 45 04	09 ♉ 57 46	16 12	23 20	07 49	14 54	04 49	13 38	19 39	26 32
28	02 25 17	08 03 47	17 ♉ 06 45	24 ♉ 11 20	17 51	24 27	08 25	14 46	04 51	13 41	19 40	26 31
29	02 29 13	09 02 08	01 ♊ 10 57	08 ♊ 05 08	19 32	25 34	09 00	14 38	04 53	13 43	19 40	26 31
30	02 33 10	10 ♉ 00 26	14 ♊ 53 33	21 ♊ 36 01	21 ♈ 15	26 ♓ 42	09 ♋ 35	14 ♏ 31	04 ♌ 56	13 ♓ 45	19 ♒ 41	26 ♐ 30

Date	Moon True ☊ ° '	Moon Mean ☊ ° '	Moon ☽ Latitude ° '
01	04 ♈ 20	04 ♈ 12	03 N 59
02	04 R 19	04 09	04 42
03	04 18	04 06	05 08
04	04 18	04 03	05 17
05	04 D 17	04 00	05 11
06	04 18	03 56	04 51
07	04 18	03 53	04 18
08	04 20	03 50	03 35
09	04 21	03 47	02 43
10	04 22	03 44	01 44
11	04 23	03 40	00 N 40
12	04 R 23	03 37	00 S 26
13	04 22	03 34	01 32
14	04 21	03 31	02 34
15	04 18	03 28	03 30
16	04 15	03 25	04 16
17	04 11	03 21	04 51
18	04 08	03 18	05 10
19	04 05	03 15	05 13
20	04 04	03 12	04 59
21	04 D 04	03 09	04 27
22	04 05	03 06	03 39
23	04 06	03 02	02 36
24	04 07	02 59	01 23
25	04 08	02 56	00 S 03
26	04 R 07	02 53	01 N 16
27	04 05	02 50	02 30
28	04 01	02 46	03 33
29	03 56	02 43	04 22
30	03 ♈ 50	02 ♈ 40	04 N 54

DECLINATIONS

Date	Sun ☉ ° '	Moon ☽ ° '	Mercury ☿ ° '	Venus ♀ ° '	Mars ♂ ° '	Jupiter ♃ ° '	Saturn ♄ ° '	Uranus ♅ ° '	Neptune ♆ ° '	Pluto ⯓ ° '
01	04 N 36	22 N 26	06 S 33	12 S 08	24 N 57	15 S 50	19 N 53	07 S 35	15 S 15	15 S 48
02	04 59	26 06	06 29	11 53	24 58	15 49	19 53	07 33	15 15	15 48
03	05 22	28 11	06 23	11 37	25 00	15 47	19 53	07 32	15 14	15 47
04	05 45	28 41	06 15	11 21	25 01	15 46	19 53	07 31	15 14	15 47
05	06 08	27 41	06 05	11 05	25 02	15 44	19 53	07 30	15 13	15 47
06	06 30	25 22	05 53	10 48	25 03	15 43	19 53	07 29	15 13	15 47
07	06 53	21 57	05 39	10 31	25 04	15 41	19 53	07 28	15 13	15 47
08	07 15	17 41	05 24	10 13	25 05	15 39	19 53	07 27	15 12	15 47
09	07 38	12 46	05 06	09 55	25 05	15 37	19 53	07 25	15 12	15 47
10	08 00	07 24	04 48	09 37	25 06	15 36	19 53	07 24	15 11	15 46
11	08 22	01 N 45	04 27	09 18	25 06	15 34	19 53	07 23	15 11	15 46
12	08 44	04 S 01	04 05	08 59	25 06	15 32	19 53	07 22	15 11	15 46
13	09 06	09 42	03 41	08 39	25 06	15 30	19 53	07 21	15 10	15 46
14	09 28	15 07	03 17	08 20	25 06	15 28	19 52	07 20	15 10	15 46
15	09 49	19 59	02 50	08 00	25 05	15 26	19 52	07 19	15 10	15 46
16	10 10	24 03	02 22	07 39	25 05	15 24	19 52	07 18	15 09	15 46
17	10 32	26 59	01 53	07 18	25 04	15 22	19 52	07 17	15 09	15 45
18	10 53	28 30	01 23	06 57	25 03	15 20	19 51	07 16	15 08	15 45
19	11 13	28 24	00 51	06 36	25 02	15 18	19 51	07 15	15 08	15 45
20	11 34	26 34	00 S 18	06 15	25 01	15 16	19 51	07 14	15 08	15 45
21	11 54	23 07	00 N 15	05 53	24 59	15 14	19 50	07 13	15 08	15 45
22	12 15	18 16	00 51	05 31	24 58	15 12	19 50	07 12	15 07	15 45
23	12 35	12 20	01 27	05 08	24 56	15 10	19 49	07 11	15 07	15 45
24	12 55	05 S 39	02 04	04 46	24 54	15 08	19 49	07 10	15 07	15 45
25	13 14	01 N 22	02 42	04 23	24 52	15 06	19 48	07 09	15 06	15 44
26	13 34	08 19	03 22	04 00	24 50	15 04	19 48	07 08	15 06	15 44
27	13 53	14 47	04 02	03 37	24 48	15 01	19 47	07 07	15 06	15 44
28	14 12	20 21	04 43	03 13	24 46	14 59	19 47	07 07	15 06	15 44
29	14 31	24 40	05 25	02 50	24 43	14 57	19 46	07 06	15 05	15 44
30	14 N 49	27 N 28	06 N 07	02 S 26	24 N 40	14 S 55	19 N 46	07 S 05	15 S 05	15 S 44

ZODIAC SIGN ENTRIES

Date	h m	Planets
01	23 49	☽ ♊
04	06 15	☽ ♋
06	01 21	♀ ♓
06	16 25	☽ ♌
09	04 58	☽ ♍
11	17 47	☽ ♎
14	00 59	♂ ♋
14	05 08	☽ ♏
16	12 20	☿ ♈
16	14 19	☽ ♐
18	21 13	☽ ♑
20	05 26	☉ ♉
21	01 56	☽ ♒
23	04 43	☽ ♓
25	06 12	☽ ♈
27	07 27	☽ ♉
29	09 58	☽ ♊

LATITUDES

Date	Mercury ☿ ° '	Venus ♀ ° '	Mars ♂ ° '	Jupiter ♃ ° '	Saturn ♄ ° '	Uranus ♅ ° '	Neptune ♆ ° '	Pluto ⯓ ° '
01	00 S 48	01 N 02	01 N 42	01 N 19	00 N 44	00 S 44	00 S 10	07 N 36
04	01 20	00 44	01 42	01 20	00 45	00 44	00 11	07 37
07	01 47	00 27	01 41	01 20	00 45	00 44	00 11	07 37
10	02 08	00 N 11	01 40	01 20	00 45	00 45	00 11	07 38
13	02 24	00 S 04	01 39	01 20	00 45	00 45	00 11	07 38
16	02 35	00 18	01 39	01 20	00 45	00 45	00 11	07 38
19	02 41	00 32	01 38	01 21	00 45	00 45	00 11	07 39
22	02 41	00 44	01 37	01 21	00 45	00 45	00 11	07 39
25	02 37	00 56	01 36	01 21	00 45	00 45	00 11	07 39
28	02 29	01 06	01 35	01 20	00 45	00 45	00 11	07 40
31	02 S 15	01 S 16	01 N 34	01 N 20	00 N 45	00 S 45	00 S 11	07 N 40

DATA

Julian Date	2453827
Delta T	+66 seconds
Ayanamsa	23° 56' 38"
Synetic vernal point	05° ♓ 10' 21"
True obliquity of ecliptic	23° 26' 28"

MOON'S PHASES, APSIDES AND POSITIONS ☽

Date	h m	Phase	Longitude ° '	Eclipse Indicator
05	12 01	☽	15 ♋ 34	
13	16 40	○	23 ♎ 37	
21	03 28	☾	00 ♒ 54	
27	19 44	●	07 ♉ 24	

Day	h m		
09	13 12	Apogee	
25	10 24	Perigee	
04	07 38	Max dec	28° N 43'
11	19 19	0S	
18	22 26	Max dec	28° S 40'
25	07 23	0N	

LONGITUDES

Date	Chiron ⚷ ° '	Ceres ⚳ ° '	Pallas ⚴ ° '	Juno ⚵ ° '	Vesta ⚶ ° '	Black Moon Lilith ⚸ ° '
01	08 ♒ 40	12 ♒ 05	15 ♑ 33	01 ♋ 36	12 ♋ 02	07 ♍ 40
11	09 ♒ 06	15 ♒ 07	17 ♑ 25	05 ♋ 43	14 ♋ 30	08 ♍ 47
21	09 ♒ 26	17 ♒ 54	18 ♑ 45	09 ♋ 58	17 ♋ 22	09 ♍ 53
31	09 ♒ 39	20 ♒ 25	19 ♑ 29	14 ♋ 20	20 ♋ 33	11 ♍ 00

All ephemeris data is given at 12.00 UT and the Moon's longitude is additionally given for 24.00 UT

ASPECTARIAN

h m	Aspects
01 Saturday	
00 40	☽ ⊥ ♂
01 45	☽ ⊥ ☉
02 39	☽ ☍ ♃
05 01	☽ □ ♆
07 49	☽ ± ⯓
10 38	☽ Q ♄
11 21	☽ ⊻ ♂
13 34	☉ ± ♃
14 10	☽ Q ♅
15 52	☽ □ ♀
18 10	☽ ⊼ ⯓
18 23	☽ ∠ ☉
19 24	☽ Q ☿
02 Sunday	
03 29	☽ ∥ ♂
07 32	☽ ✱ ♄
09 07	☉ ⊻ ♅
19 44	♂ ± ♃
22 03	☽ □ ♅
22 35	♀ ✱ ⯓
23 02	☽ ✱ ☉
03 Monday	
05 02	☽ □ ☿
07 09	☽ ⊼ ♃
08 12	♃ ∥ ⯓
10 09	☽ △ ♆
10 31	☽ ∠ ♄
18 07	☽ ± ♃
19 18	☽ ☌ ♂
22 50	☽ Q ☉
04 Tuesday	
00 09	☽ ☍ ⯓
02 24	☽ △ ♀
03 12	☽ ⊥ ♄
10 50	☽ ♇ ♃
14 11	☽ ♇ ♆
14 31	☽ ⊻ ♄
15 25	☿ △ ♃
05 Wednesday	
06 20	☽ △ ♅
07 28	☽ ± ♆
09 35	☽ ♇ ♀
09 53	☉ # ☿
12 01	☽ □ ☉
12 54	♄ St D
15 27	☽ △ ♃
17 19	☽ △ ☿
19 10	☽ ⊼ ♆
06 Thursday	
04 50	☽ ± ♀
07 38	☽ ⊻ ♂
09 56	☽ ⊼ ⯓
11 50	☽ ♇ ♅
14 27	☽ ∥ ♂
17 51	☽ ⊼ ♀
20 08	☽ ⊥ ♂
20 08	☿ ⊻ ♆
21 52	☽ ± ⯓
23 20	☿ ♇ ♄
07 Friday	
01 09	☽ ☌ ♄
01 18	☽ ♇ ☿
03 38	☉ ⊼ ♃
05 53	☽ ± ♅
15 06	☽ ∠ ♂
15 56	☽ ♇ ⯓
18 00	☽ ⊼ ♅
20 56	☽ ± ☿
08 Saturday	
00 07	☽ ∥ ♄
02 46	☽ □ ♃
05 04	☽ △ ☉
07 13	☽ ☍ ♆
08 22	♂ ☍ ⯓
10 09	☽ ⊼ ☿
19 14	☉ ⊥ ♅
21 38	☽ # ⯓
22 19	☽ △ ⯓
22 19	☽ # ♃
23 02	☽ ✱ ♂
23 18	☉ # ♅
09 Sunday	
00 28	☽ # ♆
07 33	☉ ✱ ♆
12 25	☽ ☍ ♀
13 54	☽ ⊻ ♄
14 19	☽ ♇ ☉
15 02	☽ Q ♃
10 Monday	
00 41	☽ Q ♂
01 45	☽ # ♀
02 07	☽ ⊥ ♄
04 59	♀ ⊼ ♄
07 11	☽ ☍ ♅
09 35	☽ ∥ ☉
10 16	☽ ± ☉
12 00	☽ # ♅
15 13	☽ ✱ ♃
20 17	☽ ⊼ ♆
20 24	☽ ∠ ♄
23 32	☽ ⊼ ☉
23 55	☽ # ☿
11 Tuesday	
04 56	☽ ☍ ☿
06 30	♂ ⊥ ♄
08 26	☽ ± ♆
11 09	☽ □ ⯓
14 59	☽ □ ♂
21 13	☽ ∠ ♃
12 Wednesday	
02 33	☽ ♇ ♆
02 38	☽ ✱ ♄
07 04	☽ ⊼ ♀
12 17	☽ ∥ ☿
15 03	☽ ⊥ ♃
18 28	☿ ⊥ ♆
19 45	☽ ⊼ ♅
20 10	☽ ± ♀
20 36	☉ # ♀
23 07	☽ Q ⯓
13 Thursday	
02 02	☽ ∥ ♅
02 32	☽ Q ♄
02 50	☽ ⊻ ♃
07 40	☽ ± ♅
07 47	☽ ∥ ♀
08 24	☽ △ ♆
09 14	☽ # ☉
15 39	☽ ♇ ♀
16 40	☽ ☍ ☉
20 54	☿ □ ⯓
22 42	☽ ✱ ⯓
22 53	☽ ⊼ ☿
14 Friday	
01 22	☽ ♇ ♅
04 56	♀ Q ⯓
05 21	☽ △ ♂
11 50	☽ ± ☿
12 15	☽ ∥ ♆
13 40	☽ ∥ ♃
13 46	☽ □ ♄
15 03	☽ ∥ ⯓
23 33	☽ △ ♀
15 Saturday	
03 41	☽ ∠ ⯓
06 26	☽ △ ♅
07 03	☽ ♇ ☿
11 23	☽ # ♄
11 36	☽ ♇ ♂
12 32	☽ ☌ ♃
18 29	☽ □ ♆
20 23	♀ ± ♄
20 48	☽ ⊥ ⯓
16 Sunday	
04 49	♂ ♇ ♃
05 27	☽ ± ♂
07 08	☽ ⊼ ☉
08 06	☽ ⊻ ⯓
14 33	☽ △ ☿
17 12	☽ ⊼ ♂
19 16	☽ ± ☉
19 17	☽ # ♂
19 35	☉ △ ⯓
22 41	☽ △ ♄
17 Monday	
04 13	☽ Q ♆
08 43	☿ ♇ ♃
13 09	☽ □ ♀
13 16	☽ ♇ ☉
13 50	♀ ∥ ♅
14 48	☽ □ ♅
20 00	☽ ⊻ ♃
18 Tuesday	
01 39	☉ ∠ ♀
02 15	☽ ✱ ♆
02 18	☽ ♇ ♄
06 49	☽ ⊥ ♃
09 43	♀ ☌ ♅
09 56	☿ □ ♂
10 44	☉ ∠ ♅
15 12	☽ ☌ ⯓
18 35	☽ ± ♄
18 41	☽ △ ☉
22 53	☽ ∠ ♃
23 31	☽ Q ♅
19 Wednesday	
00 42	☽ Q ♀
02 27	☽ ☍ ♂
03 33	☽ □ ☿
05 17	☽ ∠ ♆
05 20	☽ ⊼ ♄
20 32	☿ ∠ ♆
20 53	☽ ✱ ♅
21 08	☿ △ ♄
21 15	☽ ⊥ ♆
23 50	☽ ✱ ♀
20 Thursday	
01 15	☽ ✱ ♃
07 47	☽ ⊻ ♆
14 53	☽ Q ☿
15 45	♀ △ ♃
20 06	☽ ⊻ ⯓
21 58	☽ Q ♃
23 07	☽ ∠ ♅
21 Friday	
00 15	☽ # ♂
03 28	☽ □ ☉
04 10	☽ ∠ ♀
06 26	☽ ⊥ ⯓
09 14	☽ ⊼ ♂
09 50	☽ ☍ ♄
13 57	☽ ✱ ☿
14 37	☽ ⊥ ♅
19 48	☉ Q ♆
19 56	☽ ± ♂
20 52	☽ ⊥ ♀
21 48	☽ ∠ ⯓
21 56	♂ ♇ ♆
22 Saturday	
00 22	♂ ⊻ ♄
00 52	☽ ⊻ ♅
04 28	☽ □ ♃
04 54	☽ # ♄
07 54	☽ ⊻ ♀
11 15	☽ ☌ ♆
11 48	☽ ♇ ♂
12 22	☽ Q ☉
18 17	☽ ∠ ☿
22 39	☽ ∥ ⯓
23 03	☽ ✱ ⯓
23 Sunday	
00 55	☽ ∥ ♃
01 11	☽ ∥ ♆
05 00	☿ ± ♃
09 54	☽ ✱ ☉
11 03	☽ ⊥ ☿
11 06	☽ # ☉
12 29	☽ ⊼ ♄
13 55	☽ △ ♂
18 59	☽ Q ⯓
22 10	☽ ⊻ ☿
22 27	☽ ± ♄
24 Monday	
03 06	☽ ☌ ♅
03 37	♀ ⊻ ♆
05 38	♀ ♇ ♄
06 04	☽ △ ♃
06 42	☽ ∥ ♅
12 28	☽ ∠ ☉
13 09	☽ ⊻ ♆
13 18	☽ ♇ ♄
13 55	☽ ☌ ♀
15 16	☽ ∥ ♀
23 00	☽ ⊥ ♆
23 19	☽ # ☿
25 Tuesday	
00 35	☽ □ ⯓
01 33	☉ □ ♄
02 50	♃ ∥ ♆
04 17	☽ ⊥ ☉
06 29	☽ ♇ ♃
13 46	☽ ∠ ♆
13 57	☽ △ ♄
14 48	☽ ⊻ ☉
17 03	☽ ∥ ☿
17 18	☽ □ ♂
21 06	☽ ± ♃
21 09	☿ ⊻ ♅
21 47	☽ # ♀
26 Wednesday	
04 27	☽ ⊻ ♅
05 19	☽ ☌ ☿
06 50	☽ ⊼ ♃
07 54	☽ # ♅
14 19	☽ ⊥ ♅
14 22	☽ ✱ ♆
18 14	☿ ⊼ ♃
19 09	☽ ⊻ ♀
27 Thursday	
00 04	☽ Q ♂
01 44	☽ △ ⯓
02 34	☿ # ♀
05 11	☽ ∠ ♅
05 52	☽ ⊥ ♀
08 19	☽ ∥ ☉
10 10	☽ Q ♆
12 58	☽ # ♃
13 16	☽ # ♆
15 25	☽ □ ♄
15 50	☽ # ⯓
19 44	☽ ☌ ☉
20 47	☽ ✱ ♂
22 03	☽ ∠ ♀
28 Friday	
02 37	☽ ♇ ⯓
06 13	☽ ✱ ♅
08 05	☽ ☍ ♃
09 18	☽ ∥ ♄
13 25	☽ ⊻ ☿
16 18	☽ □ ♆
17 46	☽ ± ⯓
21 45	☽ Q ♄
23 08	☽ ∠ ♂
29 Saturday	
00 58	☽ ⊥ ☿
01 31	☽ ✱ ♀
02 35	☽ Q ♅
03 58	☽ ⊼ ⯓
09 32	☉ ✱ ♂
12 18	☽ ∥ ♂
13 56	☿ ✱ ♆
14 12	♀ ⊥ ♆
14 38	☿ ⊥ ♅
15 17	☽ ⊥ ♂
18 27	☽ ✱ ♄
18 38	☽ ∠ ☿
30 Sunday	
00 05	☽ Q ♀
02 12	☽ ⊻ ♂
02 43	☽ ⊻ ☉
07 47	♀ □ ⯓
09 59	☽ □ ♅
11 20	☽ ⊼ ♃
14 08	☽ ⊥ ☉
18 47	☉ # ♃
20 34	☽ △ ♆
21 02	☽ ∠ ♄
21 57	☽ ± ♃

MAY 2006

LONGITUDES

Date	Sidereal time h m s	Sun ☉ ° ' "	Moon ☽ ° ' "	Moon ☽ 24.00 ° ' "	Mercury ☿ ° '	Venus ♀ ° '	Mars ♂ ° '	Jupiter ♃ ° '	Saturn ♄ ° '	Uranus ♅ ° '	Neptune ♆ ° '	Pluto ♇ ° '
01	02 37 07	10 ♉ 58 43	28 ♊ 12 28	04 ♋ 42 58	23 ♈ 00	27 ♓ 49	10 ♋ 10	14 ♏ 23	04 ♌ 59	13 ♓ 48	19 ♒ 42	26 ♐ 29
02	02 41 03	11 56 58	11 ♋ 07 44	17 ♋ 27 02	24 46	28 ♓ 57	10 46	14 R 16	05 02	13 50	19 42	26 R 28
03	02 45 00	12 55 11	23 ♋ 41 16	29 ♋ 50 55	26 35	00 ♈ 04	11 21	14 08	05 04	13 52	19 43	26 27
04	02 48 56	13 53 21	05 ♌ 56 30	11 ♌ 58 35	28 ♈ 25	01 12	11 56	14 00	05 07	13 54	19 44	26 26
05	02 52 53	14 51 30	17 ♌ 57 49	23 ♌ 54 47	00 ♉ 17	02 20	12 32	13 53	05 10	13 56	19 44	26 25
06	02 56 49	15 49 37	29 ♌ 50 10	05 ♍ 44 35	02 10	03 28	13 07	13 45	05 14	13 58	19 45	26 23
07	03 00 46	16 47 41	11 ♍ 38 42	17 ♍ 33 06	04 06	04 37	13 43	13 38	05 17	14 00	19 45	26 22
08	03 04 42	17 45 44	23 ♍ 28 24	29 ♍ 25 10	06 03	05 45	14 18	13 30	05 20	14 02	19 46	26 21
09	03 08 39	18 43 45	05 ♎ 23 54	11 ♎ 25 05	08 02	06 53	14 54	13 23	05 24	14 04	19 46	26 20
10	03 12 36	19 41 44	17 ♎ 29 09	23 ♎ 36 26	10 03	08 02	15 29	13 15	05 27	14 06	19 47	26 19
11	03 16 32	20 39 42	29 ♎ 47 16	06 ♏ 01 52	12 05	09 10	16 05	13 07	05 31	14 08	19 47	26 18
12	03 20 29	21 37 38	12 ♏ 20 23	18 ♏ 42 55	14 09	10 19	16 40	13 00	05 35	14 10	19 47	26 17
13	03 24 25	22 35 32	25 ♏ 09 29	01 ♐ 40 01	16 15	11 27	17 16	12 52	05 39	14 11	19 48	26 15
14	03 28 22	23 33 25	08 ♐ 14 26	14 ♐ 52 34	18 22	12 36	17 51	12 45	05 42	14 13	19 48	26 14
15	03 32 18	24 31 16	21 ♐ 34 10	28 ♐ 19 02	20 30	13 45	18 27	12 37	05 46	14 15	19 48	26 13
16	03 36 15	25 29 06	05 ♑ 06 54	11 ♑ 57 28	22 39	14 54	19 03	12 30	05 50	14 16	19 48	26 12
17	03 40 11	26 26 55	18 ♑ 50 30	25 ♑ 45 45	24 49	16 03	19 39	12 23	05 55	14 18	19 49	26 10
18	03 44 08	27 24 43	02 ♒ 42 57	09 ♒ 41 56	27 00	17 12	20 14	12 16	05 59	14 19	19 49	26 09
19	03 48 05	28 22 29	16 ♒ 42 29	23 ♒ 44 27	29 ♉ 11	18 22	20 50	12 08	06 03	14 21	19 49	26 08
20	03 52 01	29 ♉ 20 15	00 ♓ 47 41	07 ♓ 52 04	01 ♊ 23	19 31	21 26	12 01	06 08	14 22	19 49	26 06
21	03 55 58	00 ♊ 17 59	14 ♓ 57 24	22 ♓ 03 34	03 34	20 40	22 02	11 54	06 12	14 24	19 49	26 05
22	03 59 54	01 15 42	29 ♓ 10 20	06 ♈ 17 28	05 45	21 50	22 38	11 48	06 17	14 25	19 49	26 04
23	04 03 51	02 13 24	13 ♈ 24 41	20 ♈ 31 37	07 56	22 59	23 13	11 41	06 21	14 26	19 R 49	26 02
24	04 07 47	03 11 05	27 ♈ 37 52	04 ♉ 42 58	10 05	24 09	23 49	11 34	06 26	14 28	19 49	26 01
25	04 11 44	04 08 45	11 ♉ 46 26	18 ♉ 47 44	12 14	25 19	24 25	11 27	06 31	14 29	19 49	25 59
26	04 15 40	05 06 24	25 ♉ 46 18	02 ♊ 41 38	14 21	26 28	25 01	11 21	06 36	14 30	19 49	25 58
27	04 19 37	06 04 02	09 ♊ 33 14	16 ♊ 20 39	16 26	27 38	25 37	11 14	06 41	14 31	19 49	25 56
28	04 23 34	07 01 39	23 ♊ 03 32	29 ♊ 41 36	18 29	28 48	26 13	11 08	06 46	14 32	19 48	25 55
29	04 27 30	07 59 14	06 ♋ 14 40	12 ♋ 42 42	20 31	29 ♈ 58	26 49	11 02	06 51	14 33	19 48	25 54
30	04 31 27	08 56 49	19 ♋ 05 42	25 ♋ 23 50	22 30	01 08	27 25	10 56	06 56	14 34	19 48	25 52
31	04 35 23	09 ♊ 54 22	01 ♌ 37 22	07 ♌ 46 36	24 ♊ 27	02 ♉ 18	28 ♋ 01	10 ♏ 50	07 ♌ 01	14 ♓ 35	19 ♒ 48	25 ♐ 51

DECLINATIONS

Date	Moon True ☊ ° '	Moon Mean ☊ ° '	Moon ☽ Latitude ° '	Sun ☉ ° '	Moon ☽ ° '	Mercury ☿ ° '	Venus ♀ ° '	Mars ♂ ° '	Jupiter ♃ ° '	Saturn ♄ ° '	Uranus ♅ ° '	Neptune ♆ ° '	Pluto ♇ ° '
01	03 ♈ 45	02 ♈ 37	05 N 10	15 N 07	28 N 35	06 N 51	02 S 02	24 N 37	14 S 53	19 N 45	07 S 04	15 S 05	15 S 44
02	03 R 40	02 34	05 08	15 25	28 06	07 35	01 38	24 34	14 50	19 44	07 03	15 05	15 44
03	03 36	02 31	04 52	15 43	26 10	08 20	01 13	24 31	14 48	19 44	07 02	15 05	15 44
04	03 35	02 27	04 23	16 01	23 02	09 05	00 49	24 28	14 46	19 43	07 02	15 05	15 43
05	03 D 34	02 24	03 43	16 18	18 59	09 51	00 S 24	24 24	14 44	19 42	07 01	15 04	15 43
06	03 35	02 21	02 53	16 35	14 10	10 37	00 00	24 20	14 42	19 41	07 00	15 04	15 43
07	03 37	02 18	01 57	16 51	09 00	11 24	00 N 25	24 16	14 39	19 41	06 59	15 04	15 43
08	03 38	02 15	00 N 55	17 08	03 N 26	12 11	00 50	24 12	14 37	19 40	06 59	15 04	15 43
09	03 R 38	02 12	00 S 10	17 24	02 S 17	12 58	01 15	24 08	14 35	19 39	06 58	15 04	15 43
10	03 37	02 08	01 14	17 40	08 00	13 44	01 40	24 04	14 33	19 38	06 57	15 04	15 43
11	03 34	02 05	02 17	17 55	13 32	14 31	02 05	23 59	14 31	19 37	06 57	15 04	15 43
12	03 29	02 02	03 13	18 10	18 36	15 18	02 30	23 55	14 29	19 36	06 56	15 03	15 43
13	03 22	01 59	04 02	18 25	22 58	16 04	02 55	23 50	14 26	19 35	06 55	15 03	15 43
14	03 14	01 56	04 38	18 40	26 26	16 49	03 20	23 45	14 24	19 34	06 55	15 03	15 43
15	03 05	01 52	05 00	18 54	28 10	17 34	03 45	23 40	14 22	19 33	06 54	15 03	15 43
16	02 57	01 49	05 06	19 08	28 26	18 18	04 10	23 34	14 20	19 32	06 54	15 03	15 42
17	02 50	01 46	04 55	19 22	26 59	19 00	04 36	23 29	14 18	19 31	06 53	15 03	15 42
18	02 46	01 43	04 26	19 35	23 52	19 42	05 01	23 23	14 16	19 30	06 52	15 03	15 42
19	02 43	01 40	03 41	19 48	19 20	20 21	05 26	23 18	14 14	19 29	06 52	15 03	15 42
20	02 D 43	01 37	02 42	20 01	13 43	20 59	05 51	23 12	14 12	19 28	06 51	15 03	15 42
21	02 43	01 33	01 34	20 13	07 22	21 35	06 16	23 06	14 10	19 27	06 51	15 03	15 42
22	02 44	01 30	00 S 19	20 25	00 S 37	22 08	06 41	22 59	14 08	19 26	06 50	15 03	15 42
23	02 R 43	01 27	00 N 57	20 36	06 N 10	22 40	07 06	22 53	14 06	19 25	06 50	15 03	15 42
24	02 41	01 24	02 09	20 48	12 38	23 08	07 31	22 47	14 04	19 24	06 49	15 03	15 42
25	02 36	01 21	03 13	20 59	18 25	23 35	07 56	22 40	14 02	19 22	06 49	15 03	15 42
26	02 29	01 17	04 04	21 09	23 09	23 58	08 20	22 33	14 00	19 21	06 49	15 03	15 42
27	02 19	01 14	04 40	21 19	26 30	24 19	08 45	22 26	13 59	19 20	06 48	15 03	15 42
28	02 09	01 11	04 59	21 29	28 24	24 38	09 09	22 19	13 57	19 19	06 48	15 03	15 42
29	01 58	01 08	05 02	21 39	28 20	24 53	09 34	22 12	13 55	19 17	06 47	15 03	15 42
30	01 48	01 05	04 50	21 48	26 52	25 06	09 58	22 04	13 53	19 16	06 47	15 03	15 42
31	01 ♈ 40	01 ♈ 02	04 N 23	21 N 56	24 N 04	25 N 16	10 N 22	21 N 57	13 S 52	19 N 15	06 S 47	15 S 04	15 S 42

ZODIAC SIGN ENTRIES

Date	h	m	Planets
01	15	17	☽ ♋
03	10	25	♀ ♈
04	00	18	☽ ♌
05	08	28	☿ ♉
06	12	20	☽ ♍
09	01	10	☽ ♎
11	12	25	☽ ♏
13	20	56	☽ ♐
16	02	59	☽ ♑
18	07	19	☽ ♒
19	20	52	☿ ♊
20	10	39	☽ ♓
21	04	32	☉ ♊
22	13	24	☽ ♈
24	16	00	☽ ♉
26	19	19	☽ ♊
29	00	34	☽ ♋
29	12	41	♀ ♉
31	08	51	☽ ♌

LATITUDES

Date	Mercury ☿ ° '	Venus ♀ ° '	Mars ♂ ° '	Jupiter ♃ ° '	Saturn ♄ ° '	Uranus ♅ ° '	Neptune ♆ ° '	Pluto ♇ ° '
01	02 S 15	01 S 16	01 N 34	01 N 20	00 N 45	00 S 45	00 S 11	07 N 40
04	01 57	01 25	01 34	01 20	00 45	00 45	00 11	07 40
07	01 35	01 33	01 33	01 20	00 45	00 45	00 11	07 40
10	01 09	01 40	01 32	01 20	00 45	00 45	00 11	07 41
13	00 40	01 45	01 31	01 19	00 45	00 45	00 11	07 41
16	00 S 09	01 50	01 30	01 19	00 45	00 46	00 11	07 41
19	00 N 23	01 55	01 29	01 19	00 45	00 46	00 11	07 41
22	00 53	01 58	01 28	01 18	00 45	00 46	00 11	07 41
25	01 20	02 00	01 27	01 18	00 45	00 46	00 11	07 41
28	01 41	02 01	01 26	01 17	00 45	00 46	00 11	07 41
31	01 N 57	02 S 02	01 N 25	01 N 17	00 N 45	00 S 46	00 S 11	07 N 41

DATA

Julian Date	2453857
Delta T	+66 seconds
Ayanamsa	23° 56' 42"
Synetic vernal point	05° ♓ 10' 17"
True obliquity of ecliptic	23° 26' 28"

MOON'S PHASES, APSIDES AND POSITIONS ☽

Date	h	m	Phase	Longitude ° '	Eclipse Indicator
05	05	13	☽	14 ♌ 35	
13	06	51	○	22 ♏ 23	
20	09	21	☾	29 ♒ 14	
27	05	26	●	05 ♊ 48	

Day	h	m		
07	06	44	Apogee	
22	15	14	Perigee	
01	16	27	Max dec	28° N 37'
09	02	26	0S	
16	03	50	Max dec	28° S 32'
22	14	11	0N	
29	01	11	Max dec	28° N 29'

LONGITUDES

Date	Chiron ⚷ ° '	Ceres ⚳ ° '	Pallas ⚴ ° '	Juno ⚵ ° '	Vesta ⚶ ° '	Black Moon Lilith ⚸ ° '
01	09 ♒ 39	20 ♒ 25	19 ♑ 29	14 ♋ 20	20 ♋ 33	11 ♍ 00
11	09 ♒ 46	22 ♒ 35	19 ♑ 34	18 ♋ 46	24 ♋ 01	12 ♍ 06
21	09 ♒ 45	24 ♒ 23	18 ♑ 55	23 ♋ 15	27 ♋ 43	13 ♍ 13
31	09 ♒ 37	25 ♒ 45	17 ♑ 33	27 ♋ 47	01 ♌ 36	14 ♍ 19

All ephemeris data is given at 12.00 UT and the Moon's longitude is additionally given for 24.00 UT

ASPECTARIAN

h m	Aspects
01 Monday	
01 05	☽ ⚹ ☿
07 36	☽ ∠ ☉
08 51	☽ ☍ ♇
09 03	☉ # ♆
11 13	☽ □ ♀
13 25	☽ ⊥ ♄
14 09	☽ ⚼ ♃
19 07	☿ # ♅
23 58	☽ ⚼ ♆
02 Tuesday	
00 07	☉ ⚼ ♇
00 32	☽ ⊻ ♄
02 31	☽ Q ☿
11 16	☽ ☌ ♂
13 41	☽ ⚹ ☉
16 53	☽ ± ♆
17 07	☽ △ ♅
17 52	☽ △ ♃
18 04	♀ ⚼ ♃
03 Wednesday	
04 20	☽ ⊼ ♆
10 16	☿ △ ♇
12 41	☉ # ♇
14 35	☽ Q ☉
17 21	☽ ⊼ ♇
18 35	☽ □ ☿
22 06	☽ ⚼ ♅
04 Thursday	
01 43	☽ △ ♀
01 51	☽ ∥ ♂
05 04	☽ ± ♇
10 22	☽ ☌ ♄
12 19	☉ ⚹ ♅
14 36	☉ ☍ ♃
15 54	☽ ± ♅
18 28	☿ ∠ ♅
22 53	☽ ⚼ ♇
05 Friday	
00 32	☽ ⊻ ♂
03 48	♃ △ ♅
03 54	☽ □ ♃
03 54	☽ ⊼ ♅
05 13	☽ □ ☉
08 01	☽ ∥ ♄
10 37	☽ ⚼ ♀
13 12	☽ ⊥ ♂
15 34	☽ ☍ ♆
16 40	☿ Q ♂
06 Saturday	
01 08	☽ ∥ ☉
04 46	☽ # ♇
05 02	☽ △ ♇
06 37	☿ Q ♆
06 42	☽ ± ♀
07 57	☽ # ♆
08 20	☽ ∠ ♂
09 46	☽ # ♃
15 51	☽ Q ♃
17 40	☽ △ ☿
20 10	☽ ⊼ ♀
23 00	☽ ⊻ ♄
07 Sunday	
02 36	☽ ∥ ☿
09 13	♂ △ ♃
11 16	☽ ⊥ ♄
13 53	♂ ± ♆
15 07	♀ ∠ ♆
15 59	☽ ⚹ ♃
16 25	☽ ⚹ ♂
16 48	☽ ☍ ♅
20 47	☽ # ♅
23 24	☽ △ ☉
08 Monday	
00 38	♂ △ ♅
02 57	♀ △ ♄
03 02	☿ □ ♄
03 08	☿ ⊻ ♀
04 29	☽ ⊼ ♆
05 37	☽ ∠ ♄
06 07	☽ ⚼ ☿
16 38	☽ ± ♆
17 49	☽ □ ♇
18 01	☽ Q ♂
22 02	☽ ∠ ♃
22 13	☽ ∥ ♀
09 Tuesday	
03 55	☽ ± ☿
07 17	☽ # ♀
08 22	☽ ⚼ ☉
10 45	☽ ⚼ ♆
12 00	☽ ⚹ ♄
15 17	☽ ☍ ♀
15 54	☽ ⊥ ♃
18 19	☽ ⊼ ☿
21 13	♀ ± ♃
10 Wednesday	
03 43	☽ ⊻ ♃
03 52	☽ ± ☉
05 18	☽ ⊼ ♅
05 45	☽ Q ♇
07 33	☽ ∥ ♅
07 51	☽ □ ♂
11 56	☽ Q ♄
14 02	☉ □ ♆
16 30	☽ △ ♆
16 43	☽ ⊼ ☉
17 09	☽ ± ♅
11 Thursday	
02 49	☿ ⚼ ♇
03 06	☉ ± ♇
05 15	☽ ⚹ ♇
10 43	☽ ⚼ ♅
11 44	☿ # ♃
16 26	☽ ∥ ♃
17 17	☽ # ☿
18 58	☽ ∥ ♆
22 00	☽ ∥ ♇
23 04	☽ □ ♄
23 20	☿ ☍ ♃
12 Friday	
04 35	☿ # ♆
07 47	☽ ⊼ ♀
09 42	☽ # ☉
09 59	☽ ∠ ♇
12 03	☿ ⚹ ♅
13 14	☽ ☌ ♃
15 27	☽ △ ♅
16 06	☽ ☍ ☿
17 05	☽ # ♄
20 14	☽ ± ♀
20 34	☽ △ ♂
13 Saturday	
00 54	☿ # ♇
02 01	☽ □ ♆
02 54	☽ ⊥ ♇
06 51	☽ ☍ ☉
14 02	☽ ⊻ ♇
14 38	☽ ⚼ ♀
17 26	☽ # ♂
14 Sunday	
01 43	☽ ⚼ ♂
04 05	☿ ⚹ ♂
07 22	☽ △ ♄
11 12	☽ Q ♆
14 40	♀ ⊼ ♃
16 00	☉ Q ♄
17 57	☿ ⊥ ♀
18 52	☽ ± ♂
20 05	☽ ⊻ ♃
20 39	☽ △ ♀
22 50	☽ □ ♅
15 Monday	
04 12	☿ □ ♆
06 10	☽ ⊼ ♂
06 47	☽ ⊥ ♃
08 51	☽ ⚹ ♆
08 51	☿ ± ♇
09 43	☽ ⊼ ☿
10 34	☽ ⚼ ♄
17 40	☽ ⊼ ☉
20 15	☽ ☌ ♇
22 26	☽ ± ☿
22 29	♀ ⊻ ♅
22 40	☽ ∠ ♃
16 Tuesday	
02 39	☽ ± ♄
05 07	☽ ± ☉
06 59	☽ Q ♅
11 27	☽ ∠ ♆
13 17	☽ ⊼ ♄
17 18	☽ ⚼ ☿
22 08	☽ ⚼ ☉
17 Wednesday	
00 50	☽ ⚹ ♃
01 37	☿ Q ♄
03 14	☽ ⊥ ♆
04 04	☽ ⚹ ♅
05 15	☉ ⊼ ♇
06 42	☽ □ ♀
08 08	☉ Q ♅
13 27	☽ ☍ ♂
13 41	☽ ⊻ ♆
18 46	♂ ⊼ ♆
21 32	☽ Q ♃
18 Thursday	
00 18	☽ △ ☿
00 41	☽ ⊻ ♇
02 10	☽ △ ☉
02 43	☿ ⊼ ♇
04 11	☉ ∥ ♄
04 27	☿ Q ♅
05 30	☿ ∥ ♄
06 08	☽ ∠ ♅
06 12	☉ ∥ ☿
11 02	☽ ⊥ ♇
14 57	☽ # ♂
16 40	☽ Q ♀
17 39	☽ ☍ ♄
20 02	☉ ☌ ☿
21 39	☽ ⊥ ♅
19 Friday	
02 28	☽ ∠ ♇
04 15	☽ □ ♃
07 43	☽ # ☿
07 57	☽ ⊻ ♅
09 55	☽ # ☉
11 18	☽ # ♄
15 05	☽ ⚹ ♀
17 18	☽ ☌ ♆
19 21	☽ ⊼ ♂
20 Saturday	
03 58	☽ ∥ ♇
04 02	☽ ⚹ ♇
06 02	☽ ± ♂
06 39	☽ ∥ ♆
09 21	☽ □ ☉
10 05	☽ ∥ ♃
13 11	☽ □ ☿
18 13	♀ ⚹ ♆
18 53	☽ ∠ ♀
21 06	☽ ⊼ ♄
21 59	☽ ⚼ ♂
21 Sunday	
00 23	☽ Q ♇
06 07	♀ ⊥ ♅
06 53	☽ △ ♃
07 19	☽ ± ♄
11 03	☽ ☌ ♅
11 29	☽ ⊥ ♀
13 53	☽ ∥ ♅
15 44	☽ # ♀
18 04	☽ Q ☉
20 13	☽ ⊻ ♆
22 31	☽ ⊻ ♀
22 37	☽ ⚼ ♄
22 Monday	
00 28	☽ △ ♂
01 13	☽ Q ☿
06 20	☽ ⊥ ♆
06 46	☽ □ ♇
08 01	☽ ⚼ ♃
13 05	♆ St R
15 47	☽ ⚹ ☉
17 58	☿ ⚹ ♄
20 36	♀ # ♅
21 31	☽ ∠ ♆
23 04	☽ ± ♃
23 Tuesday	
00 03	☽ △ ♄
01 06	☽ ⚹ ☿
09 06	☽ ⊼ ♃
13 25	☿ ∠ ♀
13 44	☽ ⊻ ♅
14 23	☽ # ♅
15 36	☽ ∥ ♀
16 30	☿ ∠ ♂
18 54	☽ ∠ ☉
20 55	☿ ∥ ♂
22 00	♀ □ ♂
22 48	☽ ⚹ ♆
23 52	☽ ⊥ ♅
24 Wednesday	
05 17	☽ □ ♂
05 36	☽ ☌ ♀
06 56	☽ ∠ ☿
09 16	☽ △ ♇
11 11	☽ ⊥ ☉
15 06	☽ ∠ ♅
17 36	☽ # ♃
19 05	☽ Q ♆
21 36	☽ # ♆
22 05	☽ ⊻ ☉
25 Thursday	
00 15	☽ # ♇
00 53	☽ ⊥ ☿
03 00	☽ □ ♄
03 44	☿ ⊼ ♃
10 40	☽ ⚼ ♇
11 28	☽ ☍ ♃
12 55	☽ ⊻ ☿
13 09	☽ Q ♂
16 22	☽ ∥ ♄
16 37	☽ ⚹ ♅
26 Friday	
00 40	☽ ∥ ☉
01 43	♀ △ ♇
01 45	☽ □ ♆
02 01	☽ ± ♇
08 40	☽ ∥ ♂
09 58	☽ Q ♄
10 39	☽ ⚹ ♂
12 20	☽ ⊼ ♇
13 16	☽ Q ♅
13 19	☽ ⊻ ♀
13 48	☿ □ ♅
17 34	☽ ∥ ☿
27 Saturday	
00 41	☽ ⊥ ♀
05 26	☽ ☌ ☉
06 56	☽ ⚹ ♄
13 58	☽ ∠ ♂
14 57	☽ ⊼ ♃
17 57	☽ ∠ ♀
20 46	☽ □ ♅
20 56	☿ ± ♃
28 Sunday	
00 19	♂ ⊼ ♇
01 29	☽ ± ♃
02 21	☽ ☌ ☿
04 48	☉ ⚹ ♄
06 10	☽ △ ♆
06 40	☽ ⊥ ♂
09 40	☽ ∠ ♄
17 08	☽ ☍ ♇
17 30	☽ ⚼ ♃
17 58	☽ ⊻ ♂
23 23	☽ ⚹ ♀
29 Monday	
02 03	☽ ⊥ ♄
03 23	♀ ∠ ♅
03 34	☿ △ ♆
09 21	☽ ⚼ ♆
13 08	☽ ⊻ ♄
15 29	☽ ⊻ ☉
17 18	☿ ⊥ ♂
20 48	☽ △ ♃
23 40	☽ Q ♀
30 Tuesday	
02 02	☽ ± ♆
03 28	☽ △ ♅
03 33	☽ ⊥ ☉
04 50	☿ ∠ ♄
13 20	☽ ⊼ ♆
19 39	☽ ⊻ ☿
21 59	☽ ∠ ☉
31 Wednesday	
00 53	☽ ⊼ ♇
01 40	♀ Q ♆
03 20	☽ ∥ ☿
04 42	☽ ☌ ♂
08 03	☽ ⚼ ♅
08 41	♃ ∠ ♇
09 18	☽ ⊥ ☿
12 26	☽ ± ♇
12 59	☉ ∥ ♂
13 27	☽ □ ♀
22 36	☽ ☌ ♄

LONGITUDES

Date	Sidereal time h m s	Sun ☉ ° ' "	Moon ☽ ° ' "	Moon ☽ 24.00 ° ' "	Mercury ☿ ° '	Venus ♀ ° '	Mars ♂ ° '	Jupiter ♃ ° '	Saturn ♄ ° '	Uranus ♅ ° '	Neptune ♆ ° '	Pluto ♇ ° '
01	04 39 20	10 ♊ 51 53	13 ♌ 52 00	19 ♌ 54 01	26 ♊ 21	03 ♉ 28	28 ♋ 38	10 ♏ 44	07 ♌ 07	14 ♓ 36	19 ♒ 47	25 ♐ 49
02	04 43 16	11 49 24	25 ♌ 53 14	01 ♍ 50 13	28 ♊ 13	04 38	29 14	10 R 38	07 12	14 37	19 R 47	25 R 48
03	04 47 13	12 46 52	07 ♍ 45 39	13 ♍ 40 09	00 ♋ 03	05 48	29 ♋ 50	10 33	07 18	14 38	19 47	25 46
04	04 51 09	13 44 20	19 ♍ 34 24	25 ♍ 29 07	01 50	06 59	00 ♌ 26	10 27	07 23	14 38	19 46	25 45
05	04 55 06	14 41 47	01 ♎ 24 56	07 ♎ 22 33	03 34	08 09	01 02	10 22	07 29	14 39	19 46	25 43
06	04 59 03	15 39 12	13 ♎ 22 34	19 ♎ 25 37	05 15	09 19	01 38	10 17	07 35	14 40	19 45	25 41
07	05 02 59	16 36 36	25 ♎ 32 13	01 ♏ 42 53	06 54	10 30	02 15	10 12	07 40	14 40	19 45	25 40
08	05 06 56	17 33 59	07 ♏ 58 01	14 ♏ 17 58	08 30	11 40	02 51	10 07	07 46	14 41	19 44	25 38
09	05 10 52	18 31 21	20 ♏ 42 58	27 ♏ 13 08	10 03	12 51	03 27	10 02	07 52	14 41	19 44	25 37
10	05 14 49	19 28 43	03 ♐ 48 30	10 ♐ 28 58	11 34	14 01	04 04	09 58	07 58	14 42	19 43	25 35
11	05 18 45	20 26 03	17 ♐ 14 19	24 ♐ 04 15	13 02	15 12	04 40	09 54	08 04	14 42	19 43	25 34
12	05 22 42	21 23 23	00 ♑ 58 19	07 ♑ 56 03	14 26	16 22	05 16	09 49	08 10	14 43	19 42	25 32
13	05 26 38	22 20 42	14 ♑ 56 53	22 ♑ 00 13	15 48	17 33	05 53	09 45	08 16	14 43	19 41	25 31
14	05 30 35	23 18 00	29 ♑ 05 28	06 ♒ 12 02	17 07	18 44	06 29	09 41	08 22	14 43	19 41	25 29
15	05 34 32	24 15 18	13 ♒ 19 22	20 ♒ 26 59	18 23	19 55	07 05	09 38	08 28	14 43	19 40	25 27
16	05 38 28	25 12 35	27 ♒ 34 27	04 ♓ 41 26	19 36	21 06	07 42	09 34	08 35	14 44	19 39	25 26
17	05 42 25	26 09 53	11 ♓ 47 37	18 ♓ 52 49	20 46	22 16	08 18	09 31	08 41	14 44	19 38	25 24
18	05 46 21	27 07 09	25 ♓ 56 51	02 ♈ 59 37	21 53	23 27	08 55	09 27	08 47	14 44	19 38	25 23
19	05 50 18	28 04 26	10 ♈ 01 01	17 ♈ 00 59	22 57	24 38	09 31	09 24	08 54	14 R 44	19 37	25 21
20	05 54 14	29 01 42	23 ♈ 59 25	00 ♉ 56 15	23 57	25 49	10 08	09 21	09 00	14 44	19 36	25 20
21	05 58 11	29 ♊ 58 58	07 ♉ 51 20	14 ♉ 44 31	24 54	27 01	10 44	09 19	09 07	14 44	19 35	25 18
22	06 02 07	00 ♋ 56 14	21 ♉ 35 37	28 ♉ 24 24	25 48	28 12	11 21	09 16	09 13	14 44	19 34	25 16
23	06 06 04	01 53 30	05 ♊ 10 37	11 ♊ 54 01	26 38	29 ♉ 23	11 58	09 14	09 20	14 43	19 33	25 15
24	06 10 01	02 50 46	18 ♊ 34 19	25 ♊ 11 16	27 24	00 ♊ 34	12 34	09 11	09 27	14 43	19 32	25 13
25	06 13 57	03 48 01	01 ♋ 44 38	08 ♋ 14 14	28 07	01 45	13 11	09 09	09 33	14 43	19 31	25 12
26	06 17 54	04 45 17	14 ♋ 39 54	21 ♋ 01 35	28 45	02 57	13 48	09 07	09 40	14 43	19 30	25 10
27	06 21 50	05 42 31	27 ♋ 19 17	03 ♌ 33 03	29 20	04 08	14 24	09 06	09 47	14 42	19 29	25 09
28	06 25 47	06 39 46	09 ♌ 43 03	15 ♌ 49 30	29 ♋ 51	05 19	15 01	09 04	09 54	14 42	19 28	25 07
29	06 29 43	07 37 00	21 ♌ 52 44	27 ♌ 53 06	00 ♌ 17	06 31	15 38	09 03	10 01	14 41	19 27	25 06
30	06 33 40	08 ♋ 34 13	03 ♍ 51 05	09 ♍ 47 08	00 ♌ 39	07 ♊ 42	16 ♌ 15	09 ♏ 02	10 ♌ 08	14 ♓ 41	19 ♒ 26	25 ♐ 04

Moon Node and Latitude

Date	Moon True ☊ ° '	Moon Mean ☊ ° '	Moon ☽ Latitude ° '
01	01 ♈ 35	00 ♈ 58	03 N 45
02	01 R 32	00 55	02 58
03	01 30	00 52	02 03
04	01 D 30	00 49	01 03
05	01 31	00 46	00 N 01
06	01 R 30	00 43	01 S 03
07	01 28	00 39	02 04
08	01 23	00 36	03 01
09	01 16	00 33	03 50
10	01 06	00 30	04 28
11	00 55	00 27	04 52
12	00 44	00 23	05 01
13	00 32	00 20	04 51
14	00 23	00 17	04 24
15	00 16	00 14	03 41
16	00 12	00 11	02 43
17	00 10	00 08	01 36
18	00 D 10	00 04	00 S 22
19	00 R 09	00 ♈ 01	00 N 52
20	00 08	29 ♓ 58	02 02
21	00 05	29 55	03 05
22	29 ♓ 59	29 52	03 56
23	29 50	29 49	04 33
24	29 39	29 45	04 55
25	29 27	29 42	05 00
26	29 14	29 39	04 50
27	29 02	29 36	04 25
28	28 53	29 33	03 48
29	28 45	29 29	03 02
30	28 ♓ 41	29 ♓ 26	02 N 07

DECLINATIONS

Date	Sun ☉ ° '	Moon ☽ ° '	Mercury ☿ ° '	Venus ♀ ° '	Mars ♂ ° '	Jupiter ♃ ° '	Saturn ♄ ° '	Uranus ♅ ° '	Neptune ♆ ° '	Pluto ♇ ° '
01	22 N 05	20 N 15	25 N 24	10 N 46	21 N 49	13 S 50	19 N 14	06 S 46	15 S 04	15 S 42
02	22 12	15 40	25 29	11 09	21 41	13 49	19 12	06 46	15 04	15 42
03	22 20	10 34	25 32	11 33	21 33	13 47	19 11	06 46	15 04	15 42
04	22 27	05 N 06	25 33	11 56	21 25	13 46	19 09	06 46	15 04	15 42
05	22 34	00 S 33	25 31	12 19	21 17	13 44	19 08	06 45	15 04	15 42
06	22 40	06 15	25 27	12 42	21 09	13 43	19 07	06 45	15 04	15 42
07	22 46	11 48	25 22	13 04	21 00	13 41	19 05	06 45	15 04	15 42
08	22 52	17 01	25 14	13 26	20 52	13 40	19 04	06 45	15 05	15 42
09	22 57	21 37	25 05	13 48	20 43	13 39	19 02	06 45	15 05	15 42
10	23 01	25 18	24 55	14 10	20 34	13 38	19 01	06 45	15 05	15 42
11	23 06	27 41	24 42	14 31	20 25	13 37	18 59	06 44	15 05	15 42
12	23 10	28 27	24 29	14 53	20 16	13 35	18 58	06 44	15 06	15 42
13	23 13	27 26	24 14	15 13	20 06	13 34	18 56	06 44	15 06	15 42
14	23 16	24 39	23 58	15 34	19 57	13 33	18 54	06 44	15 06	15 42
15	23 19	20 21	23 42	15 54	19 47	13 32	18 53	06 44	15 06	15 42
16	23 21	14 52	23 24	16 14	19 38	13 31	18 51	06 44	15 07	15 42
17	23 23	08 37	23 05	16 33	19 28	13 31	18 50	06 44	15 07	15 42
18	23 25	01 S 57	22 46	16 53	19 18	13 30	18 48	06 44	15 07	15 42
19	23 26	04 N 46	22 26	17 11	19 08	13 29	18 46	06 44	15 07	15 42
20	23 26	11 12	22 06	17 30	18 58	13 28	18 45	06 44	15 08	15 42
21	23 26	17 03	21 45	17 47	18 47	13 28	18 43	06 44	15 08	15 42
22	23 26	21 58	21 24	18 05	18 37	13 27	18 41	06 44	15 08	15 42
23	23 26	25 39	21 03	18 22	18 26	13 27	18 39	06 44	15 09	15 43
24	23 25	27 51	20 42	18 39	18 16	13 26	18 38	06 45	15 09	15 43
25	23 23	28 26	20 21	18 55	18 05	13 26	18 36	06 45	15 09	15 43
26	23 21	27 26	20 00	19 11	17 54	13 26	18 34	06 45	15 10	15 43
27	23 19	25 02	19 39	19 26	17 43	13 25	18 32	06 45	15 10	15 43
28	23 16	21 29	19 19	19 41	17 32	13 25	18 31	06 45	15 10	15 43
29	23 13	17 05	18 59	19 55	17 21	13 25	18 29	06 45	15 11	15 43
30	23 N 10	12 N 04	18 N 40	20 N 09	17 N 09	13 S 25	18 N 27	06 S 46	15 S 11	15 S 43

ZODIAC SIGN ENTRIES

Date	h	m	Planets
02	20	17	☽ ♍
03	11	21	☿ ♋
03	18	43	♂ ♌
05	09	08	☽ ♎
07	20	41	☽ ♏
10	05	05	☽ ♐
12	10	19	☽ ♑
14	13	32	☽ ♒
16	16	05	☽ ♓
18	18	54	☽ ♈
20	22	23	☽ ♉
21	12	26	☉ ♋
23	02	49	☽ ♊
24	00	31	♀ ♊
25	08	48	☽ ♋
27	17	09	☽ ♌
28	19	57	☿ ♌
30	04	15	☽ ♍

LATITUDES

Date	Mercury ☿ ° '	Venus ♀ ° '	Mars ♂ ° '	Jupiter ♃ ° '	Saturn ♄ ° '	Uranus ♅ ° '	Neptune ♆ ° '	Pluto ♇ ° '
01	02 N 01	02 S 02	01 N 25	01 N 16	00 N 45	00 S 46	00 S 11	07 N 41
04	02 07	02 02	01 24	01 16	00 45	00 46	00 12	07 41
07	02 06	02 00	01 23	01 15	00 45	00 46	00 12	07 41
10	01 59	01 58	01 22	01 14	00 46	00 47	00 12	07 40
13	01 45	01 55	01 21	01 14	00 46	00 47	00 12	07 40
16	01 24	01 52	01 20	01 13	00 46	00 47	00 12	07 40
19	00 58	01 48	01 19	01 12	00 46	00 47	00 12	07 40
22	00 N 26	01 43	01 18	01 11	00 46	00 47	00 12	07 39
25	00 S 12	01 38	01 16	01 11	00 46	00 47	00 12	07 39
28	00 53	01 32	01 15	01 10	00 46	00 47	00 12	07 38
31	01 S 38	01 S 26	01 N 14	01 N 09	00 N 46	00 S 47	00 S 12	07 N 38

DATA

Julian Date	2453888
Delta T	+66 seconds
Ayanamsa	23° 56' 47"
Synetic vernal point	05° ♓ 10' 12"
True obliquity of ecliptic	23° 26' 27"

MOON'S PHASES, APSIDES AND POSITIONS ☽

Date	h	m	Phase	Longitude ° '	Eclipse Indicator
03	23	06	☽	13 ♍ 13	
11	18	03	○	20 ♐ 41	
18	14	08	☾	27 ♓ 12	
25	16	05	●	03 ♋ 58	

Day	h	m		
04	01	40	Apogee	
16	17	08	Perigee	
05	09	40	0S	
12	10	31	Max dec	28° S 27'
18	18	57	0N	
25	08	39	Max dec	28° N 27'

LONGITUDES

Date	Chiron ⚷ ° '	Ceres ⚳ ° '	Pallas ⚴ ° '	Juno ⚵ ° '	Vesta ⚶ ° '	Black Moon Lilith ⚸ ° '
01	09 ♒ 36	25 ♒ 52	17 ♑ 23	28 ♋ 14	02 ♌ 00	14 ♍ 26
11	09 ♒ 21	26 ♒ 41	15 ♑ 18	02 ♌ 46	06 ♌ 05	15 ♍ 33
21	09 ♒ 00	26 ♒ 57	12 ♑ 44	07 ♌ 19	10 ♌ 18	16 ♍ 39
31	08 ♒ 35	26 ♒ 38	09 ♑ 55	11 ♌ 51	14 ♌ 38	17 ♍ 46

All ephemeris data is given at 12.00 UT and the Moon's longitude is additionally given for 24.00 UT

ASPECTARIAN

h m	Aspects
01 Thursday	
01 36	☽ ± ♅
01 37	☽ ∥ ☉
02 31	☽ ∥ ♂
04 30	☿ ⚼ ♃
05 15	☿ ☍ ♇
05 34	☽ ⚹ ☉
05 52	☽ □ ♃
05 59	☽ ⚼ ♇
06 08	☽ ∠ ☿
09 04	☉ ⊼ ♃
13 27	☽ ⊼ ♅
17 45	☽ ∥ ♄
23 47	☽ ☍ ♆
02 Friday	
07 30	☽ Q ☉
11 49	☽ △ ♇
11 54	☽ # ♇
14 59	☽ # ♆
17 30	☽ Q ♃
17 34	☽ ⚹ ☿
19 05	☽ ⊻ ♂
21 02	☽ # ♃
03 Saturday	
03 44	♂ ⚼ ♅
07 36	☽ △ ♀
07 40	☿ ⊻ ♂
07 50	☽ ∥ ♀
07 53	☽ ⊥ ♂
11 03	☽ ⊻ ♄
17 37	☽ ⚹ ♃
22 16	☽ Q ☿
23 06	☽ □ ☉
23 20	☽ ⊥ ♄
04 Sunday	
01 58	☽ ☍ ♅
03 08	☽ ∠ ♂
04 48	☽ # ♅
05 40	☿ ⊥ ♄
12 24	☽ ⊼ ♆
17 25	☽ ⚼ ♀
17 46	☽ ∠ ♄
21 08	♀ □ ♄
23 51	☽ ∠ ♃
05 Monday	
00 30	☽ □ ♇
00 34	☽ ± ♆
10 52	☉ □ ♅
11 12	☽ ⚹ ♂
13 39	☽ ± ♀
17 04	☽ □ ☿
17 55	☽ ⊥ ♃
18 45	☽ ⚼ ♆
06 Tuesday	
00 19	☽ ⚹ ♄
03 01	☽ ⊼ ♀
04 53	☿ ⚼ ♆
05 52	☽ ⊻ ♃
12 33	☽ Q ♂
12 38	☽ Q ♇
13 53	♂ ± ♇
14 10	☽ ∥ ♅
14 34	☽ ⊼ ♅
16 55	☽ △ ☉
07 Wednesday	
00 23	☽ Q ♄
00 39	☽ △ ♆
02 27	☽ ± ♅
02 33	☉ ± ♃
06 23	♀ ☍ ♃
12 15	☽ ⚹ ♇
15 25	♀ ⚼ ♇
18 06	☽ # ♀
20 03	☽ ⚼ ♅
20 27	☽ ∥ ♃
08 Thursday	
00 10	☿ ⊻ ♄
00 45	☉ ⊥ ♀
00 47	☽ ⚼ ☉
01 41	☽ □ ♂
02 51	☽ ∥ ♆
05 44	☽ ∥ ♇
11 37	☽ □ ♄
13 10	☽ △ ☿
16 04	☽ ☌ ♃
17 04	☽ ∠ ♇
19 24	☽ ± ☉
19 45	☽ ☍ ♀
22 08	☽ # ♄
09 Friday	
00 44	☽ △ ♅
02 12	♀ # ♃
07 06	☽ # ♂
07 21	☉ ∠ ♂
07 35	☽ ⊼ ☉
09 57	☽ ⊥ ♇
10 10	☽ □ ♆
11 46	☿ △ ♃
20 02	☽ # ☉
21 02	☽ ⊻ ♇
21 05	☽ ⚼ ☿
10 Saturday	
09 15	☽ # ☿
12 29	☽ △ ♂
15 34	☽ ± ☿
18 03	☉ △ ♆
19 03	☽ Q ♆
19 33	☽ △ ♄
23 01	☽ ⊻ ♃
11 Sunday	
01 55	♀ ⚹ ♅
03 38	☽ ⊼ ☿
07 31	☽ □ ♅
08 03	☽ ⊼ ♀
09 38	☽ ⊥ ♃
16 21	☽ ⚹ ♆
16 29	☽ ⚼ ♂
18 03	☽ ☍ ☉
19 37	☽ ± ♀
22 19	☽ ⚼ ♄
23 28	☿ ± ♆
12 Monday	
01 22	☽ ∠ ♃
02 34	☽ ☌ ♇
08 55	☽ ± ♂
12 46	☽ ⚼ ♀
14 05	☽ ± ♄
15 00	☽ Q ♅
16 44	☿ △ ♅
18 26	☽ ∠ ♆
19 46	☽ ⊼ ♂
13 Tuesday	
00 29	☽ ⊼ ♄
03 07	♀ # ♆
03 10	☽ ⚹ ♃
09 51	☽ ⊥ ♆
11 36	☽ ⚹ ♅
13 37	☽ ☍ ☿
16 50	☽ △ ♀
20 04	☽ ⊻ ♆
23 31	☽ Q ♃
14 Wednesday	
01 29	☽ ⊼ ☉
05 54	☽ ⊻ ♇
12 23	☽ ± ☉
13 04	☽ ∠ ♅
13 58	☉ ∠ ♄
16 02	☽ ⊥ ♇
16 41	☽ # ☿
20 28	☽ # ☉
21 31	♀ # ♇
15 Thursday	
01 02	☽ ☍ ♂
02 58	♀ ± ♇
03 46	☽ ☍ ♄
04 15	☽ ⊥ ♅
04 39	☽ ⚼ ☉
05 48	☽ □ ♃
07 04	♀ □ ♆
07 11	☽ ∠ ♇
14 22	☽ ⊻ ♅
14 44	☽ # ♂
18 54	☽ # ♄
20 46	☉ ⚼ ♃
21 20	☽ ⊼ ☿
22 40	☽ ☌ ♆
16 Friday	
00 06	☽ □ ♀
06 41	☽ # ♀
07 44	☽ △ ☉
08 23	☽ ± ☿
08 24	☽ ⚹ ♇
08 36	☽ ∥ ♇
11 01	☽ ∥ ♆
13 00	☿ ⊼ ♆
15 01	☉ ∥ ☿
17 22	☽ ∥ ♃
17 24	☉ ☍ ♇
17 Saturday	
00 55	☽ ⚼ ☿
04 36	☽ Q ♇
05 50	☽ ⊼ ♂
06 42	☽ ⊼ ♄
08 09	☽ △ ♃
09 12	☽ Q ♀
16 26	☽ ± ♂
16 56	☽ ± ♄
16 58	☽ ☌ ♅
18 51	☽ ∥ ♅
22 11	♄ ± ♅
18 Sunday	
01 17	☽ ⊻ ♆
04 31	☽ △ ☿
04 42	♂ ± ♅
06 04	♂ ☌ ♄
07 23	☽ ⚹ ♀
08 18	☽ ⚼ ♄
08 23	☽ ⚼ ♂
09 28	☽ ⚼ ♃
11 02	☽ □ ♇
11 27	☽ ⊥ ♆
14 08	☽ □ ☉
19 Monday	
00 44	☽ ± ♃
02 46	☽ ∠ ♆
07 40	♅ St R
07 40	♂ □ ♃
10 04	☽ △ ♄
10 57	☽ ⊼ ♃
11 07	☽ △ ♂
11 18	☽ ∠ ♀
19 11	☽ # ♅
20 04	☽ ⊻ ♅
23 09	☽ Q ☉
20 Tuesday	
02 08	♀ ⊼ ♇
04 10	☽ ⊥ ♀
04 27	☽ ⚹ ♆
06 23	☽ ⊥ ♅
11 56	☽ □ ☿
14 18	☽ △ ♇
15 27	☽ ⊻ ♀
19 22	♂ ⚼ ♇
20 57	☽ # ♃
21 20	☽ ⚹ ☉
21 54	☽ ∠ ♅
21 Wednesday	
01 08	☽ Q ♆
03 47	☽ # ♆
06 12	☽ # ♇
06 19	♀ Q ♅
14 12	☽ □ ♄
14 21	♀ Q ♄
14 31	☽ ☍ ♃
15 32	☽ ∥ ♀
16 15	☽ ⚼ ♇
17 15	☽ □ ♂
19 33	☽ ∥ ♄
19 41	☽ ∥ ♂
21 25	☽ Q ☿
22 13	☿ ⊼ ♇
23 58	☽ ⚹ ♅
22 Thursday	
00 30	♂ ∥ ♄
01 21	☽ ∠ ☉
07 56	☽ ± ♇
08 27	☽ □ ♆
09 12	☽ ∥ ☿
18 19	☽ ⊥ ☉
18 28	☽ ⊼ ♇
18 44	♃ □ ♄
19 53	☽ ⚹ ☿
20 35	☽ ∥ ☉
21 02	☽ Q ♅
22 00	☽ Q ♄
23 Friday	
00 44	☽ ☌ ♀
02 19	☽ Q ♂
05 43	☽ ⊻ ☉
15 48	♀ ∥ ♂
19 12	☽ ⊼ ♃
19 28	☽ ⚹ ♄
24 Saturday	
00 14	☽ ∠ ☿
00 41	☽ ⚹ ♂
05 04	☽ □ ♅
05 55	☽ ± ♃
10 39	♀ ∥ ♄
12 07	♀ Q ♂
13 45	☽ △ ♆
17 25	☽ ⊥ ☿
22 09	☽ ⚼ ♃
22 44	☽ ∠ ♄
25 Sunday	
00 02	☽ ☍ ♇
04 58	☽ ⊻ ☿
05 05	☉ ⊥ ♄
05 09	☽ ∠ ♂
12 01	☽ ⊻ ♀
15 22	☽ ⊥ ♄
16 05	☽ ☌ ☉
17 07	☽ ⚼ ♆
22 32	☽ ⊥ ♂
26 Monday	
00 14	☽ ⊥ ♀
01 40	☽ △ ♃
02 35	☽ ⊻ ♄
05 48	☉ ⚼ ♆
09 50	☽ ± ♆
10 17	☽ ⊻ ♂
12 05	☽ △ ♅
18 48	☽ ∠ ♀
21 06	☽ ⊼ ♆
27 Tuesday	
07 51	☽ ⊼ ♇
16 03	☽ ☌ ☿
16 34	☽ ⚼ ♅
19 20	☽ ± ♇
21 02	☿ ∥ ♀
23 33	♂ ⊼ ♅
28 Wednesday	
00 34	☽ ∥ ☉
02 32	☽ ⚹ ♀
04 41	☿ ⚼ ♅
05 32	☽ ⊻ ☉
10 00	☽ ± ♅
10 44	☽ □ ♃
12 21	☽ ☌ ♄
12 47	☽ ⚼ ♇
18 16	☽ ⊥ ☉
21 46	☽ ⊼ ♅
21 46	☽ ∥ ♀
22 57	☽ ☌ ♂
29 Thursday	
01 18	☽ ∥ ☿
04 35	☽ Q ♀
04 43	☽ ∥ ♄
07 11	☽ ☍ ♆
10 35	☽ ∥ ♂
13 36	☽ ∠ ☉
18 24	☽ △ ♇
18 46	☽ # ♇
21 23	☽ # ♆
22 18	☽ Q ♃
30 Friday	
01 39	♄ ⚼ ♇
05 22	☽ ⊻ ☿
05 48	☽ # ♃
17 49	☽ ⊥ ☿
20 39	☽ □ ♀
22 22	☽ ⚹ ☉
22 27	☽ ⚹ ♃
23 21	☉ △ ♃

JULY 2006

LONGITUDES

Date	Sidereal time h m s	Sun ☉ ° ' "	Moon ☽ ° ' "	Moon ☽ 24.00 ° ' "	Mercury ☿ ° '	Venus ♀ ° '	Mars ♂ ° '	Jupiter ♃ ° '	Saturn ♄ ° '	Uranus ♅ ° '	Neptune ♆ ° '	Pluto ♇ ° '
01	06 37 36	09 ♋ 31 26	15 ♍ 41 51	21 ♍ 35 49	00 ♌ 57	08 ♊ 54	16 ♌ 51	09 ♏ 01	10 ♌ 15	14 ♓ 40	19 ♒ 25	25 ♐ 03
02	06 41 33	10 28 39	27 ♍ 29 40	03 ♎ 24 04	01 10	10 05	17 28	09 R 00	10 22	14 R 40	19 R 24	25 R 01
03	06 45 30	11 25 51	09 ♎ 19 42	15 ♎ 17 14	01 18	11 17	18 05	08 59	10 29	14 39	19 22	25 00
04	06 49 26	12 23 03	21 ♎ 17 23	27 ♎ 20 47	01 22	12 29	18 42	08 59	10 36	14 38	19 21	24 58
05	06 53 23	13 20 15	03 ♏ 28 06	09 ♏ 39 54	01 R 21	13 40	19 19	08 59	10 43	14 38	19 20	24 57
06	06 57 19	14 17 27	15 ♏ 56 45	22 ♏ 18 04	01 17	14 52	19 56	08 D 59	10 50	14 37	19 19	24 55
07	07 01 16	15 14 38	28 ♏ 47 20	05 ♐ 21 41	01 06	16 04	20 33	08 59	10 57	14 36	19 17	24 54
08	07 05 12	16 11 50	12 ♐ 02 17	18 ♐ 49 07	00 51	17 16	21 10	08 59	11 05	14 35	19 16	24 52
09	07 09 09	17 09 01	25 ♐ 42 01	02 ♑ 40 40	00 32	18 27	21 46	09 00	11 12	14 34	19 15	24 51
10	07 13 05	18 06 12	09 ♑ 44 34	16 ♑ 53 07	00 ♌ 09	19 39	22 23	09 01	11 19	14 33	19 14	24 49
11	07 17 02	19 03 24	24 ♑ 05 36	01 ♒ 21 11	29 ♋ 42	20 51	23 01	09 01	11 26	14 32	19 12	24 48
12	07 20 59	20 00 36	08 ♒ 38 59	15 ♒ 58 06	29 12	22 03	23 38	09 02	11 34	14 31	19 11	24 47
13	07 24 55	20 57 48	23 ♒ 17 41	00 ♓ 36 54	28 38	23 15	24 15	09 03	11 41	14 31	19 10	24 45
14	07 28 52	21 55 00	07 ♓ 55 00	15 ♓ 11 21	28 02	24 27	24 52	09 05	11 48	14 29	19 08	24 44
15	07 32 48	22 52 13	22 ♓ 25 25	29 ♓ 36 47	27 24	25 39	25 29	09 06	11 56	14 28	19 07	24 43
16	07 36 45	23 49 26	06 ♈ 45 09	13 ♈ 50 17	26 45	26 51	26 06	09 08	12 03	14 26	19 05	24 41
17	07 40 41	24 46 40	20 ♈ 52 06	27 ♈ 50 31	26 04	28 04	26 43	09 10	12 10	14 25	19 04	24 40
18	07 44 38	25 43 55	04 ♉ 45 32	11 ♉ 37 10	25 24	29 ♊ 16	27 20	09 12	12 18	14 24	19 02	24 39
19	07 48 34	26 41 10	18 ♉ 25 30	25 ♉ 10 34	24 45	00 ♋ 28	27 57	09 14	12 26	14 23	19 01	24 37
20	07 52 31	27 38 27	01 ♊ 52 26	08 ♊ 31 06	24 06	01 40	28 35	09 17	12 33	14 21	18 59	24 36
21	07 56 28	28 35 44	15 ♊ 06 38	21 ♊ 39 02	23 30	02 53	29 12	09 19	12 41	14 20	18 58	24 35
22	08 00 24	29 ♋ 33 02	28 ♊ 08 18	04 ♋ 34 25	22 57	04 05	29 ♌ 49	09 22	12 48	14 18	18 56	24 34
23	08 04 21	00 ♌ 30 20	10 ♋ 57 22	17 ♋ 17 10	22 27	05 18	00 ♍ 27	09 25	12 56	14 17	18 55	24 32
24	08 08 17	01 27 40	23 ♋ 33 50	29 ♋ 47 22	22 01	06 30	01 04	09 28	13 04	14 15	18 53	24 31
25	08 12 14	02 24 59	05 ♌ 57 50	12 ♌ 05 22	21 39	07 43	01 41	09 31	13 11	14 14	18 52	24 30
26	08 16 10	03 22 20	18 ♌ 10 05	24 ♌ 12 11	21 22	08 55	02 19	09 35	13 19	14 12	18 50	24 29
27	08 20 07	04 19 41	00 ♍ 11 53	06 ♍ 09 30	21 11	10 08	02 56	09 38	13 26	14 10	18 49	24 28
28	08 24 03	05 17 02	12 ♍ 05 22	17 ♍ 59 53	21 05	11 20	03 34	09 42	13 34	14 09	18 47	24 27
29	08 28 00	06 14 24	23 ♍ 53 29	29 ♍ 46 39	21 D 04	12 33	04 11	09 46	13 42	14 07	18 46	24 26
30	08 31 57	07 11 47	05 ♎ 39 56	11 ♎ 33 54	21 10	13 46	04 49	09 50	13 49	14 05	18 44	24 25
31	08 35 53	08 ♌ 09 10	17 ♎ 29 08	23 ♎ 26 16	21 ♋ 22	14 ♋ 58	05 ♍ 26	09 ♏ 55	13 ♌ 57	14 ♓ 04	18 ♒ 42	24 ♐ 24

Date	Moon True ☊ ° '	Moon Mean ☊ ° '	Moon ☽ Latitude ° '
01	28 ♓ 39	29 ♓ 23	01 N 08
02	28 R 38	29 20	00 N 06
03	28 D 38	29 17	00 S 57
04	28 R 38	29 14	01 57
05	28 36	29 10	02 54
06	28 33	29 07	03 43
07	28 27	29 04	04 23
08	28 18	29 01	04 50
09	28 08	28 58	05 02
10	27 58	28 55	04 56
11	27 47	28 51	04 32
12	27 39	28 48	03 49
13	27 33	28 45	02 51
14	27 29	28 42	01 42
15	27 28	28 39	00 S 27
16	27 D 28	28 35	00 N 49
17	27 28	28 32	02 02
18	27 R 28	28 29	03 05
19	27 26	28 26	03 58
20	27 22	28 23	04 36
21	27 15	28 20	04 58
22	27 06	28 16	05 05
23	26 57	28 13	04 56
24	26 47	28 10	04 33
25	26 37	28 07	03 57
26	26 30	28 04	03 10
27	26 24	28 01	02 16
28	26 21	27 57	01 16
29	26 20	27 54	00 N 13
30	26 D 20	27 51	00 S 50
31	26 ♓ 21	27 ♓ 48	01 S 52

DECLINATIONS

Date	Sun ☉ ° '	Moon ☽ ° '	Mercury ☿ ° '	Venus ♀ ° '	Mars ♂ ° '	Jupiter ♃ ° '	Saturn ♄ ° '	Uranus ♅ ° '	Neptune ♆ ° '	Pluto ♇ ° '
01	23 N 06	06 N 41	18 N 21	20 N 22	16 N 58	13 S 25	18 N 25	06 S 46	15 S 11	15 S 43
02	23 02	01 N 05	18 03	20 35	16 46	13 25	18 23	06 46	15 12	15 43
03	22 57	04 S 34	17 46	20 47	16 35	13 25	18 21	06 46	15 12	15 43
04	22 52	10 07	17 30	20 59	16 23	13 25	18 19	06 47	15 13	15 44
05	22 46	15 24	17 15	21 10	16 11	13 25	18 18	06 47	15 13	15 44
06	22 40	20 10	17 02	21 21	15 59	13 25	18 16	06 47	15 13	15 44
07	22 34	24 10	16 49	21 31	15 47	13 26	18 14	06 48	15 13	15 44
08	22 27	27 02	16 38	21 40	15 35	13 26	18 12	06 48	15 14	15 44
09	22 20	28 24	16 29	21 49	15 22	13 26	18 10	06 49	15 14	15 44
10	22 13	28 00	16 20	21 57	15 10	13 27	18 08	06 49	15 15	15 44
11	22 05	25 45	16 14	22 05	14 57	13 27	18 06	06 49	15 15	15 45
12	21 57	21 47	16 09	22 12	14 45	13 28	18 04	06 50	15 15	15 45
13	21 48	16 27	16 05	22 19	14 32	13 29	18 02	06 50	15 16	15 45
14	21 39	10 11	16 04	22 25	14 19	13 29	18 00	06 51	15 17	15 45
15	21 30	03 S 25	16 03	22 30	14 07	13 30	17 58	06 51	15 17	15 45
16	21 20	03 N 26	16 05	22 35	13 54	13 31	17 56	06 52	15 18	15 45
17	21 10	10 01	16 08	22 39	13 41	13 32	17 54	06 52	15 18	15 46
18	21 00	16 01	16 12	22 42	13 27	13 32	17 52	06 53	15 19	15 46
19	20 49	21 07	16 18	22 45	13 14	13 33	17 50	06 53	15 19	15 46
20	20 38	25 02	16 25	22 47	13 01	13 34	17 48	06 54	15 20	15 46
21	20 27	27 33	16 33	22 48	12 48	13 36	17 46	06 54	15 20	15 46
22	20 15	28 30	16 43	22 49	12 34	13 37	17 44	06 55	15 21	15 47
23	20 03	27 54	16 53	22 49	12 20	13 38	17 41	06 56	15 21	15 47
24	19 50	25 52	17 04	22 49	12 07	13 39	17 39	06 56	15 21	15 47
25	19 37	22 36	17 16	22 48	11 53	13 40	17 37	06 57	15 22	15 47
26	19 24	18 24	17 28	22 46	11 39	13 42	17 35	06 57	15 22	15 47
27	19 11	13 31	17 41	22 43	11 25	13 43	17 33	06 58	15 23	15 48
28	18 57	08 12	17 53	22 40	11 12	13 45	17 31	06 59	15 23	15 48
29	18 43	02 N 38	18 06	22 36	10 58	13 46	17 29	07 00	15 24	15 48
30	18 28	03 S 01	18 19	22 32	10 43	13 48	17 27	07 00	15 24	15 48
31	18 N 14	08 S 35	18 N 31	22 N 27	10 N 29	13 S 49	17 N 24	07 S 01	15 S 25	15 S 49

ZODIAC SIGN ENTRIES

Date	h	m	Planets
02	17	06	☽ ♎
05	05	13	☽ ♏
07	14	14	☽ ♐
09	19	25	☽ ♑
10	20	18	☿ ♋
11	21	46	☽ ♒
13	22	59	☽ ♓
16	00	39	☽ ♈
18	03	44	☽ ♉
19	02	41	♀ ♋
20	08	38	☽ ♊
22	15	28	☽ ♋
22	18	53	♂ ♍
22	23	18	☉ ♌
25	00	24	☽ ♌
27	11	36	☽ ♍
30	00	27	☽ ♎

LATITUDES

Date	Mercury ☿ ° '	Venus ♀ ° '	Mars ♂ ° '	Jupiter ♃ ° '	Saturn ♄ ° '	Uranus ♅ ° '	Neptune ♆ ° '	Pluto ♇ ° '
01	01 S 38	01 S 26	01 N 14	01 N 09	00 N 46	00 S 47	00 S 12	07 N 38
04	02 25	01 19	01 13	01 08	00 46	00 47	00 12	07 37
07	03 10	01 12	01 12	01 07	00 47	00 48	00 12	07 37
10	03 52	01 05	01 11	01 07	00 47	00 48	00 12	07 36
13	04 26	00 57	01 10	01 06	00 47	00 48	00 12	07 36
16	04 49	00 50	01 09	01 05	00 47	00 48	00 12	07 35
19	04 57	00 42	01 07	01 04	00 47	00 48	00 12	07 34
22	04 51	00 34	01 06	01 04	00 47	00 48	00 12	07 34
25	04 30	00 25	01 05	01 03	00 48	00 48	00 12	07 33
28	03 57	00 17	01 04	01 02	00 48	00 48	00 12	07 32
31	03 S 15	00 S 09	01 N 03	01 N 01	00 N 48	00 S 48	00 S 12	07 N 31

DATA

Julian Date	2453918
Delta T	+66 seconds
Ayanamsa	23° 56' 53"
Synetic vernal point	05° ♓ 10' 06"
True obliquity of ecliptic	23° 26' 27"

MOON'S PHASES, APSIDES AND POSITIONS ☽

Date	h	m	Phase	Longitude ° '	Eclipse Indicator
03	16	37	☽	11 ♎ 37	
11	03	02	○	18 ♑ 42	
17	19	13	☾	25 ♈ 04	
25	04	31	●	02 ♌ 07	

Day	h	m		
01	20	13	Apogee	
13	17	42	Perigee	
29	13	07	Apogee	
02	16	37	0S	
09	18	52	Max dec	28° S 29'
15	23	55	0N	
22	14	35	Max dec	28° N 31'
29	23	10	0S	

LONGITUDES

Date	Chiron ⚷ ° '	Ceres ⚳ ° '	Pallas ⚴ ° '	Juno ⚵ ° '	Vesta ⚶ ° '	Black Moon Lilith ⚸ ° '
01	08 ♒ 35	26 ♒ 38	09 ♑ 55	11 ♌ 51	14 ♌ 38	17 ♍ 46
11	08 ♒ 05	25 ♒ 44	07 ♑ 58	16 ♌ 22	19 ♌ 05	18 ♍ 28
21	07 ♒ 33	24 ♒ 17	04 ♑ 41	20 ♌ 52	23 ♌ 38	19 ♍ 59
31	06 ♒ 59	22 ♒ 24	02 ♑ 45	25 ♌ 19	28 ♌ 16	21 ♍ 05

All ephemeris data is given at 12.00 UT and the Moon's longitude is additionally given for 24.00 UT

ASPECTARIAN

h m	Aspects
01 Saturday	
00 49	☽ ⚺ ♄
06 11	☿ ∥ ♄
09 55	☽ ☍ ♅
11 40	☽ # ♅
12 31	☽ ∠ ☿
13 07	☽ ⊥ ♄
14 19	♀ ⚻ ♃
14 29	☽ ⚺ ♂
19 33	☽ ⚻ ♆
20 32	☿ ± ♇
02 Sunday	
00 54	☽ Q ☉
03 22	☽ ⊥ ♂
04 54	☽ ∠ ♃
06 58	☽ □ ♇
07 37	☽ ∠ ♄
07 44	☽ ± ♆
08 41	☉ ⚺ ♄
18 04	♀ ✶ ♄
19 34	☽ ✶ ☿
22 40	☽ ∠ ♂
23 11	☽ ⊥ ♃
03 Monday	
01 59	☽ ⚼ ♆
11 19	☽ ⚺ ♃
14 21	☽ ✶ ♄
16 23	☽ △ ♀
16 37	☽ □ ☉
19 22	☽ Q ♇
20 05	☽ Q ☿
21 29	☽ ∥ ♅
22 43	☽ ⚻ ♅
04 Tuesday	
02 47	☉ ⚺ ♀
06 33	☽ ✶ ♂
08 09	☽ △ ♆
10 42	☽ ± ♅
14 38	☽ Q ♄
19 17	☽ ✶ ♇
19 33	☿ St R
05 Wednesday	
01 35	☽ ⚼ ♀
02 48	☽ ∥ ♃
04 29	☽ ⚼ ♅
07 34	☽ Q ♂
07 53	☽ □ ☿
10 13	☉ ⊥ ♂
11 09	☽ ∥ ♆
11 53	☉ ± ♆
12 47	♂ ☍ ♆
13 35	☽ ∥ ♇
15 37	☽ # ♂
20 34	☽ # ☿
21 01	☽ ± ♀
22 41	☽ ☌ ♃
06 Thursday	
00 31	☽ ∠ ♇
02 04	☽ # ♄
02 09	☽ □ ♄
06 57	♀ □ ♅
07 15	♃ St D
08 36	☽ △ ☉
09 28	☽ △ ♅
09 44	☽ ⚻ ♀
14 15	♀ ± ♃
17 37	☽ ⊥ ♇
18 21	☽ □ ♆
18 46	☽ # ♀
19 54	☽ □ ♂
19 59	☉ △ ♅
07 Friday	
01 57	☽ # ☉
04 49	☽ ⚺ ♇
12 33	☿ ∠ ♀
14 53	☽ ⚼ ☉
16 10	☽ △ ☿
17 30	♂ # ♇
08 Saturday	
03 28	☽ Q ♆
06 32	☽ ⚺ ♃
08 27	☽ ± ☉
10 05	☿ ± ♇
10 16	☽ △ ♄
16 31	☽ □ ♅
17 15	☽ ⊥ ♃
18 38	☽ ⚼ ☿
19 56	☽ ⚻ ☉
22 09	☽ ☍ ♀
09 Sunday	
00 46	☽ ✶ ♆
04 51	☽ △ ♂
09 02	☽ ∠ ♃
10 02	☽ ± ☿
10 31	☽ ☌ ♇
12 52	☽ ⚼ ♄
20 07	☽ ⚻ ☿
23 48	☽ Q ♅
10 Monday	
02 35	♂ # ♆
02 39	☽ ∠ ♆
03 35	♀ △ ♆
04 26	☽ ± ♄
07 50	☽ ⚼ ♂
10 45	☽ ✶ ♃
14 41	☽ ⚻ ♄
17 52	☽ ⊥ ♆
20 05	☽ ✶ ♅
23 41	☽ ± ♂
11 Tuesday	
03 02	☽ ☍ ☉
03 53	☽ ⚺ ♆
06 08	☽ ⚻ ♀
06 53	☽ Q ♃
10 07	☽ ⚻ ♂
12 03	☉ ∥ ♀
13 10	☽ ⚺ ♇
15 37	☉ ⚻ ♆
16 59	☽ ± ♀
20 19	☿ ⚼ ♅
20 58	☽ ☍ ☿
21 00	☽ ∠ ♅
23 04	☽ ⊥ ♇
12 Wednesday	
09 08	☽ ⚼ ♀
09 53	☽ # ♀
11 07	☽ # ☉
11 47	☽ ⊥ ♅
12 38	☽ □ ♃
13 51	☽ ∠ ♇
16 49	☽ ☍ ♄
21 37	☽ ⚺ ♅
13 Thursday	
03 38	☿ ⊥ ♀
05 14	☽ ☌ ♆
05 21	☽ # ♄
07 55	☽ ⚻ ☉
11 55	☽ △ ♀
13 28	☽ # ☿
13 37	☽ ☍ ♂
14 23	☽ ✶ ♇
14 50	☽ ∥ ♇
16 44	☽ ∥ ♆
18 26	☽ ± ☉
19 54	☽ # ♂
20 25	☽ ⚻ ☿
23 42	☽ ∥ ♃
14 Friday	
04 23	♀ ⚼ ♃
05 52	☽ ± ☿
07 08	♂ △ ♇
10 03	☽ Q ♇
10 14	☽ ⚼ ☉
13 55	☽ △ ♃
17 28	♀ ☍ ♇
18 28	☽ ⚻ ♄
20 05	☽ ⚼ ☿
22 49	☽ ☌ ♅
23 57	☽ ∥ ♅
15 Saturday	
04 29	☽ ± ♄
04 50	♀ ✶ ♂
06 30	☽ ⚺ ♆
12 48	☽ △ ☉
14 48	☽ ⚼ ♃
15 48	☽ □ ♇
16 28	☽ ⊥ ♆
17 19	☽ ⚻ ♂
17 52	☽ □ ♀
19 35	☽ ⚼ ♄
19 56	☽ △ ☿
16 Sunday	
03 49	☽ ± ♂
05 54	☽ ± ♃
07 31	☽ ∠ ♆
10 32	☿ ⚺ ♀
16 02	☽ ⚻ ♃
16 26	♀ ∠ ♄
19 41	☽ ⚼ ♂
21 03	☽ △ ♄
23 59	☿ ⚺ ♂
17 Monday	
00 18	☽ # ♅
01 00	☽ ⚺ ♅
03 01	☽ Q ♀
08 55	☽ ✶ ♆
09 14	☉ ⚻ ♇
11 14	☽ ⊥ ♅
18 31	☽ △ ♇
19 13	☽ □ ☉
20 32	☽ □ ☿
22 32	☽ △ ♂
18 Tuesday	
01 33	☽ ✶ ♀
01 41	☽ # ♃
01 45	☽ ∥ ♂
02 42	☽ ∠ ♅
03 24	♂ # ♃
05 33	☽ Q ♆
06 20	♂ Q ♃
07 07	☉ ☌ ☿
08 59	☽ # ♆
10 55	☽ # ♇
12 49	☽ ∥ ☿
19 47	☽ ☍ ♃
20 06	☽ ∥ ♄
20 31	☽ ⚼ ♇
19 Wednesday	
01 20	☽ □ ♄
02 26	☽ Q ☿
04 52	☽ ✶ ♅
04 54	☽ Q ☉
06 16	☽ ∠ ♀
10 31	☽ ∥ ☉
12 21	☽ ± ♇
13 03	☽ □ ♆
16 38	☿ ⚻ ♇
21 09	☽ ∥ ♀
22 43	☽ ✶ ☿
23 00	☽ ⚻ ♇
23 48	☽ ⊥ ♀
20 Thursday	
02 07	☽ Q ♅
03 49	☽ ✶ ☉
05 48	☽ □ ♂
09 36	☽ Q ♄
11 36	☽ ⚺ ♀
21 Friday	
00 29	☽ ∠ ☿
01 25	☽ ⚻ ♃
07 31	☽ ✶ ♄
09 01	☽ ∠ ☉
10 35	☽ □ ♅
12 23	☽ ± ♃
16 01	☽ Q ♂
16 12	☽ ⊥ ☿
18 07	☿ ⊥ ♂
19 03	☽ △ ♆
22 Saturday	
02 47	☽ ⚺ ☿
02 50	☽ ⊥ ☉
04 59	☽ ⚼ ♃
05 23	☽ ☍ ♇
06 00	☉ ⚼ ♅
09 11	♀ ⚼ ♆
11 23	☽ ∠ ♄
14 50	☽ ⚺ ☉
15 17	☽ ✶ ♂
22 48	☽ ⚼ ♆
23 Sunday	
00 14	☽ ☌ ♀
04 21	☽ ⊥ ♄
07 33	☉ ⚺ ♂
09 05	☽ △ ♃
12 51	☉ ± ♇
15 42	☽ ± ♆
15 47	☽ ⚺ ♄
18 17	☽ △ ♅
20 56	☽ ∠ ♂
24 Monday	
03 04	☽ ⚻ ♆
09 07	☽ ☌ ☿
13 50	☽ ⚻ ♇
15 02	☽ ⊥ ♂
22 57	☽ ⚼ ♅
25 Tuesday	
00 25	♀ ⊥ ♄
01 24	☽ ± ♇
03 14	☽ ⚺ ♂
04 31	☽ ☌ ☉
04 54	♃ ∠ ♇
10 46	☽ ∥ ♀
15 47	☽ ⚺ ♀
16 25	☽ ± ♅
18 54	☽ ⚼ ♇
19 00	☽ □ ♃
26 Wednesday	
02 19	☽ ☌ ♄
04 11	☽ ⚻ ♅
04 52	☽ ⊥ ♀
06 23	☽ ∥ ☉
13 19	☽ ☍ ♆
16 16	☽ ∥ ♄
16 38	☽ ∥ ☿
18 14	☽ ⚺ ☿
23 39	☿ ∥ ♄
27 Thursday	
00 32	☽ △ ♇
00 43	☽ ∠ ♀
01 11	☽ # ♇
01 50	♀ △ ♃
03 11	☽ # ♆
06 01	☽ ⊥ ☿
06 51	☽ Q ♃
11 06	☽ # ♃
17 49	☽ ☌ ♂
21 02	☽ ⚺ ☉
22 05	☽ ∥ ♂
23 55	☽ ∠ ☿
28 Friday	
07 09	☽ ✶ ♃
10 13	☽ ⊥ ☉
10 18	☽ ✶ ♀
15 02	☽ ⚺ ♄
16 10	☽ ☍ ♅
17 18	☽ # ♅
29 Saturday	
00 39	☿ St D
01 34	☽ ⚻ ♆
03 22	☽ ⊥ ♄
06 07	☽ ∠ ☉
06 15	☽ ✶ ☿
13 06	☽ □ ♇
13 30	☽ Q ♀
13 46	☽ ± ♆
13 48	☽ ∠ ♃
16 03	♀ ± ♆
21 54	☽ ∠ ♄
30 Sunday	
06 52	☽ Q ☿
08 04	☽ ⚼ ♆
08 15	☽ ⊥ ♃
10 10	☽ ⚺ ♂
13 22	♀ ⚺ ♄
15 24	☽ ✶ ☉
18 20	♀ △ ♅
20 21	☉ ∥ ☿
20 33	☽ ⚺ ♃
23 03	☽ ⊥ ♂
31 Monday	
01 42	☽ Q ♇
04 46	☽ ✶ ♄
05 05	☽ ⚻ ♅
05 08	☽ ∥ ♅
06 20	☽ □ ♀
09 44	☉ ± ♅
14 28	☽ △ ♆
17 11	☽ ± ♅
17 51	☽ Q ☉
18 17	☽ ∠ ♂
20 02	☽ □ ☿
20 03	☽ # ♂

AUGUST 2006

Date	Sidereal time h m s	Sun ☉ ° ' "	Moon ☽ ° ' "	Moon ☽ 24.00 ° ' "	Mercury ☿ ° '	Venus ♀ ° '	Mars ♂ ° '	Jupiter ♃ ° '	Saturn ♄ ° '	Uranus ♅ ° '	Neptune ♆ ° '	Pluto ♇ ° '
01	08 39 50	09 ♌ 06 34	29 ♎ 25 58	05 ♏ 28 51	21 ♋ 41	16 ♋ 11	06 ♍ 04	09 ♏ 59	14 ♌ 05	14 ♓ 02	18 ♒ 41	24 ♐ 23
02	08 43 46	10 03 58	11 ♏ 35 34	17 ♏ 46 46	22 05	17 24	06 41	10 04	14 12	14 R 00	18 R 39	24 R 22
03	08 47 43	11 01 24	24 ♏ 03 02	00 ♐ 24 55	22 37	18 37	07 19	10 08	14 20	13 58	18 38	24 21
04	08 51 39	11 58 49	06 ♐ 52 54	13 ♐ 27 22	23 14	19 50	07 57	10 13	14 28	13 56	18 36	24 20
05	08 55 36	12 56 16	20 ♐ 08 35	26 ♐ 56 42	23 58	21 03	08 34	10 18	14 36	13 54	18 34	24 19
06	08 59 32	13 53 43	03 ♑ 51 43	10 ♑ 53 27	24 48	22 16	09 12	10 24	14 43	13 52	18 33	24 18
07	09 03 29	14 51 11	18 ♑ 01 34	25 ♑ 15 30	25 44	23 29	09 50	10 29	14 51	13 50	18 31	24 17
08	09 07 26	15 48 40	02 ♒ 34 35	09 ♒ 57 56	26 46	24 42	10 27	10 35	14 59	13 48	18 29	24 16
09	09 11 22	16 46 10	17 ♒ 24 37	24 ♒ 53 33	27 54	25 55	11 05	10 40	15 06	13 46	18 28	24 15
10	09 15 19	17 43 41	02 ♓ 23 38	09 ♓ 53 46	29 ♋ 07	27 08	11 43	10 46	15 14	13 44	18 26	24 15
11	09 19 15	18 41 13	17 ♓ 22 55	24 ♓ 50 05	00 ♌ 26	28 21	12 21	10 52	15 22	13 42	18 25	24 14
12	09 23 12	19 38 46	02 ♈ 14 26	09 ♈ 35 13	01 51	29 ♋ 35	12 59	10 58	15 29	13 40	18 23	24 14
13	09 27 08	20 36 21	16 ♈ 51 51	24 ♈ 03 55	03 20	00 ♌ 48	13 36	11 04	15 37	13 38	18 21	24 13
14	09 31 05	21 33 57	01 ♉ 11 04	08 ♉ 13 10	04 54	02 01	14 14	11 11	15 45	13 36	18 20	24 12
15	09 35 01	22 31 35	15 ♉ 10 07	22 ♉ 01 57	06 32	03 14	14 52	11 17	15 53	13 34	18 18	24 11
16	09 38 58	23 29 14	28 ♉ 48 46	05 ♊ 30 43	08 14	04 28	15 30	11 24	16 00	13 31	18 16	24 11
17	09 42 55	24 26 55	12 ♊ 08 01	18 ♊ 40 52	10 00	05 41	16 08	11 31	16 08	13 29	18 15	24 10
18	09 46 51	25 24 38	25 ♊ 09 32	01 ♋ 34 15	11 48	06 55	16 46	11 38	16 15	13 27	18 13	24 09
19	09 50 48	26 22 22	07 ♋ 55 15	14 ♋ 12 48	13 40	08 08	17 24	11 45	16 23	13 25	18 12	24 09
20	09 54 44	27 20 07	20 ♋ 27 06	26 ♋ 38 24	15 34	09 22	18 03	11 52	16 31	13 22	18 10	24 08
21	09 58 41	28 17 55	02 ♌ 46 54	08 ♌ 52 49	17 29	10 35	18 41	11 59	16 38	13 20	18 08	24 08
22	10 02 37	29 ♌ 15 43	14 ♌ 56 21	20 ♌ 57 43	19 27	11 49	19 19	12 07	16 46	13 18	18 07	24 07
23	10 06 34	00 ♍ 13 33	26 ♌ 57 08	02 ♍ 54 49	21 25	13 03	19 57	12 14	16 54	13 16	18 05	24 07
24	10 10 30	01 11 25	08 ♍ 51 00	14 ♍ 45 58	23 24	14 16	20 35	12 22	17 01	13 13	18 04	24 07
25	10 14 27	02 09 18	20 ♍ 39 59	26 ♍ 33 22	25 23	15 30	21 14	12 30	17 09	13 11	18 02	24 06
26	10 18 24	03 07 12	02 ♎ 26 26	08 ♎ 19 35	27 23	16 44	21 52	12 38	17 16	13 09	18 00	24 06
27	10 22 20	04 05 08	14 ♎ 13 13	20 ♎ 07 45	29 ♌ 23	17 58	22 30	12 46	17 24	13 06	17 59	24 06
28	10 26 17	05 03 05	26 ♎ 03 39	02 ♏ 01 25	01 ♍ 22	19 12	23 08	12 54	17 31	13 04	17 57	24 06
29	10 30 13	06 01 03	08 ♏ 01 35	14 ♏ 04 39	03 21	20 26	23 47	13 03	17 39	13 01	17 56	24 05
30	10 34 10	06 59 03	20 ♏ 11 13	26 ♏ 21 48	05 19	21 39	24 25	13 11	17 46	12 59	17 54	24 05
31	10 38 06	07 ♍ 57 04	02 ♐ 36 59	08 ♐ 57 17	07 ♍ 16	22 ♌ 53	25 ♍ 04	13 ♏ 20	17 ♌ 53	12 ♓ 57	17 ♒ 53	24 ♐ 05

Date	Moon True Node ° '	Moon Mean Node ° '	Moon ☽ Latitude ° '	DECLINATIONS Sun ☉ ° '	Moon ☽ ° '	Mercury ☿ ° '	Venus ♀ ° '	Mars ♂ ° '	Jupiter ♃ ° '	Saturn ♄ ° '	Uranus ♅ ° '	Neptune ♆ ° '	Pluto ♇ ° '
01	26 ♓ 23	27 ♓ 45	02 S 49	17 N 59	13 S 55	18 N 43	22 N 21	10 N 15	13 S 51	17 N 22	07 S 02	15 S 25	15 S 49
02	26 R 23	27 41	03 40	17 43	18 48	18 55	22 15	10 01	13 52	17 20	07 02	15 26	15 49
03	26 22	27 38	04 22	17 28	23 01	19 05	22 08	09 46	13 54	17 18	07 03	15 27	15 49
04	26 19	27 35	04 52	17 12	26 16	19 15	22 00	09 32	13 56	17 16	07 04	15 27	15 49
05	26 15	27 32	05 08	16 56	28 12	19 23	21 52	09 17	13 58	17 14	07 05	15 28	15 50
06	26 09	27 29	05 07	16 39	28 30	19 30	21 43	09 03	14 00	17 11	07 05	15 28	15 50
07	26 03	27 26	04 47	16 23	26 58	19 36	21 33	08 48	14 02	17 09	07 06	15 29	15 50
08	25 57	27 22	04 09	16 06	23 38	19 40	21 23	08 34	14 03	17 07	07 07	15 29	15 50
09	25 52	27 19	03 14	15 49	18 41	19 42	21 12	08 19	14 05	17 05	07 08	15 30	15 51
10	25 48	27 16	02 04	15 31	12 33	19 42	21 01	08 04	14 07	17 03	07 09	15 30	15 51
11	25 47	27 13	00 S 46	15 14	05 S 41	19 39	20 49	07 49	14 10	17 00	07 09	15 31	15 51
12	25 D 46	27 10	00 N 35	14 56	01 N 26	19 35	20 36	07 34	14 12	16 58	07 10	15 31	15 52
13	25 47	27 06	01 53	14 37	08 22	19 28	20 23	07 19	14 14	16 56	07 11	15 32	15 52
14	25 49	27 03	03 01	14 19	14 43	19 18	20 09	07 04	14 16	16 54	07 12	15 32	15 52
15	25 50	27 00	03 58	14 00	20 10	19 05	19 55	06 49	14 18	16 52	07 13	15 33	15 52
16	25 R 50	26 57	04 39	13 41	24 26	18 50	19 40	06 34	14 21	16 49	07 14	15 33	15 53
17	25 49	26 54	05 04	13 22	27 16	18 32	19 24	06 19	14 23	16 47	07 14	15 34	15 53
18	25 46	26 51	05 13	13 03	28 34	18 12	19 08	06 04	14 25	16 45	07 15	15 34	15 53
19	25 42	26 47	05 06	12 44	28 18	17 48	18 52	05 49	14 28	16 43	07 16	15 35	15 54
20	25 38	26 44	04 44	12 24	26 34	17 23	18 34	05 33	14 30	16 41	07 17	15 35	15 54
21	25 33	26 41	04 10	12 04	23 35	16 54	18 17	05 18	14 32	16 38	07 18	15 36	15 54
22	25 29	26 38	03 24	11 44	19 36	16 23	17 58	05 03	14 35	16 36	07 19	15 36	15 54
23	25 25	26 35	02 30	11 24	14 53	15 51	17 40	04 47	14 37	16 34	07 20	15 37	15 55
24	25 23	26 32	01 30	11 03	09 38	15 15	17 20	04 32	14 40	16 32	07 21	15 37	15 55
25	25 22	26 28	00 N 26	10 42	04 N 06	14 39	17 01	04 16	14 43	16 30	07 22	15 38	15 55
26	25 D 22	26 25	00 S 39	10 22	01 S 34	14 00	16 40	04 01	14 45	16 27	07 22	15 38	15 56
27	25 23	26 22	01 42	10 01	07 11	13 20	16 20	03 45	14 48	16 25	07 23	15 39	15 56
28	25 25	26 19	02 42	09 40	12 34	12 38	15 59	03 30	14 51	16 23	07 24	15 39	15 56
29	25 26	26 16	03 35	09 18	17 34	11 56	15 37	03 14	14 53	16 21	07 25	15 40	15 56
30	25 28	26 12	04 19	08 57	21 57	11 12	15 15	02 59	14 56	16 19	07 26	15 40	15 57
31	25 ♓ 28	26 ♓ 09	04 S 52	08 N 35	25 S 27	10 N 28	14 N 52	02 N 43	14 S 59	16 N 17	07 S 27	15 S 41	15 S 57

ZODIAC SIGN ENTRIES

Date	h	m	Planets
01	13	08	☽ ♏
03	23	13	☽ ♐
06	05	19	☽ ♑
08	07	47	☽ ♒
10	08	10	☽ ♓
11	04	09	☿ ♌
12	08	22	☽ ♈
12	20	21	♀ ♌
14	10	00	☽ ♉
16	14	07	☽ ♊
18	21	03	☽ ♋
21	06	33	☽ ♌
23	06	23	☉ ♍
23	18	08	☽ ♍
26	07	01	☽ ♎
27	19	30	☿ ♍
28	19	56	☽ ♏
31	07	00	☽ ♐

LATITUDES

Date	Mercury ☿ ° '	Venus ♀ ° '	Mars ♂ ° '	Jupiter ♃ ° '	Saturn ♄ ° '	Uranus ♅ ° '	Neptune ♆ ° '	Pluto ♇ ° '
01	03 S 00	00 S 06	01 N 02	01 N 01	00 N 48	00 S 48	00 S 12	07 N 31
04	02 13	00 N 02	01 01	01 00	00 48	00 48	00 12	07 30
07	01 25	00 10	01 00	00 59	00 48	00 48	00 13	07 29
10	00 S 39	00 17	00 58	00 59	00 49	00 49	00 13	07 28
13	00 N 03	00 25	00 57	00 58	00 49	00 49	00 13	07 27
16	00 39	00 32	00 56	00 57	00 49	00 49	00 13	07 26
19	01 08	00 39	00 55	00 57	00 50	00 49	00 13	07 26
22	01 29	00 45	00 53	00 56	00 50	00 49	00 13	07 25
25	01 41	00 52	00 52	00 55	00 50	00 49	00 13	07 24
28	01 46	00 57	00 51	00 55	00 50	00 49	00 13	07 23
31	01 N 45	01 N 03	00 N 49	00 N 54	00 N 51	00 S 49	00 S 13	07 N 22

DATA

Julian Date	2453949
Delta T	+66 seconds
Ayanamsa	23° 56' 58"
Synetic vernal point	05° ♓ 10' 01"
True obliquity of ecliptic	23° 26' 27"

MOON'S PHASES, APSIDES AND POSITIONS ☽

Date	h	m	Phase	Longitude ° '	Eclipse Indicator
02	08	46	☽	09 ♏ 56	
09	10	54	○	16 ♒ 44	
16	01	51	☾	23 ♉ 05	
23	19	10	●	00 ♍ 31	
31	22	56	☽	08 ♐ 24	

Day	h	m		
10	18	34	Perigee	
26	01	32	Apogee	
06	04	12	Max dec	28° S 36'
12	07	10	ON	
18	19	44	Max dec	28° N 39'
26	05	22	0S	

LONGITUDES

Date	Chiron ⚷ ° '	Ceres ⚳ ° '	Pallas ⚴ ° '	Juno ⚵ ° '	Vesta ⚶ ° '	Black Moon Lilith ⚸ ° '
01	06 ♒ 56	22 ♒ 11	02 ♑ 36	25 ♌ 46	28 ♌ 44	21 ♍ 12
11	06 ♒ 23	20 ♒ 02	01 ♑ 23	00 ♍ 11	03 ♍ 26	22 ♍ 18
21	05 ♒ 52	17 ♒ 51	00 ♑ 51	04 ♍ 34	08 ♍ 12	23 ♍ 25
31	05 ♒ 24	15 ♒ 52	00 ♑ 58	08 ♍ 53	13 ♍ 02	24 ♍ 31

All ephemeris data is given at 12.00 UT and the Moon's longitude is additionally given for 24.00 UT

ASPECTARIAN

h m Aspects

01 Tuesday
01 54 ☽ ∗ ♇
04 22 ♄ ⊼ ♅
05 14 ☽ Q ♄
11 12 ☽ ⚻ ♅
11 41 ☽ ∥ ♃
18 35 ☉ ⚻ ♇
19 10 ☽ ∥ ♆
21 02 ☽ ∥ ♇

02 Wednesday
01 52 ☽ ∗ ♂
04 34 ☽ # ♄
06 45 ☽ # ☉
07 38 ☽ ∠ ♇
08 46 ☽ □ ☉
08 59 ☽ ☌ ♃
11 51 ☉ □ ♃
12 37 ☽ # ☿
16 40 ☽ △ ♅
17 09 ☽ □ ♄

03 Thursday
00 29 ☽ △ ♀
01 06 ☽ ⊥ ♇
01 39 ☽ □ ♆
02 29 ☽ Q ♂
06 42 ☽ # ♀
09 08 ☽ △ ☿
12 11 ♀ ⊼ ♆
12 33 ☽ ⊻ ♇

04 Friday
05 29 ☉ ∥ ♄
07 49 ☽ ⚻ ♀
11 29 ☽ Q ♆
14 03 ☽ □ ♂
14 37 ☽ ⚻ ☿
18 10 ☽ ⊻ ♃
22 03 ☽ △ ☉

05 Saturday
00 50 ☽ □ ♅
01 57 ☽ ± ♀
01 57 ☽ △ ♄
05 06 ☽ ⊥ ♃
07 52 ☽ ± ☿
09 12 ☽ ∗ ♆
13 46 ☽ ⊼ ♀
19 11 ☽ ⊼ ☿
19 22 ☽ ☌ ♇
21 11 ☽ ∠ ♃
22 20 ☿ ⊼ ♇

06 Sunday
02 46 ☽ ⚻ ☉
04 46 ☽ ⚻ ♄
08 35 ☽ Q ♅
11 26 ☉ ⊼ ♅
11 27 ☽ ∠ ♆
19 25 ☽ ± ☉
20 24 ☽ ± ♄
21 33 ☽ △ ♂
23 14 ☽ ∗ ♃

07 Monday
02 46 ☽ ⊥ ♆
05 00 ☽ ∗ ♅
06 18 ☽ ⊼ ☉
06 38 ☽ ⊼ ♄
11 54 ☉ ☌ ♄
12 49 ☽ ⊻ ♆
19 28 ☽ Q ♃
21 54 ☽ ☍ ♀
22 23 ☽ ⊻ ♇
23 48 ☽ ⚻ ♂

08 Tuesday
01 44 ☽ ☍ ☿
03 41 ♀ ⊼ ♇
05 51 ☽ ∠ ♅
08 14 ☽ ⊥ ♇
15 12 ☽ ± ♂
17 20 ♂ ∗ ♃
20 29 ☽ ⊥ ♅
22 52 ☽ ∠ ♇

09 Wednesday
00 11 ☽ # ♀
01 04 ☽ □ ♃
01 22 ☽ ⊼ ♂
06 09 ☽ ⊻ ♅
07 36 ☽ # ☿
08 16 ☽ ☍ ♄
09 09 ☉ # ♇
10 54 ☽ ☍ ☉
13 41 ☽ ☌ ♆
18 43 ☽ # ♄
18 54 ♀ ∠ ♂
22 58 ☽ ∗ ♇
23 35 ☽ ∥ ♇

10 Thursday
00 18 ☽ # ☉
00 56 ☽ ∥ ♆
02 51 ☽ ⊼ ♀
04 51 ☿ ⚻ ♅
06 12 ☽ ∥ ♃
06 17 ☽ ⊼ ☿
13 17 ☽ ± ♀
13 28 ☉ # ♆
16 46 ☽ ± ☿
18 09 ☽ Q ♇

11 Friday
01 29 ☽ △ ♃
03 34 ☽ ☍ ♂
04 26 ☽ # ♂
04 58 ☽ ⚻ ♀
05 14 ☉ ☍ ♆
06 07 ☽ ☌ ♅
07 00 ☽ ∥ ♅
08 21 ☿ ± ♇
08 34 ☽ ⚻ ☿
08 44 ☽ ⊼ ♄
13 39 ☽ ⊻ ♆
14 14 ☽ ⊼ ☉
18 27 ☽ ± ♄
18 39 ♀ ⚻ ♅
23 01 ☽ □ ♇
23 17 ☽ ⊥ ♆

12 Saturday
00 34 ☽ ± ☉
01 46 ☽ ⚻ ♃
07 17 ☽ △ ♀
09 08 ☽ ⚻ ♄
11 17 ☽ △ ☿
13 51 ☽ ∠ ♆
16 11 ☽ ⚻ ☉
16 28 ☽ ± ♃

13 Sunday
00 33 ♀ ± ♇
02 22 ☽ ⊼ ♃
06 22 ☽ ⊼ ♂
06 40 ☽ ⊻ ♅
07 49 ☽ # ♅
08 27 ☽ ∥ ♂
09 55 ☽ △ ♄
12 51 ♂ ☍ ♅
14 28 ☽ ∗ ♆
16 35 ☽ ⊥ ♅
16 46 ☽ ± ♂
18 40 ☽ △ ☉

14 Monday
00 14 ☽ △ ♇
00 47 ♂ # ♅
07 38 ☽ ∠ ♅
08 33 ☽ ⚻ ♂
10 11 ☽ # ♃
10 28 ☽ ∥ ☉
10 33 ☽ Q ♆
13 33 ☽ □ ♀
15 20 ☽ # ♆
15 27 ☉ # ♃
16 42 ☽ # ♇
19 08 ☽ □ ☿
20 59 ☽ ∥ ♄

15 Tuesday
01 40 ☽ ⚻ ♇
05 14 ☽ ☍ ♃
07 04 ☽ ∥ ☿
09 13 ☽ ∗ ♅
10 49 ☽ ∥ ♀
11 28 ☽ △ ♂
13 14 ☽ □ ♄
17 15 ☽ ± ♇
17 27 ☽ □ ♆
23 39 ☽ Q ♀

16 Wednesday
01 51 ☽ □ ☉
02 17 ☿ ± ♅
03 47 ☽ ⊼ ♇
06 11 ☽ Q ♅
06 45 ☽ Q ☿
21 22 ☽ Q ♄
23 08 ☽ ∗ ♀

17 Thursday
00 52 ☿ ⚻ ♇
05 02 ☉ △ ♇
07 30 ☽ ∗ ☿
10 51 ☽ ⊼ ♃
11 37 ♂ ⊻ ♄
12 37 ☽ Q ☉
14 28 ☽ □ ♅
14 59 ☿ ⊥ ♂
19 23 ☽ ∗ ♄
19 42 ☽ □ ♂
21 56 ☽ ± ♃
23 10 ☽ △ ♆

18 Friday
05 21 ☽ ∠ ♀
09 30 ☿ □ ♃
10 08 ☽ ☍ ♇
12 30 ☽ ∗ ☉
14 46 ☽ ⚻ ♃
15 35 ☽ ∠ ☿
22 12 ♀ ± ♅
23 32 ☽ ∠ ♄
23 54 ☽ ⊥ ♀

19 Saturday
03 04 ☽ ⚻ ♆
06 59 ☽ Q ♂
08 49 ☿ ⊼ ♅
11 26 ☽ ⊥ ☿
12 27 ☽ ⊻ ♀
16 44 ☽ ⊥ ♄
19 07 ☽ ∠ ☉
19 21 ☽ △ ♃
20 07 ☽ ± ♆
22 26 ☽ △ ♅

20 Sunday
00 54 ☽ ⊻ ☿
04 20 ☽ ⊻ ♄
07 06 ☽ ∗ ♂
07 36 ☽ ⊼ ♆
07 39 ♀ ⚻ ♇
13 51 ☽ ⊥ ☉
16 27 ♂ ⊼ ♆
19 08 ☽ ⊼ ♇

21 Monday
00 43 ☿ ☌ ♄
02 29 ☽ ⊻ ☉
03 20 ☽ ⚻ ♅
06 49 ☽ ± ♇
13 51 ☽ ∠ ♂
19 53 ☿ ☍ ♆
20 56 ☽ ± ♅

22 Tuesday
00 29 ☽ ⚻ ♇
01 29 ☿ ∥ ♄
05 07 ☽ ☌ ♀
06 20 ☽ □ ♃
08 36 ☽ ⊥ ♂
08 45 ☽ ⊼ ♅
09 40 ☿ ⊻ ♂
15 40 ☽ ☌ ♄
18 18 ☽ ☍ ♆
18 26 ♀ □ ♃
21 12 ☽ ⊻ ♂
21 17 ☽ ∥ ♀
22 43 ☽ ☌ ☿

23 Wednesday
03 43 ☽ ∥ ♄
06 19 ☽ △ ♇
06 45 ☽ ∥ ☿
07 00 ☽ # ♇
08 28 ☽ # ♆
08 59 ☿ # ♇
12 25 ☉ Q ♃
13 12 ☽ # ♃
16 02 ♀ ⊼ ♅
18 41 ☽ Q ♃
19 10 ☽ ☌ ☉
21 28 ☿ # ♆

24 Thursday
05 16 ☽ ∥ ☉
19 13 ☽ ∗ ♃
20 35 ☿ △ ♇
20 50 ☽ ☍ ♅
22 01 ☽ # ♅

25 Friday
00 17 ☽ ⊻ ♀
00 42 ♀ ⊥ ♂
04 45 ☽ ⊻ ♄
06 39 ☽ ⊼ ♆
09 37 ☿ # ♃
11 12 ☽ ∥ ♂
13 12 ☽ ☌ ♂
13 54 ☽ ⊥ ♀
17 06 ☽ ⊥ ♄
18 50 ☽ ± ♆
19 00 ☽ □ ♇
23 35 ☽ ⊻ ☿

26 Saturday
02 05 ☽ ∠ ♃
10 23 ☽ ∠ ♀
11 39 ☽ ∠ ♄
13 09 ☽ ⚻ ♆
13 31 ☽ ⊻ ☉
14 19 ☽ ⊥ ☿
20 39 ☽ ⊥ ♃
21 57 ☽ # ♂
23 38 ♀ ☌ ♄

27 Sunday
02 50 ☽ ⊥ ☉
04 54 ♀ ∥ ♄
07 41 ☽ Q ♇
09 01 ☽ ⊻ ♃
09 44 ☽ ⊼ ♅
12 19 ♀ ☍ ♆
12 23 ☽ ∠ ☿
12 55 ☽ ∥ ♅
18 31 ☽ ∗ ♄
19 37 ☽ △ ♆
20 29 ☽ ∗ ♀
21 53 ☽ ± ♅
22 46 ☽ ∠ ☉
23 40 ☽ # ☉

28 Monday
05 46 ☽ ⊻ ♂
06 02 ☿ Q ♃
08 02 ☽ ∗ ♇
12 17 ☽ # ☿
14 28 ♀ # ♇
16 02 ☽ ⚻ ♅
18 34 ☽ ⊥ ♂
19 02 ☽ Q ♄
22 43 ☽ ∥ ♃
23 31 ☽ Q ♀

29 Tuesday
00 48 ☽ ∗ ☿
02 32 ☽ ∥ ♆
03 00 ☽ # ♀
03 56 ☽ ∥ ♇
05 35 ♂ ⊥ ♄
05 58 ☽ # ♄
07 39 ☽ ∗ ☉
09 06 ♀ # ♆
09 13 ♃ △ ♅
13 35 ☽ ∠ ♂
14 07 ☽ ∠ ♇
17 19 ♂ ± ♆
21 53 ☽ △ ♅
22 05 ☽ ☌ ♃
23 29 ♂ □ ♇

30 Wednesday
05 17 ☽ Q ☿
07 13 ☽ □ ♄
07 32 ☽ □ ♆
07 53 ☽ ⊥ ♇
09 27 ☽ Q ☉
15 11 ☽ □ ♀
19 35 ☽ ⊻ ♇
20 41 ☽ ∗ ♂

31 Thursday
06 02 ♀ # ♃
09 54 ♄ ☍ ♆
18 11 ☽ Q ♆
20 53 ☽ Q ♂
22 25 ☽ □ ☿
22 56 ☽ □ ☉

SEPTEMBER 2006

LONGITUDES

Date	Sidereal time h	m	s	Sun ☉ °	′	″	Moon ☽ °	′	″	Moon ☽ 24.00 °	′	″	Mercury ☿ °	′	Venus ♀ °	′	Mars ♂ °	′	Jupiter ♃ °	′	Saturn ♄ °	′	Uranus ♅ °	′	Neptune ♆ °	′	Pluto ♇ °	′
01	10	42	03	08 ♍	55	07	15 ♐	23	13	21 ♐	55	13	09 ♍	12	24 ♌	07	25 ♍	42	13 ♏	28	18 ♌	01	12 ♓	54	17 ♒	51	24 ♐	05
02	10	45	59	09	53	11	28 ♐	33	39	05 ♑	18	50	11	08	25	21	26	21	13	37	18	08	12 R	52	17 R	50	24 R	05
03	10	49	56	10	51	16	12 ♑	10	55	19 ♑	09	57	13	02	26	35	26	59	13	46	18	16	12	50	17	48	24	05
04	10	53	53	11	49	23	26 ♑	15	48	03 ♒	28	10	14	56	27	50	27	38	13	55	18	23	12	47	17	47	24	05
05	10	57	49	12	47	31	10 ♒	46	37	18 ♒	10	28	16	48	29 ♌	04	28	16	14	04	18	30	12	45	17	45	24 D	05
06	11	01	46	13	45	41	25 ♒	38	55	03 ♓	10	59	18	39	00 ♍	18	28	55	14	13	18	37	12	42	17	44	24	05
07	11	05	42	14	43	52	10 ♓	45	36	18 ♓	21	35	20	29	01	32	29 ♍	34	14	23	18	45	12	40	17	42	24	05
08	11	09	39	15	42	05	25 ♓	57	43	03 ♈	32	48	22	17	02	46	00 ♎	12	14	33	18	52	12	38	17	41	24	05
09	11	13	35	16	40	20	11 ♈	05	42	18 ♈	35	20	24	05	04	00	00	51	14	42	18	59	12	35	17	40	24	05
10	11	17	32	17	38	37	26 ♈	00	48	03 ♉	21	19	25	51	05	15	01	30	14	52	19	06	12	33	17	38	24	05
11	11	21	28	18	36	56	10 ♉	36	15	17 ♉	45	11	27	37	06	29	02	09	15	01	19	13	12	30	17	37	24	06
12	11	25	25	19	35	17	24 ♉	47	48	01 ♊	44	01	29 ♍	21	07	43	02	48	15	11	19	20	12	28	17	36	24	06
13	11	29	22	20	33	40	08 ♊	33	48	15 ♊	17	18	01 ♎	04	08	58	03	26	15	21	19	27	12	26	17	34	24	06
14	11	33	18	21	32	05	21 ♊	54	45	28 ♊	26	26	02	46	10	12	04	05	15	31	19	34	12	23	17	33	24	06
15	11	37	15	22	30	32	04 ♋	52	44	11 ♋	14	02	04	26	11	26	04	44	15	42	19	41	12	21	17	32	24	06
16	11	41	11	23	29	02	17 ♋	30	46	23 ♋	43	22	06	06	12	41	05	23	15	52	19	48	12	19	17	30	24	07
17	11	45	08	24	27	34	29 ♋	52	18	05 ♌	58	00	07	45	13	55	06	02	16	02	19	55	12	16	17	29	24	07
18	11	49	04	25	26	07	12 ♌	00	53	18 ♌	01	21	09	22	15	10	06	41	16	13	20	01	12	14	17	28	24	08
19	11	53	01	26	24	43	23 ♌	59	48	29 ♌	56	37	10	59	16	25	07	20	16	23	20	08	12	12	17	27	24	08
20	11	56	57	27	23	21	05 ♍	52	07	11 ♍	46	39	12	35	17	39	08	00	16	34	20	15	12	09	17	26	24	08
21	12	00	54	28	22	01	17 ♍	40	31	23 ♍	34	01	14	09	18	54	08	39	16	44	20	21	12	07	17	25	24	09
22	12	04	51	29 ♍	20	43	29 ♍	27	27	05 ♎	21	05	15	43	20	08	09	18	16	55	20	28	12	05	17	23	24	10
23	12	08	47	00 ♎	19	27	11 ♎	15	11	17 ♎	10	03	17	16	21	23	09	57	17	06	20	35	12	03	17	22	24	10
24	12	12	44	01	18	12	23 ♎	05	57	29 ♎	03	12	18	47	22	38	10	36	17	17	20	41	12	00	17	21	24	11
25	12	16	40	02	17	00	05 ♏	02	05	11 ♏	02	57	20	18	23	52	11	16	17	28	20	48	11	58	17	20	24	11
26	12	20	37	03	15	50	17 ♏	06	07	23 ♏	11	57	21	48	25	07	11	55	17	39	20	54	11	56	17	19	24	12
27	12	24	33	04	14	41	29 ♏	20	49	05 ♐	33	08	23	17	26	22	12	35	17	50	21	00	11	54	17	18	24	13
28	12	28	30	05	13	34	11 ♐	49	16	18 ♐	09	39	24	44	27	37	13	14	18	02	21	07	11	52	17	17	24	13
29	12	32	26	06	12	29	24 ♐	34	41	01 ♑	04	44	26	11	28 ♍	51	13	53	18	13	21	13	11	49	17	16	24	14
30	12	36	23	07 ♎	11	26	07 ♑	40	11	14 ♑	21	20	27 ♎	37	00 ♎	06	14 ♎	33	18 ♏	24	21 ♌	19	11 ♓	47	17 ♒	15	24 ♐	15

Date	Moon True ☊ °	′	Moon Mean ☊ °	′	Moon ☽ Latitude °	′
01	25 ♓	28	26 ♓	06	05 S	12
02	25 R	27	26	03	05	16
03	25	26	26	00	05	03
04	25	24	25	57	04	32
05	25	23	25	53	03	43
06	25	22	25	50	02	37
07	25	21	25	47	01 S	20
08	25 D	21	25	44	00 N	03
09	25	21	25	41	01	26
10	25	21	25	38	02	42
11	25	22	25	34	03	46
12	25	22	25	31	04	34
13	25	23	25	28	05	04
14	25 R	23	25	25	05	17
15	25	23	25	22	05	13
16	25	23	25	18	04	54
17	25	22	25	15	04	22
18	25 D	22	25	12	03	38
19	25	23	25	09	02	46
20	25	23	25	06	01	46
21	25	23	25	03	00 N	43
22	25 R	23	24	59	00 S	23
23	25	23	24	56	01	27
24	25	22	24	53	02	28
25	25	21	24	50	03	23
26	25	20	24	47	04	09
27	25	19	24	44	04	45
28	25	18	24	40	05	08
29	25	17	24	37	05	17
30	25 ♓	17	24 ♓	34	05 S	10

DECLINATIONS

Date	Sun ☉ °	′	Moon ☽ °	′	Mercury ☿ °	′	Venus ♀ °	′	Mars ♂ °	′	Jupiter ♃ °	′	Saturn ♄ °	′	Uranus ♅ °	′	Neptune ♆ °	′	Pluto ♇ °	′
01	08 N	14	27 S	48	09 N	42	14 N	30	02 N	27	15 S	02	16 N	14	07 S	28	15 S	41	15 S	58
02	07	52	28	42	08	57	14	06	02	12	15	04	16	12	07	29	15	42	15	58
03	07	30	27	55	08	10	13	43	01	56	15	07	16	10	07	30	15	42	15	58
04	07	08	25	21	07	23	13	19	01	40	15	10	16	08	07	31	15	42	15	59
05	06	46	21	06	06	36	12	54	01	24	15	13	16	06	07	32	15	43	15	59
06	06	23	15	26	05	49	12	29	01	08	15	16	16	04	07	33	15	43	15	59
07	06	01	08	46	05	02	12	04	00	53	15	19	16	02	07	33	15	44	16	00
08	05	38	01 S	33	04	15	11	39	00	37	15	22	15	59	07	34	15	44	16	00
09	05	16	05 N	43	03	27	11	13	00	21	15	25	15	57	07	35	15	45	16	00
10	04	53	12	34	02	40	10	47	00 N	05	15	28	15	55	07	36	15	45	16	01
11	04	30	18	35	01	53	10	20	00 S	11	15	31	15	53	07	37	15	46	16	01
12	04	07	23	23	01	06	09	54	00	27	15	34	15	51	07	38	15	46	16	01
13	03	44	26	44	00 N	19	09	27	00	42	15	37	15	49	07	39	15	46	16	02
14	03	21	28	28	00 S	27	08	59	00	58	15	40	15	47	07	40	15	47	16	02
15	02	58	28	34	01	13	08	32	01	14	15	43	15	45	07	41	15	47	16	02
16	02	35	27	10	01	59	08	04	01	30	15	46	15	43	07	42	15	48	16	03
17	02	12	24	27	02	44	07	36	01	46	15	49	15	41	07	42	15	48	16	03
18	01	49	20	41	03	29	07	08	02	02	15	53	15	39	07	43	15	48	16	04
19	01	26	16	08	04	14	06	39	02	18	15	56	15	37	07	44	15	49	16	04
20	01	02	11	00	04	58	06	11	02	34	15	59	15	35	07	45	15	49	16	04
21	00	39	05 N	32	05	41	05	42	02	49	16	02	15	33	07	46	15	49	16	05
22	00 N	16	00 S	06	06	24	05	13	03	05	16	05	15	31	07	47	15	50	16	05
23	00 S	08	05	47	07	07	04	44	03	21	16	09	15	29	07	48	15	50	16	05
24	00	31	11	16	07	49	04	14	03	37	16	12	15	27	07	48	15	50	16	06
25	00	54	16	23	08	30	03	45	03	53	16	15	15	25	07	49	15	51	16	06
26	01	18	20	56	09	10	03	16	04	09	16	18	15	23	07	50	15	51	16	06
27	01	41	24	39	09	50	02	46	04	24	16	22	15	21	07	51	15	51	16	07
28	02	05	27	18	10	30	02	16	04	40	16	25	15	19	07	52	15	52	16	07
29	02	28	28	37	11	08	01	46	04	56	16	28	15	17	07	53	15	52	16	07
30	02 S	51	28 S	22	11 S	46	01 N	16	05 S	12	16 S	31	15 N	16	07 S	53	15 S	52	16 S	08

ZODIAC SIGN ENTRIES

Date	h	m	Planets	
02	14	34	☽	♑
04	18	15	☽	♒
06	06	15	♀	♍
06	18	56	☽	♓
08	04	18	♂	♎
08	18	23	☽	♈
10	18	30	☽	♉
12	20	59	☽	♊
12	21	08	☿	♎
15	02	53	☽	♋
17	12	15	☽	♌
20	00	07	☽	♍
22	13	06	☽	♎
23	04	03	☉	♎
25	01	54	☽	♏
27	13	16	☽	♐
29	22	01	☽	♑
30	10	02	♀	♎

LATITUDES

Date	Mercury ☿ °	′	Venus ♀ °	′	Mars ♂ °	′	Jupiter ♃ °	′	Saturn ♄ °	′	Uranus ♅ °	′	Neptune ♆ °	′	Pluto ♇ °	′
01	01 N	43	01 N	04	00 N	49	00 N	54	00 N	51	00 S	49	00 S	13	07 N	21
04	01	35	01	09	00	47	00	53	00	51	00	49	00	13	07	20
07	01	22	01	13	00	46	00	53	00	52	00	49	00	13	07	19
10	01	07	01	17	00	45	00	52	00	52	00	49	00	13	07	18
13	00	49	01	20	00	43	00	52	00	52	00	49	00	13	07	17
16	00	29	01	22	00	42	00	51	00	53	00	49	00	13	07	16
19	00 N	08	01	24	00	40	00	51	00	53	00	49	00	13	07	15
22	00 S	14	01	25	00	39	00	50	00	53	00	49	00	13	07	14
25	00	37	01	26	00	37	00	50	00	54	00	49	00	13	07	13
28	00	59	01	26	00	36	00	49	00	54	00	49	00	13	07	12
31	01 S	21	01 N	26	00 N	35	00 N	49	00 N	55	00 S	48	00 S	13	07 N	11

DATA

Julian Date	2453980
Delta T	+66 seconds
Ayanamsa	23° 57′ 02″
Synetic vernal point	05° ♓ 09′ 57″
True obliquity of ecliptic	23° 26′ 28″

MOON'S PHASES, APSIDES AND POSITIONS ☽

Date	h	m	Phase	Longitude °	′	Eclipse Indicator
07	18	42	○	15 ♓	00	partial
14	11	15	☾	21 ♊	30	
22	11	45	●	29 ♍	20	Annular
30	11	04	☽	07 ♑	09	

Day	h	m		
08	03	10	Perigee	
22	05	37	Apogee	
02	13	11	Max dec	28° S 42′
08	17	05	0N	
15	01	28	Max dec	28° N 43′
22	11	27	0S	
29	20	32	Max dec	28° S 43′

LONGITUDES

Date	Chiron ⚷ °	′	Ceres ⚳ °	′	Pallas ⚴ °	′	Juno ⚵ °	′	Vesta ⚶ °	′	Black Moon Lilith ⚸ °	′
01	05 ♒	21	15 ♒	41	01 ♑	01	09 ♍	19	13 ♍	31	24 ♍	38
11	04 ♒	58	14 ♒	09	01 ♑	46	13 ♍	35	18 ♍	24	25 ♍	44
21	04 ♒	41	13 ♒	10	03 ♑	03	17 ♍	46	23 ♍	19	26 ♍	51
31	04 ♒	29	12 ♒	46	04 ♑	45	21 ♍	52	28 ♍	16	27 ♍	57

All ephemeris data is given at 12.00 UT and the Moon's longitude is additionally given for 24.00 UT

ASPECTARIAN

h m Aspects

01 Friday
04 49 ☉ ☌ ☿
07 24 ☽ □ ♅
08 25 ☽ ⊻ ♃
11 11 ♀ △ ♇
16 33 ☽ ⚹ ♆
16 54 ☽ △ ♄
19 37 ☽ ⊥ ♃

02 Saturday
03 55 ☽ ☌ ♇
05 39 ☽ △ ♀
07 49 ☽ □ ♂
12 07 ☽ ∠ ♃
16 06 ☽ Q ♅
19 35 ☽ ∠ ♆
20 14 ☽ Ᵽ ♄

03 Sunday
09 22 ☿ ☍ ♅
09 31 ☽ △ ☉
10 52 ☽ Ᵽ ♀
11 21 ☽ ⊥ ♆
12 00 ☉ # ♅
12 08 ☽ ± ♄
13 07 ☽ ⚹ ♅
13 43 ☽ △ ☿
14 47 ☽ ⚹ ♃
21 39 ☽ ⊻ ♆
22 05 ☿ ⚹ ♃
22 33 ☽ ⊼ ♄

04 Monday
03 48 ☽ ± ♀
04 05 ♀ ⊻ ♂
08 19 ☽ ⊻ ♇
08 21 ☿ # ♅
11 25 ☽ Q ♃
13 00 ☽ Ᵽ ☉
14 24 ☽ △ ♂
14 33 ☽ ∠ ♅
14 52 ☽ ⊼ ♀
18 22 ☽ ⊥ ♇
19 02 ☽ Ᵽ ☿
23 19 ♇ St D

05 Tuesday
03 14 ☉ ∥ ☿
05 01 ☽ ± ☉
05 25 ☽ ⊥ ♅
09 13 ☽ ∠ ♇
10 54 ☉ ☍ ♅
12 02 ☽ ± ☿
15 12 ☽ ⊻ ♅
15 31 ☽ ⊼ ☉
16 15 ☽ Ᵽ ♂
17 25 ☽ □ ♃
23 10 ☽ ⊼ ☿
23 18 ☽ ☌ ♆

06 Wednesday
00 14 ☿ ⊼ ♆
00 38 ☽ ☍ ♄
07 26 ☽ ± ♂
09 29 ☽ ⚹ ♇
09 35 ☽ # ♄
09 53 ☽ ∥ ♇
10 55 ☽ ∥ ♆
11 39 ☿ ⊻ ♄
12 40 ☽ ∥ ♃
17 27 ☽ ⊼ ♂
20 04 ☽ ☍ ♀
23 42 ☽ # ♀

07 Thursday
01 45 ☉ ⚹ ♃
03 15 ♂ ∠ ♃
04 36 ☽ Q ♇
15 00 ☽ ☌ ♅
16 06 ☽ ∥ ♅
17 47 ☽ △ ♃
18 42 ☽ ☍ ☉
21 48 ☽ # ☉
22 57 ☽ ⊻ ♆

08 Friday
00 42 ☽ ⊼ ♄
02 05 ☽ # ☿
05 25 ☽ ☍ ☿
05 51 ♄ # ♇
06 56 ♀ Q ♃
08 24 ☽ ⊥ ♆
09 02 ☽ □ ♇
10 15 ☽ ± ♄
15 11 ☽ # ♂
17 43 ☽ Ᵽ ♃
18 51 ☽ ∥ ♂
19 01 ☽ ☍ ♂
22 37 ☽ ∠ ♆
23 44 ☽ ⊼ ♀

09 Saturday
00 36 ☽ Ᵽ ♄
05 12 ☽ ∥ ☿
06 24 ☿ ± ♆
08 09 ☽ ± ♃
10 07 ☽ ± ♀
10 33 ☽ ∥ ☉
12 00 ☿ □ ♇
14 22 ☽ ⊻ ♅
17 49 ☽ ⊼ ♃
18 22 ☽ # ♅
21 32 ☽ ⊼ ☉
22 30 ☽ ⚹ ♆
23 58 ☽ ⊥ ♅

10 Sunday
00 44 ☽ △ ♄
00 59 ☿ ⊥ ♄
01 49 ☽ Ᵽ ♀
05 54 ☽ ∥ ♀
07 53 ☽ ± ☉
08 52 ☽ △ ♇
11 43 ☽ ⊼ ☿
11 52 ☉ ⊼ ♆
14 29 ☽ ∠ ♅
17 54 ☽ Q ♆
21 22 ☽ ⊼ ♂
22 51 ☽ ± ☿
23 09 ☽ # ♃
23 36 ☽ Ᵽ ☉

11 Monday
00 13 ☽ # ♆
00 48 ☽ ∥ ♄
01 15 ☽ # ♇
04 31 ☽ △ ♀
07 43 ☽ ± ♂
09 29 ☽ Ᵽ ♇
15 10 ☽ ⚹ ♅
15 49 ☽ Ᵽ ☿
19 29 ☽ ☍ ♃
23 30 ☽ Ᵽ ♂
23 45 ☽ □ ♆

12 Tuesday
00 34 ☽ ± ♇
02 27 ☽ △ ☉
02 36 ☽ □ ♄
04 51 ♂ Ᵽ ♆
04 53 ☉ ⊻ ♄
10 47 ☽ ⊼ ♇
11 26 ☽ Q ♅
20 58 ☽ △ ☿

13 Wednesday
01 02 ☿ ∠ ♃
02 32 ☽ △ ♂
03 07 ☿ # ♂
10 01 ☽ Q ♄
12 47 ☽ □ ♀
18 51 ☽ □ ♅

14 Thursday
00 16 ☽ ⊼ ♃
04 05 ☽ △ ♆
07 42 ☽ ⚹ ♄
09 03 ☿ Ᵽ ♆
11 15 ☽ □ ☉
11 17 ☽ ± ♃
12 57 ♄ # ♆
16 00 ☽ ☍ ♇

15 Friday
00 47 ☽ Q ♀
04 04 ☽ Ᵽ ♃
07 36 ☽ Ᵽ ♆
09 29 ♂ ∠ ♄
11 03 ☽ □ ☿
11 38 ☽ ∠ ♄
11 43 ☽ □ ♂
12 43 ☿ ∥ ♂
15 44 ☿ ∠ ♄
19 01 ☿ ☌ ♂
19 33 ♃ # ♄
23 31 ☽ Q ☉

16 Saturday
00 32 ☽ ± ♆
01 45 ☽ ⚹ ♀
02 05 ☽ △ ♅
04 49 ☽ ⊥ ♄
05 01 ♀ ☍ ♅
08 47 ☽ △ ♃
11 59 ☽ ⊼ ♆
12 35 ☉ ± ♆
16 26 ☽ ⊻ ♄
22 53 ♃ ∥ ♆
23 59 ☽ Q ♂

17 Sunday
00 31 ☽ ⚹ ☉
00 42 ☉ # ☿
00 46 ☽ ⊼ ♇
02 42 ☽ Q ☿
03 34 ☉ □ ♇
06 37 ♀ # ♅
06 56 ☽ Ᵽ ♅
09 56 ☽ ∠ ♀
12 29 ☽ ± ♇

18 Monday
00 34 ☽ ± ♅
00 50 ☽ ⚹ ♂
04 03 ☉ # ♂
05 41 ☽ ⊥ ♀
05 56 ☽ ⚹ ☿
06 15 ☽ Ᵽ ♇
08 35 ☽ ∠ ☉
12 26 ☽ ⊼ ♅
19 01 ☽ ⊻ ♀
20 30 ☽ □ ♃
22 52 ☽ ☍ ♆

19 Tuesday
01 55 ☿ ⊥ ♃
04 09 ☽ ⊥ ☉
04 10 ☽ ☌ ♄
04 20 ☉ ⊥ ♄
08 29 ☽ ∠ ♂
11 28 ♀ ⚹ ♃
12 17 ☽ △ ♇
12 19 ☽ # ♇
12 58 ☽ # ♃
13 33 ☽ # ♆
14 33 ☽ ∥ ♄
16 38 ☽ ∠ ☿
17 18 ☽ ⊻ ☉

20 Wednesday
03 41 ☽ ⊥ ♂
05 21 ☿ Q ♇
05 46 ☿ ⊼ ♅
07 45 ♀ ⊼ ♆
09 19 ☽ Q ♃
13 40 ☽ ⊥ ☿
16 34 ☽ ⊻ ♂

21 Thursday
00 44 ☽ ☍ ♅
02 21 ☽ # ♅
03 44 ☽ ⊻ ☿
10 04 ☽ ⚹ ♃
11 12 ☽ ∥ ♀
11 23 ☽ # ☿
11 27 ☽ ⊼ ♆
12 12 ☿ # ♀
14 47 ☽ ☌ ♀
17 31 ☽ ⊻ ♄
23 00 ☽ # ♂
23 40 ☽ ± ♆

22 Friday
01 12 ☽ □ ♇
05 51 ☽ ⊥ ♄
08 40 ♃ ∥ ♇
10 14 ☽ ∥ ☉
11 45 ☽ ☌ ☉
12 31 ☽ # ☉
17 05 ☽ ∠ ♃
17 58 ☽ Ᵽ ♆
18 58 ♀ ⊻ ♄

23 Saturday
00 21 ☽ ∠ ♄
01 07 ☽ ∥ ♂
07 50 ☽ # ♀
09 10 ☿ ⊻ ♃
09 12 ☽ ☌ ♂
11 41 ☽ ⊥ ♃
13 36 ☽ ⊼ ♅
13 43 ☿ △ ♆
13 52 ☽ Q ♇
18 31 ☽ ∥ ☿
20 41 ☽ ∥ ♅
23 58 ☿ ± ♅

24 Sunday
00 03 ☽ ⊻ ♃
00 24 ☽ △ ♆
01 44 ☽ ± ♅
02 00 ☽ ☌ ☿
07 05 ☽ ⚹ ♄
10 56 ☽ ⊻ ♀
11 58 ☿ ∥ ♅
14 11 ☽ ⚹ ♇
19 51 ☽ Ᵽ ♅
20 31 ♃ □ ♆

25 Monday
00 26 ☽ ⊥ ♀
01 49 ♀ ± ♆
06 00 ☽ ⊻ ☉
07 18 ☽ # ♄
07 28 ☽ Q ♄
07 51 ♀ # ♂
09 21 ☽ ∥ ♆
10 36 ☽ ∥ ♇
11 20 ☽ ∥ ♃
13 17 ☉ Ᵽ ♆
17 31 ☉ ∠ ♃
18 10 ♀ □ ♇
19 04 ☽ ⊥ ☉
20 18 ☽ ∠ ♇
20 28 ☿ ⚹ ♄
20 33 ☽ ∠ ♀
22 20 ♂ ⊥ ♃

26 Tuesday
01 08 ☽ ⊻ ♂
01 47 ☽ △ ♅
12 26 ☽ □ ♆
12 27 ♂ ⊼ ♅
13 06 ☽ ☌ ♃
13 42 ☽ ⊥ ♂
14 10 ☽ ⊥ ♇
14 30 ☽ ∠ ☉
19 33 ☽ □ ♄
22 28 ♂ Q ♇
22 32 ☽ ⊻ ☿

27 Wednesday
01 59 ☽ ⊻ ♇
05 32 ☽ ⚹ ♀
08 22 ☽ ∠ ♂
11 51 ☽ ⊥ ☿
22 18 ☽ ⚹ ☉
23 30 ☽ Q ♆

28 Thursday
01 29 ♀ ⊥ ♄
03 27 ☿ ⚹ ♇
07 19 ☽ Q ♀
07 31 ☽ ∠ ☿
12 04 ☽ □ ♅
14 50 ☽ ⚹ ♂
17 12 ☉ # ♀
22 20 ☽ ⚹ ♆
23 06 ☽ Q ☉
23 55 ☽ ⊻ ♃

29 Friday
05 40 ☽ △ ♄
11 19 ☽ ⊥ ♃
11 22 ☽ ☌ ♇
12 08 ☉ ∠ ♄
14 34 ☽ Q ♂
15 10 ♃ ⊥ ♇
15 21 ☽ ⚹ ☿
20 45 ☽ □ ♀
21 40 ☽ Q ♅
22 25 ☿ Ᵽ ♅

30 Saturday
02 10 ☽ ∠ ♆
04 08 ☽ ∠ ♃
09 32 ☽ Ᵽ ♄
11 04 ☽ □ ☉
15 56 ☽ Q ☿
18 27 ☽ ⊥ ♆
19 24 ☽ ⚹ ♅

OCTOBER 2006

Date	Sidereal time h m s	LONGITUDES Sun ☉ ° ' "	Moon ☽ ° ' "	Moon ☽ 24.00 ° ' "	Mercury ☿ ° '	Venus ♀ ° '	Mars ♂ ° '	Jupiter ♃ ° '	Saturn ♄ ° '	Uranus ♅ ° '	Neptune ♆ ° '	Pluto ⯓ ° '
01	12 40 20	08 ♎ 10 24	21 ♑ 08 27	28 ♑ 01 43	29 ♎ 02	01 ♎ 21	15 ♎ 12	18 ♏ 36	21 ♌ 25	11 ♓ 45	17 ♒ 14	24 ♐ 16
02	12 44 16	09 09 25	05 ♒ 01 13	12 ♒ 06 55	00 ♏ 26	02 36	15 52	18 47	21 31	11 R 43	17 R 14	24 17
03	12 48 13	10 08 27	19 ♒ 18 37	26 ♒ 36 00	01 48	03 51	16 32	18 59	21 37	11 41	17 13	24 18
04	12 52 09	11 07 30	03 ♓ 58 36	11 ♓ 25 44	03 10	05 06	17 11	19 11	21 43	11 39	17 12	24 19
05	12 56 06	12 06 36	18 ♓ 56 35	26 ♓ 30 11	04 30	06 20	17 51	19 22	21 49	11 37	17 11	24 19
06	13 00 02	13 05 43	04 ♈ 05 27	11 ♈ 41 14	05 50	07 35	18 31	19 34	21 55	11 35	17 10	24 20
07	13 03 59	14 04 52	19 ♈ 16 17	26 ♈ 49 25	07 08	08 50	19 10	19 46	22 00	11 33	17 10	24 21
08	13 07 55	15 04 03	04 ♉ 19 28	11 ♉ 45 22	08 24	10 05	19 50	19 58	22 06	11 32	17 09	24 23
09	13 11 52	16 03 17	19 ♉ 06 12	26 ♉ 21 12	09 40	11 20	20 30	20 10	22 12	11 30	17 08	24 24
10	13 15 49	17 02 33	03 ♊ 29 47	10 ♊ 31 33	10 53	12 35	21 10	20 22	22 17	11 28	17 08	24 25
11	13 19 45	18 01 51	17 ♊ 26 17	24 ♊ 13 57	12 06	13 50	21 50	20 34	22 23	11 26	17 07	24 26
12	13 23 42	19 01 12	00 ♋ 54 38	07 ♋ 28 36	13 16	15 05	22 30	20 46	22 28	11 24	17 07	24 27
13	13 27 38	20 00 34	13 ♋ 56 12	20 ♋ 17 53	14 25	16 20	23 10	20 58	22 33	11 23	17 06	24 28
14	13 31 35	21 00 00	26 ♋ 34 08	02 ♌ 45 32	15 32	17 35	23 50	21 11	22 38	11 21	17 06	24 29
15	13 35 31	21 59 27	08 ♌ 52 38	14 ♌ 56 03	16 36	18 50	24 30	21 23	22 44	11 19	17 05	24 31
16	13 39 28	22 58 57	20 ♌ 56 22	26 ♌ 54 10	17 38	20 05	25 10	21 35	22 49	11 18	17 05	24 32
17	13 43 24	23 58 29	02 ♍ 50 01	08 ♍ 44 29	18 38	21 20	25 50	21 48	22 54	11 16	17 04	24 33
18	13 47 21	24 58 03	14 ♍ 38 02	20 ♍ 31 10	19 35	22 36	26 31	22 00	22 59	11 15	17 04	24 34
19	13 51 18	25 57 39	26 ♍ 24 19	02 ♎ 17 53	20 28	23 51	27 11	22 13	23 03	11 13	17 03	24 36
20	13 55 14	26 57 17	08 ♎ 12 13	14 ♎ 07 39	21 18	25 06	27 51	22 25	23 08	11 12	17 03	24 37
21	13 59 11	27 56 58	20 ♎ 04 28	26 ♎ 02 55	22 05	26 21	28 31	22 38	23 13	11 10	17 03	24 39
22	14 03 07	28 56 40	02 ♏ 03 14	08 ♏ 05 36	22 47	27 36	29 12	22 51	23 18	11 09	17 03	24 40
23	14 07 04	29 ♎ 56 25	14 ♏ 10 12	20 ♏ 17 11	23 24	28 ♎ 51	29 ♎ 52	23 03	23 22	11 07	17 02	24 41
24	14 11 00	00 ♏ 56 11	26 ♏ 26 43	02 ♐ 38 57	23 57	00 ♏ 06	00 ♏ 33	23 16	23 26	11 06	17 02	24 43
25	14 14 57	01 56 00	08 ♐ 54 02	15 ♐ 12 07	24 24	01 22	01 13	23 29	23 31	11 05	17 02	24 44
26	14 18 53	02 55 50	21 ♐ 33 22	27 ♐ 57 59	24 44	02 37	01 54	23 42	23 35	11 04	17 02	24 46
27	14 22 50	03 55 42	04 ♑ 26 07	10 ♑ 58 01	24 58	03 52	02 34	23 55	23 39	11 03	17 02	24 48
28	14 26 47	04 55 36	17 ♑ 33 53	24 ♑ 13 55	25 04	05 07	03 15	24 07	23 43	11 01	17 02	24 49
29	14 30 43	05 55 31	00 ♒ 58 19	07 ♒ 47 17	25 R 03	06 22	03 56	24 20	23 47	11 00	17 D 02	24 51
30	14 34 40	06 55 28	14 ♒ 40 57	21 ♒ 39 25	24 52	07 38	04 36	24 33	23 51	10 59	17 02	24 52
31	14 38 36	07 ♏ 55 27	28 ♒ 42 42	05 ♓ 50 44	24 ♏ 33	08 ♏ 53	05 ♏ 17	24 ♏ 46	23 ♌ 55	10 ♓ 58	17 ♒ 02	24 ♐ 54

Date	Moon True ☊ ° '	Moon Mean ☊ ° '	Moon ☽ Latitude ° '
01	25 ♓ 17	24 ♓ 31	04 S 45
02	25 D 18	24 28	04 04
03	25 19	24 24	03 07
04	25 20	24 21	01 56
05	25 21	24 18	00 S 35
06	25 R 21	24 15	00 N 48
07	25 20	24 12	02 08
08	25 18	24 09	03 19
09	25 15	24 05	04 15
10	25 12	24 02	04 53
11	25 10	23 59	05 12
12	25 08	23 56	05 14
13	25 06	23 53	04 58
14	25 D 06	23 50	04 29
15	25 07	23 46	03 48
16	25 09	23 43	02 57
17	25 11	23 40	02 00
18	25 12	23 37	00 N 58
19	25 R 13	23 34	00 S 07
20	25 12	23 30	01 11
21	25 09	23 27	02 12
22	25 05	23 24	03 08
23	24 59	23 21	03 56
24	24 53	23 18	04 33
25	24 46	23 15	04 59
26	24 40	23 11	05 10
27	24 35	23 08	05 06
28	24 32	23 05	04 46
29	24 30	23 02	04 10
30	24 D 30	22 59	03 19
31	24 ♓ 32	22 ♓ 55	02 S 15

DECLINATIONS

Date	Sun ☉ ° '	Moon ☽ ° '	Mercury ☿ ° '	Venus ♀ ° '	Mars ♂ ° '	Jupiter ♃ ° '	Saturn ♄ ° '	Uranus ♅ ° '	Neptune ♆ ° '	Pluto ⯓ ° '
01	03 S 15	26 S 29	12 S 24	00 N 46	05 S 28	16 S 35	15 N 14	07 S 54	15 S 52	16 S 08
02	03 38	22 57	13 00	00 N 16	05 43	16 38	15 12	07 55	15 53	16 09
03	04 01	17 59	13 36	00 S 14	05 59	16 41	15 10	07 56	15 53	16 09
04	04 24	11 51	14 10	00 44	06 15	16 45	15 08	07 56	15 53	16 09
05	04 47	04 S 55	14 44	01 14	06 30	16 48	15 07	07 57	15 54	16 10
06	05 10	02 N 22	15 17	01 44	06 46	16 51	15 05	07 58	15 54	16 10
07	05 33	09 31	15 49	02 14	07 01	16 55	15 03	07 59	15 54	16 10
08	05 56	16 05	16 21	02 44	07 17	16 58	15 01	07 59	15 54	16 11
09	06 19	21 35	16 51	03 14	07 32	17 01	15 00	08 00	15 54	16 11
10	06 42	25 39	17 20	03 44	07 48	17 05	14 58	08 01	15 55	16 11
11	07 04	28 02	17 48	04 14	08 03	17 08	14 57	08 01	15 55	16 12
12	07 27	28 40	18 15	04 44	08 18	17 11	14 55	08 02	15 55	16 12
13	07 49	27 39	18 41	05 13	08 34	17 15	14 53	08 02	15 55	16 12
14	08 12	25 14	19 05	05 43	08 49	17 18	14 52	08 03	15 55	16 13
15	08 34	21 42	19 29	06 13	09 04	17 21	14 50	08 04	15 55	16 13
16	08 56	17 19	19 50	06 42	09 19	17 25	14 49	08 04	15 56	16 13
17	09 18	12 20	20 11	07 11	09 35	17 28	14 47	08 05	15 56	16 14
18	09 40	06 56	20 30	07 40	09 50	17 31	14 46	08 05	15 56	16 14
19	10 02	01 N 20	20 47	08 09	10 05	17 35	14 44	08 06	15 56	16 15
20	10 23	04 S 20	21 03	08 38	10 20	17 38	14 43	08 06	15 56	16 15
21	10 45	09 53	21 16	09 07	10 34	17 41	14 42	08 07	15 56	16 15
22	11 06	15 07	21 28	09 35	10 49	17 44	14 40	08 08	15 56	16 16
23	11 27	19 50	21 38	10 03	11 04	17 48	14 39	08 08	15 56	16 16
24	11 48	23 47	21 45	10 31	11 19	17 51	14 38	08 08	15 56	16 16
25	12 09	26 42	21 51	10 59	11 33	17 54	14 36	08 09	15 56	16 17
26	12 29	28 20	21 53	11 26	11 48	17 58	14 35	08 09	15 56	16 17
27	12 50	28 28	21 53	11 53	12 02	18 01	14 34	08 10	15 56	16 17
28	13 10	27 01	21 49	12 20	12 17	18 04	14 33	08 10	15 56	16 18
29	13 30	24 00	21 42	12 47	12 31	18 07	14 32	08 11	15 56	16 18
30	13 50	19 36	21 32	13 13	12 45	18 11	14 30	08 11	15 56	16 18
31	14 S 09	14 S 02	21 S 18	13 S 39	12 S 59	18 S 14	14 N 29	08 S 11	15 S 56	16 S 19

ZODIAC SIGN ENTRIES

Date	h	m	Planets
02	03	24	☽ ♒
02	04	38	☿ ♏
04	05	33	☽ ♓
06	05	32	☽ ♈
08	05	04	☽ ♉
10	06	06	☽ ♊
12	10	21	☽ ♋
14	18	38	☽ ♌
17	06	16	☽ ♍
19	19	19	☽ ♎
22	07	54	☽ ♏
23	13	26	☉ ♏
23	16	38	♂ ♏
24	09	58	♀ ♏
24	18	53	☽ ♐
27	03	47	☽ ♑
29	10	17	☽ ♒
31	14	11	☽ ♓

LATITUDES

Date	Mercury ☿ ° '	Venus ♀ ° '	Mars ♂ ° '	Jupiter ♃ ° '	Saturn ♄ ° '	Uranus ♅ ° '	Neptune ♆ ° '	Pluto ⯓ ° '
01	01 S 21	01 N 26	00 N 35	00 N 49	00 N 55	00 S 48	00 S 13	07 N 11
04	01 42	01 25	00 33	00 48	00 55	00 48	00 13	07 10
07	02 03	01 23	00 32	00 48	00 56	00 48	00 13	07 09
10	02 21	01 21	00 30	00 48	00 56	00 48	00 13	07 08
13	02 38	01 18	00 28	00 47	00 57	00 48	00 13	07 07
16	02 52	01 15	00 27	00 47	00 57	00 48	00 13	07 07
19	03 02	01 11	00 25	00 46	00 58	00 48	00 13	07 06
22	03 06	01 07	00 24	00 46	00 58	00 48	00 13	07 05
25	03 04	01 02	00 22	00 46	00 59	00 48	00 13	07 04
28	02 52	00 57	00 20	00 45	00 59	00 48	00 13	07 03
31	02 S 28	00 N 51	00 N 19	00 N 45	01 N 00	00 S 48	00 S 13	07 N 02

DATA

Julian Date	2454010
Delta T	+66 seconds
Ayanamsa	23° 57' 06"
Synetic vernal point	05° ♓ 09' 53"
True obliquity of ecliptic	23° 26' 28"

MOON'S PHASES, APSIDES AND POSITIONS ☽

Date	h	m	Phase	Longitude ° '	Eclipse Indicator
07	03	13	○	13 ♈ 43	
14	00	26	☾	20 ♋ 31	
22	05	14	●	28 ♎ 40	
29	21	25	☽	06 ♒ 19	

Day	h	m		
06	14	19	Perigee	
19	09	49	Apogee	
06	04	15	0N	
12	08	54	Max dec	28° N 41'
19	17	38	0S	
27	02	07	Max dec	28° S 36'

LONGITUDES

Date	Chiron ⚷ ° '	Ceres ⚳ ° '	Pallas ⚴ ° '	Juno ⚵ ° '	Vesta ⚶ ° '	Black Moon Lilith ⚸ ° '
01	04 ♒ 29	12 ♒ 46	04 ♑ 45	21 ♍ 52	28 ♍ 16	27 ♍ 57
11	04 ♒ 25	12 ♒ 58	06 ♑ 50	25 ♍ 53	03 ♎ 14	29 ♍ 03
21	04 ♒ 28	13 ♒ 44	09 ♑ 14	29 ♍ 47	08 ♎ 13	00 ♎ 10
31	04 ♒ 37	15 ♒ 01	11 ♑ 53	03 ♎ 34	13 ♎ 12	01 ♎ 16

All ephemeris data is given at 12.00 UT and the Moon's longitude is additionally given for 24.00 UT

ASPECTARIAN

01 Sunday
00 59 ☽ □ ♂
01 49 ☽ ± ♄
05 08 ☽ ⊻ ♆
07 27 ☽ ⚹ ♃
12 29 ☽ ⊼ ♄
17 28 ☽ ⊻ ⯓
21 46 ☽ ∠ ♅

02 Monday
03 16 ☽ □ ☿
03 52 ☽ ⊥ ⯓
04 39 ☽ Q ♃
04 58 ♀ ♇ ♆
07 27 ☽ △ ♀
13 11 ☽ ⊥ ♅
19 14 ☽ ∠ ⯓
19 32 ☽ △ ☉
23 19 ☽ ⊻ ♅

03 Tuesday
07 09 ☽ △ ♂
08 31 ☽ ☌ ♆
11 09 ☽ ♇ ♀
11 27 ☽ □ ♃
15 07 ♀ ∠ ♃
15 50 ☽ ☍ ♄
17 22 ☽ ∥ ♃
19 36 ☽ ∥ ⯓
20 14 ☽ ⚹ ⯓
20 40 ☽ ∥ ♆
22 18 ☽ ♇ ☉
23 33 ☽ ⋕ ♄

04 Wednesday
03 20 ☽ ± ♀
04 04 ☽ ∥ ☿
08 58 ☽ ♇ ♂
10 33 ☽ △ ☿
12 23 ♂ △ ♆
13 58 ☽ ⊼ ♀
13 59 ☽ ± ☉
15 46 ☽ Q ⯓
22 36 ☿ Q ♄

05 Thursday
00 09 ☽ ± ♂
00 19 ☽ ⊼ ☉
00 20 ☽ ☌ ♅
00 29 ☉ ⊼ ♅
01 45 ☽ ∥ ♅
04 06 ♂ ± ♅
06 52 ☽ ∥ ♂
09 12 ☽ ⊻ ♆
10 11 ☽ ⊼ ♂
12 25 ☽ ∥ ☉
12 41 ☽ △ ♃
12 59 ☽ ♇ ☿
16 36 ☽ ⊼ ♄
17 19 ☉ Q ⯓
18 44 ☽ ⊥ ♆
20 33 ☽ □ ⯓
21 52 ♀ ∠ ♄
23 25 ☽ ∥ ♀

06 Friday
02 10 ☽ ± ♄
03 21 ☿ ⋕ ♄
04 37 ☽ ± ☿
08 58 ☽ ∠ ♆
09 46 ☽ ⋕ ♀
12 46 ☽ ♇ ♃
15 00 ☽ ⊼ ☿
16 29 ☽ ♇ ♄
18 01 ☽ ☍ ♀
21 48 ☽ ⋕ ☉
23 49 ☽ ⊻ ♅

07 Saturday
02 24 ☉ ⊥ ♃
03 08 ☽ ⋕ ♂
03 10 ☽ ± ♃
03 13 ☽ ☍ ☉
06 41 ☽ ⋕ ♅
08 40 ☽ ⚹ ♆
09 17 ☽ ⊥ ♅
11 50 ☽ ☍ ♂
12 48 ☽ ⊼ ♃
15 28 ☿ ∥ ♆
16 22 ☽ △ ♄
20 05 ☽ △ ⯓
23 33 ☽ ∠ ♅

08 Sunday
03 43 ☽ Q ♆
04 11 ☿ ∥ ⯓
07 55 ☽ ∥ ♄
11 18 ☽ ⋕ ♆
12 22 ☽ ⋕ ⯓
13 07 ☽ ⋕ ☿
15 33 ☽ ⋕ ♃
18 33 ♂ ⊻ ♃
19 11 ☽ ☍ ☿
20 09 ☽ ♇ ⯓
22 09 ☽ ⊼ ♀
23 36 ☽ ⚹ ♅

09 Monday
06 39 ☽ ⊼ ☉
06 47 ☿ ∠ ⯓
08 47 ☽ □ ♆
08 50 ☽ ± ♀
10 50 ☽ ± ⯓
13 46 ☽ ☍ ♃
14 25 ☽ ⊼ ♂
15 00 ♀ ⊼ ♅
17 08 ☽ □ ♄
17 13 ☽ ± ☉
19 14 ☽ Q ♅
20 45 ☽ ⊼ ⯓
21 35 ☿ ∥ ♃

10 Tuesday
00 50 ☽ ± ♂
01 06 ☽ ♇ ♀
08 37 ♀ Q ⯓
09 22 ☽ ♇ ☉
14 05 ☉ △ ♆
16 46 ☽ ♇ ♂
21 57 ☉ ± ♅
23 07 ☿ △ ♅
23 40 ☽ Q ♄

11 Wednesday
01 35 ☽ □ ♅
01 50 ☽ ⊼ ☿
05 06 ☽ △ ♀
08 55 ♂ ∥ ♅
11 27 ☽ △ ♆
13 07 ☽ △ ☉
13 15 ☽ ± ☿
17 35 ☽ ⊼ ♃
20 08 ☽ △ ♂
20 45 ☽ ⚹ ♄

12 Thursday
00 22 ☽ ☍ ⯓
04 25 ☽ ± ♃
04 47 ♀ ⊥ ♃
06 46 ☽ ♇ ☿
10 38 ♂ ⚹ ♄
14 10 ☽ ♇ ♆
21 00 ☽ ♇ ♃

13 Friday
00 04 ☽ ∠ ♄
06 43 ☽ ± ♆
07 14 ☽ △ ♅
12 59 ☽ △ ☿
16 57 ☽ ⊥ ♄
17 00 ☽ □ ♀
17 57 ☽ ⊼ ♆

14 Saturday
00 26 ☽ □ ☉
01 30 ☽ △ ♃
02 26 ☉ ∥ ♅
02 35 ♀ △ ♆
04 25 ☽ ⊻ ♄
06 27 ☽ □ ♂
07 33 ♀ ± ♅
08 00 ☽ ⊼ ⯓
11 35 ☽ ♇ ♅
17 26 ☉ ⊻ ♃
19 36 ☽ ± ⯓

15 Sunday
05 02 ☽ ± ♅
07 32 ☽ Q ♀
12 21 ♂ ⚹ ⯓
13 15 ☽ ♇ ⯓
14 24 ☽ Q ☉
16 49 ☽ ⊼ ♅
19 35 ☽ Q ♂
23 00 ☿ □ ♆
23 41 ☽ ⋕ ☿

16 Monday
04 17 ☽ ☍ ♆
04 47 ☽ □ ☿
07 29 ☉ ⚹ ♄
10 06 ☽ ⚹ ♀
11 32 ☽ ⋕ ♃
13 20 ☽ □ ♃
15 47 ☽ ☌ ♄
16 28 ☽ ⚹ ☉
17 28 ☽ ⋕ ⯓
18 57 ☽ ⋕ ♆
19 14 ☽ △ ⯓
21 01 ☽ ⚹ ♂

17 Tuesday
00 24 ☽ ∥ ♄
10 02 ☿ ⊥ ⯓
19 58 ☽ ∠ ♀
20 23 ☽ Q ☿
22 30 ♀ ⊻ ♃
23 53 ☽ ⋕ ♂

18 Wednesday
00 47 ☽ ⋕ ☉
01 37 ☽ ∠ ☉
02 17 ☉ ⚹ ⯓
02 24 ☽ Q ♃
02 50 ♂ ♇ ♅
05 06 ☽ ☍ ♅
05 15 ☽ ∠ ♂
06 59 ☽ ⋕ ♅
09 04 ☽ ⋕ ♀
16 28 ☽ ⊥ ♀
16 57 ☽ ⊼ ♆
19 54 ♀ ⚹ ♄
21 39 ☽ ⊥ ☉
22 55 ☽ ⚹ ☿

19 Thursday
00 42 ☽ ⊥ ♂
03 18 ☽ ⚹ ♃
05 08 ☽ ⊻ ♄
05 11 ☽ ± ♆
06 09 ☽ ⊻ ♀
08 18 ☽ □ ⯓
09 09 ♀ ∥ ♅
11 01 ☽ ⊻ ☉
13 40 ☽ ⊻ ♂
17 26 ☽ ⊥ ♄
18 03 ☉ ♇ ♅
22 14 ☉ ∥ ♂
23 30 ☽ ♇ ♆

20 Friday
02 42 ♀ ⚹ ⯓
07 52 ☽ ∠ ☿
10 23 ☽ ∠ ♃
11 52 ☽ ∠ ♄
18 03 ☽ ⊼ ♅
20 58 ☽ Q ⯓

21 Saturday
03 24 ☽ ⊥ ☿
04 15 ☽ ∥ ♅
04 56 ☽ ⊥ ♃
05 54 ☽ △ ♆
06 09 ☽ ± ♅
08 17 ☽ ∥ ♀
08 38 ♀ ♇ ♅
15 13 ☽ ∥ ♂
16 06 ☽ ∥ ☉
16 18 ☽ ⊻ ☿
17 14 ☽ ⊻ ♃
18 21 ☽ ⚹ ♄
21 12 ☽ ⚹ ⯓

22 Sunday
00 13 ☽ ♇ ♅
02 04 ☽ ☌ ♀
05 14 ☽ ☌ ☉
05 58 ☽ ☌ ♂
09 52 ☽ ⋕ ♄
15 19 ☿ ☌ ♃
15 56 ☽ ∥ ♆
17 31 ☽ ∥ ⯓
18 29 ☽ Q ♄

23 Monday
01 04 ☽ ∥ ♃
03 09 ☽ ∠ ⯓
06 00 ☽ △ ♅
06 46 ☉ ☌ ♂
10 06 ☿ □ ♄
17 38 ☽ □ ♆
20 54 ☽ ⊥ ⯓
22 36 ☽ ∥ ☿

24 Tuesday
05 43 ☽ ☌ ♃
06 07 ☽ □ ♄
06 56 ☽ ☌ ☿
08 38 ☽ ⊻ ⯓
19 53 ☽ ⊻ ♀
20 24 ☽ ⊻ ♂
21 27 ☽ ⊻ ☉

25 Wednesday
04 35 ☽ Q ♆
06 10 ♀ ☌ ♂
08 36 ☽ ⊥ ♂
08 43 ☽ ⊥ ♀
09 59 ☽ ⊥ ☉
16 09 ☽ □ ♅
17 26 ♃ □ ♄

26 Thursday
02 43 ☽ ∠ ♂
03 28 ☽ ⚹ ♆
03 45 ☽ ∠ ♀
04 35 ☽ ∠ ☉
14 47 ☿ ⊻ ⯓
15 50 ☽ △ ♄
16 05 ☽ ⊻ ♃
18 02 ☽ ☌ ⯓
18 06 ☽ ⊻ ☿

27 Friday
02 01 ☽ Q ♅
03 28 ☽ ⊥ ♃
05 29 ☽ ⊥ ☿
07 33 ☽ ∠ ♆
08 22 ☽ ⚹ ♂
10 50 ☽ ⚹ ♀
10 59 ☽ ⚹ ☉
17 50 ☉ ☌ ♀
19 48 ☽ ♇ ♄
20 22 ☽ ∠ ♃
22 17 ☽ ∠ ☿

28 Saturday
00 07 ☽ ⊥ ♆
00 07 ☽ ⚹ ♅
04 53 ♀ ∥ ♂
07 34 ☽ Q ♂
10 45 ☽ Q ☉
11 02 ☽ ⊻ ♆
11 07 ☽ Q ♀
12 17 ☽ ± ♄
19 15 ☿ St R
23 09 ☽ ⊼ ♄

29 Sunday
00 00 ☽ ⚹ ♃
00 11 ♀ Q ♄
01 05 ☽ ⊻ ⯓
01 30 ☽ ⚹ ☿
03 11 ☽ ∠ ♅
07 54 ♆ St D
08 29 ☉ Q ♄
11 47 ☽ ⊥ ⯓
17 29 ☽ □ ♂
19 07 ☽ ⊥ ♅
21 25 ☽ □ ☉
21 37 ☽ Q ♃
22 29 ☽ □ ♀
22 36 ☽ Q ☿

30 Monday
01 54 ☽ ∥ ☿
03 38 ☽ ∠ ⯓
05 36 ☽ ⊻ ♅
12 10 ☿ ⊻ ⯓
16 03 ☽ ☌ ♆
18 32 ☽ ∥ ♃

31 Tuesday
02 44 ☽ ∥ ⯓
03 02 ☿ ☌ ♃
03 49 ☽ ☍ ♄
04 16 ☽ ∥ ♆
05 08 ☽ □ ☿
05 13 ☽ □ ♃
05 31 ☽ ⚹ ⯓
10 12 ☽ ⋕ ♄
11 34 ☽ ∥ ☉
13 26 ☽ ∥ ♀
15 58 ☽ ∥ ♂
23 37 ☽ △ ♂

NOVEMBER 2006

Date	Sidereal time h m s	Sun ☉ ° ′ ″	Moon ☽ ° ′ ″	Moon ☽ 24.00 ° ′ ″	Mercury ☿ ° ′	Venus ♀ ° ′	Mars ♂ ° ′	Jupiter ♃ ° ′	Saturn ♄ ° ′	Uranus ♅ ° ′	Neptune ♆ ° ′	Pluto ♇ ° ′
01	14 42 33	08 ♏ 55 27	13 ♓ 03 20	20 ♓ 20 11	24 ♏ 05	10 ♏ 08	05 ♏ 58	24 ♏ 59	23 ♌ 59	10 ♓ 57	17 ♒ 02	24 ♐ 56
02	14 46 29	09 55 28	27 ♓ 40 50	05 ♈ 04 42	23 R 26	11 23	06 39	25 12	24 02	10 R 56	17 02	24 57
03	14 50 26	10 55 31	12 ♈ 31 02	19 ♈ 58 58	22 39	12 39	07 19	25 25	24 06	10 56	17 02	24 59
04	14 54 22	11 55 36	27 ♈ 27 30	04 ♉ 55 35	21 42	13 54	08 00	25 39	24 09	10 55	17 02	25 01
05	14 58 19	12 55 43	12 ♉ 22 08	19 ♉ 46 01	20 38	15 09	08 41	25 52	24 13	10 54	17 03	25 03
06	15 02 16	13 55 52	27 ♉ 06 14	04 ♊ 21 49	19 27	16 24	09 22	26 05	24 16	10 53	17 03	25 04
07	15 06 12	14 56 02	11 ♊ 31 57	18 ♊ 36 01	18 11	17 40	10 03	26 18	24 19	10 53	17 03	25 06
08	15 10 09	15 56 15	25 ♊ 33 32	02 ♋ 24 12	16 52	18 55	10 44	26 31	24 22	10 52	17 04	25 08
09	15 14 05	16 56 29	09 ♋ 07 55	15 ♋ 44 44	15 32	20 10	11 26	26 44	24 25	10 51	17 04	25 10
10	15 18 02	17 56 46	22 ♋ 14 53	28 ♋ 38 42	14 15	21 25	12 07	26 58	24 28	10 51	17 04	25 12
11	15 21 58	18 57 04	04 ♌ 56 37	11 ♌ 09 10	13 03	22 41	12 48	27 11	24 31	10 51	17 05	25 14
12	15 25 55	19 57 24	17 ♌ 16 58	23 ♌ 20 39	11 58	23 56	13 29	27 24	24 33	10 50	17 05	25 15
13	15 29 51	20 57 47	29 ♌ 20 54	05 ♍ 18 22	11 01	25 11	14 10	27 37	24 36	10 50	17 06	25 17
14	15 33 48	21 58 11	11 ♍ 13 46	17 ♍ 07 46	10 15	26 27	14 52	27 51	24 38	10 49	17 06	25 19
15	15 37 45	22 58 37	23 ♍ 01 00	28 ♍ 54 06	09 40	27 42	15 33	28 04	24 41	10 49	17 07	25 21
16	15 41 41	23 59 04	04 ♎ 47 39	10 ♎ 42 11	09 17	28 ♏ 57	16 15	28 17	24 43	10 49	17 07	25 23
17	15 45 38	24 59 34	16 ♎ 38 12	22 ♎ 36 07	09 05	00 ♐ 12	16 56	28 31	24 45	10 49	17 08	25 25
18	15 49 34	26 00 05	28 ♎ 36 20	04 ♏ 39 09	09 D 05	01 28	17 38	28 44	24 47	10 49	17 09	25 27
19	15 53 31	27 00 38	10 ♏ 44 50	16 ♏ 53 33	09 15	02 43	18 19	28 57	24 49	10 49	17 09	25 29
20	15 57 27	28 01 13	23 ♏ 05 26	29 ♏ 20 35	09 36	03 58	19 01	29 11	24 51	10 D 49	17 10	25 31
21	16 01 24	29 ♏ 01 49	05 ♐ 38 59	12 ♐ 00 37	10 06	05 14	19 42	29 24	24 52	10 49	17 11	25 33
22	16 05 20	00 ♐ 02 27	18 ♐ 25 27	24 ♐ 53 22	10 44	06 29	20 24	29 37	24 54	10 49	17 12	25 35
23	16 09 17	01 03 06	01 ♑ 24 17	07 ♑ 58 06	11 29	07 44	21 06	29 ♏ 51	24 55	10 49	17 13	25 37
24	16 13 14	02 03 47	14 ♑ 34 44	21 ♑ 14 07	12 21	09 00	21 48	00 ♐ 04	24 57	10 49	17 13	25 39
25	16 17 10	03 04 28	27 ♑ 56 11	04 ♒ 40 55	13 18	10 15	22 30	00 17	24 58	10 49	17 14	25 41
26	16 21 07	04 05 11	11 ♒ 28 19	18 ♒ 18 26	14 21	11 30	23 11	00 31	24 59	10 50	17 15	25 44
27	16 25 03	05 05 55	25 ♒ 11 18	02 ♓ 06 58	15 28	12 46	23 53	00 44	25 00	10 50	17 16	25 46
28	16 29 00	06 06 39	09 ♓ 05 28	16 ♓ 06 49	16 38	14 01	24 35	00 57	25 01	10 50	17 17	25 48
29	16 32 56	07 07 25	23 ♓ 11 01	00 ♈ 17 56	17 52	15 16	25 17	01 11	25 02	10 51	17 18	25 50
30	16 36 53	08 ♐ 08 12	07 ♈ 27 24	14 ♈ 39 10	19 ♏ 09	16 ♐ 32	25 ♏ 59	01 ♐ 24	25 ♌ 02	10 ♓ 51	17 ♒ 19	25 ♐ 52

Date	Moon True ☊ ° ′	Moon Mean ☊ ° ′	Moon ☽ Latitude ° ′
01	24 ♓ 33	22 ♓ 52	01 S 02
02	24 R 33	22 49	00 N 17
03	24 32	22 46	01 36
04	24 28	22 43	02 48
05	24 23	22 40	03 49
06	24 15	22 36	04 33
07	24 07	22 33	05 00
08	23 58	22 30	05 07
09	23 51	22 27	04 56
10	23 46	22 24	04 30
11	23 43	22 21	03 51
12	23 41	22 17	03 03
13	23 D 42	22 14	02 07
14	23 43	22 11	01 07
15	23 43	22 08	00 N 04
16	23 R 43	22 05	00 S 59
17	23 40	22 01	01 59
18	23 34	21 58	02 55
19	23 26	21 55	03 43
20	23 16	21 52	04 22
21	23 04	21 49	04 49
22	22 51	21 46	05 01
23	22 39	21 42	04 59
24	22 29	21 39	04 40
25	22 22	21 36	04 06
26	22 17	21 33	03 18
27	22 15	21 30	02 18
28	22 D 15	21 27	01 S 09
29	22 R 15	21 23	00 N 05
30	22 ♓ 15	21 ♓ 20	01 N 20

DECLINATIONS

Date	Sun ☉ ° ′	Moon ☽ ° ′	Mercury ☿ ° ′	Venus ♀ ° ′	Mars ♂ ° ′	Jupiter ♃ ° ′	Saturn ♄ ° ′	Uranus ♅ ° ′	Neptune ♆ ° ′	Pluto ♇ ° ′
01	14 S 28	07 S 37	21 S 00	14 S 05	13 S 13	18 S 17	14 N 28	08 S 12	15 S 56	16 S 19
02	14 47	00 S 40	20 38	14 30	13 27	18 20	14 27	08 12	15 56	16 19
03	15 06	06 N 25	20 11	14 55	13 41	18 24	14 26	08 12	15 56	16 20
04	15 25	13 11	19 41	15 19	13 55	18 27	14 25	08 12	15 56	16 20
05	15 43	19 11	19 07	15 44	14 09	18 30	14 24	08 13	15 56	16 20
06	16 01	23 57	18 29	16 07	14 22	18 33	14 24	08 13	15 56	16 20
07	16 19	27 07	17 49	16 31	14 36	18 36	14 23	08 13	15 56	16 21
08	16 37	28 29	17 07	16 53	14 49	18 39	14 22	08 13	15 56	16 21
09	16 54	28 03	16 25	17 16	15 02	18 43	14 21	08 13	15 56	16 21
10	17 11	26 03	15 43	17 38	15 16	18 46	14 20	08 14	15 56	16 22
11	17 27	22 46	15 02	17 59	15 29	18 49	14 20	08 14	15 56	16 22
12	17 44	18 33	14 25	18 20	15 42	18 52	14 19	08 14	15 55	16 22
13	18 00	13 41	13 52	18 41	15 55	18 55	14 18	08 14	15 55	16 23
14	18 16	08 23	13 23	19 01	16 07	18 58	14 18	08 14	15 55	16 23
15	18 31	02 N 50	13 00	19 20	16 20	19 01	14 17	08 14	15 55	16 23
16	18 46	02 S 48	12 42	19 39	16 33	19 04	14 16	08 14	15 55	16 24
17	19 01	08 23	12 30	19 58	16 45	19 07	14 16	08 14	15 55	16 24
18	19 15	13 42	12 23	20 16	16 57	19 10	14 15	08 14	15 54	16 24
19	19 29	18 35	12 21	20 33	17 09	19 13	14 15	08 14	15 54	16 24
20	19 43	22 46	12 24	20 50	17 21	19 16	14 15	08 14	15 54	16 24
21	19 57	25 59	12 32	21 06	17 33	19 19	14 14	08 14	15 54	16 25
22	20 10	27 56	12 43	21 21	17 45	19 22	14 14	08 14	15 54	16 25
23	20 22	28 25	12 57	21 36	17 57	19 25	14 14	08 14	15 53	16 25
24	20 35	27 17	13 14	21 51	18 08	19 28	14 13	08 14	15 53	16 26
25	20 46	24 36	13 34	22 04	18 19	19 31	14 13	08 14	15 53	16 26
26	20 58	20 31	13 55	22 17	18 30	19 33	14 13	08 14	15 52	16 26
27	21 09	15 18	14 19	22 30	18 42	19 36	14 13	08 13	15 52	16 26
28	21 20	09 14	14 43	22 41	18 52	19 39	14 13	08 13	15 52	16 26
29	21 30	02 38	15 09	22 52	19 03	19 42	14 13	08 13	15 52	16 27
30	21 S 40	04 N 11	15 S 35	23 S 03	19 S 14	19 S 45	14 N 13	08 S 13	15 S 51	16 S 27

ZODIAC SIGN ENTRIES

Date	h	m	Planets	
02	15	46	☽	♈
04	16	05	☽	♉
06	16	46	☽	♊
08	19	46	☽	♋
11	02	34	☽	♌
13	13	19	☽	♍
16	02	14	☽	♎
17	08	02	♀	♐
18	14	47	☽	♏
21	01	15	☽	♐
22	11	02	☉	♐
23	09	25	☽	♑
24	04	43	♃	♐
25	15	41	☽	♒
27	20	21	☽	♓
29	23	30	☽	♈

LATITUDES

Date	Mercury ☿ ° ′	Venus ♀ ° ′	Mars ♂ ° ′	Jupiter ♃ ° ′	Saturn ♄ ° ′	Uranus ♅ ° ′	Neptune ♆ ° ′	Pluto ♇ ° ′
01	02 S 17	00 N 49	00 N 18	00 N 45	01 N 00	00 S 48	00 S 13	07 N 02
04	01 33	00 43	00 17	00 45	01 01	00 48	00 13	07 01
07	00 S 36	00 37	00 15	00 45	01 01	00 47	00 13	07 01
10	00 N 26	00 30	00 13	00 44	01 02	00 47	00 13	07 00
13	01 20	00 24	00 12	00 44	01 02	00 47	00 13	06 59
16	01 59	00 17	00 10	00 44	01 03	00 47	00 13	06 59
19	02 21	00 09	00 08	00 44	01 04	00 47	00 13	06 58
22	02 27	00 N 02	00 06	00 43	01 04	00 47	00 13	06 57
25	02 23	00 S 05	00 05	00 43	01 05	00 47	00 13	06 57
28	02 11	00 12	00 03	00 43	01 05	00 47	00 13	06 56
31	01 N 54	00 S 19	00 N 01	00 N 43	01 N 06	00 S 46	00 S 13	06 N 56

DATA

Julian Date	2454041
Delta T	+66 seconds
Ayanamsa	23° 57′ 10″
Synetic vernal point	05° ♓ 09′ 49″
True obliquity of ecliptic	23° 26′ 28″

MOON'S PHASES, APSIDES AND POSITIONS ☽

Date	h	m	Phase	Longitude ° ′	Eclipse Indicator
05	12	58	○	12 ♉ 58	
12	17	45	☾	20 ♌ 12	
20	22	18	●	28 ♏ 27	
28	06	29	☽	05 ♓ 53	

Day	h	m		
03	23	56	Perigee	
15	23	28	Apogee	
02	14	15	ON	
08	18	02	Max dec	28° N 32′
16	00	03	OS	
23	07	14	Max dec	28° S 27′
29	21	18	ON	

LONGITUDES

Date	Chiron ⚷ ° ′	Ceres ⚳ ° ′	Pallas ⚴ ° ′	Juno ⚵ ° ′	Vesta ⚶ ° ′	Black Moon Lilith ⚸ ° ′
01	04 ♒ 39	15 ♒ 11	12 ♑ 09	03 ♎ 56	13 ♎ 42	01 ♎ 23
11	04 ♒ 56	16 ♒ 58	15 ♑ 07	07 ♎ 34	18 ♎ 40	02 ♎ 29
21	05 ♒ 19	19 ♒ 09	18 ♑ 05	11 ♎ 01	23 ♎ 37	03 ♎ 36
31	05 ♒ 49	21 ♒ 41	21 ♑ 16	14 ♎ 15	28 ♎ 32	04 ♎ 42

All ephemeris data is given at 12.00 UT and the Moon's longitude is additionally given for 24.00 UT

ASPECTARIAN

h m	Aspects
01 Wednesday	
01 47	☽ Q ♇
04 11	♃ ⊻ ♇
04 38	☽ △ ☉
06 42	☽ △ ♀
07 57	♀ ∠ ♇
08 31	☽ ☌ ♅
09 56	☽ ∥ ♅
11 55	☉ # ♄
12 37	♂ Q ♄
15 43	☿ □ ♄
18 34	☽ ⊻ ♆
02 Thursday	
01 40	☽ ⚼ ♂
03 32	♀ △ ♅
04 25	☽ ⊥ ♆
05 24	☽ △ ☿
06 02	☽ ⊼ ♄
07 11	☽ ⚼ ☉
07 33	☽ □ ♇
07 54	☽ △ ♃
09 34	♀ # ♄
09 42	☽ ⚼ ♀
12 46	☉ ∠ ♇
15 51	☽ ± ♄
17 03	☽ ± ♂
19 04	☽ ∠ ♆
22 52	☽ ± ☉
03 Friday	
01 39	☽ ± ♀
03 14	☽ ⊼ ♂
04 34	☽ ⚼ ☿
06 28	☽ ⚼ ♄
08 35	☽ ⚼ ♃
09 15	☽ ⊼ ☉
09 26	☽ ⊻ ♅
12 02	☉ △ ♅
12 13	☽ ⊼ ♀
18 11	☽ # ♅
18 16	☽ ± ☿
19 05	☽ ⊥ ♅
19 16	☽ ⚹ ♆
23 16	☽ ± ♃
04 Saturday	
03 20	☽ ⊼ ☿
06 41	☽ △ ♄
08 04	☽ △ ♇
09 03	☽ ⊼ ♃
09 31	☽ ∠ ♅
14 33	☽ Q ♆
14 51	☽ # ♂
16 39	☽ ∥ ♄
20 42	☽ # ♀
20 56	☽ # ☉
22 32	☽ # ♆
05 Sunday	
00 06	☽ # ♇
05 46	☽ ☍ ♂
08 14	☽ ⚼ ♇
09 01	☽ # ♃
09 38	☽ ⚹ ♅
10 43	☉ ∥ ♀
11 45	☽ # ☿
12 58	☽ ☍ ☉
16 55	☽ ☍ ♀
19 35	☽ □ ♆
22 51	☽ ± ♇
06 Monday	
00 26	☽ ☍ ☿
00 44	♀ ∥ ♆
05 06	☽ Q ♅
05 09	☉ ∥ ♆
07 20	☽ □ ♄
08 40	☽ ⊼ ♇
09 49	☿ ∥ ♃
10 18	☽ ☍ ♃
14 04	♂ # ♄
19 02	☿ ⊥ ♇
07 Tuesday	
00 23	♀ □ ♆
01 43	♀ ∥ ♇
09 23	☽ ⊼ ♂
10 54	☽ □ ♅
13 20	☽ Q ♄
13 41	♂ ∠ ♇
14 19	☉ ∥ ♇
16 51	☿ ☌ ♀
18 12	☽ ⊼ ☉
20 03	☽ ± ♂
21 22	☽ △ ♆
22 19	☽ ⊼ ☿
23 25	☽ ⊼ ♀
08 Wednesday	
05 15	☽ ± ☉
07 44	☽ ± ☿
08 24	☿ □ ♆
09 56	☽ ⚹ ♄
10 46	☽ ± ♀
11 15	☽ ☍ ♇
12 20	☽ ⚼ ♂
13 42	☽ ⊼ ♃
16 18	♀ ⊥ ♇
16 23	♂ △ ♅
17 03	☿ ∥ ♀
21 30	☉ ☌ ☿
22 04	☽ ⚼ ☿
22 10	☽ ⚼ ☉
23 24	☽ ⚼ ♆
09 Thursday	
00 14	☉ ∥ ☿
00 24	☽ ± ♃
04 11	☽ ⚼ ♀
12 31	☽ ∠ ♄
13 48	☿ ∥ ♇
14 59	☉ □ ♆
15 07	☽ △ ♅
15 29	☽ ± ♆
16 22	☽ △ ♂
16 47	☽ ⚼ ♃
22 34	☽ △ ☿
10 Friday	
02 26	☽ ⊼ ♆
03 22	☽ △ ☉
04 22	☿ ∥ ♆
04 58	☽ ⊥ ♄
10 18	☽ △ ♀
16 09	☽ ⊻ ♄
17 31	☽ ⊼ ♇
18 44	☽ ⚼ ♅
20 59	☽ △ ♃
23 56	☿ ∥ ♂
11 Saturday	
04 53	☽ ± ♇
11 48	☽ ± ♅
15 16	☿ ☌ ♂
18 45	☉ ⊥ ♇
22 13	☽ ⚼ ♇
23 23	☽ ⊼ ♅
12 Sunday	
02 23	☽ □ ☿
04 07	☽ □ ♂
10 24	☽ # ♃
11 37	☽ ☍ ♆
13 03	☽ # ♀
16 02	☽ # ☉
16 19	☿ # ♄
17 45	☽ □ ☉
23 05	☽ # ♇
13 Monday	
00 19	♀ □ ♄
01 18	☽ # ♆
01 48	☽ # ♂
02 28	☽ ☌ ♄
02 42	☽ □ ♀
03 51	☽ △ ♇
08 29	☽ □ ♃
09 04	☽ ∥ ♄
11 04	☽ # ☿
11 23	☽ Q ☿
13 16	♂ ∥ ♆
14 00	♀ ⊻ ♇
17 26	☿ △ ♅
18 02	☽ Q ♂
14 Tuesday	
07 51	♀ ∥ ♃
09 12	☽ Q ☉
09 35	☿ ∠ ♇
10 07	☽ ⚹ ☿
11 11	☽ ☍ ♅
12 38	☽ # ♅
19 18	☽ Q ♀
19 51	☽ ⚹ ♂
21 34	☽ Q ♃
23 58	☽ ⊼ ♆
15 Wednesday	
11 55	☽ ⚹ ☉
12 12	☽ ± ♆
15 14	☽ ∠ ☿
15 24	☽ ⊻ ♄
16 47	☽ □ ♇
18 03	♂ ∥ ♇
20 35	♀ ☌ ♃
22 30	☽ ⚹ ♃
22 41	☽ ⚹ ♀
16 Thursday	
03 40	☽ ⊥ ♄
04 19	☽ ∠ ♂
06 34	☽ ⚼ ♆
09 00	☽ ⊥ ☿
20 55	☽ ⊻ ☿
21 18	☽ ∠ ☉
22 01	☽ ∠ ♄
23 45	☽ ⊥ ♂
17 Friday	
00 13	☽ ⊼ ♅
05 29	☽ Q ♇
05 34	☽ ∠ ♃
05 58	☉ □ ♄
08 46	☽ ∠ ♀
11 24	☽ ∥ ♅
12 21	☽ ± ♅
12 38	☽ ⊻ ♂
13 00	☽ △ ♆
17 11	☽ ⊥ ☉
19 00	♂ □ ♆
22 28	☉ ⊻ ♇
18 Saturday	
00 02	☽ ⊥ ♃
00 26	☿ St D
00 50	☉ ∥ ♃
04 21	☽ ⚹ ♄
05 00	☽ ⊥ ♀
05 41	☽ ⚹ ♇
06 00	☽ ∥ ☿
06 20	☽ ⊻ ☉
06 25	☽ ⚼ ♅
12 15	☽ ⊻ ♃
14 36	☽ # ♄
18 20	☽ ⊻ ♀
22 30	☽ ∥ ♆
19 Sunday	
00 55	☽ ∥ ♇
04 15	☽ Q ♄
04 21	☽ ∥ ♂
09 01	☽ ☌ ☿
11 29	☽ ∠ ♇
12 07	☽ △ ♅
15 24	☽ ∥ ♃
17 07	☽ ∥ ☉
23 28	☽ ∥ ♀
20 Monday	
00 32	☽ □ ♆
03 39	☽ ☌ ♂
05 05	☽ ⊥ ♇
06 07	♅ St D
15 23	☽ □ ♄
16 41	☽ ⊻ ♇
22 18	☽ ☌ ☉
23 54	☽ ☌ ♃
21 Tuesday	
06 21	♂ ⊥ ♇
11 05	♀ Q ♆
11 07	☽ Q ♆
11 07	☽ ☌ ♀
20 48	☽ ⊻ ☿
21 45	☽ □ ♅
23 15	☉ ☌ ♃
22 Wednesday	
06 46	☿ ∠ ♇
08 39	☽ ⊥ ☿
09 42	☽ ⚹ ♆
14 50	☿ △ ♅
15 54	☽ ⊻ ♂
23 Thursday	
00 02	☽ △ ♄
01 19	☽ ☌ ♇
02 22	☽ ∠ ☿
03 38	☽ ⊥ ♂
07 14	☽ Q ♅
09 05	☽ ⊻ ♃
11 18	☽ ⊻ ☉
13 29	☽ ∠ ♆
20 16	☽ ⊥ ♃
21 04	☽ ∠ ♂
23 11	☽ ⊥ ☉
24 Friday	
00 48	☽ ⊻ ♀
03 35	☽ ⚼ ♄
05 11	☽ ⚹ ♅
05 55	☽ ⊥ ♆
07 40	☽ ⚹ ☿
12 50	☽ ⊥ ♀
12 54	☽ ∠ ♃
16 47	☽ ⊻ ♆
16 51	☽ ∠ ☉
19 54	☽ ± ♄
25 Saturday	
01 43	☽ ⚹ ♂
06 41	☽ ⊼ ♄
06 42	☽ ∠ ♀
06 56	☽ Q ☿
07 59	☽ ⊻ ♇
08 13	☽ ∠ ♅
16 16	☽ ⚹ ♃
18 42	☽ ⊥ ♇
21 53	☽ ⚹ ☉
22 55	♀ □ ♅
26 Sunday	
00 15	☽ ⊥ ♅
00 18	☽ Q ♂
02 57	☽ ∥ ♀
09 47	☽ ∥ ☉
10 41	☽ ∠ ♇
10 52	☽ ⊻ ♅
12 04	☽ ⚹ ♀
13 52	☽ Q ♃
16 45	☽ ∥ ♃
17 30	☽ □ ☿
20 45	☽ Q ☉
21 27	☽ ∥ ♂
22 10	☽ ☌ ♆
27 Monday	
06 10	☿ # ♄
07 07	☽ ∥ ♇
09 34	☽ ∥ ♆
09 37	☽ □ ♂
11 11	☽ Q ♀
11 40	☽ ☍ ♄
13 00	☽ ⚹ ♇
15 51	☽ ∥ ☿
16 08	☉ Q ♆
16 31	☽ # ♄
21 46	☽ □ ♃
28 Tuesday	
06 29	☽ □ ☉
09 46	☽ Q ♇
15 00	☽ ☌ ♅
15 47	☽ ∥ ♅
21 15	☽ □ ♀
29 Wednesday	
00 54	☿ □ ♆
02 01	☽ ⊻ ♆
02 07	☽ △ ☿
03 00	♂ □ ♄
12 12	☽ ⊥ ♆
15 07	☽ ⊼ ♄
15 44	☽ △ ♂
16 29	☽ □ ♇
30 Thursday	
01 14	☽ ± ♄
01 42	☽ △ ♃
03 23	☽ ∠ ♆
05 55	☽ ⚼ ☿
07 39	♂ ⊻ ♇
13 13	☽ △ ☉
16 19	☽ ⚼ ♄
17 41	☽ ⊻ ♅
18 12	☽ ⚼ ♂
22 26	☽ ± ☿

DECEMBER 2006

LONGITUDES

Date	Sidereal time h m s	Sun ☉ ° ' "	Moon ☽ ° ' "	Moon ☽ 24.00 ° ' "	Mercury ☿ ° '	Venus ♀ ° '	Mars ♂ ° '	Jupiter ♃ ° '	Saturn ♄ ° '	Uranus ♅ ° '	Neptune ♆ ° '	Pluto ♇ ° '
01	16 40 49	09 ♐ 08 59	21 ♈ 52 49	29 ♈ 07 53	20 ♏ 28	17 ♐ 47	26 ♏ 41	01 ♐ 37	25 ♌ 03	10 ♓ 52	17 ♒ 20	25 ♐ 54
02	16 44 46	10 09 47	06 ♉ 23 45	13 ♉ 39 42	21 49	19 02	27 23	01 51	25 04	10 52	17 21	25 56
03	16 48 43	11 10 37	20 ♉ 54 58	28 ♉ 08 43	23 11	20 18	28 06	02 04	25 04	10 53	17 23	25 59
04	16 52 39	12 11 27	05 ♊ 20 05	12 ♊ 28 15	24 35	21 33	28 48	02 17	25 04	10 54	17 24	26 01
05	16 56 36	13 12 19	19 ♊ 32 25	26 ♊ 31 56	26 01	22 48	29 ♏ 30	02 30	25 04	10 55	17 25	26 03
06	17 00 32	14 13 12	03 ♋ 26 11	10 ♋ 14 47	27 27	24 04	00 ♐ 12	02 44	25 R 04	10 55	17 26	26 05
07	17 04 29	15 14 06	16 ♋ 57 24	23 ♋ 33 56	28 ♏ 55	25 19	00 55	02 57	25 04	10 56	17 28	26 07
08	17 08 25	16 15 01	00 ♌ 04 24	06 ♌ 28 57	00 ♐ 23	26 34	01 37	03 10	25 04	10 57	17 29	26 09
09	17 12 22	17 15 57	12 ♌ 47 51	19 ♌ 01 31	01 51	27 50	02 20	03 23	25 04	10 58	17 30	26 12
10	17 16 18	18 16 54	25 ♌ 10 26	01 ♍ 15 09	03 21	29 ♐ 05	03 02	03 37	25 03	10 59	17 32	26 14
11	17 20 15	19 17 52	07 ♍ 16 17	13 ♍ 14 31	04 50	00 ♑ 20	03 45	03 50	25 03	11 00	17 33	26 16
12	17 24 12	20 18 52	19 ♍ 10 31	25 ♍ 05 00	06 21	01 36	04 27	04 03	25 02	11 01	17 34	26 18
13	17 28 08	21 19 52	00 ♎ 58 41	06 ♎ 52 16	07 51	02 51	05 10	04 16	25 01	11 02	17 36	26 21
14	17 32 05	22 20 54	12 ♎ 46 26	18 ♎ 41 51	09 22	04 06	05 52	04 29	25 00	11 04	17 37	26 23
15	17 36 01	23 21 56	24 ♎ 39 08	00 ♏ 38 51	10 53	05 21	06 35	04 42	24 59	11 05	17 39	26 25
16	17 39 58	24 23 00	06 ♏ 41 31	12 ♏ 47 35	12 25	06 37	07 18	04 55	24 58	11 06	17 40	26 27
17	17 43 54	25 24 04	18 ♏ 57 26	25 ♏ 11 21	13 57	07 52	08 01	05 08	24 57	11 07	17 42	26 29
18	17 47 51	26 25 10	01 ♐ 29 32	07 ♐ 52 06	15 28	09 07	08 43	05 21	24 56	11 09	17 43	26 32
19	17 51 47	27 26 16	14 ♐ 19 04	20 ♐ 50 21	17 01	10 23	09 26	05 34	24 54	11 10	17 45	26 34
20	17 55 44	28 27 23	27 ♐ 25 47	04 ♑ 05 09	18 33	11 38	10 09	05 47	24 53	11 12	17 47	26 36
21	17 59 41	29 ♐ 28 30	10 ♑ 48 07	17 ♑ 34 22	20 05	12 53	10 52	06 00	24 51	11 13	17 48	26 38
22	18 03 37	00 ♑ 29 38	24 ♑ 23 31	01 ♒ 15 11	21 38	14 09	11 35	06 12	24 49	11 15	17 50	26 41
23	18 07 34	01 30 46	08 ♒ 08 59	15 ♒ 04 36	23 11	15 24	12 18	06 25	24 48	11 17	17 52	26 43
24	18 11 30	02 31 54	22 ♒ 01 41	29 ♒ 00 00	24 44	16 39	13 01	06 38	24 46	11 18	17 53	26 45
25	18 15 27	03 33 02	05 ♓ 59 20	12 ♓ 59 30	26 18	17 54	13 44	06 50	24 44	11 20	17 55	26 47
26	18 19 23	04 34 11	20 ♓ 00 24	27 ♓ 01 56	27 51	19 10	14 28	07 03	24 41	11 22	17 57	26 50
27	18 23 20	05 35 19	04 ♈ 04 01	11 ♈ 06 36	29 ♐ 25	20 25	15 11	07 16	24 39	11 24	17 59	26 52
28	18 27 16	06 36 28	18 ♈ 09 34	25 ♈ 12 50	00 ♑ 59	21 40	15 54	07 28	24 37	11 25	18 00	26 54
29	18 31 13	07 37 36	02 ♉ 16 13	09 ♉ 19 30	02 34	22 55	16 37	07 41	24 34	11 27	18 02	26 56
30	18 35 10	08 38 44	16 ♉ 22 26	23 ♉ 24 39	04 08	24 11	17 21	07 53	24 32	11 29	18 04	26 58
31	18 39 06	09 ♑ 39 53	00 ♊ 25 46	07 ♊ 25 20	05 ♑ 43	25 ♑ 26	18 ♐ 04	08 ♐ 05	24 ♌ 29	11 ♓ 31	18 ♒ 06	27 ♐ 01

Date	Moon True ☊ ° '	Moon Mean ☊ ° '	Moon ☽ Latitude ° '
01	22 ♓ 12	21 ♓ 17	02 N 30
02	22 R 06	21 14	03 31
03	21 58	21 11	04 18
04	21 47	21 07	04 48
05	21 34	21 04	05 00
06	21 22	21 01	04 54
07	21 10	20 58	04 31
08	21 01	20 55	03 54
09	20 55	20 52	03 07
10	20 51	20 48	02 11
11	20 49	20 45	01 11
12	20 D 49	20 42	00 N 09
13	20 R 49	20 39	00 S 54
14	20 48	20 36	01 53
15	20 45	20 33	02 49
16	20 39	20 29	03 37
17	20 30	20 26	04 17
18	20 19	20 23	04 44
19	20 05	20 20	04 59
20	19 52	20 17	04 58
21	19 39	20 13	04 40
22	19 27	20 10	04 07
23	19 19	20 07	03 19
24	19 14	20 04	02 18
25	19 11	20 01	01 S 09
26	19 D 11	19 58	00 N 04
27	19 11	19 54	01 18
28	19 R 10	19 51	02 27
29	19 08	19 48	03 28
30	19 03	19 45	04 15
31	18 ♓ 55	19 ♓ 42	04 N 47

DECLINATIONS

Date	Sun ☉ ° '	Moon ☽ ° '	Mercury ☿ ° '	Venus ♀ ° '	Mars ♂ ° '	Jupiter ♃ ° '	Saturn ♄ ° '	Uranus ♅ ° '	Neptune ♆ ° '	Pluto ♇ ° '
01	21 S 49	10 N 51	16 S 02	23 S 12	19 S 24	19 S 47	14 N 13	08 S 13	15 S 51	16 S 27
02	21 58	16 58	16 29	23 21	19 34	19 50	14 13	08 12	15 51	16 27
03	22 07	22 08	16 56	23 29	19 44	19 53	14 13	08 12	15 50	16 28
04	22 15	25 55	17 23	23 37	19 54	19 55	14 13	08 12	15 50	16 28
05	22 23	28 01	17 50	23 44	20 04	19 58	14 13	08 11	15 50	16 28
06	22 30	28 18	18 17	23 50	20 14	20 01	14 13	08 11	15 49	16 28
07	22 37	26 51	18 43	23 55	20 23	20 03	14 13	08 11	15 49	16 28
08	22 44	23 57	19 09	24 00	20 32	20 06	14 14	08 10	15 48	16 29
09	22 50	19 57	19 34	24 03	20 42	20 08	14 14	08 10	15 48	16 29
10	22 55	15 12	19 59	24 07	20 50	20 11	14 14	08 09	15 48	16 29
11	23 01	09 56	20 23	24 09	20 59	20 13	14 15	08 09	15 47	16 29
12	23 05	04 N 25	20 46	24 11	21 08	20 16	14 15	08 09	15 47	16 29
13	23 09	01 S 13	21 08	24 12	21 16	20 18	14 15	08 08	15 46	16 30
14	23 13	06 47	21 30	24 12	21 24	20 21	14 16	08 08	15 46	16 30
15	23 16	12 10	21 50	24 11	21 32	20 23	14 17	08 07	15 45	16 30
16	23 19	17 10	22 10	24 10	21 40	20 26	14 17	08 07	15 45	16 30
17	23 22	21 34	22 28	24 08	21 48	20 28	14 18	08 06	15 44	16 30
18	23 24	25 06	22 46	24 05	21 55	20 30	14 18	08 05	15 44	16 30
19	23 25	27 28	23 02	24 01	22 02	20 33	14 19	08 05	15 43	16 30
20	23 26	28 22	23 17	23 57	22 09	20 35	14 20	08 04	15 43	16 31
21	23 26	27 39	23 31	23 52	22 16	20 37	14 20	08 04	15 43	16 31
22	23 26	25 17	23 44	23 46	22 22	20 39	14 21	08 03	15 42	16 31
23	23 26	21 26	23 56	23 39	22 29	20 42	14 22	08 02	15 41	16 31
24	23 25	16 21	24 07	23 32	22 35	20 44	14 23	08 02	15 41	16 31
25	23 24	10 23	24 16	23 24	22 41	20 46	14 24	08 01	15 40	16 31
26	23 22	03 S 53	24 24	23 15	22 47	20 48	14 25	08 00	15 40	16 31
27	23 19	02 N 49	24 31	23 06	22 52	20 50	14 26	08 00	15 39	16 32
28	23 17	09 23	24 36	22 56	22 57	20 52	14 27	07 59	15 39	16 32
29	23 13	15 30	24 40	22 45	23 03	20 54	14 28	07 58	15 38	16 32
30	23 10	20 48	24 43	22 33	23 07	20 56	14 29	07 57	15 38	16 32
31	23 S 05	24 N 55	24 S 44	22 S 21	23 S 12	20 S 58	14 N 30	07 S 57	15 S 37	16 S 32

ZODIAC SIGN ENTRIES

Date	h m	Planets
02	01 26	☽ ♉
04	03 05	☽ ♊
06	04 58	♂ ♐
06	06 00	☽ ♋
08	05 52	☿ ♐
08	11 52	☽ ♌
10	21 31	☽ ♍
11	05 33	♀ ♑
13	10 00	☽ ♎
15	22 43	☽ ♏
18	09 10	☽ ♐
20	16 39	☽ ♑
22	00 22	☉ ♑
22	21 49	☽ ♒
25	01 43	☽ ♓
27	05 04	☽ ♈
27	20 55	☿ ♑
29	08 08	☽ ♉
31	11 16	☽ ♊

LATITUDES

Date	Mercury ☿ ° '	Venus ♀ ° '	Mars ♂ ° '	Jupiter ♃ ° '	Saturn ♄ ° '	Uranus ♅ ° '	Neptune ♆ ° '	Pluto ♇ ° '
01	01 N 54	00 S 19	00 N 01	00 N 43	01 N 06	00 S 46	00 S 13	06 N 56
04	01 35	00 27	00 S 01	00 43	01 07	00 46	00 13	06 55
07	01 13	00 34	00 03	00 43	01 07	00 46	00 13	06 55
10	00 51	00 40	00 05	00 42	01 08	00 46	00 13	06 54
13	00 29	00 47	00 06	00 42	01 09	00 46	00 13	06 54
16	00 N 07	00 53	00 08	00 42	01 09	00 46	00 13	06 54
19	00 S 14	00 59	00 10	00 42	01 10	00 46	00 13	06 53
22	00 34	01 05	00 12	00 42	01 10	00 46	00 13	06 53
25	00 53	01 10	00 14	00 42	01 11	00 46	00 13	06 53
28	01 10	01 15	00 16	00 42	01 12	00 45	00 13	06 53
31	01 S 25	01 S 19	00 S 18	00 N 42	01 N 12	00 S 45	00 S 13	06 N 53

DATA

Julian Date	2454071
Delta T	+66 seconds
Ayanamsa	23° 57' 15"
Synetic vernal point	05° ♓ 09' 45"
True obliquity of ecliptic	23° 26' 27"

MOON'S PHASES, APSIDES AND POSITIONS ☽

Date	h m	Phase	Longitude ° '	Eclipse Indicator
05	00 25	○	12 ♊ 43	
12	14 32	☾	20 ♍ 25	
20	14 01	●	28 ♐ 33	
27	14 48	☽	05 ♈ 42	

Day	h m		
02	00 12	Perigee	
13	18 58	Apogee	
28	02 21	Perigee	
06	03 38	Max dec	28° N 24'
13	06 51	0S	
20	13 40	Max dec	28° S 23'
27	01 57	0N	

LONGITUDES

Date	Chiron ⚷ ° '	Ceres ⚳ ° '	Pallas ⚴ ° '	Juno ⚵ ° '	Vesta ⚶ ° '	Black Moon Lilith ⚸ ° '
01	05 ♒ 49	21 ♒ 41	21 ♑ 16	14 ♎ 15	28 ♎ 32	04 ♎ 42
11	06 ♒ 23	24 ♒ 30	24 ♑ 34	17 ♎ 16	03 ♏ 23	05 ♎ 49
21	07 ♒ 02	27 ♒ 33	27 ♑ 57	20 ♎ 01	08 ♏ 09	06 ♎ 55
31	07 ♒ 45	00 ♓ 49	01 ♒ 23	22 ♎ 26	12 ♏ 50	08 ♎ 02

All ephemeris data is given at 12.00 UT and the Moon's longitude is additionally given for 24.00 UT

ASPECTARIAN

01 Friday
01 36 ☿ ⊥ ♇
02 22 ☽ # ♅
02 31 ☿ ∥ ♆
03 08 ☽ ⚼ ♃
03 22 ♀ ⚹ ♆
03 40 ☽ ⊥ ♅
04 27 ☽ ⚹ ♆
04 34 ☽ △ ♀
09 25 ☽ ⚻ ☿
09 56 ☽ ± ♂
16 03 ☽ ⚼ ☉
17 15 ☽ △ ♄
18 18 ☽ ± ♃
18 36 ☽ ∠ ♅
18 41 ☽ △ ♇
20 22 ☽ ⚻ ♂

02 Saturday
00 22 ☽ Q ♆
00 49 ☽ ∥ ♄
04 22 ☽ ⚻ ♃
07 21 ☽ # ♆
07 45 ☽ ⚼ ♀
08 02 ☽ ± ☉
09 47 ☽ # ☿
09 52 ☽ # ♇
10 44 ☿ ∥ ♇
18 41 ☽ ⚻ ☉
19 24 ☽ ⚹ ♅
19 31 ☽ ⚼ ♇
23 49 ☽ # ♂

03 Sunday
00 01 ☽ ± ♀
00 45 ☽ # ♃
05 00 ☉ □ ♅
06 08 ☽ □ ♆
10 26 ☽ ± ♇
10 52 ☽ ⚻ ♀
11 56 ☽ # ☉
15 16 ☽ Q ♅
16 10 ☽ ☍ ☿
18 53 ☽ □ ♄
19 48 ☽ # ♀
20 25 ☽ ⚻ ♇

04 Monday
00 32 ☽ ☍ ♂
06 49 ☽ ☍ ♃
15 15 ♂ ∥ ♃
20 07 ☿ □ ♄
21 21 ☽ □ ♅

05 Tuesday
00 25 ☽ ☍ ☉
01 01 ☽ Q ♄
08 23 ☽ △ ♆
12 36 ☿ ⚺ ♇
18 08 ☽ ☍ ♀
21 29 ☽ ⚹ ♄
23 12 ☽ ☍ ♇

06 Wednesday
00 23 ☽ ⚻ ☿
04 06 ♄ St R
06 04 ☽ ⚻ ♂
10 15 ☽ ⚼ ♆
10 45 ☽ ⚻ ♃
12 02 ☽ ± ☿
17 08 ☽ ± ♂
21 28 ☽ ± ♃
23 41 ☽ ∠ ♄

07 Thursday
01 13 ☽ △ ♅
02 08 ☽ ± ♆
05 51 ☽ ⚼ ☿
07 17 ♀ △ ♄
08 39 ☽ ⚻ ☉
10 01 ☽ ⚼ ♂
12 54 ☽ ⚻ ♆
13 49 ☽ ⚼ ♃
15 49 ☽ ⊥ ♄
20 24 ☽ ± ☉

08 Friday
02 45 ☽ ⚺ ♄
03 51 ♀ ☌ ♇
04 22 ☽ ⚼ ♅
04 44 ☽ ⚻ ♇
04 50 ☽ ⚻ ♀
11 42 ☽ # ♀
12 38 ☽ △ ☿
14 23 ☽ ⚼ ☉
15 03 ☽ △ ♂
15 54 ☽ ± ♇
17 10 ☽ ± ♀
17 53 ☽ △ ♃
19 44 ☽ # ☉
21 08 ☽ ± ♅

09 Saturday
08 04 ☽ # ♂
08 30 ☽ ⚻ ♅
08 56 ☽ ⚼ ♇
11 00 ☽ # ♃
12 04 ☽ ⚼ ♀
13 52 ☽ # ☿
17 43 ☉ ⚹ ♆
21 04 ☽ ☍ ♆
21 22 ☽ △ ☉

10 Sunday
02 28 ☿ ☌ ♂
05 47 ☽ # ♇
09 07 ☽ # ♆
10 07 ♀ Q ♅
11 46 ☽ ☌ ♄
14 05 ☽ △ ♇
16 30 ☽ ∥ ♄
16 58 ☿ ☌ ♃
20 35 ☽ △ ♀

11 Monday
01 09 ☿ ∥ ♃
04 31 ☽ □ ♂
05 00 ☽ □ ♃
06 27 ☽ □ ☿
16 11 ♂ ☌ ♃
19 30 ☽ ☍ ♅
19 53 ☽ # ♅
23 28 ☿ Q ♆

12 Tuesday
08 45 ☽ ⚻ ♆
14 32 ☽ □ ☉
17 56 ☽ Q ♃
19 05 ☽ Q ♂
20 57 ☽ ± ♆
23 53 ☽ ⚺ ♄

13 Wednesday
00 02 ☽ Q ☿
02 32 ☽ □ ♇
07 05 ♀ ∠ ♆
12 05 ☽ ⊥ ♄
15 18 ☽ ⚼ ♆
16 16 ☽ □ ♀
18 49 ☽ ⚹ ♃
21 04 ☽ ⚹ ♂

14 Thursday
01 38 ☿ ∥ ♂
03 10 ♂ Q ♆
04 04 ☽ ⚹ ☿
06 23 ☽ ∠ ♄
06 36 ☽ Q ☉
08 31 ☽ ⚻ ♅
15 16 ☽ Q ♇
17 51 ☽ ∥ ♅
20 42 ☽ ± ♅
20 47 ♀ ⚺ ♃
21 50 ☽ △ ♆

15 Friday
01 50 ☽ ∠ ♃
05 26 ☽ ∠ ♂
09 06 ☽ Q ♀
09 10 ☽ ⚹ ☉
12 41 ☽ ⚹ ♄
14 51 ☽ ∠ ☿
14 52 ☽ ⚼ ♅
15 03 ☿ □ ♅
15 33 ☽ ⚹ ♇
20 16 ☽ ⊥ ♃
21 52 ☽ # ♄

16 Saturday
00 37 ☽ ⊥ ♂
04 57 ☽ ∥ ♆
08 25 ☽ ⚺ ♃
08 39 ☽ ∥ ♇
11 22 ☽ ⊥ ☿
11 50 ☽ ⚹ ♀
12 33 ☽ Q ♄
13 16 ☽ ⚺ ♂
17 47 ☽ ∠ ☉
20 42 ☽ △ ♅
21 24 ☽ ∠ ♇

17 Sunday
00 52 ☽ ⚺ ☿
01 36 ☉ △ ♄
05 32 ☽ ∥ ♃
09 33 ☽ □ ♆
12 56 ☽ ⊥ ☉
13 23 ☽ ∥ ♂
14 58 ☽ ⊥ ♇
18 00 ☽ ∥ ☿
18 17 ♀ ⚺ ♂
20 23 ☽ ∠ ♀
23 26 ☽ ∥ ☉
23 31 ☽ □ ♄

18 Monday
01 30 ☽ ⚺ ☉
02 32 ☽ ⚺ ♇
04 25 ☽ ∥ ♀
14 40 ☉ ☌ ♇
15 25 ☽ ⊥ ♀
19 24 ☽ ☌ ♃
20 00 ☽ Q ♆

19 Tuesday
02 24 ☽ ☌ ♂
03 07 ♀ ⚼ ♄
03 54 ☽ ⚺ ♀
06 09 ☽ □ ♅
17 38 ☽ ☌ ☿
18 21 ☽ ⚹ ♆
23 44 ☿ ⚹ ♆

20 Wednesday
03 30 ♀ ⚹ ♅
07 23 ☽ △ ♄
10 30 ☽ ☌ ♇
11 39 ♃ Q ♆
14 01 ☽ ☌ ☉
14 47 ♀ ⊥ ♆
15 12 ☽ Q ♅
15 21 ♀ ⊥ ♃
21 40 ☽ ∠ ♆

21 Thursday
03 02 ☉ ∥ ☿
03 17 ☽ ⚺ ♃
05 55 ☉ Q ♅
10 19 ☽ ⚼ ♄
12 08 ☽ ⚺ ♂
12 45 ☽ ⚹ ♅
13 47 ☽ ⊥ ♆
14 09 ☽ ⊥ ♃
16 05 ☽ ☌ ♀
23 21 ☽ ⊥ ♂

22 Friday
00 17 ♂ □ ♅
00 26 ☽ ⚺ ♆
02 14 ☽ ± ♄
06 19 ☽ ∠ ♃
06 33 ☽ ⚺ ☿
12 45 ☽ ⚻ ♄
14 00 ☿ ∥ ♀
15 16 ☽ ∠ ♅
16 01 ☽ ⚺ ♇
16 03 ☽ ∠ ♂
18 24 ☽ ⊥ ☿
22 09 ☽ ∥ ☿
22 50 ☽ ∥ ♀
23 32 ☽ ⚺ ☉

23 Saturday
00 35 ☽ ∥ ☉
02 31 ☽ ⊥ ♇
06 22 ☽ ∥ ♂
07 00 ☽ ⊥ ♅
08 57 ☽ ⚹ ♃
10 48 ☽ ⊥ ☉
12 04 ☽ ∠ ☿
15 48 ☽ ∥ ♃
17 26 ☽ ⚺ ♅
18 12 ☽ ∠ ♇
19 36 ☽ ⚹ ♂

24 Sunday
01 48 ☽ ⚺ ♀
03 38 ☽ ∠ ☉
04 51 ☽ ☌ ♆
06 03 ☽ Q ♃
11 16 ☽ ∥ ♇
12 19 ☿ △ ♄
13 11 ☽ ⊥ ♀
14 51 ☽ ∥ ♆
16 42 ☽ ☍ ♄
17 15 ☽ ⚹ ☿
17 26 ☽ Q ♂
20 09 ☽ ⚹ ♇
20 14 ☽ # ♄
20 38 ☉ ∠ ♆

25 Monday
06 11 ☽ ∠ ♀
07 29 ☽ ⚹ ☉
12 11 ♀ ⚺ ♆
13 29 ☽ □ ♃
13 41 ☉ ∥ ♀
16 27 ☽ Q ☿
16 49 ☽ Q ♇
19 47 ☿ ☌ ♇
20 57 ☽ ∥ ♅
21 11 ☽ ☌ ♅

26 Tuesday
02 00 ☽ □ ♂
03 13 ♀ ± ♄
05 40 ☽ Q ☉
08 28 ☽ ⚺ ♆
10 25 ☽ ⚹ ♀
18 45 ☽ ⊥ ♆
19 59 ☽ ⚻ ♄
23 41 ☽ □ ♇

27 Wednesday
03 05 ☽ □ ☿
06 11 ☽ ± ♄
08 55 ☽ Q ♀
10 08 ☽ ∠ ♆
11 37 ☿ Q ♅
14 48 ☽ □ ☉
17 32 ☽ △ ♃
21 29 ☽ ⚼ ♄

28 Thursday
00 31 ☽ ⚺ ♅
06 46 ☽ # ♅
07 57 ☽ △ ♂
09 25 ♀ ∥ ♂
10 45 ☽ ⊥ ♅
11 44 ☽ ⚹ ♆
18 33 ☽ □ ♀
19 27 ☽ ⚼ ♃
22 25 ♀ ⊥ ♂
22 57 ☽ △ ♄

29 Friday
02 05 ☽ ∠ ♅
02 54 ☽ △ ♇
06 22 ♀ ∠ ♃
07 41 ☽ ∥ ♄
08 12 ☽ Q ♆
10 50 ☽ ⚼ ♂
10 59 ☽ ± ♃
12 33 ☽ # ♆
12 33 ☽ △ ☿
13 31 ☉ ⚺ ♃
16 19 ☽ # ♇
19 25 ☿ ∠ ♆
21 20 ☽ ⚻ ♃
21 49 ☽ △ ☉

30 Saturday
02 58 ☽ ± ♂
03 40 ☽ ⚹ ♅
04 29 ☽ ⚼ ♇
12 41 ☽ # ♃
13 45 ☽ ⚻ ♂
14 54 ☽ □ ♆
17 18 ☽ ⚼ ☿
17 47 ☉ ∥ ♂
18 29 ♀ ⚻ ♄
19 52 ☽ ± ♇
20 54 ☽ # ♀

31 Sunday
00 10 ☽ Q ♅
00 33 ☽ # ☉
00 48 ☽ # ♂
01 22 ☽ ⚼ ☉
01 52 ☽ □ ♄
02 37 ☽ △ ♀
06 08 ☽ ⚻ ♇
07 55 ☉ ⚼ ♄
10 38 ☽ ± ☿
10 47 ☽ # ☿
13 05 ♂ ⚹ ♆
17 59 ☽ ± ☉
22 14 ☽ ⚻ ☿

JANUARY 2007

LONGITUDES

Date	Sidereal time h m s	Sun ☉ ° ' "	Moon ☽ ° ' "	Moon ☽ 24.00 ° ' "	Mercury ☿ ° '	Venus ♀ ° '	Mars ♂ ° '	Jupiter ♃ ° '	Saturn ♄ ° '	Uranus ♅ ° '	Neptune ♆ ° '	Pluto ♇ ° '
01	18 43 03	10 ♑ 41 01	14 ♊ 22 51	21 ♊ 17 50	07 ♑ 18	26 ♑ 41	18 ♐ 47	08 ♐ 18	24 ♌ 26	11 ♓ 33	18 ♒ 08	27 ♐ 03
02	18 46 59	11 42 09	28 ♊ 09 47	04 ♋ 58 13	08 54	27 56	19 31	08 30	24 R 23	11 35	18 10	27 05
03	18 50 56	12 43 17	11 ♋ 42 44	18 ♋ 22 58	10 30	29 ♑ 11	20 14	08 42	24 20	11 38	18 12	27 07
04	18 54 52	13 44 25	24 ♋ 58 39	01 ♌ 29 36	12 06	00 ♒ 27	20 58	08 54	24 17	11 40	18 14	27 09
05	18 58 49	14 45 34	07 ♌ 55 44	14 ♌ 17 06	13 42	01 42	21 42	09 07	24 14	11 42	18 16	27 11
06	19 02 45	15 46 42	20 ♌ 33 49	26 ♌ 46 06	15 19	02 57	22 25	09 19	24 11	11 44	18 18	27 14
07	19 06 42	16 47 50	02 ♍ 54 17	08 ♍ 58 46	16 57	04 12	23 09	09 31	24 08	11 46	18 20	27 16
08	19 10 39	17 48 58	15 ♍ 00 00	20 ♍ 58 32	18 35	05 27	23 52	09 42	24 04	11 49	18 22	27 18
09	19 14 35	18 50 06	26 ♍ 54 58	02 ♎ 49 53	20 13	06 42	24 36	09 54	24 01	11 51	18 24	27 20
10	19 18 32	19 51 15	08 ♎ 43 58	14 ♎ 37 52	21 51	07 57	25 20	10 06	23 57	11 53	18 26	27 22
11	19 22 28	20 52 23	20 ♎ 32 18	26 ♎ 27 56	23 30	09 12	26 04	10 18	23 54	11 56	18 28	27 24
12	19 26 25	21 53 31	02 ♏ 25 27	08 ♏ 25 29	25 10	10 27	26 48	10 29	23 50	11 58	18 30	27 26
13	19 30 21	22 54 39	14 ♏ 28 42	20 ♏ 35 38	26 50	11 43	27 32	10 41	23 46	12 01	18 32	27 28
14	19 34 18	23 55 47	26 ♏ 46 50	03 ♐ 02 45	28 ♑ 30	12 58	28 16	10 52	23 43	12 03	18 34	27 30
15	19 38 14	24 56 56	09 ♐ 23 44	15 ♐ 50 03	00 ♒ 11	14 13	29 00	11 04	23 39	12 06	18 36	27 32
16	19 42 11	25 58 03	22 ♐ 21 52	28 ♐ 59 14	01 52	15 28	29 ♐ 44	11 15	23 35	12 08	18 38	27 34
17	19 46 08	26 59 11	05 ♑ 42 03	12 ♑ 30 07	03 34	16 43	00 ♑ 28	11 26	23 31	12 11	18 40	27 36
18	19 50 04	28 00 18	19 ♑ 23 07	26 ♑ 20 36	05 15	17 58	01 12	11 38	23 26	12 14	18 43	27 38
19	19 54 01	29 ♑ 01 25	03 ♒ 22 03	10 ♒ 26 52	06 58	19 13	01 56	11 49	23 22	12 16	18 45	27 40
20	19 57 57	00 ♒ 02 31	17 ♒ 34 24	24 ♒ 43 58	08 40	20 28	02 40	12 00	23 18	12 19	18 47	27 42
21	20 01 54	01 03 36	01 ♓ 54 55	09 ♓ 06 37	10 23	21 42	03 24	12 11	23 14	12 22	18 49	27 44
22	20 05 50	02 04 40	16 ♓ 18 28	23 ♓ 29 56	12 05	22 57	04 09	12 22	23 09	12 25	18 51	27 46
23	20 09 47	03 05 43	00 ♈ 40 34	07 ♈ 49 58	13 48	24 12	04 53	12 32	23 05	12 28	18 53	27 48
24	20 13 43	04 06 46	14 ♈ 57 51	22 ♈ 03 56	15 31	25 27	05 37	12 43	23 01	12 30	18 56	27 50
25	20 17 40	05 07 47	29 ♈ 08 04	06 ♉ 10 04	17 13	26 42	06 22	12 54	22 56	12 33	18 58	27 52
26	20 21 37	06 08 47	13 ♉ 09 49	20 ♉ 07 15	18 55	27 57	07 06	13 04	22 52	12 36	19 00	27 54
27	20 25 33	07 09 46	27 ♉ 02 16	03 ♊ 54 46	20 36	29 ♒ 12	07 51	13 14	22 47	12 39	19 02	27 56
28	20 29 30	08 10 44	10 ♊ 44 40	17 ♊ 31 51	22 16	00 ♓ 26	08 35	13 25	22 42	12 42	19 05	27 57
29	20 33 26	09 11 41	24 ♊ 16 13	00 ♋ 57 38	23 55	01 41	09 20	13 35	22 38	12 45	19 07	27 59
30	20 37 23	10 12 37	07 ♋ 35 59	14 ♋ 11 09	25 32	02 56	10 04	13 45	22 33	12 48	19 09	28 01
31	20 41 19	11 ♒ 13 31	20 ♋ 42 59	27 ♋ 11 25	27 ♒ 07	04 ♓ 10	10 ♑ 49	13 ♐ 55	22 ♌ 28	12 ♓ 51	19 ♒ 11	28 ♐ 03

Date	Moon True ☊ ° '	Moon Mean ☊ ° '	Moon ☽ Latitude ° '
01	18 ♓ 45	19 ♓ 39	05 N 01
02	18 R 34	19 35	04 58
03	18 22	19 32	04 38
04	18 11	19 29	04 03
05	18 03	19 26	03 16
06	17 56	19 23	02 20
07	17 53	19 19	01 19
08	17 51	19 16	00 N 15
09	17 D 52	19 13	00 S 48
10	17 53	19 10	01 50
11	17 53	19 07	02 46
12	17 R 53	19 04	03 36
13	17 50	19 00	04 17
14	17 45	18 57	04 46
15	17 38	18 54	05 03
16	17 29	18 51	05 05
17	17 20	18 48	04 51
18	17 11	18 44	04 20
19	17 03	18 41	03 33
20	16 58	18 38	02 32
21	16 55	18 35	01 20
22	16 54	18 32	00 S 03
23	16 D 54	18 29	01 N 14
24	16 56	18 25	02 26
25	16 57	18 22	03 29
26	16 R 57	18 19	04 18
27	16 55	18 16	04 52
28	16 51	18 13	05 09
29	16 46	18 10	05 08
30	16 40	18 06	04 50
31	16 ♓ 33	18 ♓ 03	04 N 17

DECLINATIONS

Date	Sun ☉ ° '	Moon ☽ ° '	Mercury ☿ ° '	Venus ♀ ° '	Mars ♂ ° '	Jupiter ♃ ° '	Saturn ♄ ° '	Uranus ♅ ° '	Neptune ♆ ° '	Pluto ♇ ° '
01	23 S 01	27 N 31	24 S 44	22 S 08	23 S 16	21 S 00	14 N 31	07 S 56	15 S 37	16 S 32
02	22 56	28 24	24 43	21 55	23 21	21 02	14 32	07 55	15 36	16 32
03	22 50	27 32	24 40	21 41	23 25	21 04	14 33	07 54	15 35	16 32
04	22 44	25 07	24 36	21 26	23 28	21 06	14 34	07 53	15 35	16 32
05	22 37	21 27	24 30	21 10	23 32	21 08	14 36	07 52	15 34	16 32
06	22 30	16 51	24 23	20 54	23 35	21 10	14 37	07 51	15 34	16 32
07	22 23	11 40	24 14	20 38	23 38	21 11	14 38	07 51	15 33	16 32
08	22 15	06 04	24 04	20 20	23 41	21 13	14 39	07 50	15 32	16 33
09	22 07	00 N 29	23 52	20 02	23 43	21 15	14 41	07 49	15 32	16 33
10	21 58	05 S 08	23 39	19 44	23 46	21 17	14 42	07 48	15 31	16 33
11	21 49	10 35	23 25	19 25	23 48	21 18	14 43	07 47	15 30	16 33
12	21 40	15 41	23 08	19 05	23 50	21 20	14 45	07 46	15 30	16 33
13	21 30	20 16	22 50	18 45	23 51	21 22	14 46	07 45	15 29	16 33
14	21 19	24 05	22 31	18 25	23 53	21 23	14 48	07 44	15 29	16 33
15	21 09	26 51	22 10	18 03	23 54	21 25	14 49	07 43	15 28	16 33
16	20 57	28 18	21 48	17 42	23 55	21 26	14 51	07 42	15 27	16 33
17	20 46	28 10	21 24	17 19	23 55	21 28	14 52	07 41	15 27	16 33
18	20 34	26 20	20 58	16 57	23 56	21 30	14 54	07 40	15 26	16 33
19	20 21	22 51	20 31	16 34	23 56	21 31	14 55	07 39	15 25	16 33
20	20 09	17 58	20 02	16 10	23 56	21 32	14 57	07 38	15 25	16 33
21	19 55	12 02	19 32	15 46	23 55	21 34	14 58	07 37	15 24	16 33
22	19 42	05 S 27	19 01	15 21	23 55	21 35	15 00	07 35	15 23	16 33
23	19 28	01 N 24	18 28	14 57	23 54	21 37	15 01	07 34	15 23	16 33
24	19 14	08 09	17 53	14 31	23 53	21 38	15 03	07 33	15 22	16 33
25	18 59	14 25	17 17	14 06	23 51	21 39	15 05	07 32	15 21	16 33
26	18 44	19 54	16 40	13 40	23 50	21 41	15 06	07 31	15 21	16 33
27	18 29	24 14	16 02	13 13	23 48	21 42	15 08	07 30	15 20	16 33
28	18 13	27 09	15 23	12 46	23 46	21 43	15 09	07 29	15 19	16 33
29	17 57	28 26	14 43	12 19	23 44	21 44	15 11	07 28	15 18	16 33
30	17 41	28 03	14 02	11 52	23 41	21 46	15 13	07 26	15 18	16 33
31	17 S 25	26 N 05	13 S 21	11 S 24	23 S 38	21 S 47	15 N 14	07 S 25	15 S 17	16 S 33

ZODIAC SIGN ENTRIES

Date	h	m	Planets
02	15	14	☽ ♋
04	03	31	♀ ♒
04	21	14	☽ ♌
07	06	18	☽ ♍
09	18	15	☽ ♎
12	07	08	☽ ♏
14	18	11	☽ ♐
15	09	25	☿ ♒
16	20	54	♂ ♑
17	01	49	☽ ♑
19	06	16	☽ ♒
20	11	01	☉ ♒
21	08	48	☽ ♓
23	10	52	☽ ♈
25	13	28	☽ ♉
27	17	10	☽ ♊
28	03	32	♀ ♓
29	22	16	☽ ♋

LATITUDES

Date	Mercury ☿ ° '	Venus ♀ ° '	Mars ♂ ° '	Jupiter ♃ ° '	Saturn ♄ ° '	Uranus ♅ ° '	Neptune ♆ ° '	Pluto ♇ ° '
01	01 S 30	01 S 21	00 S 18	00 N 42	01 N 12	00 S 45	00 S 13	06 N 53
04	01 43	01 24	00 20	00 42	01 13	00 45	00 13	06 52
07	01 53	01 28	00 22	00 42	01 13	00 45	00 13	06 52
10	02 01	01 30	00 24	00 42	01 14	00 45	00 13	06 52
13	02 05	01 32	00 26	00 42	01 15	00 45	00 13	06 52
16	02 06	01 34	00 28	00 42	01 15	00 45	00 13	06 52
19	02 03	01 35	00 30	00 42	01 16	00 45	00 13	06 52
22	01 55	01 35	00 32	00 42	01 16	00 45	00 14	06 52
25	01 42	01 34	00 34	00 42	01 16	00 45	00 14	06 52
28	01 22	01 33	00 36	00 42	01 17	00 45	00 14	06 53
31	00 S 56	01 S 32	00 S 38	00 N 42	01 N 17	00 S 44	00 S 14	06 N 53

DATA

Julian Date	2454102
Delta T	+66 seconds
Ayanamsa	23° 57' 21"
Synetic vernal point	05° ♓ 09' 38"
True obliquity of ecliptic	23° 26' 26"

MOON'S PHASES, APSIDES AND POSITIONS ☽

Date	h	m	Phase	Longitude ° '	Eclipse Indicator
03	13	57	○	12 ♋ 48	
11	12	45	☾	20 ♎ 54	
19	04	01	●	28 ♑ 41	
25	23	01	☽	05 ♉ 36	

Day	h	m		
10	16	23	Apogee	
22	12	25	Perigee	
02	11	59	Max dec	28° N 24'
09	14	04	0S	
16	22	06	Max dec	28° S 27'
23	07	06	0N	
29	18	21	Max dec	28° N 30'

LONGITUDES

Date	Chiron ⚷ ° '	Ceres ⚳ ° '	Pallas ⚴ ° '	Juno ⚵ ° '	Vesta ⚶ ° '	Black Moon Lilith ⚸ ° '
01	07 ♒ 49	01 ♓ 09	01 ♒ 44	22 ♎ 39	13 ♏ 17	08 ♎ 09
11	08 ♒ 35	04 ♓ 36	05 ♒ 12	24 ♎ 40	17 ♏ 49	09 ♎ 15
21	09 ♒ 22	08 ♓ 11	08 ♒ 42	26 ♎ 24	22 ♏ 11	10 ♎ 22
31	10 ♒ 10	11 ♓ 53	12 ♒ 11	27 ♎ 19	26 ♏ 21	11 ♎ 28

All ephemeris data is given at 12.00 UT and the Moon's longitude is additionally given for 24.00 UT

ASPECTARIAN

h m Aspects

01 Monday
01 21 ☽ ☍ ♃
05 07 ☽ ⊼ ☉
06 53 ☽ ⚼ ♀
07 06 ☽ □ ♅
08 39 ☽ Q ♄
09 28 ♀ ∠ ♅
18 31 ☽ △ ♆
19 08 ♀ ⊻ ♇
20 04 ☽ ☍ ♂

02 Tuesday
00 02 ☽ ± ♀
05 11 ☿ ⊻ ♃
05 25 ☽ ⚹ ♄
09 15 ☉ ⚹ ♅
10 06 ☽ ☍ ♇
11 34 ☽ ⊼ ♀
19 12 ☿ ⚼ ♄
20 49 ☽ ⚼ ♆
23 10 ☉ ⊥ ♆

03 Wednesday
06 33 ☽ ⊼ ♃
07 47 ☽ ∠ ♄
09 32 ☽ ☍ ☿
11 51 ☽ △ ♅
12 52 ☽ ± ♆
13 57 ☽ ☍ ☉
17 27 ☽ ± ♃
23 41 ☽ ⊼ ♆
23 53 ☽ ⊥ ♄

04 Thursday
04 15 ☽ ⊼ ♂
05 19 ☿ ⚹ ♅
10 01 ☽ ⚼ ♃
10 45 ☽ ⊻ ♄
13 58 ☿ ⊥ ♆
15 06 ☽ ⚼ ♅
15 52 ☽ ± ♂
16 00 ☽ # ☿
16 00 ☽ ⊼ ♇
23 08 ☽ ☍ ♀
23 27 ☽ # ♂

05 Friday
03 07 ☽ ± ♇
04 48 ☽ # ☉
07 49 ☽ ± ♅
09 33 ☽ ⚼ ♂
13 38 ☽ # ♀
13 46 ☽ # ♃
14 15 ☽ △ ♃
15 31 ♀ ‖ ♃
19 07 ☽ ⊼ ♅
20 03 ☽ ⚼ ♇
22 16 ☉ ⊥ ♃

06 Saturday
00 30 ☽ ⊼ ☿
02 02 ☽ ⊼ ☉
07 38 ☽ ☍ ♆
11 46 ☿ ⊥ ♃
13 31 ☽ # ♇
13 41 ☽ ± ☿
14 33 ☽ ± ☉
15 48 ☽ △ ♂
17 29 ♀ ⊥ ♇
18 13 ☽ # ♆
18 57 ☽ ☌ ♄
22 36 ☽ ‖ ♄

07 Sunday
00 56 ☽ △ ♇
06 05 ☉ ☌ ☿
09 38 ☽ ⚼ ☉
09 50 ☽ ⚼ ☿
14 50 ☽ ⊼ ♀

08 Monday
01 16 ☽ □ ♃
04 06 ☽ ± ♀
04 46 ☽ # ♅
04 52 ☿ ± ♄
05 37 ☽ ☍ ♅
08 45 ☿ ⊻ ♆
17 45 ☉ ± ♄
18 05 ♂ △ ♄
18 10 ☽ △ ☉
18 45 ☽ ⊼ ♆
19 07 ♀ ⊥ ♅
20 19 ☽ △ ☿

09 Tuesday
00 14 ☽ ⚼ ♀
01 14 ☉ ⊻ ♆
06 10 ☽ ⊻ ♄
06 53 ☽ ± ♆
07 01 ☽ □ ♂
12 51 ☽ □ ♇
14 02 ☽ Q ♃
18 15 ☽ ⊥ ♄

10 Wednesday
01 11 ☽ ⚼ ♆
02 09 ☿ ‖ ♂
10 14 ☽ △ ♀
12 27 ☽ ∠ ♄
14 50 ☽ ⚹ ♃
18 27 ☽ ⊼ ♅
21 59 ☽ Q ♂
23 32 ☽ ‖ ♅

11 Thursday
01 32 ☽ Q ♇
06 41 ☽ ± ♅
07 46 ☽ △ ♆
12 45 ☽ □ ☉
17 28 ☿ ⊼ ♄
18 46 ☽ ⚹ ♄
19 00 ☽ □ ☿
21 48 ☽ ∠ ♃
23 55 ☽ ⚹ ♂

12 Friday
00 59 ☽ ⚼ ♅
01 56 ☽ ⚹ ♇
07 23 ☽ # ♄
11 04 ☽ ‖ ♆
12 45 ♀ ⚹ ♃
16 13 ☽ ⊥ ♃
16 16 ☽ ‖ ♇
17 19 ☿ ∠ ♃
18 48 ☽ Q ♄

13 Saturday
04 16 ☽ ‖ ♀
04 18 ☽ Q ☉
04 22 ☽ ⊻ ♃
05 54 ☽ □ ♀
07 07 ☽ △ ♅
07 54 ☽ ∠ ♂
08 01 ☽ ∠ ♇
10 02 ♂ ☌ ♇
12 48 ☽ Q ☿
14 42 ☿ ∠ ♅
18 03 ♀ ⊻ ♅
18 25 ☽ ‖ ♃
18 53 ☽ ‖ ☉
19 59 ☽ □ ♆
21 24 ☿ ⊻ ♇

14 Sunday
01 45 ☽ ⊥ ♇
02 20 ☽ ‖ ☿
02 42 ☽ ⊥ ♂
03 02 ♀ ∠ ♇
04 06 ☉ ‖ ♃
05 50 ☿ ⊻ ♂
06 00 ☽ ⚹ ☉
06 05 ☽ □ ♄
07 06 ☉ ⊼ ♄
10 34 ☽ ‖ ♂
13 24 ☽ ⊻ ♇
15 02 ☽ ⊻ ♂
15 50 ☽ ⚹ ☿
20 54 ☽ Q ♀

15 Monday
01 57 ♀ ∠ ♂
06 43 ☽ Q ♆
13 08 ☽ ∠ ☉
15 10 ☽ ☌ ♃
17 04 ☽ □ ♅
21 57 ☽ ⚹ ♀

16 Tuesday
00 24 ☽ ∠ ☿
05 09 ☽ ⚹ ♆
07 15 ☽ ⊥ ☉
14 12 ☽ △ ♄
19 06 ☽ ⊻ ☉
19 18 ☽ ⊥ ☿
20 16 ☉ ∠ ♃
21 28 ☽ ☌ ♇

17 Wednesday
02 06 ☽ ☌ ♂
02 07 ☽ Q ♅
02 23 ♂ Q ♅
04 09 ☽ ∠ ♀
07 39 ☽ ⊻ ☿
08 01 ☿ ‖ ♃
08 23 ☽ ∠ ♆
12 40 ☿ ⊥ ♇
16 54 ☉ ∠ ♅
16 57 ☽ ⚼ ♄
21 45 ☽ ⊥ ♀
22 17 ☽ ⊻ ♃
23 29 ☽ ⚹ ♅

18 Thursday
00 20 ☽ ⊥ ♆
03 06 ☉ ⊻ ♇
08 38 ☽ ± ♄
08 54 ☽ ⊥ ♃
09 17 ☽ ⊻ ♀
10 49 ☽ ⊻ ♆
18 58 ☽ ⊼ ♄

19 Friday
00 39 ☽ ∠ ♃
01 34 ☽ ∠ ♅
02 04 ☿ ⊥ ♅
02 15 ☽ ⊻ ♇
02 48 ♀ ☌ ♆
04 01 ☽ ☌ ☉
05 35 ☽ ‖ ♂
09 25 ☽ ⊻ ♂
12 31 ☽ ⊥ ♇
12 38 ♀ ‖ ♇
16 57 ☽ ⊥ ♅
18 56 ☽ ☌ ☿
19 11 ☽ ‖ ♃
20 10 ☽ ⊥ ♂

20 Saturday
01 33 ☽ ‖ ☿
01 37 ☽ ‖ ☉
02 30 ☽ ⚹ ♃
02 44 ☉ ‖ ☿
03 08 ☽ ⊻ ♅
03 47 ☽ ∠ ♇
12 03 ☿ ⊥ ♂
12 10 ☽ ∠ ♂
14 02 ☽ ☌ ♆
17 18 ☽ ☌ ♀
18 06 ☽ ‖ ♇
20 16 ☽ ‖ ♀
21 33 ☽ ☍ ♄
22 49 ☽ ‖ ♆
22 54 ☽ Q ♃

21 Sunday
00 38 ☽ # ♄
05 00 ☽ ⚹ ♇
10 28 ☽ ⊻ ☉
14 37 ☽ ⚹ ♂
21 14 ☽ ⊥ ☉

22 Monday
01 05 ☽ Q ♇
02 03 ♂ ∠ ♆
04 01 ☽ ⊻ ☿
04 21 ☽ ‖ ♅
05 20 ☽ □ ♃
05 29 ☽ ☌ ♅
10 12 ♀ ‖ ♆
11 43 ☽ Q ♂
13 23 ☽ ∠ ☉
15 22 ☽ ⊥ ☿
15 38 ♀ ☍ ♄
16 13 ☿ ⚹ ♃
16 16 ☽ ⊻ ♆
16 39 ☿ ⊻ ♅
21 42 ♃ □ ♅
21 43 ☿ ∠ ♇
23 22 ☽ ⊼ ♄

23 Tuesday
00 09 ☽ ⊻ ♀
02 18 ☽ ⊥ ♆
07 11 ☽ □ ♇
07 43 ♀ # ♄
08 26 ☽ ∠ ☿
09 21 ☽ ± ♄
11 08 ☽ ⊥ ♀
16 22 ☽ ⚹ ☉
17 24 ☽ ∠ ♆
19 26 ☽ □ ♂
19 29 ♀ Q ♃

24 Wednesday
00 21 ☽ ⚼ ♄
03 41 ☽ ∠ ♀
05 12 ☉ ⊥ ♇
07 51 ☽ ⊻ ♅
08 10 ☽ △ ♃
09 52 ☽ # ♅
13 03 ☽ ⚹ ☿
14 05 ☽ Q ☉
18 00 ☽ ⊥ ♅
18 43 ☽ ⚹ ♆

25 Thursday
01 32 ☽ △ ♄
07 28 ☽ ⚹ ♀
09 18 ☽ ∠ ♅
09 50 ☽ △ ♇
09 52 ☽ ⚼ ♃
10 46 ☽ # ♀
12 09 ☽ Q ☿
14 41 ☽ ‖ ♄
15 08 ☽ Q ♆
15 49 ☽ # ♆
20 51 ☽ # ♇
22 51 ☽ # ☿
23 01 ☽ □ ☉

26 Friday
01 01 ☽ △ ♂
01 24 ☽ ± ♃
05 56 ☽ Q ♀
06 48 ☽ # ☉
10 59 ♀ ⚹ ♇
11 02 ☽ ⚹ ♅
11 32 ☽ ⚼ ♇
11 50 ☽ ⊼ ♃
13 20 ☿ ☌ ♆
16 44 ☿ ‖ ♇
21 05 ☽ # ♃
22 06 ☽ □ ♆
23 17 ☽ □ ☿
23 21 ☉ ⊥ ♅

27 Saturday
03 06 ☽ ± ♇
04 18 ☽ ⚼ ♂
04 39 ☽ □ ♄
07 50 ☽ Q ♅
09 16 ☽ # ♂
10 14 ♂ ⚼ ♄
13 33 ☽ ⊼ ♇
16 08 ☽ □ ♀
20 51 ☽ ± ♂

28 Sunday
07 07 ☽ △ ☉
07 59 ☽ ⊼ ♂
11 56 ☽ Q ♄
14 27 ☿ ‖ ♆
15 28 ☽ □ ♅
16 46 ☽ ☍ ♃
18 07 ☿ ☍ ♄
19 55 ☿ # ♄

29 Monday
02 47 ☽ △ ♆
09 05 ☽ ⚹ ♄
11 16 ☽ △ ☿
11 51 ☽ ⚼ ☉
18 40 ☽ ☍ ♇
23 14 ☿ ∠ ♂
23 32 ☉ ⊻ ♂

30 Tuesday
02 41 ☽ △ ♀
05 21 ☽ ± ☉
05 44 ☽ ⚼ ♆
11 55 ☽ ∠ ♄
15 37 ☿ Q ♃
16 45 ☽ ☍ ♂
17 08 ☽ ⊼ ☉
18 04 ☽ ⚼ ☿
21 30 ☽ △ ♅
22 08 ☽ ± ♆
23 21 ☽ ⊼ ♃

31 Wednesday
04 14 ☽ ⊥ ♄
08 51 ☽ ⚼ ♀
09 11 ☽ ⊼ ♆
10 30 ☽ ± ♃
12 51 ☽ ± ☿
15 13 ☽ ⊻ ♄

FEBRUARY 2007

LONGITUDES

Date	Sidereal time h m s	Sun ☉ ° ' "	Moon ☽ ° ' "	Moon ☽ 24.00 ° ' "	Mercury ☿ ° '	Venus ♀ ° '	Mars ♂ ° '	Jupiter ♃ ° '	Saturn ♄ ° '	Uranus ♅ ° '	Neptune ♆ ° '	Pluto ♇ ° '
01	20 45 16	12 ♒ 14 25	03 ♌ 36 22	09 ♌ 57 47	28 ♒ 40	05 ♓ 25	11 ♑ 33	14 ♐ 05	22 ♌ 24	12 ♓ 54	19 ♒ 14	28 ♐ 04
02	20 49 12	13 15 17	16 ♌ 15 41	22 ♌ 30 07	00 ♓ 10	06 40	12 18	14 15	22 R 19	12 57	19 16	28 06
03	20 53 09	14 16 08	28 ♌ 41 10	04 ♍ 49 00	01 36	07 54	13 03	14 24	22 14	13 00	19 18	28 08
04	20 57 06	15 16 58	10 ♍ 53 49	16 ♍ 55 54	02 58	09 09	13 47	14 34	22 09	13 04	19 20	28 10
05	21 01 02	16 17 47	22 ♍ 55 33	28 ♍ 53 09	04 16	10 23	14 32	14 43	22 04	13 07	19 23	28 11
06	21 04 59	17 18 35	04 ♎ 49 07	10 ♎ 43 54	05 27	11 38	15 17	14 53	21 59	13 10	19 25	28 13
07	21 08 55	18 19 23	16 ♎ 38 03	22 ♎ 32 04	06 33	12 52	16 02	15 02	21 55	13 13	19 27	28 14
08	21 12 52	19 20 09	28 ♎ 26 34	04 ♏ 22 07	07 31	14 07	16 47	15 11	21 50	13 16	19 30	28 16
09	21 16 48	20 20 54	10 ♏ 19 21	16 ♏ 18 53	08 21	15 21	17 32	15 20	21 45	13 19	19 32	28 18
10	21 20 45	21 21 38	22 ♏ 21 21	28 ♏ 27 22	09 03	16 36	18 17	15 29	21 40	13 23	19 34	28 19
11	21 24 41	22 22 21	04 ♐ 37 31	10 ♐ 52 22	09 36	17 50	19 02	15 38	21 35	13 26	19 36	28 21
12	21 28 38	23 23 03	17 ♐ 12 24	23 ♐ 38 05	09 59	19 04	19 47	15 46	21 30	13 29	19 39	28 22
13	21 32 35	24 23 44	00 ♑ 09 44	06 ♑ 47 39	10 11	20 18	20 32	15 55	21 25	13 32	19 41	28 23
14	21 36 31	25 24 24	13 ♑ 31 56	20 ♑ 22 36	10 R 13	21 33	21 17	16 03	21 20	13 36	19 43	28 25
15	21 40 28	26 25 02	27 ♑ 19 30	04 ♒ 22 22	10 05	22 47	22 02	16 11	21 16	13 39	19 46	28 26
16	21 44 24	27 25 40	11 ♒ 30 44	18 ♒ 44 03	09 46	24 01	22 47	16 19	21 11	13 42	19 48	28 28
17	21 48 21	28 26 15	26 ♒ 01 34	03 ♓ 22 30	09 17	25 15	23 32	16 27	21 06	13 46	19 50	28 29
18	21 52 17	29 ♒ 26 50	10 ♓ 45 56	18 ♓ 10 56	08 39	26 29	24 17	16 35	21 01	13 49	19 52	28 30
19	21 56 14	00 ♓ 27 22	25 ♓ 36 32	03 ♈ 01 47	07 52	27 43	25 02	16 43	20 56	13 53	19 55	28 32
20	22 00 10	01 27 53	10 ♈ 25 50	17 ♈ 47 50	06 59	28 ♓ 57	25 48	16 51	20 51	13 56	19 57	28 33
21	22 04 07	02 28 22	25 ♈ 07 05	02 ♉ 22 59	06 00	00 ♈ 11	26 33	16 58	20 47	13 59	19 59	28 34
22	22 08 04	03 28 50	09 ♉ 35 04	16 ♉ 42 57	04 57	01 25	27 18	17 05	20 42	14 03	20 01	28 35
23	22 12 00	04 29 15	23 ♉ 46 23	00 ♊ 45 13	03 51	02 39	28 04	17 13	20 37	14 06	20 04	28 36
24	22 15 57	05 29 39	07 ♊ 39 22	14 ♊ 28 53	02 45	03 53	28 49	17 20	20 33	14 09	20 06	28 38
25	22 19 53	06 30 00	21 ♊ 13 49	27 ♊ 54 18	01 40	05 07	29 ♑ 34	17 26	20 28	14 13	20 08	28 39
26	22 23 50	07 30 20	04 ♋ 30 29	11 ♋ 02 33	00 ♓ 37	06 20	00 ♒ 20	17 33	20 24	14 16	20 10	28 40
27	22 27 46	08 30 38	17 ♋ 30 43	23 ♋ 55 09	29 ♒ 38	07 34	01 05	17 40	20 19	14 20	20 12	28 41
28	22 31 43	09 ♓ 30 53	00 ♌ 16 05	06 ♌ 33 43	28 ♒ 44	08 ♈ 47	01 ♒ 51	17 ♐ 46	20 ♌ 15	14 ♓ 23	20 ♒ 15	28 ♐ 42

Date	Moon True ☊ ° '	Moon Mean ☊ ° '	Moon ☽ Latitude ° '
01	16 ♓ 27	18 ♓ 00	03 N 32
02	16 R 23	17 57	02 37
03	16 20	17 54	01 35
04	16 18	17 50	00 N 30
05	16 D 18	17 47	00 S 36
06	16 19	17 44	01 40
07	16 21	17 41	02 39
08	16 23	17 38	03 31
09	16 24	17 35	04 15
10	16 R 25	17 31	04 48
11	16 24	17 28	05 08
12	16 22	17 25	05 15
13	16 20	17 22	05 06
14	16 17	17 19	04 40
15	16 14	17 16	03 58
16	16 11	17 12	03 00
17	16 09	17 09	01 49
18	16 08	17 06	00 S 30
19	16 D 09	17 03	00 N 52
20	16 09	17 00	02 11
21	16 10	16 56	03 20
22	16 11	16 53	04 15
23	16 12	16 50	04 53
24	16 R 12	16 47	05 14
25	16 12	16 44	05 16
26	16 11	16 41	05 01
27	16 10	16 37	04 31
28	16 ♓ 09	16 ♓ 34	03 N 48

DECLINATIONS

Date	Sun ☉ ° '	Moon ☽ ° '	Mercury ☿ ° '	Venus ♀ ° '	Mars ♂ ° '	Jupiter ♃ ° '	Saturn ♄ ° '	Uranus ♅ ° '	Neptune ♆ ° '	Pluto ♇ ° '
01	17 S 08	22 N 47	12 S 39	10 S 56	23 S 35	21 S 48	15 N 16	07 S 24	15 S 16	16 S 33
02	16 50	18 27	11 57	10 28	23 32	21 49	15 18	07 23	15 16	16 33
03	16 33	13 25	11 15	09 59	23 28	21 50	15 19	07 22	15 15	16 33
04	16 15	07 56	10 34	09 30	23 24	21 51	15 21	07 20	15 14	16 33
05	15 57	02 N 15	09 53	09 01	23 20	21 52	15 23	07 19	15 14	16 33
06	15 39	03 S 27	09 14	08 32	23 16	21 53	15 24	07 18	15 13	16 33
07	15 20	08 59	08 36	08 02	23 11	21 54	15 26	07 17	15 12	16 33
08	15 01	14 13	08 00	07 32	23 06	21 55	15 28	07 15	15 11	16 33
09	14 42	18 57	07 27	07 03	23 01	21 56	15 29	07 14	15 11	16 32
10	14 23	23 00	06 56	06 32	22 56	21 57	15 31	07 13	15 10	16 32
11	14 03	26 07	06 28	06 02	22 50	21 58	15 33	07 12	15 09	16 32
12	13 44	28 03	06 05	05 32	22 45	21 59	15 35	07 10	15 09	16 32
13	13 23	28 32	05 45	05 01	22 39	22 00	15 36	07 09	15 08	16 32
14	13 03	27 24	05 30	04 30	22 32	22 01	15 38	07 08	15 07	16 32
15	12 43	24 35	05 19	03 59	22 26	22 02	15 39	07 07	15 07	16 32
16	12 22	20 13	05 13	03 28	22 19	22 02	15 41	07 05	15 06	16 32
17	12 01	14 33	05 12	02 57	22 12	22 03	15 43	07 04	15 05	16 32
18	11 40	07 59	05 16	02 26	22 05	22 04	15 44	07 03	15 04	16 32
19	11 19	00 S 57	05 25	01 55	21 57	22 05	15 46	07 01	15 04	16 32
20	10 57	06 N 08	05 38	01 24	21 50	22 05	15 48	07 00	15 03	16 32
21	10 36	12 49	05 54	00 52	21 42	22 06	15 49	06 59	15 02	16 32
22	10 14	18 42	06 14	00 S 21	21 33	22 07	15 51	06 57	15 02	16 32
23	09 52	23 27	06 37	00 N 10	21 25	22 07	15 52	06 56	15 01	16 31
24	09 30	26 44	07 02	00 42	21 17	22 08	15 54	06 55	15 00	16 31
25	09 08	28 24	07 29	01 13	21 08	22 09	15 55	06 54	15 00	16 31
26	08 45	28 23	07 56	01 44	20 59	22 09	15 57	06 52	14 59	16 31
27	08 23	26 46	08 24	02 16	20 49	22 10	15 58	06 51	14 58	16 31
28	08 S 00	23 N 48	08 S 51	02 N 47	20 S 40	22 S 10	16 N 00	06 S 50	14 S 58	16 S 31

ZODIAC SIGN ENTRIES

Date	h	m	Planets
01	05	15	☽ ♌
02	09	20	☿ ♓
03	14	34	☽ ♍
06	02	15	☽ ♎
08	15	09	☽ ♏
11	03	01	☽ ♐
13	11	42	☽ ♑
15	16	34	☽ ♒
17	18	30	☽ ♓
19	01	09	☉ ♓
19	19	06	☽ ♈
21	08	21	♀ ♈
21	20	03	☽ ♉
23	22	42	☽ ♊
26	01	32	♂ ♒
26	03	48	☽ ♋
27	03	00	☿ ♒
28	11	29	☽ ♌

LATITUDES

Date	Mercury ☿ ° '	Venus ♀ ° '	Mars ♂ ° '	Jupiter ♃ ° '	Saturn ♄ ° '	Uranus ♅ ° '	Neptune ♆ ° '	Pluto ♇ ° '
01	00 S 45	01 S 31	00 S 39	00 N 42	01 N 17	00 S 44	00 S 14	06 N 53
04	00 S 10	01 28	00 41	00 42	01 18	00 44	00 14	06 53
07	00 N 33	01 25	00 43	00 42	01 18	00 44	00 14	06 53
10	01 21	01 21	00 45	00 42	01 18	00 44	00 14	06 53
13	02 09	01 17	00 47	00 42	01 19	00 44	00 14	06 54
16	02 54	01 12	00 49	00 42	01 19	00 44	00 14	06 54
19	03 27	01 06	00 51	00 42	01 19	00 44	00 14	06 54
22	03 43	01 00	00 52	00 42	01 19	00 44	00 14	06 55
25	03 38	00 53	00 54	00 42	01 19	00 44	00 14	06 55
28	03 16	00 46	00 56	00 43	01 20	00 44	00 14	06 55
31	02 N 41	00 S 38	00 S 58	00 N 43	01 N 20	00 S 44	00 S 14	06 N 56

DATA

Julian Date	2454133
Delta T	+66 seconds
Ayanamsa	23° 57' 27"
Synetic vernal point	05° ♓ 09' 32"
True obliquity of ecliptic	23° 26' 27"

MOON'S PHASES, APSIDES AND POSITIONS ☽

Date	h	m	Phase	Longitude ° '	Eclipse Indicator
02	05	45	○	12 ♌ 59	
10	09	51	☾	21 ♏ 16	
17	16	14	●	28 ♒ 37	
24	07	56	☽	05 ♊ 19	

Day	h	m		
07	12	32	Apogee	
19	09	30	Perigee	
05	21	27	0S	
13	07	30	Max dec	28° S 34'
19	15	11	0N	
25	23	37	Max dec	28° N 36'

LONGITUDES

Date	Chiron ⚷ ° '	Ceres ⚳ ° '	Pallas ⚴ ° '	Juno ⚵ ° '	Vesta ⚶ ° '	Black Moon Lilith ⚸ ° '
01	10 ♒ 14	12 ♓ 16	12 ♒ 32	27 ♎ 24	26 ♏ 46	11 ♎ 35
11	11 ♒ 02	16 ♓ 03	16 ♒ 00	27 ♎ 52	00 ♐ 39	12 ♎ 41
21	11 ♒ 48	19 ♓ 55	19 ♒ 25	27 ♎ 44	04 ♐ 14	13 ♎ 48
31	12 ♒ 33	23 ♓ 49	22 ♒ 46	27 ♎ 00	07 ♐ 27	14 ♎ 55

All ephemeris data is given at 12.00 UT and the Moon's longitude is additionally given for 24.00 UT

ASPECTARIAN

h m	Aspects
01 Thursday	
01 17	☽ bQ ♅
01 30	☽ ⊼ ☿
01 37	☽ ⊼ ♇
02 33	☿ ⚹ ♇
03 19	☽ ± ♀
03 25	☽ bQ ♃
06 49	☽ # ♂
12 53	☽ ± ♇
15 00	♄ # ♆
15 47	☽ ⊼ ♀
17 54	☽ # ♃
18 14	☽ ± ♅
02 Friday	
03 58	☽ ⊼ ♂
04 32	☉ ⊻ ♅
05 40	☽ ⊼ ♅
05 45	☽ ☍ ☉
05 57	☽ bQ ♇
08 05	☽ △ ♃
08 18	☉ ∠ ♇
16 09	☽ ± ♂
17 47	☽ ☍ ♆
20 30	☽ # ☉
21 25	☽ # ♇
23 34	☽ ☌ ♄
03 Saturday	
03 16	☽ ∥ ♄
03 33	☽ # ♆
10 40	☽ bQ ♂
10 41	♂ ⚹ ♅
10 55	☽ △ ♇
12 18	☉ ∥ ♇
15 47	☉ ⚹ ♃
18 26	☽ ☍ ☿
20 46	♂ ⊥ ♆
23 02	☽ # ☿
04 Sunday	
04 37	☽ # ♀
08 09	☽ ☍ ♀
14 33	☽ # ♅
16 18	☽ ☍ ♅
18 07	☽ △ ♂
19 23	☽ □ ♃
21 31	☽ ⊼ ☉
05 Monday	
04 52	☽ ⊼ ♆
07 58	♀ Q ♇
10 18	☽ ⊻ ♄
10 37	☽ ± ☉
16 57	☽ ± ♆
19 30	♂ ⊻ ♃
22 17	☽ ⊥ ♄
22 37	☽ □ ♇
06 Tuesday	
06 27	☽ bQ ☉
08 01	☽ Q ♃
11 11	☽ bQ ♆
13 26	☽ ⊼ ☿
16 22	☽ ∠ ♄
07 Wednesday	
02 52	☽ ± ☿
03 27	☽ ⊼ ♀
04 31	☽ ∥ ♅
05 01	☽ ⊼ ♅
05 05	☉ # ♄
08 10	☽ ∥ ♀
08 31	♂ ± ♄
08 42	☽ ⚹ ♃
10 28	☽ ∥ ☿
10 41	☽ □ ♂
11 12	☽ Q ♇
15 45	☽ △ ☉
17 05	☽ ± ♀
17 17	☽ ± ♅
17 45	☽ △ ♆
18 57	♀ ☌ ♅
22 39	☽ ⚹ ♄
22 47	☉ ∥ ♆
22 55	☽ bQ ☿
08 Thursday	
11 38	☽ ⚹ ♇
11 39	☽ bQ ♅
13 31	☽ bQ ♀
15 34	☽ ∠ ♃
15 42	☽ ∥ ☉
15 52	☉ ☌ ♆
16 45	☽ ∥ ♆
18 07	☽ # ♄
22 50	☽ Q ♄
23 28	☽ ∥ ♇
09 Friday	
01 41	☽ Q ♂
02 14	♀ ∥ ♅
07 46	☽ △ ☿
09 59	☽ ⊥ ♃
11 31	♀ □ ♃
17 58	☽ ∠ ♇
18 03	☽ △ ♅
21 44	☿ ∥ ♅
22 10	☽ ⊻ ♃
23 14	☽ △ ♀
10 Saturday	
03 22	☽ ⚹ ♂
05 20	☽ ∥ ♃
06 28	☽ □ ♆
09 51	☽ □ ☉
10 39	☽ □ ♄
11 37	☽ ∥ ♂
11 55	☽ ⊥ ♇
18 42	☉ ☍ ♄
23 45	☽ ⊻ ♇
11 Sunday	
10 46	☽ ∠ ♂
17 46	☽ Q ♆
21 54	☽ □ ☿
12 Monday	
00 01	☽ Q ☉
04 57	☽ □ ♅
05 07	☽ ⊥ ♂
07 36	♂ ⊻ ♆
09 16	☽ ☌ ♃
15 52	☽ □ ♀
16 35	☽ ⚹ ♆
17 07	☽ ⊻ ♂
19 59	☽ △ ♄
23 30	♀ ⊻ ♆
13 Tuesday	
00 31	☽ ⚹ ☉
08 21	☽ Q ☿
08 45	☽ ☌ ♇
14 31	☽ Q ♅
20 13	☽ ∠ ♆
22 47	♀ ⚹ ♂
23 16	☽ bQ ♄
14 Wednesday	
04 12	☽ Q ♀
04 38	☿ St R
06 00	☽ ∠ ☉
06 08	☽ ⚹ ☿
08 17	♀ ⊼ ♄
12 07	☽ ⚹ ♅
12 20	☽ ⊥ ♆
13 49	♂ ⊼ ♄
15 10	☽ ± ♄
16 29	☽ ⊻ ♃
22 53	☽ ⊻ ♆
23 08	☽ ⊥ ☉
15 Thursday	
01 36	☽ ⊼ ♄
02 21	☽ ☌ ♂
03 03	☽ ⊥ ♃
03 24	☽ ⚹ ♀
08 12	☽ ∠ ☿
10 19	☽ ⊻ ☉
13 55	☽ ⊻ ♇
14 17	☽ ∠ ♅
18 12	♂ ⊥ ♃
18 40	☽ ∠ ♃
23 17	☽ ⊥ ☿
16 Friday	
00 08	☽ ⊥ ♇
01 08	☽ ∥ ♂
02 55	☽ ∥ ♃
05 36	☽ ⊥ ♅
07 26	☽ ∠ ♀
09 09	☽ ⊻ ☿
15 15	☽ ∠ ♇
15 41	☽ ⊻ ♅
20 05	☽ ⚹ ♃
23 49	☽ ⊥ ♀
17 Saturday	
01 47	☽ ☌ ♆
03 57	☽ ☍ ♄
04 06	☽ ∥ ♇
07 27	☽ # ♄
07 41	☽ ⊻ ♂
09 55	☽ ∥ ♆
10 37	☽ ⊻ ♀
12 32	☉ Q ♃
13 07	☉ ⚹ ♇
16 01	☽ Q ♃
16 02	☽ ⚹ ♇
16 14	☽ ☌ ☉
18 03	☽ ⊥ ♂
22 07	☽ ∥ ☉
23 40	♀ ⊥ ♆
18 Sunday	
01 13	☿ ∠ ♂
08 43	☽ ☌ ☿
09 29	☽ ∠ ♂
11 35	☽ Q ♇
14 35	♂ ∥ ♃
15 17	☽ ∥ ♅
16 58	☽ ☌ ♅
21 13	☽ ∥ ☿
21 30	☽ □ ♃
21 41	♀ ± ♄
19 Monday	
02 46	☽ ⊻ ♆
04 30	☽ ⊼ ♄
08 29	☽ ∥ ♀
11 02	☽ ⚹ ♂
12 29	☽ ⊥ ♆
14 08	☽ ± ♄
15 43	☽ ☌ ♀
16 43	☽ □ ♇
20 24	☽ ⊻ ☉
20 59	☽ # ♀
20 Tuesday	
03 05	☽ ∠ ♆
03 57	♀ □ ♇
04 37	☽ bQ ♄
06 44	☽ ⊻ ☿
06 50	☽ ⊥ ☉
07 30	☽ Q ♂
10 12	☽ # ☿
15 01	☽ # ♅
15 54	☽ ⊥ ☿
17 43	☽ ⊻ ♅
22 32	☽ △ ♃
22 33	☽ ∠ ☉
21 Wednesday	
03 33	☽ ⊥ ♅
03 33	☽ ⚹ ♆
04 12	☽ # ☉
04 55	☽ △ ♄
05 40	☽ ∠ ☿
14 29	☽ □ ♂
17 42	☽ △ ♇
18 24	☽ ∠ ♅
19 21	☉ ⊥ ♂
20 36	☽ # ♆
21 08	☽ ⊻ ♀
23 22	☽ Q ♆
23 24	☽ bQ ♃
23 47	☽ ∥ ♄
22 Thursday	
01 04	☽ ⚹ ☉
02 38	☽ # ♇
04 47	☽ ⚹ ☿
08 02	☽ ⊥ ♀
14 33	☽ ± ♃
18 44	☽ bQ ♇
19 31	☽ ⚹ ♅
22 40	☽ Q ☉
23 30	☽ Q ☿
23 Friday	
00 35	☽ ∠ ♀
00 44	☽ ⊼ ♃
01 14	☽ # ♂
04 34	☽ # ♃
04 45	☉ ☌ ☿
05 39	☽ □ ♆
06 39	☽ □ ♄
09 18	☿ ⊥ ♂
10 00	☽ ± ♇
16 00	☽ Q ♅
19 47	☽ △ ♂
20 19	☽ ⊼ ♇
24 Saturday	
00 22	☿ ⊻ ♀
04 05	☽ □ ☿
04 47	☽ ⚹ ♀
05 16	☿ ∥ ♅
05 49	♂ ⊻ ♇
07 56	☽ □ ☉
13 33	☽ Q ♄
23 28	☽ bQ ♂
23 29	☽ □ ♅
23 42	♂ ∠ ♅
25 Sunday	
03 55	☽ Q ♀
05 11	☽ ☍ ♃
10 02	☽ △ ♆
10 39	☽ ⚹ ♄
12 29	♀ ∠ ♆
16 27	☽ ± ♂
18 37	♀ bQ ♄
26 Monday	
01 21	☽ ☍ ♇
03 56	☽ ⊼ ♂
05 26	☽ △ ☿
13 13	☽ bQ ♆
13 36	☽ ∠ ♄
15 03	☿ ⊥ ♀
15 42	☽ □ ♀
15 58	☿ ⊻ ♂
17 57	☽ △ ☉
27 Tuesday	
05 50	☽ ± ♆
06 03	☽ △ ♅
06 06	☽ ⊥ ♄
07 02	☽ bQ ☿
11 29	☉ ∥ ☿
11 31	☿ Q ♃
12 17	☽ ⊼ ♃
17 03	☽ ⊼ ♆
17 13	☽ ⊻ ♄
22 42	☽ ± ☿
23 37	☽ ± ♃
28 Wednesday	
00 11	☽ bQ ☉
09 01	☽ ⊼ ♇
09 17	☽ ⊼ ☿
10 19	☽ bQ ♅
12 01	♄ ☍ ♆
13 09	☿ ⚹ ♇
15 11	☽ ☍ ♂
16 48	☽ bQ ♃
18 43	☽ ± ☉
20 27	☽ ± ♇
22 21	☽ # ♃

MARCH 2007

LONGITUDES

Date	Sidereal time h m s	Sun ☉ ° ' "	Moon ☽ ° ' "	Moon ☽ 24.00 ° ' "	Mercury ☿ ° '	Venus ♀ ° '	Mars ♂ ° '	Jupiter ♃ ° '	Saturn ♄ ° '	Uranus ♅ ° '	Neptune ♆ ° '	Pluto ♇ ° '
01	22 35 39	10 ♓ 31 07	12 ♌ 48 15	18 ♌ 59 52	27 ♒ 56	10 ♈ 01	02 ♒ 36	17 ♐ 52	20 ♌ 10	14 ♓ 27	20 ♒ 17	28 ♐ 43
02	22 39 36	11 31 19	25 ♌ 08 47	01 ♍ 15 12	27 R 14	11 15	03 22	17 59	20 R 06	14 30	20 19	28 44
03	22 43 33	12 31 29	07 ♍ 19 17	13 ♍ 21 16	26 39	12 28	04 07	18 05	20 01	14 33	20 21	28 45
04	22 47 29	13 31 37	19 ♍ 21 21	25 ♍ 19 47	26 10	13 41	04 53	18 10	19 57	14 37	20 23	28 46
05	22 51 26	14 31 43	01 ♎ 16 48	07 ♎ 12 40	25 49	14 55	05 38	18 16	19 53	14 40	20 25	28 47
06	22 55 22	15 31 47	13 ♎ 07 41	19 ♎ 02 10	25 34	16 08	06 24	18 22	19 49	14 44	20 27	28 47
07	22 59 19	16 31 50	24 ♎ 56 28	00 ♏ 50 58	25 26	17 21	07 10	18 27	19 45	14 47	20 30	28 48
08	23 03 15	17 31 51	06 ♏ 46 05	12 ♏ 42 15	25 D 25	18 34	07 55	18 32	19 41	14 51	20 32	28 49
09	23 07 12	18 31 50	18 ♏ 39 56	24 ♏ 39 39	25 30	19 47	08 41	18 37	19 37	14 54	20 34	28 50
10	23 11 08	19 31 48	00 ♐ 41 53	06 ♐ 47 11	25 41	21 00	09 27	18 42	19 33	14 58	20 36	28 50
11	23 15 05	20 31 44	12 ♐ 56 06	19 ♐ 09 10	25 57	22 13	10 12	18 47	19 29	15 01	20 38	28 51
12	23 19 02	21 31 39	25 ♐ 26 54	01 ♑ 49 49	26 19	23 26	10 58	18 51	19 25	15 04	20 40	28 52
13	23 22 58	22 31 32	08 ♑ 18 23	14 ♑ 53 01	26 46	24 39	11 44	18 56	19 21	15 08	20 42	28 52
14	23 26 55	23 31 23	21 ♑ 34 03	28 ♑ 21 43	27 18	25 52	12 30	19 00	19 18	15 11	20 44	28 53
15	23 30 51	24 31 13	05 ♒ 16 08	12 ♒ 17 19	27 54	27 05	13 15	19 04	19 14	15 15	20 46	28 54
16	23 34 48	25 31 00	19 ♒ 25 04	26 ♒ 39 04	28 34	28 17	14 01	19 08	19 11	15 18	20 48	28 54
17	23 38 44	26 30 46	03 ♓ 58 47	11 ♓ 23 33	29 ♒ 18	29 ♈ 30	14 47	19 11	19 07	15 21	20 50	28 55
18	23 42 41	27 30 30	18 ♓ 52 31	26 ♓ 24 41	00 ♓ 05	00 ♉ 42	15 33	19 15	19 04	15 25	20 52	28 55
19	23 46 37	28 30 12	03 ♈ 58 56	11 ♈ 34 06	00 56	01 55	16 19	19 18	19 01	15 28	20 54	28 55
20	23 50 34	29 ♓ 29 52	19 ♈ 08 57	26 ♈ 42 17	01 50	03 07	17 05	19 21	18 58	15 32	20 55	28 56
21	23 54 30	00 ♈ 29 30	04 ♉ 12 59	11 ♉ 40 02	02 47	04 19	17 51	19 24	18 55	15 35	20 57	28 56
22	23 58 27	01 29 06	19 ♉ 02 31	26 ♉ 19 44	03 46	05 32	18 36	19 27	18 52	15 38	20 59	28 57
23	00 02 24	02 28 40	03 ♊ 31 08	10 ♊ 36 21	04 49	06 44	19 22	19 30	18 49	15 42	21 01	28 57
24	00 06 20	03 28 11	17 ♊ 35 10	24 ♊ 27 33	05 54	07 56	20 08	19 32	18 46	15 45	21 03	28 57
25	00 10 17	04 27 40	01 ♋ 13 34	07 ♋ 53 26	07 01	09 08	20 54	19 34	18 44	15 48	21 05	28 57
26	00 14 13	05 27 07	14 ♋ 27 27	20 ♋ 55 57	08 10	10 20	21 40	19 36	18 41	15 52	21 06	28 58
27	00 18 10	06 26 33	27 ♋ 19 22	03 ♌ 38 08	09 22	11 32	22 26	19 38	18 38	15 55	21 08	28 58
28	00 22 06	07 25 53	09 ♌ 52 42	16 ♌ 03 31	10 36	12 43	23 12	19 40	18 36	15 58	21 10	28 58
29	00 26 03	08 25 13	22 ♌ 11 03	28 ♌ 15 44	11 51	13 55	23 58	19 41	18 34	16 01	21 12	28 58
30	00 29 59	09 24 30	04 ♍ 17 57	10 ♍ 18 08	13 09	15 07	24 44	19 43	18 32	16 05	21 13	28 58
31	00 33 56	10 ♈ 23 45	16 ♍ 16 36	22 ♍ 13 43	14 ♓ 28	16 ♉ 18	25 ♒ 30	19 ♐ 44	18 ♌ 30	16 ♓ 08	21 ♒ 15	28 ♐ 58

Date	Moon True ☊ ° '	Moon Mean ☊ ° '	Moon ☽ Latitude ° '
01	16 ♓ 09	16 ♓ 31	02 N 55
02	16 R 08	16 28	01 54
03	16 08	16 25	00 N 49
04	16 D 08	16 22	00 S 18
05	16 08	16 18	01 23
06	16 08	16 15	02 24
07	16 R 08	16 12	03 19
08	16 08	16 09	04 06
09	16 07	16 06	04 42
10	16 07	16 02	05 06
11	16 07	15 59	05 17
12	16 D 07	15 56	05 13
13	16 07	15 53	04 54
14	16 08	15 50	04 19
15	16 08	15 47	03 28
16	16 09	15 43	02 23
17	16 10	15 40	01 S 07
18	16 R 10	15 37	00 N 15
19	16 10	15 34	01 37
20	16 08	15 31	02 53
21	16 07	15 28	03 56
22	16 05	15 24	04 42
23	16 03	15 21	05 09
24	16 02	15 18	05 16
25	16 01	15 15	05 05
26	16 D 01	15 12	04 38
27	16 02	15 08	03 58
28	16 04	15 05	03 07
29	16 05	15 02	02 08
30	16 07	14 59	01 05
31	16 ♓ 07	14 ♓ 56	00 S 01

DECLINATIONS

Date	Sun ☉ ° '	Moon ☽ ° '	Mercury ☿ ° '	Venus ♀ ° '	Mars ♂ ° '	Jupiter ♃ ° '	Saturn ♄ ° '	Uranus ♅ ° '	Neptune ♆ ° '	Pluto ♇ ° '
01	07 S 37	19 N 46	09 S 17	03 N 18	20 S 30	22 S 11	16 N 01	06 S 48	14 S 57	16 S 31
02	07 15	14 56	09 43	03 49	20 20	22 12	16 03	06 47	14 56	16 31
03	06 52	09 35	10 06	04 20	20 10	22 12	16 04	06 46	14 56	16 31
04	06 29	03 N 56	10 28	04 51	20 00	22 12	16 05	06 44	14 55	16 30
05	06 05	01 S 47	10 48	05 22	19 49	22 13	16 07	06 43	14 54	16 30
06	05 42	07 24	11 06	05 53	19 39	22 13	16 08	06 42	14 54	16 30
07	05 19	12 45	11 22	06 24	19 28	22 14	16 09	06 40	14 53	16 30
08	04 56	17 38	11 36	06 54	19 18	22 14	16 11	06 39	14 52	16 30
09	04 32	21 54	11 48	07 25	19 05	22 14	16 12	06 38	14 52	16 30
10	04 09	25 17	11 57	07 55	18 54	22 15	16 13	06 36	14 51	16 30
11	03 45	27 36	12 04	08 25	18 42	22 15	16 14	06 35	14 50	16 30
12	03 22	28 35	12 09	08 55	18 30	22 15	16 16	06 34	14 50	16 30
13	02 58	28 04	12 12	09 24	18 18	22 16	16 17	06 32	14 49	16 29
14	02 34	25 58	12 13	09 54	18 06	22 16	16 18	06 31	14 48	16 29
15	02 11	22 19	12 12	10 23	17 53	22 16	16 19	06 30	14 48	16 29
16	01 47	17 16	12 09	10 52	17 40	22 17	16 20	06 28	14 47	16 29
17	01 23	11 06	12 05	11 20	17 28	22 17	16 21	06 27	14 47	16 29
18	00 59	04 S 10	11 58	11 49	17 15	22 17	16 22	06 26	14 46	16 29
19	00 36	03 N 04	11 50	12 17	17 01	22 17	16 23	06 24	14 45	16 29
20	00 S 12	10 10	11 40	12 45	16 48	22 18	16 24	06 23	14 45	16 29
21	00 N 12	16 37	11 28	13 13	16 35	22 18	16 25	06 22	14 44	16 28
22	00 35	22 00	11 15	13 40	16 21	22 18	16 26	06 21	14 44	16 28
23	00 59	25 54	11 00	14 07	16 07	22 18	16 27	06 19	14 43	16 28
24	01 23	28 06	10 43	14 34	15 53	22 18	16 28	06 18	14 42	16 28
25	01 46	28 31	10 25	15 00	15 39	22 18	16 29	06 17	14 42	16 28
26	02 10	27 16	10 06	15 26	15 25	22 18	16 29	06 15	14 41	16 28
27	02 33	24 35	09 45	15 51	15 10	22 19	16 30	06 14	14 41	16 28
28	02 57	20 46	09 23	16 17	14 56	22 19	16 31	06 13	14 40	16 28
29	03 20	16 08	08 59	16 41	14 41	22 19	16 32	06 12	14 40	16 27
30	03 44	10 56	08 34	17 06	14 26	22 19	16 32	06 10	14 39	16 27
31	04 N 07	05 N 24	08 S 07	17 N 30	14 S 11	22 S 19	16 N 33	06 S 09	14 S 39	16 S 27

ZODIAC SIGN ENTRIES

Date	h	m	Planets
02	21	32	☽ ♍
05	09	25	☽ ♎
07	22	17	☽ ♏
10	10	37	☽ ♐
12	20	35	☽ ♑
15	02	52	☽ ♒
17	05	30	☽ ♓
17	22	00	♀ ♉
18	09	35	☿ ♓
19	05	42	☽ ♈
21	00	07	☉ ♈
21	05	15	☽ ♉
23	06	06	☽ ♊
25	09	49	☽ ♋
27	17	04	☽ ♌
30	03	27	☽ ♍

LATITUDES

Date	Mercury ☿ ° '	Venus ♀ ° '	Mars ♂ ° '	Jupiter ♃ ° '	Saturn ♄ ° '	Uranus ♅ ° '	Neptune ♆ ° '	Pluto ♇ ° '
01	03 N 05	00 S 43	00 S 57	00 N 43	01 N 20	00 S 44	00 S 14	06 N 55
04	02 28	00 35	00 59	00 43	01 20	00 44	00 14	06 56
07	01 46	00 27	01 01	00 43	01 20	00 44	00 14	06 56
10	01 05	00 19	01 02	00 43	01 20	00 44	00 14	06 56
13	00 N 25	00 S 10	01 04	00 43	01 20	00 44	00 14	06 57
16	00 S 12	00 00	01 06	00 43	01 20	00 44	00 14	06 57
19	00 44	00 N 09	01 08	00 43	01 20	00 44	00 14	06 58
22	01 12	00 19	01 09	00 43	01 20	00 44	00 14	06 58
25	01 36	00 29	01 11	00 44	01 20	00 44	00 14	06 58
28	01 56	00 39	01 13	00 44	01 20	00 44	00 14	06 59
31	02 S 11	00 N 49	01 S 14	00 N 44	01 N 20	00 S 44	00 S 14	06 N 59

DATA

Julian Date	2454161
Delta T	+66 seconds
Ayanamsa	23° 57' 31"
Synetic vernal point	05° ♓ 09' 28"
True obliquity of ecliptic	23° 26' 27"

MOON'S PHASES, APSIDES AND POSITIONS ☽

Date	h	m	Phase	Longitude ° '	Eclipse Indicator
03	23	17	○	13 ♍ 00	total
12	03	54	☾	21 ♐ 11	
19	02	43	●	28 ♓ 07	Partial
25	18	16	☽	04 ♋ 43	

Day	h	m		
07	03	24	Apogee	
19	18	33	Perigee	
05	04	31	0S	
12	16	10	Max dec	28° S 36'
19	01	51	0N	
25	05	40	Max dec	28° N 35'

LONGITUDES

Date	Chiron ⚷ ° '	Ceres ⚳ ° '	Pallas ⚴ ° '	Juno ⚵ ° '	Vesta ⚶ ° '	Black Moon Lilith ⚸ ° '
01	12 ♒ 24	23 ♓ 02	22 ♒ 06	27 ♎ 11	06 ♐ 51	14 ♎ 41
11	13 ♒ 06	26 ♓ 57	25 ♒ 24	25 ♎ 58	09 ♐ 43	15 ♎ 48
21	13 ♒ 44	00 ♈ 54	28 ♒ 36	24 ♎ 13	12 ♐ 04	16 ♎ 55
31	14 ♒ 18	04 ♈ 51	01 ♓ 41	22 ♎ 04	13 ♐ 48	18 ♎ 01

ASPECTARIAN

h m Aspects

01 Thursday
03 34 ☽ ± ♅
06 03 ☽ △ ♀
07 13 ☽ ⊼ ☉
07 48 ☽ # ♂
13 46 ☽ bQ ♇
15 11 ☽ ⊼ ♅
16 47 ☉ Q ♇
21 54 ☽ △ ♃
21 57 ♂ ∠ ♃

02 Friday
02 12 ☽ ☌ ♄
02 32 ☽ ☍ ♆
04 28 ☽ # ♇
06 45 ☽ ∥ ♄
11 58 ☽ # ♆
14 23 ☽ bQ ♀
15 53 ☽ ☍ ☿
19 03 ☽ △ ♇

03 Saturday
00 55 ☿ ∠ ♀
05 14 ☽ ⊼ ♂
09 51 ☽ # ☿
10 07 ☽ ± ♀
17 56 ☽ ± ♂
18 24 ☉ ⊻ ♀
18 46 ☉ ∥ ♅
23 17 ☽ ☍ ☉
23 23 ☽ ⊼ ♀

04 Sunday
00 08 ☽ # ♅
00 30 ☽ # ☉
02 28 ☽ ☍ ♅
08 13 ♂ ⊥ ♇
08 27 ☽ ∥ ♀
09 37 ☽ □ ♃
13 07 ☽ bQ ♂
13 11 ☽ ⊻ ♄
14 04 ☽ ⊼ ♆

05 Monday
01 11 ☽ ⊥ ♄
01 16 ☽ ⊼ ☿
02 10 ☽ ± ♆
06 56 ☽ □ ♇
07 06 ♀ ⊻ ♅
13 03 ☽ ± ☿
15 39 ☉ ☌ ♅
19 14 ☽ ∠ ♄
20 24 ☽ bQ ♆
21 25 ☽ △ ♂
22 10 ☽ Q ♃

06 Tuesday
04 46 ☽ # ♀
05 08 ☽ ∥ ☉
06 53 ☽ bQ ☿
07 13 ☉ # ♀
08 56 ☽ ∥ ♅
15 16 ☽ ⊼ ♅
17 20 ☽ ⊼ ☉
18 48 ☽ ☍ ♀
19 27 ☽ Q ♇
22 42 ☽ ✶ ♃

07 Wednesday
01 30 ☽ ✶ ♄
02 56 ☽ △ ♆
03 31 ☽ ± ♅
05 21 ☽ ∥ ☿
06 39 ☽ ± ☉
13 00 ☽ △ ☿
19 51 ☽ ✶ ♇
21 53 ☽ bQ ♅
22 07 ☽ ∥ ♆

08 Thursday
00 28 ♀ # ♅
01 44 ☽ Q ♄
02 37 ☽ bQ ☉
04 30 ☽ # ♄
04 45 ☿ St D
05 24 ☽ ∠ ♃
06 09 ☽ ∥ ♇
11 16 ♀ △ ♃
14 30 ☽ □ ♂
20 23 ☽ ∥ ♂
23 45 ☽ ⊥ ♃

09 Friday
02 16 ☽ ∠ ♇
04 24 ☽ △ ♅
08 41 ♀ △ ♄
11 42 ☽ △ ☉
11 54 ☽ ⊻ ♃
13 53 ☽ □ ♄
14 11 ☽ ∥ ♃
14 19 ☉ □ ♃
14 30 ☽ ⊼ ♀
15 49 ☽ □ ♆
19 30 ♂ ⊥ ♅
20 21 ☽ ⊥ ♇

10 Saturday
01 51 ☽ □ ☿
03 41 ♀ ✶ ♆
03 51 ☽ ± ♀
05 06 ☽ Q ♂
08 19 ☽ ⊻ ♇
11 02 ♀ ⊥ ♅
12 21 ☉ ⊼ ♄
23 37 ☽ bQ ♀

11 Sunday
03 35 ☽ Q ♆
06 20 ☽ ✶ ♂
11 09 ♀ Q ♂
14 02 ☽ Q ☿
14 31 ☉ ⊻ ♆
16 03 ☽ □ ♅
23 21 ☽ ☌ ♃

12 Monday
00 34 ☽ △ ♄
02 52 ☽ ✶ ♆
03 54 ☽ □ ☉
07 47 ☽ △ ♀
13 03 ☽ ∠ ♂
13 42 ☽ ✶ ☿
18 27 ☽ ☌ ♇

13 Tuesday
02 23 ☽ Q ♅
04 44 ☽ bQ ♄
06 57 ☽ ⊥ ♂
07 10 ☽ ∠ ♆
16 24 ☽ Q ☉
18 35 ☽ ∠ ☿
18 39 ☽ ⊻ ♂
21 11 ☽ ± ♄
23 42 ☽ ⊥ ♆

14 Wednesday
00 30 ☽ ✶ ♅
07 23 ☽ ⊻ ♃
07 58 ☽ ⊼ ♄
10 30 ☽ ⊻ ♆
11 30 ☽ ⊥ ☿
15 45 ☽ ✶ ☉
18 07 ☽ ⊥ ♃
20 21 ☽ □ ♀
22 35 ☽ ⊻ ☿

15 Thursday
00 55 ☽ ⊻ ♇
03 15 ☽ ∠ ♅
09 55 ☽ ∠ ♃
11 21 ☽ ⊥ ♇
12 13 ☽ ∥ ♃
18 51 ☽ ⊥ ♅
19 51 ☽ ∠ ☉

16 Friday
02 25 ☽ ☌ ♂
02 44 ☽ ∠ ♇
04 21 ☉ ± ♄
05 04 ☽ ⊻ ♅
06 16 ☽ Q ♀
08 13 ♂ ∠ ♇
10 10 ☽ ∥ ♂
11 31 ☽ ✶ ♃
11 36 ☽ ☍ ♄
12 11 ☽ ⊥ ☉
14 18 ☽ ☌ ♆
15 15 ☽ ∥ ♇
15 51 ☽ # ♄
22 06 ☽ ∥ ♆
22 42 ♃ △ ♄
22 52 ☽ ⊻ ☉
23 34 ☿ ✶ ♇

17 Saturday
00 18 ♀ △ ♇
01 23 ☿ ✶ ♀
03 43 ☽ ✶ ♇
03 56 ☽ ☌ ☿
04 01 ☽ ✶ ♀
07 26 ☽ Q ♃
08 20 ☽ ∥ ☿
11 09 ☽ # ♀
19 52 ☉ ⊥ ♆
23 14 ☽ Q ♇

18 Sunday
04 21 ☽ ∥ ♅
05 57 ♀ ∠ ♅
06 23 ☽ ⊻ ♂
06 26 ☽ ☌ ♅
06 29 ☽ ∠ ♀
07 26 ♂ ⊻ ♅
12 19 ☽ ⊼ ♄
12 36 ☽ □ ♃
15 11 ☽ ⊻ ♆
16 30 ☽ ⊥ ♂
18 12 ☿ # ♀
21 50 ☽ ± ♄
22 06 ☽ ⊥ ♀
23 12 ☽ ∥ ☉

19 Monday
00 44 ☽ ⊥ ♆
02 43 ☽ ☌ ☉
03 59 ☽ □ ♇
04 15 ☽ # ☉
06 52 ☽ ⊻ ☿
07 33 ☽ ∠ ♂
08 26 ☽ ⊻ ♀
12 03 ☽ bQ ♄
15 02 ☽ ∠ ♆
16 57 ☽ ⊥ ☿
22 14 ☉ □ ♇
22 46 ☿ Q ♃
23 06 ☽ # ♅

20 Tuesday
06 14 ☽ ⊻ ♅
08 02 ♀ Q ♆
08 05 ☽ ∠ ☿
08 33 ☽ ✶ ♂
11 42 ☽ △ ♄
12 20 ☽ △ ♃
14 49 ☽ ✶ ♆
15 47 ☽ ⊥ ♅
17 11 ☽ # ☿
21 57 ☽ ∥ ♀

21 Wednesday
03 33 ☽ △ ♇
04 38 ☽ Q ♂
04 39 ☽ # ♆
05 37 ☽ ⊻ ☉
06 10 ☽ ∠ ♅
09 32 ☽ ✶ ☿
09 59 ☽ Q ♆
11 11 ☽ ∥ ♄
11 25 ☽ # ♇
11 49 ☽ # ♂
12 11 ☽ ☌ ♀
12 18 ☽ bQ ♃
13 41 ♀ bQ ♃
15 55 ☽ ⊥ ☉
22 53 ♂ ∥ ♇

22 Thursday
02 52 ☽ ± ♃
03 32 ♂ # ♄
03 41 ☽ bQ ♇
06 16 ☽ Q ☿
06 25 ☽ ✶ ♅
07 31 ☽ ∠ ☉
11 15 ☽ □ ♂
11 43 ☽ □ ♄
12 40 ☽ ⊼ ♃
13 34 ☽ # ♃
15 12 ☽ □ ♆
18 24 ☽ ± ♇
19 34 ♂ ☍ ♄

23 Friday
02 14 ☽ Q ♅
04 21 ☽ ⊼ ♇
10 07 ☽ ✶ ☉
14 21 ☽ □ ☿
16 01 ♂ ✶ ♃
17 33 ☽ Q ♄
17 55 ☽ ⊻ ♀

24 Saturday
05 06 ☽ ⊥ ♀
08 04 ☽ Q ☉
08 49 ☽ □ ♅
14 03 ☽ ✶ ♄
15 24 ☽ ☍ ♃
16 42 ☽ △ ♂
18 02 ☽ △ ♆
18 57 ☉ bQ ♄
19 22 ♄ # ♇
19 58 ♀ # ♆
22 13 ☽ ∠ ♀

25 Sunday
07 57 ☽ ☍ ♇
16 28 ☽ ∠ ♄
17 39 ♂ ☌ ♆
18 16 ☽ □ ☉
20 44 ☽ bQ ♆
20 55 ☽ bQ ♂
23 24 ☽ △ ☿

26 Monday
03 41 ☽ ✶ ♀
08 45 ☽ ⊥ ♄
11 19 ♀ # ♂
13 12 ☽ ± ♆
14 22 ☽ ± ♂
14 36 ☽ △ ♅
19 47 ☽ ⊻ ♄
21 33 ☽ ⊼ ♃

27 Tuesday
00 21 ☽ ⊼ ♆
02 13 ☽ ⊼ ♂
04 07 ☽ Q ♀
04 21 ☉ ∠ ♆
05 51 ☽ bQ ☿
08 49 ☽ ± ♃
15 06 ☽ ⊼ ♇
18 50 ☽ bQ ♅

28 Wednesday
00 43 ☽ ± ☿
01 57 ☽ bQ ♃
02 32 ☽ ± ♇
03 01 ☽ # ♃
06 52 ☽ △ ☉
12 10 ☽ ± ♅
13 32 ☽ ⊼ ☿
18 06 ☽ □ ♀
19 06 ☿ Q ♇
19 55 ☽ bQ ♇
22 32 ♀ # ♇
23 53 ☽ ⊼ ♅

29 Thursday
02 10 ♀ ∥ ♄
04 55 ☽ ☌ ♄
07 06 ☽ △ ♃
07 20 ♀ ± ♃
09 31 ☽ ∥ ♀
10 03 ☽ ☍ ♆
10 07 ☽ ∥ ♄
10 27 ☽ # ♇
12 57 ♀ bQ ♇
13 52 ♂ ∥ ♆
14 39 ☽ bQ ☉
15 45 ☽ ☍ ♂
19 02 ☽ # ♆
19 18 ☽ # ♂

30 Friday
01 24 ☽ △ ♇
10 04 ☽ ± ☉
23 08 ☽ ⊼ ☉
23 19 ☽ # ☿

31 Saturday
07 55 ☽ ☍ ☿
08 22 ♀ ✶ ♅
08 47 ☽ # ♅
11 42 ☽ ☍ ♅
12 03 ☽ △ ♀
16 27 ☽ ⊻ ♄
17 07 ☽ ∥ ☉
18 58 ☽ □ ♃
22 03 ☽ ⊼ ♆
22 45 ♇ St R
23 06 ☉ ⊥ ♀
23 22 ☉ ∠ ♂

All ephemeris data is given at 12.00 UT and the Moon's longitude is additionally given for 24.00 UT

APRIL 2007

Date	Sidereal time h m s	LONGITUDES Sun ☉ ° ' "	Moon ☽ ° ' "	Moon ☽ 24.00 ° ' "	Mercury ☿ ° '	Venus ♀ ° '	Mars ♂ ° '	Jupiter ♃ ° '	Saturn ♄ ° '	Uranus ♅ ° '	Neptune ♆ ° '	Pluto ♇ ° '
01	00 37 53	11 ♈ 22 58	28 ♍ 09 48	04 ♎ 05 07	15 ♓ 50	17 ♉ 30	26 ♒ 16	19 ♐ 45	18 ♌ 28	16 ♓ 11	21 ♒ 17	28 ♐ 58
02	00 41 49	12 22 09	09 ♎ 59 56	15 ♎ 54 32	17 13	18 41	27 02	19 46	18 R 26	16 14	21 18	28 R 58
03	00 45 46	13 21 18	21 ♎ 49 09	27 ♎ 44 02	18 37	19 52	27 48	19 46	18 24	16 17	21 20	28 58
04	00 49 42	14 20 25	03 ♏ 39 25	09 ♏ 35 33	20 04	21 03	28 34	19 47	18 22	16 20	21 21	28 58
05	00 53 39	15 19 30	15 ♏ 32 42	21 ♏ 31 09	21 31	22 14	29 ♒ 20	19 47	18 21	16 24	21 23	28 58
06	00 57 35	16 18 33	27 ♏ 31 10	03 ♐ 33 07	23 01	23 25	00 ♓ 06	19 R 47	18 19	16 27	21 24	28 58
07	01 01 32	17 17 34	09 ♐ 37 18	15 ♐ 44 07	24 32	24 36	00 52	19 47	18 18	16 30	21 26	28 57
08	01 05 28	18 16 34	21 ♐ 53 57	28 ♐ 07 14	26 05	25 47	01 38	19 46	18 16	16 33	21 27	28 57
09	01 09 25	19 15 32	04 ♑ 24 23	10 ♑ 45 50	27 39	26 57	02 24	19 46	18 15	16 36	21 29	28 57
10	01 13 22	20 14 28	17 ♑ 12 04	23 ♑ 43 28	29 ♓ 15	28 08	03 10	19 45	18 14	16 39	21 30	28 57
11	01 17 18	21 13 22	00 ♒ 20 29	07 ♒ 03 27	00 ♈ 52	29 ♉ 18	03 56	19 44	18 13	16 42	21 31	28 56
12	01 21 15	22 12 15	13 ♒ 52 39	20 ♒ 48 19	02 31	00 ♊ 29	04 42	19 43	18 12	16 45	21 33	28 56
13	01 25 11	23 11 05	27 ♒ 50 31	04 ♓ 59 12	04 12	01 39	05 28	19 42	18 12	16 48	21 34	28 56
14	01 29 08	24 09 54	12 ♓ 14 10	19 ♓ 35 01	05 54	02 49	06 14	19 40	18 11	16 51	21 35	28 55
15	01 33 04	25 08 42	27 ♓ 01 09	04 ♈ 31 49	07 38	03 59	07 01	19 38	18 10	16 54	21 37	28 55
16	01 37 01	26 07 27	12 ♈ 06 01	19 ♈ 42 39	09 23	05 09	07 47	19 36	18 10	16 56	21 38	28 54
17	01 40 57	27 06 10	27 ♈ 20 28	04 ♉ 58 07	11 09	06 19	08 33	19 34	18 10	16 59	21 39	28 54
18	01 44 54	28 04 52	12 ♉ 34 17	20 ♉ 07 39	12 58	07 28	09 19	19 32	18 09	17 02	21 40	28 53
19	01 48 51	29 ♈ 03 32	27 ♉ 37 01	05 ♊ 01 20	14 48	08 38	10 05	19 30	18 09	17 05	21 41	28 53
20	01 52 47	00 ♉ 02 09	12 ♊ 19 42	19 ♊ 31 29	16 39	09 47	10 51	19 27	18 D 09	17 08	21 43	28 52
21	01 56 44	01 00 45	26 ♊ 36 13	03 ♋ 33 38	18 32	10 57	11 37	19 24	18 09	17 10	21 44	28 51
22	02 00 40	01 59 18	10 ♋ 23 44	17 ♋ 06 36	20 27	12 06	12 23	19 21	18 10	17 13	21 45	28 51
23	02 04 37	02 57 49	23 ♋ 42 31	00 ♌ 11 52	22 23	13 15	13 09	19 18	18 10	17 16	21 46	28 50
24	02 08 33	03 56 18	06 ♌ 35 09	12 ♌ 52 53	24 21	14 24	13 55	19 15	18 11	17 18	21 47	28 49
25	02 12 30	04 54 45	19 ♌ 05 40	25 ♌ 14 06	26 21	15 33	14 41	19 12	18 11	17 21	21 48	28 49
26	02 16 26	05 53 10	01 ♍ 18 48	07 ♍ 20 22	28 ♈ 22	16 41	15 26	19 08	18 12	17 23	21 49	28 48
27	02 20 23	06 51 32	13 ♍ 19 24	19 ♍ 16 27	00 ♉ 24	17 50	16 12	19 04	18 12	17 26	21 50	28 47
28	02 24 20	07 49 52	25 ♍ 12 02	01 ♎ 06 38	02 28	18 58	16 58	19 00	18 13	17 28	21 50	28 46
29	02 28 16	08 48 11	07 ♎ 00 42	12 ♎ 54 38	04 33	20 06	17 44	18 56	18 14	17 31	21 51	28 45
30	02 32 13	09 ♉ 46 27	18 ♎ 48 48	24 ♎ 43 31	06 ♉ 40	21 ♊ 14	18 ♓ 30	18 ♐ 52	18 ♌ 15	17 ♓ 33	21 ♒ 52	28 ♐ 45

Date	Moon True ☊ ° '	Moon Mean ☊ ° '	Moon ☽ Latitude ° '
01	16 ♓ 07	14 ♓ 53	01 S 06
02	16 R 05	14 49	02 08
03	16 01	14 46	03 04
04	15 57	14 43	03 52
05	15 52	14 40	04 30
06	15 46	14 37	04 57
07	15 41	14 33	05 10
08	15 37	14 30	05 10
09	15 35	14 27	04 56
10	15 34	14 24	04 26
11	15 D 34	14 21	03 42
12	15 35	14 18	02 44
13	15 37	14 14	01 35
14	15 38	14 11	00 S 18
15	15 R 37	14 08	01 N 02
16	15 35	14 05	02 19
17	15 31	14 02	03 27
18	15 25	13 59	04 20
19	15 18	13 55	04 54
20	15 12	13 52	05 09
21	15 06	13 49	05 03
22	15 02	13 46	04 39
23	15 00	13 43	04 02
24	15 D 00	13 39	03 12
25	15 00	13 36	02 15
26	15 02	13 33	01 13
27	15 02	13 30	00 N 09
28	15 R 02	13 27	00 S 54
29	14 59	13 24	01 55
30	14 ♓ 54	13 ♓ 20	02 S 51

DECLINATIONS

Date	Sun ☉ ° '	Moon ☽ ° '	Mercury ☿ ° '	Venus ♀ ° '	Mars ♂ ° '	Jupiter ♃ ° '	Saturn ♄ ° '	Uranus ♅ ° '	Neptune ♆ ° '	Pluto ♇ ° '
01	04 N 30	00 S 17	07 S 40	17 N 53	13 S 56	22 S 19	16 N 33	06 S 08	14 S 38	16 S 27
02	04 53	05 55	07 10	18 16	13 41	22 19	16 34	06 07	14 38	16 27
03	05 16	11 20	06 40	18 39	13 25	22 19	16 34	06 06	14 37	16 27
04	05 39	16 22	06 09	19 01	13 10	22 19	16 35	06 04	14 37	16 27
05	06 02	20 48	05 36	19 23	12 54	22 19	16 35	06 03	14 36	16 27
06	06 25	24 25	05 02	19 44	12 38	22 19	16 36	06 02	14 36	16 26
07	06 47	27 01	04 27	20 05	12 22	22 19	16 36	06 01	14 35	16 26
08	07 10	28 21	03 51	20 25	12 06	22 19	16 37	06 00	14 35	16 26
09	07 32	28 17	03 13	20 45	11 50	22 19	16 37	05 58	14 34	16 26
10	07 55	26 44	02 35	21 04	11 34	22 19	16 37	05 57	14 34	16 26
11	08 17	23 41	01 55	21 23	11 18	22 19	16 37	05 56	14 34	16 26
12	08 39	19 07	01 15	21 41	11 01	22 19	16 38	05 55	14 33	16 26
13	09 01	13 43	00 S 33	21 59	10 45	22 18	16 38	05 54	14 33	16 26
14	09 22	07 15	00 N 10	22 16	10 28	22 18	16 38	05 53	14 32	16 26
15	09 44	00 S 15	00 53	22 32	10 12	22 18	16 38	05 52	14 32	16 25
16	10 05	06 N 55	01 38	22 48	09 55	22 18	16 38	05 51	14 32	16 25
17	10 27	13 45	02 23	23 03	09 38	22 18	16 38	05 49	14 31	16 25
18	10 48	19 44	03 09	23 18	09 21	22 18	16 38	05 48	14 31	16 25
19	11 08	24 24	03 56	23 32	09 04	22 17	16 38	05 47	14 31	16 25
20	11 29	27 22	04 44	23 46	08 47	22 17	16 38	05 46	14 30	16 25
21	11 50	28 26	05 33	23 59	08 30	22 17	16 38	05 45	14 30	16 25
22	12 10	27 41	06 22	24 11	08 12	22 17	16 38	05 44	14 29	16 25
23	12 30	25 19	07 12	24 23	07 55	22 17	16 38	05 43	14 29	16 24
24	12 50	21 44	08 02	24 34	07 38	22 16	16 38	05 42	14 29	16 24
25	13 10	17 14	08 53	24 44	07 20	22 16	16 37	05 41	14 29	16 24
26	13 29	12 09	09 44	24 54	07 03	22 16	16 37	05 40	14 28	16 24
27	13 48	06 42	10 35	25 03	06 45	22 15	16 37	05 39	14 28	16 24
28	14 07	01 N 04	11 26	25 11	06 28	22 15	16 36	05 38	14 28	16 24
29	14 26	04 S 33	12 17	25 19	06 10	22 15	16 36	05 37	14 28	16 24
30	14 N 45	10 S 00	13 N 09	25 N 26	05 S 52	22 S 14	16 N 36	05 S 37	14 S 27	16 S 24

ZODIAC SIGN ENTRIES

Date	h	m	Planets
01	15	43	☽ ♎
04	04	36	☽ ♏
06	08	49	♂ ♓
06	16	57	☽ ♐
09	03	36	☽ ♑
10	23	07	☿ ♈
11	11	23	☽ ♒
12	02	15	♀ ♊
13	15	39	☽ ♓
15	16	47	☽ ♈
17	16	11	☽ ♉
19	15	51	☽ ♊
20	11	07	☉ ♉
21	17	50	☽ ♋
23	23	38	☽ ♌
26	09	24	☽ ♍
27	07	16	☿ ♉
28	21	45	☽ ♎

LATITUDES

Date	Mercury ☿ ° '	Venus ♀ ° '	Mars ♂ ° '	Jupiter ♃ ° '	Saturn ♄ ° '	Uranus ♅ ° '	Neptune ♆ ° '	Pluto ♇ ° '
01	02 S 15	00 N 52	01 S 15	00 N 44	01 N 20	00 S 44	00 S 14	06 N 59
04	02 24	01 02	01 16	00 44	01 19	00 44	00 14	07 00
07	02 29	01 12	01 17	00 44	01 19	00 44	00 14	07 00
10	02 29	01 22	01 19	00 44	01 19	00 44	00 14	07 00
13	02 25	01 31	01 20	00 44	01 19	00 45	00 15	07 01
16	02 16	01 40	01 21	00 44	01 19	00 45	00 15	07 01
19	02 03	01 49	01 23	00 44	01 19	00 45	00 15	07 01
22	01 45	01 57	01 24	00 44	01 19	00 45	00 15	07 02
25	01 23	02 05	01 25	00 44	01 18	00 45	00 15	07 02
28	00 57	02 13	01 26	00 44	01 18	00 45	00 15	07 02
31	00 S 28	02 N 19	01 S 27	00 N 44	01 N 18	00 S 45	00 S 15	07 N 02

DATA

Julian Date	2454192
Delta T	+66 seconds
Ayanamsa	23° 57' 34"
Synetic vernal point	05° ♓ 09' 25"
True obliquity of ecliptic	23° 26' 27"

MOON'S PHASES, APSIDES AND POSITIONS ☽

Date	h	m	Phase	Longitude ° '	Eclipse Indicator
02	17	15	○	12 ♎ 35	
10	18	04	☾	20 ♑ 29	
17	11	36	●	27 ♈ 05	
24	06	36	☽	03 ♌ 43	

Day	h	m		
03	08	21	Apogee	
17	05	51	Perigee	
30	10	44	Apogee	
01	10	50	0S	
08	23	00	Max dec	28° S 31'
15	12	49	0N	
21	13	39	Max dec	28° N 27'
28	16	34	0S	

LONGITUDES

Date	Chiron ⚷ ° '	Ceres ⚳ ° '	Pallas ⚴ ° '	Juno ⚵ ° '	Vesta ⚶ ° '	Black Moon Lilith ⚸ ° '
01	14 ♒ 21	05 ♈ 14	02 ♓ 00	21 ♎ 50	13 ♐ 56	18 ♎ 08
11	14 ♒ 50	09 ♈ 10	04 ♓ 55	19 ♎ 30	14 ♐ 54	19 ♎ 15
21	15 ♒ 12	13 ♈ 04	07 ♓ 41	17 ♎ 13	15 ♐ 05	20 ♎ 21
31	15 ♒ 28	16 ♈ 56	10 ♓ 15	15 ♎ 12	14 ♐ 26	21 ♎ 28

All ephemeris data is given at 12.00 UT and the Moon's longitude is additionally given for 24.00 UT

ASPECTARIAN

h m	Aspects
01 Sunday	
04 32	☽ ⊥ ♄
07 54	☽ ⊼ ♂
10 12	☽ ± ♆
13 38	☽ □ ♇
18 29	☿ ☌ ♅
20 53	☽ ± ♂
21 45	☽ ⚼ ♀
22 42	☽ ∠ ♄
02 Monday	
04 29	☽ ⚼ ♆
07 01	♀ □ ♄
07 15	☽ # ☉
07 27	☽ Q ♃
12 51	☽ ∥ ♅
16 25	☽ ⚼ ♂
17 01	☽ ∥ ☿
17 15	☽ ☍ ☉
18 03	☽ ± ♀
03 Tuesday	
00 43	☽ ⊼ ♅
02 09	☽ Q ♇
04 37	☽ ⊼ ☿
05 04	☽ ✶ ♄
07 36	☽ ⊼ ♀
07 50	☽ ✶ ♃
08 19	☿ ⊼ ♄
10 00	♀ ⊼ ♃
11 00	☽ △ ♆
12 57	☽ ± ♅
18 28	☽ ± ☿
21 11	☽ ∥ ♂
04 Wednesday	
00 59	☽ △ ♂
02 30	☽ ✶ ♇
03 22	☽ ∥ ♆
05 21	☽ Q ♄
07 17	☿ □ ♃
07 17	☽ ⚼ ♅
12 24	☽ ∥ ♇
13 06	☽ # ♄
14 16	☽ ∠ ♃
15 14	☽ ⚼ ☿
15 21	☿ ∥ ♅
18 16	♀ □ ♆
05 Thursday	
00 22	♂ ✶ ♇
00 44	☉ # ☿
03 11	☽ # ♀
08 27	☽ ⊥ ♃
08 49	☽ ∠ ♇
09 36	☿ ⊻ ♆
11 31	☽ ⊼ ☉
13 01	☉ # ♅
13 43	☽ △ ♅
17 37	☽ □ ♄
20 31	☽ ⊻ ♃
21 22	☽ ∥ ♃
23 45	☽ □ ♆
06 Friday	
00 39	☽ ± ☉
01 25	♃ St R
01 43	☽ △ ☿
02 42	♀ ± ♇
02 53	☽ ⊥ ♇
02 54	☽ ☍ ♀
14 52	☽ ⊻ ♇
15 29	☉ ⊻ ♅
17 30	☽ □ ♂
20 13	☽ ⚼ ☉
07 Saturday	
08 15	☿ ± ♄
11 37	☽ Q ♆
16 10	☿ ✶ ♀
08 Sunday	
01 33	☽ □ ♅
04 21	☽ △ ☉
04 58	☽ △ ♄
07 19	☽ Q ♂
07 52	☽ ☌ ♃
11 08	☽ ✶ ♆
11 56	☉ △ ♄
16 08	♂ Q ♃
20 17	☽ ⊼ ♀
21 14	☽ □ ☿
09 Monday	
01 35	☽ ☌ ♇
07 57	☽ ✶ ♂
08 57	☽ ± ♀
09 18	☿ ⊥ ♆
09 49	☽ ⚼ ♄
12 22	☽ Q ♅
15 56	☽ ∠ ♆
10 Tuesday	
00 07	☉ △ ♃
02 46	☽ ± ♄
03 40	☽ ⚼ ♀
07 25	☿ □ ♇
08 50	☽ ⊥ ♆
10 58	☽ ✶ ♅
12 06	☽ Q ☿
13 54	☽ ∠ ♂
13 55	☽ ⊼ ♄
16 42	☽ ⊻ ♃
18 04	☽ □ ☉
19 57	☽ ⊻ ♆
23 03	♀ Q ♅
11 Wednesday	
03 40	☽ ⊥ ♃
04 33	♀ ⊼ ♇
07 24	☽ ⊥ ♂
09 28	☽ ⊻ ♇
09 57	☽ △ ♀
13 06	☽ ✶ ☿
14 27	☽ ∠ ♅
18 50	☽ ⊻ ♂
19 33	☉ ✶ ♆
19 51	☽ ∠ ♃
20 14	☽ ⊥ ♇
20 19	☽ ∥ ♃
12 Thursday	
00 34	☽ # ♀
05 04	☽ Q ☉
06 30	☽ ⊥ ♅
12 06	☽ ∠ ♇
17 01	☽ ⊻ ♅
19 12	☽ ∠ ☿
19 31	☽ ☍ ♄
21 45	☿ ⚼ ♄
22 06	☽ ✶ ♃
13 Friday	
00 00	☽ # ♄
00 52	☽ ∥ ♇
01 18	☽ ☌ ♆
01 59	☉ ⊥ ♅
03 29	☽ ✶ ☉
08 41	☽ ∥ ♆
12 41	☽ ⊥ ☿
13 50	☽ ✶ ♇
18 29	☽ Q ♃
18 59	☽ □ ♀
23 53	☽ ∥ ♂
14 Saturday	
00 07	☽ ⊻ ☿
01 32	☽ ☌ ♂
04 49	☽ # ☉
06 34	☽ ∠ ☉
09 50	☽ Q ♇
15 39	♀ # ♃
16 50	☽ ∥ ♅
19 34	☽ ☌ ♅
20 37	☿ ⊻ ♂
21 43	☽ ⊼ ♄
21 46	☿ ∠ ♆
22 23	☽ ⊥ ☉
15 Sunday	
00 07	☽ □ ♃
03 12	☽ Q ♀
03 16	☽ ⊻ ♆
07 26	☽ ± ♄
08 47	☽ ⊻ ☉
10 02	☽ # ☿
12 57	☽ ⊥ ♆
15 02	☽ □ ♇
16 13	☽ ∥ ☿
21 50	☽ ⚼ ♄
16 Monday	
00 04	☽ ✶ ♀
03 20	☽ ∠ ♆
04 47	☽ ⊻ ♂
05 23	☉ # ♂
07 08	☽ ☌ ☿
08 23	☽ # ♅
14 47	☽ ⊥ ♂
19 40	☽ ⊻ ♅
21 34	☽ △ ♄
21 54	☽ # ♂
23 30	☽ ∥ ☉
23 49	☽ △ ♃
17 Tuesday	
01 44	☽ ∠ ♀
03 03	☽ ✶ ♆
05 08	☽ ⊥ ♅
05 43	☽ ∠ ♂
08 56	♀ Q ♄
11 36	☽ ☌ ☉
14 27	☽ △ ♇
14 54	☽ # ♆
17 03	☽ ⊥ ♀
19 20	☽ ∠ ♅
21 56	☽ Q ♆
22 11	☽ # ♇
23 03	☽ ∥ ♄
23 21	☽ ⚼ ♃
18 Wednesday	
03 17	☽ ⊻ ♀
06 34	☽ ✶ ♂
12 42	☽ ⊻ ☿
13 31	☽ ± ♃
14 05	☽ ⚼ ♇
19 06	☽ ✶ ♅
20 52	☽ □ ♄
23 02	☽ ⊼ ♃
23 33	☽ ⊥ ☿
19 Thursday	
00 09	☽ # ♃
02 29	☽ □ ♆
02 38	☽ Q ♂
04 24	☽ ± ♇
06 33	☽ ∥ ♀
07 35	☉ △ ♇
14 02	☽ ⊼ ♇
14 22	☽ Q ♅
14 29	☽ ⊻ ☉
16 01	☽ ∠ ☿
21 23	♄ St D
20 Friday	
00 55	☽ ⊥ ☉
01 51	☽ Q ♄
07 27	☽ ☌ ♀
09 25	☽ □ ♂
12 43	♂ Q ♇
16 06	☿ ⊥ ♂
16 49	☽ ∠ ☉
18 10	☿ ⊻ ♅
20 00	☽ □ ♅
20 16	☽ ✶ ☿
21 42	☽ ✶ ♄
23 51	☽ ☍ ♃
21 Saturday	
03 42	☽ △ ♆
07 09	☿ △ ♄
15 52	☽ ☍ ♇
18 02	☿ # ♅
19 50	☽ Q ☿
20 09	☽ ✶ ☉
22 39	☿ △ ♃
23 18	☽ ∠ ♄
22 Sunday	
05 33	☽ ⚼ ♆
15 08	☽ ⊥ ♄
15 18	☽ ⊻ ♀
15 44	☽ △ ♂
17 51	☉ ∠ ♅
18 54	☽ Q ☉
21 33	☽ ± ♆
23 Monday	
00 14	☽ △ ♅
01 54	☽ ⊻ ♄
03 05	☽ ⊥ ♀
04 00	☽ ⊼ ♃
04 11	☿ ✶ ♆
05 30	♀ □ ♂
08 26	☽ ⊼ ♆
09 10	☽ □ ☿
14 55	☽ ± ♃
18 43	☽ ∥ ♀
20 42	☽ ⚼ ♂
21 11	☽ ∠ ♀
21 27	☽ ⊼ ♇
22 53	☿ ⊥ ♅
24 Tuesday	
03 27	☿ # ♂
03 54	☽ ⚼ ♅
06 36	☽ □ ☉
07 37	☽ ⚼ ♃
08 01	☉ Q ♆
08 40	☽ ± ♇
08 47	☽ # ♃
14 40	☽ ± ♂
19 17	☉ ⚼ ♃
21 00	☽ ± ♅
25 Wednesday	
01 48	☽ ⚼ ♇
02 53	☽ ⊼ ♂
04 25	☽ ✶ ♀
08 36	☽ ⊼ ♅
10 14	☽ ☌ ♄
12 11	☽ △ ♃
15 03	☽ ∥ ♄
16 07	☽ # ♇
17 16	☽ ☍ ♆
26 Thursday	
01 19	☽ # ♆
05 00	☽ △ ☿
06 16	☽ Q ♀
06 16	☽ ∥ ☉
07 02	☽ △ ♇
17 06	☿ △ ♇
21 21	☽ ∥ ☿
21 54	☽ △ ☉
27 Friday	
03 20	♀ □ ♅
11 44	☽ # ♂
16 30	☽ # ♅
17 04	☽ ⚼ ☿
18 12	☽ ☍ ♂
20 07	♀ ✶ ♄
20 18	☽ ☍ ♅
21 51	☽ ⊻ ♄
22 02	☽ □ ♀
23 31	☽ □ ♃
28 Saturday	
02 52	☿ ∠ ♂
05 11	☽ ⊼ ♆
06 46	☽ ⚼ ☉
10 01	☽ ⊥ ♄
12 03	☿ ∠ ♅
12 44	♀ ☍ ♃
15 07	☽ ± ☿
17 22	☽ ± ♆
19 14	☽ □ ♇
29 Sunday	
02 40	☽ ± ☉
03 56	☿ Q ♆
04 19	☽ ∠ ♄
04 37	♂ ☌ ♅
05 07	☿ ⚼ ♃
05 55	☽ ⊼ ☿
11 41	☽ ⚼ ♆
11 51	☽ Q ♃
13 47	☉ # ♆
15 58	☽ ⊼ ☉
16 38	☽ ∥ ♅
18 40	☽ ∥ ♂
30 Monday	
01 36	☿ ∠ ♀
04 09	♂ ⊼ ♄
07 48	☽ Q ♇
09 26	☽ ⊼ ♅
10 52	☽ ✶ ♄
11 19	☽ ⊼ ♂
12 06	☽ ✶ ♃
17 27	☽ △ ♀
18 13	☽ △ ♆
21 39	☽ ± ♅
22 21	♂ □ ♃

MAY 2007

LONGITUDES

Date	Sidereal time h m s	Sun ☉ ° ' "	Moon ☽ ° ' "	Moon ☽ 24.00 ° ' "	Mercury ☿ ° '	Venus ♀ ° '	Mars ♂ ° '	Jupiter ♃ ° '	Saturn ♄ ° '	Uranus ♅ ° '	Neptune ♆ ° '	Pluto ♇ ° '
01	02 36 09	10 ♉ 44 42	00 ♏ 39 02	06 ♏ 35 38	08 ♉ 47	22 ♊ 22	19 ♓ 16	18 ♐ 47	18 ♌ 17	17 ♓ 36	21 ♒ 53	28 ♐ 44
02	02 40 06	11 42 55	12 ♏ 33 30	18 ♏ 32 50	10 55	23 30	20 02	18 R 43	18 18	17 38	21 54	28 R 43
03	02 44 02	12 41 06	24 ♏ 33 48	00 ♐ 36 33	13 05	24 37	20 48	18 38	18 19	17 40	21 54	28 42
04	02 47 59	13 39 15	06 ♐ 41 15	12 ♐ 48 03	15 14	25 45	21 33	18 33	18 21	17 43	21 55	28 41
05	02 51 55	14 37 23	18 ♐ 57 08	25 ♐ 08 40	17 24	26 52	22 19	18 28	18 22	17 45	21 56	28 40
06	02 55 52	15 35 30	01 ♑ 22 52	07 ♑ 39 58	19 34	27 59	23 05	18 23	18 24	17 47	21 56	28 39
07	02 59 49	16 33 34	14 ♑ 00 13	20 ♑ 23 56	21 44	29 ♊ 06	23 51	18 17	18 26	17 49	21 57	28 38
08	03 03 45	17 31 38	26 ♑ 51 23	03 ♒ 22 56	23 53	00 ♋ 13	24 36	18 12	18 28	17 51	21 58	28 37
09	03 07 42	18 29 40	09 ♒ 58 53	16 ♒ 39 34	26 01	01 19	25 22	18 06	18 30	17 53	21 58	28 36
10	03 11 38	19 27 41	23 ♒ 25 18	00 ♓ 16 20	28 ♉ 08	02 25	26 08	18 00	18 32	17 55	21 59	28 35
11	03 15 35	20 25 40	07 ♓ 12 52	14 ♓ 15 00	00 ♊ 14	03 32	26 54	17 55	18 34	17 57	21 59	28 34
12	03 19 31	21 23 38	21 ♓ 22 45	28 ♓ 35 57	02 18	04 38	27 39	17 49	18 37	17 59	21 59	28 32
13	03 23 28	22 21 35	05 ♈ 54 18	13 ♈ 17 19	04 21	05 43	28 25	17 42	18 39	18 01	22 00	28 31
14	03 27 24	23 19 30	20 ♈ 44 18	28 ♈ 14 25	06 21	06 49	29 10	17 36	18 41	18 03	22 00	28 30
15	03 31 21	24 17 25	05 ♉ 46 38	13 ♉ 19 48	08 18	07 54	29 ♓ 56	17 30	18 44	18 05	22 01	28 29
16	03 35 18	25 15 18	20 ♉ 52 40	28 ♉ 23 59	10 13	08 59	00 ♈ 42	17 23	18 47	18 07	22 01	28 28
17	03 39 14	26 13 09	05 ♊ 52 29	13 ♊ 17 00	12 06	10 04	01 27	17 17	18 50	18 09	22 01	28 26
18	03 43 11	27 10 59	20 ♊ 36 30	27 ♊ 50 06	13 56	11 09	02 13	17 10	18 53	18 10	22 01	28 25
19	03 47 07	28 08 48	04 ♋ 57 10	11 ♋ 57 14	15 42	12 13	02 58	17 03	18 56	18 12	22 02	28 24
20	03 51 04	29 ♉ 06 35	18 ♋ 50 03	25 ♋ 35 34	17 26	13 18	03 43	16 56	18 59	18 14	22 02	28 23
21	03 55 00	00 ♊ 04 20	02 ♌ 13 56	08 ♌ 45 24	19 07	14 21	04 29	16 49	19 02	18 15	22 02	28 21
22	03 58 57	01 02 04	15 ♌ 10 25	21 ♌ 29 27	20 44	15 25	05 14	16 42	19 05	18 17	22 02	28 20
23	04 02 53	01 59 46	27 ♌ 43 07	03 ♍ 52 02	22 18	16 29	05 59	16 35	19 08	18 18	22 02	28 19
24	04 06 50	02 57 26	09 ♍ 56 51	15 ♍ 58 15	23 49	17 32	06 45	16 28	19 12	18 20	22 02	28 17
25	04 10 47	03 55 05	21 ♍ 56 56	27 ♍ 53 31	25 17	18 35	07 30	16 21	19 16	18 21	22 R 02	28 16
26	04 14 43	04 52 42	03 ♎ 48 40	09 ♎ 42 58	26 41	19 37	08 15	16 14	19 19	18 22	22 02	28 14
27	04 18 40	05 50 18	15 ♎ 37 00	21 ♎ 31 16	28 02	20 40	09 00	16 06	19 23	18 24	22 02	28 13
28	04 22 36	06 47 53	27 ♎ 26 14	03 ♏ 22 20	29 ♊ 20	21 42	09 45	15 59	19 27	18 25	22 02	28 12
29	04 26 33	07 45 26	09 ♏ 19 55	15 ♏ 19 18	00 ♋ 34	22 43	10 30	15 51	19 31	18 26	22 02	28 10
30	04 30 29	08 42 58	21 ♏ 20 43	27 ♏ 24 24	01 44	23 45	11 15	15 44	19 35	18 27	22 02	28 09
31	04 34 26	09 ♊ 40 29	03 ♐ 30 29	09 ♐ 39 04	02 ♋ 51	24 ♋ 46	12 ♈ 00	15 ♐ 36	19 ♌ 39	18 ♓ 29	22 ♒ 01	28 ♐ 07

Date	Moon True ☊ ° '	Moon Mean ☊ ° '	Moon ☽ Latitude ° '
01	14 ♓ 47	13 ♓ 17	03 S 39
02	14 R 38	13 14	04 18
03	14 27	13 11	04 46
04	14 16	13 08	05 01
05	14 06	13 05	05 03
06	13 57	13 01	04 50
07	13 50	12 58	04 23
08	13 46	12 55	03 43
09	13 44	12 52	02 50
10	13 D 43	12 49	01 46
11	13 44	12 45	00 S 35
12	13 R 44	12 42	00 N 41
13	13 42	12 39	01 55
14	13 38	12 36	03 03
15	13 31	12 33	03 59
16	13 22	12 30	04 39
17	13 12	12 26	04 59
18	13 01	12 23	04 59
19	12 52	12 20	04 40
20	12 44	12 17	04 05
21	12 39	12 14	03 17
22	12 37	12 11	02 20
23	12 36	12 07	01 18
24	12 D 36	12 04	00 N 14
25	12 R 35	12 01	00 S 49
26	12 34	11 58	01 50
27	12 30	11 55	02 45
28	12 24	11 51	03 33
29	12 15	11 48	04 13
30	12 04	11 45	04 41
31	11 ♓ 51	11 ♓ 42	04 S 57

DECLINATIONS

Date	Sun ☉ ° '	Moon ☽ ° '	Mercury ☿ ° '	Venus ♀ ° '	Mars ♂ ° '	Jupiter ♃ ° '	Saturn ♄ ° '	Uranus ♅ ° '	Neptune ♆ ° '	Pluto ♇ ° '
01	15 N 03	15 S 07	13 N 59	25 N 32	05 S 35	22 S 14	16 N 35	05 S 36	14 S 27	16 S 24
02	15 21	19 42	14 50	25 38	05 17	22 14	16 35	05 35	14 27	16 24
03	15 39	23 32	15 39	25 43	04 59	22 13	16 34	05 34	14 27	16 24
04	15 56	26 22	16 28	25 48	04 41	22 13	16 34	05 33	14 26	16 24
05	16 14	28 01	17 15	25 51	04 24	22 12	16 33	05 32	14 26	16 23
06	16 31	28 16	18 01	25 54	04 06	22 12	16 33	05 31	14 26	16 23
07	16 47	27 04	18 46	25 57	03 48	22 11	16 32	05 30	14 26	16 23
08	17 04	24 26	19 29	25 59	03 30	22 11	16 31	05 30	14 26	16 23
09	17 20	20 28	20 10	26 00	03 12	22 11	16 31	05 29	14 26	16 23
10	17 36	15 23	20 48	26 00	02 54	22 10	16 30	05 28	14 25	16 23
11	17 51	09 24	21 25	26 00	02 36	22 10	16 29	05 27	14 25	16 23
12	18 07	02 S 48	21 59	25 59	02 18	22 09	16 29	05 27	14 25	16 23
13	18 22	04 N 06	22 31	25 57	02 00	22 09	16 28	05 26	14 25	16 23
14	18 36	10 55	23 00	25 55	01 42	22 08	16 27	05 25	14 25	16 23
15	18 51	17 12	23 27	25 52	01 24	22 08	16 26	05 24	14 25	16 23
16	19 05	22 27	23 51	25 49	01 06	22 07	16 25	05 24	14 25	16 23
17	19 18	26 12	24 13	25 45	00 48	22 07	16 24	05 23	14 25	16 23
18	19 32	28 05	24 32	25 40	00 30	22 06	16 23	05 23	14 25	16 23
19	19 45	28 01	24 48	25 35	00 S 12	22 05	16 22	05 22	14 25	16 23
20	19 58	26 09	25 02	25 29	00 N 06	22 05	16 21	05 21	14 24	16 23
21	20 10	22 51	25 14	25 22	00 23	22 04	16 20	05 21	14 24	16 23
22	20 22	18 31	25 23	25 15	00 41	22 04	16 19	05 20	14 24	16 23
23	20 34	13 29	25 30	25 08	00 59	22 03	16 18	05 20	14 24	16 23
24	20 45	08 03	25 35	24 59	01 17	22 02	16 17	05 19	14 24	16 23
25	20 56	02 N 26	25 38	24 51	01 35	22 02	16 16	05 19	14 25	16 23
26	21 07	03 S 12	25 39	24 41	01 52	22 01	16 15	05 18	14 25	16 23
27	21 17	08 41	25 38	24 31	02 10	22 00	16 13	05 18	14 25	16 23
28	21 27	13 53	25 36	24 21	02 28	22 00	16 12	05 17	14 25	16 23
29	21 36	18 35	25 32	24 10	02 45	21 59	16 11	05 17	14 25	16 23
30	21 45	22 37	25 27	23 58	03 03	21 58	16 09	05 16	14 25	16 23
31	21 N 54	25 S 42	25 N 20	23 N 46	03 N 20	21 S 58	16 N 08	05 S 16	14 S 25	16 S 23

ZODIAC SIGN ENTRIES

Date	h	m	Planets
01	10	41	☽ ♏
03	22	48	☽ ♐
06	09	21	☽ ♑
08	07	28	♀ ♋
08	17	48	☽ ♒
10	23	32	☽ ♓
11	09	17	☿ ♊
13	02	19	☽ ♈
15	02	48	☽ ♉
15	14	06	♂ ♈
17	02	34	☽ ♊
19	03	38	☽ ♋
21	07	57	☽ ♌
21	10	12	☉ ♊
23	16	26	☽ ♍
26	04	16	☽ ♎
28	17	11	☽ ♏
29	00	56	☿ ♋
31	05	07	☽ ♐

LATITUDES

Date	Mercury ☿ ° '	Venus ♀ ° '	Mars ♂ ° '	Jupiter ♃ ° '	Saturn ♄ ° '	Uranus ♅ ° '	Neptune ♆ ° '	Pluto ♇ ° '
01	00 S 28	02 N 19	01 S 27	00 N 44	01 N 18	00 S 45	00 S 15	07 N 02
04	00 N 03	02 25	01 28	00 44	01 18	00 45	00 15	07 03
07	00 35	02 31	01 28	00 44	01 18	00 45	00 15	07 03
10	01 05	02 35	01 29	00 44	01 17	00 45	00 15	07 03
13	01 32	02 39	01 30	00 44	01 17	00 45	00 15	07 03
16	01 53	02 41	01 30	00 44	01 17	00 45	00 15	07 03
19	02 08	02 43	01 31	00 43	01 17	00 46	00 15	07 03
22	02 16	02 43	01 31	00 43	01 17	00 46	00 15	07 03
25	02 17	02 43	01 31	00 43	01 17	00 46	00 15	07 03
28	02 10	02 41	01 31	00 43	01 16	00 46	00 15	07 03
31	01 N 55	02 N 38	01 S 31	00 N 42	01 N 16	00 S 46	00 S 16	07 N 03

DATA

Julian Date	2454222
Delta T	+66 seconds
Ayanamsa	23° 57' 38"
Synetic vernal point	05° ♓ 09' 22"
True obliquity of ecliptic	23° 26' 27"

MOON'S PHASES, APSIDES AND POSITIONS ☽

Date	h	m	Phase	Longitude ° '	Eclipse Indicator
02	10	09	○	11 ♏ 38	
10	04	27	☾	19 ♒ 09	
16	19	27	●	25 ♉ 33	
23	21	03	☽	02 ♍ 21	

Day	h	m		
15	14	59	Perigee	
27	21	57	Apogee	
06	04	21	Max dec	28° S 21'
12	21	47	0N	
18	23	03	Max dec	28° N 17'
25	22	21	0S	

LONGITUDES

Date	Chiron ⚷ ° '	Ceres ⚳ ° '	Pallas ⚴ ° '	Juno ⚵ ° '	Vesta ⚶ ° '	Black Moon Lilith ⚸ ° '
01	15 ♒ 28	16 ♈ 56	10 ♓ 15	15 ♎ 12	14 ♐ 26	21 ♎ 28
11	15 ♒ 38	20 ♈ 44	12 ♓ 36	13 ♎ 38	13 ♐ 41	22 ♎ 35
21	15 ♒ 41	24 ♈ 28	14 ♓ 40	12 ♎ 35	10 ♐ 57	23 ♎ 42
31	15 ♒ 38	28 ♈ 07	16 ♓ 25	12 ♎ 05	08 ♐ 35	24 ♎ 49

ASPECTARIAN

h m Aspects

01 Tuesday
00 20 ☽ ± ♂
01 34 ♀ △ ♆
05 23 ☽ # ☿
08 07 ☽ ✶ ♇
08 43 ☽ ∥ ♆
10 45 ♂ ∥ ♅
11 15 ☽ Q ♄
11 37 ☽ # ☉
15 56 ☽ ⚼ ♅
18 18 ☽ ∠ ♃
18 21 ☽ ∥ ♇
19 19 ☽ # ♄
19 48 ☽ ⚼ ♂

02 Wednesday
01 09 ☿ # ♆
02 59 ☽ ⚼ ♀
08 00 ☽ ☍ ☿
10 09 ☽ ☍ ☉
12 18 ☽ ⊥ ♃
14 19 ☽ ∠ ♇
22 12 ☽ △ ♅
22 55 ☽ ± ♀
23 31 ☽ □ ♄

03 Thursday
00 15 ☽ ⊻ ♃
03 10 ☽ ∥ ♃
03 59 ☽ △ ♂
04 05 ☉ ☌ ☿
06 42 ☽ □ ♆
07 14 ☿ ± ♃
08 17 ☽ ⊥ ♇
10 47 ☉ ± ♃
11 49 ☉ ∥ ☿
12 08 ☽ ⚻ ♀
18 52 ☿ ⚼ ♇
20 12 ☽ ⊻ ♇

04 Friday
06 04 ☽ # ♀
09 59 ☿ # ♇
12 40 ☉ ⚼ ♇
15 06 ☿ ∥ ♄
18 21 ☽ Q ♆
23 32 ♂ ⊻ ♆

05 Saturday
02 50 ☽ ⚻ ☉
08 20 ☽ ⚻ ☿
09 39 ☽ □ ♅
10 52 ☽ △ ♄
11 04 ☽ ☌ ♃
15 31 ☽ ± ☉
15 53 ☿ ✶ ♅
17 47 ☽ ✶ ♆
18 58 ☽ □ ♂
22 27 ☽ ± ☿
22 55 ☿ □ ♄
23 20 ☿ ⚻ ♃

06 Sunday
01 51 ☉ # ♇
04 50 ☽ ☍ ♀
06 46 ☽ ☌ ♇
07 11 ♃ △ ♄
10 22 ☽ ⚼ ☉
14 56 ☉ ∥ ♄
15 53 ☽ ⚼ ♄
19 22 ☽ ⚼ ☿
20 26 ☽ Q ♅
22 38 ☽ ∠ ♆

07 Monday
02 05 ♀ ☍ ♇
07 40 ☽ Q ♂
09 02 ☽ ± ♄
14 28 ☿ □ ♆
15 40 ☽ ⊥ ♆
17 12 ☽ △ ☉
19 12 ☽ ✶ ♅
20 00 ☽ ⊻ ♃
20 21 ☽ ⚻ ♄
21 58 ☿ ± ♇
23 37 ☽ # ♀

08 Tuesday
02 54 ☽ ⊻ ♆
05 23 ☽ △ ☿
07 07 ☽ ⊥ ♃
07 16 ♂ ± ♄
07 34 ☽ ✶ ♂
15 14 ☽ ⊻ ♇
18 45 ☽ ⚻ ♀
19 38 ☿ ⊥ ♀
20 25 ☉ ✶ ♅
23 04 ☽ ∠ ♅
23 35 ☽ ∠ ♃

09 Wednesday
00 39 ☿ ✶ ♂
02 14 ☽ ⊥ ♇
02 30 ☽ ∥ ♃
03 10 ☉ ⚻ ♃
06 44 ☽ ± ♀
12 05 ☉ □ ♄
12 45 ☽ ∠ ♂
13 25 ☽ # ☿
15 27 ☽ ⊥ ♅
18 30 ☽ ∠ ♇

10 Thursday
00 25 ☽ ⚼ ♀
02 14 ☽ ⊻ ♅
02 28 ☽ ✶ ♃
02 36 ☽ # ☉
03 19 ☽ ☍ ♄
04 27 ☽ □ ☉
05 50 ☽ ⊥ ♂
07 05 ☽ # ♄
07 36 ☽ ∥ ♇
09 27 ☽ ☌ ♆
16 04 ☽ ∥ ♆
16 57 ☿ ⚻ ♇
17 03 ☽ ⊻ ♂
21 02 ☽ ✶ ♇
21 47 ☽ □ ☿
23 28 ☽ Q ♃

11 Friday
03 32 ♃ □ ♅
05 06 ☽ △ ♀
08 43 ☿ Q ♅
13 00 ♀ ∠ ♄
14 14 ☽ Q ☉
17 43 ☽ Q ♇

12 Saturday
02 33 ☽ ∥ ♅
06 03 ☽ □ ♃
06 18 ☽ ☌ ♅
07 21 ☽ ⚻ ♄
09 54 ☽ Q ☿
12 02 ☽ ✶ ☉
13 01 ☽ ⊻ ♆
13 49 ☽ ∥ ♂
17 24 ☽ ± ♄
19 15 ☿ # ♃
22 44 ♂ ⊥ ♆
23 00 ☽ ⊥ ♆
23 01 ☽ ☌ ♂
23 53 ☽ □ ♇

13 Sunday
02 56 ☉ □ ♆
05 01 ☽ # ♂
08 18 ☽ ⚼ ♄
09 02 ☽ ✶ ☿
11 40 ☽ □ ♀
13 47 ☽ ∠ ♆
14 32 ☽ ∠ ☉
15 15 ♂ □ ♇
15 54 ☉ ± ♇
16 36 ☽ # ♅

14 Monday
06 08 ☽ ⊥ ☉
07 00 ☽ △ ♃
07 41 ☽ ⊻ ♅
08 42 ☽ △ ♄
13 07 ☽ ∠ ☿
14 02 ☽ ✶ ♆
16 12 ♀ ⚼ ♆
16 19 ☿ Q ♄
16 26 ☽ ⊻ ☉
17 19 ☽ ⊥ ♅
19 02 ☽ Q ♀

15 Tuesday
00 24 ☽ △ ♇
00 48 ☿ ⊻ ♀
00 57 ☽ # ♆
02 12 ☽ ⊻ ♂
05 39 ☽ ⊥ ☿
06 49 ☽ ⚼ ♃
07 42 ☽ ∠ ♅
08 39 ☽ # ♇
08 52 ☽ ∥ ♄
09 11 ☽ Q ♆
12 16 ☽ ⊥ ♂
15 38 ☽ ✶ ♀
16 37 ☽ ⊻ ☿
19 12 ☽ ∥ ☉
21 01 ☽ ± ♃

16 Wednesday
00 13 ☽ ⚼ ♇
03 19 ☽ ∠ ♂
06 29 ☽ ⚻ ♃
07 36 ☽ ✶ ♅
08 39 ☽ □ ♄
10 14 ☽ # ♃
13 49 ☽ □ ♆
14 31 ☽ ± ♇
17 20 ☽ ∠ ♀
19 27 ☽ ☌ ☉
20 33 ☽ ∥ ☿

17 Thursday
00 05 ☽ ⚻ ♇
02 46 ☽ Q ♅
04 30 ☽ ✶ ♂
08 30 ☽ ∥ ♀
08 52 ☽ ⊥ ♀
13 32 ☽ Q ♄
19 19 ☽ ⊻ ♀
23 31 ☽ ☌ ☿

18 Friday
00 56 ☽ Q ♂
06 23 ☽ ☍ ♃
07 59 ☽ □ ♅
09 08 ☽ ✶ ♄
14 20 ☽ △ ♆
18 31 ☿ Q ♂
23 42 ☽ ⊻ ☉

19 Saturday
00 03 ♄ # ♇
00 57 ☽ ☍ ♇
08 26 ☽ □ ♂
10 15 ☽ ∠ ♄
10 32 ☽ ⊥ ☉
15 32 ☽ ⚼ ♆
18 06 ☉ ⚻ ♇

20 Sunday
01 30 ☽ ☌ ♀
01 44 ☽ ⊥ ♄
03 07 ☽ ∠ ☉
04 34 ♀ ⊥ ♄
05 30 ☿ ☍ ♃
07 05 ☽ ± ♆
08 42 ☽ ⚻ ♃
09 11 ☽ ⊻ ☿
10 56 ☽ △ ♅
12 15 ☽ ⊻ ♄
17 39 ☽ ⚻ ♆
17 55 ☽ ∥ ♀
19 12 ☽ ± ♃
20 39 ☽ ∥ ☿
20 41 ♂ ⚼ ♄
21 19 ☽ ⊥ ☿
23 26 ☿ □ ♅

21 Monday
04 58 ☽ ⚻ ♇
07 46 ☽ ✶ ☉
10 48 ☿ ✶ ♄
11 16 ☽ ⚼ ♃
13 52 ☽ ⚼ ♅
15 52 ☽ ± ♇
15 56 ☽ ∠ ☿
16 22 ☽ △ ♂
16 38 ☉ Q ♅
16 44 ☽ # ♃

22 Tuesday
00 34 ☿ ∥ ♀
02 43 ☽ ∥ ☉
06 33 ☽ ± ♅
07 39 ☽ Q ☉
08 32 ☽ ⚼ ♇
12 30 ☽ ⊻ ♀
14 52 ☽ △ ♃
17 53 ☽ ⚻ ♅
19 26 ☽ ☌ ♄
22 13 ☽ ⚼ ♂
22 31 ☽ # ♇
22 51 ☽ ∥ ♄

23 Wednesday
00 05 ☽ ✶ ☿
00 57 ☽ ⊥ ♀
01 02 ☽ ☍ ♆
01 55 ♀ ± ♆
07 47 ☽ # ♆
07 47 ☿ △ ♆
13 09 ☽ △ ♇
14 17 ♀ ⚻ ♃
16 42 ☽ ± ♂
20 00 ☽ ∠ ♀
21 03 ☽ □ ☉

24 Thursday
02 42 ☽ Q ☿
05 14 ☽ ⚻ ♂
21 18 ♂ ∠ ♆
23 46 ☽ # ♅

25 Friday
00 52 ☽ □ ♃
01 08 ♆ St R
04 34 ☽ ✶ ♀
04 45 ☽ ☍ ♅
06 34 ☽ ⊻ ♄
06 41 ♀ △ ♅
12 10 ☽ ⚻ ♆
15 28 ☽ ∥ ♂
18 43 ☽ ⊥ ♄
19 38 ☽ □ ☿

26 Saturday
00 17 ☽ ± ♆
00 44 ☽ □ ♇
04 37 ♀ ⊻ ♄
06 01 ☽ # ♂
07 08 ☽ Q ♀
12 50 ☽ Q ♃
13 02 ☽ ∠ ♄
14 22 ☽ △ ☉
18 33 ☽ ⚼ ♆
21 05 ☽ ∥ ♅
21 38 ☽ ☍ ♂

27 Sunday
12 59 ☽ ✶ ♃
13 13 ☽ Q ♇
15 13 ☿ ☍ ♇
17 40 ☽ ⚻ ♅
19 42 ☽ ✶ ♄
23 14 ☽ □ ♀
23 33 ☽ ⚼ ☉

28 Monday
01 02 ☽ △ ♆
05 52 ☽ ± ♅
13 32 ☽ ✶ ♇
14 35 ☽ ∥ ♆
16 17 ☽ △ ☿
17 58 ♀ ± ♃
19 06 ☽ ∠ ♃
19 24 ☽ ± ☉
19 53 ♀ ⚻ ♆
20 09 ☽ Q ♄
23 25 ☽ # ♄

29 Tuesday
00 07 ☽ ⚼ ♅
00 21 ☽ ∥ ♇
05 22 ☉ Q ♄
08 33 ☽ ⚻ ☉
13 02 ☽ ⊥ ♃
14 30 ☽ ⚻ ♂
19 41 ☽ ∠ ♇

30 Wednesday
00 56 ☽ ⊻ ♃
01 49 ☉ ∠ ♀
01 51 ☽ ⚼ ☿
03 19 ☽ ± ♂
06 14 ☽ # ☉
06 15 ☽ △ ♅
07 50 ☽ ∥ ♃
08 28 ☽ □ ♄
13 21 ☽ □ ♆
13 35 ☽ ⊥ ♇
17 11 ☽ △ ♀
21 01 ☽ # ♀
21 36 ☽ ± ☿
22 21 ☽ ⚼ ♂

31 Thursday
01 26 ☽ ⊻ ♇
08 41 ☽ # ☿
10 35 ☽ ⚻ ☿
20 52 ☉ # ♃

All ephemeris data is given at 12.00 UT and the Moon's longitude is additionally given for 24.00 UT

JUNE 2007

Date	Sidereal time h m s	Sun ☉ ° ' "	Moon ☽ ° ' "	Moon ☽ 24.00 ° ' "	Mercury ☿ ° '	Venus ♀ ° '	Mars ♂ ° '	Jupiter ♃ ° '	Saturn ♄ ° '	Uranus ♅ ° '	Neptune ♆ ° '	Pluto ♇ ° '
01	04 38 22	10 ♊ 37 59	15 ♐ 50 15	22 ♐ 04 05	03 ♋ 54	25 ♋ 46	12 ♈ 45	15 ♐ 29	19 ♌ 43	18 ♓ 30	22 ♒ 01	28 ♐ 06
02	04 42 19	11 35 28	28 ♐ 20 36	04 ♑ 39 49	04 54	26 47	13 30	15 R 21	19 47	18 31	22 R 01	28 R 04
03	04 46 16	12 32 56	11 ♑ 01 45	17 ♑ 26 28	05 50	27 47	14 15	15 13	19 51	18 32	22 01	28 03
04	04 50 12	13 30 23	23 ♑ 53 59	00 ♒ 24 24	06 42	28 46	14 59	15 06	19 56	18 33	22 00	28 01
05	04 54 09	14 27 49	06 ♒ 57 49	13 ♒ 34 20	07 30	29 ♋ 45	15 44	14 58	20 00	18 34	22 00	28 00
06	04 58 05	15 25 14	20 ♒ 14 08	26 ♒ 57 22	08 14	00 ♌ 44	16 29	14 51	20 05	18 34	22 00	27 58
07	05 02 02	16 22 39	03 ♓ 44 13	10 ♓ 34 51	08 53	01 42	17 13	14 43	20 10	18 35	21 59	27 57
08	05 05 58	17 20 03	17 ♓ 29 24	24 ♓ 27 58	09 29	02 40	17 58	14 35	20 14	18 36	21 59	27 55
09	05 09 55	18 17 27	01 ♈ 30 36	08 ♈ 37 13	10 01	03 38	18 42	14 28	20 19	18 37	21 58	27 54
10	05 13 51	19 14 50	15 ♈ 47 42	23 ♈ 01 43	10 28	04 35	19 27	14 20	20 24	18 37	21 58	27 52
11	05 17 48	20 12 12	00 ♉ 18 52	07 ♉ 38 35	10 50	05 32	20 11	14 12	20 29	18 38	21 57	27 51
12	05 21 45	21 09 35	15 ♉ 00 10	22 ♉ 22 46	11 08	06 28	20 56	14 05	20 34	18 39	21 57	27 49
13	05 25 41	22 06 56	29 ♉ 45 27	07 ♊ 07 15	11 22	07 24	21 40	13 57	20 39	18 39	21 56	27 48
14	05 29 38	23 04 17	14 ♊ 27 07	21 ♊ 44 04	11 31	08 19	22 24	13 50	20 44	18 40	21 55	27 46
15	05 33 34	24 01 38	28 ♊ 57 10	06 ♋ 05 35	11 35	09 14	23 08	13 42	20 49	18 40	21 55	27 44
16	05 37 31	24 58 58	13 ♋ 08 39	20 ♋ 05 49	11 R 35	10 08	23 53	13 35	20 55	18 40	21 54	27 43
17	05 41 27	25 56 17	26 ♋ 56 43	03 ♌ 41 10	11 31	11 01	24 37	13 28	21 00	18 41	21 53	27 41
18	05 45 24	26 53 36	10 ♌ 19 08	16 ♌ 50 46	11 22	11 54	25 21	13 20	21 06	18 41	21 53	27 40
19	05 49 20	27 50 53	23 ♌ 16 19	29 ♌ 36 08	11 09	12 47	26 04	13 13	21 11	18 41	21 52	27 38
20	05 53 17	28 48 10	05 ♍ 50 43	12 ♍ 00 36	10 52	13 38	26 48	13 06	21 17	18 41	21 51	27 37
21	05 57 14	29 ♊ 45 26	18 ♍ 06 23	24 ♍ 08 43	10 31	14 29	27 32	12 59	21 22	18 41	21 50	27 35
22	06 01 10	00 ♋ 42 41	00 ♎ 08 14	06 ♎ 05 37	10 06	15 20	28 16	12 52	21 28	18 42	21 49	27 33
23	06 05 07	01 39 56	12 ♎ 01 32	17 ♎ 56 38	09 39	16 09	28 59	12 45	21 34	18 42	21 49	27 32
24	06 09 03	02 37 10	23 ♎ 51 34	29 ♎ 46 55	09 09	16 58	29 ♈ 43	12 38	21 40	18 R 42	21 48	27 30
25	06 13 00	03 34 23	05 ♏ 43 15	11 ♏ 41 06	08 36	17 47	00 ♉ 26	12 31	21 46	18 41	21 47	27 29
26	06 16 56	04 31 36	17 ♏ 40 55	23 ♏ 43 06	08 02	18 34	01 10	12 25	21 52	18 41	21 46	27 27
27	06 20 53	05 28 48	29 ♏ 48 02	05 ♐ 55 58	07 27	19 21	01 53	12 18	21 58	18 41	21 45	27 26
28	06 24 49	06 26 00	12 ♐ 07 08	18 ♐ 21 42	06 52	20 06	02 36	12 12	22 04	18 41	21 44	27 24
29	06 28 46	07 23 12	24 ♐ 39 44	01 ♑ 01 16	06 16	20 51	03 20	12 05	22 10	18 41	21 43	27 23
30	06 32 43	08 ♋ 20 24	07 ♑ 26 17	13 ♑ 54 43	05 ♋ 41	21 ♌ 35	04 ♉ 03	11 ♐ 59	22 ♌ 16	18 ♓ 40	21 ♒ 42	27 ♐ 21

DECLINATIONS

Date	Moon True ☊ ° '	Moon Mean ☊ ° '	Moon ☽ Latitude ° '	Sun ☉ ° '	Moon ☽ ° '	Mercury ☿ ° '	Venus ♀ ° '	Mars ♂ ° '	Jupiter ♃ ° '	Saturn ♄ ° '	Uranus ♅ ° '	Neptune ♆ ° '	Pluto ♇ ° '
01	11 ♓ 37	11 ♓ 39	04 S 59	22 N 03	27 S 38	25 N 11	23 N 34	03 N 38	21 S 57	16 N 07	05 S 15	14 S 25	16 S 23
02	11 R 24	11 36	04 47	22 11	28 13	25 02	23 21	03 55	21 56	16 05	05 15	14 25	16 23
03	11 13	11 32	04 21	22 18	27 19	24 52	23 07	04 13	21 55	16 04	05 15	14 25	16 23
04	11 05	11 29	03 41	22 25	24 57	24 40	22 53	04 30	21 55	16 02	05 14	14 25	16 23
05	10 59	11 26	02 49	22 32	21 15	24 28	22 39	04 47	21 54	16 01	05 14	14 25	16 23
06	10 56	11 23	01 47	22 39	16 26	24 14	22 24	05 04	21 53	16 00	05 14	14 26	16 23
07	10 D 55	11 20	00 S 38	22 45	10 44	24 01	22 09	05 21	21 52	15 58	05 13	14 26	16 23
08	10 R 55	11 17	00 N 35	22 50	04 S 25	23 46	21 53	05 39	21 52	15 56	05 13	14 26	16 23
09	10 54	11 13	01 47	22 55	02 N 14	23 31	21 37	05 55	21 51	15 55	05 13	14 26	16 23
10	10 53	11 10	02 53	23 00	08 53	23 15	21 21	06 12	21 50	15 53	05 13	14 26	16 23
11	10 48	11 07	03 50	23 05	15 10	23 00	21 04	06 29	21 49	15 52	05 13	14 26	16 23
12	10 42	11 04	04 32	23 09	20 40	22 43	20 47	06 46	21 49	15 50	05 12	14 27	16 23
13	10 32	11 01	04 56	23 12	24 55	22 27	20 30	07 03	21 48	15 48	05 12	14 27	16 23
14	10 22	10 57	05 00	23 16	27 30	22 11	20 12	07 19	21 47	15 47	05 12	14 27	16 23
15	10 11	10 54	04 45	23 18	28 11	21 54	19 54	07 36	21 46	15 45	05 12	14 27	16 23
16	10 01	10 51	04 13	23 21	26 59	21 38	19 36	07 52	21 46	15 43	05 12	14 27	16 23
17	09 53	10 48	03 26	23 23	24 08	21 22	19 17	08 08	21 45	15 42	05 12	14 28	16 23
18	09 47	10 45	02 29	23 24	20 03	21 06	18 59	08 24	21 44	15 40	05 12	14 28	16 23
19	09 44	10 42	01 26	23 25	15 07	20 50	18 40	08 40	21 44	15 38	05 12	14 28	16 23
20	09 43	10 38	00 N 21	23 26	09 41	20 35	18 20	08 56	21 43	15 36	05 12	14 29	16 23
21	09 D 43	10 35	00 S 45	23 26	04 N 01	20 20	18 01	09 12	21 42	15 34	05 12	14 29	16 23
22	09 43	10 32	01 46	23 26	01 S 41	20 06	17 41	09 28	21 41	15 32	05 12	14 29	16 24
23	09 R 43	10 29	02 43	23 26	07 15	19 53	17 21	09 44	21 41	15 31	05 12	14 29	16 24
24	09 40	10 26	03 32	23 25	12 33	19 41	17 01	09 59	21 40	15 29	05 12	14 30	16 24
25	09 36	10 22	04 12	23 24	17 24	19 29	16 41	10 15	21 39	15 27	05 12	14 30	16 24
26	09 29	10 19	04 42	23 22	21 37	19 19	16 21	10 30	21 39	15 25	05 12	14 30	16 24
27	09 20	10 16	04 59	23 20	24 59	19 09	16 00	10 45	21 38	15 23	05 12	14 31	16 24
28	09 10	10 13	05 03	23 17	27 15	19 01	15 40	11 00	21 37	15 21	05 12	14 31	16 24
29	08 59	10 10	04 52	23 14	28 12	18 53	15 20	11 15	21 37	15 19	05 12	14 31	16 24
30	08 ♓ 49	10 ♓ 07	04 S 27	23 N 11	27 S 40	18 N 47	14 N 59	11 N 30	21 S 36	15 N 17	05 S 12	14 S 32	16 S 24

ZODIAC SIGN ENTRIES

Date	h	m	Planets
02	15	09	☽ ♑
04	23	15	☽ ♒
05	17	59	♀ ♌
07	05	24	☽ ♓
09	09	26	☽ ♈
11	11	29	☽ ♉
13	12	24	☽ ♊
15	13	45	☽ ♋
17	17	25	☽ ♌
20	00	46	☽ ♍
21	18	06	☉ ♋
22	11	43	☽ ♎
24	21	27	♂ ♉
25	00	26	☽ ♏
27	12	24	☽ ♐
29	22	05	☽ ♑

LATITUDES

Date	Mercury ☿ ° '	Venus ♀ ° '	Mars ♂ ° '	Jupiter ♃ ° '	Saturn ♄ ° '	Uranus ♅ ° '	Neptune ♆ ° '	Pluto ♇ ° '
01	01 N 48	02 N 37	01 S 31	00 N 42	01 N 16	00 S 46	00 S 16	07 N 03
04	01 24	02 32	01 31	00 42	01 16	00 46	00 16	07 03
07	00 52	02 26	01 31	00 42	01 16	00 46	00 16	07 03
10	00 N 14	02 18	01 31	00 41	01 16	00 46	00 16	07 03
13	00 S 30	02 09	01 31	00 41	01 16	00 47	00 16	07 02
16	01 19	01 58	01 30	00 40	01 16	00 47	00 16	07 02
19	02 09	01 45	01 30	00 40	01 16	00 47	00 16	07 02
22	02 57	01 31	01 29	00 40	01 16	00 47	00 16	07 02
25	03 41	01 14	01 29	00 39	01 15	00 47	00 16	07 01
28	04 16	00 56	01 28	00 38	01 15	00 47	00 16	07 01
31	04 S 38	00 N 36	01 S 27	00 N 38	01 N 15	00 S 47	00 S 16	07 N 00

DATA

Julian Date	2454253
Delta T	+66 seconds
Ayanamsa	23° 57' 43"
Synetic vernal point	05° ♓ 09' 16"
True obliquity of ecliptic	23° 26' 26"

MOON'S PHASES, APSIDES AND POSITIONS ☽

Date	h	m	Phase	Longitude ° '	Eclipse Indicator
01	01	04	○	10 ♐ 12	
08	11	43	☾	17 ♓ 19	
15	03	13	●	23 ♊ 41	
22	13	15	☽	00 ♎ 46	
30	13	49	○	08 ♑ 25	

Day	h	m	
12	16	58	Perigee
24	14	23	Apogee

Day	h	m		
02	09	29	Max dec	28° S 13'
09	04	00	ON	
15	08	28	Max dec	28° N 13'
22	04	53	OS	
29	15	40	Max dec	28° S 13'

LONGITUDES

Date	Chiron ⚷ ° '	Ceres ⚳ ° '	Pallas ⚴ ° '	Juno ⚵ ° '	Vesta ⚶ ° '	Black Moon Lilith ⚸ ° '
01	15 ♒ 37	28 ♈ 28	16 ♓ 35	12 ♎ 04	08 ♐ 20	24 ♎ 55
11	15 ♒ 26	02 ♉ 00	17 ♓ 57	12 ♎ 12	06 ♐ 01	26 ♎ 02
21	15 ♒ 10	05 ♉ 24	18 ♓ 53	12 ♎ 50	04 ♐ 06	27 ♎ 09
31	14 ♒ 48	08 ♉ 38	19 ♓ 21	13 ♎ 55	02 ♐ 50	28 ♎ 16

All ephemeris data is given at 12.00 UT and the Moon's longitude is additionally given for 24.00 UT

ASPECTARIAN

h m	Aspects
01 Friday	
00 43	☽ Q ♆
01 04	☽ ☍ ☉
01 18	☽ bQ ♀
05 38	☽ △ ♂
11 19	☽ ☌ ♃
17 08	☽ □ ♅
19 31	☽ △ ♄
20 15	☽ ± ♀
23 54	☽ ⚹ ♆
02 Saturday	
08 45	☽ ⊼ ♀
08 56	☿ ∠ ♄
11 29	☽ ☌ ♇
03 Sunday	
00 18	☽ bQ ♄
01 27	☽ ☍ ☿
03 31	☽ Q ♅
04 26	☽ ∠ ♆
15 05	☽ ⊼ ☉
17 20	☽ ± ♄
18 23	♀ ⊼ ♇
18 24	☽ □ ♂
19 47	☽ ⊻ ♃
21 20	☽ ⊥ ♆
04 Monday	
02 03	☽ ⚹ ♅
03 12	☽ ± ☉
04 36	☽ ⊼ ♄
06 51	☽ ⊥ ♃
08 29	☽ ⊻ ♆
14 21	☽ # ☿
14 58	♂ △ ♃
19 36	☽ ⊻ ♇
21 02	☿ bQ ♆
21 11	☽ bQ ☉
21 43	☽ ☍ ♀
23 19	☽ ∠ ♃
05 Tuesday	
03 19	☽ # ♀
04 45	☽ # ☉
05 45	☽ Q ♂
05 46	☽ ∠ ♅
06 36	☽ ⊥ ♇
08 20	☽ ∥ ♃
13 02	☽ ⊼ ☿
16 39	♀ bQ ♃
19 42	☉ ∥ ♀
22 11	☽ ⊥ ♅
22 56	☽ ∠ ♇
23 13	☉ ☍ ♃
06 Wednesday	
00 34	☽ ± ☿
02 23	☽ ⚹ ♃
02 40	☽ △ ☉
04 51	☽ ⚹ ♂
09 01	☽ ⊻ ♅
11 43	☽ ☍ ♄
12 15	☽ ∥ ♇
13 59	☽ # ♄
15 09	☽ ☌ ♆
17 39	☽ bQ ☿
20 48	☽ ∥ ♆
23 41	☽ Q ♃
07 Thursday	
00 56	♂ # ♅
01 47	☽ ⚹ ♇
08 09	☽ ⊼ ♀
09 11	☽ ∠ ♂
19 30	☽ ± ♀
21 28	☽ △ ☿
22 52	☽ Q ♇
08 Friday	
01 52	☽ ⊥ ♂
07 01	☽ □ ♃
07 38	☽ # ♂
09 01	☽ ∥ ♅
11 43	☽ □ ☉
12 21	☽ bQ ♀
12 52	☽ ⊻ ♂
13 55	☽ ☌ ♅
14 42	♀ # ♃
16 46	☽ ⊼ ♄
19 44	☽ ⊻ ♆
09 Saturday	
03 07	☽ ± ♄
05 52	☽ □ ♇
05 59	☽ ⊥ ♆
08 49	♂ ⊻ ♅
11 26	♀ bQ ♅
15 51	☽ △ ♀
18 25	♀ ± ♇
18 29	☽ bQ ♄
20 08	☉ □ ♅
20 40	☽ Q ☉
21 13	☽ ∠ ♆
22 42	☽ # ♅
10 Sunday	
01 51	☽ ∥ ♂
02 49	☽ □ ☿
09 35	☽ △ ♃
16 42	☽ ⊻ ♅
18 09	☽ ⚹ ☉
18 24	☽ ☌ ♂
19 42	☽ △ ♄
22 14	☽ ⚹ ♆
11 Monday	
02 39	☽ ⊥ ♅
05 48	☉ ∥ ☿
07 57	☽ △ ♇
09 06	☽ # ♆
09 31	☽ Q ☿
10 12	☽ bQ ♃
10 27	☉ ⚹ ♂
14 49	☽ ∥ ♄
16 58	☽ # ♇
17 27	☽ ∠ ♅
17 58	☽ Q ♆
19 40	☉ ⚹ ♄
20 34	☽ ∠ ☉
21 08	☽ □ ♀
22 43	♂ △ ♄
12 Tuesday	
00 49	☽ ± ♃
05 35	☽ ⚹ ☿
08 27	☽ bQ ♇
10 31	☽ ⊼ ♃
12 16	☽ ⊥ ☉
12 35	☽ ∥ ♀
17 45	☽ # ♃
17 56	☽ ⚹ ♅
21 06	☽ □ ♄
22 00	☽ ∥ ☿
22 09	☽ ⊻ ♂
22 43	☽ ⊻ ☉
23 04	☽ ± ♇
23 17	☽ □ ♆
13 Wednesday	
01 12	☽ ∥ ☉
04 26	☽ Q ♀
06 24	☽ ∠ ☿
07 30	☉ △ ♆
08 25	☽ ⊥ ♂
08 48	☽ ⊼ ♇
13 27	☽ Q ♅
20 35	♂ ⚹ ♆
21 15	☽ ⊥ ☿
23 51	☽ ∠ ♂
14 Thursday	
01 17	☽ ⚹ ♀
02 35	☽ Q ♄
07 09	☽ ⊻ ☿
10 59	☽ ☍ ♃
18 55	☽ □ ♅
22 25	☽ ⚹ ♄
15 Friday	
00 18	☽ △ ♆
01 49	☽ ⚹ ♂
03 13	☽ ☌ ☉
03 36	☽ ∠ ♀
09 59	☽ ☍ ♇
19 39	☽ ⊥ ♀
22 57	☽ Q ♂
23 37	☽ ∠ ♄
23 40	☿ St R
23 46	☿ # ♃
16 Saturday	
01 23	☽ bQ ♆
06 30	☽ ⊻ ♀
09 21	☽ ☌ ☿
12 45	☽ ⊼ ♃
15 03	☽ ⊥ ♄
16 44	☽ ± ♆
21 32	☽ △ ♅
23 01	☽ ± ♃
17 Sunday	
01 31	☽ ⊻ ♄
03 08	☽ ⊼ ♆
07 39	☽ □ ♂
10 05	☽ ⊻ ☉
13 19	☽ ⊼ ♇
14 14	♂ ⊥ ♅
14 39	☽ bQ ♃
16 57	☽ ∥ ☉
21 33	☽ ⊥ ☉
23 40	☿ ⊻ ♀
23 59	☽ ± ♇
23 59	☽ bQ ♅
18 Monday	
02 46	☽ # ♃
06 02	☽ ∥ ☿
13 53	☽ ⊻ ☿
14 57	☉ ∠ ♀
15 06	☽ ☌ ♀
15 06	☽ ∠ ☉
16 17	☽ bQ ♇
16 19	☽ ± ♅
17 29	☽ △ ♃
17 55	☽ ∥ ♀
19 Tuesday	
00 46	☽ ⊥ ☿
03 25	☽ ⊼ ♅
06 06	☽ # ♇
06 49	☉ ☍ ♇
08 03	☽ ☌ ♄
08 16	♀ bQ ♇
09 22	☽ ☍ ♆
09 30	♀ ± ♅
09 37	☽ ∥ ♄
14 57	☽ # ♆
17 19	☽ ∠ ☿
17 37	☽ △ ♂
20 14	☽ △ ♇
21 22	☽ ⚹ ☉
22 48	♀ △ ♃
20 Wednesday	
15 03	☽ ∥ ♂
21 30	☽ ⚹ ☿
22 27	☽ Q ☉
21 Thursday	
00 20	☽ bQ ♂
02 00	☽ □ ♃
04 20	☽ ⊻ ♀
07 04	☽ # ♅
13 09	☽ ☍ ♅
13 36	♂ △ ♇
17 05	☽ ⊥ ♀
18 32	☽ ⊻ ♄
19 14	☽ ± ♂
19 24	☽ ⊼ ♆
20 28	☽ Q ☿
22 Friday	
00 42	♂ bQ ♃
06 36	☽ ⊥ ♄
06 50	☽ □ ♇
07 22	☽ ± ♆
08 00	☽ ⊼ ♂
08 45	☿ Q ♂
12 25	☽ ∠ ♀
13 15	☽ □ ☉
13 27	☽ Q ♃
23 Saturday	
00 52	☽ ∠ ♄
01 28	☽ bQ ♆
02 34	☿ ⊥ ♀
03 01	☽ ∥ ♅
07 22	☽ □ ☿
13 27	☽ ⚹ ♃
14 42	♅ St R
19 06	☽ Q ♇
21 00	☽ ⚹ ♀
23 35	☽ # ♂
24 Sunday	
01 31	☽ ⊼ ♅
07 30	☽ ⚹ ♄
07 49	☽ △ ♆
13 41	☽ ± ♅
19 23	☽ ⚹ ♇
19 35	☽ ∠ ♃
21 21	☽ ∥ ♆
23 07	☽ Q ♀
25 Monday	
00 38	☽ ☍ ♂
02 06	☽ # ♄
06 50	☽ ∥ ♇
07 17	☽ △ ☉
07 54	☽ bQ ♅
08 01	☽ Q ♄
08 34	☽ # ♀
13 36	☽ ⊥ ♃
15 54	♄ ☍ ♆
17 33	☽ △ ☿
22 55	☽ # ☿
26 Tuesday	
01 33	☽ ⊻ ♃
01 34	☽ ∠ ♇
08 22	♀ # ♇
12 12	☽ ∥ ♃
13 53	☽ □ ♀
14 00	☽ △ ♅
15 43	♀ ⊼ ♅
16 00	☽ bQ ☉
19 30	☽ ⊥ ♇
20 07	☽ □ ♆
20 23	☽ □ ♄
22 10	☽ bQ ☿
23 31	☽ # ☉
27 Wednesday	
07 21	☽ ⊻ ♇
11 19	☽ ± ☉
15 06	☽ ± ☿
16 21	☽ ⊼ ♂
28 Thursday	
00 03	☽ ⊼ ☉
02 17	☽ ⊼ ☿
04 47	☽ ± ♂
05 10	☿ ∠ ♄
07 23	☽ Q ♆
12 09	☽ ☌ ♃
17 32	☿ bQ ♆
18 40	☉ ☌ ☿
19 20	☉ bQ ♆
23 12	☽ bQ ♂
29 Friday	
00 37	☽ □ ♅
04 18	☽ △ ♀
05 41	☉ ∠ ♄
06 24	☽ ⚹ ♆
07 13	☽ △ ♄
13 00	♀ ∥ ♄
17 08	☽ ☌ ♇
19 34	☿ ∠ ♀
23 43	♂ ∠ ♅
30 Saturday	
00 37	♂ Q ♆
05 17	☽ △ ♂
08 53	☽ ☍ ☿
10 19	☽ bQ ♀
10 35	☽ Q ♅
10 37	☽ ∠ ♆
11 41	☽ bQ ♄
13 49	☽ ☍ ☉
15 30	♀ ☍ ♆
20 23	☽ ⊻ ♃

JULY 2007

Date	Sidereal time h m s	Sun ☉ ° ' "	Moon ☽ ° ' "	Moon ☽ 24.00 ° ' "	Mercury ☿ ° '	Venus ♀ ° '	Mars ♂ ° '	Jupiter ♃ ° '	Saturn ♄ ° '	Uranus ♅ ° '	Neptune ♆ ° '	Pluto ♇ ° '
01	06 36 39	09 ♋ 17 35	20 ♑ 26 27	27 ♑ 01 23	05 ♋ 08	22 ♌ 18	04 ♉ 46	11 ♐ 53	22 ♌ 22	18 ♓ 40	21 ♒ 41	27 ♐ 20
02	06 40 36	10 14 46	03 ♒ 39 20	10 ♒ 20 10	04 R 36	23 00	05 29	11 R 47	22 28	18 R 40	21 R 40	27 R 18
03	06 44 32	11 11 57	17 ♒ 03 46	23 ♒ 49 58	04 07	23 41	06 11	11 41	22 35	18 39	21 38	27 17
04	06 48 29	12 09 09	00 ♓ 38 42	07 ♓ 29 50	03 41	24 21	06 54	11 36	22 41	18 39	21 37	27 15
05	06 52 25	13 06 20	14 ♓ 23 20	21 ♓ 19 07	03 18	25 00	07 37	11 30	22 48	18 38	21 36	27 14
06	06 56 22	14 03 32	28 ♓ 17 08	05 ♈ 17 19	03 00	25 38	08 20	11 25	22 54	18 38	21 35	27 12
07	07 00 18	15 00 44	12 ♈ 19 37	19 ♈ 23 53	02 45	26 14	09 02	11 19	23 01	18 37	21 34	27 11
08	07 04 15	15 57 56	26 ♈ 29 59	03 ♉ 37 41	02 35	26 50	09 45	11 14	23 07	18 36	21 32	27 09
09	07 08 12	16 55 09	10 ♉ 46 43	17 ♉ 56 41	02 29	27 24	10 27	11 09	23 14	18 36	21 31	27 08
10	07 12 08	17 52 22	25 ♉ 07 10	02 ♊ 17 38	02 D 29	27 56	11 09	11 04	23 21	18 35	21 30	27 06
11	07 16 05	18 49 36	09 ♊ 27 32	16 ♊ 36 13	02 33	28 28	11 51	10 59	23 27	18 34	21 29	27 05
12	07 20 01	19 46 50	23 ♊ 43 03	00 ♋ 47 23	02 43	28 58	12 34	10 55	23 34	18 33	21 27	27 03
13	07 23 58	20 44 04	07 ♋ 48 34	14 ♋ 46 03	02 58	29 26	13 16	10 50	23 41	18 32	21 26	27 02
14	07 27 54	21 41 19	21 ♋ 39 19	28 ♋ 27 57	03 19	29 ♌ 53	13 57	10 46	23 48	18 31	21 25	27 01
15	07 31 51	22 38 34	05 ♌ 11 38	11 ♌ 50 09	03 45	00 ♍ 18	14 39	10 42	23 55	18 30	21 23	26 59
16	07 35 47	23 35 50	18 ♌ 23 26	24 ♌ 51 29	04 16	00 42	15 21	10 38	24 02	18 29	21 22	26 58
17	07 39 44	24 33 05	01 ♍ 14 28	07 ♍ 32 34	04 53	01 04	16 03	10 34	24 09	18 28	21 20	26 57
18	07 43 41	25 30 21	13 ♍ 46 08	19 ♍ 55 33	05 35	01 24	16 44	10 31	24 16	18 27	21 19	26 55
19	07 47 37	26 27 36	26 ♍ 01 17	02 ♎ 03 52	06 22	01 43	17 25	10 27	24 23	18 26	21 18	26 54
20	07 51 34	27 24 52	08 ♎ 03 50	14 ♎ 01 47	07 14	01 59	18 07	10 24	24 30	18 25	21 16	26 53
21	07 55 30	28 22 09	19 ♎ 58 20	25 ♎ 54 06	08 12	02 14	18 48	10 21	24 37	18 23	21 15	26 51
22	07 59 27	29 ♋ 19 25	01 ♏ 49 43	07 ♏ 45 49	09 14	02 27	19 29	10 18	24 44	18 22	21 13	26 50
23	08 03 23	00 ♌ 16 42	13 ♏ 43 00	19 ♏ 41 51	10 22	02 37	20 10	10 15	24 51	18 21	21 12	26 49
24	08 07 20	01 13 59	25 ♏ 42 55	01 ♐ 46 42	11 35	02 45	20 51	10 13	24 58	18 19	21 10	26 48
25	08 11 16	02 11 17	07 ♐ 53 42	14 ♐ 04 18	12 52	02 52	21 31	10 10	25 05	18 18	21 09	26 46
26	08 15 13	03 08 35	20 ♐ 18 51	26 ♐ 37 37	14 14	02 56	22 12	10 08	25 13	18 17	21 07	26 45
27	08 19 10	04 05 53	03 ♑ 00 48	09 ♑ 28 31	15 41	02 57	22 52	10 06	25 20	18 15	21 06	26 44
28	08 23 06	05 03 13	16 ♑ 00 48	22 ♑ 37 35	17 12	02 R 57	23 33	10 05	25 27	18 14	21 04	26 43
29	08 27 03	06 00 32	29 ♑ 18 45	06 ♒ 04 06	18 47	02 54	24 13	10 03	25 35	18 12	21 03	26 42
30	08 30 59	06 57 53	12 ♒ 53 21	19 ♒ 46 11	20 27	02 48	24 53	10 01	25 42	18 10	21 01	26 41
31	08 34 56	07 ♌ 55 14	26 ♒ 42 15	03 ♓ 41 09	22 ♋ 10	02 ♍ 41	25 ♉ 33	10 ♐ 00	25 ♌ 49	18 ♓ 09	20 ♒ 59	26 ♐ 40

Date	Moon True ☊ ° '	Moon Mean ☊ ° '	Moon ☽ Latitude ° '
01	08 ♓ 40	10 ♓ 03	03 S 47
02	08 R 33	10 00	02 55
03	08 29	09 57	01 52
04	08 27	09 54	00 S 42
05	08 D 26	09 51	00 N 32
06	08 27	09 48	01 44
07	08 28	09 44	02 52
08	08 R 27	09 41	03 49
09	08 25	09 38	04 32
10	08 21	09 35	04 59
11	08 15	09 32	05 07
12	08 08	09 28	04 56
13	08 00	09 25	04 27
14	07 53	09 22	03 42
15	07 47	09 19	02 46
16	07 44	09 16	01 43
17	07 42	09 13	00 N 35
18	07 D 42	09 09	00 S 33
19	07 43	09 06	01 38
20	07 45	09 03	02 37
21	07 46	09 00	03 29
22	07 R 46	08 57	04 12
23	07 45	08 54	04 44
24	07 43	08 50	05 04
25	07 39	08 47	05 11
26	07 34	08 44	05 03
27	07 28	08 41	04 41
28	07 23	08 38	04 04
29	07 18	08 34	03 12
30	07 15	08 31	02 09
31	07 ♓ 13	08 ♓ 28	00 S 57

DECLINATIONS

Date	Sun ☉ ° '	Moon ☽ ° '	Mercury ☿ ° '	Venus ♀ ° '	Mars ♂ ° '	Jupiter ♃ ° '	Saturn ♄ ° '	Uranus ♅ ° '	Neptune ♆ ° '	Pluto ♇ ° '
01	23 N 07	25 S 38	18 N 43	14 N 38	11 N 45	21 S 35	15 N 15	05 S 12	14 S 32	16 S 24
02	23 03	22 10	18 40	14 18	11 59	21 35	15 13	05 13	14 32	16 25
03	22 58	17 30	18 38	13 57	12 14	21 34	15 11	05 13	14 33	16 25
04	22 53	11 54	18 37	13 37	12 28	21 33	15 08	05 13	14 33	16 25
05	22 48	05 S 39	18 38	13 16	12 42	21 33	15 06	05 13	14 34	16 25
06	22 42	00 N 55	18 40	12 56	12 56	21 32	15 04	05 14	14 34	16 25
07	22 36	07 30	18 44	12 35	13 10	21 32	15 02	05 14	14 34	16 25
08	22 29	13 46	18 48	12 15	13 24	21 31	15 00	05 14	14 35	16 25
09	22 22	19 22	18 54	11 55	13 38	21 31	14 58	05 14	14 35	16 26
10	22 15	23 53	19 01	11 35	13 51	21 30	14 55	05 15	14 36	16 26
11	22 07	26 55	19 09	11 15	14 05	21 30	14 53	05 15	14 36	16 26
12	21 59	28 13	19 18	10 56	14 18	21 29	14 51	05 15	14 37	16 26
13	21 50	27 39	19 28	10 36	14 31	21 29	14 49	05 16	14 37	16 26
14	21 42	25 22	19 38	10 17	14 44	21 28	14 46	05 16	14 37	16 26
15	21 32	21 40	19 49	09 58	14 57	21 28	14 44	05 17	14 38	16 27
16	21 23	16 57	20 00	09 40	15 09	21 28	14 42	05 17	14 38	16 27
17	21 13	11 35	20 12	09 22	15 22	21 27	14 40	05 18	14 39	16 27
18	21 02	05 53	20 23	09 04	15 34	21 27	14 37	05 18	14 39	16 27
19	20 52	00 N 05	20 35	08 47	15 46	21 27	14 35	05 19	14 40	16 27
20	20 41	05 S 36	20 46	08 30	15 58	21 27	14 33	05 19	14 40	16 27
21	20 29	11 02	20 57	08 13	16 10	21 26	14 30	05 20	14 41	16 28
22	20 18	16 03	21 07	07 57	16 22	21 26	14 28	05 20	14 41	16 28
23	20 06	20 28	21 17	07 42	16 33	21 26	14 25	05 21	14 42	16 28
24	19 53	24 07	21 25	07 27	16 44	21 26	14 23	05 21	14 42	16 28
25	19 40	26 44	21 33	07 13	16 56	21 26	14 21	05 22	14 43	16 28
26	19 27	28 08	21 38	06 59	17 07	21 26	14 18	05 22	14 43	16 29
27	19 14	28 05	21 43	06 46	17 17	21 25	14 16	05 23	14 44	16 29
28	19 00	26 31	21 46	06 34	17 28	21 25	14 13	05 24	14 44	16 29
29	18 46	23 26	21 46	06 23	17 39	21 25	14 11	05 24	14 45	16 29
30	18 32	19 01	21 45	06 12	17 49	21 25	14 08	05 25	14 45	16 30
31	18 N 17	13 S 31	21 N 41	06 N 02	17 N 59	21 S 25	14 N 06	05 S 26	14 S 46	16 S 30

ZODIAC SIGN ENTRIES

Date	h m	Planets
02	05 24	☽ ♒
04	10 52	☽ ♓
06	14 57	☽ ♈
08	17 54	☽ ♉
10	20 10	☽ ♊
12	22 39	☽ ♋
14	18 23	♀ ♍
15	02 43	☽ ♌
17	09 39	☽ ♍
19	19 53	☽ ♎
22	08 18	☽ ♏
23	05 00	☉ ♌
24	20 29	☽ ♐
27	06 21	☽ ♑
29	13 14	☽ ♒
31	17 40	☽ ♓

LATITUDES

Date	Mercury ☿ ° '	Venus ♀ ° '	Mars ♂ ° '	Jupiter ♃ ° '	Saturn ♄ ° '	Uranus ♅ ° '	Neptune ♆ ° '	Pluto ♇ ° '
01	04 S 38	00 N 36	01 S 27	00 N 38	01 N 15	00 S 47	00 S 16	07 N 00
04	04 46	00 N 13	01 26	00 37	01 15	00 47	00 16	07 00
07	04 41	00 S 12	01 25	00 37	01 15	00 47	00 16	06 59
10	04 24	00 39	01 23	00 36	01 15	00 48	00 16	06 59
13	03 57	01 08	01 22	00 36	01 15	00 48	00 16	06 58
16	03 22	01 40	01 21	00 35	01 15	00 48	00 16	06 58
19	02 43	02 14	01 19	00 34	01 15	00 48	00 16	06 57
22	02 00	02 50	01 17	00 34	01 15	00 48	00 16	06 56
25	01 17	03 28	01 16	00 33	01 16	00 48	00 16	06 56
28	00 S 35	04 08	01 14	00 33	01 16	00 48	00 17	06 55
31	00 N 04	04 S 48	01 S 12	00 N 32	01 N 16	00 S 48	00 S 17	06 N 54

DATA

Julian Date	2454283
Delta T	+66 seconds
Ayanamsa	23° 57' 49"
Synetic vernal point	05° ♓ 09' 10"
True obliquity of ecliptic	23° 26' 26"

MOON'S PHASES, APSIDES AND POSITIONS ☽

Date	h	m	Phase	Longitude ° '	Eclipse Indicator
07	16	54	☾	15 ♈ 12	
14	12	04	●	21 ♋ 41	
22	06	29	☽	29 ♎ 06	
30	00	48	○	06 ♒ 31	

Day	h	m		
09	21	31	Perigee	
22	08	43	Apogee	
06	08	41	0N	
12	16	37	Max dec	28° N 15'
19	12	21	0S	
26	23	22	Max dec	28° S 18'

LONGITUDES

Date	Chiron ⚷ ° '	Ceres ⚳ ° '	Pallas ⚴ ° '	Juno ⚵ ° '	Vesta ⚶ ° '	Black Moon Lilith ⚸ ° '
01	14 ♒ 48	08 ♉ 38	19 ♓ 21	13 ♎ 55	02 ♐ 50	28 ♎ 16
11	14 ♒ 21	11 ♉ 41	19 ♓ 17	15 ♎ 25	02 ♐ 22	29 ♎ 23
21	13 ♒ 51	14 ♉ 30	18 ♓ 40	17 ♎ 15	02 ♐ 42	00 ♏ 30
31	13 ♒ 20	17 ♉ 04	17 ♓ 27	19 ♎ 23	03 ♐ 48	01 ♏ 37

ASPECTARIAN

h m	Aspects
01 Sunday	
03 16	☽ ⊥ ♆
03 58	☽ ± ♀
04 28	☽ ± ♄
07 21	☽ ⊥ ♃
08 45	☽ ✶ ♅
14 15	☽ ⊻ ♆
14 38	♀ ☌ ♄
15 33	☽ ⊼ ♄
15 36	☽ ⊼ ♀
19 08	☿ ✶ ♂
19 13	♀ # ♆
23 40	☽ ∠ ♃
02 Monday	
00 32	☽ ⊻ ♇
06 35	☽ # ☉
11 22	☽ ⊥ ♇
12 01	☽ ∠ ♅
13 39	☽ ⊼ ☿
15 23	☽ ∥ ♃
15 28	☽ □ ♂
21 09	♂ ± ♃
03 Tuesday	
00 02	☽ ± ☿
00 45	☽ ⊼ ☉
02 29	☽ ✶ ♃
03 29	☽ ∠ ♇
04 09	☽ ⊥ ♅
06 38	☽ # ☿
12 16	☽ ± ☉
14 50	☽ ⊻ ♅
15 32	☽ bQ ☿
16 56	☽ ∥ ♇
20 07	☽ ☌ ♆
21 52	☽ ☍ ♄
22 26	☽ # ♄
23 13	☉ ⊼ ♃
23 40	☽ Q ♃
04 Wednesday	
00 21	☽ ☍ ♀
01 02	☽ ∥ ♆
01 20	☽ Q ♂
04 33	☽ # ♀
05 24	☽ bQ ☉
06 03	☽ ✶ ♇
09 46	☽ # ♂
17 11	☽ △ ☿
23 34	☽ ✶ ♂
05 Thursday	
03 02	☽ Q ♇
07 01	☽ □ ♃
09 36	☽ △ ☉
13 37	☽ ∥ ♅
19 22	☽ ☌ ♅
06 Friday	
00 28	☽ ⊻ ♆
02 40	☽ ⊼ ♄
03 00	☽ ∠ ♂
07 13	☽ ⊼ ♀
10 08	☽ □ ♇
10 48	☽ ⊥ ♆
11 29	♀ ∥ ♂
13 04	☽ ± ♄
18 00	☽ ± ♀
19 18	☽ ⊥ ♂
19 55	☽ □ ☿
07 Saturday	
02 11	☽ ∠ ♆
03 39	☽ # ♅
04 35	☽ bQ ♄
06 06	☽ ⊻ ♂
10 04	☽ bQ ♀
10 18	☽ △ ♃
16 54	☽ □ ☉
22 40	☽ ⊻ ♅
08 Sunday	
01 32	☉ ± ♆
02 06	☽ Q ☿
03 38	☽ ✶ ♆
06 15	☽ △ ♄
06 18	☽ ∥ ♀
08 48	☽ ⊥ ♅
10 28	☽ ∥ ♂
11 33	☽ bQ ♃
12 34	☽ △ ♀
13 06	☽ △ ♇
15 15	☽ # ♆
16 56	☽ ∥ ♄
22 09	☽ ✶ ☿
22 55	☽ # ♇
23 50	☽ Q ♆
23 57	☽ ∠ ♅
09 Monday	
01 09	♀ △ ♇
01 28	☽ Q ☉
02 37	☽ ± ♃
09 49	☽ ∥ ☿
11 25	☽ ☌ ♂
12 37	☽ ⊼ ♃
14 16	☽ bQ ♇
17 23	☉ ± ♃
20 54	☉ ⊥ ♄
22 35	☽ # ♃
23 01	☽ ✶ ☉
23 13	☽ ∠ ☿
10 Tuesday	
01 04	☽ ✶ ♅
02 15	☿ St D
02 47	☽ ∥ ☉
05 18	☽ ± ♇
05 57	☽ □ ♆
09 00	☽ □ ♄
09 24	♂ ⊼ ♃
14 17	☽ ⊥ ☿
15 19	☽ ⊼ ♇
16 54	☽ □ ♀
21 07	☽ Q ♅
11 Wednesday	
00 22	☽ ⊻ ☿
01 54	☽ ∠ ☉
05 33	☉ △ ♅
14 33	☽ ☍ ♃
15 22	☽ Q ♄
16 14	☽ ⊻ ♂
18 03	☽ ⊥ ☉
19 23	♂ bQ ♇
12 Thursday	
00 12	☽ Q ♀
02 51	☽ ⊥ ♂
03 17	☽ □ ♅
04 52	☽ ⊻ ☉
08 11	☽ △ ♆
11 45	☽ ✶ ♄
17 39	☽ ☍ ♇
18 51	☽ ∠ ♂
21 12	☽ ✶ ♀
13 Friday	
03 33	☽ ☌ ☿
09 39	☽ bQ ♆
13 31	☽ ∠ ♄
17 11	☽ ⊼ ♃
21 53	☽ ✶ ♂
23 36	♂ # ♆
23 49	☽ ∠ ♀
14 Saturday	
01 08	☽ ± ♆
03 31	☽ ± ♃
05 10	☉ ⊼ ♆
05 12	☽ ⊥ ♄
06 32	☽ △ ♅
11 34	☽ ⊼ ♆
12 04	☽ ☌ ☉
15 47	☽ ⊻ ♄
16 02	☽ ⊥ ♀
16 12	♂ ∥ ♄
19 12	☽ bQ ♃
19 58	☽ Q ♂
21 25	☽ ⊼ ♇
15 Sunday	
03 00	☽ ⊻ ♀
08 03	☽ ± ♇
08 59	☽ bQ ♅
09 19	☽ ⊻ ☿
12 42	☽ ∥ ☉
13 04	☽ # ♃
20 32	☽ ⊥ ☿
21 31	☽ ∥ ☿
21 53	☽ △ ♃
22 57	☉ # ♃
16 Monday	
00 15	☽ bQ ♇
01 12	☽ ± ♅
06 06	☽ □ ♂
12 11	☽ ⊼ ♅
13 42	☽ ∠ ☿
14 19	☽ # ♇
17 29	☽ ☍ ♆
19 56	☽ ∥ ♂
22 23	☽ ∥ ♄
22 25	☽ ⊻ ☉
22 32	☽ ☌ ♄
22 33	☽ # ♆
17 Tuesday	
00 16	☉ ⊻ ♄
03 55	☽ △ ♇
10 35	☽ ⊥ ☉
11 40	☽ ☌ ♀
18 51	♄ # ♆
19 18	☽ ✶ ☿
21 57	☽ ∥ ♀
18 Wednesday	
05 10	☽ ∠ ☉
05 44	☽ □ ♃
08 16	☉ ⊥ ♀
12 10	☉ bQ ♃
14 23	☽ # ♅
18 06	☽ △ ♂
19 54	☽ Q ☿
21 06	☽ ☍ ♅
19 Thursday	
02 42	☽ ⊼ ♆
08 43	☽ ⊻ ♄
10 03	☿ bQ ♆
12 57	☽ ✶ ☉
13 44	☽ □ ♇
14 31	☽ ± ♆
16 48	☽ Q ♃
20 43	☽ ⊥ ♄
22 47	☉ ⊼ ♇
23 35	☽ ⊻ ♀
20 Friday	
01 29	☽ bQ ♂
06 14	☉ ∥ ☿
08 25	☽ bQ ♆
10 12	☽ □ ☿
10 45	☽ ∥ ♅
11 51	☽ ⊥ ♀
14 54	☽ ∠ ♄
14 57	☽ Q ☉
16 40	☽ ✶ ♃
20 38	☽ ± ♂
22 13	♂ ✶ ♅
23 58	☽ # ♀
21 Saturday	
01 41	☽ Q ♇
06 21	☽ ∠ ♀
08 49	☽ ⊼ ♅
09 29	☽ ⊼ ♂
14 34	☽ △ ♆
20 55	☽ ± ♅
21 29	☽ ✶ ♄
22 50	☽ ∠ ♃
22 Sunday	
01 54	☽ ✶ ♇
04 12	☽ # ♄
05 13	☽ ∥ ♆
06 29	☽ □ ☉
13 16	☽ ✶ ♀
13 38	☽ # ♂
14 06	☽ ∥ ♇
15 07	☽ bQ ♅
16 59	☽ ⊥ ♃
22 01	☽ Q ♄
23 57	☿ ∠ ♄
23 Monday	
01 18	♂ # ♇
04 32	☽ △ ☿
05 03	☽ ⊻ ♃
08 11	☽ ∠ ♇
09 46	☿ ⊼ ♃
09 52	☽ # ☉
13 50	☽ Q ♀
17 02	☽ # ☿
17 48	☽ ∥ ♃
21 17	☽ △ ♅
24 Tuesday	
01 43	☽ ☍ ♂
02 13	☽ ⊥ ♇
02 58	☽ □ ♆
10 18	♂ ± ♇
10 30	☽ □ ♄
13 50	☿ # ♃
13 55	☽ bQ ☿
14 08	☽ ⊻ ♇
23 10	♂ □ ♆
23 51	☽ △ ☉
25 Wednesday	
02 03	☽ □ ♀
09 45	☽ ± ☿
14 26	☽ Q ♆
16 25	☽ ☌ ♃
22 51	☽ ⊼ ☿
26 Thursday	
02 25	☉ ± ♇
06 16	☉ ⊻ ♀
07 30	☽ bQ ☉
08 06	☽ □ ♅
13 32	☽ ✶ ♆
15 17	☉ bQ ♅
15 48	☽ ⊼ ♂
21 25	☽ △ ♄
27 Friday	
00 13	☽ ☌ ♇
02 02	☽ ± ☉
02 36	☿ ± ♆
03 48	☽ ± ♂
11 54	☽ △ ♀
14 11	☽ ⊼ ☉
17 28	♀ St R
17 44	☽ ∠ ♆
18 01	☽ Q ♅
18 39	☿ ± ♃
21 32	☽ bQ ♂
28 Saturday	
01 07	☽ ⊻ ♃
01 43	☽ bQ ♄
10 17	☽ ⊥ ♆
12 06	☽ ⊥ ♃
14 27	☽ ☍ ☿
15 31	☽ bQ ♀
16 01	☽ ✶ ♅
18 19	☽ ± ♄
21 10	☽ ⊻ ♆
23 09	☿ ∠ ♀
29 Sunday	
02 23	☽ △ ♂
03 25	☿ △ ♅
04 22	☽ ∠ ♃
05 15	☽ ⊼ ♄
07 20	☽ ⊻ ♇
07 42	☽ ± ♀
18 01	☽ ⊥ ♇
18 21	☽ ⊼ ♀
18 54	☽ ∠ ♅
21 50	☽ # ☿
23 44	☽ ∥ ♃
30 Monday	
00 31	☿ ⊥ ♄
00 48	☽ ☍ ☉
05 25	☉ Q ♂
06 58	☽ ✶ ♃
09 53	☽ ∠ ♇
10 45	☽ ⊥ ♅
14 22	☽ # ☉
17 24	☽ # ♂
20 00	☿ ⊼ ♆
21 13	☽ ⊻ ♅
23 29	☽ ∥ ♇
31 Tuesday	
02 08	☽ ☌ ♆
03 01	☽ ⊼ ☿
03 53	☽ Q ♃
06 52	☽ ∥ ♆
09 35	☽ # ♄
09 55	☽ □ ♂
10 28	☽ ☍ ♄
11 56	☽ ✶ ♇
14 52	☽ ± ☿
22 09	☽ ☍ ♀
23 59	♂ □ ♄

All ephemeris data is given at 12.00 UT and the Moon's longitude is additionally given for 24.00 UT

AUGUST 2007

LONGITUDES

Date	Sidereal time h m s	Sun ☉ ° ' "	Moon ☽ ° ' "	Moon ☽ 24.00 ° ' "	Mercury ☿ ° '	Venus ♀ ° '	Mars ♂ ° '	Jupiter ♃ ° '	Saturn ♄ ° '	Uranus ♅ ° '	Neptune ♆ ° '	Pluto ♇ ° '
01	08 38 52	08 ♌ 52 36	10 ♓ 42 27	17 ♓ 45 46	23 ♋ 56	02 ♍ 30	26 ♉ 13	09 ♐ 59	25 ♌ 57	18 ♓ 07	20 ♒ 58	26 ♐ 39
02	08 42 49	09 49 59	24 ♓ 50 41	01 ♈ 56 49	25 46	02 R 18	26 53	09 R 58	26 04	18 R 05	20 R 56	26 R 38
03	08 46 45	10 47 23	09 ♈ 03 46	16 ♈ 11 12	27 39	02 03	27 32	09 57	26 12	18 04	20 55	26 37
04	08 50 42	11 44 48	23 ♈ 18 47	00 ♉ 26 13	29 ♋ 34	01 46	28 12	09 56	26 19	18 02	20 53	26 36
05	08 54 39	12 42 14	07 ♉ 33 11	14 ♉ 39 27	01 ♌ 32	01 26	28 51	09 56	26 27	18 00	20 51	26 35
06	08 58 35	13 39 42	21 ♉ 44 42	28 ♉ 48 42	03 32	01 05	29 ♉ 31	09 56	26 34	17 58	20 50	26 34
07	09 02 32	14 37 11	05 ♊ 51 09	12 ♊ 51 49	05 33	00 41	00 ♊ 10	09 D 56	26 42	17 57	20 48	26 33
08	09 06 28	15 34 42	19 ♊ 50 24	26 ♊ 46 38	07 35	00 ♍ 15	00 49	09 56	26 49	17 55	20 47	26 32
09	09 10 25	16 32 14	03 ♋ 40 14	10 ♋ 30 58	09 38	29 ♌ 47	01 28	09 56	26 57	17 53	20 45	26 31
10	09 14 21	17 29 47	17 ♋ 18 33	24 ♋ 02 45	11 41	29 18	02 06	09 57	27 04	17 51	20 43	26 30
11	09 18 18	18 27 21	00 ♌ 43 24	07 ♌ 20 17	13 45	28 47	02 45	09 58	27 12	17 49	20 42	26 29
12	09 22 14	19 24 57	13 ♌ 53 18	20 ♌ 22 22	15 49	28 14	03 23	09 59	27 19	17 47	20 40	26 29
13	09 26 11	20 22 33	26 ♌ 47 28	03 ♍ 08 37	17 52	27 40	04 01	10 00	27 27	17 45	20 38	26 28
14	09 30 08	21 20 11	09 ♍ 25 54	15 ♍ 39 29	19 55	27 05	04 40	10 01	27 35	17 43	20 37	26 27
15	09 34 04	22 17 50	21 ♍ 49 35	27 ♍ 56 25	21 57	26 29	05 17	10 02	27 42	17 41	20 35	26 26
16	09 38 01	23 15 30	04 ♎ 00 21	10 ♎ 01 44	23 58	25 52	05 55	10 04	27 50	17 39	20 34	26 26
17	09 41 57	24 13 11	16 ♎ 00 58	21 ♎ 58 31	25 58	25 15	06 33	10 06	27 58	17 36	20 32	26 25
18	09 45 54	25 10 53	27 ♎ 54 53	03 ♏ 50 35	27 57	24 38	07 10	10 08	28 05	17 34	20 30	26 24
19	09 49 50	26 08 36	09 ♏ 46 09	15 ♏ 42 11	29 ♌ 55	24 01	07 48	10 10	28 13	17 32	20 29	26 24
20	09 53 47	27 06 21	21 ♏ 39 14	27 ♏ 37 53	01 ♍ 52	23 24	08 25	10 12	28 20	17 30	20 27	26 23
21	09 57 43	28 04 06	03 ♐ 38 44	09 ♐ 42 21	03 47	22 47	09 02	10 15	28 28	17 28	20 25	26 23
22	10 01 40	29 01 53	15 ♐ 49 17	22 ♐ 00 03	05 41	22 12	09 38	10 18	28 36	17 25	20 24	26 22
23	10 05 37	29 ♌ 59 41	28 ♐ 15 08	04 ♑ 34 57	07 34	21 37	10 15	10 21	28 43	17 23	20 22	26 22
24	10 09 33	00 ♍ 57 30	10 ♑ 59 53	17 ♑ 30 12	09 25	21 03	10 51	10 24	28 51	17 21	20 21	26 21
25	10 13 30	01 55 20	24 ♑ 06 06	00 ♒ 47 41	11 15	20 31	11 28	10 27	28 59	17 19	20 19	26 21
26	10 17 26	02 53 12	07 ♒ 34 55	14 ♒ 27 42	13 04	20 00	12 04	10 30	29 06	17 16	20 17	26 20
27	10 21 23	03 51 05	21 ♒ 25 46	28 ♒ 28 46	14 52	19 31	12 39	10 34	29 14	17 14	20 16	26 20
28	10 25 19	04 48 59	05 ♓ 36 12	12 ♓ 47 29	16 38	19 04	13 15	10 38	29 21	17 12	20 14	26 20
29	10 29 16	05 46 55	20 ♓ 01 59	27 ♓ 18 57	18 22	18 39	13 51	10 42	29 29	17 09	20 13	26 19
30	10 33 12	06 44 52	04 ♈ 37 36	11 ♈ 57 10	20 06	18 16	14 26	10 46	29 37	17 07	20 11	26 19
31	10 37 09	07 ♍ 42 52	19 ♈ 16 51	26 ♈ 35 54	21 ♍ 48	17 ♌ 56	15 ♊ 01	10 ♐ 50	29 ♌ 44	17 ♓ 05	20 ♒ 10	26 ♐ 19

Date	Moon True ☊ ° '	Moon Mean ☊ ° '	Moon ☽ Latitude ° '
01	07 ♓ 13	08 ♓ 25	00 N 19
02	07 D 14	08 22	01 35
03	07 15	08 19	02 46
04	07 16	08 15	03 47
05	07 17	08 12	04 33
06	07 R 17	08 09	05 03
07	07 16	08 06	05 14
08	07 14	08 03	05 07
09	07 11	08 00	04 42
10	07 08	07 56	04 01
11	07 06	07 53	03 07
12	07 04	07 50	02 04
13	07 03	07 47	00 N 56
14	07 D 03	07 44	00 S 13
15	07 03	07 40	01 21
16	07 04	07 37	02 24
17	07 05	07 34	03 19
18	07 06	07 31	04 06
19	07 07	07 28	04 41
20	07 08	07 25	05 05
21	07 R 08	07 21	05 16
22	07 07	07 18	05 13
23	07 07	07 15	04 56
24	07 06	07 12	04 23
25	07 05	07 09	03 36
26	07 05	07 06	02 36
27	07 04	07 02	01 26
28	07 04	06 59	00 S 08
29	07 D 04	06 56	01 N 11
30	07 04	06 53	02 27
31	07 ♓ 04	06 ♓ 50	03 N 33

DECLINATIONS

Date	Sun ☉ ° '	Moon ☽ ° '	Mercury ☿ ° '	Venus ♀ ° '	Mars ♂ ° '	Jupiter ♃ ° '	Saturn ♄ ° '	Uranus ♅ ° '	Neptune ♆ ° '	Pluto ♇ ° '
01	18 N 02	07 S 15	21 N 35	05 N 53	18 N 09	21 S 25	14 N 03	05 S 26	14 S 46	16 S 30
02	17 47	00 S 35	21 27	05 45	18 19	21 25	14 01	05 27	14 47	16 30
03	17 32	06 N 08	21 16	05 38	18 29	21 26	13 59	05 28	14 47	16 30
04	17 16	12 34	21 02	05 31	18 38	21 26	13 56	05 28	14 48	16 31
05	17 00	18 20	20 46	05 26	18 48	21 26	13 53	05 29	14 48	16 31
06	16 44	23 04	20 26	05 22	18 57	21 26	13 51	05 30	14 49	16 31
07	16 27	26 26	20 05	05 18	19 06	21 26	13 48	05 31	14 49	16 31
08	16 10	28 09	19 41	05 16	19 15	21 26	13 46	05 31	14 50	16 32
09	15 53	28 05	19 14	05 15	19 23	21 27	13 43	05 32	14 50	16 32
10	15 35	26 18	18 45	05 14	19 32	21 27	13 41	05 33	14 51	16 32
11	15 18	23 02	18 14	05 15	19 40	21 27	13 38	05 34	14 52	16 32
12	15 00	18 39	17 41	05 16	19 48	21 28	13 36	05 34	14 52	16 33
13	14 42	13 28	17 06	05 19	19 56	21 28	13 33	05 35	14 53	16 33
14	14 23	07 50	16 29	05 22	20 04	21 28	13 31	05 36	14 53	16 33
15	14 05	02 N 00	15 51	05 27	20 12	21 29	13 28	05 37	14 54	16 34
16	13 46	03 S 47	15 12	05 32	20 19	21 29	13 25	05 38	14 54	16 34
17	13 27	09 22	14 31	05 38	20 27	21 30	13 23	05 39	14 55	16 34
18	13 08	14 33	13 49	05 45	20 34	21 30	13 20	05 39	14 55	16 34
19	12 48	19 11	13 07	05 52	20 41	21 31	13 18	05 40	14 56	16 35
20	12 29	23 05	12 23	06 00	20 47	21 31	13 15	05 41	14 56	16 35
21	12 09	26 03	11 39	06 09	20 54	21 32	13 12	05 42	14 57	16 35
22	11 49	27 52	10 54	06 18	21 00	21 32	13 10	05 43	14 57	16 36
23	11 28	28 21	10 09	06 28	21 07	21 33	13 07	05 44	14 58	16 36
24	11 08	27 21	09 23	06 38	21 13	21 34	13 05	05 45	14 58	16 36
25	10 47	24 50	08 38	06 49	21 19	21 34	13 02	05 46	14 59	16 36
26	10 27	20 54	07 52	06 59	21 25	21 35	12 59	05 46	14 59	16 37
27	10 06	15 43	07 05	07 10	21 30	21 36	12 57	05 47	15 00	16 37
28	09 45	09 35	06 19	07 21	21 36	21 36	12 54	05 48	15 00	16 37
29	09 24	02 S 51	05 33	07 32	21 41	21 37	12 52	05 49	15 01	16 38
30	09 02	04 N 05	04 47	07 43	21 46	21 38	12 49	05 50	15 01	16 38
31	08 N 41	10 N 50	04 N 01	07 N 54	21 N 51	21 S 39	12 N 46	05 S 51	15 S 02	16 S 38

ZODIAC SIGN ENTRIES

Date	h	m	Planets
02	20	43	☽ ♈
04	17	15	☿ ♌
04	23	16	☽ ♉
07	02	01	☽ ♊
07	06	01	♂ ♊
09	01	10	♀ ♌
09	05	36	☽ ♋
11	10	42	☽ ♌
13	18	03	☽ ♍
16	04	04	☽ ♎
18	16	13	☽ ♏
19	13	01	☿ ♍
21	04	44	☽ ♐
23	12	08	☉ ♍
23	15	20	☽ ♑
25	22	35	☽ ♒
28	02	34	☽ ♓
30	04	25	☽ ♈

LATITUDES

Date	Mercury ☿ ° '	Venus ♀ ° '	Mars ♂ ° '	Jupiter ♃ ° '	Saturn ♄ ° '	Uranus ♅ ° '	Neptune ♆ ° '	Pluto ♇ ° '
01	00 N 16	05 S 02	01 S 11	00 N 32	01 N 16	00 S 48	00 S 17	06 N 54
04	00 49	05 42	01 09	00 31	01 16	00 48	00 17	06 53
07	01 14	06 20	01 07	00 31	01 16	00 48	00 17	06 52
10	01 32	06 55	01 04	00 30	01 16	00 48	00 17	06 52
13	01 42	07 25	01 02	00 29	01 16	00 48	00 17	06 51
16	01 46	07 49	00 59	00 29	01 16	00 49	00 17	06 50
19	01 43	08 05	00 57	00 28	01 17	00 49	00 17	06 49
22	01 35	08 14	00 54	00 28	01 17	00 49	00 17	06 48
25	01 23	08 15	00 51	00 27	01 17	00 49	00 17	06 47
28	01 08	08 09	00 48	00 26	01 17	00 49	00 17	06 46
31	00 N 49	07 S 56	00 S 45	00 N 26	01 N 18	00 S 49	00 S 17	06 N 45

DATA

Julian Date	2454314
Delta T	+66 seconds
Ayanamsa	23° 57' 54"
Synetic vernal point	05° ♓ 09' 05"
True obliquity of ecliptic	23° 26' 26"

MOON'S PHASES, APSIDES AND POSITIONS ☽

Date	h	m	Phase	Longitude ° '	Eclipse Indicator
05	21	20	☾	13 ♉ 05	
12	23	03	●	19 ♌ 51	
20	23	54	☽	27 ♏ 35	
28	10	35	○	04 ♓ 46	total

Day	h	m		
04	00	01	Perigee	
19	03	29	Apogee	
31	00	22	Perigee	
02	14	06	0N	
08	23	03	Max dec	28° N 20'
15	20	15	0S	
23	08	02	Max dec	28° S 22'
29	21	53	0N	

LONGITUDES

Date	Chiron ⚷ ° '	Ceres ⚳ ° '	Pallas ⚴ ° '	Juno ⚵ ° '	Vesta ⚶ ° '	Black Moon Lilith ⚸ ° '
01	13 ♒ 16	17 ♉ 18	17 ♓ 18	19 ♎ 37	03 ♐ 57	01 ♏ 43
11	12 ♒ 44	19 ♉ 31	15 ♓ 31	22 ♎ 02	05 ♐ 47	02 ♏ 50
21	12 ♒ 12	21 ♉ 21	13 ♓ 17	24 ♎ 39	08 ♐ 12	03 ♏ 57
31	11 ♒ 43	22 ♉ 44	10 ♓ 47	27 ♎ 27	11 ♐ 06	05 ♏ 04

ASPECTARIAN

h m	Aspects
01 Wednesday	
05 26	☉ ∥ ♂
08 29	☽ Q ♇
08 32	☽ ⅊ ☿
08 39	☽ ⊼ ☉
10 45	☽ □ ♃
17 08	☽ # ♀
18 16	☽ Q ♂
18 38	☽ ∥ ♅
19 37	☽ ± ☉
02 Thursday	
00 35	☽ ☌ ♅
01 35	☿ ⅊ ♃
03 00	♂ ⊼ ♇
05 24	☽ ⊻ ♆
11 59	☽ ⅊ ☉
13 48	☽ △ ☿
14 05	☽ ⊼ ♄
15 01	☽ □ ♇
15 12	☉ △ ♃
15 16	☿ # ♃
15 32	☽ ⊥ ♆
15 37	☽ ⚹ ♂
16 09	☿ ⊻ ♄
18 03	☿ ⊥ ♀
22 54	☿ ⊼ ♇
03 Friday	
00 19	☽ ± ♄
00 23	☽ ⊼ ♀
06 42	☽ ∠ ♆
09 34	☽ # ♅
09 53	☿ ⚹ ♂
10 13	☽ ∥ ♀
10 20	☽ ± ♀
13 30	☽ △ ♃
15 07	☽ △ ☉
15 37	☽ ⅊ ♄
18 09	☽ ∠ ♂
04 Saturday	
01 12	☽ ⅊ ♀
03 08	☽ ⊻ ♅
07 55	☽ ⚹ ♆
08 12	☉ ⅊ ♇
10 02	☽ ⊥ ♂
13 13	☽ ⊥ ♅
14 44	☽ ⅊ ♃
17 06	☽ △ ♄
17 22	☽ ∥ ♄
17 31	☽ △ ♇
18 58	☉ ± ♅
20 38	☽ ⊻ ♂
20 55	☽ # ♆
05 Sunday	
00 09	♀ # ♅
00 13	☽ □ ☿
01 56	☽ △ ♀
04 03	☽ # ♇
04 06	☽ Q ♆
04 21	☽ ∠ ♅
05 54	☽ ± ♃
06 23	☽ ∥ ☉
11 00	☿ ⊻ ♀
14 10	☽ ∥ ♂
16 01	☽ ⊼ ♃
18 47	☽ ⅊ ♇
21 20	☽ □ ☉
22 50	☽ ∥ ☿
06 Monday	
00 29	☿ ± ♇
02 58	☽ # ♃
05 27	☿ ⅊ ♅
05 37	☽ ⚹ ♅
10 00	☽ ± ♇
10 27	☽ □ ♆
10 35	♄ △ ♇
11 34	☽ Q ☿
20 10	☽ ⊼ ♇
20 16	☽ □ ♄
07 Tuesday	
01 50	☽ ☌ ♂
01 57	☽ Q ♅
02 04	♃ St D
03 26	☽ □ ♀
04 15	♂ Q ♅
05 33	☉ # ♇
06 05	☽ Q ☉
11 23	☽ ⚹ ☿
18 58	☽ ☍ ♃
23 36	♀ □ ♂
08 Wednesday	
03 17	☽ Q ♄
04 07	☽ ⚹ ☉
08 41	☽ □ ♅
09 21	☽ Q ♀
13 37	☽ △ ♆
17 33	☽ ∠ ☿
23 34	☽ ☍ ♇
09 Thursday	
00 11	☽ ⚹ ♄
05 27	☽ ⚹ ♀
05 48	☿ ∥ ♂
07 57	☽ ⊻ ♂
08 00	☽ ∠ ☉
11 55	☽ ⊥ ☿
15 36	☿ △ ♃
15 38	☽ ⅊ ♆
18 58	☽ ⊥ ♂
23 00	☽ ⊼ ♃
10 Friday	
00 18	☽ ⊻ ☿
00 57	☽ ⊥ ☉
02 39	☽ ∠ ♄
06 51	☽ ∠ ♀
07 26	☽ ± ♆
09 35	☽ ± ♃
09 50	☿ ⅊ ♇
11 37	☽ ∠ ♂
12 21	☽ ⊻ ☉
12 57	☽ △ ♅
13 48	☿ ± ♅
18 03	☽ ⊼ ♆
18 45	☽ ⊥ ♄
20 28	☉ ⊼ ♅
22 16	☽ ⊥ ♀
11 Saturday	
01 38	☽ ⅊ ♃
04 23	☽ ⊼ ♇
05 35	☽ ⊻ ♄
08 37	☽ ⊻ ♀
15 11	☽ ± ♇
15 46	☽ ⅊ ♅
15 51	☽ ⚹ ♂
21 16	☽ # ♃
12 Sunday	
04 48	☽ △ ♃
04 49	☿ Q ♂
06 17	☽ ∥ ♂
07 34	☽ ⅊ ♇
08 08	☽ ± ♅
14 54	☽ Q ♂
16 13	☽ ☌ ☿
17 18	☽ ∥ ☿
19 10	☽ ⊼ ♅
22 04	☽ # ♇
22 07	☉ # ♆
23 03	☽ ☌ ☉
13 Monday	
00 31	☽ ☍ ♆
05 44	☽ # ♆
06 12	☽ ∥ ☉
10 37	☿ ⊼ ♅
11 23	☽ △ ♇
11 37	☽ ∥ ♄
13 15	☽ ☌ ♄
13 34	☽ ☌ ♀
18 25	☉ ☍ ♆
19 16	♀ ☌ ♄
14 Tuesday	
02 24	☽ □ ♂
09 31	☿ # ♇
13 07	☽ □ ♃
20 09	☿ ☍ ♆
21 12	☽ # ♅
22 03	☽ ∥ ♀
15 Wednesday	
03 56	☽ ☍ ♅
09 35	☽ ⊼ ♆
12 17	☽ ⊻ ☿
13 00	☽ ⊻ ☉
13 44	♀ △ ♇
19 56	☉ ☌ ☿
20 42	☽ ⊻ ♀
21 02	☽ □ ♇
21 19	☽ ± ♆
23 39	☽ ⊻ ♄
16 Thursday	
00 13	☽ Q ♃
01 48	☽ ⊥ ☉
02 24	☽ ⊥ ☿
07 58	☽ ⊥ ♀
11 39	☽ ⊥ ♄
15 05	☽ ⅊ ♆
16 01	☽ △ ♂
19 30	☽ # ♀
19 48	☽ ∥ ♅
21 12	☽ ∠ ☉
22 17	☿ # ♆
23 51	☽ ∠ ☿
17 Friday	
00 06	☽ ⚹ ♃
01 01	☽ ∠ ♀
05 27	☿ ☌ ♀
05 48	☽ ∠ ♄
08 47	☽ Q ♇
14 26	♀ # ♅
15 11	☽ ⊼ ♅
17 23	☿ △ ♇
18 00	☉ ∥ ♄
21 04	☽ △ ♆
23 45	☽ ⅊ ♂
18 Saturday	
03 15	☽ ± ♅
03 41	☉ ☌ ♀
05 35	☽ # ☉
05 42	☽ ⚹ ♀
05 59	☽ ⚹ ☉
06 13	☽ # ♄
06 21	☽ ∠ ♃
08 54	☽ # ☿
08 57	☽ ⚹ ♇
12 05	☽ ⚹ ☿
12 21	☽ ⚹ ♄
13 44	☿ ☌ ♄
13 48	☽ ∥ ♆
18 57	☽ ± ♂
21 24	☽ ⅊ ♅
22 04	☽ ∥ ♇
19 Sunday	
00 37	☽ ⊥ ♃
04 46	☽ Q ♀
05 25	☿ ∥ ♄
07 47	☽ ⊼ ♂
08 25	☽ Q ☉
12 48	☽ ⊻ ♃
12 55	☽ Q ♄
15 17	☽ ∠ ♇
17 12	☽ Q ☿
18 14	☉ △ ♇
20 50	☽ # ♂
20 Monday	
01 42	☽ ∥ ♃
03 39	☽ △ ♅
06 28	☉ ∥ ☿
09 27	☽ ⊥ ♇
09 35	☽ □ ♆
15 20	☽ □ ♀
21 30	☽ ⊻ ♇
23 54	☽ □ ☉
21 Tuesday	
01 34	☽ □ ♄
12 20	☽ □ ☿
21 27	☽ Q ♆
23 14	☽ ☍ ♂
23 28	☉ ☌ ♄
22 Wednesday	
01 07	☽ ☌ ♃
15 07	☽ □ ♅
20 53	☽ ⚹ ♆
23 08	♀ Q ♂
23 49	☽ △ ♀
23 Thursday	
08 23	☽ ☌ ♇
12 54	☽ △ ♄
15 36	☽ △ ☉
16 03	♂ ☍ ♃
24 Friday	
01 27	☽ ∠ ♆
01 28	☽ Q ♅
03 09	☽ ⅊ ♀
08 35	☽ △ ☿
10 52	☽ ⊻ ♃
11 43	☽ ⊼ ♂
17 20	☽ ⅊ ♄
18 11	☽ ⊥ ♆
19 12	☽ ± ♀
21 54	☽ ⅊ ☉
22 00	☽ ⊥ ♃
23 20	☽ ± ♂
23 41	☽ ⚹ ♅
25 Saturday	
01 03	☿ □ ♃
05 09	☽ ⊻ ♆
05 45	☽ ⊼ ♀
09 57	☽ ± ♄
14 26	☽ ∠ ♃
15 32	☽ ± ☉
15 59	☿ □ ♂
16 03	☽ ⊻ ♇
16 27	☽ ⅊ ♂
16 30	☽ ⅊ ☿
20 50	☽ ⊼ ♄
21 48	♀ ☍ ♆
26 Sunday	
02 39	☽ ∠ ♅
02 45	☽ ⊥ ♇
03 05	☽ ⊼ ☉
08 21	☽ ∥ ♃
09 19	☽ # ♂
10 58	☽ ± ☿
17 08	☽ ⚹ ♃
18 26	☽ ⊥ ♅
18 34	☽ ∠ ♇
20 11	☽ △ ♂
23 01	☽ ⊼ ☿
27 Monday	
04 49	☽ ⊻ ♅
08 09	☽ ∥ ♇
08 50	☽ ☍ ♀
09 55	☿ ∥ ♀
10 00	☽ ☌ ♆
13 57	☽ Q ♃
14 58	☽ ∥ ♆
20 22	☽ ⚹ ♇
23 17	☽ # ♄
28 Tuesday	
01 23	☽ ☍ ♄
10 35	☽ ☍ ☉
11 22	☽ # ☉
14 11	♂ # ♃
16 34	☽ Q ♇
19 39	☿ ☍ ♅
19 55	☽ # ♀
20 26	☽ □ ♃
29 Wednesday	
01 19	☽ □ ♂
01 21	☽ # ☿
01 38	☽ ∥ ♅
03 42	☿ # ♅
07 15	☽ ☌ ♅
08 53	☽ ☍ ☿
09 47	☽ ⊼ ♀
12 18	☽ ⊻ ♆
15 12	☿ ⊻ ♀
19 25	☽ ± ♀
22 10	☽ ⊥ ♆
22 22	☽ □ ♇
30 Thursday	
03 42	☽ ⊼ ♄
08 15	☽ Q ♂
09 50	☽ ⅊ ♀
12 55	☽ ∠ ♆
13 14	☿ ⊼ ♆
13 38	☽ ± ♄
14 10	☽ ∥ ☿
15 43	☽ ⊼ ☉
18 06	☽ # ♅
22 06	☽ △ ♃
31 Friday	
01 05	☽ ∥ ♀
02 14	☽ ± ☉
04 30	☽ ⅊ ♄
04 33	☽ ∥ ☉
04 44	☽ ⚹ ♂
08 24	☽ ⊻ ♅
09 50	☽ △ ♀
13 26	☽ ⚹ ♆
16 40	☽ ⊼ ☿
18 02	☽ ⅊ ☉
18 13	☽ ⊥ ♅
19 12	☽ ∥ ♄
22 48	☽ ⅊ ♃
23 32	☽ △ ♇

All ephemeris data is given at 12.00 UT and the Moon's longitude is additionally given for 24.00 UT

SEPTEMBER 2007

LONGITUDES

Date	Sidereal time h m s	Sun ☉ ° ' "	Moon ☽ ° ' "	Moon ☽ 24.00 ° ' "	Mercury ☿ ° '	Venus ♀ ° '	Mars ♂ ° '	Jupiter ♃ ° '	Saturn ♄ ° '	Uranus ♅ ° '	Neptune ♆ ° '	Pluto ♇ ° '
01	10 41 06	08 ♍ 40 53	03 ♉ 53 37	11 ♉ 09 23	23 ♍ 29	17 ♌ 37	15 ♊ 36	10 ♐ 54	29 ♌ 52	17 ♓ 02	20 ♒ 08	26 ♐ 19
02	10 45 02	09 38 56	18 ♉ 22 37	25 ♉ 32 53	25 08	17 R 21	16 11	10 59	29 ♌ 59	17 R 00	20 R 06	26 19
03	10 48 59	10 37 01	02 ♊ 39 49	09 ♊ 43 07	26 47	17 08	16 45	11 04	00 ♍ 07	16 58	20 05	26 18
04	10 52 55	11 35 08	16 ♊ 42 37	23 ♊ 38 10	28 ♍ 24	16 56	17 19	11 09	00 15	16 55	20 03	26 18
05	10 56 52	12 33 17	00 ♋ 29 43	07 ♋ 17 17	00 ♎ 00	16 47	17 54	11 14	00 22	16 53	20 02	26 18
06	11 00 48	13 31 28	14 ♋ 00 54	20 ♋ 40 33	01 35	16 41	18 27	11 19	00 30	16 51	20 01	26 18
07	11 04 45	14 29 41	27 ♋ 16 33	03 ♌ 48 47	03 08	16 37	19 01	11 24	00 37	16 48	19 59	26 18
08	11 08 41	15 27 56	10 ♌ 17 28	16 ♌ 42 43	04 41	16 35	19 34	11 30	00 45	16 46	19 58	26 D 18
09	11 12 38	16 26 13	23 ♌ 04 40	29 ♌ 23 25	06 12	16 D 36	20 08	11 36	00 52	16 43	19 56	26 18
10	11 16 35	17 24 32	05 ♍ 39 08	11 ♍ 51 57	07 42	16 39	20 40	11 42	01 00	16 41	19 55	26 18
11	11 20 31	18 22 52	18 ♍ 02 01	24 ♍ 09 30	09 10	16 45	21 13	11 48	01 07	16 39	19 53	26 18
12	11 24 28	19 21 15	00 ♎ 14 34	06 ♎ 17 24	10 38	16 52	21 46	11 54	01 14	16 36	19 52	26 18
13	11 28 24	20 19 39	12 ♎ 18 15	18 ♎ 17 21	12 04	17 02	22 18	12 00	01 22	16 34	19 51	26 19
14	11 32 21	21 18 05	24 ♎ 14 59	00 ♏ 11 28	13 30	17 14	22 50	12 07	01 29	16 31	19 49	26 19
15	11 36 17	22 16 32	06 ♏ 07 08	12 ♏ 02 21	14 53	17 28	23 21	12 13	01 36	16 29	19 48	26 19
16	11 40 14	23 15 02	17 ♏ 57 33	23 ♏ 53 11	16 16	17 44	23 53	12 20	01 44	16 27	19 47	26 19
17	11 44 10	24 13 33	29 ♏ 49 42	05 ♐ 47 38	17 37	18 02	24 24	12 27	01 51	16 24	19 45	26 20
18	11 48 07	25 12 06	11 ♐ 47 31	17 ♐ 49 53	18 57	18 21	24 55	12 34	01 58	16 22	19 44	26 20
19	11 52 04	26 10 40	23 ♐ 55 18	00 ♑ 04 19	20 16	18 43	25 25	12 41	02 05	16 19	19 42	26 21
20	11 56 00	27 09 16	06 ♑ 17 21	12 ♑ 35 27	21 33	19 06	25 55	12 49	02 12	16 17	19 41	26 21
21	11 59 57	28 07 54	18 ♑ 58 36	25 ♑ 27 26	22 49	19 31	26 25	12 56	02 20	16 15	19 40	26 21
22	12 03 53	29 ♍ 06 34	02 ♒ 02 22	08 ♒ 43 41	24 03	19 58	26 55	13 04	02 27	16 12	19 39	26 22
23	12 07 50	00 ♎ 05 15	15 ♒ 31 37	22 ♒ 26 14	25 15	20 26	27 24	13 11	02 34	16 10	19 38	26 22
24	12 11 46	01 03 58	29 ♒ 27 29	06 ♓ 35 08	26 26	20 56	27 53	13 19	02 41	16 08	19 37	26 23
25	12 15 43	02 02 43	13 ♓ 48 48	21 ♓ 07 54	27 35	21 28	28 22	13 27	02 48	16 06	19 36	26 23
26	12 19 39	03 01 29	28 ♓ 31 42	05 ♈ 59 17	28 41	22 00	28 50	13 36	02 55	16 03	19 35	26 24
27	12 23 36	04 00 17	13 ♈ 29 39	21 ♈ 01 38	29 ♎ 46	22 35	29 19	13 44	03 02	16 01	19 34	26 24
28	12 27 33	04 59 08	28 ♈ 34 06	06 ♉ 05 51	00 ♏ 48	23 10	29 ♊ 46	13 52	03 09	15 59	19 33	26 25
29	12 31 29	05 58 01	13 ♉ 35 45	21 ♉ 02 45	01 48	23 47	00 ♋ 14	14 01	03 16	15 57	19 32	26 26
30	12 35 26	06 ♎ 56 56	28 ♉ 25 56	05 ♊ 44 33	02 ♏ 46	24 ♌ 25	00 ♋ 41	14 ♐ 09	03 ♍ 23	15 ♓ 54	19 ♒ 31	26 ♐ 26

Date	Moon True ☊ ° '	Moon Mean ☊ ° '	Moon ☽ Latitude ° '
01	07 ♓ 04	06 ♓ 46	04 N 26
02	07 R 04	06 43	05 00
03	07 04	06 40	05 16
04	07 D 04	06 37	05 13
05	07 04	06 34	04 51
06	07 05	06 31	04 14
07	07 05	06 27	03 23
08	07 06	06 24	02 23
09	07 07	06 21	01 17
10	07 06	06 18	00 N 08
11	07 R 07	06 15	01 S 00
12	07 06	06 11	02 04
13	07 04	06 08	03 02
14	07 02	06 05	03 52
15	07 00	06 02	04 31
16	06 58	05 59	04 58
17	06 56	05 56	05 13
18	06 54	05 52	05 14
19	06 54	05 49	05 01
20	06 D 54	05 46	04 35
21	06 55	05 43	03 54
22	06 56	05 40	03 01
23	06 58	05 37	01 55
24	06 59	05 33	00 S 41
25	06 R 59	05 30	00 N 37
26	06 58	05 27	01 56
27	06 55	05 24	03 07
28	06 52	05 21	04 06
29	06 48	05 17	04 47
30	06 ♓ 44	05 ♓ 14	05 N 09

DECLINATIONS

Date	Sun ☉ ° '	Moon ☽ ° '	Mercury ☿ ° '	Venus ♀ ° '	Mars ♂ ° '	Jupiter ♃ ° '	Saturn ♄ ° '	Uranus ♅ ° '	Neptune ♆ ° '	Pluto ♇ ° '
01	08 N 19	16 N 59	03 N 15	08 N 05	21 N 56	21 S 39	12 N 44	05 S 52	15 S 02	16 S 38
02	07 57	22 06	02 29	08 15	22 01	21 40	12 41	05 53	15 03	16 39
03	07 35	25 51	01 44	08 26	22 06	21 41	12 39	05 54	15 03	16 39
04	07 13	27 58	00 58	08 36	22 10	21 42	12 36	05 55	15 04	16 39
05	06 51	28 17	00 N 13	08 45	22 14	21 43	12 34	05 56	15 04	16 40
06	06 29	26 54	00 S 31	08 55	22 18	21 44	12 31	05 57	15 05	16 40
07	06 06	24 02	01 15	09 04	22 22	21 45	12 28	05 58	15 05	16 40
08	05 44	19 58	01 59	09 12	22 26	21 46	12 26	05 58	15 06	16 41
09	05 21	15 02	02 42	09 21	22 30	21 47	12 23	05 59	15 06	16 41
10	04 58	09 34	03 25	09 28	22 34	21 47	12 21	06 00	15 07	16 41
11	04 36	03 N 49	04 07	09 35	22 37	21 48	12 18	06 01	15 07	16 42
12	04 13	02 S 00	04 49	09 42	22 40	21 49	12 16	06 02	15 08	16 42
13	03 50	07 40	05 30	09 48	22 44	21 50	12 13	06 03	15 08	16 42
14	03 27	13 00	06 11	09 54	22 47	21 52	12 11	06 04	15 09	16 43
15	03 04	17 49	06 51	09 59	22 50	21 53	12 08	06 05	15 09	16 43
16	02 41	21 57	07 30	10 04	22 52	21 54	12 06	06 06	15 09	16 43
17	02 18	25 12	08 09	10 08	22 55	21 55	12 03	06 07	15 10	16 44
18	01 54	27 23	08 47	10 11	22 58	21 56	12 01	06 08	15 10	16 44
19	01 31	28 19	09 24	10 14	23 00	21 57	11 58	06 09	15 10	16 44
20	01 08	27 52	10 00	10 16	23 03	21 58	11 56	06 10	15 11	16 45
21	00 45	25 58	10 36	10 18	23 05	21 59	11 53	06 10	15 11	16 45
22	00 N 21	22 38	11 10	10 19	23 07	22 00	11 51	06 11	15 12	16 45
23	00 S 02	18 01	11 44	10 20	23 09	22 01	11 48	06 12	15 12	16 46
24	00 25	12 18	12 17	10 20	23 11	22 02	11 46	06 13	15 12	16 46
25	00 49	05 S 47	12 48	10 19	23 13	22 04	11 44	06 14	15 13	16 46
26	01 12	01 N 11	13 19	10 18	23 15	22 05	11 41	06 15	15 13	16 47
27	01 36	08 12	13 49	10 16	23 17	22 06	11 39	06 15	15 13	16 47
28	01 59	14 47	14 17	10 14	23 18	22 08	11 36	06 16	15 14	16 47
29	02 22	20 29	14 44	10 11	23 20	22 08	11 34	06 17	15 14	16 47
30	02 S 46	24 N 50	15 S 09	10 N 07	23 N 21	22 S 09	11 N 32	06 S 18	15 S 14	16 S 48

ZODIAC SIGN ENTRIES

Date	h	m	Planets
01	05	35	☽ ♉
02	13	49	♄ ♍
03	07	30	☽ ♊
05	11	08	☽ ♋
05	12	02	☿ ♎
07	16	59	☽ ♌
10	01	10	☽ ♍
12	11	31	☽ ♎
14	23	37	☽ ♏
17	12	21	☽ ♐
19	23	52	☽ ♑
22	08	18	☽ ♒
23	09	51	☉ ♎
24	12	55	☽ ♓
26	14	22	☽ ♈
27	17	17	☿ ♏
28	14	17	☽ ♉
28	23	55	♂ ♋
30	14	34	☽ ♊

LATITUDES

Date	Mercury ☿ ° '	Venus ♀ ° '	Mars ♂ ° '	Jupiter ♃ ° '	Saturn ♄ ° '	Uranus ♅ ° '	Neptune ♆ ° '	Pluto ♇ ° '
01	00 N 43	07 S 50	00 S 44	00 N 26	01 N 18	00 S 49	00 S 17	06 N 45
04	00 N 22	07 30	00 40	00 25	01 18	00 49	00 17	06 44
07	00 00	07 07	00 37	00 25	01 18	00 49	00 17	06 43
10	00 S 24	06 41	00 33	00 24	01 19	00 49	00 17	06 42
13	00 48	06 13	00 29	00 24	01 19	00 49	00 17	06 41
16	01 12	05 44	00 25	00 23	01 19	00 49	00 17	06 40
19	01 36	05 14	00 21	00 23	01 20	00 49	00 17	06 39
22	01 59	04 45	00 17	00 22	01 20	00 49	00 17	06 38
25	02 21	04 15	00 12	00 22	01 20	00 49	00 17	06 37
28	02 42	03 47	00 08	00 21	01 21	00 49	00 17	06 37
31	03 S 00	03 S 19	00 S 03	00 N 21	01 N 21	00 S 49	00 S 17	06 N 36

DATA

Julian Date	2454345
Delta T	+66 seconds
Ayanamsa	23° 57' 58"
Synetic vernal point	05° ♓ 09' 01"
True obliquity of ecliptic	23° 26' 27"

LONGITUDES

Date	Chiron ⚷ ° '	Ceres ⚳ ° '	Pallas ⚴ ° '	Juno ⚵ ° '	Vesta ⚶ ° '	Black Moon Lilith ⚸ ° '
01	11 ♒ 40	22 ♉ 51	10 ♓ 31	27 ♎ 45	11 ♐ 25	05 ♏ 11
11	11 ♒ 15	23 ♉ 41	07 ♓ 57	00 ♏ 41	14 ♐ 45	06 ♏ 18
21	10 ♒ 54	23 ♉ 56	05 ♓ 34	03 ♏ 48	18 ♐ 17	07 ♏ 25
31	10 ♒ 39	23 ♉ 33	03 ♓ 34	06 ♏ 59	22 ♐ 24	08 ♏ 32

MOON'S PHASES, APSIDES AND POSITIONS ☽

Date	h	m	Phase	Longitude ° '	Eclipse Indicator
04	02	32	☾	11 ♊ 12	
11	12	44	●	18 ♍ 25	Partial
19	16	48	☽	26 ♐ 22	
26	19	45	○	03 ♈ 20	

Day	h	m		
15	21	11	Apogee	
28	02	02	Perigee	
05	04	27	Max dec	28° N 23'
12	03	42	0S	
19	16	23	Max dec	28° S 21'
26	07	59	0N	

ASPECTARIAN

h m Aspects

01 Saturday
03 47 ☽ ± ☿
04 04 ☽ # ♆
05 19 ☽ △ ♄
06 21 ☽ ∠ ♂
08 57 ☽ ∠ ♅
09 06 ☽ Q ♆
10 36 ☽ # ♇
13 41 ☽ ± ♃
13 45 ☿ ⊥ ♀
20 28 ☽ △ ☉
20 33 ☽ ⚼ ☿
21 49 ☽ ⊥ ♂
22 28 ☉ ∥ ♀
23 39 ☽ ⊼ ♃

02 Sunday
00 15 ☽ ⚼ ♇
08 11 ☽ ⊻ ♂
09 42 ☽ # ♃
09 43 ☽ ✱ ♅
10 19 ☽ □ ♀
11 32 ☽ ∥ ♂
14 53 ☽ □ ♆
15 13 ☽ ± ♇

03 Monday
00 47 ☽ △ ☿
01 17 ☽ ⊼ ♇
01 56 ☿ ± ♆
05 04 ☿ □ ♇
05 46 ☽ Q ♅
07 39 ☽ □ ♄
16 07 ☽ Q ♀
20 09 ♂ □ ♅
23 36 ♀ ✱ ♂

04 Tuesday
00 04 ☉ □ ♃
02 23 ☽ ☍ ♃
02 32 ☽ □ ☉
12 22 ☽ □ ♅
12 23 ☽ ✱ ♀
13 06 ☽ ☌ ♂
14 40 ☽ Q ♄
15 12 ♀ ⊼ ♅
17 47 ☽ △ ♆
23 48 ☿ Q ♃

05 Wednesday
04 39 ☽ ☍ ♇
11 01 ☽ □ ☿
11 46 ☽ ✱ ♄
12 07 ☽ Q ☉
14 16 ☽ ∠ ♀
18 05 ☿ ⊻ ♄
19 59 ☽ ⚼ ♆

06 Thursday
06 05 ☽ ⊥ ♀
07 09 ☽ ⊼ ♃
11 03 ☽ ✱ ☉
11 59 ☽ ± ♆
13 34 ☿ ∠ ♀
14 01 ♂ Q ♄
14 41 ☽ ∠ ♄
16 46 ☽ ⊻ ♀
17 04 ☽ △ ♅
17 58 ☽ ± ♃
20 20 ☽ ⊻ ♂
22 46 ☽ ⊼ ♆
23 21 ☽ Q ☿

07 Friday
07 06 ☽ ⊥ ♄
07 42 ☽ ⊥ ♂
10 13 ☽ ⊼ ♇
10 24 ☽ ⚼ ♃
14 54 ♇ St D
16 23 ☽ ∠ ☉
18 11 ☽ ⊻ ♄
20 16 ☽ ⚼ ♅
20 58 ☉ # ♅
21 13 ☽ ± ♇
22 20 ☽ ∥ ♂

08 Saturday
00 12 ☽ ✱ ☿
00 56 ☽ ∠ ♂
02 09 ☽ # ♃
10 20 ☽ ⊥ ☉
12 52 ☽ ± ♅
13 53 ☽ ⚼ ♇
14 16 ☽ △ ♃
16 14 ♀ St D
16 25 ☿ ⚼ ♆
22 27 ☽ ⊻ ☉
23 47 ☽ ☌ ♀

09 Sunday
00 03 ☽ ⊼ ♅
04 05 ♂ △ ♆
04 21 ☽ # ♇
06 05 ☽ ☍ ♆
06 10 ☽ ✱ ♂
07 58 ☽ ∠ ☿
11 42 ☽ # ♆
16 17 ☉ ⊻ ♀
18 07 ☽ △ ♇
18 46 ☉ ☍ ♅
23 42 ☿ ⊥ ♄
23 56 ☽ ∥ ♄

10 Monday
02 58 ☽ ☌ ♄
03 23 ☽ ⊥ ☿
06 01 ☽ Q ♂
12 24 ☽ ∥ ♀
16 28 ☽ ⊻ ☿
17 36 ♀ ⊼ ♅
23 46 ☽ □ ♃

11 Tuesday
02 53 ☽ # ♅
08 32 ☽ ∥ ☉
09 18 ☽ ☍ ♅
09 28 ☽ ⊻ ♀
10 51 ☽ # ☿
12 44 ☽ ☌ ☉
15 37 ☽ ⊼ ♆
18 31 ☽ □ ♂
21 19 ☽ ⊥ ♀
22 31 ☉ # ☿

12 Wednesday
03 23 ☽ ± ♆
04 14 ☽ □ ♇
11 19 ☽ Q ♃
14 00 ☽ ⊻ ♄
15 16 ☽ ∠ ♀
20 41 ☽ # ☉
21 09 ☽ ⚼ ♆

13 Thursday
00 21 ☉ ⊼ ♆
01 27 ☽ ∥ ☿
02 02 ☽ ⊥ ♄
05 03 ☽ ∥ ♅
10 43 ☿ ✱ ♃
11 24 ☽ ✱ ♃
11 29 ☽ ☌ ☿
16 01 ☽ Q ♇
20 13 ☽ ∠ ♄
20 30 ☽ ⊼ ♅
21 36 ☽ # ♀
21 38 ☽ ✱ ♀

14 Friday
03 06 ☽ △ ♆
05 32 ☽ ⊻ ☉
07 50 ☿ ∥ ♅
08 13 ☽ # ♄
08 32 ☽ ± ♅
09 00 ☽ △ ♂
16 10 ☽ ✱ ♇
17 50 ☽ ∠ ♃
18 43 ☽ ⊥ ☉
22 15 ☽ Q ♀
22 20 ☽ ∥ ♆

15 Saturday
02 06 ☿ Q ♇
02 39 ☽ ⚼ ♅
02 46 ☽ ✱ ♄
06 14 ☽ ∥ ♇
12 13 ☽ ⊥ ♃
14 33 ☽ ∠ ☉
16 44 ☽ ⚼ ♂
22 33 ☽ ∠ ♇

16 Sunday
00 29 ☽ ⊻ ♃
03 20 ☽ Q ♄
08 08 ☽ ⊻ ☿
08 56 ☽ △ ♅
11 31 ☽ □ ♀
11 38 ☽ ∥ ♃
11 50 ☽ ± ♂
14 59 ☿ ⊼ ♅
15 41 ☽ □ ♆
16 47 ☽ ⊥ ♇
18 14 ☽ # ♂
20 54 ☿ ∠ ♄
21 51 ☽ ⊥ ☿
23 40 ☽ ✱ ☉

17 Monday
00 32 ☽ ⊼ ♂
04 47 ☉ ⊥ ♀
04 56 ☽ ⊻ ♇
16 07 ☽ □ ♄
18 21 ☽ ∠ ☿
20 49 ☉ □ ♂
21 28 ☿ ✱ ♀

18 Tuesday
02 00 ☽ Q ☉
03 54 ☽ Q ♆
13 34 ☽ ☌ ♃
21 04 ☽ □ ♅

19 Wednesday
00 51 ☉ ± ♆
01 26 ☽ △ ♀
02 00 ☿ △ ♆
03 44 ☽ ✱ ♆
03 57 ☽ ✱ ☿
15 03 ☽ ☍ ♂
15 59 ☉ □ ♇
16 44 ☽ ☌ ♇
16 48 ☽ □ ☉

20 Thursday
04 04 ☽ △ ♄
06 08 ☽ Q ☿
07 39 ☽ ⚼ ♀
08 09 ☽ Q ♅
08 56 ☽ ∠ ♆
23 26 ☿ # ♀

21 Friday
00 32 ☽ ⊻ ♃
01 24 ☽ ± ♀
01 30 ☿ ± ♅
02 04 ☽ ⊥ ♆
06 54 ☽ ✱ ♅
08 40 ♂ ☍ ♇
08 54 ☽ ⚼ ♄
11 55 ☽ ⊥ ♃
13 03 ☽ ⊼ ♀
13 18 ☽ ⊻ ♆
19 53 ☽ □ ☿
20 03 ♀ ☍ ♆

22 Saturday
01 39 ☽ ⊻ ♇
01 44 ☽ ± ♄
02 19 ☽ ⊼ ♂
04 42 ☽ ∠ ♃
06 15 ☽ △ ☉
09 05 ☽ # ♂
10 30 ☽ ∠ ♅
12 35 ☽ ⊥ ♇
12 45 ☽ ⊼ ♄
13 39 ☽ ± ♂
15 42 ☽ ∥ ♃

23 Sunday
02 35 ☽ ⊥ ♅
04 41 ☽ ∠ ♇
06 19 ☽ ⚼ ♂
07 52 ☽ ✱ ♃
11 10 ☽ ⚼ ☉
13 07 ☽ ⊻ ♅
14 55 ☿ # ♄
17 39 ☽ ∥ ♇
19 09 ☽ ☌ ♆
20 52 ☽ ☍ ♀

24 Monday
00 20 ☽ ∥ ♆
03 57 ☽ ± ☉
04 53 ☽ Q ♃
06 23 ☽ △ ☿
06 45 ☽ ✱ ♇
09 14 ☽ △ ♂
10 55 ☿ ✱ ♇
12 06 ☽ ∥ ☿
14 07 ☽ # ♄
14 55 ☽ ⊼ ☉
17 30 ☽ ☍ ♄
19 17 ☉ Q ♃
19 34 ☽ # ♀

25 Tuesday
03 00 ☽ Q ♇
09 47 ☽ ⚼ ☿
10 26 ☽ ∥ ♅
11 24 ☽ □ ♃
15 45 ☽ ☌ ♅
21 29 ☽ ⊻ ♆

26 Wednesday
01 01 ☽ ⊼ ♀
01 46 ☽ ± ☿
04 19 ☽ ∥ ☉
07 14 ☽ ⊥ ♆
08 33 ☽ □ ♇
09 00 ☉ ⊻ ♄
09 35 ☿ ∠ ♃
11 07 ☽ ± ♀
12 05 ☽ # ☉
12 17 ☽ ⊼ ☿
12 31 ☽ □ ♂
17 53 ☿ △ ♂
19 08 ☽ ⊼ ♄
19 45 ☽ ☍ ☉
21 44 ☽ ∠ ♆

27 Thursday
02 10 ☽ ⚼ ♀
04 49 ☽ ± ♄
05 18 ☽ # ♅
12 23 ☽ △ ♃
16 01 ☽ ⊻ ♅
18 17 ☽ Q ♂
19 15 ☽ ∥ ♀
19 17 ☽ ⚼ ♄
21 39 ☽ ✱ ♆

28 Friday
00 11 ☽ ∥ ♄
01 25 ☉ ⚼ ♆
01 33 ☽ ⊥ ♅
03 03 ☽ △ ♀
08 34 ☽ △ ♇
09 53 ☽ # ☿
12 29 ☽ ⚼ ♃
13 42 ☽ # ♆
13 59 ☽ ✱ ♂
15 49 ☽ ☍ ☿
15 50 ☽ ∠ ♅
15 56 ☿ ⚼ ♅
16 44 ☽ Q ♆
19 21 ☽ △ ♄
19 54 ☽ # ♇
22 56 ☽ ⊼ ☉

29 Saturday
02 58 ☽ ± ♃
08 31 ☽ ⚼ ♇
09 12 ☽ ± ☉
12 40 ☽ ⊼ ♃
14 42 ☽ ∠ ♂
15 46 ☽ ✱ ♅
20 12 ☽ # ♃
21 32 ☽ □ ♆
23 01 ☽ ± ♇

30 Sunday
00 43 ☽ ⚼ ☉
02 47 ☽ ∥ ♂
05 10 ☽ □ ♀
05 42 ☽ ⊥ ♂
08 45 ☽ ⊼ ♇
11 09 ☽ Q ♅
15 48 ☽ ⊻ ♂
16 55 ☿ ∥ ♆
19 35 ☽ ⊼ ☿
20 10 ☽ □ ♄

All ephemeris data is given at 12.00 UT and the Moon's longitude is additionally given for 24.00 UT

OCTOBER 2007

LONGITUDES

Date	Sidereal time h m s	Sun ☉ ° ' "	Moon ☽ ° ' "	Moon ☽ 24.00 ° ' "	Mercury ☿ ° '	Venus ♀ ° '	Mars ♂ ° '	Jupiter ♃ ° '	Saturn ♄ ° '	Uranus ♅ ° '	Neptune ♆ ° '	Pluto ♇ ° '
01	12 39 22	07 ♎ 55 53	12 ♊ 57 58	20 ♊ 05 47	03 ♏ 40	25 ♌ 04	01 ♋ 08	14 ♐ 18	03 ♍ 29	15 ♓ 52	19 ♒ 30	26 ♐ 27
02	12 43 19	08 54 53	27 ♊ 07 45	04 ♋ 03 46	04 32	25 45	01 34	14 27	03 36	15 R 50	19 R 29	26 28
03	12 47 15	09 53 55	10 ♋ 53 52	17 ♋ 38 12	05 20	26 27	02 00	14 36	03 43	15 48	19 28	26 29
04	12 51 12	10 52 59	24 ♋ 17 00	00 ♌ 50 36	06 05	27 09	02 26	14 45	03 49	15 46	19 27	26 30
05	12 55 08	11 52 06	07 ♌ 19 21	13 ♌ 43 37	06 45	27 53	02 51	14 54	03 56	15 44	19 26	26 30
06	12 59 05	12 51 15	20 ♌ 03 49	26 ♌ 20 20	07 21	28 38	03 16	15 04	04 02	15 42	19 25	26 31
07	13 03 02	13 50 26	02 ♍ 33 33	08 ♍ 43 49	07 53	29 ♌ 23	03 40	15 13	04 09	15 40	19 25	26 32
08	13 06 58	14 49 39	14 ♍ 51 29	20 ♍ 56 52	08 19	00 ♍ 10	04 04	15 23	04 15	15 38	19 24	26 33
09	13 10 55	15 48 55	27 ♍ 00 14	03 ♎ 01 52	08 40	00 58	04 28	15 33	04 22	15 36	19 23	26 34
10	13 14 51	16 48 12	09 ♎ 01 58	15 ♎ 00 46	08 55	01 46	04 51	15 42	04 28	15 34	19 22	26 35
11	13 18 48	17 47 32	20 ♎ 58 28	26 ♎ 55 17	09 03	02 35	05 14	15 52	04 34	15 32	19 22	26 36
12	13 22 44	18 46 54	02 ♏ 51 24	08 ♏ 47 01	09 R 04	03 25	05 36	16 02	04 41	15 30	19 21	26 37
13	13 26 41	19 46 18	14 ♏ 42 22	20 ♏ 37 41	08 58	04 16	05 58	16 12	04 47	15 28	19 20	26 38
14	13 30 37	20 45 43	26 ♏ 33 15	02 ♐ 29 20	08 44	05 07	06 19	16 23	04 53	15 26	19 20	26 40
15	13 34 34	21 45 11	08 ♐ 26 17	14 ♐ 24 28	08 21	05 59	06 40	16 33	04 59	15 24	19 19	26 41
16	13 38 31	22 44 41	20 ♐ 24 17	26 ♐ 26 11	07 50	06 52	07 00	16 43	05 05	15 23	19 19	26 42
17	13 42 27	23 44 12	02 ♑ 30 38	08 ♑ 38 09	07 11	07 46	07 20	16 54	05 11	15 21	19 18	26 43
18	13 46 24	24 43 45	14 ♑ 49 17	21 ♑ 04 34	06 23	08 40	07 39	17 05	05 17	15 19	19 18	26 44
19	13 50 20	25 43 20	27 ♑ 24 33	03 ♒ 49 48	05 28	09 35	07 58	17 15	05 23	15 18	19 17	26 46
20	13 54 17	26 42 57	10 ♒ 20 51	16 ♒ 58 10	04 25	10 30	08 16	17 26	05 28	15 16	19 17	26 47
21	13 58 13	27 42 35	23 ♒ 42 10	00 ♓ 33 09	03 17	11 26	08 34	17 37	05 34	15 14	19 17	26 48
22	14 02 10	28 42 15	07 ♓ 31 19	14 ♓ 36 42	02 04	12 23	08 51	17 48	05 39	15 13	19 16	26 50
23	14 06 06	29 ♎ 41 57	21 ♓ 49 10	29 ♓ 08 20	00 ♏ 49	13 20	09 08	17 59	05 45	15 11	19 16	26 51
24	14 10 03	00 ♏ 41 40	06 ♈ 33 39	14 ♈ 04 18	29 ♎ 34	14 17	09 24	18 10	05 50	15 10	19 16	26 52
25	14 14 00	01 41 26	21 ♈ 39 16	29 ♈ 17 21	28 21	15 16	09 39	18 21	05 56	15 08	19 16	26 54
26	14 17 56	02 41 13	06 ♉ 57 13	14 ♉ 37 24	27 12	16 14	09 54	18 33	06 01	15 07	19 15	26 55
27	14 21 53	03 41 02	22 ♉ 16 27	29 ♉ 52 58	26 09	17 13	10 09	18 44	06 06	15 06	19 15	26 57
28	14 25 49	04 40 54	07 ♊ 25 39	14 ♊ 53 21	25 15	18 13	10 22	18 56	06 12	15 04	19 15	26 58
29	14 29 46	05 40 47	22 ♊ 15 09	29 ♊ 30 22	24 30	19 13	10 35	19 07	06 17	15 03	19 15	27 00
30	14 33 42	06 40 43	06 ♋ 38 31	13 ♋ 39 23	23 56	20 13	10 48	19 19	06 22	15 02	19 15	27 01
31	14 37 39	07 ♏ 40 41	20 ♋ 32 56	27 ♋ 19 18	23 ♎ 34	21 ♍ 14	10 ♋ 59	19 ♐ 31	06 ♍ 27	15 ♓ 01	19 ♒ 15	27 ♐ 03

MOON

Date	Moon True ☊ ° '	Moon Mean ☊ ° '	Moon ☽ Latitude ° '
01	06 ♓ 40	05 ♓ 11	05 N 11
02	06 R 38	05 08	04 53
03	06 D 38	05 05	04 19
04	06 38	05 02	03 31
05	06 40	04 58	02 33
06	06 42	04 55	01 30
07	06 43	04 52	00 N 23
08	06 R 42	04 49	00 S 44
09	06 40	04 46	01 48
10	06 36	04 43	02 46
11	06 30	04 39	03 37
12	06 23	04 36	04 17
13	06 14	04 33	04 46
14	06 06	04 30	05 03
15	05 58	04 27	05 07
16	05 51	04 23	04 58
17	05 47	04 20	04 35
18	05 44	04 17	03 59
19	05 D 44	04 14	03 13
20	05 44	04 11	02 12
21	05 45	04 08	01 S 04
22	05 R 46	04 04	00 N 09
23	05 45	04 01	01 25
24	05 42	03 58	02 37
25	05 36	03 55	03 40
26	05 28	03 52	04 28
27	05 19	03 49	04 57
28	05 10	03 45	05 04
29	05 02	03 42	04 51
30	04 56	03 39	04 19
31	04 ♓ 53	03 ♓ 36	03 N 33

DECLINATIONS

Date	Sun ☉ ° '	Moon ☽ ° '	Mercury ☿ ° '	Venus ♀ ° '	Mars ♂ ° '	Jupiter ♃ ° '	Saturn ♄ ° '	Uranus ♅ ° '	Neptune ♆ ° '	Pluto ♇ ° '
01	03 S 09	27 N 29	15 S 33	10 N 03	23 N 23	22 S 11	11 N 29	06 S 19	15 S 15	16 S 48
02	03 32	28 17	15 56	09 58	23 24	22 12	11 27	06 20	15 15	16 48
03	03 55	27 17	16 17	09 52	23 26	22 12	11 25	06 21	15 15	16 49
04	04 18	24 43	16 36	09 46	23 27	22 14	11 22	06 22	15 15	16 49
05	04 42	20 55	16 53	09 40	23 28	22 15	11 20	06 22	15 16	16 49
06	05 05	16 13	17 08	09 33	23 29	22 16	11 18	06 23	15 16	16 50
07	05 28	10 55	17 21	09 25	23 30	22 18	11 16	06 24	15 16	16 50
08	05 51	05 N 17	17 31	09 17	23 31	22 19	11 14	06 25	15 17	16 50
09	06 13	00 S 28	17 39	09 08	23 32	22 20	11 11	06 25	15 17	16 51
10	06 36	06 08	17 44	08 59	23 34	22 21	11 09	06 26	15 17	16 51
11	06 59	11 32	17 45	08 49	23 34	22 22	11 07	06 27	15 17	16 51
12	07 22	16 29	17 44	08 38	23 35	22 24	11 05	06 28	15 17	16 52
13	07 44	20 48	17 38	08 27	23 36	22 25	11 03	06 28	15 18	16 52
14	08 06	24 18	17 29	08 16	23 37	22 26	11 01	06 29	15 18	16 52
15	08 29	26 46	17 15	08 04	23 38	22 27	10 59	06 30	15 18	16 53
16	08 51	28 03	16 57	07 52	23 39	22 28	10 57	06 30	15 18	16 53
17	09 13	28 00	16 35	07 39	23 40	22 29	10 55	06 31	15 18	16 53
18	09 35	26 35	16 08	07 25	23 41	22 31	10 53	06 32	15 18	16 54
19	09 57	23 48	15 37	07 11	23 42	22 32	10 51	06 32	15 19	16 54
20	10 18	19 46	15 01	06 57	23 43	22 33	10 49	06 33	15 19	16 54
21	10 40	14 38	14 22	06 42	23 44	22 34	10 47	06 33	15 19	16 55
22	11 01	08 36	13 40	06 27	23 45	22 35	10 45	06 34	15 19	16 55
23	11 22	01 S 57	12 55	06 11	23 46	22 36	10 43	06 34	15 19	16 55
24	11 43	05 N 01	12 10	05 55	23 47	22 37	10 41	06 35	15 19	16 55
25	12 04	11 50	11 24	05 39	23 48	22 39	10 39	06 36	15 19	16 56
26	12 24	18 03	10 40	05 22	23 50	22 40	10 38	06 36	15 19	16 56
27	12 45	23 07	09 59	05 04	23 51	22 41	10 36	06 37	15 19	16 56
28	13 05	26 33	09 21	04 46	23 52	22 42	10 34	06 37	15 19	16 57
29	13 25	28 03	08 48	04 28	23 53	22 43	10 32	06 37	15 19	16 57
30	13 45	27 35	08 20	04 10	23 55	22 44	10 31	06 38	15 19	16 57
31	14 S 04	25 N 23	07 S 58	03 N 51	23 N 56	22 S 45	10 N 29	06 S 38	15 S 19	16 S 58

ZODIAC SIGN ENTRIES

Date	h	m	Planets
02	16	57	☽ ♋
04	22	27	☽ ♌
07	07	03	☽ ♍
08	06	53	♀ ♍
09	17	58	☽ ♎
12	06	13	☽ ♏
14	18	58	☽ ♐
17	07	03	☽ ♑
19	16	52	☽ ♒
21	23	02	☽ ♓
23	19	15	☉ ♏
24	01	24	☽ ♈
24	03	36	☿ ♎
26	01	07	☽ ♉
28	00	11	☽ ♊
30	00	49	☽ ♋

LATITUDES

Date	Mercury ☿ ° '	Venus ♀ ° '	Mars ♂ ° '	Jupiter ♃ ° '	Saturn ♄ ° '	Uranus ♅ ° '	Neptune ♆ ° '	Pluto ♇ ° '
01	03 S 00	03 S 19	00 S 03	00 N 21	01 N 21	00 S 49	00 S 17	06 N 36
04	03 14	02 51	00 N 02	00 21	01 22	00 48	00 17	06 35
07	03 24	02 25	00 07	00 20	01 22	00 48	00 17	06 34
10	03 26	02 00	00 12	00 19	01 22	00 48	00 17	06 33
13	03 20	01 36	00 18	00 19	01 23	00 48	00 17	06 32
16	03 00	01 13	00 24	00 18	01 24	00 48	00 17	06 31
19	02 24	00 51	00 30	00 18	01 24	00 48	00 17	06 30
22	01 34	00 30	00 36	00 18	01 25	00 48	00 17	06 29
25	00 S 33	00 S 11	00 43	00 17	01 25	00 48	00 17	06 29
28	00 N 27	00 N 07	00 50	00 17	01 26	00 48	00 17	06 28
31	01 N 17	00 N 25	00 N 57	00 N 17	01 N 26	00 S 48	00 S 17	06 N 27

DATA

Julian Date	2454375
Delta T	+66 seconds
Ayanamsa	23° 58' 02"
Synetic vernal point	05° ♓ 08' 58"
True obliquity of ecliptic	23° 26' 27"

MOON'S PHASES, APSIDES AND POSITIONS ☽

Date	h	m	Phase	Longitude ° '	Eclipse Indicator
03	10	06	☾	09 ♋ 49	
11	05	01	●	17 ♎ 30	
19	08	33	☽	25 ♑ 35	
26	04	52	○	02 ♉ 23	

Day	h	m		
13	10	02	Apogee	
26	11	57	Perigee	
02	10	21	Max dec	28° N 18'
09	10	04	0S	
16	23	16	Max dec	28° S 12'
23	18	45	0N	
29	18	10	Max dec	28° N 07'

LONGITUDES

Date	Chiron ⚷ ° '	Ceres ⚳ ° '	Pallas ⚴ ° '	Juno ⚵ ° '	Vesta ⚶ ° '	Black Moon Lilith ⚸ ° '
01	10 ♒ 39	23 ♉ 33	03 ♓ 34	06 ♏ 59	22 ♐ 24	08 ♏ 32
11	10 ♒ 30	22 ♉ 33	02 ♓ 04	10 ♏ 16	26 ♐ 35	09 ♏ 39
21	10 ♒ 28	20 ♉ 57	01 ♓ 10	13 ♏ 36	00 ♑ 58	10 ♏ 46
31	10 ♒ 32	18 ♉ 54	00 ♓ 52	16 ♏ 58	05 ♑ 31	11 ♏ 53

All ephemeris data is given at 12.00 UT and the Moon's longitude is additionally given for 24.00 UT

ASPECTARIAN

h m	Aspects
01 Monday	
03 01	☽ △ ☉
06 09	☽ ± ☿
06 16	☿ ⚹ ♄
12 11	☽ Q ♀
14 16	☽ ☍ ♃
16 51	☽ □ ♅
22 13	☽ bQ ☿
22 58	☽ △ ♆
02 Tuesday	
02 29	☽ Q ♄
09 31	☽ ⚹ ♀
10 52	☽ ☍ ♇
19 55	☽ ☌ ♂
23 17	☽ ⚹ ♄
03 Wednesday	
00 43	☽ bQ ♆
01 38	☽ △ ☿
06 52	☉ ⊥ ♄
10 06	☽ □ ☉
13 01	☽ ∠ ♀
13 16	♀ △ ♇
16 33	☽ ± ♆
18 39	☽ ⊼ ♃
20 41	☽ △ ♅
04 Thursday	
02 03	☽ ∠ ♄
03 16	☽ ⊼ ♆
05 32	☽ ± ♃
06 00	☽ ⊥ ♀
16 02	☽ ⊼ ♇
17 32	☽ ⊻ ♀
18 31	☽ ⊥ ♄
20 45	☽ ∥ ♂
21 05	☽ Q ☉
22 07	☽ bQ ♃
23 49	☽ bQ ♅
05 Friday	
03 04	☽ ± ♇
03 25	☽ ⊻ ♂
04 18	☽ # ♃
05 39	☽ ⊻ ♄
06 26	☿ ∥ ♇
10 53	☽ □ ☿
14 56	☽ ⊥ ♂
16 29	☽ ± ♅
19 50	☽ bQ ♇
21 13	☽ ⚹ ☉
06 Saturday	
02 24	☽ △ ♃
03 44	☽ ⊼ ♅
07 45	☽ # ☿
08 28	☽ ∠ ♂
09 02	☽ # ♇
10 47	☽ ☍ ♆
16 26	☽ # ♆
22 35	☽ Q ☿
07 Sunday	
00 22	☽ △ ♇
04 12	☽ ∠ ☉
05 28	☽ ☌ ♀
10 28	☽ ∥ ♄
14 13	☽ ⚹ ♂
15 07	☽ ☌ ♄
18 38	☽ ∥ ♀
22 45	☽ ⚹ ☿
23 10	☽ ⊥ ☉
08 Monday	
05 14	☉ Q ♇
07 17	☽ # ♅
09 48	☽ # ☉
11 56	☽ ⊻ ☉
13 02	☽ □ ♃
13 30	☽ ☍ ♅
14 28	☽ Q ♂
20 56	☽ ⊼ ♆
09 Tuesday	
03 44	♂ ⚹ ♄
04 04	☉ ⚹ ♃
05 13	☽ ∠ ☿
06 49	☉ ⊼ ♅
07 27	♂ bQ ♆
08 47	☽ ± ♆
11 08	☽ □ ♇
18 23	♃ □ ♅
20 26	☽ ⊻ ♀
23 33	☽ ⊥ ☿
10 Wednesday	
00 58	☉ ∥ ♅
01 12	☽ Q ♃
02 41	☽ bQ ♆
02 47	☽ ⊻ ♄
03 21	☽ □ ♂
06 42	☉ ∠ ♀
09 17	☽ ⊥ ♀
11 46	☽ ⊻ ☿
13 20	☽ ∥ ♅
14 13	☽ ∥ ☉
14 54	☽ ⊥ ♄
23 10	☽ Q ♇
11 Thursday	
00 06	☽ # ♀
01 04	☽ ⊼ ♅
01 35	☽ ⚹ ♃
04 40	☽ ∠ ♀
05 01	☽ ☌ ☉
08 45	☽ △ ♆
09 09	☽ ∠ ♄
10 07	☽ # ♄
13 07	☽ ± ♅
23 23	☽ ⚹ ♇
12 Friday	
03 59	☿ St R
05 57	☽ ∥ ♆
07 15	☽ bQ ♅
08 16	☽ ∠ ♃
13 13	☽ ⚹ ♀
13 57	☽ ∥ ♇
15 43	☽ ⚹ ♄
17 43	☽ △ ♂
18 24	☽ ∥ ☿
13 Saturday	
00 31	☽ ☌ ☿
01 40	☉ △ ♆
02 45	☽ ⊥ ♃
05 47	☽ ∠ ♇
12 13	☉ ∠ ♄
13 32	☽ △ ♅
15 05	☽ ⊻ ♃
15 24	☽ Q ♀
16 14	☽ Q ♄
21 23	☽ □ ♆
22 19	☽ ∥ ♃
23 12	☽ ⊻ ☉
14 Sunday	
00 03	☽ ⊥ ♇
01 04	☽ bQ ♂
04 28	♀ ☌ ♄
06 47	☽ # ♂
12 13	☽ ⊻ ♇
12 28	☽ ⊥ ☉
18 47	☉ # ♀
19 50	☽ ± ♂
15 Monday	
03 52	☉ ± ♅
04 58	☽ □ ♄
06 40	☽ □ ♀
08 18	☽ ∠ ☉
08 19	☽ ⊼ ♂
09 45	☽ Q ♆
11 51	☽ ⊻ ☿
23 26	☽ ⊥ ☿
16 Tuesday	
01 58	☽ □ ♅
04 32	☽ ☌ ♃
09 49	☽ ⚹ ♆
16 38	☽ ∠ ☿
17 05	☿ ∥ ♇
17 05	☽ ⚹ ☉
17 37	♀ ⚹ ♂
17 Wednesday	
00 32	☽ ☌ ♇
03 18	☿ ⚹ ♀
08 36	☿ △ ♂
13 39	☽ Q ♅
15 31	☽ ∠ ♆
17 17	☽ △ ♄
18 54	☽ Q ☉
20 38	☽ ⚹ ☿
21 43	☽ ☍ ♂
23 07	☽ △ ♀
18 Thursday	
09 03	☽ ⊥ ♆
12 58	☽ ⚹ ♅
16 24	☽ ⊻ ♃
18 25	☽ Q ☿
20 36	☽ ⊻ ♆
22 34	☽ bQ ♄
19 Friday	
04 02	☽ ⊥ ♃
06 14	☽ bQ ♀
08 33	☽ □ ☉
10 47	☽ ⊻ ♇
12 43	☽ # ♂
13 54	☿ ⚹ ♄
15 43	☽ ± ♄
17 24	☽ ∠ ♅
20 21	☽ ∥ ♃
21 12	☽ ∠ ♃
22 02	☽ ⊥ ♇
20 Saturday	
00 25	☽ ± ♀
00 32	☿ ∥ ♆
01 56	☽ □ ☿
02 58	☽ ⊼ ♄
08 07	☽ ⊼ ♂
10 02	☽ ⊥ ♅
12 18	☽ ⊼ ♀
13 39	☉ ⚹ ♇
14 37	☽ ∠ ♇
19 18	☽ ± ♂
20 55	☽ ⊻ ♅
21 Sunday	
01 01	☽ ⚹ ♃
01 57	☽ ∥ ♇
04 08	☽ ☌ ♆
09 04	☽ ∥ ♆
11 45	☽ bQ ♂
13 18	☽ ∥ ☿
17 28	☽ ⚹ ♇
19 30	☉ # ♄
19 36	☽ △ ☉
22 31	☽ Q ♃
23 32	☿ ∠ ♃
22 Monday	
01 32	♀ # ♅
03 16	☽ ∥ ☉
03 23	☽ △ ☿
03 45	☽ # ♄
08 47	☽ ☍ ♄
14 14	☽ Q ♇
14 19	☽ △ ♂
19 31	☽ ∥ ♅
20 17	☽ # ♀
20 50	☽ ☍ ♀
23 16	☽ bQ ☉
23 Tuesday	
00 59	☽ ☌ ♅
02 49	☽ bQ ☿
05 33	☽ □ ♃
07 46	☽ ⊻ ♆
15 20	☽ ± ☉
16 33	☽ ± ☿
17 40	☽ ⊥ ♆
20 17	☽ □ ♇
23 31	☉ bQ ♅
23 55	☉ ☌ ☿
24 Wednesday	
00 15	☿ bQ ♅
01 34	☽ ⊼ ☿
01 51	☽ ⊼ ☉
08 18	☽ ∠ ♆
10 50	☽ ⊼ ♄
14 58	☿ ∠ ♀
15 01	☽ ∥ ♀
16 38	☽ □ ♂
17 26	☽ # ♅
20 30	☽ ± ♄
21 36	☉ ∥ ☿
25 Thursday	
01 12	☽ ⊼ ♀
01 43	☽ ⊻ ♅
06 44	☽ △ ♃
07 46	☽ ∥ ♄
08 13	☽ ⚹ ♆
09 08	♀ ☍ ♅
10 35	☽ # ☿
10 51	☽ bQ ♄
11 11	☽ ⊥ ♅
11 20	☽ ± ♀
12 51	☽ # ☉
20 16	☽ △ ♇
21 36	☽ Q ♂
21 46	☽ ☍ ☿
26 Friday	
01 02	☽ # ♆
01 19	☽ ∠ ♅
02 26	☽ bQ ♀
03 05	☽ Q ♆
04 52	☽ ☍ ☉
06 36	☽ bQ ♃
07 25	☽ # ♇
10 32	☽ △ ♄
13 27	☿ # ♄
16 41	☽ ⚹ ♂
17 51	☿ ⚹ ♇
19 47	☽ bQ ♇
20 51	☽ ± ♃
27 Saturday	
00 45	☽ ⚹ ♅
03 31	☽ △ ♀
06 22	☽ ⊼ ♃
07 15	☽ □ ♆
09 39	☽ # ♃
09 55	☽ ± ♇
13 32	☉ ∠ ♃
16 15	☽ ∥ ♂
16 35	☽ ∠ ♂
17 44	☽ ⊼ ☿
19 22	☽ ⊼ ♇
19 35	☽ Q ♅
28 Sunday	
02 41	☽ ± ☿
07 03	☽ ⊥ ♂
07 18	☽ ⊼ ☉
10 01	☽ □ ♄
16 17	☽ bQ ☿
16 47	☽ ⊻ ♂
17 35	☽ ± ☉
29 Monday	
00 17	☽ □ ♅
01 53	☿ ⊥ ♀
06 40	☽ □ ♀
06 48	☽ ☍ ♃
07 05	☽ △ ♆
09 08	♀ □ ♃
09 14	☽ bQ ☉
12 47	♀ ⊼ ♆
15 21	☽ Q ♄
15 33	☽ △ ☿
19 50	☽ ☍ ♇
30 Tuesday	
03 42	☉ ⚹ ♄
03 58	♃ ⚹ ♆
07 57	☽ bQ ♆
11 31	☽ ⚹ ♄
12 04	☽ △ ☉
14 54	☽ Q ♀
19 11	☽ ☌ ♂
23 18	☽ ± ♆
31 Wednesday	
02 21	☽ △ ♅
09 43	☽ ⊼ ♆
10 09	☽ ⊼ ♃
13 19	☽ ⚹ ♀
13 35	☽ ∠ ♄
17 13	☽ □ ☿
20 06	♆ St D
20 54	☽ ± ♃
22 36	☽ ∥ ♂
23 32	☽ ⊼ ♇

NOVEMBER 2007

LONGITUDES

Date	Sidereal time h m s	Sun ☉ ° ' "	Moon ☽ ° ' "	Moon ☽ 24.00 ° ' "	Mercury ☿ ° '	Venus ♀ ° '	Mars ♂ ° '	Jupiter ♃ ° '	Saturn ♄ ° '	Uranus ♅ ° '	Neptune ♆ ° '	Pluto ♇ ° '
01	14 41 35	08 ♏ 40 41	03 ♌ 58 46	10 ♌ 31 44	23 ♎ 23	22 ♍ 16	11 ♋ 10	19 ♐ 42	06 ♍ 32	15 ♓ 00	19 ♒ 15	27 ♐ 05
02	14 45 32	09 40 43	16 ♌ 58 43	23 ♌ 20 13	23 D 23	23 18	11 21	19 54	06 36	14 R 58	19 D 15	27 06
03	14 49 29	10 40 48	29 ♌ 36 50	05 ♍ 49 09	23 35	24 20	11 30	20 06	06 41	14 57	19 15	27 08
04	14 53 25	11 40 54	11 ♍ 57 43	18 ♍ 03 07	23 56	25 22	11 39	20 18	06 46	14 56	19 15	27 09
05	14 57 22	12 41 03	24 ♍ 05 52	00 ♎ 06 27	24 28	26 25	11 48	20 30	06 50	14 55	19 15	27 11
06	15 01 18	13 41 13	06 ♎ 05 19	12 ♎ 02 52	25 08	27 28	11 55	20 42	06 55	14 55	19 15	27 13
07	15 05 15	14 41 26	17 ♎ 59 29	23 ♎ 55 27	25 56	28 32	12 02	20 55	06 59	14 54	19 16	27 15
08	15 09 11	15 41 40	29 ♎ 51 03	05 ♏ 46 31	26 51	29 ♍ 36	12 08	21 07	07 03	14 53	19 16	27 16
09	15 13 08	16 41 57	11 ♏ 42 04	17 ♏ 37 51	27 52	00 ♎ 40	12 13	21 19	07 07	14 52	19 16	27 18
10	15 17 04	17 42 15	23 ♏ 34 04	29 ♏ 30 51	28 ♎ 59	01 44	12 17	21 32	07 12	14 51	19 16	27 20
11	15 21 01	18 42 35	05 ♐ 28 23	11 ♐ 26 49	00 ♏ 10	02 49	12 21	21 44	07 16	14 51	19 17	27 22
12	15 24 58	19 42 56	17 ♐ 26 20	23 ♐ 27 09	01 25	03 54	12 24	21 57	07 20	14 50	19 17	27 24
13	15 28 54	20 43 19	29 ♐ 29 31	05 ♑ 33 43	02 44	05 00	12 26	22 09	07 23	14 49	19 18	27 25
14	15 32 51	21 43 44	11 ♑ 40 03	17 ♑ 48 55	04 05	06 05	12 27	22 22	07 27	14 49	19 18	27 27
15	15 36 47	22 44 10	24 ♑ 00 42	00 ♒ 15 50	05 29	07 11	12 R 27	22 35	07 31	14 48	19 19	27 29
16	15 40 44	23 44 37	06 ♒ 34 49	12 ♒ 58 08	06 55	08 18	12 26	22 47	07 34	14 48	19 19	27 31
17	15 44 40	24 45 06	19 ♒ 26 17	25 ♒ 59 46	08 22	09 24	12 25	23 00	07 38	14 48	19 20	27 33
18	15 48 37	25 45 36	02 ♓ 39 04	09 ♓ 24 35	09 51	10 31	12 23	23 13	07 41	14 47	19 20	27 35
19	15 52 33	26 46 07	16 ♓ 16 40	23 ♓ 15 34	11 21	11 38	12 20	23 26	07 44	14 47	19 21	27 37
20	15 56 30	27 46 40	00 ♈ 21 20	07 ♈ 33 54	12 52	12 45	12 16	23 39	07 48	14 47	19 21	27 39
21	16 00 27	28 47 13	14 ♈ 52 59	22 ♈ 18 04	14 24	13 52	12 11	23 52	07 51	14 47	19 22	27 41
22	16 04 23	29 ♏ 47 48	29 ♈ 48 23	07 ♉ 22 58	15 56	15 00	12 05	24 05	07 54	14 47	19 23	27 43
23	16 08 20	00 ♐ 48 24	15 ♉ 00 39	22 ♉ 40 04	17 29	16 07	11 59	24 18	07 57	14 46	19 24	27 45
24	16 12 16	01 49 02	00 ♊ 19 47	07 ♊ 58 19	19 02	17 15	11 51	24 31	07 59	14 D 46	19 24	27 47
25	16 16 13	02 49 41	15 ♊ 34 12	23 ♊ 06 05	20 35	18 24	11 43	24 44	08 02	14 46	19 25	27 49
26	16 20 09	03 50 22	00 ♋ 32 48	07 ♋ 53 22	22 09	19 32	11 34	24 57	08 05	14 47	19 26	27 51
27	16 24 06	04 51 04	15 ♋ 07 03	22 ♋ 13 21	23 43	20 41	11 24	25 11	08 07	14 47	19 27	27 53
28	16 28 02	05 51 47	29 ♋ 12 01	06 ♌ 03 01	25 17	21 49	11 13	25 24	08 09	14 47	19 28	27 55
29	16 31 59	06 52 33	12 ♌ 46 29	19 ♌ 22 45	26 51	22 58	11 02	25 37	08 12	14 47	19 29	27 57
30	16 35 56	07 ♐ 53 19	25 ♌ 52 14	02 ♍ 15 28	28 ♏ 25	24 ♎ 07	10 ♋ 49	25 ♐ 51	08 ♍ 14	14 ♓ 47	19 ♒ 30	27 ♐ 59

MOON NODES AND LATITUDE

Date	Moon True ☊ ° '	Moon Mean ☊ ° '	Moon ☽ Latitude ° '
01	04 ♓ 51	03 ♓ 33	02 N 37
02	04 D 52	03 29	01 34
03	04 52	03 26	00 N 28
04	04 R 52	03 23	00 S 38
05	04 50	03 20	01 40
06	04 46	03 17	02 38
07	04 38	03 14	03 28
08	04 28	03 10	04 08
09	04 16	03 07	04 38
10	04 02	03 04	04 55
11	03 48	03 01	05 00
12	03 34	02 58	04 52
13	03 23	02 55	04 30
14	03 14	02 51	03 56
15	03 08	02 48	03 11
16	03 05	02 45	02 15
17	03 04	02 42	01 11
18	03 D 04	02 39	00 S 02
19	03 R 04	02 35	01 N 09
20	03 02	02 32	02 18
21	02 57	02 29	03 21
22	02 50	02 26	04 12
23	02 40	02 23	04 46
24	02 29	02 20	05 00
25	02 17	02 16	04 52
26	02 06	02 13	04 24
27	01 58	02 10	03 40
28	01 52	02 07	02 43
29	01 50	02 04	01 39
30	01 ♓ 49	02 ♓ 00	00 N 31

DECLINATIONS

Date	Sun ☉ ° '	Moon ☽ ° '	Mercury ☿ ° '	Venus ♀ ° '	Mars ♂ ° '	Jupiter ♃ ° '	Saturn ♄ ° '	Uranus ♅ ° '	Neptune ♆ ° '	Pluto ♇ ° '
01	14 S 24	21 N 48	07 S 41	03 N 32	23 N 58	22 S 46	10 N 28	06 S 39	15 S 19	16 S 58
02	14 43	17 14	07 31	03 12	23 59	22 47	10 26	06 39	15 19	16 58
03	15 02	12 03	07 26	02 52	24 01	22 48	10 24	06 39	15 19	16 58
04	15 20	06 30	07 27	02 32	24 03	22 49	10 23	06 40	15 19	16 59
05	15 39	00 N 48	07 33	02 12	24 05	22 50	10 21	06 40	15 19	16 59
06	15 57	04 S 50	07 43	01 51	24 06	22 51	10 20	06 41	15 19	16 59
07	16 15	10 15	07 57	01 30	24 08	22 52	10 18	06 41	15 19	16 59
08	16 32	15 17	08 15	01 08	24 11	22 53	10 17	06 41	15 19	17 00
09	16 50	19 45	08 36	00 47	24 13	22 54	10 16	06 41	15 19	17 00
10	17 07	23 26	09 00	00 25	24 15	22 55	10 14	06 42	15 19	17 00
11	17 23	26 08	09 26	00 N 03	24 17	22 56	10 13	06 42	15 19	17 01
12	17 40	27 41	09 54	00 S 19	24 20	22 57	10 12	06 42	15 19	17 01
13	17 56	27 56	10 24	00 42	24 23	22 58	10 11	06 42	15 19	17 01
14	18 12	26 51	10 54	01 04	24 25	22 58	10 09	06 42	15 18	17 01
15	18 27	24 26	11 26	01 27	24 28	22 59	10 08	06 43	15 18	17 02
16	18 43	20 48	11 59	01 50	24 31	23 00	10 07	06 43	15 18	17 02
17	18 57	16 07	12 31	02 13	24 34	23 01	10 06	06 43	15 18	17 02
18	19 12	10 34	13 05	02 36	24 37	23 02	10 05	06 43	15 18	17 02
19	19 26	04 S 21	13 38	03 00	24 40	23 02	10 04	06 43	15 17	17 03
20	19 40	02 N 16	14 11	03 23	24 44	23 03	10 03	06 43	15 17	17 03
21	19 53	08 57	14 45	03 47	24 47	23 04	10 02	06 43	15 17	17 03
22	20 06	15 20	15 17	04 10	24 51	23 05	10 01	06 43	15 17	17 03
23	20 19	20 53	15 50	04 34	24 54	23 05	10 00	06 43	15 17	17 04
24	20 32	25 06	16 22	04 58	24 58	23 06	10 00	06 43	15 16	17 04
25	20 44	27 30	16 54	05 22	25 02	23 06	09 59	06 43	15 16	17 04
26	20 55	27 51	17 25	05 46	25 05	23 07	09 58	06 43	15 16	17 04
27	21 06	26 13	17 55	06 10	25 09	23 08	09 57	06 43	15 16	17 04
28	21 17	22 59	18 25	06 34	25 13	23 08	09 57	06 43	15 16	17 04
29	21 28	18 33	18 54	06 58	25 17	23 09	09 56	06 43	15 15	17 05
30	21 S 38	13 N 23	19 S 22	07 S 22	25 N 21	23 S 09	09 N 55	06 S 42	15 S 15	17 S 05

ZODIAC SIGN ENTRIES

Date	h	m	Planets
01	04	48	☽ ♌
03	12	45	☽ ♍
05	23	47	☽ ♎
08	12	18	☽ ♏
08	21	05	♀ ♎
11	00	59	☽ ♐
11	08	41	☿ ♏
13	13	00	☽ ♑
15	23	30	☽ ♒
18	07	14	☽ ♓
20	11	24	☽ ♈
22	12	18	☽ ♉
22	16	50	☉ ♐
24	11	29	☽ ♊
26	11	07	☽ ♋
28	13	23	☽ ♌
30	19	44	☽ ♍

LATITUDES

Date	Mercury ☿ ° '	Venus ♀ ° '	Mars ♂ ° '	Jupiter ♃ ° '	Saturn ♄ ° '	Uranus ♅ ° '	Neptune ♆ ° '	Pluto ♇ ° '
01	01 N 30	00 N 30	01 N 00	00 N 16	01 N 27	00 S 48	00 S 17	06 N 27
04	01 59	00 45	01 07	00 16	01 27	00 48	00 17	06 26
07	02 13	01 00	01 15	00 16	01 28	00 47	00 17	06 25
10	02 15	01 13	01 23	00 15	01 29	00 47	00 17	06 24
13	02 09	01 24	01 31	00 15	01 29	00 47	00 17	06 24
16	01 57	01 35	01 40	00 15	01 30	00 47	00 17	06 23
19	01 41	01 44	01 49	00 14	01 31	00 47	00 17	06 23
22	01 23	01 53	01 58	00 14	01 31	00 47	00 17	06 22
25	01 02	02 00	02 07	00 14	01 32	00 47	00 17	06 22
28	00 41	02 06	02 16	00 13	01 33	00 47	00 17	06 22
31	00 N 20	02 N 10	02 N 25	00 N 13	01 N 33	00 S 47	00 S 17	06 N 21

DATA

Julian Date	2454406
Delta T	+66 seconds
Ayanamsa	23° 58' 06"
Synetic vernal point	05° ♓ 08' 53"
True obliquity of ecliptic	23° 26' 26"

MOON'S PHASES, APSIDES AND POSITIONS ☽

Date	h	m	Phase	Longitude ° '	Eclipse Indicator
01	21	18	☾	09 ♌ 04	
09	23	03	●	17 ♏ 10	
17	22	33	☽	25 ♒ 12	
24	14	30	○	01 ♊ 55	

Day	h	m		
09	12	49	Apogee	
24	00	19	Perigee	
05	15	25	0S	
13	04	37	Max dec	28° S 00'
20	03	25	0N	
26	04	00	Max dec	27° N 57'

LONGITUDES

Date	Chiron ⚷ ° '	Ceres ⚳ ° '	Pallas ⚴ ° '	Juno ⚵ ° '	Vesta ⚶ ° '	Black Moon Lilith ⚸ ° '
01	10 ♒ 33	18 ♉ 41	00 ♓ 52	17 ♏ 18	05 ♑ 59	12 ♏ 00
11	10 ♒ 45	16 ♉ 23	01 ♓ 12	20 ♏ 43	10 ♑ 41	12 ♏ 07
21	11 ♒ 03	14 ♉ 08	02 ♓ 04	24 ♏ 07	15 ♑ 29	14 ♏ 14
31	11 ♒ 28	12 ♉ 10	03 ♓ 25	27 ♏ 31	20 ♑ 23	15 ♏ 22

All ephemeris data is given at 12.00 UT and the Moon's longitude is additionally given for 24.00 UT

ASPECTARIAN

h m Aspects

01 Thursday
04 48 ☽ bQ ♅
05 43 ☽ ⊥ ♄
06 14 ☽ # ♃
10 21 ☽ ± ♇
13 21 ☽ bQ ♃
16 40 ☽ ⊻ ♄
18 30 ☽ ∠ ♀
21 09 ☽ ± ♅
21 18 ☽ □ ☉
22 59 ☿ St D

02 Friday
01 22 ☽ ⊻ ♂
01 32 ☽ Q ☿
02 54 ☽ bQ ♇
08 15 ☽ ⊼ ♅
12 38 ☽ ⊥ ♀
12 42 ☽ ⊥ ♂
13 19 ☽ # ♇
13 32 ♀ Q ♂
14 27 ☿ ⊻ ♀
16 16 ☽ ☍ ♆
17 35 ☽ △ ♃
21 07 ☽ # ♆
23 16 ☽ # ☉

03 Saturday
00 14 ☽ ⚹ ☿
00 59 ☽ ⊻ ♀
05 58 ☽ ∠ ♂
07 13 ☽ △ ♇
10 03 ☽ Q ☉
19 14 ☽ ∥ ♄

04 Sunday
01 46 ☽ ☌ ♄
05 52 ☽ ∠ ☿
07 58 ☽ # ☿
09 17 ♀ ± ♆
10 38 ☉ ∥ ♆
11 17 ☽ # ♅
11 19 ☉ △ ♂
11 24 ☽ ⚹ ♂
11 24 ☽ ⚹ ☉
17 51 ☽ ☍ ♅
23 44 ☉ ∠ ♇

05 Monday
00 16 ☽ ⊥ ☿
02 23 ☽ ⊼ ♆
04 44 ☽ □ ♃
05 47 ☽ ∥ ♀
11 23 ☽ Q ♂
12 46 ☽ ⊻ ☿
14 18 ☽ ± ♆
17 04 ☽ ☌ ♀
18 10 ☽ □ ♇
19 48 ☽ ∠ ☉
23 58 ☽ # ♀

06 Tuesday
06 01 ♀ □ ♇
08 19 ☽ bQ ♆
13 40 ☽ ⊻ ♄
15 31 ☽ ⊥ ☉
17 22 ☽ Q ♃
20 02 ☽ ∥ ♅
23 51 ☽ □ ♂

07 Wednesday
01 09 ☽ ∥ ☿
01 50 ☽ ⊥ ♄
04 43 ☽ ⊻ ☉
05 45 ☽ ⊼ ♅
06 26 ☽ Q ♇
12 15 ☽ # ♄
14 34 ☽ △ ♆
16 49 ☉ △ ♅
17 52 ☽ ± ♅
18 00 ☽ ⚹ ♃
18 36 ☉ ⊥ ♃
20 07 ☽ ∠ ♄

08 Thursday
05 24 ☽ ☌ ☿
06 46 ☽ ⚹ ♇
11 26 ☽ ⊻ ♀
12 04 ☽ bQ ♅
12 10 ☽ ∥ ♆
18 50 ☽ ∥ ☉
20 48 ☽ ∥ ♇
22 27 ☿ ⚹ ♇

09 Friday
00 47 ☽ ⊥ ♀
00 55 ☽ ∠ ♃
02 41 ☽ ⚹ ♄
13 03 ☽ △ ♂
13 13 ☽ ∠ ♇
18 24 ☽ △ ♅
19 27 ☽ ⊥ ♃
20 49 ☽ ∠ ♀
23 03 ☽ ☌ ☉

10 Saturday
02 43 ☉ ∥ ♇
03 07 ☽ Q ♄
03 19 ☽ □ ♆
07 28 ☽ ⊥ ♇
07 48 ☽ ⊻ ♃
08 16 ☽ ∥ ♃
18 29 ☽ # ♂
19 33 ☽ bQ ♂
19 37 ☽ ⊻ ♇

11 Sunday
00 07 ☽ ⊻ ☿
05 39 ☿ bQ ♅
06 08 ☽ ⚹ ♀
13 33 ☽ ⊥ ☿
13 46 ☽ ± ♂
15 37 ☽ □ ♄
15 38 ☽ Q ♆

12 Monday
01 42 ☉ □ ♆
01 52 ☽ ⊼ ♂
02 04 ☉ Q ♄
06 48 ☽ □ ♅
08 38 ☽ Q ♀
09 43 ☽ ∠ ☿
15 42 ☽ ⚹ ♆
16 58 ☽ ⊻ ☉
20 26 ♀ bQ ♆
21 09 ☽ ☌ ♃

13 Tuesday
01 57 ☿ # ♄
06 00 ☽ ⊥ ☉
07 53 ☽ ☌ ♇
18 36 ☽ Q ♅
19 12 ☽ ⚹ ☿
21 30 ☽ ∠ ♆
23 58 ☽ □ ♀

14 Wednesday
01 26 ☽ ∠ ☉
03 41 ☽ △ ♄
05 16 ☉ ⊥ ♇
13 32 ☽ ☍ ♂
15 12 ☽ ⊥ ♆
18 09 ☽ ⚹ ♅
21 44 ☽ Q ☿

15 Thursday
02 54 ☽ ⊻ ♆
07 12 ☉ ⊻ ♃
08 24 ♂ St R
09 06 ☽ bQ ♄
09 11 ☽ ⊻ ♃
09 19 ☽ ⚹ ☉
11 44 ☽ # ♂
18 42 ☽ ⊻ ♇
19 28 ♀ ⊻ ♄
20 55 ☽ ⊥ ♃
22 28 ☽ ∥ ♃
23 07 ☽ ∠ ♅

16 Friday
02 27 ☽ ± ♄
06 11 ☽ ⊥ ♇
10 17 ☽ Q ☉
12 42 ☽ □ ☿
13 53 ☽ ⊼ ♄
14 19 ☽ ∠ ♃
15 32 ☽ △ ♀
16 11 ☽ ⊥ ♅
22 47 ☽ ∥ ☉
23 00 ☽ ⊼ ♂
23 11 ☽ ∠ ♇
23 22 ☿ ⚹ ♄

17 Saturday
03 24 ☽ ⊻ ♅
04 57 ☿ ∠ ♃
07 41 ☽ ∥ ♇
10 07 ☽ ± ♂
11 48 ☽ ☌ ♆
15 47 ☽ ∥ ♆
18 39 ☽ ⚹ ♃
21 56 ☽ bQ ♀
22 33 ☽ □ ☉

18 Sunday
02 29 ☽ ∥ ☿
02 33 ☽ bQ ♂
02 51 ☽ ⚹ ♇
13 57 ☽ # ♄
15 37 ☽ ± ♀
16 39 ☽ Q ♃
21 00 ☽ ☍ ♄

19 Monday
00 20 ☽ Q ♇
02 21 ☽ △ ☿
03 06 ☽ ∥ ♅
03 10 ☽ ⊼ ♀
05 09 ☽ △ ♂
06 51 ♀ Q ♃
09 25 ☽ ☌ ♅
16 44 ☽ ∥ ♀
17 18 ☽ ⊻ ♆

20 Tuesday
00 29 ☽ □ ♃
00 33 ☉ bQ ♂
02 18 ♀ □ ♂
02 52 ☿ △ ♂
03 34 ☽ ⊥ ♆
04 34 ☿ ⊻ ♀
07 19 ☽ bQ ☿
07 20 ☽ △ ☉
07 26 ☽ □ ♇
08 27 ☿ ∠ ♇
08 49 ☉ ⊻ ♇
16 16 ☽ # ♀
18 41 ☽ ∠ ♆
19 13 ☉ ∠ ♀

21 Wednesday
00 07 ☽ ± ☿
00 25 ☽ ⊼ ♄
03 55 ☽ # ♅
07 37 ☽ □ ♂
10 05 ☽ bQ ☉
10 12 ☽ ☍ ♀
10 18 ☽ ± ♄
11 07 ☽ ⊼ ☿
11 31 ♀ ⊥ ♄
11 50 ☽ ⊻ ♅
15 55 ☽ ∥ ♄
17 57 ☿ △ ♅
19 17 ☽ ⚹ ♆
21 33 ☽ ⊥ ♅

22 Thursday
00 55 ☽ bQ ♄
01 42 ☽ ± ☉
02 43 ☽ △ ♃
07 23 ♀ ⊼ ♅
08 40 ☽ △ ♇
11 33 ☿ ∥ ♆
11 48 ☽ # ♆
11 50 ☽ # ☿
11 57 ☽ ∠ ♅
11 59 ☽ ⊼ ☉
12 27 ☽ Q ♂
14 30 ☽ Q ♆
18 58 ☽ # ♇

23 Friday
00 51 ☽ △ ♄
02 53 ☽ bQ ♃
03 49 ♀ Q ♇
07 16 ☽ ⚹ ♂
08 26 ☽ bQ ♇
09 10 ☽ # ☉
11 38 ☽ ⚹ ♅
13 53 ☽ ⊼ ♀
16 19 ☽ ☍ ☿
17 14 ☽ ± ♃
18 53 ☽ □ ♆
22 35 ☽ ± ♇
23 25 ☽ # ♃

24 Saturday
00 02 ☽ ± ♀
02 46 ☿ ⊥ ♃
02 46 ☽ ⊼ ♃
06 26 ☽ Q ♅
06 36 ☽ ∠ ♂
08 00 ☽ ⊼ ♇
10 16 ♅ St D
11 02 ☽ ∥ ♂
14 30 ☽ ☍ ☉
15 16 ☽ bQ ♀
17 49 ☿ □ ♆
20 36 ☽ ⊥ ♂

25 Sunday
00 04 ☽ □ ♄
03 11 ☿ Q ♄
05 58 ☽ ⊻ ♂
10 44 ☽ □ ♅
16 51 ☽ △ ♀
18 08 ☽ △ ♆
19 45 ☿ ∥ ♇
20 54 ☽ ⊼ ☿

26 Monday
02 51 ☽ ☍ ♃
04 45 ☽ Q ♄
07 17 ☿ ⊥ ♇
07 38 ☽ ☍ ♇
07 40 ☽ ± ☿
09 55 ♀ △ ♆
17 45 ☽ ⊼ ☉
18 20 ☽ bQ ♆

27 Tuesday
00 05 ☽ bQ ☿
00 21 ☽ ⚹ ♄
04 22 ☽ ± ☉
05 53 ☽ ☌ ♂
09 13 ☽ ± ♆
11 26 ☽ △ ♅
14 09 ♀ ± ♅
19 18 ☽ ⊼ ♆
20 35 ☽ bQ ☉
20 57 ☽ ∥ ♂
22 12 ☽ □ ♀
23 06 ☉ ± ♂

28 Wednesday
01 34 ☽ ∠ ♄
04 22 ☽ △ ☿
05 20 ☽ ⊼ ♃
09 46 ☽ ⊼ ♇
11 00 ☽ # ♃
13 01 ☽ bQ ♅
14 11 ☿ ⊻ ♃
15 54 ☽ ± ♃
17 10 ☽ ⊥ ♄
20 16 ☽ ± ♇
20 57 ♀ ∥ ♅
21 26 ☽ # ☉

29 Thursday
00 37 ☽ △ ☉
00 52 ☿ bQ ♂
03 47 ☽ ⊻ ♄
04 51 ☽ ± ♅
08 04 ☽ bQ ♃
08 27 ☽ Q ♀
08 54 ☽ ⊻ ♂
10 28 ☽ # ☿
12 19 ☽ bQ ♇
15 38 ☽ ⊼ ♅
16 47 ♀ ∠ ♄
19 09 ☽ # ♇
19 35 ☽ ⊥ ♂

30 Friday
00 12 ☽ ☍ ♆
02 34 ☉ Q ♆
03 39 ☽ # ♆
05 24 ☿ ⊻ ♇
08 26 ☽ ⚹ ♀
11 54 ☽ ∠ ♂
11 57 ☽ △ ♃
15 58 ☽ △ ♇
17 25 ☽ □ ☿
20 23 ☉ □ ♄

DECEMBER 2007

LONGITUDES

Date	Sidereal time h m s	Sun ☉ ° ' "	Moon ☽ ° ' "	Moon ☽ 24.00 ° ' "	Mercury ☿ ° '	Venus ♀ ° '	Mars ♂ ° '	Jupiter ♃ ° '	Saturn ♄ ° '	Uranus ♅ ° '	Neptune ♆ ° '	Pluto ♇ ° '
01	16 39 52	08 ♐ 54 07	08 ♍ 33 03	14 ♍ 45 38	29 ♏ 59	25 ♎ 17	10 ♋ 36	26 ♐ 04	08 ♍ 16	14 ♓ 48	19 ♒ 31	28 ♐ 01
02	16 43 49	09 54 56	20 ♍ 53 52	26 ♍ 58 25	01 ♐ 33	26 26	10 R 22	26 17	08 18	14 48	19 32	28 04
03	16 47 45	10 55 47	02 ♎ 59 55	08 ♎ 59 00	03 07	27 36	10 07	26 31	08 20	14 49	19 33	28 06
04	16 51 42	11 56 40	14 ♎ 56 16	20 ♎ 52 14	04 41	28 46	09 51	26 44	08 21	14 49	19 34	28 08
05	16 55 38	12 57 33	26 ♎ 47 26	02 ♏ 42 18	06 15	29 ♎ 56	09 34	26 58	08 23	14 50	19 35	28 10
06	16 59 35	13 58 28	08 ♏ 37 15	14 ♏ 32 36	07 49	01 ♏ 06	09 17	27 11	08 24	14 50	19 36	28 12
07	17 03 31	14 59 24	20 ♏ 28 41	26 ♏ 25 43	09 23	02 16	08 59	27 25	08 26	14 51	19 38	28 14
08	17 07 28	16 00 22	02 ♐ 23 56	08 ♐ 23 29	10 57	03 26	08 41	27 39	08 27	14 51	19 39	28 16
09	17 11 25	17 01 20	14 ♐ 24 31	20 ♐ 27 08	12 31	04 37	08 22	27 52	08 28	14 52	19 40	28 19
10	17 15 21	18 02 19	26 ♐ 31 28	02 ♑ 37 36	14 05	05 48	08 02	28 06	08 29	14 53	19 41	28 21
11	17 19 18	19 03 20	08 ♑ 45 38	14 ♑ 55 43	15 39	06 58	07 41	28 19	08 30	14 54	19 43	28 23
12	17 23 14	20 04 21	21 ♑ 07 57	27 ♑ 22 33	17 13	08 09	07 20	28 33	08 31	14 55	19 44	28 25
13	17 27 11	21 05 22	03 ♒ 39 42	09 ♒ 59 38	18 47	09 20	06 59	28 47	08 32	14 56	19 45	28 27
14	17 31 07	22 06 24	16 ♒ 22 39	22 ♒ 49 03	20 21	10 31	06 37	29 00	08 33	14 57	19 47	28 30
15	17 35 04	23 07 27	29 ♒ 19 09	05 ♓ 53 20	21 56	11 42	06 15	29 14	08 33	14 58	19 48	28 32
16	17 39 00	24 08 30	12 ♓ 31 56	19 ♓ 15 16	23 30	12 54	05 52	29 28	08 34	14 59	19 50	28 34
17	17 42 57	25 09 34	26 ♓ 03 40	02 ♈ 57 21	25 05	14 05	05 29	29 42	08 34	15 00	19 51	28 36
18	17 46 54	26 10 38	09 ♈ 56 29	17 ♈ 01 05	26 39	15 17	05 06	29 ♐ 55	08 34	15 01	19 53	28 38
19	17 50 50	27 11 42	24 ♈ 11 04	01 ♉ 26 09	28 14	16 28	04 43	00 ♑ 09	08 34	15 02	19 54	28 41
20	17 54 47	28 12 46	08 ♉ 45 56	16 ♉ 09 45	29 ♐ 49	17 40	04 19	00 23	08 R 34	15 04	19 56	28 43
21	17 58 43	29 ♐ 13 51	23 ♉ 36 49	01 ♊ 06 08	01 ♑ 24	18 52	03 56	00 37	08 34	15 05	19 57	28 45
22	18 02 40	00 ♑ 14 57	08 ♊ 36 38	16 ♊ 07 07	03 00	20 04	03 32	00 50	08 34	15 07	19 59	28 47
23	18 06 36	01 16 02	23 ♊ 36 20	01 ♋ 03 05	04 35	21 16	03 08	01 04	08 33	15 08	20 01	28 50
24	18 10 33	02 17 08	08 ♋ 26 14	15 ♋ 44 46	06 11	22 28	02 45	01 18	08 33	15 09	20 02	28 52
25	18 14 29	03 18 15	22 ♋ 57 50	00 ♌ 04 44	07 47	23 40	02 21	01 32	08 32	15 11	20 04	28 54
26	18 18 26	04 19 22	07 ♌ 05 02	13 ♌ 58 25	09 23	24 52	01 58	01 45	08 31	15 13	20 06	28 56
27	18 22 23	05 20 29	20 ♌ 44 50	27 ♌ 24 20	11 00	26 04	01 34	01 59	08 31	15 14	20 07	28 58
28	18 26 19	06 21 37	03 ♍ 57 10	10 ♍ 23 41	12 36	27 17	01 11	02 13	08 30	15 16	20 09	29 01
29	18 30 16	07 22 45	16 ♍ 44 20	22 ♍ 59 40	14 13	28 29	00 48	02 27	08 29	15 18	20 11	29 03
30	18 34 12	08 23 54	29 ♍ 10 16	05 ♎ 16 44	15 50	29 ♏ 42	00 26	02 40	08 27	15 19	20 13	29 05
31	18 38 09	09 ♑ 25 03	11 ♎ 19 46	17 ♎ 19 58	17 ♑ 28	00 ♐ 54	00 ♋ 04	02 ♑ 54	08 ♍ 26	15 ♓ 21	20 ♒ 14	29 ♐ 07

Date	Moon True ☊ ° '	Moon Mean ☊ ° '	Moon ☽ Latitude ° '
01	01 ♓ 49	01 ♓ 57	00 S 36
02	01 R 48	01 54	01 39
03	01 46	01 51	02 37
04	01 42	01 48	03 27
05	01 35	01 45	04 07
06	01 25	01 41	04 37
07	01 12	01 38	04 55
08	00 58	01 35	05 00
09	00 44	01 32	04 52
10	00 30	01 29	04 30
11	00 19	01 26	03 56
12	00 10	01 22	03 10
13	00 04	01 19	02 15
14	00 01	01 16	01 12
15	00 00	01 13	00 S 04
16	00 D 00	01 10	01 N 06
17	00 R 00	01 06	02 14
18	00 00	01 03	03 16
19	29 ♒ 57	01 00	04 07
20	29 51	00 57	04 44
21	29 44	00 54	05 02
22	29 34	00 51	05 00
23	29 25	00 47	04 38
24	29 16	00 44	03 56
25	29 09	00 41	03 01
26	29 04	00 38	01 55
27	29 02	00 35	00 N 44
28	29 D 02	00 32	00 S 26
29	29 03	00 28	01 34
30	29 04	00 25	02 35
31	29 ♒ 04	00 ♓ 22	03 S 27

DECLINATIONS

Date	Sun ☉ ° '	Moon ☽ ° '	Mercury ☿ ° '	Venus ♀ ° '	Mars ♂ ° '	Jupiter ♃ ° '	Saturn ♄ ° '	Uranus ♅ ° '	Neptune ♆ ° '	Pluto ♇ ° '
01	21 S 47	07 N 49	19 S 49	07 S 45	25 N 25	23 S 10	09 N 55	06 S 42	15 S 14	17 S 05
02	21 56	02 N 05	20 15	08 09	25 29	23 10	09 54	06 42	15 14	17 05
03	22 05	03 S 35	20 41	08 33	25 33	23 11	09 54	06 42	15 14	17 05
04	22 13	09 04	21 05	08 57	25 37	23 11	09 54	06 42	15 13	17 06
05	22 21	14 10	21 28	09 20	25 42	23 12	09 53	06 41	15 13	17 06
06	22 29	18 45	21 51	09 44	25 46	23 12	09 53	06 41	15 13	17 06
07	22 36	22 36	22 12	10 07	25 50	23 12	09 52	06 41	15 12	17 06
08	22 42	25 32	22 32	10 31	25 54	23 13	09 52	06 41	15 12	17 06
09	22 48	27 21	22 51	10 54	25 58	23 13	09 52	06 40	15 12	17 06
10	22 54	27 54	23 09	11 17	26 02	23 13	09 52	06 40	15 11	17 07
11	22 59	27 05	23 26	11 40	26 06	23 14	09 52	06 39	15 11	17 07
12	23 04	24 55	23 42	12 02	26 09	23 14	09 52	06 39	15 10	17 07
13	23 08	21 31	23 56	12 25	26 13	23 14	09 52	06 39	15 10	17 07
14	23 12	17 04	24 09	12 47	26 17	23 14	09 52	06 38	15 09	17 07
15	23 16	11 46	24 21	13 09	26 20	23 15	09 52	06 38	15 09	17 07
16	23 19	05 S 50	24 32	13 31	26 24	23 15	09 52	06 37	15 09	17 08
17	23 21	00 N 29	24 41	13 52	26 27	23 15	09 52	06 37	15 08	17 08
18	23 23	06 56	24 49	14 14	26 30	23 15	09 52	06 36	15 08	17 08
19	23 25	13 12	24 56	14 35	26 33	23 15	09 52	06 36	15 07	17 08
20	23 26	18 54	25 01	14 55	26 36	23 15	09 52	06 35	15 07	17 08
21	23 26	23 33	25 05	15 16	26 38	23 15	09 53	06 35	15 06	17 08
22	23 26	26 41	25 08	15 36	26 41	23 15	09 53	06 34	15 06	17 08
23	23 26	27 54	25 09	15 56	26 43	23 15	09 53	06 34	15 05	17 08
24	23 25	27 06	25 09	16 15	26 45	23 15	09 54	06 33	15 05	17 08
25	23 24	24 27	25 07	16 34	26 47	23 15	09 54	06 32	15 04	17 09
26	23 22	20 21	25 04	16 53	26 49	23 15	09 55	06 32	15 04	17 09
27	23 20	15 17	24 59	17 11	26 51	23 15	09 55	06 31	15 03	17 09
28	23 17	09 39	24 53	17 29	26 52	23 15	09 56	06 30	15 02	17 09
29	23 14	03 N 48	24 45	17 46	26 54	23 15	09 57	06 30	15 02	17 09
30	23 10	02 S 02	24 35	18 03	26 55	23 15	09 57	06 29	15 01	17 09
31	23 S 06	07 S 40	24 S 25	18 S 20	26 N 56	23 S 14	09 N 58	06 S 28	15 S 01	17 S 09

ZODIAC SIGN ENTRIES

Date	h	m	Planets
01	12	21	☿ ♐
03	06	01	☽ ♎
05	13	29	♀ ♏
05	18	31	☽ ♏
08	07	11	☽ ♐
10	18	51	☽ ♑
13	05	01	☽ ♒
15	13	15	☽ ♓
17	18	52	☽ ♈
18	20	11	♃ ♑
19	21	38	☽ ♉
20	14	43	☿ ♑
21	22	14	☽ ♊
22	06	08	☉ ♑
23	22	18	☽ ♋
25	23	52	☽ ♌
28	04	44	☽ ♍
30	13	37	☽ ♎
30	18	02	♀ ♐
31	16	00	♂ ♊

LATITUDES

Date	Mercury ☿ ° '	Venus ♀ ° '	Mars ♂ ° '	Jupiter ♃ ° '	Saturn ♄ ° '	Uranus ♅ ° '	Neptune ♆ ° '	Pluto ♇ ° '
01	00 N 20	02 N 10	02 N 25	00 N 13	01 N 33	00 S 47	00 S 17	06 N 21
04	00 S 01	02 14	02 33	00 13	01 34	00 46	00 17	06 20
07	00 21	02 17	02 42	00 12	01 35	00 46	00 17	06 20
10	00 40	02 18	02 50	00 12	01 36	00 46	00 17	06 19
13	00 58	02 19	02 58	00 12	01 36	00 46	00 17	06 19
16	01 15	02 18	03 05	00 12	01 37	00 46	00 17	06 18
19	01 30	02 17	03 12	00 11	01 38	00 46	00 17	06 18
22	01 44	02 15	03 17	00 11	01 39	00 46	00 17	06 18
25	01 54	02 12	03 22	00 11	01 39	00 46	00 17	06 18
28	02 03	02 08	03 26	00 10	01 40	00 45	00 17	06 17
31	02 S 08	02 N 03	03 N 30	00 N 10	01 N 41	00 S 45	00 S 17	06 N 17

DATA

Julian Date	2454436
Delta T	+66 seconds
Ayanamsa	23° 58' 11"
Synetic vernal point	05° ♓ 08' 48"
True obliquity of ecliptic	23° 26' 25"

MOON'S PHASES, APSIDES AND POSITIONS ☽

Date	h	m	Phase	Longitude ° '	Eclipse Indicator
01	12	44	☾	08 ♍ 56	
09	17	40	●	17 ♐ 16	
17	10	18	☽	25 ♓ 05	
24	01	16	○	01 ♋ 50	
31	07	51	☾	09 ♎ 14	

Day	h	m		
06	17	07	Apogee	
22	10	20	Perigee	
02	20	46	0S	
10	09	38	Max dec	27° S 54'
17	10	11	0N	
23	14	26	Max dec	27° N 55'
30	03	33	0S	

LONGITUDES

Date	Chiron ⚷ ° '	Ceres ⚳ ° '	Pallas ⚴ ° '	Juno ⚵ ° '	Vesta ⚶ ° '	Black Moon Lilith ⚸ ° '
01	11 ♒ 28	12 ♉ 10	03 ♓ 25	27 ♏ 31	20 ♑ 23	15 ♏ 22
11	11 ♒ 58	10 ♉ 42	05 ♓ 10	00 ♐ 54	25 ♑ 20	16 ♏ 29
21	12 ♒ 32	09 ♉ 52	07 ♓ 18	04 ♐ 13	00 ♒ 21	17 ♏ 36
31	13 ♒ 11	09 ♉ 42	09 ♓ 43	07 ♐ 28	05 ♒ 24	18 ♏ 43

All ephemeris data is given at 12.00 UT and the Moon's longitude is additionally given for 24.00 UT

ASPECTARIAN

h m Aspects

01 Saturday
03 04 ☽ ∥ ♄
11 27 ☽ ☌ ♄
12 13 ☽ # ♀
12 44 ☽ □ ☉
15 40 ☽ ∠ ♀
15 52 ☽ ⚹ ♂
16 41 ☽ # ♅

02 Sunday
00 04 ☽ ☍ ♅
08 14 ♀ ⚹ ♃
08 57 ☽ Q ☿
09 19 ☽ ⊼ ♆
11 00 ☽ ⊥ ♀
14 49 ☽ Q ♂
20 27 ☉ ⊼ ♂
21 09 ☽ ± ♆
22 51 ☽ □ ♃

03 Monday
00 06 ☽ ⊻ ♀
02 12 ☽ □ ♇
03 09 ☽ Q ☉
12 16 ☽ ⚹ ☿
15 06 ☽ ⚻ ♆
22 34 ♀ ⚹ ♇
22 42 ☽ ⊻ ♄

04 Tuesday
01 07 ☿ ± ♂
01 28 ☽ ∥ ♅
01 58 ☽ □ ♂
05 24 ☽ ⚹ ☉
10 49 ☽ ⊥ ♄
11 26 ☽ ∥ ♀
11 36 ☽ Q ♃
11 45 ☽ ⊼ ♅
14 25 ☽ Q ♇
15 46 ☽ # ♄
16 58 ☿ ⊥ ♀
21 23 ☽ △ ♆
23 03 ☽ ∠ ☿
23 54 ☽ ± ♅

05 Wednesday
05 04 ☽ ∠ ♄
09 53 ♀ ⚻ ♅
12 22 ☽ ⚹ ♃
14 36 ☽ ∠ ☉
14 48 ☽ ⚹ ♇
17 13 ☽ ∥ ♆
18 10 ☽ ⚻ ♅
19 04 ☽ ☌ ♀
20 05 ☽ ⊥ ☿

06 Thursday
02 59 ☽ ∥ ♇
08 47 ☿ Q ♆
10 07 ☽ ⊻ ☿
10 34 ☽ ⊥ ☉
11 34 ☽ ⚹ ♄
13 19 ☽ △ ♂
19 23 ☽ ∠ ♃
20 53 ♀ # ♄
21 16 ☿ □ ♄
21 19 ☽ ∠ ♇
23 52 ☽ ⊻ ☉

07 Friday
00 36 ☽ △ ♅
06 59 ☿ ⊼ ♂
08 34 ☉ □ ♅
08 57 ☽ ∥ ☿
10 16 ☽ □ ♆
11 54 ☽ Q ♄
11 57 ☽ ∥ ☉
13 56 ☽ ⊥ ♃
15 34 ☽ ⊥ ♇
16 21 ☽ ∥ ♃
18 54 ☽ ⚻ ♂

08 Saturday
02 16 ☽ ⊻ ♃
03 41 ☽ ⊻ ♇
12 33 ☽ ± ♂
14 19 ☽ ⊻ ♀
15 48 ☽ # ♂
22 32 ☽ Q ♆

09 Sunday
00 09 ☽ □ ♄
00 15 ☽ ⊼ ♂
03 37 ☽ ⊥ ♀
03 59 ♂ ⚹ ♄
06 18 ☉ ∥ ☿
07 40 ☽ ☌ ☿
12 55 ☽ □ ♅
17 40 ☽ ☌ ☉
22 28 ☽ ⚹ ♆
23 27 ☽ ∠ ♀

10 Monday
15 09 ☽ ☌ ♃
15 36 ☽ ☌ ♇
17 40 ☿ ∥ ♃

11 Tuesday
00 25 ☿ □ ♅
00 31 ☽ Q ♅
04 04 ☽ ∠ ♆
08 08 ☽ ⚹ ♀
09 58 ☽ ☍ ♂
11 30 ☽ △ ♄
19 36 ♃ ☌ ♇
21 39 ☽ ⊥ ♆
23 16 ♀ △ ♂
23 57 ☽ ⚹ ♅

12 Wednesday
00 23 ☽ # ♂
03 20 ☽ ⊻ ☿
03 48 ☉ ⚹ ♆
09 18 ☽ ⊻ ♆
09 46 ☽ ⊻ ☉
09 55 ☽ Q ♀
16 36 ☽ ⊥ ☿
16 36 ☽ ⚻ ♄
19 33 ♀ ⚹ ♄
20 55 ☽ ∥ ☿
22 21 ☽ ⊥ ☉

13 Thursday
00 54 ☽ ∥ ♃
01 45 ☽ ∥ ☉
02 03 ☽ ⊻ ♇
02 31 ☽ ⊻ ♃
04 53 ☽ ∠ ♅
09 51 ☽ ± ♄
12 16 ☽ ∠ ☿
13 31 ☽ ⊥ ♇
14 10 ☽ ⊥ ♃
17 01 ☽ ∠ ☉
18 08 ☽ ⊼ ♂
21 15 ☽ ⊼ ♄
22 00 ☽ ⊥ ♅
23 52 ☽ □ ♀

14 Friday
03 02 ☿ ⚹ ♆
05 09 ☽ ± ♂
06 35 ☽ ∠ ♇
07 29 ☽ ∠ ♃
09 19 ☽ ⊻ ♅
11 45 ☽ ∥ ♇
18 22 ☽ ☌ ♆
20 28 ☽ ⚹ ☿
21 04 ☽ ∥ ♆
21 30 ☽ ⚻ ♂
23 36 ☽ ⚹ ☉

15 Saturday
03 27 ☉ ∥ ♃
06 25 ☽ ∥ ♀
10 33 ☽ ⚹ ♇
11 51 ☽ ⚹ ♃
19 57 ☽ # ♄
21 35 ☽ Q ☿
23 30 ☽ Q ☉

16 Sunday
00 18 ☽ △ ♂
04 50 ☽ ☍ ♄
08 27 ☽ Q ♇
08 56 ☽ ∥ ♅
10 03 ☽ Q ♃
12 43 ☽ △ ♀
16 24 ☽ ☌ ♅

17 Monday
01 02 ☽ ⊻ ♆
01 58 ♀ ∠ ♇
10 03 ☽ □ ☿
10 18 ☽ □ ☉
11 38 ☽ ⊥ ♆
15 27 ☉ ☌ ☿
16 27 ☽ □ ♇
17 47 ☽ ⚻ ♀
18 27 ☽ □ ♃

18 Tuesday
03 07 ♀ ∠ ♃
03 18 ☽ ∠ ♆
03 56 ☽ □ ♂
06 43 ♀ △ ♅
09 39 ☽ ⊼ ♄
10 46 ☽ ± ♀
10 46 ☽ # ♅
19 52 ☽ ± ♄
20 38 ☽ ⊻ ♅
21 54 ☽ ⊼ ♀
23 03 ☽ ∥ ♄

19 Wednesday
01 10 ♂ ⚻ ♆
04 50 ☽ ⚹ ♆
06 45 ☽ ⊥ ♅
09 37 ☽ Q ♂
10 58 ☽ ⚻ ♄
14 10 ♄ St R
17 23 ☽ △ ☉
17 51 ☽ # ♀
18 50 ☿ ☌ ♇
19 28 ☽ △ ♇
19 33 ☽ △ ☿
19 42 ☽ # ♆
21 43 ☽ ∠ ♅
22 02 ☽ △ ♃

20 Thursday
00 47 ☽ Q ♆
04 11 ☽ # ♇
04 56 ☽ ⚹ ♂
11 41 ☽ △ ♄
19 46 ☽ ⚻ ☉
20 04 ☽ ⚻ ♇
21 54 ☿ ☌ ♃
22 14 ☽ ⚹ ♅
22 54 ☽ ⚻ ♃
23 00 ☽ ⚻ ☿

21 Friday
00 17 ☉ ☌ ♇
00 58 ♀ ∥ ♆
03 41 ☽ ☍ ♀
04 39 ☽ ∠ ♂
06 06 ☽ □ ♆
10 15 ☽ # ♃
10 37 ☽ ± ♇
11 20 ☽ # ☉
11 20 ☽ ± ☉
12 59 ♀ ⚻ ♂
13 37 ☽ ± ♃
15 13 ☽ ± ☿
17 35 ☽ Q ♅
18 44 ☽ ⊥ ♂
20 16 ☽ ⊼ ♇
21 40 ☽ ⊼ ☉
22 21 ☽ # ☿
23 23 ☽ ⊼ ♃

22 Saturday
01 58 ☽ ⊼ ☿
04 06 ☽ ⊻ ♂
10 22 ♀ □ ♆
11 55 ☽ □ ♄
12 02 ☽ ∥ ♂
13 43 ☿ Q ♅
18 29 ☿ ☍ ♂
21 56 ♀ Q ♄
22 24 ☽ □ ♅

23 Sunday
05 56 ☉ ☌ ♃
06 13 ☽ △ ♆
07 54 ☽ ⊼ ♀
16 44 ☽ Q ♄
18 24 ☽ ± ♀
18 25 ☿ ∠ ♆
20 26 ☽ ☍ ♇

24 Monday
00 13 ☽ ☍ ♃
01 16 ☽ ☍ ☉
02 59 ☽ ☌ ♂
06 27 ☽ ⚻ ♆
07 53 ☽ ☍ ☿
10 16 ☽ ⚻ ♀
12 11 ☽ ⚹ ♄
16 12 ☽ ∥ ♂
19 47 ☉ ☍ ♂
20 16 ♀ ⊥ ♇
21 12 ☽ ± ♆
23 03 ☽ △ ♅

25 Tuesday
07 07 ☽ # ☿
07 09 ☽ ⊼ ♆
09 04 ☉ Q ♅
12 57 ☽ ∠ ♄
13 17 ☽ △ ♀
18 57 ☽ # ☉
19 49 ☽ # ♃
22 01 ☽ ⊼ ♇
23 10 ☿ △ ♄

26 Wednesday
00 12 ☽ ⚻ ♅
02 42 ☽ ⊼ ♃
03 26 ☽ ⊻ ♂
04 10 ☽ ⊥ ♄
06 52 ☽ ⊼ ☉
08 17 ☽ ± ♇
13 11 ☽ ± ♃
13 28 ☽ ⊥ ♂
14 29 ☽ ⊻ ♄
15 41 ☽ ± ♅
16 31 ☽ ⊼ ☿
18 04 ☽ ± ☉
19 52 ♂ ☍ ♃
23 58 ☽ ⚻ ♇

27 Thursday
02 12 ☽ ⊼ ♅
03 34 ☽ # ♇
03 52 ☽ # ♀
04 26 ♀ ± ♂
04 26 ☽ ± ☿
04 47 ☽ ∠ ♂
05 12 ☽ ⚻ ♃
06 40 ☉ ∠ ♆
09 07 ♀ ∥ ♇
09 50 ♀ ⊥ ♃
10 53 ☽ ☍ ♆
11 13 ☽ ⚻ ☉
13 01 ☽ # ♆
16 35 ☿ ∠ ♀
22 33 ☽ □ ♀
22 45 ☽ ⚻ ☿

28 Friday
02 54 ☽ △ ♇
07 03 ☽ ⚹ ♂
08 44 ☽ △ ♃
10 50 ☽ ∥ ♄
16 51 ☽ △ ☉
20 26 ☽ ☌ ♄

29 Saturday
00 57 ☽ # ♅
04 45 ☽ Q ♂
06 31 ☽ △ ☿
07 27 ☉ ∥ ♃
09 15 ☽ ☍ ♅
11 22 ☿ ⊥ ♆
11 28 ☽ Q ♀
18 36 ☽ ⊼ ♆
23 27 ♀ ⊻ ♇

30 Sunday
04 10 ☿ ⚹ ♅
06 13 ☽ ± ♆
11 50 ☽ □ ♇
13 08 ☽ ⚹ ♀
13 23 ☉ △ ♄
14 23 ☽ □ ♂
19 00 ☽ □ ♃
23 09 ♀ ⊼ ♂
23 54 ☽ ⚻ ♆

31 Monday
06 16 ☽ ⊻ ♄
06 49 ☽ ∥ ♅
07 51 ☽ □ ☉
18 11 ☽ ⊥ ♄
20 03 ☽ ⊼ ♅
22 10 ☽ ∠ ♀
22 17 ☽ # ♄
23 36 ☽ Q ♇

JANUARY 2008

LONGITUDES

Date	Sidereal time h m s	Sun ☉ ° ' "	Moon ☽ ° ' "	Moon ☽ 24.00 ° ' "	Mercury ☿ ° '	Venus ♀ ° '	Mars ♂ ° '	Jupiter ♃ ° '	Saturn ♄ ° '	Uranus ♅ ° '	Neptune ♆ ° '	Pluto ♇ ° '
01	18 42 05	10 ♑ 26 13	23 ♎ 18 00	29 ♎ 14 30	19 ♑ 05	02 ♐ 07	29 ♊ 42	03 ♑ 08	08 ♍ 25	15 ♓ 23	20 ♒ 16	29 ♐ 09
02	18 46 02	11 27 23	05 ♏ 10 03	11 ♏ 05 15	20 43	03 20	29 R 21	03 22	08 R 23	15 25	20 18	29 12
03	18 49 58	12 28 33	17 ♏ 00 36	22 ♏ 56 37	22 21	04 33	29 00	03 35	08 22	15 27	20 20	29 14
04	18 53 55	13 29 43	28 ♏ 53 43	04 ♐ 52 18	23 59	05 45	28 39	03 49	08 20	15 29	20 22	29 16
05	18 57 52	14 30 54	10 ♐ 52 41	16 ♐ 55 10	25 36	06 58	28 20	04 03	08 18	15 31	20 24	29 18
06	19 01 48	15 32 05	22 ♐ 59 59	29 ♐ 07 18	27 14	08 11	28 00	04 16	08 16	15 33	20 26	29 20
07	19 05 45	16 33 16	05 ♑ 17 15	11 ♑ 29 57	28 ♑ 52	09 24	27 42	04 30	08 14	15 35	20 28	29 22
08	19 09 41	17 34 27	17 ♑ 45 27	24 ♑ 03 46	00 ♒ 29	10 37	27 24	04 44	08 12	15 37	20 30	29 24
09	19 13 38	18 35 38	00 ♒ 24 57	06 ♒ 48 59	02 06	11 51	27 07	04 57	08 10	15 39	20 32	29 27
10	19 17 34	19 36 48	13 ♒ 15 53	19 ♒ 45 39	03 43	13 04	26 50	05 11	08 08	15 42	20 34	29 29
11	19 21 31	20 37 58	26 ♒ 18 20	02 ♓ 53 56	05 19	14 17	26 34	05 24	08 05	15 44	20 36	29 31
12	19 25 27	21 39 07	09 ♓ 32 31	16 ♓ 14 08	06 54	15 30	26 19	05 38	08 03	15 46	20 38	29 33
13	19 29 24	22 40 16	22 ♓ 58 53	29 ♓ 46 49	08 27	16 43	26 05	05 51	08 00	15 48	20 40	29 35
14	19 33 21	23 41 24	06 ♈ 38 01	13 ♈ 32 31	09 59	17 57	25 52	06 05	07 58	15 51	20 42	29 37
15	19 37 17	24 42 32	20 ♈ 30 20	27 ♈ 31 25	11 30	19 10	25 39	06 18	07 55	15 53	20 44	29 39
16	19 41 14	25 43 39	04 ♉ 35 40	11 ♉ 42 55	12 58	20 24	25 27	06 32	07 52	15 56	20 46	29 41
17	19 45 10	26 44 45	18 ♉ 52 52	26 ♉ 05 10	14 23	21 37	25 16	06 45	07 49	15 58	20 48	29 43
18	19 49 07	27 45 50	03 ♊ 19 20	10 ♊ 34 47	15 45	22 50	25 06	06 58	07 46	16 01	20 50	29 45
19	19 53 03	28 46 55	17 ♊ 50 53	25 ♊ 06 54	17 03	24 04	24 56	07 12	07 43	16 03	20 52	29 47
20	19 57 00	29 ♑ 47 59	02 ♋ 22 04	09 ♋ 35 34	18 17	25 18	24 47	07 25	07 40	16 06	20 55	29 49
21	20 00 56	00 ♒ 49 02	16 ♋ 46 39	23 ♋ 54 34	19 26	26 31	24 39	07 38	07 36	16 08	20 57	29 51
22	20 04 53	01 50 04	00 ♌ 58 39	07 ♌ 58 17	20 28	27 45	24 32	07 52	07 33	16 11	20 59	29 53
23	20 08 50	02 51 06	14 ♌ 53 02	21 ♌ 42 30	21 24	28 ♐ 58	24 26	08 05	07 30	16 14	21 01	29 55
24	20 12 46	03 52 06	28 ♌ 26 29	05 ♍ 04 52	22 12	00 ♑ 12	24 21	08 18	07 26	16 17	21 03	29 57
25	20 16 43	04 53 06	11 ♍ 37 39	18 ♍ 04 59	22 52	01 26	24 16	08 31	07 22	16 19	21 06	29 ♐ 59
26	20 20 39	05 54 06	24 ♍ 27 05	00 ♎ 44 16	23 22	02 39	24 12	08 44	07 19	16 22	21 08	00 ♑ 01
27	20 24 36	06 55 05	06 ♎ 56 55	13 ♎ 05 32	23 43	03 53	24 09	08 57	07 15	16 25	21 10	00 03
28	20 28 32	07 56 03	19 ♎ 10 36	25 ♎ 12 40	23 52	05 07	24 07	09 10	07 11	16 28	21 12	00 04
29	20 32 29	08 57 01	01 ♏ 12 20	07 ♏ 10 11	23 R 50	06 21	24 05	09 23	07 07	16 31	21 14	00 06
30	20 36 25	09 57 57	13 ♏ 06 49	19 ♏ 02 51	23 37	07 35	24 05	09 36	07 03	16 33	21 17	00 08
31	20 40 22	10 ♒ 58 54	24 ♏ 58 53	00 ♐ 55 29	23 ♒ 13	08 ♑ 48	24 ♊ 05	09 ♑ 48	06 ♍ 59	16 ♓ 36	21 ♒ 19	00 ♑ 10

Date	Moon True ☊ ° '	Moon Mean ☊ ° '	Moon ☽ Latitude ° '
01	29 ♒ 03	00 ♓ 19	04 S 10
02	29 R 00	00 16	04 42
03	28 55	00 12	05 01
04	28 48	00 09	05 08
05	28 39	00 06	05 01
06	28 31	00 03	04 40
07	28 22	00 ♓ 00	04 07
08	28 15	29 ♒ 57	03 21
09	28 10	29 53	02 24
10	28 07	29 50	01 20
11	28 06	29 47	00 S 10
12	28 D 06	29 44	01 N 02
13	28 08	29 41	02 11
14	28 09	29 38	03 15
15	28 10	29 34	04 07
16	28 R 10	29 31	04 46
17	28 08	29 28	05 08
18	28 05	29 25	05 11
19	28 01	29 22	04 54
20	27 57	29 18	04 18
21	27 53	29 15	03 26
22	27 49	29 12	02 22
23	27 48	29 09	01 N 10
24	27 D 47	29 06	00 S 04
25	27 48	29 03	01 15
26	27 49	28 59	02 22
27	27 51	28 56	03 19
28	27 52	28 53	04 06
29	27 53	28 50	04 42
30	27 R 53	28 47	05 05
31	27 ♒ 53	28 ♒ 44	05 S 15

DECLINATIONS

Date	Sun ☉ ° '	Moon ☽ ° '	Mercury ☿ ° '	Venus ♀ ° '	Mars ♂ ° '	Jupiter ♃ ° '	Saturn ♄ ° '	Uranus ♅ ° '	Neptune ♆ ° '	Pluto ♇ ° '
01	23 S 02	12 S 55	24 S 12	18 S 36	26 N 57	23 S 14	09 N 59	06 S 27	15 S 00	17 S 09
02	22 57	17 40	23 58	18 52	26 58	23 14	09 59	06 27	15 00	17 09
03	22 51	21 43	23 43	19 07	26 58	23 14	10 00	06 26	14 59	17 09
04	22 45	24 55	23 26	19 22	26 59	23 13	10 01	06 25	14 58	17 09
05	22 39	27 02	23 07	19 36	26 59	23 13	10 02	06 24	14 58	17 09
06	22 32	27 55	22 47	19 50	26 59	23 13	10 03	06 23	14 57	17 09
07	22 25	27 27	22 25	20 03	26 59	23 12	10 04	06 23	14 57	17 10
08	22 17	25 35	22 02	20 16	26 59	23 12	10 05	06 22	14 56	17 10
09	22 09	22 25	21 37	20 28	26 59	23 11	10 06	06 21	14 55	17 10
10	22 00	18 07	21 11	20 40	26 58	23 11	10 07	06 20	14 55	17 10
11	21 51	12 54	20 44	20 51	26 58	23 11	10 08	06 19	14 54	17 10
12	21 42	07 02	20 15	21 01	26 57	23 10	10 09	06 18	14 53	17 10
13	21 32	00 S 46	19 45	21 11	26 57	23 10	10 10	06 17	14 53	17 10
14	21 22	05 N 37	19 14	21 21	26 56	23 09	10 12	06 16	14 52	17 10
15	21 11	11 50	18 42	21 29	26 56	23 09	10 13	06 15	14 51	17 10
16	21 00	17 33	18 10	21 37	26 55	23 08	10 14	06 14	14 51	17 10
17	20 48	22 22	17 36	21 45	26 54	23 07	10 15	06 13	14 50	17 10
18	20 37	25 54	17 02	21 52	26 53	23 07	10 17	06 12	14 49	17 10
19	20 24	27 46	16 28	21 58	26 52	23 06	10 18	06 11	14 49	17 10
20	20 12	27 43	15 54	22 04	26 51	23 06	10 20	06 10	14 48	17 10
21	19 59	25 48	15 21	22 09	26 50	23 05	10 21	06 09	14 47	17 10
22	19 45	22 15	14 48	22 13	26 50	23 04	10 22	06 08	14 47	17 10
23	19 31	17 29	14 16	22 17	26 49	23 03	10 24	06 07	14 46	17 10
24	19 17	11 58	13 46	22 20	26 48	23 03	10 25	06 06	14 45	17 10
25	19 03	06 02	13 18	22 23	26 47	23 02	10 27	06 05	14 45	17 10
26	18 48	00 N 02	12 52	22 24	26 46	23 01	10 28	06 04	14 44	17 10
27	18 33	05 S 48	12 29	22 25	26 45	23 00	10 30	06 03	14 43	17 10
28	18 17	11 18	12 09	22 26	26 44	23 00	10 32	06 02	14 42	17 10
29	18 01	16 18	11 53	22 26	26 43	22 59	10 33	06 00	14 42	17 10
30	17 45	20 37	11 41	22 25	26 42	22 58	10 35	05 59	14 41	17 10
31	17 S 29	24 S 06	11 S 33	22 S 23	26 N 41	22 S 57	10 N 37	05 S 58	14 S 40	17 S 10

ZODIAC SIGN ENTRIES

Date	h	m	Planets
02	01	32	☽ ♏
04	14	13	☽ ♐
07	01	43	☽ ♑
08	04	46	☿ ♒
09	11	13	☽ ♒
11	18	44	☽ ♓
14	00	23	☽ ♈
16	04	13	☽ ♉
18	06	30	☽ ♊
20	08	05	☽ ♋
20	16	43	☉ ♒
22	10	20	☽ ♌
24	08	06	♀ ♑
24	14	48	☽ ♍
26	02	37	♇ ♑
26	22	35	☽ ♎
29	09	35	☽ ♏
31	22	08	☽ ♐

LATITUDES

Date	Mercury ☿ ° '	Venus ♀ ° '	Mars ♂ ° '	Jupiter ♃ ° '	Saturn ♄ ° '	Uranus ♅ ° '	Neptune ♆ ° '	Pluto ♇ ° '
01	02 S 09	02 N 01	03 N 30	00 N 10	01 N 41	00 S 45	00 S 17	06 N 17
04	02 09	01 56	03 33	00 10	01 42	00 45	00 17	06 17
07	02 05	01 50	03 34	00 10	01 43	00 45	00 17	06 17
10	01 55	01 43	03 34	00 09	01 43	00 45	00 17	06 17
13	01 40	01 36	03 34	00 09	01 44	00 45	00 17	06 17
16	01 17	01 28	03 33	00 09	01 45	00 45	00 17	06 17
19	00 47	01 20	03 32	00 09	01 45	00 45	00 17	06 17
22	00 S 09	01 12	03 30	00 08	01 46	00 45	00 17	06 17
25	00 N 38	01 03	03 28	00 08	01 46	00 45	00 17	06 17
28	01 30	00 55	03 25	00 08	01 47	00 45	00 17	06 17
31	02 N 22	00 N 46	03 N 22	00 N 07	01 N 47	00 S 44	00 S 17	06 N 17

DATA

Julian Date	2454467
Delta T	+67 seconds
Ayanamsa	23° 58' 16"
Synetic vernal point	05° ♓ 08' 43"
True obliquity of ecliptic	23° 26' 25"

MOON'S PHASES, APSIDES AND POSITIONS ☽

Date	h	m	Phase	Longitude ° '	Eclipse Indicator
08	11	37	●	17 ♑ 33	
15	19	46	☽	25 ♈ 02	
22	13	35	○	01 ♌ 54	
30	05	03	☾	09 ♏ 40	

Day	h	m		
03	08	13	Apogee	
19	08	39	Perigee	
31	04	28	Apogee	
06	15	46	Max dec	27° S 56'
13	14	55	0N	
19	23	28	Max dec	27° N 59'
26	12	09	0S	

LONGITUDES

Date	Chiron ⚷ ° '	Ceres ⚳ ° '	Pallas ⚴ ° '	Juno ⚵ ° '	Vesta ⚶ ° '	Black Moon Lilith ⚸ ° '
01	13 ♒ 15	09 ♉ 43	09 ♓ 59	07 ♐ 47	05 ♒ 55	18 ♏ 50
11	13 ♒ 57	10 ♉ 17	12 ♓ 42	10 ♐ 56	10 ♒ 59	19 ♏ 57
21	14 ♒ 41	11 ♉ 26	15 ♓ 39	13 ♐ 58	16 ♒ 05	21 ♏ 04
31	15 ♒ 27	13 ♉ 07	18 ♓ 47	16 ♐ 50	21 ♒ 10	22 ♏ 12

ASPECTARIAN

h m	Aspects
01 Tuesday	
02 11	☽ □ ☿
05 53	☽ △ ♆
07 33	☽ Q ♃
08 08	☽ ± ♅
12 14	☽ ∠ ♄
18 20	☽ ⊥ ♀
22 09	☽ ∥ ♆
23 21	☽ Q ☉
23 52	☽ ✶ ♇
02 Wednesday	
00 33	☽ △ ♂
02 21	☽ ⚼ ♅
05 48	☿ ⊻ ♆
07 51	☽ ⊻ ♀
08 16	☽ ✶ ♃
09 16	☽ ∥ ♇
12 43	♀ ⊻ ♃
18 31	☽ ✶ ♄
18 46	♃ Q ♅
19 05	☽ ∥ ♀
20 20	☽ Q ☿
21 23	♂ ☍ ♇
03 Thursday	
01 57	☽ ✶ ☉
06 04	☽ ⚼ ♂
06 21	☽ ∠ ♇
08 49	☽ △ ♅
15 15	☽ ∠ ♃
18 45	☽ □ ♆
18 46	☽ Q ♄
19 32	☽ ∥ ☉
22 25	☽ ∥ ♃
23 46	☽ ± ♂
04 Friday	
00 30	☽ ✶ ☿
00 37	☽ ⊥ ♇
00 56	☽ ∥ ☿
02 45	☿ ⚼ ♄
09 47	☽ ⊥ ♃
11 07	☽ ∠ ☉
11 32	☽ ⊼ ♂
12 45	☽ ⊻ ♇
22 05	☽ ⊻ ♃
05 Saturday	
03 20	☽ ☌ ♀
04 26	☿ ∥ ♃
06 51	☽ ⊥ ☉
06 53	☽ □ ♄
07 02	☽ Q ♆
09 08	☉ ⊥ ♆
11 07	☽ # ♂
11 23	☽ ∠ ☿
19 54	☽ ⊻ ☉
21 14	☽ □ ♅
06 Sunday	
06 55	☽ ✶ ♆
08 00	☽ ⊥ ☿
12 19	☉ ✶ ♅
13 39	♀ □ ♄
16 53	♀ Q ♆
21 32	☿ ⊼ ♂
21 35	☽ ☍ ♂
21 35	☽ ⊻ ☿
07 Monday	
00 27	☽ ☌ ♇
08 41	☽ Q ♅
10 26	☽ ☌ ♃
12 20	☽ ∠ ♆
12 32	☉ ∥ ☿
17 42	☽ △ ♄
19 39	☿ ⊻ ♇
19 54	☽ # ♂
20 50	☽ ⊻ ♀
08 Tuesday	
05 44	☽ ⊥ ♆
07 54	☽ ✶ ♅
09 36	☽ ⊥ ♀
11 37	☽ ☌ ☉
13 57	☿ ∠ ♅
17 14	☽ ⊻ ♆
22 21	☽ ⚼ ♄
09 Wednesday	
04 33	☽ ∠ ♀
05 55	☽ ⊼ ♂
06 52	☽ ∥ ♃
10 10	☽ ⊻ ♇
12 27	☽ ∠ ♅
12 55	☿ ± ♄
13 44	☽ ∥ ☉
15 17	☽ ± ♄
15 38	☽ ☌ ☿
16 58	☽ ± ♂
17 28	☽ ∥ ☿
20 40	☽ ⊻ ♃
21 28	☽ ⊥ ♇
23 04	☽ ∥ ♀
10 Thursday	
00 47	☿ ± ♂
02 29	☽ ⊼ ♄
05 21	☽ ⊥ ♅
08 04	☽ ⊥ ♃
09 24	☽ ⚼ ♂
11 35	☽ ✶ ♀
14 15	☽ ∠ ♇
16 30	☽ ⊻ ♅
16 40	☽ ∥ ♇
11 Friday	
00 43	☽ ⊻ ☉
01 00	☽ ∠ ♃
01 31	☽ ☌ ♆
03 11	☽ ∥ ♆
07 45	☿ ∥ ♀
11 06	☉ ⊻ ♆
11 57	☽ Q ♀
12 29	☽ △ ♂
12 39	☽ ⊥ ☉
13 39	☿ ⊻ ♃
15 07	☿ ⊥ ♇
17 52	☽ ✶ ♇
23 33	☽ # ♄
12 Saturday	
04 49	☽ ✶ ♃
06 21	☽ ∠ ☉
06 35	☽ ⊻ ☿
09 19	☽ ☍ ♄
11 55	♃ ∠ ♆
14 52	☽ ∥ ♅
15 37	☽ Q ♇
17 23	♀ □ ♅
18 49	☽ ⊥ ☿
23 12	☽ ☌ ♅
23 46	☽ □ ♀
13 Sunday	
02 44	☽ Q ♃
05 17	☿ ⊼ ♄
05 42	☉ ⊥ ♀
07 53	☽ ⊻ ♆
11 24	☽ ✶ ☉
12 57	☽ ∠ ☿
17 24	☽ □ ♂
18 32	☽ ⊥ ♆
19 34	☉ ⚼ ♄
22 14	♂ Q ♄
23 41	☽ □ ♇
14 Monday	
09 42	☿ ⊥ ♅
10 13	☽ Q ☉
10 22	☽ ∠ ♆
11 01	☽ □ ♃
13 26	☉ ∥ ♀
14 18	☽ ⊼ ♄
14 29	☽ # ♅
18 34	☽ ✶ ☿
15 Tuesday	
00 06	☿ ⚼ ♂
00 21	☽ Q ♂
00 41	☽ ± ♄
04 02	☽ ⊻ ♅
05 37	☽ ∥ ♄
09 29	☽ △ ♀
12 24	☽ ✶ ♆
14 23	☽ ⊥ ♅
16 07	☽ ⚼ ♄
17 44	☽ Q ☿
19 46	☽ □ ☉
20 41	☽ ✶ ♂
16 Wednesday	
00 20	☽ # ♆
03 33	☿ ⊥ ♃
03 39	☽ △ ♇
05 46	☽ ∠ ♅
06 29	☉ ⊼ ♂
08 54	☽ Q ♆
10 18	☽ # ♇
13 29	☽ ⚼ ♀
14 32	☽ # ☿
15 19	☽ △ ♃
17 30	☽ △ ♄
19 35	♀ ✶ ♆
21 45	☽ ∠ ♂
17 Thursday	
03 40	☽ □ ☿
03 56	☽ # ☉
05 01	☽ ⚼ ♇
06 02	☽ ± ♀
07 07	☽ ✶ ♅
08 28	☽ # ♀
12 38	☽ ⊥ ♂
15 13	☽ □ ♆
16 21	☽ # ♃
16 52	☽ ⚼ ♃
16 59	☽ ⊼ ♀
17 56	☿ ∠ ♇
20 05	☽ ± ♇
22 30	☽ ⊻ ♂
18 Friday	
02 05	☽ △ ☉
03 10	☽ Q ♅
06 04	☽ ⊼ ♇
06 45	☿ ∥ ♇
08 03	☽ ± ♃
16 49	☿ ⊻ ♅
18 08	☽ ⊼ ♃
19 20	☽ □ ♄
21 47	☽ ∥ ♂
19 Saturday	
04 47	☽ ⚼ ☉
09 02	☽ □ ♅
10 34	☽ △ ☿
17 01	☽ △ ♆
20 46	☽ ± ☉
23 13	☽ ☍ ♀
23 35	☽ ☌ ♂
20 Sunday	
00 57	☽ Q ♄
03 10	♀ ☍ ♂
07 26	☽ ⊼ ☉
07 46	☽ ☍ ♇
12 29	☉ ⊻ ♇
13 40	☽ ⚼ ☿
17 54	☽ ⚼ ♆
20 31	☽ ☍ ♃
20 45	☽ ✶ ♄
21 Monday	
01 35	☽ ∥ ♂
05 56	☽ ± ☿
08 39	☉ ± ♂
08 56	☽ ± ♆
09 14	♃ △ ♄
10 56	☽ △ ♅
16 49	☽ ⊼ ☿
19 01	☽ ⊼ ♆
19 59	☉ ∠ ♅
21 46	☽ ∠ ♄
22 Tuesday	
01 09	☽ ⊻ ♂
05 39	☉ ± ♄
05 58	☽ ⊼ ♀
07 10	☽ # ♃
10 08	☽ ⊼ ♇
11 16	☽ ⊥ ♂
12 09	☽ # ♀
12 21	☽ ⚼ ♅
12 58	☽ ⊥ ♄
12 59	☿ ∥ ♆
13 35	☽ ☍ ☉
17 11	☽ ± ♀
20 26	☽ ± ♇
23 14	☽ ⊻ ♄
23 Wednesday	
00 00	☽ ⊼ ♃
01 15	☿ ☌ ♆
01 51	☽ # ☉
02 36	☽ ∠ ♂
03 53	☽ ± ♅
10 15	☽ ⚼ ♀
10 34	☽ ± ♃
12 03	☽ ⚼ ♇
13 30	☽ # ♇
14 22	☽ ⊼ ♅
22 49	☽ ☍ ♆
24 Thursday	
00 09	☽ # ♆
00 13	☽ ☍ ☿
02 40	☽ ⚼ ♃
03 41	☽ # ☿
04 44	☽ ✶ ♂
06 58	♀ ☌ ♇
14 43	☽ △ ♇
15 29	☽ △ ♀
18 18	☽ ∥ ♄
22 37	☽ ⊼ ☉
25 Friday	
02 13	☽ Q ♂
04 13	☽ ☌ ♄
06 11	☽ △ ♃
10 31	☽ ± ☉
11 50	☽ # ♅
20 44	☽ ☍ ♅
26 Saturday	
04 43	☽ ⚼ ☉
05 42	☽ ⊼ ♆
09 53	☽ ⊼ ☿
11 32	☽ □ ♂
14 42	☉ ⊥ ♇
17 07	☽ ± ♆
21 41	☽ ± ☿
22 38	☽ □ ♇
27 Sunday	
05 25	☽ □ ♀
10 29	☽ ⚼ ♆
11 56	☽ △ ☉
12 35	☽ ⊻ ♄
13 00	☽ ∥ ♅
15 30	☽ ⚼ ☿
15 58	☽ □ ♃
19 20	☉ ⊼ ♄
22 44	♀ Q ♅
28 Monday	
00 15	☽ ⊥ ♄
06 37	☽ ⊼ ♅
08 28	☽ # ♄
09 49	☽ Q ♇
15 41	☽ ∥ ☿
16 02	☽ △ ♆
17 56	☽ ∠ ♄
18 33	☽ ± ♅
20 31	☿ St R
20 43	☽ Q ♀
21 21	☽ △ ☿
21 48	☽ △ ♂
29 Tuesday	
04 00	☽ ∥ ♆
04 12	☽ Q ♃
09 47	☽ ✶ ♇
09 53	♀ ∠ ♆
12 37	☽ ⚼ ♅
15 16	☉ ⚼ ♂
16 30	☽ ∥ ♇
20 35	☽ ∥ ☉
23 32	☽ ✶ ♀
23 50	☽ ✶ ♄
30 Wednesday	
00 52	☉ ⊻ ♃
02 19	♀ △ ♄
03 51	☽ ⚼ ♂
04 46	☽ ✶ ♃
05 03	☽ □ ☉
16 06	☽ ∠ ♇
19 00	☽ △ ♅
22 02	☽ ± ♂
22 34	♂ St D
23 30	☽ ∥ ♀
23 57	☽ Q ♄
31 Thursday	
02 42	☉ ⊥ ♅
03 23	☽ ∥ ♃
03 40	☿ ∠ ♀
04 34	☽ □ ♆
08 35	☽ □ ☿
09 21	☽ ∠ ♀
10 11	☽ ⊼ ♂
10 21	☽ ⊥ ♇
11 39	☽ ∠ ♃
20 50	☽ Q ☉
22 30	☽ ⊻ ♇

All ephemeris data is given at 12.00 UT and the Moon's longitude is additionally given for 24.00 UT

FEBRUARY 2008

Date	Sidereal time h m s	LONGITUDES Sun ☉ ° ' "	Moon ☽ ° ' "	Moon ☽ 24.00 ° ' "	Mercury ☿ ° '	Venus ♀ ° '	Mars ♂ ° '	Jupiter ♃ ° '	Saturn ♄ ° '	Uranus ♅ ° '	Neptune ♆ ° '	Pluto ♇ ° '
01	20 44 19	11 ♒ 59 49	06 ♐ 53 14	12 ♐ 52 39	22 ♒ 38	10 ♑ 02	24 ♊ 06	10 ♑ 01	06 ♍ 55	16 ♓ 39	21 ♒ 21	00 ♑ 12
02	20 48 15	13 00 44	18 ♐ 54 12	24 ♐ 58 22	21 R 53	11 16	24 D 07	10 14	06 R 51	16 42	21 23	00 13
03	20 52 12	14 01 38	01 ♑ 05 32	07 ♑ 16 02	20 59	12 30	24 09	10 27	06 47	16 45	21 26	00 15
04	20 56 08	15 02 31	13 ♑ 30 09	19 ♑ 48 06	19 57	13 44	24 12	10 39	06 42	16 48	21 28	00 17
05	21 00 05	16 03 23	26 ♑ 10 04	02 ♒ 36 06	18 50	14 58	24 16	10 52	06 38	16 51	21 30	00 19
06	21 04 01	17 04 14	09 ♒ 06 14	15 ♒ 40 25	17 39	16 12	24 21	11 04	06 34	16 55	21 33	00 20
07	21 07 58	18 05 04	22 ♒ 18 34	29 ♒ 00 31	16 27	17 26	24 26	11 17	06 29	16 58	21 35	00 22
08	21 11 54	19 05 53	05 ♓ 46 03	12 ♓ 34 56	15 15	18 40	24 32	11 29	06 25	17 01	21 37	00 23
09	21 15 51	20 06 40	19 ♓ 26 54	26 ♓ 21 39	14 05	19 54	24 38	11 41	06 20	17 04	21 39	00 25
10	21 19 48	21 07 25	03 ♈ 18 53	10 ♈ 18 16	12 59	21 08	24 45	11 53	06 16	17 07	21 42	00 27
11	21 23 44	22 08 09	17 ♈ 19 29	24 ♈ 22 14	11 59	22 22	24 53	12 05	06 11	17 10	21 44	00 28
12	21 27 41	23 08 52	01 ♉ 26 13	08 ♉ 31 08	11 05	23 36	25 01	12 18	06 06	17 13	21 46	00 30
13	21 31 37	24 09 33	15 ♉ 36 41	22 ♉ 42 35	10 18	24 50	25 10	12 30	06 02	17 17	21 48	00 31
14	21 35 34	25 10 12	29 ♉ 48 34	06 ♊ 54 20	09 39	26 04	25 20	12 41	05 57	17 20	21 51	00 33
15	21 39 30	26 10 50	13 ♊ 59 36	21 ♊ 04 04	09 08	27 18	25 30	12 53	05 52	17 23	21 53	00 34
16	21 43 27	27 11 26	28 ♊ 07 27	05 ♋ 09 25	08 44	28 32	25 41	13 05	05 48	17 26	21 55	00 36
17	21 47 23	28 12 00	12 ♋ 09 39	19 ♋ 07 50	08 29	29 ♑ 47	25 52	13 17	05 43	17 30	21 58	00 37
18	21 51 20	29 ♒ 12 32	26 ♋ 03 39	02 ♌ 56 45	08 21	01 ♒ 01	26 04	13 28	05 38	17 33	22 00	00 38
19	21 55 17	00 ♓ 13 03	09 ♌ 46 52	16 ♌ 33 41	08 D 20	02 15	26 17	13 40	05 33	17 36	22 02	00 40
20	21 59 13	01 13 32	23 ♌ 16 59	29 ♌ 56 31	08 26	03 29	26 30	13 51	05 29	17 40	22 04	00 41
21	22 03 10	02 13 59	06 ♍ 32 09	13 ♍ 03 46	08 38	04 43	26 43	14 03	05 24	17 43	22 07	00 42
22	22 07 06	03 14 24	19 ♍ 31 17	25 ♍ 54 44	08 56	05 57	26 57	14 14	05 19	17 46	22 09	00 44
23	22 11 03	04 14 48	02 ♎ 14 10	08 ♎ 29 44	09 20	07 11	27 12	14 25	05 14	17 50	22 11	00 45
24	22 14 59	05 15 11	14 ♎ 41 38	20 ♎ 50 08	09 49	08 25	27 27	14 36	05 09	17 53	22 13	00 46
25	22 18 56	06 15 32	26 ♎ 55 32	02 ♏ 58 13	10 23	09 39	27 42	14 47	05 04	17 56	22 16	00 47
26	22 22 52	07 15 52	08 ♏ 58 36	14 ♏ 57 11	11 02	10 54	27 58	14 58	05 00	18 00	22 18	00 48
27	22 26 49	08 16 10	20 ♏ 54 26	26 ♏ 50 54	11 44	12 08	28 15	15 09	04 55	18 03	22 20	00 49
28	22 30 46	09 16 26	02 ♐ 47 09	08 ♐ 43 46	12 31	13 22	28 31	15 20	04 50	18 07	22 22	00 50
29	22 34 42	10 ♓ 16 42	14 ♐ 41 20	20 ♐ 40 27	13 ♒ 20	14 ♒ 36	28 ♊ 49	15 ♑ 31	04 ♍ 45	18 ♓ 10	22 ♒ 24	00 ♑ 52

Date	Moon True ☊ ° '	Moon Mean ☊ ° '	Moon ☽ Latitude ° '	DECLINATIONS Sun ☉ ° '	Moon ☽ ° '	Mercury ☿ ° '	Venus ♀ ° '	Mars ♂ ° '	Jupiter ♃ ° '	Saturn ♄ ° '	Uranus ♅ ° '	Neptune ♆ ° '	Pluto ♇ ° '
01	27 ♒ 51	28 ♒ 40	05 S 12	17 S 12	26 S 35	11 S 29	22 S 21	26 N 40	22 S 56	10 N 38	05 S 57	14 S 40	17 S 10
02	27 R 49	28 37	04 54	16 55	27 52	11 29	22 18	26 39	22 55	10 40	05 56	14 39	17 10
03	27 47	28 34	04 24	16 37	27 50	11 34	22 15	26 38	22 55	10 42	05 55	14 38	17 10
04	27 45	28 31	03 41	16 19	26 25	11 42	22 10	26 37	22 54	10 43	05 53	14 37	17 09
05	27 43	28 28	02 46	16 01	23 38	11 54	22 06	26 36	22 53	10 45	05 52	14 37	17 09
06	27 42	28 24	01 41	15 43	19 36	12 09	22 00	26 35	22 52	10 47	05 51	14 36	17 09
07	27 42	28 21	00 S 30	15 25	14 33	12 26	21 54	26 34	22 51	10 49	05 50	14 35	17 09
08	27 D 42	28 18	00 N 45	15 06	08 42	12 45	21 47	26 34	22 50	10 51	05 49	14 35	17 09
09	27 42	28 15	01 58	14 47	02 S 22	13 05	21 39	26 33	22 49	10 52	05 47	14 34	17 09
10	27 43	28 12	03 05	14 27	04 N 09	13 26	21 31	26 32	22 48	10 54	05 46	14 33	17 09
11	27 43	28 09	04 02	14 08	10 31	13 47	21 23	26 31	22 47	10 56	05 45	14 32	17 09
12	27 44	28 05	04 44	13 48	16 25	14 08	21 13	26 30	22 46	10 58	05 44	14 32	17 09
13	27 44	28 02	05 10	13 28	21 27	14 28	21 03	26 29	22 45	11 00	05 42	14 31	17 09
14	27 R 44	27 59	05 17	13 08	25 16	14 47	20 52	26 29	22 44	11 02	05 41	14 30	17 09
15	27 44	27 56	05 04	12 47	27 31	15 05	20 41	26 28	22 43	11 03	05 40	14 30	17 09
16	27 44	27 53	04 34	12 27	27 59	15 22	20 29	26 27	22 42	11 05	05 38	14 29	17 09
17	27 D 44	27 49	03 47	12 06	26 39	15 38	20 17	26 26	22 41	11 07	05 37	14 28	17 09
18	27 44	27 46	02 47	11 45	23 40	15 52	20 04	26 26	22 40	11 09	05 36	14 27	17 09
19	27 44	27 43	01 38	11 24	19 22	16 04	19 50	26 25	22 39	11 11	05 35	14 27	17 09
20	27 44	27 40	00 N 25	11 02	14 09	16 15	19 36	26 24	22 37	11 13	05 33	14 26	17 08
21	27 R 44	27 37	00 S 49	10 41	08 22	16 24	19 21	26 23	22 36	11 15	05 32	14 25	17 08
22	27 44	27 34	01 58	10 19	02 N 21	16 32	19 05	26 22	22 35	11 16	05 31	14 24	17 08
23	27 43	27 30	03 00	09 57	03 S 38	16 38	18 49	26 22	22 34	11 18	05 29	14 24	17 08
24	27 42	27 27	03 52	09 35	09 21	16 42	18 33	26 21	22 33	11 20	05 28	14 23	17 08
25	27 41	27 24	04 32	09 13	14 36	16 45	18 16	26 20	22 32	11 22	05 27	14 22	17 08
26	27 40	27 21	05 00	08 51	19 13	16 46	17 58	26 19	22 31	11 24	05 25	14 22	17 08
27	27 39	27 18	05 14	08 28	23 02	16 45	17 40	26 18	22 30	11 26	05 24	14 21	17 08
28	27 38	27 15	05 15	08 06	25 52	16 43	17 22	26 17	22 29	11 28	05 23	14 20	17 08
29	27 ♒ 38	27 ♒ 11	05 S 02	07 S 43	27 S 34	16 S 40	17 S 03	26 N 16	22 S 27	11 N 29	05 S 21	14 S 19	17 S 08

ZODIAC SIGN ENTRIES

Date	h	m	Planets
03	09	52	☽ ♑
05	19	10	☽ ♒
08	01	46	☽ ♓
10	06	17	☽ ♈
12	09	34	☽ ♉
14	12	19	☽ ♊
16	15	12	☽ ♋
17	16	22	♀ ♒
18	18	51	☽ ♌
19	06	50	☉ ♓
21	00	06	☽ ♍
23	07	45	☽ ♎
25	18	06	☽ ♏
28	06	22	☽ ♐

LATITUDES

Date	Mercury ☿ ° '	Venus ♀ ° '	Mars ♂ ° '	Jupiter ♃ ° '	Saturn ♄ ° '	Uranus ♅ ° '	Neptune ♆ ° '	Pluto ♇ ° '
01	02 N 38	00 N 43	03 N 21	00 N 07	01 N 48	00 S 44	00 S 17	06 N 17
04	03 18	00 34	03 18	00 07	01 48	00 44	00 18	06 17
07	03 39	00 25	03 15	00 07	01 49	00 44	00 18	06 17
10	03 38	00 16	03 12	00 07	01 49	00 44	00 18	06 17
13	03 19	00 N 07	03 08	00 06	01 49	00 44	00 18	06 17
16	02 48	00 S 02	03 05	00 06	01 50	00 44	00 18	06 18
19	02 11	00 11	03 02	00 06	01 50	00 44	00 18	06 18
22	01 33	00 19	02 58	00 06	01 50	00 44	00 18	06 18
25	00 56	00 27	02 55	00 05	01 51	00 44	00 18	06 18
28	00 N 21	00 35	02 51	00 05	01 51	00 44	00 18	06 19
31	00 S 12	00 S 42	02 N 48	00 N 05	01 N 51	00 S 44	00 S 18	06 N 19

DATA

Julian Date	2454498
Delta T	+67 seconds
Ayanamsa	23° 58' 22"
Synetic vernal point	05° ♓ 08' 37"
True obliquity of ecliptic	23° 26' 25"

MOON'S PHASES, APSIDES AND POSITIONS ☽

Date	h	m	Phase	Longitude ° '	Eclipse Indicator
07	03	44	●	17 ♒ 44	Annular
14	03	33	☽	24 ♉ 49	
21	03	30	○	01 ♍ 53	total
29	02	18	☾	09 ♐ 52	

Day	h	m		
14	00	56	Perigee	
28	01	25	Apogee	
02	23	29	Max dec	28° S 01'
09	20	45	ON	
16	06	09	Max dec	28° N 02'
22	21	20	OS	

LONGITUDES

Date	Chiron ⚷ ° '	Ceres ⚳ ° '	Pallas ⚴ ° '	Juno ⚵ ° '	Vesta ⚶ ° '	Black Moon Lilith ⚸ ° '
01	15 ♒ 31	13 ♉ 19	19 ♓ 06	17 ♐ 07	21 ♒ 40	22 ♏ 18
11	16 ♒ 17	15 ♉ 30	22 ♓ 25	19 ♐ 48	26 ♒ 45	23 ♏ 26
21	17 ♒ 02	18 ♉ 04	25 ♓ 52	22 ♐ 14	01 ♓ 48	24 ♏ 33
31	17 ♒ 46	20 ♉ 58	29 ♓ 25	24 ♐ 26	06 ♓ 49	25 ♏ 40

All ephemeris data is given at 12.00 UT and the Moon's longitude is additionally given for 24.00 UT

ASPECTARIAN

h m Aspects

01 Friday
05 37 ☽ ⊥ ♀
06 08 ☽ ⊥ ♃
11 33 ♀ ☌ ♃
12 04 ☽ □ ♄
13 05 ☽ # ♂
15 00 ☉ ∥ ♇
16 58 ☽ Q ♆
18 24 ☽ ⊻ ♃
19 03 ☽ ⊻ ♀
19 06 ☽ Q ☿
23 11 ☽ ⚹ ☉
02 Saturday
07 37 ☽ □ ♅
16 57 ☽ ⚹ ♆
17 31 ☽ ⚹ ☿
22 21 ☽ ☍ ♂
03 Sunday
00 56 ☿ ☌ ♆
07 36 ☽ ∠ ☉
10 21 ☽ ☌ ♇
19 10 ☽ Q ♅
20 48 ☽ ∠ ☿
22 25 ☽ ∠ ♆
22 59 ☽ △ ♄
04 Monday
02 40 ☽ ⊥ ☉
06 26 ☽ ☌ ♃
09 36 ☽ # ♂
12 30 ☽ ☌ ♀
12 48 ☽ ⊥ ☿
14 16 ☿ ⊥ ♀
15 12 ☽ ⊻ ☉
15 46 ☽ ⊥ ♆
17 49 ☉ ∠ ♇
18 20 ☽ ⚹ ♅
23 18 ☽ ⊻ ☿
05 Tuesday
03 12 ☽ ⊻ ♆
03 31 ☽ bQ ♄
08 25 ☽ ⊼ ♂
17 04 ☽ ∥ ♃
19 42 ☽ ± ♂
19 45 ☽ ⊻ ♇
20 18 ☽ ± ♄
22 14 ☽ ∥ ♀
22 40 ☽ ∠ ♅
22 47 ♀ ⊥ ♆
06 Wednesday
06 54 ☽ ⊥ ♇
07 21 ☽ ⊼ ♄
07 57 ☉ ⊻ ♅
11 56 ☉ ⊥ ♃
12 26 ☽ bQ ♂
15 19 ☽ ⊥ ♅
15 40 ☽ ⊻ ♃
18 19 ☉ ☌ ☿
21 57 ☿ ⊥ ♃
23 25 ☽ ∠ ♇
07 Thursday
00 10 ☽ ∥ ♇
02 14 ☿ ⊻ ♅
02 17 ☽ ⊻ ♀
02 18 ☽ ☌ ☿
02 18 ☽ ⊻ ♅
02 18 ☿ ⊻ ♀
02 23 ♀ ⚹ ♅
02 46 ☽ ⊥ ♃
03 44 ☽ ☌ ☉
07 56 ☽ ∥ ☉
10 41 ☽ ☌ ♆
11 49 ☽ ∥ ♆
14 14 ☽ ⊥ ♀
15 50 ☽ △ ♂
19 14 ☽ ∠ ♃
20 27 ♂ Q ♄
20 33 ☽ ∥ ☿
08 Friday
02 27 ☽ ⚹ ♇
03 33 ☽ # ♄
07 55 ☽ ∠ ♀
09 09 ☿ ∠ ♇
13 08 ☽ ☍ ♄
22 13 ☽ ⚹ ♃
23 11 ☽ ∥ ♅
23 41 ☽ Q ♇
09 Saturday
03 21 ☽ ⊻ ☿
07 50 ☽ ☌ ♅
12 52 ☽ ⚹ ♀
13 01 ☽ ⊥ ☿
13 15 ☽ ⊻ ☉
15 51 ☽ ⊻ ♆
19 28 ☽ Q ♃
21 05 ☽ □ ♂
10 Sunday
00 29 ☽ ⊥ ☉
02 17 ☽ ⊥ ♆
03 28 ☽ ∠ ☿
04 44 ☉ ∥ ♆
07 03 ☽ □ ♇
10 49 ☉ ⊻ ♀
11 40 ☽ Q ♀
14 19 ♀ bQ ♄

17 02 ☽ ⊼ ♄
17 12 ☽ ∠ ☉
17 50 ☽ ∠ ♆
18 00 ☽ # ♅
23 14 ♀ ⊻ ♆
11 Monday
02 03 ☉ ☌ ♆
02 56 ☽ □ ♃
03 16 ☽ ± ♄
03 28 ☽ ⚹ ☿
04 20 ☽ Q ♂
09 45 ☿ ⊻ ♃
11 44 ☽ ⊻ ♅
13 37 ☽ ∥ ♄
18 33 ☽ bQ ♄
19 32 ☽ ⚹ ♆
20 50 ☽ ⚹ ☉
21 25 ☽ □ ♀
22 00 ☽ ⊥ ♅
22 38 ☽ Q ☿
12 Tuesday
00 27 ☉ ∥ ☿
01 00 ☽ ⚹ ♂
01 37 ☽ # ☉
01 45 ☽ # ☿
04 02 ☽ # ♆
08 16 ☿ ⊥ ♅
10 24 ☽ △ ♇
13 20 ☽ ∠ ♅
15 14 ☽ # ♇
15 58 ☽ Q ♆
18 46 ☽ Q ☉
19 52 ☽ △ ♄
13 Wednesday
02 42 ☽ ∠ ♂
03 28 ☽ □ ☿
06 39 ☽ △ ♃
09 57 ☽ # ♀
11 51 ☽ bQ ♇
14 50 ☽ ⚹ ♅
15 36 ☿ bQ ♂
15 39 ☿ ∥ ♆
18 05 ☽ ⊥ ♂
19 13 ☽ # ♃
19 29 ♀ ⊼ ♂
22 30 ☽ □ ♆
14 Thursday
03 05 ☽ ± ♇
03 33 ☽ □ ☉
04 21 ☽ ⊻ ♂
05 05 ☽ △ ♀
08 22 ☽ bQ ♃
11 11 ☽ Q ♅
13 15 ☽ ⊼ ♇
16 39 ☉ △ ♂
22 20 ☽ □ ♄
22 39 ☽ ∥ ♂
23 48 ☽ ± ♃
15 Friday
04 02 ☽ △ ☿
08 52 ☽ bQ ♀
10 06 ☽ ⊼ ♃
17 46 ☽ □ ♅
16 Saturday
01 25 ☽ △ ♆
01 35 ☽ ± ♀
04 40 ☽ Q ♄
04 43 ☽ bQ ☿
07 48 ☽ ☌ ♂
10 17 ☽ △ ☉
12 47 ☽ ⊼ ♀
16 13 ☽ ☍ ♇
19 42 ☽ ± ☿
17 Sunday
01 01 ☽ ⚹ ♄
03 04 ☽ bQ ♆
05 47 ☽ ⊼ ☿
10 54 ♀ ± ♄
13 55 ☽ bQ ☉
13 57 ☽ ☍ ♃
14 11 ☽ ∥ ♂
14 21 ☉ ∠ ♃
18 33 ☽ ± ♆
21 13 ☽ △ ♅
18 Monday
02 39 ☽ ∠ ♄
04 39 ♀ ⊻ ♇
04 56 ☽ ⊼ ♆
06 40 ☽ ± ☉
12 01 ☽ ⊻ ♂
17 55 ☽ ⊼ ☉
18 11 ☽ ⊥ ♄
18 17 ☽ # ♃
19 59 ☽ ⊼ ♇
21 28 ☽ ☍ ♀
22 38 ☽ ⊥ ♂
23 21 ☽ bQ ♅
19 Tuesday
02 58 ☿ St D
04 37 ☽ ⊻ ♄
06 30 ☽ ± ♇
09 26 ☽ ☍ ☿
09 35 ☽ # ♀
12 51 ♀ ± ♂

14 41 ☽ ∠ ♂
15 14 ☽ ± ♅
18 58 ☽ ⊼ ♃
19 18 ♀ ∠ ♅
22 25 ☽ bQ ♇
22 40 ☽ # ♇
22 48 ☉ ⚹ ♇
20 Wednesday
01 19 ☉ # ♄
01 55 ☽ ⊼ ♅
03 02 ☽ # ☿
05 47 ☽ ± ♃
09 50 ☽ ☍ ♆
10 46 ☽ # ♆
17 53 ☽ ⚹ ♂
22 11 ☽ bQ ♃
21 Thursday
00 18 ☽ ∥ ♄
01 22 ☽ △ ♇
01 57 ☽ # ☉
03 30 ☽ ☍ ☉
08 20 ☽ ⊼ ♀
09 56 ☽ ☌ ♄
15 55 ☽ ⊼ ☿
16 05 ☽ Q ♂
20 28 ☽ ± ♀
23 22 ☽ # ♅
22 Friday
00 23 ♀ ⊼ ♄
02 01 ☽ △ ♃
03 15 ☽ ± ☿
08 43 ☽ ☍ ♅
14 58 ☽ bQ ♀
16 56 ☽ ⊼ ♆
20 32 ☽ bQ ☿
23 Saturday
02 15 ☽ □ ♂
03 18 ♀ ⊥ ♇
04 17 ☽ ± ♆
09 09 ☽ □ ♇
16 11 ☽ ⊼ ☉
17 42 ☽ ⊻ ♄
19 35 ☽ ∥ ♅
21 30 ☽ bQ ♆
22 31 ☽ △ ♀
24 Sunday
02 10 ☽ △ ☿
04 44 ☽ ± ☉
05 11 ☽ ⊥ ♄
09 48 ☉ ☍ ♄
11 50 ☽ □ ♃
12 58 ☽ ∥ ☉
18 15 ☽ ⊼ ♅
19 57 ☽ Q ♇
20 50 ☽ # ♄
22 36 ☽ ∠ ♄
23 50 ☽ bQ ☉
25 Monday
02 46 ☽ △ ♆
06 05 ☽ ± ♅
10 54 ☽ ∥ ♆
13 35 ☽ △ ♂
19 40 ☽ ⚹ ♇
22 44 ☽ ∥ ☿
23 49 ☽ Q ♃
26 Tuesday
00 00 ☽ bQ ♅
00 41 ☽ ∥ ♇
04 05 ☽ ⚹ ♄
05 34 ☽ ∥ ♀
08 16 ☽ △ ☉
16 17 ☽ □ ♀
16 22 ☽ □ ☿
17 52 ☿ ☌ ♀
20 12 ☽ bQ ♂
27 Wednesday
00 13 ☽ ⚹ ♃
01 44 ☽ ∠ ♇
04 00 ☽ Q ♄
06 13 ☽ △ ♅
08 17 ☽ ∥ ♃
10 25 ♀ ⊥ ♅
14 46 ☽ ± ♂
14 53 ☽ □ ♆
19 55 ☽ ⊥ ♇
22 49 ☿ ⊥ ♅
28 Thursday
03 10 ☽ ⊼ ♂
06 58 ☽ ∠ ♃
07 04 ☽ Q ☿
08 04 ☽ ⊻ ♇
08 48 ☽ Q ♀
15 56 ♀ bQ ♂
16 06 ☽ □ ♄
16 36 ☽ # ♂
29 Friday
01 25 ☽ ⊥ ♃
02 18 ☽ □ ☉
03 21 ☽ Q ♆
05 43 ♀ ∥ ♇
09 05 ☽ ⚹ ☿
11 48 ☽ ⚹ ♀
13 40 ☽ ⊻ ♃
19 01 ☽ □ ♅

MARCH 2008

Date	Sidereal time h m s	Sun ☉ ° ′ ″	Moon ☽ ° ′ ″	Moon ☽ 24.00 ° ′ ″	Mercury ☿ ° ′	Venus ♀ ° ′	Mars ♂ ° ′	Jupiter ♃ ° ′	Saturn ♄ ° ′	Uranus ♅ ° ′	Neptune ♆ ° ′	Pluto ♇ ° ′
01	22 38 39	11 ♓ 16 56	26 ♐ 41 43	02 ♑ 45 41	14 ♒ 14	15 ♒ 50	29 ♊ 06	15 ♑ 41	04 ♍ 40	18 ♓ 13	22 ♒ 27	00 ♑ 53
02	22 42 35	12 17 08	08 ♑ 52 55	15 ♑ 03 57	15 10	17 05	29 24	15 52	04 R 35	18 17	22 29	00 54
03	22 46 32	13 17 19	21 ♑ 19 14	27 ♑ 39 11	16 09	18 19	29 ♊ 43	16 02	04 31	18 20	22 31	00 55
04	22 50 28	14 17 28	04 ♒ 04 10	10 ♒ 34 26	17 06	19 33	00 ♋ 02	16 12	04 26	18 24	22 33	00 55
05	22 54 25	15 17 35	17 ♒ 10 10	23 ♒ 51 26	18 15	20 47	00 21	16 22	04 21	18 27	22 35	00 56
06	22 58 21	16 17 41	00 ♓ 38 11	07 ♓ 30 15	19 21	22 01	00 40	16 32	04 17	18 31	22 37	00 57
07	23 02 18	17 17 44	14 ♓ 27 21	21 ♓ 29 05	20 30	23 16	01 00	16 42	04 12	18 34	22 40	00 58
08	23 06 15	18 17 46	28 ♓ 34 56	05 ♈ 44 19	21 40	24 30	01 21	16 52	04 07	18 37	22 42	00 59
09	23 10 11	19 17 46	12 ♈ 56 31	20 ♈ 10 49	22 53	25 44	01 41	17 02	04 03	18 41	22 44	01 00
10	23 14 08	20 17 44	27 ♈ 26 26	04 ♉ 42 37	24 08	26 58	02 02	17 12	03 58	18 44	22 46	01 00
11	23 18 04	21 17 40	11 ♉ 58 37	19 ♉ 13 45	25 24	28 12	02 24	17 21	03 54	18 48	22 48	01 01
12	23 22 01	22 17 34	26 ♉ 27 23	03 ♊ 39 00	26 42	29 ♒ 26	02 45	17 30	03 49	18 51	22 50	01 02
13	23 25 57	23 17 25	10 ♊ 48 10	17 ♊ 54 30	28 02	00 ♓ 41	03 08	17 40	03 45	18 55	22 52	01 03
14	23 29 54	24 17 15	24 ♊ 57 46	01 ♋ 57 48	29 ♒ 23	01 55	03 30	17 49	03 40	18 58	22 54	01 03
15	23 33 50	25 17 02	08 ♋ 54 29	15 ♋ 47 46	00 ♓ 46	03 09	03 52	17 58	03 36	19 01	22 56	01 04
16	23 37 47	26 16 47	22 ♋ 37 40	29 ♋ 24 14	02 10	04 23	04 15	18 07	03 32	19 05	22 58	01 04
17	23 41 44	27 16 29	06 ♌ 07 29	12 ♌ 47 31	03 36	05 37	04 39	18 15	03 28	19 08	23 00	01 05
18	23 45 40	28 16 09	19 ♌ 24 22	25 ♌ 58 08	05 03	06 51	05 02	18 24	03 23	19 12	23 02	01 05
19	23 49 37	29 ♓ 15 47	02 ♍ 28 52	08 ♍ 56 35	06 32	08 06	05 26	18 33	03 19	19 15	23 04	01 06
20	23 53 33	00 ♈ 15 23	15 ♍ 21 22	21 ♍ 43 13	08 02	09 20	05 50	18 41	03 15	19 18	23 06	01 06
21	23 57 30	01 14 57	28 ♍ 02 11	04 ♎ 18 21	09 33	10 34	06 14	18 49	03 11	19 22	23 08	01 07
22	00 01 26	02 14 28	10 ♎ 31 43	16 ♎ 42 23	11 06	11 48	06 39	18 57	03 07	19 25	23 10	01 07
23	00 05 23	03 13 58	22 ♎ 50 28	28 ♎ 56 07	12 40	13 02	07 03	19 05	03 03	19 29	23 11	01 07
24	00 09 19	04 13 25	04 ♏ 59 28	11 ♏ 00 47	14 15	14 16	07 28	19 13	03 00	19 32	23 13	01 08
25	00 13 16	05 12 51	17 ♏ 00 18	22 ♏ 58 20	15 52	15 30	07 54	19 21	02 56	19 35	23 15	01 08
26	00 17 13	06 12 15	28 ♏ 55 15	04 ♐ 51 27	17 30	16 45	08 19	19 28	02 52	19 39	23 17	01 08
27	00 21 09	07 11 38	10 ♐ 47 23	16 ♐ 43 33	19 09	17 59	08 45	19 36	02 49	19 42	23 19	01 08
28	00 25 06	08 10 58	22 ♐ 40 28	28 ♐ 38 42	20 50	19 13	09 11	19 43	02 45	19 45	23 20	01 09
29	00 29 02	09 10 17	04 ♑ 38 51	10 ♑ 41 32	22 32	20 27	09 37	19 50	02 42	19 49	23 22	01 09
30	00 32 59	10 09 34	16 ♑ 47 20	22 ♑ 56 54	24 15	21 41	10 03	19 57	02 39	19 52	23 24	01 09
31	00 36 55	11 ♈ 08 49	29 ♑ 10 50	05 ♒ 29 42	26 ♓ 00	22 ♓ 55	10 ♋ 30	20 ♑ 04	02 ♍ 35	19 ♓ 55	23 ♒ 26	01 ♑ 09

Date	Moon True ☊ ° ′	Moon Mean ☊ ° ′	Moon ☽ Latitude ° ′
01	27 ♒ 39	27 ♒ 08	04 S 37
02	27 D 40	27 05	03 58
03	27 41	27 02	03 08
04	27 43	26 59	02 07
05	27 44	26 55	00 S 58
06	27 R 44	26 52	00 N 16
07	27 43	26 49	01 31
08	27 41	26 46	02 42
09	27 39	26 43	03 44
10	27 36	26 40	04 32
11	27 32	26 36	05 02
12	27 30	26 33	05 14
13	27 28	26 30	05 05
14	27 D 27	26 27	04 39
15	27 28	26 24	03 56
16	27 29	26 21	03 00
17	27 31	26 17	01 55
18	27 32	26 14	00 N 44
19	27 R 32	26 11	00 S 27
20	27 31	26 08	01 36
21	27 28	26 05	02 39
22	27 23	26 01	03 33
23	27 17	25 58	04 16
24	27 10	25 55	04 47
25	27 03	25 52	05 05
26	26 57	25 49	05 09
27	26 52	25 46	05 01
28	26 49	25 42	04 39
29	26 48	25 39	04 05
30	26 D 48	25 36	03 20
31	26 ♒ 49	25 ♒ 33	02 S 24

DECLINATIONS

Date	Sun ☉ ° ′	Moon ☽ ° ′	Mercury ☿ ° ′	Venus ♀ ° ′	Mars ♂ ° ′	Jupiter ♃ ° ′	Saturn ♄ ° ′	Uranus ♅ ° ′	Neptune ♆ ° ′	Pluto ♇ ° ′
01	07 S 20	28 S 01	16 S 35	16 S 43	26 N 15	22 S 26	11 N 31	05 S 20	14 S 19	17 S 07
02	06 57	27 06	16 29	16 23	26 14	22 25	11 33	05 19	14 18	17 07
03	06 34	24 50	16 21	16 02	26 13	22 24	11 35	05 17	14 17	17 07
04	06 11	21 18	16 11	15 42	26 12	22 23	11 37	05 16	14 17	17 07
05	05 48	16 37	16 00	15 20	26 11	22 22	11 38	05 15	14 16	17 07
06	05 24	11 00	15 48	14 58	26 10	22 21	11 40	05 13	14 15	17 07
07	05 01	04 S 43	15 34	14 36	26 09	22 19	11 42	05 12	14 15	17 07
08	04 38	01 N 55	15 19	14 13	26 07	22 18	11 44	05 11	14 14	17 07
09	04 14	08 33	15 03	13 50	26 06	22 17	11 45	05 09	14 13	17 07
10	03 51	14 47	14 45	13 27	26 05	22 16	11 47	05 08	14 13	17 07
11	03 27	20 13	14 26	13 03	26 03	22 15	11 49	05 06	14 12	17 06
12	03 03	24 27	14 05	12 39	26 02	22 14	11 50	05 05	14 11	17 06
13	02 40	27 06	13 44	12 15	26 00	22 13	11 52	05 04	14 11	17 06
14	02 16	27 59	13 21	11 50	25 58	22 12	11 53	05 02	14 10	17 06
15	01 52	27 03	12 56	11 25	25 57	22 11	11 55	05 01	14 09	17 06
16	01 29	24 30	12 30	10 59	25 55	22 10	11 57	05 00	14 09	17 06
17	01 05	20 36	12 04	10 34	25 53	22 08	11 58	04 58	14 08	17 06
18	00 41	15 42	11 35	10 08	25 51	22 07	12 00	04 57	14 08	17 06
19	00 S 18	10 10	11 06	09 41	25 49	22 06	12 01	04 56	14 07	17 06
20	00 N 06	04 N 18	10 35	09 15	25 47	22 05	12 02	04 54	14 06	17 05
21	00 30	01 S 39	10 03	08 48	25 45	22 04	12 04	04 53	14 06	17 05
22	00 53	07 26	09 30	08 21	25 43	22 03	12 05	04 52	14 05	17 05
23	01 17	12 50	08 56	07 54	25 41	22 02	12 07	04 51	14 04	17 05
24	01 41	17 41	08 20	07 26	25 38	22 01	12 08	04 49	14 04	17 05
25	02 04	21 47	07 44	06 58	25 36	22 00	12 09	04 48	14 03	17 05
26	02 28	24 57	07 06	06 31	25 33	21 59	12 11	04 47	14 03	17 05
27	02 51	27 01	06 27	06 02	25 30	21 58	12 12	04 45	14 02	17 05
28	03 15	27 53	05 47	05 34	25 28	21 57	12 13	04 44	14 02	17 05
29	03 38	27 26	05 05	05 06	25 25	21 56	12 14	04 43	14 01	17 04
30	04 01	25 41	04 23	04 37	25 22	21 55	12 15	04 41	14 00	17 04
31	04 N 25	22 S 40	03 S 40	04 S 09	25 N 19	21 S 55	12 N 17	04 S 40	14 S 00	17 S 04

ZODIAC SIGN ENTRIES

Date	h	m	Planets
01	18	33	☽ ♑
04	04	24	☽ ♒
04	10	01	♂ ♋
06	10	53	☽ ♓
08	14	23	☽ ♈
10	16	14	☽ ♉
12	17	54	☽ ♊
12	22	51	♀ ♓
14	20	37	☽ ♋
14	22	46	☿ ♓
17	01	04	☽ ♌
19	07	25	☽ ♍
20	05	48	☉ ♈
21	15	45	☽ ♎
24	02	06	☽ ♏
26	14	11	☽ ♐
29	02	43	☽ ♑
31	13	34	☽ ♒

LATITUDES

Date	Mercury ☿ ° ′	Venus ♀ ° ′	Mars ♂ ° ′	Jupiter ♃ ° ′	Saturn ♄ ° ′	Uranus ♅ ° ′	Neptune ♆ ° ′	Pluto ♇ ° ′
01	00 S 01	00 S 39	02 N 49	00 N 05	01 N 51	00 S 44	00 S 18	06 N 19
04	00 31	00 46	02 46	00 05	01 51	00 44	00 18	06 19
07	00 58	00 53	02 42	00 04	01 51	00 44	00 18	06 19
10	01 21	00 59	02 39	00 04	01 51	00 44	00 18	06 20
13	01 40	01 05	02 36	00 04	01 51	00 44	00 18	06 20
16	01 56	01 10	02 33	00 04	01 51	00 44	00 18	06 20
19	02 08	01 15	02 30	00 03	01 51	00 44	00 18	06 21
22	02 16	01 19	02 27	00 03	01 51	00 44	00 18	06 21
25	02 21	01 22	02 24	00 03	01 51	00 44	00 18	06 21
28	02 20	01 25	02 21	00 02	01 51	00 44	00 18	06 22
31	02 S 15	01 S 27	02 N 18	00 N 02	01 N 51	00 S 44	00 S 18	06 N 22

DATA

Julian Date	2454527
Delta T	+67 seconds
Ayanamsa	23° 58′ 26″
Synetic vernal point	05° ♓ 08′ 33″
True obliquity of ecliptic	23° 26′ 26″

MOON'S PHASES, APSIDES AND POSITIONS ☽

Date	h	m	Phase	Longitude ° ′	Eclipse Indicator
07	17	14	●	17 ♓ 31	
14	10	46	☽	24 ♊ 14	
21	18	40	○	01 ♎ 31	
29	21	47	☾	09 ♑ 34	

Day	h	m	
10	21	33	Perigee
26	20	05	Apogee

01	07	59	Max dec	28° S 02′
08	05	09	0N	
14	11	32	Max dec	27° N 59′
21	05	20	0S	
28	15	57	Max dec	27° S 54′

LONGITUDES

Date	Chiron ⚷ ° ′	Ceres ⚳ ° ′	Pallas ⚴ ° ′	Juno ⚵ ° ′	Vesta ⚶ ° ′	Black Moon Lilith ⚸ ° ′
01	17 ♒ 41	20 ♉ 40	29 ♓ 04	24 ♐ 13	06 ♓ 19	25 ♏ 33
11	18 ♒ 23	23 ♉ 49	02 ♈ 43	26 ♐ 08	11 ♓ 17	26 ♏ 41
21	19 ♒ 01	27 ♉ 12	06 ♈ 27	27 ♐ 41	16 ♓ 12	27 ♏ 48
31	19 ♒ 36	00 ♊ 46	10 ♈ 15	28 ♐ 49	21 ♓ 04	28 ♏ 55

ASPECTARIAN

h m Aspects

01 Saturday
03 31 ☽ ✶ ♆
07 13 ☿ ⚼ ♂
08 32 ♀ ⊻ ♃
12 44 ♀ ∠ ♇
16 54 ☽ ☍ ♂
17 26 ☽ ∠ ☿
17 36 ☽ Q ☉
20 18 ☽ ☌ ♇
21 08 ☽ ∠ ♀

02 Sunday
01 59 ☿ ∥ ♀
03 39 ☽ △ ♄
06 53 ☽ Q ♅
09 15 ☽ ∠ ♆
12 35 ☽ ⊥ ☿
16 45 ☽ ⊥ ♀
19 12 ☽ ✶ ☉
23 02 ☽ # ♂

03 Monday
01 13 ☽ ⊻ ☿
01 43 ☽ ☌ ♃
02 46 ☽ ⊥ ♆
02 46 ☉ Q ♇
05 37 ☽ ⊻ ♀
06 15 ☿ ∠ ♇
06 16 ☽ ✶ ♅
08 34 ☽ ⚼ ♄
08 46 ☿ ⊻ ♃
12 30 ♀ ⊻ ♅
14 17 ☽ ⊻ ♆

04 Tuesday
01 32 ☽ ± ♄
02 19 ☽ ∠ ☉
04 16 ☽ ⊼ ♂
05 26 ☽ ∥ ♃
06 08 ☽ ⊻ ♇
10 44 ☽ ∠ ♅
12 40 ☽ ⊼ ♄
15 43 ☽ ± ♂
17 18 ☽ ⊥ ♇
20 27 ☽ ⊥ ☉

05 Wednesday
03 24 ☽ ⊥ ♅
08 19 ☽ ⊻ ☉
08 37 ☽ ⚼ ♂
09 38 ☽ ∥ ♇
09 46 ☽ ∠ ♇
10 32 ☽ ⊻ ♃
14 06 ☽ ☌ ☿
14 19 ☽ ⊻ ♅
14 53 ☽ ∥ ☿
16 51 ☿ ⊻ ♅
18 11 ☽ ∥ ♀
19 10 ☽ ☌ ♀
21 28 ☽ ⊥ ♃
21 46 ☽ ☌ ♆
22 28 ☽ ∥ ♆

06 Thursday
09 19 ☽ # ♄
12 04 ☽ △ ♂
12 34 ☽ ✶ ♇
13 36 ☽ ∠ ♃
18 21 ☽ ☍ ♄
19 04 ☉ ✶ ♃
23 38 ♀ ⊥ ♃

07 Friday
00 02 ♀ ☌ ♆
00 17 ☉ ∥ ♅
03 24 ♃ ⊥ ♆
09 12 ♂ ☍ ♇
09 27 ☽ Q ♇
10 14 ☽ ∥ ♅
10 50 ☽ ∥ ☉
15 54 ☽ ✶ ♃
17 14 ☽ ☌ ☉
19 04 ☽ ☌ ♅
23 15 ☽ ⊻ ☿

08 Saturday
02 02 ☽ ⊻ ♆
04 27 ☽ ⊻ ♀
10 20 ☽ ⊥ ☿
11 27 ♀ ∥ ♆
12 11 ☽ ⊥ ♆
12 29 ☽ Q ♃
15 31 ☽ ⊥ ♀
16 02 ☽ □ ♇
16 45 ☽ □ ♂
20 19 ☉ ☌ ♅
21 13 ☽ # ☉
21 15 ☽ ⊼ ♄
23 42 ☽ # ♅

09 Sunday
02 48 ☽ ∠ ☿
03 18 ☽ ∠ ♆
07 12 ☽ ± ♄
07 59 ☽ ∠ ♀
08 52 ☿ ☌ ♆
15 18 ☿ ⊥ ♃
18 52 ☽ □ ♃
21 33 ☽ ⊻ ♅
22 04 ☽ ⚼ ♄
23 19 ☽ ⊻ ☉
23 28 ☽ Q ♂

10 Monday
00 07 ☽ ∥ ♄
04 15 ☽ ✶ ♆
06 01 ☽ ✶ ☿
06 59 ☽ # ♀
07 31 ☽ ⊥ ♅
09 42 ☽ # ♆
09 58 ☽ ⊥ ☉
11 09 ☽ ✶ ♀
11 53 ☽ # ☿
17 54 ☽ △ ♇
19 47 ☽ ✶ ♂
21 44 ☽ # ♇
22 26 ☽ ∠ ♅
22 43 ☽ △ ♄

11 Tuesday
00 07 ☽ Q ♆
01 55 ☽ ∠ ☉
03 43 ☽ Q ☿
08 48 ☽ Q ♀
18 41 ☽ ⚼ ♇
20 59 ☽ △ ♃
21 12 ☽ ∠ ♂
22 31 ☽ # ♃
23 20 ☽ ✶ ♅

12 Wednesday
04 34 ☽ ✶ ☉
05 03 ☿ ∥ ♆
05 58 ☽ □ ♆
09 38 ☽ ± ♇
12 27 ☽ □ ☿
12 31 ☽ ⊥ ♂
17 26 ☽ □ ♀
19 21 ☽ Q ♅
19 38 ☽ ⊼ ♇
22 12 ☽ ⚼ ♃
22 47 ☽ ⊻ ♂

13 Thursday
00 13 ☽ □ ♄
00 15 ☽ ∥ ♂
01 27 ☉ ⊻ ♆
02 03 ☽ Q ☉
13 28 ☽ ± ♃
19 09 ♀ ✶ ♇
23 42 ☽ ⊼ ♃

14 Friday
01 45 ☽ □ ♅
06 25 ☽ Q ♄
08 28 ☽ △ ♆
08 49 ♀ # ♄
10 46 ☽ □ ☉
20 23 ☽ △ ☿
21 24 ♂ ✶ ♄
22 26 ☽ ☍ ♇

15 Saturday
01 04 ☽ △ ♀
02 52 ☽ ✶ ♄
03 03 ☽ ☌ ♂
07 52 ♀ ∠ ♃
10 18 ☽ ⚼ ♆
17 10 ☿ ✶ ♇
20 16 ♀ ☍ ♄

16 Sunday
00 18 ☽ ∥ ♂
01 18 ☽ ⚼ ☿
02 02 ☽ ± ♆
03 58 ☽ ☍ ♃
04 50 ☽ ∠ ♄
05 44 ☽ ⚼ ♀
05 44 ☽ △ ♅
08 21 ♀ △ ♂
12 36 ☽ ⊼ ♆
18 58 ☽ △ ☉
19 00 ☽ ± ☿
20 38 ☽ ⊥ ♄
23 13 ☽ ± ♀

17 Monday
02 59 ☽ ⊼ ♇
03 15 ☽ # ♃
05 40 ☿ ∠ ♃
06 56 ☽ ⊼ ☿
07 15 ☽ ⊻ ♄
08 26 ☽ ⚼ ♅
09 16 ☽ ⊻ ♂
09 47 ☿ ☍ ♄
11 00 ☽ ⊼ ♀
13 43 ☽ ± ♇
16 29 ☿ # ♄
20 22 ☽ ⊥ ♂
23 58 ☽ ⚼ ☉

18 Tuesday
00 41 ☽ ± ♅
05 33 ☽ # ♇
05 58 ☽ ⚼ ♇
10 09 ☽ ⊼ ♃
10 41 ♃ ⚼ ♄
11 37 ☽ ⊼ ♅
11 37 ☿ △ ♂
13 11 ☽ ∠ ♂
17 39 ☽ ± ☉
18 38 ☽ ☍ ♆
19 05 ☽ # ♆
21 13 ☽ ± ♃

19 Wednesday
04 15 ☽ ∥ ♄
05 34 ☽ ⊼ ☉
07 03 ☉ ⊥ ♆
07 45 ☽ # ☿
09 26 ☽ △ ♇
13 33 ☽ ☌ ♄
13 59 ☽ ⚼ ♃
14 11 ☽ # ♀
17 38 ☽ ✶ ♂
20 29 ☽ ☍ ☿
23 31 ☽ ☍ ♀

20 Thursday
09 32 ☽ # ♅
16 48 ☽ Q ♂
18 20 ☽ △ ♃
19 28 ☽ ☍ ♅
23 58 ☉ Q ♃

21 Friday
02 39 ☽ ⊼ ♆
03 52 ☽ ∥ ☉
07 01 ☽ # ☉
08 38 ☉ □ ♇
14 05 ☽ ± ♆
17 53 ☽ □ ♇
18 40 ☽ ☍ ☉
21 48 ☽ ⊻ ♄

22 Saturday
01 16 ☽ ∥ ♅
04 15 ☽ □ ♂
07 25 ☽ ⚼ ♆
09 18 ☽ ⊥ ♄
13 15 ☽ ⊼ ☿
14 44 ☽ ⊼ ♀
15 39 ☽ ∥ ♀
20 07 ☽ ∥ ☿

23 Sunday
02 37 ☽ ± ☿
02 41 ☽ ∠ ♄
03 43 ☽ ± ♀
04 34 ☽ □ ♃
04 43 ☽ Q ♇
05 23 ☽ ⊼ ♅
08 01 ☉ ⊼ ♄
08 37 ☽ # ♄
12 41 ☽ △ ♆
13 41 ♀ Q ♇
17 12 ☽ ± ♅
17 48 ☽ ∥ ♆
19 02 ☿ Q ♇
22 54 ☽ ⚼ ☿
23 23 ☽ ⚼ ♀

24 Monday
04 20 ☽ ✶ ♇
08 03 ☽ ✶ ♄
08 49 ☽ ∥ ♇
10 20 ☽ ⊼ ☉
11 05 ☽ ⚼ ♅
13 29 ☿ ☌ ♀
16 29 ☽ Q ♃
17 07 ☽ △ ♂
23 22 ☽ ± ☉

25 Tuesday
07 52 ☽ Q ♄
08 39 ☽ △ ♀
09 21 ☽ △ ☿
10 15 ☽ ∠ ♇
13 27 ☽ ∥ ♃
16 45 ☽ ✶ ♃
17 13 ☽ △ ♅
19 02 ☽ ⚼ ☉

26 Wednesday
00 17 ☽ ⚼ ♂
00 36 ☽ □ ♆
04 22 ☽ ⊥ ♇
09 48 ♂ ⚼ ♆
16 29 ☽ ⊻ ♇
17 40 ☽ # ♂
19 07 ☽ ± ♂
19 57 ☽ □ ♄
23 21 ☽ ∠ ♃

27 Thursday
04 04 ☽ △ ☉
07 43 ☽ ⊼ ♂
13 03 ☽ Q ♆
17 44 ☽ ⊥ ♃
18 56 ☿ ✶ ♃
20 07 ☿ ☌ ♅

28 Friday
04 13 ☽ □ ♀
05 59 ☽ ⊻ ♃
06 05 ☽ □ ♅
07 40 ☽ □ ☿
13 21 ☽ ✶ ♆
15 57 ☉ ∠ ♆
22 54 ♀ ✶ ♃
22 58 ♀ ☌ ♅

29 Saturday
00 16 ♃ ✶ ♅
01 07 ☉ ± ♄
05 00 ☽ ☌ ♇
08 08 ☽ △ ♄
11 05 ☿ ∥ ♀
18 19 ☽ Q ♅
19 25 ☽ ∠ ♆
20 25 ☽ Q ♀
21 47 ☽ □ ☉
22 14 ☽ ☍ ♂
23 54 ☿ ⊻ ♆

30 Sunday
01 18 ☿ ∥ ♅
01 37 ☽ Q ☿
07 21 ☉ □ ♂
08 22 ♀ ∥ ♅
13 12 ☽ ⊥ ♆
13 40 ☽ ⚼ ♄
15 08 ☽ # ♂
18 02 ☽ ✶ ♅
18 15 ☽ ☌ ♃
19 49 ☉ # ☿
22 37 ☽ ✶ ♀

31 Monday
00 54 ☽ ⊻ ♆
04 37 ☉ # ♀
04 54 ☽ ✶ ☿
07 03 ☽ ± ♄
11 56 ☽ Q ☉
15 45 ☽ ⊻ ♇
16 55 ☽ ∥ ♃
18 28 ☽ ⊼ ♄
22 04 ♀ ⊻ ♆
22 57 ☽ ∠ ♅

All ephemeris data is given at 12.00 UT and the Moon's longitude is additionally given for 24.00 UT

APRIL 2008

LONGITUDES

Date	Sidereal time h m s	Sun ☉ ° ' "	Moon ☽ ° ' "	Moon ☽ 24.00 ° ' "	Mercury ☿ ° '	Venus ♀ ° '	Mars ♂ ° '	Jupiter ♃ ° '	Saturn ♄ ° '	Uranus ♅ ° '	Neptune ♆ ° '	Pluto ♇ ° '
01	00 40 52	12 ♈ 08 02	11 ♒ 54 03	18 ♒ 24 21	27 ♓ 46	24 ♓ 09	10 ♋ 56	20 ♑ 11	02 ♍ 32	19 ♓ 58	23 ♒ 27	01 ♑ 09
02	00 44 48	13 07 13	25 ♒ 00 58	01 ♓ 44 12	29 ♓ 34	25 23	11 23	20 18	02 R 29	20 02	23 29	01 R 09
03	00 48 45	14 06 23	08 ♓ 34 11	15 ♓ 30 54	01 ♈ 23	26 38	11 51	20 24	02 26	20 05	23 31	01 09
04	00 52 42	15 05 30	22 ♓ 34 13	29 ♓ 43 44	03 14	27 52	12 18	20 30	02 23	20 08	23 32	01 09
05	00 56 38	16 04 36	06 ♈ 58 55	14 ♈ 19 03	05 05	29 ♓ 06	12 45	20 36	02 20	20 11	23 34	01 09
06	01 00 35	17 03 40	21 ♈ 43 14	29 ♈ 10 25	06 59	00 ♈ 20	13 13	20 42	02 17	20 14	23 35	01 09
07	01 04 31	18 02 42	06 ♉ 39 29	14 ♉ 09 15	08 53	01 34	13 41	20 48	02 15	20 17	23 37	01 09
08	01 08 28	19 01 41	21 ♉ 38 33	29 ♉ 06 15	10 50	02 48	14 09	20 54	02 12	20 21	23 38	01 08
09	01 12 24	20 00 39	06 ♊ 31 20	13 ♊ 52 54	12 47	04 02	14 37	20 59	02 10	20 24	23 40	01 08
10	01 16 21	20 59 34	21 ♊ 10 14	28 ♊ 22 48	14 46	05 16	15 06	21 04	02 07	20 27	23 41	01 08
11	01 20 17	21 58 28	05 ♋ 30 12	12 ♋ 32 16	16 46	06 30	15 34	21 10	02 05	20 30	23 43	01 08
12	01 24 14	22 57 18	19 ♋ 28 53	26 ♋ 20 10	18 48	07 44	16 03	21 15	02 03	20 33	23 44	01 07
13	01 28 11	23 56 07	03 ♌ 06 09	09 ♌ 47 25	20 51	08 58	16 32	21 19	02 01	20 36	23 45	01 07
14	01 32 07	24 54 53	16 ♌ 23 55	22 ♌ 56 07	22 54	10 12	17 01	21 24	01 59	20 39	23 47	01 07
15	01 36 04	25 53 36	29 ♌ 24 19	05 ♍ 48 53	24 59	11 26	17 30	21 28	01 57	20 42	23 48	01 06
16	01 40 00	26 52 18	12 ♍ 10 08	18 ♍ 28 22	27 05	12 40	18 00	21 33	01 55	20 45	23 49	01 06
17	01 43 57	27 50 57	24 ♍ 43 50	00 ♎ 56 49	29 ♈ 12	13 54	18 29	21 37	01 54	20 48	23 51	01 05
18	01 47 53	28 49 34	07 ♎ 07 29	13 ♎ 16 02	01 ♉ 19	15 08	18 59	21 41	01 52	20 51	23 52	01 05
19	01 51 50	29 ♈ 48 10	19 ♎ 22 38	25 ♎ 27 25	03 26	16 22	19 29	21 45	01 51	20 53	23 53	01 04
20	01 55 46	00 ♉ 46 43	01 ♏ 30 30	07 ♏ 32 03	05 34	17 36	19 58	21 48	01 49	20 56	23 54	01 04
21	01 59 43	01 45 14	13 ♏ 32 10	19 ♏ 31 00	07 41	18 50	20 28	21 52	01 48	20 59	23 55	01 03
22	02 03 40	02 43 44	25 ♏ 28 45	01 ♐ 25 36	09 48	20 04	20 59	21 55	01 47	21 02	23 56	01 03
23	02 07 36	03 42 12	07 ♐ 21 47	13 ♐ 17 36	11 54	21 18	21 29	21 58	01 46	21 05	23 57	01 02
24	02 11 33	04 40 38	19 ♐ 13 20	25 ♐ 09 23	13 59	22 32	21 59	22 01	01 45	21 07	23 58	01 01
25	02 15 29	05 39 02	01 ♑ 06 10	07 ♑ 04 07	16 03	23 46	22 30	22 04	01 44	21 10	23 59	01 01
26	02 19 26	06 37 25	13 ♑ 03 46	19 ♑ 05 39	18 05	25 00	23 00	22 06	01 43	21 13	24 00	01 00
27	02 23 22	07 35 46	25 ♑ 10 21	01 ♒ 18 29	20 05	26 13	23 31	22 08	01 42	21 15	24 01	00 59
28	02 27 19	08 34 06	07 ♒ 30 39	13 ♒ 47 31	22 03	27 27	24 02	22 11	01 42	21 18	24 02	00 58
29	02 31 15	09 32 24	20 ♒ 09 39	26 ♒ 37 41	23 58	28 41	24 33	22 13	01 41	21 20	24 03	00 58
30	02 35 12	10 ♉ 30 40	03 ♓ 12 06	09 ♓ 53 22	25 ♉ 50	29 ♈ 55	25 ♋ 04	22 ♑ 14	01 ♍ 41	21 ♓ 23	24 ♒ 04	00 ♑ 57

Date	Moon True ☊ ° '	Moon Mean ☊ ° '	Moon ☽ Latitude ° '
01	26 ♒ 50	25 ♒ 30	01 S 20
02	26 D 51	25 27	00 S 10
03	26 R 50	25 23	01 N 03
04	26 48	25 20	02 14
05	26 43	25 17	03 19
06	26 36	25 14	04 12
07	26 28	25 11	04 48
08	26 20	25 07	05 05
09	26 13	25 04	05 01
10	26 08	25 01	04 37
11	26 04	24 58	03 57
12	26 03	24 55	03 03
13	26 D 03	24 52	02 00
14	26 04	24 48	00 N 52
15	26 R 04	24 45	00 S 18
16	26 03	24 42	01 25
17	25 59	24 39	02 27
18	25 53	24 36	03 20
19	25 44	24 33	04 04
20	25 33	24 29	04 36
21	25 21	24 26	04 55
22	25 09	24 23	05 02
23	24 57	24 20	04 55
24	24 47	24 17	04 35
25	24 39	24 13	04 04
26	24 35	24 10	03 21
27	24 32	24 07	02 29
28	24 31	24 04	01 29
29	24 D 32	24 01	00 23
30	24 ♒ 31	23 ♒ 58	00 N 46

DECLINATIONS

Date	Sun ☉ ° '	Moon ☽ ° '	Mercury ☿ ° '	Venus ♀ ° '	Mars ♂ ° '	Jupiter ♃ ° '	Saturn ♄ ° '	Uranus ♅ ° '	Neptune ♆ ° '	Pluto ♇ ° '
01	04 N 48	18 S 30	02 S 55	03 S 40	25 N 16	21 S 54	12 N 18	04 S 39	13 S 59	17 S 04
02	05 11	13 20	02 10	03 11	25 13	21 53	12 19	04 38	13 59	17 04
03	05 34	07 23	01 23	02 42	25 09	21 52	12 20	04 36	13 58	17 04
04	05 57	00 S 54	00 S 36	02 13	25 06	21 51	12 21	04 35	13 58	17 04
05	06 19	05 N 49	00 N 13	01 44	25 02	21 50	12 22	04 34	13 57	17 04
06	06 42	12 21	01 02	01 15	24 59	21 49	12 23	04 33	13 57	17 04
07	07 05	18 16	01 53	00 45	24 55	21 49	12 23	04 31	13 56	17 04
08	07 27	23 05	02 44	00 S 16	24 51	21 48	12 24	04 30	13 56	17 03
09	07 49	26 20	03 35	00 N 13	24 47	21 47	12 25	04 29	13 55	17 03
10	08 12	27 46	04 28	00 43	24 43	21 47	12 26	04 28	13 55	17 03
11	08 34	27 16	05 21	01 12	24 39	21 46	12 27	04 27	13 55	17 03
12	08 56	25 02	06 15	01 41	24 35	21 45	12 27	04 25	13 54	17 03
13	09 17	21 24	07 09	02 11	24 30	21 45	12 28	04 24	13 54	17 03
14	09 39	16 44	08 03	02 40	24 26	21 44	12 29	04 23	13 53	17 03
15	10 00	11 24	08 57	03 09	24 21	21 43	12 29	04 22	13 53	17 03
16	10 21	05 N 41	09 52	03 38	24 16	21 43	12 30	04 21	13 52	17 03
17	10 43	00 S 09	10 46	04 08	24 11	21 42	12 30	04 20	13 52	17 03
18	11 03	05 54	11 40	04 37	24 06	21 42	12 31	04 19	13 52	17 03
19	11 24	11 21	12 34	05 06	24 01	21 41	12 31	04 18	13 51	17 02
20	11 45	16 19	13 27	05 34	23 56	21 41	12 32	04 16	13 51	17 02
21	12 05	20 36	14 19	06 03	23 51	21 40	12 32	04 15	13 51	17 02
22	12 25	24 01	15 10	06 32	23 45	21 40	12 32	04 14	13 50	17 02
23	12 45	26 23	16 00	07 00	23 40	21 40	12 33	04 13	13 50	17 02
24	13 05	27 34	16 48	07 29	23 34	21 39	12 33	04 12	13 50	17 02
25	13 24	27 30	17 34	07 57	23 28	21 39	12 33	04 11	13 49	17 02
26	13 44	26 08	18 19	08 25	23 22	21 39	12 33	04 10	13 49	17 02
27	14 03	23 32	19 02	08 52	23 16	21 38	12 33	04 09	13 49	17 02
28	14 22	19 50	19 42	09 20	23 10	21 38	12 33	04 08	13 48	17 02
29	14 40	15 08	20 20	09 47	23 04	21 38	12 33	04 07	13 48	17 02
30	14 N 59	09 S 37	20 N 56	10 N 15	22 N 57	21 S 38	12 N 33	04 S 06	13 S 48	17 S 02

ZODIAC SIGN ENTRIES

Date	h	m	Planets
02	17	45	☿ ♈
02	20	55	☽ ♓
05	00	27	☽ ♈
06	05	35	♀ ♈
07	01	20	☽ ♉
09	01	27	☽ ♊
11	02	43	☽ ♋
13	06	29	☽ ♌
15	13	07	☽ ♍
17	21	07	☿ ♉
17	22	10	☽ ♎
19	16	51	☉ ♉
20	09	00	☽ ♏
22	21	07	☽ ♐
25	09	47	☽ ♑
27	21	27	☽ ♒
30	06	11	☽ ♓
30	13	34	♀ ♉

LATITUDES

Date	Mercury ☿ ° '	Venus ♀ ° '	Mars ♂ ° '	Jupiter ♃ ° '	Saturn ♄ ° '	Uranus ♅ ° '	Neptune ♆ ° '	Pluto ♇ ° '
01	02 S 13	01 S 28	02 N 17	00 N 02	01 N 51	00 S 44	00 S 18	06 N 22
04	02 03	01 29	02 14	00 01	01 51	00 44	00 18	06 22
07	01 48	01 30	02 11	00 01	01 50	00 44	00 18	06 23
10	01 28	01 30	02 09	00 N 01	01 50	00 44	00 18	06 23
13	01 04	01 30	02 06	00 00	01 50	00 44	00 19	06 23
16	00 37	01 29	02 03	00 00	01 50	00 44	00 19	06 23
19	00 S 06	01 27	02 01	00 00	01 49	00 44	00 19	06 24
22	00 N 26	01 25	01 58	00 S 01	01 49	00 45	00 19	06 24
25	00 58	01 23	01 56	00 01	01 49	00 45	00 19	06 24
28	01 28	01 19	01 53	00 01	01 48	00 45	00 19	06 24
31	01 N 54	01 S 16	01 N 51	00 S 02	01 N 48	00 S 45	00 S 19	06 N 25

DATA

Julian Date	2454558
Delta T	+67 seconds
Ayanamsa	23° 58' 30"
Synetic vernal point	05° ♓ 08' 30"
True obliquity of ecliptic	23° 26' 26"

MOON'S PHASES, APSIDES AND POSITIONS ☽

Date	h	m	Phase	Longitude ° '	Eclipse Indicator
06	03	55	●	16 ♈ 44	
12	18	32	☽	23 ♋ 13	
20	10	25	○	00 ♏ 43	
28	14	12	☾	08 ♒ 39	

Day	h	m		
07	19	24	Perigee	
23	09	21	Apogee	
04	15	13	0N	
10	17	41	Max dec	27° N 49'
17	11	23	0S	
24	22	33	Max dec	27° S 42'

LONGITUDES

Date	Chiron ⚷ ° '	Ceres ⚳ ° '	Pallas ⚴ ° '	Juno ⚵ ° '	Vesta ⚶ ° '	Black Moon Lilith ⚸ ° '
01	19 ♒ 39	01 ♊ 08	10 ♈ 38	28 ♐ 54	21 ♓ 33	29 ♏ 02
11	20 ♒ 08	04 ♊ 53	14 ♈ 30	29 ♐ 32	26 ♓ 19	00 ♐ 09
21	20 ♒ 32	08 ♊ 46	18 ♈ 25	29 ♐ 39	01 ♈ 00	01 ♐ 16
31	20 ♒ 51	12 ♊ 46	22 ♈ 23	29 ♐ 12	05 ♈ 36	02 ♐ 23

All ephemeris data is given at 12.00 UT and the Moon's longitude is additionally given for 24.00 UT

ASPECTARIAN

01 Tuesday
03 07 ☽ ⊥ ♇
03 16 ☉ # ♅
06 20 ☽ ∠ ♀
10 09 ☽ ⊼ ♂
12 28 ☽ ∗ ☉
13 53 ☽ ∠ ☿
15 51 ☽ ⊥ ♅
19 06 ☽ ∥ ♇
19 52 ☽ ∠ ♇
21 39 ☽ ± ♂

02 Wednesday
00 44 ☽ ⊥ ♀
02 55 ☽ ⊻ ♅
03 22 ☽ ⊻ ♃
08 59 ☽ ⊥ ☿
09 13 ☽ ∥ ♆
09 14 ☽ ☌ ♆
09 23 ♇ St R
10 51 ☿ ⊥ ♆
12 45 ☽ ⊻ ♀
14 19 ☽ ⊥ ♃
14 33 ☽ Ꝑ ♂
16 20 ☽ # ♄
18 01 ☽ ∠ ☉
21 25 ☽ ⊻ ☿
22 58 ☽ ∗ ♇

03 Thursday
01 16 ☽ ☍ ♄
06 25 ☽ ∠ ♃
08 54 ☿ □ ♇
11 08 ☽ ⊥ ☉
17 52 ☽ △ ♂
18 30 ☽ # ☉
19 56 ☽ Q ♇
22 19 ☽ ⊻ ☉
22 30 ☽ ∥ ♅

04 Friday
01 21 ☿ ⊼ ♄
02 04 ☿ Q ♃
06 51 ☽ ∥ ♀
07 52 ☽ ☌ ♅
08 29 ☽ ∗ ♃
13 15 ☽ ∥ ☿
13 38 ☽ ⊻ ♆
16 47 ☽ # ☿
21 43 ☽ ☌ ♀
22 24 ☽ # ♀
23 42 ☽ ⊥ ♆

05 Saturday
02 22 ☽ □ ♇
04 21 ☽ ⊼ ♄
04 44 ☽ Q ♃
07 33 ☽ # ♅
08 25 ☽ ☌ ☿
13 57 ☽ ∥ ☉
14 13 ☽ ± ♄
14 36 ☽ ∠ ♆
21 16 ♀ ⊥ ♆
21 46 ☽ □ ♂

06 Sunday
03 55 ☽ ☌ ☉
04 51 ☽ Ꝑ ♄
09 36 ☽ ⊻ ♅
10 21 ☽ □ ♃
12 06 ☽ ∥ ♄
15 01 ☽ ∗ ♆
15 42 ☿ # ♀
17 22 ☉ Ꝑ ♄
18 09 ☽ # ♆
19 18 ☽ ⊥ ♅

07 Monday
03 06 ☽ ⊻ ♀
03 10 ☽ △ ♇
03 46 ☽ Q ♂
03 48 ♀ □ ♇
04 08 ☿ ± ♄
04 57 ☽ △ ♄
06 47 ☽ # ♇
08 29 ☿ ∠ ♆
09 48 ☽ ∠ ♅
10 19 ☽ Q ♆
13 35 ☽ ⊥ ♀
16 06 ☽ ⊻ ☿
23 37 ☽ ∗ ♂

08 Tuesday
00 50 ♀ ⊼ ♄
03 08 ☽ ⊥ ☿
03 11 ☽ Ꝑ ♇
04 56 ☽ # ♃
05 17 ☽ ∠ ♀
07 31 ☽ ⊻ ☉
09 54 ☽ ∗ ♅
10 48 ☽ △ ♃
14 01 ♀ Q ♃
15 13 ☽ □ ♆
17 37 ☽ ± ♇
17 49 ☽ ⊥ ☉
19 44 ☽ ∠ ☿
23 14 ☽ ∥ ♂

09 Wednesday
00 28 ☽ ∠ ♂
03 17 ☽ ⊼ ♇
04 58 ☽ □ ♄
05 17 ☽ Q ♅
07 36 ☽ ∗ ♀
09 22 ☽ ∠ ☉
11 07 ☽ Ꝑ ♃
15 32 ☽ ⊥ ♂
21 54 ☉ ⊻ ♅
23 48 ☽ ∗ ☿

10 Thursday
01 39 ☽ ⊻ ♂
01 54 ☽ ± ♃
04 58 ☽ Q ♀
10 16 ☽ Q ♄
10 48 ☽ □ ♅
11 41 ☽ ∗ ☉
11 50 ☽ ⊼ ♃
12 00 ☿ # ♅
14 11 ☉ □ ♃
16 11 ☽ △ ♆
17 13 ☿ □ ♂
22 48 ☽ Q ☿

11 Friday
04 37 ☽ ☍ ♇
06 14 ☽ ∗ ♄
09 13 ☽ Q ☉
13 51 ☽ □ ♀
15 41 ☿ Ꝑ ♄
17 28 ☽ Ꝑ ♆

12 Saturday
05 51 ☽ ☌ ♂
07 47 ☽ ∠ ♄
08 58 ☽ ± ♆
10 36 ☽ □ ☿
13 52 ☽ △ ♅
15 05 ☽ ☍ ♃
15 41 ☽ ∥ ♂
17 58 ♀ ± ♄
18 32 ☽ □ ☉
19 26 ☽ ⊼ ♆
23 28 ☽ ⊥ ♄

13 Sunday
07 30 ☉ ∗ ♆
07 47 ♀ ∠ ♆
08 28 ☽ ⊼ ♇
09 04 ☿ ⊻ ♅
10 03 ☽ # ♃
10 04 ☽ ⊻ ♄
16 29 ☽ Ꝑ ♅
17 49 ☿ □ ♃
19 11 ☽ ± ♇
23 36 ☽ △ ♀

14 Monday
08 48 ☽ ± ♅
10 18 ♂ ∠ ♄
10 32 ☽ # ♇
11 28 ☽ Ꝑ ♇
13 11 ☽ ⊻ ♂
19 49 ☽ ⊼ ♅
21 13 ☽ ⊼ ♃
22 09 ☿ ∗ ♆

15 Tuesday
00 38 ☽ ⊥ ♂
01 08 ☽ # ♆
01 35 ☽ ☍ ♆
02 14 ☽ △ ☿
04 56 ☽ △ ☉
05 54 ☽ Ꝑ ♀
07 18 ☽ ∥ ♄
08 23 ☽ ± ♃
15 10 ☽ △ ♇
16 45 ☽ ☌ ♄
17 38 ☽ ∥ ☉
18 01 ☽ ∠ ♂
20 58 ☽ ∥ ☿

16 Wednesday
00 30 ☽ ± ♀
01 19 ☽ Ꝑ ♃
03 06 ♂ ± ♆
07 24 ☉ ☌ ☿
08 01 ☿ ⊥ ♅
08 45 ☉ ⊥ ♅
11 23 ☽ Ꝑ ☉
11 49 ☽ Ꝑ ☿
13 03 ☽ ⊼ ♀
17 33 ☽ # ♅
19 47 ☽ ∥ ♀
23 32 ☽ ∗ ♂

17 Thursday
04 25 ☽ ☍ ♅
05 59 ☽ △ ♃
06 00 ☽ ± ☉
08 27 ☽ ± ☿
09 16 ☉ ∥ ☿
10 17 ☽ ⊼ ♆
18 31 ☽ ⊼ ☉
21 41 ♀ # ♅
21 53 ☽ ± ♆
22 23 ☽ ⊼ ☿
23 34 ☽ Q ♂

18 Friday
00 16 ☽ □ ♇
01 48 ☽ ⊻ ♄
05 19 ☽ ∥ ♅
06 03 ☽ # ♀
09 22 ☿ △ ♇
13 27 ☽ ⊥ ♄
15 24 ☽ Ꝑ ♆
18 11 ☿ △ ♄

19 Saturday
05 25 ☽ ☍ ♀
07 01 ☽ ∠ ♄
10 39 ☿ ∥ ♄
11 24 ☽ Q ♇
12 12 ☽ □ ♂
12 17 ☽ # ☉
15 00 ☽ ⊼ ♅
16 41 ☽ □ ♃
17 27 ☽ # ♄
18 52 ☽ # ☿
20 54 ☽ △ ♆
21 06 ♀ Ꝑ ♄
23 46 ☽ ∥ ♆

20 Sunday
02 54 ☽ ± ♅
10 25 ☽ ☍ ☉
11 07 ☽ ∗ ♇
12 37 ☽ ∗ ♄
15 49 ☽ ∥ ♇
15 51 ☿ Q ♆
16 19 ☿ ∠ ♅
18 56 ☉ △ ♇
20 51 ☽ Ꝑ ♅
21 48 ☽ ☍ ☿
22 51 ☿ # ♆

21 Monday
04 37 ☽ Q ♃
12 32 ☽ Q ♄
13 05 ☉ △ ♄
17 03 ☽ ∠ ♇
18 53 ☽ ∥ ♃
23 42 ☿ Q ♂
23 51 ☽ ⊼ ♀

22 Tuesday
02 32 ☽ △ ♂
03 00 ☽ △ ♅
04 47 ☽ ∗ ♃
08 54 ☽ □ ♆
09 59 ☽ # ♂
11 07 ☽ ⊥ ♇
13 19 ☽ ± ♀
14 48 ♂ △ ♅
20 36 ☉ ∥ ♄
23 13 ☽ ⊻ ♇

23 Wednesday
00 42 ☽ □ ♄
03 56 ☽ ⊼ ☉
07 33 ♀ ⊻ ♅
09 36 ☽ Ꝑ ♀
10 08 ☽ Ꝑ ♂
11 12 ☽ ∠ ♃
17 09 ☽ ± ☉
18 07 ♀ □ ♂
21 19 ☽ Q ♆
23 09 ☽ ⊼ ☿

24 Thursday
01 35 ♀ □ ♃
05 10 ☽ ± ♂
05 29 ☽ ⊥ ♃
13 00 ☽ Ꝑ ☉
13 23 ♂ ☍ ♃
13 53 ☽ ± ☿
15 51 ☽ □ ♅
17 40 ☽ ⊻ ♃
17 51 ☽ ⊼ ♂
19 10 ☿ # ♇
19 28 ☽ △ ♀
21 37 ☽ ∗ ♆

25 Friday
11 33 ☿ Ꝑ ♇
11 49 ☽ ☌ ♇
11 52 ☽ Ꝑ ☿
13 16 ☽ △ ♄
16 33 ♀ ∗ ♆
20 33 ☉ Q ♆
21 58 ☽ △ ☉

26 Saturday
01 19 ☉ ∠ ♅
03 53 ☽ ∠ ♆
04 16 ☽ Q ♅
18 29 ☉ # ♆
19 17 ☽ Ꝑ ♄
21 51 ☽ ⊥ ♆
23 59 ☽ △ ☿

27 Sunday
04 15 ☽ ∗ ♅
06 01 ☽ ☌ ♃
08 36 ☽ ☍ ♂
09 44 ☽ ⊻ ♆
13 03 ☽ ± ♄
14 06 ☽ # ♂
14 18 ☽ □ ♀
23 22 ☽ ⊻ ♇

28 Monday
00 46 ☽ ⊼ ♄
01 08 ☽ ∥ ♃
02 36 ☿ ∗ ♅
08 48 ♀ ⊥ ♅
09 39 ☽ ∠ ♅
10 58 ☽ ⊥ ♇
12 14 ♂ ⊼ ♆
12 37 ☽ # ☿
13 41 ☿ △ ♃
14 12 ☽ □ ☉

29 Tuesday
02 50 ☽ ∥ ♇
02 54 ☽ ⊥ ♅
04 07 ☽ ∠ ♇
04 47 ☽ Q ♀
13 10 ☿ □ ♆
14 01 ☽ # ☉
14 13 ☽ ⊻ ♅
15 50 ☽ ⊻ ♃
18 06 ☽ ∥ ♆
19 15 ☽ ☌ ♆
20 17 ☽ □ ☿
20 30 ☽ ⊼ ♂
22 19 ☿ ∗ ♂
23 36 ☽ # ♄

30 Wednesday
00 38 ☿ ± ♇
02 46 ☽ Q ☉
02 57 ☽ ⊥ ♃
05 25 ☽ ∗ ♀
07 55 ☽ ∗ ♇
07 58 ☽ ± ♂
09 15 ☽ ☍ ♄
09 37 ☽ # ♀
19 17 ☽ ∠ ♃

MAY 2008

LONGITUDES

Date	Sidereal time h m s	Sun ☉ ° ′ ″	Moon ☽ ° ′ ″	Moon ☽ 24.00 ° ′ ″	Mercury ☿ ° ′	Venus ♀ ° ′	Mars ♂ ° ′	Jupiter ♃ ° ′	Saturn ♄ ° ′	Uranus ♅ ° ′	Neptune ♆ ° ′	Pluto ⯓ ° ′
01	02 39 09	11 ♉ 28 55	16 ♓ 41 49	23 ♓ 57 40	27 ♉ 39	01 ♉ 09	25 ♋ 35	22 ♑ 16	01 ♍ 41	21 ♓ 25	24 ♒ 05	00 ♑ 56
02	02 43 05	12 27 09	00 ♈ 40 55	07 ♈ 51 25	29 ♉ 25	02 23	26 07	22 17	01 R 41	21 28	24 06	00 R 55
03	02 47 02	13 25 21	15 ♈ 08 45	22 ♈ 32 19	01 ♊ 08	03 37	26 38	22 19	01 D 41	21 30	24 07	00 54
04	02 50 58	14 23 31	00 ♉ 01 16	07 ♉ 34 31	02 47	04 51	27 10	22 20	01 41	21 33	24 07	00 53
05	02 54 55	15 21 40	15 ♉ 10 50	22 ♉ 48 52	04 23	06 05	27 41	22 21	01 41	21 35	24 08	00 52
06	02 58 51	16 19 47	00 ♊ 27 09	08 ♊ 04 16	05 54	07 18	28 13	22 21	01 41	21 37	24 09	00 51
07	03 02 48	17 17 53	15 ♊ 38 53	23 ♊ 09 46	07 22	08 32	28 45	22 22	01 42	21 40	24 09	00 50
08	03 06 44	18 15 57	00 ♋ 35 53	07 ♋ 56 24	08 46	09 46	29 17	22 22	01 42	21 42	24 10	00 49
09	03 10 41	19 13 59	15 ♋ 10 44	22 ♋ 18 28	10 06	11 00	29 ♋ 49	22 22	01 43	21 44	24 10	00 48
10	03 14 38	20 11 59	29 ♋ 19 28	06 ♌ 13 43	11 22	12 14	00 ♌ 21	22 R 22	01 44	21 46	24 11	00 47
11	03 18 34	21 09 57	13 ♌ 01 24	19 ♌ 42 48	12 34	13 28	00 53	22 22	01 45	21 48	24 12	00 46
12	03 22 31	22 07 53	26 ♌ 18 17	02 ♍ 48 18	13 41	14 41	01 26	22 21	01 45	21 51	24 12	00 45
13	03 26 27	23 05 47	09 ♍ 13 20	15 ♍ 33 52	14 45	15 55	01 58	22 21	01 47	21 53	24 12	00 44
14	03 30 24	24 03 40	21 ♍ 50 24	28 ♍ 03 25	15 43	17 09	02 30	22 20	01 48	21 55	24 13	00 43
15	03 34 20	25 01 30	04 ♎ 13 22	10 ♎ 20 41	16 38	18 23	03 03	22 19	01 49	21 57	24 13	00 42
16	03 38 17	25 59 20	16 ♎ 25 43	22 ♎ 28 50	17 28	19 37	03 36	22 17	01 50	21 58	24 14	00 40
17	03 42 13	26 57 07	28 ♎ 30 19	04 ♏ 30 28	18 14	20 50	04 08	22 16	01 52	22 00	24 14	00 39
18	03 46 10	27 54 53	10 ♏ 29 28	16 ♏ 27 34	18 55	22 04	04 41	22 14	01 53	22 02	24 14	00 38
19	03 50 07	28 52 38	22 ♏ 24 55	28 ♏ 21 43	19 31	23 18	05 14	22 13	01 55	22 04	24 14	00 37
20	03 54 03	29 ♉ 50 21	04 ♐ 18 06	10 ♐ 14 16	20 03	24 32	05 47	22 11	01 57	22 06	24 15	00 35
21	03 58 00	00 ♊ 48 03	16 ♐ 10 23	22 ♐ 06 39	20 29	25 45	06 20	22 09	01 59	22 08	24 15	00 34
22	04 01 56	01 45 43	28 ♐ 03 18	04 ♑ 00 35	20 52	26 59	06 53	22 06	02 01	22 09	24 15	00 33
23	04 05 53	02 43 23	09 ♑ 58 48	15 ♑ 58 18	21 09	28 13	07 27	22 04	02 03	22 11	24 15	00 32
24	04 09 49	03 41 01	21 ♑ 59 27	28 ♑ 02 40	21 21	29 ♉ 27	08 00	22 01	02 05	22 12	24 15	00 30
25	04 13 46	04 38 39	04 ♒ 08 27	10 ♒ 17 17	21 29	00 ♊ 40	08 33	21 58	02 07	22 14	24 15	00 29
26	04 17 42	05 36 15	16 ♒ 29 42	22 ♒ 46 15	21 32	01 54	09 07	21 55	02 09	22 16	24 15	00 28
27	04 21 39	06 33 50	29 ♒ 07 31	05 ♓ 34 02	21 R 31	03 08	09 40	21 52	02 12	22 17	24 R 15	00 26
28	04 25 36	07 31 25	12 ♓ 06 21	18 ♓ 44 58	21 25	04 22	10 14	21 48	02 14	22 18	24 15	00 25
29	04 29 32	08 28 58	25 ♓ 30 16	02 ♈ 22 33	21 14	05 35	10 47	21 45	02 17	22 20	24 15	00 23
30	04 33 29	09 26 31	09 ♈ 22 01	16 ♈ 28 39	21 00	06 49	11 21	21 41	02 20	22 21	24 15	00 22
31	04 37 25	10 ♊ 24 03	23 ♈ 42 16	01 ♉ 02 26	20 ♊ 42	08 ♊ 03	11 ♌ 55	21 ♑ 37	02 ♍ 23	22 ♓ 22	24 ♒ 15	00 ♑ 21

Date	Moon True ☊ ° ′	Moon Mean ☊ ° ′	Moon ☽ Latitude ° ′
01	24 ♒ 30	23 ♒ 54	01 N 54
02	24 R 26	23 51	02 59
03	24 19	23 48	03 54
04	24 11	23 45	04 34
05	24 00	23 42	04 57
06	23 49	23 38	04 58
07	23 39	23 35	04 39
08	23 31	23 32	04 00
09	23 26	23 29	03 07
10	23 23	23 26	02 03
11	23 D 22	23 23	00 N 54
12	23 R 22	23 19	00 S 16
13	23 21	23 16	01 23
14	23 19	23 13	02 24
15	23 15	23 10	03 18
16	23 08	23 07	04 01
17	22 58	23 04	04 33
18	22 46	23 00	04 53
19	22 32	22 57	04 59
20	22 18	22 54	04 53
21	22 05	22 51	04 34
22	21 54	22 48	04 03
23	21 45	22 44	03 21
24	21 39	22 41	02 30
25	21 36	22 38	01 31
26	21 35	22 35	00 S 27
27	21 D 35	22 32	00 N 40
28	21 R 35	22 29	01 47
29	21 34	22 25	02 50
30	21 31	22 22	03 46
31	21 ♒ 26	22 ♒ 19	04 N 28

DECLINATIONS

Date	Sun ☉ ° ′	Moon ☽ ° ′	Mercury ☿ ° ′	Venus ♀ ° ′	Mars ♂ ° ′	Jupiter ♃ ° ′	Saturn ♄ ° ′	Uranus ♅ ° ′	Neptune ♆ ° ′	Pluto ⯓ ° ′
01	15 N 17	03 S 30	21 N 29	10 N 42	22 N 51	21 S 38	12 N 33	04 S 05	13 S 48	17 S 02
02	15 35	03 N 00	22 11	11 08	22 44	21 38	12 33	04 04	13 47	17 02
03	15 52	09 33	22 28	11 35	22 37	21 38	12 33	04 03	13 47	17 02
04	16 09	15 45	22 54	12 01	22 30	21 37	12 33	04 02	13 47	17 01
05	16 27	21 07	23 17	12 27	22 23	21 37	12 33	04 01	13 47	17 01
06	16 43	25 06	23 38	12 52	22 16	21 37	12 33	04 01	13 46	17 01
07	17 00	27 17	23 56	13 18	22 09	21 38	12 32	04 00	13 46	17 01
08	17 16	27 24	24 12	13 43	22 01	21 38	12 32	03 59	13 46	17 01
09	17 32	25 40	24 26	14 07	21 54	21 38	12 32	03 58	13 46	17 01
10	17 48	22 18	24 37	14 32	21 46	21 38	12 31	03 57	13 46	17 01
11	18 03	17 47	24 47	14 56	21 39	21 38	12 31	03 56	13 46	17 01
12	18 18	12 30	24 54	15 19	21 31	21 38	12 31	03 56	13 45	17 01
13	18 33	06 50	24 59	15 42	21 23	21 39	12 30	03 55	13 45	17 01
14	18 47	01 N 02	25 02	16 05	21 14	21 39	12 29	03 54	13 45	17 01
15	19 01	04 S 42	25 04	16 28	21 06	21 39	12 29	03 53	13 45	17 01
16	19 15	10 10	25 04	16 50	20 58	21 39	12 28	03 53	13 45	17 01
17	19 29	15 12	25 02	17 11	20 49	21 40	12 28	03 52	13 45	17 01
18	19 42	19 36	24 58	17 32	20 41	21 40	12 27	03 51	13 45	17 01
19	19 55	23 12	24 53	17 53	20 32	21 41	12 26	03 50	13 45	17 01
20	20 07	25 48	24 47	18 13	20 23	21 41	12 26	03 50	13 45	17 01
21	20 19	27 16	24 39	18 33	20 14	21 42	12 25	03 49	13 45	17 01
22	20 31	27 28	24 29	18 52	20 05	21 42	12 24	03 48	13 45	17 01
23	20 42	26 24	24 18	19 11	19 56	21 43	12 23	03 48	13 45	17 01
24	20 53	24 06	24 06	19 29	19 47	21 43	12 22	03 47	13 45	17 01
25	21 04	20 42	23 53	19 47	19 37	21 44	12 21	03 47	13 45	17 01
26	21 14	16 19	23 39	20 04	19 28	21 44	12 20	03 46	13 45	17 01
27	21 24	11 09	23 23	20 21	19 18	21 45	12 19	03 46	13 45	17 01
28	21 34	05 S 23	23 07	20 37	19 08	21 46	12 18	03 45	13 45	17 01
29	21 43	00 N 48	22 50	20 52	18 59	21 47	12 17	03 45	13 45	17 01
30	21 52	07 09	22 32	21 07	18 49	21 47	12 16	03 44	13 45	17 01
31	22 N 01	13 N 21	22 N 14	21 N 22	18 N 38	21 S 48	12 N 15	03 S 44	13 S 45	17 S 01

ZODIAC SIGN ENTRIES

Date	h	m	Planets
02	10	51	☽ ♈
02	20	00	☿ ♊
04	11	58	☽ ♉
06	11	17	☽ ♊
08	11	02	☽ ♋
09	20	20	♂ ♌
10	13	10	☽ ♌
12	18	48	☽ ♍
15	03	46	☽ ♎
17	14	59	☽ ♏
20	03	18	☽ ♐
20	16	01	☉ ♊
22	15	55	☽ ♑
24	22	52	♀ ♊
25	03	52	☽ ♒
27	13	38	☽ ♓
29	19	52	☽ ♈
31	22	18	☽ ♉

LATITUDES

Date	Mercury ☿ ° ′	Venus ♀ ° ′	Mars ♂ ° ′	Jupiter ♃ ° ′	Saturn ♄ ° ′	Uranus ♅ ° ′	Neptune ♆ ° ′	Pluto ⯓ ° ′
01	01 N 54	01 S 16	01 N 51	00 S 02	01 N 48	00 S 45	00 S 19	06 N 25
04	02 13	01 11	01 49	00 02	01 48	00 45	00 19	06 25
07	02 26	01 07	01 46	00 02	01 47	00 45	00 19	06 25
10	02 30	01 02	01 44	00 03	01 47	00 45	00 19	06 25
13	02 26	00 56	01 42	00 03	01 46	00 45	00 19	06 25
16	02 14	00 51	01 40	00 04	01 46	00 45	00 19	06 25
19	01 52	00 44	01 37	00 04	01 46	00 45	00 19	06 25
22	01 22	00 38	01 35	00 05	01 45	00 45	00 19	06 25
25	00 N 43	00 31	01 33	00 05	01 45	00 45	00 19	06 25
28	00 S 02	00 25	01 31	00 05	01 45	00 46	00 20	06 25
31	00 S 53	00 S 18	01 N 29	00 S 06	01 N 44	00 S 46	00 S 20	06 N 25

DATA

Julian Date	2454588
Delta T	+67 seconds
Ayanamsa	23° 58′ 33″
Synetic vernal point	05° ♓ 08′ 26″
True obliquity of ecliptic	23° 26′ 25″

MOON'S PHASES, APSIDES AND POSITIONS ☽

Date	h	m	Phase	Longitude ° ′	Eclipse Indicator
05	12	18	●	15 ♉ 22	
12	03	47	☽	21 ♌ 48	
20	02	11	○	29 ♏ 27	
28	02	57	☾	07 ♓ 10	

Day	h	m	
06	03	12	Perigee
20	14	10	Apogee

02	01	00	ON	
08	01	50	Max dec	27° N 38′
14	16	15	OS	
22	03	57	Max dec	27° S 33′
29	08	56	ON	

LONGITUDES

Date	Chiron ⚷ ° ′	Ceres ⚳ ° ′	Pallas ⚴ ° ′	Juno ⚵ ° ′	Vesta ⚶ ° ′	Black Moon Lilith ⚸ ° ′
01	20 ♒ 51	12 ♊ 46	22 ♈ 23	29 ♐ 12	05 ♈ 36	02 ♐ 23
11	21 ♒ 03	16 ♊ 51	26 ♈ 22	28 ♐ 12	10 ♈ 04	03 ♐ 31
21	21 ♒ 09	21 ♊ 01	00 ♉ 22	26 ♐ 42	14 ♈ 25	04 ♐ 38
31	21 ♒ 08	25 ♊ 14	04 ♉ 23	24 ♐ 46	18 ♈ 37	05 ♐ 45

All ephemeris data is given at 12.00 UT and the Moon's longitude is additionally given for 24.00 UT

ASPECTARIAN

h m	Aspects
01 Thursday	
00 49	☽ ♇ ♂
02 07	☽ ✶ ☉
05 24	☽ Q ⯓
07 48	♀ △ ⯓
09 45	☽ ∥ ♅
09 54	☽ Q ☿
10 57	☽ ∠ ♀
16 16	♂ ⊥ ♄
18 36	☿ # ♃
20 14	☽ ☌ ♅
21 41	☽ ✶ ♃
22 20	♀ △ ♄
02 Friday	
00 48	☽ ⊻ ♆
03 57	☽ △ ♂
04 01	☽ ⊥ ♀
06 08	☽ ∠ ☉
09 35	☽ ✶ ☿
11 01	☽ ⊥ ♆
12 24	☽ □ ⯓
13 41	☽ ⊼ ♄
15 08	☽ ⊻ ♀
15 53	☽ # ♅
18 04	☽ Q ♃
22 22	☽ ⊥ ☉
23 42	☽ ± ♄
03 Saturday	
02 04	☽ ∠ ♆
03 06	♄ St D
08 45	☿ ⊼ ⯓
08 59	☽ ⊻ ☉
13 49	☽ ∠ ☿
14 30	☽ ♇ ♄
18 39	☿ ∥ ♂
19 50	☿ □ ♄
20 10	☽ ∥ ♀
22 22	☽ ⊻ ♅
23 19	☽ ∥ ♄
23 39	☽ □ ♃
04 Sunday	
02 32	☽ ✶ ♆
04 06	☽ # ♆
06 12	☽ ⊥ ☿
07 16	☽ □ ♂
08 02	☽ ⊥ ♅
13 23	☽ △ ⯓
13 44	☽ ∥ ☉
14 39	☽ △ ♄
16 56	☽ ⊻ ☿
17 16	☽ # ⯓
20 21	☽ ☌ ♀
21 42	☽ Q ♆
22 24	☽ ∠ ♅
23 36	☿ Q ♅
05 Monday	
12 18	☽ ☌ ☉
12 50	☽ Q ♂
13 05	☽ ♇ ⯓
13 07	♀ Q ♆
14 37	☽ # ♃
17 36	♀ ∥ ♄
18 31	☽ ∥ ♂
22 06	☽ ✶ ♅
22 14	♀ ∠ ♅
23 16	☽ △ ♃
06 Tuesday	
00 27	☉ ♇ ⯓
00 58	☽ ∥ ☿
02 05	☽ □ ♆
03 13	☽ ± ⯓
05 50	☉ Q ♂
08 22	☽ ✶ ♂
12 38	☽ ⊼ ⯓
13 57	☽ □ ♄
17 00	☽ Q ♅
21 31	☽ ☌ ☿
22 52	☽ ♇ ♃
23 45	☽ ⊻ ♀
07 Wednesday	
08 52	☽ ∠ ♂
10 05	☽ ⊥ ♀
11 50	☿ ♇ ♃
13 08	☽ ± ♃
14 09	☉ # ⯓
14 48	☽ ⊻ ☉
18 27	☽ Q ♄
21 37	☽ □ ♅
22 43	☽ ⊼ ♃
23 45	☽ ⊥ ♂
08 Thursday	
01 04	☽ ⊥ ☉
01 36	☽ △ ♆
01 44	☽ ∠ ♀
09 47	☽ ⊻ ♂
12 22	☽ ☍ ⯓
13 48	☽ ✶ ♄
15 11	♀ # ♆
16 39	☽ ∠ ☉
09 Friday	
02 02	☽ ♇ ♆
02 44	☽ ⊻ ☿
04 24	☽ ✶ ♀
12 09	♃ St R
13 42	☽ ⊥ ☿
14 34	☽ ∠ ♄
17 01	☽ ± ♆
19 18	☽ ✶ ☉
21 27	☽ ∥ ☿
23 03	☽ △ ♅
10 Saturday	
00 06	☽ ☍ ♃
02 26	☽ Q ♀
03 11	☽ ⊼ ♆
05 49	☽ ⊥ ♄
06 26	☽ ∠ ☿
13 50	☽ ☌ ♂
14 31	☽ ⊼ ⯓
15 13	☽ ∥ ♂
15 55	☽ # ♃
16 10	☽ ⊻ ♄
17 21	☽ Q ☉
11 Sunday	
00 58	☽ ± ⯓
00 59	☽ ♇ ♅
06 56	♂ ⊼ ⯓
10 44	☽ ∥ ☉
11 06	☽ ✶ ☿
12 51	☽ □ ♀
13 18	♂ # ♃
15 37	☽ # ⯓
16 54	☽ ♇ ⯓
16 59	☽ ± ♅
12 Monday	
00 22	☽ ∥ ♀
03 47	☽ □ ☉
03 50	☽ ⊼ ♅
04 32	☉ ✶ ♅
04 48	☽ ⊼ ♃
06 31	☽ # ♆
08 09	☽ ☍ ♆
10 46	☽ Q ☿
11 59	☽ ∥ ♄
15 46	☽ ± ♃
17 29	☉ △ ♃
20 11	☽ △ ⯓
21 51	☽ ⊻ ♂
22 04	☽ ☌ ♄
13 Tuesday	
03 16	♂ ⊻ ♄
08 23	♀ ♇ ⯓
08 28	☽ ♇ ♃
09 32	☽ ⊥ ♂
23 20	☽ □ ☿
14 Wednesday	
00 07	☽ # ♅
02 03	☽ △ ♀
03 20	☽ ∠ ♂
12 08	☽ ☍ ♅
12 56	☽ △ ♃
15 50	☉ □ ♆
16 34	☽ ⊼ ♆
16 38	☽ △ ☉
15 Thursday	
03 26	☿ ± ♃
03 54	☉ ± ⯓
04 12	☽ ± ♆
05 08	☽ □ ⯓
07 18	☽ ⊻ ♄
08 33	☽ ∥ ♅
09 36	☽ ✶ ♂
10 10	☽ ♇ ♀
19 03	☽ ⊥ ♄
21 48	☽ ♇ ♆
16 Friday	
00 21	☽ ♇ ☉
05 48	☽ ± ♀
10 16	☽ Q ♂
12 49	☽ ∠ ♄
14 12	☽ △ ☿
16 26	☽ Q ⯓
19 01	☽ ⊼ ♀
19 40	☽ ± ☉
22 41	☽ # ♄
23 01	☽ ⊼ ♅
23 36	☽ □ ♃
17 Saturday	
00 47	♀ # ⯓
03 29	☽ △ ♆
04 50	☽ ∥ ♆
08 38	☽ ⊼ ☉
11 00	☽ ± ♅
16 17	☽ ✶ ⯓
18 43	☽ ✶ ♄
21 29	☽ ∥ ⯓
22 02	☽ ♇ ☿
23 17	☽ # ♀
23 48	☽ □ ♂
18 Sunday	
05 03	☽ ♇ ♅
11 23	♀ ✶ ♅
11 30	☽ Q ♃
12 37	☽ # ☉
15 18	♀ △ ♃
17 08	☽ ± ☿
18 23	☽ # ♂
18 51	☽ Q ♄
22 19	☽ ∠ ⯓
19 Monday	
01 07	☽ ∥ ♃
05 52	☽ ⊼ ☿
09 48	♀ Q ♂
11 18	☽ △ ♅
11 35	☽ ✶ ♃
13 59	☽ ☍ ♀
15 41	☽ □ ♆
16 25	☽ ⊥ ⯓
20 Tuesday	
01 49	☽ # ☿
02 11	☽ ☍ ☉
04 31	☽ ⊻ ⯓
06 28	♀ □ ♆
07 00	☿ Q ♄
07 14	☽ □ ♄
13 14	♀ ± ⯓
15 09	☽ △ ♂
17 48	☽ ∠ ♃
21 Wednesday	
04 04	☽ Q ♆
06 22	☉ ⊼ ⯓
06 29	☉ ∥ ♂
11 56	☽ ⊥ ♃
18 05	♃ ✶ ♅
21 02	☽ ☍ ☿
21 47	♂ ± ⯓
22 57	☽ ♇ ♂
22 Thursday	
00 02	☽ ⊻ ♃
00 04	☽ □ ♅
04 19	☽ ✶ ♆
09 36	☽ ⊼ ♀
17 01	☽ ☌ ⯓
17 59	☽ ± ♂
18 25	☉ □ ♄
20 00	☽ △ ♄
20 08	☽ ⊼ ☉
23 05	☽ ± ♀
23 Friday	
00 08	♂ ♇ ♅
06 39	☽ ⊼ ♂
09 15	☽ ± ☉
10 32	☽ ∠ ♆
12 24	☽ Q ♅
19 14	☽ ♇ ♀
24 Saturday	
02 11	☽ ♇ ♄
04 33	☽ ⊥ ♆
04 51	☽ ♇ ☉
10 43	☽ ⊼ ☿
12 01	☽ # ☿
12 03	☽ ☌ ♃
12 26	☽ ✶ ♅
16 30	☽ ⊻ ♆
20 08	☽ ± ♄
22 47	☽ ± ☿
25 Sunday	
01 28	☉ Q ♅
03 35	♀ ∥ ♂
04 25	☽ △ ♀
04 49	☽ ⊻ ⯓
05 28	☽ ∥ ♃
08 01	☽ ⊼ ♄
08 21	♀ ⊼ ⯓
09 48	☽ # ☉
13 04	☽ △ ☉
16 34	☽ ⊥ ⯓
16 37	☽ ♇ ☿
17 06	☽ # ♀
18 04	☽ ∠ ♅
18 37	☽ # ♂
21 02	☽ ☍ ♂
26 Monday	
08 28	☽ ∥ ⯓
10 01	☽ ∠ ⯓
11 33	☽ ⊥ ♅
15 48	☿ St R
16 16	♆ St R
17 10	♀ □ ♄
21 39	☽ △ ☿
22 20	☽ ⊻ ♃
23 03	☽ ⊻ ♅
27 Tuesday	
00 23	☽ ∥ ♆
02 49	☽ ☌ ♆
06 50	☽ # ♄
09 38	☽ ⊥ ♃
14 27	☽ ✶ ⯓
17 46	☽ ☍ ♄
19 06	☉ ♇ ♃
20 16	☽ □ ♀
28 Wednesday	
02 20	☽ ∠ ♃
02 57	☽ □ ☉
08 25	☽ ⊼ ♂
10 58	♀ Q ♅
12 33	☽ Q ⯓
18 27	☽ ∥ ♅
19 48	☽ ± ♂
29 Thursday	
04 33	☽ □ ☿
05 23	☽ ✶ ♃
06 23	☽ ☌ ♅
08 17	☽ Q ♀
09 48	☽ ⊻ ♆
12 31	☽ ♇ ♂
13 51	☽ Q ☉
20 19	☽ ⊥ ♆
20 32	☽ □ ⯓
21 41	☉ # ♃
23 06	☽ # ♅
23 53	☽ ⊼ ♄
30 Friday	
02 18	☽ Q ♃
07 14	☽ ✶ ♀
09 33	♀ ♇ ♃
10 14	☽ ± ♄
11 23	☽ Q ☿
11 48	☽ ∠ ♆
12 08	☽ ✶ ☉
15 30	☽ △ ♂
31 Saturday	
01 28	☽ ♇ ♄
07 08	☽ ✶ ☿
07 39	☽ ∥ ♄
08 34	☽ □ ♃
09 48	☽ ⊻ ♅
10 49	☽ ∠ ♀
12 54	☽ ✶ ♆
13 36	☽ # ♆
14 59	☽ ∠ ☉
19 40	☽ ⊥ ♅
22 51	☽ △ ⯓
23 59	☉ ∥ ☿

JUNE 2008

LONGITUDES

Date	Sidereal time h m s	Sun ☉ ° ' "	Moon ☽ ° ' "	Moon ☽ 24.00 ° ' "	Mercury ☿ ° '	Venus ♀ ° '	Mars ♂ ° '	Jupiter ♃ ° '	Saturn ♄ ° '	Uranus ♅ ° '	Neptune ♆ ° '	Pluto ♇ ° '
01	04 41 22	11 ♊ 21 34	08 ♉ 28 33	15 ♉ 59 42	20 ♊ 20	09 ♊ 17	12 ♌ 29	21 ♑ 33	02 ♍ 26	22 ♓ 24	24 ♒ 15	00 ♑ 19
02	04 45 18	12 19 04	23 ♉ 34 51	01 ♊ 12 43	19 R 55	10 30	13 03	21 R 29	02 29	22 25	24 R 15	00 R 18
03	04 49 15	13 16 34	08 ♊ 51 56	16 ♊ 31 03	19 27	11 44	13 37	21 24	02 32	22 26	24 14	00 16
04	04 53 11	14 14 03	24 ♊ 08 39	01 ♋ 43 21	18 57	12 58	14 11	21 20	02 35	22 27	24 14	00 15
05	04 57 08	15 11 30	09 ♋ 13 56	16 ♋ 39 23	18 26	14 11	14 45	21 15	02 38	22 28	24 14	00 13
06	05 01 05	16 08 57	23 ♋ 58 50	01 ♌ 11 42	17 53	15 25	15 19	21 10	02 42	22 29	24 13	00 12
07	05 05 01	17 06 22	08 ♌ 17 35	15 ♌ 16 19	17 19	16 39	15 53	21 05	02 45	22 30	24 13	00 10
08	05 08 58	18 03 47	22 ♌ 07 56	28 ♌ 52 35	16 46	17 53	16 28	21 00	02 49	22 31	24 13	00 09
09	05 12 54	19 01 10	05 ♍ 30 35	12 ♍ 02 20	16 13	19 06	17 02	20 55	02 52	22 32	24 12	00 07
10	05 16 51	19 58 32	18 ♍ 28 18	24 ♍ 49 00	15 41	20 20	17 37	20 49	02 56	22 33	24 12	00 06
11	05 20 47	20 55 53	01 ♎ 05 01	07 ♎ 16 53	15 11	21 34	18 11	20 43	03 00	22 34	24 11	00 04
12	05 24 44	21 53 13	13 ♎ 25 09	19 ♎ 30 21	14 43	22 48	18 46	20 38	03 04	22 34	24 11	00 03
13	05 28 40	22 50 32	25 ♎ 33 00	01 ♏ 33 35	14 17	24 01	19 20	20 32	03 08	22 35	24 10	00 01
14	05 32 37	23 47 50	07 ♏ 32 32	13 ♏ 30 14	13 54	25 15	19 55	20 26	03 12	22 36	24 10	00 ♑ 00
15	05 36 34	24 45 07	19 ♏ 27 04	25 ♏ 23 22	13 35	26 29	20 30	20 20	03 16	22 36	24 09	29 ♐ 58
16	05 40 30	25 42 24	01 ♐ 19 24	07 ♐ 15 26	13 20	27 42	21 05	20 13	03 21	22 37	24 08	29 56
17	05 44 27	26 39 40	13 ♐ 11 42	19 ♐ 08 25	13 09	28 ♊ 56	21 39	20 07	03 25	22 37	24 08	29 55
18	05 48 23	27 36 55	25 ♐ 05 47	01 ♑ 03 58	13 02	00 ♋ 10	22 14	20 01	03 29	22 38	24 07	29 53
19	05 52 20	28 34 10	07 ♑ 03 12	13 ♑ 03 38	12 59	01 24	22 49	19 54	03 34	22 38	24 06	29 52
20	05 56 16	29 ♊ 31 25	19 ♑ 05 31	25 ♑ 09 04	13 D 01	02 37	23 24	19 47	03 38	22 38	24 06	29 50
21	06 00 13	00 ♋ 28 39	01 ♒ 14 33	07 ♒ 22 14	13 07	03 51	23 59	19 41	03 43	22 39	24 05	29 49
22	06 04 09	01 25 52	13 ♒ 32 26	19 ♒ 45 30	13 18	05 05	24 35	19 34	03 48	22 39	24 04	29 47
23	06 08 06	02 23 06	26 ♒ 01 48	02 ♓ 21 42	13 34	06 18	25 10	19 27	03 53	22 39	24 03	29 46
24	06 12 03	03 20 19	08 ♓ 45 39	15 ♓ 14 01	13 55	07 32	25 45	19 20	03 57	22 39	24 02	29 44
25	06 15 59	04 17 32	21 ♓ 47 14	28 ♓ 25 40	14 20	08 46	26 20	19 12	04 02	22 39	24 01	29 43
26	06 19 56	05 14 45	05 ♈ 09 38	11 ♈ 59 24	14 50	10 00	26 56	19 05	04 07	22 39	24 00	29 41
27	06 23 52	06 11 59	18 ♈ 55 10	25 ♈ 56 57	15 24	11 13	27 31	18 58	04 12	22 R 39	24 00	29 39
28	06 27 49	07 09 12	03 ♉ 04 41	10 ♉ 18 08	16 03	12 27	28 07	18 51	04 18	22 39	23 59	29 38
29	06 31 45	08 06 25	17 ♉ 36 51	25 ♉ 00 16	16 47	13 41	28 42	18 43	04 23	22 39	23 58	29 36
30	06 35 42	09 ♋ 03 39	02 ♊ 27 35	09 ♊ 57 50	17 ♊ 35	14 ♋ 54	29 ♌ 18	18 ♑ 36	04 ♍ 28	22 ♓ 39	23 ♒ 57	29 ♐ 35

Date	Moon True ☊ ° '	Moon Mean ☊ ° '	Moon ☽ Latitude ° '
01	21 ♒ 18	22 ♒ 16	04 N 55
02	21 R 09	22 13	05 02
03	20 59	22 10	04 48
04	20 50	22 06	04 13
05	20 42	22 03	03 21
06	20 37	22 00	02 17
07	20 35	21 57	01 N 05
08	20 D 34	21 54	00 S 08
09	20 35	21 50	01 19
10	20 35	21 47	02 23
11	20 R 35	21 44	03 19
12	20 32	21 41	04 04
13	20 28	21 38	04 37
14	20 21	21 35	04 57
15	20 12	21 31	05 05
16	20 03	21 28	04 59
17	19 52	21 25	04 40
18	19 43	21 22	04 09
19	19 35	21 19	03 27
20	19 29	21 16	02 36
21	19 25	21 12	01 36
22	19 23	21 09	00 S 31
23	19 D 23	21 06	00 N 36
24	19 24	21 03	01 43
25	19 25	21 00	02 46
26	19 26	20 56	03 42
27	19 R 25	20 53	04 27
28	19 23	20 50	04 57
29	19 19	20 47	05 09
30	19 ♒ 14	20 ♒ 44	05 N 01

DECLINATIONS

Date	Sun ☉ ° '	Moon ☽ ° '	Mercury ☿ ° '	Venus ♀ ° '	Mars ♂ ° '	Jupiter ♃ ° '	Saturn ♄ ° '	Uranus ♅ ° '	Neptune ♆ ° '	Pluto ♇ ° '
01	22 N 09	18 N 59	21 N 55	21 N 35	18 N 28	21 S 49	12 N 14	03 S 43	13 S 45	17 S 01
02	22 16	23 32	21 36	21 49	18 18	21 50	12 13	03 43	13 45	17 01
03	22 24	26 31	21 16	22 01	18 07	21 50	12 11	03 42	13 45	17 01
04	22 31	27 32	20 57	22 13	17 57	21 51	12 10	03 42	13 45	17 01
05	22 37	26 28	20 38	22 24	17 46	21 52	12 09	03 41	13 45	17 01
06	22 43	23 33	20 19	22 35	17 36	21 53	12 07	03 41	13 45	17 01
07	22 49	19 14	20 00	22 45	17 25	21 54	12 06	03 41	13 46	17 02
08	22 54	14 00	19 43	22 55	17 14	21 55	12 05	03 40	13 46	17 02
09	22 59	08 16	19 26	23 03	17 03	21 56	12 03	03 40	13 46	17 02
10	23 04	02 N 22	19 10	23 11	16 51	21 57	12 02	03 40	13 46	17 02
11	23 08	03 S 28	18 56	23 19	16 40	21 58	12 00	03 40	13 46	17 02
12	23 11	09 02	18 43	23 25	16 29	21 59	11 59	03 39	13 46	17 02
13	23 15	14 10	18 31	23 31	16 17	22 00	11 57	03 39	13 47	17 02
14	23 18	18 42	18 21	23 37	16 06	22 01	11 56	03 39	13 47	17 02
15	23 20	22 29	18 13	23 41	15 54	22 02	11 54	03 39	13 47	17 02
16	23 22	25 18	18 07	23 45	15 42	22 03	11 52	03 38	13 47	17 02
17	23 24	27 01	18 02	23 49	15 30	22 04	11 51	03 38	13 48	17 02
18	23 25	27 30	17 59	23 51	15 18	22 05	11 49	03 38	13 48	17 02
19	23 26	26 42	17 59	23 53	15 06	22 07	11 47	03 38	13 48	17 02
20	23 26	24 39	17 59	23 54	14 54	22 08	11 46	03 38	13 48	17 03
21	23 26	21 27	18 02	23 55	14 42	22 09	11 44	03 38	13 49	17 03
22	23 26	17 16	18 06	23 54	14 30	22 10	11 42	03 38	13 49	17 03
23	23 25	12 17	18 12	23 53	14 17	22 11	11 40	03 38	13 49	17 03
24	23 24	06 42	18 19	23 52	14 05	22 12	11 38	03 38	13 49	17 03
25	23 22	00 S 43	18 28	23 49	13 52	22 13	11 36	03 38	13 50	17 03
26	23 20	05 N 27	18 38	23 46	13 39	22 15	11 35	03 38	13 50	17 03
27	23 18	11 31	18 50	23 42	13 26	22 16	11 33	03 38	13 50	17 03
28	23 15	17 11	19 02	23 38	13 14	22 17	11 31	03 38	13 51	17 03
29	23 12	22 02	19 15	23 33	13 01	22 18	11 29	03 38	13 51	17 04
30	23 N 08	25 N 34	19 N 29	23 N 27	12 N 48	22 S 19	11 N 27	03 S 38	13 S 51	17 S 04

ZODIAC SIGN ENTRIES

Date	h	m	Planets
02	22	06	☽ ♊
04	21	16	☽ ♋
06	22	00	☽ ♌
09	02	01	☽ ♍
11	09	55	☽ ♎
13	20	53	☽ ♏
14	05	13	♇ ♐
16	09	19	☽ ♐
18	08	48	♀ ♋
18	21	52	☽ ♑
20	23	59	☉ ♋
21	09	33	☽ ♒
23	19	32	☽ ♓
26	02	49	☽ ♈
28	06	50	☽ ♉
30	08	03	☽ ♊

LATITUDES

Date	Mercury ☿ ° '	Venus ♀ ° '	Mars ♂ ° '	Jupiter ♃ ° '	Saturn ♄ ° '	Uranus ♅ ° '	Neptune ♆ ° '	Pluto ♇ ° '
01	01 S 10	00 S 15	01 N 28	00 S 06	01 N 44	00 S 46	00 S 20	06 N 25
04	02 02	00 08	01 26	00 06	01 44	00 46	00 20	06 25
07	02 50	00 S 01	01 24	00 07	01 44	00 46	00 20	06 25
10	03 31	00 N 06	01 22	00 07	01 43	00 46	00 20	06 25
13	04 01	00 13	01 20	00 08	01 43	00 46	00 20	06 24
16	04 19	00 20	01 18	00 08	01 43	00 46	00 20	06 24
19	04 25	00 27	01 16	00 09	01 42	00 46	00 20	06 24
22	04 20	00 34	01 14	00 09	01 42	00 47	00 20	06 24
25	04 05	00 40	01 12	00 10	01 42	00 47	00 20	06 23
28	03 42	00 47	01 10	00 10	01 41	00 47	00 20	06 23
31	03 S 13	00 N 52	01 N 08	00 S 11	01 N 41	00 S 47	00 S 20	06 N 23

DATA

Julian Date	2454619
Delta T	+67 seconds
Ayanamsa	23° 58' 38"
Synetic vernal point	05° ♓ 08' 21"
True obliquity of ecliptic	23° 26' 24"

MOON'S PHASES, APSIDES AND POSITIONS ☽

Date	h	m	Phase	Longitude ° '	Eclipse Indicator
03	19	23	●	13 ♊ 34	
10	15	04	☽	20 ♍ 06	
18	17	30	○	27 ♐ 50	
26	12	10	☾	05 ♈ 15	

Day	h	m		
03	13	05	Perigee	
16	17	22	Apogee	
04	11	34	Max dec	27° N 32'
10	21	38	0S	
18	09	06	Max dec	27° S 31'
25	14	48	0N	

LONGITUDES

Date	Chiron ⚷ ° '	Ceres ⚳ ° '	Pallas ⚴ ° '	Juno ⚵ ° '	Vesta ⚶ ° '	Black Moon Lilith ⚸ ° '
01	21 ♒ 07	25 ♊ 39	04 ♉ 48	24 ♐ 33	19 ♈ 02	05 ♐ 52
11	21 ♒ 00	29 ♊ 56	08 ♉ 49	22 ♐ 20	23 ♈ 03	06 ♐ 59
21	20 ♒ 46	04 ♋ 14	12 ♉ 50	20 ♐ 04	26 ♈ 52	08 ♐ 06
31	20 ♒ 27	08 ♋ 34	16 ♉ 50	17 ♐ 58	00 ♉ 27	09 ♐ 13

All ephemeris data is given at 12.00 UT and the Moon's longitude is additionally given for 24.00 UT

ASPECTARIAN

h m	Aspects
01 Sunday	
02 13	☽ △ ♄
02 52	☽ ⊥ ♀
03 15	☽ # ♇
06 33	☿ Q ♄
06 39	☽ ⊥ ☉
07 04	☽ ∠ ☿
08 25	☽ Q ♆
09 45	☽ ∥ ♂
10 16	☽ ∠ ♅
13 24	☽ ⊻ ♀
16 56	☽ ⊻ ☉
18 39	☽ □ ♂
19 34	☿ # ♃
21 07	☽ ⊥ ☿
22 54	☽ ⚼ ♇
02 Monday	
01 33	☽ ∥ ♀
01 40	☽ ∥ ☿
02 07	☽ # ♃
02 34	☿ ∥ ♀
04 21	☽ ∥ ☉
06 22	☽ ⊻ ☿
08 42	☽ △ ♃
10 10	☽ ⚹ ♅
13 03	☽ □ ♆
13 07	☽ ± ♇
14 07	♀ # ♃
22 33	☽ ⊼ ♇
03 Tuesday	
00 11	☽ Q ♂
02 02	☽ □ ♄
05 03	☽ Q ♅
08 10	☽ ⚼ ♃
16 53	☽ ☌ ♀
19 23	☽ ☌ ☉
19 43	☽ ⚹ ♂
22 12	☽ ± ♃
04 Wednesday	
04 06	☽ ☌ ☿
06 22	☽ Q ♄
07 35	☽ ⊼ ♃
08 34	☉ ⚹ ♂
09 20	☽ □ ♅
12 08	☽ △ ♆
20 16	☽ ∠ ♂
21 38	☽ ☍ ♇
05 Thursday	
01 25	☽ ⚹ ♄
11 11	☽ ⊥ ♂
12 00	☽ ⚼ ♆
13 23	☉ ± ♃
20 43	☽ ⊻ ♀
21 15	☽ ⊻ ♂
22 17	☽ ⊻ ☉
06 Friday	
01 39	☽ ∠ ♄
02 21	☽ ⊻ ☿
02 33	☽ ± ♆
07 05	♂ ⚼ ♇
07 24	☽ ⊥ ♀
07 24	☽ ☍ ♃
07 25	♀ ± ♃
08 16	♀ ⚹ ♂
08 46	☽ ⊥ ☉
09 32	☽ △ ♅
11 51	☽ ⊥ ☿
12 24	☽ ⊼ ♆
16 31	☽ ⊥ ♄
17 05	☽ ∥ ☉
17 46	☽ ∥ ♀
22 02	☽ # ♃
22 19	☽ ⊼ ♇
23 42	☽ ∠ ♀
07 Saturday	
00 47	☽ ∠ ☉
02 17	☽ ∠ ☿
02 35	☽ ⊻ ♄
07 52	☽ ∥ ☿
08 24	☽ ± ♇
10 39	☽ ⚼ ♅
15 27	☉ ☌ ☿
21 03	☿ ☌ ♀
21 04	☽ ∥ ♂
22 32	☽ # ♇
23 48	☽ ⚼ ♇
08 Sunday	
01 38	☽ ☌ ♂
02 09	☽ ± ♅
02 58	☽ ⚹ ☿
03 48	☽ ⚹ ♀
04 19	☽ ⚹ ☉
10 01	☽ ⊼ ♃
10 21	☉ ∥ ♀
12 41	☽ ⊼ ♅
13 02	☽ # ♆
14 27	♂ ± ♅
15 40	☽ ☍ ♆
18 28	☿ ⚹ ♂
20 15	☽ ∥ ♄
20 35	☽ ± ♃
23 20	☽ Q ☿
09 Monday	
02 15	☽ △ ♇
03 12	☽ Q ♀
03 13	☽ Q ☉
04 18	☉ ☌ ♀
07 11	☽ ☌ ♄
12 44	☽ ⚼ ♃
14 02	♂ # ♇
10 Tuesday	
06 43	☽ # ♅
06 59	☽ □ ☿
10 18	☽ ⊻ ♂
15 04	☽ □ ☉
15 53	☽ □ ♀
16 23	☽ △ ♃
19 42	☽ ☍ ♅
20 46	♀ ⊼ ♃
22 10	☽ ⊥ ♂
22 49	☽ ⊼ ♆
11 Wednesday	
00 24	♀ Q ♄
07 17	☉ ⊼ ♃
10 03	☽ □ ♇
10 17	☽ ± ♆
12 48	☽ ∥ ♅
13 53	☉ Q ♄
15 43	☽ ⊻ ♄
16 15	☽ ∠ ♂
12 Thursday	
03 26	☽ ⊥ ♄
03 42	☽ ⚼ ♆
07 38	♀ □ ♅
14 27	☽ △ ☿
17 34	☿ ± ♃
21 05	☽ Q ♇
21 12	☽ ∠ ♄
23 03	☽ ⚹ ♂
13 Friday	
01 24	☽ # ♄
02 07	☽ □ ♃
05 24	☉ □ ♅
06 06	☽ ⊼ ♅
06 09	☽ △ ☉
08 37	☽ △ ♀
09 15	☽ △ ♆
10 04	☽ ∥ ♆
14 53	♀ △ ♆
18 03	☽ ± ♅
19 13	☽ ⚼ ☿
20 54	☽ ⚹ ♇
22 20	☽ # ♂
14 Saturday	
00 09	☽ Q ♂
02 43	☽ ∥ ♇
03 14	☽ ⚹ ♄
10 05	☽ # ☿
12 06	☽ ⚼ ♅
12 43	☽ ± ☿
13 46	☽ Q ♃
14 45	☽ ⚼ ☉
18 04	☽ ⚼ ♀
21 02	☉ △ ♆
15 Sunday	
00 28	☽ ⊼ ☿
02 58	☽ ∠ ♇
03 31	☽ Q ♄
06 07	♂ ⊼ ♃
08 52	☽ ∥ ♃
10 28	☽ ± ☉
13 45	☽ ⚹ ♃
14 13	☽ □ ♂
14 19	☽ ± ♀
18 22	☽ △ ♅
18 32	☽ # ☉
21 06	☽ ⊥ ♇
21 29	☽ □ ♆
21 31	☽ # ♀
23 39	☽ ⊼ ☉
16 Monday	
03 50	☽ ⊼ ♀
09 13	☽ ⊻ ♇
16 06	☽ □ ♄
19 49	☽ ∠ ♃
17 Tuesday	
09 51	☽ Q ♆
11 54	☽ ☍ ☿
13 51	☽ ⊥ ♃
18 Wednesday	
01 51	☽ ⊻ ♃
05 57	☽ △ ♂
06 46	♀ ☍ ♇
07 02	☽ □ ♅
10 02	☽ ⚹ ♆
17 30	☽ ☍ ☉
21 37	☽ ☌ ♇
23 21	☽ ☍ ♀
19 Thursday	
04 08	♂ ⊼ ♅
04 58	☽ △ ♄
13 37	☽ ⚼ ♂
14 31	☿ St D
16 06	☽ ∠ ♆
19 10	☽ Q ♅
23 51	☽ ⊼ ☿
20 Friday	
08 29	☽ ± ♂
10 01	☽ ⊥ ♆
11 06	☽ ⚼ ♄
11 50	☽ ± ☿
13 22	☽ ☌ ♃
18 21	☽ # ♀
19 02	☽ ⚹ ♅
19 42	☉ ☍ ♇
20 59	☽ ⊼ ♂
21 54	☽ ⊻ ♆
22 02	☽ # ☉
21 Saturday	
05 01	☽ ± ♄
05 47	☽ ⚼ ☿
07 24	☽ ∥ ♃
09 12	☽ ⊻ ♇
09 14	♀ ⚹ ♄
10 22	☽ ⊼ ☉
15 32	♂ ☍ ♆
16 53	☽ ⊼ ♄
17 41	☽ ⊼ ♀
20 56	☽ ⊥ ♇
23 07	☽ ± ☉
22 Sunday	
00 32	☽ ∠ ♅
06 42	☽ ± ♀
07 38	☽ # ☿
11 32	☽ △ ☿
13 07	☽ ∥ ♇
14 24	☽ ∠ ♇
18 01	☽ ⊥ ♅
18 03	☽ ⚼ ☉
23 31	☽ ⊻ ♃
23 Monday	
01 59	☽ ⚼ ♀
02 21	☽ # ♂
04 50	☿ ± ♃
04 56	☽ ∥ ♆
05 33	☽ ⊻ ♅
08 14	☽ ☌ ♆
10 16	☽ ☍ ♂
10 54	☽ ⊥ ♃
14 45	☽ # ♄
19 04	☽ ⚹ ♇
21 33	♂ ± ♃
24 Tuesday	
01 01	☽ △ ☉
02 57	☽ ☍ ♄
03 47	☽ ∠ ♃
09 28	☽ △ ♀
17 31	☽ Q ♇
21 51	☽ □ ☿
25 Wednesday	
00 27	☽ ∥ ♅
05 01	☉ ⚹ ♄
07 20	☽ ⚹ ♃
10 04	☿ Q ♂
13 35	☽ ☌ ♅
15 56	♂ # ♆
16 03	☽ ⊻ ♆
17 01	♀ ⚼ ♆
20 38	☽ ⊼ ♂
26 Thursday	
02 16	☽ □ ♇
02 51	☽ ⊥ ♆
04 50	☽ Q ♃
04 56	☽ # ♅
07 42	☽ Q ☿
07 52	☽ ± ♂
07 57	♃ ⚼ ♄
10 09	☽ ⊼ ♄
12 10	☽ □ ☉
18 46	☽ ∠ ♆
20 47	☽ ± ♄
21 21	☽ □ ♀
27 Friday	
00 02	♅ St R
00 26	☽ ⚼ ♂
05 39	☽ ⚹ ☿
12 05	☽ □ ♃
12 05	☽ ∥ ♄
12 30	☽ ⚼ ♄
18 24	☽ ⊻ ♅
19 34	☽ ∥ ♂
20 40	☽ ⚹ ♆
21 32	☽ # ♆
21 41	☽ Q ☉
28 Saturday	
03 18	☽ △ ♂
04 35	☽ ⊥ ♅
06 14	☽ △ ♇
07 11	☽ Q ♀
08 27	☽ ∠ ☿
11 25	☽ # ♇
14 02	☽ △ ♄
16 50	☽ Q ♆
19 16	☽ ⚹ ☉
19 37	☽ ∠ ♅
20 57	☽ ∥ ☿
23 37	☿ Q ♂
29 Sunday	
00 11	☽ ⊥ ☿
04 58	☽ ⚹ ♀
07 05	☽ ⚼ ♇
10 34	☽ ⊻ ☿
12 57	♀ ∠ ♂
13 33	☽ # ♃
13 47	☽ △ ♃
18 44	☽ ∥ ☉
20 12	☽ ⚹ ♅
20 49	☽ ∥ ♀
21 33	☽ ∠ ☉
21 43	☽ ± ♇
22 18	☽ □ ♆
30 Monday	
06 43	☽ □ ♂
07 23	☽ ⊼ ♇
07 32	☽ ∠ ♀
09 05	☉ ⚼ ♆
13 02	☽ ⊥ ☉
13 48	☽ ⚼ ♃
15 14	☽ □ ♄
15 31	☽ Q ♅
22 57	♂ △ ♇
23 14	☽ ⊥ ♀
23 16	☽ ⊻ ☉

JULY 2008

LONGITUDES

Date	Sidereal time h	m	s	Sun ☉ °	'	"	Moon ☽ °	'	"	Moon ☽ 24.00 °	'	"	Mercury ☿ °	'	Venus ♀ °	'	Mars ♂ °	'	Jupiter ♃ °	'	Saturn ♄ °	'	Uranus ♅ °	'	Neptune ♆ °	'	Pluto ♇ °	'
01	06	39	38	10 ♋	00	53	17 ♊	29	59	25 ♊	02	50	18 ♊	28	16 ♋	08	29 ♌	54	18 ♑	28	04 ♍	33	22 ♓	39	23 ♒	56	29 ♐	33
02	06	43	35	10	58	07	02 ♋	35	11	10 ♋	05	48	19	24	17	22	00 ♍	29	18 R	21	04	39	22 R	39	23 R	54	29 R	32
03	06	47	32	11	55	20	17 ♋	33	34	24 ♋	57	24	20	25	18	36	01	05	18	13	04	44	22	38	23	53	29	30
04	06	51	28	12	52	34	02 ♌	16	25	09 ♌	29	52	21	31	19	50	01	41	18	05	04	50	22	38	23	52	29	29
05	06	55	25	13	49	48	16 ♌	37	11	23 ♌	38	01	22	40	21	03	02	17	17	58	04	56	22	38	23	51	29	27
06	06	59	21	14	47	01	00 ♍	32	09	07 ♍	19	34	23	54	22	17	02	53	17	50	05	01	22	37	23	50	29	26
07	07	03	18	15	44	14	14 ♍	00	23	20 ♍	34	50	25	12	23	31	03	29	17	42	05	07	22	37	23	49	29	24
08	07	07	14	16	41	27	27 ♍	03	15	03 ♎	26	04	26	33	24	45	04	05	17	35	05	13	22	36	23	48	29	23
09	07	11	11	17	38	40	09 ♎	43	46	15 ♎	56	52	27	59	25	58	04	41	17	27	05	19	22	36	23	46	29	21
10	07	15	07	18	35	52	22 ♎	05	55	28 ♎	11	30	29 ♊	28	27	12	05	17	17	19	05	25	22	35	23	45	29	20
11	07	19	04	19	33	05	04 ♏	14	10	10 ♏	14	28	01 ♋	02	28	26	05	53	17	11	05	31	22	34	23	44	29	18
12	07	23	01	20	30	17	16 ♏	12	56	22 ♏	10	06	02	38	29 ♋	40	06	29	17	04	05	37	22	34	23	43	29	17
13	07	26	57	21	27	30	28 ♏	06	25	04 ♐	02	23	04	19	00 ♌	53	07	06	16	56	05	43	22	33	23	41	29	16
14	07	30	54	22	24	43	09 ♐	58	23	15 ♐	54	49	06	03	02	07	07	42	16	48	05	49	22	32	23	40	29	14
15	07	34	50	23	21	56	21 ♐	52	02	27 ♐	50	21	07	50	03	21	08	18	16	41	05	55	22	31	23	39	29	13
16	07	38	47	24	19	09	03 ♑	50	02	09 ♑	51	22	09	41	04	35	08	55	16	33	06	01	22	30	23	37	29	11
17	07	42	43	25	16	22	15 ♑	54	32	21 ♑	59	47	11	34	05	48	09	31	16	26	06	08	22	29	23	36	29	10
18	07	46	40	26	13	36	28 ♑	07	15	04 ♒	17	09	13	30	07	02	10	08	16	18	06	14	22	28	23	34	29	09
19	07	50	36	27	10	50	10 ♒	29	37	16 ♒	44	49	15	29	08	16	10	44	16	11	06	20	22	27	23	33	29	07
20	07	54	33	28	08	05	23 ♒	02	55	29 ♒	24	03	17	29	09	30	11	21	16	03	06	27	22	26	23	32	29	06
21	07	58	30	29 ♋	05	20	05 ♓	48	23	12 ♓	16	05	19	32	10	43	11	58	15	56	06	33	22	25	23	30	29	05
22	08	02	26	00 ♌	02	36	18 ♓	47	19	25 ♓	22	15	21	36	11	57	12	34	15	49	06	40	22	24	23	29	29	04
23	08	06	23	00	59	52	02 ♈	01	02	08 ♈	43	48	23	41	13	11	13	11	15	41	06	47	22	23	23	27	29	02
24	08	10	19	01	57	10	15 ♈	30	41	22 ♈	21	45	25	47	14	25	13	48	15	34	06	53	22	22	23	26	29	01
25	08	14	16	02	54	28	29 ♈	17	02	06 ♉	16	30	27 ♋	54	15	38	14	25	15	27	07	00	22	20	23	24	29	00
26	08	18	12	03	51	48	13 ♉	20	04	20 ♉	27	30	00 ♌	01	16	52	15	02	15	20	07	07	22	19	23	23	28	59
27	08	22	09	04	49	08	27 ♉	38	33	04 ♊	52	47	02	08	18	06	15	39	15	13	07	13	22	18	23	21	28	57
28	08	26	05	05	46	30	12 ♊	09	44	19 ♊	28	46	04	15	19	20	16	16	15	06	07	20	22	16	23	20	28	56
29	08	30	02	06	43	53	26 ♊	49	12	04 ♋	10	16	06	21	20	33	16	53	15	00	07	27	22	15	23	18	28	55
30	08	33	59	07	41	16	11 ♋	31	08	18 ♋	50	59	08	26	21	47	17	30	14	53	07	34	22	13	23	17	28	54
31	08	37	55	08 ♌	38	41	26 ♋	08	57	03 ♌	24	13	10 ♌	31	23 ♌	01	18 ♍	07	14 ♑	47	07 ♍	41	22 ♓	12	23 ♒	15	28 ♐	53

Date	Moon True ☊ °	'	Moon Mean ☊ °	'	Moon ☽ Latitude °	'
01	19 ♒	08	20 ♒	41	04 N	33
02	19 R	03	20	37	03	45
03	18	58	20	34	02	42
04	18	56	20	31	01	29
05	18	55	20	28	00 N	12
06	18 D	55	20	25	01 S	03
07	18	56	20	21	02	13
08	18	58	20	18	03	13
09	18	59	20	15	04	02
10	18 R	59	20	12	04	39
11	18	58	20	09	05	03
12	18	56	20	06	05	12
13	18	53	20	02	05	09
14	18	48	19	59	04	52
15	18	44	19	56	04	22
16	18	40	19	53	03	41
17	18	37	19	50	02	50
18	18	34	19	47	01	50
19	18	33	19	43	00 S	44
20	18 D	33	19	40	00 N	25
21	18	34	19	37	01	34
22	18	35	19	34	02	39
23	18	36	19	31	03	37
24	18	37	19	27	04	25
25	18	38	19	24	04	58
26	18 R	38	19	21	05	15
27	18	37	19	18	05	12
28	18	36	19	15	04	50
29	18	34	19	12	04	08
30	18	33	19	08	03	11
31	18 ♒	32	19 ♒	05	02 N	01

DECLINATIONS

Date	Sun ☉ °	'	Moon ☽ °	'	Mercury ☿ °	'	Venus ♀ °	'	Mars ♂ °	'	Jupiter ♃ °	'	Saturn ♄ °	'	Uranus ♅ °	'	Neptune ♆ °	'	Pluto ♇ °	'
01	23 N	04	27 N	23	19 N	44	23 N	20	12 N	34	22 S	20	11 N	25	03 S	38	13 S	52	17 S	04
02	22	59	27	10	20	00	23	13	12	21	22	22	11	23	03	38	13	52	17	04
03	22	54	24	58	20	15	23	05	12	08	22	23	11	20	03	38	13	53	17	04
04	22	49	21	06	20	32	22	56	11	54	22	24	11	18	03	39	13	53	17	04
05	22	43	16	03	20	48	22	47	11	41	22	25	11	16	03	39	13	53	17	04
06	22	37	10	18	21	04	22	37	11	27	22	26	11	14	03	39	13	54	17	05
07	22	31	04 N	15	21	19	22	26	11	14	22	27	11	12	03	39	13	54	17	05
08	22	24	01 S	47	21	35	22	15	11	00	22	29	11	10	03	39	13	55	17	05
09	22	17	07	34	21	49	22	03	10	46	22	30	11	07	03	40	13	55	17	05
10	22	09	12	55	22	03	21	50	10	33	22	31	11	05	03	40	13	55	17	05
11	22	01	17	40	22	16	21	37	10	19	22	32	11	03	03	40	13	56	17	05
12	21	53	21	40	22	28	21	23	10	05	22	33	11	01	03	41	13	56	17	06
13	21	44	24	45	22	38	21	08	09	51	22	34	10	58	03	41	13	57	17	06
14	21	35	26	45	22	47	20	53	09	36	22	35	10	56	03	41	13	57	17	06
15	21	25	27	33	22	54	20	38	09	22	22	36	10	54	03	42	13	58	17	06
16	21	15	27	04	22	59	20	21	09	08	22	37	10	51	03	42	13	58	17	06
17	21	05	25	18	23	02	20	05	08	54	22	38	10	49	03	42	13	59	17	06
18	20	54	22	20	23	02	19	47	08	39	22	40	10	47	03	43	13	59	17	07
19	20	43	18	19	23	00	19	29	08	25	22	41	10	44	03	43	14	00	17	07
20	20	32	13	27	22	56	19	11	08	10	22	42	10	42	03	44	14	00	17	07
21	20	20	07	56	22	49	18	52	07	56	22	43	10	39	03	44	14	01	17	07
22	20	09	02 S	00	22	39	18	32	07	41	22	44	10	37	03	45	14	01	17	07
23	19	56	04 N	08	22	26	18	12	07	26	22	45	10	34	03	45	14	02	17	08
24	19	44	10	11	22	11	17	51	07	12	22	46	10	32	03	46	14	02	17	08
25	19	31	15	52	21	53	17	30	06	57	22	46	10	29	03	46	14	03	17	08
26	19	17	20	50	21	33	17	09	06	42	22	47	10	27	03	47	14	03	17	08
27	19	04	24	42	21	10	16	47	06	27	22	48	10	24	03	47	14	04	17	08
28	18	50	27	02	20	45	16	24	06	12	22	49	10	22	03	48	14	04	17	09
29	18	35	27	32	20	17	16	01	05	57	22	50	10	19	03	48	14	05	17	09
30	18	21	26	06	19	48	15	38	05	42	22	51	10	17	03	49	14	05	17	09
31	18 N	06	22 N	54	19 N	16	15 N	14	05 N	27	22 S	52	10 N	14	03 S	50	14 S	06	17 S	09

ZODIAC SIGN ENTRIES

Date	h	m	Planets
01	16	21	♂ ♍
02	07	53	☽ ♋
04	08	15	☽ ♌
06	11	04	☽ ♍
08	17	31	☽ ♎
10	20	17	☿ ♋
11	03	35	☽ ♏
12	18	39	♀ ♌
13	15	50	☽ ♐
16	04	20	☽ ♑
18	15	40	☽ ♒
21	01	08	☽ ♓
22	10	55	☉ ♌
23	08	22	☽ ♈
25	13	14	☽ ♉
26	11	48	☿ ♌
27	15	55	☽ ♊
29	17	12	☽ ♋
31	18	22	☽ ♌

LATITUDES

Date	Mercury ☿ °	'	Venus ♀ °	'	Mars ♂ °	'	Jupiter ♃ °	'	Saturn ♄ °	'	Uranus ♅ °	'	Neptune ♆ °	'	Pluto ♇ °	'
01	03 S	13	00 N	52	01 N	08	00 S	11	01 N	41	00 S	47	00 S	20	06 N	23
04	02	39	00	58	01	06	00	11	01	41	00	47	00	20	06	22
07	02	02	01	03	01	04	00	11	01	41	00	47	00	20	06	22
10	01	23	01	08	01	02	00	12	01	41	00	47	00	20	06	21
13	00	44	01	13	01	00	00	12	01	40	00	47	00	20	06	21
16	00 S	06	01	17	00	59	00	13	01	40	00	47	00	20	06	20
19	00 N	28	01	20	00	57	00	13	01	40	00	47	00	21	06	19
22	00	57	01	23	00	55	00	13	01	40	00	48	00	21	06	19
25	01	20	01	25	00	53	00	14	01	40	00	48	00	21	06	18
28	01	36	01	27	00	51	00	14	01	40	00	48	00	21	06	18
31	01 N	44	01 N	28	00 N	49	00 S	15	01 N	40	00 S	48	00 S	21	06 N	17

DATA

Julian Date	2454649
Delta T	+67 seconds
Ayanamsa	23° 58' 44"
Synetic vernal point	05° ♓ 08' 15"
True obliquity of ecliptic	23° 26' 24"

MOON'S PHASES, APSIDES AND POSITIONS ☽

Date	h	m	Phase	Longitude °	'	Eclipse Indicator
03	02	19	●	11 ♋	32	
10	04	35	☽	18 ♎	18	
18	07	59	○	26 ♑	04	
25	18	42	☾	03 ♉	10	

Day	h	m		
01	21	20	Perigee	
14	04	09	Apogee	
29	23	14	Perigee	
01	21	32	Max dec	27° N 32'
08	04	50	0S	
15	15	02	Max dec	27° S 34'
22	19	51	0N	
29	06	16	Max dec	27° N 36'

LONGITUDES

Date	Chiron ⚷ °	'	Ceres ⚳ °	'	Pallas ⚴ °	'	Juno ⚵ °	'	Vesta ⚶ °	'	Black Moon Lilith ⚸ °	'
01	20 ♒	27	08 ♋	34	16 ♉	50	17 ♐	58	00 ♉	27	09 ♐	13
11	20 ♒	03	12 ♋	55	20 ♉	48	16 ♐	12	03 ♉	47	10 ♐	21
21	19 ♒	36	17 ♋	16	24 ♉	43	14 ♐	55	06 ♉	47	11 ♐	28
31	19 ♒	06	21 ♋	37	28 ♉	34	14 ♐	10	09 ♉	26	12 ♐	35

All ephemeris data is given at 12.00 UT and the Moon's longitude is additionally given for 24.00 UT

ASPECTARIAN

h m	Aspects
01 Tuesday	
04 04	☽ ± ♃
09 38	☽ ⊻ ♀
12 15	☿ ⊼ ♃
12 39	☽ Q ♂
13 32	☽ ⊼ ♃
13 37	☽ ☌ ☿
20 05	☽ Q ♄
20 11	☽ □ ♅
22 12	☽ △ ♆
02 Wednesday	
07 08	☽ ☍ ♇
08 31	☽ ⚹ ♂
15 19	☽ ⚹ ♄
22 05	☽ ꝑ ♆
22 25	♀ ± ♆
03 Thursday	
02 19	☽ ☌ ☉
05 18	♀ ☍ ♃
09 31	☽ ∠ ♂
12 32	☽ ± ♆
13 03	☽ ☍ ♃
13 50	☽ ☌ ♀
15 33	☽ ∠ ♄
17 00	☽ ⊻ ☿
20 14	☽ △ ♅
22 15	☽ ⊼ ♆
04 Friday	
00 43	☽ ⊥ ♂
01 20	☽ ∥ ♀
02 12	☽ ∥ ☉
03 32	☽ ⊥ ☿
04 54	☽ # ♃
06 18	☽ ⊥ ♄
07 24	☽ ⊼ ♇
10 59	☽ ⊻ ♂
12 10	♀ ∠ ♄
14 51	☽ ∥ ☿
16 16	☽ ⊻ ♄
17 18	☽ ± ♇
19 37	☽ ∠ ☿
20 53	☽ ꝑ ♅
05 Saturday	
06 56	☽ ⊻ ☉
07 29	☽ # ♇
08 20	☽ ꝑ ♇
11 07	☿ □ ♅
12 01	☽ ± ♅
14 16	☽ ⊼ ♃
17 32	☿ Q ♄
17 52	☽ ⊥ ☉
20 18	☽ ⊻ ♀
21 16	☽ # ♆
22 16	☽ ⊼ ♅
23 20	☽ ⚹ ☿
06 Sunday	
00 22	☽ ☍ ♆
00 27	☽ ± ♃
07 08	☽ ∥ ♂
07 41	☽ ⊥ ♀
08 12	☽ ∥ ♄
08 39	☉ ∥ ♀
10 04	☽ △ ♇
10 34	♂ ꝑ ♃
10 35	☽ ∠ ☉
10 46	☿ △ ♆
12 17	♃ ⊥ ♆
16 00	☽ ꝑ ♃
16 18	☽ ☌ ♂
18 32	♀ △ ♅
19 58	☽ ☌ ♄
22 27	☽ Q ☿
07 Monday	
01 07	☽ ∠ ♀
09 09	♀ # ♃
14 21	☽ # ♅
15 23	☽ ⚹ ☉
16 00	♂ ∥ ♄
17 46	♀ ⊼ ♆
18 40	☽ △ ♃
22 02	☉ # ♃
08 Tuesday	
03 44	☽ ☍ ♅
05 57	☽ ⊼ ♆
07 15	☽ ⚹ ♀
10 57	☽ □ ☿
15 19	☽ Q ☉
16 21	☽ □ ♇
17 07	☽ ± ♆
19 37	☽ ∥ ♅
09 Wednesday	
01 53	☽ ⊻ ♂
03 30	☽ ⊻ ♄
07 39	☉ ☍ ♃
08 16	☽ Q ♀
10 10	☽ ꝑ ♆
13 55	☽ ⊥ ♂
15 04	☽ ⊥ ♄
15 10	☉ ± ♆
23 55	☿ ∥ ♀
10 Thursday	
01 30	☽ # ♂
02 42	☽ Q ♇
02 46	☽ □ ♃
03 34	☽ # ♄
04 35	☽ □ ☉
08 15	☽ ∠ ♂
08 40	☽ ∠ ♄
09 50	☿ ☍ ♇
12 57	☽ ⊼ ♅
15 14	☽ △ ♆
16 49	☽ ∥ ♆
18 11	♂ ☌ ♄
18 17	☉ ∥ ☿
23 10	☽ □ ♀
11 Friday	
00 46	☽ ± ♅
02 14	☽ ⚹ ♇
04 39	☽ △ ☿
08 51	☽ ∥ ♇
13 53	☽ Q ♃
14 34	☽ ⚹ ♄
15 28	☽ ⚹ ♂
18 39	☽ ꝑ ♅
12 Saturday	
04 48	♀ ⊼ ♇
08 07	☽ ∠ ♇
10 12	☽ # ♀
10 58	♀ ⊥ ♄
13 20	☽ # ☉
13 41	☽ ⚹ ♃
14 50	☽ Q ♄
15 00	☉ ∠ ♄
15 20	☽ ꝑ ☿
16 49	☽ Q ♂
17 51	☽ # ☿
18 10	☽ ∥ ♃
21 24	☽ △ ☉
13 Sunday	
00 45	☿ # ♃
00 47	☽ △ ♅
02 13	☽ ⊥ ♇
03 05	☽ □ ♆
12 30	☽ ± ☿
14 20	☽ ⊻ ♇
18 17	☽ △ ♀
19 40	☽ ∠ ♃
19 51	♀ ⊥ ♂
14 Monday	
02 42	☽ ⊼ ☿
03 31	☽ □ ♄
06 22	☽ ꝑ ☉
07 09	☽ □ ♂
08 36	☿ ⚹ ♄
13 40	☽ ⊥ ♃
15 03	☉ △ ♅
15 25	☽ Q ♆
15 Tuesday	
01 39	☽ ⊻ ♃
02 08	☽ ± ☉
04 06	☽ ꝑ ♀
07 48	☉ ∠ ♂
13 19	☽ □ ♅
15 17	☽ ⊼ ☉
15 34	☽ ⚹ ♆
18 50	☉ ⊼ ♆
21 11	☿ ⚹ ♂
22 27	☿ ꝑ ♆
16 Wednesday	
00 16	☽ ± ♀
02 44	☽ ☌ ♇
13 39	☽ ⊼ ♀
16 24	☽ △ ♄
21 31	☽ ∠ ♆
22 40	☽ △ ♂
23 47	♀ ± ♇
17 Thursday	
01 17	☽ Q ♅
01 48	☽ ☍ ☿
13 01	☽ ☌ ♃
15 20	☽ ⊥ ♆
18 53	♀ ⊻ ♄
22 23	☽ ꝑ ♄
18 Friday	
00 57	☽ ⚹ ♅
03 07	☽ ⊻ ♆
05 50	☽ ꝑ ♂
07 03	☽ # ☿
07 59	☽ ☍ ☉
09 46	☽ ∥ ♃
14 00	☽ ⊻ ♇
16 10	☽ ± ♄
20 28	♀ ꝑ ♅
21 40	☽ # ☉
19 Saturday	
00 18	☽ ± ♂
01 39	☽ ⊥ ♇
03 55	☽ ⊼ ♄
05 04	☽ # ♀
06 09	☽ ∠ ♅
07 14	☽ ☍ ♀
12 30	☽ ⊼ ♂
18 18	☽ ∥ ♇
18 58	☽ ∠ ♇
19 54	☿ ☍ ♃
22 48	☽ ⊻ ♃
23 23	☽ ⊼ ☿
23 26	☽ ⊥ ♅
20 Sunday	
09 26	☽ ∥ ♆
10 08	☽ ⊥ ♃
10 51	☽ ⊻ ♅
12 27	☿ ± ♆
12 54	☽ ☌ ♆
13 00	☽ ± ☿
22 24	☽ ⊼ ☉
23 25	☽ ⚹ ♇
21 Monday	
00 20	☽ # ♄
02 58	☽ ∠ ♃
09 10	☽ ꝑ ☿
10 33	☽ ± ☉
11 48	☉ ⊼ ♇
12 00	☽ # ♂
13 25	☽ ☍ ♄
21 47	☽ Q ♇
22 06	☽ ⊼ ♀
22 Tuesday	
00 00	☽ ☍ ♂
02 05	☿ # ♃
04 35	☽ ꝑ ☉
05 03	☽ ∥ ♅
06 35	☽ ⚹ ♃
10 18	☽ ± ♀
12 49	☿ ∠ ♄
18 06	☽ △ ☿
18 36	☽ ☌ ♅
20 33	☽ ⊻ ♆
21 10	☿ △ ♅
23 Wednesday	
04 16	☽ Q ♃
04 23	☽ ꝑ ♀
05 41	☉ ⊥ ♄
06 39	☽ □ ♇
07 24	☽ ⊥ ♆
09 23	☿ ⊼ ♆
10 02	☽ △ ☉
10 32	☽ # ♅
12 09	♀ ⊻ ♂
20 35	☽ ⊼ ♄
23 29	☽ ∠ ♆
24 Thursday	
00 33	☽ ∥ ♂
04 27	♀ ꝑ ♇
07 20	☽ ± ♄
08 50	☽ ⊼ ♂
09 52	☽ △ ♀
12 06	☽ □ ♃
13 26	☽ ∥ ♄
19 53	☽ ± ♂
23 16	☽ ꝑ ♄
23 59	☽ ⊻ ♅
25 Friday	
01 50	☽ ⚹ ♆
04 03	☽ # ♆
08 39	♀ ⊼ ♃
09 11	☽ □ ☿
10 22	☽ ⊥ ♅
11 30	☽ △ ♇
12 14	☽ ꝑ ♂
17 46	☽ # ♇
18 42	☽ □ ☉
19 00	☽ ∥ ♀
22 30	☽ Q ♆
26 Saturday	
00 19	☿ ⊼ ♇
01 20	☽ △ ♄
01 27	♀ ± ♅
01 48	☽ ∠ ♅
04 24	☽ ∥ ☉
12 11	☿ ∠ ♂
12 33	♀ # ♇
13 05	☽ ꝑ ♇
15 00	☽ △ ♂
15 21	☽ △ ♃
15 35	☽ ∥ ☿
18 32	☽ □ ♀
21 17	☽ Q ☿
22 04	♂ △ ♃
23 11	☽ # ♃
27 Sunday	
01 05	☿ ⊥ ♄
03 06	☽ ⚹ ♅
03 23	☽ Q ☉
04 12	☽ ± ♇
04 52	☽ □ ♆
14 11	☽ ⊼ ♇
15 24	☉ ± ♇
16 15	☽ ꝑ ♃
20 44	☽ ⚹ ☿
23 01	☽ Q ♅
28 Monday	
00 44	☽ ⚹ ☉
03 18	☽ Q ♀
04 00	☽ □ ♄
07 01	☽ ± ♃
16 48	☽ ⊼ ♃
19 02	☽ □ ♂
19 49	☿ ± ♇
29 Tuesday	
00 50	☽ ⚹ ♀
01 34	☽ ∠ ☿
03 06	☽ ∠ ☉
04 33	☽ □ ♅
06 16	☽ △ ♆
09 45	☽ Q ♄
15 25	☽ ☍ ♇
18 43	☽ ⊥ ☿
18 50	☽ ⊥ ☉
19 50	♀ ± ♃
20 04	☉ ☌ ☿
22 12	☿ ꝑ ♅
30 Wednesday	
00 40	☉ ꝑ ♅
01 22	☿ ⊻ ♄
01 44	☽ Q ♂
03 34	☽ ∠ ♀
05 18	☽ ⊻ ☉
05 29	☽ ⚹ ♄
06 08	☽ ⊻ ☿
06 43	☽ ꝑ ♆
08 29	☉ ⊻ ♄
17 28	☽ ☍ ♃
19 37	☽ ⊥ ♀
20 22	♀ ⊼ ♅
21 24	☽ ± ♆
22 13	☽ ⚹ ♂
31 Thursday	
05 31	☽ △ ♅
06 15	☽ ∠ ♄
06 22	☽ ⊻ ♀
07 14	☽ ⊼ ♆
12 12	☽ # ♃
16 29	♀ ☍ ♆
16 30	☽ ⊼ ♇
21 13	☽ ⊥ ♄

AUGUST 2008

LONGITUDES

Date	Sidereal time h m s	Sun ☉ ° ' "	Moon ☽ ° ' "	Moon ☽ 24.00 ° ' "	Mercury ☿ ° '
01	08 41 52	09 ♌ 36 06	10 ♌ 36 02	17 ♌ 43 43	12 ♌ 34
02	08 45 48	10 33 33	24 ♌ 46 42	01 ♍ 44 32	14 37
03	08 49 45	11 31 00	08 ♍ 36 53	15 ♍ 23 32	16 38
04	08 53 41	12 28 27	22 ♍ 04 24	28 ♍ 39 33	18 38
05	08 57 38	13 25 56	05 ♎ 09 05	11 ♎ 33 17	20 36
06	09 01 34	14 23 25	17 ♎ 52 26	24 ♎ 06 57	22 33
07	09 05 31	15 20 55	00 ♏ 17 18	06 ♏ 23 57	24 28
08	09 09 28	16 18 26	12 ♏ 27 27	18 ♏ 28 21	26 22
09	09 13 24	17 15 58	24 ♏ 27 13	00 ♐ 24 37	28 ♌ 14
10	09 17 21	18 13 30	06 ♐ 21 08	12 ♐ 17 18	00 ♍ 05
11	09 21 17	19 11 04	18 ♐ 13 40	24 ♐ 10 45	01 55
12	09 25 14	20 08 38	00 ♑ 09 02	06 ♑ 08 58	03 42
13	09 29 10	21 06 14	12 ♑ 11 00	18 ♑ 15 28	05 29
14	09 33 07	22 03 50	24 ♑ 22 44	00 ♒ 33 03	07 13
15	09 37 03	23 01 28	06 ♒ 46 42	13 ♒ 03 50	08 57
16	09 41 00	23 59 07	19 ♒ 24 35	25 ♒ 49 04	10 39
17	09 44 57	24 56 46	02 ♓ 17 18	08 ♓ 49 16	12 19
18	09 48 53	25 54 28	15 ♓ 24 57	22 ♓ 04 14	13 58
19	09 52 50	26 52 10	28 ♓ 47 00	05 ♈ 33 07	15 35
20	09 56 46	27 49 54	12 ♈ 22 23	19 ♈ 14 38	17 11
21	10 00 43	28 47 40	26 ♈ 09 38	03 ♉ 07 10	18 46
22	10 04 39	29 ♌ 45 28	10 ♉ 07 01	17 ♉ 08 57	20 19
23	10 08 36	00 ♍ 43 17	24 ♉ 12 41	01 ♊ 17 59	21 51
24	10 12 32	01 41 08	08 ♊ 24 34	15 ♊ 32 08	23 21
25	10 16 29	02 39 01	22 ♊ 40 23	29 ♊ 48 59	24 50
26	10 20 26	03 36 56	06 ♋ 57 33	14 ♋ 05 44	26 18
27	10 24 22	04 34 52	21 ♋ 13 06	28 ♋ 19 14	27 44
28	10 28 19	05 32 50	05 ♌ 23 41	12 ♌ 25 59	29 ♍ 08
29	10 32 15	06 30 50	19 ♌ 25 44	26 ♌ 22 28	00 ♎ 32
30	10 36 12	07 28 52	03 ♍ 15 47	10 ♍ 05 20	01 53
31	10 40 08	08 ♍ 26 55	16 ♍ 50 48	23 ♍ 31 55	03 ♎ 13

Date	Venus ♀ ° '	Mars ♂ ° '	Jupiter ♃ ° '	Saturn ♄ ° '	Uranus ♅ ° '	Neptune ♆ ° '	Pluto ♇ ° '
01	24 ♌ 15	18 ♍ 44	14 ♑ 40	07 ♍ 48	22 ♓ 10	23 ♒ 14	28 ♐ 52
02	25 29	19 22	14 R 34	07 55	22 R 09	23 R 12	28 R 51
03	26 42	19 59	14 28	08 02	22 07	23 10	28 50
04	27 56	20 36	14 22	08 09	22 06	23 09	28 49
05	29 ♌ 10	21 14	14 16	08 16	22 04	23 07	28 48
06	00 ♍ 24	21 51	14 10	08 23	22 02	23 06	28 47
07	01 37	22 29	14 04	08 30	22 00	23 04	28 46
08	02 51	23 06	13 59	08 37	21 59	23 02	28 45
09	04 05	23 44	13 54	08 45	21 57	23 01	28 44
10	05 19	24 22	13 48	08 52	21 55	22 59	28 43
11	06 32	25 00	13 43	08 59	21 53	22 57	28 42
12	07 46	25 37	13 39	09 06	21 51	22 56	28 41
13	09 00	26 15	13 34	09 14	21 49	22 54	28 41
14	10 13	26 53	13 29	09 21	21 47	22 53	28 40
15	11 27	27 31	13 25	09 28	21 45	22 51	28 39
16	12 41	28 09	13 21	09 36	21 43	22 49	28 38
17	13 54	28 47	13 16	09 43	21 41	22 48	28 38
18	15 08	29 ♍ 25	13 13	09 51	21 39	22 46	28 37
19	16 22	00 ♎ 03	13 09	09 58	21 37	22 44	28 36
20	17 35	00 41	13 05	10 05	21 35	22 43	28 36
21	18 49	01 19	13 02	10 13	21 33	22 41	28 35
22	20 03	01 58	12 59	10 20	21 31	22 39	28 35
23	21 16	02 36	12 56	10 28	21 28	22 38	28 34
24	22 30	03 14	12 53	10 35	21 26	22 36	28 34
25	23 44	03 53	12 50	10 43	21 24	22 35	28 33
26	24 57	04 31	12 47	10 50	21 22	22 33	28 33
27	26 11	05 10	12 45	10 58	21 20	22 31	28 32
28	27 25	05 48	12 43	11 05	21 17	22 30	28 32
29	28 38	06 27	12 41	11 13	21 15	22 28	28 32
30	29 ♍ 52	07 06	12 39	11 20	21 13	22 27	28 31
31	01 ♎ 05	07 ♎ 45	12 ♑ 38	11 ♍ 28	21 ♓ 10	22 ♒ 25	28 ♐ 31

Date	Moon True ☊ ° '	Moon Mean ☊ ° '	Moon ☽ Latitude ° '
01	18 ♒ 32	19 ♒ 02	00 N 44
02	18 D 32	18 59	00 S 35
03	18 32	18 56	01 49
04	18 32	18 53	02 55
05	18 33	18 49	03 51
06	18 33	18 46	04 33
07	18 34	18 43	05 01
08	18 34	18 40	05 15
09	18 R 34	18 37	05 15
10	18 34	18 33	05 02
11	18 D 34	18 30	04 36
12	18 34	18 27	03 57
13	18 34	18 24	03 08
14	18 34	18 21	02 10
15	18 35	18 18	01 S 05
16	18 R 35	18 14	00 N 05
17	18 35	18 11	01 15
18	18 34	18 08	02 23
19	18 33	18 05	03 25
20	18 32	18 02	04 16
21	18 31	17 59	04 53
22	18 30	17 55	05 13
23	18 30	17 52	05 15
24	18 D 30	17 49	04 57
25	18 30	17 46	04 22
26	18 31	17 43	03 30
27	18 32	17 39	02 25
28	18 33	17 36	01 N 12
29	18 R 34	17 33	00 S 05
30	18 33	17 30	01 20
31	18 ♒ 31	17 ♒ 27	02 S 30

DECLINATIONS

Date	Sun ☉ ° '	Moon ☽ ° '	Mercury ☿ ° '	Venus ♀ ° '	Mars ♂ ° '	Jupiter ♃ ° '	Saturn ♄ ° '	Uranus ♅ ° '	Neptune ♆ ° '	Pluto ♇ ° '
01	17 N 51	18 N 17	18 N 43	14 N 50	05 N 12	22 S 53	10 N 11	03 S 50	14 S 06	17 S 10
02	17 35	12 43	18 09	14 25	04 57	22 53	10 09	03 51	14 07	17 10
03	17 20	06 39	17 32	14 00	04 42	22 54	10 06	03 52	14 07	17 10
04	17 04	00 N 28	16 55	13 35	04 26	22 55	10 03	03 52	14 08	17 10
05	16 47	05 S 34	16 16	13 09	04 11	22 56	10 01	03 53	14 08	17 10
06	16 31	11 13	15 37	12 43	03 56	22 56	09 58	03 54	14 09	17 11
07	16 14	16 16	14 56	12 17	03 40	22 57	09 55	03 54	14 09	17 11
08	15 57	20 34	14 15	11 50	03 25	22 58	09 53	03 55	14 10	17 11
09	15 40	23 58	13 33	11 23	03 09	22 58	09 50	03 56	14 11	17 11
10	15 22	26 19	12 50	10 56	02 54	22 59	09 47	03 57	14 11	17 12
11	15 04	27 30	12 07	10 28	02 38	23 00	09 44	03 57	14 12	17 12
12	14 46	27 24	11 24	10 00	02 23	23 00	09 42	03 58	14 12	17 12
13	14 28	26 00	10 40	09 32	02 07	23 01	09 39	03 59	14 13	17 12
14	14 09	23 23	09 56	09 04	01 52	23 01	09 36	04 00	14 13	17 13
15	13 51	19 38	09 12	08 35	01 36	23 02	09 33	04 01	14 14	17 13
16	13 32	14 56	08 28	08 06	01 20	23 02	09 31	04 01	14 14	17 13
17	13 12	09 29	07 43	07 37	01 05	23 03	09 28	04 02	14 15	17 14
18	12 53	03 S 33	06 59	07 08	00 49	23 03	09 25	04 03	14 15	17 14
19	12 33	02 N 39	06 15	06 38	00 33	23 04	09 22	04 04	14 16	17 14
20	12 14	08 49	05 31	06 09	00 17	23 04	09 20	04 05	14 17	17 14
21	11 54	14 38	04 47	05 39	00 N 02	23 05	09 17	04 06	14 17	17 15
22	11 33	19 48	04 03	05 09	00 S 14	23 05	09 14	04 06	14 18	17 15
23	11 13	23 54	03 20	04 39	00 30	23 06	09 11	04 07	14 18	17 15
24	10 53	26 36	02 37	04 08	00 46	23 06	09 08	04 08	14 19	17 15
25	10 32	27 36	01 54	03 38	01 02	23 06	09 05	04 09	14 19	17 16
26	10 11	26 45	01 12	03 08	01 17	23 07	09 03	04 10	14 20	17 16
27	09 50	24 09	00 N 30	02 37	01 33	23 07	09 00	04 11	14 20	17 16
28	09 29	20 05	00 S 12	02 06	01 49	23 07	08 57	04 12	14 21	17 17
29	09 07	14 55	00 53	01 36	02 05	23 07	08 54	04 13	14 21	17 17
30	08 46	09 04	01 33	01 05	02 21	23 08	08 51	04 14	14 22	17 17
31	08 N 24	02 N 54	02 S 13	00 N 34	02 S 37	23 S 08	08 N 48	04 S 14	14 S 22	17 S 17

ZODIAC SIGN ENTRIES

Date	h	m	Planets
02	20	59	☽ ♍
05	02	28	☽ ♎
06	04	20	♀ ♍
07	11	26	☽ ♏
09	23	10	☽ ♐
10	10	51	☿ ♍
12	11	42	☽ ♑
14	22	56	☽ ♒
17	07	46	☽ ♓
19	10	03	♂ ♎
19	14	10	☽ ♈
21	18	38	☽ ♉
22	18	02	☉ ♍
23	21	48	☽ ♊
26	00	19	☽ ♋
28	02	51	☽ ♌
29	02	50	☿ ♎
30	06	18	☽ ♍
30	14	41	♀ ♎

LATITUDES

Date	Mercury ☿ ° '	Venus ♀ ° '	Mars ♂ ° '	Jupiter ♃ ° '	Saturn ♄ ° '	Uranus ♅ ° '	Neptune ♆ ° '	Pluto ♇ ° '
01	01 N 46	01 N 29	00 N 49	00 S 15	01 N 40	00 S 48	00 S 21	06 N 17
04	01 46	01 29	00 47	00 15	01 40	00 48	00 21	06 16
07	01 40	01 29	00 45	00 15	01 40	00 48	00 21	06 15
10	01 30	01 28	00 43	00 16	01 40	00 48	00 21	06 14
13	01 15	01 27	00 41	00 16	01 40	00 48	00 21	06 14
16	00 57	01 25	00 39	00 16	01 40	00 48	00 21	06 13
19	00 37	01 22	00 37	00 17	01 40	00 48	00 21	06 12
22	00 N 15	01 19	00 36	00 17	01 40	00 48	00 21	06 11
25	00 S 10	01 15	00 34	00 17	01 40	00 48	00 21	06 10
28	00 35	01 10	00 32	00 17	01 40	00 48	00 21	06 09
31	01 S 01	01 N 05	00 N 30	00 S 18	01 N 40	00 S 48	00 S 21	06 N 09

DATA

Julian Date	2454680
Delta T	+67 seconds
Ayanamsa	23° 58' 50"
Synetic vernal point	05° ♓ 08' 09"
True obliquity of ecliptic	23° 26' 24"

MOON'S PHASES, APSIDES AND POSITIONS ☽

Date	h	m	Phase	Longitude ° '	Eclipse Indicator
01	10	13	●	09 ♌ 32	Total
08	20	20	☽	16 ♏ 38	
16	21	16	○	24 ♒ 21	partial
23	23	50	☾	01 ♊ 12	
30	19	58	●	07 ♍ 48	

Day	h	m		
10	20	16	Apogee	
26	03	49	Perigee	
04	13	47	0S	
11	22	12	Max dec	27° S 37'
19	01	48	0N	
25	13	03	Max dec	27° N 36'
31	23	13	0S	

LONGITUDES

Date	Chiron ⚷ ° '	Ceres ⚳ ° '	Pallas ⚴ ° '	Juno ⚵ ° '	Vesta ⚶ ° '	Black Moon Lilith ⚸ ° '
01	19 ♒ 03	22 ♋ 03	28 ♉ 56	14 ♐ 08	09 ♉ 41	12 ♐ 42
11	18 ♒ 31	26 ♋ 23	02 ♊ 40	14 ♐ 01	11 ♉ 51	13 ♐ 49
21	18 ♒ 00	00 ♌ 41	06 ♊ 14	14 ♐ 28	13 ♉ 31	14 ♐ 56
31	17 ♒ 30	04 ♌ 57	09 ♊ 37	15 ♐ 25	14 ♉ 35	16 ♐ 03

All ephemeris data is given at 12.00 UT and the Moon's longitude is additionally given for 24.00 UT

ASPECTARIAN

h m Aspects

01 Friday
00 03 ☽ ∠ ♂
02 26 ☽ ± ♇
06 17 ☽ ₽ ♅
07 16 ☽ ⊻ ♄
09 40 ☽ ∥ ☿
10 13 ☽ ☌ ☉
14 06 ☽ ∥ ☉
14 47 ☿ ⊥ ♂
15 45 ☽ ⊥ ♂
15 52 ☽ ☌ ☿
17 07 ☽ ⋕ ♇
17 28 ☽ ₽ ♇
18 47 ☽ ⊼ ♃
21 21 ☽ ± ♅

02 Saturday
02 21 ☽ ⊻ ♂
03 00 ☿ ₽ ♇
04 25 ☽ ∥ ♀
04 52 ☽ ± ♃
06 16 ☽ ⋕ ♆
07 31 ☽ ⊼ ♅
09 18 ☽ ☍ ♆
11 27 ☿ ⊼ ♃
13 19 ☽ ☌ ♀
18 59 ☽ △ ♇
20 10 ☽ ₽ ♃
22 25 ☽ ∥ ♄

03 Sunday
05 30 ♀ ⋕ ♆
05 58 ☿ ± ♅
10 58 ☽ ☌ ♄
17 30 ☽ ⊻ ☉
19 55 ☽ ∥ ♂
22 16 ☽ △ ♃
22 48 ☽ ⋕ ♅

04 Monday
02 22 ☉ ⋕ ♇
02 23 ☿ ⋕ ♇
02 23 ☉ ∥ ☿
04 42 ☽ ⊻ ☿
05 01 ☽ ⊥ ☉
09 13 ☽ ☌ ♂
12 02 ☽ ☍ ♅
13 56 ☽ ⊼ ♆
17 27 ☽ ⊥ ☿
22 36 ☽ ∠ ☉
23 47 ☽ ⊻ ♀

05 Tuesday
00 16 ☽ □ ♇
00 52 ☽ ± ♆
04 53 ♀ △ ♇
05 08 ☽ ∥ ♅
06 35 ☽ ⋕ ♂
08 05 ☿ ± ♃
12 02 ☽ ⊥ ♀
12 59 ☽ ∠ ☿
13 49 ♀ ₽ ♃
17 32 ☽ ₽ ♆
17 52 ☽ ⊻ ♄
20 55 ☉ ₽ ♇
23 23 ☿ ⊻ ♂

06 Wednesday
04 49 ☽ ✶ ☉
05 00 ☽ □ ♃
05 17 ☽ ⊥ ♄
05 46 ☿ ⊼ ♅
06 33 ☽ ⋕ ♄
06 46 ☽ ∠ ♀
06 56 ☉ ⊼ ♃
09 55 ☽ Q ♇
14 54 ♂ ⋕ ♅
18 19 ☽ ⋕ ♀
18 38 ♂ ☍ ♅
18 40 ☿ ☍ ♆
19 58 ☽ ⊼ ♅
20 03 ☽ ⊻ ♂
22 00 ☽ △ ♆
22 37 ☽ ✶ ☿
22 41 ☽ ∠ ♄

07 Thursday
01 35 ☽ ∥ ♆
05 47 ☽ Q ☉
06 11 ☽ ⋕ ☿
07 34 ☽ ± ♅
08 17 ☽ ⊥ ♂
09 02 ☽ ✶ ♇
11 51 ☽ ⋕ ☉
14 54 ☽ ✶ ♀
15 28 ☽ Q ♃
16 45 ☽ ∥ ♇

08 Friday
01 10 ☽ ₽ ♅
02 24 ☽ Q ☿
02 54 ☽ ∠ ♂
03 59 ☉ ± ♅
04 19 ☽ ✶ ♄
09 27 ♂ ⊼ ♆
14 34 ☽ ∠ ♇
14 46 ☿ ⋕ ♆
15 01 ☽ ✶ ♃
17 18 ☽ Q ♀
20 20 ☽ □ ☉

09 Saturday
04 09 ☽ ∥ ♃
04 28 ☽ Q ♄
06 59 ☽ △ ♅
08 33 ☽ ⊥ ♇
09 06 ☽ □ ♆
10 28 ☽ ✶ ♂
18 18 ☿ △ ♇
20 04 ☿ ₽ ♃
20 36 ☽ ⊻ ♇
20 53 ☽ ∠ ♃
21 02 ☽ □ ☿

10 Sunday
09 39 ☽ □ ♀
12 01 ☽ Q ♂
14 55 ☽ ⊥ ♃
17 08 ☽ □ ♄
21 21 ☽ Q ♆
22 00 ☉ ⊥ ♂

11 Monday
02 58 ☽ ⊻ ♃
14 06 ☽ △ ☉
19 21 ☽ □ ♅
21 31 ☽ ✶ ♆

12 Tuesday
00 25 ☉ ± ♃
02 24 ☽ □ ♂
09 04 ☽ ☌ ♇
20 22 ☽ △ ☿
22 52 ☽ ₽ ☉

13 Wednesday
03 31 ☽ ∠ ♆
04 57 ☽ △ ♀
05 29 ♀ ∥ ♄
06 04 ☽ △ ♄
07 19 ☽ Q ♅
14 43 ☽ ☌ ♃
17 03 ♀ ☌ ♄
18 17 ☽ ± ☉
21 19 ☽ ⊥ ♆

14 Thursday
05 20 ☉ ⊼ ♅
06 57 ☽ ✶ ♅
07 05 ☉ ⋕ ♆
07 05 ☽ ⊼ ☉
07 05 ☽ ₽ ☿
09 04 ☽ ⊻ ♆
11 57 ☽ ₽ ♄
13 50 ☽ ₽ ♀
14 35 ☽ ∥ ♃
17 09 ☽ △ ♂
20 20 ☽ ⊻ ♇
23 33 ☿ ∥ ♄

15 Friday
03 27 ☽ ± ☿
05 35 ☽ ± ♄
07 43 ☉ ☍ ♆
07 55 ☽ ⊥ ♇
09 10 ☽ ± ♀
11 57 ☽ ∠ ♅
16 48 ☽ ⊼ ☿
17 12 ☽ ⊼ ♄
19 59 ☿ ☌ ♄
21 54 ☽ ⊼ ♀
23 32 ☽ ₽ ♂

16 Saturday
00 36 ☽ ⊻ ♃
00 50 ☽ ∥ ♇
01 06 ☽ ∠ ♇
05 03 ☽ ⊥ ♅
11 52 ☽ ⊥ ♃
15 12 ☽ ∥ ♆
16 20 ☽ ⊻ ♅
17 25 ☽ ± ♂
18 23 ☽ ☌ ♆
18 52 ☽ ⋕ ☉
21 16 ☽ ☍ ☉

17 Sunday
00 16 ♀ △ ♃
04 36 ☽ ∠ ♃
05 11 ☽ ⊼ ♂
05 14 ☽ ✶ ♇
06 15 ♂ □ ♇
12 06 ☽ ⋕ ♄
12 25 ♂ ± ♆
20 21 ☽ ⋕ ☿
20 26 ☽ ⋕ ♀
22 15 ☿ ∥ ♀

18 Monday
01 23 ☿ △ ♃
01 46 ☽ ☍ ♄
03 17 ☽ Q ♇
08 01 ☽ ✶ ♃
09 00 ☽ ☍ ☿
10 02 ☽ ∥ ♅
11 26 ☽ ☍ ♀
23 07 ☽ ⋕ ♂
23 13 ☽ ☌ ♅

19 Tuesday
01 13 ☽ ⊻ ♆
04 16 ☽ ∥ ♂
05 32 ☽ Q ♃
08 20 ☽ ⊼ ☉
11 41 ☽ □ ♇
11 55 ☽ ⊥ ♆
14 22 ☽ ☍ ♂
17 29 ☽ ⋕ ♅
19 49 ☽ ± ☉

20 Wednesday
00 28 ☽ ∥ ☿
02 19 ☽ ∥ ♀
03 50 ☽ ∠ ♆
07 57 ☽ ⊼ ♄
12 52 ☽ ₽ ☉
13 15 ☽ □ ♃
14 02 ☽ ∥ ♄
18 00 ☉ ₽ ♃
18 34 ☽ ± ♄
21 31 ☽ ⊼ ☿
22 01 ☽ ⊼ ♀

21 Thursday
01 04 ☽ ∥ ☉
04 01 ☽ ⊻ ♅
05 59 ☽ ✶ ♆
06 51 ☉ △ ♇
09 17 ☽ ± ☿
09 27 ☽ ± ♀
10 21 ☽ ₽ ♄
10 28 ☽ ⋕ ♆
14 23 ☽ ⊥ ♅
15 44 ☿ ☌ ♀
16 11 ☽ △ ♇
16 53 ☽ △ ☉
21 20 ☽ ⊼ ♂
23 39 ☽ ⋕ ♇

22 Friday
02 28 ☽ ₽ ♀
02 40 ☽ Q ♆
02 45 ☽ ₽ ☿
05 50 ☽ ∠ ♅
08 08 ☽ ± ♂
10 25 ☿ ⋕ ♅
12 23 ☽ △ ♄
16 52 ☽ △ ♃
17 54 ☽ ₽ ♇

23 Saturday
00 14 ☽ ₽ ♂
06 13 ☿ ☍ ♅
06 32 ☽ △ ♀
06 37 ☽ ⋕ ♃
07 22 ☽ ✶ ♅
07 31 ☽ △ ☿
09 13 ☽ ± ♇
09 19 ☽ □ ♆
15 49 ♀ ☍ ♅
18 16 ☽ ₽ ♃
19 23 ☽ ⊼ ♇
23 50 ☽ □ ☉

24 Sunday
00 11 ☿ ⊼ ♆
02 52 ☽ △ ♂
03 38 ☽ Q ♅
09 26 ☽ ± ♃
12 15 ♀ ⋕ ♅
13 58 ♀ ⊼ ♆
15 42 ☽ □ ♄
19 30 ☽ ⊼ ♃

25 Monday
08 21 ☽ Q ☉
09 52 ☽ □ ♅
11 50 ☽ △ ♆
13 56 ☽ □ ♀
16 03 ☽ □ ☿
21 52 ☽ ☍ ♇
22 14 ☽ Q ♄

26 Tuesday
05 58 ☽ ✶ ☉
07 43 ☽ □ ♂
09 41 ☿ ⋕ ♂
12 59 ☽ ₽ ♆
18 35 ☽ ✶ ♄
21 47 ☽ ☍ ♃
23 02 ☽ Q ♀

27 Wednesday
01 43 ☽ Q ☿
04 06 ☽ ± ♆
09 02 ☽ ∠ ☉
12 11 ☽ △ ♅
14 12 ☽ ⊼ ♆
15 26 ☽ Q ♂
18 57 ☽ ⋕ ♃
20 05 ☽ ∠ ♄
21 11 ☽ ✶ ♀

28 Thursday
00 13 ☽ ✶ ☿
00 22 ☽ ⊼ ♇
01 11 ☿ ± ♆
01 21 ☽ ⊥ ☉
01 37 ☿ □ ♇
10 32 ☽ ± ♇
11 29 ☽ ⊥ ♄
12 17 ☽ ⊻ ☉
12 44 ☽ ✶ ♂
13 31 ☽ ₽ ♅
20 53 ♀ ⋕ ♂
21 48 ☽ ⊻ ♄

29 Friday
00 27 ☽ ⊼ ♃
01 06 ☽ ∠ ♀
01 32 ☽ ⋕ ♇
01 52 ☽ ₽ ♇
04 34 ☽ ∠ ☿
04 50 ☽ ± ♅
07 24 ☉ ⊻ ♂
08 50 ♀ ± ♆
09 51 ♀ □ ♇
10 43 ☽ ± ♃
14 24 ☽ ⋕ ♆
15 08 ☽ ⊼ ♅
15 39 ☽ ∠ ♂
17 14 ☽ ☍ ♆
18 04 ☽ ⊥ ♀
21 46 ☽ ⊥ ☿

30 Saturday
02 15 ☽ ₽ ♃
02 26 ☿ ⋕ ♀
03 44 ☽ △ ♇
04 54 ☉ ∥ ♄
05 29 ☽ ⊻ ♀
08 02 ☽ ⊥ ♂
09 20 ☽ ⊻ ☿
12 49 ☽ ∥ ♄
13 15 ☽ ∥ ☉
19 04 ☽ ⊻ ♂
19 58 ☽ ☌ ☉

31 Sunday
00 25 ♂ ₽ ♆
02 21 ☽ ☌ ♄
04 31 ☽ △ ♃
06 49 ☽ ⋕ ♅
13 03 ☽ ⋕ ♂
14 22 ☽ ⋕ ☿
19 44 ☽ ☍ ♅
21 51 ☽ ∥ ♀
21 58 ☽ ⊼ ♆

SEPTEMBER 2008

LONGITUDES

Date	Sidereal time h m s	Sun ☉ ° ' "	Moon ☽ ° ' "	Moon ☽ 24.00 ° ' "	Mercury ☿ ° '	Venus ♀ ° '	Mars ♂ ° '	Jupiter ♃ ° '	Saturn ♄ ° '	Uranus ♅ ° '	Neptune ♆ ° '	Pluto ♇ ° '
01	10 44 05	09 ♍ 24 59	00 ♎ 08 33	06 ♎ 40 33	04 ♎ 32	02 ♎ 19	08 ♎ 23	12 ♑ 36	11 ♍ 36	21 ♓ 08	22 ♒ 24	28 ♐ 31
02	10 48 01	10 23 05	13 ♎ 07 56	19 ♎ 30 46	05 48	03 33	09 02	12 R 35	11 43	21 R 06	22 R 22	28 R 30
03	10 51 58	11 21 13	25 ♎ 49 10	02 ♏ 03 22	07 03	04 46	09 41	12 34	11 51	21 03	22 20	28 30
04	10 55 55	12 19 22	08 ♏ 13 40	14 ♏ 20 26	08 17	06 00	10 20	12 33	11 58	21 01	22 19	28 30
05	10 59 51	13 17 33	20 ♏ 24 06	26 ♏ 25 07	09 28	07 13	10 59	12 33	12 06	20 59	22 17	28 30
06	11 03 48	14 15 46	02 ♐ 24 00	08 ♐ 21 20	10 38	08 27	11 38	12 32	12 13	20 56	22 16	28 30
07	11 07 44	15 13 59	14 ♐ 17 40	20 ♐ 13 36	11 45	09 40	12 17	12 32	12 21	20 54	22 14	28 30
08	11 11 41	16 12 15	26 ♐ 09 45	02 ♑ 06 44	12 50	10 54	12 56	12 D 32	12 29	20 51	22 13	28 30
09	11 15 37	17 10 32	08 ♑ 05 08	14 ♑ 05 34	13 53	12 07	13 36	12 32	12 36	20 49	22 12	28 D 30
10	11 19 34	18 08 50	20 ♑ 08 35	26 ♑ 14 42	14 54	13 21	14 15	12 33	12 44	20 47	22 10	28 30
11	11 23 30	19 07 10	02 ♒ 24 27	08 ♒ 38 13	15 52	14 34	14 54	12 33	12 51	20 44	22 09	28 30
12	11 27 27	20 05 32	14 ♒ 56 25	21 ♒ 19 20	16 48	15 48	15 33	12 34	12 59	20 42	22 07	28 30
13	11 31 24	21 03 55	27 ♒ 47 11	04 ♓ 20 05	17 40	17 01	16 13	12 35	13 06	20 39	22 06	28 30
14	11 35 20	22 02 20	10 ♓ 58 04	17 ♓ 41 02	18 29	18 14	16 52	12 36	13 14	20 37	22 05	28 30
15	11 39 17	23 00 46	24 ♓ 28 48	01 ♈ 21 05	19 15	19 28	17 32	12 37	13 21	20 35	22 03	28 30
16	11 43 13	23 59 15	08 ♈ 17 30	15 ♈ 17 34	19 58	20 41	18 11	12 39	13 29	20 32	22 02	28 31
17	11 47 10	24 57 45	22 ♈ 20 45	29 ♈ 26 29	20 36	21 54	18 51	12 40	13 36	20 30	22 01	28 31
18	11 51 06	25 56 18	06 ♉ 34 08	13 ♉ 43 05	21 11	23 08	19 31	12 42	13 44	20 28	21 59	28 31
19	11 55 03	26 54 53	20 ♉ 52 45	28 ♉ 02 35	21 41	24 21	20 10	12 44	13 51	20 25	21 58	28 31
20	11 58 59	27 53 30	05 ♊ 12 03	12 ♊ 20 44	22 06	25 34	20 50	12 46	13 58	20 23	21 57	28 32
21	12 02 56	28 52 09	19 ♊ 28 16	26 ♊ 34 19	22 25	26 48	21 30	12 49	14 06	20 20	21 55	28 32
22	12 06 53	29 ♍ 50 51	03 ♋ 38 41	10 ♋ 41 09	22 40	28 01	22 10	12 52	14 13	20 18	21 54	28 33
23	12 10 49	00 ♎ 49 35	17 ♋ 41 36	24 ♋ 39 56	22 48	29 ♎ 14	22 50	12 54	14 21	20 16	21 53	28 33
24	12 14 46	01 48 21	01 ♌ 36 03	08 ♌ 29 55	22 R 50	00 ♏ 28	23 30	12 57	14 28	20 13	21 52	28 33
25	12 18 42	02 47 09	15 ♌ 21 33	22 ♌ 10 29	22 45	01 41	24 10	13 00	14 35	20 11	21 51	28 34
26	12 22 39	03 46 00	28 ♌ 57 01	05 ♍ 40 55	22 34	02 54	24 50	13 04	14 43	20 09	21 50	28 35
27	12 26 35	04 44 52	12 ♍ 22 02	19 ♍ 00 15	22 15	04 07	25 30	13 07	14 50	20 06	21 49	28 35
28	12 30 32	05 43 47	25 ♍ 35 24	02 ♎ 07 21	21 48	05 20	26 10	13 11	14 57	20 04	21 47	28 36
29	12 34 28	06 42 44	08 ♎ 35 59	15 ♎ 01 12	21 14	06 34	26 50	13 15	15 04	20 02	21 46	28 36
30	12 38 25	07 ♎ 41 43	21 ♎ 22 55	27 ♎ 41 07	20 ♎ 32	07 ♏ 47	27 ♎ 31	13 ♑ 19	15 ♍ 12	20 ♓ 00	21 ♒ 45	28 ♐ 37

DECLINATIONS

Date	Moon True ☊ ° '	Moon Mean ☊ ° '	Moon ☽ Latitude ° '	Sun ☉ ° '	Moon ☽ ° '	Mercury ☿ ° '	Venus ♀ ° '	Mars ♂ ° '	Jupiter ♃ ° '	Saturn ♄ ° '	Uranus ♅ ° '	Neptune ♆ ° '	Pluto ♇ ° '
01	18 ♒ 29	17 ♒ 24	03 S 29	08 N 02	03 S 15	02 S 52	00 N 03	02 S 53	23 S 08	08 N 46	04 S 15	14 S 23	17 S 18
02	18 R 26	17 20	04 17	07 40	09 07	03 31	00 S 28	03 08	23 08	08 43	04 16	14 23	17 18
03	18 22	17 17	04 50	07 18	14 28	04 09	00 59	03 24	23 08	08 40	04 17	14 24	17 18
04	18 19	17 14	05 09	06 56	19 07	04 46	01 30	03 40	23 09	08 37	04 18	14 24	17 19
05	18 16	17 11	05 14	06 34	22 53	05 22	02 01	03 56	23 09	08 34	04 19	14 25	17 19
06	18 14	17 08	05 04	06 12	25 37	05 58	02 32	04 12	23 09	08 31	04 20	14 25	17 19
07	18 14	17 05	04 42	05 49	27 11	06 32	03 02	04 28	23 09	08 29	04 21	14 26	17 19
08	18 D 14	17 01	04 08	05 27	27 31	07 06	03 33	04 43	23 09	08 26	04 22	14 26	17 20
09	18 15	16 58	03 23	05 04	26 34	07 38	04 04	04 59	23 09	08 23	04 23	14 27	17 20
10	18 17	16 55	02 28	04 41	24 22	08 09	04 35	05 15	23 09	08 20	04 24	14 27	17 20
11	18 19	16 52	01 26	04 18	21 01	08 39	05 05	05 31	23 09	08 17	04 25	14 28	17 21
12	18 20	16 49	00 S 19	03 55	16 39	09 08	05 36	05 46	23 09	08 14	04 26	14 28	17 21
13	18 R 19	16 45	00 N 51	03 33	11 26	09 35	06 06	06 02	23 09	08 12	04 27	14 29	17 21
14	18 17	16 42	02 00	03 10	05 S 36	10 01	06 36	06 18	23 09	08 09	04 28	14 29	17 22
15	18 14	16 39	03 04	02 46	00 N 38	10 26	07 06	06 34	23 09	08 06	04 29	14 30	17 22
16	18 09	16 36	03 59	02 23	06 56	10 48	07 36	06 49	23 09	08 03	04 29	14 30	17 22
17	18 03	16 33	04 40	02 00	13 01	11 09	08 06	07 05	23 09	08 00	04 30	14 30	17 22
18	17 58	16 30	05 04	01 37	18 29	11 27	08 36	07 20	23 09	07 58	04 31	14 31	17 23
19	17 53	16 26	05 09	01 14	22 57	11 44	09 05	07 36	23 09	07 55	04 32	14 31	17 23
20	17 49	16 23	04 56	00 50	26 01	11 58	09 35	07 51	23 08	07 52	04 33	14 32	17 23
21	17 47	16 20	04 24	00 27	27 25	12 09	10 04	08 07	23 08	07 49	04 34	14 32	17 24
22	17 D 47	16 17	03 36	00 N 04	27 00	12 17	10 33	08 22	23 08	07 47	04 35	14 32	17 24
23	17 48	16 14	02 36	00 S 20	24 51	12 23	11 01	08 38	23 08	07 44	04 36	14 33	17 24
24	17 49	16 10	01 27	00 43	21 13	12 25	11 30	08 53	23 08	07 41	04 37	14 33	17 25
25	17 50	16 07	00 N 13	01 06	16 27	12 24	11 58	09 08	23 07	07 38	04 38	14 34	17 25
26	17 R 49	16 04	01 S 00	01 30	10 54	12 19	12 26	09 24	23 07	07 36	04 39	14 34	17 25
27	17 47	16 01	02 08	01 53	04 N 57	12 09	12 54	09 39	23 07	07 33	04 40	14 34	17 25
28	17 42	15 58	03 09	02 17	01 S 08	11 56	13 21	09 54	23 07	07 30	04 40	14 35	17 26
29	17 35	15 55	03 59	02 40	07 04	11 38	13 48	10 09	23 06	07 27	04 41	14 35	17 26
30	17 ♒ 26	15 ♒ 51	04 S 35	03 S 03	12 S 35	11 S 16	14 S 15	10 S 24	23 S 06	07 N 25	04 S 42	14 S 35	17 S 26

ZODIAC SIGN ENTRIES

Date	h m	Planets
01	11 44	☽ ♎
03	20 02	☽ ♏
06	07 11	☽ ♐
08	19 45	☽ ♑
11	07 20	☽ ♒
13	16 04	☽ ♓
15	21 39	☽ ♈
18	00 57	☽ ♉
20	03 17	☽ ♊
22	05 49	☽ ♋
22	15 44	☉ ♎
24	02 59	♀ ♏
24	09 13	☽ ♌
26	13 52	☽ ♍
28	20 05	☽ ♎

LATITUDES

Date	Mercury ☿ ° '	Venus ♀ ° '	Mars ♂ ° '	Jupiter ♃ ° '	Saturn ♄ ° '	Uranus ♅ ° '	Neptune ♆ ° '	Pluto ♇ ° '
01	01 S 10	01 N 04	00 N 30	00 S 18	01 N 40	00 S 48	00 S 21	06 N 08
04	01 37	00 58	00 28	00 18	01 40	00 48	00 21	06 07
07	02 03	00 52	00 26	00 18	01 41	00 48	00 21	06 06
10	02 29	00 45	00 24	00 18	01 41	00 48	00 21	06 06
13	02 53	00 38	00 22	00 19	01 41	00 48	00 21	06 05
16	03 14	00 30	00 20	00 19	01 41	00 48	00 21	06 04
19	03 32	00 23	00 19	00 19	01 42	00 48	00 21	06 03
22	03 45	00 14	00 17	00 19	01 42	00 48	00 21	06 02
25	03 49	00 N 06	00 15	00 19	01 42	00 48	00 21	06 01
28	03 43	00 S 03	00 13	00 19	01 43	00 48	00 21	06 00
31	03 S 21	00 S 12	00 N 11	00 S 20	01 N 43	00 S 48	00 S 21	05 N 59

DATA

Julian Date	2454711
Delta T	+67 seconds
Ayanamsa	23° 58' 54"
Synetic vernal point	05° ♓ 08' 06"
True obliquity of ecliptic	23° 26' 25"

MOON'S PHASES, APSIDES AND POSITIONS ☽

Date	h m	Phase	Longitude ° '	Eclipse Indicator
07	14 04	☽	15 ♐ 19	
15	09 13	○	22 ♓ 54	
22	05 04	☾	29 ♊ 34	
29	08 12	●	06 ♎ 33	

Day	h m		
07	14 57	Apogee	
20	03 38	Perigee	
08	06 12	Max dec	27° S 33'
15	09 37	0N	
21	18 28	Max dec	27° N 28'
28	07 29	0S	

LONGITUDES

Date	Chiron ⚷ ° '	Ceres ⚳ ° '	Pallas ⚴ ° '	Juno ⚵ ° '	Vesta ⚶ ° '	Black Moon Lilith ⚸ ° '
01	17 ♒ 27	05 ♌ 23	09 ♊ 56	15 ♐ 33	14 ♉ 39	16 ♐ 09
11	17 ♒ 01	09 ♌ 34	13 ♊ 01	17 ♐ 01	14 ♉ 58	17 ♐ 17
21	16 ♒ 38	13 ♌ 42	15 ♊ 45	18 ♐ 53	14 ♉ 34	18 ♐ 24
31	16 ♒ 20	17 ♌ 43	18 ♊ 00	21 ♐ 07	13 ♉ 24	19 ♐ 31

ASPECTARIAN

01 Monday
00 23 ☽ # ♀
08 49 ☽ ± ♆
09 02 ☽ □ ♇
10 18 ☽ II ☿
10 25 ☽ II ♂
12 14 ☿ II ♂
16 00 ☽ II ♅
16 23 ☽ ☌ ♀
20 55 ☽ ☌ ☿

02 Tuesday
01 18 ☽ ⚼ ♆
03 58 ☽ ☌ ♂
06 17 ☽ # ☉
06 28 ☽ ⊻ ☉
09 20 ☽ ⊻ ♄
10 17 ☽ # ♄
10 59 ☽ □ ♃
18 20 ☽ Q ♇
18 36 ☽ ⊥ ☉
20 42 ☽ ⊥ ♄

03 Wednesday
02 57 ☽ ⊼ ♅
05 23 ☽ △ ♆
11 38 ☽ II ♆
13 07 ☽ ∠ ☉
13 59 ☽ ∠ ♄
14 22 ☽ ± ♅
17 09 ☽ ⚹ ♇
17 25 ☿ ⚼ ♆
17 37 ☿ II ♅
21 07 ☽ Q ♃

04 Thursday
02 00 ☉ ☌ ♄
02 09 ☽ II ♇
07 10 ☽ ⊻ ♀
07 42 ☽ ⚼ ♅
12 07 ☽ ⊻ ☿
16 21 ☽ ⊻ ♂
17 41 ☉ △ ♃
19 25 ☽ ⚹ ♄
20 12 ☽ ⊥ ♀
20 29 ☽ ⚹ ♃
20 43 ☽ ⚹ ☉
22 21 ☽ ∠ ♇

05 Friday
01 10 ☽ ⊥ ☿
04 50 ☽ ⊥ ♂
13 08 ☽ △ ♅
13 23 ♀ ⚼ ♆
13 59 ☽ II ♃
15 45 ☽ □ ♆
16 02 ☽ ∠ ♀
16 10 ☽ ⊥ ♇
19 26 ☽ Q ♄
20 59 ☽ ∠ ☿
22 36 ☽ Q ☉
23 46 ☽ ∠ ♂

06 Saturday
02 15 ☽ ∠ ♃
04 10 ☽ ⊻ ♇
17 54 ☉ # ☿
20 20 ☽ ⊥ ♃

07 Sunday
01 26 ♂ II ♅
01 35 ☽ ⚹ ♀
03 49 ☽ Q ♆
06 20 ☽ ⚹ ☿
07 42 ☽ ⚹ ♂
08 02 ☽ □ ♄
08 27 ☽ ⊻ ♃
14 04 ☽ □ ☉
14 52 ♂ ⊻ ♄
21 05 ♂ □ ♃

08 Monday
01 19 ☽ □ ♅
02 47 ☿ ⊻ ♄
04 03 ☽ ⚹ ♆
04 13 ♃ St D
04 38 ☽ Q ♀
05 09 ☿ □ ♃
09 04 ☽ Q ☿
09 23 ☽ Q ♂
16 43 ☽ ☌ ♇
17 44 ☿ ☌ ♂
23 18 ♃ △ ♄

09 Tuesday
03 13 ♇ St D
10 13 ☽ ∠ ♆
13 28 ☽ Q ♅
14 58 ☉ # ♂
20 12 ♀ □ ♃
20 54 ☽ ☌ ♃
20 59 ☽ □ ♀
21 07 ☽ △ ♄
22 31 ♀ ⊻ ♄
23 38 ☽ □ ♂

10 Wednesday
00 40 ☽ □ ☿
03 18 ♀ II ♅
04 09 ☽ ⊥ ♆
07 42 ☽ △ ☉
13 15 ☽ ⚹ ♅
14 58 ☉ # ♀
15 59 ☽ ⊻ ♆
19 51 ☿ # ♄
21 40 ☽ II ♃

11 Thursday
03 03 ☽ ⚼ ♄
04 24 ☽ ⊻ ♇
05 32 ☉ # ♅
15 36 ☽ ⚼ ☉
16 02 ☽ ⊥ ♇
18 25 ☽ ∠ ♅
20 40 ☽ ± ♄

12 Friday
02 06 ♀ ☌ ♂
04 11 ☿ Q ♇
07 29 ☽ ⊻ ♃
08 15 ☽ ⊼ ♄
08 29 ☽ II ♇
09 16 ☽ ∠ ♇
10 16 ☽ ± ☉
11 33 ☽ ⊥ ♅
13 14 ☽ △ ♂
13 47 ☽ △ ♀
15 46 ☽ △ ☿
18 50 ☽ ⊥ ♃
22 27 ☽ II ♆
22 30 ☽ ⊼ ☉
22 48 ☽ ⊻ ♅

13 Saturday
01 28 ☽ ☌ ♆
01 54 ♀ Q ♇
02 21 ☉ ☍ ♅
05 34 ♀ II ♂
11 37 ☽ ∠ ♃
13 19 ☽ ⚹ ♇
18 38 ☽ ⚼ ♂
19 16 ☽ II ☿
20 34 ☽ ⚼ ♀
21 34 ☽ ⚼ ☿
22 31 ♂ Q ♇

14 Sunday
01 41 ☽ # ♄
08 18 ☽ II ♀
09 20 ☽ II ♂
11 10 ☽ Q ♇
11 49 ☽ ± ♂
12 54 ☉ ⊼ ♆
14 31 ☽ ± ♀
14 54 ☽ ± ☿
14 56 ☽ ⚹ ♃
16 06 ☽ ☍ ♄
16 27 ☽ II ♅
22 09 ☽ # ☉
23 06 ☽ ⊼ ♂

15 Monday
01 37 ☿ ☌ ♀
02 15 ☽ ⊼ ☿
02 16 ☽ ⊼ ♀
05 09 ☽ ☌ ♅
07 44 ☽ ⊻ ♆
09 13 ☽ ☍ ☉
09 38 ♀ ⊥ ♄
12 15 ☽ Q ♃
15 47 ☿ ⊥ ♄
18 15 ☽ ⊥ ♆
19 03 ☽ □ ♇
19 41 ☽ II ☉

16 Tuesday
02 38 ☽ # ♅
09 13 ♀ ⊼ ♅
09 50 ☽ ∠ ♆
11 31 ☽ # ♂
14 47 ☽ # ♀
16 15 ☽ II ♄
19 29 ☽ □ ♃
20 59 ☽ ⊼ ♄

17 Wednesday
03 58 ☽ # ☿
05 46 ☽ ☍ ♂
07 18 ☽ ± ♄
07 41 ♀ # ♄
08 02 ☿ ⊼ ♅
08 52 ☽ ⊻ ♅
08 55 ☽ ☍ ☿
11 11 ☽ ☍ ♀
11 26 ☽ ⚹ ♆
13 58 ♀ △ ♆
16 46 ☽ ⊼ ☉
18 13 ☽ # ♆
19 01 ☽ ⊥ ♅
22 26 ☽ △ ♇
22 41 ☽ ⚼ ♄

18 Thursday
03 38 ☽ ± ☉
06 49 ☽ # ♇
07 40 ☽ Q ♆
10 08 ☽ ∠ ♅
19 52 ☽ ⚼ ☉
21 40 ♂ ⊥ ♄
22 19 ☽ △ ♃
23 40 ☽ ⚼ ♇

19 Friday
00 07 ☽ △ ♄
10 45 ☽ ⊼ ♂
11 14 ☽ ⚹ ♅
13 15 ☽ # ♃
13 23 ☽ ⊼ ☿
13 49 ☽ □ ♆
14 45 ☽ ± ♇
18 21 ☽ ⊼ ♀
20 26 ♂ ⊼ ♅
21 18 ☽ ± ♂
22 51 ☽ △ ☉
23 31 ☽ ⚼ ♃
23 45 ☽ ± ☿

20 Saturday
00 49 ☽ ⊼ ♇
03 21 ☿ △ ♆
05 21 ☽ ± ♀
07 17 ☽ Q ♅
12 54 ♂ # ♄
13 07 ☽ ⚼ ♂
13 18 ☉ ± ♆
14 39 ☽ ± ♃
15 16 ☽ ⚼ ☿
21 52 ☽ ⚼ ♀

21 Sunday
00 45 ☽ ⊼ ♃
02 52 ☽ □ ♄
03 20 ♀ ± ♅
03 46 ☉ □ ♇
13 28 ☽ □ ♅
15 35 ☽ △ ♂
16 08 ☽ △ ♆
17 05 ☽ △ ☿

22 Monday
01 33 ☽ △ ♀
02 57 ♂ △ ♆
03 20 ☽ ☍ ♇
05 04 ☽ □ ☉
09 34 ☽ Q ♄
17 32 ☽ ⚼ ♆
22 24 ♀ ⚹ ♇

23 Tuesday
03 46 ☽ ☍ ♃
06 12 ☽ ⚹ ♄
08 54 ☽ ± ♆
10 51 ☿ ☌ ♂
14 06 ☽ Q ☉
14 20 ♀ ∠ ♄
16 24 ☽ △ ♅
19 12 ☽ ⊼ ♆
20 50 ☽ □ ☿
21 17 ☽ □ ♂

24 Wednesday
00 25 ☽ # ♃
06 43 ☽ ⊼ ♇
07 16 ☿ St R
08 16 ☽ ∠ ♄
09 50 ☽ □ ♀
12 23 ☽ ⚹ ☉
17 08 ☽ ± ♇
18 16 ☽ ⚼ ♅
22 12 ♀ Q ♃

25 Thursday
00 03 ☽ ⊥ ♄
04 01 ☽ Q ☿
06 07 ☽ Q ♂
07 29 ☽ # ♇
07 52 ☽ ⊼ ♃
08 52 ☽ ⚼ ♇
09 57 ☽ ± ♅
10 38 ☽ ⊻ ♄
16 36 ☽ ∠ ☉
18 27 ☽ ± ♃
20 21 ☽ Q ♀
20 27 ☽ # ♆
20 28 ☽ ⊼ ♅
23 24 ☽ ☍ ♆

26 Friday
00 52 ☽ ⚹ ☿
04 19 ☽ ⚹ ♂
06 01 ☽ # ☿
06 06 ☽ # ♀
06 55 ☿ II ♀
09 44 ☽ ⊥ ☉
10 25 ☽ ⚼ ♃
11 20 ☽ △ ♇
17 56 ☽ # ♂
19 44 ☽ ⚹ ♀
21 15 ☽ ⊻ ☉

27 Saturday
01 34 ☽ II ♄
03 02 ☽ ∠ ☿
08 28 ☽ ∠ ♂
13 07 ☽ # ♅
13 22 ☽ △ ♃
16 29 ☽ ☌ ♄
18 47 ☽ ⊥ ☿
23 19 ☽ # ☉

28 Sunday
01 27 ☽ ∠ ♀
01 35 ☽ ⊥ ♂
01 58 ☽ ☍ ♅
05 05 ☽ ⊼ ♆
05 20 ☽ ⊻ ☿
06 48 ♀ ⚼ ♅
08 33 ♂ ± ♅
12 25 ☿ △ ♆
13 07 ☽ ⊻ ♂
16 02 ☽ ± ♆
16 50 ☽ II ☉
17 31 ☽ □ ♇
19 35 ☽ ⊥ ♀

29 Monday
02 13 ☽ II ♅
07 49 ☽ ⊻ ♀
08 12 ☽ ☌ ☉
08 37 ☽ ⚼ ♆
13 29 ☉ ⚼ ♆
13 38 ☽ # ♄
17 00 ☿ ⊥ ♄
20 44 ☽ □ ♃

30 Tuesday
00 13 ☽ ⊻ ♄
01 47 ☽ II ♂
03 00 ☽ Q ♇
03 25 ☉ ⊻ ♀
06 28 ☽ II ☿
09 23 ☽ ⊼ ♅
10 30 ☽ ☌ ☿
11 38 ☽ ⊥ ♄
12 43 ☽ △ ♆
20 29 ☽ II ♀
20 44 ☽ ⊥ ♅
21 24 ☽ II ♅

All ephemeris data is given at 12.00 UT and the Moon's longitude is additionally given for 24.00 UT

OCTOBER 2008

LONGITUDES

Date	Sidereal time h m s	Sun ☉ ° ' "	Moon ☽ ° ' "	Moon ☽ 24.00 ° ' "	Mercury ☿ ° '	Venus ♀ ° '	Mars ♂ ° '	Jupiter ♃ ° '	Saturn ♄ ° '	Uranus ♅ ° '	Neptune ♆ ° '	Pluto ♇ ° '
01	12 42 22	08 ♎ 40 43	03 ♏ 55 51	10 ♏ 07 11	19 ♎ 44	09 ♏ 00	28 ♎ 11	13 ♑ 23	15 ♍ 19	19 ♓ 57	21 ♒ 44	28 ♐ 38
02	12 46 18	09 39 46	16 ♏ 15 17	22 ♏ 20 21	18 R 49	10 13	28 52	13 28	15 26	19 R 55	21 R 43	28 38
03	12 50 15	10 38 51	28 ♏ 22 42	04 ♐ 22 38	17 48	11 26	29 ♎ 32	13 32	15 33	19 53	21 43	28 39
04	12 54 11	11 37 57	10 ♐ 20 36	16 ♐ 17 01	16 43	12 39	00 ♏ 13	13 37	15 40	19 51	21 42	28 40
05	12 58 08	12 37 06	22 ♐ 12 26	28 ♐ 07 24	15 34	13 52	00 53	13 42	15 47	19 49	21 41	28 41
06	13 02 04	13 36 16	04 ♑ 02 30	09 ♑ 58 22	14 24	15 05	01 34	13 47	15 54	19 46	21 40	28 42
07	13 06 01	14 35 28	15 ♑ 55 39	21 ♑ 55 01	13 14	16 18	02 14	13 52	16 01	19 44	21 39	28 42
08	13 09 57	15 34 42	27 ♑ 57 07	04 ♒ 02 38	12 07	17 31	02 55	13 58	16 08	19 42	21 38	28 43
09	13 13 54	16 33 57	10 ♒ 12 12	16 ♒ 26 24	11 03	18 44	03 36	14 04	16 15	19 40	21 37	28 44
10	13 17 51	17 33 14	22 ♒ 45 48	29 ♒ 10 51	10 06	19 57	04 17	14 09	16 22	19 38	21 37	28 45
11	13 21 47	18 32 33	05 ♓ 41 58	12 ♓ 19 24	09 16	21 10	04 58	14 15	16 29	19 36	21 36	28 46
12	13 25 44	19 31 54	19 ♓ 03 19	25 ♓ 53 41	08 35	22 23	05 39	14 21	16 35	19 34	21 35	28 47
13	13 29 40	20 31 17	02 ♈ 50 21	09 ♈ 52 59	08 04	23 36	06 20	14 28	16 42	19 32	21 35	28 48
14	13 33 37	21 30 41	17 ♈ 01 03	24 ♈ 13 54	07 44	24 49	07 01	14 34	16 49	19 30	21 34	28 49
15	13 37 33	22 30 08	01 ♉ 30 43	08 ♉ 50 33	07 35	26 02	07 42	14 41	16 55	19 28	21 33	28 50
16	13 41 30	23 29 37	16 ♉ 12 26	23 ♉ 35 19	07 D 36	27 15	08 23	14 47	17 02	19 26	21 33	28 52
17	13 45 26	24 29 08	00 ♊ 58 11	08 ♊ 20 04	07 49	28 27	09 04	14 54	17 09	19 25	21 32	28 53
18	13 49 23	25 28 41	15 ♊ 40 08	22 ♊ 57 36	08 12	29 ♏ 40	09 45	15 01	17 15	19 23	21 32	28 54
19	13 53 20	26 28 17	00 ♋ 11 55	07 ♋ 22 36	08 45	00 ♐ 53	10 27	15 09	17 22	19 21	21 31	28 55
20	13 57 16	27 27 55	14 ♋ 29 21	21 ♋ 32 00	09 27	02 06	11 08	15 16	17 28	19 19	21 31	28 56
21	14 01 13	28 27 35	28 ♋ 30 29	05 ♌ 24 51	10 17	03 18	11 50	15 23	17 34	19 18	21 30	28 58
22	14 05 09	29 ♎ 27 17	12 ♌ 15 12	19 ♌ 01 42	11 14	04 31	12 31	15 31	17 41	19 16	21 30	28 59
23	14 09 06	00 ♏ 27 02	25 ♌ 44 32	02 ♍ 23 54	12 18	05 44	13 13	15 39	17 47	19 14	21 30	29 00
24	14 13 02	01 26 49	08 ♍ 59 59	15 ♍ 32 59	13 28	06 56	13 54	15 47	17 53	19 13	21 29	29 02
25	14 16 59	02 26 38	22 ♍ 03 03	28 ♍ 30 18	14 43	08 09	14 36	15 55	17 59	19 11	21 29	29 03
26	14 20 55	03 26 29	04 ♎ 54 48	11 ♎ 16 40	16 02	09 22	15 18	16 03	18 05	19 10	21 29	29 04
27	14 24 52	04 26 22	17 ♎ 35 54	23 ♎ 52 32	17 25	10 34	15 59	16 11	18 11	19 08	21 29	29 06
28	14 28 48	05 26 17	00 ♏ 06 35	06 ♏ 18 06	18 51	11 47	16 41	16 20	18 17	19 07	21 28	29 07
29	14 32 45	06 26 15	12 ♏ 27 04	18 ♏ 33 35	20 20	12 59	17 23	16 28	18 23	19 05	21 28	29 09
30	14 36 42	07 26 14	24 ♏ 37 42	00 ♐ 39 34	21 50	14 11	18 05	16 37	18 29	19 04	21 28	29 10
31	14 40 38	08 ♏ 26 15	06 ♐ 39 21	12 ♐ 37 17	23 ♎ 23	15 ♐ 24	18 ♏ 47	16 ♑ 46	18 ♍ 35	19 ♓ 03	21 ♒ 28	29 ♐ 12

Moon

Date	Moon True ☊ ° '	Moon Mean ☊ ° '	Moon ☽ Latitude ° '
01	17 ♒ 16	15 ♒ 48	04 S 58
02	17 R 07	15 45	05 06
03	16 58	15 42	05 00
04	16 52	15 39	04 41
05	16 47	15 36	04 10
06	16 45	15 32	03 28
07	16 D 44	15 29	02 37
08	16 45	15 26	01 39
09	16 46	15 23	00 S 35
10	16 R 46	15 20	00 N 32
11	16 44	15 16	01 39
12	16 40	15 13	02 43
13	16 33	15 10	03 39
14	16 24	15 07	04 24
15	16 14	15 04	04 52
16	16 04	15 01	05 02
17	15 54	14 57	04 52
18	15 47	14 54	04 23
19	15 42	14 51	03 36
20	15 39	14 48	02 37
21	15 D 39	14 45	01 31
22	15 39	14 42	00 N 18
23	15 R 39	14 38	00 S 53
24	15 37	14 35	02 00
25	15 32	14 32	03 00
26	15 25	14 29	03 49
27	15 14	14 26	04 26
28	15 01	14 22	04 50
29	14 47	14 19	04 59
30	14 33	14 16	04 55
31	14 ♒ 20	14 ♒ 13	04 S 38

DECLINATIONS

Date	Sun ☉ ° '	Moon ☽ ° '	Mercury ☿ ° '	Venus ♀ ° '	Mars ♂ ° '	Jupiter ♃ ° '	Saturn ♄ ° '	Uranus ♅ ° '	Neptune ♆ ° '	Pluto ♇ ° '
01	03 S 26	17 S 29	10 S 49	14 S 41	10 S 39	23 S 05	07 N 22	04 S 43	14 S 36	17 S 27
02	03 50	21 34	10 18	15 07	10 54	23 05	07 19	04 44	14 36	17 27
03	04 13	24 40	09 44	15 33	11 09	23 05	07 17	04 45	14 36	17 27
04	04 36	26 38	09 05	15 58	11 24	23 04	07 14	04 46	14 37	17 27
05	04 59	27 22	08 24	16 23	11 39	23 04	07 12	04 47	14 37	17 28
06	05 22	26 51	07 40	16 47	11 53	23 03	07 09	04 47	14 37	17 28
07	05 45	25 05	06 55	17 12	12 08	23 03	07 06	04 48	14 37	17 28
08	06 08	22 11	06 11	17 35	12 23	23 02	07 04	04 49	14 38	17 29
09	06 31	18 15	05 27	17 59	12 37	23 02	07 01	04 50	14 38	17 29
10	06 53	13 25	04 46	18 21	12 51	23 01	06 59	04 51	14 38	17 29
11	07 16	07 53	04 07	18 44	13 06	23 01	06 56	04 51	14 38	17 30
12	07 39	01 S 50	03 33	19 06	13 20	23 00	06 54	04 52	14 39	17 30
13	08 01	04 N 29	03 03	19 27	13 34	22 59	06 51	04 53	14 39	17 30
14	08 23	10 45	02 39	19 48	13 48	22 59	06 49	04 54	14 39	17 30
15	08 45	16 34	02 20	20 08	14 02	22 58	06 46	04 54	14 39	17 31
16	09 07	21 30	02 07	20 28	14 16	22 57	06 44	04 55	14 39	17 31
17	09 29	25 07	02 00	20 48	14 30	22 57	06 41	04 56	14 40	17 31
18	09 51	27 01	01 58	21 06	14 43	22 56	06 39	04 56	14 40	17 32
19	10 13	27 03	02 01	21 25	14 57	22 55	06 37	04 57	14 40	17 32
20	10 34	25 15	02 10	21 42	15 10	22 54	06 34	04 58	14 40	17 32
21	10 56	21 55	02 23	22 00	15 24	22 54	06 32	04 58	14 40	17 32
22	11 17	17 25	02 40	22 16	15 37	22 53	06 30	04 59	14 40	17 33
23	11 38	12 06	03 01	22 32	15 50	22 52	06 27	05 00	14 40	17 33
24	11 59	06 20	03 25	22 47	16 03	22 51	06 25	05 00	14 41	17 33
25	12 19	00 N 24	03 52	23 02	16 16	22 50	06 23	05 01	14 41	17 33
26	12 40	05 S 27	04 22	23 16	16 29	22 49	06 21	05 01	14 41	17 34
27	13 00	11 00	04 54	23 30	16 42	22 48	06 19	05 02	14 41	17 34
28	13 20	16 02	05 27	23 42	16 54	22 47	06 16	05 02	14 41	17 34
29	13 40	20 19	06 02	23 55	17 07	22 46	06 14	05 03	14 41	17 34
30	14 00	23 42	06 38	24 06	17 19	22 45	06 12	05 03	14 41	17 35
31	14 S 19	25 S 59	07 S 15	24 S 17	17 S 31	22 S 44	06 N 10	05 S 04	14 S 41	17 S 35

ZODIAC SIGN ENTRIES

Date	h	m	Planets
01	04	26	☽ ♏
03	15	14	☽ ♐
04	04	34	♂ ♏
06	03	48	☽ ♑
08	16	03	☽ ♒
11	01	31	☽ ♓
13	07	07	☽ ♈
15	09	31	☽ ♉
17	10	25	☽ ♊
18	18	31	♀ ♐
19	11	40	☽ ♋
21	14	35	☽ ♌
23	01	09	☉ ♏
23	19	40	☽ ♍
26	02	48	☽ ♎
28	11	47	☽ ♏
30	22	41	☽ ♐

LATITUDES

Date	Mercury ☿ ° '	Venus ♀ ° '	Mars ♂ ° '	Jupiter ♃ ° '	Saturn ♄ ° '	Uranus ♅ ° '	Neptune ♆ ° '	Pluto ♇ ° '
01	03 S 21	00 S 12	00 N 11	00 S 20	01 N 43	00 S 48	00 S 21	05 N 59
04	02 43	00 21	00 09	00 20	01 43	00 48	00 21	05 59
07	01 51	00 30	00 08	00 20	01 44	00 48	00 21	05 58
10	00 S 50	00 39	00 06	00 20	01 44	00 48	00 21	05 57
13	00 N 09	00 48	00 04	00 20	01 45	00 48	00 21	05 56
16	00 59	00 57	00 N 02	00 20	01 45	00 48	00 21	05 55
19	01 34	01 06	00 00	00 21	01 45	00 48	00 21	05 54
22	01 56	01 15	00 S 01	00 21	01 46	00 48	00 21	05 54
25	02 05	01 23	00 03	00 21	01 47	00 48	00 21	05 53
28	02 05	01 31	00 05	00 21	01 47	00 48	00 21	05 52
31	01 N 58	01 S 39	00 S 07	00 S 21	01 N 48	00 S 48	00 S 21	05 N 51

DATA

Julian Date	2454741
Delta T	+67 seconds
Ayanamsa	23° 58' 57"
Synetic vernal point	05° ♓ 08' 03"
True obliquity of ecliptic	23° 26' 25"

MOON'S PHASES, APSIDES AND POSITIONS ☽

Date	h	m	Phase	Longitude ° '	Eclipse Indicator
07	09	04	☽	14 ♑ 28	
14	20	02	○	21 ♈ 51	
21	11	55	☾	28 ♋ 27	
28	23	14	●	05 ♏ 54	

Day	h	m		
05	10	36	Apogee	
17	06	19	Perigee	
05	14	04	Max dec	27° S 22'
12	19	00	0N	
19	00	19	Max dec	27° N 16'
25	13	37	0S	

LONGITUDES

Date	Chiron ⚷ ° '	Ceres ⚳ ° '	Pallas ⚴ ° '	Juno ⚵ ° '	Vesta ⚶ ° '	Black Moon Lilith ⚸ ° '
01	16 ♒ 20	17 ♌ 43	18 ♊ 00	21 ♐ 07	13 ♉ 24	19 ♐ 31
11	16 ♒ 08	21 ♌ 37	19 ♊ 39	23 ♐ 39	11 ♉ 34	20 ♐ 38
21	16 ♒ 03	25 ♌ 21	20 ♊ 32	26 ♐ 28	09 ♉ 13	21 ♐ 44
31	16 ♒ 03	28 ♌ 54	20 ♊ 31	29 ♐ 31	06 ♉ 38	22 ♐ 51

ASPECTARIAN

h m	Aspects
01 Wednesday	
00 20	☽ ☌ ♂
01 48	☽ ⚹ ♇
04 58	☽ ∠ ♄
05 26	☿ ⊼ ♅
07 04	☽ Q ♃
07 06	♀ ‖ ♆
11 46	☽ ‖ ♇
13 58	☽ ⚼ ♅
17 31	☿ ‖ ♂
22 00	☽ ⊻ ☉
22 54	☽ ☌ ♀
02 Thursday	
04 01	♂ ⚹ ♇
06 30	☽ ⚹ ♃
06 52	☽ ∠ ♇
10 22	☽ ⚹ ♄
10 44	☽ ⊥ ☉
16 40	☽ ⊻ ☿
19 11	☽ △ ♅
22 41	☽ ‖ ♃
22 46	☽ □ ♆
03 Friday	
00 36	☽ ⊥ ♇
03 37	☽ ⊥ ☿
06 05	☽ ∠ ☉
10 20	☽ Q ♄
12 19	☽ ∠ ♃
12 33	☽ ⊻ ♇
14 27	☽ ⊻ ♂
20 07	☽ ∠ ☿
04 Saturday	
03 11	☽ ⊥ ♂
06 29	☽ ⊥ ♃
10 41	☽ Q ♆
13 00	☿ Q ♇
14 50	☽ ⚹ ☉
17 12	☽ ⊻ ♀
18 39	☽ ⊻ ♃
22 23	☉ ‖ ♅
22 25	☽ ∠ ♂
22 52	☽ □ ♄
23 45	☽ ⚹ ☿
05 Sunday	
06 44	☽ ⊥ ♀
07 09	☽ □ ♅
07 43	♂ ∠ ♄
07 56	☿ ⊻ ♄
08 07	♀ ∠ ♇
08 22	♀ ⚹ ♃
10 56	☽ ⚹ ♆
17 20	☽ Q ☉
21 54	☽ Q ☿
06 Monday	
01 09	☽ ☌ ♇
03 04	☽ ∠ ♀
05 05	☿ ⊻ ♀
06 40	☽ ⚹ ♂
16 52	☉ □ ♃
17 18	☽ ∠ ♆
19 32	☽ Q ♅
20 53	☉ ☌ ☿
21 06	♂ Q ♃
23 46	☿ □ ♃
07 Tuesday	
05 42	♀ ⚹ ♄
05 48	☿ # ♄
07 04	☽ □ ☿
07 50	☽ ☌ ♃
08 24	☽ Q ♂
09 04	☽ □ ☉
11 27	☽ ⊥ ♆
12 11	☽ △ ♄
12 51	☽ ⚹ ♀
19 37	☽ ⚹ ♅
23 27	☽ ⊻ ♆
08 Wednesday	
05 10	♀ ‖ ♇
05 48	☽ ‖ ♃
13 02	☉ ‖ ☿
13 32	☽ ⊻ ♇
15 27	☽ Q ♀
18 08	☿ ⊥ ♀
18 21	☽ ⚼ ♄
22 22	☽ □ ♂
09 Thursday	
01 15	☽ ∠ ♅
01 21	☽ ⊥ ♇
03 18	☉ ⊻ ♄
12 05	☽ ± ♄
13 22	☽ ‖ ♀
13 32	☽ △ ☿
16 06	☽ ‖ ♇
16 15	☉ Q ♇
18 40	☽ ⊥ ♅
18 50	☽ ∠ ♇
19 30	☽ ⊻ ♃
23 45	☽ ⊼ ♄
10 Friday	
01 17	☽ △ ☉
05 48	♀ △ ♅
06 06	☽ ⊻ ♅
06 08	☽ □ ♀
06 20	☽ ‖ ♆
07 02	☽ ⊥ ♃
09 11	☿ ‖ ♅
09 50	☽ ☌ ♆
14 29	☽ ‖ ♂
16 06	☽ ⚼ ☿
17 04	☉ # ♄
23 13	☽ ⚹ ♇
23 53	♂ ⚼ ♅
11 Saturday	
00 03	☽ ∠ ♃
07 44	☽ ⚼ ☉
07 48	☽ ± ☿
10 35	☽ △ ♂
14 22	☽ ‖ ☉
15 53	☽ # ♄
18 09	☽ ⊼ ☿
20 20	♀ □ ♆
21 13	☽ Q ♇
12 Sunday	
00 10	☽ ‖ ♅
01 23	☽ ± ☉
03 35	☽ ⚹ ♃
04 41	☽ ‖ ☿
07 36	☽ ☍ ♄
12 51	☉ ⊼ ♅
12 54	☽ ☌ ♅
12 54	☽ ⊼ ☉
14 57	☽ ⚼ ♂
16 28	☽ ⊻ ♆
18 27	☽ △ ♀
20 00	♀ ⊥ ♇
13 Monday	
00 54	☽ Q ♃
02 56	☽ ⊥ ♆
04 20	☿ ∠ ♀
05 02	☽ □ ♇
06 58	☽ # ☿
07 27	☽ ± ♂
13 30	☽ # ♅
18 16	☽ ⊼ ♂
18 23	☽ ∠ ♆
20 41	☽ ☍ ☿
20 56	☽ ‖ ♄
22 45	☽ ⚼ ♀
14 Tuesday	
02 16	☽ # ☉
07 52	☽ □ ♃
11 39	☽ ⊼ ♄
13 19	☉ △ ♆
15 17	☽ ± ♀
16 08	☽ ⊻ ♅
19 35	☽ ⚹ ♆
20 02	☽ ☍ ☉
21 44	☽ ± ♄
15 Wednesday	
00 47	☽ # ♂
02 04	☽ ⊥ ♅
02 09	☽ ⊼ ♀
03 48	☽ # ♆
07 36	☽ △ ♇
08 12	☿ ⊻ ♂
12 41	☽ ⚼ ♄
15 21	☽ Q ♆
16 13	☽ # ♇
16 51	☽ ∠ ♅
20 06	☿ St D
21 55	☽ ⊼ ☿
22 38	☽ ☍ ♂
23 31	☉ ⊥ ♄
16 Thursday	
06 03	☽ # ♀
07 44	☽ ± ☿
08 10	☽ ⚼ ♇
09 41	☽ △ ♃
13 21	☽ △ ♄
17 15	☽ ⚹ ♅
20 28	☽ # ♃
20 41	☽ □ ♆
22 31	☽ ⚼ ☿
22 50	☽ ± ♇
17 Friday	
00 42	☽ ⊼ ☉
07 33	☽ ☍ ♀
08 36	☽ ⊼ ♇
10 15	☽ ⚼ ♃
11 09	☽ ± ☉
12 43	☽ Q ♅
20 28	♀ ⊻ ♇
23 25	☽ △ ☿
18 Saturday	
01 02	☽ ± ♃
01 51	☽ ⊼ ♂
02 53	☽ ⚼ ☉
02 54	♀ Q ♄
05 30	♂ ‖ ♆
09 41	☉ ± ♅
10 56	☽ ⊼ ♃
12 09	☽ ± ♂
14 37	☽ □ ♄
18 05	☽ □ ♅
19 43	♀ ∠ ♃
21 38	☽ △ ♆
19 Sunday	
03 43	☽ ⚼ ♂
05 21	☽ △ ☉
09 52	☽ ☍ ♇
13 15	☽ ⊼ ♀
20 41	☽ Q ♄
22 34	☽ ⚼ ♆
20 Monday	
00 12	☽ ± ♀
03 00	☽ □ ☿
06 02	☽ △ ♂
13 20	☽ ☍ ♃
13 44	☽ ± ♆
16 51	☽ ⚼ ♀
17 06	☽ ⚹ ♄
20 12	☽ △ ♅
23 58	☽ ⊼ ♆
21 Tuesday	
05 54	☽ # ♃
11 34	☽ Q ☿
11 36	☽ # ♀
11 55	☽ □ ☉
12 47	☽ ⊼ ♇
19 06	☽ ∠ ♄
21 08	☽ △ ♀
22 01	☽ ⚼ ♅
23 14	☽ ± ♇
22 Wednesday	
00 23	☉ ⚹ ♇
09 16	♃ ⊥ ♆
10 04	☽ ⚹ ☿
10 59	☽ ⊥ ♄
11 22	☽ # ♇
12 30	☽ □ ♂
13 47	☽ ± ♅
15 04	☽ ⚼ ♇
17 49	☽ ⊼ ♃
20 06	☽ # ♂
21 40	☽ ⊻ ♄
21 56	☽ Q ☉
23 Thursday	
00 24	☽ ⊼ ♅
00 44	☽ # ♆
04 24	☽ ☍ ♆
04 36	☽ ± ♃
13 55	☽ # ☉
15 04	☽ ∠ ☿
17 53	☽ △ ♇
20 55	☽ ⚼ ♃
21 10	☽ ⚹ ☉
22 24	☽ Q ♂
24 Friday	
07 52	☽ □ ♀
08 56	☽ ⊥ ☿
11 40	☽ ‖ ♄
16 25	♂ ∠ ♇
17 24	☽ # ♅
17 36	♀ ‖ ♃
21 01	☽ ⊻ ☿
21 29	☽ ⚹ ♂
23 00	☽ # ☿
25 Saturday	
00 33	☽ △ ♃
02 47	☽ ∠ ☉
04 26	☽ ☌ ♄
06 43	☽ ☍ ♅
07 11	☿ ⊻ ♂
10 57	☽ ⊼ ♆
20 24	☽ Q ♀
20 51	☽ ⊥ ☉
22 06	☽ ± ♆
26 Sunday	
01 03	☽ □ ♇
02 34	☉ ∠ ♄
02 51	☽ ∠ ♂
07 02	☽ ‖ ☿
09 00	☽ ⊻ ☉
10 11	☽ ‖ ♅
12 14	☿ □ ♃
14 26	♀ Q ♆
14 57	☽ ⚼ ♆
15 43	☽ # ♄
20 44	☽ ⊥ ♂
21 15	☽ ⚹ ♀
21 26	☉ ⊥ ♀
27 Monday	
03 29	♀ ⊥ ♃
04 51	☉ ⚼ ♅
04 58	☉ Q ♃
06 26	☿ Q ♇
08 46	☽ ⊻ ♂
09 17	☽ □ ♃
11 03	☽ Q ♇
11 37	☽ ☌ ☿
13 08	☽ ⊻ ♄
14 55	☽ ⊼ ♅
18 08	☿ ‖ ♅
19 24	☽ △ ♆
20 29	♂ ⚹ ♃
21 50	☽ ‖ ☉
28 Tuesday	
00 42	☽ ⊥ ♄
01 58	☿ ⊻ ♄
02 23	☽ ± ♅
04 53	☽ ∠ ♀
05 15	☽ ‖ ♆
10 05	☽ ⚹ ♇
16 11	☿ ⊼ ♅
16 47	☽ ‖ ♂
18 12	☽ ∠ ♄
19 44	☽ ⚼ ♅
20 07	☽ ‖ ♇
20 16	☽ Q ♃
23 14	☽ ☌ ☉
29 Wednesday	
00 10	☽ ⊥ ♀
13 09	☽ ⊻ ♀
15 20	☽ ∠ ♇
19 48	☿ # ♄
19 59	☽ ⚹ ♃
22 17	☽ ☌ ♂
23 45	☽ ⚹ ♄
30 Thursday	
01 01	☽ △ ♅
04 35	☽ ‖ ♃
05 41	☽ ⊻ ☿
05 45	☽ □ ♆
06 11	☿ △ ♆
09 07	☽ ⊥ ♇
15 47	☽ ‖ ♀
19 19	☽ ⊥ ☿
21 03	☽ ⊻ ♇
23 45	☽ Q ♄
31 Friday	
02 05	☽ ∠ ♃
03 51	♂ ⚹ ♄
15 54	☽ ⊻ ☉
15 59	☽ ∠ ☿
17 39	☽ Q ♆
19 17	♂ ‖ ♇
20 22	☽ ⊥ ♃
20 32	♂ △ ♅

All ephemeris data is given at 12.00 UT and the Moon's longitude is additionally given for 24.00 UT

NOVEMBER 2008

LONGITUDES

Date	Sidereal time h m s	Sun ☉ ° ' "	Moon ☽ ° ' "	Moon ☽ 24.00 ° ' "	Mercury ☿ ° '	Venus ♀ ° '	Mars ♂ ° '	Jupiter ♃ ° '	Saturn ♄ ° '	Uranus ♅ ° '	Neptune ♆ ° '	Pluto ♇ ° '
01	14 44 35	09 ♏ 26 18	18 ♐ 33 38	24 ♐ 28 45	24 ♎ 57	16 ♐ 36	19 ♏ 29	16 ♑ 55	18 ♍ 41	19 ♓ 01	21 ♒ 28	29 ♐ 13
02	14 48 31	10 26 23	00 ♑ 23 01	06 ♑ 16 54	26 32	17 49	20 11	17 04	18 46	19 R 00	21 D 28	29 15
03	14 52 28	11 26 29	12 ♑ 10 53	18 ♑ 05 32	28 07	19 01	20 53	17 13	18 52	18 59	21 28	29 17
04	14 56 24	12 26 37	24 ♑ 01 27	29 ♑ 59 16	29 ♎ 44	20 13	21 36	17 23	18 57	18 58	21 28	29 18
05	15 00 21	13 26 46	05 ♒ 59 39	12 ♒ 03 16	01 ♏ 21	21 25	22 18	17 32	19 03	18 57	21 28	29 20
06	15 04 17	14 26 57	18 ♒ 12 36	24 ♒ 26 03	02 58	22 38	23 00	17 42	19 08	18 56	21 28	29 22
07	15 08 14	15 27 09	00 ♓ 40 30	07 ♓ 03 54	04 36	23 50	23 43	17 52	19 13	18 55	21 29	29 23
08	15 12 11	16 27 23	13 ♓ 33 45	20 ♓ 10 32	06 13	25 02	24 25	18 01	19 19	18 54	21 29	29 25
09	15 16 07	17 27 39	26 ♓ 54 34	03 ♈ 46 04	07 51	26 14	25 08	18 11	19 24	18 53	21 29	29 27
10	15 20 04	18 27 55	10 ♈ 45 03	17 ♈ 51 18	09 28	27 26	25 50	18 21	19 29	18 52	21 29	29 28
11	15 24 00	19 28 14	25 ♈ 04 26	02 ♉ 23 50	11 06	28 38	26 33	18 32	19 34	18 51	21 30	29 30
12	15 27 57	20 28 34	09 ♉ 48 38	17 ♉ 17 49	12 43	29 ♐ 50	27 15	18 42	19 39	18 50	21 30	29 32
13	15 31 53	21 28 55	24 ♉ 50 11	02 ♊ 24 26	14 20	01 ♑ 02	27 58	18 52	19 44	18 49	21 30	29 34
14	15 35 50	22 29 19	09 ♊ 59 14	17 ♊ 33 14	15 57	02 13	28 41	19 03	19 48	18 49	21 30	29 36
15	15 39 46	23 29 44	25 ♊ 05 13	02 ♋ 34 04	17 33	03 25	29 ♏ 24	19 13	19 53	18 48	21 31	29 37
16	15 43 43	24 30 11	09 ♋ 58 50	17 ♋ 18 47	19 10	04 37	00 ♐ 06	19 24	19 58	18 48	21 32	29 39
17	15 47 40	25 30 40	24 ♋ 33 22	01 ♌ 42 14	20 46	05 48	00 49	19 35	20 02	18 47	21 32	29 41
18	15 51 36	26 31 10	08 ♌ 45 13	15 ♌ 42 18	22 22	07 00	01 32	19 46	20 07	18 46	21 33	29 43
19	15 55 33	27 31 43	22 ♌ 33 38	29 ♌ 19 25	23 58	08 12	02 15	19 57	20 11	18 46	21 33	29 45
20	15 59 29	28 32 17	05 ♍ 59 57	12 ♍ 35 35	25 34	09 23	02 58	20 08	20 15	18 46	21 34	29 47
21	16 03 26	29 ♏ 32 53	19 ♍ 06 41	25 ♍ 33 39	27 09	10 34	03 41	20 19	20 19	18 45	21 34	29 49
22	16 07 22	00 ♐ 33 30	01 ♎ 56 49	08 ♎ 16 34	28 ♏ 44	11 46	04 25	20 31	20 23	18 45	21 35	29 51
23	16 11 19	01 34 09	14 ♎ 33 13	20 ♎ 47 02	00 ♐ 19	12 57	05 08	20 42	20 27	18 45	21 36	29 53
24	16 15 15	02 34 50	26 ♎ 58 18	03 ♏ 07 14	01 54	14 08	05 51	20 54	20 31	18 45	21 36	29 55
25	16 19 12	03 35 33	09 ♏ 14 01	15 ♏ 18 49	03 29	15 19	06 35	21 05	20 35	18 44	21 37	29 57
26	16 23 09	04 36 17	21 ♏ 21 47	27 ♏ 23 02	05 03	16 30	07 18	21 17	20 39	18 44	21 38	29 ♐ 59
27	16 27 05	05 37 02	03 ♐ 22 44	09 ♐ 21 00	06 38	17 41	08 01	21 29	20 43	18 44	21 39	00 ♑ 01
28	16 31 02	06 37 49	15 ♐ 17 59	21 ♐ 13 52	08 12	18 52	08 45	21 40	20 46	18 D 44	21 40	00 03
29	16 34 58	07 38 37	27 ♐ 08 50	03 ♑ 03 09	09 46	20 03	09 28	21 52	20 50	18 44	21 41	00 05
30	16 38 55	08 ♐ 39 26	08 ♑ 57 05	14 ♑ 50 57	11 ♐ 20	21 ♑ 14	10 ♐ 12	22 ♑ 04	20 ♍ 53	18 ♓ 45	21 ♒ 42	00 ♑ 07

Date	Moon True ☊ ° '	Moon Mean ☊ ° '	Moon ☽ Latitude ° '
01	14 ♒ 09	14 ♒ 10	04 S 08
02	14 R 01	14 07	03 28
03	13 55	14 03	02 39
04	13 53	14 00	01 43
05	13 D 52	13 57	00 S 42
06	13 R 52	13 54	00 N 23
07	13 51	13 51	01 28
08	13 49	13 48	02 30
09	13 45	13 44	03 27
10	13 38	13 41	04 13
11	13 28	13 38	04 45
12	13 17	13 35	05 00
13	13 05	13 32	04 54
14	12 54	13 28	04 28
15	12 45	13 25	03 43
16	12 40	13 22	02 43
17	12 36	13 19	01 34
18	12 35	13 16	00 N 20
19	12 D 35	13 13	00 S 53
20	12 R 35	13 09	02 01
21	12 34	13 06	03 01
22	12 30	13 03	03 50
23	12 23	13 00	04 27
24	12 13	12 57	04 51
25	12 01	12 54	05 01
26	11 48	12 50	04 58
27	11 35	12 47	04 41
28	11 23	12 44	04 11
29	11 12	12 41	03 31
30	11 ♒ 05	12 ♒ 38	02 S 42

DECLINATIONS

Date	Sun ☉ ° '	Moon ☽ ° '	Mercury ☿ ° '	Venus ♀ ° '	Mars ♂ ° '	Jupiter ♃ ° '	Saturn ♄ ° '	Uranus ♅ ° '	Neptune ♆ ° '	Pluto ♇ ° '
01	14 S 38	27 S 04	07 S 53	24 S 27	17 S 43	22 S 43	06 N 08	05 S 04	14 S 41	17 S 35
02	14 57	26 54	08 31	24 36	17 55	22 42	06 06	05 05	14 41	17 35
03	15 16	25 31	09 09	24 45	18 07	22 41	06 04	05 05	14 41	17 36
04	15 34	23 00	09 48	24 53	18 19	22 40	06 02	05 06	14 41	17 36
05	15 53	19 27	10 27	25 00	18 30	22 39	06 00	05 06	14 41	17 36
06	16 10	15 01	11 05	25 07	18 42	22 37	05 58	05 06	14 41	17 36
07	16 28	09 52	11 43	25 13	18 53	22 36	05 56	05 07	14 41	17 37
08	16 46	04 S 09	12 21	25 18	19 04	22 35	05 54	05 07	14 41	17 37
09	17 03	01 N 56	12 59	25 22	19 15	22 34	05 53	05 08	14 41	17 37
10	17 19	08 08	13 36	25 26	19 26	22 32	05 51	05 08	14 41	17 37
11	17 36	14 08	14 13	25 29	19 36	22 31	05 49	05 08	14 40	17 38
12	17 52	19 29	14 49	25 31	19 46	22 30	05 47	05 08	14 40	17 38
13	18 08	23 44	15 24	25 33	19 57	22 28	05 46	05 09	14 40	17 38
14	18 24	26 22	15 59	25 34	20 07	22 27	05 44	05 09	14 40	17 38
15	18 39	27 04	16 33	25 34	20 17	22 25	05 42	05 09	14 40	17 38
16	18 54	25 47	17 06	25 33	20 26	22 24	05 41	05 09	14 40	17 39
17	19 08	22 45	17 38	25 32	20 36	22 22	05 39	05 09	14 40	17 39
18	19 23	18 24	18 10	25 29	20 45	22 21	05 38	05 10	14 39	17 39
19	19 37	13 10	18 41	25 27	20 55	22 19	05 36	05 10	14 39	17 39
20	19 50	07 26	19 10	25 23	21 04	22 18	05 35	05 10	14 39	17 39
21	20 03	01 N 32	19 39	25 19	21 12	22 16	05 33	05 10	14 39	17 40
22	20 16	04 S 18	20 07	25 13	21 21	22 14	05 32	05 10	14 39	17 40
23	20 29	09 51	20 34	25 08	21 30	22 13	05 30	05 10	14 38	17 40
24	20 41	14 55	21 00	25 01	21 38	22 11	05 29	05 10	14 38	17 40
25	20 52	19 20	21 25	24 54	21 46	22 09	05 28	05 10	14 38	17 40
26	21 04	22 53	21 49	24 46	21 54	22 08	05 27	05 10	14 38	17 41
27	21 15	25 25	22 11	24 37	22 01	22 06	05 25	05 10	14 37	17 41
28	21 25	26 47	22 33	24 28	22 09	22 04	05 24	05 10	14 37	17 41
29	21 35	26 56	22 53	24 18	22 16	22 02	05 23	05 10	14 37	17 41
30	21 S 45	25 S 50	23 S 13	24 S 07	22 S 23	22 S 00	05 N 22	05 S 10	14 S 37	17 S 41

ZODIAC SIGN ENTRIES

Date	h	m	Planets
02	11	13	☽ ♑
04	16	00	☿ ♏
05	00	01	☽ ♒
07	10	43	☽ ♓
09	17	26	☽ ♈
11	20	05	☽ ♉
12	15	25	♀ ♑
13	20	11	☽ ♊
15	19	52	☽ ♋
16	08	26	♂ ♐
17	21	07	☽ ♌
20	01	13	☽ ♍
21	22	44	☉ ♐
22	08	20	☽ ♎
23	07	09	☿ ♐
24	17	54	☽ ♏
27	01	03	♇ ♑
27	05	14	☽ ♐
29	17	48	☽ ♑

LATITUDES

Date	Mercury ☿ ° '	Venus ♀ ° '	Mars ♂ ° '	Jupiter ♃ ° '	Saturn ♄ ° '	Uranus ♅ ° '	Neptune ♆ ° '	Pluto ♇ ° '
01	01 N 55	01 S 41	00 S 08	00 S 21	01 N 48	00 S 48	00 S 21	05 N 51
04	01 41	01 49	00 09	00 21	01 49	00 47	00 21	05 50
07	01 25	01 55	00 11	00 21	01 49	00 47	00 21	05 50
10	01 06	02 01	00 13	00 21	01 50	00 47	00 21	05 49
13	00 46	02 07	00 15	00 22	01 50	00 47	00 21	05 48
16	00 26	02 12	00 16	00 22	01 51	00 47	00 21	05 48
19	00 N 05	02 16	00 18	00 22	01 52	00 47	00 21	05 47
22	00 S 15	02 19	00 20	00 22	01 52	00 47	00 21	05 47
25	00 34	02 21	00 22	00 22	01 53	00 47	00 21	05 46
28	00 53	02 23	00 23	00 23	01 54	00 46	00 21	05 45
31	01 S 11	02 S 23	00 S 25	00 S 23	01 N 55	00 S 46	00 S 21	05 N 45

DATA

Julian Date	2454772
Delta T	+67 seconds
Ayanamsa	23° 59' 00"
Synetic vernal point	05° ♓ 07' 59"
True obliquity of ecliptic	23° 26' 24"

MOON'S PHASES, APSIDES AND POSITIONS ☽

Date	h	m	Phase	Longitude ° '	Eclipse Indicator
06	04	03	☽	14 ♒ 07	
13	06	17	○	21 ♉ 15	
19	21	31	☾	27 ♌ 56	
27	16	55	●	05 ♐ 49	

Day	h	m		
02	05	01	Apogee	
14	10	08	Perigee	
29	17	06	Apogee	
01	20	55	Max dec	27° S 09'
09	04	29	ON	
15	08	18	Max dec	27° N 05'
21	18	16	OS	
29	02	38	Max dec	27° S 01'

LONGITUDES

Date	Chiron ⚷ ° '	Ceres ⚳ ° '	Pallas ⚴ ° '	Juno ⚵ ° '	Vesta ⚶ ° '	Black Moon Lilith ⚸ ° '
01	16 ♒ 04	29 ♌ 15	20 ♊ 27	29 ♐ 50	06 ♉ 22	22 ♐ 58
11	16 ♒ 12	02 ♍ 32	19 ♊ 16	03 ♑ 06	03 ♉ 53	24 ♐ 05
21	16 ♒ 26	05 ♍ 32	17 ♊ 32	06 ♑ 32	01 ♉ 49	25 ♐ 12
31	16 ♒ 47	08 ♍ 10	14 ♊ 01	10 ♑ 08	00 ♉ 21	26 ♐ 19

All ephemeris data is given at 12.00 UT and the Moon's longitude is additionally given for 24.00 UT

ASPECTARIAN

h m Aspects

01 Saturday
05 06 ☽ ⊥ ☉
07 36 ☽ ☌ ♀
07 38 ☿ ⊥ ♄
08 38 ☽ ⊻ ♃
12 14 ☽ □ ♄
12 56 ☽ □ ♅
13 09 ☿ ± ♅
14 00 ☽ ⊻ ♂
15 25 ☉ ∥ ♆
17 53 ☽ ⚹ ♆
19 05 ♀ ⊻ ♃

02 Sunday
01 01 ☽ ∠ ☉
02 56 ☽ ⊥ ♂
02 57 ☽ ⚹ ☿
06 36 ♆ St D
09 41 ☽ ☌ ♇
22 24 ☽ ∠ ♂

03 Monday
00 23 ☽ ∠ ♆
01 26 ☽ Q ♅
07 10 ☽ Q ☿
08 41 ♀ □ ♄
10 21 ☽ ⚹ ☉
11 18 ♀ □ ♅
18 41 ☽ ⊥ ♆
20 02 ☽ ∥ ♀
22 22 ☽ ☌ ♃

04 Tuesday
01 40 ☽ △ ♄
01 47 ☽ ⚹ ♅
03 27 ☽ ⊻ ♀
05 31 ☿ ⚹ ♇
06 47 ☽ ⚹ ♂
06 50 ☽ ⊻ ♆
07 44 ♂ □ ♆
12 55 ☽ Q ☉
13 36 ♄ ☍ ♅
14 34 ☽ ∥ ♃
16 56 ☽ ⊥ ♀
22 39 ☽ ⊻ ♇

05 Wednesday
01 17 ☽ □ ☿
07 55 ☽ ∠ ♅
08 05 ☽ bQ ♄
08 25 ☽ Q ♂
10 41 ☽ ⊥ ♇
12 56 ♀ ⚹ ♆
12 57 ☽ ∠ ♀
17 15 ☽ ∥ ♂
22 31 ☽ ∥ ♇

06 Thursday
01 44 ☽ ⊥ ♅
02 03 ☽ ± ♄
04 03 ☽ □ ☉
04 31 ☽ ∠ ♇
06 29 ☽ ∥ ☉
09 47 ☉ ∠ ♇
11 03 ☽ ⊻ ♃
13 27 ☽ ⊻ ♅
13 41 ☽ ∥ ♆
13 52 ☽ ⊼ ♄
18 23 ☽ ☌ ♆
21 32 ☽ ⚹ ♀
21 54 ☽ □ ♂
22 49 ☽ ⊥ ♃

07 Friday
00 35 ♂ ⊥ ♇
02 01 ☿ bQ ♅
04 35 ☽ ∥ ☿
06 13 ♀ ⊻ ♂
06 14 ☿ ∠ ♄
09 33 ☽ ⚹ ♇
16 11 ☽ ∠ ♃
20 27 ☽ △ ☿
22 42 ☽ Q ♀

08 Saturday
04 48 ☽ # ♄
08 03 ☽ Q ♇
08 04 ☽ ∥ ♅
08 49 ☿ Q ♃
17 43 ☽ △ ☉
20 13 ☽ ⚹ ♃
21 41 ☽ ☌ ♅
22 31 ☽ ☍ ♄

09 Sunday
02 21 ☽ ⊻ ♆
03 47 ☽ bQ ☿
08 40 ☽ △ ♂
10 41 ☽ □ ♀
13 01 ☽ ⊥ ♆
16 28 ☽ □ ♇
17 50 ☽ Q ♃
21 49 ☽ ± ☿
22 30 ☽ bQ ☉

10 Monday
00 22 ☽ # ♅
03 11 ☽ ∥ ♄
04 42 ☽ ∠ ♆
08 54 ☉ ⚹ ♃
09 32 ☽ ⊼ ☿
12 09 ☽ bQ ♂
15 08 ☽ ± ☉
21 21 ☉ △ ♅

11 Tuesday
01 00 ☽ □ ♃
01 40 ☽ ⊻ ♅
02 00 ☽ ⊼ ☉
02 48 ☽ ⊼ ♄
04 06 ☽ ± ♂
06 04 ☽ ⚹ ♆
11 38 ☽ ⊥ ♅
12 25 ☽ # ☿
12 49 ☽ ± ♄
14 18 ☽ # ♆
14 24 ☉ ⚹ ♄
14 27 ☉ ∥ ♇
14 33 ☽ ⊼ ♂
18 22 ☽ △ ♀
19 18 ☽ △ ♇

12 Wednesday
01 47 ☽ Q ♆
02 21 ☽ ∠ ♅
03 13 ☽ # ♇
03 36 ☽ bQ ♄
03 54 ☽ # ☉
05 55 ♀ ☌ ♇
06 20 ☿ ∥ ♆
13 28 ☽ # ♂
17 14 ☽ ☍ ☿
19 36 ☽ bQ ♇
20 45 ☽ bQ ♀

13 Thursday
02 24 ☽ △ ♃
02 27 ☽ ⚹ ♅
03 50 ☽ △ ♄
04 02 ☽ # ♃
05 40 ♃ ⚹ ♅
06 17 ☽ ☍ ☉
06 42 ☽ □ ♆
09 58 ☽ ± ♇
12 20 ☽ ± ♀
12 31 ☉ □ ♆
15 31 ☿ ∠ ♇
17 13 ☽ ☍ ♂
19 31 ☽ ⊼ ♇
21 29 ☽ Q ♅
22 39 ☽ ⊼ ♀

14 Friday
02 29 ☽ bQ ♃
02 41 ☽ # ♀
16 54 ☽ ± ♃
22 35 ☽ ⊼ ☿

15 Saturday
01 59 ☽ □ ♅
02 33 ☽ ⊼ ♃
03 40 ☽ □ ♄
06 18 ☽ △ ♆
09 16 ☽ ± ☿
09 17 ☽ ⊼ ☉
15 11 ☉ ⊥ ♇
19 14 ☽ ⊼ ♂
19 17 ☽ ☍ ♇
19 34 ☽ ± ☉
20 12 ♂ ⊻ ♇

16 Sunday
01 26 ☽ bQ ☿
02 32 ☽ ☍ ♀
05 24 ☽ ± ♂
06 23 ☽ bQ ♆
06 26 ☿ △ ♅
08 42 ☽ Q ♄
11 10 ☽ bQ ☉
14 24 ☽ # ♀
16 00 ☿ ⚹ ♃
20 48 ☽ bQ ♂
21 04 ☽ ± ♆

17 Monday
00 28 ☿ ⚹ ♄
02 26 ☽ △ ♅
03 39 ☽ ☍ ♃
04 27 ☽ ⚹ ♄
04 56 ☽ △ ☿
06 58 ☽ ⊼ ♆
12 16 ☿ ∥ ♇
13 43 ☽ △ ☉
14 12 ☿ ∠ ♀
14 24 ☽ # ♃
20 37 ☽ ⊼ ♇
23 04 ☽ △ ♂
23 31 ☿ □ ♆

18 Tuesday
00 11 ☽ # ♂
02 43 ♀ ∠ ♆
03 31 ☽ bQ ♅
05 45 ☽ ∠ ♄
06 48 ☽ ± ♇
07 19 ☽ # ☉
07 29 ♀ Q ♅
08 43 ☽ ⊼ ♀
13 02 ☽ # ☿
15 38 ☽ # ♇
18 55 ☽ ± ♅
20 00 ☽ ± ♀
21 17 ☽ ⊥ ♄
22 19 ☽ bQ ♇

19 Wednesday
05 21 ☽ ⊼ ♅
05 27 ☽ # ♆
07 21 ☽ ⊼ ♃
07 48 ☽ ⊻ ♄
08 41 ☿ ⊥ ♇
09 20 ♂ Q ♄
10 13 ☽ ☍ ♆
13 13 ☽ bQ ♀
14 49 ☽ □ ☿
15 08 ♀ ⊥ ♂
18 05 ☽ ± ♃
21 31 ☽ □ ☉

20 Thursday
00 48 ☽ △ ♇
06 14 ☽ □ ♂
10 25 ☽ bQ ♃
18 45 ☽ △ ♀
19 38 ☽ ∥ ♄
21 17 ☽ # ♅

21 Friday
03 41 ☽ Q ☿
08 52 ☽ Q ☉
11 20 ☽ ☍ ♅
12 13 ♃ △ ♄
14 15 ☽ ☌ ♄
14 17 ☽ △ ♃
16 34 ☽ ⊼ ♆
17 04 ☽ Q ♂
18 34 ☉ ⊻ ♇

22 Saturday
03 47 ☽ ± ♆
05 06 ☽ ⚹ ☿
08 02 ☽ □ ♇
09 09 ☽ ⚹ ☉
15 41 ☽ ∥ ♅
16 56 ☽ ⚹ ♂
17 12 ☽ # ♄
20 47 ☽ bQ ♆

23 Sunday
02 41 ☉ ∥ ☿
05 12 ☿ ⊻ ♇
08 36 ☽ □ ♀
13 41 ☽ ∠ ☿
16 13 ☽ ∠ ☉
18 25 ☽ Q ♇
20 04 ☽ ⊼ ♅
23 24 ☽ ∠ ♂
23 26 ☽ ⊻ ♄

24 Monday
00 01 ☽ □ ♃
01 35 ☽ △ ♆
07 40 ☽ ± ♅
09 37 ☽ ⊥ ☿
10 33 ☉ Q ♄
10 36 ☽ ∥ ♆
11 07 ☽ ⊥ ♄
11 10 ☽ ⊥ ☉
13 47 ♂ ∠ ♃
17 45 ☽ ⚹ ♇
17 58 ☽ ⊥ ♂
21 53 ☿ Q ♄
23 02 ☽ ⊻ ☿
23 09 ☽ Q ♀
23 56 ☽ ⊻ ☉

25 Tuesday
01 13 ☽ bQ ♅
02 30 ☽ ∥ ♇
04 48 ☽ ∠ ♄
06 27 ☽ ⊻ ♂
11 42 ☽ Q ♃
16 52 ☉ ☌ ☿
18 09 ♀ ⊥ ♆
22 12 ☽ ∥ ☉
23 18 ☽ ∠ ♇

26 Wednesday
01 19 ☽ ⚹ ♀
03 02 ☽ ∥ ☿
04 22 ☽ ∥ ♂
06 21 ☽ ∥ ♃
06 47 ☽ △ ♅
10 34 ☽ ⚹ ♄
11 50 ☽ ⚹ ♃
12 32 ☽ □ ♆
17 14 ☽ ⊥ ♇
20 01 ☿ ∥ ♂

27 Thursday
03 46 ☽ ∥ ♀
05 15 ☽ ⊻ ♇
06 25 ☿ ∥ ♃
09 22 ☿ ∠ ♃
10 28 ☽ ∠ ♀
10 39 ☽ Q ♄
16 07 ♅ St D
16 55 ☽ ☌ ☉
18 19 ☽ ∠ ♃
19 31 ☽ ☌ ☿
21 56 ☽ ☌ ♂
23 09 ♂ ∥ ♃

28 Friday
00 37 ☽ Q ♆
06 33 ☽ ⊥ ♀
09 20 ♀ ⚹ ♅
10 35 ♃ ⊻ ♆
12 46 ☽ ⊥ ♃
13 16 ☉ ∠ ♃
18 57 ☽ □ ♅
20 01 ☽ ⊻ ♀
23 07 ☽ □ ♄

29 Saturday
00 53 ☽ ⚹ ♆
01 07 ☽ ⊻ ♃
03 39 ☿ ☌ ♂
10 37 ☿ Q ♆
17 59 ☽ ☌ ♇
18 52 ♂ Q ♆

30 Sunday
04 36 ♀ △ ♄
07 24 ☽ ∠ ♆
07 30 ☽ Q ♅
11 21 ☽ ⊻ ☉
14 43 ☽ ⊻ ♂
17 36 ☽ ⊻ ☿
21 35 ♀ ⊻ ♆

DECEMBER 2008

Date	Sidereal time h m s	Sun ☉ ° ' "	Moon ☽ ° ' "	Moon ☽ 24.00 ° ' "	Mercury ☿ ° '	Venus ♀ ° '	Mars ♂ ° '	Jupiter ♃ ° '	Saturn ♄ ° '	Uranus ♅ ° '	Neptune ♆ ° '	Pluto ♇ ° '
01	16 42 51	09 ♐ 40 16	20 ♑ 45 08	26 ♑ 40 03	12 ♐ 54	22 ♑ 24	10 ♐ 56	22 ♑ 16	20 ♍ 56	18 ♓ 45	21 ♒ 43	00 ♑ 09
02	16 46 48	10 41 07	02 ♒ 36 10	08 ♒ 33 59	14 28	23 35	11 39	22 29	20 59	18 45	21 44	00 11
03	16 50 44	11 42 00	14 ♒ 34 04	20 ♒ 37 00	16 02	24 45	12 23	22 41	21 02	18 45	21 45	00 13
04	16 54 41	12 42 52	26 ♒ 43 24	02 ♓ 53 55	17 36	25 56	13 07	22 53	21 05	18 46	21 46	00 15
05	16 58 38	13 43 46	09 ♓ 09 10	15 ♓ 29 48	19 10	27 06	13 51	23 06	21 08	18 46	21 47	00 18
06	17 02 34	14 44 41	21 ♓ 56 24	28 ♓ 29 31	20 44	28 16	14 35	23 18	21 11	18 46	21 48	00 20
07	17 06 31	15 45 36	05 ♈ 09 37	11 ♈ 57 04	22 17	29 ♑ 26	15 19	23 31	21 14	18 47	21 49	00 22
08	17 10 27	16 46 32	18 ♈ 52 03	25 ♈ 54 39	23 51	00 ♒ 36	16 03	23 43	21 16	18 47	21 50	00 24
09	17 14 24	17 47 28	03 ♉ 04 42	10 ♉ 21 49	25 25	01 46	16 47	23 56	21 19	18 48	21 51	00 26
10	17 18 20	18 48 25	17 ♉ 45 25	25 ♉ 14 41	26 59	02 56	17 31	24 09	21 21	18 49	21 53	00 28
11	17 22 17	19 49 23	02 ♊ 48 33	10 ♊ 25 48	28 ♐ 33	04 05	18 15	24 21	21 23	18 49	21 54	00 30
12	17 26 13	20 50 22	18 ♊ 05 05	25 ♊ 45 00	00 ♑ 07	05 14	18 59	24 34	21 26	18 50	21 55	00 33
13	17 30 10	21 51 22	03 ♋ 24 06	11 ♋ 01 02	01 41	06 24	19 44	24 47	21 28	18 51	21 57	00 35
14	17 34 07	22 52 23	18 ♋ 34 34	26 ♋ 03 36	03 15	07 33	20 28	25 00	21 30	18 52	21 58	00 37
15	17 38 03	23 53 24	03 ♌ 27 15	10 ♌ 44 52	04 49	08 42	21 12	25 13	21 31	18 52	21 59	00 39
16	17 42 00	24 54 26	17 ♌ 55 58	25 ♌ 00 19	06 23	09 51	21 57	25 26	21 33	18 53	22 01	00 41
17	17 45 56	25 55 29	01 ♍ 57 50	08 ♍ 48 35	07 56	10 59	22 41	25 39	21 35	18 54	22 02	00 44
18	17 49 53	26 56 33	15 ♍ 32 48	22 ♍ 10 47	09 30	12 08	23 26	25 53	21 36	18 55	22 04	00 46
19	17 53 49	27 57 38	28 ♍ 42 55	05 ♎ 09 39	11 04	13 16	24 10	26 06	21 38	18 57	22 05	00 48
20	17 57 46	28 58 44	11 ♎ 31 27	17 ♎ 48 48	12 37	14 24	24 55	26 19	21 39	18 58	22 07	00 50
21	18 01 42	29 ♐ 59 51	24 ♎ 02 10	00 ♏ 12 02	14 10	15 32	25 39	26 33	21 40	18 59	22 08	00 52
22	18 05 39	01 ♑ 00 58	06 ♏ 18 50	12 ♏ 23 01	15 42	16 40	26 24	26 46	21 41	19 00	22 10	00 55
23	18 09 36	02 02 06	18 ♏ 24 56	24 ♏ 24 59	17 14	17 48	27 09	26 59	21 42	19 01	22 11	00 57
24	18 13 32	03 03 15	00 ♐ 23 27	06 ♐ 20 40	18 46	18 55	27 54	27 13	21 43	19 03	22 13	00 59
25	18 17 29	04 04 24	12 ♐ 16 54	18 ♐ 12 23	20 16	20 03	28 39	27 26	21 44	19 04	22 15	01 01
26	18 21 25	05 05 33	24 ♐ 07 21	00 ♑ 02 03	21 46	21 10	29 ♐ 23	27 40	21 44	19 06	22 16	01 03
27	18 25 22	06 06 43	05 ♑ 56 39	11 ♑ 51 25	23 14	22 17	00 ♑ 08	27 54	21 45	19 07	22 18	01 06
28	18 29 18	07 07 54	17 ♑ 46 34	23 ♑ 42 19	24 41	23 23	00 53	28 07	21 45	19 09	22 20	01 08
29	18 33 15	08 09 04	29 ♑ 38 56	05 ♒ 36 44	26 06	24 30	01 38	28 21	21 46	19 10	22 21	01 10
30	18 37 11	09 10 15	11 ♒ 35 59	17 ♒ 37 04	27 29	25 36	02 24	28 35	21 46	19 12	22 23	01 12
31	18 41 08	10 ♑ 11 25	23 ♒ 40 20	29 ♒ 46 11	28 ♑ 49	26 ♒ 42	03 ♑ 09	28 ♑ 49	21 ♍ 46	19 ♓ 13	22 ♒ 25	01 ♑ 14

Date	Moon True ☊ ° '	Moon Mean ☊ ° '	Moon ☽ Latitude ° '
01	11 ♒ 00	12 ♒ 34	01 S 46
02	10 R 57	12 31	00 S 45
03	10 D 57	12 28	00 N 19
04	10 58	12 25	01 24
05	10 59	12 22	02 25
06	10 R 58	12 19	03 22
07	10 56	12 15	04 09
08	10 52	12 12	04 44
09	10 46	12 09	05 04
10	10 38	12 06	05 04
11	10 30	12 03	04 44
12	10 22	11 59	04 03
13	10 16	11 56	03 05
14	10 12	11 53	01 54
15	10 10	11 50	00 N 36
16	10 D 10	11 47	00 S 42
17	10 11	11 44	01 55
18	10 13	11 40	03 00
19	10 R 13	11 37	03 53
20	10 12	11 34	04 32
21	10 10	11 31	04 58
22	10 05	11 28	05 09
23	09 59	11 25	05 07
24	09 52	11 21	04 51
25	09 44	11 18	04 22
26	09 38	11 15	03 42
27	09 32	11 12	02 53
28	09 28	11 09	01 56
29	09 26	11 05	00 S 53
30	09 D 26	11 02	00 N 12
31	09 ♒ 26	10 ♒ 59	01 N 17

DECLINATIONS

Date	Sun ☉ ° '	Moon ☽ ° '	Mercury ☿ ° '	Venus ♀ ° '	Mars ♂ ° '	Jupiter ♃ ° '	Saturn ♄ ° '	Uranus ♅ ° '	Neptune ♆ ° '	Pluto ♇ ° '
01	21 S 54	23 S 35	23 S 31	23 S 56	22 S 30	21 S 58	05 N 21	05 S 10	14 S 36	17 S 41
02	22 03	20 18	23 48	23 44	22 37	21 56	05 20	05 10	14 36	17 42
03	22 11	16 09	24 04	23 32	22 43	21 54	05 19	05 10	14 36	17 42
04	22 19	11 18	24 18	23 18	22 49	21 52	05 18	05 09	14 35	17 42
05	22 27	05 54	24 31	23 04	22 55	21 50	05 17	05 09	14 35	17 42
06	22 34	00 S 07	24 43	22 50	23 01	21 48	05 16	05 09	14 35	17 42
07	22 41	05 N 52	24 54	22 35	23 06	21 46	05 15	05 09	14 34	17 42
08	22 47	11 46	25 03	22 19	23 12	21 44	05 15	05 09	14 34	17 42
09	22 53	17 18	25 11	22 03	23 17	21 42	05 14	05 08	14 33	17 43
10	22 58	21 59	25 18	21 46	23 21	21 40	05 13	05 08	14 33	17 43
11	23 03	25 22	25 23	21 29	23 26	21 38	05 13	05 08	14 33	17 43
12	23 07	26 57	25 27	21 11	23 30	21 35	05 12	05 07	14 32	17 43
13	23 11	26 29	25 29	20 52	23 34	21 33	05 11	05 07	14 32	17 43
14	23 15	24 02	25 30	20 33	23 38	21 31	05 11	05 07	14 31	17 43
15	23 18	19 58	25 29	20 14	23 42	21 29	05 10	05 06	14 31	17 43
16	23 21	14 47	25 27	19 53	23 45	21 26	05 10	05 06	14 30	17 43
17	23 23	08 59	25 24	19 33	23 48	21 24	05 10	05 05	14 30	17 43
18	23 24	02 N 56	25 18	19 12	23 51	21 22	05 09	05 05	14 29	17 44
19	23 25	03 S 03	25 12	18 51	23 54	21 19	05 09	05 05	14 29	17 44
20	23 26	08 44	25 04	18 29	23 56	21 17	05 09	05 04	14 28	17 44
21	23 26	13 56	24 54	18 06	23 59	21 14	05 09	05 04	14 28	17 44
22	23 26	18 29	24 43	17 44	24 00	21 12	05 08	05 03	14 27	17 44
23	23 25	22 13	24 30	17 20	24 02	21 09	05 08	05 02	14 27	17 44
24	23 24	24 58	24 16	16 57	24 03	21 07	05 08	05 02	14 26	17 44
25	23 23	26 36	24 01	16 33	24 05	21 04	05 08	05 01	14 26	17 44
26	23 21	27 01	23 44	16 09	24 05	21 01	05 08	05 01	14 25	17 44
27	23 18	26 11	23 26	15 44	24 06	20 59	05 08	05 00	14 25	17 44
28	23 15	24 10	23 06	15 19	24 06	20 56	05 08	04 59	14 24	17 44
29	23 11	21 06	22 45	14 54	24 06	20 53	05 08	04 59	14 24	17 44
30	23 07	17 07	22 23	14 28	24 06	20 51	05 09	04 58	14 23	17 44
31	23 S 03	12 S 25	22 S 00	14 S 02	24 S 06	20 S 48	05 N 09	04 S 57	14 S 22	17 S 45

ZODIAC SIGN ENTRIES

Date	h	m	Planets
02	06	45	☽ ♒
04	18	23	☽ ♓
07	02	44	☽ ♈
07	23	37	♀ ♒
09	06	52	☽ ♉
11	07	33	☽ ♊
12	10	13	☿ ♑
13	06	40	☽ ♋
15	06	22	☽ ♌
17	08	36	☽ ♍
19	14	23	☽ ♎
21	12	04	☉ ♑
21	23	36	☽ ♏
24	11	13	☽ ♐
26	23	56	☽ ♑
27	07	30	♂ ♑
29	12	42	☽ ♒

LATITUDES

Date	Mercury ☿ ° '	Venus ♀ ° '	Mars ♂ ° '	Jupiter ♃ ° '	Saturn ♄ ° '	Uranus ♅ ° '	Neptune ♆ ° '	Pluto ♇ ° '
01	01 S 11	02 S 23	00 S 25	00 S 23	01 N 55	00 S 46	00 S 21	05 N 45
04	01 27	02 23	00 27	00 23	01 55	00 46	00 21	05 45
07	01 41	02 22	00 29	00 23	01 56	00 46	00 21	05 44
10	01 53	02 19	00 30	00 23	01 57	00 46	00 21	05 44
13	02 03	02 16	00 32	00 23	01 58	00 46	00 21	05 43
16	02 10	02 11	00 34	00 24	01 59	00 46	00 21	05 43
19	02 14	02 06	00 35	00 24	01 59	00 46	00 21	05 43
22	02 13	01 59	00 37	00 24	02 00	00 46	00 21	05 42
25	02 08	01 51	00 39	00 24	02 00	00 45	00 21	05 42
28	01 56	01 41	00 40	00 24	02 02	00 45	00 21	05 42
31	01 S 38	01 S 31	00 S 42	00 S 25	02 N 03	00 S 45	00 S 21	05 N 42

DATA

Julian Date	2454802
Delta T	+67 seconds
Ayanamsa	23° 59' 06"
Synetic vernal point	05° ♓ 07' 54"
True obliquity of ecliptic	23° 26' 23"

MOON'S PHASES, APSIDES AND POSITIONS ☽

Date	h	m	Phase	Longitude ° '	Eclipse Indicator
05	21	26	☽	14 ♓ 08	
12	16	37	○	21 ♊ 02	
19	10	29	☾	27 ♍ 54	
27	12	22	●	06 ♑ 08	

Day	h	m		
12	21	45	Perigee	
26	18	04	Apogee	
06	12	26	ON	
12	18	37	Max dec	27° N 01'
18	23	41	OS	
26	07	59	Max dec	27° S 02'

LONGITUDES

Date	Chiron ⚷ ° '	Ceres ⚳ ° '	Pallas ⚴ ° '	Juno ⚵ ° '	Vesta ⚶ ° '	Black Moon Lilith ⚸ ° '
01	16 ♒ 47	08 ♍ 10	14 ♊ 01	10 ♑ 08	00 ♉ 21	26 ♐ 19
11	17 ♒ 13	10 ♍ 22	10 ♊ 38	13 ♑ 51	29 ♈ 38	27 ♐ 26
21	17 ♒ 44	12 ♍ 04	07 ♊ 32	17 ♑ 40	29 ♈ 39	28 ♐ 32
31	18 ♒ 19	13 ♍ 10	05 ♊ 15	21 ♑ 34	00 ♉ 23	29 ♐ 39

ASPECTARIAN

h m	Aspects
01 Monday	
00 42	☽ ⊥ ☉
01 44	☽ ⊥ ♆
03 43	☽ ⊥ ♂
07 40	☽ ⊥ ☿
07 55	☽ ✶ ♅
08 37	☽ ∥ ♀
08 45	♀ ☌ ♃
12 23	☽ △ ♄
12 31	☽ ∥ ☿
12 55	☉ Q ♆
13 57	☽ ⊻ ♆
15 09	☽ ☌ ♃
15 44	☽ ☌ ♀
20 26	☽ ∥ ♂
20 42	☽ ∠ ☉
21 16	☉ ∥ ♃
23 12	☽ ∠ ♂
02 Tuesday	
00 35	☽ ∥ ☉
00 46	☽ ∥ ♃
04 42	☽ ∠ ☿
07 07	☽ ⊻ ♇
08 56	☿ ∥ ♀
14 19	☽ ∠ ♅
18 51	☽ ₽ ♄
19 15	☽ ⊥ ♇
03 Wednesday	
03 36	☽ ∥ ♇
05 45	☽ ✶ ☉
07 22	☽ ✶ ♂
08 23	☽ ⊥ ♅
12 57	☽ ± ♄
13 19	☽ ∠ ♇
15 21	☽ ✶ ☿
20 04	☽ ∥ ♆
20 19	☽ ⊻ ♅
23 26	☿ ⊥ ♃
04 Thursday	
00 53	☽ ⊼ ♄
02 15	☽ ☌ ♆
04 21	☽ ⊻ ♃
07 43	☽ Q ☉
08 39	☽ Q ♂
10 17	☽ ⊻ ♀
16 17	☽ ⊥ ♃
18 25	☽ Q ☿
18 54	☽ ✶ ♇
23 11	☽ ⊥ ♀
05 Friday	
05 52	☿ □ ♅
09 57	☽ ∠ ♃
14 35	☽ # ♄
15 09	☽ ∥ ♅
17 59	☽ Q ♇
18 10	☽ ∠ ♀
21 26	☽ □ ☉
21 26	☽ □ ♂
22 04	☉ ☌ ♂
23 06	♀ ∥ ♂
06 Saturday	
06 07	☽ ☌ ♅
09 27	☽ □ ☿
10 36	☽ ☍ ♄
11 44	☽ ⊻ ♆
14 33	☽ ✶ ♃
19 14	☿ □ ♄
22 45	☽ ⊥ ♆
07 Sunday	
00 43	☽ ✶ ♀
03 22	☽ □ ♇
04 38	☿ ✶ ♆
05 29	☉ ∥ ♀
09 08	☽ # ♅
09 35	☽ ∥ ♄
12 38	☽ Q ♃
14 57	☽ ∠ ♆
08 Monday	
00 08	☽ Q ♀
06 52	☽ △ ♂
07 44	♀ ⊻ ♇
08 07	☽ △ ☉
09 35	☿ ⊻ ♃
11 52	☽ ⊻ ♅
16 08	☽ ⊼ ♄
17 06	☽ ✶ ♆
20 25	☽ □ ♃
21 35	☽ △ ☿
22 07	☽ ⊥ ♅
23 48	☽ # ♆
09 Tuesday	
02 20	☽ ± ♄
07 35	☽ △ ♇
09 37	☽ □ ♀
09 44	☽ ₽ ♂
11 29	☽ ₽ ☉
12 54	♀ ∠ ♂
13 12	☽ ∠ ♅
13 18	☽ Q ♆
13 57	☽ # ♇
16 10	☉ ⊥ ♃
17 22	☽ ₽ ♄
10 Wednesday	
01 21	☽ ± ♂
01 32	☽ ₽ ☿
03 23	☽ ± ☉
08 18	☽ ₽ ♇
08 26	☿ ⊥ ♀
10 09	☽ # ♃
10 48	☽ # ♀
11 35	☽ ⊼ ♂
12 03	☉ □ ♅
13 42	☽ ✶ ♅
13 49	☽ ⊼ ☉
17 48	☽ △ ♄
17 48	☽ ± ☿
18 05	☽ # ☉
18 38	☽ □ ♆
20 40	☽ # ♂
21 40	♀ ∥ ♃
22 23	☽ △ ♃
22 48	☽ ± ♇
11 Thursday	
04 29	☽ ⊼ ☿
06 28	♀ ∠ ♅
08 21	☽ ⊼ ♇
08 51	☽ Q ♅
12 12	☽ # ☿
14 11	☽ △ ♀
16 47	♂ ⊥ ♃
22 28	☽ ₽ ♃
12 Friday	
06 50	♂ □ ♅
12 46	☽ ± ♃
13 10	☽ □ ♅
13 29	☽ ☍ ♂
15 39	☽ ₽ ♀
16 37	☽ ☍ ☉
17 15	☽ □ ♄
18 01	☽ △ ♆
18 43	☿ ☌ ♇
22 18	☽ ⊼ ♃
13 Saturday	
02 20	☉ □ ♄
06 54	☽ ± ♀
07 34	☽ ☍ ♇
09 00	☽ ☍ ☿
13 23	♀ ₽ ♄
14 05	☉ ✶ ♆
15 59	♀ ⊥ ♇
17 06	☽ ⊼ ♀
17 35	☽ ₽ ♆
21 34	☽ Q ♄
23 56	☽ # ☿
14 Sunday	
07 50	☽ ± ♆
12 27	☽ △ ♅
14 43	☽ # ♂
15 10	☽ ⊼ ♂
16 40	☽ ✶ ♄
17 17	☽ # ☉
17 26	☽ ⊼ ♆
19 23	☽ ⊼ ☉
22 27	☽ ☍ ♃
15 Monday	
01 19	☽ ± ♂
03 55	☽ # ♃
05 46	☽ ± ☉
07 26	☽ ⊼ ♇
10 35	☽ # ♀
12 41	☽ ₽ ♅
14 29	☽ ⊼ ☿
16 45	☽ ₽ ♂
17 02	☽ ∠ ♄
17 15	☽ ± ♇
21 21	☽ ☍ ♀
21 36	☽ ₽ ☉
22 46	♂ □ ♄
22 55	☽ # ♇
16 Tuesday	
01 35	☽ ± ☿
03 33	☽ ± ♅
08 00	☽ ⊥ ♄
08 13	☽ ₽ ♇
13 14	☽ # ♆
13 37	☽ ⊼ ♅
14 16	♂ ✶ ♆
18 08	☽ ⊻ ♄
18 33	☽ ₽ ☿
18 55	☽ ☍ ♆
19 09	☽ △ ♂
19 57	☿ Q ♅
21 53	☿ ∠ ♆
23 48	☉ ∠ ♀
17 Wednesday	
00 45	☽ △ ☉
00 57	☽ ⊼ ♃
03 56	☉ ⊻ ♃
09 51	☽ △ ♇
11 27	☽ ± ♃
23 49	☽ △ ☿
18 Thursday	
03 12	☽ ∥ ♄
03 29	☽ # ♅
03 31	☽ ₽ ♃
05 20	☽ ⊼ ♀
17 05	☽ ± ♀
18 06	☽ ☍ ♅
22 58	☽ ☌ ♄
23 48	☽ ⊼ ♆
19 Friday	
03 08	☽ □ ♂
04 58	♀ ⊥ ♅
07 05	☽ △ ♃
10 29	☽ □ ☉
10 50	☽ ± ♆
11 06	☽ ₽ ♀
15 52	☽ □ ♇
20 22	☽ ∥ ♅
20 41	☽ # ♄
20 Saturday	
03 39	☽ ₽ ♆
14 22	☽ □ ☿
14 48	☽ Q ♂
18 02	☽ △ ♀
23 19	☽ Q ☉
21 Sunday	
02 00	☽ Q ♇
02 14	☽ ⊼ ♅
07 25	☽ ⊻ ♄
08 19	☽ △ ♆
13 50	☽ ± ♅
14 36	☽ ∥ ♆
14 46	♀ ± ♄
15 21	☽ ✶ ♂
16 57	☽ □ ♃
19 04	☽ ⊥ ♄
19 16	♀ ∠ ♇
22 Monday	
00 39	☽ ✶ ☉
01 21	☽ ✶ ♇
06 08	☽ Q ☿
07 27	☽ ₽ ♅
07 44	☽ ∥ ♇
08 03	☽ ∥ ♀
09 24	☉ ☌ ♇
11 40	♀ ∥ ♇
12 44	☽ ∠ ♄
19 16	☿ ⊥ ♆
22 43	☽ ∠ ♂
23 Tuesday	
04 37	☽ ∥ ♃
04 41	♂ ⊻ ♃
05 03	☽ Q ♃
07 04	☽ ∠ ♇
09 00	☽ ∠ ☉
09 19	☽ ✶ ☿
10 39	☽ □ ♀
13 13	☽ △ ♅
17 49	☽ ⊥ ♂
18 35	☽ ✶ ♄
19 33	☽ □ ♆
21 23	☽ ∥ ☉
24 Wednesday	
01 06	☽ ⊥ ♇
02 54	☽ ∥ ♂
04 40	☽ ⊥ ☉
05 30	☽ ✶ ♃
05 36	☽ ∥ ☿
06 39	☽ ⊻ ♂
13 12	☽ ⊻ ♇
14 38	♀ ⊻ ♅
16 33	☿ ✶ ♅
17 52	☽ ⊻ ☉
18 43	☽ Q ♄
19 47	☽ ∠ ☿
21 57	☿ ⊻ ♀
25 Thursday	
02 33	☽ Q ♀
06 47	☿ ∥ ♂
07 52	☽ Q ♆
12 20	☽ ∠ ♃
16 37	☽ ⊥ ☿
26 Friday	
01 46	☽ □ ♅
05 22	☽ ✶ ♀
06 32	☽ ⊻ ☿
06 55	☽ ⊥ ♃
07 10	☽ □ ♄
08 14	☽ ✶ ♆
11 38	☿ △ ♄
19 20	☽ ⊻ ♃
20 24	☿ ⊻ ♆
23 25	☽ ☌ ♂
27 Saturday	
00 31	♀ ⊼ ♄
02 07	☽ ☌ ♇
12 22	☽ ☌ ☉
12 29	♀ ☌ ♆
14 23	☽ Q ♅
14 46	☽ ∠ ♆
14 59	☽ ∠ ♀
23 08	☉ ∥ ☿
28 Sunday	
09 04	☽ ⊥ ♆
11 08	☽ ⊥ ♀
12 15	☉ Q ♅
12 36	☽ ∥ ♂
14 46	☽ ✶ ♅
16 47	☉ ∠ ♆
20 02	♂ ☌ ♇
20 04	☽ △ ♄
20 14	☽ ∥ ☉
21 14	☽ ⊻ ♆
22 29	☽ ∥ ☿
29 Monday	
00 32	☽ ⊻ ♀
03 53	☽ ☌ ☿
09 20	☽ ☌ ♃
13 23	☽ ∥ ♃
15 04	☽ ⊻ ♇
16 17	☽ ⊻ ♂
21 07	☽ ∠ ♅
30 Tuesday	
02 19	☽ ₽ ♄
03 10	☽ ⊥ ♇
05 09	☽ ⊥ ♂
06 41	☽ ⊻ ☉
08 31	☽ ∥ ♇
15 12	☽ ⊥ ♅
16 47	♀ ∥ ♆
19 47	☽ ⊥ ☉
20 19	☽ ± ♄
21 13	☽ ∠ ♇
31 Wednesday	
00 19	☽ ∠ ♂
02 21	☽ ∥ ♆
03 10	☽ ⊻ ♅
03 16	☽ ∥ ♀
08 14	☽ ⊼ ♄
09 31	☽ ☌ ♆
11 40	☿ ☌ ♃
15 16	☽ ∠ ☉
18 06	♄ St R
18 34	☽ ☌ ♀
22 19	☽ ⊻ ♃
23 21	☽ ⊻ ☿

All ephemeris data is given at 12.00 UT and the Moon's longitude is additionally given for 24.00 UT

JANUARY 2009

LONGITUDES

Date	Sidereal time h m s	Sun ☉ ° ′ ″	Moon ☽ ° ′ ″	Moon ☽ 24.00 ° ′ ″	Mercury ☿ ° ′	Venus ♀ ° ′	Mars ♂ ° ′	Jupiter ♃ ° ′	Saturn ♄ ° ′	Uranus ♅ ° ′	Neptune ♆ ° ′	Pluto ♇ ° ′
01	18 45 05	11 ♑ 12 35	05 ♓ 55 04	12 ♓ 07 26	00 ♒ 07	27 ♒ 48	03 ♑ 54	29 ♑ 02	21 ♍ 46	19 ♓ 15	22 ♒ 27	01 ♑ 16
02	18 49 01	12 13 45	18 ♓ 23 46	24 ♓ 44 31	01 21	28 53	04 39	29 16	21 R 46	19 17	22 29	01 19
03	18 52 58	13 14 55	01 ♈ 10 11	07 ♈ 41 13	02 30	29 ♒ 58	05 24	29 30	21 46	19 19	22 31	01 21
04	18 56 54	14 16 05	14 ♈ 18 02	21 ♈ 00 58	03 35	01 ♓ 03	06 10	29 44	21 45	19 20	22 32	01 23
05	19 00 51	15 17 14	27 ♈ 50 20	04 ♉ 46 15	04 34	02 08	06 55	29 ♑ 58	21 45	19 22	22 34	01 25
06	19 04 47	16 18 23	11 ♉ 48 48	18 ♉ 57 50	05 27	03 12	07 40	00 ♒ 12	21 44	19 24	22 36	01 27
07	19 08 44	17 19 32	26 ♉ 13 05	03 ♊ 34 03	06 13	04 16	08 26	00 26	21 44	19 26	22 38	01 29
08	19 12 40	18 20 40	11 ♊ 00 05	18 ♊ 30 19	06 50	05 20	09 11	00 40	21 43	19 28	22 40	01 32
09	19 16 37	19 21 48	26 ♊ 03 46	03 ♋ 39 17	07 19	06 23	09 57	00 54	21 42	19 30	22 42	01 34
10	19 20 34	20 22 56	11 ♋ 15 39	18 ♋ 51 36	07 37	07 26	10 42	01 08	21 41	19 32	22 44	01 36
11	19 24 30	21 24 03	26 ♋ 25 52	03 ♌ 57 16	07 45	08 29	11 28	01 22	21 40	19 35	22 46	01 38
12	19 28 27	22 25 10	11 ♌ 24 44	18 ♌ 47 20	07 R 42	09 31	12 13	01 36	21 38	19 37	22 48	01 40
13	19 32 23	23 26 16	26 ♌ 04 16	03 ♍ 14 59	07 26	10 33	12 59	01 50	21 37	19 39	22 50	01 42
14	19 36 20	24 27 22	10 ♍ 19 03	17 ♍ 16 17	06 59	11 35	13 45	02 04	21 36	19 41	22 52	01 44
15	19 40 16	25 28 28	24 ♍ 06 36	00 ♎ 50 05	06 21	12 36	14 30	02 18	21 34	19 43	22 54	01 46
16	19 44 13	26 29 34	07 ♎ 26 59	13 ♎ 57 36	05 32	13 36	15 16	02 32	21 33	19 46	22 56	01 48
17	19 48 09	27 30 40	20 ♎ 22 20	26 ♎ 41 39	04 33	14 37	16 02	02 47	21 31	19 48	22 58	01 50
18	19 52 06	28 31 45	02 ♏ 56 04	09 ♏ 06 08	03 26	15 36	16 48	03 01	21 29	19 51	23 01	01 52
19	19 56 03	29 ♑ 32 50	15 ♏ 12 23	21 ♏ 15 23	02 13	16 36	17 34	03 15	21 27	19 53	23 03	01 54
20	19 59 59	00 ♒ 33 55	27 ♏ 15 39	03 ♐ 13 45	00 ♒ 57	17 35	18 20	03 29	21 25	19 56	23 05	01 56
21	20 03 56	01 35 00	09 ♐ 10 10	15 ♐ 05 23	29 ♑ 39	18 33	19 06	03 43	21 23	19 58	23 07	01 58
22	20 07 52	02 36 04	20 ♐ 59 51	26 ♐ 54 00	28 23	19 31	19 52	03 58	21 21	20 01	23 09	02 00
23	20 11 49	03 37 07	02 ♑ 48 13	08 ♑ 42 51	27 10	20 28	20 38	04 12	21 18	20 03	23 11	02 02
24	20 15 45	04 38 10	14 ♑ 38 14	20 ♑ 34 40	26 01	21 25	21 24	04 26	21 16	20 06	23 13	02 04
25	20 19 42	05 39 12	26 ♑ 32 25	02 ♒ 31 43	25 00	22 21	22 10	04 40	21 13	20 08	23 16	02 06
26	20 23 38	06 40 13	08 ♒ 32 49	14 ♒ 35 54	24 06	23 17	22 56	04 54	21 11	20 11	23 18	02 08
27	20 27 35	07 41 14	20 ♒ 41 13	26 ♒ 48 56	23 21	24 12	23 42	05 09	21 08	20 14	23 20	02 10
28	20 31 32	08 42 13	02 ♓ 59 15	09 ♓ 12 22	22 44	25 06	24 28	05 23	21 05	20 17	23 22	02 12
29	20 35 28	09 43 12	15 ♓ 28 28	21 ♓ 47 47	22 17	26 00	25 14	05 37	21 02	20 19	23 24	02 13
30	20 39 25	10 44 09	28 ♓ 10 30	04 ♈ 36 51	21 58	26 53	26 01	05 51	20 59	20 22	23 27	02 15
31	20 43 21	11 ♒ 45 05	11 ♈ 07 03	17 ♈ 41 17	21 ♑ 47	27 ♓ 45	26 ♑ 47	06 ♒ 05	20 ♍ 56	20 ♓ 25	23 ♒ 29	02 ♑ 17

Date	Moon True ☊ ° ′	Moon Mean ☊ ° ′	Moon ☽ Latitude ° ′
01	09 ♒ 28	10 ♒ 56	02 N 20
02	09 D 30	10 53	03 18
03	09 31	10 50	04 07
04	09 R 32	10 46	04 45
05	09 31	10 43	05 09
06	09 30	10 40	05 15
07	09 27	10 37	05 02
08	09 24	10 34	04 29
09	09 21	10 31	03 37
10	09 19	10 27	02 29
11	09 18	10 24	01 N 11
12	09 D 17	10 21	00 S 12
13	09 18	10 18	01 32
14	09 19	10 15	02 44
15	09 20	10 11	03 44
16	09 21	10 08	04 29
17	09 22	10 05	05 00
18	09 R 22	10 02	05 15
19	09 21	09 59	05 15
20	09 21	09 56	05 02
21	09 19	09 52	04 35
22	09 18	09 49	03 57
23	09 17	09 46	03 09
24	09 17	09 43	02 13
25	09 16	09 40	01 10
26	09 16	09 37	00 S 04
27	09 D 16	09 33	01 N 03
28	09 17	09 30	02 08
29	09 R 17	09 27	03 08
30	09 16	09 24	04 00
31	09 ♒ 16	09 ♒ 21	04 N 40

DECLINATIONS

Date	Sun ☉ ° ′	Moon ☽ ° ′	Mercury ☿ ° ′	Venus ♀ ° ′	Mars ♂ ° ′	Jupiter ♃ ° ′	Saturn ♄ ° ′	Uranus ♅ ° ′	Neptune ♆ ° ′	Pluto ♇ ° ′
01	22 S 58	07 S 10	21 S 36	13 S 36	24 S 05	20 S 45	05 N 09	04 S 57	14 S 22	17 S 45
02	22 53	01 S 33	21 11	13 10	24 04	20 42	05 09	04 56	14 21	17 45
03	22 47	04 N 15	20 46	12 43	24 03	20 40	05 10	04 55	14 21	17 45
04	22 41	10 01	20 21	12 16	24 02	20 37	05 10	04 55	14 20	17 45
05	22 34	15 30	19 55	11 49	24 00	20 34	05 10	04 54	14 19	17 45
06	22 27	20 22	19 30	11 21	23 58	20 31	05 11	04 53	14 19	17 45
07	22 19	24 11	19 05	10 54	23 56	20 28	05 11	04 52	14 18	17 45
08	22 11	26 31	18 41	10 26	23 53	20 25	05 12	04 51	14 18	17 45
09	22 03	26 59	18 18	09 58	23 50	20 22	05 13	04 51	14 17	17 45
10	21 54	25 26	17 57	09 30	23 47	20 19	05 13	04 50	14 16	17 45
11	21 44	22 01	17 38	09 02	23 44	20 16	05 14	04 49	14 16	17 45
12	21 35	17 10	17 21	08 33	23 40	20 13	05 15	04 48	14 15	17 45
13	21 24	11 23	17 07	08 05	23 36	20 10	05 15	04 47	14 14	17 45
14	21 14	05 N 11	16 55	07 36	23 32	20 07	05 16	04 46	14 14	17 45
15	21 03	01 S 05	16 47	07 08	23 28	20 04	05 17	04 45	14 13	17 45
16	20 51	07 05	16 42	06 39	23 23	20 01	05 18	04 44	14 12	17 45
17	20 40	12 35	16 39	06 10	23 18	19 58	05 19	04 43	14 12	17 45
18	20 27	17 25	16 40	05 41	23 13	19 54	05 20	04 42	14 11	17 45
19	20 15	21 25	16 43	05 13	23 08	19 51	05 21	04 41	14 10	17 45
20	20 02	24 26	16 49	04 44	23 02	19 48	05 22	04 40	14 10	17 45
21	19 48	26 21	16 57	04 15	22 56	19 45	05 23	04 39	14 09	17 45
22	19 35	27 05	17 06	03 46	22 50	19 41	05 24	04 38	14 08	17 45
23	19 21	26 34	17 16	03 17	22 44	19 38	05 25	04 37	14 07	17 45
24	19 06	24 50	17 28	02 48	22 37	19 35	05 27	04 36	14 07	17 45
25	18 51	22 00	17 40	02 20	22 30	19 32	05 28	04 35	14 06	17 45
26	18 36	18 11	17 52	01 51	22 23	19 28	05 29	04 34	14 05	17 45
27	18 21	13 36	18 05	01 22	22 15	19 25	05 30	04 33	14 05	17 45
28	18 05	08 25	18 17	00 54	22 08	19 22	05 32	04 32	14 04	17 45
29	17 49	02 S 50	18 29	00 S 25	22 00	19 18	05 33	04 31	14 03	17 45
30	17 33	02 N 56	18 41	00 N 03	21 52	19 15	05 34	04 30	14 02	17 45
31	17 S 16	08 N 42	18 S 52	00 N 31	21 S 43	19 S 11	05 N 36	04 S 28	14 S 02	17 S 44

ZODIAC SIGN ENTRIES

Date	h m	Planets
01	00 27	☽ ♓
01	09 51	☿ ♒
03	09 50	☽ ♈
03	12 35	♀ ♓
05	15 41	♃ ♒
05	15 46	☽ ♉
07	18 12	☽ ♊
09	18 14	☽ ♋
11	17 41	☽ ♌
13	18 33	☽ ♍
15	22 30	☽ ♎
18	06 20	☽ ♏
19	22 40	☉ ♒
20	17 30	☽ ♐
21	05 36	☿ ♑
23	06 18	☽ ♑
25	18 56	☽ ♒
28	06 12	☽ ♓
30	15 25	☽ ♈

LATITUDES

Date	Mercury ☿ ° ′	Venus ♀ ° ′	Mars ♂ ° ′	Jupiter ♃ ° ′	Saturn ♄ ° ′	Uranus ♅ ° ′	Neptune ♆ ° ′	Pluto ♇ ° ′
01	01 S 30	01 S 27	00 S 42	00 S 25	02 N 03	00 S 45	00 S 21	05 N 41
04	01 01	01 15	00 44	00 25	02 04	00 45	00 21	05 41
07	00 S 22	01 01	00 45	00 25	02 05	00 45	00 21	05 41
10	00 N 26	00 47	00 47	00 25	02 06	00 45	00 21	05 41
13	01 21	00 31	00 48	00 26	02 06	00 45	00 21	05 41
16	02 16	00 S 13	00 50	00 26	02 07	00 45	00 21	05 41
19	03 01	00 N 05	00 51	00 26	02 08	00 45	00 21	05 41
22	03 27	00 25	00 52	00 26	02 09	00 44	00 21	05 41
25	03 32	00 46	00 54	00 27	02 09	00 44	00 21	05 41
28	03 17	01 09	00 55	00 27	02 10	00 44	00 21	05 41
31	02 N 51	01 N 32	00 S 56	00 S 27	02 N 11	00 S 44	00 S 21	05 N 41

DATA

Julian Date	2454833
Delta T	+67 seconds
Ayanamsa	23° 59′ 12″
Synetic vernal point	05° ♓ 07′ 48″
True obliquity of ecliptic	23° 26′ 23″

MOON'S PHASES, APSIDES AND POSITIONS ☽

Date	h m	Phase	Longitude ° ′	Eclipse Indicator
04	11 56	☽	14 ♈ 16	
11	03 27	○	21 ♋ 02	
18	02 46	☾	28 ♎ 08	
26	07 55	●	06 ♒ 30	Annular

Day	h m		
10	10 56	Perigee	
23	00 23	Apogee	
02	18 28	0N	
09	05 34	Max dec	27° N 04′
15	07 48	0S	
22	13 58	Max dec	27° S 05′
29	23 50	0N	

LONGITUDES

Date	Chiron ⚷ ° ′	Ceres ⚳ ° ′	Pallas ⚴ ° ′	Juno ⚵ ° ′	Vesta ⚶ ° ′	Black Moon Lilith ⚸ ° ′
01	18 ♒ 23	13 ♍ 14	05 ♊ 05	21 ♑ 58	00 ♉ 29	29 ♐ 46
11	19 ♒ 02	13 ♍ 36	04 ♊ 04	25 ♑ 57	01 ♉ 54	00 ♑ 53
21	19 ♒ 44	13 ♍ 14	04 ♊ 16	00 ♒ 00	03 ♉ 51	02 ♑ 00
31	20 ♒ 27	12 ♍ 09	05 ♊ 34	04 ♒ 05	06 ♉ 14	03 ♑ 06

All ephemeris data is given at 12.00 UT and the Moon's longitude is additionally given for 24.00 UT

ASPECTARIAN

h m	Aspects
01 Thursday	
02 55	☽ ⚹ ♇
07 49	☽ ⚹ ♂
10 15	☽ ⊥ ♃
12 25	☽ ⊥ ☿
20 45	☽ # ♄
21 39	☽ ∥ ♅
23 09	☽ ⚹ ☉
02 Friday	
02 15	☽ Q ♇
03 58	☽ ∠ ♃
07 40	☽ ∠ ☿
08 28	☽ Q ♂
11 19	☿ ⊻ ♇
13 41	☽ ☌ ♅
18 23	☽ ☍ ♄
19 45	☽ ⊻ ♆
22 43	♀ ⊻ ♃
03 Saturday	
00 00	☽ Q ☉
07 02	☽ ⊥ ♆
08 51	☽ ⚹ ♃
09 34	☽ ⊻ ♀
12 20	☽ □ ♇
14 42	☽ ⚹ ☿
14 47	☽ # ♅
15 47	☽ ∥ ♄
18 59	☿ ∥ ♃
20 18	☽ □ ♂
21 40	☽ ⊥ ♀
23 42	☽ ∠ ♆
04 Sunday	
07 17	☽ Q ♃
11 56	☽ □ ☉
14 30	☽ Q ☿
15 26	☽ ∠ ♀
19 32	♀ ⚹ ♇
20 53	☽ # ♀
21 03	☽ ⊻ ♅
05 Monday	
01 18	☽ ⊼ ♄
02 44	☽ ⚹ ♆
06 41	☽ # ♆
06 45	☿ ∠ ♅
07 40	☽ ⊥ ♅
11 50	☽ ± ♄
15 46	☽ □ ♃
18 14	☽ △ ♇
20 05	☽ ⚹ ♀
22 35	☽ # ♇
23 20	☽ ∠ ♅
23 41	☽ Q ♆
06 Tuesday	
00 28	☽ □ ☿
03 11	♂ Q ♅
03 22	☽ P ♄
04 34	☽ △ ♂
07 46	☽ # ☿
09 47	♂ ∠ ♆
12 49	☽ # ♃
18 10	☽ Q ♀
19 13	☉ ⊥ ♆
19 50	☽ P ♇
20 08	☽ △ ☉
23 41	☽ # ☉
07 Wednesday	
00 46	☽ ⚹ ♅
04 36	☽ △ ♄
06 05	☽ □ ♆
07 09	☽ P ♂
10 03	☽ # ♂
10 48	☽ ± ♇
19 00	☽ △ ♃
20 33	☽ Q ♅
20 39	☽ ⊼ ♇
22 42	☽ ± ♂
22 43	☽ P ☉
08 Thursday	
02 09	☽ □ ♀
05 02	☽ △ ☿
06 43	☿ P ♄
08 55	☽ ⊼ ♂
14 19	☽ ± ☉
19 35	☽ P ♃
09 Friday	
00 36	☽ ⊼ ☉
01 34	☽ □ ♅
05 05	☽ □ ♄
05 53	☽ P ☿
06 40	☽ △ ♆
10 07	☽ ± ♃
15 28	☉ ⚹ ♅
19 46	☽ ⊼ ♃
20 30	☽ ± ☿
20 43	☽ ☍ ♇
10 Saturday	
02 56	♀ ⊥ ♃
05 31	☽ △ ♀
06 10	☽ ⊼ ☿
06 25	☽ P ♆
08 45	☿ ⊥ ♇
09 31	☽ Q ♄
11 04	☽ ☍ ♂
17 14	☿ ⊻ ♀
20 40	☽ ± ♆
11 Sunday	
01 06	☽ △ ♅
01 15	☽ # ♂
02 58	☿ ∥ ♇
03 27	☽ ☍ ☉
04 27	☽ ⚹ ♄
06 10	☽ ⊼ ♆
06 58	☽ P ♀
13 39	☽ # ☉
16 43	☿ St R
18 02	☉ △ ♄
19 59	☽ ☍ ♃
20 18	☽ ⊼ ♇
21 29	☽ # ♃
22 22	☽ ± ♀
12 Monday	
01 02	☽ P ♅
04 19	☽ ∠ ♄
05 57	☽ ± ♇
06 03	☽ ☍ ☿
08 43	☽ ⊼ ♀
09 24	☽ # ♇
11 09	☽ # ☿
13 23	☽ ⊼ ♂
15 24	☿ ⊥ ♇
15 34	☽ ± ♅
18 51	☽ ⊥ ♄
19 51	♃ ⊻ ♇
20 33	☽ P ♇
21 19	☉ ⊻ ♆
23 41	☽ ± ♂
13 Tuesday	
00 29	☽ # ♆
01 23	☽ ⊼ ♅
04 39	☽ ⊻ ♄
06 38	☽ ☍ ♆
07 19	☽ ⊼ ☉
15 21	☽ P ♂
18 02	☽ ± ☉
21 25	☽ △ ♇
21 47	☽ ⊼ ♃
14 Wednesday	
01 57	☽ # ♀
06 32	☽ ⊼ ☿
08 06	☽ ± ♃
10 25	☽ P ☉
11 38	☽ ∥ ♄
13 33	☽ # ♅
14 20	☽ ☍ ♀
16 25	☽ ± ☿
18 14	☽ △ ♂
23 51	☽ P ♃
15 Thursday	
04 12	☿ P ♄
04 16	☽ ☍ ♅
07 22	☽ P ☿
07 32	☽ ☌ ♄
08 36	☿ ⊥ ♀
09 52	☽ ⊼ ♆
14 37	☽ △ ☉
20 33	☽ ± ♆
16 Friday	
01 43	☽ □ ♇
02 28	☽ ∥ ♅
02 55	☽ △ ♃
04 42	☽ # ♄
08 43	☽ △ ☿
10 22	☽ ∥ ♀
12 54	☽ P ♆
16 51	♀ Q ♇
17 Saturday	
00 18	☽ ⊼ ♀
03 21	☽ □ ♂
06 17	☿ ∠ ♅
10 55	☽ ⊼ ♅
10 59	☽ Q ♇
12 29	☽ ± ♀
14 09	☽ ⊻ ♄
16 56	☽ △ ♆
19 35	☽ ∥ ♆
22 20	☽ ± ♅
18 Sunday	
01 32	☽ ⊥ ♄
02 46	☽ □ ☉
07 07	☽ P ♀
07 59	☽ ∥ ☿
09 56	☽ ⚹ ♇
12 09	☽ □ ♃
12 53	☽ □ ☿
13 48	☽ ∥ ♇
15 43	☽ P ♅
15 51	☽ Q ♂
18 52	☽ ∠ ♄
18 59	♂ ⊥ ♆
19 07	☿ ☌ ♃
19 Monday	
02 05	☽ ∥ ♃
04 47	☽ ∥ ☉
05 16	♀ # ♄
14 59	☽ △ ♀
15 22	☽ ∠ ♇
16 58	☽ ⚹ ♂
17 03	☽ Q ☉
17 56	☿ ⊻ ♇
18 42	☿ ∠ ♀
21 00	☽ Q ☿
21 18	☽ △ ♅
20 Tuesday	
00 11	☽ ∥ ♂
00 14	☽ Q ♃
00 21	☽ ⚹ ♄
03 37	☽ □ ♆
09 20	☽ ⊥ ♇
14 53	♀ ∥ ♅
15 59	☉ ☌ ☿
18 41	☽ ⚹ ☿
19 15	☽ ⚹ ☉
21 25	☽ ⊻ ♇
21 Wednesday	
00 20	☽ Q ♄
00 46	☽ ⚹ ♃
01 02	☽ ∠ ♂
15 57	☽ Q ♆
17 41	♀ ∠ ♃
20 30	☽ ⊥ ♂
20 45	☉ ∥ ♃
21 25	☉ ⊻ ♇
22 02	☽ ∠ ☿
22 Thursday	
04 27	☽ ∠ ☉
07 46	☽ ∠ ♃
08 43	☽ □ ♀
09 32	☽ ⊻ ♂
09 59	☽ □ ♅
12 42	☽ □ ♄
14 32	☽ ⊥ ☿
16 23	☽ ⚹ ♆
16 59	♂ ⚹ ♅
23 Friday	
00 28	☽ ⊥ ☉
01 01	♀ ☌ ♅
01 35	☽ ⊻ ☿
02 27	☽ ⊥ ♃
10 26	☽ ☌ ♇
13 49	☽ ⊻ ☉
14 53	☽ ⊻ ♃
22 42	☽ Q ♅
22 58	☽ ∠ ♆
24 Saturday	
00 30	☽ Q ♀
05 44	☉ ☌ ♃
08 04	♂ △ ♄
08 15	♀ ☍ ♄
09 05	♀ ⚹ ♂
17 15	☽ ⊥ ♆
23 04	☽ ⚹ ♅
23 21	☉ ∠ ♅
25 Sunday	
01 20	☽ △ ♄
02 35	☽ ☌ ♂
02 52	☽ ⚹ ♀
05 23	☽ ⊻ ♆
08 09	☽ ∥ ♂
09 08	☽ ☌ ☿
21 35	☿ ∥ ♇
23 10	☽ ⊻ ♇
26 Monday	
00 48	☉ P ♄
04 24	☽ ∥ ♃
04 36	☽ ☌ ♃
05 17	☽ ∠ ♅
07 18	☽ P ♄
07 55	☽ ☌ ☉
09 28	☽ ∥ ☉
11 10	☽ ⊥ ♇
11 25	☽ ∠ ♀
12 29	♀ ⊻ ♆
13 44	☽ ∥ ☿
14 31	☽ ∥ ♇
23 14	☽ ⊥ ♅
23 31	☿ ⚹ ♀
23 58	♂ ⊻ ♆
27 Tuesday	
01 06	☽ ± ♄
05 03	☽ ∠ ♇
06 11	☿ ☌ ♂
06 42	☽ ⊥ ♀
09 39	☽ ∥ ♆
11 06	☽ ⊻ ♅
12 22	☿ ⊻ ♆
12 52	☽ ⊼ ♄
16 57	☽ ⊻ ☿
17 12	☽ ☌ ♆
18 18	☽ ⊻ ♂
19 26	☽ ⊻ ♀
23 05	♃ ∠ ♅
23 34	☉ ⊥ ♇
28 Wednesday	
01 59	☉ ∥ ☿
04 06	☽ ⊥ ☿
06 48	☽ ⊥ ♂
10 27	☽ ⚹ ♇
16 43	☽ ⊻ ♃
20 49	☽ ∠ ☿
29 Thursday	
00 01	☽ ⊻ ☉
00 32	☽ # ♄
01 20	☽ ∠ ♂
04 29	☽ ⊥ ♃
04 54	☽ ∥ ♅
09 36	☽ Q ♇
12 31	☽ ⊥ ☉
18 31	☉ ∥ ♇
21 15	☽ ☌ ♅
21 57	☽ ∠ ♃
22 31	☽ ☍ ♄
22 59	☽ ∥ ♀
30 Friday	
00 34	☽ ⚹ ☿
00 34	☽ # ♀
03 05	☽ ⊻ ♆
07 02	☽ ∠ ☉
07 41	☽ ⚹ ♂
09 24	☽ ☌ ♀
14 23	☽ ⊥ ♆
18 25	☽ # ♅
19 38	☽ □ ♇
22 37	☽ Q ☿
22 58	☽ ∥ ♄
23 03	♃ P ♄
31 Saturday	
02 34	☽ ⚹ ♃
07 08	☽ ∠ ♆
07 26	☽ Q ♂
13 16	☽ ⚹ ☉

FEBRUARY 2009

LONGITUDES

Date	Sidereal time h m s	Sun ☉ ° ′ ″	Moon ☽ ° ′ ″	Moon ☽ 24.00 ° ′ ″	Mercury ☿ ° ′	Venus ♀ ° ′	Mars ♂ ° ′	Jupiter ♃ ° ′	Saturn ♄ ° ′	Uranus ♅ ° ′	Neptune ♆ ° ′	Pluto ♇ ° ′
01	20 47 18	12 ♒ 46 00	24 ♈ 19 46	01 ♉ 02 41	21 ♑ 45	28 ♓ 37	27 ♑ 33	06 ♒ 19	20 ♍ 53	20 ♓ 28	23 ♒ 31	02 ♑ 19
02	20 51 14	13 46 53	07 ♉ 50 11	14 ♉ 42 20	21 D 50	29 ♓ 27	28 20	06 34	20 R 50	20 31	23 33	02 21
03	20 55 11	14 47 45	21 ♉ 39 13	28 ♉ 40 46	22 02	00 ♈ 17	29 06	06 48	20 46	20 34	23 36	02 22
04	20 59 07	15 48 36	05 ♊ 46 54	12 ♊ 57 22	22 21	01 06	29 ♑ 52	07 02	20 43	20 37	23 38	02 24
05	21 03 04	16 49 25	20 ♊ 11 50	27 ♊ 29 53	22 46	01 54	00 ♒ 39	07 16	20 39	20 40	23 40	02 26
06	21 07 01	17 50 13	04 ♋ 50 56	12 ♋ 14 17	23 17	02 42	01 25	07 30	20 36	20 43	23 42	02 27
07	21 10 57	18 51 00	19 ♋ 39 11	27 ♋ 04 45	23 52	03 28	02 12	07 44	20 32	20 46	23 45	02 29
08	21 14 54	19 51 45	04 ♌ 30 05	11 ♌ 54 12	24 32	04 13	02 58	07 58	20 28	20 49	23 47	02 31
09	21 18 50	20 52 28	19 ♌ 16 11	26 ♌ 35 06	25 17	04 58	03 45	08 12	20 25	20 52	23 49	02 32
10	21 22 47	21 53 10	03 ♍ 50 08	11 ♍ 00 32	26 05	05 41	04 31	08 26	20 21	20 55	23 52	02 34
11	21 26 43	22 53 51	18 ♍ 05 41	25 ♍ 05 05	26 57	06 23	05 18	08 40	20 17	20 58	23 54	02 36
12	21 30 40	23 54 31	01 ♎ 58 23	08 ♎ 45 24	27 52	07 04	06 04	08 54	20 13	21 01	23 56	02 37
13	21 34 36	24 55 09	15 ♎ 26 06	22 ♎ 00 31	28 50	07 44	06 51	09 08	20 09	21 04	23 58	02 39
14	21 38 33	25 55 46	28 ♎ 28 53	04 ♏ 51 29	29 ♑ 51	08 22	07 38	09 22	20 05	21 07	24 01	02 40
15	21 42 30	26 56 23	11 ♏ 08 44	17 ♏ 21 05	00 ♒ 54	09 00	08 24	09 36	20 01	21 11	24 03	02 42
16	21 46 26	27 56 57	23 ♏ 29 03	29 ♏ 33 13	02 00	09 35	09 11	09 50	19 56	21 14	24 05	02 43
17	21 50 23	28 57 31	05 ♐ 34 10	11 ♐ 32 31	03 08	10 10	09 58	10 04	19 52	21 17	24 08	02 44
18	21 54 19	29 ♒ 58 04	17 ♐ 28 52	23 ♐ 23 52	04 18	10 43	10 45	10 18	19 48	21 20	24 10	02 46
19	21 58 16	00 ♓ 58 35	29 ♐ 18 05	05 ♑ 12 06	05 30	11 15	11 31	10 31	19 43	21 24	24 12	02 47
20	22 02 12	01 59 05	11 ♑ 06 30	17 ♑ 01 47	06 43	11 45	12 18	10 45	19 39	21 27	24 14	02 49
21	22 06 09	02 59 33	22 ♑ 58 27	28 ♑ 56 57	07 59	12 13	13 05	10 59	19 35	21 30	24 17	02 50
22	22 10 05	04 00 00	04 ♒ 57 40	11 ♒ 00 57	09 16	12 40	13 52	11 13	19 30	21 33	24 19	02 51
23	22 14 02	05 00 26	17 ♒ 07 08	23 ♒ 16 25	10 34	13 05	14 38	11 26	19 26	21 37	24 21	02 52
24	22 17 59	06 00 50	29 ♒ 29 02	05 ♓ 45 06	11 54	13 28	15 25	11 40	19 21	21 40	24 23	02 54
25	22 21 55	07 01 12	12 ♓ 04 42	18 ♓ 27 54	13 16	13 49	16 12	11 53	19 16	21 43	24 26	02 55
26	22 25 52	08 01 32	24 ♓ 54 41	01 ♈ 25 00	14 38	14 09	16 59	12 07	19 12	21 47	24 28	02 56
27	22 29 48	09 01 51	07 ♈ 58 48	14 ♈ 35 57	16 02	14 26	17 46	12 20	19 07	21 50	24 30	02 57
28	22 33 45	10 ♓ 02 08	21 ♈ 16 22	27 ♈ 59 54	17 ♒ 28	14 ♈ 42	18 ♒ 33	12 ♒ 34	19 ♍ 02	21 ♓ 53	24 ♒ 32	02 ♑ 58

Date	Moon True ☊ ° ′	Moon Mean ☊ ° ′	Moon ☽ Latitude ° ′
01	09 ♒ 16	09 ♒ 17	05 N 07
02	09 R 16	09 14	05 17
03	09 D 16	09 11	05 10
04	09 16	09 08	04 44
05	09 17	09 05	04 00
06	09 17	09 02	03 00
07	09 18	08 58	01 47
08	09 18	08 55	00 N 27
09	09 R 18	08 52	00 S 55
10	09 18	08 49	02 12
11	09 16	08 46	03 19
12	09 14	08 43	04 12
13	09 12	08 39	04 50
14	09 10	08 36	05 11
15	09 08	08 33	05 16
16	09 08	08 30	05 06
17	09 D 08	08 27	04 43
18	09 09	08 23	04 08
19	09 10	08 20	03 23
20	09 12	08 17	02 29
21	09 14	08 14	01 28
22	09 14	08 11	00 S 24
23	09 R 14	08 08	00 N 43
24	09 13	08 04	01 49
25	09 10	08 01	02 51
26	09 06	07 58	03 45
27	09 01	07 55	04 28
28	08 ♒ 56	07 ♒ 52	04 N 58

DECLINATIONS

Date	Sun ☉ ° ′	Moon ☽ ° ′	Mercury ☿ ° ′	Venus ♀ ° ′	Mars ♂ ° ′	Jupiter ♃ ° ′	Saturn ♄ ° ′	Uranus ♅ ° ′	Neptune ♆ ° ′	Pluto ♇ ° ′
01	16 S 59	14 N 11	19 S 02	00 N 59	21 S 35	19 S 08	05 N 37	04 S 27	14 S 01	17 S 44
02	16 41	19 07	19 12	01 27	21 26	19 04	05 39	04 26	14 00	17 44
03	16 24	23 10	19 21	01 54	21 17	19 01	05 40	04 25	14 00	17 44
04	16 06	25 56	19 29	02 22	21 07	18 58	05 42	04 24	13 59	17 44
05	15 48	27 04	19 37	02 49	20 58	18 54	05 43	04 23	13 58	17 44
06	15 29	26 21	19 43	03 16	20 48	18 51	05 45	04 21	13 57	17 44
07	15 10	23 46	19 48	03 42	20 38	18 47	05 47	04 20	13 57	17 44
08	14 51	19 34	19 52	04 09	20 27	18 43	05 48	04 19	13 56	17 44
09	14 32	14 10	19 56	04 35	20 17	18 40	05 50	04 18	13 55	17 44
10	14 13	08 03	19 58	05 00	20 06	18 36	05 52	04 17	13 54	17 44
11	13 53	01 N 39	19 59	05 25	19 55	18 33	05 53	04 15	13 54	17 44
12	13 33	04 S 38	19 58	05 50	19 44	18 29	05 55	04 14	13 53	17 44
13	13 13	10 32	19 57	06 15	19 33	18 26	05 57	04 13	13 52	17 44
14	12 53	15 46	19 54	06 39	19 21	18 22	05 59	04 12	13 51	17 44
15	12 32	20 10	19 50	07 03	19 09	18 18	06 00	04 10	13 51	17 44
16	12 11	23 35	19 45	07 26	18 57	18 15	06 02	04 09	13 50	17 43
17	11 50	25 53	19 39	07 49	18 45	18 11	06 04	04 08	13 49	17 43
18	11 29	26 58	19 31	08 11	18 33	18 08	06 06	04 06	13 48	17 43
19	11 08	26 49	19 22	08 33	18 20	18 04	06 08	04 05	13 48	17 43
20	10 46	25 27	19 12	08 54	18 07	18 00	06 09	04 04	13 47	17 43
21	10 24	22 56	19 01	09 15	17 54	17 57	06 11	04 02	13 46	17 43
22	10 03	19 24	18 48	09 35	17 41	17 53	06 13	04 01	13 45	17 43
23	09 41	15 01	18 34	09 54	17 28	17 49	06 15	04 00	13 45	17 43
24	09 18	09 57	18 19	10 12	17 14	17 46	06 17	03 59	13 44	17 43
25	08 56	04 S 24	18 02	10 30	17 01	17 42	06 19	03 57	13 43	17 43
26	08 34	01 N 25	17 44	10 47	16 47	17 38	06 21	03 56	13 42	17 43
27	08 11	07 16	17 25	11 04	16 33	17 35	06 23	03 55	13 42	17 43
28	07 S 48	12 N 54	17 S 05	11 N 19	16 S 19	17 S 31	06 N 25	03 S 53	13 S 41	17 S 42

ZODIAC SIGN ENTRIES

Date	h m	Planets
01	22 09	☽ ♉
03	03 41	♀ ♈
04	02 14	☽ ♊
04	15 55	♂ ♒
06	04 06	☽ ♋
08	04 43	☽ ♌
10	05 38	☽ ♍
12	08 33	☽ ♎
14	14 51	☽ ♏
14	15 39	☿ ♒
17	00 53	☽ ♐
18	12 46	☉ ♓
19	13 25	☽ ♑
22	02 06	☽ ♒
24	13 00	☽ ♓
26	21 24	☽ ♈

LATITUDES

Date	Mercury ☿ ° ′	Venus ♀ ° ′	Mars ♂ ° ′	Jupiter ♃ ° ′	Saturn ♄ ° ′	Uranus ♅ ° ′	Neptune ♆ ° ′	Pluto ♇ ° ′
01	02 N 41	01 N 40	00 S 57	00 S 27	02 N 11	00 S 44	00 S 21	05 N 41
04	02 07	02 06	00 58	00 28	02 12	00 44	00 21	05 41
07	01 33	02 32	00 59	00 28	02 12	00 44	00 21	05 41
10	00 59	02 59	01 00	00 28	02 13	00 44	00 21	05 41
13	00 N 27	03 28	01 01	00 28	02 13	00 44	00 21	05 41
16	00 S 02	03 57	01 02	00 29	02 14	00 44	00 22	05 41
19	00 29	04 27	01 03	00 29	02 14	00 44	00 22	05 41
22	00 54	04 58	01 04	00 29	02 15	00 44	00 22	05 42
25	01 15	05 29	01 05	00 30	02 15	00 44	00 22	05 42
28	01 33	06 00	01 06	00 30	02 15	00 44	00 22	05 42
31	01 S 48	06 N 30	01 S 06	00 S 31	02 N 16	00 S 44	00 S 22	05 N 42

DATA

Julian Date	2454864
Delta T	+67 seconds
Ayanamsa	23° 59′ 17″
Synetic vernal point	05° ♓ 07′ 42″
True obliquity of ecliptic	23° 26′ 23″

MOON'S PHASES, APSIDES AND POSITIONS ☽

Date	h m	Phase	Longitude ° ′	Eclipse Indicator
02	23 13	☽	14 ♉ 15	
09	14 49	○	21 ♌ 00	
16	21 37	☾	28 ♏ 21	
25	01 35	●	06 ♓ 35	

Day	h m		
07	20 14	Perigee	
19	17 06	Apogee	
05	14 53	Max dec	27° N 05′
11	18 14	0S	
18	21 06	Max dec	27° S 03′
26	06 13	0N	

LONGITUDES

Date	Chiron ⚷ ° ′	Ceres ⚳ ° ′	Pallas ⚴ ° ′	Juno ⚵ ° ′	Vesta ⚶ ° ′	Black Moon Lilith ⚸ ° ′
01	20 ♒ 31	12 ♍ 00	05 ♊ 46	04 ♒ 30	06 ♉ 30	03 ♑ 13
11	21 ♒ 15	10 ♍ 14	08 ♊ 06	08 ♒ 36	09 ♉ 19	04 ♑ 20
21	21 ♒ 59	08 ♍ 02	11 ♊ 11	12 ♒ 44	12 ♉ 26	05 ♑ 26
31	22 ♒ 41	05 ♍ 43	14 ♊ 52	16 ♒ 51	15 ♉ 48	06 ♑ 33

All ephemeris data is given at 12.00 UT and the Moon's longitude is additionally given for 24.00 UT

ASPECTARIAN

h m	Aspects
01 Sunday	
00 58	☽ Q ♃
05 00	☽ ⊻ ♅
05 49	☽ ⊼ ♄
07 11	☿ St D
07 21	☽ □ ☿
10 32	☽ ⚹ ♆
11 14	☽ ⋕ ♆
12 51	☽ Q ☉
15 51	☽ ⊥ ♅
16 33	☽ ± ♄
18 08	☽ □ ♂
20 11	☽ ⊻ ♀
22 02	☿ ∥ ♃
02 Monday	
00 26	☽ ⋕ ☉
02 17	☽ △ ♇
04 55	☽ ⋕ ♇
07 32	☽ ⊥ ♀
07 54	☽ ∠ ♅
07 59	☽ Q ♆
08 29	☽ Ᵽ ♄
09 43	☽ □ ♃
11 45	☽ ⋕ ♃
12 26	☽ ⋕ ☿
15 04	♀ ⊥ ♆
23 13	☽ □ ☉
03 Tuesday	
00 19	☽ ∠ ♀
00 25	☽ ⋕ ♂
04 36	☽ Ᵽ ♇
06 10	☉ ⊥ ♅
10 07	☽ ⚹ ♅
10 29	☽ △ ♄
11 25	☉ ± ♄
12 41	☽ △ ☿
15 20	☽ □ ♆
20 05	☽ ± ♇
04 Wednesday	
01 27	☽ △ ♂
03 38	☽ ⚹ ♀
06 18	☽ ⊼ ♇
06 38	☽ Q ♅
14 08	☽ △ ♃
14 43	☽ Ᵽ ☿
05 Thursday	
00 58	☽ Q ♀
04 03	☽ Ᵽ ♂
06 00	☽ △ ☉
06 09	☽ ± ☿
10 56	♄ ☍ ♅
12 45	☽ □ ♄
12 46	☽ □ ♅
15 28	☽ Ᵽ ♃
16 23	☽ ⊼ ☿
17 44	☽ △ ♆
19 44	☽ ± ♂
21 10	☉ ∠ ♀
06 Friday	
02 45	☉ ∠ ♇
04 25	♀ □ ♇
06 06	☽ ⊼ ♂
06 27	☽ ± ♃
08 06	☽ ☍ ♇
08 18	☽ □ ♀
08 29	☽ Ᵽ ☉
16 23	☽ ⊼ ♃
18 04	☽ Q ♄
18 17	☽ Ᵽ ♆
07 Saturday	
00 11	☽ ± ☉
07 02	☿ ⊻ ♆
08 55	☽ ± ♆
10 36	☽ ⊼ ☉
13 25	☽ ⚹ ♄
13 48	☽ △ ♅
18 38	☽ ⊼ ♆
19 07	☽ ☍ ☿
21 17	♂ ⊻ ♇
08 Sunday	
07 20	☽ ⋕ ♂
08 47	☽ ⊼ ♇
09 23	☽ ☍ ♂
10 30	☽ ⋕ ☿
11 32	☽ △ ♀
13 34	☽ ∠ ♄
14 08	☽ Ᵽ ♅
16 07	☽ ⋕ ♃
17 43	☽ ☍ ♃
18 31	☽ ± ♇
20 40	☽ ⋕ ♇
21 12	♀ ⋕ ♅
09 Monday	
01 39	☉ ⊼ ♄
04 07	☽ ⊥ ♄
04 47	☽ ± ♅
09 10	☽ Ᵽ ♇
10 24	☽ ⋕ ☉
11 42	☉ ⊻ ♅
13 03	☽ ⋕ ♆
13 11	☽ Ᵽ ♀
13 51	☽ ⊻ ♄
14 37	☽ ⊼ ♅
14 49	☽ ☍ ☉
19 29	☽ ☍ ♆
22 24	☽ ⊼ ☿
10 Tuesday	
04 45	☽ ± ♀
08 54	☽ ± ☿
09 53	☽ △ ♇
13 12	☽ ⊼ ♂
15 14	☽ ⊼ ♀
19 49	☽ ⊼ ♃
20 14	☽ ∥ ♄
22 46	☽ ∥ ♀
23 49	☽ ± ♂
11 Wednesday	
00 53	☽ Ᵽ ☿
02 15	☽ ⋕ ♅
02 38	♃ ⊥ ♇
05 04	☿ ∥ ♂
06 05	☽ ± ♃
11 21	☉ ∥ ♆
11 31	♂ Ᵽ ♄
15 43	☽ ☌ ♄
15 59	☽ Ᵽ ♂
16 56	☽ ☍ ♅
20 52	☽ ⊼ ☉
21 44	☽ Ᵽ ♃
21 59	☽ ⊼ ♆
12 Thursday	
04 17	☽ △ ☿
08 06	☽ ± ☉
08 25	☽ ± ♆
10 07	♂ ∠ ♅
10 26	☽ ∥ ♅
12 41	☉ ☌ ♆
13 08	☽ □ ♇
16 46	♀ ∥ ♄
17 04	☽ ⋕ ♄
17 05	☽ ⋕ ♀
19 40	☽ △ ♂
21 28	☽ ☍ ♀
13 Friday	
00 21	☽ Ᵽ ♆
00 29	☽ △ ♃
01 16	☽ Ᵽ ☉
20 32	☽ ⊻ ♄
21 31	☽ Q ♇
22 19	☽ ⊼ ♅
23 10	☽ ∥ ☉
14 Saturday	
02 52	☽ ∥ ♆
03 40	☽ △ ♆
06 51	☽ △ ☉
07 33	☽ ⊥ ♄
09 27	☽ ± ♅
14 46	☽ □ ☿
19 52	☽ ⚹ ♇
22 04	☽ ∥ ♇
15 Sunday	
00 21	☽ ∠ ♄
01 22	☽ ∥ ♃
02 28	☽ Ᵽ ♅
06 16	☽ ∥ ♂
06 24	☽ □ ♂
07 40	☽ ⊼ ♀
08 59	☽ □ ♃
10 02	☽ ∥ ☿
14 26	♀ ∠ ♆
19 48	☽ ± ♀
21 06	♂ ⊥ ♇
16 Monday	
00 41	☽ ∠ ♇
04 29	☽ Q ☿
05 05	☽ ⚹ ♄
07 33	☽ △ ♅
13 12	☽ □ ♆
14 17	☽ Ᵽ ♀
18 23	☽ ⊥ ♇
19 48	☽ Q ♂
20 45	☽ Q ♃
21 37	☽ □ ☉
17 Tuesday	
03 43	☿ ⊻ ♇
04 39	☽ Q ♄
04 47	♀ ⚹ ♃
06 20	☽ ⊻ ♇
06 36	☽ ⚹ ☿
16 27	♂ ☌ ♃
21 12	☽ ⚹ ♃
21 26	☽ ⚹ ♂
21 41	☽ △ ♀
18 Wednesday	
01 13	☽ Q ♆
09 31	♀ ⚹ ♂
13 05	☽ Q ☉
16 05	☽ ∠ ☿
16 40	☽ □ ♄
19 51	☽ □ ♅
21 34	☿ Ᵽ ♄
19 Thursday	
01 36	☽ ⚹ ♆
04 10	☽ ∠ ♃
05 57	☽ ∠ ♂
12 26	☽ ⊥ ☿
15 44	☽ ⚹ ☉
19 06	☽ ☌ ♇
22 50	☽ ⊥ ♃
20 Friday	
01 33	☽ ⊥ ♂
02 03	☽ ⊻ ☿
06 26	☿ ∠ ♅
08 12	☽ ∠ ♆
08 37	☽ Q ♅
11 16	☽ ⊻ ♃
13 21	☽ □ ♀
14 35	☽ ⊻ ♂
21 Saturday	
01 01	☽ ∠ ☉
02 30	☽ ⊥ ♆
05 11	☽ △ ♄
06 16	♂ ∥ ♃
08 03	☉ ⚹ ♇
09 01	☽ ⚹ ♅
14 38	☽ ⊻ ♆
20 49	☽ ⊥ ☉
22 Sunday	
01 56	♂ ± ♄
03 06	☽ Q ♀
04 16	☿ ⊥ ♇
07 48	☽ ⊻ ♇
09 01	♂ ∥ ♇
09 55	☽ ⊻ ☉
11 05	☽ Ᵽ ♄
15 11	☽ ∠ ♅
15 48	☽ ∥ ☿
19 44	☽ ⊥ ♇
20 57	☽ ∥ ♃
21 33	☽ ☌ ☿
21 46	☽ ∥ ♇
22 27	☽ ∥ ♂
23 Monday	
00 37	☽ ☌ ♃
03 48	☽ ⚹ ♀
04 16	☉ ⋕ ♀
04 48	☽ ± ♄
06 48	☽ ☌ ♂
09 02	☽ ⊥ ♅
13 29	☽ ∠ ♇
16 29	☽ ⊼ ♄
18 19	☽ ∥ ♆
20 49	☽ ⊻ ♅
24 Tuesday	
02 08	☽ ☌ ♆
06 53	☿ ☌ ♃
09 59	☽ ∠ ♀
10 55	☽ ⋕ ♀
15 06	☽ ∥ ☉
18 33	☽ ⚹ ♇
20 08	♂ ⊥ ♅
25 Wednesday	
01 35	☽ ☌ ☉
03 43	☽ ⊥ ♀
03 56	☽ ⋕ ♄
08 13	♃ ∥ ♇
11 38	☽ ⊻ ♃
12 13	☿ ± ♄
13 53	☽ ∥ ♅
14 30	☽ ⊻ ☿
15 23	☽ ⊻ ♀
17 21	☽ Q ♇
20 16	☽ ⊻ ♂
23 07	☽ ⊥ ♃
26 Thursday	
00 59	☿ ⚹ ♀
01 26	☽ ☍ ♄
03 06	☽ ⊥ ☿
06 09	☽ ☌ ♅
08 12	☽ ⊥ ♂
11 10	☽ ⊻ ♆
14 20	☿ ∥ ♇
16 09	☽ ∠ ♃
16 11	☉ ⊥ ♀
21 27	☿ ∥ ♃
21 47	☽ ∠ ☿
22 14	☽ ⋕ ♅
22 17	☽ ⊥ ♆
27 Friday	
01 52	☽ ∠ ♂
02 48	☽ □ ♇
08 19	☽ ∥ ♄
08 23	☿ ⊥ ♅
14 04	☽ ⊻ ☉
14 47	☽ ∠ ♆
15 35	☽ ⋕ ☉
17 54	♂ ∠ ♇
20 03	☽ ⚹ ♃
23 57	☽ ☌ ♀
28 Saturday	
01 49	☽ ⊥ ☉
04 20	☽ ⚹ ☿
04 47	☽ ∥ ♀
06 48	☽ ⚹ ♂
08 01	☽ ⊼ ♄
13 07	☽ ⊻ ♅
15 31	☽ ⋕ ♆
17 51	☽ ⚹ ♆
17 59	☽ Q ♃
18 42	☽ ± ♄
19 16	☽ ∠ ☉
20 39	☿ ∠ ♇
23 52	☽ ⊥ ♅

MARCH 2009

LONGITUDES

Date	Sidereal time h m s	Sun ☉ ° ' "	Moon ☽ ° ' "	Moon ☽ 24.00 ° ' "	Mercury ☿ ° '	Venus ♀ ° '	Mars ♂ ° '	Jupiter ♃ ° '	Saturn ♄ ° '	Uranus ♅ ° '	Neptune ♆ ° '	Pluto ♇ ° '
01	22 37 41	11 ♓ 02 22	04 ♉ 46 26	11 ♉ 35 50	18 ♒ 54	14 ♈ 55	19 ♒ 20	12 ♒ 47	18 ♍ 58	21 ♓ 57	24 ♒ 35	02 ♑ 59
02	22 41 38	12 02 35	18 ♉ 27 59	25 ♉ 22 44	20 22	15 06	20 07	13 01	18 R 53	22 00	24 37	03 00
03	22 45 34	13 02 46	02 ♊ 19 59	09 ♊ 19 37	21 51	15 15	20 53	13 14	18 48	22 04	24 39	03 01
04	22 49 31	14 02 55	16 ♊ 21 29	23 ♊ 25 28	23 21	15 21	21 40	13 27	18 43	22 07	24 41	03 02
05	22 53 28	15 03 02	00 ♋ 31 23	07 ♋ 39 01	24 52	15 26	22 27	13 40	18 39	22 10	24 43	03 03
06	22 57 24	16 03 06	14 ♋ 48 08	21 ♋ 58 26	26 24	15 27	23 14	13 53	18 34	22 14	24 45	03 04
07	23 01 21	17 03 08	29 ♋ 09 32	06 ♌ 21 01	27 58	15 R 27	24 01	14 07	18 29	22 17	24 48	03 05
08	23 05 17	18 03 08	13 ♌ 32 24	20 ♌ 43 09	29 ♒ 32	15 24	24 48	14 20	18 24	22 21	24 50	03 06
09	23 09 14	19 03 06	27 ♌ 52 39	05 ♍ 00 18	01 ♓ 08	15 18	25 35	14 33	18 20	22 24	24 52	03 07
10	23 13 10	20 03 02	12 ♍ 05 28	19 ♍ 07 33	02 45	15 10	26 22	14 45	18 15	22 28	24 54	03 08
11	23 17 07	21 02 56	26 ♍ 05 58	03 ♎ 00 12	04 23	14 59	27 09	14 58	18 10	22 31	24 56	03 09
12	23 21 03	22 02 48	09 ♎ 49 47	16 ♎ 34 24	06 02	14 46	27 56	15 11	18 05	22 34	24 58	03 09
13	23 25 00	23 02 38	23 ♎ 13 47	29 ♎ 47 49	07 43	14 30	28 43	15 24	18 01	22 38	25 00	03 10
14	23 28 57	24 02 27	06 ♏ 16 28	12 ♏ 39 52	09 24	14 12	29 ♒ 30	15 36	17 56	22 41	25 02	03 11
15	23 32 53	25 02 13	18 ♏ 58 12	25 ♏ 11 46	11 07	13 51	00 ♓ 17	15 49	17 51	22 45	25 04	03 11
16	23 36 50	26 01 58	01 ♐ 20 58	07 ♐ 26 15	12 51	13 28	01 04	16 02	17 46	22 48	25 06	03 12
17	23 40 46	27 01 41	13 ♐ 28 09	19 ♐ 27 16	14 36	13 03	01 51	16 14	17 42	22 52	25 08	03 13
18	23 44 43	28 01 23	25 ♐ 24 11	01 ♑ 19 34	16 22	12 36	02 38	16 26	17 37	22 55	25 10	03 13
19	23 48 39	29 ♓ 01 03	07 ♑ 14 04	13 ♑ 08 21	18 10	12 07	03 25	16 39	17 32	22 58	25 12	03 14
20	23 52 36	00 ♈ 00 41	19 ♑ 03 06	24 ♑ 58 56	19 59	11 37	04 12	16 51	17 28	23 02	25 14	03 14
21	23 56 32	01 00 17	00 ♒ 56 32	06 ♒ 56 27	21 49	11 04	04 59	17 03	17 23	23 05	25 16	03 15
22	00 00 29	01 59 51	12 ♒ 59 15	19 ♒ 05 28	23 40	10 30	05 46	17 15	17 19	23 09	25 18	03 15
23	00 04 26	02 59 24	25 ♒ 15 31	01 ♓ 29 46	25 33	09 55	06 33	17 27	17 14	23 12	25 20	03 16
24	00 08 22	03 58 55	07 ♓ 48 32	14 ♓ 11 59	27 27	09 19	07 20	17 39	17 10	23 15	25 22	03 16
25	00 12 19	04 58 23	20 ♓ 40 15	27 ♓ 13 19	29 ♓ 22	08 42	08 07	17 51	17 05	23 19	25 24	03 16
26	00 16 15	05 57 50	03 ♈ 51 05	10 ♈ 33 22	01 ♈ 18	08 05	08 54	18 03	17 01	23 22	25 26	03 17
27	00 20 12	06 57 15	17 ♈ 19 52	24 ♈ 10 14	03 16	07 27	09 41	18 15	16 56	23 26	25 28	03 17
28	00 24 08	07 56 37	01 ♉ 04 02	08 ♉ 00 46	05 14	06 49	10 28	18 26	16 52	23 29	25 29	03 17
29	00 28 05	08 55 58	14 ♉ 59 58	22 ♉ 01 07	07 14	06 12	11 15	18 38	16 48	23 32	25 31	03 17
30	00 32 01	09 55 16	29 ♉ 03 45	06 ♊ 07 23	09 15	05 35	12 02	18 49	16 44	23 36	25 33	03 18
31	00 35 58	10 ♈ 54 33	13 ♊ 11 39	20 ♊ 16 12	11 ♈ 17	04 ♈ 59	12 ♓ 49	19 ♒ 00	16 ♍ 39	23 ♓ 39	25 ♒ 35	03 ♑ 18

DECLINATIONS

Date	Moon True ☊ ° '	Moon Mean ☊ ° '	Moon ☽ Latitude ° '	Sun ☉ ° '	Moon ☽ ° '	Mercury ☿ ° '	Venus ♀ ° '	Mars ♂ ° '	Jupiter ♃ ° '	Saturn ♄ ° '	Uranus ♅ ° '	Neptune ♆ ° '	Pluto ♇ ° '
01	08 ♒ 51	07 ♒ 48	05 N 11	07 S 26	18 N 00	16 S 43	11 N 33	16 S 04	17 S 27	06 N 27	03 S 52	13 S 40	17 S 42
02	08 R 47	07 45	05 08	07 03	22 15	16 20	11 47	15 50	17 24	06 29	03 50	13 40	17 42
03	08 45	07 42	04 46	06 40	25 18	15 56	12 00	15 35	17 20	06 31	03 49	13 39	17 42
04	08 D 45	07 39	04 08	06 17	26 51	15 31	12 11	15 20	17 16	06 33	03 48	13 38	17 42
05	08 46	07 36	03 14	05 53	26 40	15 04	12 21	15 05	17 13	06 35	03 46	13 37	17 42
06	08 47	07 33	02 07	05 30	24 43	14 36	12 31	14 50	17 09	06 36	03 45	13 37	17 42
07	08 48	07 29	00 N 52	05 07	21 11	14 06	12 39	14 34	17 06	06 38	03 44	13 36	17 42
08	08 R 49	07 26	00 S 26	04 43	16 21	13 36	12 46	14 19	17 02	06 40	03 42	13 35	17 42
09	08 47	07 23	01 42	04 20	10 37	13 04	12 51	14 03	16 58	06 42	03 41	13 35	17 42
10	08 44	07 20	02 51	03 56	04 N 24	12 31	12 56	13 47	16 55	06 44	03 40	13 34	17 41
11	08 38	07 17	03 48	03 33	01 S 56	11 57	12 58	13 32	16 51	06 46	03 38	13 33	17 41
12	08 31	07 14	04 31	03 09	08 03	11 21	13 00	13 16	16 47	06 48	03 37	13 33	17 41
13	08 23	07 10	04 58	02 46	13 38	10 45	13 00	12 59	16 44	06 50	03 36	13 32	17 41
14	08 16	07 07	05 08	02 22	18 27	10 07	12 59	12 43	16 40	06 52	03 34	13 31	17 41
15	08 09	07 04	05 03	01 58	22 19	09 27	12 56	12 27	16 37	06 54	03 33	13 30	17 41
16	08 04	07 01	04 44	01 35	25 03	08 47	12 51	12 10	16 33	06 56	03 32	13 30	17 41
17	08 01	06 58	04 12	01 11	26 35	08 05	12 45	11 53	16 29	06 57	03 30	13 29	17 41
18	08 00	06 54	03 29	00 47	26 51	07 23	12 38	11 37	16 26	06 59	03 29	13 29	17 41
19	08 D 00	06 51	02 38	00 S 23	25 53	06 39	12 29	11 20	16 22	07 01	03 27	13 28	17 41
20	08 01	06 48	01 41	00 00	23 45	05 53	12 18	11 03	16 19	07 03	03 26	13 27	17 40
21	08 02	06 45	00 S 38	00 N 24	20 34	05 07	12 06	10 46	16 15	07 05	03 25	13 27	17 40
22	08 R 02	06 42	00 N 27	00 48	16 29	04 20	11 53	10 29	16 12	07 06	03 23	13 26	17 40
23	08 01	06 39	01 31	01 11	11 40	03 31	11 39	10 11	16 08	07 08	03 22	13 25	17 40
24	07 57	06 35	02 33	01 35	06 17	02 42	11 23	09 54	16 05	07 10	03 21	13 25	17 40
25	07 50	06 32	03 28	01 59	00 S 31	01 51	11 06	09 36	16 01	07 12	03 19	13 24	17 40
26	07 42	06 29	04 13	02 22	05 N 24	01 00	10 48	09 19	15 58	07 13	03 18	13 23	17 40
27	07 32	06 26	04 46	02 46	11 12	00 S 08	10 28	09 01	15 54	07 15	03 17	13 23	17 40
28	07 21	06 23	05 02	03 09	16 34	00 N 46	10 08	08 44	15 51	07 17	03 15	13 22	17 40
29	07 12	06 20	05 01	03 32	21 08	01 40	09 48	08 26	15 47	07 18	03 14	13 22	17 40
30	07 04	06 16	04 42	03 56	24 32	02 34	09 26	08 08	15 44	07 20	03 13	13 21	17 40
31	06 ♒ 59	06 ♒ 13	04 N 06	04 N 19	26 N 27	03 N 30	09 N 05	07 S 50	15 S 41	07 N 22	03 S 11	13 S 21	17 S 39

ZODIAC SIGN ENTRIES

Date	h	m	Planets
01	03	33	☽ ♉
03	07	59	☽ ♊
05	11	07	☽ ♋
07	13	24	☽ ♌
08	18	56	☿ ♓
09	15	34	☽ ♍
11	18	46	☽ ♎
14	00	22	☽ ♏
15	03	20	♂ ♓
16	09	21	☽ ♐
18	21	19	☽ ♑
20	11	44	☉ ♈
21	10	06	☽ ♒
23	21	08	☽ ♓
25	19	55	☿ ♈
26	05	03	☽ ♈
28	10	09	☽ ♉
30	13	36	☽ ♊

LATITUDES

Date	Mercury ☿ ° '	Venus ♀ ° '	Mars ♂ ° '	Jupiter ♃ ° '	Saturn ♄ ° '	Uranus ♅ ° '	Neptune ♆ ° '	Pluto ♇ ° '
01	01 S 39	06 N 10	01 S 06	00 S 30	02 N 16	00 S 44	00 S 22	05 N 42
04	01 53	06 40	01 07	00 31	02 16	00 44	00 22	05 42
07	02 03	07 08	01 07	00 31	02 16	00 44	00 22	05 43
10	02 10	07 33	01 08	00 32	02 16	00 44	00 22	05 43
13	02 14	07 55	01 08	00 32	02 16	00 44	00 22	05 43
16	02 13	08 11	01 09	00 32	02 16	00 44	00 22	05 43
19	02 08	08 22	01 09	00 33	02 16	00 44	00 22	05 44
22	01 59	08 25	01 10	00 33	02 16	00 44	00 22	05 44
25	01 45	08 15	01 10	00 34	02 16	00 44	00 22	05 44
28	01 26	08 06	01 10	00 34	02 16	00 44	00 22	05 44
31	01 S 03	07 N 44	01 S 10	00 S 35	02 N 16	00 S 44	00 S 22	05 N 45

DATA

Julian Date	2454892
Delta T	+67 seconds
Ayanamsa	23° 59' 20"
Synetic vernal point	05° ♓ 07' 39"
True obliquity of ecliptic	23° 26' 23"

MOON'S PHASES, APSIDES AND POSITIONS ☽

Date	h	m	Phase	Longitude ° '	Eclipse Indicator
04	07	46	☽	13 ♊ 52	
11	02	38	○	20 ♍ 40	
18	17	47	☾	28 ♐ 16	
26	16	06	●	06 ♈ 08	

Day	h	m		
07	15	13	Perigee	
19	13	17	Apogee	
04	21	34	Max dec	26° N 59'
11	04	37	0S	
18	05	05	Max dec	26° S 54'
25	14	05	0N	

LONGITUDES

Date	Chiron ⚷ ° '	Ceres ⚳ ° '	Pallas ⚴ ° '	Juno ⚵ ° '	Vesta ⚶ ° '	Black Moon Lilith ☾ ° '
01	22 ♒ 33	06 ♍ 11	14 ♊ 05	16 ♒ 02	15 ♉ 07	06 ♑ 20
11	23 ♒ 14	03 ♍ 58	18 ♊ 08	20 ♒ 08	18 ♉ 39	07 ♑ 26
21	23 ♒ 52	02 ♍ 09	22 ♊ 33	24 ♒ 13	22 ♉ 08	08 ♑ 33
31	24 ♒ 28	00 ♍ 56	27 ♊ 13	28 ♒ 16	26 ♉ 14	09 ♑ 40

ASPECTARIAN

h m	Aspects
01 Sunday	
01 47	♂ ⊼ ♄
02 59	☽ # ♂
04 20	☽ Q ☿
05 32	☽ Q ♂
06 07	☽ # ☿
08 51	☽ △ ♇
09 20	☽ # ♃
10 32	☽ # ♇
10 34	☽ ⚼ ♄
12 56	☿ ⊼ ♄
15 11	☽ Q ♆
15 51	☽ ∠ ♅
23 54	☽ ⚹ ☉
02 Monday	
01 50	♃ ± ♄
02 19	☽ □ ♃
03 05	☿ ☌ ♂
06 03	☽ ⊻ ♀
11 12	☽ ⚼ ♇
12 43	☽ △ ♄
15 02	☽ □ ♂
15 42	☽ □ ☿
16 38	☽ ⊥ ♀
18 11	☽ ⚹ ♅
22 27	☽ Q ☉
22 42	☽ □ ♆
03 Tuesday	
02 50	☽ ± ♇
08 23	☽ ∠ ♀
13 11	☽ ⊼ ♇
14 59	☽ Q ♅
15 35	☿ ⊻ ♅
17 43	☉ ⊻ ♃
20 40	☉ # ♄
04 Wednesday	
06 58	☽ △ ♃
07 46	☽ □ ☉
10 17	☽ ⚹ ♀
16 00	☽ □ ♄
21 34	☽ △ ♂
21 49	☽ □ ♅
05 Thursday	
01 17	☽ △ ☿
02 10	☽ △ ♆
02 41	♂ ⊻ ♅
06 45	☽ Q ♀
08 50	☽ ⚼ ♃
09 40	☿ ☌ ♆
09 50	☿ ∥ ♂
12 09	☉ Q ♇
16 17	☽ ☍ ♇
21 25	☉ ⊻ ♀
22 15	☽ Q ♄
06 Friday	
00 13	☽ ± ♃
00 21	☽ ⚼ ♂
03 31	☽ ⚼ ♆
05 37	☽ ⚼ ☿
10 27	☽ ⊼ ♃
13 06	☽ □ ♀
14 15	☽ △ ☉
16 19	☽ ± ♂
17 17	♀ St R
18 16	☽ ⚹ ♄
18 38	☽ ± ♆
22 31	☽ ± ☿
07 Saturday	
00 29	☽ △ ♅
02 56	☽ ⊼ ♂
04 42	☽ ⊼ ♆
09 46	☽ ⊼ ☿
17 11	☽ ⚼ ☉
18 34	☽ ⊼ ♇
19 11	☽ ∠ ♄
08 Sunday	
01 37	☽ ⚼ ♅
04 35	☽ ± ♇
05 49	☽ # ♇
08 52	☽ # ♃
09 20	☽ ± ☉
10 07	☽ ⊥ ♄
12 30	☿ ∥ ♆
12 50	♂ ☌ ♆
13 20	☽ ☍ ♃
15 05	☽ △ ♀
16 42	☽ ± ♅
19 38	☽ ⚼ ♇
19 53	☉ ☍ ♄
20 05	☽ ⊻ ♄
20 06	☽ ⊼ ☉
21 18	☽ # ♂
23 57	☽ # ♆
09 Monday	
00 11	☿ ∠ ♀
01 04	☽ # ☿
02 47	☽ ⊼ ♅
03 06	☽ ∥ ♀
06 56	☽ ☍ ♆
07 56	☽ ☍ ♂
16 02	☽ ⚼ ♀
18 11	☽ ☍ ☿
20 14	☿ # ♀
20 49	☽ △ ♇
10 Tuesday	
03 08	☽ ∥ ♄
07 05	☽ ± ♀
13 50	☽ # ☉
14 47	☽ # ♅
16 36	☽ ⊼ ♃
17 10	☽ ⊼ ♀
17 38	☿ ⚹ ♇
22 26	☽ ☌ ♄
11 Wednesday	
02 38	☽ ☍ ☉
03 02	☽ ± ♃
05 48	☽ ☍ ♅
06 12	☉ ∥ ♅
09 27	♂ ∥ ♆
09 38	☉ ⊥ ♃
09 59	☽ ⊼ ♆
12 35	♀ ⚹ ♃
13 56	☽ ⊼ ♂
17 49	☽ ∥ ☉
18 31	☽ ∥ ♅
18 49	☽ ⚼ ♃
20 25	☽ ± ♆
12 Thursday	
00 15	☽ □ ♇
01 00	☽ ± ♂
04 24	☽ ⊼ ☿
06 57	☽ # ♄
12 15	☽ ⚼ ♆
16 28	☽ ± ☿
17 51	☽ ⚼ ♂
20 36	☽ ☍ ♀
21 40	☽ △ ♃
13 Friday	
00 30	☽ ∥ ☿
01 27	☉ ☌ ♅
02 38	☽ ⊻ ♄
08 16	☽ Q ♇
09 08	☽ # ♀
09 13	☽ ∥ ♂
10 55	☽ ⊼ ♅
10 55	☽ ⚼ ☿
11 07	♀ # ♂
11 32	☽ ∥ ♆
11 38	☽ ⊼ ☉
13 24	☽ ⊥ ♄
15 14	☽ △ ♆
21 34	☿ ⊥ ♀
21 54	☽ ± ♅
22 39	☽ △ ♂
23 30	☽ ± ☉
14 Saturday	
02 42	☽ ∥ ♃
05 23	♀ ∠ ♂
05 49	☽ ∠ ♄
06 14	☽ ⚹ ♇
07 51	☽ ∥ ♇
14 39	☽ ⚼ ♅
17 37	☽ ⚼ ☉
18 45	☽ △ ☿
15 Sunday	
02 31	☽ ⊼ ♀
05 53	☽ □ ♃
09 52	☽ ⚹ ♄
10 30	☽ ∠ ♇
12 53	☉ ⊻ ♆
13 39	☽ ± ♀
19 18	☽ △ ♅
23 48	☽ □ ♆
16 Monday	
00 43	☽ △ ☉
03 53	☽ ⊥ ♇
06 33	☽ ⚼ ♀
08 56	☽ Q ♄
11 24	☽ □ ♂
15 38	☽ ⊻ ♇
17 21	☽ Q ♃
19 00	☿ ⊻ ♀
17 Tuesday	
11 12	☽ △ ♀
11 20	☽ Q ♆
14 39	☽ □ ☿
17 38	☽ ⚹ ♃
20 22	☿ Q ♇
20 24	☽ □ ♄
18 Wednesday	
01 42	☽ Q ♂
06 57	☽ □ ♅
11 32	☽ ⚹ ♆
13 03	☿ ⊻ ♃
17 47	☽ □ ☉
19 Thursday	
00 16	☿ # ♄
00 27	☽ ∠ ♃
03 41	☽ ⚹ ♂
03 52	☽ ☌ ♇
04 00	☿ ☍ ♄
06 15	♂ ⚹ ♇
09 26	☽ Q ☿
18 03	☽ ∠ ♆
19 03	☽ ⊥ ♃
19 38	☽ Q ♅
21 32	☽ □ ♀
20 Friday	
07 27	☽ ⊻ ♃
08 48	☽ △ ♄
09 42	☽ Q ☉
12 19	☽ ∠ ♂
12 23	☽ ⊥ ♆
14 13	☽ ⚹ ☿
20 06	☽ ⚹ ♅
21 Saturday	
00 33	☽ ⊻ ♆
07 47	☽ ⊥ ♂
08 24	☽ Q ♀
12 08	☽ ⚹ ☉
13 34	♀ ⊥ ♂
14 53	☽ ⚼ ♄
16 37	☽ ⊻ ♇
17 46	♃ ⊥ ♅
18 40	☉ ⊥ ♆
20 39	☽ ⊻ ♂
22 Sunday	
01 53	☽ ∠ ☿
02 21	☽ ∠ ♅
04 36	☽ ⊥ ♇
05 02	☿ ☌ ♅
05 31	☽ ∥ ♇
06 01	☿ ⊥ ♃
07 18	☽ ⚹ ♀
08 42	☽ ± ♄
13 37	☽ ∥ ♃
16 46	♃ ⊼ ♄
19 47	☉ ∠ ♃
19 54	♀ ∠ ♆
20 13	☽ ⊥ ♅
20 27	☽ ⊼ ♄
20 33	☽ ☌ ♃
20 36	☽ ∠ ☉
22 22	☽ ∠ ♇
22 53	☽ ⊥ ☿
23 Monday	
03 38	☽ ∥ ♆
08 00	☽ ⊻ ♅
09 17	☿ ⊻ ♆
11 22	☽ ∠ ♀
12 08	☽ # ♀
12 09	☽ ☌ ♆
12 39	☽ ⊻ ☿
15 38	☽ ⊥ ☉
16 43	☿ ∥ ♅
18 35	☉ □ ♇
19 13	☽ ∥ ♂
24 Tuesday	
03 23	☽ ⚹ ♇
03 52	☽ ⊥ ♀
04 07	☽ ⊻ ☉
08 12	☽ # ♄
11 02	☽ ☌ ♂
14 43	☽ ⊻ ♀
25 Wednesday	
00 24	☽ ∥ ♅
02 00	☽ Q ♇
05 24	☽ ☍ ♄
05 33	☽ ∥ ☿
06 23	☽ # ☉
06 43	☽ ⊻ ♃
09 43	☉ # ☿
16 53	☽ ☌ ♅
17 56	☽ ⊥ ♃
20 25	☽ # ☿
20 42	☽ ⊻ ♆
22 05	♀ ⊻ ♂
22 50	☽ ∥ ☉
26 Thursday	
03 30	☽ # ♅
06 37	☽ ☌ ☿
07 37	☽ ⊥ ♆
10 32	☽ ∠ ♃
10 58	☽ □ ♇
13 36	☿ ⊥ ♆
16 06	☽ ☌ ☉
19 15	☽ ☌ ♀
19 28	☽ ∥ ♄
21 36	☽ ⊻ ♂
23 48	☽ ∠ ♆
27 Friday	
03 17	☽ # ♂
08 54	☽ ⊥ ♂
09 06	☽ ∥ ♀
11 19	☽ ⊼ ♄
11 45	☿ ∠ ♃
12 16	☿ □ ♇
13 38	☽ ⚹ ♃
17 03	♃ ∠ ♇
19 24	☉ ☌ ♀
21 26	☽ # ♆
21 48	☽ ± ♄
22 45	☽ ⊻ ♅
28 Saturday	
01 40	☽ ∠ ♂
02 17	☽ ⚹ ♆
08 40	☽ # ♃
09 14	☽ ⊥ ♅
10 53	☽ Q ♃
13 23	☽ ⚼ ♄
15 51	☽ △ ♇
17 22	☽ # ♇
18 08	☉ # ♅
20 25	☽ ⊻ ☿
21 31	☽ ⊻ ♀
23 07	☽ Q ♆
29 Sunday	
00 47	☽ ⊻ ☉
00 52	☽ ∠ ♅
02 31	☿ ☌ ♀
05 11	☽ ⚹ ♂
07 24	☽ ⊥ ♀
08 28	☽ ⊥ ☿
11 53	☽ ⊥ ☉
15 04	☽ △ ♄
17 38	☽ ⚼ ♇
18 18	☽ □ ♃
22 09	☽ ∠ ♀
30 Monday	
02 26	☽ ∠ ☿
02 39	☽ ⚹ ♅
02 55	☽ Q ♂
04 25	☽ ∠ ☉
06 00	☽ □ ♆
08 59	☽ ± ♇
19 12	☽ ⊼ ♇
22 37	☽ ⚹ ♀
23 09	☽ Q ♅
31 Tuesday	
03 29	☉ ☌ ☿
03 36	☿ ∠ ♆
03 43	☉ ∠ ♆
04 21	☿ # ♅
07 50	☽ ⚹ ☉
08 12	☽ ⚹ ☿
11 18	☽ □ ♂
17 51	☽ □ ♄
18 09	☽ Q ♀
22 00	☽ △ ♃

All ephemeris data is given at 12.00 UT and the Moon's longitude is additionally given for 24.00 UT

APRIL 2009

LONGITUDES

Date	Sidereal time h m s	Sun ☉ ° ' "	Moon ☽ ° ' "	Moon ☽ 24.00 ° ' "	Mercury ☿ ° '	Venus ♀ ° '	Mars ♂ ° '	Jupiter ♃ ° '	Saturn ♄ ° '	Uranus ♅ ° '	Neptune ♆ ° '	Pluto ♇ ° '
01	00 39 55	11 ♈ 53 46	27 ♊ 20 43	04 ♋ 25 00	13 ♈ 19	04 ♈ 23	13 ♓ 35	19 ♒ 12	16 ♍ 35	23 ♓ 42	25 ♒ 36	03 ♑ 18
02	00 43 51	12 52 58	11 ♋ 28 52	18 ♋ 32 11	15 23	03 R 49	14 22	19 23	16 R 31	23 46	25 38	03 18
03	00 47 48	13 52 07	25 ♋ 34 48	02 ♌ 36 40	17 27	03 16	15 09	19 34	16 27	23 49	25 40	03 18
04	00 51 44	14 51 14	09 ♌ 37 39	16 ♌ 37 37	19 31	02 44	15 56	19 45	16 23	23 52	25 41	03 18
05	00 55 41	15 50 18	23 ♌ 36 27	00 ♍ 33 57	21 35	02 15	16 43	19 56	16 20	23 55	25 43	03 R 18
06	00 59 37	16 49 20	07 ♍ 29 54	14 ♍ 24 01	23 39	01 47	17 30	20 07	16 16	23 59	25 45	03 18
07	01 03 34	17 48 20	21 ♍ 16 02	28 ♍ 05 36	25 43	01 21	18 17	20 17	16 12	24 02	25 46	03 18
08	01 07 30	18 47 17	04 ♎ 52 22	11 ♎ 36 02	27 46	00 58	19 03	20 28	16 08	24 05	25 48	03 18
09	01 11 27	19 46 13	18 ♎ 16 15	24 ♎ 52 44	29 ♈ 48	00 36	19 50	20 38	16 05	24 08	25 49	03 18
10	01 15 24	20 45 06	01 ♏ 25 15	07 ♏ 53 38	01 ♉ 49	00 17	20 37	20 49	16 01	24 11	25 51	03 18
11	01 19 20	21 43 58	14 ♏ 17 46	20 ♏ 37 40	03 48	00 ♈ 01	21 24	20 59	15 58	24 14	25 52	03 17
12	01 23 17	22 42 48	26 ♏ 53 22	03 ♐ 05 03	05 44	29 ♓ 46	22 10	21 09	15 55	24 18	25 54	03 17
13	01 27 13	23 41 36	09 ♐ 12 58	15 ♐ 17 25	07 39	29 35	22 57	21 19	15 51	24 21	25 55	03 17
14	01 31 10	24 40 22	21 ♐ 18 48	27 ♐ 17 36	09 30	29 25	23 44	21 29	15 48	24 24	25 57	03 17
15	01 35 06	25 39 06	03 ♑ 14 21	09 ♑ 09 37	11 18	29 18	24 30	21 39	15 45	24 27	25 58	03 16
16	01 39 03	26 37 49	15 ♑ 04 01	20 ♑ 58 13	13 03	29 14	25 17	21 48	15 42	24 30	25 59	03 16
17	01 42 59	27 36 30	26 ♑ 52 53	02 ♒ 48 43	14 45	29 12	26 04	21 58	15 39	24 33	26 01	03 15
18	01 46 56	28 35 09	08 ♒ 46 24	14 ♒ 46 38	16 22	29 D 13	26 50	22 08	15 36	24 36	26 02	03 15
19	01 50 53	29 ♈ 33 47	20 ♒ 50 04	26 ♒ 57 20	17 55	29 15	27 37	22 17	15 33	24 39	26 03	03 15
20	01 54 49	00 ♉ 32 22	03 ♓ 09 01	09 ♓ 25 38	19 24	29 20	28 24	22 26	15 31	24 42	26 04	03 14
21	01 58 46	01 30 56	15 ♓ 47 38	22 ♓ 15 20	20 48	29 28	29 10	22 35	15 28	24 45	26 06	03 14
22	02 02 42	02 29 29	28 ♓ 48 58	05 ♈ 28 39	22 08	29 37	29 ♓ 57	22 44	15 26	24 48	26 07	03 13
23	02 06 39	03 27 59	12 ♈ 14 18	19 ♈ 05 44	23 23	29 ♓ 49	00 ♈ 43	22 53	15 23	24 50	26 08	03 13
24	02 10 35	04 26 28	26 ♈ 02 37	03 ♉ 04 27	24 33	00 ♈ 03	01 30	23 02	15 21	24 53	26 09	03 12
25	02 14 32	05 24 55	10 ♉ 10 37	17 ♉ 20 24	25 38	00 19	02 16	23 10	15 19	24 56	26 10	03 11
26	02 18 28	06 23 21	24 ♉ 32 59	01 ♊ 47 33	26 38	00 36	03 02	23 19	15 16	24 59	26 11	03 11
27	02 22 25	07 21 44	09 ♊ 03 16	16 ♊ 19 18	27 33	00 56	03 49	23 27	15 14	25 02	26 12	03 10
28	02 26 22	08 20 06	23 ♊ 34 54	00 ♋ 49 26	28 22	01 17	04 35	23 35	15 12	25 04	26 13	03 09
29	02 30 18	09 18 25	08 ♋ 02 18	15 ♋ 13 06	29 06	01 41	05 21	23 43	15 11	25 07	26 14	03 09
30	02 34 15	10 ♉ 16 43	22 ♋ 21 27	29 ♋ 27 10	29 ♉ 45	02 ♈ 05	06 ♈ 08	23 ♒ 51	15 ♍ 09	25 ♓ 10	26 ♒ 15	03 ♑ 08

Date	Moon True ☊ ° '	Moon Mean ☊ ° '	Moon ☽ Latitude ° '
01	06 ♒ 56	06 ♒ 10	03 N 14
02	06 R 55	06 07	02 11
03	06 D 55	06 04	01 N 00
04	06 R 55	06 00	00 S 14
05	06 54	05 57	01 28
06	06 51	05 54	02 35
07	06 45	05 51	03 32
08	06 36	05 48	04 17
09	06 25	05 45	04 47
10	06 13	05 41	05 00
11	06 00	05 38	04 58
12	05 49	05 35	04 41
13	05 40	05 32	04 12
14	05 33	05 29	03 31
15	05 29	05 26	02 42
16	05 27	05 22	01 46
17	05 D 26	05 19	00 S 45
18	05 R 26	05 16	00 N 18
19	05 26	05 13	01 21
20	05 23	05 10	02 21
21	05 19	05 06	03 16
22	05 11	05 03	04 03
23	05 01	05 00	04 37
24	04 50	04 57	04 57
25	04 38	04 54	04 59
26	04 26	04 51	04 42
27	04 17	04 47	04 07
28	04 10	04 44	03 16
29	04 06	04 41	02 13
30	04 ♒ 04	04 ♒ 38	01 N 02

DECLINATIONS

Date	Sun ☉ ° '	Moon ☽ ° '	Mercury ☿ ° '	Venus ♀ ° '	Mars ♂ ° '	Jupiter ♃ ° '	Saturn ♄ ° '	Uranus ♅ ° '	Neptune ♆ ° '	Pluto ♇ ° '
01	04 N 42	26 N 39	04 N 25	08 N 42	07 S 32	15 S 37	07 N 23	03 S 10	13 S 20	17 S 39
02	05 05	25 07	05 21	08 20	07 14	15 34	07 25	03 09	13 19	17 39
03	05 28	22 01	06 18	07 58	06 56	15 31	07 26	03 08	13 19	17 39
04	05 51	17 37	07 14	07 35	06 38	15 27	07 28	03 06	13 18	17 39
05	06 14	12 16	08 11	07 13	06 19	15 24	07 29	03 05	13 18	17 39
06	06 37	06 22	09 07	06 51	06 01	15 21	07 30	03 04	13 17	17 39
07	06 59	00 N 13	10 02	06 30	05 43	15 18	07 32	03 02	13 17	17 39
08	07 22	05 S 52	10 57	06 09	05 25	15 14	07 33	03 01	13 16	17 39
09	07 44	11 35	11 51	05 48	05 06	15 11	07 34	03 00	13 16	17 39
10	08 06	16 39	12 44	05 28	04 48	15 08	07 36	02 59	13 15	17 39
11	08 28	20 52	13 35	05 10	04 29	15 05	07 37	02 58	13 15	17 39
12	08 50	24 01	14 25	04 51	04 11	15 02	07 38	02 56	13 14	17 38
13	09 12	25 59	15 13	04 34	03 52	14 59	07 39	02 55	13 14	17 38
14	09 33	26 40	15 59	04 18	03 34	14 56	07 41	02 54	13 13	17 38
15	09 55	26 06	16 43	04 03	03 15	14 53	07 42	02 53	13 13	17 38
16	10 16	24 20	17 25	03 48	02 56	14 50	07 43	02 51	13 12	17 38
17	10 37	21 31	18 04	03 35	02 38	14 47	07 44	02 50	13 12	17 38
18	10 58	17 47	18 41	03 23	02 19	14 44	07 45	02 49	13 12	17 38
19	11 19	13 17	19 16	03 11	02 01	14 41	07 46	02 48	13 11	17 38
20	11 40	08 09	19 47	03 01	01 42	14 38	07 47	02 47	13 11	17 38
21	12 00	02 S 35	20 17	02 52	01 23	14 36	07 48	02 46	13 10	17 38
22	12 20	03 N 14	20 43	02 44	01 05	14 33	07 48	02 45	13 10	17 38
23	12 40	09 05	21 07	02 37	00 46	14 30	07 49	02 44	13 10	17 38
24	13 00	14 40	21 29	02 31	00 28	14 28	07 50	02 42	13 09	17 38
25	13 20	19 36	21 48	02 26	00 S 09	14 25	07 51	02 41	13 09	17 38
26	13 39	23 28	22 04	02 22	00 N 10	14 22	07 51	02 40	13 09	17 38
27	13 58	25 53	22 18	02 19	00 28	14 20	07 52	02 39	13 08	17 38
28	14 17	26 33	22 29	02 17	00 47	14 17	07 53	02 38	13 08	17 38
29	14 36	25 24	22 38	02 16	01 05	14 15	07 53	02 37	13 08	17 38
30	14 N 54	22 N 36	22 N 44	02 N 16	01 N 24	14 S 12	07 N 54	02 S 36	13 S 07	17 S 38

ZODIAC SIGN ENTRIES

Date	h	m	Planets
01	16	30	☽ ♋
03	19	32	☽ ♌
05	23	01	☽ ♍
08	03	22	☽ ♎
09	14	21	☿ ♉
10	09	23	☽ ♏
11	12	47	♀ ♓
12	18	01	☽ ♐
15	05	27	☽ ♑
17	18	19	☽ ♒
19	22	44	☉ ♉
20	05	55	☽ ♓
22	13	44	♂ ♈
22	14	09	☽ ♈
24	07	18	♀ ♈
24	18	46	☽ ♉
26	21	02	☽ ♊
28	22	38	☽ ♋
30	22	29	☿ ♊

LATITUDES

Date	Mercury ☿ ° '	Venus ♀ ° '	Mars ♂ ° '	Jupiter ♃ ° '	Saturn ♄ ° '	Uranus ♅ ° '	Neptune ♆ ° '	Pluto ♇ ° '
01	00 S 55	07 N 36	01 S 10	00 S 35	02 N 16	00 S 44	00 S 22	05 N 45
04	00 S 26	07 05	01 10	00 35	02 16	00 44	00 22	05 45
07	00 N 06	06 30	01 10	00 36	02 16	00 44	00 22	05 45
10	00 40	05 51	01 10	00 37	02 16	00 44	00 22	05 45
13	01 13	05 10	01 10	00 37	02 15	00 44	00 22	05 46
16	01 44	04 29	01 10	00 38	02 15	00 44	00 23	05 46
19	02 10	03 48	01 09	00 38	02 15	00 44	00 23	05 46
22	02 30	03 09	01 09	00 39	02 14	00 44	00 23	05 46
25	02 42	02 31	01 09	00 39	02 14	00 44	00 23	05 46
28	02 45	01 56	01 08	00 40	02 14	00 44	00 23	05 47
31	02 N 38	01 N 23	01 S 08	00 S 41	02 N 13	00 S 44	00 S 23	05 N 47

DATA

Julian Date	2454923
Delta T	+67 seconds
Ayanamsa	23° 59' 24"
Synetic vernal point	05° ♓ 07' 35"
True obliquity of ecliptic	23° 26' 23"

MOON'S PHASES, APSIDES AND POSITIONS ☽

Date	h	m	Phase	Longitude ° '	Eclipse Indicator
02	14	34	☽	12 ♋ 59	
09	14	56	○	19 ♎ 53	
17	13	36	☾	27 ♑ 40	
25	03	23	●	05 ♉ 04	

Day	h	m		
02	02	16	Perigee	
16	09	13	Apogee	
28	06	19	Perigee	
01	02	49	Max dec	26° N 47'
07	12	49	0S	
14	12	59	Max dec	26° S 40'
21	22	44	0N	
28	08	49	Max dec	26° N 34'

LONGITUDES

Date	Chiron ⚷ ° '	Ceres ⚳ ° '	Pallas ⚴ ° '	Juno ⚵ ° '	Vesta ⚶ ° '	Black Moon Lilith ⚸ ° '
01	24 ♒ 31	00 ♍ 51	27 ♊ 42	28 ♒ 40	26 ♉ 37	09 ♑ 46
11	25 ♒ 01	00 ♍ 25	02 ♋ 34	02 ♓ 39	00 ♊ 37	10 ♑ 53
21	25 ♒ 27	00 ♍ 40	07 ♋ 35	06 ♓ 33	04 ♊ 42	11 ♑ 59
31	25 ♒ 48	01 ♍ 36	12 ♋ 40	10 ♓ 21	08 ♊ 52	13 ♑ 06

All ephemeris data is given at 12.00 UT and the Moon's longitude is additionally given for 24.00 UT

ASPECTARIAN

h m	Aspects
01 Wednesday	
05 43	☽ Q ☉
05 48	☽ □ ♅
07 59	☽ Q ☿
09 03	☽ △ ♆
17 02	☿ ⊻ ♂
17 56	♀ ∠ ♃
22 06	☽ ☍ ♇
22 58	♂ ⋕ ♄
23 29	☽ □ ♀
23 47	☽ bQ ♃
02 Thursday	
00 14	☽ Q ♄
00 16	☉ ∥ ☿
10 34	☽ bQ ♆
14 34	☽ □ ☉
15 16	☽ ± ♃
17 12	☽ △ ♂
19 46	☽ □ ☿
20 32	☽ ✶ ♄
03 Friday	
00 52	☿ ⚻ ♄
01 37	☽ ⚻ ♃
01 54	☽ ± ♆
08 59	☽ △ ♅
10 23	♀ □ ♇
12 09	☽ ⚻ ♆
16 31	♂ Q ♇
20 16	☽ bQ ♂
21 59	☽ ∠ ♄
04 Saturday	
00 13	☿ ⋕ ♂
00 38	☽ △ ♀
01 11	☽ ⚻ ♇
10 42	☽ bQ ♅
11 26	☽ ± ♇
11 48	☽ ⋕ ♇
12 33	☽ ± ♂
13 18	☽ ⊥ ♄
14 58	☿ ✶ ♃
17 34	♇ St R
17 47	☿ ∥ ♄
18 24	☿ ∥ ♀
19 50	♀ ∥ ♄
21 38	☽ △ ☉
22 13	☽ ⋕ ♃
23 27	☽ ⚻ ♂
23 32	☽ ⊻ ♄
05 Sunday	
00 55	♂ ☍ ♄
01 26	☽ bQ ♀
02 11	☽ ± ♅
02 52	☽ bQ ♇
05 35	☽ ☍ ♃
07 38	☽ ⋕ ♆
07 55	☽ △ ☿
12 33	☽ ⚻ ♅
15 14	☉ ⋕ ♂
15 39	☽ ☍ ♆
16 24	☽ ± ♀
20 19	☿ ± ♄
23 09	☉ ⚻ ♄
06 Monday	
01 25	☽ bQ ☉
02 25	☽ ⚻ ♀
02 31	☽ ∥ ☿
04 44	☽ △ ♇
07 29	☽ ∥ ♄
09 02	☿ ⊥ ♂
09 56	☽ ∥ ♀
11 04	☽ ∥ ☉
13 24	☽ ⋕ ♂
13 57	♀ ⊥ ♆
14 22	☽ bQ ☿
15 50	☿ ⊻ ♅
18 13	☽ ± ☉
19 52	☉ ∥ ♀
07 Tuesday	
00 57	☽ ⋕ ♅
03 11	☽ ☌ ♄
05 29	☽ ⚻ ☉
06 27	☽ ☍ ♂
08 48	☽ ± ☿
10 16	☽ ⚻ ♃
12 38	☿ ✶ ♆
16 52	☽ ☍ ♅
19 55	☽ ⚻ ♆
20 56	☽ ± ♃
21 12	☽ ⚻ ☿
08 Wednesday	
00 41	☽ ∥ ♅
05 15	☽ ☍ ♀
06 32	☽ ± ♆
09 12	☽ □ ♇
10 15	☽ ∥ ♂
13 04	☽ ⋕ ♀
13 04	☽ bQ ♃
18 30	☽ ⋕ ☉
18 54	☽ ⋕ ♄
22 35	☽ bQ ♆
09 Thursday	
01 07	☉ ∥ ♄
08 04	☽ ⊻ ♄
13 27	☽ ⋕ ☿
14 56	☽ ☍ ☉
15 01	☽ ⚻ ♂
16 05	☿ ⊥ ♅
16 20	☽ △ ♃
17 29	☽ Q ♇
18 52	☽ ⊥ ♄
19 35	☽ ∥ ♆
19 47	☉ ⊻ ♂
20 12	☿ ⊻ ♀
22 41	☽ ⚻ ♅
10 Friday	
01 45	☽ △ ♆
02 37	☽ ± ♂
02 48	☿ bQ ♄
04 30	☽ ∥ ♃
09 43	☽ ± ♅
09 57	☽ ⚻ ♀
11 16	☽ ∠ ♄
12 51	☽ ☍ ☿
13 43	☉ ✶ ♃
15 27	☽ ✶ ♇
17 11	☽ ∥ ♇
19 41	♂ ⊻ ♃
20 16	☽ bQ ♂
20 49	☽ ± ♀
11 Saturday	
01 11	☿ Q ♃
02 29	☽ bQ ♅
02 34	☿ ⋕ ♆
05 52	☿ △ ♇
13 19	☽ bQ ♀
15 08	☽ ✶ ♄
17 21	☉ ± ♄
19 33	☽ ∠ ♇
12 Sunday	
00 51	☽ □ ♃
02 21	☽ △ ♂
03 18	☽ ⚻ ☉
06 59	☽ △ ♅
10 05	☽ □ ♆
12 22	☿ ⊥ ♀
12 46	☽ ⊥ ♇
13 57	☽ Q ♄
15 49	☽ ± ☉
17 28	☽ △ ♀
13 Monday	
00 23	☽ ⊻ ♇
05 19	☿ ⋕ ♃
08 21	☽ ⚻ ☿
10 53	☽ bQ ☉
12 12	☽ Q ♃
15 35	☿ Q ♆
18 46	☿ ∠ ♂
21 18	☽ Q ♆
22 20	☽ ± ☿
14 Tuesday	
01 04	☽ □ ♄
04 52	☉ ⊻ ♅
10 36	☿ ∠ ♅
12 21	☽ ✶ ♃
17 11	☽ □ ♂
18 12	☽ □ ♅
19 20	☽ △ ☉
19 32	☽ bQ ☿
21 18	☽ ✶ ♆
15 Wednesday	
04 07	☽ □ ♀
10 00	♂ ☌ ♅
12 04	☽ ☌ ♇
19 00	☽ ∠ ♃
19 55	☉ ✶ ♆
16 Thursday	
03 42	☽ ∠ ♆
06 45	☽ Q ♅
07 13	☽ △ ☿
08 07	☽ Q ♂
13 17	☽ △ ♄
13 32	☽ ⊥ ♃
16 23	☽ Q ♀
18 55	♂ ∥ ♅
20 02	☿ ⋕ ♇
22 02	☽ ⊥ ♆
17 Friday	
01 53	☽ ⊻ ♃
04 15	☿ ∠ ♀
07 15	☽ ✶ ♅
10 13	☽ ✶ ♂
10 14	☽ ⊻ ♆
10 21	♂ ⊻ ♆
13 36	☽ □ ☉
16 42	☽ ✶ ♀
19 24	♀ St D
19 36	☽ bQ ♄
18 Saturday	
00 52	☿ △ ♄
00 54	☽ ⊻ ♇
07 22	☽ ⋕ ☿
12 51	☽ ∥ ♇
12 57	☽ ⊥ ♇
13 39	☽ ± ♄
13 40	☽ ∠ ♅
18 34	☽ ∠ ♂
22 54	☽ ∠ ♀
19 Sunday	
01 35	☽ ⚻ ♄
03 57	☉ ⊻ ♀
04 47	☽ ∥ ♃
04 58	☽ Q ☉
05 25	☽ □ ☿
06 53	☽ ∠ ♇
07 40	☽ ⊥ ♅
12 26	☽ ∥ ♆
13 39	☽ ⊥ ♂
14 53	☽ ☌ ♃
16 47	☽ ⊥ ♀
17 06	☿ bQ ♇
19 31	☽ ⊻ ♅
20 52	☽ ⋕ ☉
22 16	☽ ☌ ♆
20 Monday	
02 11	☽ ⊻ ♂
04 35	☽ ⊻ ♀
06 32	☽ ✶ ☉
11 19	☉ bQ ♄
12 10	☽ ✶ ♇
13 37	♂ ⊥ ♃
13 41	☽ ⋕ ♄
16 05	☉ ⊥ ♅
21 11	☽ Q ☿
21 Tuesday	
10 47	☽ ⋕ ♀
10 56	☽ Q ♇
11 16	☽ ∥ ♅
11 23	☽ ☍ ♄
13 28	☽ ∠ ☉
17 16	☽ ∥ ♂
22 25	☽ ✶ ☿
23 19	♀ ☌ ♂
22 Wednesday	
00 45	☽ ⊻ ♃
03 37	☽ ⋕ ♂
04 38	☽ ☌ ♅
07 04	☽ ⊻ ♆
07 26	☽ ⊥ ☉
09 58	☽ ⋕ ♅
09 59	☽ ∥ ♀
10 30	♀ ⋕ ♅
11 51	☽ ⊥ ♃
13 29	☽ ☌ ♀
14 10	☽ ☌ ♂
17 58	☽ ⊥ ♆
19 10	☽ ⊻ ☉
19 57	☽ □ ♇
23 Thursday	
00 53	☿ □ ♃
04 12	☽ ∠ ♃
04 30	☽ ∠ ☿
05 44	☉ △ ♇
06 44	☽ ∥ ♄
10 03	☽ ∠ ♆
17 31	☽ ⚻ ♄
21 52	☽ ⊥ ☿
24 Friday	
03 56	☽ ± ♄
04 10	☽ ∥ ☉
05 19	☽ ⋕ ♆
06 45	☽ ✶ ♃
09 12	☽ ⊻ ☿
10 01	☽ ⊻ ♅
11 05	☽ ⋕ ♃
12 11	☽ ✶ ♆
18 58	☽ ⊻ ♀
19 20	☽ bQ ♄
19 38	☿ ✶ ♅
20 19	☽ ⊥ ♅
21 51	☽ ⊻ ♂
23 10	☉ ⋕ ♆
25 Saturday	
00 12	☽ △ ♇
01 56	☽ ⋕ ♇
03 23	☽ ☌ ☉
03 28	☽ Q ♃
04 59	☉ Q ♃
05 21	☽ ⊥ ♀
08 36	☽ ⊥ ♂
08 37	☽ Q ♆
08 56	♂ ⊥ ♆
11 36	☽ ∠ ♅
20 35	☽ △ ♄
20 47	☽ ∠ ♀
26 Sunday	
00 33	☽ ∠ ♂
00 54	☿ □ ♆
01 24	☽ bQ ♇
01 39	☽ ∥ ☿
09 56	☽ □ ♃
12 43	☽ ✶ ♅
14 43	☽ □ ♆
15 42	☽ ☌ ☿
16 14	♂ □ ♇
16 21	☽ ± ♇
19 55	☉ ⊥ ♀
22 16	☽ ✶ ♀
27 Monday	
01 59	☿ ± ♇
02 17	☽ ⚻ ♇
02 51	☽ ✶ ♂
08 39	☽ Q ♅
09 00	☽ ⊻ ☉
18 34	☽ Q ♀
19 37	☽ ⊥ ☉
22 11	☽ □ ♄
23 47	☽ Q ♂
28 Tuesday	
09 10	☉ Q ♆
11 34	☽ ∠ ☉
12 01	☽ △ ♃
12 14	☉ ⋕ ♃
14 29	☽ □ ♅
16 23	☽ △ ♆
20 22	☽ ⊻ ☿
29 Wednesday	
01 07	☽ □ ♀
03 52	☽ ☍ ♇
03 55	☽ Q ♄
06 51	☽ ⊥ ☿
07 17	☽ □ ♂
13 09	☽ bQ ♃
14 16	☽ ✶ ☉
17 21	☽ bQ ♆
22 38	☽ ∠ ☿
23 54	☽ ✶ ♄
30 Thursday	
04 21	☽ ± ♃
08 27	☽ ± ♆
09 00	☉ ∠ ♅
11 06	☽ ∥ ☿
11 51	☽ Q ☉
14 33	☽ ⚻ ♃
16 45	☽ △ ♅
18 35	☽ ⚻ ♆
21 27	♀ ⊥ ♆

MAY 2009

LONGITUDES

Date	Sidereal time h m s	Sun ☉ ° ' "	Moon ☽ ° ' "	Moon ☽ 24.00 ° ' "	Mercury ☿ ° '	Venus ♀ ° '	Mars ♂ ° '	Jupiter ♃ ° '	Saturn ♄ ° '	Uranus ♅ ° '	Neptune ♆ ° '	Pluto ♇ ° '
01	02 38 11	11 ♉ 14 58	06 ♌ 30 04	13 ♌ 30 06	00 ♊ 18	02 ♈ 32	06 ♈ 54	23 ♒ 59	15 ♍ 07	25 ♓ 12	26 ♒ 16	03 ♑ 07
02	02 42 08	12 13 11	20 ♌ 27 15	27 ♌ 21 32	00 46	03 00	07 40	24 07	15 R 06	25 15	26 17	03 R 06
03	02 46 04	13 11 22	04 ♍ 13 00	11 ♍ 01 42	01 08	03 29	08 26	24 14	15 04	25 18	26 18	03 05
04	02 50 01	14 09 31	17 ♍ 47 40	24 ♍ 30 54	01 25	04 00	09 12	24 22	15 03	25 20	26 19	03 05
05	02 53 57	15 07 38	01 ♎ 11 25	07 ♎ 49 11	01 37	04 33	09 59	24 29	15 02	25 23	26 19	03 04
06	02 57 54	16 05 43	14 ♎ 24 08	20 ♎ 56 12	01 43	05 07	10 45	24 36	15 00	25 25	26 20	03 03
07	03 01 51	17 03 47	27 ♎ 25 18	03 ♏ 51 20	01 R 44	05 42	11 31	24 43	14 59	25 28	26 21	03 02
08	03 05 47	18 01 48	10 ♏ 14 14	16 ♏ 33 57	01 40	06 18	12 17	24 50	14 58	25 30	26 22	03 01
09	03 09 44	18 59 48	22 ♏ 50 25	29 ♏ 03 41	01 32	06 56	13 03	24 56	14 58	25 32	26 22	03 00
10	03 13 40	19 57 47	05 ♐ 13 47	11 ♐ 20 50	01 18	07 34	13 48	25 03	14 57	25 35	26 23	02 59
11	03 17 37	20 55 44	17 ♐ 25 00	23 ♐ 26 31	01 00	08 14	14 34	25 09	14 56	25 37	26 24	02 58
12	03 21 33	21 53 39	29 ♐ 25 40	05 ♑ 22 49	00 39	08 55	15 20	25 15	14 56	25 39	26 24	02 57
13	03 25 30	22 51 34	11 ♑ 18 22	17 ♑ 12 48	00 ♊ 14	09 37	16 06	25 21	14 55	25 41	26 25	02 56
14	03 29 26	23 49 26	23 ♑ 06 37	29 ♑ 00 23	29 ♉ 45	10 20	16 52	25 27	14 55	25 44	26 25	02 55
15	03 33 23	24 47 18	04 ♒ 54 42	10 ♒ 50 13	29 15	11 04	17 37	25 33	14 55	25 46	26 26	02 54
16	03 37 20	25 45 08	16 ♒ 47 35	22 ♒ 47 28	28 42	11 48	18 23	25 38	14 55	25 48	26 26	02 52
17	03 41 16	26 42 57	28 ♒ 50 33	04 ♓ 57 30	28 08	12 34	19 09	25 43	14 D 55	25 50	26 26	02 51
18	03 45 13	27 40 45	11 ♓ 08 58	17 ♓ 25 34	27 33	13 20	19 54	25 49	14 55	25 52	26 27	02 50
19	03 49 09	28 38 32	23 ♓ 47 50	00 ♈ 16 14	26 58	14 08	20 40	25 54	14 55	25 54	26 27	02 49
20	03 53 06	29 ♉ 36 18	06 ♈ 51 09	13 ♈ 32 50	26 24	14 56	21 25	25 58	14 55	25 56	26 27	02 48
21	03 57 02	00 ♊ 34 02	20 ♈ 21 22	27 ♈ 16 42	25 50	15 45	22 11	26 03	14 56	25 58	26 28	02 47
22	04 00 59	01 31 46	04 ♉ 18 36	11 ♉ 26 38	25 18	16 34	22 56	26 08	14 56	26 00	26 28	02 45
23	04 04 55	02 29 28	18 ♉ 40 13	25 ♉ 58 35	24 49	17 24	23 42	26 12	14 57	26 01	26 28	02 44
24	04 08 52	03 27 09	03 ♊ 20 49	10 ♊ 45 55	24 21	18 15	24 27	26 16	14 58	26 03	26 28	02 43
25	04 12 49	04 24 49	18 ♊ 12 48	25 ♊ 40 23	23 57	19 07	25 12	26 20	14 58	26 05	26 28	02 41
26	04 16 45	05 22 28	03 ♋ 07 36	10 ♋ 33 28	23 36	19 59	25 57	26 24	14 59	26 07	26 29	02 40
27	04 20 42	06 20 05	17 ♋ 57 05	25 ♋ 17 42	23 19	20 52	26 42	26 27	15 00	26 08	26 29	02 39
28	04 24 38	07 17 41	02 ♌ 34 43	09 ♌ 47 39	23 06	21 45	27 28	26 31	15 01	26 10	26 29	02 37
29	04 28 35	08 15 16	16 ♌ 56 12	24 ♌ 00 09	22 57	22 39	28 13	26 34	15 03	26 12	26 R 29	02 36
30	04 32 31	09 12 49	00 ♍ 59 26	07 ♍ 54 03	22 53	23 33	28 58	26 37	15 04	26 13	26 29	02 35
31	04 36 28	10 ♊ 10 20	14 ♍ 44 06	21 ♍ 29 42	22 ♉ 53	24 ♈ 28	29 ♈ 43	26 ♒ 40	15 ♍ 06	26 ♓ 15	26 ♒ 29	02 ♑ 33

Date	Moon True ☊ ° '	Moon Mean ☊ ° '	Moon ☽ Latitude ° '
01	04 ♒ 04	04 ♒ 35	00 S 13
02	04 R 04	04 32	01 26
03	04 02	04 28	02 32
04	03 59	04 25	03 29
05	03 53	04 22	04 14
06	03 44	04 19	04 44
07	03 33	04 16	04 59
08	03 20	04 12	04 59
09	03 08	04 09	04 44
10	02 56	04 06	04 15
11	02 47	04 03	03 35
12	02 40	04 00	02 46
13	02 35	03 57	01 51
14	02 33	03 53	00 S 50
15	02 D 33	03 50	00 N 13
16	02 33	03 47	01 15
17	02 R 33	03 44	02 16
18	02 32	03 41	03 11
19	02 30	03 37	03 58
20	02 24	03 34	04 35
21	02 17	03 31	04 58
22	02 09	03 28	05 04
23	01 59	03 25	04 52
24	01 50	03 22	04 20
25	01 43	03 18	03 31
26	01 38	03 15	02 26
27	01 35	03 12	01 N 12
28	01 D 34	03 09	00 S 05
29	01 35	03 06	01 22
30	01 36	03 03	02 31
31	01 ♒ 36	02 ♒ 59	03 S 31

DECLINATIONS

Date	Sun ☉ ° '	Moon ☽ ° '	Mercury ☿ ° '	Venus ♀ ° '	Mars ♂ ° '	Jupiter ♃ ° '	Saturn ♄ ° '	Uranus ♅ ° '	Neptune ♆ ° '	Pluto ♇ ° '
01	15 N 12	18 N 26	22 N 48	02 N 16	01 N 42	14 S 10	07 N 54	02 S 35	13 S 07	17 S 37
02	15 30	13 19	22 49	02 18	02 01	14 08	07 55	02 34	13 07	17 37
03	15 48	07 36	22 48	02 20	02 19	14 05	07 55	02 33	13 07	17 37
04	16 05	01 N 37	22 45	02 23	02 37	14 03	07 56	02 32	13 06	17 37
05	16 22	04 S 21	22 39	02 27	02 56	14 01	07 56	02 31	13 06	17 37
06	16 39	10 03	22 31	02 32	03 14	13 59	07 56	02 30	13 06	17 37
07	16 56	15 12	22 21	02 37	03 32	13 57	07 56	02 29	13 06	17 37
08	17 12	19 37	22 09	02 44	03 50	13 55	07 57	02 28	13 05	17 37
09	17 28	23 03	21 55	02 50	04 09	13 53	07 57	02 27	13 05	17 37
10	17 44	25 21	21 39	02 58	04 27	13 51	07 57	02 26	13 05	17 37
11	17 59	26 25	21 21	03 06	04 45	13 49	07 57	02 25	13 05	17 37
12	18 14	26 13	21 01	03 15	05 03	13 47	07 57	02 25	13 05	17 37
13	18 29	24 48	20 40	03 25	05 21	13 45	07 57	02 24	13 05	17 37
14	18 44	22 17	20 18	03 35	05 38	13 43	07 57	02 23	13 04	17 37
15	18 58	18 50	19 55	03 45	05 56	13 42	07 57	02 22	13 04	17 37
16	19 12	14 36	19 30	03 56	06 14	13 40	07 57	02 21	13 04	17 37
17	19 25	09 45	19 06	04 08	06 31	13 38	07 57	02 20	13 04	17 37
18	19 39	04 S 26	18 41	04 20	06 49	13 37	07 57	02 20	13 04	17 37
19	19 51	01 N 11	18 16	04 33	07 06	13 35	07 56	02 19	13 04	17 37
20	20 04	06 56	17 52	04 46	07 24	13 34	07 56	02 18	13 04	17 37
21	20 16	12 33	17 28	04 59	07 41	13 33	07 56	02 17	13 04	17 37
22	20 28	17 44	17 05	05 13	07 58	13 31	07 55	02 17	13 04	17 37
23	20 40	22 03	16 43	05 27	08 15	13 30	07 55	02 16	13 04	17 37
24	20 51	25 05	16 23	05 42	08 32	13 29	07 55	02 15	13 04	17 37
25	21 01	26 25	16 04	05 57	08 49	13 28	07 54	02 15	13 03	17 38
26	21 12	25 51	15 47	06 12	09 06	13 27	07 54	02 14	13 03	17 38
27	21 22	23 26	15 32	06 28	09 23	13 26	07 53	02 13	13 03	17 38
28	21 32	19 30	15 20	06 43	09 39	13 25	07 52	02 13	13 03	17 38
29	21 41	14 28	15 09	07 00	09 56	13 24	07 52	02 12	13 04	17 38
30	21 50	08 46	15 00	07 16	10 12	13 23	07 51	02 12	13 04	17 38
31	21 N 58	02 N 46	14 N 54	07 N 32	10 N 28	13 S 22	07 N 50	02 S 11	13 S 04	17 S 38

ZODIAC SIGN ENTRIES

Date	h	m	Planets
01	00	56	☽ ♌
03	04	37	☽ ♍
05	09	51	☽ ♎
07	16	48	☽ ♏
10	01	49	☽ ♐
12	13	09	☽ ♑
13	23	53	☿ ♉
15	02	01	☽ ♒
17	14	17	☽ ♓
19	23	30	☽ ♈
20	21	51	☉ ♊
22	04	40	☽ ♉
24	06	34	☽ ♊
26	06	58	☽ ♋
28	07	44	☽ ♌
30	10	17	☽ ♍
31	21	18	♂ ♉

LATITUDES

Date	Mercury ☿ ° '	Venus ♀ ° '	Mars ♂ ° '	Jupiter ♃ ° '	Saturn ♄ ° '	Uranus ♅ ° '	Neptune ♆ ° '	Pluto ♇ ° '
01	02 N 38	01 N 23	01 S 08	00 S 41	02 N 13	00 S 44	00 S 23	05 N 47
04	02 21	00 52	01 07	00 41	02 12	00 44	00 23	05 47
07	01 53	00 N 24	01 06	00 42	02 12	00 44	00 23	05 47
10	01 15	00 S 02	01 05	00 43	02 11	00 44	00 23	05 47
13	00 N 29	00 26	01 04	00 43	02 11	00 45	00 23	05 47
16	00 S 22	00 48	01 03	00 44	02 10	00 45	00 23	05 47
19	01 15	01 07	01 02	00 45	02 10	00 45	00 23	05 47
22	02 04	01 24	01 01	00 46	02 09	00 45	00 23	05 47
25	02 47	01 40	01 00	00 46	02 09	00 45	00 23	05 47
28	03 20	01 53	00 59	00 47	02 08	00 45	00 23	05 47
31	03 S 43	02 S 05	00 S 58	00 S 48	02 N 08	00 S 45	00 S 24	05 N 47

DATA

Julian Date	2454953
Delta T	+67 seconds
Ayanamsa	23° 59' 28"
Synetic vernal point	05° ♓ 07' 31"
True obliquity of ecliptic	23° 26' 22"

MOON'S PHASES, APSIDES AND POSITIONS ☽

Date	h	m	Phase	Longitude ° '	Eclipse Indicator
01	20	44	☽	11 ♌ 36	
09	04	01	○	18 ♏ 41	
17	07	26	☾	26 ♒ 32	
24	12	11	●	03 ♊ 28	
31	03	22	☽	09 ♍ 50	

Day	h	m		
14	02	50	Apogee	
26	03	39	Perigee	
04	18	25	0S	
11	20	02	Max dec	26° S 29'
19	07	01	0N	
25	16	53	Max dec	26° N 27'
31	23	02	0S	

LONGITUDES

Date	Chiron ⚷ ° '	Ceres ⚳ ° '	Pallas ⚴ ° '	Juno ⚵ ° '	Vesta ⚶ ° '	Black Moon Lilith ⚸ ° '
01	25 ♒ 48	01 ♍ 36	12 ♋ 40	10 ♓ 21	08 ♊ 52	13 ♑ 06
11	26 ♒ 03	03 ♍ 07	17 ♋ 49	14 ♓ 02	13 ♊ 05	14 ♑ 12
21	26 ♒ 11	05 ♍ 09	22 ♋ 58	17 ♓ 34	17 ♊ 21	15 ♑ 19
31	26 ♒ 14	07 ♍ 37	28 ♋ 08	20 ♓ 55	21 ♊ 39	16 ♑ 26

All ephemeris data is given at 12.00 UT and the Moon's longitude is additionally given for 24.00 UT

ASPECTARIAN

h m	Aspects
01 Friday	
01 02	☽ ⚹ ☿
01 09	☽ ∠ ♄
05 00	☽ △ ♀
06 14	☽ ⊼ ♇
12 43	☽ △ ♂
16 05	☽ # ♇
16 28	☽ ⊥ ♄
16 28	☽ ± ♇
18 22	☽ ⚼ ♅
20 44	☽ □ ☉
22 18	☽ Q ☿
02 Saturday	
02 43	☽ ∥ ☉
02 46	☽ ⊻ ♄
07 36	☽ ⚼ ♀
07 56	☽ ⚼ ♇
08 24	☽ # ♃
09 55	☽ ± ♅
12 53	☽ # ♆
16 04	☽ ⚼ ♂
17 10	♀ □ ♇
18 24	☽ ☍ ♃
20 21	☽ ⊼ ♅
22 08	☽ ☍ ♆
23 47	☽ ± ♀
03 Sunday	
06 28	☽ □ ☿
08 42	☽ ± ♂
10 02	☽ △ ♇
10 41	☽ ⊼ ♀
10 41	☽ ∥ ♄
13 25	♀ ∥ ♂
19 52	☽ ⊼ ♂
04 Monday	
05 03	☽ △ ☉
05 04	♂ # ♅
07 08	☽ ☌ ♄
08 09	☽ ∥ ♂
08 19	☽ # ♅
08 56	☽ ∥ ♀
17 39	♂ ∠ ♃
23 50	☽ ⊼ ♃
05 Tuesday	
01 31	☽ ☍ ♅
03 14	☽ ⊼ ♆
04 10	☽ # ♀
04 32	☽ ∥ ♅
05 53	☽ # ♂
09 34	☉ △ ♄
09 56	☽ ⚼ ☉
10 42	☽ ± ♃
12 47	☽ △ ☿
14 03	☽ ± ♆
15 23	☽ □ ♇
18 20	☽ ☍ ♀
06 Wednesday	
02 54	☽ # ♄
03 09	☽ ⚼ ♃
03 31	☽ ± ☉
04 19	♀ # ♅
04 54	☽ ☍ ♂
06 24	☽ ⚼ ♆
13 06	☽ ⊻ ♄
15 21	☽ ⊼ ☉
16 16	☽ ⚼ ☿
07 Thursday	
00 07	☽ ⊥ ♄
00 11	☽ Q ♇
01 49	☽ ∥ ♆
05 01	☿ St R
05 53	☽ ∥ ♃
06 53	♂ ∠ ♆
06 56	☽ △ ♃
08 21	☽ ⊼ ♅
08 53	☽ ± ☿
10 00	☽ △ ♆
16 46	☽ ∠ ♄
19 32	☽ ± ♅
20 01	☽ ⊼ ☿
21 26	☽ # ☉
22 27	☽ ⚹ ♇
08 Friday	
00 35	☽ ∥ ♇
04 13	☽ ⊼ ♀
11 39	☉ ⚼ ♇
12 30	☽ ⚼ ♅
16 06	☽ ± ♀
16 06	☽ ⊼ ♂
20 58	☽ ⚹ ♄
09 Saturday	
02 45	☽ ∠ ♇
03 50	☽ # ☿
04 01	☽ ☍ ☉
04 15	☽ ± ♂
10 09	☽ ⚼ ♀
16 04	☽ □ ♃
17 12	☽ △ ♅
17 25	☉ ⊥ ♂
18 48	☽ □ ♆
19 56	☽ Q ♄
20 00	☽ ⊥ ♇
22 41	☽ ⚼ ♂
10 Sunday	
01 57	☉ # ♇
04 30	☽ ☍ ☿
07 37	☽ ⊻ ♇
16 50	☽ △ ♀
11 Monday	
03 29	☽ Q ♃
05 59	☽ △ ♂
06 00	☽ Q ♆
07 06	☽ □ ♄
19 36	☽ ⊼ ☉
23 23	♂ ⊼ ♄
12 Tuesday	
03 33	☽ ⚹ ♃
04 24	☽ □ ♅
05 55	☽ ⚹ ♆
08 39	☽ ± ☉
14 22	☽ ⊼ ☿
18 24	☿ ∠ ♂
19 05	☽ ☌ ♇
13 Wednesday	
02 05	☽ ± ☿
04 24	☽ ⚼ ☉
08 21	☽ □ ♀
10 03	☽ ∠ ♃
12 13	☽ ∠ ♆
16 51	☽ Q ♅
19 20	☽ △ ♄
19 40	☽ ⚼ ☿
22 24	☽ □ ♂
14 Thursday	
04 29	☽ ⊥ ♃
06 31	☽ ⊥ ♆
13 35	☽ △ ☉
16 33	♀ ∠ ♃
16 48	☽ ⊻ ♃
17 20	☽ ⚹ ♅
18 44	☽ ⊻ ♆
23 19	☽ Q ♀
15 Friday	
00 58	☽ △ ☿
01 51	☽ ⚼ ♄
04 16	☽ # ☿
07 55	☽ ⊻ ♇
11 13	☽ # ☉
13 33	☽ Q ♂
19 18	☽ ∥ ♇
20 03	☽ ⊥ ♇
20 07	☽ ± ♄
23 53	☽ ∠ ♅
23 59	♀ ∠ ♆
16 Saturday	
01 17	☽ ⚹ ♀
08 13	☽ ⊼ ♄
08 45	☉ □ ♃
13 09	☉ ⚹ ♅
14 10	☽ ∠ ♇
15 25	☽ ⚹ ♂
16 54	☽ ∥ ♃
18 02	☽ ⊥ ♅
19 55	☽ ∥ ♆
23 45	☉ ∥ ☿
17 Sunday	
02 06	♄ St D
05 05	☉ □ ♆
05 47	☽ ☌ ♃
06 02	☽ ⊻ ♅
07 15	☽ ☌ ♆
07 26	☽ □ ☉
09 19	☽ ∠ ♀
10 40	☽ □ ☿
15 24	☉ ± ♇
19 53	☽ ⚹ ♇
20 23	☽ # ♄
22 05	☿ ∠ ♀
23 06	☽ ∠ ♂
18 Monday	
02 02	☽ # ♂
04 09	☽ ⊥ ♀
10 02	☉ ☌ ☿
12 27	☽ # ♀
16 29	☽ ⊻ ♀
17 38	☽ ⊥ ♂
19 03	☽ Q ♇
19 13	☽ ☍ ♄
20 04	☽ Q ☿
21 09	☽ ∥ ♅
21 24	☽ Q ☉
19 Tuesday	
05 45	☽ ⊻ ♂
14 25	♃ ⊻ ♅
15 56	☽ ☌ ♅
15 56	☽ ⊻ ♃
16 43	☽ # ♅
16 57	☽ ⊻ ♆
17 29	☿ ⊥ ♂
17 39	☽ ⚹ ☿
18 35	☿ ± ♇
20 03	♂ ± ♄
21 43	☽ ⚹ ☉
20 Wednesday	
02 35	☽ ∥ ♀
03 04	☽ ⊥ ♃
04 00	☽ ⊥ ♆
04 38	☽ □ ♇
09 21	☿ □ ♆
11 43	♀ ⊼ ♄
14 03	☽ ∥ ♂
16 13	☽ ∥ ♄
19 27	☽ ∠ ♃
19 50	☽ ∠ ☿
20 17	☽ ∠ ♆
21 Thursday	
02 14	☿ # ♇
02 27	☽ ⊼ ♄
02 56	☽ ∠ ☉
03 23	☽ ☌ ♀
03 44	☿ □ ♃
06 46	☿ ⚹ ♅
11 07	☽ ⊥ ☿
13 00	☽ ± ♄
14 15	☽ # ♆
15 22	☽ ☌ ♂
16 23	☽ # ♃
19 52	☽ ⊥ ☉
21 10	☽ ⊻ ☿
21 45	☽ ⊻ ♅
21 56	☽ ⚹ ♃
22 36	☽ ⚹ ♆
22 Friday	
04 33	☽ ⚼ ♄
06 56	☽ ⊻ ☉
08 03	☽ ⊥ ♅
08 11	♂ ∥ ♄
09 03	☽ ∥ ☿
09 22	☽ △ ♇
11 30	☽ # ♇
18 28	☽ Q ♃
19 01	☽ Q ♆
19 44	☉ ∠ ♀
23 16	☽ ∠ ♅
23 Saturday	
03 11	☽ ∥ ☉
05 50	☽ △ ♄
09 47	☽ ⊻ ♀
10 27	☽ ⚼ ♇
17 55	☉ ⊼ ♇
20 16	☽ ⊥ ♀
20 43	☽ ⊻ ♂
21 47	☽ ☌ ☿
24 Sunday	
00 06	☽ ⚹ ♅
00 25	☽ □ ♃
00 48	☽ □ ♆
01 13	☽ ± ♇
07 02	☽ ⊥ ♂
10 09	☿ ⊻ ♂
10 58	☽ ⊼ ♇
11 51	☽ ∠ ♀
12 11	☽ ☌ ☉
13 52	☿ ⊥ ♀
19 38	☽ Q ♅
22 24	☽ ∠ ♂
25 Monday	
06 47	☽ □ ♄
13 32	☽ ⚹ ♀
21 01	☽ ⊻ ☿
23 50	☽ ⚹ ♂
26 Tuesday	
00 41	☽ □ ♅
01 07	☽ △ ♃
01 17	☽ △ ♆
06 27	☽ ⊥ ☿
10 03	☽ Q ♀
11 16	☽ ☍ ♇
11 47	☽ Q ♄
15 53	☽ ⊻ ☉
17 09	♂ ⊻ ♅
20 13	☽ Q ♂
20 40	☽ ∠ ☿
27 Wednesday	
01 25	☽ ⚼ ♃
01 29	☽ ⚼ ♆
02 15	☽ ⊥ ☉
03 19	♂ ⚹ ♃
04 37	♂ ⚹ ♆
07 12	☽ ⚹ ♄
15 58	♀ ± ♄
16 06	☽ ± ♃
16 07	☽ ± ♆
17 03	☽ □ ♀
17 54	☽ ∠ ☉
20 06	♃ ☌ ♆
20 38	☽ ⚹ ☿
28 Thursday	
01 03	☽ ∥ ☉
01 25	☽ △ ♅
01 56	☽ ⊼ ♆
01 58	☽ ⊼ ♃
03 06	☽ □ ♂
07 46	☽ ∠ ♄
12 04	☽ ⊼ ♇
16 08	☽ Q ☿
20 23	☽ ⚹ ☉
21 25	☽ # ♇
22 02	☽ ± ♇
22 44	☽ ⊥ ♄
29 Friday	
02 19	☽ ⚼ ♅
04 33	♆ St R
08 48	☽ ⊻ ♄
08 51	☽ ∥ ☿
10 23	☉ Q ♅
13 07	☽ ⚼ ♇
16 39	☽ # ♃
17 31	☽ ± ♅
18 02	☽ Q ☉
18 06	☽ # ♆
19 23	☿ ⊻ ♀
22 09	☽ □ ☿
22 21	☽ △ ♀
30 Saturday	
03 46	☽ ⊼ ♅
04 14	☽ ☍ ♆
04 27	☽ ☍ ♃
06 22	☽ ∥ ♂
08 18	☽ △ ♂
14 44	☽ △ ♇
15 42	☽ ∥ ♄
17 47	☽ ∥ ♀
31 Sunday	
01 22	☿ St D
02 05	☽ ⚼ ♀
03 22	☽ □ ☉
11 57	☽ ⚼ ♂
12 38	☽ ☌ ♄
14 19	☽ # ♅
19 06	☽ ± ♀

JUNE 2009

LONGITUDES

Date	Sidereal time h m s	Sun ☉ ° ' "	Moon ☽ ° ' "	Moon ☽ 24.00 ° ' "	Mercury ☿ ° '	Venus ♀ ° '	Mars ♂ ° '	Jupiter ♃ ° '	Saturn ♄ ° '	Uranus ♅ ° '	Neptune ♆ ° '	Pluto ⯓ ° '
01	04 40 24	11 ♊ 07 50	28 ♍ 11 03	04 ♎ 48 18	22 ♉ 57	25 ♈ 23	00 ♉ 27	26 ♒ 43	15 ♍ 07	26 ♓ 16	26 ♒ 28	02 ♑ 32
02	04 44 21	12 05 19	11 ♎ 21 42	17 ♎ 51 24	23 D 05	26 19	01 12	26 45	15 09	26 17	26 R 28	02 R 30
03	04 48 18	13 02 47	24 ♎ 17 37	00 ♏ 40 31	23 19	27 15	01 57	26 48	15 10	26 19	26 28	02 29
04	04 52 14	14 00 13	07 ♏ 00 16	13 ♏ 17 00	23 36	28 12	02 42	26 50	15 12	26 20	26 28	02 28
05	04 56 11	14 57 39	19 ♏ 30 53	25 ♏ 42 01	23 58	29 ♈ 09	03 26	26 52	15 14	26 21	26 28	02 26
06	05 00 07	15 55 03	01 ♐ 50 32	07 ♐ 56 36	24 25	00 ♉ 07	04 11	26 54	15 16	26 22	26 28	02 25
07	05 04 04	16 52 27	14 ♐ 00 20	20 ♐ 01 54	24 55	01 05	04 56	26 55	15 18	26 24	26 27	02 23
08	05 08 00	17 49 49	26 ♐ 01 30	01 ♑ 59 20	25 30	02 03	05 40	26 57	15 21	26 25	26 27	02 22
09	05 11 57	18 47 11	07 ♑ 55 40	13 ♑ 50 45	26 09	03 02	06 25	26 58	15 23	26 26	26 27	02 20
10	05 15 53	19 44 32	19 ♑ 44 57	25 ♑ 38 35	26 52	04 01	07 09	26 59	15 25	26 27	26 26	02 19
11	05 19 50	20 41 53	01 ♒ 32 06	07 ♒ 25 54	27 39	05 01	07 53	27 00	15 28	26 28	26 26	02 17
12	05 23 47	21 39 13	13 ♒ 20 28	19 ♒ 16 21	28 30	06 00	08 38	27 00	15 31	26 29	26 25	02 16
13	05 27 43	22 36 32	25 ♒ 14 04	01 ♓ 14 13	29 ♉ 25	07 01	09 22	27 01	15 33	26 30	26 25	02 14
14	05 31 40	23 33 51	07 ♓ 17 22	13 ♓ 24 09	00 ♊ 23	08 01	10 06	27 01	15 36	26 30	26 24	02 13
15	05 35 36	24 31 09	19 ♓ 35 09	25 ♓ 50 57	01 25	09 02	10 50	27 R 01	15 39	26 31	26 24	02 11
16	05 39 33	25 28 27	02 ♈ 12 09	08 ♈ 39 13	02 30	10 03	11 34	27 01	15 42	26 32	26 23	02 10
17	05 43 29	26 25 45	15 ♈ 12 39	21 ♈ 52 46	03 39	11 04	12 18	27 01	15 45	26 33	26 23	02 08
18	05 47 26	27 23 03	28 ♈ 39 49	05 ♉ 33 56	04 52	12 06	13 02	27 00	15 48	26 33	26 22	02 07
19	05 51 22	28 20 20	12 ♉ 35 03	19 ♉ 42 57	06 07	13 08	13 46	26 59	15 51	26 34	26 21	02 05
20	05 55 19	29 ♊ 17 37	26 ♉ 57 14	04 ♊ 17 17	07 27	14 10	14 30	26 58	15 55	26 34	26 21	02 04
21	05 59 16	00 ♋ 14 54	11 ♊ 42 20	19 ♊ 11 27	08 49	15 13	15 13	26 57	15 58	26 35	26 20	02 02
22	06 03 12	01 12 10	26 ♊ 43 34	04 ♋ 17 31	10 15	16 15	15 57	26 56	16 02	26 35	26 19	02 00
23	06 07 09	02 09 27	11 ♋ 52 07	19 ♋ 26 09	11 44	17 18	16 41	26 55	16 05	26 36	26 19	01 59
24	06 11 05	03 06 42	26 ♋ 58 29	04 ♌ 28 05	13 16	18 21	17 24	26 53	16 09	26 36	26 18	01 57
25	06 15 02	04 03 58	11 ♌ 54 00	19 ♌ 15 29	14 52	19 25	18 08	26 51	16 13	26 37	26 17	01 56
26	06 18 58	05 01 12	26 ♌ 31 53	03 ♍ 42 48	16 31	20 28	18 51	26 49	16 17	26 37	26 16	01 54
27	06 22 55	05 58 26	10 ♍ 47 54	17 ♍ 47 03	18 12	21 32	19 35	26 47	16 21	26 37	26 15	01 53
28	06 26 51	06 55 40	24 ♍ 40 14	01 ♎ 27 32	19 57	22 36	20 18	26 44	16 25	26 37	26 14	01 51
29	06 30 48	07 52 53	08 ♎ 09 08	14 ♎ 45 17	21 45	23 40	21 01	26 41	16 29	26 37	26 13	01 50
30	06 34 45	08 ♋ 50 06	21 ♎ 16 17	27 ♎ 42 27	23 ♊ 35	24 ♉ 45	21 ♉ 44	26 ♒ 39	16 ♍ 33	26 ♓ 37	26 ♒ 12	01 ♑ 48

Date	Moon True ☊ ° '	Moon Mean ☊ ° '	Moon ☽ Latitude ° '
01	01 ♒ 34	02 ♒ 56	04 S 17
02	01 R 31	02 53	04 49
03	01 25	02 50	05 05
04	01 18	02 47	05 06
05	01 10	02 43	04 52
06	01 02	02 40	04 25
07	00 54	02 37	03 46
08	00 48	02 34	02 57
09	00 44	02 31	02 00
10	00 41	02 28	00 S 59
11	00 D 41	02 24	00 N 05
12	00 41	02 21	01 08
13	00 43	02 18	02 10
14	00 45	02 15	03 06
15	00 45	02 12	03 56
16	00 R 45	02 09	04 35
17	00 44	02 05	05 01
18	00 41	02 02	05 12
19	00 37	01 59	05 06
20	00 33	01 56	04 40
21	00 28	01 53	03 56
22	00 25	01 49	02 54
23	00 22	01 46	01 40
24	00 22	01 43	00 N 19
25	00 D 22	01 40	01 S 03
26	00 23	01 37	02 19
27	00 24	01 34	03 25
28	00 26	01 30	04 16
29	00 R 26	01 27	04 52
30	00 ♒ 25	01 ♒ 24	05 S 11

DECLINATIONS

Date	Sun ☉ ° '	Moon ☽ ° '	Mercury ☿ ° '	Venus ♀ ° '	Mars ♂ ° '	Jupiter ♃ ° '	Saturn ♄ ° '	Uranus ♅ ° '	Neptune ♆ ° '	Pluto ⯓ ° '
01	22 N 07	03 S 13	14 N 50	07 N 49	10 N 44	13 S 22	07 N 50	02 S 11	13 S 04	17 S 38
02	22 14	08 55	14 48	08 06	11 00	13 21	07 49	02 10	13 04	17 38
03	22 22	14 09	14 48	08 23	11 16	13 21	07 48	02 09	13 04	17 38
04	22 29	18 40	14 51	08 40	11 32	13 20	07 47	02 09	13 04	17 38
05	22 35	22 18	14 55	08 58	11 48	13 20	07 46	02 09	13 04	17 38
06	22 42	24 51	15 02	09 16	12 03	13 19	07 45	02 08	13 04	17 38
07	22 48	26 13	15 10	09 34	12 19	13 19	07 44	02 08	13 04	17 38
08	22 53	26 19	15 20	09 52	12 34	13 19	07 43	02 07	13 04	17 38
09	22 58	25 12	15 32	10 10	12 49	13 19	07 42	02 07	13 04	17 38
10	23 03	22 58	15 45	10 28	13 04	13 19	07 41	02 06	13 05	17 38
11	23 07	19 45	16 00	10 46	13 19	13 19	07 40	02 06	13 05	17 39
12	23 11	15 43	16 16	11 04	13 34	13 19	07 39	02 06	13 05	17 39
13	23 14	11 04	16 34	11 22	13 48	13 19	07 37	02 05	13 05	17 39
14	23 17	05 57	16 52	11 40	14 03	13 19	07 36	02 05	13 05	17 39
15	23 20	00 S 31	17 12	11 58	14 17	13 19	07 35	02 05	13 06	17 39
16	23 22	05 N 05	17 32	12 16	14 31	13 19	07 34	02 05	13 06	17 39
17	23 23	10 37	17 53	12 34	14 45	13 20	07 32	02 04	13 06	17 39
18	23 25	15 52	18 15	12 52	14 59	13 20	07 31	02 04	13 06	17 39
19	23 26	20 28	18 38	13 10	15 13	13 21	07 29	02 04	13 06	17 39
20	23 26	24 01	19 01	13 28	15 26	13 21	07 28	02 04	13 07	17 40
21	23 26	26 05	19 24	13 46	15 40	13 22	07 27	02 04	13 07	17 40
22	23 26	26 18	19 47	14 03	15 53	13 22	07 25	02 03	13 07	17 40
23	23 25	24 34	20 10	14 21	16 06	13 23	07 23	02 03	13 07	17 40
24	23 24	21 04	20 33	14 38	16 19	13 24	07 22	02 03	13 08	17 40
25	23 23	16 13	20 56	14 56	16 31	13 25	07 20	02 03	13 08	17 40
26	23 21	10 29	21 18	15 13	16 44	13 26	07 19	02 03	13 08	17 40
27	23 18	04 N 22	21 39	15 29	16 56	13 27	07 17	02 03	13 09	17 40
28	23 15	01 S 48	22 00	15 46	17 08	13 28	07 15	02 03	13 09	17 41
29	23 12	07 42	22 19	16 03	17 20	13 29	07 14	02 03	13 09	17 41
30	23 N 09	13 S 06	22 N 37	16 N 19	17 N 32	13 S 30	07 N 12	02 S 03	13 S 10	17 S 41

ZODIAC SIGN ENTRIES

Date	h	m	Planets
01	15	17	☽ ♎
03	22	44	☽ ♏
06	08	24	☽ ♐
06	09	07	♀ ♉
08	20	00	☽ ♑
11	08	52	☽ ♒
13	21	32	☽ ♓
14	02	47	☿ ♊
16	07	52	☽ ♈
18	14	20	☽ ♉
20	17	00	☽ ♊
21	05	45	☉ ♋
22	17	12	☽ ♋
24	16	50	☽ ♌
26	17	47	☽ ♍
28	21	24	☽ ♎

LATITUDES

Date	Mercury ☿ ° '	Venus ♀ ° '	Mars ♂ ° '	Jupiter ♃ ° '	Saturn ♄ ° '	Uranus ♅ ° '	Neptune ♆ ° '	Pluto ⯓ ° '
01	03 S 48	02 S 09	00 S 57	00 S 48	02 N 08	00 S 45	00 S 24	05 N 47
04	03 57	02 18	00 56	00 49	02 07	00 45	00 24	05 47
07	03 57	02 26	00 54	00 50	02 07	00 45	00 24	05 47
10	03 48	02 33	00 53	00 50	02 06	00 45	00 24	05 47
13	03 33	02 38	00 51	00 51	02 06	00 46	00 24	05 47
16	03 11	02 42	00 50	00 52	02 05	00 46	00 24	05 46
19	02 44	02 44	00 48	00 53	02 05	00 46	00 24	05 46
22	02 14	02 45	00 46	00 54	02 04	00 46	00 24	05 46
25	01 40	02 46	00 44	00 54	02 04	00 46	00 24	05 45
28	01 04	02 45	00 42	00 55	02 03	00 46	00 24	05 45
31	00 S 28	02 S 42	00 S 40	00 S 56	02 N 03	00 S 46	00 S 24	05 N 45

DATA

Julian Date	2454984
Delta T	+67 seconds
Ayanamsa	23° 59' 33"
Synetic vernal point	05° ♓ 07' 27"
True obliquity of ecliptic	23° 26' 22"

MOON'S PHASES, APSIDES AND POSITIONS ☽

Date	h	m	Phase	Longitude ° '	Eclipse Indicator
07	18	12	○	17 ♐ 07	
15	22	15	☾	24 ♓ 56	
22	19	35	●	01 ♋ 30	
29	11	28	☽	07 ♎ 52	

Day	h	m		
10	15	51	Apogee	
23	10	33	Perigee	
08	02	04	Max dec	26° S 26'
15	14	12	0N	
22	02	44	Max dec	26° N 27'
28	04	55	0S	

LONGITUDES

Date	Chiron ⚷ ° '	Ceres ⚳ ° '	Pallas ⚴ ° '	Juno ⚵ ° '	Vesta ⚶ ° '	Black Moon Lilith ⚸ ° '
01	26 ♒ 14	07 ♍ 53	28 ♋ 39	21 ♓ 14	22 ♊ 05	16 ♑ 32
11	26 ♒ 09	10 ♍ 45	03 ♌ 49	24 ♓ 21	26 ♊ 25	17 ♑ 39
21	25 ♒ 59	13 ♍ 54	08 ♌ 57	27 ♓ 10	00 ♋ 45	18 ♑ 45
31	25 ♒ 43	17 ♍ 19	14 ♌ 03	29 ♓ 39	05 ♋ 06	19 ♑ 52

All ephemeris data is given at 12.00 UT and the Moon's longitude is additionally given for 24.00 UT

ASPECTARIAN

h m	Aspects
01 Monday	
00 41	♂ ♇ ♄
02 32	☽ △ ☿
04 54	☽ ± ♂
06 36	☽ ⚻ ♀
07 48	☽ ∥ ♅
08 32	☽ ☍ ♅
08 55	☽ ⚻ ♆
09 21	☽ ⚻ ♃
12 20	♀ ∥ ♄
16 21	☽ ⚻ ♂
19 46	☽ ± ♆
19 51	☽ □ ⯓
20 13	☽ ± ♃
02 Tuesday	
05 55	☽ ♇ ☿
07 12	☽ ⋕ ♄
08 16	☽ ⋕ ♀
11 10	♀ ⊻ ♅
12 12	☽ ♇ ♆
12 44	☽ ♇ ♃
13 27	☽ △ ☉
15 54	♀ ⚹ ♆
18 59	☽ ⊻ ♄
21 45	☽ ⋕ ♂
22 45	☽ ± ☿
23 39	♀ ⚹ ♃
03 Wednesday	
04 54	☽ Q ⯓
06 09	☽ ⊥ ♄
06 47	☽ ∥ ♆
08 08	☽ ∥ ♃
10 07	☽ ⚻ ☿
15 17	☽ ⋕ ☿
15 47	☽ ⚻ ♅
16 05	☽ △ ♆
16 42	☽ △ ♃
18 00	☽ ☍ ♀
19 37	☽ ♇ ☉
23 05	☽ ∠ ♄
23 56	♂ ⊥ ♅
04 Thursday	
03 07	☽ ± ♅
03 19	☽ ☍ ♂
03 23	☽ ⚹ ⯓
04 37	♂ △ ⯓
06 07	☽ ∥ ⯓
14 04	☽ ± ☉
20 17	☽ ♇ ♅
05 Friday	
02 30	☽ ⚻ ☉
03 44	☽ ⚹ ♄
08 00	☽ ∠ ⯓
14 25	☽ ⋕ ☉
19 10	☉ □ ♄
20 56	☽ ☍ ☿
06 Saturday	
01 18	☽ △ ♅
01 24	☽ ⊥ ⯓
01 29	☽ □ ♆
02 18	☽ □ ♃
03 02	☽ Q ♄
08 20	☽ ⚻ ♀
13 07	☽ ⊻ ⯓
16 01	♀ ♇ ♄
16 54	☽ ⚻ ♂
21 07	☽ ± ♀
07 Sunday	
05 30	☽ ± ♂
12 53	☽ Q ♆
13 49	☽ Q ♃
14 36	☽ □ ♄
16 29	☽ ♇ ♀
18 12	☽ ☍ ☉
08 Monday	
00 34	☽ ♇ ♂
10 54	☽ ⚻ ☿
12 47	☽ □ ♅
12 51	☽ ⚹ ♆
13 51	☽ ⚹ ♃
19 23	♀ △ ⯓
20 57	♀ ⊥ ♅
23 38	☽ ± ☿
09 Tuesday	
00 44	☽ ☌ ⯓
01 13	☽ △ ♀
08 44	☽ △ ♂
18 10	☿ ± ⯓
18 57	☽ ♇ ☿
19 07	☽ ∠ ♆
20 11	☽ ∠ ♃
21 46	☿ ⚹ ♅
21 53	☿ □ ♆
10 Wednesday	
01 12	☽ Q ♅
01 44	♅ ⊻ ♆
03 10	☽ △ ♄
11 18	☽ ⋕ ☉
11 59	☽ ⚻ ☉
12 49	♂ ⋕ ♆
13 24	☽ ⊥ ♆
14 31	☽ ⊥ ♃
15 34	☿ □ ♃
11 Thursday	
01 17	☽ ± ☉
01 37	☽ ⊻ ♆
01 39	☽ ⚹ ♅
02 44	☽ ⊻ ♃
03 31	☽ △ ☿
09 49	☽ ♇ ♄
11 15	♂ ⋕ ♃
13 32	☽ ⊻ ⯓
19 44	☽ □ ♀
21 13	☽ ♇ ☉
12 Friday	
01 03	☽ ∥ ⯓
01 43	☽ ⊥ ⯓
01 48	☽ □ ♂
04 12	☽ ± ♄
05 26	♂ Q ♆
08 13	☽ ∠ ♅
09 09	☽ ⋕ ☿
16 24	☽ ⚻ ♄
19 55	☽ ∠ ⯓
22 55	☽ ⋕ ♂
13 Saturday	
00 26	♂ Q ♃
00 48	☽ ∥ ♃
01 57	☽ ∥ ♆
02 27	☽ ⊥ ♅
06 16	☽ △ ☉
10 38	☽ ⋕ ♀
11 30	☽ Q ♀
14 22	☽ ☌ ♆
14 32	☽ ⊻ ♅
15 34	☽ ☌ ♃
16 33	☽ Q ♂
21 04	☽ □ ☿
14 Sunday	
01 58	☽ ⚹ ⯓
04 24	☽ ⋕ ♄
13 34	☽ ⚹ ♀
17 53	☽ ⚹ ♂
21 10	♀ Q ♆
15 Monday	
01 33	☽ Q ⯓
04 21	☽ ☍ ♄
05 09	☽ ∥ ♅
07 48	♃ St R
11 38	☽ Q ☿
11 41	♀ Q ♃
21 17	☽ ∠ ♀
22 15	☽ □ ☉
23 09	☽ ⋕ ♅
16 Tuesday	
00 43	☽ ∠ ♂
01 02	☽ ⊻ ♆
01 17	☽ ☌ ♅
02 13	☽ ⊻ ♃
04 48	☿ ⚻ ⯓
10 48	♂ ∠ ♅
11 55	☽ □ ⯓
12 21	☽ ⊥ ♆
12 37	☽ ⚹ ☿
13 31	☽ ⊥ ♃
15 45	☽ ⊥ ♀
18 40	☽ ⊥ ♂
19 53	☿ ⋕ ⯓
22 38	☽ ∥ ♄
17 Wednesday	
03 48	☽ ⊻ ♀
05 01	☽ ∠ ♆
06 10	☽ ∠ ♃
06 23	☽ ⊻ ♂
10 28	☽ Q ☉
10 45	☉ △ ♆
12 59	☽ ⚻ ♄
14 56	☉ □ ♅
18 50	☽ ∠ ☿
21 14	☽ ∥ ♀
23 07	☽ ⋕ ♆
23 11	♀ ∠ ♅
23 49	☽ ± ♄
18 Thursday	
00 10	☽ ⋕ ♃
02 29	☉ △ ♃
07 38	☽ ∥ ♂
07 58	☽ ⚹ ♆
08 17	☽ ⊻ ♅
09 05	☽ ⚹ ♃
09 35	☽ ⚹ ☉
12 23	☽ ⊥ ☿
15 45	☽ ♇ ♄
18 00	☽ △ ⯓
18 48	☽ ⊥ ♅
20 52	☽ ⋕ ⯓
23 51	☽ ⊻ ☿
19 Friday	
01 00	☽ ∥ ☿
04 48	☽ Q ♆
05 53	☽ Q ♃
06 52	♀ ⋕ ♆
10 16	☽ ∠ ♅
13 00	☽ ☌ ♀
13 22	☽ ∠ ☉
14 07	☽ ☌ ♂
17 33	☽ △ ♄
19 35	☽ ♇ ⯓
20 Saturday	
02 23	♀ ⋕ ♃
05 32	☽ ⊥ ☉
07 22	☽ ∥ ☉
10 32	☽ ± ⯓
11 00	☽ □ ♆
11 23	☽ ⚹ ♅
12 02	☽ □ ♃
16 07	☽ ⊻ ☉
20 21	☽ ⚻ ⯓
21 Sunday	
06 52	☽ ☌ ☿
06 58	☽ Q ♅
07 56	☿ Q ♅
09 30	☉ ∠ ♂
13 09	♀ ☌ ♂
17 57	☽ ⊻ ♂
18 03	☽ ⊻ ♀
18 52	☽ □ ♄
22 24	☉ ∠ ♀
22 Monday	
04 01	☽ ⊥ ♂
04 21	☽ ⊥ ♀
06 30	♀ △ ♄
11 22	☽ △ ♆
11 47	☽ □ ♅
12 20	☽ △ ♃
14 44	♂ △ ♄
19 03	☽ ∠ ♂
19 35	☽ ☌ ☉
19 43	☽ ∠ ♀
20 22	☽ ☍ ⯓
23 38	☽ Q ♄
23 Tuesday	
04 49	♀ ♇ ⯓
07 42	☉ ☍ ⯓
11 07	☽ ♇ ♆
11 46	☽ ⊻ ☿
12 04	☽ ♇ ♃
18 43	☽ ⚹ ♄
20 01	☽ ⚹ ♂
21 12	☽ ∥ ☉
21 16	☽ ⚹ ♀
21 39	♂ ♇ ⯓
22 20	☽ ⊥ ☿
24 Wednesday	
01 23	☽ ± ♆
02 19	☽ ± ♃
10 55	☽ ⚻ ♆
11 24	☽ △ ♅
11 51	☽ ⚻ ♃
14 19	☽ ∠ ☿
14 37	☽ ∥ ☿
16 05	☽ Q ♂
17 49	☽ Q ♀
18 42	☽ ∠ ♄
19 57	☽ ⚻ ⯓
22 29	☽ ⊻ ☉
25 Thursday	
05 20	☽ ⋕ ⯓
05 35	☽ ± ⯓
08 49	☽ ⊥ ☉
09 15	☽ ⊥ ♄
10 39	☽ ∥ ♂
11 32	☽ ♇ ♅
15 59	☉ Q ♄
17 22	☽ ∥ ♀
17 25	☽ ⚹ ☿
19 03	☽ ⊻ ♄
20 10	☽ ♇ ⯓
22 41	☽ □ ♂
26 Friday	
00 01	☽ ⋕ ♃
00 30	☽ ∠ ☉
01 13	☽ □ ♀
01 13	☽ ⋕ ♆
02 13	☽ ± ♅
08 32	☿ □ ♄
11 34	☽ ☍ ♆
12 08	☽ ⚻ ♅
12 28	☽ ☍ ♃
15 43	☽ Q ☿
20 57	☽ △ ⯓
27 Saturday	
00 36	☽ ∥ ♄
03 13	☽ ⚹ ☉
20 57	☽ ⋕ ♅
21 34	☽ ☌ ♄
28 Sunday	
01 14	☽ Q ☉
02 33	☽ □ ☿
03 56	☽ △ ♂
08 05	☽ △ ♀
12 59	☽ ∥ ♅
14 45	☽ ⚻ ♆
15 26	☽ ☍ ♅
15 38	☽ ⚻ ♃
19 48	☿ ⊻ ♂
29 Monday	
00 41	☽ □ ⯓
01 23	☽ ± ♆
02 14	☽ ± ♃
07 56	☽ ♇ ♂
10 01	☽ ⋕ ♄
11 28	☽ □ ☉
13 02	☽ ♇ ♀
17 33	☽ ♇ ♆
18 24	☽ ♇ ♃
30 Tuesday	
01 12	☽ ± ♂
03 15	☽ ⊻ ♄
06 55	☽ ± ♀
09 17	☽ Q ⯓
12 17	☽ ∥ ♆
12 55	☽ ⚻ ♂
13 56	☽ ∥ ♃
14 23	☽ ⊥ ♄
17 02	☽ △ ☿
19 03	☽ ⚻ ♀
21 11	☽ △ ♆
21 58	☽ ⚻ ♅
21 59	☽ △ ♃

JULY 2009

LONGITUDES

Date	Sidereal time h m s	Sun ☉ ° ' "	Moon ☽ ° ' "	Moon ☽ 24.00 ° ' "	Mercury ☿ ° '	Venus ♀ ° '	Mars ♂ ° '	Jupiter ♃ ° '	Saturn ♄ ° '	Uranus ♅ ° '	Neptune ♆ ° '	Pluto ♇ ° '
01	06 38 41	09 ♋ 47 18	04 ♏ 04 09	10 ♏ 21 46	25 ♊ 29	25 ♉ 50	22 ♉ 27	26 ♒ 36	16 ♍ 37	26 ♓ 37	26 ♒ 11	01 ♑ 47
02	06 42 38	10 44 30	16 ♏ 35 39	22 ♏ 46 09	27 25	26 54	23 10	26 R 33	16 42	26 R 37	26 R 10	01 R 45
03	06 46 34	11 41 41	28 ♏ 53 38	04 ♐ 58 25	29 ♊ 23	27 59	23 53	26 29	16 46	26 37	26 09	01 44
04	06 50 31	12 38 53	11 ♐ 00 51	17 ♐ 01 13	01 ♋ 24	29 ♉ 05	24 36	26 26	16 51	26 37	26 08	01 42
05	06 54 27	13 36 04	22 ♐ 59 49	28 ♐ 56 57	03 27	00 ♊ 10	25 19	26 22	16 55	26 37	26 07	01 41
06	06 58 24	14 33 15	04 ♑ 52 51	10 ♑ 47 50	05 32	01 15	26 01	26 18	17 00	26 37	26 06	01 39
07	07 02 20	15 30 26	16 ♑ 42 09	22 ♑ 36 04	07 38	02 21	26 44	26 14	17 05	26 36	26 05	01 38
08	07 06 17	16 27 38	28 ♑ 29 53	04 ♒ 23 54	09 45	03 27	27 26	26 10	17 09	26 36	26 04	01 36
09	07 10 14	17 24 49	10 ♒ 18 24	16 ♒ 13 45	11 54	04 33	28 09	26 06	17 14	26 36	26 03	01 35
10	07 14 10	18 22 01	22 ♒ 10 17	28 ♒ 08 23	14 03	05 39	28 51	26 01	17 19	26 35	26 02	01 33
11	07 18 07	19 19 13	04 ♓ 08 26	10 ♓ 10 51	16 12	06 46	29 ♉ 34	25 57	17 24	26 35	26 00	01 32
12	07 22 03	20 16 25	16 ♓ 16 06	22 ♓ 24 37	18 22	07 52	00 ♊ 16	25 52	17 29	26 34	25 59	01 30
13	07 26 00	21 13 38	28 ♓ 36 52	04 ♈ 53 20	20 31	08 59	00 58	25 47	17 34	26 34	25 58	01 29
14	07 29 56	22 10 51	11 ♈ 14 28	17 ♈ 40 43	22 40	10 06	01 40	25 41	17 40	26 33	25 57	01 27
15	07 33 53	23 08 05	24 ♈ 12 29	00 ♉ 50 10	24 48	11 13	02 22	25 36	17 45	26 33	25 55	01 26
16	07 37 49	24 05 19	07 ♉ 34 01	14 ♉ 24 17	26 56	12 20	03 04	25 31	17 50	26 32	25 54	01 25
17	07 41 46	25 02 35	21 ♉ 21 03	28 ♉ 24 17	29 ♋ 02	13 27	03 46	25 25	17 56	26 31	25 53	01 23
18	07 45 43	25 59 51	05 ♊ 33 49	12 ♊ 49 19	01 ♌ 07	14 34	04 28	25 19	18 01	26 30	25 51	01 22
19	07 49 39	26 57 08	20 ♊ 10 18	27 ♊ 36 04	03 11	15 42	05 10	25 13	18 07	26 30	25 50	01 20
20	07 53 36	27 54 25	05 ♋ 05 49	12 ♋ 38 35	05 13	16 49	05 51	25 07	18 12	26 29	25 49	01 19
21	07 57 32	28 51 43	20 ♋ 13 17	27 ♋ 48 45	07 14	17 57	06 33	25 01	18 18	26 28	25 47	01 18
22	08 01 29	29 ♋ 49 02	05 ♌ 23 48	12 ♌ 57 16	09 13	19 05	07 14	24 55	18 24	26 27	25 46	01 16
23	08 05 25	00 ♌ 46 21	20 ♌ 28 02	27 ♌ 55 03	11 11	20 13	07 56	24 48	18 30	26 26	25 44	01 15
24	08 09 22	01 43 40	05 ♍ 17 27	12 ♍ 34 27	13 07	21 21	08 37	24 42	18 36	26 25	25 43	01 14
25	08 13 18	02 41 00	19 ♍ 45 30	26 ♍ 50 11	15 01	22 30	09 18	24 35	18 41	26 24	25 41	01 13
26	08 17 15	03 38 20	03 ♎ 48 15	10 ♎ 39 38	16 54	23 38	10 00	24 28	18 47	26 22	25 40	01 11
27	08 21 12	04 35 41	17 ♎ 24 22	24 ♎ 02 39	18 44	24 46	10 41	24 22	18 53	26 21	25 38	01 10
28	08 25 08	05 33 02	00 ♏ 34 45	07 ♏ 01 02	20 34	25 55	11 22	24 15	19 00	26 20	25 37	01 09
29	08 29 05	06 30 24	13 ♏ 21 56	19 ♏ 37 54	22 21	27 04	12 02	24 08	19 06	26 19	25 35	01 08
30	08 33 01	07 27 46	25 ♏ 49 27	01 ♐ 57 05	24 07	28 12	12 43	24 00	19 12	26 18	25 34	01 07
31	08 36 58	08 ♌ 25 09	08 ♐ 01 19	14 ♐ 02 40	25 ♌ 51	29 ♊ 21	13 ♊ 24	23 ♒ 53	19 ♍ 18	26 ♓ 16	25 ♒ 32	01 ♑ 05

Date	Moon True ☊ ° '	Moon Mean ☊ ° '	Moon ☽ Latitude ° '
01	00 ♒ 24	01 ♒ 21	05 S 15
02	00 R 22	01 18	05 03
03	00 19	01 15	04 37
04	00 16	01 11	03 59
05	00 14	01 08	03 11
06	00 12	01 05	02 15
07	00 11	01 02	01 14
08	00 10	00 59	00 S 09
09	00 D 10	00 55	00 N 56
10	00 11	00 52	01 59
11	00 12	00 49	02 58
12	00 13	00 46	03 49
13	00 14	00 43	04 31
14	00 14	00 40	05 00
15	00 R 15	00 36	05 16
16	00 14	00 33	05 15
17	00 14	00 30	04 56
18	00 14	00 27	04 19
19	00 13	00 24	03 25
20	00 13	00 21	02 15
21	00 13	00 17	00 N 55
22	00 D 13	00 14	00 S 29
23	00 R 13	00 11	01 50
24	00 13	00 08	03 02
25	00 13	00 05	04 02
26	00 12	00 ♒ 01	04 45
27	00 12	29 ♑ 58	05 10
28	00 12	29 55	05 18
29	00 D 12	29 52	05 09
30	00 12	29 49	04 47
31	00 ♒ 13	29 ♑ 46	04 S 11

DECLINATIONS

Date	Sun ☉ ° '	Moon ☽ ° '	Mercury ☿ ° '	Venus ♀ ° '	Mars ♂ ° '	Jupiter ♃ ° '	Saturn ♄ ° '	Uranus ♅ ° '	Neptune ♆ ° '	Pluto ♇ ° '
01	23 N 05	17 S 48	22 N 54	16 N 35	17 N 44	13 S 32	07 N 10	02 S 03	13 S 10	17 S 41
02	23 00	21 38	23 09	16 51	17 55	13 33	07 08	02 03	13 10	17 41
03	22 55	24 25	23 22	17 06	18 07	13 34	07 06	02 03	13 11	17 41
04	22 50	26 03	23 33	17 21	18 18	13 36	07 04	02 03	13 11	17 41
05	22 45	26 26	23 41	17 36	18 29	13 37	07 02	02 03	13 12	17 42
06	22 39	25 36	23 48	17 51	18 40	13 39	07 01	02 03	13 12	17 42
07	22 32	23 37	23 51	18 05	18 50	13 40	06 59	02 04	13 12	17 42
08	22 25	20 37	23 53	18 19	19 01	13 42	06 57	02 04	13 13	17 42
09	22 18	16 46	23 51	18 33	19 11	13 44	06 55	02 04	13 13	17 42
10	22 11	12 15	23 47	18 46	19 21	13 46	06 52	02 04	13 14	17 42
11	22 03	07 14	23 40	18 59	19 31	13 47	06 50	02 04	13 14	17 43
12	21 55	01 S 54	23 30	19 12	19 40	13 49	06 48	02 05	13 14	17 43
13	21 46	03 N 35	23 17	19 24	19 50	13 51	06 46	02 05	13 15	17 43
14	21 37	09 03	23 02	19 36	19 59	13 53	06 44	02 05	13 15	17 43
15	21 27	14 17	22 45	19 47	20 08	13 55	06 42	02 05	13 16	17 43
16	21 18	19 00	22 24	19 58	20 17	13 57	06 40	02 06	13 16	17 44
17	21 07	22 52	22 02	20 09	20 26	13 59	06 37	02 06	13 17	17 44
18	20 57	25 29	21 38	20 19	20 34	14 01	06 35	02 06	13 17	17 44
19	20 46	26 29	21 11	20 29	20 43	14 04	06 33	02 07	13 18	17 44
20	20 35	25 36	20 43	20 38	20 51	14 06	06 31	02 07	13 18	17 44
21	20 23	22 50	20 13	20 47	20 59	14 08	06 28	02 07	13 19	17 44
22	20 11	18 27	19 41	20 55	21 06	14 11	06 26	02 08	13 19	17 45
23	19 59	12 55	19 08	21 03	21 14	14 13	06 24	02 08	13 20	17 45
24	19 47	06 44	18 34	21 10	21 21	14 15	06 21	02 09	13 20	17 45
25	19 34	00 N 21	17 58	21 17	21 28	14 18	06 19	02 09	13 21	17 45
26	19 20	05 S 52	17 22	21 23	21 35	14 20	06 16	02 10	13 21	17 46
27	19 07	11 36	16 44	21 29	21 42	14 23	06 14	02 10	13 22	17 46
28	18 53	16 38	16 06	21 34	21 48	14 25	06 11	02 11	13 22	17 46
29	18 39	20 46	15 27	21 39	21 55	14 28	06 09	02 11	13 23	17 46
30	18 24	23 51	14 48	21 43	22 01	14 30	06 06	02 12	13 23	17 46
31	18 N 10	25 S 47	14 N 08	21 N 47	22 N 07	14 S 33	06 N 04	02 S 12	13 S 24	17 S 47

ZODIAC SIGN ENTRIES

Date	h m	Planets
01	04 19	☽ ♏
03	14 11	☽ ♐
03	19 20	☿ ♋
05	08 23	♀ ♊
06	02 07	☽ ♑
08	15 03	☽ ♒
11	03 44	☽ ♓
12	02 56	♂ ♊
13	14 40	☽ ♈
15	22 30	☽ ♉
17	23 07	☿ ♌
18	02 41	☽ ♊
20	03 51	☽ ♋
22	03 28	☽ ♌
22	16 36	☉ ♌
24	03 23	☽ ♍
26	05 25	☽ ♎
28	10 56	☽ ♏
30	20 10	☽ ♐

LATITUDES

Date	Mercury ☿ ° '	Venus ♀ ° '	Mars ♂ ° '	Jupiter ♃ ° '	Saturn ♄ ° '	Uranus ♅ ° '	Neptune ♆ ° '	Pluto ♇ ° '
01	00 S 28	02 S 42	00 S 40	00 S 56	02 N 03	00 S 46	00 S 24	05 N 45
04	00 N 07	02 39	00 38	00 57	02 02	00 46	00 24	05 44
07	00 38	02 36	00 36	00 58	02 02	00 46	00 24	05 44
10	01 05	02 31	00 34	00 58	02 02	00 47	00 24	05 43
13	01 26	02 25	00 32	00 59	02 01	00 47	00 25	05 43
16	01 40	02 19	00 30	01 00	02 01	00 47	00 25	05 43
19	01 47	02 12	00 27	01 00	02 00	00 47	00 25	05 42
22	01 48	02 05	00 25	01 01	02 00	00 47	00 25	05 41
25	01 43	01 57	00 23	01 02	02 00	00 47	00 25	05 41
28	01 33	01 48	00 20	01 02	02 00	00 47	00 25	05 40
31	01 N 18	01 S 40	00 S 18	01 S 03	01 N 59	00 S 47	00 S 25	05 N 39

DATA

Julian Date	2455014
Delta T	+67 seconds
Ayanamsa	23° 59' 38"
Synetic vernal point	05° ♓ 07' 21"
True obliquity of ecliptic	23° 26' 21"

MOON'S PHASES, APSIDES AND POSITIONS ☽

Date	h m	Phase	Longitude ° '	Eclipse Indicator
07	09 21	○	15 ♑ 24	
15	09 53	☾	23 ♈ 03	
22	02 35	●	29 ♋ 27	Total
28	22 00	☽	05 ♏ 57	

Day	h m		
07	21 22	Apogee	
21	20 08	Perigee	
05	07 38	Max dec	26° S 28'
12	20 21	0N	
19	12 58	Max dec	26° N 29'
25	13 18	0S	

LONGITUDES

Date	Chiron ⚷ ° '	Ceres ⚳ ° '	Pallas ⚴ ° '	Juno ⚵ ° '	Vesta ⚶ ° '	Black Moon Lilith ⚸ ° '
01	25 ♒ 43	17 ♍ 19	14 ♌ 03	29 ♓ 39	05 ♋ 06	19 ♑ 52
11	25 ♒ 22	20 ♍ 56	19 ♌ 07	01 ♈ 43	09 ♋ 27	20 ♑ 58
21	24 ♒ 57	24 ♍ 44	24 ♌ 09	03 ♈ 18	13 ♋ 47	22 ♑ 05
31	24 ♒ 29	28 ♍ 40	29 ♌ 09	04 ♈ 17	18 ♋ 06	23 ♑ 11

ASPECTARIAN

h m Aspects

01 Wednesday
01 03 ♃ ⊻ ♅
04 56 ☽ # ♀
05 41 ♂ # ♇
07 20 ☽ ∠ ♄
07 38 ♅ St R
07 40 ☽ ⚹ ♇
09 16 ☽ ± ♅
10 56 ♀ ± ♇
11 20 ☽ ∥ ♇
11 35 ☽ # ♂
20 01 ♀ □ ♆
20 48 ☿ △ ♆
21 49 ☿ ⊻ ♀
23 48 ☽ △ ☉

02 Thursday
01 26 ☉ ∥ ☿
01 34 ☿ △ ♃
02 13 ☿ □ ♅
02 25 ☽ ⅊ ♅
02 27 ☽ ⅊ ☿
04 24 ♀ □ ♃
05 42 ♀ ⚹ ♅
12 12 ☽ ⚹ ♄
12 18 ☽ ∠ ♇
22 24 ☽ # ☉
22 43 ☉ ⅊ ♆
23 08 ☽ ± ☿

03 Friday
00 54 ☽ # ☿
01 34 ☽ ☍ ♂
05 48 ☽ ⊥ ♇
06 38 ☽ □ ♆
07 09 ☉ ⅊ ♃
07 18 ☽ □ ♃
07 19 ☽ ⅊ ☉
07 32 ☽ △ ♅
10 03 ☽ ☍ ♀
11 45 ☽ Q ♄
13 10 ☽ ⊼ ☿
17 34 ☽ ⊻ ♇
21 12 ☿ ⊥ ♂

04 Saturday
02 35 ☽ ± ☉
15 29 ☿ ☍ ♇
15 32 ☽ ⊼ ☉
18 14 ☽ Q ♆
18 47 ☽ Q ♃
23 43 ☽ □ ♄

05 Sunday
16 57 ☽ ⊼ ♂
18 17 ☽ ⚹ ♆
18 46 ☽ ⚹ ♃
19 17 ☽ □ ♅
20 39 ♀ # ♇
23 54 ♂ ± ♇

06 Monday
03 55 ☽ ⊼ ♀
05 28 ☽ ☌ ♇
05 41 ☿ Q ♄
05 51 ☽ ± ♂
13 35 ☽ ☍ ☿
14 43 ♂ □ ♆
17 18 ☽ ± ♀
20 26 ♀ ⊼ ♇
20 48 ♂ □ ♃

07 Tuesday
00 36 ☽ ∠ ♆
00 58 ☽ ∠ ♃
01 15 ☽ ⅊ ♂
07 45 ☽ Q ♅
07 48 ♂ ⚹ ♅
09 21 ☽ ☍ ☉
09 43 ☽ # ☿
12 46 ☽ △ ♄
13 27 ☽ ⅊ ♀
18 52 ☽ ⊥ ♆
19 09 ☽ ⊥ ♃
21 54 ☽ # ☉

08 Wednesday
04 58 ☿ ⊥ ♀
07 04 ☽ ⊻ ♆
07 17 ☽ ⊻ ♃
08 09 ☽ ⚹ ♅
09 43 ☽ △ ♂
18 18 ☽ ⊻ ♇
19 30 ☽ ⅊ ♄
22 07 ☽ # ♂
23 06 ☽ △ ♀

09 Thursday
02 00 ☽ # ♀
02 37 ☿ ⅊ ♆
03 23 ☿ ⅊ ♃
06 28 ☽ ⊥ ♇
06 32 ☽ ∥ ♇
07 09 ☉ ⚹ ♄
13 54 ☽ ± ♄
14 37 ☽ ∠ ♅
15 56 ☽ ⊼ ☿

10 Friday
00 41 ☽ ∠ ♇
02 08 ☽ ⊼ ♄
03 39 ☽ ⊼ ☉
04 19 ☽ ∥ ♃
06 45 ☽ ± ☿
07 01 ☽ ∥ ♆
08 49 ☽ ⊥ ♅
08 54 ☿ ∠ ♂
09 13 ♃ ☌ ♆
16 48 ☽ ± ☉
19 42 ☽ ☌ ♃
19 45 ☽ ☌ ♆
20 53 ☽ ⊻ ♅

11 Saturday
02 17 ☽ □ ♂
04 51 ☽ ⅊ ☿
06 48 ☽ ⚹ ♇
12 23 ☽ ⅊ ☉
13 49 ☽ # ♄
17 44 ☽ □ ♀

12 Sunday
01 55 ☿ ⚹ ♄
02 26 ☉ ± ♃
04 55 ☉ ± ♆
06 34 ☽ Q ♇
11 14 ☽ ∥ ♅
14 25 ☽ ☍ ♄
16 09 ☽ Q ♂
16 58 ☽ △ ☿
20 30 ☽ △ ☉

13 Monday
03 06 ♀ Q ♅
04 05 ☿ ± ♃
05 26 ☽ # ♅
05 55 ☿ ± ♆
06 34 ☽ ⊻ ♃
06 54 ☽ ⊻ ♆
08 03 ☽ ☌ ♅
08 32 ☽ Q ♀
16 47 ☽ ⚹ ♂
17 29 ☽ □ ♇
18 02 ☽ ⊥ ♃
18 25 ☽ ⊥ ♆

14 Tuesday
01 49 ☽ ∥ ♄
02 15 ☉ ☌ ☿
04 52 ♂ ⊼ ♇
09 38 ☽ ⚹ ♀
10 58 ☽ ∠ ♃
11 27 ☽ ∠ ♆
22 43 ☽ ∠ ♂

15 Wednesday
00 03 ☽ ⊼ ♄
07 12 ☽ # ♆
09 53 ☽ □ ☉
10 17 ☽ # ♃
11 10 ☽ ± ♄
13 18 ☽ □ ☿
14 31 ☽ ⚹ ♃
15 07 ☽ ⚹ ♆
15 59 ☽ ∠ ♀
16 09 ☽ ⊥ ♂
16 15 ☽ ⊻ ♅
20 39 ☿ ⊼ ♃

16 Thursday
00 31 ☿ ⊼ ♆
01 03 ☽ △ ♇
03 03 ☽ ⊥ ♅
03 32 ☽ ⅊ ♄
03 34 ☽ ⊻ ♂
05 11 ☽ # ♇
07 33 ☿ △ ♅
09 36 ☽ ⊥ ♀
11 54 ☽ Q ♃
12 35 ☽ Q ♆
17 47 ☽ ∥ ♀
18 58 ☽ ∠ ♅
19 37 ☽ ∥ ♂
20 33 ☽ Q ☉
21 07 ☽ ⊻ ♀
21 45 ☿ ∠ ♀

17 Friday
00 54 ☽ ∥ ☉
03 14 ☽ Q ☿
03 27 ☽ ⅊ ♇
06 05 ☽ △ ♄
06 49 ☽ ∥ ☿
18 46 ☽ ⚹ ☉
18 52 ☽ ± ♇
18 54 ☽ □ ♃
19 43 ☽ □ ♆
20 33 ☉ ⊼ ♃
20 48 ☽ ⚹ ♅

18 Saturday
03 18 ☽ ⚹ ☿
04 59 ☽ ⊼ ♇
08 31 ☉ ⊼ ♆
10 05 ☽ ☌ ♂
14 49 ☿ ⊼ ♇
16 53 ☽ Q ♅
21 38 ☽ ∠ ☉

19 Sunday
00 36 ☉ △ ♅
04 06 ☽ ☌ ♀
08 14 ☽ ∠ ☿
08 38 ☽ □ ♄
11 12 ☿ ∠ ♄
13 21 ☽ ⊥ ☉
16 18 ☉ ∥ ♂
20 07 ☽ △ ♃
21 08 ☽ △ ♆
22 12 ☽ □ ♅
23 42 ☽ ⊻ ☉

20 Monday
01 07 ☽ ⊥ ☿
05 58 ☽ ☍ ♇
06 46 ☿ ∥ ♂
08 15 ☉ ∥ ♀
12 13 ☽ ⊻ ☿
13 16 ☽ ⊻ ♂
13 47 ☽ Q ♄
14 53 ☿ ∥ ♀
19 57 ☽ ⅊ ♃
21 05 ☽ ⅊ ♆
22 19 ☉ ∥ ☿
23 16 ☽ ⊥ ♂
23 32 ☿ ⚹ ♂

21 Tuesday
08 07 ☽ ⊻ ♀
08 57 ☽ ⚹ ♄
10 07 ☽ ± ♃
11 19 ☽ ± ♆
12 45 ☿ ± ♇
14 12 ☽ ∠ ♂
18 23 ☽ ⊥ ♀
19 32 ☽ ⊼ ♃
20 05 ♀ □ ♄
20 47 ☽ ⊼ ♆
21 51 ☽ △ ♅
22 48 ☽ ∥ ♂
23 50 ☽ ∥ ♀

22 Wednesday
02 35 ☽ ☌ ☉
02 58 ☽ ∥ ☉
05 12 ☽ ∥ ☿
05 29 ☽ ⊼ ♇
08 49 ☽ ∠ ♄
09 45 ☽ ∠ ♀
14 58 ☽ ± ♇
15 04 ☽ ⚹ ♂
15 20 ☽ # ♇
18 58 ☽ ☌ ☿
21 35 ☽ ⅊ ♅
23 11 ☽ ⊥ ♄

23 Thursday
05 16 ☽ ⅊ ♇
06 45 ☽ # ♃
08 49 ☽ ⊻ ♄
10 22 ☽ # ♆
11 06 ☽ Q ♂
11 34 ☽ ⚹ ♀
11 56 ☽ ± ♅
15 03 ☿ ⅊ ♅
18 56 ☽ ☍ ♃
20 28 ☽ ☍ ♆
21 35 ☽ ⊼ ♅
23 47 ☉ ⊼ ♇

24 Friday
04 57 ♂ Q ♅
05 10 ☿ ⊥ ♄
05 23 ☽ △ ♇
05 47 ☽ ⊻ ☉
08 34 ☽ Q ♀
13 27 ☽ ∥ ♄
16 17 ☽ ⊥ ☉
17 44 ☽ □ ♂

25 Saturday
02 52 ☽ ⊻ ☿
05 12 ☽ # ♅
08 16 ☽ ∠ ☉
10 12 ☽ ☌ ♄
14 27 ☽ ⊥ ☿
17 01 ☽ □ ♀
20 06 ☽ ⊼ ♃
20 25 ☿ # ♇
21 31 ☽ ∥ ♅
22 02 ☽ ⊼ ♆
23 14 ☽ ☍ ♅

26 Sunday
03 02 ☿ ⅊ ♇
06 17 ☽ ± ♃
07 29 ☽ □ ♇
08 11 ☽ ∠ ☿
08 18 ☽ ± ♆
11 42 ☽ ⚹ ☉
13 37 ☽ # ♄
16 16 ☉ ∠ ♄
21 50 ☽ ⅊ ♃
23 24 ☽ △ ♂
23 59 ☽ ⅊ ♆

27 Monday
04 07 ♀ △ ♃
10 26 ☽ Q ☉
14 05 ☿ ⊻ ♄
14 41 ☽ ⊻ ♄
14 47 ☽ ⚹ ☿
15 10 ☽ Q ♇
20 00 ☽ ∥ ♆

28 Tuesday
00 28 ☽ △ ♃
00 53 ☽ ∥ ♃
01 39 ☽ ⊥ ♄
02 36 ☽ △ ♀
02 53 ☽ △ ♆
03 48 ☽ ⅊ ♂
04 12 ☽ ⊼ ♅
05 52 ♀ △ ♆
09 02 ☿ ± ♅
09 37 ☽ # ☿
13 03 ☽ ⚹ ♇
15 15 ☽ ± ♅
16 16 ☽ Q ☿
18 05 ☽ ∥ ♇
18 23 ☽ ∠ ♄
20 38 ♀ □ ♅
21 24 ☽ ± ♂
22 00 ☽ □ ☉
23 43 ☽ # ☉

29 Wednesday
08 07 ☽ ⅊ ♅
09 16 ☽ ⅊ ♀
09 20 ☽ ⊼ ♂
17 16 ☽ ∠ ♇
18 13 ☽ # ♀
20 15 ☽ # ♂
23 03 ☽ ⚹ ♄

30 Thursday
03 17 ☉ ± ♇
04 15 ☽ ± ♀
08 07 ☽ □ ☿
08 30 ☽ □ ♃
10 37 ☽ ⊥ ♇
10 38 ☿ ☍ ♃
11 30 ☽ □ ♆
12 55 ☽ △ ♅
17 08 ☽ ⊼ ♀
21 55 ☿ # ♃
22 20 ☽ ⊻ ♇
22 37 ☽ Q ♄

31 Friday
01 44 ☿ Q ♂
07 47 ☿ ☍ ♆
12 51 ☽ △ ☉
17 49 ☿ ⊼ ♅
19 37 ☽ Q ♃
22 58 ☽ Q ♆
23 21 ☽ ☍ ♂

All ephemeris data is given at 12.00 UT and the Moon's longitude is additionally given for 24.00 UT

AUGUST 2009

Date	Sidereal time h m s	Sun ☉ ° ' "	Moon ☽ ° ' "	Moon ☽ 24.00 ° ' "	Mercury ☿ ° '	Venus ♀ ° '	Mars ♂ ° '	Jupiter ♃ ° '	Saturn ♄ ° '	Uranus ♅ ° '	Neptune ♆ ° '	Pluto ♇ ° '
01	08 40 54	09 ♌ 22 32	20 ♐ 01 38	25 ♐ 58 41	27 ♌ 32	00 ♋ 30	14 ♊ 05	23 ♒ 46	19 ♍ 24	26 ♓ 15	25 ♒ 31	01 ♑ 04
02	08 44 51	10 19 56	01 ♑ 54 18	07 ♑ 48 53	29 ♌ 14	01 39	14 45	23 R 38	19 31	26 R 13	25 R 29	01 R 03
03	08 48 47	11 17 21	13 ♑ 42 52	19 ♑ 36 36	00 ♍ 53	02 49	15 26	23 31	19 37	26 12	25 28	01 02
04	08 52 44	12 14 46	25 ♑ 30 28	01 ♒ 24 47	02 30	03 58	16 06	23 24	19 44	26 10	25 26	01 01
05	08 56 41	13 12 13	07 ♒ 19 51	13 ♒ 15 57	04 06	05 07	16 46	23 16	19 50	26 09	25 25	01 00
06	09 00 37	14 09 40	19 ♒ 13 22	25 ♒ 12 21	05 40	06 17	17 26	23 08	19 57	26 07	25 23	00 59
07	09 04 34	15 07 08	01 ♓ 13 08	07 ♓ 15 59	07 12	07 27	18 06	23 01	20 03	26 06	25 21	00 58
08	09 08 30	16 04 38	13 ♓ 21 08	19 ♓ 28 50	08 43	08 36	18 46	22 53	20 10	26 04	25 20	00 57
09	09 12 27	17 02 08	25 ♓ 39 18	01 ♈ 52 50	10 12	09 46	19 26	22 45	20 16	26 02	25 18	00 56
10	09 16 23	17 59 40	08 ♈ 09 39	14 ♈ 30 02	11 40	10 56	20 06	22 37	20 23	26 01	25 17	00 55
11	09 20 20	18 57 13	20 ♈ 54 16	27 ♈ 22 36	13 05	12 06	20 46	22 30	20 30	25 59	25 15	00 54
12	09 24 16	19 54 48	03 ♉ 55 18	10 ♉ 32 38	14 29	13 16	21 26	22 22	20 37	25 57	25 13	00 53
13	09 28 13	20 52 24	17 ♉ 14 49	24 ♉ 02 02	15 51	14 26	22 05	22 14	20 43	25 55	25 12	00 52
14	09 32 10	21 50 02	00 ♊ 54 26	07 ♊ 52 06	17 12	15 37	22 45	22 06	20 50	25 54	25 10	00 52
15	09 36 06	22 47 41	14 ♊ 54 59	22 ♊ 03 00	18 30	16 47	23 24	21 58	20 57	25 52	25 08	00 51
16	09 40 03	23 45 22	29 ♊ 15 54	06 ♋ 33 22	19 47	17 58	24 03	21 50	21 04	25 50	25 07	00 50
17	09 43 59	24 43 04	13 ♋ 54 52	21 ♋ 19 47	21 02	19 08	24 42	21 43	21 11	25 48	25 05	00 49
18	09 47 56	25 40 48	28 ♋ 47 21	06 ♌ 16 42	22 15	20 19	25 21	21 35	21 18	25 46	25 03	00 48
19	09 51 52	26 38 33	13 ♌ 46 49	21 ♌ 16 41	23 26	21 30	26 00	21 27	21 25	25 44	25 02	00 48
20	09 55 49	27 36 20	28 ♌ 45 13	06 ♍ 11 20	24 34	22 41	26 39	21 19	21 32	25 42	25 00	00 47
21	09 59 45	28 34 07	13 ♍ 34 01	20 ♍ 52 22	25 40	23 52	27 18	21 11	21 39	25 40	24 59	00 46
22	10 03 42	29 ♌ 31 57	28 ♍ 05 34	05 ♎ 12 57	26 44	25 03	27 56	21 04	21 46	25 38	24 57	00 46
23	10 07 39	00 ♍ 29 47	12 ♎ 14 03	19 ♎ 08 32	27 46	26 14	28 35	20 56	21 53	25 36	24 55	00 45
24	10 11 35	01 27 39	25 ♎ 56 16	02 ♏ 37 14	28 45	27 25	29 13	20 48	22 00	25 34	24 54	00 45
25	10 15 32	02 25 32	09 ♏ 11 35	15 ♏ 39 37	29 ♍ 41	28 36	29 ♊ 52	20 41	22 08	25 31	24 52	00 44
26	10 19 28	03 23 26	22 ♏ 01 41	28 ♏ 18 16	00 ♎ 35	29 ♋ 48	00 ♋ 30	20 33	22 15	25 29	24 50	00 44
27	10 23 25	04 21 22	04 ♐ 29 54	10 ♐ 37 08	01 25	00 ♌ 59	01 08	20 26	22 22	25 27	24 49	00 44
28	10 27 21	05 19 19	16 ♐ 40 35	22 ♐ 40 53	02 12	02 10	01 46	20 19	22 29	25 25	24 47	00 43
29	10 31 18	06 17 17	28 ♐ 38 39	04 ♑ 34 30	02 56	03 22	02 24	20 11	22 37	25 23	24 46	00 43
30	10 35 14	07 15 16	10 ♑ 29 02	16 ♑ 22 49	03 36	04 34	03 01	20 04	22 44	25 20	24 44	00 42
31	10 39 11	08 ♍ 13 17	22 ♑ 16 26	28 ♑ 10 21	04 ♎ 12	05 ♌ 45	03 ♋ 39	19 ♒ 57	22 ♍ 51	25 ♓ 18	24 ♒ 42	00 ♑ 42

Date	Moon True ☊ ° '	Moon Mean ☊ ° '	Moon ☽ Latitude ° '
01	00 ♒ 14	29 ♑ 42	03 S 25
02	00 D 15	29 39	02 31
03	00 15	29 36	01 31
04	00 16	29 33	00 S 26
05	00 R 16	29 30	00 N 39
06	00 15	29 26	01 43
07	00 14	29 23	02 43
08	00 12	29 20	03 36
09	00 09	29 17	04 20
10	00 07	29 14	04 53
11	00 04	29 11	05 12
12	00 03	29 07	05 15
13	00 02	29 04	05 01
14	00 D 02	29 01	04 31
15	00 03	28 58	03 44
16	00 05	28 55	02 42
17	00 06	28 52	01 28
18	00 06	28 48	00 N 07
19	00 R 06	28 45	01 S 15
20	00 04	28 42	02 31
21	00 01	28 39	03 36
22	29 ♑ 57	28 36	04 26
23	29 52	28 32	04 58
24	29 48	28 29	05 12
25	29 45	28 26	05 09
26	29 43	28 23	04 50
27	29 D 42	28 20	04 17
28	29 43	28 17	03 34
29	29 44	28 13	02 42
30	29 46	28 10	01 43
31	29 ♑ 47	28 ♑ 07	00 S 41

DECLINATIONS

Date	Sun ☉ ° '	Moon ☽ ° '	Mercury ☿ ° '	Venus ♀ ° '	Mars ♂ ° '	Jupiter ♃ ° '	Saturn ♄ ° '	Uranus ♅ ° '	Neptune ♆ ° '	Pluto ♇ ° '
01	17 N 54	26 S 29	13 N 27	21 N 50	22 N 12	14 S 35	06 N 01	02 S 13	13 S 24	17 S 47
02	17 39	25 57	12 47	21 52	22 18	14 38	05 59	02 14	13 25	17 47
03	17 23	24 14	12 06	21 54	22 23	14 41	05 56	02 14	13 25	17 47
04	17 08	21 28	11 24	21 56	22 28	14 43	05 54	02 15	13 26	17 48
05	16 51	17 49	10 43	21 56	22 33	14 46	05 51	02 15	13 27	17 48
06	16 35	13 25	10 02	21 57	22 38	14 48	05 48	02 16	13 27	17 48
07	16 18	08 30	09 20	21 56	22 43	14 51	05 46	02 17	13 28	17 48
08	16 01	03 S 13	08 39	21 55	22 47	14 54	05 43	02 17	13 28	17 49
09	15 44	02 N 15	07 58	21 54	22 51	14 57	05 40	02 18	13 29	17 49
10	15 26	07 43	07 17	21 51	22 55	14 59	05 38	02 19	13 29	17 49
11	15 09	12 58	06 36	21 49	22 59	15 02	05 35	02 19	13 30	17 49
12	14 51	17 45	05 56	21 45	23 02	15 05	05 32	02 20	13 30	17 49
13	14 32	21 48	05 16	21 42	23 06	15 07	05 30	02 21	13 31	17 50
14	14 14	24 45	04 36	21 37	23 09	15 10	05 27	02 22	13 32	17 50
15	13 55	26 18	03 57	21 32	23 12	15 13	05 24	02 22	13 32	17 50
16	13 36	26 08	03 18	21 26	23 15	15 15	05 21	02 23	13 33	17 50
17	13 17	24 10	02 39	21 20	23 17	15 18	05 18	02 24	13 33	17 51
18	12 58	20 31	02 02	21 13	23 20	15 20	05 16	02 25	13 34	17 51
19	12 38	15 30	01 25	21 05	23 22	15 23	05 13	02 26	13 34	17 51
20	12 18	09 33	00 49	20 57	23 24	15 26	05 10	02 26	13 35	17 52
21	11 58	03 N 08	00 N 13	20 49	23 26	15 28	05 07	02 27	13 35	17 52
22	11 38	03 S 19	00 S 21	20 39	23 27	15 31	05 04	02 28	13 36	17 52
23	11 18	09 25	00 55	20 30	23 29	15 33	05 02	02 29	13 37	17 52
24	10 57	14 52	01 27	20 19	23 30	15 36	04 59	02 30	13 37	17 53
25	10 37	19 26	01 59	20 08	23 31	15 38	04 56	02 31	13 38	17 53
26	10 16	22 56	02 29	19 57	23 32	15 41	04 53	02 31	13 38	17 53
27	09 55	25 15	02 58	19 44	23 33	15 43	04 50	02 32	13 39	17 53
28	09 34	26 19	03 26	19 32	23 34	15 45	04 47	02 33	13 39	17 54
29	09 12	26 08	03 52	19 18	23 34	15 48	04 44	02 34	13 40	17 54
30	08 51	24 45	04 16	19 05	23 34	15 50	04 41	02 35	13 40	17 54
31	08 N 29	22 S 16	04 S 38	18 N 50	23 N 34	15 S 52	04 N 39	02 S 36	13 S 41	17 S 54

ZODIAC SIGN ENTRIES

Date	h	m	Planets	
01	01	28	♀	♋
02	08	08	☽	♑
02	23	07	☿	♍
04	21	08	☽	♒
07	09	34	☽	♓
09	20	23	☽	♈
12	04	50	☽	♉
14	10	25	☽	♊
16	13	13	☽	♋
18	13	57	☽	♌
20	14	00	☽	♍
22	15	12	☽	♎
22	23	39	☉	♍
24	19	16	☽	♏
25	17	15	♂	♋
25	20	18	☿	♎
26	16	12	♀	♌
27	03	16	☽	♐
29	14	44	☽	♑

LATITUDES

Date	Mercury ☿ ° '	Venus ♀ ° '	Mars ♂ ° '	Jupiter ♃ ° '	Saturn ♄ ° '	Uranus ♅ ° '	Neptune ♆ ° '	Pluto ♇ ° '
01	01 N 12	01 S 37	00 S 17	01 S 03	01 N 59	00 S 47	00 S 25	05 N 39
04	00 53	01 27	00 14	01 03	01 59	00 47	00 25	05 39
07	00 31	01 18	00 12	01 04	01 59	00 47	00 25	05 38
10	00 N 06	01 08	00 09	01 04	01 59	00 48	00 25	05 37
13	00 S 21	00 58	00 06	01 05	01 58	00 48	00 25	05 36
16	00 49	00 48	00 04	01 05	01 58	00 48	00 25	05 36
19	01 18	00 38	00 S 01	01 05	01 58	00 48	00 25	05 35
22	01 48	00 28	00 N 02	01 06	01 58	00 48	00 25	05 34
25	02 18	00 19	00 05	01 06	01 58	00 48	00 25	05 34
28	02 47	00 09	00 08	01 06	01 58	00 48	00 25	05 33
31	03 S 14	00 00	00 N 11	01 S 06	01 N 58	00 S 48	00 S 25	05 N 32

DATA

Julian Date	2455045
Delta T	+67 seconds
Ayanamsa	23° 59' 43"
Synetic vernal point	05° ♓ 07' 16"
True obliquity of ecliptic	23° 26' 21"

MOON'S PHASES, APSIDES AND POSITIONS ☽

Date	h	m	Phase	Longitude ° '	Eclipse Indicator
06	00	55	○	13 ♒ 43	
13	18	55	☾	21 ♉ 09	
20	10	02	●	27 ♌ 32	
27	11	42	☽	04 ♐ 21	

Day	h	m		
04	00	30	Apogee	
19	04	50	Perigee	
31	10	57	Apogee	
01	13	31	Max dec	26° S 29'
09	02	08	0N	
15	21	54	Max dec	26° N 27'
21	23	35	0S	
28	20	18	Max dec	26° S 24'

LONGITUDES

Date	Chiron ⚷ ° '	Ceres ⚳ ° '	Pallas ⚴ ° '	Juno ⚵ ° '	Vesta ⚶ ° '	Black Moon Lilith ⚸ ° '
01	24 ♒ 26	29 ♍ 04	29 ♌ 38	04 ♈ 21	18 ♋ 32	23 ♑ 18
11	23 ♒ 56	03 ♎ 08	04 ♍ 35	04 ♈ 36	22 ♋ 49	24 ♑ 24
21	23 ♒ 26	07 ♎ 18	09 ♍ 29	04 ♈ 06	27 ♋ 03	25 ♑ 31
31	22 ♒ 56	11 ♎ 32	14 ♍ 20	02 ♈ 52	01 ♌ 14	26 ♑ 37

ASPECTARIAN

h m	Aspects
01 Saturday	
10 44	☽ □ ♄
13 45	☿ # ♆
19 27	☽ ✶ ♃
21 32	☽ ♇ ☉
23 02	☽ ✶ ♆
23 36	♀ ☍ ⯓
23 44	☉ # ⯓
02 Sunday	
00 31	☽ □ ♅
05 42	☽ △ ☿
10 16	☽ ☌ ⯓
11 27	☽ ☍ ♀
17 22	☽ ± ☉
03 Monday	
01 32	☽ ∠ ♃
05 24	☽ ∠ ♆
06 38	☽ ⊼ ☉
09 47	☉ ♇ ♅
12 59	☽ Q ♅
14 12	☿ △ ⯓
15 42	☽ ⊼ ♂
17 07	☽ ♇ ☿
19 40	☽ ⊥ ♃
23 40	☽ ⊥ ♆
04 Tuesday	
00 08	☽ △ ♄
04 26	☽ # ♂
04 39	☽ ± ♂
07 45	☽ ⊻ ♃
08 35	☽ # ♀
11 51	☽ ⊻ ♆
13 21	☽ ✶ ♅
14 21	☽ ± ☿
23 11	☽ ⊻ ⯓
05 Wednesday	
00 03	☽ ♇ ♂
04 26	☽ ⊼ ☿
06 54	☽ ♇ ♄
07 03	☽ ⊼ ♀
11 20	☽ ⊥ ⯓
12 04	☽ ∥ ⯓
17 55	☽ # ☉
19 42	☽ ∠ ♅
20 30	☽ ± ♀
06 Thursday	
00 55	☽ ☍ ☉
01 16	☽ ± ♄
04 51	☽ ∥ ♃
05 29	☽ ∠ ⯓
05 49	☉ ⊥ ♄
08 12	☽ △ ♂
11 52	☽ ∥ ♆
13 28	☽ ⊼ ♄
13 48	☽ ⊥ ♅
16 35	☽ ♇ ♀
19 47	☽ ☌ ♃
07 Friday	
00 20	☽ ☌ ♆
01 48	☽ ⊻ ♅
07 25	☽ # ☿
11 30	☽ ✶ ⯓
22 34	♀ ♇ ♃
08 Saturday	
00 41	☽ # ♄
01 35	☽ ☍ ☿
01 39	☽ △ ♀
01 54	♀ Q ♄
03 52	☿ ✶ ♀
08 50	☉ ♇ ⯓
11 12	☽ Q ⯓
16 04	☽ ∥ ♅
17 48	☽ ⊼ ☉
23 14	☽ □ ♂
09 Sunday	
01 27	☽ ☍ ♄
06 26	☽ ⊻ ♃
06 30	☽ ± ☉
11 19	☽ ⊻ ♆
12 11	☽ # ♅
12 45	☽ ☌ ♅
17 55	☽ ⊥ ♃
22 10	☽ □ ⯓
22 44	♀ ♇ ♆
22 52	☽ ⊥ ♆
10 Monday	
01 19	☽ ♇ ☉
02 49	☽ ∥ ♄
10 17	☽ ∥ ☿
10 59	☽ ∠ ♃
11 53	☽ Q ♂
16 00	☽ ∠ ♆
17 48	☽ □ ♀
19 29	☽ ⊼ ☿
11 Tuesday	
00 16	♂ □ ♄
08 04	☽ △ ☉
08 11	☽ ± ☿
11 14	☽ ⊼ ♄
11 44	☽ ✶ ♂
14 32	☽ # ♆
14 56	☽ ✶ ♃
19 46	☉ # ♃
20 03	☽ ✶ ♆
21 24	☽ ⊻ ♅
21 56	☽ ∥ ☉
22 05	☽ # ♃
22 28	☽ ± ♄
12 Wednesday	
02 55	☽ ♇ ☿
06 28	☽ △ ⯓
06 41	☽ Q ♀
08 25	☽ ⊥ ♅
12 22	☽ # ⯓
12 48	☽ Q ♃
12 58	☉ ± ♅
15 06	☽ ♇ ♄
16 48	☽ ∠ ♂
17 59	☽ Q ♆
13 Thursday	
00 42	☽ ∠ ♅
03 03	☿ ∥ ♄
06 31	☽ ✶ ♀
07 44	☉ ⊻ ♄
09 15	☽ △ ☿
09 33	☽ ♇ ⯓
09 49	☽ ⊥ ♂
11 18	☽ ∥ ♀
16 30	♂ △ ♃
18 13	☽ △ ♄
18 55	☽ □ ☉
20 45	☽ □ ♃
21 00	☽ ⊻ ♂
21 30	☽ ∥ ♂
14 Friday	
01 28	☽ ± ⯓
02 00	☽ □ ♆
03 17	☽ ✶ ♅
11 27	☽ ∠ ♀
11 55	☽ ⊼ ⯓
17 53	☉ ☍ ♃
20 59	♀ ± ♃
15 Saturday	
00 01	☽ Q ♅
04 21	☽ ⊥ ♀
04 29	☽ Q ☉
12 51	☉ ⊥ ♀
15 27	☽ ⊻ ♀
18 40	☽ □ ☿
22 15	☽ □ ♄
23 46	☽ △ ♃
16 Sunday	
02 12	☽ ✶ ☉
02 56	☽ ☌ ♂
05 07	☽ △ ♆
06 19	☽ □ ♅
14 35	☽ ☍ ⯓
16 16	☉ # ♆
17 Monday	
00 21	☽ ♇ ♃
03 19	☽ Q ☿
04 14	☽ Q ♄
04 41	☽ ∠ ☉
05 47	☽ ♇ ♆
10 56	♀ ± ♆
10 58	☉ ✶ ♂
14 53	☽ ± ♃
15 14	☿ ☌ ♄
18 47	☽ ∥ ♂
20 19	☽ ⊥ ☉
20 22	☽ ± ♆
20 55	☉ ☍ ♆
21 12	☽ ☌ ♀
21 37	☿ # ♅
23 51	☽ ✶ ♄
18 Tuesday	
00 00	☿ ⊼ ♃
00 30	☽ ⊼ ♃
00 33	☽ ✶ ☿
01 28	♂ △ ♆
06 01	☽ ⊼ ♆
06 14	☽ ⊻ ♂
06 40	☽ ⊻ ☉
07 09	☽ △ ♅
08 00	☽ ∥ ♀
14 03	☉ ⊼ ♅
15 14	☽ ⊼ ⯓
16 18	☽ ⊥ ♂
19 Wednesday	
00 08	☽ ∠ ♄
00 50	☽ ± ⯓
01 26	☽ # ⯓
02 24	♂ □ ♅
02 43	☽ ∠ ☿
07 08	☽ ♇ ♅
07 22	☽ ∠ ♂
10 11	♀ ✶ ♄
11 07	♀ ⊼ ♃
12 30	☽ # ♃
14 38	☽ ⊥ ♄
15 07	♃ ⊼ ♄
15 13	☽ ♇ ⯓
18 19	☽ ⊥ ☿
20 05	☽ # ♆
21 30	☽ ± ♅
20 Thursday	
00 10	☽ ☍ ♃
00 19	☽ ⊻ ♄
00 36	☽ ∥ ☉
01 24	☽ ⊻ ♀
04 44	☽ ⊻ ☿
05 59	☽ ☍ ♆
07 06	☽ ⊼ ♅
08 28	☽ ✶ ♂
10 02	☽ ☌ ☉
11 52	☽ ⊥ ♀
15 16	☽ △ ⯓
21 07	☿ ⊼ ♆
21 Friday	
03 40	☽ ∠ ♀
04 35	☽ ∥ ♄
04 43	☽ Q ♂
11 47	☿ ☍ ♅
14 30	☽ # ♅
23 23	☽ # ☿
23 50	☽ ∥ ☿
22 Saturday	
00 25	☽ ⊼ ♃
01 24	☽ ☌ ♄
06 28	☽ ✶ ♀
06 46	☽ ⊼ ♆
07 54	☽ ☍ ♅
08 48	☽ ∥ ♅
09 34	☽ ☌ ☿
10 07	♀ ⊼ ♆
10 17	☽ ± ♃
11 44	☽ □ ♂
14 35	☽ ⊻ ☉
16 29	☽ □ ⯓
16 47	☽ ± ♆
18 34	☿ ± ♃
18 41	☽ # ♄
23 31	♀ △ ♅
23 Sunday	
01 19	☽ ♇ ♃
01 27	☽ ⊥ ☉
04 30	☽ Q ♀
08 02	☽ ♇ ♆
18 04	☽ ∠ ☉
18 20	☉ △ ⯓
19 29	☽ # ☉
23 19	☽ Q ⯓
24 Monday	
03 00	☽ △ ♃
04 58	☽ ⊻ ♄
06 12	☽ ∥ ♆
10 09	☽ △ ♆
11 20	☽ ⊼ ♅
14 54	☽ □ ♀
15 35	☽ ∥ ♃
15 44	☽ ⊥ ♄
17 25	☽ ⊻ ☿
18 10	☽ △ ♂
20 36	☽ ✶ ⯓
22 03	☽ ± ♅
22 41	☽ ✶ ☉
25 Tuesday	
03 15	☽ ∥ ⯓
05 05	☽ ⊥ ☿
08 10	☽ ∠ ♄
14 27	☽ ♇ ♅
16 05	☽ # ♀
22 29	☽ Q ☉
22 57	☽ ∠ ☿
23 03	☽ ♇ ♂
26 Wednesday	
00 08	☽ ∠ ⯓
03 57	☿ □ ♂
09 14	☽ □ ♃
12 25	☽ ✶ ♄
13 56	☿ ∥ ♅
16 09	☿ □ ⯓
16 57	☽ ± ♂
17 08	☽ ⊥ ⯓
17 12	☽ # ♂
17 21	☽ □ ♆
18 35	☽ △ ♅
19 10	☿ ± ♆
20 34	♂ ☍ ⯓
27 Thursday	
04 27	☽ △ ♀
04 40	☽ ⊻ ⯓
05 06	☽ ⊼ ♂
05 35	☽ ✶ ☿
06 43	♀ ⊼ ⯓
11 42	☽ □ ☉
11 45	☽ Q ♄
18 24	♀ ⊻ ♂
19 37	☽ Q ♃
28 Friday	
04 18	☽ Q ♆
06 45	☽ Q ☿
13 06	☽ ♇ ♀
13 26	☿ ✶ ♀
19 10	☽ ✶ ♃
23 44	☽ □ ♄
29 Saturday	
04 12	☽ ✶ ♆
05 26	☽ □ ♅
09 08	☽ ± ♀
16 10	☽ ☌ ⯓
20 00	☽ ☍ ♂
21 12	☽ □ ☿
22 37	☽ ⊼ ♀
30 Sunday	
01 07	☽ ∠ ♃
04 51	☽ △ ☉
10 29	☽ ∠ ♆
17 48	☽ Q ♅
19 13	☽ ⊥ ♃
31 Monday	
00 36	☽ # ♂
04 45	☽ ⊥ ♆
07 19	☽ ⊻ ♃
12 05	☿ # ♄
13 12	☽ △ ♄
14 06	☽ ♇ ☉
16 56	☽ ⊻ ♆
18 09	☽ ✶ ♅

All ephemeris data is given at 12.00 UT and the Moon's longitude is additionally given for 24.00 UT

SEPTEMBER 2009

Date	Sidereal time h m s	Sun ☉ LONGITUDES	Moon ☽	Moon ☽ 24.00	Mercury ☿	Venus ♀	Mars ♂	Jupiter ♃	Saturn ♄	Uranus ♅	Neptune ♆	Pluto ♇
01	10 43 08	09 ♍ 11 19	04 ♒ 05 05	10 ♒ 01 02	04 ♎ 44	06 ♌ 57	04 ♋ 16	19 ♒ 50	22 ♍ 59	25 ♓ 16	24 ♒ 41	00 ♑ 41
02	10 47 04	10 09 23	15 ♒ 58 36	21 ♒ 58 07	05 11	08 09	04 54	19 R 43	23 06	25 R 14	24 R 39	00 R 41
03	10 51 01	11 07 28	27 ♒ 59 53	04 ♓ 04 09	05 34	09 21	05 31	19 36	23 13	25 11	24 38	00 41
04	10 54 57	12 05 35	10 ♓ 11 06	16 ♓ 20 55	05 52	10 33	06 08	19 30	23 21	25 09	24 36	00 40
05	10 58 54	13 03 44	22 ♓ 33 41	28 ♓ 49 31	06 05	11 45	06 45	19 23	23 28	25 07	24 35	00 40
06	11 02 50	14 01 54	05 ♈ 08 28	11 ♈ 30 34	06 12	12 57	07 22	19 16	23 35	25 04	24 33	00 40
07	11 06 47	15 00 06	17 ♈ 55 49	24 ♈ 24 16	06 R 13	14 10	07 59	19 10	23 43	25 02	24 32	00 40
08	11 10 43	15 58 20	00 ♉ 55 57	07 ♉ 30 46	06 08	15 22	08 36	19 04	23 50	25 00	24 30	00 40
09	11 14 40	16 56 36	14 ♉ 08 52	20 ♉ 50 14	05 57	16 34	09 12	18 58	23 58	24 57	24 29	00 40
10	11 18 37	17 54 54	27 ♉ 34 55	04 ♊ 22 58	05 39	17 47	09 49	18 52	24 05	24 55	24 27	00 40
11	11 22 33	18 53 14	11 ♊ 14 25	18 ♊ 09 20	05 14	18 59	10 25	18 46	24 13	24 53	24 26	00 40
12	11 26 30	19 51 37	25 ♊ 07 42	02 ♋ 09 31	04 43	20 12	11 01	18 40	24 20	24 50	24 24	00 D 40
13	11 30 26	20 50 01	09 ♋ 14 43	16 ♋ 23 09	04 06	21 25	11 37	18 35	24 28	24 48	24 23	00 40
14	11 34 23	21 48 28	23 ♋ 34 37	00 ♌ 48 47	03 22	22 37	12 13	18 29	24 35	24 45	24 21	00 40
15	11 38 19	22 46 57	08 ♌ 05 15	15 ♌ 23 28	02 33	23 50	12 49	18 24	24 43	24 43	24 20	00 40
16	11 42 16	23 45 28	22 ♌ 42 48	00 ♍ 02 31	01 38	25 03	13 24	18 19	24 50	24 41	24 19	00 40
17	11 46 12	24 44 00	07 ♍ 21 48	14 ♍ 39 47	00 ♎ 39	26 16	14 00	18 14	24 58	24 38	24 17	00 40
18	11 50 09	25 42 35	21 ♍ 55 36	29 ♍ 08 22	29 ♍ 38	27 29	14 35	18 09	25 05	24 36	24 16	00 40
19	11 54 06	26 41 12	06 ♎ 17 16	13 ♎ 21 34	28 34	28 42	15 10	18 05	25 13	24 33	24 15	00 40
20	11 58 02	27 39 51	20 ♎ 20 40	27 ♎ 14 03	27 30	29 ♌ 55	15 45	18 00	25 20	24 31	24 13	00 41
21	12 01 59	28 38 31	04 ♏ 01 24	10 ♏ 42 33	26 27	01 ♍ 09	16 20	17 56	25 27	24 29	24 12	00 41
22	12 05 55	29 ♍ 37 13	17 ♏ 17 26	23 ♏ 46 12	25 27	02 22	16 55	17 52	25 35	24 26	24 11	00 41
23	12 09 52	00 ♎ 35 57	00 ♐ 09 05	06 ♐ 26 25	24 31	03 35	17 29	17 48	25 42	24 24	24 09	00 42
24	12 13 48	01 34 43	12 ♐ 38 41	18 ♐ 46 23	23 41	04 48	18 04	17 44	25 50	24 21	24 08	00 42
25	12 17 45	02 33 31	24 ♐ 50 07	00 ♑ 50 32	22 58	06 02	18 38	17 41	25 57	24 19	24 07	00 43
26	12 21 41	03 32 20	06 ♑ 48 17	12 ♑ 44 03	22 24	07 15	19 12	17 37	26 05	24 17	24 06	00 43
27	12 25 38	04 31 11	18 ♑ 38 31	24 ♑ 32 23	21 58	08 29	19 46	17 34	26 12	24 14	24 05	00 43
28	12 29 35	05 30 04	00 ♒ 26 16	06 ♒ 20 51	21 42	09 42	20 20	17 31	26 20	24 12	24 04	00 44
29	12 33 31	06 28 58	12 ♒ 16 43	18 ♒ 14 26	21 37	10 56	20 53	17 29	26 27	24 10	24 02	00 45
30	12 37 28	07 ♎ 27 54	24 ♒ 14 30	00 ♓ 17 22	21 ♍ 41	12 ♍ 10	21 ♋ 27	17 ♒ 26	26 ♍ 34	24 ♓ 07	24 ♒ 01	00 ♑ 45

Date	Moon True ☊	Moon Mean ☊	Moon ☽ Latitude
01	29 ♑ 47	28 ♑ 04	00 N 23
02	29 R 46	28 01	01 27
03	29 43	27 58	02 27
04	29 37	27 54	03 21
05	29 31	27 51	04 07
06	29 23	27 48	04 41
07	29 15	27 45	05 02
08	29 08	27 42	05 08
09	29 02	27 38	04 57
10	28 58	27 35	04 31
11	28 56	27 32	03 48
12	28 D 56	27 29	02 52
13	28 57	27 26	01 44
14	28 57	27 23	00 N 29
15	28 R 57	27 19	00 S 49
16	28 55	27 16	02 04
17	28 50	27 13	03 11
18	28 43	27 10	04 05
19	28 35	27 07	04 42
20	28 25	27 04	05 02
21	28 15	27 00	05 03
22	28 07	26 57	04 48
23	28 01	26 54	04 19
24	27 57	26 51	03 37
25	27 55	26 48	02 47
26	27 D 55	26 44	01 50
27	27 55	26 41	00 S 49
28	27 R 55	26 38	00 N 13
29	27 54	26 35	01 16
30	27 ♑ 51	26 ♑ 32	02 N 15

DECLINATIONS

Date	Sun ☉	Moon ☽	Mercury ☿	Venus ♀	Mars ♂	Jupiter ♃	Saturn ♄	Uranus ♅	Neptune ♆	Pluto ♇
01	08 N 07	18 S 51	04 S 59	18 N 35	23 N 34	15 S 55	04 N 36	02 S 37	13 S 41	17 S 55
02	07 46	14 40	05 17	18 20	23 34	15 57	04 33	02 38	13 42	17 55
03	07 24	09 52	05 34	18 04	23 34	15 59	04 30	02 39	13 43	17 55
04	07 02	04 S 38	05 47	17 48	23 33	16 01	04 27	02 40	13 43	17 56
05	06 39	00 N 50	05 58	17 31	23 32	16 03	04 24	02 41	13 44	17 56
06	06 17	06 21	06 06	17 13	23 31	16 05	04 21	02 41	13 44	17 56
07	05 55	11 41	06 12	16 55	23 30	16 07	04 18	02 42	13 45	17 56
08	05 32	16 36	06 13	16 37	23 29	16 09	04 15	02 43	13 45	17 57
09	05 09	20 49	06 12	16 18	23 28	16 11	04 12	02 44	13 46	17 57
10	04 47	24 01	06 06	15 59	23 26	16 13	04 09	02 45	13 46	17 57
11	04 24	25 54	05 57	15 39	23 25	16 15	04 06	02 46	13 47	17 57
12	04 01	26 13	05 44	15 19	23 23	16 16	04 03	02 47	13 47	17 58
13	03 38	24 51	05 26	14 58	23 21	16 18	04 00	02 48	13 48	17 58
14	03 15	21 51	05 05	14 37	23 19	16 20	03 58	02 49	13 48	17 58
15	02 52	17 27	04 39	14 15	23 16	16 21	03 55	02 50	13 48	17 59
16	02 29	11 59	04 10	13 53	23 14	16 23	03 52	02 51	13 49	17 59
17	02 06	05 N 51	03 38	13 31	23 12	16 24	03 49	02 52	13 49	17 59
18	01 42	00 S 33	03 02	13 08	23 09	16 26	03 46	02 53	13 50	17 59
19	01 19	06 49	02 24	12 45	23 06	16 27	03 43	02 54	13 50	18 00
20	00 56	12 36	01 45	12 21	23 03	16 29	03 40	02 55	13 51	18 00
21	00 32	17 36	01 04	11 58	23 00	16 30	03 37	02 56	13 51	18 00
22	00 N 09	21 36	00 S 23	11 33	22 57	16 31	03 34	02 57	13 52	18 00
23	00 S 14	24 23	00 N 16	11 09	22 54	16 32	03 31	02 58	13 52	18 01
24	00 38	25 54	00 54	10 44	22 51	16 33	03 28	02 59	13 52	18 01
25	01 01	26 07	01 30	10 19	22 47	16 34	03 25	02 59	13 53	18 01
26	01 24	25 06	02 02	09 53	22 44	16 35	03 22	03 00	13 53	18 02
27	01 48	22 57	02 30	09 28	22 40	16 36	03 20	03 01	13 54	18 02
28	02 11	19 50	02 54	09 01	22 36	16 37	03 17	03 02	13 54	18 02
29	02 34	15 54	03 14	08 35	22 32	16 38	03 14	03 03	13 54	18 02
30	02 S 58	11 S 19	03 N 28	08 N 09	22 N 28	16 S 38	03 N 11	03 S 04	13 S 55	18 S 03

ZODIAC SIGN ENTRIES

Date	h	m	Planets	
01	03	43	☽	♒
03	15	58	☽	♓
06	02	14	☽	♈
08	10	18	☽	♉
10	16	17	☽	♊
12	20	20	☽	♋
14	22	39	☽	♌
16	23	56	☽	♍
18	03	26	☿	♍
19	01	26	☽	♎
20	13	32	♀	♍
21	04	52	☽	♏
22	21	19	☉	♎
23	11	43	☽	♐
25	22	19	☽	♑
28	11	07	☽	♒
30	23	26	☽	♓

LATITUDES

Date	Mercury ☿	Venus ♀	Mars ♂	Jupiter ♃	Saturn ♄	Uranus ♅	Neptune ♆	Pluto ♇
01	03 S 23	00 N 04	00 N 12	01 S 06	01 N 58	00 S 48	00 S 25	05 N 32
04	03 46	00 13	00 15	01 06	01 58	00 48	00 25	05 31
07	04 03	00 21	00 18	01 06	01 58	00 48	00 25	05 30
10	04 12	00 30	00 22	01 06	01 58	00 48	00 25	05 29
13	04 09	00 38	00 25	01 06	01 58	00 48	00 25	05 28
16	03 50	00 46	00 28	01 06	01 58	00 48	00 25	05 27
19	03 14	00 53	00 32	01 06	01 58	00 48	00 25	05 27
22	02 24	00 59	00 35	01 06	01 58	00 48	00 25	05 26
25	01 24	01 05	00 39	01 05	01 59	00 48	00 25	05 25
28	00 S 25	01 11	00 43	01 05	01 59	00 48	00 25	05 24
31	00 N 27	01 N 16	00 N 46	01 S 05	01 N 59	00 S 48	00 S 25	05 N 23

DATA

Julian Date	2455076
Delta T	+67 seconds
Ayanamsa	23° 59' 48"
Synetic vernal point	05° ♓ 07' 12"
True obliquity of ecliptic	23° 26' 22"

MOON'S PHASES, APSIDES AND POSITIONS ☽

Date	h	m	Phase	Longitude	Eclipse Indicator
04	16	03	○	12 ♓ 15	
12	02	16	☾	19 ♊ 28	
18	18	44	●	25 ♍ 59	
26	04	50	☽	03 ♑ 15	

Day	h	m		
16	07	47	Perigee	
28	03	31	Apogee	
05	08	24	0N	
12	04	39	Max dec	26° N 17'
18	09	57	0S	
25	04	02	Max dec	26° S 11'

LONGITUDES

Date	Chiron ⚷	Ceres ⚳	Pallas ⚴	Juno ⚵	Vesta ⚶	Black Moon Lilith ⚸
01	22 ♒ 53	11 ♎ 58	14 ♍ 49	02 ♈ 43	01 ♌ 39	26 ♑ 44
11	22 ♒ 25	16 ♎ 17	19 ♍ 37	00 ♈ 17	05 ♌ 45	27 ♑ 50
21	22 ♒ 00	20 ♎ 38	24 ♍ 23	28 ♓ 26	09 ♌ 46	28 ♑ 57
31	21 ♒ 40	25 ♎ 02	29 ♍ 04	26 ♓ 00	13 ♌ 40	00 ♒ 03

All ephemeris data is given at 12.00 UT and the Moon's longitude is additionally given for 24.00 UT

ASPECTARIAN

h m Aspects

01 Tuesday
05 07 ☽ ⊻ ♇
06 40 ♀ ± ♇
10 01 ☽ ± ☉
12 24 ☽ ⊼ ♂
13 22 ☽ △ ☿
13 46 ☽ # ♀
16 04 ☿ Ᵽ ♃
17 16 ☽ ⊥ ♇
17 46 ☽ ∥ ♇
18 28 ☽ ☍ ♀
19 58 ☽ Ᵽ ♄
23 15 ☽ ⊼ ☉

02 Wednesday
00 28 ☽ ∠ ♅
01 13 ☽ ± ♂
05 07 ☽ ∥ ♃
06 07 ♂ Ᵽ ♃
10 50 ♀ ∠ ♄
11 24 ☽ ∠ ♇
14 17 ☽ ± ♄
17 02 ☽ ∥ ♆
18 30 ☽ ⊥ ♅
19 26 ☽ ☌ ♃
20 17 ☽ Ᵽ ♂
20 44 ☽ Ᵽ ☿

03 Thursday
02 24 ☽ ⊼ ♄
05 19 ☽ ☌ ♆
06 26 ☽ ⊻ ♅
15 12 ☽ ± ☿
16 19 ☿ □ ♂
17 18 ☽ ✶ ♇

04 Friday
00 25 ☽ # ☉
00 55 ♀ # ♇
03 21 ☽ ⊼ ☿
03 39 ☽ △ ♂
04 15 ♀ Ᵽ ♅
07 03 ☽ ∥ ☿
12 48 ☽ ⊼ ♀
12 52 ☽ # ♄
16 03 ☽ ☍ ☉
16 51 ☽ Q ♇
20 45 ☽ ∥ ♅

05 Saturday
01 43 ☽ ± ♀
05 55 ☽ ⊻ ♃
13 46 ☽ ☍ ♄
15 52 ☽ ⊻ ♆
16 53 ☽ ☌ ♅
17 22 ☽ ⊥ ♃
20 02 ☽ # ♅
20 54 ☽ Ᵽ ♀

06 Sunday
03 18 ☽ ⊥ ♆
03 22 ☽ ∥ ♄
03 31 ☽ □ ♇
10 22 ☽ ∠ ♃
10 57 ☽ # ☿
11 45 ☽ ∥ ☉
14 00 ☽ ☍ ☿
16 25 ☽ □ ♂
20 18 ☽ ∠ ♆
20 51 ☉ # ☿

07 Monday
04 14 ☽ △ ♀
04 44 ☿ St R
04 50 ♀ ⊥ ♂
06 06 ☽ ⊼ ☉
14 17 ☽ ✶ ♃
18 10 ☽ ± ☉
21 45 ☽ # ♆
22 50 ☽ ⊼ ♄

08 Tuesday
00 12 ☽ ✶ ♆
01 07 ☽ ⊻ ♅
03 39 ☽ Q ♂
09 41 ☽ # ♃
09 59 ☽ ± ♄
11 30 ☽ △ ♇
12 05 ☽ ∥ ♀
12 05 ☽ Ᵽ ☉
12 07 ☽ ⊥ ♅
12 14 ☽ Q ♃
17 53 ♀ Ᵽ ♇
19 10 ☽ # ♇
21 23 ☽ ⊼ ☿
22 09 ☽ Q ♆

09 Wednesday
02 33 ☽ Ᵽ ♄
02 38 ☽ ✶ ♂
04 27 ☽ ∠ ♅
08 06 ☽ ± ☿
14 43 ☽ Ᵽ ♇
15 12 ♃ ⊥ ♅
16 48 ☽ □ ♀
17 25 ☽ △ ☉
19 55 ♀ # ♃
20 35 ☽ □ ♃
22 24 ♂ Ᵽ ♆
23 57 ☽ Ᵽ ☿

10 Thursday
05 44 ☽ △ ♄
06 27 ☽ □ ♆
06 49 ☽ ± ♇
06 51 ☽ ∠ ♂
06 59 ☽ ∥ ♂
07 17 ☽ ✶ ♅
17 27 ☽ ⊼ ♇
18 48 ♀ ⊥ ♄
23 30 ☽ ⊥ ♂

11 Friday
01 42 ☉ ⊻ ♀
01 50 ☽ △ ☿
03 51 ☽ Q ♀
04 24 ☽ Q ♅
07 54 ♀ ☍ ♃
09 17 ☉ ⊼ ♃
09 49 ♀ ± ♅
10 30 ☽ ⊻ ♂
16 55 ♇ St D

12 Saturday
00 58 ☽ △ ♃
02 16 ☽ □ ☉
02 43 ☽ ✶ ♀
05 32 ☿ ∠ ♀
09 03 ☉ ∥ ♄
10 38 ☽ □ ♄
10 46 ☽ △ ♆
11 30 ☽ □ ♅
21 27 ☽ ☍ ♇
22 58 ♄ ⊼ ♆

13 Sunday
02 28 ☽ Ᵽ ♃
03 41 ☽ □ ☿
06 46 ☽ ∠ ♀
11 15 ☽ Q ☉
12 14 ☽ Ᵽ ♆
16 10 ☽ ☌ ♂
17 28 ☽ Q ♄
17 35 ☽ ± ♃
23 20 ☽ ⊥ ♀

14 Monday
01 14 ♃ ± ♄
01 43 ☽ ∥ ♂
03 19 ☽ ± ♆
03 34 ☽ ⊼ ♃
07 46 ☿ Ᵽ ♃
08 31 ☽ Q ☿
08 50 ☽ ✶ ☉
10 16 ☽ ⊻ ♀
13 18 ☽ ⊼ ♆
13 41 ☽ ✶ ♄
13 57 ☽ △ ♅
21 37 ♂ ± ♃
23 45 ☽ ⊼ ♇

15 Tuesday
03 22 ☽ ✶ ☿
06 52 ♂ Q ♄
09 29 ☽ # ♇
09 39 ☽ ± ♇
11 28 ☽ ∠ ☉
12 51 ♄ ☍ ♅
13 53 ☉ # ♅
14 40 ☽ Ᵽ ♅
14 42 ☽ ∠ ♄
17 09 ☽ # ♃
20 06 ☽ ⊻ ♂
21 37 ♀ ☍ ♆

16 Wednesday
00 27 ☽ Ᵽ ♇
02 38 ☽ ∠ ☿
03 18 ☽ ⊥ ☉
03 32 ☽ ∥ ♀
04 24 ☽ # ♆
04 48 ♀ ⊼ ♅
04 50 ☽ ☍ ♃
05 24 ☽ ± ♅
05 36 ☽ ⊥ ♄
06 21 ☽ ⊥ ♂
07 13 ♀ ⊻ ♄
13 50 ☽ ⊻ ☉
14 37 ☽ ☍ ♆
15 12 ☽ ⊼ ♅
15 30 ☽ ⊻ ♄
16 11 ☽ ☌ ♀
16 29 ♀ # ♆
16 30 ☽ ⊥ ☿
18 27 ☿ ⊥ ♀
21 43 ☽ ∠ ♂

17 Thursday
00 41 ☉ ± ♃
01 01 ☽ △ ♇
01 17 ☉ ⊼ ♆
01 42 ☽ ⊻ ☿
03 25 ☿ # ♄
09 41 ☉ ☍ ♅
11 46 ☿ □ ♇
18 22 ☉ ☌ ♄
19 45 ☽ ∥ ♄
20 56 ☿ ± ♆
21 12 ☽ # ☿
23 12 ☽ # ♅
23 21 ☽ ✶ ♂

18 Friday
03 02 ☽ ∥ ☉
05 48 ☽ ⊼ ♃
15 41 ☽ ± ♃
15 52 ☽ ⊼ ♆
16 06 ☽ # ☉
16 25 ☽ ☍ ♅
17 17 ☽ ☌ ♄
17 53 ☿ ∥ ♅
18 44 ☽ ☌ ☉
20 04 ☽ Q ♂
20 33 ☽ ∥ ☿
20 50 ☽ ∥ ♅
22 06 ☽ ⊻ ♀
23 56 ☽ ☌ ☿

19 Saturday
00 04 ☽ # ♄
01 52 ☽ ± ♆
02 34 ☽ □ ♇
06 38 ☽ Ᵽ ♃
09 05 ☽ ⊥ ♀
10 35 ☿ ⊻ ♀
16 59 ☽ Ᵽ ♆

20 Sunday
01 47 ☽ ∠ ♀
03 45 ☽ □ ♂
07 59 ☽ △ ♃
08 20 ☿ Q ♂
09 07 ☽ Q ♇
10 05 ☉ ☌ ☿
11 00 ☽ # ♀
17 22 ☉ Q ♂
17 35 ☽ ∥ ♆
18 43 ☽ △ ♆
19 13 ☽ ⊼ ♅
20 45 ☽ ⊻ ♄
23 34 ☽ ⊻ ☿

21 Monday
01 44 ☽ ⊻ ☉
02 57 ♀ △ ♇
05 43 ☽ ± ♅
06 04 ☽ ✶ ♇
06 15 ☽ ∥ ♃
06 23 ☽ ✶ ♀
07 24 ☽ ⊥ ♄
09 24 ☽ ⊥ ☿
13 11 ☽ ⊥ ☉
14 07 ☽ ∥ ♇
19 10 ♀ ∠ ♂
21 45 ☽ Ᵽ ♅
23 39 ☽ ∠ ♄

22 Tuesday
00 24 ☽ ∠ ☿
06 05 ☽ Q ♀
06 43 ☽ ∠ ☉
09 04 ☽ ∠ ♇
09 08 ☿ ☌ ♄
11 16 ☽ △ ♂
13 03 ☽ □ ♃
22 10 ☽ # ♂

23 Wednesday
00 17 ☉ ∥ ☿
00 45 ☽ □ ♆
01 12 ☽ △ ♅
01 24 ☉ ± ♆
01 43 ☽ ⊥ ♇
02 06 ☽ ✶ ☿
03 33 ☽ ✶ ♄
08 55 ☉ # ☿
12 55 ☽ ✶ ☉
13 02 ☽ ⊻ ♇
14 22 ☉ □ ♇
15 36 ☿ ☍ ♅
16 39 ☽ Ᵽ ♂
19 14 ☽ □ ♀
22 20 ☿ ⊼ ♆
22 43 ☽ Q ♃
23 22 ☽ Q ☿
23 50 ♂ ⊼ ♃

24 Thursday
02 35 ☽ Q ♄
10 17 ☿ ± ♃
10 49 ☽ ± ♂
11 01 ☽ Q ♆
13 59 ☽ Q ☉
15 07 ♂ ± ♆
21 55 ☽ ✶ ♃
23 07 ☽ ⊼ ♂

25 Friday
08 29 ☽ □ ☿
10 34 ☽ ✶ ♆
10 58 ☽ □ ♅
14 15 ☽ □ ♄
14 49 ☉ Ᵽ ♃
23 44 ☽ ☌ ♇

26 Saturday
03 37 ☽ ∠ ♃
04 50 ☽ □ ☉
13 01 ☽ △ ♀
16 38 ☽ ∠ ♆
21 42 ☽ ⊥ ♃
23 02 ☽ Q ♅

27 Sunday
09 50 ☽ ⊻ ♃
10 51 ☽ ⊥ ♆
14 24 ☽ ☍ ♂
14 38 ☽ # ♂
18 35 ☽ △ ☿
22 59 ☽ Ᵽ ♀
23 03 ☽ ⊻ ♆
23 21 ☽ ✶ ♅

28 Monday
03 33 ☽ △ ♄
12 36 ☽ ⊻ ♇
19 25 ☽ ± ♀
21 32 ☿ # ♅
23 13 ☽ △ ☉
23 33 ☽ ∥ ♇

29 Tuesday
00 35 ☽ Ᵽ ☿
00 47 ☽ ⊥ ♇
05 43 ☽ ∠ ♅
07 54 ☽ ∥ ♃
08 58 ☽ ⊼ ♀
10 19 ☽ Ᵽ ♄
12 18 ☿ ∥ ♄
13 13 ☿ St D
18 43 ☽ ± ☿
18 59 ☽ ∠ ♇
22 26 ☽ ☌ ♃
22 48 ☽ ∥ ♆
23 48 ☽ ⊥ ♅

30 Wednesday
04 36 ☽ ± ♄
06 08 ☽ ⊼ ♂
06 51 ☽ ⊼ ☿
08 08 ☽ Ᵽ ☉
11 34 ☽ ☌ ♆
11 46 ☽ ⊻ ♅
16 41 ☽ ⊼ ♄
18 40 ☽ ± ♂
18 46 ☉ ∥ ♅

OCTOBER 2009

LONGITUDES

Date	Sidereal time h m s	Sun ☉ ° ' "	Moon ☽ ° ' "	Moon ☽ 24.00 ° ' "	Mercury ☿ ° '	Venus ♀ ° '	Mars ♂ ° '	Jupiter ♃ ° '	Saturn ♄ ° '	Uranus ♅ ° '	Neptune ♆ ° '	Pluto ♇ ° '
01	12 41 24	08 ♎ 26 52	06 ♓ 23 26	12 ♓ 33 00	21 ♍ 56	13 ♍ 24	22 ♋ 00	17 ♒ 24	26 ♍ 42	24 ♓ 05	24 ♒ 00	00 ♑ 46
02	12 45 21	09 25 52	18 ♓ 46 20	25 ♓ 03 34	22 D 21	14 37	22 33	17 R 21	26 49	24 R 03	23 R 59	00 46
03	12 49 17	10 24 54	01 ♈ 24 48	07 ♈ 50 01	22 55	15 51	23 06	17 19	26 57	24 00	23 58	00 47
04	12 53 14	11 23 57	14 ♈ 19 10	20 ♈ 52 07	23 39	17 05	23 38	17 17	27 04	23 58	23 57	00 48
05	12 57 10	12 23 03	27 ♈ 28 39	04 ♉ 08 33	24 30	18 19	24 11	17 16	27 11	23 56	23 56	00 48
06	13 01 07	13 22 11	10 ♉ 51 32	17 ♉ 37 21	25 30	19 33	24 43	17 14	27 19	23 54	23 55	00 49
07	13 05 04	14 21 21	24 ♉ 25 42	01 ♊ 16 20	26 36	20 47	25 15	17 13	27 26	23 51	23 54	00 50
08	13 09 00	15 20 34	08 ♊ 09 00	15 ♊ 03 30	27 48	22 01	25 47	17 12	27 33	23 49	23 54	00 51
09	13 12 57	16 19 48	21 ♊ 59 40	28 ♊ 57 22	29 ♍ 06	23 15	26 19	17 11	27 40	23 47	23 53	00 52
10	13 16 53	17 19 06	05 ♋ 56 31	12 ♋ 57 02	00 ♎ 29	24 29	26 51	17 11	27 48	23 45	23 52	00 53
11	13 20 50	18 18 25	19 ♋ 58 52	27 ♋ 01 55	01 56	25 44	27 22	17 10	27 55	23 43	23 51	00 53
12	13 24 46	19 17 47	04 ♌ 06 07	11 ♌ 11 21	03 26	26 58	27 53	17 10	28 02	23 41	23 50	00 54
13	13 28 43	20 17 11	18 ♌ 17 26	25 ♌ 24 07	04 59	28 12	28 24	17 D 10	28 09	23 39	23 50	00 55
14	13 32 39	21 16 37	02 ♍ 31 06	09 ♍ 38 00	06 34	29 ♍ 27	28 55	17 10	28 16	23 37	23 49	00 56
15	13 36 36	22 16 06	16 ♍ 44 20	23 ♍ 49 36	08 11	00 ♎ 41	29 25	17 11	28 23	23 35	23 48	00 57
16	13 40 33	23 15 36	00 ♎ 53 12	07 ♎ 54 32	09 50	01 55	29 ♋ 56	17 11	28 30	23 33	23 48	00 58
17	13 44 29	24 15 09	14 ♎ 53 01	21 ♎ 48 03	11 30	03 10	00 ♌ 26	17 12	28 37	23 31	23 47	01 00
18	13 48 26	25 14 44	28 ♎ 39 06	05 ♏ 25 42	13 11	04 24	00 55	17 13	28 44	23 29	23 46	01 01
19	13 52 22	26 14 21	12 ♏ 07 30	18 ♏ 44 13	14 52	05 39	01 25	17 14	28 51	23 27	23 46	01 02
20	13 56 19	27 14 00	25 ♏ 15 43	01 ♐ 41 58	16 34	06 54	01 54	17 15	28 58	23 25	23 45	01 03
21	14 00 15	28 13 41	08 ♐ 03 05	14 ♐ 19 14	18 16	08 08	02 23	17 17	29 05	23 23	23 45	01 04
22	14 04 12	29 ♎ 13 24	20 ♐ 30 46	26 ♐ 38 03	19 58	09 23	02 52	17 19	29 12	23 21	23 44	01 05
23	14 08 08	00 ♏ 13 08	02 ♑ 41 34	08 ♑ 41 53	21 41	10 38	03 21	17 21	29 19	23 19	23 44	01 07
24	14 12 05	01 12 54	14 ♑ 39 35	20 ♑ 35 20	23 23	11 52	03 49	17 23	29 25	23 18	23 44	01 08
25	14 16 02	02 12 42	26 ♑ 29 47	02 ♒ 23 38	25 04	13 07	04 17	17 25	29 32	23 16	23 43	01 09
26	14 19 58	03 12 32	08 ♒ 17 35	14 ♒ 12 20	26 46	14 22	04 45	17 28	29 39	23 14	23 43	01 11
27	14 23 55	04 12 23	20 ♒ 08 34	26 ♒ 06 58	28 ♎ 27	15 37	05 12	17 30	29 45	23 13	23 43	01 12
28	14 27 51	05 12 16	02 ♓ 08 09	08 ♓ 12 41	00 ♏ 08	16 51	05 40	17 33	29 52	23 11	23 42	01 13
29	14 31 48	06 12 10	14 ♓ 21 07	20 ♓ 33 52	01 48	18 06	06 07	17 37	29 ♍ 59	23 09	23 42	01 15
30	14 35 44	07 12 06	26 ♓ 51 20	03 ♈ 13 46	03 28	19 21	06 33	17 40	00 ♎ 05	23 08	23 42	01 16
31	14 39 41	08 ♏ 12 04	09 ♈ 41 21	16 ♈ 14 08	05 ♏ 08	20 ♎ 36	07 ♌ 00	17 ♒ 43	00 ♎ 12	23 ♓ 06	23 ♒ 42	01 ♑ 18

Date	Moon True ☊ ° '	Moon Mean ☊ ° '	Moon ☽ Latitude ° '
01	27 ♑ 45	26 ♑ 29	03 N 09
02	27 R 37	26 25	03 56
03	27 26	26 22	04 31
04	27 14	26 19	04 54
05	27 02	26 16	05 01
06	26 50	26 13	04 52
07	26 40	26 10	04 27
08	26 33	26 06	03 46
09	26 29	26 03	02 52
10	26 27	26 00	01 46
11	26 D 27	25 57	00 N 34
12	26 R 27	25 54	00 S 40
13	26 26	25 50	01 53
14	26 22	25 47	02 58
15	26 16	25 44	03 53
16	26 07	25 41	04 32
17	25 55	25 38	04 55
18	25 43	25 35	05 00
19	25 30	25 31	04 48
20	25 18	25 28	04 21
21	25 09	25 25	03 41
22	25 03	25 22	02 52
23	24 59	25 19	01 55
24	24 57	25 15	00 S 54
25	24 D 57	25 12	00 N 08
26	24 R 57	25 09	01 10
27	24 56	25 06	02 09
28	24 53	25 03	03 04
29	24 48	25 00	03 50
30	24 40	24 56	04 27
31	24 ♑ 30	24 ♑ 53	04 N 52

DECLINATIONS

Date	Sun ☉ ° '	Moon ☽ ° '	Mercury ☿ ° '	Venus ♀ ° '	Mars ♂ ° '	Jupiter ♃ ° '	Saturn ♄ ° '	Uranus ♅ ° '	Neptune ♆ ° '	Pluto ♇ ° '
01	03 S 21	06 S 14	03 N 37	07 N 42	22 N 24	16 S 39	03 N 08	03 S 05	13 S 55	18 S 03
02	03 44	00 S 50	03 41	07 15	22 20	16 40	03 05	03 06	13 55	18 03
03	04 07	04 N 42	03 40	06 47	22 16	16 40	03 02	03 07	13 56	18 03
04	04 31	10 09	03 34	06 20	22 12	16 41	03 00	03 08	13 56	18 04
05	04 54	15 15	03 23	05 52	22 07	16 41	02 57	03 08	13 56	18 04
06	05 17	19 42	03 08	05 24	22 03	16 41	02 54	03 09	13 57	18 04
07	05 40	23 11	02 49	04 56	21 58	16 42	02 51	03 10	13 57	18 04
08	06 02	25 23	02 27	04 28	21 54	16 42	02 48	03 11	13 57	18 05
09	06 25	26 03	02 01	04 00	21 49	16 42	02 45	03 12	13 58	18 05
10	06 48	25 05	01 32	03 31	21 44	16 42	02 43	03 13	13 58	18 05
11	07 11	22 31	01 00	03 03	21 40	16 42	02 40	03 14	13 58	18 06
12	07 33	18 35	00 N 26	02 34	21 35	16 42	02 37	03 14	13 58	18 06
13	07 56	13 33	00 S 10	02 05	21 30	16 42	02 34	03 15	13 59	18 06
14	08 18	07 48	00 47	01 36	21 25	16 42	02 32	03 16	13 59	18 06
15	08 40	01 N 40	01 26	01 07	21 20	16 42	02 29	03 17	13 59	18 07
16	09 02	04 S 31	02 06	00 38	21 15	16 41	02 26	03 18	13 59	18 07
17	09 24	10 23	02 47	00 N 09	21 10	16 41	02 24	03 18	13 59	18 07
18	09 46	15 40	03 29	00 S 21	21 05	16 41	02 21	03 19	14 00	18 07
19	10 08	20 02	04 11	00 50	21 00	16 40	02 18	03 20	14 00	18 08
20	10 29	23 18	04 53	01 19	20 55	16 40	02 16	03 21	14 00	18 08
21	10 51	25 17	05 36	01 48	20 50	16 39	02 13	03 21	14 00	18 08
22	11 12	25 57	06 19	02 18	20 45	16 39	02 11	03 22	14 00	18 08
23	11 33	25 20	07 02	02 47	20 39	16 38	02 08	03 23	14 00	18 08
24	11 54	23 32	07 44	03 16	20 34	16 37	02 05	03 23	14 01	18 09
25	12 14	20 43	08 26	03 45	20 29	16 36	02 03	03 24	14 01	18 09
26	12 35	17 04	09 09	04 15	20 24	16 35	02 00	03 25	14 01	18 09
27	12 55	12 43	09 50	04 44	20 19	16 34	01 58	03 25	14 01	18 09
28	13 15	07 51	10 31	05 13	20 14	16 33	01 55	03 26	14 01	18 10
29	13 35	02 S 37	11 12	05 41	20 09	16 32	01 53	03 26	14 01	18 10
30	13 55	02 N 50	11 52	06 10	20 04	16 31	01 50	03 27	14 01	18 10
31	14 S 14	08 N 18	12 S 32	06 S 39	19 N 58	16 S 30	01 N 48	03 S 28	14 S 01	18 S 10

ZODIAC SIGN ENTRIES

Date	h m	Planets
03	09 21	☽ ♈
05	16 33	☽ ♉
07	21 46	☽ ♊
10	01 48	☽ ♋
10	03 46	☿ ♎
12	05 02	☽ ♌
14	07 45	☽ ♍
14	22 46	♀ ♎
16	10 29	☽ ♎
16	15 32	♂ ♌
18	14 23	☽ ♏
20	20 49	☽ ♐
23	06 39	☽ ♑
23	06 43	☉ ♏
25	19 08	☽ ♒
28	07 45	☽ ♓
28	10 09	☿ ♏
29	17 09	♄ ♎
30	17 56	☽ ♈

LATITUDES

Date	Mercury ☿ ° '	Venus ♀ ° '	Mars ♂ ° '	Jupiter ♃ ° '	Saturn ♄ ° '	Uranus ♅ ° '	Neptune ♆ ° '	Pluto ♇ ° '
01	00 N 27	01 N 16	00 N 46	01 S 05	01 N 59	00 S 48	00 S 25	05 N 23
04	01 08	01 20	00 50	01 05	01 59	00 48	00 25	05 23
07	01 36	01 24	00 54	01 04	02 00	00 48	00 25	05 22
10	01 53	01 27	00 58	01 04	02 00	00 48	00 25	05 21
13	01 59	01 29	01 02	01 04	02 00	00 48	00 25	05 20
16	01 57	01 31	01 07	01 03	02 01	00 48	00 25	05 19
19	01 49	01 32	01 11	01 03	02 01	00 47	00 25	05 19
22	01 37	01 33	01 16	01 03	02 01	00 47	00 25	05 18
25	01 21	01 33	01 20	01 02	02 02	00 47	00 25	05 17
28	01 04	01 32	01 25	01 02	02 02	00 47	00 25	05 16
31	00 N 45	01 N 30	01 N 30	01 S 02	02 N 03	00 S 47	00 S 25	05 N 16

DATA

Julian Date	2455106
Delta T	+67 seconds
Ayanamsa	23° 59' 51"
Synetic vernal point	05° ♓ 07' 09"
True obliquity of ecliptic	23° 26' 22"

MOON'S PHASES, APSIDES AND POSITIONS ☽

Date	h m	Phase	Longitude ° '	Eclipse Indicator
04	06 10	○	11 ♈ 10	
11	08 56	☾	18 ♋ 11	
18	05 33	●	24 ♎ 59	
26	00 42	☽	02 ♒ 44	

Day	h m		
13	12 13	Perigee	
25	23 18	Apogee	
02	15 36	ON	
09	09 49	Max dec	26° N 03'
15	18 25	OS	
22	12 09	Max dec	25° S 57'
29	23 35	ON	

LONGITUDES

Date	Chiron ⚷ ° '	Ceres ⚳ ° '	Pallas ⚴ ° '	Juno ⚵ ° '	Vesta ⚶ ° '	Black Moon Lilith ⚸ ° '
01	21 ♒ 40	25 ♎ 02	29 ♍ 04	26 ♓ 00	13 ♌ 40	00 ♒ 03
11	21 ♒ 25	29 ♎ 26	03 ♎ 43	23 ♓ 52	17 ♌ 25	01 ♒ 09
21	21 ♒ 16	03 ♏ 52	08 ♎ 17	22 ♓ 20	21 ♌ 00	02 ♒ 16
31	21 ♒ 13	08 ♏ 17	12 ♎ 46	21 ♓ 35	24 ♌ 22	03 ♒ 22

All ephemeris data is given at 12.00 UT and the Moon's longitude is additionally given for 24.00 UT

ASPECTARIAN

h m	Aspects
01 Thursday	
00 07	☉ # ♄
00 55	☽ ⚹ ♇
03 35	☽ ± ☉
04 38	☽ # ♀
06 00	☿ ⚹ ♂
13 14	☽ bQ ♂
16 22	☽ ⊼ ☉
23 34	☽ # ☿
02 Friday	
00 04	☽ ∥ ☉
00 25	☽ Q ♇
01 22	☉ bQ ♆
02 00	☽ # ♄
02 04	☽ ∥ ♅
03 08	☽ ☍ ♀
07 57	♄ # ♅
08 12	☉ # ☿
09 17	☽ ⊻ ♃
19 08	☽ ☍ ☿
19 33	☽ △ ♂
20 44	☽ ⊥ ♃
21 57	☽ ⊻ ♆
22 02	☽ ☌ ♅
03 Saturday	
03 29	☽ ☍ ♄
04 50	☽ ∥ ♄
05 04	☽ # ♅
07 31	☽ ∥ ☿
09 17	☽ # ☉
09 17	☽ ⊥ ♆
10 49	☽ □ ♇
13 42	☽ ∠ ♃
20 23	☽ ∥ ♀
04 Sunday	
01 21	☿ ± ♃
02 06	☽ ∠ ♆
06 10	☽ ☍ ☉
11 37	☿ ⚹ ♂
15 58	♀ ⊼ ♃
17 27	☽ ⚹ ♃
17 36	☽ ⊼ ♀
20 57	☿ ⊼ ♆
21 06	☿ ☍ ♅
05 Monday	
01 33	♂ ⊼ ♆
01 37	♂ △ ♅
03 08	♅ ⊻ ♆
05 34	☽ # ♆
05 35	☽ ⊻ ♅
05 36	☽ ⚹ ♆
05 41	☽ ± ♀
05 46	☽ □ ♂
06 13	☽ ⊼ ☿
11 28	☽ ⊼ ♄
15 13	☽ Q ♃
16 25	☽ ⊥ ♅
17 53	☽ ± ☿
18 01	☽ △ ♇
19 17	☽ # ♃
22 23	☽ ± ♄
23 35	☽ bQ ♀
06 Tuesday	
02 39	☽ # ♇
03 12	☽ Q ♆
08 30	☽ ∠ ♅
10 39	☿ # ♅
11 18	☽ bQ ☿
14 36	☽ bQ ♄
15 27	☽ Q ♂
15 39	☉ # ♀
16 49	☽ ⊼ ☉
20 49	☽ bQ ♇
23 18	☽ □ ♃
07 Wednesday	
02 54	☽ ∥ ♂
04 16	☽ ± ☉
04 56	☽ △ ♀
09 46	☿ ∥ ♄
11 00	☽ ⚹ ♅
11 05	☽ □ ♆
12 43	☽ ± ♇
13 31	☽ ⚹ ♂
16 10	☽ △ ☿
17 19	☽ △ ♄
21 19	☽ bQ ☉
23 15	☽ ⊼ ♇
08 Thursday	
06 34	☿ ☌ ♄
07 57	☽ Q ♅
16 46	☽ ∠ ♂
09 Friday	
01 27	☽ △ ☉
03 42	☽ △ ♃
08 59	☽ ⊥ ♂
10 45	♀ ± ♃
14 23	☽ □ ♀
15 05	☽ □ ♅
15 15	☽ △ ♆
19 45	☽ ⊻ ♂
21 53	☽ □ ♄
22 01	♀ ☍ ♅
10 Saturday	
00 01	♀ ⊼ ♆
01 32	☿ ± ♆
01 35	☽ □ ☿
03 18	☽ ☍ ♇
05 33	☽ bQ ♃
08 38	☉ △ ♃
17 00	☽ bQ ♆
18 44	☿ □ ♇
20 58	☽ ± ♃
11 Sunday	
00 18	☽ Q ♀
03 03	♀ # ♅
05 00	☽ Q ♄
07 13	☽ ⊼ ♃
08 22	☽ ± ♆
08 56	☽ □ ☉
11 54	☽ Q ☿
15 58	☿ bQ ♃
18 03	☽ ∥ ♂
18 20	☽ △ ♅
18 35	☽ ⊼ ♆
22 43	☽ ⚹ ♀
12 Monday	
01 03	☽ ☌ ♂
01 37	☽ ⚹ ♄
02 24	☉ Q ♇
06 34	☽ ⊼ ♇
08 51	♀ ∥ ♄
10 43	☽ ⚹ ☿
14 31	☽ # ♇
16 45	☽ ± ♇
17 49	☽ Q ☉
19 44	☽ bQ ♅
20 51	♂ ⚹ ♄
21 29	☽ # ♃
13 Tuesday	
02 35	☽ ∠ ♀
03 15	☽ ∠ ♄
04 34	♃ St D
08 00	☽ bQ ♇
10 06	☽ ☍ ♃
10 09	☽ # ♆
10 52	♀ ☌ ♄
10 55	☽ ± ♅
15 12	☽ ∠ ☿
15 37	☽ ⚹ ☉
18 31	♀ ⚹ ♂
18 34	☽ ⊥ ♄
19 14	☽ ⊥ ♀
21 01	☽ ⊼ ♅
21 20	☽ ☍ ♆
14 Wednesday	
04 47	☽ ⊻ ♄
05 42	☽ ⊻ ♂
06 19	☽ ⊻ ♀
08 17	☽ ⊥ ☿
09 20	☽ △ ♇
10 08	☽ # ☉
16 11	☽ ⊥ ♂
18 49	☽ ∠ ☉
19 09	♀ ± ♆
19 42	☽ ⊻ ☿
15 Thursday	
05 18	☿ # ♀
05 46	☽ # ♅
07 56	☽ ∠ ♂
08 48	☽ ∥ ♄
11 09	☽ ⊥ ☉
12 44	☽ ⊼ ♃
12 48	☽ # ☿
14 18	☽ ∥ ♀
17 21	♀ □ ♇
20 58	☿ bQ ♆
21 57	☽ # ♀
22 04	☽ ⊻ ☉
22 54	☽ ± ♃
23 33	☽ ☍ ♅
23 57	☽ ⊼ ♆
16 Friday	
01 24	☽ ∥ ☿
03 56	☽ # ♄
07 12	☽ ∥ ♅
07 55	☽ ☌ ♄
10 08	☽ ± ♆
10 18	☽ ⚹ ♂
12 09	☽ □ ♇
13 56	☽ ☌ ♀
14 13	☽ bQ ♃
17 05	♀ bQ ♃
18 36	☉ ⊼ ♅
23 14	☿ # ♄
17 Saturday	
00 46	☉ △ ♆
01 30	☽ bQ ♆
05 22	☽ ☌ ☿
07 32	☽ ∥ ☉
07 36	☽ Q ♂
16 00	☽ △ ♃
19 07	☽ Q ♇
18 Sunday	
02 57	☽ ⊼ ♅
03 27	☽ △ ♆
04 01	☽ ∥ ♆
05 33	☽ ☌ ☉
06 25	☿ ∥ ♅
06 49	☿ Q ♂
12 09	☽ ⊻ ♄
13 27	☽ ± ♅
16 10	☽ □ ♂
16 10	☽ ⚹ ♇
16 24	♂ ⊼ ♇
17 07	☽ ∥ ♃
22 52	☽ ⊥ ♄
23 13	☽ ⊻ ♀
19 Monday	
00 49	☽ ∥ ♇
05 24	☽ bQ ♅
11 03	☽ ⊥ ♀
15 09	☽ ∠ ♄
17 42	☽ ⊻ ☿
18 06	☽ # ♂
19 05	☽ ∠ ♇
21 16	☽ □ ♃
20 Tuesday	
05 08	☽ ∠ ♀
06 17	☽ ⊥ ☿
08 36	☽ △ ♅
09 13	☽ □ ♆
11 36	☽ ⊥ ♇
15 58	☽ ⊻ ☉
18 57	☽ ⚹ ♄
21 49	☿ △ ♃
22 48	☽ ⊻ ♇
21 Wednesday	
00 53	☽ △ ♂
01 34	☽ ∠ ☿
04 09	☽ ⊥ ☉
06 44	☽ Q ♃
12 11	☽ ⚹ ♀
17 50	☽ Q ♄
19 03	☽ Q ♆
22 45	☽ ∠ ☉
23 23	☿ Q ♇
23 42	♀ bQ ♆
22 Thursday	
05 46	☽ ⚹ ♃
06 34	♀ # ♄
06 39	☽ bQ ♂
10 47	☽ ⚹ ☿
11 19	☉ ⊻ ♄
13 53	☽ Q ♀
15 01	☉ ± ♅
17 32	☽ □ ♅
18 18	☽ ⚹ ♆
23 Friday	
00 58	☽ ± ♂
03 48	♃ ⊥ ♅
05 14	☽ □ ♄
06 39	☽ ⚹ ☉
08 51	☽ ☌ ♇
11 18	☽ ∠ ♃
13 21	☽ ⊼ ♂
14 17	☽ Q ☿
24 Saturday	
00 04	☽ ∠ ♆
05 14	☽ Q ♅
05 22	☽ ⊥ ♃
05 43	☽ □ ♀
08 49	☽ Q ☉
09 57	☉ ⚹ ♇
10 50	☿ ⊼ ♅
16 55	☿ △ ♆
17 31	☽ ⊻ ♃
17 55	♀ ∥ ♅
18 12	☽ ⊥ ♆
25 Sunday	
05 27	☽ ⚹ ♅
06 22	☽ ⊻ ♆
08 37	☽ □ ☿
13 45	☽ # ♂
18 15	☽ △ ♄
21 30	☽ ⊻ ♇
26 Monday	
00 42	☽ □ ☉
04 30	☽ ☍ ♂
05 20	☽ ∥ ♇
09 44	☽ ⊥ ♇
11 53	☽ ∠ ♅
14 48	☽ ∥ ♃
27 Tuesday	
01 01	☽ bQ ♄
01 46	☽ △ ♀
02 02	♄ ± ♆
04 01	☽ ∠ ♇
05 11	☽ ∥ ♆
06 06	☽ ⊥ ♅
06 40	☽ ☌ ♃
11 02	☽ ∥ ☉
18 09	☽ ⊻ ♅
19 10	☽ ☌ ♆
19 20	☽ ± ♄
22 40	☿ ± ♅
28 Wednesday	
00 43	☽ ∥ ☿
07 22	☽ △ ☿
07 27	☽ ⊼ ♄
07 59	☿ ⊻ ♄
10 11	☽ ⚹ ♇
11 23	☽ bQ ♀
18 37	☽ △ ☉
19 15	☽ ⊼ ♂
23 15	☽ ∥ ♀
29 Thursday	
02 02	♀ △ ♃
03 53	☿ ⚹ ♇
05 54	☉ ⊥ ♄
07 08	☽ ± ♀
07 28	☽ ± ♂
07 56	☉ □ ♂
08 19	☽ ∥ ♅
09 51	☽ Q ♇
12 06	♀ Q ♂
15 18	☽ # ♄
17 29	☽ bQ ☿
18 20	☽ ⊻ ♃
20 05	☽ ⊼ ♀
30 Friday	
01 32	☽ bQ ♂
02 22	☽ bQ ☉
04 56	☽ ☌ ♅
05 54	☽ ⊥ ♃
06 00	☽ ⊻ ♆
07 41	☽ ∥ ♄
10 20	♀ Q ♇
13 20	☽ ± ☿
14 41	☽ # ♅
17 22	☽ ⊥ ♆
18 09	☽ ☍ ♄
19 32	☉ ∥ ♆
20 21	☽ □ ♇
20 54	☽ ± ☉
23 00	☽ ∠ ♃
31 Saturday	
02 17	☽ ⊼ ☿
03 59	☽ # ♀
06 50	☽ △ ♂
09 01	☽ ⊼ ☉
09 46	☉ bQ ♅
10 10	☽ ∠ ♆

NOVEMBER 2009

LONGITUDES

Date	Sidereal time h m s	Sun ☉ ° ′ ″	Moon ☽ ° ′ ″	Moon ☽ 24.00 ° ′ ″	Mercury ☿ ° ′	Venus ♀ ° ′	Mars ♂ ° ′	Jupiter ♃ ° ′	Saturn ♄ ° ′	Uranus ♅ ° ′	Neptune ♆ ° ′	Pluto ♇ ° ′
01	14 43 37	09 ♏ 12 04	22 ♈ 52 03	29 ♈ 34 56	06 ♏ 47	21 ♎ 51	07 ♌ 26	17 ♒ 47	00 ♎ 18	23 ♓ 05	23 ♒ 42	01 ♑ 19
02	14 47 34	10 12 05	06 ♉ 22 29	13 ♉ 14 19	08 25	23 06	07 51	17 51	00 24	23 R 03	23 R 42	01 21
03	14 51 31	11 12 09	20 ♉ 09 58	27 ♉ 08 54	10 03	24 21	08 17	17 55	00 31	23 02	23 41	01 22
04	14 55 27	12 12 14	04 ♊ 10 32	11 ♊ 14 17	11 41	25 36	08 42	17 59	00 37	23 01	23 41	01 24
05	14 59 24	13 12 21	18 ♊ 19 35	25 ♊ 25 53	13 18	26 51	09 07	18 04	00 43	23 00	23 D 41	01 25
06	15 03 20	14 12 31	02 ♋ 32 41	09 ♋ 39 34	14 55	28 06	09 31	18 08	00 49	22 58	23 41	01 27
07	15 07 17	15 12 42	16 ♋ 46 09	23 ♋ 52 10	16 32	29 ♎ 21	09 55	18 13	00 55	22 57	23 42	01 28
08	15 11 13	16 12 55	00 ♌ 57 22	08 ♌ 01 36	18 08	00 ♏ 36	10 19	18 18	01 01	22 56	23 42	01 30
09	15 15 10	17 13 11	15 ♌ 04 44	22 ♌ 06 40	19 44	01 51	10 43	18 23	01 07	22 55	23 42	01 32
10	15 19 06	18 13 28	29 ♌ 07 19	06 ♍ 06 37	21 19	03 07	11 06	18 29	01 13	22 54	23 42	01 33
11	15 23 03	19 13 47	13 ♍ 04 26	20 ♍ 00 39	22 54	04 22	11 28	18 34	01 19	22 53	23 42	01 35
12	15 27 00	20 14 09	26 ♍ 55 08	03 ♎ 47 40	24 29	05 37	11 51	18 40	01 25	22 52	23 42	01 37
13	15 30 56	21 14 32	10 ♎ 38 01	17 ♎ 25 58	26 03	06 52	12 13	18 46	01 30	22 51	23 43	01 39
14	15 34 53	22 14 57	24 ♎ 11 13	00 ♏ 53 30	27 37	08 07	12 34	18 52	01 36	22 50	23 43	01 40
15	15 38 49	23 15 24	07 ♏ 32 32	14 ♏ 08 05	29 ♏ 11	09 23	12 55	18 58	01 42	22 49	23 43	01 42
16	15 42 46	24 15 52	20 ♏ 39 55	27 ♏ 07 53	00 ♐ 45	10 38	13 16	19 04	01 47	22 48	23 44	01 44
17	15 46 42	25 16 23	03 ♐ 31 51	09 ♐ 51 48	02 18	11 53	13 36	19 11	01 53	22 48	23 44	01 46
18	15 50 39	26 16 55	16 ♐ 07 45	22 ♐ 19 49	03 51	13 09	13 56	19 18	01 58	22 47	23 45	01 48
19	15 54 35	27 17 28	28 ♐ 28 10	04 ♑ 33 06	05 24	14 24	14 15	19 25	02 03	22 46	23 45	01 50
20	15 58 32	28 18 02	10 ♑ 34 56	16 ♑ 34 04	06 57	15 39	14 34	19 32	02 09	22 46	23 46	01 51
21	16 02 29	29 ♏ 18 38	22 ♑ 30 58	28 ♑ 26 08	08 29	16 55	14 53	19 39	02 14	22 45	23 46	01 53
22	16 06 25	00 ♐ 19 16	04 ♒ 20 10	10 ♒ 13 39	10 02	18 10	15 11	19 46	02 19	22 45	23 47	01 55
23	16 10 22	01 19 54	16 ♒ 07 13	22 ♒ 01 33	11 34	19 25	15 28	19 54	02 24	22 44	23 47	01 57
24	16 14 18	02 20 33	27 ♒ 57 17	03 ♓ 55 08	13 06	20 41	15 45	20 01	02 29	22 44	23 48	01 59
25	16 18 15	03 21 14	09 ♓ 55 44	15 ♓ 59 46	14 37	21 56	16 02	20 09	02 34	22 43	23 49	02 01
26	16 22 11	04 21 55	22 ♓ 07 51	28 ♓ 20 33	16 09	23 11	16 18	20 17	02 39	22 43	23 49	02 03
27	16 26 08	05 22 38	04 ♈ 38 23	11 ♈ 01 47	17 40	24 27	16 34	20 25	02 43	22 43	23 50	02 05
28	16 30 04	06 23 22	17 ♈ 31 07	24 ♈ 06 35	19 11	25 42	16 49	20 33	02 48	22 43	23 51	02 07
29	16 34 01	07 24 07	00 ♉ 48 18	07 ♉ 36 15	20 42	26 58	17 03	20 42	02 52	22 42	23 52	02 09
30	16 37 58	08 ♐ 24 53	14 ♉ 30 13	21 ♉ 29 54	22 ♐ 13	28 ♏ 13	17 ♌ 17	20 ♒ 50	02 ♎ 57	22 ♓ 42	23 ♒ 53	02 ♑ 11

MOON / DECLINATIONS

Date	Moon True ☊ ° ′	Moon Mean ☊ ° ′	Moon ☽ Latitude ° ′	Sun ☉ ° ′	Moon ☽ ° ′	Mercury ☿ ° ′	Venus ♀ ° ′	Mars ♂ ° ′	Jupiter ♃ ° ′	Saturn ♄ ° ′	Uranus ♅ ° ′	Neptune ♆ ° ′	Pluto ♇ ° ′
01	24 ♑ 18	24 ♑ 50	05 N 01	14 S 34	13 N 33	13 S 11	07 S 07	19 N 53	16 S 29	01 N 46	03 S 28	14 S 01	18 S 10
02	24 R 05	24 47	04 55	14 53	18 16	13 49	07 36	19 48	16 27	01 43	03 29	14 01	18 11
03	23 54	24 44	04 31	15 11	22 08	14 26	08 04	19 43	16 26	01 41	03 29	14 01	18 11
04	23 44	24 41	03 50	15 30	24 45	15 03	08 32	19 38	16 25	01 39	03 30	14 01	18 11
05	23 37	24 37	02 55	15 48	25 50	15 39	09 00	19 33	16 23	01 36	03 30	14 01	18 11
06	23 33	24 34	01 49	16 06	25 14	16 15	09 28	19 29	16 22	01 34	03 31	14 01	18 12
07	23 31	24 31	00 N 36	16 24	22 59	16 49	09 55	19 24	16 20	01 32	03 31	14 01	18 12
08	23 D 31	24 28	00 S 40	16 41	19 18	17 22	10 23	19 19	16 18	01 30	03 31	14 01	18 12
09	23 31	24 25	01 52	16 58	14 31	17 55	10 50	19 14	16 17	01 27	03 32	14 01	18 12
10	23 R 31	24 21	02 58	17 15	09 00	18 27	11 16	19 10	16 15	01 25	03 32	14 01	18 12
11	23 28	24 18	03 52	17 32	03 N 04	18 58	11 43	19 05	16 13	01 23	03 33	14 01	18 12
12	23 24	24 15	04 32	17 48	02 S 57	19 28	12 09	19 01	16 11	01 21	03 33	14 01	18 13
13	23 16	24 12	04 57	18 04	08 45	19 57	12 35	18 56	16 09	01 19	03 33	14 01	18 13
14	23 06	24 09	05 04	18 20	14 05	20 25	13 01	18 52	16 07	01 17	03 34	14 01	18 13
15	22 56	24 06	04 54	18 35	18 40	20 52	13 26	18 48	16 05	01 15	03 34	14 01	18 13
16	22 45	24 02	04 29	18 50	22 14	21 18	13 51	18 44	16 03	01 13	03 34	14 00	18 13
17	22 35	23 59	03 51	19 05	24 38	21 42	14 16	18 40	16 01	01 11	03 34	14 00	18 14
18	22 27	23 56	03 01	19 19	25 43	22 06	14 40	18 36	15 59	01 09	03 35	14 00	18 14
19	22 22	23 53	02 01	19 33	25 30	22 29	15 04	18 32	15 57	01 07	03 35	14 00	18 14
20	22 19	23 50	01 S 03	19 47	24 03	22 51	15 28	18 29	15 54	01 05	03 35	14 00	18 14
21	22 D 18	23 47	00 N 01	20 00	21 32	23 11	15 51	18 25	15 52	01 03	03 35	14 00	18 14
22	22 18	23 43	01 05	20 13	18 08	23 30	16 14	18 22	15 49	01 01	03 35	13 59	18 14
23	22 20	23 40	02 05	20 26	14 01	23 49	16 36	18 19	15 47	01 00	03 36	13 59	18 15
24	22 21	23 37	03 00	20 38	09 22	24 05	16 58	18 15	15 44	00 58	03 36	13 59	18 15
25	22 R 20	23 34	03 49	20 50	04 S 19	24 21	17 19	18 13	15 42	00 56	03 36	13 59	18 15
26	22 19	23 31	04 27	21 01	00 N 58	24 35	17 40	18 10	15 39	00 54	03 36	13 58	18 15
27	22 15	23 27	04 55	21 12	06 21	24 49	18 01	18 07	15 37	00 53	03 36	13 58	18 15
28	22 09	23 24	05 08	21 22	11 37	25 00	18 21	18 05	15 34	00 51	03 36	13 58	18 15
29	22 02	23 21	05 08	21 33	16 31	25 11	18 40	18 02	15 31	00 49	03 36	13 58	18 15
30	21 ♑ 54	23 ♑ 18	04 N 45	21 S 42	20 N 44	25 S 20	19 S 00	18 N 00	15 S 28	00 N 48	03 S 36	13 S 57	18 S 16

ZODIAC SIGN ENTRIES

Date	h	m	Planets
02	00	45	☽ ♉
04	04	53	☽ ♊
06	07	42	☽ ♋
08	00	23	♀ ♏
08	10	23	☽ ♌
10	13	30	☽ ♍
12	17	22	☽ ♎
14	22	24	☽ ♏
16	00	28	☿ ♐
17	05	22	☽ ♐
19	15	01	☽ ♑
22	03	11	☽ ♒
22	04	23	☉ ♐
24	16	07	☽ ♓
27	03	11	☽ ♈
29	10	34	☽ ♉

LATITUDES

Date	Mercury ☿ ° ′	Venus ♀ ° ′	Mars ♂ ° ′	Jupiter ♃ ° ′	Saturn ♄ ° ′	Uranus ♅ ° ′	Neptune ♆ ° ′	Pluto ♇ ° ′
01	00 N 38	01 N 30	01 N 32	01 S 02	02 N 03	00 S 47	00 S 25	05 N 15
04	00 N 18	01 28	01 37	01 02	02 03	00 47	00 25	05 15
07	00 S 02	01 25	01 42	01 01	02 04	00 47	00 25	05 14
10	00 22	01 21	01 47	01 01	02 05	00 47	00 25	05 14
13	00 42	01 18	01 53	01 00	02 05	00 47	00 25	05 13
16	01 00	01 13	01 59	01 00	02 06	00 47	00 25	05 12
19	01 18	01 08	02 05	01 00	02 06	00 46	00 25	05 12
22	01 34	01 03	02 11	00 59	02 07	00 46	00 25	05 11
25	01 49	00 58	02 17	00 59	02 08	00 46	00 25	05 11
28	02 01	00 52	02 23	00 59	02 08	00 46	00 25	05 10
31	02 S 11	00 N 45	02 N 30	00 S 58	02 N 09	00 S 46	00 S 25	05 N 10

DATA

Julian Date	2455137
Delta T	+67 seconds
Ayanamsa	23° 59′ 54″
Synetic vernal point	05° ♓ 07′ 05″
True obliquity of ecliptic	23° 26′ 21″

MOON'S PHASES, APSIDES AND POSITIONS ☽

Date	h	m	Phase	Longitude ° ′	Eclipse Indicator
02	19	14	○	10 ♉ 30	
09	15	56	☾	17 ♌ 23	
16	19	14	●	24 ♏ 34	
24	21	39	☽	02 ♓ 45	

Day	h	m		
07	07	36	Perigee	
22	20	09	Apogee	
05	15	30	Max dec	25° N 51′
12	00	14	0S	
18	19	52	Max dec	25° S 48′
26	07	38	0N	

LONGITUDES

Date	Chiron ⚷ ° ′	Ceres ⚳ ° ′	Pallas ⚴ ° ′	Juno ⚵ ° ′	Vesta ⚶ ° ′	Black Moon Lilith ⚸ ° ′
01	21 ♒ 13	08 ♏ 43	13 ♎ 13	21 ♓ 34	24 ♌ 41	03 ♒ 29
11	21 ♒ 17	13 ♏ 07	17 ♎ 37	21 ♓ 48	27 ♌ 45	04 ♒ 36
21	21 ♒ 27	17 ♏ 30	21 ♎ 54	22 ♓ 53	00 ♍ 30	05 ♒ 42
31	21 ♒ 44	21 ♏ 49	26 ♎ 04	24 ♓ 46	02 ♍ 51	06 ♒ 48

All ephemeris data is given at 12.00 UT and the Moon's longitude is additionally given for 24.00 UT

ASPECTARIAN

h m Aspects

01 Sunday
02 46 ☽ ⚹ ♃
04 34 ☿ ⊥ ♄
05 35 ♂ ± ♇
09 59 ☽ ☍ ♀
09 59 ☽ ⋕ ☿
12 23 ☽ ⊻ ♅
13 29 ☽ ⚹ ♆
14 15 ☽ ⋕ ♆
17 12 ☽ ⋕ ☉
23 05 ☽ ⊥ ♅

02 Monday
00 25 ☽ Q ♃
00 50 ☿ □ ♂
01 23 ☽ ⊼ ♄
02 23 ☽ ⋕ ♃
03 06 ☽ △ ♇
06 47 ☿ ⚼ ♅
10 48 ☽ Q ♆
11 12 ♀ ⊼ ♅
11 28 ☽ ⋕ ♇
12 03 ☽ ± ♄
14 41 ☽ □ ♂
14 57 ☽ ∠ ♅
16 05 ☽ ☍ ☿
19 14 ☽ ☍ ☉
19 48 ☿ ∥ ♆
20 37 ☽ ∥ ♂
22 49 ♂ ⚼ ♅
23 21 ♀ △ ♆

03 Tuesday
03 53 ☽ ⚼ ♄
05 25 ☽ ⚼ ♇
08 06 ☽ □ ♃
16 56 ☽ ⚹ ♅
18 04 ☽ □ ♆
19 55 ☽ ⊼ ♀
20 58 ☽ ± ♇
22 50 ☽ Q ♂

04 Wednesday
05 53 ☽ △ ♄
07 11 ☽ ± ♀
07 15 ☽ ⊼ ♇
13 25 ☽ Q ♅
18 10 ♆ St D
19 55 ☽ ⚹ ♂
23 59 ☽ ⚼ ♀

05 Thursday
02 25 ☽ ⊼ ☿
02 41 ☽ ⊼ ☉
08 02 ☉ ☌ ☿
11 33 ☽ △ ♃
13 36 ☽ ± ☉
13 52 ☽ ± ☿
19 52 ☽ □ ♅
21 04 ☽ △ ♆
22 04 ☽ ∠ ♂

06 Friday
00 16 ☉ ∥ ☿
03 47 ☽ △ ♀
05 57 ☽ ⚼ ☉
07 01 ☽ ⚼ ☿
09 04 ☽ □ ♄
10 09 ☽ ☍ ♇
13 01 ☽ ⚼ ♃
13 42 ☽ ⊥ ♂
16 41 ☿ ∥ ♃
22 22 ☽ ⚼ ♆

07 Saturday
00 07 ☽ ⊻ ♂
02 16 ☿ ∠ ♄
04 17 ☽ ± ♃
04 24 ♀ ± ♅
07 09 ☉ ∥ ♃
09 10 ☽ △ ☉
11 09 ☿ ∠ ♇
11 33 ☽ △ ☿
13 34 ☽ ± ♆
14 28 ☽ ⊼ ♃
15 40 ☽ Q ♄
22 26 ☽ △ ♅
23 42 ☽ ⊼ ♆

08 Sunday
06 50 ☉ ∠ ♄
11 21 ☽ □ ♀
11 54 ☽ ∥ ♂
12 07 ☽ ⚹ ♄
12 56 ☽ ⊼ ♇
14 42 ☿ □ ♃
17 59 ☽ ⋕ ♇
19 03 ☉ ∠ ♇
20 39 ♀ ⊻ ♄
21 11 ☽ ⋕ ☿
23 08 ☽ ± ♇
23 49 ☽ ⚼ ♅

09 Monday
00 55 ☽ ⋕ ☉
03 39 ☽ ⋕ ♃
04 21 ☽ ☌ ♂
05 34 ♀ ⚹ ♇
13 47 ☽ ∠ ♄
14 19 ☽ ⋕ ♆
14 29 ☽ ⚼ ♇
15 07 ☽ ± ♅
15 56 ☽ □ ☉
17 41 ☽ ☍ ♃
20 57 ☽ □ ☿
20 57 ☽ Q ♀

10 Tuesday
00 49 ☿ ∥ ♇
01 21 ☽ ⊼ ♅
02 43 ☽ ☍ ♆
03 07 ☽ ⋕ ♀
05 16 ☽ ⊥ ♄
15 37 ☽ ⊻ ♄
16 11 ☽ △ ♇
18 41 ☉ □ ♃
19 31 ☽ ⚹ ♀

11 Wednesday
01 09 ☽ Q ☉
07 46 ☽ Q ☿
09 10 ☽ ⊻ ♂
10 08 ☽ ⋕ ♅
11 38 ☿ △ ♅
17 09 ☿ ⋕ ♂
18 46 ☽ ∥ ♄
19 49 ☽ ⊥ ♂
21 34 ☽ ⊼ ♃
23 29 ☽ ⚹ ☉
23 58 ☽ ∠ ♀

12 Thursday
00 11 ☿ □ ♆
04 57 ☽ ☍ ♅
05 39 ☽ ⋕ ♄
06 25 ☽ ⊼ ♆
07 13 ☽ ⚹ ☿
08 03 ☽ ± ♃
11 52 ☽ ∠ ♂
14 27 ☽ ∥ ♅
16 52 ☽ ± ♆
17 10 ☽ ⊥ ♀
19 53 ☽ ☌ ♄
20 12 ☽ □ ♇
23 52 ☽ ⚼ ♃

13 Friday
03 41 ☽ ∠ ☉
04 43 ☽ ⊻ ♀
05 36 ☿ ⊥ ♇
08 37 ☽ ⚼ ♆
12 50 ☽ ∠ ☿
14 51 ☽ ⚹ ♂
20 47 ☽ ⊥ ☉

14 Saturday
01 11 ♀ ⊥ ♄
01 22 ☉ ∥ ♇
02 28 ☽ △ ♃
03 57 ☽ Q ♇
06 27 ♀ ⚼ ♅
06 29 ☽ ∥ ♀
06 50 ☽ ⊥ ☿
08 16 ☽ ⊻ ☉
09 35 ☽ ⊼ ♅
11 10 ☽ △ ♆
11 39 ☽ ∥ ♆
12 42 ☽ Q ♂
18 57 ☽ ⊻ ☿
20 18 ☽ ± ♅
22 03 ☽ ∥ ♃

15 Sunday
01 22 ☽ ⊻ ♄
01 26 ☽ ⚹ ♇
01 41 ☉ △ ♅
09 28 ☽ ∥ ♇
11 33 ☽ ∥ ☉
12 17 ☽ ⊥ ♄
12 30 ☽ ⚼ ♅
12 49 ☽ ⋕ ♂
15 20 ♄ □ ♇
15 41 ☽ ☌ ♀
22 03 ☽ □ ♂
23 11 ☉ □ ♆

16 Monday
03 51 ☽ ∥ ☿
03 58 ☉ ⋕ ♂
04 45 ☽ ∠ ♇
04 49 ☽ ∠ ♄
09 02 ☽ □ ♃
15 57 ☽ △ ♅
17 40 ☽ □ ♆
19 14 ☽ ☌ ☉
20 54 ♀ ∥ ♆
21 25 ☽ ⊥ ♇

17 Tuesday
03 29 ☿ ⊻ ♇
05 00 ☿ ⚹ ♄
08 40 ☽ ⊻ ♇
08 52 ☽ ⚹ ♄
09 22 ☽ ☌ ☿
18 58 ☽ Q ♃

18 Wednesday
00 04 ☉ ⊥ ♇
03 35 ☽ Q ♆
05 38 ☽ ⊻ ♀
07 40 ☽ △ ♂
07 49 ☽ Q ♄
18 10 ☽ ⚹ ♃
18 28 ☽ ⊥ ♀

19 Thursday
00 52 ☽ □ ♅
02 46 ☽ ⚹ ♆
08 22 ♀ □ ♂
09 29 ☽ ⊻ ☉
13 35 ☽ ⚼ ♂
14 02 ☽ ∠ ♀
18 38 ☽ ☌ ♇
19 07 ☽ □ ♄
22 22 ☽ ⊥ ☉
23 50 ☽ ∠ ♃

20 Friday
03 42 ☽ ⊻ ☿
07 53 ☽ ± ♂
08 22 ☽ ∠ ♆
12 21 ☽ Q ♅
17 26 ☽ ⊥ ☿
17 56 ☽ ∠ ☉
17 57 ☽ ⊥ ♃
20 12 ☽ ⊼ ♂
21 43 ☿ Q ♃
23 12 ☽ ∥ ☿
23 21 ☽ ⚹ ♀

21 Saturday
02 25 ☽ ⊥ ♆
06 09 ☽ ⊻ ♃
11 36 ♀ ∠ ♇
12 28 ☽ ⚹ ♅
12 54 ♀ ∥ ♃
14 16 ☽ ∠ ☿
14 32 ☽ ⊻ ♆
18 35 ♀ ∠ ♄
22 50 ☽ ∥ ☉

22 Sunday
02 30 ☽ Q ♀
03 04 ☽ ⚹ ☉
07 04 ☽ ⊻ ♇
07 51 ☽ △ ♄
10 30 ☽ ⋕ ♂
11 18 ☽ ∥ ♇
18 56 ☽ ∠ ♅
19 19 ☽ ⊥ ♇
22 35 ☽ ∥ ♀

23 Monday
01 20 ☽ ⚹ ☿
02 02 ☽ ∥ ♃
05 47 ☽ Q ☉
10 39 ☽ ☍ ♂
12 09 ☽ ∥ ♆
13 15 ☽ ⊥ ♅
13 42 ☽ ∠ ♇
14 37 ☽ ⚼ ♄
15 34 ☿ Q ♆
19 31 ☽ □ ♀
19 45 ☽ ☌ ♃
22 02 ♀ □ ♃

24 Tuesday
01 26 ☽ ⊻ ♅
03 15 ☉ ⊻ ♇
03 36 ☽ ☌ ♆
05 22 ☽ Q ☿
09 00 ☽ ± ♄
15 35 ☉ ⚹ ♄
17 54 ♂ ⋕ ♇
20 08 ☽ ⚹ ♇
21 11 ☽ ⊼ ♄
21 39 ☽ □ ☉

25 Wednesday
10 59 ☿ Q ♄
15 19 ☽ ∥ ♅
20 07 ☽ Q ♇
22 38 ☽ □ ☿

26 Thursday
00 21 ☽ ⊼ ♂
03 00 ♀ △ ♅
03 30 ☽ ⋕ ♄
08 22 ☽ ⊻ ♃
11 42 ☽ ∥ ♄
12 20 ☽ ± ♂
13 08 ☽ ☌ ♅
14 17 ☽ △ ♀
14 53 ☿ △ ♂
15 17 ☽ ⊻ ♆
20 08 ☽ ⊥ ♃
23 43 ☽ ⋕ ♅

27 Friday
00 15 ♀ □ ♆
02 51 ☽ ⊥ ♆
06 02 ☽ ⚼ ♂
07 08 ☽ □ ♇
08 20 ☽ ☍ ♄
13 29 ☽ ∠ ♃
13 31 ☽ △ ☉
18 32 ♀ ⋕ ♂
19 55 ☽ ∠ ♆
22 01 ☽ ⚼ ♀

28 Saturday
02 27 ♂ ± ♅
05 07 ♀ ∥ ♇
10 40 ☽ △ ♂
15 27 ☽ △ ☿
16 25 ☽ ± ♀
17 37 ☽ ⚹ ♃
19 39 ☽ ⚼ ☉
20 08 ♀ ⊥ ♇
21 28 ☽ ⊻ ♅
23 12 ☽ ⋕ ♆
23 33 ☽ ⚹ ♆

29 Sunday
04 25 ☽ ⊼ ♀
06 56 ☽ ⋕ ♃
08 16 ☽ ⊥ ♅
11 54 ☿ ⚹ ♃
13 09 ☽ ± ☉
14 24 ☽ △ ♇
15 23 ☽ Q ♃
15 41 ☽ ⊼ ♄
20 04 ☽ ∥ ♂
20 57 ☽ Q ♆
21 20 ☽ ⋕ ♇
21 44 ☽ ⚼ ☿

30 Monday
00 05 ♂ ⚼ ♇
00 11 ☽ ∠ ♅
00 34 ☽ ⊼ ☉
00 38 ☽ ⋕ ♀
02 18 ☽ ± ♄
15 18 ☽ ± ☿
16 38 ☽ ⚼ ♇
16 52 ☽ □ ♂
17 58 ☽ ⚼ ♄
18 51 ☽ ⋕ ☉
19 52 ☿ □ ♅
22 59 ☽ □ ♃
23 41 ☉ Q ♃

DECEMBER 2009

LONGITUDES

Date	Sidereal time h m s	Sun ☉ ° ' "	Moon ☽ ° ' "	Moon ☽ 24.00 ° ' "	Mercury ☿ ° '	Venus ♀ ° '	Mars ♂ ° '	Jupiter ♃ ° '	Saturn ♄ ° '	Uranus ♅ ° '	Neptune ♆ ° '	Pluto ♇ ° '
01	16 41 54	09 ♐ 25 40	28 ♉ 34 48	05 ♊ 44 20	23 ♐ 43	29 ♏ 28	17 ♌ 30	20 ♒ 59	03 ♎ 01	22 ♓ 42	23 ♒ 54	02 ♑ 13
02	16 45 51	10 26 28	12 ♊ 57 46	20 ♊ 14 18	25 13	00 ♐ 44	17 43	21 08	03 06	22 D 42	23 55	02 15
03	16 49 47	11 27 17	27 ♊ 33 04	04 ♋ 53 12	26 42	01 59	17 55	21 17	03 10	22 42	23 55	02 17
04	16 53 44	12 28 08	12 ♋ 13 50	19 ♋ 34 09	28 12	03 15	18 07	21 26	03 14	22 42	23 56	02 19
05	16 57 40	13 29 00	26 ♋ 53 23	04 ♌ 10 53	29 ♐ 40	04 30	18 18	21 35	03 18	22 43	23 57	02 21
06	17 01 37	14 29 53	11 ♌ 26 05	18 ♌ 38 32	01 ♑ 08	05 46	18 28	21 44	03 22	22 43	23 59	02 23
07	17 05 33	15 30 47	25 ♌ 47 52	02 ♍ 53 50	02 35	07 01	18 38	21 54	03 26	22 43	24 00	02 26
08	17 09 30	16 31 43	09 ♍ 56 15	16 ♍ 55 02	04 02	08 17	18 47	22 03	03 30	22 43	24 01	02 28
09	17 13 27	17 32 39	23 ♍ 50 08	00 ♎ 41 32	05 28	09 32	18 56	22 13	03 33	22 44	24 02	02 30
10	17 17 23	18 33 37	07 ♎ 29 17	14 ♎ 13 27	06 52	10 48	19 03	22 23	03 37	22 44	24 03	02 32
11	17 21 20	19 34 37	20 ♎ 54 03	27 ♎ 31 12	08 15	12 03	19 10	22 33	03 40	22 45	24 04	02 34
12	17 25 16	20 35 37	04 ♏ 04 55	10 ♏ 35 18	09 37	13 19	19 17	22 43	03 44	22 45	24 05	02 36
13	17 29 13	21 36 38	17 ♏ 02 22	23 ♏ 26 12	10 56	14 34	19 23	22 53	03 47	22 46	24 07	02 38
14	17 33 09	22 37 41	29 ♏ 46 50	06 ♐ 04 20	12 14	15 50	19 28	23 04	03 50	22 46	24 08	02 40
15	17 37 06	23 38 44	12 ♐ 18 47	18 ♐ 30 14	13 29	17 05	19 32	23 14	03 53	22 47	24 09	02 43
16	17 41 02	24 39 48	24 ♐ 38 50	00 ♑ 44 42	14 41	18 21	19 35	23 25	03 56	22 48	24 11	02 45
17	17 44 59	25 40 53	06 ♑ 47 59	12 ♑ 48 55	15 50	19 36	19 38	23 35	03 59	22 49	24 12	02 47
18	17 48 56	26 41 58	18 ♑ 47 43	24 ♑ 44 41	16 55	20 52	19 40	23 46	04 02	22 49	24 13	02 49
19	17 52 52	27 43 04	00 ♒ 40 07	06 ♒ 34 25	17 56	22 07	19 41	23 57	04 05	22 50	24 15	02 51
20	17 56 49	28 44 10	12 ♒ 27 58	18 ♒ 21 13	18 51	23 23	19 42	24 08	04 08	22 51	24 16	02 53
21	18 00 45	29 ♐ 45 17	24 ♒ 14 40	00 ♓ 08 50	19 40	24 38	19 R 41	24 19	04 10	22 52	24 18	02 56
22	18 04 42	00 ♑ 46 24	06 ♓ 04 17	12 ♓ 01 33	20 23	25 54	19 40	24 30	04 13	22 53	24 19	02 58
23	18 08 38	01 47 31	18 ♓ 01 17	24 ♓ 04 04	20 58	27 09	19 38	24 42	04 15	22 54	24 21	03 00
24	18 12 35	02 48 38	00 ♈ 10 30	06 ♈ 21 11	21 24	28 25	19 36	24 53	04 17	22 55	24 22	03 02
25	18 16 31	03 49 45	12 ♈ 36 43	18 ♈ 57 38	21 41	29 ♐ 40	19 32	25 04	04 19	22 57	24 24	03 04
26	18 20 28	04 50 52	25 ♈ 24 24	01 ♉ 57 27	21 48	00 ♑ 56	19 28	25 16	04 21	22 58	24 26	03 07
27	18 24 25	05 52 00	08 ♉ 37 05	15 ♉ 23 30	21 R 43	02 11	19 22	25 28	04 23	22 59	24 27	03 09
28	18 28 21	06 53 07	22 ♉ 16 48	29 ♉ 16 53	21 27	03 27	19 16	25 40	04 25	23 00	24 29	03 11
29	18 32 18	07 54 15	06 ♊ 23 31	13 ♊ 36 17	20 59	04 42	19 10	25 51	04 27	23 02	24 31	03 13
30	18 36 14	08 55 23	20 ♊ 54 35	28 ♊ 17 42	20 20	05 58	19 02	26 03	04 28	23 03	24 32	03 15
31	18 40 11	09 ♑ 56 30	05 ♋ 44 44	13 ♋ 14 40	19 ♑ 29	07 ♑ 13	18 ♌ 54	26 ♒ 15	04 ♎ 30	23 ♓ 05	24 ♒ 34	03 ♑ 17

MOON NODES AND LATITUDE

Date	Moon True ☊ ° '	Moon Mean ☊ ° '	Moon ☽ Latitude ° '
01	21 ♑ 47	23 ♑ 15	04 N 08
02	21 R 41	23 12	03 14
03	21 37	23 08	02 07
04	21 35	23 05	00 N 51
05	21 D 34	23 02	00 S 29
06	21 35	22 59	01 46
07	21 37	22 56	02 56
08	21 38	22 53	03 54
09	21 R 38	22 49	04 37
10	21 37	22 46	05 03
11	21 33	22 43	05 12
12	21 29	22 40	05 05
13	21 24	22 37	04 42
14	21 19	22 33	04 05
15	21 14	22 30	03 17
16	21 10	22 27	02 21
17	21 08	22 24	01 18
18	21 07	22 21	00 S 13
19	21 D 08	22 18	00 N 52
20	21 09	22 14	01 55
21	21 11	22 11	02 53
22	21 12	22 08	03 44
23	21 14	22 05	04 25
24	21 14	22 02	04 56
25	21 R 14	21 59	05 13
26	21 13	21 55	05 16
27	21 11	21 52	05 02
28	21 09	21 49	04 31
29	21 08	21 46	03 43
30	21 06	21 43	02 40
31	21 ♑ 05	21 ♑ 39	01 N 24

DECLINATIONS

Date	Sun ☉	Moon ☽	Mercury ☿	Venus ♀	Mars ♂	Jupiter ♃	Saturn ♄	Uranus ♅	Neptune ♆	Pluto ♇
01	21 S 52	23 N 52	25 S 28	19 S 18	17 N 58	15 S 25	00 N 46	03 S 36	13 S 57	18 S 16
02	22 01	25 34	25 34	19 36	17 56	15 22	00 45	03 36	13 57	18 16
03	22 09	25 32	25 39	19 53	17 55	15 20	00 43	03 36	13 56	18 16
04	22 17	23 43	25 43	20 10	17 53	15 17	00 42	03 36	13 56	18 16
05	22 25	20 18	25 45	20 27	17 52	15 13	00 41	03 36	13 56	18 16
06	22 32	15 39	25 46	20 42	17 51	15 10	00 39	03 36	13 55	18 16
07	22 39	10 10	25 45	20 57	17 50	15 07	00 38	03 35	13 55	18 16
08	22 45	04 N 14	25 43	21 12	17 50	15 04	00 37	03 35	13 55	18 16
09	22 51	01 S 47	25 40	21 26	17 49	15 01	00 36	03 35	13 54	18 17
10	22 57	07 37	25 35	21 39	17 49	14 57	00 34	03 35	13 54	18 17
11	23 02	12 59	25 28	21 52	17 49	14 54	00 33	03 35	13 54	18 17
12	23 06	17 39	25 20	22 04	17 50	14 51	00 32	03 34	13 53	18 17
13	23 10	21 26	25 11	22 15	17 50	14 47	00 31	03 34	13 53	18 17
14	23 14	24 06	25 00	22 26	17 51	14 44	00 30	03 34	13 52	18 17
15	23 17	25 32	24 49	22 36	17 52	14 40	00 29	03 34	13 52	18 17
16	23 20	25 40	24 35	22 45	17 53	14 37	00 28	03 33	13 51	18 17
17	23 22	24 34	24 21	22 54	17 55	14 33	00 27	03 33	13 51	18 17
18	23 24	22 34	24 06	23 02	17 57	14 30	00 26	03 33	13 50	18 17
19	23 25	19 09	23 50	23 09	17 59	14 26	00 25	03 32	13 50	18 17
20	23 26	15 13	23 33	23 15	18 01	14 22	00 25	03 32	13 50	18 17
21	23 26	10 43	23 16	23 21	18 03	14 18	00 24	03 31	13 49	18 18
22	23 26	05 49	22 58	23 26	18 06	14 15	00 23	03 31	13 49	18 18
23	23 26	00 S 40	22 40	23 31	18 09	14 11	00 23	03 30	13 48	18 18
24	23 25	04 N 35	22 22	23 34	18 13	14 07	00 22	03 30	13 47	18 18
25	23 23	09 47	22 04	23 37	18 16	14 03	00 21	03 29	13 47	18 18
26	23 21	14 43	21 46	23 40	18 20	13 59	00 21	03 29	13 46	18 18
27	23 19	19 08	21 30	23 41	18 24	13 55	00 20	03 28	13 46	18 18
28	23 16	22 42	21 14	23 42	18 28	13 51	00 20	03 28	13 45	18 18
29	23 12	25 02	20 59	23 42	18 33	13 47	00 19	03 27	13 45	18 18
30	23 08	25 47	20 46	23 41	18 38	13 43	00 19	03 27	13 44	18 18
31	23 S 04	24 N 43	20 S 34	23 S 40	18 N 43	13 S 39	00 N 19	03 S 26	13 S 44	18 S 18

ZODIAC SIGN ENTRIES

Date	h	m	Planets
01	14	23	☽ ♊
01	22	04	♀ ♐
03	16	01	☽ ♋
05	17	07	☽ ♌
05	17	24	☿ ♑
07	19	05	☽ ♍
09	22	47	☽ ♎
12	04	31	☽ ♏
14	12	25	☽ ♐
16	22	32	☽ ♑
19	10	39	☽ ♒
21	17	47	☉ ♑
21	23	42	☽ ♓
24	11	39	☽ ♈
25	18	17	♀ ♑
26	20	26	☽ ♉
29	01	13	☽ ♊
31	02	45	☽ ♋

LATITUDES

Date	Mercury ☿	Venus ♀	Mars ♂	Jupiter ♃	Saturn ♄	Uranus ♅	Neptune ♆	Pluto ♇
01	02 S 11	00 N 45	02 N 30	00 S 58	02 N 09	00 S 46	00 S 25	05 N 10
04	02 17	00 39	02 37	00 58	02 10	00 46	00 25	05 09
07	02 21	00 32	02 44	00 58	02 11	00 46	00 25	05 09
10	02 19	00 25	02 51	00 58	02 11	00 46	00 25	05 08
13	02 12	00 18	02 59	00 57	02 12	00 46	00 25	05 08
16	01 58	00 11	03 06	00 57	02 13	00 45	00 25	05 07
19	01 37	00 N 03	03 14	00 57	02 14	00 45	00 25	05 07
22	01 05	00 S 04	03 21	00 57	02 15	00 45	00 25	05 07
25	00 S 22	00 11	03 29	00 56	02 15	00 45	00 25	05 07
28	00 N 30	00 18	03 37	00 56	02 16	00 45	00 25	05 06
31	01 N 28	00 S 25	03 N 44	00 S 56	02 N 17	00 S 45	00 S 25	05 N 06

DATA

Julian Date	2455167
Delta T	+67 seconds
Ayanamsa	23° 59' 59"
Synetic vernal point	05° ♓ 07' 01"
True obliquity of ecliptic	23° 26' 20"

MOON'S PHASES, APSIDES AND POSITIONS ☽

Date	h	m	Phase	Longitude ° '	Eclipse Indicator
02	07	30	○	10 ♊ 15	
09	00	13	☾	17 ♍ 03	
16	12	02	●	24 ♐ 40	
24	17	36	☽	03 ♈ 03	
31	19	13	○	10 ♋ 15	partial

Day	h	m		
04	14	27	Perigee	
20	14	59	Apogee	
02	23	36	Max dec	25° N 46'
09	04	51	0S	
16	02	38	Max dec	25° S 46'
23	15	03	0N	
30	10	05	Max dec	25° N 47'

LONGITUDES

Date	Chiron ⚷ ° '	Ceres ⚳ ° '	Pallas ⚴ ° '	Juno ⚵ ° '	Vesta ⚶ ° '	Black Moon Lilith ⚸ ° '
01	21 ♒ 44	21 ♏ 49	26 ♎ 04	24 ♓ 46	02 ♍ 51	06 ♒ 48
11	22 ♒ 06	26 ♏ 05	00 ♏ 06	27 ♓ 21	04 ♍ 42	07 ♒ 55
21	22 ♒ 33	00 ♐ 16	03 ♏ 57	00 ♈ 32	06 ♍ 00	09 ♒ 02
31	23 ♒ 05	04 ♐ 22	07 ♏ 36	04 ♈ 14	06 ♍ 37	10 ♒ 08

All ephemeris data is given at 12.00 UT and the Moon's longitude is additionally given for 24.00 UT

ASPECTARIAN

h m Aspects

01 Tuesday
02 03 ☽ ⚹ ♅
02 48 ☽ ⊼ ☿
04 04 ☽ □ ♆
08 00 ☽ ± ♇
13 39 ☽ ☍ ♀
14 52 ☿ ⚹ ♆
18 08 ☽ ⊼ ♇
19 30 ☽ △ ♄
20 26 ♅ St D
22 16 ☽ Q ♅
23 47 ☽ Q ♂

02 Wednesday
07 30 ☽ ☍ ☉
12 19 ☽ # ☿
19 58 ☽ ⚹ ♂

03 Thursday
01 36 ☽ △ ♃
04 03 ☽ □ ♅
06 03 ☽ △ ♆
08 34 ☽ # ☿
10 28 ☽ ☍ ☿
17 53 ♀ ⊻ ♇
19 46 ☽ ☍ ♇
19 56 ☽ ⊼ ♀
20 54 ☽ ∠ ♂
21 14 ☽ □ ♄
23 18 ☉ Q ♆

04 Friday
02 25 ☽ ⚼ ♃
06 37 ☽ ⚼ ♆
06 40 ☽ ± ♀
11 45 ♀ ⚹ ♄
11 49 ☽ ⊥ ♂
12 25 ☽ ⊼ ☉
17 17 ☽ ± ♃
21 21 ☽ ± ♆
21 45 ☽ ⊻ ♂
22 45 ☽ ⚼ ♀
22 53 ☽ # ☉
22 57 ☽ ± ☉

05 Saturday
02 47 ☽ Q ♄
03 12 ☽ ⊼ ♃
05 09 ☽ △ ♅
07 11 ☽ ⊼ ♆
11 14 ☽ # ♀
12 19 ♂ ∠ ♄
14 49 ☽ ⚼ ☉
17 05 ☽ ⊼ ☿
21 01 ☽ ⊼ ♇
22 36 ☽ ⚹ ♄
23 09 ☽ # ♇

06 Sunday
01 16 ☽ ∥ ♂
01 43 ☽ △ ♀
04 05 ☽ ± ☿
05 50 ☽ ⚼ ♅
06 57 ☽ ± ♇
14 13 ☽ # ♃
17 29 ☽ △ ☉
19 51 ☽ # ♆
20 42 ☽ ⚼ ☿
20 47 ☽ ± ♅
21 56 ☽ ⚼ ♇
23 35 ☽ ∠ ♄
23 51 ☽ ☌ ♂

07 Monday
05 22 ☽ ☍ ♃
06 49 ☽ ⊼ ♅
08 58 ☽ ☍ ♆
09 12 ☿ ☌ ♇
09 55 ☉ Q ♄
14 46 ☽ ⊥ ♄
23 14 ☽ △ ♇

08 Tuesday
00 47 ☽ △ ☿
00 58 ☽ ⊻ ♄
02 35 ☿ □ ♄
07 24 ☿ ⚼ ♂
08 53 ☽ □ ♀
14 34 ☽ # ♅

09 Wednesday
00 13 ☽ □ ☉
02 27 ☽ ∥ ♄
03 23 ☽ ⊻ ♂
07 13 ☽ # ♄
09 09 ☽ ⊼ ♃
10 04 ☽ ☍ ♅
12 20 ☽ ⊼ ♆
13 55 ☽ ⊥ ♂
19 07 ☽ Q ♀
19 16 ☽ ∥ ♅
19 45 ☽ ± ♃
22 51 ☽ ± ♆

10 Thursday
03 01 ♀ Q ♃
03 13 ☽ □ ♇
05 07 ☽ ☌ ♄
05 52 ☽ ∠ ♂
10 13 ☽ Q ☉
10 46 ☽ □ ☿
11 49 ☽ ⚼ ♃
14 47 ☽ ⚼ ♆
18 29 ☽ ⚹ ♀
22 07 ☿ ∠ ♃

11 Friday
01 18 ☉ △ ♂
08 52 ☽ ⚹ ♂
09 25 ☽ ⚹ ☉
11 24 ☽ Q ♇
12 23 ♀ Q ♆
15 01 ☽ △ ♃
15 20 ☽ ⊼ ♅
16 23 ☽ ∥ ♆
17 44 ☽ △ ♆
21 16 ☽ ∥ ♃
22 49 ☽ Q ☿

12 Saturday
00 19 ☽ ∠ ♀
02 14 ☽ ± ♅
02 38 ☿ ∠ ♆
06 50 ☽ Q ♂
09 17 ☽ ⚹ ♇
11 21 ☽ ⊻ ♄
12 57 ☽ # ♂
15 01 ☽ ∠ ☉
15 34 ☽ ∥ ♇
17 19 ♃ ⊻ ♅
18 34 ☽ ⊥ ♀
18 46 ☽ ⚼ ♅
22 27 ☽ ⊥ ♄
23 22 ☽ ⚹ ☿

13 Sunday
06 53 ☽ ⊻ ♀
08 45 ☿ Q ♅
09 06 ☽ ⊥ ☉
13 07 ☽ ∠ ♇
15 17 ☽ ∠ ♄
16 24 ☽ □ ♂
18 48 ☽ ∥ ♀
21 18 ☽ ⊻ ☉
22 44 ☽ △ ♅
23 07 ☽ □ ♃

14 Monday
01 18 ☽ □ ♆
02 47 ☽ ∥ ☉
06 06 ☽ ⊥ ♇
06 38 ☽ ∠ ☿
12 16 ♀ Q ♄
15 27 ☉ □ ♅
17 31 ☽ ⊻ ♇
19 46 ☽ ⚹ ♄
23 06 ☽ ∥ ☿

15 Tuesday
00 17 ☉ ⚹ ♃
01 41 ☽ ⊥ ☿
09 53 ☽ Q ♃
11 42 ☽ Q ♆
12 56 ☿ ± ♂
14 31 ☽ ⊻ ☿
18 57 ☽ Q ♄
22 17 ☽ ☌ ♀

16 Wednesday
00 17 ☉ ⚹ ♆
02 04 ☽ △ ♂
08 22 ☽ □ ♅
09 32 ☽ ⚹ ♃
11 05 ☽ ⚹ ♆
12 02 ☽ ☌ ☉

17 Thursday
04 00 ☽ ☌ ♇
06 24 ☽ □ ♄
07 41 ☽ ⚼ ♂
12 39 ♀ △ ♂
15 17 ☽ ∥ ☿
15 37 ☽ ∠ ♃
16 47 ☽ ∠ ♆
20 00 ☽ Q ♅

18 Friday
01 41 ☽ ± ♂
02 01 ☽ ∥ ☉
05 55 ☽ ∥ ♀
07 52 ☽ ☌ ☿
09 54 ☽ ⊥ ♃
10 51 ☽ ⊥ ♆
13 46 ☽ ⊼ ♂
16 39 ☽ ⊻ ♀
20 08 ☽ ⚹ ♅
22 11 ☽ ⊻ ♃
22 58 ☽ ⊻ ♆

19 Saturday
05 27 ☽ ⊻ ☉
06 13 ☽ ⊥ ♀
12 39 ☿ ⊥ ♃
16 27 ☽ ⊻ ♇
17 38 ☽ ∥ ♇
18 47 ☽ ⊥ ☉
18 58 ☽ △ ♄
19 33 ☽ # ♂
20 16 ☿ ⊥ ♆

20 Sunday
01 52 ♀ □ ♅
02 35 ☽ ∠ ♅
02 41 ☽ ∠ ♀
04 42 ☽ ⊥ ♇
13 26 ♂ St R
14 50 ☽ ∠ ☉
16 49 ☽ ∥ ♃
19 44 ☽ ∥ ♆
20 57 ☽ ⊥ ♅
21 35 ☉ ∥ ☿
23 06 ☽ ∠ ♇

21 Monday
01 37 ☽ ⚼ ♄
02 01 ☽ ⊻ ☿
02 44 ☽ ☍ ♂
04 53 ♀ ⚹ ♃
05 24 ♀ ⚹ ♆
06 12 ☿ ∥ ♀
08 50 ♃ ☌ ♆
09 12 ☽ ⊻ ♅
12 06 ☽ ☌ ♆
12 09 ☽ ☌ ♃
12 33 ☿ ⊼ ♂
12 54 ☽ ⚹ ♀
15 06 ☽ ⊥ ☿
20 01 ☽ ± ♄

22 Tuesday
00 16 ☽ ⚹ ☉
05 42 ☽ ⚹ ♇
08 13 ☽ ⊼ ♄
10 31 ☽ ∠ ☿
11 23 ☉ ∥ ♀
16 07 ☽ Q ♀
22 51 ☽ ∥ ♅

23 Wednesday
02 45 ☽ Q ☉
05 57 ☽ Q ♇
13 20 ☽ # ♄
15 13 ☽ ⊼ ♂
16 46 ☽ ∥ ♄
18 06 ☽ ⚹ ☿
21 43 ☽ ☌ ♅

24 Thursday
00 35 ☽ ⊻ ♆
01 27 ☽ ⊻ ♃
03 03 ☽ ± ♂
07 02 ☽ # ♅
08 09 ☽ □ ♀
12 23 ☽ ⊥ ♆
13 24 ☽ ⊥ ♃
17 32 ☉ ☌ ♇
17 36 ☽ □ ♇
17 36 ☽ □ ☉
18 28 ☽ Q ☿
20 01 ☽ ☍ ♄
20 33 ☽ ⚼ ♂

25 Friday
05 51 ☽ ∠ ♆
07 05 ☽ ∠ ♃
23 05 ♂ # ♇
23 58 ☉ □ ♄

26 Saturday
01 00 ☽ △ ♂
03 31 ☉ ⚼ ♂
05 16 ☽ □ ☿
07 15 ☽ # ♆
07 28 ☽ ⊻ ♅
08 20 ☽ # ♃
10 11 ☽ ⚹ ♆
11 44 ☽ ⚹ ♃
14 37 ☿ St R
18 33 ☽ ⊥ ♅
23 12 ☽ △ ♀

27 Sunday
02 08 ☽ △ ♇
04 23 ☽ ⊼ ♄
06 40 ☽ △ ☉
07 07 ☽ # ♇
07 38 ☽ ∥ ♂
08 07 ☽ Q ♆
09 51 ♂ ∠ ♄
09 54 ☽ Q ♃
10 52 ☽ ∠ ♅
15 09 ☽ ± ♄

28 Monday
01 57 ☽ # ☿
04 40 ☽ ⚼ ♀
04 52 ☽ ⚼ ♇
06 49 ♀ ☌ ♇
06 50 ☽ □ ♂
07 01 ☽ ⚼ ♄
10 36 ☽ △ ☿
11 16 ☽ ⚼ ☉
13 15 ☽ ⚹ ♅
15 48 ☽ □ ♆
16 30 ☽ # ☉
17 54 ☽ □ ♃
20 27 ☽ ± ♇
20 31 ☽ # ♀
21 45 ☽ ± ♀

29 Tuesday
02 31 ♀ ⚼ ♂
03 52 ☽ ± ☉
06 39 ☽ ⊼ ♇
06 55 ♀ □ ♄
08 43 ☽ △ ♄
08 54 ☽ ⊼ ♀
09 43 ☽ Q ♅
11 21 ☽ ⚼ ☿
13 17 ☽ Q ♂
14 43 ☽ ⊼ ☉

30 Wednesday
01 42 ☽ ± ☿
02 23 ♃ ∥ ♆
08 58 ☽ ⚹ ♂
11 06 ☽ ⊼ ☿
15 30 ☽ □ ♅
17 55 ☽ △ ♆
18 40 ☿ ⊥ ♃
20 29 ☽ △ ♃

31 Thursday
02 55 ☉ ∠ ♆
08 03 ☽ ☍ ♇
09 03 ☽ ∠ ♂
09 59 ☽ □ ♄
14 35 ☽ ☍ ♀
18 08 ☽ ⚼ ♆
19 13 ☽ ☍ ☉
20 57 ☽ ⚼ ♃
22 46 ☽ # ♀
23 20 ☽ ⊥ ♂

JANUARY 2010

LONGITUDES

Date	Sidereal time h m s	Sun ☉ ° ' "	Moon ☽ ° ' "	Moon ☽ 24.00 ° ' "	Mercury ☿ ° '	Venus ♀ ° '	Mars ♂ ° '	Jupiter ♃ ° '	Saturn ♄ ° '	Uranus ♅ ° '	Neptune ♆ ° '	Pluto ♇ ° '
01	18 44 07	10 ♑ 57 38	20 ♋ 46 26	28 ♋ 18 54	18 ♑ 28	08 ♑ 29	18 ♌ 44	26 ♒ 28	04 ♎ 31	23 ♓ 06	24 ♒ 36	03 ♑ 20
02	18 48 04	11 58 46	05 ♌ 50 56	13 ♌ 21 28	17 R 19	09 44	18 R 34	26 40	04 32	23 08	24 38	03 22
03	18 52 00	12 59 54	20 ♌ 49 30	28 ♌ 14 07	16 03	11 00	18 23	26 52	04 34	23 09	24 39	03 24
04	18 55 57	14 01 02	05 ♍ 34 35	12 ♍ 50 18	14 43	12 15	18 12	27 05	04 35	23 11	24 41	03 26
05	18 59 54	15 02 11	20 ♍ 00 47	27 ♍ 05 45	13 22	13 31	17 59	27 17	04 36	23 13	24 43	03 28
06	19 03 50	16 03 19	04 ♎ 05 01	10 ♎ 58 32	12 02	14 46	17 46	27 30	04 36	23 14	24 45	03 30
07	19 07 47	17 04 28	17 ♎ 46 24	24 ♎ 28 43	10 46	16 02	17 32	27 42	04 37	23 16	24 47	03 33
08	19 11 43	18 05 37	01 ♏ 05 45	07 ♏ 37 44	09 36	17 17	17 17	27 55	04 38	23 18	24 49	03 35
09	19 15 40	19 06 46	14 ♏ 05 01	20 ♏ 27 55	08 33	18 33	17 01	28 08	04 38	23 20	24 51	03 37
10	19 19 36	20 07 55	26 ♏ 46 47	03 ♐ 01 57	07 39	19 48	16 45	28 20	04 39	23 22	24 53	03 39
11	19 23 33	21 09 04	09 ♐ 13 45	15 ♐ 22 32	06 54	21 04	16 28	28 33	04 39	23 24	24 55	03 41
12	19 27 29	22 10 13	21 ♐ 28 37	27 ♐ 32 16	06 20	22 19	16 10	28 46	04 39	23 26	24 57	03 43
13	19 31 26	23 11 22	03 ♑ 33 48	09 ♑ 33 27	05 55	23 35	15 52	28 59	04 39	23 28	24 59	03 45
14	19 35 23	24 12 31	15 ♑ 31 31	21 ♑ 28 14	05 40	24 50	15 33	29 12	04 R 39	23 30	25 01	03 47
15	19 39 19	25 13 39	27 ♑ 23 51	03 ♒ 18 36	05 33	26 06	15 13	29 26	04 39	23 32	25 03	03 49
16	19 43 16	26 14 47	09 ♒ 12 46	15 ♒ 06 35	05 D 36	27 21	14 53	29 39	04 39	23 34	25 05	03 51
17	19 47 12	27 15 54	21 ♒ 00 22	26 ♒ 54 24	05 46	28 36	14 32	29 ♒ 52	04 38	23 36	25 07	03 53
18	19 51 09	28 17 00	02 ♓ 49 00	08 ♓ 44 31	06 04	29 ♑ 52	14 11	00 ♓ 05	04 38	23 39	25 09	03 55
19	19 55 05	29 ♑ 18 06	14 ♓ 41 19	20 ♓ 39 49	06 29	01 ♒ 07	13 49	00 19	04 37	23 41	25 11	03 57
20	19 59 02	00 ♒ 19 11	26 ♓ 40 27	02 ♈ 43 38	06 59	02 23	13 27	00 32	04 37	23 43	25 13	03 59
21	20 02 58	01 20 15	08 ♈ 49 52	14 ♈ 59 39	07 35	03 38	13 04	00 46	04 36	23 46	25 15	04 01
22	20 06 55	02 21 18	21 ♈ 13 28	27 ♈ 31 50	08 17	04 54	12 42	00 59	04 35	23 48	25 17	04 03
23	20 10 52	03 22 21	03 ♉ 55 13	10 ♉ 24 07	09 02	06 09	12 18	01 13	04 34	23 51	25 19	04 05
24	20 14 48	04 23 22	16 ♉ 58 57	23 ♉ 40 04	09 52	07 24	11 55	01 27	04 33	23 53	25 21	04 07
25	20 18 45	05 24 22	00 ♊ 27 47	07 ♊ 22 15	10 46	08 40	11 31	01 40	04 32	23 55	25 24	04 09
26	20 22 41	06 25 21	14 ♊ 23 34	21 ♊ 31 36	11 43	09 55	11 07	01 54	04 30	23 58	25 26	04 11
27	20 26 38	07 26 20	28 ♊ 46 08	06 ♋ 06 43	12 42	11 10	10 44	02 08	04 29	24 01	25 28	04 13
28	20 30 34	08 27 17	13 ♋ 32 44	21 ♋ 03 23	13 45	12 26	10 20	02 22	04 27	24 03	25 30	04 15
29	20 34 31	09 28 13	28 ♋ 37 40	06 ♌ 14 29	14 50	13 41	09 56	02 36	04 26	24 06	25 32	04 17
30	20 38 27	10 29 08	13 ♌ 52 36	21 ♌ 30 43	15 58	14 56	09 32	02 50	04 24	24 09	25 35	04 19
31	20 42 24	11 ♒ 30 02	29 ♌ 07 32	06 ♍ 41 48	17 ♑ 07	16 ♒ 12	09 ♌ 08	03 ♓ 04	04 ♎ 22	24 ♓ 11	25 ♒ 37	04 ♑ 21

Date	Moon True ☊ ° '	Moon Mean ☊ ° '	Moon ☽ Latitude ° '
01	21 ♑ 05	21 ♑ 36	00 N 02
02	21 D 05	21 33	01 S 21
03	21 06	21 30	02 38
04	21 06	21 27	03 43
05	21 07	21 24	04 32
06	21 07	21 20	05 04
07	21 R 07	21 17	05 17
08	21 07	21 14	05 13
09	21 07	21 11	04 53
10	21 07	21 08	04 19
11	21 D 07	21 04	03 33
12	21 07	21 01	02 38
13	21 07	20 58	01 36
14	21 07	20 55	00 S 31
15	21 R 07	20 52	00 N 35
16	21 07	20 49	01 39
17	21 06	20 45	02 39
18	21 05	20 42	03 32
19	21 04	20 39	04 16
20	21 03	20 36	04 49
21	21 02	20 33	05 10
22	21 01	20 30	05 17
23	21 D 01	20 26	05 09
24	21 01	20 23	04 45
25	21 02	20 20	04 05
26	21 03	20 17	03 10
27	21 04	20 14	02 01
28	21 05	20 10	00 N 42
29	21 R 05	20 07	00 S 42
30	21 04	20 04	02 03
31	21 ♑ 02	20 ♑ 01	03 S 15

DECLINATIONS

Sun ☉ ° '	Moon ☽ ° '	Mercury ☿ ° '	Venus ♀ ° '	Mars ♂ ° '	Jupiter ♃ ° '	Saturn ♄ ° '	Uranus ♅ ° '	Neptune ♆ ° '	Pluto ♇ ° '
22 S 59	21 N 52	20 S 23	23 S 37	18 N 48	13 S 34	00 N 18	03 S 25	13 S 43	18 S 18
22 54	17 30	20 14	23 34	18 53	13 30	00 18	03 25	13 42	18 18
22 48	12 03	20 07	23 31	18 59	13 26	00 18	03 24	13 42	18 18
22 42	06 N 00	20 00	23 26	19 05	13 22	00 18	03 23	13 41	18 18
22 35	00 S 13	19 56	23 21	19 11	13 17	00 18	03 23	13 41	18 18
22 28	06 16	19 53	23 15	19 17	13 13	00 18	03 22	13 40	18 18
22 21	11 51	19 51	23 09	19 24	13 09	00 18	03 21	13 39	18 18
22 13	16 44	19 50	23 02	19 31	13 04	00 18	03 20	13 39	18 18
22 05	20 43	19 51	22 54	19 37	13 00	00 18	03 20	13 38	18 18
21 56	23 38	19 54	22 45	19 44	12 55	00 18	03 19	13 37	18 18
21 47	25 20	19 57	22 35	19 52	12 51	00 18	03 18	13 37	18 18
21 37	25 47	20 02	22 25	19 59	12 46	00 18	03 17	13 36	18 18
21 27	25 00	20 07	22 15	20 06	12 42	00 18	03 16	13 35	18 18
21 16	23 03	20 14	22 03	20 14	12 37	00 19	03 16	13 35	18 18
21 05	20 07	20 21	21 51	20 21	12 32	00 19	03 15	13 34	18 18
20 54	16 22	20 29	21 38	20 29	12 28	00 19	03 14	13 33	18 18
20 42	11 59	20 37	21 25	20 37	12 23	00 20	03 13	13 33	18 18
20 30	07 11	20 45	21 11	20 44	12 18	00 20	03 12	13 32	18 18
20 18	02 S 06	20 54	20 56	20 52	12 14	00 21	03 11	13 31	18 18
20 05	03 N 06	21 02	20 41	21 00	12 09	00 21	03 10	13 31	18 18
19 52	08 15	21 10	20 25	21 08	12 04	00 22	03 09	13 30	18 18
19 38	13 10	21 18	20 08	21 15	11 59	00 22	03 08	13 29	18 18
19 24	17 40	21 25	19 51	21 23	11 54	00 23	03 07	13 29	18 18
19 10	21 28	21 32	19 34	21 31	11 49	00 24	03 06	13 28	18 18
18 55	24 15	21 38	19 15	21 38	11 44	00 24	03 05	13 27	18 18
18 40	25 40	21 43	18 57	21 45	11 40	00 25	03 04	13 26	18 18
18 25	25 27	21 48	18 37	21 53	11 35	00 26	03 03	13 26	18 18
18 09	23 26	21 52	18 17	22 00	11 30	00 27	03 02	13 25	18 18
17 53	19 45	21 54	17 57	22 07	11 25	00 28	03 01	13 24	18 18
17 37	14 42	21 56	17 36	22 14	11 20	00 29	03 00	13 23	18 17
17 S 20	08 N 44	21 S 57	17 S 15	22 N 20	11 S 15	00 N 30	02 S 59	13 S 23	18 S 17

ZODIAC SIGN ENTRIES

Date	h	m	Planets
02	02	41	☽ ♌
04	02	52	☽ ♍
06	04	58	☽ ♎
08	10	00	☽ ♏
10	18	10	☽ ♐
13	04	54	☽ ♑
15	17	17	☽ ♒
18	02	10	♃ ♓
18	06	17	☽ ♓
18	14	35	♀ ♒
20	04	28	☉ ♒
20	18	36	☽ ♈
23	04	39	☽ ♉
25	11	11	☽ ♊
27	14	01	☽ ♋
29	14	10	☽ ♌
31	13	23	☽ ♍

LATITUDES

Date	Mercury ☿ ° '	Venus ♀ ° '	Mars ♂ ° '	Jupiter ♃ ° '	Saturn ♄ ° '	Uranus ♅ ° '	Neptune ♆ ° '	Pluto ♇ ° '
01	01 N 48	00 S 27	03 N 47	00 S 56	02 N 17	00 S 45	00 S 25	05 N 06
04	02 38	00 34	03 54	00 56	02 18	00 45	00 25	05 06
07	03 10	00 41	04 01	00 56	02 19	00 45	00 25	05 05
10	03 20	00 47	04 07	00 56	02 20	00 44	00 25	05 05
13	03 11	00 53	04 13	00 56	02 21	00 44	00 25	05 05
16	02 51	00 58	04 18	00 55	02 22	00 44	00 25	05 05
19	02 23	01 03	04 23	00 55	02 23	00 44	00 25	05 05
22	01 53	01 08	04 27	00 55	02 23	00 44	00 25	05 05
25	01 23	01 12	04 29	00 55	02 24	00 44	00 25	05 05
28	00 52	01 16	04 31	00 55	02 25	00 44	00 25	05 05
31	00 N 24	01 S 19	04 N 32	00 S 55	02 N 26	00 S 44	00 S 25	05 N 05

DATA

Julian Date	2455198
Delta T	+67 seconds
Ayanamsa	24° 00' 05"
Synetic vernal point	05° ♓ 06' 54"
True obliquity of ecliptic	23° 26' 20"

MOON'S PHASES, APSIDES AND POSITIONS ☽

Date	h	m	Phase	Longitude ° '	Eclipse Indicator
07	10	39	☾	17 ♎ 01	
15	07	11	●	25 ♑ 01	Annular
23	10	53	☽	03 ♉ 20	
30	06	18	○	10 ♌ 15	

Day	h	m		
01	20	40	Perigee	
17	01	53	Apogee	
30	09	11	Perigee	
05	11	09	0S	
12	08	33	Max dec	25° S 48'
19	21	42	0N	
26	21	00	Max dec	25° N 47'

LONGITUDES

Date	Chiron ⚷ ° '	Ceres ⚳ ° '	Pallas ⚴ ° '	Juno ⚵ ° '	Vesta ⚶ ° '	Black Moon Lilith ⚸ ° '
01	23 ♒ 09	04 ♐ 46	07 ♏ 57	04 ♈ 38	06 ♍ 38	10 ♒ 15
11	23 ♒ 45	08 ♐ 43	11 ♏ 20	08 ♈ 47	06 ♍ 27	11 ♒ 21
21	24 ♒ 23	12 ♐ 32	14 ♏ 26	13 ♈ 18	05 ♍ 28	12 ♒ 28
31	25 ♒ 04	16 ♐ 11	17 ♏ 09	18 ♈ 06	03 ♍ 45	13 ♒ 34

All ephemeris data is given at 12.00 UT and the Moon's longitude is additionally given for 24.00 UT

ASPECTARIAN

h m	Aspects
01 Friday	
03 53	☽ ⋕ ☉
04 53	☿ ⊼ ♂
08 32	☽ ± ♆
08 35	☽ ☍ ☿
08 48	☽ ⊻ ♂
09 14	☿ ⊥ ♆
11 30	☽ ± ♃
14 47	☽ Q ♄
15 26	☉ Q ♅
15 43	☽ △ ♅
18 06	☽ ⊼ ♆
21 10	☽ ⊼ ♃
21 18	☽ ⋕ ☿
02 Saturday	
02 42	☉ ∠ ♃
05 13	☽ ∥ ♂
08 02	☽ ⊼ ♇
08 04	☽ ⋕ ♇
09 50	♀ ∠ ♆
09 55	☽ ✶ ♄
15 39	☽ ⅊ ♅
17 37	☽ ± ♇
18 46	☽ ⊼ ♀
22 30	☽ ⊼ ☉
23 51	☉ ± ♂
03 Sunday	
04 56	☽ ⊼ ☿
05 07	☽ ⋕ ♆
05 16	☽ ± ♀
06 05	☽ ± ♅
06 10	☽ ⋕ ♃
08 05	☽ ⅊ ♇
08 08	☽ ☌ ♂
08 50	☽ ± ☉
09 57	☽ ∠ ♄
10 57	♂ ⅊ ♇
13 49	☽ ± ☿
15 07	♀ Q ♅
15 46	☽ ⊼ ♅
18 12	☽ ☍ ♆
21 08	☽ ⅊ ♀
21 55	☽ ☍ ♃
04 Monday	
00 29	☽ ⅊ ☉
00 32	☽ ⊥ ♄
03 13	☽ ⅊ ☿
07 55	♀ ∠ ♃
08 29	☽ △ ♇
10 21	☽ ⊻ ♄
11 01	♀ ± ♂
19 06	☉ ☌ ☿
22 06	☽ ⋕ ♅
05 Tuesday	
00 05	☽ △ ♀
01 50	☽ △ ☿
03 01	☽ △ ☉
08 39	☽ ⊻ ♂
10 00	☽ ∥ ♄
10 39	☿ ☌ ♀
12 17	☽ ⋕ ♄
17 25	☽ ☍ ♅
18 37	☽ ⊥ ♂
19 58	☽ ⊼ ♆
06 Wednesday	
00 20	☽ ∥ ♅
00 30	☽ ⊼ ♃
04 48	☿ ∠ ♃
06 14	☽ ± ♆
09 46	☽ ∠ ♂
10 58	☽ ± ♃
11 00	☽ □ ♇
12 54	☽ ☌ ♄
18 04	☿ ± ♂
21 52	☽ ⅊ ♆
07 Thursday	
00 40	☽ □ ☿
02 29	☿ Q ♅
02 53	☽ ⅊ ♃
08 35	☽ □ ♀
10 39	☽ □ ☉
11 34	☽ ✶ ♂
17 52	☽ ∥ ♃
18 45	☽ Q ♇
20 22	☽ ∥ ♆
20 39	☉ ⊼ ♂
21 51	☽ ⊼ ♅
08 Friday	
00 34	☽ △ ♆
06 07	☽ △ ♃
06 08	☽ Q ☿
07 29	☿ ∠ ♆
08 43	☽ ± ♅
08 45	☽ Q ♂
10 27	♂ ± ♅
11 56	♀ ⊼ ♂
16 33	☽ ✶ ♇
18 28	☽ ⊻ ♄
20 30	☽ Q ♀
20 43	☽ ∥ ♇
21 57	☽ Q ☉
09 Saturday	
01 16	☽ ⅊ ♅
02 26	☽ ✶ ☿
04 31	☽ ⋕ ♂
05 27	☉ ⊥ ♆
05 34	☽ ⊥ ♄
06 11	☽ ∥ ☿
17 24	☽ □ ♂
17 52	♀ ⊥ ♆
20 32	☽ ∠ ♇
21 18	☽ ✶ ♀
21 36	☽ ∥ ☉
22 16	☽ ✶ ☉
22 26	☽ ∠ ♄
10 Sunday	
04 06	☽ ∥ ♀
04 38	☽ ∠ ☿
05 29	☽ △ ♅
08 22	☽ □ ♆
13 40	☽ ⊥ ♇
15 02	☽ □ ♃
20 47	☽ ⊥ ☿
11 Monday	
01 13	☽ ⊻ ♇
03 07	☽ ✶ ♄
05 09	☽ ∠ ♀
05 29	☽ ∠ ☉
07 43	☽ ⊻ ☿
19 12	☽ Q ♆
21 06	☉ ☌ ♀
22 31	♃ ± ♄
12 Tuesday	
00 37	☽ ⊥ ☉
00 41	☽ ⊥ ♀
01 48	☽ △ ♂
02 30	☽ Q ♄
02 34	☽ Q ♃
13 30	☽ ⊻ ☉
13 51	☽ ⊻ ♀
15 52	☽ □ ♅
18 52	☽ ✶ ♆
22 26	♀ ⊥ ♃
13 Wednesday	
02 43	☽ ✶ ♃
05 59	☉ ⊥ ♃
06 45	☽ ⅊ ♂
09 48	♀ ✶ ♅
12 23	☽ ☌ ♇
14 11	☽ □ ♄
15 56	♄ St R
16 35	☽ ☌ ☿
18 42	☉ ✶ ♅
14 Thursday	
00 17	☽ ± ♂
00 52	☽ ∠ ♆
03 52	☽ Q ♅
09 18	☽ ∠ ♃
12 02	☽ ⊼ ♂
14 42	☿ ⋕ ♂
15 25	♀ ⊻ ♆
19 03	☽ ⊥ ♆
21 41	☽ ∥ ♀
15 Friday	
03 48	☽ ⊥ ♃
04 09	☽ ✶ ♅
04 19	☽ ∥ ☉
07 11	☽ ☌ ☉
07 13	☽ ⊻ ♆
07 29	☉ ⊻ ♆
09 02	☽ ☌ ♀
10 18	☽ ⋕ ♂
10 20	☽ ∥ ☿
16 12	☽ ⊻ ♃
16 53	☿ St D
16 Saturday	
00 09	☽ ∥ ♇
01 05	☽ ⊻ ♇
02 43	☽ △ ♄
04 36	☽ ⊻ ☿
10 41	☽ ∠ ♅
13 19	☽ ⊥ ♇
16 54	☽ ⊥ ☿
23 13	☽ ☍ ♂
17 Sunday	
01 50	☿ ⋕ ♂
03 45	☽ ∥ ♆
05 04	☽ ⊥ ♅
07 41	☽ ∠ ♇
09 13	☽ ⅊ ♄
09 54	☽ ∥ ♃
11 31	☽ ∠ ☿
17 19	☽ ⊻ ♅
18 26	☉ ∥ ☿
18 52	☉ ⋕ ♂
20 22	☽ ☌ ♆
18 Monday	
01 56	☽ ⊻ ☉
03 31	☽ ± ♄
05 18	☽ ⊻ ♀
06 22	☽ ☌ ♃
14 15	☽ ✶ ♇
15 15	☽ ⊥ ☉
15 41	☽ ⊼ ♄
17 15	♀ ⊻ ♃
18 48	☽ ✶ ☿
18 55	☽ ⊥ ♀
19 Tuesday	
06 55	☽ ∥ ♅
10 18	☽ ⊼ ♂
11 09	☽ ∠ ☉
14 34	☽ Q ♇
14 36	☿ ∥ ♀
15 13	☽ ∠ ♀
16 11	♀ ⋕ ♂
19 56	☽ Q ☿
20 06	☽ ⋕ ♄
22 00	☽ ± ♂
23 19	☽ ∥ ♄
20 Wednesday	
06 06	☽ ☌ ♅
09 05	☽ ⊻ ♆
12 19	☽ ⋕ ♅
15 25	☽ ⅊ ♂
18 38	☉ ⊻ ♃
19 49	☽ ⊻ ♃
19 54	☽ ✶ ☉
21 02	☽ ⊥ ♆
23 36	☿ ± ♂
21 Thursday	
00 37	☽ ✶ ♀
02 32	☽ □ ♇
03 42	☽ ☍ ♄
07 52	☽ ⊥ ♃
09 26	☽ □ ☿
14 47	☽ ∠ ♆
19 36	♀ ⊻ ♇
20 02	☽ △ ♂
21 34	☽ Q ☉
22 Friday	
01 44	☽ ∠ ♃
02 44	☽ Q ♀
06 07	♀ △ ♄
06 09	☽ ⋕ ♃
13 35	☽ ⋕ ♆
16 56	☽ ⊻ ♅
19 46	☽ ✶ ♆
23 Saturday	
04 20	☽ ⊥ ♅
06 51	☽ ✶ ♃
10 53	☽ □ ☉
12 19	☽ △ ♇
13 12	☽ ⊼ ♄
15 39	☽ ⋕ ♇
16 36	☽ □ ♀
18 20	☽ Q ♆
21 09	☽ ∠ ♅
21 43	☽ ⋕ ☉
22 07	☽ △ ☿
24 Sunday	
00 14	☽ ⋕ ♀
00 17	☽ ± ♄
03 03	☽ □ ♂
05 28	☽ Q ♃
05 29	☉ ⊻ ♇
12 21	☽ ∥ ♂
12 30	☽ ⋕ ☿
15 37	☉ △ ♄
15 52	☽ ⅊ ♇
16 37	☽ ⅊ ♄
25 Monday	
00 25	☽ ✶ ♅
01 52	☿ ∠ ♆
03 03	☽ □ ♆
03 07	☽ ⅊ ☿
07 56	☽ ± ♇
09 03	☿ ⋕ ♂
10 24	☽ Q ♂
14 09	☽ □ ♃
17 11	♀ ∠ ♅
18 27	☽ ⊼ ♇
19 04	☽ △ ♄
20 02	☽ ± ☿
21 17	☽ △ ☉
21 32	☽ Q ♅
26 Tuesday	
01 40	☿ ⊼ ♂
03 38	☽ △ ♀
06 35	☽ ✶ ♂
07 06	☽ ⊼ ☿
17 15	♀ ⊥ ♇
18 35	☿ Q ♅
27 Wednesday	
00 44	☽ ⅊ ☉
04 07	☽ □ ♅
05 29	♀ ☍ ♂
06 32	☽ △ ♆
07 07	☽ ∠ ♂
07 19	☽ ⅊ ♀
16 42	☽ ± ☉
17 36	☽ △ ♃
20 56	☽ ☍ ♇
21 20	☽ □ ♄
21 29	☽ ⊥ ♂
22 50	☉ ∥ ♇
23 27	☽ ± ♀
28 Thursday	
03 12	☽ ⊼ ☉
06 57	☽ ⊻ ♂
07 06	☽ ⅊ ♆
10 03	☽ ⊼ ♀
11 43	♀ ∥ ♇
12 21	☽ ☍ ☿
16 13	♂ ± ♇
18 13	☽ ⅊ ♃
21 33	☽ ± ♆
22 19	☽ ∥ ♂
23 25	☽ ⋕ ☿
29 Friday	
02 12	☽ Q ♄
02 48	☉ ∠ ♅
04 49	☽ △ ♅
07 06	☽ ⊼ ♆
08 44	☽ ± ♃
18 21	☽ ⊼ ♃
19 34	☽ ⋕ ♇
19 43	☉ ☍ ♂
20 56	☽ ⊼ ♇
21 08	☽ ✶ ♄
21 56	☽ ⋕ ♀
22 07	☽ ⋕ ☉
30 Saturday	
04 32	☽ ⅊ ♅
05 20	☽ ☌ ♂
06 18	☽ ☍ ☉
06 23	☽ ± ♇
07 46	☉ ⊥ ♇
09 01	☉ ∥ ♀
13 49	☽ ☍ ♀
15 33	☽ ⊼ ☿
17 32	☽ ⋕ ♆
18 43	☽ ± ♅
20 33	☽ ⅊ ♇
20 39	☽ ∠ ♄
31 Sunday	
01 45	☽ ± ☿
02 05	☽ ⋕ ♃
04 11	☽ ⊼ ♅
06 27	☽ ☍ ♆
08 41	♂ ⅊ ♅
10 48	☽ ⊥ ♄
17 08	☽ ⅊ ☿
18 19	☽ ☍ ♃
20 17	☽ △ ♇
20 17	☽ ⊻ ♄
21 27	♄ □ ♇

FEBRUARY 2010

LONGITUDES

Date	Sidereal time h m s	Sun ☉ ° ' "	Moon ☽ ° ' "	Moon ☽ 24.00 ° ' "	Mercury ☿ ° '	Venus ♀ ° '	Mars ♂ ° '	Jupiter ♃ ° '	Saturn ♄ ° '	Uranus ♅ ° '	Neptune ♆ ° '	Pluto ♇ ° '
01	20 46 21	12 ♒ 30 54	14 ♍ 12 20	21 ♍ 38 07	18 ♑ 19	17 ♒ 27	08 ♌ 44	03 ♓ 18	04 ♎ 20	24 ♓ 14	25 ♒ 39	04 ♑ 22
02	20 50 17	13 31 46	28 ♍ 58 18	06 ♎ 12 12	19 32	18 42	08 R 20	03 32	04 R 18	24 17	25 41	04 24
03	20 54 14	14 32 38	13 ♎ 19 22	20 ♎ 19 31	20 47	19 58	07 57	03 46	04 16	24 20	25 44	04 26
04	20 58 10	15 33 28	27 ♎ 12 34	03 ♏ 58 34	22 04	21 13	07 33	04 00	04 14	24 22	25 46	04 28
05	21 02 07	16 34 18	10 ♏ 37 46	17 ♏ 10 29	23 22	22 28	07 10	04 14	04 11	24 25	25 48	04 29
06	21 06 03	17 35 06	23 ♏ 37 07	29 ♏ 58 09	24 41	23 43	06 48	04 28	04 09	24 28	25 50	04 31
07	21 10 00	18 35 54	06 ♐ 14 07	12 ♐ 25 33	26 01	24 59	06 25	04 42	04 06	24 31	25 53	04 33
08	21 13 56	19 36 41	18 ♐ 33 02	24 ♐ 37 05	27 23	26 14	06 04	04 57	04 04	24 34	25 55	04 35
09	21 17 53	20 37 27	00 ♑ 38 16	06 ♑ 37 04	28 ♑ 46	27 29	05 42	05 11	04 01	24 37	25 57	04 36
10	21 21 50	21 38 12	12 ♑ 34 00	18 ♑ 29 30	00 ♒ 10	28 44	05 21	05 25	03 58	24 40	25 59	04 38
11	21 25 46	22 38 56	24 ♑ 23 59	00 ♒ 17 50	01 35	29 ♒ 59	05 01	05 39	03 55	24 43	26 02	04 39
12	21 29 43	23 39 38	06 ♒ 11 25	12 ♒ 05 01	03 02	01 ♓ 15	04 41	05 54	03 52	24 46	26 04	04 41
13	21 33 39	24 40 19	17 ♒ 58 56	23 ♒ 53 25	04 29	02 30	04 22	06 08	03 49	24 49	26 06	04 43
14	21 37 36	25 40 59	29 ♒ 48 43	05 ♓ 45 02	05 57	03 45	04 03	06 22	03 46	24 52	26 09	04 44
15	21 41 32	26 41 37	11 ♓ 42 35	17 ♓ 41 35	07 26	05 00	03 45	06 37	03 43	24 55	26 11	04 46
16	21 45 29	27 42 14	23 ♓ 42 14	29 ♓ 44 44	08 56	06 15	03 28	06 51	03 40	24 58	26 13	04 47
17	21 49 25	28 42 49	05 ♈ 49 21	11 ♈ 56 18	10 27	07 30	03 11	07 06	03 36	25 01	26 15	04 49
18	21 53 22	29 ♒ 43 22	18 ♈ 05 51	24 ♈ 18 20	11 59	08 45	02 55	07 20	03 33	25 05	26 18	04 50
19	21 57 19	00 ♓ 43 54	00 ♉ 34 02	06 ♉ 53 18	13 32	10 00	02 40	07 34	03 29	25 08	26 20	04 51
20	22 01 15	01 44 24	13 ♉ 16 29	19 ♉ 43 59	15 06	11 15	02 25	07 49	03 26	25 11	26 22	04 53
21	22 05 12	02 44 52	26 ♉ 16 08	02 ♊ 53 20	16 41	12 30	02 12	08 03	03 22	25 14	26 24	04 54
22	22 09 08	03 45 18	09 ♊ 35 54	16 ♊ 24 08	18 16	13 45	01 59	08 18	03 18	25 18	26 27	04 56
23	22 13 05	04 45 43	23 ♊ 18 15	00 ♋ 18 25	19 53	15 00	01 46	08 32	03 15	25 21	26 29	04 57
24	22 17 01	05 46 05	07 ♋ 24 38	14 ♋ 36 49	21 31	16 15	01 35	08 47	03 11	25 24	26 31	04 58
25	22 20 58	06 46 25	21 ♋ 54 40	29 ♋ 17 44	23 09	17 30	01 24	09 01	03 07	25 27	26 33	04 59
26	22 24 54	07 46 44	06 ♌ 45 24	14 ♌ 16 50	24 49	18 45	01 15	09 16	03 03	25 31	26 36	05 01
27	22 28 51	08 47 01	21 ♌ 51 02	29 ♌ 26 50	26 30	20 00	01 06	09 30	02 59	25 34	26 38	05 02
28	22 32 48	09 ♓ 47 15	07 ♍ 03 01	14 ♍ 38 14	28 ♒ 11	21 ♓ 15	00 ♌ 57	09 ♓ 45	02 ♎ 55	25 ♓ 37	26 ♒ 40	05 ♑ 03

Date	Moon True ☊ ° '	Moon Mean ☊ ° '	Moon ☽ Latitude ° '
01	20 ♑ 59	19 ♑ 58	04 S 13
02	20 R 56	19 55	04 52
03	20 53	19 51	05 12
04	20 51	19 48	05 13
05	20 50	19 45	04 57
06	20 D 50	19 42	04 25
07	20 51	19 39	03 42
08	20 53	19 36	02 49
09	20 54	19 32	01 49
10	20 56	19 29	00 S 46
11	20 R 56	19 26	00 N 19
12	20 55	19 23	01 23
13	20 52	19 20	02 23
14	20 47	19 16	03 16
15	20 41	19 13	04 02
16	20 34	19 10	04 37
17	20 26	19 07	05 00
18	20 20	19 04	05 10
19	20 14	19 01	05 05
20	20 11	18 57	04 45
21	20 09	18 54	04 11
22	20 D 09	18 51	03 22
23	20 10	18 48	02 20
24	20 11	18 45	01 N 09
25	20 R 12	18 42	00 S 09
26	20 10	18 38	01 28
27	20 07	18 35	02 42
28	20 ♑ 02	18 ♑ 32	03 S 45

DECLINATIONS

Date	Sun ☉ ° '	Moon ☽ ° '	Mercury ☿ ° '	Venus ♀ ° '	Mars ♂ ° '	Jupiter ♃ ° '	Saturn ♄ ° '	Uranus ♅ ° '	Neptune ♆ ° '	Pluto ♇ ° '
01	17 S 03	02 N 19	21 S 57	16 S 53	22 N 27	11 S 09	00 N 31	02 S 58	13 S 22	18 S 17
02	16 46	04 S 04	21 55	16 30	22 33	11 04	00 32	02 56	13 21	18 17
03	16 28	10 03	21 53	16 07	22 39	10 59	00 33	02 55	13 20	18 17
04	16 10	15 21	21 50	15 44	22 45	10 54	00 34	02 54	13 20	18 17
05	15 52	19 42	21 45	15 20	22 50	10 49	00 35	02 53	13 19	18 17
06	15 34	22 57	21 39	14 56	22 56	10 44	00 36	02 52	13 18	18 17
07	15 15	24 59	21 32	14 32	23 01	10 39	00 38	02 51	13 17	18 17
08	14 56	25 45	21 23	14 07	23 06	10 33	00 39	02 50	13 17	18 17
09	14 37	25 15	21 14	13 41	23 11	10 28	00 40	02 48	13 16	18 17
10	14 18	23 36	21 03	13 16	23 15	10 23	00 41	02 47	13 15	18 17
11	13 58	20 56	20 51	12 50	23 19	10 18	00 43	02 46	13 14	18 17
12	13 38	17 23	20 38	12 23	23 23	10 13	00 44	02 45	13 14	18 17
13	13 18	13 11	20 23	11 57	23 26	10 07	00 46	02 43	13 13	18 17
14	12 58	08 28	20 07	11 30	23 30	10 02	00 47	02 42	13 12	18 16
15	12 37	03 S 26	19 50	11 02	23 33	09 57	00 48	02 41	13 11	18 16
16	12 16	01 N 46	19 31	10 34	23 36	09 51	00 50	02 40	13 11	18 16
17	11 55	06 54	19 11	10 07	23 38	09 46	00 52	02 38	13 10	18 16
18	11 34	11 52	18 50	09 38	23 41	09 41	00 53	02 37	13 09	18 16
19	11 13	16 26	18 28	09 10	23 43	09 35	00 55	02 36	13 08	18 16
20	10 51	20 21	18 04	08 41	23 44	09 30	00 56	02 35	13 07	18 16
21	10 30	23 23	17 39	08 12	23 46	09 25	00 58	02 33	13 07	18 16
22	10 08	25 13	17 13	07 43	23 47	09 19	01 00	02 32	13 06	18 16
23	09 46	25 36	16 45	07 14	23 48	09 14	01 01	02 31	13 05	18 16
24	09 24	24 22	16 16	06 44	23 49	09 08	01 03	02 29	13 04	18 16
25	09 02	21 30	15 46	06 15	23 50	09 03	01 05	02 28	13 04	18 16
26	08 39	17 09	15 14	05 45	23 50	08 58	01 06	02 27	13 03	18 15
27	08 17	11 40	14 41	05 15	23 50	08 52	01 08	02 25	13 02	18 15
28	07 S 54	05 N 26	14 S 07	04 S 45	23 N 50	08 S 47	01 N 10	02 S 24	13 S 01	18 S 15

ZODIAC SIGN ENTRIES

Date	h	m	Planets
02	13	42	☽ ♎
04	16	55	☽ ♏
07	00	04	☽ ♐
09	10	43	☽ ♑
10	09	06	☿ ♒
11	12	10	♀ ♓
11	23	24	☽ ♒
14	12	23	☽ ♓
17	00	30	☽ ♈
18	18	36	☉ ♓
19	10	55	☽ ♉
21	18	47	☽ ♊
23	23	29	☽ ♋
26	01	08	☽ ♌
28	00	52	☽ ♍

LATITUDES

Date	Mercury ☿ ° '	Venus ♀ ° '	Mars ♂ ° '	Jupiter ♃ ° '	Saturn ♄ ° '	Uranus ♅ ° '	Neptune ♆ ° '	Pluto ♇ ° '
01	00 N 15	01 S 20	04 N 32	00 S 55	02 N 26	00 S 44	00 S 25	05 N 05
04	00 S 12	01 23	04 31	00 55	02 27	00 44	00 25	05 05
07	00 36	01 25	04 30	00 55	02 28	00 44	00 25	05 05
10	00 57	01 26	04 27	00 55	02 28	00 44	00 25	05 05
13	01 17	01 27	04 24	00 55	02 29	00 43	00 25	05 05
16	01 33	01 28	04 20	00 56	02 30	00 43	00 25	05 05
19	01 47	01 27	04 16	00 56	02 30	00 43	00 25	05 05
22	01 57	01 27	04 11	00 56	02 31	00 43	00 25	05 05
25	02 05	01 25	04 05	00 56	02 31	00 43	00 25	05 05
28	02 09	01 23	03 59	00 56	02 32	00 43	00 25	05 05
31	02 S 09	01 S 21	03 N 53	00 S 56	02 N 32	00 S 43	00 S 26	05 N 06

DATA

Julian Date	2455229
Delta T	+67 seconds
Ayanamsa	24° 00' 10"
Synetic vernal point	05° ♓ 06' 49"
True obliquity of ecliptic	23° 26' 20"

MOON'S PHASES, APSIDES AND POSITIONS ☽

Date	h	m	Phase	Longitude ° '	Eclipse Indicator
05	23	48	☾	17 ♏ 04	
14	02	51	●	25 ♒ 18	
22	00	42	☽	03 ♊ 17	
28	16	38	○	09 ♍ 59	

Day	h	m		
13	02	24	Apogee	
27	21	45	Perigee	
01	20	39	0S	
08	14	27	Max dec	25° S 45'
16	03	58	0N	
23	05	59	Max dec	25° N 39'

LONGITUDES

Date	Chiron ⚷ ° '	Ceres ⚳ ° '	Pallas ⚴ ° '	Juno ⚵ ° '	Vesta ⚶ ° '	Black Moon Lilith ⚸ ° '
01	25 ♒ 08	16 ♐ 32	17 ♏ 25	18 ♈ 36	03 ♍ 33	13 ♒ 41
11	25 ♒ 50	19 ♐ 58	19 ♏ 40	23 ♈ 40	01 ♍ 12	14 ♒ 48
21	26 ♒ 33	23 ♐ 09	21 ♏ 25	28 ♈ 56	28 ♌ 35	15 ♒ 54
31	27 ♒ 14	26 ♐ 03	22 ♏ 34	04 ♉ 21	26 ♌ 03	17 ♒ 01

All ephemeris data is given at 12.00 UT and the Moon's longitude is additionally given for 24.00 UT

ASPECTARIAN

h m Aspects

01 Monday
03 28 ☽ ⊻ ♂
09 05 ☽ ⊼ ☉
09 38 ☽ # ♅
11 29 ☿ ∠ ♃
12 49 ☽ ⊥ ♂
17 42 ☽ ⊼ ♀
18 43 ☽ ∥ ♄
19 12 ☽ △ ☿
19 27 ☽ ± ☉
22 35 ☽ # ♄

02 Tuesday
03 00 ☽ ∠ ♂
03 31 ♀ ⊥ ♅
04 17 ☽ ☍ ♅
04 21 ☽ ± ♀
06 36 ☽ ⊼ ♆
07 44 ☽ ∥ ♅
11 13 ☽ Ᵽ ☉
15 03 ☿ ⊥ ♆
16 30 ☽ ± ♆
19 40 ☽ ⊼ ♃
20 35 ☽ Ᵽ ♀
20 48 ☽ ☌ ♄
21 01 ☽ □ ♇
23 04 ♀ Ᵽ ♄

03 Wednesday
01 39 ♀ ∠ ♇
03 09 ☽ ✶ ♂
05 52 ☽ ± ♃
07 35 ☽ Ᵽ ♆
14 14 ☽ △ ☉
15 56 ☽ ∥ ♃
21 28 ☽ Ᵽ ♃
23 02 ☽ Q ♂

04 Thursday
00 30 ☽ △ ♀
02 05 ☽ □ ☿
02 26 ☽ ∥ ♆
03 41 ☽ Q ♇
07 01 ☽ ⊼ ♅
09 27 ☽ △ ♆
13 49 ☽ ∥ ♀
15 55 ☽ ∥ ☉
17 36 ☽ ± ♅

05 Friday
00 15 ☽ △ ♃
00 25 ☽ ⊻ ♄
00 54 ☽ ✶ ♇
03 33 ☽ ∥ ♇
05 55 ☽ □ ♂
08 14 ♃ ⊼ ♄
09 48 ☽ Ᵽ ♅
11 12 ☽ ⊥ ♄
13 28 ☽ Q ☿
23 48 ☽ □ ☉

06 Saturday
01 38 ☽ ∥ ☿
03 41 ☽ ∠ ♄
04 20 ☽ ∠ ♇
08 02 ☿ ✶ ♅
11 48 ☽ # ♂
12 13 ☽ □ ♀
13 36 ☽ △ ♅
14 14 ☽ ✶ ☿
16 11 ☽ □ ♆
17 51 ♃ ✶ ♇
21 16 ☽ ⊥ ♇

07 Sunday
02 50 ♀ ⊻ ♅
07 55 ☽ ✶ ♄
08 45 ☽ ⊻ ♇
09 00 ☽ □ ♃
09 18 ☿ ⊻ ♆
09 59 ☉ ⊥ ♅
12 21 ☽ △ ♂
12 46 ☽ Q ☉
22 25 ☽ ∠ ☿
23 33 ☉ Ᵽ ♄

08 Monday
02 34 ☽ Q ♀
02 53 ☽ Q ♆
05 44 ♀ ☌ ♆
07 08 ☽ Q ♄
11 07 ☉ ∠ ♇
14 17 ☽ ✶ ☉
16 49 ☽ Ᵽ ♂
18 19 ☽ ⊥ ☿
20 51 ☽ Q ♃
23 57 ☽ □ ♅

09 Tuesday
02 37 ☽ ✶ ♆
04 58 ☽ ✶ ♀
07 47 ☽ ⊻ ☿
10 11 ☽ ± ♂
18 45 ☽ □ ♄
19 58 ☽ ☌ ♇
20 31 ☿ ⊥ ♃
21 18 ☽ ✶ ♃
21 51 ♀ ± ♄
21 52 ☽ ⊼ ♂
22 56 ☽ ∠ ☉

10 Wednesday
08 48 ☽ ∠ ♆
09 24 ♂ ⊼ ♃
12 12 ☽ Q ♅
12 38 ♀ ∥ ♆
14 39 ☽ ∠ ♀
15 39 ☽ # ♂
18 48 ☽ ⊥ ☉

11 Thursday
03 05 ☽ ⊥ ♆
04 14 ☽ ∠ ♃
08 06 ☽ ⊻ ☉
11 04 ☽ ⊥ ♀
12 38 ☽ ∥ ☿
12 39 ☽ ✶ ♅
15 19 ☽ ⊻ ♆
22 55 ☽ ⊥ ♃

12 Friday
00 44 ☽ ⊻ ♀
04 39 ☽ ☌ ☿
06 26 ☽ ∥ ♇
07 18 ☽ △ ♄
08 55 ☽ ⊻ ♇
09 01 ☽ ☍ ♂
11 23 ☽ ⊻ ♃
12 00 ♂ ⊼ ♇
19 19 ☽ ∠ ♅
21 10 ☽ ⊥ ♇

13 Saturday
01 34 ☿ △ ♄
10 25 ☿ ☍ ♂
11 19 ☽ ∥ ☉
11 49 ☽ ∥ ♆
13 42 ☽ Ᵽ ♄
13 42 ☽ ⊥ ♅
15 31 ☽ ∠ ♇
15 38 ☉ ⊻ ♅
15 51 ☿ ⊻ ♇
18 13 ☉ ∥ ♆
19 13 ☽ ∥ ♀

14 Sunday
01 56 ☽ ⊻ ♅
02 51 ☽ ☌ ☉
04 07 ☽ ∥ ♃
04 33 ☽ ☌ ♆
07 53 ☽ ± ♄
12 24 ♀ ⊼ ♄
16 39 ♀ ⊼ ♂
19 58 ☽ ⊼ ♄
20 13 ☿ ⊻ ♃
20 21 ☽ ⊼ ♂
20 54 ☽ ☌ ♀
21 58 ☽ ✶ ♇
23 19 ☉ ☌ ♆

15 Monday
01 32 ☽ ☌ ♃
02 10 ☽ ⊻ ☿
07 17 ♀ ✶ ♇
08 09 ☽ ± ♂
15 20 ♂ ✶ ♄
15 33 ☽ ∥ ♅
15 58 ☽ ⊥ ☿
22 09 ☽ Q ♇

16 Tuesday
00 10 ☽ # ♄
01 47 ☽ Ᵽ ♂
07 47 ☽ ∥ ♄
11 04 ☉ ± ♄
12 32 ☽ ∠ ☿
14 32 ☽ ☌ ♅
16 14 ☽ # ♅
17 01 ☽ ⊻ ♆
20 41 ☽ ⊻ ☉

17 Wednesday
02 14 ♀ ☌ ♃
04 57 ☽ ⊥ ♆
05 00 ☿ ∠ ♅
06 55 ☽ △ ♂
07 39 ☽ ☍ ♄
09 37 ☽ ⊥ ☉
10 00 ☽ □ ♇
14 33 ☽ ⊻ ♃
15 41 ☽ ⊻ ♀
17 41 ☿ ⊥ ♇
22 23 ☽ ✶ ☿
22 42 ☽ ∠ ♆

18 Thursday
01 25 ☽ # ♃
02 00 ☽ # ♀
02 33 ☽ ⊥ ♃
04 46 ☽ ⊥ ♀
04 51 ☽ ∠ ☉
09 36 ♀ ∥ ♃
10 37 ☽ # ☉
14 34 ♀ ± ♂
18 29 ☽ # ♆
20 22 ☽ ∠ ♃

19 Friday
00 10 ☽ ∠ ♀
01 00 ☽ Q ☿
01 33 ☽ ⊻ ♅
03 52 ☽ ✶ ♆
12 20 ☽ ✶ ☉
13 05 ☽ ⊥ ♅
15 55 ☽ □ ♂
17 32 ☽ ⊼ ♄
20 10 ☽ △ ♇
22 40 ☽ # ♇
22 48 ☽ # ☿

20 Saturday
00 12 ☿ ∥ ♇
01 33 ☽ ✶ ♃
02 46 ☽ Q ♆
04 49 ☽ ± ♄
06 11 ☽ ∠ ♅
07 49 ☽ ✶ ♀
12 57 ☽ Q ☉
15 53 ☽ □ ☿
21 33 ☽ Ᵽ ♄

21 Sunday
00 18 ☽ Ᵽ ♇
00 23 ☽ Q ♃
01 02 ☽ Q ♂
01 13 ☉ ⊼ ♂
08 27 ☽ Q ♀
10 07 ☽ ✶ ♅
12 15 ☽ □ ♆
15 55 ☽ ∥ ♂
16 48 ☽ ± ♇
19 05 ♂ ± ♃
22 34 ☽ ✶ ♂

22 Monday
00 42 ☽ □ ☉
00 48 ☽ △ ♄
01 56 ☉ ⊼ ♄
03 39 ☽ ⊼ ♇
07 53 ☽ Q ♅
09 39 ☽ □ ♃
12 27 ☿ Ᵽ ♄
20 06 ☽ □ ♀

23 Tuesday
00 49 ☽ ∠ ♂
03 43 ☿ ⊥ ♅
05 18 ☽ △ ☿
12 55 ☿ ∠ ♇
15 32 ☽ □ ♅
16 12 ☽ ⊥ ♂
16 31 ☉ ✶ ♇
17 29 ☽ △ ♆

24 Wednesday
02 18 ☽ ⊻ ♂
04 54 ☽ □ ♄
07 53 ☽ ☍ ♇
09 02 ☽ △ ☉
10 18 ☽ Ᵽ ☿
14 20 ☽ △ ♃
17 30 ♀ Ᵽ ♂
17 53 ☽ ∥ ♂
18 53 ☽ Ᵽ ♆

25 Thursday
01 55 ♀ Q ♇
03 12 ☽ ± ☿
04 06 ☽ △ ♀
09 47 ☽ ± ♆
09 53 ☉ ∥ ♃
10 42 ☽ Q ♄
11 46 ☽ Ᵽ ☉
14 17 ☽ ⊼ ☿
15 30 ☽ Ᵽ ♃
17 48 ☽ △ ♅
19 35 ☽ ⊼ ♆

26 Friday
00 59 ☉ ± ♂
03 14 ☽ ☌ ♂
03 26 ☽ ± ☉
06 04 ☽ ✶ ♄
06 18 ☽ ± ♃
06 34 ☽ # ♇
06 45 ☽ Ᵽ ♀
09 12 ☽ ⊼ ♇
13 45 ☽ ⊼ ☉
16 04 ☽ ⊼ ♃
18 01 ☽ Ᵽ ♅
18 48 ☽ ± ♇
21 53 ☽ # ☿
22 19 ☿ ⊻ ♅
22 26 ☽ ± ♀

27 Saturday
05 54 ☽ ∠ ♄
06 21 ☽ # ♆
08 22 ☽ ± ♅
08 49 ☽ ⊼ ♀
09 07 ☽ Ᵽ ♇
14 03 ☿ ☌ ♆
17 54 ☽ ⊼ ♅
18 39 ☿ ± ♄
19 35 ☽ ☍ ♆
20 04 ☽ ⊥ ♄
20 15 ☽ ☍ ☿
23 10 ☽ # ♃

28 Sunday
02 08 ☽ # ☉
02 28 ☽ ⊻ ♂
05 30 ☽ ⊻ ♄
08 50 ☽ △ ♇
10 44 ☉ ☌ ♃
11 51 ☽ ⊥ ♂
14 47 ☽ # ♀
16 20 ☽ ☍ ♃
16 38 ☽ ☍ ☉
23 16 ☽ # ♅

MARCH 2010

LONGITUDES

Date	Sidereal time h m s	Sun ☉ ° ' "	Moon ☽ ° ' "	Moon ☽ 24.00 ° ' "	Mercury ☿ ° '	Venus ♀ ° '	Mars ♂ ° '	Jupiter ♃ ° '	Saturn ♄ ° '	Uranus ♅ ° '	Neptune ♆ ° '	Pluto ♇ ° '
01	22 36 44	10 ♓ 47 28	22 ♍ 11 12	29 ♍ 40 39	29 ♒ 54	22 ♓ 30	00 ♌ 50	09 ♓ 59	02 ♎ 50	25 ♓ 41	26 ♒ 42	05 ♑ 04
02	22 40 41	11 47 39	07 ♎ 05 28	14 ♎ 24 40	01 ♓ 37	23 45	00 R 43	10 14	02 R 46	25 44	26 45	05 05
03	22 44 37	12 47 49	21 ♎ 37 27	28 ♎ 43 16	03 22	24 59	00 37	10 28	02 42	25 47	26 47	05 06
04	22 48 34	13 47 57	05 ♏ 41 45	12 ♏ 32 46	05 08	26 14	00 32	10 43	02 38	25 51	26 49	05 07
05	22 52 30	14 48 03	19 ♏ 16 20	25 ♏ 52 40	06 55	27 29	00 28	10 57	02 33	25 54	26 51	05 08
06	22 56 27	15 48 08	02 ♐ 22 08	08 ♐ 45 11	08 43	28 44	00 24	11 12	02 29	25 57	26 53	05 09
07	23 00 23	16 48 11	15 ♐ 02 21	21 ♐ 14 16	10 32	29 ♓ 58	00 22	11 26	02 25	26 01	26 56	05 10
08	23 04 20	17 48 13	27 ♐ 21 33	03 ♑ 24 54	12 22	01 ♈ 13	00 20	11 41	02 20	26 04	26 58	05 11
09	23 08 17	18 48 13	09 ♑ 24 58	15 ♑ 22 25	14 13	02 28	00 18	11 55	02 16	26 08	27 00	05 12
10	23 12 13	19 48 12	21 ♑ 17 53	27 ♑ 11 59	16 05	03 42	00 18	12 10	02 11	26 11	27 02	05 13
11	23 16 10	20 48 09	03 ♒ 05 17	08 ♒ 58 20	17 59	04 57	00 D 18	12 24	02 06	26 14	27 04	05 14
12	23 20 06	21 48 04	14 ♒ 51 36	20 ♒ 45 31	19 53	06 11	00 19	12 38	02 02	26 18	27 06	05 15
13	23 24 03	22 47 57	26 ♒ 40 29	02 ♓ 36 49	21 48	07 26	00 21	12 53	01 57	26 21	27 09	05 16
14	23 27 59	23 47 48	08 ♓ 34 48	14 ♓ 34 41	23 45	08 40	00 23	13 07	01 53	26 25	27 11	05 16
15	23 31 56	24 47 38	20 ♓ 36 38	26 ♓ 40 49	25 42	09 55	00 26	13 22	01 48	26 28	27 13	05 17
16	23 35 52	25 47 25	02 ♈ 47 20	08 ♈ 56 17	27 40	11 09	00 30	13 36	01 43	26 32	27 15	05 18
17	23 39 49	26 47 11	15 ♈ 07 45	21 ♈ 21 47	29 ♓ 39	12 24	00 34	13 50	01 39	26 35	27 17	05 18
18	23 43 46	27 46 54	27 ♈ 38 27	03 ♉ 57 49	01 ♈ 39	13 38	00 39	14 05	01 34	26 38	27 19	05 19
19	23 47 42	28 46 35	10 ♉ 20 00	16 ♉ 45 05	03 39	14 53	00 45	14 19	01 29	26 42	27 21	05 20
20	23 51 39	29 ♓ 46 15	23 ♉ 13 14	29 ♉ 44 35	05 39	16 07	00 52	14 33	01 24	26 45	27 23	05 20
21	23 55 35	00 ♈ 45 52	06 ♊ 19 21	12 ♊ 57 45	07 39	17 21	00 59	14 47	01 20	26 49	27 25	05 21
22	23 59 32	01 45 26	19 ♊ 39 59	26 ♊ 26 17	09 40	18 36	01 06	15 02	01 15	26 52	27 27	05 21
23	00 03 28	02 44 59	03 ♋ 16 51	10 ♋ 11 53	11 39	19 50	01 15	15 16	01 10	26 56	27 29	05 22
24	00 07 25	03 44 29	17 ♋ 11 29	24 ♋ 15 41	13 38	21 04	01 24	15 30	01 06	26 59	27 31	05 22
25	00 11 21	04 43 57	01 ♌ 24 26	08 ♌ 37 32	15 36	22 18	01 33	15 44	01 01	27 02	27 33	05 23
26	00 15 18	05 43 22	15 ♌ 54 40	23 ♌ 15 19	17 33	23 33	01 43	15 58	00 56	27 06	27 35	05 23
27	00 19 15	06 42 45	00 ♍ 38 52	08 ♍ 04 31	19 28	24 47	01 54	16 12	00 51	27 09	27 36	05 23
28	00 23 11	07 42 06	15 ♍ 31 17	22 ♍ 58 10	21 20	26 01	02 05	16 26	00 47	27 13	27 38	05 24
29	00 27 08	08 41 25	00 ♎ 24 03	07 ♎ 47 47	23 10	27 15	02 17	16 40	00 42	27 16	27 40	05 24
30	00 31 04	09 40 41	15 ♎ 08 17	22 ♎ 24 34	24 57	28 29	02 29	16 54	00 37	27 19	27 42	05 24
31	00 35 01	10 ♈ 39 56	29 ♎ 35 43	06 ♏ 41 01	26 ♈ 41	29 ♈ 43	02 ♌ 42	17 ♓ 08	00 ♎ 33	27 ♓ 23	27 ♒ 44	05 ♑ 25

DECLINATIONS

Date	Moon True ☊ ° '	Moon Mean ☊ ° '	Moon ☽ Latitude ° '	Sun ☉ ° '	Moon ☽ ° '	Mercury ☿ ° '	Venus ♀ ° '	Mars ♂ ° '	Jupiter ♃ ° '	Saturn ♄ ° '	Uranus ♅ ° '	Neptune ♆ ° '	Pluto ♇ ° '
01	19 ♑ 54	18 ♑ 29	04 S 32	07 S 31	01 S 04	13 S 31	04 S 14	23 N 50	08 S 41	01 N 12	02 S 23	13 S 01	18 S 15
02	19 R 46	18 26	04 59	07 08	07 24	12 54	03 44	23 49	08 36	01 14	02 21	13 00	18 15
03	19 38	18 22	05 06	06 45	13 10	12 16	03 13	23 48	08 30	01 15	02 20	12 59	18 15
04	19 31	18 19	04 55	06 22	18 03	11 37	02 43	23 48	08 25	01 17	02 19	12 58	18 15
05	19 25	18 16	04 26	05 59	21 49	10 56	02 12	23 47	08 20	01 19	02 17	12 58	18 15
06	19 22	18 13	03 45	05 36	24 18	10 14	01 41	23 45	08 14	01 21	02 16	12 57	18 15
07	19 21	18 10	02 54	05 13	25 28	09 30	01 11	23 44	08 09	01 23	02 15	12 56	18 15
08	19 D 22	18 07	01 55	04 49	25 20	08 46	00 40	23 42	08 03	01 25	02 13	12 56	18 15
09	19 23	18 03	00 S 53	04 26	23 59	08 00	00 S 09	23 40	07 58	01 27	02 12	12 55	18 14
10	19 R 23	18 00	00 N 10	04 02	21 35	07 13	00 N 22	23 38	07 52	01 28	02 11	12 54	18 14
11	19 22	17 57	01 13	03 39	18 17	06 25	00 53	23 36	07 47	01 30	02 09	12 53	18 14
12	19 18	17 54	02 12	03 15	14 17	05 35	01 24	23 34	07 41	01 32	02 08	12 53	18 14
13	19 12	17 51	03 05	02 51	09 43	04 45	01 55	23 32	07 36	01 34	02 07	12 52	18 14
14	19 04	17 48	03 51	02 28	04 S 47	03 54	02 25	23 29	07 30	01 36	02 05	12 51	18 14
15	18 53	17 44	04 27	02 04	00 N 22	03 01	02 56	23 26	07 25	01 38	02 04	12 51	18 14
16	18 40	17 41	04 51	01 40	05 33	02 08	03 27	23 23	07 20	01 40	02 02	12 50	18 14
17	18 28	17 38	05 01	01 17	10 35	01 13	03 58	23 20	07 14	01 42	02 01	12 49	18 14
18	18 16	17 35	04 58	00 53	15 16	00 S 18	04 28	23 17	07 09	01 44	02 00	12 48	18 14
19	18 05	17 32	04 40	00 29	19 21	00 N 37	04 59	23 14	07 03	01 46	01 58	12 48	18 13
20	17 58	17 28	04 07	00 S 05	22 34	01 33	05 29	23 10	06 58	01 48	01 57	12 47	18 13
21	17 53	17 25	03 21	00 N 18	24 24	02 30	05 59	23 07	06 52	01 50	01 56	12 46	18 13
22	17 50	17 22	02 23	00 42	25 25	03 27	06 29	23 03	06 47	01 52	01 54	12 46	18 13
23	17 D 50	17 19	01 16	01 06	24 40	04 23	06 59	22 59	06 42	01 53	01 53	12 45	18 13
24	17 50	17 16	00 N 03	01 29	22 23	05 20	07 29	22 55	06 36	01 55	01 52	12 44	18 13
25	17 R 49	17 13	01 S 11	01 53	18 41	06 16	07 59	22 51	06 31	01 57	01 50	12 44	18 13
26	17 47	17 09	02 23	02 16	13 48	07 11	08 28	22 47	06 25	01 59	01 49	12 43	18 13
27	17 42	17 06	03 26	02 40	08 02	08 06	08 57	22 42	06 20	02 01	01 47	12 43	18 13
28	17 34	17 03	04 15	03 03	01 N 47	08 59	09 26	22 38	06 15	02 03	01 46	12 42	18 13
29	17 24	17 00	04 48	03 27	04 S 33	09 51	09 55	22 33	06 09	02 05	01 45	12 41	18 13
30	17 13	16 57	05 00	03 50	10 35	10 42	10 24	22 29	06 04	02 07	01 43	12 41	18 13
31	17 ♑ 01	16 ♑ 53	04 S 54	04 N 13	15 S 54	11 N 31	10 N 52	22 N 24	05 S 59	02 N 08	01 S 42	12 S 40	18 S 13

ZODIAC SIGN ENTRIES

Date	h m	Planets
01	13 28	☿ ♓
02	00 31	☽ ♎
04	02 11	☽ ♏
06	07 36	☽ ♐
07	12 33	♀ ♈
08	17 13	☽ ♑
11	05 42	☽ ♒
13	18 44	☽ ♓
16	06 32	☽ ♈
17	16 12	☿ ♈
18	16 29	☽ ♉
20	17 32	☉ ♈
21	00 28	☽ ♊
23	06 16	☽ ♋
25	09 39	☽ ♌
27	10 57	☽ ♍
29	11 21	☽ ♎
31	12 41	☽ ♏
31	17 35	♀ ♉

LATITUDES

Date	Mercury ☿ ° '	Venus ♀ ° '	Mars ♂ ° '	Jupiter ♃ ° '	Saturn ♄ ° '	Uranus ♅ ° '	Neptune ♆ ° '	Pluto ♇ ° '
01	02 S 09	01 S 22	03 N 57	00 S 56	02 N 32	00 S 43	00 S 26	05 N 06
04	02 08	01 20	03 51	00 56	02 32	00 43	00 26	05 06
07	02 02	01 16	03 45	00 56	02 33	00 43	00 26	05 06
10	01 53	01 12	03 38	00 56	02 33	00 43	00 26	05 06
13	01 38	01 08	03 32	00 57	02 34	00 43	00 26	05 06
16	01 19	01 03	03 26	00 57	02 34	00 43	00 26	05 06
19	00 54	00 58	03 19	00 57	02 34	00 43	00 26	05 07
22	00 S 25	00 52	03 13	00 58	02 34	00 43	00 26	05 07
25	00 N 08	00 46	03 07	00 58	02 34	00 43	00 26	05 07
28	00 43	00 39	03 01	00 58	02 34	00 43	00 26	05 07
31	01 N 19	00 S 32	02 N 55	00 S 58	02 N 34	00 S 43	00 S 26	05 N 07

DATA

Julian Date	2455257
Delta T	+67 seconds
Ayanamsa	24° 00' 14"
Synetic vernal point	05° ♓ 06' 45"
True obliquity of ecliptic	23° 26' 20"

MOON'S PHASES, APSIDES AND POSITIONS ☽

Date	h m	Phase	Longitude ° '	Eclipse Indicator
07	15 42	☾	16 ♐ 57	
15	21 01	●	25 ♓ 10	
23	11 00	☽	02 ♋ 43	
30	02 25	○	09 ♎ 17	

Day	h m		
12	10 17	Apogee	
28	05 05	Perigee	
01	08 04	OS	
07	21 19	Max dec	25° S 34'
15	10 19	ON	
22	12 16	Max dec	25° N 25'
28	18 42	OS	

LONGITUDES

Date	Chiron ⚷ ° '	Ceres ⚳ ° '	Pallas ⚴ ° '	Juno ⚵ ° '	Vesta ⚶ ° '	Black Moon Lilith ⚸ ° '
01	27 ♒ 06	25 ♐ 29	22 ♏ 24	03 ♉ 15	26 ♌ 32	16 ♒ 47
11	27 ♒ 47	28 ♐ 08	23 ♏ 00	08 ♉ 47	24 ♌ 17	17 ♒ 54
21	28 ♒ 25	00 ♑ 24	22 ♏ 50	14 ♉ 24	22 ♌ 38	19 ♒ 01
31	29 ♒ 01	02 ♑ 15	21 ♏ 52	20 ♉ 06	21 ♌ 46	20 ♒ 07

ASPECTARIAN

h m	Aspects
01 Monday	
01 58	☽ ∠ ♂
03 42	☽ ∥ ♄
12 29	☽ # ♄
12 32	☽ ☍ ♀
16 52	☽ ∥ ♅
17 36	☽ ☍ ♅
19 15	☽ ⚻ ♆
22 59	☽ ∥ ♀
02 Tuesday	
00 14	☿ ⚻ ♂
01 45	☽ ⚹ ♂
01 58	☽ ⚻ ☿
04 56	☽ ± ♆
05 01	☽ ☌ ♄
08 22	☿ ∥ ♆
08 44	☽ □ ♇
11 04	☽ ∥ ☉
12 59	☽ ± ☿
16 42	☽ ∥ ♃
17 13	☽ ⚻ ♃
17 58	♄ ± ♆
19 38	☽ bQ ♆
20 15	☽ ⚻ ☉
21 09	☽ Q ♂
03 Wednesday	
03 13	☿ ⚻ ♄
03 16	☽ ± ♃
05 48	☽ bQ ☿
06 55	☽ ± ☉
08 29	☽ ∥ ☿
11 13	☽ ∥ ♆
14 29	☽ Q ♇
18 13	☽ ⚻ ♀
18 36	☽ bQ ♃
19 03	☽ ⚻ ♅
20 43	☽ △ ♆
23 13	☽ bQ ☉
04 Thursday	
03 09	☽ □ ♂
04 07	♀ ☌ ♅
05 19	☽ ± ♅
05 26	☽ ± ♀
06 43	☽ ⊻ ♄
10 52	☽ △ ☿
11 00	☽ ⚹ ♇
11 55	☿ ⚹ ♇
13 07	☽ ∥ ♇
17 05	☽ ⊥ ♄
20 56	☽ △ ♃
21 02	☽ bQ ♅
22 40	☽ bQ ♀
23 34	♀ ⊻ ♆
05 Friday	
03 21	☽ △ ☉
06 14	☿ ± ♂
07 42	♀ ∥ ♅
08 56	☽ ∠ ♄
13 34	☽ ∠ ♇
06 Saturday	
00 06	☽ △ ♅
01 50	☽ □ ♆
02 59	☉ bQ ♂
04 31	☽ △ ♀
05 29	☽ # ♂
06 02	☽ ⊥ ♇
08 22	☽ △ ♂
12 13	☽ ⚹ ♄
17 13	☽ ⊻ ♇
07 Sunday	
01 55	☽ □ ☿
03 03	♀ # ♄
04 58	☽ □ ♃
10 48	☽ Q ♄
11 47	☽ Q ♆
12 37	☽ bQ ♂
15 42	☽ □ ☉
19 15	♀ △ ♂
21 02	☉ Q ♇
08 Monday	
01 45	☿ ☌ ♃
06 03	☽ ± ♂
09 27	☽ □ ♅
11 13	☽ ⚹ ♆
16 40	☽ Q ♃
17 51	☽ ⚻ ♂
19 00	☽ Q ☿
20 30	☽ □ ♀
21 47	☽ □ ♄
09 Tuesday	
03 33	☽ ☌ ♇
06 17	☽ Q ☉
08 21	♀ ☍ ♄
13 21	☿ ∥ ♃
15 53	☽ # ♂
17 08	☽ ⚹ ♃
17 13	☽ ∠ ♆
21 32	☽ Q ♅
22 43	♀ ⊥ ♆
23 28	☽ ⚹ ☿
10 Wednesday	
01 55	☿ bQ ♂
08 41	☽ ⚹ ☉
11 28	☽ ⊥ ♆
12 55	☽ Q ♀
17 09	♂ St D
21 59	☽ ⚹ ♅
23 42	☽ ⊻ ♆
11 Thursday	
00 10	☽ ∠ ♃
02 32	☿ Q ♇
06 19	☽ ☍ ♂
10 01	☽ △ ♄
11 44	☽ ∠ ☿
12 19	☽ ∥ ♇
16 14	☽ ⚹ ♀
16 23	☽ ⊻ ♇
17 36	♀ □ ♇
18 03	☽ ∠ ☉
18 54	☽ ⊥ ♃
12 Friday	
04 38	☽ ⊥ ♇
04 42	☽ ∠ ♅
07 23	☽ ⊻ ♃
09 37	☽ ⊥ ☿
14 06	☽ ⊥ ☉
16 23	☽ bQ ♄
19 07	♀ ∥ ♄
19 39	☽ ∥ ♆
22 58	☽ ∠ ♇
23 07	☽ ⊥ ♅
13 Saturday	
00 12	☽ ⊻ ☿
02 23	☽ ∠ ♀
03 25	☽ ⊻ ☉
10 33	☽ ± ♄
11 21	☽ ⊻ ♅
12 57	☽ ☌ ♆
19 26	☽ ⚻ ♂
20 54	♀ # ♅
22 36	☽ ⚻ ♄
22 41	☽ ∥ ♃
22 44	☽ ⊥ ♀
14 Sunday	
05 21	☽ ⚹ ♇
07 34	☽ ± ♂
12 13	☽ ⊻ ♀
13 04	☉ # ♀
13 16	☉ ☌ ☿
17 04	☽ ∥ ☿
21 17	☽ ☌ ♃
22 04	☽ # ♀
23 47	☽ ∥ ☉
15 Monday	
00 42	☽ ∥ ♅
01 39	☽ bQ ♂
02 48	☽ # ♄
05 23	☽ Q ♇
12 21	☉ ∥ ♅
13 24	☿ # ♀
17 53	☽ ∥ ♄
19 18	☽ # ☉
19 48	☽ # ♅
21 01	☽ ☌ ☉
21 40	☿ ☌ ♅
22 27	☽ # ☿
23 38	☽ ☌ ♅
16 Tuesday	
00 01	☽ ☌ ☿
01 05	☽ ⊻ ♆
01 10	☽ ∥ ♀
06 44	☿ ⊻ ♆
07 29	☽ △ ♂
09 55	☽ ☍ ♄
12 21	☉ # ♄
12 54	☽ ⊥ ♆
14 22	☿ ∥ ♅
16 55	☽ □ ♇
20 11	☽ # ♃
23 50	☿ # ♄
17 Wednesday	
06 08	☽ ☌ ♀
06 28	☽ ∠ ♆
06 50	☉ ☌ ♅
09 25	☉ ∥ ☿
09 27	☽ ⊻ ♃
09 38	♀ ∠ ♆
21 15	☽ ⊥ ♃
23 07	☽ # ♆
23 33	☿ △ ♂
18 Thursday	
00 19	☉ ⊻ ♆
10 05	☽ ⊻ ♅
11 04	☿ ☍ ♄
11 23	☽ ⚹ ♆
12 17	☽ ⊻ ☉
14 47	☽ ∠ ♃
17 47	☽ □ ♂
19 25	☽ ⚻ ♄
21 02	☽ ⊻ ☿
21 32	☽ ⊥ ♅
22 29	♀ ⊻ ♃
19 Friday	
00 39	☽ ⊥ ☉
02 34	☽ △ ♇
04 59	☽ # ♇
06 41	☽ ± ♄
08 23	☿ ⊥ ♆
09 34	☉ # ☿
10 09	☽ Q ♆
10 28	☽ ⊥ ☿
14 34	☽ ∠ ♅
18 59	☽ ∠ ☉
19 36	☽ ⚹ ♃
21 25	☽ ⊻ ♀
23 26	☽ bQ ♄
20 Saturday	
03 51	☽ Q ♂
06 22	☽ ∠ ☿
06 40	☽ bQ ♇
08 16	☿ □ ♇
09 45	☽ ⊥ ♀
17 33	☽ ∥ ♂
18 15	☽ Q ♃
18 19	☿ ∥ ♄
18 33	☽ ⚹ ♅
19 41	☽ □ ♆
21 47	☿ # ♅
23 16	☽ ± ♇
21 Sunday	
01 02	☽ ⚹ ☉
02 10	☽ ⚹ ♂
02 57	☽ △ ♄
04 02	☽ ∠ ♀
10 14	☽ ⚻ ♇
14 51	☽ ⚹ ☿
16 32	☽ Q ♅
17 54	☉ △ ♂
22 Monday	
00 35	☽ Q ☉
00 37	☉ ☍ ♄
03 33	☽ □ ♃
05 35	☽ ∠ ♂
09 54	☽ ⚹ ♀
16 10	☽ Q ☿
21 45	☽ ⊥ ♂
23 Tuesday	
00 02	♀ # ♃
00 49	☽ □ ♅
01 49	☽ △ ♆
03 53	♂ ⚹ ♄
07 32	♄ # ♅
08 20	☽ □ ♄
08 24	☽ ⊻ ♂
09 13	☽ Q ♀
11 00	☽ □ ☉
15 38	☽ ☍ ♇
22 07	☿ ∠ ♆
24 Wednesday	
03 58	☽ bQ ♆
04 55	☽ □ ☿
06 15	☉ ⊥ ♆
07 29	☽ ∥ ♂
09 03	☽ △ ♃
15 13	☽ Q ♄
19 14	☽ □ ♀
19 22	☽ ± ♆
22 16	♀ ⊥ ♃
25 Thursday	
04 39	☽ △ ♅
05 31	☽ ⚻ ♆
09 25	☉ # ♅
10 51	☽ bQ ♃
11 21	☽ ⚹ ♄
12 15	☽ ☌ ♂
13 46	☿ ⊻ ♃
14 33	☽ # ♇
16 55	☉ ∥ ♄
17 55	☿ # ♃
17 57	☽ △ ☉
18 37	☽ ⚻ ♇
26 Friday	
02 04	☽ ± ♃
03 45	☉ □ ♇
04 33	☽ ± ♇
05 42	☽ bQ ♅
12 02	☽ ∠ ♄
12 06	☽ ⚻ ♃
15 05	☽ △ ☿
16 43	☽ # ♆
19 19	☽ bQ ♇
20 26	☽ bQ ☉
20 31	☽ ± ♅
27 Saturday	
01 36	☽ △ ♀
02 39	☽ ⊥ ♄
06 19	☽ ⚻ ♅
07 04	☽ ☍ ♆
08 36	☽ ∥ ♀
11 46	☽ ∥ ☿
12 07	☽ ± ☉
12 20	☽ ⊻ ♄
14 03	☽ ⊻ ♂
18 44	☽ # ♃
19 04	☽ bQ ☿
19 40	☽ △ ♇
22 30	☽ ⚻ ☉
23 52	☽ ⊥ ♂
28 Sunday	
04 05	☽ bQ ♀
07 28	☽ ∥ ☉
10 59	☽ ∥ ♄
11 40	☽ ± ☿
12 03	☽ # ♅
13 30	☽ ☍ ♃
14 33	☽ ∠ ♂
19 53	☽ ± ♀
22 42	☽ ⚻ ☿
29 Monday	
01 20	☽ ∥ ♅
02 30	☽ # ♄
04 23	☿ ⊥ ♃
06 26	☽ ⚻ ♀
06 55	☽ ☍ ♅
07 28	☽ # ☉
07 34	☽ ⚻ ♆
12 24	♀ ⊻ ♅
12 29	☽ ☌ ♄
15 05	☽ ⚹ ♂
16 05	☿ ∥ ♀
17 18	☽ ± ♆
18 06	☽ ∥ ♃
20 06	☽ □ ♇
20 25	♀ ⚹ ♆
30 Tuesday	
02 25	☽ ☍ ☉
08 00	☽ bQ ♆
10 55	☽ Q ♂
11 11	☽ # ♀
12 36	☽ # ☿
14 57	☽ ⚻ ♃
21 00	☽ ∥ ♆
31 Wednesday	
01 02	☽ ± ♃
01 39	☽ Q ♇
06 27	☽ ☍ ☿
08 16	☽ ⚻ ♅
08 52	☽ △ ♆
12 13	☽ ☍ ♀
13 36	☽ ⊻ ♄
16 21	☽ bQ ♃
17 19	☽ □ ♂
18 25	☽ ± ♅
21 50	☽ ⚹ ♇
22 22	☿ ⊻ ♅
23 42	☽ ⊥ ♄

All ephemeris data is given at 12.00 UT and the Moon's longitude is additionally given for 24.00 UT

APRIL 2010

LONGITUDES

Date	Sidereal time h m s	Sun ☉ ° ' "	Moon ☽ ° ' "	Moon ☽ 24.00 ° ' "	Mercury ☿ ° '	Venus ♀ ° '	Mars ♂ ° '	Jupiter ♃ ° '	Saturn ♄ ° '	Uranus ♅ ° '	Neptune ♆ ° '	Pluto ♇ ° '
01	00 38 57	11 ♈ 39 08	13 ♏ 39 56	20 ♏ 32 06	28 ♈ 20	00 ♉ 57	02 ♌ 55	17 ♓ 22	00 ♎ 28	27 ♓ 26	27 ♒ 45	05 ♑ 25
02	00 42 54	12 38 19	27 ♏ 17 22	03 ♐ 55 44	29 ♈ 56	02 11	03 09	17 36	00 R 24	27 29	27 47	05 25
03	00 46 50	13 37 28	10 ♐ 27 22	16 ♐ 52 36	01 ♉ 27	03 25	03 23	17 49	00 19	27 33	27 49	05 25
04	00 50 47	14 36 35	23 ♐ 11 51	29 ♐ 25 38	02 53	04 38	03 38	18 03	00 15	27 36	27 51	05 25
05	00 54 44	15 35 41	05 ♑ 34 34	11 ♑ 39 17	04 14	05 52	03 53	18 17	00 10	27 39	27 52	05 25
06	00 58 40	16 34 44	17 ♑ 40 27	23 ♑ 38 47	05 29	07 06	04 09	18 31	00 06	27 43	27 54	05 25
07	01 02 37	17 33 46	29 ♑ 34 57	05 ♒ 29 40	06 39	08 20	04 25	18 44	00 ♎ 01	27 46	27 56	05 25
08	01 06 33	18 32 46	11 ♒ 23 36	17 ♒ 17 22	07 43	09 33	04 41	18 58	29 ♍ 57	27 49	27 57	05 R 25
09	01 10 30	19 31 45	23 ♒ 11 34	29 ♒ 06 47	08 41	10 47	04 58	19 11	29 53	27 53	27 59	05 25
10	01 14 26	20 30 41	05 ♓ 03 30	11 ♓ 02 11	09 32	12 01	05 16	19 25	29 48	27 56	28 00	05 25
11	01 18 23	21 29 36	17 ♓ 03 12	23 ♓ 06 53	10 18	13 14	05 33	19 38	29 44	27 59	28 02	05 25
12	01 22 19	22 28 28	29 ♓ 13 30	05 ♈ 23 13	10 57	14 28	05 51	19 51	29 40	28 02	28 04	05 25
13	01 26 16	23 27 19	11 ♈ 36 10	17 ♈ 52 26	11 30	15 42	06 10	20 05	29 36	28 05	28 05	05 25
14	01 30 13	24 26 08	24 ♈ 12 01	00 ♉ 34 51	11 56	16 55	06 29	20 18	29 32	28 09	28 07	05 24
15	01 34 09	25 24 55	07 ♉ 00 54	13 ♉ 30 03	12 16	18 09	06 48	20 31	29 28	28 12	28 08	05 24
16	01 38 06	26 23 40	20 ♉ 02 10	26 ♉ 37 10	12 29	19 22	07 08	20 44	29 24	28 15	28 09	05 24
17	01 42 02	27 22 23	03 ♊ 14 55	09 ♊ 55 20	12 37	20 35	07 28	20 57	29 20	28 18	28 11	05 24
18	01 45 59	28 21 04	16 ♊ 38 21	23 ♊ 23 55	12 R 38	21 49	07 48	21 10	29 16	28 21	28 12	05 23
19	01 49 55	29 ♈ 19 43	00 ♋ 12 03	07 ♋ 02 44	12 33	23 02	08 09	21 23	29 12	28 24	28 13	05 23
20	01 53 52	00 ♉ 18 19	13 ♋ 56 02	20 ♋ 51 56	12 22	24 15	08 30	21 36	29 08	28 27	28 15	05 22
21	01 57 48	01 16 54	27 ♋ 50 30	04 ♌ 51 43	12 06	25 29	08 52	21 49	29 05	28 30	28 16	05 22
22	02 01 45	02 15 26	11 ♌ 55 32	19 ♌ 01 50	11 45	26 42	09 13	22 02	29 01	28 33	28 17	05 22
23	02 05 42	03 13 55	26 ♌ 10 27	03 ♍ 21 05	11 20	27 55	09 36	22 14	28 58	28 36	28 19	05 21
24	02 09 38	04 12 23	10 ♍ 33 22	17 ♍ 46 48	10 50	29 ♉ 08	09 58	22 27	28 54	28 39	28 20	05 21
25	02 13 35	05 10 48	25 ♍ 00 47	02 ♎ 14 41	10 17	00 ♊ 21	10 21	22 39	28 51	28 42	28 21	05 20
26	02 17 31	06 09 11	09 ♎ 27 43	16 ♎ 39 08	09 42	01 34	10 44	22 52	28 48	28 45	28 22	05 19
27	02 21 28	07 07 33	23 ♎ 48 07	00 ♏ 53 56	09 04	02 47	11 07	23 04	28 45	28 48	28 23	05 19
28	02 25 24	08 05 52	07 ♏ 55 52	14 ♏ 53 18	08 25	04 00	11 30	23 17	28 41	28 51	28 24	05 18
29	02 29 21	09 04 10	21 ♏ 45 42	28 ♏ 32 43	07 45	05 13	11 54	23 29	28 38	28 53	28 25	05 18
30	02 33 17	10 ♉ 02 26	05 ♐ 14 06	11 ♐ 49 43	07 ♉ 06	06 ♊ 26	12 ♌ 18	23 ♓ 41	28 ♍ 35	28 ♓ 56	28 ♒ 26	05 ♑ 17

Date	Moon True ☊ ° '	Moon Mean ☊ ° '	Moon ☽ Latitude ° '
01	16 ♑ 51	16 ♑ 50	04 S 29
02	16 R 42	16 47	03 49
03	16 37	16 44	02 59
04	16 34	16 41	02 00
05	16 D 32	16 38	00 S 58
06	16 R 32	16 34	00 N 06
07	16 32	16 31	01 09
08	16 31	16 28	02 08
09	16 27	16 25	03 01
10	16 21	16 22	03 47
11	16 12	16 19	04 23
12	16 00	16 15	04 48
13	15 47	16 12	04 59
14	15 34	16 09	04 57
15	15 21	16 06	04 39
16	15 10	16 03	04 07
17	15 01	15 59	03 21
18	14 56	15 56	02 24
19	14 53	15 53	01 17
20	14 53	15 50	00 N 05
21	14 D 53	15 47	01 S 08
22	14 R 53	15 44	02 18
23	14 51	15 40	03 21
24	14 47	15 37	04 11
25	14 40	15 34	04 45
26	14 31	15 31	05 01
27	14 20	15 28	04 58
28	14 10	15 25	04 37
29	14 00	15 21	04 00
30	13 ♑ 53	15 ♑ 18	03 S 10

DECLINATIONS

Date	Sun ☉ ° '	Moon ☽ ° '	Mercury ☿ ° '	Venus ♀ ° '	Mars ♂ ° '	Jupiter ♃ ° '	Saturn ♄ ° '	Uranus ♅ ° '	Neptune ♆ ° '	Pluto ♇ ° '
01	04 N 36	20 S 13	12 N 18	11 N 20	22 N 19	05 S 53	02 N 10	01 S 41	12 S 40	18 S 13
02	05 00	23 16	13 02	11 48	22 14	05 48	02 12	01 39	12 39	18 12
03	05 23	24 58	13 44	12 16	22 09	05 43	02 14	01 38	12 38	18 12
04	05 45	25 16	14 24	12 43	22 03	05 38	02 16	01 37	12 38	18 12
05	06 08	24 17	15 01	13 10	21 58	05 32	02 17	01 36	12 37	18 12
06	06 31	22 10	15 35	13 37	21 52	05 27	02 19	01 34	12 37	18 12
07	06 54	19 07	16 06	14 03	21 47	05 22	02 21	01 33	12 36	18 12
08	07 16	15 19	16 34	14 29	21 41	05 17	02 22	01 32	12 36	18 12
09	07 38	10 56	17 00	14 55	21 35	05 12	02 24	01 30	12 35	18 12
10	08 01	06 08	17 22	15 20	21 29	05 06	02 26	01 29	12 35	18 12
11	08 23	01 S 05	17 41	15 45	21 23	05 01	02 27	01 28	12 34	18 12
12	08 45	04 N 05	17 57	16 09	21 17	04 56	02 29	01 27	12 34	18 12
13	09 07	09 11	18 09	16 33	21 11	04 51	02 31	01 25	12 33	18 12
14	09 28	13 59	18 19	16 57	21 04	04 46	02 32	01 24	12 33	18 12
15	09 50	18 15	18 25	17 20	20 58	04 41	02 34	01 23	12 32	18 12
16	10 11	21 43	18 28	17 43	20 51	04 36	02 35	01 22	12 32	18 12
17	10 32	24 06	18 27	18 05	20 44	04 31	02 37	01 20	12 31	18 12
18	10 53	25 09	18 24	18 27	20 37	04 26	02 38	01 19	12 31	18 12
19	11 14	24 43	18 17	18 49	20 30	04 21	02 39	01 18	12 30	18 12
20	11 35	22 48	18 07	19 10	20 23	04 16	02 41	01 17	12 30	18 11
21	11 55	19 29	17 54	19 30	20 16	04 11	02 42	01 16	12 29	18 11
22	12 15	15 00	17 38	19 50	20 09	04 06	02 43	01 14	12 29	18 11
23	12 35	09 39	17 20	20 09	20 01	04 01	02 45	01 13	12 29	18 11
24	12 55	03 N 44	16 59	20 28	19 54	03 56	02 46	01 12	12 28	18 11
25	13 15	02 S 23	16 36	20 47	19 46	03 52	02 47	01 11	12 28	18 11
26	13 34	08 22	16 11	21 04	19 38	03 47	02 48	01 10	12 27	18 11
27	13 53	13 51	15 44	21 21	19 30	03 42	02 49	01 09	12 27	18 11
28	14 12	18 31	15 17	21 38	19 22	03 37	02 51	01 08	12 27	18 11
29	14 31	22 04	14 48	21 54	19 14	03 33	02 52	01 06	12 26	18 11
30	14 N 50	24 S 17	14 N 19	22 N 09	19 N 06	03 S 28	02 N 53	01 S 05	12 S 26	18 S 11

ZODIAC SIGN ENTRIES

Date	h	m	Planets
02	13	06	☿ ♉
02	16	52	☽ ♐
05	01	07	☽ ♑
07	12	51	☽ ♒
07	18	51	♄ ♍
10	01	48	☽ ♓
12	13	31	☽ ♈
14	22	55	☽ ♉
17	06	08	☽ ♊
19	11	39	☽ ♋
20	04	30	☉ ♉
21	15	42	☽ ♌
23	18	24	☽ ♍
25	05	05	♀ ♊
25	20	16	☽ ♎
27	22	28	☽ ♏
30	02	36	☽ ♐

LATITUDES

Date	Mercury ☿ ° '	Venus ♀ ° '	Mars ♂ ° '	Jupiter ♃ ° '	Saturn ♄ ° '	Uranus ♅ ° '	Neptune ♆ ° '	Pluto ♇ ° '
01	01 N 31	00 S 30	02 N 53	00 S 59	02 N 34	00 S 43	00 S 26	05 N 07
04	02 03	00 22	02 48	00 59	02 34	00 43	00 26	05 08
07	02 31	00 15	02 42	00 59	02 34	00 43	00 26	05 08
10	02 51	00 S 07	02 37	01 00	02 34	00 43	00 26	05 08
13	03 01	00 N 01	02 32	01 00	02 34	00 43	00 26	05 08
16	03 01	00 09	02 26	01 00	02 33	00 43	00 26	05 08
19	02 49	00 17	02 21	01 01	02 33	00 43	00 26	05 08
22	02 24	00 26	02 17	01 01	02 33	00 43	00 26	05 09
25	01 47	00 34	02 12	01 02	02 32	00 43	00 27	05 09
28	01 01	00 42	02 08	01 02	02 32	00 43	00 27	05 09
31	00 N 10	00 N 50	02 N 03	01 S 03	02 N 31	00 S 44	00 S 27	05 N 09

DATA

Julian Date	2455288
Delta T	+67 seconds
Ayanamsa	24° 00' 17"
Synetic vernal point	05° ♓ 06' 42"
True obliquity of ecliptic	23° 26' 20"

MOON'S PHASES, APSIDES AND POSITIONS ☽

Date	h	m	Phase	Longitude ° '	Eclipse Indicator
06	09	37	☾	16 ♑ 29	
14	12	29	●	24 ♈ 27	
21	18	20	☽	01 ♌ 32	
28	12	18	○	08 ♏ 07	

Day	h	m		
09	02	50	Apogee	
24	21	06	Perigee	
04	05	26	Max dec	25° S 19'
11	17	01	0N	
18	17	14	Max dec	25° N 11'
25	02	41	0S	

LONGITUDES

Date	Chiron ⚷ ° '	Ceres ⚳ ° '	Pallas ⚴ ° '	Juno ⚵ ° '	Vesta ⚶ ° '	Black Moon Lilith ⚸ ° '
01	29 ♒ 04	02 ♑ 25	21 ♏ 44	20 ♉ 40	21 ♌ 43	20 ♒ 14
11	29 ♒ 36	03 ♑ 43	19 ♏ 53	26 ♉ 25	21 ♌ 40	21 ♒ 21
21	00 ♓ 03	04 ♑ 28	17 ♏ 22	02 ♊ 11	22 ♌ 28	22 ♒ 27
31	00 ♓ 25	04 ♑ 37	14 ♏ 26	07 ♊ 59	23 ♌ 55	23 ♒ 34

All ephemeris data is given at 12.00 UT and the Moon's longitude is additionally given for 24.00 UT

ASPECTARIAN

h m	Aspects
01 Thursday	
00 03	☽ ∥ ♇
03 16	♀ ⊼ ♄
03 21	☿ ⚹ ♆
08 15	☽ ⊼ ☉
09 52	☽ Ᵽ ♅
15 07	☽ ∠ ♄
17 06	♃ Q ♇
18 33	☽ △ ♃
19 29	☽ ± ☉
23 33	☿ # ♆
23 47	☽ ∠ ♇
02 Friday	
02 52	☽ # ♂
12 22	☽ △ ♅
12 41	☽ Ᵽ ☉
12 54	☽ □ ♆
15 44	☉ ∠ ♆
15 49	☽ ⊥ ♇
17 22	☽ ⊼ ☿
17 33	☽ ⚹ ♄
18 53	☿ ⊼ ♄
21 43	☽ ⊼ ♀
21 58	♀ ∠ ♃
22 46	☽ △ ♂
03 Saturday	
02 43	☽ ⊻ ♇
05 43	☽ ± ☿
09 52	☽ ± ♀
11 25	♀ □ ♂
14 48	♀ ⊥ ♅
15 26	☽ Q ♄
18 23	☽ △ ☉
22 02	☽ Q ♆
04 Sunday	
00 38	☽ Ᵽ ☿
02 02	☽ □ ♃
03 08	☽ Ᵽ ♂
04 30	☽ Ᵽ ♀
05 17	☉ # ♃
07 26	♀ # ♆
15 38	☿ ∠ ♃
20 30	☽ □ ♅
20 42	☽ ± ♂
20 57	☽ ⚹ ♆
05 Monday	
01 15	☿ ⊥ ♅
01 30	☽ □ ♄
03 12	♀ △ ♇
04 17	☿ □ ♂
08 37	☽ ⊼ ♂
09 02	☽ △ ☿
11 42	☽ ☌ ♇
12 39	☽ △ ♀
13 25	☽ Q ♃
17 29	♀ ± ♄
06 Tuesday	
02 27	☽ ∠ ♆
08 04	☽ Q ♅
09 37	☽ □ ☉
10 46	☿ △ ♇
13 42	☽ ⚹ ♃
14 47	☽ # ♂
20 30	☽ ⊥ ♆
23 39	☿ ± ♄
07 Wednesday	
02 36	♇ St R
08 18	☽ ⚹ ♅
08 39	☽ ⊻ ♆
12 53	☽ △ ♄
18 12	☽ ∥ ♇
20 35	☽ ∠ ♃
22 02	☽ ☍ ♂
23 51	☽ ⊻ ♇
08 Thursday	
01 15	☽ Q ☉
03 47	☽ □ ☿
05 17	☽ # ☿
07 50	☽ □ ♀
12 03	☽ ⊥ ♇
14 55	☽ ∠ ♅
15 15	☽ ⊥ ♃
16 21	☽ # ♀
19 11	☽ Ᵽ ♄
19 56	♀ Q ♆
09 Friday	
01 09	☉ ⊻ ♃
03 16	☽ ∥ ♆
03 42	☽ ⊻ ♃
03 53	☽ ⚹ ☉
06 22	☽ ∠ ♇
09 19	☽ ⊥ ♅
13 23	☽ ± ♄
19 39	☽ Q ☿
21 32	☽ ⊻ ♅
21 44	☽ ☌ ♆
10 Saturday	
00 39	☽ Q ♀
01 28	☽ ⊼ ♄
03 29	☽ # ☉
12 25	☽ ⊼ ♂
12 43	☽ ⚹ ♇
13 00	☽ ∠ ☉
17 03	☽ ∥ ♃
21 38	☽ ⚹ ☿
11 Sunday	
00 45	☽ ± ♂
00 51	♂ ⊼ ♇
02 58	☿ Q ♆
03 33	☽ ⚹ ♀
05 34	☽ # ♄
06 44	♀ ∠ ♅
08 37	☽ ⊥ ☉
10 11	☽ ∥ ♅
12 43	☽ Q ♇
17 13	☽ ☌ ♃
19 07	☽ Ᵽ ♂
21 34	☽ ⊻ ☉
23 46	☽ # ♅
12 Monday	
04 30	☽ ∥ ♄
05 14	☽ ∠ ☿
09 40	☽ ☌ ♅
09 43	☽ ⊻ ♆
12 32	☽ ∠ ♀
12 51	☽ ☍ ♄
15 40	♀ Ᵽ ♄
15 53	☽ # ♃
21 27	☽ ⊥ ♆
23 42	☽ ⊥ ☿
13 Tuesday	
00 03	☽ □ ♇
01 14	☽ △ ♂
07 43	♅ ⊻ ♆
07 55	☽ ⊥ ♀
11 39	☽ ∥ ☉
11 47	☽ ⊻ ☿
14 51	☽ ∠ ♆
17 37	☿ # ♇
20 41	☽ ⊻ ♀
14 Wednesday	
04 29	☽ ⊻ ♃
04 37	☽ # ♆
12 29	☽ ☌ ☉
16 02	☽ ⊥ ♃
19 23	☽ ⚹ ♆
19 27	☽ ⊻ ♅
21 58	☽ ⊼ ♄
15 Thursday	
05 59	☽ ∥ ♀
06 44	☽ ⊥ ♅
09 00	☽ △ ♇
09 08	☽ ± ♄
09 10	☽ ∠ ♃
11 36	☽ □ ♂
11 40	☽ # ♇
13 01	☽ ∥ ☿
17 47	☽ Q ♆
21 55	☽ ☌ ☿
23 29	☽ ∠ ♅
16 Friday	
01 42	☽ Ᵽ ♄
05 33	☽ ∥ ♂
10 39	☽ ☌ ♀
12 40	☽ Ᵽ ♇
13 18	☽ ⚹ ♃
21 32	☽ Q ♂
22 49	☉ ⊥ ♃
17 Saturday	
00 31	☽ ⊻ ☉
02 49	☽ □ ♆
03 01	☽ ⚹ ♅
04 57	☽ △ ♄
05 02	☽ ± ♇
08 08	♀ Ᵽ ♇
11 28	☽ Q ♃
12 15	☽ ⊥ ☉
15 52	☽ ⊼ ♇
18 40	♀ # ♇
19 48	☽ ⚹ ♂
20 43	♀ ⚹ ♃
18 Sunday	
00 43	☽ Q ♅
04 07	☿ St R
04 51	☽ ⊻ ☿
05 41	☽ ∠ ☉
08 16	☉ ⚹ ♆
08 35	☿ ∥ ♀
11 58	☉ ⊻ ♅
15 32	☽ ⊥ ☿
20 11	☽ □ ♃
22 06	☽ ⊻ ♀
23 14	☽ ∠ ♂
19 Monday	
07 22	☽ ∠ ☿
08 31	☽ △ ♆
08 49	☽ □ ♅
09 06	☉ ⊼ ♄
09 45	☽ ⊥ ♀
10 15	☽ □ ♄
10 21	☽ ⚹ ☉
15 31	☽ ⊥ ♂
21 05	☽ ☍ ♇
20 Tuesday	
01 39	☿ # ♇
02 18	☽ ⊻ ♂
03 04	☽ ∠ ♀
08 57	☽ Q ☉
09 19	☽ ⚹ ☿
10 48	☽ Ᵽ ♆
17 32	☽ Q ♄
21 Wednesday	
01 29	☽ △ ♃
02 24	☽ ± ♆
05 43	☽ Q ☿
06 53	☽ ∥ ♂
07 33	☽ ⚹ ♀
11 52	☽ ∥ ♀
12 44	☽ ⊼ ♆
13 08	☽ △ ♅
14 07	☽ ⚹ ♄
18 20	☽ □ ☉
19 30	☽ # ♇
21 40	☽ ∥ ☿
22 Thursday	
00 51	☽ ⊼ ♇
03 34	☽ Ᵽ ♃
06 00	☽ Q ♀
07 18	☽ ☌ ♂
11 02	☽ ± ♇
11 43	☽ □ ☿
14 46	☽ Ᵽ ♅
15 32	☽ ∠ ♄
19 02	☽ ± ♃
23 42	☽ # ♆
23 57	☽ ∥ ☉
23 Friday	
02 14	☽ Ᵽ ♇
02 54	♀ Q ♂
03 54	☉ # ♆
04 38	♀ ∥ ♂
05 18	☽ ⊼ ♃
05 59	☽ ± ♅
06 38	☽ ⊥ ♄
15 11	☽ □ ♀
15 35	☽ ☍ ♆
16 05	☽ ⊼ ♅
16 39	☽ ⊻ ♄
19 54	♀ □ ♆
24 Saturday	
00 39	☽ △ ☉
02 06	♀ ⚹ ♅
03 19	☽ △ ♇
07 44	♀ △ ♄
10 59	☽ ⊻ ♂
11 12	☽ # ♃
12 27	☽ △ ☿
15 50	☽ ∥ ♄
16 05	♀ ± ♇
21 14	☽ ⊥ ♂
22 01	☽ # ♅
23 32	☉ ⊥ ♅
25 Sunday	
03 24	☽ Ᵽ ☉
04 19	☉ ± ♄
07 19	☽ ∥ ♅
08 02	☽ ☍ ♃
09 38	☿ Q ♆
10 41	☿ □ ♂
12 27	☽ Ᵽ ☿
12 34	☽ ∠ ♂
13 36	☽ # ♄
15 44	☉ △ ♇
17 32	☽ ⊼ ♆
17 45	☽ ∥ ♃
18 08	☽ ☍ ♅
18 21	☽ ☌ ♄
19 25	☽ ± ☉
21 40	☽ △ ♀
26 Monday	
02 48	☽ ± ☿
03 31	☽ ± ♆
05 07	☽ □ ♇
06 06	☽ ⊼ ☉
12 23	☽ ⊼ ☿
14 10	☽ ⚹ ♂
18 31	☽ Ᵽ ♆
23 23	♄ ☍ ♅
27 Tuesday	
00 57	☽ Ᵽ ♀
05 35	☽ ∥ ♆
10 45	☽ ⊼ ♃
10 48	☽ Q ♂
11 11	☽ Q ♇
12 11	☽ # ☉
17 30	☽ ± ♀
19 45	☽ △ ♆
20 19	☽ ⊻ ♄
20 19	☽ # ☿
20 28	☽ ⊼ ♅
21 02	☽ ± ♃
28 Wednesday	
00 02	♂ ± ♇
04 38	☽ ⊼ ♀
06 29	☽ ⊥ ♄
06 42	☽ ± ♅
07 30	☽ ⚹ ♇
10 06	☽ ∥ ♇
12 18	☽ ☍ ☉
12 36	☽ Ᵽ ♃
12 48	☽ ☍ ☿
15 58	☿ ∠ ♃
16 44	☉ ☌ ☿
16 55	☽ # ♂
17 35	☉ ∠ ♃
18 20	☽ □ ♂
21 53	☽ ∠ ♄
22 13	☽ Ᵽ ♅
29 Thursday	
09 25	☽ ∠ ♇
10 32	☽ # ♀
13 32	♀ ⊼ ♇
15 04	☽ △ ♃
18 20	♀ Q ♃
20 32	☉ ∥ ☿
23 48	☽ □ ♆
30 Friday	
00 07	☽ ⚹ ♄
00 39	☽ △ ♅
01 19	☽ ⊥ ♇
12 05	☽ ⊻ ♇
14 22	☽ ☍ ♀
15 13	☽ ⊼ ☿
20 39	☿ ⊻ ♀
21 25	☽ ⊼ ☉
21 42	☽ Q ♄
22 00	☉ Q ♆

MAY 2010

LONGITUDES

Date	Sidereal time h m s	Sun ☉ ° ′ ″	Moon ☽ ° ′ ″	Moon ☽ 24.00 ° ′ ″	Mercury ☿ ° ′	Venus ♀ ° ′	Mars ♂ ° ′	Jupiter ♃ ° ′	Saturn ♄ ° ′	Uranus ♅ ° ′	Neptune ♆ ° ′	Pluto ♇ ° ′
01	02 37 14	11 ♉ 00 41	18 ♐ 19 37	24 ♐ 43 57	06 ♉ 27	07 ♊ 38	12 ♌ 42	23 ♓ 53	28 ♍ 33	28 ♓ 59	28 ♒ 27	05 ♑ 16
02	02 41 11	11 58 53	01 ♑ 02 57	07 ♑ 17 02	05 R 50	08 51	13 07	24 05	28 R 30	29 02	28 28	05 R 15
03	02 45 07	12 57 05	13 ♑ 26 37	19 ♑ 32 13	05 15	10 04	13 32	24 17	28 27	29 04	28 29	05 15
04	02 49 04	13 55 15	25 ♑ 34 26	01 ♒ 33 52	04 42	11 17	13 57	24 29	28 25	29 07	28 30	05 14
05	02 53 00	14 53 23	07 ♒ 31 10	13 ♒ 26 59	04 13	12 29	14 22	24 40	28 22	29 10	28 31	05 13
06	02 56 57	15 51 30	19 ♒ 22 00	25 ♒ 16 52	03 47	13 42	14 48	24 52	28 20	29 12	28 32	05 12
07	03 00 53	16 49 35	01 ♓ 12 14	07 ♓ 08 43	03 25	14 54	15 13	25 04	28 17	29 15	28 33	05 11
08	03 04 50	17 47 39	13 ♓ 06 56	19 ♓ 07 24	03 07	16 07	15 39	25 15	28 15	29 17	28 33	05 10
09	03 08 46	18 45 42	25 ♓ 10 39	01 ♈ 17 06	02 53	17 19	16 06	25 27	28 13	29 20	28 34	05 09
10	03 12 43	19 43 43	07 ♈ 27 09	13 ♈ 41 05	02 44	18 31	16 32	25 38	28 11	29 22	28 35	05 09
11	03 16 40	20 41 42	19 ♈ 59 08	26 ♈ 21 27	02 40	19 44	16 59	25 49	28 09	29 25	28 35	05 08
12	03 20 36	21 39 41	02 ♉ 48 05	09 ♉ 19 01	02 D 40	20 56	17 26	26 00	28 07	29 27	28 36	05 07
13	03 24 33	22 37 38	15 ♉ 54 09	22 ♉ 33 18	02 45	22 08	17 53	26 11	28 05	29 30	28 37	05 06
14	03 28 29	23 35 33	29 ♉ 16 16	06 ♊ 02 44	02 55	23 21	18 20	26 22	28 03	29 32	28 37	05 05
15	03 32 26	24 33 27	12 ♊ 52 25	19 ♊ 44 58	03 09	24 33	18 48	26 33	28 02	29 34	28 38	05 03
16	03 36 22	25 31 20	26 ♊ 40 02	03 ♋ 37 17	03 28	25 45	19 15	26 44	28 00	29 37	28 38	05 02
17	03 40 19	26 29 11	10 ♋ 36 25	17 ♋ 37 07	03 51	26 57	19 43	26 54	27 59	29 39	28 39	05 01
18	03 44 15	27 27 00	24 ♋ 39 08	01 ♌ 42 12	04 18	28 09	20 11	27 05	27 58	29 41	28 39	05 00
19	03 48 12	28 24 48	08 ♌ 46 07	15 ♌ 50 41	04 49	29 ♊ 21	20 40	27 15	27 56	29 43	28 40	04 59
20	03 52 09	29 ♉ 22 33	22 ♌ 55 42	00 ♍ 00 59	05 25	00 ♋ 33	21 08	27 25	27 55	29 45	28 40	04 58
21	03 56 05	00 ♊ 20 17	07 ♍ 06 19	14 ♍ 11 28	06 04	01 45	21 37	27 35	27 54	29 47	28 40	04 57
22	04 00 02	01 18 00	21 ♍ 16 12	28 ♍ 20 13	06 47	02 56	22 05	27 46	27 53	29 49	28 41	04 56
23	04 03 58	02 15 40	05 ♎ 23 13	12 ♎ 24 49	07 34	04 08	22 34	27 55	27 53	29 51	28 41	04 54
24	04 07 55	03 13 20	19 ♎ 24 39	26 ♎ 22 20	08 24	05 20	23 04	28 05	27 52	29 53	28 41	04 53
25	04 11 51	04 10 57	03 ♏ 17 26	10 ♏ 09 35	09 18	06 31	23 33	28 15	27 51	29 55	28 42	04 52
26	04 15 48	05 08 34	16 ♏ 58 24	23 ♏ 43 33	10 15	07 43	24 02	28 25	27 51	29 57	28 42	04 51
27	04 19 44	06 06 09	00 ♐ 24 44	07 ♐ 01 45	11 15	08 54	24 32	28 34	27 50	29 ♓ 59	28 42	04 49
28	04 23 41	07 03 43	13 ♐ 34 25	20 ♐ 02 40	12 18	10 06	25 02	28 43	27 50	00 ♈ 01	28 42	04 48
29	04 27 38	08 01 16	26 ♐ 26 31	02 ♑ 46 01	13 25	11 17	25 32	28 53	27 50	00 03	28 42	04 47
30	04 31 34	08 58 48	09 ♑ 01 22	15 ♑ 12 46	14 35	12 28	26 02	29 02	27 50	00 04	28 42	04 45
31	04 35 31	09 ♊ 56 19	21 ♑ 20 33	27 ♑ 25 05	15 ♉ 47	13 ♋ 39	26 ♌ 32	29 ♓ 11	27 ♍ 50	00 ♈ 06	28 ♒ 42	04 ♑ 44

Date	Moon True ☊ ° ′	Moon Mean ☊ ° ′	Moon ☽ Latitude ° ′
01	13 ♑ 48	15 ♑ 15	02 S 11
02	13 R 45	15 12	01 07
03	13 44	15 09	00 S 02
04	13 D 44	15 05	01 N 03
05	13 45	15 02	02 04
06	13 R 45	14 59	02 59
07	13 44	14 56	03 46
08	13 41	14 53	04 24
09	13 35	14 50	04 51
10	13 28	14 46	05 04
11	13 19	14 43	05 04
12	13 10	14 40	04 48
13	13 01	14 37	04 18
14	12 54	14 34	03 32
15	12 48	14 31	02 34
16	12 45	14 27	01 26
17	12 44	14 24	00 N 11
18	12 D 44	14 21	01 S 04
19	12 46	14 18	02 16
20	12 47	14 15	03 20
21	12 R 47	14 11	04 12
22	12 45	14 08	04 48
23	12 42	14 05	05 07
24	12 37	14 02	05 08
25	12 31	13 59	04 50
26	12 25	13 56	04 16
27	12 19	13 52	03 28
28	12 15	13 49	02 30
29	12 12	13 46	01 25
30	12 11	13 43	00 17
31	12 ♑ 10	13 ♑ 40	00 N 50

DECLINATIONS

Date	Sun ☉ ° ′	Moon ☽ ° ′	Mercury ☿ ° ′	Venus ♀ ° ′	Mars ♂ ° ′	Jupiter ♃ ° ′	Saturn ♄ ° ′	Uranus ♅ ° ′	Neptune ♆ ° ′	Pluto ♇ ° ′
01	15 N 08	25 S 06	13 N 50	22 N 24	18 N 58	03 S 23	02 N 54	01 S 04	12 S 26	18 S 11
02	15 26	24 33	13 21	22 38	18 49	03 19	02 55	01 03	12 25	18 11
03	15 44	22 47	12 53	22 52	18 41	03 14	02 56	01 02	12 25	18 11
04	16 01	20 00	12 27	23 05	18 32	03 10	02 56	01 01	12 25	18 11
05	16 18	16 24	12 01	23 17	18 24	03 05	02 57	01 00	12 25	18 11
06	16 35	12 11	11 37	23 29	18 15	03 01	02 58	00 59	12 24	18 11
07	16 52	07 31	11 16	23 40	18 06	02 56	02 59	00 58	12 24	18 11
08	17 08	02 S 34	10 56	23 50	17 57	02 52	03 00	00 57	12 24	18 11
09	17 24	02 N 32	10 38	24 00	17 48	02 48	03 00	00 56	12 24	18 11
10	17 40	07 37	10 23	24 09	17 38	02 43	03 01	00 55	12 23	18 11
11	17 56	12 30	10 10	24 17	17 29	02 39	03 02	00 54	12 23	18 11
12	18 11	16 57	10 00	24 25	17 20	02 35	03 02	00 53	12 23	18 11
13	18 26	20 42	09 52	24 32	17 10	02 31	03 03	00 52	12 23	18 11
14	18 40	23 27	09 47	24 38	17 00	02 26	03 03	00 51	12 23	18 11
15	18 55	24 53	09 44	24 43	16 50	02 22	03 04	00 50	12 22	18 11
16	19 08	24 49	09 44	24 48	16 41	02 18	03 04	00 50	12 22	18 11
17	19 22	23 12	09 46	24 52	16 31	02 14	03 04	00 49	12 22	18 11
18	19 35	20 09	09 50	24 56	16 21	02 10	03 05	00 48	12 22	18 11
19	19 48	15 53	09 56	24 58	16 10	02 06	03 05	00 47	12 22	18 12
20	20 01	10 43	10 05	25 00	16 00	02 02	03 05	00 46	12 22	18 12
21	20 13	05 N 00	10 15	25 02	15 50	01 59	03 06	00 45	12 22	18 12
22	20 25	00 S 58	10 28	25 02	15 39	01 55	03 06	00 45	12 22	18 12
23	20 37	06 50	10 42	25 02	15 29	01 51	03 06	00 44	12 21	18 12
24	20 48	12 20	10 58	25 01	15 18	01 47	03 06	00 43	12 21	18 12
25	20 59	17 09	11 16	24 59	15 07	01 44	03 06	00 42	12 21	18 12
26	21 09	20 59	11 35	24 57	14 56	01 40	03 06	00 42	12 21	18 12
27	21 20	23 37	11 55	24 54	14 45	01 37	03 06	00 41	12 21	18 12
28	21 29	24 54	12 17	24 50	14 34	01 33	03 06	00 40	12 21	18 12
29	21 39	24 48	12 41	24 46	14 23	01 30	03 06	00 40	12 21	18 12
30	21 48	23 25	13 05	24 41	14 12	01 26	03 06	00 39	12 21	18 12
31	21 N 56	20 S 55	13 N 31	24 N 35	14 N 00	01 S 23	03 N 05	00 S 38	12 S 21	18 S 12

ZODIAC SIGN ENTRIES

Date	h	m	Planets
02	10	00	☽ ♑
04	20	52	☽ ♒
07	09	34	☽ ♓
09	21	29	☽ ♈
12	06	48	☽ ♉
14	13	18	☽ ♊
16	17	45	☽ ♋
18	21	06	☽ ♌
20	01	05	♀ ♋
20	23	58	☽ ♍
21	03	34	☉ ♊
23	02	50	☽ ♎
25	06	17	☽ ♏
27	11	15	☽ ♐
28	01	44	♅ ♈
29	18	44	☽ ♑

LATITUDES

Date	Mercury ☿ ° ′	Venus ♀ ° ′	Mars ♂ ° ′	Jupiter ♃ ° ′	Saturn ♄ ° ′	Uranus ♅ ° ′	Neptune ♆ ° ′	Pluto ♇ ° ′
01	00 N 10	00 N 50	02 N 03	01 S 03	02 N 31	00 S 44	00 S 27	05 N 09
04	00 S 41	00 58	01 59	01 03	02 31	00 44	00 27	05 09
07	01 29	01 05	01 55	01 04	02 30	00 44	00 27	05 09
10	02 10	01 12	01 51	01 04	02 30	00 44	00 27	05 09
13	02 43	01 19	01 47	01 05	02 29	00 44	00 27	05 09
16	03 07	01 26	01 43	01 06	02 29	00 44	00 27	05 09
19	03 23	01 32	01 39	01 06	02 28	00 44	00 27	05 09
22	03 31	01 38	01 36	01 07	02 28	00 44	00 27	05 09
25	03 31	01 43	01 32	01 08	02 27	00 44	00 27	05 09
28	03 24	01 47	01 28	01 08	02 26	00 44	00 27	05 09
31	03 S 11	01 N 51	01 N 25	01 S 09	02 N 26	00 S 44	00 S 27	05 N 09

DATA

Julian Date	2455318
Delta T	+67 seconds
Ayanamsa	24° 00′ 20″
Synetic vernal point	05° ♓ 06′ 39″
True obliquity of ecliptic	23° 26′ 19″

MOON'S PHASES, APSIDES AND POSITIONS ☽

Date	h	m	Phase	Longitude ° ′	Eclipse Indicator
06	04	15	☾	15 ♒ 33	
14	01	04	●	23 ♉ 09	
20	23	43	☽	29 ♌ 51	
27	23	07	○	06 ♐ 33	

Day	h	m		
06	21	54	Apogee	
20	08	43	Perigee	
01	14	04	Max dec	25° S 07′
09	00	08	0N	
15	23	05	Max dec	25° N 03′
22	08	09	0S	
28	22	13	Max dec	25° S 02′

LONGITUDES

Date	Chiron ⚷ ° ′	Ceres ⚳ ° ′	Pallas ⚴ ° ′	Juno ⚵ ° ′	Vesta ⚶ ° ′	Black Moon Lilith ⚸ ° ′
01	00 ♓ 25	04 ♑ 37	14 ♏ 26	07 ♊ 59	23 ♌ 55	23 ♒ 34
11	00 ♓ 42	04 ♑ 09	11 ♏ 24	13 ♊ 45	25 ♌ 58	24 ♒ 41
21	00 ♓ 54	03 ♑ 03	08 ♏ 37	19 ♊ 31	28 ♌ 31	25 ♒ 48
31	00 ♓ 59	01 ♑ 26	06 ♏ 24	25 ♊ 14	01 ♍ 31	26 ♒ 55

All ephemeris data is given at 12.00 UT and the Moon's longitude is additionally given for 24.00 UT

ASPECTARIAN

h m Aspects

01 Saturday
01 17 ☽ △ ♂
01 39 ☽ ± ☿
08 31 ☽ Q ♆
09 21 ☽ ± ☉
17 34 ☽ ⚼ ☿
22 34 ☽ □ ♃

02 Sunday
03 37 ☽ ⚼ ☉
06 13 ☽ ⚼ ♂
07 05 ☽ ✶ ♆
07 09 ☽ □ ♄
08 08 ☽ □ ♅
20 04 ☽ ☌ ♇
20 46 ☽ △ ☿
22 22 ♄ ⊼ ♆

03 Monday
00 05 ☽ ± ♂
04 41 ☽ ⊼ ♀
07 57 ♂ ∠ ♄
09 41 ☽ Q ♃
10 57 ☽ △ ☉
11 13 ☽ # ♀
12 01 ☿ △ ♇
12 05 ☽ ∠ ♆
12 10 ☽ ⊼ ♂
17 43 ☽ ± ♀
18 47 ☿ ⊥ ♅
19 10 ☽ Q ♅
23 51 ☉ ⚼ ♄

04 Tuesday
04 04 ☿ ⊥ ♀
05 52 ☽ ⊥ ♆
08 45 ♀ Q ♅
09 47 ☽ ✶ ♃
13 09 ☉ □ ♂
13 33 ☽ ⚼ ♀
13 39 ☿ # ♆
17 07 ☉ ∠ ♅
17 39 ☽ △ ♄
17 52 ☽ ⊻ ♆
19 07 ☽ ✶ ♅
22 43 ☽ # ♂
22 53 ♂ ⚼ ♅

05 Wednesday
00 38 ☽ ∥ ♇
03 19 ☿ ± ♄
05 35 ☽ □ ☿
07 22 ☽ ⊻ ♇
12 31 ☽ # ☉
16 26 ☽ ∠ ♃
19 28 ☽ ⊥ ♇
23 11 ☽ △ ♀
23 47 ☽ ⚼ ♄

06 Thursday
01 30 ☽ ∠ ♅
02 23 ☽ ☍ ♂
04 15 ☽ □ ☉
10 47 ☽ ∥ ♆
10 58 ☽ ⊥ ♃
13 42 ☽ ∠ ♇
15 14 ☽ # ☿
16 44 ☽ Q ☿
17 59 ☽ ± ♄
19 49 ☽ ⊥ ♅
21 28 ♂ # ♇
23 21 ☽ ⊻ ♃
23 59 ♃ # ♄

07 Friday
06 07 ☽ ⊼ ♄
06 36 ☽ ☌ ♆
08 02 ☽ ⊻ ♅
16 21 ☽ ✶ ☿
19 58 ☽ Q ☉
20 03 ☽ ✶ ♇
21 55 ♀ ✶ ♂

08 Saturday
09 59 ☽ # ♄
10 34 ☽ ∥ ♃
17 17 ☽ ⊼ ♂
18 40 ☽ □ ♀
19 41 ☽ ∥ ♅
20 06 ☽ Q ♇
21 47 ☽ ∠ ☿
22 10 ☽ ✶ ☉

09 Sunday
04 33 ☽ # ♅
05 40 ☽ ± ♂
12 32 ☽ ☌ ♃
13 13 ☽ # ♃
14 15 ☽ ∥ ♄
15 19 ☽ ⊥ ☿
17 58 ☽ ☍ ♄
18 41 ☽ ⊻ ♆
20 12 ☽ ☌ ♅
21 56 ☿ ∠ ♀

10 Monday
00 04 ☽ ⚼ ♂
02 56 ☽ ⊻ ☿
06 16 ☽ ∠ ☉
06 25 ☽ ⊥ ♆
07 32 ☽ □ ♇
10 01 ☽ Q ♀
10 22 ☉ ∥ ♂
22 07 ☉ ⚼ ♇
23 49 ☽ ∠ ♆

11 Tuesday
00 51 ☽ ∥ ☿
01 06 ☽ ⊥ ☉
06 05 ☽ △ ♂
11 25 ☽ # ♆
11 28 ☽ ✶ ♀
13 27 ☽ ⊻ ☉
22 27 ☿ St D
23 09 ☽ ⊻ ♃

12 Wednesday
03 18 ☽ ⊼ ♄
04 11 ☽ ✶ ♆
05 46 ☽ ⊻ ♅
10 30 ☽ ⊥ ♃
11 46 ☽ ☌ ☿
12 53 ☉ # ♇
14 05 ☽ ∥ ♂
14 25 ☽ ± ♄
16 16 ☽ △ ♇
16 55 ☽ ⊥ ♅
18 23 ☽ ∠ ♀
19 20 ☽ # ♇
19 46 ☽ ∥ ☉

13 Thursday
02 22 ☽ Q ♆
03 18 ☽ ∠ ♃
06 53 ☽ ⚼ ♄
09 26 ☽ ∠ ♅
12 28 ☽ ⊥ ♀
15 42 ☽ □ ♂
19 34 ☽ ⚼ ♇

14 Friday
00 22 ☽ ⊻ ♀
01 04 ☽ ☌ ☉
06 45 ☽ ✶ ♃
09 50 ☽ △ ♄
10 51 ☽ □ ♆
11 39 ☽ ± ♇
12 28 ☽ ✶ ♅
18 34 ☽ ⊻ ☿
22 16 ☽ ⊼ ♇

15 Saturday
00 57 ☽ Q ♂
04 18 ☽ Q ♃
05 20 ☽ ⊥ ☿
07 00 ☽ ∥ ♀
09 43 ☽ Q ♅
13 13 ☉ ⊻ ♀
21 25 ☽ ∠ ☿
22 42 ☽ ✶ ♂

16 Sunday
09 52 ☽ ⊻ ☉
10 15 ☽ ☌ ♀
12 06 ☽ □ ♃
12 33 ☽ ∥ ♀
14 19 ☽ □ ♄
15 25 ☽ △ ♆
17 06 ☽ □ ♅
21 00 ☽ ⊥ ☉

17 Monday
00 02 ☽ ✶ ☿
01 32 ☽ ∠ ♂
02 25 ☽ ☍ ♇
10 57 ♀ □ ♃
13 37 ☽ ∠ ☉
17 13 ☽ ⚼ ♆
17 31 ☽ ⊥ ♂
19 24 ☿ ± ♄
21 11 ☽ Q ♄
21 15 ☽ Q ☿

18 Tuesday
00 41 ☉ ✶ ♃
02 57 ♂ ⚼ ♇
04 07 ☽ ⊻ ♂
08 18 ♀ □ ♄
08 36 ☽ ± ♆
15 18 ☽ ∥ ☉
16 11 ☽ △ ♃
17 07 ☽ ✶ ☉
17 37 ☽ ✶ ♄
18 30 ☽ ⊻ ♀
18 49 ☽ ⊼ ♆
20 35 ☽ △ ♅
22 12 ♀ △ ♆
23 43 ☽ # ♇

19 Wednesday
00 27 ☉ △ ♄
05 01 ☽ □ ☿
05 35 ☽ ⊼ ♇
05 39 ☽ ⊥ ♀
10 27 ☽ ∥ ♂
15 00 ☽ Q ☉
15 45 ☽ ± ♇
17 59 ☽ ⚼ ♃
18 14 ☉ □ ♆
18 47 ☿ △ ♇
19 04 ☽ ∠ ♄
19 41 ♀ □ ♅
22 07 ☽ ⚼ ♅
22 20 ☽ ∠ ♀

20 Thursday
01 58 ☉ ± ♇
04 42 ☽ # ♆
06 59 ☽ ⚼ ♇
08 51 ☽ ☌ ♂
09 25 ☽ ± ♃
10 18 ☽ ⊥ ♄
13 24 ☽ ± ♅
14 42 ☽ ∥ ☿
19 42 ☽ ⊼ ♃
20 27 ☽ ⊻ ♄
21 43 ☽ ☍ ♆
21 48 ☉ ✶ ♅
23 35 ☽ ⊼ ♅
23 43 ☽ □ ☉

21 Friday
01 41 ☿ ⊥ ♅
02 05 ☽ ✶ ♀
08 21 ☽ △ ♇
10 09 ☽ △ ☿
10 34 ♂ ± ♃
19 43 ☽ ∥ ♄

22 Saturday
00 17 ☽ Q ♀
00 20 ☽ # ♃
02 19 ♂ ⊥ ♄
05 09 ☽ # ♅
11 08 ☽ ∥ ♅
12 55 ☽ ⚼ ☿
13 26 ☽ ⊻ ♂
15 48 ☽ ∥ ♃
20 37 ☽ # ♄
23 09 ☽ ☍ ♃
23 14 ☽ ☌ ♄
23 59 ☽ ⊥ ♂

23 Sunday
00 35 ☽ ⊼ ♆
02 34 ☽ ☍ ♅
05 05 ☽ ± ☿
05 36 ♃ ☍ ♄
06 17 ☽ △ ☉
09 40 ☽ □ ♀
10 48 ☽ ± ♆
11 11 ☽ □ ♇
15 52 ☽ ∠ ♂
15 56 ☽ ⊼ ☿

24 Monday
02 11 ☽ ⚼ ♆
03 15 ♀ ☍ ♇
05 26 ☽ # ☿
09 48 ☽ ⚼ ☉
12 05 ☽ ∥ ♆
17 58 ☽ Q ♇
18 30 ☽ ✶ ♂

25 Tuesday
01 44 ☽ # ♂
02 28 ☽ ± ☉
02 34 ☽ ⊻ ♄
03 09 ☽ ⊼ ♃
04 01 ☽ △ ♆
06 08 ☽ ⊼ ♅
12 59 ☽ ⊥ ♄
13 40 ☽ ⊼ ☉
13 41 ☽ ± ♃
14 44 ☽ ✶ ♇
16 05 ☽ Q ♂
16 36 ☽ ± ♅
17 56 ☽ ∥ ♇
18 10 ☽ △ ♀
23 15 ☽ ☍ ☿

26 Wednesday
04 39 ☉ ⊼ ♇
04 44 ☽ ∠ ♄
05 38 ☽ ⚼ ♃
07 30 ♂ ± ♅
08 25 ☽ ⚼ ♅
13 19 ☽ # ☉
17 05 ☽ ∠ ♇
22 56 ☿ Q ♆
23 11 ☽ ⚼ ♀

27 Thursday
01 02 ☽ □ ♂
07 22 ☽ ✶ ♄
08 38 ☽ △ ♃
08 55 ☽ □ ♆
09 08 ☽ ⊥ ♇
11 13 ☽ △ ♅
16 57 ☽ ± ♀
19 58 ☽ ⊻ ♇
23 07 ☽ ☍ ☉

28 Friday
04 58 ☽ ⊼ ♀
05 08 ☽ Q ♄
08 14 ♃ ⊻ ♆
09 28 ☽ ⊼ ☿
09 46 ♀ ∠ ♂
09 49 ☽ # ♀
16 03 ☿ # ♆
17 47 ☽ Q ♆
21 35 ☽ ± ☿
23 32 ☿ ⚼ ♄

29 Saturday
10 12 ☽ △ ♂
13 21 ☽ # ♀
14 38 ☽ □ ♄
16 06 ☽ ⚼ ☿
16 16 ☽ ✶ ♆
16 40 ☽ □ ♃
18 50 ☽ □ ♅
23 06 ☿ ∠ ♃

30 Sunday
03 49 ☽ ☌ ♇
11 55 ☽ ⊼ ☉
16 02 ☽ ⚼ ♂
18 07 ♄ St D
19 23 ☽ ☍ ♀
21 04 ☽ ∠ ♆
22 08 ☿ ∠ ♅
23 55 ☽ △ ☿

31 Monday
00 31 ☽ ± ☉
03 44 ☽ Q ♃
03 46 ☽ # ☉
05 37 ☽ Q ♅
10 20 ☽ ± ♂
12 56 ♀ ⚼ ♆
14 41 ☽ ⊥ ♆
18 52 ♆ St R
19 42 ☽ ⚼ ☉
22 41 ☽ ⊼ ♂

JUNE 2010

Date	Sidereal time h m s	LONGITUDES Sun ☉ ° ' "	Moon ☽ ° ' "	Moon ☽ 24.00 ° ' "	Mercury ☿ ° '	Venus ♀ ° '	Mars ♂ ° '	Jupiter ♃ ° '	Saturn ♄ ° '	Uranus ♅ ° '	Neptune ♆ ° '	Pluto ♇ ° '
01	04 39 27	10 ♊ 53 49	03 ♒ 26 47	09 ♒ 26 07	17 ♉ 03	14 ♋ 51	27 ♌ 02	29 ♓ 20	27 ♍ 50	00 ♈ 08	28 ♒ 42	04 ♑ 43
02	04 43 24	11 51 18	15 ♒ 23 38	21 ♒ 19 52	18 21	16 02	27 33	29 28	27 D 50	00 09	28 R 42	04 R 41
03	04 47 20	12 48 46	27 ♒ 15 23	03 ♓ 10 49	19 42	17 13	28 04	29 37	27 51	00 11	28 42	04 40
04	04 51 17	13 46 13	09 ♓ 06 45	15 ♓ 03 47	21 06	18 24	28 34	29 45	27 51	00 12	28 42	04 38
05	04 55 13	14 43 40	21 ♓ 02 34	27 ♓ 03 39	22 33	19 34	29 05	29 ♓ 54	27 52	00 14	28 42	04 37
06	04 59 10	15 41 06	03 ♈ 07 38	09 ♈ 15 01	24 02	20 45	29 ♌ 36	00 ♈ 02	27 52	00 15	28 42	04 36
07	05 03 07	16 38 32	15 ♈ 26 20	21 ♈ 42 00	25 34	21 56	00 ♍ 08	00 10	27 53	00 16	28 41	04 34
08	05 07 03	17 35 57	28 ♈ 02 22	04 ♉ 27 46	27 09	23 07	00 39	00 18	27 54	00 18	28 41	04 33
09	05 11 00	18 33 21	10 ♉ 58 22	17 ♉ 34 18	28 ♉ 46	24 17	01 10	00 26	27 55	00 19	28 41	04 31
10	05 14 56	19 30 45	24 ♉ 15 33	01 ♊ 02 03	00 ♊ 27	25 28	01 42	00 33	27 56	00 20	28 41	04 30
11	05 18 53	20 28 08	07 ♊ 53 34	14 ♊ 49 48	02 09	26 38	02 14	00 41	27 57	00 21	28 40	04 28
12	05 22 49	21 25 30	21 ♊ 50 21	28 ♊ 54 44	03 55	27 49	02 46	00 48	27 58	00 23	28 40	04 27
13	05 26 46	22 22 52	06 ♋ 02 23	13 ♋ 12 43	05 43	28 ♋ 59	03 18	00 55	28 00	00 24	28 40	04 25
14	05 30 42	23 20 13	20 ♋ 25 04	27 ♋ 38 50	07 33	00 ♌ 09	03 50	01 02	28 01	00 25	28 39	04 24
15	05 34 39	24 17 33	04 ♌ 53 20	12 ♌ 07 59	09 26	01 19	04 22	01 09	28 03	00 26	28 39	04 22
16	05 38 35	25 14 52	19 ♌ 22 12	26 ♌ 35 28	11 22	02 30	04 54	01 16	28 04	00 27	28 38	04 21
17	05 42 32	26 12 11	03 ♍ 47 17	10 ♍ 57 16	13 20	03 40	05 27	01 22	28 06	00 27	28 38	04 19
18	05 46 29	27 09 28	18 ♍ 05 03	25 ♍ 10 21	15 20	04 49	06 00	01 29	28 08	00 28	28 37	04 18
19	05 50 25	28 06 45	02 ♎ 12 54	09 ♎ 12 31	17 22	05 59	06 32	01 35	28 10	00 29	28 37	04 16
20	05 54 22	29 ♊ 04 00	16 ♎ 09 02	23 ♎ 02 21	19 26	07 09	07 05	01 41	28 12	00 30	28 36	04 15
21	05 58 18	00 ♋ 01 15	29 ♎ 52 19	06 ♏ 38 55	21 32	08 19	07 38	01 47	28 14	00 31	28 35	04 13
22	06 02 15	00 58 30	13 ♏ 22 03	20 ♏ 01 42	23 39	09 28	08 11	01 53	28 16	00 31	28 35	04 12
23	06 06 11	01 55 44	26 ♏ 37 50	03 ♐ 10 26	25 48	10 38	08 44	01 59	28 18	00 32	28 34	04 10
24	06 10 08	02 52 57	09 ♐ 39 32	16 ♐ 05 08	27 ♊ 57	11 47	09 18	02 04	28 21	00 32	28 33	04 09
25	06 14 04	03 50 10	22 ♐ 27 18	28 ♐ 46 05	00 ♋ 08	12 56	09 51	02 09	28 23	00 33	28 32	04 07
26	06 18 01	04 47 22	05 ♑ 01 36	11 ♑ 13 57	02 19	14 05	10 24	02 14	28 26	00 33	28 32	04 06
27	06 21 58	05 44 34	17 ♑ 23 19	23 ♑ 29 52	04 30	15 14	10 58	02 19	28 29	00 34	28 31	04 04
28	06 25 54	06 41 46	29 ♑ 33 51	05 ♒ 35 31	06 41	16 23	11 32	02 24	28 31	00 34	28 30	04 02
29	06 29 51	07 38 58	11 ♒ 35 11	17 ♒ 33 10	08 52	17 32	12 05	02 29	28 34	00 35	28 29	04 01
30	06 33 47	08 ♋ 36 10	23 ♒ 29 52	29 ♒ 25 40	11 ♋ 03	18 ♌ 41	12 ♍ 39	02 ♈ 33	28 ♍ 37	00 ♈ 35	28 ♒ 28	03 ♑ 59

Date	Moon True ☊ ° '	Moon Mean ☊ ° '	Moon ☽ Latitude ° '
01	12 ♑ 13	13 ♑ 37	01 N 54
02	12 D 15	13 33	02 52
03	12 16	13 30	03 42
04	12 17	13 27	04 23
05	12 R 17	13 24	04 53
06	12 16	13 21	05 10
07	12 13	13 17	05 13
08	12 10	13 14	05 02
09	12 07	13 11	04 35
10	12 03	13 08	03 53
11	12 01	13 05	02 57
12	11 59	13 02	01 50
13	11 58	12 58	00 N 33
14	11 D 58	12 55	00 S 47
15	11 59	12 52	02 03
16	12 00	12 49	03 12
17	12 01	12 46	04 09
18	12 02	12 42	04 49
19	12 R 02	12 39	05 12
20	12 01	12 36	05 15
21	12 00	12 33	05 01
22	11 59	12 30	04 30
23	11 57	12 27	03 45
24	11 56	12 23	02 49
25	11 56	12 20	01 46
26	11 55	12 17	00 S 38
27	11 D 55	12 14	00 N 30
28	11 55	12 11	01 36
29	11 56	12 08	02 37
30	11 ♑ 56	12 ♑ 04	03 N 31

DECLINATIONS

Date	Sun ☉ ° '	Moon ☽ ° '	Mercury ☿ ° '	Venus ♀ ° '	Mars ♂ ° '	Jupiter ♃ ° '	Saturn ♄ ° '	Uranus ♅ ° '	Neptune ♆ ° '	Pluto ♇ ° '
01	22 N 05	17 S 32	13 N 57	24 N 28	13 N 49	01 S 20	03 N 05	00 S 38	12 S 21	18 S 12
02	22 13	13 29	14 25	24 21	13 37	01 16	03 05	00 37	12 21	18 12
03	22 20	08 57	14 53	24 13	13 26	01 13	03 04	00 37	12 21	18 12
04	22 27	04 S 05	15 22	24 05	13 14	01 10	03 04	00 36	12 21	18 13
05	22 34	00 N 56	15 51	23 56	13 02	01 07	03 04	00 36	12 22	18 13
06	22 40	05 59	16 21	23 46	12 50	01 04	03 03	00 35	12 22	18 13
07	22 46	10 54	16 52	23 35	12 38	01 01	03 03	00 34	12 22	18 13
08	22 52	15 28	17 23	23 24	12 26	00 58	03 02	00 34	12 22	18 13
09	22 57	19 28	17 53	23 12	12 14	00 55	03 02	00 33	12 22	18 13
10	23 01	22 36	18 24	23 00	12 02	00 52	03 01	00 33	12 22	18 13
11	23 06	24 32	18 55	22 47	11 49	00 50	03 00	00 33	12 22	18 13
12	23 10	25 00	19 25	22 33	11 37	00 47	03 00	00 32	12 22	18 13
13	23 13	23 51	19 55	22 19	11 24	00 44	02 59	00 32	12 22	18 14
14	23 16	21 07	20 25	22 04	11 12	00 42	02 58	00 31	12 23	18 14
15	23 19	17 03	20 53	21 49	10 59	00 39	02 57	00 31	12 23	18 14
16	23 21	11 58	21 21	21 33	10 46	00 37	02 57	00 31	12 23	18 14
17	23 23	06 15	21 47	21 17	10 33	00 34	02 56	00 30	12 23	18 14
18	23 24	00 N 16	22 12	21 00	10 20	00 32	02 55	00 30	12 23	18 14
19	23 25	05 S 39	22 36	20 42	10 07	00 30	02 54	00 30	12 24	18 14
20	23 26	11 12	22 58	20 24	09 54	00 28	02 53	00 29	12 24	18 14
21	23 26	16 07	23 18	20 05	09 41	00 26	02 52	00 29	12 24	18 15
22	23 26	20 08	23 36	19 46	09 28	00 23	02 51	00 29	12 24	18 15
23	23 25	23 03	23 51	19 27	09 14	00 22	02 49	00 29	12 25	18 15
24	23 24	24 41	24 04	19 07	09 01	00 20	02 48	00 29	12 25	18 15
25	23 23	24 59	24 15	18 46	08 47	00 18	02 47	00 28	12 25	18 15
26	23 21	23 59	24 22	18 25	08 34	00 16	02 46	00 28	12 25	18 15
27	23 19	21 48	24 27	18 04	08 20	00 14	02 45	00 28	12 26	18 15
28	23 16	18 40	24 30	17 42	08 07	00 13	02 43	00 28	12 26	18 16
29	23 13	14 47	24 29	17 19	07 53	00 11	02 42	00 28	12 26	18 16
30	23 N 10	10 S 22	24 N 25	16 N 57	07 N 39	00 S 10	02 N 41	00 S 28	12 S 27	18 S 16

ZODIAC SIGN ENTRIES

Date	h	m	Planets
01	05	08	☽ ♒
03	17	34	☽ ♓
06	05	50	☽ ♈
06	06	28	♃ ♈
07	06	11	♂ ♍
08	15	41	☽ ♉
10	05	41	☿ ♊
10	22	11	☽ ♊
13	01	50	☽ ♋
14	08	50	♀ ♌
15	03	54	☽ ♌
17	05	41	☽ ♍
19	08	13	☽ ♎
21	11	28	☉ ♋
21	12	14	☽ ♏
23	18	10	☽ ♐
25	10	32	☿ ♋
26	02	21	☽ ♑
28	12	52	☽ ♒

LATITUDES

Date	Mercury ☿ ° '	Venus ♀ ° '	Mars ♂ ° '	Jupiter ♃ ° '	Saturn ♄ ° '	Uranus ♅ ° '	Neptune ♆ ° '	Pluto ♇ ° '
01	03 S 06	01 N 52	01 N 24	01 S 09	02 N 25	00 S 44	00 S 28	05 N 09
04	02 46	01 55	01 21	01 10	02 25	00 44	00 28	05 09
07	02 21	01 58	01 17	01 11	02 24	00 45	00 28	05 09
10	01 53	01 59	01 14	01 11	02 23	00 45	00 28	05 09
13	01 21	02 00	01 11	01 12	02 23	00 45	00 28	05 09
16	00 48	02 00	01 08	01 13	02 22	00 45	00 28	05 08
19	00 S 14	01 59	01 05	01 13	02 22	00 45	00 28	05 08
22	00 N 19	01 58	01 02	01 15	02 21	00 45	00 28	05 08
25	00 48	01 55	00 59	01 15	02 20	00 45	00 28	05 08
28	01 13	01 51	00 56	01 16	02 20	00 45	00 28	05 07
31	01 N 33	01 N 47	00 N 54	01 S 17	02 N 19	00 S 45	00 S 28	05 N 07

DATA

Julian Date	2455349
Delta T	+67 seconds
Ayanamsa	24° 00' 25"
Synetic vernal point	05° ♓ 06' 34"
True obliquity of ecliptic	23° 26' 18"

MOON'S PHASES, APSIDES AND POSITIONS ☽

Date	h	m	Phase	Longitude ° '	Eclipse Indicator
04	22	13	☾	14 ♓ 11	
12	11	15	●	21 ♊ 24	
19	04	29	☽	27 ♍ 49	
26	11	30	○	04 ♑ 46	partial

Day	h	m		
03	16	48	Apogee	
15	14	53	Perigee	
05	07	34	0N	
12	07	03	Max dec	25° N 02'
18	13	05	0S	
25	05	16	Max dec	25° S 02'

LONGITUDES

Date	Chiron ⚷ ° '	Ceres ⚳ ° '	Pallas ⚴ ° '	Juno ⚵ ° '	Vesta ⚶ ° '	Black Moon Lilith ⚸ ° '
01	00 ♓ 59	01 ♑ 15	06 ♏ 13	25 ♊ 48	01 ♍ 50	27 ♒ 01
11	00 ♓ 58	29 ♐ 12	04 ♏ 48	01 ♋ 29	05 ♍ 13	28 ♒ 08
21	00 ♓ 51	27 ♐ 00	04 ♏ 11	07 ♋ 06	08 ♍ 55	29 ♒ 15
31	00 ♓ 38	24 ♐ 53	04 ♏ 19	12 ♋ 39	12 ♍ 52	00 ♓ 22

All ephemeris data is given at 12.00 UT and the Moon's longitude is additionally given for 24.00 UT

ASPECTARIAN

h m Aspects

01 Tuesday
- 00 50 ☽ △ ♄
- 02 33 ☽ ⊻ ♆
- 03 41 ☽ ⚹ ♃
- 05 22 ☽ ⚹ ♅
- 06 41 ☿ ∥ ♂
- 07 40 ☽ ∥ ♇
- 14 31 ☽ ⊻ ♇

02 Wednesday
- 00 42 ☉ Q ♃
- 02 32 ☽ ⊥ ♇
- 04 15 ☽ △ ☉
- 06 51 ☽ ⚼ ♄
- 07 17 ☽ ⋕ ☿
- 08 10 ♀ Q ♄
- 10 07 ☽ ∠ ♃
- 11 11 ☽ ⋕ ♂
- 11 31 ☽ ∠ ♅
- 13 25 ☽ ⊼ ♀
- 18 09 ☽ ∥ ♆
- 18 44 ☽ □ ☿
- 19 39 ☉ Q ♅
- 20 39 ☽ ∠ ♇

03 Thursday
- 01 02 ☽ ± ♄
- 01 50 ♂ ⊻ ♄
- 02 54 ☽ ± ♀
- 04 32 ☽ ⊥ ♃
- 05 45 ☽ ⊥ ♅
- 11 21 ☿ ⚼ ♇
- 13 12 ☽ ⊼ ♄
- 13 42 ☽ ☍ ♂
- 14 56 ☽ ☌ ♆
- 16 50 ☽ ⊻ ♃
- 17 56 ☽ ⊻ ♅
- 23 09 ☽ ⚼ ♀

04 Friday
- 02 58 ☽ ⚹ ♇
- 11 58 ☽ Q ☿
- 16 55 ☽ ⋕ ♄
- 17 52 ♂ ☍ ♆
- 22 13 ☽ □ ☉

05 Saturday
- 02 10 ☽ ∥ ♃
- 03 09 ☽ Q ♇
- 04 44 ☽ ∥ ♅
- 08 44 ☽ △ ♀
- 10 22 ☽ ⋕ ♅
- 12 50 ☽ ⋕ ♃
- 15 25 ☽ ⚹ ☿
- 22 04 ☽ ∥ ♄

06 Sunday
- 01 36 ☽ ☍ ♄
- 03 15 ☽ ⊻ ♆
- 04 44 ☽ ⊼ ♂
- 05 49 ☽ ☌ ♃
- 06 19 ☽ ☌ ♅
- 13 12 ☽ Q ☉
- 14 53 ☽ □ ♇
- 15 05 ☽ ⊥ ♆
- 17 05 ☽ ± ♂

07 Monday
- 01 13 ☽ ∠ ☿
- 08 37 ☽ ∠ ♆
- 11 22 ☽ ⚼ ♂
- 14 26 ♂ ⊼ ♃
- 14 31 ☽ ⚹ ☉
- 19 03 ♂ ⊼ ♅
- 19 28 ☽ ⋕ ♆
- 20 30 ☽ ∥ ♂
- 21 04 ☽ ⊥ ☿

08 Tuesday
- 01 44 ☽ □ ♀
- 03 22 ♀ ± ♆
- 10 05 ☽ ⊻ ☿
- 11 27 ♃ ☌ ♅
- 11 44 ☽ ⊼ ♄
- 13 13 ☽ ⚹ ♆
- 16 14 ☽ ⊻ ♅
- 16 17 ☽ ⊻ ♃
- 17 06 ☽ △ ♂
- 20 09 ♂ ⋕ ♆
- 21 13 ☽ ∠ ☉
- 22 58 ☽ ± ♄
- 23 14 ☿ △ ♄

09 Wednesday
- 00 08 ☽ △ ♇
- 00 29 ☽ ∥ ☿
- 03 25 ☽ ⊥ ♅
- 03 33 ☽ ⊥ ♃
- 03 59 ☽ ⋕ ♇
- 08 21 ☿ ± ♇
- 10 39 ☿ □ ♆
- 11 28 ☽ Q ♆
- 14 39 ☽ Q ♀
- 15 07 ☽ ⊥ ☉
- 15 33 ☽ ⚼ ♄
- 19 56 ☽ ∠ ♅
- 20 12 ☽ ∠ ♃

10 Thursday
- 02 51 ☽ ⊻ ☉
- 03 24 ☿ ⋕ ♇
- 03 29 ☽ ⚼ ♇
- 10 05 ☉ ∥ ♀
- 10 28 ☿ ⚹ ♅
- 13 42 ☿ ⚹ ♃
- 14 21 ☽ ⚹ ♀
- 15 34 ☽ ∥ ♀
- 16 15 ☽ ∥ ☉
- 17 19 ☉ ⊥ ♀
- 18 32 ☽ △ ♄
- 19 31 ☽ ± ♇
- 19 50 ☽ □ ♆
- 20 47 ♀ ⊥ ♂
- 22 33 ☉ Q ♂
- 22 47 ☽ ⚹ ♅
- 23 16 ☽ ⚹ ♃

11 Friday
- 00 31 ☽ ☌ ☿
- 01 42 ☽ □ ♂
- 06 03 ☽ ⊼ ♇
- 13 27 ☿ □ ♂
- 19 06 ☽ ∠ ♀
- 19 45 ☽ Q ♅
- 20 22 ☽ Q ♃

12 Saturday
- 10 05 ☽ Q ♂
- 11 15 ☽ ☌ ☉
- 11 57 ☽ ⊥ ♀
- 15 18 ♀ ⚹ ♄
- 19 05 ☿ ⊼ ♇
- 22 25 ☽ □ ♄
- 23 03 ☽ ⊻ ♀
- 23 35 ☽ △ ♆

13 Sunday
- 02 29 ☽ □ ♅
- 03 19 ☽ □ ♃
- 05 23 ♀ ⊼ ♆
- 07 13 ☽ ⚹ ♂
- 09 17 ☽ ☍ ♇
- 11 22 ☽ ⊻ ☿
- 18 42 ☽ ∥ ☉
- 22 53 ☽ ⊥ ☿

14 Monday
- 00 44 ☽ ⚼ ♆
- 04 18 ☽ ∥ ♀
- 04 40 ☽ Q ♄
- 09 15 ☽ ∠ ♂
- 15 43 ☽ ± ♆
- 16 05 ☽ ∠ ☿
- 16 12 ☽ ∥ ☿
- 17 12 ☽ ⊻ ☉
- 17 20 ♀ △ ♅

15 Tuesday
- 00 38 ☽ ⚹ ♄
- 00 47 ☽ ⊥ ♂
- 01 40 ☽ ⊼ ♆
- 03 51 ☽ ⊥ ☉
- 04 36 ☽ △ ♅
- 05 35 ☽ ☌ ♀
- 05 42 ☽ ⋕ ♇
- 05 46 ☽ △ ♃
- 08 07 ♀ △ ♃
- 11 06 ☽ ⊻ ♂
- 11 09 ☽ ⊼ ♇
- 12 12 ♂ △ ♇
- 19 49 ☽ ∠ ☉
- 20 41 ☽ ⚹ ☿
- 21 04 ☽ ± ♇

16 Wednesday
- 01 32 ☽ ∠ ♄
- 05 29 ☽ ⚼ ♅
- 06 49 ☽ ⚼ ♃
- 10 09 ☽ ⋕ ♆
- 11 58 ☽ ⚼ ♇
- 16 30 ☽ ⊥ ♄
- 17 25 ☽ ∥ ♂
- 18 48 ☿ ∥ ♀
- 19 40 ☽ Q ☿
- 20 26 ☽ ± ♅
- 21 52 ☽ ± ♃
- 22 27 ☽ ⚹ ☉

17 Thursday
- 01 21 ☿ Q ♅
- 02 29 ☽ ⊻ ♄
- 03 24 ☽ ☍ ♆
- 06 26 ☽ ⊼ ♅
- 07 56 ☽ ⊼ ♃
- 11 46 ☽ ⊻ ♀
- 12 37 ☿ Q ♃
- 12 53 ☽ △ ♇
- 14 53 ☽ ☌ ♂
- 19 55 ☽ Q ☉
- 22 42 ☽ ⊥ ♀

18 Friday
- 01 23 ♀ ⊼ ♇
- 01 26 ☽ ∥ ♄
- 06 35 ☽ □ ☿
- 10 57 ☽ ⋕ ♃
- 11 05 ☽ ⋕ ♅
- 15 05 ☽ ∥ ♅
- 15 12 ☽ ∠ ♀
- 15 13 ☽ ∥ ♃

19 Saturday
- 00 46 ☽ ⋕ ♄
- 04 29 ☽ □ ☉
- 05 04 ☽ ☌ ♄
- 05 51 ☽ ⊼ ♆
- 09 03 ☽ ☍ ♅
- 10 55 ☽ ☍ ♃
- 13 18 ☉ □ ♄
- 13 23 ♃ ∥ ♅
- 15 31 ☽ □ ♇
- 16 06 ☽ ± ♆
- 19 03 ☽ ⚹ ♀
- 19 42 ☽ ⊻ ♂

20 Sunday
- 00 21 ☉ △ ♆
- 06 24 ☽ ⋕ ♂
- 06 28 ☽ ⊥ ♂
- 07 35 ☽ ⚼ ♆
- 09 29 ♀ ⊻ ♂
- 17 32 ☽ ∥ ♆
- 17 41 ☽ Q ♀
- 18 43 ☽ △ ☿
- 22 36 ☽ Q ♇
- 22 46 ☽ ∠ ♂

21 Monday
- 09 06 ☽ ⊻ ♄
- 09 44 ☽ △ ♆
- 12 17 ☽ △ ☉
- 13 08 ☽ ⊼ ♅
- 15 24 ☽ ⊼ ♃
- 19 40 ☽ ⚹ ♇
- 19 44 ☽ ⊥ ♄
- 22 55 ☉ ∥ ☿
- 23 46 ☽ ± ♅
- 23 56 ☽ ∥ ♇

22 Tuesday
- 00 27 ☉ □ ♅
- 01 59 ☽ ⚼ ☿
- 02 08 ☽ ± ♃
- 02 20 ☽ ⚹ ♂
- 04 22 ☽ □ ♀
- 09 42 ☽ ⋕ ♀
- 11 49 ☽ ∠ ♄
- 15 52 ☽ ⚼ ♅
- 17 03 ☽ ⚼ ☉
- 18 22 ☽ ⚼ ♃
- 21 11 ☽ ± ☿
- 22 28 ☽ ∠ ♇

23 Wednesday
- 00 49 ☽ Q ♂
- 02 44 ♀ ± ♇
- 07 54 ☿ ∠ ♀
- 10 11 ☽ ⊼ ☿
- 10 37 ☽ ± ☉
- 13 21 ☉ □ ♃
- 14 48 ☽ ⊥ ♇
- 15 04 ☽ ⚹ ♄
- 15 32 ☽ □ ♆
- 16 04 ☽ ⋕ ☉
- 19 09 ☽ △ ♅
- 21 52 ☽ △ ♃
- 22 28 ☽ ⊼ ☉
- 22 51 ☽ ⋕ ☿

24 Thursday
- 01 48 ☽ ⊻ ♇
- 02 05 ☿ Q ♂
- 11 17 ☽ □ ♂
- 13 17 ☽ Q ♄
- 16 20 ☽ △ ♀
- 16 23 ☿ □ ♄
- 18 33 ☿ △ ♆

25 Friday
- 00 52 ☽ Q ♆
- 16 36 ☿ □ ♅
- 18 55 ☉ ☍ ♇
- 21 50 ♀ ∠ ♄
- 23 19 ☽ □ ♄
- 23 28 ☽ ⚼ ♀
- 23 33 ☽ ⚹ ♆

26 Saturday
- 03 25 ☽ □ ♅
- 05 41 ☽ ☍ ☿
- 06 01 ☽ ⋕ ☿
- 06 37 ☽ □ ♃
- 10 12 ☽ ☌ ♇
- 11 08 ☿ □ ♃
- 11 30 ☽ ☍ ☉
- 18 31 ☽ ± ♀
- 20 26 ☽ ⋕ ☉
- 22 53 ☽ △ ♂
- 22 54 ♀ ⋕ ♇

27 Sunday
- 04 27 ☽ ∠ ♆
- 07 15 ☿ ☍ ♇
- 07 22 ☽ ⊼ ♀
- 14 18 ☽ Q ♅
- 17 48 ☽ Q ♃
- 18 52 ♀ ⚼ ♅
- 22 03 ☽ ⊥ ♆

28 Monday
- 02 46 ♄ ⊼ ♆
- 05 41 ☽ ⚼ ♂
- 09 54 ☽ ⊻ ♆
- 09 56 ☽ △ ♄
- 12 07 ☉ ☌ ☿
- 14 00 ☽ ⚹ ♅
- 14 45 ☽ ∥ ♇
- 17 41 ☽ ⚹ ♃
- 19 08 ☽ ⋕ ♀
- 20 53 ☽ ⊻ ♇

29 Tuesday
- 00 27 ☽ ± ♂
- 03 26 ☽ ⊼ ☉
- 05 21 ☽ ⊼ ☿
- 08 51 ☽ ⊥ ♇
- 10 49 ♀ ⚼ ♃
- 13 04 ☽ ⊼ ♂
- 16 00 ☽ ⚼ ♄
- 16 30 ☽ ± ☉
- 20 01 ☽ ∠ ♅
- 20 04 ☽ ± ☿
- 23 56 ☽ ∠ ♃

30 Wednesday
- 01 01 ☽ ∥ ♆
- 01 14 ☽ ☍ ♀
- 02 55 ☽ ∠ ♇
- 10 13 ☽ ± ♄
- 12 14 ☽ ⚼ ☉
- 14 11 ☽ ⊥ ♅
- 18 13 ☽ ⊥ ♃
- 18 18 ☽ ⚼ ☿
- 18 24 ♀ ⚼ ♇
- 22 03 ☽ ☌ ♆
- 22 25 ☽ ⊼ ♄

JULY 2010

LONGITUDES

Date	Sidereal time h m s	Sun ☉ ° ' "	Moon ☽ ° ' "	Moon ☽ 24.00 ° ' "	Mercury ☿ ° '	Venus ♀ ° '	Mars ♂ ° '	Jupiter ♃ ° '	Saturn ♄ ° '	Uranus ♅ ° '	Neptune ♆ ° '	Pluto ♇ ° '
01	06 37 44	09 ♋ 33 22	05 ♓ 21 03	11 ♓ 16 27	13 ♋ 12	19 ♌ 49	13 ♍ 13	02 ♈ 38	28 ♍ 40	00 ♈ 35	28 ♒ 27	03 ♑ 58
02	06 41 40	10 30 33	17 ♓ 12 24	23 ♓ 09 25	15 20	20 58	13 47	02 42	28 44	00 35	28 R 26	03 R 56
03	06 45 37	11 27 45	29 ♓ 08 02	05 ♈ 08 50	17 28	22 06	14 21	02 45	28 47	00 35	28 26	03 55
04	06 49 33	12 24 57	11 ♈ 12 21	17 ♈ 19 09	19 33	23 14	14 56	02 48	28 50	00 35	28 25	03 53
05	06 53 30	13 22 10	23 ♈ 29 48	29 ♈ 44 48	21 38	24 23	15 30	02 53	28 54	00 35	28 24	03 52
06	06 57 27	14 19 23	06 ♉ 04 39	12 ♉ 29 47	23 40	25 31	16 05	02 56	28 57	00 R 35	28 22	03 50
07	07 01 23	15 16 36	19 ♉ 00 35	25 ♉ 37 20	25 42	26 38	16 39	02 59	29 01	00 35	28 21	03 49
08	07 05 20	16 13 49	02 ♊ 20 13	09 ♊ 09 22	27 41	27 46	17 14	03 02	29 04	00 35	28 20	03 47
09	07 09 16	17 11 03	16 ♊ 04 42	23 ♊ 06 03	29 ♋ 38	28 ♌ 54	17 49	03 05	29 08	00 35	28 19	03 46
10	07 13 13	18 08 17	00 ♋ 13 07	07 ♋ 25 24	01 ♌ 34	00 ♍ 01	18 23	03 08	29 12	00 35	28 18	03 44
11	07 17 09	19 05 31	14 ♋ 42 19	22 ♋ 03 08	03 28	01 09	18 58	03 10	29 16	00 35	28 17	03 43
12	07 21 06	20 02 46	29 ♋ 26 59	06 ♌ 52 57	05 19	02 16	19 33	03 12	29 20	00 34	28 16	03 41
13	07 25 02	21 00 01	14 ♌ 20 03	21 ♌ 47 17	07 09	03 23	20 08	03 14	29 24	00 34	28 15	03 40
14	07 28 59	21 57 15	29 ♌ 13 40	06 ♍ 38 17	08 57	04 30	20 44	03 16	29 28	00 34	28 13	03 38
15	07 32 56	22 54 30	14 ♍ 00 16	21 ♍ 18 54	10 43	05 37	21 19	03 18	29 32	00 33	28 12	03 37
16	07 36 52	23 51 45	28 ♍ 33 33	05 ♎ 43 44	12 28	06 43	21 54	03 19	29 37	00 33	28 11	03 36
17	07 40 49	24 49 00	12 ♎ 49 08	19 ♎ 49 29	14 10	07 50	22 30	03 21	29 41	00 32	28 10	03 34
18	07 44 45	25 46 15	26 ♎ 44 42	03 ♏ 34 46	15 50	08 56	23 05	03 22	29 45	00 32	28 08	03 33
19	07 48 42	26 43 30	10 ♏ 19 47	16 ♏ 59 54	17 29	10 02	23 41	03 23	29 50	00 31	28 07	03 31
20	07 52 38	27 40 46	23 ♏ 35 19	00 ♐ 06 16	19 05	11 08	24 17	03 23	29 55	00 30	28 06	03 30
21	07 56 35	28 38 02	06 ♐ 33 03	12 ♐ 55 55	20 40	12 14	24 53	03 24	29 ♍ 59	00 30	28 04	03 29
22	08 00 31	29 ♋ 35 18	19 ♐ 15 10	25 ♐ 31 06	22 13	13 20	25 29	03 24	00 ♎ 04	00 29	28 03	03 27
23	08 04 28	00 ♌ 32 34	01 ♑ 43 57	07 ♑ 54 01	23 43	14 25	26 05	03 24	00 09	00 28	28 01	03 26
24	08 08 25	01 29 51	14 ♑ 01 33	20 ♑ 06 47	25 12	15 31	26 41	03 R 24	00 14	00 27	28 00	03 25
25	08 12 21	02 27 08	26 ♑ 09 58	02 ♒ 11 18	26 39	16 36	27 17	03 24	00 19	00 26	27 59	03 23
26	08 16 18	03 24 26	08 ♒ 11 03	14 ♒ 09 25	28 04	17 40	27 53	03 23	00 24	00 25	27 57	03 22
27	08 20 14	04 21 45	20 ♒ 06 38	26 ♒ 02 58	29 ♌ 27	18 45	28 29	03 23	00 29	00 24	27 56	03 21
28	08 24 11	05 19 04	01 ♓ 58 40	07 ♓ 54 02	00 ♍ 48	19 50	29 06	03 22	00 34	00 23	27 54	03 20
29	08 28 07	06 16 24	13 ♓ 49 22	19 ♓ 44 59	02 07	20 54	29 ♍ 42	03 21	00 40	00 22	27 53	03 18
30	08 32 04	07 13 45	25 ♓ 41 16	01 ♈ 38 37	03 23	21 58	00 ♎ 19	03 19	00 45	00 21	27 51	03 17
31	08 36 00	08 ♌ 11 07	07 ♈ 37 26	13 ♈ 38 11	04 ♍ 38	23 ♍ 02	00 ♎ 55	03 ♈ 18	00 ♎ 50	00 ♈ 20	27 ♒ 50	03 ♑ 16

Date	Moon True ☊ ° '	Moon Mean ☊ ° '	Moon ☽ Latitude ° '
01	11 ♑ 57	12 ♑ 01	04 N 15
02	11 D 57	11 58	04 48
03	11 57	11 55	05 09
04	11 R 57	11 52	05 17
05	11 D 57	11 48	05 11
06	11 57	11 45	04 50
07	11 57	11 42	04 13
08	11 58	11 39	03 23
09	11 58	11 36	02 19
10	11 58	11 33	01 N 05
11	11 R 58	11 29	00 S 15
12	11 58	11 26	01 35
13	11 58	11 23	02 50
14	11 57	11 20	03 53
15	11 55	11 17	04 40
16	11 54	11 14	05 08
17	11 53	11 10	05 16
18	11 D 53	11 07	05 06
19	11 54	11 04	04 38
20	11 54	11 01	03 57
21	11 56	10 58	03 03
22	11 57	10 54	02 02
23	11 58	10 51	00 S 56
24	11 R 58	10 48	00 N 11
25	11 58	10 45	01 17
26	11 56	10 42	02 19
27	11 53	10 39	03 15
28	11 50	10 35	04 01
29	11 46	10 32	04 37
30	11 42	10 29	05 01
31	11 ♑ 38	10 ♑ 26	05 N 13

DECLINATIONS

Date	Sun ☉ ° '	Moon ☽ ° '	Mercury ☿ ° '	Venus ♀ ° '	Mars ♂ ° '	Jupiter ♃ ° '	Saturn ♄ ° '	Uranus ♅ ° '	Neptune ♆ ° '	Pluto ♇ ° '
01	23 N 06	05 S 36	24 N 19	16 N 34	07 N 25	00 S 08	02 N 39	00 S 28	12 S 27	18 S 16
02	23 01	00 S 38	24 10	16 10	07 11	00 07	02 38	00 28	12 27	18 16
03	22 57	04 N 23	23 59	15 46	06 57	00 06	02 36	00 28	12 28	18 16
04	22 52	09 18	23 45	15 22	06 43	00 04	02 35	00 28	12 28	18 17
05	22 46	13 56	23 29	14 58	06 29	00 03	02 33	00 28	12 29	18 17
06	22 40	18 06	23 10	14 33	06 15	00 02	02 32	00 28	12 29	18 17
07	22 34	21 32	22 49	14 08	06 01	00 S 01	02 30	00 28	12 29	18 17
08	22 27	23 56	22 27	13 42	05 46	00 00	02 28	00 28	12 30	18 17
09	22 20	25 01	22 02	13 16	05 32	00 N 01	02 27	00 28	12 30	18 17
10	22 13	24 31	21 36	12 50	05 17	00 01	02 25	00 28	12 31	18 18
11	22 05	22 23	21 09	12 24	05 03	00 02	02 23	00 28	12 31	18 18
12	21 57	18 43	20 40	11 57	04 48	00 03	02 22	00 28	12 32	18 18
13	21 48	13 49	20 10	11 30	04 34	00 03	02 20	00 29	12 32	18 18
14	21 39	08 06	19 38	11 03	04 19	00 04	02 18	00 29	12 32	18 18
15	21 30	01 N 59	19 06	10 36	04 04	00 04	02 16	00 29	12 33	18 19
16	21 20	04 S 08	18 33	10 08	03 50	00 04	02 14	00 29	12 33	18 19
17	21 10	09 55	17 59	09 40	03 35	00 05	02 12	00 29	12 34	18 19
18	21 00	15 04	17 24	09 12	03 20	00 05	02 10	00 30	12 34	18 19
19	20 49	19 19	16 48	08 44	03 05	00 05	02 08	00 30	12 35	18 19
20	20 37	22 29	16 12	08 16	02 50	00 05	02 06	00 30	12 35	18 20
21	20 26	24 25	15 36	07 47	02 35	00 05	02 04	00 31	12 36	18 20
22	20 14	24 59	14 59	07 18	02 20	00 05	02 02	00 31	12 36	18 20
23	20 02	24 22	14 22	06 50	02 05	00 04	02 00	00 31	12 37	18 20
24	19 50	22 31	13 45	06 21	01 50	00 04	01 58	00 32	12 37	18 20
25	19 37	19 39	13 08	05 52	01 35	00 04	01 56	00 32	12 38	18 21
26	19 24	15 58	12 31	05 23	01 20	00 03	01 54	00 32	12 38	18 21
27	19 10	11 42	11 54	04 53	01 05	00 03	01 52	00 33	12 39	18 21
28	18 56	07 01	11 16	04 24	00 49	00 02	01 49	00 33	12 39	18 21
29	18 42	02 S 06	10 40	03 55	00 34	00 01	01 47	00 34	12 40	18 21
30	18 28	02 N 54	10 03	03 25	00 19	00 01	01 45	00 34	12 40	18 22
31	18 N 13	07 N 49	09 N 26	02 N 56	00 N 04	00 00	01 N 43	00 S 35	12 S 41	18 S 22

ZODIAC SIGN ENTRIES

Date	h m	Planets
01	01 10	☽ ♓
03	13 44	☽ ♈
06	00 29	☽ ♉
08	07 51	☽ ♊
09	16 29	☿ ♌
10	11 32	♀ ♍
10	11 38	☽ ♋
12	12 53	☽ ♌
14	13 15	☽ ♍
16	14 24	☽ ♎
18	17 42	☽ ♏
20	23 48	☽ ♐
21	15 10	♄ ♎
22	22 21	☉ ♌
23	08 39	☽ ♑
25	19 38	☽ ♒
27	21 43	☿ ♍
28	08 00	☽ ♓
29	23 46	♂ ♎
30	20 42	☽ ♈

LATITUDES

Date	Mercury ☿ ° '	Venus ♀ ° '	Mars ♂ ° '	Jupiter ♃ ° '	Saturn ♄ ° '	Uranus ♅ ° '	Neptune ♆ ° '	Pluto ♇ ° '
01	01 N 33	01 N 47	00 N 54	01 S 17	02 N 19	00 S 45	00 S 28	05 N 07
04	01 45	01 42	00 51	01 18	02 19	00 46	00 28	05 06
07	01 51	01 35	00 48	01 19	02 18	00 46	00 28	05 06
10	01 51	01 28	00 46	01 20	02 17	00 46	00 28	05 06
13	01 44	01 20	00 43	01 21	02 17	00 46	00 29	05 05
16	01 33	01 10	00 40	01 22	02 16	00 46	00 29	05 05
19	01 16	01 00	00 38	01 23	02 16	00 46	00 29	05 04
22	00 56	00 49	00 35	01 23	02 15	00 46	00 29	05 04
25	00 32	00 37	00 33	01 24	02 15	00 46	00 29	05 03
28	00 N 05	00 24	00 30	01 25	02 14	00 46	00 29	05 03
31	00 S 24	00 N 10	00 N 28	01 S 26	02 N 14	00 S 46	00 S 29	05 N 02

DATA

Julian Date	2455379
Delta T	+67 seconds
Ayanamsa	24° 00' 31"
Synetic vernal point	05° ♓ 06' 28"
True obliquity of ecliptic	23° 26' 18"

MOON'S PHASES, APSIDES AND POSITIONS ☽

Date	h m	Phase	Longitude ° '	Eclipse Indicator
04	14 35	☾	12 ♈ 31	
11	19 40	●	19 ♋ 24	Total
18	10 11	☽	25 ♎ 42	
26	01 37	○	03 ♒ 00	

Day	h m		
01	10 04	Apogee	
13	11 16	Perigee	
28	23 35	Apogee	
02	15 00	0N	
09	16 46	Max dec	25° N 03'
15	19 43	0S	
22	11 21	Max dec	25° S 02'
29	22 05	0N	

LONGITUDES

Date	Chiron ⚷ ° '	Ceres ⚳ ° '	Pallas ⚴ ° '	Juno ⚵ ° '	Vesta ⚶ ° '	Black Moon Lilith ⚸ ° '
01	00 ♓ 38	24 ♐ 53	04 ♏ 19	12 ♋ 39	12 ♍ 52	00 ♓ 22
11	00 ♓ 19	23 ♐ 03	05 ♏ 08	18 ♋ 08	17 ♍ 00	01 ♓ 29
21	29 ♒ 57	21 ♐ 41	06 ♏ 34	23 ♋ 32	21 ♍ 24	02 ♓ 36
31	29 ♒ 31	20 ♐ 53	08 ♏ 30	28 ♋ 50	25 ♍ 55	03 ♓ 43

ASPECTARIAN

h m	Aspects
01 Thursday	
02 20	☽ ⊻ ♅
02 31	☽ # ♂
06 27	☽ ⊻ ♃
09 12	☽ ⚹ ♇
12 19	☿ ⚹ ♂
14 51	☿ ⚼ ♆
21 16	☽ △ ☉
02 Friday	
02 20	☽ # ♄
02 54	☿ ⊥ ♀
04 44	☽ ☍ ♂
07 24	☽ △ ☿
09 27	☽ Q ♇
12 47	☽ ∥ ♅
14 28	☽ ∥ ♃
15 31	☽ # ♃
17 12	☽ # ♅
20 23	☽ ⊼ ♀
03 Saturday	
03 30	☽ ∥ ♄
04 04	☿ Q ♄
09 43	☽ ± ♀
10 35	☽ ⊻ ♆
11 17	☽ ☍ ♄
14 55	☽ ☌ ♅
19 17	☽ ☌ ♃
21 32	☽ □ ♇
22 33	☽ ⊥ ♆
23 53	☽ ∥ ♂
04 Sunday	
03 01	♀ ⊥ ♄
05 32	☽ ⚼ ♀
14 35	☽ □ ☉
16 20	☽ ∠ ♆
19 41	☽ ⊼ ♂
05 Monday	
04 14	☽ # ♆
07 40	☽ □ ☿
07 57	☽ ± ♂
12 33	☉ ⚼ ♆
13 52	☽ △ ♀
16 35	♀ ± ♅
16 51	♅ St R
17 08	☽ ∥ ♀
20 50	☿ ± ♆
21 24	☽ ⚹ ♆
22 25	☽ ⊼ ♄
06 Tuesday	
01 37	☽ ⊻ ♅
02 05	☽ ⚼ ♂
04 19	☽ Q ☉
06 02	☽ ⊻ ♃
07 47	☽ △ ♇
09 52	☽ ± ♄
12 58	☽ ⊥ ♅
13 09	☽ # ♇
17 23	☽ ⊥ ♃
20 02	☽ Q ♆
07 Wednesday	
00 25	☽ Q ☿
02 46	☽ ⚼ ♄
04 36	☽ ⚹ ☉
05 43	☽ ∠ ♅
07 29	☽ △ ♂
10 07	☽ ∠ ♃
11 38	☽ ⚼ ♇
19 44	♀ ± ♃
20 34	☽ ∥ ☉
21 57	☽ ∥ ☿
08 Thursday	
02 15	☽ ⚹ ☿
03 06	☽ □ ♀
03 54	☽ ± ♇
04 53	☽ □ ♆
06 10	☽ △ ♄
08 54	☽ ⚹ ♅
09 53	☽ ∠ ☉
11 31	☉ ∥ ☿
13 15	☽ ⚹ ♃
14 33	☿ ⊻ ♀
14 34	☽ ⊼ ♇
19 58	☿ ⊼ ♆
23 55	♀ ☍ ♆
09 Friday	
02 54	☽ ⊥ ☉
05 35	☿ ⚹ ♄
05 58	☽ Q ♅
09 07	☽ ∠ ☿
10 17	☽ Q ♃
10 38	☉ Q ♄
13 32	☽ Q ♀
14 02	☽ ⊻ ☉
15 06	☽ □ ♂
17 21	♀ ⊻ ♄
23 45	☿ △ ♅
10 Saturday	
02 57	☽ ⊥ ☿
08 47	☽ △ ♆
10 17	☽ □ ♄
11 39	☽ ⚹ ♀
12 37	☽ □ ♅
14 36	☽ ⊻ ☿
16 53	☽ □ ♃
17 52	☽ ☍ ♇
22 43	☽ Q ♂
23 55	♀ ⊼ ♅
11 Sunday	
04 13	☉ ⚹ ♂
05 30	♀ # ♆
08 13	☿ △ ♃
09 40	☽ ⚼ ♆
14 28	☽ ∥ ☉
14 33	☽ ∠ ♀
15 13	☿ ⊼ ♇
16 12	☽ Q ♄
19 16	☽ ⚹ ♂
19 40	☽ ☌ ☉
21 32	☿ ∠ ♂
22 33	☽ ∥ ☿
12 Monday	
00 21	☽ ± ♆
06 25	☽ ⊥ ♀
10 05	☽ ⊼ ♆
11 48	☽ ⚹ ♄
13 49	☽ △ ♅
14 14	☽ # ♇
16 55	☽ ⊻ ♀
18 05	☽ △ ♃
18 50	☽ ⊼ ♇
20 35	☽ ∠ ♂
22 49	☽ ☌ ☿
13 Tuesday	
04 30	☽ ± ♇
08 48	♀ ⊼ ♃
11 41	☽ ⊥ ♂
12 06	☽ ∠ ♄
13 59	☽ ⚼ ♅
17 38	☽ # ♆
17 55	♀ △ ♇
18 18	☽ ⚼ ♃
18 58	☽ ⚼ ♇
21 44	☽ ⊻ ♂
22 54	☽ ∥ ♀
23 28	☽ ⊻ ☉
14 Wednesday	
02 40	☽ ⊥ ♄
04 28	☽ ± ♅
08 50	☽ ± ♃
09 48	☽ ⊥ ☉
10 23	☽ ☍ ♆
12 23	☽ ⊻ ♄
14 09	☽ ⊼ ♅
18 33	☽ ⊼ ♃
18 35	☉ ± ♆
19 08	☽ △ ♇
21 08	☿ ± ♇
21 14	☽ ☌ ♀
15 Thursday	
01 23	☽ ∠ ☉
03 33	☽ ∥ ♂
05 55	☽ ⊻ ☿
10 54	☽ ∥ ♄
17 03	☽ ⊥ ☿
17 50	☽ # ♅
19 27	☽ ∥ ♃
19 59	☽ # ♃
21 36	☽ ∥ ♅
16 Friday	
00 31	☽ ☌ ♂
03 40	☽ ⚹ ☉
04 31	☽ # ♄
09 55	☽ ∠ ☿
10 49	☽ # ♂
11 22	☽ ⊼ ♆
13 46	☽ ☌ ♄
15 19	☽ ☍ ♅
19 58	☽ ☍ ♃
20 24	☽ □ ♇
21 23	☽ ± ♆
21 49	☿ # ♇
17 Saturday	
01 06	☽ Q ☉
02 50	☽ ⊻ ♀
11 01	☽ # ♀
12 35	☽ ⚼ ♆
13 53	☽ ⊥ ♀
14 36	☽ ⚹ ☿
19 44	☿ ∠ ♄
23 55	☽ ∥ ♆
18 Sunday	
02 59	☽ Q ♇
05 22	☽ ⊻ ♂
06 41	☽ ∠ ♀
07 32	☿ ⚼ ♅
10 11	☽ □ ☉
14 10	☽ Q ☿
14 26	☽ △ ♆
16 17	☽ ⊥ ♂
17 18	☽ ⊻ ♄
18 37	☽ ⊼ ♅
22 58	☽ # ☿
23 38	☽ ⊼ ♃
23 55	☽ ⚹ ♇
19 Monday	
03 57	☽ ⊥ ♄
05 12	☽ ± ♅
05 50	☽ ∥ ♇
08 56	☽ ∠ ♂
10 18	☽ ± ♃
11 26	☽ ⚹ ♀
16 49	☿ ⊥ ♂
20 08	☽ ∠ ♄
21 19	☽ ⚼ ♅
21 41	☽ # ☉
20 Tuesday	
01 26	☿ ⚼ ♃
02 31	☽ ⚼ ♃
02 39	☽ □ ☿
02 44	☽ ∠ ♇
03 19	☿ ⚼ ♇
11 06	☽ Q ♀
13 20	☽ ⚹ ♂
19 11	☽ ⊥ ♇
20 07	☽ △ ☉
20 16	☽ □ ♆
22 09	☉ ⊼ ♆
23 43	☽ ⚹ ♄
21 Wednesday	
00 44	☽ △ ♅
06 06	☽ △ ♃
06 17	☽ ⊻ ♇
12 38	☽ Q ♂
22 17	☽ Q ♄
23 42	☽ □ ♀
22 Thursday	
02 25	☽ ⚼ ☉
05 55	☽ Q ♆
18 26	☽ △ ☿
20 59	☽ ± ☉
23 Friday	
00 31	☽ □ ♂
01 12	☉ ⚹ ♄
04 50	☽ ⚹ ♆
08 55	☽ □ ♄
09 30	☽ ⊼ ☉
09 33	☽ □ ♅
10 06	☉ △ ♅
12 08	♃ St R
15 14	☽ □ ♃
15 18	☽ ☌ ♇
19 15	☿ ⊥ ♄
21 13	♂ ∥ ♄
23 51	☿ ± ♅
24 Saturday	
03 29	☽ ⚼ ☿
09 59	☽ ∠ ♆
15 12	☽ △ ♀
20 42	☽ Q ♅
23 36	☽ ± ☿
25 Sunday	
02 32	☽ Q ♃
03 42	☽ ⊥ ♆
04 20	♃ □ ♇
12 16	☽ # ☉
13 06	☽ ⊼ ☿
14 20	☽ △ ♂
15 36	☽ ⊻ ♆
20 19	☽ △ ♄
20 30	☽ ⚹ ♅
21 03	☽ ∥ ♇
23 53	☽ ⚼ ♀
26 Monday	
00 27	☿ ± ♃
01 37	☽ ☍ ☉
02 23	☽ ⊻ ♇
02 24	☽ ⚹ ♃
06 25	☿ ⊻ ♂
07 23	☿ # ♆
10 03	☿ ☍ ♆
11 04	☉ ⊼ ♇
11 30	☉ △ ♃
14 22	☽ ⊥ ♇
14 41	♂ ⊼ ♆
17 07	♄ ☍ ♅
19 42	☽ ± ♀
21 56	☽ ⚼ ♂
27 Tuesday	
02 31	☽ ∠ ♅
02 36	☽ ⚼ ♄
06 56	☽ ∥ ♆
08 27	☽ ∠ ♇
08 30	☽ ∠ ♃
08 59	☽ ⊼ ♀
10 47	☽ # ☿
17 03	☽ ± ♂
20 40	☽ ⊥ ♅
20 54	☽ ± ♄
28 Wednesday	
02 40	☽ ⊥ ♃
03 46	☽ ☌ ♆
04 42	☿ ⊼ ♅
05 51	☽ ⊼ ♂
07 37	☿ ⊻ ♄
08 47	☽ ⊻ ♅
09 08	☽ ⊼ ♄
09 19	☽ ☍ ☿
14 44	☽ ⚹ ♇
14 48	☽ ⊻ ♃
19 22	☽ ⊼ ☉
29 Thursday	
02 16	☽ # ♀
08 35	☽ ± ☉
12 46	♂ # ♅
13 30	☽ # ♄
15 00	☽ Q ♇
19 22	☽ ∥ ♅
19 44	☽ # ♂
22 00	☽ # ♃
22 09	☽ ∥ ♃
30 Friday	
00 12	☽ ∥ ♂
00 48	☽ # ♅
03 44	☽ ☍ ♀
04 24	☽ ⚼ ☉
06 32	☽ ∥ ♄
10 04	☿ △ ♇
10 44	☿ ⊼ ♃
13 31	♂ ☍ ♅
14 18	☽ ∥ ♀
16 22	☽ ⊻ ♆
21 23	☽ ☌ ♅
21 49	☽ ☍ ♂
21 57	☉ # ♇
22 16	☽ ☍ ♄
31 Saturday	
03 17	☽ □ ♇
03 21	☽ ☌ ♃
04 24	☽ ⊥ ♆
05 19	☽ ⊼ ☿
08 07	♂ ☌ ♄
13 13	☽ △ ☉
16 55	♂ # ♃
18 41	☽ ± ☿
18 42	♂ ∥ ♃
19 16	☽ ∥ ☿
22 22	☽ ∠ ♆

All ephemeris data is given at 12.00 UT and the Moon's longitude is additionally given for 24.00 UT

AUGUST 2010

Date	Sidereal time h m s	LONGITUDES Sun ☉ ° ' "	Moon ☽ ° ' "	Moon ☽ 24.00 ° ' "	Mercury ☿ ° '	Venus ♀ ° '	Mars ♂ ° '	Jupiter ♃ ° '	Saturn ♄ ° '	Uranus ♅ ° '	Neptune ♆ ° '	Pluto ♇ ° '
01	08 39 57	09 ♌ 08 31	19 ♈ 41 21	25 ♈ 47 25	05 ♍ 50	24 ♍ 05	01 ♎ 32	03 ♈ 16	00 ♎ 56	00 ♈ 19	27 ♒ 48	03 ♑ 15
02	08 43 54	10 05 55	01 ♉ 56 55	08 ♉ 10 22	07 00	25 08	02 09	03 R 14	01 01	00 R 17	27 R 47	03 R 14
03	08 47 50	11 03 20	14 ♉ 28 18	20 ♉ 51 13	08 08	26 11	02 46	03 12	01 07	00 16	27 45	03 13
04	08 51 47	12 00 47	27 ♉ 19 37	03 ♊ 53 56	09 13	27 14	03 23	03 10	01 12	00 15	27 44	03 11
05	08 55 43	12 58 15	10 ♊ 34 33	17 ♊ 21 45	10 15	28 17	04 00	03 08	01 18	00 13	27 42	03 10
06	08 59 40	13 55 44	24 ♊ 15 43	01 ♋ 16 30	11 15	29 ♍ 19	04 37	03 05	01 24	00 12	27 40	03 09
07	09 03 36	14 53 15	08 ♋ 24 00	15 ♋ 37 56	12 12	00 ♎ 21	05 14	03 02	01 30	00 11	27 39	03 08
08	09 07 33	15 50 46	22 ♋ 57 50	00 ♌ 23 02	13 06	01 23	05 51	02 59	01 35	00 09	27 37	03 07
09	09 11 29	16 48 19	07 ♌ 52 42	15 ♌ 25 48	13 57	02 24	06 28	02 56	01 41	00 08	27 36	03 06
10	09 15 26	17 45 53	23 ♌ 01 11	00 ♍ 37 35	14 45	03 26	07 06	02 52	01 47	00 06	27 34	03 05
11	09 19 23	18 43 28	08 ♍ 13 43	15 ♍ 48 17	15 30	04 27	07 43	02 49	01 53	00 04	27 32	03 04
12	09 23 19	19 41 04	23 ♍ 20 04	00 ♎ 47 59	16 10	05 27	08 21	02 45	01 59	00 03	27 31	03 03
13	09 27 16	20 38 41	08 ♎ 11 03	15 ♎ 28 31	16 47	06 27	08 59	02 41	02 06	00 ♈ 01	27 29	03 02
14	09 31 12	21 36 19	22 ♎ 39 50	29 ♎ 44 37	17 21	07 27	09 36	02 37	02 12	29 ♓ 59	27 28	03 01
15	09 35 09	22 33 58	06 ♏ 42 41	13 ♏ 34 03	17 49	08 27	10 14	02 33	02 18	29 58	27 26	03 01
16	09 39 05	23 31 37	20 ♏ 18 50	26 ♏ 57 19	18 14	09 26	10 52	02 28	02 24	29 56	27 24	03 00
17	09 43 02	24 29 18	03 ♐ 29 50	09 ♐ 56 49	18 34	10 25	11 30	02 24	02 31	29 54	27 23	02 59
18	09 46 58	25 27 00	16 ♐ 18 44	22 ♐ 36 05	18 49	11 23	12 08	02 19	02 37	29 52	27 21	02 58
19	09 50 55	26 24 43	28 ♐ 49 22	04 ♑ 59 05	18 59	12 21	12 46	02 14	02 43	29 50	27 19	02 57
20	09 54 52	27 22 27	11 ♑ 05 42	17 ♑ 09 41	19 03	13 19	13 24	02 09	02 50	29 49	27 18	02 57
21	09 58 48	28 20 13	23 ♑ 11 28	29 ♑ 11 26	19 R 02	14 16	14 03	02 03	02 56	29 47	27 16	02 56
22	10 02 45	29 ♌ 17 59	05 ♒ 09 57	11 ♒ 07 19	18 56	15 13	14 41	01 58	03 03	29 45	27 15	02 55
23	10 06 41	00 ♍ 15 47	17 ♒ 03 51	22 ♒ 59 49	18 43	16 09	15 19	01 52	03 10	29 43	27 13	02 55
24	10 10 38	01 13 36	28 ♒ 55 27	04 ♓ 50 58	18 25	17 05	15 58	01 47	03 16	29 41	27 11	02 54
25	10 14 34	02 11 26	10 ♓ 46 34	16 ♓ 42 28	18 01	18 00	16 36	01 41	03 23	29 39	27 10	02 53
26	10 18 31	03 09 18	22 ♓ 38 52	28 ♓ 35 59	17 31	18 55	17 15	01 35	03 30	29 37	27 08	02 53
27	10 22 27	04 07 12	04 ♈ 34 02	10 ♈ 33 17	16 55	19 49	17 54	01 29	03 36	29 34	27 06	02 52
28	10 26 24	05 05 07	16 ♈ 33 59	22 ♈ 36 28	16 15	20 43	18 32	01 22	03 43	29 32	27 05	02 52
29	10 30 21	06 03 04	28 ♈ 41 04	04 ♉ 48 08	15 29	21 36	19 11	01 16	03 50	29 30	27 03	02 51
30	10 34 17	07 01 03	10 ♉ 58 05	17 ♉ 11 22	14 39	22 29	19 50	01 09	03 57	29 28	27 01	02 51
31	10 38 14	07 ♍ 59 04	23 ♉ 28 27	29 ♉ 49 47	13 ♍ 46	23 ♎ 21	20 ♎ 29	01 ♈ 03	04 ♎ 04	29 ♓ 26	27 ♒ 00	02 ♑ 50

Date	Moon True ☊ ° '	Moon Mean ☊ ° '	Moon ☽ Latitude ° '
01	11 ♑ 35	10 ♑ 23	05 N 10
02	11 R 34	10 20	04 54
03	11 D 34	10 16	04 23
04	11 34	10 13	03 39
05	11 36	10 10	02 42
06	11 37	10 07	01 34
07	11 38	10 04	00 N 18
08	11 R 38	10 00	01 S 01
09	11 36	09 57	02 18
10	11 32	09 54	03 26
11	11 27	09 51	04 20
12	11 22	09 48	04 55
13	11 17	09 45	05 10
14	11 12	09 41	05 04
15	11 09	09 38	04 40
16	11 08	09 35	04 01
17	11 D 09	09 32	03 10
18	11 10	09 29	02 11
19	11 11	09 26	01 06
20	11 12	09 22	00 S 01
21	11 R 11	09 19	01 N 04
22	11 08	09 16	02 06
23	11 03	09 13	03 01
24	10 56	09 10	03 48
25	10 47	09 06	04 25
26	10 37	09 03	04 51
27	10 27	09 00	05 04
28	10 18	08 57	05 03
29	10 10	08 54	04 50
30	10 05	08 51	04 22
31	10 ♑ 02	08 ♑ 47	03 N 42

DECLINATIONS

Date	Sun ☉ ° '	Moon ☽ ° '	Mercury ☿ ° '	Venus ♀ ° '	Mars ♂ ° '	Jupiter ♃ ° '	Saturn ♄ ° '	Uranus ♅ ° '	Neptune ♆ ° '	Pluto ♇ ° '
01	17 N 58	12 N 29	08 N 50	02 N 26	00 S 12	00 S 01	01 N 40	00 S 35	12 S 41	18 S 22
02	17 43	16 44	08 15	01 56	00 27	00 02	01 38	00 36	12 42	18 22
03	17 27	20 22	07 40	01 27	00 42	00 03	01 36	00 36	12 42	18 23
04	17 11	23 07	07 06	00 57	00 58	00 05	01 33	00 37	12 43	18 23
05	16 55	24 42	06 32	00 N 27	01 13	00 06	01 31	00 37	12 43	18 23
06	16 39	24 52	05 59	00 S 02	01 29	00 07	01 29	00 38	12 44	18 23
07	16 22	23 28	05 27	00 32	01 44	00 08	01 26	00 39	12 45	18 24
08	16 05	20 28	04 56	01 01	02 00	00 10	01 24	00 39	12 45	18 24
09	15 48	16 04	04 26	01 31	02 15	00 11	01 21	00 40	12 46	18 24
10	15 30	10 36	03 57	02 00	02 31	00 13	01 19	00 40	12 46	18 24
11	15 13	04 N 28	03 29	02 30	02 46	00 15	01 16	00 41	12 47	18 24
12	14 55	01 S 52	03 03	02 59	03 02	00 16	01 14	00 42	12 47	18 25
13	14 37	07 59	02 38	03 29	03 17	00 18	01 11	00 42	12 48	18 25
14	14 18	13 31	02 15	03 58	03 33	00 20	01 09	00 43	12 49	18 25
15	13 59	18 10	01 54	04 27	03 48	00 22	01 06	00 44	12 49	18 25
16	13 41	21 41	01 35	04 56	04 04	00 24	01 03	00 45	12 50	18 26
17	13 22	23 58	01 18	05 25	04 19	00 26	01 01	00 45	12 50	18 26
18	13 02	24 54	01 03	05 54	04 35	00 28	00 58	00 46	12 51	18 26
19	12 43	24 32	00 50	06 22	04 50	00 30	00 56	00 47	12 51	18 26
20	12 23	22 59	00 40	06 51	05 06	00 33	00 53	00 48	12 52	18 27
21	12 03	20 23	00 33	07 19	05 21	00 35	00 50	00 48	12 53	18 27
22	11 43	16 57	00 29	07 47	05 37	00 37	00 48	00 49	12 53	18 27
23	11 23	12 51	00 28	08 15	05 52	00 40	00 45	00 50	12 54	18 27
24	11 02	08 17	00 30	08 43	06 08	00 42	00 42	00 51	12 54	18 28
25	10 42	03 S 26	00 36	09 10	06 23	00 45	00 39	00 52	12 55	18 28
26	10 21	01 N 32	00 45	09 37	06 39	00 47	00 37	00 52	12 55	18 28
27	10 00	06 27	00 57	10 04	06 54	00 50	00 34	00 53	12 56	18 28
28	09 39	11 11	01 13	10 31	07 10	00 53	00 31	00 54	12 57	18 29
29	09 18	15 31	01 32	10 58	07 25	00 55	00 28	00 55	12 57	18 29
30	08 56	19 16	01 55	11 24	07 40	00 58	00 26	00 56	12 58	18 29
31	08 N 35	22 N 14	02 N 21	11 S 50	07 S 56	01 S 01	00 N 23	00 S 57	12 S 58	18 S 29

ZODIAC SIGN ENTRIES

Date	h	m	Planets
02	08	13	☽ ♉
04	16	54	☽ ♊
06	21	50	☽ ♋
07	03	47	♀ ♎
08	23	23	☽ ♌
10	23	01	☽ ♍
12	22	43	☽ ♎
14	03	36	♅ ♓
15	00	26	☽ ♏
17	05	34	☽ ♐
19	14	17	☽ ♑
22	01	37	☽ ♒
23	05	27	☉ ♍
24	14	11	☽ ♓
27	02	49	☽ ♈
29	14	35	☽ ♉

LATITUDES

Date	Mercury ☿ ° '	Venus ♀ ° '	Mars ♂ ° '	Jupiter ♃ ° '	Saturn ♄ ° '	Uranus ♅ ° '	Neptune ♆ ° '	Pluto ♇ ° '
01	00 S 34	00 N 05	00 N 27	01 S 26	02 N 14	00 S 46	00 S 29	05 N 02
04	01 06	00 S 10	00 25	01 27	02 13	00 47	00 29	05 01
07	01 40	00 25	00 22	01 28	02 13	00 47	00 29	05 01
10	02 14	00 42	00 20	01 29	02 12	00 47	00 29	05 00
13	02 48	01 00	00 18	01 30	02 12	00 47	00 29	04 59
16	03 20	01 18	00 16	01 30	02 12	00 47	00 29	04 59
19	03 50	01 37	00 13	01 31	02 11	00 47	00 29	04 58
22	04 14	01 57	00 11	01 32	02 11	00 47	00 29	04 57
25	04 30	02 17	00 09	01 33	02 11	00 47	00 29	04 57
28	04 34	02 38	00 07	01 33	02 11	00 47	00 29	04 56
31	04 S 22	02 S 59	00 N 05	01 S 34	02 N 10	00 S 47	00 S 29	04 N 55

DATA

Julian Date	2455410
Delta T	+67 seconds
Ayanamsa	24° 00' 35"
Synetic vernal point	05° ♓ 06' 24"
True obliquity of ecliptic	23° 26' 18"

MOON'S PHASES, APSIDES AND POSITIONS ☽

Date	h	m	Phase	Longitude ° '	Eclipse Indicator
03	04	59	☾	10 ♉ 47	
10	03	08	●	17 ♌ 25	
16	18	14	☽	23 ♏ 47	
24	17	05	○	01 ♓ 26	

Day	h	m		
10	17	51	Perigee	
25	05	34	Apogee	
06	02	48	Max dec	24° N 59'
12	04	53	0S	
18	17	11	Max dec	24° S 56'
26	04	37	0N	

LONGITUDES

Date	Chiron ⚷ ° '	Ceres ⚳ ° '	Pallas ⚴ ° '	Juno ⚵ ° '	Vesta ⚶ ° '	Black Moon Lilith ⚸ ° '
01	29 ♒ 29	20 ♐ 50	08 ♏ 43	29 ♋ 22	26 ♍ 23	03 ♓ 49
11	29 ♒ 00	20 ♐ 42	11 ♏ 07	04 ♌ 34	01 ♎ 04	04 ♓ 56
21	28 ♒ 30	21 ♐ 09	13 ♏ 52	09 ♌ 40	05 ♎ 52	06 ♓ 03
31	28 ♒ 00	22 ♐ 09	16 ♏ 55	14 ♌ 39	10 ♎ 47	07 ♓ 10

All ephemeris data is given at 12.00 UT and the Moon's longitude is additionally given for 24.00 UT

ASPECTARIAN

h m Aspects

01 Sunday
13 06 ☽ # ♆
14 30 ☽ Ᵽ ☿
14 35 ☉ ± ♇
21 28 ☽ ⊼ ♀

02 Monday
01 49 ☉ ∠ ♀
03 54 ☽ ⚹ ♆
08 47 ☽ ⊻ ♅
10 11 ☽ ⊼ ♄
10 17 ☽ ± ♀
12 24 ☽ ⊼ ♂
14 28 ☽ △ ♇
14 29 ☽ ⊻ ♃
17 37 ☽ ∥ ☉
20 22 ☽ ⊥ ♅
21 52 ☽ ± ♄
22 13 ☽ # ♇
22 44 ☽ △ ☿

03 Tuesday
00 34 ☽ ± ♂
02 00 ☽ ⊥ ♃
02 07 ♂ ∥ ♅
03 02 ☽ Q ♆
04 04 ♀ ∥ ♄
04 59 ☽ □ ☉
05 12 ☽ Ᵽ ♀
05 33 ♃ □ ♇
13 30 ☽ ∠ ♅
15 07 ☽ Ᵽ ♄
18 31 ☽ Ᵽ ♂
19 01 ☽ ∠ ♃
19 02 ☽ Ᵽ ♇

04 Wednesday
04 20 ♂ ☍ ♃
04 58 ♂ □ ♇
11 39 ♀ # ♂
11 45 ☽ ± ♇
11 49 ☽ △ ♀
12 44 ☽ □ ♆
17 19 ☽ Q ☉
17 21 ☽ ⚹ ♅
19 09 ☽ △ ♄
22 38 ☽ ⚹ ♃
22 42 ☽ ⊼ ♇
22 57 ♀ ⊼ ♆
23 36 ☽ △ ♂

05 Thursday
01 05 ♂ ± ♆
04 07 ♀ # ♅
11 23 ☽ □ ☿
14 56 ☽ Q ♅
16 35 ☽ ⚹ ☉
20 03 ☽ Q ♃

06 Friday
04 51 ♀ # ♃
11 58 ♂ # ♄
16 09 ♀ ∥ ♃
17 51 ☽ △ ♆
20 36 ☽ ∠ ☉
21 11 ☽ Q ☿
21 22 ☽ □ ♀
22 09 ☽ □ ♅

07 Saturday
00 17 ☽ □ ♄
03 01 ☽ □ ♃
03 10 ☽ ☍ ♇
06 27 ☽ □ ♂
07 59 ♀ ☍ ♅
12 52 ☽ ⊥ ☉
17 37 ♀ ∥ ♅
18 46 ☽ ⚹ ☿
19 03 ☉ Ᵽ ♅
19 03 ☽ Ᵽ ♆
23 32 ☽ ⊻ ☉

08 Sunday
05 43 ☽ Q ♀
06 28 ☽ Q ♄
09 49 ☽ ± ♆
13 30 ☽ Q ♂
17 24 ♀ ☌ ♄
19 32 ☽ ⊼ ♆
20 51 ☽ ∠ ☿
23 36 ☽ △ ♅

09 Monday
00 11 ☽ # ♇
02 02 ☽ ⚹ ♄
02 36 ☽ ⚹ ♀
04 07 ☽ △ ♃
04 22 ☽ ⊼ ♇
04 45 ♀ # ♄
08 47 ☉ ∠ ♄
09 40 ☽ ⚹ ♂
12 08 ☽ ⊥ ☿
13 23 ☽ ∥ ☉
13 57 ☽ ± ♇
22 13 ☽ ⊻ ☿
23 30 ☽ Ᵽ ♅
23 40 ♀ ☍ ♃

10 Tuesday
02 05 ☽ ∠ ♄
02 55 ☽ # ♆
03 08 ☽ ☌ ☉
03 54 ☽ Ᵽ ♃
04 05 ♀ □ ♇
04 13 ☽ Ᵽ ♇
04 13 ☽ ∠ ♀
10 29 ☽ ∠ ♂
13 42 ☽ ± ♅
14 35 ☉ Ᵽ ♃
15 14 ♀ ± ♆
16 24 ☽ ⊥ ♄
18 04 ☽ ± ♃
19 10 ☽ ☍ ♆
19 27 ☽ ⊥ ♀
19 54 ☉ Ᵽ ♇
23 09 ☽ ⊼ ♅

11 Wednesday
01 17 ☽ ⊥ ♂
01 56 ☽ ⊻ ♄
03 29 ☽ ⊼ ♃
03 52 ☽ △ ♇
05 36 ☽ ⊻ ♀
11 10 ☽ ⊻ ♂
15 58 ☽ ∥ ☿
18 10 ☽ # ♂
18 54 ☽ # ♀

12 Thursday
00 04 ☽ ☌ ☿
00 10 ☽ ∥ ♄
02 16 ☽ # ♅
03 53 ☽ # ♃
05 46 ☽ ⊻ ☉
05 54 ☽ ∥ ♃
07 31 ☽ ∥ ♅
09 34 ☽ # ♄
12 52 ☿ # ♂
13 40 ☿ # ♀
16 01 ☽ ⊥ ☉
16 02 ♀ ∥ ♂
16 14 ☽ # ☿
16 37 ☽ ∥ ♂
16 38 ☽ ∥ ♀
18 41 ☽ ⊼ ♆
22 46 ☽ ☍ ♅

13 Friday
02 01 ☽ ☌ ♄
03 06 ☽ ☍ ♃
03 38 ☽ □ ♇
04 22 ☽ ± ♆
07 34 ☽ ∠ ☉
08 58 ☽ ☌ ♀
13 21 ☽ ☌ ♂
19 03 ☽ Ᵽ ♆

14 Saturday
02 46 ☽ ⊻ ☿
08 43 ☽ ∥ ♆
09 15 ☽ Q ♇
10 06 ☽ ⚹ ☉
13 11 ☽ ⊥ ☿
15 31 ☽ # ☉
20 06 ☽ △ ♆

15 Sunday
00 24 ☽ ⊼ ♅
04 19 ☽ ⊻ ♄
04 50 ☽ ⊼ ♃
05 03 ☽ ∠ ☿
05 36 ☽ ⚹ ♇
08 01 ☽ Q ☉
10 42 ☽ ± ♅
13 33 ☽ ∥ ♇
14 47 ☽ ⊥ ♄
15 11 ☽ ± ♃
15 15 ☽ ⊻ ♀
18 27 ☽ ⊻ ♂

16 Monday
02 26 ☽ Ᵽ ♅
02 37 ☽ ⊥ ♀
05 33 ☽ ⊥ ♂
06 46 ☽ ∠ ♄
06 57 ☽ Ᵽ ♃
07 52 ☽ ∠ ♇
08 10 ☽ ⚹ ☿
18 14 ☽ □ ☉
20 01 ☽ ∠ ♀
20 45 ♃ ☍ ♄
21 48 ☉ ± ♅
22 32 ☽ ∠ ♂

17 Tuesday
00 04 ☽ ⊥ ♇
00 48 ☽ □ ♆
05 24 ☽ △ ♅
06 28 ☽ Q ☿
09 59 ☽ △ ♃
10 10 ☽ ⚹ ♄
11 03 ☽ ⊻ ♇

18 Wednesday
01 56 ☽ ⚹ ♀
03 42 ☽ ⚹ ♂
08 46 ☽ Q ♄
10 11 ☽ Q ♆
16 50 ☽ □ ☿
19 51 ♂ Ᵽ ♆
21 59 ☿ ∥ ♄

19 Thursday
01 37 ☉ # ♆
02 39 ☽ Q ♀
03 45 ☽ Q ♂
06 57 ☽ △ ☉
07 51 ☉ ± ♃
09 06 ☽ ⚹ ♆
11 15 ♀ Ᵽ ♆
13 58 ☽ □ ♅
18 35 ☽ □ ♃
18 48 ☿ # ♅
19 39 ☽ □ ♄
20 02 ☽ ☌ ♇
20 46 ☉ ⊥ ♄

20 Friday
10 07 ☉ ☍ ♆
14 22 ☽ ∠ ♆
14 44 ☽ Ᵽ ☉
16 45 ☽ □ ♀
16 49 ☽ □ ♂
18 49 ♀ ☌ ♂
19 58 ☿ St R

21 Saturday
01 15 ☽ Q ♅
03 46 ☽ △ ☿
05 48 ☽ Q ♃
06 34 ☿ # ♃
08 11 ☽ ⊥ ♆
10 09 ☽ ± ☉
10 16 ♄ □ ♇
20 08 ☽ ⊻ ♆
23 11 ☽ ⊼ ☉

22 Sunday
01 00 ♄ # ♅
01 08 ☽ ⚹ ♅
02 07 ☽ ∥ ♇
05 37 ☽ ⚹ ♃
07 29 ☽ ⊻ ♇
07 42 ☽ △ ♄
09 33 ☽ Ᵽ ☿
19 33 ☽ ⊥ ♇
22 42 ☉ ⊼ ♅

23 Monday
03 24 ☽ ± ☿
07 15 ☽ ∠ ♅
08 17 ☽ △ ♂
10 00 ☽ △ ♀
11 37 ☽ ∠ ♃
11 44 ☽ ∥ ♆
13 42 ☽ ∠ ♇
14 14 ☽ Ᵽ ♄
15 17 ☽ ⊼ ☿
16 21 ☉ ∠ ♂
20 35 ☽ # ☉
21 34 ♄ ± ♆

24 Tuesday
01 24 ☽ ⊥ ♅
05 41 ☽ ⊥ ♃
08 29 ☽ ☌ ♆
08 37 ☽ ± ♄
10 02 ☽ ∥ ♀
11 38 ♃ # ♄
13 31 ☽ ⊻ ♅
16 22 ☽ Ᵽ ♂
17 05 ☽ ☍ ☉
17 44 ☽ ⊻ ♃
18 56 ☽ Ᵽ ♀
20 03 ☽ ⚹ ♇
20 53 ☽ ⊼ ♄
22 14 ☽ ∥ ♂

25 Wednesday
00 28 ☉ ⊼ ♃
11 38 ☽ ± ♂
12 09 ☿ ⊻ ♀
14 42 ☽ ± ♀
20 19 ☽ Q ♇
20 45 ☿ ∥ ♄

26 Thursday
00 26 ☽ ∥ ♅
00 28 ☽ ⊼ ♂
00 54 ☽ ∥ ♃
01 23 ☽ # ☿
01 34 ☽ # ♄
02 04 ☽ ☍ ☿
03 51 ☽ ⊼ ♀
05 12 ☉ △ ♇
07 36 ☽ ∥ ♄
08 03 ☽ ∥ ☿
08 23 ☽ # ♃
08 49 ☽ # ♅
17 18 ☿ ⊻ ♂
19 41 ☿ # ♃
21 01 ☽ ⊻ ♆
21 31 ☉ ⊻ ♄

27 Friday
02 00 ☽ ☌ ♅
04 56 ☿ # ♅
05 51 ☽ ☌ ♃
08 36 ☽ □ ♇
09 04 ☽ ⊥ ♆
09 44 ☉ # ♀
10 03 ☽ ☍ ♄
11 01 ☽ ⊼ ☉
14 20 ☽ # ♂

28 Saturday
00 06 ☽ ± ☉
03 04 ☽ ∠ ♆
04 36 ☽ ∥ ☉
08 12 ☽ # ♀
11 23 ☽ ⊼ ☿
15 48 ♀ Q ♇
16 09 ☽ ☍ ♂
19 36 ☽ Ᵽ ☉
20 55 ☽ ☍ ♀
21 29 ☽ # ♆
22 38 ☽ ± ☿

29 Sunday
06 40 ♃ ∥ ♅
08 47 ☽ ⚹ ♆
10 14 ☿ ⊥ ♀
13 36 ☽ ⊻ ♅
15 19 ☽ Ᵽ ☿
17 02 ☽ ⊻ ♃
20 11 ☽ △ ♇
22 12 ☽ ⊼ ♄

30 Monday
01 20 ☽ ⊥ ♅
03 40 ☽ △ ☉
04 39 ☽ ⊥ ♃
06 36 ☽ # ♇
08 14 ☽ Q ♆
10 00 ☽ ± ♄
18 40 ☽ △ ☿
18 45 ☽ ∠ ♅
21 55 ☽ ∠ ♃

31 Tuesday
00 55 ☿ ⊥ ♂
01 15 ☽ Ᵽ ♇
03 31 ☽ Ᵽ ♄
06 00 ☽ ⊼ ♂
11 45 ☽ ⊼ ♀
18 00 ☽ ± ♂
18 22 ☽ ± ♇
18 39 ☽ □ ♆
23 13 ☽ ⚹ ♅
23 54 ☽ ± ♀

SEPTEMBER 2010

LONGITUDES

Date	Sidereal time h m s	Sun ☉ ° ' "	Moon ☽ ° ' "	Moon ☽ 24.00 ° ' "	Mercury ☿ ° '	Venus ♀ ° '	Mars ♂ ° '	Jupiter ♃ ° '	Saturn ♄ ° '	Uranus ♅ ° '	Neptune ♆ ° '	Pluto ♇ ° '
01	10 42 10	08 ♍ 57 06	06 ♊ 15 53	12 ♊ 47 12	12 ♍ 50	24 ♎ 12	21 ♎ 08	00 ♈ 56	04 ♎ 11	29 ♓ 24	26 ♒ 58	02 ♑ 50
02	10 46 07	09 55 11	19 ♊ 24 13	26 ♊ 07 20	11 R 53	25 03	21 47	00 R 49	04 18	29 R 22	26 R 57	02 R 50
03	10 50 03	10 53 18	02 ♋ 56 53	09 ♋ 53 06	10 56	25 53	22 26	00 42	04 25	29 19	26 55	02 49
04	10 54 00	11 51 26	16 ♋ 56 08	24 ♋ 05 54	10 00	26 43	23 06	00 35	04 32	29 17	26 54	02 49
05	10 57 56	12 49 37	01 ♌ 22 12	08 ♌ 44 36	09 06	27 31	23 45	00 28	04 39	29 15	26 52	02 49
06	11 01 53	13 47 49	16 ♌ 12 26	23 ♌ 44 52	08 15	28 19	24 24	00 20	04 46	29 12	26 50	02 48
07	11 05 50	14 46 03	01 ♍ 20 49	08 ♍ 59 03	07 29	29 06	25 04	00 13	04 53	29 10	26 49	02 48
08	11 09 46	15 44 19	16 ♍ 38 11	24 ♍ 16 49	06 49	29 ♎ 53	25 43	00 05	05 00	29 08	26 47	02 48
09	11 13 43	16 42 37	01 ♎ 53 29	09 ♎ 26 52	06 15	00 ♏ 38	26 23	29 ♓ 58	05 07	29 06	26 46	02 48
10	11 17 39	17 40 57	16 ♎ 55 43	24 ♎ 19 01	05 49	01 23	27 03	29 50	05 14	29 03	26 44	02 48
11	11 21 36	18 39 18	01 ♏ 35 55	08 ♏ 45 49	05 32	02 07	27 43	29 42	05 21	29 01	26 43	02 48
12	11 25 32	19 37 40	15 ♏ 48 22	22 ♏ 43 25	05 23	02 50	28 22	29 35	05 29	28 58	26 41	02 47
13	11 29 29	20 36 05	29 ♏ 31 02	06 ♐ 11 25	05 D 23	03 32	29 02	29 27	05 36	28 56	26 40	02 47
14	11 33 25	21 34 31	12 ♐ 44 56	19 ♐ 12 03	05 33	04 12	29 ♎ 42	29 19	05 43	28 54	26 38	02 D 47
15	11 37 22	22 32 58	25 ♐ 33 19	01 ♑ 49 19	05 52	04 52	00 ♏ 22	29 11	05 50	28 51	26 37	02 47
16	11 41 19	23 31 28	08 ♑ 00 40	14 ♑ 08 01	06 20	05 31	01 02	29 03	05 58	28 49	26 36	02 47
17	11 45 15	24 29 58	20 ♑ 11 58	26 ♑ 13 08	06 57	06 08	01 43	28 55	06 05	28 47	26 34	02 48
18	11 49 12	25 28 31	02 ♒ 12 06	08 ♒ 09 24	07 42	06 45	02 23	28 47	06 12	28 44	26 33	02 48
19	11 53 08	26 27 05	14 ♒ 05 31	20 ♒ 00 55	08 36	07 20	03 03	28 39	06 20	28 42	26 31	02 48
20	11 57 05	27 25 41	25 ♒ 55 59	01 ♓ 51 04	09 37	07 53	03 44	28 31	06 27	28 39	26 30	02 48
21	12 01 01	28 24 18	07 ♓ 46 30	13 ♓ 42 32	10 45	08 26	04 24	28 23	06 34	28 37	26 29	02 48
22	12 04 58	29 ♍ 22 57	19 ♓ 39 23	25 ♓ 37 15	11 59	08 57	05 05	28 15	06 42	28 35	26 27	02 49
23	12 08 54	00 ♎ 21 39	01 ♈ 36 18	07 ♈ 36 40	13 20	09 26	05 45	28 07	06 49	28 32	26 26	02 49
24	12 12 51	01 20 22	13 ♈ 38 30	19 ♈ 41 55	14 45	09 54	06 26	27 59	06 56	28 30	26 25	02 49
25	12 16 48	02 19 07	25 ♈ 47 04	01 ♉ 54 06	16 14	10 20	07 07	27 51	07 04	28 27	26 24	02 49
26	12 20 44	03 17 55	08 ♉ 03 12	14 ♉ 14 33	17 48	10 45	07 47	27 43	07 11	28 25	26 22	02 50
27	12 24 41	04 16 44	20 ♉ 28 25	26 ♉ 45 02	19 25	11 08	08 28	27 35	07 19	28 23	26 21	02 50
28	12 28 37	05 15 36	03 ♊ 04 45	09 ♊ 27 53	21 04	11 30	09 09	27 27	07 26	28 20	26 20	02 51
29	12 32 34	06 14 30	15 ♊ 54 48	22 ♊ 25 54	22 46	11 49	09 50	27 20	07 33	28 18	26 19	02 51
30	12 36 30	07 ♎ 13 26	29 ♊ 01 34	05 ♋ 42 10	24 ♍ 29	12 ♏ 07	10 ♏ 31	27 ♓ 12	07 ♎ 41	28 ♓ 15	26 ♒ 18	02 ♑ 52

Moon Nodes and Latitude

Date	Moon True ☊ ° '	Moon Mean ☊ ° '	Moon ☽ Latitude ° '
01	10 ♑ 00	08 ♑ 44	02 N 50
02	10 D 01	08 41	01 48
03	10 01	08 38	00 N 38
04	10 R 01	08 35	00 S 37
05	10 00	08 31	01 51
06	09 56	08 28	03 00
07	09 49	08 25	03 58
08	09 40	08 22	04 39
09	09 30	08 19	05 00
10	09 21	08 16	05 00
11	09 12	08 12	04 40
12	09 06	08 09	04 00
13	09 02	08 06	03 31
14	09 00	08 03	02 15
15	09 D 00	08 00	01 11
16	09 00	07 57	00 S 05
17	09 R 00	07 53	00 N 59
18	08 57	07 50	02 00
19	08 53	07 47	02 54
20	08 45	07 44	03 41
21	08 35	07 41	04 18
22	08 23	07 37	04 44
23	08 09	07 34	04 58
24	07 55	07 31	04 58
25	07 42	07 28	04 45
26	07 30	07 25	04 19
27	07 22	07 22	03 40
28	07 16	07 18	02 50
29	07 14	07 15	01 50
30	07 ♑ 13	07 ♑ 12	00 N 43

DECLINATIONS

Date	Sun ☉ ° '	Moon ☽ ° '	Mercury ☿ ° '	Venus ♀ ° '	Mars ♂ ° '	Jupiter ♃ ° '	Saturn ♄ ° '	Uranus ♅ ° '	Neptune ♆ ° '	Pluto ♇ ° '
01	08 N 13	24 N 09	02 N 49	12 S 16	08 S 11	01 S 04	00 N 20	00 S 58	12 S 59	18 S 30
02	07 51	24 48	03 19	12 41	08 26	01 07	00 17	00 59	12 59	18 30
03	07 29	24 02	03 52	13 07	08 42	01 10	00 14	00 59	13 00	18 30
04	07 07	21 45	04 25	13 32	08 57	01 13	00 12	01 00	13 00	18 31
05	06 45	18 03	04 59	13 56	09 12	01 16	00 09	01 01	13 01	18 31
06	06 22	13 07	05 33	14 20	09 27	01 19	00 06	01 02	13 02	18 31
07	06 00	07 17	06 06	14 44	09 42	01 22	00 N 03	01 03	13 02	18 31
08	05 37	00 N 59	06 38	15 08	09 57	01 25	00 00	01 04	13 03	18 32
09	05 15	05 S 20	07 08	15 31	10 13	01 28	00 S 03	01 05	13 03	18 32
10	04 52	11 16	07 35	15 53	10 27	01 31	00 06	01 06	13 04	18 32
11	04 29	16 24	08 00	16 16	10 42	01 34	00 08	01 07	13 04	18 32
12	04 06	20 27	08 21	16 38	10 57	01 38	00 11	01 08	13 05	18 33
13	03 43	23 12	08 39	16 59	11 12	01 41	00 14	01 09	13 05	18 33
14	03 20	24 33	08 53	17 20	11 27	01 44	00 17	01 10	13 06	18 33
15	02 57	24 33	09 02	17 41	11 42	01 47	00 20	01 11	13 06	18 33
16	02 34	23 17	09 08	18 01	11 56	01 50	00 23	01 12	13 07	18 34
17	02 11	20 57	09 09	18 20	12 11	01 54	00 26	01 13	13 07	18 34
18	01 48	17 43	09 06	18 40	12 25	01 57	00 29	01 13	13 08	18 34
19	01 25	13 49	08 58	18 58	12 40	02 00	00 32	01 14	13 08	18 34
20	01 01	09 24	08 47	19 16	12 54	02 03	00 34	01 15	13 09	18 35
21	00 38	04 S 39	08 32	19 33	13 08	02 06	00 37	01 16	13 09	18 35
22	00 N 15	00 N 16	08 12	19 50	13 23	02 10	00 40	01 17	13 10	18 35
23	00 S 09	05 11	07 50	20 06	13 37	02 13	00 43	01 18	13 10	18 35
24	00 32	09 58	07 24	20 22	13 51	02 16	00 46	01 19	13 10	18 36
25	00 55	14 23	06 55	20 37	14 05	02 19	00 49	01 20	13 11	18 36
26	01 19	18 16	06 24	20 51	14 19	02 22	00 52	01 21	13 11	18 36
27	01 42	21 24	05 49	21 05	14 33	02 25	00 55	01 22	13 12	18 36
28	02 05	23 33	05 13	21 17	14 46	02 29	00 58	01 23	13 12	18 37
29	02 29	24 31	04 35	21 29	15 00	02 32	01 01	01 24	13 12	18 37
30	02 S 52	24 N 09	03 N 55	21 S 41	15 S 14	02 S 35	01 S 03	01 S 25	13 S 13	18 S 37

ZODIAC SIGN ENTRIES

Date	h	m	Planets
01	00	19	☽ ♊
03	06	50	☽ ♋
05	09	45	☽ ♌
07	09	53	☽ ♍
08	15	44	♀ ♏
09	04	49	♃ ♓
09	09	01	☽ ♎
11	09	21	☽ ♏
13	12	52	☽ ♐
14	22	38	♂ ♏
15	20	30	☽ ♑
18	07	35	☽ ♒
20	20	15	☽ ♓
23	03	09	☉ ♎
23	08	47	☽ ♈
25	20	17	☽ ♉
28	06	10	☽ ♊
30	13	46	☽ ♋

LATITUDES

Date	Mercury ☿ ° '	Venus ♀ ° '	Mars ♂ ° '	Jupiter ♃ ° '	Saturn ♄ ° '	Uranus ♅ ° '	Neptune ♆ ° '	Pluto ♇ ° '
01	04 S 15	03 S 06	00 N 04	01 S 34	02 N 10	00 S 47	00 S 29	04 N 55
04	03 40	03 28	00 N 02	01 35	02 10	00 47	00 29	04 54
07	02 52	03 50	00 00	01 35	02 10	00 47	00 29	04 53
10	01 55	04 12	00 S 02	01 35	02 10	00 47	00 29	04 53
13	00 57	04 34	00 04	01 35	02 10	00 47	00 29	04 52
16	00 S 04	04 57	00 07	01 36	02 10	00 47	00 29	04 51
19	00 N 40	05 19	00 09	01 36	02 10	00 47	00 29	04 50
22	01 14	05 40	00 10	01 36	02 10	00 47	00 29	04 50
25	01 37	06 00	00 12	01 36	02 10	00 47	00 29	04 49
28	01 50	06 20	00 14	01 36	02 10	00 47	00 29	04 48
31	01 N 54	06 S 38	00 S 16	01 S 36	02 N 10	00 S 47	00 S 29	04 N 47

DATA

Julian Date	2455441
Delta T	+67 seconds
Ayanamsa	24° 00' 39"
Synetic vernal point	05° ♓ 06' 20"
True obliquity of ecliptic	23° 26' 18"

MOON'S PHASES, APSIDES AND POSITIONS ☽

Date	h	m	Phase	Longitude ° '	Eclipse Indicator
01	17	22	☾	09 ♊ 10	
08	10	30	●	15 ♍ 41	
15	05	50	☽	22 ♐ 18	
23	09	17	○	00 ♈ 15	

Day	h	m		
08	03	52	Perigee	
21	07	51	Apogee	
02	11	24	Max dec	24° N 49'
08	15	43	0S	
14	23	48	Max dec	24° S 43'
22	10	44	0N	
29	17	42	Max dec	24° N 33'

LONGITUDES

Date	Chiron ⚷ ° '	Ceres ⚳ ° '	Pallas ⚴ ° '	Juno ⚵ ° '	Vesta ⚶ ° '	Black Moon Lilith ⚸ ° '
01	27 ♒ 57	22 ♐ 17	17 ♏ 14	15 ♌ 08	11 ♎ 17	07 ♓ 17
11	27 ♒ 29	23 ♐ 48	20 ♏ 32	19 ♌ 58	16 ♎ 18	08 ♓ 24
21	27 ♒ 03	25 ♐ 45	24 ♏ 02	24 ♌ 40	21 ♎ 23	09 ♓ 31
31	26 ♒ 41	28 ♐ 03	27 ♏ 42	29 ♌ 13	26 ♎ 34	10 ♓ 38

ASPECTARIAN

h m	Aspects
01 Wednesday	
00 58	♂ Q ♆
02 09	☽ ⚹ ♃
05 37	☽ ⊼ ♇
08 05	☽ △ ♄
11 45	☽ ⚼ ♂
13 04	☉ # ♂
17 22	☽ □ ☉
17 49	☽ ⚼ ♀
21 26	☽ Q ♅
23 17	☽ □ ☿
02 Thursday	
00 09	☽ Q ♃
16 30	☽ △ ♂
22 46	☽ △ ♀
03 Friday	
01 26	☽ △ ♆
04 20	☽ Q ☉
05 22	♀ ∥ ♆
05 26	☽ Q ☿
05 40	☽ □ ♅
08 06	☽ □ ♃
11 45	☉ ∠ ♀
11 47	☽ ☍ ♇
12 35	☉ ☌ ☿
12 39	☿ ∠ ♀
14 34	☽ □ ♄
04 Saturday	
00 56	☽ ⚹ ☿
02 44	☽ ⚹ ☉
03 27	☽ ⚼ ♆
17 11	♀ △ ♆
18 38	☽ ± ♆
21 28	☽ Q ♄
22 49	☽ □ ♂
05 Sunday	
00 41	☽ ∠ ☿
04 36	☽ ⊼ ♆
05 18	☽ □ ♀
05 46	☽ ∠ ☉
08 31	☽ △ ♅
09 23	☽ # ♇
10 31	☽ △ ♃
14 21	☽ ⊼ ♇
14 40	☽ ⊥ ☿
17 23	☽ ⚹ ♄
17 23	☿ ∠ ♂
21 31	☽ ⊥ ☉
23 52	☽ ⊻ ☿
06 Monday	
00 06	☽ ± ♇
05 37	☽ Q ♂
06 52	☽ # ♀
07 52	☽ ⊻ ☉
08 48	☽ ⚼ ♅
10 37	☽ ⚼ ♃
12 12	☽ Q ♀
12 23	☽ # ♆
14 33	☽ ⚼ ♇
17 43	☽ ∠ ♄
23 07	☽ ± ♅
07 Tuesday	
00 50	☽ ± ♃
01 38	☽ ⚹ ♂
02 46	☽ # ♂
04 52	☽ ☍ ♆
08 05	☽ ⊥ ♄
08 17	☽ ⚹ ♀
08 35	☽ ⊼ ♅
09 17	☉ ∥ ☿
10 14	☽ ⊼ ♃
13 49	♀ ⊼ ♅
14 17	☽ △ ♇
16 15	☽ ∥ ☿
17 20	☽ ∥ ☉
17 36	☽ ⊻ ♄
21 13	☽ ☌ ☿
08 Wednesday	
02 19	☽ ∠ ♂
09 06	☽ ∠ ♀
10 25	☽ # ♃
10 30	☽ ☌ ☉
11 43	☽ # ♅
15 43	☽ # ♄
15 44	☽ ∥ ♄
17 04	☽ ⊥ ♂
17 36	♀ ⊼ ♃
19 45	☽ ∥ ♅
21 07	☽ ∥ ♃
23 58	☽ ⊥ ♀
09 Thursday	
02 55	☽ ⊻ ♂
03 55	☽ ⊼ ♆
07 35	☽ ☍ ♅
08 59	☽ ☍ ♃
09 55	☽ ⊻ ♀
11 40	☽ # ☉
13 23	☽ ± ♆
13 26	☽ □ ♇
17 09	☽ ☌ ♄
18 42	☽ ⊻ ☿
19 37	☽ # ☿
10 Friday	
01 16	♂ △ ♆
03 40	☽ ⚼ ♆
04 00	☽ ⊥ ☿
08 25	☽ ∥ ♂
13 18	☽ ⊻ ☉
18 09	☽ ∠ ☿
18 15	☽ Q ♇
19 55	☽ ∥ ♆
23 44	☽ ⊥ ☉
11 Saturday	
03 57	☽ △ ♆
05 16	☽ ☌ ♂
07 44	☽ ⊼ ♅
08 54	☽ ⊼ ♃
11 13	☽ ∥ ♀
12 54	☽ ☌ ♀
13 59	☽ ⚹ ♇
15 40	☽ ∠ ☉
17 41	☽ ± ♅
18 19	☽ ⊻ ♄
18 28	☽ ⚹ ☿
18 48	☽ ± ♃
23 47	☽ ∥ ♇
12 Sunday	
01 54	☿ ⊻ ♄
04 32	☽ ⊥ ♄
08 52	☽ ⚼ ♅
09 55	☽ ⚼ ♃
10 43	♀ ⚹ ♇
14 42	☽ Q ☿
15 25	☽ ∠ ♇
19 06	☽ ⚹ ☉
20 09	☽ ∠ ♄
23 08	☿ St D
13 Monday	
06 57	☽ □ ♆
07 09	☽ ⊥ ♇
08 30	♂ ⊼ ♅
10 58	☽ △ ♅
11 06	☽ ⊻ ♂
11 53	☽ △ ♃
17 52	☽ ⊻ ♇
17 57	☽ Q ☉
19 34	☽ ⊻ ♀
22 26	☽ ⊥ ♂
22 38	☽ □ ☿
23 02	☽ ⚹ ♄
14 Tuesday	
00 19	♂ ⊼ ♃
04 33	♇ St D
07 04	☽ ⊥ ♀
15 29	☽ Q ♆
15 49	☽ ∠ ♂
21 19	☽ Q ♄
15 Wednesday	
00 40	☽ ∠ ♀
05 50	☽ □ ☉
09 48	☿ ⊻ ♄
11 32	♀ ± ♅
14 01	☽ ⚹ ♆
18 17	☽ □ ♅
18 52	☽ □ ♃
21 41	♀ ± ♃
21 44	☽ ⚹ ♂
16 Thursday	
01 52	☽ ☌ ♇
06 53	☽ ⚹ ♀
07 58	☽ □ ♄
08 35	☽ △ ☿
18 59	☽ ∠ ♆
22 25	☽ Q ♂
17 Friday	
05 14	☽ Q ♅
05 34	☽ Q ♃
07 41	☽ Q ♀
09 17	♀ ⊻ ♄
12 44	☽ ⊥ ♆
15 41	☽ ⚼ ☿
21 19	☽ △ ☉
18 Saturday	
00 41	☽ ⊻ ♆
03 57	☉ # ♃
04 57	♀ ∥ ♇
05 04	☽ ⚹ ♅
05 13	☽ ⚹ ♃
06 08	☽ ∥ ♀
06 15	☽ ∥ ♇
10 55	☽ ± ☿
12 23	☽ □ ♂
13 12	☽ ⊻ ♇
20 09	☽ △ ♄
21 37	☽ □ ♀
23 57	☽ ⊼ ☿
19 Sunday	
01 04	♃ ☌ ♅
01 17	☽ ⊥ ♇
02 48	♂ ⚹ ♇
06 11	☽ ⚼ ☉
11 07	☽ ∠ ♃
11 12	☽ ∠ ♅
13 43	☉ ⊼ ♆
15 52	☽ ∥ ♆
18 09	☽ ∥ ♂
19 30	☽ ∠ ♇
22 07	☉ # ♅
20 Monday	
02 03	☽ ± ☉
02 49	☽ ⚼ ♄
05 09	☽ ⊥ ♃
05 23	☽ ⊥ ♅
13 09	☽ ☌ ♆
15 18	☽ ⊼ ☉
15 23	☽ # ☿
17 11	☽ ⊻ ♃
17 30	☽ ⊻ ♅
21 15	☽ ± ♄
21 Tuesday	
01 56	☽ ⚹ ♇
04 45	☽ △ ♂
09 32	☽ ⊼ ♄
11 33	♂ ± ♃
11 36	☉ ☍ ♃
11 43	☉ ⊥ ♂
12 40	☉ # ♄
12 51	♂ ∥ ♆
13 23	☽ △ ♀
16 58	☉ ☍ ♅
18 42	☽ ☍ ☿
19 11	♂ ± ♅
22 Wednesday	
00 22	☽ ∥ ♃
02 13	☽ Q ♇
04 30	☽ ∥ ♅
07 31	☽ ∥ ♄
09 20	☽ # ☉
11 56	☽ ∥ ☉
12 54	☽ ⚼ ♂
14 00	☽ # ♄
17 00	☽ # ♅
21 00	☽ ⚼ ♀
21 19	☽ # ♃
23 Thursday	
01 39	☽ ⊻ ♆
05 06	☽ ☌ ♃
05 52	☽ ☌ ♅
08 04	☽ ± ♂
09 17	☽ ☍ ☉
13 39	☽ ⊥ ♆
14 25	☽ □ ♇
15 49	☽ ± ♀
20 47	☽ ⊼ ♂
22 31	☽ ☍ ♄
24 Friday	
00 05	☽ ∥ ☿
04 16	☽ ⊼ ♀
07 35	☽ ∠ ♆
14 30	☽ ⊼ ☿
25 Saturday	
04 02	☽ ± ☿
04 34	☉ ∥ ♄
05 12	☽ # ♆
09 55	♂ ⊻ ♄
10 10	☽ # ♂
13 12	☽ ⚹ ♆
13 46	☉ ± ♆
16 01	☽ ⊻ ♃
17 14	☽ ⊻ ♅
26 Sunday	
00 15	☽ ⚼ ☿
00 26	☉ □ ♇
01 48	☽ △ ♇
01 56	☽ ⊼ ☉
03 39	☽ ⊥ ♃
04 56	☽ ⊥ ♅
10 18	☽ ⊼ ♄
11 28	☽ ☍ ♂
12 37	☽ Q ♆
14 16	☽ # ♇
14 37	☉ ∥ ♅
14 38	☽ ± ☉
17 25	☽ ☍ ♀
20 58	☽ ∠ ♃
22 03	☽ ± ♄
22 22	☽ ∠ ♅
27 Monday	
06 56	☽ ⚼ ♇
08 57	☽ # ♀
09 31	☽ ⚼ ☉
09 39	☽ △ ☿
15 33	☽ ⚼ ♄
23 13	☽ □ ♆
28 Tuesday	
00 10	☽ ± ♇
01 27	☽ ⚹ ♃
03 03	☽ ⚹ ♅
11 33	☽ ⊼ ♇
16 27	☽ △ ☉
20 16	☽ △ ♄
20 40	☉ ⊥ ♀
23 52	☽ Q ♃
29 Wednesday	
00 04	☽ ⊼ ♂
01 35	☽ Q ♅
04 12	☽ ⊼ ♀
11 51	☽ ± ♂
15 33	☉ ∥ ♃
15 36	☽ ± ♀
30 Thursday	
02 30	☽ □ ☿
05 18	☽ ⚼ ♂
07 03	☽ △ ♆
08 28	☽ ⚼ ♀
08 43	☽ □ ♃
10 37	☽ □ ♅
16 58	♀ ⚼ ♃
18 55	☽ ☍ ♇

All ephemeris data is given at 12.00 UT and the Moon's longitude is additionally given for 24.00 UT

OCTOBER 2010

LONGITUDES

Date	Sidereal time h m s	Sun ☉ ° ' "	Moon ☽ ° ' "	Moon ☽ 24.00 ° ' "	Mercury ☿ ° '	Venus ♀ ° '	Mars ♂ ° '	Jupiter ♃ ° '	Saturn ♄ ° '	Uranus ♅ ° '	Neptune ♆ ° '	Pluto ♇ ° '
01	12 40 27	08 ♎ 12 25	12 ♋ 28 05	19 ♋ 19 35	26 ♍ 14	12 ♏ 23	11 ♏ 13	27 ♓ 04	07 ♎ 48	28 ♓ 13	26 ♒ 16	02 ♑ 52
02	12 44 23	09 11 26	26 ♋ 16 53	03 ♌ 20 06	28 00	12 36	11 54	26 R 56	07 56	28 R 11	26 R 15	02 53
03	12 48 20	10 10 30	10 ♌ 29 12	17 ♌ 43 59	29 ♍ 46	12 48	12 35	26 49	08 03	28 08	26 14	02 53
04	12 52 17	11 09 35	25 ♌ 04 03	02 ♍ 28 50	01 ♎ 33	12 58	13 16	26 41	08 10	28 06	26 13	02 54
05	12 56 13	12 08 43	09 ♍ 57 32	17 ♍ 29 11	03 20	13 05	13 58	26 34	08 18	28 04	26 12	02 54
06	13 00 10	13 07 53	25 ♍ 02 39	02 ♎ 36 40	05 08	13 10	14 39	26 26	08 25	28 01	26 11	02 55
07	13 04 06	14 07 05	10 ♎ 09 56	17 ♎ 41 06	06 55	13 13	15 21	26 19	08 33	27 59	26 10	02 56
08	13 08 03	15 06 19	25 ♎ 08 55	02 ♏ 32 15	08 42	13 R 14	16 03	26 12	08 40	27 57	26 09	02 57
09	13 11 59	16 05 35	09 ♏ 50 06	17 ♏ 01 42	10 28	13 12	16 44	26 05	08 47	27 55	26 08	02 57
10	13 15 56	17 04 53	24 ♏ 06 28	01 ♐ 04 04	12 14	13 08	17 26	25 58	08 55	27 52	26 07	02 58
11	13 19 52	18 04 14	07 ♐ 54 21	14 ♐ 37 23	14 00	13 02	18 08	25 51	09 02	27 50	26 07	02 59
12	13 23 49	19 03 35	21 ♐ 13 22	27 ♐ 42 40	15 45	12 53	18 50	25 44	09 09	27 48	26 06	03 00
13	13 27 46	20 02 59	04 ♑ 05 46	10 ♑ 23 11	17 29	12 42	19 32	25 38	09 17	27 46	26 05	03 01
14	13 31 42	21 02 25	16 ♑ 35 32	22 ♑ 43 29	19 13	12 28	20 14	25 31	09 24	27 44	26 04	03 02
15	13 35 39	22 01 52	28 ♑ 47 42	04 ♒ 48 50	20 56	12 12	20 56	25 25	09 31	27 41	26 03	03 03
16	13 39 35	23 01 21	10 ♒ 47 33	16 ♒ 44 29	22 38	11 54	21 38	25 18	09 39	27 39	26 03	03 04
17	13 43 32	24 00 51	22 ♒ 40 15	28 ♒ 35 24	24 20	11 33	22 20	25 12	09 46	27 37	26 02	03 05
18	13 47 28	25 00 23	04 ♓ 30 27	10 ♓ 25 54	26 01	11 10	23 03	25 06	09 53	27 35	26 01	03 06
19	13 51 25	25 59 58	16 ♓ 22 10	22 ♓ 19 36	27 42	10 45	23 45	25 01	10 00	27 33	26 01	03 07
20	13 55 21	26 59 34	28 ♓ 18 32	04 ♈ 19 13	29 ♎ 21	10 18	24 28	24 55	10 07	27 31	26 00	03 08
21	13 59 18	27 59 11	10 ♈ 21 51	16 ♈ 26 37	01 ♏ 01	09 49	25 10	24 49	10 15	27 29	26 00	03 09
22	14 03 15	28 58 51	22 ♈ 33 39	28 ♈ 43 00	02 39	09 18	25 53	24 44	10 22	27 27	25 59	03 10
23	14 07 11	29 ♎ 58 33	04 ♉ 54 46	11 ♉ 08 59	04 17	08 46	26 35	24 39	10 29	27 25	25 59	03 11
24	14 11 08	00 ♏ 58 17	17 ♉ 25 41	23 ♉ 44 54	05 55	08 13	27 18	24 34	10 36	27 23	25 58	03 12
25	14 15 04	01 58 03	00 ♊ 06 42	06 ♊ 31 09	07 31	07 38	28 01	24 29	10 43	27 22	25 58	03 14
26	14 19 01	02 57 51	12 ♊ 58 20	19 ♊ 28 22	09 08	07 03	28 43	24 24	10 50	27 20	25 57	03 15
27	14 22 57	03 57 41	26 ♊ 01 25	02 ♋ 37 37	10 43	06 27	29 ♏ 26	24 20	10 57	27 18	25 57	03 16
28	14 26 54	04 57 33	09 ♋ 17 10	16 ♋ 00 17	12 19	05 51	00 ♐ 09	24 16	11 04	27 16	25 56	03 18
29	14 30 50	05 57 28	22 ♋ 47 08	29 ♋ 37 55	13 53	05 14	00 52	24 11	11 11	27 14	25 56	03 19
30	14 34 47	06 57 24	06 ♌ 32 45	13 ♌ 31 44	15 28	04 37	01 35	24 07	11 18	27 13	25 56	03 20
31	14 38 44	07 ♏ 57 23	20 ♌ 34 51	27 ♌ 42 02	17 ♏ 01	04 ♏ 01	02 ♐ 19	24 ♓ 04	11 ♎ 25	27 ♓ 11	25 ♒ 56	03 ♑ 22

Date	Moon True ☊ ° '	Moon Mean ☊ ° '	Moon ☽ Latitude ° '
01	07 ♑ 13	07 ♑ 09	00 S 28
02	07 R 12	07 06	01 39
03	07 10	07 03	02 46
04	07 05	06 59	03 44
05	06 58	06 56	04 28
06	06 48	06 53	04 54
07	06 37	06 50	05 00
08	06 25	06 47	04 45
09	06 15	06 43	04 11
10	06 07	06 40	03 22
11	06 02	06 37	02 23
12	06 00	06 34	01 17
13	05 59	06 31	00 S 10
14	05 D 59	06 28	00 N 56
15	05 R 59	06 24	01 58
16	05 57	06 21	02 53
17	05 54	06 18	03 41
18	05 47	06 15	04 19
19	05 38	06 12	04 45
20	05 27	06 09	04 59
21	05 15	06 05	05 01
22	05 02	06 02	04 48
23	04 50	05 59	04 22
24	04 40	05 56	03 43
25	04 33	05 53	02 52
26	04 28	05 49	01 52
27	04 26	05 46	00 N 45
28	04 D 26	05 43	00 S 26
29	04 26	05 40	01 37
30	04 27	05 37	02 43
31	04 ♑ 26	05 ♑ 34	03 S 41

DECLINATIONS

Date	Sun ☉ ° '	Moon ☽ ° '	Mercury ☿ ° '	Venus ♀ ° '	Mars ♂ ° '	Jupiter ♃ ° '	Saturn ♄ ° '	Uranus ♅ ° '	Neptune ♆ ° '	Pluto ♇ ° '
01	03 S 15	22 N 24	03 N 14	21 S 51	15 S 27	02 S 38	01 S 06	01 S 26	13 S 13	18 S 37
02	03 39	19 16	02 32	22 00	15 40	02 41	01 09	01 27	13 14	18 38
03	04 02	14 57	01 48	22 09	15 54	02 44	01 12	01 28	13 14	18 38
04	04 25	09 39	01 04	22 17	16 07	02 47	01 15	01 29	13 14	18 38
05	04 48	03 N 42	00 N 20	22 23	16 20	02 50	01 18	01 29	13 15	18 38
06	05 11	02 S 32	00 S 26	22 29	16 33	02 52	01 21	01 30	13 15	18 38
07	05 34	08 37	01 11	22 33	16 45	02 55	01 24	01 31	13 15	18 39
08	05 57	14 09	01 57	22 37	16 58	02 58	01 27	01 32	13 16	18 39
09	06 20	18 44	02 42	22 39	17 11	03 01	01 29	01 33	13 16	18 39
10	06 43	22 04	03 28	22 40	17 23	03 03	01 32	01 34	13 16	18 39
11	07 05	23 58	04 13	22 40	17 35	03 06	01 35	01 35	13 16	18 40
12	07 28	24 26	04 59	22 39	17 48	03 09	01 38	01 36	13 17	18 40
13	07 50	23 32	05 44	22 36	18 00	03 11	01 41	01 37	13 17	18 40
14	08 13	21 29	06 28	22 32	18 12	03 14	01 44	01 37	13 17	18 40
15	08 35	18 29	07 12	22 27	18 23	03 16	01 46	01 38	13 18	18 41
16	08 57	14 45	07 56	22 20	18 35	03 18	01 49	01 39	13 18	18 41
17	09 19	10 28	08 40	22 12	18 46	03 21	01 52	01 40	13 18	18 41
18	09 41	05 50	09 22	22 02	18 58	03 23	01 55	01 41	13 18	18 41
19	10 02	01 S 00	10 05	21 51	19 09	03 25	01 57	01 41	13 18	18 41
20	10 24	03 N 54	10 46	21 38	19 20	03 27	02 00	01 42	13 19	18 42
21	10 45	08 43	11 27	21 24	19 31	03 29	02 03	01 43	13 19	18 42
22	11 07	13 14	12 07	21 09	19 42	03 31	02 06	01 44	13 19	18 42
23	11 28	17 16	12 47	20 52	19 52	03 33	02 08	01 44	13 19	18 42
24	11 49	20 36	13 26	20 34	20 03	03 35	02 11	01 45	13 19	18 42
25	12 09	22 59	14 04	20 15	20 13	03 37	02 14	01 46	13 19	18 43
26	12 30	24 12	14 42	19 55	20 23	03 38	02 16	01 47	13 20	18 43
27	12 50	24 07	15 18	19 34	20 33	03 40	02 19	01 47	13 20	18 43
28	13 10	22 41	15 54	19 12	20 43	03 41	02 22	01 48	13 20	18 43
29	13 30	19 56	16 29	18 49	20 52	03 43	02 24	01 49	13 20	18 43
30	13 50	16 00	17 03	18 25	21 01	03 44	02 27	01 49	13 20	18 44
31	14 S 10	11 N 08	17 S 37	18 S 01	21 S 11	03 S 45	02 S 29	01 S 50	13 S 20	18 S 44

ZODIAC SIGN ENTRIES

Date	h m	Planets
02	18 21	☽ ♌
03	15 04	☿ ♎
04	20 00	☽ ♍
06	19 52	☽ ♎
08	19 52	☽ ♏
10	22 09	☽ ♐
13	04 17	☽ ♑
15	14 24	☽ ♒
18	02 52	☽ ♓
20	15 23	☽ ♈
20	21 19	☿ ♏
23	02 30	☽ ♉
23	12 35	☉ ♏
25	11 47	☽ ♊
27	19 14	☽ ♋
28	06 47	♂ ♐
30	00 39	☽ ♌

LATITUDES

Date	Mercury ☿ ° '	Venus ♀ ° '	Mars ♂ ° '	Jupiter ♃ ° '	Saturn ♄ ° '	Uranus ♅ ° '	Neptune ♆ ° '	Pluto ♇ ° '
01	01 N 54	06 S 38	00 S 16	01 S 36	02 N 10	00 S 47	00 S 29	04 N 47
04	01 51	06 53	00 18	01 36	02 10	00 47	00 29	04 46
07	01 42	07 06	00 20	01 35	02 10	00 47	00 29	04 46
10	01 29	07 14	00 22	01 35	02 10	00 47	00 29	04 45
13	01 14	07 19	00 24	01 35	02 10	00 47	00 29	04 44
16	00 56	07 18	00 25	01 34	02 11	00 47	00 29	04 44
19	00 37	07 10	00 27	01 34	02 11	00 47	00 29	04 43
22	00 N 17	06 56	00 29	01 33	02 11	00 47	00 29	04 42
25	00 S 03	06 34	00 31	01 33	02 11	00 47	00 29	04 41
28	00 23	06 05	00 32	01 32	02 12	00 47	00 29	04 41
31	00 S 43	05 S 29	00 S 34	01 S 31	02 N 12	00 S 47	00 S 29	04 N 40

DATA

Julian Date	2455471
Delta T	+67 seconds
Ayanamsa	24° 00' 43"
Synetic vernal point	05° ♓ 06' 17"
True obliquity of ecliptic	23° 26' 18"

MOON'S PHASES, APSIDES AND POSITIONS ☽

Date	h m	Phase	Longitude ° '	Eclipse Indicator
01	03 52	☾	07 ♋ 52	
07	18 44	●	14 ♎ 24	
14	21 27	☽	21 ♑ 26	
23	01 37	○	29 ♈ 33	
30	12 46	☾	06 ♌ 59	

Day	h m		
06	13 31	Perigee	
18	18 12	Apogee	
06	02 16	0S	
12	07 48	Max dec	24° S 27'
19	16 53	0N	
26	22 38	Max dec	24° N 20'

LONGITUDES

Date	Chiron ⚷ ° '	Ceres ⚳ ° '	Pallas ⚴ ° '	Juno ⚵ ° '	Vesta ⚶ ° '	Black Moon Lilith ⚸ ° '
01	26 ♒ 41	28 ♐ 03	27 ♏ 42	29 ♌ 13	26 ♎ 34	10 ♓ 38
11	26 ♒ 23	00 ♑ 41	01 ♐ 30	03 ♍ 34	01 ♏ 48	11 ♓ 45
21	26 ♒ 11	03 ♑ 34	05 ♐ 24	07 ♍ 44	07 ♏ 04	12 ♓ 52
31	26 ♒ 05	06 ♑ 41	09 ♐ 23	11 ♍ 41	12 ♏ 24	13 ♓ 59

All ephemeris data is given at 12.00 UT and the Moon's longitude is additionally given for 24.00 UT

ASPECTARIAN

h m	Aspects
01 Friday	
00 42	☉ ☌ ♄
03 40	☽ □ ♄
03 52	☽ □ ☉
09 40	☽ △ ♂
09 54	☽ ⚼ ♆
11 25	☿ ∠ ♂
11 33	☉ # ☿
11 50	☽ △ ♀
12 32	☿ ⊼ ♆
15 34	☽ Q ☿
16 45	☽ # ♀
22 36	☿ ☍ ♃
02 Saturday	
01 38	☽ ± ♆
05 55	☿ ∠ ♀
07 14	☿ # ♃
11 23	☽ Q ♄
11 57	☽ ⊼ ♆
13 07	☽ △ ♃
13 18	♂ ⚼ ♃
13 40	☽ Q ☉
14 23	☿ ☍ ♅
15 14	☽ △ ♅
15 22	☽ ⚹ ☿
16 02	☽ # ♇
23 14	☽ ⊼ ♇
03 Sunday	
07 26	☽ # ♂
07 54	☽ ⚹ ♄
09 20	☽ ± ♇
11 27	☽ ⚹ ☉
14 11	☽ ⚼ ♃
15 40	☽ □ ♂
15 54	☽ □ ♀
16 24	☽ ⚼ ♅
20 07	☽ ∠ ☿
20 11	☽ # ♆
21 58	♀ ☌ ♂
23 06	☿ # ♅
04 Monday	
00 16	☽ ⚼ ♇
04 55	☽ ± ♃
06 17	♂ ⚼ ♅
06 32	☿ # ♄
07 10	☽ ± ♅
08 53	☽ ∠ ♄
12 54	☽ ⊥ ☿
13 27	☉ ⚼ ♆
13 52	☽ ☍ ♆
13 54	☽ ∠ ☉
14 37	☽ ⊼ ♃
16 55	☽ ⊼ ♅
20 51	☿ ± ♆
21 38	☽ Q ♀
22 33	☽ Q ♂
23 36	☽ ⊥ ♄
23 56	☽ ⊻ ☿
05 Tuesday	
00 41	☽ △ ♇
05 28	☽ ⊥ ☉
06 08	☿ □ ♇
07 55	☽ # ☉
08 24	♀ ⚼ ♅
09 19	☽ ⊻ ♄
15 21	☽ # ♃
15 44	☽ ⊻ ☉
17 01	☽ ⚹ ♀
18 42	☽ ⚹ ♂
20 31	☽ # ♅
21 12	☽ # ♄
06 Wednesday	
01 51	☽ # ☿
02 01	♂ ⊥ ♄
02 47	☽ ‖ ☿
07 24	☽ ‖ ♄
08 03	☽ ‖ ♅
13 02	☉ ⊻ ♀
13 20	☽ ‖ ♃
13 49	☽ ⊼ ♆
14 12	☽ ☍ ♃
16 43	☽ ☍ ♅
16 59	☽ ∠ ♀
19 40	☽ ∠ ♂
22 59	☽ ‖ ☉
23 19	☽ ± ♆
07 Thursday	
00 30	☽ □ ♇
06 08	☽ ☌ ☿
07 18	☽ ⊥ ♀
09 24	☽ ☌ ♄
10 38	☽ ⊥ ♂
13 36	☽ ⚼ ♆
16 12	☿ ⊥ ♀
16 53	☽ ⊻ ♀
18 44	☽ ☌ ☉
19 08	☿ ‖ ♄
20 40	☽ ⊻ ♂
22 55	☿ ‖ ♅
08 Friday	
05 13	☽ Q ♇
07 05	♀ St R
07 55	☽ ‖ ♆
11 35	☿ ☌ ♄
13 37	☽ △ ♆
13 41	☽ ⊼ ♃
16 31	☽ ⊼ ♅
22 14	♃ ⊻ ♆
23 21	☽ ± ♃
09 Saturday	
00 40	☽ ⚹ ♇
02 16	☽ ± ♅
02 46	☽ ‖ ♂
10 15	☽ ⊻ ♄
11 34	☽ ‖ ♇
13 12	☽ ⊻ ☿
14 03	☽ ⚼ ♃
17 06	☽ ⚼ ♅
17 35	☽ ☌ ♀
18 02	☿ ⊥ ♂
20 19	☽ ⊥ ♄
21 01	☿ ⚼ ♆
22 23	☿ ‖ ♃
23 12	☽ ⊻ ☉
10 Sunday	
00 06	☽ ☌ ♂
00 37	☽ ⊥ ☿
01 34	☽ ∠ ♇
10 07	☽ ⊥ ☉
11 40	☽ ∠ ♄
15 09	☽ △ ♃
15 27	☽ □ ♆
16 55	☽ ⊥ ♇
17 58	☽ ‖ ♀
18 09	☽ ∠ ☿
18 27	☽ △ ♅
23 39	☿ ⊻ ♀
11 Monday	
02 50	☽ ∠ ☉
03 20	☽ ⊻ ♇
06 47	♂ ∠ ♇
08 58	♄ ‖ ♅
14 01	☽ ⚹ ♄
17 04	☉ ⊻ ♂
21 03	☽ ⊻ ♀
23 04	☽ Q ♆
12 Tuesday	
00 31	☽ ⚹ ☿
07 23	☽ ⊻ ♂
07 43	☽ ⚹ ☉
07 46	☽ ⊥ ♀
11 53	☽ Q ♄
19 02	☽ ⊥ ♂
20 15	☽ □ ♃
20 59	☽ ⚹ ♆
13 Wednesday	
00 08	☽ □ ♅
00 09	☽ ∠ ♀
01 57	☽ Q ☿
02 09	♀ ⚼ ♅
07 48	☽ Q ☉
09 57	☽ ☌ ♇
12 52	☽ ∠ ♂
21 58	☽ □ ♄
14 Thursday	
00 52	☽ ‖ ♀
01 19	☽ ∠ ♆
04 09	☽ ⚹ ♀
06 05	☽ Q ♃
10 19	☽ Q ♅
11 42	☉ Q ♇
17 57	☽ □ ☿
18 47	☽ ⊥ ♆
19 32	☽ ⚹ ♂
21 27	☽ □ ☉
15 Friday	
03 06	☽ Q ♀
05 21	☽ ⚹ ♃
06 35	☽ ⊻ ♆
09 49	☽ ⚹ ♅
10 35	☽ ‖ ♇
12 05	☿ ⊻ ♂
12 35	☽ ‖ ♂
13 37	☿ Q ♇
20 28	☽ ⊻ ♇
20 45	☽ Q ♂
16 Saturday	
08 31	☽ ⊥ ♇
09 40	☽ △ ♄
11 02	☽ ∠ ♃
14 09	☽ □ ♀
15 44	☽ ∠ ♅
20 25	☽ ‖ ♆
17 Sunday	
00 22	♂ ‖ ♇
01 05	☉ ☌ ☿
02 41	☽ ∠ ♇
05 03	☽ ⊥ ♃
09 53	☽ ⊥ ♅
11 17	☽ □ ♂
14 58	☽ △ ☉
15 56	☽ △ ☿
16 17	☽ ⚼ ♄
17 06	☽ ⊻ ♃
17 41	☽ ‖ ☉
18 49	☽ ☌ ♆
20 17	☽ ‖ ☿
22 00	☽ ⊻ ♅
23 42	☿ ⊼ ♃
18 Monday	
09 08	☽ ⚹ ♇
10 43	☽ ± ♄
12 03	☿ △ ♆
14 12	☉ ⊼ ♃
23 00	☽ ⊼ ♄
19 Tuesday	
00 09	☽ ⚼ ☉
00 11	☽ ‖ ♃
01 02	☽ △ ♀
03 21	☽ ⚼ ☿
07 20	☽ ‖ ♄
08 36	☽ ‖ ♅
09 27	☽ Q ♇
09 35	☉ ‖ ☿
10 00	☿ ⊼ ♅
12 18	☉ △ ♆
19 59	☽ ± ☉
20 Wednesday	
00 28	☽ ± ☿
01 10	☽ # ♅
02 35	☽ # ♄
03 48	☽ △ ♂
05 15	☽ ☌ ♃
06 12	☽ ⚼ ♀
07 23	☽ ⊻ ♆
09 08	☽ ⊼ ☉
09 45	☽ # ♃
10 25	☽ ☌ ♅
14 26	☽ ⊼ ☿
19 03	♀ ⊻ ♄
19 22	☽ ⊥ ♆
21 39	☽ □ ♇
23 30	☽ ± ♀
21 Thursday	
00 17	☉ ⊼ ♅
01 42	♂ △ ♃
09 25	☿ ± ♃
10 58	☽ ⊼ ♀
11 35	☽ ⚼ ♂
11 38	♀ ⚼ ♃
11 46	☽ ☍ ♄
13 15	☽ ∠ ♆
15 10	♂ ∠ ♄
23 32	☽ # ☉
22 Friday	
04 51	☽ # ☿
06 25	☽ ± ♂
12 29	☽ # ♆
15 35	♂ □ ♆
16 13	☽ ⊻ ♃
18 41	☽ ⚹ ♆
18 52	☽ ⊼ ♂
19 38	☿ ⚹ ♇
21 31	☽ ⊻ ♅
23 30	☿ ± ♅
23 Saturday	
01 37	☽ ☍ ☉
03 48	☽ ⊥ ♃
08 40	☽ △ ♇
09 08	☽ ⊥ ♅
10 37	☽ ☍ ☿
17 54	☽ Q ♆
19 07	☽ ☍ ♀
21 03	☽ ∠ ♃
21 40	☽ # ♇
22 49	☽ ⊼ ♄
24 Sunday	
02 24	☽ ∠ ♅
02 57	☉ ± ♃
07 18	☽ # ♂
07 51	☿ ‖ ♆
08 51	♂ ⊥ ♇
10 25	☽ ± ♄
11 51	☽ # ♀
13 29	☽ ⚼ ♇
14 56	♂ △ ♅
25 Monday	
01 27	☽ ⚹ ♃
03 39	☽ ⚼ ♄
04 11	☽ □ ♆
06 34	☽ ± ♇
06 50	☽ ⚹ ♅
07 49	☽ ☍ ♂
13 17	☿ ☌ ♀
14 01	♀ ‖ ♂
15 47	☽ ⊼ ☉
17 51	☽ ⊼ ♇
23 52	☽ Q ♃
26 Tuesday	
01 29	☽ ⊼ ♀
03 51	☽ ⊼ ☿
03 56	☽ ± ☉
05 15	☽ Q ♅
08 00	☽ △ ♄
12 08	☽ ± ♀
15 59	☿ ⚼ ♃
16 33	☽ ± ☿
19 01	☉ ⚹ ♇
20 30	☉ ± ♅
21 59	☽ ⚼ ☉
27 Wednesday	
04 00	☽ ⚼ ♀
08 56	☽ □ ♃
10 30	♄ ⚼ ♆
11 23	☽ ⚼ ☿
11 52	☽ △ ♆
14 19	☽ □ ♅
15 45	☿ ⊻ ♄
18 35	☽ ⊼ ♂
28 Thursday	
01 11	☽ ☍ ♇
03 35	☽ △ ☉
06 03	☽ ± ♂
06 05	☽ △ ♀
11 22	☿ ⚼ ♅
14 58	☽ ⚼ ♆
15 14	☽ □ ♄
18 08	☽ △ ☿
23 05	☽ ⚼ ♂
23 19	☉ ‖ ♆
29 Friday	
01 10	☉ ☌ ♀
05 16	☽ # ♂
06 58	☽ ± ♆
14 28	☽ △ ♃
17 29	♀ ‖ ♇
17 32	☽ ⊼ ♆
19 48	☽ △ ♅
20 05	☽ # ♇
20 23	☽ # ♀
23 19	☽ Q ♄
30 Saturday	
02 56	☽ △ ♂
06 26	☽ ⊼ ♇
06 50	☽ # ☿
08 49	☽ □ ♀
12 46	☽ □ ☉
16 25	☽ ⚼ ♃
16 49	☽ ± ♇
20 15	☽ ⚹ ♄
21 44	☽ ⚼ ♅
22 29	☽ # ☉
31 Sunday	
01 39	☽ # ♆
05 13	☽ □ ☿
07 45	☽ ± ♃
08 14	☽ ⚼ ♇
13 01	☽ ± ♅
14 20	☽ Q ♀
17 51	☽ ⊼ ♃
18 35	☿ ⊥ ♄
21 01	☽ ☍ ♆
21 45	☽ Q ☉
21 56	☽ ∠ ♄
22 15	☿ ‖ ♀
23 07	☽ ⊼ ♅

LONGITUDES

Date	Sidereal time h m s	Sun ☉ ° ' "	Moon ☽ ° ' "	Moon ☽ 24.00 ° ' "	Mercury ☿ ° '	Venus ♀ ° '	Mars ♂ ° '	Jupiter ♃ ° '	Saturn ♄ ° '	Uranus ♅ ° '	Neptune ♆ ° '	Pluto ♇ ° '
01	14 42 40	08 ♏ 57 24	04 ♍ 53 01	12 ♍ 07 30	18 ♏ 35	03 ♏ 26	03 ♐ 02	24 ♓ 00	11 ♎ 32	27 ♓ 09	25 ≈ 55	03 ♑ 23
02	14 46 37	09 57 27	19 ♍ 24 57	26 ♍ 44 46	20 08	02 R 51	03 45	23 R 57	11 39	27 R 08	25 R 55	03 24
03	14 50 33	10 57 32	04 ♎ 06 11	11 ♎ 28 20	21 40	02 17	04 28	23 53	11 46	27 06	25 55	03 26
04	14 54 30	11 57 40	18 ♎ 50 15	26 ♎ 10 58	23 12	01 44	05 12	23 50	11 52	27 05	25 55	03 27
05	14 58 26	12 57 49	03 ♏ 29 28	10 ♏ 44 48	24 44	01 13	05 55	23 48	11 59	27 03	25 55	03 29
06	15 02 23	13 58 00	17 ♏ 56 06	25 ♏ 02 37	26 15	00 44	06 39	23 45	12 06	27 02	25 55	03 30
07	15 06 19	14 58 13	02 ♐ 03 43	08 ♐ 58 58	27 46	00 ♏ 16	07 22	23 43	12 12	27 01	25 D 55	03 32
08	15 10 16	15 58 28	15 ♐ 48 03	22 ♐ 30 50	29 ♏ 16	29 ♎ 51	08 06	23 40	12 19	26 59	25 55	03 34
09	15 14 13	16 58 45	29 ♐ 07 21	05 ♑ 37 44	00 ♐ 46	29 27	08 50	23 38	12 26	26 58	25 55	03 35
10	15 18 09	17 59 03	12 ♑ 02 17	18 ♑ 21 22	02 16	29 06	09 34	23 37	12 32	26 57	25 55	03 37
11	15 22 06	18 59 22	24 ♑ 35 27	00 ≈ 45 05	03 45	28 46	10 18	23 35	12 39	26 55	25 55	03 38
12	15 26 02	19 59 43	06 ≈ 50 49	12 ≈ 53 17	05 13	28 30	11 01	23 34	12 45	26 54	25 55	03 40
13	15 29 59	21 00 05	18 ≈ 53 07	24 ≈ 50 58	06 42	28 15	11 45	23 32	12 51	26 53	25 56	03 42
14	15 33 55	22 00 28	00 ♓ 47 27	06 ♓ 43 13	08 09	28 03	12 29	23 32	12 58	26 52	25 56	03 44
15	15 37 52	23 00 53	12 ♓ 38 52	18 ♓ 34 58	09 37	27 54	13 13	23 31	13 04	26 51	25 56	03 45
16	15 41 48	24 01 19	24 ♓ 32 04	00 ♈ 30 40	11 03	27 46	13 58	23 30	13 10	26 50	25 56	03 47
17	15 45 45	25 01 47	06 ♈ 31 13	12 ♈ 34 07	12 29	27 42	14 42	23 30	13 16	26 49	25 57	03 49
18	15 49 42	26 02 16	18 ♈ 39 41	24 ♈ 48 14	13 55	27 40	15 26	23 30	13 22	26 48	25 57	03 51
19	15 53 38	27 02 46	00 ♉ 59 59	07 ♉ 15 05	15 20	27 D 40	16 10	23 D 30	13 28	26 47	25 57	03 52
20	15 57 35	28 03 18	13 ♉ 33 37	19 ♉ 55 40	16 44	27 43	16 55	23 30	13 34	26 46	25 58	03 54
21	16 01 31	29 ♏ 03 51	26 ♉ 21 12	02 ♊ 50 12	18 07	27 48	17 39	23 30	13 40	26 46	25 58	03 56
22	16 05 28	00 ♐ 04 26	09 ♊ 22 33	15 ♊ 58 09	19 29	27 55	18 24	23 31	13 46	26 45	25 59	03 58
23	16 09 24	01 05 03	22 ♊ 36 52	29 ♊ 18 34	20 49	28 05	19 08	23 32	13 52	26 44	25 59	04 00
24	16 13 21	02 05 40	06 ♋ 03 06	12 ♋ 50 20	22 09	28 17	19 53	23 33	13 58	26 44	26 00	04 02
25	16 17 17	03 06 20	19 ♋ 40 08	26 ♋ 32 22	23 27	28 31	20 37	23 34	14 04	26 43	26 01	04 04
26	16 21 14	04 07 01	03 ♌ 26 56	10 ♌ 23 42	24 44	28 47	21 22	23 36	14 09	26 43	26 01	04 05
27	16 25 11	05 07 43	17 ♌ 22 34	24 ♌ 23 24	25 58	29 05	22 07	23 38	14 15	26 42	26 02	04 07
28	16 29 07	06 08 27	01 ♍ 26 03	08 ♍ 30 23	27 10	29 25	22 52	23 40	14 20	26 42	26 03	04 09
29	16 33 04	07 09 13	15 ♍ 38 30	22 ♍ 43 06	28 20	29 ♎ 48	23 37	23 42	14 26	26 41	26 03	04 11
30	16 37 00	08 ♐ 10 00	29 ♍ 50 57	06 ♎ 59 19	29 ♐ 27	00 ♏ 12	24 ♐ 22	23 ♓ 44	14 ♎ 31	26 ♓ 41	26 ≈ 04	04 ♑ 13

Moon True ☊ / Mean ☊ / Latitude

Date	Moon True ☊ ° '	Moon Mean ☊ ° '	Moon ☽ Latitude ° '
01	04 ♑ 23	05 ♑ 30	04 S 27
02	04 R 18	05 27	04 56
03	04 11	05 24	05 06
04	04 03	05 21	04 56
05	03 55	05 18	04 26
06	03 47	05 15	03 40
07	03 41	05 11	02 41
08	03 37	05 08	01 35
09	03 36	05 05	00 S 24
10	03 D 36	05 02	00 N 46
11	03 37	04 59	01 51
12	03 39	04 55	02 50
13	03 40	04 52	03 41
14	03 R 39	04 49	04 21
15	03 37	04 46	04 50
16	03 33	04 43	05 06
17	03 27	04 40	05 10
18	03 21	04 36	04 59
19	03 14	04 33	04 35
20	03 07	04 30	03 57
21	03 02	04 27	03 06
22	02 58	04 24	02 05
23	02 56	04 20	00 N 56
24	02 D 56	04 17	00 S 17
25	02 57	04 14	01 30
26	02 58	04 11	02 39
27	03 00	04 08	03 40
28	03 01	04 05	04 28
29	03 R 00	04 01	05 00
30	02 ♑ 59	03 ♑ 58	05 S 14

DECLINATIONS

Date	Sun ☉ ° '	Moon ☽ ° '	Mercury ☿ ° '	Venus ♀ ° '	Mars ♂ ° '	Jupiter ♃ ° '	Saturn ♄ ° '	Uranus ♅ ° '	Neptune ♆ ° '	Pluto ♇ ° '
01	14 S 29	05 N 35	18 S 09	17 S 37	21 S 20	03 S 47	02 S 32	01 S 50	13 S 20	18 S 44
02	14 48	00 S 21	18 41	17 12	21 28	03 48	02 35	01 51	13 20	18 44
03	15 07	06 18	19 11	16 47	21 37	03 49	02 37	01 52	13 20	18 44
04	15 25	11 56	19 41	16 22	21 45	03 50	02 40	01 52	13 20	18 44
05	15 44	16 51	20 10	15 58	21 54	03 51	02 42	01 53	13 20	18 45
06	16 02	20 42	20 37	15 33	22 02	03 52	02 45	01 53	13 20	18 45
07	16 20	23 13	21 04	15 09	22 10	03 52	02 47	01 54	13 20	18 45
08	16 37	24 15	21 30	14 46	22 17	03 53	02 49	01 54	13 20	18 45
09	16 54	23 50	21 54	14 23	22 25	03 54	02 52	01 55	13 20	18 45
10	17 11	22 08	22 18	14 01	22 32	03 54	02 54	01 55	13 20	18 45
11	17 28	19 23	22 40	13 40	22 39	03 54	02 57	01 56	13 20	18 46
12	17 44	15 49	23 02	13 20	22 45	03 55	02 59	01 56	13 20	18 46
13	18 00	11 40	23 22	13 01	22 52	03 55	03 01	01 57	13 20	18 46
14	18 16	07 07	23 41	12 43	22 58	03 55	03 04	01 57	13 20	18 46
15	18 32	02 S 21	23 59	12 25	23 04	03 55	03 06	01 57	13 20	18 46
16	18 47	02 N 31	24 15	12 09	23 10	03 55	03 08	01 58	13 20	18 46
17	19 01	07 19	24 31	11 55	23 16	03 55	03 10	01 58	13 20	18 47
18	19 16	11 55	24 45	11 41	23 21	03 55	03 13	01 58	13 20	18 47
19	19 30	16 06	24 58	11 28	23 27	03 55	03 15	01 59	13 19	18 47
20	19 44	19 40	25 09	11 17	23 32	03 54	03 17	01 59	13 19	18 47
21	19 57	22 21	25 19	11 07	23 36	03 54	03 19	01 59	13 19	18 47
22	20 10	23 55	25 28	10 58	23 41	03 53	03 21	02 00	13 19	18 47
23	20 23	24 10	25 35	10 50	23 45	03 53	03 23	02 00	13 19	18 47
24	20 35	23 01	25 41	10 44	23 49	03 52	03 25	02 00	13 19	18 47
25	20 47	20 30	25 46	10 38	23 53	03 51	03 27	02 00	13 18	18 48
26	20 58	16 48	25 49	10 33	23 56	03 50	03 29	02 00	13 18	18 48
27	21 09	12 08	25 51	10 30	23 59	03 50	03 31	02 00	13 18	18 48
28	21 20	06 47	25 51	10 28	24 02	03 49	03 33	02 01	13 18	18 48
29	21 30	01 04	25 50	10 26	24 05	03 47	03 35	02 01	13 17	18 48
30	21 S 40	04 S 44	25 S 48	10 S 26	24 S 08	03 S 46	03 S 37	02 S 01	13 S 17	18 S 48

ZODIAC SIGN ENTRIES

Date	h m	Planets
01	03 51	☽ ♍
03	05 19	☽ ♎
05	06 16	☽ ♏
07	08 27	☽ ♐
08	03 06	♀ ♎
08	23 43	☿ ♐
09	13 36	☽ ♑
11	22 32	☽ ≈
14	10 24	☽ ♓
16	22 59	☽ ♈
19	10 04	☽ ♉
21	18 46	☽ ♊
22	10 15	☉ ♐
24	01 14	☽ ♋
26	06 01	☽ ♌
28	09 34	☽ ♍
30	00 33	♀ ♏
30	12 15	☽ ♎

LATITUDES

Date	Mercury ☿ ° '	Venus ♀ ° '	Mars ♂ ° '	Jupiter ♃ ° '	Saturn ♄ ° '	Uranus ♅ ° '	Neptune ♆ ° '	Pluto ♇ ° '
01	00 S 50	05 S 16	00 S 34	01 S 31	02 N 12	00 S 47	00 S 29	04 N 40
04	01 09	04 35	00 36	01 31	02 13	00 46	00 29	04 39
07	01 27	03 50	00 38	01 30	02 13	00 46	00 29	04 39
10	01 43	03 04	00 39	01 29	02 13	00 46	00 29	04 38
13	01 58	02 19	00 41	01 28	02 14	00 46	00 29	04 37
16	02 11	01 35	00 42	01 28	02 14	00 46	00 29	04 37
19	02 21	00 53	00 44	01 27	02 15	00 46	00 29	04 36
22	02 27	00 S 15	00 45	01 26	02 15	00 46	00 29	04 36
25	02 30	00 N 20	00 46	01 25	02 16	00 46	00 29	04 35
28	02 27	00 52	00 48	01 24	02 17	00 46	00 29	04 35
31	02 S 18	01 N 20	00 S 49	01 S 24	02 N 17	00 S 45	00 S 29	04 N 34

DATA

Julian Date	2455502
Delta T	+67 seconds
Ayanamsa	24° 00' 46"
Synetic vernal point	05° ♓ 06' 13"
True obliquity of ecliptic	23° 26' 18"

LONGITUDES

Date	Chiron ⚷ ° '	Ceres ⚳ ° '	Pallas ⚴ ° '	Juno ⚵ ° '	Vesta ⚶ ° '	Black Moon Lilith ⚸ ° '
01	26 ≈ 05	07 ♑ 00	09 ♐ 47	12 ♍ 04	12 ♏ 56	14 ♓ 06
11	26 ≈ 05	10 ♑ 20	13 ♐ 50	15 ♍ 43	18 ♏ 17	15 ♓ 13
21	26 ≈ 12	13 ♑ 48	17 ♐ 55	19 ♍ 04	23 ♏ 39	16 ♓ 20
31	26 ≈ 25	17 ♑ 25	22 ♐ 01	22 ♍ 05	29 ♏ 02	17 ♓ 28

MOON'S PHASES, APSIDES AND POSITIONS ☽

Date	h m	Phase	Longitude ° '	Eclipse Indicator
06	04 52	●	13 ♏ 40	
13	16 39	☽	21 ≈ 12	
21	17 27	○	29 ♉ 18	
28	20 36	☾	06 ♍ 30	

Day	h m		
03	17 16	Perigee	
15	11 44	Apogee	
30	18 45	Perigee	
02	10 37	0S	
08	16 57	Max dec	24° S 17'
15	23 36	0N	
23	04 22	Max dec	24° N 14'
29	16 24	0S	

ASPECTARIAN

h m Aspects

01 Monday
08 45 ☽ □ ♂
08 57 ☿ ∠ ♇
09 30 ☽ △ ♇
09 40 ☽ ⚹ ♀
13 01 ☉ ⚼ ♃
13 05 ☽ ⊥ ♄
13 37 ♀ ⚹ ♇
15 09 ☽ Q ☿
19 16 ♀ ⊻ ♂
19 16 ☽ ⚹ ☉
19 23 ☽ # ♃
23 07 ☽ ⊻ ♄
23 32 ♀ ± ♅

02 Tuesday
00 11 ♂ ⊻ ♇
00 22 ☽ # ♄
03 13 ☽ # ♅
09 31 ☽ ∠ ♀
13 18 ☽ ⚹ ☿
14 42 ☿ ∥ ♇
16 02 ☽ Q ♂
18 02 ☽ ∥ ♅
19 24 ☽ ☍ ♃
20 59 ☽ ∥ ♄
21 44 ☽ ∠ ☉
22 39 ☽ ⊼ ♆
23 43 ☽ ⊥ ♀

03 Wednesday
00 36 ☽ ☍ ♅
01 52 ☽ ∥ ♃
08 26 ☽ ± ♆
09 08 ☽ ⊻ ♀
10 54 ☽ □ ♇
12 38 ☽ ⚹ ♂
13 30 ☽ ⊥ ☉
16 40 ☽ ∠ ☿
23 06 ☽ ⚼ ♆
23 59 ☽ ⊻ ☉

04 Thursday
00 34 ☽ ☌ ♄
09 01 ☽ ⊥ ☿
09 37 ☉ ⊻ ♄
14 20 ☽ ∠ ♂
14 47 ☉ ⚼ ♅
16 17 ☽ Q ♇
18 26 ☽ ∥ ♆
19 57 ☽ ⊻ ☿
20 09 ☽ ⊼ ♃
21 46 ☿ △ ♃
23 34 ☽ △ ♆

05 Friday
01 27 ☽ ⊼ ♅
05 42 ☽ ∥ ☉
05 49 ☽ ⊥ ♂
05 56 ☽ ± ♃
07 40 ☽ ∥ ♀
08 24 ☽ ☌ ♀
11 17 ☽ ± ♅
11 59 ☽ ⚹ ♇
16 13 ☽ ⊻ ♂
19 46 ☉ ∥ ♀
20 44 ☽ ⚼ ♃
22 53 ☽ ∥ ♇

06 Saturday
02 09 ☽ ⚼ ♅
02 10 ☽ ⊻ ♄
04 52 ☽ ☌ ☉
06 44 ☿ □ ♆
11 21 ☽ ∥ ☿
12 16 ☽ ⊥ ♄
12 58 ☽ ∠ ♇
13 43 ♀ ⊥ ♂
21 47 ☽ △ ♃
23 36 ☽ ∥ ♂

07 Sunday
00 15 ☿ △ ♅
01 29 ☽ □ ♆
02 31 ☿ ∠ ♄
03 21 ☽ △ ♅
03 37 ☽ ∠ ♄
03 44 ☽ ☌ ☿
04 14 ☽ ⊥ ♇
06 05 ♆ St D
08 20 ☿ ⊥ ♇
09 01 ☽ ⊻ ♀
14 33 ☽ ⊻ ♇
19 03 ☽ ⊥ ♀
21 42 ☽ ☌ ♂

08 Monday
05 48 ☽ ⚹ ♄
08 40 ☽ Q ♆
10 21 ☽ ∠ ♀
12 20 ☽ ⊻ ☉
19 16 ☿ ⊻ ♀
23 07 ♀ ± ♃
23 56 ☽ ⊥ ☉

09 Tuesday
02 03 ☽ □ ♃
03 23 ☽ Q ♄
06 09 ☽ ⚹ ♆
08 04 ☽ □ ♅
12 35 ☽ ⚹ ♀
15 24 ☽ ⊻ ☿
17 41 ☽ ∠ ☉
20 14 ☽ ☌ ♇

10 Wednesday
03 58 ☽ ⊥ ☿
07 04 ☽ ⊻ ♂
07 55 ☽ ∥ ♂
09 53 ☽ ∠ ♆
10 16 ☽ Q ♀
10 31 ☽ ∥ ☿
11 12 ☽ Q ♃
12 57 ☽ □ ♄
17 29 ☽ Q ♅
19 05 ☽ ⊥ ♂
23 14 ☽ ∠ ☿

11 Thursday
00 16 ☽ ⚹ ☉
02 44 ☉ ⊥ ♄
03 00 ☽ ⊥ ♆
03 28 ☉ ∠ ♇
09 17 ☿ ∥ ♂
10 03 ☽ ⚹ ♃
10 17 ☿ ⊻ ♇
13 27 ☽ ∠ ♂
14 35 ☽ ⊻ ♆
16 31 ☽ ⚹ ♅
16 32 ☽ ∥ ♇
19 57 ☽ □ ♀

12 Friday
00 30 ☽ ∥ ☉
01 35 ☽ Q ☉
01 57 ☿ ⊥ ♀
05 43 ☽ ⊻ ♇
08 21 ☽ ⚹ ☿
11 48 ♀ ∥ ♆
15 23 ☽ ∠ ♃
17 36 ☽ ⊥ ♇
20 49 ☽ ⚹ ♂
22 01 ☽ ∠ ♅
23 50 ☽ △ ♄

13 Saturday
02 40 ☽ ∥ ♆
03 54 ☽ ∥ ♀
09 19 ☽ ⊥ ♃
11 34 ☽ Q ☿
11 37 ☽ ∠ ♇
16 01 ☽ ⊥ ♅
16 39 ☽ □ ☉
21 21 ☽ ⊻ ♃
22 26 ☽ Q ♂

14 Sunday
02 10 ☽ ☌ ♆
04 05 ☽ ⊻ ♅
06 14 ☽ ⚼ ♄
06 33 ☽ △ ♀
17 57 ☽ ⚹ ♇

15 Monday
00 36 ☽ ± ♄
02 57 ♀ ∠ ♂
04 12 ☽ ∥ ♃
05 00 ☽ □ ☿
05 58 ♂ ⚹ ♄
08 19 ☽ ∥ ♄
12 29 ☽ ⚼ ♀
12 51 ☽ ⊼ ♄
13 15 ☽ □ ♂
13 57 ☽ ∥ ♅
18 18 ☽ Q ♇
23 44 ☉ △ ♃

16 Tuesday
06 29 ☽ ± ♀
09 16 ☽ # ♅
09 55 ☽ ☌ ♃
10 52 ☽ △ ☉
11 17 ♂ Q ♆
11 40 ☉ ∥ ♇
14 50 ☽ ⊻ ♆
15 06 ☽ # ♄
16 37 ☽ ☌ ♅
18 28 ☽ ⊼ ♀
18 57 ☽ # ♃

17 Wednesday
02 52 ☽ ⊥ ♆
06 35 ☽ □ ♇
15 19 ☿ ∠ ♀
19 36 ☽ ⚼ ☉
20 47 ☽ ∠ ♆

18 Thursday
01 25 ☽ △ ☿
01 30 ☽ ☍ ♄
02 09 ☿ ⚹ ♄
05 15 ☽ △ ♂
09 54 ☉ □ ♆
10 47 ☽ # ♀
12 35 ☿ Q ♆
14 56 ☽ ± ☉
16 53 ♃ St D
19 45 ☽ # ♆
21 18 ♀ St D
21 27 ☽ ⊻ ♃

19 Friday
02 14 ☽ ⚹ ♆
03 41 ☽ ⊼ ☉
03 52 ☽ ⊻ ♅
05 33 ☽ ☍ ♀
05 57 ☉ △ ♅
09 06 ☽ ⊥ ♃
10 32 ☽ ⚼ ☿
12 21 ☽ ⚼ ♂
15 26 ☽ ⊥ ♅
17 33 ☽ △ ♇

20 Saturday
01 21 ☽ Q ♆
02 23 ☽ ∠ ♃
03 17 ☉ ⊻ ♀
05 30 ☽ # ♇
05 57 ☽ ± ☿
06 40 ☽ ± ♂
08 17 ☉ ⊥ ♇
08 37 ☽ ∠ ♅
12 02 ☽ ⊼ ♄
12 28 ☽ # ☉
18 43 ☽ ⊼ ☿
18 43 ☽ ⊼ ♂
18 53 ☿ ☌ ♂
22 06 ☽ ⚼ ♇
23 26 ☽ ± ♄

21 Sunday
01 43 ☉ ∠ ♄
06 42 ☽ ⚹ ♃
11 17 ☽ □ ♆
12 46 ☽ ⚹ ♅
14 42 ☽ ⊼ ♀
14 57 ☽ ± ♇
16 20 ☽ ⚼ ♄
17 27 ☽ ☍ ☉

22 Monday
01 53 ☽ ± ♀
02 03 ☽ ⊼ ♇
04 56 ☽ Q ♃
06 21 ☽ # ♂
10 51 ☽ Q ♅
18 32 ☽ ⚼ ♀
20 04 ☽ △ ♄

23 Tuesday
05 22 ☽ ☍ ♂
08 25 ☽ ☍ ☿
13 39 ☽ □ ♃
18 04 ☽ △ ♆
19 24 ☽ □ ♅
21 56 ☽ △ ♀
23 41 ☽ # ♂

24 Wednesday
04 24 ☽ ⊼ ☉
08 24 ☽ ☍ ♇
15 55 ☽ ± ☉
20 46 ☽ ⚼ ♆

25 Thursday
02 05 ☽ □ ♄
09 02 ☽ ⚼ ☉
10 01 ☽ # ☉
12 36 ☽ ± ♆
13 46 ☽ ⊼ ♂
14 18 ☿ □ ♃
18 50 ☽ △ ♃
19 18 ☽ ⊼ ☿
23 05 ☽ ⊼ ♆
23 58 ☽ # ♇

26 Friday
00 18 ☽ △ ♅
00 50 ☽ ± ♂
03 44 ☽ □ ♀
06 48 ☽ ± ☿
09 44 ☽ Q ♄
11 23 ☉ ⊻ ♇
13 07 ☽ ⊼ ♇
13 15 ☽ △ ☉
17 20 ☽ ⚼ ♂
20 55 ☽ ⚼ ♃
23 30 ☽ ± ♇
23 55 ☽ ⚼ ☿

27 Saturday
02 16 ☽ ⚼ ♅
06 22 ☽ # ♆
06 36 ☽ ⚹ ♄
10 33 ☉ ⊥ ♀
11 30 ☽ Q ♀
12 26 ☽ ± ♃
13 13 ☿ ⚹ ♆
15 00 ☽ ⚼ ♇
17 42 ☽ ± ♅
17 55 ☿ Q ♄
19 40 ☽ # ♀
20 34 ☽ △ ♂
22 43 ☽ ⊼ ♃

28 Sunday
02 27 ☿ □ ♅
02 49 ☽ ☍ ♆
03 56 ☽ ⊼ ♅
04 05 ☽ △ ☿
08 25 ☽ ∠ ♄
08 30 ☽ ⚹ ♀
16 38 ☽ △ ♇
20 36 ☽ □ ☉
23 48 ☽ ⊥ ♄

29 Monday
00 40 ☽ # ♃
01 37 ☽ # ♄
05 05 ♀ ± ♃
08 06 ☽ # ♅
10 00 ☽ ⊻ ♄
10 36 ☽ ∠ ♀
14 51 ♂ □ ♃

30 Tuesday
00 42 ☽ ∥ ♅
01 41 ☽ ☍ ♃
02 12 ☽ ⊥ ♀
02 15 ☽ □ ♂
05 20 ☽ Q ☉
05 38 ☽ ⊼ ♆
06 41 ☽ ☍ ♅
07 20 ☽ ∥ ♄
08 00 ☽ ∥ ♃
11 17 ☽ □ ☿
12 36 ☽ ⊻ ♀
15 44 ☽ ± ♆
19 22 ☽ □ ♇

All ephemeris data is given at 12.00 UT and the Moon's longitude is additionally given for 24.00 UT

DECEMBER 2010

LONGITUDES

Date	Sidereal time h m s	Sun ☉ ° ' "	Moon ☽ ° ' "	Moon ☽ 24.00 ° ' "	Mercury ☿ ° '	Venus ♀ ° '	Mars ♂ ° '	Jupiter ♃ ° '	Saturn ♄ ° '	Uranus ♅ ° '	Neptune ♆ ° '	Pluto ♇ ° '
01	16 40 57	09 ♐ 10 49	14 ♎ 07 47	21 ♎ 15 54	00 ♑ 31	00 ♏ 38	25 ♐ 07	23 ♓ 47	14 ♎ 36	26 ♓ 41	26 ♒ 05	04 ♑ 15
02	16 44 53	10 11 39	28 ♎ 23 09	05 ♏ 29 00	01 31	01 05	25 52	23 49	14 42	26 R 41	26 06	04 17
03	16 48 50	11 12 31	12 ♏ 32 53	19 ♏ 34 16	02 26	01 34	26 37	23 52	14 47	26 40	26 07	04 19
04	16 52 46	12 13 23	26 ♏ 32 37	03 ♐ 27 27	03 17	02 05	27 22	23 55	14 52	26 40	26 07	04 21
05	16 56 43	13 14 18	10 ♐ 18 22	17 ♐ 04 58	04 02	02 37	28 07	23 59	14 57	26 40	26 08	04 23
06	17 00 40	14 15 13	23 ♐ 47 01	00 ♑ 24 19	04 41	03 11	28 52	24 02	15 02	26 D 40	26 09	04 25
07	17 04 36	15 16 09	06 ♑ 56 46	13 ♑ 24 22	05 13	03 47	29 ♐ 38	24 06	15 07	26 40	26 10	04 28
08	17 08 33	16 17 06	19 ♑ 47 15	26 ♑ 05 33	05 36	04 23	00 ♑ 23	24 10	15 12	26 40	26 11	04 30
09	17 12 29	17 18 04	02 ♒ 19 34	08 ♒ 29 37	05 51	05 01	01 08	24 14	15 16	26 41	26 12	04 32
10	17 16 26	18 19 03	14 ♒ 36 06	20 ♒ 39 28	05 56	05 40	01 54	24 18	15 21	26 41	26 13	04 34
11	17 20 22	19 20 02	26 ♒ 40 14	02 ♓ 38 56	05 R 51	06 21	02 39	24 23	15 25	26 41	26 15	04 36
12	17 24 19	20 21 02	08 ♓ 36 08	14 ♓ 32 25	05 34	07 02	03 25	24 27	15 30	26 41	26 16	04 38
13	17 28 15	21 22 02	20 ♓ 28 22	26 ♓ 24 37	05 07	07 45	04 11	24 32	15 34	26 42	26 17	04 40
14	17 32 12	22 23 03	02 ♈ 21 45	08 ♈ 20 21	04 27	08 29	04 56	24 37	15 39	26 42	26 18	04 42
15	17 36 09	23 24 04	14 ♈ 20 59	20 ♈ 24 13	03 37	09 14	05 42	24 42	15 43	26 43	26 20	04 44
16	17 40 05	24 25 06	26 ♈ 30 31	02 ♉ 40 21	02 36	10 00	06 28	24 48	15 47	26 43	26 21	04 46
17	17 44 02	25 26 08	08 ♉ 54 08	15 ♉ 12 11	01 26	10 46	07 13	24 53	15 51	26 44	26 22	04 49
18	17 47 58	26 27 11	21 ♉ 34 48	28 ♉ 02 08	00 ♑ 10	11 34	07 59	24 59	15 55	26 44	26 23	04 51
19	17 51 55	27 28 14	04 ♊ 34 19	11 ♊ 11 22	28 ♐ 48	12 23	08 45	25 05	15 59	26 45	26 25	04 53
20	17 55 51	28 29 18	17 ♊ 53 12	24 ♊ 39 39	27 26	13 12	09 31	25 11	16 03	26 46	26 26	04 55
21	17 59 48	29 ♐ 30 23	01 ♋ 30 30	08 ♋ 25 23	26 04	14 02	10 17	25 17	16 06	26 46	26 28	04 57
22	18 03 44	00 ♑ 31 27	15 ♋ 23 56	22 ♋ 25 41	24 46	14 54	11 03	25 24	16 10	26 47	26 29	04 59
23	18 07 41	01 32 33	29 ♋ 30 08	06 ♌ 36 45	23 33	15 46	11 49	25 30	16 14	26 48	26 31	05 02
24	18 11 38	02 33 39	13 ♌ 44 59	20 ♌ 54 18	22 29	16 38	12 35	25 37	16 17	26 49	26 32	05 04
25	18 15 34	03 34 45	28 ♌ 04 11	05 ♍ 14 08	21 35	17 31	13 21	25 44	16 20	26 50	26 34	05 06
26	18 19 31	04 35 52	12 ♍ 23 41	19 ♍ 32 25	20 50	18 26	14 08	25 51	16 24	26 51	26 35	05 08
27	18 23 27	05 37 00	26 ♍ 39 59	03 ♎ 46 02	20 16	19 20	14 54	25 58	16 27	26 52	26 37	05 10
28	18 27 24	06 38 08	10 ♎ 50 19	17 ♎ 52 35	19 53	20 16	15 40	26 06	16 30	26 53	26 38	05 12
29	18 31 20	07 39 17	24 ♎ 52 38	01 ♏ 50 19	19 41	21 11	16 26	26 13	16 33	26 54	26 40	05 14
30	18 35 17	08 40 26	08 ♏ 45 27	15 ♏ 37 56	19 D 38	22 08	17 13	26 21	16 36	26 56	26 42	05 17
31	18 39 13	09 ♑ 41 36	22 ♏ 27 39	29 ♏ 14 26	19 ♐ 44	23 ♏ 05	17 ♑ 59	26 ♓ 29	16 ♎ 38	26 ♓ 57	26 ♒ 43	05 ♑ 19

Date	Moon True ☊ ° '	Moon Mean ☊ ° '	Moon ☽ Latitude ° '
01	02 ♑ 56	03 ♑ 55	05 S 08
02	02 R 53	03 52	04 44
03	02 50	03 49	04 02
04	02 47	03 46	03 06
05	02 45	03 42	02 01
06	02 44	03 39	00 S 49
07	02 D 43	03 36	00 N 23
08	02 44	03 33	01 33
09	02 45	03 30	02 36
10	02 47	03 26	03 31
11	02 48	03 23	04 16
12	02 49	03 20	04 49
13	02 49	03 17	05 09
14	02 R 49	03 14	05 17
15	02 48	03 11	05 10
16	02 47	03 07	04 50
17	02 46	03 04	04 16
18	02 45	03 01	03 29
19	02 45	02 58	02 30
20	02 44	02 55	01 21
21	02 44	02 52	00 N 07
22	02 D 44	02 48	01 S 10
23	02 44	02 45	02 23
24	02 44	02 42	03 28
25	02 R 44	02 39	04 21
26	02 44	02 36	04 57
27	02 44	02 32	05 15
28	02 D 44	02 29	05 14
29	02 44	02 26	04 54
30	02 45	02 23	04 17
31	02 ♑ 45	02 ♑ 20	03 S 25

DECLINATIONS

Date	Sun ☉ ° '	Moon ☽ ° '	Mercury ☿ ° '	Venus ♀ ° '	Mars ♂ ° '	Jupiter ♃ ° '	Saturn ♄ ° '	Uranus ♅ ° '	Neptune ♆ ° '	Pluto ♇ ° '
01	21 S 50	10 S 18	25 S 44	10 S 26	24 S 10	03 S 45	03 S 39	02 S 01	13 S 17	18 S 48
02	21 59	15 19	25 39	10 28	24 12	03 44	03 41	02 01	13 17	18 48
03	22 07	19 26	25 32	10 30	24 13	03 42	03 43	02 01	13 16	18 48
04	22 15	22 24	25 25	10 33	24 15	03 41	03 44	02 01	13 16	18 48
05	22 23	23 59	25 15	10 37	24 16	03 39	03 46	02 01	13 16	18 49
06	22 31	24 07	25 05	10 42	24 17	03 38	03 48	02 01	13 15	18 49
07	22 37	22 52	24 54	10 47	24 18	03 36	03 49	02 01	13 15	18 49
08	22 44	20 27	24 41	10 53	24 18	03 34	03 51	02 01	13 15	18 49
09	22 50	17 06	24 28	11 00	24 18	03 32	03 53	02 01	13 14	18 49
10	22 55	13 05	24 13	11 07	24 18	03 30	03 54	02 01	13 14	18 49
11	23 01	08 37	23 58	11 15	24 18	03 28	03 56	02 00	13 13	18 49
12	23 05	03 S 53	23 41	11 23	24 17	03 26	03 57	02 00	13 13	18 49
13	23 09	00 N 58	23 24	11 32	24 16	03 24	03 59	02 00	13 13	18 49
14	23 13	05 47	23 07	11 42	24 15	03 22	04 00	02 00	13 12	18 49
15	23 16	10 25	22 49	11 52	24 13	03 20	04 02	02 00	13 12	18 49
16	23 19	14 44	22 30	12 02	24 11	03 17	04 03	02 00	13 12	18 49
17	23 22	18 30	22 11	12 13	24 09	03 15	04 04	01 59	13 11	18 49
18	23 23	21 31	21 52	12 24	24 07	03 12	04 06	01 59	13 11	18 49
19	23 25	23 31	21 33	12 35	24 05	03 10	04 07	01 59	13 10	18 49
20	23 26	24 14	21 15	12 47	24 02	03 07	04 08	01 58	13 09	18 49
21	23 26	23 33	20 58	12 59	23 59	03 04	04 09	01 58	13 09	18 49
22	23 26	21 24	20 43	13 12	23 55	03 02	04 10	01 58	13 08	18 49
23	23 26	17 55	20 30	13 24	23 52	02 59	04 11	01 57	13 08	18 49
24	23 25	13 22	20 19	13 37	23 48	02 56	04 13	01 57	13 07	18 50
25	23 23	08 03	20 10	13 50	23 43	02 53	04 14	01 56	13 07	18 50
26	23 21	02 N 19	20 04	14 03	23 39	02 50	04 15	01 56	13 06	18 50
27	23 19	03 S 30	20 00	14 16	23 34	02 47	04 16	01 55	13 06	18 50
28	23 16	09 06	19 59	14 29	23 29	02 44	04 16	01 55	13 05	18 50
29	23 13	14 11	20 01	14 43	23 24	02 40	04 17	01 54	13 05	18 50
30	23 09	18 28	20 04	14 56	23 18	02 37	04 18	01 54	13 04	18 50
31	23 S 05	21 S 42	20 S 10	15 S 10	23 S 13	02 S 34	04 S 19	01 S 53	13 S 04	18 S 50

ZODIAC SIGN ENTRIES

Date	h m	Planets
01	00 10	☿ ♑
02	14 44	☽ ♏
04	17 59	☽ ♐
06	23 16	☽ ♑
07	23 49	♂ ♑
09	07 30	☽ ♒
11	18 41	☽ ♓
14	07 15	☽ ♈
16	18 49	☽ ♉
18	14 53	☿ ♐
19	03 37	☽ ♊
21	09 22	☽ ♋
21	23 38	☉ ♑
23	12 51	☽ ♌
25	15 14	☽ ♍
27	17 38	☽ ♎
29	20 49	☽ ♏

LATITUDES

Date	Mercury ☿ ° '	Venus ♀ ° '	Mars ♂ ° '	Jupiter ♃ ° '	Saturn ♄ ° '	Uranus ♅ ° '	Neptune ♆ ° '	Pluto ♇ ° '
01	02 S 18	01 N 20	00 S 49	01 S 24	02 N 17	00 S 45	00 S 29	04 N 34
04	02 01	01 45	00 50	01 23	02 18	00 45	00 29	04 34
07	01 34	02 07	00 51	01 22	02 18	00 45	00 29	04 33
10	00 55	02 26	00 53	01 21	02 19	00 45	00 29	04 33
13	00 S 04	02 42	00 54	01 20	02 20	00 45	00 29	04 32
16	00 N 55	02 55	00 55	01 20	02 21	00 45	00 29	04 32
19	01 53	03 06	00 56	01 19	02 21	00 45	00 29	04 32
22	02 37	03 15	00 57	01 18	02 22	00 45	00 29	04 31
25	03 01	03 22	00 58	01 17	02 23	00 44	00 29	04 31
28	03 04	03 27	00 58	01 17	02 24	00 44	00 29	04 30
31	02 N 53	03 N 29	00 S 59	01 S 16	02 N 24	00 S 44	00 S 29	04 N 30

DATA

Julian Date	2455532
Delta T	+67 seconds
Ayanamsa	24° 00' 50"
Synetic vernal point	05° ♓ 06' 09"
True obliquity of ecliptic	23° 26' 17"

MOON'S PHASES, APSIDES AND POSITIONS ☽

Date	h m	Phase	Longitude ° '	Eclipse Indicator
05	17 36	●	13 ♐ 28	
13	13 59	☽	21 ♓ 27	
21	08 13	○	29 ♊ 21	total
28	04 18	☾	06 ♎ 19	

Day	h m		
13	08 35	Apogee	
25	12 28	Perigee	
06	02 08	Max dec	24° S 14'
13	07 13	0N	
20	12 33	Max dec	24° N 14'
26	21 33	0S	

LONGITUDES

Date	Chiron ⚷ ° '	Ceres ⚳ ° '	Pallas ⚴ ° '	Juno ⚵ ° '	Vesta ⚶ ° '	Black Moon Lilith ⚸ ° '
01	26 ♒ 25	17 ♑ 25	22 ♐ 01	22 ♍ 05	29 ♏ 02	17 ♓ 28
11	26 ♒ 43	21 ♑ 08	26 ♐ 08	24 ♍ 41	04 ♐ 24	18 ♓ 35
21	27 ♒ 07	24 ♑ 56	00 ♑ 14	26 ♍ 49	09 ♐ 46	19 ♓ 42
31	27 ♒ 36	28 ♑ 47	04 ♑ 18	28 ♍ 24	15 ♐ 06	20 ♓ 49

All ephemeris data is given at 12.00 UT and the Moon's longitude is additionally given for 24.00 UT

ASPECTARIAN

h m	Aspects
01 Wednesday	
03 03	☽ ✶ ☉
06 52	☽ ᵽ ♆
10 11	☽ Q ♂
12 38	☽ ‖ ♀
12 48	☽ ☌ ♄
16 34	☿ ✶ ♀
19 57	☽ Q ☿
02 Thursday	
01 42	☽ Q ♇
01 50	☽ ‖ ♆
04 17	☽ ⊼ ♃
06 12	☽ ∠ ☉
07 30	☽ ✶ ♂
08 08	☽ △ ♆
09 07	☽ ⊼ ♅
14 26	☽ ± ♃
16 43	☽ ☌ ♀
17 40	☽ ✶ ☿
19 15	☽ ± ♅
19 39	♂ ✶ ♆
22 00	☽ ✶ ♇
22 34	☽ ⊥ ☉
03 Friday	
05 43	☽ ᵽ ♃
07 52	☽ ‖ ♇
09 33	☽ ⊻ ☉
10 12	♃ ‖ ♄
10 19	☽ ∠ ♂
10 31	☽ ᵽ ♅
14 01	♂ □ ♅
15 50	☽ ⊻ ♄
18 05	♂ Q ♄
20 54	☽ ∠ ☿
23 36	☽ ∠ ♇
04 Saturday	
02 10	☽ ⊥ ♄
02 34	☽ ⊥ ♂
07 27	☽ △ ♃
10 27	☽ ‖ ☉
11 17	☽ □ ♆
12 13	☽ △ ♅
13 22	☽ ⊥ ☿
13 30	☽ ⊻ ♂
15 09	☽ ⊥ ♇
17 47	☽ ∠ ♄
22 00	☽ ⊻ ♀
05 Sunday	
00 24	☽ ⊻ ☿
01 36	☽ ⊻ ♇
08 55	☽ ⊥ ♀
14 03	♀ ± ♅
17 36	☽ ☌ ☉
18 47	☽ Q ♆
20 16	☽ ✶ ♄
06 Monday	
01 20	☿ ☌ ♇
01 32	☽ ∠ ♀
01 50	♅ St D
09 40	☉ Q ♆
12 27	☽ □ ♃
16 17	☽ ✶ ♆
17 13	☽ □ ♅
17 54	☽ Q ♄
21 46	☽ ☌ ♂
07 Tuesday	
05 53	☽ ✶ ♀
07 24	☽ ☌ ♇
08 00	☉ ✶ ♄
08 41	☽ ☌ ☿
14 51	☽ ‖ ☉
19 51	☽ ∠ ♆
21 36	☽ Q ♃
08 Wednesday	
02 22	☽ Q ♅
03 17	☽ □ ♄
04 49	☽ ⊻ ☉
05 16	☽ Q ♀
12 46	☽ ⊥ ♆
16 19	♀ ✶ ♇
17 09	☽ ⊥ ☉
20 21	☽ ✶ ♃
09 Thursday	
00 12	☽ ⊻ ♆
00 25	☽ ‖ ♇
01 07	☽ ✶ ♅
09 34	☽ ⊻ ♂
11 57	☽ ∠ ☉
16 17	☽ ⊻ ♇
17 31	☽ □ ♀
18 55	☽ ⊻ ☿
21 58	☽ ⊥ ♂
10 Friday	
01 31	☽ ∠ ♃
03 51	☿ ‖ ♂
04 02	☽ ⊥ ♇
06 15	☽ ∠ ♅
06 44	☽ ⊥ ☿
11 09	☽ ‖ ♆
12 03	☿ St R
13 29	☽ △ ♄
16 51	☽ ∠ ♂
19 22	☽ ⊥ ♃
20 02	☽ ✶ ☉
21 01	☿ ✶ ♀
21 51	☽ ∠ ♇
22 28	☽ ‖ ♀
11 Saturday	
00 03	☽ ⊥ ♅
00 30	☽ ∠ ☿
07 23	☽ ⊻ ♃
11 09	☽ ☌ ♆
12 02	☽ ⊻ ♅
19 35	☽ ᵽ ♄
22 14	☽ Q ☉
12 Sunday	
00 50	☽ ✶ ♂
03 58	☽ ✶ ♇
06 04	☽ ✶ ☿
08 39	☽ △ ♀
11 36	☽ ‖ ♄
13 49	☽ ± ♄
14 12	☽ ‖ ♃
21 18	☽ ‖ ♅
13 Monday	
02 02	☽ ⊼ ♄
02 43	☽ Q ♂
04 17	☽ Q ♇
05 30	☽ Q ☿
13 59	☽ □ ☉
16 54	☽ ᵽ ♀
17 07	☽ # ♅
20 16	☽ ☌ ♃
23 46	☽ ⊻ ♆
23 59	☽ # ♃
14 Tuesday	
00 35	☽ ☌ ♅
03 02	☽ # ♄
04 05	☿ ☌ ♇
04 10	☿ ☌ ♂
04 15	♂ ☌ ♇
04 52	☉ ‖ ☿
11 53	☽ ⊥ ♆
12 15	☽ ± ♀
15 57	☽ □ ☿
16 43	☽ □ ♇
17 32	☽ □ ♂
15 Wednesday	
01 06	☽ ⊼ ♀
05 58	☽ ∠ ♆
14 44	☽ ☍ ♄
20 06	☽ # ♀
16 Thursday	
03 10	☽ # ♆
05 02	♀ ᵽ ♃
07 32	☽ △ ☉
08 37	☽ ⊻ ♃
11 41	☽ ✶ ♆
12 25	☽ ⊻ ♅
20 25	☽ ⊥ ♃
21 47	☉ □ ♃
22 52	☽ △ ☿
17 Friday	
00 06	☽ ⊥ ♅
04 07	☽ △ ♇
08 34	☽ △ ♂
10 58	☽ Q ♆
13 54	☽ ∠ ♃
14 14	☽ # ♇
15 12	☽ ᵽ ☉
15 49	☽ ☍ ♀
17 24	☽ ∠ ♅
18 Saturday	
01 03	☽ ᵽ ☿
01 18	☽ ⊼ ♄
08 45	☽ ᵽ ♇
09 42	☽ ± ☉
10 29	☉ ✶ ♆
12 38	☽ ± ♄
14 48	☽ ᵽ ♂
14 59	☽ # ☿
16 22	☽ ± ☿
17 09	♀ ᵽ ♅
18 24	☽ ✶ ♃
18 48	☉ □ ♅
20 58	☽ □ ♆
21 37	☽ ✶ ♅
21 51	☽ ⊼ ☉
19 Sunday	
01 05	☉ ∠ ♀
01 32	☽ ± ♇
02 26	☽ ⊼ ☿
05 24	☽ ᵽ ♄
08 28	☽ ± ♂
10 23	☽ # ☉
12 34	☽ ⊼ ♇
16 36	☽ Q ♃
19 36	☽ Q ♅
20 04	☽ ⊼ ♂
20 Monday	
00 16	☽ # ♂
00 51	☉ Q ♄
01 23	☉ ☌ ☿
01 44	☿ Q ♄
03 04	☽ ⊼ ♀
03 34	☿ ∠ ♀
08 42	☽ △ ♄
14 30	☽ ± ♀
23 30	☿ □ ♅
21 Tuesday	
01 01	☽ □ ♃
02 23	☽ # ♂
03 09	☽ △ ♆
03 19	☽ ☍ ☿
03 43	☽ □ ♅
05 03	☿ ✶ ♆
07 25	☽ ᵽ ♀
08 13	☽ ☍ ☉
13 43	☽ # ☉
18 00	☽ ☍ ♇
22 Wednesday	
00 59	☿ □ ♃
04 06	☽ ☍ ♂
05 16	☽ ᵽ ♆
06 22	♀ ‖ ♆
11 05	☽ △ ♀
13 19	☽ □ ♄
17 52	☽ # ☿
20 42	☽ ± ♆
23 Thursday	
01 58	♂ ∠ ♆
02 41	☽ ⊼ ☿
05 11	☽ △ ♃
06 29	☽ # ♇
06 55	☽ ⊼ ♆
07 25	☽ △ ♅
12 05	☽ ± ☿
15 43	☽ ⊼ ☉
20 01	☽ Q ♄
21 21	☽ ⊼ ♇
24 Friday	
01 43	♀ ⊻ ♄
02 11	☽ ᵽ ☿
02 37	☽ ± ☉
06 42	☽ ᵽ ♃
07 28	☽ ± ♇
08 45	☽ ᵽ ♅
09 56	☽ ⊼ ♂
10 08	☿ ⊥ ♀
10 52	☽ # ♀
13 10	☽ # ♆
16 16	☽ ✶ ♄
17 10	☽ □ ♀
18 53	☽ ᵽ ☉
20 35	☽ ± ♂
21 56	☽ ± ♃
22 37	☽ ᵽ ♇
23 52	☽ ± ♅
25 Saturday	
01 45	☽ △ ☿
08 04	☽ ⊼ ♃
09 28	☽ ☍ ♆
09 56	☽ ⊼ ♅
12 30	☽ ᵽ ♂
17 30	☽ ∠ ♄
21 56	☽ △ ☉
23 48	☽ △ ♇
26 Sunday	
01 19	☽ Q ♀
01 57	♂ Q ♃
04 06	☽ # ♄
08 38	☽ ⊥ ♄
09 54	☽ # ♃
13 37	☽ # ♅
15 04	☽ △ ♂
18 44	☽ ⊻ ♄
22 49	☽ ✶ ♀
27 Monday	
01 04	☉ ☌ ♇
01 36	☽ □ ☿
05 29	☽ ‖ ♅
09 02	☽ ‖ ♃
10 49	☽ ☍ ♃
11 06	♂ Q ♅
11 54	☽ ⊼ ♆
12 20	☽ ☍ ♅
15 11	☽ ‖ ♄
22 04	☽ ± ♆
28 Tuesday	
01 52	☽ ∠ ♀
02 25	☽ □ ♇
04 18	☽ □ ☉
04 53	☿ ⊻ ♀
07 06	☽ Q ☿
10 33	♀ ∠ ♇
13 22	☽ ᵽ ♆
18 14	☽ ⊥ ♀
20 42	☽ □ ♂
21 40	☽ ☌ ♄
29 Wednesday	
03 11	☽ ✶ ☿
05 13	☽ ⊻ ♀
06 30	☽ ‖ ♆
09 11	☽ Q ♇
13 27	☽ Q ☉
14 20	☽ ⊼ ♃
14 51	☽ ‖ ♀
15 05	☽ △ ♆
15 29	♂ □ ♄
15 30	☽ ⊼ ♅
30 Thursday	
00 47	☽ ± ♃
01 52	☽ ± ♅
04 49	☽ ∠ ☿
05 29	☽ Q ♂
05 56	☽ ✶ ♇
07 22	☿ St D
11 51	☽ ✶ ☉
14 18	☽ ‖ ♇
16 34	☽ ᵽ ♃
17 32	☽ ᵽ ♅
20 32	☽ ⊥ ☿
23 08	☽ ‖ ☿
31 Friday	
00 11	♀ ⊥ ♄
01 44	☽ ⊻ ♄
03 39	☽ ✶ ♂
07 08	☽ ⊻ ☿
08 12	☽ ∠ ♇
12 19	☽ ⊥ ♄
13 11	☽ ☌ ♀
16 16	☽ ∠ ☉
19 11	☽ △ ♃
19 33	☽ □ ♆
19 57	☽ △ ♅

JANUARY 2011

LONGITUDES

Date	Sidereal time h m s	Sun ☉ ° ' "	Moon ☽ ° ' "	Moon ☽ 24.00 ° ' "	Mercury ☿ ° '	Venus ♀ ° '	Mars ♂ ° '	Jupiter ♃ ° '	Saturn ♄ ° '	Uranus ♅ ° '	Neptune ♆ ° '	Pluto ♇ ° '
01	18 43 10	10 ♑ 42 46	05 ♐ 58 14	12 ♐ 38 56	19 ♐ 59	24 ♏ 03	18 ♑ 46	26 ♓ 37	16 ♎ 41	26 ♓ 58	26 ♒ 45	05 ♑ 21
02	18 47 07	11 43 57	19 ♐ 16 26	25 ♐ 50 39	20 21	25 01	19 32	26 45	16 43	27 00	26 47	05 23
03	18 51 03	12 45 08	02 ♑ 21 32	08 ♑ 49 01	20 51	26 00	20 19	26 54	16 46	27 01	26 48	05 25
04	18 55 00	13 46 18	15 ♑ 13 05	21 ♑ 33 45	21 26	26 59	21 05	27 02	16 48	27 02	26 50	05 27
05	18 58 56	14 47 29	27 ♑ 51 03	04 ♒ 05 03	22 08	27 58	21 52	27 11	16 51	27 04	26 52	05 30
06	19 02 53	15 48 40	10 ♒ 15 53	16 ♒ 23 44	22 54	28 58	22 38	27 20	16 53	27 05	26 54	05 32
07	19 06 49	16 49 50	22 ♒ 28 49	28 ♒ 31 23	23 45	29 ♏ 59	23 25	27 28	16 55	27 07	26 56	05 34
08	19 10 46	17 51 00	04 ♓ 31 46	10 ♓ 30 20	24 40	01 ♐ 00	24 12	27 38	16 57	27 09	26 57	05 36
09	19 14 42	18 52 09	16 ♓ 27 30	22 ♓ 23 42	25 39	02 01	24 58	27 47	16 59	27 10	26 59	05 38
10	19 18 39	19 53 19	28 ♓ 19 27	04 ♈ 15 15	26 40	03 02	25 45	27 56	17 00	27 12	27 01	05 40
11	19 22 36	20 54 27	10 ♈ 11 40	16 ♈ 09 17	27 45	04 04	26 32	28 06	17 02	27 14	27 03	05 42
12	19 26 32	21 55 35	22 ♈ 08 42	28 ♈ 10 29	28 ♐ 52	05 07	27 19	28 15	17 03	27 16	27 05	05 44
13	19 30 29	22 56 43	04 ♉ 15 17	10 ♉ 23 39	00 ♑ 02	06 10	28 05	28 25	17 05	27 18	27 07	05 46
14	19 34 25	23 57 50	16 ♉ 36 10	22 ♉ 53 21	01 13	07 13	28 52	28 35	17 06	27 20	27 09	05 49
15	19 38 22	24 58 56	29 ♉ 15 42	05 ♊ 43 38	02 27	08 16	29 ♑ 39	28 45	17 07	27 21	27 11	05 51
16	19 42 18	26 00 02	12 ♊ 17 28	18 ♊ 57 27	03 42	09 20	00 ♒ 26	28 55	17 08	27 23	27 13	05 53
17	19 46 15	27 01 07	25 ♊ 43 41	02 ♋ 36 10	04 58	10 24	01 13	29 05	17 09	27 26	27 15	05 55
18	19 50 11	28 02 11	09 ♋ 34 45	16 ♋ 39 06	06 16	11 28	02 00	29 15	17 10	27 28	27 17	05 57
19	19 54 08	29 ♑ 03 15	23 ♋ 48 46	01 ♌ 03 07	07 36	12 32	02 47	29 26	17 11	27 30	27 19	05 59
20	19 58 05	00 ♒ 04 18	08 ♌ 21 24	15 ♌ 42 45	08 56	13 37	03 34	29 36	17 12	27 32	27 21	06 01
21	20 02 01	01 05 21	23 ♌ 06 13	00 ♍ 30 48	10 18	14 42	04 21	29 47	17 12	27 34	27 23	06 03
22	20 05 58	02 06 22	07 ♍ 55 29	15 ♍ 19 18	11 41	15 48	05 08	29 ♓ 58	17 13	27 36	27 25	06 05
23	20 09 54	03 07 24	22 ♍ 41 20	00 ♎ 00 46	13 04	16 53	05 55	00 ♈ 09	17 13	27 39	27 27	06 07
24	20 13 51	04 08 25	07 ♎ 16 55	14 ♎ 29 14	14 29	17 59	06 42	00 20	17 13	27 41	27 30	06 09
25	20 17 47	05 09 25	21 ♎ 37 18	28 ♎ 40 52	15 55	19 05	07 29	00 31	17 14	27 43	27 32	06 11
26	20 21 44	06 10 25	05 ♏ 39 46	12 ♏ 33 58	17 21	20 11	08 16	00 42	17 R 14	27 46	27 34	06 13
27	20 25 40	07 11 24	19 ♏ 23 32	26 ♏ 08 36	18 48	21 18	09 04	00 53	17 14	27 48	27 36	06 15
28	20 29 37	08 12 23	02 ♐ 49 22	09 ♐ 26 02	20 16	22 25	09 51	01 05	17 13	27 51	27 38	06 17
29	20 33 34	09 13 22	15 ♐ 58 51	22 ♐ 28 04	21 45	23 32	10 38	01 16	17 13	27 53	27 40	06 18
30	20 37 30	10 14 19	28 ♐ 53 55	05 ♑ 16 38	23 14	24 39	11 25	01 28	17 13	27 56	27 43	06 20
31	20 41 27	11 ♒ 15 16	11 ♑ 36 25	17 ♑ 53 28	24 ♑ 45	25 ♐ 46	12 ♒ 12	01 ♈ 39	17 ♎ 12	27 ♓ 58	27 ♒ 45	06 ♑ 22

Date	Moon True ☊ ° '	Moon Mean ☊ ° '	Moon ☽ Latitude ° '
01	02 ♑ 46	02 ♑ 17	02 S 23
02	02 D 47	02 13	01 14
03	02 47	02 10	00 S 02
04	02 R 47	02 07	01 N 08
05	02 46	02 04	02 14
06	02 44	02 01	03 13
07	02 42	01 58	04 01
08	02 40	01 54	04 38
09	02 37	01 51	05 03
10	02 35	01 48	05 15
11	02 34	01 45	05 13
12	02 D 34	01 42	04 57
13	02 34	01 38	04 29
14	02 35	01 35	03 47
15	02 37	01 32	02 54
16	02 38	01 29	01 50
17	02 39	01 26	00 N 38
18	02 R 39	01 23	00 S 38
19	02 38	01 19	01 54
20	02 35	01 16	03 03
21	02 32	01 13	04 02
22	02 27	01 10	04 44
23	02 23	01 07	05 08
24	02 19	01 04	05 11
25	02 17	01 00	04 55
26	02 16	00 57	04 21
27	02 D 17	00 54	03 33
28	02 18	00 51	02 34
29	02 20	00 48	01 28
30	02 21	00 44	00 19
31	02 ♑ 20	00 ♑ 41	00 N 50

DECLINATIONS

Date	Sun ☉ ° '	Moon ☽ ° '	Mercury ☿ ° '	Venus ♀ ° '	Mars ♂ ° '	Jupiter ♃ ° '	Saturn ♄ ° '	Uranus ♅ ° '	Neptune ♆ ° '	Pluto ♇ ° '
01	23 S 00	23 S 39	20 S 17	15 S 23	23 S 06	02 S 30	04 S 20	01 S 53	13 S 03	18 S 50
02	22 55	24 14	20 25	15 37	23 00	02 27	04 20	01 52	13 02	18 50
03	22 50	23 27	20 34	15 50	22 54	02 23	04 21	01 52	13 02	18 50
04	22 44	21 26	20 45	16 04	22 47	02 20	04 22	01 51	13 01	18 50
05	22 37	18 24	20 55	16 17	22 40	02 16	04 22	01 50	13 01	18 50
06	22 30	14 34	21 06	16 30	22 32	02 12	04 23	01 50	13 00	18 50
07	22 23	10 13	21 18	16 43	22 25	02 09	04 24	01 49	12 59	18 50
08	22 15	05 31	21 29	16 56	22 17	02 05	04 24	01 48	12 59	18 50
09	22 07	00 S 41	21 40	17 09	22 09	02 01	04 24	01 48	12 58	18 50
10	21 58	04 N 09	21 51	17 22	22 00	01 57	04 25	01 47	12 57	18 50
11	21 49	08 50	22 02	17 35	21 52	01 53	04 25	01 46	12 57	18 49
12	21 39	13 13	22 12	17 47	21 43	01 49	04 26	01 45	12 56	18 49
13	21 29	17 09	22 21	17 59	21 34	01 45	04 26	01 45	12 55	18 49
14	21 19	20 25	22 30	18 11	21 24	01 41	04 26	01 44	12 55	18 49
15	21 08	22 49	22 38	18 22	21 15	01 37	04 26	01 43	12 54	18 49
16	20 57	24 05	22 45	18 34	21 05	01 33	04 26	01 42	12 53	18 49
17	20 45	24 00	22 51	18 45	20 55	01 28	04 27	01 41	12 53	18 49
18	20 33	22 28	22 57	18 55	20 44	01 24	04 27	01 41	12 52	18 49
19	20 21	19 29	23 01	19 06	20 34	01 20	04 27	01 40	12 51	18 49
20	20 08	15 13	23 04	19 16	20 23	01 16	04 27	01 39	12 51	18 49
21	19 55	10 00	23 06	19 26	20 12	01 11	04 27	01 38	12 50	18 49
22	19 41	04 N 12	23 07	19 35	20 01	01 07	04 26	01 37	12 49	18 49
23	19 27	01 S 49	23 07	19 44	19 49	01 02	04 26	01 36	12 48	18 49
24	19 13	07 39	23 06	19 53	19 38	00 58	04 26	01 35	12 48	18 49
25	18 59	12 59	23 03	20 01	19 26	00 53	04 26	01 34	12 47	18 49
26	18 44	17 30	23 00	20 09	19 14	00 48	04 26	01 33	12 46	18 49
27	18 28	20 59	22 55	20 16	19 01	00 44	04 25	01 32	12 46	18 49
28	18 13	23 14	22 49	20 23	18 49	00 39	04 25	01 31	12 45	18 49
29	17 57	24 10	22 41	20 29	18 36	00 34	04 25	01 30	12 44	18 49
30	17 41	23 45	22 33	20 35	18 23	00 30	04 24	01 29	12 43	18 49
31	17 S 24	22 S 06	22 S 23	20 S 41	18 S 10	00 S 25	04 S 24	01 S 28	12 S 43	18 S 48

ZODIAC SIGN ENTRIES

Date	h	m	Planets
01	01	21	☽ ♐
03	07	39	☽ ♑
05	16	08	☽ ♒
07	12	30	♀ ♐
08	02	57	☽ ♓
10	15	24	☽ ♈
13	03	37	☽ ♉
13	11	25	☿ ♑
15	13	23	☽ ♊
15	22	41	♂ ♒
17	19	29	☽ ♋
19	22	16	☽ ♌
20	10	18	☉ ♒
21	23	10	☽ ♍
22	17	11	♃ ♈
23	23	59	☽ ♎
26	02	15	☽ ♏
28	06	55	☽ ♐
30	14	04	☽ ♑

LATITUDES

Date	Mercury ☿ ° '	Venus ♀ ° '	Mars ♂ ° '	Jupiter ♃ ° '	Saturn ♄ ° '	Uranus ♅ ° '	Neptune ♆ ° '	Pluto ♇ ° '
01	02 N 47	03 N 30	01 S 00	01 S 16	02 N 25	00 S 44	00 S 29	04 N 30
04	02 25	03 31	01 00	01 15	02 25	00 44	00 29	04 30
07	02 00	03 30	01 01	01 15	02 26	00 44	00 29	04 30
10	01 32	03 28	01 02	01 14	02 27	00 44	00 29	04 30
13	01 05	03 24	01 02	01 13	02 28	00 44	00 29	04 30
16	00 38	03 19	01 03	01 13	02 29	00 44	00 29	04 29
19	00 N 12	03 14	01 03	01 12	02 30	00 44	00 29	04 29
22	00 S 12	03 07	01 04	01 12	02 31	00 43	00 29	04 29
25	00 34	02 59	01 04	01 11	02 31	00 43	00 29	04 29
28	00 55	02 51	01 04	01 11	02 32	00 43	00 29	04 29
31	01 S 13	02 N 41	01 S 05	01 S 10	02 N 33	00 S 43	00 S 29	04 N 29

DATA

Julian Date	2455563
Delta T	+68 seconds
Ayanamsa	24° 00' 56"
Synetic vernal point	05° ♓ 06' 03"
True obliquity of ecliptic	23° 26' 16"

MOON'S PHASES, APSIDES AND POSITIONS ☽

Date	h	m	Phase	Longitude ° '	Eclipse Indicator
04	09	03	●	13 ♑ 39	Partial
12	11	31	☽	21 ♈ 54	
19	21	21	○	29 ♋ 27	
26	12	57	☾	06 ♏ 13	

Day	h	m		
10	05	40	Apogee	
22	00	18	Perigee	
02	10	05	Max dec	24° S 14'
09	15	24	0N	
16	22	46	Max dec	24° N 13'
23	04	45	0S	
29	16	25	Max dec	24° S 11'

LONGITUDES

Date	Chiron ⚷ ° '	Ceres ⚳ ° '	Pallas ⚴ ° '	Juno ⚵ ° '	Vesta ⚶ ° '	Black Moon Lilith ⚸ ° '
01	27 ♒ 39	29 ♑ 11	04 ♑ 42	28 ♍ 32	15 ♐ 38	20 ♓ 56
11	28 ♒ 12	03 ♒ 05	08 ♑ 43	29 ♍ 27	20 ♐ 55	22 ♓ 03
21	28 ♒ 48	07 ♒ 02	12 ♑ 40	29 ♍ 41	26 ♐ 08	23 ♓ 10
31	29 ♒ 27	10 ♒ 58	16 ♑ 32	29 ♍ 12	01 ♑ 18	24 ♓ 18

All ephemeris data is given at 12.00 UT and the Moon's longitude is additionally given for 24.00 UT

ASPECTARIAN

h m	Aspects
01 Saturday	
00 10	☽ ⊥ ♇
02 30	☽ ∥ ☉
03 55	☽ ∥ ♂
04 19	☽ ∠ ♄
07 48	☽ ∠ ♂
09 34	☽ ⊥ ☉
10 53	☽ ⊻ ♇
21 13	☽ ⊻ ☉
02 Sunday	
00 57	☽ ⊥ ♂
03 50	☽ Q ♆
07 21	☽ ⚹ ♄
12 30	☽ ⊻ ♂
13 05	☉ ∠ ♆
14 02	☽ ☌ ☿
16 51	♃ ⊻ ♆
23 19	☽ ⊻ ♀
03 Monday	
01 45	☽ ⚹ ♆
01 49	☽ □ ♃
02 08	☽ □ ♅
05 21	☽ Q ♄
11 16	☽ ⊥ ♀
17 42	☽ ☌ ♇
20 47	☽ ∥ ♂
21 36	☽ ∥ ☉
04 Tuesday	
04 01	♂ ⊥ ♆
05 24	☽ ∠ ♀
05 38	☽ ∠ ♆
08 27	♀ □ ♆
09 03	☽ ☌ ☉
11 39	☽ Q ♃
11 40	☽ Q ♅
12 53	♃ ☌ ♅
13 33	♀ △ ♅
13 39	♀ △ ♃
15 00	☽ □ ♄
17 49	☽ ∥ ☿
22 39	☽ ⊥ ♆
23 49	☽ ☌ ♂
05 Wednesday	
00 26	☽ ⊻ ☿
08 58	☽ ∥ ♇
10 07	☽ ⊻ ♆
10 30	☽ ⚹ ♅
10 42	☽ ⚹ ♃
12 15	☽ ⚹ ♀
12 34	☽ ⊥ ☿
18 37	☉ Q ♅
22 40	☉ Q ♃
06 Thursday	
01 01	☽ ∥ ♀
02 46	☽ ⊻ ♇
07 05	☽ ∠ ☿
13 30	☽ Q ♀
14 28	☽ ⊥ ♇
15 34	☽ ∠ ♅
16 04	☽ ∠ ♃
20 57	☽ ∥ ♆
23 50	☽ ⊻ ☉
07 Friday	
00 59	☽ △ ♄
01 47	♀ ⊥ ♇
08 12	☽ ∠ ♇
09 18	☽ ⊥ ♅
09 59	☽ ⊥ ♃
12 45	☽ ⊥ ☉
13 59	☽ ⊻ ♂
14 00	☉ □ ♄
14 43	☽ ⚹ ☿
20 51	☽ ☌ ♆
21 13	☽ ⊻ ♅
22 02	☽ ⊻ ♃
08 Saturday	
02 44	☽ ⊥ ♂
04 16	☽ □ ♀
06 49	☽ ⚼ ♄
08 20	☽ ∠ ☉
14 09	☽ ⚹ ♇
16 40	☽ Q ☿
17 36	☽ ∥ ♄
22 01	☽ ∠ ♂
09 Sunday	
00 55	☽ ± ♄
05 20	☽ ∥ ♃
06 31	☽ ∥ ♅
11 05	♀ ∠ ♄
13 03	☽ ⊼ ♄
14 23	☽ Q ♇
17 20	☽ ⚹ ☉
10 Monday	
00 15	☽ # ♅
01 13	☽ # ♃
06 25	☽ ⚹ ♂
08 20	☽ □ ☿
09 21	☽ ⊻ ♆
09 43	☽ ☌ ♅
11 12	☽ ☌ ♃
13 22	☽ # ♄
19 54	☽ Q ☉
20 00	☉ ∥ ☿
20 03	☿ ⚹ ♆
21 32	☽ ⊥ ♆
22 27	☽ △ ♀
23 05	☿ ∥ ♂
11 Tuesday	
00 13	☿ □ ♅
02 54	☽ □ ♇
08 24	☽ Q ♂
15 32	☉ ⊥ ♆
15 45	☽ ∠ ♆
20 37	☿ □ ♃
12 Wednesday	
01 47	☽ ☍ ♄
04 47	♂ ⊻ ♆
07 33	☽ ⚼ ♀
10 24	☽ # ♆
10 30	♂ ⚹ ♅
11 31	☽ □ ☉
16 00	☿ Q ♄
21 52	☽ ⚹ ♆
22 13	☽ ⊻ ♅
23 00	☽ □ ♂
13 Thursday	
00 19	☽ ⊻ ♃
02 47	☽ △ ☿
02 54	♀ ⊻ ♇
03 10	☽ ± ♀
10 06	☽ ⊥ ♅
12 19	☽ ⊥ ♃
14 59	☽ △ ♇
15 22	♃ ∥ ♅
16 05	☽ ⊼ ♀
17 58	☽ # ♀
21 32	☽ Q ♆
23 35	☽ # ♇
14 Friday	
00 35	♂ ⚹ ♃
03 44	☽ ∠ ♅
06 06	☽ ∠ ♃
11 11	☽ ⚼ ☿
12 58	☽ ⊼ ♄
19 20	☽ # ☉
20 04	☽ ⚼ ♇
20 10	☽ # ♂
15 Saturday	
00 25	☽ ± ♄
03 16	☽ △ ☉
06 09	☽ ± ☿
08 06	☽ □ ♆
08 25	☽ ⚹ ♅
09 34	☽ # ☿
11 01	☽ ⚹ ♃
12 47	☽ △ ♂
13 05	☽ ± ♇
17 20	☽ ⚼ ♄
18 34	☽ ⊼ ☿
16 Sunday	
00 15	☽ ⊼ ♇
06 08	☽ ☍ ♀
06 42	☽ Q ♅
09 28	☽ ⚼ ☉
09 28	☽ Q ♃
18 02	☽ ⚼ ♂
20 46	☽ △ ♄
17 Monday	
03 00	☽ ± ☉
11 03	☽ ± ♂
14 27	☽ ⊼ ☉
14 41	☽ △ ♆
14 59	☽ □ ♅
17 39	☉ ⊻ ♆
17 57	☽ □ ♃
21 56	☉ ⚹ ♅
21 59	♀ ∥ ♇
22 10	☽ ⊼ ♂
18 Tuesday	
05 45	☽ ☍ ☿
05 45	☽ ☍ ♇
05 50	☿ ☌ ♇
06 45	☽ # ☿
15 29	☽ ⊼ ♀
16 37	☽ ⚼ ♆
19 Wednesday	
00 53	☽ □ ♄
02 27	☽ ± ♀
04 01	☽ # ♂
05 37	☽ # ☉
07 50	☽ ± ♆
14 19	☽ # ♀
16 10	☽ # ♇
17 50	☽ ⊼ ♆
18 08	☽ △ ♅
18 41	☽ ⚼ ♀
21 21	☽ ☍ ☉
21 26	☽ △ ♃
22 39	☉ ⚹ ♃
20 Thursday	
03 42	☽ ☍ ♂
06 49	☽ Q ♄
08 09	☽ ⊼ ♇
13 03	☽ ⊼ ☿
17 59	☽ ± ♇
18 50	☽ ⚼ ♅
21 17	☽ △ ♀
22 19	☽ ⚼ ♃
23 25	☽ # ♆
23 50	☽ ± ☿
21 Friday	
02 25	☽ ⚹ ♄
08 40	☽ ⚼ ♇
09 30	☽ ± ♅
13 07	☽ ± ♃
15 55	☽ ⚼ ☿
18 57	☽ ☍ ♆
19 15	☽ ⊼ ♅
22 57	☽ ⊼ ♃
22 Saturday	
01 53	☽ ⊼ ☉
02 45	☽ ∠ ♄
03 34	♀ Q ♆
07 14	☽ ⊼ ♂
09 00	☽ △ ♇
11 01	☽ # ♄
12 19	☽ ± ☉
17 20	☽ ⊥ ♄
17 29	☽ ± ♂
18 34	☉ ∥ ♀
18 43	☽ △ ☿
22 20	☽ # ♅
23 Sunday	
00 29	☽ # ♃
01 10	☿ ∠ ♆
01 47	☽ □ ♀
03 05	☽ ⊻ ♄
04 00	☽ ⚼ ☉
08 55	☽ ∥ ♃
08 57	☽ ⚼ ♂
11 09	☽ ∥ ♅
18 08	♀ ∥ ♂
18 16	♂ ⊻ ♇
19 19	♀ ⚹ ♄
19 49	☽ ⊼ ♆
20 08	☽ ☍ ♅
22 37	☽ ∥ ♄
24 Monday	
00 22	☽ ☍ ♃
05 43	☽ ± ♆
06 25	☽ △ ☉
09 41	☽ Q ♀
10 07	☽ □ ♇
10 59	☽ △ ♂
20 41	☽ ⚼ ♆
25 Tuesday	
01 19	☽ □ ☿
04 35	☽ ☌ ♄
07 22	☽ ⚹ ♀
08 45	☿ Q ♅
11 03	☽ ∥ ♆
16 21	☽ Q ♇
22 03	☽ △ ♆
22 23	☽ ⊼ ♅
26 Wednesday	
03 20	☽ ⊼ ♃
03 49	☉ ∥ ♇
06 10	♄ St R
08 43	☽ ± ♅
09 58	☿ □ ♄
11 07	☽ ∠ ♀
11 24	☽ Q ☿
12 55	☉ ⊻ ♇
12 57	☽ ⚹ ♇
12 57	☽ □ ☉
13 49	☽ ± ♃
16 48	☽ □ ♂
19 05	☽ ∥ ☉
20 07	☽ ∥ ♇
22 17	☽ ∥ ♂
27 Thursday	
00 23	☽ ⚼ ♅
04 09	☽ ⊥ ♀
05 44	☽ ⚼ ♃
06 05	☽ ∥ ♀
08 11	☽ ⊻ ♄
10 50	☽ ⚹ ☿
13 34	☿ Q ♃
15 17	☽ ∠ ♇
15 41	☽ ⊻ ♀
18 48	☽ ⊥ ♄
23 08	☽ Q ☉
28 Friday	
02 29	☽ Q ♂
02 39	☽ □ ♆
03 01	☽ △ ♅
06 30	☽ ∥ ☿
07 24	☽ ⊥ ♇
08 48	☽ △ ♃
10 55	☽ ∠ ♄
12 15	♂ ∥ ♇
16 59	☽ ∠ ☿
18 16	☽ ⊻ ♇
22 35	☽ ⚹ ☉
29 Saturday	
01 34	☽ ⚹ ♂
10 45	☿ ⊥ ♆
11 26	☽ Q ♆
11 31	☽ ⊥ ☿
14 17	☽ ⚹ ♄
30 Sunday	
00 03	☽ ⊻ ☿
03 18	☽ ☌ ♀
04 34	☽ ∠ ☉
07 04	☽ ∠ ♂
09 46	☽ ⚹ ♆
10 10	☽ □ ♅
12 35	☽ Q ♄
16 53	☽ □ ♃
22 55	☽ ⊥ ☉
31 Monday	
01 05	☽ ⊥ ♂
02 02	☽ ☌ ♇
08 36	☽ ∥ ☿
11 16	☽ ⊻ ☉
13 13	☽ ⊻ ♂
14 11	☽ ∠ ♆
17 08	♂ ⊥ ♇
20 21	☽ Q ♅
22 40	☽ □ ♄

FEBRUARY 2011

LONGITUDES

Date	Sidereal time h m s	Sun ☉ ° ' "	Moon ☽ ° ' "	Moon ☽ 24.00 ° ' "	Mercury ☿ ° '	Venus ♀ ° '	Mars ♂ ° '	Jupiter ♃ ° '	Saturn ♄ ° '	Uranus ♅ ° '	Neptune ♆ ° '	Pluto ♇ ° '
01	20 45 23	12 ♒ 16 12	24 ♑ 07 57	00 ♒ 19 59	26 ♑ 16	26 ♐ 54	13 ♒ 00	01 ♈ 51	17 ♎ 12	28 ♓ 01	27 ♒ 47	06 ♑ 24
02	20 49 20	13 17 07	06 ♒ 29 44	12 ♒ 37 19	27 48	28 01	13 47	02 03	17 R 11	28 03	27 49	06 26
03	20 53 16	14 18 01	18 ♒ 42 50	24 ♒ 46 25	29 ♑ 20	29 ♐ 09	14 34	02 15	17 10	28 06	27 51	06 28
04	20 57 13	15 18 54	00 ♓ 48 13	06 ♓ 48 22	00 ♒ 53	00 ♑ 17	15 22	02 27	17 09	28 09	27 54	06 29
05	21 01 09	16 19 46	12 ♓ 47 05	18 ♓ 44 33	02 27	01 25	16 09	02 39	17 08	28 12	27 56	06 31
06	21 05 06	17 20 36	24 ♓ 41 02	00 ♈ 36 49	04 02	02 34	16 56	02 51	17 07	28 14	27 58	06 33
07	21 09 03	18 21 25	06 ♈ 32 15	12 ♈ 27 43	05 37	03 42	17 44	03 04	17 06	28 17	28 00	06 35
08	21 12 59	19 22 12	18 ♈ 23 38	24 ♈ 20 29	07 14	04 51	18 31	03 16	17 04	28 20	28 03	06 36
09	21 16 56	20 22 58	00 ♉ 18 46	06 ♉ 19 03	08 51	05 59	19 18	03 29	17 03	28 23	28 05	06 38
10	21 20 52	21 23 43	12 ♉ 21 54	18 ♉ 27 56	10 28	07 08	20 06	03 41	17 01	28 26	28 07	06 40
11	21 24 49	22 24 26	24 ♉ 37 47	00 ♊ 52 02	12 07	08 17	20 53	03 54	17 00	28 29	28 09	06 41
12	21 28 45	23 25 07	07 ♊ 11 21	13 ♊ 36 16	13 46	09 26	21 40	04 06	16 58	28 31	28 12	06 43
13	21 32 42	24 25 47	20 ♊ 07 21	26 ♊ 45 02	15 27	10 35	22 28	04 19	16 56	28 34	28 14	06 45
14	21 36 38	25 26 25	03 ♋ 29 42	10 ♋ 21 35	17 08	11 45	23 15	04 32	16 54	28 37	28 16	06 46
15	21 40 35	26 27 02	17 ♋ 20 46	24 ♋ 27 10	18 50	12 54	24 02	04 45	16 52	28 40	28 19	06 48
16	21 44 32	27 27 36	01 ♌ 40 27	09 ♌ 00 08	20 32	14 04	24 50	04 58	16 50	28 43	28 21	06 49
17	21 48 28	28 28 10	16 ♌ 25 28	23 ♌ 55 31	22 16	15 13	25 37	05 11	16 48	28 46	28 23	06 51
18	21 52 25	29 ♒ 28 41	01 ♍ 29 09	09 ♍ 05 05	24 01	16 23	26 25	05 24	16 45	28 50	28 25	06 52
19	21 56 21	00 ♓ 29 11	16 ♍ 41 58	24 ♍ 18 24	25 46	17 33	27 12	05 37	16 43	28 53	28 28	06 54
20	22 00 18	01 29 40	01 ♎ 53 03	09 ♎ 24 40	27 33	18 43	27 59	05 51	16 40	28 56	28 30	06 55
21	22 04 14	02 30 07	16 ♎ 52 09	24 ♎ 14 36	29 ♒ 20	19 53	28 47	06 04	16 38	28 59	28 32	06 57
22	22 08 11	03 30 32	01 ♏ 31 19	08 ♏ 41 48	01 ♓ 08	21 03	29 ♒ 34	06 17	16 35	29 02	28 35	06 58
23	22 12 07	04 30 57	15 ♏ 45 49	22 ♏ 43 15	02 57	22 13	00 ♓ 22	06 31	16 32	29 05	28 37	06 59
24	22 16 04	05 31 20	29 ♏ 34 13	06 ♐ 18 54	04 48	23 24	01 09	06 44	16 29	29 08	28 39	07 01
25	22 20 01	06 31 42	12 ♐ 57 39	19 ♐ 30 52	06 39	24 34	01 56	06 58	16 26	29 12	28 41	07 02
26	22 23 57	07 32 02	25 ♐ 58 58	02 ♑ 22 27	08 30	25 45	02 44	07 11	16 23	29 15	28 44	07 03
27	22 27 54	08 32 21	08 ♑ 41 47	14 ♑ 57 26	10 23	26 55	03 31	07 25	16 20	29 18	28 46	07 04
28	22 31 50	09 ♓ 32 39	21 ♑ 09 52	27 ♑ 19 27	12 ♓ 16	28 ♑ 06	04 ♓ 18	07 ♈ 38	16 ♎ 17	29 ♓ 21	28 ♒ 48	07 ♑ 06

Date	Moon True ☊ ° '	Moon Mean ☊ ° '	Moon ☽ Latitude ° '
01	02 ♑ 18	00 ♑ 38	01 N 56
02	02 R 14	00 35	02 54
03	02 07	00 32	03 44
04	01 59	00 29	04 24
05	01 50	00 25	04 51
06	01 41	00 22	05 05
07	01 33	00 19	05 06
08	01 27	00 16	04 54
09	01 22	00 13	04 29
10	01 20	00 09	03 52
11	01 D 19	00 06	03 04
12	01 20	00 03	02 06
13	01 21	00 ♑ 00	01 N 00
14	01 R 21	29 ♐ 57	00 S 11
15	01 20	29 54	01 25
16	01 17	29 50	02 35
17	01 11	29 47	03 37
18	01 03	29 44	04 25
19	00 54	29 41	04 55
20	00 44	29 38	05 04
21	00 36	29 35	04 52
22	00 30	29 31	04 21
23	00 26	29 28	03 34
24	00 24	29 25	02 36
25	00 D 24	29 22	01 32
26	00 25	29 19	00 S 24
27	00 R 25	29 15	00 N 44
28	00 ♑ 23	29 ♐ 12	01 N 48

DECLINATIONS

Date	Sun ☉ ° '	Moon ☽ ° '	Mercury ☿ ° '	Venus ♀ ° '	Mars ♂ ° '	Jupiter ♃ ° '	Saturn ♄ ° '	Uranus ♅ ° '	Neptune ♆ ° '	Pluto ♇ ° '
01	17 S 07	19 S 23	22 S 11	20 S 46	17 S 57	00 S 20	04 S 23	01 S 27	12 S 42	18 S 48
02	16 50	15 50	21 59	20 51	17 43	00 15	04 23	01 26	12 41	18 48
03	16 32	11 39	21 45	20 55	17 30	00 10	04 22	01 25	12 40	18 48
04	16 15	07 04	21 29	20 58	17 16	00 S 05	04 22	01 24	12 40	18 48
05	15 56	02 S 17	21 13	21 01	17 02	00 00	04 21	01 23	12 39	18 48
06	15 38	02 N 33	20 55	21 04	16 47	00 N 05	04 20	01 21	12 38	18 48
07	15 20	07 17	20 35	21 06	16 33	00 10	04 20	01 20	12 37	18 48
08	15 01	11 44	20 14	21 07	16 18	00 15	04 19	01 19	12 36	18 48
09	14 42	15 47	19 52	21 08	16 04	00 20	04 18	01 18	12 36	18 48
10	14 22	19 14	19 29	21 08	15 49	00 25	04 17	01 17	12 35	18 48
11	14 03	21 54	19 04	21 08	15 33	00 30	04 16	01 16	12 34	18 48
12	13 43	23 35	18 37	21 07	15 18	00 35	04 15	01 15	12 33	18 48
13	13 23	24 04	18 10	21 06	15 03	00 40	04 14	01 13	12 33	18 47
14	13 02	23 12	17 40	21 04	14 47	00 46	04 14	01 12	12 32	18 47
15	12 42	20 55	17 10	21 02	14 32	00 51	04 12	01 11	12 31	18 47
16	12 21	17 16	16 38	20 59	14 16	00 56	04 11	01 10	12 30	18 47
17	12 00	12 28	16 05	20 55	14 00	01 01	04 10	01 08	12 29	18 47
18	11 39	06 49	15 30	20 51	13 44	01 07	04 09	01 07	12 29	18 47
19	11 18	00 N 43	14 54	20 46	13 27	01 12	04 08	01 06	12 28	18 47
20	10 57	05 S 24	14 16	20 41	13 11	01 17	04 07	01 05	12 27	18 47
21	10 35	11 07	13 37	20 35	12 54	01 23	04 06	01 03	12 26	18 47
22	10 13	16 05	12 57	20 28	12 38	01 28	04 04	01 02	12 26	18 47
23	09 51	19 58	12 15	20 21	12 21	01 33	04 03	01 01	12 25	18 47
24	09 29	22 36	11 32	20 14	12 04	01 39	04 02	01 00	12 24	18 47
25	09 07	23 52	10 48	20 05	11 47	01 44	04 01	00 58	12 23	18 46
26	08 45	23 46	10 03	19 57	11 30	01 50	03 59	00 57	12 22	18 46
27	08 22	22 25	09 16	19 47	11 12	01 55	03 58	00 56	12 22	18 46
28	07 S 59	20 S 00	08 S 28	19 S 37	10 S 55	02 N 01	03 S 56	00 S 54	12 S 21	18 S 46

ZODIAC SIGN ENTRIES

Date	h	m	Planets
01	23	21	☽ ♒
03	22	19	☿ ♒
04	05	58	♀ ♑
04	10	24	☽ ♓
06	22	45	☽ ♈
09	11	22	☽ ♉
11	22	20	☽ ♊
14	05	48	☽ ♋
16	09	14	☽ ♌
18	09	39	☽ ♍
19	00	25	☉ ♓
20	09	01	☽ ♎
21	20	53	☿ ♓
22	09	29	☽ ♏
23	01	06	♂ ♓
24	12	46	☽ ♐
26	19	32	☽ ♑

LATITUDES

Date	Mercury ☿ ° '	Venus ♀ ° '	Mars ♂ ° '	Jupiter ♃ ° '	Saturn ♄ ° '	Uranus ♅ ° '	Neptune ♆ ° '	Pluto ♇ ° '
01	01 S 19	02 N 38	01 S 05	01 S 10	02 N 33	00 S 43	00 S 29	04 N 29
04	01 34	02 28	01 05	01 10	02 34	00 43	00 29	04 29
07	01 47	02 18	01 05	01 09	02 35	00 43	00 29	04 29
10	01 56	02 07	01 05	01 09	02 36	00 43	00 29	04 29
13	02 03	01 55	01 05	01 08	02 37	00 43	00 29	04 29
16	02 06	01 44	01 05	01 08	02 37	00 43	00 29	04 29
19	02 05	01 32	01 05	01 08	02 38	00 43	00 29	04 29
22	02 01	01 20	01 04	01 07	02 39	00 43	00 29	04 29
25	01 52	01 08	01 04	01 07	02 39	00 43	00 29	04 29
28	01 38	00 56	01 04	01 07	02 40	00 43	00 29	04 29
31	01 S 19	00 N 44	01 S 03	01 S 06	02 N 41	00 S 42	00 S 29	04 N 29

DATA

Julian Date	2455594
Delta T	+68 seconds
Ayanamsa	24° 01' 02"
Synetic vernal point	05° ♓ 05' 57"
True obliquity of ecliptic	23° 26' 16"

MOON'S PHASES, APSIDES AND POSITIONS ☽

Date	h	m	Phase	Longitude ° '	Eclipse Indicator
03	02	31	●	13 ♒ 54	
11	07	18	☽	22 ♉ 13	
18	08	36	○	29 ♌ 20	
24	23	26	☾	06 ♐ 00	

Day	h	m		
06	23	18	Apogee	
19	07	31	Perigee	
05	23	17	ON	
13	08	57	Max dec	24° N 05'
19	14	48	OS	
25	22	12	Max dec	23° S 59'

LONGITUDES

Date	Chiron ⚷ ° '	Ceres ⚳ ° '	Pallas ⚴ ° '	Juno ⚵ ° '	Vesta ⚶ ° '	Black Moon Lilith ⚸ ° '
01	29 ♒ 31	11 ♒ 22	16 ♑ 55	29 ♍ 07	01 ♑ 49	24 ♓ 24
11	00 ♓ 11	15 ♒ 18	20 ♑ 40	27 ♍ 52	06 ♑ 52	25 ♓ 32
21	00 ♓ 52	19 ♒ 13	24 ♑ 16	26 ♍ 00	11 ♑ 49	26 ♓ 39
31	01 ♓ 33	23 ♒ 06	27 ♑ 44	23 ♍ 42	16 ♑ 37	27 ♓ 46

All ephemeris data is given at 12.00 UT and the Moon's longitude is additionally given for 24.00 UT

ASPECTARIAN

01 Tuesday
01 10 ☽ ∥ ♀
03 38 ☽ Q ♃
07 28 ☽ ⊥ ♆
12 35 ♂ ∠ ♅
15 10 ☉ ⊥ ♇
16 18 ☽ ∥ ♇
16 41 ☽ ☌ ☿
17 52 ☽ ⊻ ♀
19 05 ☽ ⊻ ♆
19 32 ☽ ✱ ♅
23 03 ☽ ∥ ♂

02 Wednesday
03 12 ☽ ✱ ♃
05 12 ☽ ∥ ☉
06 21 ☉ ∠ ♅
06 41 ☽ ⊥ ♀
07 35 ♀ ✱ ♆
11 52 ☽ ⊻ ♇
12 27 ☿ ⊻ ♆
12 47 ♀ □ ♅
16 15 ☿ ✱ ♅
23 39 ☽ ⊥ ♇

03 Thursday
00 54 ☽ ∠ ♅
01 27 ☿ ⊻ ♀
02 05 ☽ ∠ ♀
02 31 ☽ ☌ ☉
03 16 ☽ ☌ ♂
06 24 ☽ ∥ ♆
08 57 ☽ △ ♄
09 04 ☽ ∠ ♃
12 17 ♀ Q ♄
17 27 ☽ ∠ ♇
18 44 ☽ ⊥ ♅

04 Friday
03 11 ☽ ⊥ ♃
06 11 ☽ ☌ ♆
06 41 ☽ ⊻ ♅
10 51 ☽ ✱ ♀
12 12 ☽ ⊻ ☿
14 41 ☽ ⚼ ♄
15 21 ☽ ⊻ ♃
16 40 ☉ ☌ ♂
17 07 ♀ ∠ ♂
17 59 ☉ ∠ ♀
23 24 ☽ ✱ ♇

05 Saturday
01 41 ☽ ∥ ♄
01 59 ☽ ⊥ ☿
08 41 ☽ ± ♄
13 25 ☽ Q ♀
15 30 ☿ ✱ ♃
16 29 ☽ ∥ ♅
19 15 ☽ ⊻ ♂
19 48 ☽ ⊻ ☉
20 44 ☽ ⊼ ♄
22 50 ☽ ∠ ☿
23 08 ☽ # ♃
23 27 ☽ ∥ ♃
23 35 ☽ Q ♇

06 Sunday
01 23 ☿ ∥ ♀
06 03 ☽ # ♅
06 40 ☉ △ ♄
08 13 ☽ ⊥ ♂
09 02 ☽ ⊥ ☉
17 16 ♂ △ ♄
18 40 ☽ ⊻ ♆
19 13 ☽ ☌ ♅
19 39 ♀ □ ♃
20 55 ☽ # ♄

07 Monday
03 15 ☉ ∠ ♃
03 44 ☽ ∠ ♂
04 50 ☽ ☌ ♃
04 57 ☽ ∠ ☉
05 38 ☽ □ ♀
06 51 ☽ ⊥ ♆
09 52 ☽ ✱ ☿
12 05 ☽ □ ♇

08 Tuesday
01 09 ☽ ∠ ♆
01 53 ♂ ∠ ♃
02 33 ☿ ⊻ ♇
09 20 ☽ ☍ ♄
12 16 ☽ ✱ ♂
13 57 ☽ Q ☿
14 09 ☽ ✱ ☉
16 54 ☽ # ♆

09 Wednesday
05 44 ☽ # ☉
07 31 ☽ ✱ ♆
08 06 ☽ ⊻ ♅
13 40 ☽ # ♂
14 07 ☽ Q ♂
16 32 ☽ Q ☉
18 27 ☽ ⊻ ♃
20 10 ☽ ⊥ ♅

10 Thursday
00 32 ☽ △ ♀
00 40 ☽ △ ♇
01 52 ♀ ☌ ♇
06 36 ☽ ⊥ ♃
07 33 ☽ Q ♆
07 40 ☽ □ ☿
08 40 ☽ # ♇
13 41 ☽ # ☿
14 06 ☽ ∠ ♅
18 30 ☉ ∠ ♇
21 09 ☽ ⊼ ♄

11 Friday
00 39 ☽ ∠ ♃
04 13 ☽ □ ♂
04 15 ☽ # ♀
06 17 ☽ ⚼ ♇
07 18 ☽ □ ☉
08 50 ☽ ± ♄
09 07 ☽ ⚼ ♀
13 42 ☉ ⊥ ♅
18 50 ☽ □ ♆
19 27 ☽ ✱ ♅
20 27 ☿ ⊥ ♇
23 41 ☽ ± ♇

12 Saturday
02 07 ☽ ⚼ ♄
02 46 ☿ ∥ ♇
04 11 ☽ ± ♀
06 04 ☽ ✱ ♃
08 17 ☿ ∠ ♅
11 06 ☽ ⊼ ♇
13 25 ♂ ∠ ♇
16 38 ☽ ⊼ ♀
18 17 ☽ Q ♅

13 Sunday
00 03 ♃ ⊥ ♆
02 08 ☽ △ ☿
04 55 ☽ Q ♃
06 10 ☽ △ ♄
15 39 ♂ ⊥ ♅
16 32 ☽ △ ♂
20 28 ☽ △ ☉

14 Monday
02 42 ☽ △ ♆
03 19 ☽ □ ♅
08 48 ☿ △ ♄
09 15 ☽ ⚼ ☿
13 52 ☽ □ ♃
17 46 ☽ ☍ ♇
20 51 ☽ ⚼ ♂

15 Tuesday
01 06 ☽ ⚼ ☉
03 11 ☽ ± ☿
03 42 ☽ ☍ ♀
05 04 ☽ ⚼ ♆
11 04 ☽ # ♀
11 11 ☽ □ ♄
13 15 ☽ ± ♂
14 52 ☽ ⊼ ☿
15 14 ☿ ⊥ ♀
17 40 ☽ ± ☉
20 25 ☽ ± ♆
20 44 ♀ ∠ ♆
23 58 ☽ ⊼ ♂

16 Wednesday
01 13 ☉ ∥ ♆
02 48 ☿ ∠ ♃
02 59 ☽ # ♇
04 30 ☽ ⊼ ☉
06 29 ☽ ⊼ ♆
07 06 ☽ △ ♅
15 59 ☽ # ☿
17 10 ☽ Q ♄
17 30 ☽ △ ♃
20 28 ☽ ⊼ ♇

17 Thursday
04 28 ☽ # ♂
06 03 ☿ ∠ ♇
06 13 ☽ ± ♇
07 43 ☽ ⚼ ♅
09 54 ☽ ⊼ ♀
09 56 ☉ ☌ ♆
11 53 ☽ # ♆
12 35 ☽ ✱ ♄
14 13 ☽ # ☉
18 07 ☽ ⚼ ♃
19 11 ☿ ⊥ ♅
19 39 ☉ ⊻ ♅
20 20 ☽ ± ♀
20 42 ☽ ⚼ ♇
22 12 ☽ ± ♅
22 35 ☽ ☍ ☿

18 Friday
03 30 ☽ ☍ ♂
07 08 ☽ ☍ ♆
07 46 ☽ ⊼ ♅
08 36 ☽ ☍ ☉
08 39 ☽ ± ♃
09 41 ☉ ⊥ ♃
11 50 ☽ ⚼ ♀
12 25 ☽ ∠ ♄
14 09 ♃ # ♅
18 17 ☽ ⊼ ♃
19 21 ♀ □ ♄
20 31 ☽ △ ♇
21 31 ♀ Q ♅
22 38 ☽ # ♄

19 Saturday
02 36 ☽ ⊥ ♄
10 10 ☽ ∥ ♃
10 32 ☽ # ♅
12 01 ☽ ⊻ ♄
13 27 ☽ △ ♀
19 03 ☽ ∥ ♅
19 35 ☽ # ♃

20 Sunday
04 12 ☽ ⊼ ☿
05 29 ☽ ⊼ ♂
06 37 ☽ ⊼ ♆
06 55 ☽ ∥ ♄
07 18 ☽ ☍ ♅
11 20 ☽ ⊼ ☉
15 00 ☽ ± ☿
15 32 ☽ ± ♂
16 02 ☉ ⚼ ♄
16 10 ☽ ± ♆
18 23 ☽ ☍ ♃
20 02 ☽ □ ♇
21 35 ☽ ± ☉
22 44 ☿ ☌ ♂

21 Monday
01 07 ☿ ☌ ♆
04 17 ♂ ☌ ♆
06 37 ☽ ⚼ ♆
06 44 ☽ ⚼ ♂
07 10 ☿ ⊻ ♅
07 21 ☽ ⚼ ☿
09 47 ☽ ∥ ☉
11 37 ☽ ☌ ♄
13 06 ☽ ⚼ ☉
17 18 ☽ □ ♀
17 58 ☽ ∥ ♆
18 36 ♂ ⊻ ♅
19 44 ☽ ∥ ♂
22 13 ☽ ∥ ☿
23 06 ☿ ⊥ ♃

22 Tuesday
01 10 ☽ Q ♇
07 07 ☽ △ ♆
07 52 ☽ ⊼ ♅
08 35 ☽ △ ♂
11 16 ☽ △ ☿
15 33 ☽ △ ☉
17 45 ☿ ⚼ ♄
17 52 ☽ ± ♅
20 04 ☽ ⊼ ♃
21 06 ☽ ✱ ♇

23 Wednesday
01 44 ☽ Q ♀
03 48 ☽ ∥ ♇
06 06 ♂ ∥ ♆
06 22 ☽ ± ♃
06 37 ☿ ∥ ♆
06 56 ☿ ∥ ♂
09 08 ☽ ⚼ ♅
13 19 ☽ ⊻ ♄
14 44 ☽ ∥ ♀
18 22 ♂ ⊥ ♃
20 14 ♀ ⊥ ♆
22 04 ☽ ⚼ ♃
22 45 ☽ ∠ ♇
23 38 ☽ ⊥ ♄

24 Thursday
00 10 ☽ ✱ ♀
10 22 ☽ □ ♆
11 14 ☽ △ ♅
14 33 ☽ ⊥ ♇
14 58 ☽ □ ♂
15 23 ☽ ∠ ♄
21 44 ♂ ⚼ ♄
22 45 ☽ □ ☿
23 26 ☽ □ ☉

25 Friday
00 58 ☽ △ ♃
01 16 ☽ ⊻ ♇
05 16 ☽ ∠ ♀
08 48 ☉ ☌ ☿
16 39 ☿ ⊻ ♃
17 06 ☿ ✱ ♇
18 20 ☽ ✱ ♄
18 50 ☽ Q ♆
20 40 ♃ □ ♇
21 49 ♀ Q ♃
23 17 ☽ ⊥ ♀

26 Saturday
00 17 ☉ ✱ ♇
01 14 ☉ ⊻ ♃
01 37 ☽ Q ♂
11 05 ☽ Q ☉
11 31 ☽ ⊻ ♀
13 09 ☽ Q ☿
16 29 ☽ Q ♄
17 09 ☽ ✱ ♆
18 08 ☽ □ ♅

27 Sunday
01 31 ☽ ✱ ♂
08 54 ☽ ☌ ♇
09 30 ☽ □ ♃
11 26 ☿ ± ♄
11 40 ☽ ✱ ☉
15 48 ☽ ✱ ☿
21 44 ☽ ∠ ♆

28 Monday
02 36 ☽ □ ♄
04 36 ☽ Q ♅
08 09 ☽ ∠ ♂
15 11 ☽ ⊥ ♆
15 13 ☽ ∥ ♀
19 09 ☽ ∠ ☉
20 52 ☽ Q ♃
21 36 ☽ ∥ ♇

MARCH 2011

LONGITUDES

Date	Sidereal time h m s	Sun ☉ ° ' "	Moon ☽ ° ' "	Moon ☽ 24.00 ° ' "	Mercury ☿ ° '	Venus ♀ ° '	Mars ♂ ° '	Jupiter ♃ ° '	Saturn ♄ ° '	Uranus ♅ ° '	Neptune ♆ ° '	Pluto ♇ ° '
01	22 35 47	10 ♓ 32 55	03 ♒ 26 36	09 ♒ 31 39	14 ♓ 11	29 ♑ 17	05 ♓ 06	07 ♈ 52	16 ♎ 14	29 ♓ 25	28 ♒ 50	07 ♑ 07
02	22 39 43	11 33 09	15 ♒ 34 53	21 ♒ 36 34	16 05	00 ♒ 28	05 53	08 06	16 R 10	29 28	28 53	07 08
03	22 43 40	12 33 22	27 ♒ 36 55	03 ♓ 36 08	18 00	01 39	06 41	08 20	16 07	29 31	28 55	07 09
04	22 47 36	13 33 33	09 ♓ 34 23	15 ♓ 31 50	19 56	02 49	07 28	08 33	16 03	29 35	28 57	07 10
05	22 51 33	14 33 42	21 ♓ 28 37	27 ♓ 24 54	21 52	04 00	08 15	08 47	16 00	29 38	28 59	07 11
06	22 55 30	15 33 49	03 ♈ 20 51	09 ♈ 16 38	23 48	05 12	09 03	09 01	15 56	29 41	29 01	07 13
07	22 59 26	16 33 54	15 ♈ 12 28	21 ♈ 08 36	25 43	06 23	09 50	09 15	15 53	29 45	29 04	07 14
08	23 03 23	17 33 58	27 ♈ 05 20	03 ♉ 02 58	27 38	07 34	10 37	09 29	15 49	29 48	29 06	07 15
09	23 07 19	18 33 59	09 ♉ 01 54	15 ♉ 02 33	29 ♓ 33	08 45	11 25	09 43	15 45	29 51	29 08	07 16
10	23 11 16	19 33 58	21 ♉ 05 22	27 ♉ 10 53	01 ♈ 26	09 56	12 12	09 57	15 41	29 55	29 10	07 16
11	23 15 12	20 33 55	03 ♊ 19 39	09 ♊ 32 13	03 18	11 08	12 59	10 12	15 37	29 ♓ 58	29 12	07 17
12	23 19 09	21 33 50	15 ♊ 49 12	22 ♊ 11 12	05 07	12 19	13 46	10 26	15 33	00 ♈ 02	29 15	07 18
13	23 23 05	22 33 43	28 ♊ 38 48	05 ♋ 12 33	06 55	13 31	14 34	10 40	15 29	00 05	29 17	07 19
14	23 27 02	23 33 33	11 ♋ 52 58	18 ♋ 40 26	08 40	14 42	15 21	10 54	15 25	00 08	29 19	07 20
15	23 30 59	24 33 22	25 ♋ 35 15	02 ♌ 37 33	10 21	15 54	16 08	11 08	15 21	00 12	29 21	07 21
16	23 34 55	25 33 08	09 ♌ 47 17	17 ♌ 04 12	11 59	17 05	16 55	11 23	15 17	00 15	29 23	07 21
17	23 38 52	26 32 51	24 ♌ 27 46	01 ♍ 57 16	13 32	18 17	17 42	11 37	15 12	00 19	29 25	07 22
18	23 42 48	27 32 33	09 ♍ 31 42	17 ♍ 09 53	15 01	19 28	18 29	11 51	15 08	00 22	29 27	07 23
19	23 46 45	28 32 12	24 ♍ 50 25	02 ♎ 31 49	16 24	20 40	19 17	12 05	15 04	00 26	29 29	07 24
20	23 50 41	29 ♓ 31 50	10 ♎ 12 35	17 ♎ 51 12	17 42	21 52	20 04	12 20	14 59	00 29	29 31	07 24
21	23 54 38	00 ♈ 31 25	25 ♎ 26 18	02 ♏ 56 39	18 54	23 04	20 51	12 34	14 55	00 32	29 33	07 25
22	23 58 34	01 30 59	10 ♏ 21 14	17 ♏ 39 18	19 59	24 16	21 38	12 49	14 50	00 36	29 35	07 25
23	00 02 31	02 30 31	24 ♏ 50 18	01 ♐ 53 58	20 57	25 27	22 25	13 03	14 46	00 39	29 37	07 26
24	00 06 28	03 30 01	08 ♐ 50 12	15 ♐ 39 08	21 49	26 39	23 12	13 17	14 41	00 43	29 39	07 26
25	00 10 24	04 29 29	22 ♐ 21 02	28 ♐ 56 16	22 33	27 51	23 59	13 32	14 37	00 46	29 41	07 27
26	00 14 21	05 28 56	05 ♑ 25 19	11 ♑ 48 44	23 10	29 ♒ 03	24 46	13 46	14 32	00 49	29 43	07 27
27	00 18 17	06 28 21	18 ♑ 07 04	24 ♑ 20 56	23 39	00 ♓ 15	25 33	14 01	14 28	00 53	29 45	07 28
28	00 22 14	07 27 44	00 ♒ 30 53	06 ♒ 37 31	24 01	01 27	26 20	14 15	14 23	00 56	29 47	07 28
29	00 26 10	08 27 05	12 ♒ 41 20	18 ♒ 42 50	24 14	02 40	27 07	14 30	14 18	01 00	29 49	07 29
30	00 30 07	09 26 25	24 ♒ 42 31	00 ♓ 40 45	24 21	03 52	27 53	14 44	14 14	01 03	29 51	07 29
31	00 34 03	10 ♈ 25 42	06 ♓ 37 55	12 ♓ 34 22	24 ♈ 20	05 ♓ 04	28 ♓ 40	14 ♈ 59	14 ♎ 09	01 ♈ 07	29 ♒ 53	07 ♑ 29

DECLINATIONS

Date	Moon True ☊ ° '	Moon Mean ☊ ° '	Moon ☽ Latitude ° '	Sun ☉ ° '	Moon ☽ ° '	Mercury ☿ ° '	Venus ♀ ° '	Mars ♂ ° '	Jupiter ♃ ° '	Saturn ♄ ° '	Uranus ♅ ° '	Neptune ♆ ° '	Pluto ♇ ° '
01	00 ♑ 18	29 ♐ 09	02 N 46	07 S 37	16 S 42	07 S 39	19 S 27	10 S 38	02 N 06	03 S 55	00 S 53	12 S 20	18 S 46
02	00 R 10	29 06	03 35	07 14	12 44	06 48	19 16	10 20	02 12	03 54	00 52	12 19	18 46
03	00 00	29 03	04 14	06 51	08 19	05 57	19 04	10 02	02 17	03 52	00 50	12 19	18 46
04	29 ♐ 47	29 00	04 42	06 28	03 S 37	05 05	18 52	09 45	02 23	03 51	00 49	12 18	18 46
05	29 33	28 56	04 57	06 05	01 N 10	04 12	18 39	09 27	02 28	03 49	00 48	12 17	18 46
06	29 19	28 53	04 59	05 41	05 54	03 18	18 26	09 09	02 34	03 47	00 46	12 16	18 46
07	29 06	28 50	04 49	05 18	10 25	02 23	18 13	08 51	02 39	03 46	00 45	12 16	18 46
08	28 54	28 47	04 25	04 55	14 33	01 28	17 58	08 33	02 45	03 44	00 44	12 15	18 45
09	28 46	28 44	03 50	04 31	18 08	00 S 33	17 44	08 15	02 50	03 43	00 42	12 14	18 45
10	28 40	28 41	03 04	04 08	20 59	00 N 22	17 28	07 56	02 56	03 41	00 41	12 13	18 45
11	28 37	28 37	02 09	03 44	22 56	01 17	17 13	07 38	03 02	03 39	00 40	12 13	18 45
12	28 36	28 34	01 N 07	03 21	23 48	02 11	16 56	07 20	03 07	03 38	00 38	12 12	18 45
13	28 D 36	28 31	00 00	02 57	23 26	03 05	16 40	07 01	03 13	03 36	00 37	12 11	18 45
14	28 R 36	28 28	01 S 10	02 33	21 45	03 58	16 23	06 43	03 18	03 34	00 36	12 10	18 45
15	28 34	28 25	02 17	02 10	18 46	04 49	16 05	06 24	03 24	03 33	00 34	12 10	18 45
16	28 29	28 21	03 19	01 46	14 36	05 39	15 47	06 06	03 30	03 31	00 33	12 09	18 45
17	28 22	28 18	04 10	01 22	09 26	06 28	15 28	05 47	03 35	03 29	00 31	12 08	18 45
18	28 13	28 15	04 44	00 59	03 N 36	07 14	15 09	05 29	03 41	03 27	00 30	12 07	18 45
19	28 02	28 12	05 00	00 35	02 S 32	07 58	14 50	05 10	03 47	03 25	00 29	12 07	18 45
20	27 51	28 09	04 53	00 S 11	08 32	08 39	14 30	04 51	03 52	03 24	00 27	12 06	18 44
21	27 40	28 06	04 26	00 N 12	13 58	09 18	14 10	04 32	03 58	03 22	00 26	12 05	18 44
22	27 32	28 02	03 41	00 36	18 25	09 54	13 49	04 14	04 04	03 20	00 25	12 05	18 44
23	27 27	27 59	02 43	01 00	21 36	10 26	13 28	03 55	04 09	03 18	00 23	12 04	18 44
24	27 24	27 56	01 36	01 23	23 22	10 55	13 07	03 36	04 15	03 16	00 22	12 03	18 44
25	27 23	27 53	00 S 27	01 47	23 40	11 20	12 45	03 17	04 20	03 15	00 21	12 03	18 44
26	27 D 23	27 50	00 N 42	02 11	22 37	11 42	12 23	02 58	04 26	03 13	00 19	12 02	18 44
27	27 R 23	27 47	01 47	02 34	20 26	12 00	12 00	02 39	04 32	03 11	00 18	12 01	18 44
28	27 21	27 43	02 46	02 58	17 21	12 14	11 37	02 20	04 37	03 09	00 16	12 01	18 44
29	27 16	27 40	03 35	03 21	13 34	12 23	11 14	02 01	04 43	03 07	00 15	12 00	18 44
30	27 09	27 37	04 14	03 44	09 17	12 29	10 51	01 42	04 49	03 05	00 14	11 59	18 44
31	26 ♐ 59	27 ♐ 34	04 N 42	04 N 08	04 S 42	12 N 31	10 S 27	01 S 23	04 N 54	03 S 04	00 S 12	11 S 59	18 S 44

ZODIAC SIGN ENTRIES

Date	h m	Planets
01	05 14	☽ ♒
02	02 39	♀ ♒
03	16 47	☽ ♓
06	05 14	☽ ♈
08	17 52	☽ ♉
09	17 47	☿ ♈
11	05 31	☽ ♊
12	00 49	♅ ♈
13	14 29	☽ ♋
15	19 33	☽ ♌
17	20 53	☽ ♍
19	20 03	☽ ♎
20	23 21	☉ ♈
21	19 17	☽ ♏
23	20 45	☽ ♐
26	01 57	☽ ♑
27	06 53	♀ ♓
28	11 00	☽ ♒
30	22 38	☽ ♓

LATITUDES

Date	Mercury ☿ ° '	Venus ♀ ° '	Mars ♂ ° '	Jupiter ♃ ° '	Saturn ♄ ° '	Uranus ♅ ° '	Neptune ♆ ° '	Pluto ♇ ° '
01	01 S 32	00 N 52	01 S 04	01 S 07	02 N 40	00 S 43	00 S 29	04 N 29
04	01 11	00 41	01 03	01 06	02 41	00 42	00 29	04 29
07	00 45	00 29	01 03	01 06	02 41	00 42	00 29	04 29
10	00 S 14	00 18	01 02	01 06	02 42	00 42	00 29	04 29
13	00 N 22	00 N 07	01 01	01 06	02 42	00 42	00 29	04 30
16	01 00	00 S 04	01 01	01 05	02 43	00 42	00 30	04 30
19	01 39	00 15	01 00	01 05	02 43	00 42	00 30	04 30
22	02 15	00 24	00 59	01 05	02 43	00 42	00 30	04 30
25	02 46	00 34	00 58	01 05	02 44	00 42	00 30	04 30
28	03 08	00 43	00 57	01 05	02 44	00 42	00 30	04 30
31	03 N 19	00 S 51	00 S 56	01 S 05	02 N 44	00 S 42	00 S 30	04 N 30

DATA

Julian Date	2455622
Delta T	+68 seconds
Ayanamsa	24° 01' 05"
Synetic vernal point	05° ♓ 05' 54"
True obliquity of ecliptic	23° 26' 17"

MOON'S PHASES, APSIDES AND POSITIONS ☽

Date	h m	Phase	Longitude ° '	Eclipse Indicator
04	20 46	●	13 ♓ 56	
12	23 45	☽	22 ♊ 03	
19	18 10	○	28 ♍ 48	
26	12 07	☾	05 ♑ 29	

Day	h m		
06	08 03	Apogee	
19	19 15	Perigee	
05	06 08	0N	
12	17 06	Max dec	23° N 49'
19	02 07	0S	
25	05 04	Max dec	23° S 43'

LONGITUDES

Date	Chiron ⚷ ° '	Ceres ⚳ ° '	Pallas ⚴ ° '	Juno ⚵ ° '	Vesta ⚶ ° '	Black Moon Lilith ⚸ ° '
01	01 ♓ 24	22 ♒ 20	27 ♑ 03	24 ♍ 11	15 ♑ 40	27 ♓ 33
11	02 ♓ 04	26 ♒ 10	00 ♒ 22	21 ♍ 42	20 ♑ 22	28 ♓ 40
21	02 ♓ 43	29 ♒ 56	03 ♒ 28	19 ♍ 16	24 ♑ 52	29 ♓ 47
31	03 ♓ 19	03 ♓ 36	06 ♒ 20	17 ♍ 08	29 ♑ 10	00 ♈ 54

All ephemeris data is given at 12.00 UT and the Moon's longitude is additionally given for 24.00 UT

ASPECTARIAN

h m	Aspects
01 Tuesday	
00 47	♀ ⊥ ♂
02 05	☽ ∠ ☿
02 45	♀ ⊻ ♆
02 53	☽ ⊥ ♂
02 56	☽ ⊻ ♆
02 57	☽ ☌ ♀
04 03	☽ ⚹ ♅
04 46	☉ ± ♄
13 37	☉ ∥ ☿
14 22	☽ ⊥ ☉
14 49	♀ ⚹ ♅
15 27	☿ ∠ ♀
15 29	☽ ⊻ ♂
19 15	☽ ⊻ ♇
20 53	☽ ⚹ ♃
23 04	☽ ⊥ ☿
02 Wednesday	
03 17	☽ ⊻ ☉
07 08	☽ ⊥ ♇
09 47	☽ ∠ ♅
13 03	☿ ⊼ ♄
13 10	☽ △ ♄
13 12	☽ ⊻ ☿
14 21	☽ ∥ ♆
03 Thursday	
01 04	☽ ∠ ♇
02 13	☽ ∥ ♂
03 16	☽ ∠ ♃
03 47	☽ ⊥ ♅
14 36	☽ ☌ ♆
15 50	☽ ⊻ ♅
18 59	☽ bQ ♄
20 18	☽ ∥ ☉
20 57	☽ ⊻ ♀
21 38	☽ ⊥ ♃
04 Friday	
02 26	☿ Q ♇
02 52	♂ ⚹ ♇
02 59	☽ ∥ ☿
07 10	☽ ⚹ ♇
07 28	☽ ☌ ♂
09 55	☽ ⊻ ♃
10 20	☽ ⊥ ♀
10 53	☽ ∥ ♄
12 58	☽ ± ♄
18 07	☽ ǂ ♃
20 46	☽ ☌ ☉
05 Saturday	
00 13	♀ ∥ ♇
01 00	☽ ⊼ ♄
02 07	☽ ∥ ♅
06 28	☽ ∠ ♀
07 23	☽ Q ♇
10 07	☽ ǂ ♅
12 56	☽ ☌ ☿
18 38	☽ ∥ ♃
22 22	☿ ∥ ♄
06 Sunday	
00 48	☽ ǂ ☿
01 14	☽ ǂ ♄
03 14	☽ ⊻ ♆
04 34	☽ ☌ ♅
10 58	☽ ǂ ☉
11 03	♂ ⊻ ♃
15 24	☽ ⊥ ♆
16 09	☽ ⚹ ♀
19 49	☽ □ ♇
20 26	☉ ⊼ ♄
23 43	☽ ☌ ♃
07 Monday	
00 21	☽ ⊻ ♂
03 59	☽ ǂ ♂
05 37	☿ ǂ ♃
09 40	☽ ∠ ♆
13 13	♂ ± ♄
13 21	☽ ☍ ♄
13 21	☽ ⊥ ♂
15 00	☽ ⊻ ☉
19 07	☽ Q ♀
22 19	☽ ǂ ♆
08 Tuesday	
04 14	☽ ⊥ ☉
05 23	♀ ⊻ ♇
08 50	☽ ∠ ♂
13 19	☽ ⊻ ☿
16 04	☽ ⚹ ♆
17 29	☽ ⊻ ♅
09 Wednesday	
00 02	☽ ∠ ☉
03 41	☽ ⊥ ☿
05 36	☽ ⊥ ♅
06 44	☿ ⊻ ♆
08 02	☿ ∥ ♅
08 27	☽ △ ♇
09 14	☽ ǂ ♀
11 23	☽ □ ♀
13 25	☽ ⊻ ♃
16 05	☿ ☌ ♅
16 13	☽ Q ♆
16 41	☽ ǂ ♇
17 05	☽ ⚹ ♂
23 41	☽ ∠ ♅
10 Thursday	
01 03	☽ ∠ ☿
01 20	☽ ⊼ ♄
01 37	☽ ⊥ ♃
04 53	☉ Q ♇
08 43	☽ ⚹ ☉
12 25	♀ ⚹ ♃
13 10	☽ ± ♄
14 21	☽ bQ ♇
18 33	☽ Q ♂
19 47	☽ ∠ ♃
20 14	☿ ǂ ♅
11 Friday	
03 57	☽ □ ♆
05 26	☽ ⚹ ♅
06 45	☽ bQ ♄
08 02	☽ ± ♇
10 23	☽ Q ☉
11 55	☽ ⚹ ☿
17 28	☉ ∥ ♄
19 41	☽ ⊼ ♇
12 Saturday	
01 31	☽ ⚹ ♃
04 38	☽ △ ♀
04 44	☽ Q ♅
07 51	☽ □ ♂
11 30	☽ △ ♄
13 37	☿ ⊥ ♆
14 53	☽ Q ☿
23 03	☉ ǂ ♃
23 45	☽ □ ☉
13 Sunday	
00 41	☽ Q ♃
08 06	♀ ⊥ ♇
09 36	☉ ǂ ☿
11 43	☽ bQ ♀
13 10	☽ △ ♆
14 39	☽ □ ♅
16 04	☿ ∥ ♃
17 32	☿ □ ♇
14 Monday	
01 39	☿ ǂ ♄
03 50	☽ ☍ ♇
05 22	☽ □ ☿
05 45	☽ ± ♀
10 13	☽ □ ♃
13 54	♂ ⊼ ♄
16 20	☽ bQ ♆
17 29	☽ ⊼ ♀
18 14	☽ □ ♄
18 32	☽ △ ♂
21 16	♀ ∠ ♅
15 Tuesday	
01 34	♀ △ ♄
08 08	☽ ± ♆
10 05	☽ △ ☉
12 10	☽ ǂ ♇
18 27	☽ ⊼ ♆
19 55	☽ △ ♅
22 02	☽ bQ ♂
16 Wednesday	
01 09	☽ Q ♄
01 26	☿ ☌ ♃
02 10	♀ ⊻ ♂
05 21	☽ ǂ ♀
07 57	☽ ⊼ ♇
13 22	☽ bQ ☉
13 59	☽ ± ♂
14 41	☽ △ ♃
15 26	♃ ǂ ♄
16 05	☽ △ ☿
17 55	☽ ± ♇
21 01	☽ ⚹ ♄
21 04	☽ bQ ♅
21 23	☿ ǂ ♂
23 55	☽ ǂ ♆
17 Thursday	
00 25	☽ ⊼ ♂
01 05	☽ ☍ ♀
05 13	☽ ± ☉
08 37	☽ bQ ♇
11 45	☽ ± ♅
15 31	☽ bQ ♃
15 36	☽ ⊼ ☉
19 16	☽ bQ ☿
19 58	☽ ☍ ♆
21 10	☽ ∠ ♄
21 25	☽ ⊼ ♅
23 01	☽ ∥ ☿
18 Friday	
02 32	☿ ∠ ♆
04 05	☽ ǂ ♂
06 06	☽ ± ♃
08 37	☽ △ ♇
11 06	☽ ± ☿
11 23	☽ ⊥ ♄
11 41	☽ ∥ ♃
12 35	☽ ǂ ♄
13 55	☿ ☍ ♄
15 43	☽ ⊼ ♃
20 46	☽ ⊻ ♄
21 30	☽ ⊼ ☿
23 01	☽ ǂ ☉
19 Saturday	
00 13	☽ ǂ ♅
02 50	☽ ☍ ♂
04 01	☽ ∥ ♅
04 51	☽ ∥ ☉
04 56	☽ ⊼ ♀
15 06	☽ ± ♀
15 28	☽ ∥ ♄
15 37	♂ Q ♇
16 57	☽ ǂ ♃
18 10	☽ ☍ ☉
18 41	☉ ∥ ♅
19 16	☽ ⊼ ♆
20 45	☽ ☍ ♅
21 51	☽ ∥ ♂
20 Sunday	
04 39	☽ ± ♆
06 20	☽ bQ ♀
07 36	☽ □ ♇
11 47	☉ ⊻ ♆
12 35	☽ ǂ ☿
15 22	☽ ☍ ♃
18 47	☽ bQ ♆
19 27	☽ ☌ ♄
22 53	♀ ∠ ♇
21 Monday	
00 47	☽ ☍ ☿
03 20	☽ ∥ ♆
04 20	☽ ⊼ ♂
07 54	☽ △ ♀
11 58	☽ Q ♇
12 24	☉ ☌ ♅
12 55	☽ ∥ ♀
14 22	☽ ± ♂
18 35	☽ △ ♆
20 10	☽ ⊼ ♅
20 41	☽ ⊼ ☉
22 Tuesday	
00 55	☉ ǂ ♅
05 37	☽ bQ ♂
05 52	☽ ± ♅
07 03	☽ ± ☉
07 14	☽ ⚹ ♇
14 01	☽ ∥ ♇
16 05	☽ ⊼ ♃
19 05	♀ ⊥ ♅
19 19	☽ ⊻ ♄
20 38	☽ bQ ♅
21 48	♂ ǂ ♃
22 52	☽ bQ ☉
23 Wednesday	
02 09	☽ ± ♃
05 02	☽ ⊼ ☿
05 12	☽ ⊥ ♄
07 42	☽ △ ♂
07 57	☽ ∠ ♇
13 08	☽ □ ♀
15 49	☽ ± ☿
17 31	☽ bQ ♃
20 08	☽ □ ♆
20 18	☽ ∠ ♄
21 54	☽ △ ♅
23 12	☽ ⊥ ♇
24 Thursday	
02 03	☽ △ ☉
08 16	☽ bQ ☿
09 34	☽ ⊻ ♇
19 30	♄ bQ ♆
19 57	☽ △ ♃
22 14	☽ ⚹ ♄
23 14	☽ Q ♀
25 Friday	
03 36	☽ Q ♆
12 23	☽ △ ☿
15 08	☽ □ ♂
15 12	♂ ∥ ♄
19 42	☽ Q ♄
23 02	☽ ⚹ ♀
26 Saturday	
01 25	☽ ⚹ ♆
03 26	☽ □ ♅
04 52	♀ ∠ ♃
12 07	☽ □ ☉
15 48	☽ ☌ ♇
17 58	☉ ⊥ ♆
21 03	♀ bQ ♄
27 Sunday	
01 39	♀ ☌ ♆
02 43	☽ Q ♂
04 01	☽ □ ♃
05 04	☽ □ ♄
05 34	☽ ∠ ♆
05 58	☽ ∠ ♀
10 43	♀ ∥ ♆
12 19	☿ ǂ ♀
13 28	☽ Q ♅
14 28	☿ ǂ ♆
14 47	☉ ǂ ♂
22 52	☽ ⊥ ♆
23 00	☽ □ ☿
28 Monday	
01 05	☽ ⊥ ♀
01 07	♀ ⊻ ♅
01 18	☽ Q ☉
01 59	☽ ∥ ♇
03 17	☽ ⚹ ♂
10 34	☽ ⊻ ♆
12 13	☉ □ ♇
12 33	☉ ⊥ ♀
12 50	☽ ⚹ ♅
14 03	☽ ⊻ ♀
15 28	☽ Q ♃
21 55	♃ ☍ ♄
22 55	☉ ǂ ♄
29 Tuesday	
01 41	☽ ⊻ ♇
02 52	☽ ⚹ ☉
10 46	☽ ∠ ♂
11 06	☽ Q ☿
13 34	☽ ⊥ ♇
15 12	☽ △ ♄
15 40	☽ ⚹ ♃
18 36	☽ ǂ ☿
18 37	☽ ∠ ♅
21 02	☽ ∥ ♆
30 Wednesday	
02 39	☽ ∥ ♀
05 58	☽ ⊥ ♂
07 32	☽ ∠ ♇
11 16	☽ ⚹ ☿
11 25	☽ ∠ ☉
12 42	☽ ⊥ ♅
18 50	☽ ⊻ ♂
20 48	☿ St R
21 01	☽ bQ ♄
22 19	☽ ∠ ♃
22 21	☽ ☌ ♆
31 Thursday	
00 41	♃ ∠ ♆
00 49	☽ ⊻ ♅
07 09	☽ ⊥ ☉
08 29	☽ ☌ ♀
11 01	☽ ǂ ♃
13 44	☽ ⚹ ♇
14 44	☽ ǂ ☉
15 03	☽ ± ♄
16 50	☽ ⊥ ♃
17 25	☽ ∠ ☿
20 22	☽ ⊻ ☉
20 28	☽ ∥ ♄

APRIL 2011

Date	Sidereal time h m s	Sun ☉	Moon ☽	Moon ☽ 24.00	Mercury ☿	Venus ♀	Mars ♂	Jupiter ♃	Saturn ♄	Uranus ♅	Neptune ♆	Pluto ♇
01	00 38 00	11 ♈ 24 58	18 ♓ 30 21	24 ♓ 26 07	24 ♈ 12	06 ♓ 16	29 ♓ 27	15 ♈ 13	14 ♎ 04	01 ♈ 10	29 ♒ 55	07 ♑ 29
02	00 41 57	12 24 12	00 ♈ 21 54	06 ♈ 17 52	23 R 57	07 28	00 ♈ 14	15 28	14 R 00	01 13	29 56	07 30
03	00 45 53	13 23 23	12 ♈ 14 13	18 ♈ 11 05	23 36	08 40	01 01	15 42	13 55	01 17	29 ♒ 58	07 30
04	00 49 50	14 22 33	24 ♈ 08 39	00 ♉ 07 04	23 09	09 53	01 47	15 57	13 50	01 20	00 ♓ 00	07 30
05	00 53 46	15 21 41	06 ♉ 06 33	12 ♉ 07 17	22 37	11 05	02 34	16 11	13 46	01 23	00 02	07 30
06	00 57 43	16 20 46	18 ♉ 09 31	24 ♉ 13 32	22 00	12 17	03 21	16 26	13 41	01 27	00 03	07 30
07	01 01 39	17 19 50	00 ♊ 19 39	06 ♊ 28 12	21 19	13 30	04 07	16 40	13 36	01 30	00 05	07 30
08	01 05 36	18 18 51	12 ♊ 39 35	18 ♊ 54 15	20 36	14 42	04 54	16 55	13 32	01 33	00 07	07 30
09	01 09 32	19 17 50	25 ♊ 12 37	01 ♋ 35 12	19 51	15 54	05 41	17 09	13 27	01 37	00 08	07 R 30
10	01 13 29	20 16 47	08 ♋ 02 27	14 ♋ 34 53	19 05	17 07	06 27	17 24	13 23	01 40	00 10	07 30
11	01 17 26	21 15 41	21 ♋ 12 55	27 ♋ 56 58	18 18	18 19	07 14	17 38	13 18	01 43	00 12	07 30
12	01 21 22	22 14 33	04 ♌ 47 22	11 ♌ 44 20	17 33	19 31	08 00	17 53	13 13	01 47	00 13	07 30
13	01 25 19	23 13 23	18 ♌ 47 55	25 ♌ 58 05	16 49	20 44	08 46	18 07	13 09	01 50	00 15	07 30
14	01 29 15	24 12 10	03 ♍ 14 32	10 ♍ 36 48	16 07	21 56	09 33	18 22	13 04	01 53	00 16	07 30
15	01 33 12	25 10 55	18 ♍ 04 12	25 ♍ 35 48	15 28	23 09	10 19	18 36	13 00	01 56	00 18	07 30
16	01 37 08	26 09 38	03 ♎ 10 33	10 ♎ 47 10	14 53	24 21	11 05	18 51	12 55	02 00	00 19	07 30
17	01 41 05	27 08 19	18 ♎ 24 20	26 ♎ 00 40	14 21	25 34	11 52	19 05	12 51	02 03	00 21	07 29
18	01 45 01	28 06 58	03 ♏ 34 49	11 ♏ 05 30	13 54	26 46	12 38	19 19	12 46	02 06	00 22	07 29
19	01 48 58	29 ♈ 05 35	18 ♏ 31 36	25 ♏ 52 10	13 32	27 59	13 24	19 34	12 42	02 09	00 24	07 29
20	01 52 55	00 ♉ 04 10	03 ♐ 06 28	10 ♐ 13 58	13 15	29 ♓ 11	14 10	19 48	12 38	02 12	00 25	07 29
21	01 56 51	01 02 44	17 ♐ 14 22	24 ♐ 07 33	13 03	00 ♈ 24	14 56	20 03	12 33	02 15	00 26	07 28
22	02 00 48	02 01 16	00 ♑ 53 34	07 ♑ 32 40	12 55	01 36	15 42	20 17	12 29	02 18	00 28	07 28
23	02 04 44	02 59 46	14 ♑ 05 10	20 ♑ 31 31	12 D 53	02 49	16 28	20 31	12 25	02 21	00 29	07 27
24	02 08 41	03 58 15	26 ♑ 52 14	03 ♒ 07 52	12 56	04 02	17 14	20 46	12 21	02 24	00 30	07 27
25	02 12 37	04 56 42	09 ♒ 19 00	15 ♒ 26 14	13 04	05 14	18 00	21 00	12 17	02 28	00 31	07 26
26	02 16 34	05 55 07	21 ♒ 30 11	27 ♒ 31 25	13 17	06 27	18 46	21 14	12 12	02 31	00 33	07 26
27	02 20 30	06 53 31	03 ♓ 30 30	09 ♓ 27 58	13 34	07 40	19 32	21 28	12 08	02 34	00 34	07 25
28	02 24 27	07 51 53	15 ♓ 24 19	21 ♓ 20 00	13 56	08 52	20 18	21 43	12 04	02 36	00 35	07 25
29	02 28 24	08 50 13	27 ♓ 15 26	03 ♈ 11 01	14 23	10 05	21 04	21 57	12 01	02 39	00 36	07 24
30	02 32 20	09 ♉ 48 32	09 ♈ 07 03	15 ♈ 03 51	14 ♈ 53	11 ♈ 18	21 ♈ 49	22 ♈ 11	11 ♎ 57	02 ♈ 42	00 ♓ 37	07 ♑ 24

Date	Moon True ☊	Moon Mean ☊	Moon ☽ Latitude
01	26 ♐ 47	27 ♐ 31	04 N 57
02	26 R 33	27 27	05 00
03	26 20	27 24	04 49
04	26 07	27 21	04 26
05	25 55	27 18	03 51
06	25 47	27 15	03 05
07	25 41	27 12	02 10
08	25 39	27 08	01 08
09	25 38	27 05	00 N 02
10	25 D 38	27 02	01 S 06
11	25 39	26 59	02 12
12	25 R 38	26 56	03 13
13	25 35	26 53	04 04
14	25 30	26 49	04 42
15	25 23	26 46	05 02
16	25 15	26 43	05 01
17	25 06	26 40	04 40
18	24 58	26 37	03 58
19	24 51	26 33	03 01
20	24 47	26 30	01 53
21	24 45	26 27	00 S 40
22	24 D 45	26 24	00 N 33
23	24 46	26 21	01 42
24	24 47	26 18	02 44
25	24 R 47	26 14	03 36
26	24 46	26 11	04 18
27	24 42	26 08	04 47
28	24 37	26 05	05 04
29	24 30	26 02	05 07
30	24 ♐ 22	25 ♐ 58	04 N 58

DECLINATIONS

Date	Sun ☉	Moon ☽	Mercury ☿	Venus ♀	Mars ♂	Jupiter ♃	Saturn ♄	Uranus ♅	Neptune ♆	Pluto ♇
01	04 N 31	00 N 01	12 N 29	10 S 03	01 S 04	05 N 00	03 S 02	00 S 11	11 S 58	18 S 44
02	04 54	04 44	12 22	09 38	00 46	05 05	03 00	00 10	11 58	18 44
03	05 17	09 16	12 12	09 14	00 27	05 11	02 58	00 08	11 57	18 44
04	05 40	13 29	11 58	08 49	00 S 08	05 17	02 56	00 07	11 56	18 43
05	06 03	17 11	11 41	08 23	00 N 11	05 22	02 55	00 06	11 56	18 43
06	06 26	20 12	11 20	07 58	00 30	05 28	02 53	00 04	11 55	18 43
07	06 48	22 20	10 57	07 32	00 49	05 33	02 51	00 03	11 55	18 43
08	07 11	23 27	10 31	07 06	01 08	05 39	02 49	00 S 02	11 54	18 43
09	07 33	23 23	10 03	06 40	01 27	05 44	02 47	00 00	11 53	18 43
10	07 55	22 06	09 33	06 14	01 45	05 50	02 46	00 N 01	11 53	18 43
11	08 18	19 36	09 03	05 48	02 04	05 56	02 44	00 02	11 52	18 43
12	08 40	15 57	08 31	05 21	02 23	06 01	02 42	00 03	11 52	18 43
13	09 01	11 19	08 00	04 54	02 41	06 07	02 40	00 05	11 51	18 43
14	09 23	05 56	07 29	04 27	03 00	06 12	02 39	00 06	11 51	18 43
15	09 45	00 N 05	06 58	04 00	03 19	06 18	02 37	00 07	11 50	18 43
16	10 06	05 S 52	06 29	03 33	03 37	06 23	02 35	00 09	11 50	18 43
17	10 27	11 31	06 02	03 05	03 56	06 28	02 33	00 10	11 49	18 43
18	10 48	16 27	05 36	02 38	04 14	06 34	02 32	00 11	11 49	18 43
19	11 09	20 14	05 13	02 10	04 32	06 39	02 30	00 12	11 48	18 43
20	11 30	22 38	04 51	01 43	04 51	06 45	02 29	00 14	11 48	18 43
21	11 50	23 30	04 33	01 15	05 09	06 50	02 27	00 15	11 47	18 43
22	12 11	22 53	04 16	00 47	05 27	06 56	02 25	00 16	11 47	18 43
23	12 31	21 00	04 03	00 S 19	05 45	07 01	02 24	00 17	11 46	18 43
24	12 50	18 06	03 51	00 N 09	06 03	07 06	02 22	00 18	11 46	18 43
25	13 10	14 27	03 43	00 37	06 21	07 12	02 21	00 20	11 46	18 43
26	13 30	10 16	03 37	01 05	06 39	07 17	02 19	00 21	11 45	18 43
27	13 49	05 45	03 34	01 33	06 57	07 22	02 18	00 22	11 45	18 43
28	14 08	01 S 05	03 33	02 01	07 15	07 27	02 16	00 23	11 44	18 43
29	14 27	03 N 36	03 34	02 29	07 32	07 33	02 15	00 24	11 44	18 43
30	14 N 45	08 N 10	03 N 38	02 N 57	07 N 50	07 N 38	02 S 14	00 N 25	11 S 44	18 S 43

ZODIAC SIGN ENTRIES

Date	h	m	Planets
02	04	51	♂ ♈
02	11	16	☽ ♈
04	13	50	♆ ♓
04	23	46	☽ ♉
07	11	21	☽ ♊
09	21	02	☽ ♋
12	03	37	☽ ♌
14	06	40	☽ ♍
16	06	59	☽ ♎
18	06	19	☽ ♏
20	06	50	☽ ♐
20	10	17	☉ ♉
21	04	06	♀ ♈
22	10	24	☽ ♑
24	17	59	☽ ♒
27	04	57	☽ ♓
29	17	33	☽ ♈

LATITUDES

Date	Mercury ☿	Venus ♀	Mars ♂	Jupiter ♃	Saturn ♄	Uranus ♅	Neptune ♆	Pluto ♇
01	03 N 20	00 S 54	00 S 56	01 S 05	02 N 44	00 S 42	00 S 30	04 N 30
04	03 13	01 01	00 55	01 05	02 44	00 42	00 30	04 30
07	02 50	01 08	00 54	01 05	02 44	00 42	00 30	04 31
10	02 15	01 15	00 53	01 05	02 44	00 42	00 30	04 31
13	01 30	01 20	00 51	01 05	02 44	00 42	00 30	04 31
16	00 N 41	01 25	00 50	01 05	02 44	00 42	00 30	04 31
19	00 S 08	01 29	00 49	01 05	02 44	00 42	00 30	04 31
22	00 54	01 33	00 47	01 05	02 44	00 43	00 30	04 31
25	01 34	01 36	00 46	01 05	02 43	00 43	00 30	04 31
28	02 07	01 38	00 45	01 05	02 43	00 43	00 30	04 31
31	02 S 33	01 S 40	00 S 43	01 S 05	02 N 43	00 S 43	00 S 31	04 N 31

DATA

Julian Date	2455653
Delta T	+68 seconds
Ayanamsa	24° 01' 08"
Synetic vernal point	05° ♓ 05' 51"
True obliquity of ecliptic	23° 26' 16"

MOON'S PHASES, APSIDES AND POSITIONS ☽

Date	h	m	Phase	Longitude	Eclipse Indicator
03	14	32	●	13 ♈ 30	
11	12	05	☽	21 ♋ 16	
18	02	44	○	27 ♎ 44	
25	02	47	☾	04 ♒ 34	

Day	h	m		
02	09	17	Apogee	
17	06	06	Perigee	
29	18	12	Apogee	
01	11	55	0N	
08	22	55	Max dec	23° N 34'
15	12	20	0S	
21	13	43	Max dec	23° S 30'
28	17	32	0N	

LONGITUDES

Date	Chiron ⚷	Ceres ⚳	Pallas ⚴	Juno ⚵	Vesta ⚶	Black Moon Lilith ⚸
01	03 ♓ 22	03 ♓ 58	06 ♒ 37	16 ♍ 56	29 ♑ 35	01 ♈ 01
11	03 ♓ 54	07 ♓ 32	09 ♒ 10	15 ♍ 20	03 ♒ 36	02 ♈ 08
21	04 ♓ 23	10 ♓ 58	11 ♒ 23	14 ♍ 21	07 ♒ 19	03 ♈ 16
31	04 ♓ 47	14 ♓ 14	13 ♒ 13	14 ♍ 00	10 ♒ 41	04 ♈ 23

All ephemeris data is given at 12.00 UT and the Moon's longitude is additionally given for 24.00 UT

ASPECTARIAN

h m	Aspects
01 Friday	
03 06	☽ ⊼ ♄
05 13	☽ ⊻ ♃
06 06	☽ ∥ ♂
10 59	☽ ∥ ♅
11 23	☽ ⊥ ☿
12 51	☽ ⋕ ♅
14 00	☽ Q ♇
17 02	☽ ⋕ ♂
23 19	☽ ⊻ ☿
02 Saturday	
02 36	♂ ⊻ ♆
03 12	☽ ⋕ ♄
11 08	☽ ⊻ ♆
11 43	☽ ☌ ♂
12 30	♀ ⚹ ♇
12 59	☽ ∥ ☉
13 45	☽ ☌ ♅
13 55	☽ ∥ ♃
21 52	♀ ± ♄
23 18	☽ ⊥ ♆
03 Sunday	
02 26	☽ □ ♇
03 37	☉ ∥ ♃
03 59	☽ ⊻ ♀
10 50	☿ ∠ ♀
11 47	☽ ⋕ ♀
14 32	☽ ☌ ☉
15 22	☽ ☍ ♄
17 28	☽ ⊥ ♀
17 32	☽ ∠ ♆
19 08	☽ ☌ ♃
20 52	♂ ☌ ♅
23 56	☉ ☍ ♄
04 Monday	
02 57	☽ ⋕ ♆
03 39	☽ ∥ ☿
10 04	☽ ☌ ☿
12 58	♂ ∥ ♅
13 39	☽ ∠ ♀
13 40	♀ ⊥ ♃
15 18	☿ ⋕ ♆
23 47	☽ ⚹ ♆
05 Tuesday	
02 30	☽ ⊻ ♅
03 36	☉ ∠ ♆
04 25	☽ ⊻ ♂
05 29	♂ ⋕ ♅
14 34	☽ ⊥ ♅
14 47	☽ △ ♇
17 15	☽ ⊥ ♂
23 02	☽ ⚹ ♀
23 30	☽ ⋕ ♇
23 50	☽ Q ♆
06 Wednesday	
03 10	☽ ⊼ ♄
08 05	☽ ⊻ ☉
08 30	☽ ⊻ ♃
08 35	☽ ∠ ♅
12 24	☽ ∠ ♂
14 40	☉ ☌ ♃
15 00	☽ ± ♄
19 13	☽ ⊻ ☿
20 36	☽ Ꝑ ♇
20 37	☽ ⊥ ♃
21 01	☽ ⊥ ☉
07 Thursday	
01 27	☽ Q ♀
06 25	☽ ⊥ ☿
08 39	☽ Ꝑ ♄
11 31	☽ □ ♆
14 09	♀ ⊼ ♄
14 19	☽ ± ♇
14 19	☽ ⚹ ♅
14 41	☽ ∠ ♃
16 16	☽ ∠ ☉
19 56	☽ ⚹ ♂
23 05	☽ ∠ ☿
08 Friday	
02 01	☽ ⊼ ♇
09 52	☉ ⋕ ♀
10 50	☿ ⊥ ♀
13 40	☽ △ ♄
13 44	☽ Q ♅
16 21	☽ □ ♀
20 21	☽ ⚹ ♃
20 42	☽ Q ♂
23 48	☽ ⚹ ☉
09 Saturday	
02 24	☽ ⚹ ☿
08 52	♇ St R
19 35	☽ Q ♃
19 36	☉ ☌ ☿
21 19	☽ △ ♆
23 47	☽ Q ☿
10 Sunday	
00 06	☽ □ ♅
00 25	☽ Q ☉
02 52	♂ ⊥ ♆
08 52	☽ □ ♂
11 01	☽ ☍ ♇
19 06	♀ ⊻ ♃
21 45	☽ □ ♄
11 Monday	
01 06	☽ Ꝑ ♆
05 26	☽ □ ♃
06 03	♀ ⋕ ♃
06 15	☽ △ ♀
07 03	☽ □ ☿
11 53	☿ ⊻ ♀
12 05	☽ □ ☉
17 20	☽ ± ♆
18 27	☽ ⋕ ♇
20 41	♂ □ ♇
12 Tuesday	
03 59	☿ ☌ ♃
03 59	☽ ⊼ ♆
05 48	☽ Q ♄
06 43	☽ △ ♅
08 15	☉ ∥ ☿
11 30	☽ Ꝑ ♀
11 37	♀ Q ♇
16 43	☽ ⊼ ♇
17 54	☽ △ ♂
13 Wednesday	
02 28	☽ ⚹ ♄
03 01	☽ ± ♇
04 28	☽ ± ♀
08 40	☽ Ꝑ ♅
08 48	☽ △ ☿
09 25	☽ ⋕ ♆
10 43	♂ ⋕ ♄
10 50	☽ △ ♃
15 33	☽ ⊼ ♀
18 13	☽ Ꝑ ♇
19 58	☽ △ ☉
20 49	☽ Ꝑ ♂
21 54	☽ ∥ ☉
23 49	☽ ± ♅
14 Thursday	
03 32	☽ ∠ ♄
04 42	☽ ∥ ☿
07 07	☽ ☍ ♆
08 40	☽ Ꝑ ☿
09 46	☽ ⊼ ♅
10 53	☽ ∥ ♃
12 12	☽ Ꝑ ♃
12 32	☽ ± ♂
18 13	☽ ⊥ ♄
18 44	☽ ⋕ ♀
18 57	☽ △ ♇
22 24	☽ Ꝑ ☉
22 43	☽ ± ☿
22 50	☽ ⊼ ♂
23 35	☽ ∥ ♂
15 Friday	
01 43	☽ ⋕ ♄
03 04	☽ ± ♃
03 53	☽ ⊻ ♄
08 00	☽ ⊼ ☿
11 51	☽ ∥ ♅
12 50	☽ ⋕ ♅
12 52	☽ ⊼ ♃
13 54	☽ ± ☉
18 24	☿ ∠ ♆
20 49	☽ ☍ ♀
22 47	☽ ∥ ♄
16 Saturday	
00 08	☽ ⊼ ☉
02 23	☽ ⋕ ♂
03 15	☽ ∥ ♀
07 29	☽ ⊼ ♆
09 44	♀ ⋕ ♂
10 07	☽ ☍ ♅
14 08	☽ ⋕ ♃
14 21	☽ ⋕ ☿
16 33	☿ ∥ ♃
16 58	☽ ± ♆
18 49	☽ □ ♇
17 Sunday	
01 09	☽ ☍ ♂
03 17	☽ ☌ ♄
05 49	☽ ☍ ☿
06 57	☽ ⋕ ☉
07 10	☽ Ꝑ ♆
13 05	☽ ☍ ♃
13 21	☽ ∥ ♆
23 10	☽ Q ♇
18 Monday	
00 16	☽ ⊼ ♀
02 44	☽ ☍ ☉
06 54	☽ △ ♆
09 38	☽ ⊼ ♅
10 36	☽ ± ♀
16 02	♂ ☍ ♄
17 36	♀ ∥ ♄
18 13	☽ ⚹ ♇
19 14	☽ ± ♅
19 Tuesday	
01 21	☽ ∥ ♇
02 15	☽ Ꝑ ♀
02 38	☽ ⊻ ♄
03 16	☽ ⊼ ♂
04 07	☽ ⊼ ☿
09 46	☽ Ꝑ ♅
12 17	☽ ⊥ ♄
13 30	☽ ± ♂
13 36	☽ ± ☿
13 43	☽ ⊼ ♃
14 58	☿ ☌ ♂
18 26	☽ ∠ ♇
23 41	☽ ± ♃
20 Wednesday	
02 57	☽ ∠ ♄
04 04	☽ Ꝑ ☿
04 53	☽ △ ♀
05 05	☽ Ꝑ ♂
06 34	☽ ⊼ ☉
07 30	☽ □ ♆
09 17	☽ ⊥ ♇
10 29	☽ △ ♅
12 30	☿ ∥ ♂
14 53	☽ Ꝑ ♃
17 20	☽ ± ☉
19 20	☽ ⊻ ♇
20 42	☉ ⚹ ♆
21 Thursday	
04 00	☽ ⚹ ♄
04 53	☽ △ ☿
07 48	☽ △ ♂
08 43	☉ ⋕ ♆
09 47	☽ Ꝑ ☉
12 49	♀ ⊻ ♆
14 05	☽ Q ♆
16 57	☽ △ ♃
22 Friday	
00 41	☽ Q ♄
04 05	♂ ∠ ♆
11 14	☽ ⚹ ♆
13 24	☽ □ ♀
14 11	☽ △ ☉
14 33	☽ □ ♅
19 24	☉ ⊻ ♅
23 51	☽ ☌ ♇
23 Saturday	
02 28	♀ ☌ ♅
08 56	☽ □ ♄
09 47	☽ □ ☿
10 04	☿ St D
13 33	♀ ⋕ ♅
14 35	☽ ∠ ♆
16 42	☽ □ ♂
23 44	☽ Q ♅
24 Sunday	
00 13	☽ □ ♃
01 52	☽ Q ♀
06 12	☉ ⊻ ♀
07 28	☽ ∥ ♇
07 30	☽ ⊥ ♆
18 57	☽ ⊻ ♆
19 50	☽ Q ☿
20 34	♀ ∥ ♅
22 39	☽ ⚹ ♅
25 Monday	
02 47	☽ □ ☉
03 13	☽ ⚹ ♀
05 08	☽ Q ♂
08 21	☽ ⊻ ♇
11 22	☽ Q ♃
17 45	☽ △ ♄
19 02	☽ ⋕ ☉
19 27	☽ ⚹ ☿
20 04	☽ ⊥ ♇
26 Tuesday	
03 42	☽ ∥ ♆
04 03	☽ ∠ ♅
06 13	☽ ⚹ ♂
11 28	☽ ⚹ ♃
11 53	☽ ∠ ♀
13 51	☽ ∠ ♇
13 55	♀ ⊥ ♆
17 14	☽ Q ☉
22 01	☽ ⊥ ♅
23 18	☽ Ꝑ ♄
27 Wednesday	
01 50	☽ ∠ ☿
03 43	☽ ⋕ ♃
06 05	☽ ☌ ♆
06 10	☽ ⋕ ♂
07 21	♀ □ ♇
07 52	☽ ⊥ ♀
10 05	☽ ⊻ ♅
14 12	☽ ∠ ♂
17 16	☽ ± ♄
18 05	☽ ∠ ♃
19 25	☽ ⚹ ☉
19 52	☽ ⚹ ♇
20 25	☽ ⊥ ☿
21 18	☽ ⊻ ♀
23 24	☽ ⋕ ☿
28 Thursday	
01 00	☉ △ ♇
05 18	☽ ⊼ ♄
05 54	☽ ∥ ♄
07 41	☽ ⋕ ♀
08 55	☽ ⊻ ☿
09 36	☽ ⊥ ♂
12 38	☽ ⊥ ♃
15 33	☽ ⋕ ♅
19 31	☽ ∥ ♅
20 06	☽ Q ♇
22 35	☽ ⊻ ♂
29 Friday	
00 39	♀ ⋕ ♄
01 01	☽ ⊻ ♃
04 27	☽ ∠ ☉
05 03	☽ ⋕ ♄
05 34	☽ ∥ ♀
07 19	☉ ⊥ ♅
11 49	☽ ∥ ☿
13 01	♂ ∥ ♃
18 47	☽ ⊻ ♆
22 59	☽ ☌ ♅
30 Saturday	
00 18	☽ ⊥ ☉
06 57	☽ ⊥ ♆
08 31	☽ □ ♇
09 03	☽ ∥ ♃
10 03	☽ ∥ ♂
13 31	☽ ⊻ ☉
16 54	☽ ☌ ♀
17 41	☽ ☍ ♄

MAY 2011

LONGITUDES

Date	Sidereal time h m s	Sun ☉ ° ' "	Moon ☽ ° ' "	Moon ☽ 24.00 ° ' "	Mercury ☿ ° '	Venus ♀ ° '	Mars ♂ ° '	Jupiter ♃ ° '	Saturn ♄ ° '	Uranus ♅ ° '	Neptune ♆ ° '	Pluto ♇ ° '
01	02 36 17	10 ♉ 46 49	21 ♈ 01 40	27 ♈ 00 45	15 ♈ 28	12 ♈ 30	22 ♈ 35	22 ♈ 25	11 ♎ 53	02 ♈ 45	00 ♓ 38	07 ♑ 23
02	02 40 13	11 45 04	03 ♉ 01 17	09 ♉ 03 26	16 06	13 43	23 21	22 39	11 R 49	02 48	00 39	07 R 22
03	02 44 10	12 43 18	15 ♉ 07 24	21 ♉ 13 19	16 49	14 56	24 06	22 53	11 45	02 51	00 40	07 22
04	02 48 06	13 41 30	27 ♉ 21 21	03 ♊ 31 39	17 35	16 09	24 52	23 07	11 42	02 54	00 41	07 21
05	02 52 03	14 39 41	09 ♊ 44 25	15 ♊ 59 49	18 24	17 21	25 37	23 21	11 38	02 56	00 42	07 20
06	02 55 59	15 37 49	22 ♊ 18 03	28 ♊ 39 22	19 17	18 34	26 23	23 35	11 35	02 59	00 43	07 19
07	02 59 56	16 35 56	05 ♋ 03 59	11 ♋ 32 10	20 13	19 47	27 08	23 49	11 31	03 02	00 44	07 19
08	03 03 53	17 34 01	18 ♋ 04 12	24 ♋ 40 20	21 12	21 00	27 54	24 03	11 28	03 05	00 45	07 18
09	03 07 49	18 32 04	01 ♌ 20 50	08 ♌ 05 56	22 14	22 12	28 39	24 17	11 25	03 07	00 46	07 17
10	03 11 46	19 30 05	14 ♌ 55 48	21 ♌ 50 34	23 19	23 25	29 ♈ 24	24 31	11 22	03 10	00 46	07 16
11	03 15 42	20 28 04	28 ♌ 50 16	05 ♍ 54 52	24 26	24 38	00 ♉ 09	24 45	11 19	03 12	00 47	07 15
12	03 19 39	21 26 01	13 ♍ 04 11	20 ♍ 17 53	25 36	25 51	00 54	24 58	11 16	03 15	00 48	07 14
13	03 23 35	22 23 56	27 ♍ 35 32	04 ♎ 56 33	26 49	27 03	01 40	25 12	11 13	03 17	00 48	07 13
14	03 27 32	23 21 50	12 ♎ 20 10	19 ♎ 45 33	28 05	28 16	02 25	25 26	11 10	03 20	00 49	07 12
15	03 31 28	24 19 42	27 ♎ 11 45	04 ♏ 37 46	29 ♈ 22	29 ♈ 29	03 10	25 39	11 07	03 22	00 50	07 11
16	03 35 25	25 17 32	12 ♏ 02 34	19 ♏ 25 08	00 ♉ 43	00 ♉ 42	03 55	25 53	11 04	03 25	00 50	07 10
17	03 39 22	26 15 20	26 ♏ 44 31	03 ♐ 59 51	02 06	01 55	04 39	26 06	11 02	03 27	00 51	07 09
18	03 43 18	27 13 08	11 ♐ 10 24	18 ♐ 15 33	03 31	03 08	05 24	26 20	10 59	03 30	00 52	07 08
19	03 47 15	28 10 54	25 ♐ 14 53	02 ♑ 08 06	04 58	04 20	06 09	26 33	10 57	03 32	00 52	07 07
20	03 51 11	29 ♉ 08 39	08 ♑ 55 02	15 ♑ 35 43	06 28	05 33	06 54	26 46	10 54	03 34	00 53	07 06
21	03 55 08	00 ♊ 06 22	22 ♑ 10 16	28 ♑ 38 56	08 00	06 46	07 38	27 00	10 52	03 36	00 53	07 05
22	03 59 04	01 04 04	05 ♒ 02 03	11 ♒ 20 02	09 34	07 59	08 23	27 13	10 50	03 39	00 53	07 04
23	04 03 01	02 01 46	17 ♒ 33 20	23 ♒ 42 30	11 11	09 12	09 08	27 26	10 48	03 41	00 54	07 03
24	04 06 57	02 59 26	29 ♒ 48 04	05 ♓ 50 37	12 50	10 25	09 52	27 39	10 46	03 43	00 54	07 02
25	04 10 54	03 57 05	11 ♓ 50 43	17 ♓ 48 57	14 31	11 38	10 37	27 52	10 44	03 45	00 54	07 00
26	04 14 51	04 54 43	23 ♓ 45 53	29 ♓ 42 03	16 14	12 51	11 21	28 05	10 42	03 47	00 55	06 59
27	04 18 47	05 52 20	05 ♈ 38 00	11 ♈ 34 14	18 00	14 04	12 06	28 18	10 41	03 49	00 55	06 58
28	04 22 44	06 49 56	17 ♈ 31 12	23 ♈ 29 20	19 47	15 17	12 50	28 31	10 39	03 51	00 55	06 56
29	04 26 40	07 47 32	29 ♈ 29 03	05 ♉ 30 40	21 38	16 30	13 34	28 44	10 37	03 53	00 55	06 55
30	04 30 37	08 45 06	11 ♉ 34 32	17 ♉ 40 54	23 30	17 42	14 18	28 56	10 36	03 55	00 55	06 54
31	04 34 33	09 ♊ 42 39	23 ♉ 49 59	00 ♊ 02 01	25 ♉ 24	18 ♉ 55	15 ♉ 02	29 ♈ 09	10 ♎ 35	03 ♈ 57	00 ♓ 56	06 ♑ 53

Moon Nodes and Latitude; DECLINATIONS

Date	Moon True ☊ ° '	Moon Mean ☊ ° '	Moon ☽ Latitude ° '	Sun ☉	Moon ☽	Mercury ☿	Venus ♀	Mars ♂	Jupiter ♃	Saturn ♄	Uranus ♅	Neptune ♆	Pluto ♇
01	24 ♐ 13	25 ♐ 55	04 N 35	15 N 03	12 N 27	03 N 44	03 N 25	08 N 07	07 N 43	02 S 12	00 N 27	11 S 43	18 S 43
02	24 R 05	25 52	04 00	15 22	16 17	03 53	03 53	08 25	07 48	02 11	00 28	11 43	18 43
03	23 59	25 49	03 14	15 39	19 28	04 03	04 21	08 42	07 54	02 10	00 29	11 42	18 43
04	23 54	25 46	02 19	15 57	21 49	04 16	04 48	08 59	07 59	02 08	00 30	11 42	18 43
05	23 51	25 43	01 16	16 14	23 10	04 30	05 16	09 16	08 04	02 07	00 31	11 42	18 43
06	23 50	25 39	00 N 08	16 31	23 21	04 46	05 44	09 33	08 09	02 06	00 32	11 42	18 43
07	23 D 50	25 36	01 S 01	16 48	22 20	05 04	06 11	09 50	08 14	02 05	00 33	11 41	18 43
08	23 52	25 33	02 08	17 04	20 06	05 24	06 38	10 07	08 19	02 03	00 34	11 41	18 43
09	23 53	25 30	03 10	17 20	16 46	05 45	07 06	10 23	08 24	02 02	00 35	11 41	18 43
10	23 54	25 27	04 03	17 36	12 29	06 08	07 33	10 40	08 29	02 01	00 36	11 41	18 43
11	23 R 54	25 24	04 43	17 52	07 27	06 33	08 00	10 56	08 34	02 00	00 37	11 40	18 43
12	23 52	25 20	05 07	18 07	01 N 56	06 58	08 27	11 12	08 39	01 59	00 38	11 40	18 43
13	23 49	25 17	05 11	18 22	03 S 48	07 25	08 53	11 29	08 44	01 58	00 39	11 40	18 43
14	23 46	25 14	04 56	18 37	09 25	07 54	09 20	11 45	08 49	01 57	00 40	11 40	18 43
15	23 41	25 11	04 21	18 51	14 31	08 23	09 46	12 01	08 54	01 56	00 41	11 40	18 43
16	23 38	25 08	03 28	19 05	18 45	08 54	10 12	12 16	08 59	01 56	00 42	11 39	18 43
17	23 35	25 04	02 22	19 19	21 43	09 25	10 38	12 32	09 03	01 55	00 43	11 39	18 43
18	23 33	25 01	01 S 07	19 32	23 14	09 58	11 04	12 48	09 08	01 54	00 44	11 39	18 44
19	23 D 32	24 58	00 N 09	19 45	23 12	10 31	11 29	13 03	09 13	01 53	00 45	11 39	18 44
20	23 33	24 55	01 24	19 58	21 45	11 06	11 54	13 18	09 18	01 52	00 46	11 39	18 44
21	23 34	24 52	02 31	20 10	19 08	11 41	12 19	13 33	09 22	01 52	00 46	11 39	18 44
22	23 36	24 49	03 29	20 22	15 38	12 16	12 44	13 48	09 27	01 51	00 47	11 39	18 44
23	23 37	24 45	04 15	20 34	11 31	12 53	13 08	14 03	09 32	01 50	00 48	11 38	18 44
24	23 38	24 42	04 48	20 45	07 02	13 29	13 32	14 18	09 36	01 50	00 49	11 38	18 44
25	23 R 38	24 39	05 08	20 56	02 S 22	14 07	13 56	14 32	09 41	01 49	00 50	11 38	18 44
26	23 37	24 36	05 15	21 07	02 N 21	14 44	14 19	14 47	09 45	01 49	00 51	11 38	18 44
27	23 35	24 33	05 08	21 17	06 57	15 22	14 42	15 01	09 50	01 48	00 51	11 38	18 44
28	23 32	24 30	04 48	21 27	11 19	16 00	15 05	15 15	09 54	01 48	00 52	11 38	18 44
29	23 30	24 26	04 15	21 36	15 16	16 38	15 27	15 29	09 59	01 48	00 53	11 38	18 44
30	23 28	24 23	03 31	21 46	18 39	17 15	15 49	15 43	10 03	01 47	00 54	11 38	18 45
31	23 ♐ 26	24 ♐ 20	02 N 36	21 N 54	21 N 15	17 N 53	16 N 11	15 N 56	10 N 08	01 S 47	00 N 54	11 S 38	18 S 45

ZODIAC SIGN ENTRIES

Date	h	m	Planets
02	05	58	☽ ♉
04	17	09	☽ ♊
07	02	31	☽ ♋
09	09	35	☽ ♌
11	07	03	♂ ♉
11	13	59	☽ ♍
13	15	56	☽ ♎
15	16	31	☽ ♏
15	22	12	♀ ♉
15	23	18	☿ ♉
17	17	22	☽ ♐
19	20	16	☽ ♑
21	09	21	☉ ♊
22	02	31	☽ ♒
24	12	24	☽ ♓
27	00	36	☽ ♈
29	13	02	☽ ♉
31	23	56	☽ ♊

LATITUDES

Date	Mercury ☿ ° '	Venus ♀ ° '	Mars ♂ ° '	Jupiter ♃ ° '	Saturn ♄ ° '	Uranus ♅ ° '	Neptune ♆ ° '	Pluto ♇ ° '
01	02 S 33	01 S 40	00 S 43	01 S 05	02 N 43	00 S 43	00 S 31	04 N 31
04	02 51	01 40	00 42	01 05	02 42	00 43	00 31	04 31
07	03 03	01 40	00 40	01 05	02 42	00 43	00 31	04 32
10	03 08	01 40	00 38	01 05	02 41	00 43	00 31	04 32
13	03 07	01 39	00 37	01 06	02 41	00 43	00 31	04 32
16	03 01	01 37	00 35	01 06	02 40	00 43	00 31	04 32
19	02 49	01 34	00 33	01 06	02 39	00 43	00 31	04 32
22	02 32	01 31	00 31	01 06	02 39	00 43	00 31	04 31
25	02 11	01 28	00 30	01 06	02 38	00 43	00 31	04 31
28	01 45	01 24	00 28	01 07	02 38	00 43	00 31	04 31
31	01 S 16	01 S 19	00 S 26	01 S 07	02 N 37	00 S 43	00 S 31	04 N 31

DATA

Julian Date	2455683
Delta T	+68 seconds
Ayanamsa	24° 01' 11"
Synetic vernal point	05° ♓ 05' 48"
True obliquity of ecliptic	23° 26' 16"

MOON'S PHASES, APSIDES AND POSITIONS ☽

Date	h	m	Phase	Longitude ° '	Eclipse Indicator
03	06	51	●	12 ♉ 31	
10	20	33	☽	19 ♌ 51	
17	11	09	○	26 ♏ 13	
24	18	52	☾	03 ♓ 16	

Day	h	m		
15	11	31	Perigee	
27	10	04	Apogee	
06	03	54	Max dec	23° N 25'
12	20	06	0S	
18	23	26	Max dec	23° S 24'
26	00	01	0N	

LONGITUDES

Date	Chiron ⚷ ° '	Ceres ⚳ ° '	Pallas ⚴ ° '	Juno ⚵ ° '	Vesta ⚶ ° '	Black Moon Lilith ⚸ ° '
01	04 ♓ 47	14 ♓ 14	13 ♒ 13	14 ♍ 00	10 ♒ 41	04 ♈ 23
11	05 ♓ 06	17 ♓ 20	14 ♒ 37	14 ♍ 15	13 ♒ 38	05 ♈ 30
21	05 ♓ 19	20 ♓ 13	15 ♒ 30	15 ♍ 06	15 ♒ 06	06 ♈ 37
31	05 ♓ 27	22 ♓ 51	15 ♒ 50	16 ♍ 20	17 ♒ 58	07 ♈ 44

All ephemeris data is given at 12.00 UT and the Moon's longitude is additionally given for 24.00 UT

ASPECTARIAN

h m Aspects

01 Sunday
00 13 ☽ ☌ ☿
00 14 ♀ ☍ ♄
01 08 ☽ ∠ ♆
04 26 ♂ ☌ ♃
07 47 ☽ # ♆
14 51 ☽ ☌ ♃
15 20 ☽ ☌ ♂
18 54 ☿ ∠ ♆

02 Monday
05 23 ☽ ∥ ☉
07 16 ☽ ✶ ♆
11 33 ☽ ⊻ ♅
12 05 ☿ ∥ ♀
13 34 ☉ ⊼ ♄
20 39 ☽ △ ♇
23 32 ☽ ⊥ ♅

03 Tuesday
05 23 ☽ ⊼ ♄
05 50 ☽ # ♇
06 51 ☽ ☌ ☉
07 09 ☽ Q ♆
10 43 ☉ Q ♆
11 35 ☽ ⊻ ♀
15 33 ☽ ⊻ ☿
17 10 ☽ ± ♄
17 23 ☽ ∠ ♅

04 Wednesday
00 41 ☽ ⊥ ♀
02 13 ☽ BQ ♇
02 53 ♀ ∠ ♆
03 34 ☽ ⊻ ♃
04 07 ☽ ⊥ ☿
06 49 ☽ ⊻ ♂
10 43 ☽ BQ ♄
15 30 ☽ ⊥ ♃
18 30 ☽ □ ♆
19 17 ☽ ⊥ ♂
19 46 ☽ ± ♇
20 10 ☽ ∠ ♀
22 49 ☽ ✶ ♅
22 52 ☽ ∠ ☿

05 Thursday
07 22 ☽ ⊼ ♇
09 17 ☽ ∠ ♃
13 48 ☽ ∠ ♂
15 38 ☽ △ ♄
22 01 ☽ Q ♅
22 14 ☽ ⊻ ☉

06 Friday
04 09 ☽ ✶ ♀
05 49 ☽ ✶ ☿
10 37 ☽ ⊥ ☉
14 29 ☽ ✶ ♃
20 12 ☽ ✶ ♂

07 Saturday
03 53 ☽ △ ♆
05 00 ☽ ∠ ☉
05 13 ☽ Q ♀
06 15 ☽ Q ☿
08 11 ☽ □ ♅
13 26 ☽ Q ♃
16 10 ☽ ☍ ♇
20 02 ☽ Q ♂
23 56 ☽ □ ♄

08 Sunday
07 45 ☽ BQ ♆
09 41 ☉ ± ♄
11 00 ☽ ✶ ☉
17 52 ☽ □ ♀
18 10 ☽ □ ☿
22 52 ☽ # ♇
23 04 ☽ □ ♃

09 Monday
00 09 ☽ ± ♆
01 14 ☉ ∠ ♅
06 52 ☽ □ ♂
08 33 ☽ Q ♄
08 35 ☽ ∥ ☉
10 26 ☽ Q ☉
10 57 ☽ ⊼ ♆
15 10 ☽ △ ♅
15 44 ☿ ☌ ♀
22 33 ☽ ⊼ ♇

10 Tuesday
05 46 ☽ ✶ ♄
09 06 ☽ ± ♇
16 03 ☽ # ♆
17 39 ☽ BQ ♅
20 33 ☽ □ ☉
20 33 ☽ ∥ ♂

11 Wednesday
00 43 ☽ BQ ♇
03 48 ☽ △ ☿
04 07 ☽ △ ♀
04 52 ☽ △ ♃
07 00 ☽ ∥ ♃
07 42 ☽ ∠ ♄
09 12 ☽ ± ♅
09 43 ☽ ∥ ♀
14 22 ☽ △ ♂
14 42 ♀ ☌ ♃
15 19 ☽ ☍ ♆
15 46 ☽ ∥ ☿
19 27 ☽ ⊼ ♅
19 57 ☿ ☌ ♃
22 56 ☽ ⊥ ♄

12 Thursday
02 14 ☽ △ ♇
06 44 ☽ BQ ♃
07 31 ☽ BQ ☿
07 56 ☽ BQ ♀
08 28 ♂ ✶ ♆
08 59 ☽ ⊻ ♄
11 45 ☽ # ♄
16 59 ☽ BQ ♂
17 25 ☽ ∥ ♅
21 58 ☽ ± ♃
22 47 ☽ # ♅
23 50 ☽ ± ☿

13 Friday
00 17 ☽ ± ♀
01 36 ♀ ∥ ♃
02 52 ☽ △ ☉
04 21 ☽ ∥ ♄
07 38 ☉ BQ ♇
08 01 ☽ ⊼ ♃
08 39 ☽ ± ♂
10 37 ☽ ⊼ ☿
11 03 ☽ ⊼ ♀
17 16 ☽ ⊼ ♆
19 00 ☽ ⊼ ♂
21 20 ☽ ☍ ♅

14 Saturday
03 03 ☽ ± ♆
03 41 ☽ □ ♇
04 44 ☽ # ☿
04 47 ♂ # ♆
05 07 ☽ BQ ☉
09 19 ☽ # ♃
10 06 ☽ ☌ ♄
11 36 ☽ # ♀
17 39 ☽ BQ ♆
20 42 ☽ ± ☉
22 10 ☽ ∥ ♆
22 59 ☉ # ♇
23 07 ☽ # ♂

15 Sunday
07 03 ☽ ⊼ ☉
08 46 ☽ Q ♇
09 28 ☽ ☍ ♃
15 51 ☽ ☍ ☿
16 01 ☽ ☍ ♀
17 52 ☽ △ ♆
19 15 ♂ ⊻ ♅
22 00 ☽ ⊼ ♅
22 08 ☽ ☍ ♂

16 Monday
04 07 ☽ ✶ ♇
07 44 ☽ ± ♅
09 25 ☿ ☌ ♀
10 26 ☽ ⊻ ♄
11 53 ☽ ∥ ♇
14 16 ☿ ✶ ♆
14 27 ☽ # ☉
14 52 ♀ ✶ ♆
16 26 ☿ ∥ ♃
20 09 ☽ ⊥ ♄
22 23 ☽ BQ ♅

17 Tuesday
04 29 ☽ ∠ ♇
06 35 ☉ BQ ♄
07 00 ☉ ⊻ ♃
10 50 ☽ ∠ ♄
10 56 ☽ ⊼ ♃
11 09 ☽ ☍ ☉
18 47 ☽ □ ♆
19 16 ☽ ⊥ ♇
21 00 ☽ ± ♃
21 19 ☽ ⊼ ♀
21 47 ☽ ⊼ ☿
23 08 ☽ △ ♅

18 Wednesday
01 49 ☽ ⊼ ♂
05 14 ☽ ⊻ ♇
08 15 ☽ ± ♀
08 54 ☽ ± ☿
11 41 ☽ ✶ ♄
11 42 ☿ ⊻ ♅
12 16 ☽ BQ ♃
12 25 ☽ ± ♂
19 30 ♀ ⊻ ♅

19 Thursday
00 53 ☽ BQ ♀
01 02 ☽ Q ♆
01 51 ☽ BQ ☿
04 33 ☽ BQ ♂
08 02 ☽ Q ♄
14 17 ☽ △ ♃
17 28 ☽ ⊼ ☉
21 23 ♀ # ♆
21 47 ☽ ✶ ♆

20 Friday
02 30 ☽ □ ♅
04 47 ☽ ± ☉
05 26 ☽ △ ♀
07 05 ☽ △ ☿
08 12 ☽ △ ♂
08 46 ☽ ☌ ♇
15 33 ☽ □ ♄
18 22 ♂ △ ♇
21 54 ☿ △ ♇
22 06 ☽ BQ ☉

21 Saturday
00 31 ☽ ∠ ♆
01 20 ☿ ☌ ♂
04 06 ☽ # ☉
10 42 ☿ # ♆
10 58 ☽ Q ♅
15 02 ☽ ∥ ♇
17 00 ☽ ⊥ ♆
18 04 ♀ △ ♇
21 04 ☽ □ ♃

22 Sunday
03 55 ☽ △ ☉
04 11 ☽ ⊻ ♆
07 31 ☉ □ ♆
09 22 ☽ ✶ ♅
11 50 ☉ ± ♇
13 09 ☿ ⊥ ♅
15 50 ☽ ⊻ ♇
18 12 ☽ □ ♀
18 46 ☽ □ ♂
21 53 ☽ □ ☿
22 23 ☽ # ♂
23 01 ☽ △ ♄

23 Monday
03 18 ☽ ⊥ ♇
03 43 ☽ # ♀
05 21 ☽ # ☿
06 30 ☿ ⊼ ♄
07 49 ☽ Q ♃
08 25 ♀ ☌ ♂
11 20 ☽ ∥ ♆
14 11 ☽ ∠ ♅
20 43 ☽ ∠ ♇
21 48 ♀ ⊥ ♅
22 40 ☽ # ♃

24 Tuesday
04 04 ☽ BQ ♄
06 44 ♂ ⊥ ♅
07 40 ☽ ✶ ♃
07 52 ☽ ⊥ ♅
07 56 ☽ Q ♂
08 57 ☽ Q ♀
13 05 ☿ Q ♆
14 11 ☽ ☌ ♆
14 21 ☽ Q ☿
16 44 ☿ ∥ ♀
18 47 ♀ ⊼ ♄
18 52 ☽ □ ☉
19 47 ☽ ⊻ ♅
21 50 ☽ ± ♄

25 Wednesday
02 20 ☽ ✶ ♇
06 47 ☉ ✶ ♅
09 18 ☉ ⊥ ♃
09 22 ☽ ✶ ♂
09 47 ☽ ⊼ ♄
11 31 ☽ ✶ ♀
14 05 ☽ ∠ ♃
14 46 ☽ ∥ ♄
15 49 ♂ ⊼ ♄
18 15 ☽ ✶ ☿
19 47 ☽ # ♅

26 Thursday
02 22 ☽ Q ♇
04 18 ☽ ∥ ♅
08 33 ☽ ⊥ ♃
09 17 ☽ # ♄
10 08 ☽ Q ☉
13 19 ♀ Q ♆
14 34 ☿ ∥ ♂
17 35 ☽ ∠ ♂
18 22 ☿ ± ♄
20 54 ☽ ⊻ ♃
21 11 ☽ ∠ ♀

27 Friday
02 27 ☽ ⊻ ♆
05 43 ☽ ∠ ☿
08 19 ☽ ☌ ♅
12 32 ☽ ✶ ☉
12 59 ☽ ⊥ ♂
14 36 ☽ ⊥ ♆
14 41 ☽ □ ♇
17 28 ☽ ⊥ ♀
22 10 ☽ ☍ ♄
23 16 ☿ ∠ ♅

28 Saturday
01 55 ☽ ⊻ ♂
03 08 ☽ ⊥ ☿
03 56 ☽ ∥ ♃
06 58 ☽ ⊻ ♀
08 46 ☽ ∠ ♆
13 52 ☽ # ♆
14 38 ☉ ⊼ ♇
14 52 ♂ Q ♆
17 24 ☽ ⊻ ☿
21 26 ☽ ∠ ☉

29 Sunday
10 28 ☽ ☌ ♃
13 21 ☽ ∥ ♀
13 29 ☽ ∥ ♂
14 33 ♀ ± ♄
14 52 ☽ ✶ ♆
15 44 ☿ BQ ♇
16 40 ♀ ∥ ♂
17 00 ☽ ⊥ ☉
20 47 ☽ ⊻ ♅
23 08 ☽ ∥ ☿

30 Monday
02 46 ☽ △ ♇
05 57 ☽ ⊻ ☉
08 43 ☽ ⊥ ♅
10 05 ☽ ⊼ ♄
12 46 ☽ # ♇
14 39 ☽ Q ♆
17 43 ☽ ☌ ♂
21 52 ☽ ± ♄

31 Tuesday
01 23 ☽ ☌ ♀
02 27 ☽ ∠ ♅
08 12 ☽ BQ ♇
12 25 ♀ ∠ ♅
14 07 ☿ BQ ♄
15 23 ☽ BQ ♄
15 37 ☽ ☌ ☿
20 34 ☽ ∥ ☉
22 28 ☽ ⊻ ♃

JUNE 2011

Date	Sidereal time h m s	LONGITUDES Sun ⊙ ° ' "	Moon ☽ ° ' "	Moon ☽ 24.00 ° ' "	Mercury ☿ ° '	Venus ♀ ° '	Mars ♂ ° '	Jupiter ♃ ° '	Saturn ♄ ° '	Uranus ♅ ° '	Neptune ♆ ° '	Pluto ♇ ° '
01	04 38 30	10 ♊ 40 12	06 ♊ 17 06	12 ♊ 35 23	27 ♉ 21	20 ♉ 08	15 ♉ 47	29 ♈ 22	10 ♎ 33	03 ♈ 59	00 ♓ 56	06 ♑ 51
02	04 42 26	11 37 43	18 ♊ 56 58	25 ♊ 21 52	29 ♉ 20	21 21	16 31	29 34	10 R 32	04 00	00 56	06 R 50
03	04 46 23	12 35 14	01 ♋ 50 10	08 ♋ 21 51	01 ♊ 21	22 34	17 15	29 47	10 31	04 02	00 R 56	06 49
04	04 50 20	13 32 43	14 ♋ 56 57	21 ♋ 35 26	03 23	23 48	17 59	29 ♈ 59	10 30	04 04	00 56	06 47
05	04 54 16	14 30 11	28 ♋ 17 16	05 ♌ 02 27	05 28	25 01	18 42	00 ♉ 11	10 30	04 05	00 56	06 46
06	04 58 13	15 27 38	11 ♌ 50 55	18 ♌ 42 35	07 34	26 14	19 26	00 24	10 29	04 07	00 55	06 44
07	05 02 09	16 25 04	25 ♌ 37 23	02 ♍ 35 13	09 42	27 27	20 10	00 36	10 28	04 09	00 55	06 43
08	05 06 06	17 22 28	09 ♍ 35 56	16 ♍ 39 20	11 51	28 40	20 54	00 48	10 28	04 10	00 55	06 42
09	05 10 02	18 19 51	23 ♍ 45 12	00 ♎ 53 17	14 01	29 ♉ 53	21 37	01 00	10 27	04 11	00 55	06 40
10	05 13 59	19 17 14	08 ♎ 03 13	15 ♎ 14 38	16 12	01 ♊ 06	22 21	01 12	10 27	04 13	00 55	06 39
11	05 17 55	20 14 35	22 ♎ 27 04	29 ♎ 40 02	18 23	02 19	23 04	01 23	10 27	04 14	00 55	06 37
12	05 21 52	21 11 55	06 ♏ 53 00	14 ♏ 05 22	20 35	03 32	23 48	01 35	10 27	04 16	00 54	06 36
13	05 25 49	22 09 14	21 ♏ 16 32	28 ♏ 25 55	22 47	04 45	24 31	01 47	10 D 27	04 17	00 54	06 34
14	05 29 45	23 06 33	05 ♐ 32 54	12 ♐ 36 56	24 59	05 58	25 15	01 58	10 27	04 18	00 54	06 33
15	05 33 42	24 03 50	19 ♐ 37 28	26 ♐ 34 02	27 11	07 11	25 58	02 10	10 27	04 19	00 53	06 31
16	05 37 38	25 01 07	03 ♑ 26 16	10 ♑ 13 51	29 ♊ 21	08 24	26 41	02 21	10 27	04 20	00 53	06 30
17	05 41 35	25 58 24	16 ♑ 56 34	23 ♑ 34 16	01 ♋ 31	09 38	27 24	02 33	10 28	04 22	00 52	06 28
18	05 45 31	26 55 40	00 ♒ 06 56	06 ♒ 34 38	03 39	10 51	28 07	02 44	10 28	04 23	00 52	06 27
19	05 49 28	27 52 55	12 ♒ 57 30	19 ♒ 15 46	05 47	12 04	28 50	02 55	10 29	04 24	00 51	06 25
20	05 53 24	28 50 10	25 ♒ 29 45	01 ♓ 39 47	07 52	13 17	29 ♉ 33	03 06	10 29	04 25	00 51	06 24
21	05 57 21	29 ♊ 47 25	07 ♓ 46 19	13 ♓ 49 48	09 56	14 30	00 ♊ 16	03 17	10 30	04 26	00 50	06 22
22	06 01 18	00 ♋ 44 40	19 ♓ 50 46	25 ♓ 49 44	11 58	15 44	00 59	03 28	10 31	04 26	00 50	06 21
23	06 05 14	01 41 54	01 ♈ 47 16	07 ♈ 43 57	13 58	16 57	01 42	03 38	10 32	04 27	00 49	06 19
24	06 09 11	02 39 09	13 ♈ 40 21	19 ♈ 37 03	15 57	18 10	02 25	03 49	10 33	04 28	00 48	06 18
25	06 13 07	03 36 23	25 ♈ 34 36	01 ♉ 33 35	17 53	19 23	03 08	04 00	10 34	04 29	00 48	06 16
26	06 17 04	04 33 37	07 ♉ 34 30	13 ♉ 37 52	19 47	20 37	03 50	04 10	10 36	04 29	00 47	06 15
27	06 21 00	05 30 51	19 ♉ 44 08	25 ♉ 53 43	21 39	21 50	04 33	04 20	10 37	04 30	00 46	06 13
28	06 24 57	06 28 05	02 ♊ 06 58	08 ♊ 24 13	23 29	23 03	05 15	04 31	10 38	04 31	00 46	06 12
29	06 28 53	07 25 19	14 ♊ 45 41	21 ♊ 11 32	25 16	24 17	05 58	04 41	10 40	04 31	00 45	06 10
30	06 32 50	08 ♋ 22 33	27 ♊ 41 53	04 ♋ 16 44	27 ♋ 02	25 ♊ 30	06 ♊ 40	04 ♉ 51	10 ♎ 42	04 ♈ 32	00 ♓ 44	06 ♑ 09

Date	Moon True ☊ ° '	Moon Mean ☊ ° '	Moon ☽ Latitude ° '
01	23 ♐ 25	24 ♐ 17	01 N 33
02	23 R 24	24 14	00 N 25
03	23 D 24	24 10	00 S 47
04	23 25	24 07	01 56
05	23 26	24 04	03 01
06	23 26	24 01	03 57
07	23 27	23 58	04 40
08	23 27	23 55	05 08
09	23 R 27	23 51	05 17
10	23 27	23 48	05 07
11	23 27	23 45	04 37
12	23 27	23 42	03 50
13	23 26	23 39	02 49
14	23 26	23 36	01 38
15	23 D 26	23 32	00 S 21
16	23 R 26	23 29	00 N 55
17	23 26	23 26	02 07
18	23 26	23 23	03 10
19	23 25	23 20	04 01
20	23 25	23 16	04 40
21	23 24	23 13	05 05
22	23 24	23 10	05 16
23	23 D 24	23 07	05 13
24	23 24	23 04	04 57
25	23 24	23 01	04 28
26	23 25	22 57	03 47
27	23 26	22 54	02 56
28	23 27	22 51	01 55
29	23 28	22 48	00 48
30	23 ♐ 28	22 ♐ 45	00 S 23

DECLINATIONS

Date	Sun ⊙ ° '	Moon ☽ ° '	Mercury ☿ ° '	Venus ♀ ° '	Mars ♂ ° '	Jupiter ♃ ° '	Saturn ♄ ° '	Uranus ♅ ° '	Neptune ♆ ° '	Pluto ♇ ° '
01	22 N 03	22 N 53	18 N 30	16 N 32	16 N 10	10 N 12	01 S 47	00 N 55	11 S 38	18 S 45
02	22 11	23 23	19 06	16 53	16 23	10 16	01 46	00 56	11 38	18 45
03	22 18	22 39	19 42	17 13	16 36	10 21	01 46	00 56	11 38	18 45
04	22 25	20 40	20 16	17 33	16 49	10 25	01 46	00 57	11 38	18 45
05	22 32	17 32	20 50	17 53	17 02	10 29	01 46	00 58	11 38	18 45
06	22 39	13 26	21 22	18 12	17 14	10 33	01 46	00 58	11 38	18 45
07	22 45	08 34	21 52	18 30	17 27	10 37	01 46	00 59	11 38	18 45
08	22 50	03 N 13	22 21	18 48	17 39	10 41	01 46	00 59	11 38	18 46
09	22 55	02 S 22	22 48	19 06	17 51	10 45	01 46	01 00	11 38	18 46
10	23 00	07 53	23 13	19 23	18 03	10 50	01 46	01 00	11 39	18 46
11	23 05	13 01	23 35	19 40	18 14	10 53	01 46	01 01	11 39	18 46
12	23 09	17 26	23 55	19 56	18 26	10 57	01 46	01 01	11 39	18 46
13	23 12	20 48	24 13	20 12	18 37	11 01	01 47	01 02	11 39	18 46
14	23 15	22 50	24 27	20 27	18 48	11 05	01 47	01 02	11 39	18 46
15	23 18	23 23	24 39	20 41	18 59	11 09	01 47	01 03	11 39	18 46
16	23 21	22 28	24 48	20 55	19 10	11 13	01 47	01 03	11 39	18 47
17	23 23	20 16	24 55	21 08	19 20	11 17	01 48	01 04	11 40	18 47
18	23 24	17 02	24 58	21 21	19 31	11 20	01 48	01 04	11 40	18 47
19	23 25	13 04	24 59	21 33	19 41	11 24	01 49	01 04	11 40	18 47
20	23 26	08 37	24 57	21 45	19 51	11 28	01 49	01 05	11 40	18 47
21	23 26	03 S 56	24 52	21 56	20 01	11 31	01 50	01 05	11 40	18 47
22	23 26	00 N 50	24 45	22 07	20 10	11 35	01 50	01 05	11 41	18 47
23	23 26	05 30	24 35	22 16	20 20	11 38	01 51	01 06	11 41	18 48
24	23 25	09 57	24 24	22 26	20 29	11 42	01 52	01 06	11 41	18 48
25	23 23	14 03	24 10	22 34	20 38	11 45	01 52	01 06	11 41	18 48
26	23 22	17 37	23 54	22 42	20 47	11 48	01 53	01 06	11 42	18 48
27	23 19	20 29	23 36	22 49	20 55	11 52	01 54	01 07	11 42	18 48
28	23 17	22 28	23 16	22 56	21 04	11 55	01 55	01 07	11 42	18 48
29	23 14	23 22	22 55	23 02	21 12	11 58	01 55	01 07	11 43	18 49
30	23 N 10	23 N 02	22 N 33	23 N 07	21 N 20	12 N 01	01 S 56	01 N 07	11 S 43	18 S 49

ZODIAC SIGN ENTRIES

Date	h m	Planets
02	20 02	☿ ♊
03	08 36	☽ ♋
04	13 56	♃ ♉
05	15 03	☽ ♌
07	19 33	☽ ♍
09	14 23	♀ ♊
09	22 31	☽ ♎
12	00 33	☽ ♏
14	02 38	☽ ♐
16	05 59	☽ ♑
16	19 09	☿ ♋
18	11 47	☽ ♒
20	20 45	☽ ♓
21	02 50	♂ ♊
21	17 16	⊙ ♋
23	08 24	☽ ♈
25	20 53	☽ ♉
28	07 56	☽ ♊
30	16 13	☽ ♋

LATITUDES

Date	Mercury ☿ ° '	Venus ♀ ° '	Mars ♂ ° '	Jupiter ♃ ° '	Saturn ♄ ° '	Uranus ♅ ° '	Neptune ♆ ° '	Pluto ♇ ° '
01	01 S 06	01 S 17	00 S 25	01 S 07	02 N 37	00 S 43	00 S 31	04 N 31
04	00 34	01 12	00 23	01 07	02 36	00 44	00 31	04 31
07	00 S 02	01 07	00 21	01 08	02 35	00 44	00 32	04 31
10	00 N 30	01 01	00 19	01 08	02 35	00 44	00 32	04 31
13	00 58	00 54	00 17	01 08	02 34	00 44	00 32	04 31
16	01 22	00 48	00 15	01 09	02 33	00 44	00 32	04 30
19	01 40	00 41	00 13	01 09	02 32	00 44	00 32	04 30
22	01 51	00 34	00 11	01 10	02 32	00 44	00 32	04 30
25	01 56	00 27	00 09	01 10	02 31	00 44	00 32	04 30
28	01 54	00 20	00 07	01 10	02 30	00 44	00 32	04 29
31	01 N 47	00 S 12	00 S 05	01 S 11	02 N 29	00 S 44	00 S 32	04 N 29

DATA

Julian Date	2455714
Delta T	+68 seconds
Ayanamsa	24° 01' 16"
Synetic vernal point	05° ♓ 05' 43"
True obliquity of ecliptic	23° 26' 15"

MOON'S PHASES, APSIDES AND POSITIONS ☽

Date	h m	Phase	Longitude ° '	Eclipse Indicator
01	21 03	●	11 ♊ 02	Partial
09	02 11	☽	17 ♍ 56	
15	20 14	○	24 ♐ 23	total
23	11 48	☾	01 ♈ 41	

Day	h m		
12	01 47	Perigee	
24	04 14	Apogee	
02	09 52	Max dec	23° N 23'
09	01 52	0S	
15	08 53	Max dec	23° S 24'
22	07 49	0N	
29	17 44	Max dec	23° N 24'

LONGITUDES

Date	Chiron ⚷ ° '	Ceres ⚳ ° '	Pallas ⚴ ° '	Juno ⚵ ° '	Vesta ⚶ ° '	Black Moon Lilith ⚸ ° '
01	05 ♓ 28	23 ♓ 06	15 ♒ 50	16 ♍ 29	18 ♒ 08	07 ♈ 51
11	05 ♓ 29	25 ♓ 24	15 ♒ 29	18 ♍ 14	19 ♒ 16	08 ♈ 58
21	05 ♓ 24	27 ♓ 21	14 ♒ 30	20 ♍ 18	19 ♒ 40	10 ♈ 05
31	05 ♓ 14	28 ♓ 55	12 ♒ 54	22 ♍ 40	19 ♒ 16	11 ♈ 13

All ephemeris data is given at 12.00 UT and the Moon's longitude is additionally given for 24.00 UT

ASPECTARIAN

h m Aspects

01 Wednesday
01 36 ☽ ± ♇
01 43 ☽ □ ♆
07 34 ☽ ✶ ♅
09 15 ⊙ △ ♄
10 12 ☽ ⊥ ♃
13 05 ☽ ⚻ ♇
20 08 ☽ △ ♄
21 03 ☽ ☌ ⊙
21 55 ☿ ǂ ♇

02 Thursday
03 37 ☽ ∠ ♃
06 27 ☽ Q ♅
07 08 ☽ ⚺ ♂
12 56 ♂ ± ♄
15 13 ☿ ⚺ ♃
16 59 ☽ ⚺ ♀
19 05 ☽ ⊥ ♂
21 10 ♀ ꝑ ♇

03 Friday
05 20 ☽ ⊥ ♀
05 45 ☿ ± ♇
07 05 ☿ □ ♆
07 30 ♆ St R
08 08 ☽ ✶ ♃
10 19 ☽ △ ♆
10 55 ☽ ⚺ ☿
12 48 ☽ ∠ ♂
16 04 ☽ □ ♅
17 07 ☽ ∥ ⊙
21 08 ☽ ☍ ♇
23 38 ☽ ∠ ♀

04 Saturday
00 00 ☽ ⊥ ☿
03 56 ☽ □ ♄
06 31 ☽ Q ♃
09 15 ☽ ⚺ ⊙
13 46 ☽ ꝑ ♆
14 59 ☽ ∥ ☿
17 48 ☽ ✶ ♂
19 22 ☽ ∠ ☿
19 57 ☿ ✶ ♅
20 58 ☽ ⊥ ⊙

05 Sunday
03 38 ☽ ǂ ♇
05 33 ☽ ✶ ♀
05 59 ☽ ± ♆
09 57 ☽ ∥ ♀
12 22 ☽ Q ♄
14 20 ☽ ∠ ⊙
15 09 ☽ ∥ ♂
15 26 ☽ □ ♃
16 33 ☽ Q ♂
16 42 ☽ ⚻ ♆
21 13 ☿ ⊥ ♃
21 27 ♀ ꝑ ♄
22 20 ☽ △ ♅

06 Monday
01 02 ♂ ∠ ♅
02 43 ☿ ⚻ ♇
03 01 ☽ ⚻ ♇
03 04 ☽ ✶ ☿
05 00 ☽ Q ♀
09 36 ☽ ✶ ♄
09 50 ⊙ ∠ ♃
13 34 ☽ ± ♇
18 48 ☽ ✶ ⊙
21 15 ☽ ǂ ♆

07 Tuesday
00 44 ☽ ꝑ ♅
02 00 ☽ □ ♂
02 23 ☽ ∥ ♃
03 57 ☽ Q ☿
04 54 ⊙ Q ♅
05 15 ☽ ꝑ ♇
11 44 ☽ ∠ ♄
15 27 ☽ □ ♀
16 21 ☽ ± ♅
17 11 ☽ Q ⊙
20 40 ☿ △ ♄
20 42 ☽ △ ♃
21 08 ☽ ☍ ♆

08 Wednesday
02 42 ☽ ⚻ ♅
03 13 ☽ ⊥ ♄
07 03 ☽ △ ♇
08 05 ♀ ǂ ♇
13 28 ☽ ⚺ ♄
16 31 ☽ □ ☿
18 17 ☽ ǂ ♄
21 36 ☽ ∥ ♅
22 42 ☽ ꝑ ♃

09 Thursday
02 11 ☽ □ ⊙
02 39 ♃ ✶ ♆
06 07 ☽ ǂ ♅
08 13 ☽ △ ♂
09 25 ☽ ∥ ♄
13 31 ♂ ꝑ ♇
14 07 ☽ ± ♃
20 35 ⊙ ∥ ☿
23 16 ☽ △ ♀

10 Friday
00 03 ☽ ⚻ ♆
00 21 ☽ ⚻ ♃
03 17 ♀ ± ♇
05 34 ☽ ☍ ♅
08 24 ♀ □ ♆
09 39 ☽ □ ♇
10 06 ☽ ± ♆
10 45 ☽ ꝑ ♂
11 59 ☿ ∠ ♃
12 13 ☿ Q ♅
14 18 ♀ ⚺ ♃
16 00 ☽ ☌ ♄

11 Saturday
01 07 ☽ ꝑ ♆
01 35 ☽ ǂ ♃
02 34 ☽ ± ♂
02 40 ☽ ꝑ ♀
04 02 ☽ △ ☿
05 17 ☽ ∥ ♆
08 04 ☽ △ ⊙
13 05 ☽ ⚻ ♂
15 36 ☽ Q ♇
19 01 ☽ ± ♀

12 Sunday
02 04 ☽ △ ♆
03 04 ☽ ☍ ♃
05 55 ☽ ⚻ ♀
07 38 ☽ ⚻ ♅
09 28 ☽ ꝑ ☿
10 47 ☽ ꝑ ⊙
11 31 ☽ ✶ ♇
17 38 ☽ ± ♅
17 56 ☽ ⚺ ♄
18 38 ☽ ǂ ♂
20 33 ☽ ∥ ♇
23 44 ⊙ ☌ ☿

13 Monday
02 36 ♀ ✶ ♅
02 50 ☽ ± ⊙
03 09 ☽ ± ☿
03 52 ♄ St D
03 56 ☽ ⊥ ♄
06 30 ☽ ǂ ♀
08 40 ☽ ꝑ ♅
12 30 ☽ ∠ ♇
13 34 ☽ ⚻ ⊙
15 00 ☽ ⚻ ☿
17 43 ☽ ☍ ♂
18 59 ☽ ∠ ♄

14 Tuesday
03 35 ☽ ⊥ ♇
04 09 ☽ □ ♆
05 53 ☽ ⚻ ♃
07 45 ♂ ǂ ♇
09 53 ☽ △ ♅
12 47 ☽ ☍ ♀
13 41 ☽ ⚺ ♇
16 10 ☽ ± ♃
16 11 ☿ ⚺ ♂
18 44 ♂ ꝑ ♄
20 18 ☽ ✶ ♄
23 10 ♀ ⚻ ♇
23 28 ☽ ǂ ⊙

15 Wednesday
07 43 ☽ ꝑ ♃
10 44 ☽ Q ♆
16 52 ☽ Q ♄
16 54 ☽ ǂ ⊙
20 14 ☽ ☍ ⊙
23 33 ☽ ⚻ ♂

16 Thursday
03 31 ☽ ☍ ☿
07 31 ☽ ✶ ♆
10 04 ☽ △ ♃
10 36 ☽ ± ♂
10 48 ♀ ⊥ ♃
13 35 ☽ □ ♅
17 23 ☽ ☌ ♇
21 38 ☽ ⚻ ♀

17 Friday
00 24 ☽ □ ♄
03 25 ☽ ꝑ ♂
04 35 ☽ ǂ ♀
04 52 ☿ △ ♆
09 24 ☽ ± ♀
10 05 ☽ ∠ ♆
19 12 ☽ ǂ ♂
21 49 ☽ Q ♅
23 52 ☽ ∥ ♇

18 Saturday
00 35 ☿ ✶ ♃
02 22 ☽ ⊥ ♆
03 21 ☽ ꝑ ♀
04 31 ♀ △ ♄
05 40 ☽ ⚻ ⊙
08 07 ☽ △ ♂
13 23 ☽ ⚺ ♆
16 55 ☽ □ ♃
17 37 ☽ ± ⊙
19 51 ☽ ⚻ ☿
19 55 ☽ ✶ ♅
19 56 ☿ ⊥ ♂
20 11 ☿ □ ♅
23 44 ☽ ⚺ ♇

19 Sunday
07 19 ☽ △ ♄
09 19 ☽ ± ☿
10 08 ☽ △ ♀
10 59 ☽ ⊥ ♇
11 51 ☽ ꝑ ⊙
19 18 ☿ ☍ ♇
19 46 ☽ ∥ ♆
21 06 ☽ ǂ ♃

20 Monday
00 16 ☽ ∠ ♅
03 24 ☽ Q ♃
04 07 ☽ ∠ ♇
05 55 ☽ ꝑ ☿
11 59 ☽ ꝑ ♄
17 40 ☽ ⊥ ♅
19 02 ☽ △ ⊙
20 23 ☽ □ ♂
22 24 ☽ ☌ ♆

21 Tuesday
03 02 ☽ ✶ ♃
05 24 ☽ ⚺ ♅
05 33 ☽ ± ♄
09 15 ☽ ✶ ♇
15 15 ♂ ± ♇
17 09 ☽ △ ☿
17 24 ☽ ⚻ ♄
18 42 ☿ □ ♄
22 35 ☽ ∥ ♄

22 Wednesday
02 20 ☽ ǂ ♅
02 51 ☽ □ ♀
06 45 ♂ □ ♆
09 01 ☽ Q ♇
09 11 ☽ ∠ ♃
10 11 ☽ Q ♂
13 20 ☽ ∥ ♅
14 07 ⊙ △ ♆
17 09 ☽ ǂ ♄

23 Thursday
02 12 ♀ Q ♅
03 31 ☽ ⊥ ♃
10 03 ☽ ⚺ ♆
11 48 ☽ □ ⊙
11 49 ☽ ✶ ♂
12 19 ⊙ ⚺ ♂
15 48 ☽ ⚺ ♃
17 23 ☽ ☌ ♅
19 06 ☽ Q ♀
21 08 ☽ □ ♇
22 09 ☽ ⊥ ♆

24 Friday
05 41 ☽ ☍ ♄
08 08 ♃ ǂ ♆
10 18 ☿ Q ♃
10 21 ☿ ꝑ ♆
16 19 ☽ ∠ ♆
17 29 ☽ □ ☿
20 02 ☽ ∠ ♂
21 51 ☽ ǂ ♆
22 02 ☽ ∥ ♃
22 07 ☽ ✶ ♀

25 Saturday
02 55 ♀ ∠ ♃
03 19 ☽ Q ⊙
15 19 ☽ ⊥ ♂
16 54 ☿ ∠ ♂
22 28 ☽ ✶ ♆
23 54 ⊙ ✶ ♃

26 Sunday
04 05 ☽ ⚺ ♂
05 07 ☽ ☌ ♃
05 29 ☽ ✶ ⊙
05 51 ☽ ⚺ ♅
07 39 ☽ ∠ ♀
09 22 ☽ △ ♇
10 13 ⊙ □ ♅
12 29 ☽ Q ☿
17 48 ☽ ⊥ ♅
18 00 ☽ ⚻ ♄
21 08 ☽ ǂ ♇
22 19 ☽ Q ♆

27 Monday
02 43 ♂ ⚺ ♃
03 29 ☽ ⊥ ♀
05 52 ☽ ± ♄
10 26 ♂ ✶ ♅
11 32 ☽ ∠ ♅
13 39 ☽ ∠ ⊙
14 54 ☽ ꝑ ♇
16 24 ☽ ✶ ☿
16 33 ☽ ⚺ ♀
16 38 ☽ ∥ ♂
19 07 ☿ ⚺ ♀
23 29 ☽ ꝑ ♄

28 Tuesday
05 19 ⊙ ☍ ♇
08 19 ☽ ± ♇
08 34 ☽ ⊥ ⊙
09 24 ☽ □ ♆
11 38 ⊙ ∥ ☿
12 20 ♃ ⚺ ♅
16 35 ☽ ✶ ♅
16 39 ☽ ⚺ ♃
18 22 ☽ ☌ ♂
19 47 ☽ ⚻ ♇
21 00 ☽ ⚺ ⊙
21 47 ☽ ∥ ♀

29 Wednesday
00 49 ☽ ∥ ☿
02 10 ☽ ∠ ☿
04 12 ☽ ⊥ ♃
04 16 ☽ △ ♄
04 57 ☿ ± ♆
05 47 ☽ ∥ ⊙
06 23 ☿ ∥ ♀
15 18 ☽ Q ♅
18 47 ♂ ⚻ ♇
21 18 ☽ ∠ ♃
21 47 ☽ ⊥ ☿

30 Thursday
07 33 ☽ ☌ ♀
07 40 ☽ ∥ ⊙
09 52 ☽ ∥ ♀
10 36 ☽ ⚺ ☿
17 33 ☽ △ ♆
21 23 ⊙ ∥ ♀

JULY 2011

LONGITUDES

Date	Sidereal time h m s	Sun ☉ ° ' "	Moon ☽ ° ' "	Moon ☽ 24.00 ° ' "	Mercury ☿ ° '	Venus ♀ ° '	Mars ♂ ° '	Jupiter ♃ ° '	Saturn ♄ ° '	Uranus ♅ ° '	Neptune ♆ ° '	Pluto ♇ ° '
01	06 36 47	09 ♋ 19 47	10 ♋ 56 01	17 ♋ 39 37	28 ♋ 45	26 ♊ 43	07 ♊ 23	05 ♉ 00	10 ♎ 43	04 ♈ 32	00 ♓ 43	06 ♑ 07
02	06 40 43	10 17 01	24 ♋ 27 19	01 ♌ 18 49	00 ♌ 27	27 57	08 05	05 10	10 45	04 33	00 R 42	06 R 06
03	06 44 40	11 14 15	08 ♌ 13 49	15 ♌ 11 55	02 06	29 ♊ 10	08 47	05 20	10 47	04 33	00 41	06 04
04	06 48 36	12 11 28	22 ♌ 12 42	29 ♌ 15 43	03 43	00 ♋ 24	09 29	05 29	10 49	04 33	00 40	06 03
05	06 52 33	13 08 41	06 ♍ 20 31	13 ♍ 26 39	05 17	01 37	10 11	05 39	10 51	04 33	00 40	06 01
06	06 56 29	14 05 54	20 ♍ 33 42	27 ♍ 41 14	06 50	02 51	10 53	05 48	10 54	04 34	00 39	06 00
07	07 00 26	15 03 06	04 ♎ 48 52	11 ♎ 56 16	08 20	04 04	11 35	05 57	10 56	04 34	00 38	05 58
08	07 04 22	16 00 18	19 ♎ 03 03	26 ♎ 09 03	09 48	05 17	12 17	06 06	10 59	04 34	00 37	05 57
09	07 08 19	16 57 30	03 ♏ 13 53	10 ♏ 17 21	11 14	06 31	12 59	06 15	11 01	04 34	00 35	05 55
10	07 12 16	17 54 42	17 ♏ 19 12	24 ♏ 19 15	12 37	07 44	13 41	06 24	11 04	04 R 34	00 34	05 54
11	07 16 12	18 51 54	01 ♐ 17 16	08 ♐ 13 02	13 58	08 58	14 22	06 32	11 06	04 34	00 33	05 52
12	07 20 09	19 49 06	15 ♐ 06 21	21 ♐ 56 59	15 17	10 12	15 04	06 41	11 09	04 34	00 32	05 51
13	07 24 05	20 46 18	28 ♐ 44 44	05 ♑ 29 24	16 33	11 25	15 46	06 49	11 12	04 34	00 31	05 49
14	07 28 02	21 43 30	12 ♑ 10 46	18 ♑ 48 41	17 47	12 39	16 27	06 58	11 15	04 33	00 30	05 48
15	07 31 58	22 40 42	25 ♑ 23 00	01 ♒ 53 35	18 59	13 52	17 09	07 06	11 18	04 33	00 29	05 46
16	07 35 55	23 37 55	08 ♒ 20 23	14 ♒ 43 21	20 07	15 06	17 50	07 14	11 21	04 33	00 27	05 45
17	07 39 51	24 35 08	21 ♒ 02 33	27 ♒ 18 04	21 14	16 20	18 31	07 21	11 25	04 33	00 26	05 43
18	07 43 48	25 32 21	03 ♓ 30 02	09 ♓ 38 40	22 17	17 33	19 13	07 29	11 28	04 32	00 25	05 42
19	07 47 45	26 29 35	15 ♓ 44 16	21 ♓ 47 08	23 17	18 47	19 54	07 37	11 31	04 32	00 24	05 41
20	07 51 41	27 26 49	27 ♓ 47 41	03 ♈ 46 20	24 15	20 01	20 35	07 44	11 35	04 31	00 22	05 39
21	07 55 38	28 24 05	09 ♈ 43 36	15 ♈ 39 58	25 09	21 14	21 16	07 51	11 38	04 31	00 21	05 38
22	07 59 34	29 ♋ 21 21	21 ♈ 36 02	27 ♈ 32 21	26 01	22 28	21 57	07 58	11 42	04 30	00 20	05 36
23	08 03 31	00 ♌ 18 38	03 ♉ 29 33	09 ♉ 28 13	26 49	23 42	22 38	08 05	11 46	04 30	00 18	05 35
24	08 07 27	01 15 55	15 ♉ 29 00	21 ♉ 32 30	27 33	24 56	23 19	08 12	11 50	04 29	00 17	05 34
25	08 11 24	02 13 14	27 ♉ 39 19	03 ♊ 49 59	28 14	26 09	24 00	08 19	11 54	04 28	00 16	05 32
26	08 15 20	03 10 34	10 ♊ 05 04	16 ♊ 25 00	28 51	27 23	24 41	08 25	11 58	04 27	00 14	05 31
27	08 19 17	04 07 54	22 ♊ 50 11	29 ♊ 20 56	29 24	28 37	25 21	08 31	12 02	04 27	00 13	05 30
28	08 23 14	05 05 15	05 ♋ 57 27	12 ♋ 39 50	29 ♌ 53	29 ♋ 51	26 02	08 38	12 06	04 26	00 11	05 28
29	08 27 10	06 02 38	19 ♋ 28 02	26 ♋ 21 53	00 ♍ 18	01 ♌ 05	26 43	08 43	12 10	04 25	00 10	05 27
30	08 31 07	07 00 01	03 ♌ 21 04	10 ♌ 25 07	00 39	02 19	27 23	08 49	12 14	04 24	00 08	05 26
31	08 35 03	07 ♌ 57 24	17 ♌ 33 27	24 ♌ 45 22	00 ♍ 54	03 ♌ 33	28 ♊ 04	08 ♉ 55	12 ♎ 19	04 ♈ 23	00 ♓ 07	05 ♑ 25

MOON NODES AND LATITUDE / DECLINATIONS

Date	Moon True ☊ ° '	Moon Mean ☊ ° '	Moon ☽ Latitude ° '	Sun ☉ ° '	Moon ☽ ° '	Mercury ☿ ° '	Venus ♀ ° '	Mars ♂ ° '	Jupiter ♃ ° '	Saturn ♄ ° '	Uranus ♅ ° '	Neptune ♆ ° '	Pluto ♇ ° '
01	23 ♐ 28	22 ♐ 42	01 S 35	23 N 07	21 N 25	22 N 09	23 N 12	21 N 27	12 N 05	01 S 57	01 N 07	11 S 43	18 S 49
02	23 R 26	22 38	02 43	23 02	18 33	21 44	23 16	21 35	12 08	01 58	01 07	11 44	18 49
03	23 24	22 35	03 42	22 58	14 38	21 17	23 19	21 42	12 11	01 59	01 08	11 44	18 49
04	23 22	22 32	04 30	22 53	09 51	20 50	23 21	21 50	12 14	02 00	01 08	11 44	18 50
05	23 19	22 29	05 01	22 47	04 N 31	20 22	23 23	21 56	12 17	02 01	01 08	11 45	18 50
06	23 17	22 26	05 14	22 41	01 S 05	19 53	23 25	22 03	12 20	02 02	01 08	11 45	18 50
07	23 16	22 22	05 08	22 35	06 38	19 23	23 25	22 10	12 22	02 04	01 08	11 45	18 50
08	23 D 16	22 19	04 43	22 29	11 50	18 53	23 25	22 16	12 25	02 05	01 08	11 46	18 50
09	23 16	22 16	04 01	22 22	16 22	18 22	23 24	22 22	12 28	02 06	01 08	11 46	18 50
10	23 18	22 13	03 05	22 14	19 58	17 51	23 22	22 28	12 31	02 07	01 08	11 47	18 51
11	23 19	22 10	01 58	22 07	22 21	17 19	23 20	22 34	12 33	02 08	01 08	11 47	18 51
12	23 20	22 07	00 S 45	21 58	23 21	16 48	23 17	22 39	12 36	02 10	01 08	11 47	18 51
13	23 R 20	22 03	00 N 30	21 50	22 56	16 16	23 14	22 44	12 39	02 11	01 08	11 48	18 51
14	23 19	22 00	01 41	21 41	21 12	15 44	23 09	22 49	12 41	02 12	01 07	11 48	18 51
15	23 17	21 57	02 46	21 32	18 20	15 12	23 04	22 54	12 44	02 14	01 07	11 49	18 52
16	23 13	21 54	03 41	21 22	14 37	14 41	22 59	22 59	12 46	02 15	01 07	11 49	18 52
17	23 08	21 51	04 24	21 12	10 18	14 09	22 52	23 03	12 48	02 17	01 07	11 50	18 52
18	23 02	21 47	04 54	21 02	05 39	13 38	22 45	23 07	12 51	02 18	01 07	11 50	18 52
19	22 57	21 44	05 09	20 51	00 S 53	13 08	22 38	23 11	12 53	02 20	01 07	11 50	18 53
20	22 53	21 41	05 10	20 40	03 N 52	12 38	22 29	23 15	12 55	02 21	01 06	11 51	18 53
21	22 50	21 38	04 58	20 29	08 25	12 08	22 20	23 19	12 57	02 23	01 06	11 51	18 53
22	22 48	21 35	04 33	20 17	12 38	11 39	22 11	23 22	12 59	02 25	01 06	11 52	18 53
23	22 D 47	21 32	03 56	20 05	16 23	11 11	22 00	23 25	13 01	02 26	01 06	11 52	18 54
24	22 48	21 28	03 09	19 53	19 29	10 44	21 49	23 28	13 03	02 28	01 05	11 53	18 54
25	22 50	21 25	02 13	19 40	21 47	10 18	21 38	23 31	13 05	02 30	01 05	11 53	18 54
26	22 51	21 22	01 N 09	19 27	23 06	09 53	21 26	23 33	13 07	02 31	01 05	11 54	18 54
27	22 52	21 19	00 00	19 13	23 15	09 29	21 13	23 35	13 09	02 33	01 04	11 54	18 55
28	22 R 51	21 16	01 S 10	19 00	22 08	09 07	20 59	23 38	13 11	02 35	01 04	11 55	18 55
29	22 48	21 13	02 19	18 46	19 44	08 46	20 45	23 39	13 13	02 37	01 04	11 56	18 55
30	22 44	21 09	03 21	18 31	16 09	08 27	20 30	23 41	13 14	02 39	01 03	11 56	18 55
31	22 ♐ 38	21 ♐ 06	04 S 12	18 N 17	11 N 34	08 N 10	20 N 15	23 N 43	13 N 16	02 S 41	01 N 03	11 S 57	18 S 55

ZODIAC SIGN ENTRIES

Date	h	m	Planets
02	05	38	☿ ♌
02	21	43	☽ ♌
04	04	17	♀ ♋
05	01	15	☽ ♍
07	03	54	☽ ♎
09	06	31	☽ ♏
11	09	47	☽ ♐
13	14	13	☽ ♑
15	20	30	☽ ♒
18	05	13	☽ ♓
20	16	25	☽ ♈
23	04	12	☉ ♌
23	04	58	☽ ♉
25	16	34	☽ ♊
28	01	11	☽ ♋
28	14	59	♀ ♌
28	17	59	☿ ♍
30	06	16	☽ ♌

LATITUDES

Date	Mercury ☿ ° '	Venus ♀ ° '	Mars ♂ ° '	Jupiter ♃ ° '	Saturn ♄ ° '	Uranus ♅ ° '	Neptune ♆ ° '	Pluto ♇ ° '
01	01 N 47	00 S 12	00 S 05	01 S 11	02 N 29	00 S 44	00 S 32	04 N 29
04	01 33	00 S 05	00 03	01 11	02 29	00 45	00 32	04 29
07	01 15	00 N 03	00 S 01	01 12	02 28	00 45	00 32	04 28
10	00 52	00 10	00 N 02	01 12	02 27	00 45	00 32	04 28
13	00 N 25	00 17	00 04	01 13	02 26	00 45	00 33	04 28
16	00 S 06	00 24	00 06	01 13	02 26	00 45	00 33	04 27
19	00 40	00 31	00 08	01 14	02 25	00 45	00 33	04 27
22	01 16	00 37	00 10	01 14	02 24	00 45	00 33	04 26
25	01 54	00 44	00 13	01 15	02 24	00 45	00 33	04 26
28	02 33	00 50	00 15	01 15	02 23	00 45	00 33	04 25
31	03 S 11	00 N 55	00 N 17	01 S 16	02 N 23	00 S 45	00 S 33	04 N 25

DATA

Julian Date	2455744
Delta T	+68 seconds
Ayanamsa	24° 01' 21"
Synetic vernal point	05° ♓ 05' 38"
True obliquity of ecliptic	23° 26' 14"

MOON'S PHASES, APSIDES AND POSITIONS ☽

Date	h	m	Phase	Longitude ° '	Eclipse Indicator
01	08	54	●	09 ♋ 12	Partial
08	06	29	☽	15 ♎ 47	
15	06	40	○	22 ♑ 28	
23	05	02	☾	00 ♉ 02	
30	18	40	●	07 ♌ 16	

Day	h	m		
07	13	51	Perigee	
21	22	43	Apogee	
06	07	24	0S	
12	16	55	Max dec	23° S 23'
19	16	23	0N	
27	03	00	Max dec	23° N 20'

LONGITUDES

Date	Chiron ⚷ ° '	Ceres ⚳ ° '	Pallas ⚴ ° '	Juno ⚵ ° '	Vesta ⚶ ° '	Black Moon Lilith ⚸ ° '
01	05 ♓ 14	28 ♓ 55	12 ♒ 54	22 ♍ 40	19 ♒ 16	11 ♈ 13
11	04 ♓ 59	00 ♈ 00	10 ♒ 46	25 ♍ 17	18 ♒ 05	12 ♈ 20
21	04 ♓ 39	00 ♈ 35	08 ♒ 16	28 ♍ 05	16 ♒ 13	13 ♈ 27
31	04 ♓ 15	00 ♈ 35	05 ♒ 36	01 ♎ 04	13 ♒ 54	14 ♈ 34

All ephemeris data is given at 12.00 UT and the Moon's longitude is additionally given for 24.00 UT

ASPECTARIAN

h m	Aspects
01 Friday	
00 28	☽ □ ♅
00 36	☽ ∥ ☿
01 11	☽ ⚹ ♃
03 21	☽ ☍ ♇
05 15	☽ ⊻ ♂
08 54	☽ ☌ ☉
11 31	☽ ∥ ♂
11 31	☿ Q ♄
11 37	☽ □ ♄
16 37	☽ ⊥ ♂
20 32	☽ ⚼ ♆
22 58	☽ Q ♃
02 Saturday	
09 27	☽ ∠ ♂
10 08	☽ # ♇
12 26	☽ ± ♆
15 43	☿ ⊼ ♆
18 07	☿ ∥ ♂
18 43	☽ ⊻ ♀
19 33	☽ Q ♄
22 56	☽ ⊼ ♆
23 55	☽ ☌ ☿
03 Sunday	
00 17	☉ □ ♄
05 37	☽ △ ♅
06 11	☽ ⊥ ♀
06 55	☽ □ ♃
08 16	☽ ⊼ ♇
13 00	☽ ⚹ ♂
16 25	☽ ⚹ ♄
17 34	☽ ⊻ ☉
18 36	☽ ± ♇
23 13	☽ ∠ ♀
04 Monday	
00 37	☽ ∥ ♃
02 57	☽ # ♆
04 38	☽ ⊥ ☉
07 27	☽ ⚼ ♅
10 01	☽ ⚼ ♇
10 42	☽ Q ♂
17 26	♀ △ ♆
18 10	☽ ∠ ♄
21 06	☽ ∠ ☉
22 48	☽ ± ♅
05 Tuesday	
00 47	☿ △ ♅
02 23	☽ ☍ ♆
03 14	☽ ⚹ ♀
08 59	☽ ⊼ ♅
09 29	☽ ⊥ ♄
10 00	☽ ⊻ ☿
10 48	☽ △ ♃
11 27	☽ △ ♇
18 09	☿ □ ♃
18 50	☽ □ ♂
19 39	☽ ⊻ ♄
21 23	☽ ⊥ ☿
22 44	☽ # ♄
23 08	☿ ⊼ ♇
06 Wednesday	
00 19	☽ ⚹ ☉
01 27	☽ Q ♀
02 35	☽ ∥ ♅
12 13	☽ # ♅
12 16	♂ △ ♄
12 24	☽ ⚼ ♃
14 23	☽ ∠ ☿
16 07	☽ ∥ ♄
22 00	☽ Q ☉
07 Thursday	
03 43	☽ ± ♃
04 57	☽ ⊼ ♆
10 37	☽ □ ♀
11 34	☽ ☍ ♅
13 56	☽ ⊼ ♃
13 56	☽ □ ♇
14 07	♃ △ ♇
15 03	☽ ± ♆
18 37	☽ ⚹ ☿
21 44	♀ □ ♅
22 20	☽ ☌ ♄
08 Friday	
00 00	☽ △ ♂
02 12	☉ ⚼ ♆
03 57	♂ ⊥ ♃
06 12	☽ ⚼ ♆
06 29	☽ □ ☉
11 41	☽ ∥ ♆
13 55	☿ # ♇
14 57	☽ # ♃
17 10	☽ Q ☿
20 15	☽ Q ♇
09 Saturday	
00 31	♀ ☍ ♇
02 39	☽ ⚼ ♂
06 05	♀ ⚹ ♃
07 32	☽ △ ♆
08 19	☿ ⚹ ♄
11 17	☉ ∥ ♂
14 16	☽ ⊼ ♅
16 33	☽ ⚹ ♇
17 11	☽ ☍ ♃
18 07	☽ △ ♀
18 42	☽ ± ♂
22 53	☽ # ☿
23 36	☿ ± ♇
10 Sunday	
00 28	☽ ± ♅
00 34	♅ St R
01 17	☽ ⊻ ♄
03 06	☽ □ ☿
03 43	☽ ∥ ♇
05 27	☽ ⊼ ♂
11 33	☽ ⊥ ♄
13 05	☽ △ ☉
15 50	☽ ⚼ ♅
18 06	☽ ∠ ♇
22 11	☽ ⚼ ♀
11 Monday	
02 23	☉ Q ♃
03 02	☽ ∠ ♄
09 05	☽ # ☉
09 33	☽ ⊥ ♇
10 44	☽ □ ♆
15 11	☽ ± ♀
15 25	☽ # ♂
16 47	☽ ⚼ ☉
17 40	☽ △ ♅
19 55	☽ ⊻ ♇
21 11	☽ ⊼ ♃
12 Tuesday	
02 36	☽ ⊼ ♀
03 28	☿ ⚹ ♂
05 05	☽ ⚹ ♄
07 44	☽ ± ♃
08 37	☽ # ♀
09 35	☽ ± ☉
11 56	☽ ☍ ♂
12 21	☽ △ ☿
18 00	☽ Q ♆
20 52	☽ ⊼ ☉
23 39	☽ ⚼ ♃
13 Wednesday	
02 10	☽ Q ♄
03 11	☽ # ♀
07 35	♀ □ ♄
15 08	☽ ⚹ ♆
15 42	☽ # ♂
17 30	☽ ⚼ ☿
22 20	☽ □ ♅
14 Thursday	
00 34	☽ ☌ ♇
02 32	☽ △ ♃
06 18	☽ # ☉
10 19	☽ □ ♄
11 14	☽ ± ☿
12 55	☽ ☍ ♀
15 33	♂ Q ♅
17 59	☽ ∠ ♆
20 09	☽ ⊼ ♂
23 09	☽ ⊼ ☿
15 Friday	
06 40	☽ ☍ ☉
06 49	☽ Q ♅
07 40	☽ ± ♂
08 07	☽ ∥ ♇
10 21	☽ ⊥ ♆
21 22	☽ ⊻ ♆
23 52	☿ ⚼ ♅
16 Saturday	
01 10	☽ ⚼ ♂
04 56	☽ ⚹ ♅
07 10	☽ ⊻ ♇
09 54	☽ □ ♃
11 32	☽ # ☿
12 03	♀ ∥ ♂
17 41	☽ △ ♄
18 23	☽ ⊥ ♇
18 55	♀ ⚼ ♆
22 30	☽ # ♃
17 Sunday	
01 09	☿ ⚼ ♇
02 04	☽ ⊼ ♀
03 49	☽ ∥ ♆
06 24	☉ ⊥ ♂
06 56	☽ △ ♂
08 21	☉ ± ♆
09 09	☽ ∠ ♅
11 23	☽ ∠ ♇
12 23	☽ ☍ ☿
14 43	☽ ± ♀
19 20	☽ ⊼ ☉
20 21	☽ Q ♃
22 20	☽ ⚼ ♄
18 Monday	
02 23	☽ ⊥ ♅
06 02	☽ ☌ ♆
07 52	☽ ± ☉
09 57	☽ ⚼ ♀
14 01	☽ ⊻ ♅
15 51	☽ ± ♄
16 16	☽ ⚹ ♇
19 51	☽ ⚹ ♃
19 Tuesday	
02 55	☽ ⚼ ☉
03 39	☽ ⊼ ♄
04 45	☽ ∥ ♄
10 49	☽ # ♅
15 50	☽ Q ♇
18 43	☽ △ ♀
20 45	☽ □ ♂
21 59	☽ ∥ ♅
22 58	☿ ∥ ♃
20 Wednesday	
01 47	☽ ∠ ♃
04 15	☽ # ♄
04 19	☽ ⊼ ☿
06 01	♀ Q ♃
11 15	☽ △ ☉
17 10	☽ ⊻ ♆
17 20	☽ ± ☿
19 59	☽ ⊥ ♃
21 Thursday	
01 30	☽ ☌ ♅
03 45	☽ □ ♇
05 12	☽ ⊥ ♆
08 11	☽ ⊻ ♃
11 01	☽ Q ♂
12 56	☽ ⚼ ☿
13 23	♀ ⊻ ♂
15 53	☽ ☍ ♄
23 21	☽ ∠ ♆
22 Friday	
01 33	☿ # ♆
06 48	☽ ∥ ☿
07 25	☽ # ♆
12 45	☽ ⚹ ♂
13 57	☽ □ ♀
14 09	☽ ∥ ♃
21 18	☉ Q ♄
21 34	☽ △ ☿
23 Saturday	
05 02	☽ □ ☉
05 36	☽ ⚹ ♆
10 24	☿ ∠ ♄
11 54	☉ ⊼ ♆
14 01	☽ ⊻ ♅
16 12	☽ △ ♇
20 50	☽ ∠ ♂
21 19	☽ ☌ ♃
23 43	♀ ± ♆
24 Sunday	
02 02	☽ ⊥ ♅
04 41	☽ ⊼ ♄
05 38	☽ Q ♆
06 19	☽ Q ♀
06 55	☽ # ♇
07 06	♂ ∠ ♃
15 13	☽ ∥ ☉
15 52	☽ ⊥ ♂
16 41	☽ ± ♄
19 55	☽ ∠ ♅
20 09	☽ Q ☉
22 03	☽ ⚼ ♇
25 Monday	
04 25	☽ ⊻ ♂
08 44	☽ ⚹ ♀
10 10	☽ ∥ ♀
10 30	☽ ⚼ ♄
13 12	☽ □ ☿
15 40	☽ ± ♇
17 04	☽ □ ♆
20 39	☿ ± ♅
21 37	☽ ⚹ ☉
26 Tuesday	
01 13	☽ ⚹ ♅
03 16	☽ ⊼ ♇
08 47	☽ ⊻ ♃
15 35	☽ △ ♄
16 51	☽ ∠ ♀
20 18	☽ ⊥ ♃
27 Wednesday	
00 04	☽ Q ♅
01 25	☽ Q ☿
04 32	☽ ∠ ☉
11 33	☽ ⊥ ♀
13 17	☽ ∠ ♃
16 55	☽ ☌ ♂
19 43	☉ △ ♅
22 32	☽ ⊥ ☉
23 46	☽ ⊻ ♀
28 Thursday	
00 35	☽ ⚹ ☿
01 33	☽ △ ♆
09 15	☽ □ ♅
10 19	☽ ⊻ ☉
11 08	☽ ☍ ♇
13 19	☿ ⊻ ♀
16 50	☽ ⚹ ♃
17 11	♀ Q ♄
18 33	♀ ⊼ ♆
20 40	☉ # ♇
21 30	☉ ⊼ ♇
23 03	☽ □ ♄
29 Friday	
02 21	☽ ∥ ♀
04 00	☿ ☍ ♆
04 27	☽ ⚼ ♆
04 28	☽ ∠ ☿
14 13	☽ Q ♃
18 12	☽ # ♇
19 51	☽ ∥ ☉
20 11	☽ ± ♆
20 39	☽ ⊥ ☿
30 Saturday	
01 14	☽ ⊻ ♂
06 31	☽ ⊼ ♆
06 39	☽ Q ♄
07 16	☽ ⊻ ☿
10 03	☽ ☌ ♀
12 04	☽ ⊥ ♂
13 47	☽ △ ♅
15 33	☽ ⊼ ♇
18 40	☽ ☌ ☉
21 22	☽ □ ♃
31 Sunday	
01 41	☽ ± ♇
03 09	☽ ⚹ ♄
03 39	☽ ∥ ♃
04 05	☽ ∠ ♂
10 10	☽ # ♆
15 03	☽ ⚼ ♅
16 46	☽ ⚼ ♇

AUGUST 2011

LONGITUDES

Date	Sidereal time h m s	Sun ☉ ° ′ ″	Moon ☽ ° ′ ″	Moon ☽ 24.00 ° ′ ″	Mercury ☿ ° ′	Venus ♀ ° ′	Mars ♂ ° ′	Jupiter ♃ ° ′	Saturn ♄ ° ′	Uranus ♅ ° ′	Neptune ♆ ° ′	Pluto ♇ ° ′
01	08 39 00	08 ♌ 54 49	02 ♍ 00 07	09 ♍ 16 50	01 ♍ 05	04 ♌ 46	28 ♊ 44	09 ♉ 01	12 ♎ 23	04 ♈ 22	00 ♓ 06	05 ♑ 23
02	08 42 56	09 52 14	16 ♍ 34 39	23 ♍ 52 44	01 11	06 00	29 ♊ 24	09 06	12 28	04 R 21	00 R 04	05 R 22
03	08 46 53	10 49 40	01 ♎ 10 17	08 ♎ 26 33	01 R 12	07 14	00 ♋ 04	09 11	12 32	04 20	00 02	05 21
04	08 50 49	11 47 07	15 ♎ 40 53	22 ♎ 52 47	01 07	08 28	00 45	09 16	12 37	04 19	00 ♓ 01	05 20
05	08 54 46	12 44 34	00 ♏ 01 49	07 ♏ 07 42	00 58	09 42	01 25	09 21	12 42	04 17	29 ♒ 59	05 19
06	08 58 43	13 42 02	14 ♏ 10 12	21 ♏ 09 14	00 43	10 56	02 05	09 26	12 47	04 16	29 58	05 18
07	09 02 39	14 39 30	28 ♏ 04 46	04 ♐ 56 49	00 ♍ 23	12 10	02 45	09 30	12 52	04 15	29 56	05 17
08	09 06 36	15 37 00	11 ♐ 45 28	18 ♐ 30 47	29 ♌ 57	13 24	03 25	09 34	12 57	04 14	29 55	05 16
09	09 10 32	16 34 30	25 ♐ 12 54	01 ♑ 51 54	29 27	14 39	04 04	09 38	13 02	04 12	29 53	05 14
10	09 14 29	17 32 01	08 ♑ 27 53	15 ♑ 00 54	28 52	15 53	04 44	09 42	13 07	04 11	29 52	05 13
11	09 18 25	18 29 33	21 ♑ 31 01	27 ♑ 58 16	28 13	17 07	05 24	09 46	13 12	04 09	29 50	05 12
12	09 22 22	19 27 06	04 ♒ 22 41	10 ♒ 44 16	27 31	18 21	06 03	09 50	13 17	04 08	29 48	05 11
13	09 26 18	20 24 40	17 ♒ 03 02	23 ♒ 18 59	26 45	19 35	06 43	09 53	13 22	04 06	29 47	05 10
14	09 30 15	21 22 15	29 ♒ 32 09	05 ♓ 42 35	25 57	20 49	07 23	09 56	13 28	04 05	29 45	05 10
15	09 34 12	22 19 51	11 ♓ 50 22	17 ♓ 55 38	25 07	22 03	08 02	09 59	13 33	04 03	29 44	05 09
16	09 38 08	23 17 29	23 ♓ 58 32	29 ♓ 59 17	24 16	23 17	08 41	10 02	13 39	04 02	29 42	05 08
17	09 42 05	24 15 08	05 ♈ 58 10	11 ♈ 55 29	23 26	24 32	09 21	10 05	13 44	04 00	29 40	05 07
18	09 46 01	25 12 48	17 ♈ 51 38	23 ♈ 47 01	22 37	25 46	10 00	10 07	13 50	03 58	29 39	05 06
19	09 49 58	26 10 30	29 ♈ 42 08	05 ♉ 37 30	21 51	27 00	10 39	10 09	13 56	03 57	29 37	05 05
20	09 53 54	27 08 14	11 ♉ 33 40	17 ♉ 31 16	21 07	28 14	11 18	10 12	14 01	03 55	29 35	05 04
21	09 57 51	28 05 59	23 ♉ 30 54	29 ♉ 33 15	20 28	29 ♌ 29	11 57	10 13	14 07	03 53	29 34	05 04
22	10 01 47	29 ♌ 03 46	05 ♊ 38 57	11 ♊ 48 40	19 54	00 ♍ 43	12 36	10 15	14 13	03 51	29 32	05 03
23	10 05 44	00 ♍ 01 35	18 ♊ 03 02	24 ♊ 22 39	19 26	01 57	13 15	10 17	14 19	03 49	29 30	05 02
24	10 09 41	00 59 25	00 ♋ 48 06	07 ♋ 19 50	19 04	03 11	13 54	10 18	14 25	03 47	29 29	05 01
25	10 13 37	01 57 17	13 ♋ 58 13	20 ♋ 43 32	18 50	04 26	14 32	10 19	14 31	03 46	29 27	05 01
26	10 17 34	02 55 11	27 ♋ 35 53	04 ♌ 35 10	18 42	05 40	15 11	10 20	14 37	03 44	29 26	05 00
27	10 21 30	03 53 07	11 ♌ 41 09	18 ♌ 53 22	18 D 43	06 54	15 50	10 20	14 43	03 42	29 24	04 59
28	10 25 27	04 51 04	26 ♌ 11 10	03 ♍ 33 41	18 52	08 09	16 28	10 21	14 49	03 40	29 22	04 59
29	10 29 23	05 49 03	10 ♍ 59 56	18 ♍ 28 47	19 09	09 23	17 07	10 21	14 55	03 38	29 21	04 58
30	10 33 20	06 47 03	25 ♍ 59 00	03 ♎ 29 23	19 34	10 38	17 45	10 R 21	15 01	03 36	29 19	04 58
31	10 37 16	07 ♍ 45 04	10 ♎ 58 44	18 ♎ 25 58	20 ♌ 07	11 ♍ 52	18 ♋ 23	10 ♉ 21	15 ♎ 08	03 ♈ 33	29 ♒ 17	04 ♑ 57

Date	Moon True ☊ ° ′	Moon Mean ☊ ° ′	Moon ☽ Latitude ° ′
01	22 ♐ 31	21 ♐ 03	04 S 48
02	22 R 24	21 00	05 06
03	22 18	20 57	05 04
04	22 13	20 53	04 42
05	22 10	20 50	04 03
06	22 D 10	20 47	03 10
07	22 10	20 44	02 06
08	22 11	20 41	00 S 56
09	22 R 12	20 38	00 N 16
10	22 10	20 34	01 26
11	22 07	20 31	02 30
12	22 01	20 28	03 26
13	21 53	20 25	04 10
14	21 43	20 22	04 41
15	21 32	20 19	04 59
16	21 21	20 15	05 03
17	21 11	20 12	04 53
18	21 03	20 09	04 31
19	20 57	20 06	03 57
20	20 54	20 03	03 13
21	20 52	19 59	02 20
22	20 D 52	19 56	01 20
23	20 53	19 53	00 N 15
24	20 R 52	19 50	00 S 53
25	20 50	19 47	01 59
26	20 46	19 44	03 02
27	20 39	19 40	03 55
28	20 30	19 37	04 35
29	20 19	19 34	04 57
30	20 09	19 31	05 00
31	19 ♐ 59	19 ♐ 28	04 S 41

DECLINATIONS

Date	Sun ☉ ° ′	Moon ☽ ° ′	Mercury ☿ ° ′	Venus ♀ ° ′	Mars ♂ ° ′	Jupiter ♃ ° ′	Saturn ♄ ° ′	Uranus ♅ ° ′	Neptune ♆ ° ′	Pluto ♇ ° ′
01	18 N 02	06 N 16	07 N 55	19 N 59	23 N 44	13 N 17	02 S 43	01 N 02	11 S 57	18 S 55
02	17 46	00 N 36	07 41	19 43	23 45	13 19	02 45	01 02	11 58	18 56
03	17 31	05 S 07	07 30	19 26	23 46	13 20	02 46	01 01	11 58	18 56
04	17 15	10 30	07 21	19 08	23 46	13 22	02 48	01 01	11 59	18 56
05	16 59	15 16	07 15	18 50	23 47	13 23	02 51	01 00	11 59	18 56
06	16 43	19 06	07 11	18 32	23 47	13 24	02 53	01 00	12 00	18 57
07	16 26	21 46	07 10	18 13	23 47	13 26	02 55	00 59	12 00	18 57
08	16 09	23 07	07 12	17 53	23 47	13 27	02 57	00 59	12 01	18 57
09	15 52	23 05	07 17	17 33	23 47	13 28	02 59	00 58	12 02	18 57
10	15 35	21 44	07 24	17 12	23 46	13 29	03 01	00 58	12 02	18 58
11	15 17	19 15	07 34	16 51	23 45	13 30	03 03	00 57	12 03	18 58
12	14 59	15 50	07 47	16 29	23 44	13 31	03 05	00 57	12 03	18 58
13	14 41	11 45	08 03	16 07	23 43	13 32	03 08	00 56	12 04	18 58
14	14 23	07 14	08 20	15 45	23 42	13 32	03 10	00 55	12 04	18 58
15	14 04	02 S 31	08 40	15 22	23 40	13 33	03 12	00 55	12 05	18 59
16	13 45	02 N 15	09 02	14 58	23 39	13 34	03 14	00 54	12 06	18 59
17	13 26	06 52	09 25	14 35	23 37	13 34	03 17	00 53	12 06	18 59
18	13 07	11 11	09 50	14 11	23 35	13 35	03 19	00 53	12 07	18 59
19	12 48	15 04	10 15	13 46	23 32	13 35	03 21	00 52	12 07	19 00
20	12 28	18 22	10 40	13 21	23 30	13 36	03 24	00 51	12 08	19 00
21	12 08	20 55	11 06	12 56	23 27	13 36	03 26	00 50	12 09	19 00
22	11 48	22 34	11 31	12 30	23 25	13 37	03 28	00 50	12 09	19 00
23	11 28	23 09	11 55	12 04	23 22	13 37	03 31	00 49	12 10	19 01
24	11 07	22 34	12 18	11 38	23 18	13 37	03 33	00 48	12 10	19 01
25	10 47	20 44	12 39	11 11	23 15	13 37	03 36	00 47	12 11	19 01
26	10 26	17 40	12 59	10 44	23 12	13 37	03 38	00 47	12 11	19 01
27	10 05	13 31	13 16	10 17	23 08	13 37	03 41	00 46	12 12	19 02
28	09 44	08 28	13 31	09 49	23 04	13 37	03 43	00 45	12 13	19 02
29	09 23	02 N 51	13 43	09 22	23 00	13 37	03 46	00 44	12 13	19 02
30	09 01	02 S 59	13 52	08 54	22 56	13 37	03 48	00 43	12 14	19 02
31	08 N 40	08 S 39	13 N 59	08 N 25	22 N 52	13 N 37	03 S 51	00 N 43	12 S 14	19 S 03

ZODIAC SIGN ENTRIES

Date	h m	Planets
01	08 41	☽ ♍
03	09 22	♂ ♋
03	10 04	☽ ♎
05	02 54	♆ ♒
05	11 57	☽ ♏
07	15 21	☽ ♐
08	09 46	☿ ♌
09	20 38	☽ ♑
12	03 47	☽ ♒
14	12 54	☽ ♓
17	00 01	☽ ♈
19	12 36	☽ ♉
21	22 11	♀ ♍
22	00 53	☽ ♊
23	11 21	☉ ♍
24	10 31	☽ ♋
26	16 09	☽ ♌
28	18 13	☽ ♍
30	18 25	☽ ♎

LATITUDES

Date	Mercury ☿ ° ′	Venus ♀ ° ′	Mars ♂ ° ′	Jupiter ♃ ° ′	Saturn ♄ ° ′	Uranus ♅ ° ′	Neptune ♆ ° ′	Pluto ♇ ° ′
01	03 S 24	00 N 57	00 N 18	01 S 17	02 N 22	00 S 45	00 S 33	04 N 24
04	03 58	01 02	00 20	01 17	02 22	00 46	00 33	04 24
07	04 27	01 07	00 23	01 18	02 21	00 46	00 33	04 23
10	04 45	01 11	00 25	01 18	02 21	00 46	00 33	04 23
13	04 50	01 14	00 27	01 19	02 20	00 46	00 33	04 22
16	04 39	01 18	00 30	01 20	02 20	00 46	00 33	04 22
19	04 12	01 20	00 32	01 20	02 19	00 46	00 33	04 21
22	03 30	01 22	00 34	01 21	02 19	00 46	00 33	04 20
25	02 39	01 24	00 37	01 22	02 18	00 46	00 33	04 20
28	01 44	01 25	00 39	01 22	02 18	00 46	00 33	04 19
31	00 S 51	01 N 25	00 N 41	01 S 23	02 N 18	00 S 46	00 S 33	04 N 18

DATA

Julian Date	2455775
Delta T	+68 seconds
Ayanamsa	24° 01′ 26″
Synetic vernal point	05° ♓ 05′ 33″
True obliquity of ecliptic	23° 26′ 15″

MOON'S PHASES, APSIDES AND POSITIONS ☽

Date	h m	Phase	Longitude ° ′	Eclipse Indicator
06	11 08	☽	13 ♏ 40	
13	18 57	○	20 ♒ 41	
21	21 54	☾	28 ♉ 30	
29	03 04	●	05 ♍ 27	

Day	h m		
02	20 58	Perigee	
18	16 14	Apogee	
30	17 30	Perigee	
02	14 30	0S	
08	23 21	Max dec	23° S 16′
16	00 38	0N	
23	12 19	Max dec	23° N 09′
29	23 45	0S	

LONGITUDES

Date	Chiron ⚷ ° ′	Ceres ⚳ ° ′	Pallas ⚴ ° ′	Juno ⚵ ° ′	Vesta ⚶ ° ′	Black Moon Lilith ⚸ ° ′
01	04 ♓ 13	00 ♈ 33	05 ♒ 20	01 ♎ 22	13 ♒ 39	14 ♈ 41
11	03 ♓ 45	29 ♓ 54	02 ♒ 48	04 ♎ 30	11 ♒ 12	15 ♈ 48
21	03 ♓ 17	28 ♓ 40	00 ♒ 35	07 ♎ 44	09 ♒ 00	16 ♈ 55
31	02 ♓ 47	26 ♓ 56	28 ♑ 52	11 ♎ 03	07 ♒ 19	18 ♈ 02

All ephemeris data is given at 12.00 UT and the Moon's longitude is additionally given for 24.00 UT

ASPECTARIAN

h m Aspects

01 Monday
04 09 ♀ △ ♅
04 20 ☽ ∠ ♄
04 29 ☽ ∥ ☿
06 00 ☽ ± ♅
06 20 ☽ ⚹ ♂
08 51 ☽ ☍ ♆
10 09 ♀ ⊥ ♂
10 28 ☽ ☌ ☿
14 41 ☉ □ ♃
15 54 ☽ ⊼ ♅
17 00 ☽ ⊻ ♀
17 35 ☽ △ ♇
19 16 ☽ ⊥ ♄
23 38 ☽ △ ♃
23 49 ♀ ⊼ ♇

02 Tuesday
00 12 ☽ ⊻ ☉
03 05 ☽ Q ♂
03 05 ☽ ⋕ ♄
03 48 ☽ ⊥ ♀
05 12 ☽ ⊻ ♄
10 11 ☽ ∥ ♅
10 45 ☽ ⊥ ☉
18 48 ☽ ⋕ ♅
19 57 ☽ ∠ ♀

03 Wednesday
00 26 ☽ ₽ ♃
02 03 ☽ ∥ ♄
02 35 ☽ ∠ ☉
03 50 ☿ St R
10 06 ☽ □ ♂
10 09 ☽ ⊼ ♆
10 53 ♂ △ ♆
12 02 ☽ ⊻ ☿
15 20 ☽ ± ♃
17 12 ☽ ☍ ♅
18 53 ☽ □ ♇
20 01 ☽ ± ♆
21 54 ☽ ⊥ ☿
22 05 ☽ ⋕ ☿
22 56 ☽ ⚹ ♀

04 Thursday
00 53 ☉ ± ♇
01 18 ☽ ⊼ ♃
05 05 ☽ ⚹ ☉
06 53 ☽ ☌ ♄
10 54 ☽ ₽ ♆
12 44 ☽ ∠ ☿
19 02 ☽ ∥ ♆
20 44 ☽ Q ♀
23 19 ☿ ⚹ ♂

05 Friday
00 44 ☽ Q ♇
01 57 ☽ ⋕ ♃
02 29 ☽ Q ☉
04 10 ♀ ⋕ ♇
04 34 ♀ □ ♃
10 47 ☉ ⚹ ♄
11 56 ☽ △ ♆
13 33 ☽ ⚹ ☿
14 27 ☽ △ ♂
19 11 ☽ ⊼ ♅
20 55 ☽ ⚹ ♇
21 17 ☽ ⋕ ☉

06 Saturday
03 52 ☽ ☍ ♃
05 21 ☽ ± ♅
05 57 ☽ □ ♀
08 14 ☽ ⋕ ♀
09 34 ☽ Q ☿
09 36 ☽ ⊻ ♄
10 49 ☽ ∥ ♇
11 08 ☽ □ ☉
17 14 ☽ ₽ ♂
18 48 ♀ ± ♇
19 57 ☽ ⊥ ♄
20 44 ☽ ₽ ♅
22 30 ☽ ∠ ♇

07 Sunday
09 34 ☽ ± ♂
11 37 ☽ ∠ ♄
14 05 ☽ ⊥ ♇
15 14 ☽ □ ♆
15 54 ☽ □ ☿
20 33 ☽ ⊼ ♂
22 45 ☽ △ ♅

08 Monday
00 34 ☽ ⊻ ♇
02 19 ♀ ⚹ ♄
08 07 ☽ ⊼ ♃
14 07 ☽ ⚹ ♄
14 27 ☿ ☍ ♆
15 13 ☽ △ ♀
18 48 ☽ ± ♃
19 22 ☽ △ ☉
22 54 ☽ Q ♆

09 Tuesday
10 58 ☽ ₽ ♃
11 40 ☽ Q ♄
16 33 ♂ □ ♅
19 20 ☽ △ ☿
20 24 ☽ ⚹ ♆
20 48 ☽ ₽ ♀

10 Wednesday
00 22 ☽ ₽ ☉
04 13 ☽ □ ♅
04 51 ☽ ☍ ♂
06 06 ☽ ☌ ♇
14 17 ☽ △ ♃
14 51 ☽ ± ♀
18 03 ☽ ± ☉
20 34 ☽ □ ♄
21 27 ☽ ₽ ☿
23 41 ☽ ∠ ♆

11 Thursday
03 00 ☽ ⊼ ♀
05 14 ♂ ☍ ♇
05 58 ☽ ⊼ ☉
12 46 ☿ ∠ ♄
13 11 ☽ Q ♅
13 15 ☽ ± ☿
14 13 ☽ ∥ ♇
14 25 ☿ ± ♅
16 17 ☽ ⊥ ♆
23 50 ☽ ⊼ ☿

12 Friday
03 27 ☽ ⊻ ♆
04 12 ☉ ₽ ♅
07 18 ☽ ⋕ ♀
11 32 ☽ ⚹ ♅
13 32 ☽ ⊻ ♇
15 20 ☽ ⊼ ♂
17 42 ☽ ⋕ ☉
22 20 ☽ □ ♃

13 Saturday
00 50 ☽ ⊥ ♇
01 59 ☽ ⋕ ♃
02 57 ♀ ₽ ♅
03 18 ☽ ± ♂
04 57 ☽ △ ♄
06 10 ☉ ₽ ♇
10 16 ☽ ∥ ♆
15 55 ☽ ∠ ♅
17 22 ☽ ☍ ♀
17 58 ☽ ∠ ♇
18 57 ☽ ☍ ☉
21 26 ☽ ₽ ♂
23 21 ♀ ₽ ♇

14 Sunday
05 29 ☽ ☍ ☿
06 39 ☽ ⋕ ☿
08 54 ☽ Q ♃
09 11 ☽ ⊥ ♅
09 55 ☽ ₽ ♄
12 25 ☽ ☌ ♆
20 48 ☽ ⊻ ♅
22 55 ☽ ⚹ ♇

15 Monday
03 32 ☽ ± ♄
04 07 ☽ △ ♂
08 21 ☽ ⚹ ♃
08 33 ☽ ∥ ♄
15 24 ☽ ⊼ ♄
20 03 ☽ ⋕ ♅
22 26 ☽ Q ♇

16 Tuesday
05 11 ☽ ∥ ♅
09 27 ♃ ⊥ ♅
10 29 ☽ ⊼ ♀
10 31 ☽ ⊼ ☉
12 08 ☉ ☌ ♀
12 33 ☽ ⊼ ☿
14 07 ☽ ∠ ♃
17 07 ☽ ⋕ ♄
21 21 ☿ ∠ ♂
23 21 ☿ ☌ ♀
23 24 ☽ ⊻ ♆
23 32 ☽ ± ☉
23 45 ☽ ± ☿
23 49 ☽ ± ♀

17 Wednesday
01 04 ☉ ☌ ☿
02 02 ☉ ∥ ♃
04 27 ♀ ∠ ♂
08 03 ☽ ☌ ♅
08 11 ☽ ⊥ ♃
10 17 ☽ □ ♇
11 24 ☽ ⊥ ♆
16 38 ☽ ₽ ☿
19 06 ☉ ∠ ♂
19 11 ☽ □ ♂
19 11 ☽ ₽ ☉
20 00 ☽ ₽ ♀
20 18 ☽ ⊻ ♃

18 Thursday
03 26 ☽ ∥ ☿
03 47 ☽ ☍ ♄
05 30 ☽ ∠ ♆
16 49 ♂ ⚹ ♃
17 30 ☽ ⋕ ♆
21 02 ☽ △ ☿
22 41 ☽ ∥ ☉

19 Friday
02 29 ☽ ∥ ♃
04 13 ☽ △ ☉
04 26 ☽ ∥ ♀
05 53 ☽ △ ♀
09 44 ☽ Q ♂
11 50 ☽ ⚹ ♆
20 34 ☽ ⊻ ♅
22 00 ♀ ∥ ♃
22 54 ☽ △ ♇

20 Saturday
05 52 ♀ ± ♅
08 41 ☽ ⊥ ♅
09 14 ☽ ☌ ♃
11 27 ☽ ⚹ ♂
12 03 ☽ Q ♆
17 00 ☽ ⊼ ♄
17 21 ☽ ⋕ ♇

21 Sunday
02 46 ☽ ∠ ♅
04 29 ♀ ∠ ♄
05 06 ☽ ₽ ♇
05 09 ☽ ± ♄
06 13 ☽ □ ☿
06 47 ☉ ± ♅
11 20 ☉ ⋕ ♆
13 39 ♀ ☍ ♆
19 14 ☽ ∠ ♂
21 54 ☽ □ ☉
23 01 ☽ ± ♇
23 14 ☽ ₽ ♄
23 59 ☽ □ ♆

22 Monday
01 11 ☽ □ ♀
05 25 ☿ ₽ ♇
08 29 ☽ ⚹ ♅
10 49 ☽ ⊼ ♇
13 58 ☽ ⊥ ♂
16 13 ☉ ∠ ♄
16 14 ☽ Q ☿
21 00 ☽ ⊻ ♃
21 07 ☉ ∥ ☿
23 26 ☉ ☍ ♆

23 Tuesday
02 16 ☽ ⊻ ♂
04 46 ☽ △ ♄
06 55 ♀ ⋕ ♆
07 45 ☽ Q ♅
08 36 ☽ ⊥ ♃
11 57 ☽ Q ☉
14 33 ☽ ⚹ ☿
16 01 ☽ Q ♀
16 14 ☿ ⊥ ♂
16 19 ☿ ∥ ♀

24 Wednesday
01 43 ☽ ∠ ♃
03 30 ☿ ⋕ ♆
09 33 ☽ △ ♆
12 23 ☽ ⚹ ☉
16 53 ☽ ⚹ ♀
17 30 ☽ □ ♅
17 54 ☽ ∠ ☿
19 47 ☽ ☍ ♇
23 22 ♀ ⊼ ♅

25 Thursday
00 49 ♄ ₽ ♆
05 25 ☽ ⚹ ♃
08 53 ♂ ₽ ♆
09 59 ☽ ⊥ ☿
10 46 ♂ □ ♄
12 52 ☽ ₽ ♆
12 59 ☽ □ ♄
13 04 ☽ ☌ ♂
17 44 ☽ ∠ ☉
20 33 ☽ ⊻ ☿
22 41 ☽ ∠ ♀
23 12 ♀ △ ♇

26 Friday
02 30 ☽ ⋕ ♇
02 38 ☿ ₽ ♅
02 49 ☽ Q ♃
04 45 ☽ ± ♆
10 44 ☽ ⊥ ☉
15 09 ☽ ⊼ ♆
15 55 ☽ ⊥ ♀
20 42 ☽ Q ♄
21 50 ☽ ⊻ ☉
22 02 ☿ St D
22 31 ☽ △ ♅

27 Saturday
00 42 ☽ ⊼ ♇
03 10 ☽ ⊻ ♀
05 27 ☿ ₽ ♅
07 24 ☉ ⊼ ♅
09 44 ☽ □ ♃
10 50 ☽ ± ♇
11 27 ☽ ∥ ♃
13 12 ☽ ∥ ☿
17 06 ☽ ⚹ ♄
18 36 ☽ ⋕ ♆
19 15 ☽ ⊻ ♂
23 39 ☽ ₽ ♅
23 48 ☽ ☌ ☿

28 Sunday
01 49 ☽ ₽ ♇
05 19 ☽ ∥ ♀
05 38 ☽ ⊥ ♂
05 55 ☽ ∥ ☉
14 24 ☽ ± ♅
15 12 ☉ △ ♇
17 11 ☽ ☍ ♆
17 58 ☽ ∠ ♄
21 00 ☽ ∠ ♂
23 32 ☿ ∥ ♃

29 Monday
00 08 ☽ ⊼ ♅
02 10 ♀ ⊥ ♄
02 17 ☽ △ ♇
03 04 ☽ ☌ ☉
08 16 ☽ ⋕ ♄
08 17 ☉ ∥ ♀
08 38 ☽ ⊥ ♄
09 10 ☽ ☌ ♀
10 58 ☽ △ ♃
18 21 ☽ ⊻ ♄
20 45 ☽ ∥ ♅
22 15 ☽ ⚹ ♂

30 Tuesday
01 25 ☽ ⊻ ☿
02 44 ☽ ⋕ ♅
06 46 ♀ △ ♃
09 17 ♃ St R
11 00 ☽ ₽ ♃
11 18 ☽ ⊥ ☿
15 24 ☽ ∥ ♄
17 19 ☽ ⊼ ♆
18 17 ☽ Q ♂

31 Wednesday
00 08 ☽ ☍ ♅
01 23 ☽ ± ♃
02 13 ☽ ∠ ☿
02 21 ☽ □ ♇
02 54 ☽ ± ♆
05 46 ☿ ₽ ♇
06 28 ☽ ⊻ ☉
11 00 ☽ ⊼ ♃
11 03 ☽ ⋕ ♀
12 01 ☽ ⋕ ☉
13 33 ☽ ⊻ ♀
16 46 ☽ ⊥ ☉
17 19 ☽ ₽ ♆
18 43 ☽ ☌ ♄

SEPTEMBER 2011

LONGITUDES

Date	Sidereal time h m s	Sun ☉ ° ' "	Moon ☽ ° ' "	Moon ☽ 24.00 ° ' "	Mercury ☿ ° '	Venus ♀ ° '	Mars ♂ ° '	Jupiter ♃ ° '	Saturn ♄ ° '	Uranus ♅ ° '	Neptune ♆ ° '	Pluto ♇ ° '
01	10 41 13	08 ♍ 43 07	25 ♎ 50 06	03 ♏ 10 21	20 ♌ 48	13 ♍ 06	19 ♋ 01	10 ♉ 21	15 ♎ 14	03 ♈ 31	29 ♒ 16	04 ♑ 57
02	10 45 10	09 41 12	10 ♏ 26 04	17 ♏ 36 50	21 37	14 21	19 40	10 R 20	15 21	03 R 29	29 R 14	04 R 56
03	10 49 06	10 39 18	24 ♏ 42 22	01 ♐ 42 33	22 33	15 35	20 18	10 20	15 27	03 27	29 13	04 56
04	10 53 03	11 37 25	08 ♐ 37 26	15 ♐ 27 07	23 36	16 50	20 56	10 19	15 33	03 25	29 11	04 56
05	10 56 59	12 35 34	22 ♐ 11 51	28 ♐ 51 54	24 46	18 04	21 33	10 18	15 40	03 23	29 09	04 55
06	11 00 56	13 33 45	05 ♑ 27 33	11 ♑ 59 10	26 02	19 19	22 11	10 16	15 47	03 21	29 08	04 55
07	11 04 52	14 31 56	18 ♑ 27 02	24 ♑ 51 29	27 24	20 33	22 49	10 15	15 53	03 18	29 06	04 54
08	11 08 49	15 30 09	01 ♒ 12 46	07 ♒ 31 11	28 ♌ 52	21 48	23 27	10 13	16 00	03 16	29 05	04 54
09	11 12 45	16 28 24	13 ♒ 46 56	20 ♒ 00 11	00 ♍ 23	23 02	24 04	10 11	16 06	03 14	29 03	04 54
10	11 16 42	17 26 40	26 ♒ 11 08	02 ♓ 19 54	02 00	24 16	24 42	10 09	16 13	03 11	29 02	04 54
11	11 20 39	18 24 58	08 ♓ 26 36	14 ♓ 31 20	03 39	25 31	25 19	10 07	16 20	03 09	29 00	04 54
12	11 24 35	19 23 18	20 ♓ 34 14	26 ♓ 35 24	05 22	26 45	25 56	10 04	16 27	03 07	28 59	04 54
13	11 28 32	20 21 40	02 ♈ 34 59	08 ♈ 33 06	07 07	28 00	26 34	10 01	16 33	03 05	28 57	04 53
14	11 32 28	21 20 03	14 ♈ 29 59	20 ♈ 25 49	08 55	29 ♍ 14	27 11	09 59	16 40	03 02	28 56	04 53
15	11 36 25	22 18 28	26 ♈ 20 55	02 ♉ 15 34	10 44	00 ♎ 29	27 48	09 55	16 47	03 00	28 54	04 53
16	11 40 21	23 16 56	08 ♉ 10 09	14 ♉ 05 04	12 34	01 44	28 25	09 52	16 54	02 57	28 53	04 53
17	11 44 18	24 15 26	20 ♉ 00 49	25 ♉ 57 53	14 25	02 58	29 02	09 49	17 01	02 55	28 51	04 D 53
18	11 48 14	25 13 57	01 ♊ 56 49	07 ♊ 58 15	16 17	04 13	29 ♋ 39	09 45	17 08	02 53	28 50	04 53
19	11 52 11	26 12 31	14 ♊ 02 46	20 ♊ 11 03	18 09	05 27	00 ♌ 16	09 41	17 15	02 50	28 48	04 53
20	11 56 08	27 11 07	26 ♊ 23 45	02 ♋ 41 29	20 02	06 42	00 52	09 37	17 22	02 48	28 47	04 54
21	12 00 04	28 09 46	09 ♋ 04 55	15 ♋ 34 36	21 54	07 56	01 29	09 33	17 29	02 46	28 45	04 54
22	12 04 01	29 ♍ 08 26	22 ♋ 11 03	28 ♋ 54 41	23 46	09 11	02 05	09 29	17 36	02 43	28 44	04 54
23	12 07 57	00 ♎ 07 09	05 ♌ 45 47	12 ♌ 44 26	25 37	10 25	02 42	09 24	17 43	02 41	28 43	04 54
24	12 11 54	01 05 54	19 ♌ 50 35	27 ♌ 03 57	27 28	11 40	03 18	09 19	17 50	02 38	28 41	04 54
25	12 15 50	02 04 41	04 ♍ 24 00	11 ♍ 50 00	29 ♍ 18	12 55	03 54	09 14	17 57	02 36	28 40	04 54
26	12 19 47	03 03 31	19 ♍ 20 57	26 ♍ 55 42	01 ♎ 08	14 09	04 30	09 09	18 05	02 34	28 39	04 55
27	12 23 43	04 02 22	04 ♎ 32 56	12 ♎ 11 14	02 57	15 24	05 06	09 04	18 12	02 31	28 38	04 55
28	12 27 40	05 01 15	19 ♎ 49 12	27 ♎ 25 25	04 45	16 38	05 42	08 59	18 19	02 29	28 36	04 55
29	12 31 37	06 00 10	04 ♏ 58 39	12 ♏ 27 47	06 32	17 53	06 18	08 53	18 26	02 26	28 35	04 56
30	12 35 33	06 ♎ 59 08	19 ♏ 51 52	27 ♏ 10 14	08 ♎ 18	19 ♎ 08	06 ♌ 54	08 ♉ 47	18 ♎ 33	02 ♈ 24	28 ♒ 34	04 ♑ 56

DECLINATIONS

Date	Moon True ☊ ° '	Moon Mean ☊ ° '	Moon ☽ Latitude ° '	Sun ☉ ° '	Moon ☽ ° '	Mercury ☿ ° '	Venus ♀ ° '	Mars ♂ ° '	Jupiter ♃ ° '	Saturn ♄ ° '	Uranus ♅ ° '	Neptune ♆ ° '	Pluto ♇ ° '
01	19 ♐ 51	19 ♐ 25	04 S 04	08 N 18	13 S 46	14 N 02	07 N 57	22 N 47	13 N 36	03 S 53	00 N 42	12 S 15	19 S 03
02	19 R 46	19 21	03 12	07 56	17 59	14 02	07 28	22 42	13 36	03 56	00 41	12 16	19 03
03	19 44	19 18	02 08	07 34	21 01	13 58	06 59	22 38	13 36	03 58	00 40	12 16	19 03
04	19 D 43	19 15	00 S 58	07 12	22 42	13 51	06 30	22 33	13 35	04 01	00 39	12 17	19 04
05	19 R 43	19 12	00 N 13	06 50	22 59	13 41	06 01	22 28	13 35	04 04	00 38	12 17	19 04
06	19 43	19 09	01 22	06 28	21 57	13 27	05 31	22 22	13 34	04 06	00 37	12 18	19 04
07	19 40	19 05	02 26	06 05	19 46	13 10	05 02	22 17	13 33	04 09	00 36	12 18	19 05
08	19 36	19 02	03 20	05 43	16 38	12 50	04 32	22 11	13 33	04 11	00 36	12 19	19 05
09	19 28	18 59	04 04	05 20	12 47	12 26	04 02	22 06	13 32	04 14	00 35	12 19	19 05
10	19 17	18 56	04 36	04 58	08 27	12 00	03 32	22 00	13 31	04 17	00 34	12 20	19 05
11	19 04	18 53	04 55	04 35	03 S 51	11 30	03 02	21 54	13 30	04 19	00 33	12 21	19 05
12	18 50	18 50	04 59	04 12	00 N 51	10 59	02 31	21 48	13 29	04 22	00 32	12 21	19 05
13	18 36	18 46	04 51	03 49	05 28	10 24	02 01	21 42	13 28	04 25	00 31	12 22	19 06
14	18 24	18 43	04 30	03 26	09 51	09 48	01 31	21 35	13 27	04 27	00 30	12 22	19 06
15	18 13	18 40	03 57	03 03	13 51	09 10	01 00	21 29	13 26	04 30	00 29	12 23	19 06
16	18 05	18 37	03 14	02 40	17 17	08 30	00 N 30	21 22	13 25	04 33	00 28	12 23	19 06
17	18 00	18 34	02 22	02 17	20 01	07 48	00 S 01	21 15	13 23	04 36	00 27	12 24	19 07
18	17 57	18 31	01 24	01 54	21 55	07 06	00 32	21 09	13 22	04 38	00 26	12 24	19 07
19	17 56	18 27	00 N 21	01 30	22 50	06 22	01 02	21 02	13 21	04 41	00 25	12 25	19 07
20	17 D 56	18 24	00 S 45	01 07	22 39	05 37	01 33	20 54	13 19	04 44	00 24	12 25	19 07
21	17 R 56	18 21	01 50	00 44	21 18	04 51	02 03	20 47	13 18	04 46	00 23	12 26	19 08
22	17 54	18 18	02 51	00 N 21	18 48	04 05	02 34	20 40	13 16	04 49	00 22	12 26	19 08
23	17 50	18 15	03 45	00 S 03	15 12	03 19	03 04	20 32	13 15	04 52	00 21	12 27	19 08
24	17 44	18 11	04 27	00 26	10 38	02 32	03 35	20 25	13 13	04 55	00 21	12 27	19 08
25	17 35	18 08	04 54	00 50	05 N 20	01 45	04 05	20 17	13 11	04 57	00 20	12 27	19 08
26	17 24	18 05	05 01	01 13	00 S 24	00 58	04 36	20 09	13 10	05 00	00 19	12 28	19 09
27	17 14	18 02	04 48	01 36	06 13	00 N 10	05 06	20 02	13 08	05 03	00 18	12 28	19 09
28	17 04	17 59	04 14	02 00	11 40	00 S 37	05 36	19 54	13 06	05 06	00 17	12 29	19 09
29	16 56	17 56	03 23	02 23	16 22	01 24	06 06	19 46	13 04	05 08	00 16	12 29	19 09
30	16 ♐ 51	17 ♐ 52	02 S 18	02 S 46	19 S 55	02 S 11	06 S 36	19 N 37	13 N 02	05 S 11	00 N 15	12 S 30	19 S 10

ZODIAC SIGN ENTRIES

Date	h m	Planets
01	18 48	☽ ♏
03	21 03	☽ ♐
06	02 03	☽ ♑
08	09 42	☽ ♒
09	05 58	☿ ♍
10	19 26	☽ ♓
13	06 49	☽ ♈
15	02 40	♀ ♎
15	19 25	☽ ♉
18	08 06	☽ ♊
19	01 51	♂ ♌
20	18 53	☽ ♋
23	01 55	☽ ♌
23	09 05	☉ ♎
25	04 49	☽ ♍
25	21 09	☿ ♎
27	04 51	☽ ♎
29	04 05	☽ ♏

LATITUDES

Date	Mercury ☿ ° '	Venus ♀ ° '	Mars ♂ ° '	Jupiter ♃ ° '	Saturn ♄ ° '	Uranus ♅ ° '	Neptune ♆ ° '	Pluto ♇ ° '
01	00 S 34	01 N 25	00 N 42	01 S 23	02 N 17	00 S 46	00 S 33	04 N 18
04	00 N 13	01 25	00 45	01 24	02 17	00 46	00 33	04 17
07	00 51	01 24	00 47	01 24	02 17	00 46	00 33	04 17
10	01 19	01 22	00 50	01 25	02 16	00 46	00 33	04 16
13	01 38	01 20	00 52	01 26	02 16	00 46	00 33	04 16
16	01 47	01 17	00 55	01 26	02 16	00 46	00 33	04 15
19	01 50	01 14	00 57	01 27	02 16	00 46	00 33	04 14
22	01 45	01 10	01 00	01 27	02 16	00 46	00 33	04 13
25	01 36	01 06	01 02	01 27	02 15	00 46	00 33	04 12
28	01 23	01 01	01 05	01 28	02 15	00 46	00 33	04 12
31	01 N 07	00 N 56	01 N 08	01 S 28	02 N 15	00 S 46	00 S 33	04 N 11

DATA

Julian Date	2455806
Delta T	+68 seconds
Ayanamsa	24° 01' 30"
Synetic vernal point	05° ♓ 05' 29"
True obliquity of ecliptic	23° 26' 15"

MOON'S PHASES, APSIDES AND POSITIONS ☽

Date	h m	Phase	Longitude ° '	Eclipse Indicator
04	17 39	☽	11 ♐ 51	
12	09 27	○	19 ♓ 17	
20	13 39	☾	27 ♊ 15	
27	11 09	●	04 ♎ 00	

Day	h m		
15	06 10	Apogee	
28	00 57	Perigee	
05	05 01	Max dec	23° S 03'
12	07 38	0N	
19	20 11	Max dec	22° N 53'
26	10 20	0S	

LONGITUDES

Date	Chiron ⚷ ° '	Ceres ⚳ ° '	Pallas ⚴ ° '	Juno ⚵ ° '	Vesta ⚶ ° '	Black Moon Lilith ⚸ ° '
01	02 ♓ 44	26 ♓ 45	28 ♑ 44	11 ♎ 23	07 ♒ 11	18 ♈ 09
11	02 ♓ 16	24 ♓ 38	27 ♑ 40	14 ♎ 47	06 ♒ 18	19 ♈ 16
21	01 ♓ 49	22 ♓ 24	27 ♑ 12	18 ♎ 14	06 ♒ 11	20 ♈ 23
31	01 ♓ 25	20 ♓ 16	27 ♑ 20	21 ♎ 44	06 ♒ 50	21 ♈ 30

All ephemeris data is given at 12.00 UT and the Moon's longitude is additionally given for 24.00 UT

ASPECTARIAN

h m	Aspects
01 Thursday	
00 06	☽ ⊥ ♀
00 28	☽ □ ♂
03 24	☽ ✶ ☿
04 28	☽ ∥ ♆
07 19	☽ Q ♇
08 19	☽ ∠ ☉
11 10	☽ # ♃
13 20	☽ # ☿
16 03	☽ ∠ ♀
17 35	☽ △ ♆
02 Friday	
00 02	☽ Q ☿
00 33	☽ ⚻ ♅
02 24	☉ ⊥ ♄
02 55	☽ ✶ ♇
10 26	☽ ± ♅
10 40	☽ ✶ ☉
11 51	☽ ☍ ♃
19 08	☽ ✶ ♀
19 24	☽ ∥ ♇
20 15	☽ ⊻ ♄
03 Saturday	
01 26	☽ ⚼ ♅
03 55	☽ ∠ ♇
03 59	☉ △ ♃
04 10	☽ △ ♂
06 26	☽ ⊥ ♄
08 03	☽ □ ☿
08 15	☽ Q ☉
09 05	♀ ⊻ ♄
17 24	☽ Q ♀
19 13	☽ ⊥ ♇
19 41	☽ □ ♆
21 54	☽ ∠ ♄
04 Sunday	
02 58	☽ △ ♅
05 34	☽ ⊻ ♇
07 05	☽ ⚼ ♂
08 48	☽ # ♂
14 57	☽ ⚻ ♃
17 39	☽ □ ☉
23 37	☽ ± ♂
05 Monday	
00 17	☽ ✶ ♄
01 30	☽ ± ♃
03 02	☽ Q ♆
03 54	☽ □ ♀
10 48	☽ ⚻ ♂
17 05	☽ △ ☿
17 33	☽ ⚼ ♃
21 55	☽ Q ♄
23 57	☿ ∥ ♃
06 Tuesday	
00 30	☽ ✶ ♆
04 44	☽ # ♂
08 09	☽ □ ♅
11 00	☽ ☌ ♇
15 03	♂ Q ♃
20 49	☽ △ ♃
23 26	☽ ⚼ ☿
07 Wednesday	
03 56	☽ ∠ ♆
04 07	☽ △ ☉
07 11	☽ □ ♄
10 19	☿ ± ♅
16 21	☽ △ ♀
17 19	☽ Q ♅
17 53	☽ ∥ ♇
18 13	☽ ± ☿
20 35	☽ ☍ ♂
20 41	☽ ⊥ ♆
22 31	♂ ± ♆
08 Thursday	
06 57	☽ ⚻ ☿
07 58	☽ ⊻ ♆
10 33	☽ ⚼ ☉
15 25	☿ ☍ ♆
15 53	☽ ✶ ♅
19 01	☽ ⊻ ♇
23 46	☽ ⚼ ♀
09 Friday	
01 46	☉ ⊻ ♄
03 00	♀ # ♄
03 43	☿ ⊥ ♂
05 07	☽ □ ♃
05 07	☽ ± ☉
06 28	☽ ⊥ ♇
07 36	☽ # ♃
14 16	☽ # ☿
14 40	☽ ∥ ♆
16 31	☽ △ ♄
17 37	☽ ⚻ ☉
18 08	☿ # ♆
18 57	☽ ± ♀
20 33	☽ ∠ ♅
23 40	☿ ∠ ♄
23 48	☽ ∠ ♇
10 Saturday	
07 52	☽ ⚻ ♀
08 57	☽ ⚻ ♂
13 57	☽ ⊥ ♅
15 49	☽ Q ♃
17 32	☽ ☌ ♆
21 16	☽ ± ♂
21 55	☽ ⚼ ♄
11 Sunday	
01 06	☽ ☍ ☿
01 38	☽ ⊻ ♅
04 22	♀ ✶ ♂
04 25	♀ ⚼ ♃
05 01	☿ ⚻ ♅
05 02	☽ ✶ ♇
07 54	☽ # ☉
09 35	☽ ∥ ♄
15 16	☽ ✶ ♃
15 45	☽ ± ♄
15 54	☽ ⚼ ♂
16 42	☽ # ♀
12 Monday	
02 36	☉ # ♄
03 44	☽ ⚻ ♄
04 42	☽ Q ♇
04 54	☽ # ♅
05 27	☿ △ ♇
09 27	☽ ☍ ☉
10 22	☽ ∥ ♅
19 44	☽ ∥ ♀
20 56	☽ ∠ ♃
23 17	☽ △ ♂
13 Tuesday	
01 45	☽ ☍ ♀
03 58	☽ ∥ ☉
04 44	☽ ⊻ ♆
06 21	☽ # ♄
12 59	☽ ☌ ♅
14 53	☽ ⊥ ♃
16 38	☽ □ ♇
16 45	☽ ⊥ ♆
22 43	☽ ⚻ ☿
14 Wednesday	
02 54	☽ ⊻ ♃
06 01	♀ ⚻ ♆
10 50	☽ ∠ ♆
11 44	☽ ∥ ☿
12 59	☽ ± ☿
16 26	☽ ☍ ♄
15 Thursday	
01 43	☿ △ ♃
02 49	☽ # ♆
03 04	☽ ⚻ ☉
09 24	☽ ∥ ♃
10 31	☽ ⚼ ☿
12 48	☿ ⊥ ♄
15 06	☽ □ ♂
16 20	☽ ± ☉
17 10	☽ ✶ ♆
21 23	☽ ⚻ ♀
16 Friday	
01 27	☽ ⊻ ♅
05 20	☽ △ ♇
11 00	☽ ± ♀
12 15	☽ ⚼ ☉
13 12	♀ ∥ ♅
13 36	☽ ⊥ ♅
15 26	☽ ☌ ♃
17 29	☽ Q ♆
18 27	♇ St D
22 35	☽ △ ☿
17 Saturday	
03 13	☽ # ♇
04 29	☿ ∠ ♂
05 14	♂ ⚻ ♆
05 38	☽ Q ♂
05 53	☽ ⚻ ♄
07 23	☽ ⚼ ♀
07 47	☽ ∠ ♅
11 05	♀ ☍ ♅
11 45	☽ ⚼ ♇
18 07	☽ ± ♄
21 20	☽ △ ☉
18 Sunday	
00 51	☉ ⚼ ♃
01 24	☽ ∥ ♂
03 33	♀ ± ♃
05 46	☽ □ ♆
05 52	☽ ± ♇
07 09	☽ ✶ ♂
07 58	♀ # ♅
12 22	☽ ⚼ ♄
13 51	☽ ✶ ♅
17 02	☽ △ ♀
17 52	☽ ⚻ ♇
23 36	☿ ⊻ ♄
23 43	♀ ± ♆
19 Monday	
01 08	♀ □ ♇
03 27	☽ ⊻ ♃
13 33	☽ Q ♅
14 30	☽ ∠ ♂
15 12	☽ ⊥ ♃
18 20	☽ △ ♄
21 30	☽ □ ☿
20 Tuesday	
00 36	☉ # ♀
08 36	☽ ∠ ♃
08 55	☽ ⊥ ♂
13 39	☽ □ ☉
16 33	☽ △ ♆
20 59	☽ ⊻ ♂
21 Wednesday	
00 10	☽ □ ♅
04 09	☽ ☍ ♇
09 38	☽ □ ♀
12 52	☽ ✶ ♃
13 46	☽ Q ☿
14 29	☿ # ♄
18 21	☽ ∥ ♂
20 39	☽ ⚼ ♆
22 Thursday	
02 07	☽ Q ☉
02 17	☉ ⚻ ♆
03 37	☽ □ ♄
09 23	☽ # ♇
09 56	☉ ∥ ♅
10 44	☽ Q ♃
12 59	☽ ± ♆
15 17	☽ ✶ ☿
17 23	♀ ⚻ ♃
20 53	☿ ⚼ ♃
21 50	☽ Q ♀
23 40	☽ ⚻ ♆
23 Friday	
01 22	☽ ✶ ☉
06 24	☽ ☌ ♂
06 39	☽ △ ♅
10 30	☽ ⚻ ♇
11 24	♂ △ ♅
11 55	☽ Q ♄
16 28	☿ # ♀
18 15	☽ □ ♃
20 49	☽ ✶ ♀
20 51	☽ ± ♇
21 39	☽ ∠ ☿
22 54	☽ ∥ ♃
24 Saturday	
02 59	☽ # ♆
05 14	☽ ∠ ☉
06 22	☉ # ♅
08 19	☽ ⚼ ♅
08 36	☽ ✶ ♄
12 06	☽ ⚼ ♇
15 07	☽ ⊥ ☿
21 23	☽ ⊥ ☉
23 16	☽ ± ♅
25 Sunday	
00 24	☽ ∠ ♀
02 29	☽ ⊻ ☿
02 39	☽ ☍ ♆
03 48	☿ ⚻ ♆
07 57	☽ ⊻ ☉
09 05	☽ ⚻ ♅
09 38	☽ ∠ ♄
11 10	☽ ⊻ ♂
12 49	☽ △ ♇
13 36	☽ # ♄
16 26	☽ ⊥ ♀
16 52	☽ # ♀
19 47	☽ △ ♃
21 17	☽ ⊥ ♂
26 Monday	
00 15	☉ ☍ ♅
00 18	☽ ⊥ ♄
02 23	♀ ⚼ ♆
02 58	☽ ⊻ ♀
05 29	☽ ∥ ☿
05 44	☽ # ☉
06 47	☉ # ☿
09 03	☽ ∥ ♅
09 58	☽ ⊻ ♄
11 36	☽ # ♅
12 16	☽ ∠ ♂
14 00	☽ # ☿
14 09	☉ ± ♃
15 34	☽ ∥ ☉
19 34	☽ ⚼ ♃
27 Tuesday	
02 41	☽ ⚻ ♆
04 20	♂ ⚻ ♇
06 30	☿ ☍ ♅
06 54	☽ ∥ ♀
07 05	☽ ∥ ♄
08 14	☿ ∥ ♅
08 49	☽ ☍ ♅
09 08	☽ ☌ ☿
09 23	♀ ∥ ♄
09 41	☽ ± ♃
11 09	☽ ☌ ☉
12 07	☽ ± ♆
12 35	☽ □ ♇
12 55	☽ ✶ ♂
13 34	☿ ± ♃
19 03	☽ ⚻ ♃
28 Wednesday	
01 59	☿ # ♅
02 03	☉ ± ♆
02 14	☽ ⚼ ♆
06 33	☽ ☌ ♀
08 32	☽ Q ♂
09 37	☉ □ ♇
09 37	☽ ☌ ♄
10 11	☿ ± ♆
14 26	☿ □ ♇
15 50	☽ ∥ ♆
16 54	☽ Q ♇
18 47	☽ # ♃
20 16	☉ ☌ ☿
29 Thursday	
01 51	☽ △ ♆
07 27	☿ ✶ ♂
07 58	☽ ⚻ ♅
11 55	☽ ✶ ♇
13 45	☽ ⊻ ☉
14 12	☽ □ ♂
14 49	☽ ⊻ ☿
17 31	☽ ± ♅
18 12	☽ ☍ ♃
18 38	♂ Q ♄
23 48	♀ ☌ ♄
30 Friday	
00 03	☽ ⊥ ☉
01 45	☽ ⊥ ☿
03 35	♀ Q ♂
06 04	☽ ∥ ♇
06 43	☉ ✶ ♂
08 00	☽ ⚼ ♅
09 44	☽ # ♂
09 51	☽ ⊻ ♄
10 41	☽ ⊻ ♀
12 07	☽ ∠ ♇
15 43	☽ ∠ ☉
18 16	☿ ⚻ ♃
18 24	☽ ∠ ☿
19 45	☽ ⊥ ♄
21 26	☽ ⊥ ♀

Date	Sidereal time h m s	LONGITUDES Sun ☉ ° ' "	Moon ☽ ° ' "	Moon ☽ 24.00 ° ' "	Mercury ☿ ° '	Venus ♀ ° '	Mars ♂ ° '	Jupiter ♃ ° '	Saturn ♄ ° '	Uranus ♅ ° '	Neptune ♆ ° '	Pluto ♇ ° '
01	12 39 30	07 ♎ 58 07	04 ♐ 22 24	11 ♐ 28 03	10 ♎ 04	20 ♎ 22	07 ♌ 30	08 ♉ 41	18 ♎ 41	02 ♈ 21	28 ♒ 33	04 ♑ 57
02	12 43 26	08 57 08	18 ♐ 27 08	25 ♐ 19 43	11 48	21 37	08 05	08 R 35	18 48	02 R 19	28 R 32	04 57
03	12 47 23	09 56 10	02 ♑ 05 59	08 ♑ 46 15	13 32	22 52	08 41	08 29	18 55	02 17	28 30	04 58
04	12 51 19	10 55 14	15 ♑ 20 53	21 ♑ 50 19	15 15	24 06	09 16	08 23	19 02	02 14	28 29	04 58
05	12 55 16	11 54 20	28 ♑ 15 01	04 ♒ 35 25	16 57	25 21	09 51	08 16	19 10	02 12	28 28	04 59
06	12 59 12	12 53 28	10 ♒ 51 58	17 ♒ 05 07	18 38	26 35	10 26	08 10	19 17	02 10	28 27	04 59
07	13 03 09	13 52 38	23 ♒ 15 16	29 ♒ 22 48	20 18	27 50	11 01	08 03	19 24	02 07	28 26	05 00
08	13 07 06	14 51 49	05 ♓ 28 01	11 ♓ 31 16	21 58	29 ♎ 05	11 36	07 56	19 32	02 05	28 25	05 01
09	13 11 02	15 51 02	17 ♓ 32 47	23 ♓ 32 50	23 37	00 ♏ 19	12 11	07 49	19 39	02 03	28 24	05 01
10	13 14 59	16 50 17	29 ♓ 31 37	05 ♈ 29 20	25 15	01 34	12 46	07 42	19 46	02 00	28 23	05 02
11	13 18 55	17 49 34	11 ♈ 26 10	17 ♈ 22 17	26 52	02 48	13 21	07 35	19 54	01 58	28 22	05 03
12	13 22 52	18 48 53	23 ♈ 17 53	29 ♈ 13 10	28 ♎ 29	04 03	13 55	07 28	20 01	01 56	28 21	05 04
13	13 26 48	19 48 14	05 ♉ 08 19	11 ♉ 03 36	00 ♏ 05	05 18	14 29	07 20	20 08	01 53	28 20	05 04
14	13 30 45	20 47 38	16 ♉ 59 15	22 ♉ 55 35	01 40	06 32	15 04	07 13	20 16	01 51	28 20	05 05
15	13 34 41	21 47 03	28 ♉ 52 56	04 ♊ 51 41	03 14	07 47	15 38	07 05	20 23	01 49	28 19	05 06
16	13 38 38	22 46 31	10 ♊ 52 15	16 ♊ 55 05	04 48	09 01	16 12	06 58	20 30	01 47	28 18	05 07
17	13 42 35	23 46 01	23 ♊ 00 42	29 ♊ 09 36	06 21	10 16	16 46	06 50	20 38	01 45	28 17	05 08
18	13 46 31	24 45 33	05 ♋ 22 21	11 ♋ 39 30	07 54	11 31	17 20	06 42	20 45	01 42	28 16	05 09
19	13 50 28	25 45 08	18 ♋ 01 38	24 ♋ 29 16	09 25	12 45	17 53	06 34	20 52	01 40	28 16	05 10
20	13 54 24	26 44 44	01 ♌ 02 55	07 ♌ 43 02	10 57	14 00	18 27	06 26	20 59	01 38	28 15	05 11
21	13 58 21	27 44 24	14 ♌ 29 58	21 ♌ 23 58	12 27	15 14	19 00	06 19	21 07	01 36	28 14	05 12
22	14 02 17	28 44 06	28 ♌ 25 07	05 ♍ 33 22	13 57	16 29	19 34	06 11	21 14	01 34	28 14	05 13
23	14 06 14	29 ♎ 43 48	12 ♍ 48 25	20 ♍ 09 47	15 27	17 44	20 07	06 02	21 21	01 32	28 13	05 14
24	14 10 10	00 ♏ 43 34	27 ♍ 36 47	05 ♎ 08 28	16 55	18 58	20 40	05 54	21 29	01 30	28 13	05 15
25	14 14 07	01 43 22	12 ♎ 43 46	20 ♎ 21 24	18 23	20 13	21 13	05 46	21 36	01 28	28 12	05 16
26	14 18 04	02 43 12	28 ♎ 00 02	05 ♏ 38 18	19 51	21 28	21 46	05 38	21 43	01 26	28 12	05 18
27	14 22 00	03 43 04	13 ♏ 14 49	20 ♏ 48 19	21 18	22 42	22 18	05 30	21 50	01 24	28 11	05 19
28	14 25 57	04 42 58	28 ♏ 17 40	05 ♐ 41 54	22 44	23 57	22 51	05 22	21 58	01 22	28 11	05 20
29	14 29 53	05 42 54	13 ♐ 00 14	20 ♐ 12 09	24 09	25 11	23 23	05 14	22 05	01 20	28 11	05 21
30	14 33 50	06 42 52	27 ♐ 17 16	04 ♑ 15 26	25 34	26 26	23 55	05 06	22 12	01 18	28 10	05 23
31	14 37 46	07 ♏ 42 51	11 ♑ 06 39	17 ♑ 51 04	26 ♏ 58	27 ♏ 41	24 ♌ 27	04 ♉ 57	22 ♎ 19	01 ♈ 17	28 ♒ 10	05 ♑ 24

Date	Moon True ☊ ° '	Moon Mean ☊ ° '	Moon ☽ Latitude ° '	DECLINATIONS Sun ☉ ° '	Moon ☽ ° '	Mercury ☿ ° '	Venus ♀ ° '	Mars ♂ ° '	Jupiter ♃ ° '	Saturn ♄ ° '	Uranus ♅ ° '	Neptune ♆ ° '	Pluto ♇ ° '
01	16 ♐ 49	17 ♐ 49	01 S 06	03 S 10	22 S 05	02 S 57	07 S 06	19 N 29	13 N 00	05 S 14	00 N 14	12 S 30	19 S 10
02	16 D 48	17 46	00 N 09	03 33	22 47	03 43	07 36	19 21	12 58	05 17	00 13	12 30	19 10
03	16 49	17 43	01 21	03 56	22 05	04 29	08 05	19 12	12 56	05 20	00 12	12 31	19 10
04	16 R 49	17 40	02 26	04 19	20 08	05 15	08 34	19 04	12 54	05 22	00 11	12 31	19 10
05	16 47	17 36	03 22	04 42	17 13	06 00	09 03	18 55	12 52	05 25	00 10	12 32	19 11
06	16 44	17 33	04 07	05 05	13 33	06 44	09 32	18 47	12 50	05 28	00 09	12 32	19 11
07	16 38	17 30	04 39	05 28	09 22	07 28	10 01	18 38	12 47	05 31	00 08	12 32	19 11
08	16 29	17 27	04 58	05 51	04 53	08 12	10 30	18 29	12 45	05 33	00 07	12 33	19 11
09	16 19	17 24	05 04	06 14	00 S 16	08 55	10 58	18 20	12 43	05 36	00 06	12 33	19 11
10	16 07	17 21	04 55	06 37	04 N 20	09 37	11 26	18 11	12 41	05 39	00 06	12 33	19 12
11	15 56	17 17	04 35	07 00	08 44	10 19	11 54	18 02	12 38	05 42	00 05	12 34	19 12
12	15 45	17 14	04 02	07 22	12 47	11 00	12 21	17 53	12 36	05 44	00 04	12 34	19 12
13	15 36	17 11	03 18	07 45	16 21	11 40	12 48	17 44	12 33	05 47	00 03	12 34	19 12
14	15 30	17 08	02 26	08 07	19 15	12 20	13 15	17 35	12 31	05 50	00 02	12 35	19 12
15	15 26	17 05	01 28	08 29	21 20	12 59	13 42	17 26	12 28	05 53	00 N 01	12 35	19 13
16	15 24	17 02	00 N 24	08 51	22 29	13 37	14 08	17 16	12 26	05 55	00 00	12 35	19 13
17	15 D 25	16 58	00 S 41	09 13	22 34	14 15	14 34	17 07	12 23	05 58	00 S 01	12 35	19 13
18	15 26	16 55	01 46	09 35	21 34	14 51	14 59	16 58	12 21	06 01	00 01	12 36	19 13
19	15 27	16 52	02 47	09 57	19 28	15 27	15 24	16 48	12 18	06 03	00 02	12 36	19 13
20	15 R 27	16 49	03 41	10 19	16 20	16 03	15 49	16 39	12 16	06 06	00 03	12 36	19 14
21	15 26	16 46	04 25	10 40	12 15	16 37	16 13	16 29	12 13	06 09	00 04	12 36	19 14
22	15 22	16 42	04 55	11 01	07 24	17 10	16 37	16 20	12 10	06 12	00 05	12 37	19 14
23	15 17	16 39	05 08	11 23	02 N 00	17 43	17 01	16 10	12 08	06 14	00 06	12 37	19 14
24	15 11	16 36	05 02	11 44	03 S 40	18 15	17 24	16 01	12 05	06 17	00 06	12 37	19 14
25	15 04	16 33	04 34	12 04	09 14	18 46	17 47	15 51	12 03	06 20	00 07	12 37	19 14
26	14 57	16 30	03 46	12 25	14 17	19 16	18 09	15 41	12 00	06 22	00 08	12 37	19 15
27	14 53	16 27	02 43	12 45	18 24	19 45	18 30	15 32	11 57	06 25	00 09	12 37	19 15
28	14 49	16 23	01 29	13 06	21 13	20 12	18 52	15 22	11 55	06 28	00 09	12 38	19 15
29	14 48	16 20	00 S 10	13 26	22 31	20 39	19 12	15 12	11 52	06 30	00 10	12 38	19 15
30	14 D 49	16 17	01 N 07	13 45	22 17	21 05	19 33	15 03	11 49	06 33	00 11	12 38	19 15
31	14 ♐ 50	16 ♐ 14	02 N 18	14 S 05	20 S 40	21 S 30	19 S 52	14 N 53	11 N 47	06 S 35	00 S 11	12 S 38	19 S 15

ZODIAC SIGN ENTRIES

Date	h	m	Planets
01	04	42	☽ ♐
03	08	16	☽ ♑
05	15	18	☽ ♒
08	01	13	☽ ♓
09	05	50	♀ ♏
10	12	57	☽ ♈
13	01	35	☽ ♉
13	10	52	☿ ♏
15	14	15	☽ ♊
18	01	38	☽ ♋
20	10	06	☽ ♌
22	14	41	☽ ♍
23	18	30	☉ ♏
24	15	49	☽ ♎
26	15	08	☽ ♏
28	14	45	☽ ♐
30	16	39	☽ ♑

LATITUDES

Date	Mercury ☿ ° '	Venus ♀ ° '	Mars ♂ ° '	Jupiter ♃ ° '	Saturn ♄ ° '	Uranus ♅ ° '	Neptune ♆ ° '	Pluto ♇ ° '
01	01 N 07	00 N 56	01 N 08	01 S 28	02 N 15	00 S 46	00 S 33	04 N 11
04	00 49	00 50	01 10	01 29	02 15	00 46	00 33	04 10
07	00 30	00 44	01 13	01 29	02 15	00 46	00 33	04 10
10	00 N 10	00 38	01 16	01 29	02 15	00 46	00 33	04 09
13	00 S 11	00 31	01 19	01 29	02 15	00 46	00 33	04 08
16	00 32	00 24	01 21	01 29	02 15	00 46	00 33	04 08
19	00 52	00 16	01 24	01 29	02 15	00 46	00 33	04 07
22	01 12	00 09	01 27	01 29	02 15	00 46	00 33	04 06
25	01 31	00 N 01	01 30	01 29	02 15	00 46	00 33	04 06
28	01 49	00 S 06	01 33	01 29	02 15	00 46	00 33	04 05
31	02 S 05	00 S 14	01 N 36	01 S 29	02 N 16	00 S 46	00 S 33	04 N 04

DATA

Julian Date	2455836
Delta T	+68 seconds
Ayanamsa	24° 01' 33"
Synetic vernal point	05° ♓ 05' 27"
True obliquity of ecliptic	23° 26' 15"

MOON'S PHASES, APSIDES AND POSITIONS ☽

Date	h	m	Phase	Longitude ° '	Eclipse Indicator
04	03	15	☽	10 ♑ 34	
12	02	06	○	18 ♈ 24	
20	03	30	☾	26 ♋ 24	
26	19	56	●	03 ♏ 03	

Day	h	m		
12	11	25	Apogee	
26	12	19	Perigee	
02	11	33	Max dec	22° S 47'
09	13	21	0N	
17	02	11	Max dec	22° N 40'
23	20	32	0S	
29	20	10	Max dec	22° S 36'

LONGITUDES

Date	Chiron ⚷ ° '	Ceres ⚳ ° '	Pallas ⚴ ° '	Juno ⚵ ° '	Vesta ⚶ ° '	Black Moon Lilith ⚸ ° '
01	01 ♓ 25	20 ♓ 16	27 ♑ 20	21 ♎ 44	06 ♒ 50	21 ♈ 30
11	01 ♓ 06	18 ♓ 29	28 ♑ 01	25 ♎ 14	08 ♒ 10	22 ♈ 37
21	00 ♓ 51	17 ♓ 12	29 ♑ 10	28 ♎ 45	10 ♒ 07	23 ♈ 44
31	00 ♓ 42	16 ♓ 31	00 ♒ 44	02 ♏ 15	12 ♒ 34	24 ♈ 51

All ephemeris data is given at 12.00 UT and the Moon's longitude is additionally given for 24.00 UT

ASPECTARIAN

h m	Aspects
01 Saturday	
02 17	☽ □ ♆
02 56	☽ ⊥ ♇
08 38	☽ △ ♅
10 49	☽ ∠ ♄
12 58	☽ ⊻ ♇
13 50	☽ ∠ ♀
17 29	☽ △ ♂
18 31	☽ ⚹ ☉
19 14	☽ ⊼ ♃
22 58	☽ ⚹ ☿
02 Sunday	
01 02	☉ ∥ ☿
03 57	☉ ⊼ ♃
05 23	☽ ± ♃
08 41	☽ Q ♆
12 36	☽ ⚹ ♄
16 40	☽ Q ☉
18 03	☽ ⚹ ♀
20 26	☽ ⚼ ♂
20 53	☽ ⚼ ♃
22 40	☽ Q ☿
03 Monday	
05 21	♂ □ ♃
05 37	☽ ⚹ ♆
09 53	☽ Q ♄
11 39	☿ ⚼ ♆
12 19	☽ □ ♅
13 05	☽ ± ♂
13 59	♀ Q ♇
17 08	☽ ☌ ♇
17 27	☽ Q ♀
18 03	♂ # ♇
23 24	☽ △ ♃
04 Tuesday	
00 23	☽ ⊼ ♂
03 15	☽ □ ☉
08 36	☽ ∠ ♆
11 47	☽ □ ☿
16 19	☿ ∥ ♄
18 52	☽ □ ♄
20 44	☽ ∥ ♇
21 00	☽ Q ♅
22 10	☽ # ♂
05 Wednesday	
01 11	☽ ⊥ ♆
05 58	☽ □ ♀
12 25	☽ ⊻ ♆
19 26	☽ ⚹ ♅
06 Thursday	
00 45	☽ ⊻ ♇
06 52	☽ □ ♃
11 08	☽ ☍ ♂
12 14	☽ ⊥ ♇
12 58	♃ ⊥ ♅
16 14	☽ △ ☉
16 21	☽ # ♃
18 02	☽ ∥ ♆
22 01	☿ ☌ ♄
07 Friday	
00 06	☽ ∠ ♅
01 25	☉ ⚼ ♆
04 26	☽ △ ♄
05 21	☽ △ ☿
05 39	☽ ∠ ♇
08 46	☽ ∥ ♀
11 00	♂ ± ♇
14 36	☉ ∥ ♄
17 25	☽ Q ♃
17 35	☽ ⊥ ♅
20 51	☽ ∥ ☿
21 59	☽ △ ♀
22 08	☽ ☌ ♆
23 28	♀ △ ♆
23 59	☽ ⚼ ☉
08 Saturday	
05 20	☽ ⊻ ♅
07 16	☽ ∥ ☉
08 29	☽ ∥ ♄
10 07	☽ ⚼ ♄
11 06	☽ ⚹ ♇
15 26	☽ ⚼ ☿
16 51	☽ ⚹ ♃
19 19	☽ ± ☉
09 Sunday	
00 47	☽ ⊼ ♂
03 18	☿ Q ♇
04 09	☽ ± ♄
07 03	☽ ⚼ ♀
08 19	☽ ⊼ ☉
10 57	☽ Q ♇
12 09	☽ ± ☿
12 48	☽ # ♅
13 21	☽ ± ♂
13 54	☽ ∥ ♅
16 15	☽ ⊼ ♄
22 27	☽ ∠ ♃
10 Monday	
00 59	☿ Q ♂
02 03	☽ ⊼ ☿
03 07	☽ ± ♀
08 17	☽ ⚼ ♂
09 43	☽ ⊻ ♆
16 20	☽ ⊥ ♃
16 34	☽ ⊼ ♀
16 58	☽ ☌ ♅
19 07	☽ # ♄
20 15	♀ ⊼ ♅
21 46	☽ ⊥ ♆
23 06	☽ □ ♇
11 Tuesday	
01 26	☽ # ☉
04 18	☽ ⊻ ♃
15 54	☽ ∠ ♆
16 03	☽ △ ♂
22 52	☽ # ☿
12 Wednesday	
02 06	☽ ☍ ☉
05 17	☽ ☍ ♄
08 53	☽ # ♀
10 10	☿ △ ♆
10 36	☽ # ♆
10 48	☽ ∥ ♃
22 14	☽ ⚹ ♆
23 36	♀ ∥ ♆
23 59	♀ # ♃
13 Thursday	
00 08	☽ ☍ ☿
03 42	♃ # ♆
05 26	☽ ⊻ ♅
07 43	♀ ⚹ ♇
11 52	☽ △ ♇
12 21	☽ ☍ ♀
16 25	☽ ☌ ♃
17 34	☽ ⊥ ♅
21 13	☉ ☌ ♄
22 13	☽ ∥ ♂
22 32	☽ Q ♆
14 Friday	
07 54	☽ □ ♂
11 37	☽ # ♇
11 44	☽ ∠ ♅
14 51	☿ ⊼ ♅
18 17	☽ ⚼ ♇
18 22	☿ # ♃
18 41	☽ ⊼ ♄
20 24	☽ ⊼ ☉
21 06	☿ ∥ ♆
23 53	♀ ☍ ♃
15 Saturday	
06 55	☽ ± ♄
09 35	☽ ± ☉
10 51	☽ □ ♆
12 27	☽ ± ♇
12 40	♀ ± ♅
17 53	☽ ⚹ ♅
22 01	☽ Q ♂
22 03	☽ ⊼ ☿
16 Sunday	
00 30	☽ ⊼ ♇
01 11	☽ ⚼ ♄
04 17	☽ ⊻ ♃
05 16	☽ ⚼ ☉
07 54	☽ ⊼ ♀
11 50	☽ ± ☿
16 07	☽ ⊥ ♃
16 58	☿ ⚹ ♇
17 46	☽ Q ♅
20 25	☉ Q ♇
21 12	☽ ± ♀
23 06	☽ ⚹ ♂
17 Monday	
07 16	☽ △ ♄
08 16	☽ ⚼ ☿
09 43	☽ ∠ ♃
11 08	♂ ⚼ ♅
13 37	☽ △ ☉
16 55	☽ ⚼ ♀
18 55	☿ ☍ ♃
22 18	☽ △ ♆
18 Tuesday	
04 57	☽ □ ♅
05 51	☽ ∠ ♂
09 10	☿ ± ♅
11 34	☽ ☍ ♇
14 32	☽ ⚹ ♃
17 30	☽ △ ☿
23 54	☽ ⊥ ♂
19 Wednesday	
01 00	☽ △ ♀
03 03	☽ ⚼ ♆
04 56	☿ ∥ ♀
11 44	☽ ⊻ ♂
13 01	☽ Q ♃
14 12	☽ # ♇
17 21	☽ □ ♄
19 53	☽ ± ♆
20 Thursday	
03 30	☽ □ ☉
06 55	☽ ⊼ ♆
09 45	☽ ∥ ♂
13 04	☽ △ ♅
13 36	☽ # ☿
14 59	☽ # ♀
19 28	☽ ⊼ ♇
21 37	☽ □ ♃
21 Friday	
02 24	☽ Q ♄
06 11	☽ ± ♇
07 50	☿ # ♂
07 57	☽ □ ☿
10 05	☽ # ♆
12 10	☽ ∥ ♃
13 26	☽ □ ♀
14 21	☽ Q ☉
15 40	☽ ⚼ ♅
19 37	☽ # ☉
20 11	☽ ☌ ♂
21 56	☽ ⚼ ♇
23 26	♀ # ♂
23 36	☽ ⚹ ♄
23 57	☉ △ ♆
22 Saturday	
07 10	☽ ± ♅
11 41	☽ ☍ ♆
12 35	☽ ⚹ ☉
13 31	♀ ⚼ ♅
17 18	☽ ⊼ ♅
17 29	☽ # ♄
18 40	☽ Q ☿
23 11	☽ Q ♀
23 27	☽ △ ♇
23 Sunday	
00 55	☽ △ ♃
01 15	☽ ∠ ♄
15 23	☽ ∠ ☉
16 13	☽ ⊥ ♄
16 48	☽ ⚹ ☿
17 26	♂ ⚼ ♇
20 07	☽ # ♅
20 47	☽ ⚹ ♀
20 57	☽ ∥ ♅
24 Monday	
00 23	☽ ⊻ ♂
01 18	☽ ⚼ ♃
02 03	☽ ⊻ ♄
05 15	☿ ⚼ ♅
07 02	☽ ⊥ ☉
10 25	☽ ⊥ ♂
12 57	☽ ⊼ ♆
15 38	☽ ± ♃
17 20	☽ ⊻ ☉
18 12	☽ ☍ ♅
19 38	☽ ∠ ☿
22 31	☽ ± ♆
23 03	☽ ∠ ♀
23 14	☽ ∥ ♄
25 Tuesday	
00 12	☽ □ ♇
01 06	☽ ⊼ ♃
01 19	☽ ∠ ♂
05 59	☉ ⊼ ♅
10 15	☉ # ♃
11 25	☽ ⊥ ☿
12 45	☽ ⚼ ♆
13 09	♀ ∠ ♇
14 33	☽ ⊥ ♀
21 51	☽ ⊻ ☿
26 Wednesday	
00 49	☽ ⊻ ♀
00 49	☽ # ♃
01 51	☽ ⚹ ♂
02 02	☽ ∥ ☉
02 04	☽ ☌ ♄
03 42	☽ ∥ ♆
04 36	☽ Q ♇
09 37	♂ ⚹ ♄
11 14	☿ ∥ ♇
12 18	☽ △ ♆
17 23	☽ ⊼ ♅
17 34	♀ ⊻ ♄
19 14	☽ # ♂
19 28	☿ ∠ ♇
19 56	☽ ☌ ☉
21 23	☽ Q ♂
22 23	♀ □ ♂
23 28	☽ ⚹ ♇
23 53	☽ ☍ ♃
27 Thursday	
02 35	☉ ∥ ♆
02 47	☽ ± ♅
12 49	☽ ∥ ♀
16 59	☽ ⚼ ♅
18 04	☽ ∥ ♇
21 55	☿ ⊻ ♄
23 14	☽ ∠ ♇
23 53	☽ ∥ ☿
28 Friday	
01 46	☽ ⊻ ♄
02 08	☽ ☌ ☿
02 56	☽ □ ♂
04 23	☽ ☌ ♀
11 27	☽ ⊥ ♄
11 49	☽ □ ♆
13 41	☽ ⊥ ♇
15 08	☿ □ ♂
16 29	♃ △ ♇
16 57	☽ △ ♅
18 52	☉ Q ♂
23 09	☽ ⊻ ☉
23 21	☽ ⊼ ♃
23 25	☽ ⊻ ♇
29 Saturday	
01 42	☉ ☍ ♃
02 11	☽ ∠ ♄
03 11	☉ ⚹ ♇
09 06	☽ ± ♃
09 43	☽ ⊥ ☉
15 15	♀ ∥ ♇
17 15	☽ Q ♆
23 56	☽ ⚼ ♃
30 Sunday	
01 50	☽ ∠ ☉
03 17	☽ ⚹ ♄
06 03	☽ △ ♂
08 45	☽ ⊻ ☿
10 24	☽ ⊻ ♀
13 30	☽ ⚹ ♆
18 53	☽ □ ♅
20 10	☽ ⊥ ☿
21 43	☽ ⊥ ♀
31 Monday	
00 00	☽ Q ♄
01 19	☽ △ ♃
01 47	☉ ± ♅
01 58	☽ ☌ ♇
03 47	☽ ∥ ☿
05 33	☽ ⚹ ☉
08 58	☽ ⚼ ♂
13 41	☽ ∠ ☿
15 03	☽ ∠ ♀
15 38	☽ ∠ ♆
19 08	☽ ∥ ♀
21 19	♀ □ ♆

NOVEMBER 2011

LONGITUDES

Date	Sidereal time h m s	Sun ☉ ° ' "	Moon ☽ ° ' "	Moon ☽ 24.00 ° ' "	Mercury ☿ ° '	Venus ♀ ° '	Mars ♂ ° '	Jupiter ♃ ° '	Saturn ♄ ° '	Uranus ♅ ° '	Neptune ♆ ° '	Pluto ♇ ° '
01	14 41 43	08 ♏ 42 52	24 ♑ 28 59	01 ♒ 00 45	28 ♏ 21	28 ♏ 55	24 ♌ 59	04 ♉ 49	22 ♎ 26	01 ♈ 15	28 ♒ 09	05 ♑ 25
02	14 45 39	09 42 54	07 ♒ 26 48	13 ♒ 47 38	29 ♏ 43	00 ♐ 10	25 31	04 R 41	22 33	01 R 13	28 R 09	05 27
03	14 49 36	10 42 58	20 ♒ 03 45	26 ♒ 15 42	01 ♐ 05	01 24	26 03	04 33	22 41	01 11	28 09	05 28
04	14 53 33	11 43 04	02 ♓ 23 59	08 ♓ 29 07	02 25	02 39	26 34	04 25	22 48	01 10	28 09	05 29
05	14 57 29	12 43 11	14 ♓ 31 36	20 ♓ 31 55	03 44	03 54	27 05	04 17	22 55	01 08	28 09	05 31
06	15 01 26	13 43 19	26 ♓ 30 29	02 ♈ 27 43	05 02	05 08	27 37	04 09	23 02	01 06	28 09	05 32
07	15 05 22	14 43 30	08 ♈ 23 59	14 ♈ 19 38	06 18	06 23	28 08	04 01	23 09	01 05	28 08	05 34
08	15 09 19	15 43 42	20 ♈ 14 58	26 ♈ 10 17	07 34	07 37	28 38	03 53	23 16	01 03	28 08	05 35
09	15 13 15	16 43 55	02 ♉ 05 51	08 ♉ 01 53	08 47	08 52	29 09	03 46	23 23	01 02	28 08	05 37
10	15 17 12	17 44 10	13 ♉ 58 37	19 ♉ 56 17	09 58	10 06	29 ♌ 39	03 38	23 30	01 00	28 D 08	05 38
11	15 21 08	18 44 27	25 ♉ 55 06	01 ♊ 55 16	11 08	11 21	00 ♍ 10	03 30	23 36	00 59	28 08	05 40
12	15 25 05	19 44 46	07 ♊ 57 03	14 ♊ 00 39	12 15	12 35	00 40	03 23	23 43	00 58	28 08	05 42
13	15 29 02	20 45 07	20 ♊ 06 22	26 ♊ 14 27	13 20	13 50	01 10	03 15	23 50	00 56	28 08	05 43
14	15 32 58	21 45 29	02 ♋ 25 12	08 ♋ 38 57	14 21	15 04	01 40	03 08	23 57	00 55	28 09	05 45
15	15 36 55	22 45 54	14 ♋ 56 02	21 ♋ 16 48	15 20	16 19	02 09	03 01	24 04	00 54	28 09	05 47
16	15 40 51	23 46 20	27 ♋ 41 37	04 ♌ 10 49	16 14	17 33	02 38	02 54	24 10	00 53	28 09	05 48
17	15 44 48	24 46 47	10 ♌ 44 47	17 ♌ 23 48	17 05	18 48	03 08	02 47	24 17	00 52	28 09	05 50
18	15 48 44	25 47 17	24 ♌ 08 09	00 ♍ 58 03	17 50	20 02	03 37	02 40	24 24	00 50	28 10	05 52
19	15 52 41	26 47 49	07 ♍ 53 38	14 ♍ 54 54	18 30	21 17	04 05	02 33	24 30	00 49	28 10	05 53
20	15 56 37	27 48 22	21 ♍ 01 46	29 ♍ 13 59	19 05	22 31	04 34	02 26	24 37	00 48	28 10	05 55
21	16 00 34	28 48 57	06 ♎ 31 10	13 ♎ 52 46	19 32	23 46	05 02	02 20	24 43	00 47	28 11	05 57
22	16 04 31	29 ♏ 49 34	21 ♎ 18 04	28 ♎ 46 14	19 52	25 00	05 30	02 14	24 50	00 47	28 11	05 59
23	16 08 27	00 ♐ 50 13	06 ♏ 16 18	13 ♏ 47 12	20 04	26 15	05 58	02 07	24 56	00 46	28 11	06 01
24	16 12 24	01 50 53	21 ♏ 17 49	28 ♏ 47 03	20 R 07	27 29	06 26	02 01	25 03	00 45	28 12	06 02
25	16 16 20	02 51 35	06 ♐ 13 49	13 ♐ 37 06	20 00	28 44	06 53	01 55	25 09	00 44	28 12	06 04
26	16 20 17	03 52 18	20 ♐ 56 01	28 ♐ 09 46	19 43	29 ♐ 58	07 21	01 49	25 15	00 43	28 13	06 06
27	16 24 13	04 53 02	05 ♑ 17 47	12 ♑ 19 37	19 15	01 ♑ 13	07 48	01 44	25 21	00 43	28 14	06 08
28	16 28 10	05 53 48	19 ♑ 14 58	26 ♑ 03 44	18 36	02 27	08 14	01 38	25 28	00 42	28 14	06 10
29	16 32 06	06 54 34	02 ♒ 45 57	09 ♒ 21 44	17 47	03 41	08 41	01 33	25 34	00 42	28 15	06 12
30	16 36 03	07 ♐ 55 22	15 ♒ 51 24	22 ♒ 15 16	16 ♐ 47	04 ♑ 56	09 ♍ 07	01 ♉ 28	25 ♎ 40	00 ♈ 41	28 ♒ 16	06 ♑ 14

Moon

Date	Moon True ☊ ° '	Moon Mean ☊ ° '	Moon ☽ Latitude ° '
01	14 ♐ 52	16 ♐ 11	03 N 19
02	14 D 53	16 08	04 08
03	14 R 52	16 04	04 44
04	14 50	16 01	05 05
05	14 47	15 58	05 12
06	14 42	15 55	05 06
07	14 37	15 52	04 46
08	14 31	15 48	04 14
09	14 26	15 45	03 31
10	14 22	15 42	02 39
11	14 19	15 39	01 39
12	14 18	15 36	00 N 35
13	14 D 18	15 33	00 S 32
14	14 19	15 29	01 38
15	14 21	15 26	02 41
16	14 22	15 23	03 37
17	14 23	15 20	04 23
18	14 24	15 17	04 57
19	14 R 24	15 14	05 14
20	14 22	15 10	05 13
21	14 21	15 07	04 53
22	14 19	15 04	04 13
23	14 17	15 01	03 15
24	14 16	14 58	02 04
25	14 15	14 54	00 S 45
26	14 D 15	14 51	00 N 37
27	14 16	14 48	01 54
28	14 16	14 45	03 02
29	14 17	14 42	03 58
30	14 ♐ 18	14 ♐ 39	04 N 39

DECLINATIONS

Date	Sun ☉ ° '	Moon ☽ ° '	Mercury ☿ ° '	Venus ♀ ° '	Mars ♂ ° '	Jupiter ♃ ° '	Saturn ♄ ° '	Uranus ♅ ° '	Neptune ♆ ° '	Pluto ♇ ° '
01	14 S 24	17 S 57	21 S 54	20 S 11	14 N 43	11 N 44	06 S 38	00 S 12	12 S 38	19 S 16
02	14 43	14 24	22 17	20 30	14 33	11 42	06 40	00 13	12 38	19 16
03	15 02	10 18	22 38	20 48	14 24	11 39	06 43	00 13	12 38	19 16
04	15 21	05 52	22 59	21 05	14 14	11 36	06 46	00 14	12 38	19 16
05	15 39	01 S 17	23 18	21 22	14 04	11 34	06 48	00 15	12 38	19 16
06	15 57	03 N 17	23 36	21 39	13 54	11 31	06 51	00 15	12 38	19 16
07	16 15	07 43	23 52	21 54	13 45	11 29	06 53	00 16	12 38	19 16
08	16 33	11 50	24 08	22 09	13 35	11 26	06 56	00 16	12 38	19 17
09	16 50	15 30	24 22	22 24	13 25	11 24	06 58	00 17	12 38	19 17
10	17 07	18 34	24 34	22 37	13 15	11 21	07 01	00 18	12 38	19 17
11	17 24	20 51	24 46	22 50	13 06	11 19	07 03	00 18	12 38	19 17
12	17 40	22 12	24 56	23 03	12 56	11 17	07 05	00 19	12 38	19 17
13	17 56	22 32	25 04	23 15	12 47	11 14	07 08	00 19	12 38	19 17
14	18 12	21 47	25 11	23 26	12 37	11 12	07 10	00 20	12 38	19 17
15	18 28	19 56	25 16	23 36	12 27	11 10	07 13	00 20	12 38	19 17
16	18 43	17 04	25 20	23 46	12 18	11 07	07 15	00 20	12 38	19 18
17	18 58	13 19	25 22	23 55	12 08	11 05	07 17	00 21	12 38	19 18
18	19 12	08 48	25 22	24 03	11 59	11 03	07 20	00 21	12 38	19 18
19	19 26	03 N 45	25 21	24 11	11 50	11 01	07 22	00 22	12 38	19 18
20	19 40	01 S 38	25 17	24 18	11 40	10 59	07 24	00 22	12 38	19 18
21	19 54	07 04	25 12	24 24	11 31	10 57	07 26	00 22	12 37	19 18
22	20 07	12 13	25 05	24 29	11 22	10 55	07 29	00 23	12 37	19 18
23	20 19	16 41	24 56	24 34	11 12	10 53	07 31	00 23	12 37	19 18
24	20 32	20 04	24 45	24 38	11 03	10 51	07 33	00 23	12 37	19 18
25	20 44	22 04	24 31	24 41	10 54	10 49	07 35	00 23	12 37	19 18
26	20 55	22 31	24 15	24 43	10 45	10 47	07 37	00 24	12 37	19 18
27	21 06	21 26	23 57	24 45	10 36	10 46	07 39	00 24	12 36	19 19
28	21 17	19 03	23 37	24 46	10 27	10 44	07 42	00 24	12 36	19 19
29	21 28	15 40	23 14	24 46	10 19	10 42	07 44	00 24	12 36	19 19
30	21 S 38	11 S 38	22 S 49	24 S 45	10 N 10	10 N 41	07 S 46	00 S 25	12 S 36	19 S 19

ZODIAC SIGN ENTRIES

Date	h	m	Planets	
01	22	08	☽	♒
02	08	51	♀	♐
02	16	54	☿	♐
04	07	18	☽	♓
06	19	02	☽	♈
09	07	45	☽	♉
11	04	15	♂	♍
11	20	10	☽	♊
14	07	19	☽	♋
16	16	17	☽	♌
18	22	19	☽	♍
21	01	16	☽	♎
22	16	08	☉	♐
23	01	58	☽	♏
25	01	57	☽	♐
26	12	36	♀	♑
27	03	04	☽	♑
29	07	02	☽	♒

LATITUDES

Date	Mercury ☿ ° '	Venus ♀ ° '	Mars ♂ ° '	Jupiter ♃ ° '	Saturn ♄ ° '	Uranus ♅ ° '	Neptune ♆ ° '	Pluto ♇ ° '
01	02 S 10	00 S 17	01 N 37	01 S 29	02 N 16	00 S 46	00 S 33	04 N 04
04	02 23	00 25	01 40	01 28	02 16	00 46	00 33	04 04
07	02 33	00 32	01 44	01 28	02 16	00 45	00 33	04 03
10	02 40	00 40	01 47	01 27	02 16	00 45	00 33	04 02
13	02 41	00 48	01 50	01 27	02 17	00 45	00 33	04 02
16	02 37	00 55	01 54	01 26	02 17	00 45	00 33	04 01
19	02 25	01 02	01 57	01 26	02 17	00 45	00 33	04 01
22	02 02	01 09	02 01	01 25	02 18	00 45	00 33	04 00
25	01 28	01 15	02 04	01 24	02 18	00 45	00 33	04 00
28	00 S 40	01 21	02 08	01 24	02 19	00 45	00 33	03 59
31	00 N 18	01 S 27	02 N 12	01 S 23	02 N 19	00 S 45	00 S 33	03 N 59

DATA

Julian Date	2455867
Delta T	+68 seconds
Ayanamsa	24° 01' 36"
Synetic vernal point	05° ♓ 05' 23"
True obliquity of ecliptic	23° 26' 14"

MOON'S PHASES, APSIDES AND POSITIONS ☽

Date	h	m	Phase	Longitude ° '	Eclipse Indicator
02	16	38	☽	09 ♒ 55	
10	20	16	○	18 ♉ 05	
18	15	09	☾	25 ♌ 55	
25	06	10	●	02 ♐ 37	Partial

Day	h	m		
08	13	09	Apogee	
23	23	13	Perigee	
05	18	43	0N	
13	07	22	Max dec	22° N 33'
20	04	47	0S	
26	06	39	Max dec	22° S 33'

LONGITUDES

Date	Chiron ⚷ ° '	Ceres ⚳ ° '	Pallas ⚴ ° '	Juno ⚵ ° '	Vesta ⚶ ° '	Black Moon Lilith ⚸ ° '
01	00 ♓ 41	16 ♓ 29	00 ♒ 54	02 ♏ 36	12 ♒ 50	24 ♈ 57
11	00 ♓ 39	16 ♓ 29	02 ♒ 52	06 ♏ 04	15 ♒ 46	26 ♈ 04
21	00 ♓ 42	17 ♓ 05	05 ♒ 06	09 ♏ 29	19 ♒ 03	27 ♈ 11
31	00 ♓ 51	18 ♓ 14	07 ♒ 36	12 ♏ 50	22 ♒ 38	28 ♈ 18

ASPECTARIAN

h m	Aspects
01 Tuesday	
01 30	☽ ∥ ♇
01 38	☽ ± ♂
01 44	♀ ⊥ ♄
02 32	☽ Q ♅
04 36	☽ Q ☉
07 46	☽ ⊥ ♆
08 15	☽ □ ♄
08 37	☿ □ ♆
12 58	☽ ⊼ ♂
13 40	☿ ⊥ ♄
18 44	☽ ⊻ ♆
19 55	☽ ⚹ ☿
21 00	☽ ⚹ ♀
21 52	♀ ⊥ ♇
23 00	♂ ± ♅
02 Wednesday	
00 24	☽ ⚹ ♅
03 37	☉ # ♂
06 53	☽ □ ♃
07 02	☿ ⊥ ♇
08 14	☽ ⊻ ♇
10 10	☽ ∥ ☉
11 01	☽ # ♂
16 38	☽ □ ☉
19 33	☽ ⊥ ♇
21 02	☽ Q ☿
21 52	☽ Q ♀
22 40	☽ ∥ ♆
03 Thursday	
04 19	☽ # ♃
04 35	☽ ∠ ♅
07 54	♀ △ ♅
12 47	☽ ∠ ♇
13 58	☿ △ ♅
16 45	☽ Q ♃
17 06	☽ △ ♄
21 54	☽ ⊥ ♅
04 Friday	
00 06	☽ ☍ ♂
03 40	☽ ☌ ♆
07 19	☽ ∥ ♄
09 35	☽ ⊻ ♅
12 02	☽ □ ☿
12 33	☽ □ ♀
15 56	☽ ⚹ ♃
18 06	☽ ⚹ ♇
22 44	☽ ⚼ ♄
05 Saturday	
08 05	☽ △ ☉
16 48	☽ ± ♄
17 25	☽ ∥ ♅
17 59	☽ Q ♇
18 52	♀ ⊼ ♃
20 01	☽ # ♅
21 13	☿ ⊼ ♃
21 24	☽ ∠ ♃
06 Sunday	
04 56	☽ ⊼ ♄
14 19	☽ ⊼ ♂
15 16	☽ ⊥ ♃
15 17	☽ ⊻ ♆
16 52	☽ ⚼ ☉
19 59	♀ ⊻ ♇
21 15	☽ ☌ ♅
21 42	☿ ⊻ ♇
07 Monday	
02 58	☽ ± ♂
03 15	☽ ⊻ ♃
03 23	☽ ⊥ ♆
06 15	☽ □ ♇
07 16	☽ △ ☿
07 23	☽ # ♄
07 26	☽ △ ♀
12 34	♂ ☍ ♆
12 43	☽ ± ☉
21 35	☽ ∠ ♆
22 00	☽ ⚼ ♂
08 Tuesday	
01 59	☽ ⊼ ☉
09 38	☽ ∥ ♃
17 00	☽ # ♆
17 13	☽ ⚼ ☿
17 22	☽ ⚼ ♀
18 10	☽ ☍ ♄
19 39	☉ ⚼ ♅
22 33	☽ ∥ ♂
09 Wednesday	
01 41	♀ ∠ ♄
03 10	☿ ∠ ♄
03 59	☽ ⚹ ♆
05 46	☽ △ ♂
09 51	☽ ⊻ ♅
13 33	☽ ± ☿
13 44	☽ ± ♀
15 20	☽ ☌ ♃
18 55	♆ St D
19 08	☽ △ ♇
21 58	☽ ⊥ ♅
22 48	☽ # ☉
10 Thursday	
03 02	☽ ⊼ ☿
03 16	☽ ⊼ ♀
03 45	♀ ± ♃
04 15	☽ Q ♆
05 44	☿ ± ♃
16 05	☽ ∠ ♅
18 41	☽ # ♇
20 16	☽ ☍ ☉
11 Friday	
01 27	☽ ⚼ ♇
07 19	☽ ⊼ ♄
16 27	☽ □ ♆
19 27	☽ ± ♄
19 31	☽ ± ♇
20 52	☽ □ ♂
22 07	☽ ⚹ ♅
12 Saturday	
03 00	☽ ⊻ ♃
04 22	♄ Q ♇
07 30	☽ ⊼ ♇
13 33	☽ ⚼ ♄
14 49	☽ ⊥ ♃
21 22	☽ ☍ ☿
21 55	☽ Q ♅
22 14	☽ ☍ ♀
13 Sunday	
01 41	♂ ⊼ ♅
08 24	☽ ∠ ♃
10 04	☽ Q ♂
11 15	☉ ∠ ♇
13 23	☽ ⊼ ☉
19 22	☽ △ ♄
14 Monday	
02 09	☽ ± ☉
03 42	☽ △ ♆
08 58	♂ # ♆
09 06	☽ □ ♅
10 28	☽ ⚹ ♂
13 22	☽ ⚹ ♃
18 26	☽ ☍ ♇
21 06	☽ ⚼ ☉
15 Tuesday	
08 36	☽ ⚼ ♆
08 45	♀ Q ♆
12 09	☽ Q ♃
12 48	☽ ⊼ ☿
14 54	☽ ⊼ ♀
16 23	☽ ∠ ♂
18 10	☽ # ♇
16 Wednesday	
00 15	☽ # ☉
01 02	☽ ± ☿
01 38	☽ ± ♆
03 27	☽ ± ♀
04 03	☽ △ ☉
05 22	☽ □ ♄
09 37	☿ Q ♆
09 58	☽ ⊥ ♂
12 51	☽ ⊼ ♆
17 54	☽ △ ♅
18 01	♀ ⚼ ♃
19 03	☽ ⚼ ☿
21 31	☽ ⊻ ♂
21 33	☽ □ ♃
21 57	☽ ⚼ ♀
22 08	♂ △ ♃
22 45	☉ ⊻ ♄
17 Thursday	
03 01	☽ ⊼ ♇
13 59	☽ ± ♇
14 49	☽ Q ♄
15 51	☽ # ♆
18 50	☽ ∥ ♂
21 14	☽ ⚼ ♅
18 Friday	
00 08	☽ △ ☿
00 23	☽ ∥ ♃
03 59	☽ △ ♀
06 11	☽ ⚼ ♇
07 05	☿ ⚼ ♃
12 28	☽ ⚹ ♄
13 15	☽ ± ♅
15 09	☽ □ ☉
19 05	☽ ☍ ♆
19 11	☽ # ♄
21 25	☉ ∥ ♇
23 46	☽ ⊼ ♅
19 Saturday	
02 50	☽ △ ♃
05 12	☽ ☌ ♂
08 32	☽ △ ♇
14 47	☽ ∠ ♄
20 Sunday	
00 42	☽ Q ☉
03 10	☽ # ♅
04 20	☽ ⚼ ♃
06 13	☽ ⊥ ♄
06 24	☽ ∥ ♅
06 51	☽ □ ☿
12 54	☽ □ ♀
16 21	☽ ⊻ ♄
19 19	☽ ± ♃
20 42	☉ □ ♆
22 15	☽ ⊼ ♆
22 21	☽ ⚹ ☉
21 Monday	
02 35	☽ ☍ ♅
05 10	☽ ⊼ ♃
08 09	☽ ± ♆
09 29	☽ ⊻ ♂
11 04	☽ □ ♇
13 40	☽ ∥ ♄
13 43	☽ Q ☿
19 37	☽ ⊥ ♂
21 21	☽ Q ♀
22 52	☽ ⚼ ♆
22 Tuesday	
00 46	☽ ∠ ☉
05 45	☽ # ♃
07 58	☽ # ♂
08 19	♀ ⚹ ♄
09 39	☽ ⚹ ☿
10 41	☽ ∠ ♂
14 01	☽ ∥ ♆
15 46	☉ ⊥ ♇
16 19	☽ Q ♇
16 22	☽ ⊥ ☉
17 43	☽ ☌ ♄
18 30	☽ ⚹ ♀
23 04	☽ △ ♆
23 Wednesday	
02 41	☽ ⊻ ☉
03 12	☽ ⊼ ♅
05 25	☽ ☍ ♃
10 04	☽ ∠ ☿
10 14	☉ △ ♅
11 30	☽ ⚹ ♂
11 35	☽ ⚹ ♇
12 47	☽ ± ♅
14 07	♂ △ ♇
14 41	☉ ⊥ ♄
20 40	☽ ∠ ♀
24 Thursday	
00 31	☽ ⊥ ☿
03 08	☽ ⚼ ♅
05 41	☽ ∥ ♇
07 17	☽ Q ♂
07 18	☿ St R
10 07	☽ ⊻ ☿
11 35	☽ ∠ ♇
12 20	☽ ⊥ ♀
15 43	☉ ⊼ ♃
16 32	☽ ∥ ☉
18 03	☽ ⊻ ♄
22 25	☿ ∥ ♀
22 49	☽ ⊻ ♀
23 04	☽ □ ♆
25 Friday	
01 52	♀ ⚹ ♆
02 03	☽ ⊥ ♇
03 09	☽ △ ♅
03 45	☽ ⊥ ♄
05 05	☽ ⊼ ♃
06 10	☽ ☌ ☉
11 45	☽ ⊻ ♇
13 06	☽ □ ♂
14 43	☽ ± ♃
18 24	☽ ∠ ♄
26 Saturday	
04 15	☽ Q ♆
04 41	♂ ∥ ♃
05 17	☽ ⚼ ♃
10 03	☽ ☌ ☿
19 12	☽ ⚹ ♄
27 Sunday	
00 06	☽ ⚹ ♆
02 27	♀ □ ♅
04 16	☽ □ ♅
04 27	☽ ☌ ♀
06 01	☽ △ ♃
11 15	☽ ⊻ ☉
13 25	☽ ☌ ♇
15 32	☽ Q ♄
15 44	☽ ∥ ☉
16 23	☽ △ ♂
21 23	♀ △ ♃
22 16	☽ ⊥ ☉
28 Monday	
01 34	☽ ∠ ♆
09 48	☽ ∥ ♇
10 56	☽ ⊻ ☿
11 03	☽ Q ♅
15 07	☽ ∠ ☉
17 15	☽ ⊥ ♆
18 36	☉ ⊻ ♇
19 14	☽ ⚼ ♂
20 55	☽ ⊥ ☿
23 01	☽ □ ♄
29 Tuesday	
03 53	☽ ⊻ ♆
08 16	☽ ⚹ ♅
09 50	☽ □ ♃
11 50	☽ ± ♂
12 01	☽ ∠ ☿
13 51	☽ ⊻ ♀
18 14	☽ ⊻ ♇
20 09	☽ ⚹ ☉
23 07	☽ ⊼ ♂
30 Wednesday	
01 55	☽ ⊥ ♀
02 01	☉ ± ♃
05 16	☽ ⊥ ♇
06 32	☽ ∥ ♆
11 41	☽ ∠ ♅
13 36	☽ ⚹ ☿
17 17	☽ # ♃
18 42	☽ Q ♃
19 34	☿ ⚼ ♃
20 16	☽ Q ☉
20 22	☽ # ♂
20 26	☽ ∠ ♀
22 06	☽ ∠ ♇
23 17	☿ Q ♆

All ephemeris data is given at 12.00 UT and the Moon's longitude is additionally given for 24.00 UT

DECEMBER 2011

LONGITUDES

Date	Sidereal time h m s	Sun ☉ ° ' "	Moon ☽ ° ' "	Moon ☽ 24.00 ° ' "	Mercury ☿ ° '	Venus ♀ ° '	Mars ♂ ° '	Jupiter ♃ ° '	Saturn ♄ ° '	Uranus ♅ ° '	Neptune ♆ ° '	Pluto ♇ ° '
01	16 40 00	08 ♐ 56 10	28 ♒ 33 48	04 ♓ 47 29	15 ♐ 39	06 ♑ 10	09 ♍ 33	01 ♉ 23	25 ♎ 46	00 ♈ 41	28 ♒ 16	06 ♑ 16
02	16 43 56	09 56 59	10 ♓ 56 51	17 ♓ 02 27	14 R 23	07 25	09 59	01 R 18	25 52	00 R 40	28 17	06 18
03	16 47 53	10 57 49	23 ♓ 04 52	29 ♓ 04 40	13 03	08 39	10 24	01 14	25 58	00 40	28 18	06 20
04	16 51 49	11 58 40	05 ♈ 02 26	10 ♈ 58 42	11 40	09 53	10 49	01 09	26 04	00 39	28 19	06 22
05	16 55 46	12 59 32	16 ♈ 54 01	22 ♈ 48 54	10 18	11 08	11 14	01 05	26 09	00 39	28 20	06 24
06	16 59 42	14 00 25	28 ♈ 43 49	04 ♉ 39 12	08 59	12 22	11 39	01 01	26 15	00 39	28 20	06 26
07	17 03 39	15 01 19	10 ♉ 35 30	16 ♉ 33 03	07 46	13 36	12 03	00 57	26 21	00 39	28 21	06 28
08	17 07 35	16 02 13	22 ♉ 32 13	28 ♉ 33 17	06 41	14 50	12 27	00 53	26 26	00 39	28 22	06 30
09	17 11 32	17 03 09	04 ♊ 36 31	10 ♊ 42 09	05 45	16 05	12 51	00 50	26 32	00 39	28 23	06 32
10	17 15 29	18 04 05	16 ♊ 50 23	23 ♊ 01 22	05 00	17 19	13 14	00 47	26 37	00 D 39	28 24	06 34
11	17 19 25	19 05 02	29 ♊ 15 15	05 ♋ 32 09	04 27	18 33	13 37	00 44	26 43	00 39	28 25	06 36
12	17 23 22	20 06 00	11 ♋ 52 11	18 ♋ 15 24	04 04	19 47	14 00	00 41	26 48	00 39	28 26	06 38
13	17 27 18	21 07 00	24 ♋ 41 54	01 ♌ 11 44	03 53	21 02	14 22	00 38	26 53	00 39	28 28	06 40
14	17 31 15	22 08 00	07 ♌ 44 58	14 ♌ 21 39	03 D 52	22 16	14 44	00 36	26 59	00 39	28 29	06 42
15	17 35 11	23 09 01	21 ♌ 01 49	27 ♌ 45 31	04 01	23 30	15 06	00 33	27 04	00 39	28 30	06 44
16	17 39 08	24 10 02	04 ♍ 32 47	11 ♍ 23 37	04 19	24 44	15 27	00 31	27 09	00 40	28 31	06 46
17	17 43 04	25 11 05	18 ♍ 17 59	25 ♍ 15 52	04 46	25 58	15 48	00 29	27 14	00 40	28 32	06 48
18	17 47 01	26 12 09	02 ♎ 17 09	09 ♎ 21 42	05 19	27 12	16 09	00 28	27 19	00 40	28 34	06 51
19	17 50 58	27 13 14	16 ♎ 29 19	23 ♎ 39 44	06 00	28 26	16 29	00 26	27 24	00 41	28 35	06 53
20	17 54 54	28 14 19	00 ♏ 52 34	08 ♏ 07 24	06 46	29 ♑ 40	16 49	00 25	27 28	00 41	28 36	06 55
21	17 58 51	29 ♐ 15 26	15 ♏ 23 44	22 ♏ 40 57	07 37	00 ♒ 54	17 09	00 24	27 33	00 42	28 38	06 57
22	18 02 47	00 ♑ 16 33	29 ♏ 58 26	07 ♐ 15 26	08 33	02 08	17 28	00 23	27 38	00 42	28 39	06 59
23	18 06 44	01 17 41	14 ♐ 31 15	21 ♐ 45 06	09 33	03 22	17 46	00 22	27 42	00 43	28 40	07 01
24	18 10 40	02 18 50	28 ♐ 56 16	06 ♑ 04 01	10 37	04 36	18 05	00 22	27 47	00 44	28 42	07 03
25	18 14 37	03 19 59	13 ♑ 07 45	20 ♑ 06 53	11 43	05 50	18 22	00 22	27 51	00 44	28 43	07 05
26	18 18 33	04 21 08	27 ♑ 00 58	03 ♒ 49 40	12 52	07 04	18 40	00 D 22	27 55	00 45	28 45	07 08
27	18 22 30	05 22 18	10 ♒ 32 44	17 ♒ 10 06	14 04	08 18	18 57	00 22	28 00	00 46	28 46	07 10
28	18 26 27	06 23 27	23 ♒ 41 46	00 ♓ 07 53	15 18	09 31	19 13	00 23	28 04	00 47	28 48	07 12
29	18 30 23	07 24 37	06 ♓ 28 40	12 ♓ 44 27	16 34	10 45	19 29	00 23	28 08	00 48	28 49	07 14
30	18 34 20	08 25 46	18 ♓ 55 40	25 ♓ 02 46	17 51	11 59	19 44	00 24	28 12	00 49	28 51	07 16
31	18 38 16	09 ♑ 26 55	01 ♈ 06 18	07 ♈ 06 50	19 ♐ 10	13 ♒ 13	19 ♍ 59	00 ♉ 25	28 ♎ 15	00 ♈ 50	28 ♒ 53	07 ♑ 18

MOON NODES, LATITUDE AND DECLINATIONS

Date	Moon True ☊ ° '	Moon Mean ☊ ° '	Moon ☽ Latitude ° '	Sun ☉ ° '	Moon ☽ ° '	Mercury ☿ ° '	Venus ♀ ° '	Mars ♂ ° '	Jupiter ♃ ° '	Saturn ♄ ° '	Uranus ♅ ° '	Neptune ♆ ° '	Pluto ♇ ° '
01	14 ♐ 18	14 ♐ 35	05 N 06	21 S 47	07 S 11	22 S 22	24 S 44	10 N 01	10 N 39	07 S 48	00 S 25	12 S 35	19 S 19
02	14 R 18	14 32	05 17	21 56	02 S 34	21 53	24 42	09 53	10 38	07 50	00 25	12 35	19 19
03	14 18	14 29	05 14	22 05	02 N 04	21 53	24 39	09 44	10 37	07 52	00 25	12 35	19 19
04	14 18	14 26	04 57	22 13	06 33	20 53	24 36	09 36	10 35	07 54	00 25	12 35	19 19
05	14 18	14 23	04 28	22 21	10 46	20 24	24 31	09 28	10 34	07 56	00 25	12 34	19 19
06	14 18	14 19	03 47	22 29	14 33	19 56	24 26	09 20	10 33	07 57	00 25	12 34	19 19
07	14 D 18	14 16	02 57	22 36	17 48	19 30	24 20	09 12	10 32	07 59	00 25	12 34	19 19
08	14 18	14 13	01 58	22 42	20 18	19 07	24 14	09 04	10 31	08 01	00 25	12 33	19 19
09	14 18	14 10	00 N 54	22 48	21 56	18 48	24 06	08 56	10 30	08 03	00 25	12 33	19 19
10	14 R 18	14 07	00 S 14	22 54	22 33	18 32	23 58	08 48	10 29	08 05	00 25	12 33	19 19
11	14 18	14 04	01 22	22 59	22 04	18 21	23 50	08 40	10 28	08 07	00 25	12 32	19 19
12	14 17	14 00	02 27	23 04	20 28	18 13	23 40	08 33	10 28	08 08	00 25	12 32	19 19
13	14 17	13 57	03 26	23 08	17 48	18 09	23 30	08 26	10 27	08 10	00 25	12 31	19 19
14	14 16	13 54	04 15	23 12	14 13	18 09	23 19	08 18	10 27	08 12	00 25	12 31	19 19
15	14 14	13 51	04 51	23 16	09 53	18 12	23 08	08 11	10 26	08 13	00 25	12 31	19 19
16	14 14	13 48	05 12	23 19	04 N 59	18 18	22 56	08 04	10 26	08 15	00 25	12 30	19 19
17	14 13	13 45	05 16	23 21	00 S 14	18 26	22 43	07 58	10 25	08 17	00 24	12 30	19 19
18	14 D 13	13 41	05 01	23 23	05 31	18 37	22 29	07 51	10 25	08 18	00 24	12 29	19 19
19	14 14	13 38	04 28	23 24	10 36	18 49	22 15	07 44	10 25	08 20	00 24	12 29	19 19
20	14 15	13 35	03 38	23 26	15 11	19 03	22 00	07 38	10 25	08 21	00 24	12 28	19 20
21	14 16	13 32	02 33	23 26	18 54	19 18	21 44	07 32	10 24	08 23	00 24	12 28	19 20
22	14 17	13 29	01 S 18	23 26	21 25	19 34	21 28	07 26	10 24	08 24	00 23	12 27	19 20
23	14 R 17	13 25	00 N 01	23 26	22 31	19 50	21 12	07 20	10 25	08 25	00 23	12 27	19 20
24	14 17	13 22	01 20	23 25	22 06	20 07	20 54	07 14	10 25	08 27	00 23	12 26	19 20
25	14 15	13 19	02 33	23 24	20 15	20 24	20 36	07 09	10 25	08 28	00 22	12 26	19 20
26	14 13	13 16	03 35	23 22	17 15	20 41	20 18	07 03	10 25	08 30	00 22	12 25	19 19
27	14 09	13 13	04 23	23 20	13 13	20 57	19 59	06 58	10 26	08 31	00 22	12 25	19 19
28	14 06	13 10	04 55	23 17	08 58	21 14	19 39	06 53	10 26	08 32	00 21	12 24	19 19
29	14 03	13 06	05 12	23 14	04 S 18	21 30	19 19	06 48	10 27	08 33	00 21	12 24	19 19
30	14 00	13 03	05 14	23 10	00 N 26	21 46	18 58	06 44	10 27	08 34	00 20	12 23	19 19
31	13 ♐ 59	13 ♐ 00	05 N 01	23 S 06	05 N 02	22 S 01	18 S 37	06 N 39	10 N 28	08 S 36	00 S 20	12 S 23	19 S 19

ZODIAC SIGN ENTRIES

Date	h	m	Planets
01	14	45	☽ ♓
04	01	51	☽ ♈
06	14	34	☽ ♉
09	02	52	☽ ♊
11	13	26	☽ ♋
13	21	48	☽ ♌
16	03	58	☽ ♍
18	08	06	☽ ♎
20	10	33	☽ ♏
20	18	26	♀ ♒
22	05	30	☉ ♑
22	12	03	☽ ♐
24	13	47	☽ ♑
26	17	14	☽ ♒
28	23	45	☽ ♓
31	09	48	☽ ♈

LATITUDES

Date	Mercury ☿ ° '	Venus ♀ ° '	Mars ♂ ° '	Jupiter ♃ ° '	Saturn ♄ ° '	Uranus ♅ ° '	Neptune ♆ ° '	Pluto ♇ ° '
01	00 N 18	01 S 27	02 N 12	01 S 23	02 N 19	00 S 45	00 S 33	03 N 59
04	01 18	01 32	02 16	01 22	02 20	00 44	00 33	03 58
07	02 08	01 36	02 20	01 21	02 20	00 44	00 33	03 58
10	02 38	01 40	02 24	01 20	02 21	00 44	00 33	03 57
13	02 49	01 44	02 28	01 19	02 21	00 44	00 33	03 57
16	02 45	01 47	02 32	01 18	02 22	00 44	00 33	03 56
19	02 31	01 49	02 37	01 18	02 22	00 44	00 33	03 56
22	02 12	01 50	02 41	01 17	02 23	00 44	00 33	03 56
25	01 49	01 51	02 46	01 16	02 24	00 44	00 33	03 56
28	01 24	01 51	02 51	01 15	02 24	00 44	00 33	03 55
31	00 N 59	01 S 50	02 N 56	01 S 14	02 N 25	00 S 43	00 S 33	03 N 55

DATA

Julian Date	2455897
Delta T	+68 seconds
Ayanamsa	24° 01' 41"
Synetic vernal point	05° ♓ 05' 18"
True obliquity of ecliptic	23° 26' 13"

MOON'S PHASES, APSIDES AND POSITIONS ☽

Date	h	m	Phase	Longitude ° '	Eclipse Indicator
02	09	52	☽	09 ♓ 52	
10	14	36	○	18 ♊ 11	total
18	00	48	☾	25 ♍ 44	
24	18	06	●	02 ♑ 34	

Day	h	m		
06	01	08	Apogee	
22	02	49	Perigee	
03	01	16	0N	
10	13	35	Max dec	22° N 33'
17	10	58	0S	
23	17	18	Max dec	22° S 33'
30	09	48	0N	

LONGITUDES

Date	Chiron ⚷ ° '	Ceres ⚳ ° '	Pallas ⚴ ° '	Juno ⚵ ° '	Vesta ⚶ ° '	Black Moon Lilith ⚸ ° '
01	00 ♓ 51	18 ♓ 14	07 ♒ 36	12 ♏ 50	22 ♒ 38	28 ♈ 18
11	01 ♓ 07	19 ♓ 54	10 ♒ 19	16 ♏ 06	26 ♒ 28	29 ♈ 25
21	01 ♓ 27	21 ♓ 59	13 ♒ 11	19 ♏ 16	00 ♓ 49	00 ♉ 32
31	01 ♓ 53	24 ♓ 26	16 ♒ 12	22 ♏ 17	04 ♓ 40	01 ♉ 39

ASPECTARIAN

h m Aspects

01 Thursday
04 35 ☽ ⊥ ♅
06 37 ☽ △ ♄
08 50 ☽ ∥ ♄
10 24 ☽ Q ☿
11 26 ☽ ☌ ♆
13 51 ♀ ☌ ♇
16 03 ☽ ⊻ ♅
17 23 ☽ ⚹ ♃

02 Friday
02 54 ☽ ⚹ ♇
04 19 ☽ ⚹ ♀
09 48 ☉ ∥ ☿
09 52 ☽ □ ☉
10 02 ☽ ☍ ♂
11 50 ☽ ⚼ ♄
13 08 ☉ □ ♂
18 06 ☽ □ ☿
21 33 ♀ Q ♄
22 29 ☽ ∠ ♃
23 07 ☽ ∥ ♅

03 Saturday
02 31 ☽ Q ♇
03 26 ☽ ⫲ ♅
05 44 ☽ ± ♄
06 36 ☽ Q ♀
11 58 ☉ ∠ ♄
16 15 ☽ ⊥ ♃
17 48 ☽ ⚻ ♄
22 27 ☽ ⊻ ♆

04 Sunday
03 10 ☽ ☌ ♅
04 13 ☽ ⊻ ♃
08 52 ☉ ☌ ☿
10 32 ☽ ⊥ ♆
14 40 ☽ □ ♇
19 32 ☽ ⫲ ♄
21 51 ☿ ∠ ♄
22 56 ☽ □ ♀
23 19 ☿ □ ♂

05 Monday
00 00 ☽ △ ☿
00 06 ☽ ⚻ ♂
03 20 ☽ △ ☉
04 19 ☿ ⊻ ♀
04 39 ☽ ∥ ♂
04 45 ☽ ∠ ♆
06 06 ♂ ∠ ♄
10 54 ☽ ∥ ♃
12 42 ☽ ± ♂
15 07 ♀ △ ♂
23 03 ☽ ⫲ ♆

06 Tuesday
03 19 ☽ ⚼ ☿
06 56 ☽ ☍ ♄
07 37 ☽ ⚼ ♂
11 13 ☽ ⚹ ♆
12 37 ☽ ⚼ ☉
15 53 ☽ ⊻ ♅
16 36 ☽ ☌ ♃
19 47 ☽ ± ☿

07 Wednesday
03 38 ☽ △ ♇
04 02 ☽ ⊥ ♅
06 48 ☽ ⚻ ☿
07 08 ♀ ∠ ♆
08 32 ☽ ± ☉
11 31 ☽ Q ♆
13 35 ☿ ⊥ ♀
15 02 ☽ △ ♂
18 46 ☽ △ ♀
21 45 ☽ ⚻ ☉
22 11 ☽ ∠ ♅
22 51 ☿ ∥ ♇

08 Thursday
01 16 ☽ ⫲ ☿
01 39 ☽ ⫲ ♇
06 42 ☿ ± ♃
08 43 ☉ ⚼ ♃
09 55 ☽ ⚼ ♇
16 13 ☿ ⊻ ♇
19 51 ☽ ⚻ ♄
20 02 ☉ Q ♆
23 39 ☽ □ ♆

09 Friday
03 54 ☽ ± ♇
04 09 ☽ ⚹ ♅
04 13 ☽ ⚼ ♀
04 33 ☽ ⊻ ♃
07 52 ☽ ± ♄
14 07 ☽ ☍ ☿
15 48 ☽ ⚻ ♇
16 22 ☽ ⊥ ♃
23 59 ☽ ± ♀

10 Saturday
01 44 ☽ ⚼ ♄
03 48 ☽ Q ♅
04 44 ☽ □ ♂
07 03 ♅ St D
09 56 ☽ ∠ ♃
13 02 ☽ ⚻ ♀
14 36 ☽ ☍ ☉

11 Sunday
07 05 ☽ △ ♄
10 24 ☽ △ ♆
13 44 ♀ Q ♅
14 40 ☽ □ ♅
14 49 ☽ ⚹ ♃
16 40 ☽ Q ♂
21 36 ☽ ⚻ ☿

12 Monday
00 56 ☿ ∠ ♀
02 03 ☽ ☍ ♇
08 41 ☽ ± ☿
13 31 ☽ Q ♃
14 58 ☽ ⚼ ♆
16 08 ☽ ⚹ ♂
23 25 ☽ ⫲ ♇

13 Tuesday
01 17 ☽ ⚼ ☿
04 27 ☽ ☍ ♀
04 46 ☽ ⚻ ☉
05 23 ♃ ⊻ ♅
07 50 ☽ ± ♆
09 16 ☽ ⫲ ☿
16 05 ☽ □ ♄
16 52 ☽ ± ☉
18 58 ☽ ⚻ ♆
20 53 ☽ ∠ ♂
21 55 ☉ ⊻ ♀
22 56 ☽ □ ♃
23 00 ☽ △ ♅

14 Wednesday
01 44 ☿ St D
04 53 ☽ △ ☿
10 05 ☽ ⚻ ♇
10 47 ☽ ⚼ ☉
13 51 ☽ ⊥ ♂
16 16 ♀ ⊥ ♆
21 01 ☽ ± ♇
21 52 ☽ ⫲ ♆
23 26 ☉ ∥ ♀

15 Thursday
01 02 ☽ ⊻ ♂
01 12 ☽ Q ♄
02 20 ☽ ⚼ ♅
06 26 ♂ ⫲ ♄
09 06 ☽ ∥ ♃
13 16 ☽ ⚼ ♇
16 06 ☽ △ ☉
16 52 ☽ ⚻ ♀
18 29 ☽ ± ♅
20 20 ☽ ⫲ ♄
20 45 ☽ ∥ ♂
22 50 ☽ ⚹ ♄

16 Friday
01 20 ☽ ☍ ♆
04 36 ☽ ± ♀
04 55 ☽ △ ♃
05 08 ☽ ⚻ ♅
11 36 ☽ □ ☿
15 55 ☽ △ ♇
16 01 ♂ ⚼ ♃
22 00 ☽ ⚼ ♀

17 Saturday
01 24 ☽ ∠ ♄
07 08 ☽ ⚼ ♃
07 34 ☽ ☌ ♂
09 07 ☽ ⫲ ♅
12 49 ☽ ∥ ♅
17 05 ☽ ⊥ ♄
20 00 ☽ Q ☿
22 39 ☽ ± ♃

18 Sunday
00 48 ☽ □ ☉
02 29 ☽ △ ♀
03 28 ☽ ⊻ ♄
05 38 ☽ ⚻ ♆
08 54 ☽ ⚻ ♃
09 15 ☽ ☍ ♅
14 18 ♀ □ ♄
15 52 ☽ ± ♆
17 24 ☽ ⚹ ☿
19 45 ☽ □ ♇
22 34 ☽ ⫲ ♂

19 Monday
00 59 ☽ ∥ ♄
07 06 ☽ ⚼ ♆
09 42 ☽ Q ☉
11 03 ☽ ⫲ ♃
12 00 ☽ ⊻ ♂
14 52 ♀ ⊻ ♆
16 24 ☉ ⚹ ♄
19 57 ☽ ∠ ☿
21 24 ☽ ∥ ♆
22 17 ☽ ⊥ ♂

20 Tuesday
01 45 ☿ ± ♃
02 04 ☽ Q ♇
06 19 ☽ ☌ ♄
07 17 ☽ ⚹ ☉
08 13 ☽ △ ♆
09 49 ☽ □ ♀
11 14 ☽ ☍ ♃
11 41 ☽ ⚻ ♅
11 48 ☽ ⊥ ☿
13 36 ☽ ∠ ♂
16 32 ☿ ⊻ ♇
20 46 ☉ ⚹ ♆
21 38 ☽ ± ♅
22 01 ☽ ⚹ ♇
22 21 ☽ ⊻ ☿

21 Wednesday
02 19 ♀ □ ♃
07 56 ♀ ⚹ ♅
09 59 ☽ ∠ ☉
12 30 ☽ ⚼ ♅
14 37 ☿ ∥ ♇
14 57 ☽ ⚹ ♂
15 22 ☽ ∥ ♇
15 27 ☽ ∥ ☿
18 19 ☽ Q ♀
22 49 ☽ ∠ ♇

22 Thursday
01 55 ☽ ⊥ ☉
08 07 ☽ ⊻ ♄
09 49 ☽ □ ♆
11 08 ☽ Q ♂
12 32 ☽ ⊻ ☉
12 35 ☽ ∥ ♀
12 41 ☽ ⚻ ♃
13 12 ☽ △ ♅
13 40 ☽ ⊥ ♇
14 32 ☉ △ ♃
15 53 ☽ ⚹ ♀
18 03 ☽ ⊥ ♄
20 30 ♀ ⚼ ♂
22 14 ☉ □ ♅
22 33 ☽ ± ♃
23 35 ☽ ⊻ ♇

23 Friday
03 10 ☽ ☌ ☿
08 59 ☽ ∠ ♄
13 25 ☽ ⚼ ♃
15 34 ☽ Q ♆
17 30 ☽ □ ♂
18 58 ☽ ∠ ♀

24 Saturday
10 03 ☽ ⚹ ♄
11 23 ☽ ⊥ ♀
11 36 ☽ ⚹ ♆
14 24 ☽ △ ♃
15 00 ☽ □ ♅
18 06 ☽ ☌ ☉
22 26 ☽ ⊻ ♀

25 Sunday
01 42 ☽ ☌ ♇
06 23 ☽ Q ♄
07 59 ☽ ∥ ♀
09 23 ☽ ⊻ ☿
10 47 ☽ ∥ ☿
13 01 ☽ ∠ ♆
20 21 ☽ ∥ ♇
20 32 ☿ ∥ ♀
20 34 ☽ ⊥ ☿
21 11 ☽ △ ♂
21 38 ☽ Q ♅
22 08 ♃ St D

26 Monday
04 33 ☽ ⊥ ♆
13 02 ☿ ∠ ♄
13 17 ♀ ⊻ ♇
13 36 ☽ □ ♄
13 39 ☽ ∠ ☿
15 02 ☽ ⊻ ♆
17 53 ☽ □ ♃
18 34 ☽ ⚹ ♅
23 57 ☽ ⚼ ♂

27 Tuesday
01 59 ☽ ⊻ ☉
05 55 ☽ ⊻ ♇
07 33 ☽ ☌ ♀
13 37 ☽ ⊥ ☉
16 25 ☽ ± ♂
16 44 ☽ ⊥ ♇
17 28 ☽ ∥ ♆
19 01 ☽ ⚹ ☿
21 28 ☽ ∠ ♅

28 Wednesday
02 12 ☽ Q ♃
03 34 ☽ ⚻ ♂
04 16 ☽ ⫲ ♃
07 23 ☽ ∠ ☉
09 13 ☽ ∠ ♇
13 28 ☿ ⚼ ♃
14 01 ☽ ⊥ ♅
14 17 ☽ ∥ ♄
19 25 ☽ Q ☿
20 10 ☽ △ ♄
21 31 ☽ ☌ ♆
23 00 ☽ ⫲ ♂

29 Thursday
00 28 ☽ ⚹ ♃
01 15 ☽ ⊻ ♅
07 43 ☉ ☌ ♇
11 05 ♀ ∥ ♇
13 27 ☽ ⚹ ♇
13 56 ☽ ⚹ ☉
16 57 ☿ Q ♆
21 04 ☽ ⊻ ♀

30 Friday
00 49 ☽ ⚼ ♄
05 08 ☽ ∠ ♃
08 04 ☽ ∥ ♅
09 40 ☽ □ ☿
09 57 ☽ ⊥ ♀
11 32 ☽ ⫲ ♅
12 40 ☽ Q ♇
13 37 ☽ ☍ ♂
15 12 ☽ Q ☉
18 25 ☽ ± ♄
22 45 ☽ ⊥ ♃

31 Saturday
05 36 ☽ ∠ ♀
06 19 ☽ ⚻ ♄
07 34 ☽ ⊻ ♆
10 38 ☽ ⊻ ♃
11 28 ☽ ☌ ♅
13 57 ♀ ⊥ ♇
19 32 ☽ ⊥ ♆
20 35 ☽ ∥ ♂

All ephemeris data is given at 12.00 UT and the Moon's longitude is additionally given for 24.00 UT

JANUARY 2012

LONGITUDES

Date	Sidereal time h m s	Sun ☉ ° ' "	Moon ☽ ° ' "	Moon ☽ 24.00 ° ' "	Mercury ☿ ° '	Venus ♀ ° '	Mars ♂ ° '	Jupiter ♃ ° '	Saturn ♄ ° '	Uranus ♅ ° '	Neptune ♆ ° '	Pluto ♇ ° '
01	18 42 13	10 ♑ 28 05	13 ♈ 04 59	19 ♈ 01 21	20 ♐ 30	14 ♒ 26	20 ♍ 14	00 ♉ 26	28 ♎ 19	00 ♈ 51	28 ♒ 54	07 ♑ 20
02	18 46 09	11 29 14	24 ♈ 56 34	00 ♉ 51 17	21 52	15 40	20 28	00 28	28 23	00 52	28 56	07 23
03	18 50 06	12 30 23	06 ♉ 46 06	12 ♉ 41 38	23 14	16 53	20 41	00 30	28 27	00 54	28 58	07 25
04	18 54 02	13 31 31	18 ♉ 38 27	24 ♉ 37 07	24 38	18 07	20 54	00 31	28 30	00 55	28 59	07 27
05	18 57 59	14 32 40	00 ♊ 38 20	06 ♊ 41 56	26 02	19 20	21 07	00 34	28 33	00 56	29 01	07 29
06	19 01 56	15 33 48	12 ♊ 48 56	18 ♊ 59 30	27 27	20 34	21 19	00 36	28 37	00 57	29 03	07 31
07	19 05 52	16 34 56	25 ♊ 13 54	01 ♋ 32 20	28 ♐ 53	21 47	21 30	00 38	28 40	00 59	29 05	07 33
08	19 09 49	17 36 04	07 ♋ 54 57	14 ♋ 21 47	00 ♑ 20	23 00	21 41	00 41	28 43	01 00	29 06	07 35
09	19 13 45	18 37 12	20 ♋ 52 50	27 ♋ 27 59	01 47	24 14	21 51	00 44	28 46	01 02	29 08	07 38
10	19 17 42	19 38 19	04 ♌ 07 06	10 ♌ 49 57	03 15	25 27	22 00	00 47	28 49	01 03	29 10	07 40
11	19 21 38	20 39 26	17 ♌ 36 16	24 ♌ 25 44	04 43	26 40	22 09	00 50	28 52	01 05	29 12	07 42
12	19 25 35	21 40 33	01 ♍ 18 01	08 ♍ 12 46	06 12	27 53	22 18	00 54	28 55	01 07	29 14	07 44
13	19 29 31	22 41 40	15 ♍ 09 39	22 ♍ 08 20	07 42	29 ♒ 06	22 25	00 57	28 57	01 08	29 16	07 46
14	19 33 28	23 42 47	29 ♍ 08 30	06 ♎ 09 53	09 12	00 ♓ 19	22 33	01 01	29 00	01 10	29 18	07 48
15	19 37 25	24 43 53	13 ♎ 12 15	20 ♎ 15 21	10 43	01 32	22 39	01 05	29 02	01 12	29 20	07 50
16	19 41 21	25 45 00	27 ♎ 19 00	04 ♏ 23 03	12 14	02 45	22 45	01 10	29 05	01 14	29 21	07 52
17	19 45 18	26 46 06	11 ♏ 27 18	18 ♏ 31 37	13 46	03 57	22 50	01 14	29 07	01 15	29 23	07 54
18	19 49 14	27 47 12	25 ♏ 35 48	02 ♐ 39 39	15 18	05 10	22 54	01 19	29 09	01 17	29 25	07 56
19	19 53 11	28 48 18	09 ♐ 42 56	16 ♐ 45 22	16 51	06 23	22 58	01 23	29 11	01 19	29 27	07 58
20	19 57 07	29 ♑ 49 24	23 ♐ 46 38	00 ♑ 46 23	18 25	07 35	23 01	01 28	29 13	01 21	29 29	08 00
21	20 01 04	00 ♒ 50 29	07 ♑ 44 13	14 ♑ 39 44	19 59	08 48	23 03	01 34	29 15	01 23	29 32	08 02
22	20 05 00	01 51 34	21 ♑ 32 31	28 ♑ 22 08	21 33	10 00	23 05	01 39	29 17	01 25	29 34	08 04
23	20 08 57	02 52 38	05 ♒ 08 12	11 ♒ 50 22	23 08	11 13	23 05	01 45	29 18	01 27	29 36	08 06
24	20 12 54	03 53 42	18 ♒ 28 20	25 ♒ 01 53	24 44	12 25	23 R 05	01 50	29 20	01 30	29 38	08 08
25	20 16 50	04 54 44	01 ♓ 30 53	07 ♓ 55 16	26 20	13 37	23 05	01 56	29 21	01 32	29 40	08 10
26	20 20 47	05 55 45	14 ♓ 15 05	20 ♓ 30 29	27 57	14 49	23 03	02 02	29 23	01 34	29 42	08 12
27	20 24 43	06 56 46	26 ♓ 41 41	02 ♈ 49 00	29 ♑ 35	16 01	23 01	02 08	29 24	01 36	29 44	08 14
28	20 28 40	07 57 45	08 ♈ 52 50	14 ♈ 53 39	01 ♒ 13	17 13	22 58	02 15	29 25	01 39	29 46	08 16
29	20 32 36	08 58 43	20 ♈ 51 58	26 ♈ 48 21	02 52	18 25	22 54	02 21	29 26	01 41	29 48	08 18
30	20 36 33	09 59 41	02 ♉ 43 26	08 ♉ 37 52	04 31	19 37	22 50	02 28	29 27	01 43	29 51	08 20
31	20 40 29	11 ♒ 00 36	14 ♉ 32 18	20 ♉ 27 26	06 ♒ 11	20 ♓ 48	22 ♍ 44	02 ♉ 35	29 ♎ 28	01 ♈ 46	29 ♒ 53	08 ♑ 22

DECLINATIONS

Date	Moon True ☊ ° '	Moon Mean ☊ ° '	Moon ☽ Latitude ° '	Sun ☉ ° '	Moon ☽ ° '	Mercury ☿ ° '	Venus ♀ ° '	Mars ♂ ° '	Jupiter ♃ ° '	Saturn ♄ ° '	Uranus ♅ ° '	Neptune ♆ ° '	Pluto ♇ ° '
01	13 ♐ 58	12 ♐ 57	04 N 35	23 S 01	09 N 23	22 S 15	18 S 15	06 N 35	10 N 29	08 S 37	00 S 19	12 S 22	19 S 19
02	13 D 59	12 54	03 58	22 56	13 21	22 29	17 53	06 31	10 29	08 38	00 19	12 21	19 19
03	14 01	12 51	03 11	22 51	16 46	22 41	17 30	06 27	10 30	08 39	00 18	12 21	19 19
04	14 03	12 47	02 15	22 45	19 32	22 53	17 07	06 24	10 31	08 40	00 18	12 20	19 19
05	14 04	12 44	01 13	22 39	21 28	23 04	16 43	06 20	10 32	08 41	00 17	12 20	19 19
06	14 05	12 41	00 N 07	22 32	22 27	23 14	16 19	06 17	10 33	08 42	00 17	12 19	19 19
07	14 R 04	12 38	01 S 01	22 24	22 20	23 23	15 54	06 14	10 34	08 43	00 16	12 18	19 19
08	14 02	12 35	02 07	22 17	21 06	23 31	15 29	06 12	10 36	08 44	00 16	12 18	19 19
09	13 58	12 31	03 07	22 09	18 44	23 38	15 04	06 09	10 37	08 44	00 15	12 17	19 19
10	13 53	12 28	03 59	22 00	15 21	23 43	14 38	06 07	10 38	08 45	00 14	12 16	19 19
11	13 47	12 25	04 39	21 51	11 08	23 48	14 12	06 05	10 40	08 46	00 14	12 16	19 19
12	13 40	12 22	05 03	21 42	06 17	23 51	13 46	06 03	10 41	08 47	00 13	12 15	19 19
13	13 35	12 19	05 10	21 32	01 N 05	23 53	13 19	06 02	10 43	08 47	00 12	12 14	19 19
14	13 30	12 16	04 59	21 21	04 S 14	23 54	12 52	06 01	10 44	08 48	00 12	12 14	19 19
15	13 28	12 12	04 30	21 11	09 21	23 53	12 24	06 00	10 46	08 49	00 11	12 13	19 19
16	13 D 27	12 09	03 44	21 00	14 00	23 51	11 56	05 59	10 48	08 49	00 10	12 12	19 19
17	13 28	12 06	02 45	20 48	17 53	23 48	11 28	05 59	10 50	08 50	00 09	12 12	19 19
18	13 29	12 03	01 35	20 36	20 42	23 44	11 00	05 59	10 52	08 50	00 08	12 11	19 19
19	13 30	12 00	00 S 21	20 24	22 15	23 38	10 31	05 59	10 53	08 51	00 08	12 10	19 19
20	13 R 30	11 57	00 N 55	20 11	22 22	23 31	10 02	05 59	10 55	08 51	00 07	12 10	19 19
21	13 27	11 53	02 07	19 58	21 06	23 23	09 33	06 00	10 58	08 52	00 06	12 09	19 18
22	13 22	11 50	03 10	19 45	18 35	23 13	09 04	06 01	11 00	08 52	00 05	12 08	19 18
23	13 15	11 47	04 02	19 31	15 04	23 02	08 34	06 02	11 02	08 52	00 04	12 07	19 18
24	13 06	11 44	04 39	19 17	10 52	22 49	08 04	06 04	11 04	08 53	00 03	12 07	19 18
25	12 56	11 41	05 00	19 02	06 15	22 35	07 34	06 05	11 06	08 53	00 02	12 06	19 18
26	12 47	11 37	05 06	18 47	01 S 29	22 19	07 04	06 08	11 09	08 53	00 02	12 05	19 18
27	12 38	11 34	04 57	18 32	03 N 14	22 03	06 34	06 10	11 11	08 53	00 S 01	12 05	19 18
28	12 32	11 31	04 35	18 17	07 44	21 44	06 03	06 13	11 14	08 53	00 00	12 04	19 18
29	12 29	11 28	04 01	18 01	11 51	21 24	05 32	06 16	11 16	08 54	00 N 01	12 03	19 18
30	12 26	11 25	03 16	17 44	15 29	21 03	05 01	06 20	11 19	08 54	00 02	12 02	19 18
31	12 ♐ 25	11 ♐ 22	02 N 24	17 S 28	18 N 29	20 S 40	04 S 31	06 N 23	11 N 21	08 S 54	00 N 03	12 S 01	19 S 18

ZODIAC SIGN ENTRIES

Date	h	m	Planets
02	22	16	☽ ♉
05	10	44	☽ ♊
07	21	05	☽ ♋
08	06	34	☿ ♑
10	04	35	☽ ♌
12	09	44	☽ ♍
14	05	47	♀ ♓
14	13	28	☽ ♎
16	16	33	☽ ♏
18	19	29	☽ ♐
20	16	10	☉ ♒
20	22	40	☽ ♑
23	02	53	☽ ♒
25	09	11	☽ ♓
27	18	12	☿ ♒
27	18	28	☽ ♈
30	06	28	☽ ♉

LATITUDES

Date	Mercury ☿ ° '	Venus ♀ ° '	Mars ♂ ° '	Jupiter ♃ ° '	Saturn ♄ ° '	Uranus ♅ ° '	Neptune ♆ ° '	Pluto ♇ ° '
01	00 N 51	01 S 50	02 N 57	01 S 14	02 N 25	00 S 43	00 S 33	03 N 55
04	00 26	01 48	03 02	01 13	02 26	00 43	00 33	03 55
07	00 N 03	01 45	03 07	01 13	02 27	00 43	00 33	03 54
10	00 S 19	01 42	03 13	01 11	02 28	00 43	00 33	03 54
13	00 40	01 38	03 18	01 10	02 28	00 43	00 33	03 54
16	00 59	01 33	03 23	01 09	02 29	00 43	00 33	03 54
19	01 16	01 27	03 28	01 08	02 30	00 43	00 33	03 54
22	01 31	01 21	03 34	01 07	02 31	00 43	00 33	03 54
25	01 44	01 13	03 39	01 06	02 31	00 42	00 33	03 53
28	01 54	01 05	03 44	01 06	02 32	00 42	00 33	03 53
31	02 S 01	00 S 56	03 N 49	01 S 05	02 N 33	00 S 42	00 S 33	03 N 53

DATA

Julian Date	2455928
Delta T	+68 seconds
Ayanamsa	24° 01' 46"
Synetic vernal point	05° ♓ 05' 13"
True obliquity of ecliptic	23° 26' 13"

MOON'S PHASES, APSIDES AND POSITIONS ☽

Date	h	m	Phase	Longitude ° '	Eclipse Indicator
01	06	15	☽	10 ♈ 13	
09	07	30	○	18 ♋ 26	
16	09	08	☾	25 ♎ 38	
23	07	39	●	02 ♒ 42	
31	04	10	☽	10 ♉ 41	

Day	h	m		
02	20	17	Apogee	
17	21	13	Perigee	
30	17	42	Apogee	
06	21	45	Max dec	22° N 32'
13	16	52	0S	
20	02	07	Max dec	22° S 29'
26	19	30	0N	

LONGITUDES

Date	Chiron ⚷ ° '	Ceres ⚳ ° '	Pallas ⚴ ° '	Juno ⚵ ° '	Vesta ⚶ ° '	Black Moon Lilith ⚸ ° '
01	01 ♓ 56	24 ♓ 42	16 ♒ 30	22 ♏ 25	05 ♓ 06	01 ♉ 45
11	02 ♓ 16	27 ♓ 29	19 ♒ 38	25 ♏ 35	09 ♓ 11	02 ♉ 52
21	03 ♓ 00	00 ♈ 33	22 ♒ 51	28 ♏ 03	13 ♓ 51	03 ♉ 59
31	03 ♓ 36	03 ♈ 49	26 ♒ 07	00 ♐ 26	18 ♓ 22	05 ♉ 06

ASPECTARIAN

h m	Aspects
01 Sunday	
00 25	☽ □ ♇
06 06	☿ □ ♂
06 15	☽ □ ☉
07 04	♀ ± ♂
07 34	☽ ⋕ ♄
08 18	☉ Q ♄
13 39	☽ ∠ ♆
15 03	☽ ⚹ ♀
18 22	☽ ∥ ♃
02 Monday	
02 45	☽ ⊼ ♂
04 56	☽ △ ☿
05 46	☽ ⋕ ♆
15 09	☽ ± ♂
16 10	♀ ∠ ♅
18 10	☽ Q ♀
19 01	☽ ☍ ♄
20 07	☽ ⚹ ♆
23 14	☽ ☌ ♃
03 Tuesday	
00 03	☽ ⊻ ♅
09 46	☽ ⚼ ♂
12 15	☽ ⊥ ♅
13 19	☽ △ ♇
15 22	☽ ⚼ ☿
17 04	☽ ⋕ ♀
20 31	☽ Q ♆
04 Wednesday	
00 43	☽ △ ☉
00 50	☉ ∥ ☿
06 30	☽ ∠ ♅
09 54	☽ ⋕ ♇
10 49	☽ □ ♀
11 58	☽ ± ☿
16 38	☽ △ ♂
19 40	☽ ⚼ ♇
20 17	♀ Q ♃
23 12	☉ ∠ ♆
05 Thursday	
01 37	☽ ⊼ ☿
07 51	☽ ⊼ ♄
08 46	☽ □ ♆
09 38	☽ ⚼ ☉
11 51	☽ ⊻ ♃
12 36	☽ ⚹ ♅
13 41	☽ ± ♇
19 49	☽ ± ♄
23 46	☽ ⊥ ♃
06 Friday	
01 35	☽ ⊼ ♇
05 03	☽ ± ☉
12 17	☽ Q ♅
13 34	☽ ⚼ ♄
16 08	☽ ⋕ ☉
17 26	☽ ∠ ♃
17 50	☽ ⊼ ☉
07 Saturday	
04 40	☽ △ ♀
04 44	☽ □ ♂
05 27	♀ ⊼ ♂
08 13	☿ ⚹ ♄
08 33	☽ ⋕ ☉
15 15	☿ ⚹ ♆
18 35	☽ △ ♄
19 21	☽ △ ♆
19 52	☽ ☍ ☿
22 20	☽ ⚹ ♃
22 58	☽ □ ♅
08 Sunday	
03 37	♀ ∠ ♇
11 23	☽ ☍ ♇
12 11	☽ ⚼ ♀
15 20	☽ Q ♂
18 05	☿ △ ♃
20 55	☽ Q ♃
23 23	☿ □ ♅
23 33	☽ ⚼ ♆
09 Monday	
06 37	☽ ± ♀
06 55	☽ ⋕ ♇
07 30	☽ ☍ ☉
13 48	☽ ⚹ ♂
16 08	☽ ± ♆
18 44	☽ ⊼ ♀
21 55	☉ Q ♅
10 Tuesday	
02 25	☽ □ ♄
03 03	☽ ⊼ ♆
04 10	♀ ⊥ ♅
05 58	☽ □ ♃
06 29	☽ △ ♅
10 15	☽ ⊼ ☿
16 53	☽ ⋕ ♀
17 14	☽ ∠ ♂
18 22	☽ ⊼ ♇
22 18	☽ ± ☿
11 Wednesday	
05 04	☽ ± ♇
05 54	☽ ⋕ ♆
09 18	☽ ⚼ ♅
09 25	☽ ⊥ ♂
10 42	☽ Q ♄
11 10	☉ ⊥ ♀
14 24	☽ ∥ ♃
16 11	☽ ⚼ ☿
17 49	☽ ⊼ ☉
20 06	☽ ⊻ ♂
20 59	☽ ⚼ ♇
23 58	☽ ⋕ ♄
12 Thursday	
01 10	☽ ± ♅
05 10	☽ ± ☉
05 28	☽ ☍ ♀
07 49	☽ ⚹ ♄
08 23	☽ ☍ ♆
11 18	☽ △ ♃
11 40	☽ ⊼ ♅
13 05	☽ ∥ ♂
21 33	☽ △ ☿
22 05	☽ ⚼ ☉
23 12	☽ △ ♇
13 Friday	
04 46	☉ △ ♂
09 05	♀ △ ♄
09 55	☽ ∠ ♄
13 03	☿ ☌ ♇
13 23	☽ ⚼ ♃
15 17	♀ ☌ ♆
15 57	☽ ⋕ ♅
17 46	☽ ∥ ♅
14 Saturday	
00 36	☽ ☌ ♂
01 27	☽ ⊥ ♄
01 47	☉ ⊥ ♆
01 58	☽ △ ☉
04 55	☽ ± ♃
11 45	☽ ⊻ ♄
12 16	☽ ⊼ ♆
14 12	☽ ⊼ ♀
15 14	☽ ⊼ ♃
15 28	☽ ☍ ♅
20 10	☽ ⋕ ♂
22 32	☽ ± ♆
15 Sunday	
01 25	☽ ± ♀
02 47	♀ ⚹ ♃
02 50	☽ □ ♇
05 15	♀ ⊻ ♅
07 15	☽ □ ☿
09 23	☽ ∥ ♄
13 55	☽ ⚼ ♆
17 14	☿ Q ♄
18 12	☽ ⚼ ♀
19 02	☽ ⋕ ♃
21 49	♀ ∥ ♆
16 Monday	
01 58	☽ ∥ ♀
02 23	☽ ∥ ♆
04 11	☽ ⊻ ♂
09 08	☽ □ ☉
09 32	☽ Q ♇
14 27	☽ ⊥ ♂
15 00	☽ ☌ ♄
15 28	☽ △ ♆
17 34	☽ Q ☿
18 34	☽ ☍ ♃
18 39	☽ ⊼ ♅
22 05	☽ △ ♀
17 Tuesday	
04 52	☽ ± ♅
05 49	☽ ∠ ♂
05 58	☽ ⚹ ♇
16 24	☽ ⚹ ☿
18 03	☽ Q ☉
20 10	☽ ⚼ ♅
21 56	☿ ∠ ♆
22 56	☽ ∥ ♇
18 Wednesday	
00 12	♃ ⊻ ♅
07 24	☽ ⚹ ♂
07 29	☽ ∠ ♇
11 00	☽ ∥ ☉
16 00	☽ ⚹ ☉
18 03	☽ ⊻ ♄
18 31	☽ □ ♆
18 33	♀ ⋕ ♃
20 59	☽ ∠ ☿
21 41	☽ △ ♅
21 46	☽ ⊼ ♃
22 48	☽ ⊥ ♇
19 Thursday	
03 53	☽ Q ♂
04 17	☽ ⊥ ♄
05 47	☽ □ ♀
08 01	☽ ± ♃
09 02	☽ ⊻ ♇
14 11	☽ ⊥ ☿
19 31	☽ ∠ ☉
19 38	☽ ∠ ♄
21 18	☉ □ ♄
23 27	☽ ⚼ ♃
20 Friday	
01 14	☽ Q ♆
01 41	☽ ⊻ ☿
03 55	☉ ⊻ ♆
10 41	☽ □ ♂
12 05	☽ ⊥ ☉
15 24	☽ Q ♀
20 32	♀ ⚹ ♇
21 21	☽ ⚹ ♄
21 49	☽ ⚹ ♆
23 11	☽ ⊻ ☉
21 Saturday	
01 02	☽ □ ♅
01 17	☽ △ ♃
02 45	☿ Q ♅
12 31	☽ ☌ ♇
14 00	☽ ⚹ ♀
18 06	☽ Q ♄
23 48	☽ ∠ ♆
22 Sunday	
01 15	☽ ∥ ☉
01 20	☉ ⚹ ♅
06 03	☽ ∥ ♇
06 35	☉ □ ♃
08 17	☽ Q ♅
12 02	☽ ☌ ☿
14 42	☽ △ ♂
15 33	☽ ⊥ ♆
18 40	☽ ∠ ♀
21 20	♀ ∥ ♄
23 Monday	
01 38	☽ □ ♄
02 08	☽ ⊻ ♆
05 27	☽ ⚹ ♅
05 56	☽ □ ♃
07 39	☽ ☌ ☉
11 15	☿ △ ♂
12 09	☽ ⊥ ♀
17 17	☽ ⚼ ♂
17 19	☽ ⊻ ♇
19 01	☿ ⊥ ♆
23 57	☽ ⊻ ♀
24 Tuesday	
00 53	♂ St R
04 08	☽ ⊥ ♇
05 10	☽ ∥ ♆
08 24	☽ ∠ ♅
09 20	☉ ∥ ♇
09 30	☽ ± ♂
10 56	☽ ⋕ ♃
14 30	☽ Q ♃
20 26	☽ ⊼ ♂
20 33	☽ ∠ ♇
22 32	☽ ∥ ♄
25 Wednesday	
00 53	☽ ⊥ ♅
01 04	☽ ⊻ ☿
04 30	☽ ∥ ♀
07 59	☽ △ ♄
08 33	☽ ☌ ♆
12 02	☽ ⊻ ♅
12 47	☽ ⚹ ♃
12 48	☽ ⋕ ♂
13 45	☽ ⊥ ☿
18 54	☽ ⊻ ☉
26 Thursday	
00 30	☽ ⚹ ♇
03 03	♀ ⚼ ♄
07 12	☽ ⊥ ☉
09 10	☽ ∠ ☿
12 15	☽ ⚼ ♄
13 12	☽ ☌ ♀
17 22	☽ ∠ ♃
19 23	☽ ∥ ♅
19 36	☽ ⋕ ♅
23 27	☽ Q ♇
27 Friday	
01 57	☽ ∠ ☉
04 52	☽ ☍ ♂
05 35	☽ ± ♄
09 19	☿ □ ♄
10 55	☽ ⊥ ♃
14 21	☿ ⊻ ♆
17 18	☽ ⊼ ♄
17 58	☽ ⊻ ♆
18 30	☽ ⚹ ☿
21 39	☽ ☌ ♅
22 46	☽ ⊻ ♃
28 Saturday	
03 41	☽ ∥ ♂
03 48	☽ ⋕ ♀
04 41	♀ ⋕ ♂
05 49	☽ ⊥ ♆
10 01	☽ ⚹ ☉
10 47	☽ □ ♇
12 05	☉ ⚼ ♂
12 37	♀ ∠ ♃
18 26	☿ ⚹ ♅
18 31	☽ ⋕ ♄
19 28	☉ ⊻ ♇
22 01	☽ Q ☿
23 47	☽ ∠ ♆
29 Sunday	
04 08	☿ □ ♃
06 31	☽ ⊻ ♀
08 22	☽ ∥ ♃
12 15	☽ Q ☉
13 11	☽ ⋕ ♆
16 05	☽ ⊼ ♂
19 58	☽ ⊥ ♀
30 Monday	
04 09	☽ ± ♂
05 21	☽ ☍ ♄
06 08	☽ ⚹ ♆
09 58	☽ ⊻ ♅
11 29	☽ ☌ ♃
16 14	☽ □ ☿
16 16	☽ ∠ ♀
16 47	☿ ∠ ♀
22 11	☽ ⊥ ♅
22 17	☽ ⚼ ♂
23 25	☽ △ ♇
31 Tuesday	
02 52	♀ Q ♇
03 56	☽ ⋕ ☉
04 10	☽ □ ☉
06 35	☽ Q ♆
16 32	☽ ∠ ♅
19 42	☽ ⋕ ♇

All ephemeris data is given at 12.00 UT and the Moon's longitude is additionally given for 24.00 UT

FEBRUARY 2012

Date	Sidereal time h m s	Sun ☉ ° ' "	Moon ☽ ° ' "	Moon ☽ 24.00 ° ' "	Mercury ☿ ° '	Venus ♀ ° '	Mars ♂ ° '	Jupiter ♃ ° '	Saturn ♄ ° '	Uranus ♅ ° '	Neptune ♆ ° '	Pluto ♇ ° '
01	20 44 26	12 ♒ 01 31	26 ♉ 23 56	02 ♊ 22 30	07 ♒ 52	22 ♓ 00	22 ♍ 38	02 ♉ 42	29 ♎ 29	01 ♈ 48	29 ♒ 55	08 ♑ 24
02	20 48 22	13 02 25	08 ♊ 23 45	14 ♊ 28 21	09 34	23 11	22 R 31	02 49	29 29	01 51	29 57	08 26
03	20 52 19	14 03 17	20 ♊ 36 51	26 ♊ 49 46	11 16	24 23	22 23	02 57	29 30	01 53	29 ♒ 59	08 27
04	20 56 16	15 04 08	03 ♋ 07 35	09 ♋ 30 37	12 59	25 34	22 15	03 04	29 30	01 56	00 ♓ 02	08 29
05	21 00 12	16 04 57	15 ♋ 59 09	22 ♋ 33 20	14 42	26 45	22 05	03 12	29 30	01 58	00 04	08 31
06	21 04 09	17 05 45	29 ♋ 13 10	05 ♌ 58 33	16 27	27 56	21 55	03 20	29 30	02 01	00 06	08 33
07	21 08 05	18 06 32	12 ♌ 49 14	19 ♌ 44 49	18 12	29 ♓ 07	21 44	03 28	29 30	02 04	00 08	08 35
08	21 12 02	19 07 18	26 ♌ 44 48	03 ♍ 48 36	19 58	00 ♈ 18	21 33	03 36	29 R 30	02 06	00 10	08 36
09	21 15 58	20 08 02	10 ♍ 55 31	18 ♍ 04 49	21 45	01 28	21 20	03 44	29 30	02 09	00 13	08 38
10	21 19 55	21 08 45	25 ♍ 15 44	02 ♎ 27 31	23 32	02 39	21 07	03 52	29 30	02 12	00 15	08 40
11	21 23 51	22 09 27	09 ♎ 39 30	16 ♎ 51 02	25 20	03 49	20 53	04 01	29 30	02 15	00 17	08 41
12	21 27 48	23 10 08	24 ♎ 01 35	01 ♏ 10 41	27 09	04 59	20 39	04 10	29 29	02 18	00 20	08 43
13	21 31 45	24 10 48	08 ♏ 18 00	15 ♏ 23 17	28 ♒ 58	06 09	20 23	04 18	29 29	02 20	00 22	08 45
14	21 35 41	25 11 27	22 ♏ 26 23	29 ♏ 27 11	00 ♓ 47	07 19	20 07	04 27	29 28	02 23	00 24	08 46
15	21 39 38	26 12 04	06 ♐ 25 40	13 ♐ 21 51	02 38	08 29	19 50	04 36	29 27	02 26	00 26	08 48
16	21 43 34	27 12 41	20 ♐ 15 43	27 ♐ 07 20	04 28	09 39	19 33	04 46	29 26	02 29	00 29	08 49
17	21 47 31	28 13 17	03 ♑ 56 41	10 ♑ 43 46	06 19	10 49	19 15	04 55	29 25	02 32	00 31	08 51
18	21 51 27	29 ♒ 13 51	17 ♑ 28 32	24 ♑ 10 55	08 10	11 58	18 56	05 04	29 24	02 35	00 33	08 53
19	21 55 24	00 ♓ 14 24	00 ♒ 50 48	07 ♒ 28 03	10 00	13 07	18 37	05 14	29 23	02 38	00 35	08 54
20	21 59 20	01 14 55	14 ♒ 02 32	20 ♒ 34 03	11 51	14 16	18 17	05 24	29 22	02 41	00 38	08 55
21	22 03 17	02 15 25	27 ♒ 02 27	03 ♓ 27 36	13 41	15 25	17 56	05 34	29 20	02 44	00 40	08 57
22	22 07 14	03 15 53	09 ♓ 49 22	16 ♓ 07 42	15 30	16 34	17 36	05 44	29 19	02 47	00 42	08 58
23	22 11 10	04 16 20	22 ♓ 22 34	28 ♓ 34 00	17 17	17 43	17 14	05 54	29 17	02 50	00 45	09 00
24	22 15 07	05 16 45	04 ♈ 42 07	10 ♈ 47 06	19 03	18 51	16 52	06 04	29 16	02 53	00 47	09 01
25	22 19 03	06 17 08	16 ♈ 49 12	22 ♈ 48 44	20 47	19 59	16 30	06 14	29 14	02 56	00 49	09 03
26	22 23 00	07 17 29	28 ♈ 46 05	04 ♉ 41 43	22 29	21 07	16 08	06 25	29 12	03 00	00 51	09 04
27	22 26 56	08 17 49	10 ♉ 36 08	16 ♉ 29 55	24 08	22 15	15 45	06 35	29 10	03 03	00 54	09 05
28	22 30 53	09 18 06	22 ♉ 23 39	28 ♉ 18 00	25 43	23 23	15 22	06 46	29 08	03 06	00 56	09 06
29	22 34 49	10 ♓ 18 22	04 ♊ 13 39	10 ♊ 11 16	27 ♓ 14	24 ♈ 30	14 ♍ 58	06 ♉ 56	29 ♎ 06	03 ♈ 09	00 ♓ 58	09 ♑ 08

Date	Moon True ☊ ° '	Moon Mean ☊ ° '	Moon ☽ Latitude ° '
01	12 ♐ 26	11 ♐ 18	01 N 25
02	12 D 27	11 15	00 N 22
03	12 R 27	11 12	00 S 44
04	12 24	11 09	01 48
05	12 20	11 06	02 49
06	12 12	11 03	03 42
07	12 03	10 59	04 24
08	11 52	10 56	04 52
09	11 41	10 53	05 02
10	11 30	10 50	04 54
11	11 22	10 47	04 27
12	11 16	10 43	03 43
13	11 13	10 40	02 45
14	11 12	10 37	01 38
15	11 D 12	10 34	00 S 25
16	11 R 12	10 31	00 N 48
17	11 10	10 28	01 57
18	11 06	10 24	02 59
19	10 59	10 21	03 51
20	10 48	10 18	04 29
21	10 36	10 15	04 52
22	10 22	10 12	05 00
23	10 08	10 08	04 53
24	09 55	10 05	04 33
25	09 45	10 02	04 00
26	09 37	09 59	03 18
27	09 32	09 56	02 27
28	09 30	09 53	01 29
29	09 ♐ 29	09 ♐ 49	00 N 28

DECLINATIONS

Date	Sun ☉ ° '	Moon ☽ ° '	Mercury ☿ ° '	Venus ♀ ° '	Mars ♂ ° '	Jupiter ♃ ° '	Saturn ♄ ° '	Uranus ♅ ° '	Neptune ♆ ° '	Pluto ♇ ° '
01	17 S 11	20 N 43	20 S 16	03 S 59	06 N 27	11 N 24	08 S 54	00 N 04	12 S 01	19 S 18
02	16 54	22 04	19 51	03 28	06 31	11 27	08 54	00 05	12 00	19 18
03	16 37	22 23	19 23	02 57	06 36	11 29	08 53	00 06	11 59	19 17
04	16 19	21 36	18 55	02 26	06 40	11 32	08 53	00 07	11 58	19 17
05	16 01	19 41	18 25	01 54	06 45	11 35	08 53	00 08	11 58	19 17
06	15 43	16 41	17 53	01 23	06 51	11 38	08 53	00 09	11 57	19 17
07	15 24	12 44	17 20	00 52	06 56	11 41	08 53	00 11	11 56	19 17
08	15 05	08 01	16 46	00 S 20	07 02	11 44	08 53	00 12	11 55	19 17
09	14 46	02 N 48	16 10	00 N 11	07 08	11 47	08 52	00 13	11 54	19 17
10	14 27	02 S 37	15 32	00 43	07 15	11 50	08 52	00 14	11 54	19 17
11	14 07	07 55	14 53	01 14	07 21	11 53	08 52	00 15	11 53	19 17
12	13 48	12 46	14 13	01 46	07 28	11 56	08 51	00 16	11 52	19 17
13	13 28	16 52	13 31	02 17	07 35	11 59	08 51	00 17	11 51	19 17
14	13 07	19 57	12 49	02 48	07 42	12 03	08 50	00 18	11 51	19 16
15	12 47	21 48	12 04	03 19	07 50	12 06	08 50	00 20	11 50	19 16
16	12 26	22 17	11 19	03 51	07 57	12 09	08 49	00 21	11 49	19 16
17	12 05	21 25	10 32	04 22	08 05	12 13	08 49	00 22	11 48	19 16
18	11 44	19 20	09 44	04 53	08 13	12 16	08 48	00 23	11 47	19 16
19	11 23	16 13	08 56	05 24	08 21	12 19	08 47	00 24	11 47	19 16
20	11 02	12 19	08 06	05 55	08 30	12 23	08 47	00 26	11 46	19 16
21	10 40	07 55	07 16	06 25	08 38	12 26	08 46	00 27	11 45	19 16
22	10 18	03 S 15	06 25	06 56	08 47	12 30	08 45	00 28	11 44	19 16
23	09 57	01 N 28	05 34	07 26	08 55	12 33	08 44	00 29	11 43	19 16
24	09 34	06 03	04 42	07 56	09 04	12 37	08 44	00 31	11 43	19 16
25	09 12	10 18	03 51	08 27	09 13	12 40	08 43	00 32	11 42	19 15
26	08 50	14 07	03 00	08 56	09 21	12 44	08 42	00 33	11 41	19 15
27	08 27	17 19	02 09	09 26	09 30	12 48	08 41	00 34	11 40	19 15
28	08 05	19 48	01 19	09 56	09 39	12 51	08 40	00 36	11 39	19 15
29	07 S 42	21 N 27	00 S 31	10 N 25	09 N 48	12 N 55	08 S 39	00 N 37	11 S 39	19 S 15

ZODIAC SIGN ENTRIES

Date	h m	Planets
01	19 14	☽ ♊
03	19 03	♆ ♓
04	06 04	☽ ♋
06	13 24	☽ ♌
08	06 01	♀ ♈
08	17 32	☽ ♍
10	19 54	☽ ♎
12	22 01	☽ ♏
14	01 38	☿ ♓
15	00 56	☽ ♐
17	05 03	☽ ♑
19	06 18	☉ ♓
19	10 28	☽ ♒
21	17 31	☽ ♓
24	02 48	☽ ♈
26	14 29	☽ ♉
29	03 27	☽ ♊

LATITUDES

Date	Mercury ☿ ° '	Venus ♀ ° '	Mars ♂ ° '	Jupiter ♃ ° '	Saturn ♄ ° '	Uranus ♅ ° '	Neptune ♆ ° '	Pluto ♇ ° '
01	02 S 02	00 S 53	03 N 50	01 S 05	02 N 33	00 S 42	00 S 33	03 N 53
04	02 05	00 44	03 55	01 04	02 34	00 42	00 33	03 53
07	02 04	00 33	03 59	01 03	02 35	00 42	00 33	03 53
10	01 58	00 22	04 03	01 02	02 36	00 42	00 33	03 53
13	01 48	00 S 11	04 07	01 01	02 37	00 42	00 33	03 53
16	01 33	00 N 01	04 09	01 01	02 37	00 42	00 33	03 53
19	01 12	00 14	04 11	01 00	02 38	00 42	00 33	03 53
22	00 45	00 27	04 13	00 59	02 39	00 42	00 33	03 53
25	00 S 13	00 41	04 13	00 59	02 40	00 42	00 33	03 53
28	00 N 25	00 55	04 13	00 58	02 40	00 42	00 33	03 53
31	01 N 06	01 N 09	04 N 11	00 S 57	02 N 41	00 S 42	00 S 33	03 N 53

DATA

Julian Date	2455959
Delta T	+68 seconds
Ayanamsa	24° 01' 51"
Synetic vernal point	05° ♓ 05' 08"
True obliquity of ecliptic	23° 26' 13"

MOON'S PHASES, APSIDES AND POSITIONS ☽

Date	h m	Phase	Longitude ° '	Eclipse Indicator
07	21 54	○	18 ♌ 32	
14	17 04	☾	25 ♏ 24	
21	22 35	●	02 ♓ 42	

Day	h m		
11	18 47	Perigee	
27	14 04	Apogee	
03	07 10	Max dec	22° N 24'
10	00 27	0S	
16	08 32	Max dec	22° S 18'
23	04 30	0N	

LONGITUDES

Date	Chiron ⚷ ° '	Ceres ⚳ ° '	Pallas ⚴ ° '	Juno ⚵ ° '	Vesta ⚶ ° '	Black Moon Lilith ⚸ ° '
01	03 ♓ 40	04 ♈ 09	26 ♒ 27	00 ♐ 39	18 ♓ 49	05 ♉ 12
11	04 ♓ 18	07 ♈ 37	29 ♒ 46	02 ♐ 44	23 ♓ 24	06 ♉ 19
21	04 ♓ 58	11 ♈ 14	03 ♓ 06	04 ♐ 28	28 ♓ 01	07 ♉ 26
31	05 ♓ 38	14 ♈ 58	06 ♓ 25	05 ♐ 48	02 ♈ 39	08 ♉ 32

All ephemeris data is given at 12.00 UT and the Moon's longitude is additionally given for 24.00 UT

ASPECTARIAN

h m	Aspects	h m	Aspects	h m	Aspects
01 Wednesday		20 08	☽ ± ☿	07 20	☽ ⊻ ☿
02 08	☽ ✶ ♀	20 20	☽ ⊼ ♆	08 51	☽ ± ♂
04 28	☽ △ ♂	23 36	☽ ☍ ♅	11 41	☽ # ♃
05 55	☽ ⚼ ♇	**11 Saturday**		12 28	☽ ✶ ♀
07 18	☽ # ☿	01 24	☽ ☍ ♀	13 37	☽ ⊥ ♇
08 52	☿ ⚼ ♂	02 30	☽ ⊼ ♃	15 12	☽ ∥ ♆
18 12	☽ ⊼ ♄	06 22	☽ ± ♆	18 43	☽ ∠ ♅
19 06	☽ □ ♆	07 31	☽ ⚼ ☉	19 35	☽ ⊼ ♂
19 38	☿ ⊻ ♇	08 03	☉ Q ♃	19 57	☽ ∥ ☉
22 54	☽ ✶ ♅	09 21	☽ # ♂	**21 Tuesday**	
23 41	♀ ☍ ♂	10 23	☽ □ ♇	05 27	☽ Q ♃
02 Thursday		11 16	♃ # ♆	06 15	☽ ∠ ♇
00 04	☽ ± ♇	13 17	☽ ⚼ ☿	07 32	☽ ∥ ♄
00 47	☽ ⊻ ♃	16 28	☽ ∥ ♄	08 21	☽ # ♂
04 55	☽ Q ♀	16 34	♀ ⊻ ♃	11 26	☽ ⊥ ♅
06 12	☽ ± ♄	21 25	☽ ⚼ ♆	16 12	☽ ∥ ☿
12 04	☽ ⊼ ♇	**12 Sunday**		16 17	☽ △ ♄
12 51	☽ ⊥ ♃	00 27	☿ ⊥ ♅	17 20	♀ ∠ ♆
14 41	☽ △ ☿	06 26	☽ ⊻ ♂	18 47	☽ ☌ ♆
18 01	♀ ± ♄	07 20	☽ ∥ ♆	18 55	☽ ∠ ♀
22 01	☽ △ ☉	07 37	☽ # ♃	19 02	☽ # ♀
22 48	☽ Q ♅	10 27	☽ △ ☉	20 36	☿ ⚼ ♄
03 Friday		16 19	☽ ⊥ ♂	22 35	☽ ☌ ☉
00 02	☽ ⚼ ♄	16 31	☽ Q ♇	22 41	☽ ⊻ ♅
06 45	☽ ∠ ♃	17 09	☽ ∥ ☉	23 58	☉ ⊻ ♅
15 24	☽ □ ♂	17 59	☽ △ ☿	**22 Wednesday**	
17 09	☿ ∥ ♇	18 49	☽ ∥ ☿	02 55	☿ # ♀
20 03	☽ □ ♀	21 09	☽ ☌ ♄	04 09	☽ ✶ ♃
21 48	☉ ⊥ ♇	22 36	☽ △ ♆	07 59	♂ # ♄
04 Saturday		**13 Monday**		10 23	☽ ✶ ♇
00 39	☽ ⚼ ☿	01 22	☉ ∠ ♇	13 33	☽ ⊥ ♀
05 06	☽ △ ♄	01 55	☽ ⊼ ♅	20 31	☽ ⚼ ♄
05 41	☽ ⚼ ☉	05 12	☽ ☍ ♃	**23 Thursday**	
06 06	☽ △ ♆	07 10	☽ ∠ ♂	00 35	☽ ☌ ☿
09 44	☽ □ ♅	08 04	☽ ⊼ ♀	02 03	☽ # ♅
11 53	☽ ✶ ♃	12 04	☽ ± ♅	02 08	☽ ⊻ ♀
20 24	☽ ± ☿	12 45	☽ ✶ ♇	02 24	☽ ☍ ♂
22 07	☽ ☍ ♇	16 10	☉ ∥ ☿	04 25	♀ ⊼ ♂
05 Sunday		16 22	♀ ⊥ ♆	06 58	☽ ∥ ♅
00 08	☽ ± ☉	18 44	☿ △ ♄	09 06	☽ ∠ ♃
01 13	☽ Q ♂	19 07	☽ ± ♀	09 20	☽ Q ♇
09 17	☽ ⊼ ☿	**14 Tuesday**		11 25	☿ ☍ ♂
09 18	☿ ⊥ ♇	03 22	☽ ⚼ ♅	13 45	☽ ± ♄
10 18	☽ ⚼ ♆	05 50	☽ ∥ ♇	**24 Friday**	
10 32	☽ Q ♃	06 47	☿ ☌ ♆	01 22	☽ ⊼ ♄
12 10	☉ ± ♂	08 07	☽ ✶ ♂	02 47	☽ ⊥ ♃
12 12	☽ ⊼ ☉	11 47	☽ ⚼ ♀	04 01	☿ ⊻ ♀
15 44	☽ # ♇	14 17	☽ ∠ ♇	04 18	☽ ⊻ ♆
22 09	♀ ⊥ ♃	17 04	☽ □ ☉	05 59	☽ # ☿
23 01	☽ ✶ ♂	**15 Wednesday**		08 26	☽ ☌ ♅
06 Monday		00 01	☽ ⊻ ♄	13 14	☽ ⊻ ☉
01 34	☽ # ☿	01 40	☽ □ ♆	14 43	☽ ⊻ ♃
02 46	☽ ± ♆	04 15	☽ Q ♂	16 06	☽ ⊥ ♆
05 25	☿ ± ♂	04 28	☽ □ ☿	20 31	☽ □ ♇
09 28	☽ △ ♀	05 06	☽ △ ♅	23 48	☽ ∥ ♀
10 04	☉ ∠ ♅	05 44	☽ ⊥ ♇	**25 Saturday**	
12 31	☽ □ ♄	07 02	☿ ⊥ ♀	02 09	☽ ⊥ ☉
13 35	☽ ⊼ ♆	08 49	☽ ⊼ ♃	02 47	☽ # ♄
17 00	☽ △ ♅	09 25	☿ ⊻ ♅	05 23	☽ ∥ ♂
19 02	☽ # ☉	10 19	☽ ⊥ ♄	06 06	☽ # ☉
19 23	☽ □ ♃	11 13	☿ # ♃	10 00	☽ ∠ ♆
20 01	☿ ∠ ♅	15 53	☽ △ ♀	10 32	☉ ✶ ♃
07 Tuesday		16 06	☽ ⊻ ♇	11 23	☽ ⊼ ♂
01 30	☽ ∠ ♂	17 50	☉ ⊥ ♅	11 39	☉ # ♂
04 34	☽ ⊼ ♇	18 32	♀ □ ♇	15 34	☿ Q ♇
09 02	☉ ☌ ☿	19 18	☽ ± ♃	18 56	☿ ∠ ♃
14 03	♄ St R	19 53	☿ ∥ ♆	19 00	☽ ☌ ♀
14 28	☽ ⚼ ♀	**16 Thursday**		20 21	☽ # ♆
15 04	☽ ± ♇	00 42	♂ ⚼ ♃	21 16	☽ ⊻ ☿
16 19	☽ # ♆	01 53	☽ ∠ ♄	21 45	☽ ∠ ☉
17 00	☽ ⊥ ♂	02 31	☽ Q ☉	23 02	☽ ± ♂
17 35	☽ ∥ ♃	08 53	☽ Q ♆	**26 Sunday**	
19 23	☽ ⚼ ♅	10 47	☽ □ ♂	00 35	♀ # ♄
20 00	♀ ⊼ ♄	11 07	☽ ⚼ ♃	02 46	☽ ∥ ♃
20 08	☽ Q ♄	16 07	☿ ✶ ♃	09 02	☉ # ♀
21 54	☽ ☍ ☉	16 27	☽ Q ☿	11 20	☽ ⊥ ☿
22 41	☽ ☍ ☿	**17 Friday**		12 52	☽ ☍ ♄
08 Wednesday		01 08	☽ ✶ ☉	16 14	☽ ✶ ♆
03 13	☽ ⊻ ♂	04 03	☽ ✶ ♄	16 37	☽ ⚼ ♂
06 37	☽ ⚼ ♇	05 01	☉ # ♃	20 35	☽ ⊻ ♅
07 25	☽ ± ♀	05 57	☽ ✶ ♆	21 06	☉ ∥ ♄
07 51	☽ # ♄	09 30	☽ □ ♅	22 07	☿ ± ♄
09 30	♀ ⊻ ♆	13 44	☽ △ ♃	**27 Monday**	
10 54	☽ ± ♅	16 51	☽ ✶ ☿	03 43	☽ ☌ ♃
16 34	☽ ∥ ♂	20 41	☽ ☌ ♇	03 57	♀ ± ♂
16 42	☽ ✶ ♄	**18 Saturday**		06 53	☽ ✶ ☉
17 51	☽ ☍ ♆	01 13	☽ Q ♄	08 32	☽ ∠ ☿
18 19	♀ # ♅	01 17	☽ □ ♀	08 49	☽ ⊥ ♅
18 35	☽ ⊼ ♀	05 45	☽ ∠ ☉	08 55	☽ △ ♇
21 09	☽ ⊼ ♅	08 32	☉ ∥ ♆	16 41	☽ Q ♆
23 45	☽ △ ♃	08 34	☽ ∠ ♆	16 53	♀ ∥ ♂
09 Thursday		12 34	☽ ∥ ♇	22 08	☽ △ ♂
07 09	☿ ⊼ ♂	14 33	☽ △ ♂	**28 Tuesday**	
08 08	☽ △ ♇	16 02	☉ △ ♄	03 13	☽ ∠ ♅
11 51	☿ Q ♃	17 34	☽ Q ♅	05 57	☽ # ♇
13 14	♀ ∥ ♅	21 23	☿ ✶ ♇	07 15	☉ ✶ ♇
18 01	☽ ∠ ♄	23 08	☽ ⊥ ☉	09 34	☽ Q ☉
22 36	☽ ∥ ♀	23 48	☽ ∠ ☿	14 13	☽ ⊻ ♀
23 28	☽ ∥ ♅	**19 Sunday**		15 29	☽ ⚼ ♇
10 Friday		00 42	☽ ⊥ ♆	19 46	☽ ✶ ☿
01 13	☽ ⚼ ♃	09 22	☽ □ ♄	**29 Wednesday**	
01 26	☽ # ♅	10 49	☽ ⊻ ☉	01 39	☽ ⊼ ♄
02 29	♀ ☌ ♅	11 32	☽ ⊻ ♆	03 41	☽ ⊥ ♀
02 42	☽ # ♀	12 32	☽ Q ♀	05 23	☽ □ ♆
04 36	☽ ⊼ ☉	15 15	☽ ✶ ♅	09 03	☿ # ♅
05 11	☽ ☌ ♂	16 10	☿ ∥ ♄	09 46	☽ ± ♇
08 42	☽ ⊼ ☿	16 53	☽ ⚼ ♂	09 49	☽ ✶ ♅
09 04	☽ ⊥ ♄	18 39	☽ ⊥ ☿	13 45	☽ ± ♄
11 30	☉ ⊼ ♂	20 02	☽ □ ♃	17 33	☽ ⊻ ♃
13 46	☿ ∠ ♇	20 41	☉ ☌ ♆	21 53	☽ ⊼ ♇
15 23	☽ ± ☉	**20 Monday**		23 29	☽ Q ☿
16 24	☽ ± ♃	02 16	☿ # ♂	23 44	☽ ∠ ♀
19 04	☽ ⊻ ♄	02 38	☽ ⊻ ♇		

MARCH 2012

LONGITUDES

Date	Sidereal time h m s	Sun ☉ ° ' "	Moon ☽ ° ' "	Moon ☽ 24.00 ° ' "	Mercury ☿ ° '	Venus ♀ ° '	Mars ♂ ° '	Jupiter ♃ ° '	Saturn ♄ ° '	Uranus ♅ ° '	Neptune ♆ ° '	Pluto ♇ ° '
01	22 38 46	11 ♓ 18 35	16 ♊ 11 35	22 ♊ 15 17	28 ♓ 40	25 ♈ 38	14 ♍ 35	07 ♉ 07	29 ♎ 03	03 ♈ 12	01 ♓ 00	09 ♑ 09
02	22 42 43	12 18 47	28 ♊ 23 03	04 ♋ 35 33	00 ♈ 01	26 45	14 R 11	07 18	29 R 01	03 16	01 03	09 10
03	22 46 39	13 18 56	10 ♋ 53 23	17 ♋ 17 03	01 16	27 51	13 47	07 29	28 59	03 19	01 05	09 11
04	22 50 36	14 19 04	23 ♋ 47 02	00 ♌ 23 37	02 25	28 ♈ 58	13 24	07 40	28 56	03 22	01 07	09 12
05	22 54 32	15 19 09	07 ♌ 07 00	13 ♌ 57 13	03 26	00 ♉ 04	13 00	07 52	28 53	03 26	01 09	09 14
06	22 58 29	16 19 12	20 ♌ 54 07	27 ♌ 57 22	04 21	01 11	12 36	08 03	28 51	03 29	01 12	09 15
07	23 02 25	17 19 13	05 ♍ 06 28	12 ♍ 20 43	05 07	02 16	12 13	08 14	28 48	03 32	01 14	09 16
08	23 06 22	18 19 12	19 ♍ 39 15	27 ♍ 01 07	05 45	03 22	11 49	08 26	28 45	03 35	01 16	09 17
09	23 10 18	19 19 10	04 ♎ 25 16	11 ♎ 50 36	06 14	04 27	11 26	08 38	28 42	03 39	01 18	09 18
10	23 14 15	20 19 05	19 ♎ 16 02	26 ♎ 40 33	06 35	05 32	11 03	08 49	28 39	03 42	01 20	09 19
11	23 18 12	21 18 59	04 ♏ 03 15	11 ♏ 23 19	06 46	06 37	10 40	09 01	28 36	03 46	01 23	09 20
12	23 22 08	22 18 51	18 ♏ 40 07	25 ♏ 53 11	06 R 49	07 42	10 18	09 13	28 33	03 49	01 25	09 21
13	23 26 05	23 18 42	03 ♐ 02 08	10 ♐ 06 48	06 43	08 46	09 56	09 25	28 29	03 52	01 27	09 21
14	23 30 01	24 18 30	17 ♐ 07 06	24 ♐ 03 03	06 29	09 50	09 34	09 37	28 26	03 56	01 29	09 22
15	23 33 58	25 18 18	00 ♑ 54 45	07 ♑ 42 20	06 07	10 54	09 12	09 49	28 23	03 59	01 31	09 23
16	23 37 54	26 18 03	14 ♑ 26 00	21 ♑ 05 57	05 38	11 57	08 52	10 01	28 19	04 02	01 33	09 24
17	23 41 51	27 17 47	27 ♑ 42 22	04 ♒ 15 27	05 02	13 00	08 31	10 13	28 16	04 06	01 35	09 25
18	23 45 47	28 17 29	10 ♒ 45 21	17 ♒ 12 13	04 20	14 03	08 11	10 26	28 12	04 09	01 38	09 26
19	23 49 44	29 ♓ 17 09	23 ♒ 36 10	29 ♒ 57 18	03 34	15 05	07 52	10 38	28 08	04 13	01 40	09 27
20	23 53 41	00 ♈ 16 48	06 ♓ 15 39	12 ♓ 31 19	02 44	16 07	07 33	10 51	28 05	04 16	01 42	09 27
21	23 57 37	01 16 24	18 ♓ 44 18	24 ♓ 54 42	01 51	17 09	07 15	11 03	28 01	04 20	01 44	09 27
22	00 01 34	02 15 59	01 ♈ 02 33	07 ♈ 07 56	00 57	18 10	06 58	11 16	27 57	04 23	01 46	09 28
23	00 05 30	03 15 31	13 ♈ 10 59	19 ♈ 11 48	00 ♈ 03	19 11	06 41	11 28	27 53	04 26	01 48	09 28
24	00 09 27	04 15 01	25 ♈ 10 37	01 ♉ 07 39	29 ♓ 10	20 12	06 25	11 41	27 49	04 30	01 50	09 29
25	00 13 23	05 14 30	07 ♉ 03 11	12 ♉ 57 32	28 19	21 12	06 09	11 54	27 45	04 33	01 52	09 30
26	00 17 20	06 13 56	18 ♉ 51 07	24 ♉ 44 21	27 30	22 12	05 54	12 07	27 41	04 37	01 54	09 30
27	00 21 16	07 13 20	00 ♊ 37 44	06 ♊ 31 47	26 46	23 11	05 40	12 20	27 37	04 40	01 56	09 31
28	00 25 13	08 12 42	12 ♊ 27 06	18 ♊ 24 16	26 05	24 10	05 27	12 33	27 33	04 44	01 58	09 31
29	00 29 10	09 12 01	24 ♊ 23 56	00 ♋ 26 45	25 30	25 08	05 15	12 46	27 28	04 47	02 00	09 31
30	00 33 06	10 11 18	06 ♋ 33 21	12 ♋ 44 31	24 59	26 06	05 03	12 59	27 24	04 50	02 01	09 32
31	00 37 03	11 ♈ 10 33	19 ♋ 00 45	25 ♋ 22 44	24 ♓ 34	27 ♉ 03	04 ♍ 52	13 ♉ 12	27 ♎ 20	04 ♈ 54	02 ♓ 03	09 ♑ 32

Date	Moon True ☊ ° '	Moon Mean ☊ ° '	Moon ☽ Latitude ° '
01	09 ♐ 29	09 ♐ 46	00 S 35
02	09 R 29	09 43	01 38
03	09 26	09 40	02 38
04	09 22	09 37	03 32
05	09 14	09 34	04 15
06	09 04	09 30	04 46
07	08 53	09 27	05 00
08	08 41	09 24	04 55
09	08 30	09 21	04 31
10	08 21	09 18	03 48
11	08 14	09 14	02 50
12	08 11	09 11	01 41
13	08 10	09 08	00 S 27
14	08 D 10	09 05	00 N 47
15	08 R 10	09 02	01 58
16	08 09	08 59	03 00
17	08 05	08 55	03 51
18	07 59	08 52	04 29
19	07 50	08 49	04 53
20	07 38	08 46	05 02
21	07 26	08 43	04 57
22	07 14	08 40	04 37
23	07 02	08 36	04 05
24	06 53	08 33	03 23
25	06 46	08 30	02 31
26	06 42	08 27	01 34
27	06 40	08 24	00 N 32
28	06 D 40	08 20	00 S 31
29	06 41	08 17	01 34
30	06 42	08 14	02 33
31	06 ♐ 41	08 ♐ 11	03 S 27

DECLINATIONS

Date	Sun ☉ ° '	Moon ☽ ° '	Mercury ☿ ° '	Venus ♀ ° '	Mars ♂ ° '	Jupiter ♃ ° '	Saturn ♄ ° '	Uranus ♅ ° '	Neptune ♆ ° '	Pluto ♇ ° '
01	07 S 19	22 N 08	00 N 16	10 N 54	09 N 57	12 N 59	08 S 38	00 N 38	11 S 38	19 S 15
02	06 56	21 47	01 01	11 23	10 05	13 03	08 37	00 40	11 37	19 15
03	06 33	20 22	01 44	11 51	10 14	13 06	08 36	00 41	11 36	19 15
04	06 10	17 52	02 24	12 19	10 22	13 10	08 35	00 42	11 35	19 15
05	05 47	14 22	03 02	12 47	10 31	13 14	08 34	00 44	11 35	19 15
06	05 24	10 00	03 36	13 15	10 39	13 18	08 32	00 45	11 34	19 15
07	05 01	04 N 59	04 06	13 43	10 47	13 22	08 31	00 46	11 33	19 14
08	04 37	00 S 26	04 33	14 10	10 55	13 26	08 30	00 48	11 32	19 14
09	04 14	05 54	04 56	14 37	11 03	13 29	08 29	00 49	11 32	19 14
10	03 50	11 04	05 15	15 03	11 10	13 33	08 28	00 50	11 31	19 14
11	03 27	15 32	05 29	15 29	11 18	13 37	08 26	00 52	11 30	19 14
12	03 03	19 00	05 38	15 55	11 25	13 41	08 25	00 53	11 29	19 14
13	02 39	21 12	05 43	16 21	11 32	13 45	08 24	00 54	11 28	19 14
14	02 16	22 02	05 43	16 46	11 39	13 49	08 22	00 56	11 28	19 14
15	01 52	21 28	05 39	17 11	11 45	13 53	08 21	00 57	11 27	19 14
16	01 28	19 40	05 29	17 35	11 51	13 57	08 20	00 58	11 26	19 14
17	01 04	16 50	05 16	17 59	11 57	14 01	08 18	01 00	11 26	19 14
18	00 41	13 13	04 59	18 23	12 03	14 05	08 17	01 01	11 25	19 14
19	00 S 17	09 02	04 38	18 46	12 09	14 09	08 15	01 02	11 24	19 13
20	00 N 07	04 S 32	04 14	19 09	12 14	14 13	08 14	01 04	11 23	19 13
21	00 30	00 N 06	03 47	19 31	12 19	14 17	08 12	01 05	11 23	19 13
22	00 54	04 39	03 18	19 53	12 23	14 21	08 11	01 07	11 22	19 13
23	01 18	08 58	02 47	20 14	12 27	14 25	08 09	01 08	11 21	19 13
24	01 41	12 53	02 16	20 35	12 31	14 29	08 08	01 09	11 21	19 13
25	02 05	16 15	01 44	20 56	12 35	14 33	08 06	01 11	11 20	19 13
26	02 28	18 56	01 12	21 16	12 38	14 38	08 05	01 12	11 19	19 13
27	02 52	20 48	00 41	21 36	12 41	14 42	08 03	01 13	11 19	19 13
28	03 15	21 46	00 N 11	21 55	12 44	14 46	08 02	01 15	11 18	19 13
29	03 39	21 45	00 S 17	22 14	12 46	14 50	08 00	01 16	11 17	19 13
30	04 02	20 43	00 44	22 32	12 48	14 54	07 58	01 17	11 17	19 13
31	04 N 25	18 N 40	01 S 08	22 N 50	12 N 50	14 N 58	07 S 57	01 N 19	11 S 16	19 S 13

ZODIAC SIGN ENTRIES

Date	h	m	Planets
02	11	41	☿ ♈
02	15	08	☽ ♋
04	23	17	☽ ♌
05	10	25	♀ ♉
07	03	27	☽ ♍
09	04	50	☽ ♎
11	05	24	☽ ♏
13	06	53	☽ ♐
15	10	24	☽ ♑
17	16	11	☽ ♒
20	00	05	☽ ♓
20	05	14	☉ ♈
22	09	57	☽ ♈
23	13	22	☿ ♓
24	21	43	☽ ♉
27	10	43	☽ ♊
29	23	07	☽ ♋

LATITUDES

Date	Mercury ☿ ° '	Venus ♀ ° '	Mars ♂ ° '	Jupiter ♃ ° '	Saturn ♄ ° '	Uranus ♅ ° '	Neptune ♆ ° '	Pluto ♇ ° '
01	00 N 52	01 N 04	04 N 12	00 S 58	02 N 41	00 S 42	00 S 33	03 N 53
04	01 34	01 18	04 10	00 57	02 41	00 42	00 33	03 53
07	02 16	01 33	04 07	00 56	02 42	00 42	00 33	03 53
10	02 52	01 47	04 04	00 56	02 42	00 41	00 33	03 53
13	03 19	02 02	03 59	00 55	02 43	00 41	00 33	03 53
16	03 33	02 16	03 54	00 55	02 44	00 41	00 33	03 53
19	03 30	02 31	03 48	00 54	02 44	00 41	00 33	03 53
22	03 11	02 45	03 42	00 54	02 45	00 41	00 33	03 53
25	02 37	02 59	03 35	00 53	02 45	00 41	00 33	03 53
28	01 54	03 13	03 28	00 53	02 45	00 41	00 34	03 53
31	01 N 07	03 N 26	03 N 20	00 S 52	02 N 46	00 S 41	00 S 34	03 N 53

DATA

Julian Date	2455988
Delta T	+68 seconds
Ayanamsa	24° 01' 54"
Synetic vernal point	05° ♓ 05' 05"
True obliquity of ecliptic	23° 26' 13"

MOON'S PHASES, APSIDES AND POSITIONS ☽

Date	h	m	Phase	Longitude ° '	Eclipse Indicator
01	01	21	☽	10 ♊ 52	
08	09	39	○	18 ♍ 13	
15	01	25	☾	24 ♐ 52	
22	14	37	●	02 ♈ 22	
30	19	41	☽	10 ♋ 30	

Day	h	m		
10	10	09	Perigee	
26	06	10	Apogee	
01	16	12	Max dec	22° N 09'
08	10	06	0S	
14	14	06	Max dec	22° S 02'
21	11	30	0N	
28	23	38	Max dec	21° N 53'

LONGITUDES

Date	Chiron ⚷ ° '	Ceres ⚳ ° '	Pallas ⚴ ° '	Juno ⚵ ° '	Vesta ⚶ ° '	Black Moon Lilith ⚸ ° '
01	05 ♓ 34	14 ♈ 36	06 ♓ 06	05 ♐ 41	02 ♈ 11	08 ♉ 26
11	06 ♓ 13	18 ♈ 25	09 ♓ 24	06 ♐ 38	06 ♈ 50	09 ♉ 32
21	06 ♓ 51	22 ♈ 18	12 ♓ 41	07 ♐ 05	11 ♈ 29	10 ♉ 39
31	07 ♓ 27	26 ♈ 16	15 ♓ 55	07 ♐ 00	16 ♈ 08	11 ♉ 45

All ephemeris data is given at 12.00 UT and the Moon's longitude is additionally given for 24.00 UT

ASPECTARIAN

h m Aspects

01 Thursday
01 21 ☽ □ ☉
05 47 ☽ ⊥ ♃
07 45 ☽ ⅊ ♄
08 53 ☽ □ ♂
10 02 ☽ Q ♅
18 33 ☿ ⊼ ♄
23 55 ☽ ∠ ♃

02 Friday
00 08 ☿ ∥ ♅
08 29 ☽ ⚹ ♀
13 14 ☽ △ ♄
15 33 ☽ □ ☿
17 11 ☽ △ ♆
19 09 ☽ Q ♂
21 29 ☽ □ ♅
23 31 ♂ ∠ ♄
23 45 ♀ # ♆

03 Saturday
05 27 ☽ ⚹ ♃
08 10 ☿ ⊻ ♆
08 46 ☽ ☍ ♇
09 51 ☽ Q ♀
16 58 ☽ △ ☉
17 13 ☿ ⊥ ♃
17 18 ☽ ⚹ ♂
20 10 ☉ ☍ ♂
21 47 ☽ ⅊ ♆

04 Sunday
00 01 ☽ # ♇
02 53 ♀ ⅊ ♂
03 11 ☉ ⅊ ♄
04 19 ☽ Q ♃
11 18 ♀ ☍ ♄
14 27 ☽ ± ♆
20 09 ☽ ∠ ♂
21 20 ☽ □ ♄
22 17 ☽ □ ♀
22 53 ☽ ⅊ ☉

05 Monday
01 21 ☽ ⊼ ♆
04 57 ☽ △ ☿
05 25 ☽ △ ♅
11 35 ☿ ☌ ♅
11 48 ☽ ⊥ ♂
13 20 ☽ □ ♃
15 44 ☽ ⊼ ♇
16 12 ☽ ± ☉
18 35 ☽ ∥ ♃
20 17 ☽ ∥ ♀
22 03 ☽ ⊻ ♂

06 Tuesday
02 14 ☽ ± ♇
03 30 ☽ ⊼ ☉
03 52 ☽ # ♆
05 02 ☽ Q ♄
07 49 ☽ ⅊ ♅
08 48 ☽ ∥ ♂
09 09 ☽ ⅊ ☿
12 25 ♀ ⚹ ♆
14 38 ♀ ∥ ♃
17 43 ☽ ⅊ ♇
19 18 ☽ # ♄
23 14 ☽ ± ♅

07 Wednesday
01 25 ☽ ± ☿
01 27 ☽ ⚹ ♄
05 30 ☽ ☍ ♆
06 52 ☽ △ ♀
09 22 ☽ ⊼ ♅
11 51 ☽ # ☉
12 01 ☽ ⊼ ☿
15 38 ☽ ∥ ☿
17 17 ☽ △ ♃
18 55 ☽ △ ♇
23 28 ☽ ☌ ♂

08 Thursday
00 02 ☉ ∠ ♀
02 21 ☽ ∠ ♄
06 40 ☽ ∥ ♅
09 39 ☽ ☍ ☉
09 43 ☽ ⅊ ♀
13 35 ☽ # ♅
13 56 ☉ # ☿
17 02 ☽ ⊥ ♄
17 13 ♀ ⊻ ♅
18 15 ☽ ⅊ ♃

09 Friday
01 34 ☽ ± ♀
02 46 ☽ ⊻ ♄
05 05 ☽ ∥ ☉
06 56 ☽ ⊼ ♆
07 24 ☽ # ☿
09 03 ☽ ± ♃
10 45 ☽ ☍ ♅
12 04 ☽ ⊼ ♀
15 01 ☽ ☍ ☿
16 40 ☽ ± ♆
18 53 ☽ ⊼ ♃
19 53 ☽ □ ♇
23 03 ☽ ⊻ ♂
23 40 ☽ ∥ ♄

10 Saturday
07 15 ☽ ⅊ ♆
08 30 ☽ ⊥ ♂
12 35 ☽ # ♂
13 49 ☽ ⊼ ☉
14 15 ☽ ∥ ♆
22 43 ☽ ∠ ♂

11 Sunday
00 15 ☽ ± ☉
01 00 ☽ # ♃
01 03 ☽ Q ♇
03 09 ☽ ☌ ♄
07 38 ☽ △ ♆
11 31 ☽ ⊼ ♅
11 41 ☽ # ♀
12 16 ☉ Q ♇
15 46 ☿ ⊻ ♀
15 58 ☽ ⅊ ☉
16 28 ☽ ⊼ ☿
16 31 ☽ ☍ ♀
20 13 ☽ ☍ ♃
20 38 ☽ ⚹ ♇
21 22 ☽ ± ♅
22 33 ☽ ⚹ ♂

12 Monday
02 21 ☽ ± ☿
07 49 ☿ St R
12 15 ☽ ⅊ ♅
13 59 ☽ ∥ ♇
17 13 ☽ ⅊ ☿
17 16 ☉ ± ♄
17 52 ☽ Q ♂
18 30 ☽ △ ☉
21 26 ☽ ∠ ♇

13 Tuesday
01 03 ♂ # ♆
04 23 ☽ ⊻ ♄
04 43 ♃ △ ♇
09 19 ☽ □ ♆
12 33 ☽ ⊥ ♇
13 25 ☽ △ ♅
14 27 ☽ ⊥ ♄
18 09 ☽ △ ☿
22 30 ☽ ⊼ ♀
22 43 ☽ ⊻ ♇
22 58 ☽ ⊼ ♃
23 23 ☽ □ ♂

14 Wednesday
01 27 ♀ △ ♇
05 42 ☽ ∠ ♄
05 54 ♀ ☌ ♃
07 27 ♀ △ ♂
09 22 ☽ ± ♃
09 36 ☽ ± ♀
09 51 ♂ △ ♃
14 15 ♀ ⊥ ♅
16 06 ☽ Q ♆
21 13 ☉ ∠ ♃

15 Thursday
00 24 ♂ △ ♇
01 10 ☽ ⅊ ♃
01 25 ☽ □ ☉
02 29 ☽ ⅊ ♀
07 34 ☽ ⚹ ♄
13 04 ☽ ⚹ ♆
17 26 ☽ □ ♅
20 53 ☽ □ ☿

16 Friday
02 18 ☽ △ ♂
03 00 ☽ ☌ ♇
04 00 ☽ △ ♃
04 41 ☽ Q ♄
07 09 ☿ ⊥ ♀
07 11 ☽ △ ♀
11 45 ☽ Q ☉
15 38 ♃ ⊥ ♅
15 49 ☽ ∠ ♆
16 22 ☽ ∥ ♇

17 Saturday
01 46 ☽ Q ♅
03 53 ☽ Q ☿
04 15 ☽ # ♀
04 34 ☽ ⅊ ♂
08 08 ☽ ⊥ ♆
11 12 ☽ ⚹ ☉
13 00 ☽ □ ♄
16 34 ☉ # ♅
19 07 ☽ ⊻ ♆
20 35 ☽ ± ♂
23 45 ☽ ⚹ ♅

18 Sunday
00 46 ☽ ⚹ ☿
02 00 ♀ Q ♆
06 41 ☽ # ♃
07 22 ☽ ⊼ ♂
09 32 ☽ ⊻ ♇
09 35 ☿ ⊥ ♃
09 56 ☉ ⊼ ♄
11 23 ☽ □ ♃
17 06 ☽ ∠ ☉
17 22 ☿ ☌ ♅
18 39 ☽ □ ♀
18 46 ☽ # ♂
20 41 ☽ ⊥ ♇
22 40 ☽ ∥ ♆

19 Monday
03 05 ☽ ∠ ☿
03 43 ☽ ∠ ♅
11 21 ☽ ⊥ ☉
13 34 ☽ ∠ ♇
16 15 ☽ ∥ ♄
19 01 ☽ ⊥ ☿
20 31 ☽ △ ♄
20 44 ☽ ⊥ ♅
21 40 ☽ Q ♃
23 39 ☽ ⊻ ☉

20 Tuesday
03 17 ☽ ☌ ♆
05 41 ☽ ⊻ ☿
07 33 ☽ Q ♀
08 11 ☽ ⊻ ♅
13 44 ☽ # ☿
14 25 ☽ ☍ ♂
17 13 ♀ # ♇
18 06 ☽ ⚹ ♇
20 55 ☽ ⚹ ♃

21 Wednesday
01 00 ☽ ⅊ ♄
05 54 ☽ # ♅
08 16 ☿ ∠ ♀
08 39 ☽ ⚹ ♀
09 07 ☽ # ☉
14 20 ☽ ∥ ☉
15 09 ☿ ⊻ ♆
17 11 ☽ ∥ ♅
17 17 ☽ Q ♇
18 19 ☽ ± ♄
19 21 ☉ ☌ ☿
23 24 ☉ ⊻ ♆

22 Thursday
02 28 ☽ ∠ ♃
05 29 ☽ ∥ ☿
05 58 ☽ ⊼ ♄
11 50 ☽ ☌ ☿
13 25 ☽ ⊻ ♆
14 37 ☽ ☌ ☉
16 34 ☽ ∠ ♀
18 36 ☽ ☌ ♅
20 27 ☽ ⊥ ♃
23 23 ☽ ⊼ ♂

23 Friday
01 17 ☽ ⊥ ♆
01 24 ☉ ∥ ♅
04 38 ☽ □ ♇
07 23 ☽ # ♄
08 33 ☽ ⊻ ♃
11 01 ☽ ± ♂
12 00 ☽ ⊥ ♀
18 25 ♀ ∠ ♅
19 13 ☽ ∠ ♆

24 Saturday
01 05 ☽ ⊻ ♀
02 14 ☽ # ♆
04 36 ☽ ⅊ ♂
09 36 ☽ ∥ ♂
17 17 ☽ ☍ ♄
18 20 ☉ ☌ ♅
19 29 ☽ ⊻ ☿
23 08 ☽ ∥ ♃

25 Sunday
01 27 ☽ ⚹ ♆
02 50 ☉ ∥ ☿
06 49 ☽ ⊥ ☿
06 55 ☽ ⊻ ♅
07 59 ☽ ⊻ ☉
10 13 ☽ △ ♂
16 58 ☽ △ ♇
19 09 ☽ ⊥ ♅
21 17 ☽ ⊥ ☉
22 02 ☽ ☌ ♃
23 53 ☽ ∠ ☿

26 Monday
01 53 ☽ Q ♆
05 39 ☉ ⊼ ♂
06 09 ☿ ⊼ ♄
08 23 ☉ ⊥ ♃
12 04 ☿ ∥ ♅
13 33 ☽ ∠ ♅
15 03 ☽ # ♇
17 18 ☽ ∠ ☉
19 26 ☽ ☌ ♀
21 34 ☿ ∠ ♃
23 32 ☽ ⅊ ♇

27 Tuesday
04 34 ☽ ⚹ ☿
05 53 ☽ ⊼ ♄
14 39 ☽ □ ♆
17 52 ☽ ± ♇
18 02 ☽ ± ♄
20 16 ☽ ⚹ ♅
22 04 ☽ □ ♂

28 Wednesday
02 38 ☽ ⚹ ☉
03 37 ☽ Q ☿
05 42 ☉ ⊥ ♆
06 04 ☽ ⊼ ♇
12 11 ☽ ⅊ ♄
12 12 ☽ ⊻ ♃
19 24 ♄ Q ♇
20 40 ☽ Q ♅
20 51 ♀ ⅊ ♇

29 Thursday
00 31 ☽ ⊥ ♃
05 02 ☽ Q ☉
09 44 ☽ Q ♂
13 36 ☽ ⊻ ♀
14 05 ☽ □ ☿
17 43 ☿ ⚹ ♀
18 05 ☽ △ ♄
18 49 ☽ ∠ ♃
19 55 ☉ □ ♇

30 Friday
02 31 ☽ ⊥ ♀
03 05 ☽ △ ♆
08 38 ☽ □ ♅
09 06 ☽ ⚹ ♂
17 48 ☽ ☍ ♇
19 41 ☽ □ ☉
21 34 ☽ ∠ ♀

31 Saturday
00 41 ☽ ⚹ ♃
05 39 ☉ ± ♂
06 37 ☽ # ♇
08 16 ☽ ⅊ ♆
09 03 ♂ ⊼ ♅
13 36 ☽ ∠ ♂
18 24 ♀ ⊼ ♄
22 12 ☽ △ ☿
23 52 ☽ Q ♃
23 56 ☿ # ♅

APRIL 2012

Date	Sidereal time h m s	Sun ☉	Moon ☽	Moon ☽ 24.00	Mercury ☿	Venus ♀	Mars ♂	Jupiter ♃	Saturn ♄	Uranus ♅	Neptune ♆	Pluto ♇
01	00 40 59	12 ♈ 09 46	01 ♌ 50 58	08 ♌ 25 56	24 ♓ 15	28 ♉ 00	04 ♍ 42	13 ♉ 25	27 ♎ 15	04 ♈ 57	02 ♓ 05	09 ♑ 32
02	00 44 56	13 08 56	15 ♌ 07 59	21 ♌ 57 19	24 R 01	28 57	04 R 33	13 39	27 R 11	05 01	02 07	09 33
03	00 48 52	14 08 04	28 ♌ 54 00	05 ♍ 57 54	23 53	29 ♉ 52	04 24	13 52	27 07	05 04	02 09	09 33
04	00 52 49	15 07 09	13 ♍ 08 42	20 ♍ 25 50	23 D 51	00 ♊ 48	04 16	14 05	27 02	05 07	02 11	09 33
05	00 56 45	16 06 13	27 ♍ 48 36	05 ♎ 16 04	23 54	01 42	04 09	14 19	26 58	05 11	02 12	09 33
06	01 00 42	17 05 14	12 ♎ 47 08	20 ♎ 20 37	24 03	02 36	04 03	14 32	26 53	05 14	02 14	09 33
07	01 04 39	18 04 13	27 ♎ 55 17	05 ♏ 29 50	24 16	03 30	03 58	14 46	26 49	05 18	02 16	09 33
08	01 08 35	19 03 10	13 ♏ 03 04	20 ♏ 33 53	24 35	04 22	03 53	14 59	26 44	05 21	02 18	09 34
09	01 12 32	20 02 06	28 ♏ 01 16	05 ♐ 24 25	24 58	05 14	03 49	15 13	26 40	05 24	02 19	09 34
10	01 16 28	21 00 59	12 ♐ 42 40	19 ♐ 55 33	25 26	06 06	03 46	15 26	26 35	05 28	02 21	09 34
11	01 20 25	21 59 51	27 ♐ 02 46	04 ♑ 04 10	25 58	06 56	03 44	15 40	26 30	05 31	02 23	09 R 34
12	01 24 21	22 58 42	10 ♑ 59 45	17 ♑ 49 37	26 34	07 46	03 42	15 54	26 26	05 34	02 24	09 34
13	01 28 18	23 57 30	24 ♑ 33 56	01 ♒ 12 58	27 14	08 35	03 41	16 07	26 21	05 38	02 26	09 34
14	01 32 14	24 56 17	07 ♒ 47 01	14 ♒ 16 24	27 58	09 24	03 D 41	16 21	26 17	05 41	02 27	09 34
15	01 36 11	25 55 02	20 ♒ 41 27	27 ♒ 02 32	28 46	10 11	03 42	16 35	26 12	05 44	02 29	09 33
16	01 40 08	26 53 45	03 ♓ 19 58	09 ♓ 34 03	29 ♓ 36	10 58	03 43	16 49	26 07	05 47	02 30	09 33
17	01 44 04	27 52 27	15 ♓ 45 06	21 ♓ 53 24	00 ♈ 30	11 44	03 45	17 03	26 03	05 51	02 32	09 33
18	01 48 01	28 51 07	27 ♓ 59 12	04 ♈ 02 45	01 27	12 29	03 48	17 17	25 58	05 54	02 33	09 33
19	01 51 57	29 ♈ 49 44	10 ♈ 04 16	16 ♈ 03 59	02 27	13 12	03 51	17 31	25 54	05 57	02 35	09 33
20	01 55 54	00 ♉ 48 21	22 ♈ 02 06	27 ♈ 58 50	03 30	13 55	03 55	17 45	25 49	06 00	02 36	09 33
21	01 59 50	01 46 55	03 ♉ 54 25	09 ♉ 49 05	04 35	14 37	04 00	17 58	25 44	06 03	02 38	09 32
22	02 03 47	02 45 27	15 ♉ 43 04	21 ♉ 36 39	05 43	15 18	04 06	18 12	25 40	06 07	02 39	09 32
23	02 07 43	03 43 58	27 ♉ 30 09	03 ♊ 23 52	06 53	15 58	04 12	18 27	25 35	06 10	02 40	09 31
24	02 11 40	04 42 26	09 ♊ 18 11	15 ♊ 13 29	08 06	16 36	04 19	18 41	25 31	06 13	02 42	09 31
25	02 15 37	05 40 53	21 ♊ 10 13	27 ♊ 08 50	09 21	17 13	04 26	18 55	25 26	06 16	02 43	09 30
26	02 19 33	06 39 17	03 ♋ 09 49	09 ♋ 13 42	10 39	17 49	04 34	19 09	25 22	06 19	02 44	09 30
27	02 23 30	07 37 40	15 ♋ 21 01	21 ♋ 32 19	11 58	18 24	04 43	19 23	25 17	06 22	02 45	09 29
28	02 27 26	08 36 00	27 ♋ 48 09	04 ♌ 09 03	13 20	18 57	04 53	19 37	25 13	06 25	02 47	09 29
29	02 31 23	09 34 18	10 ♌ 35 32	17 ♌ 08 04	14 43	19 29	05 02	19 51	25 08	06 28	02 48	09 28
30	02 35 19	10 ♉ 32 34	23 ♌ 47 03	00 ♍ 32 46	16 ♈ 09	20 ♊ 00	05 ♍ 13	20 ♉ 05	25 ♎ 04	06 ♈ 31	02 ♓ 49	09 ♑ 28

Date	Moon True ☊	Moon Mean ☊	Moon ☽ Latitude
01	06 ♐ 39	08 ♐ 08	04 S 13
02	06 R 35	08 05	04 46
03	06 29	08 01	05 04
04	06 22	07 58	05 05
05	06 14	07 55	04 46
06	06 06	07 52	04 07
07	06 00	07 49	03 10
08	05 56	07 46	02 00
09	05 54	07 42	00 S 43
10	05 D 54	07 39	00 N 37
11	05 55	07 36	01 52
12	05 56	07 33	02 59
13	05 R 57	07 30	03 53
14	05 56	07 26	04 34
15	05 53	07 23	05 00
16	05 49	07 20	05 10
17	05 43	07 17	05 06
18	05 37	07 14	04 47
19	05 30	07 11	04 16
20	05 24	07 07	03 34
21	05 19	07 04	02 43
22	05 16	07 01	01 45
23	05 14	06 58	00 N 42
24	05 D 14	06 55	00 S 22
25	05 15	06 52	01 26
26	05 16	06 48	02 28
27	05 18	06 45	03 23
28	05 19	06 42	04 10
29	05 R 20	06 39	04 46
30	05 ♐ 19	06 ♐ 36	05 S 09

DECLINATIONS

Date	Sun ☉	Moon ☽	Mercury ☿	Venus ♀	Mars ♂	Jupiter ♃	Saturn ♄	Uranus ♅	Neptune ♆	Pluto ♇
01	04 N 48	15 N 38	01 S 30	23 N 07	12 N 51	15 N 02	07 S 55	01 N 20	11 S 15	19 S 13
02	05 11	11 44	01 50	23 24	12 52	15 06	07 53	01 21	11 15	19 13
03	05 34	07 06	02 07	23 40	12 53	15 10	07 52	01 23	11 14	19 13
04	05 57	01 N 56	02 22	23 56	12 53	15 14	07 50	01 24	11 13	19 13
05	06 20	03 S 30	02 34	24 11	12 53	15 18	07 48	01 25	11 13	19 13
06	06 43	08 50	02 43	24 26	12 53	15 23	07 47	01 27	11 12	19 13
07	07 05	13 42	02 50	24 40	12 53	15 27	07 45	01 28	11 12	19 13
08	07 28	17 40	02 55	24 54	12 52	15 31	07 44	01 29	11 11	19 12
09	07 50	20 24	02 57	25 07	12 51	15 35	07 42	01 31	11 10	19 12
10	08 12	21 43	02 56	25 20	12 50	15 39	07 40	01 32	11 10	19 12
11	08 34	21 32	02 53	25 32	12 48	15 43	07 38	01 33	11 09	19 12
12	08 56	20 01	02 48	25 44	12 47	15 47	07 37	01 35	11 09	19 12
13	09 18	17 23	02 41	25 55	12 45	15 51	07 35	01 36	11 08	19 12
14	09 39	13 54	02 32	26 06	12 42	15 55	07 33	01 37	11 08	19 12
15	10 01	09 51	02 21	26 16	12 40	15 59	07 32	01 39	11 07	19 12
16	10 22	05 28	02 07	26 25	12 37	16 03	07 30	01 40	11 06	19 12
17	10 43	00 S 55	01 52	26 34	12 34	16 07	07 29	01 41	11 06	19 12
18	11 04	03 N 36	01 35	26 43	12 30	16 11	07 27	01 42	11 05	19 12
19	11 25	07 55	01 17	26 51	12 27	16 15	07 25	01 44	11 05	19 12
20	11 45	11 53	00 56	26 58	12 23	16 19	07 24	01 45	11 04	19 12
21	12 06	15 22	00 34	27 05	12 19	16 23	07 22	01 46	11 04	19 12
22	12 26	18 13	00 S 11	27 12	12 15	16 27	07 20	01 47	11 04	19 12
23	12 46	20 17	00 N 14	27 18	12 10	16 31	07 19	01 49	11 03	19 12
24	13 05	21 28	00 40	27 23	12 06	16 35	07 17	01 50	11 03	19 12
25	13 25	21 42	01 08	27 28	12 01	16 39	07 16	01 51	11 02	19 12
26	13 44	20 56	01 37	27 32	11 55	16 43	07 14	01 52	11 02	19 12
27	14 03	19 11	02 07	27 36	11 50	16 47	07 12	01 54	11 01	19 12
28	14 22	16 31	02 39	27 40	11 45	16 51	07 11	01 55	11 01	19 12
29	14 41	12 59	03 11	27 42	11 39	16 55	07 09	01 56	11 01	19 13
30	14 N 59	08 N 44	03 N 45	27 N 45	11 N 33	16 N 59	07 S 08	01 N 57	11 S 00	19 S 13

ZODIAC SIGN ENTRIES

Date	h m	Planets
01	08 35	☽ ♌
03	13 53	☽ ♍
03	15 18	♀ ♊
05	15 32	☽ ♎
07	15 17	☽ ♏
09	15 12	☽ ♐
11	17 02	☽ ♑
13	21 48	☽ ♒
16	05 38	☽ ♓
16	22 42	☿ ♈
18	15 59	☽ ♈
19	16 12	☉ ♉
21	04 05	☽ ♉
23	17 05	☽ ♊
26	05 42	☽ ♋
28	16 10	☽ ♌
30	23 02	☽ ♍

LATITUDES

Date	Mercury ☿	Venus ♀	Mars ♂	Jupiter ♃	Saturn ♄	Uranus ♅	Neptune ♆	Pluto ♇
01	00 N 51	03 N 30	03 N 18	00 S 52	02 N 46	00 S 41	00 S 34	03 N 53
04	00 N 05	03 42	03 10	00 52	02 46	00 41	00 34	03 54
07	00 S 37	03 54	03 02	00 51	02 46	00 41	00 34	03 54
10	01 13	04 05	02 54	00 51	02 46	00 41	00 34	03 54
13	01 44	04 14	02 47	00 51	02 46	00 41	00 34	03 54
16	02 08	04 23	02 39	00 50	02 46	00 41	00 34	03 54
19	02 27	04 30	02 32	00 50	02 46	00 41	00 34	03 54
22	02 40	04 36	02 24	00 50	02 46	00 41	00 34	03 54
25	02 48	04 40	02 17	00 49	02 46	00 42	00 34	03 54
28	02 50	04 42	02 10	00 49	02 46	00 42	00 34	03 54
31	02 S 47	04 N 42	02 N 04	00 S 49	02 N 46	00 S 42	00 S 34	03 N 54

DATA

Julian Date	2456019
Delta T	+68 seconds
Ayanamsa	24° 01' 57"
Synetic vernal point	05° ♓ 05' 02"
True obliquity of ecliptic	23° 26' 13"

MOON'S PHASES, APSIDES AND POSITIONS ☽

Date	h m	Phase	Longitude	Eclipse Indicator
06	19 19	○	17 ♎ 23	
13	10 50	☾	23 ♑ 55	
21	07 18	●	01 ♉ 35	
29	09 57	☽	09 ♌ 29	

Day	h m		
07	17 06	Perigee	
22	14 02	Apogee	
04	20 34	0S	
10	21 02	Max dec	21° S 49'
17	16 51	0N	
25	05 36	Max dec	21° N 44'

LONGITUDES

Date	Chiron ⚷	Ceres ⚳	Pallas ⚴	Juno ⚵	Vesta ⚶	Black Moon Lilith ⚸
01	07 ♓ 30	26 ♈ 40	16 ♓ 14	06 ♐ 57	16 ♈ 36	11 ♉ 52
11	08 ♓ 06	00 ♉ 39	19 ♓ 33	06 ♐ 16	21 ♈ 13	12 ♉ 59
21	08 ♓ 32	04 ♉ 41	22 ♓ 27	05 ♐ 02	25 ♈ 49	14 ♉ 05
31	08 ♓ 57	08 ♉ 43	25 ♓ 24	03 ♐ 20	00 ♉ 22	15 ♉ 12

All ephemeris data is given at 12.00 UT and the Moon's longitude is additionally given for 24.00 UT

ASPECTARIAN

h m	Aspects
01 Sunday	
01 18	☽ ± ♆
03 33	☽ □ ♄
04 20	☽ ✶ ♀
06 15	☽ ⊥ ♂
12 26	☽ ⊼ ♆
16 00	☽ ∥ ♃
17 09	☽ ⊻ ♂
17 43	☽ △ ♅
02 Monday	
01 13	☽ Ᵽ ☿
02 00	☽ ⊼ ♇
03 57	☽ Q ♀
05 33	☽ ∥ ♂
08 11	☽ △ ☉
09 18	☽ □ ♃
12 05	☽ Q ♄
12 44	☽ ± ♇
14 43	☽ ⋕ ♆
17 03	☽ ± ☿
20 38	☽ Ᵽ ♅
03 Tuesday	
03 25	☽ ⊼ ☿
03 34	☉ ⊻ ♃
04 30	☽ Ᵽ ♇
08 13	☽ ⋕ ♄
08 57	☽ ✶ ♄
12 17	☽ ± ♅
12 26	☽ Ᵽ ☉
13 47	☽ □ ♀
17 33	☽ ☍ ♆
18 46	☽ ∥ ☉
21 16	☽ ☌ ♂
22 32	☽ ⊼ ♅
04 Wednesday	
04 48	☽ ± ☉
06 01	☽ △ ♇
10 06	☽ ⋕ ☿
10 10	☽ ∠ ♄
10 12	☿ St D
13 35	☽ △ ♃
14 20	☽ ∥ ♅
15 30	☽ ⊼ ☉
23 00	♃ Q ♆
05 Thursday	
00 55	☽ ⊥ ♄
02 50	☽ ⋕ ♅
05 37	☽ ☍ ☿
07 44	☽ ∥ ☿
10 38	☽ ⊻ ♄
14 28	☽ Ᵽ ♃
18 41	☽ △ ♀
19 06	☽ ⊼ ♆
22 09	☽ ⊻ ♂
23 54	☽ ☍ ♅
06 Friday	
01 32	☽ ⋕ ☉
01 49	♀ □ ♆
04 44	☽ ± ♆
05 07	☽ ± ♃
06 51	☽ □ ♇
07 09	☽ ∥ ♄
07 40	☽ ⊥ ♂
14 50	☽ ⊼ ♃
15 45	☉ ∠ ♆
18 59	♀ ± ♄
19 05	☽ Ᵽ ♆
19 19	☽ ☍ ☉
20 08	☽ Ᵽ ♀
21 53	☽ ∠ ♂
23 14	☽ ∥ ♆
07 Saturday	
06 07	☽ ⊼ ☿
07 44	☽ ⋕ ♂
10 15	☽ ☌ ♄
11 17	☽ ± ♀
11 26	☽ Q ♇
13 48	♀ ± ♇
15 47	☽ ± ☿
18 53	☽ △ ♆
21 22	☽ ⊼ ♀
21 31	☽ ✶ ♂
21 58	☽ ⋕ ♃
23 36	♀ □ ♂
23 43	☽ ⊼ ♅
08 Sunday	
06 21	☽ Ᵽ ☿
06 27	☽ ✶ ♇
08 07	☉ Ᵽ ♂
09 17	☽ ± ♅
15 08	☽ ☍ ♃
16 29	☽ Q ♂
22 15	☽ ⊼ ☉
23 42	☽ Ᵽ ♅
09 Monday	
00 00	☽ ∥ ♇
03 54	☉ ⋕ ♄
06 25	☽ ∠ ♇
06 56	☽ △ ☿
08 34	☽ ± ☉
09 49	☽ ⊻ ♄
16 51	♀ ✶ ♅
18 59	☽ □ ♆
19 29	☽ ⊥ ♄
21 00	☽ ⊥ ♇
21 22	☽ □ ♂
10 Tuesday	
00 02	☽ △ ♅
00 12	☽ Ᵽ ☉
00 27	☽ ☍ ♀
06 49	☽ ⊻ ♇
10 09	☽ ∠ ♄
16 24	♇ St R
16 36	☽ ⊼ ♃
11 Wednesday	
00 44	☽ Q ♆
02 21	☉ ∠ ♀
02 46	☽ ± ♃
02 51	☽ △ ☉
10 06	☽ □ ☿
11 05	☽ ✶ ♄
18 16	☽ Ᵽ ♃
21 07	☽ ✶ ♆
23 23	☽ △ ♂
12 Thursday	
02 33	☽ □ ♅
06 02	☽ ⊼ ♀
07 16	☿ ⊼ ♄
07 34	☽ Q ♄
09 30	☽ ☌ ♇
17 10	☽ ± ♀
18 34	☽ Q ☿
20 22	☽ ∥ ♇
20 44	☽ △ ♃
23 16	☽ ∠ ♆
13 Friday	
01 32	☽ Ᵽ ♂
10 09	☽ Ᵽ ♀
10 19	☽ Q ♅
10 50	☽ □ ☉
15 12	☽ □ ♄
15 21	☽ ⊥ ♆
17 05	☽ ✶ ☿
17 36	☽ ± ♂
22 57	☽ ⋕ ♃
14 Saturday	
02 14	☽ ⊻ ♆
03 53	♂ St D
04 29	☽ ⊼ ♂
08 07	☽ ✶ ♅
15 10	☽ △ ♀
15 16	☽ ⊻ ♇
16 52	♀ ⊼ ♇
19 29	☽ ⋕ ♂
22 11	☽ ∠ ☿
22 18	☽ Q ☉
15 Sunday	
02 23	☽ ⊥ ♇
04 10	☽ □ ♃
04 47	☽ ∥ ♆
11 11	☽ ⋕ ☉
12 05	☽ ∠ ♅
16 10	☽ ⊥ ☿
18 26	☉ ☍ ♄
19 17	☽ ∠ ♇
22 20	☽ △ ♄
22 42	☽ ✶ ☉
16 Monday	
00 56	☽ ∥ ♄
04 21	☽ ⊻ ☿
05 12	☽ ⊥ ♅
10 25	☽ ☌ ♆
12 44	☽ ☍ ♂
14 54	☽ Q ♃
16 28	♀ Ᵽ ♄
16 44	☽ ⊻ ♅
23 58	☽ ✶ ♇
17 Tuesday	
02 55	☽ Ᵽ ♄
03 40	☽ □ ♀
05 55	☽ ∠ ☉
06 41	☽ ∥ ☿
07 58	☽ ⋕ ♅
14 34	☽ ✶ ♃
20 20	☽ ± ♄
23 20	☽ Q ♇
18 Wednesday	
01 01	☽ ⊥ ☉
01 51	☽ ∥ ♅
01 54	☽ ⋕ ☿
02 47	☿ ⋕ ♅
08 03	☽ ⊼ ♄
13 44	☉ ⋕ ♆
13 52	☽ ⊻ ☉
17 14	☽ Q ♀
19 28	☽ ☌ ☿
20 39	☽ ∠ ♃
21 04	☽ ⊻ ♆
23 33	☽ ⊼ ♂
19 Thursday	
03 45	☽ ☌ ♅
09 01	☽ ⊥ ♆
09 12	☽ ⋕ ♄
10 57	☽ □ ♇
11 34	☽ ± ♂
13 43	☿ ∠ ♃
14 56	☽ ⊥ ♃
15 05	☿ ⊻ ♆
18 40	☽ ✶ ♀
20 Friday	
03 04	☽ ∠ ♆
03 12	☽ ⊻ ♃
05 42	☽ Ᵽ ♂
06 52	☽ ⋕ ♆
11 02	☽ ∥ ☉
15 08	☽ ∥ ♂
19 35	☽ ☍ ♄
22 15	☿ ⊼ ♂
21 Saturday	
02 46	☽ ∠ ♀
07 18	☽ ☌ ☉
09 24	☽ ✶ ♆
12 12	☽ △ ♂
13 31	☽ ⊻ ☿
16 23	☽ ⊻ ♅
20 06	☽ ∥ ♃
22 09	☽ ⊥ ♀
23 25	☽ △ ♇
22 Sunday	
01 18	☉ ∥ ♂
02 59	☽ ⊥ ☿
04 38	☽ ⊥ ♅
09 18	☉ ✶ ♆
09 49	☽ Q ♆
11 06	☽ ⊻ ♀
17 10	☽ ☌ ♃
20 29	☿ ☌ ♅
22 19	☽ ⋕ ♇
23 02	☽ ∠ ♅
23 18	☽ ∠ ☿
23 Monday	
05 56	☽ Ᵽ ♇
08 08	☽ ⊼ ♄
20 16	☽ ± ♄
22 33	☽ □ ♆
24 Tuesday	
00 15	☽ ± ♇
00 59	☉ △ ♂
01 46	☽ □ ♂
01 49	☽ ⊻ ☉
05 42	☽ ✶ ♅
09 17	☽ ✶ ☿
12 26	☽ ⊼ ♇
14 26	☽ Ᵽ ♄
15 06	☽ ⊥ ☉
23 38	☿ ⊥ ♆
25 Wednesday	
03 36	☽ ☌ ♀
06 07	☽ Q ♅
07 21	☽ ⊻ ♃
10 56	☽ ∠ ☉
12 25	☽ Q ☿
14 35	☽ Q ♂
14 51	☿ □ ♇
19 40	☽ ⊥ ♃
20 31	☽ △ ♄
26 Thursday	
03 12	☉ ⊻ ♅
10 35	☿ ± ♂
11 09	☽ △ ♆
13 59	☽ ∠ ♃
14 50	☽ ✶ ♂
18 17	☽ □ ♅
19 31	☽ ✶ ☉
27 Friday	
00 31	☽ ☍ ♇
00 53	☿ ∥ ♅
04 34	☽ □ ☿
10 33	♀ Q ♅
11 50	☽ ⋕ ♇
16 41	☽ Ᵽ ♆
18 13	☽ ⊻ ♀
19 59	☽ ✶ ♃
20 36	☽ ∠ ♂
21 01	☽ Q ☉
28 Saturday	
06 19	☽ ⊥ ♀
07 05	☽ □ ♄
09 26	☽ ∥ ♃
10 03	☽ ± ♆
14 04	☽ ⊥ ♂
18 02	☿ ⊥ ♃
19 22	☽ Q ♃
21 26	☽ ⊼ ♆
29 Sunday	
00 08	☽ ∠ ♀
01 32	☽ ⊻ ♂
02 00	☽ ∥ ☉
04 18	☽ △ ♅
09 34	☉ △ ♇
09 56	☽ ⊼ ♇
09 57	☽ □ ☉
16 40	☽ Q ♄
20 08	☽ ∥ ♂
20 31	☽ △ ☿
20 58	☽ ± ♇
23 37	☽ ⋕ ♆
30 Monday	
04 55	☽ ✶ ♀
05 14	☽ □ ♃
07 55	☽ Ᵽ ♅
13 13	☽ Ᵽ ♇
14 17	☽ ✶ ♄
20 16	☽ ⋕ ♄
20 54	♀ ⊻ ♃

MAY 2012

LONGITUDES

Date	Sidereal time h m s	Sun ☉ ° ' "	Moon ☽ ° ' "	Moon ☽ 24.00 ° ' "	Mercury ☿ ° '	Venus ♀ ° '	Mars ♂ ° '	Jupiter ♃ ° '	Saturn ♄ ° '	Uranus ♅ ° '	Neptune ♆ ° '	Pluto ♇ ° '
01	02 39 16	11 ♉ 30 49	07 ♍ 25 27	14 ♍ 25 08	17 ♈ 37	20 ♊ 28	05 ♍ 24	20 ♉ 19	25 ♎ 00	06 ♈ 34	02 ♓ 50	09 ♑ 27
02	02 43 12	12 29 00	21 ♍ 31 43	28 ♍ 44 55	19 07	20 56	05 36	20 34	24 R 55	06 37	02 51	09 R 27
03	02 47 09	13 27 10	06 ♎ 04 17	13 ♎ 29 08	20 38	21 21	05 48	20 48	24 51	06 40	02 52	09 26
04	02 51 06	14 25 19	20 ♎ 58 39	28 ♎ 31 50	22 12	21 45	06 01	21 02	24 47	06 43	02 53	09 25
05	02 55 02	15 23 25	06 ♏ 07 32	13 ♏ 44 33	23 47	22 08	06 14	21 16	24 43	06 46	02 54	09 24
06	02 58 59	16 21 30	21 ♏ 21 38	28 ♏ 57 31	25 25	22 28	06 28	21 30	24 39	06 48	02 55	09 24
07	03 02 55	17 19 33	06 ♐ 31 02	14 ♐ 01 05	27 05	22 46	06 42	21 45	24 35	06 51	02 56	09 23
08	03 06 52	18 17 34	21 ♐ 26 46	28 ♐ 47 11	28 ♈ 46	23 03	06 57	21 59	24 31	06 54	02 57	09 22
09	03 10 48	19 15 34	06 ♑ 01 50	13 ♑ 10 14	00 ♉ 29	23 17	07 12	22 13	24 27	06 57	02 58	09 21
10	03 14 45	20 13 33	20 ♑ 12 09	27 ♑ 07 26	02 15	23 30	07 28	22 27	24 23	07 00	02 59	09 21
11	03 18 41	21 11 30	03 ♒ 56 10	10 ♒ 38 27	04 02	23 40	07 44	22 41	24 19	07 02	02 59	09 20
12	03 22 38	22 09 26	17 ♒ 14 35	23 ♒ 44 52	05 51	23 48	08 00	22 56	24 15	07 05	03 00	09 19
13	03 26 35	23 07 21	00 ♓ 09 43	06 ♓ 29 32	07 42	23 54	08 18	23 10	24 11	07 07	03 01	09 18
14	03 30 31	24 05 15	12 ♓ 44 48	18 ♓ 55 58	09 35	23 58	08 35	23 24	24 08	07 10	03 02	09 17
15	03 34 28	25 03 07	25 ♓ 03 30	01 ♈ 07 53	11 30	24 00	08 53	23 38	24 04	07 13	03 02	09 16
16	03 38 24	26 00 58	07 ♈ 09 33	13 ♈ 08 57	13 27	23 R 59	09 11	23 53	24 00	07 15	03 03	09 15
17	03 42 21	26 58 48	19 ♈ 06 28	25 ♈ 02 31	15 26	23 55	09 30	24 07	23 57	07 18	03 04	09 14
18	03 46 17	27 56 36	00 ♉ 57 27	06 ♉ 51 37	17 27	23 50	09 49	24 21	23 53	07 20	03 04	09 13
19	03 50 14	28 54 24	12 ♉ 45 21	18 ♉ 38 57	19 29	23 41	10 09	24 35	23 50	07 23	03 05	09 12
20	03 54 10	29 ♉ 52 10	24 ♉ 32 42	00 ♊ 26 55	21 33	23 31	10 29	24 49	23 47	07 25	03 05	09 11
21	03 58 07	00 ♊ 49 54	06 ♊ 21 51	12 ♊ 17 47	23 39	23 18	10 49	25 04	23 44	07 27	03 06	09 10
22	04 02 04	01 47 38	18 ♊ 15 01	24 ♊ 13 48	25 46	23 03	11 10	25 18	23 40	07 30	03 06	09 09
23	04 06 00	02 45 20	00 ♋ 14 28	06 ♋ 17 17	27 ♉ 55	22 45	11 31	25 32	23 37	07 32	03 07	09 07
24	04 09 57	03 43 00	12 ♋ 22 35	18 ♋ 30 43	00 ♊ 04	22 25	11 53	25 46	23 34	07 34	03 07	09 06
25	04 13 53	04 40 40	24 ♋ 42 00	00 ♌ 56 49	02 15	22 03	12 14	26 00	23 32	07 36	03 07	09 05
26	04 17 50	05 38 17	07 ♌ 15 30	13 ♌ 38 26	04 26	21 38	12 37	26 14	23 29	07 39	03 08	09 04
27	04 21 46	06 35 54	20 ♌ 05 58	26 ♌ 38 26	06 38	21 12	12 59	26 28	23 26	07 41	03 08	09 03
28	04 25 43	07 33 28	03 ♍ 16 08	09 ♍ 59 22	08 50	20 43	13 22	26 42	23 23	07 43	03 08	09 02
29	04 29 39	08 31 02	16 ♍ 48 17	23 ♍ 43 02	11 02	20 13	13 45	26 56	23 21	07 45	03 08	09 00
30	04 33 36	09 28 34	00 ♎ 43 39	07 ♎ 50 02	13 13	19 41	14 09	27 11	23 18	07 47	03 09	08 59
31	04 37 33	10 ♊ 26 04	15 ♎ 01 57	22 ♎ 19 04	15 ♊ 24	19 ♊ 08	14 ♍ 32	27 ♉ 25	23 ♎ 16	07 ♈ 49	03 ♓ 09	08 ♑ 58

Date	Moon True ☊ ° '	Moon Mean ☊ ° '	Moon ☽ Latitude ° '
01	05 ♐ 17	06 ♐ 32	05 S 15
02	05 R 15	06 29	05 03
03	05 12	06 26	04 31
04	05 09	06 23	03 40
05	05 07	06 20	02 33
06	05 06	06 17	01 S 15
07	05 D 05	06 13	00 N 08
08	05 06	06 10	01 29
09	05 07	06 07	02 43
10	05 08	06 04	03 45
11	05 09	06 01	04 31
12	05 09	05 57	05 02
13	05 R 09	05 54	05 16
14	05 09	05 51	05 14
15	05 08	05 48	04 58
16	05 06	05 45	04 29
17	05 05	05 42	03 48
18	05 04	05 38	02 58
19	05 04	05 35	02 01
20	05 03	05 32	00 N 58
21	05 D 03	05 29	00 S 07
22	05 03	05 26	01 10
23	05 03	05 23	02 16
24	05 03	05 19	03 13
25	05 04	05 16	04 02
26	05 R 04	05 13	04 41
27	05 03	05 10	05 07
28	05 03	05 07	05 18
29	05 D 03	05 03	05 11
30	05 04	05 00	04 47
31	05 ♐ 04	04 ♐ 57	04 S 04

DECLINATIONS

Date	Sun ☉ ° '	Moon ☽ ° '	Mercury ☿ ° '	Venus ♀ ° '	Mars ♂ ° '	Jupiter ♃ ° '	Saturn ♄ ° '	Uranus ♅ ° '	Neptune ♆ ° '	Pluto ♇ ° '
01	15 N 17	03 N 54	04 N 20	27 N 47	11 N 27	17 N 02	07 S 06	01 N 58	11 S 00	19 S 13
02	15 35	01 S 17	04 56	27 48	11 20	17 06	07 05	01 59	10 59	19 13
03	15 53	06 33	05 33	27 49	11 14	17 10	07 03	02 00	10 59	19 13
04	16 10	11 35	06 11	27 49	11 07	17 14	07 02	02 02	10 59	19 13
05	16 27	15 58	06 49	27 49	11 00	17 18	07 01	02 03	10 58	19 13
06	16 44	19 19	07 29	27 49	10 53	17 21	06 59	02 04	10 58	19 13
07	17 00	21 16	08 09	27 48	10 46	17 25	06 58	02 05	10 58	19 13
08	17 16	21 40	08 50	27 46	10 39	17 29	06 56	02 06	10 57	19 13
09	17 32	20 35	09 32	27 44	10 31	17 33	06 55	02 07	10 57	19 13
10	17 48	18 13	10 14	27 41	10 23	17 36	06 54	02 08	10 57	19 13
11	18 03	14 52	10 57	27 38	10 15	17 40	06 53	02 09	10 57	19 13
12	18 18	10 52	11 40	27 34	10 07	17 44	06 51	02 10	10 56	19 13
13	18 33	06 29	12 23	27 30	09 59	17 47	06 50	02 11	10 56	19 13
14	18 48	01 S 56	13 07	27 25	09 51	17 51	06 49	02 12	10 56	19 13
15	19 02	02 N 36	13 51	27 19	09 42	17 55	06 48	02 13	10 56	19 13
16	19 15	06 57	14 35	27 12	09 34	17 58	06 47	02 14	10 56	19 13
17	19 29	11 00	15 19	27 07	09 25	18 02	06 45	02 15	10 55	19 14
18	19 42	14 35	16 02	27 00	09 16	18 05	06 44	02 16	10 55	19 14
19	19 55	17 35	16 45	26 52	09 07	18 09	06 43	02 17	10 55	19 14
20	20 07	19 51	17 28	26 43	08 57	18 12	06 42	02 18	10 55	19 14
21	20 19	21 15	18 10	26 34	08 48	18 16	06 41	02 19	10 55	19 14
22	20 31	21 43	18 51	26 24	08 38	18 19	06 40	02 20	10 55	19 14
23	20 42	21 11	19 31	26 14	08 29	18 23	06 39	02 21	10 54	19 14
24	20 54	19 39	20 09	26 02	08 19	18 26	06 38	02 22	10 54	19 14
25	21 04	17 12	20 46	25 51	08 09	18 29	06 37	02 22	10 54	19 14
26	21 15	13 55	21 22	25 38	07 59	18 33	06 37	02 23	10 54	19 15
27	21 25	09 55	21 55	25 25	07 49	18 36	06 36	02 24	10 54	19 15
28	21 34	05 22	22 27	25 11	07 39	18 39	06 35	02 25	10 54	19 15
29	21 43	00 N 26	22 56	24 56	07 28	18 43	06 34	02 26	10 54	19 15
30	21 52	04 S 40	23 22	24 41	07 18	18 46	06 34	02 26	10 54	19 15
31	22 N 01	09 S 40	23 N 46	24 N 25	07 N 07	18 N 49	06 S 33	02 N 27	10 S 54	19 S 15

ZODIAC SIGN ENTRIES

Date	h	m	Planets
03	02	04	☽ ♎
05	02	20	☽ ♏
07	01	39	☽ ♐
09	02	00	☽ ♑
09	05	14	☿ ♉
11	05	03	☽ ♒
13	11	42	☽ ♓
15	21	45	☽ ♈
18	10	30	☽ ♉
20	15	15	☉ ♊
20	23	05	☽ ♊
23	11	31	☽ ♋
24	11	12	☿ ♊
25	22	11	☽ ♌
28	06	06	☽ ♍
30	10	46	☽ ♎

LATITUDES

Date	Mercury ☿ ° '	Venus ♀ ° '	Mars ♂ ° '	Jupiter ♃ ° '	Saturn ♄ ° '	Uranus ♅ ° '	Neptune ♆ ° '	Pluto ♇ ° '
01	02 S 47	04 N 42	02 N 04	00 S 49	02 N 46	00 S 42	00 S 34	03 N 54
04	02 39	04 39	01 57	00 49	02 45	00 42	00 34	03 54
07	02 27	04 34	01 51	00 48	02 45	00 42	00 35	03 54
10	02 09	04 25	01 45	00 48	02 45	00 42	00 35	03 54
13	01 47	04 12	01 39	00 48	02 44	00 42	00 35	03 54
16	01 22	03 56	01 33	00 48	02 44	00 42	00 35	03 54
19	00 53	03 35	01 27	00 48	02 43	00 42	00 35	03 54
22	00 S 22	03 09	01 22	00 47	02 43	00 42	00 35	03 54
25	00 N 10	02 39	01 17	00 47	02 42	00 42	00 35	03 54
28	00 41	02 05	01 12	00 47	02 41	00 42	00 35	03 54
31	01 N 09	01 N 26	01 N 07	00 S 47	02 N 41	00 S 42	00 S 35	03 N 54

DATA

Julian Date	2456049
Delta T	+68 seconds
Ayanamsa	24° 02' 00"
Synetic vernal point	05° ♓ 04' 59"
True obliquity of ecliptic	23° 26' 12"

MOON'S PHASES, APSIDES AND POSITIONS ☽

Date	h	m	Phase	Longitude ° '	Eclipse Indicator
06	03	35	○	16 ♏ 01	
12	21	47	☾	22 ♒ 33	
20	23	47	●	00 ♊ 21	Annular
28	20	16	☽	07 ♍ 53	

Day	h	m		
06	03	40	Perigee	
19	16	30	Apogee	
02	06	11	0S	
08	06	18	Max dec	21° S 43'
14	22	10	0N	
22	11	13	Max dec	21° N 43'
29	14	02	0S	

LONGITUDES

Date	Chiron ⚷ ° '	Ceres ⚳ ° '	Pallas ⚴ ° '	Juno ⚵ ° '	Vesta ⚶ ° '	Black Moon Lilith ⚸ ° '
01	08 ♓ 57	08 ♉ 43	25 ♓ 24	03 ♐ 20	00 ♉ 22	15 ♉ 12
11	09 ♓ 17	12 ♉ 45	28 ♓ 13	01 ♐ 17	04 ♉ 52	16 ♉ 18
21	09 ♓ 32	16 ♉ 47	00 ♈ 57	29 ♏ 03	09 ♉ 18	17 ♉ 25
31	09 ♓ 41	20 ♉ 47	03 ♈ 19	26 ♏ 50	13 ♉ 41	18 ♉ 31

All ephemeris data is given at 12.00 UT and the Moon's longitude is additionally given for 24.00 UT

ASPECTARIAN

h m Aspects

01 Tuesday
00 00 ☽ ± ♅
02 37 ☽ BQ ☿
03 05 ☽ Q ♀
04 00 ☽ ☍ ♆
08 27 ☽ ☌ ♂
10 11 ☽ ∥ ☿
10 31 ☽ ⊼ ♅
15 30 ☽ △ ♇
15 37 ☿ ∠ ♆
16 25 ☽ ∠ ♄
19 34 ☽ △ ☉
20 04 ☽ ± ☿
21 04 ☽ ∥ ♅

02 Wednesday
07 27 ☽ ⊼ ☿
07 39 ☽ ⊥ ♄
10 21 ☽ △ ♃
10 58 ☽ □ ♀
15 15 ☽ # ♅
15 27 ☉ ⊥ ♅
17 38 ☽ ⊻ ♄
22 37 ☽ BQ ☉

03 Thursday
06 46 ☽ ⊼ ♆
06 46 ☽ # ☿
11 33 ☽ BQ ♃
11 33 ☽ ⊻ ♂
12 58 ☽ ☍ ♅
14 19 ☽ ∥ ♄
14 24 ☽ ± ☉
14 54 ☿ BQ ♂
14 54 ☿ ⊻ ♃
16 33 ☽ ± ♆
17 27 ☽ □ ♇
21 25 ☽ ⊥ ♂

04 Friday
00 47 ☽ ⊼ ☉
02 20 ☽ ± ♃
03 01 ☿ ⚹ ♀
07 03 ☽ BQ ♆
08 59 ☽ ∥ ♆
09 44 ☽ # ♂
12 03 ☽ ∠ ♂
12 05 ☽ ⊼ ♃
13 16 ☽ △ ♀
14 10 ☽ ☍ ☿
18 02 ☽ ☌ ♄
22 14 ☽ Q ♇
23 43 ☉ Q ♆

05 Saturday
06 55 ☽ △ ♆
12 10 ☽ ⚹ ♂
13 00 ☽ ⊼ ♅
13 37 ☽ BQ ♀
15 12 ☽ # ☉
17 10 ☽ ⚹ ♇
18 28 ♂ # ♆
18 43 ☿ # ♄
20 40 ☽ # ♃
22 29 ☽ ± ♅

06 Sunday
01 07 ☿ ☍ ♄
03 35 ☽ ☍ ☉
04 07 ☽ ± ♀
07 22 ☽ Q ♂
11 09 ☽ ∥ ♇
12 14 ☽ ☍ ♃
12 42 ☽ BQ ♅
13 47 ☽ ⊼ ♀
15 57 ☉ ⊥ ♀
16 47 ☽ ∠ ♇
17 10 ☽ ⊻ ♄
19 11 ☽ ⊼ ☿

07 Monday
02 36 ☽ ⊥ ♄
05 51 ☽ ± ☿
06 18 ☽ □ ♆
07 01 ☽ ⊥ ♇
12 18 ☽ □ ♂
12 32 ☽ △ ♅
16 34 ☽ ⊻ ♇
16 52 ☽ ∠ ♄
22 00 ☽ BQ ☿

08 Tuesday
02 09 ♃ ∠ ♅
06 32 ☽ ⊼ ☉
06 45 ♂ ⊼ ♅
11 12 ☽ Q ♆
12 53 ☽ ⊼ ♃
14 39 ☽ ☍ ♀
16 58 ☽ ± ☉
16 58 ☽ ⚹ ♄
22 51 ☽ ± ♃

09 Wednesday
01 34 ☽ △ ☿
06 54 ☽ ⚹ ♆
08 50 ☽ BQ ☉
12 37 ☉ ∥ ♃
12 41 ☽ Q ♄
13 32 ☽ □ ♅
13 59 ☽ △ ♂
14 01 ☽ BQ ♃
17 34 ☽ ☌ ♇

10 Thursday
03 09 ☽ ∥ ♇
08 11 ☽ ∠ ♆
12 03 ☽ △ ☉
15 04 ☽ # ☉
15 57 ☽ △ ♃
15 58 ☽ BQ ♂
16 28 ☿ ∥ ♂
16 45 ☽ # ♃
17 46 ☽ ⊼ ♀
19 11 ☽ □ ♄
20 19 ☽ Q ♅
21 59 ☿ ⚹ ♆
23 45 ☽ ⊥ ♆

11 Friday
04 22 ☽ ± ♀
08 01 ☽ ± ♂
10 19 ☽ ⊻ ♆
12 03 ☿ # ♆
12 12 ☽ □ ☿
17 33 ☽ ⚹ ♅
18 55 ☽ ⊼ ♂
20 33 ☽ BQ ♀
21 37 ☽ ⊻ ♇

12 Saturday
08 08 ☽ # ☿
08 29 ☽ ⊥ ♇
10 01 ☉ ∠ ♅
11 35 ☽ ∥ ♆
16 20 ☽ # ♂
20 56 ☽ ∠ ♅
21 47 ☽ □ ☉
22 40 ☽ □ ♃

13 Sunday
00 13 ☽ △ ♀
00 52 ☽ △ ♄
01 02 ☽ ∠ ♇
02 14 ☽ Q ☿
04 22 ☿ ⊻ ♅
10 06 ☽ ∥ ♄
13 23 ☉ ☌ ♃
13 49 ☽ ⊥ ♅
17 24 ☽ ☌ ♆
20 54 ☿ △ ♂

14 Monday
01 15 ☽ ⊻ ♅
03 48 ☽ ☍ ♂
03 56 ☿ ∠ ♀
04 50 ☽ ⚹ ☿
05 04 ☽ BQ ♄
05 21 ☽ ⚹ ♇
08 09 ☿ △ ♇
08 54 ☉ ⊻ ♀
09 21 ☽ Q ♃
10 35 ☽ # ♅
10 37 ☽ Q ☉
12 55 ☉ ⊼ ♄
16 46 ☉ BQ ♇
22 23 ☽ ± ♄

15 Tuesday
04 34 ☽ Q ♇
09 09 ☽ ⚹ ♃
09 54 ☽ □ ♀
09 58 ☽ ∥ ♅
10 04 ☽ ⊼ ♄
11 59 ☽ ⚹ ☉
14 33 ♀ St R
15 23 ☽ ∠ ☿

16 Wednesday
00 06 ♂ ∠ ♄
03 48 ☽ ⊻ ♆
08 34 ☉ # ♇
09 28 ☿ ⊥ ♅
10 58 ☽ # ♄
12 11 ☽ ☌ ♅
12 42 ☽ ⊥ ☿
15 30 ☽ ∠ ♃
15 47 ☽ ⊥ ♆
16 10 ☽ ⊼ ♂
16 10 ☽ □ ♇
16 26 ♂ △ ♇
20 24 ☽ ∠ ☉
20 41 ♀ ⊻ ♃
21 37 ☽ Q ♀
22 42 ♃ ⊼ ♄

17 Thursday
02 38 ☽ ∥ ♂
03 07 ☽ ⊻ ☿
04 32 ☽ ± ♂
07 27 ☿ Q ♆
09 53 ☽ ∠ ♆
09 57 ☽ ⊥ ♃
11 31 ☽ # ♆
16 07 ☽ ⊥ ☉
21 40 ☽ ⚹ ♀
21 44 ☽ ☍ ♄
22 19 ☽ ⊻ ♃
23 12 ☽ BQ ♂
23 22 ♃ BQ ♇

18 Friday
05 20 ☽ ⊻ ☉
16 15 ☿ ⊥ ♀
16 18 ☽ ⚹ ♆

19 Saturday
01 01 ☽ ⊻ ♅
02 27 ☽ ∥ ☿
03 50 ☽ ∠ ♀
04 46 ☽ △ ♇
06 32 ☽ △ ♂
13 16 ☽ ⊥ ♅
16 44 ☽ Q ♆
17 24 ☽ ∥ ♃
21 55 ☽ ⊥ ♀

20 Sunday
04 35 ☽ # ♇
04 37 ☽ ☌ ☿
07 39 ☽ ∠ ♅
09 55 ☉ ± ♄
09 56 ☽ ⊻ ♀
10 27 ☽ ⊼ ♄
11 16 ☽ BQ ♇
12 35 ☽ ☌ ♃
16 15 ☽ ∥ ☉
22 05 ☿ ∠ ♅
22 36 ☽ ± ♄
23 47 ☽ ☌ ☉

21 Monday
05 22 ☽ □ ♆
05 31 ☽ ± ♇
08 24 ☿ ⊻ ♀
12 51 ☿ ⊼ ♄
14 13 ☽ ⚹ ♅
15 38 ☿ ∥ ♃
16 46 ☽ BQ ♄
17 39 ☽ ⊼ ♇
17 46 ☿ BQ ♇
21 17 ☽ □ ♂

22 Tuesday
05 58 ☿ ☌ ♃
14 31 ☽ Q ♅
21 24 ☽ ☌ ♀
22 51 ☽ △ ♄

23 Wednesday
01 49 ☿ # ♇
02 25 ☽ ⊻ ♃
06 21 ☽ ⊻ ☿
10 31 ☽ Q ♂
14 37 ☽ ⊥ ♃
17 26 ☽ ⊻ ☉
17 42 ☽ △ ♆
20 07 ☽ ∥ ☉
20 52 ☽ ⊥ ☿
20 55 ☉ □ ♆
21 02 ☉ ± ♇

24 Thursday
02 30 ☽ □ ♅
05 34 ☽ ☍ ♇
06 19 ☽ ⊥ ☉
06 36 ☿ ± ♄
07 16 ☽ ∥ ☿
08 46 ☽ ∠ ♃
10 59 ☽ ⚹ ♂
16 49 ☽ # ♇
18 25 ☽ ∠ ☿
23 14 ☽ BQ ♆

25 Friday
00 46 ☽ ∥ ♃
01 27 ☽ ∠ ☉
07 01 ☽ ⊻ ♀
09 44 ☽ □ ♄
14 34 ☽ ⚹ ♃
16 40 ☽ ± ♆
17 03 ☽ ∠ ♂
18 15 ☽ ⊥ ♀
21 06 ☿ ± ♇
21 37 ☿ □ ♆

26 Saturday
04 10 ☽ ⊼ ♆
04 51 ☉ ∥ ☿
05 32 ☽ ⚹ ☿
08 41 ☽ ⚹ ☉
10 44 ☽ ⊥ ♂
10 52 ☽ ∠ ♀
12 44 ☽ △ ♅
13 53 ☽ Q ♃
15 24 ☽ ⊼ ♇
19 55 ☽ Q ♄
22 22 ☽ ⊻ ♂

27 Sunday
02 38 ☽ ± ♇
06 26 ☽ # ♆
08 44 ☽ Q ☿
09 00 ☽ Q ☉
11 19 ☉ ☌ ☿
13 57 ☽ ⚹ ♀
16 46 ☽ BQ ♅
18 07 ☽ ⚹ ♄
19 15 ☽ BQ ♇
23 36 ☿ ⚹ ♅
23 51 ☽ ∥ ♂
23 54 ☽ □ ♃

28 Monday
05 47 ☽ # ♄
07 15 ☿ BQ ♄
09 12 ☽ ± ♅
11 03 ☽ Q ♀
11 46 ☽ ☍ ♆
14 05 ☿ ⊼ ♇
16 04 ☉ ⚹ ♅
19 59 ☽ ⊼ ♅
20 16 ☽ □ ☉
21 08 ☽ ∠ ♄
22 16 ☽ △ ♇
23 53 ☽ □ ☿

29 Tuesday
02 28 ☽ ∥ ♅
06 29 ☽ ☌ ♂
07 53 ☉ BQ ♄
12 57 ☽ ⊥ ♄
17 44 ☽ □ ♀
23 19 ☽ ⊻ ♄
23 56 ☉ ⊼ ♇

30 Wednesday
01 30 ☽ # ♅
05 50 ☽ △ ♃
07 52 ♀ Q ♅
16 06 ☽ ⊼ ♆
20 55 ☽ ∥ ♄
23 56 ☽ ☍ ♅

31 Thursday
00 01 ☽ # ♂
00 20 ☿ □ ♂
01 54 ☽ □ ♇
02 12 ☽ ± ♆
03 48 ☽ △ ☉
07 34 ☽ BQ ♃
11 10 ☽ ⊻ ♂
12 44 ☽ △ ☿
17 09 ☽ BQ ♆
18 11 ☽ ∥ ♆
18 30 ☽ △ ♀
21 20 ☽ ⊥ ♂
22 41 ☽ ± ♃

JUNE 2012

LONGITUDES

Date	Sidereal time h m s	Sun ☉ ° ' "	Moon ☽ ° ' "	Moon ☽ 24.00 ° ' "	Mercury ☿ ° '	Venus ♀ ° '	Mars ♂ ° '	Jupiter ♃ ° '	Saturn ♄ ° '	Uranus ♅ ° '	Neptune ♆ ° '	Pluto ♇ ° '
01	04 41 29	11 ♊ 23 33	29 ♎ 40 51	07 ♏ 06 40	17 ♊ 34	18 ♊ 33	14 ♍ 56	27 ♉ 39	23 ♎ 14	07 ♈ 51	03 ♓ 09	08 ♑ 56
02	04 45 26	12 21 01	14 ♏ 35 42	22 ♏ 07 02	19 43	17 R 57	15 21	27 52	23 R 11	07 53	03 09	08 R 55
03	04 49 22	13 18 28	29 ♏ 39 39	07 ♐ 12 27	21 51	17 20	15 46	28 06	23 09	07 55	03 09	08 54
04	04 53 19	14 15 54	14 ♐ 44 20	22 ♐ 14 11	23 57	16 43	16 11	28 20	23 07	07 56	03 09	08 52
05	04 57 15	15 13 19	29 ♐ 40 57	07 ♑ 03 40	26 01	16 05	16 36	28 34	23 05	07 58	03 R 09	08 51
06	05 01 12	16 10 43	14 ♑ 21 30	21 ♑ 33 45	28 ♊ 03	15 28	17 01	28 48	23 03	08 00	03 09	08 50
07	05 05 08	17 08 07	28 ♑ 39 53	05 ♒ 39 31	00 ♋ 04	14 50	17 27	29 02	23 01	08 02	03 09	08 48
08	05 09 05	18 05 30	12 ♒ 32 27	19 ♒ 18 37	02 02	14 13	17 53	29 16	23 00	08 03	03 09	08 47
09	05 13 02	19 02 52	25 ♒ 58 08	02 ♓ 31 12	03 58	13 36	18 19	29 29	22 58	08 05	03 09	08 46
10	05 16 58	20 00 13	08 ♓ 58 08	15 ♓ 19 20	05 51	13 00	18 46	29 43	22 57	08 07	03 09	08 44
11	05 20 55	20 57 34	21 ♓ 35 16	27 ♓ 46 28	07 42	12 26	19 13	29 ♉ 57	22 55	08 08	03 08	08 43
12	05 24 51	21 54 55	03 ♈ 53 28	09 ♈ 56 51	09 31	11 52	19 40	00 ♊ 11	22 54	08 10	03 08	08 41
13	05 28 48	22 52 15	15 ♈ 57 11	21 ♈ 55 04	11 18	11 20	20 07	00 24	22 53	08 11	03 08	08 40
14	05 32 44	23 49 34	27 ♈ 51 01	03 ♉ 45 38	13 01	10 50	20 35	00 38	22 52	08 12	03 08	08 38
15	05 36 41	24 46 54	09 ♉ 39 23	15 ♉ 32 48	14 43	10 21	21 03	00 51	22 51	08 14	03 07	08 37
16	05 40 37	25 44 13	21 ♉ 26 20	27 ♉ 20 25	16 22	09 54	21 31	01 05	22 50	08 15	03 07	08 36
17	05 44 34	26 41 31	03 ♊ 15 25	09 ♊ 11 43	17 58	09 30	21 59	01 18	22 49	08 16	03 06	08 34
18	05 48 31	27 38 49	15 ♊ 09 38	21 ♊ 09 28	19 32	09 07	22 27	01 32	22 48	08 18	03 06	08 33
19	05 52 27	28 36 07	27 ♊ 11 27	03 ♋ 15 49	21 03	08 47	22 56	01 45	22 47	08 19	03 06	08 31
20	05 56 24	29 ♊ 33 24	09 ♋ 22 46	15 ♋ 32 29	22 32	08 29	23 25	01 59	22 47	08 20	03 05	08 30
21	06 00 20	00 ♋ 30 40	21 ♋ 45 08	28 ♋ 00 50	23 58	08 13	23 54	02 12	22 47	08 21	03 05	08 28
22	06 04 17	01 27 57	04 ♌ 19 43	10 ♌ 41 56	25 22	08 00	24 23	02 25	22 46	08 22	03 04	08 27
23	06 08 13	02 25 12	17 ♌ 07 34	23 ♌ 36 46	26 43	07 49	24 53	02 38	22 46	08 23	03 04	08 25
24	06 12 10	03 22 27	00 ♍ 09 38	06 ♍ 46 16	28 01	07 41	25 23	02 52	22 46	08 24	03 03	08 24
25	06 16 06	04 19 41	13 ♍ 26 47	20 ♍ 11 17	29 ♋ 16	07 35	25 53	03 05	22 D 46	08 25	03 02	08 22
26	06 20 03	05 16 55	26 ♍ 59 51	03 ♎ 52 32	00 ♌ 29	07 31	26 23	03 18	22 46	08 26	03 02	08 21
27	06 24 00	06 14 08	10 ♎ 49 22	17 ♎ 50 18	01 38	07 29	26 53	03 31	22 46	08 26	03 01	08 19
28	06 27 56	07 11 21	24 ♎ 55 15	02 ♏ 04 04	02 45	07 D 30	27 24	03 44	22 46	08 27	03 00	08 17
29	06 31 53	08 08 33	09 ♏ 16 30	16 ♏ 32 11	03 49	07 33	27 55	03 56	22 47	08 28	02 59	08 16
30	06 35 49	09 ♋ 05 45	23 ♏ 50 41	01 ♐ 11 26	04 ♌ 50	07 ♊ 39	28 ♍ 25	04 ♊ 09	22 ♎ 47	08 ♈ 28	02 ♓ 59	08 ♑ 14

Date	Moon True ☊ ° '	Moon Mean ☊ ° '	Moon ☽ Latitude ° '
01	05 ♐ 05	04 ♐ 54	03 S 04
02	05 D 05	04 51	01 51
03	05 05	04 48	00 S 30
04	05 R 05	04 44	00 N 53
05	05 05	04 41	02 12
06	05 04	04 38	03 21
07	05 02	04 35	04 15
08	05 01	04 32	04 53
09	05 00	04 29	05 13
10	04 59	04 25	05 16
11	04 D 58	04 22	05 04
12	04 59	04 19	04 38
13	05 00	04 16	04 00
14	05 01	04 13	03 12
15	05 03	04 09	02 16
16	05 04	04 06	01 15
17	05 04	04 03	00 N 10
18	05 R 04	04 00	00 S 55
19	05 03	03 57	01 59
20	05 00	03 54	02 58
21	04 56	03 50	03 49
22	04 52	03 47	04 30
23	04 48	03 44	04 59
24	04 44	03 41	05 12
25	04 42	03 38	05 10
26	04 40	03 35	04 50
27	04 D 40	03 31	04 13
28	04 41	03 28	03 21
29	04 42	03 25	02 15
30	04 ♐ 43	03 ♐ 22	00 S 59

DECLINATIONS

Date	Sun ☉ ° '	Moon ☽ ° '	Mercury ☿ ° '	Venus ♀ ° '	Mars ♂ ° '	Jupiter ♃ ° '	Saturn ♄ ° '	Uranus ♅ ° '	Neptune ♆ ° '	Pluto ♇ ° '
01	22 N 09	14 S 14	24 N 08	24 N 09	06 N 56	18 N 52	06 S 32	02 N 28	10 S 54	19 S 15
02	22 16	17 59	24 27	23 52	06 45	18 55	06 32	02 29	10 54	19 15
03	22 24	20 34	24 43	23 35	06 34	18 59	06 31	02 29	10 54	19 15
04	22 31	21 41	24 56	23 18	06 23	19 02	06 31	02 30	10 54	19 16
05	22 37	21 14	25 06	23 00	06 12	19 05	06 30	02 31	10 54	19 16
06	22 43	19 20	25 14	22 42	06 01	19 08	06 30	02 31	10 54	19 16
07	22 49	16 15	25 19	22 23	05 49	19 11	06 29	02 32	10 54	19 16
08	22 54	12 21	25 22	22 05	05 38	19 14	06 29	02 33	10 54	19 16
09	22 59	07 57	25 22	21 47	05 26	19 17	06 28	02 33	10 54	19 16
10	23 04	03 S 19	25 19	21 29	05 14	19 20	06 28	02 34	10 54	19 16
11	23 08	01 N 19	25 15	21 11	05 03	19 23	06 28	02 34	10 54	19 17
12	23 11	05 47	25 08	20 53	04 51	19 26	06 27	02 35	10 54	19 17
13	23 15	09 57	25 00	20 36	04 39	19 29	06 27	02 36	10 55	19 17
14	23 18	13 41	24 49	20 19	04 27	19 31	06 27	02 36	10 55	19 17
15	23 20	16 51	24 37	20 03	04 14	19 34	06 27	02 37	10 55	19 17
16	23 22	19 19	24 22	19 47	04 02	19 37	06 27	02 37	10 55	19 17
17	23 24	20 58	24 07	19 33	03 50	19 40	06 27	02 38	10 55	19 18
18	23 25	21 42	23 50	19 18	03 37	19 43	06 27	02 38	10 55	19 18
19	23 26	21 25	23 32	19 05	03 25	19 45	06 27	02 38	10 56	19 18
20	23 26	20 09	23 12	18 52	03 12	19 48	06 27	02 39	10 56	19 18
21	23 26	17 54	22 52	18 41	02 59	19 51	06 27	02 39	10 56	19 18
22	23 26	14 48	22 30	18 30	02 46	19 53	06 27	02 40	10 56	19 18
23	23 25	10 57	22 08	18 20	02 34	19 56	06 27	02 40	10 56	19 19
24	23 24	06 32	21 45	18 10	02 21	19 58	06 27	02 40	10 57	19 19
25	23 22	01 N 44	21 22	18 02	02 08	20 01	06 28	02 41	10 57	19 19
26	23 20	03 S 15	20 58	17 55	01 54	20 03	06 28	02 41	10 57	19 19
27	23 17	08 10	20 33	17 48	01 41	20 06	06 28	02 41	10 57	19 19
28	23 14	12 46	20 08	17 42	01 28	20 08	06 28	02 41	10 58	19 20
29	23 11	16 43	19 43	17 37	01 15	20 11	06 29	02 42	10 58	19 20
30	23 N 07	19 S 41	19 N 18	17 N 33	01 N 01	20 N 13	06 S 29	02 N 42	10 S 58	19 S 20

ZODIAC SIGN ENTRIES

Date	h	m	Planets
01	12	31	☽ ♏
03	12	32	☽ ♐
05	12	31	☽ ♑
07	11	16	☿ ♋
07	14	17	☽ ♒
09	19	22	☽ ♓
11	17	22	♃ ♊
12	04	21	☽ ♈
14	16	22	☽ ♉
17	05	24	☽ ♊
19	17	34	☽ ♋
20	23	09	☉ ♋
22	03	47	☽ ♌
24	11	42	☽ ♍
26	02	24	☿ ♌
26	17	15	☽ ♎
28	20	32	☽ ♏
30	22	04	☽ ♐

LATITUDES

Date	Mercury ☿ ° '	Venus ♀ ° '	Mars ♂ ° '	Jupiter ♃ ° '	Saturn ♄ ° '	Uranus ♅ ° '	Neptune ♆ ° '	Pluto ♇ ° '
01	01 N 17	01 N 13	01 N 05	00 S 47	02 N 40	00 S 42	00 S 35	03 N 53
04	01 38	00 N 31	01 01	00 47	02 40	00 42	00 35	03 53
07	01 53	00 S 11	00 56	00 47	02 39	00 42	00 35	03 53
10	02 01	00 53	00 52	00 47	02 38	00 43	00 36	03 53
13	02 03	01 33	00 48	00 46	02 38	00 43	00 36	03 53
16	01 57	02 10	00 44	00 46	02 37	00 43	00 36	03 53
19	01 46	02 43	00 40	00 46	02 36	00 43	00 36	03 52
22	01 28	03 11	00 36	00 46	02 35	00 43	00 36	03 52
25	01 05	03 35	00 32	00 46	02 35	00 43	00 36	03 52
28	00 37	03 54	00 28	00 46	02 34	00 43	00 36	03 52
31	00 N 04	04 S 10	00 N 25	00 S 46	02 N 33	00 S 43	00 S 36	03 N 51

DATA

Julian Date	2456080
Delta T	+68 seconds
Ayanamsa	24° 02' 05"
Synetic vernal point	05° ♓ 04' 54"
True obliquity of ecliptic	23° 26' 11"

MOON'S PHASES, APSIDES AND POSITIONS ☽

Date	h	m	Phase	Longitude ° '	Eclipse Indicator
04	11	12	○	14 ♐ 14	partial
11	10	41	☾	20 ♓ 54	
19	15	02	●	28 ♊ 43	
27	03	30	☽	05 ♎ 54	

Day	h	m		
03	13	22	Perigee	
16	01	35	Apogee	
04	17	02	Max dec	21° S 43'
11	05	08	0N	
18	17	38	Max dec	21° N 43'
25	20	25	0S	

LONGITUDES

Date	Chiron ⚷ ° '	Ceres ⚳ ° '	Pallas ⚴ ° '	Juno ⚵ ° '	Vesta ⚶ ° '	Black Moon Lilith ⚸ ° '
01	09 ♓ 42	21 ♉ 11	03 ♈ 33	26 ♏ 37	14 ♉ 07	18 ♉ 38
11	09 ♓ 45	25 ♉ 08	05 ♈ 45	24 ♏ 40	18 ♉ 24	19 ♉ 45
21	09 ♓ 43	29 ♉ 03	07 ♈ 41	23 ♏ 06	22 ♉ 36	20 ♉ 51
31	09 ♓ 35	02 ♊ 54	09 ♈ 16	22 ♏ 01	26 ♉ 40	21 ♉ 58

All ephemeris data is given at 12.00 UT and the Moon's longitude is additionally given for 24.00 UT

ASPECTARIAN

h m	Aspects
01 Friday	
01 31	☽ ☌ ♄
06 17	☽ bQ ☉
07 33	☽ Q ♇
08 38	☽ ⊼ ♃
12 26	☽ ∠ ♂
12 50	☿ ∥ ♀
17 29	☽ bQ ☿
17 37	☽ △ ♆
18 01	☽ bQ ♀
20 31	☿ ☌ ♀
21 52	☽ ± ☉
02 Saturday	
01 13	☽ ⊼ ♅
02 55	☽ ⚹ ♇
07 56	☽ ± ♀
08 10	☽ ⊼ ☉
10 22	☽ ± ☿
10 51	☽ ± ♅
13 14	☽ ⚹ ♂
13 48	☿ Q ♅
17 09	☽ ⊼ ♀
19 31	☽ # ♃
21 32	☽ ⊼ ☿
22 16	☽ ∥ ♇
03 Sunday	
01 14	☽ bQ ♅
01 40	☽ ⊻ ♄
02 51	☽ ∠ ♇
08 53	☽ Q ♂
09 29	☽ ☍ ♃
11 12	☽ ⊥ ♄
17 08	☽ ⊥ ♇
17 33	☽ □ ♆
19 24	♂ # ♄
04 Monday	
01 09	☽ △ ♅
01 28	☽ ∠ ♄
02 38	☿ △ ♄
02 40	☽ ⊻ ♇
11 12	☽ ☍ ☉
14 22	☽ □ ♂
15 02	☽ ☍ ♀
21 03	♆ St R
22 16	☽ Q ♆
05 Tuesday	
00 28	♀ □ ♂
01 23	☽ ⚹ ♄
05 08	☽ ☍ ☿
10 10	☽ ⊼ ♃
17 38	☽ ⚹ ♆
20 04	☽ ± ♃
20 45	☽ Q ♄
06 Wednesday	
01 09	☉ ☌ ♀
01 31	☽ □ ♅
02 55	☽ ☌ ♇
10 22	☉ ∥ ♀
11 04	☽ bQ ♃
12 37	☽ ∥ ♇
13 43	♂ ⊥ ♄
13 45	☽ ⊼ ♀
13 48	☽ # ♃
15 14	☽ ⊼ ☉
16 33	☽ △ ♂
18 18	☽ ∠ ♆
22 01	☿ ⊻ ♃
23 20	☽ ± ♀
07 Thursday	
01 58	☽ ± ☉
02 29	☽ □ ♄
07 31	☽ Q ♅
09 26	☽ ⊥ ♆
11 19	♃ ± ♄
12 38	☽ △ ♃
13 55	☽ bQ ♀
14 46	☽ ⊼ ☿
18 22	☽ bQ ☉
18 41	☽ bQ ♂
19 40	☽ ⊻ ♆
08 Friday	
02 30	☉ □ ♂
02 49	☽ ± ☿
04 09	☽ ⚹ ♅
05 26	☽ ⊻ ♇
10 49	☽ ± ♂
14 49	☽ △ ♀
15 57	☽ ⊥ ♇
20 07	☽ ∥ ♆
21 16	☽ bQ ☿
21 46	☽ ⊼ ♂
22 35	☽ △ ☉
09 Saturday	
01 49	☿ △ ♆
06 20	♃ # ♇
06 35	☽ △ ♄
06 46	☽ ∠ ♅
08 00	☽ ∠ ♇
18 33	☽ □ ♃
19 43	☽ ∥ ♄
23 13	☽ ⊥ ♅
10 Sunday	
01 09	☽ ☌ ♆
01 40	☽ # ♂
05 11	☽ △ ☿
10 03	☿ ⊥ ♃
10 05	☽ bQ ♄
10 23	☽ ⊻ ♅
11 34	☽ ⚹ ♇
14 43	☉ Q ♅
15 52	☽ # ♅
19 16	☽ □ ♀
11 Monday	
03 04	☽ ± ♄
03 30	☿ Q ♂
04 53	☽ Q ♃
07 16	☽ ☍ ♂
10 19	☽ Q ♇
10 41	☽ □ ☉
14 34	☽ ⊼ ♄
17 41	☿ □ ♅
18 38	☽ ∥ ♅
12 Tuesday	
01 04	☿ ☍ ♇
04 26	☽ Q ♀
04 34	☽ ⚹ ♃
07 02	☽ ∥ ♂
10 31	☽ ⊻ ♆
15 42	☽ # ♄
20 28	☽ ☌ ♅
21 29	☽ □ ♇
22 23	☽ ⊥ ♆
13 Wednesday	
00 58	☽ Q ☉
01 05	☽ □ ☿
03 09	☽ ⚹ ♀
10 53	☽ ∠ ♃
12 12	☉ △ ♄
12 27	☿ ⊻ ♀
16 22	☽ ∠ ♆
17 50	☽ # ♆
20 43	☽ ⊼ ♂
14 Thursday	
01 55	☽ ☍ ♄
03 09	☽ ⚹ ☉
05 21	☽ ⊥ ♃
08 04	☽ ∠ ♀
09 19	☽ ± ♂
17 45	☽ ⊻ ♃
19 31	☽ Q ☿
22 42	☽ ⚹ ♆
15 Friday	
01 37	☽ ⊥ ♀
04 20	☽ bQ ♂
09 05	☽ ⊻ ♅
09 53	☽ △ ♇
12 17	☽ ∠ ☉
13 22	☽ ⊻ ♀
21 20	☽ ⊥ ♅
23 08	☽ Q ♆
16 Saturday	
00 00	☽ ⚹ ☿
06 43	☿ ⊥ ♀
07 13	☿ ∠ ♃
08 14	☽ ⊥ ☉
11 38	☽ # ♇
12 09	☽ △ ♂
14 49	☽ ⊼ ♄
15 37	☽ ∥ ♃
15 42	☽ ∠ ♅
16 22	☽ bQ ♇
17 02	☽ ∥ ♀
21 31	☽ ⊻ ☉
17 Sunday	
02 02	♀ ∥ ♃
03 00	☽ ± ♄
07 58	☽ ☌ ♃
10 36	☽ ± ♇
11 20	☽ ∠ ☿
11 42	☽ □ ♆
14 06	☿ bQ ♆
21 12	☽ bQ ♄
22 09	☽ ⚹ ♅
22 43	☽ ⊼ ♇
18 Monday	
00 13	☽ ☌ ♀
08 15	☽ ⊥ ☿
13 16	♀ # ♇
22 02	☽ ⊻ ☿
22 17	☽ Q ♅
19 Tuesday	
03 11	☽ □ ♂
03 16	☽ △ ♄
05 03	♂ ⊻ ♄
15 02	☽ ☌ ☉
19 21	☉ ∥ ☿
21 12	☽ ⊻ ♃
23 39	☽ △ ♆
20 Wednesday	
09 12	☽ ⊥ ♃
09 57	☽ □ ♅
10 16	☽ ☍ ♇
10 17	☽ ⊻ ♀
11 18	♀ ⊼ ♇
16 05	☿ □ ♄
16 08	☽ Q ♂
16 24	☽ ∥ ♃
21 44	☽ ⊥ ♀
22 17	☽ # ♇
21 Thursday	
00 55	♀ ⚹ ♅
01 19	☿ ∠ ♀
03 03	☽ ∠ ♃
04 10	☽ ∥ ♀
04 55	☽ bQ ♆
10 14	☿ ⚹ ♂
13 58	☽ □ ♄
14 47	☽ ∠ ♀
16 18	☽ ⚹ ♂
16 48	☽ ☌ ☿
22 12	☽ ± ♆
22 Friday	
06 08	☽ ⊻ ☉
08 19	☽ ⚹ ♃
09 37	☽ ⊼ ♆
14 19	♃ ± ♇
18 24	☽ ⊥ ☉
18 50	☽ ⚹ ♀
19 38	☽ △ ♅
19 45	☽ ⊼ ♇
21 56	☽ ∠ ♂
23 Saturday	
00 08	☽ Q ♄
00 29	♂ ∥ ♅
06 58	☽ ± ♇
07 17	☽ Q ♃
12 04	☽ # ♆
12 35	☽ ∠ ☉
15 23	☽ ⊥ ♂
16 56	☽ Q ♀
18 19	☿ ± ♆
19 10	☉ ⊻ ♃
20 35	♀ bQ ♄
22 26	☽ ⚹ ♄
23 35	☽ bQ ♅
23 37	☽ bQ ♇
24 Sunday	
02 54	☽ ⊻ ♂
03 54	☉ △ ♆
07 39	☽ ⊻ ☿
09 12	♅ □ ♇
12 25	☽ # ♄
16 05	☽ ± ♅
17 00	☽ □ ♃
17 15	☽ ☍ ♆
18 18	☽ ⚹ ☉
19 45	☽ ⊥ ☿
25 Monday	
01 31	☽ □ ♀
01 47	☽ ∠ ♄
02 54	☽ △ ♇
02 57	☽ ⊼ ♅
07 24	☽ ∥ ♅
07 56	♃ □ ♆
08 00	♄ St D
10 00	☽ ∥ ♂
13 37	☽ ∠ ☿
17 32	☽ Q ☉
17 55	☽ ⊥ ♄
26 Tuesday	
04 33	☽ ⊻ ♄
05 52	☽ # ♂
09 18	☽ # ♅
10 53	☽ ☌ ♂
18 40	☽ ⚹ ☿
22 31	☽ ⊼ ♆
23 10	☽ △ ♃
27 Wednesday	
03 30	☽ □ ☉
03 37	☽ ∥ ♄
06 16	☽ △ ♀
07 42	☽ □ ♇
07 54	☽ ☍ ♅
08 54	☽ ± ♆
15 07	♀ St D
17 16	☽ Q ☿
28 Thursday	
00 17	☽ bQ ♆
01 21	☽ bQ ♃
02 18	☽ ∥ ♆
07 54	☽ bQ ♀
08 22	☽ ☌ ♄
11 58	☿ ∥ ♃
14 18	☽ Q ♇
16 19	☽ ⊻ ♂
16 48	☽ ± ♃
17 28	☿ ⊼ ♆
20 14	☉ ⊻ ♀
23 05	☽ ± ♀
29 Friday	
01 33	☽ △ ♆
02 13	☽ □ ☿
02 45	☽ ⊥ ♂
03 00	☽ ⊼ ♃
09 08	☽ ⊼ ♀
09 59	☽ △ ☉
10 20	☽ ⚹ ♇
10 39	☽ ⊼ ♅
15 02	☉ ☍ ♇
15 37	☿ ⚹ ♃
18 14	☽ ∠ ♂
18 18	☽ # ♀
20 10	☉ □ ♅
20 35	☽ ± ♅
30 Saturday	
08 33	☽ ∥ ♇
08 48	☽ # ☿
10 16	☽ ⊻ ♄
10 20	☿ # ♇
11 01	☽ ∠ ♇
11 24	☽ bQ ♅
12 26	☽ bQ ☉
17 51	☽ # ♃
19 46	☽ ⚹ ♂
20 05	☽ ⊥ ♄

JULY 2012

Date	Sidereal time h m s	Sun ☉ ° ' "	Moon ☽ ° ' "	Moon ☽ 24.00 ° ' "	Mercury ☿ ° '	Venus ♀ ° '	Mars ♂ ° '	Jupiter ♃ ° '	Saturn ♄ ° '	Uranus ♅ ° '	Neptune ♆ ° '	Pluto ♇ ° '
01	06 39 46	10 ♋ 02 56	08 ♐ 33 46	15 ♐ 56 56	05 ♌ 47	07 ♊ 46	28 ♍ 57	04 ♊ 22	22 ♎ 48	08 ♈ 29	02 ♓ 58	08 ♑ 13
02	06 43 42	11 00 07	23 ♐ 20 06	00 ♑ 42 22	06 41	07 56	29 28	04 35	22 48	08 30	02 R 57	08 R 11
03	06 47 39	11 57 18	08 ♑ 02 51	15 ♑ 20 38	07 32	08 08	29 ♍ 59	04 47	22 49	08 30	02 56	08 10
04	06 51 35	12 54 29	22 ♑ 34 51	29 ♑ 44 45	08 20	08 21	00 ♎ 31	05 00	22 50	08 31	02 55	08 08
05	06 55 32	13 51 40	06 ♒ 49 38	13 ♒ 48 58	09 03	08 37	01 03	05 13	22 51	08 31	02 54	08 07
06	06 59 29	14 48 51	20 ♒ 42 21	27 ♒ 29 30	09 43	08 55	01 35	05 25	22 52	08 31	02 53	08 05
07	07 03 25	15 46 02	04 ♓ 10 19	10 ♓ 44 51	10 19	09 14	02 07	05 37	22 53	08 32	02 52	08 04
08	07 07 22	16 43 14	17 ♓ 13 15	23 ♓ 35 48	10 51	09 36	02 39	05 50	22 54	08 32	02 52	08 02
09	07 11 18	17 40 26	29 ♓ 52 53	06 ♈ 04 59	11 19	09 59	03 12	06 02	22 56	08 32	02 50	08 01
10	07 15 15	18 37 38	12 ♈ 12 36	18 ♈ 16 20	11 43	10 24	03 44	06 14	22 57	08 32	02 49	07 59
11	07 19 11	19 34 51	24 ♈ 16 49	00 ♉ 14 39	12 02	10 50	04 17	06 26	22 59	08 32	02 48	07 58
12	07 23 08	20 32 04	06 ♉ 10 31	12 ♉ 05 04	12 16	11 18	04 50	06 38	23 00	08 32	02 47	07 56
13	07 27 04	21 29 18	17 ♉ 58 55	23 ♉ 52 41	12 26	11 47	05 23	06 50	23 02	08 R 32	02 46	07 55
14	07 31 01	22 26 32	29 ♉ 46 58	05 ♊ 42 20	12 32	12 18	05 56	07 02	23 04	08 32	02 45	07 53
15	07 34 58	23 23 47	11 ♊ 39 18	17 ♊ 38 19	12 R 32	12 51	06 30	07 14	23 06	08 32	02 44	07 52
16	07 38 54	24 21 02	23 ♊ 39 50	29 ♊ 44 11	12 28	13 24	07 03	07 26	23 08	08 32	02 43	07 51
17	07 42 51	25 18 18	05 ♋ 51 41	12 ♋ 02 34	12 19	13 59	07 37	07 37	23 10	08 32	02 42	07 49
18	07 46 47	26 15 35	18 ♋ 17 01	24 ♋ 35 08	12 05	14 35	08 11	07 49	23 13	08 32	02 40	07 48
19	07 50 44	27 12 52	00 ♌ 56 59	07 ♌ 22 32	11 46	15 12	08 45	08 00	23 15	08 32	02 39	07 46
20	07 54 40	28 10 09	13 ♌ 51 44	20 ♌ 24 29	11 23	15 51	09 19	08 12	23 17	08 31	02 38	07 45
21	07 58 37	29 ♋ 07 26	27 ♌ 00 39	03 ♍ 40 04	10 56	16 30	09 53	08 23	23 19	08 31	02 37	07 43
22	08 02 33	00 ♌ 04 45	10 ♍ 22 34	17 ♍ 07 57	10 24	17 11	10 27	08 34	23 22	08 30	02 35	07 42
23	08 06 30	01 02 03	23 ♍ 56 05	00 ♎ 46 46	09 50	17 53	11 02	08 45	23 25	08 30	02 34	07 41
24	08 10 27	01 59 22	07 ♎ 39 54	14 ♎ 35 21	09 12	18 35	11 37	08 56	23 27	08 29	02 33	07 39
25	08 14 23	02 56 41	21 ♎ 33 00	28 ♎ 32 47	08 31	19 19	12 11	09 07	23 30	08 29	02 31	07 38
26	08 18 20	03 54 00	05 ♏ 34 34	12 ♏ 38 18	07 49	20 03	12 46	09 18	23 33	08 28	02 30	07 37
27	08 22 16	04 51 20	19 ♏ 43 50	26 ♏ 51 00	07 06	20 48	13 22	09 29	23 36	08 28	02 28	07 35
28	08 26 13	05 48 41	03 ♐ 59 37	11 ♐ 09 23	06 22	21 34	13 57	09 40	23 39	08 27	02 27	07 34
29	08 30 09	06 46 02	18 ♐ 20 00	25 ♐ 31 00	05 39	22 21	14 32	09 50	23 43	08 26	02 26	07 33
30	08 34 06	07 43 23	02 ♑ 41 56	09 ♑ 52 14	04 57	23 09	15 08	10 01	23 46	08 25	02 24	07 32
31	08 38 02	08 ♌ 40 45	17 ♑ 01 16	24 ♑ 08 25	04 ♌ 16	23 ♊ 58	15 ♎ 43	10 ♊ 11	23 ♎ 49	08 ♈ 25	02 ♓ 23	07 ♑ 30

Date	Moon True ☊ ° '	Moon Mean ☊ ° '	Moon ☽ Latitude ° '
01	04 ♐ 44	03 ♐ 19	00 N 21
02	04 R 42	03 15	01 40
03	04 40	03 12	02 52
04	04 35	03 09	03 51
05	04 29	03 06	04 35
06	04 23	03 03	05 02
07	04 17	03 00	05 10
08	04 12	02 56	05 02
09	04 09	02 53	04 40
10	04 07	02 50	04 05
11	04 D 07	02 47	03 19
12	04 08	02 44	02 26
13	04 09	02 41	01 27
14	04 10	02 37	00 N 24
15	04 R 10	02 34	00 S 40
16	04 08	02 31	01 43
17	04 04	02 28	02 42
18	03 58	02 25	03 35
19	03 49	02 21	04 17
20	03 40	02 18	04 48
21	03 31	02 15	05 04
22	03 22	02 12	05 03
23	03 15	02 09	04 46
24	03 11	02 06	04 12
25	03 08	02 02	03 23
26	03 D 08	01 59	02 22
27	03 08	01 56	01 S 11
28	03 R 09	01 53	00 N 05
29	03 08	01 50	01 20
30	03 05	01 46	02 31
31	02 ♐ 59	01 ♐ 43	03 N 31

DECLINATIONS

Date	Sun ☉ ° '	Moon ☽ ° '	Mercury ☿ ° '	Venus ♀ ° '	Mars ♂ ° '	Jupiter ♃ ° '	Saturn ♄ ° '	Uranus ♅ ° '	Neptune ♆ ° '	Pluto ♇ ° '
01	23 N 03	21 S 23	18 N 53	17 N 30	00 N 48	20 N 15	06 S 30	02 N 42	10 S 59	19 S 20
02	22 59	21 36	18 28	17 27	00 34	20 18	06 30	02 42	10 59	19 20
03	22 54	20 20	18 03	17 25	00 21	20 20	06 31	02 42	10 59	19 20
04	22 49	17 45	17 39	17 24	00 N 07	20 22	06 31	02 43	11 00	19 21
05	22 43	14 07	17 15	17 23	00 S 06	20 25	06 32	02 43	11 00	19 21
06	22 37	09 49	16 51	17 23	00 20	20 27	06 33	02 43	11 00	19 21
07	22 30	05 09	16 28	17 24	00 34	20 29	06 33	02 43	11 01	19 21
08	22 23	00 S 24	16 06	17 25	00 48	20 31	06 34	02 43	11 01	19 22
09	22 16	04 N 14	15 44	17 26	01 02	20 33	06 35	02 43	11 02	19 22
10	22 08	08 35	15 24	17 29	01 16	20 35	06 35	02 43	11 02	19 22
11	22 00	12 30	15 05	17 31	01 30	20 37	06 36	02 43	11 02	19 22
12	21 52	15 52	14 47	17 34	01 44	20 39	06 37	02 43	11 03	19 22
13	21 43	18 34	14 30	17 38	01 58	20 41	06 38	02 43	11 03	19 23
14	21 34	20 29	14 14	17 41	02 12	20 43	06 39	02 43	11 04	19 23
15	21 25	21 31	14 01	17 45	02 26	20 45	06 40	02 43	11 04	19 23
16	21 15	21 34	13 48	17 49	02 40	20 47	06 41	02 43	11 04	19 23
17	21 04	20 36	13 38	17 54	02 54	20 49	06 42	02 43	11 05	19 23
18	20 54	18 39	13 29	17 59	03 09	20 51	06 43	02 43	11 05	19 24
19	20 43	15 45	13 23	18 04	03 23	20 53	06 44	02 43	11 06	19 24
20	20 32	12 04	13 18	18 09	03 37	20 55	06 45	02 42	11 06	19 24
21	20 20	07 45	13 15	18 14	03 52	20 56	06 46	02 42	11 07	19 24
22	20 08	03 N 00	13 15	18 19	04 06	20 58	06 47	02 42	11 07	19 25
23	19 56	01 S 58	13 16	18 25	04 21	21 00	06 49	02 42	11 08	19 25
24	19 43	06 54	13 19	18 30	04 35	21 01	06 50	02 42	11 08	19 25
25	19 30	11 33	13 25	18 36	04 50	21 03	06 51	02 41	11 09	19 25
26	19 17	15 36	13 32	18 41	05 04	21 05	06 52	02 41	11 09	19 26
27	19 03	18 48	13 41	18 47	05 19	21 06	06 54	02 41	11 10	19 26
28	18 49	20 52	13 52	18 52	05 33	21 08	06 55	02 41	11 10	19 26
29	18 35	21 36	14 04	18 57	05 48	21 09	06 57	02 40	11 11	19 26
30	18 20	20 54	14 17	19 03	06 02	21 11	06 58	02 40	11 11	19 26
31	18 N 05	18 S 51	14 N 32	19 N 08	06 S 17	21 N 12	06 S 59	02 N 39	11 S 12	19 S 27

ZODIAC SIGN ENTRIES

Date	h m	Planets
02	22 51	☽ ♑
03	12 31	♂ ♎
05	00 26	☽ ♒
07	04 29	☽ ♓
09	12 14	☽ ♈
11	23 30	☽ ♉
14	12 26	☽ ♊
17	00 31	☽ ♋
19	10 13	☽ ♌
21	17 24	☽ ♍
22	10 01	☉ ♌
23	22 38	☽ ♎
26	02 29	☽ ♏
28	05 18	☽ ♐
30	07 29	☽ ♑

LATITUDES

Date	Mercury ☿ ° '	Venus ♀ ° '	Mars ♂ ° '	Jupiter ♃ ° '	Saturn ♄ ° '	Uranus ♅ ° '	Neptune ♆ ° '	Pluto ♇ ° '
01	00 N 04	04 S 10	00 N 25	00 S 46	02 N 33	00 S 43	00 S 36	03 N 51
04	00 S 33	04 21	00 21	00 46	02 32	00 43	00 36	03 51
07	01 14	04 29	00 18	00 46	02 31	00 43	00 36	03 50
10	01 57	04 34	00 15	00 46	02 31	00 44	00 36	03 50
13	02 41	04 37	00 11	00 46	02 30	00 44	00 36	03 50
16	03 23	04 37	00 08	00 46	02 29	00 44	00 37	03 49
19	04 02	04 35	00 05	00 46	02 28	00 44	00 37	03 49
22	04 33	04 31	00 N 02	00 46	02 28	00 44	00 37	03 49
25	04 53	04 25	00 S 01	00 46	02 27	00 44	00 37	03 48
28	04 58	04 18	00 03	00 46	02 26	00 44	00 37	03 48
31	04 S 47	04 S 10	00 S 06	00 S 46	02 N 25	00 S 44	00 S 37	03 N 47

DATA

Julian Date	2456110
Delta T	+68 seconds
Ayanamsa	24° 02' 10"
Synetic vernal point	05° ♓ 04' 49"
True obliquity of ecliptic	23° 26' 11"

MOON'S PHASES, APSIDES AND POSITIONS ☽

Date	h	m	Phase	Longitude ° '	Eclipse Indicator
03	18	52	○	12 ♑ 14	
11	01	48	☾	19 ♈ 11	
19	04	24	●	26 ♋ 55	
26	08	56	☽	03 ♏ 47	

Day	h	m		
01	18	09	Perigee	
13	16	52	Apogee	
29	08	34	Perigee	
02	03	31	Max dec	21° S 42'
08	14	03	ON	
16	01	17	Max dec	21° N 40'
23	02	32	OS	
29	12	14	Max dec	21° S 36'

LONGITUDES

Date	Chiron ⚷ ° '	Ceres ⚳ ° '	Pallas ⚴ ° '	Juno ⚵ ° '	Vesta ⚶ ° '	Black Moon Lilith ⚸ ° '
01	09 ♓ 35	02 ♊ 54	09 ♈ 16	22 ♏ 01	26 ♉ 40	21 ♉ 58
11	09 ♓ 21	06 ♊ 39	10 ♈ 29	21 ♏ 29	00 ♊ 38	23 ♉ 04
21	09 ♓ 03	10 ♊ 19	11 ♈ 15	21 ♏ 29	04 ♊ 26	24 ♉ 11
31	08 ♓ 41	13 ♊ 51	11 ♈ 29	22 ♏ 00	08 ♊ 04	25 ♉ 17

All ephemeris data is given at 12.00 UT and the Moon's longitude is additionally given for 24.00 UT

ASPECTARIAN

h m	Aspects
01 Sunday	
01 41	☽ ⊥ ♇
02 54	☽ □ ♆
04 09	☽ ± ☉
05 05	☽ ☍ ♃
07 11	☽ △ ☿
10 42	☽ ☍ ♀
10 45	☽ ∠ ♄
11 26	☽ ⊻ ♇
11 52	☽ △ ♅
14 35	☽ ⊼ ☉
16 01	☽ Q ♂
16 31	♀ ⚼ ♄
22 20	☉ ⊥ ♃
02 Monday	
08 08	☽ Q ♆
09 10	☽ ⚼ ☿
11 08	☽ ⚹ ♄
22 21	☽ □ ♂
03 Tuesday	
00 43	☽ ± ☿
03 39	☽ ⚹ ♆
06 35	☽ ⊼ ♃
06 43	☽ Q ♄
11 07	☽ ⊼ ☿
12 01	☽ # ♃
12 08	☽ ⊼ ♀
12 12	☽ ☌ ♇
12 45	☽ □ ♅
15 45	♀ ⊼ ♇
16 34	☽ ± ♃
18 52	☽ ☍ ☉
22 09	☽ ± ♀
22 33	☽ ∥ ♇
04 Wednesday	
04 16	☽ ∠ ♆
06 19	☿ ⊼ ♇
07 39	☽ ⚼ ♃
12 25	☽ □ ♄
12 51	☽ # ☿
13 19	☽ ⚼ ♀
13 26	☿ ⚹ ♀
14 35	☽ # ♀
17 53	☿ △ ♅
18 34	☽ Q ♅
19 15	☽ ⊥ ♆
05 Thursday	
01 49	☽ △ ♂
02 36	♀ ⚹ ♅
03 13	☿ ∥ ♀
05 21	☽ ⊻ ♆
09 12	☽ △ ♃
14 12	☽ ⊻ ♇
14 53	☽ ⚹ ♅
15 08	☽ △ ♀
16 00	☽ ☍ ☿
06 Friday	
00 30	☽ ⊥ ♇
00 58	☽ ⊼ ☉
04 30	☽ ⚼ ♂
05 39	☽ ∥ ♆
12 12	☽ ± ☉
15 48	☉ ⊥ ♀
15 48	☽ △ ♄
16 11	☽ ∠ ♇
16 58	☽ ∠ ♅
20 57	☽ ± ♂
07 Saturday	
04 56	☽ ∥ ♄
05 23	☽ ⚼ ☉
08 08	☽ ⊼ ♂
09 02	☽ ⊥ ♅
09 39	☽ ☌ ♆
14 40	☽ □ ♃
18 46	☽ ⚼ ♄
19 04	☽ ⚹ ♇
19 56	☽ ⊻ ♅
21 29	☽ □ ♀
23 43	☽ ⊼ ☿
08 Sunday	
00 19	☽ # ♅
10 06	☽ ∥ ♂
10 59	☽ △ ☉
11 17	☽ ± ☿
11 25	☽ ± ♄
14 40	☿ Q ♄
17 16	☽ Q ♇
18 26	☽ # ♂
20 57	♂ ⊼ ♆
22 43	☽ ⊼ ♄
09 Monday	
00 39	☽ Q ♃
04 03	☽ ∥ ♅
04 56	☽ ⚼ ☿
08 14	☽ Q ♀
16 09	☉ ⚼ ♆
17 42	☽ ⊻ ♆
18 41	☽ ☍ ♂
10 Tuesday	
00 06	☽ ⚹ ♃
00 45	☽ # ♄
03 44	☽ □ ♇
04 47	☽ ☌ ♅
05 22	☽ ⊥ ♆
08 18	☽ ⚹ ♀
10 59	☽ △ ☿
23 06	☽ ∠ ♆
11 Wednesday	
01 48	☽ □ ☉
02 42	☽ # ♆
06 13	☽ ∠ ♃
09 23	☽ ☍ ♄
15 15	☽ ∠ ♀
12 Thursday	
00 36	☽ ⊥ ♃
04 24	☽ ∥ ☿
05 09	☽ ⚹ ♆
09 08	☽ ⊼ ♂
10 09	☽ ⊥ ♀
12 57	☽ ⊻ ♃
15 34	☽ △ ♇
16 48	☽ ⊻ ♅
17 12	☽ Q ☉
21 55	☽ ± ♂
22 51	☽ ⊻ ♀
13 Friday	
00 35	☽ □ ☿
02 37	☽ ∥ ♀
05 00	☽ ⊥ ♅
05 28	☽ Q ♆
09 49	♅ St R
17 07	☽ ⚼ ♂
19 46	☽ ⚹ ☉
20 52	☽ # ♇
22 01	☽ ⚼ ♇
22 19	☽ ⊼ ♄
23 05	☉ ∠ ♃
23 19	☽ ∠ ♅
14 Saturday	
10 32	☽ ± ♄
13 31	☽ Q ☿
16 06	☽ ∥ ♃
16 16	☽ ± ♇
18 01	☽ □ ♆
22 41	☿ ⚹ ♀
15 Sunday	
01 05	☽ △ ♂
02 16	☿ St R
02 56	☽ ☌ ♃
04 14	☉ □ ♄
04 23	☽ ⊼ ♇
04 49	☽ ⚼ ♄
04 52	☽ ∠ ☉
05 43	☽ ⚹ ♅
08 46	☽ ∥ ☉
13 46	☽ ⚹ ☿
14 30	☽ ☌ ♀
16 Monday	
00 31	☽ ⊥ ☉
05 47	☽ Q ♅
10 56	☽ △ ♄
13 29	☽ ⊻ ☉
16 29	♂ # ♅
19 27	☽ ∠ ☿
17 Tuesday	
01 33	☽ ∥ ☉
05 49	☽ △ ♆
08 23	☽ ∥ ♃
12 38	♂ △ ♃
12 52	☽ ⊥ ☿
15 29	☽ ⊻ ♃
15 34	☽ □ ♂
15 48	☽ ☍ ♇
17 12	☽ □ ♅
20 26	♂ □ ♇
18 Wednesday	
00 18	☽ ⊻ ☿
03 17	☽ ⊥ ♃
04 08	☽ # ♇
04 32	☽ ⊻ ♀
06 52	☉ Q ♂
09 46	♃ ⊼ ♇
10 50	☽ ⚼ ♆
16 37	☽ ⊥ ♀
17 15	☉ ∥ ♃
18 03	☽ ∥ ♀
20 46	☽ ∠ ♃
21 24	☽ □ ♄
22 08	☉ ± ♆
19 Thursday	
02 55	♂ ☍ ♅
03 43	☽ Q ♂
03 55	☽ ± ♆
04 24	☽ ☌ ☉
08 15	♂ ± ♆
10 32	☽ ∠ ♀
15 11	☽ ⊼ ♆
20 Friday	
00 43	☽ ⊼ ♇
01 22	☽ ⚹ ♃
02 08	☽ △ ♅
03 12	☽ ⚹ ♂
04 22	☽ ∥ ☿
07 14	☽ Q ♄
07 34	☽ ☌ ☿
11 47	☽ ± ♇
15 51	☽ ⚹ ♀
17 06	☿ Q ♄
17 37	☽ # ♆
23 47	☽ Q ♃
21 Saturday	
02 00	♃ ⚼ ♄
04 14	☽ ⚼ ♇
05 17	☽ ⚹ ♄
05 40	☽ ⚼ ♅
07 58	☽ ∠ ♂
14 51	☽ Q ♀
16 07	☽ ⊻ ☉
17 03	☽ # ♄
21 55	☽ ± ♅
22 05	☽ ☍ ♆
22 Sunday	
00 56	☽ ⊥ ♂
03 44	☽ ⊥ ☉
04 04	♃ ⚹ ♅
06 48	☽ # ♂
07 14	☽ △ ♇
08 24	☽ ∠ ♄
08 40	☽ ⊼ ♅
08 44	☽ □ ♃
10 55	☿ ⚹ ♂
12 03	☽ ⊻ ☿
12 09	☽ ⊻ ♂
13 26	☽ ∥ ♅
21 00	☽ ∠ ☉
22 17	☽ ⊥ ☿
23 Monday	
00 27	☽ ⊥ ♄
00 44	☽ □ ♀
11 05	☽ ⊻ ♄
13 30	☽ ∠ ☿
15 30	☽ # ♅
24 Tuesday	
00 03	☽ ∥ ♂
01 22	☽ ⚹ ☉
03 06	☽ ⊼ ♆
11 39	☽ ∥ ♄
11 59	☽ □ ♇
13 26	☽ ☍ ♅
13 31	☽ ± ♆
14 15	☽ △ ♃
14 32	☽ ⚹ ☿
19 09	☽ ☌ ♂
19 14	☿ ⚹ ♃
23 46	☽ Q ☉
25 Wednesday	
01 34	☉ ⊼ ♆
05 04	☽ ⚼ ♆
07 56	☽ △ ♀
09 51	☽ ∥ ♆
10 19	☽ Q ☿
13 27	☿ △ ♅
15 22	☽ ☌ ♄
16 29	☽ ⚼ ♃
19 00	☽ Q ♇
20 13	☉ # ♇
22 49	☽ # ☿
26 Thursday	
06 46	☽ △ ♆
08 05	☽ ± ♃
08 56	☽ □ ☉
11 03	☽ ⚼ ♀
15 28	☽ ⚹ ♇
15 38	☽ □ ☿
16 55	☽ ⊼ ♅
18 25	☽ ⊼ ♃
19 12	☿ ⊼ ♇
27 Friday	
00 45	☽ ⊻ ♂
01 18	♀ Q ♅
03 06	☽ ± ♅
03 12	☽ ± ♀
11 21	☽ ⊥ ♂
11 44	☽ # ♀
13 55	☽ ⊼ ♀
14 02	☽ # ☉
16 49	☽ ∠ ♇
17 54	☽ ∥ ♇
18 17	☽ ⚼ ♅
18 34	☽ ⊻ ♄
28 Saturday	
03 10	☽ ∠ ♂
04 42	☽ ⊥ ♄
07 56	☽ ⊥ ♇
08 08	☉ ∥ ♀
08 45	☿ ∠ ♀
09 25	☽ □ ♆
11 58	♅ ⊥ ♆
15 16	☽ △ ☉
15 47	☽ △ ☿
17 03	☽ # ♃
17 59	☽ ⊻ ♇
19 28	☽ △ ♅
19 51	☽ ∠ ♄
19 57	☉ ☌ ☿
21 37	☽ ☍ ♃
29 Sunday	
05 23	☽ ⚹ ♂
15 29	☽ Q ♆
15 41	☽ ⚼ ☿
18 09	☽ ⚼ ☉
19 07	☽ ☍ ♀
21 01	☽ ⚹ ♄
30 Monday	
02 17	☽ Q ♂
06 01	☽ ± ☿
06 33	☽ # ♃
07 09	☉ ⊼ ♇
10 15	☽ ± ☉
11 30	☽ ⚹ ♆
15 35	☽ ⊼ ☿
17 09	☽ Q ♄
20 04	☽ ☌ ♇
21 00	☽ ⊼ ☉
21 34	☽ □ ♅
31 Tuesday	
00 23	☽ ⊼ ♃
05 21	☉ △ ♅
06 24	☽ ∥ ♇
07 38	♀ △ ♄
09 32	☽ # ♀
09 43	☽ □ ♂
10 35	☽ ± ♃
12 36	☽ ∠ ♆
19 10	☽ # ☉
22 43	☿ Q ♂
23 30	☽ □ ♄

AUGUST 2012

LONGITUDES

Date	Sidereal time h m s	Sun ☉ ° ' "	Moon ☽ ° ' "	Moon ☽ 24.00 ° ' "	Mercury ☿ ° '	Venus ♀ ° '	Mars ♂ ° '	Jupiter ♃ ° '	Saturn ♄ ° '	Uranus ♅ ° '	Neptune ♆ ° '	Pluto ♇ ° '
01	08 41 59	09 ♌ 38 08	01 ♒ 13 00	08 ♒ 14 22	03 ♌ 39	24 ♊ 47	16 ♎ 19	10 ♊ 21	23 ♎ 53	08 ♈ 24	02 ♓ 21	07 ♑ 29
02	08 45 56	10 35 31	15 ♒ 11 56	22 ♒ 05 09	03 R 05	25 37	16 55	10 31	23 56	08 R 23	02 R 20	07 R 28
03	08 49 52	11 32 56	28 ♒ 53 34	05 ♓ 36 50	02 35	26 27	17 31	10 41	24 00	08 22	02 18	07 27
04	08 53 49	12 30 21	12 ♓ 14 44	18 ♓ 47 09	02 09	27 19	18 07	10 51	24 04	08 21	02 17	07 25
05	08 57 45	13 27 48	25 ♓ 14 06	01 ♈ 35 44	01 50	28 10	18 43	11 01	24 07	08 20	02 15	07 24
06	09 01 42	14 25 15	07 ♈ 52 17	14 ♈ 04 06	01 35	29 03	19 19	11 11	24 11	08 19	02 14	07 23
07	09 05 38	15 22 44	20 ♈ 11 38	26 ♈ 15 23	01 27	29 ♊ 56	19 56	11 21	24 15	08 18	02 12	07 22
08	09 09 35	16 20 14	02 ♉ 15 55	08 ♉ 13 50	01 D 26	00 ♋ 50	20 32	11 30	24 19	08 16	02 11	07 21
09	09 13 31	17 17 46	14 ♉ 09 48	20 ♉ 04 30	01 31	01 44	21 09	11 39	24 23	08 15	02 09	07 20
10	09 17 28	18 15 19	25 ♉ 58 35	01 ♊ 52 46	01 43	02 39	21 46	11 49	24 28	08 14	02 07	07 19
11	09 21 25	19 12 53	07 ♊ 47 44	13 ♊ 44 06	02 02	03 34	22 23	11 58	24 32	08 13	02 06	07 18
12	09 25 21	20 10 29	19 ♊ 42 32	25 ♊ 43 37	02 28	04 30	23 00	12 07	24 36	08 11	02 04	07 17
13	09 29 18	21 08 06	01 ♋ 47 53	07 ♋ 55 49	03 01	05 27	23 37	12 16	24 41	08 10	02 03	07 16
14	09 33 14	22 05 45	14 ♋ 07 49	20 ♋ 24 13	03 41	06 23	24 14	12 24	24 45	08 08	02 01	07 15
15	09 37 11	23 03 25	26 ♋ 45 15	03 ♌ 11 04	04 28	07 21	24 52	12 33	24 50	08 07	01 59	07 14
16	09 41 07	24 01 06	09 ♌ 41 42	16 ♌ 17 05	05 21	08 18	25 29	12 42	24 54	08 05	01 58	07 13
17	09 45 04	24 58 49	22 ♌ 57 03	29 ♌ 41 20	06 21	09 17	26 07	12 50	24 59	08 04	01 56	07 12
18	09 49 00	25 56 33	06 ♍ 29 35	13 ♍ 21 25	07 28	10 15	26 44	12 58	25 04	08 02	01 55	07 11
19	09 52 57	26 54 18	20 ♍ 16 21	27 ♍ 13 56	08 40	11 14	27 22	13 06	25 09	08 01	01 53	07 10
20	09 56 54	27 52 04	04 ♎ 13 38	11 ♎ 15 01	09 58	12 13	28 00	13 14	25 13	07 59	01 51	07 09
21	10 00 50	28 49 52	18 ♎ 17 38	25 ♎ 21 05	11 22	13 13	28 38	13 22	25 18	07 57	01 50	07 09
22	10 04 47	29 ♌ 47 41	02 ♏ 25 03	09 ♏ 29 14	12 51	14 13	29 16	13 30	25 24	07 56	01 48	07 08
23	10 08 43	00 ♍ 45 31	16 ♏ 33 25	23 ♏ 37 26	14 25	15 14	29 ♎ 55	13 37	25 29	07 54	01 46	07 07
24	10 12 40	01 43 22	00 ♐ 41 09	07 ♐ 44 27	16 02	16 14	00 ♏ 33	13 45	25 34	07 52	01 45	07 06
25	10 16 36	02 41 15	14 ♐ 47 13	21 ♐ 49 23	17 44	17 15	01 11	13 52	25 39	07 50	01 43	07 06
26	10 20 33	03 39 08	28 ♐ 50 47	05 ♑ 51 16	19 29	18 17	01 50	13 59	25 44	07 48	01 41	07 05
27	10 24 29	04 37 03	12 ♑ 50 38	19 ♑ 48 39	21 17	19 19	02 29	14 06	25 50	07 47	01 40	07 04
28	10 28 26	05 34 59	26 ♑ 45 00	03 ♒ 39 22	23 08	20 21	03 07	14 13	25 55	07 45	01 38	07 04
29	10 32 23	06 32 57	10 ♒ 31 23	17 ♒ 20 41	25 00	21 23	03 46	14 19	26 01	07 43	01 36	07 03
30	10 36 19	07 30 56	24 ♒ 06 54	00 ♓ 49 39	26 54	22 26	04 25	14 26	26 06	07 41	01 35	07 03
31	10 40 16	08 ♍ 28 56	07 ♓ 28 38	14 ♓ 03 35	28 ♌ 50	23 ♋ 29	05 ♏ 04	14 ♊ 32	26 ♎ 12	07 ♈ 39	01 ♓ 33	07 ♑ 02

Date	Moon True ☊ ° '	Moon Mean ☊ ° '	Moon ☽ Latitude ° '
01	02 ♐ 51	01 ♐ 40	04 N 18
02	02 R 41	01 37	04 49
03	02 30	01 34	05 02
04	02 20	01 31	04 58
05	02 10	01 27	04 39
06	02 03	01 24	04 06
07	01 58	01 21	03 23
08	01 55	01 18	02 31
09	01 55	01 15	01 33
10	01 D 55	01 12	00 N 32
11	01 R 55	01 08	00 S 31
12	01 53	01 05	01 33
13	01 50	01 02	02 32
14	01 45	00 59	03 24
15	01 36	00 56	04 08
16	01 25	00 52	04 40
17	01 13	00 49	04 58
18	01 01	00 46	04 59
19	00 50	00 43	04 44
20	00 40	00 40	04 11
21	00 34	00 37	03 23
22	00 30	00 33	02 23
23	00 28	00 30	01 S 13
24	00 D 28	00 27	00 N 01
25	00 R 28	00 24	01 15
26	00 26	00 21	02 24
27	00 23	00 18	03 24
28	00 16	00 14	04 11
29	00 07	00 11	04 44
30	29 ♏ 56	00 08	04 59
31	29 ♏ 43	00 ♐ 05	04 N 58

DECLINATIONS

Date	Sun ☉ ° '	Moon ☽ ° '	Mercury ☿ ° '	Venus ♀ ° '	Mars ♂ ° '	Jupiter ♃ ° '	Saturn ♄ ° '	Uranus ♅ ° '	Neptune ♆ ° '	Pluto ♇ ° '
01	17 N 50	15 S 41	14 N 47	19 N 13	06 S 31	21 N 14	07 S 01	02 N 39	11 S 13	19 S 27
02	17 35	11 40	15 04	19 18	06 46	21 15	07 03	02 39	11 13	19 27
03	17 19	07 08	15 20	19 22	07 01	21 17	07 04	02 38	11 14	19 27
04	17 03	02 S 22	15 37	19 27	07 15	21 18	07 06	02 38	11 14	19 28
05	16 47	02 N 22	15 54	19 31	07 30	21 19	07 07	02 37	11 15	19 28
06	16 30	06 53	16 10	19 35	07 44	21 21	07 09	02 37	11 15	19 28
07	16 13	11 01	16 26	19 39	07 59	21 22	07 10	02 37	11 16	19 28
08	15 56	14 37	16 41	19 43	08 13	21 23	07 12	02 36	11 16	19 29
09	15 39	17 34	16 56	19 46	08 28	21 25	07 14	02 36	11 17	19 29
10	15 21	19 45	17 09	19 49	08 43	21 26	07 16	02 35	11 18	19 29
11	15 04	21 06	17 22	19 51	08 57	21 27	07 17	02 34	11 18	19 29
12	14 45	21 29	17 32	19 54	09 12	21 28	07 19	02 34	11 19	19 30
13	14 27	20 54	17 41	19 56	09 26	21 29	07 21	02 33	11 19	19 30
14	14 09	18 17	17 49	19 57	09 41	21 30	07 23	02 33	11 20	19 30
15	13 50	16 45	17 54	19 59	09 55	21 31	07 25	02 32	11 21	19 30
16	13 31	13 19	17 57	20 00	10 10	21 32	07 27	02 32	11 21	19 30
17	13 12	09 10	17 58	20 00	10 24	21 33	07 28	02 31	11 22	19 31
18	12 52	04 N 29	17 56	20 00	10 38	21 34	07 30	02 30	11 22	19 31
19	12 33	00 S 30	17 52	20 00	10 53	21 35	07 32	02 30	11 23	19 31
20	12 13	05 31	17 45	19 59	11 07	21 36	07 34	02 29	11 24	19 31
21	11 53	10 18	17 35	19 58	11 21	21 37	07 36	02 28	11 24	19 32
22	11 33	14 32	17 22	19 57	11 36	21 38	07 38	02 28	11 25	19 32
23	11 12	17 57	17 07	19 55	11 50	21 39	07 40	02 27	11 25	19 32
24	10 52	20 16	16 48	19 52	12 04	21 40	07 42	02 26	11 26	19 32
25	10 31	21 20	16 27	19 50	12 18	21 41	07 44	02 25	11 27	19 33
26	10 10	21 02	16 03	19 46	12 32	21 41	07 47	02 25	11 27	19 33
27	09 49	19 26	15 36	19 42	12 46	21 42	07 49	02 24	11 28	19 33
28	09 28	16 41	15 07	19 38	13 00	21 43	07 51	02 23	11 28	19 33
29	09 06	13 03	14 35	19 33	13 14	21 44	07 53	02 22	11 29	19 34
30	08 45	08 46	14 01	19 28	13 28	21 44	07 55	02 22	11 30	19 33
31	08 N 23	04 S 09	13 N 25	19 N 23	13 S 42	21 N 45	07 S 57	02 N 21	11 S 30	19 S 34

ZODIAC SIGN ENTRIES

Date	h	m	Planets
01	09	56	☽ ♒
03	13	58	☽ ♓
05	20	58	☽ ♈
07	13	43	♀ ♋
08	07	28	☽ ♉
10	20	11	☽ ♊
13	08	27	☽ ♋
15	18	05	☽ ♌
18	00	33	☽ ♍
20	04	45	☽ ♎
22	07	54	☽ ♏
22	17	07	☉ ♍
23	15	24	♂ ♏
24	10	50	☽ ♐
26	13	58	☽ ♑
28	17	38	☽ ♒
30	22	31	☽ ♓

LATITUDES

Date	Mercury ☿ ° '	Venus ♀ ° '	Mars ♂ ° '	Jupiter ♃ ° '	Saturn ♄ ° '	Uranus ♅ ° '	Neptune ♆ ° '	Pluto ♇ ° '
01	04 S 40	04 S 07	00 S 07	00 S 46	02 N 25	00 S 44	00 S 37	03 N 47
04	04 10	03 58	00 10	00 47	02 24	00 44	00 37	03 46
07	03 29	03 47	00 12	00 47	02 24	00 44	00 37	03 46
10	02 41	03 36	00 15	00 47	02 23	00 44	00 37	03 45
13	01 50	03 24	00 17	00 47	02 22	00 45	00 37	03 45
16	01 01	03 11	00 20	00 47	02 22	00 45	00 37	03 44
19	00 S 14	02 58	00 22	00 47	02 21	00 45	00 37	03 44
22	00 N 26	02 45	00 24	00 47	02 21	00 45	00 37	03 43
25	00 59	02 31	00 27	00 47	02 20	00 45	00 37	03 43
28	01 23	02 17	00 29	00 47	02 19	00 45	00 37	03 42
31	01 N 39	02 S 03	00 S 31	00 S 48	02 N 19	00 S 45	00 S 37	03 N 41

DATA

Julian Date	2456141
Delta T	+68 seconds
Ayanamsa	24° 02' 15"
Synetic vernal point	05° ♓ 04' 44"
True obliquity of ecliptic	23° 26' 11"

MOON'S PHASES, APSIDES AND POSITIONS ☽

Date	h	m	Phase	Longitude ° '	Eclipse Indicator
02	03	27	○	10 ♒ 15	
09	18	55	☾	17 ♉ 34	
17	15	54	●	25 ♌ 08	
24	13	54	☽	01 ♐ 48	
31	13	58	○	08 ♓ 34	

Day	h	m		
10	10	53	Apogee	
23	19	25	Perigee	
04	23	54	0N	
12	09	47	Max dec	21° N 30'
19	09	38	0S	
25	18	44	Max dec	21° S 23'

LONGITUDES

Date	Chiron ⚷ ° '	Ceres ⚳ ° '	Pallas ⚴ ° '	Juno ⚵ ° '	Vesta ⚶ ° '	Black Moon Lilith ⚸ ° '
01	08 ♓ 39	14 ♊ 12	11 ♈ 29	22 ♏ 05	08 ♊ 25	25 ♉ 24
11	08 ♓ 13	17 ♊ 34	11 ♈ 06	23 ♏ 07	11 ♊ 50	26 ♉ 30
21	07 ♓ 45	20 ♊ 46	10 ♈ 05	24 ♏ 34	14 ♊ 59	27 ♉ 37
31	07 ♓ 16	23 ♊ 44	08 ♈ 28	26 ♏ 23	17 ♊ 51	28 ♉ 43

All ephemeris data is given at 12.00 UT and the Moon's longitude is additionally given for 24.00 UT

ASPECTARIAN

h m	Aspects
01 Wednesday	
00 24	☽ ⊼ ♀
01 56	☽ ⚼ ♃
03 46	☽ ⊥ ♆
03 50	☽ Q ♅
11 13	☽ ± ♀
13 56	☽ ⊻ ♆
15 58	☽ ☍ ☿
17 21	☽ # ☿
22 41	☽ ⊻ ♇
02 Thursday	
00 15	☽ ✶ ♅
03 27	☽ ☍ ☉
03 34	☽ ⚼ ♀
03 50	☽ △ ♃
09 00	☽ ⊥ ♇
09 56	☉ ✶ ♃
14 29	☽ ∥ ♆
15 06	☽ △ ♂
15 53	☉ ∠ ♀
03 Friday	
00 39	☽ ∠ ♇
02 15	☽ ∠ ♅
03 19	☽ △ ♄
04 00	♂ ⚼ ♆
07 24	☽ △ ♀
12 20	☽ ∥ ♄
12 36	☽ ∥ ♂
14 15	☿ ⊥ ♀
18 04	☽ ☌ ♆
18 10	☽ ⊥ ♅
18 21	☽ ⊼ ☿
18 28	♂ ∥ ♄
18 45	☽ ⚼ ♂
04 Saturday	
00 03	☉ Q ♄
03 16	☽ ✶ ♇
04 06	☿ ⊼ ♆
04 48	☽ ± ☿
04 56	☽ ⊻ ♅
06 11	☽ ⚼ ♄
09 27	☽ □ ♃
10 41	☽ # ♅
11 45	☽ ± ♂
12 31	☽ ⊼ ☉
15 58	♀ # ♇
20 45	☽ ⚼ ☿
22 43	☽ ± ♄
23 17	☽ ⊼ ♂
05 Sunday	
00 24	☽ ± ☉
01 10	☽ Q ♇
09 55	☽ ⊼ ♄
10 34	☉ ± ♇
13 17	☽ ∥ ♅
17 56	☽ □ ♀
18 34	☽ ⚼ ☉
19 13	☽ Q ♃
06 Monday	
00 11	☽ △ ☿
01 13	☽ ⊻ ♆
11 04	☽ □ ♇
12 41	☽ ⊥ ♆
12 51	☽ ☌ ♅
13 26	☽ # ♄
17 00	☽ # ♂
18 29	☽ ✶ ♃
07 Tuesday	
01 45	☽ △ ☉
02 41	☉ ∥ ☿
06 08	☽ ∠ ♆
07 13	☽ Q ♀
11 27	☽ ☍ ♂
13 32	☽ # ♆
20 04	☽ ☍ ♄
08 Wednesday	
00 20	☽ ∠ ♃
05 40	☿ St D
08 54	☽ ✶ ♀
10 19	☽ □ ☿
11 49	☽ ✶ ♆
18 35	☽ ⊥ ♃
21 11	☽ ∥ ☉
22 12	☽ △ ♇
09 Thursday	
00 04	☽ ⊻ ♅
05 14	☿ ⊻ ♀
05 48	☽ ∥ ☿
06 51	☽ ⊻ ♃
11 58	☽ Q ♆
12 11	☽ ⊥ ♅
17 39	☽ ∠ ♀
18 55	☽ □ ☉
22 34	♀ △ ♆
23 01	☽ Q ☿
10 Friday	
02 58	☽ ⊼ ♂
04 34	☽ ⚼ ♇
06 26	☽ ∠ ♅
08 25	☽ # ♇
08 54	☽ ⊼ ♄
12 47	☽ ∥ ♀
13 29	☽ ⊥ ♀
15 50	☽ ± ♂
21 10	☽ ± ♄
22 50	☽ ± ♇
23 58	☽ ✶ ☿
11 Saturday	
00 28	☽ □ ♆
02 43	☽ ⊻ ♀
10 43	☽ Q ☉
10 59	☽ ⊼ ♇
11 07	☽ ⚼ ♂
12 50	☽ ✶ ♅
15 32	☽ ⚼ ♄
15 47	☿ ⊼ ♆
20 32	☽ ☌ ♃
12 Sunday	
03 26	☽ ∥ ♃
07 19	☽ ∠ ☿
12 57	☽ Q ♅
13 01	☽ ✶ ☉
15 13	☽ ∥ ♃
18 55	☽ △ ♂
21 49	☽ △ ♄
13 Monday	
02 05	☽ ⊥ ☿
12 29	☽ △ ♆
14 31	☽ ⊻ ☿
19 45	☽ ☌ ♀
21 13	☽ ∠ ☉
22 41	☽ ☍ ♇
14 Tuesday	
00 26	☽ □ ♅
03 58	☽ ∥ ♀
08 38	☽ ⊻ ♃
09 44	☽ # ♇
15 41	☉ ⚼ ♇
16 05	☽ ⊥ ☉
17 32	☽ ⚼ ♆
20 17	☽ ⊥ ♃
15 Wednesday	
02 32	☽ ∥ ☿
04 28	☽ ⊻ ☉
08 15	☽ □ ♂
08 21	☽ □ ♄
09 11	♀ ☍ ♇
10 34	☽ ± ♆
10 35	♂ ☌ ♄
13 27	☉ ⚼ ♅
13 31	☽ ∠ ♃
21 46	☽ ⊼ ♆
16 Thursday	
01 55	♂ Q ♇
03 24	☽ ☌ ☿
06 48	♀ □ ♅
07 27	☽ ⊼ ♇
09 04	☽ △ ♅
09 15	☽ ⊻ ♀
10 42	☽ ∥ ☉
17 32	☽ ✶ ♃
17 54	☽ Q ♄
18 25	☽ ± ♇
19 16	☽ Q ♂
21 05	☽ ⊥ ♀
23 50	☽ # ♆
17 Friday	
05 38	☽ # ♂
07 40	☉ Q ♃
10 39	☽ ⚼ ♇
12 04	☉ ✶ ♄
12 12	☽ ⚼ ♅
14 33	☽ ∠ ♀
15 24	☽ Q ♃
15 39	☽ ✶ ♄
15 54	☽ ☌ ☉
17 55	☽ ✶ ♂
20 54	☽ # ♄
18 Saturday	
03 57	☽ ☍ ♆
04 10	☽ ± ♅
06 17	☿ ⊼ ♇
13 13	☽ △ ♇
13 51	☽ ⊻ ☿
14 42	☽ ⊼ ♅
18 17	☽ ∠ ♄
19 05	☽ ✶ ♀
21 37	☽ ∠ ♂
21 41	☽ ∥ ♅
23 26	☽ □ ♃
23 29	☿ △ ♅
19 Sunday	
01 20	☽ ⊥ ☿
10 02	☽ ⊥ ♄
13 59	☽ ⊥ ♂
17 30	☽ Q ♀
18 27	☽ ∠ ☿
18 57	♄ Q ♇
20 27	☽ ⊻ ♄
21 28	☽ # ♅
20 Monday	
00 17	☽ ⊻ ☉
00 49	☽ ⊻ ♂
07 57	☽ ⊼ ♆
11 20	☽ ⊥ ☉
17 00	☽ □ ♇
18 11	☽ ± ♆
18 25	☽ ☍ ♅
21 45	☉ ✶ ♂
22 07	☽ ∥ ♄
22 53	☽ ✶ ☿
23 10	♂ ⚼ ♃
21 Tuesday	
02 42	☽ □ ♀
03 32	☽ △ ♃
03 51	☽ ∠ ☉
09 30	☽ ⚼ ♆
16 06	♀ ⊻ ♃
16 53	♂ ∥ ♆
17 54	☽ ∥ ♆
17 57	☽ ∥ ♂
19 53	☽ # ☉
21 38	☽ Q ☿
23 38	☽ Q ♇
22 Wednesday	
00 00	☽ ☌ ♄
05 17	☽ ⚼ ♃
06 24	☽ ☌ ♂
07 13	☽ ✶ ☉
09 52	☉ # ♂
10 57	☽ △ ♆
16 21	☿ ± ♇
20 00	☽ ✶ ♇
20 42	☽ ± ♃
20 57	☿ Q ♄
20 58	☉ # ♆
21 20	☽ ⊼ ♅
22 56	☿ ✶ ♃
23 Thursday	
05 05	☽ Q ☉
05 56	☽ # ☿
06 58	☽ ⊼ ♃
07 30	☽ ± ♅
07 54	☽ □ ☿
09 34	☽ △ ♀
21 26	☽ ∠ ♇
22 45	☽ ⚼ ♅
24 Friday	
02 57	☽ ∥ ♇
03 15	☽ ⊻ ♄
06 56	☽ # ♀
11 45	☽ ⊻ ♂
12 32	☉ ☍ ♆
12 43	☽ ⊥ ♇
13 01	☽ ⚼ ♀
13 30	☽ ⊥ ♄
13 48	☽ □ ♆
13 54	☽ □ ☉
15 32	☉ ± ♅
19 14	☿ ⊻ ♀
22 27	☽ ⊥ ♂
22 55	☽ ⊻ ♇
23 38	♀ ⚼ ♆
25 Saturday	
00 12	☽ △ ♅
04 55	☽ ∠ ♄
05 31	☽ ± ♀
10 25	☽ ☍ ♃
14 30	☽ ∠ ♂
16 32	☽ ⊼ ♀
17 44	☽ △ ☿
20 23	☽ Q ♆
26 Sunday	
06 39	☽ ✶ ♄
06 53	♂ △ ♆
16 51	☽ ✶ ♆
17 21	☽ ✶ ♂
19 17	☿ Q ♂
20 50	☽ △ ☉
23 04	☽ ⚼ ☿
27 Monday	
02 06	☽ ☌ ♇
03 19	☽ □ ♅
03 20	☽ Q ♄
08 46	☽ # ♀
10 40	☽ ∥ ♇
14 10	☽ ⊼ ♃
14 57	☽ Q ♂
16 50	☽ ± ☿
18 34	☽ ∠ ♆
22 14	☿ ⚼ ♇
28 Tuesday	
00 02	☽ ☍ ♀
00 32	☽ ⚼ ☉
00 36	☽ ± ♃
04 46	☽ ⊼ ☿
07 08	☿ ⚼ ♅
08 31	♀ ⊥ ♃
10 04	☽ ⊥ ♆
10 16	☽ Q ♅
10 33	☽ □ ♄
16 18	☽ ⚼ ♃
17 17	☽ ± ☉
20 28	☽ ⊻ ♆
23 37	☽ □ ♂
29 Wednesday	
00 50	☽ # ☿
04 31	☽ ⊼ ☉
05 56	☽ ⊻ ♇
07 06	☽ ✶ ♅
10 54	☽ ∥ ♂
11 51	♀ # ♇
16 26	☽ ⊥ ♇
18 43	☽ △ ♃
21 05	☽ ∥ ♆
30 Thursday	
00 21	☉ △ ♇
01 25	☿ ✶ ♄
05 40	☿ Q ♃
08 19	☽ ∠ ♇
08 45	☽ ⊼ ♀
09 27	☽ ∠ ♅
12 08	☽ # ☉
15 34	☽ △ ♄
15 57	☉ ⊼ ♅
16 30	☽ ∥ ♄
17 48	☽ ☍ ☿
20 21	☽ ± ♀
31 Friday	
01 20	☽ ☌ ♆
01 30	☽ ⊥ ♅
04 05	☿ # ♂
07 25	☽ △ ♂
11 12	☽ ✶ ♇
12 18	☽ ⊻ ♅
12 51	☉ ∠ ♀
13 58	☽ ☍ ☉
13 59	☽ ⚼ ♀
18 49	☽ ⚼ ♄
21 11	☽ # ♅

SEPTEMBER 2012

Date	Sidereal time h m s	Sun ☉ ° ' "	Moon ☽ ° ' "	Moon ☽ 24.00 ° ' "	Mercury ☿ ° '	Venus ♀ ° '	Mars ♂ ° '	Jupiter ♃ ° '	Saturn ♄ ° '	Uranus ♅ ° '	Neptune ♆ ° '	Pluto ♇ ° '
01	10 44 12	09 ♍ 26 58	20 ♓ 34 16	27 ♓ 00 36	00 ♍ 46	24 ♋ 32	05 ♏ 43	14 ♊ 38	26 ♎ 18	07 ♈ 37	01 ♓ 32	07 ♑ 01
02	10 48 09	10 25 02	03 ♈ 22 31	09 ♈ 40 04	02 43	25 36	06 23	14 44	26 23	07 R 35	01 R 30	07 R 01
03	10 52 05	11 23 07	15 ♈ 53 24	22 ♈ 02 45	04 40	26 39	07 02	14 50	26 29	07 33	01 28	07 01
04	10 56 02	12 21 15	28 ♈ 08 24	04 ♉ 10 47	06 37	27 43	07 42	14 56	26 35	07 31	01 27	07 00
05	10 59 58	13 19 24	10 ♉ 10 18	16 ♉ 07 31	08 34	28 48	08 21	15 01	26 41	07 28	01 25	07 00
06	11 03 55	14 17 35	22 ♉ 02 59	27 ♉ 57 19	10 30	29 ♋ 52	09 01	15 07	26 47	07 26	01 24	06 59
07	11 07 52	15 15 49	03 ♊ 51 09	09 ♊ 45 10	12 26	00 ♌ 57	09 40	15 12	26 53	07 24	01 22	06 59
08	11 11 48	16 14 04	15 ♊ 40 03	21 ♊ 36 30	14 21	02 02	10 20	15 17	26 59	07 22	01 20	06 59
09	11 15 45	17 12 22	27 ♊ 35 11	03 ♋ 36 46	16 15	03 08	11 00	15 22	27 05	07 20	01 19	06 58
10	11 19 41	18 10 41	09 ♋ 41 54	15 ♋ 51 09	18 09	04 13	11 40	15 26	27 11	07 17	01 17	06 58
11	11 23 38	19 09 03	22 ♋ 05 05	28 ♋ 24 08	20 02	05 19	12 20	15 31	27 17	07 15	01 16	06 58
12	11 27 34	20 07 26	04 ♌ 48 41	11 ♌ 18 59	21 53	06 25	13 01	15 35	27 23	07 13	01 14	06 58
13	11 31 31	21 05 52	17 ♌ 55 11	24 ♌ 37 18	23 44	07 31	13 41	15 39	27 30	07 11	01 13	06 58
14	11 35 27	22 04 19	01 ♍ 25 11	08 ♍ 18 35	25 33	08 38	14 21	15 43	27 36	07 08	01 11	06 57
15	11 39 24	23 02 49	15 ♍ 17 06	22 ♍ 20 11	27 22	09 45	15 02	15 47	27 42	07 06	01 10	06 57
16	11 43 21	24 01 20	29 ♍ 27 13	06 ♎ 37 27	29 ♍ 09	10 51	15 42	15 51	27 49	07 04	01 08	06 57
17	11 47 17	24 59 53	13 ♎ 50 08	21 ♎ 04 30	00 ♎ 56	11 58	16 23	15 54	27 55	07 01	01 07	06 57
18	11 51 14	25 58 29	28 ♎ 19 44	05 ♏ 35 08	02 41	13 06	17 04	15 57	28 02	06 59	01 05	06 D 57
19	11 55 10	26 57 06	12 ♏ 50 01	20 ♏ 03 49	04 26	14 13	17 45	16 00	28 08	06 57	01 04	06 57
20	11 59 07	27 55 44	27 ♏ 16 03	04 ♐ 26 18	06 09	15 21	18 26	16 03	28 15	06 54	01 02	06 57
21	12 03 03	28 54 25	11 ♐ 34 17	18 ♐ 39 48	07 51	16 28	19 07	16 06	28 22	06 52	01 01	06 57
22	12 07 00	29 ♍ 53 07	25 ♐ 42 41	02 ♑ 42 51	09 32	17 36	19 48	16 08	28 28	06 50	01 00	06 58
23	12 10 56	00 ♎ 51 50	09 ♑ 40 14	16 ♑ 34 51	11 13	18 45	20 29	16 11	28 35	06 47	00 58	06 58
24	12 14 53	01 50 36	23 ♑ 26 38	00 ♒ 15 36	12 52	19 53	21 11	16 13	28 42	06 45	00 57	06 58
25	12 18 50	02 49 23	07 ♒ 01 43	13 ♒ 44 57	14 30	21 01	21 52	16 15	28 48	06 42	00 55	06 58
26	12 22 46	03 48 11	20 ♒ 25 14	27 ♒ 02 31	16 08	22 10	22 33	16 16	28 55	06 40	00 54	06 58
27	12 26 43	04 47 02	03 ♓ 36 42	10 ♓ 07 44	17 44	23 19	23 15	16 18	29 02	06 37	00 53	06 59
28	12 30 39	05 45 54	16 ♓ 35 30	22 ♓ 59 58	19 20	24 28	23 57	16 19	29 09	06 35	00 52	06 59
29	12 34 36	06 44 48	29 ♓ 21 04	05 ♈ 38 48	20 55	25 37	24 38	16 20	29 16	06 33	00 50	06 59
30	12 38 32	07 ♎ 43 45	11 ♈ 53 11	18 ♈ 04 16	22 ♎ 29	26 ♌ 46	25 ♏ 20	16 ♊ 21	29 ♎ 23	06 ♈ 30	00 ♓ 49	07 ♑ 00

Date	Moon True ☊ ° '	Moon Mean ☊ ° '	Moon ☽ Latitude ° '
01	29 ♏ 31	00 ♐ 02	04 N 41
02	29 R 20	29 ♏ 58	04 10
03	29 11	29 55	03 28
04	29 05	29 52	02 36
05	29 01	29 49	01 38
06	29 00	29 46	00 N 37
07	29 D 00	29 43	00 S 26
08	29 00	29 39	01 28
09	28 R 59	29 36	02 26
10	28 57	29 33	03 19
11	28 53	29 30	04 04
12	28 46	29 27	04 38
13	28 37	29 24	04 58
14	28 27	29 20	05 03
15	28 16	29 17	04 50
16	28 06	29 14	04 20
17	27 58	29 11	03 32
18	27 52	29 08	02 31
19	27 49	29 04	01 19
20	27 48	29 01	00 S 03
21	27 D 49	28 58	01 N 13
22	27 49	28 55	02 24
23	27 R 49	28 52	03 25
24	27 47	28 49	04 13
25	27 42	28 45	04 47
26	27 36	28 42	05 04
27	27 28	28 39	05 05
28	27 18	28 36	04 49
29	27 09	28 33	04 20
30	27 ♏ 01	28 ♏ 30	03 N 38

DECLINATIONS

Date	Sun ☉ ° '	Moon ☽ ° '	Mercury ☿ ° '	Venus ♀ ° '	Mars ♂ ° '	Jupiter ♃ ° '	Saturn ♄ ° '	Uranus ♅ ° '	Neptune ♆ ° '	Pluto ♇ ° '
01	08 N 02	00 N 34	12 N 47	19 N 16	13 S 56	21 N 46	07 S 59	02 N 20	11 S 31	19 S 34
02	07 40	05 10	12 08	19 10	14 09	21 46	08 02	02 19	11 31	19 34
03	07 18	09 27	11 27	19 02	14 23	21 47	08 04	02 18	11 32	19 35
04	06 55	13 14	10 44	18 55	14 36	21 48	08 06	02 17	11 32	19 35
05	06 33	16 25	10 01	18 47	14 50	21 48	08 08	02 17	11 33	19 35
06	06 11	18 52	09 16	18 38	15 03	21 49	08 11	02 16	11 34	19 35
07	05 48	20 30	08 31	18 29	15 16	21 49	08 13	02 15	11 34	19 36
08	05 26	21 13	07 45	18 19	15 30	21 50	08 15	02 14	11 35	19 36
09	05 03	20 59	06 59	18 09	15 43	21 50	08 18	02 13	11 35	19 36
10	04 40	19 47	06 12	17 58	15 56	21 51	08 20	02 12	11 36	19 36
11	04 18	17 37	05 24	17 47	16 09	21 51	08 22	02 11	11 36	19 36
12	03 55	14 34	04 37	17 35	16 22	21 51	08 25	02 10	11 37	19 37
13	03 32	10 43	03 49	17 23	16 34	21 52	08 27	02 10	11 37	19 37
14	03 09	06 15	03 02	17 11	16 47	21 52	08 29	02 09	11 38	19 37
15	02 46	01 N 20	02 14	16 57	17 00	21 53	08 32	02 08	11 39	19 37
16	02 22	03 S 45	01 27	16 44	17 12	21 53	08 34	02 07	11 39	19 38
17	01 59	08 43	00 N 39	16 30	17 24	21 53	08 37	02 06	11 40	19 38
18	01 36	13 13	00 S 08	16 15	17 37	21 53	08 39	02 05	11 40	19 38
19	01 13	16 57	00 55	16 00	17 49	21 54	08 41	02 04	11 41	19 38
20	00 49	19 36	01 41	15 45	18 01	21 54	08 44	02 03	11 41	19 38
21	00 26	20 58	02 27	15 29	18 13	21 54	08 46	02 02	11 42	19 39
22	00 N 03	20 58	03 13	15 12	18 24	21 54	08 49	02 01	11 42	19 39
23	00 S 21	19 41	03 59	14 55	18 36	21 55	08 51	02 00	11 43	19 39
24	00 44	17 14	04 44	14 38	18 47	21 55	08 54	01 59	11 43	19 39
25	01 07	13 53	05 28	14 20	18 59	21 55	08 56	01 58	11 44	19 39
26	01 31	09 52	06 12	14 02	19 10	21 55	08 59	01 57	11 44	19 40
27	01 54	05 27	06 56	13 43	19 21	21 55	09 01	01 56	11 45	19 40
28	02 17	00 S 51	07 39	13 24	19 32	21 55	09 04	01 55	11 45	19 40
29	02 41	03 N 43	08 21	13 05	19 43	21 55	09 06	01 54	11 45	19 40
30	03 S 04	08 N 03	09 S 03	12 N 45	19 S 54	21 N 55	09 S 09	01 N 54	11 S 46	19 S 40

ZODIAC SIGN ENTRIES

Date	h m	Planets
01	02 32	☿ ♍
02	05 37	☽ ♈
04	15 41	☽ ♉
06	14 48	♀ ♌
07	04 10	☽ ♊
09	16 49	☽ ♋
12	03 00	☽ ♌
14	09 30	☽ ♍
16	12 55	☽ ♎
16	23 22	☿ ♎
18	14 46	☽ ♏
20	16 34	☽ ♐
22	14 49	☉ ♎
22	19 20	☽ ♑
24	23 32	☽ ♒
27	05 23	☽ ♓
29	13 14	☽ ♈

LATITUDES

Date	Mercury ☿ ° '	Venus ♀ ° '	Mars ♂ ° '	Jupiter ♃ ° '	Saturn ♄ ° '	Uranus ♅ ° '	Neptune ♆ ° '	Pluto ♇ ° '
01	01 N 42	01 S 58	00 S 32	00 S 48	02 N 19	00 S 45	00 S 37	03 N 41
04	01 47	01 44	00 34	00 48	02 18	00 45	00 37	03 40
07	01 45	01 30	00 36	00 48	02 18	00 45	00 37	03 40
10	01 38	01 16	00 38	00 48	02 17	00 45	00 37	03 39
13	01 27	01 02	00 40	00 48	02 17	00 45	00 37	03 39
16	01 13	00 48	00 41	00 49	02 17	00 45	00 37	03 38
19	00 55	00 35	00 43	00 49	02 16	00 45	00 37	03 37
22	00 36	00 22	00 45	00 49	02 16	00 45	00 37	03 37
25	00 N 16	00 S 10	00 47	00 49	02 16	00 45	00 37	03 36
28	00 S 05	00 N 03	00 48	00 49	02 15	00 45	00 37	03 35
31	00 S 27	00 N 14	00 S 50	00 S 49	02 N 15	00 S 45	00 S 37	03 N 35

DATA

Julian Date	2456172
Delta T	+68 seconds
Ayanamsa	24° 02' 18"
Synetic vernal point	05° ♓ 04' 41"
True obliquity of ecliptic	23° 26' 11"

MOON'S PHASES, APSIDES AND POSITIONS ☽

Date	h m	Phase	Longitude ° '	Eclipse Indicator
08	13 15	☾	16 ♊ 17	
16	02 11	●	23 ♍ 37	
22	19 41	☽	00 ♑ 12	
30	03 19	○	07 ♈ 22	

Day	h m		
07	05 57	Apogee	
19	02 43	Perigee	
01	09 05	0N	
08	18 13	Max dec	21° N 15'
15	18 19	0S	
22	00 09	Max dec	21° S 08'
28	16 24	0N	

LONGITUDES

Date	Chiron ⚷ ° '	Ceres ⚳ ° '	Pallas ⚴ ° '	Juno ⚵ ° '	Vesta ⚶ ° '	Black Moon Lilith ⚸ ° '
01	07 ♓ 13	24 ♊ 00	08 ♈ 16	26 ♏ 35	18 ♊ 07	28 ♉ 50
11	06 ♓ 45	26 ♊ 41	06 ♈ 03	28 ♏ 45	20 ♊ 37	29 ♉ 56
21	06 ♓ 18	29 ♊ 01	03 ♈ 26	01 ♐ 12	22 ♊ 40	01 ♊ 02
31	05 ♓ 53	00 ♋ 58	00 ♈ 40	03 ♐ 52	24 ♊ 15	02 ♊ 09

All ephemeris data is given at 12.00 UT and the Moon's longitude is additionally given for 24.00 UT

ASPECTARIAN

h m	Aspects
01 Saturday	
00 58	☽ □ ♃
05 43	☿ ⊥ ♀
09 08	☽ Q ♇
11 29	☽ ± ♄
12 18	☽ ₽ ♂
14 05	☉ # ♄
20 02	☽ △ ♀
21 02	☽ ॥ ♅
21 16	☿ ☍ ♆
22 17	☿ ± ♅
22 44	☽ ⊼ ♄
02 Sunday	
06 02	☽ ± ♂
08 27	☽ ⊻ ♆
09 56	♀ ± ♆
10 31	☽ ⊼ ☿
10 47	☽ Q ♃
18 01	☽ ⊼ ♂
18 55	☽ □ ♇
19 50	☽ ⊥ ♆
19 59	☽ ☌ ♅
03 Monday	
00 03	☽ ± ☿
00 38	☽ ॥ ☉
02 34	☽ ⊼ ☉
03 57	☽ # ♄
07 43	♀ □ ♄
09 05	☿ # ♆
09 56	☽ ✶ ♃
11 05	♂ ✶ ♇
13 08	☽ ∠ ♆
14 43	☉ ∠ ♄
15 09	☽ ± ☉
20 43	☽ ₽ ☿
22 20	☽ ॥ ☿
04 Tuesday	
00 48	☽ # ♆
05 41	♂ ⊼ ♅
08 54	☽ ☍ ♄
10 19	☽ ₽ ☉
11 06	☽ □ ♀
15 34	☽ ∠ ♃
16 46	☿ △ ♇
18 32	☽ ✶ ♆
22 25	☽ # ♂
22 51	☿ ⊼ ♅
05 Wednesday	
05 38	☽ △ ♇
06 36	☽ ⊻ ♅
08 06	☿ ✶ ♂
08 08	☽ ☍ ♂
08 08	☽ △ ☿
09 40	☽ ⊥ ♃
18 31	☽ Q ♆
18 38	☽ ⊥ ♅
18 54	☽ △ ☉
21 51	☽ ⊻ ♃
06 Thursday	
02 41	☽ Q ♀
09 26	☽ ॥ ♀
11 53	☽ ₽ ♇
12 47	☽ ∠ ♅
16 12	♂ ± ♃
17 46	♀ ∠ ♃
21 07	☽ # ♇
21 41	☽ ⊼ ♄
07 Friday	
04 44	☿ ∠ ♄
05 30	☽ ✶ ♀
06 10	☽ ± ♇
06 57	☽ □ ♆
10 00	☽ ± ♄
10 15	☉ □ ♃
18 22	☽ ⊼ ♇
19 12	☽ ✶ ♅
20 52	♀ ⊼ ♆
21 03	☿ # ♄
08 Saturday	
00 33	☽ ⊼ ♂
04 27	☽ ₽ ♄
08 49	☽ □ ☿
11 13	☽ ☌ ♃
13 15	☽ □ ☉
13 26	☽ ± ♂
15 04	☽ ∠ ♀
19 27	☽ Q ♅
09 Sunday	
00 13	☿ □ ♃
08 39	☽ ₽ ♂
10 59	☽ △ ♄
11 00	☽ ⊥ ♀
19 25	☽ △ ♆
10 Monday	
00 08	☽ ⊻ ♀
03 44	☽ Q ☿
04 28	☽ Q ☉
06 38	☽ ☍ ♇
07 17	☽ □ ♅
12 44	☉ ☌ ☿
14 22	☽ # ♇
16 05	☽ △ ♂
20 21	♅ ⊥ ♆
23 16	☽ ⊻ ♃
11 Tuesday	
00 49	☽ ₽ ♆
05 54	☽ ✶ ☉
07 22	☽ ✶ ☿
10 20	☽ ॥ ♀
10 54	☽ ⊥ ♃
18 03	☽ ± ♆
21 09	☿ ∠ ♀
21 58	☽ □ ♄
23 36	☽ # ♂
12 Wednesday	
04 04	☽ ∠ ♃
05 11	☿ ⊥ ♄
05 20	☽ ⊼ ♆
12 38	☽ ∠ ☉
15 16	☽ ☌ ♀
15 59	☽ ⊼ ♇
16 27	☽ △ ♅
16 29	☽ ∠ ☿
18 58	♂ ± ♅
23 47	♀ ⊼ ♇
13 Thursday	
03 00	☽ ± ♇
03 54	☽ □ ♂
04 45	♀ △ ♅
06 29	☽ ⊥ ☉
06 46	☽ # ♆
07 35	☽ Q ♄
07 53	☽ ✶ ♃
11 36	☽ ⊥ ☿
18 10	☽ ⊻ ☉
19 15	☽ ₽ ♇
19 37	☽ ₽ ♅
22 59	☉ ⊥ ♄
14 Friday	
00 03	☽ ⊻ ☿
00 27	☽ # ♄
05 14	☽ ✶ ♄
05 18	☉ ॥ ☿
05 28	☽ Q ♃
11 31	☽ ± ♅
11 35	☽ ☍ ♆
13 43	☽ Q ♂
21 39	☽ △ ♇
21 57	☽ ⊼ ♅
15 Saturday	
01 39	☽ ⊻ ♀
04 37	☽ ॥ ☉
06 53	☽ ॥ ☿
07 33	☽ ∠ ♄
08 12	☽ ॥ ♅
09 59	♀ # ♂
11 33	☽ ✶ ♂
12 51	☽ ⊥ ♀
12 52	☽ □ ♃
15 21	☿ ॥ ♅
16 50	☿ ⊻ ♄
23 01	☽ ⊥ ♄
16 Sunday	
02 11	☽ ☌ ☉
02 36	☽ # ☿
04 18	☽ # ♅
05 26	☽ ∠ ♀
05 57	☽ # ☉
09 13	☽ ⊻ ♄
11 26	☽ ☌ ☿
14 13	☽ ∠ ♂
14 49	☽ ⊼ ♆
17 24	♂ ⊼ ♃
17 Monday	
00 33	☽ □ ♇
00 42	☽ ☍ ♅
00 50	☽ ± ♆
04 52	☉ ॥ ♅
05 59	☽ ⊥ ♂
08 39	☽ ✶ ♀
11 28	☽ ॥ ♄
14 24	☿ ⊼ ♆
15 27	☽ △ ♃
15 46	☽ ₽ ♆
16 26	☽ ⊻ ♂
22 05	☿ ∠ ♂
18 Tuesday	
03 19	☽ ॥ ♆
05 06	♇ St D
06 12	☽ Q ♀
06 25	☽ Q ♇
07 50	☽ ⊻ ☉
09 00	♀ ± ♇
11 30	☽ ☌ ♄
16 22	☽ ₽ ♃
16 33	☽ △ ♆
18 28	☽ ⊥ ☉
20 12	☽ ⊻ ☿
19 Wednesday	
02 16	☽ ✶ ♇
02 16	☽ ⊼ ♅
05 44	☽ # ♀
05 57	♅ □ ♇
07 18	☽ ± ♃
07 28	☽ ⊥ ☿
10 26	☽ ∠ ☉
12 11	☽ ± ♅
14 29	☽ □ ♀
17 17	☽ ⊼ ♃
18 12	☉ # ☿
19 16	☽ ॥ ♂
20 33	☽ ☌ ♂
20 Thursday	
00 26	☽ ∠ ☿
03 05	☽ ₽ ♅
03 09	☽ ∠ ♇
12 34	☽ ॥ ♇
13 11	☽ ✶ ☉
13 39	☽ ⊻ ♄
18 10	☽ ⊥ ♇
18 18	☽ □ ♆
20 50	☉ ⊻ ♄
22 23	☿ ☍ ♅
23 06	☿ # ♅
23 21	☿ □ ♇
23 46	☽ ⊥ ♄
21 Friday	
00 20	☿ ± ♆
03 45	♀ ✶ ♃
04 06	☽ △ ♅
04 14	☽ ⊻ ♇
04 53	☽ ✶ ☿
09 17	♀ Q ♄
10 48	☽ Q ☉
15 02	☽ ∠ ♄
19 41	☽ ☍ ♃
21 00	☽ △ ♀
22 Saturday	
00 35	☽ Q ♆
01 25	☽ ⊻ ♂
03 56	☽ Q ☿
12 10	☽ ⊥ ♂
16 45	☽ ✶ ♄
19 41	☽ □ ☉
21 02	☽ ✶ ♆
23 Sunday	
00 52	☽ ₽ ♀
04 24	☽ ∠ ♂
07 02	☽ □ ♅
07 19	☽ ☌ ♇
12 19	☽ ॥ ♇
13 35	☽ Q ♄
14 31	☉ ⊼ ♆
15 02	☽ □ ☿
17 48	☽ ± ♀
22 55	☽ ∠ ♆
23 00	☽ ॥ ♂
23 20	☽ ⊼ ♃
24 Monday	
05 12	☽ ⊼ ♀
07 49	☽ ✶ ♂
09 50	☽ ± ♃
14 17	☽ Q ♅
14 38	☽ ⊥ ♆
21 19	☽ □ ♄
25 Tuesday	
01 11	☽ ⊻ ♆
01 43	☽ ₽ ♃
03 57	☽ △ ☉
06 05	☽ Q ♂
06 43	♂ ₽ ♅
08 47	☽ # ♀
11 26	☽ ✶ ♅
11 54	☽ ⊻ ♇
15 34	♂ ∠ ♇
22 36	☽ ⊥ ♇
26 Wednesday	
01 16	☽ ॥ ♆
01 53	♀ ₽ ♅
03 13	☽ △ ☿
04 31	☽ △ ♃
07 57	♀ ₽ ♇
08 38	☿ ₽ ♆
08 51	☽ ₽ ☉
14 09	☿ △ ♃
14 15	☽ ∠ ♅
14 48	☽ ∠ ♇
15 27	☽ ☍ ♀
16 05	☽ □ ♂
16 56	☽ ॥ ♄
23 06	☿ ⊥ ♂
27 Thursday	
02 28	☽ ± ☉
03 33	☽ △ ♄
05 13	☽ ॥ ☿
06 33	☽ ⊥ ♅
07 00	☽ ☌ ♆
08 48	♀ □ ♂
10 11	☽ ₽ ☿
14 20	☽ ⊼ ☉
14 20	☉ # ♅
17 31	☽ ⊻ ♅
18 11	☽ ✶ ♇
28 Friday	
05 05	☽ ॥ ☉
05 05	☽ ± ☿
06 23	☽ # ♅
07 24	☽ ₽ ♄
11 30	☽ □ ♃
16 28	☽ Q ♇
17 51	☽ ⊼ ☿
29 Saturday	
00 23	☽ ± ♄
02 26	☽ ॥ ♅
02 34	☽ △ ♂
04 13	☽ ⊼ ♀
05 57	☽ # ☉
06 00	♂ ॥ ♇
07 15	☉ ☍ ♅
11 50	☽ ⊼ ♄
14 11	☉ ± ♆
14 49	☽ ⊻ ♆
16 44	☽ ± ♀
17 55	☉ □ ♇
21 31	☽ Q ♃
30 Sunday	
01 41	☽ ☌ ♅
02 16	☽ ⊥ ♆
02 35	☽ □ ♇
03 19	☽ ☍ ☉
08 50	☽ ₽ ♂
11 45	☽ ₽ ♀
15 35	☿ ॥ ♄
18 30	☽ # ♄
19 04	☽ # ☿
19 36	☽ ∠ ♆
20 40	☽ ✶ ♃

OCTOBER 2012

LONGITUDES

Date	Sidereal time h m s	Sun ☉ ° ' "	Moon ☽ ° ' "	Moon ☽ 24.00 ° ' "	Mercury ☿ ° '	Venus ♀ ° '	Mars ♂ ° '	Jupiter ♃ ° '	Saturn ♄ ° '	Uranus ♅ ° '	Neptune ♆ ° '	Pluto ♇ ° '
01	12 42 29	08 ♎ 42 43	24 ♈ 12 11	00 ♉ 17 07	24 ♎ 02	27 ♌ 55	26 ♏ 02	16 ♊ 22	29 ♎ 30	06 ♈ 28	00 ♓ 48	07 ♑ 00
02	12 46 25	09 41 43	06 ♉ 19 16	12 ♉ 18 56	25 34	29 ♌ 05	26 44	16 22	29 36	06 R 25	00 R 47	07 00
03	12 50 22	10 40 46	18 ♉ 16 26	24 ♉ 12 11	27 05	00 ♍ 15	27 26	16 23	29 43	06 23	00 45	07 01
04	12 54 19	11 39 51	00 ♊ 06 37	06 ♊ 00 14	28 ♎ 36	01 24	28 08	16 23	29 50	06 21	00 44	07 01
05	12 58 15	12 38 58	11 ♊ 53 32	17 ♊ 47 07	00 ♏ 05	02 34	28 51	16 R 23	29 ♎ 57	06 18	00 43	07 02
06	13 02 12	13 38 07	23 ♊ 41 36	29 ♊ 37 34	01 34	03 45	29 ♏ 33	16 23	00 ♏ 05	06 16	00 42	07 02
07	13 06 08	14 37 19	05 ♋ 35 42	11 ♋ 36 37	03 02	04 55	00 ♐ 15	16 22	00 12	06 13	00 41	07 03
08	13 10 05	15 36 33	17 ♋ 41 00	23 ♋ 49 29	04 29	06 05	00 58	16 21	00 19	06 11	00 40	07 04
09	13 14 01	16 35 49	00 ♌ 02 38	06 ♌ 21 03	05 55	07 16	01 40	16 20	00 26	06 09	00 39	07 04
10	13 17 58	17 35 07	12 ♌ 45 12	19 ♌ 15 31	07 21	08 26	02 23	16 19	00 33	06 06	00 38	07 05
11	13 21 54	18 34 28	25 ♌ 52 19	02 ♍ 35 47	08 45	09 37	03 06	16 18	00 40	06 04	00 37	07 06
12	13 25 51	19 33 51	09 ♍ 25 58	16 ♍ 22 45	10 09	10 48	03 49	16 17	00 47	06 02	00 36	07 06
13	13 29 48	20 33 17	23 ♍ 25 53	00 ♎ 34 56	11 31	11 59	04 31	16 15	00 54	05 59	00 35	07 07
14	13 33 44	21 32 44	07 ♎ 49 17	15 ♎ 08 12	12 53	13 10	05 14	16 13	01 02	05 57	00 34	07 08
15	13 37 41	22 32 14	22 ♎ 30 47	29 ♎ 56 06	14 13	14 21	05 57	16 11	01 09	05 55	00 33	07 09
16	13 41 37	23 31 45	07 ♏ 23 06	14 ♏ 50 46	15 33	15 33	06 41	16 08	01 16	05 52	00 32	07 10
17	13 45 34	24 31 19	22 ♏ 18 05	29 ♏ 44 05	16 51	16 44	07 24	16 06	01 23	05 50	00 32	07 11
18	13 49 30	25 30 55	07 ♐ 07 55	14 ♐ 28 50	18 08	17 56	08 07	16 03	01 30	05 48	00 31	07 11
19	13 53 27	26 30 32	21 ♐ 46 13	28 ♐ 59 35	19 23	19 08	08 50	16 00	01 38	05 46	00 30	07 12
20	13 57 23	27 30 11	06 ♑ 08 33	13 ♑ 12 54	20 38	20 19	09 34	15 57	01 45	05 43	00 29	07 13
21	14 01 20	28 29 52	20 ♑ 12 30	27 ♑ 07 18	21 50	21 31	10 17	15 54	01 52	05 41	00 29	07 14
22	14 05 17	29 ♎ 29 34	03 ♒ 57 21	10 ♒ 42 45	23 01	22 43	11 01	15 50	02 00	05 39	00 28	07 15
23	14 09 13	00 ♏ 29 19	17 ♒ 23 38	24 ♒ 00 11	24 10	23 55	11 45	15 47	02 07	05 37	00 27	07 16
24	14 13 10	01 29 04	00 ♓ 32 35	07 ♓ 01 04	25 17	25 07	12 28	15 43	02 14	05 35	00 27	07 18
25	14 17 06	02 28 52	13 ♓ 25 48	19 ♓ 47 02	26 22	26 20	13 12	15 39	02 21	05 33	00 26	07 19
26	14 21 03	03 28 41	26 ♓ 04 55	02 ♈ 19 41	27 24	27 32	13 56	15 35	02 29	05 31	00 26	07 20
27	14 24 59	04 28 32	08 ♈ 31 30	14 ♈ 40 33	28 23	28 44	14 40	15 30	02 36	05 29	00 25	07 21
28	14 28 56	05 28 25	20 ♈ 47 02	26 ♈ 51 07	29 ♏ 19	29 ♍ 57	15 24	15 26	02 43	05 27	00 25	07 22
29	14 32 52	06 28 20	02 ♉ 53 00	08 ♉ 52 54	00 ♐ 12	01 ♎ 09	16 08	15 21	02 50	05 25	00 24	07 23
30	14 36 49	07 28 16	14 ♉ 51 01	20 ♉ 47 36	01 01	02 22	16 52	15 16	02 58	05 23	00 24	07 25
31	14 40 46	08 ♏ 28 15	26 ♉ 42 55	02 ♊ 37 16	01 ♐ 46	03 ♎ 35	17 ♐ 36	15 ♊ 11	03 ♏ 05	05 ♈ 21	00 ♓ 23	07 ♑ 26

Date	Moon True ☊ ° '	Moon Mean ☊ ° '	Moon ☽ Latitude ° '
01	26 ♏ 54	28 ♏ 26	02 N 47
02	26 R 50	28 23	01 49
03	26 48	28 20	00 N 46
04	26 D 47	28 17	00 S 18
05	26 48	28 14	01 21
06	26 50	28 10	02 21
07	26 51	28 07	03 15
08	26 R 52	28 04	04 02
09	26 51	28 01	04 38
10	26 48	27 58	05 02
11	26 44	27 55	05 11
12	26 38	27 51	05 04
13	26 32	27 48	04 38
14	26 27	27 45	03 54
15	26 22	27 42	02 54
16	26 19	27 39	01 42
17	26 18	27 35	00 S 22
18	26 D 19	27 32	00 N 59
19	26 20	27 29	02 15
20	26 21	27 26	03 22
21	26 22	27 23	04 14
22	26 R 23	27 20	04 51
23	26 22	27 16	05 11
24	26 19	27 13	05 14
25	26 16	27 10	05 01
26	26 12	27 07	04 33
27	26 09	27 04	03 53
28	26 05	27 01	03 03
29	26 03	26 57	02 05
30	26 02	26 54	01 01
31	26 ♏ 01	26 ♏ 51	00 S 04

DECLINATIONS

Date	Sun ☉ ° '	Moon ☽ ° '	Mercury ☿ ° '	Venus ♀ ° '	Mars ♂ ° '	Jupiter ♃ ° '	Saturn ♄ ° '	Uranus ♅ ° '	Neptune ♆ ° '	Pluto ♇ ° '
01	03 S 27	11 N 58	09 S 44	12 N 25	20 S 04	21 N 55	09 S 11	01 N 53	11 S 46	19 S 41
02	03 50	15 20	10 24	12 04	20 14	21 55	09 14	01 52	11 47	19 41
03	04 14	18 00	11 04	11 43	20 25	21 55	09 16	01 51	11 47	19 41
04	04 37	19 53	11 44	11 22	20 35	21 55	09 19	01 50	11 48	19 41
05	05 00	20 52	12 22	11 00	20 44	21 55	09 21	01 49	11 48	19 41
06	05 23	20 56	13 00	10 38	20 54	21 55	09 24	01 48	11 48	19 42
07	05 46	20 04	13 37	10 16	21 04	21 55	09 26	01 47	11 49	19 42
08	06 09	18 16	14 13	09 53	21 13	21 55	09 29	01 46	11 49	19 42
09	06 31	15 36	14 48	09 30	21 22	21 55	09 31	01 45	11 49	19 42
10	06 54	12 09	15 23	09 07	21 31	21 55	09 34	01 44	11 50	19 42
11	07 17	08 00	15 57	08 43	21 40	21 55	09 36	01 43	11 50	19 42
12	07 39	03 N 20	16 30	08 19	21 49	21 54	09 39	01 42	11 50	19 43
13	08 02	01 S 39	17 02	07 55	21 57	21 54	09 41	01 41	11 51	19 43
14	08 24	06 41	17 33	07 30	22 05	21 54	09 44	01 41	11 51	19 43
15	08 46	11 27	18 03	07 05	22 13	21 54	09 46	01 40	11 51	19 43
16	09 08	15 35	18 32	06 40	22 21	21 53	09 49	01 39	11 52	19 43
17	09 30	18 42	19 01	06 15	22 29	21 53	09 52	01 38	11 52	19 43
18	09 52	20 32	19 28	05 49	22 36	21 53	09 54	01 37	11 52	19 44
19	10 14	20 56	19 54	05 24	22 44	21 52	09 57	01 36	11 52	19 44
20	10 35	19 56	20 19	04 58	22 51	21 52	09 59	01 35	11 53	19 44
21	10 56	17 43	20 42	04 32	22 58	21 52	10 02	01 34	11 53	19 44
22	11 18	14 33	21 05	04 05	23 04	21 51	10 04	01 34	11 53	19 44
23	11 39	10 41	21 26	03 39	23 11	21 51	10 07	01 33	11 53	19 44
24	11 59	06 23	21 46	03 12	23 17	21 51	10 09	01 32	11 54	19 44
25	12 20	01 S 53	22 05	02 45	23 23	21 50	10 12	01 31	11 54	19 45
26	12 40	02 N 37	22 22	02 18	23 28	21 50	10 14	01 30	11 54	19 45
27	13 01	06 57	22 37	01 51	23 34	21 49	10 16	01 30	11 54	19 45
28	13 21	10 56	22 51	01 24	23 39	21 49	10 19	01 29	11 54	19 45
29	13 41	14 25	23 04	00 57	23 44	21 48	10 21	01 28	11 54	19 45
30	14 00	17 16	23 14	00 29	23 49	21 48	10 24	01 27	11 55	19 45
31	14 S 20	19 N 21	23 S 23	00 N 02	23 S 54	21 N 47	10 S 26	01 N 27	11 S 55	19 S 45

ZODIAC SIGN ENTRIES

Date	h m	Planets
01	23 26	☽ ♉
03	06 59	♀ ♍
04	11 47	☽ ♊
05	10 35	☿ ♏
05	20 34	♄ ♏
07	00 45	☽ ♋
07	03 21	♂ ♐
09	11 55	☽ ♌
11	19 23	☽ ♍
13	23 02	☽ ♎
16	00 06	☽ ♏
18	00 26	☽ ♐
20	01 41	☽ ♑
22	05 02	☽ ♒
23	00 14	☉ ♏
24	11 00	☽ ♓
26	19 31	☽ ♈
28	13 04	♀ ♎
29	06 15	☽ ♉
29	06 18	☿ ♐
31	18 40	☽ ♊

LATITUDES

Date	Mercury ☿ ° '	Venus ♀ ° '	Mars ♂ ° '	Jupiter ♃ ° '	Saturn ♄ ° '	Uranus ♅ ° '	Neptune ♆ ° '	Pluto ♇ ° '
01	00 S 27	00 N 14	00 S 50	00 S 49	02 N 15	00 S 45	00 S 37	03 N 35
04	00 48	00 25	00 51	00 49	02 15	00 45	00 37	03 34
07	01 10	00 36	00 53	00 50	02 15	00 45	00 37	03 33
10	01 30	00 46	00 54	00 50	02 14	00 45	00 37	03 33
13	01 50	00 55	00 56	00 50	02 14	00 45	00 37	03 32
16	02 08	01 03	00 57	00 50	02 14	00 45	00 37	03 32
19	02 25	01 11	00 58	00 50	02 14	00 45	00 37	03 31
22	02 39	01 18	00 59	00 50	02 14	00 45	00 37	03 30
25	02 49	01 25	01 00	00 50	02 14	00 45	00 37	03 30
28	02 55	01 30	01 01	00 50	02 14	00 45	00 37	03 29
31	02 S 56	01 N 35	01 S 02	00 S 50	02 N 14	00 S 45	00 S 37	03 N 29

DATA

Julian Date	2456202
Delta T	+68 seconds
Ayanamsa	24° 02' 21"
Synetic vernal point	05° ♓ 04' 38"
True obliquity of ecliptic	23° 26' 11"

MOON'S PHASES, APSIDES AND POSITIONS ☽

Date	h m	Phase	Longitude ° '	Eclipse Indicator
08	07 33	☾	15 ♋ 26	
15	12 02	●	22 ♎ 32	
22	03 32	☽	29 ♑ 09	
29	19 49	○	06 ♉ 48	

Day	h m		
05	00 35	Apogee	
17	00 54	Perigee	
06	01 44	Max dec	21° N 02'
13	04 10	0S	
19	06 43	Max dec	20° S 58'
25	21 59	0N	

LONGITUDES

Date	Chiron ⚷ ° '	Ceres ⚳ ° '	Pallas ⚴ ° '	Juno ⚵ ° '	Vesta ⚶ ° '	Black Moon Lilith ⚸ ° '
01	05 ♓ 53	00 ♋ 58	00 ♈ 40	03 ♐ 52	24 ♊ 15	02 ♊ 09
11	05 ♓ 32	02 ♋ 27	28 ♓ 00	06 ♐ 45	25 ♊ 14	03 ♊ 15
21	05 ♓ 16	03 ♋ 23	25 ♓ 41	09 ♐ 49	25 ♊ 34	04 ♊ 22
31	05 ♓ 05	03 ♋ 44	23 ♓ 56	13 ♐ 01	25 ♊ 11	05 ♊ 28

All ephemeris data is given at 12.00 UT and the Moon's longitude is additionally given for 24.00 UT

ASPECTARIAN

h m	Aspects
01 Monday	
03 20	☽ ± ♂
10 43	☽ # ♆
11 36	☽ ☍ ☿
14 42	☽ ∥ ♀
15 50	☽ ⊼ ♂
20 06	☽ △ ♀
21 16	♀ Q ♃
22 32	☽ ☍ ♄
02 Tuesday	
01 00	☽ ⚹ ♆
02 09	☽ ∠ ♃
03 14	☿ Q ♇
12 12	☽ ⊻ ♅
13 22	☽ △ ♇
19 21	☽ ⊼ ☉
20 07	☽ ⊥ ♃
03 Wednesday	
00 05	♀ ⚹ ♄
00 11	☽ ⊥ ♅
00 54	☽ Q ♆
07 48	♀ # ♆
08 11	☽ ⊻ ♃
08 30	☽ ± ☉
14 48	♀ ± ♅
18 16	☽ ∠ ♅
19 34	☽ bQ ♇
22 24	☿ ⊻ ♂
22 26	♀ ☍ ♆
04 Thursday	
03 27	☿ # ♀
04 21	☽ bQ ☉
07 44	☽ ☍ ♂
08 28	☽ ⊼ ☿
08 57	☽ # ♇
11 27	☽ ⊼ ♄
13 17	☽ □ ♆
13 17	♃ St R
13 52	☽ ± ♇
14 31	☿ ∥ ♆
14 56	☽ □ ♀
22 27	☽ ± ☿
23 47	☽ ± ♄
05 Friday	
00 39	☽ ⚹ ♅
02 05	☽ ⊼ ♇
05 41	☽ # ♂
09 44	☿ ☌ ♄
13 41	☽ △ ☉
18 18	☽ bQ ♄
19 27	☽ bQ ☿
21 08	☽ ☌ ♃
22 06	☿ △ ♆
06 Saturday	
01 01	☽ Q ♅
07 36	☽ Q ♀
08 53	☿ bQ ♃
13 28	☽ # ♂
07 Sunday	
00 35	☽ ⊼ ♂
01 02	☽ △ ♄
02 09	☽ △ ♆
06 09	☽ △ ☿
09 30	♂ ⊻ ♄
10 29	☽ ⚹ ♀
13 15	☽ □ ♅
13 24	☽ ± ♂
14 55	☽ ☍ ♇
18 05	☽ # ♇
08 Monday	
02 10	♂ □ ♆
07 33	☽ □ ☉
08 02	☽ bQ ♆
08 24	☽ bQ ♂
09 23	☽ ⊻ ♃
13 21	☉ bQ ♆
13 55	♀ ⊼ ♅
15 22	♂ ⊥ ♇
19 22	☽ ∠ ♀
21 08	☽ ⊥ ♃
09 Tuesday	
01 37	☽ ± ♆
05 52	☉ △ ♃
08 04	♀ △ ♇
10 45	♀ # ♄
12 45	☽ □ ♄
13 09	☽ ⊼ ♆
14 29	☽ ∠ ♃
14 34	☽ ⊥ ♀
15 18	☽ △ ♂
15 38	☿ ⊼ ♅
17 06	☽ # ☿
18 31	☉ ∠ ♂
21 25	☽ Q ☉
23 34	☽ △ ♅
10 Wednesday	
00 36	☽ □ ☿
01 22	☽ ⊼ ♇
03 07	☽ ⊻ ♀
07 31	☿ ⚹ ♇
12 37	☽ ± ♇
13 58	☽ # ♆
18 36	☽ ⚹ ♃
21 40	☽ ⚹ ☉
22 48	☽ Q ♄
11 Thursday	
02 38	♄ △ ♆
03 13	☽ # ♄
03 19	☽ bQ ♅
05 10	☽ bQ ♇
07 43	☽ ∥ ♀
13 46	☽ Q ☿
15 36	☽ # ☉
16 21	☽ Q ♃
19 29	☽ ± ♅
20 29	☽ ☍ ♆
20 39	☽ ⚹ ♄
12 Friday	
01 36	☽ □ ♂
02 48	☽ ∠ ☉
06 04	☽ ⊼ ♅
07 56	☽ △ ♇
13 23	☽ ⚹ ☿
14 13	☿ ± ♃
14 36	☽ ☌ ♀
19 42	☽ ⊥ ☉
20 00	☽ ∥ ♅
23 05	☽ ∠ ♄
23 48	☽ □ ♃
13 Saturday	
03 35	♂ # ♃
06 45	☽ ⊻ ☉
08 25	☉ # ♀
10 23	☽ Q ♂
12 12	☽ # ♅
14 31	☽ ⊥ ♄
17 45	☽ ∠ ☿
20 00	☿ ± ♅
23 59	☽ ⊼ ♆
14 Sunday	
00 39	☽ ⊻ ♄
07 31	☽ ⚹ ♂
08 55	☽ ☍ ♅
09 56	☽ ± ♆
10 17	☽ ⊥ ☿
10 52	☽ □ ♇
15 38	☽ # ♀
21 03	☽ ∥ ☉
21 09	☽ ⊻ ☿
21 34	☽ ⊻ ♀
15 Monday	
00 42	☽ bQ ♆
01 44	☽ △ ♃
03 11	☽ ∥ ♄
08 12	☽ ⊥ ♀
09 21	☽ ∠ ♂
10 33	♂ △ ♅
12 02	☽ ☌ ☉
14 09	☽ ∥ ♆
16 16	☽ Q ♇
16 56	♃ bQ ♄
16 Tuesday	
00 02	☽ ∠ ♀
00 39	☽ ⊥ ♂
00 59	☽ △ ♆
01 58	☽ bQ ♃
02 04	☽ ☌ ♄
09 34	☽ ⊼ ♅
10 48	☽ ⊻ ♂
11 38	☽ ⚹ ♇
12 31	☿ ⚹ ♀
16 25	☽ ± ♃
19 12	☽ ± ♅
22 36	☿ ⊼ ♃
23 34	♀ □ ♃
17 Wednesday	
02 03	☽ ⊼ ♃
02 16	☽ ⚹ ♀
02 23	☽ ☌ ☿
04 09	♀ ∠ ♄
04 30	♂ ⊻ ♇
09 39	☽ bQ ♅
11 41	♂ ⊥ ♄
11 48	☽ ∠ ♇
15 50	☽ ⊻ ☉
15 51	☽ ∥ ☿
23 14	☽ ∥ ♇
23 18	☽ Q ♀
18 Thursday	
01 16	☽ □ ♆
02 14	☽ ⊥ ☉
02 21	☽ ⊥ ♇
02 48	☽ ⊻ ♄
04 02	☉ Q ♇
09 50	☽ △ ♅
12 06	☽ ⊻ ♇
12 37	☽ ⊥ ♄
13 41	☽ ☌ ♂
14 42	☉ ∥ ♄
17 55	☽ ∠ ☉
19 Friday	
02 32	☽ ☍ ♃
02 40	☿ ∥ ♇
03 27	☽ ∠ ♄
06 37	☽ Q ♆
07 15	☽ □ ♀
07 42	☽ ⊻ ☿
18 34	☽ ⊥ ☿
20 26	☽ ⚹ ☉
20 Saturday	
02 31	☽ ⚹ ♆
04 33	☽ ⚹ ♄
07 22	☽ ∥ ☿
11 03	☽ ∠ ☿
11 18	☽ □ ♅
13 50	☽ ☌ ♇
13 53	☿ bQ ♅
14 53	☽ ∥ ♇
18 06	☽ ⊻ ♂
18 07	☽ Q ☉
21 Sunday	
01 01	☽ Q ♄
03 53	☽ ∠ ♆
04 37	☽ ⊼ ♃
04 53	☽ ⊥ ♂
14 29	☽ △ ♀
14 55	☽ ± ♃
15 05	☽ ⚹ ☿
18 00	☽ Q ♅
19 23	☽ ⊥ ♆
20 15	☿ ∠ ♇
21 18	☽ ∠ ♂
22 Monday	
03 32	☽ □ ☉
05 52	☽ ⊻ ♆
06 32	☽ bQ ♃
08 30	☽ □ ♄
14 03	☽ Q ☿
15 00	☽ ⚹ ♅
17 51	☽ ⊻ ♇
19 19	☽ bQ ♀
23 Tuesday	
01 16	☽ ⚹ ♂
04 35	☽ ⊥ ♇
04 51	☽ ∥ ♆
06 47	☽ ∥ ☉
09 06	☽ △ ♃
11 15	☉ △ ♆
13 03	☽ ± ♀
15 16	☽ ∥ ♄
17 49	☽ ∠ ♅
18 36	☉ bQ ♃
20 52	☽ ∠ ♇
24 Wednesday	
00 12	☽ Q ♂
01 02	☽ ⊼ ♀
01 27	☽ □ ☿
05 20	☉ ∥ ♆
10 14	☽ ⊥ ♅
11 49	☽ ☌ ♆
13 53	☽ △ ☉
15 09	☽ △ ♄
17 38	☿ # ♃
21 18	☽ ⊻ ♅
25 Thursday	
00 32	☽ ⚹ ♇
06 52	☽ # ♀
08 32	☉ ☌ ♄
11 33	☽ □ ♂
12 38	♀ ⊥ ♄
13 56	☽ # ♅
16 09	☽ □ ♃
17 10	☿ ⚹ ♀
19 28	☽ bQ ♄
20 17	☽ bQ ☉
23 07	☽ Q ♇
26 Friday	
05 54	☉ ⊥ ♀
06 01	☽ ∥ ♅
10 28	☽ ∥ ♀
12 46	☽ ± ♄
14 44	☽ △ ☿
14 54	☽ ± ☉
15 04	☽ ☍ ♀
20 20	☽ ⊻ ♆
27 Saturday	
00 24	☽ ⊼ ♄
02 20	☽ Q ♃
03 28	☽ ⊼ ☉
06 06	☽ ☌ ♅
07 55	☽ ⊥ ♆
09 43	☽ □ ♇
22 16	☽ bQ ☿
28 Sunday	
00 44	☽ △ ♂
01 27	☽ ∠ ♆
01 32	☽ ⚹ ♃
07 44	♀ ∥ ♅
08 04	☽ # ♄
11 21	☉ ⊼ ♅
12 54	♂ ☍ ♃
17 25	☽ ± ☿
18 20	☽ # ♆
21 11	♀ ⊼ ♆
29 Monday	
05 56	☽ # ☉
06 15	☽ ⊼ ☿
06 59	☽ ∠ ♃
07 04	☽ ⚹ ♆
08 10	☽ ⊼ ♀
08 17	☽ bQ ♂
11 55	☽ ☍ ♄
17 02	☽ ⊻ ♅
17 40	☿ □ ♆
19 49	☽ ☍ ☉
21 01	☽ △ ♇
21 30	☽ ± ♀
30 Tuesday	
00 51	☽ ⊥ ♃
03 28	☽ ± ♂
05 02	☽ ⊥ ♅
07 04	☽ Q ♆
10 30	☉ ⚹ ♇
12 50	☽ ⊻ ♃
16 20	☽ ⊼ ♂
17 39	☽ bQ ♀
23 08	☽ ∠ ♅
31 Wednesday	
00 29	☿ ⊥ ♇
01 01	♀ ⊻ ♄
03 18	☽ bQ ♇
18 08	☽ # ♇
19 28	☽ □ ♆
21 36	☽ ± ♇
22 55	☽ ☍ ☿

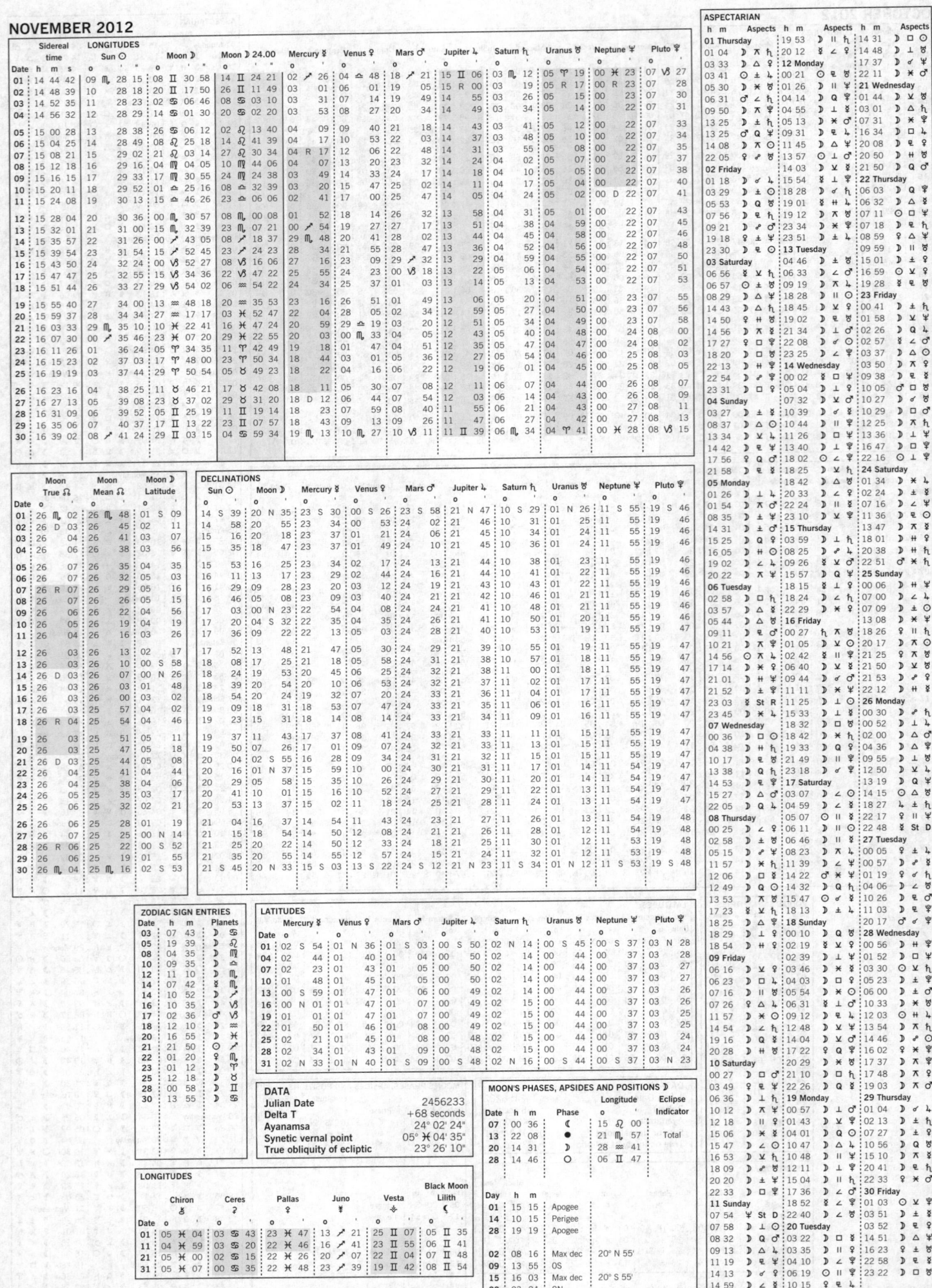

NOVEMBER 2012

LONGITUDES

Date	Sidereal time h m s	Sun ☉ ° ' "	Moon ☽ ° ' "	Moon ☽ 24.00 ° ' "	Mercury ☿ ° '	Venus ♀ ° '	Mars ♂ ° '	Jupiter ♃ ° '	Saturn ♄ ° '	Uranus ♅ ° '	Neptune ♆ ° '	Pluto ♇ ° '
01	14 44 42	09 ♏ 28 15	08 ♊ 30 58	14 ♊ 24 21	02 ♐ 26	04 ♎ 48	18 ♐ 21	15 ♊ 06	03 ♏ 12	05 ♈ 19	00 ♓ 23	07 ♑ 27
02	14 48 39	10 28 18	20 ♊ 17 50	26 ♊ 11 49	03 01	06 01	19 05	15 R 00	03 19	05 R 17	00 R 23	07 28
03	14 52 35	11 28 23	02 ♋ 06 46	08 ♋ 03 10	03 31	07 14	19 49	14 55	03 26	05 15	00 23	07 30
04	14 56 32	12 28 29	14 ♋ 01 30	20 ♋ 02 20	03 53	08 27	20 34	14 49	03 34	05 14	00 22	07 31
05	15 00 28	13 28 38	26 ♋ 06 12	02 ♌ 13 40	04 09	09 40	21 18	14 43	03 41	05 12	00 22	07 33
06	15 04 25	14 28 49	08 ♌ 25 18	14 ♌ 41 39	04 17	10 53	22 03	14 37	03 48	05 10	00 22	07 34
07	15 08 21	15 29 02	21 ♌ 03 14	27 ♌ 30 34	04 R 17	12 06	22 48	14 31	03 55	05 08	00 22	07 35
08	15 12 18	16 29 16	04 ♍ 04 05	10 ♍ 44 06	04 07	13 20	23 32	14 24	04 02	05 07	00 22	07 37
09	15 16 15	17 29 33	17 ♍ 30 55	24 ♍ 24 38	03 49	14 33	24 17	14 18	04 10	05 05	00 22	07 38
10	15 20 11	18 29 52	01 ♎ 25 16	08 ♎ 32 39	03 20	15 47	25 02	14 11	04 17	05 04	00 22	07 40
11	15 24 08	19 30 13	15 ♎ 46 26	23 ♎ 06 06	02 41	17 00	25 47	14 05	04 24	05 02	00 D 22	07 41
12	15 28 04	20 30 36	00 ♏ 30 57	08 ♏ 00 08	01 52	18 14	26 32	13 58	04 31	05 01	00 22	07 43
13	15 32 01	21 31 00	15 ♏ 32 39	23 ♏ 07 21	00 ♐ 54	19 27	27 17	13 51	04 38	04 59	00 22	07 45
14	15 35 57	22 31 26	00 ♐ 43 05	08 ♐ 18 37	29 ♏ 48	20 41	28 02	13 44	04 45	04 58	00 22	07 46
15	15 39 54	23 31 54	15 ♐ 52 45	23 ♐ 24 23	28 34	21 55	28 47	13 36	04 52	04 56	00 22	07 48
16	15 43 50	24 32 24	00 ♑ 52 27	08 ♑ 16 06	27 16	23 09	29 ♐ 32	13 29	04 59	04 55	00 22	07 50
17	15 47 47	25 32 55	15 ♑ 34 36	22 ♑ 47 22	25 55	24 23	00 ♑ 18	13 22	05 06	04 54	00 22	07 51
18	15 51 44	26 33 27	29 ♑ 54 02	06 ♒ 54 22	24 34	25 37	01 03	13 14	05 13	04 53	00 22	07 53
19	15 55 40	27 34 00	13 ♒ 48 18	20 ♒ 35 53	23 16	26 51	01 49	13 06	05 20	04 51	00 23	07 55
20	15 59 37	28 34 34	27 ♒ 17 17	03 ♓ 52 47	22 04	28 05	02 34	12 59	05 27	04 50	00 23	07 56
21	16 03 33	29 ♏ 35 10	10 ♓ 22 41	16 ♓ 47 24	20 59	29 ♎ 19	03 20	12 51	05 34	04 49	00 23	07 58
22	16 07 30	00 ♐ 35 46	23 ♓ 07 20	29 ♓ 22 55	20 03	00 ♏ 33	04 05	12 43	05 40	04 48	00 24	08 00
23	16 11 26	01 36 24	05 ♈ 34 35	11 ♈ 42 49	19 18	01 47	04 51	12 35	05 47	04 47	00 24	08 02
24	16 15 23	02 37 03	17 ♈ 48 00	23 ♈ 50 34	18 44	03 01	05 36	12 27	05 54	04 46	00 25	08 03
25	16 19 19	03 37 44	29 ♈ 50 54	05 ♉ 49 23	18 22	04 16	06 22	12 19	06 01	04 45	00 25	08 05
26	16 23 16	04 38 25	11 ♉ 46 21	17 ♉ 42 08	18 11	05 30	07 08	12 11	06 07	04 44	00 26	08 07
27	16 27 13	05 39 08	23 ♉ 37 02	29 ♉ 31 20	18 D 12	06 44	07 54	12 03	06 14	04 43	00 26	08 09
28	16 31 09	06 39 52	05 ♊ 25 19	11 ♊ 19 14	18 23	07 59	08 40	11 55	06 21	04 43	00 27	08 11
29	16 35 06	07 40 37	17 ♊ 13 22	23 ♊ 07 57	18 43	09 13	09 26	11 47	06 27	04 42	00 27	08 13
30	16 39 02	08 ♐ 41 24	29 ♊ 03 15	04 ♋ 59 34	19 ♏ 13	10 ♏ 27	10 ♑ 11	11 ♊ 39	06 ♏ 34	04 ♈ 41	00 ♓ 28	08 ♑ 15

Date	Moon True ☊ ° '	Moon Mean ☊ ° '	Moon ☽ Latitude ° '
01	26 ♏ 02	26 ♏ 48	01 S 09
02	26 D 03	26 45	02 11
03	26 04	26 41	03 07
04	26 06	26 38	03 56
05	26 07	26 35	04 35
06	26 07	26 32	05 03
07	26 R 07	26 29	05 16
08	26 07	26 26	05 15
09	26 06	26 22	04 56
10	26 05	26 19	04 19
11	26 04	26 16	03 26
12	26 03	26 13	02 17
13	26 03	26 10	00 S 58
14	26 D 03	26 07	00 N 26
15	26 03	26 03	01 48
16	26 03	26 00	03 02
17	26 03	25 57	04 02
18	26 R 04	25 54	04 46
19	26 03	25 51	05 11
20	26 03	25 47	05 18
21	26 D 03	25 44	05 08
22	26 04	25 41	04 44
23	26 04	25 38	04 06
24	26 05	25 35	03 17
25	26 06	25 32	02 21
26	26 06	25 28	01 19
27	26 07	25 25	00 N 14
28	26 R 06	25 22	00 S 52
29	26 06	25 19	01 55
30	26 ♏ 04	25 ♏ 16	02 S 53

DECLINATIONS

Date	Sun ☉ ° '	Moon ☽ ° '	Mercury ☿ ° '	Venus ♀ ° '	Mars ♂ ° '	Jupiter ♃ ° '	Saturn ♄ ° '	Uranus ♅ ° '	Neptune ♆ ° '	Pluto ♇ ° '
01	14 S 39	20 N 35	23 S 30	00 S 26	23 S 58	21 N 47	10 S 29	01 N 26	11 S 55	19 S 46
02	14 58	20 55	23 34	00 53	24 02	21 46	10 31	01 25	11 55	19 46
03	15 16	20 18	23 37	01 21	24 06	21 45	10 34	01 24	11 55	19 46
04	15 35	18 47	23 37	01 49	24 10	21 45	10 36	01 24	11 55	19 46
05	15 53	16 25	23 34	02 17	24 13	21 44	10 38	01 23	11 55	19 46
06	16 11	13 17	23 29	02 44	24 16	21 44	10 41	01 22	11 55	19 46
07	16 29	09 28	23 20	03 12	24 19	21 43	10 43	01 22	11 55	19 46
08	16 46	05 08	23 09	03 40	24 21	21 42	10 46	01 21	11 55	19 46
09	17 03	00 N 23	22 54	04 08	24 24	21 41	10 48	01 21	11 55	19 46
10	17 20	04 S 32	22 35	04 35	24 26	21 41	10 50	01 20	11 55	19 46
11	17 36	09 22	22 13	05 03	24 28	21 40	10 53	01 19	11 55	19 47
12	17 52	13 48	21 47	05 30	24 29	21 39	10 55	01 19	11 55	19 47
13	18 08	17 25	21 18	05 58	24 31	21 38	10 57	01 18	11 55	19 47
14	18 24	19 53	20 45	06 25	24 32	21 38	11 00	01 18	11 55	19 47
15	18 39	20 54	20 10	06 53	24 32	21 37	11 02	01 17	11 55	19 47
16	18 54	20 24	19 32	07 20	24 33	21 36	11 04	01 17	11 55	19 47
17	19 09	18 31	18 53	07 47	24 33	21 35	11 06	01 16	11 55	19 47
18	19 23	15 31	18 14	08 14	24 33	21 34	11 09	01 16	11 55	19 47
19	19 37	11 43	17 37	08 41	24 33	21 33	11 11	01 15	11 55	19 47
20	19 50	07 26	17 01	09 07	24 32	21 33	11 13	01 15	11 55	19 47
21	20 04	02 S 55	16 28	09 34	24 31	21 32	11 15	01 15	11 55	19 47
22	20 16	01 N 37	15 59	10 00	24 30	21 31	11 17	01 14	11 54	19 47
23	20 29	05 58	15 35	10 26	24 29	21 30	11 20	01 14	11 54	19 47
24	20 41	10 01	15 16	10 52	24 27	21 29	11 22	01 13	11 54	19 47
25	20 53	13 37	15 02	11 18	24 25	21 28	11 24	01 13	11 54	19 47
26	21 04	16 37	14 54	11 43	24 23	21 27	11 26	01 13	11 54	19 48
27	21 15	18 54	14 50	12 08	24 21	21 26	11 28	01 12	11 54	19 48
28	21 25	20 22	14 50	12 33	24 18	21 25	11 30	01 12	11 53	19 48
29	21 35	20 55	14 55	12 57	24 15	21 24	11 32	01 12	11 53	19 48
30	21 S 45	20 N 33	15 S 03	13 S 22	24 S 12	21 N 23	11 S 34	01 N 12	11 S 53	19 S 48

ZODIAC SIGN ENTRIES

Date	h	m	Planets
03	07	43	☽ ♋
05	19	39	☽ ♌
08	04	35	☽ ♍
10	09	35	☽ ♎
12	11	10	☽ ♏
14	07	42	☿ ♏
14	10	52	☽ ♐
16	10	35	☽ ♑
17	02	36	♂ ♑
18	12	10	☽ ♒
20	16	55	☽ ♓
21	21	50	☉ ♐
22	01	20	♀ ♏
23	01	12	☽ ♈
25	12	18	☽ ♉
28	00	58	☽ ♊
30	13	55	☽ ♋

LATITUDES

Date	Mercury ☿ ° '	Venus ♀ ° '	Mars ♂ ° '	Jupiter ♃ ° '	Saturn ♄ ° '	Uranus ♅ ° '	Neptune ♆ ° '	Pluto ♇ ° '
01	02 S 54	01 N 36	01 S 03	00 S 50	02 N 14	00 S 45	00 S 37	03 N 28
04	02 44	01 40	01 04	00 50	02 14	00 44	00 37	03 28
07	02 23	01 43	01 05	00 50	02 14	00 44	00 37	03 27
10	01 48	01 45	01 05	00 50	02 14	00 44	00 37	03 27
13	00 S 59	01 47	01 06	00 49	02 14	00 44	00 37	03 26
16	00 N 01	01 47	01 07	00 49	02 15	00 44	00 37	03 26
19	01 01	01 47	01 07	00 49	02 15	00 44	00 37	03 25
22	01 50	01 46	01 08	00 49	02 15	00 44	00 37	03 25
25	02 21	01 45	01 08	00 48	02 15	00 44	00 37	03 24
28	02 34	01 43	01 09	00 48	02 15	00 44	00 37	03 24
31	02 N 33	01 N 40	01 S 09	00 S 48	02 N 16	00 S 44	00 S 37	03 N 23

DATA

Julian Date	2456233
Delta T	+68 seconds
Ayanamsa	24° 02' 24"
Synetic vernal point	05° ♓ 04' 35"
True obliquity of ecliptic	23° 26' 10"

MOON'S PHASES, APSIDES AND POSITIONS ☽

Date	h	m	Phase	Longitude ° '	Eclipse Indicator
07	00	36	☾	15 ♌ 00	
13	22	08	●	21 ♏ 57	Total
20	14	31	☽	28 ♒ 41	
28	14	46	○	06 ♊ 47	

Day	h	m		
01	15	15	Apogee	
14	10	15	Perigee	
28	19	19	Apogee	
02	08	16	Max dec	20° N 55'
09	13	55	0S	
15	16	03	Max dec	20° S 55'
22	03	24	0N	
29	14	31	Max dec	20° N 56'

LONGITUDES

Date	Chiron ⚷ ° '	Ceres ⚳ ° '	Pallas ⚴ ° '	Juno ⚵ ° '	Vesta ⚶ ° '	Black Moon Lilith ⚸ ° '
01	05 ♓ 04	03 ♋ 43	23 ♓ 47	13 ♐ 21	25 ♊ 07	05 ♊ 35
11	04 ♓ 59	03 ♋ 20	22 ♓ 46	16 ♐ 41	23 ♊ 55	06 ♊ 41
21	05 ♓ 00	02 ♋ 15	22 ♓ 26	20 ♐ 07	22 ♊ 04	07 ♊ 48
31	05 ♓ 07	00 ♋ 35	22 ♓ 48	23 ♐ 39	19 ♊ 42	08 ♊ 54

All ephemeris data is given at 12.00 UT and the Moon's longitude is additionally given for 24.00 UT

ASPECTARIAN

h m	Aspects
01 Thursday	
01 04	☽ ⊼ ♄
03 33	☽ △ ♀
03 41	☉ ± ♃
05 30	☽ ⚹ ♅
06 31	♂ ∠ ♄
09 50	☽ ⊼ ♇
13 25	☽ ± ♄
13 25	♂ Q ♆
14 08	☽ ⊼ ☉
22 05	♀ ☍ ♅
02 Friday	
01 18	☽ ☌ ♃
03 29	☽ ± ☉
05 53	☽ Q ♅
07 56	☽ ⚼ ♄
09 21	☽ ☍ ♂
19 18	♀ ± ♆
23 30	☽ ⚼ ☉
03 Saturday	
06 56	☿ ⊻ ♄
06 57	☉ ± ♅
08 29	☽ △ ♆
14 43	☽ △ ♄
14 50	♀ # ♅
14 56	☽ ⊼ ☿
17 27	♀ □ ♇
18 20	☽ □ ♅
22 13	☽ # ♇
22 54	☽ ☍ ♇
23 31	☽ □ ♀
04 Sunday	
03 27	☽ ± ☿
08 37	☽ △ ☉
13 34	☽ ⊻ ♃
14 42	☽ ⚼ ♆
17 56	♀ Q ♂
21 58	☽ ⚼ ☿
05 Monday	
01 26	☽ ⊥ ♃
01 54	☽ ⊼ ♂
08 35	☽ ± ♆
14 31	☽ ± ♂
15 25	☽ Q ♀
16 05	☽ # ☉
19 02	☽ ∠ ♃
20 22	☽ ⊼ ♆
06 Tuesday	
02 58	☽ □ ♄
03 57	☽ △ ☿
05 44	☽ △ ♅
09 11	☽ ⚼ ♂
10 21	☽ ⊼ ♇
14 56	☉ ⊼ ♃
17 14	☽ ⚹ ♀
21 01	☽ # ♆
21 52	☽ ± ♇
23 03	☿ St R
23 45	☽ ⚹ ♃
07 Wednesday	
00 36	☽ □ ☉
04 38	☽ # ♄
10 17	☽ ⚼ ♅
13 38	☽ Q ♄
14 53	☽ ⚼ ♇
15 27	☽ △ ♂
22 05	☽ Q ♃
08 Thursday	
00 25	☽ ∠ ♀
02 58	☽ ± ♅
05 15	☽ ☍ ♆
11 57	☽ ⚹ ♄
12 06	☽ □ ☿
12 49	☽ Q ☉
13 53	☽ ⊼ ♅
17 23	☿ ⊻ ♄
18 25	☽ △ ♇
18 29	☽ ⊥ ♀
18 54	☽ # ♀
09 Friday	
06 16	☽ ⊻ ♀
06 23	☽ □ ♃
07 16	☽ ∥ ♅
07 26	♀ △ ♃
11 57	☽ ⚹ ☉
14 54	☽ ∠ ♄
19 16	☽ Q ☿
20 28	☽ # ♅
10 Saturday	
00 27	☽ □ ♂
03 49	♀ ⚼ ♆
06 36	☽ ⊥ ♄
10 12	☽ ⊼ ♆
12 18	☽ ∥ ♀
15 06	☽ ⚹ ☿
15 47	☽ ∠ ☉
16 53	☽ ⊻ ♄
18 09	☽ ☍ ♅
20 20	☽ ± ♆
22 33	☽ □ ♇
11 Sunday	
07 54	♆ St D
07 58	☽ ⊥ ☉
08 32	☽ Q ♂
09 13	☽ △ ♃
11 19	☽ ⚼ ♆
14 13	☽ ☌ ♀
14 59	☽ ∠ ☿
18 35	☽ ⊻ ☉
19 53	☽ ∥ ♄
20 12	☿ ∠ ♀
12 Monday	
00 21	☉ ⚼ ♅
01 26	☽ ∥ ♆
04 14	☽ Q ♇
04 55	☽ ⊥ ☿
05 13	☽ ⚹ ♂
09 31	☽ ⚼ ♃
11 45	☽ △ ♆
13 57	☉ ⊥ ♂
14 03	☽ ⊻ ☿
15 54	☿ ⊥ ♇
18 28	☽ ☌ ♄
19 01	☿ # ♃
19 12	☽ ⊼ ♅
23 34	☽ ⚹ ♇
23 51	☽ ± ♃
13 Tuesday	
04 46	☽ ± ♅
06 33	☽ ∠ ♂
09 19	☽ ⊼ ♃
18 28	☽ ∥ ☉
18 45	☽ ⊻ ♀
19 02	☽ ⚼ ♅
21 34	☽ ⊥ ♂
22 08	☽ ☌ ☉
23 25	☽ ∠ ♇
14 Wednesday	
00 02	☿ □ ♆
05 04	☽ ⊥ ♀
07 32	☽ ⊻ ♂
10 39	☽ ☌ ☿
10 44	☽ ∥ ♇
11 26	☽ □ ♆
13 40	☽ ⊥ ♇
18 02	☉ ∠ ♇
18 25	☽ ⊻ ♄
18 42	☽ △ ♅
20 33	☽ ∠ ♀
22 24	☽ ∥ ☿
23 10	☽ ⊻ ♇
15 Thursday	
03 59	☽ ⊥ ♄
08 25	☽ ☍ ♃
09 26	☿ ⊻ ♂
15 57	☽ Q ♆
18 15	☿ ⊥ ♀
18 24	☽ ∠ ♄
22 29	☽ ⚹ ♀
16 Friday	
00 27	♄ ⊼ ♅
01 05	☽ ⊻ ☉
02 42	☿ ∥ ♇
06 40	☽ ⊻ ☿
09 44	☽ ☌ ♂
11 11	☽ ⚹ ♆
11 25	☽ ⊥ ☉
15 33	☽ ⊥ ☿
18 32	☽ □ ♅
18 42	☽ ⚹ ♄
19 33	☽ Q ♀
21 49	☽ ∥ ♇
23 18	☽ ☌ ♇
17 Saturday	
03 07	☽ ∠ ☉
04 59	☽ ∠ ☿
05 07	☉ ∥ ☿
06 11	☽ ∥ ☉
06 46	☽ ∥ ☿
08 23	☽ ⊼ ♃
11 39	☽ ∠ ♆
14 22	♂ ⚹ ♆
14 32	☽ Q ♄
15 47	☉ ☌ ☿
18 13	☽ ± ♃
18 Sunday	
00 10	☽ Q ♅
02 19	☿ ⊻ ♀
02 39	☽ ⊥ ♆
03 46	☽ ⚹ ☿
04 03	☽ □ ♀
05 54	☽ ⚹ ☉
06 31	☿ ⊥ ♂
09 12	☽ ⚼ ♃
12 48	☽ ⊻ ♆
14 04	☽ ⊻ ♂
17 22	♀ Q ♇
20 29	☽ ⚹ ♅
21 10	☽ □ ♄
22 26	☽ Q ☿
19 Monday	
00 57	☽ ⊥ ♂
01 43	☽ ⊻ ♇
04 01	☽ Q ☉
10 47	☽ △ ♃
10 48	☽ ∥ ♆
12 11	☽ ⊥ ♇
15 04	☽ ∥ ♄
17 36	☽ ∠ ♂
18 52	☿ ∠ ♇
22 40	☽ ∠ ♅
20 Tuesday	
03 22	☽ □ ☿
03 35	☽ ∥ ♀
04 10	☽ ∠ ♇
06 19	☉ ∥ ♇
10 15	♀ ⚼ ♃
13 35	☽ △ ♀
14 31	☽ □ ☉
14 48	☽ ⊥ ♅
17 37	☽ ☌ ♆
22 11	☽ ⚹ ♂
21 Wednesday	
01 44	☽ ⊻ ♅
03 01	☽ △ ♄
07 31	☽ ⚹ ♇
16 34	☽ □ ♃
20 08	☽ ⚼ ♀
20 50	☽ # ♅
21 50	☽ Q ♂
22 Thursday	
06 03	☽ Q ♇
06 32	☽ △ ☿
07 11	☉ □ ♆
07 18	☽ ⚼ ♄
08 59	♀ △ ♆
09 59	☽ ∥ ♅
15 01	☽ ± ♀
16 59	☉ ⊻ ♀
19 28	☿ ⚼ ♅
23 Friday	
00 41	☽ ± ♄
01 58	☽ ⊻ ♆
02 26	☽ Q ♃
02 57	☿ ∠ ♂
03 37	☽ △ ☉
03 50	☽ ⊼ ♀
09 38	☽ ⚼ ☿
10 05	♂ □ ♅
10 27	☽ ☌ ♅
10 29	☽ □ ♂
12 25	☽ ⊼ ♄
13 36	☽ ⊥ ♆
16 47	☽ □ ♇
22 16	☉ ⊥ ♇
24 Saturday	
01 34	☽ ⚹ ♃
02 24	☽ ± ☿
07 16	☽ ∠ ♆
11 36	☽ ⚼ ☉
13 47	☽ ⊼ ☿
18 01	☽ # ♀
20 38	☽ # ♄
22 51	♂ ⚹ ♄
25 Sunday	
00 06	☽ # ♆
07 00	☽ ∠ ♃
07 09	☽ ± ☉
13 08	☽ ⚹ ♆
18 26	♀ ∥ ♄
20 17	☽ ⊼ ☉
21 25	♀ ⊼ ♅
21 50	☽ ⊻ ♅
21 53	☽ ☍ ♀
22 12	☽ # ☿
26 Monday	
00 30	☽ ☍ ♄
00 52	☽ ⊥ ♃
02 00	☽ △ ♂
04 36	☽ △ ♇
09 55	☽ ⊥ ♅
12 50	☽ ⊻ ♃
13 19	☽ Q ♆
14 15	☉ △ ♅
18 27	♃ ± ♄
22 17	♀ ∥ ♆
22 48	☿ St D
27 Tuesday	
00 05	♀ ± ♃
00 57	☽ ☍ ☿
01 19	♀ ☌ ♄
04 06	☽ ∠ ♅
10 26	☽ ⚼ ♂
11 03	☽ ⚼ ♇
20 17	♂ ☌ ♇
28 Wednesday	
00 56	☽ # ♇
01 52	☽ □ ♆
03 30	☉ ⊻ ♄
05 23	☽ ± ♇
06 00	☽ ± ♂
10 33	☽ ⚹ ♅
12 03	☉ # ♃
13 54	☽ ⊼ ♄
14 46	☽ ☍ ☉
16 02	♀ ⚹ ♇
17 37	☽ ⊼ ♇
17 48	☽ ⊼ ♀
19 03	☽ ⊼ ♂
29 Thursday	
01 04	☽ ☌ ♃
02 13	☽ ± ♄
07 27	☽ ± ♀
10 56	☽ Q ♅
15 10	☽ ⊼ ☿
20 41	☽ ⚼ ♄
22 33	♀ ⚹ ♂
30 Friday	
01 03	☉ ⊻ ♇
03 51	☽ ± ☿
03 52	☽ ⚼ ♀
14 51	☽ △ ♆
16 23	♀ ± ♅
22 58	☽ ⚼ ☿
23 22	☽ □ ♅

DECEMBER 2012

LONGITUDES

Date	Sidereal time h m s	Sun ☉ ° ' "	Moon ☽ ° ' "	Moon ☽ 24.00 ° ' "	Mercury ☿ ° '	Venus ♀ ° '	Mars ♂ ° '	Jupiter ♃ ° '	Saturn ♄ ° '	Uranus ♅ ° '	Neptune ♆ ° '	Pluto ♇ ° '
01	16 42 59	09 ♐ 42 12	10 ♋ 57 09	16 ♋ 56 20	19 ♏ 50	11 ♏ 42	10 ♑ 58	11 ♊ 31	06 ♏ 40	04 ♈ 41	00 ♓ 29	08 ♑ 16
02	16 46 55	10 43 01	22 ♋ 57 24	29 ♋ 00 44	20 34	12 56	11 44	11 R 23	06 47	04 R 40	00 29	08 18
03	16 50 52	11 43 51	05 ♌ 06 41	11 ♌ 15 38	21 24	14 11	12 30	11 14	06 53	04 39	00 30	08 20
04	16 54 48	12 44 43	17 ♌ 27 59	23 ♌ 44 09	22 20	15 26	13 16	11 06	07 00	04 39	00 31	08 22
05	16 58 45	13 45 36	00 ♍ 04 35	06 ♍ 29 40	23 21	16 40	14 02	10 58	07 06	04 39	00 32	08 24
06	17 02 42	14 46 31	12 ♍ 59 51	19 ♍ 35 31	24 25	17 55	14 48	10 50	07 12	04 38	00 32	08 26
07	17 06 38	15 47 26	26 ♍ 16 59	03 ♎ 04 34	25 34	19 10	15 35	10 42	07 18	04 38	00 33	08 28
08	17 10 35	16 48 23	09 ♎ 58 27	16 ♎ 58 44	26 45	20 24	16 21	10 34	07 25	04 37	00 34	08 30
09	17 14 31	17 49 21	24 ♎ 05 23	01 ♏ 18 13	27 59	21 39	17 07	10 26	07 31	04 37	00 35	08 32
10	17 18 28	18 50 21	08 ♏ 36 55	16 ♏ 00 57	29 ♏ 16	22 54	17 54	10 18	07 37	04 37	00 36	08 34
11	17 22 24	19 51 21	23 ♏ 29 37	01 ♐ 02 02	00 ♐ 34	24 09	18 40	10 10	07 43	04 37	00 37	08 36
12	17 26 21	20 52 23	08 ♐ 37 10	16 ♐ 13 53	01 54	25 23	19 27	10 02	07 49	04 37	00 38	08 38
13	17 30 17	21 53 25	23 ♐ 50 54	01 ♑ 26 58	03 16	26 38	20 13	09 54	07 55	04 37	00 39	08 40
14	17 34 14	22 54 28	09 ♑ 00 48	16 ♑ 31 11	04 39	27 53	21 00	09 46	08 01	04 D 37	00 40	08 42
15	17 38 11	23 55 32	23 ♑ 57 03	01 ♒ 17 27	06 03	29 ♏ 08	21 47	09 38	08 06	04 37	00 41	08 44
16	17 42 07	24 56 37	08 ♒ 31 39	15 ♒ 39 05	07 28	00 ♐ 23	22 33	09 31	08 12	04 37	00 43	08 47
17	17 46 04	25 57 41	22 ♒ 39 26	29 ♒ 32 31	08 54	01 38	23 20	09 23	08 18	04 37	00 44	08 49
18	17 50 00	26 58 46	06 ♓ 18 22	12 ♓ 57 10	10 21	02 53	24 07	09 16	08 23	04 37	00 45	08 51
19	17 53 57	27 59 52	19 ♓ 29 12	25 ♓ 54 55	11 48	04 08	24 54	09 08	08 29	04 38	00 46	08 53
20	17 57 53	29 ♐ 00 57	02 ♈ 14 48	08 ♈ 29 24	13 16	05 23	25 40	09 01	08 34	04 38	00 48	08 55
21	18 01 50	00 ♑ 02 03	14 ♈ 39 18	20 ♈ 45 07	14 45	06 38	26 27	08 54	08 40	04 38	00 49	08 57
22	18 05 46	01 03 09	26 ♈ 47 27	02 ♉ 46 55	16 14	07 53	27 14	08 47	08 45	04 39	00 50	08 59
23	18 09 43	02 04 16	08 ♉ 44 06	14 ♉ 39 33	17 43	09 08	28 01	08 40	08 50	04 39	00 52	09 01
24	18 13 40	03 05 22	20 ♉ 33 49	26 ♉ 27 22	19 13	10 23	28 48	08 33	08 56	04 40	00 53	09 03
25	18 17 36	04 06 29	02 ♊ 20 39	08 ♊ 14 06	20 44	11 38	29 ♑ 35	08 26	09 01	04 41	00 54	09 06
26	18 21 33	05 07 36	14 ♊ 08 03	20 ♊ 02 52	22 14	12 53	00 ♒ 22	08 20	09 06	04 41	00 56	09 08
27	18 25 29	06 08 43	25 ♊ 58 48	01 ♋ 56 06	23 45	14 08	01 09	08 13	09 11	04 42	00 57	09 10
28	18 29 26	07 09 50	07 ♋ 55 01	13 ♋ 55 43	25 16	15 23	01 56	08 07	09 16	04 43	00 59	09 12
29	18 33 22	08 10 58	19 ♋ 58 21	26 ♋ 03 07	26 48	16 38	02 43	08 01	09 21	04 43	01 00	09 14
30	18 37 19	09 12 06	02 ♌ 10 07	08 ♌ 19 30	28 20	17 53	03 30	07 55	09 26	04 44	01 02	09 16
31	18 41 15	10 ♑ 13 14	14 ♌ 31 25	20 ♌ 46 02	29 ♐ 52	19 ♐ 08	04 ♒ 17	07 ♊ 49	09 ♏ 30	04 ♈ 45	01 ♓ 03	09 ♑ 18

Date	Moon True ☊ ° '	Moon Mean ☊ ° '	Moon ☽ Latitude ° '
01	26 ♏ 02	25 ♏ 13	03 S 44
02	26 R 00	25 09	04 25
03	25 58	25 06	04 56
04	25 56	25 03	05 13
05	25 54	25 00	05 15
06	25 54	24 57	05 02
07	25 D 54	24 53	04 33
08	25 55	24 50	03 48
09	25 56	24 47	02 47
10	25 58	24 44	01 34
11	25 59	24 41	00 S 14
12	25 R 58	24 38	01 N 09
13	25 56	24 34	02 28
14	25 53	24 31	03 35
15	25 49	24 28	04 27
16	25 45	24 25	05 00
17	25 41	24 22	05 13
18	25 38	24 19	05 08
19	25 36	24 15	04 47
20	25 D 36	24 12	04 12
21	25 37	24 09	03 26
22	25 38	24 06	02 32
23	25 40	24 03	01 31
24	25 41	23 59	00 N 28
25	25 R 41	23 56	00 S 36
26	25 39	23 53	01 39
27	25 35	23 50	02 37
28	25 29	23 47	03 29
29	25 22	23 44	04 12
30	25 13	23 40	04 44
31	25 ♏ 04	23 ♏ 37	05 S 03

DECLINATIONS

Date	Sun ☉ ° '	Moon ☽ ° '	Mercury ☿ ° '	Venus ♀ ° '	Mars ♂ ° '	Jupiter ♃ ° '	Saturn ♄ ° '	Uranus ♅ ° '	Neptune ♆ ° '	Pluto ♇ ° '
01	21 S 54	19 N 16	15 S 14	13 S 46	24 S 08	21 N 22	11 S 36	01 N 11	11 S 53	19 S 48
02	22 03	17 07	15 28	14 09	24 04	21 21	11 38	01 11	11 52	19 48
03	22 11	14 12	15 44	14 32	24 00	21 20	11 40	01 11	11 52	19 48
04	22 19	10 37	16 03	14 55	23 56	21 19	11 42	01 11	11 52	19 48
05	22 27	06 31	16 23	15 18	23 51	21 18	11 44	01 11	11 52	19 48
06	22 34	02 N 01	16 44	15 40	23 46	21 17	11 46	01 11	11 51	19 48
07	22 41	02 S 42	17 06	16 02	23 41	21 16	11 48	01 11	11 51	19 48
08	22 47	07 26	17 28	16 23	23 35	21 15	11 50	01 11	11 51	19 48
09	22 53	11 56	17 51	16 44	23 30	21 14	11 52	01 10	11 50	19 48
10	22 58	15 52	18 15	17 05	23 24	21 13	11 53	01 10	11 50	19 48
11	23 03	18 52	18 38	17 25	23 17	21 12	11 55	01 10	11 50	19 48
12	23 07	20 36	19 02	17 45	23 11	21 11	11 57	01 10	11 49	19 48
13	23 11	20 50	19 25	18 04	23 04	21 10	11 59	01 10	11 49	19 48
14	23 15	19 33	19 48	18 23	22 57	21 09	12 00	01 11	11 48	19 48
15	23 18	16 56	20 10	18 41	22 50	21 08	12 02	01 11	11 48	19 48
16	23 20	13 18	20 32	18 59	22 42	21 07	12 04	01 11	11 48	19 48
17	23 22	09 01	20 53	19 16	22 34	21 06	12 06	01 11	11 47	19 48
18	23 24	04 S 25	21 13	19 33	22 26	21 06	12 07	01 11	11 47	19 48
19	23 25	00 N 14	21 33	19 49	22 18	21 05	12 09	01 11	11 46	19 48
20	23 26	04 45	21 52	20 04	22 09	21 04	12 10	01 11	11 46	19 48
21	23 26	08 56	22 10	20 19	22 00	21 03	12 12	01 11	11 45	19 48
22	23 26	12 41	22 27	20 34	21 51	21 02	12 14	01 12	11 45	19 48
23	23 25	15 51	22 43	20 48	21 42	21 01	12 15	01 12	11 44	19 48
24	23 24	18 20	22 58	21 01	21 32	21 00	12 17	01 12	11 44	19 48
25	23 22	20 02	23 12	21 14	21 22	20 59	12 18	01 12	11 43	19 48
26	23 20	20 51	23 25	21 26	21 12	20 58	12 19	01 13	11 43	19 48
27	23 18	20 45	23 37	21 37	21 02	20 58	12 21	01 13	11 42	19 48
28	23 15	19 43	23 47	21 48	20 51	20 57	12 22	01 13	11 42	19 48
29	23 11	17 48	23 57	21 58	20 40	20 56	12 24	01 14	11 41	19 47
30	23 07	15 04	24 05	22 08	20 29	20 55	12 25	01 14	11 41	19 47
31	23 S 03	11 N 38	24 S 12	22 S 16	20 S 18	20 N 55	12 S 26	01 N 14	11 S 40	19 S 47

ZODIAC SIGN ENTRIES

Date	h	m	Planets
03	01	57	☽ ♌
05	11	51	☽ ♍
07	18	35	☽ ♎
09	21	51	☽ ♏
11	01	40	☿ ♐
11	22	22	☽ ♐
13	21	42	☽ ♑
15	21	53	☽ ♒
16	04	38	♀ ♐
18	00	48	☽ ♓
20	07	43	☽ ♈
21	11	12	☉ ♑
22	18	25	☽ ♉
25	07	13	☽ ♊
26	00	49	♂ ♒
27	20	06	☽ ♋
30	07	45	☽ ♌
31	14	03	☿ ♑

LATITUDES

Date	Mercury ☿ ° '	Venus ♀ ° '	Mars ♂ ° '	Jupiter ♃ ° '	Saturn ♄ ° '	Uranus ♅ ° '	Neptune ♆ ° '	Pluto ♇ ° '
01	02 N 33	01 N 40	01 S 09	00 S 48	02 N 16	00 S 44	00 S 37	03 N 23
04	02 23	01 36	01 09	00 47	02 16	00 43	00 37	03 23
07	02 07	01 32	01 10	00 47	02 16	00 43	00 37	03 22
10	01 47	01 27	01 10	00 47	02 17	00 43	00 37	03 22
13	01 25	01 22	01 10	00 46	02 17	00 43	00 37	03 22
16	01 02	01 17	01 10	00 45	02 18	00 43	00 37	03 21
19	00 39	01 11	01 10	00 45	02 18	00 43	00 37	03 21
22	00 N 17	01 04	01 10	00 44	02 19	00 43	00 37	03 21
25	00 S 05	00 57	01 10	00 44	02 19	00 43	00 37	03 20
28	00 26	00 50	01 09	00 43	02 20	00 42	00 37	03 20
31	00 S 46	00 N 43	01 S 09	00 S 43	02 N 20	00 S 42	00 S 37	03 N 20

DATA

Julian Date	2456263
Delta T	+68 seconds
Ayanamsa	24° 02' 29"
Synetic vernal point	05° ♓ 04' 31"
True obliquity of ecliptic	23° 26' 10"

MOON'S PHASES, APSIDES AND POSITIONS ☽

Date	h	m	Phase	Longitude ° '	Eclipse Indicator
06	15	31	☾	14 ♍ 55	
13	08	42	●	21 ♐ 45	
20	05	19	☽	28 ♓ 44	
28	10	21	○	07 ♋ 06	

Day	h	m		
12	23	07	Perigee	
25	21	08	Apogee	
06	22	21	0S	
13	03	41	Max dec	20° S 56'
19	10	45	0N	
26	21	23	Max dec	20° N 56'

LONGITUDES

Date	Chiron ⚷ ° '	Ceres ⚳ ° '	Pallas ⚴ ° '	Juno ⚵ ° '	Vesta ⚶ ° '	Black Moon Lilith ⚸ ° '
01	05 ♓ 07	00 ♋ 35	22 ♓ 48	23 ♐ 39	19 ♊ 42	08 ♊ 54
11	05 ♓ 20	28 ♊ 27	23 ♓ 48	27 ♐ 14	17 ♊ 05	10 ♊ 01
21	05 ♓ 38	26 ♊ 06	25 ♓ 22	00 ♑ 52	14 ♊ 33	11 ♊ 07
31	06 ♓ 01	23 ♊ 49	27 ♓ 26	04 ♑ 31	12 ♊ 23	12 ♊ 14

All ephemeris data is given at 12.00 UT and the Moon's longitude is additionally given for 24.00 UT

ASPECTARIAN

h m Aspects

01 Saturday
03 18 ☽ △ ♄
03 58 ☽ ⋕ ♇
06 32 ☿ ⚼ ♅
06 36 ☽ ☍ ♇
08 45 ♀ ⊼ ♃
09 15 ☽ ⊼ ☉
12 01 ☽ ☍ ♂
13 07 ☽ ⊻ ♃
13 40 ☽ △ ♀
21 05 ☽ ⚼ ♆
22 24 ☽ ± ☉

02 Sunday
01 00 ☽ ⊥ ♃
02 42 ♂ ⊼ ♃
06 55 ☽ △ ☿
15 03 ☽ ± ♆
17 59 ☽ ⚼ ☉
18 43 ☽ ∠ ♃

03 Monday
01 07 ☽ ⋕ ☿
01 45 ☉ ☍ ♃
02 56 ☽ ⊼ ♆
09 44 ☽ ⋕ ♀
11 07 ☽ △ ♅
15 30 ☽ □ ♄
18 20 ☽ ⊼ ♇
23 50 ☽ ⚹ ♃

04 Tuesday
02 04 ☽ △ ☉
03 21 ☽ ⊼ ♂
04 07 ☽ ⋕ ♆
05 14 ☽ ⋕ ♄
06 01 ☽ ± ♇
07 38 ☽ □ ♀
15 41 ☽ ± ♂
16 11 ☽ ⚼ ♅
18 32 ☉ ⊥ ♄
22 07 ☽ □ ☿
22 41 ☽ Q ♃
23 20 ☽ ⚼ ♇

05 Wednesday
02 31 ☽ Q ♄
09 18 ☽ ± ♅
09 55 ☽ ⚼ ♂
12 51 ☽ ☍ ♆
13 25 ☿ ∠ ♇
20 33 ☽ ⊼ ♅
21 32 ☽ Q ♀

06 Thursday
01 14 ☽ ⚹ ♄
03 35 ☽ △ ♇
08 04 ☽ □ ♃
10 51 ☽ Q ☿
15 01 ☉ ⊻ ♂
15 31 ☽ △ ♂
15 31 ☽ □ ☉
16 20 ☽ ∥ ♅
21 54 ☽ ⚹ ♀

07 Friday
04 19 ☽ ⋕ ♅
04 50 ☽ ∠ ♄
10 35 ☽ ⚹ ☿
11 17 ♂ ∠ ♆
19 35 ☽ ⊼ ♆
20 58 ☽ ⊥ ♄
21 01 ♀ ⚼ ♅

08 Saturday
02 19 ☽ Q ☉
02 43 ☽ ☍ ♅
03 17 ☽ ∠ ♀
06 06 ☽ ± ♆
07 32 ☽ ⊻ ♄
09 27 ☽ □ ♇
13 00 ☽ △ ♃
15 21 ☽ ∠ ☿
17 34 ♂ ± ♃
20 21 ☽ ⊥ ♀
21 16 ♄ ∥ ♆
21 37 ☽ ⚼ ♆
23 34 ☽ □ ♂

09 Sunday
00 37 ☽ ⚹ ☉
07 31 ☽ ⊻ ♀
08 08 ☽ ⊥ ☿
11 28 ☽ ∥ ♆
11 36 ☽ ∥ ♄
14 13 ☽ ⚼ ♃
16 06 ☽ Q ♇
19 07 ☽ ⊻ ☿
22 49 ☽ △ ♆

10 Monday
03 36 ☽ ∠ ☉
04 59 ☽ ± ♃
05 28 ☽ ⊼ ♅
06 19 ☉ Q ♆
07 19 ☽ Q ♂
10 21 ☽ ☌ ♄
11 56 ☽ ⚹ ♇
14 42 ☽ ⊼ ♃
15 16 ☽ ± ♅
19 22 ☽ ⊥ ☉
21 48 ☽ ∥ ♀

11 Tuesday
01 19 ♀ ∠ ♇
03 51 ☽ ⚹ ♂
05 45 ☽ ⊻ ☉
05 48 ☽ ⚼ ♅
09 24 ☽ ∥ ☿
12 11 ☽ ∠ ♇
12 57 ☿ □ ♆
13 08 ☽ ☌ ♀
22 31 ☽ ∥ ♇
23 21 ☽ □ ♆

12 Wednesday
00 21 ☽ ☌ ☿
02 31 ☽ ⊥ ♇
05 03 ☽ ∠ ♂
05 40 ☽ △ ♅
10 43 ☽ ⊻ ♄
12 02 ☽ ⊻ ♇
14 12 ☽ ☍ ♃
19 39 ☉ ∥ ♂
20 01 ☽ ⊥ ♂
20 15 ☽ ⊥ ♄

13 Thursday
00 57 ♂ Q ♄
01 20 ☿ ⊥ ♇
03 48 ☽ Q ♆
05 59 ☽ ⊻ ♂
08 42 ☽ ☌ ☉
10 31 ☽ ∠ ♄
12 01 ♅ St D
16 48 ☽ ⊻ ♀
22 45 ☽ ⚹ ♆

14 Friday
03 08 ☽ ⊥ ♀
04 22 ☽ ⊻ ☿
05 01 ☽ □ ♅
08 55 ☽ ∥ ♇
09 30 ☽ ∥ ☿
10 24 ☽ ⚹ ♄
11 24 ☿ △ ♅
11 31 ☽ ☌ ♇
12 21 ☿ ∥ ♇
13 11 ☽ ⊼ ♃
14 38 ☉ ∠ ♄
14 52 ☽ ⊥ ☿
18 44 ☽ ∠ ♀
22 39 ☽ ∠ ♆
22 42 ☽ ± ♃
22 51 ☽ ∥ ♀

15 Saturday
05 44 ☽ Q ♄
06 49 ☽ ∠ ☿
08 17 ☽ ☌ ♂
09 50 ☽ Q ♅
11 57 ☽ ⊻ ☉
13 06 ☽ ⚼ ♃
13 12 ☽ ⊥ ♆
21 15 ☽ ⚹ ♀
22 29 ☽ ⊥ ☉
23 02 ☽ ⊻ ♆

16 Sunday
05 30 ☽ ⚹ ♅
10 03 ☽ ⚹ ☿
11 27 ☽ □ ♄
12 25 ☽ ⊻ ♇
13 38 ☽ △ ♃
13 55 ♂ Q ♅
14 33 ☽ ∠ ☉
15 10 ☿ ∠ ♂
18 23 ♀ □ ♆
19 06 ☽ Q ♀
19 11 ☽ ∥ ♄
20 48 ☽ ∥ ♆
22 32 ☽ ⊥ ♇

17 Monday
01 08 ☿ ⊻ ♄
06 46 ☽ ∠ ♅
08 38 ☽ Q ☿
10 25 ☿ ⊻ ♇
13 14 ☽ ⊻ ♂
14 00 ☽ ∠ ♇
18 11 ☽ ⚹ ☉
19 22 ☿ ☍ ♃
22 23 ☽ ⊥ ♅

18 Tuesday
00 20 ☽ ⊥ ♂
02 07 ☽ ☌ ♆
03 11 ☿ ⋕ ♃
05 17 ☽ □ ♀
09 00 ☽ ⊻ ♅
11 19 ♀ ⊥ ♇
15 46 ☽ △ ♄
15 53 ♂ ⚼ ♃
16 34 ☽ ⚹ ♇
17 12 ☽ Q ☉
17 16 ☽ □ ♃
17 21 ☽ ∠ ♂
20 10 ☽ □ ☿

19 Wednesday
04 38 ☽ ⋕ ♅
08 08 ♂ ⊥ ♆
10 38 ♀ ∥ ♇
14 36 ☽ Q ♇
16 56 ☽ ∥ ♅
19 29 ☽ ⚼ ♄
21 38 ♀ △ ♅
22 44 ☽ ⚹ ♂

20 Thursday
02 10 ☽ Q ♃
05 19 ☽ □ ☉
09 13 ☽ ⊻ ♆
12 38 ☽ ± ♄
16 34 ☽ ☌ ♅
18 40 ☽ △ ♀
20 44 ☽ ⊥ ♆
23 07 ☽ Q ♂

21 Friday
00 15 ☽ ⊼ ♄
00 52 ☽ □ ♇
00 54 ☽ ⚹ ♃
03 16 ☿ ∥ ♂
03 36 ♃ ⊼ ♇
10 32 ☿ ⊥ ♄
12 13 ☽ △ ☿
14 17 ☽ ∠ ♆

22 Saturday
03 19 ☽ ⚼ ♀
05 42 ☽ ⋕ ♆
06 04 ☽ ∠ ♃
06 48 ☉ ⚹ ♆
08 52 ☽ ⋕ ♄
12 57 ☽ □ ♂
15 03 ♃ ⊼ ♄
20 07 ☽ ⚹ ♆
21 19 ☽ △ ☉
22 09 ☽ ⚼ ☿
23 23 ☽ ± ♀
23 53 ☽ ⊥ ♃

23 Sunday
03 46 ☽ ⊻ ♅
03 50 ♀ ☍ ♃
06 05 ♀ ⊻ ♄
09 53 ♀ ⊻ ♇
11 51 ☽ ⊻ ♃
12 13 ☽ ☍ ♄
12 35 ☽ △ ♇
12 53 ☽ ⊼ ♀
15 53 ☽ ⊥ ♅
18 55 ☽ ± ☿
20 22 ☽ Q ♆

24 Monday
06 30 ☽ ⚼ ☉
06 30 ☿ Q ♆
08 52 ☽ ⊼ ☿
10 10 ☽ ∠ ♅
10 39 ♀ ⋕ ♃
19 08 ☽ ⚼ ♇

25 Tuesday
02 33 ☽ ± ☉
05 58 ☽ △ ♂
07 43 ☽ ⋕ ♇
09 04 ☽ □ ♆
13 32 ☽ ± ♇
15 56 ☽ ⊼ ☉
16 45 ☽ ⚹ ♅
21 17 ♀ ∥ ♂

26 Wednesday
00 18 ☽ ☌ ♃
01 31 ☉ □ ♅
01 41 ☽ ⊼ ♄
01 47 ☽ ⊼ ♇
04 24 ☉ ∥ ☿
09 09 ☽ ☍ ♀
13 58 ☽ ± ♄
14 41 ☽ ⚼ ♂
17 11 ☽ Q ♅

27 Thursday
01 41 ♄ ⚹ ♇
05 52 ♂ ⊻ ♆
06 50 ☽ ☍ ☿
08 21 ☽ ⚼ ♄
10 12 ☽ ± ♂
19 13 ☿ ∠ ♄
22 03 ☽ △ ♆
22 13 ♂ ⋕ ♃
23 09 ☽ ⊼ ♂

28 Friday
05 34 ☽ □ ♅
09 37 ♀ ⊥ ♄
10 21 ☽ ☍ ☉
10 50 ☽ ⋕ ♇
12 24 ☽ ⊻ ♃
14 34 ☽ ☍ ♇
14 43 ☽ △ ♄

29 Saturday
00 17 ☽ ⊥ ♃
04 07 ☽ ⚼ ♆
04 37 ☽ ⊼ ♀
08 27 ☉ ⊼ ♃
09 22 ☿ ⊥ ♂
17 52 ☽ ± ♀
17 58 ☽ ∠ ♃
21 58 ☽ ± ♆

30 Sunday
03 24 ☽ ⊼ ☿
09 46 ☽ ⊼ ♆
13 34 ☽ ⚼ ♀
13 39 ☉ ☌ ♇
14 47 ☽ ☍ ♂
16 50 ☽ ± ☿
17 01 ☽ △ ♅
17 45 ☉ ⚹ ♄
23 07 ☽ ⚹ ♃

31 Monday
01 52 ☽ ⊼ ♇
02 14 ☽ □ ♄
02 56 ☽ ⊼ ☉
06 46 ☽ ⋕ ♄
10 28 ♀ Q ♆
11 48 ☽ ⋕ ♆
12 46 ☽ ⚼ ☿
13 31 ☽ ± ♇
15 34 ☽ ± ☉
19 46 ♀ ∠ ♂
21 52 ☽ △ ♀
22 04 ☽ ⚼ ♅
22 07 ☽ Q ♃

JANUARY 2013

LONGITUDES

Date	Sidereal time h m s	Sun ☉ ° ' "	Moon ☽ ° ' "	Moon ☽ 24.00 ° ' "	Mercury ☿ ° '	Venus ♀ ° '	Mars ♂ ° '	Jupiter ♃ ° '	Saturn ♄ ° '	Uranus ♅ ° '	Neptune ♆ ° '	Pluto ♇ ° '
01	18 45 12	11 ♑ 14 22	27 ♌ 03 29	03 ♍ 24 00	01 ♑ 25	20 ♐ 23	05 ♒ 04	07 ♊ 44	09 ♏ 35	04 ♈ 46	01 ♓ 05	09 ♑ 20
02	18 49 09	12 15 31	09 ♍ 47 47	16 ♍ 15 02	02 58	21 38	05 52	07 R 38	09 40	04 47	01 07	09 23
03	18 53 05	13 16 39	22 ♍ 46 02	29 ♍ 21 03	04 31	22 54	06 39	07 33	09 44	04 48	01 08	09 25
04	18 57 02	14 17 48	06 ♎ 00 19	12 ♎ 44 06	06 04	24 09	07 26	07 28	09 48	04 49	01 10	09 27
05	19 00 58	15 18 57	19 ♎ 32 40	26 ♎ 26 11	07 38	25 24	08 13	07 23	09 53	04 50	01 12	09 29
06	19 04 55	16 20 07	03 ♏ 24 47	10 ♏ 28 32	09 13	26 39	09 00	07 18	09 57	04 52	01 13	09 31
07	19 08 51	17 21 17	17 ♏ 37 22	24 ♏ 51 07	10 48	27 54	09 48	07 13	10 01	04 53	01 15	09 33
08	19 12 48	18 22 27	02 ♐ 09 26	09 ♐ 31 50	12 23	29 ♐ 09	10 35	07 09	10 05	04 54	01 17	09 35
09	19 16 44	19 23 37	16 ♐ 57 40	24 ♐ 26 06	13 58	00 ♑ 24	11 22	07 04	10 09	04 56	01 19	09 37
10	19 20 41	20 24 47	01 ♑ 56 09	09 ♑ 26 46	15 34	01 40	12 10	07 00	10 13	04 57	01 21	09 39
11	19 24 38	21 25 56	16 ♑ 56 46	24 ♑ 24 57	17 11	02 55	12 57	06 56	10 17	04 58	01 22	09 42
12	19 28 34	22 27 06	01 ♒ 50 10	09 ♒ 11 18	18 47	04 10	13 44	06 53	10 21	05 00	01 24	09 44
13	19 32 31	23 28 15	16 ♒ 27 23	23 ♒ 37 38	20 25	05 25	14 32	06 49	10 24	05 01	01 26	09 46
14	19 36 27	24 29 23	00 ♓ 41 23	07 ♓ 38 14	22 03	06 40	15 19	06 46	10 28	05 03	01 28	09 48
15	19 40 24	25 30 31	14 ♓ 27 56	21 ♓ 10 27	23 41	07 56	16 06	06 43	10 31	05 05	01 30	09 50
16	19 44 20	26 31 38	27 ♓ 45 56	04 ♈ 14 38	25 20	09 11	16 54	06 40	10 35	05 06	01 32	09 52
17	19 48 17	27 32 44	10 ♈ 36 59	16 ♈ 53 29	26 59	10 26	17 41	06 37	10 38	05 08	01 34	09 54
18	19 52 13	28 33 50	23 ♈ 04 43	29 ♈ 11 19	28 ♑ 39	11 41	18 29	06 34	10 41	05 10	01 36	09 56
19	19 56 10	29 ♑ 34 54	05 ♉ 13 59	11 ♉ 13 23	00 ♒ 19	12 56	19 16	06 32	10 44	05 12	01 38	09 58
20	20 00 07	00 ♒ 35 58	17 ♉ 10 14	23 ♉ 05 13	02 00	14 12	20 03	06 30	10 47	05 13	01 40	10 00
21	20 04 03	01 37 01	28 ♉ 58 59	04 ♊ 52 11	03 42	15 27	20 51	06 28	10 50	05 15	01 42	10 02
22	20 08 00	02 38 03	10 ♊ 45 25	16 ♊ 39 15	05 24	16 42	21 38	06 26	10 53	05 17	01 44	10 04
23	20 11 56	03 39 04	22 ♊ 34 10	28 ♊ 30 38	07 06	17 57	22 26	06 25	10 56	05 19	01 46	10 06
24	20 15 53	04 40 05	04 ♋ 29 02	10 ♋ 29 43	08 49	19 12	23 13	06 23	10 59	05 21	01 48	10 08
25	20 19 49	05 41 04	16 ♋ 32 57	22 ♋ 38 56	10 33	20 28	24 01	06 22	11 01	05 23	01 50	10 10
26	20 23 46	06 42 02	28 ♋ 47 49	04 ♌ 59 43	12 17	21 43	24 48	06 21	11 04	05 25	01 52	10 12
27	20 27 42	07 43 00	11 ♌ 14 39	17 ♌ 32 37	14 01	22 58	25 35	06 21	11 06	05 28	01 54	10 14
28	20 31 39	08 43 56	23 ♌ 53 37	00 ♍ 17 34	15 46	24 13	26 23	06 20	11 08	05 30	01 56	10 16
29	20 35 36	09 44 52	06 ♍ 44 25	13 ♍ 14 05	17 31	25 28	27 10	06 20	11 10	05 32	01 59	10 18
30	20 39 32	10 45 46	19 ♍ 46 32	26 ♍ 21 43	19 17	26 43	27 58	06 D 20	11 12	05 34	02 01	10 20
31	20 43 29	11 ♒ 46 40	02 ♎ 59 37	09 ♎ 40 14	21 ♒ 02	27 ♑ 59	28 ♒ 45	06 ♊ 20	11 ♏ 14	05 ♈ 37	02 ♓ 03	10 ♑ 22

Date	Moon True ☊ ° '	Moon Mean ☊ ° '	Moon ☽ Latitude ° '
01	24 ♏ 57	23 ♏ 34	05 S 08
02	24 R 50	23 31	04 58
03	24 46	23 28	04 32
04	24 44	23 24	03 52
05	24 D 43	23 21	02 58
06	24 44	23 18	01 52
07	24 45	23 15	00 S 38
08	24 R 45	23 12	00 N 40
09	24 43	23 09	01 56
10	24 38	23 05	03 06
11	24 31	23 02	04 02
12	24 22	22 59	04 42
13	24 13	22 56	05 02
14	24 03	22 53	05 03
15	23 54	22 50	04 46
16	23 48	22 46	04 13
17	23 45	22 43	03 29
18	23 43	22 40	02 36
19	23 D 43	22 37	01 37
20	23 44	22 34	00 N 35
21	23 R 44	22 30	00 S 28
22	23 42	22 27	01 29
23	23 38	22 24	02 27
24	23 31	22 21	03 19
25	23 22	22 18	04 02
26	23 10	22 15	04 35
27	22 57	22 11	04 55
28	22 43	22 08	05 01
29	22 30	22 05	04 52
30	22 20	22 02	04 28
31	22 ♏ 12	21 ♏ 59	03 S 49

DECLINATIONS

Date	Sun ☉ ° '	Moon ☽ ° '	Mercury ☿ ° '	Venus ♀ ° '	Mars ♂ ° '	Jupiter ♃ ° '	Saturn ♄ ° '	Uranus ♅ ° '	Neptune ♆ ° '	Pluto ♇ ° '
01	22 S 58	07 N 40	24 S 18	22 S 25	20 S 07	20 N 54	12 S 28	01 N 15	11 S 39	19 S 47
02	22 52	03 N 18	24 22	22 32	19 55	20 53	12 29	01 15	11 39	19 47
03	22 46	01 S 18	24 25	22 39	19 43	20 53	12 30	01 16	11 38	19 47
04	22 40	05 56	24 27	22 45	19 31	20 52	12 31	01 16	11 38	19 47
05	22 33	10 24	24 28	22 51	19 19	20 51	12 32	01 17	11 37	19 47
06	22 26	14 25	24 27	22 56	19 06	20 51	12 34	01 17	11 36	19 47
07	22 19	17 42	24 25	23 00	18 53	20 50	12 35	01 18	11 36	19 47
08	22 11	19 56	24 21	23 03	18 40	20 50	12 36	01 18	11 35	19 47
09	22 02	20 52	24 16	23 06	18 27	20 49	12 37	01 19	11 35	19 47
10	21 53	20 20	24 10	23 08	18 14	20 49	12 38	01 19	11 34	19 47
11	21 44	18 21	24 02	23 09	18 00	20 48	12 39	01 20	11 33	19 47
12	21 34	15 10	23 52	23 10	17 46	20 48	12 40	01 21	11 33	19 47
13	21 24	11 06	23 41	23 09	17 32	20 48	12 41	01 21	11 32	19 47
14	21 13	06 30	23 29	23 09	17 18	20 47	12 42	01 22	11 31	19 46
15	21 02	01 S 43	23 15	23 07	17 04	20 47	12 42	01 23	11 31	19 46
16	20 51	02 N 59	23 00	23 05	16 50	20 47	12 43	01 23	11 30	19 46
17	20 39	07 24	22 43	23 02	16 35	20 47	12 44	01 24	11 29	19 46
18	20 27	11 23	22 25	22 58	16 20	20 46	12 45	01 25	11 28	19 46
19	20 14	14 47	22 05	22 54	16 05	20 46	12 46	01 26	11 28	19 46
20	20 01	17 31	21 44	22 49	15 50	20 46	12 46	01 26	11 27	19 46
21	19 48	19 28	21 21	22 43	15 34	20 46	12 47	01 27	11 26	19 46
22	19 34	20 35	20 57	22 36	15 19	20 46	12 48	01 28	11 26	19 46
23	19 20	20 47	20 31	22 29	15 03	20 46	12 48	01 29	11 25	19 46
24	19 06	20 03	20 03	22 21	14 47	20 46	12 49	01 30	11 24	19 46
25	18 51	18 25	19 34	22 13	14 32	20 46	12 49	01 30	11 23	19 46
26	18 36	15 55	19 04	22 03	14 15	20 46	12 50	01 31	11 23	19 45
27	18 20	12 40	18 32	21 54	13 59	20 46	12 50	01 32	11 22	19 45
28	18 04	08 49	17 58	21 43	13 43	20 46	12 51	01 33	11 21	19 45
29	17 48	04 N 31	17 23	21 32	13 26	20 46	12 51	01 34	11 20	19 45
30	17 32	00 S 03	16 47	21 20	13 10	20 47	12 52	01 35	11 20	19 45
31	17 S 15	04 S 41	16 S 09	21 S 07	12 S 53	20 N 47	12 S 52	01 N 36	11 S 19	19 S 45

ZODIAC SIGN ENTRIES

Date	h	m	Planets
01	17	35	☽ ♍
04	01	11	☽ ♎
06	06	09	☽ ♏
08	08	28	☽ ♐
09	04	11	♀ ♑
10	08	54	☽ ♑
12	09	01	☽ ♒
14	10	49	☽ ♓
16	16	07	☽ ♈
19	01	36	☽ ♉
19	07	25	☿ ♒
19	21	52	☉ ♒
21	14	04	☽ ♊
24	03	00	☽ ♋
26	14	20	☽ ♌
28	23	27	☽ ♍
31	06	36	☽ ♎

LATITUDES

Date	Mercury ☿ ° '	Venus ♀ ° '	Mars ♂ ° '	Jupiter ♃ ° '	Saturn ♄ ° '	Uranus ♅ ° '	Neptune ♆ ° '	Pluto ♇ ° '
01	00 S 52	00 N 41	01 S 09	00 S 42	02 N 20	00 S 42	00 S 37	03 N 20
04	01 09	00 33	01 09	00 42	02 21	00 42	00 37	03 19
07	01 25	00 26	01 08	00 41	02 22	00 42	00 37	03 19
10	01 39	00 18	01 08	00 40	02 22	00 42	00 37	03 19
13	01 50	00 10	01 07	00 40	02 23	00 42	00 37	03 19
16	01 58	00 N 02	01 06	00 39	02 24	00 42	00 37	03 18
19	02 03	00 S 05	01 06	00 38	02 24	00 42	00 37	03 18
22	02 05	00 13	01 05	00 37	02 25	00 42	00 37	03 18
25	02 03	00 20	01 04	00 37	02 26	00 41	00 37	03 18
28	01 57	00 27	01 03	00 36	02 26	00 41	00 37	03 18
31	01 S 46	00 S 34	01 S 02	00 S 35	02 N 27	00 S 41	00 S 37	03 N 18

DATA

Julian Date	2456294
Delta T	+68 seconds
Ayanamsa	24° 02' 34"
Synetic vernal point	05° ♓ 04' 25"
True obliquity of ecliptic	23° 26' 09"

MOON'S PHASES, APSIDES AND POSITIONS ☽

Date	h	m	Phase	Longitude ° '	Eclipse Indicator
05	03	58	☾	14 ♎ 58	
11	19	44	●	21 ♑ 46	
18	23	45	☽	29 ♈ 04	
27	04	38	○	07 ♌ 24	

Day	h	m		
10	10	19	Perigee	
22	10	47	Apogee	
03	05	16	0S	
09	15	16	Max dec	20° S 53'
15	20	40	0N	
23	05	12	Max dec	20° N 49'
30	11	43	0S	

LONGITUDES

Date	Chiron ⚷ ° '	Ceres ⚳ ° '	Pallas ⚴ ° '	Juno ⚵ ° '	Vesta ⚶ ° '	Black Moon Lilith ⚸ ° '
01	06 ♓ 04	23 ♊ 37	27 ♓ 40	04 ♑ 53	12 ♊ 12	12 ♊ 21
11	06 ♓ 32	21 ♊ 44	00 ♈ 13	08 ♑ 33	10 ♊ 43	13 ♊ 27
21	07 ♓ 04	20 ♊ 25	03 ♈ 08	12 ♑ 13	09 ♊ 57	14 ♊ 34
31	07 ♓ 38	19 ♊ 48	06 ♈ 23	15 ♑ 51	09 ♊ 56	15 ♊ 40

All ephemeris data is given at 12.00 UT and the Moon's longitude is additionally given for 24.00 UT

ASPECTARIAN

h m	Aspects
01 Tuesday	
02 31	♂ ⚹ ♅
06 49	☽ ⚼ ♇
06 50	☿ ⚹ ♆
10 18	☽ ⚼ ☉
13 00	☽ Q ♄
15 15	☽ ± ♅
19 39	☽ ☍ ♆
21 24	☽ △ ☿
02 Wednesday	
02 36	☽ ⚻ ♅
04 08	☽ ⚻ ♂
07 59	☽ □ ♃
11 13	☽ △ ♇
11 45	☽ ⚹ ♄
16 06	☽ ± ♂
16 59	☽ △ ☉
22 43	☽ ∥ ♅
03 Thursday	
03 46	♂ ∥ ♇
09 49	☽ ⚼ ♂
11 48	☽ ⋕ ♅
12 15	☽ □ ♀
15 37	☽ ∠ ♄
16 29	☿ □ ♅
17 49	☉ ± ♃
04 Friday	
02 03	☉ ∥ ♀
03 16	☽ ⚻ ♆
08 02	☽ ⊥ ♄
09 52	☽ ☍ ♅
12 08	☽ □ ☿
12 46	♂ △ ♃
14 05	☽ ± ♆
14 35	☽ △ ♃
14 43	☽ △ ♂
18 10	☽ □ ♇
18 50	☽ ⚺ ♄
05 Saturday	
00 04	☽ Q ♀
01 29	♀ ∠ ♄
03 58	☽ □ ☉
06 06	☽ ⚼ ♆
08 09	☿ ⚻ ♃
16 55	☽ ⚼ ♃
18 57	☽ ∥ ♆
23 13	☽ ⚹ ♀
23 59	☽ Q ☿
06 Sunday	
00 27	☽ ∥ ♄
01 51	☽ Q ♇
05 41	☿ ⚺ ♂
08 15	☽ △ ♆
08 24	☽ ± ♃
09 19	☉ ∠ ♆
13 42	☽ Q ☉
14 28	☽ ⚻ ♅
16 43	☿ ☌ ♇
18 35	☽ ⚻ ♃
22 04	☽ □ ♂
22 24	☽ ⚹ ♇
23 06	☽ ⚹ ☿
23 10	☽ ☌ ♄
23 44	☿ ⚹ ♄
07 Monday	
00 40	☽ ± ♅
03 20	☽ ∠ ♀
04 17	♂ ⚺ ♇
11 31	☽ ⚹ ☉
15 46	☽ ⚼ ♅
19 33	♂ □ ♄
19 48	☽ ⊥ ♀
22 12	☽ ∥ ♂
23 32	☽ ∠ ♇
08 Tuesday	
03 12	☽ ∠ ☿
05 48	☽ Q ♂
06 37	☽ ⚺ ♀
09 44	☽ ∥ ♇
10 34	☽ □ ♆
14 08	☽ ∠ ☉
14 20	☽ ⊥ ♇
16 29	☽ △ ♅
19 43	☽ ⊥ ☿
20 05	☽ ☍ ♃
23 03	☿ ± ♃
09 Wednesday	
00 07	☽ ⚺ ♇
00 58	☽ ⚺ ♄
02 28	☽ ⚹ ♂
05 50	☽ ⊥ ☉
06 36	☽ ⚺ ☿
08 39	☽ ⋕ ♃
10 42	☽ ⊥ ♄
15 47	☽ Q ♆
16 12	☽ ⚺ ☉
22 09	☽ ⋕ ♃
10 Thursday	
01 13	☽ ∠ ♄
03 56	☽ ∠ ♂
05 45	♀ ⚹ ♆
11 03	☽ ⚹ ♆
11 31	☽ ☌ ♀
16 49	☽ □ ♅
19 07	☽ ⊥ ♂
20 04	☽ ⚻ ♃
20 33	☽ ∥ ♇
23 48	☿ ∠ ♆
11 Friday	
00 22	☽ ☌ ♇
01 18	☽ ⚹ ♄
05 15	☽ ⚺ ♂
05 37	☽ ± ♃
11 05	☽ ∠ ♆
12 25	☽ ☌ ☿
15 23	☽ ∥ ♂
19 44	☽ ☌ ☉
19 59	☽ ⚼ ♃
20 36	☽ Q ♄
21 42	☽ Q ♅
23 13	☉ ⚼ ♃
12 Saturday	
01 34	☽ ⊥ ♆
09 22	☉ Q ♄
11 18	☽ ⚺ ♆
16 09	☽ ⚺ ♀
17 09	☽ ⚹ ♅
20 11	☽ △ ♃
13 Sunday	
00 55	☽ ⚺ ♇
01 11	☉ Q ♅
01 58	☽ □ ♄
02 53	☽ ⊥ ♀
03 10	☽ ∥ ♄
04 13	♀ □ ♅
08 37	☽ ☌ ♂
09 36	☽ ∥ ♆
10 51	☽ ⊥ ♇
17 57	☽ ∠ ♅
19 15	☽ ∠ ♀
19 27	☽ ⚺ ☿
14 Monday	
00 38	☽ ⚺ ☉
01 57	☽ ∠ ♇
06 53	☽ ⊥ ☿
08 00	☿ ⚼ ♃
09 11	☽ ⊥ ♅
11 38	☽ ⊥ ☉
13 20	☽ ☌ ♆
13 36	♀ ⚻ ♃
18 28	☿ Q ♄
19 31	☽ ⚺ ♅
22 26	☽ □ ♃
23 21	☽ ⚹ ♀
15 Tuesday	
00 26	☽ ∠ ☿
03 01	☿ Q ♅
03 16	♂ ⊥ ♇
03 49	☽ ⚹ ♇
04 28	☽ ∠ ☉
05 01	☽ △ ♄
11 47	☉ ⊥ ♆
13 43	☽ ⋕ ♅
15 06	☽ ⚺ ♂
22 46	☽ Q ♀
16 Wednesday	
01 13	☽ Q ♇
02 33	☽ ⊥ ♂
03 43	☽ ∥ ♅
03 44	☿ ∥ ♀
06 20	☽ Q ♃
06 54	☽ ⚹ ☿
07 59	☽ ⚼ ♄
09 32	☽ ⚹ ☉
15 02	☿ ⊥ ♆
18 58	☽ ⚺ ♆
20 07	☽ ∠ ♂
20 17	☉ ⋕ ♃
17 Thursday	
00 41	☽ ± ♄
01 29	♀ ☌ ♇
01 38	☽ ☌ ♅
04 28	☽ ⚹ ♃
06 13	☽ ⊥ ♆
08 26	☽ Q ☿
09 48	☽ Q ☉
10 38	☽ □ ♇
11 37	☽ □ ♀
12 02	☽ ⚻ ♄
16 02	♀ ⚹ ♄
23 24	☽ ∠ ♆
18 Friday	
02 27	☽ ⚹ ♂
08 56	☉ ☌ ☿
09 04	☽ ∠ ♃
12 35	☽ ⋕ ♆
21 07	☽ ⋕ ♄
23 45	☽ □ ☉
19 Saturday	
00 40	☽ □ ☿
02 41	☽ ⊥ ♃
03 34	☽ Q ♂
04 25	♀ ± ♃
04 49	☽ ⚹ ♆
11 55	☽ ⚺ ♅
14 35	☽ ⚺ ♃
21 30	☽ △ ♇
21 43	☽ ⋕ ♂
23 05	☽ ☍ ♄
23 58	☽ ⊥ ♅
20 Sunday	
04 54	☽ Q ♆
04 55	♀ ⊥ ♂
05 17	☽ △ ♀
07 05	☿ ⚺ ♆
17 15	♂ ∠ ♅
18 12	☽ ∠ ♅
18 16	☽ □ ♂
21 Monday	
03 56	☽ ⚼ ♇
13 58	☉ ⚺ ♆
15 20	☽ ⚼ ♀
15 28	☉ ∥ ♇
16 37	☽ ⋕ ☉
16 48	☽ ⋕ ♇
17 33	☽ □ ♆
17 53	☽ △ ☉
22 20	☽ ± ♇
23 13	☽ △ ☿
22 Tuesday	
00 49	☽ ⚹ ♅
03 13	☽ ☌ ♃
10 28	☿ ⚹ ♅
10 36	☽ ⚻ ♇
11 52	☽ ± ♀
12 16	☽ ⚻ ♄
12 38	♀ ∠ ♆
21 11	☽ ∥ ♃
21 47	☽ ⋕ ☿
22 12	☿ ⋕ ♃
23 Wednesday	
00 31	☽ ± ♄
01 19	☽ Q ♅
01 32	☽ ⚻ ♀
02 26	☿ △ ♃
03 19	☽ ⚼ ☉
10 54	☽ ⚼ ☿
11 42	☽ △ ♂
13 13	☽ ∥ ♃
18 50	☽ ⚼ ♄
23 14	☽ ± ☉
24 Thursday	
06 36	☽ △ ♆
08 06	☽ ± ☿
11 50	☽ ⋕ ☿
12 24	☽ ⚻ ☉
13 45	☽ □ ♅
15 48	☽ ⚺ ♃
17 23	☽ ⋕ ♇
19 59	☽ ⚼ ♂
22 07	☽ ⚻ ☿
23 19	☽ ☍ ♇
25 Friday	
01 00	☽ △ ♄
02 49	☿ ∥ ♇
03 44	☽ ⊥ ♃
04 46	☉ ⚹ ♅
05 53	☽ ⋕ ☉
06 38	☿ ⚺ ♇
12 34	☽ ⚼ ♆
15 05	☽ ± ♂
18 42	☿ □ ♄
20 35	☽ ☍ ♀
21 29	☽ ∠ ♃
26 Saturday	
03 40	☽ ⚻ ♂
03 56	☉ △ ♃
05 14	♀ ⚼ ♃
06 17	☽ ± ♆
17 59	☽ ⚻ ♆
27 Sunday	
00 40	♂ ∠ ♇
00 52	☽ △ ♅
02 04	☽ ⋕ ♂
02 36	☽ ⚹ ♃
04 38	☽ ☍ ☉
10 04	☽ ⚻ ♇
10 52	☽ ⋕ ♄
11 43	☽ □ ♄
14 38	♀ Q ♄
18 09	☽ ☍ ☿
20 34	☽ ⋕ ♆
21 32	☽ ± ♇
21 45	♀ Q ♅
28 Monday	
01 30	☽ Q ♃
03 42	☉ ∥ ☿
05 34	☽ ⚼ ♅
12 41	☽ ⚻ ♀
14 35	☽ ⚼ ♇
16 59	☽ ☍ ♂
18 54	☿ ⊥ ♇
21 52	☽ Q ♄
22 32	☽ ± ♅
29 Tuesday	
01 08	☽ ± ♀
03 07	☽ ☍ ♆
09 45	☽ ⚻ ♅
11 14	☽ □ ♃
18 02	☽ ⚻ ☉
18 36	☽ △ ♇
19 38	☽ ⚼ ♀
20 13	☽ ⚹ ♄
21 59	♀ ⊥ ♆
30 Wednesday	
01 23	☉ ⚺ ♇
03 32	☽ ∥ ♅
06 01	☽ ± ☉
10 57	☽ ⚻ ☿
11 35	♃ St D
19 55	☽ ⋕ ♅
22 49	☉ □ ♄
23 35	☽ ± ☿
23 45	☽ ∠ ♄
23 49	☽ ⚼ ☉
31 Thursday	
01 59	☽ △ ♀
03 51	☽ ⚻ ♂
05 59	☿ ∠ ♅
10 17	☽ ⚻ ♆
13 02	♂ ∥ ♄
15 22	☽ ± ♂
16 03	☽ ⊥ ♄
16 43	☽ ☍ ♅
18 00	☽ △ ♃
18 19	☽ ⚼ ☿
21 07	☽ ± ♆

FEBRUARY 2013

LONGITUDES

Date	Sidereal time h m s	Sun ☉ ° ' "	Moon ☽ ° ' "	Moon ☽ 24.00 ° ' "	Mercury ☿ ° '	Venus ♀ ° '	Mars ♂ ° '	Jupiter ♃ ° '	Saturn ♄ ° '	Uranus ♅ ° '	Neptune ♆ ° '	Pluto ♇ ° '
01	20 47 25	12 ♒ 47 33	16 ♎ 23 39	23 ♎ 09 55	22 ♒ 48	29 ♑ 14	29 ♒ 33	06 ♊ 20	11 ♏ 16	05 ♈ 39	02 ♓ 05	10 ♑ 23
02	20 51 22	13 48 26	29 ♎ 59 08	06 ♏ 51 24	24 34	00 ♒ 29	00 ♓ 20	06 21	11 18	05 41	02 07	10 25
03	20 55 18	14 49 17	13 ♏ 46 49	20 ♏ 45 27	26 19	01 44	01 07	06 21	11 19	05 44	02 09	10 27
04	20 59 15	15 50 08	27 ♏ 47 20	04 ♐ 52 26	28 03	02 59	01 55	06 22	11 21	05 46	02 12	10 29
05	21 03 11	16 50 58	12 ♐ 00 40	19 ♐ 11 46	29 ♒ 47	04 14	02 42	06 23	11 22	05 49	02 14	10 31
06	21 07 08	17 51 48	26 ♐ 25 27	03 ♑ 41 13	01 ♓ 30	05 30	03 30	06 25	11 24	05 51	02 16	10 33
07	21 11 05	18 52 36	10 ♑ 58 31	18 ♑ 16 38	03 12	06 45	04 17	06 26	11 25	05 54	02 18	10 34
08	21 15 01	19 53 23	25 ♑ 34 45	02 ♒ 52 01	04 52	08 00	05 04	06 28	11 26	05 56	02 21	10 36
09	21 18 58	20 54 09	10 ♒ 07 29	17 ♒ 20 16	06 29	09 15	05 52	06 30	11 27	05 59	02 23	10 38
10	21 22 54	21 54 54	24 ♒ 29 28	01 ♓ 34 18	08 04	10 30	06 39	06 32	11 28	06 02	02 25	10 39
11	21 26 51	22 55 37	08 ♓ 34 05	15 ♓ 28 16	09 36	11 45	07 26	06 34	11 29	06 04	02 27	10 41
12	21 30 47	23 56 19	22 ♓ 16 38	28 ♓ 58 27	11 04	13 00	08 14	06 37	11 30	06 07	02 30	10 43
13	21 34 44	24 56 59	05 ♈ 34 09	12 ♈ 03 39	12 28	14 16	09 01	06 40	11 30	06 10	02 32	10 44
14	21 38 40	25 57 38	18 ♈ 27 11	24 ♈ 45 05	13 46	15 31	09 48	06 42	11 31	06 12	02 34	10 46
15	21 42 37	26 58 15	00 ♉ 57 48	07 ♉ 05 52	14 59	16 46	10 36	06 46	11 31	06 15	02 36	10 48
16	21 46 34	27 58 51	13 ♉ 09 53	19 ♉ 10 29	16 05	18 01	11 23	06 49	11 31	06 18	02 39	10 49
17	21 50 30	28 59 24	25 ♉ 08 21	01 ♊ 04 12	17 04	19 16	12 10	06 52	11 32	06 21	02 41	10 51
18	21 54 27	29 ♒ 59 56	06 ♊ 58 44	12 ♊ 52 38	17 55	20 31	12 58	06 56	11 32	06 24	02 43	10 52
19	21 58 23	01 ♓ 00 26	18 ♊ 46 35	24 ♊ 41 16	18 37	21 46	13 45	07 00	11 R 32	06 27	02 45	10 54
20	22 02 20	02 00 54	00 ♋ 37 17	06 ♋ 35 13	19 10	23 01	14 32	07 04	11 31	06 30	02 48	10 55
21	22 06 16	03 01 21	12 ♋ 35 37	18 ♋ 38 55	19 34	24 16	15 19	07 08	11 31	06 33	02 50	10 57
22	22 10 13	04 01 46	24 ♋ 45 32	00 ♌ 55 48	19 48	25 31	16 06	07 13	11 31	06 36	02 52	10 58
23	22 14 09	05 02 08	07 ♌ 09 57	13 ♌ 28 08	19 R 52	26 46	16 53	07 17	11 30	06 39	02 55	11 00
24	22 18 06	06 02 29	19 ♌ 50 26	26 ♌ 16 50	19 46	28 01	17 41	07 22	11 30	06 42	02 57	11 01
25	22 22 03	07 02 48	02 ♍ 47 16	09 ♍ 21 33	19 31	29 ♒ 16	18 28	07 27	11 29	06 45	02 59	11 03
26	22 25 59	08 03 06	15 ♍ 59 30	22 ♍ 40 49	19 06	00 ♓ 31	19 15	07 32	11 28	06 48	03 01	11 04
27	22 29 56	09 03 22	29 ♍ 25 14	06 ♎ 12 26	18 32	01 46	20 02	07 37	11 28	06 51	03 04	11 05
28	22 33 52	10 ♓ 03 36	13 ♎ 02 07	19 ♎ 54 00	17 ♓ 50	03 ♓ 01	20 ♓ 49	07 ♊ 43	11 ♏ 27	06 ♈ 54	03 ♓ 06	11 ♑ 07

DECLINATIONS

Date	Moon True ☊ ° '	Moon Mean ☊ ° '	Moon ☽ Latitude ° '	Sun ☉ ° '	Moon ☽ ° '	Mercury ☿ ° '	Venus ♀ ° '	Mars ♂ ° '	Jupiter ♃ ° '	Saturn ♄ ° '	Uranus ♅ ° '	Neptune ♆ ° '	Pluto ♇ ° '
01	22 ♏ 07	21 ♏ 56	02 S 57	16 S 58	09 S 10	15 S 30	20 S 54	12 S 36	20 N 47	12 S 52	01 N 37	11 S 18	19 S 45
02	22 R 05	21 52	01 54	16 41	13 15	14 50	20 40	12 19	20 47	12 53	01 38	11 17	19 45
03	22 D 04	21 49	00 S 44	16 23	16 40	14 08	20 26	12 02	20 48	12 53	01 39	11 16	19 45
04	22 R 04	21 46	00 N 30	16 05	19 10	13 26	20 11	11 45	20 48	12 53	01 40	11 16	19 45
05	22 03	21 43	01 43	15 47	20 31	12 42	19 56	11 27	20 49	12 53	01 41	11 15	19 44
06	22 00	21 40	02 51	15 29	20 32	11 57	19 39	11 10	20 49	12 54	01 42	11 14	19 44
07	21 54	21 36	03 48	15 10	19 12	11 12	19 23	10 52	20 50	12 54	01 43	11 13	19 44
08	21 45	21 33	04 30	14 51	16 36	10 27	19 05	10 34	20 50	12 54	01 44	11 12	19 44
09	21 34	21 30	04 54	14 32	12 58	09 41	18 47	10 17	20 51	12 54	01 45	11 12	19 44
10	21 21	21 27	05 00	14 12	08 38	08 55	18 29	09 59	20 51	12 54	01 46	11 11	19 44
11	21 09	21 24	04 47	13 52	03 S 55	08 09	18 10	09 41	20 52	12 54	01 47	11 10	19 44
12	20 58	21 21	04 17	13 32	00 N 53	07 24	17 51	09 23	20 52	12 54	01 48	11 09	19 44
13	20 49	21 17	03 34	13 12	05 29	06 40	17 31	09 05	20 53	12 54	01 49	11 08	19 44
14	20 43	21 14	02 42	12 52	09 43	05 57	17 10	08 47	20 54	12 54	01 50	11 08	19 44
15	20 39	21 11	01 42	12 31	13 24	05 15	16 49	08 28	20 55	12 54	01 52	11 07	19 43
16	20 38	21 08	00 N 40	12 10	16 25	04 36	16 28	08 10	20 55	12 54	01 53	11 06	19 43
17	20 D 38	21 05	00 S 24	11 49	18 40	03 59	16 06	07 51	20 56	12 54	01 54	11 05	19 43
18	20 R 38	21 02	01 26	11 28	20 04	03 25	15 44	07 33	20 57	12 53	01 55	11 04	19 43
19	20 37	20 58	02 23	11 07	20 35	02 54	15 21	07 14	20 58	12 53	01 56	11 04	19 43
20	20 33	20 55	03 15	10 45	20 11	02 27	14 58	06 56	20 59	12 53	01 57	11 03	19 43
21	20 27	20 52	03 59	10 24	18 53	02 04	14 34	06 37	21 00	12 53	01 59	11 02	19 43
22	20 19	20 49	04 32	10 02	16 42	01 45	14 10	06 19	21 01	12 52	02 00	11 01	19 43
23	20 08	20 46	04 54	09 40	13 44	01 30	13 46	06 00	21 02	12 52	02 01	11 00	19 43
24	19 55	20 42	05 01	09 18	10 05	01 21	13 21	05 41	21 03	12 51	02 02	11 00	19 43
25	19 42	20 39	04 54	08 55	05 55	01 16	12 56	05 22	21 04	12 51	02 03	10 59	19 42
26	19 30	20 36	04 30	08 33	01 N 22	01 17	12 30	05 03	21 05	12 51	02 05	10 58	19 42
27	19 20	20 33	03 52	08 10	03 S 19	01 22	12 05	04 44	21 06	12 50	02 06	10 57	19 42
28	19 ♏ 12	20 ♏ 30	02 S 59	07 S 48	07 S 54	01 S 32	11 S 38	04 S 25	21 N 07	12 S 50	02 N 07	10 S 56	19 S 42

ZODIAC SIGN ENTRIES

Date	h	m	Planets
02	01	54	♂ ♓
02	02	47	♀ ♒
02	12	02	☽ ♏
04	15	45	☽ ♐
05	14	55	☿ ♓
06	17	55	☽ ♑
08	19	16	☽ ♒
10	21	19	☽ ♓
13	01	51	☽ ♈
15	10	08	☽ ♉
17	21	50	☽ ♊
18	12	02	☉ ♓
20	10	45	☽ ♋
22	22	12	☽ ♌
25	06	52	☽ ♍
26	02	03	♀ ♓
27	13	02	☽ ♎

LATITUDES

Date	Mercury ☿ ° '	Venus ♀ ° '	Mars ♂ ° '	Jupiter ♃ ° '	Saturn ♄ ° '	Uranus ♅ ° '	Neptune ♆ ° '	Pluto ♇ ° '
01	01 S 41	00 S 36	01 S 02	00 S 35	02 N 27	00 S 41	00 S 37	03 N 18
04	01 22	00 43	01 01	00 34	02 28	00 41	00 37	03 17
07	00 56	00 49	01 00	00 34	02 29	00 41	00 37	03 17
10	00 S 24	00 55	00 59	00 33	02 29	00 41	00 37	03 17
13	00 N 14	01 00	00 58	00 32	02 30	00 41	00 37	03 17
16	00 58	01 05	00 56	00 32	02 31	00 41	00 37	03 17
19	01 44	01 10	00 55	00 31	02 32	00 41	00 37	03 17
22	02 29	01 14	00 54	00 30	02 32	00 41	00 37	03 17
25	03 08	01 17	00 53	00 30	02 33	00 41	00 37	03 17
28	03 34	01 20	00 51	00 29	02 33	00 41	00 37	03 17
31	03 N 42	01 S 22	00 S 50	00 S 29	02 N 34	00 S 40	00 S 37	03 N 17

DATA

Julian Date	2456325
Delta T	+68 seconds
Ayanamsa	24° 02' 39"
Synetic vernal point	05° ♓ 04' 21"
True obliquity of ecliptic	23° 26' 10"

MOON'S PHASES, APSIDES AND POSITIONS ☽

Date	h	m	Phase	Longitude ° '	Eclipse Indicator
03	13	56	☾	14 ♏ 54	
10	07	20	●	21 ♒ 43	
17	20	31	☽	29 ♉ 21	
25	20	26	○	07 ♍ 24	

Date	h	m		
07	12	03	Perigee	
19	06	28	Apogee	
06	00	24	Max dec	20° S 42'
12	07	35	0N	
19	13	34	Max dec	20° N 35'
26	19	03	0S	

LONGITUDES

Date	Chiron ⚷ ° '	Ceres ⚳ ° '	Pallas ⚴ ° '	Juno ⚵ ° '	Vesta ⚶ ° '	Black Moon Lilith ⚸ ° '
01	07 ♓ 42	19 ♊ 46	06 ♈ 44	16 ♑ 13	09 ♊ 58	15 ♊ 47
11	08 ♓ 19	19 ♊ 55	10 ♈ 18	19 ♑ 48	10 ♊ 43	16 ♊ 53
21	08 ♓ 57	20 ♊ 44	14 ♈ 07	23 ♑ 18	12 ♊ 06	18 ♊ 00
31	09 ♓ 36	22 ♊ 10	18 ♈ 09	26 ♑ 45	14 ♊ 00	19 ♊ 07

All ephemeris data is given at 12.00 UT and the Moon's longitude is additionally given for 24.00 UT

ASPECTARIAN

h m	Aspects
01 Friday	
01 16	☽ □ ♇
02 50	☽ ⊻ ♄
05 04	☽ △ ☉
08 30	☽ ꝑ ♂
13 14	☽ ꝑ ♆
15 22	♂ ⊥ ♅
20 46	☽ ꝑ ♃
02 Saturday	
00 08	☽ ∥ ♆
00 12	♀ # ♃
01 03	☽ △ ☿
04 17	♀ ⊻ ♂
06 38	☽ ∥ ♂
09 15	☽ Q ♇
09 43	☽ ∥ ♄
12 38	☽ ± ♃
12 39	☽ △ ♂
12 57	☽ □ ♀
15 45	☽ △ ♆
20 43	☽ ∥ ☿
22 00	☽ ⊼ ♅
23 07	☽ ⊼ ♃
03 Sunday	
00 00	☿ ∠ ♇
06 14	☽ ⚹ ♇
07 45	☽ ☌ ♄
08 26	☽ ± ♅
09 58	☽ ∥ ☉
13 56	☽ □ ☉
20 21	♀ ⊻ ♆
23 15	☽ Q ♀
23 59	☽ ꝑ ♅
04 Monday	
08 04	☽ ∠ ♇
12 31	☽ □ ☿
19 25	☽ □ ♂
19 29	☽ □ ♆
19 43	☽ ∥ ♇
20 56	♂ ☌ ♆
21 40	☽ ⚹ ♀
23 02	☽ Q ☉
23 22	☽ ⊥ ♇
05 Tuesday	
00 48	☽ ∥ ♀
01 33	☽ △ ♅
02 32	☽ ☍ ♃
03 46	☉ ⊥ ♇
05 42	☿ ∥ ♄
09 29	☽ ⊻ ♇
10 56	☽ ⊻ ♄
12 17	☿ ⊥ ♅
20 42	☽ ⚹ ☉
20 59	☽ ⊥ ♄
22 58	☽ Q ☿
06 Wednesday	
01 13	☽ ∠ ♀
01 45	☽ Q ♆
03 21	☽ Q ♂
04 38	♀ ∥ ♇
05 46	♂ ∥ ♆
11 57	☽ ∠ ♄
17 33	☽ ⊥ ♀
19 08	♀ ⚹ ♅
21 31	☽ ⚹ ☿
21 41	☽ ⚹ ♆
23 00	☿ ☌ ♆
23 26	☽ ∠ ☉
07 Thursday	
00 21	☽ ⚹ ♂
03 37	☽ □ ♅
04 23	☽ ⊻ ♀
04 31	☽ ⊼ ♃
04 50	☽ ∥ ♇
05 57	♀ △ ♃
09 24	☽ ∥ ♀
11 20	☽ ☌ ♇
11 31	☿ ∥ ♆
12 44	☽ ⚹ ♄
14 24	☽ ± ♃
15 22	☽ ⊥ ☉
22 26	☽ ∠ ♆
08 Friday	
01 24	☽ ∠ ☿
01 57	☽ ⊻ ☉
02 26	☽ ∠ ♂
05 13	☽ ꝑ ♃
05 25	☿ ∥ ♂
08 28	☽ Q ♄
09 18	☽ Q ♅
13 15	☽ ⊥ ♆
17 57	☿ ☌ ♂
18 05	☽ ⊥ ♂
18 05	☽ ⊥ ☿
23 10	☽ ⊻ ♆
09 Saturday	
01 30	☽ ∥ ☉
04 14	☿ ⊻ ♅
04 32	☽ ⊻ ♂
05 07	☽ ⚹ ♅
05 14	☽ ⊻ ☿
05 59	☽ △ ♃
10 25	☽ ☌ ♀
12 08	☿ □ ♃
12 27	☽ ∥ ♄
12 50	☽ ⊻ ♇
13 57	☉ ∠ ♅
14 12	☽ □ ♄
15 49	♂ ⊻ ♅
22 16	☽ ∥ ♆
22 50	☽ ⊥ ♇
10 Sunday	
04 22	☽ ∥ ♂
06 09	☽ ∠ ♅
07 20	☽ ☌ ☉
08 12	♂ □ ♃
10 15	☽ ∥ ☿
13 58	☽ ∠ ♇
15 03	♀ ⊻ ♇
21 24	☽ ⊥ ♅
11 Monday	
01 29	☽ ☌ ♆
06 42	♀ □ ♄
07 41	☽ ⊻ ♅
08 33	☽ □ ♃
09 56	☽ ☌ ♂
14 00	☽ ☌ ☿
15 40	☽ ⚹ ♇
17 03	☽ △ ♄
18 04	☽ ⊻ ♀
22 37	☽ # ♅
12 Tuesday	
05 38	☽ ⊥ ♀
06 00	☿ ⚹ ♇
12 47	☽ Q ♇
15 12	☽ ⊻ ☉
16 11	☽ Q ♃
16 44	☽ ∥ ♅
19 15	☿ △ ♄
19 32	☽ ꝑ ♄
23 19	☽ ∠ ♀
13 Wednesday	
02 53	☽ ⊥ ☉
06 26	☽ ⊻ ♆
11 53	☽ ± ♄
13 05	☽ ☌ ♅
14 01	☽ ⚹ ♃
17 28	☽ ⊥ ♆
17 29	☽ # ☿
18 46	☽ ⊻ ♂
20 46	☽ ∠ ☉
21 34	☽ □ ♇
22 58	☽ ⊼ ♄
14 Thursday	
02 12	☽ ⊻ ☿
05 51	☽ ⚹ ♀
06 41	☽ ⊥ ♂
06 46	☽ # ♂
07 19	☉ ∠ ♇
09 30	☉ ∥ ♄
10 19	☽ ∠ ♆
14 46	☽ ⊥ ☿
18 12	☽ ∠ ♃
20 40	☽ # ♆
15 Friday	
00 55	☽ ∠ ♂
03 35	☽ ⚹ ☉
06 24	☽ # ☉
07 15	☽ Q ♀
08 26	☽ # ♄
09 53	☽ ∠ ☿
11 36	☽ ⊥ ♃
12 39	♀ ⊥ ♇
15 12	☽ ⚹ ♆
18 17	♂ ⚹ ♇
22 23	☽ ⊻ ♅
23 23	☽ ⊻ ♃
16 Saturday	
05 07	☽ Q ☉
07 20	☽ △ ♇
08 13	☽ ⚹ ♂
08 44	☽ ☍ ♄
10 17	☽ ⊥ ♅
12 23	☽ # ♀
14 57	☽ Q ♆
16 14	♂ △ ♄
18 21	☽ ⚹ ☿
22 48	☽ □ ♀
17 Sunday	
04 20	☽ ∠ ♅
09 55	☽ Q ♂
13 26	☽ ꝑ ♇
20 31	☽ □ ☉
20 34	☽ Q ☿
18 Monday	
03 19	☽ □ ♆
04 29	☽ # ♇
07 43	☽ ± ♇
10 49	☽ ⚹ ♅
11 54	☽ ☌ ♃
17 02	♄ St R
19 56	☽ ⊼ ♇
21 15	☽ ⊼ ♄
21 59	☉ ⊥ ♅
19 Tuesday	
01 02	☽ □ ♂
05 37	♀ ∠ ♅
09 27	☽ ± ♄
11 20	☽ Q ♅
11 39	☽ □ ☿
15 51	☉ ∥ ♆
18 48	☽ △ ♀
20 Wednesday	
03 43	☽ ꝑ ♄
15 04	☽ △ ☉
16 24	☽ △ ♆
22 38	☽ # ♇
23 52	☽ □ ♅
21 Thursday	
01 02	☽ ⊻ ♃
04 36	☽ ꝑ ♀
07 18	☉ ☌ ♆
08 43	☽ ☍ ♇
09 52	☽ △ ♄
13 05	☽ ⊥ ♃
17 33	☿ # ♅
17 47	☽ △ ♂
22 25	☽ ꝑ ♆
23 44	☽ ꝑ ☉
22 Friday	
00 32	☽ ± ♀
02 08	☽ △ ☿
06 59	☽ ∠ ♃
13 39	☽ ⊼ ♀
16 08	☽ ± ♆
18 56	☽ ± ☉
20 53	♀ ∠ ♇
23 Saturday	
01 10	☽ ꝑ ♂
03 48	☽ ⊼ ♆
07 33	☽ ⊼ ☉
07 36	☽ ꝑ ☿
09 41	☿ St R
11 00	☽ △ ♅
11 48	☽ # ♀
12 14	☽ ⚹ ♃
18 10	☽ # ♄
19 19	☽ ⊼ ♇
19 35	☽ ± ♂
20 16	☽ □ ♄
24 Sunday	
00 42	☽ ± ☿
06 24	☽ # ♆
06 42	☽ ± ♇
07 40	☽ ⊼ ♂
11 06	☽ Q ♃
11 53	☽ ⊼ ☿
15 29	☽ ꝑ ♅
17 16	☽ # ☉
23 32	☽ ꝑ ♇
25 Monday	
04 28	☉ ⊻ ♅
04 50	☽ ☍ ♀
05 56	☽ Q ♄
08 14	☽ ± ♅
12 22	☽ ☍ ♆
15 09	☽ # ♂
16 39	♀ ∥ ♄
19 16	☽ ⊼ ♅
20 26	☽ ☍ ☉
20 34	☽ □ ♃
22 26	☉ □ ♃
26 Tuesday	
03 05	☽ △ ♇
03 51	☽ ⚹ ♄
08 21	☽ ∥ ♅
09 09	☿ ☌ ♂
12 29	☽ # ☿
17 22	☽ ☍ ☿
17 38	♀ ⊥ ♅
18 13	☽ ☍ ♂
27 Wednesday	
01 47	☽ ∥ ☿
05 46	☽ # ♅
06 45	☽ ∠ ♄
16 35	☽ ⊼ ♀
18 28	☽ ⊼ ♆
18 54	☽ ∥ ♂
22 40	☽ ⊥ ♄
28 Thursday	
01 11	☽ ☍ ♅
02 35	☽ △ ♃
04 14	☽ ± ♀
05 04	☽ ± ♆
06 22	☽ ⊼ ☉
08 37	☽ □ ♇
09 13	☽ ⊻ ♄
11 30	☽ ∥ ☉
13 37	♀ ☌ ♆
17 43	☽ ± ☉
19 57	☽ ⊼ ☿
20 53	☽ ꝑ ♆
21 35	☽ ꝑ ♀

MARCH 2013

LONGITUDES

Date	Sidereal time h m s	Sun ☉ ° ' "	Moon ☽ ° ' "	Moon ☽ 24.00 ° ' "	Mercury ☿ ° '	Venus ♀ ° '	Mars ♂ ° '	Jupiter ♃ ° '	Saturn ♄ ° '	Uranus ♅ ° '	Neptune ♆ ° '	Pluto ♇ ° '
01	22 37 49	11 ♓ 03 48	26 ♎ 47 48	03 ♏ 43 19	17 ♓ 02	04 ♓ 16	21 ♓ 36	07 ♊ 48	11 ♏ 26	06 ♈ 57	03 ♓ 08	11 ♑ 08
02	22 41 45	12 03 59	10 ♏ 40 20	17 ♏ 38 43	16 R 08	05 31	22 23	07 54	11 R 24	07 01	03 10	11 09
03	22 45 42	13 04 09	24 ♏ 38 21	01 ♐ 39 07	15 10	06 46	23 10	08 00	11 23	07 04	03 13	11 10
04	22 49 38	14 04 17	08 ♐ 40 58	15 ♐ 43 48	14 09	08 01	23 57	08 06	11 22	07 07	03 15	11 11
05	22 53 35	15 04 23	22 ♐ 47 32	29 ♐ 52 00	13 07	09 16	24 43	08 12	11 20	07 10	03 17	11 13
06	22 57 32	16 04 28	06 ♑ 57 03	14 ♑ 02 26	12 06	10 31	25 30	08 19	11 19	07 13	03 19	11 14
07	23 01 28	17 04 32	21 ♑ 07 51	28 ♑ 12 57	11 06	11 45	26 17	08 25	11 17	07 17	03 22	11 15
08	23 05 25	18 04 34	05 ♒ 17 18	12 ♒ 20 24	10 09	13 00	27 04	08 32	11 15	07 20	03 24	11 16
09	23 09 21	19 04 34	19 ♒ 21 44	26 ♒ 20 47	09 16	14 15	27 51	08 39	11 14	07 23	03 26	11 17
10	23 13 18	20 04 32	03 ♓ 16 58	10 ♓ 09 48	08 27	15 30	28 37	08 46	11 12	07 26	03 28	11 18
11	23 17 14	21 04 29	16 ♓ 58 48	23 ♓ 43 33	07 45	16 45	29 ♓ 24	08 53	11 10	07 30	03 31	11 19
12	23 21 11	22 04 23	00 ♈ 23 45	06 ♈ 59 08	07 08	18 00	00 ♈ 11	09 01	11 07	07 33	03 33	11 20
13	23 25 07	23 04 16	13 ♈ 29 37	19 ♈ 55 09	06 37	19 15	00 57	09 08	11 05	07 36	03 35	11 21
14	23 29 04	24 04 06	26 ♈ 15 50	02 ♉ 31 51	06 13	20 29	01 44	09 16	11 03	07 40	03 37	11 22
15	23 33 01	25 03 55	08 ♉ 43 29	14 ♉ 51 05	05 55	21 44	02 31	09 24	11 01	07 43	03 39	11 23
16	23 36 57	26 03 41	20 ♉ 55 07	26 ♉ 56 04	05 43	22 59	03 17	09 32	10 58	07 47	03 41	11 24
17	23 40 54	27 03 25	02 ♊ 54 30	08 ♊ 51 00	05 38	24 14	04 04	09 40	10 55	07 50	03 44	11 24
18	23 44 50	28 03 07	14 ♊ 46 12	20 ♊ 40 45	05 D 39	25 28	04 50	09 48	10 53	07 53	03 46	11 25
19	23 48 47	29 ♓ 02 47	26 ♊ 35 19	02 ♋ 30 33	05 46	26 43	05 37	09 56	10 50	07 57	03 48	11 26
20	23 52 43	00 ♈ 02 24	08 ♋ 27 08	14 ♋ 25 40	05 59	27 58	06 23	10 05	10 47	08 00	03 50	11 27
21	23 56 40	01 01 59	20 ♋ 26 48	26 ♋ 31 06	06 16	29 ♓ 13	07 09	10 13	10 44	08 04	03 52	11 27
22	00 00 36	02 01 32	02 ♌ 39 04	08 ♌ 51 12	06 39	00 ♈ 27	07 56	10 22	10 41	08 07	03 54	11 28
23	00 04 33	03 01 03	15 ♌ 07 52	21 ♌ 29 23	07 07	01 42	08 42	10 31	10 38	08 10	03 56	11 29
24	00 08 30	04 00 31	27 ♌ 55 58	04 ♍ 27 43	07 39	02 56	09 28	10 40	10 35	08 14	03 58	11 29
25	00 12 26	04 59 57	11 ♍ 04 41	17 ♍ 46 44	08 15	04 11	10 14	10 49	10 32	08 17	04 00	11 30
26	00 16 23	05 59 21	24 ♍ 33 40	01 ♎ 25 12	08 55	05 26	11 01	10 58	10 29	08 21	04 02	11 31
27	00 20 19	06 58 43	08 ♎ 20 55	15 ♎ 20 22	09 39	06 40	11 47	11 07	10 25	08 24	04 04	11 31
28	00 24 16	07 58 03	22 ♎ 23 01	29 ♎ 28 19	10 27	07 55	12 33	11 17	10 22	08 27	04 06	11 32
29	00 28 12	08 57 21	06 ♏ 35 41	13 ♏ 44 32	11 18	09 09	13 19	11 27	10 18	08 31	04 08	11 32
30	00 32 09	09 56 37	20 ♏ 54 20	28 ♏ 04 34	12 12	10 24	14 05	11 36	10 15	08 34	04 10	11 32
31	00 36 05	10 ♈ 55 51	05 ♐ 14 46	12 ♐ 24 31	13 ♓ 09	11 ♈ 38	14 ♈ 51	11 ♊ 46	10 ♏ 11	08 ♈ 38	04 ♓ 12	11 ♑ 33

Date	Moon True ☊ ° '	Moon Mean ☊ ° '	Moon ☽ Latitude ° '
01	19 ♏ 08	20 ♏ 27	01 S 56
02	19 R 06	20 23	00 S 45
03	19 D 06	20 20	00 N 30
04	19 06	20 17	01 42
05	19 R 06	20 14	02 50
06	19 04	20 11	03 46
07	19 00	20 08	04 29
08	18 53	20 04	04 56
09	18 44	20 01	05 04
10	18 34	19 58	04 54
11	18 24	19 55	04 28
12	18 15	19 52	03 47
13	18 08	19 48	02 54
14	18 03	19 45	01 54
15	18 01	19 42	00 N 50
16	18 D 00	19 39	00 S 16
17	18 01	19 36	01 20
18	18 02	19 33	02 19
19	18 03	19 29	03 13
20	18 R 03	19 26	03 58
21	18 01	19 23	04 34
22	17 57	19 20	04 58
23	17 51	19 17	05 08
24	17 44	19 13	05 04
25	17 36	19 10	04 44
26	17 29	19 07	04 07
27	17 23	19 04	03 16
28	17 19	19 01	02 11
29	17 17	18 58	00 S 58
30	17 D 16	18 54	00 N 20
31	17 ♏ 17	18 ♏ 51	01 N 37

DECLINATIONS

Date	Sun ☉ ° '	Moon ☽ ° '	Mercury ☿ ° '	Venus ♀ ° '	Mars ♂ ° '	Jupiter ♃ ° '	Saturn ♄ ° '	Uranus ♅ ° '	Neptune ♆ ° '	Pluto ♇ ° '
01	07 S 25	12 S 08	01 S 46	11 S 12	04 S 07	21 N 08	12 S 49	02 N 08	10 S 55	19 S 42
02	07 02	15 44	02 04	10 45	03 48	21 09	12 48	02 10	10 55	19 42
03	06 39	18 27	02 26	10 18	03 29	21 10	12 48	02 11	10 54	19 42
04	06 16	20 04	02 51	09 51	03 10	21 11	12 47	02 12	10 53	19 42
05	05 53	20 25	03 18	09 23	02 50	21 12	12 47	02 14	10 52	19 42
06	05 30	19 29	03 46	08 55	02 31	21 14	12 46	02 15	10 51	19 42
07	05 06	17 20	04 16	08 27	02 12	21 15	12 45	02 16	10 51	19 41
08	04 43	14 09	04 46	07 59	01 53	21 16	12 44	02 17	10 50	19 41
09	04 19	10 11	05 15	07 31	01 34	21 17	12 44	02 19	10 49	19 41
10	03 56	05 43	05 44	07 02	01 15	21 19	12 43	02 20	10 48	19 41
11	03 32	01 S 02	06 12	06 33	00 56	21 20	12 42	02 21	10 48	19 41
12	03 09	03 N 37	06 39	06 04	00 37	21 21	12 41	02 23	10 47	19 41
13	02 45	08 00	07 03	05 35	00 S 18	21 23	12 40	02 24	10 46	19 41
14	02 21	11 55	07 25	05 05	00 N 01	21 24	12 39	02 25	10 45	19 41
15	01 58	15 12	07 46	04 36	00 20	21 25	12 39	02 27	10 44	19 41
16	01 34	17 44	08 04	04 06	00 39	21 27	12 38	02 28	10 44	19 41
17	01 10	19 26	08 19	03 36	00 58	21 28	12 37	02 29	10 43	19 41
18	00 46	20 16	08 32	03 07	01 17	21 29	12 36	02 31	10 42	19 41
19	00 S 23	20 11	08 43	02 37	01 36	21 31	12 35	02 32	10 41	19 41
20	00 N 01	19 12	08 52	02 07	01 55	21 32	12 34	02 33	10 41	19 41
21	00 25	17 22	08 58	01 37	02 13	21 34	12 33	02 35	10 40	19 40
22	00 48	14 43	09 01	01 06	02 32	21 35	12 31	02 36	10 39	19 40
23	01 12	11 23	09 03	00 36	02 51	21 37	12 30	02 38	10 38	19 40
24	01 36	07 26	09 02	00 S 06	03 10	21 38	12 29	02 39	10 38	19 40
25	01 59	03 N 02	09 00	00 N 24	03 28	21 40	12 28	02 40	10 37	19 40
26	02 23	01 S 37	08 55	00 54	03 47	21 41	12 27	02 42	10 36	19 40
27	02 46	06 18	08 48	01 25	04 05	21 42	12 26	02 43	10 36	19 40
28	03 10	10 44	08 40	01 55	04 24	21 44	12 25	02 44	10 35	19 40
29	03 33	14 38	08 29	02 25	04 42	21 45	12 23	02 46	10 34	19 40
30	03 56	17 40	08 17	02 55	05 01	21 47	12 22	02 47	10 34	19 40
31	04 N 20	19 S 35	08 S 03	03 N 25	05 N 19	21 N 48	12 S 21	02 N 48	10 S 33	19 S 40

ZODIAC SIGN ENTRIES

Date	h	m	Planets
01	17	33	☽ ♏
03	21	11	☽ ♐
06	00	14	☽ ♑
08	03	01	☽ ♒
10	06	19	☽ ♓
12	06	26	♂ ♈
12	11	17	☽ ♈
14	19	08	☽ ♉
17	06	09	☽ ♊
19	18	55	☽ ♋
20	11	02	☉ ♈
22	03	15	♀ ♈
22	06	50	☽ ♌
24	15	49	☽ ♍
26	21	32	☽ ♎
29	00	53	☽ ♏
31	03	13	☽ ♐

LATITUDES

Date	Mercury ☿ ° '	Venus ♀ ° '	Mars ♂ ° '	Jupiter ♃ ° '	Saturn ♄ ° '	Uranus ♅ ° '	Neptune ♆ ° '	Pluto ♇ ° '
01	03 N 39	01 S 21	00 S 51	00 S 29	02 N 34	00 S 41	00 S 37	03 N 17
04	03 40	01 23	00 49	00 28	02 35	00 40	00 37	03 17
07	03 23	01 25	00 48	00 28	02 35	00 40	00 37	03 17
10	02 52	01 26	00 46	00 27	02 36	00 40	00 37	03 17
13	02 11	01 26	00 45	00 27	02 36	00 40	00 37	03 17
16	01 27	01 26	00 43	00 26	02 37	00 40	00 37	03 17
19	00 44	01 25	00 41	00 26	02 38	00 40	00 37	03 17
22	00 N 03	01 24	00 40	00 25	02 38	00 40	00 37	03 17
25	00 S 34	01 22	00 38	00 25	02 39	00 40	00 37	03 17
28	01 06	01 20	00 36	00 24	02 39	00 40	00 37	03 17
31	01 S 33	01 S 17	00 S 35	00 S 24	02 N 39	00 S 40	00 S 37	03 N 17

DATA

Julian Date	2456353
Delta T	+68 seconds
Ayanamsa	24° 02' 42"
Synetic vernal point	05° ♓ 04' 17"
True obliquity of ecliptic	23° 26' 10"

MOON'S PHASES, APSIDES AND POSITIONS ☽

Date	h	m	Phase	Longitude ° '	Eclipse Indicator
04	21	53	☾	14 ♐ 29	
11	19	51	●	21 ♓ 24	
19	17	27	☽	29 ♊ 16	
27	09	27	○	06 ♎ 52	

Day	h	m		
05	23	11	Perigee	
19	03	13	Apogee	
31	04	02	Perigee	
05	06	39	Max dec	20° S 27'
11	17	14	0N	
18	21	49	Max dec	20° N 20'
26	03	43	0S	

LONGITUDES

Date	Chiron ⚷ ° '	Ceres ⚳ ° '	Pallas ⚴ ° '	Juno ⚵ ° '	Vesta ⚶ ° '	Black Moon Lilith ⚸ ° '
01	09 ♓ 28	21 ♊ 50	17 ♈ 20	26 ♑ 04	13 ♊ 35	18 ♊ 53
11	10 ♓ 07	23 ♊ 41	21 ♈ 32	29 ♑ 25	15 ♊ 52	20 ♊ 00
21	10 ♓ 44	25 ♊ 59	25 ♈ 55	02 ♒ 37	18 ♊ 32	21 ♊ 07
31	11 ♓ 20	28 ♊ 42	00 ♉ 28	05 ♒ 40	21 ♊ 33	22 ♊ 13

All ephemeris data is given at 12.00 UT and the Moon's longitude is additionally given for 24.00 UT

ASPECTARIAN

h m	Aspects
01 Friday	
02 25	☽ ⊼ ♂
04 54	☽ ∥ ♆
05 01	☽ bQ ♃
05 50	☽ ± ☿
07 01	☽ ∥ ♀
10 38	☽ bQ ☉
13 28	☽ ± ♂
13 38	☉ ⚹ ♇
16 03	☽ Q ♇
16 14	☽ ∥ ♄
20 31	☉ △ ♄
20 32	☽ bQ ☿
20 45	☽ ± ♃
23 01	☽ △ ♆
02 Saturday	
02 13	☽ △ ♀
03 18	♀ ∥ ♆
05 39	☽ ⊼ ♅
05 59	☽ bQ ♂
07 11	☽ ⊼ ♃
12 50	☽ ⚹ ♇
13 16	☽ ☌ ♄
14 35	☽ △ ☉
16 02	☽ ± ♅
18 51	☿ # ♅
20 48	☽ △ ☿
03 Sunday	
07 34	☽ bQ ♅
09 19	☽ △ ♂
12 21	♂ Q ♇
14 38	☽ ∠ ♇
17 58	♀ ⊻ ♅
04 Monday	
02 42	☽ □ ♆
04 41	☽ ∥ ♇
06 02	☽ ⊥ ♇
09 19	☽ △ ♅
10 45	☽ □ ♀
11 00	☽ ☍ ♃
12 58	☉ ☌ ☿
13 52	♀ □ ♃
16 17	☽ ⊻ ♇
16 34	☽ ⊻ ♄
20 41	☽ □ ☿
21 53	☽ □ ☉
21 58	☿ ∥ ♂
05 Tuesday	
02 45	☽ ⊥ ♄
09 26	☽ Q ♆
15 28	☽ □ ♂
18 00	☽ ∠ ♄
20 19	☽ Q ♀
06 Wednesday	
01 10	☽ Q ☿
05 50	☽ ⚹ ♆
06 46	☽ Q ☉
08 38	☽ ∥ ♇
12 28	☽ □ ♅
14 20	☽ ⊼ ♃
18 36	☽ ⚹ ♀
19 15	☽ ☌ ♇
19 22	☽ ⚹ ♄
20 08	☽ ⚹ ☿
23 44	☽ Q ♂
07 Thursday	
00 34	☽ ± ♃
02 03	♀ ⚹ ♇
03 07	♀ △ ♄
04 37	☽ ⚹ ☉
04 54	☿ ☌ ♀
07 17	☿ △ ♄
07 18	☽ ∠ ♆
07 35	♂ # ♅
08 23	☿ ⚹ ♇
12 01	♂ bQ ♄
15 38	☽ Q ♄
15 55	☽ bQ ♃
16 59	♂ Q ♃
19 03	☽ Q ♅
19 52	☽ ∠ ☿
21 14	☽ ⚹ ♂
22 27	☽ ∠ ♀
22 35	☽ ⊥ ♆
08 Friday	
07 06	♄ ⚹ ♇
07 57	☽ ∠ ☉
08 47	☽ ⊻ ♆
10 11	☽ ⊥ ☿
10 40	☉ ∥ ☿
15 12	☽ ⊥ ♀
15 29	☽ ⚹ ♅
17 34	☽ △ ♃
19 46	☽ ⊻ ☿
21 04	☽ ∥ ♄
22 08	☽ □ ♄
22 11	☽ ⊻ ♇
09 Saturday	
00 12	☽ ∠ ♂
00 26	☽ ⊥ ☉
02 25	☽ ⊻ ♀
08 25	☽ ∥ ♆
08 26	☽ ⊥ ♇
11 28	☽ ⊻ ☉
16 31	☽ ⊥ ♂
17 12	☽ ∠ ♅
23 54	☽ ∠ ♇
10 Sunday	
03 27	☽ ⊻ ♂
03 37	☿ □ ♃
04 20	☽ ∥ ♀
08 48	☽ ⊥ ♅
11 54	☽ ∥ ☿
12 20	☽ ☌ ♆
19 16	☽ ⊻ ♅
20 33	☽ ☌ ☿
21 39	☽ □ ♃
22 03	☽ ∥ ☉
11 Monday	
01 46	☽ △ ♄
02 01	☽ ⚹ ♇
05 16	☽ # ♅
11 33	☽ ☌ ♀
12 30	☽ ∥ ♂
19 51	☽ ☌ ☉
20 21	☿ ⊻ ♅
20 55	☿ ∥ ♀
21 24	☽ # ♂
23 17	☽ Q ♇
12 Tuesday	
04 19	☽ bQ ♄
05 27	☽ ∥ ♅
05 50	☽ Q ♃
09 41	☽ # ☉
11 35	☽ ☌ ♂
17 44	☽ ⊻ ♆
20 34	☽ ± ♄
23 46	☽ ⊻ ☿
23 49	☽ # ♀
13 Wednesday	
01 06	☽ ☌ ♅
03 52	☽ ⚹ ♃
04 45	☽ ⊥ ♆
06 03	☽ # ☿
07 33	☽ ⊼ ♄
08 02	☽ □ ♇
10 26	☽ ⊥ ☿
18 48	☉ Q ♇
21 31	☽ ∠ ♆
23 53	☽ ⊻ ♀
14 Thursday	
02 42	☽ ∠ ☿
04 33	☽ # ♆
07 28	☽ ⊻ ☉
08 06	☉ # ♅
08 10	☽ ∠ ♃
12 29	☽ ⊥ ♀
17 02	☽ # ♄
19 54	☽ ⊥ ☉
23 10	☽ ⊻ ♂
15 Friday	
01 33	☽ ⊥ ♃
02 08	☽ ⚹ ♆
06 39	☽ ⚹ ☿
07 42	☽ ∠ ♀
10 02	☽ ⊻ ♅
11 33	☽ ⊥ ♂
13 19	☽ ⊻ ♃
14 51	☽ ∠ ☉
16 27	☽ ☍ ♄
17 11	☽ △ ♇
21 49	☽ ⊥ ♅
16 Saturday	
01 37	☽ Q ♆
05 45	☽ Q ☿
06 25	☽ ∠ ♂
09 50	☉ bQ ♄
15 43	☽ ∠ ♅
16 35	☽ ⚹ ♀
20 01	♀ Q ♇
22 56	☽ bQ ♇
23 11	☽ ⚹ ☉
17 Sunday	
01 04	♂ ⊻ ♆
13 39	☽ □ ♆
14 29	☽ ⚹ ♂
16 55	☽ # ♇
17 03	☽ ± ♇
17 30	☽ □ ☿
18 56	☉ # ♂
19 29	☽ Q ♀
20 03	☿ St D
21 59	☽ ⚹ ♅
18 Monday	
01 33	☽ Q ☉
01 48	☽ ☌ ♃
04 08	☽ ⊼ ♄
04 52	☉ Q ♃
05 12	☽ ⊼ ♇
13 17	♂ ± ♄
16 16	☽ ± ♄
16 29	☽ Q ♂
19 33	♀ bQ ♄
22 27	☽ Q ♅
19 Tuesday	
10 29	☽ bQ ♄
12 18	☽ □ ♀
15 31	♀ # ♅
17 27	☽ □ ☉
18 20	☿ ⊻ ♂
20 Wednesday	
02 39	☽ △ ♆
02 52	☽ # ♇
06 54	☽ △ ☿
07 32	☽ □ ♂
11 05	☽ □ ♅
14 28	♀ Q ♃
15 19	☽ ⊻ ♃
16 41	☽ △ ♄
17 56	♀ # ♂
18 02	☽ ☍ ♇
21 Thursday	
03 29	☽ ⊥ ♃
08 51	☽ bQ ♆
13 41	☽ bQ ☿
21 34	☽ ∠ ♃
22 Friday	
02 41	☽ ± ♆
07 14	☽ △ ♀
07 58	☽ ± ☿
10 41	☽ △ ☉
14 26	☽ ⊼ ♆
17 28	♂ ∥ ♅
18 17	♂ ☌ ♅
20 02	☽ ⊼ ☿
20 05	☉ # ♀
22 38	☽ △ ♅
22 54	☽ △ ♂
23 Saturday	
03 05	☽ ⚹ ♃
03 28	☽ □ ♄
04 22	☽ # ♄
05 02	☽ ⊼ ♇
15 18	☽ bQ ♀
16 27	☽ ± ♇
16 45	☽ # ♆
17 56	☽ bQ ☉
24 Sunday	
02 05	☽ Q ♃
02 32	☽ # ☿
02 51	♃ ⊼ ♄
03 13	☽ bQ ♅
05 10	☽ bQ ♂
09 19	☽ bQ ♇
09 58	☽ ± ♀
10 59	☉ ⊻ ♆
12 09	☽ ± ☉
13 12	☽ Q ♄
19 57	☽ ± ♅
22 11	☽ ⊼ ♀
22 49	☽ ± ♂
23 08	☽ ☍ ♆
25 Monday	
00 05	☽ ⊼ ☉
01 17	☉ ± ♄
04 15	♂ ⊥ ♆
06 38	☽ ☍ ☿
06 56	☽ ⊼ ♅
08 22	♀ ⊻ ♆
09 50	☽ ∥ ♂
10 24	☽ ⊼ ♂
11 01	☽ ⚹ ♄
11 31	☽ □ ♃
12 46	☽ △ ♇
13 34	☿ ⊻ ♅
13 54	☽ ∥ ♅
17 03	☽ ∥ ☉
18 26	♀ ± ♄
20 32	♂ ⊼ ♄
26 Tuesday	
00 20	☽ ∥ ♀
07 55	☽ # ♀
10 26	♂ ⚹ ♃
13 36	☽ ∠ ♄
16 12	☽ # ☉
17 29	☽ # ♅
23 47	☽ # ♂
27 Wednesday	
03 47	♂ □ ♇
04 35	☽ ⊼ ♆
05 14	☽ ⊥ ♄
08 27	☉ ∥ ♅
08 49	☽ ☍ ♀
09 27	☽ ☍ ☉
12 05	☽ ☍ ♅
14 23	☽ ⊼ ☿
14 58	☽ ± ♆
15 33	☽ ⊻ ♄
16 50	☽ △ ♃
17 27	☽ □ ♇
18 14	☽ ☍ ♂
28 Thursday	
00 52	☽ ∥ ☿
01 16	☽ ± ☿
06 24	☽ bQ ♆
09 48	☿ △ ♄
11 06	☽ ∥ ♆
17 05	☉ ☌ ♀
17 31	☽ bQ ☿
18 41	☽ bQ ♃
21 45	☽ ∥ ♄
23 01	♀ ☌ ♅
29 Friday	
00 06	☽ Q ♇
00 38	☉ ☌ ♅
07 51	☽ △ ♆
10 02	☽ ± ♃
15 14	☽ ⊼ ♅
16 16	☽ ⊼ ☉
16 43	☽ ⊼ ♀
16 56	☿ □ ♃
18 13	☽ ☌ ♄
18 34	☿ ⚹ ♇
20 14	☽ ⊼ ♃
20 18	☽ ⚹ ♇
20 25	☽ △ ☿
23 55	☽ ⊼ ♂
30 Saturday	
01 21	☽ ± ♅
02 19	♃ ⊼ ♇
03 04	☽ ± ☉
03 44	☽ ± ♀
05 15	♀ ∥ ♅
07 24	♀ ⊥ ♆
09 14	♀ ⊼ ♄
10 32	☽ ± ♂
16 29	☽ bQ ♅
17 36	☉ ⊥ ♆
18 58	☉ ⊼ ♄
19 15	☽ bQ ☉
20 14	☽ bQ ♀
21 26	☽ ∠ ♇
31 Sunday	
02 27	☽ bQ ♂
10 12	♀ □ ♇
10 15	☽ □ ♆
12 30	☽ ⊥ ♇
13 26	☽ ∥ ♇
14 49	♀ ⚹ ♃
17 41	☽ △ ♅
20 14	☽ ⊻ ♄
22 14	☽ △ ☉
22 34	☽ ⊻ ♇
23 03	☽ ☍ ♃
23 44	☽ △ ♀

APRIL 2013

LONGITUDES

Date	Sidereal time h m s	Sun ☉ ° ' "	Moon ☽ ° ' "	Moon ☽ 24.00 ° ' "	Mercury ☿ ° '	Venus ♀ ° '	Mars ♂ ° '	Jupiter ♃ ° '	Saturn ♄ ° '	Uranus ♅ ° '	Neptune ♆ ° '	Pluto ♇ ° '
01	00 40 02	11 ♈ 55 04	19 ♐ 33 27	26 ♐ 41 16	14 ♓ 09	12 ♈ 53	15 ♈ 37	11 ♊ 56	10 ♏ 08	08 ♈ 41	04 ♓ 14	11 ♑ 33
02	00 43 59	12 54 14	03 ♑ 47 41	10 ♑ 52 29	15 11	14 07	16 22	12 06	10 R 04	08 45	04 16	11 34
03	00 47 55	13 53 23	17 ♑ 55 29	24 ♑ 56 29	16 16	15 22	17 08	12 16	10 00	08 48	04 18	11 34
04	00 51 52	14 52 31	01 ♒ 55 19	08 ♒ 51 51	17 24	16 36	17 54	12 26	09 56	08 51	04 19	11 34
05	00 55 48	15 51 36	15 ♒ 45 55	22 ♒ 37 22	18 34	17 51	18 40	12 37	09 52	08 55	04 21	11 34
06	00 59 45	16 50 40	29 ♒ 26 00	06 ♓ 11 42	19 46	19 05	19 25	12 47	09 48	08 58	04 23	11 35
07	01 03 41	17 49 42	12 ♓ 54 16	19 ♓ 33 34	21 00	20 20	20 11	12 58	09 44	09 02	04 25	11 35
08	01 07 38	18 48 42	26 ♓ 09 28	02 ♈ 41 50	22 16	21 34	20 57	13 08	09 40	09 05	04 27	11 35
09	01 11 34	19 47 40	09 ♈ 10 35	15 ♈ 35 39	23 35	22 48	21 42	13 19	09 36	09 08	04 28	11 35
10	01 15 31	20 46 36	21 ♈ 57 02	28 ♈ 14 47	24 55	24 03	22 28	13 30	09 32	09 12	04 30	11 35
11	01 19 28	21 45 30	04 ♉ 28 57	10 ♉ 39 43	26 17	25 17	23 13	13 41	09 28	09 15	04 32	11 35
12	01 23 24	22 44 22	16 ♉ 47 14	22 ♉ 51 46	27 41	26 31	23 59	13 52	09 23	09 19	04 33	11 35
13	01 27 21	23 43 12	28 ♉ 53 38	04 ♊ 53 10	29 ♓ 07	27 46	24 44	14 03	09 19	09 22	04 35	11 R 35
14	01 31 17	24 42 00	10 ♊ 50 46	16 ♊ 46 53	00 ♈ 34	29 ♈ 00	25 29	14 14	09 15	09 25	04 37	11 35
15	01 35 14	25 40 46	22 ♊ 42 00	28 ♊ 36 38	02 04	00 ♉ 14	26 15	14 25	09 11	09 29	04 38	11 35
16	01 39 10	26 39 30	04 ♋ 31 21	10 ♋ 26 42	03 35	01 28	27 00	14 36	09 06	09 32	04 40	11 35
17	01 43 07	27 38 11	16 ♋ 23 17	22 ♋ 21 42	05 07	02 43	27 45	14 48	09 02	09 35	04 42	11 35
18	01 47 03	28 36 50	28 ♋ 22 34	04 ♌ 26 28	06 42	03 57	28 30	14 59	08 57	09 38	04 43	11 35
19	01 51 00	29 ♈ 35 27	10 ♌ 33 59	16 ♌ 45 39	08 18	05 11	29 ♈ 15	15 11	08 53	09 42	04 45	11 35
20	01 54 57	00 ♉ 34 02	23 ♌ 02 00	29 ♌ 23 28	09 56	06 25	00 ♉ 00	15 23	08 48	09 45	04 46	11 34
21	01 58 53	01 32 35	05 ♍ 50 27	12 ♍ 23 14	11 35	07 39	00 45	15 34	08 44	09 48	04 47	11 34
22	02 02 50	02 31 05	19 ♍ 02 02	25 ♍ 46 56	13 16	08 53	01 30	15 46	08 39	09 51	04 49	11 34
23	02 06 46	03 29 33	02 ♎ 37 54	09 ♎ 34 46	14 59	10 08	02 15	15 58	08 35	09 55	04 50	11 33
24	02 10 43	04 27 59	16 ♎ 37 14	23 ♎ 44 52	16 43	11 22	03 00	16 10	08 30	09 58	04 52	11 33
25	02 14 39	05 26 24	00 ♏ 57 07	08 ♏ 13 17	18 29	12 36	03 45	16 22	08 26	10 01	04 53	11 33
26	02 18 36	06 24 46	15 ♏ 32 37	22 ♏ 54 17	20 17	13 50	04 29	16 34	08 21	10 04	04 54	11 32
27	02 22 32	07 23 07	00 ♐ 17 24	07 ♐ 41 06	22 07	15 04	05 14	16 46	08 17	10 07	04 56	11 32
28	02 26 29	08 21 26	15 ♐ 04 29	22 ♐ 26 44	23 58	16 18	05 59	16 58	08 12	10 11	04 57	11 31
29	02 30 26	09 19 44	29 ♐ 47 05	07 ♑ 04 54	25 51	17 32	06 43	17 11	08 08	10 14	04 58	11 31
30	02 34 22	10 ♉ 18 00	14 ♑ 19 34	21 ♑ 30 38	27 ♈ 46	18 ♉ 46	07 ♉ 28	17 ♊ 23	08 ♏ 03	10 ♈ 17	04 ♓ 59	11 ♑ 30

Date	Moon True ☊ ° '	Moon Mean ☊ ° '	Moon ☽ Latitude ° '
01	17 ♏ 19	18 ♏ 48	02 N 47
02	17 D 20	18 45	03 47
03	17 R 20	18 42	04 32
04	17 19	18 39	05 01
05	17 16	18 35	05 12
06	17 12	18 32	05 06
07	17 08	18 29	04 42
08	17 03	18 26	04 03
09	16 59	18 23	03 12
10	16 56	18 19	02 13
11	16 54	18 16	01 08
12	16 53	18 13	00 N 01
13	16 D 54	18 10	01 S 06
14	16 55	18 07	02 08
15	16 57	18 04	03 05
16	16 58	18 00	03 54
17	16 59	17 57	04 32
18	17 00	17 54	05 00
19	16 R 59	17 51	05 14
20	16 58	17 48	05 14
21	16 57	17 45	04 59
22	16 55	17 41	04 28
23	16 53	17 38	03 41
24	16 51	17 35	02 40
25	16 50	17 32	01 27
26	16 50	17 29	00 S 07
27	16 D 50	17 25	01 N 14
28	16 51	17 22	02 30
29	16 51	17 19	03 37
30	16 ♏ 52	17 ♏ 16	04 N 28

DECLINATIONS

Date	Sun ☉ ° '	Moon ☽ ° '	Mercury ☿ ° '	Venus ♀ ° '	Mars ♂ ° '	Jupiter ♃ ° '	Saturn ♄ ° '	Uranus ♅ ° '	Neptune ♆ ° '	Pluto ♇ ° '
01	04 N 43	20 S 15	07 S 47	03 N 55	05 N 37	21 N 50	12 S 20	02 N 50	10 S 32	19 S 40
02	05 06	19 36	07 30	04 25	05 55	21 51	12 18	02 51	10 31	19 40
03	05 29	17 44	07 11	04 55	06 14	21 53	12 17	02 52	10 31	19 40
04	05 52	14 50	06 50	05 25	06 32	21 54	12 16	02 54	10 30	19 40
05	06 14	11 08	06 28	05 54	06 50	21 56	12 14	02 55	10 30	19 40
06	06 37	06 54	06 05	06 24	07 07	21 57	12 13	02 56	10 29	19 40
07	07 00	02 S 22	05 40	06 53	07 25	21 59	12 12	02 58	10 28	19 40
08	07 22	02 N 12	05 14	07 22	07 43	22 00	12 10	02 59	10 28	19 40
09	07 44	06 35	04 46	07 51	08 00	22 02	12 09	03 00	10 27	19 40
10	08 07	10 36	04 17	08 20	08 18	22 03	12 08	03 02	10 26	19 40
11	08 29	14 05	03 47	08 49	08 35	22 05	12 06	03 03	10 26	19 40
12	08 51	16 52	03 16	09 17	08 53	22 06	12 05	03 04	10 25	19 40
13	09 12	18 50	02 43	09 45	09 10	22 08	12 03	03 06	10 25	19 40
14	09 34	19 57	02 09	10 14	09 27	22 09	12 02	03 07	10 24	19 40
15	09 55	20 09	01 34	10 41	09 44	22 10	12 01	03 08	10 24	19 40
16	10 17	19 28	00 58	11 09	10 01	22 12	11 59	03 10	10 23	19 40
17	10 38	17 55	00 S 21	11 36	10 18	22 13	11 58	03 11	10 22	19 40
18	10 59	15 35	00 N 18	12 03	10 34	22 15	11 56	03 12	10 22	19 40
19	11 20	12 32	00 57	12 30	10 51	22 16	11 55	03 13	10 21	19 40
20	11 40	08 53	01 37	12 56	11 07	22 18	11 53	03 15	10 21	19 40
21	12 01	04 44	02 18	13 22	11 24	22 19	11 52	03 16	10 20	19 40
22	12 21	00 N 14	03 00	13 48	11 40	22 20	11 51	03 17	10 20	19 40
23	12 41	04 S 26	03 43	14 13	11 56	22 22	11 49	03 19	10 19	19 40
24	13 00	08 59	04 27	14 39	12 12	22 23	11 48	03 20	10 19	19 40
25	13 20	13 10	05 12	15 03	12 28	22 25	11 46	03 21	10 18	19 40
26	13 39	16 36	05 57	15 28	12 43	22 26	11 45	03 22	10 18	19 40
27	13 58	19 00	06 44	15 52	12 59	22 27	11 43	03 23	10 18	19 40
28	14 17	20 07	07 30	16 15	13 14	22 29	11 42	03 25	10 17	19 40
29	14 36	19 50	08 18	16 38	13 30	22 30	11 40	03 26	10 17	19 40
30	14 N 54	18 S 14	09 N 05	17 N 01	13 N 45	22 N 31	11 S 39	03 N 27	10 S 16	19 S 40

ZODIAC SIGN ENTRIES

Date	h m	Planets
02	05 35	☽ ♑
04	08 41	☽ ♒
06	13 00	☽ ♓
08	19 02	☽ ♈
11	03 22	☽ ♉
13	14 13	☽ ♊
14	02 37	☿ ♈
15	07 25	♀ ♉
16	02 49	☽ ♋
18	15 13	☽ ♌
19	22 03	☉ ♉
20	11 48	♂ ♉
21	01 08	☽ ♍
23	07 25	☽ ♎
25	10 25	☽ ♏
27	11 32	☽ ♐
29	12 21	☽ ♑

LATITUDES

Date	Mercury ☿ ° '	Venus ♀ ° '	Mars ♂ ° '	Jupiter ♃ ° '	Saturn ♄ ° '	Uranus ♅ ° '	Neptune ♆ ° '	Pluto ♇ ° '
01	01 S 41	01 S 16	00 S 34	00 S 23	02 N 40	00 S 40	00 S 37	03 N 17
04	02 01	01 13	00 32	00 23	02 40	00 40	00 37	03 17
07	02 17	01 08	00 30	00 22	02 40	00 40	00 37	03 17
10	02 28	01 04	00 29	00 22	02 40	00 40	00 37	03 17
13	02 34	00 59	00 27	00 22	02 41	00 40	00 38	03 17
16	02 36	00 54	00 25	00 21	02 41	00 40	00 38	03 17
19	02 33	00 48	00 23	00 21	02 41	00 40	00 38	03 17
22	02 25	00 42	00 21	00 20	02 41	00 40	00 38	03 17
25	02 13	00 35	00 19	00 20	02 41	00 40	00 38	03 17
28	01 56	00 29	00 17	00 19	02 41	00 40	00 38	03 17
31	01 S 34	00 S 22	00 S 15	00 S 19	02 N 41	00 S 40	00 S 38	03 N 17

DATA

Julian Date	2456384
Delta T	+68 seconds
Ayanamsa	24° 02' 44"
Synetic vernal point	05° ♓ 04' 15"
True obliquity of ecliptic	23° 26' 10"

MOON'S PHASES, APSIDES AND POSITIONS ☽

Date	h m	Phase	Longitude ° '	Eclipse Indicator
03	04 37	☾	13 ♑ 35	
10	09 35	●	20 ♈ 41	
18	12 31	☽	28 ♋ 38	
25	19 57	○	05 ♏ 46	partial

Day	h m		
15	22 23	Apogee	
27	20 01	Perigee	
01	12 02	Max dec	20° S 15'
08	00 26	0N	
15	05 29	Max dec	20° N 11'
22	13 10	0S	
28	19 06	Max dec	20° S 10'

LONGITUDES

Date	Chiron ⚷ ° '	Ceres ⚳ ° '	Pallas ⚴ ° '	Juno ⚵ ° '	Vesta ⚶ ° '	Black Moon Lilith ⚸ ° '
01	11 ♓ 24	28 ♊ 59	00 ♉ 56	05 ♒ 58	21 ♊ 52	22 ♊ 20
11	11 ♓ 57	02 ♋ 03	05 ♉ 40	08 ♒ 48	25 ♊ 10	23 ♊ 27
21	12 ♓ 27	05 ♋ 23	10 ♉ 32	11 ♒ 24	28 ♊ 42	24 ♊ 33
31	12 ♓ 53	08 ♋ 58	15 ♉ 32	13 ♒ 42	02 ♋ 26	25 ♊ 40

All ephemeris data is given at 12.00 UT and the Moon's longitude is additionally given for 24.00 UT

ASPECTARIAN

h m Aspects

01 Monday
02 13 ☽ □ ☿
03 05 ☉ □ ♇
05 00 ☽ △ ♂
06 16 ☽ ⊥ ♄
12 24 ☉ ⚹ ♃
16 30 ☽ Q ♆
21 20 ☽ ∠ ♄

02 Tuesday
10 47 ☽ ∥ ♇
10 53 ☽ Q ☿
12 48 ☽ ⚹ ♆
20 25 ☽ □ ♅
22 35 ☽ ⚹ ♄

03 Wednesday
01 10 ☽ ☌ ♇
02 15 ☽ ⊼ ♃
04 37 ☽ □ ☉
07 13 ☽ □ ♀
08 57 ☽ ⚹ ☿
10 35 ☽ □ ♂
12 35 ☽ ± ♃
14 20 ☽ ∠ ♆
18 56 ☽ Q ♄

04 Thursday
03 15 ☽ Q ♅
04 11 ☽ ᵽ ♃
05 48 ☽ ⊥ ♆
12 54 ☽ ∠ ☿
13 46 ☽ Q ☉
16 09 ☽ ⊻ ♆
17 05 ☽ Q ♀
19 16 ☽ Q ♂
23 25 ☿ # ♂

05 Friday
00 02 ☽ ⚹ ♅
01 48 ☽ □ ♄
04 42 ☽ ⊻ ♇
05 11 ☽ ∥ ♄
05 54 ☽ ⊥ ☿
06 26 ☽ △ ♃
12 11 ☽ ⚹ ☉
15 09 ☽ ⊥ ♇
15 47 ☽ ∥ ♆
16 00 ☽ ⚹ ♀
17 21 ☽ ⊻ ☿
17 22 ☽ ⚹ ♂
17 41 ☿ ⊻ ♂
19 20 ☉ # ☿

06 Saturday
02 20 ☽ ∠ ♅
03 33 ☿ # ♀
06 57 ☽ ∠ ♇
10 41 ♂ ∠ ♆
10 50 ☽ # ♂
13 23 ☽ # ☉
14 26 ☽ # ♀
16 36 ☽ ∠ ☉
16 49 ☽ ∥ ☿
17 55 ♀ ∠ ♆
18 18 ☽ ⊥ ♅
20 48 ☽ ☌ ♆
21 05 ☽ ∠ ♀
21 23 ☽ ∠ ♂

07 Sunday
04 58 ♀ ☌ ♂
05 01 ☽ ⊻ ♅
06 21 ☽ △ ♄
08 56 ☽ # ♅
09 37 ☽ ⚹ ♇
09 55 ☽ ⊥ ☉
12 06 ☽ □ ♃
14 26 ☽ ⊥ ♂
14 49 ☽ ⊥ ♀
21 35 ☽ ⊻ ☉

08 Monday
01 56 ☽ ⊻ ♂
02 46 ☽ ⊻ ♀
04 10 ☽ ☌ ☿
07 18 ☽ Q ♇
09 18 ☽ ᵽ ♄
11 33 ☉ ∥ ♀
16 15 ☽ ∥ ♅
21 15 ☽ Q ♃

09 Tuesday
01 43 ☽ ± ♄
02 52 ☽ # ☿
03 16 ☽ ⊻ ♆
03 55 ☉ ∠ ♆
11 56 ☽ ☌ ♅
12 05 ☿ Q ♇
12 47 ☽ ⊼ ♄
14 25 ☽ ⊥ ♆
16 29 ☽ □ ♇
19 18 ☽ ∥ ☉
19 50 ☽ ⚹ ♃
20 16 ☽ ∥ ♀
20 50 ☽ ∥ ♂

10 Wednesday
05 29 ☿ ᵽ ♄
07 21 ☽ ∠ ♆
07 24 ♀ ∥ ♂
09 35 ☽ ☌ ☉
10 58 ☽ # ♆
13 02 ☽ ☌ ♂
16 25 ☽ ☌ ♀
18 19 ☽ ⊻ ☿
21 57 ☽ # ♄

11 Thursday
00 40 ☽ ∠ ♃
07 14 ☽ ⊥ ☿
12 06 ☽ ⚹ ♆
18 17 ☽ ⊥ ♃
21 18 ☽ ⊻ ♅
21 36 ☽ ☍ ♄

12 Friday
01 48 ☽ △ ♇
02 55 ☽ ∠ ☿
06 10 ☽ ⊻ ♃
09 05 ☽ ⊥ ♅
11 33 ☽ Q ♆
19 31 ♇ St R
19 58 ☿ # ♅
22 55 ☉ ∥ ♂

13 Saturday
00 48 ☽ ⊻ ☉
02 56 ☽ ∠ ♅
03 10 ☽ ⊻ ♂
03 29 ♄ ⊼ ♅
07 24 ☽ ᵽ ♇
09 29 ☽ ⊻ ♀
12 30 ☽ ⚹ ☿
13 48 ☽ ⊥ ☉
15 56 ☽ ⊥ ♂
22 52 ☽ ⊥ ♀
23 25 ☽ □ ♆

14 Sunday
01 24 ☽ ± ♇
03 33 ☽ # ♇
08 48 ☽ ⊼ ♄
09 07 ☽ ⚹ ♅
09 29 ☽ ∠ ☉
11 14 ☽ ∠ ♂
13 30 ☽ ⊼ ♇
15 59 ☽ Q ☿
17 18 ♀ ∠ ♃
18 57 ☽ ☌ ♃
19 07 ☽ ∠ ♀
20 51 ☽ ± ♄
20 55 ♀ # ♆

15 Monday
09 30 ☽ Q ♅
14 59 ☽ ᵽ ♄
18 30 ☿ Q ♃
18 36 ☽ ⚹ ☉
19 41 ☽ ⚹ ♂

16 Tuesday
04 51 ☿ ± ♄
05 05 ☽ ⚹ ♀
07 33 ☽ # ♇
09 48 ☽ □ ☿
12 18 ☽ △ ♆
18 52 ☉ # ♆
21 08 ☽ Q ☉
21 14 ☽ △ ♄
21 41 ☽ Q ♂
22 12 ☽ □ ♅

17 Wednesday
02 18 ☽ ☍ ♇
05 14 ☿ ⊻ ♆
08 14 ☽ Q ♀
08 45 ☽ ⊻ ♃
18 32 ♂ # ♆
18 40 ☽ ᵽ ♆
21 01 ☽ ⊥ ♃

18 Thursday
00 20 ☉ ☌ ♂
06 14 ♀ # ♄
08 53 ♃ ± ♄
12 16 ☽ □ ♂
12 31 ☽ □ ☉
12 41 ☽ ± ♆
15 15 ☽ ∠ ♃

19 Friday
00 16 ☽ □ ♀
00 34 ☽ ⊼ ♆
03 15 ♀ ⚹ ♆
06 54 ☽ △ ☿
08 44 ☽ □ ♄
10 18 ☽ △ ♅
12 16 ☽ ∥ ♀
13 58 ☽ ⊼ ♇
16 25 ☽ # ♄
19 39 ☽ ∥ ☉
20 16 ☿ ⊼ ♄
21 06 ☽ ⚹ ♃
22 44 ☽ ∥ ♂

20 Saturday
01 34 ☽ ± ♇
02 47 ☽ # ♆
06 10 ☉ ∠ ♃
09 20 ☿ ☌ ♅
15 16 ☽ ᵽ ♅
16 08 ☽ ᵽ ☿
18 42 ☽ ᵽ ♇
19 06 ☽ Q ♄
20 21 ☽ Q ♃

21 Sunday
00 25 ☿ ⊥ ♆
01 58 ☽ △ ♂
02 33 ☉ # ♄
03 22 ☽ △ ☉
04 04 ♂ ∠ ♃
08 13 ☽ ± ♅
10 03 ☽ ☍ ♆
11 27 ☽ ± ☿
11 46 ☿ □ ♇
15 42 ☽ △ ♀
17 17 ☽ ⚹ ♄
19 19 ☽ ⊼ ♅
19 54 ☽ ∥ ♅
22 30 ☽ △ ♇
23 19 ☽ ∥ ☿

22 Monday
00 04 ☽ ⊼ ☿
06 02 ☽ □ ♃
07 11 ☽ ᵽ ♂
07 43 ♀ ☍ ♄
09 04 ☽ ᵽ ☉
20 12 ☽ ∠ ♄
21 31 ☽ ᵽ ♀
21 42 ☿ ∥ ♅

23 Tuesday
00 11 ☽ ± ♂
02 19 ☽ ± ☉
02 28 ♂ # ♄
06 13 ☽ # ♅
07 40 ♀ ⊻ ♅
07 43 ☽ # ☿
08 20 ♀ ⊥ ♃
11 18 ☽ ⊼ ♂
11 55 ☽ ⊥ ♄
13 37 ☽ ⊼ ☉
14 51 ☽ ± ♀
15 50 ☽ ⊼ ♆
22 14 ☽ ⊻ ♄

24 Wednesday
00 37 ☽ ☍ ♅
02 11 ☽ ± ♆
02 11 ☽ ⊼ ♀
03 23 ☽ □ ♇
03 25 ☿ ⚹ ♃
11 13 ☽ △ ♃
12 12 ☽ ☍ ☿
15 43 ♀ △ ♇
17 29 ☽ ᵽ ♆
19 17 ☽ ∥ ♆
21 58 ☉ ⚹ ♆

25 Thursday
03 40 ☽ ∥ ♄
07 26 ☽ # ♂
09 40 ☽ Q ♇
12 42 ☽ ᵽ ♃
13 10 ☽ # ☉
16 52 ☽ ☍ ♂
18 31 ☽ △ ♆
19 57 ☽ ☍ ☉

26 Friday
00 17 ☽ ☌ ♄
02 05 ☽ # ♀
03 00 ☽ ⊼ ♅
03 44 ☽ ± ♃
05 27 ☽ ⚹ ♇
06 52 ☿ ∠ ♆
08 56 ☽ ☍ ♀
12 52 ☽ ± ♅
13 42 ☽ ⊼ ♃
20 50 ☽ ⊼ ☿

27 Saturday
01 46 ♂ ⚹ ♆
03 35 ☽ ᵽ ♅
05 54 ☽ ∠ ♇
07 58 ☽ ± ☿
19 32 ☽ □ ♆
20 27 ☽ ⊼ ♂
20 30 ☽ ⊥ ♇
22 45 ☽ ∥ ♇

28 Sunday
00 19 ☽ ⊼ ☉
00 39 ☽ ᵽ ☿
00 54 ☽ ⊻ ♄
04 01 ☽ △ ♅
06 14 ☽ ⊻ ♇
06 42 ☽ ± ♂
08 27 ☉ ☍ ♄
09 34 ♀ ⊥ ♅
10 35 ☽ ⊥ ♄
10 45 ☽ ± ☉
14 10 ☽ ⊼ ♀
15 08 ☽ ☍ ♃
22 07 ☽ ᵽ ♂

29 Monday
00 50 ☽ ± ♀
00 50 ☽ Q ♆
00 56 ♀ Q ♆
01 10 ☽ ∠ ♄
02 27 ☽ ᵽ ☉
03 51 ♀ ⊻ ♃
04 37 ☽ △ ☿
15 35 ☽ ∥ ♇
16 55 ☽ ᵽ ♀
20 32 ☽ ⚹ ♆

30 Tuesday
00 01 ☽ △ ♂
01 39 ☽ ⚹ ♄
04 50 ☽ △ ☉
05 16 ☽ □ ♅
07 19 ☽ ☌ ♇
11 27 ☉ ⊻ ♅
17 10 ☽ ⊼ ♃
20 05 ☽ △ ♀
21 28 ☽ ∠ ♆
21 30 ☽ Q ♄
22 20 ☽ # ♀

MAY 2013

LONGITUDES

Date	Sidereal time h m s	Sun ☉ ° ′ ″	Moon ☽ ° ′ ″	Moon ☽ 24.00 ° ′ ″	Mercury ☿ ° ′	Venus ♀ ° ′
01	02 38 19	11 ♉ 16 14	28 ♑ 37 45	05 ♒ 40 37	29 ♈ 42	20 ♉ 00
02	02 42 15	12 14 27	12 ♒ 39 07	19 ♒ 33 08	01 ♉ 40	21 14
03	02 46 12	13 12 39	26 ♒ 22 39	03 ♓ 07 45	03 40	22 27
04	02 50 08	14 10 49	09 ♓ 48 29	16 ♓ 25 01	05 42	23 41
05	02 54 05	15 08 57	22 ♓ 57 29	29 ♓ 26 04	07 45	24 55
06	02 58 01	16 07 05	05 ♈ 50 57	12 ♈ 12 18	09 49	26 09
07	03 01 58	17 05 10	18 ♈ 30 19	24 ♈ 45 12	11 55	27 23
08	03 05 55	18 03 14	00 ♉ 57 07	07 ♉ 06 17	14 02	28 37
09	03 09 51	19 01 17	13 ♉ 12 53	19 ♉ 17 08	16 10	29 ♉ 51
10	03 13 48	19 59 18	25 ♉ 19 14	01 ♊ 19 24	18 20	01 ♊ 04
11	03 17 44	20 57 18	07 ♊ 17 54	13 ♊ 14 59	20 30	02 18
12	03 21 41	21 55 16	19 ♊ 10 56	25 ♊ 06 05	22 40	03 32
13	03 25 37	22 53 12	01 ♋ 00 46	06 ♋ 55 21	24 51	04 46
14	03 29 34	23 51 07	12 ♋ 50 14	18 ♋ 45 51	27 02	05 59
15	03 33 30	24 49 00	24 ♋ 42 39	00 ♌ 41 09	29 ♉ 13	07 13
16	03 37 27	25 46 51	06 ♌ 41 49	12 ♌ 45 12	01 ♊ 23	08 27
17	03 41 24	26 44 40	18 ♌ 51 50	25 ♌ 02 15	03 32	09 41
18	03 45 20	27 42 28	01 ♍ 16 59	07 ♍ 36 33	05 41	10 54
19	03 49 17	28 40 14	14 ♍ 01 26	20 ♍ 32 05	07 48	12 08
20	03 53 13	29 ♉ 37 58	27 ♍ 08 51	03 ♎ 52 03	09 53	13 21
21	03 57 10	00 ♊ 35 41	10 ♎ 41 52	17 ♎ 38 21	11 57	14 35
22	04 01 06	01 33 22	24 ♎ 41 26	01 ♏ 50 54	13 59	15 49
23	04 05 03	02 31 02	09 ♏ 06 21	16 ♏ 27 12	15 58	17 02
24	04 08 59	03 28 41	23 ♏ 52 43	01 ♐ 22 02	17 55	18 16
25	04 12 56	04 26 18	08 ♐ 54 05	16 ♐ 27 47	19 50	19 29
26	04 16 53	05 23 54	24 ♐ 01 56	01 ♑ 35 22	21 41	20 43
27	04 20 49	06 21 29	09 ♑ 06 55	16 ♑ 35 32	23 31	21 56
28	04 24 46	07 19 03	24 ♑ 00 14	01 ♒ 20 13	25 17	23 10
29	04 28 42	08 16 36	08 ♒ 34 51	15 ♒ 43 40	27 01	24 23
30	04 32 39	09 14 08	22 ♒ 46 22	29 ♒ 42 48	28 ♊ 42	25 36
31	04 36 35	10 ♊ 11 39	06 ♓ 32 59	13 ♓ 17 02	00 ♋ 20	26 ♊ 50

Date	Mars ♂ ° ′	Jupiter ♃ ° ′	Saturn ♄ ° ′	Uranus ♅ ° ′	Neptune ♆ ° ′	Pluto ♇ ° ′
01	08 ♉ 12	17 ♊ 35	07 ♏ 58	10 ♈ 20	05 ♓ 01	11 ♑ 30
02	08 57	17 48	07 R 54	10 23	05 02	11 R 29
03	09 41	18 00	07 49	10 26	05 03	11 29
04	10 25	18 13	07 45	10 29	05 04	11 28
05	11 10	18 26	07 40	10 32	05 05	11 28
06	11 54	18 38	07 36	10 35	05 06	11 27
07	12 38	18 51	07 31	10 38	05 07	11 26
08	13 22	19 04	07 27	10 41	05 08	11 25
09	14 06	19 17	07 23	10 43	05 09	11 25
10	14 50	19 30	07 18	10 46	05 10	11 24
11	15 34	19 43	07 14	10 49	05 11	11 23
12	16 18	19 56	07 09	10 52	05 12	11 22
13	17 02	20 09	07 05	10 55	05 12	11 21
14	17 46	20 22	07 01	10 57	05 13	11 21
15	18 30	20 35	06 57	11 00	05 14	11 20
16	19 13	20 48	06 52	11 03	05 15	11 19
17	19 57	21 01	06 48	11 05	05 15	11 18
18	20 41	21 14	06 44	11 08	05 16	11 17
19	21 24	21 28	06 40	11 11	05 17	11 16
20	22 08	21 41	06 36	11 13	05 17	11 15
21	22 51	21 54	06 32	11 16	05 18	11 14
22	23 34	22 07	06 28	11 18	05 18	11 13
23	24 18	22 21	06 24	11 21	05 19	11 12
24	25 01	22 34	06 20	11 23	05 19	11 10
25	25 44	22 48	06 17	11 25	05 20	11 09
26	26 27	23 01	06 13	11 28	05 20	11 08
27	27 11	23 15	06 09	11 30	05 21	11 07
28	27 54	23 28	06 06	11 32	05 21	11 06
29	28 37	23 42	06 02	11 34	05 21	11 05
30	29 ♉ 20	23 55	05 59	11 37	05 21	11 03
31	00 ♊ 02	24 ♊ 09	05 ♏ 56	11 ♈ 39	05 ♓ 22	11 ♑ 02

Date	Moon True ☊ ° ′	Moon Mean ☊ ° ′	Moon ☽ Latitude ° ′
01	16 ♏ 52	17 ♏ 13	05 N 02
02	16 R 53	17 10	05 17
03	16 52	17 06	05 13
04	16 52	17 03	04 53
05	16 52	17 00	04 17
06	16 52	16 57	03 29
07	16 D 52	16 54	02 31
08	16 52	16 51	01 27
09	16 52	16 47	00 N 20
10	16 R 52	16 44	00 S 47
11	16 52	16 41	01 51
12	16 51	16 38	02 50
13	16 51	16 35	03 41
14	16 50	16 31	04 23
15	16 49	16 28	04 54
16	16 48	16 25	05 12
17	16 47	16 22	05 17
18	16 D 47	16 19	05 07
19	16 48	16 16	04 42
20	16 49	16 12	04 02
21	16 50	16 09	03 07
22	16 51	16 06	02 00
23	16 51	16 03	00 S 43
24	16 R 51	16 00	00 N 39
25	16 51	15 57	01 59
26	16 49	15 53	03 11
27	16 47	15 50	04 10
28	16 44	15 47	04 51
29	16 42	15 44	05 12
30	16 40	15 41	05 13
31	16 ♏ 39	15 ♏ 37	04 N 57

DECLINATIONS

Date	Sun ☉ ° ′	Moon ☽ ° ′	Mercury ☿ ° ′	Venus ♀ ° ′	Mars ♂ ° ′	Jupiter ♃ ° ′	Saturn ♄ ° ′	Uranus ♅ ° ′	Neptune ♆ ° ′	Pluto ♇ ° ′
01	15 N 13	15 S 30	09 N 54	17 N 23	14 N 00	22 N 32	11 S 38	03 N 28	10 S 16	19 S 40
02	15 30	11 56	10 42	17 45	14 15	22 34	11 36	03 29	10 15	19 40
03	15 48	07 48	11 31	18 06	14 29	22 35	11 35	03 31	10 15	19 40
04	16 06	03 S 22	12 20	18 27	14 44	22 36	11 33	03 32	10 15	19 40
05	16 23	01 N 08	13 09	18 48	14 58	22 37	11 32	03 33	10 14	19 40
06	16 40	05 31	13 58	19 08	15 13	22 39	11 31	03 34	10 14	19 41
07	16 56	09 35	14 46	19 27	15 27	22 40	11 29	03 35	10 14	19 41
08	17 12	13 10	15 34	19 46	15 41	22 41	11 28	03 36	10 13	19 41
09	17 28	16 08	16 21	20 04	15 54	22 42	11 26	03 37	10 13	19 41
10	17 44	18 20	17 08	20 22	16 08	22 43	11 25	03 39	10 13	19 41
11	18 00	19 42	17 53	20 39	16 21	22 44	11 24	03 40	10 12	19 41
12	18 15	20 10	18 37	20 56	16 35	22 46	11 22	03 41	10 12	19 41
13	18 29	19 45	19 20	21 12	16 48	22 47	11 21	03 42	10 12	19 41
14	18 44	18 27	20 01	21 28	17 01	22 48	11 20	03 43	10 12	19 41
15	18 58	16 21	20 40	21 43	17 14	22 49	11 19	03 44	10 11	19 41
16	19 12	13 33	21 17	21 57	17 26	22 50	11 17	03 45	10 11	19 41
17	19 26	10 09	21 52	22 11	17 39	22 51	11 16	03 46	10 11	19 41
18	19 39	06 14	22 25	22 24	17 51	22 52	11 15	03 47	10 11	19 42
19	19 52	01 N 57	22 55	22 36	18 03	22 53	11 14	03 48	10 10	19 42
20	20 04	02 S 34	23 23	22 48	18 15	22 54	11 13	03 49	10 10	19 42
21	20 16	07 06	23 48	22 59	18 27	22 55	11 11	03 50	10 10	19 42
22	20 28	11 25	24 10	23 10	18 38	22 56	11 10	03 51	10 10	19 42
23	20 40	15 12	24 30	23 20	18 49	22 56	11 09	03 52	10 10	19 42
24	20 51	18 07	24 47	23 29	19 01	22 57	11 08	03 53	10 10	19 42
25	21 02	19 49	25 01	23 37	19 12	22 58	11 07	03 54	10 09	19 42
26	21 12	20 07	25 13	23 45	19 22	22 59	11 06	03 54	10 09	19 43
27	21 22	18 58	25 23	23 52	19 33	23 00	11 05	03 55	10 09	19 43
28	21 32	16 32	25 30	23 59	19 43	23 01	11 04	03 56	10 09	19 43
29	21 41	13 05	25 35	24 05	19 53	23 01	11 03	03 57	10 09	19 43
30	21 50	08 59	25 37	24 10	20 03	23 02	11 02	03 58	10 09	19 43
31	21 N 58	04 S 31	25 N 38	24 N 14	20 N 13	23 N 03	11 S 01	03 N 59	10 S 09	19 S 43

ZODIAC SIGN ENTRIES

Date	h	m	Planets
01	14	19	☽ ♒
01	15	37	☿ ♉
03	18	25	☽ ♓
06	01	03	☽ ♈
08	10	09	☽ ♉
09	15	03	♀ ♊
10	21	21	☽ ♊
13	09	57	☽ ♋
15	20	41	☿ ♊
15	22	38	☽ ♌
18	09	33	☽ ♍
20	17	07	☽ ♎
20	21	09	☉ ♊
22	20	55	☽ ♏
24	21	49	☽ ♐
26	21	28	☽ ♑
28	21	48	☽ ♒
31	00	30	☽ ♓
31	07	07	☿ ♋
31	10	39	♂ ♊

LATITUDES

Date	Mercury ☿ ° ′	Venus ♀ ° ′	Mars ♂ ° ′	Jupiter ♃ ° ′	Saturn ♄ ° ′	Uranus ♅ ° ′	Neptune ♆ ° ′	Pluto ♇ ° ′
01	01 S 34	00 S 22	00 S 15	00 S 19	02 N 41	00 S 40	00 S 38	03 N 17
04	01 09	00 15	00 13	00 19	02 41	00 40	00 38	03 17
07	00 40	00 S 07	00 12	00 18	02 41	00 40	00 38	03 17
10	00 S 10	00 00	00 10	00 18	02 40	00 40	00 38	03 17
13	00 N 22	00 N 07	00 08	00 18	02 40	00 41	00 38	03 17
16	00 52	00 15	00 06	00 17	02 40	00 41	00 39	03 17
19	01 20	00 22	00 04	00 17	02 39	00 41	00 39	03 16
22	01 43	00 29	00 S 02	00 17	02 39	00 41	00 39	03 16
25	01 59	00 36	00 00	00 16	02 39	00 41	00 39	03 16
28	02 09	00 43	00 N 02	00 16	02 38	00 41	00 39	03 16
31	02 N 11	00 N 50	00 N 04	00 S 16	02 N 37	00 S 41	00 S 39	03 N 16

DATA

Julian Date	2456414
Delta T	+68 seconds
Ayanamsa	24° 02′ 48″
Synetic vernal point	05° ♓ 04′ 11″
True obliquity of ecliptic	23° 26′ 09″

MOON'S PHASES, APSIDES AND POSITIONS ☽

Date	h	m	Phase	Longitude ° ′	Eclipse Indicator
02	11	14	☾	12 ♒ 13	
10	00	28	●	19 ♉ 31	Annular
18	04	35	☽	27 ♌ 25	
25	04	25	○	04 ♐ 08	
31	18	58	☾	10 ♓ 28	

Day	h	m		
13	13	38	Apogee	
26	01	50	Perigee	
05	05	55	ON	
12	12	32	Max dec	20° N 10′
19	22	26	OS	
26	04	49	Max dec	20° S 11′

LONGITUDES

Date	Chiron ⚷ ° ′	Ceres ⚳ ° ′	Pallas ⚴ ° ′	Juno ⚵ ° ′	Vesta ⚶ ° ′	Black Moon Lilith ⚸ ° ′
01	12 ♓ 53	08 ♋ 58	15 ♉ 32	13 ♒ 42	02 ♋ 26	25 ♊ 40
11	13 ♓ 15	12 ♋ 43	20 ♉ 40	15 ♒ 41	06 ♋ 19	26 ♊ 47
21	13 ♓ 31	16 ♋ 39	25 ♉ 55	17 ♒ 15	10 ♋ 20	27 ♊ 54
31	13 ♓ 43	20 ♋ 42	01 ♊ 17	18 ♒ 21	14 ♋ 28	29 ♊ 00

All ephemeris data is given at 12.00 UT and the Moon's longitude is additionally given for 24.00 UT

ASPECTARIAN

h m Aspects

01 Wednesday
03 22 ☽ ± ♃
05 12 ♂ ☍ ♄
11 29 ☽ Q ♅
12 39 ☽ ⊥ ♆
14 02 ☽ # ☉
14 07 ☽ □ ☿
17 36 ☉ △ ♇
18 50 ☽ bQ ♃
22 02 ☽ # ♂
22 06 ☉ ⊥ ♃
22 52 ☿ # ♆
22 52 ☽ ⊻ ♆

02 Thursday
03 51 ☽ □ ♄
05 15 ☽ □ ♂
08 04 ☽ ⚹ ♅
10 00 ☽ ⊻ ♇
11 14 ☽ □ ☉
14 05 ☽ ‖ ♄
18 13 ☽ # ☿
20 24 ☽ ⊥ ♇
21 05 ☽ △ ♃
22 06 ☽ ‖ ♆

03 Friday
02 17 ☽ Q ☿
03 09 ☿ ∠ ♃
04 24 ☽ □ ♀
10 19 ☽ ∠ ♅
12 11 ☽ ∠ ♇
13 46 ☿ # ♄
14 27 ☽ Q ♂
21 15 ☽ Q ☉

04 Saturday
02 23 ☽ ⊥ ♅
03 16 ☽ ⚹ ☿
03 27 ☽ ☌ ♆
04 31 ☿ ⚹ ♆
08 18 ☽ △ ♄
11 10 ☽ # ♅
13 11 ☽ ⚹ ♂
13 13 ☽ ⊻ ♅
13 56 ♂ ⊻ ♅
15 00 ☽ ⚹ ♇
15 45 ☽ Q ♀
20 33 ☽ ⚹ ☉

05 Sunday
03 32 ☽ □ ♃
11 12 ☿ ☍ ♄
11 29 ☽ bQ ♄
11 32 ☽ ∠ ☿
12 55 ☽ Q ♇
16 00 ☽ ⚹ ♀
18 17 ☽ ∠ ♂
21 29 ♂ △ ♇

06 Monday
00 22 ♀ ∠ ♅
01 07 ☽ ‖ ♅
02 25 ☽ ∠ ☉
04 05 ☽ ± ♄
07 27 ☽ ⊥ ☿
10 36 ☽ ⊻ ♆
12 06 ☽ ⊥ ♂
13 31 ☽ Q ♃
15 16 ☽ ⊼ ♄
17 42 ♀ bQ ♇
20 43 ☽ ⊥ ☉
20 57 ☿ ⊻ ♅
20 57 ☽ ☌ ♅
20 57 ☽ ⊻ ☿
21 55 ☽ ⊥ ♆
22 33 ☽ □ ♇
23 05 ☽ ∠ ♀

07 Tuesday
00 02 ☉ ⊥ ♅
00 08 ☽ ⊻ ♂
06 34 ☿ △ ♇
09 04 ☽ ⊻ ☉
12 40 ☽ ⚹ ♃
12 47 ☉ Q ♆
15 05 ☽ ∠ ♆
16 04 ☽ # ♆
18 07 ☽ ⊥ ♀
21 53 ♂ ⊥ ♃
23 49 ☿ ⊥ ♃

08 Wednesday
00 13 ☽ # ♄
00 33 ☿ ☌ ♂
05 10 ♀ # ♇
06 58 ☽ ⊻ ♀
16 47 ☿ ‖ ♂
18 10 ☽ ∠ ♃
20 09 ☽ ⚹ ♆

09 Thursday
00 36 ☽ ☍ ♄
07 05 ☽ ⊻ ♅
08 27 ☽ △ ♇
09 49 ☽ ‖ ♂
12 08 ☽ ⊥ ♃
13 52 ☽ ☌ ♂
15 06 ☽ ‖ ☿
18 18 ☿ ⊥ ♅
18 57 ☽ ⊥ ♅
19 06 ☽ ☌ ☿
19 47 ☽ Q ♆
20 12 ☉ ⊻ ♃
22 59 ☿ Q ♆

10 Friday
00 12 ☽ ⊻ ♃
00 28 ☽ ☌ ☉
03 29 ☽ ‖ ☉
12 54 ☽ ∠ ♅
14 09 ☽ bQ ♇

11 Saturday
00 49 ☽ ☌ ♀
02 22 ☿ ⊻ ♃
07 44 ☽ □ ♆
08 09 ☽ ± ♇
11 34 ☽ # ♇
11 52 ☽ ⊼ ♄
17 15 ☉ ‖ ☿
19 07 ☽ ⚹ ♅
20 14 ☽ ⊼ ♇
21 10 ☉ ☌ ☿
23 53 ☽ ± ♄

12 Sunday
05 48 ☽ ⊻ ♂
13 32 ☽ ☌ ♃
17 59 ☽ bQ ♄
18 03 ☽ ⊻ ☉
18 45 ☽ ⊥ ♂
19 30 ☽ Q ♅
20 40 ☽ ⊻ ☿

13 Monday
07 18 ☽ ⊥ ☉
07 37 ♂ ⊥ ♅
11 36 ☽ ⊥ ☿
13 34 ☽ # ♇
14 13 ☽ ∠ ♂
17 44 ♂ Q ♆
17 54 ☽ ‖ ☿
20 30 ☽ ⊻ ♀
20 32 ☽ △ ♆
20 47 ♀ □ ♆
23 28 ♀ ± ♇
23 53 ☿ ∠ ♅

14 Tuesday
00 14 ☿ # ♇
00 15 ☽ △ ♄
03 12 ☽ ∠ ☉
04 26 ☿ bQ ♇
08 10 ☽ □ ♅
08 24 ☽ ‖ ☉
08 58 ☽ ☍ ♇
10 00 ☽ ∠ ☿
10 05 ☽ ⊥ ♀
22 38 ☽ ⚹ ♂

15 Wednesday
02 57 ☽ bQ ♆
03 31 ☽ ⊻ ♃
03 50 ☽ ‖ ♂
06 24 ☽ ∠ ♀
06 54 ♀ ⊼ ♄
12 14 ☽ ⚹ ☉
15 50 ☽ ⊥ ♃
21 06 ☽ ± ♆
23 03 ☽ ⚹ ☿

16 Thursday
00 22 ☽ Q ♂
09 06 ☽ ⊼ ♆
10 11 ☽ ∠ ♃
12 21 ☽ □ ♄
14 20 ☽ Q ☉
15 52 ☽ ⚹ ♀
18 55 ☉ ∠ ♅
20 39 ☽ △ ♅
21 08 ☽ ⊼ ♇

17 Friday
01 00 ☉ bQ ♇
04 06 ☽ Q ☿
04 27 ☽ # ♄
08 56 ☽ ± ♇
11 45 ☽ # ♆
14 15 ☽ □ ♂
16 17 ☽ ⚹ ♃
18 05 ☽ Q ♀
23 29 ☽ Q ♄

18 Saturday
02 05 ☽ bQ ♅
02 25 ☽ bQ ♇
04 35 ☽ □ ☉
07 19 ☿ □ ♆
07 31 ☿ ± ♇
10 31 ☿ ‖ ♀
15 47 ☽ Q ♃
16 40 ♀ ⚹ ♅
17 17 ☉ # ♇
19 16 ♀ ⊼ ♇
19 21 ☽ ± ♅
19 35 ☽ ☍ ♆
22 02 ☽ □ ☿
22 18 ☽ ⚹ ♄
23 33 ☿ ⊼ ♄

19 Sunday
01 52 ☽ ‖ ♅
06 41 ☽ ⊼ ♅
06 52 ☽ △ ♇
08 06 ☽ □ ♀
09 54 ☿ ‖ ♃
14 40 ♂ ⊻ ♃
21 58 ♀ ± ♄

20 Monday
01 55 ☽ □ ♃
02 00 ☽ ∠ ♄
02 23 ☽ △ ♂
05 21 ♃ bQ ♄
16 48 ☽ △ ☉
18 10 ☽ ⊥ ♄
18 36 ☽ # ♅
23 02 ♅ □ ♇

21 Tuesday
00 47 ♀ ‖ ♃
02 31 ☽ ⊼ ♆
03 38 ☿ ⊼ ♇
03 46 ☿ ⚹ ♅
04 45 ☽ ⊻ ♄
06 45 ☽ bQ ♂
12 55 ☽ □ ♇
12 59 ☽ ☍ ♅
13 03 ☽ ± ♆
14 33 ☽ △ ☿
18 41 ☿ ± ♄
19 24 ☽ △ ♀
21 07 ☽ bQ ☉
23 14 ☽ ± ♂

22 Wednesday
04 33 ☽ bQ ♆
04 48 ☽ ‖ ♆
07 35 ☽ △ ♃
10 01 ☽ ⊼ ♂
10 34 ☽ ‖ ♄
13 34 ☽ ± ☉
15 16 ♅ ⊥ ♆
19 35 ☽ Q ♇
20 22 ☽ bQ ☿
23 13 ☽ bQ ♀

23 Thursday
00 20 ☽ ⊼ ☉
05 45 ☽ △ ♆
07 34 ☽ ☌ ♄
09 04 ☽ bQ ♃
13 38 ☽ ± ☿
15 25 ☽ ⚹ ♇
15 27 ☽ ± ♀
15 41 ☽ ⊼ ♅

24 Friday
00 01 ☽ ± ♃
00 55 ☽ ⊼ ☿
01 29 ☽ ± ♅
02 07 ☽ ⊼ ♀
09 52 ☽ ⊼ ♃
13 55 ☽ ☍ ♂
15 41 ☽ ∠ ♇
16 02 ☽ bQ ♅
23 28 ☽ # ♂
23 54 ☿ ☌ ♀

25 Saturday
04 25 ☽ ☍ ☉
06 03 ☽ ⊥ ♇
06 19 ☽ □ ♆
07 51 ☽ ⊻ ♄
09 29 ☽ ‖ ♇
15 35 ☽ ⊻ ♇
16 01 ☽ △ ♅
17 20 ☽ ⊥ ♄

26 Sunday
01 36 ♂ bQ ♇
05 35 ☉ ± ♇
06 03 ☿ bQ ♄
06 16 ☽ ☍ ♀
07 33 ☽ ∠ ♄
07 46 ☽ ☍ ☿
10 22 ☽ ☍ ♃
10 26 ☉ □ ♆
10 54 ☽ Q ♆
12 09 ♂ ∠ ♅
16 02 ☽ ⊼ ♂
21 28 ♀ bQ ♄

27 Monday
00 07 ☽ ‖ ♇
02 03 ☽ ± ♂
04 13 ☽ # ♂
05 58 ☽ ⚹ ♆
07 16 ☉ ⊼ ♄
07 18 ☽ ⚹ ♄
07 18 ☽ ⊼ ☉
07 56 ☿ ☌ ♃
11 50 ☿ Q ♅
15 12 ☽ ☌ ♇
15 50 ☽ □ ♅
17 09 ☽ bQ ♂
17 33 ☽ ± ☉

28 Tuesday
02 28 ☽ Q ♄
06 04 ☽ ∠ ♆
09 04 ☽ bQ ☉
10 30 ☽ ⊼ ♀
11 00 ♂ # ♇
11 07 ☽ ⊼ ♃
14 22 ☽ ⊼ ☿
18 40 ☽ △ ♂
19 29 ♀ ☌ ♃
19 39 ♀ Q ♅
20 33 ♃ Q ♅
20 44 ☽ ⊥ ♆
21 04 ☽ Q ♅
21 04 ☽ ± ♃
21 11 ☽ ± ♀

29 Wednesday
01 32 ☽ ± ☿
06 38 ☽ ⊻ ♆
07 47 ☽ □ ♄
11 27 ☽ △ ☉
12 12 ☽ bQ ♃
13 28 ☽ bQ ♀
16 10 ☽ ⊻ ♇
17 01 ☽ ⚹ ♅
18 31 ☽ bQ ☿

30 Thursday
00 20 ☽ ‖ ♄
02 16 ☽ ⊥ ♇
05 26 ☽ ‖ ♆
14 00 ☽ △ ♃
17 21 ☽ △ ♀
17 39 ☽ ∠ ♇
18 38 ☽ ∠ ♅
23 37 ☽ △ ☿
23 57 ☽ □ ♂

31 Friday
04 20 ☿ ⊻ ♂
09 54 ☽ ☌ ♆
10 24 ☽ ⊥ ♅
10 54 ☽ △ ♄
14 48 ☽ # ♅
18 58 ☽ □ ☉
19 58 ☽ ⚹ ♇
21 05 ☽ ⊻ ♅

JUNE 2013

LONGITUDES

Date	Sidereal time h m s	Sun ☉ ° ' "	Moon ☽ ° ' "	Moon ☽ 24.00 ° ' "	Mercury ☿ ° '	Venus ♀ ° '	Mars ♂ ° '	Jupiter ♃ ° '	Saturn ♄ ° '	Uranus ♅ ° '	Neptune ♆ ° '	Pluto ♇ ° '
01	04 40 32	11 ♊ 09 10	19 ♓ 55 11	26 ♓ 27 45	01 ♋ 55	28 ♊ 03	00 ♊ 45	24 ♊ 23	05 ♏ 52	11 ♈ 41	05 ♓ 22	11 ♑ 01
02	04 44 28	12 06 39	02 ♈ 55 06	09 ♈ 17 37	03 27	29 ♊ 17	01 28	24 36	05 R 49	11 43	05 22	11 R 00
03	04 48 25	13 04 08	15 ♈ 35 45	21 ♈ 49 55	04 56	00 ♋ 30	02 11	24 50	05 46	11 45	05 22	10 58
04	04 52 22	14 01 37	28 ♈ 00 33	04 ♉ 08 04	06 22	01 43	02 54	25 03	05 43	11 47	05 22	10 57
05	04 56 18	14 59 04	10 ♉ 12 52	16 ♉ 15 21	07 45	02 57	03 36	25 17	05 40	11 49	05 22	10 56
06	05 00 15	15 56 31	22 ♉ 15 50	28 ♉ 14 40	09 05	04 10	04 19	25 31	05 37	11 51	05 22	10 55
07	05 04 11	16 53 57	04 ♊ 12 10	10 ♊ 08 35	10 21	05 23	05 01	25 45	05 34	11 53	05 R 22	10 53
08	05 08 08	17 51 22	16 ♊ 04 12	21 ♊ 59 16	11 35	06 36	05 44	25 58	05 31	11 55	05 22	10 52
09	05 12 04	18 48 47	27 ♊ 54 02	03 ♋ 48 45	12 45	07 50	06 26	26 12	05 28	11 56	05 22	10 50
10	05 16 01	19 46 10	09 ♋ 43 38	15 ♋ 38 58	13 52	09 03	07 09	26 26	05 26	11 58	05 22	10 49
11	05 19 57	20 43 33	21 ♋ 35 00	27 ♋ 32 02	14 56	10 16	07 51	26 39	05 23	12 00	05 22	10 48
12	05 23 54	21 40 55	03 ♌ 30 21	09 ♌ 30 19	15 56	11 29	08 33	26 53	05 21	12 02	05 22	10 46
13	05 27 51	22 38 16	15 ♌ 32 16	21 ♌ 36 36	16 53	12 43	09 15	27 07	05 18	12 03	05 22	10 45
14	05 31 47	23 35 36	27 ♌ 43 44	03 ♍ 54 06	17 46	13 56	09 58	27 21	05 16	12 05	05 22	10 43
15	05 35 44	24 32 55	10 ♍ 08 10	16 ♍ 26 25	18 35	15 09	10 40	27 35	05 14	12 06	05 21	10 42
16	05 39 40	25 30 13	22 ♍ 49 19	29 ♍ 17 21	19 21	16 22	11 22	27 48	05 12	12 08	05 21	10 41
17	05 43 37	26 27 31	05 ♎ 50 56	12 ♎ 30 30	20 03	17 35	12 04	28 02	05 10	12 09	05 21	10 39
18	05 47 33	27 24 47	19 ♎ 16 23	26 ♎ 08 50	20 40	18 48	12 46	28 16	05 08	12 11	05 20	10 38
19	05 51 30	28 22 03	03 ♏ 08 01	10 ♏ 13 56	21 14	20 01	13 28	28 30	05 06	12 12	05 20	10 36
20	05 55 26	29 ♊ 19 18	17 ♏ 26 26	24 ♏ 45 11	21 44	21 14	14 09	28 43	05 04	12 13	05 20	10 35
21	05 59 23	00 ♋ 16 32	02 ♐ 09 41	09 ♐ 39 11	22 09	22 27	14 51	28 57	05 03	12 15	05 19	10 33
22	06 03 20	01 13 46	17 ♐ 12 45	24 ♐ 49 18	22 29	23 40	15 33	29 11	05 01	12 16	05 19	10 32
23	06 07 16	02 11 00	02 ♑ 27 34	10 ♑ 06 15	22 46	24 53	16 15	29 25	05 00	12 17	05 18	10 30
24	06 11 13	03 08 13	17 ♑ 43 58	25 ♑ 19 22	22 57	26 06	16 56	29 38	04 58	12 18	05 18	10 29
25	06 15 09	04 05 25	02 ♒ 51 11	10 ♒ 18 18	23 04	27 19	17 38	29 ♊ 52	04 57	12 19	05 17	10 27
26	06 19 06	05 02 38	17 ♒ 39 46	24 ♒ 54 51	23 07	28 32	18 19	00 ♋ 06	04 56	12 20	05 17	10 26
27	06 23 02	05 59 50	02 ♓ 03 02	09 ♓ 04 00	23 R 05	29 ♋ 45	19 01	00 20	04 55	12 21	05 16	10 24
28	06 26 59	06 57 03	15 ♓ 57 40	22 ♓ 44 04	22 58	00 ♌ 57	19 42	00 33	04 54	12 22	05 16	10 23
29	06 30 55	07 54 15	29 ♓ 23 29	05 ♈ 56 13	22 47	02 10	20 24	00 47	04 53	12 23	05 15	10 21
30	06 34 52	08 ♋ 51 28	12 ♈ 22 45	18 ♈ 43 36	22 ♋ 32	03 ♌ 23	21 ♊ 05	01 ♋ 01	04 ♏ 52	12 ♈ 24	05 ♓ 14	10 ♑ 20

Moon

Date	Moon True ☊ ° '	Moon Mean ☊ ° '	Moon Latitude ° '
01	16 ♏ 39	15 ♏ 34	04 N 24
02	16 D 40	15 31	03 38
03	16 42	15 28	02 43
04	16 44	15 25	01 41
05	16 45	15 22	00 N 36
06	16 R 45	15 18	00 S 30
07	16 43	15 15	01 34
08	16 40	15 12	02 34
09	16 36	15 09	03 26
10	16 31	15 06	04 10
11	16 25	15 02	04 43
12	16 19	14 59	05 03
13	16 13	14 56	05 11
14	16 09	14 53	05 05
15	16 06	14 50	04 44
16	16 05	14 47	04 10
17	16 D 05	14 43	03 21
18	16 07	14 40	02 21
19	16 08	14 37	01 S 10
20	16 R 08	14 34	00 N 07
21	16 07	14 31	01 26
22	16 04	14 28	02 40
23	16 00	14 24	03 44
24	15 53	14 21	04 32
25	15 46	14 18	05 00
26	15 40	14 15	05 08
27	15 34	14 12	04 55
28	15 30	14 08	04 26
29	15 28	14 05	03 42
30	15 ♏ 28	14 ♏ 02	02 N 48

DECLINATIONS

Date	Sun ☉	Moon ☽	Mercury ☿	Venus ♀	Mars ♂	Jupiter ♃	Saturn ♄	Uranus ♅	Neptune ♆	Pluto ♇
01	22 N 07	00 N 03	25 N 36	24 N 18	20 N 23	23 N 03	11 S 00	03 N 59	10 S 09	19 S 43
02	22 14	04 30	25 33	24 21	20 32	23 04	10 59	04 00	10 09	19 44
03	22 22	08 39	25 27	24 23	20 41	23 05	10 58	04 01	10 09	19 44
04	22 29	12 20	25 21	24 24	20 50	23 05	10 58	04 02	10 09	19 44
05	22 35	15 27	25 12	24 25	20 59	23 06	10 57	04 03	10 09	19 44
06	22 42	17 51	25 02	24 25	21 08	23 07	10 56	04 03	10 09	19 44
07	22 47	19 26	24 51	24 24	21 16	23 07	10 55	04 04	10 09	19 44
08	22 53	20 10	24 39	24 23	21 24	23 08	10 54	04 05	10 09	19 45
09	22 58	19 59	24 26	24 21	21 32	23 08	10 54	04 05	10 09	19 45
10	23 02	18 56	24 11	24 18	21 40	23 09	10 53	04 06	10 09	19 45
11	23 07	17 03	23 56	24 14	21 48	23 09	10 53	04 07	10 09	19 45
12	23 10	14 27	23 40	24 10	21 55	23 10	10 52	04 07	10 09	19 45
13	23 14	11 13	23 23	24 05	22 02	23 10	10 51	04 08	10 09	19 46
14	23 17	07 29	23 06	23 59	22 09	23 10	10 51	04 08	10 09	19 46
15	23 19	03 N 23	22 48	23 53	22 16	23 11	10 50	04 09	10 10	19 46
16	23 22	00 S 58	22 30	23 46	22 22	23 11	10 50	04 10	10 10	19 46
17	23 23	05 24	22 12	23 38	22 29	23 11	10 49	04 10	10 10	19 46
18	23 25	09 43	21 53	23 30	22 35	23 12	10 49	04 11	10 10	19 46
19	23 26	13 39	21 35	23 21	22 41	23 12	10 49	04 11	10 10	19 47
20	23 26	16 55	21 16	23 11	22 46	23 12	10 48	04 12	10 10	19 47
21	23 26	19 11	20 58	23 00	22 52	23 12	10 48	04 12	10 10	19 47
22	23 26	20 10	20 40	22 49	22 57	23 13	10 48	04 13	10 11	19 47
23	23 25	19 41	20 22	22 37	23 02	23 13	10 47	04 13	10 11	19 47
24	23 24	17 46	20 04	22 25	23 07	23 13	10 47	04 13	10 11	19 47
25	23 22	14 39	19 48	22 12	23 11	23 13	10 47	04 14	10 11	19 48
26	23 20	10 39	19 31	21 58	23 16	23 13	10 47	04 14	10 12	19 48
27	23 18	06 09	19 16	21 44	23 20	23 13	10 47	04 15	10 12	19 48
28	23 15	01 S 27	19 01	21 29	23 24	23 13	10 47	04 15	10 12	19 48
29	23 12	03 N 09	18 47	21 14	23 28	23 13	10 47	04 15	10 12	19 49
30	23 N 08	07 N 29	18 N 34	20 N 57	23 N 31	23 N 13	10 S 47	04 N 16	10 S 13	19 S 49

ZODIAC SIGN ENTRIES

Date	h	m	Planets	
02	06	33	☽	♈
03	02	13	♀	♋
04	15	53	☽	♉
07	03	32	☽	♊
09	16	16	☽	♋
12	04	58	☽	♌
14	16	26	☽	♍
17	01	19	☽	♎
19	06	38	☽	♏
21	05	04	☉	♋
21	08	31	☽	♐
23	08	08	☽	♑
25	07	26	☽	♒
26	01	40	♃	♋
27	08	32	☽	♓
27	17	03	♀	♌
29	13	06	☽	♈

LATITUDES

Date	Mercury ☿	Venus ♀	Mars ♂	Jupiter ♃	Saturn ♄	Uranus ♅	Neptune ♆	Pluto ♇
01	02 N 11	00 N 52	00 N 05	00 S 15	02 N 37	00 S 41	00 S 39	03 N 16
04	02 04	00 59	00 07	00 15	02 37	00 41	00 39	03 16
07	01 50	01 05	00 08	00 15	02 36	00 41	00 39	03 16
10	01 29	01 10	00 10	00 15	02 35	00 41	00 39	03 15
13	01 02	01 16	00 12	00 14	02 35	00 41	00 39	03 15
16	00 N 28	01 21	00 14	00 14	02 34	00 41	00 40	03 15
19	00 S 11	01 25	00 16	00 14	02 33	00 41	00 40	03 15
22	00 55	01 29	00 18	00 13	02 33	00 42	00 40	03 15
25	01 42	01 32	00 20	00 13	02 32	00 42	00 40	03 14
28	02 30	01 35	00 22	00 13	02 31	00 42	00 40	03 14
31	03 S 16	01 N 37	00 N 24	00 S 13	02 N 30	00 S 42	00 S 40	03 N 14

DATA

Julian Date	2456445
Delta T	+68 seconds
Ayanamsa	24° 02' 52"
Synetic vernal point	05° ♓ 04' 07"
True obliquity of ecliptic	23° 26' 08"

MOON'S PHASES, APSIDES AND POSITIONS ☽

Date	h	m	Phase	Longitude ° '	Eclipse Indicator
08	15	56	●	18 ♊ 01	
16	17	24	☽	25 ♍ 43	
23	11	32	○	02 ♑ 10	
30	04	54	☾	08 ♈ 35	

Day	h	m		
09	21	53	Apogee	
23	11	18	Perigee	
01	11	43	ON	
08	19	16	Max dec	20° N 12'
16	06	42	OS	
22	16	11	Max dec	20° S 11'
28	19	28	ON	

LONGITUDES

Date	Chiron ⚷ ° '	Ceres ⚳ ° '	Pallas ⚴ ° '	Juno ⚵ ° '	Vesta ⚶ ° '	Black Moon Lilith ⚸ ° '
01	13 ♓ 44	21 ♋ 07	01 ♊ 49	18 ♒ 26	14 ♋ 53	29 ♊ 07
11	13 ♓ 49	25 ♋ 18	07 ♊ 18	18 ♒ 58	19 ♋ 07	00 ♋ 14
21	13 ♓ 49	29 ♋ 34	12 ♊ 52	18 ♒ 53	23 ♋ 26	01 ♋ 21
31	13 ♓ 44	03 ♌ 55	18 ♊ 30	18 ♒ 11	27 ♋ 49	02 ♋ 28

ASPECTARIAN

h m Aspects

01 Saturday
08 40 ☉ ⚻ ♇
09 46 ☽ Q ♂
13 44 ☽ bQ ♄
17 39 ☽ Q ♇
20 18 ☽ □ ♃

02 Sunday
01 46 ☉ ⚹ ♅
04 30 ☽ □ ♀
05 01 ☉ ± ♄
06 14 ☽ ± ♄
06 21 ☽ Q ☉
09 08 ☽ ⚹ ♂
09 16 ☽ ∥ ♅
13 07 ☽ □ ☿
16 36 ☽ ⚺ ♆
17 25 ☽ ⚻ ♄

03 Monday
03 12 ☽ □ ♇
03 56 ☽ ⊥ ♆
04 39 ☽ ☌ ♅
06 37 ☽ Q ♃
06 47 ☽ ⚹ ☉
15 13 ☽ ∠ ♂
18 11 ☽ Q ♀
19 17 ☿ △ ♆
21 11 ☽ ∠ ♆
21 22 ☽ # ♆

04 Tuesday
01 22 ☿ △ ♄
02 40 ☽ # ♄
03 59 ☽ Q ☿
06 09 ☽ ⚹ ♃
09 41 ☽ ⊥ ♂
14 09 ☽ ∠ ☉
20 04 ☽ ⚹ ♀
22 09 ☽ ⚺ ♂

05 Wednesday
02 26 ☽ ⚹ ♆
03 03 ☽ ☍ ♄
06 30 ☽ ⚹ ☿
09 21 ☽ ⊥ ☉
12 09 ☽ ∠ ♃
13 25 ☽ △ ♇
15 11 ☽ ⚺ ♅
22 17 ☽ ⚺ ☉

06 Thursday
02 14 ☽ Q ♆
03 09 ☽ ⊥ ♅
05 06 ☽ ∠ ♀
06 24 ☽ ⊥ ♃
16 05 ☽ ∠ ☿
18 38 ☽ ⚺ ♃
18 54 ♀ ⚺ ♂
19 18 ☽ bQ ♇
21 13 ☽ ∠ ♅

07 Friday
01 11 ☽ ⊥ ♀
07 34 ♂ ± ♇
08 24 ♆ St R
11 46 ♀ △ ♆
12 21 ☽ ⊥ ☿
13 23 ☽ ± ♇
13 45 ☽ ☌ ♂
14 22 ☽ □ ♆
14 40 ☽ ⚺ ♀
14 44 ☽ ⚻ ♄
15 23 ♀ △ ♄
19 00 ☽ # ♇
22 03 ☿ ☍ ♇
23 57 ♂ □ ♆

08 Saturday
01 29 ☽ ⚻ ♇
01 53 ☽ ⚺ ☿
02 49 ☽ ± ♄
03 33 ☽ ⚹ ♅
05 18 ♂ ⚻ ♄
15 56 ☽ ☌ ☉
18 47 ☿ □ ♅
19 16 ☿ ⊥ ♂
20 59 ☽ bQ ♄

09 Sunday
03 57 ☽ Q ♅
08 29 ☽ ☌ ♃
19 30 ☽ # ♇
22 07 ☿ ∥ ♀

10 Monday
03 10 ☽ △ ♆
03 19 ☽ △ ♄
06 26 ☽ ⚺ ♂
10 28 ☽ ☌ ♀
14 12 ☽ ☍ ♇
16 34 ☽ □ ♅
19 22 ☽ ⊥ ♂
21 15 ☽ ☌ ☿

11 Tuesday
03 53 ☉ bQ ♄
09 33 ☽ bQ ♆
10 07 ☽ ⚺ ☉
14 43 ☽ ∠ ♂
22 08 ♀ ☍ ♇
22 26 ☽ ⚺ ♃
23 17 ☽ ⊥ ☉
23 26 ♄ △ ♆

12 Wednesday
03 41 ☽ ± ♆
05 21 ☉ ∥ ♃
10 44 ☽ ⊥ ♃
15 41 ☽ □ ♄
15 44 ☽ ⚻ ♆
18 55 ☽ ∠ ☉
22 44 ☽ ⚹ ♂
22 49 ♀ □ ♅

13 Thursday
02 30 ☽ ⚻ ♇
05 04 ☽ △ ♅
05 04 ☽ ∠ ♃
05 45 ☽ ⚺ ♀
14 24 ☽ ± ♇
14 29 ☽ # ♄
14 53 ☽ ⚺ ☿
18 59 ☽ ⊥ ♀
19 09 ☽ # ♆
23 10 ☉ ∥ ☿

14 Friday
00 00 ☽ Q ♂
03 13 ☽ ⚹ ☉
03 17 ☽ Q ♄
03 39 ☽ ⊥ ☿
06 07 ☿ ∥ ♃
08 05 ☽ bQ ♇
10 44 ☽ bQ ♅
11 14 ☽ ⚹ ♃
14 36 ☽ ∠ ♀
22 31 ☽ ∠ ☿

15 Saturday
00 35 ☉ Q ♅
02 36 ☽ ⚹ ♄
02 49 ☽ ☍ ♆
04 15 ☽ ± ♅
04 32 ☽ Q ☉
07 38 ☽ ∥ ♅
10 54 ☽ Q ♃
13 04 ☽ □ ♂
13 05 ☽ △ ♇
13 19 ♂ ⚻ ♇
15 46 ☽ ⚻ ♅
22 34 ☽ ⚹ ♀

16 Sunday
05 06 ☽ ⚹ ☿
06 39 ♂ ± ♄
07 06 ☽ ∠ ♄
17 24 ☽ □ ☉
19 32 ☿ ∥ ♂
21 26 ☽ □ ♃
23 22 ☽ Q ♀
23 48 ☽ ⊥ ♄

17 Monday
04 41 ☽ Q ☿
05 17 ☽ # ♅
10 46 ☽ ⚺ ♄
11 05 ☽ ⚻ ♆
15 20 ♂ ⚹ ♅
20 39 ☽ □ ♇
21 55 ☽ ± ♆
23 02 ☿ bQ ♆
23 23 ☽ ☍ ♅
23 49 ☽ △ ♂

18 Tuesday
10 06 ♀ ⊥ ♂
11 05 ☽ □ ♀
13 53 ☽ bQ ♆
14 34 ☽ □ ☿
14 38 ☽ ∥ ♆
18 25 ☽ ∥ ♄

19 Wednesday
00 27 ☉ ∥ ♀
03 14 ☽ △ ☉
03 34 ☽ bQ ♂
03 55 ☽ △ ♃
04 15 ☽ Q ♇
15 20 ☽ ☌ ♄
15 44 ☽ △ ♆
16 11 ☉ ☌ ♃
18 11 ♀ bQ ♆
19 43 ☽ ± ♂

20 Thursday
00 36 ☽ ⚹ ♇
03 19 ☽ ⚻ ♅
05 44 ☽ bQ ♃
06 17 ☽ ⚻ ♂
06 28 ☽ bQ ☉
08 58 ♀ ∥ ♃
13 18 ☽ ± ♅
18 49 ☽ △ ♀
19 16 ☽ △ ☿
20 49 ☽ ± ♃
22 20 ☽ ± ☉

21 Friday
01 19 ☽ ∠ ♇
02 57 ☿ ☌ ♀
04 02 ☽ bQ ♅
06 44 ☽ ⚻ ♃
08 45 ☽ ⚻ ☉
15 50 ☽ ⊥ ♇
16 38 ☽ ⚺ ♄
17 04 ☽ □ ♆
20 12 ☽ bQ ☿
21 14 ☽ bQ ♀
22 22 ☽ ∥ ♇

22 Saturday
00 39 ♀ ∥ ♂
01 25 ☽ ⚺ ♇
02 12 ☽ ⊥ ♄
04 09 ☽ △ ♅
09 14 ☽ ☍ ♂
10 50 ☽ ± ☿
12 47 ☽ ± ♀
16 26 ☽ ∠ ♄
20 29 ☽ ⚻ ☿
21 37 ☽ Q ♆
23 04 ☽ ⚻ ♀

23 Sunday
07 08 ☽ ☍ ♃
09 48 ☽ ∥ ♇
11 32 ☽ ☍ ☉
15 58 ☽ ⚹ ♄
16 28 ☽ ⚹ ♆

24 Monday
00 37 ☽ ☌ ♇
03 27 ☽ □ ♅
10 41 ☽ ⚻ ♂
10 48 ☽ Q ♄
12 45 ☿ ⊥ ♂
16 02 ☽ ∠ ♆
20 19 ☽ ☍ ☿
20 37 ☽ ± ♂

25 Tuesday
02 24 ☽ ☍ ♀
06 18 ☽ ⊥ ♆
07 10 ☽ ⚻ ♃
07 57 ☽ Q ♅
11 38 ☽ bQ ♂
11 46 ☿ # ♇
14 07 ☽ ⚻ ☉
15 22 ☽ □ ♄
15 54 ☽ ⚺ ♆
16 55 ☽ ± ♃
20 36 ♂ ∥ ♃

26 Wednesday
00 13 ☽ ⚺ ♇
00 27 ☽ ± ☉
03 18 ☽ ⚹ ♅
07 44 ☽ bQ ♃
09 12 ☉ △ ♄
09 59 ☽ ⊥ ♇
11 16 ☽ ∥ ♄
13 08 ☿ St R
13 08 ☽ △ ♂
14 32 ☽ ∥ ♆
16 12 ☽ bQ ☉
17 48 ☉ △ ♆
21 00 ☽ ⚻ ☿

27 Thursday
00 50 ☽ ∠ ♇
02 38 ♀ ± ♆
04 05 ☽ ∠ ♅
04 47 ☉ ∥ ♂
07 00 ☽ ± ☿
07 44 ☽ ⚻ ♀
09 02 ☽ △ ♃
16 52 ☽ △ ♄
17 28 ☽ ☌ ♆
18 53 ☽ ± ♀
19 13 ☽ △ ☉
19 21 ☽ ⊥ ♅
21 45 ☽ # ♅
22 14 ☽ bQ ☿

28 Friday
02 12 ♀ ⚺ ♃
02 17 ☽ ⚹ ♇
05 43 ☽ ⚺ ♅
12 00 ☽ bQ ♀
18 32 ♂ bQ ♄
18 56 ☽ bQ ♄
18 58 ☽ □ ♂
23 21 ☽ Q ♇

29 Saturday
00 16 ☽ △ ☿
03 02 ☉ ∥ ♃
11 04 ☽ ± ♄
14 35 ☽ □ ♃
17 35 ☽ △ ♀
17 55 ☽ ∥ ♅
22 02 ☽ ⚻ ♄
22 43 ☽ ⚺ ♆

30 Sunday
04 54 ☽ □ ☉
05 29 ☽ Q ♂
08 10 ☽ □ ♇
09 51 ☽ ⊥ ♆
12 03 ☽ ☌ ♅

All ephemeris data is given at 12.00 UT and the Moon's longitude is additionally given for 24.00 UT

JULY 2013

LONGITUDES

Date	Sidereal time h m s	Sun ☉ ° ' "	Moon ☽ ° ' "	Moon ☽ 24.00 ° ' "	Mercury ☿ ° '	Venus ♀ ° '	Mars ♂ ° '	Jupiter ♃ ° '	Saturn ♄ ° '	Uranus ♅ ° '	Neptune ♆ ° '	Pluto ♇ ° '
01	06 38 49	09 ♋ 48 40	24 ♈ 59 18	01 ♉ 10 28	22 ♋ 12	04 ♌ 36	21 ♊ 46	01 ♋ 14	04 ♏ 51	12 ♈ 25	05 ♓ 13	10 ♑ 18
02	06 42 45	10 45 53	07 ♉ 17 41	13 ♉ 21 33	21 R 49	05 49	22 27	01 28	04 R 51	12 26	05 R 12	10 R 17
03	06 46 42	11 43 06	19 ♉ 22 37	25 ♉ 21 25	21 22	07 01	23 09	01 42	04 50	12 26	05 12	10 15
04	06 50 38	12 40 20	01 ♊ 18 29	07 ♊ 14 15	20 52	08 14	23 50	01 55	04 50	12 27	05 11	10 14
05	06 54 35	13 37 33	13 ♊ 09 10	19 ♊ 03 37	20 20	09 27	24 31	02 09	04 49	12 28	05 10	10 12
06	06 58 31	14 34 47	24 ♊ 57 56	00 ♋ 52 26	19 45	10 39	25 12	02 23	04 49	12 28	05 09	10 11
07	07 02 28	15 32 01	06 ♋ 47 22	12 ♋ 42 58	19 08	11 52	25 53	02 36	04 49	12 29	05 08	10 09
08	07 06 24	16 29 14	18 ♋ 39 29	24 ♋ 37 04	18 31	13 05	26 34	02 50	04 D 49	12 29	05 07	10 08
09	07 10 21	17 26 28	00 ♌ 35 55	06 ♌ 36 12	17 53	14 17	27 15	03 03	04 49	12 30	05 06	10 06
10	07 14 18	18 23 42	12 ♌ 38 06	18 ♌ 41 48	17 15	15 30	27 56	03 17	04 49	12 30	05 05	10 05
11	07 18 14	19 20 57	24 ♌ 47 31	00 ♍ 55 28	16 38	16 42	28 36	03 30	04 50	12 30	05 04	10 03
12	07 22 11	20 18 11	07 ♍ 05 54	13 ♍ 19 07	16 03	17 55	29 17	03 44	04 50	12 31	05 03	10 02
13	07 26 07	21 15 25	19 ♍ 35 25	25 ♍ 55 09	15 30	19 07	29 ♊ 58	03 57	04 50	12 31	05 02	10 00
14	07 30 04	22 12 39	02 ♎ 18 41	08 ♎ 46 25	14 59	20 20	00 ♋ 38	04 11	04 51	12 31	05 01	09 59
15	07 34 00	23 09 53	15 ♎ 18 45	21 ♎ 56 02	14 32	21 32	01 19	04 24	04 52	12 31	05 00	09 57
16	07 37 57	24 07 07	28 ♎ 38 41	05 ♏ 27 00	14 09	22 44	01 59	04 37	04 52	12 31	04 59	09 56
17	07 41 53	25 04 22	12 ♏ 21 14	19 ♏ 21 34	13 50	23 57	02 40	04 50	04 53	12 31	04 58	09 54
18	07 45 50	26 01 36	26 ♏ 28 01	03 ♐ 40 29	13 35	25 09	03 20	05 04	04 54	12 R 31	04 57	09 53
19	07 49 47	26 58 51	10 ♐ 58 41	18 ♐ 22 09	13 26	26 21	04 01	05 17	04 55	12 31	04 55	09 52
20	07 53 43	27 56 06	25 ♐ 50 11	03 ♑ 21 56	13 22	27 34	04 41	05 30	04 57	12 31	04 54	09 50
21	07 57 40	28 53 21	10 ♑ 56 20	18 ♑ 32 10	13 D 23	28 46	05 21	05 43	04 58	12 31	04 53	09 49
22	08 01 36	29 ♋ 50 37	26 ♑ 08 08	03 ♒ 42 53	13 30	29 ♌ 58	06 01	05 56	04 59	12 31	04 52	09 47
23	08 05 33	00 ♌ 47 53	11 ♒ 15 05	18 ♒ 43 28	13 43	01 ♍ 10	06 42	06 10	05 01	12 31	04 51	09 46
24	08 09 29	01 45 10	26 ♒ 06 57	03 ♓ 24 34	14 01	02 22	07 22	06 23	05 02	12 30	04 49	09 45
25	08 13 26	02 42 27	10 ♓ 35 36	17 ♓ 39 32	14 26	03 34	08 02	06 36	05 04	12 30	04 48	09 43
26	08 17 22	03 39 46	24 ♓ 36 07	01 ♈ 25 14	14 56	04 46	08 42	06 49	05 05	12 29	04 47	09 42
27	08 21 19	04 37 05	08 ♈ 07 01	14 ♈ 41 43	15 33	05 58	09 22	07 01	05 07	12 29	04 45	09 40
28	08 25 16	05 34 25	21 ♈ 09 45	27 ♈ 31 36	16 15	07 10	10 02	07 14	05 09	12 29	04 44	09 39
29	08 29 12	06 31 46	03 ♉ 47 50	09 ♉ 59 05	17 03	08 22	10 42	07 27	05 11	12 28	04 42	09 38
30	08 33 09	07 29 09	16 ♉ 06 01	22 ♉ 09 17	17 57	09 34	11 21	07 40	05 13	12 27	04 41	09 37
31	08 37 05	08 ♌ 26 32	28 ♉ 09 33	04 ♊ 07 27	18 ♋ 56	10 ♍ 46	12 ♋ 01	07 ♋ 53	05 ♏ 16	12 ♈ 27	04 ♓ 40	09 ♑ 35

DECLINATIONS

Date	Moon True ☊ ° '	Moon Mean ☊ ° '	Moon ☽ Latitude ° '	Sun ☉ ° '	Moon ☽ ° '	Mercury ☿ ° '	Venus ♀ ° '	Mars ♂ ° '	Jupiter ♃ ° '	Saturn ♄ ° '	Uranus ♅ ° '	Neptune ♆ ° '	Pluto ♇ ° '
01	15 ♏ 28	13 ♏ 59	01 N 48	23 N 04	11 N 21	18 N 22	20 N 41	23 N 34	23 N 13	10 S 47	04 N 16	10 S 13	19 S 49
02	15 D 29	13 56	00 N 44	23 00	14 38	18 12	20 24	23 38	23 13	10 47	04 16	10 13	19 49
03	15 R 30	13 53	00 S 21	22 55	17 14	18 02	20 06	23 40	23 13	10 47	04 16	10 14	19 49
04	15 28	13 49	01 24	22 50	19 03	17 54	19 47	23 43	23 13	10 47	04 17	10 14	19 50
05	15 25	13 46	02 23	22 44	20 01	17 47	19 28	23 45	23 13	10 47	04 17	10 14	19 50
06	15 19	13 43	03 15	22 38	20 06	17 41	19 09	23 48	23 13	10 47	04 17	10 15	19 50
07	15 10	13 40	03 59	22 32	19 17	17 37	18 49	23 50	23 13	10 47	04 17	10 15	19 50
08	15 00	13 37	04 32	22 25	17 39	17 34	18 29	23 51	23 12	10 48	04 17	10 15	19 50
09	14 48	13 34	04 54	22 18	15 14	17 33	18 08	23 53	23 12	10 48	04 17	10 16	19 51
10	14 37	13 30	05 03	22 10	12 10	17 33	17 46	23 54	23 12	10 48	04 18	10 16	19 51
11	14 26	13 27	04 58	22 02	08 34	17 34	17 24	23 56	23 12	10 49	04 18	10 16	19 51
12	14 17	13 24	04 40	21 54	04 34	17 36	17 02	23 56	23 11	10 49	04 18	10 17	19 51
13	14 11	13 21	04 08	21 45	00 N 19	17 40	16 39	23 57	23 11	10 49	04 18	10 17	19 52
14	14 07	13 18	03 23	21 36	04 S 01	17 45	16 16	23 58	23 11	10 50	04 18	10 18	19 52
15	14 05	13 14	02 27	21 27	08 17	17 51	15 52	23 58	23 10	10 50	04 18	10 18	19 52
16	14 D 05	13 11	01 21	21 17	12 15	17 58	15 28	23 58	23 10	10 51	04 18	10 19	19 52
17	14 05	13 08	00 S 09	21 07	15 41	18 06	15 04	23 58	23 10	10 51	04 18	10 19	19 53
18	14 R 05	13 05	01 N 05	20 56	18 18	18 15	14 39	23 58	23 09	10 52	04 18	10 19	19 53
19	14 02	13 02	02 18	20 46	19 49	18 25	14 14	23 57	23 09	10 52	04 18	10 20	19 53
20	13 58	12 59	03 22	20 34	20 00	18 35	13 48	23 56	23 08	10 53	04 18	10 20	19 53
21	13 50	12 55	04 14	20 23	18 46	18 46	13 22	23 55	23 08	10 54	04 18	10 21	19 54
22	13 41	12 52	04 48	20 11	16 13	18 57	12 56	23 54	23 07	10 54	04 17	10 21	19 54
23	13 31	12 49	05 01	19 59	12 34	19 08	12 30	23 53	23 07	10 55	04 17	10 22	19 54
24	13 20	12 46	04 54	19 46	08 12	19 20	12 03	23 51	23 06	10 56	04 17	10 22	19 54
25	13 11	12 43	04 28	19 33	03 S 27	19 31	11 36	23 50	23 06	10 57	04 17	10 23	19 54
26	13 04	12 40	03 46	19 20	01 N 19	19 42	11 08	23 48	23 05	10 58	04 17	10 23	19 55
27	13 00	12 36	02 53	19 06	05 52	19 53	10 40	23 46	23 05	10 58	04 17	10 24	19 55
28	12 58	12 33	01 53	18 52	10 00	20 03	10 12	23 43	23 04	10 59	04 16	10 24	19 55
29	12 D 57	12 30	00 N 48	18 38	13 32	20 12	09 44	23 41	23 03	11 00	04 16	10 25	19 55
30	12 R 57	12 27	00 S 17	18 24	16 23	20 20	09 16	23 38	23 03	11 01	04 16	10 25	19 56
31	12 ♏ 57	12 ♏ 24	01 S 20	18 N 09	18 N 27	20 N 28	08 N 47	23 N 35	23 N 02	11 S 02	04 N 16	10 S 26	19 S 56

ZODIAC SIGN ENTRIES

Date	h	m	Planets
01	21	43	☽ ♉
04	09	22	☽ ♊
06	22	14	☽ ♋
09	10	48	☽ ♌
11	22	12	☽ ♍
13	13	22	♂ ♋
14	07	41	☽ ♎
16	14	24	☽ ♏
18	17	54	☽ ♐
20	18	39	☽ ♑
22	12	41	♀ ♍
22	15	56	☉ ♌
22	18	07	☽ ♒
24	18	22	☽ ♓
26	21	29	☽ ♈
29	04	43	☽ ♉
31	15	42	☽ ♊

LATITUDES

Date	Mercury ☿ ° '	Venus ♀ ° '	Mars ♂ ° '	Jupiter ♃ ° '	Saturn ♄ ° '	Uranus ♅ ° '	Neptune ♆ ° '	Pluto ♇ ° '
01	03 S 16	01 N 37	00 N 24	00 S 13	02 N 30	00 S 42	00 S 40	03 N 14
04	03 58	01 39	00 26	00 12	02 30	00 42	00 40	03 13
07	04 30	01 39	00 27	00 12	02 29	00 42	00 40	03 13
10	04 49	01 40	00 29	00 12	02 28	00 42	00 40	03 13
13	04 54	01 39	00 31	00 12	02 27	00 42	00 40	03 12
16	04 45	01 38	00 33	00 11	02 26	00 42	00 40	03 12
19	04 22	01 36	00 35	00 11	02 26	00 42	00 40	03 12
22	03 49	01 33	00 36	00 11	02 25	00 42	00 41	03 11
25	03 09	01 30	00 38	00 10	02 24	00 43	00 41	03 11
28	02 25	01 26	00 40	00 10	02 23	00 43	00 41	03 10
31	01 S 39	01 N 21	00 N 42	00 S 10	02 N 22	00 S 43	00 S 41	03 N 10

DATA

Julian Date	2456475
Delta T	+68 seconds
Ayanamsa	24° 02' 57"
Synetic vernal point	05° ♓ 04' 02"
True obliquity of ecliptic	23° 26' 08"

MOON'S PHASES, APSIDES AND POSITIONS ☽

Date	h	m	Phase	Longitude ° '	Eclipse Indicator
08	07	14	●	16 ♋ 18	
16	03	18	☽	23 ♎ 46	
22	18	16	○	00 ♒ 06	
29	17	43	☾	06 ♉ 45	

Day	h	m		
07	00	51	Apogee	
21	20	29	Perigee	
06	02	07	Max dec	20° N 10'
13	13	48	OS	
20	03	14	Max dec	20° S 06'
26	05	18	ON	

LONGITUDES

Date	Chiron ⚷ ° '	Ceres ⚳ ° '	Pallas ⚴ ° '	Juno ⚵ ° '	Vesta ⚶ ° '	Black Moon Lilith ⚸ ° '
01	13 ♓ 44	03 ♌ 55	18 ♊ 30	18 ♒ 11	27 ♋ 49	02 ♋ 28
11	13 ♓ 33	08 ♌ 20	24 ♊ 13	16 ♒ 51	02 ♌ 16	03 ♋ 35
21	13 ♓ 17	12 ♌ 48	29 ♊ 59	14 ♒ 58	06 ♌ 46	04 ♋ 42
31	12 ♓ 57	17 ♌ 18	05 ♋ 46	12 ♒ 42	11 ♌ 17	05 ♋ 48

All ephemeris data is given at 12.00 UT and the Moon's longitude is additionally given for 24.00 UT

ASPECTARIAN

h m	Aspects
01 Monday	
00 47	☽ Q ♃
02 51	☽ ∠ ♆
04 38	☽ # ♆
05 27	☽ ⚹ ♂
06 48	☽ □ ☿
08 14	☽ # ♄
17 03	♀ □ ♄
17 55	☽ Q ☉
21 54	☿ ⊻ ♂
02 Tuesday	
00 06	☉ ☍ ♇
00 12	♀ ⊼ ♆
00 21	☽ ⚹ ♃
07 11	☽ ☍ ♄
07 54	☽ ⚹ ♆
08 45	☽ □ ♀
12 20	☽ ∠ ♂
16 49	☽ Q ☿
17 53	☽ △ ♇
19 26	☽ ⚹ ☉
22 10	☽ ⊻ ♅
03 Wednesday	
06 32	☽ ∠ ♃
07 16	☽ ⊥ ♂
07 39	☽ Q ♆
10 07	☽ ⊥ ♅
15 51	☽ ⚹ ☿
20 01	☽ ⊻ ♂
20 41	☽ ∥ ☿
23 46	☽ ⚼ ♇
04 Thursday	
00 36	☽ Q ♀
00 56	☽ ⊥ ♃
04 01	☽ ∠ ☉
04 13	☽ ∠ ♅
04 25	♀ ⊥ ♃
06 22	☉ □ ♅
09 09	♀ # ♇
13 16	☽ ⊻ ♃
17 54	☽ ± ♇
19 07	☽ ⊼ ♄
19 49	☽ □ ♆
20 50	☽ ∠ ☿
23 48	☽ ⊥ ☉
23 49	☽ ∥ ♀
05 Friday	
03 37	☽ ⚹ ♀
05 08	☽ # ♇
06 02	☽ ⊼ ♇
07 17	☽ ± ♄
10 08	♂ Q ♅
10 36	☽ ⚹ ♅
13 03	☽ ⊻ ☉
14 17	☽ ⊥ ☿
15 08	♀ ∠ ♂
19 03	☿ ⚼ ♆
06 Saturday	
01 33	☽ ⚼ ♄
01 54	☽ ⊻ ☿
02 43	♀ ⊼ ♇
11 00	☽ Q ♅
12 30	☽ ☌ ♂
13 34	☽ ∠ ♀
22 56	☽ # ♇
07 Sunday	
03 21	☽ ☌ ♃
08 00	☽ △ ♄
08 39	☽ △ ♆
09 55	☽ ⊥ ♀
18 48	☽ ☍ ♇
22 28	☽ ∥ ♀
23 27	☽ ⊻ ♀
23 32	☽ □ ♅
08 Monday	
00 14	♀ △ ♅
05 11	♄ St D
07 14	☽ ☌ ☉
11 44	☽ ☌ ☿
12 51	☽ ∥ ☿
14 57	☽ ⚼ ♆
09 Tuesday	
04 52	☽ ⊻ ♂
09 01	☽ ± ♆
17 00	☽ ⊻ ♃
17 37	☽ ⊥ ♂
18 41	☉ ☌ ☿
20 27	☽ □ ♄
21 00	☽ ⊼ ♆
10 Wednesday	
05 12	☽ ⊥ ♃
06 56	☽ ⊼ ♇
11 44	☽ △ ♅
12 37	☽ ∠ ♂
18 18	☽ ☌ ♀
18 49	☽ ± ♇
20 42	☽ ⊻ ☿
21 27	☽ # ♄
23 20	♀ ± ♇
23 23	☽ ∠ ♃
11 Thursday	
00 23	☽ ⊻ ☉
01 01	☽ # ♆
02 31	☿ ∥ ♀
07 58	☽ ⊥ ☿
08 08	☽ Q ♄
11 08	☿ ⊻ ♀
12 31	☽ ⚼ ♇
13 11	☽ ⊥ ☉
17 19	☽ ⚼ ♅
19 54	☽ ⚹ ♂
12 Friday	
00 47	☽ ∠ ☿
05 20	☽ ⚹ ♃
05 54	☉ ⚼ ♆
07 36	☽ ⚹ ♄
08 03	☽ ☍ ♆
08 14	☽ ∠ ☉
10 52	☽ ± ♅
13 36	☽ ∥ ♅
17 39	☽ △ ♇
20 33	☽ Q ♂
22 27	☽ ⊼ ♅
13 Saturday	
04 30	☽ ⚹ ☿
04 56	☽ Q ♃
07 53	♀ ∠ ♃
11 01	☽ ⊻ ♀
12 29	☽ ∠ ♄
15 26	☽ ⚹ ☉
23 36	☽ ⊥ ♀
14 Sunday	
02 23	☽ Q ☿
05 31	☽ ⊥ ♄
08 42	☽ □ ♂
13 32	☽ # ♅
15 32	☽ □ ♃
15 49	☽ Q ☉
16 44	☽ ⊻ ♄
17 02	☽ ⊼ ♆
18 12	☽ ∠ ♀
21 28	☿ ⊥ ♀
15 Monday	
02 12	☽ □ ♇
04 07	☽ ± ♆
06 54	☽ ☍ ♅
10 38	☽ □ ☿
20 30	☽ ⚼ ♆
23 55	☽ ∥ ♆
16 Tuesday	
00 24	☽ ⚹ ♀
03 10	☽ ∥ ♄
03 18	☽ □ ☉
10 44	☽ Q ♇
14 41	♀ Q ♄
18 14	☽ △ ♂
22 43	☽ △ ♃
23 00	☽ ☌ ♄
23 10	☽ △ ♆
23 48	☽ Q ♀
17 Wednesday	
07 46	☽ ⚹ ♇
07 48	☽ # ♀
12 17	☽ ⊼ ♅
14 30	☽ △ ☿
17 20	♅ St R
17 31	♃ △ ♄
21 34	☽ ⚼ ♂
22 34	☽ ± ♅
18 Thursday	
00 14	♃ △ ♆
01 01	☽ ⚼ ♃
06 45	♀ ⚼ ♇
09 21	☽ ∠ ♇
09 35	☽ □ ♀
11 13	☽ △ ☉
11 22	☽ # ☿
13 32	☽ ± ♂
13 46	☽ ⚼ ♅
15 30	☽ ⚼ ☿
16 24	☽ ± ♃
19 Friday	
00 00	☽ ⊼ ♂
00 19	☽ ⊥ ♇
02 03	☽ ⊻ ♄
02 05	☽ □ ♆
02 31	☽ ⊼ ♃
06 15	☽ ± ☿
10 10	☽ ⊻ ♇
11 55	☽ ⊥ ♄
13 20	♄ △ ♆
13 45	☽ ⚼ ☉
14 00	☽ ∥ ♇
14 31	☽ △ ♅
15 58	☽ ⊼ ☿
20 Saturday	
02 31	☽ ∠ ♄
05 19	☽ ± ☉
07 19	☽ Q ♆
11 11	♀ ⚼ ♅
15 00	☽ △ ♀
15 35	☽ ⊼ ☉
16 14	☽ ∥ ♇
18 22	☿ St D
19 42	♂ △ ♆
21 34	♂ △ ♄
21 Sunday	
02 25	☽ ⚹ ♆
02 31	☽ ⚹ ♄
02 45	☽ ☍ ♂
03 37	☽ ☍ ♃
04 09	☿ ∠ ♀
10 13	☽ ☌ ♇
11 53	☉ ± ♆
12 07	☽ # ☿
14 30	☽ □ ♅
15 53	☽ ☍ ☿
16 51	☽ ⚼ ♀
21 32	☽ Q ♄
22 Monday	
00 11	☉ ⊻ ♀
02 07	☽ ∠ ♆
07 35	♂ ☌ ♃
08 17	☽ ± ♀
16 18	☽ ⊥ ♆
18 16	☽ ☍ ☉
18 35	☽ ⊼ ♀
18 55	☽ Q ♅
23 Tuesday	
01 48	☽ ⊻ ♆
02 02	☽ □ ♄
03 46	☽ ⊼ ♃
04 24	☽ ⊼ ♂
09 38	☽ ⊻ ♇
12 30	☽ # ♀
13 28	☽ ± ♃
14 01	☽ ⚹ ♅
14 25	☽ ± ♂
16 01	☽ ⊼ ☿
19 13	☽ ⊥ ♇
20 41	☉ # ♇
21 23	☽ ∥ ♄
24 Wednesday	
00 27	☽ ∥ ♆
01 52	☽ ± ☿
04 10	☽ ⚼ ♃
05 36	☽ ⚼ ♂
09 46	☽ ∠ ♇
14 16	☽ ∠ ♅
16 53	☽ ⚼ ☿
21 55	☽ ⊼ ☉
23 12	☽ ☍ ♀
25 Thursday	
02 19	☽ ☌ ♆
02 44	☽ △ ♄
05 08	☽ ⊥ ♅
05 12	☽ △ ♃
07 29	☽ △ ♂
07 52	☽ # ♅
08 36	☽ ± ☉
10 32	☽ ⚹ ♇
14 13	☉ ∥ ☿
15 13	☽ ⊻ ♅
18 43	☽ △ ☿
26 Friday	
00 58	☽ ⚼ ☉
04 10	☽ ⚼ ♄
06 58	☽ Q ♇
12 08	♀ ☍ ♆
18 36	♀ ⚹ ♄
19 54	☽ ± ♄
20 51	♀ # ♄
27 Saturday	
03 23	☽ ∥ ♅
05 13	☽ △ ☉
05 57	☽ ⊻ ♆
06 36	☽ ⊼ ♄
07 45	☽ ⊼ ♀
10 00	☽ □ ♃
14 23	☽ □ ♂
14 49	☽ □ ♇
15 20	☉ ⊼ ♆
16 47	☽ ⊥ ♆
17 38	☿ # ♇
19 42	☽ ± ♀
19 56	☽ ☌ ♅
22 13	♀ ± ♅
22 51	♂ ☍ ♇
28 Sunday	
01 05	☉ □ ♄
01 55	♀ # ♆
02 19	☽ □ ☿
09 20	☽ ∠ ♆
13 10	☽ ∥ ♀
13 42	♀ ⚹ ♃
14 05	☽ ⚼ ♀
14 35	☽ # ♆
18 20	☽ # ♄
19 48	☽ Q ♃
29 Monday	
01 41	☽ Q ♂
13 45	☽ ⚹ ♆
14 36	☽ Q ☿
14 41	☽ ☍ ♄
17 43	☽ □ ☉
19 12	☽ ⚹ ♃
21 48	☽ △ ♀
23 17	☽ △ ♇
30 Tuesday	
02 09	☽ ⚹ ♂
04 50	☽ ⊻ ♅
12 52	♀ △ ♇
13 09	☽ Q ♆
15 58	☽ ⚹ ☿
16 39	☽ ⊥ ♅
17 46	☉ ⊻ ♃
31 Wednesday	
01 15	☽ ∠ ♃
04 52	☽ ⚼ ♇
08 16	☽ Q ☉
08 18	☽ ∥ ☉
09 35	☽ ∠ ♂
10 34	☽ ∠ ♅
19 36	☽ ⊥ ♃
22 54	☽ ± ♇

LONGITUDES

Date	Sidereal time h m s	Sun ☉ ° ' "	Moon ☽ ° ' "	Moon ☽ 24.00 ° ' "	Mercury ☿ ° '	Venus ♀ ° '	Mars ♂ ° '	Jupiter ♃ ° '	Saturn ♄ ° '	Uranus ♅ ° '	Neptune ♆ ° '	Pluto ♇ ° '
01	08 41 02	09 ♌ 23 57	10 ♊ 03 37	15 ♊ 58 38	20 ♋ 01	11 ♍ 58	12 ♋ 41	08 ♋ 05	05 ♏ 18	12 ♈ 26	04 ♓ 38	09 ♑ 34
02	08 44 58	10 21 22	21 ♊ 53 03	27 ♊ 47 20	21 12	13 09	13 21	08 18	05 20	12 R 25	04 R 37	09 R 33
03	08 48 55	11 18 49	03 ♋ 41 57	09 ♋ 37 18	22 28	14 21	14 00	08 30	05 23	12 25	04 35	09 31
04	08 52 51	12 16 17	15 ♋ 33 44	21 ♋ 31 31	23 49	15 33	14 40	08 43	05 25	12 24	04 34	09 30
05	08 56 48	13 13 46	27 ♋ 30 56	03 ♌ 32 08	25 14	16 44	15 20	08 55	05 28	12 23	04 32	09 29
06	09 00 45	14 11 16	09 ♌ 35 20	15 ♌ 40 37	26 45	17 56	15 59	09 08	05 31	12 22	04 31	09 28
07	09 04 41	15 08 46	21 ♌ 48 06	27 ♌ 57 52	28 20	19 07	16 39	09 20	05 34	12 21	04 29	09 27
08	09 08 38	16 06 18	04 ♍ 10 00	10 ♍ 24 33	29 ♋ 59	20 19	17 18	09 32	05 36	12 20	04 28	09 26
09	09 12 34	17 03 51	16 ♍ 41 38	23 ♍ 01 20	01 ♌ 42	21 30	17 57	09 45	05 39	12 19	04 26	09 24
10	09 16 31	18 01 25	29 ♍ 23 48	05 ♎ 49 09	03 29	22 42	18 37	09 57	05 43	12 18	04 25	09 23
11	09 20 27	18 58 59	12 ♎ 17 36	18 ♎ 49 21	05 18	23 53	19 16	10 09	05 46	12 17	04 23	09 22
12	09 24 24	19 56 35	25 ♎ 24 38	02 ♏ 03 41	07 11	25 05	19 55	10 21	05 49	12 16	04 22	09 21
13	09 28 20	20 54 12	08 ♏ 46 48	15 ♏ 34 11	09 06	26 16	20 34	10 33	05 52	12 14	04 20	09 20
14	09 32 17	21 51 49	22 ♏ 26 03	29 ♏ 22 35	11 03	27 27	21 13	10 45	05 56	12 13	04 18	09 19
15	09 36 14	22 49 28	06 ♐ 23 51	13 ♐ 29 51	13 01	28 38	21 52	10 57	05 59	12 12	04 17	09 18
16	09 40 10	23 47 07	20 ♐ 40 26	27 ♐ 55 20	15 01	29 ♍ 49	22 31	11 08	06 03	12 11	04 15	09 17
17	09 44 07	24 44 47	05 ♑ 14 07	12 ♑ 36 12	17 01	01 ♎ 00	23 10	11 20	06 07	12 09	04 14	09 16
18	09 48 03	25 42 29	20 ♑ 00 48	27 ♑ 27 02	19 03	02 11	23 49	11 32	06 10	12 08	04 12	09 15
19	09 52 00	26 40 12	04 ♒ 53 53	12 ♒ 20 14	21 04	03 22	24 28	11 43	06 14	12 06	04 10	09 14
20	09 55 56	27 37 55	19 ♒ 45 00	27 ♒ 07 03	23 05	04 33	25 07	11 54	06 18	12 05	04 09	09 14
21	09 59 53	28 35 40	04 ♓ 25 22	11 ♓ 39 00	25 06	05 44	25 46	12 06	06 22	12 03	04 07	09 13
22	10 03 49	29 ♌ 33 27	18 ♓ 47 11	25 ♓ 49 20	27 07	06 55	26 24	12 17	06 26	12 02	04 05	09 12
23	10 07 46	00 ♍ 31 15	02 ♈ 45 00	09 ♈ 33 58	29 ♌ 07	08 05	27 03	12 28	06 31	12 00	04 04	09 11
24	10 11 43	01 29 04	16 ♈ 16 11	22 ♈ 51 44	01 ♍ 06	09 16	27 42	12 39	06 35	11 58	04 02	09 10
25	10 15 39	02 26 55	29 ♈ 20 52	05 ♉ 43 58	03 05	10 27	28 20	12 50	06 39	11 57	04 00	09 09
26	10 19 36	03 24 48	12 ♉ 01 29	18 ♉ 13 58	05 02	11 37	28 59	13 01	06 43	11 55	03 59	09 09
27	10 23 32	04 22 43	24 ♉ 22 01	00 ♊ 26 15	06 58	12 48	29 ♋ 37	13 12	06 48	11 53	03 57	09 08
28	10 27 29	05 20 40	06 ♊ 27 16	12 ♊ 25 59	08 54	13 58	00 ♌ 16	13 23	06 52	11 52	03 56	09 07
29	10 31 25	06 18 38	18 ♊ 22 47	24 ♊ 18 25	10 47	15 08	00 54	13 34	06 57	11 50	03 54	09 07
30	10 35 22	07 16 39	00 ♋ 13 30	06 ♋ 08 37	12 40	16 19	01 33	13 44	07 02	11 48	03 52	09 06
31	10 39 18	08 ♍ 14 41	12 ♋ 04 18	18 ♋ 01 05	14 ♍ 32	17 ♎ 29	02 ♌ 11	13 ♋ 55	07 ♏ 07	11 ♈ 46	03 ♓ 51	09 ♑ 05

Date	Moon True ☊ ° '	Moon Mean ☊ ° '	Moon ☽ Latitude ° '
01	12 ♏ 55	12 ♏ 20	02 S 18
02	12 R 50	12 17	03 10
03	12 43	12 14	03 54
04	12 33	12 11	04 28
05	12 20	12 08	04 50
06	12 07	12 05	05 00
07	11 53	12 01	04 55
08	11 41	11 58	04 37
09	11 30	11 55	04 06
10	11 22	11 52	03 22
11	11 17	11 49	02 27
12	11 14	11 46	01 23
13	11 D 14	11 42	00 S 13
14	11 R 14	11 39	00 N 59
15	11 13	11 36	02 09
16	11 11	11 33	03 13
17	11 06	11 30	04 05
18	10 59	11 26	04 42
19	10 50	11 23	05 00
20	10 40	11 20	04 58
21	10 29	11 17	04 36
22	10 20	11 14	03 57
23	10 13	11 11	03 05
24	10 08	11 07	02 03
25	10 06	11 04	00 N 57
26	10 D 05	11 01	00 S 10
27	10 06	10 58	01 15
28	10 R 06	10 55	02 16
29	10 05	10 51	03 09
30	10 03	10 48	03 54
31	09 ♏ 58	10 ♏ 45	04 S 29

DECLINATIONS

Date	Sun ☉ ° '	Moon ☽ ° '	Mercury ☿ ° '	Venus ♀ ° '	Mars ♂ ° '	Jupiter ♃ ° '	Saturn ♄ ° '	Uranus ♅ ° '	Neptune ♆ ° '	Pluto ♇ ° '
01	17 N 54	19 N 41	20 N 33	08 N 18	23 N 32	23 N 01	11 S 03	04 N 15	10 S 27	19 S 56
02	17 39	20 01	20 38	07 49	23 29	23 01	11 04	04 15	10 27	19 56
03	17 23	19 29	20 41	07 19	23 25	23 00	11 05	04 15	10 28	19 56
04	17 07	18 06	20 41	06 50	23 22	22 59	11 06	04 14	10 28	19 57
05	16 51	15 55	20 40	06 20	23 18	22 59	11 07	04 14	10 29	19 57
06	16 34	13 02	20 37	05 50	23 14	22 58	11 09	04 14	10 29	19 57
07	16 17	09 34	20 31	05 20	23 10	22 57	11 10	04 13	10 30	19 58
08	16 00	05 40	20 23	04 50	23 05	22 56	11 11	04 13	10 30	19 58
09	15 43	01 N 28	20 12	04 20	23 01	22 56	11 12	04 12	10 31	19 58
10	15 26	02 S 51	19 59	03 49	22 56	22 55	11 13	04 12	10 32	19 58
11	15 08	07 06	19 42	03 19	22 51	22 54	11 15	04 12	10 32	19 59
12	14 50	11 07	19 24	02 48	22 46	22 53	11 16	04 11	10 33	19 59
13	14 32	14 38	19 02	02 17	22 40	22 52	11 17	04 11	10 33	19 59
14	14 13	17 25	18 38	01 47	22 35	22 51	11 19	04 10	10 34	19 59
15	13 54	19 15	18 12	01 16	22 29	22 50	11 20	04 10	10 35	20 00
16	13 35	19 54	17 43	00 45	22 24	22 50	11 22	04 09	10 35	20 00
17	13 16	19 15	17 12	00 N 14	22 18	22 49	11 23	04 08	10 36	20 00
18	12 57	17 18	16 38	00 S 17	22 12	22 48	11 24	04 08	10 36	20 00
19	12 37	14 11	16 03	00 48	22 05	22 47	11 26	04 07	10 37	20 00
20	12 18	10 10	15 26	01 19	21 59	22 46	11 27	04 07	10 38	20 01
21	11 58	05 36	14 47	01 50	21 52	22 45	11 29	04 06	10 38	20 01
22	11 38	00 S 48	14 07	02 21	21 46	22 44	11 31	04 05	10 39	20 01
23	11 17	03 N 55	13 26	02 52	21 39	22 43	11 32	04 05	10 39	20 01
24	10 57	08 18	12 43	03 22	21 32	22 42	11 34	04 04	10 40	20 02
25	10 36	12 08	12 00	03 53	21 24	22 41	11 35	04 03	10 41	20 02
26	10 15	15 17	11 16	04 24	21 17	22 40	11 37	04 03	10 41	20 02
27	09 54	17 38	10 31	04 55	21 10	22 39	11 39	04 02	10 42	20 02
28	09 33	19 19	09 45	05 25	21 02	22 38	11 40	04 01	10 42	20 03
29	09 12	19 47	08 59	05 56	20 54	22 37	11 42	04 01	10 43	20 03
30	08 50	19 32	08 13	06 26	20 46	22 36	11 44	04 00	10 44	20 03
31	08 N 29	18 N 25	07 N 26	06 S 56	20 N 38	22 N 35	11 S 45	03 N 59	10 S 44	20 S 03

ZODIAC SIGN ENTRIES

Date	h	m	Planets
03	04	29	☽ ♋
05	16	58	☽ ♌
08	03	57	☽ ♍
08	12	13	☿ ♌
10	13	08	☽ ♎
12	20	18	☽ ♏
15	01	04	☽ ♐
16	15	37	♀ ♎
17	03	25	☽ ♑
19	04	07	☽ ♒
21	04	43	☽ ♓
22	23	02	☉ ♍
23	07	13	☽ ♈
23	22	36	☿ ♍
25	13	13	☽ ♉
27	23	03	☽ ♊
28	02	05	♂ ♌
30	11	33	☽ ♋

LATITUDES

Date	Mercury ☿ ° '	Venus ♀ ° '	Mars ♂ ° '	Jupiter ♃ ° '	Saturn ♄ ° '	Uranus ♅ ° '	Neptune ♆ ° '	Pluto ♇ ° '
01	01 S 24	01 N 19	00 N 42	00 S 10	02 N 22	00 S 43	00 S 41	03 N 10
04	00 S 39	01 14	00 44	00 10	02 21	00 43	00 41	03 10
07	00 N 02	01 07	00 46	00 09	02 21	00 43	00 41	03 09
10	00 37	01 00	00 48	00 09	02 20	00 43	00 41	03 08
13	01 06	00 53	00 49	00 09	02 19	00 43	00 41	03 08
16	01 27	00 44	00 51	00 09	02 18	00 43	00 41	03 07
19	01 40	00 35	00 53	00 08	02 18	00 43	00 41	03 07
22	01 45	00 26	00 55	00 08	02 17	00 43	00 41	03 06
25	01 45	00 16	00 56	00 08	02 16	00 43	00 41	03 06
28	01 38	00 N 06	00 58	00 08	02 16	00 43	00 41	03 05
31	01 N 28	00 S 05	01 N 00	00 S 07	02 N 15	00 S 43	00 S 41	03 N 05

DATA

Julian Date	2456506
Delta T	+68 seconds
Ayanamsa	24° 03' 02"
Synetic vernal point	05° ♓ 03' 58"
True obliquity of ecliptic	23° 26' 08"

MOON'S PHASES, APSIDES AND POSITIONS ☽

Date	h	m	Phase	Longitude ° '	Eclipse Indicator
06	21	51	●	14 ♌ 35	
14	10	56	☽	21 ♏ 49	
21	01	45	○	28 ♒ 11	
28	09	35	☾	05 ♊ 15	

Day	h	m		
03	09	04	Apogee	
19	01	32	Perigee	
30	23	51	Apogee	
02	09	21	Max dec	20° N 02'
09	20	13	0S	
16	12	11	Max dec	19° S 54'
22	16	00	0N	
29	17	02	Max dec	19° N 48'

LONGITUDES

Date	Chiron ⚷ ° '	Ceres ⚳ ° '	Pallas ⚴ ° '	Juno ⚵ ° '	Vesta ⚶ ° '	Black Moon Lilith ⚸ ° '
01	12 ♓ 55	17 ♌ 45	06 ♋ 21	12 ♒ 28	11 ♌ 45	05 ♋ 55
11	12 ♓ 31	22 ♌ 17	12 ♋ 10	10 ♒ 02	16 ♌ 19	07 ♋ 02
21	12 ♓ 04	26 ♌ 50	17 ♋ 57	07 ♒ 44	20 ♌ 54	08 ♋ 09
31	11 ♓ 36	01 ♍ 24	23 ♋ 43	05 ♒ 51	25 ♌ 31	09 ♋ 16

All ephemeris data is given at 12.00 UT and the Moon's longitude is additionally given for 24.00 UT

ASPECTARIAN

h m	Aspects
01 Thursday	
00 46	☽ ∠ ☿
01 03	☽ □ ♆
02 20	☽ ⚻ ♄
03 08	♂ □ ♅
03 52	☿ ⚼ ♆
04 46	☽ ⊥ ♂
07 56	☽ ⚺ ♃
10 33	☽ ⚹ ☉
11 00	☽ ⚻ ♇
14 31	☽ ± ♄
16 06	☉ ⚻ ♇
16 17	☽ □ ♀
16 48	☽ ⚹ ♅
17 38	☽ ⚺ ♂
20 54	☽ ⊥ ☿
21 28	♀ ⚻ ♅
22 34	☽ # ♇
02 Friday	
08 51	☽ ⚼ ♄
10 27	☽ ⚺ ☿
17 09	☽ Q ♅
19 41	☽ ∠ ☉
19 55	☽ # ♇
20 35	♀ ⚹ ♂
03 Saturday	
08 57	☽ Q ♀
13 48	☽ △ ♆
15 25	☽ △ ♄
15 34	☽ ⊥ ☉
21 55	☽ ☌ ♃
23 47	☽ ☍ ♇
04 Sunday	
04 46	☽ ⚺ ☉
05 37	☽ □ ♅
10 05	☽ ☌ ♂
11 58	☽ ⚹ ♀
15 05	☉ △ ♅
20 02	☽ ⚼ ♆
05 Monday	
01 26	☽ ∥ ☉
06 49	☽ ☌ ☿
14 02	☽ ± ♆
21 21	☽ ∠ ♀
06 Tuesday	
01 58	☽ ⚻ ♆
03 54	☽ □ ♄
11 05	☽ ⚺ ♃
11 45	☽ ⚻ ♇
17 08	☽ ⊥ ♀
17 29	☽ △ ♅
21 51	☽ ☌ ☉
23 07	☽ ⊥ ♃
23 34	☽ ± ♇
07 Wednesday	
01 19	☽ ⚺ ♂
01 28	☽ # ♄
05 56	☽ # ♆
06 12	☽ ⚺ ♀
13 44	☽ ⊥ ♂
14 16	☿ ± ♆
15 27	☽ Q ♄
17 02	☽ ∠ ♃
17 09	☽ ⚼ ♇
18 02	☉ ⊥ ♃
19 21	☉ ± ♇
22 48	☽ ⚼ ♅
23 46	♃ ☍ ♇
08 Thursday	
02 39	☽ ⚺ ☿
08 12	☽ ∠ ♂
12 34	☽ ☍ ♆
14 47	☽ ⚹ ♄
16 03	☽ ⊥ ☿
16 10	☽ ± ♅
17 34	☽ ∥ ♀
18 08	♀ ∠ ♄
20 29	☽ ∥ ♅
22 06	☽ △ ♇
22 30	☽ ⚹ ♃
09 Friday	
03 40	☽ ⚻ ♅
12 01	☽ ∠ ☿
12 46	☽ ⚺ ☉
14 32	☽ ⚹ ♂
17 44	♀ ∥ ♅
17 45	♀ Q ♃
19 33	☽ ∠ ♄
21 44	☽ Q ♃
22 05	☽ ☌ ♀
10 Saturday	
01 04	☽ ⊥ ☉
12 25	☿ # ♇
12 35	☽ ⊥ ♄
14 24	☽ Q ♂
16 51	☽ # ♀
18 29	♂ ∥ ♃
19 20	☽ ∠ ☉
19 32	☽ # ♅
20 53	☽ ⚹ ☿
21 22	☽ ⚻ ♆
23 51	☽ ⚺ ♄
11 Sunday	
00 10	☿ ⚻ ♆
06 36	☽ □ ♇
07 58	☽ □ ♃
08 29	☽ ± ♆
11 59	☽ ☍ ♅
16 16	♂ ⚼ ♆
18 04	☿ □ ♄
22 45	☽ Q ☿
12 Monday	
01 00	☽ ⚼ ♆
01 16	☽ ⚹ ☉
01 29	☽ □ ♂
08 28	☽ ∥ ♆
10 01	☉ ⚺ ♂
11 20	☽ ⚺ ♀
13 01	☽ ∥ ♄
15 31	☽ Q ♇
23 14	☽ ⊥ ♀
13 Tuesday	
00 42	☽ Q ☉
04 05	☽ △ ♆
06 48	☽ ☌ ♄
11 18	☽ # ☉
12 39	☽ □ ☿
12 59	☽ ⚹ ♇
14 58	☿ ⚻ ♇
15 11	☽ △ ♃
16 50	☽ ∠ ♀
18 07	☽ ⚻ ♅
14 Wednesday	
04 40	☽ ± ♅
07 57	☿ ⚺ ♃
09 47	☽ △ ♂
10 56	☽ □ ☉
15 16	☽ ∠ ♇
17 49	☽ ⚼ ♃
20 16	☽ ⚼ ♅
21 30	☽ ⚹ ♀
23 26	☽ # ☿
15 Thursday	
02 10	☿ △ ♅
06 44	☽ ⊥ ♇
08 24	☽ □ ♆
09 29	☽ ± ♃
11 18	☽ ⚺ ♄
12 51	☽ ⚼ ♂
16 55	☽ ⚺ ♇
19 48	☽ ⚻ ♃
19 50	☽ Q ♀
21 30	☽ ⊥ ♄
21 48	☽ △ ♅
16 Friday	
01 00	☽ △ ☿
04 45	☽ ± ♂
06 23	☿ ∠ ♀
08 05	☉ ⊥ ♀
12 38	☽ ∠ ♄
14 37	☽ Q ♆
15 13	☽ ⚻ ♂
15 15	☿ ± ♇
16 52	♀ ⊥ ♄
17 32	☽ △ ☉
19 03	☉ Q ♄
17 Saturday	
00 18	☉ ⚼ ♇
04 28	☽ □ ♀
05 54	☽ ⚼ ☿
10 21	☽ ⚹ ♆
13 26	☽ ⚹ ♄
16 04	☿ ⊥ ♃
18 34	☽ ☌ ♇
19 52	☽ ⚼ ☉
22 04	☽ ☍ ♃
22 56	☽ ± ☿
23 15	☽ □ ♅
18 Sunday	
09 01	☽ Q ♄
10 11	☽ ⚻ ☿
10 41	☽ ∠ ♆
11 28	☽ ± ☉
18 26	☽ ☍ ♂
19 17	☽ # ☿
21 50	☽ ⚻ ☉
19 Monday	
01 11	☽ ⊥ ♆
04 17	☽ Q ♅
09 20	☽ △ ♀
10 50	☽ ⚺ ♆
13 28	☉ ∠ ♃
14 10	☽ □ ♄
19 00	☽ ⚺ ♇
22 35	☉ ⚼ ♅
22 46	☽ # ☉
23 08	☽ ⚻ ♃
23 36	☽ ⚹ ♅
20 Tuesday	
03 53	♀ ⚻ ♆
04 41	☽ ⊥ ♇
04 48	☽ ∥ ♄
08 58	☽ ± ♃
09 29	☽ ∥ ♆
11 39	☽ ⚼ ♀
18 17	☽ ☍ ☿
19 16	☽ ∠ ♇
21 08	☽ ⚻ ♂
23 49	☽ ⚼ ♃
23 55	☽ ∠ ♅
21 Wednesday	
01 24	☿ ⚼ ♇
01 45	☽ ☍ ☉
02 56	☿ Q ♄
03 36	☽ ± ♀
07 15	♃ □ ♅
07 25	☽ ± ♂
11 30	☽ ☌ ♆
14 21	☽ ⚻ ♀
14 42	☽ ⊥ ♅
15 14	☽ △ ♄
19 33	☽ # ♅
19 56	☽ ⚹ ♇
23 01	☽ ⚼ ♂
23 29	☿ ⚺ ♂
22 Thursday	
00 39	☽ ⚺ ♅
00 55	☽ △ ♃
01 46	♀ ⚺ ♄
05 02	☽ ∥ ♀
10 56	☿ ⚼ ♅
14 11	☿ ∠ ♃
16 05	☽ Q ♇
16 32	☽ ⚼ ♄
19 44	☉ # ♄
23 Friday	
01 38	☽ △ ♂
04 38	☽ ⚻ ☿
05 50	☽ # ♀
07 50	☽ ⚻ ☉
08 05	☽ ± ♄
12 51	☽ ∥ ♅
14 17	☽ ⚺ ♆
16 51	☽ ± ☿
18 38	☽ ⚻ ♄
19 07	☽ ± ☉
22 17	☽ ☍ ♀
23 19	☽ □ ♇
24 Saturday	
00 51	☽ ⊥ ♆
04 18	☽ ☌ ♅
05 25	☽ □ ♃
10 01	♀ □ ♇
11 39	☽ ⚼ ☿
12 25	☽ ⚼ ☉
17 00	☽ ∠ ♆
20 56	☉ ☌ ☿
25 Sunday	
00 07	♂ ± ♆
02 26	☽ # ♆
02 49	☽ ∥ ☉
03 17	♀ ± ♆
06 51	☉ # ♆
07 17	♀ Q ♂
08 20	☽ # ♄
10 01	☽ □ ♂
11 17	☽ ∥ ☿
14 50	☽ Q ♃
18 17	☽ △ ☉
19 46	♀ # ♅
20 16	☽ △ ☿
20 43	☽ ⚹ ♆
23 12	☿ ☍ ♆
26 Monday	
00 59	☿ # ♄
01 49	☽ ☍ ♄
06 30	☽ △ ♇
11 01	☿ ⊥ ♂
11 09	☽ ⚻ ♀
11 48	☽ ⚺ ♅
13 57	☽ ⚹ ♃
17 56	♀ ☍ ♅
19 37	☽ Q ♆
22 06	☽ Q ♂
22 44	☿ ± ♅
23 22	☽ ⊥ ♅
23 57	☽ ± ♀
27 Tuesday	
01 43	☉ ☍ ♆
06 13	☿ # ♆
06 19	☿ ⊥ ♀
09 44	☿ ⚹ ♄
11 32	☽ ⚼ ♇
16 58	☽ ∠ ♅
19 29	☽ ⚼ ♀
19 41	☽ ∠ ♃
21 53	♀ □ ♃
22 58	☽ ⚹ ♂
28 Wednesday	
05 21	☽ ± ♇
06 57	☽ □ ♆
09 35	☽ □ ☉
12 51	☽ ⚻ ♄
13 53	☽ ⊥ ♃
14 52	☿ △ ♇
17 20	☽ ⚻ ♇
17 49	☽ □ ☿
22 49	☽ ⚹ ♅
23 55	☉ ∥ ☿
29 Thursday	
00 24	☉ ± ♅
00 58	☽ ± ♄
02 08	☽ ⚺ ♃
04 44	☽ △ ♀
06 43	☽ ∠ ♂
19 16	☽ ⚼ ♄
23 00	☽ Q ♅
30 Friday	
01 00	☿ ⚻ ♅
01 04	☽ Q ☉
01 58	☽ ⊥ ♂
05 18	☉ ⚹ ♄
13 04	☽ Q ☿
14 50	☽ ⚺ ♂
19 23	☽ △ ♆
31 Saturday	
01 53	☽ △ ♄
03 10	☿ ⚹ ♃
03 34	☽ ⚹ ☉
05 58	☽ ☍ ♇
07 28	☉ ⊥ ♂
11 23	☽ □ ♅
15 46	☽ ☌ ♃
17 53	☽ ⚹ ☿
21 17	☿ # ♀

SEPTEMBER 2013

LONGITUDES

Date	Sidereal time h m s	Sun ☉ ° ' "	Moon ☽ ° ' "	Moon ☽ 24.00 ° ' "	Mercury ☿ ° '	Venus ♀ ° '	Mars ♂ ° '	Jupiter ♃ ° '	Saturn ♄ ° '	Uranus ♅ ° '	Neptune ♆ ° '	Pluto ⯓ ° '
01	10 43 15	09 ♍ 12 45	23 ♋ 59 25	29 ♋ 59 40	16 ♍ 22	18 ♎ 39	02 ♌ 49	14 ♋ 05	07 ♏ 11	11 ♈ 44	03 ♓ 49	09 ♑ 05
02	10 47 12	10 10 51	06 ♌ 02 13	12 ♌ 07 19	18 11	19 49	03 27	14 15	07 16	11 R 42	03 R 47	09 R 04
03	10 51 08	11 08 59	18 ♌ 15 11	24 ♌ 26 01	19 59	20 59	04 06	14 26	07 21	11 40	03 46	09 04
04	10 55 05	12 07 08	00 ♍ 39 54	06 ♍ 56 53	21 46	22 09	04 44	14 36	07 26	11 38	03 44	09 03
05	10 59 01	13 05 19	13 ♍ 17 00	19 ♍ 40 14	23 31	23 19	05 22	14 46	07 31	11 36	03 42	09 03
06	11 02 58	14 03 32	26 ♍ 06 32	02 ♎ 35 50	25 15	24 29	06 00	14 56	07 36	11 34	03 41	09 02
07	11 06 54	15 01 46	09 ♎ 08 05	15 ♎ 43 13	26 58	25 38	06 38	15 05	07 42	11 32	03 39	09 02
08	11 10 51	16 00 02	22 ♎ 21 09	29 ♎ 01 53	28 ♍ 40	26 48	07 16	15 15	07 47	11 30	03 38	09 02
09	11 14 47	16 58 20	05 ♏ 45 24	12 ♏ 31 40	00 ♎ 20	27 58	07 54	15 25	07 52	11 28	03 36	09 01
10	11 18 44	17 56 40	19 ♏ 20 43	26 ♏ 12 34	02 00	29 ♎ 08	08 31	15 34	07 58	11 26	03 34	09 01
11	11 22 41	18 55 00	03 ♐ 07 14	10 ♐ 04 44	03 38	00 ♏ 17	09 09	15 43	08 03	11 24	03 33	09 01
12	11 26 37	19 53 23	17 ♐ 05 17	24 ♐ 08 03	05 15	01 26	09 47	15 53	08 09	11 21	03 31	09 00
13	11 30 34	20 51 47	01 ♑ 13 41	08 ♑ 21 42	06 52	02 35	10 24	16 02	08 14	11 19	03 30	09 00
14	11 34 30	21 50 13	15 ♑ 31 50	22 ♑ 43 42	08 27	03 44	11 02	16 11	08 20	11 17	03 28	09 00
15	11 38 27	22 48 40	29 ♑ 56 49	07 ♒ 10 38	10 01	04 53	11 40	16 20	08 26	11 15	03 27	09 00
16	11 42 23	23 47 08	14 ♒ 24 29	21 ♒ 37 41	11 33	06 02	12 17	16 28	08 31	11 12	03 25	09 00
17	11 46 20	24 45 39	28 ♒ 49 28	05 ♓ 59 07	13 05	07 11	12 55	16 37	08 37	11 10	03 24	09 00
18	11 50 16	25 44 11	13 ♓ 05 53	20 ♓ 09 06	14 36	08 20	13 32	16 45	08 43	11 08	03 22	08 59
19	11 54 13	26 42 45	27 ♓ 08 36	04 ♈ 02 31	16 06	09 28	14 09	16 54	08 49	11 05	03 21	08 59
20	11 58 09	27 41 21	10 ♈ 51 49	17 ♈ 35 46	17 34	10 37	14 47	17 02	08 55	11 03	03 19	08 59
21	12 02 06	28 39 58	24 ♈ 14 14	00 ♉ 47 10	19 02	11 45	15 24	17 10	09 01	11 01	03 18	08 D 59
22	12 06 03	29 ♍ 38 38	07 ♉ 14 40	13 ♉ 36 55	20 28	12 54	16 01	17 18	09 07	10 58	03 16	08 59
23	12 09 59	00 ♎ 37 20	19 ♉ 54 15	26 ♉ 07 00	21 54	14 02	16 38	17 26	09 13	10 56	03 15	09 00
24	12 13 56	01 36 05	02 ♊ 15 39	08 ♊ 20 41	23 18	15 10	17 15	17 34	09 19	10 54	03 13	09 00
25	12 17 52	02 34 51	14 ♊ 22 41	20 ♊ 22 12	24 41	16 18	17 52	17 41	09 26	10 51	03 12	09 00
26	12 21 49	03 33 40	26 ♊ 19 53	02 ♋ 16 19	26 03	17 26	18 29	17 49	09 32	10 49	03 11	09 00
27	12 25 45	04 32 31	08 ♋ 12 08	14 ♋ 07 58	27 24	18 34	19 06	17 56	09 38	10 47	03 09	09 00
28	12 29 42	05 31 25	20 ♋ 04 26	26 ♋ 02 05	28 ♎ 43	19 41	19 43	18 03	09 44	10 44	03 08	09 00
29	12 33 38	06 30 20	02 ♌ 01 29	08 ♌ 03 09	00 ♏ 01	20 49	20 20	18 10	09 51	10 42	03 07	09 01
30	12 37 35	07 ♎ 29 18	14 ♌ 07 32	20 ♌ 15 05	01 ♏ 18	21 ♏ 56	20 ♌ 57	18 ♋ 17	09 ♏ 57	10 ♈ 39	03 ♓ 05	09 ♑ 01

Moon Node / Latitude and DECLINATIONS

Date	Moon True ☊ ° '	Moon Mean ☊ ° '	Moon Latitude ° '	Sun ☉ ° '	Moon ☽ ° '	Mercury ☿ ° '	Venus ♀ ° '	Mars ♂ ° '	Jupiter ♃ ° '	Saturn ♄ ° '	Uranus ♅ ° '	Neptune ♆ ° '	Pluto ⯓ ° '
01	09 ♏ 50	10 ♏ 42	04 S 53	08 N 07	16 N 30	06 N 39	07 S 27	20 N 30	22 N 34	11 S 47	03 N 58	10 S 45	20 S 03
02	09 R 41	10 39	05 04	07 45	13 51	05 52	07 57	20 22	22 33	11 49	03 58	10 45	20 04
03	09 31	10 36	05 01	07 23	10 35	05 05	08 26	20 13	22 32	11 51	03 57	10 46	20 04
04	09 20	10 32	04 44	07 01	06 49	04 18	08 56	20 05	22 31	11 53	03 56	10 47	20 04
05	09 10	10 29	04 13	06 39	02 N 41	03 31	09 26	19 56	22 30	11 54	03 55	10 47	20 04
06	09 02	10 26	03 28	06 16	01 S 38	02 45	09 55	19 47	22 29	11 56	03 54	10 48	20 05
07	08 56	10 23	02 32	05 54	05 57	01 58	10 24	19 38	22 28	11 58	03 54	10 48	20 05
08	08 52	10 20	01 27	05 31	10 03	01 12	10 53	19 29	22 27	12 00	03 53	10 49	20 05
09	08 51	10 17	00 S 17	05 09	13 42	00 N 26	11 22	19 20	22 26	12 02	03 52	10 50	20 05
10	08 D 51	10 13	00 N 56	04 46	16 40	00 S 20	11 50	19 11	22 25	12 04	03 51	10 50	20 05
11	08 52	10 10	02 07	04 23	18 42	01 06	12 19	19 01	22 24	12 06	03 50	10 51	20 06
12	08 53	10 07	03 11	04 00	19 38	01 51	12 47	18 52	22 23	12 08	03 49	10 51	20 06
13	08 R 52	10 04	04 05	03 37	19 21	02 35	13 14	18 42	22 22	12 10	03 48	10 52	20 06
14	08 50	10 01	04 43	03 14	17 50	03 20	13 42	18 32	22 21	12 12	03 48	10 52	20 06
15	08 46	09 57	05 05	02 51	15 11	04 03	14 09	18 22	22 20	12 14	03 47	10 53	20 06
16	08 40	09 54	05 07	02 28	11 36	04 47	14 36	18 12	22 19	12 16	03 46	10 54	20 07
17	08 33	09 51	04 49	02 05	07 21	05 30	15 02	18 02	22 18	12 18	03 45	10 54	20 07
18	08 26	09 48	04 14	01 42	02 S 43	06 12	15 29	17 52	22 18	12 20	03 44	10 55	20 07
19	08 20	09 45	03 24	01 18	01 N 59	06 54	15 55	17 42	22 17	12 22	03 43	10 55	20 07
20	08 16	09 42	02 23	00 55	06 29	07 35	16 20	17 31	22 16	12 24	03 42	10 56	20 07
21	08 13	09 38	01 15	00 32	10 34	08 15	16 45	17 21	22 15	12 26	03 41	10 56	20 08
22	08 12	09 35	00 N 05	00 N 08	14 01	08 55	17 10	17 10	22 14	12 28	03 40	10 57	20 08
23	08 D 13	09 32	01 S 03	00 S 15	16 42	09 34	17 35	17 00	22 13	12 30	03 39	10 57	20 08
24	08 14	09 29	02 07	00 38	18 32	10 13	17 59	16 49	22 12	12 32	03 38	10 58	20 08
25	08 16	09 26	03 04	01 02	19 28	10 50	18 23	16 38	22 11	12 34	03 38	10 58	20 08
26	08 17	09 23	03 53	01 25	19 30	11 27	18 46	16 27	22 10	12 36	03 37	10 59	20 09
27	08 R 17	09 19	04 31	01 48	18 41	12 04	19 09	16 16	22 09	12 38	03 36	10 59	20 09
28	08 16	09 16	04 57	02 12	17 02	12 39	19 31	16 05	22 08	12 40	03 35	11 00	20 09
29	08 14	09 13	05 11	02 35	14 39	13 13	19 53	15 54	22 08	12 42	03 34	11 00	20 09
30	08 ♏ 10	09 ♏ 10	05 S 11	02 S 58	11 N 37	13 S 47	20 S 15	15 N 43	22 N 07	12 S 45	03 N 33	11 S 01	20 S 09

ZODIAC SIGN ENTRIES

Date	h	m	Planets
02	00	01	☽ ♌
04	10	43	☽ ♍
06	19	12	☽ ♎
09	01	44	☽ ♏
09	07	07	☿ ♎
11	06	16	♀ ♏
11	06	36	☽ ♐
13	09	56	☽ ♑
15	12	05	☽ ♒
17	13	58	☽ ♓
19	16	58	☽ ♈
21	22	33	☽ ♉
22	20	44	☉ ♎
24	07	34	☽ ♊
26	19	24	☽ ♋
29	07	57	☽ ♌
29	11	38	☿ ♏

LATITUDES

Date	Mercury ☿ ° '	Venus ♀ ° '	Mars ♂ ° '	Jupiter ♃ ° '	Saturn ♄ ° '	Uranus ♅ ° '	Neptune ♆ ° '	Pluto ⯓ ° '
01	01 N 23	00 S 09	01 N 00	00 S 07	02 N 15	00 S 43	00 S 41	03 N 04
04	01 08	00 20	01 02	00 07	02 14	00 44	00 41	03 04
07	00 50	00 32	01 03	00 07	02 14	00 44	00 41	03 03
10	00 30	00 43	01 05	00 06	02 13	00 44	00 41	03 03
13	00 N 09	00 56	01 07	00 06	02 12	00 44	00 41	03 02
16	00 S 14	01 08	01 08	00 06	02 12	00 44	00 41	03 02
19	00 37	01 20	01 10	00 06	02 11	00 44	00 41	03 01
22	01 00	01 32	01 12	00 05	02 11	00 44	00 41	03 00
25	01 23	01 44	01 13	00 05	02 10	00 44	00 41	03 00
28	01 45	01 57	01 15	00 05	02 10	00 44	00 41	03 00
31	02 S 06	02 S 08	01 N 17	00 S 04	02 N 10	00 S 44	00 S 41	02 N 59

DATA

Julian Date	2456537
Delta T	+68 seconds
Ayanamsa	24° 03' 05"
Synetic vernal point	05° ♓ 03' 54"
True obliquity of ecliptic	23° 26' 08"

MOON'S PHASES, APSIDES AND POSITIONS ☽

Date	h	m	Phase	Longitude ° '	Eclipse Indicator
05	11	36	●	13 ♍ 04	
12	17	08	☽	20 ♐ 06	
19	11	13	○	26 ♓ 41	
27	03	55	☾	04 ♋ 13	

Day	h	m		
15	16	37	Perigee	
27	18	17	Apogee	
06	02	56	0S	
12	18	33	Max dec	19° S 41'
19	01	51	0N	
26	01	04	Max dec	19° N 36'

LONGITUDES

Date	Chiron ⚷ ° '	Ceres ⚳ ° '	Pallas ⚴ ° '	Juno ⚵ ° '	Vesta ⚶ ° '	Black Moon Lilith ⚸ ° '
01	11 ♓ 33	01 ♍ 51	24 ♋ 17	05 ♒ 41	25 ♌ 59	09 ♋ 23
11	11 ♓ 05	06 ♍ 25	29 ♋ 57	04 ♒ 26	00 ♍ 36	10 ♋ 50
21	10 ♓ 37	10 ♍ 57	05 ♌ 31	03 ♒ 52	05 ♍ 12	11 ♋ 37
31	10 ♓ 12	15 ♍ 28	10 ♌ 56	04 ♒ 02	09 ♍ 48	12 ♋ 44

All ephemeris data is given at 12.00 UT and the Moon's longitude is additionally given for 24.00 UT

ASPECTARIAN

h m	Aspects
01 Sunday	
00 06	☽ □ ♀
01 38	☽ ♇ ♆
08 46	☉ △ ⯓
12 29	☽ ∠ ☉
15 20	♀ ♇ ♆
19 38	☽ ± ♆
02 Monday	
05 21	☽ ∠ ☿
06 36	☽ ☌ ♂
06 39	☉ ⫲ ♀
07 33	☽ ⊼ ♆
08 00	☽ ⊥ ☉
14 27	☽ □ ♄
15 54	☽ Q ♀
17 35	☿ ∠ ♂
17 59	☽ ⊼ ⯓
20 53	☽ ⊻ ☉
23 09	☽ △ ♅
03 Tuesday	
00 00	♂ ⊼ ♆
02 13	☽ ⊥ ☿
03 18	☽ ⫲ ♄
04 25	☽ ⊻ ♃
05 46	☽ ± ⯓
10 45	☽ ⫲ ♆
15 56	☽ ⊻ ☿
16 17	☽ ⊥ ♃
17 52	☽ ⚹ ♀
23 17	☽ ♇ ⯓
04 Wednesday	
00 21	☽ ⫲ ♀
00 28	☉ ⊼ ♅
01 52	☽ Q ♄
04 17	☽ ♇ ♅
09 55	☽ ∠ ♃
10 38	☽ ∥ ☉
13 40	♂ ⫲ ⯓
17 52	☽ ☍ ♆
20 11	☽ ⊻ ♂
21 29	☽ ± ♅
21 42	☿ ∠ ♄
23 38	☿ ∥ ♅
05 Thursday	
01 01	☽ ⚹ ♄
01 39	☽ ∠ ♀
03 51	☿ ⊻ ♀
03 59	☽ △ ⯓
04 55	☽ ∥ ♅
06 06	☽ ∥ ☿
08 11	☽ ⊥ ♂
08 50	☽ ⊼ ♅
11 36	☽ ☌ ☉
14 49	☽ ⚹ ♃
20 21	☽ ⊥ ♀
06 Friday	
01 59	☽ ∠ ♂
05 27	☽ ∠ ♄
08 40	☽ ⊻ ♀
10 10	☽ ☌ ☿
13 32	☽ Q ♃
17 10	☽ ⫲ ☿
22 15	☽ ⊥ ♄
07 Saturday	
00 29	☽ ⫲ ♅
01 58	☽ ⊼ ♆
07 11	☽ ⚹ ♂
09 21	☽ ⊻ ♄
11 42	☽ ⫲ ☉
11 49	☽ □ ⯓
12 57	☽ ± ♆
13 46	☉ ⚹ ♃
13 55	☿ Q ♃
16 22	☽ ☍ ♅
22 59	☽ □ ♃
23 36	☽ ⊻ ☉
08 Sunday	
05 17	☽ ♇ ♆
06 08	☽ Q ♂
08 36	♀ ∥ ♆
11 19	☽ ⊥ ☉
16 37	♀ Q ⯓
16 46	☽ ∥ ♆
17 55	☽ ∥ ♀
20 24	☽ Q ⯓
20 46	☽ ☌ ♀
09 Monday	
00 27	☽ ∥ ♄
00 58	☽ ⊻ ☿
04 44	☽ ∠ ☉
08 10	☽ △ ♆
11 06	♂ □ ♄
13 11	☽ ⊥ ☿
15 47	☽ ☌ ♄
15 59	☽ □ ♂
17 48	☽ ⚹ ⯓
22 06	☽ ⊼ ♅
10 Tuesday	
05 17	☽ △ ♃
07 19	☽ ∠ ☿
08 39	☽ ± ♅
09 21	☽ ⚹ ☉
11 27	☿ ⊥ ♄
20 10	☽ ∠ ⯓
11 Wednesday	
00 16	♀ ∥ ♄
00 21	☽ ♇ ♅
06 37	☽ ⊻ ♀
06 39	♂ ⊼ ⯓
07 48	☽ ♇ ♃
07 54	☽ Q ☉
10 43	☿ ⊼ ♆
11 49	☽ ⊥ ⯓
12 44	☽ □ ♆
13 01	☽ ⚹ ☿
16 55	☽ ⫲ ♂
17 57	☽ ⊥ ♀
20 34	☽ ⊻ ♄
22 10	☽ ⊻ ⯓
22 54	☽ △ ♂
23 31	☽ ± ♃
12 Thursday	
02 13	☽ △ ♅
06 57	☽ ⊥ ♄
09 55	☽ ⊼ ♃
10 47	☽ ∠ ♀
12 20	☽ Q ☿
17 08	☽ □ ☉
19 33	☽ Q ♆
22 23	☽ ∠ ♄
23 49	☉ ∥ ♅
13 Friday	
01 42	☽ ♇ ♂
14 29	☽ ⚹ ♀
15 49	☽ ⚹ ♆
17 36	☽ ± ♂
22 40	☽ □ ☿
23 52	☽ ⚹ ♄
14 Saturday	
01 04	☽ ☌ ⯓
02 11	☽ ⫲ ♂
04 08	☽ ⊼ ♂
04 55	☽ □ ♅
06 33	♀ △ ♆
10 03	☉ ⫲ ☿
10 13	☿ ⊻ ♄
12 23	☽ Q ♀
13 05	☽ ☍ ♃
16 54	☽ ∠ ♆
20 04	☽ Q ♄
20 29	☿ □ ⯓
20 57	♂ △ ♅
23 17	☽ △ ☉
15 Sunday	
02 57	☿ ⫲ ♅
03 28	☿ ± ♆
07 51	☽ ⊥ ♆
10 50	☽ Q ♅
17 48	☽ ⊻ ♆
18 42	☽ ∥ ♀
20 55	☽ □ ♀
16 Monday	
02 00	☽ ♇ ☉
02 10	☽ □ ♄
03 01	☽ ⊻ ⯓
04 50	☉ ∠ ♄
06 41	☿ ☍ ♅
06 42	☽ ⚹ ♅
06 42	☽ △ ☿
08 03	☽ ∥ ♄
08 19	☽ ☍ ♂
12 58	☽ ⊥ ⯓
15 28	☽ ⊼ ♃
16 15	☽ ∥ ♆
18 01	☽ ± ☉
17 Tuesday	
01 32	☽ ± ♃
03 56	☽ ∠ ⯓
04 44	☽ ⊼ ☉
07 17	☿ ⚹ ♂
07 35	☽ ∠ ♅
10 37	☽ ♇ ☿
16 43	☽ ♇ ♃
19 38	☽ ☌ ♆
20 29	☽ ∥ ☿
22 36	☽ ⊥ ♅
18 Wednesday	
03 15	☽ △ ♀
03 30	☽ ± ☿
04 33	☽ △ ♄
05 04	☽ ⚹ ⯓
06 50	☽ ⫲ ♅
08 41	☽ ⊻ ♅
12 46	☽ ⊼ ♂
14 51	☽ ⊼ ☿
17 42	☽ ⫲ ☉
18 16	☽ △ ♃
20 53	♀ ☌ ♄
23 27	☽ ± ♂
19 Thursday	
01 26	☽ Q ⯓
01 51	♀ ⚹ ⯓
06 14	☽ ♇ ♄
07 00	☽ ♇ ♀
08 47	☽ ∥ ☉
11 13	☽ ☍ ☉
15 40	☽ ♇ ♂
21 01	☽ ∥ ♅
21 56	☽ ± ♄
22 46	☽ ⊻ ♆
20 Friday	
00 00	☽ ± ♀
02 20	☿ □ ♃
08 32	☽ ⊼ ♄
08 41	☽ □ ⯓
09 17	☽ ⊥ ♆
11 31	☽ ⊼ ♀
12 20	☽ ☌ ♅
15 28	⯓ St D
19 17	☽ △ ♂
19 18	☽ ⫲ ☿
20 14	♂ ± ⯓
20 53	♀ ⊼ ♅
23 06	☽ □ ♃
21 Saturday	
00 06	☿ ♇ ♆
01 17	☽ ∠ ♆
01 25	☽ ☍ ☿
05 45	♄ ⚹ ⯓
14 24	☽ ⫲ ♆
20 45	☽ ⊼ ☉
22 Sunday	
00 30	☽ ⫲ ♄
02 14	☉ Q ♃
04 37	☽ ⚹ ♆
08 20	☽ Q ♃
08 46	☽ ± ☉
12 05	♀ ⫲ ♂
15 16	☽ △ ⯓
15 32	☽ ☍ ♄
18 59	☽ ⊻ ♅
23 41	☽ ☍ ♀
23 Monday	
03 07	☽ Q ♆
03 07	☽ ♇ ☉
05 25	☽ □ ♂
06 20	☽ ⊥ ♅
07 13	☽ ⚹ ♃
15 01	☽ ∥ ♂
16 19	☽ ⊼ ☿
19 53	☽ ♇ ⯓
23 37	☽ ∠ ♅
24 Tuesday	
01 00	☽ ⫲ ♀
05 27	☽ ± ☿
10 36	☽ △ ☉
12 36	☽ ∠ ♃
13 26	☽ ± ⯓
13 53	☽ □ ♆
18 12	☽ Q ♂
25 Wednesday	
01 17	☽ ⊼ ⯓
01 27	☽ ♇ ☿
02 03	☽ ⊼ ♄
02 52	♂ ⊻ ♃
05 00	☽ ⚹ ♅
06 35	☽ ⊥ ♃
14 07	☽ ± ♄
16 14	☽ ⊼ ♀
17 12	☿ ∥ ♆
18 41	☽ ⊻ ♃
19 22	☽ ⚹ ♂
23 24	♀ ± ♅
26 Thursday	
02 51	☉ ⊼ ♆
04 56	☽ Q ♅
05 33	☽ ± ♀
07 14	☉ ∠ ♂
08 20	☽ ♇ ♄
11 08	☉ ⊥ ♄
11 21	☽ △ ☿
21 01	♀ △ ♃
27 Friday	
01 37	☽ ♇ ♀
01 48	☽ △ ♆
03 15	☽ ∠ ♂
03 55	☽ □ ☉
04 29	☽ ⫲ ♀
04 59	☿ Q ⯓
13 37	☽ ☍ ⯓
14 55	☽ △ ♄
17 11	☽ □ ♅
22 28	☽ ⊥ ♂
28 Saturday	
07 52	☽ ☌ ♃
08 06	☽ ♇ ♆
11 08	☽ △ ♀
11 15	☽ ⊻ ♂
12 59	☿ ∥ ♄
13 32	♀ □ ♂
19 34	☽ Q ☉
23 21	☽ ∥ ♂
29 Sunday	
02 11	☽ ± ♆
02 38	♃ ♇ ♆
07 30	☽ □ ☿
14 10	☽ ⊼ ♆
21 43	☽ ⚹ ☉
22 03	☽ ⫲ ☿
30 Monday	
01 54	☽ ⊼ ⯓
03 42	☽ □ ♄
03 47	☽ ⫲ ♄
05 11	☽ △ ♅
05 27	♀ ∥ ⯓
13 45	☽ ± ⯓
16 21	☽ ⫲ ♆
20 13	☽ ⊻ ♃
23 18	☽ Q ☿

OCTOBER 2013

LONGITUDES

Date	Sidereal time h m s	Sun ☉ ° ′ ″	Moon ☽ ° ′ ″	Moon ☽ 24.00 ° ′ ″	Mercury ☿ ° ′	Venus ♀ ° ′	Mars ♂ ° ′	Jupiter ♃ ° ′	Saturn ♄ ° ′	Uranus ♅ ° ′	Neptune ♆ ° ′	Pluto ♇ ° ′
01	12 41 32	08 ♎ 28 18	26 ♌ 26 09	02 ♍ 41 00	02 ♏ 33	23 ♏ 03	21 ♌ 33	18 ♋ 23	10 ♏ 04	10 ♈ 37	03 ♓ 04	09 ♑ 01
02	12 45 28	09 27 20	08 ♍ 59 53	15 ♍ 22 56	03 47	24 10	22 10	18 30	10 10	10 R 34	03 R 03	09 02
03	12 49 25	10 26 24	21 ♍ 50 13	28 ♍ 21 46	05 00	25 17	22 47	18 36	10 17	10 32	03 02	09 02
04	12 53 21	11 25 31	04 ♎ 57 29	11 ♎ 37 14	06 10	26 24	23 23	18 43	10 23	10 30	03 00	09 02
05	12 57 18	12 24 39	18 ♎ 20 51	25 ♎ 08 03	07 19	27 31	24 00	18 49	10 30	10 27	02 59	09 03
06	13 01 14	13 23 50	01 ♏ 58 35	08 ♏ 52 07	08 26	28 37	24 36	18 55	10 37	10 25	02 58	09 03
07	13 05 11	14 23 02	15 ♏ 48 20	22 ♏ 46 53	09 30	29 ♏ 44	25 13	19 00	10 43	10 22	02 57	09 04
08	13 09 07	15 22 17	29 ♏ 47 27	06 ♐ 49 42	10 33	00 ♐ 50	25 49	19 06	10 50	10 20	02 56	09 04
09	13 13 04	16 21 33	13 ♐ 53 19	20 ♐ 57 59	11 33	01 56	26 25	19 11	10 57	10 18	02 55	09 05
10	13 17 01	17 20 51	28 ♐ 03 25	05 ♑ 09 22	12 30	03 02	27 01	19 17	11 04	10 15	02 54	09 05
11	13 20 57	18 20 11	12 ♑ 15 31	19 ♑ 21 38	13 24	04 07	27 37	19 22	11 10	10 13	02 53	09 06
12	13 24 54	19 19 33	26 ♑ 27 25	03 ♒ 32 37	14 15	05 13	28 14	19 27	11 17	10 10	02 52	09 07
13	13 28 50	20 18 56	10 ♒ 36 56	17 ♒ 40 04	15 02	06 18	28 50	19 31	11 24	10 08	02 51	09 07
14	13 32 47	21 18 21	24 ♒ 41 42	01 ♓ 41 32	15 46	07 23	29 ♌ 26	19 36	11 31	10 06	02 50	09 08
15	13 36 43	22 17 48	08 ♓ 39 12	15 ♓ 34 25	16 25	08 28	00 ♍ 01	19 40	11 38	10 03	02 49	09 09
16	13 40 40	23 17 16	22 ♓ 26 50	29 ♓ 16 08	16 59	09 33	00 37	19 45	11 45	10 01	02 49	09 10
17	13 44 36	24 16 47	06 ♈ 02 05	12 ♈ 44 23	17 29	10 37	01 13	19 49	11 52	09 59	02 47	09 11
18	13 48 33	25 16 19	19 ♈ 22 52	25 ♈ 57 21	17 52	11 41	01 49	19 53	11 59	09 56	02 46	09 11
19	13 52 30	26 15 53	02 ♉ 27 46	08 ♉ 54 04	18 10	12 45	02 24	19 56	12 06	09 54	02 45	09 12
20	13 56 26	27 15 30	15 ♉ 16 16	21 ♉ 34 28	18 20	13 49	03 00	20 00	12 13	09 52	02 44	09 13
21	14 00 23	28 15 08	27 ♉ 48 50	03 ♊ 59 34	18 R 24	14 53	03 35	20 03	12 20	09 49	02 44	09 14
22	14 04 19	29 ♎ 14 49	10 ♊ 06 57	16 ♊ 11 20	18 19	15 56	04 11	20 06	12 27	09 47	02 43	09 15
23	14 08 16	00 ♏ 14 32	22 ♊ 13 05	28 ♊ 12 39	18 06	16 59	04 46	20 09	12 34	09 45	02 42	09 16
24	14 12 12	01 14 17	04 ♋ 10 29	10 ♋ 07 08	17 45	18 01	05 21	20 12	12 41	09 43	02 42	09 17
25	14 16 09	02 14 04	16 ♋ 03 06	21 ♋ 58 58	17 14	19 04	05 57	20 15	12 48	09 41	02 41	09 18
26	14 20 05	03 13 53	27 ♋ 55 18	03 ♌ 52 43	16 35	20 06	06 32	20 17	12 55	09 38	02 40	09 19
27	14 24 02	04 13 45	09 ♌ 51 46	15 ♌ 53 04	15 46	21 08	07 07	20 19	13 02	09 36	02 40	09 20
28	14 27 59	05 13 39	21 ♌ 57 11	28 ♌ 04 38	14 49	22 09	07 42	20 21	13 10	09 34	02 39	09 21
29	14 31 55	06 13 35	04 ♍ 15 57	10 ♍ 31 36	13 45	23 10	08 17	20 23	13 17	09 32	02 39	09 22
30	14 35 52	07 13 33	16 ♍ 51 59	23 ♍ 17 26	12 34	24 11	08 52	20 25	13 24	09 30	02 38	09 24
31	14 39 48	08 ♏ 13 33	29 ♍ 48 12	06 ♎ 24 28	11 ♏ 19	25 ♐ 11	09 ♍ 26	20 ♋ 26	13 ♏ 31	09 ♈ 28	02 ♓ 38	09 ♑ 25

Date	Moon True ☊ ° ′	Moon Mean ☊ ° ′	Moon ☽ Latitude ° ′
01	08 ♏ 05	09 ♏ 07	04 S 57
02	08 R 00	09 03	04 28
03	07 56	09 00	03 46
04	07 52	08 57	02 51
05	07 49	08 54	01 45
06	07 48	08 51	00 S 32
07	07 D 48	08 48	00 N 44
08	07 49	08 44	01 58
09	07 51	08 41	03 06
10	07 52	08 38	04 03
11	07 53	08 35	04 45
12	07 R 53	08 32	05 09
13	07 52	08 29	05 15
14	07 50	08 25	05 02
15	07 48	08 22	04 31
16	07 46	08 19	03 45
17	07 44	08 16	02 46
18	07 43	08 13	01 40
19	07 42	08 09	00 N 29
20	07 D 42	08 06	00 S 42
21	07 43	08 03	01 49
22	07 44	08 00	02 50
23	07 45	07 57	03 43
24	07 45	07 54	04 25
25	07 46	07 50	04 55
26	07 46	07 47	05 13
27	07 R 46	07 44	05 17
28	07 46	07 41	05 08
29	07 46	07 38	04 44
30	07 46	07 35	04 07
31	07 ♏ 46	07 ♏ 31	03 S 16

DECLINATIONS

Date	Sun ☉ ° ′	Moon ☽ ° ′	Mercury ☿ ° ′	Venus ♀ ° ′	Mars ♂ ° ′	Jupiter ♃ ° ′	Saturn ♄ ° ′	Uranus ♅ ° ′	Neptune ♆ ° ′	Pluto ♇ ° ′
01	03 S 22	08 N 03	14 S 20	20 S 36	15 N 31	22 N 06	12 S 47	03 N 32	11 S 01	20 S 09
02	03 45	04 N 03	14 52	20 57	15 20	22 05	12 49	03 31	11 02	20 10
03	04 08	00 S 13	15 22	21 17	15 09	22 04	12 51	03 30	11 02	20 10
04	04 31	04 35	15 52	21 37	14 57	22 04	12 53	03 29	11 02	20 10
05	04 54	08 49	16 21	21 56	14 46	22 03	12 55	03 28	11 03	20 10
06	05 17	12 40	16 48	22 15	14 34	22 02	12 57	03 27	11 03	20 10
07	05 40	15 52	17 14	22 33	14 22	22 02	13 00	03 26	11 04	20 10
08	06 03	18 11	17 39	22 51	14 10	22 01	13 02	03 25	11 04	20 11
09	06 26	19 23	18 03	23 08	13 59	22 00	13 04	03 24	11 04	20 11
10	06 49	19 23	18 25	23 24	13 47	22 00	13 06	03 23	11 05	20 11
11	07 11	18 09	18 45	23 41	13 35	21 59	13 08	03 22	11 05	20 11
12	07 34	15 48	19 04	23 56	13 23	21 58	13 11	03 22	11 06	20 11
13	07 56	12 31	19 21	24 11	13 11	21 58	13 13	03 21	11 06	20 11
14	08 19	08 32	19 37	24 25	12 59	21 57	13 15	03 20	11 06	20 11
15	08 41	04 S 08	19 50	24 39	12 47	21 57	13 17	03 19	11 07	20 12
16	09 03	00 N 27	20 01	24 52	12 34	21 56	13 19	03 18	11 07	20 12
17	09 25	04 57	20 09	25 05	12 22	21 56	13 22	03 17	11 07	20 12
18	09 47	09 07	20 15	25 17	12 10	21 55	13 24	03 16	11 08	20 12
19	10 08	12 47	20 19	25 29	11 58	21 55	13 26	03 15	11 08	20 12
20	10 30	15 45	20 19	25 39	11 45	21 54	13 28	03 14	11 08	20 12
21	10 51	17 54	20 16	25 50	11 33	21 54	13 30	03 13	11 08	20 12
22	11 12	19 09	20 10	25 59	11 21	21 54	13 33	03 13	11 09	20 12
23	11 33	19 30	19 59	26 08	11 08	21 53	13 35	03 12	11 09	20 13
24	11 54	18 58	19 45	26 17	10 56	21 53	13 37	03 11	11 09	20 13
25	12 15	17 35	19 27	26 24	10 43	21 53	13 39	03 11	11 09	20 13
26	12 35	15 28	19 04	26 32	10 31	21 53	13 41	03 09	11 10	20 13
27	12 56	12 41	18 37	26 38	10 18	21 52	13 44	03 08	11 10	20 13
28	13 16	09 20	18 05	26 44	10 06	21 52	13 46	03 08	11 10	20 13
29	13 36	05 32	17 30	26 49	09 53	21 52	13 48	03 07	11 10	20 13
30	13 55	01 24	16 51	26 54	09 41	21 52	13 50	03 06	11 10	20 13
31	14 S 15	02 S 55	16 S 09	26 S 58	09 N 28	21 N 52	13 S 52	03 N 05	11 S 10	20 S 13

ZODIAC SIGN ENTRIES

Date	h m	Planets
01	18 52	☽ ♍
04	02 59	☽ ♎
06	08 33	☽ ♏
07	17 54	♀ ♐
08	12 21	☽ ♐
10	15 17	☽ ♑
12	18 00	☽ ♒
14	21 06	☽ ♓
15	11 05	♂ ♍
17	01 17	☽ ♈
19	07 27	☽ ♉
21	16 14	☽ ♊
23	06 10	☉ ♏
24	03 36	☽ ♋
26	16 12	☽ ♌
29	03 45	☽ ♍
31	12 22	☽ ♎

LATITUDES

Date	Mercury ☿ ° ′	Venus ♀ ° ′	Mars ♂ ° ′	Jupiter ♃ ° ′	Saturn ♄ ° ′	Uranus ♅ ° ′	Neptune ♆ ° ′	Pluto ♇ ° ′
01	02 S 06	02 S 08	01 N 17	00 S 04	02 N 10	00 S 44	00 S 41	02 N 59
04	02 26	02 20	01 18	00 04	02 09	00 44	00 41	02 58
07	02 44	02 31	01 20	00 04	02 09	00 44	00 41	02 58
10	02 58	02 42	01 22	00 03	02 09	00 44	00 41	02 57
13	03 09	02 52	01 23	00 03	02 08	00 44	00 41	02 56
16	03 14	03 02	01 25	00 03	02 08	00 44	00 41	02 56
19	03 12	03 11	01 26	00 02	02 08	00 44	00 41	02 55
22	03 00	03 19	01 28	00 02	02 08	00 43	00 41	02 55
25	02 34	03 26	01 30	00 02	02 08	00 43	00 41	02 54
28	01 53	03 32	01 31	00 01	02 07	00 43	00 41	02 54
31	00 S 59	03 S 37	01 N 33	00 S 01	02 N 07	00 S 43	00 S 41	02 N 53

DATA

Julian Date	2456567
Delta T	+68 seconds
Ayanamsa	24° 03' 08"
Synetic vernal point	05° ♓ 03' 52"
True obliquity of ecliptic	23° 26' 08"

MOON'S PHASES, APSIDES AND POSITIONS ☽

Date	h m	Phase	Longitude ° ′	Eclipse Indicator
05	00 35	●	11 ♎ 56	
11	23 02	☽	18 ♑ 47	
18	23 38	○	25 ♈ 45	
26	23 40	☾	03 ♌ 46	

Day	h m		
10	23 10	Perigee	
25	14 21	Apogee	
03	10 46	0S	
09	23 49	Max dec	19° S 32'
16	09 39	0N	
23	09 14	Max dec	19° N 31'
30	19 51	0S	

LONGITUDES

Date	Chiron ⚷ ° ′	Ceres ⚳ ° ′	Pallas ⚴ ° ′	Juno ⚵ ° ′	Vesta ⚶ ° ′	Black Moon Lilith ⚸ ° ′
01	10 ♓ 12	15 ♍ 28	10 ♌ 56	04 ♒ 02	09 ♍ 48	12 ♋ 44
11	09 ♓ 49	19 ♍ 56	16 ♌ 10	04 ♒ 52	14 ♍ 23	13 ♋ 51
21	09 ♓ 31	24 ♍ 21	21 ♌ 10	06 ♒ 21	18 ♍ 55	14 ♋ 58
31	09 ♓ 18	28 ♍ 41	25 ♌ 52	08 ♒ 24	23 ♍ 24	16 ♋ 05

ASPECTARIAN

h m Aspects

01 Tuesday
02 03 ☽ ☌ ♂
04 48 ☽ □ ♀
05 46 ☽ ∠ ☉
07 20 ☽ ꝑ ♇
08 01 ☽ ⊥ ♃
10 25 ☽ ꝑ ♅
15 10 ☽ Q ♄
21 46 ☿ △ ♆
22 16 ☉ # ♅

02 Wednesday
00 35 ☽ ⊥ ☉
00 43 ☽ ☍ ♆
01 03 ☽ ⚹ ☿
01 28 ☉ □ ♇
01 28 ☽ ∠ ♃
02 16 ☉ ± ♆
03 38 ☽ ± ♅
08 50 ♀ ∠ ♇
12 03 ☽ △ ♇
12 56 ☽ ⊻ ☉
13 35 ☽ # ☉
14 14 ☽ ⚹ ♄
14 58 ☽ ⊼ ♅
15 04 ☽ ‖ ♅
18 33 ☽ Q ♀

03 Thursday
03 16 ☿ Q ♂
04 07 ☿ # ♂
05 58 ☽ ⚹ ♃
07 34 ☉ ⊻ ♄
08 14 ☽ ∠ ☿
13 50 ☽ ⊻ ♂
14 12 ☉ ☍ ♅
17 05 ♀ ꝑ ♅
18 24 ☽ ∠ ♄
18 57 ☽ ⚹ ♀

04 Friday
01 23 ☽ ⊥ ♂
02 27 ☽ ⊥ ☿
04 14 ☽ Q ♃
05 58 ☽ # ♅
08 28 ☽ ⊼ ♆
10 58 ☽ ⊥ ♄
11 38 ☽ ‖ ☉
14 24 ☽ ⊻ ☿
16 16 ☉ ∠ ♀
18 29 ☽ ∠ ♂
19 18 ☽ ± ♆
19 22 ☽ □ ♇
21 52 ☽ ⊻ ♄
21 57 ☽ ☍ ♅

05 Saturday
00 35 ☽ ☌ ☉
00 40 ☽ ∠ ♀
04 45 ♄ ⊼ ♅
11 22 ☽ ꝑ ♆
12 50 ☽ □ ♃
14 00 ♂ ꝑ ♇
18 07 ☽ ⊥ ♀
20 30 ♀ # ♃
22 28 ☽ ⚹ ♂

06 Sunday
01 35 ☽ ‖ ♆
03 22 ☽ Q ♇
05 37 ☽ ⊻ ♀
13 44 ☽ △ ♆
14 03 ☽ ‖ ♄
20 26 ☽ Q ♂

07 Monday
00 12 ☽ ☌ ☿
00 20 ☽ ⚹ ♇
00 46 ☽ # ♂
01 57 ☿ ⚹ ♇
02 25 ♂ ⊥ ♃
02 38 ☽ ⊼ ♅
03 08 ☽ ☌ ♄
09 22 ☽ ⊻ ☉
12 58 ☽ ± ♅
17 33 ☽ △ ♃
18 01 ♂ ꝑ ♅
20 29 ☽ ⊥ ☉

08 Tuesday
02 12 ☽ ∠ ♇
03 58 ☽ ‖ ☿
04 24 ☽ ꝑ ♅
04 54 ☽ □ ♂
07 11 ☿ ⊼ ♅
13 04 ☽ ∠ ☉
13 56 ☽ ☌ ♀
17 21 ☽ □ ♆
17 36 ☽ ⊥ ♇
19 24 ☽ ꝑ ♃
19 39 ☿ ☌ ♄

09 Wednesday
03 50 ☽ ⊻ ♇
05 55 ☽ △ ♅
06 58 ☽ ⊻ ♄
07 43 ☽ ⊻ ☿
10 48 ☽ ± ♃
16 30 ☽ ⚹ ☉
17 14 ☽ ⊥ ♄
18 39 ☽ ⊥ ☿
21 03 ☽ ⊼ ♃
23 54 ☽ Q ♆

10 Thursday
08 36 ☽ ∠ ♄
09 05 ♀ □ ♆
10 10 ☽ △ ♂
10 59 ☽ ∠ ☿
13 21 ♀ ⊥ ♇
14 21 ☽ Q ☉
20 10 ☽ ⚹ ♆
21 07 ☽ ⊻ ♀

11 Friday
01 03 ☉ ꝑ ♆
04 34 ☽ ‖ ☿
06 40 ☽ ☌ ♇
08 06 ☽ ⊥ ♀
08 33 ☽ □ ♅
10 09 ☽ ⚹ ♄
12 39 ☽ ꝑ ♂
14 03 ☽ ⚹ ☿
17 40 ♀ ꝑ ♃
21 29 ☽ ∠ ♆
23 02 ☽ □ ☉

12 Saturday
00 04 ☽ ☍ ♃
00 34 ☽ ∠ ♀
04 32 ☽ ± ♂
06 36 ☽ Q ♄
11 38 ☽ Q ☿
12 41 ☽ ⊥ ♆
14 54 ☽ Q ♅
15 08 ☽ ⊼ ♂
15 09 ☉ □ ♃
22 50 ☽ ⊻ ♆

13 Sunday
04 04 ☽ ⚹ ♀
07 17 ☽ # ♂
07 22 ☽ ‖ ♄
08 39 ♂ # ♄
09 28 ☽ ⊻ ♇
11 11 ☽ ⚹ ♅
13 21 ☽ □ ♄
19 40 ☽ ⊥ ♇
19 57 ☽ □ ☿
20 55 ☽ ‖ ♆

14 Monday
02 11 ☽ Q ♀
03 15 ☽ ⊼ ♃
05 46 ☽ △ ☉
11 03 ☽ ∠ ♇
12 41 ☽ ∠ ♅
13 11 ☽ ‖ ☉
13 33 ☽ ± ♃
16 29 ♂ Q ♄
20 28 ☽ ☍ ♂
23 06 ☿ ± ♅

15 Tuesday
01 56 ☽ ☌ ♆
04 05 ☽ ⊥ ♅
05 06 ☽ ꝑ ♃
09 29 ☽ ꝑ ☉
11 39 ☽ □ ♀
12 51 ☽ ⚹ ♇
14 25 ☽ ⊻ ♅
16 19 ☽ # ♅
17 12 ☽ △ ♄

16 Wednesday
02 04 ☽ △ ☿
02 17 ☽ ± ☉
03 18 ♀ ⊻ ♇
07 15 ☽ △ ♃
09 45 ☽ Q ♇
13 35 ☽ ⊼ ☉
19 37 ☽ ꝑ ♄
22 04 ♀ △ ♅

17 Thursday
03 02 ☽ ‖ ♅
03 03 ☽ ⊼ ♂
05 28 ☽ ꝑ ☿
06 14 ☽ ⊻ ♆
11 41 ☽ ± ♄
14 12 ☽ ± ♂
16 54 ☽ ⊥ ♆
17 37 ☽ □ ♇
19 01 ☽ ☌ ♅
20 35 ☿ ‖ ♇
20 55 ☽ △ ♀
22 03 ☽ ± ☿
22 31 ☽ ⊼ ♄

18 Friday
07 07 ☽ ꝑ ♂
09 05 ☽ ∠ ♆
09 11 ☽ ⊼ ☿
12 54 ☽ □ ♃
16 26 ☽ # ☉
19 16 ♀ ⊻ ♄
23 38 ☽ ☍ ☉

19 Saturday
00 40 ☽ # ♆
02 32 ☽ ꝑ ♀
06 34 ☽ ‖ ♂
11 53 ☽ △ ♂
12 32 ☽ ⚹ ♆
16 51 ☽ # ♄
20 43 ☽ ± ♀
22 15 ☽ Q ♃

20 Sunday
00 35 ☽ △ ♇
01 50 ☽ ⊻ ♅
01 53 ♂ ☍ ♆
06 10 ☽ ☍ ♄
09 00 ☽ ⊼ ♀
11 00 ☽ Q ♆
11 02 ☉ Q ♇
13 07 ☽ ⊥ ♅
16 15 ♀ ± ♃
17 52 ☽ ☍ ☿
21 02 ☽ ⚹ ♃

21 Monday
05 05 ☽ ꝑ ♇
06 15 ☽ ∠ ♅
10 28 ☿ St R
12 55 ☽ ⊼ ☉
20 59 ♂ ± ♅
21 32 ☽ □ ♆
22 32 ☽ ± ♇
23 47 ☽ □ ♂

22 Tuesday
01 36 ☽ ± ☉
02 08 ☽ ∠ ♃
03 04 ☿ ‖ ♇
07 50 ☉ ‖ ♆
10 18 ☽ ⊼ ♇
11 21 ☽ ⚹ ♅
16 39 ☽ ⊼ ♄
17 57 ☉ # ♂
19 54 ☽ ⊥ ♃
20 53 ☽ ꝑ ☉

23 Wednesday
00 35 ☽ ☍ ♀
03 59 ☽ ⊼ ☿
04 39 ☽ ± ♄
07 52 ☽ ⊻ ♃
10 36 ♂ # ♆
11 04 ☽ Q ♅
13 09 ☽ Q ♂
15 41 ☽ ± ☿
22 49 ☽ ꝑ ♄

24 Thursday
05 08 ♂ ∠ ♃
05 33 ☽ △ ☉
07 27 ☿ ⊻ ♀
09 01 ☽ △ ♆
09 13 ☽ ꝑ ☿
14 30 ☽ ⚹ ♂
20 55 ☿ Q ♂
22 20 ☽ ☍ ♇
23 09 ☽ □ ♅

25 Friday
05 13 ♀ ⊥ ♄
05 22 ☽ △ ♄
14 17 ☽ △ ☿
15 18 ☽ ꝑ ♆
18 41 ☽ ⊼ ♀
20 31 ☽ ☌ ♃
22 25 ☽ ∠ ♂
22 40 ☉ △ ♆

26 Saturday
07 58 ☽ ± ♀
09 29 ☽ ± ♆
16 31 ♀ ⊼ ♃
17 32 ☽ ⊥ ♂
21 34 ☽ ⊼ ♆
23 40 ☽ □ ☉

27 Sunday
01 16 ♀ Q ♆
03 41 ☽ # ♄
03 49 ☽ ꝑ ♀
06 13 ☽ ⊻ ♂
10 14 ☽ # ☉
10 57 ☽ ⊼ ♇
11 29 ☽ △ ♅
16 33 ☿ ± ♅
18 24 ☽ □ ♄
19 57 ☿ ⊥ ♀
22 56 ☽ ± ♇
22 56 ☽ □ ☿
23 19 ☽ # ♆

28 Monday
06 30 ☽ ‖ ♂
08 50 ☽ ⊻ ♃
12 26 ☽ △ ♀
14 44 ☽ Q ☉
16 44 ☽ ꝑ ♇
17 07 ☽ ꝑ ♅
20 40 ☽ ⊥ ♃

29 Tuesday
06 10 ☽ Q ♄
07 32 ☽ Q ☿
08 52 ☽ ☍ ♆
10 35 ☽ ± ♅
14 10 ☽ ∠ ♃
16 06 ☽ ⚹ ☉
20 05 ☽ ☌ ♂
20 48 ☿ ☌ ♄
21 49 ☽ △ ♇
22 05 ☽ ⊼ ♅

30 Wednesday
02 16 ☽ ‖ ♅
04 35 ☽ ⚹ ☿
04 45 ☉ ‖ ♄
05 24 ☽ ⚹ ♄
18 39 ☽ ⚹ ♃
22 52 ☽ ∠ ☉

31 Thursday
02 48 ☽ □ ♀
06 10 ☽ ∠ ☿
09 37 ☽ ∠ ♄
10 49 ♂ △ ♇
12 55 ♂ ⊼ ♅
12 58 ☽ # ♅
16 47 ☽ ⊥ ☉
16 49 ☽ Q ♃
17 09 ☽ ⊼ ♆
21 08 ☽ ⊥ ☿
23 47 ☿ ∠ ♀

All ephemeris data is given at 12.00 UT and the Moon's longitude is additionally given for 24.00 UT

NOVEMBER 2013

LONGITUDES

Date	Sidereal time h m s	Sun ☉ ° ' "	Moon ☽ ° ' "	Moon ☽ 24.00 ° ' "	Mercury ☿ ° '	Venus ♀ ° '	Mars ♂ ° '	Jupiter ♃ ° '	Saturn ♄ ° '	Uranus ♅ ° '	Neptune ♆ ° '	Pluto ⯓ ° '
01	14 43 45	09 ♏ 13 35	13 ♎ 06 17	19 ♎ 53 36	10 ♏ 01	26 ♐ 12	10 ♍ 01	20 ♋ 27	13 ♏ 38	09 ♈ 26	02 ♓ 37	09 ♑ 26
02	14 47 41	10 13 39	26 ♎ 46 56	03 ♏ 44 00	08 R 44	27 11	10 36	20 28	13 45	09 R 24	02 R 37	09 27
03	14 51 38	11 13 46	10 ♏ 46 23	17 ♏ 52 57	07 29	28 11	11 10	20 29	13 53	09 22	02 37	09 29
04	14 55 34	12 13 54	25 ♏ 03 05	02 ♐ 16 08	06 18	29 ♐ 09	11 45	20 30	14 00	09 20	02 36	09 30
05	14 59 31	13 14 04	09 ♐ 31 22	16 ♐ 48 02	05 15	00 ♑ 08	12 19	20 30	14 07	09 18	02 36	09 31
06	15 03 28	14 14 15	24 ♐ 05 23	01 ♑ 22 40	04 20	01 06	12 54	20 31	14 14	09 16	02 36	09 33
07	15 07 24	15 14 29	08 ♑ 39 11	15 ♑ 54 18	03 36	02 04	13 28	20 R 31	14 21	09 14	02 35	09 34
08	15 11 21	16 14 43	23 ♑ 07 26	00 ♒ 18 08	03 02	03 01	14 02	20 30	14 29	09 13	02 35	09 35
09	15 15 17	17 15 00	07 ♒ 25 58	14 ♒ 30 40	02 40	03 57	14 36	20 30	14 36	09 11	02 35	09 37
10	15 19 14	18 15 17	21 ♒ 31 58	28 ♒ 29 46	02 30	04 54	15 10	20 30	14 43	09 09	02 35	09 38
11	15 23 10	19 15 36	05 ♓ 23 56	12 ♓ 14 28	02 D 32	05 49	15 44	20 29	14 50	09 07	02 35	09 40
12	15 27 07	20 15 56	19 ♓ 01 22	25 ♓ 44 40	02 44	06 44	16 18	20 28	14 57	09 06	02 35	09 41
13	15 31 03	21 16 18	02 ♈ 24 26	09 ♈ 00 44	03 06	07 39	16 51	20 27	15 05	09 04	02 35	09 43
14	15 35 00	22 16 41	15 ♈ 33 40	22 ♈ 03 18	03 37	08 32	17 25	20 25	15 12	09 03	02 D 35	09 44
15	15 38 57	23 17 06	28 ♈ 29 43	04 ♉ 53 00	04 17	09 25	17 58	20 24	15 19	09 01	02 35	09 46
16	15 42 53	24 17 32	11 ♉ 13 14	17 ♉ 30 31	05 05	10 18	18 32	20 22	15 26	08 59	02 35	09 47
17	15 46 50	25 18 00	23 ♉ 44 55	29 ♉ 56 34	05 59	11 10	19 05	20 20	15 33	08 58	02 35	09 49
18	15 50 46	26 18 29	06 ♊ 05 35	12 ♊ 12 06	06 58	12 01	19 38	20 18	15 40	08 57	02 35	09 51
19	15 54 43	27 19 00	18 ♊ 16 17	24 ♊ 18 21	08 03	12 51	20 11	20 16	15 47	08 55	02 35	09 52
20	15 58 39	28 19 32	00 ♋ 18 31	06 ♋ 17 03	09 12	13 41	20 44	20 13	15 55	08 54	02 36	09 54
21	16 02 36	29 ♏ 20 06	12 ♋ 14 17	18 ♋ 10 32	10 25	14 30	21 17	20 10	16 02	08 53	02 36	09 56
22	16 06 32	00 ♐ 20 42	24 ♋ 06 13	00 ♌ 01 45	11 42	15 18	21 50	20 07	16 09	08 51	02 36	09 57
23	16 10 29	01 21 20	05 ♌ 57 36	11 ♌ 54 16	13 00	16 05	22 23	20 04	16 16	08 50	02 36	09 59
24	16 14 26	02 21 59	17 ♌ 52 18	23 ♌ 52 14	14 22	16 51	22 56	20 01	16 23	08 49	02 37	10 01
25	16 18 22	03 22 39	29 ♌ 54 39	06 ♍ 00 09	15 45	17 36	23 28	19 57	16 30	08 48	02 37	10 03
26	16 22 19	04 23 21	12 ♍ 09 19	18 ♍ 22 45	17 10	18 20	24 00	19 54	16 37	08 47	02 38	10 04
27	16 26 15	05 24 05	24 ♍ 40 59	01 ♎ 04 35	18 36	19 04	24 33	19 50	16 44	08 46	02 38	10 06
28	16 30 12	06 24 51	07 ♎ 33 59	14 ♎ 09 36	20 03	19 46	25 05	19 46	16 51	08 45	02 38	10 08
29	16 34 08	07 25 37	20 ♎ 51 45	27 ♎ 40 35	21 32	20 27	25 37	19 42	16 58	08 44	02 39	10 10
30	16 38 05	08 ♐ 26 26	04 ♏ 36 12	11 ♏ 38 27	23 ♏ 01	21 ♑ 07	26 ♍ 09	19 ♋ 37	17 ♏ 05	08 ♈ 43	02 ♓ 40	10 ♑ 12

DECLINATIONS

Date	Moon True ☊ ° '	Moon Mean ☊ ° '	Moon ☽ Latitude ° '	Sun ☉ ° '	Moon ☽ ° '	Mercury ☿ ° '	Venus ♀ ° '	Mars ♂ ° '	Jupiter ♃ ° '	Saturn ♄ ° '	Uranus ♅ ° '	Neptune ♆ ° '	Pluto ⯓ ° '
01	07 ♏ 46	07 ♏ 28	02 S 13	14 S 34	07 S 13	15 S 25	27 S 01	09 N 15	21 N 52	13 S 54	03 N 04	11 S 11	20 S 13
02	07 D 46	07 25	01 S 01	14 53	11 16	14 41	27 04	09 03	21 52	13 57	03 04	11 11	20 14
03	07 R 46	07 22	00 N 17	15 12	14 47	13 57	27 07	08 50	21 52	13 59	03 03	11 11	20 14
04	07 46	07 19	01 34	15 30	17 30	13 15	27 08	08 37	21 52	14 01	03 02	11 11	20 14
05	07 45	07 15	02 47	15 49	19 07	12 36	27 09	08 25	21 52	14 03	03 01	11 11	20 14
06	07 45	07 12	03 50	16 07	19 29	12 01	27 10	08 12	21 52	14 05	03 01	11 11	20 14
07	07 44	07 09	04 37	16 24	18 33	11 31	27 09	08 00	21 52	14 07	03 00	11 11	20 14
08	07 43	07 06	05 06	16 42	16 25	11 06	27 07	07 47	21 52	14 09	02 59	11 11	20 14
09	07 43	07 03	05 17	16 59	13 18	10 47	27 07	07 34	21 52	14 12	02 59	11 11	20 14
10	07 D 43	07 00	05 08	17 16	09 28	10 33	27 05	07 22	21 53	14 14	02 58	11 11	20 14
11	07 43	06 56	04 41	17 32	05 10	10 25	27 03	07 09	21 53	14 16	02 57	11 11	20 14
12	07 44	06 53	03 58	17 49	00 S 41	10 23	27 00	06 56	21 53	14 18	02 57	11 11	20 14
13	07 45	06 50	03 04	18 05	03 N 46	10 26	26 56	06 44	21 53	14 20	02 56	11 11	20 14
14	07 46	06 47	02 00	18 20	07 58	10 33	26 52	06 31	21 54	14 22	02 56	11 11	20 14
15	07 47	06 44	00 N 51	18 35	11 44	10 44	26 48	06 19	21 54	14 24	02 55	11 11	20 14
16	07 R 47	06 40	00 S 19	18 50	14 54	10 59	26 43	06 06	21 54	14 26	02 54	11 11	20 14
17	07 46	06 37	01 27	19 05	17 18	11 17	26 37	05 54	21 55	14 28	02 54	11 11	20 15
18	07 45	06 34	02 30	19 19	18 52	11 37	26 31	05 41	21 55	14 30	02 53	11 11	20 15
19	07 42	06 31	03 25	19 33	19 31	12 00	26 24	05 29	21 56	14 32	02 53	11 11	20 15
20	07 39	06 28	04 10	19 47	19 16	12 25	26 17	05 16	21 56	14 34	02 52	11 11	20 15
21	07 35	06 25	04 44	20 00	18 09	12 51	26 10	05 04	21 57	14 36	02 52	11 11	20 15
22	07 31	06 21	05 06	20 13	16 13	13 19	26 02	04 51	21 57	14 38	02 51	11 11	20 15
23	07 28	06 18	05 14	20 26	13 42	13 47	25 53	04 39	21 58	14 40	02 51	11 11	20 15
24	07 26	06 15	05 09	20 38	10 34	14 16	25 45	04 27	21 59	14 42	02 51	11 11	20 15
25	07 25	06 12	04 50	20 50	06 58	14 46	25 35	04 14	21 59	14 44	02 50	11 10	20 15
26	07 D 25	06 09	04 18	21 01	03 N 01	15 16	25 26	04 02	22 00	14 46	02 50	11 10	20 15
27	07 27	06 06	03 34	21 12	01 S 09	15 46	25 16	03 50	22 01	14 48	02 49	11 10	20 15
28	07 28	06 02	02 37	21 23	05 24	16 16	25 06	03 37	22 01	14 50	02 49	11 10	20 15
29	07 30	05 59	01 30	21 33	09 32	16 46	24 55	03 25	22 02	14 52	02 49	11 10	20 15
30	07 ♏ 31	05 ♏ 56	00 S 16	21 S 43	13 S 18	17 S 14	24 S 44	03 N 13	22 N 03	14 S 54	02 N 48	11 S 09	20 S 15

ZODIAC SIGN ENTRIES

Date	h	m	Planets
02	17	35	☽ ♏
04	20	14	☽ ♐
05	08	43	♀ ♑
06	21	44	☽ ♑
08	23	30	☽ ♒
11	02	36	☽ ♓
13	07	39	☽ ♈
15	14	49	☽ ♉
18	00	07	☽ ♊
20	11	23	☽ ♋
22	03	48	☉ ♐
22	23	56	☽ ♌
25	12	11	☽ ♍
27	22	00	☽ ♎
30	04	03	☽ ♏

LATITUDES

Date	Mercury ☿ ° '	Venus ♀ ° '	Mars ♂ ° '	Jupiter ♃ ° '	Saturn ♄ ° '	Uranus ♅ ° '	Neptune ♆ ° '	Pluto ⯓ ° '
01	00 S 38	03 S 39	01 N 34	00 S 01	02 N 07	00 S 43	00 S 41	02 N 53
04	00 N 23	03 42	01 35	00 S 01	02 07	00 43	00 41	02 52
07	01 16	03 44	01 37	00 00	02 07	00 43	00 41	02 52
10	01 54	03 45	01 39	00 00	02 07	00 43	00 41	02 51
13	02 15	03 44	01 40	00 N 01	02 07	00 43	00 41	02 51
16	02 22	03 41	01 42	00 01	02 07	00 43	00 41	02 50
19	02 18	03 36	01 44	00 01	02 07	00 43	00 41	02 50
22	02 08	03 30	01 46	00 02	02 07	00 43	00 41	02 50
25	01 52	03 21	01 47	00 02	02 07	00 43	00 41	02 49
28	01 33	03 09	01 49	00 03	02 07	00 42	00 41	02 49
31	01 N 13	02 S 55	01 N 51	00 N 03	02 N 07	00 S 42	00 S 41	02 N 48

DATA

Julian Date	2456598
Delta T	+68 seconds
Ayanamsa	24° 03' 10"
Synetic vernal point	05° ♓ 03' 49"
True obliquity of ecliptic	23° 26' 08"

MOON'S PHASES, APSIDES AND POSITIONS ☽

Date	h	m	Phase	Longitude ° '	Eclipse Indicator
03	12	50	●	11 ♏ 16	Ann-Total
10	05	57	☽	18 ♒ 00	
17	15	16	○	25 ♉ 26	
25	19	28	☾	03 ♍ 42	

Day	h	m		
06	09	16	Perigee	
22	09	41	Apogee	
06	06	38	Max dec	19° S 31'
12	15	40	0N	
19	17	16	Max dec	19° N 32'
27	05	26	0S	

LONGITUDES

Date	Chiron ⚷ ° '	Ceres ⚳ ° '	Pallas ⚴ ° '	Juno ⚵ ° '	Vesta ⚶ ° '	Black Moon Lilith ⚸ ° '
01	09 ♓ 17	29 ♍ 06	26 ♌ 19	08 ♒ 38	23 ♍ 51	16 ♋ 12
11	09 ♓ 09	03 ♎ 20	00 ♍ 37	11 ♒ 14	28 ♍ 14	17 ♋ 19
21	09 ♓ 07	07 ♎ 25	04 ♍ 29	14 ♒ 15	02 ♎ 32	18 ♋ 26
31	09 ♓ 11	11 ♎ 22	07 ♍ 49	17 ♒ 39	06 ♎ 42	19 ♋ 33

ASPECTARIAN

h m	Aspects
01 Friday	
02 08	☽ ⊥ ♄
03 59	☽ ± ♆
04 31	☽ ⊻ ☉
05 26	☽ □ ⯓
05 27	☽ ☍ ♅
06 15	☽ ⊻ ♂
06 59	☽ ⊻ ☿
10 30	♅ □ ⯓
12 03	☿ ⚹ ♂
12 57	☽ ⊻ ♄
14 05	☽ Q ♀
16 44	☉ ⊼ ♅
17 05	☉ ⚹ ⯓
17 24	☽ ⊥ ♂
20 00	☽ ♇ ♆
20 19	☉ ☌ ☿
22 44	☿ ⚹ ⯓
23 15	☽ # ♂
23 15	☿ ⊼ ♅
02 Saturday	
01 00	☽ □ ♃
07 28	☉ ∥ ☿
09 52	☽ ∠ ♂
11 29	☽ ∥ ♆
12 47	☽ ⚹ ♀
13 11	☽ Q ⯓
22 05	☽ △ ♆
03 Sunday	
05 58	☽ ∥ ♄
06 51	☽ ☌ ☿
06 57	☽ ∥ ☿
08 49	☉ ⚹ ♂
09 37	☽ ⊼ ♅
09 48	☽ ⚹ ⯓
11 14	☿ ∥ ♄
12 42	☽ ⚹ ♂
12 50	☽ ☌ ☉
15 29	☽ ∥ ☉
16 22	☽ ∠ ♀
17 18	☽ ☌ ♄
19 45	☽ ± ♅
04 Monday	
04 23	☽ △ ♃
07 31	♀ ∠ ♄
08 37	☽ ⊥ ♀
09 44	☽ Q ♂
10 48	☽ ♇ ♅
11 05	☽ ∠ ⯓
19 20	☽ ⊻ ♀
05 Tuesday	
00 33	☽ □ ♆
02 03	☽ ⊥ ⯓
05 22	☽ ♇ ♃
05 23	☽ ⊻ ☿
11 38	☽ △ ♅
12 00	☽ ⊻ ⯓
14 40	☽ ⊥ ☿
16 48	☽ □ ♂
18 35	☽ ⊻ ☉
19 39	☽ ⊻ ♄
20 13	☽ ± ♃
06 Wednesday	
04 36	☽ ∠ ☿
05 11	☽ ⊥ ☉
05 36	☽ ⊥ ♄
06 06	☽ ⊼ ♃
06 15	☽ Q ♆
12 01	☉ ☌ ♄
20 33	☽ ∠ ♄
21 06	☽ ∠ ☉
07 Thursday	
00 22	☽ ☌ ♀
02 00	☽ ⚹ ♆
04 01	☽ ⚹ ☿
05 02	♃ St R
11 57	☉ ± ♅
12 58	☽ □ ♅
13 31	☽ ☌ ⯓
20 17	☽ △ ♂
21 31	☽ ⚹ ♄
23 01	☽ Q ☿
23 43	☽ ⚹ ☉
08 Friday	
01 16	♀ ⚹ ♆
02 48	☽ ∠ ♆
06 21	☿ ∥ ♆
07 39	☽ ☍ ♃
09 40	☽ ∥ ☉
12 23	☿ ⚹ ♀
17 39	☽ Q ♄
17 47	☽ ⊥ ♆
18 48	☽ Q ♅
21 12	☽ Q ☉
22 17	☽ ♇ ♂
09 Saturday	
03 50	☽ ⊻ ♆
04 09	☽ □ ☿
05 44	☽ ⊻ ♀
05 48	☽ ∥ ♄
11 57	♂ ⚹ ♄
14 03	☽ ± ♂
14 57	☽ ⚹ ♅
15 42	☽ ⊻ ⯓
16 34	☽ ⊥ ♀
21 35	☿ △ ♆
10 Sunday	
00 15	☽ □ ♄
00 40	☽ ⊼ ♂
01 38	☽ ∥ ♆
01 54	☽ ⊥ ⯓
05 12	☽ ∥ ☿
05 57	☽ □ ☉
08 59	☽ ∠ ♀
10 13	☽ ⊼ ♃
16 29	☽ ∠ ♅
17 21	☽ ∠ ⯓
20 32	☽ ± ♃
21 13	☿ St D
11 Monday	
00 36	☽ # ♂
06 58	☽ △ ☿
07 05	☽ ☌ ♆
08 02	☽ ⊥ ♅
12 08	☽ ♇ ♃
12 47	☽ ⚹ ♀
18 30	☽ ⊻ ♅
19 29	☽ ⚹ ⯓
20 46	☿ △ ♆
23 56	☽ # ♅
12 Tuesday	
04 44	☽ △ ♄
06 57	☽ ☍ ♂
09 39	☽ ♇ ☿
11 27	☽ Q ♀
14 23	☽ △ ☉
14 34	☽ △ ♃
16 39	☉ △ ♃
16 45	☽ Q ⯓
13 Wednesday	
02 07	☽ ± ☿
07 29	☽ ∥ ♅
07 45	☽ ♇ ♄
10 35	♄ ± ♅
12 19	☽ ⊻ ♆
13 18	☽ ⊼ ☿
18 40	♆ St D
19 35	☽ ♇ ☉
22 12	☽ □ ♀
23 13	☽ ⊥ ♆
14 Thursday	
00 05	☽ ☌ ♅
00 14	☽ ± ♄
01 18	☽ □ ⯓
03 55	☽ ∥ ♂
11 19	☽ ⊼ ♄
13 26	☽ ± ☉
15 34	☽ ⊼ ♂
15 43	☽ ∠ ♆
20 57	☽ □ ♃
15 Friday	
01 13	♀ □ ♅
01 28	☽ ⊼ ☉
03 11	☽ ± ♂
04 51	☽ # ☿
08 18	☽ # ♆
19 40	☽ ⚹ ♆
20 47	☽ ♇ ♂
21 33	♀ ☌ ⯓
23 34	☽ ☍ ☿
16 Saturday	
05 00	☉ ♇ ♅
06 36	☽ Q ♃
07 47	☽ ⊻ ♅
08 07	☽ # ♄
09 17	☽ △ ⯓
10 07	☽ △ ♀
18 24	☽ Q ♆
19 10	☽ ⊥ ♅
20 07	☽ ☍ ♄
17 Sunday	
00 11	☉ ∠ ⯓
02 36	☽ △ ♂
05 03	☿ ∥ ♆
05 26	☽ ⚹ ♃
12 25	☽ ∠ ♅
14 04	☽ ♇ ⯓
15 16	☽ ☍ ☉
17 01	☽ ♇ ♀
18 Monday	
05 09	☽ □ ♆
07 36	☽ ± ⯓
10 27	☽ ∠ ♃
11 50	☽ ± ♀
13 53	☽ ⊼ ☿
17 35	☽ ⚹ ♅
19 23	☽ ⊼ ⯓
19 Tuesday	
00 30	☽ ⊼ ♀
02 49	☽ ± ☿
04 05	☽ ⊥ ♃
07 02	☽ ⊼ ♄
14 48	♂ ⚹ ♃
15 56	☽ ⊻ ♃
15 59	☽ □ ♂
17 15	☽ Q ♅
19 04	☽ ± ♄
22 29	☽ ♇ ☿
20 Wednesday	
05 50	☿ ⊼ ♅
07 40	☽ ⊼ ☉
13 13	☽ ♇ ♄
16 35	☽ △ ♆
20 48	☽ ± ☉
21 Thursday	
02 09	☿ ⚹ ⯓
05 14	☽ □ ♅
05 46	☽ Q ♂
07 20	☽ ☍ ⯓
07 55	☽ △ ☿
16 38	☽ ♇ ☉
16 53	☽ ☍ ♀
19 44	☽ △ ♄
22 50	☽ ♇ ♆
22 Friday	
03 58	☽ ☌ ♃
06 15	☉ ∠ ♀
07 11	☽ ⚹ ♂
14 47	☉ ∥ ⯓
17 04	☽ ± ♆
23 Saturday	
01 49	☽ △ ☉
03 43	☽ # ♄
05 13	☽ ⊼ ♆
11 23	☽ # ☿
15 01	☽ ∠ ♂
17 48	☽ △ ♅
18 45	♀ ⚹ ♄
20 09	☽ ⊼ ⯓
24 Sunday	
04 02	☽ □ ☿
07 37	☽ # ♆
08 16	☽ ± ⯓
08 59	☽ □ ♄
09 48	☽ ⊼ ♀
10 01	☽ ⊥ ♂
16 17	☽ ⊻ ♃
17 52	☉ □ ♆
19 49	☿ ± ♅
22 36	☽ ⊻ ♂
22 38	☽ ± ♀
23 52	☽ ♇ ♅
25 Monday	
02 19	☽ ♇ ⯓
04 12	☽ ⊥ ♃
10 25	☿ ∥ ♄
12 33	♀ ∠ ♆
17 21	☽ ☍ ♆
17 40	☽ ♇ ♀
17 41	☽ ± ♅
19 28	☽ □ ☉
20 34	☽ Q ☿
21 08	☽ Q ♄
21 54	☽ ∠ ♃
26 Tuesday	
01 54	☿ ☌ ♄
04 16	☉ ⊥ ⯓
05 26	☽ ⊼ ♅
05 40	☽ ∥ ♂
07 57	☽ △ ⯓
13 06	☽ ∥ ♅
20 42	☽ ⚹ ♄
22 55	☽ ⚹ ☿
23 18	☉ ♇ ♃
27 Wednesday	
00 39	☽ △ ♀
02 50	☽ ⚹ ♃
09 22	☽ Q ☉
11 44	☽ ☌ ♂
21 24	☽ # ♅
28 Thursday	
01 20	☽ Q ♃
01 20	☽ ∠ ♄
02 24	☽ # ♂
02 45	☿ ⚹ ♀
02 54	☽ ⊼ ♆
06 48	☽ ∠ ☿
07 23	☿ △ ♃
09 42	☽ ⚹ ☉
12 00	♀ ☍ ♃
13 58	☽ ± ♆
14 09	☽ ☍ ♅
16 42	☽ □ ⯓
18 03	☽ ⊥ ♄
29 Friday	
01 17	☽ ⊥ ☿
04 59	☽ ⊻ ♄
06 16	☽ ♇ ♆
09 56	☽ □ ♃
11 14	☽ □ ♀
13 20	☽ ⊻ ☿
15 00	☽ ∠ ☉
20 44	☽ ⊻ ♂
21 26	☉ Q ♂
21 59	☽ ∥ ♆
30 Saturday	
00 53	☽ Q ⯓
07 36	☽ ⊥ ♂
07 59	☽ ⊥ ☉
08 39	☽ △ ♆
18 25	☉ △ ♅
19 02	☽ ⊼ ♅
19 05	☽ ⊻ ☉
20 05	☽ Q ♀
21 34	☽ ⚹ ⯓
23 04	☿ ♇ ♅
23 35	☽ ∥ ♄
23 36	☽ ∠ ♂

All ephemeris data is given at 12.00 UT and the Moon's longitude is additionally given for 24.00 UT

DECEMBER 2013

Date	Sidereal time h m s	Sun ☉ ° ' "	Moon ☽ ° ' "	Moon ☽ 24.00 ° ' "	Mercury ☿ ° '	Venus ♀ ° '	Mars ♂ ° '	Jupiter ♃ ° '	Saturn ♄ ° '	Uranus ♅ ° '	Neptune ♆ ° '	Pluto ♇ ° '
01	16 42 01	09 ♐ 27 16	18 ♏ 47 04	26 ♏ 01 36	24 ♏ 31	21 ♑ 46	26 ♍ 41	19 ♋ 33	17 ♏ 11	08 ♈ 42	02 ♓ 40	10 ♑ 14
02	16 45 58	10 28 07	03 ♐ 21 24	10 ♐ 45 39	26 02	22 23	27 12	19 R 28	17 18	08 R 41	02 41	10 15
03	16 49 55	11 28 59	18 ♐ 13 22	25 ♐ 43 28	27 33	22 59	27 44	19 23	17 25	08 41	02 41	10 17
04	16 53 51	12 29 53	03 ♑ 14 48	10 ♑ 46 09	29 ♏ 04	23 34	28 15	19 18	17 32	08 40	02 42	10 19
05	16 57 48	13 30 47	18 ♑ 16 20	25 ♑ 44 17	00 ♐ 36	24 08	28 47	19 12	17 39	08 39	02 43	10 21
06	17 01 44	14 31 43	03 ♒ 08 58	10 ♒ 29 33	02 07	24 40	29 18	19 07	17 45	08 39	02 44	10 23
07	17 05 41	15 32 39	17 ♒ 45 22	24 ♒ 55 55	03 40	25 10	29 ♍ 49	19 01	17 52	08 38	02 44	10 25
08	17 09 37	16 33 35	02 ♓ 00 52	09 ♓ 00 05	05 12	25 39	00 ♎ 20	18 56	17 59	08 38	02 45	10 27
09	17 13 34	17 34 33	15 ♓ 53 32	22 ♓ 41 19	06 44	26 06	00 50	18 50	18 05	08 37	02 46	10 29
10	17 17 30	18 35 31	29 ♓ 23 41	06 ♈ 00 53	08 17	26 32	01 21	18 44	18 12	08 37	02 47	10 31
11	17 21 27	19 36 29	12 ♈ 33 15	19 ♈ 01 11	09 50	26 55	01 51	18 38	18 18	08 36	02 48	10 33
12	17 25 24	20 37 28	25 ♈ 25 02	01 ♉ 45 11	11 23	27 17	02 22	18 31	18 25	08 36	02 49	10 35
13	17 29 20	21 38 28	08 ♉ 02 00	14 ♉ 15 51	12 56	27 37	02 52	18 25	18 31	08 36	02 50	10 37
14	17 33 17	22 39 28	20 ♉ 27 01	26 ♉ 35 48	14 29	27 55	03 22	18 18	18 38	08 36	02 51	10 39
15	17 37 13	23 40 29	02 ♊ 42 27	08 ♊ 47 13	16 02	28 11	03 52	18 11	18 44	08 35	02 52	10 41
16	17 41 10	24 41 31	14 ♊ 50 17	20 ♊ 51 49	17 35	28 24	04 21	18 05	18 50	08 35	02 53	10 43
17	17 45 06	25 42 33	26 ♊ 52 01	02 ♋ 51 02	19 09	28 36	04 51	17 58	18 57	08 35	02 54	10 45
18	17 49 03	26 43 36	08 ♋ 49 00	14 ♋ 46 08	20 42	28 45	05 20	17 51	19 03	08 D 35	02 55	10 47
19	17 52 59	27 44 39	20 ♋ 42 35	26 ♋ 38 34	22 16	28 52	05 49	17 43	19 09	08 35	02 56	10 49
20	17 56 56	28 45 44	02 ♌ 34 19	08 ♌ 30 07	23 50	28 57	06 18	17 36	19 15	08 36	02 58	10 51
21	18 00 53	29 ♐ 46 48	14 ♌ 26 16	20 ♌ 23 08	25 24	28 59	06 47	17 29	19 21	08 36	02 59	10 53
22	18 04 49	00 ♑ 47 54	26 ♌ 21 05	02 ♍ 20 36	26 58	28 R 59	07 16	17 21	19 27	08 36	03 00	10 55
23	18 08 46	01 49 00	08 ♍ 22 07	14 ♍ 26 11	28 ♐ 32	28 56	07 44	17 14	19 33	08 36	03 01	10 58
24	18 12 42	02 50 07	20 ♍ 33 21	26 ♍ 44 12	00 ♑ 07	28 51	08 13	17 06	19 39	08 37	03 03	11 00
25	18 16 39	03 51 14	02 ♎ 59 19	09 ♎ 19 18	01 42	28 43	08 41	16 58	19 45	08 37	03 04	11 02
26	18 20 35	04 52 22	15 ♎ 44 45	22 ♎ 16 13	03 17	28 33	09 09	16 51	19 51	08 37	03 06	11 04
27	18 24 32	05 53 31	28 ♎ 54 12	05 ♏ 39 06	04 52	28 20	09 37	16 43	19 56	08 38	03 07	11 06
28	18 28 28	06 54 40	12 ♏ 31 15	19 ♏ 30 47	06 28	28 05	10 04	16 35	20 02	08 38	03 08	11 08
29	18 32 25	07 55 50	26 ♏ 37 43	03 ♐ 51 50	08 04	27 48	10 31	16 27	20 08	08 39	03 10	11 10
30	18 36 22	08 57 00	11 ♐ 12 41	18 ♐ 39 36	09 40	27 28	10 58	16 19	20 13	08 40	03 11	11 12
31	18 40 18	09 ♑ 58 11	26 ♐ 11 42	03 ♑ 47 50	11 ♑ 17	27 ♑ 06	11 ♎ 25	16 ♋ 11	20 ♏ 19	08 ♈ 40	03 ♓ 13	11 ♑ 14

Date	Moon True ☊ ° '	Moon Mean ☊ ° '	Moon ☽ Latitude ° '
01	07 ♏ 30	05 ♏ 53	01 N 01
02	07 R 28	05 50	02 16
03	07 24	05 46	03 24
04	07 19	05 43	04 18
05	07 14	05 40	04 54
06	07 09	05 37	05 10
07	07 04	05 34	05 05
08	07 01	05 31	04 42
09	07 00	05 27	04 02
10	07 D 01	05 24	03 10
11	07 02	05 21	02 09
12	07 04	05 18	01 N 03
13	07 R 04	05 15	00 S 05
14	07 03	05 12	01 12
15	07 00	05 08	02 14
16	06 54	05 05	03 09
17	06 47	05 02	03 55
18	06 37	04 59	04 31
19	06 27	04 56	04 54
20	06 16	04 52	05 05
21	06 07	04 49	05 02
22	05 59	04 46	04 47
23	05 53	04 43	04 18
24	05 50	04 40	03 38
25	05 49	04 37	02 46
26	05 D 50	04 33	01 45
27	05 51	04 30	00 S 37
28	05 R 51	04 27	00 N 36
29	05 49	04 24	01 49
30	05 44	04 21	02 57
31	05 ♏ 37	04 ♏ 18	03 N 54

DECLINATIONS

Date	Sun ☉ ° '	Moon ☽ ° '	Mercury ☿ ° '	Venus ♀ ° '	Mars ♂ ° '	Jupiter ♃ ° '	Saturn ♄ ° '	Uranus ♅ ° '	Neptune ♆ ° '	Pluto ♇ ° '
01	21 S 52	16 S 26	17 S 43	24 S 33	03 N 01	22 N 04	14 S 56	02 N 48	11 S 09	20 S 15
02	22 01	18 35	18 12	24 22	02 49	22 05	14 57	02 48	11 09	20 15
03	22 09	19 32	18 40	24 10	02 37	22 05	14 59	02 48	11 09	20 15
04	22 17	19 06	19 07	23 58	02 25	22 06	15 01	02 47	11 08	20 15
05	22 25	17 20	19 33	23 46	02 13	22 07	15 03	02 47	11 08	20 15
06	22 32	14 25	19 59	23 34	02 01	22 08	15 05	02 47	11 08	20 15
07	22 39	10 39	20 24	23 21	01 50	22 09	15 07	02 47	11 08	20 15
08	22 45	06 22	20 48	23 08	01 38	22 10	15 08	02 47	11 07	20 15
09	22 51	01 S 50	21 11	22 56	01 26	22 11	15 10	02 46	11 07	20 15
10	22 57	02 N 40	21 33	22 43	01 15	22 12	15 12	02 46	11 07	20 15
11	23 02	06 56	21 54	22 30	01 03	22 13	15 13	02 46	11 06	20 15
12	23 06	10 48	22 14	22 16	00 52	22 14	15 15	02 46	11 06	20 15
13	23 10	14 06	22 33	22 03	00 40	22 15	15 17	02 46	11 06	20 15
14	23 14	16 42	22 51	21 50	00 29	22 16	15 19	02 46	11 05	20 15
15	23 17	18 30	23 08	21 37	00 18	22 17	15 20	02 46	11 05	20 14
16	23 20	19 26	23 24	21 24	00 N 06	22 18	15 22	02 46	11 04	20 14
17	23 22	19 29	23 39	21 11	00 S 05	22 19	15 23	02 46	11 04	20 14
18	23 24	18 38	23 52	20 57	00 16	22 20	15 25	02 46	11 04	20 14
19	23 25	16 59	24 04	20 44	00 27	22 21	15 27	02 46	11 03	20 14
20	23 26	14 38	24 15	20 31	00 38	22 22	15 28	02 46	11 03	20 14
21	23 26	11 40	24 25	20 18	00 48	22 23	15 30	02 46	11 02	20 14
22	23 26	08 14	24 33	20 06	00 59	22 25	15 31	02 46	11 02	20 14
23	23 25	04 26	24 40	19 53	01 10	22 26	15 33	02 47	11 01	20 14
24	23 24	00 N 24	24 46	19 40	01 20	22 27	15 34	02 47	11 01	20 14
25	23 23	03 S 44	24 50	19 28	01 31	22 28	15 36	02 47	11 00	20 14
26	23 21	07 49	24 53	19 16	01 41	22 29	15 37	02 47	11 00	20 14
27	23 18	11 40	24 55	19 04	01 51	22 30	15 38	02 47	10 59	20 14
28	23 15	15 02	24 55	18 52	02 01	22 31	15 40	02 48	10 59	20 14
29	23 12	17 38	24 54	18 41	02 12	22 32	15 41	02 48	10 58	20 14
30	23 08	19 12	24 51	18 29	02 21	22 33	15 43	02 48	10 58	20 14
31	23 S 04	19 S 28	24 S 47	18 S 18	02 S 31	22 N 35	15 S 44	02 N 48	10 S 57	20 S 14

ZODIAC SIGN ENTRIES

Date	h	m	Planets
02	06	31	☽ ♐
04	06	49	☽ ♑
05	02	42	☿ ♐
06	06	53	☽ ♒
07	20	41	♂ ♎
08	08	34	☽ ♓
10	13	05	☽ ♈
12	20	40	☽ ♉
15	06	41	☽ ♊
17	18	17	☽ ♋
20	06	48	☽ ♌
21	17	11	☉ ♑
22	19	19	☽ ♍
24	10	12	☿ ♑
25	06	17	☽ ♎
27	13	58	☽ ♏
29	17	37	☽ ♐
31	18	01	☽ ♑

LATITUDES

Date	Mercury ☿ ° '	Venus ♀ ° '	Mars ♂ ° '	Jupiter ♃ ° '	Saturn ♄ ° '	Uranus ♅ ° '	Neptune ♆ ° '	Pluto ♇ ° '
01	01 N 13	02 S 55	01 N 51	00 N 03	02 N 07	00 S 42	00 S 41	02 N 48
04	00 51	02 38	01 53	00 03	02 08	00 42	00 41	02 48
07	00 29	02 17	01 55	00 04	02 08	00 42	00 41	02 47
10	00 N 08	01 54	01 56	00 04	02 08	00 42	00 41	02 47
13	00 S 13	01 27	01 58	00 05	02 08	00 42	00 41	02 47
16	00 33	00 56	02 00	00 05	02 08	00 42	00 40	02 46
19	00 52	00 S 22	02 02	00 06	02 09	00 42	00 40	02 46
22	01 09	00 N 16	02 04	00 06	02 09	00 41	00 40	02 46
25	01 25	00 58	02 06	00 06	02 09	00 41	00 40	02 45
28	01 38	01 42	02 08	00 07	02 10	00 41	00 40	02 45
31	01 S 50	02 N 29	02 N 10	00 N 07	02 N 10	00 S 41	00 S 40	02 N 45

DATA

Julian Date	2456628
Delta T	+68 seconds
Ayanamsa	24° 03' 14"
Synetic vernal point	05° ♓ 03' 45"
True obliquity of ecliptic	23° 26' 07"

LONGITUDES

Date	Chiron ⚷ ° '	Ceres ⚳ ° '	Pallas ⚴ ° '	Juno ⚵ ° '	Vesta ⚶ ° '	Black Moon Lilith ⚸ ° '
01	09 ♓ 11	11 ♎ 22	07 ♍ 49	17 ♒ 39	06 ♎ 42	19 ♋ 33
11	09 ♓ 21	15 ♎ 06	10 ♍ 32	21 ♒ 30	10 ♎ 42	20 ♋ 41
21	09 ♓ 37	18 ♎ 37	12 ♍ 29	25 ♒ 23	14 ♎ 30	21 ♋ 48
31	09 ♓ 57	21 ♎ 51	13 ♍ 34	29 ♒ 39	18 ♎ 03	22 ♋ 55

MOON'S PHASES, APSIDES AND POSITIONS ☽

Date	h	m	Phase	Longitude ° '	Eclipse Indicator
03	00	22	●	10 ♐ 59	
09	15	12	☽	17 ♓ 43	
17	09	28	○	25 ♊ 36	
25	13	48	☾	03 ♎ 56	

Day	h	m		
04	10	03	Perigee	
19	23	34	Apogee	
03	16	35	Max dec	19° S 33'
09	21	43	ON	
17	00	57	Max dec	19° N 34'
24	14	21	OS	
31	04	46	Max dec	19° S 32'

ASPECTARIAN

h m Aspects

01 Sunday
05 10 ☽ ± ♅
09 19 ☽ ☌ ♄
13 15 ☽ △ ♃
17 11 ☽ ✶ ♀
20 09 ☽ bQ ♅
22 37 ☽ ☌ ☿
22 42 ☽ ∠ ♇
23 31 ☿ ∠ ♇

02 Monday
01 34 ☽ ✶ ♂
04 45 ☽ ∥ ☿
06 50 ☉ ⊻ ♇
10 54 ☽ □ ♆
13 28 ☽ ⊥ ♇
13 47 ☽ bQ ♃
14 27 ♂ ∥ ♅
18 49 ☽ ∠ ♀
20 39 ☽ △ ♅
20 42 ♅ ⊥ ♆
21 50 ☽ Q ♂
23 13 ☽ ⊻ ♇
23 25 ☉ # ♃

03 Tuesday
00 22 ☽ ☌ ☉
04 16 ☽ ± ♃
09 56 ☽ ⊥ ♀
10 42 ☽ ⊻ ♄
13 51 ☽ ⚻ ♃
15 57 ☽ Q ♆
16 35 ☿ ✶ ♂
19 56 ☽ ⊻ ♀
20 23 ☽ ⊥ ♄

04 Wednesday
03 45 ☽ □ ♂
04 35 ☽ ⊻ ☿
10 51 ☽ ∠ ♄
11 08 ☽ ✶ ♆
11 46 ☽ ∥ ☿
15 14 ☽ ⊥ ☿
20 38 ☽ □ ♅
23 18 ☽ ☌ ♇

05 Thursday
03 50 ☽ ⊻ ☉
05 21 ☉ ± ♃
07 13 ☽ ∠ ☿
10 59 ☽ ✶ ♄
11 06 ☽ ∠ ♆
13 29 ☽ ☍ ♃
14 08 ☽ ⊥ ☉
21 46 ☽ ☌ ♀

06 Friday
01 28 ☽ Q ♅
01 35 ☽ ⊥ ♆
05 31 ☽ △ ♂
05 42 ☽ ∠ ☉
06 27 ☽ Q ♄
07 15 ☽ ∥ ♄
10 08 ☽ ✶ ☿
11 19 ☽ ⊻ ♆
20 58 ☽ ✶ ♅
21 31 ☿ □ ♆
23 51 ☽ ⊻ ♇

07 Saturday
03 04 ☿ ∥ ♇
06 57 ☽ bQ ♂
08 03 ☽ ✶ ☉
08 07 ☽ Q ☿
09 13 ☽ ∥ ♆
09 46 ☽ ⊥ ♇
12 11 ☽ □ ♄
14 06 ☽ ⚻ ♃
17 21 ☿ bQ ♃
21 49 ☽ ∠ ♅
22 30 ☽ ± ♂

08 Sunday
00 04 ☿ ⊥ ♇
00 04 ☽ ± ♃
00 50 ☽ ⊻ ♀
00 51 ☽ ∠ ♇
05 40 ☽ Q ☉
09 01 ☽ ⚻ ♂
11 21 ☽ ⊥ ♀
13 03 ☽ ⊥ ♅
13 16 ☽ ☌ ♆
15 15 ☽ bQ ♃
18 07 ☽ □ ☿
23 21 ☽ ⊻ ♅

09 Monday
02 33 ☽ ✶ ♇
03 22 ☽ ∠ ♀
07 05 ☽ # ♅
11 34 ♂ Q ♃
14 13 ☽ # ♂
15 12 ☽ □ ☉
15 53 ☽ △ ♄
17 08 ☽ △ ♃
17 28 ☉ ∥ ♀
23 40 ☽ Q ♇

10 Tuesday
01 31 ☉ ⊻ ♄
04 40 ☽ ∥ ♂
06 41 ☽ ✶ ♀
12 35 ☽ ∥ ♅
14 56 ☉ ⚻ ♃
15 40 ☽ ☍ ♂
17 01 ♀ Q ♅
17 05 ☿ △ ♅
18 08 ☽ ⊻ ♆
18 56 ☽ bQ ♄

11 Wednesday
04 06 ♀ ⊥ ♆
04 44 ☽ ☌ ♅
05 05 ☽ ⊥ ♆
05 07 ☽ Q ♀
06 19 ☽ △ ☿
08 18 ☽ □ ♇
11 32 ☽ ± ♄
21 44 ☽ ∠ ♆
22 45 ☽ ⚻ ♄
23 10 ☽ □ ♃
23 24 ☿ ⊻ ♇

12 Thursday
02 13 ☽ △ ☉
11 18 ☿ # ♃
13 31 ☿ ∥ ♀
13 59 ☽ # ♆
14 04 ☽ bQ ☿
15 37 ☽ □ ♀
16 25 ♀ # ♃
16 34 ☉ Q ♆

13 Friday
00 01 ♃ △ ♄
01 42 ☽ ⚻ ♂
02 02 ☽ ✶ ♆
04 33 ☿ ± ♃
05 54 ☿ ∠ ♀
08 55 ☽ Q ♃
09 06 ☽ bQ ☉
09 35 ☽ ± ☿
10 20 ♂ ⚻ ♆
11 05 ☉ ⊥ ♀
13 05 ☽ ⊻ ♅
13 40 ☽ ± ♂
16 59 ☽ △ ♇
22 08 ☽ # ♄
22 46 ☽ ⚻ ☿

14 Saturday
00 38 ☽ ⊥ ♅
01 07 ☽ Q ♆
03 58 ☽ ± ☉
07 46 ☽ bQ ♂
07 52 ☽ ✶ ♃
08 25 ☽ ☍ ♄
16 41 ☽ ⚻ ☉
18 08 ☽ ∠ ♅
22 10 ☽ bQ ♇

15 Sunday
02 54 ☽ △ ♀
04 03 ♂ ∠ ♄
08 09 ☿ Q ♂
12 19 ☽ □ ♆
12 57 ☽ ∠ ♃
14 22 ☽ △ ♂
15 54 ☽ ± ♇
23 37 ☽ ✶ ♅

16 Monday
03 48 ☽ ⚻ ♇
03 58 ☉ ∥ ☿
06 34 ☽ ⊥ ♃
09 06 ☽ bQ ♀
15 50 ☉ ⊥ ♄
18 17 ☽ ☍ ☿
18 23 ☽ ⊻ ♃
19 02 ☿ ⚻ ♃
20 02 ☽ ⚻ ♄
23 27 ☽ Q ♅

17 Tuesday
03 20 ☽ ± ♀
08 07 ☽ ± ♄
08 40 ☿ ⊻ ♄
09 28 ☽ ☍ ☉
15 31 ☽ ⚻ ♀
17 40 ♅ St D
22 19 ♃ bQ ♆

18 Wednesday
00 07 ☽ △ ♆
02 19 ☽ bQ ♄
04 42 ☽ □ ♂
11 33 ☽ □ ♅
15 24 ☿ Q ♆
15 59 ☽ ☍ ♇

19 Thursday
06 02 ☽ ☌ ♃
06 24 ☽ bQ ♆
08 49 ☽ △ ♄
15 37 ☽ ⚻ ☿
18 34 ☽ Q ♂
21 45 ☿ ⊥ ♀

20 Friday
00 37 ☽ ± ♆
03 34 ☽ ⚻ ☉
04 18 ☽ # ♄
04 37 ☽ ☍ ♀
05 36 ☽ ± ☿
12 47 ☽ ⚻ ♆
16 29 ☉ ⊻ ♀
16 51 ☽ ± ☉
19 53 ☽ ✶ ♂

21 Saturday
00 11 ☽ △ ♅
02 35 ☽ bQ ☿
04 48 ☽ ⚻ ♇
11 15 ☿ ⊥ ♄
12 45 ☽ bQ ☉
16 41 ☽ # ♆
16 58 ☽ ± ♇
18 05 ☽ ⊻ ♃
19 58 ♀ ∥ ♇
21 53 ♀ St R
22 00 ☽ □ ♄

22 Sunday
03 27 ☽ ∠ ♂
06 03 ☽ ⊥ ♃
06 28 ☽ bQ ♅
11 08 ☽ bQ ♇
13 25 ☽ △ ☿
17 15 ☽ ⚻ ♀
21 44 ☽ △ ☉
22 15 ☽ ⊥ ♂
23 54 ☽ ∠ ♃

23 Monday
00 31 ☽ ± ♅
01 20 ☽ ☍ ♆
05 12 ☽ ± ♀
10 22 ☽ Q ♄
10 42 ☽ ⊻ ♂
12 28 ☽ ⚻ ♅
17 09 ☽ △ ♇
17 41 ☿ ⊻ ♀
22 00 ☽ ∥ ♅
22 56 ☽ bQ ♀

24 Tuesday
05 19 ☽ ✶ ♃
06 45 ☽ # ♂
10 13 ☽ ✶ ♄
17 05 ☉ ✶ ♆
22 34 ☽ ∥ ♂

25 Wednesday
03 55 ☽ △ ♀
04 23 ☽ Q ♃
06 29 ☽ # ♅
08 33 ♂ ☍ ♅
09 11 ☽ □ ☿
12 09 ☽ ⚻ ♆
13 48 ☽ □ ☉
15 22 ☽ ∠ ♄
22 41 ☽ ☍ ♅
23 12 ☽ ☌ ♂
23 33 ☽ ± ♆

26 Thursday
03 15 ☽ □ ♇
08 26 ☽ ⊥ ♄
09 01 ♂ ± ♆
09 03 ☿ ✶ ♆
11 14 ☉ ∠ ♄
14 01 ☽ □ ♃
16 21 ☽ bQ ♆
19 37 ☽ ⊻ ♄
23 36 ☽ Q ☿

27 Friday
02 12 ☽ Q ☉
07 38 ☽ ∥ ♆
11 00 ☽ □ ♀
12 21 ☽ Q ♇
13 04 ☿ ∠ ♄
19 32 ☽ △ ♆

28 Saturday
00 02 ☽ ✶ ☿
01 26 ☽ ✶ ☉
05 14 ☽ ⚻ ♅
07 35 ☽ ⊻ ♂
09 35 ☽ ✶ ♇
15 40 ☽ ± ♅
17 12 ☽ ∥ ♄
18 02 ☽ Q ♀
18 19 ☽ ⊥ ♂
18 56 ☽ △ ♃

29 Sunday
00 58 ☽ ☌ ♄
05 15 ☽ ∠ ☿
05 19 ☽ ∠ ☉
06 27 ☉ ☌ ☿
07 00 ☽ bQ ♅
10 05 ☽ ∠ ♂
11 14 ☽ ∠ ♇
13 54 ☽ ✶ ♀
19 57 ☽ bQ ♃
20 48 ☿ □ ♅
21 29 ☽ ⊥ ☉
22 09 ☽ ⊥ ☿
22 52 ☽ □ ♆

30 Monday
00 12 ☽ ∥ ♀
02 11 ☽ ⊥ ♇
05 05 ☉ □ ♅
07 51 ☽ △ ♅
08 03 ☽ ⊻ ☉
09 11 ☽ ⊻ ☿
10 34 ☽ ± ♃
11 36 ☽ ✶ ♂
11 59 ☽ ⊻ ♇
13 59 ☽ ∠ ♀
20 10 ☽ ⚻ ♃

31 Tuesday
01 22 ♂ □ ♇
02 35 ☽ ⊻ ♄
04 04 ☽ Q ♆
04 06 ☽ ⊥ ♀
05 04 ♀ ⊥ ♆
07 28 ☽ Q ♂
11 26 ☿ ☌ ♇
12 11 ☽ ⊥ ♄
13 24 ☽ ⊻ ♀
14 59 ☿ □ ♂
23 06 ☽ ✶ ♆

All ephemeris data is given at 12.00 UT and the Moon's longitude is additionally given for 24.00 UT

JANUARY 2014

LONGITUDES

Date	Sidereal time h m s	Sun ☉ ° ' "	Moon ☽ ° ' "	Moon ☽ 24.00 ° ' "	Mercury ☿ ° '	Venus ♀ ° '	Mars ♂ ° '	Jupiter ♃ ° '	Saturn ♄ ° '	Uranus ♅ ° '	Neptune ♆ ° '	Pluto ♇ ° '
01	18 44 15	10 ♑ 59 22	11 ♑ 26 42	19 ♑ 06 53	12 ♑ 53	26 ♑ 41	11 ♎ 52	16 ♋ 03	20 ♏ 24	08 ♈ 41	03 ♓ 14	11 ♑ 16
02	18 48 11	12 00 33	26 ♑ 46 53	04 ♒ 25 15	14 31	26 R 15	12 18	15 R 55	20 30	08 42	03 16	11 19
03	18 52 08	13 01 43	12 ♒ 00 34	19 ♒ 31 35	16 08	25 47	12 45	15 47	20 35	08 43	03 18	11 21
04	18 56 04	14 02 54	26 ♒ 57 15	04 ♓ 16 45	17 46	25 17	13 10	15 39	20 40	08 44	03 19	11 23
05	19 00 01	15 04 04	11 ♓ 29 29	18 ♓ 35 06	19 24	24 45	13 36	15 31	20 45	08 44	03 21	11 25
06	19 03 57	16 05 14	25 ♓ 33 28	02 ♈ 24 39	21 03	24 12	14 02	15 22	20 50	08 45	03 22	11 27
07	19 07 54	17 06 24	09 ♈ 08 52	15 ♈ 46 27	22 41	23 38	14 27	15 14	20 55	08 46	03 24	11 29
08	19 11 51	18 07 33	22 ♈ 17 52	28 ♈ 43 38	24 21	23 02	14 52	15 06	21 00	08 48	03 26	11 31
09	19 15 47	19 08 41	05 ♉ 04 16	11 ♉ 20 22	26 00	22 26	15 16	14 58	21 05	08 49	03 28	11 33
10	19 19 44	20 09 49	17 ♉ 32 29	23 ♉ 41 11	27 40	21 49	15 41	14 50	21 10	08 50	03 29	11 35
11	19 23 40	21 10 57	29 ♉ 46 58	05 ♊ 50 19	29 ♑ 20	21 13	16 05	14 42	21 15	08 51	03 31	11 38
12	19 27 37	22 12 04	11 ♊ 51 42	17 ♊ 51 30	01 ♒ 00	20 36	16 29	14 34	21 19	08 52	03 33	11 40
13	19 31 33	23 13 11	23 ♊ 50 04	29 ♊ 47 42	02 41	19 59	16 52	14 26	21 24	08 54	03 35	11 42
14	19 35 30	24 14 17	05 ♋ 44 41	11 ♋ 41 13	04 21	19 23	17 16	14 18	21 29	08 55	03 37	11 44
15	19 39 26	25 15 23	17 ♋ 37 32	23 ♋ 33 45	06 02	18 48	17 39	14 10	21 33	08 57	03 39	11 46
16	19 43 23	26 16 28	29 ♋ 30 04	05 ♌ 26 37	07 42	18 14	18 01	14 03	21 37	08 58	03 40	11 48
17	19 47 20	27 17 32	11 ♌ 23 33	17 ♌ 21 02	09 23	17 41	18 24	13 55	21 42	09 00	03 42	11 50
18	19 51 16	28 18 37	23 ♌ 19 14	29 ♌ 18 21	11 02	17 09	18 46	13 47	21 46	09 01	03 44	11 52
19	19 55 13	29 ♑ 19 40	05 ♍ 18 39	11 ♍ 20 24	12 42	16 39	19 08	13 40	21 50	09 03	03 46	11 54
20	19 59 09	00 ♒ 20 43	17 ♍ 23 55	23 ♍ 29 34	14 21	16 11	19 29	13 32	21 54	09 04	03 48	11 56
21	20 03 06	01 21 46	29 ♍ 37 47	05 ♎ 48 59	15 58	15 45	19 50	13 25	21 58	09 06	03 50	11 58
22	20 07 02	02 22 48	12 ♎ 03 42	18 ♎ 22 25	17 35	15 21	20 11	13 18	22 02	09 08	03 52	12 00
23	20 10 59	03 23 50	24 ♎ 45 42	01 ♏ 14 04	19 10	15 00	20 31	13 10	22 06	09 10	03 54	12 02
24	20 14 55	04 24 52	07 ♏ 48 03	14 ♏ 28 09	20 43	14 40	20 51	13 03	22 09	09 12	03 56	12 04
25	20 18 52	05 25 53	21 ♏ 14 45	28 ♏ 08 13	22 13	14 23	21 11	12 56	22 13	09 13	03 58	12 06
26	20 22 49	06 26 54	05 ♐ 08 43	12 ♐ 16 17	23 41	14 09	21 30	12 49	22 16	09 15	04 00	12 08
27	20 26 45	07 27 54	19 ♐ 30 47	26 ♐ 51 49	25 05	13 57	21 49	12 43	22 20	09 17	04 02	12 10
28	20 30 42	08 28 53	04 ♑ 18 47	11 ♑ 50 50	26 25	13 47	22 08	12 36	22 23	09 19	04 04	12 12
29	20 34 38	09 29 52	19 ♑ 26 52	27 ♑ 05 38	27 40	13 40	22 26	12 29	22 26	09 21	04 07	12 14
30	20 38 35	10 30 50	04 ♒ 45 43	12 ♒ 25 37	28 50	13 36	22 44	12 23	22 30	09 24	04 09	12 16
31	20 42 31	11 ♒ 31 47	20 ♒ 03 50	27 ♒ 38 57	29 ♒ 54	13 ♑ 34	23 ♎ 01	12 ♋ 17	22 ♏ 33	09 ♈ 26	04 ♓ 11	12 ♑ 18

Date	Moon True ☊ ° '	Moon Mean ☊ ° '	Moon ☽ Latitude ° '
01	05 ♏ 28	04 ♏ 14	04 N 37
02	05 R 18	04 11	04 59
03	05 07	04 08	05 00
04	04 58	04 05	04 41
05	04 52	04 02	04 03
06	04 48	03 58	03 12
07	04 46	03 55	02 12
08	04 D 46	03 52	01 N 06
09	04 R 46	03 49	00 S 02
10	04 45	03 46	01 07
11	04 43	03 43	02 08
12	04 37	03 39	03 03
13	04 29	03 36	03 48
14	04 18	03 33	04 24
15	04 04	03 30	04 47
16	03 49	03 27	04 58
17	03 34	03 24	04 57
18	03 20	03 20	04 42
19	03 08	03 17	04 14
20	03 00	03 14	03 35
21	02 54	03 11	02 46
22	02 51	03 08	01 47
23	02 D 50	03 04	00 S 43
24	02 R 50	03 01	00 N 26
25	02 49	02 58	01 36
26	02 47	02 55	02 42
27	02 42	02 52	03 40
28	02 35	02 49	04 25
29	02 25	02 45	04 53
30	02 13	02 42	05 00
31	02 ♏ 02	02 ♏ 39	04 N 46

DECLINATIONS

Date	Sun ☉ ° '	Moon ☽ ° '	Mercury ☿ ° '	Venus ♀ ° '	Mars ♂ ° '	Jupiter ♃ ° '	Saturn ♄ ° '	Uranus ♅ ° '	Neptune ♆ ° '	Pluto ♇ ° '
01	22 S 59	18 S 21	24 S 41	18 S 08	02 S 41	22 N 36	15 S 45	02 N 49	10 S 57	20 S 13
02	22 54	15 54	24 34	17 57	02 51	22 37	15 46	02 49	10 56	20 13
03	22 48	12 23	24 25	17 47	03 00	22 38	15 48	02 49	10 55	20 13
04	22 42	08 07	24 15	17 37	03 10	22 39	15 49	02 50	10 55	20 13
05	22 35	03 S 30	24 04	17 28	03 19	22 40	15 50	02 50	10 54	20 13
06	22 28	01 N 11	23 51	17 18	03 28	22 41	15 51	02 51	10 54	20 13
07	22 20	05 38	23 36	17 10	03 37	22 42	15 52	02 51	10 53	20 13
08	22 13	09 42	23 20	17 01	03 46	22 43	15 54	02 52	10 52	20 13
09	22 04	13 11	23 02	16 53	03 55	22 44	15 55	02 52	10 52	20 13
10	21 55	15 59	22 42	16 46	04 04	22 45	15 56	02 53	10 51	20 13
11	21 46	18 01	22 21	16 38	04 13	22 46	15 57	02 53	10 50	20 13
12	21 36	19 11	21 59	16 32	04 21	22 47	15 58	02 54	10 50	20 12
13	21 26	19 29	21 35	16 25	04 29	22 48	15 59	02 54	10 49	20 12
14	21 16	18 55	21 09	16 19	04 38	22 49	16 00	02 55	10 48	20 12
15	21 05	17 31	20 42	16 14	04 46	22 50	16 01	02 55	10 48	20 12
16	20 54	15 23	20 14	16 09	04 54	22 51	16 02	02 56	10 47	20 12
17	20 42	12 36	19 44	16 05	05 02	22 52	16 03	02 57	10 46	20 12
18	20 30	09 18	19 12	16 01	05 09	22 53	16 04	02 57	10 46	20 12
19	20 17	05 37	18 40	15 57	05 17	22 54	16 05	02 58	10 45	20 12
20	20 04	01 N 40	18 06	15 54	05 24	22 55	16 06	02 59	10 44	20 12
21	19 51	02 S 23	17 31	15 52	05 31	22 56	16 06	02 59	10 43	20 12
22	19 38	06 25	16 54	15 50	05 39	22 57	16 07	03 00	10 43	20 11
23	19 24	10 15	16 17	15 48	05 46	22 57	16 08	03 01	10 42	20 11
24	19 09	13 42	15 40	15 47	05 52	22 58	16 09	03 02	10 41	20 11
25	18 55	16 32	15 01	15 47	05 59	22 59	16 10	03 02	10 41	20 11
26	18 39	18 30	14 23	15 46	06 06	23 00	16 10	03 03	10 40	20 11
27	18 24	19 22	13 44	15 46	06 12	23 01	16 11	03 04	10 39	20 11
28	18 08	18 57	13 05	15 47	06 18	23 01	16 12	03 05	10 38	20 11
29	17 52	17 12	12 27	15 48	06 24	23 02	16 12	03 06	10 37	20 11
30	17 36	14 13	11 50	15 49	06 30	23 03	16 13	03 06	10 37	20 11
31	17 S 19	10 S 16	11 S 14	15 S 50	06 S 36	23 N 03	16 S 13	03 N 07	10 S 36	20 S 10

ZODIAC SIGN ENTRIES

Date	h	m	Planets
02	17	03	☽ ♒
04	16	58	☽ ♓
06	19	45	☽ ♈
09	02	24	☽ ♉
11	12	26	☽ ♊
11	21	35	☿ ♒
14	00	25	☽ ♋
16	13	00	☽ ♌
19	01	23	☽ ♍
20	03	51	☉ ♒
21	12	43	☽ ♎
23	21	43	☽ ♏
26	03	13	☽ ♐
28	05	04	☽ ♑
30	04	33	☽ ♒
31	14	29	☿ ♓

LATITUDES

Date	Mercury ☿ ° '	Venus ♀ ° '	Mars ♂ ° '	Jupiter ♃ ° '	Saturn ♄ ° '	Uranus ♅ ° '	Neptune ♆ ° '	Pluto ♇ ° '
01	01 S 53	02 N 44	02 N 11	00 N 07	02 N 10	00 S 41	00 S 40	02 N 45
04	02 01	03 31	02 13	00 08	02 11	00 41	00 40	02 44
07	02 06	04 16	02 15	00 08	02 11	00 41	00 40	02 44
10	02 07	04 58	02 17	00 09	02 12	00 41	00 40	02 44
13	02 05	05 35	02 19	00 09	02 12	00 41	00 40	02 43
16	01 57	06 06	02 21	00 10	02 13	00 41	00 40	02 43
19	01 44	06 29	02 23	00 10	02 13	00 40	00 40	02 43
22	01 25	06 46	02 25	00 10	02 14	00 40	00 40	02 43
25	00 58	06 55	02 27	00 11	02 14	00 40	00 40	02 43
28	00 S 24	06 59	02 30	00 11	02 15	00 40	00 40	02 42
31	00 N 17	06 N 57	02 N 32	00 N 11	02 N 15	00 S 40	00 S 40	02 N 42

DATA

Julian Date	2456659
Delta T	+69 seconds
Ayanamsa	24° 03' 20"
Synetic vernal point	05° ♓ 03' 39"
True obliquity of ecliptic	23° 26' 07"

MOON'S PHASES, APSIDES AND POSITIONS ☽

Date	h	m	Phase	Longitude ° '	Eclipse Indicator
01	11	14	●	10 ♑ 57	
08	03	39	☽	17 ♈ 46	
16	04	52	○	25 ♋ 58	
24	05	19	☾	04 ♏ 08	
30	21	39	●	10 ♒ 55	

Day	h	m		
01	20	53	Perigee	
16	01	35	Apogee	
30	09	52	Perigee	
06	05	55	ON	
13	08	12	Max dec	19° N 30'
20	21	56	OS	
27	16	30	Max dec	19° S 23'

LONGITUDES

Date	Chiron ⚷ ° '	Ceres ⚳ ° '	Pallas ⚴ ° '	Juno ⚵ ° '	Vesta ⚶ ° '	Black Moon Lilith ⚸ ° '
01	10 ♓ 00	22 ♎ 10	13 ♍ 38	00 ♓ 05	18 ♎ 23	23 ♋ 02
11	10 ♓ 25	25 ♎ 02	13 ♍ 37	04 ♓ 35	21 ♎ 35	24 ♋ 09
21	10 ♓ 55	27 ♎ 29	12 ♍ 33	09 ♓ 16	24 ♎ 23	25 ♋ 16
31	11 ♓ 27	29 ♎ 28	10 ♍ 29	14 ♓ 08	26 ♎ 44	26 ♋ 23

All ephemeris data is given at 12.00 UT and the Moon's longitude is additionally given for 24.00 UT

ASPECTARIAN

h m	Aspects
01 Wednesday	
02 28	☽ ∠ ♄
07 40	☽ □ ♅
11 14	☽ ☌ ☉
11 44	☽ ☌ ♇
12 28	♀ Q ♅
12 41	☽ □ ♂
14 32	☽ ☌ ☿
15 05	☽ ∥ ♀
18 57	☉ ☌ ♇
19 09	☽ ☍ ♃
22 39	☽ ∠ ♆
02 Thursday	
02 06	☽ ⚹ ♄
07 42	♂ # ♅
11 12	☽ ☌ ♀
11 52	☽ Q ♅
12 46	☽ ⊥ ♆
12 59	☽ ∥ ♄
21 01	☽ Q ♄
22 12	☽ ⊻ ♆
03 Friday	
00 14	☉ □ ♂
06 46	☽ ⚹ ♅
07 10	☿ ☍ ♃
10 57	☽ ⊻ ♇
13 12	☽ △ ♂
13 44	☽ ⊻ ☉
17 57	☽ ⊼ ♃
19 22	☽ ⊻ ☿
20 31	☽ ⊥ ♇
20 35	☽ ∥ ♆
04 Saturday	
00 01	☽ ⊥ ☉
01 47	☽ □ ♄
03 29	☽ ± ♃
06 11	☽ ⊥ ☿
06 46	☽ ∠ ♅
09 22	☽ ⊻ ♀
11 04	☽ ∠ ♇
14 03	☽ ⚼ ♂
15 40	☽ ∠ ☉
17 58	☽ ⚼ ♃
18 49	☽ ⊥ ♀
20 16	☿ ∠ ♆
20 36	☉ # ♃
21 27	☽ ⊥ ♅
22 26	☽ ☌ ♆
22 42	☽ ∠ ☿
05 Sunday	
05 19	☽ ± ♂
07 24	☽ ⊻ ♅
09 11	☽ ∠ ♀
11 52	☽ ⚹ ♇
12 54	☽ ∥ ♂
15 22	☽ # ♅
15 40	☽ ⊼ ♂
18 29	☽ ⚹ ☉
18 43	☽ △ ♃
21 11	☉ ☍ ♃
06 Monday	
03 10	☽ ⚹ ☿
03 48	☽ △ ♄
08 20	☽ Q ♇
08 51	☿ ⚹ ♄
09 44	☽ ⚹ ♀
16 45	☽ Q ☉
20 48	☽ ∥ ♅
07 Tuesday	
00 33	☽ # ♂
01 44	☽ ⊻ ♆
02 56	☽ Q ☿
05 58	☽ Q ♀
06 12	☽ ⚼ ♄
11 20	☽ ☌ ♅
12 28	☽ ⊥ ♆
16 13	☽ □ ♇
21 54	☽ ☍ ♂
22 02	☿ ☌ ♀
22 31	☽ ± ♄
22 55	☽ □ ♃
08 Wednesday	
03 39	☽ □ ☉
04 51	☽ ∠ ♆
09 36	☽ ⊼ ♄
13 19	☽ □ ♀
16 22	☽ □ ☿
19 25	☉ ∠ ♆
19 36	☽ # ♆
22 24	♂ ⊥ ♄
22 36	♂ □ ♃
09 Thursday	
08 03	☽ Q ♃
08 56	☽ ⚹ ♆
19 09	☽ ⊻ ♅
23 49	☿ Q ♅
10 Friday	
00 27	☽ △ ♇
06 44	☽ ⊥ ♅
06 48	☽ ⚹ ♃
08 00	☽ Q ♆
08 16	☽ ⊼ ♂
08 34	☿ # ♃
09 25	☿ ⊥ ♆
11 26	☽ # ♄
17 34	☽ △ ☉
19 07	☽ ☍ ♄
19 39	☽ # ♀
19 57	☽ △ ♀
20 21	☽ ± ♂
11 Saturday	
00 18	☽ ∠ ♅
05 45	☽ ⚼ ♇
10 44	♀ ⚹ ♄
10 58	☽ △ ☿
11 51	☽ ∠ ♃
12 24	☉ ☌ ♀
13 39	☉ ⚹ ♄
14 40	☽ ⚼ ♂
19 25	☽ □ ♆
23 37	☽ ± ♇
12 Sunday	
00 07	☽ ⚼ ♀
01 51	☽ ⚼ ☉
05 30	☽ ⊥ ♃
06 02	☽ ⚹ ♅
11 36	☽ ⊼ ♇
17 12	☽ ± ♀
17 21	☽ ⊻ ♃
21 29	☽ ± ☉
21 33	☽ △ ♂
21 38	☽ ⚼ ☿
13 Monday	
04 39	☽ ⊼ ♀
06 05	☽ Q ♅
07 05	☽ ⊼ ♄
10 39	☽ ⊼ ☉
18 39	☽ ± ☿
19 13	☽ ± ♄
22 51	☿ Q ♄
14 Tuesday	
01 09	☿ ⊻ ♆
01 39	☉ ∥ ☿
07 41	☽ △ ♆
08 44	☽ ⊼ ☿
13 29	☽ ⚼ ♄
18 25	☽ □ ♅
15 Wednesday	
00 07	☽ ☍ ♇
05 06	☽ ☌ ♃
12 03	☽ □ ♂
14 04	☽ ⚼ ♆
14 16	☽ ☍ ♀
18 09	♀ ∠ ♆
19 59	☽ △ ♄
16 Thursday	
03 56	☽ # ♀
04 52	☽ ☍ ☉
05 32	☽ # ♄
08 18	☽ ± ♆
13 11	☿ ∥ ♇
17 13	♀ □ ♂
20 27	☽ ⊼ ♆
17 Friday	
01 36	☽ Q ♂
04 46	☉ Q ♅
06 25	☿ ⚹ ♅
07 09	☽ △ ♅
07 16	☽ ☍ ☿
12 53	☽ ⊼ ♇
17 02	☽ ⊻ ♃
21 03	♀ ∥ ♄
22 03	☉ ⊥ ♆
18 Saturday	
00 07	☽ ⊼ ♀
01 00	☽ ± ♇
01 46	☽ # ♆
02 33	☽ ⚹ ♂
04 59	☽ ⊥ ♃
08 51	☽ □ ♄
09 56	♂ ⚼ ♆
11 41	☽ ± ♀
13 24	☽ ⚼ ♅
19 08	☽ ⚼ ♇
22 51	☽ ∠ ♃
22 56	☽ ⊼ ☉
19 Sunday	
00 11	☿ ⊻ ♇
04 59	☽ ⚼ ♀
07 28	☽ ± ♅
08 55	☽ ☍ ♆
09 34	☽ ∠ ♂
12 02	☽ ± ☉
14 02	☽ # ♂
19 27	☽ ⊼ ♅
21 03	☽ Q ♄
22 26	☉ ∥ ♇
20 Monday	
01 03	☿ ⊼ ♃
01 09	☽ △ ♇
04 01	☽ ⊥ ♂
04 12	☽ ∥ ♅
04 26	☽ ⚹ ♃
05 01	☽ ⊼ ☿
07 34	☽ ⚼ ☉
09 42	☽ △ ♀
16 15	☽ ⊻ ♂
18 42	☽ ± ☿
20 55	☽ ⚹ ♄
21 Tuesday	
03 51	☽ Q ♃
09 26	☿ ⊻ ♀
15 00	☽ ⚼ ☿
15 34	☽ # ♅
15 40	☽ △ ☉
20 11	☽ ⊼ ♆
22 Wednesday	
02 18	☽ ∠ ♄
06 22	☽ ☍ ♅
07 12	☽ ∥ ♂
07 47	☽ ± ♆
11 53	☽ □ ♇
14 20	☽ □ ♃
18 06	☽ □ ♀
18 29	☿ ⊥ ♇
19 36	☽ ⊥ ♄
23 Thursday	
00 00	☽ △ ☿
00 58	☽ ⚼ ♆
03 50	☽ ☌ ♂
06 59	☽ ⊻ ♄
12 08	☿ ± ♃
14 58	☽ ∥ ♆
17 52	☿ ∥ ♄
21 49	☽ Q ♇
24 Friday	
00 20	☉ ⊻ ♆
02 52	☽ Q ♀
04 56	☽ △ ♆
05 19	☽ □ ☉
05 30	☉ Q ♄
07 03	☿ ∥ ♀
11 30	☿ ⊥ ♀
14 32	☽ ⊼ ♅
14 57	☿ △ ♂
19 43	☽ ⚹ ♇
21 23	☽ △ ♃
25 Saturday	
00 06	☽ ⚹ ♀
00 57	☽ ∥ ☿
01 19	☽ ± ♅
04 59	☽ ∥ ♀
08 26	☽ ∥ ♄
11 54	☽ ⊻ ♂
11 56	☿ □ ♄
13 42	☽ ☌ ♄
13 55	☽ □ ☿
16 08	☽ Q ☉
17 13	☽ ⚼ ♅
22 14	☽ ∠ ♇
22 36	☽ ⊥ ♂
23 33	☽ ⚼ ♃
26 Sunday	
01 55	☽ ∠ ♀
10 03	☽ □ ♆
13 41	☽ ⊥ ♇
14 16	☽ ∥ ☉
14 22	☽ ∠ ♂
14 23	☽ ⚹ ☉
14 49	☽ ± ♃
17 00	☽ ⊥ ♀
18 58	☽ △ ♅
21 57	☿ ∠ ♅
23 48	☽ ⊻ ♇
27 Monday	
00 13	☽ Q ☿
00 49	☽ ⊼ ♃
02 55	☽ ⊻ ♀
15 52	☽ ⚹ ♂
16 09	☽ Q ♆
16 38	☽ ⊻ ♄
17 12	☽ ∠ ☉
22 02	☽ ⚹ ☿
28 Tuesday	
02 26	☽ ⊥ ♄
08 51	☽ ⊥ ☉
11 37	☽ ⚹ ♆
11 42	☽ Q ♂
16 56	☽ ∠ ♄
19 08	☽ ⊻ ☉
20 01	☽ □ ♅
29 Wednesday	
00 22	☽ ∠ ☿
00 35	☽ ☌ ♇
01 06	☽ ☍ ♃
02 57	☽ ☌ ♀
03 06	☿ ∠ ♇
03 29	☽ ∥ ☉
08 33	☉ ⚹ ♅
08 39	☿ ⚼ ♃
11 28	☽ ∠ ♆
12 54	♂ ⊻ ♄
15 48	☽ ⊥ ☿
16 43	☽ ⚹ ♄
16 47	☽ □ ♂
21 01	☽ ∥ ♄
30 Thursday	
00 20	☽ ∥ ♀
00 26	☽ Q ♅
01 37	☽ ⊥ ♆
01 59	☽ ⊻ ☿
07 04	☿ ∠ ♀
11 02	☽ ⊻ ♆
11 35	☽ Q ♄
19 16	☽ ⚹ ♅
21 39	☽ ☌ ☉
23 46	☽ ⊻ ♇
23 51	☽ ⊼ ♃
31 Friday	
01 47	☽ ⊻ ♀
05 36	☽ ∥ ☿
09 12	☽ ⊥ ♇
09 12	☽ ± ♃
09 16	♃ ☍ ♇
10 08	☽ ∥ ♆
11 12	☽ ⊥ ♀
15 56	☽ □ ♄
16 45	☽ △ ♂
18 54	☽ ∠ ♅
20 49	♀ St D
23 20	☽ ⚼ ♃
23 28	☽ ∠ ♇

FEBRUARY 2014

Date	Sidereal time h m s	LONGITUDES Sun ☉ ° ' "	Moon ☽ ° ' "	Moon ☽ 24.00 ° ' "	Mercury ☿ ° '	Venus ♀ ° '	Mars ♂ ° '	Jupiter ♃ ° '	Saturn ♄ ° '	Uranus ♅ ° '	Neptune ♆ ° '	Pluto ⯓ ° '
01	20 46 28	12 ♒ 32 42	05 ♓ 09 40	12 ♓ 34 52	00 ♓ 50	13 ♑ 34	23 ♎ 18	12 ♋ 11	22 ♏ 36	09 ♈ 28	04 ♓ 13	12 ♑ 20
02	20 50 24	13 33 37	19 ♓ 53 42	27 ♓ 05 30	01 39	13 D 37	23 34	12 R 05	22 39	09 30	04 15	12 21
03	20 54 21	14 34 30	04 ♈ 09 52	11 ♈ 06 40	02 19	13 42	23 50	11 59	22 41	09 32	04 17	12 23
04	20 58 18	15 35 22	17 ♈ 55 54	24 ♈ 37 49	02 49	13 49	24 05	11 53	22 44	09 35	04 19	12 25
05	21 02 14	16 36 12	01 ♉ 12 44	07 ♉ 41 09	03 10	13 59	24 20	11 48	22 47	09 37	04 22	12 27
06	21 06 11	17 37 01	14 ♉ 03 36	20 ♉ 20 40	03 19	14 10	24 35	11 42	22 49	09 39	04 24	12 29
07	21 10 07	18 37 49	26 ♉ 33 00	02 ♊ 41 14	03 R 18	14 24	24 49	11 37	22 52	09 42	04 26	12 31
08	21 14 04	19 38 35	08 ♊ 45 59	14 ♊ 47 53	03 06	14 40	25 02	11 32	22 54	09 44	04 28	12 32
09	21 18 00	20 39 20	20 ♊ 47 30	26 ♊ 45 23	02 44	14 58	25 15	11 27	22 56	09 47	04 30	12 34
10	21 21 57	21 40 03	02 ♋ 42 02	08 ♋ 37 55	02 10	15 18	25 28	11 23	22 58	09 49	04 33	12 36
11	21 25 53	22 40 45	14 ♋ 33 25	20 ♋ 28 54	01 28	15 40	25 40	11 18	23 00	09 52	04 35	12 38
12	21 29 50	23 41 25	26 ♋ 24 42	02 ♌ 21 03	00 ♓ 37	16 04	25 51	11 14	23 02	09 54	04 37	12 39
13	21 33 47	24 42 03	08 ♌ 18 12	14 ♌ 16 20	29 ♒ 39	16 29	26 02	11 10	23 04	09 57	04 39	12 41
14	21 37 43	25 42 40	20 ♌ 15 37	26 ♌ 15 58	28 35	16 56	26 12	11 06	23 06	10 00	04 42	12 43
15	21 41 40	26 43 16	02 ♍ 18 11	08 ♍ 21 44	27 28	17 25	26 22	11 02	23 07	10 02	04 44	12 44
16	21 45 36	27 43 50	14 ♍ 26 58	20 ♍ 34 01	26 19	17 55	26 31	10 58	23 09	10 05	04 46	12 46
17	21 49 33	28 44 22	26 ♍ 43 04	02 ♎ 54 19	25 11	18 27	26 40	10 55	23 10	10 08	04 48	12 47
18	21 53 29	29 ♒ 44 54	09 ♎ 07 59	15 ♎ 24 20	24 04	19 01	26 48	10 52	23 11	10 10	04 51	12 49
19	21 57 26	00 ♓ 45 24	21 ♎ 43 38	28 ♎ 06 15	23 00	19 35	26 55	10 49	23 13	10 13	04 53	12 51
20	22 01 22	01 45 52	04 ♏ 32 30	11 ♏ 02 45	22 01	20 11	27 02	10 46	23 14	10 16	04 55	12 52
21	22 05 19	02 46 20	17 ♏ 37 24	24 ♏ 16 47	21 08	20 49	27 08	10 43	23 15	10 19	04 58	12 54
22	22 09 16	03 46 46	01 ♐ 01 15	07 ♐ 51 03	20 21	21 27	27 13	10 41	23 16	10 22	05 00	12 55
23	22 13 12	04 47 11	14 ♐ 46 26	21 ♐ 47 28	19 41	22 07	27 18	10 38	23 16	10 25	05 02	12 57
24	22 17 09	05 47 34	28 ♐ 54 08	06 ♑ 06 15	19 08	22 48	27 22	10 36	23 17	10 28	05 04	12 58
25	22 21 05	06 47 56	13 ♑ 23 28	20 ♑ 45 16	18 43	23 30	27 26	10 34	23 18	10 31	05 07	12 59
26	22 25 02	07 48 17	28 ♑ 10 55	05 ♒ 39 31	18 24	24 13	27 28	10 33	23 18	10 34	05 09	13 01
27	22 28 58	08 48 36	13 ♒ 10 03	20 ♒ 41 21	18 14	24 58	27 30	10 31	23 19	10 37	05 11	13 02
28	22 32 55	09 ♓ 48 53	28 ♒ 12 12	05 ♓ 41 23	18 ♒ 10	25 ♑ 43	27 ♎ 31	10 ♋ 30	23 ♏ 19	10 ♈ 40	05 ♓ 14	13 ♑ 04

Date	Moon True ☊ ° '	Moon Mean ☊ ° '	Moon ☽ Latitude ° '
01	01 ♏ 52	02 ♏ 36	04 N 12
02	01 R 44	02 33	03 21
03	01 38	02 29	02 20
04	01 36	02 26	01 12
05	01 35	02 23	00 N 02
06	01 D 35	02 20	01 S 06
07	01 R 35	02 17	02 08
08	01 33	02 14	03 04
09	01 29	02 10	03 50
10	01 22	02 07	04 25
11	01 13	02 04	04 49
12	01 01	02 01	05 01
13	00 48	01 58	04 59
14	00 35	01 55	04 45
15	00 23	01 51	04 17
16	00 13	01 48	03 38
17	00 05	01 45	02 48
18	00 00	01 42	01 49
19	29 ♎ 58	01 39	00 S 44
20	29 D 58	01 35	00 N 24
21	29 59	01 32	01 33
22	29 59	01 29	02 38
23	29 R 59	01 26	03 36
24	29 57	01 23	04 23
25	29 52	01 20	04 54
26	29 46	01 16	05 06
27	29 38	01 13	04 58
28	29 ♎ 30	01 ♏ 10	04 N 29

DECLINATIONS

Date	Sun ☉ ° '	Moon ☽ ° '	Mercury ☿ ° '	Venus ♀ ° '	Mars ♂ ° '	Jupiter ♃ ° '	Saturn ♄ ° '	Uranus ♅ ° '	Neptune ♆ ° '	Pluto ⯓ ° '
01	17 S 02	05 S 43	10 S 40	15 S 52	06 S 41	23 N 04	16 S 14	03 N 08	10 S 35	20 S 10
02	16 45	00 S 55	10 08	15 54	06 47	23 05	16 15	03 09	10 34	20 10
03	16 27	03 N 48	09 38	15 56	06 52	23 05	16 15	03 10	10 34	20 10
04	16 10	08 08	09 12	15 58	06 57	23 06	16 16	03 11	10 33	20 10
05	15 51	11 55	08 49	16 01	07 02	23 07	16 16	03 12	10 32	20 10
06	15 33	15 01	08 29	16 03	07 06	23 07	16 16	03 13	10 31	20 10
07	15 14	17 18	08 14	16 06	07 11	23 08	16 17	03 14	10 30	20 10
08	14 55	18 44	08 04	16 08	07 15	23 08	16 17	03 15	10 30	20 10
09	14 36	19 18	07 57	16 11	07 19	23 09	16 18	03 16	10 29	20 09
10	14 17	18 59	07 56	16 14	07 23	23 09	16 18	03 17	10 28	20 09
11	13 57	17 51	07 59	16 17	07 27	23 10	16 18	03 18	10 27	20 09
12	13 37	15 57	08 07	16 19	07 30	23 10	16 18	03 19	10 26	20 09
13	13 17	13 22	08 19	16 22	07 33	23 11	16 19	03 20	10 26	20 09
14	12 57	10 14	08 34	16 24	07 37	23 11	16 19	03 21	10 25	20 09
15	12 36	06 39	08 53	16 27	07 39	23 12	16 19	03 22	10 24	20 09
16	12 16	02 N 46	09 14	16 29	07 42	23 12	16 19	03 23	10 23	20 09
17	11 55	01 S 15	09 38	16 31	07 45	23 12	16 19	03 24	10 22	20 08
18	11 34	05 17	10 02	16 33	07 47	23 13	16 20	03 25	10 21	20 08
19	11 12	09 09	10 27	16 34	07 49	23 13	16 20	03 27	10 21	20 08
20	10 51	12 39	10 52	16 36	07 51	23 13	16 20	03 28	10 20	20 08
21	10 29	15 36	11 17	16 37	07 52	23 14	16 20	03 29	10 19	20 08
22	10 07	17 46	11 41	16 38	07 54	23 14	16 20	03 30	10 18	20 08
23	09 45	18 59	12 04	16 38	07 55	23 14	16 20	03 31	10 17	20 08
24	09 23	19 03	12 25	16 39	07 56	23 14	16 20	03 32	10 16	20 08
25	09 01	17 53	12 44	16 38	07 57	23 15	16 20	03 33	10 16	20 08
26	08 38	15 31	13 02	16 38	07 57	23 15	16 20	03 35	10 15	20 08
27	08 16	12 06	13 18	16 37	07 58	23 15	16 20	03 36	10 14	20 07
28	07 S 53	07 S 53	13 S 32	16 S 36	07 S 58	23 N 15	16 S 19	03 N 37	10 S 13	20 S 07

ZODIAC SIGN ENTRIES

Date	h	m	Planets
01	03	44	☽ ♓
03	04	55	☽ ♈
05	09	46	☽ ♉
07	18	44	☽ ♊
10	06	33	☽ ♋
12	19	15	☽ ♌
13	03	30	☿ ♒
15	07	25	☽ ♍
17	18	22	☽ ♎
18	17	59	☉ ♓
20	03	33	☽ ♏
22	10	12	☽ ♐
24	13	50	☽ ♑
26	14	55	☽ ♒
28	14	52	☽ ♓

LATITUDES

Date	Mercury ☿ ° '	Venus ♀ ° '	Mars ♂ ° '	Jupiter ♃ ° '	Saturn ♄ ° '	Uranus ♅ ° '	Neptune ♆ ° '	Pluto ⯓ ° '
01	00 N 33	06 N 55	02 N 32	00 N 12	02 N 15	00 S 40	00 S 40	02 N 42
04	01 22	06 47	02 35	00 12	02 16	00 40	00 40	02 42
07	02 12	06 36	02 37	00 12	02 17	00 40	00 40	02 42
10	02 58	06 22	02 39	00 12	02 17	00 40	00 40	02 42
13	03 30	06 06	02 41	00 13	02 18	00 40	00 40	02 42
16	03 43	05 48	02 43	00 13	02 18	00 40	00 40	02 41
19	03 35	05 29	02 44	00 13	02 19	00 39	00 40	02 41
22	03 11	05 09	02 46	00 14	02 20	00 39	00 40	02 41
25	02 36	04 49	02 48	00 14	02 20	00 39	00 40	02 41
28	01 56	04 28	02 49	00 14	02 21	00 39	00 40	02 41
31	01 N 16	04 N 07	02 N 50	00 N 14	02 N 22	00 S 39	00 S 40	02 N 41

DATA

Julian Date	2456690
Delta T	+69 seconds
Ayanamsa	24° 03' 25"
Synetic vernal point	05° ♓ 03' 34"
True obliquity of ecliptic	23° 26' 07"

MOON'S PHASES, APSIDES AND POSITIONS ☽

Date	h	m	Phase	Longitude ° '	Eclipse Indicator
06	19	22	☽	17 ♉ 56	
14	23	53	○	26 ♌ 13	
22	17	15	☾	04 ♐ 00	

Day	h	m		
12	05	00	Apogee	
27	19	43	Perigee	
02	16	35	0N	
09	15	21	Max dec	19° N 18'
17	04	33	0S	
24	01	24	Max dec	19° S 10'

LONGITUDES

Date	Chiron ⚷ ° '	Ceres ⚳ ° '	Pallas ⚴ ° '	Juno ⚵ ° '	Vesta ⚶ ° '	Black Moon Lilith ⚸ ° '
01	11 ♓ 31	29 ♎ 39	10 ♍ 13	14 ♓ 37	26 ♎ 56	26 ♋ 30
11	12 ♓ 06	01 ♏ 01	07 ♍ 17	19 ♓ 38	28 ♎ 39	27 ♋ 37
21	12 ♓ 43	01 ♏ 46	03 ♍ 57	24 ♓ 47	29 ♎ 42	28 ♋ 45
31	13 ♓ 21	01 ♏ 49	00 ♍ 44	00 ♈ 03	29 ♎ 59	29 ♋ 52

All ephemeris data is given at 12.00 UT and the Moon's longitude is additionally given for 24.00 UT

ASPECTARIAN

h m	Aspects
01 Saturday	
01 27	☽ ∠ ♀
04 05	☉ ⊼ ♃
04 38	☽ ☌ ☿
06 39	☉ ⊻ ⯓
07 07	☽ ∥ ♂
09 16	☽ ⊥ ♅
10 29	☽ ☌ ♆
15 37	☿ ∥ ♆
17 09	☽ ♇ ♂
18 57	☽ ⊻ ♅
23 16	☽ △ ♃
23 37	☽ ⚹ ⯓
02 Sunday	
00 49	☽ ⊻ ☉
00 52	☽ # ♅
01 38	☽ ⚹ ♀
08 05	☽ ± ♂
11 24	☽ ⊥ ☉
13 15	☉ ⊻ ♀
16 34	☽ △ ♄
18 13	☽ ⊼ ♂
19 26	☽ Q ⯓
21 34	☽ Q ♀
03 Monday	
03 35	☽ ∠ ☉
08 42	☽ ⊻ ☿
08 43	☽ ∥ ♅
12 13	☽ ⊻ ♆
18 05	☽ ♇ ♄
19 26	☽ ⊥ ☿
21 17	☽ ☌ ♅
22 36	☽ ⊥ ♆
04 Tuesday	
01 26	☽ □ ♃
02 16	☽ □ ⯓
04 16	☉ ∥ ♄
04 40	☽ □ ♀
05 00	☽ # ♂
07 31	☽ ⚹ ☉
09 52	☽ ± ♄
11 48	☽ ∠ ☿
14 29	☽ ∠ ♆
17 44	☽ # ☿
20 37	☽ ⊼ ♄
23 14	☽ ☍ ♂
05 Wednesday	
01 29	☉ ∥ ♀
02 44	☽ # ♆
06 49	☽ Q ☉
09 25	☽ Q ♃
15 39	☽ ⚹ ☿
17 50	☽ ⚹ ♆
06 Thursday	
03 40	☽ ⊻ ♅
07 35	☽ ⚹ ♃
09 00	☽ △ ⯓
12 13	☽ △ ♀
13 56	☉ ± ♃
14 25	☽ Q ☿
15 02	☽ ⊥ ♅
16 25	☽ # ☉
16 27	☽ Q ♆
19 22	☽ □ ☉
21 43	☿ St R
22 03	☽ # ♀
07 Friday	
00 11	☽ # ♄
04 49	☽ ☍ ♄
08 23	☽ ∠ ♅
08 34	☽ ⊼ ♂
09 02	☉ ⊥ ⯓
12 08	☽ ∠ ♃
13 52	☽ ♇ ⯓
17 41	☽ ♇ ♀
20 29	☽ ± ♂
08 Saturday	
01 03	☽ □ ☿
03 29	☽ □ ♆
05 39	☽ ⊥ ♃
07 35	☽ ± ⯓
11 48	☽ ± ♀
13 56	☽ ⚹ ♅
14 34	☽ ♇ ♂
17 28	☽ ⊻ ♃
19 31	☽ ⊼ ⯓
09 Sunday	
00 02	☽ ⊼ ♀
11 42	☽ △ ☉
13 59	☽ Q ♅
16 19	☽ ⊼ ♄
21 08	☽ △ ♂
23 01	☉ ⊥ ♀
10 Monday	
04 27	☽ ± ♄
10 59	☽ △ ☿
15 45	☽ △ ♆
20 46	☽ ♇ ☉
22 41	☽ ♇ ♄
11 Tuesday	
02 27	☽ □ ♅
05 27	☽ ☌ ♃
08 05	☽ ☍ ⯓
14 20	☽ ☍ ♀
15 37	☽ ♇ ☿
16 42	☽ ± ☉
19 57	☉ □ ♄
22 13	☽ ♇ ♆
12 Wednesday	
03 40	☿ ∠ ♀
03 56	♀ ∥ ♄
05 09	☽ △ ♄
05 59	☽ ⊼ ☉
07 57	☽ # ♀
08 01	☽ # ♄
08 37	☽ ± ☿
10 51	☽ □ ♂
16 29	☽ ± ♆
19 52	☽ ⊼ ☿
13 Thursday	
04 38	☽ ⊼ ♆
12 45	☽ # ☉
15 19	☽ △ ♅
17 43	☽ ⊻ ♃
18 09	☉ ∠ ♅
20 50	☽ ⊼ ⯓
23 42	☽ Q ♂
14 Friday	
05 05	☽ ⊼ ♀
05 42	☽ ⊥ ♃
08 53	☽ ± ⯓
10 40	☽ # ♆
17 34	☽ ± ♀
17 41	☽ □ ♄
20 35	☉ ♇ ♃
21 29	☽ ♇ ♅
22 31	☽ # ☿
23 35	☽ ∠ ♃
23 53	☽ ☍ ☉
15 Saturday	
00 02	☽ ⚹ ♂
02 05	☉ △ ♂
02 54	☽ ♇ ⯓
03 13	☽ ☍ ☿
05 34	☽ # ♂
06 30	☿ ∠ ⯓
12 14	☽ ♇ ♀
15 27	☽ ± ♅
16 50	☽ ☍ ♆
20 22	☉ ☌ ☿
16 Sunday	
03 22	☽ ⊼ ♅
05 11	☽ ⚹ ♃
05 29	☽ Q ♄
06 10	☽ ∠ ♂
08 18	☽ ∥ ♅
08 19	☿ △ ♂
08 41	☽ △ ⯓
12 49	☉ ∠ ⯓
19 08	☽ △ ♀
19 43	☿ ♇ ♃
17 Monday	
00 03	☽ ⊥ ♂
04 37	☽ Q ♃
05 04	☽ ⚹ ♄
09 15	☽ ⊼ ☿
11 54	☽ ⊻ ♂
13 00	☿ ∠ ♅
16 17	☽ ⊼ ☉
19 56	☽ ± ☿
22 21	☿ ⊥ ♀
18 Tuesday	
00 47	☽ # ♅
03 43	☽ ⊼ ♆
04 55	☽ ± ☉
10 11	☽ ∠ ♄
11 52	☽ ♇ ☿
14 00	☽ ☍ ♅
15 18	☽ ± ♆
15 18	☽ □ ♃
19 04	☽ □ ⯓
23 41	☽ ♇ ☉
19 Wednesday	
03 25	☽ ⊥ ♄
03 28	☽ ∥ ♂
06 00	☿ ∥ ♆
07 11	☿ □ ♄
07 45	☽ □ ♀
08 30	☽ ♇ ♆
14 13	☽ △ ☿
14 48	☽ ⊻ ♄
19 51	☽ ∥ ♆
21 45	☽ ∥ ☿
21 52	☽ ☌ ♂
20 Thursday	
00 27	☽ ∥ ☉
00 41	♀ ∠ ♆
05 09	☽ Q ⯓
06 24	☽ △ ☉
11 12	☉ ∥ ☿
12 42	☽ △ ♆
19 05	☽ Q ♀
22 37	☽ ⊼ ♅
23 27	☽ △ ♃
21 Friday	
03 22	☽ ⚹ ⯓
09 37	☽ ± ♅
17 08	☿ ⊻ ♀
17 58	☽ □ ☿
18 04	☽ ⚹ ♀
19 07	☽ ∥ ♄
22 08	☽ ∥ ♀
22 10	☽ ☌ ♄
23 33	☉ ∥ ♆
22 Saturday	
01 54	☽ ♇ ♅
02 32	☽ ♇ ♃
05 13	☽ ⊻ ♂
06 29	☽ ∠ ⯓
15 55	☽ ⊥ ♂
17 15	☽ □ ☉
18 26	☽ ± ♃
19 02	☽ □ ♆
22 03	☽ ∠ ♀
22 23	☽ ⊥ ⯓
23 Sunday	
00 14	☽ Q ☿
02 38	☉ ⊥ ♅
04 26	☽ △ ♅
04 52	☽ ⊼ ♃
07 43	☽ ∠ ♂
08 50	☽ ⊻ ⯓
14 26	☽ ⊥ ♀
18 11	☉ ☌ ♆
20 04	☽ ⚹ ☿
24 Monday	
01 12	☽ ⊻ ♀
02 09	☽ Q ♆
02 32	☽ ⊻ ♄
02 44	☽ Q ☉
09 25	☽ ⚹ ♂
12 39	☽ ⊥ ♄
20 06	☿ ⊥ ⯓
20 27	☽ ∠ ☿
22 19	☽ ⚹ ♆
25 Tuesday	
00 20	☽ ⚹ ☉
03 37	☽ ∠ ♄
04 50	♀ ⚹ ♄
05 28	☽ Q ♂
07 15	☽ □ ♅
07 23	☽ ☍ ♃
10 54	☽ ⊥ ☿
11 21	☽ ☌ ⯓
20 29	☽ ⊻ ☿
22 59	☽ ∠ ♆
26 Wednesday	
02 05	☽ ∥ ♀
02 41	☽ ∠ ☉
04 08	☽ ⚹ ♄
05 00	☽ ∥ ♄
05 17	☽ ☌ ♀
07 29	♃ □ ♅
10 51	☽ □ ♂
12 37	☽ Q ♅
13 34	☽ ⊥ ♆
18 15	☽ ⊥ ☉
23 13	☽ ⊻ ♆
23 26	☽ Q ♄
27 Thursday	
04 32	☽ ⊻ ☉
04 45	☽ ∥ ☿
07 47	☽ ⊼ ♃
07 54	☽ ⚹ ♅
11 47	☽ ⊻ ⯓
17 21	☽ ± ♃
20 01	☽ ☌ ☿
21 23	☽ ⊥ ⯓
23 07	☽ ∥ ♆
28 Friday	
04 11	☽ □ ♄
07 20	☉ ∥ ♂
07 41	☽ ♇ ♃
07 48	☽ ⊻ ♀
07 55	☽ ∠ ♅
10 55	☽ △ ♂
11 35	☽ ∥ ♂
11 46	☽ ∠ ⯓
11 59	☽ ∥ ☉
14 01	☿ St D
17 55	☽ ⊥ ♀
22 23	☽ ⊥ ♅
23 17	☽ ☌ ♆

LONGITUDES

Date	Sidereal time h m s	Sun ☉ ° ' "	Moon ☽ ° ' "	Moon ☽ 24.00 ° ' "	Mercury ☿ ° '	Venus ♀ ° '	Mars ♂ ° '	Jupiter ♃ ° '	Saturn ♄ ° '	Uranus ♅ ° '	Neptune ♆ ° '	Pluto ⯓ ° '
01	22 36 51	10 ♓ 49 09	13 ♓ 07 43	20 ♓ 30 06	18 ♒ 12	26 ♑ 29	27 ♎ 32	10 ♋ 29	23 ♏ 19	10 ♈ 43	05 ♓ 16	13 ♑ 05
02	22 40 48	11 49 23	27 ♓ 47 37	04 ♈ 59 27	18 D 21	27 16	27 R 32	10 R 28	23 19	10 46	05 18	13 06
03	22 44 45	12 49 35	12 ♈ 05 02	19 ♈ 03 58	18 36	28 03	27 31	10 27	23 R 19	10 49	05 20	13 07
04	22 48 41	13 49 45	25 ♈ 56 03	02 ♉ 41 14	18 57	28 52	27 29	10 27	23 19	10 52	05 23	13 09
05	22 52 38	14 49 53	09 ♉ 19 41	15 ♉ 51 37	19 22	29 ♑ 41	27 27	10 27	23 19	10 55	05 25	13 10
06	22 56 34	15 49 59	22 ♉ 17 27	28 ♉ 37 38	19 53	00 ♒ 31	27 23	10 D 27	23 18	10 58	05 27	13 11
07	23 00 31	16 50 03	04 ♊ 52 42	11 ♊ 03 13	20 28	01 22	27 19	10 27	23 18	11 02	05 29	13 12
08	23 04 27	17 50 05	17 ♊ 09 47	23 ♊ 13 03	21 07	02 14	27 15	10 27	23 17	11 05	05 32	13 13
09	23 08 24	18 50 05	29 ♊ 13 36	05 ♋ 12 04	21 51	03 06	27 09	10 27	23 17	11 08	05 34	13 14
10	23 12 20	19 50 02	11 ♋ 09 03	17 ♋ 05 05	22 37	03 59	27 03	10 28	23 16	11 11	05 36	13 16
11	23 16 17	20 49 58	23 ♋ 00 43	28 ♋ 56 27	23 28	04 52	26 56	10 29	23 16	11 15	05 38	13 17
12	23 20 14	21 49 51	04 ♌ 52 43	10 ♌ 49 56	24 21	05 46	26 48	10 30	23 14	11 18	05 41	13 18
13	23 24 10	22 49 42	16 ♌ 48 27	22 ♌ 48 37	25 18	06 41	26 39	10 31	23 13	11 21	05 43	13 19
14	23 28 07	23 49 31	28 ♌ 50 40	04 ♍ 54 51	26 17	07 36	26 30	10 33	23 12	11 24	05 45	13 20
15	23 32 03	24 49 18	11 ♍ 01 22	17 ♍ 10 20	27 19	08 32	26 20	10 35	23 11	11 28	05 47	13 20
16	23 36 00	25 49 03	23 ♍ 21 54	29 ♍ 36 10	28 24	09 28	26 09	10 36	23 09	11 31	05 49	13 21
17	23 39 56	26 48 45	05 ♎ 53 11	12 ♎ 13 03	29 ♒ 30	10 25	25 57	10 38	23 08	11 34	05 51	13 22
18	23 43 53	27 48 26	18 ♎ 35 49	25 ♎ 01 33	00 ♓ 39	11 22	25 45	10 41	23 07	11 38	05 54	13 23
19	23 47 49	28 48 05	01 ♏ 30 19	08 ♏ 02 10	01 51	12 20	25 32	10 43	23 05	11 41	05 56	13 24
20	23 51 46	29 ♓ 47 42	14 ♏ 37 13	21 ♏ 15 32	03 04	13 18	25 18	10 46	23 03	11 44	05 58	13 25
21	23 55 43	00 ♈ 47 18	27 ♏ 57 12	04 ♐ 42 20	04 19	14 16	25 03	10 48	23 01	11 48	06 00	13 26
22	23 59 39	01 46 51	11 ♐ 30 59	18 ♐ 23 13	05 36	15 15	24 48	10 51	22 59	11 51	06 02	13 26
23	00 03 36	02 46 23	25 ♐ 19 03	02 ♑ 18 27	06 55	16 15	24 32	10 55	22 57	11 55	06 04	13 27
24	00 07 32	03 45 54	09 ♑ 21 21	16 ♑ 27 33	08 15	17 15	24 15	10 58	22 55	11 58	06 06	13 28
25	00 11 29	04 45 22	23 ♑ 36 49	00 ♒ 48 49	09 37	18 15	23 58	11 01	22 53	12 01	06 08	13 28
26	00 15 25	05 44 49	08 ♒ 03 05	15 ♒ 19 05	11 01	19 15	23 40	11 05	22 51	12 05	06 10	13 29
27	00 19 22	06 44 14	22 ♒ 36 12	29 ♒ 53 43	12 27	20 16	23 22	11 09	22 49	12 08	06 12	13 30
28	00 23 18	07 43 37	07 ♓ 10 52	14 ♓ 26 52	13 54	21 17	23 03	11 13	22 46	12 12	06 14	13 30
29	00 27 15	08 42 58	21 ♓ 40 56	28 ♓ 52 17	15 22	22 19	22 43	11 17	22 44	12 15	06 16	13 31
30	00 31 12	09 42 18	06 ♈ 00 12	13 ♈ 04 04	16 52	23 21	22 23	11 22	22 41	12 18	06 18	13 31
31	00 35 08	10 ♈ 41 35	20 ♈ 03 20	26 ♈ 57 33	18 ♓ 24	24 ♒ 23	22 ♎ 02	11 ♋ 26	22 ♏ 38	12 ♈ 22	06 ♓ 20	13 ♑ 32

Moon Nodes and Latitude; DECLINATIONS

Date	Moon True ☊ ° '	Moon Mean ☊ ° '	Moon ☽ Latitude ° '	Sun ☉ ° '	Moon ☽ ° '	Mercury ☿ ° '	Venus ♀ ° '	Mars ♂ ° '	Jupiter ♃ ° '	Saturn ♄ ° '	Uranus ♅ ° '	Neptune ♆ ° '	Pluto ⯓ ° '
01	29 ♎ 22	01 ♏ 07	03 N 42	07 S 30	03 S 12	13 S 44	16 S 35	07 S 57	23 N 15	16 S 19	03 N 38	10 S 12	20 S 07
02	29 R 17	01 04	02 41	07 08	01 N 35	13 55	16 33	07 57	23 16	16 19	03 40	10 12	20 07
03	29 13	01 01	01 31	06 45	06 10	14 03	16 31	07 56	23 16	16 19	03 41	10 11	20 07
04	29 12	00 57	00 N 18	06 22	10 17	14 09	16 28	07 55	23 16	16 19	03 42	10 10	20 07
05	29 D 12	00 54	00 S 55	05 58	13 44	14 13	16 25	07 54	23 16	16 18	03 43	10 09	20 07
06	29 13	00 51	02 02	05 35	16 23	14 16	16 21	07 53	23 16	16 18	03 45	10 08	20 07
07	29 15	00 48	03 01	05 12	18 08	14 16	16 17	07 51	23 16	16 18	03 46	10 07	20 07
08	29 16	00 45	03 51	04 48	18 59	14 15	16 13	07 49	23 16	16 18	03 47	10 07	20 07
09	29 R 15	00 41	04 29	04 25	18 57	14 12	16 08	07 47	23 16	16 17	03 48	10 06	20 06
10	29 12	00 38	04 55	04 02	18 04	14 08	16 03	07 45	23 16	16 17	03 50	10 05	20 06
11	29 08	00 35	05 08	03 38	16 24	14 01	15 57	07 42	23 16	16 16	03 51	10 04	20 06
12	29 03	00 32	05 09	03 14	14 03	13 53	15 51	07 40	23 16	16 16	03 52	10 03	20 06
13	28 56	00 29	04 55	02 51	11 06	13 44	15 44	07 37	23 16	16 15	03 53	10 02	20 06
14	28 50	00 26	04 29	02 27	07 40	13 33	15 37	07 33	23 16	16 15	03 55	10 02	20 06
15	28 43	00 22	03 50	02 03	03 N 53	13 20	15 30	07 30	23 16	16 15	03 56	10 01	20 06
16	28 38	00 19	03 00	01 40	00 S 07	13 06	15 22	07 26	23 16	16 14	03 57	10 00	20 06
17	28 35	00 16	02 01	01 16	04 11	12 50	15 13	07 22	23 16	16 14	03 59	09 59	20 06
18	28 33	00 13	00 S 54	00 52	08 07	12 33	15 04	07 18	23 16	16 13	04 00	09 59	20 06
19	28 D 32	00 10	00 N 16	00 29	11 44	12 14	14 55	07 14	23 16	16 12	04 01	09 58	20 06
20	28 33	00 07	01 27	00 S 05	14 50	11 54	14 45	07 09	23 16	16 12	04 03	09 57	20 06
21	28 34	00 03	02 34	00 N 19	17 12	11 33	14 35	07 04	23 15	16 11	04 04	09 56	20 06
22	28 36	00 ♏ 00	03 34	00 42	18 37	11 10	14 24	06 59	23 15	16 11	04 05	09 56	20 06
23	28 37	29 ♎ 57	04 23	01 06	18 58	10 46	14 13	06 54	23 15	16 10	04 07	09 55	20 06
24	28 R 37	29 54	04 57	01 30	18 10	10 20	14 01	06 48	23 15	16 09	04 08	09 54	20 05
25	28 36	29 51	05 13	01 53	16 14	09 54	13 49	06 43	23 15	16 09	04 09	09 53	20 05
26	28 34	29 47	05 10	02 17	13 15	09 26	13 36	06 37	23 14	16 08	04 11	09 53	20 05
27	28 31	29 44	04 47	02 40	09 27	08 56	13 23	06 31	23 14	16 07	04 12	09 52	20 05
28	28 28	29 41	04 06	03 04	05 04	08 26	13 09	06 25	23 14	16 06	04 13	09 51	20 05
29	28 26	29 38	03 09	03 27	00 S 24	07 54	12 55	06 19	23 14	16 05	04 15	09 50	20 05
30	28 24	29 35	02 00	03 51	04 N 14	07 21	12 41	06 12	23 13	16 05	04 16	09 50	20 05
31	28 ♎ 23	29 ♎ 32	00 N 46	04 N 14	08 N 33	06 S 47	12 S 26	06 S 06	23 N 13	16 S 04	04 N 17	09 S 49	20 S 05

ZODIAC SIGN ENTRIES

Date	h	m	Planets
02	15	40	☽ ♈
04	19	12	☽ ♉
05	21	03	♀ ♒
07	02	37	☽ ♊
09	13	33	☽ ♋
12	02	09	☽ ♌
14	14	17	☽ ♍
17	00	46	☽ ♎
17	22	24	☿ ♓
19	09	13	☽ ♏
20	16	57	☉ ♈
21	15	39	☽ ♐
23	20	03	☽ ♑
25	22	39	☽ ♒
28	00	10	☽ ♓
30	01	54	☽ ♈

LATITUDES

Date	Mercury ☿ ° '	Venus ♀ ° '	Mars ♂ ° '	Jupiter ♃ ° '	Saturn ♄ ° '	Uranus ♅ ° '	Neptune ♆ ° '	Pluto ⯓ ° '
01	01 N 43	04 N 21	02 N 49	00 N 14	02 N 21	00 S 39	00 S 40	02 N 41
04	01 03	04 00	02 50	00 14	02 22	00 39	00 40	02 41
07	00 N 25	03 39	02 51	00 15	02 22	00 39	00 40	02 41
10	00 S 10	03 18	02 51	00 15	02 23	00 39	00 41	02 41
13	00 41	02 57	02 51	00 15	02 23	00 39	00 41	02 41
16	01 08	02 37	02 51	00 15	02 24	00 39	00 41	02 41
19	01 31	02 16	02 50	00 16	02 25	00 39	00 41	02 41
22	01 50	01 57	02 49	00 16	02 25	00 39	00 41	02 41
25	02 05	01 38	02 47	00 16	02 26	00 39	00 41	02 41
28	02 16	01 19	02 44	00 16	02 26	00 39	00 41	02 41
31	02 S 23	01 N 01	02 N 41	00 N 16	02 N 27	00 S 39	00 S 41	02 N 41

DATA

Julian Date	2456718
Delta T	+69 seconds
Ayanamsa	24° 03' 28"
Synetic vernal point	05° ♓ 03' 31"
True obliquity of ecliptic	23° 26' 07"

MOON'S PHASES, APSIDES AND POSITIONS ☽

Date	h	m	Phase	Longitude ° '	Eclipse Indicator
01	08	00	●	10 ♓ 39	
08	13	27	☽	17 ♊ 54	
16	17	08	○	26 ♍ 02	
24	01	46	☾	03 ♑ 21	
30	18	45	●	09 ♈ 59	

Day	h	m		
11	19	43	Apogee	
27	18	24	Perigee	
02	04	01	0N	
08	22	55	Max dec	19° N 05'
16	11	16	0S	
23	07	26	Max dec	19° S 00'
29	14	05	0N	

LONGITUDES

Date	Chiron ⚷ ° '	Ceres ⚳ ° '	Pallas ⚴ ° '	Juno ⚵ ° '	Vesta ⚶ ° '	Black Moon Lilith ⚸ ° '
01	13 ♓ 14	01 ♏ 52	01 ♍ 21	28 ♓ 59	00 ♏ 00	29 ♋ 38
11	13 ♓ 51	01 ♏ 22	28 ♌ 31	04 ♈ 20	29 ♎ 38	00 ♌ 46
21	14 ♓ 29	00 ♏ 10	26 ♌ 26	09 ♈ 46	28 ♎ 23	01 ♌ 53
31	15 ♓ 05	28 ♎ 24	25 ♌ 16	15 ♈ 17	26 ♎ 36	03 ♌ 00

All ephemeris data is given at 12.00 UT and the Moon's longitude is additionally given for 24.00 UT

ASPECTARIAN

h m	Aspects
01 Saturday	
04 05	☉ △ ♃
07 44	☽ △ ♃
08 00	☽ ☌ ☉
08 05	☽ ⊻ ♅
09 11	☽ ∠ ♀
09 20	☉ ⊻ ♅
09 51	☽ # ♅
11 02	☽ ♇ ♂
11 55	☽ ⚹ ⯓
16 23	♂ St R
20 19	☽ ⊻ ☿
02 Sunday	
01 41	☽ ± ♂
04 37	☽ △ ♄
06 15	☽ ⊥ ☿
07 33	☽ Q ⯓
11 04	☽ ⚹ ♀
11 34	☽ ⊼ ♂
16 19	♄ St R
20 03	♀ □ ♂
21 24	☽ ∠ ☿
22 40	☽ ∥ ♅
03 Monday	
00 33	☽ ⊻ ♆
04 38	☉ ♇ ♂
05 36	☽ ♇ ♄
08 20	☽ Q ♀
09 14	☽ □ ♃
09 50	☽ ☌ ♅
10 44	☽ ⊥ ♆
13 22	☽ ⊻ ☉
13 47	☽ □ ⯓
14 53	☽ # ☉
19 15	☉ ⚹ ⯓
20 59	☽ ± ♄
21 51	☽ # ♂
23 28	☽ ⚹ ☿
04 Tuesday	
00 29	☽ ⊥ ☉
02 15	☽ ∠ ♆
07 24	☽ ⊼ ♄
11 13	☽ # ♆
12 08	♀ Q ♅
14 44	☽ ☍ ♂
16 26	☽ Q ♃
16 31	☉ ∠ ♀
17 31	☽ □ ♀
17 32	☽ ∠ ☉
21 10	☽ Q ☿
05 Wednesday	
00 15	☿ ⊥ ⯓
03 43	♀ ⊥ ♆
04 53	☽ ⚹ ♆
14 02	☽ ⚹ ♃
14 55	☽ ⊻ ♅
15 58	☽ # ☿
19 02	☽ △ ⯓
22 56	☽ ⚹ ☉
06 Thursday	
02 01	☽ ⊥ ♅
02 56	☽ Q ♆
07 17	☽ □ ☿
10 42	♃ St D
11 12	☽ # ♄
11 46	☽ # ♀
13 55	☽ ☍ ♄
17 57	☽ ∠ ♃
18 59	☽ ∠ ♅
21 36	☽ ⊼ ♂
23 10	☽ ♇ ⯓
23 24	☽ Q ☉
07 Friday	
04 45	☽ △ ♀
09 01	☽ ± ♂
09 08	♀ ∥ ♄
11 10	☽ ⊥ ♃
13 11	☽ □ ♆
16 31	☽ ± ⯓
22 49	☽ ⊻ ♃
08 Saturday	
00 00	☽ ⚹ ♅
02 23	☽ ♇ ♂
04 14	☽ ⊼ ⯓
12 08	☽ ♇ ♀
13 27	☽ □ ☉
20 19	☽ △ ☿
23 47	☽ Q ♅
09 Sunday	
00 08	☽ ⊼ ♄
07 24	☽ ± ♀
07 53	☽ △ ♂
12 06	☽ ± ♄
20 23	☽ ⊼ ♀
10 Monday	
00 46	☽ △ ♆
04 22	☽ ♇ ☿
06 11	☽ ♇ ♄
10 37	☽ ☌ ♃
12 05	☽ □ ♅
16 16	☽ ☍ ⯓
23 32	☉ # ♅
23 54	☽ ± ☿
11 Tuesday	
06 14	☿ □ ♄
07 11	☽ ♇ ♆
07 11	☽ △ ☉
12 29	☽ △ ♄
12 59	☽ ⊼ ☿
13 32	☽ # ♄
14 00	☉ ± ♂
17 23	☽ # ♀
19 50	☽ □ ♂
22 07	♀ Q ♄
12 Wednesday	
01 27	☽ ± ♆
09 25	♀ ⊻ ♆
13 29	☽ # ☿
13 37	☽ ⊼ ♆
13 57	☽ ☍ ♀
16 18	☽ ♇ ☉
23 21	☽ ⊻ ♃
13 Thursday	
01 00	☽ △ ♅
04 58	☽ ⊼ ⯓
07 44	☽ Q ♂
11 26	☽ ⊥ ♃
12 03	☽ ± ☉
17 01	☽ ± ⯓
17 44	☿ ♇ ♃
19 44	☽ # ♆
21 15	☉ △ ♄
14 Friday	
00 48	☽ □ ♄
01 07	☽ ⊼ ☉
05 26	☽ ∠ ♃
06 27	☽ ☍ ☿
07 08	☽ ♇ ♅
07 24	☽ ⚹ ♂
10 58	☽ ♇ ⯓
12 44	☽ # ♂
15 02	☿ ∠ ♅
16 17	☿ △ ♂
15 Saturday	
01 02	☽ ± ♅
01 41	☽ ☍ ♆
06 42	☽ ⊼ ♀
11 07	☽ ⚹ ♃
11 39	☽ ∥ ♅
12 18	☽ Q ♄
12 35	☽ ∠ ♂
12 52	☽ ⊼ ♅
16 32	☽ △ ⯓
19 25	☽ ± ♀
16 Sunday	
00 14	☽ # ☉
00 44	☉ Q ⯓
05 52	☽ ⊥ ♂
10 32	☽ Q ♃
11 13	☿ ∠ ⯓
11 36	☽ ⚹ ♄
14 18	☽ ♇ ♀
17 08	☽ ☍ ☉
17 17	☽ ⊻ ♂
18 38	☉ ⊼ ♂
20 16	☽ ∥ ☉
22 37	☽ ⊼ ☿
17 Monday	
10 46	☽ # ♅
11 12	☽ ± ☿
11 57	☽ ⊼ ♆
16 16	☽ ∠ ♄
18 01	♀ ⊼ ♃
21 03	☽ □ ♃
21 17	☽ △ ♀
22 50	☽ ☍ ♅
23 21	☽ ± ♆
18 Tuesday	
02 12	☽ □ ⯓
05 56	☽ ♇ ☿
06 57	☽ ∥ ♂
09 13	☽ ⊥ ♄
16 18	☽ ♇ ♆
19 00	♀ ⚹ ♅
20 25	☽ ⊻ ♄
23 55	☽ ∥ ♆
19 Wednesday	
01 07	☽ ☌ ♂
06 35	☽ ⊼ ☉
11 48	☽ Q ⯓
12 41	☽ △ ☿
15 14	☽ ∥ ☿
18 34	☽ ± ☉
20 10	☽ △ ♆
20 Thursday	
04 57	☽ △ ♃
06 44	☽ ⊼ ♅
09 24	☽ □ ♀
09 48	☽ ⚹ ⯓
11 19	☽ ∥ ♀
12 21	☽ ♇ ☉
14 56	♀ ⊻ ⯓
17 41	☽ ± ♅
21 Friday	
00 45	☽ ∥ ♄
03 11	☽ ☌ ♄
06 54	☽ ⊻ ♂
08 09	☽ ♇ ♃
09 55	☽ ♇ ♅
12 51	☽ ∠ ⯓
17 25	☽ ⊥ ♂
17 27	☽ △ ☉
20 17	☽ Q ♀
22 Saturday	
00 14	☽ ± ♃
00 29	☽ □ ☿
02 19	☽ □ ♆
04 49	☽ ⊥ ⯓
09 02	☽ ∠ ♂
10 50	☽ ⊼ ♃
12 36	☽ △ ♅
15 22	☽ ⊻ ⯓
16 55	☿ ⊥ ♅
19 03	☽ ⚹ ♀
20 16	☿ ☌ ♆
23 Sunday	
07 56	☽ ⊻ ♄
09 50	☽ Q ♆
10 40	☽ ⚹ ♂
11 14	☽ Q ☿
18 15	☽ ⊥ ♄
22 58	☽ ∠ ♀
24 Monday	
01 46	☽ □ ☉
04 58	♀ ± ♃
06 28	☽ ⚹ ♆
06 50	☽ Q ♂
09 34	☽ ∠ ♄
09 56	☽ ⚹ ☿
14 44	☽ ☍ ♃
15 26	☽ ⊥ ♀
16 26	☽ □ ♅
18 57	☽ ☌ ⯓
25 Tuesday	
02 19	☽ ⊻ ♀
02 33	☿ ♇ ♂
07 51	☽ ∠ ♆
10 28	☽ Q ☉
10 47	☽ ⚹ ♄
12 13	☿ ∥ ♆
12 35	☽ □ ♂
12 50	☽ ∥ ♄
13 52	☽ ∠ ☿
22 44	☽ Q ♅
22 54	☽ ⊥ ♆
26 Wednesday	
06 26	☽ ⊥ ☿
06 43	☽ Q ♄
07 54	☽ ⚹ ☉
08 53	☽ ⊻ ♆
09 24	☽ ∥ ♀
13 11	☿ △ ♃
17 02	☽ ⊼ ♃
17 26	☽ ⊻ ☿
17 29	♀ ⊥ ⯓
18 41	☽ ⚹ ♅
20 59	☽ ⊻ ⯓
22 41	☉ ⊻ ♆
27 Thursday	
02 59	☽ ± ♃
06 40	☿ ⊻ ♅
06 52	☽ ⊥ ⯓
07 52	☽ ☌ ♀
09 34	☽ ∥ ♆
10 28	☽ ∠ ☉
12 20	☽ □ ♄
13 13	☽ △ ♂
15 20	☽ ∥ ☿
17 52	☽ ♇ ♃
19 29	☽ ∠ ♅
21 42	☽ ∠ ⯓
28 Friday	
02 22	☽ ⊥ ☉
04 42	☽ ∥ ♂
05 32	☿ ⚹ ⯓
10 22	☽ ⊥ ♅
10 27	☽ ☌ ♆
12 58	☉ ♇ ♄
12 58	☽ ⊻ ☉
13 23	☽ ♇ ♂
16 24	☽ # ♅
18 42	☽ △ ♃
20 18	☽ ⊻ ♅
21 36	☽ # ☉
22 26	☽ ⚹ ⯓
29 Saturday	
00 20	☽ ☌ ☿
03 56	☽ ± ♂
11 04	♂ ⊻ ♄
13 08	☽ ⊻ ♀
13 41	☽ ⊼ ♂
13 44	☽ △ ♄
18 23	☽ Q ⯓
19 05	♀ △ ♂
21 14	♀ □ ♄
23 55	☽ ⊥ ♀
30 Sunday	
05 40	☿ ± ♂
09 48	☽ ∥ ☉
10 48	♅ ⊥ ♆
12 14	☽ ∥ ♅
12 31	☽ ⊻ ♆
14 50	☽ ♇ ♄
16 16	☽ ∠ ♀
18 45	☽ ☌ ☉
21 08	☽ □ ♃
22 25	☽ # ♂
22 44	☽ ⊥ ♆
22 45	☽ ☌ ♅
31 Monday	
00 47	☽ □ ⯓
03 06	☽ # ☿
06 08	☽ ± ♄
08 47	☽ ⊻ ☿
14 14	☽ ∠ ♆
15 21	☽ ☍ ♂
15 48	☉ ∥ ♅
16 27	☽ ⊼ ♄
19 38	☽ # ♆
20 07	☽ ⚹ ♀
20 28	☽ ⊥ ☿

APRIL 2014

LONGITUDES

Date	Sidereal time h m s	Sun ☉ ° ' "	Moon ☽ ° ' "	Moon ☽ 24.00 ° ' "	Mercury ☿ ° '	Venus ♀ ° '	Mars ♂ ° '	Jupiter ♃ ° '	Saturn ♄ ° '	Uranus ♅ ° '	Neptune ♆ ° '	Pluto ♇ ° '
01	00 39 05	11 ♈ 40 50	03 ♉ 46 26	10 ♉ 29 46	19 ♓ 56	25 ♒ 25	21 ♎ 41	11 ♋ 31	22 ♏ 35	12 ♈ 25	06 ♓ 22	13 ♑ 32
02	00 43 01	12 40 03	17 ♉ 07 31	23 ♉ 39 42	21 31	26 28	21 R 20	11 36	22 R 33	12 29	06 24	13 32
03	00 46 58	13 39 14	00 ♊ 06 30	06 ♊ 28 09	23 07	27 31	20 58	11 41	22 30	12 32	06 26	13 33
04	00 50 54	14 38 23	12 ♊ 44 59	18 ♊ 57 25	24 44	28 34	20 36	11 46	22 27	12 36	06 28	13 33
05	00 54 51	15 37 29	25 ♊ 05 54	01 ♋ 10 57	26 23	29 ♒ 37	20 14	11 52	22 24	12 39	06 30	13 33
06	00 58 47	16 36 33	07 ♋ 13 06	13 ♋ 12 55	28 03	00 ♓ 41	19 51	11 57	22 20	12 42	06 32	13 34
07	01 02 44	17 35 35	19 ♋ 11 00	25 ♋ 07 55	29 ♓ 45	01 45	19 28	12 03	22 17	12 46	06 34	13 34
08	01 06 41	18 34 35	01 ♌ 04 16	07 ♌ 00 36	01 ♈ 28	02 49	19 05	12 09	22 14	12 49	06 35	13 34
09	01 10 37	19 33 32	12 ♌ 57 29	18 ♌ 55 28	03 12	03 54	18 43	12 15	22 10	12 53	06 37	13 34
10	01 14 34	20 32 27	24 ♌ 55 02	00 ♍ 56 39	04 59	04 58	18 19	12 21	22 07	12 56	06 39	13 35
11	01 18 30	21 31 20	07 ♍ 00 44	13 ♍ 07 42	06 46	06 03	17 56	12 27	22 04	13 00	06 41	13 35
12	01 22 27	22 30 10	19 ♍ 17 50	25 ♍ 31 27	08 35	07 08	17 33	12 34	22 00	13 03	06 43	13 35
13	01 26 23	23 28 58	01 ♎ 48 45	08 ♎ 09 53	10 26	08 13	17 11	12 40	21 56	13 06	06 44	13 35
14	01 30 20	24 27 45	14 ♎ 34 59	21 ♎ 04 04	12 18	09 18	16 48	12 47	21 53	13 10	06 46	13 35
15	01 34 16	25 26 29	27 ♎ 37 07	04 ♏ 14 05	14 12	10 24	16 25	12 54	21 49	13 13	06 48	13 R 35
16	01 38 13	26 25 11	10 ♏ 54 51	17 ♏ 39 14	16 07	11 30	16 03	13 01	21 45	13 16	06 49	13 35
17	01 42 10	27 23 52	24 ♏ 27 02	01 ♐ 18 02	18 04	12 36	15 41	13 08	21 41	13 20	06 51	13 35
18	01 46 06	28 22 30	08 ♐ 11 59	15 ♐ 08 34	20 02	13 42	15 19	13 16	21 37	13 23	06 52	13 35
19	01 50 03	29 ♈ 21 07	22 ♐ 07 31	29 ♐ 08 32	22 02	14 48	14 57	13 23	21 33	13 26	06 54	13 35
20	01 53 59	00 ♉ 19 43	06 ♑ 11 19	13 ♑ 15 33	24 03	15 54	14 36	13 31	21 29	13 30	06 56	13 34
21	01 57 56	01 18 16	20 ♑ 20 56	27 ♑ 27 10	26 05	17 01	14 15	13 38	21 25	13 33	06 57	13 34
22	02 01 52	02 16 48	04 ♒ 33 57	11 ♒ 40 58	28 ♈ 09	18 08	13 55	13 46	21 21	13 36	06 59	13 34
23	02 05 49	03 15 19	18 ♒ 47 54	25 ♒ 54 26	00 ♉ 14	19 15	13 35	13 54	21 17	13 40	07 00	13 34
24	02 09 45	04 13 47	03 ♓ 00 15	10 ♓ 05 00	02 20	20 22	13 16	14 02	21 13	13 43	07 01	13 33
25	02 13 42	05 12 15	17 ♓ 08 20	24 ♓ 09 55	04 28	21 29	12 57	14 11	21 09	13 46	07 03	13 33
26	02 17 39	06 10 40	01 ♈ 09 23	08 ♈ 06 23	06 36	22 36	12 39	14 19	21 04	13 49	07 04	13 33
27	02 21 35	07 09 04	15 ♈ 00 34	21 ♈ 51 39	08 44	23 43	12 21	14 27	21 00	13 53	07 06	13 32
28	02 25 32	08 07 26	28 ♈ 39 18	05 ♉ 23 17	10 53	24 51	12 04	14 36	20 56	13 56	07 07	13 32
29	02 29 28	09 05 47	12 ♉ 03 22	18 ♉ 39 23	13 02	25 59	11 48	14 45	20 51	13 59	07 08	13 32
30	02 33 25	10 ♉ 04 05	25 ♉ 11 15	01 ♊ 38 54	15 ♉ 11	27 ♓ 06	11 ♎ 32	14 ♋ 54	20 ♏ 47	14 ♈ 02	07 ♓ 10	13 ♑ 31

Date	Moon True ☊ ° '	Moon Mean ☊ ° '	Moon ☽ Latitude ° '
01	28 ♎ 23	29 ♎ 28	00 S 30
02	28 D 23	29 25	01 42
03	28 24	29 22	02 47
04	28 26	29 19	03 41
05	28 27	29 16	04 25
06	28 28	29 13	04 55
07	28 R 28	29 09	05 12
08	28 27	29 06	05 16
09	28 27	29 03	05 06
10	28 26	29 00	04 44
11	28 25	28 57	04 08
12	28 24	28 53	03 20
13	28 23	28 50	02 22
14	28 23	28 47	01 16
15	28 23	28 44	00 S 04
16	28 D 23	28 41	01 N 09
17	28 23	28 38	02 20
18	28 23	28 34	03 23
19	28 R 23	28 31	04 16
20	28 23	28 28	04 54
21	28 23	28 25	05 14
22	28 D 23	28 22	05 15
23	28 23	28 18	04 57
24	28 23	28 15	04 21
25	28 24	28 12	03 29
26	28 24	28 09	02 25
27	28 25	28 06	01 N 14
28	28 R 25	28 03	00 S 01
29	28 25	27 59	01 15
30	28 ♎ 24	27 ♎ 56	02 S 23

DECLINATIONS

Date	Sun ☉ ° '	Moon ☽ ° '	Mercury ☿ ° '	Venus ♀ ° '	Mars ♂ ° '	Jupiter ♃ ° '	Saturn ♄ ° '	Uranus ♅ ° '	Neptune ♆ ° '	Pluto ♇ ° '
01	04 N 37	12 N 18	06 S 12	12 S 11	05 S 59	23 N 13	16 S 03	04 N 19	09 S 48	20 S 05
02	05 00	15 19	05 35	11 55	05 52	23 12	16 02	04 20	09 48	20 05
03	05 23	17 27	04 58	11 39	05 45	23 12	16 01	04 21	09 47	20 05
04	05 46	18 40	04 19	11 22	05 39	23 11	16 00	04 23	09 46	20 05
05	06 09	18 56	03 39	11 05	05 32	23 11	16 00	04 24	09 46	20 05
06	06 32	18 20	02 58	10 48	05 25	23 11	15 59	04 25	09 45	20 05
07	06 54	16 54	02 16	10 30	05 18	23 10	15 58	04 27	09 44	20 05
08	07 17	14 46	01 33	10 12	05 11	23 10	15 57	04 28	09 44	20 05
09	07 39	12 01	00 49	09 54	05 04	23 09	15 56	04 29	09 43	20 05
10	08 01	08 45	00 S 04	09 35	04 57	23 09	15 55	04 31	09 42	20 05
11	08 23	05 06	00 N 42	09 16	04 50	23 08	15 54	04 32	09 42	20 05
12	08 45	01 N 10	01 28	08 57	04 43	23 08	15 53	04 33	09 41	20 05
13	09 07	02 S 54	02 16	08 37	04 36	23 07	15 52	04 35	09 41	20 05
14	09 29	06 55	03 04	08 17	04 29	23 06	15 51	04 36	09 40	20 05
15	09 50	10 41	03 54	07 56	04 23	23 06	15 50	04 37	09 39	20 05
16	10 12	14 00	04 44	07 35	04 16	23 05	15 49	04 39	09 39	20 05
17	10 33	16 37	05 34	07 14	04 10	23 04	15 48	04 40	09 38	20 05
18	10 54	18 19	06 26	06 53	04 04	23 04	15 46	04 41	09 38	20 05
19	11 15	18 57	07 17	06 31	03 57	23 03	15 45	04 42	09 37	20 05
20	11 35	18 24	08 10	06 10	03 52	23 02	15 44	04 44	09 36	20 05
21	11 56	16 43	09 02	05 47	03 46	23 02	15 43	04 45	09 36	20 05
22	12 16	14 01	09 55	05 25	03 40	23 01	15 42	04 46	09 35	20 05
23	12 36	10 28	10 48	05 02	03 35	23 00	15 41	04 48	09 35	20 05
24	12 56	06 20	11 41	04 39	03 30	22 59	15 40	04 49	09 34	20 05
25	13 15	01 S 52	12 33	04 16	03 25	22 59	15 39	04 50	09 34	20 05
26	13 35	02 N 41	13 25	03 53	03 20	22 58	15 38	04 51	09 33	20 05
27	13 54	07 03	14 17	03 29	03 16	22 57	15 36	04 53	09 33	20 06
28	14 13	10 58	15 07	03 06	03 11	22 56	15 35	04 54	09 32	20 06
29	14 32	14 16	15 57	02 42	03 08	22 55	15 34	04 55	09 32	20 06
30	14 N 50	16 N 45	16 N 46	02 S 18	03 S 04	22 N 54	15 S 33	04 N 56	09 S 32	20 S 06

ZODIAC SIGN ENTRIES

Date	h	m	Planets
01	05	20	☽ ♉
03	11	48	☽ ♊
05	20	31	♀ ♓
05	21	40	☽ ♋
07	15	35	☿ ♈
08	09	50	☽ ♌
10	22	08	☽ ♍
13	08	33	☽ ♎
15	16	20	☽ ♏
17	21	44	☽ ♐
20	01	28	☽ ♑
20	03	55	☉ ♉
22	04	18	☽ ♒
23	09	16	☿ ♉
24	06	55	☽ ♓
26	10	01	☽ ♈
28	14	23	☽ ♉
30	20	56	☽ ♊

LATITUDES

Date	Mercury ☿ ° '	Venus ♀ ° '	Mars ♂ ° '	Jupiter ♃ ° '	Saturn ♄ ° '	Uranus ♅ ° '	Neptune ♆ ° '	Pluto ♇ ° '
01	02 S 24	00 N 55	02 N 40	00 N 16	02 N 27	00 S 39	00 S 41	02 N 40
04	02 25	00 38	02 35	00 17	02 27	00 39	00 41	02 40
07	02 22	00 22	02 31	00 17	02 27	00 39	00 41	02 40
10	02 14	00 N 07	02 25	00 17	02 28	00 39	00 41	02 40
13	02 02	00 S 08	02 19	00 17	02 28	00 39	00 41	02 40
16	01 45	00 22	02 13	00 17	02 28	00 39	00 41	02 40
19	01 23	00 35	02 06	00 17	02 29	00 39	00 41	02 40
22	00 58	00 47	01 58	00 18	02 29	00 39	00 41	02 40
25	00 S 29	00 58	01 51	00 18	02 29	00 39	00 42	02 40
28	00 N 02	01 09	01 43	00 18	02 29	00 39	00 42	02 40
31	00 N 34	01 S 18	01 N 35	00 N 18	02 N 29	00 S 39	00 S 42	02 N 40

DATA

Julian Date	2456749
Delta T	+69 seconds
Ayanamsa	24° 03' 30"
Synetic vernal point	05° ♓ 03' 29"
True obliquity of ecliptic	23° 26' 07"

LONGITUDES

Date	Chiron ⚷ ° '	Ceres ⚳ ° '	Pallas ⚴ ° '	Juno ⚵ ° '	Vesta ⚶ ° '	Black Moon Lilith ⚸ ° '
01	15 ♓ 08	28 ♎ 12	25 ♌ 12	15 ♈ 51	26 ♎ 22	03 ♌ 07
11	15 ♓ 42	26 ♎ 02	25 ♌ 00	21 ♈ 26	23 ♎ 59	04 ♌ 14
21	16 ♓ 12	23 ♎ 46	25 ♌ 37	27 ♈ 06	21 ♎ 29	05 ♌ 21
31	16 ♓ 39	21 ♎ 41	26 ♌ 55	02 ♉ 48	19 ♎ 13	06 ♌ 29

MOON'S PHASES, APSIDES AND POSITIONS ☽

Date	h	m	Phase	Longitude ° '	Eclipse Indicator
07	08	31	☽	17 ♋ 27	
15	07	42	○	25 ♎ 16	total
22	07	52	☾	02 ♒ 07	
29	06	14	●	08 ♉ 52	Annular

Day	h	m		
08	14	51	Apogee	
23	00	18	Perigee	
05	07	14	Max dec	18° N 57'
12	18	56	OS	
19	12	54	Max dec	18° S 57'
25	21	48	ON	

ASPECTARIAN

h m Aspects

01 Tuesday
04 26 ☽ Q ♃
07 39 ☉ □ ♃
11 10 ☽ # ♀
14 21 ☽ ∠ ☿
16 38 ☽ ⚹ ♆
19 02 ☽ Q ♀
22 15 ☿ ∥ ♂

02 Wednesday
01 55 ☽ ⚹ ♃
03 16 ☽ ⊻ ☉
03 32 ☽ ⊻ ♅
05 23 ☉ ⊥ ♆
05 29 ☽ △ ♇
07 08 ☉ ☌ ♅
07 28 ♂ bQ ♆
09 44 ☿ ⊼ ♂
14 20 ☽ Q ♆
14 29 ☽ ⊥ ♅
15 03 ☽ ⊥ ☉
15 20 ♀ bQ ♃
18 58 ☽ # ♄
19 30 ☽ ⊼ ♂
21 09 ☽ ⚹ ☿
21 54 ☽ ☍ ♄

03 Thursday
01 52 ☉ # ☿
03 03 ☿ △ ♄
05 34 ☽ ∠ ♃
06 18 ☽ ± ♂
06 43 ☽ □ ♀
07 10 ☽ ∠ ♅
09 03 ☽ ∠ ☉
09 05 ☽ bQ ♇
09 22 ☉ □ ♇
12 34 ♀ ∠ ♅
22 35 ☽ ⊥ ♃
22 44 ☽ bQ ♂
22 48 ☽ Q ☿
23 58 ☽ □ ♆

04 Friday
02 03 ☽ ± ♇
05 55 ☉ # ♂
09 40 ☿ # ♅
10 06 ☽ ⊻ ♃
11 40 ♀ ∠ ♇
11 42 ☽ ⚹ ♅
13 33 ☽ ⊼ ♇
15 57 ☽ ⚹ ☉

05 Saturday
00 00 ☿ Q ♇
02 45 ☽ △ ♂
06 43 ☽ ⊼ ♄
11 07 ☽ Q ♅
14 55 ☽ □ ☿
17 24 ☽ Q ☉
18 27 ☽ ± ♄
21 46 ☽ △ ♀

06 Sunday
05 46 ☉ ± ♄
10 37 ☽ △ ♆
12 15 ☽ bQ ♄
21 33 ☽ ☌ ♃
23 02 ☽ □ ♅

07 Monday
00 42 ☽ ☍ ♇
06 37 ☽ bQ ♀
08 31 ☽ □ ☉
12 34 ☽ □ ♂
16 48 ☽ bQ ♆
18 14 ☽ △ ♄
23 39 ☽ # ♄

08 Tuesday
02 34 ☽ ± ♀
11 02 ☽ ± ♆
12 56 ☽ △ ☿
15 53 ☽ ⊼ ♀
21 03 ☉ ☍ ♂
23 11 ☽ ⊼ ♆
23 47 ☽ Q ♂

09 Wednesday
08 58 ♀ bQ ♂
10 33 ☽ ⊻ ♃
11 50 ☽ △ ♅
13 14 ☽ ⊼ ♇
22 44 ☽ ⊥ ♃
23 12 ☽ ⚹ ♂

10 Thursday
00 23 ☽ bQ ☿
01 18 ☽ ± ♇
02 28 ☽ △ ☉
02 35 ♀ ∥ ♆
05 19 ☽ # ♆
05 35 ☽ # ♀
06 26 ☽ □ ♄
11 43 ☿ ⊻ ♀
16 32 ☽ ∥ ☉
16 54 ☽ ∠ ♃
18 03 ☽ bQ ♅
19 18 ☽ bQ ♇
21 30 ☽ ± ☿

11 Friday
04 12 ☽ ∠ ♂
09 55 ☽ ☍ ♀
10 47 ☿ ⊻ ♆
10 57 ☽ bQ ☉
11 21 ☽ ☍ ♆
11 26 ☽ ⊼ ☿
11 58 ☽ ± ♅
13 44 ☽ # ♂
15 30 ☽ ∥ ♅
15 42 ☿ bQ ♄
15 59 ☉ ∠ ♆
17 58 ☽ Q ♄
21 23 ☽ ⊥ ♂
22 47 ☽ ⚹ ♃
23 47 ☽ ⊼ ♅

12 Saturday
00 23 ☉ ⊼ ♄
00 53 ☽ △ ♇
02 23 ♀ ☌ ♆
06 07 ☽ ± ☉
08 44 ☽ ⊻ ♂
10 05 ♀ ⊥ ♅
10 29 ☽ ∥ ☿
17 12 ☽ ⚹ ♄
18 32 ☉ # ♀
18 43 ☽ ⊼ ☉
22 15 ☽ Q ♃

13 Sunday
07 24 ☽ # ☿
16 12 ♂ # ♅
21 20 ☽ ⊼ ♆
21 39 ☽ ∠ ♄
21 50 ☽ ∥ ♂
22 01 ☽ # ♅

14 Monday
01 14 ☽ ⊼ ♀
07 01 ☽ ☍ ☿
08 36 ☽ ± ♆
08 37 ☽ □ ♃
09 21 ☽ ☍ ♅
10 08 ☽ □ ♇
13 28 ☽ ± ♀
14 24 ☽ ⊥ ♄
16 00 ☽ ☌ ♂
17 59 ☿ ⊥ ♆
18 32 ☿ □ ♃
19 45 ☽ ∥ ♀
23 15 ☿ ☌ ♅
23 48 ♇ St R

15 Tuesday
00 07 ☉ # ♆
01 19 ☽ bQ ♆
01 26 ☽ ⊻ ♄
04 13 ☿ □ ♇
05 13 ☽ ∥ ♆
05 46 ☽ # ☉
07 35 ☽ bQ ♀
07 42 ☽ ☍ ☉
12 19 ♀ ± ♂
19 12 ☽ Q ♇
20 38 ☉ ∠ ♀

16 Wednesday
00 20 ☿ # ♂
04 39 ☽ △ ♆
07 35 ☿ ± ♄
09 29 ☿ ∥ ♅
11 15 ☿ ☍ ♂
13 08 ☽ △ ♀
15 47 ☽ △ ♃
16 14 ☽ ⊼ ♅
16 45 ☽ ⚹ ♇
20 54 ☽ ⊻ ♂
22 50 ☽ ⊼ ☿

17 Thursday
02 56 ☽ ± ♅
03 35 ☽ ∥ ♄
07 09 ☽ ☌ ♄
07 15 ☽ ⊥ ♂
11 09 ♂ ⊥ ♄
11 12 ☽ ± ☿
17 34 ☽ ⊼ ☉
18 32 ☽ bQ ♃
18 50 ☽ bQ ♅
19 14 ☽ ∠ ♇
22 38 ☽ ∠ ♂

18 Friday
01 19 ♀ △ ♃
02 43 ☿ ⊥ ♀
04 51 ☽ ± ☉
04 54 ♀ ⊻ ♅
05 35 ☽ bQ ☿
09 26 ♀ ⚹ ♇
09 42 ☽ □ ♆
10 21 ☽ ± ♃
10 55 ☽ ⊥ ♇
20 50 ☽ ⊼ ♃
20 59 ☿ # ♀
21 00 ☽ △ ♅
21 18 ☽ ⊻ ♇
21 38 ☽ bQ ☉
22 19 ☽ □ ♀
23 59 ☽ ⚹ ♂

19 Saturday
06 31 ☿ ⊼ ♄
10 26 ☿ ∠ ♆
11 02 ☽ ⊻ ♄
11 48 ☽ △ ☿
14 31 ♀ ⊼ ♂
16 46 ☽ Q ♆
20 03 ☽ Q ♂
21 15 ☽ ⊥ ♄

20 Sunday
01 17 ☽ △ ☉
07 29 ♃ □ ♅
07 47 ☽ Q ♀
12 30 ☽ ∠ ♄
13 15 ☽ ⚹ ♆
23 26 ♃ ☍ ♇

21 Monday
00 27 ☽ □ ♅
00 32 ☽ ☌ ♇
00 32 ☽ ☍ ♃
01 56 ☽ □ ♂
05 53 ☽ ⚹ ♀
13 48 ☽ ⚹ ♄
14 43 ☽ ∠ ♆
19 21 ♅ □ ♇
21 30 ☉ Q ♃
21 56 ☽ ∥ ♄
23 21 ☽ □ ☿

22 Tuesday
03 13 ☿ # ♆
05 56 ☽ ⊥ ♆
06 59 ☽ Q ♅
07 52 ☽ □ ☉
09 22 ☽ ∠ ♀
09 58 ☽ Q ♄
16 04 ☽ ⊻ ♆
19 28 ♂ □ ♃
23 24 ☽ # ☉

23 Wednesday
01 50 ☽ ⊥ ♀
03 10 ☽ ⊻ ♇
03 18 ☽ ⚹ ♅
03 24 ☽ △ ♂
03 40 ☽ ⊼ ♃
07 08 ♂ ☍ ♅
10 22 ☽ # ☿
10 54 ☽ Q ☿
12 49 ☽ ⊻ ♀
13 17 ☽ ⊥ ♇
13 38 ♂ □ ♇
13 53 ☽ ± ♃
16 10 ☽ □ ♄
16 27 ☽ Q ☉
17 26 ☽ ∥ ♆

24 Thursday
02 37 ♀ # ♅
04 09 ☽ bQ ♂
04 29 ☽ ∠ ♇
04 43 ☽ ∠ ♅
05 14 ☽ bQ ♃
08 19 ☿ Q ♃
10 41 ☽ ⚹ ☿
14 14 ☽ ⚹ ☉
18 49 ☽ ☌ ♆
19 03 ☽ ± ♂
20 01 ☽ ⊥ ♅
20 16 ☽ # ♅
22 01 ☽ ∥ ♀

25 Friday
03 39 ☽ ∥ ♂
04 49 ♂ ± ♆
05 01 ☽ ⊼ ♂
05 15 ♀ △ ♄
05 54 ☽ ⚹ ♇
06 15 ☽ ⊻ ♅
06 54 ☽ △ ♃
16 40 ☽ ∠ ☿
17 37 ☽ ∠ ☉
18 48 ☽ △ ♄
20 03 ☽ ☌ ♀

26 Saturday
02 22 ☽ Q ♇
03 27 ☉ ☌ ☿
10 11 ☽ ⊥ ☉
10 51 ☽ ⊥ ☿
15 26 ☽ # ♂
17 27 ☿ ⚹ ♆
17 55 ☽ # ♀
19 03 ☉ ∥ ☿
20 26 ☽ bQ ♄
21 19 ☽ ⊻ ☉
22 14 ☽ ⊻ ♆
23 05 ☽ ⊻ ☿
23 48 ☽ ∥ ♅

27 Sunday
07 28 ☽ ☍ ♂
08 39 ☽ ⊥ ♆
09 26 ☽ □ ♇
10 01 ☽ ☌ ♅
10 34 ☉ ⚹ ♆
11 02 ☽ □ ♃
11 45 ☿ ∠ ♀
11 59 ☽ ± ♄
22 26 ☽ ⊼ ♄

28 Monday
00 26 ☽ ∠ ♆
02 52 ☽ # ♆
04 39 ☽ ⊻ ♀
05 01 ♀ ∥ ♂
16 15 ☽ ⊥ ♀
19 06 ☽ Q ♃
23 44 ☿ ⊼ ♂

29 Tuesday
01 04 ☿ # ♄
02 32 ♀ Q ♇
03 07 ☽ ⚹ ♆
06 14 ☽ ☌ ☉
09 52 ☽ ∠ ♀
11 32 ☽ ⊼ ♂
14 07 ☽ ☌ ☿
14 27 ☽ ∥ ☉
14 40 ☽ △ ♇
15 30 ☽ ⊻ ♅
16 56 ☽ ⚹ ♃
17 30 ☿ △ ♇
22 13 ☽ ± ♂
22 53 ☿ ⊻ ♅
23 28 ☽ # ♄

30 Wednesday
00 54 ☽ Q ♆
02 29 ☽ ⊥ ♅
03 56 ☽ ☍ ♄
08 32 ☿ ⚹ ♃
12 18 ☽ ∥ ☿
14 27 ☽ bQ ♂
15 53 ☽ ⚹ ♀
18 10 ☽ bQ ♇
19 10 ☽ ∠ ♅
20 50 ☽ ∠ ♃

All ephemeris data is given at 12.00 UT and the Moon's longitude is additionally given for 24.00 UT

MAY 2014

LONGITUDES

Date	Sidereal time h m s	Sun ☉ ° ' "	Moon ☽ ° ' "	Moon ☽ 24.00 ° ' "	Mercury ☿ ° '	Venus ♀ ° '	Mars ♂ ° '	Jupiter ♃ ° '	Saturn ♄ ° '	Uranus ♅ ° '	Neptune ♆ ° '	Pluto ♇ ° '
01	02 37 21	11 ♉ 02 22	08 ♊ 02 22	14 ♊ 21 44	17 ♉ 19	28 ♓ 14	11 ♎ 17	15 ♋ 03	20 ♏ 43	14 ♈ 05	07 ♓ 11	13 ♑ 31
02	02 41 18	12 00 37	20 ♊ 37 10	26 ♊ 48 52	19 27	29 ♓ 22	11 R 03	15 12	20 R 38	14 09	07 12	13 R 30
03	02 45 14	12 58 50	02 ♋ 57 09	09 ♋ 02 19	21 34	00 ♈ 30	10 50	15 21	20 34	14 12	07 13	13 30
04	02 49 11	13 57 02	15 ♋ 04 49	21 ♋ 05 03	23 40	01 38	10 37	15 30	20 29	14 15	07 14	13 29
05	02 53 08	14 55 11	27 ♋ 03 31	03 ♌ 00 45	25 44	02 47	10 25	15 40	20 25	14 18	07 16	13 29
06	02 57 04	15 53 18	08 ♌ 57 17	14 ♌ 53 42	27 46	03 55	10 14	15 49	20 20	14 21	07 17	13 28
07	03 01 01	16 51 23	20 ♌ 50 34	26 ♌ 48 30	29 ♉ 45	05 03	10 04	15 59	20 16	14 24	07 18	13 27
08	03 04 57	17 49 27	02 ♍ 48 04	08 ♍ 49 51	01 ♊ 43	06 12	09 54	16 08	20 11	14 27	07 19	13 27
09	03 08 54	18 47 28	14 ♍ 54 25	21 ♍ 02 18	03 38	07 21	09 45	16 18	20 07	14 30	07 20	13 26
10	03 12 50	19 45 28	27 ♍ 14 00	03 ♎ 29 57	05 30	08 29	09 37	16 28	20 02	14 33	07 21	13 25
11	03 16 47	20 43 25	09 ♎ 50 33	16 ♎ 16 06	07 19	09 38	09 30	16 38	19 58	14 36	07 22	13 25
12	03 20 43	21 41 21	22 ♎ 46 51	29 ♎ 22 56	09 05	10 47	09 24	16 48	19 53	14 39	07 23	13 24
13	03 24 40	22 39 16	06 ♏ 04 23	12 ♏ 51 08	10 48	11 56	09 18	16 58	19 49	14 42	07 24	13 23
14	03 28 37	23 37 09	19 ♏ 42 58	26 ♏ 39 37	12 28	13 05	09 14	17 09	19 44	14 44	07 24	13 22
15	03 32 33	24 35 00	03 ♐ 40 40	10 ♐ 45 35	14 04	14 14	09 10	17 19	19 40	14 47	07 25	13 21
16	03 36 30	25 32 50	17 ♐ 53 47	25 ♐ 04 37	15 37	15 24	09 06	17 30	19 35	14 50	07 26	13 20
17	03 40 26	26 30 39	02 ♑ 17 22	09 ♑ 31 20	17 06	16 33	09 04	17 40	19 31	14 53	07 27	13 20
18	03 44 23	27 28 26	16 ♑ 45 46	24 ♑ 00 02	18 32	17 42	09 02	17 51	19 26	14 56	07 28	13 19
19	03 48 19	28 26 12	01 ♒ 13 28	08 ♒ 25 33	19 54	18 52	09 02	18 02	19 22	14 58	07 29	13 18
20	03 52 16	29 ♉ 23 57	15 ♒ 35 47	22 ♒ 43 46	21 13	20 01	09 D 02	18 12	19 18	15 01	07 29	13 17
21	03 56 12	00 ♊ 21 41	29 ♒ 49 14	06 ♓ 51 55	22 27	21 11	09 02	18 23	19 13	15 04	07 30	13 16
22	04 00 09	01 19 24	13 ♓ 51 40	20 ♓ 48 24	23 38	22 20	09 04	18 34	19 09	15 06	07 30	13 15
23	04 04 06	02 17 06	27 ♓ 42 04	04 ♈ 32 38	24 45	23 30	09 06	18 45	19 04	15 09	07 31	13 14
24	04 08 02	03 14 47	11 ♈ 20 06	18 ♈ 04 30	25 49	24 40	09 09	18 56	19 00	15 11	07 31	13 13
25	04 11 59	04 12 27	24 ♈ 45 49	01 ♉ 24 06	26 48	25 50	09 13	19 08	18 56	15 14	07 32	13 12
26	04 15 55	05 10 06	07 ♉ 59 20	14 ♉ 31 32	27 43	27 00	09 17	19 19	18 52	15 16	07 32	13 11
27	04 19 52	06 07 43	21 ♉ 00 40	27 ♉ 26 45	28 35	28 10	09 22	19 30	18 47	15 19	07 33	13 09
28	04 23 48	07 05 20	03 ♊ 49 46	10 ♊ 09 44	29 ♊ 22	29 ♈ 20	09 28	19 42	18 43	15 21	07 33	13 08
29	04 27 45	08 02 56	16 ♊ 26 40	22 ♊ 40 36	00 ♋ 05	00 ♉ 30	09 35	19 53	18 39	15 24	07 34	13 07
30	04 31 41	09 00 30	28 ♊ 51 37	04 ♋ 59 51	00 43	01 40	09 42	20 05	18 35	15 26	07 34	13 06
31	04 35 38	09 ♊ 58 03	11 ♋ 05 27	17 ♋ 08 38	01 ♋ 18	02 ♉ 50	09 ♎ 50	20 ♋ 16	18 ♏ 31	15 ♈ 28	07 ♓ 34	13 ♑ 05

DECLINATIONS

Date	Moon True ☊ ° '	Moon Mean ☊ ° '	Moon ☽ Latitude ° '	Sun ☉ ° '	Moon ☽ ° '	Mercury ☿ ° '	Venus ♀ ° '	Mars ♂ ° '	Jupiter ♃ ° '	Saturn ♄ ° '	Uranus ♅ ° '	Neptune ♆ ° '	Pluto ♇ ° '
01	28 ♎ 23	27 ♎ 53	03 S 22	15 N 08	18 N 19	17 N 33	01 S 54	03 S 01	22 N 53	15 S 32	04 N 58	09 S 31	20 S 06
02	28 R 21	27 50	04 10	15 26	18 57	18 18	01 29	02 57	22 52	15 31	04 59	09 31	20 06
03	28 19	27 47	04 45	15 44	18 39	19 02	01 05	02 55	22 51	15 29	05 00	09 30	20 06
04	28 17	27 44	05 07	16 01	17 30	19 44	00 40	02 52	22 50	15 28	05 01	09 30	20 06
05	28 15	27 40	05 16	16 19	15 35	20 24	00 S 15	02 50	22 49	15 27	05 02	09 29	20 06
06	28 14	27 37	05 10	16 35	13 01	21 01	00 N 09	02 48	22 48	15 26	05 03	09 29	20 06
07	28 D 14	27 34	04 51	16 52	09 56	21 36	00 34	02 46	22 47	15 25	05 05	09 29	20 06
08	28 15	27 31	04 20	17 09	06 26	22 09	00 59	02 45	22 46	15 24	05 06	09 28	20 06
09	28 16	27 28	03 37	17 25	02 N 36	22 39	01 24	02 44	22 45	15 22	05 07	09 28	20 06
10	28 18	27 24	02 43	17 40	01 S 24	23 06	01 50	02 43	22 44	15 21	05 08	09 28	20 07
11	28 19	27 21	01 40	17 56	05 26	23 31	02 15	02 43	22 43	15 20	05 09	09 27	20 07
12	28 20	27 18	00 S 31	18 11	09 20	23 54	02 40	02 43	22 41	15 19	05 10	09 27	20 07
13	28 R 20	27 15	00 N 42	18 26	12 53	24 14	03 05	02 43	22 40	15 18	05 11	09 27	20 07
14	28 19	27 12	01 55	18 40	15 49	24 31	03 31	02 43	22 39	15 17	05 12	09 26	20 07
15	28 16	27 09	03 02	18 55	17 54	24 47	03 56	02 44	22 38	15 16	05 14	09 26	20 07
16	28 13	27 05	03 59	19 09	18 55	24 59	04 21	02 45	22 36	15 14	05 15	09 26	20 07
17	28 09	27 02	04 42	19 22	18 43	25 10	04 47	02 46	22 35	15 13	05 16	09 26	20 07
18	28 05	26 59	05 07	19 36	17 19	25 18	05 12	02 48	22 34	15 12	05 17	09 25	20 08
19	28 02	26 56	05 12	19 49	14 48	25 25	05 37	02 50	22 32	15 11	05 18	09 25	20 08
20	27 59	26 53	04 58	20 01	11 24	25 29	06 02	02 52	22 31	15 10	05 19	09 25	20 08
21	27 59	26 50	04 26	20 13	07 23	25 32	06 28	02 55	22 29	15 09	05 20	09 25	20 08
22	27 D 59	26 46	03 38	20 25	03 S 00	25 32	06 53	02 57	22 28	15 08	05 21	09 25	20 08
23	28 01	26 43	02 38	20 37	01 N 30	25 31	07 18	03 00	22 27	15 07	05 22	09 24	20 08
24	28 02	26 40	01 30	20 48	05 52	25 29	07 43	03 04	22 25	15 06	05 23	09 24	20 08
25	28 03	26 37	00 N 18	20 59	09 52	25 25	08 07	03 07	22 24	15 05	05 24	09 24	20 08
26	28 R 02	26 34	00 S 54	21 09	13 19	25 19	08 32	03 11	22 22	15 03	05 25	09 24	20 09
27	28 00	26 30	02 02	21 19	16 03	25 12	08 57	03 15	22 20	15 02	05 25	09 24	20 09
28	27 56	26 27	03 02	21 29	17 56	25 04	09 21	03 19	22 19	15 01	05 26	09 24	20 09
29	27 51	26 24	03 52	21 39	18 54	24 55	09 45	03 24	22 17	15 00	05 27	09 23	20 09
30	27 44	26 21	04 31	21 48	18 55	24 45	10 10	03 29	22 16	14 59	05 28	09 23	20 09
31	27 ♎ 36	26 ♎ 18	04 S 56	21 N 56	18 N 03	24 N 34	10 N 34	03 S 34	22 N 14	14 S 58	05 N 29	09 S 23	20 S 09

ZODIAC SIGN ENTRIES

Date	h	m	Planets
03	01	21	♀ ♈
03	06	13	☽ ♋
05	17	55	☽ ♌
07	14	57	☿ ♊
08	06	24	☽ ♍
10	17	19	☽ ♎
13	01	07	☽ ♏
15	05	44	☽ ♐
17	08	12	☽ ♑
19	09	58	☽ ♒
21	02	59	☉ ♊
21	12	18	☽ ♓
23	16	01	☽ ♈
25	21	28	☽ ♉
28	04	47	☽ ♊
29	01	45	♀ ♉
29	09	12	☿ ♋
30	14	13	☽ ♋

LATITUDES

Date	Mercury ☿ ° '	Venus ♀ ° '	Mars ♂ ° '	Jupiter ♃ ° '	Saturn ♄ ° '	Uranus ♅ ° '	Neptune ♆ ° '	Pluto ♇ ° '
01	00 N 34	01 S 18	01 N 35	00 N 18	02 N 29	00 S 39	00 S 42	02 N 40
04	01 05	01 26	01 27	00 18	02 29	00 39	00 42	02 40
07	01 33	01 34	01 19	00 18	02 29	00 39	00 42	02 40
10	01 55	01 40	01 11	00 19	02 29	00 39	00 42	02 40
13	02 12	01 46	01 03	00 19	02 29	00 39	00 42	02 40
16	02 21	01 51	00 56	00 19	02 29	00 39	00 42	02 40
19	02 22	01 55	00 49	00 19	02 28	00 39	00 42	02 39
22	02 16	01 57	00 42	00 19	02 28	00 39	00 42	02 39
25	02 01	01 59	00 35	00 20	02 28	00 39	00 43	02 39
28	01 38	02 00	00 28	00 20	02 28	00 39	00 43	02 39
31	01 N 08	02 S 01	00 N 22	00 N 20	02 N 27	00 S 39	00 S 43	02 N 39

DATA

Julian Date	2456779
Delta T	+69 seconds
Ayanamsa	24° 03' 33"
Synetic vernal point	05° ♓ 03' 26"
True obliquity of ecliptic	23° 26' 07"

MOON'S PHASES, APSIDES AND POSITIONS ☽

Date	h	m	Phase	Longitude ° '	Eclipse Indicator
07	03	15	☽	16 ♌ 30	
14	19	16	○	23 ♏ 55	
21	12	59	☾	00 ♓ 24	
28	18	40	●	07 ♊ 21	

Day	h	m		
06	10	23	Apogee	
18	12	07	Perigee	
02	16	02	Max dec	18° N 58'
10	03	43	0S	
16	20	13	Max dec	18° S 59'
23	03	57	0N	
30	00	39	Max dec	19° N 01'

LONGITUDES

Date	Chiron ⚷ ° '	Ceres ⚳ ° '	Pallas ⚴ ° '	Juno ⚵ ° '	Vesta ⚶ ° '	Black Moon Lilith ⚸ ° '
01	16 ♓ 39	21 ♎ 41	26 ♌ 55	02 ♉ 48	19 ♎ 13	06 ♌ 29
11	17 ♓ 03	20 ♎ 02	28 ♌ 48	08 ♉ 34	17 ♎ 32	07 ♌ 36
21	17 ♓ 21	18 ♎ 58	01 ♍ 09	14 ♉ 20	16 ♎ 35	08 ♌ 43
31	17 ♓ 35	18 ♎ 33	03 ♍ 53	20 ♉ 09	16 ♎ 28	09 ♌ 50

All ephemeris data is given at 12.00 UT and the Moon's longitude is additionally given for 24.00 UT

ASPECTARIAN

h m Aspects

01 Thursday
10 23 ☽ □ ♆
11 00 ☽ ± ♇
11 38 ☿ ± ♂
13 55 ☽ ⊥ ♃
16 34 ☽ Q ♀
16 55 ☉ ⊼ ♂
18 02 ☽ △ ♂
18 09 ☽ ⊻ ☉
22 22 ☽ ⊼ ♇
23 32 ☽ ✶ ♅

02 Friday
01 28 ☽ ⊻ ♃
06 34 ☽ ⊥ ☉
09 06 ☿ Q ♆
09 18 ☽ ⊻ ☿
12 02 ☽ ⊼ ♄
17 35 ☉ ⫲ ♄
19 59 ☿ ⊥ ♅
22 44 ☽ Q ♅
23 18 ☽ ⊥ ☿
23 35 ☽ ± ♄

03 Saturday
00 56 ☿ ☍ ♄
01 26 ☽ ∠ ☉
04 58 ☽ ∥ ☿
06 43 ☽ □ ♀
17 06 ☽ ⚼ ♄
20 25 ☽ △ ♆
20 36 ☽ ∠ ☿

04 Sunday
00 37 ☉ △ ♇
03 16 ☽ □ ♂
08 50 ☽ ☍ ♇
09 33 ☽ ✶ ☉
10 20 ☽ □ ♅
12 51 ☽ ☌ ♃
19 43 ☉ ⊻ ♅
22 44 ☽ △ ♄

05 Monday
01 09 ☿ ⫲ ♇
02 20 ☽ ⚼ ♆
04 50 ☽ ∥ ☉
08 42 ☿ ⚼ ♂
08 46 ☽ ✶ ☿
11 42 ☽ Q ☉
13 23 ☽ ⫲ ♄
14 42 ☽ Q ♂
20 29 ☽ ± ♆

06 Tuesday
00 45 ☽ △ ♀
08 36 ☽ ⊼ ♆
09 54 ☉ ✶ ♃
10 17 ♄ ± ♅
13 57 ☽ Q ☿
14 33 ☽ ✶ ♂
19 12 ☉ ± ♂
20 23 ☿ ⚼ ♇
21 06 ☽ ⊼ ♇
22 56 ☽ △ ♅

07 Wednesday
02 03 ☽ ⊻ ♃
03 15 ☽ □ ☉
07 32 ☿ ∠ ♅
09 12 ☽ ± ♇
10 15 ☽ ⚼ ♀
10 50 ☽ □ ♄
14 19 ☽ ⊥ ♃
15 17 ☽ ⫲ ♆
16 03 ♀ ⚼ ♄
20 22 ☽ ∠ ♂

08 Thursday
03 17 ☽ ⚼ ♇
04 14 ☿ ∠ ♃
05 16 ☽ ⚼ ♅
06 15 ☽ ± ♀
08 38 ☽ ∠ ♃
09 25 ☽ □ ☿
14 10 ☽ ⊥ ♂
19 29 ☽ ⊼ ♀
20 30 ☽ ∥ ♅
21 00 ☽ ☍ ♆
22 40 ☽ Q ♄
23 17 ☽ ± ♅

09 Friday
01 57 ☽ ⊻ ♂
09 06 ☽ △ ♇
11 12 ☽ ⊼ ♅
11 14 ☽ ⫲ ♂
11 42 ♀ ⊻ ♆
14 47 ☽ ✶ ♃
16 50 ☿ ∥ ♃
18 35 ☽ ∥ ♀
20 16 ☽ △ ☉
22 08 ☽ ✶ ♄

10 Saturday
01 37 ☉ Q ♆
14 25 ☽ Q ♃
14 52 ☽ ⫲ ♀
18 28 ☉ ☍ ♄
19 51 ☽ ∥ ♂

11 Sunday
02 50 ☽ ∠ ♄
03 35 ☽ ⚼ ☉
06 28 ☽ △ ☿
07 19 ☽ ⊼ ♆
08 39 ☉ ⊥ ♅
09 27 ♀ ☍ ♂
10 19 ☽ ⫲ ♅
11 22 ☽ ☌ ♂
11 34 ☽ ☍ ♀
12 34 ☿ □ ♆
13 10 ☿ ± ♆
18 36 ☽ ± ♆
18 40 ☽ □ ♇
19 40 ☽ ⊥ ♄
20 56 ☽ ☍ ♅
21 52 ☽ ± ☉

12 Monday
00 51 ☽ □ ♃
06 43 ☽ ⊻ ♄
09 50 ☽ ⊼ ☉
11 16 ☽ ⚼ ♆
12 46 ☽ ∥ ♆
14 27 ♀ ⫲ ♂
14 45 ☽ ⚼ ☿
16 02 ☿ △ ♂

13 Tuesday
03 37 ☽ Q ♇
09 25 ☽ ± ☿
14 21 ☽ △ ♆
14 41 ☿ ⊥ ♃
17 42 ☽ ⊻ ♂
21 34 ☽ ⊼ ☿
23 21 ☽ ⊼ ♀

14 Wednesday
00 55 ☽ ✶ ♇
03 17 ☽ ⊼ ♅
04 12 ☽ ⊥ ♂
07 07 ☽ ∥ ♄
07 28 ☽ △ ♃
10 48 ☽ ± ♀
12 02 ☽ ☌ ♄
13 47 ☽ ± ♅
17 50 ♀ □ ♇
18 47 ♀ ⊥ ♆
19 16 ☽ ☍ ☉
19 46 ☽ ∠ ♂

15 Thursday
00 45 ♀ ± ♄
01 18 ☿ ⊼ ♇
02 05 ☉ ⚼ ♂
02 55 ☽ ∠ ♇
03 45 ☽ ⚼ ♀
05 20 ☽ ⚼ ♅
09 39 ☽ ⚼ ♃
18 14 ☽ ⊥ ♇
18 22 ☽ □ ♆
21 16 ☽ ✶ ♂
21 54 ☿ ✶ ♀
23 22 ☿ ✶ ♅
23 54 ♀ ☌ ♅

16 Friday
01 06 ☽ ± ♃
04 21 ☽ ⊻ ♇
06 51 ☽ △ ♅
07 26 ☽ △ ♀
07 43 ☽ ☍ ☿
11 19 ☽ ⊼ ♃
14 49 ☽ ⊻ ♄
17 21 ☽ Q ♂

17 Saturday
00 36 ☽ Q ♆
00 47 ☽ ⊥ ♄
01 42 ☽ ⊼ ☉
12 24 ☽ ± ☉
15 40 ☽ ∠ ♄
20 34 ☽ ✶ ♆
22 42 ☿ ⊻ ♃
23 13 ☽ □ ♂

18 Sunday
04 23 ☽ ⚼ ☉
06 17 ☽ ☌ ♇
08 57 ☽ □ ♅
13 42 ☽ □ ♀
13 49 ☽ ☍ ♃
15 15 ☽ ⊼ ☿
15 31 ♀ □ ♃
16 25 ☽ ✶ ♄
16 41 ♀ ∥ ♅
21 27 ☽ ∠ ♆

19 Monday
02 15 ☽ ± ☿
02 58 ☿ ⊼ ♄
07 02 ☽ △ ☉
08 31 ☉ ⚼ ♇
08 52 ☽ ∥ ♄
12 14 ☽ Q ♄
12 25 ☽ ⊥ ♆
14 55 ☽ Q ♅
18 45 ☽ ⚼ ☿
21 50 ♀ ⊼ ♄
22 13 ☽ Q ♀
22 25 ☽ ⊻ ♆

20 Tuesday
01 00 ☽ △ ♂
01 31 ♂ St D
08 07 ☽ ⊻ ♇
11 01 ☽ ✶ ♅
16 26 ☽ ⊼ ♃
18 11 ☽ □ ♄
18 11 ☽ ⊥ ♇
20 05 ☽ ✶ ♀
22 21 ☽ △ ☿

21 Wednesday
00 15 ☽ ∥ ♆
01 06 ☉ ⫲ ♇
02 12 ☽ ⚼ ♂
02 41 ☽ ± ♃
04 06 ☉ ∠ ♅
09 22 ☽ ∠ ♇
12 24 ☽ ∠ ♅
12 59 ☽ □ ☉
16 42 ☽ ⫲ ♀
17 29 ☽ ± ♂
18 09 ☽ ⚼ ♃
23 20 ☽ ⫲ ♅
23 48 ☽ ∠ ♀

22 Thursday
01 05 ☽ ☌ ♆
03 45 ☽ ⊼ ♂
03 48 ☽ ⊥ ♅
10 57 ☽ ✶ ♇
12 12 ☽ ∥ ♂
14 09 ☽ ⊻ ♅
15 24 ♀ ∠ ♆
16 40 ☽ ⊥ ♀
20 14 ☽ △ ♃
21 05 ☽ △ ♄
22 08 ☽ Q ☉

23 Friday
04 01 ☽ ⊻ ♀
06 25 ☽ □ ☿
07 41 ☽ Q ♇
18 36 ☿ ± ♄
20 14 ☽ ⫲ ♂
20 38 ☽ ✶ ☉
23 07 ☽ ⚼ ♄

24 Saturday
05 15 ☽ ⊻ ♆
08 07 ☽ ☍ ♂
09 15 ☽ ∥ ♅
14 57 ☽ ± ♄
15 20 ☽ □ ♇
15 53 ☽ ⊥ ♆
16 46 ☽ Q ☿
17 47 ♃ △ ♄
18 52 ☽ ☌ ♅
23 58 ☽ ∥ ♀

25 Sunday
01 15 ☽ ∠ ☉
01 35 ☽ ⊼ ♄
01 44 ☽ □ ♃
07 59 ☽ ∠ ♆
09 03 ☽ ⫲ ♆
09 30 ☉ ∠ ♃
14 06 ☽ ☌ ♀
15 57 ☽ ✶ ☿
18 42 ☽ ⊥ ☉
23 27 ☿ Q ♅

26 Monday
06 27 ☽ ⊻ ☉
10 45 ☽ Q ♃
11 11 ☽ ✶ ♆
14 23 ☽ ⊼ ♂
21 19 ☽ ∠ ☿
21 30 ☽ △ ♇

27 Tuesday
01 25 ☽ ⊻ ♅
01 29 ☽ ± ♂
02 21 ☽ ⫲ ♄
07 54 ☽ ☍ ♄
09 10 ☽ ✶ ♃
09 17 ☽ Q ♆
12 34 ☽ ⊥ ♅
15 07 ☽ ⊥ ☿
18 18 ☽ ⚼ ♂

28 Wednesday
01 19 ☽ ⚼ ♇
02 41 ☽ ⊻ ♀
03 04 ☽ ⊻ ☿
05 26 ☽ ∠ ♅
13 12 ☉ ± ♇
13 40 ☽ ∠ ♃
13 58 ☿ ✶ ♀
14 26 ♀ ⫲ ♆
15 07 ☽ ⊥ ♀
18 15 ☽ ± ♇
18 40 ☽ ☌ ☉
19 03 ☽ □ ♆
22 46 ☽ △ ♂
23 43 ☉ □ ♆

29 Thursday
05 39 ☽ ⊼ ♇
07 02 ☽ ⊥ ♃
09 59 ☽ ✶ ♅
10 00 ☽ ∠ ♀
16 13 ☽ ⊼ ♄
18 43 ☽ ⊻ ♃

30 Friday
03 44 ☽ ± ♄
09 13 ☽ Q ♅
15 49 ☽ ☌ ☿
18 03 ☽ ✶ ♀
21 11 ☽ ⚼ ♄

31 Saturday
05 03 ☽ △ ♆
08 01 ☉ △ ♂
09 29 ☽ □ ♂
09 35 ☽ ⊻ ☉
15 56 ☽ ☍ ♇
20 13 ☽ Q ♀
20 42 ☽ □ ♅
22 30 ☽ ⊥ ☉

JUNE 2014

LONGITUDES

Date	Sidereal time h m s	Sun ☉ ° ' "	Moon ☽ ° ' "	Moon ☽ 24.00 ° ' "	Mercury ☿ ° '	Venus ♀ ° '	Mars ♂ ° '	Jupiter ♃ ° '	Saturn ♄ ° '	Uranus ♅ ° '	Neptune ♆ ° '	Pluto ♇ ° '
01	04 39 35	10 ♊ 55 35	23 ♋ 09 38	29 ♋ 08 47	01 ♋ 47	04 ♉ 01	09 ♎ 58	20 ♋ 28	18 ♏ 27	15 ♈ 31	07 ♓ 35	13 ♑ 04
02	04 43 31	11 53 06	05 ♌ 06 25	11 ♌ 02 57	02 13	05 11	10 08	20 40	18 R 23	15 33	07 35	13 R 02
03	04 47 28	12 50 36	16 ♌ 58 51	22 ♌ 54 36	02 33	06 21	10 18	20 52	18 19	15 35	07 35	13 01
04	04 51 24	13 48 04	28 ♌ 50 44	04 ♍ 47 50	02 49	07 32	10 28	21 04	18 15	15 37	07 35	13 00
05	04 55 21	14 45 31	10 ♍ 46 30	16 ♍ 47 19	03 01	08 42	10 39	21 16	18 11	15 39	07 35	12 59
06	04 59 17	15 42 57	22 ♍ 50 57	28 ♍ 58 00	03 08	09 53	10 51	21 28	18 08	15 41	07 36	12 57
07	05 03 14	16 40 22	05 ♎ 09 06	11 ♎ 24 50	03 R 10	11 03	11 04	21 40	18 04	15 43	07 36	12 56
08	05 07 10	17 37 45	17 ♎ 45 45	24 ♎ 12 20	03 08	12 14	11 16	21 52	18 00	15 45	07 36	12 55
09	05 11 07	18 35 08	00 ♏ 45 00	07 ♏ 24 03	03 01	13 24	11 30	22 04	17 57	15 47	07 36	12 53
10	05 15 04	19 32 29	14 ♏ 09 42	21 ♏ 01 58	02 50	14 35	11 44	22 16	17 53	15 49	07 R 36	12 52
11	05 19 00	20 29 50	28 ♏ 00 46	05 ♐ 05 48	02 35	15 46	11 59	22 28	17 50	15 51	07 36	12 51
12	05 22 57	21 27 10	12 ♐ 16 35	19 ♐ 32 31	02 16	16 57	12 14	22 41	17 46	15 53	07 36	12 49
13	05 26 53	22 24 29	26 ♐ 52 47	04 ♑ 16 25	01 54	18 07	12 30	22 53	17 43	15 55	07 36	12 48
14	05 30 50	23 21 47	11 ♑ 42 24	19 ♑ 09 36	01 28	19 18	12 46	23 05	17 40	15 57	07 35	12 47
15	05 34 46	24 19 05	26 ♑ 36 56	04 ♒ 03 17	01 00	20 29	13 03	23 18	17 37	15 58	07 35	12 45
16	05 38 43	25 16 22	11 ♒ 27 40	18 ♒ 49 13	00 ♋ 30	21 40	13 20	23 30	17 34	16 00	07 35	12 44
17	05 42 39	26 13 39	26 ♒ 07 10	03 ♓ 20 57	29 ♊ 57	22 51	13 38	23 43	17 31	16 02	07 35	12 42
18	05 46 36	27 10 56	10 ♓ 30 09	17 ♓ 34 32	29 24	24 02	13 56	23 56	17 28	16 03	07 35	12 41
19	05 50 33	28 08 12	24 ♓ 33 56	01 ♈ 28 24	28 50	25 13	14 15	24 08	17 25	16 05	07 34	12 39
20	05 54 29	29 ♊ 05 28	08 ♈ 18 01	15 ♈ 02 59	28 15	26 25	14 34	24 21	17 22	16 07	07 34	12 38
21	05 58 26	00 ♋ 02 44	21 ♈ 43 30	28 ♈ 19 51	27 41	27 36	14 54	24 34	17 19	16 08	07 34	12 37
22	06 02 22	01 00 00	04 ♉ 52 20	11 ♉ 21 11	27 09	28 47	15 14	24 46	17 17	16 09	07 33	12 35
23	06 06 19	01 57 16	17 ♉ 46 42	24 ♉ 09 07	26 38	29 ♉ 58	15 34	24 59	17 14	16 11	07 33	12 34
24	06 10 15	02 54 31	00 ♊ 28 40	06 ♊ 45 30	26 09	01 ♊ 10	15 55	25 12	17 12	16 12	07 32	12 32
25	06 14 12	03 51 47	12 ♊ 59 50	19 ♊ 11 45	25 43	02 21	16 16	25 25	17 09	16 13	07 32	12 31
26	06 18 08	04 49 02	25 ♊ 21 25	01 ♋ 28 56	25 19	03 32	16 38	25 38	17 07	16 15	07 31	12 29
27	06 22 05	05 46 17	07 ♋ 34 23	13 ♋ 37 54	25 00	04 44	17 00	25 51	17 05	16 16	07 31	12 28
28	06 26 02	06 43 31	19 ♋ 39 36	25 ♋ 39 38	24 44	05 55	17 23	26 04	17 02	16 17	07 30	12 26
29	06 29 58	07 40 46	01 ♌ 38 10	07 ♌ 35 25	24 33	07 07	17 46	26 17	17 00	16 18	07 30	12 25
30	06 33 55	08 ♋ 38 00	13 ♌ 31 37	19 ♌ 27 05	24 ♊ 25	08 ♊ 19	18 ♎ 10	26 ♋ 30	16 ♏ 58	16 ♈ 19	07 ♓ 29	12 ♑ 23

Date	Moon True ☊ ° '	Moon Mean ☊ ° '	Moon ☽ Latitude ° '
01	27 ♎ 29	26 ♎ 15	05 S 08
02	27 R 22	26 11	05 06
03	27 17	26 08	04 51
04	27 14	26 05	04 23
05	27 12	26 02	03 44
06	27 D 13	25 59	02 55
07	27 14	25 56	01 56
08	27 15	25 52	00 S 51
09	27 R 15	25 49	00 N 19
10	27 14	25 46	01 30
11	27 10	25 43	02 37
12	27 05	25 40	03 37
13	26 58	25 36	04 24
14	26 49	25 33	04 54
15	26 41	25 30	05 05
16	26 34	25 27	04 55
17	26 28	25 24	04 25
18	26 25	25 21	03 40
19	26 24	25 17	02 41
20	26 D 24	25 14	01 35
21	26 24	25 11	00 N 25
22	26 R 24	25 08	00 S 45
23	26 22	25 05	01 51
24	26 18	25 02	02 51
25	26 11	24 58	03 41
26	26 01	24 55	04 20
27	25 49	24 52	04 46
28	25 37	24 49	05 00
29	25 24	24 46	05 00
30	25 ♎ 13	24 ♎ 42	04 S 46

DECLINATIONS

Date	Sun ☉ ° '	Moon ☽ ° '	Mercury ☿ ° '	Venus ♀ ° '	Mars ♂ ° '	Jupiter ♃ ° '	Saturn ♄ ° '	Uranus ♅ ° '	Neptune ♆ ° '	Pluto ♇ ° '
01	22 N 05	16 N 24	24 N 21	10 N 57	03 S 39	22 N 12	14 S 58	05 N 30	09 S 23	20 S 10
02	22 13	14 02	24 08	11 21	03 44	22 11	14 57	05 31	09 23	20 10
03	22 20	11 07	23 55	11 44	03 50	22 09	14 56	05 32	09 23	20 10
04	22 27	07 46	23 40	12 07	03 56	22 07	14 55	05 32	09 23	20 10
05	22 34	04 04	23 25	12 30	04 02	22 05	14 54	05 33	09 23	20 10
06	22 40	00 N 10	23 10	12 53	04 09	22 03	14 53	05 34	09 23	20 11
07	22 46	03 S 50	22 54	13 16	04 15	22 02	14 52	05 35	09 23	20 11
08	22 52	07 45	22 37	13 38	04 22	22 00	14 51	05 36	09 23	20 11
09	22 57	11 26	22 21	14 00	04 29	21 58	14 50	05 36	09 23	20 11
10	23 01	14 40	22 04	14 21	04 36	21 56	14 50	05 37	09 23	20 11
11	23 06	17 09	21 47	14 43	04 43	21 54	14 49	05 38	09 23	20 11
12	23 10	18 40	21 30	15 04	04 51	21 52	14 48	05 38	09 23	20 12
13	23 13	19 00	21 14	15 24	04 59	21 50	14 47	05 39	09 23	20 12
14	23 16	18 02	20 58	15 45	05 07	21 48	14 47	05 40	09 23	20 12
15	23 19	15 50	20 42	16 05	05 15	21 46	14 46	05 40	09 23	20 12
16	23 21	12 37	20 26	16 24	05 23	21 44	14 45	05 41	09 24	20 12
17	23 23	08 39	20 11	16 44	05 31	21 42	14 45	05 42	09 24	20 13
18	23 24	04 S 14	19 57	17 03	05 40	21 40	14 44	05 42	09 24	20 13
19	23 25	00 N 19	19 44	17 21	05 49	21 38	14 43	05 43	09 24	20 13
20	23 26	04 45	19 32	17 39	05 57	21 35	14 43	05 43	09 24	20 13
21	23 26	08 51	19 21	17 57	06 06	21 33	14 42	05 44	09 24	20 14
22	23 26	12 26	19 11	18 15	06 16	21 31	14 42	05 44	09 24	20 14
23	23 25	15 21	19 02	18 31	06 25	21 29	14 41	05 45	09 25	20 14
24	23 24	17 28	18 55	18 48	06 34	21 27	14 41	05 45	09 25	20 14
25	23 23	18 42	18 50	19 04	06 44	21 24	14 40	05 46	09 25	20 14
26	23 21	19 01	18 45	19 19	06 54	21 22	14 40	05 46	09 25	20 15
27	23 19	18 27	18 43	19 34	07 03	21 20	14 40	05 47	09 25	20 15
28	23 16	17 03	18 42	19 49	07 13	21 17	14 39	05 47	09 26	20 15
29	23 13	14 55	18 42	20 03	07 23	21 15	14 39	05 48	09 26	20 15
30	23 N 09	12 N 11	18 N 44	20 N 17	07 S 33	21 N 12	14 S 38	05 N 48	09 S 26	20 S 16

ZODIAC SIGN ENTRIES

Date	h	m	Planets
02	01	43	☽ ♌
04	14	20	☽ ♍
07	02	01	☽ ♎
09	10	38	☽ ♏
11	15	23	☽ ♐
13	17	04	☽ ♑
15	17	27	☽ ♒
17	10	04	☿ ♊
17	18	26	☽ ♓
19	21	26	☽ ♈
21	10	51	☉ ♋
22	03	03	☽ ♉
23	12	33	♀ ♊
24	11	05	☽ ♊
26	21	05	☽ ♋
29	08	43	☽ ♌

LATITUDES

Date	Mercury ☿ ° '	Venus ♀ ° '	Mars ♂ ° '	Jupiter ♃ ° '	Saturn ♄ ° '	Uranus ♅ ° '	Neptune ♆ ° '	Pluto ♇ ° '
01	00 N 56	02 S 01	00 N 20	00 N 20	02 N 27	00 S 39	00 S 43	02 N 39
04	00 N 16	02 00	00 14	00 20	02 27	00 39	00 43	02 39
07	00 S 30	01 59	00 08	00 20	02 26	00 40	00 43	02 38
10	01 20	01 56	00 N 02	00 20	02 26	00 40	00 43	02 38
13	02 11	01 54	00 S 03	00 21	02 25	00 40	00 43	02 38
16	03 00	01 50	00 08	00 21	02 24	00 40	00 43	02 38
19	03 42	01 46	00 13	00 21	02 23	00 40	00 43	02 38
22	04 13	01 41	00 17	00 21	02 23	00 40	00 43	02 37
25	04 32	01 36	00 22	00 21	02 22	00 40	00 44	02 37
28	04 38	01 30	00 26	00 22	02 22	00 40	00 44	02 37
31	04 S 32	01 S 24	00 S 30	00 N 22	02 N 21	00 S 40	00 S 44	02 N 36

DATA

Julian Date	2456810
Delta T	+69 seconds
Ayanamsa	24° 03' 38"
Synetic vernal point	05° ♓ 03' 22"
True obliquity of ecliptic	23° 26' 06"

MOON'S PHASES, APSIDES AND POSITIONS ☽

Date	h	m	Phase	Longitude ° '	Eclipse Indicator
05	20	39	☽	15 ♍ 06	
13	04	11	○	22 ♐ 06	
19	18	39	☾	28 ♓ 24	
27	08	08	●	05 ♋ 37	

Day	h	m		
03	04	28	Apogee	
15	03	38	Perigee	
30	19	17	Apogee	
06	13	00	0S	
13	06	07	Max dec	19° S 02'
19	10	21	0N	
26	08	31	Max dec	19° N 02'

LONGITUDES

Date	Chiron ⚷ ° '	Ceres ⚳ ° '	Pallas ⚴ ° '	Juno ⚵ ° '	Vesta ⚶ ° '	Black Moon Lilith ⚸ ° '
01	17 ♓ 36	18 ♎ 32	04 ♍ 10	20 ♉ 44	16 ♎ 30	09 ♌ 57
11	17 ♓ 43	18 ♎ 50	07 ♍ 15	26 ♉ 32	17 ♎ 16	11 ♌ 04
21	17 ♓ 46	19 ♎ 26	10 ♍ 35	02 ♊ 21	18 ♎ 46	12 ♌ 11
31	17 ♓ 46	21 ♎ 11	14 ♍ 07	08 ♊ 08	20 ♎ 53	13 ♌ 18

All ephemeris data is given at 12.00 UT and the Moon's longitude is additionally given for 24.00 UT

ASPECTARIAN

01 Sunday
02 39 ☽ △ ♄
06 32 ☽ ☌ ♃
10 50 ☽ ⚼ ♆
18 01 ☽ ∠ ☉
21 46 ☽ Q ♂

02 Monday
03 27 ☽ # ♄
04 54 ☽ ± ♆
05 58 ☽ ⊻ ☿
06 51 ☉ ∥ ♃
12 10 ☽ □ ♀
17 00 ☽ ⊼ ♆
18 28 ☽ ⊥ ☿
22 17 ☽ ⚹ ♂

03 Tuesday
02 54 ☽ ⚹ ☉
04 00 ☽ ⊼ ♇
07 47 ☽ ∥ ♀
09 10 ☽ △ ♅
13 12 ☽ ∠ ☿
14 41 ☽ □ ♄
16 07 ☽ ± ♇
16 18 ☉ ⊼ ♇
19 59 ☽ ⊻ ♃

04 Wednesday
00 44 ☽ # ♆
05 04 ☽ ∠ ♂
05 18 ☽ Q ☉
08 20 ☽ ⊥ ♃
10 17 ☽ ⚼ ♇
13 15 ♀ ⚹ ♆
15 36 ☽ ⚼ ♅
20 11 ☽ ⚹ ☿
23 31 ☽ ⊥ ♂

05 Thursday
02 35 ☽ ∥ ♅
02 47 ☽ ∠ ♃
02 51 ☽ Q ♄
05 37 ☽ ☍ ♆
07 24 ☽ △ ♀
09 45 ☽ ± ♅
11 45 ☽ ⊻ ♂
12 11 ☽ # ♂
16 24 ☽ △ ♇
18 20 ☉ ⊥ ♀
20 34 ☽ Q ☿
20 39 ☽ □ ☉
21 46 ☽ ⊼ ♅

06 Friday
01 43 ♀ Q ♃
02 42 ☽ ⚹ ♄
03 51 ☉ ⊥ ♃
09 13 ☽ ⚹ ♃
11 20 ☉ ⚹ ♅
11 35 ☿ ⚼ ♄
16 25 ☽ ⚼ ♀

07 Saturday
07 59 ☽ ∠ ♄
08 10 ☽ □ ☿
09 04 ☽ Q ♃
11 47 ☽ ± ♀
11 57 ☿ St R
12 10 ♀ ⊼ ♂
14 39 ☽ ∥ ♂
16 42 ☽ ⊼ ♆
20 17 ☉ ∥ ☿
22 39 ☽ # ♅
23 31 ☽ ☌ ♂

08 Sunday
00 29 ☽ ⊼ ♀
01 11 ☽ ⊥ ♄
02 52 ☽ □ ♇
04 09 ☽ ± ♆
08 13 ☽ ☍ ♅
11 44 ☽ △ ☉
12 27 ☽ ⊻ ♄
19 47 ☽ □ ♃
20 53 ☉ ⊼ ♄
21 01 ☽ ⚼ ♆
22 21 ☽ ∥ ♆

09 Monday
01 40 ♀ △ ♇
12 15 ☽ Q ♇
16 04 ☽ △ ☿
17 32 ☽ ⚼ ☉
19 49 ♆ St R

10 Tuesday
00 21 ☽ △ ♆
03 36 ☿ ⚼ ♄
07 38 ☽ ⊻ ♂
09 11 ☽ # ♀
09 43 ☽ ⚹ ♇
10 49 ☽ ± ☉
12 49 ☽ ☍ ♀
13 24 ☽ ∥ ♄
14 56 ☽ ⊼ ♅
18 20 ☽ ⚼ ☿
18 23 ☽ ⊥ ♂
18 30 ☽ ☌ ♄
22 07 ☽ ⊼ ☉

11 Wednesday
00 09 ♂ ⊥ ♄
01 04 ☿ ∥ ♃
01 24 ☽ ± ♅
02 21 ☽ △ ♃
09 37 ☽ ± ☿
10 12 ☽ ∠ ♂
11 43 ☽ ∠ ♇
13 55 ♀ ⊻ ♅
16 51 ☽ ⚼ ♅
18 52 ♀ # ♄
19 37 ☽ ⊼ ☿

12 Thursday
02 11 ♃ ⚼ ♆
02 55 ☽ ⊥ ♇
04 12 ☽ □ ♆
04 13 ☽ ⚼ ♃
11 56 ☽ ⚹ ♂
12 54 ☽ ⊻ ♇
17 10 ☿ ∠ ♀
17 59 ☽ △ ♅
19 23 ☽ ± ♃
20 24 ☽ ⊼ ♀
21 03 ☽ ⊻ ♄

13 Friday
04 09 ♀ ☍ ♄
04 11 ☽ ☍ ☉
05 23 ☽ ⊼ ♃
06 52 ☽ ⊥ ♄
07 07 ☽ ± ♀
08 03 ☽ Q ♂
09 54 ☽ Q ♆
19 56 ☽ ☍ ☿
21 27 ☽ ∠ ♄
21 48 ♀ ± ♂
23 01 ☽ ⚼ ♀

14 Saturday
03 17 ☉ ⊻ ♃
05 22 ☽ ⚹ ♆
12 44 ♂ □ ♇
13 43 ☽ ☌ ♇
13 44 ☽ □ ♂
17 47 ♀ Q ♆
18 50 ☽ □ ♅
19 11 ☉ ± ♄
21 34 ☽ ⚹ ♄

15 Sunday
01 17 ☽ △ ♀
05 31 ☽ ∠ ♆
06 35 ☽ ☍ ♃
08 03 ☽ ⊼ ☉
10 05 ☽ # ♀
16 49 ☽ Q ♄
18 23 ☽ ± ☉
18 51 ☽ ⊼ ☿
20 01 ☽ ⊥ ♆
20 47 ☽ ∥ ♄
23 54 ☽ Q ♅

16 Monday
04 13 ☽ ± ☿
05 43 ☽ ⊻ ♆
09 56 ☽ ⚼ ☉
14 03 ☽ ⊻ ♇
15 06 ☽ △ ♂
18 20 ☽ ⚼ ☿
18 55 ♀ ⊥ ♅
19 25 ☽ ⚹ ♅
21 54 ☽ □ ♄
23 50 ☽ ⊥ ♇

17 Tuesday
06 09 ☽ □ ♀
07 43 ☽ ∥ ♆
07 59 ☽ ⊼ ♃
08 04 ♂ ± ♆
09 42 ☿ # ♇
12 11 ☽ △ ☉
14 37 ☽ ∠ ♇
16 14 ☽ ⚼ ♂
18 03 ☽ ± ♃
18 07 ☽ △ ☿
20 09 ☽ ∠ ♅

18 Wednesday
03 12 ☿ ⊥ ♀
04 12 ☽ # ♅
04 37 ☽ ∥ ♂
07 05 ☽ ☌ ♆
07 35 ☽ ± ♂
09 16 ♀ ⚹ ♃
09 18 ☽ ⚼ ♃
11 15 ☽ ⊥ ♅
14 50 ☽ Q ♀
15 41 ☽ ⚹ ♇
17 56 ☽ ⊼ ♂
18 52 ♂ # ♅
21 26 ☽ ⊻ ♅
23 46 ☽ △ ♄

19 Thursday
10 37 ☉ Q ♅
11 15 ☽ △ ♃
12 10 ☽ Q ♇
13 15 ☽ ⚹ ♀
18 39 ☽ □ ☉
19 05 ☽ □ ☿
22 50 ☉ ☌ ☿

20 Friday
01 36 ☽ ⚼ ♄
10 42 ☽ ⊻ ♆
17 25 ☽ ± ♄
17 30 ☽ ∥ ♅
17 55 ☿ Q ♅
18 03 ☽ ∠ ♀
19 04 ☽ # ♂
19 41 ☽ □ ♇
21 21 ☽ ⊥ ♆
23 25 ☽ ☍ ♂

21 Saturday
01 35 ☽ Q ☿
01 55 ☽ ☌ ♅
04 06 ☽ ⊼ ♄
04 52 ☽ Q ☉
11 45 ☽ ⊥ ♀
12 15 ♀ ⚼ ♇
13 18 ☿ ⊻ ♀
13 30 ☽ ∠ ♆
15 28 ☽ # ♆
17 13 ☽ □ ♃
22 24 ☽ ⚹ ☿
23 43 ☽ ⊻ ♀

22 Sunday
04 19 ☽ ⚹ ☉
16 57 ☽ ⚹ ♆

23 Monday
00 57 ☽ ∠ ☿
02 16 ☽ △ ♇
02 53 ☽ Q ♃
06 00 ☽ # ♄
07 45 ☽ ⊼ ♂
09 00 ☽ ⊻ ♅
10 20 ☽ ∠ ☉
10 59 ☽ ☍ ♄
15 19 ☽ Q ♆
17 09 ☽ ⊥ ☿
18 44 ☉ ⚼ ♄
19 19 ☽ ± ♂
20 17 ☽ ⊥ ♅

24 Tuesday
01 49 ☽ ⚹ ♃
04 04 ☽ ⊻ ☿
04 40 ☽ ⊥ ☉
05 01 ♀ ⚼ ♂
06 25 ☽ ⚼ ♇
12 50 ♀ ∠ ♅
12 52 ☽ ⚼ ♂
13 23 ☽ ∠ ♅
13 26 ☽ ☌ ♀
17 01 ☽ ⊻ ☉
19 53 ☿ ∥ ♀
23 33 ☽ ± ♇

25 Wednesday
01 29 ☽ □ ♆
06 56 ☽ ∠ ♃
08 26 ♂ ☍ ♅
11 04 ☽ ⊼ ♇
15 50 ☽ ∥ ☿
18 15 ☽ ⚹ ♅
18 31 ☽ △ ♂
20 01 ☽ ⊼ ♄
23 28 ☿ ⊻ ♃

26 Thursday
00 39 ☽ ⊥ ♃
07 38 ☽ ± ♄
11 56 ☽ ☌ ☿
12 33 ☽ ⊻ ♃
17 39 ☽ Q ♅

27 Friday
01 12 ☽ ⚼ ♄
04 31 ☽ ∥ ☿
05 47 ☽ ⊻ ♀
08 08 ☽ ☌ ☉
11 53 ☽ △ ♆
16 03 ♂ ⊻ ♄
18 56 ☽ ⊥ ♀
21 39 ☽ ☍ ♇

28 Saturday
05 16 ☽ □ ♅
06 48 ☽ △ ♄
07 19 ☽ □ ♂
14 48 ☽ ∠ ♀
17 40 ☽ ⚼ ♆
21 58 ☽ ⊻ ☿
22 06 ♀ ± ♇

29 Sunday
01 02 ☽ ☌ ♃
07 22 ☉ △ ♆
09 50 ☽ ⊥ ☿
11 43 ☽ ± ♆
14 39 ☽ # ♄
19 31 ♀ □ ♆
20 36 ☽ Q ♂
23 48 ☽ ⊼ ♆

30 Monday
00 17 ☽ ⚹ ♀
01 15 ☽ ⊻ ☉
03 46 ☽ ∠ ☿
09 42 ☽ ⊼ ♇
09 55 ♀ # ♇
14 26 ☽ ⊥ ☉
17 40 ☽ △ ♅
18 58 ☽ □ ♄
21 42 ☽ ⚹ ♂
21 49 ☽ ± ♇

JULY 2014

LONGITUDES

Date	Sidereal time h m s	Sun ☉ ° ′ ″	Moon ☽ ° ′ ″	Moon ☽ 24.00 ° ′ ″	Mercury ☿ ° ′	Venus ♀ ° ′	Mars ♂ ° ′	Jupiter ♃ ° ′	Saturn ♄ ° ′	Uranus ♅ ° ′	Neptune ♆ ° ′	Pluto ♇ ° ′
01	06 37 51	09 ♋ 35 13	25 ♌ 22 09	01 ♍ 17 12	24 ♊ 23	09 ♊ 30	18 ♎ 33	26 ♋ 43	16 ♏ 57	16 ♈ 20	07 ♓ 28	12 ♑ 22
02	06 41 48	10 32 27	07 ♍ 12 41	13 ♍ 09 05	24 D 25	10 42	18 57	26 56	16 R 55	16 21	07 R 28	12 R 20
03	06 45 44	11 29 40	19 ♍ 06 55	25 ♍ 06 47	24 32	11 54	19 22	27 09	16 53	16 22	07 27	12 19
04	06 49 41	12 26 52	01 ♎ 09 17	07 ♎ 15 01	24 44	13 05	19 47	27 22	16 51	16 23	07 26	12 17
05	06 53 37	13 24 05	13 ♎ 24 39	19 ♎ 38 50	25 01	14 17	20 12	27 35	16 50	16 24	07 25	12 16
06	06 57 34	14 21 17	25 ♎ 58 11	02 ♏ 23 19	25 24	15 29	20 38	27 48	16 49	16 25	07 25	12 14
07	07 01 31	15 18 29	08 ♏ 54 45	15 ♏ 32 59	25 51	16 41	21 04	28 02	16 47	16 25	07 24	12 13
08	07 05 27	16 15 41	22 ♏ 18 21	29 ♏ 11 06	26 23	17 53	21 30	28 15	16 46	16 26	07 23	12 11
09	07 09 24	17 12 52	06 ♐ 11 15	13 ♐ 18 42	27 00	19 05	21 57	28 28	16 45	16 27	07 22	12 10
10	07 13 20	18 10 04	20 ♐ 33 05	27 ♐ 53 50	27 42	20 16	22 23	28 41	16 44	16 27	07 21	12 08
11	07 17 17	19 07 16	05 ♑ 20 09	12 ♑ 51 02	28 29	21 28	22 51	28 54	16 43	16 28	07 20	12 07
12	07 21 13	20 04 28	20 ♑ 25 17	28 ♑ 01 35	29 ♊ 21	22 41	23 18	29 08	16 42	16 28	07 19	12 05
13	07 25 10	21 01 40	05 ♒ 38 35	13 ♒ 14 52	00 ♋ 18	23 53	23 46	29 21	16 41	16 29	07 18	12 04
14	07 29 06	21 58 52	20 ♒ 49 08	28 ♒ 20 10	01 19	25 05	24 14	29 34	16 41	16 29	07 17	12 02
15	07 33 03	22 56 05	05 ♓ 46 55	13 ♓ 08 34	02 25	26 17	24 42	29 ♋ 48	16 40	16 29	07 16	12 01
16	07 37 00	23 53 18	20 ♓ 24 29	27 ♓ 34 15	03 36	27 29	25 11	00 ♌ 01	16 40	16 30	07 15	11 59
17	07 40 56	24 50 32	04 ♈ 37 39	11 ♈ 34 39	04 51	28 41	25 40	00 14	16 39	16 30	07 14	11 58
18	07 44 53	25 47 46	18 ♈ 25 21	25 ♈ 09 59	06 11	29 ♊ 54	26 09	00 27	16 39	16 30	07 13	11 56
19	07 48 49	26 45 01	01 ♉ 48 54	08 ♉ 22 27	07 35	01 ♋ 06	26 39	00 41	16 39	16 30	07 12	11 55
20	07 52 46	27 42 18	14 ♉ 51 05	21 ♉ 15 14	09 03	02 18	27 08	00 54	16 39	16 30	07 11	11 54
21	07 56 42	28 39 34	27 ♉ 35 20	03 ♊ 51 50	10 36	03 31	27 39	01 07	16 D 39	16 31	07 09	11 52
22	08 00 39	29 ♋ 36 52	10 ♊ 05 07	16 ♊ 15 34	12 12	04 43	28 09	01 21	16 39	16 R 31	07 08	11 51
23	08 04 35	00 ♌ 34 10	22 ♊ 23 30	28 ♊ 29 15	13 53	05 56	28 39	01 34	16 39	16 30	07 07	11 49
24	08 08 32	01 31 30	04 ♋ 33 04	10 ♋ 35 11	15 36	07 08	29 10	01 47	16 39	16 30	07 06	11 48
25	08 12 29	02 28 49	16 ♋ 35 49	22 ♋ 35 08	17 24	08 21	29 ♎ 41	02 01	16 40	16 30	07 04	11 47
26	08 16 25	03 26 10	28 ♋ 33 18	04 ♌ 30 30	19 14	09 34	00 ♏ 13	02 14	16 40	16 30	07 03	11 45
27	08 20 22	04 23 31	10 ♌ 26 54	16 ♌ 22 40	21 08	10 46	00 44	02 28	16 41	16 30	07 02	11 44
28	08 24 18	05 20 53	22 ♌ 17 59	28 ♌ 13 05	23 04	11 59	01 16	02 41	16 41	16 30	07 01	11 42
29	08 28 15	06 18 16	04 ♍ 08 13	10 ♍ 03 40	25 02	13 12	01 48	02 54	16 42	16 29	06 59	11 41
30	08 32 11	07 15 39	15 ♍ 59 47	21 ♍ 56 55	27 03	14 24	02 20	03 07	16 43	16 29	06 58	11 40
31	08 36 08	08 ♌ 13 02	27 ♍ 55 30	03 ♎ 55 59	29 ♋ 05	15 ♋ 37	02 ♏ 53	03 ♌ 21	16 ♏ 44	16 ♈ 28	06 ♓ 56	11 ♑ 38

Date	Moon True ☊ ° ′	Moon Mean ☊ ° ′	Moon ☽ Latitude ° ′
01	25 ♎ 04	24 ♎ 39	04 S 21
02	24 R 57	24 36	03 44
03	24 52	24 33	02 57
04	24 51	24 30	02 02
05	24 D 50	24 27	01 S 00
06	24 R 50	24 23	00 N 06
07	24 50	24 20	01 14
08	24 47	24 17	02 20
09	24 43	24 14	03 20
10	24 36	24 11	04 10
11	24 27	24 07	04 44
12	24 16	24 04	05 00
13	24 05	24 01	04 54
14	23 56	23 58	04 28
15	23 48	23 55	03 44
16	23 44	23 52	02 46
17	23 41	23 48	01 39
18	23 D 41	23 45	00 N 28
19	23 R 41	23 42	00 S 43
20	23 40	23 39	01 49
21	23 37	23 36	02 49
22	23 33	23 33	03 39
23	23 25	23 29	04 18
24	23 14	23 26	04 45
25	23 02	23 23	04 58
26	22 48	23 20	04 59
27	22 35	23 17	04 46
28	22 23	23 13	04 21
29	22 12	23 10	03 45
30	22 05	23 07	02 58
31	22 ♎ 00	23 ♎ 04	02 S 04

DECLINATIONS

Date	Sun ☉ ° ′	Moon ☽ ° ′	Mercury ☿ ° ′	Venus ♀ ° ′	Mars ♂ ° ′	Jupiter ♃ ° ′	Saturn ♄ ° ′	Uranus ♅ ° ′	Neptune ♆ ° ′	Pluto ♇ ° ′
01	23 N 05	08 N 58	18 N 47	20 N 30	07 S 44	21 N 10	14 S 38	05 N 48	09 S 27	20 S 16
02	23 01	05 24	18 52	20 42	07 54	21 08	14 38	05 49	09 27	20 16
03	22 56	01 N 35	18 58	20 54	08 05	21 05	14 38	05 49	09 27	20 16
04	22 51	02 S 19	19 06	21 06	08 15	21 03	14 37	05 49	09 27	20 16
05	22 46	06 13	19 14	21 17	08 26	21 00	14 37	05 50	09 28	20 17
06	22 40	09 56	19 24	21 27	08 37	20 58	14 37	05 50	09 28	20 17
07	22 33	13 18	19 34	21 37	08 47	20 55	14 37	05 50	09 28	20 17
08	22 27	16 05	19 45	21 46	08 58	20 52	14 37	05 50	09 29	20 17
09	22 20	18 03	19 57	21 54	09 09	20 50	14 37	05 51	09 29	20 18
10	22 12	18 57	20 10	22 02	09 21	20 47	14 37	05 51	09 29	20 18
11	22 04	18 36	20 22	22 10	09 32	20 45	14 37	05 51	09 30	20 18
12	21 56	16 56	20 35	22 17	09 43	20 42	14 37	05 51	09 30	20 18
13	21 47	14 06	20 48	22 23	09 54	20 39	14 37	05 51	09 31	20 19
14	21 38	10 18	21 01	22 28	10 06	20 37	14 37	05 51	09 31	20 19
15	21 29	05 55	21 14	22 33	10 17	20 34	14 37	05 51	09 31	20 19
16	21 19	01 S 15	21 26	22 38	10 29	20 31	14 37	05 52	09 32	20 19
17	21 09	03 N 21	21 37	22 41	10 40	20 28	14 37	05 52	09 32	20 20
18	20 59	07 39	21 48	22 44	10 52	20 26	14 37	05 52	09 33	20 20
19	20 48	11 26	21 57	22 47	11 03	20 23	14 38	05 52	09 33	20 20
20	20 37	14 33	22 05	22 49	11 15	20 20	14 38	05 52	09 34	20 20
21	20 26	16 53	22 12	22 50	11 27	20 17	14 38	05 52	09 34	20 21
22	20 14	18 21	22 17	22 50	11 38	20 14	14 38	05 52	09 35	20 21
23	20 01	18 55	22 21	22 50	11 50	20 11	14 39	05 52	09 35	20 21
24	19 49	18 37	22 22	22 49	12 02	20 09	14 39	05 52	09 36	20 21
25	19 36	17 28	22 21	22 48	12 14	20 06	14 39	05 51	09 36	20 22
26	19 23	15 34	22 18	22 45	12 26	20 03	14 40	05 51	09 37	20 22
27	19 10	13 01	22 12	22 43	12 38	20 00	14 40	05 51	09 37	20 22
28	18 56	09 58	22 04	22 39	12 50	19 57	14 41	05 51	09 38	20 22
29	18 42	06 30	21 53	22 35	13 02	19 54	14 41	05 51	09 38	20 23
30	18 27	02 47	21 40	22 30	13 13	19 51	14 42	05 51	09 39	20 23
31	18 N 12	01 S 04	21 N 24	22 N 25	13 S 25	19 N 48	14 S 42	05 N 51	09 S 39	20 S 23

ZODIAC SIGN ENTRIES

Date	h	m	Planets
01	21	23	☽ ♍
04	09	43	☽ ♎
06	19	33	☽ ♏
09	01	24	☽ ♐
11	03	24	☽ ♑
13	03	07	☽ ♒
13	04	45	☿ ♋
15	02	40	☽ ♓
16	10	30	♃ ♌
17	04	07	☽ ♈
18	14	06	♀ ♋
19	08	42	☽ ♉
21	16	36	☽ ♊
22	21	41	☉ ♌
24	02	59	☽ ♋
26	02	25	♂ ♏
26	14	55	☽ ♌
29	03	37	☽ ♍
31	16	09	☽ ♎
31	22	46	☿ ♌

LATITUDES

Date	Mercury ☿ ° ′	Venus ♀ ° ′	Mars ♂ ° ′	Jupiter ♃ ° ′	Saturn ♄ ° ′	Uranus ♅ ° ′	Neptune ♆ ° ′	Pluto ♇ ° ′
01	04 S 32	01 S 24	00 S 30	00 N 22	02 N 21	00 S 40	00 S 44	02 N 36
04	04 14	01 17	00 34	00 22	02 20	00 40	00 44	02 36
07	03 48	01 10	00 37	00 22	02 20	00 40	00 44	02 36
10	03 15	01 02	00 41	00 23	02 19	00 40	00 44	02 36
13	02 38	00 55	00 44	00 23	02 18	00 41	00 44	02 35
16	01 57	00 47	00 47	00 23	02 17	00 41	00 44	02 35
19	01 16	00 39	00 50	00 23	02 17	00 41	00 44	02 35
22	00 S 35	00 31	00 53	00 24	02 16	00 41	00 44	02 34
25	00 N 03	00 23	00 56	00 24	02 15	00 41	00 44	02 34
28	00 37	00 15	00 59	00 24	02 14	00 41	00 45	02 33
31	01 N 05	00 S 07	01 S 01	00 N 24	02 N 13	00 S 41	00 S 45	02 N 33

DATA

Julian Date	2456840
Delta T	+69 seconds
Ayanamsa	24° 03′ 42″
Synetic vernal point	05° ♓ 03′ 17″
True obliquity of ecliptic	23° 26′ 06″

MOON'S PHASES, APSIDES AND POSITIONS ☽

Date	h	m	Phase	Longitude ° ′	Eclipse Indicator
05	11	59	☽	13 ♎ 24	
12	11	25	○	20 ♑ 03	
19	02	08	☾	26 ♈ 21	
26	22	42	●	03 ♌ 52	

Day	h	m		
13	08	33	Perigee	
28	03	41	Apogee	
03	21	47	0S	
10	17	30	Max dec	18° S 59′
16	18	27	0N	
23	15	31	Max dec	18° N 56′
31	05	22	0S	

LONGITUDES

Date	Chiron ⚷ ° ′	Ceres ⚳ ° ′	Pallas ⚴ ° ′	Juno ⚵ ° ′	Vesta ⚶ ° ′	Black Moon Lilith ⚸ ° ′
01	17 ♓ 42	21 ♎ 11	14 ♍ 07	08 ♊ 08	20 ♎ 53	13 ♌ 18
11	17 ♓ 34	23 ♎ 06	17 ♍ 50	13 ♊ 54	23 ♎ 33	14 ♌ 26
21	17 ♓ 20	25 ♎ 25	21 ♍ 41	19 ♊ 36	26 ♎ 41	15 ♌ 33
31	17 ♓ 02	28 ♎ 04	25 ♍ 39	25 ♊ 14	00 ♏ 11	16 ♌ 40

ASPECTARIAN

h m	Aspects
01 Tuesday	
03 17	☽ Q ♀
08 38	☽ # ♆
10 00	☽ ✶ ☿
10 16	☽ ∠ ☉
12 50	☿ St D
14 47	☽ ⚺ ♃
16 02	☽ ⚼ ♇
20 08	☽ # ♂
20 20	☉ ⚺ ♀
02 Wednesday	
00 07	☽ ⚼ ♅
03 10	☽ ⊥ ♃
05 11	☽ ∠ ♂
07 22	☽ Q ♄
09 18	☽ ∥ ♅
10 23	☽ Q ☿
12 30	☽ ☍ ♆
18 22	☽ ± ♅
19 19	☽ ✶ ☉
19 50	☽ □ ♀
21 43	☽ ∠ ♃
22 20	☽ △ ♇
03 Thursday	
00 01	☽ ⊥ ♂
06 28	☽ ⚻ ♅
07 32	☽ ✶ ♄
12 31	☽ ⚺ ♂
18 18	♀ ∠ ♃
20 15	♀ ⚻ ♇
21 31	☽ Q ☉
23 01	☽ □ ☿
04 Friday	
04 21	☽ ✶ ♃
06 41	♀ ∥ ♃
08 03	☉ ☍ ♇
13 23	☽ ∠ ♄
05 Saturday	
00 21	☽ ⚻ ♆
04 26	☽ Q ♃
07 00	☽ ⊥ ♄
09 33	☽ # ♅
09 47	☽ □ ♇
11 59	☽ □ ☉
12 01	☽ ± ♆
13 52	☽ △ ♀
17 46	☽ ☍ ♅
18 35	☽ ⚺ ♄
06 Sunday	
01 31	☽ ☌ ♂
02 48	☽ ∥ ♂
05 16	☽ ⚼ ♆
08 52	☽ ∥ ♆
10 52	☽ △ ☿
15 31	☽ □ ♃
19 59	☽ Q ♇
21 19	☽ ⚼ ♀
07 Monday	
06 49	♀ ✶ ♅
09 14	☽ △ ♆
14 09	♀ ⚻ ♄
15 32	☽ ± ♀
15 39	☽ ⚼ ☿
17 59	☽ ✶ ♇
22 35	☽ ∥ ♄
08 Tuesday	
00 27	☽ △ ☉
01 34	☽ ⚻ ♅
02 12	☽ ☌ ♄
03 23	☽ ⚻ ♀
08 27	☽ ± ☿
10 32	☽ ⚺ ♂
12 13	☽ ± ♅
16 23	☉ □ ♅
19 27	☽ ⚻ ☿
20 31	☽ ∠ ♇
21 23	☽ ⊥ ♂
22 32	☽ △ ♃
09 Wednesday	
00 28	☉ △ ♄
03 53	☽ ⚼ ♅
04 43	☽ ⚼ ☉
11 57	☽ ⊥ ♇
13 19	☽ ∠ ♂
14 00	☽ □ ♆
21 06	☽ ± ☉
22 04	☽ ⚺ ♇
10 Thursday	
00 27	☽ ⚼ ♃
05 14	☽ △ ♅
05 42	☽ ⚺ ♄
07 48	☽ ⚻ ☉
10 01	♂ ⚼ ♆
11 30	☽ ☍ ♀
15 07	☽ ✶ ♂
15 33	☽ ± ♃
15 34	☽ ⊥ ♄
19 51	☽ Q ♆
11 Friday	
00 19	☽ ☍ ☿
01 29	☽ ⚻ ♃
03 13	☉ ∥ ♀
03 51	☿ # ♇
06 11	☽ ∠ ♄
07 49	♂ ∥ ♆
11 11	☽ Q ♂
11 21	☿ Q ♅
15 12	☽ ✶ ♆
22 48	☽ ☌ ♇
12 Saturday	
04 04	☿ ⚺ ♃
05 45	☽ □ ♅
06 07	☽ ✶ ♄
11 25	☽ ☍ ☉
12 30	♀ ± ♄
15 00	☽ ∠ ♆
15 52	☽ ⚻ ♀
16 42	☽ □ ♂
22 07	☿ ∥ ♃
23 05	♀ ⊥ ♃
13 Sunday	
01 03	☽ Q ♄
01 56	☽ ☍ ♃
02 08	☽ ± ♀
03 00	☽ ⚻ ☿
05 10	☽ ⊥ ♆
08 14	☽ ∥ ♄
08 22	♀ △ ♂
10 10	☽ Q ♅
13 06	☽ ± ☿
14 37	☽ ⚺ ♆
17 32	☽ ⚼ ♀
22 07	☽ ⚺ ♇
14 Monday	
04 20	☽ ⚼ ☿
05 07	☽ ✶ ♅
05 26	☽ □ ♄
07 35	☽ ⊥ ♇
13 11	☽ ∥ ♂
13 58	☽ ⚻ ☉
16 30	☽ ∥ ♆
17 37	☽ △ ♂
19 23	☽ △ ♀
19 33	☉ ⚼ ♆
19 56	☿ ⚼ ♄
21 54	☽ ∠ ♇
15 Tuesday	
00 13	☽ ± ☉
02 11	☽ ⚻ ♃
05 04	☽ ∠ ♅
06 07	☽ △ ☿
12 01	☽ ± ♃
12 17	☽ # ♅
14 25	☽ ☌ ♆
15 44	☽ ⚼ ☉
18 36	☽ ⚼ ♂
19 40	☽ ⊥ ♅
22 08	☽ ✶ ♇
16 Wednesday	
02 56	☽ ⚼ ♃
04 56	☉ ∥ ☿
05 31	☽ ⚺ ♅
05 47	☽ △ ♄
09 54	☽ ± ♂
17 58	☽ Q ♇
18 13	☽ △ ☉
20 16	☽ ⚻ ♂
17 Thursday	
00 57	☽ □ ♀
04 23	☽ △ ♃
06 55	☽ ⚼ ♄
08 13	♀ Q ♅
12 25	☽ □ ☿
16 28	☽ ⚺ ♆
22 23	☽ ± ♄
18 Friday	
00 39	☽ □ ♇
01 43	☽ ∥ ♅
02 52	☽ ⊥ ♆
08 37	☽ ☌ ♅
08 52	☽ ⚻ ♄
10 59	☽ Q ♀
18 43	☽ ∠ ♆
23 25	☽ Q ☿
23 35	☽ # ♆
19 Saturday	
01 45	♀ ⚺ ♃
02 08	☽ □ ☉
02 18	☽ ☍ ♂
05 36	☿ △ ♆
06 32	☉ □ ♂
09 16	☽ # ♂
09 54	☽ □ ♃
10 34	☽ ✶ ♀
21 49	☽ ✶ ♆
22 50	♀ ⚼ ♄
23 52	☽ ✶ ☿
20 Sunday	
06 31	☽ △ ♇
09 01	♃ # ♇
12 44	☽ # ♄
13 43	☽ Q ☉
15 05	☽ ⚺ ♅
15 21	☽ ☍ ♄
17 04	☽ ∠ ♀
19 43	☽ Q ♃
20 05	☽ Q ♆
20 35	♄ St D
21 Monday	
02 22	☽ ⊥ ♅
07 40	☽ ∠ ☿
10 38	☽ ⚼ ♇
11 50	☽ ⊥ ♀
12 06	☽ ⚻ ♂
14 12	☽ ✶ ☉
15 14	♃ ± ♆
18 52	☽ ✶ ♃
19 29	☽ ∠ ♅
21 45	☉ # ♇
22 Tuesday	
00 04	☽ ± ♂
00 32	☽ ⚺ ♀
02 54	♅ St R
03 22	☽ ⊥ ☿
03 50	☽ ± ♇
06 19	☽ □ ♆
06 49	☿ ☍ ♇
10 17	☉ ∥ ♃
15 24	☽ ⚻ ♇
16 44	☽ ⚺ ☿
18 12	☽ ⚼ ♂
21 32	☽ ∠ ☉
23 Wednesday	
00 24	☽ ∠ ♃
00 29	☽ ✶ ♅
00 45	☽ ⚻ ♄
12 30	☽ ± ♄
16 39	☽ ⊥ ☉
18 22	☽ ⊥ ♃
24 Thursday	
00 02	☽ Q ♅
00 53	☽ △ ♂
01 27	☉ ± ♆
05 29	☽ ⚺ ☉
06 15	☽ ⚼ ♄
06 26	☽ ⚺ ♃
11 09	♀ △ ♆
17 03	☽ △ ♆
17 43	☽ ☌ ♀
20 44	☉ ☌ ♃
25 Friday	
00 07	☿ □ ♅
02 11	☿ △ ♄
02 23	☽ ☍ ♇
11 49	☽ □ ♅
12 08	☽ △ ♄
13 53	☽ ☌ ☿
15 54	♂ Q ♇
22 57	☽ ⚼ ♆
26 Saturday	
15 29	☽ □ ♂
17 01	☽ ± ♆
19 33	☽ ☌ ♃
21 12	☽ # ♄
22 42	☽ ☌ ☉
27 Sunday	
05 06	☽ ⚻ ♆
12 44	☽ ⚺ ♀
14 35	☽ ⚻ ♇
15 07	☽ # ♂
23 05	☿ ⚼ ♆
28 Monday	
00 14	☽ △ ♅
00 37	☽ □ ♄
02 15	☽ ⊥ ♀
02 43	☽ ± ♇
05 34	☽ Q ♂
06 39	♀ ☍ ♇
13 52	☽ ⚺ ☿
14 25	☽ # ♆
20 55	☽ ⚼ ♇
22 35	☽ ∠ ♀
29 Tuesday	
04 28	☽ ⊥ ☿
06 38	☽ ⚼ ♅
07 02	☽ ✶ ♂
09 27	☽ ⚺ ♃
13 09	☽ Q ♄
16 19	☽ ∥ ♅
16 47	☽ ⚺ ☉
17 46	☽ ☍ ♆
21 50	☽ ⊥ ♃
30 Wednesday	
00 51	☽ ± ♅
02 23	☽ ∠ ☿
03 15	☽ △ ♇
04 44	☉ ⚻ ♆
05 59	☽ ⊥ ☉
08 26	☽ ✶ ♀
12 59	☽ ⚻ ♅
13 28	☽ ✶ ♄
14 50	☽ ∠ ♂
16 23	☽ ∠ ♃
31 Thursday	
01 44	☽ ∠ ☉
09 48	☽ ⊥ ♂
11 19	☽ Q ♀
14 47	☽ ✶ ☿
19 38	☽ ∠ ♄
22 22	☽ ⚺ ♂
23 02	☽ ✶ ♃

All ephemeris data is given at 12.00 UT and the Moon's longitude is additionally given for 24.00 UT

AUGUST 2014

LONGITUDES

Date	Sidereal time h m s	Sun ☉ ° ' "	Moon ☽ ° ' "	Moon ☽ 24.00 ° ' "	Mercury ☿ ° '	Venus ♀ ° '	Mars ♂ ° '	Jupiter ♃ ° '	Saturn ♄ ° '	Uranus ♅ ° '	Neptune ♆ ° '	Pluto ♇ ° '
01	08 40 04	09 ♌ 10 27	09 ♎ 58 53	16 ♎ 04 44	01 ♌ 08	16 ♋ 50	03 ♏ 25	03 ♌ 34	16 ♏ 45	16 ♈ 28	06 ♓ 55	11 ♑ 37
02	08 44 01	10 07 52	22 ♎ 14 07	28 ♎ 27 36	03 12	18 03	03 58	03 47	16 46	16 R 27	06 R 54	11 R 36
03	08 47 58	11 05 17	04 ♏ 45 49	11 ♏ 09 20	05 17	19 16	04 31	04 01	16 48	16 27	06 52	11 35
04	08 51 54	12 02 44	17 ♏ 38 43	24 ♏ 14 30	07 22	20 29	05 05	04 14	16 49	16 26	06 51	11 33
05	08 55 51	13 00 11	00 ♐ 57 07	07 ♐ 46 54	09 28	21 42	05 38	04 27	16 51	16 25	06 49	11 32
06	08 59 47	13 57 38	14 ♐ 44 04	21 ♐ 48 37	11 32	22 55	06 12	04 40	16 52	16 25	06 48	11 31
07	09 03 44	14 55 07	29 ♐ 00 26	06 ♑ 19 06	13 37	24 08	06 46	04 54	16 54	16 24	06 46	11 30
08	09 07 40	15 52 36	13 ♑ 44 01	21 ♑ 14 21	15 41	25 21	07 20	05 07	16 56	16 23	06 45	11 29
09	09 11 37	16 50 06	28 ♑ 49 03	06 ♒ 26 51	17 44	26 34	07 55	05 20	16 57	16 22	06 43	11 27
10	09 15 33	17 47 37	14 ♒ 06 25	21 ♒ 46 17	19 46	27 48	08 29	05 33	16 59	16 21	06 42	11 26
11	09 19 30	18 45 09	29 ♒ 25 02	07 ♓ 01 17	21 46	29 ♋ 01	09 04	05 46	17 01	16 21	06 40	11 25
12	09 23 27	19 42 42	14 ♓ 33 46	22 ♓ 01 27	23 46	00 ♌ 14	09 39	06 00	17 04	16 20	06 39	11 24
13	09 27 23	20 40 17	29 ♓ 23 25	06 ♈ 39 02	25 45	01 27	10 14	06 13	17 06	16 19	06 37	11 23
14	09 31 20	21 37 53	13 ♈ 47 54	20 ♈ 49 45	27 42	02 41	10 49	06 26	17 08	16 17	06 36	11 22
15	09 35 16	22 35 30	27 ♈ 44 36	04 ♉ 32 32	29 ♌ 37	03 54	11 25	06 39	17 11	16 16	06 34	11 21
16	09 39 13	23 33 09	11 ♉ 13 50	17 ♉ 48 52	01 ♍ 32	05 08	12 00	06 52	17 13	16 15	06 32	11 21
17	09 43 09	24 30 50	24 ♉ 18 02	00 ♊ 41 52	03 25	06 21	12 36	07 05	17 16	16 14	06 31	11 19
18	09 47 06	25 28 32	07 ♊ 00 50	13 ♊ 15 30	05 16	07 35	13 12	07 18	17 18	16 13	06 29	11 18
19	09 51 02	26 26 15	19 ♊ 26 21	25 ♊ 33 54	07 06	08 48	13 48	07 31	17 21	16 11	06 28	11 17
20	09 54 59	27 24 01	01 ♋ 38 38	07 ♋ 40 59	08 55	10 02	14 25	07 44	17 24	16 10	06 26	11 16
21	09 58 56	28 21 48	13 ♋ 41 23	19 ♋ 40 13	10 42	11 15	15 01	07 57	17 27	16 09	06 24	11 15
22	10 02 52	29 ♌ 19 36	25 ♋ 37 48	01 ♌ 34 28	12 28	12 29	15 38	08 09	17 30	16 07	06 23	11 14
23	10 06 49	00 ♍ 17 26	07 ♌ 30 30	13 ♌ 26 07	14 13	13 43	16 15	08 22	17 33	16 06	06 21	11 13
24	10 10 45	01 15 18	19 ♌ 21 35	25 ♌ 17 06	15 56	14 57	16 52	08 35	17 36	16 04	06 20	11 12
25	10 14 42	02 13 11	01 ♍ 12 53	07 ♍ 09 07	17 37	16 10	17 29	08 48	17 39	16 03	06 18	11 12
26	10 18 38	03 11 05	13 ♍ 06 32	19 ♍ 03 50	19 18	17 24	18 06	09 01	17 43	16 01	06 16	11 11
27	10 22 35	04 09 01	25 ♍ 02 47	01 ♎ 03 08	20 57	18 38	18 44	09 13	17 46	16 00	06 15	11 10
28	10 26 31	05 06 58	07 ♎ 05 10	13 ♎ 09 13	22 35	19 52	19 21	09 26	17 50	15 58	06 13	11 09
29	10 30 28	06 04 57	19 ♎ 15 38	25 ♎ 24 48	24 11	21 06	19 59	09 38	17 53	15 56	06 11	11 09
30	10 34 25	07 02 57	01 ♏ 37 08	07 ♏ 53 04	25 46	22 20	20 37	09 51	17 57	15 55	06 10	11 08
31	10 38 21	08 ♍ 00 59	14 ♏ 13 03	20 ♏ 37 33	27 ♍ 20	23 ♌ 34	21 ♏ 15	10 ♌ 03	18 ♏ 01	15 ♈ 53	06 ♓ 08	11 ♑ 07

Date	Moon True ☊ ° '	Moon Mean ☊ ° '	Moon ☽ Latitude ° '
01	21 ♎ 57	23 ♎ 01	01 S 03
02	21 D 57	22 58	00 N 02
03	21 57	22 54	01 08
04	21 R 57	22 51	02 12
05	21 56	22 48	03 12
06	21 52	22 45	04 02
07	21 47	22 42	04 40
08	21 39	22 39	05 00
09	21 30	22 35	05 01
10	21 21	22 32	04 40
11	21 12	22 29	03 59
12	21 06	22 26	03 02
13	21 02	22 23	01 53
14	21 00	22 19	00 N 38
15	21 D 00	22 16	00 S 36
16	21 01	22 13	01 46
17	21 01	22 10	02 49
18	21 R 01	22 07	03 41
19	20 58	22 04	04 21
20	20 53	22 00	04 49
21	20 46	21 57	05 04
22	20 38	21 54	05 05
23	20 28	21 51	04 53
24	20 18	21 48	04 28
25	20 10	21 45	03 52
26	20 03	21 41	03 05
27	19 57	21 38	02 10
28	19 55	21 35	01 09
29	19 54	21 32	00 S 03
30	19 D 54	21 29	01 N 03
31	19 ♎ 56	21 ♎ 25	02 N 08

DECLINATIONS

Date	Sun ☉ ° '	Moon ☽ ° '	Mercury ☿ ° '	Venus ♀ ° '	Mars ♂ ° '	Jupiter ♃ ° '	Saturn ♄ ° '	Uranus ♅ ° '	Neptune ♆ ° '	Pluto ♇ ° '
01	17 N 57	04 S 55	21 N 05	22 N 19	13 S 37	19 N 45	14 S 43	05 N 50	09 S 40	20 S 23
02	17 42	08 38	20 43	22 12	13 49	19 42	14 43	05 50	09 40	20 24
03	17 27	12 03	20 19	22 05	14 01	19 39	14 44	05 50	09 41	20 24
04	17 11	14 59	19 53	21 56	14 13	19 36	14 44	05 50	09 41	20 24
05	16 55	17 13	19 25	21 48	14 25	19 33	14 45	05 49	09 42	20 24
06	16 38	18 33	18 54	21 39	14 37	19 30	14 46	05 49	09 42	20 25
07	16 21	18 46	18 21	21 29	14 49	19 27	14 47	05 49	09 43	20 25
08	16 04	17 45	17 47	21 18	15 01	19 23	14 47	05 48	09 44	20 25
09	15 47	15 29	17 11	21 07	15 13	19 20	14 48	05 48	09 44	20 25
10	15 30	12 07	16 34	20 55	15 25	19 17	14 49	05 48	09 45	20 26
11	15 12	07 56	15 55	20 43	15 37	19 14	14 50	05 47	09 45	20 26
12	14 54	03 S 17	15 15	20 30	15 48	19 11	14 50	05 47	09 46	20 26
13	14 36	01 N 29	14 34	20 16	16 00	19 08	14 51	05 47	09 47	20 26
14	14 18	06 02	13 52	20 02	16 12	19 05	14 52	05 46	09 47	20 27
15	13 59	10 06	13 09	19 47	16 24	19 01	14 53	05 46	09 48	20 27
16	13 40	13 31	12 26	19 32	16 35	18 58	14 54	05 45	09 48	20 27
17	13 21	16 07	11 42	19 16	16 47	18 55	14 55	05 45	09 49	20 27
18	13 02	17 51	10 57	19 00	16 58	18 52	14 56	05 44	09 50	20 28
19	12 42	18 40	10 13	18 43	17 10	18 49	14 57	05 44	09 50	20 28
20	12 22	18 36	09 28	18 25	17 21	18 46	14 58	05 43	09 51	20 28
21	12 02	17 42	08 42	18 07	17 33	18 42	14 59	05 43	09 51	20 28
22	11 42	16 01	07 57	17 49	17 44	18 39	15 00	05 42	09 52	20 28
23	11 22	13 40	07 11	17 30	17 55	18 36	15 01	05 41	09 53	20 29
24	11 02	10 46	06 26	17 10	18 06	18 33	15 02	05 41	09 53	20 29
25	10 41	07 25	05 40	16 50	18 17	18 29	15 04	05 40	09 54	20 29
26	10 20	03 N 47	04 55	16 30	18 28	18 26	15 05	05 40	09 54	20 29
27	09 59	00 S 02	04 10	16 09	18 39	18 23	15 06	05 39	09 55	20 30
28	09 38	03 52	03 25	15 47	18 50	18 20	15 07	05 38	09 56	20 30
29	09 17	07 35	02 40	15 25	19 01	18 16	15 08	05 38	09 56	20 30
30	08 55	11 03	01 55	15 03	19 11	18 13	15 10	05 37	09 57	20 30
31	08 N 34	14 S 04	01 N 11	14 N 40	19 S 22	18 N 10	15 S 11	05 N 36	09 S 58	20 S 30

ZODIAC SIGN ENTRIES

Date	h	m	Planets
03	02	57	☽ ♏
05	10	19	☽ ♐
07	13	38	☽ ♑
09	13	52	☽ ♒
11	12	55	☽ ♓
12	07	24	♀ ♌
13	13	00	☽ ♈
15	15	58	☽ ♉
15	16	44	☿ ♍
17	22	41	☽ ♊
20	08	45	☽ ♋
22	20	49	☽ ♌
23	04	46	☉ ♍
25	09	33	☽ ♍
27	21	54	☽ ♎
30	08	53	☽ ♏

LATITUDES

Date	Mercury ☿ ° '	Venus ♀ ° '	Mars ♂ ° '	Jupiter ♃ ° '	Saturn ♄ ° '	Uranus ♅ ° '	Neptune ♆ ° '	Pluto ♇ ° '
01	01 N 12	00 S 04	01 S 02	00 N 24	02 N 13	00 S 41	00 S 45	02 N 33
04	01 31	00 N 04	01 04	00 25	02 12	00 41	00 45	02 32
07	01 42	00 12	01 06	00 25	02 12	00 41	00 45	02 32
10	01 46	00 20	01 09	00 25	02 11	00 41	00 45	02 32
13	01 44	00 27	01 10	00 26	02 10	00 41	00 45	02 31
16	01 36	00 34	01 13	00 26	02 09	00 41	00 45	02 31
19	01 25	00 41	01 14	00 26	02 09	00 42	00 45	02 30
22	01 09	00 47	01 16	00 27	02 08	00 42	00 45	02 30
25	00 51	00 53	01 18	00 27	02 07	00 42	00 45	02 29
28	00 30	00 59	01 19	00 27	02 06	00 42	00 45	02 29
31	00 08	01 N 04	01 S 21	00 N 28	02 N 06	00 S 42	00 S 45	02 N 28

DATA

Julian Date	2456871
Delta T	+69 seconds
Ayanamsa	24° 03' 47"
Synetic vernal point	05° ♓ 03' 12"
True obliquity of ecliptic	23° 26' 06"

MOON'S PHASES, APSIDES AND POSITIONS ☽

Date	h	m	Phase	Longitude ° '	Eclipse Indicator
04	00	50	☽	11 ♏ 36	
10	18	09	○	18 ♒ 02	
17	12	26	☾	24 ♉ 32	
25	14	13	●	02 ♍ 19	

Day	h	m		
10	17	49	Perigee	
24	06	25	Apogee	
07	04	26	Max dec	18° S 50'
13	04	29	0N	
19	22	08	Max dec	18° N 45'
27	11	50	0S	

LONGITUDES

Date	Chiron ⚷ ° '	Ceres ⚳ ° '	Pallas ⚴ ° '	Juno ⚵ ° '	Vesta ⚶ ° '	Black Moon Lilith ⚸ ° '
01	17 ♓ 00	28 ♎ 21	26 ♍ 04	25 ♊ 48	00 ♏ 34	16 ♌ 47
11	16 ♓ 37	01 ♏ 19	00 ♎ 09	01 ♋ 20	04 ♏ 25	17 ♌ 54
21	16 ♓ 12	04 ♏ 32	04 ♎ 19	06 ♋ 46	08 ♏ 33	19 ♌ 01
31	15 ♓ 45	07 ♏ 56	08 ♎ 33	12 ♋ 03	12 ♏ 54	20 ♌ 08

All ephemeris data is given at 12.00 UT and the Moon's longitude is additionally given for 24.00 UT

ASPECTARIAN

h m	Aspects
01 Friday	
04 43	♀ □ ♅
05 57	☽ ⊼ ♆
09 30	☿ ± ♆
10 16	☽ ⚹ ☉
10 22	♀ △ ♄
13 32	☽ ⊥ ♄
15 14	☽ □ ♇
17 47	☽ ± ♆
17 49	☽ ⋕ ♅
19 29	☽ Q ☿
22 46	♂ □ ♃
23 12	☽ Q ♃
02 Saturday	
00 45	☽ ☍ ♅
01 21	☽ ⊻ ♄
02 58	☽ □ ♀
11 20	☽ ⚼ ♆
11 47	☽ Q ☉
19 03	☽ ∥ ♆
19 33	☿ ☌ ♃
03 Sunday	
00 01	☿ □ ♂
02 09	☽ Q ♇
07 46	☿ ⋕ ♇
10 33	☽ □ ♃
11 32	☽ ☌ ♂
13 11	☽ □ ☿
15 58	☽ △ ♆
23 59	☉ ⊼ ♇
04 Monday	
00 46	☽ ⚹ ♇
00 50	☽ □ ☉
04 49	☽ ∥ ♂
06 01	☿ ⊼ ♆
09 47	☽ ⊼ ♅
09 52	☽ ∥ ♄
10 29	☽ ☌ ♄
17 43	☽ △ ♀
20 44	☽ ± ♅
05 Tuesday	
04 08	☽ ∠ ♇
04 37	☿ ∥ ♃
08 35	☽ ⋕ ☉
12 50	☽ ⚼ ♅
14 23	♀ ⚼ ♆
18 17	☽ △ ♃
20 04	☽ ⊥ ♇
20 37	☽ ⊻ ♂
22 19	☽ □ ♆
23 06	☽ ⚼ ♀
06 Wednesday	
05 33	☽ △ ☿
06 29	☽ ⊻ ♇
07 28	☽ ⊥ ♂
10 35	☽ △ ☉
11 43	☿ ⊼ ♇
14 52	☽ △ ♅
15 39	☽ ⊻ ♄
16 04	☽ ± ♀
19 20	☽ ⋕ ☿
20 32	☽ ⚼ ♃
23 26	☽ ∠ ♂
07 Thursday	
01 48	☽ ⊥ ♄
03 09	☽ ⊼ ♀
04 59	☽ Q ♆
06 32	♂ ∥ ♄
11 15	☽ ⚼ ☿
11 49	☽ ± ♃
12 13	♂ △ ♆
13 37	☽ ⚼ ☉
16 46	☽ ∠ ♄
21 50	☽ ⊼ ♃
08 Friday	
00 43	☽ ⚹ ♆
01 15	☽ ⚹ ♂
04 24	☽ ± ☿
05 21	☽ ± ☉
08 22	☽ ☌ ♇
11 03	☽ ⋕ ☿
15 37	☽ ⊼ ☿
15 40	☽ ⊼ ☉
16 15	☽ □ ♅
16 21	☉ ☌ ☿
17 08	☽ ⚹ ♄
20 14	☿ △ ♅
21 20	☽ Q ♂
09 Saturday	
00 36	☉ △ ♅
00 47	☽ ∠ ♆
02 50	☿ □ ♄
08 09	☽ ☍ ♀
08 51	☿ ± ♇
09 02	☽ ⋕ ☉
12 13	☽ Q ♄
14 04	☽ ∥ ♂
15 00	☽ ⊥ ♆
15 10	☉ □ ♄
17 28	☽ ∥ ♄
20 44	☽ Q ♅
22 24	☽ ☍ ♃
10 Sunday	
00 25	☽ ⊻ ♆
02 51	☽ □ ♂
03 15	☉ ± ♇
07 50	☽ ⊻ ♇
15 31	☽ ⚹ ♅
16 08	☉ ⋕ ♂
16 31	☽ □ ♄
17 12	☽ ⊥ ♇
18 09	☽ ☍ ☉
22 12	☽ ☍ ☿
11 Monday	
02 02	☽ ∥ ♆
07 18	☽ ∠ ♇
11 19	☽ ⊼ ♀
15 02	☽ ∠ ♅
20 32	☿ ⋕ ♂
21 36	☽ ± ♀
22 10	☽ ⊼ ♃
23 17	☽ ⋕ ♅
23 26	☽ ☌ ♆
12 Tuesday	
03 51	☽ △ ♂
05 15	☽ ⊥ ♅
06 58	☽ ⚹ ♇
07 50	☽ ± ♃
13 10	☽ ⚼ ♀
14 49	☽ ⊻ ♅
16 01	☽ △ ♄
16 37	☉ ⋕ ♄
18 30	♀ ⋕ ♇
19 55	♀ ± ♆
20 50	☽ ⊼ ☉
22 29	☽ ⚼ ♃
13 Wednesday	
02 01	☿ ⋕ ♄
02 13	☽ Q ♇
04 55	☽ ⚼ ♂
05 07	☽ ⊼ ☿
07 14	☽ ± ☉
09 47	☉ ∥ ☿
15 42	☽ △ ♀
16 28	☽ ⚼ ♄
16 28	☽ ± ☿
19 46	☿ ⚼ ♇
20 19	☽ ± ♂
23 07	☽ ⚼ ☉
23 26	☽ △ ♃
23 56	☽ ⊻ ♆
14 Thursday	
06 46	☽ ⊼ ♂
07 30	☽ ± ♄
07 54	☽ □ ♇
09 50	☽ ⚼ ☿
09 58	☽ ⊥ ♆
10 33	☽ ∥ ♅
16 14	☽ ☌ ♅
17 41	☽ ⊼ ♄
15 Friday	
01 18	☽ ∠ ♆
02 22	☽ △ ☉
04 12	♃ ⊼ ♆
08 10	☿ Q ♂
09 29	♂ ⚹ ♇
10 01	☽ ⋕ ♆
15 50	☽ △ ☿
23 57	☽ □ ♀
16 Saturday	
03 35	☽ ⚹ ♆
04 01	☽ □ ♃
05 16	☽ ∥ ☿
08 33	☿ ⚼ ♅
12 11	☽ △ ♇
13 07	☽ ∥ ☉
13 28	☽ ☍ ♂
21 07	☽ ⊻ ♅
22 56	☽ ☍ ♄
23 54	☽ ⋕ ♄
17 Sunday	
01 19	☽ Q ♆
08 10	☽ ⊥ ♅
12 06	☽ Q ♀
12 26	☽ □ ☉
13 29	☽ Q ♃
15 08	♀ ⊼ ♆
15 45	☽ ⚼ ♇
20 43	☽ ⋕ ♂
18 Monday	
00 59	☽ ∠ ♅
05 21	♀ ☌ ♃
08 06	☽ □ ☿
08 44	☽ ± ♇
11 00	☽ □ ♆
12 28	☿ Q ♄
12 33	☽ ⚹ ♃
13 12	☽ ⚹ ♀
20 12	☽ ⊼ ♇
19 Tuesday	
00 30	☽ ⊼ ♂
01 28	☽ Q ☉
01 42	♀ ∥ ♃
03 40	☿ ☍ ♆
05 41	☽ ⚹ ♅
07 55	☽ ⊼ ♄
08 09	☉ ⚼ ♇
12 45	☽ ± ♂
13 33	☽ ∥ ♀
18 07	☿ ⊻ ♃
18 07	☽ ∠ ♃
19 41	☽ ± ♄
21 29	☽ ∠ ♀
23 50	☿ ⋕ ♆
20 Wednesday	
01 02	☽ Q ☿
02 54	☽ ⚹ ☉
05 09	☽ Q ♅
07 21	☽ ⚼ ♂
12 10	☽ ⊥ ♃
13 30	☽ ⚼ ♄
17 16	☽ ⊥ ♀
21 29	☽ △ ♆
23 10	☽ ∥ ♀
21 Thursday	
00 19	☽ ⊻ ♃
04 35	☿ ± ♅
04 59	☽ ⚹ ☿
06 35	☽ ⊻ ♀
07 07	☽ ☍ ♇
11 17	☽ ∠ ☉
11 52	♀ ⊼ ♇
14 22	☽ ⋕ ♂
14 48	☽ △ ♂
16 55	☽ □ ♅
19 20	☿ △ ♇
19 34	☽ △ ♄
22 Friday	
03 27	☽ ⚼ ♆
06 57	☽ ⊥ ☉
12 45	☿ ⊻ ♀
15 51	♀ ⋕ ♂
16 21	☽ ∠ ☿
20 07	☽ ⊻ ☉
21 34	☽ ± ♆
23 02	☽ ⋕ ♄
23 Saturday	
06 32	♂ ⊼ ♅
09 40	☽ ⊼ ♆
13 40	☽ ⊥ ☿
13 47	☽ ☌ ♃
14 33	☿ ⊥ ♃
19 30	☽ ⊼ ♇
24 Sunday	
02 01	☽ ☌ ♀
03 53	☽ ⊻ ☿
05 22	☽ △ ♅
06 40	☽ □ ♂
07 37	☉ ⚼ ♅
07 39	☽ ± ♇
08 26	☽ □ ♄
09 43	☽ ∥ ☉
14 01	☿ ⊼ ♅
18 33	☽ ⋕ ♆
25 Monday	
01 51	☽ ⚼ ♇
08 46	☿ ⚹ ♂
09 38	♀ △ ♅
11 40	☽ ⚼ ♅
12 02	☿ ∥ ♅
12 29	☿ ⚹ ♄
14 13	☽ ☌ ☉
19 30	♂ ☌ ♄
21 01	☽ Q ♄
21 06	☽ Q ♂
22 15	☽ ☍ ♆
23 47	☽ ∥ ♅
26 Tuesday	
02 49	☽ ∥ ☿
03 36	☽ ⊻ ♃
05 49	☽ ± ♅
07 42	♀ ± ♇
08 08	☽ △ ♇
08 20	♂ ⋕ ♃
15 55	☽ ⊥ ♃
17 52	☽ ⊼ ♅
18 22	♀ □ ♄
21 20	☽ ⚹ ♄
21 39	☽ ⊻ ♀
22 38	☽ ⚹ ♂
27 Wednesday	
02 29	☽ ☌ ☿
10 19	☽ ∠ ♃
11 05	☽ ⊥ ♀
15 46	♀ □ ♂
16 32	☉ ⋕ ♆
28 Thursday	
03 30	☽ ∠ ♄
06 17	☽ ∠ ♂
07 05	☽ ∠ ♀
07 45	☽ ⊻ ☉
09 36	☽ ⋕ ☿
10 17	☽ ⊼ ♆
16 44	☽ ⚹ ♃
20 03	☽ □ ♇
20 40	☽ ⊥ ☉
21 26	☽ ⊥ ♄
22 08	☽ ± ♆
23 14	☽ ⋕ ♅
29 Friday	
01 05	☽ ⊥ ♂
05 30	☽ ☍ ♅
06 55	☉ Q ♄
09 18	☽ ⊻ ♄
13 30	☽ ⊻ ♂
14 33	☉ ☍ ♆
15 46	☽ ⚼ ♆
15 52	☽ ∠ ☉
16 00	☽ ⚹ ♀
16 44	☽ Q ♃
19 51	☿ ∠ ♃
22 21	☽ ⋕ ☉
23 02	☽ ⊻ ☿
30 Saturday	
04 05	☽ ∥ ♆
05 17	♀ ⋕ ♄
07 12	☽ Q ♇
12 20	☽ ⊥ ☿
17 47	☽ Q ♀
20 41	☽ △ ♆
23 16	☽ ⚹ ☉
31 Sunday	
04 00	☽ □ ♃
06 09	☽ ⚹ ♇
07 57	☽ ∠ ☿
15 08	☽ ⊼ ♅
16 48	☽ ⋕ ♀
19 10	☽ ☌ ♄
22 27	☽ ∥ ♄
23 45	☽ Q ☉

SEPTEMBER 2014

LONGITUDES

Date	Sidereal time h m s	Sun ☉ ° ' "	Moon ☽ ° ' "	Moon ☽ 24.00 ° ' "	Mercury ☿ ° '	Venus ♀ ° '	Mars ♂ ° '	Jupiter ♃ ° '	Saturn ♄ ° '	Uranus ♅ ° '	Neptune ♆ ° '	Pluto ♇ ° '
01	10 42 18	08 ♍ 59 02	27 ♏ 07 01	03 ♐ 41 52	28 ♍ 53	24 ♌ 48	21 ♏ 54	10 ♌ 16	18 ♏ 05	15 ♈ 51	06 ♓ 06	11 ♑ 07
02	10 46 14	09 57 06	10 ♐ 22 30	17 ♐ 09 14	00 ♎ 24	26 02	22 32	10 28	18 09	15 R 49	06 R 05	11 R 06
03	10 50 11	10 55 12	24 ♐ 02 17	01 ♑ 01 46	01 54	27 16	23 11	10 41	18 13	15 48	06 03	11 05
04	10 54 07	11 53 19	08 ♑ 07 38	15 ♑ 19 43	03 23	28 30	23 49	10 53	18 17	15 46	06 01	11 05
05	10 58 04	12 51 28	22 ♑ 37 38	00 ♒ 00 48	04 50	29 ♌ 44	24 28	11 05	18 21	15 44	06 00	11 04
06	11 02 00	13 49 38	07 ♒ 28 29	14 ♒ 59 43	06 16	00 ♍ 58	25 07	11 17	18 25	15 42	05 58	11 04
07	11 05 57	14 47 49	22 ♒ 33 28	00 ♓ 08 29	07 41	02 13	25 46	11 30	18 30	15 40	05 57	11 03
08	11 09 54	15 46 01	07 ♓ 43 34	15 ♓ 17 25	09 04	03 27	26 25	11 42	18 34	15 38	05 55	11 03
09	11 13 50	16 44 17	22 ♓ 48 49	00 ♈ 16 40	10 26	04 41	27 05	11 54	18 39	15 36	05 53	11 02
10	11 17 47	17 42 33	07 ♈ 39 57	14 ♈ 57 51	11 47	05 55	27 44	12 06	18 43	15 34	05 52	11 02
11	11 21 43	18 40 52	22 ♈ 09 43	29 ♈ 15 07	13 06	07 10	28 24	12 17	18 48	15 32	05 50	11 02
12	11 25 40	19 39 14	06 ♉ 13 45	13 ♉ 05 34	14 24	08 24	29 04	12 29	18 52	15 30	05 48	11 01
13	11 29 36	20 37 35	19 ♉ 50 35	26 ♉ 29 00	15 40	09 38	29 ♏ 43	12 41	18 57	15 28	05 47	11 01
14	11 33 33	21 35 59	03 ♊ 01 08	09 ♊ 27 23	16 54	10 53	00 ♐ 23	12 53	19 02	15 25	05 45	11 01
15	11 37 29	22 34 26	15 ♊ 48 10	22 ♊ 04 01	18 07	12 07	01 04	13 04	19 07	15 23	05 44	11 01
16	11 41 26	23 32 55	28 ♊ 15 28	04 ♋ 23 04	19 18	13 22	01 44	13 16	19 12	15 21	05 42	11 00
17	11 45 23	24 31 26	10 ♋ 27 22	16 ♋ 28 56	20 27	14 36	02 24	13 28	19 17	15 19	05 41	11 00
18	11 49 19	25 30 00	22 ♋ 28 16	28 ♋ 25 53	21 34	15 51	03 05	13 39	19 22	15 17	05 39	11 00
19	11 53 16	26 28 35	04 ♌ 22 18	10 ♌ 17 56	22 39	17 05	03 45	13 50	19 27	15 14	05 38	11 00
20	11 57 12	27 27 13	16 ♌ 13 14	22 ♌ 08 34	23 42	18 20	04 26	14 02	19 32	15 12	05 36	11 00
21	12 01 09	28 25 52	28 ♌ 04 18	04 ♍ 00 46	24 42	19 35	05 07	14 13	19 38	15 10	05 35	11 00
22	12 05 05	29 ♍ 24 34	09 ♍ 58 14	15 ♍ 56 59	25 40	20 49	05 48	14 24	19 43	15 08	05 33	11 D 00
23	12 09 02	00 ♎ 23 17	21 ♍ 57 15	27 ♍ 59 15	26 35	22 04	06 29	14 35	19 48	15 06	05 32	11 00
24	12 12 58	01 22 03	04 ♎ 03 12	10 ♎ 09 18	27 27	23 19	07 10	14 46	19 54	15 03	05 30	11 00
25	12 16 55	02 20 51	16 ♎ 17 43	22 ♎ 28 39	28 16	24 33	07 52	14 57	19 59	15 01	05 29	11 00
26	12 20 52	03 19 40	28 ♎ 42 17	04 ♏ 58 49	29 01	25 48	08 33	15 08	20 05	14 58	05 27	11 00
27	12 24 48	04 18 32	11 ♏ 18 27	17 ♏ 41 23	29 ♎ 43	27 03	09 15	15 19	20 11	14 56	05 26	11 00
28	12 28 45	05 17 25	24 ♏ 07 50	00 ♐ 38 00	00 ♏ 20	28 18	09 56	15 29	20 16	14 53	05 25	11 00
29	12 32 41	06 16 20	07 ♐ 12 08	13 ♐ 50 24	00 53	29 ♍ 32	10 38	15 40	20 22	14 51	05 23	11 00
30	12 36 38	07 ♎ 15 17	20 ♐ 33 00	27 ♐ 20 04	01 ♏ 22	00 ♎ 47	11 ♐ 20	15 ♌ 50	20 ♏ 28	14 ♈ 49	05 ♓ 22	11 ♑ 01

Moon

Date	Moon True ☊ ° '	Moon Mean ☊ ° '	Moon ☽ Latitude ° '
01	19 ♎ 57	21 ♎ 22	03 N 08
02	19 R 57	21 19	04 00
03	19 57	21 16	04 40
04	19 54	21 13	05 05
05	19 50	21 10	05 11
06	19 46	21 06	04 57
07	19 40	21 03	04 22
08	19 36	21 00	03 28
09	19 32	20 57	02 21
10	19 30	20 54	01 N 04
11	19 D 30	20 51	00 S 15
12	19 31	20 47	01 30
13	19 32	20 44	02 39
14	19 34	20 41	03 36
15	19 35	20 38	04 21
16	19 R 35	20 35	04 53
17	19 34	20 31	05 10
18	19 32	20 28	05 14
19	19 29	20 25	05 04
20	19 25	20 22	04 41
21	19 21	20 19	04 06
22	19 18	20 16	03 20
23	19 16	20 12	02 25
24	19 14	20 09	01 23
25	19 13	20 06	00 S 16
26	19 D 13	20 03	00 N 52
27	19 14	20 00	01 59
28	19 15	19 56	03 01
29	19 17	19 53	03 56
30	19 ♎ 18	19 ♎ 50	04 N 38

DECLINATIONS

Date	Sun ☉ ° '	Moon ☽ ° '	Mercury ☿ ° '	Venus ♀ ° '	Mars ♂ ° '	Jupiter ♃ ° '	Saturn ♄ ° '	Uranus ♅ ° '	Neptune ♆ ° '	Pluto ♇ ° '
01	08 N 12	16 S 27	00 N 27	14 N 17	19 S 33	18 N 07	15 S 12	05 N 36	09 S 58	20 S 31
02	07 50	18 02	00 S 16	13 54	19 43	18 03	15 13	05 35	09 59	20 31
03	07 28	18 38	01 00	13 30	19 53	18 00	15 15	05 34	09 59	20 31
04	07 06	18 07	01 42	13 05	20 03	17 57	15 16	05 33	10 00	20 31
05	06 44	16 26	02 24	12 41	20 13	17 54	15 17	05 33	10 01	20 32
06	06 22	13 37	03 06	12 16	20 23	17 51	15 19	05 32	10 01	20 32
07	05 59	09 52	03 47	11 50	20 33	17 47	15 20	05 31	10 02	20 32
08	05 37	05 27	04 28	11 25	20 43	17 44	15 22	05 30	10 02	20 32
09	05 14	00 S 42	05 08	10 59	20 52	17 41	15 23	05 30	10 03	20 32
10	04 51	04 N 02	05 47	10 32	21 02	17 38	15 25	05 29	10 04	20 33
11	04 29	08 26	06 26	10 06	21 11	17 34	15 26	05 28	10 04	20 33
12	04 06	12 10	07 03	09 39	21 20	17 31	15 27	05 27	10 05	20 33
13	03 43	15 09	07 41	09 12	21 29	17 28	15 29	05 26	10 05	20 33
14	03 20	17 08	08 17	08 44	21 38	17 25	15 30	05 26	10 06	20 33
15	02 57	18 21	08 52	08 17	21 47	17 22	15 32	05 25	10 07	20 34
16	02 34	18 33	09 27	07 49	21 55	17 18	15 33	05 24	10 07	20 34
17	02 10	17 52	10 00	07 21	22 04	17 15	15 35	05 23	10 08	20 34
18	01 47	16 24	10 33	06 52	22 12	17 12	15 37	05 22	10 08	20 34
19	01 24	14 15	11 04	06 24	22 20	17 09	15 38	05 21	10 09	20 34
20	01 01	11 30	11 35	05 55	22 28	17 06	15 40	05 20	10 09	20 34
21	00 37	08 18	12 05	05 26	22 36	17 03	15 41	05 20	10 10	20 35
22	00 N 14	04 44	12 31	04 57	22 43	17 00	15 43	05 19	10 10	20 35
23	00 S 09	00 N 58	12 58	04 28	22 51	16 57	15 45	05 18	10 11	20 35
24	00 33	02 S 53	13 23	03 58	22 58	16 54	15 46	05 17	10 12	20 35
25	00 56	06 39	13 46	03 29	23 05	16 50	15 48	05 16	10 12	20 35
26	01 19	10 12	14 08	02 59	23 12	16 47	15 49	05 15	10 13	20 36
27	01 43	13 20	14 27	02 29	23 18	16 44	15 51	05 14	10 13	20 36
28	02 06	15 52	14 45	02 00	23 25	16 41	15 53	05 13	10 14	20 36
29	02 29	17 38	15 01	01 30	23 31	16 38	15 54	05 12	10 14	20 36
30	02 S 53	18 S 28	15 S 14	01 N 00	23 S 37	16 N 35	15 S 56	05 N 11	10 S 15	20 S 36

ZODIAC SIGN ENTRIES

Date	h	m	Planets
01	17	17	☽ ♐
02	05	38	☿ ♎
03	22	15	☽ ♑
05	17	07	♀ ♍
05	23	59	☽ ♒
07	23	47	☽ ♓
09	23	23	☽ ♈
12	01	17	☽ ♉
13	21	57	♂ ♐
14	06	26	☽ ♊
16	15	24	☽ ♋
19	03	10	☽ ♌
21	15	54	☽ ♍
23	02	29	☉ ♎
24	03	59	☽ ♎
26	14	29	☽ ♏
27	22	39	☿ ♏
28	22	50	☽ ♐
29	20	52	♀ ♎

LATITUDES

Date	Mercury ☿ ° '	Venus ♀ ° '	Mars ♂ ° '	Jupiter ♃ ° '	Saturn ♄ ° '	Uranus ♅ ° '	Neptune ♆ ° '	Pluto ♇ ° '
01	00 00	01 N 06	01 S 21	00 N 28	02 N 06	00 S 42	00 S 45	02 N 28
04	00 S 24	01 10	01 22	00 28	02 05	00 42	00 45	02 28
07	00 48	01 14	01 24	00 29	02 04	00 42	00 45	02 27
10	01 13	01 18	01 25	00 29	02 03	00 42	00 45	02 27
13	01 38	01 20	01 26	00 29	02 03	00 42	00 45	02 26
16	02 03	01 23	01 27	00 30	02 02	00 42	00 45	02 25
19	02 26	01 24	01 28	00 30	02 02	00 42	00 45	02 25
22	02 48	01 26	01 28	00 31	02 01	00 42	00 45	02 24
25	03 07	01 26	01 29	00 31	02 01	00 42	00 45	02 24
28	03 23	01 26	01 30	00 31	02 00	00 42	00 45	02 23
31	03 S 33	01 N 25	01 S 30	00 N 32	02 N 00	00 S 42	00 S 45	02 N 23

DATA

Julian Date	2456902
Delta T	+69 seconds
Ayanamsa	24° 03' 50"
Synetic vernal point	05° ♓ 03' 09"
True obliquity of ecliptic	23° 26' 06"

LONGITUDES

Date	Chiron ⚷ ° '	Ceres ⚳ ° '	Pallas ⚴ ° '	Juno ⚵ ° '	Vesta ⚶ ° '	Black Moon Lilith ⚸ ° '
01	15 ♓ 42	08 ♏ 17	08 ♎ 59	12 ♋ 35	13 ♏ 21	20 ♌ 14
11	15 ♓ 14	11 ♏ 53	13 ♎ 17	17 ♋ 40	17 ♏ 54	21 ♌ 22
21	14 ♓ 46	15 ♏ 36	17 ♎ 38	22 ♋ 34	22 ♏ 37	22 ♌ 29
31	14 ♓ 20	19 ♏ 27	22 ♎ 02	27 ♋ 12	27 ♏ 29	23 ♌ 36

MOON'S PHASES, APSIDES AND POSITIONS ☽

Date	h	m	Phase	Longitude ° '	Eclipse Indicator
02	11	11	☽	09 ♐ 55	
09	01	38	○	16 ♓ 19	
16	02	05	☾	23 ♊ 09	
24	06	14	●	01 ♎ 08	

Day	h	m		
08	03	37	Perigee	
20	14	32	Apogee	
03	13	11	Max dec	18° S 38'
09	15	29	0N	
16	05	13	Max dec	18° N 35'
23	18	04	0S	
30	19	28	Max dec	18° S 32'

ASPECTARIAN

h m	Aspects
01 Monday	
01 51	☽ ☌ ♂
02 18	☽ ± ♅
07 17	☽ □ ♀
10 09	☽ ∠ ♇
10 38	♂ ± ♅
15 40	☽ ✱ ☿
18 49	☽ ⚼ ♅
02 Tuesday	
02 32	☽ ⊥ ♇
04 19	☽ □ ♆
08 56	☉ ± ♅
11 11	☽ □ ☉
12 10	☽ △ ♃
12 24	☽ # ♃
13 17	☽ ⊻ ♇
13 19	♀ ⚼ ♇
16 04	☽ Q ☿
21 38	☽ △ ♅
03 Wednesday	
01 49	☽ ⊻ ♄
04 22	☉ ⊻ ♃
10 26	☽ ⊻ ♂
12 01	☽ Q ♆
12 18	☽ ⊥ ♄
14 52	☽ ⚼ ♃
16 10	☉ △ ♇
18 06	☽ △ ♀
21 16	☽ ⊥ ♂
04 Thursday	
03 04	☽ □ ☿
03 47	☽ ∠ ♄
06 27	☽ ± ♃
06 54	☉ Q ♂
08 28	☽ ✱ ♆
10 19	☿ ∠ ♄
13 13	☽ ∠ ♂
15 33	☽ # ♃
16 41	☽ ⊼ ♃
16 56	☽ ☌ ♇
18 45	☽ △ ☉
21 48	☽ ⚼ ♀
05 Friday	
00 41	☽ □ ♅
04 57	☽ ✱ ♄
09 20	☽ ∠ ♆
10 23	♃ ⊼ ♇
13 59	☽ ± ♀
15 08	☽ ✱ ♂
21 06	☽ ⚼ ☉
22 43	☽ ∥ ♄
23 57	☽ ⊥ ♆
06 Saturday	
00 36	☽ ⊼ ♀
00 36	☽ Q ♄
05 57	☽ Q ♅
06 49	♀ ⚼ ♅
07 01	☿ ⊼ ♆
09 36	☽ ⊻ ♆
09 52	☽ △ ☿
11 24	☽ Q ♂
12 36	☽ ± ☉
17 44	☽ ⊻ ♇
18 11	☽ ☍ ♃
22 29	☽ # ♀
22 50	☽ ⊼ ☉
07 Sunday	
01 05	☽ ✱ ♅
03 17	☽ ⊥ ♇
05 32	☽ □ ♄
09 19	♂ ∥ ♇
11 02	☽ ∥ ♆
12 13	☽ ⚼ ☿
17 19	☽ □ ♂
17 32	☽ ∠ ♇
22 27	♂ ∠ ♇
08 Monday	
00 48	☽ ∠ ♅
03 54	☽ ± ☿
04 38	☽ ☍ ♀
08 47	☉ ⊼ ♅
09 08	☽ ☌ ♆
11 03	☽ # ☉
11 40	☽ # ♅
14 21	☽ ⊼ ☿
15 01	☽ ⊥ ♅
16 24	☽ ∥ ☿
17 16	☽ ✱ ♇
18 22	☽ ⊼ ♃
18 53	☉ ∥ ♅
09 Tuesday	
00 31	☽ ⊻ ♅
01 38	☽ ☍ ☉
04 02	☽ ± ♃
05 18	☽ △ ♄
12 22	☽ Q ♇
14 27	☉ # ☿
18 38	☽ ⚼ ♃
19 09	☽ △ ♂
22 40	☿ □ ♇
10 Wednesday	
01 05	☿ # ♅
05 32	☽ ⚼ ♄
08 54	☽ ⊼ ♀
09 04	☽ ⊻ ♆
10 51	♀ ☍ ♆
13 25	☿ ± ♆
16 02	☽ ∥ ☉
17 31	☽ □ ♇
18 37	☿ ✱ ♃
18 52	☽ ⊥ ♆
19 22	☽ △ ♃
19 25	☽ ☍ ☿
19 38	☽ ± ♀
19 40	☽ ∥ ♅
20 20	☽ ± ♄
20 43	☽ ⚼ ♂
22 55	☽ # ☿
23 55	☉ ⊥ ♃
11 Thursday	
00 58	☽ ☌ ♅
04 26	♀ Q ♄
05 45	☽ ⊼ ☉
06 03	☿ ⊥ ♄
06 20	☽ ⊼ ♄
09 47	☽ ∠ ♆
12 00	☽ ⚼ ♀
12 25	☽ ± ♂
13 20	♀ # ♆
15 03	☉ ✱ ♄
16 33	☽ ± ☉
21 10	☽ ∥ ♀
22 06	☽ # ♆
23 04	☽ ⊼ ♂
23 09	☿ ∠ ♂
12 Friday	
09 04	☽ ⚼ ☉
11 16	☽ ✱ ♆
16 09	☽ △ ♀
16 19	☿ ⊥ ♀
20 21	☽ △ ♇
23 06	☽ □ ♃
13 Saturday	
03 47	☽ ⊼ ☿
04 12	☽ ⊻ ♅
08 16	☿ ☍ ♅
08 19	☽ Q ♆
08 37	♀ ± ♅
10 24	☽ ☍ ♄
13 31	☽ △ ☉
14 54	☽ ⊥ ♅
15 18	☽ # ♄
15 36	☽ ± ☿
23 09	☽ ⚼ ♇
14 Sunday	
06 53	☽ ☍ ♂
07 14	☽ ∠ ♅
08 00	☽ Q ♃
09 43	☽ ⚼ ☿
13 09	♂ ⚼ ♅
14 35	♀ △ ♇
14 57	☽ ∥ ♃
15 42	☽ ± ♇
17 04	☽ □ ♆
22 25	☿ # ♀
15 Monday	
02 56	☽ ⊼ ♇
04 16	☽ □ ♀
06 44	☽ ✱ ♃
07 11	☿ ⊥ ♀
11 13	☽ ✱ ♅
16 52	☽ △ ☿
18 22	☽ ⊼ ♄
16 Tuesday	
02 05	☽ □ ☉
06 01	☽ ± ♄
09 49	☿ ⊻ ♄
09 50	♀ ⊻ ♃
10 14	☽ Q ♅
12 01	☽ ∠ ♃
18 45	☽ Q ♀
19 11	☽ ⊼ ♂
23 43	☽ ⚼ ♄
17 Wednesday	
02 34	☽ △ ♆
05 58	☽ ⊥ ♃
07 42	☽ ± ♂
13 05	☽ ☍ ♇
16 28	☽ Q ☉
16 48	☿ ⚼ ♆
17 31	☿ ∥ ♆
18 04	☽ ⊻ ♃
21 12	☽ ✱ ♀
21 38	☽ □ ♅
23 58	☽ ∥ ♃
18 Thursday	
01 20	♀ ⊼ ♅
02 40	☽ ⚼ ♂
05 44	☽ △ ♄
08 21	☽ ⚼ ♆
10 00	☽ □ ☿
18 38	☽ ✱ ☉
21 32	☽ # ♄
19 Friday	
02 26	☽ ± ♆
06 51	☽ ∠ ♀
10 41	☽ △ ♂
14 32	☽ ⊼ ♆
20 Saturday	
01 25	☽ ⊼ ♇
01 57	☽ Q ☿
03 12	☽ ⊥ ♀
03 41	☽ ∠ ☉
07 29	☽ ☌ ♃
09 57	☽ △ ♅
11 29	☽ # ☿
13 34	☽ ± ♇
16 47	☽ ⊻ ♀
18 46	☽ □ ♄
22 25	☽ # ♆
23 33	☽ ⊥ ☉
21 Sunday	
04 33	☽ ✱ ☿
07 48	☽ ⚼ ♇
07 48	♂ ⊥ ♇
12 48	☽ ⊻ ☉
13 03	♀ ✱ ♄
16 13	☽ ⚼ ♅
17 29	♀ ∥ ♅
22 Monday	
02 31	♀ ⊥ ♃
03 05	☽ □ ♂
03 07	☽ ☍ ♆
03 39	♂ □ ♆
07 26	☽ Q ♄
08 16	☽ ∥ ♅
10 18	☽ ± ♅
10 27	☽ ∥ ♀
11 46	☉ ∠ ♃
13 31	☽ ∠ ☿
14 04	☽ △ ♇
21 02	☽ ⊻ ♃
22 19	☽ ⊼ ♅
23 Tuesday	
00 34	♇ St D
07 41	☽ ✱ ♄
09 02	☽ ⊥ ☿
09 14	☽ ⊥ ♃
12 11	☿ Q ♃
12 15	☽ ☌ ♀
16 38	☽ # ☉
17 21	☽ Q ♂
19 49	☽ ∥ ☉
21 56	☽ ⊻ ☿
24 Wednesday	
03 24	☽ ∠ ♃
06 14	☽ ☌ ☉
13 41	☽ ∠ ♄
14 51	☽ ⊼ ♆
18 04	☽ # ♀
18 30	☽ ✱ ♂
25 Thursday	
01 39	☽ □ ♇
02 37	☽ ± ♆
03 05	☽ # ♅
07 28	☽ ⊥ ♄
09 21	☽ ✱ ♃
09 30	☽ ☍ ♅
18 19	♃ △ ♅
19 14	☽ ⊻ ♄
20 07	☽ ⚼ ♆
26 Friday	
01 29	☽ ∠ ♂
05 48	☽ ⊻ ♀
08 56	☽ Q ♃
11 19	☿ Q ♇
12 05	☽ ∥ ♆
12 34	☽ Q ♇
12 39	☽ ☌ ☿
18 35	☽ ⊥ ♀
19 48	☽ ⊥ ♂
21 36	☽ ⊻ ☉
27 Saturday	
00 53	☽ △ ♆
07 52	☽ ⊻ ♂
09 57	☽ ⊥ ☉
11 25	☽ ✱ ♇
13 33	☽ ∠ ♀
18 48	☽ ⊼ ♅
19 39	☽ □ ♃
20 35	♀ Q ♂
23 12	☽ ∥ ☿
28 Sunday	
04 16	☽ ∠ ☉
04 46	☽ ☌ ♄
06 00	☽ ± ♅
09 05	☉ # ♀
11 30	☉ ∠ ♄
12 08	☽ ∥ ♄
14 52	☉ ⊼ ♆
15 28	☽ ∠ ♇
20 30	☽ ✱ ♀
21 31	☽ # ♃
22 36	☽ ⚼ ♅
23 59	☽ ⊻ ☿
29 Monday	
08 00	☽ ⊥ ♇
08 42	☽ □ ♆
10 10	☽ ✱ ☉
11 25	☽ ⊥ ☿
18 34	☽ ☌ ♂
18 54	☽ ⊻ ♇
20 40	☽ Q ♀
30 Tuesday	
00 52	♂ ⊻ ♇
01 47	☽ △ ♅
03 29	☽ △ ♃
04 16	☽ ∠ ☿
09 31	☽ Q ☉
11 51	☽ ⊻ ♄
13 12	♀ ∠ ♃
16 59	☽ Q ♆
22 33	☽ ⊥ ♄

All ephemeris data is given at 12.00 UT and the Moon's longitude is additionally given for 24.00 UT

OCTOBER 2014

LONGITUDES

Date	Sidereal time h m s	Sun ☉ ° ' "	Moon ☽ ° ' "	Moon ☽ 24.00 ° ' "	Mercury ☿ ° '	Venus ♀ ° '	Mars ♂ ° '	Jupiter ♃ ° '	Saturn ♄ ° '	Uranus ♅ ° '	Neptune ♆ ° '	Pluto ♇ ° '
01	12 40 34	08 ♎ 14 16	04 ♑ 11 42	11 ♑ 07 58	01 ♏ 45	02 ♎ 02	12 ♐ 02	16 ♌ 01	20 ♏ 34	14 ♈ 46	05 ♓ 21	11 ♑ 01
02	12 44 31	09 13 16	18 ♑ 08 49	25 ♑ 14 08	02 08	03 17	12 44	16 11	20 40	14 R 44	05 R 19	11 01
03	12 48 27	10 12 18	02 ♒ 23 42	09 ♒ 37 10	02 14	04 32	13 26	16 21	20 46	14 41	05 18	11 01
04	12 52 24	11 11 22	16 ♒ 54 05	24 ♒ 13 53	02 19	05 47	14 09	16 31	20 52	14 39	05 17	11 02
05	12 56 21	12 10 27	01 ♓ 35 53	08 ♓ 59 19	02 R 17	07 01	14 51	16 41	20 58	14 37	05 15	11 02
06	13 00 17	13 09 35	16 ♓ 23 20	23 ♓ 47 02	02 07	08 16	15 33	16 51	21 04	14 34	05 14	11 03
07	13 04 14	14 08 44	01 ♈ 09 29	08 ♈ 29 47	01 50	09 31	16 16	17 01	21 10	14 32	05 13	11 03
08	13 08 10	15 07 55	15 ♈ 47 04	23 ♈ 00 33	01 25	10 46	16 59	17 11	21 16	14 29	05 12	11 03
09	13 12 07	16 07 08	00 ♉ 09 33	07 ♉ 13 27	00 52	12 01	17 42	17 20	21 22	14 27	05 11	11 04
10	13 16 03	17 06 23	14 ♉ 11 50	21 ♉ 04 22	00 ♏ 10	13 16	18 24	17 30	21 29	14 24	05 10	11 04
11	13 20 00	18 05 41	27 ♉ 50 53	04 ♊ 31 20	29 ♎ 21	14 31	19 07	17 39	21 35	14 22	05 08	11 05
12	13 23 56	19 05 01	11 ♊ 05 48	17 ♊ 34 28	28 25	15 46	19 50	17 49	21 41	14 20	05 07	11 06
13	13 27 53	20 04 23	23 ♊ 57 38	00 ♋ 15 38	27 23	17 01	20 34	17 58	21 48	14 17	05 06	11 06
14	13 31 50	21 03 47	06 ♋ 28 55	12 ♋ 38 00	26 15	18 16	21 17	18 07	21 54	14 15	05 05	11 07
15	13 35 46	22 03 14	18 ♋ 43 23	24 ♋ 45 38	25 04	19 31	22 00	18 16	22 01	14 12	05 04	11 08
16	13 39 43	23 02 43	00 ♌ 45 19	06 ♌ 43 03	23 51	20 46	22 44	18 24	22 07	14 10	05 03	11 08
17	13 43 39	24 02 14	12 ♌ 39 24	18 ♌ 34 56	22 38	22 01	23 27	18 33	22 14	14 08	05 02	11 09
18	13 47 36	25 01 48	24 ♌ 30 13	00 ♍ 25 47	21 27	23 17	24 11	18 42	22 20	14 05	05 01	11 10
19	13 51 32	26 01 23	06 ♍ 22 09	12 ♍ 19 48	20 21	24 32	24 54	18 50	22 27	14 03	05 01	11 11
20	13 55 29	27 01 01	18 ♍ 19 09	24 ♍ 20 36	19 21	25 47	25 38	18 59	22 34	14 00	05 00	11 11
21	13 59 25	28 00 41	00 ♎ 24 32	06 ♎ 31 14	18 29	27 02	26 22	19 07	22 40	13 58	04 59	11 12
22	14 03 22	29 ♎ 00 23	12 ♎ 40 58	18 ♎ 53 56	17 47	28 17	27 06	19 15	22 47	13 56	04 58	11 13
23	14 07 19	00 ♏ 00 07	25 ♎ 10 20	01 ♏ 30 14	17 16	29 ♎ 32	27 50	19 23	22 54	13 53	04 57	11 14
24	14 11 15	00 59 54	07 ♏ 53 44	14 ♏ 20 50	16 55	00 ♏ 47	28 34	19 31	23 01	13 51	04 57	11 15
25	14 15 12	01 59 42	20 ♏ 51 32	27 ♏ 25 47	16 46	02 03	29 ♐ 18	19 38	23 08	13 49	04 56	11 16
26	14 19 08	02 59 32	04 ♐ 03 30	10 ♐ 44 35	16 D 48	03 18	00 ♑ 02	19 46	23 14	13 47	04 55	11 17
27	14 23 05	03 59 24	17 ♐ 28 54	24 ♐ 16 19	17 01	04 33	00 47	19 53	23 21	13 44	04 54	11 18
28	14 27 01	04 59 18	01 ♑ 06 41	07 ♑ 59 51	17 25	05 48	01 31	20 00	23 28	13 42	04 54	11 19
29	14 30 58	05 59 13	14 ♑ 55 39	21 ♑ 53 54	17 58	07 03	02 16	20 08	23 35	13 40	04 53	11 20
30	14 34 54	06 59 10	28 ♑ 54 27	05 ♒ 57 06	18 40	08 19	03 00	20 15	23 42	13 38	04 53	11 21
31	14 38 51	07 ♏ 59 09	13 ♒ 01 38	20 ♒ 07 50	19 ♎ 30	09 ♏ 34	03 ♑ 45	20 ♌ 21	23 ♏ 49	13 ♈ 36	04 ♓ 52	11 ♑ 22

Date	Moon True ☊ ° '	Moon Mean ☊ ° '	Moon ☽ Latitude ° '
01	19 ♎ 18	19 ♎ 47	05 N 06
02	19 R 18	19 44	05 17
03	19 17	19 41	05 09
04	19 17	19 37	04 41
05	19 16	19 34	03 55
06	19 15	19 31	02 52
07	19 15	19 28	01 39
08	19 14	19 25	00 N 19
09	19 D 14	19 22	01 S 00
10	19 15	19 18	02 14
11	19 15	19 15	03 18
12	19 15	19 12	04 10
13	19 R 15	19 09	04 47
14	19 15	19 06	05 10
15	19 15	19 02	05 17
16	19 D 15	18 59	05 11
17	19 15	18 56	04 51
18	19 15	18 53	04 20
19	19 16	18 50	03 36
20	19 17	18 47	02 44
21	19 18	18 43	01 43
22	19 18	18 40	00 S 37
23	19 R 18	18 37	00 N 32
24	19 17	18 34	01 41
25	19 16	18 31	02 46
26	19 15	18 28	03 43
27	19 13	18 24	04 29
28	19 11	18 21	05 00
29	19 09	18 18	05 15
30	19 08	18 15	05 11
31	19 ♎ 08	18 ♎ 12	04 N 49

DECLINATIONS

Date	Sun ☉	Moon ☽	Mercury ☿	Venus ♀	Mars ♂	Jupiter ♃	Saturn ♄	Uranus ♅	Neptune ♆	Pluto ♇
01	03 S 16	18 S 16	15 S 25	00 N 30	23 S 43	16 N 32	15 S 58	05 N 10	10 S 15	20 S 36
02	03 39	16 58	15 32	00 00	23 49	16 30	16 00	05 09	10 16	20 36
03	04 02	14 36	15 37	00 S 30	23 54	16 27	16 01	05 09	10 16	20 37
04	04 26	11 18	15 39	01 00	23 59	16 24	16 03	05 08	10 16	20 37
05	04 49	07 15	15 37	01 31	24 05	16 21	16 05	05 07	10 17	20 37
06	05 12	02 S 43	15 31	02 01	24 09	16 18	16 06	05 06	10 17	20 37
07	05 35	01 N 58	15 22	02 31	24 14	16 15	16 08	05 05	10 18	20 37
08	05 58	06 30	15 08	03 01	24 18	16 12	16 10	05 04	10 18	20 37
09	06 20	10 35	14 49	03 31	24 23	16 10	16 12	05 03	10 19	20 37
10	06 43	13 58	14 26	04 01	24 26	16 07	16 13	05 02	10 19	20 38
11	07 06	16 27	13 59	04 31	24 30	16 04	16 15	05 01	10 19	20 38
12	07 28	17 59	13 27	05 00	24 34	16 02	16 17	05 00	10 20	20 38
13	07 51	18 31	12 51	05 30	24 37	15 59	16 19	04 59	10 20	20 38
14	08 13	18 07	12 12	06 00	24 40	15 56	16 21	04 58	10 20	20 38
15	08 35	16 53	11 29	06 29	24 43	15 54	16 22	04 57	10 21	20 38
16	08 57	14 55	10 44	06 58	24 45	15 51	16 24	04 56	10 21	20 38
17	09 19	12 20	09 59	07 28	24 48	15 49	16 26	04 56	10 22	20 38
18	09 41	09 16	09 13	07 57	24 50	15 46	16 28	04 55	10 22	20 38
19	10 03	05 49	08 29	08 25	24 51	15 44	16 29	04 54	10 22	20 39
20	10 25	02 N 06	07 48	08 54	24 53	15 41	16 31	04 53	10 23	20 39
21	10 46	01 S 44	07 10	09 23	24 54	15 39	16 33	04 52	10 23	20 39
22	11 07	05 34	06 36	09 51	24 55	15 37	16 35	04 51	10 23	20 39
23	11 28	09 14	06 08	10 19	24 56	15 34	16 37	04 50	10 23	20 39
24	11 49	12 33	05 45	10 47	24 57	15 32	16 38	04 49	10 24	20 39
25	12 10	15 18	05 28	11 14	24 57	15 30	16 40	04 48	10 24	20 39
26	12 30	17 08	05 17	11 42	24 57	15 28	16 42	04 48	10 24	20 39
27	12 51	18 23	05 12	12 09	24 57	15 25	16 44	04 47	10 24	20 39
28	13 11	18 25	05 12	12 35	24 56	15 23	16 46	04 46	10 25	20 39
29	13 31	17 23	05 17	13 02	24 55	15 21	16 47	04 45	10 25	20 39
30	13 51	15 18	05 28	13 28	24 55	15 19	16 49	04 44	10 25	20 39
31	14 S 10	12 S 17	05 S 42	13 S 54	24 S 53	15 N 17	16 S 51	04 N 43	10 S 25	20 S 40

ZODIAC SIGN ENTRIES

Date	h	m	Planets
01	04	41	☽ ♑
03	08	00	☽ ♒
05	09	24	☽ ♓
07	10	07	☽ ♈
09	11	44	☽ ♉
10	17	27	☿ ♎
11	15	51	☽ ♊
13	23	30	☽ ♋
16	10	29	☽ ♌
18	23	08	☽ ♍
21	11	12	☽ ♎
23	11	57	☉ ♏
23	20	52	♀ ♏
23	21	10	☽ ♏
26	04	40	☽ ♐
26	10	43	♂ ♑
28	10	03	☽ ♑
30	13	52	☽ ♒

LATITUDES

Date	Mercury ☿	Venus ♀	Mars ♂	Jupiter ♃	Saturn ♄	Uranus ♅	Neptune ♆	Pluto ♇
01	03 S 33	01 N 25	01 S 30	00 N 32	02 N 00	00 S 42	00 S 45	02 N 23
04	03 36	01 24	01 30	00 32	01 59	00 42	00 45	02 22
07	03 28	01 22	01 31	00 33	01 59	00 42	00 45	02 22
10	03 07	01 20	01 31	00 33	01 58	00 42	00 45	02 21
13	02 29	01 17	01 31	00 34	01 58	00 42	00 45	02 21
16	01 36	01 14	01 31	00 34	01 58	00 42	00 45	02 20
19	00 S 35	01 10	01 31	00 35	01 57	00 42	00 45	02 20
22	00 N 24	01 05	01 31	00 36	01 57	00 42	00 45	02 19
25	01 13	01 01	01 31	00 36	01 57	00 42	00 45	02 19
28	01 46	00 55	01 31	00 37	01 56	00 42	00 45	02 18
31	02 N 05	00 N 49	01 S 30	00 N 37	01 N 56	00 S 42	00 S 45	02 N 18

DATA

Julian Date	2456932
Delta T	+69 seconds
Ayanamsa	24° 03' 53"
Synetic vernal point	05° ♓ 03' 06"
True obliquity of ecliptic	23° 26' 06"

MOON'S PHASES, APSIDES AND POSITIONS ☽

Date	h	m	Phase	Longitude ° '	Eclipse Indicator
01	19	33	☽	08 ♑ 33	
08	10	51	○	15 ♈ 05	total
15	19	12	☾	22 ♋ 21	
23	21	57	●	00 ♏ 25	Partial
31	02	48	☽	07 ♒ 36	

Day	h	m		
06	09	45	Perigee	
18	06	11	Apogee	
07	01	55	0N	
13	13	34	Max dec	18° N 31'
21	01	13	0S	
28	00	58	Max dec	18° S 32'

LONGITUDES

Date	Chiron ⚷ ° '	Ceres ⚳ ° '	Pallas ⚴ ° '	Juno ⚵ ° '	Vesta ⚶ ° '	Black Moon Lilith ⚸ ° '
01	14 ♓ 20	19 ♏ 27	22 ♎ 02	27 ♋ 12	27 ♏ 29	23 ♌ 36
11	13 ♓ 57	23 ♏ 23	26 ♎ 26	01 ♌ 32	02 ♐ 27	24 ♌ 43
21	13 ♓ 37	27 ♏ 24	00 ♏ 52	05 ♌ 30	07 ♐ 30	25 ♌ 49
31	13 ♓ 21	01 ♐ 28	05 ♏ 18	09 ♌ 03	12 ♐ 39	26 ♌ 56

All ephemeris data is given at 12.00 UT and the Moon's longitude is additionally given for 24.00 UT

ASPECTARIAN

h m Aspects

01 Wednesday
04 20 ☿ ⊻ ♀
06 23 ☽ ⚼ ♃
07 37 ☽ ⚹ ☿
07 51 ☽ □ ♀
13 59 ☽ ⚹ ♆
14 24 ☽ ∠ ♄
19 33 ☽ □ ☉
22 12 ☽ ± ♃
23 48 ☽ ☌ ♇

02 Thursday
02 16 ☽ ⊻ ♂
04 51 ☽ Q ☿
06 11 ☽ □ ♅
08 37 ☽ ⊼ ♃
13 03 ☽ ⊥ ♂
15 41 ☽ ∠ ♆
16 18 ☽ ⚹ ♄
17 56 ☽ ⋕ ♃
23 04 ☽ ∥ ♄

03 Friday
00 05 ♄ ± ♅
03 08 ☽ ∥ ☿
05 02 ☽ ∠ ♂
06 50 ☽ ⊥ ♆
11 43 ☽ □ ☿
12 30 ☽ Q ♅
12 37 ☽ Q ♄
15 53 ☽ △ ♀
16 50 ☽ ⊻ ♆

04 Saturday
01 55 ☽ △ ☉
02 20 ☽ ⊻ ♇
02 35 ♀ ⊼ ♆
07 14 ☽ ⚹ ♂
08 06 ☉ □ ♇
08 19 ☽ ⚹ ♅
11 22 ☽ ☍ ♃
12 13 ☽ ⊥ ♇
13 45 ♀ ∠ ♄
14 08 ☉ ± ♆
17 01 ☿ St R
18 25 ☽ ∥ ♆
18 32 ☽ □ ♄
18 57 ☽ ⚼ ♀

05 Sunday
02 56 ☽ ∠ ♇
03 53 ☽ Q ♂
04 17 ☽ ⚼ ☉
04 18 ♂ △ ♅
08 46 ☽ ∠ ♅
10 59 ☽ ± ♀
13 06 ☽ △ ☿
17 56 ☽ ☌ ♆
19 58 ☽ ± ☉
21 38 ☽ ⊼ ♀
23 21 ☽ ⊥ ♅
23 35 ☽ ⋕ ♅

06 Monday
00 06 ☽ ∥ ☉
03 20 ☽ ⚹ ♇
06 01 ☉ ⋕ ♅
06 23 ☽ ⊼ ☉
09 03 ☽ ⊻ ♅
10 35 ☽ □ ♂
12 46 ☽ ⊼ ♃
13 10 ☽ ⚼ ☿
15 18 ☽ ∥ ♀
19 38 ☽ △ ♄
22 36 ☽ ± ♃
22 48 ☽ Q ♇

07 Tuesday
03 31 ☽ ± ☿
13 04 ☽ ⊼ ☿
13 25 ☽ ⚼ ♃
15 07 ☽ ⋕ ♀
17 01 ♃ ± ♇
18 37 ☽ ⊻ ♆
20 14 ☽ ⚼ ♄
20 58 ☉ ☍ ♅

08 Wednesday
00 22 ☿ ∠ ♂
02 58 ☽ ☍ ♀
04 12 ☽ □ ♇
04 12 ☽ ∥ ♅
04 27 ☽ ⊥ ♆
08 43 ☽ ⋕ ☉
09 52 ☽ ☌ ♅
10 51 ☽ ☍ ☉
11 08 ☽ ± ♄
14 05 ☽ △ ♂
14 20 ☽ △ ♃
15 43 ☉ ⊥ ♄
17 33 ♀ □ ♇
19 18 ☽ ∠ ♆
20 05 ♀ ± ♆
20 43 ♂ △ ♃
21 10 ☽ ⊼ ♄

09 Thursday
01 31 ♃ ⋕ ♄
10 16 ☽ ⋕ ♆
13 08 ☽ ☍ ☿
16 31 ☽ ⚼ ♂
20 30 ☽ ⚹ ♆

10 Friday
06 36 ☽ △ ♇
08 44 ☽ ± ♂
10 14 ☽ ⊼ ♀
12 22 ☽ ⊻ ♅
15 29 ☽ ⋕ ☿
17 09 ☽ Q ♆
17 27 ☽ ⊼ ☉
17 48 ☽ □ ♃
19 44 ☽ ⊼ ♂
21 43 ☽ ± ♀
22 48 ☽ ⊥ ♅
23 17 ☉ ⚹ ♃

11 Saturday
00 49 ☽ ☍ ♄
04 48 ☽ ± ☉
05 00 ☿ Q ♃
07 44 ☽ ∥ ♃
08 51 ☽ ⚼ ♇
09 10 ♀ ☍ ♅
09 39 ☽ ⋕ ♄
14 32 ☽ ⊼ ☿
14 42 ☽ ∠ ♅
15 18 ☽ ⚼ ♀
19 16 ☿ Q ♇
22 11 ☽ ⚼ ☉

12 Sunday
00 39 ☽ ± ☿
01 02 ☽ ± ♇
01 06 ☽ □ ♆
02 13 ☽ Q ♃
10 19 ♀ ⊥ ♄
11 52 ♀ ⋕ ♅
12 00 ☽ ⊼ ♇
15 59 ☽ ⚼ ☿
17 57 ☽ ⚹ ♅
21 34 ☽ △ ♀

13 Monday
00 35 ☽ ⚹ ♃
04 04 ☽ △ ☉
05 12 ☽ ☍ ♂
07 53 ☽ ⊼ ♄
12 47 ☉ ⚼ ♆
16 24 ☽ Q ♅
17 58 ☽ △ ☿
19 21 ☽ ± ♄

14 Tuesday
05 24 ☽ ∠ ♃
08 33 ♀ ⚹ ♃
09 18 ☽ △ ♆
12 50 ☽ ⚼ ♄
21 02 ☽ ☍ ♇
23 07 ☽ ⊥ ♃

15 Wednesday
03 07 ☽ □ ♅
07 19 ☉ ⚹ ♂
10 51 ☉ ⊻ ♄
11 04 ☽ ⊻ ♃
12 24 ♂ ⊻ ♄
13 46 ☽ □ ♀
14 40 ☽ ⚼ ♆
18 35 ☽ △ ♄
18 55 ☽ ⊼ ♂
19 01 ☽ ⋕ ♄
19 12 ☽ □ ☉
22 26 ♀ ⚼ ♆
23 27 ☽ □ ☿

16 Thursday
01 22 ☽ ∥ ♃
07 40 ☽ ± ♂
08 36 ☽ ± ♆
20 38 ☽ ⊼ ♆
20 40 ☉ ☌ ☿
22 43 ♂ Q ♆

17 Friday
00 06 ☿ ∥ ♆
01 49 ☿ ⚹ ♂
02 57 ☽ ⚼ ♂
06 03 ☽ Q ♀
08 17 ☽ Q ☿
08 57 ☽ ⊼ ♇
10 38 ☽ Q ☉
14 58 ☽ △ ♅
16 20 ♀ ⊻ ♄
17 55 ☿ ☌ ♀
19 19 ☿ ⊻ ♄
21 06 ☽ ± ♇

18 Saturday
00 05 ☽ ☌ ♃
02 01 ☉ ∥ ☿
03 50 ☽ ⋕ ♆
06 22 ☽ ⚹ ☿
07 34 ☽ □ ♄
09 13 ☽ ⚹ ♀
09 14 ☽ ⋕ ☉
11 18 ☽ △ ♂
12 27 ☽ ⋕ ☿
13 10 ☽ ⚹ ☉
15 22 ☽ ⚼ ♇
20 19 ☽ ⋕ ♀
21 15 ☽ ⚼ ♅

19 Sunday
09 16 ☽ ☍ ♆
10 07 ☽ ∠ ☿
13 18 ☿ ∥ ♀
15 22 ☽ ± ♅
18 08 ☽ ∥ ♅
19 07 ☽ ∠ ♀
19 57 ☿ ⚼ ♆
20 18 ☽ Q ♄
21 41 ☽ △ ♇
22 13 ☽ ∠ ☉

20 Monday
02 47 ☽ ⊥ ☿
03 24 ☽ ⊼ ♅
05 18 ♀ ⚹ ♂
09 49 ☉ ∥ ♆
13 20 ☽ ⊻ ♃
13 55 ☽ ⊻ ☿
15 15 ☽ ⊥ ♀
17 52 ☽ ⊥ ☉
20 32 ☽ ⚹ ♄
20 37 ☿ ⚹ ♃

21 Tuesday
01 24 ☽ ⊥ ♃
03 30 ☽ □ ♂
04 34 ☽ ⊻ ♀
06 51 ☽ ⊻ ☉
19 22 ☽ ∠ ♃
20 59 ☽ ⊼ ♆

22 Wednesday
02 23 ☽ ∠ ♄
07 27 ☽ ⋕ ♅
08 40 ☽ ± ♆
09 09 ☽ □ ♇
14 25 ☽ ☍ ♅
16 58 ☽ Q ♂
17 11 ☉ Q ♇
17 49 ☽ ∥ ☿
20 00 ☽ ⊥ ♄
21 25 ☽ ☌ ☿

23 Thursday
00 48 ☽ ⚹ ♃
02 02 ☽ ⚼ ♆
06 06 ♀ Q ♇
07 37 ☽ ⊻ ♄
15 58 ♀ ∥ ♆
17 22 ☽ ⚹ ♂
19 43 ☽ Q ♇
20 02 ☽ ∥ ♆
20 41 ☽ ∥ ♀
21 12 ☽ ☌ ♀
21 57 ☽ ☌ ☉
23 53 ☽ Q ♃

24 Friday
05 46 ☽ ∥ ☉
06 17 ☿ ⊥ ♄
06 28 ☽ △ ♆
18 16 ☽ ⚹ ♇
20 59 ☿ Q ♂
23 03 ☽ ⊼ ♅
23 11 ☽ ∠ ♂

25 Saturday
02 09 ☉ Q ♃
03 22 ♀ Q ♃
04 31 ☽ ⊻ ☿
05 11 ♂ ⊥ ♄
07 31 ☉ ☌ ♀
09 44 ☽ □ ♃
10 05 ☽ ± ♅
13 59 ☽ ⋕ ♃
15 29 ☽ ⊥ ☿
16 11 ☽ ☌ ♄
16 44 ☽ ⊥ ♂
19 17 ☿ St D
21 54 ☽ ∠ ♇

26 Sunday
02 29 ☽ ⚼ ♅
03 31 ☽ ∥ ♄
04 19 ☽ ⊻ ♂
07 54 ☽ ∠ ☿
09 55 ☽ ⊻ ☉
10 29 ☽ ⊻ ♀
13 33 ☽ □ ♆
14 12 ☽ ⊥ ♇
21 35 ☽ ⊥ ☉
22 23 ☽ ⊥ ♀

27 Monday
00 59 ☽ ⊻ ♇
05 22 ☽ △ ♅
11 10 ☽ ⚹ ☿
14 53 ☽ ∠ ☉
16 02 ☽ ∠ ♀
16 18 ☽ △ ♃
18 48 ♀ △ ♆
21 35 ☽ Q ♆
22 28 ☽ ⊻ ♄

28 Tuesday
08 56 ☽ Q ☿
09 06 ☽ ⊥ ♄
09 49 ☉ △ ♆
12 45 ☽ ☌ ♂
15 32 ☿ ⊥ ♄
18 36 ☽ ⚹ ♆
18 52 ☽ ⚼ ♃
19 18 ☽ ⚹ ☉
21 00 ☽ ⚹ ♀

29 Wednesday
00 56 ☽ ∠ ♄
05 47 ☽ ☌ ♇
09 50 ☽ □ ♅
10 36 ☽ ± ♃
17 30 ☽ □ ☿
17 41 ☽ Q ☉
19 49 ☽ Q ♀
19 58 ☽ ∥ ♄
20 32 ☽ ∠ ♆
21 02 ☽ ⊼ ♃

30 Thursday
03 01 ☽ ⚹ ♄
11 45 ☽ ⋕ ♃
11 57 ☽ ⊥ ♆
16 38 ☽ Q ♅
19 22 ☽ ⊻ ♂
22 10 ☽ ⊻ ♆
23 11 ☽ ∥ ☉
23 40 ☽ Q ♄

31 Friday
01 29 ☽ ∥ ♀
02 48 ☽ □ ☉
05 34 ☽ □ ♀
06 08 ☽ ⊥ ♂
09 12 ☽ ⊻ ♇
12 57 ☽ ⚹ ♅
19 21 ☽ ⊥ ♇
21 44 ☿ ⚼ ♆
22 12 ☽ ∠ ♂
23 41 ☽ △ ☿

NOVEMBER 2014

LONGITUDES

Date	Sidereal time h m s	Sun ☉ ° ' "	Moon ☽ ° ' "	Moon ☽ 24.00 ° ' "	Mercury ☿ ° '	Venus ♀ ° '	Mars ♂ ° '	Jupiter ♃ ° '	Saturn ♄ ° '	Uranus ♅ ° '	Neptune ♆ ° '	Pluto ♇ ° '
01	14 42 48	08 ♏ 59 09	27 ♒ 15 26	04 ♓ 24 09	20 ♎ 27	10 ♏ 49	04 ♑ 30	20 ♌ 28	23 ♏ 56	13 ♈ 34	04 ♓ 52	11 ♑ 23
02	14 46 44	09 59 11	11 ♓ 33 38	18 ♓ 43 33	21 30	12 04	05 14	20 35	24 03	13 R 31	04 R 51	11 25
03	14 50 41	10 59 14	25 ♓ 53 28	03 ♈ 02 55	22 38	13 19	05 59	20 41	24 10	13 29	04 51	11 26
04	14 54 37	11 59 19	10 ♈ 11 25	17 ♈ 18 28	23 52	14 35	06 44	20 47	24 17	13 27	04 50	11 27
05	14 58 34	12 59 25	24 ♈ 23 32	01 ♉ 26 03	25 09	15 50	07 29	20 53	24 24	13 25	04 50	11 28
06	15 02 30	13 59 33	08 ♉ 25 33	15 ♉ 21 30	26 30	17 05	08 14	20 59	24 31	13 23	04 50	11 30
07	15 06 27	14 59 43	22 ♉ 13 30	29 ♉ 01 10	27 53	18 20	08 59	21 05	24 38	13 21	04 49	11 31
08	15 10 23	15 59 55	05 ♊ 44 11	12 ♊ 22 21	29 ♎ 19	19 36	09 44	21 11	24 45	13 19	04 49	11 32
09	15 14 20	17 00 09	18 ♊ 55 34	25 ♊ 23 47	00 ♏ 47	20 51	10 29	21 16	24 52	13 17	04 49	11 34
10	15 18 17	18 00 25	01 ♋ 47 06	08 ♋ 05 40	02 17	22 06	11 15	21 22	24 59	13 16	04 48	11 35
11	15 22 13	19 00 42	14 ♋ 19 45	20 ♋ 29 42	03 48	23 21	12 00	21 27	25 07	13 14	04 48	11 36
12	15 26 10	20 01 02	26 ♋ 35 54	02 ♌ 38 49	05 20	24 37	12 45	21 32	25 14	13 12	04 48	11 38
13	15 30 06	21 01 23	08 ♌ 39 00	14 ♌ 37 00	06 54	25 52	13 31	21 36	25 21	13 10	04 48	11 39
14	15 34 03	22 01 47	20 ♌ 33 24	26 ♌ 28 51	08 27	27 07	14 16	21 41	25 28	13 08	04 48	11 41
15	15 37 59	23 02 12	02 ♍ 23 57	08 ♍ 19 22	10 02	28 23	15 02	21 46	25 35	13 07	04 48	11 42
16	15 41 56	24 02 39	14 ♍ 15 42	20 ♍ 13 37	11 36	29 ♏ 38	15 48	21 50	25 42	13 05	04 D 48	11 44
17	15 45 52	25 03 08	26 ♍ 13 41	02 ♎ 16 29	13 11	00 ♐ 53	16 33	21 54	25 49	13 03	04 48	11 45
18	15 49 49	26 03 38	08 ♎ 22 31	14 ♎ 32 17	14 46	02 08	17 19	21 58	25 57	13 02	04 48	11 47
19	15 53 46	27 04 11	20 ♎ 46 11	27 ♎ 04 34	16 22	03 24	18 05	22 02	26 04	13 00	04 48	11 48
20	15 57 42	28 04 45	03 ♏ 27 41	09 ♏ 55 41	17 57	04 39	18 51	22 05	26 11	12 59	04 48	11 50
21	16 01 39	29 ♏ 05 21	16 ♏ 28 38	23 ♏ 06 32	19 32	05 54	19 36	22 09	26 18	12 57	04 48	11 52
22	16 05 35	00 ♐ 05 58	29 ♏ 49 12	06 ♐ 36 24	21 08	07 10	20 22	22 12	26 25	12 56	04 49	11 53
23	16 09 32	01 06 37	13 ♐ 27 49	20 ♐ 23 01	22 43	08 25	21 08	22 15	26 32	12 54	04 49	11 55
24	16 13 28	02 07 18	27 ♐ 21 31	04 ♑ 22 47	24 18	09 40	21 54	22 18	26 39	12 53	04 49	11 57
25	16 17 25	03 07 59	11 ♑ 26 15	18 ♑ 31 22	25 53	10 56	22 40	22 20	26 47	12 52	04 49	11 58
26	16 21 21	04 08 42	25 ♑ 37 34	02 ♒ 44 21	27 28	12 11	23 27	22 23	26 54	12 50	04 50	12 00
27	16 25 18	05 09 26	09 ♒ 51 15	16 ♒ 57 53	29 ♏ 03	13 26	24 13	22 25	27 01	12 49	04 50	12 02
28	16 29 15	06 10 10	24 ♒ 03 53	01 ♓ 09 00	00 ♐ 38	14 42	24 59	22 27	27 08	12 48	04 51	12 04
29	16 33 11	07 10 56	08 ♓ 13 01	15 ♓ 15 47	02 12	15 57	25 45	22 29	27 15	12 47	04 51	12 05
30	16 37 08	08 ♐ 11 43	22 ♓ 17 10	29 ♓ 17 04	03 ♐ 47	17 ♐ 12	26 ♑ 31	22 ♌ 31	27 ♏ 22	12 ♈ 46	04 ♓ 51	12 ♑ 07

Date	Moon True ☊ ° '	Moon Mean ☊ ° '	Moon ☽ Latitude ° '
01	19 ♎ 09	18 ♎ 08	04 N 09
02	19 D 10	18 05	03 13
03	19 11	18 02	02 05
04	19 12	17 59	00 N 50
05	19 R 12	17 56	00 S 29
06	19 11	17 53	01 44
07	19 09	17 49	02 52
08	19 05	17 46	03 48
09	19 01	17 43	04 31
10	18 56	17 40	04 59
11	18 51	17 37	05 12
12	18 48	17 34	05 10
13	18 46	17 30	04 54
14	18 45	17 27	04 26
15	18 D 45	17 24	03 47
16	18 46	17 21	02 58
17	18 48	17 18	02 00
18	18 50	17 14	00 S 57
19	18 R 50	17 11	00 N 11
20	18 49	17 08	01 19
21	18 46	17 05	02 24
22	18 41	17 02	03 24
23	18 35	16 59	04 13
24	18 27	16 55	04 48
25	18 20	16 52	05 06
26	18 13	16 49	05 06
27	18 08	16 46	04 47
28	18 06	16 43	04 10
29	18 D 05	16 40	03 18
30	18 ♎ 06	16 ♎ 36	02 N 15

DECLINATIONS

Date	Sun ☉ ° '	Moon ☽ ° '	Mercury ☿ ° '	Venus ♀ ° '	Mars ♂ ° '	Jupiter ♃ ° '	Saturn ♄ ° '	Uranus ♅ ° '	Neptune ♆ ° '	Pluto ♇ ° '
01	14 S 29	08 S 31	06 S 00	14 S 19	24 S 52	15 N 15	16 S 53	04 N 43	10 S 25	20 S 40
02	14 48	04 S 15	06 22	14 44	24 50	15 13	16 55	04 42	10 26	20 40
03	15 07	00 N 17	06 46	15 09	24 48	15 12	16 56	04 41	10 26	20 40
04	15 26	04 48	07 13	15 33	24 46	15 10	16 58	04 40	10 26	20 40
05	15 44	09 01	07 42	15 57	24 43	15 08	17 00	04 39	10 26	20 40
06	16 02	12 40	08 13	16 20	24 40	15 06	17 02	04 39	10 26	20 40
07	16 20	15 33	08 45	16 44	24 37	15 05	17 03	04 38	10 26	20 40
08	16 37	17 31	09 19	17 06	24 34	15 03	17 05	04 37	10 26	20 40
09	16 55	18 28	09 53	17 28	24 30	15 02	17 07	04 36	10 26	20 40
10	17 12	18 26	10 28	17 50	24 26	15 00	17 09	04 36	10 26	20 40
11	17 28	17 30	11 03	18 11	24 22	14 59	17 11	04 35	10 26	20 40
12	17 45	15 45	11 39	18 32	24 18	14 57	17 12	04 35	10 26	20 40
13	18 01	13 21	12 15	18 52	24 13	14 56	17 14	04 34	10 27	20 40
14	18 16	10 26	12 50	19 12	24 08	14 55	17 16	04 34	10 27	20 40
15	18 32	07 05	13 26	19 31	24 03	14 53	17 17	04 33	10 27	20 40
16	18 47	03 N 28	14 01	19 50	23 57	14 52	17 19	04 32	10 27	20 40
17	19 02	00 S 20	14 36	20 08	23 51	14 51	17 21	04 31	10 27	20 40
18	19 16	04 11	15 10	20 26	23 45	14 50	17 23	04 31	10 27	20 40
19	19 30	07 57	15 44	20 43	23 39	14 49	17 24	04 30	10 26	20 40
20	19 44	11 26	16 17	20 59	23 32	14 48	17 26	04 29	10 26	20 40
21	19 57	14 27	16 50	21 15	23 25	14 47	17 28	04 29	10 26	20 40
22	20 10	16 47	17 22	21 30	23 18	14 46	17 29	04 28	10 26	20 40
23	20 23	18 14	17 53	21 45	23 11	14 46	17 31	04 28	10 26	20 40
24	20 35	18 37	18 23	21 59	23 03	14 45	17 33	04 27	10 26	20 40
25	20 47	17 52	18 53	22 12	22 55	14 44	17 34	04 27	10 26	20 40
26	20 58	16 00	19 22	22 24	22 47	14 44	17 36	04 26	10 26	20 40
27	21 09	13 10	19 50	22 36	22 39	14 43	17 38	04 26	10 26	20 40
28	21 20	09 34	20 17	22 48	22 30	14 43	17 39	04 26	10 25	20 40
29	21 30	05 25	20 43	22 58	22 21	14 42	17 41	04 25	10 25	20 40
30	21 S 40	01 S 00	21 S 07	23 S 08	22 S 12	14 N 42	17 S 42	04 N 25	10 S 25	20 S 40

ZODIAC SIGN ENTRIES

Date	h	m	Planets
01	16	37	☽ ♓
03	18	53	☽ ♈
05	21	33	☽ ♉
08	01	45	☽ ♊
08	23	09	☿ ♏
10	08	38	☽ ♋
12	18	44	☽ ♌
15	07	08	☽ ♍
16	19	03	♀ ♐
17	19	30	☽ ♎
20	05	31	☽ ♏
22	09	38	☉ ♐
22	12	19	☽ ♐
24	16	31	☽ ♑
26	19	23	☽ ♒
28	02	26	☿ ♐
28	22	03	☽ ♓

LATITUDES

Date	Mercury ☿ ° '	Venus ♀ ° '	Mars ♂ ° '	Jupiter ♃ ° '	Saturn ♄ ° '	Uranus ♅ ° '	Neptune ♆ ° '	Pluto ♇ ° '
01	02 N 08	00 N 47	01 S 30	00 N 37	01 N 56	00 S 42	00 S 45	02 N 18
04	02 11	00 41	01 30	00 38	01 56	00 42	00 45	02 17
07	02 06	00 35	01 29	00 39	01 55	00 42	00 45	02 17
10	01 55	00 28	01 29	00 39	01 55	00 42	00 45	02 16
13	01 40	00 21	01 28	00 40	01 55	00 41	00 45	02 16
16	01 22	00 14	01 27	00 41	01 55	00 41	00 45	02 15
19	01 01	00 N 07	01 27	00 41	01 55	00 41	00 45	02 15
22	00 42	00 00	01 26	00 42	01 55	00 41	00 45	02 15
25	00 N 21	00 S 07	01 25	00 43	01 55	00 41	00 45	02 14
28	00 00	00 14	01 24	00 43	01 55	00 41	00 45	02 14
31	00 S 20	00 S 22	01 S 23	00 N 44	01 N 55	00 S 41	00 S 44	02 N 13

DATA

Julian Date	2456963
Delta T	+69 seconds
Ayanamsa	24° 03' 56"
Synetic vernal point	05° ♓ 03' 03"
True obliquity of ecliptic	23° 26' 06"

MOON'S PHASES, APSIDES AND POSITIONS ☽

Date	h	m	Phase	Longitude ° '	Eclipse Indicator
06	22	23	○	14 ♉ 26	
14	15	15	☾	22 ♌ 10	
22	12	32	●	00 ♐ 07	
29	10	06	☽	07 ♓ 06	

Day	h	m		
03	00	35	Perigee	
15	01	57	Apogee	
27	23	07	Perigee	
03	10	32	0N	
09	23	10	Max dec	18° N 35'
17	09	53	0S	
24	08	12	Max dec	18° S 38'
30	17	21	0N	

LONGITUDES

Date	Chiron ⚷ ° '	Ceres ⚳ ° '	Pallas ⚴ ° '	Juno ⚵ ° '	Vesta ⚶ ° '	Black Moon Lilith ⚸ ° '
01	13 ♓ 20	01 ♐ 52	05 ♏ 44	09 ♌ 22	13 ♐ 10	27 ♌ 03
11	13 ♓ 10	05 ♐ 59	10 ♏ 09	12 ♌ 21	18 ♐ 23	28 ♌ 10
21	13 ♓ 06	10 ♐ 08	14 ♏ 33	14 ♌ 43	23 ♐ 38	29 ♌ 17
31	13 ♓ 07	14 ♐ 17	18 ♏ 55	16 ♌ 30	28 ♐ 57	00 ♍ 24

All ephemeris data is given at 12.00 UT and the Moon's longitude is additionally given for 24.00 UT

ASPECTARIAN

h m Aspects

01 Saturday
00 23 ☽ ∥ ♆
00 29 ☽ ☍ ♃
06 22 ☽ □ ♄
10 32 ☽ ∠ ♇
12 44 ☿ ⚹ ♃
14 11 ☽ ∠ ♅
23 09 ♀ ⚹ ♇
23 43 ♂ ⚹ ♆

02 Sunday
00 46 ☽ ☌ ♆
00 49 ☽ ⚹ ♂
01 20 ☽ ∥ ☿
02 49 ☽ Ꝑ ☿
05 15 ☽ ⊥ ♅
09 10 ☽ △ ☉
09 35 ☽ # ♅
11 45 ☽ ⚹ ♇
12 56 ☽ △ ♀
15 17 ☽ ⊻ ♅
19 09 ☽ ± ☿
22 02 ☽ Q ♂

03 Monday
00 46 ♂ Ꝑ ♃
03 13 ☽ ⊼ ♃
05 52 ☉ ∥ ♀
06 04 ☽ ⊼ ☿
07 52 ☽ Q ♇
09 05 ☽ △ ♄
12 10 ☽ Ꝑ ☉
13 20 ☽ ± ♃
14 36 ♀ # ♃
15 04 ♀ ⊼ ♅
16 28 ☽ Ꝑ ♀
17 06 ☉ # ♃
22 51 ☉ ⚹ ♇

04 Tuesday
03 00 ☽ ⊻ ♆
04 24 ☽ ± ☉
04 33 ☽ Ꝑ ♃
05 52 ☽ □ ♂
09 02 ☽ ± ♀
10 28 ☽ Ꝑ ♄
11 20 ☽ ∥ ♅
13 05 ☽ ⊥ ♆
14 07 ☽ □ ♇
15 15 ☽ ⊼ ☉
17 29 ☽ ☌ ♅
20 06 ☽ ⊼ ♀
20 51 ☿ ⊻ ♄

05 Wednesday
01 46 ☽ ± ♄
03 15 ☽ # ☿
04 16 ☽ ∠ ♆
06 01 ☽ △ ♃
12 01 ☽ ⊼ ♄
13 25 ☽ ☍ ☿
20 49 ☽ # ♆
21 59 ☉ ⊼ ♅

06 Thursday
01 46 ☿ Q ♂
05 49 ☽ ⚹ ♆
11 39 ☽ △ ♂
17 18 ☽ △ ♇
20 33 ☽ ⊻ ♅
22 23 ☽ ☍ ☉

07 Friday
02 33 ☽ Q ♆
04 31 ☽ ☍ ♀
06 59 ☽ ⊥ ♅
07 30 ☽ ∥ ♃
09 59 ☽ □ ♃
15 17 ☽ Ꝑ ♂
16 17 ☽ ☍ ♄
19 34 ☽ Ꝑ ♇
21 30 ☽ # ☉
22 48 ☽ ∠ ♅
23 10 ☽ ⊼ ☿

08 Saturday
04 08 ☽ # ♀
05 29 ☽ # ♄
06 56 ♀ ± ♅
08 12 ☽ ± ♂
10 21 ☽ □ ♆
10 55 ♀ ∥ ♄
11 10 ☽ ± ☿
11 39 ☽ ± ♇
12 40 ♂ ∠ ♄
15 37 ☿ Q ♇
18 15 ☽ Q ♃
19 39 ☽ ⊼ ♂
22 30 ☽ ⊼ ♇

09 Sunday
01 42 ☽ ⚹ ♅
05 30 ☽ Ꝑ ☿
08 10 ☽ ⊼ ☉
15 56 ☽ ⊼ ♀
16 22 ☽ ⚹ ♃
20 11 ☽ ± ☉
20 41 ♀ □ ♃
23 08 ☽ ⊼ ♄
23 46 ☽ Q ♅

10 Monday
04 19 ☽ ± ♀
07 32 ☉ ∥ ♄
10 29 ☽ ± ♄
10 57 ☿ ∥ ♆
13 05 ☽ △ ☿
14 31 ☽ Ꝑ ☉
17 44 ☽ △ ♆
20 45 ☽ ∠ ♃
23 06 ♂ ☌ ♇
23 14 ☽ Ꝑ ♀

11 Tuesday
01 07 ☽ # ♀
03 47 ☽ Ꝑ ♄
06 00 ☿ Q ♃
06 44 ☽ ☍ ♇
07 13 ☽ ☍ ♂
09 53 ☽ □ ♅
12 22 ☽ # ☉
14 10 ☽ ⊥ ♃
17 02 ☉ ± ♅
17 12 ☽ # ♄
21 55 ☽ △ ☉
22 39 ☽ Ꝑ ♆

12 Wednesday
01 57 ☽ ⊻ ♃
03 38 ☿ △ ♆
07 38 ☽ △ ♀
09 16 ☽ △ ♄
16 22 ☽ ± ♆
20 49 ☽ ∥ ♃

13 Thursday
01 02 ♀ ☌ ♄
01 29 ♂ □ ♅
04 18 ☽ ⊼ ♆
07 57 ☽ □ ☿
18 03 ☽ ⊼ ♇
19 56 ☽ # ☿
21 04 ☽ △ ♅
22 27 ☽ ⊼ ♂

14 Friday
03 05 ☉ □ ♃
03 23 ♀ ∠ ♇
06 10 ☽ ± ♇
11 23 ☽ ± ♂
11 52 ☽ # ♆
14 18 ☽ ☌ ♃
15 15 ☽ □ ☉
22 03 ☽ □ ♄

15 Saturday
00 26 ☽ Ꝑ ♇
01 46 ☽ Q ☿
02 53 ☽ □ ♀
03 20 ☽ Ꝑ ♅
06 52 ☽ Ꝑ ♂
07 03 ♀ Ꝑ ♅
16 52 ☽ ☍ ♆
21 32 ☽ ± ♅

16 Sunday
05 03 ☽ ∥ ♅
05 49 ☽ ⚹ ☿
06 53 ☽ △ ♇
07 05 ♆ St D
07 07 ☽ Q ☉
09 38 ☽ ⊼ ♅
10 52 ☽ Q ♄
13 20 ♂ ± ♃
13 57 ☿ ⚹ ♇
15 18 ☽ △ ♂
19 35 ☽ Q ♀

17 Monday
03 18 ☽ ⊻ ♃
09 26 ☽ ⚹ ☉
10 04 ☿ ⊼ ♅
11 11 ☽ ⚹ ♄
15 21 ☽ ⊥ ♃
16 29 ☽ ∠ ☿
22 13 ☿ # ♃
22 19 ☽ ⚹ ♀

18 Tuesday
04 59 ☽ ⊼ ♆
08 50 ☉ ☌ ♄
09 13 ☽ ∠ ♃
12 53 ☽ ⊥ ☿
14 00 ☽ # ♅
16 44 ☽ ± ♆
17 04 ☽ ∠ ♄
17 43 ☽ ∠ ☉
18 40 ☽ □ ♇
20 29 ♀ ∠ ♂
21 03 ☽ ☍ ♅

19 Wednesday
02 17 ☽ ⊻ ☿
05 37 ☉ ∠ ♇
06 30 ☽ □ ♂
06 56 ☽ ∠ ♀
08 27 ♀ ∥ ♇
10 09 ☽ Ꝑ ♆
10 38 ☽ ⊥ ♄
12 37 ☽ ⊥ ☉
14 25 ☽ ⚹ ♃
22 11 ☽ ⊻ ♄

20 Thursday
01 01 ☽ ⊻ ☉
01 59 ☽ ⊥ ♀
04 54 ☽ ∥ ♆
05 11 ☽ Q ♇
09 40 ☉ Ꝑ ♅
13 10 ☽ Q ♃
14 28 ☽ ⊻ ♀
14 30 ☽ △ ♆
14 55 ♀ □ ♆
18 41 ☽ Q ♂

21 Friday
03 19 ☿ ± ♅
03 33 ☽ ⚹ ♇
05 34 ☽ ⊼ ♅
11 08 ♀ ⊥ ♇
14 00 ☿ ⚹ ♂
14 58 ☽ # ♃
16 29 ☽ ± ♅
18 02 ☽ ⚹ ♂
18 18 ♂ ∠ ♆
18 19 ☽ ☌ ☿
22 18 ☽ □ ♃

22 Saturday
05 53 ☽ ☌ ♄
06 46 ☽ ∠ ♇
08 38 ☽ Ꝑ ♅
12 32 ☽ ☌ ☉
17 52 ☿ ∥ ♄
20 50 ☽ □ ♆
21 53 ☽ ∥ ♄
22 25 ☽ ∠ ♂
22 46 ☽ ⊥ ♇
23 56 ☽ ∥ ☿

23 Sunday
02 17 ☽ ☌ ♀
04 43 ☿ □ ♃
09 18 ☽ ⊻ ♇
11 02 ☽ △ ♅
15 05 ☽ ⊥ ♂

24 Monday
02 05 ☽ ⊻ ♂
03 16 ☽ △ ♃
04 12 ☽ Q ♆
06 05 ☽ ⊻ ☿
10 47 ☽ ⊻ ♄
17 41 ☽ ⊥ ☿
18 17 ☽ ∥ ☿
20 47 ☽ ⊻ ☉
21 08 ☽ ⊥ ♄
22 33 ☉ ∥ ♇

25 Tuesday
00 45 ☽ ⚹ ♆
00 55 ♂ ⊼ ♃
05 01 ☽ Ꝑ ♃
07 47 ☽ ⊥ ☉
10 57 ☽ ∠ ☿
11 03 ☽ ⊻ ♀
12 35 ☽ ∠ ♄
12 55 ☽ ☌ ♇
14 25 ☽ □ ♅
16 45 ☽ ∥ ♄
20 20 ☽ ± ♃
22 12 ☽ ⊥ ♀

26 Wednesday
00 13 ☽ ∠ ☉
02 12 ☽ ∠ ♆
02 37 ☿ ☌ ♄
04 49 ☿ ∠ ♇
06 30 ☽ ⊼ ♃
08 06 ☽ ☌ ♂
08 28 ♀ ⊻ ♇
14 10 ☽ ⚹ ♄
14 53 ☽ ∠ ♀
15 30 ☽ ⚹ ☿
17 24 ☽ ⊥ ♆
17 35 ☿ Ꝑ ♅
20 47 ☽ Q ♅
23 48 ☽ # ♃

27 Thursday
00 22 ♀ △ ♅
03 28 ☽ ⚹ ☉
03 32 ☽ ⊻ ♆
04 19 ☉ □ ♆
10 34 ☽ Q ♄
14 16 ☽ Q ☿
14 46 ♀ ∥ ♂
15 41 ☽ ⊻ ♇
16 47 ♄ ∠ ♇
17 00 ☽ ⚹ ♅
18 38 ☽ ⚹ ♀

28 Friday
01 16 ☽ Q ☉
01 50 ☽ ⊥ ♇
06 38 ☽ ∥ ♆
09 16 ☽ ☍ ♃
09 22 ☉ ⊥ ♇
13 38 ☽ ⊻ ♂
16 53 ☽ Q ♀
17 05 ☽ ∠ ♇
17 14 ☽ □ ♄
18 19 ☽ ∠ ♅

29 Saturday
00 23 ☽ ⊥ ♂
00 31 ☽ □ ☿
06 16 ☽ ☌ ♆
09 34 ☽ ⊥ ♅
09 44 ☿ ∥ ♇
10 06 ☽ □ ☉
16 34 ☽ ∠ ♂
17 32 ☽ # ♅
18 36 ☽ ⚹ ♇
19 46 ☽ ⊻ ♅

30 Sunday
02 28 ☽ □ ♀
12 23 ☽ ⊼ ♃
15 09 ☽ Q ♇
19 41 ☽ ⚹ ♂
20 47 ☽ △ ♄
22 42 ☽ ± ♃

DECEMBER 2014

LONGITUDES

Date	Sidereal time h m s	Sun ☉ ° ' "	Moon ☽ ° ' "	Moon ☽ 24.00 ° ' "	Mercury ☿ ° '	Venus ♀ ° '	Mars ♂ ° '	Jupiter ♃ ° '	Saturn ♄ ° '	Uranus ♅ ° '	Neptune ♆ ° '	Pluto ♇ ° '
01	16 41 04	09 ♐ 12 30	06 ♈ 15 23	13 ♈ 12 04	05 ♐ 22	18 ♐ 28	27 ♑ 18	22 ♌ 32	27 ♏ 29	12 ♈ 45	04 ♓ 52	12 ♑ 09
02	16 45 01	10 13 19	20 ♈ 06 59	27 ♈ 00 02	06 56	19 43	28 04	22 34	27 36	12 R 44	04 52	12 11
03	16 48 57	11 14 08	03 ♉ 51 03	10 ♉ 39 51	08 30	20 58	28 50	22 35	27 43	12 43	04 53	12 13
04	16 52 54	12 14 58	17 ♉ 26 16	24 ♉ 10 02	10 05	22 13	29 ♑ 37	22 36	27 50	12 42	04 54	12 15
05	16 56 50	13 15 50	00 ♊ 50 57	07 ♊ 28 46	11 39	23 29	00 ♒ 23	22 37	27 57	12 41	04 54	12 16
06	17 00 47	14 16 42	14 ♊ 03 15	20 ♊ 34 13	13 13	24 44	01 10	22 37	28 04	12 40	04 55	12 18
07	17 04 44	15 17 35	27 ♊ 01 31	03 ♋ 25 01	14 47	25 59	01 56	22 37	28 11	12 40	04 56	12 20
08	17 08 40	16 18 30	09 ♋ 44 40	16 ♋ 00 31	16 21	27 15	02 43	22 38	28 18	12 39	04 56	12 22
09	17 12 37	17 19 25	22 ♋ 12 39	28 ♋ 21 14	17 56	28 30	03 29	22 R 38	28 25	12 38	04 57	12 24
10	17 16 33	18 20 22	04 ♌ 26 31	10 ♌ 28 49	19 30	29 ♐ 45	04 16	22 37	28 32	12 38	04 58	12 26
11	17 20 30	19 21 20	16 ♌ 28 30	22 ♌ 26 04	21 04	01 ♑ 01	05 03	22 37	28 39	12 37	04 59	12 28
12	17 24 26	20 22 18	28 ♌ 22 00	04 ♍ 16 52	22 39	02 16	05 49	22 36	28 46	12 37	05 00	12 30
13	17 28 23	21 23 18	10 ♍ 11 16	16 ♍ 05 51	24 13	03 31	06 36	22 36	28 52	12 36	05 01	12 32
14	17 32 19	22 24 19	22 ♍ 01 17	27 ♍ 58 15	25 48	04 46	07 23	22 34	28 59	12 36	05 02	12 34
15	17 36 16	23 25 21	03 ♎ 57 27	09 ♎ 59 34	27 22	06 02	08 09	22 33	29 06	12 35	05 03	12 36
16	17 40 13	24 26 24	16 ♎ 05 15	22 ♎ 15 10	28 ♐ 57	07 17	08 56	22 32	29 13	12 35	05 04	12 38
17	17 44 09	25 27 28	28 ♎ 29 54	04 ♏ 49 57	00 ♑ 32	08 32	09 43	22 30	29 19	12 35	05 05	12 40
18	17 48 06	26 28 32	11 ♏ 15 47	17 ♏ 47 44	02 07	09 48	10 30	22 28	29 26	12 34	05 06	12 42
19	17 52 02	27 29 38	24 ♏ 26 01	01 ♐ 10 41	03 42	11 03	11 17	22 26	29 33	12 34	05 07	12 44
20	17 55 59	28 30 44	08 ♐ 01 41	14 ♐ 58 44	05 18	12 18	12 03	22 24	29 39	12 34	05 08	12 46
21	17 59 55	29 ♐ 31 51	22 ♐ 01 26	29 ♐ 09 13	06 53	13 33	12 50	22 22	29 46	12 34	05 09	12 48
22	18 03 52	00 ♑ 32 59	06 ♑ 21 19	13 ♑ 36 55	08 29	14 49	13 37	22 19	29 52	12 D 34	05 10	12 50
23	18 07 48	01 34 07	20 ♑ 55 05	28 ♑ 14 49	10 05	16 04	14 24	22 17	29 ♏ 59	12 34	05 12	12 52
24	18 11 45	02 35 16	05 ♒ 35 09	12 ♒ 55 10	11 40	17 19	15 11	22 14	00 ♐ 05	12 34	05 13	12 54
25	18 15 42	03 36 24	20 ♒ 13 58	27 ♒ 30 51	13 16	18 34	15 58	22 10	00 12	12 35	05 14	12 56
26	18 19 38	04 37 33	04 ♓ 45 10	11 ♓ 56 27	14 52	19 50	16 45	22 07	00 18	12 35	05 15	12 58
27	18 23 35	05 38 42	19 ♓ 04 22	26 ♓ 08 42	16 28	21 05	17 32	22 04	00 24	12 35	05 17	13 01
28	18 27 31	06 39 50	03 ♈ 09 21	10 ♈ 06 19	18 04	22 20	18 19	22 00	00 30	12 35	05 18	13 03
29	18 31 28	07 40 59	16 ♈ 59 40	23 ♈ 49 31	19 40	23 35	19 06	21 56	00 37	12 36	05 19	13 05
30	18 35 24	08 42 07	00 ♉ 36 02	07 ♉ 19 21	21 16	24 51	19 53	21 52	00 43	12 36	05 21	13 07
31	18 39 21	09 ♑ 43 16	13 ♉ 59 39	20 ♉ 37 02	22 ♑ 52	26 ♑ 06	20 ♒ 40	21 ♌ 48	00 ♐ 49	12 ♈ 37	05 ♓ 22	13 ♑ 09

Moon

Date	Moon True ☊ ° '	Moon Mean ☊ ° '	Moon ☽ Latitude ° '
01	18 ♎ 07	16 ♎ 33	01 N 03
02	18 R 08	16 30	00 S 11
03	18 06	16 27	01 24
04	18 03	16 24	02 31
05	17 56	16 20	03 28
06	17 48	16 17	04 14
07	17 37	16 14	04 45
08	17 26	16 11	05 01
09	17 15	16 08	05 03
10	17 06	16 05	04 50
11	16 58	16 01	04 25
12	16 53	15 58	03 49
13	16 50	15 55	03 02
14	16 49	15 52	02 08
15	16 D 50	15 49	01 08
16	16 50	15 45	00 S 04
17	16 R 50	15 42	01 N 02
18	16 47	15 39	02 06
19	16 42	15 36	03 06
20	16 34	15 33	03 57
21	16 23	15 30	04 35
22	16 12	15 26	04 57
23	16 00	15 23	05 00
24	15 49	15 20	04 44
25	15 41	15 17	04 09
26	15 35	15 14	03 18
27	15 32	15 11	02 15
28	15 D 32	15 07	01 N 05
29	15 R 32	15 04	00 S 08
30	15 31	15 01	01 19
31	15 ♎ 29	14 ♎ 58	02 S 24

DECLINATIONS

Date	Sun ☉ ° '	Moon ☽ ° '	Mercury ☿ ° '	Venus ♀ ° '	Mars ♂ ° '	Jupiter ♃ ° '	Saturn ♄ ° '	Uranus ♅ ° '	Neptune ♆ ° '	Pluto ♇ ° '
01	21 S 50	03 N 27	21 S 31	23 S 18	22 S 03	14 N 42	17 S 44	04 N 24	10 S 25	20 S 40
02	21 59	07 42	21 54	23 26	21 53	14 41	17 46	04 24	10 25	20 40
03	22 07	11 29	22 16	23 34	21 43	14 41	17 47	04 24	10 24	20 40
04	22 15	14 37	22 36	23 41	21 33	14 41	17 49	04 23	10 24	20 40
05	22 23	16 56	22 56	23 48	21 23	14 41	17 50	04 23	10 24	20 40
06	22 31	18 17	23 14	23 53	21 12	14 41	17 52	04 23	10 24	20 40
07	22 37	18 39	23 31	23 58	21 02	14 41	17 53	04 23	10 24	20 40
08	22 44	18 04	23 47	24 02	20 51	14 41	17 55	04 22	10 23	20 40
09	22 50	16 37	24 02	24 06	20 39	14 42	17 56	04 22	10 23	20 40
10	22 55	14 26	24 16	24 09	20 28	14 42	17 58	04 22	10 23	20 40
11	23 00	11 40	24 28	24 11	20 16	14 42	17 59	04 22	10 22	20 40
12	23 05	08 28	24 39	24 12	20 04	14 43	18 01	04 22	10 22	20 40
13	23 09	04 56	24 49	24 12	19 52	14 43	18 02	04 22	10 21	20 40
14	23 13	01 N 12	24 57	24 12	19 40	14 44	18 04	04 21	10 21	20 39
15	23 16	02 S 37	25 04	24 11	19 27	14 44	18 05	04 21	10 21	20 39
16	23 19	06 23	25 09	24 09	19 15	14 45	18 06	04 21	10 20	20 39
17	23 21	09 59	25 14	24 07	19 02	14 46	18 08	04 21	10 20	20 39
18	23 23	13 12	25 16	24 03	18 49	14 47	18 09	04 21	10 20	20 39
19	23 25	15 52	25 18	23 59	18 35	14 48	18 11	04 21	10 19	20 39
20	23 26	17 45	25 18	23 55	18 22	14 48	18 12	04 21	10 19	20 39
21	23 26	18 37	25 16	23 49	18 08	14 49	18 13	04 21	10 18	20 39
22	23 26	18 20	25 13	23 43	17 54	14 51	18 15	04 21	10 18	20 39
23	23 26	16 52	25 08	23 36	17 40	14 52	18 16	04 21	10 17	20 39
24	23 25	14 17	25 02	23 28	17 26	14 53	18 17	04 21	10 17	20 39
25	23 23	10 48	24 55	23 20	17 11	14 54	18 19	04 21	10 16	20 39
26	23 21	06 42	24 46	23 11	16 57	14 55	18 20	04 21	10 16	20 38
27	23 19	02 S 15	24 35	23 01	16 42	14 57	18 21	04 21	10 15	20 38
28	23 16	02 N 15	24 23	22 51	16 27	14 58	18 22	04 22	10 15	20 38
29	23 13	06 33	24 09	22 39	16 12	15 00	18 23	04 22	10 14	20 38
30	23 09	10 27	23 54	22 27	15 56	15 01	18 25	04 22	10 14	20 38
31	23 S 05	13 N 44	23 S 37	22 S 15	15 S 41	15 N 03	18 S 26	04 N 22	10 S 13	20 S 38

ZODIAC SIGN ENTRIES

Date	h m	Planets
01	01 14	☽ ♈
03	05 15	☽ ♉
04	23 57	♂ ♒
05	10 28	☽ ♊
07	17 34	☽ ♋
10	03 14	☽ ♌
10	16 42	♀ ♑
12	15 19	☽ ♍
15	04 05	☽ ♎
17	03 53	☿ ♑
17	14 52	☽ ♏
19	21 55	☽ ♐
21	23 03	☉ ♑
22	01 25	☽ ♑
23	16 34	♄ ♐
24	02 52	☽ ♒
26	04 07	☽ ♓
28	06 35	☽ ♈
30	10 56	☽ ♉

LATITUDES

Date	Mercury ☿ ° '	Venus ♀ ° '	Mars ♂ ° '	Jupiter ♃ ° '	Saturn ♄ ° '	Uranus ♅ ° '	Neptune ♆ ° '	Pluto ♇ ° '
01	00 S 20	00 S 22	01 S 23	00 N 44	01 N 55	00 S 41	00 S 44	02 N 13
04	00 39	00 29	01 21	00 45	01 55	00 41	00 44	02 13
07	00 58	00 36	01 20	00 46	01 55	00 41	00 44	02 12
10	01 15	00 42	01 19	00 46	01 55	00 41	00 44	02 12
13	01 30	00 49	01 18	00 47	01 55	00 40	00 44	02 12
16	01 44	00 55	01 16	00 48	01 55	00 40	00 44	02 11
19	01 55	01 01	01 15	00 48	01 55	00 40	00 44	02 11
22	02 03	01 07	01 13	00 49	01 55	00 40	00 44	02 11
25	02 09	01 12	01 12	00 50	01 55	00 40	00 44	02 10
28	02 11	01 17	01 10	00 50	01 56	00 40	00 44	02 10
31	02 S 09	01 S 21	01 S 08	00 N 51	01 N 56	00 S 40	00 S 44	02 N 10

DATA

Julian Date	2456993
Delta T	+69 seconds
Ayanamsa	24° 04' 00"
Synetic vernal point	05° ♓ 02' 59"
True obliquity of ecliptic	23° 26' 05"

MOON'S PHASES, APSIDES AND POSITIONS ☽

Date	h m	Phase	Longitude ° '	Eclipse Indicator
06	12 27	○	14 ♊ 18	
14	12 51	☾	22 ♍ 26	
22	01 36	●	00 ♑ 06	
28	18 31	☽	06 ♈ 56	

Day	h m		
12	22 59	Apogee	
24	16 36	Perigee	
07	09 03	Max dec	18° N 40'
14	19 34	0S	
21	18 22	Max dec	18° S 40'
27	23 57	0N	

LONGITUDES

Date	Chiron ⚷ ° '	Ceres ⚳ ° '	Pallas ⚴ ° '	Juno ⚵ ° '	Vesta ⚶ ° '	Black Moon Lilith ⚸ ° '
01	13 ♓ 07	14 ♐ 17	18 ♏ 55	16 ♌ 22	28 ♐ 57	00 ♍ 24
11	13 ♓ 15	18 ♐ 27	23 ♏ 13	17 ♌ 12	04 ♑ 16	01 ♍ 31
21	13 ♓ 27	22 ♐ 35	27 ♏ 27	17 ♌ 09	09 ♑ 36	02 ♍ 37
31	13 ♓ 45	26 ♐ 41	01 ♐ 35	16 ♌ 13	14 ♑ 57	03 ♍ 44

All ephemeris data is given at 12.00 UT and the Moon's longitude is additionally given for 24.00 UT

ASPECTARIAN

h m Aspects

01 Monday
04 26 ☿ □ ♆
09 36 ☽ ⊻ ♆
10 15 ☽ △ ☿
14 13 ☽ ⚼ ♃
17 14 ☽ ∥ ♅
17 30 ☽ △ ☉
17 33 ☽ Q ♂
19 00 ♂ ⚹ ♄
19 58 ☽ ⊥ ♆
22 12 ☽ □ ♇
22 51 ☽ ⚼ ♄
23 12 ☽ ☌ ♅

02 Tuesday
00 19 ☿ ⊥ ♇
04 59 ☉ ∥ ♂
11 14 ☽ △ ♀
11 22 ☿ ∥ ♂
11 35 ☽ ∠ ♆
14 37 ☽ ± ♄
15 34 ☽ ⚼ ☿
16 16 ☽ △ ♃
20 16 ☉ ∥ ☿
21 36 ☽ ⚼ ☉

03 Wednesday
01 10 ☽ ⊼ ♄
02 42 ☽ □ ♂
04 47 ☽ # ♆
09 20 ☽ ± ☿
10 46 ♄ ⚼ ♅
13 21 ♂ ⊥ ♆
13 49 ☽ ⚹ ♆
14 38 ☽ ± ☉
16 06 ☽ ⚼ ♀
21 16 ☽ ⊼ ☿

04 Thursday
02 04 ☽ ⊼ ☉
02 46 ☽ △ ♇
03 36 ☽ ⊻ ♅
09 38 ☽ ± ♀
11 02 ☽ Q ♆
11 51 ☉ ⊻ ♇
12 33 ☽ ∥ ♃
14 14 ☽ ⊥ ♅
19 11 ♀ △ ♃
21 12 ☽ □ ♃
21 24 ☽ ⊼ ♀
22 30 ☉ △ ♅

05 Friday
00 55 ♀ Q ♆
05 33 ☽ ⚼ ♇
06 19 ☽ ∠ ♅
06 45 ☽ ☍ ♄
11 07 ☽ △ ♂
19 20 ☽ □ ♆
21 01 ♂ Q ♅
21 47 ☿ ⊻ ♇
21 50 ☽ ± ♇

06 Saturday
02 18 ☽ # ♄
03 44 ☿ △ ♅
05 43 ☽ Q ♃
08 48 ☽ ⊼ ♇
09 28 ☽ ⚹ ♅
10 15 ☽ ☍ ☿
12 27 ☽ ☍ ☉
16 07 ☽ ⚼ ♂

07 Sunday
03 48 ☽ ⚹ ♃
07 36 ☽ Q ♅
09 27 ♀ ⊥ ♂
09 51 ☽ ± ♂
09 52 ☽ ☍ ♀
14 12 ☽ ⊼ ♄
21 49 ☽ ⊼ ♂

08 Monday
01 35 ☽ ± ♄
02 52 ☽ △ ♆
07 58 ☽ ∠ ♃
09 51 ☉ ☌ ☿
15 16 ☽ # ♄
17 02 ☽ ☍ ♇
17 32 ☽ □ ♅
18 52 ☽ ⚼ ♄
20 43 ♃ St R

09 Tuesday
01 11 ☽ ⊥ ♃
01 42 ☽ ⊼ ☉
02 30 ☽ ⊼ ☿
07 37 ☽ ⚼ ♆
10 17 ♀ ⊻ ♄
11 27 ♂ ∥ ♇
12 48 ☽ ⊻ ♃
14 22 ☽ ± ☉
15 50 ☽ ± ☿
19 58 ☿ ∥ ♀

10 Wednesday
00 14 ☽ △ ♄
01 12 ☽ ± ♆
01 41 ☽ ⊼ ♀
04 59 ☿ ∠ ♂
09 29 ☽ ∥ ♃
09 37 ☽ ⚼ ☉
11 38 ☽ ☍ ♂
12 08 ☽ ⚼ ☿
13 03 ☽ ⊼ ♆
14 54 ☽ ± ♀

11 Thursday
03 57 ☽ ⊼ ♇
04 16 ☽ △ ♅
10 02 ♂ ⊻ ♆
10 57 ☽ ⚼ ♀
16 01 ☽ ± ♇
18 20 ☽ △ ☉
22 07 ☽ # ♆
22 40 ☽ △ ☿

12 Friday
00 21 ☽ ☌ ♃
10 14 ☽ ⚼ ♇
10 28 ☽ ⚼ ♅
11 24 ☿ △ ♃
12 48 ☽ □ ♄
17 24 ☿ Q ♆
20 51 ☽ △ ♀

13 Saturday
01 28 ☽ ☍ ♆
04 12 ☽ ⊼ ♂
04 43 ☽ ± ♅
15 45 ☽ ∥ ♅
16 47 ☽ △ ♇
16 54 ☽ ⊼ ♅
17 15 ☽ ± ♂

14 Sunday
01 42 ☽ Q ♄
09 15 ☉ ∠ ♂
12 46 ☽ ⚼ ♂
12 51 ☽ □ ☉
13 07 ☽ ⊻ ♃
15 56 ☉ △ ♃
16 29 ♀ ⊥ ♄
16 54 ♀ ⚹ ♆
20 47 ☽ □ ☿

15 Monday
01 12 ☽ ⊥ ♃
02 11 ☽ ⚹ ♄
02 54 ☉ Q ♆
05 14 ♅ □ ♇
14 10 ☽ ⊼ ♆
16 37 ☽ □ ♀
19 09 ☽ ∠ ♃
20 56 ☽ △ ♂
22 58 ☽ # ♅

16 Tuesday
02 06 ☽ ± ♆
04 11 ☽ Q ☉
05 07 ☽ ☍ ♅
05 12 ☽ □ ♇
08 17 ☽ ∠ ♄
13 57 ☽ Q ☿
16 13 ☿ ⊻ ♄
16 39 ♀ ⚼ ♃
19 46 ☽ ⚼ ♆

17 Wednesday
00 31 ☽ ⚹ ♃
01 59 ☽ ⊥ ♄
05 40 ☽ ⚹ ☉
07 50 ☽ Q ♀
13 35 ☽ ⊻ ♄
14 29 ☽ ∥ ♆
16 08 ☽ Q ♇
16 26 ☽ ⚹ ☿
23 21 ☽ Q ♃

18 Thursday
00 29 ☽ △ ♆
08 59 ☽ ⚹ ♀
10 29 ☽ □ ♂
12 26 ☽ ∠ ☉
14 25 ☽ ⊼ ♅
14 40 ☽ ⚹ ♇

19 Friday
00 15 ☽ ∠ ☿
01 25 ☽ ± ♅
01 27 ☽ # ♃
06 16 ☽ ⊥ ☉
08 26 ☽ □ ♃
15 11 ☽ ∠ ♀
17 36 ☽ ⚼ ♅
17 55 ☽ ∠ ♇
17 55 ☽ ⊻ ☉
18 38 ☽ ⊥ ☿
21 10 ☽ Q ♂
21 11 ☽ ☌ ♄
21 37 ♂ Q ♄
23 34 ♀ ⊻ ♂

20 Saturday
06 37 ☽ ⊻ ☿
06 57 ☽ □ ♆
08 41 ☽ ⊥ ♀
09 31 ☿ ⚹ ♆
09 48 ☽ ⊥ ♇
17 08 ♀ □ ♅
17 49 ☿ ⊥ ♄
19 23 ☽ ⚹ ♂
19 52 ☽ △ ♅
20 08 ☽ ⊻ ♀
20 13 ☽ ⊻ ♇
21 03 ☽ ∥ ♄
21 09 ♀ ☌ ♇
22 27 ☽ ∥ ♂

21 Sunday
03 47 ♂ ∥ ♄
03 47 ♂ ⚹ ♅
10 33 ☿ ⊥ ♂
10 51 ♂ ⊻ ♇
12 35 ☽ △ ♃
13 55 ☽ Q ♆
18 08 ☉ ⊻ ♄
19 01 ☿ ⚼ ♃
22 22 ☽ ∠ ♂
22 46 ♅ St D

22 Monday
01 07 ☽ ⊻ ♄
01 36 ☽ ☌ ☉
10 02 ☽ ⚹ ♆
11 11 ☽ ⊥ ♄
13 16 ♀ ∠ ♄
13 36 ☽ ⚼ ♃
14 13 ☽ ⊥ ♂
14 26 ☽ ∥ ♄
15 57 ☽ ☌ ☿
22 17 ☽ □ ♅
22 44 ☽ ☌ ♇
23 17 ☽ ∥ ♂

23 Tuesday
00 41 ☽ ⊻ ♂
02 10 ☽ ∠ ♄
03 17 ☽ ☌ ♀
04 24 ☽ ± ♃
10 48 ☽ ∠ ♆
14 13 ☽ ⊼ ♃
15 53 ♀ ± ♃

24 Wednesday
01 34 ☽ ⊥ ♆
02 57 ☽ ⚹ ♄
03 48 ☽ Q ♅
06 44 ☽ ⊻ ☉
07 13 ☽ # ♃
11 23 ☽ ⊻ ♆
17 17 ☽ ⊥ ☉
22 43 ☽ Q ♄
23 11 ☽ ⊻ ☿
23 26 ☽ ⚹ ♅

25 Thursday
00 00 ☽ ⊻ ♇
01 13 ☉ ∥ ♀
01 31 ☿ □ ♅
04 36 ☽ ☌ ♂
06 53 ☿ ☌ ♇
09 01 ☽ ⊻ ♀
09 08 ☽ ∠ ☉
09 52 ☽ ⊥ ♇
10 14 ☽ ⊥ ☿
15 11 ☽ ☍ ♃
15 16 ☽ ∥ ♆
19 49 ☽ ⊥ ♀

26 Friday
00 06 ☽ ∠ ♅
00 44 ☽ ∠ ♇
02 54 ☽ ∠ ☿
04 33 ☽ □ ♄
11 46 ☽ ⚹ ☉
12 08 ☽ ∠ ♀
12 50 ☽ ☌ ♆
15 02 ☽ ⊥ ♅
18 51 ☿ ∠ ♄
20 20 ♀ ∠ ♆

27 Saturday
00 45 ☽ # ♅
01 04 ☽ ⊻ ♅
01 46 ☽ ⚹ ♇
03 10 ☉ ⚹ ♆
06 02 ☿ ± ♃
07 04 ☽ ⚹ ☿
09 15 ☽ ⊻ ♂
09 24 ☽ Q ☉
15 44 ☽ ⚹ ♀
17 02 ☽ ⊼ ♃
19 59 ☽ ⊥ ♂
22 05 ☽ Q ♇

28 Sunday
03 12 ☽ ± ♃
05 51 ♀ ⊼ ♃
06 02 ☽ Q ☿
07 25 ☽ △ ♄
07 55 ☉ ⊥ ♄
12 17 ☽ ∠ ♂
14 14 ☽ Q ♀
15 42 ☽ ⊻ ♆
18 31 ☽ □ ☉
18 35 ☽ ⚼ ♃
19 00 ☿ ⊻ ♂
19 24 ☉ ⚼ ♃
23 34 ☽ ∥ ♅

29 Monday
02 06 ☽ ⊥ ♆
04 19 ☽ ☌ ♅
05 09 ☽ □ ♇
09 34 ☽ ⚼ ♄
11 29 ♂ ⊥ ♇
15 54 ☽ ⚹ ♂
17 19 ☽ □ ☿
17 51 ☽ ∠ ♆
20 37 ☽ △ ♃
21 57 ☿ ∠ ♆

30 Tuesday
00 46 ☽ □ ♀
01 29 ☽ ± ♄
10 34 ☽ # ♆
12 12 ☽ ⊼ ♄
14 25 ☽ Q ♂
20 29 ☽ ⚹ ♆
20 37 ☿ ⊼ ♃

31 Wednesday
03 40 ☽ △ ☉
09 30 ☽ ⊻ ♅
10 28 ☽ △ ♇
18 07 ☽ Q ♆
20 21 ☽ ⊥ ♅
23 34 ☽ ∥ ♃

JANUARY 2015

Date	Sidereal time h m s	Sun ☉ ° ' "	Moon ☽ ° ' "	Moon ☽ 24.00 ° ' "	Mercury ☿ ° '	Venus ♀ ° '	Mars ♂ ° '	Jupiter ♃ ° '	Saturn ♄ ° '	Uranus ♅ ° '	Neptune ♆ ° '	Pluto ♇ ° '
01	18 43 17	10 ♑ 44 24	27 ♉ 11 38	03 ♊ 43 32	24 ♑ 27	27 ♑ 21	21 ♒ 27	21 ♌ 43	00 ♐ 55	12 ♈ 37	05 ♓ 24	13 ♑ 11
02	18 47 14	11 45 32	10 ♊ 12 45	16 ♊ 39 18	26 02	28 36	22 14	21 R 39	01 01	12 38	05 25	13 13
03	18 51 11	12 46 40	23 ♊ 03 12	29 ♊ 24 24	27 36	29 ♑ 51	23 00	21 34	01 07	12 38	05 27	13 15
04	18 55 07	13 47 48	05 ♋ 42 53	11 ♋ 58 36	29 ♑ 09	01 ♒ 06	23 47	21 29	01 13	12 39	05 28	13 17
05	18 59 04	14 48 56	18 ♋ 11 32	24 ♋ 21 43	00 ♒ 42	02 21	24 34	21 24	01 19	12 40	05 30	13 19
06	19 03 00	15 50 04	00 ♌ 29 11	06 ♌ 34 01	02 13	03 37	25 21	21 19	01 25	12 40	05 32	13 21
07	19 06 57	16 51 12	12 ♌ 36 22	18 ♌ 36 25	03 43	04 52	26 08	21 13	01 30	12 41	05 33	13 24
08	19 10 53	17 52 20	24 ♌ 34 27	00 ♍ 30 45	05 10	06 07	26 55	21 08	01 36	12 42	05 35	13 26
09	19 14 50	18 53 28	06 ♍ 25 43	12 ♍ 19 46	06 36	07 22	27 42	21 02	01 42	12 43	05 37	13 28
10	19 18 46	19 54 36	18 ♍ 13 25	24 ♍ 07 12	07 59	08 37	28 29	20 56	01 47	12 44	05 38	13 30
11	19 22 43	20 55 43	00 ♎ 01 43	05 ♎ 57 36	09 19	09 52	29 ♒ 16	20 50	01 53	12 45	05 40	13 32
12	19 26 40	21 56 51	11 ♎ 55 30	17 ♎ 56 06	10 35	11 07	00 ♓ 03	20 44	01 58	12 46	05 42	13 34
13	19 30 36	22 57 59	24 ♎ 00 08	00 ♏ 08 17	11 46	12 22	00 50	20 38	02 03	12 47	05 44	13 36
14	19 34 33	23 59 07	06 ♏ 21 13	12 ♏ 39 37	12 53	13 37	01 37	20 32	02 09	12 48	05 45	13 38
15	19 38 29	25 00 14	19 ♏ 04 03	25 ♏ 35 02	13 53	14 52	02 24	20 25	02 14	12 50	05 47	13 40
16	19 42 26	26 01 21	02 ♐ 13 00	08 ♐ 58 13	14 47	16 07	03 11	20 18	02 19	12 51	05 49	13 42
17	19 46 22	27 02 29	15 ♐ 50 48	22 ♐ 50 41	15 33	17 22	03 58	20 12	02 24	12 52	05 51	13 44
18	19 50 19	28 03 36	29 ♐ 57 36	07 ♑ 11 04	16 11	18 37	04 45	20 05	02 29	12 54	05 53	13 46
19	19 54 15	29 ♑ 04 42	14 ♑ 30 24	21 ♑ 54 42	16 39	19 52	05 32	19 58	02 34	12 55	05 55	13 48
20	19 58 12	00 ♒ 05 48	29 ♑ 22 54	06 ♒ 53 48	16 58	21 07	06 19	19 51	02 39	12 57	05 57	13 50
21	20 02 09	01 06 54	14 ♒ 26 10	21 ♒ 58 42	17 05	22 22	07 06	19 44	02 44	12 58	05 58	13 52
22	20 06 05	02 07 58	29 ♒ 30 10	06 ♓ 59 26	17 R 01	23 37	07 53	19 36	02 49	13 00	06 00	13 54
23	20 10 02	03 09 02	14 ♓ 25 30	21 ♓ 47 33	16 46	24 51	08 39	19 29	02 53	13 01	06 02	13 56
24	20 13 58	04 10 05	29 ♓ 04 56	06 ♈ 17 12	16 19	26 06	09 26	19 22	02 58	13 03	06 04	13 58
25	20 17 55	05 11 06	13 ♈ 24 05	20 ♈ 25 27	15 41	27 21	10 13	19 14	03 03	13 05	06 06	14 00
26	20 21 51	06 12 07	27 ♈ 21 20	04 ♉ 11 52	14 52	28 36	11 00	19 06	03 07	13 06	06 08	14 02
27	20 25 48	07 13 06	10 ♉ 57 15	17 ♉ 37 47	13 55	29 ♒ 51	11 47	18 59	03 11	13 08	06 10	14 04
28	20 29 44	08 14 05	24 ♉ 13 45	00 ♊ 45 29	12 50	01 ♓ 05	12 34	18 51	03 16	13 10	06 13	14 06
29	20 33 41	09 15 02	07 ♊ 13 19	13 ♊ 37 34	11 40	02 20	13 20	18 43	03 20	13 12	06 15	14 08
30	20 37 38	10 15 58	19 ♊ 58 30	26 ♊ 16 24	10 26	03 35	14 07	18 36	03 24	13 14	06 17	14 10
31	20 41 34	11 ♒ 16 53	02 ♋ 31 30	08 ♋ 44 00	09 ♒ 11	04 ♓ 49	14 ♓ 54	18 ♌ 28	03 ♐ 28	13 ♈ 16	06 ♓ 19	14 ♑ 12

Date	Moon True ☊ ° '	Moon Mean ☊ ° '	Moon ☽ Latitude ° '
01	15 ♎ 23	14 ♎ 55	03 S 21
02	15 R 15	14 51	04 06
03	15 04	14 48	04 38
04	14 50	14 45	04 56
05	14 36	14 42	04 59
06	14 21	14 39	04 48
07	14 08	14 36	04 24
08	13 58	14 32	03 49
09	13 50	14 29	03 04
10	13 45	14 26	02 11
11	13 42	14 23	01 12
12	13 D 42	14 20	00 S 09
13	13 R 42	14 17	00 N 54
14	13 41	14 13	01 57
15	13 39	14 10	02 56
16	13 35	14 07	03 47
17	13 27	14 04	04 28
18	13 18	14 01	04 54
19	13 06	13 57	05 02
20	12 55	13 54	04 49
21	12 44	13 51	04 17
22	12 36	13 48	03 27
23	12 30	13 45	02 23
24	12 28	13 42	01 N 11
25	12 D 27	13 38	00 S 05
26	12 27	13 35	01 19
27	12 R 28	13 32	02 26
28	12 26	13 29	03 23
29	12 23	13 26	04 09
30	12 16	13 23	04 42
31	12 ♎ 08	13 ♎ 19	05 S 00

DECLINATIONS

Date	Sun ☉ ° '	Moon ☽ ° '	Mercury ☿ ° '	Venus ♀ ° '	Mars ♂ ° '	Jupiter ♃ ° '	Saturn ♄ ° '	Uranus ♅ ° '	Neptune ♆ ° '	Pluto ♇ ° '
01	23 S 00	16 N 16	23 S 19	22 S 02	15 S 25	15 N 04	18 S 27	04 N 22	10 S 13	20 S 38
02	22 55	17 55	22 59	21 48	15 10	15 06	18 28	04 23	10 12	20 38
03	22 49	18 37	22 38	21 33	14 54	15 08	18 29	04 23	10 12	20 38
04	22 43	18 23	22 15	21 18	14 38	15 10	18 30	04 23	10 11	20 38
05	22 37	17 16	21 51	21 02	14 22	15 11	18 31	04 24	10 10	20 37
06	22 30	15 21	21 26	20 46	14 05	15 13	18 32	04 24	10 10	20 37
07	22 22	12 48	20 59	20 29	13 49	15 15	18 34	04 24	10 09	20 37
08	22 15	09 44	20 32	20 11	13 32	15 17	18 35	04 25	10 09	20 37
09	22 06	06 18	20 03	19 53	13 15	15 19	18 36	04 25	10 08	20 37
10	21 58	02 N 39	19 34	19 34	12 59	15 21	18 37	04 26	10 07	20 37
11	21 48	01 S 07	19 04	19 15	12 42	15 23	18 38	04 26	10 07	20 37
12	21 39	04 51	18 34	18 55	12 25	15 26	18 39	04 26	10 06	20 37
13	21 29	08 28	18 03	18 35	12 07	15 28	18 40	04 27	10 05	20 36
14	21 18	11 48	17 33	18 14	11 50	15 30	18 40	04 27	10 05	20 36
15	21 08	14 40	17 03	17 52	11 33	15 32	18 41	04 28	10 04	20 36
16	20 56	16 53	16 34	17 30	11 15	15 35	18 42	04 28	10 03	20 36
17	20 45	18 15	16 06	17 08	10 57	15 37	18 43	04 29	10 03	20 36
18	20 33	18 32	15 39	16 45	10 40	15 39	18 44	04 30	10 02	20 36
19	20 20	17 39	15 15	16 21	10 22	15 42	18 45	04 30	10 01	20 36
20	20 08	15 33	14 53	15 58	10 04	15 44	18 46	04 31	10 00	20 36
21	19 54	12 23	14 33	15 33	09 46	15 46	18 47	04 31	10 00	20 35
22	19 41	08 24	14 17	15 09	09 28	15 49	18 47	04 32	09 59	20 35
23	19 27	03 S 56	14 04	14 43	09 10	15 51	18 48	04 33	09 58	20 35
24	19 13	00 N 43	13 55	14 18	08 52	15 54	18 49	04 33	09 58	20 35
25	18 58	05 13	13 50	13 52	08 33	15 56	18 50	04 34	09 57	20 35
26	18 43	09 18	13 49	13 26	08 15	15 59	18 50	04 35	09 56	20 35
27	18 28	12 48	13 51	12 59	07 56	16 01	18 51	04 35	09 55	20 35
28	18 12	15 32	13 57	12 32	07 38	16 04	18 52	04 36	09 55	20 35
29	17 56	17 25	14 06	12 05	07 19	16 07	18 52	04 37	09 54	20 34
30	17 40	18 23	14 18	11 37	07 01	16 09	18 53	04 38	09 53	20 34
31	17 S 23	18 N 25	14 S 31	11 S 09	06 S 42	16 N 12	18 S 54	04 N 39	09 S 52	20 S 34

ZODIAC SIGN ENTRIES

Date	h	m	Planets
01	17	09	☽ ♊
03	14	48	♀ ♒
04	01	07	☽ ♋
05	01	08	☿ ♒
06	11	03	☽ ♌
08	22	58	☽ ♍
11	11	57	☽ ♎
12	10	20	♂ ♓
13	23	44	☽ ♏
16	08	01	☽ ♐
18	12	04	☽ ♑
20	09	43	☉ ♒
20	12	59	☽ ♒
22	12	48	☽ ♓
24	13	31	☽ ♈
26	16	37	☽ ♉
27	15	00	♀ ♓
28	22	36	☽ ♊
31	07	09	☽ ♋

LATITUDES

Date	Mercury ☿ ° '	Venus ♀ ° '	Mars ♂ ° '	Jupiter ♃ ° '	Saturn ♄ ° '	Uranus ♅ ° '	Neptune ♆ ° '	Pluto ♇ ° '
01	02 S 07	01 S 22	01 S 08	00 N 51	01 N 56	00 S 40	00 S 44	02 N 10
04	01 58	01 26	01 06	00 52	01 56	00 40	00 44	02 09
07	01 43	01 29	01 04	00 53	01 56	00 39	00 44	02 09
10	01 21	01 31	01 03	00 53	01 57	00 39	00 44	02 09
13	00 50	01 33	01 01	00 54	01 57	00 39	00 44	02 09
16	00 S 10	01 34	00 59	00 55	01 57	00 39	00 44	02 08
19	00 N 37	01 35	00 57	00 55	01 58	00 39	00 44	02 08
22	01 31	01 35	00 55	00 56	01 58	00 39	00 44	02 08
25	02 24	01 34	00 53	00 56	01 58	00 39	00 44	02 08
28	03 08	01 33	00 51	00 57	01 59	00 39	00 44	02 07
31	03 N 33	01 S 31	00 S 49	00 N 57	01 N 59	00 S 39	00 S 44	02 N 07

DATA

Julian Date	2457024
Delta T	+69 seconds
Ayanamsa	24° 04' 05"
Synetic vernal point	05° ♓ 02' 54"
True obliquity of ecliptic	23° 26' 05"

MOON'S PHASES, APSIDES AND POSITIONS ☽

Date	h	m	Phase	Longitude ° '	Eclipse Indicator
05	04	53	○	14 ♋ 31	
13	09	46	☾	22 ♎ 52	
20	13	14	●	00 ♒ 09	
27	04	48	☽	06 ♉ 55	

Day	h	m		
09	18	09	Apogee	
21	20	01	Perigee	
03	17	51	Max dec	18° N 39'
11	04	57	0S	
18	06	14	Max dec	18° S 35'
24	08	18	0N	
31	01	00	Max dec	18° N 31'

LONGITUDES

Date	Chiron ⚷ ° '	Ceres ⚳ ° '	Pallas ⚴ ° '	Juno ⚵ ° '	Vesta ⚶ ° '	Black Moon Lilith ⚸ ° '
01	13 ♓ 47	27 ♐ 06	02 ♐ 02	16 ♌ 04	15 ♑ 29	03 ♍ 51
11	14 ♓ 11	01 ♑ 09	06 ♐ 27	14 ♌ 16	20 ♑ 49	04 ♍ 58
21	14 ♓ 38	05 ♑ 33	09 ♐ 52	11 ♌ 53	26 ♑ 04	06 ♍ 04
31	15 ♓ 09	09 ♑ 03	13 ♐ 32	09 ♌ 18	01 ♒ 25	07 ♍ 11

ASPECTARIAN

h m	Aspects
01 Thursday	
00 50	☽ □ ♂
02 04	☽ □ ♃
03 58	☽ # ♂
06 18	☽ △ ☿
09 07	☽ bQ ☉
12 19	☽ △ ♀
12 47	☽ ∠ ♅
13 49	☽ bQ ♇
18 53	☽ ☍ ♄
19 49	♂ ☍ ♃
02 Friday	
03 03	☽ ± ☉
03 07	☽ □ ♆
06 26	☽ ± ♇
10 57	☽ Q ♃
13 44	☽ bQ ☿
15 07	☽ ⊼ ☉
16 30	☽ ✶ ♅
16 52	♂ # ♃
17 36	☽ ⊼ ♇
18 08	☉ ∥ ☿
18 59	☽ bQ ♀
03 Saturday	
03 29	☽ # ♄
04 03	♀ ⊥ ♆
08 40	☉ □ ♅
08 53	☽ ± ☿
09 13	☽ ✶ ♃
11 55	☽ △ ♂
13 40	☽ ± ♀
14 59	☽ Q ♅
21 47	☽ ⊼ ☿
23 34	☉ ☌ ♇
04 Sunday	
02 16	☽ ⊼ ♀
03 10	♀ Q ♅
03 22	☽ ⊼ ♄
07 22	☽ # ♄
11 32	☽ △ ♆
13 28	☽ ∠ ♃
14 15	♀ ✶ ♄
14 53	☽ ± ♄
17 01	☿ ⊥ ♆
18 17	☽ bQ ♂
05 Monday	
01 18	☽ □ ♅
02 34	☽ ☍ ♇
04 53	☽ ☍ ☉
06 38	☽ ⊥ ♃
08 20	☽ bQ ♄
11 28	☿ Q ♅
12 47	☽ ± ♂
16 29	☽ bQ ♆
18 11	☽ ⊻ ♃
22 21	☿ ✶ ♄
06 Tuesday	
00 41	☉ ± ♃
01 16	☽ ⊼ ♂
10 07	☽ ± ♆
13 20	☽ ∥ ♃
13 50	☽ △ ♄
15 53	☽ ☍ ☿
18 52	☽ ☍ ♀
21 58	☽ ⊼ ♆
07 Wednesday	
00 18	♀ ∥ ♇
02 01	☽ # ♂
02 55	☉ ∠ ♄
12 10	☽ △ ♅
13 34	☽ ⊼ ♇
21 16	☽ ⊼ ☉
08 Thursday	
01 34	♀ ⊻ ♆
01 37	☽ ± ♇
05 07	☽ ☌ ♃
07 20	☿ ∥ ♇
08 56	☽ # ♆
10 27	☽ ± ☉
17 04	☽ ☍ ♂
18 19	☽ bQ ♅
18 56	☿ ⊻ ♆
19 48	☽ bQ ♇
09 Friday	
02 19	☽ □ ♄
06 22	☽ bQ ☉
10 20	☽ ☍ ♆
12 23	♂ ∠ ♅
12 24	☽ ⊼ ☿
12 35	☽ ± ♅
14 08	☽ ⊼ ♀
10 Saturday	
00 31	☽ ∥ ♅
00 49	☽ ⊼ ♅
02 14	☽ ± ☿
02 21	☽ △ ♇
03 47	☽ ± ♀
11 09	☿ ∥ ♀
12 14	♂ ∠ ♇
15 12	☽ Q ♄
15 45	☽ △ ☉
17 29	☽ ⊻ ♃
22 56	☽ bQ ☿
11 Sunday	
00 17	☽ bQ ♀
05 35	☽ ⊥ ♃
05 39	☉ ∠ ♆
10 05	☉ ⊼ ♃
10 21	☽ ⊼ ♂
15 46	☽ ✶ ♄
23 22	☽ ± ♂
23 26	☽ ⊼ ♆
23 39	☽ ∠ ♃
12 Monday	
08 14	☿ ∥ ♄
08 59	☽ △ ☿
09 17	☽ # ♅
10 11	☽ △ ♀
11 32	☽ ± ♆
13 42	☽ ☍ ♅
15 18	☽ □ ♇
18 42	☽ bQ ♂
22 09	☽ ∠ ♄
13 Tuesday	
05 25	☽ ✶ ♃
05 31	☽ bQ ♆
06 29	♀ ∥ ♄
09 46	☽ □ ☉
16 04	☽ ⊥ ♄
20 12	♀ ✶ ♅
23 21	☽ ∥ ♆
14 Wednesday	
02 16	☽ △ ♂
02 53	☽ Q ♇
03 50	☽ ⊻ ♄
04 42	☽ Q ♃
10 25	☿ ✶ ♅
10 51	☽ △ ♆
12 17	☽ ∥ ♂
12 20	♀ ⊻ ♇
22 54	♀ Q ♄
23 40	☽ Q ☉
15 Thursday	
00 18	☽ ⊼ ♅
01 30	☽ □ ☿
01 53	☽ ✶ ♇
03 18	☽ □ ♀
06 10	♂ □ ♄
06 25	☿ ⊻ ♇
11 33	☽ ± ♅
14 29	☽ □ ♃
20 39	☽ # ♃
21 51	☿ Q ♄
23 52	☽ ✶ ☉
16 Friday	
04 07	☽ bQ ♅
05 39	☽ ∠ ♇
08 42	☽ ∥ ☿
12 11	☽ ☌ ♄
13 05	☽ Q ☿
13 50	☽ □ ♂
15 45	☽ Q ♀
18 26	☽ □ ♆
19 10	☽ ∥ ♀
21 47	☽ ⊥ ♇
17 Saturday	
04 51	☽ ∠ ☉
06 50	☽ △ ♅
08 20	☽ ⊻ ♇
11 28	☽ ✶ ☿
14 53	☽ ✶ ♀
19 25	☽ △ ♃
21 37	☽ ⊥ ☉
23 07	☽ Q ♂
18 Sunday	
01 44	☽ Q ♆
05 49	☉ ∥ ♇
08 34	☽ ⊻ ☉
11 59	☿ # ♃
14 07	☽ ∠ ☿
16 15	☽ ⊻ ♄
18 40	☽ ∠ ♀
20 26	☽ ✶ ♂
20 27	☽ bQ ♃
21 52	☽ ✶ ♆
19 Monday	
02 14	☽ ⊥ ♄
05 32	☽ ⊥ ☿
09 24	☽ □ ♅
10 49	♀ ⊥ ♇
10 51	☽ ☌ ♇
10 52	☽ ⊥ ♀
11 07	☽ ± ♃
13 44	♀ ☍ ♃
15 35	☽ ⊻ ☿
17 01	☽ ∠ ♄
20 47	☽ ⊼ ♃
21 30	☽ ⊻ ♀
22 19	☽ ∠ ♂
22 24	☽ ∠ ♆
20 Tuesday	
00 08	♂ ☌ ♆
07 30	☽ ∥ ♀
08 15	☉ ⊥ ♆
10 23	☽ # ♃
12 54	☽ ⊥ ♆
13 04	♃ ± ♇
13 14	☽ ☌ ☉
13 34	☽ ⊥ ♂
14 30	☽ Q ♅
16 44	♂ ∥ ♆
17 16	☽ ✶ ♄
18 35	☽ ∥ ☿
22 30	☽ ⊻ ♆
23 41	☽ ⊻ ♂
21 Wednesday	
00 13	♀ # ♃
08 00	♂ ⊥ ♅
08 28	☉ Q ♅
09 40	☽ ✶ ♅
09 57	☉ ⊥ ♂
11 06	☽ ⊻ ♇
12 29	☽ Q ♄
15 54	☿ St R
16 13	☽ ☌ ☿
20 21	☽ ☍ ♃
20 40	☽ ⊥ ♇
22 Thursday	
01 45	☽ ☌ ♀
02 55	☽ ∥ ♆
05 31	☽ ∥ ♂
09 35	☽ ∠ ♅
11 03	☽ ∠ ♇
16 31	☽ ⊻ ☉
17 19	☽ □ ♄
22 27	☽ ☌ ♆
23 Friday	
00 02	☽ ⊥ ♅
02 10	☽ ☌ ♂
02 51	☽ ⊥ ☉
05 23	☉ ✶ ♄
08 48	☽ # ♅
09 43	☽ ⊻ ♅
11 13	☽ ✶ ♇
15 42	☽ ⊻ ☿
18 30	☽ ∠ ☉
20 09	☽ ⊼ ♃
24 Saturday	
01 13	☽ ⊥ ☿
05 54	☽ ± ♃
06 37	☽ ⊻ ♀
06 51	☽ Q ♇
15 34	☽ ∠ ☿
17 30	☽ ⊥ ♀
18 29	☽ △ ♄
20 42	☽ bQ ♃
21 06	☽ ✶ ☉
23 40	☽ ⊻ ♆
25 Sunday	
06 18	☽ ⊻ ♂
08 27	☽ ∥ ♅
09 48	☽ ⊥ ♆
10 03	☽ ∠ ♀
11 27	☽ ☌ ♅
13 02	☽ □ ♇
13 53	☿ ∥ ♀
15 41	☽ ✶ ☿
17 05	☽ ⊥ ♂
18 57	☽ Q ☉
19 57	☽ bQ ♄
21 52	☽ △ ♃
26 Monday	
00 59	☉ ∥ ♄
01 12	☽ ∠ ♆
02 18	♀ ∠ ♅
05 42	☿ Q ♄
05 57	☽ # ♂
09 30	☽ ∠ ♂
10 30	☉ ⊻ ♆
11 13	☽ Q ☿
11 35	☽ ± ♄
14 23	☽ ✶ ♀
15 59	☽ # ♆
20 43	♀ ∠ ♇
22 09	☽ ⊼ ♄
27 Tuesday	
03 28	☽ ✶ ♆
04 48	☽ □ ☉
08 23	☿ ⊻ ♇
13 15	☽ # ♀
13 34	☽ ✶ ♂
13 45	☽ Q ♀
15 55	☽ ⊻ ♅
16 55	☽ □ ☿
17 36	☽ △ ♇
20 40	☽ # ☿
28 Wednesday	
01 01	☽ Q ♆
02 18	☽ □ ♃
02 46	☽ ⊥ ♅
05 02	☿ ✶ ♅
12 39	☽ Q ♂
15 26	☿ ⊻ ♂
17 51	☽ ∥ ♃
19 14	☽ ∠ ♅
20 58	☽ bQ ♇
29 Thursday	
01 57	☽ □ ♀
04 44	☽ ☍ ♄
07 30	♂ ⊻ ♅
10 10	☽ □ ♆
11 05	☽ Q ♃
13 43	☽ ± ♇
16 07	☽ △ ☉
19 35	☽ △ ☿
20 09	☽ # ☉
23 13	☽ ✶ ♅
30 Friday	
00 12	☽ □ ♂
01 00	☽ ⊼ ♇
08 23	♀ □ ♄
09 24	☽ ✶ ♃
13 40	♂ ✶ ♇
13 45	☉ ☌ ☿
21 27	☽ bQ ☿
22 02	☽ Q ♅
22 58	☽ bQ ☉
31 Saturday	
13 09	☽ ± ☿
13 47	☽ ∠ ♃
13 50	☽ ⊼ ♄
15 24	☿ ⊥ ♂
16 56	☽ △ ♀
17 47	☽ ± ☉
19 20	☽ △ ♆
23 42	☽ ⊼ ☿

All ephemeris data is given at 12.00 UT and the Moon's longitude is additionally given for 24.00 UT

FEBRUARY 2015

LONGITUDES

Date	Sidereal time h m s	Sun ☉ ° ' "	Moon ☽ ° ' "	Moon ☽ 24.00 ° ' "	Mercury ☿ ° '	Venus ♀ ° '	Mars ♂ ° '	Jupiter ♃ ° '	Saturn ♄ ° '	Uranus ♅ ° '	Neptune ♆ ° '	Pluto ♇ ° '
01	20 45 31	12 ♒ 17 47	14 ♋ 54 05	21 ♋ 01 54	07 ♒ 57	06 ♓ 04	15 ♓ 40	18 ♌ 20	03 ♐ 32	13 ♈ 18	06 ♓ 21	14 ♑ 14
02	20 49 27	13 18 39	27 ♋ 07 36	03 ♌ 11 19	06 R 46	07 19	16 27	18 R 12	03 36	13 20	06 23	14 16
03	20 53 24	14 19 30	09 ♌ 13 09	15 ♌ 13 16	05 40	08 33	17 14	18 04	03 40	13 22	06 25	14 18
04	20 57 20	15 20 21	21 ♌ 11 48	27 ♌ 08 55	04 40	09 48	18 00	17 56	03 44	13 24	06 27	14 20
05	21 01 17	16 21 10	03 ♍ 04 49	08 ♍ 59 45	03 46	11 02	18 47	17 48	03 47	13 26	06 29	14 21
06	21 05 13	17 21 58	14 ♍ 53 57	20 ♍ 47 46	03 01	12 17	19 34	17 40	03 51	13 28	06 32	14 23
07	21 09 10	18 22 44	26 ♍ 41 33	02 ♎ 35 42	02 24	13 31	20 20	17 32	03 54	13 31	06 34	14 25
08	21 13 07	19 23 30	08 ♎ 30 40	14 ♎ 26 57	01 55	14 45	21 07	17 24	03 58	13 33	06 36	14 27
09	21 17 03	20 24 15	20 ♎ 25 06	26 ♎ 25 40	01 35	16 00	21 53	17 16	04 01	13 35	06 38	14 29
10	21 21 00	21 24 59	02 ♏ 29 15	08 ♏ 36 30	01 23	17 14	22 40	17 08	04 04	13 38	06 40	14 30
11	21 24 56	22 25 41	14 ♏ 48 01	21 ♏ 04 27	01 18	18 28	23 26	17 00	04 07	13 40	06 43	14 32
12	21 28 53	23 26 23	27 ♏ 26 23	03 ♐ 54 25	01 D 21	19 43	24 12	16 52	04 10	13 43	06 45	14 34
13	21 32 49	24 27 04	10 ♐ 29 01	17 ♐ 10 37	01 30	20 57	24 59	16 45	04 13	13 45	06 47	14 36
14	21 36 46	25 27 43	23 ♐ 59 30	00 ♑ 55 49	01 46	22 11	25 45	16 37	04 16	13 48	06 49	14 37
15	21 40 42	26 28 22	07 ♑ 59 32	15 ♑ 10 27	02 08	23 25	26 32	16 29	04 19	13 50	06 52	14 39
16	21 44 39	27 28 59	22 ♑ 28 07	29 ♑ 51 53	02 36	24 39	27 18	16 21	04 21	13 53	06 54	14 41
17	21 48 36	28 29 35	07 ♒ 20 52	14 ♒ 54 03	03 08	25 53	28 04	16 14	04 24	13 55	06 56	14 42
18	21 52 32	29 ♒ 30 09	22 ♒ 30 12	00 ♓ 08 01	03 45	27 07	28 50	16 06	04 26	13 58	06 58	14 44
19	21 56 29	00 ♓ 30 42	07 ♓ 46 08	15 ♓ 23 14	04 27	28 21	29 ♓ 37	15 58	04 29	14 01	07 01	14 45
20	22 00 25	01 31 13	22 ♓ 58 02	00 ♈ 29 22	05 12	29 ♓ 35	00 ♈ 23	15 51	04 31	14 03	07 03	14 47
21	22 04 22	02 31 43	07 ♈ 56 17	15 ♈ 17 57	06 02	00 ♈ 49	01 09	15 44	04 33	14 06	07 05	14 49
22	22 08 18	03 32 11	22 ♈ 33 47	29 ♈ 43 22	06 54	02 03	01 55	15 36	04 35	14 09	07 07	14 50
23	22 12 15	04 32 37	06 ♉ 46 28	13 ♉ 43 02	07 50	03 17	02 41	15 29	04 37	14 12	07 10	14 52
24	22 16 11	05 33 01	20 ♉ 33 08	27 ♉ 16 58	08 48	04 30	03 27	15 22	04 39	14 14	07 12	14 53
25	22 20 08	06 33 23	03 ♊ 54 50	10 ♊ 27 04	09 50	05 44	04 13	15 15	04 41	14 17	07 14	14 55
26	22 24 05	07 33 43	16 ♊ 54 04	23 ♊ 16 17	10 53	06 58	04 59	15 08	04 43	14 20	07 17	14 56
27	22 28 01	08 34 01	29 ♊ 34 08	05 ♋ 48 04	11 59	08 11	05 45	15 01	04 44	14 23	07 19	14 57
28	22 31 58	09 ♓ 34 17	11 ♋ 58 32	18 ♋ 05 54	13 ♒ 07	09 ♈ 25	06 ♈ 31	14 ♌ 54	04 ♐ 46	14 ♈ 26	07 ♓ 21	14 ♑ 59

MOON NODES AND LATITUDE / DECLINATIONS

Date	Moon True ☊ ° '	Moon Mean ☊ ° '	Moon ☽ Latitude ° '	Sun ☉ ° '	Moon ☽ ° '	Mercury ☿ ° '	Venus ♀ ° '	Mars ♂ ° '	Jupiter ♃ ° '	Saturn ♄ ° '	Uranus ♅ ° '	Neptune ♆ ° '	Pluto ♇ ° '
01	11 ♎ 57	13 ♎ 16	05 S 04	17 S 06	17 N 34	14 S 47	10 S 41	06 S 23	16 N 14	18 S 54	04 N 39	09 S 52	20 S 34
02	11 R 45	13 13	04 53	16 49	15 56	15 04	10 13	06 05	16 17	18 55	04 40	09 51	20 34
03	11 34	13 10	04 30	16 32	13 36	15 21	09 44	05 46	16 19	18 55	04 41	09 50	20 34
04	11 23	13 07	03 55	16 14	10 43	15 38	09 15	05 27	16 22	18 56	04 42	09 49	20 34
05	11 15	13 03	03 09	15 56	07 26	15 56	08 46	05 08	16 24	18 57	04 43	09 48	20 33
06	11 09	13 00	02 16	15 38	03 51	16 13	08 16	04 49	16 27	18 57	04 44	09 48	20 33
07	11 05	12 57	01 17	15 19	00 N 09	16 29	07 47	04 30	16 30	18 58	04 45	09 47	20 33
08	11 04	12 54	00 S 14	15 00	03 S 35	16 45	07 17	04 12	16 32	18 58	04 45	09 46	20 33
09	11 D 04	12 51	00 N 50	14 41	07 12	16 59	06 47	03 53	16 35	18 59	04 46	09 45	20 33
10	11 05	12 48	01 53	14 22	10 34	17 13	06 16	03 34	16 37	18 59	04 47	09 44	20 33
11	11 06	12 44	02 52	14 02	13 32	17 25	05 46	03 15	16 40	18 59	04 48	09 44	20 33
12	11 R 07	12 41	03 44	13 42	15 57	17 36	05 16	02 56	16 42	19 00	04 49	09 43	20 33
13	11 05	12 38	04 27	13 22	17 37	17 46	04 45	02 37	16 45	19 00	04 50	09 42	20 32
14	11 02	12 35	04 56	13 02	18 22	17 54	04 14	02 17	16 47	19 01	04 51	09 41	20 32
15	10 57	12 32	05 09	12 41	18 03	18 01	03 43	01 58	16 49	19 01	04 52	09 40	20 32
16	10 50	12 29	05 03	12 21	16 35	18 07	03 12	01 39	16 52	19 01	04 53	09 39	20 32
17	10 44	12 25	04 37	12 00	13 58	18 11	02 41	01 20	16 54	19 02	04 54	09 39	20 32
18	10 37	12 22	03 51	11 39	10 22	18 14	02 10	01 01	16 56	19 02	04 55	09 38	20 32
19	10 32	12 19	02 48	11 17	06 03	18 16	01 39	00 42	16 59	19 02	04 56	09 37	20 32
20	10 29	12 16	01 34	10 56	01 S 21	18 16	01 07	00 23	17 01	19 02	04 58	09 36	20 32
21	10 28	12 13	00 N 14	10 34	03 N 22	18 15	00 36	00 S 04	17 03	19 03	04 59	09 35	20 31
22	10 D 28	12 09	01 S 06	10 13	07 46	18 13	00 S 05	00 N 15	17 06	19 03	05 00	09 34	20 31
23	10 30	12 06	02 19	09 51	11 35	18 09	00 N 27	00 34	17 08	19 03	05 01	09 34	20 31
24	10 31	12 03	03 21	09 28	14 39	18 03	00 58	00 53	17 10	19 03	05 02	09 33	20 31
25	10 32	12 00	04 11	09 06	16 49	17 57	01 30	01 11	17 12	19 03	05 03	09 32	20 31
26	10 R 32	11 57	04 47	08 44	18 02	17 49	02 01	01 30	17 14	19 03	05 04	09 31	20 31
27	10 30	11 54	05 07	08 21	18 19	17 40	02 32	01 49	17 16	19 04	05 05	09 30	20 31
28	10 ♎ 26	11 ♎ 50	05 S 13	07 S 59	17 N 42	17 S 29	03 N 03	02 N 08	17 N 18	19 S 04	05 N 07	09 S 29	20 S 31

ZODIAC SIGN ENTRIES

Date	h m	Planets
02	17 41	☽ ♌
05	05 46	☽ ♍
07	18 44	☽ ♎
10	07 05	☽ ♏
12	16 46	☽ ♐
14	22 24	☽ ♑
17	00 13	☽ ♒
18	23 47	☽ ♓
18	23 50	☉ ♓
20	00 11	♂ ♈
20	20 05	♀ ♈
20	23 13	☽ ♈
23	00 28	☽ ♉
25	04 54	☽ ♊
27	12 50	☽ ♋

LATITUDES

Date	Mercury ☿ ° '	Venus ♀ ° '	Mars ♂ ° '	Jupiter ♃ ° '	Saturn ♄ ° '	Uranus ♅ ° '	Neptune ♆ ° '	Pluto ♇ ° '
01	03 N 37	01 S 31	00 S 48	00 N 57	01 N 59	00 S 39	00 S 44	02 N 07
04	03 33	01 28	00 46	00 57	02 00	00 38	00 44	02 07
07	03 13	01 24	00 44	00 58	02 00	00 38	00 44	02 07
10	02 42	01 20	00 42	00 58	02 01	00 38	00 44	02 07
13	02 07	01 16	00 40	00 58	02 01	00 38	00 44	02 06
16	01 30	01 10	00 38	00 59	02 01	00 38	00 44	02 06
19	00 54	01 05	00 36	00 59	02 02	00 38	00 44	02 06
22	00 N 21	00 58	00 34	00 59	02 02	00 38	00 44	02 06
25	00 S 10	00 51	00 32	00 59	02 03	00 38	00 44	02 06
28	00 38	00 44	00 30	00 59	02 03	00 38	00 44	02 06
31	01 S 03	00 S 36	00 S 28	00 N 59	02 N 04	00 S 38	00 S 44	02 N 06

DATA

Julian Date	2457055
Delta T	+69 seconds
Ayanamsa	24° 04' 10"
Synetic vernal point	05° ♓ 02' 49"
True obliquity of ecliptic	23° 26' 05"

MOON'S PHASES, APSIDES AND POSITIONS ☽

Date	h m	Phase	Longitude ° '	Eclipse Indicator
03	23 09	○	14 ♌ 48	
12	03 50	☾	23 ♏ 06	
18	23 47	●	00 ♓ 00	
25	17 14	☽	06 ♊ 47	

Day	h m		
06	06 11	Apogee	
19	07 22	Perigee	
07	12 55	0S	
14	17 15	Max dec	18° S 24'
20	18 50	0N	
27	07 21	Max dec	18° N 20'

LONGITUDES

Date	Chiron ⚷ ° '	Ceres ⚳ ° '	Pallas ⚴ ° '	Juno ⚵ ° '	Vesta ⚶ ° '	Black Moon Lilith ⚸ ° '
01	15 ♓ 12	09 ♑ 26	13 ♐ 54	09 ♌ 03	01 ♒ 56	07 ♍ 18
11	15 ♓ 46	13 ♑ 13	17 ♐ 20	06 ♌ 40	07 ♒ 10	08 ♍ 25
21	16 ♓ 22	16 ♑ 53	20 ♐ 30	04 ♌ 48	12 ♒ 21	09 ♍ 31
31	16 ♓ 59	20 ♑ 24	23 ♐ 21	03 ♌ 38	17 ♒ 28	10 ♍ 38

ASPECTARIAN

h m	Aspects
01 Sunday	
01 30	☽ ± ♄
06 28	☽ ⊼ ☉
07 03	☽ ⊥ ♃
08 52	☽ □ ♅
10 42	☽ ☍ ♇
13 37	☽ △ ♂
17 36	♀ ☌ ♆
18 38	☽ ⊻ ♃
19 09	☽ P ♄
21 51	☽ # ☉
02 Monday	
00 40	☽ P ♆
01 26	☽ P ♀
06 34	☿ ⊻ ♀
07 48	☽ ∥ ♃
12 25	♀ ⊥ ♅
12 30	☉ ⚹ ♅
18 28	☽ ± ♆
19 52	☿ ⊻ ♆
20 41	☽ # ☿
21 09	☽ P ♂
21 13	☽ ± ♀
03 Tuesday	
00 53	☽ △ ♄
05 30	☽ ☍ ☿
06 24	☽ ⊼ ♆
06 53	♀ ∥ ♆
10 31	☽ ⊼ ♀
11 17	☉ ⊻ ♇
16 18	☽ ± ♂
20 19	☽ △ ♅
22 10	☽ ⊼ ♇
23 09	☽ ☍ ☉
04 Wednesday	
02 44	☉ # ♃
05 09	☽ ⊼ ♂
05 31	☽ ☌ ♃
10 04	♂ ⊼ ♃
10 15	☽ ± ♇
18 52	☽ # ♆
21 36	☿ ⊥ ♀
21 46	☉ Q ♄
05 Thursday	
00 57	☽ # ♀
02 35	☽ P ♅
04 27	☽ P ♇
11 39	☿ ⚹ ♄
11 52	☿ ∠ ♂
12 05	☉ ∥ ☿
13 19	☽ ⊼ ☿
13 26	☽ □ ♄
18 56	☽ ☍ ♆
20 52	☽ ± ♅
06 Friday	
00 44	☽ ± ☿
05 02	☽ # ♂
06 02	☽ ☍ ♀
06 17	☽ ∥ ♅
09 06	☽ ⊼ ♅
10 57	☽ △ ♇
17 29	☽ ⊼ ☉
17 34	☽ ⊻ ♃
18 01	☽ P ☿
18 20	☉ ☍ ♃
18 56	♂ # ♅
22 09	☽ ☍ ♂
07 Saturday	
02 13	☽ Q ♄
05 39	☽ ⊥ ♃
06 51	☽ ± ☉
11 54	♀ ⊻ ♅
12 36	☿ # ♃
23 07	☽ △ ☿
23 45	☽ ∠ ♃
08 Sunday	
02 44	☽ ⚹ ♄
02 52	☽ P ☉
05 53	♀ ⚹ ♇
08 07	☽ ⊼ ♆
15 38	☽ ∥ ♂
19 42	☽ # ♅
20 18	☽ ± ♆
22 13	☽ ☍ ♅
09 Monday	
00 02	☽ □ ♇
02 05	☽ ⊼ ♀
05 45	☽ ⚹ ♃
09 10	☽ ∠ ♄
09 30	☽ ∥ ♀
11 39	☿ Q ♅
11 58	☽ △ ☉
13 48	☉ ⊥ ♇
14 27	☽ P ♆
15 08	☽ ⊼ ♂
15 31	☽ ± ♀
21 32	☿ ∠ ♀
10 Tuesday	
03 13	☽ ⊥ ♄
03 55	☽ ± ♂
05 27	☽ Q ♃
05 56	☽ ∥ ♆
09 50	☽ □ ☿
10 19	♀ ⊼ ♃
11 27	☽ P ♀
12 02	☽ Q ♇
15 08	☽ ⊻ ♄
20 15	☽ △ ♆
22 50	☽ P ♂
11 Wednesday	
00 38	♂ ± ♃
09 49	☽ ⊼ ♅
11 29	☽ ⚹ ♇
14 58	☿ St D
16 01	☽ ∥ ☉
16 12	☽ □ ♃
19 49	☽ △ ♀
20 38	☽ Q ☿
21 21	☽ ± ♅
12 Thursday	
03 50	☽ □ ☉
05 32	☽ △ ♂
14 23	☽ P ♅
15 58	☽ ∠ ♇
19 20	☽ ⚹ ☿
21 42	☽ # ♃
13 Friday	
00 32	☽ ☌ ♄
05 15	☽ □ ♆
07 54	♀ # ♅
08 34	☽ ⊥ ♇
15 26	☽ ∥ ☿
15 50	☽ Q ☉
17 54	☽ △ ♅
19 25	☽ ⊻ ♇
23 00	☽ ∠ ☿
23 07	☽ △ ♃
14 Saturday	
08 31	☽ □ ♀
13 27	☽ Q ♆
13 43	☿ Q ♅
14 46	☽ ⚹ ☉
15 10	☽ ⊥ ☿
15 15	☽ □ ♂
19 34	♀ ± ♃
15 Sunday	
01 03	☽ P ♃
01 47	☽ ⊻ ☿
05 45	☽ ⊻ ♄
10 05	☽ ⚹ ♆
12 46	☽ ∥ ☿
15 55	☽ ⊥ ♄
16 00	♂ Q ♇
16 09	☽ ± ♃
17 19	☉ ⊻ ♂
18 17	☽ ∠ ☉
18 17	☽ Q ♀
21 48	☽ □ ♅
23 09	☽ ☌ ♇
23 32	☽ Q ♂
16 Monday	
02 02	☽ ⊼ ♃
06 53	☽ ∠ ♄
08 36	☽ # ♃
10 16	☽ ⊥ ☉
11 04	☽ ∠ ♆
15 53	☽ ⚹ ♀
20 17	☽ ⚹ ♂
20 45	☽ ⊻ ☉
17 Tuesday	
01 42	☽ ⊥ ♆
03 17	☽ Q ♅
05 00	☽ ☌ ☿
07 17	☽ ⚹ ♄
11 20	☽ ⊻ ♆
18 08	☽ ∠ ♀
21 35	☽ ∠ ♂
22 29	☽ ⚹ ♅
22 38	☉ ∠ ♅
23 43	☽ ⊻ ♇
18 Wednesday	
01 59	☽ ☍ ♃
02 25	☽ Q ♄
03 17	☽ ∥ ☉
04 16	♀ Q ♇
09 12	☽ ⊥ ♇
09 38	☽ ⊥ ♀
12 33	☽ ⊥ ♂
16 21	☽ ∥ ♆
17 35	☉ ∠ ♇
19 54	☽ ⊻ ♀
22 12	☽ ∠ ♅
22 30	☽ ⊻ ♂
23 23	☽ ∠ ♇
23 47	☽ ☌ ☉
19 Thursday	
06 31	☽ ⊻ ☿
06 49	☽ □ ♄
10 48	☽ ☌ ♆
12 23	☽ ⊥ ♅
13 06	☿ ⚹ ♄
16 26	☽ ⊥ ☿
17 45	☽ # ♅
21 51	☽ ⊻ ♅
23 02	☽ ⚹ ♇
20 Friday	
00 49	☽ ⊼ ♃
07 23	☽ ∠ ☿
10 14	☽ ± ♃
13 19	☽ ∥ ♀
17 13	☽ ∥ ♂
18 05	☽ Q ♇
20 14	☽ # ♂
23 15	☽ # ♀
23 30	☽ ☌ ♀
21 Saturday	
00 28	☽ ☌ ♂
00 28	☽ P ♃
00 39	♂ P ♃
02 38	☽ ⊻ ☉
06 31	☽ △ ♄
08 43	☽ ⚹ ☿
10 24	♀ P ♃
10 37	☽ ⊻ ♆
13 02	☽ ⊥ ☉
20 24	☽ ⊥ ♆
20 34	☽ ∥ ♅
22 04	☽ ☌ ♅
23 13	☽ □ ♇
22 Sunday	
00 36	☽ △ ♃
04 50	☽ ∠ ☉
05 13	♀ ☌ ♂
05 32	☽ Q ☿
07 03	☽ P ♄
07 11	♀ # ♂
11 16	☽ ∠ ♆
18 08	☿ ⊻ ♆
22 07	☽ ± ♄
22 47	☽ # ♆
23 Monday	
01 31	☽ # ☉
04 37	☽ ⊻ ♂
05 27	☽ ⊻ ♀
07 53	☽ ⚹ ☉
08 19	☽ ⊼ ♄
12 40	☽ ⚹ ♆
13 56	☉ □ ♄
13 57	☽ □ ☿
15 28	☽ ⊥ ♂
16 43	☽ ⊥ ♀
24 Tuesday	
00 52	☽ ⊻ ♅
01 02	♀ ∥ ♂
02 01	☽ △ ♇
02 57	☽ □ ♃
06 17	☽ Q ☉
07 14	☉ ∥ ♆
08 04	☽ ∠ ♂
09 36	☽ Q ♆
09 58	☽ ∠ ♀
11 27	☽ ⊥ ♅
14 58	♀ △ ♄
25 Wednesday	
03 35	☽ ∠ ♅
04 43	☽ P ♇
10 48	☽ Q ♃
12 35	☽ ⚹ ♂
13 24	☽ ☍ ♄
15 40	☽ ⚹ ♀
17 14	☽ □ ☉
18 02	☽ ∥ ♃
18 06	☽ □ ♆
21 10	☽ ± ♇
23 48	☽ △ ☿
26 Thursday	
03 08	♂ △ ♄
04 55	☉ ☌ ♆
06 42	☽ # ☿
07 11	☽ ⚹ ♅
08 19	☽ ⊼ ♇
08 43	☽ ⚹ ♃
12 10	☽ Q ♂
16 16	☽ Q ♀
18 22	♀ ⊻ ♆
27 Friday	
05 54	☽ Q ♅
06 35	☽ P ☿
07 25	☉ ⊥ ♅
12 51	☽ ∠ ♃
21 58	☽ ⊼ ♄
23 11	♃ ⊼ ♇
28 Saturday	
00 41	☽ □ ♂
01 36	☽ ± ☿
02 59	☽ △ ♆
06 05	☽ ⊥ ♃
06 27	☽ □ ♀
06 54	☽ △ ☉
09 38	☽ ± ♄
14 29	☽ ⊼ ☿
16 50	☽ □ ♅
17 41	☽ ⊻ ♃
17 46	☽ # ☿
17 53	☽ ☍ ♇
20 06	☽ ∥ ♃

All ephemeris data is given at 12.00 UT and the Moon's longitude is additionally given for 24.00 UT

MARCH 2015

Date	Sidereal time h m s	Sun ☉ ° ' "	Moon ☽ ° ' "	Moon ☽ 24.00 ° ' "	Mercury ☿ ° '	Venus ♀ ° '	Mars ♂ ° '	Jupiter ♃ ° '	Saturn ♄ ° '	Uranus ♅ ° '	Neptune ♆ ° '	Pluto ⯓ ° '
01	22 35 54	10 ♓ 34 32	24 ♋ 10 36	00 ♌ 12 59	14 ♒ 18	10 ♈ 38	07 ♈ 17	14 ♌ 48	04 ♐ 47	14 ♈ 29	07 ♓ 23	15 ♑ 00
02	22 39 51	11 34 44	06 ♌ 13 23	12 ♌ 12 08	15 30	11 52	08 02	14 R 41	04 48	14 32	07 26	15 01
03	22 43 47	12 34 54	18 ♌ 09 30	24 ♌ 05 47	16 44	13 05	08 48	14 35	04 50	14 35	07 28	15 03
04	22 47 44	13 35 02	00 ♍ 01 13	05 ♍ 56 04	18 00	14 18	09 34	14 29	04 51	14 38	07 30	15 04
05	22 51 40	14 35 09	11 ♍ 50 34	17 ♍ 44 56	19 17	15 31	10 20	14 23	04 52	14 41	07 32	15 05
06	22 55 37	15 35 13	23 ♍ 39 27	29 ♍ 34 21	20 36	16 45	11 05	14 17	04 52	14 44	07 35	15 07
07	22 59 34	16 35 16	05 ♎ 29 54	11 ♎ 26 24	21 57	17 58	11 51	14 11	04 53	14 47	07 37	15 08
08	23 03 30	17 35 17	17 ♎ 24 10	23 ♎ 23 32	23 19	19 11	12 36	14 06	04 54	14 50	07 39	15 09
09	23 07 27	18 35 16	29 ♎ 24 53	05 ♏ 28 34	24 43	20 24	13 22	14 00	04 54	14 54	07 42	15 10
10	23 11 23	19 35 14	11 ♏ 35 03	17 ♏ 44 45	26 08	21 37	14 07	13 55	04 55	14 57	07 44	15 11
11	23 15 20	20 35 10	23 ♏ 58 07	00 ♐ 15 38	27 34	22 50	14 53	13 50	04 55	15 00	07 46	15 12
12	23 19 16	21 35 04	06 ♐ 37 45	13 ♐ 04 56	29 ♒ 01	24 02	15 38	13 44	04 56	15 03	07 48	15 13
13	23 23 13	22 34 56	19 ♐ 37 36	26 ♐ 16 07	00 ♓ 30	25 15	16 23	13 40	04 56	15 06	07 50	15 14
14	23 27 09	23 34 47	03 ♑ 00 48	09 ♑ 51 53	02 00	26 28	17 09	13 35	04 56	15 10	07 53	15 16
15	23 31 06	24 34 37	16 ♑ 49 29	23 ♑ 53 33	03 32	27 40	17 54	13 30	04 R 56	15 13	07 55	15 17
16	23 35 03	25 34 24	01 ♒ 03 56	08 ♒ 20 17	05 05	28 ♈ 53	18 39	13 26	04 56	15 16	07 57	15 17
17	23 38 59	26 34 10	15 ♒ 42 05	23 ♒ 08 38	06 39	00 ♉ 05	19 24	13 22	04 55	15 19	07 59	15 18
18	23 42 56	27 33 54	00 ♓ 39 04	08 ♓ 12 24	08 14	01 18	20 09	13 18	04 55	15 23	08 01	15 19
19	23 46 52	28 33 37	15 ♓ 47 28	23 ♓ 23 07	09 50	02 30	20 54	13 14	04 55	15 26	08 04	15 20
20	23 50 49	29 ♓ 33 17	00 ♈ 58 06	08 ♈ 31 14	11 28	03 42	21 40	13 10	04 54	15 29	08 06	15 21
21	23 54 45	00 ♈ 32 55	16 ♈ 01 23	23 ♈ 27 31	13 07	04 54	22 25	13 07	04 53	15 33	08 08	15 22
22	23 58 42	01 32 31	00 ♉ 48 44	08 ♉ 04 20	14 47	06 07	23 09	13 03	04 53	15 36	08 10	15 23
23	00 02 38	02 32 05	15 ♉ 13 46	22 ♉ 16 40	16 29	07 19	23 54	13 00	04 52	15 39	08 12	15 23
24	00 06 35	03 31 37	29 ♉ 12 48	06 ♊ 02 10	18 12	08 31	24 39	12 57	04 51	15 43	08 14	15 24
25	00 10 32	04 31 06	12 ♊ 44 50	19 ♊ 21 01	19 56	09 42	25 24	12 54	04 50	15 46	08 16	15 25
26	00 14 28	05 30 34	25 ♊ 51 03	02 ♋ 15 19	21 41	10 54	26 09	12 52	04 49	15 49	08 18	15 26
27	00 18 25	06 29 59	08 ♋ 34 16	14 ♋ 48 24	23 28	12 06	26 53	12 50	04 47	15 53	08 21	15 26
28	00 22 21	07 29 21	20 ♋ 58 15	27 ♋ 04 20	25 16	13 17	27 38	12 47	04 46	15 56	08 23	15 27
29	00 26 18	08 28 41	03 ♌ 07 12	09 ♌ 07 23	27 05	14 29	28 23	12 45	04 45	16 00	08 25	15 27
30	00 30 14	09 27 59	15 ♌ 05 23	21 ♌ 01 42	28 ♓ 56	15 40	29 07	12 43	04 43	16 03	08 27	15 28
31	00 34 11	10 ♈ 27 15	26 ♌ 56 48	02 ♍ 51 08	00 ♈ 48	16 ♉ 52	29 ♈ 52	12 ♌ 42	04 ♐ 42	16 ♈ 06	08 ♓ 29	15 ♑ 28

Date	Moon True ☊ ° '	Moon Mean ☊ ° '	Moon ☽ Latitude ° '
01	10 ♎ 22	11 ♎ 47	05 S 04
02	10 R 17	11 44	04 42
03	10 11	11 41	04 07
04	10 07	11 38	03 22
05	10 03	11 34	02 29
06	10 01	11 31	01 29
07	10 00	11 28	00 S 25
08	10 D 00	11 25	00 N 41
09	10 01	11 22	01 45
10	10 02	11 19	02 46
11	10 05	11 15	03 40
12	10 05	11 12	04 24
13	10 06	11 09	04 57
14	10 R 06	11 06	05 14
15	10 05	11 03	05 15
16	10 03	11 00	04 56
17	10 01	10 56	04 18
18	10 00	10 53	03 21
19	09 59	10 50	02 10
20	09 58	10 47	00 N 50
21	09 D 58	10 44	00 S 34
22	09 58	10 40	01 53
23	09 59	10 37	03 04
24	10 00	10 34	04 01
25	10 00	10 31	04 43
26	10 00	10 28	05 08
27	10 R 01	10 25	05 18
28	10 00	10 21	05 12
29	10 00	10 18	04 53
30	10 00	10 15	04 20
31	10 ♎ 00	10 ♎ 12	03 S 37

DECLINATIONS

Date	Sun ☉	Moon ☽	Mercury ☿	Venus ♀	Mars ♂	Jupiter ♃	Saturn ♄	Uranus ♅	Neptune ♆	Pluto ⯓
01	07 S 36	16 N 17	17 S 17	03 N 35	02 N 27	17 N 20	19 S 04	05 N 08	09 S 29	20 S 31
02	07 13	14 10	17 03	04 06	02 45	17 22	19 04	05 09	09 28	20 30
03	06 50	11 28	16 49	04 37	03 04	17 24	19 04	05 10	09 27	20 30
04	06 27	08 18	16 33	05 08	03 23	17 26	19 04	05 11	09 26	20 30
05	06 04	04 50	16 15	05 39	03 41	17 27	19 04	05 12	09 25	20 30
06	05 41	01 N 09	15 57	06 09	04 00	17 29	19 04	05 14	09 24	20 30
07	05 18	02 S 34	15 37	06 40	04 18	17 31	19 04	05 15	09 24	20 30
08	04 54	06 12	15 15	07 10	04 37	17 32	19 04	05 16	09 23	20 30
09	04 31	09 37	14 53	07 41	04 55	17 34	19 04	05 17	09 22	20 30
10	04 07	12 40	14 29	08 11	05 13	17 35	19 04	05 19	09 21	20 30
11	03 44	15 12	14 04	08 41	05 32	17 37	19 04	05 20	09 20	20 30
12	03 20	17 04	13 38	09 10	05 50	17 38	19 04	05 21	09 19	20 30
13	02 57	18 06	13 10	09 40	06 08	17 40	19 03	05 22	09 19	20 29
14	02 33	18 10	12 41	10 09	06 26	17 41	19 03	05 24	09 18	20 29
15	02 09	17 10	12 11	10 38	06 44	17 42	19 03	05 25	09 17	20 29
16	01 46	15 06	11 40	11 07	07 01	17 44	19 03	05 26	09 16	20 29
17	01 22	12 01	11 07	11 36	07 19	17 45	19 03	05 27	09 15	20 29
18	00 58	08 06	10 33	12 04	07 37	17 46	19 02	05 29	09 15	20 29
19	00 34	03 S 36	09 58	12 32	07 55	17 47	19 02	05 30	09 14	20 29
20	00 S 11	01 N 09	09 22	13 00	08 12	17 48	19 02	05 31	09 13	20 29
21	00 N 13	05 47	08 45	13 27	08 29	17 49	19 02	05 33	09 12	20 29
22	00 37	09 59	08 06	13 54	08 47	17 50	19 01	05 34	09 11	20 29
23	01 00	13 28	07 26	14 21	09 04	17 51	19 01	05 35	09 11	20 29
24	01 24	16 03	06 45	14 48	09 21	17 51	19 01	05 36	09 10	20 29
25	01 48	17 39	06 03	15 14	09 38	17 52	19 01	05 38	09 09	20 29
26	02 11	18 14	05 20	15 39	09 55	17 53	19 00	05 39	09 08	20 29
27	02 35	17 52	04 36	16 05	10 12	17 53	19 00	05 40	09 08	20 29
28	02 58	16 39	03 51	16 30	10 29	17 54	18 59	05 42	09 07	20 29
29	03 22	14 43	03 04	16 54	10 45	17 54	18 59	05 43	09 06	20 29
30	03 45	12 09	02 17	17 19	11 02	17 55	18 59	05 44	09 05	20 29
31	04 N 08	09 N 07	01 S 28	17 N 42	11 N 18	17 N 55	18 S 58	05 N 46	09 S 05	20 S 28

ZODIAC SIGN ENTRIES

Date	h	m	Planets
01	23	34	☽ ♌
04	11	58	☽ ♍
07	00	52	☽ ♎
09	13	10	☽ ♏
11	23	30	☽ ♐
13	03	52	☿ ♓
14	06	40	☽ ♑
16	10	14	☽ ♒
17	10	15	♀ ♉
18	10	58	☽ ♓
20	10	28	☽ ♈
20	22	45	☉ ♈
22	10	40	☽ ♉
24	13	22	☽ ♊
26	19	45	☽ ♋
29	05	48	☽ ♌
31	01	44	☿ ♈
31	16	26	♂ ♉
31	18	12	☽ ♍

LATITUDES

Date	Mercury ☿	Venus ♀	Mars ♂	Jupiter ♃	Saturn ♄	Uranus ♅	Neptune ♆	Pluto ⯓
01	00 S 47	00 S 41	00 S 29	00 N 59	02 N 04	00 S 38	00 S 44	02 N 06
04	01 10	00 33	00 27	00 59	02 04	00 38	00 44	02 06
07	01 30	00 25	00 25	00 59	02 05	00 38	00 44	02 05
10	01 47	00 16	00 23	00 59	02 05	00 38	00 44	02 05
13	02 00	00 S 07	00 20	00 59	02 06	00 38	00 44	02 05
16	02 10	00 N 03	00 18	00 59	02 06	00 37	00 44	02 05
19	02 15	00 12	00 16	00 59	02 06	00 37	00 44	02 05
22	02 17	00 22	00 14	00 59	02 07	00 37	00 44	02 05
25	02 15	00 32	00 12	00 58	02 07	00 37	00 44	02 05
28	02 08	00 42	00 10	00 58	02 07	00 37	00 44	02 05
31	01 S 57	00 N 52	00 S 08	00 N 58	02 N 08	00 S 37	00 S 44	02 N 05

DATA

Julian Date	2457083
Delta T	+69 seconds
Ayanamsa	24° 04' 13"
Synetic vernal point	05° ♓ 02' 46"
True obliquity of ecliptic	23° 26' 06"

MOON'S PHASES, APSIDES AND POSITIONS ☽

Date	h	m	Phase	Longitude ° '	Eclipse Indicator
05	18	05	○	14 ♍ 50	
13	17	48	☾	22 ♐ 49	
20	09	36	●	29 ♓ 27	Total
27	07	43	☽	06 ♋ 19	

Day	h	m		
05	07	17	Apogee	
19	19	31	Perigee	
06	19	28	0S	
14	01	37	Max dec	18° S 16'
20	06	13	0N	
26	14	32	Max dec	18° N 14'

LONGITUDES

Date	Chiron ⚷	Ceres ⚳	Pallas ⚴	Juno ⚵	Vesta ⚶	Black Moon Lilith ⚸
01	16 ♓ 51	19 ♑ 42	22 ♐ 49	03 ♌ 48	16 ♒ 27	10 ♍ 25
11	17 ♓ 29	23 ♑ 04	25 ♐ 22	03 ♌ 15	21 ♒ 30	11 ♍ 31
21	18 ♓ 05	26 ♑ 15	27 ♐ 29	03 ♌ 26	26 ♒ 28	12 ♍ 38
31	18 ♓ 41	29 ♑ 10	29 ♐ 05	04 ♌ 19	01 ♓ 20	13 ♍ 44

All ephemeris data is given at 12.00 UT and the Moon's longitude is additionally given for 24.00 UT

ASPECTARIAN

h m Aspects

01 Sunday
03 18 ☽ ♇ ♄
05 20 ☉ ⊻ ♀
06 45 ☿ # ♃
08 27 ☽ ♇ ♆
15 01 ☽ ♇ ☉
15 44 ♂ ⊻ ♆
15 55 ☿ ⚹ ♅
21 15 ☿ ☍ ♃

02 Monday
02 23 ☽ ± ♆
02 25 ☿ ⊻ ⯓
09 10 ☽ △ ♄
10 36 ☽ ± ☉
14 25 ☽ ⊼ ♆
15 53 ☽ △ ♂
23 44 ☽ ⊼ ☉

03 Tuesday
00 36 ☽ △ ♀
04 46 ☽ △ ♅
04 52 ☽ ☌ ♃
05 43 ☽ ⊼ ⯓
08 47 ☽ ☍ ☿
12 25 ♃ △ ♅
13 46 ☿ Q ♄
17 51 ☽ ± ⯓
19 46 ♀ ⊥ ♆

04 Wednesday
00 11 ☽ ♇ ♂
03 42 ☽ # ♆
10 23 ☽ ♇ ♀
11 13 ☽ ♇ ♅
12 06 ☽ ♇ ⯓
14 49 ♀ ∥ ♅
15 14 ♀ △ ♃
18 46 ♀ ☌ ♅
19 41 ☽ ± ♂
21 48 ☽ □ ♄

05 Thursday
02 35 ☽ # ☉
03 14 ☽ ☍ ♆
03 17 ♀ □ ⯓
05 33 ☽ ± ♅
06 45 ☽ ± ♀
07 12 ☽ ∥ ♀
07 33 ☉ ⊼ ♃
08 42 ☽ ⊼ ♂
09 27 ☽ ∥ ♅
14 29 ☉ ⊻ ♅
17 07 ☽ ⊻ ♃
17 48 ☽ ⊼ ♅
18 05 ☽ ☍ ☉
18 36 ☽ △ ⯓
18 57 ☽ ∥ ♂
20 20 ☽ ⊼ ♀
23 21 ☉ # ♀

06 Friday
00 19 ☉ ⚹ ⯓
05 01 ☽ ⊼ ☿
05 12 ☽ ⊥ ♃
10 24 ☽ Q ♄
18 45 ☽ ± ☿
21 10 ☿ ⊥ ⯓
23 19 ☽ ∠ ♃

07 Saturday
10 46 ☽ ⚹ ♄
14 40 ☉ # ♅
15 19 ☽ ♇ ☿
16 18 ☽ ⊼ ♆

08 Sunday
00 24 ☽ # ♂
01 41 ☽ ☍ ♂
04 10 ☽ ∥ ☉
04 26 ☽ ± ♆
05 24 ☽ ⚹ ♃
05 42 ☽ # ♅
06 49 ☽ ☍ ♅
07 28 ☽ □ ⯓
12 24 ☽ ⊼ ☉
15 58 ☽ ☍ ♀
17 01 ☽ ∠ ♄
19 43 ☽ # ♀
22 06 ☉ # ♂
22 33 ☽ ♇ ♆

09 Monday
01 24 ☽ △ ☿
01 31 ☽ ± ☉
02 18 ♀ ♇ ♄
05 15 ☽ Q ♃
10 07 ☽ ∥ ♆
11 00 ☽ ⊥ ♄
19 27 ☽ Q ⯓
21 01 ☽ ♇ ☉
22 53 ☽ ⊻ ♄
22 57 ♂ ⊥ ♆

10 Tuesday
04 25 ☽ △ ♆
06 04 ♂ △ ♃
16 31 ☽ □ ♃
17 17 ☽ ⊼ ♂
18 35 ☽ ⊼ ♅
19 03 ☽ ⚹ ⯓
19 10 ☉ ± ♃
19 25 ♂ ∥ ♅

11 Wednesday
02 09 ☽ ∥ ☿
04 55 ☽ △ ☉
05 40 ☽ ± ♂
06 16 ☽ ± ♅
09 34 ☽ ⊼ ♀
10 48 ♀ ∠ ♆
16 06 ♂ ☌ ♅
19 46 ☽ □ ☿
22 16 ☽ ± ♀
22 41 ♂ □ ⯓
23 33 ☽ ♇ ♅
23 55 ☽ ∠ ⯓
23 59 ☽ ♇ ♂

12 Thursday
08 48 ☽ ☌ ♄
14 12 ☽ □ ♆
16 51 ☽ ⊥ ⯓
16 58 ☽ ♇ ♀
19 12 ♀ # ♆
22 58 ☽ # ♃

13 Friday
01 08 ☽ △ ♃
03 42 ☽ △ ♅
03 58 ☽ ⊻ ⯓
05 19 ☿ ∠ ♅
05 43 ☽ △ ♂
07 43 ☿ ∠ ⯓
09 42 ☽ Q ☿
17 48 ☽ □ ☉
23 11 ☽ △ ♀
23 16 ☽ Q ♆

14 Saturday
04 11 ☽ ♇ ♃
10 00 ☽ ⚹ ☿
15 03 ♄ St R
15 23 ☽ ⊻ ♄
16 18 ☿ ∠ ♂
19 59 ☽ ± ♃
20 34 ☽ ⚹ ♆

15 Sunday
01 51 ☽ ⊥ ♄
02 27 ☽ # ♃
04 08 ☽ Q ☉
06 20 ☽ ⊼ ♃
09 14 ☽ □ ♅
09 20 ☽ ☌ ⯓
13 56 ☽ □ ♂
15 16 ☽ ∠ ☿
17 18 ☽ ∠ ♄
22 23 ☽ ∠ ♆

16 Monday
02 08 ☽ ⚹ ☉
08 02 ☽ □ ♀
08 17 ☽ ⊥ ☿
09 41 ☿ □ ♄
12 55 ♀ ± ♄
13 28 ☽ ⊥ ♆
15 40 ☽ Q ♅
18 23 ☽ ⚹ ♄
19 26 ☽ ⊻ ☿
21 44 ☽ Q ♂
23 24 ☽ ⊻ ♆

17 Tuesday
00 49 ☿ # ♀
02 54 ♅ □ ⯓
04 48 ☽ ∠ ☉
08 13 ☽ ☍ ♃
11 22 ☽ ⊻ ⯓
11 23 ☽ ⚹ ♅
13 59 ☽ Q ♄
14 32 ☽ # ♀
16 12 ☽ Q ♀
18 18 ☽ ⚹ ♂
19 00 ☽ ∥ ☿
20 25 ☽ ⊥ ☉
21 04 ☽ ⊥ ⯓

18 Wednesday
04 25 ♂ ♇ ♄
05 25 ☽ ∥ ♆
06 03 ☉ Q ⯓
06 44 ☽ ⊻ ☉
08 49 ☿ ☌ ♆
11 29 ☽ ∠ ⯓
11 34 ☽ ∠ ♅
13 07 ☽ ⚹ ♀
14 32 ☽ # ♂
18 47 ☽ □ ♄
19 32 ☽ ∠ ♂
23 44 ☽ ☌ ♆

19 Thursday
01 28 ☽ ☌ ☿
01 54 ☽ ⊥ ♅
02 11 ☽ # ♅
04 32 ☉ ♇ ♃
05 44 ☿ ⊥ ♅
07 58 ☽ ⊼ ♃
10 32 ☽ ⊥ ♂
11 17 ☽ ⚹ ⯓
11 26 ☽ ⊻ ♅
14 56 ☽ ∠ ♀
17 25 ☽ ± ♃
20 30 ☽ ⊻ ♂

20 Friday
04 44 ☽ ∥ ☉
06 16 ☽ Q ⯓
06 23 ☽ ⊥ ♀
07 29 ☽ # ☉
07 35 ☽ ♇ ♃
09 36 ☽ ☌ ☉
16 43 ☽ ⊻ ♀
17 58 ☿ ∥ ♆
18 14 ☽ △ ♄
23 21 ☽ ⊻ ♆

21 Saturday
06 46 ☽ ⊻ ☿
07 21 ☽ △ ♃
08 58 ☽ ⊥ ♆
10 42 ☽ ∥ ♅
10 57 ☽ □ ⯓
11 14 ☽ ☌ ♅
11 39 ♀ ⊼ ♄
11 54 ☿ ⊼ ♃
17 36 ☽ ⊥ ☿
18 13 ☽ ♇ ♄
18 32 ☿ # ♂
22 51 ☽ ☌ ♂
23 30 ☽ ∠ ♆

22 Sunday
02 19 ☽ # ☿
04 12 ☽ ∥ ♂
07 12 ☽ # ♆
08 50 ☽ ± ♄
10 06 ☽ ∠ ☿
12 21 ♂ ∠ ♆
13 17 ☽ ⊻ ☉
18 42 ☽ ⊼ ♄
20 27 ☿ ⚹ ⯓
21 32 ☽ ☌ ♀
23 55 ☿ ⊻ ♅
23 56 ☽ ⊥ ☉

23 Monday
00 11 ☽ ⚹ ♆
08 16 ☽ □ ♃
12 16 ☽ △ ⯓
12 43 ☽ ⊻ ♅
14 24 ☽ ⚹ ☿
16 12 ☽ ∠ ☉
20 28 ☽ Q ♆
20 39 ☽ ∥ ♀
20 54 ♂ # ♆
22 59 ☽ ⊥ ♅

24 Tuesday
03 38 ☽ ⊻ ♂
06 25 ♀ ⚹ ♆
13 58 ☽ Q ☿
14 05 ☽ ♇ ⯓
14 38 ☽ ∠ ♅
14 40 ☽ ⊥ ♂
15 02 ☽ Q ♃
20 09 ☽ ⚹ ☉
21 53 ☽ ☍ ♄
22 16 ☿ ± ♃
23 12 ☿ ⊥ ♂

25 Wednesday
03 57 ☽ □ ♆
06 00 ☽ ⊻ ♀
06 01 ☽ ± ⯓
07 32 ☽ ∠ ♂
12 17 ☽ ⚹ ♃
16 50 ☽ ⊼ ⯓
17 29 ☽ ⚹ ♅
17 33 ☽ ∥ ♃
17 53 ☽ ⊥ ♀
19 23 ☽ Q ☉
19 25 ☉ △ ♄

26 Thursday
01 53 ☿ # ♅
03 05 ☽ □ ☿
12 06 ☽ ∠ ♀
12 35 ☽ ⚹ ♂
15 42 ☽ Q ♅
15 45 ☽ ∠ ♃

27 Friday
04 49 ☽ ⊼ ♄
07 43 ☽ □ ☉
08 41 ☽ ⊥ ♃
11 28 ☽ ∥ ♃
11 34 ☽ △ ♆
12 39 ☽ Q ♂
16 15 ☽ ± ♄
19 29 ☽ ⚹ ♀
20 09 ☽ ⊻ ♃

28 Saturday
01 14 ☽ ☍ ⯓
02 09 ☽ □ ♅
02 11 ♀ □ ♃
09 39 ☽ ♇ ♄
13 53 ☽ ∥ ♀
16 44 ☽ ♇ ♆
21 24 ☽ Q ♀
21 55 ☽ △ ☿

29 Sunday
01 58 ☽ □ ♂
06 02 ☉ # ☿
10 19 ☉ ⊻ ♆
10 35 ☽ ± ♆
11 24 ☉ ⊥ ♀
15 14 ☽ △ ♄
16 50 ☿ Q ⯓
20 33 ☿ ♇ ♃
22 36 ☽ ⊼ ♆
23 26 ♂ ± ♄
23 40 ☽ △ ☉

30 Monday
07 15 ☽ ☌ ♃
07 47 ♀ △ ⯓
09 15 ☽ ♇ ☿
12 46 ☽ ⊼ ⯓
13 19 ☽ □ ♀
13 57 ☽ △ ♅
16 01 ☿ ⊻ ♂
19 59 ♀ ⊻ ♅
20 33 ☽ ∥ ♂

31 Tuesday
00 54 ☽ ± ⯓
06 50 ☽ ± ☿
08 42 ☽ ♇ ☉
12 19 ☽ # ♆
18 19 ☽ △ ♂
19 10 ☽ ♇ ⯓
20 30 ☽ ♇ ♅
21 19 ☽ ⊼ ☿

APRIL 2015

LONGITUDES

Date	Sidereal time h m s	Sun ☉ ° ' "	Moon ☽ ° ' "	Moon ☽ 24.00 ° ' "	Mercury ☿ ° '	Venus ♀ ° '	Mars ♂ ° '	Jupiter ♃ ° '	Saturn ♄ ° '	Uranus ♅ ° '	Neptune ♆ ° '	Pluto ♇ ° '
01	00 38 07	11 ♈ 26 28	08 ♍ 45 08	14 ♍ 39 09	02 ♈ 42	18 ♉ 03	00 ♉ 36	12 ♌ 40	04 ♐ 40	16 ♈ 10	08 ♓ 31	15 ♑ 29
02	00 42 04	12 25 40	20 ♍ 33 34	26 ♍ 28 42	04 36	19 14	01 21	12 R 39	04 R 38	16 13	08 33	15 29
03	00 46 01	13 24 49	02 ♎ 24 52	08 ♎ 22 21	06 32	20 25	02 05	12 38	04 36	16 17	08 35	15 30
04	00 49 57	14 23 56	14 ♎ 21 23	20 ♎ 22 14	08 30	21 36	02 49	12 37	04 34	16 20	08 37	15 30
05	00 53 54	15 23 01	26 ♎ 25 08	02 ♏ 30 18	10 29	22 47	03 34	12 36	04 32	16 24	08 38	15 31
06	00 57 50	16 22 04	08 ♏ 37 57	14 ♏ 48 17	12 29	23 58	04 18	12 36	04 30	16 27	08 40	15 31
07	01 01 47	17 21 05	21 ♏ 01 32	27 ♏ 17 54	14 30	25 08	05 02	12 36	04 28	16 30	08 42	15 31
08	01 05 43	18 20 04	03 ♐ 37 36	10 ♐ 00 52	16 32	26 19	05 46	12 35	04 25	16 34	08 44	15 32
09	01 09 40	19 19 01	16 ♐ 27 55	22 ♐ 58 58	18 36	27 29	06 30	12 D 35	04 23	16 37	08 46	15 32
10	01 13 36	20 17 57	29 ♐ 34 13	06 ♑ 13 52	20 40	28 40	07 14	12 36	04 20	16 41	08 48	15 32
11	01 17 33	21 16 51	12 ♑ 58 06	19 ♑ 47 01	22 45	29 ♉ 50	07 58	12 36	04 18	16 44	08 50	15 32
12	01 21 30	22 15 43	26 ♑ 40 44	03 ♒ 39 14	24 51	01 ♊ 00	08 42	12 37	04 15	16 48	08 51	15 32
13	01 25 26	23 14 34	10 ♒ 42 30	17 ♒ 50 21	26 57	02 10	09 26	12 38	04 12	16 51	08 53	15 33
14	01 29 23	24 13 23	25 ♒ 02 32	02 ♓ 18 43	29 ♈ 03	03 20	10 10	12 39	04 10	16 54	08 55	15 33
15	01 33 19	25 12 10	09 ♓ 38 23	17 ♓ 00 56	01 ♉ 09	04 30	10 53	12 40	04 07	16 58	08 57	15 33
16	01 37 16	26 10 55	24 ♓ 25 40	01 ♈ 51 45	03 15	05 39	11 37	12 41	04 04	17 01	08 58	15 33
17	01 41 12	27 09 38	09 ♈ 18 18	16 ♈ 44 22	05 20	06 49	12 21	12 43	04 01	17 05	09 00	15 R 33
18	01 45 09	28 08 20	24 ♈ 08 58	01 ♉ 31 09	07 24	07 58	13 04	12 44	03 57	17 08	09 02	15 33
19	01 49 05	29 ♈ 07 00	08 ♉ 50 01	16 ♉ 04 43	09 27	09 08	13 48	12 46	03 54	17 11	09 03	15 33
20	01 53 02	00 ♉ 05 37	23 ♉ 14 31	00 ♊ 18 51	11 29	10 17	14 32	12 48	03 51	17 15	09 05	15 33
21	01 56 59	01 04 13	07 ♊ 17 13	14 ♊ 09 21	13 28	11 26	15 15	12 51	03 48	17 18	09 06	15 32
22	02 00 55	02 02 47	20 ♊ 55 05	27 ♊ 34 24	15 25	12 35	15 59	12 53	03 44	17 21	09 08	15 32
23	02 04 52	03 01 19	04 ♋ 07 25	10 ♋ 34 23	17 20	13 43	16 42	12 56	03 41	17 25	09 10	15 32
24	02 08 48	03 59 48	16 ♋ 55 39	23 ♋ 11 37	19 12	14 52	17 25	12 59	03 37	17 28	09 11	15 32
25	02 12 45	04 58 15	29 ♋ 22 48	05 ♌ 29 44	21 00	16 01	18 09	13 02	03 34	17 31	09 13	15 32
26	02 16 41	05 56 40	11 ♌ 33 00	17 ♌ 33 11	22 46	17 09	18 52	13 05	03 30	17 35	09 14	15 31
27	02 20 38	06 55 03	23 ♌ 30 56	29 ♌ 26 50	24 27	18 17	19 35	13 08	03 26	17 38	09 15	15 31
28	02 24 34	07 53 24	05 ♍ 21 30	11 ♍ 15 31	26 05	19 25	20 18	13 12	03 23	17 41	09 17	15 31
29	02 28 31	08 51 43	17 ♍ 09 28	23 ♍ 03 53	27 39	20 33	21 01	13 15	03 19	17 45	09 18	15 30
30	02 32 28	09 ♉ 50 00	28 ♍ 59 15	04 ♎ 56 02	29 ♉ 09	21 ♊ 41	21 ♉ 44	13 ♌ 19	03 ♐ 15	17 ♈ 48	09 ♓ 20	15 ♑ 30

Date	Moon True ☊ ° '	Moon Mean ☊ ° '	Moon ☽ Latitude ° '
01	10 ♎ 00	10 ♎ 09	02 S 45
02	10 D 00	10 06	01 46
03	10 01	10 02	00 S 42
04	10 R 01	09 59	00 N 24
05	10 00	09 56	01 30
06	10 00	09 53	02 32
07	09 59	09 50	03 29
08	09 58	09 46	04 16
09	09 57	09 43	04 51
10	09 56	09 40	05 12
11	09 55	09 37	05 17
12	09 D 55	09 34	05 04
13	09 56	09 31	04 33
14	09 57	09 27	03 44
15	09 58	09 24	02 40
16	09 59	09 21	01 25
17	09 59	09 18	00 N 04
18	09 R 59	09 15	01 S 18
19	09 57	09 12	02 33
20	09 55	09 08	03 37
21	09 52	09 05	04 26
22	09 49	09 02	04 58
23	09 46	08 59	05 14
24	09 44	08 56	05 13
25	09 43	08 52	04 57
26	09 D 43	08 49	04 28
27	09 44	08 46	03 48
28	09 46	08 43	02 58
29	09 48	08 40	02 01
30	09 ♎ 49	08 ♎ 37	00 S 59

DECLINATIONS

Date	Sun ☉ ° '	Moon ☽ ° '	Mercury ☿ ° '	Venus ♀ ° '	Mars ♂ ° '	Jupiter ♃ ° '	Saturn ♄ ° '	Uranus ♅ ° '	Neptune ♆ ° '	Pluto ♇ ° '
01	04 N 31	05 N 44	00 S 39	18 N 06	11 N 34	17 N 56	18 S 58	05 N 47	09 S 04	20 S 28
02	04 55	02 N 07	00 N 11	18 29	11 50	17 56	18 57	05 48	09 03	20 28
03	05 18	01 S 36	01 03	18 51	12 06	17 56	18 57	05 50	09 02	20 28
04	05 41	05 17	01 55	19 13	12 22	17 56	18 56	05 51	09 02	20 28
05	06 03	08 48	02 47	19 35	12 38	17 56	18 56	05 52	09 01	20 28
06	06 26	11 58	03 41	19 56	12 53	17 57	18 55	05 54	09 00	20 28
07	06 49	14 39	04 35	20 16	13 09	17 57	18 55	05 55	09 00	20 28
08	07 11	16 41	05 30	20 36	13 24	17 57	18 54	05 56	08 59	20 28
09	07 34	17 55	06 25	20 56	13 39	17 56	18 54	05 58	08 58	20 28
10	07 56	18 14	07 20	21 15	13 54	17 56	18 53	05 59	08 58	20 28
11	08 18	17 33	08 16	21 33	14 09	17 56	18 53	06 00	08 57	20 28
12	08 40	15 51	09 11	21 51	14 24	17 56	18 52	06 01	08 56	20 28
13	09 02	13 10	10 06	22 08	14 39	17 56	18 51	06 03	08 56	20 28
14	09 23	09 39	11 01	22 25	14 53	17 55	18 51	06 04	08 55	20 28
15	09 45	05 29	11 56	22 41	15 07	17 55	18 50	06 05	08 54	20 29
16	10 06	00 S 55	12 49	22 57	15 22	17 54	18 49	06 07	08 54	20 29
17	10 28	03 N 45	13 42	23 12	15 36	17 54	18 49	06 08	08 53	20 29
18	10 49	08 10	14 33	23 26	15 50	17 53	18 48	06 09	08 53	20 29
19	11 09	12 02	15 23	23 40	16 03	17 53	18 47	06 11	08 52	20 29
20	11 30	15 05	16 12	23 53	16 17	17 52	18 47	06 12	08 51	20 29
21	11 51	17 09	16 58	24 06	16 30	17 51	18 46	06 13	08 51	20 29
22	12 11	18 10	17 43	24 18	16 43	17 50	18 45	06 14	08 50	20 29
23	12 31	18 09	18 26	24 29	16 56	17 50	18 45	06 16	08 50	20 29
24	12 51	17 11	19 06	24 40	17 09	17 49	18 44	06 17	08 49	20 29
25	13 11	15 26	19 44	24 50	17 22	17 48	18 43	06 18	08 49	20 29
26	13 30	13 01	20 19	25 00	17 35	17 47	18 42	06 19	08 48	20 29
27	13 49	10 05	20 53	25 08	17 47	17 46	18 42	06 21	08 48	20 29
28	14 08	06 47	21 23	25 16	17 59	17 45	18 41	06 22	08 47	20 29
29	14 27	03 12	21 51	25 24	18 11	17 43	18 40	06 23	08 47	20 29
30	14 N 46	00 S 30	22 N 17	25 N 31	18 N 23	17 N 42	18 S 39	06 N 24	08 S 46	20 S 29

ZODIAC SIGN ENTRIES

Date	h	m	Planets
03	07	07	☽ ♎
05	19	04	☽ ♏
08	05	08	☽ ♐
10	12	47	☽ ♑
11	15	28	♀ ♊
12	17	44	☽ ♒
14	20	12	☽ ♓
14	22	51	☿ ♉
16	21	00	☽ ♈
18	21	31	☽ ♉
20	09	42	☉ ♉
20	23	28	☽ ♊
23	04	25	☽ ♋
25	13	13	☽ ♌
28	01	07	☽ ♍
30	14	03	☽ ♎

LATITUDES

Date	Mercury ☿ ° '	Venus ♀ ° '	Mars ♂ ° '	Jupiter ♃ ° '	Saturn ♄ ° '	Uranus ♅ ° '	Neptune ♆ ° '	Pluto ♇ ° '
01	01 S 52	00 N 56	00 S 07	00 N 58	02 N 08	00 S 37	00 S 44	02 N 05
04	01 35	01 06	00 05	00 58	02 09	00 37	00 45	02 04
07	01 13	01 16	00 03	00 57	02 09	00 37	00 45	02 04
10	00 47	01 25	00 S 01	00 57	02 10	00 37	00 45	02 04
13	00 S 18	01 35	00 N 01	00 57	02 10	00 37	00 45	02 04
16	00 N 15	01 44	00 03	00 57	02 10	00 37	00 45	02 04
19	00 47	01 53	00 05	00 56	02 10	00 37	00 45	02 04
22	01 19	02 01	00 07	00 56	02 11	00 37	00 45	02 04
25	01 47	02 09	00 09	00 56	02 11	00 37	00 45	02 04
28	02 11	02 16	00 10	00 56	02 11	00 37	00 45	02 04
31	02 N 27	02 N 23	00 N 12	00 N 55	02 N 11	00 S 37	00 S 45	02 N 03

DATA

Julian Date	2457114
Delta T	+69 seconds
Ayanamsa	24° 04' 15"
Synetic vernal point	05° ♓ 02' 44"
True obliquity of ecliptic	23° 26' 06"

MOON'S PHASES, APSIDES AND POSITIONS ☽

Date	h	m	Phase	Longitude ° '	Eclipse Indicator
04	12	06	○	14 ♎ 24	total
12	03	44	☾	21 ♑ 55	
18	18	57	●	28 ♈ 25	
25	23	55	☽	05 ♌ 27	

Day	h	m		
01	12	52	Apogee	
17	03	40	Perigee	
29	03	51	Apogee	
03	01	40	0S	
10	07	44	Max dec	18° S 15'
16	16	41	0N	
22	23	30	Max dec	18° N 17'
30	08	46	0S	

LONGITUDES

Date	Chiron ⚷ ° '	Ceres ⚳ ° '	Pallas ⚴ ° '	Juno ⚵ ° '	Vesta ⚶ ° '	Black Moon Lilith ⚸ ° '
01	18 ♓ 45	29 ♑ 27	29 ♐ 13	04 ♌ 26	01 ♓ 49	13 ♍ 51
11	19 ♓ 10	02 ♒ 04	00 ♑ 10	05 ♌ 58	06 ♓ 32	14 ♍ 58
21	19 ♓ 50	04 ♒ 22	00 ♑ 26	07 ♌ 58	11 ♓ 08	16 ♍ 04
31	20 ♓ 18	06 ♒ 17	29 ♐ 56	10 ♌ 23	15 ♓ 35	17 ♍ 11

All ephemeris data is given at 12.00 UT and the Moon's longitude is additionally given for 24.00 UT

ASPECTARIAN

h m Aspects

01 Wednesday
01 25 ♀ ∥ ♃
03 42 ☽ □ ♄
04 39 ☽ ± ☉
11 31 ☽ ☍ ♆
11 39 ☽ ∥ ♅
14 53 ☽ ± ♅
17 58 ☽ ⊼ ☉
19 21 ☽ ∥ ☉
19 58 ☽ ⊻ ♃
23 52 ☿ ∠ ♀

02 Thursday
01 42 ☽ △ ♇
02 52 ☽ bQ ♂
03 09 ☽ ⊼ ♅
08 08 ☽ ⊥ ♃
09 01 ☽ △ ♀
12 20 ☿ △ ♄
16 12 ☽ Q ♄
17 20 ☉ △ ♃
22 08 ☽ ∥ ☿
22 21 ☽ ± ♂

03 Friday
02 21 ☽ ∠ ♃
07 18 ☽ # ☿
11 17 ☽ ⊼ ♂
15 17 ♀ Q ♆
16 24 ☽ ⚹ ♄
18 02 ♀ # ♄
18 44 ☽ bQ ♀
21 57 ☽ ☍ ☿

04 Saturday
00 27 ☽ ⊼ ♆
08 31 ☽ ⚹ ♃
12 06 ☽ ☍ ☉
12 30 ☽ ± ♆
13 22 ☿ ⊻ ♆
14 18 ☽ □ ♇
14 46 ☽ ± ♀
14 53 ☽ # ☉
15 45 ☽ # ♅
15 58 ☽ ☍ ♅
17 18 ☉ ⊥ ♆
22 22 ☽ ∠ ♄
23 35 ☉ ∥ ♅

05 Sunday
03 40 ♀ ⊥ ♅
04 01 ☽ ⊼ ♀
06 29 ☽ bQ ♆
08 25 ☽ Q ♃
13 36 ☽ ∥ ♆
15 08 ☉ □ ♇
16 10 ☽ ⊥ ♄

06 Monday
01 59 ☽ Q ♇
02 58 ☽ ☍ ♂
03 56 ☽ ⊻ ♄
12 05 ☽ △ ♆
13 26 ☿ △ ♃
14 08 ☉ ☌ ♅
18 17 ♂ ⊼ ♄
19 43 ☽ □ ♃
20 29 ☽ # ♂
20 57 ☽ ⊼ ☿

07 Tuesday
01 23 ☽ ⚹ ♇
03 15 ☽ ⊼ ♅
04 19 ☽ ⊼ ☉
10 47 ☽ ± ☿
14 29 ☿ ⊥ ♆
14 51 ☽ ± ♅
16 50 ☽ ± ☉
20 42 ☽ ☍ ♀

08 Wednesday
00 07 ☿ □ ♇
02 37 ♀ # ♇
06 08 ☽ ∠ ♇
07 17 ☽ bQ ☿
08 05 ☽ bQ ♅
11 24 ☽ bQ ☉
12 20 ☿ ☌ ♅
13 30 ☽ ☌ ♄
16 17 ☽ ⊼ ♂
16 53 ♃ St D
21 38 ☽ □ ♆
23 05 ☽ ⊥ ♇
23 49 ☿ ∥ ♅

09 Thursday
04 12 ☽ ± ♂
04 48 ☽ △ ♃
10 16 ☽ ⊻ ♇
12 17 ☽ △ ♅
12 39 ☽ # ♃
13 31 ☉ bQ ♄
16 40 ☽ △ ☿
17 42 ☽ △ ☉
20 59 ☿ bQ ♄
21 50 ☽ bQ ♂

10 Friday
04 00 ☉ ☌ ☿
06 57 ☽ Q ♆
08 25 ☽ bQ ♃
10 11 ☽ ⊼ ♀
20 35 ☽ ⊻ ♄
22 04 ☽ ± ♀

11 Saturday
00 40 ☽ ± ♃
02 36 ☽ △ ♂
02 39 ☽ # ♃
04 37 ☽ ⚹ ♆
07 16 ☽ ⊥ ♄
11 21 ☽ ⊼ ♃
13 38 ☉ ∥ ☿
15 36 ☽ bQ ♀
16 32 ☽ ☌ ♇
18 41 ☽ □ ♅
23 07 ☽ ∠ ♄

12 Sunday
00 33 ☿ ∠ ♆
02 33 ♀ bQ ♇
03 44 ☽ □ ☉
03 59 ♀ Q ♃
05 40 ☿ # ♆
07 06 ☽ ∠ ♆
08 15 ☽ □ ☿
16 03 ☿ ⊥ ♀
17 22 ♂ ⚹ ♆
20 08 ☽ △ ♀
22 39 ☽ ⊥ ♆

13 Monday
00 48 ☽ # ♂
00 59 ☽ ⚹ ♄
02 00 ☽ Q ♅
05 09 ♀ ∠ ♅
05 30 ☉ # ♆
08 54 ☽ ⊻ ♆
09 43 ☽ □ ♂
12 58 ☽ Q ☉
15 15 ☽ ☍ ♃
20 09 ☽ ⊻ ♇
20 23 ☽ Q ☿
21 14 ☽ Q ♄
22 23 ☽ ⚹ ♅

14 Tuesday
02 05 ☿ ± ♄
04 14 ☉ ∠ ♆
04 59 ☽ # ☿
06 11 ☽ ⊥ ♇
10 33 ☽ ⚹ ☉
13 27 ☽ # ☉
16 27 ☽ ∥ ♆
17 26 ☽ Q ♂
19 45 ☽ ⚹ ☿
21 06 ☽ ∠ ♇
23 23 ☽ ∠ ♅

15 Wednesday
02 51 ☽ □ ♀
02 59 ☽ □ ♄
04 25 ♀ ☍ ♄
08 40 ☽ # ♅
10 52 ☽ ☌ ♆
12 59 ☽ ∠ ☉
14 09 ☽ ⚹ ♂
14 10 ☽ ⊥ ♅
16 56 ☽ ⊼ ♃
21 37 ☽ ⚹ ♇
23 58 ☽ ⊻ ♅

16 Thursday
00 21 ☽ ∠ ☿
02 42 ☽ ± ♃
04 39 ☽ ⊥ ☉
10 39 ☽ Q ♀
15 02 ☽ ⊻ ☉
15 43 ☽ ∠ ♂
17 02 ☽ Q ♇
17 16 ☽ bQ ♃
17 18 ☽ ⊥ ☿
21 06 ☿ ⊼ ♄

17 Friday
03 30 ☽ △ ♄
03 55 ♇ St R
04 34 ☽ ⊻ ☿
06 59 ☽ ⊥ ♂
07 39 ☽ ⚹ ♀
11 30 ☽ ⊻ ♆
17 10 ☽ ⊻ ♂
17 30 ☽ △ ♃
21 12 ☽ ⊥ ♆
22 04 ☽ □ ♇

18 Saturday
00 29 ♂ □ ♃
00 36 ☽ ☌ ♅
00 46 ☽ ∥ ♅
03 37 ☽ bQ ♄
07 47 ☉ ± ♄
09 56 ☽ ∠ ♀
11 48 ☽ ∠ ♆
16 09 ☽ # ♆
18 10 ☽ ± ♄
18 57 ☽ ☌ ☉

19 Sunday
01 50 ☽ ⊥ ♀
03 03 ☿ ⊻ ♀
03 56 ☽ ⊼ ♄
05 36 ☽ ∥ ☉
07 12 ☿ ⚹ ♆
10 29 ♀ □ ♆
12 22 ☽ ⚹ ♆
12 31 ☽ ⊻ ♀
13 12 ☽ ☌ ☿
18 31 ☽ □ ♃
20 39 ☽ ☌ ♂
20 42 ♀ ± ♇
23 07 ☽ △ ♇

20 Monday
01 54 ☽ ⊻ ♅
08 22 ☽ Q ♆
12 00 ☽ ⊥ ♅
15 31 ☿ ∥ ♂

21 Tuesday
00 23 ☽ bQ ♇
00 30 ☽ ⊻ ☉
00 53 ☽ Q ♃
01 57 ☽ ∥ ♂
03 22 ☽ ∠ ♅
04 15 ☿ □ ♃
06 00 ☽ ☍ ♄
07 15 ☽ ∥ ☿
11 36 ☽ ⊥ ☉
15 10 ☽ □ ♆
15 55 ☽ ± ♇
19 53 ☽ ☌ ♀
21 35 ♂ △ ♇
21 43 ☽ ⚹ ♃

22 Wednesday
00 37 ☽ ⊻ ☿
01 23 ☽ ∥ ♃
02 26 ☽ ⊼ ♇
02 43 ☽ ⊻ ♂
04 34 ☽ ∠ ☉
05 38 ☽ ⚹ ♅
13 03 ☽ ⊥ ☿
13 26 ☿ △ ♇
14 00 ☽ ⊥ ♂
15 56 ☿ ∥ ♃
18 41 ♀ ⚹ ♃
23 03 ☿ ☌ ♂

23 Thursday
00 37 ☽ ∠ ♃
03 19 ☽ Q ♅
05 44 ☽ ∥ ☿
07 16 ☽ ∠ ♂
08 09 ☽ ∠ ☿
09 48 ☽ ⚹ ☉
11 11 ☽ ⊼ ♄
13 03 ☿ ⊻ ♅
17 13 ☽ ⊥ ♃
21 22 ☽ △ ♆
22 17 ☽ ± ♄
22 56 ☿ # ♄
22 58 ☽ ∥ ♃

24 Friday
03 17 ☉ ⊼ ♄
04 29 ☽ ⊻ ♃
07 42 ☽ ⊻ ♀
09 21 ☽ ☍ ♇
10 05 ☽ Q ☉
12 32 ☽ ∥ ♂
13 00 ☽ ⚹ ♂
13 02 ☽ □ ♅
13 44 ♂ ⊻ ♅
15 13 ☽ bQ ♄
17 04 ☽ ⚹ ☿
20 17 ☽ ⊥ ♀

25 Saturday
01 54 ♀ ⊼ ♇
01 56 ☽ bQ ♆
13 35 ☽ Q ♂
14 48 ☿ Q ♆
15 31 ☽ ∠ ♀
19 31 ☽ ± ♆
20 09 ☽ △ ♄
20 18 ☽ Q ☿
23 55 ☽ □ ☉

26 Sunday
07 23 ☽ ⊼ ♆
08 07 ☽ ∥ ☉
15 04 ☽ ☌ ♃
18 45 ☿ # ♇
19 56 ☽ ⊼ ♇
21 31 ♀ ⚹ ♅
23 52 ☿ ⊥ ♅

27 Monday
00 06 ☽ △ ♅
00 22 ☽ ⚹ ♀
03 34 ☽ □ ♂
07 59 ☽ ± ♇
09 37 ♂ ∥ ♃
14 12 ☽ □ ☿
21 42 ☽ # ♆

28 Tuesday
02 10 ☽ bQ ♇
03 09 ☽ Q ♀
06 33 ☽ bQ ♅
08 00 ☽ □ ♄
14 50 ☽ ∥ ♅
17 37 ☽ △ ☉
20 00 ☽ ☍ ♆

29 Wednesday
00 56 ☽ ± ♅
04 01 ☽ ⊻ ♃
08 39 ☽ △ ♇
13 12 ☽ ⊼ ♅
16 17 ☽ ⊥ ♃
19 38 ☽ □ ♀
20 21 ☽ △ ♂
20 24 ☽ Q ♄
21 52 ♂ Q ♆
23 11 ☉ ⚹ ♆

30 Thursday
02 50 ☽ bQ ☉
10 38 ☽ ∠ ♃
12 23 ☽ △ ☿
15 14 ♀ ⊻ ♂
20 33 ☽ ⚹ ♄
22 39 ☽ ± ☉

MAY 2015

Date	Sidereal time h m s	Sun ☉ ° ′ ″	Moon ☽ ° ′ ″	Moon ☽ 24.00 ° ′ ″	Mercury ☿ ° ′	Venus ♀ ° ′	Mars ♂ ° ′	Jupiter ♃ ° ′	Saturn ♄ ° ′	Uranus ♅ ° ′	Neptune ♆ ° ′	Pluto ♇ ° ′
01	02 36 24	10 ♉ 48 15	10 ♎ 54 41	16 ♎ 55 33	00 ♊ 35	22 ♊ 48	22 ♉ 27	13 ♌ 23	03 ♐ 11	17 ♈ 51	09 ♓ 21	15 ♑ 30
02	02 40 21	11 46 28	22 ♎ 58 59	29 ♎ 05 14	01 57	23 56	23 10	13 27	03 R 07	17 54	09 22	15 R 29
03	02 44 17	12 44 39	05 ♏ 14 34	11 ♏ 27 07	03 14	25 03	23 53	13 32	03 03	17 57	09 23	15 29
04	02 48 14	13 42 48	17 ♏ 43 03	24 ♏ 02 26	04 27	26 10	24 36	13 36	02 59	18 01	09 25	15 28
05	02 52 10	14 40 56	00 ♐ 25 18	06 ♐ 51 41	05 35	27 17	25 18	13 41	02 55	18 04	09 26	15 28
06	02 56 07	15 39 02	13 ♐ 21 32	19 ♐ 54 48	06 39	28 23	26 01	13 46	02 51	18 07	09 27	15 27
07	03 00 03	16 37 07	26 ♐ 31 25	03 ♑ 11 18	07 38	29 ♊ 30	26 44	13 51	02 46	18 10	09 28	15 27
08	03 04 00	17 35 10	09 ♑ 54 22	16 ♑ 40 33	08 33	00 ♋ 36	27 26	13 56	02 42	18 13	09 29	15 26
09	03 07 56	18 33 11	23 ♑ 29 44	00 ♒ 21 50	09 23	01 42	28 09	14 01	02 38	18 16	09 30	15 25
10	03 11 53	19 31 12	07 ♒ 16 48	14 ♒ 14 31	10 07	02 48	28 52	14 07	02 34	18 19	09 32	15 25
11	03 15 50	20 29 11	21 ♒ 14 55	28 ♒ 17 52	10 47	03 54	29 ♉ 34	14 12	02 29	18 22	09 33	15 24
12	03 19 46	21 27 09	05 ♓ 23 14	12 ♓ 30 51	11 22	04 59	00 ♊ 16	14 18	02 25	18 25	09 34	15 23
13	03 23 43	22 25 05	19 ♓ 40 28	26 ♓ 51 47	11 53	06 05	00 59	14 24	02 21	18 28	09 35	15 23
14	03 27 39	23 23 00	04 ♈ 04 28	11 ♈ 18 05	12 18	07 10	01 41	14 30	02 16	18 31	09 35	15 22
15	03 31 36	24 20 54	18 ♈ 32 06	25 ♈ 45 57	12 38	08 15	02 24	14 36	02 12	18 34	09 36	15 21
16	03 35 32	25 18 47	02 ♉ 59 00	10 ♉ 10 34	12 53	09 19	03 06	14 42	02 08	18 37	09 37	15 20
17	03 39 29	26 16 38	17 ♉ 19 59	24 ♉ 26 31	13 03	10 24	03 48	14 49	02 03	18 40	09 38	15 19
18	03 43 25	27 14 28	01 ♊ 29 33	08 ♊ 28 27	13 08	11 28	04 30	14 55	01 59	18 43	09 39	15 19
19	03 47 22	28 12 17	15 ♊ 22 42	22 ♊ 11 53	13 R 08	12 32	05 12	15 02	01 54	18 46	09 40	15 18
20	03 51 19	29 ♉ 10 04	28 ♊ 55 41	05 ♋ 33 54	13 04	13 35	05 54	15 09	01 50	18 49	09 41	15 17
21	03 55 15	00 ♊ 07 50	12 ♋ 06 31	18 ♋ 33 33	12 55	14 39	06 36	15 16	01 45	18 51	09 41	15 16
22	03 59 12	01 05 34	24 ♋ 55 12	01 ♌ 11 44	12 42	15 42	07 18	15 23	01 41	18 54	09 42	15 15
23	04 03 08	02 03 16	07 ♌ 23 32	13 ♌ 31 04	12 25	16 45	08 00	15 30	01 36	18 57	09 43	15 14
24	04 07 05	03 00 57	19 ♌ 34 50	25 ♌ 35 26	12 04	17 47	08 42	15 38	01 32	18 59	09 43	15 13
25	04 11 01	03 58 37	01 ♍ 33 29	07 ♍ 29 36	11 40	18 49	09 24	15 45	01 27	19 02	09 44	15 12
26	04 14 58	04 56 15	13 ♍ 24 28	19 ♍ 18 44	11 13	19 51	10 06	15 53	01 23	19 05	09 44	15 11
27	04 18 54	05 53 51	25 ♍ 13 04	01 ♎ 08 08	10 43	20 53	10 47	16 01	01 18	19 07	09 45	15 10
28	04 22 51	06 51 26	07 ♎ 04 33	13 ♎ 02 54	10 12	21 54	11 29	16 09	01 14	19 10	09 45	15 09
29	04 26 48	07 48 59	19 ♎ 03 45	25 ♎ 07 37	09 39	22 55	12 11	16 17	01 10	19 12	09 46	15 08
30	04 30 44	08 46 32	01 ♏ 14 55	07 ♏ 26 03	09 05	23 56	12 52	16 25	01 05	19 15	09 46	15 07
31	04 34 41	09 ♊ 44 03	13 ♏ 41 19	20 ♏ 00 57	08 ♊ 31	24 ♋ 56	13 ♊ 34	16 ♌ 33	01 ♐ 01	19 ♈ 17	09 ♓ 47	15 ♑ 05

Date	Moon True ☊ ° ′	Moon Mean ☊ ° ′	Moon ☽ Latitude ° ′
01	09 ♎ 49	08 ♎ 33	00 N 06
02	09 R 49	08 30	01 12
03	09 46	08 27	02 15
04	09 42	08 24	03 12
05	09 37	08 21	04 02
06	09 31	08 18	04 39
07	09 25	08 14	05 03
08	09 20	08 11	05 11
09	09 16	08 08	05 01
10	09 13	08 05	04 35
11	09 D 13	08 02	03 51
12	09 13	07 58	02 53
13	09 15	07 55	01 44
14	09 15	07 52	00 N 28
15	09 R 15	07 49	00 S 50
16	09 13	07 46	02 05
17	09 09	07 43	03 11
18	09 03	07 39	04 04
19	08 55	07 36	04 42
20	08 47	07 33	05 03
21	08 39	07 30	05 07
22	08 32	07 27	04 55
23	08 27	07 23	04 29
24	08 24	07 20	03 52
25	08 23	07 17	03 05
26	08 D 23	07 14	02 10
27	08 24	07 11	01 10
28	08 24	07 08	00 S 07
29	08 R 24	07 04	00 N 57
30	08 21	07 01	01 59
31	08 ♎ 16	06 ♎ 58	02 N 57

DECLINATIONS

Date	Sun ☉ ° ′	Moon ☽ ° ′	Mercury ☿ ° ′	Venus ♀ ° ′	Mars ♂ ° ′	Jupiter ♃ ° ′	Saturn ♄ ° ′	Uranus ♅ ° ′	Neptune ♆ ° ′	Pluto ♇ ° ′
01	15 N 04	04 S 14	22 N 40	25 N 37	18 N 35	17 N 41	18 S 39	06 N 26	08 S 46	20 S 29
02	15 22	07 50	23 00	25 42	18 46	17 40	18 38	06 27	08 45	20 30
03	15 40	11 09	23 19	25 47	18 58	17 38	18 37	06 28	08 45	20 30
04	15 57	14 02	23 34	25 51	19 09	17 37	18 36	06 29	08 44	20 30
05	16 14	16 18	23 48	25 55	19 20	17 36	18 35	06 30	08 44	20 30
06	16 31	17 47	23 59	25 58	19 30	17 34	18 35	06 32	08 44	20 30
07	16 48	18 20	24 08	26 00	19 41	17 33	18 34	06 33	08 43	20 30
08	17 05	17 54	24 15	26 01	19 51	17 31	18 33	06 34	08 43	20 30
09	17 21	16 26	24 19	26 02	20 01	17 29	18 32	06 35	08 42	20 30
10	17 37	14 01	24 22	26 02	20 11	17 28	18 31	06 36	08 42	20 30
11	17 52	10 46	24 23	26 02	20 21	17 26	18 30	06 37	08 42	20 30
12	18 07	06 51	24 21	26 01	20 31	17 24	18 30	06 39	08 41	20 31
13	18 22	02 S 29	24 18	25 59	20 40	17 23	18 29	06 40	08 41	20 31
14	18 37	02 N 03	24 13	25 57	20 50	17 21	18 28	06 41	08 41	20 31
15	18 51	06 29	24 06	25 54	20 59	17 19	18 27	06 42	08 40	20 31
16	19 05	10 33	23 58	25 50	21 07	17 17	18 26	06 43	08 40	20 31
17	19 19	13 57	23 48	25 46	21 16	17 15	18 25	06 44	08 40	20 31
18	19 32	16 28	23 36	25 41	21 24	17 13	18 25	06 45	08 40	20 31
19	19 45	17 58	23 23	25 35	21 33	17 11	18 24	06 46	08 39	20 32
20	19 58	18 23	23 08	25 29	21 41	17 09	18 23	06 47	08 39	20 32
21	20 10	17 48	22 52	25 22	21 48	17 07	18 22	06 48	08 39	20 32
22	20 22	16 18	22 34	25 15	21 56	17 04	18 21	06 49	08 39	20 32
23	20 34	14 05	22 16	25 07	22 03	17 02	18 20	06 50	08 38	20 32
24	20 45	11 16	21 56	24 59	22 11	17 00	18 19	06 51	08 38	20 32
25	20 56	08 02	21 36	24 50	22 18	16 58	18 19	06 52	08 38	20 33
26	21 07	04 31	21 15	24 40	22 24	16 55	18 18	06 53	08 38	20 33
27	21 17	00 N 49	20 53	24 30	22 31	16 53	18 17	06 54	08 38	20 33
28	21 27	02 S 55	20 32	24 20	22 37	16 50	18 16	06 55	08 37	20 33
29	21 37	06 35	20 10	24 09	22 43	16 48	18 15	06 56	08 37	20 33
30	21 46	10 03	19 48	23 57	22 49	16 45	18 15	06 57	08 37	20 33
31	21 N 54	13 S 08	19 N 27	23 N 45	22 N 55	16 N 43	18 S 14	06 N 58	08 S 37	20 S 34

ZODIAC SIGN ENTRIES

Date	h	m	Planets
01	02	00	☿ ♊
03	01	47	☽ ♏
05	11	13	☽ ♐
07	18	16	☽ ♑
07	22	52	♀ ♋
09	23	22	☽ ♒
12	02	40	♂ ♊
12	02	53	☽ ♓
14	05	13	☽ ♈
16	07	02	☽ ♉
18	09	27	☽ ♊
20	13	56	☽ ♋
21	08	45	☉ ♊
22	21	42	☽ ♌
25	08	52	☽ ♍
27	21	42	☽ ♎
30	09	34	☽ ♏

LATITUDES

Date	Mercury ☿ ° ′	Venus ♀ ° ′	Mars ♂ ° ′	Jupiter ♃ ° ′	Saturn ♄ ° ′	Uranus ♅ ° ′	Neptune ♆ ° ′	Pluto ♇ ° ′
01	02 N 27	02 N 23	00 N 12	00 N 55	02 N 11	00 S 37	00 S 45	02 N 03
04	02 35	02 29	00 14	00 55	02 11	00 37	00 45	02 03
07	02 35	02 34	00 16	00 55	02 11	00 37	00 46	02 03
10	02 26	02 38	00 18	00 55	02 11	00 37	00 46	02 03
13	02 07	02 41	00 20	00 54	02 11	00 37	00 46	02 03
16	01 38	02 44	00 21	00 54	02 11	00 37	00 46	02 03
19	01 01	02 45	00 23	00 54	02 11	00 37	00 46	02 03
22	00 N 16	02 46	00 25	00 54	02 11	00 38	00 46	02 03
25	00 S 35	02 45	00 26	00 53	02 11	00 38	00 46	02 02
28	01 28	02 43	00 28	00 53	02 11	00 38	00 46	02 02
31	02 S 18	02 N 39	00 N 30	00 N 53	02 N 10	00 S 38	00 S 46	02 N 02

DATA

Julian Date	2457144
Delta T	+69 seconds
Ayanamsa	24° 04' 18"
Synetic vernal point	05° ♓ 02' 41"
True obliquity of ecliptic	23° 26' 05"

MOON'S PHASES, APSIDES AND POSITIONS ☽

Date	h	m	Phase	Longitude ° ′	Eclipse Indicator
04	03	42	○	13 ♏ 23	
11	10	36	☾	20 ♒ 26	
18	04	13	●	26 ♉ 56	
25	17	19	☽	04 ♍ 11	

Day	h	m		
15	00	08	Perigee	
26	22	11	Apogee	
07	13	38	Max dec	18° S 21'
14	01	11	ON	
20	09	43	Max dec	18° N 23'
27	17	17	OS	

LONGITUDES

Date	Chiron ⚷ ° ′	Ceres ⚳ ° ′	Pallas ⚴ ° ′	Juno ⚵ ° ′	Vesta ⚶ ° ′	Black Moon Lilith ⚸ ° ′
01	20 ♓ 18	06 ♒ 17	29 ♐ 56	10 ♌ 23	15 ♓ 35	17 ♍ 11
11	20 ♓ 42	07 ♒ 46	28 ♐ 41	13 ♌ 07	19 ♓ 50	18 ♍ 17
21	21 ♓ 03	08 ♒ 45	26 ♐ 41	16 ♌ 07	23 ♓ 54	19 ♍ 24
31	21 ♓ 18	09 ♒ 12	24 ♐ 08	19 ♌ 20	27 ♓ 43	20 ♍ 30

All ephemeris data is given at 12.00 UT and the Moon's longitude is additionally given for 24.00 UT

ASPECTARIAN

h m	Aspects
01 Friday	
04 37	☽ ⚼ ♂
08 52	☽ ⚻ ♆
10 25	☿ ⚼ ♇
11 46	☽ ⚻ ☉
16 59	☽ ⚹ ♃
19 25	♂ # ♄
20 53	☽ ± ♆
21 09	☽ □ ♇
22 33	☽ ⚼ ☿
23 45	☽ ± ♂
02 Saturday	
01 54	☽ ☍ ♅
02 25	☽ ∠ ♄
02 36	☽ # ♅
02 40	☿ Q ♃
12 23	☽ ⚻ ♂
14 03	☽ △ ♀
14 44	☽ ⚼ ♆
16 54	☽ Q ♃
18 27	☽ ∥ ♆
18 33	☽ ± ☿
20 05	☽ ⊥ ♄
03 Sunday	
06 26	☿ ∠ ♅
07 39	☽ ⚻ ☿
07 45	☽ ⚺ ♄
08 35	☽ Q ♇
08 35	☿ ☍ ♄
14 46	♂ ⊥ ♅
20 03	☽ △ ♆
22 13	☽ ⚼ ♀
04 Monday	
03 42	☽ ☍ ☉
04 05	☽ □ ♃
06 12	♄ ⚼ ♅
07 43	☽ ⚹ ♇
09 02	☉ □ ♃
12 34	☽ ⚻ ♅
17 07	☽ ± ♀
05 Tuesday	
00 00	☽ ± ♅
01 49	☽ ☍ ♂
05 33	☽ ⚻ ♀
11 10	☽ # ☉
12 05	☽ ∠ ♇
16 38	☽ ☌ ♄
16 57	☽ ⚼ ♅
22 32	☽ ☍ ☿
06 Wednesday	
04 47	☽ □ ♆
04 48	☽ ⊥ ♇
07 11	☉ △ ♇
07 39	☽ # ♃
12 45	☽ △ ♃
15 51	☽ ⚺ ♇
16 33	☽ ⚻ ☉
20 38	♀ ∠ ♃
20 45	☽ △ ♅
07 Thursday	
04 22	☽ ± ☉
12 24	☽ ⚻ ♂
13 43	☽ Q ♆
16 13	☽ ⚼ ♃
17 51	☽ ☍ ♀
21 54	☽ ⚼ ☉
23 12	☽ ⚺ ♄
23 48	☽ ± ♂
08 Friday	
03 13	♀ Q ♅
08 27	☽ ± ♃
09 25	☽ ⚻ ☿
09 52	☽ ⊥ ♄
11 15	☽ ⚹ ♆
16 45	☽ ⚼ ♂
19 11	☽ ⚻ ♃
20 18	☽ # ♃
20 47	☽ ± ☿
21 48	☽ ☌ ♇
09 Saturday	
01 09	☽ # ☉
01 45	☽ ∠ ♄
02 39	☽ △ ☉
02 47	☽ □ ♅
04 35	☉ ⚺ ♅
13 27	☿ ± ♇
13 38	☽ ⚼ ☿
13 47	☽ ∠ ♆
16 08	☿ □ ♆
20 35	☽ △ ♂
10 Sunday	
00 02	☉ ∥ ♃
03 34	☽ ⚻ ♀
03 52	☽ ⚹ ♄
05 29	☽ ⊥ ♆
07 01	♀ ⚻ ♄
10 20	☽ Q ♅
14 52	☽ ± ♀
15 53	☽ ⚺ ♆
17 10	☽ △ ☿
23 51	☽ ☍ ♃
11 Monday	
00 29	☽ Q ♄
02 00	☽ ⚺ ♇
07 04	☽ ⚹ ♅
07 39	☽ ⚼ ♀
10 36	☽ □ ☉
12 16	☽ ⊥ ♇
12 Tuesday	
01 06	☽ ∥ ♆
02 54	☽ □ ♂
03 33	☽ ∠ ♇
07 00	☽ □ ♄
08 40	☽ ∠ ♅
11 16	☽ △ ♀
11 18	♂ # ♇
13 08	☽ # ♅
14 42	☉ Q ♆
15 51	♂ ⚼ ♇
19 02	☽ ☌ ♆
19 21	☽ Q ☉
22 28	☽ □ ☿
23 53	☽ ⊥ ♅
13 Wednesday	
03 06	☽ ⚻ ♃
04 49	☽ ⚹ ♇
09 59	☽ ⚺ ♅
10 47	☽ Q ♂
13 13	☽ ± ♃
16 55	☽ ⚹ ☉
21 56	☉ # ♄
14 Thursday	
00 51	☽ Q ♇
04 20	☽ ⚼ ♃
05 32	☽ Q ☿
07 50	☽ ⚹ ♂
09 01	☽ △ ♄
17 33	☽ □ ♀
19 40	☽ ∠ ☉
21 10	☽ ⚺ ♆
15 Friday	
01 59	☽ ⚹ ☿
05 25	☽ △ ♃
06 03	♂ ☍ ♄
06 44	☽ □ ♇
07 08	☽ ⊥ ♆
09 48	☽ ⚼ ♄
10 00	☽ ∠ ♂
11 04	♀ ± ♄
11 40	☽ ⊥ ☉
12 04	☽ ☌ ♅
13 10	☽ ∥ ♅
17 49	☉ ⊥ ♅
20 19	♂ Q ♃
20 47	♀ ⊥ ♃
21 31	♀ ⊥ ♂
22 05	☽ ∠ ♆
22 20	☽ ⚺ ☉
16 Saturday	
00 30	☽ # ♆
00 39	☽ ± ♄
01 42	☽ ⊥ ♂
01 49	☽ Q ♀
03 23	☽ ∠ ☿
10 35	☽ ⚻ ♄
12 12	☽ ⚺ ♂
18 35	☽ ⊥ ☿
18 49	♀ △ ♆
23 05	☽ ⚹ ♆
23 25	☽ ⚹ ♀
17 Sunday	
04 44	☽ ⚺ ☿
07 09	♂ ∠ ♅
07 44	☽ □ ♃
08 38	☽ △ ♇
14 15	☽ ⚺ ♅
19 16	☽ Q ♆
18 Monday	
00 25	☽ ⊥ ♅
02 44	☽ ∠ ♀
04 13	☽ ☌ ☉
09 59	☽ ⚼ ♇
12 49	☽ ☍ ♄
14 28	☽ Q ♃
15 49	☽ ∠ ♅
17 26	☽ ☌ ♂
19 22	☽ ⊥ ♀
21 46	☽ ∥ ♃
19 Tuesday	
01 26	☽ ± ♇
01 50	☿ St R
02 03	☽ □ ♆
06 37	☽ ⚺ ♀
08 06	☽ ☌ ☿
11 24	☽ ⚹ ♃
11 51	☽ ⚻ ♇
17 57	☽ ⚹ ♅
20 Wednesday	
01 11	☿ ⚺ ♀
07 10	☽ # ♄
12 28	☽ ⚺ ☉
12 53	☽ # ♄
14 13	☽ ∠ ♃
15 24	☽ Q ♅
17 12	☽ ⚻ ♄
21 Thursday	
00 10	☽ ⊥ ☉
01 20	☽ ⚺ ♂
04 02	☽ ± ♄
06 43	☽ ⊥ ♃
07 32	☽ △ ♆
11 39	♃ ⚻ ♇
12 58	☽ ⊥ ♂
13 29	☽ ⚺ ☿
15 19	☉ ⚼ ♇
17 07	☽ ☌ ♀
17 51	☽ ☍ ♇
17 55	☽ ⚺ ♃
18 03	☽ ∠ ☉
20 34	☽ ⚼ ♄
22 Friday	
00 29	☽ ⊥ ☿
00 36	☽ □ ♅
00 58	☽ ∥ ♃
01 57	♀ ☍ ♇
04 00	♀ ⚺ ♃
06 45	☽ ∠ ♂
11 35	☽ ⚼ ♆
17 12	☽ ∠ ☿
23 Saturday	
00 47	☽ ⚹ ☉
00 51	☽ △ ♄
01 35	☉ ☍ ♄
04 51	☽ ± ♆
07 59	☉ # ♇
09 01	♀ ⚼ ♄
13 16	☽ ⚹ ♂
16 32	☽ ⚻ ♆
21 34	☽ ⚹ ☿
23 20	☿ ∥ ♂
24 Sunday	
02 11	☽ Q ☉
03 22	☽ ⚻ ♇
04 05	☽ ☌ ♃
08 06	☽ ⚺ ♀
10 49	☽ △ ♅
14 22	☽ Q ♂
15 15	☽ ± ♇
16 45	☿ ⊥ ♀
20 40	☽ Q ☿
21 11	☽ ⊥ ♀
25 Monday	
05 22	♂ ± ♇
05 41	☉ Q ♃
07 47	☽ # ♆
09 16	☽ ⚼ ♇
11 48	☽ □ ♄
13 33	☉ ∠ ♅
17 01	☽ ∠ ♀
17 01	☽ ⚼ ♅
17 09	♀ □ ♅
17 19	☽ □ ☉
20 05	☽ ∥ ♅
23 38	♂ □ ♆
26 Tuesday	
04 33	☽ ☍ ♆
04 51	☽ □ ♂
07 43	☽ □ ☿
11 20	☽ ± ♅
15 36	☽ △ ♇
17 05	☽ ⚺ ♃
18 04	☉ ∥ ☿
23 34	☽ ⚻ ♅
27 Wednesday	
00 04	☽ Q ♄
02 21	☽ ⚹ ♀
05 25	☽ ⊥ ♃
10 37	☿ ☌ ♂
17 44	☉ ∠ ♀
23 53	☽ ∠ ♃
28 Thursday	
00 16	☽ ⚹ ♄
05 00	☽ Q ♀
10 31	☿ # ♇
11 31	☽ △ ☉
17 24	☽ ⚻ ♆
18 00	☽ △ ☿
21 25	☽ △ ♂
29 Friday	
04 10	☽ □ ♇
05 26	☽ ± ♆
06 15	☽ ∠ ♄
06 24	☽ ⚹ ♃
07 00	☿ □ ♆
12 17	☽ ☍ ♅
14 22	☽ # ♅
20 05	☽ ⚼ ☉
20 20	☽ □ ♀
22 34	☽ ⚼ ☿
23 18	☽ ⚼ ♆
23 59	☽ ⊥ ♄
30 Saturday	
01 52	☽ ∥ ♆
05 00	☽ ⚼ ♂
06 24	☽ Q ♃
11 02	☿ ± ♇
11 41	☽ ⚺ ♄
14 26	☿ ∠ ♀
15 13	☽ ± ☉
15 26	☽ ± ☿
15 37	☽ Q ♇
16 56	☉ ☌ ☿
20 13	☉ ± ♇
23 33	☽ ± ♂
31 Sunday	
02 32	☽ ⚻ ☿
03 48	☽ ⚻ ☉
04 31	☽ △ ♆
08 16	♀ ⚼ ♆
11 45	☽ ⚻ ♂
13 08	☉ □ ♆
14 40	☽ ⚹ ♇
17 31	☽ □ ♃
22 40	☽ ⚻ ♅

LONGITUDES

Date	Sidereal time h m s	Sun ☉ ° ' "	Moon ☽ ° ' "	Moon ☽ 24.00 ° ' "	Mercury ☿ ° '	Venus ♀ ° '	Mars ♂ ° '	Jupiter ♃ ° '	Saturn ♄ ° '	Uranus ♅ ° '	Neptune ♆ ° '	Pluto ♇ ° '
01	04 38 37	10 ♊ 41 33	26 ♏ 25 04	02 ♐ 53 43	07 ♊ 58	25 ♋ 56	14 ♊ 15	16 ♌ 42	00 ♐ 56	19 ♈ 20	09 ♓ 47	15 ♑ 04
02	04 42 34	11 39 01	09 ♐ 26 52	16 ♐ 04 21	07 R 26	26 55	14 57	16 50	00 R 52	19 22	09 47	15 R 03
03	04 46 30	12 36 29	22 ♐ 45 59	29 ♐ 31 29	06 55	27 54	15 38	16 59	00 48	19 25	09 48	15 02
04	04 50 27	13 33 56	06 ♑ 20 28	13 ♑ 12 35	06 26	28 53	16 19	17 08	00 43	19 27	09 48	15 01
05	04 54 23	14 31 22	20 ♑ 07 25	27 ♑ 04 33	05 59	29 ♋ 51	17 01	17 16	00 39	19 29	09 48	15 00
06	04 58 20	15 28 47	04 ♒ 03 35	11 ♒ 04 07	05 36	00 ♌ 49	17 42	17 25	00 35	19 32	09 48	14 58
07	05 02 17	16 26 11	18 ♒ 05 50	25 ♒ 08 26	05 16	01 47	18 23	17 34	00 31	19 34	09 49	14 57
08	05 06 13	17 23 35	02 ♓ 11 40	09 ♓ 15 19	04 59	02 44	19 04	17 44	00 26	19 36	09 49	14 56
09	05 10 10	18 20 58	16 ♓ 19 14	23 ♓ 23 15	04 47	03 40	19 45	17 53	00 22	19 38	09 49	14 54
10	05 14 06	19 18 21	00 ♈ 27 15	07 ♈ 31 07	04 38	04 36	20 27	18 02	00 18	19 40	09 49	14 53
11	05 18 03	20 15 43	14 ♈ 34 41	21 ♈ 37 48	04 34	05 32	21 08	18 12	00 14	19 42	09 49	14 52
12	05 21 59	21 13 05	28 ♈ 40 16	05 ♉ 41 48	04 D 34	06 27	21 49	18 21	00 10	19 44	09 R 49	14 51
13	05 25 56	22 10 26	12 ♉ 42 08	19 ♉ 40 55	04 39	07 21	22 30	18 31	00 06	19 46	09 49	14 49
14	05 29 52	23 07 47	26 ♉ 37 44	03 ♊ 32 13	04 49	08 15	23 10	18 41	00 ♐ 02	19 48	09 49	14 48
15	05 33 49	24 05 08	10 ♊ 23 56	17 ♊ 12 26	05 02	09 08	23 51	18 50	29 ♏ 58	19 50	09 49	14 46
16	05 37 46	25 02 28	23 ♊ 57 22	00 ♋ 38 21	05 21	10 01	24 32	19 00	29 54	19 52	09 49	14 45
17	05 41 42	25 59 47	07 ♋ 15 06	13 ♋ 47 25	05 44	10 53	25 13	19 10	29 51	19 54	09 49	14 44
18	05 45 39	26 57 05	20 ♋ 15 10	26 ♋ 38 19	06 11	11 45	25 54	19 20	29 47	19 56	09 48	14 42
19	05 49 35	27 54 23	02 ♌ 56 55	09 ♌ 11 07	06 43	12 36	26 34	19 31	29 43	19 57	09 48	14 41
20	05 53 32	28 51 40	15 ♌ 21 11	21 ♌ 27 27	07 19	13 26	27 15	19 41	29 39	19 59	09 48	14 39
21	05 57 28	29 ♊ 48 57	27 ♌ 30 19	03 ♍ 30 16	08 00	14 15	27 56	19 51	29 36	20 01	09 48	14 38
22	06 01 25	00 ♋ 46 13	09 ♍ 27 50	15 ♍ 23 38	08 45	15 04	28 36	20 02	29 32	20 02	09 47	14 37
23	06 05 21	01 43 28	21 ♍ 18 16	27 ♍ 12 24	09 34	15 52	29 17	20 12	29 29	20 04	09 47	14 35
24	06 09 18	02 40 42	03 ♎ 06 44	09 ♎ 01 55	10 27	16 39	29 ♊ 57	20 23	29 26	20 06	09 47	14 34
25	06 13 15	03 37 56	14 ♎ 58 39	20 ♎ 57 38	11 24	17 26	00 ♋ 38	20 34	29 22	20 07	09 46	14 32
26	06 17 11	04 35 09	26 ♎ 59 30	03 ♏ 04 51	12 25	18 11	01 18	20 44	29 19	20 09	09 46	14 31
27	06 21 08	05 32 22	09 ♏ 14 17	15 ♏ 28 17	13 30	18 56	01 59	20 55	29 16	20 10	09 45	14 29
28	06 25 04	06 29 34	21 ♏ 47 18	28 ♏ 11 39	14 39	19 39	02 39	21 06	29 13	20 11	09 45	14 28
29	06 29 01	07 26 46	04 ♐ 41 34	11 ♐ 17 09	15 52	20 22	03 19	21 17	29 10	20 13	09 44	14 26
30	06 32 57	08 ♋ 23 58	17 ♐ 58 25	24 ♐ 45 11	17 ♊ 08	21 ♌ 04	04 ♋ 00	21 ♌ 28	29 ♏ 07	20 ♈ 14	09 ♓ 44	14 ♑ 25

Date	Moon True ☊ ° '	Moon Mean ☊ ° '	Moon ☽ Latitude ° '
01	08 ♎ 09	06 ♎ 55	03 N 47
02	08 R 00	06 52	04 27
03	07 50	06 49	04 53
04	07 39	06 45	05 03
05	07 29	06 42	04 56
06	07 22	06 39	04 31
07	07 16	06 36	03 50
08	07 14	06 33	02 55
09	07 13	06 29	01 49
10	07 D 13	06 26	00 N 36
11	07 R 13	06 23	00 S 39
12	07 11	06 20	01 51
13	07 08	06 17	02 56
14	07 01	06 14	03 50
15	06 52	06 10	04 30
16	06 41	06 07	04 54
17	06 28	06 04	05 01
18	06 16	06 01	04 52
19	06 06	05 58	04 29
20	05 57	05 55	03 54
21	05 51	05 51	03 08
22	05 48	05 48	02 15
23	05 47	05 45	01 16
24	05 D 46	05 42	00 S 14
25	05 R 46	05 39	00 N 49
26	05 45	05 35	01 50
27	05 42	05 32	02 47
28	05 37	05 29	03 38
29	05 29	05 26	04 19
30	05 ♎ 19	05 ♎ 23	04 N 47

DECLINATIONS

Date	Sun ☉ ° '	Moon ☽ ° '	Mercury ☿ ° '	Venus ♀ ° '	Mars ♂ ° '	Jupiter ♃ ° '	Saturn ♄ ° '	Uranus ♅ ° '	Neptune ♆ ° '	Pluto ♇ ° '
01	22 N 03	15 S 40	19 N 06	23 N 33	23 N 00	16 N 40	18 S 13	06 N 59	08 S 37	20 S 34
02	22 11	17 28	18 47	23 19	23 06	16 38	18 12	07 00	08 37	20 34
03	22 18	18 22	18 28	23 06	23 11	16 35	18 11	07 01	08 37	20 34
04	22 25	18 14	18 11	22 52	23 16	16 32	18 11	07 02	08 37	20 34
05	22 32	17 03	17 55	22 38	23 20	16 30	18 10	07 03	08 37	20 35
06	22 39	14 50	17 41	22 23	23 25	16 27	18 09	07 03	08 37	20 35
07	22 45	11 45	17 28	22 08	23 29	16 24	18 08	07 04	08 37	20 35
08	22 50	07 58	17 18	21 52	23 33	16 21	18 08	07 05	08 36	20 35
09	22 55	03 S 44	17 09	21 36	23 37	16 18	18 07	07 06	08 36	20 35
10	23 00	00 N 44	17 03	21 20	23 40	16 15	18 06	07 07	08 36	20 36
11	23 05	05 09	16 58	21 04	23 44	16 12	18 05	07 07	08 37	20 36
12	23 09	09 16	16 55	20 47	23 47	16 09	18 05	07 08	08 37	20 36
13	23 12	12 51	16 55	20 29	23 50	16 06	18 04	07 09	08 37	20 36
14	23 15	15 40	16 56	20 12	23 53	16 03	18 03	07 09	08 37	20 37
15	23 18	17 33	16 59	19 54	23 55	16 00	18 03	07 10	08 37	20 37
16	23 20	18 24	17 04	19 36	23 57	15 57	18 02	07 11	08 37	20 37
17	23 22	18 14	17 11	19 17	23 59	15 54	18 01	07 12	08 37	20 37
18	23 24	17 05	17 19	18 59	24 01	15 51	18 01	07 12	08 37	20 37
19	23 25	15 08	17 29	18 40	24 03	15 47	18 00	07 13	08 37	20 38
20	23 26	12 30	17 40	18 21	24 04	15 44	17 59	07 13	08 37	20 38
21	23 26	09 23	17 53	18 02	24 06	15 41	17 59	07 14	08 37	20 38
22	23 26	05 56	18 07	17 42	24 07	15 37	17 58	07 15	08 37	20 38
23	23 25	02 N 17	18 22	17 23	24 07	15 34	17 58	07 15	08 38	20 39
24	23 24	01 S 27	18 38	17 03	24 08	15 31	17 57	07 16	08 38	20 39
25	23 23	05 09	18 55	16 44	24 08	15 27	17 57	07 16	08 38	20 39
26	23 21	08 41	19 12	16 24	24 09	15 24	17 56	07 17	08 38	20 39
27	23 19	11 55	19 30	16 04	24 09	15 20	17 56	07 17	08 38	20 40
28	23 17	14 42	19 49	15 44	24 08	15 17	17 55	07 18	08 39	20 40
29	23 14	16 49	20 08	15 24	24 08	15 13	17 55	07 18	08 39	20 40
30	23 N 10	18 S 07	20 N 27	15 N 04	24 N 07	15 N 09	17 S 54	07 N 19	08 S 39	20 S 40

ZODIAC SIGN ENTRIES

Date	h	m	Planets
01	18	39	☽ ♐
04	00	50	☽ ♑
05	15	33	♀ ♌
06	05	02	☽ ♒
08	08	16	☽ ♓
10	11	14	☽ ♈
12	14	16	☽ ♉
14	17	51	☽ ♊
15	00	36	♄ ♏
16	22	51	☽ ♋
19	06	22	☽ ♌
21	16	38	☉ ♋
21	16	59	☽ ♍
24	05	41	☽ ♎
24	13	33	♂ ♋
26	17	57	☽ ♏
29	03	21	☽ ♐

LATITUDES

Date	Mercury ☿ ° '	Venus ♀ ° '	Mars ♂ ° '	Jupiter ♃ ° '	Saturn ♄ ° '	Uranus ♅ ° '	Neptune ♆ ° '	Pluto ♇ ° '
01	02 S 33	02 N 38	00 N 30	00 N 53	02 N 10	00 S 38	00 S 46	02 N 02
04	03 15	02 33	00 32	00 53	02 10	00 38	00 47	02 02
07	03 46	02 26	00 34	00 53	02 10	00 38	00 47	02 02
10	04 05	02 18	00 35	00 52	02 09	00 38	00 47	02 01
13	04 13	02 08	00 37	00 52	02 09	00 38	00 47	02 01
16	04 11	01 56	00 38	00 52	02 09	00 38	00 47	02 01
19	04 00	01 43	00 40	00 52	02 08	00 38	00 47	02 01
22	03 41	01 28	00 41	00 52	02 08	00 38	00 47	02 00
25	03 16	01 10	00 42	00 52	02 07	00 38	00 47	02 00
28	02 45	00 51	00 44	00 52	02 06	00 38	00 47	02 00
31	02 S 11	00 N 30	00 N 45	00 N 52	02 N 06	00 S 38	00 S 47	02 N 00

DATA

Julian Date	2457175
Delta T	+69 seconds
Ayanamsa	24° 04' 22"
Synetic vernal point	05° ♓ 02' 37"
True obliquity of ecliptic	23° 26' 05"

MOON'S PHASES, APSIDES AND POSITIONS ☽

Date	h	m	Phase	Longitude ° '	Eclipse Indicator
02	16	19	○	11 ♐ 49	
09	15	42	☾	18 ♓ 30	
16	14	05	●	25 ♊ 07	
24	11	03	☽	02 ♎ 38	

Day	h	m		
10	04	40	Perigee	
23	17	01	Apogee	
03	21	12	Max dec	18° S 26'
10	08	05	0N	
16	19	44	Max dec	18° N 27'
24	02	41	0S	

LONGITUDES

Date	Chiron ⚷ ° '	Ceres ⚳ ° '	Pallas ⚴ ° '	Juno ⚵ ° '	Vesta ⚶ ° '	Black Moon Lilith ⚸ ° '
01	21 ♓ 19	09 ♒ 12	23 ♐ 51	19 ♌ 39	28 ♓ 05	20 ♍ 37
11	21 ♓ 29	09 ♒ 01	20 ♐ 57	23 ♌ 03	01 ♈ 35	21 ♍ 44
21	21 ♓ 33	08 ♒ 13	18 ♐ 04	26 ♌ 34	04 ♈ 46	22 ♍ 50
31	21 ♓ 32	06 ♒ 51	15 ♐ 31	00 ♍ 12	07 ♈ 33	23 ♍ 57

ASPECTARIAN

h m	Aspects
01 Monday	
09 58	☽ ± ♅
11 01	☽ △ ♀
18 47	☽ ∠ ♇
20 21	☽ ☌ ♄
23 47	☽ ⋕ ♃
02 Tuesday	
02 41	☽ bQ ♅
08 27	☽ ☍ ☿
11 17	☽ ⊥ ♇
12 37	☽ □ ♆
15 40	♂ ⚻ ♇
16 19	☽ ☍ ☉
16 52	☽ bQ ♀
22 09	☽ ⊻ ♇
22 31	☽ ☍ ♂
03 Wednesday	
01 31	☽ △ ♃
04 42	☽ ∥ ♄
05 48	♀ ∥ ♂
05 59	☽ △ ♅
10 21	☽ ± ♀
16 18	☽ ⋕ ☿
20 57	☽ Q ♆
21 51	☽ ⚻ ♀
04 Thursday	
02 10	☽ ⊻ ♄
04 31	☽ bQ ♃
12 09	☽ ⚻ ☿
12 33	☿ ⋕ ♄
12 40	☽ ⊥ ♄
14 07	☽ ∥ ♄
14 56	☽ ⋕ ☿
18 03	☽ ✶ ♆
20 27	☽ ± ♃
22 18	☽ ± ☿
05 Friday	
01 34	☽ ⚻ ☉
03 07	☽ ☌ ♇
04 17	☽ ∠ ♄
06 19	☽ ⚻ ♂
07 00	☽ ⚻ ♃
10 54	☽ □ ♅
12 44	☽ ± ☉
13 27	☽ bQ ☿
17 15	☽ ± ♂
18 18	☉ ∥ ♀
19 18	☽ ⋕ ♃
20 05	☽ ∠ ♆
23 32	☉ ⚻ ♇
23 38	♂ ✶ ♃
06 Saturday	
05 24	☽ bQ ☉
06 02	☽ ☍ ♀
06 03	☽ ✶ ♄
06 23	♀ △ ♄
09 33	☽ bQ ♂
11 34	☽ ⊥ ♆
14 34	☽ △ ☿
17 57	☽ Q ♅
20 28	☿ Q ♃
21 51	☽ ⊻ ♆
07 Sunday	
02 31	☽ Q ♄
06 38	☽ ⊻ ♇
08 58	☽ △ ☉
11 06	☽ ☍ ♃
12 31	☽ △ ♂
14 30	☽ ✶ ♅
16 51	☽ ⊥ ♇
08 Monday	
08 09	☽ ∠ ♇
08 11	☽ ∥ ♆
09 02	☽ □ ♄
12 58	☽ ⚻ ♀
16 06	☽ ∠ ♅
16 40	☽ □ ☿
17 10	☽ ⋕ ♅
21 56	☉ ✶ ♃
23 54	☽ ± ♀
09 Tuesday	
00 57	☽ ☌ ♆
07 26	☽ ⊥ ♅
07 30	♂ ✶ ♅
09 36	☽ ✶ ♇
14 41	☽ ⚻ ♃
15 42	☽ □ ☉
15 45	♀ ± ♆
16 16	☽ bQ ♀
17 39	☽ ⊻ ♅
18 08	☽ □ ♂
22 51	☽ Q ☿
10 Wednesday	
00 59	☽ ± ♃
05 57	☽ Q ♇
06 52	☿ ∠ ♅
11 45	☽ △ ♄
12 52	☿ ✶ ♀
16 26	☽ bQ ♃
19 04	☽ ✶ ☿
19 32	☽ △ ♀
21 30	☉ ✶ ♅
11 Thursday	
00 29	☽ Q ☉
02 16	☽ Q ♂
03 54	☽ ⊻ ♆
12 29	☽ □ ♇
13 07	☽ bQ ♄
14 06	☽ ⊥ ♆
18 13	☽ △ ♃
20 29	☽ ∠ ☿
20 45	☽ ☌ ♅
22 22	☽ ✶ ☉
22 33	☿ St D
23 15	☽ ∥ ♅
23 43	☽ ✶ ♂
12 Friday	
04 21	☽ ± ♄
05 26	☽ ∠ ♆
08 00	☽ ⋕ ♆
09 05	♆ St R
11 50	☽ ⊥ ☿
14 32	☽ ⚻ ♄
22 07	☽ ⊻ ☿
13 Saturday	
01 50	☽ ∠ ☉
02 12	☽ □ ♀
02 32	♀ ⋕ ♇
02 37	☽ ∠ ♂
07 03	☽ ✶ ♆
15 38	☽ △ ♇
18 24	☽ ⊥ ☉
18 51	☽ ⊥ ♂
22 06	☽ □ ♃
14 Sunday	
00 11	☽ ⊻ ♅
03 29	♀ ∠ ♂
03 41	☽ Q ♆
05 30	☽ ⊻ ☉
05 43	☽ ⊻ ♂
10 34	☽ ⊥ ♅
11 18	☽ Q ♀
11 20	☿ ∠ ♅
15 56	☉ ☌ ♂
16 00	☽ ∥ ♃
17 29	☽ bQ ♇
17 53	☽ ☍ ♄
15 Monday	
02 15	☽ ∠ ♅
02 27	☽ ☌ ☿
03 03	☽ ∥ ☿
05 41	☽ Q ♃
09 09	☽ ± ♇
09 38	☽ ✶ ♀
10 59	☽ □ ♆
19 41	☽ ⚻ ♇
22 10	☽ ⋕ ♄
16 Tuesday	
03 05	☽ ✶ ♃
04 42	☽ ✶ ♅
05 31	☉ ∠ ♀
06 21	♀ ⚻ ♆
13 06	☽ ☌ ♂
14 02	☽ ∠ ♀
14 05	☽ ☌ ☉
22 38	☽ ⚻ ♄
17 Wednesday	
02 15	☽ Q ♅
06 19	☽ ∠ ♃
07 24	☽ ⊥ ♀
09 08	☽ ⊻ ☿
09 27	☽ ± ♄
16 41	☽ △ ♆
18 12	☽ ⋕ ♄
19 08	☽ ⊻ ♀
20 29	☽ ⊥ ☿
23 00	☽ ⊥ ♃
18 Thursday	
01 43	☽ ☍ ♇
01 53	☽ bQ ♄
08 40	☽ ∥ ☿
10 17	☽ ⊻ ♃
11 24	☽ □ ♅
13 49	☽ ∠ ☿
20 33	☽ bQ ♆
23 12	☽ ⊻ ♂
19 Friday	
01 37	☽ ⊻ ☉
04 39	☽ ∥ ♃
05 52	☽ △ ♄
11 15	☽ ⊥ ♂
13 38	☽ ± ♆
13 59	☽ ⊥ ☉
19 36	☽ ✶ ☿
20 Saturday	
01 12	☽ ⚻ ♆
05 36	☽ ∠ ♂
07 59	☽ ☌ ♀
08 50	☽ ∠ ☉
10 39	☽ ⚻ ♇
11 45	♄ ∠ ♇
20 14	☽ Q ☿
20 37	☽ ☌ ♃
21 07	☽ △ ♅
22 24	☽ ± ♇
21 Sunday	
05 26	☿ Q ♃
06 51	☉ ⚻ ♄
12 54	☽ ✶ ♂
15 52	♀ ⋕ ♄
16 09	☽ □ ♄
16 14	☽ bQ ♇
17 01	☽ ✶ ☉
17 31	☽ ⋕ ♆
18 21	☿ ∥ ♀
21 36	☿ ⋕ ♄
22 48	♀ ⚻ ♇
22 Monday	
03 04	☽ bQ ♅
03 10	☽ ∥ ♅
07 54	☿ ± ♇
10 27	☽ □ ☿
12 39	☽ ☍ ♆
13 46	♃ △ ♅
14 27	☽ Q ♂
19 16	☽ Q ☉
21 17	☽ ± ♅
22 23	☽ △ ♇
23 Tuesday	
00 10	☽ ⊻ ♀
04 16	☽ Q ♄
09 29	☽ ⚻ ♅
09 44	☽ ⊻ ♃
13 14	☽ ⊥ ♀
18 08	☿ □ ♆
18 36	♂ ⚻ ♄
20 52	☉ Q ♅
22 07	☽ ⊥ ♃
24 Wednesday	
04 33	☽ ✶ ♄
05 12	☽ □ ♂
08 50	☽ ∠ ♀
11 03	☽ □ ☉
16 41	☽ ∠ ♃
25 Thursday	
01 30	☽ ⚻ ♆
04 08	☽ △ ☿
09 16	♃ ± ♇
10 47	☽ ∠ ♄
11 07	☽ □ ♇
13 36	☽ ± ♆
17 16	☽ ✶ ♀
22 20	☽ ☍ ♅
23 22	☽ ✶ ♃
26 Friday	
02 14	☽ ⋕ ♅
04 44	☽ ⊥ ♄
07 35	☽ bQ ♆
11 38	☽ ∥ ♆
12 55	☽ bQ ☿
16 35	☽ ⊻ ♄
18 43	☽ Q ♀
21 01	☽ △ ♂
22 52	☽ Q ♇
23 30	☽ Q ♃
27 Saturday	
04 12	☽ △ ☉
05 26	☉ ± ♄
08 18	☽ ± ☿
13 00	☽ △ ♆
18 57	♂ Q ♅
21 02	☽ ⚻ ☿
22 06	☽ ✶ ♇
23 53	☉ ∠ ♃
28 Sunday	
03 43	☽ bQ ♂
07 44	☽ □ ♀
08 20	☿ ⚻ ♇
08 58	☽ ⚻ ♅
10 41	☽ □ ♃
11 24	☽ bQ ☉
17 40	☽ ⋕ ♃
20 16	☽ ± ♅
21 13	☽ ⋕ ♀
21 37	☽ ± ♂
29 Monday	
01 50	☽ ☌ ♄
02 20	☽ ∠ ♇
05 33	☽ ± ☉
06 26	♀ △ ♅
09 21	☽ ⚻ ♂
12 57	☽ bQ ♅
14 22	♀ ± ♇
17 26	☽ ⚻ ☉
18 50	☽ ⊥ ♇
21 12	☽ □ ♆
30 Tuesday	
03 19	♀ ∥ ♃
05 39	☽ ⊻ ♇
06 31	☽ ∥ ♄
10 20	☽ ☍ ☿
16 01	☽ △ ♅
17 47	☽ △ ♀
18 18	☽ △ ♃

All ephemeris data is given at 12.00 UT and the Moon's longitude is additionally given for 24.00 UT

JULY 2015

Date	Sidereal time h m s	Sun ☉ ° ' "	Moon ☽ ° ' "	Moon ☽ 24.00 ° ' "	Mercury ☿ ° '	Venus ♀ ° '	Mars ♂ ° '	Jupiter ♃ ° '	Saturn ♄ ° '	Uranus ♅ ° '	Neptune ♆ ° '	Pluto ♇ ° '
01	06 36 54	09 ♋ 21 09	01 ♑ 37 10	08 ♑ 33 59	18 ♊ 28	21 ♌ 44	04 ♋ 40	21 ♌ 39	29 ♏ 04	20 ♈ 15	09 ♓ 43	14 ♑ 23
02	06 40 50	10 18 21	15 ♑ 35 05	22 ♑ 39 52	19 52	22 24	05 20	21 51	29 R 01	20 16	09 R 43	14 R 22
03	06 44 47	11 15 32	29 ♑ 47 38	06 ♒ 57 40	21 20	23 03	06 00	22 02	28 58	20 17	09 42	14 20
04	06 48 44	12 12 43	14 ♒ 09 14	21 ♒ 21 37	22 51	23 40	06 40	22 13	28 56	20 18	09 41	14 19
05	06 52 40	13 09 54	28 ♒ 34 11	05 ♓ 46 18	24 25	24 16	07 20	22 25	28 53	20 19	09 41	14 17
06	06 56 37	14 07 05	12 ♓ 57 30	20 ♓ 07 21	26 03	24 51	08 00	22 36	28 51	20 20	09 40	14 16
07	07 00 33	15 04 17	27 ♓ 15 31	04 ♈ 21 46	27 45	25 24	08 40	22 48	28 48	20 21	09 39	14 14
08	07 04 30	16 01 29	11 ♈ 25 55	18 ♈ 27 54	29 ♊ 29	25 57	09 20	22 59	28 46	20 22	09 38	14 13
09	07 08 26	16 58 41	25 ♈ 27 36	02 ♉ 25 00	01 ♋ 17	26 27	10 00	23 11	28 44	20 23	09 37	14 12
10	07 12 23	17 55 54	09 ♉ 20 04	16 ♉ 12 47	03 08	26 57	10 40	23 23	28 41	20 24	09 37	14 10
11	07 16 19	18 53 08	23 ♉ 03 04	29 ♉ 50 53	05 02	27 25	11 20	23 34	28 39	20 25	09 36	14 09
12	07 20 16	19 50 21	06 ♊ 36 07	13 ♊ 18 40	06 59	27 51	12 00	23 46	28 37	20 25	09 35	14 07
13	07 24 13	20 47 36	19 ♊ 58 24	26 ♊ 35 08	08 57	28 16	12 39	23 58	28 36	20 26	09 34	14 06
14	07 28 09	21 44 50	03 ♋ 08 45	09 ♋ 39 05	10 58	28 39	13 19	24 10	28 34	20 27	09 33	14 04
15	07 32 06	22 42 05	16 ♋ 06 00	22 ♋ 29 24	13 02	29 00	13 59	24 22	28 32	20 27	09 32	14 03
16	07 36 02	23 39 21	28 ♋ 49 14	05 ♌ 05 29	15 06	29 20	14 38	24 34	28 30	20 28	09 31	14 01
17	07 39 59	24 36 37	11 ♌ 18 11	17 ♌ 27 29	17 12	29 38	15 18	24 46	28 29	20 28	09 30	14 00
18	07 43 55	25 33 53	23 ♌ 33 31	29 ♌ 36 32	19 19	29 ♌ 54	15 58	24 58	28 27	20 29	09 29	13 58
19	07 47 52	26 31 09	05 ♍ 36 51	11 ♍ 34 51	21 27	00 ♍ 07	16 37	25 10	28 26	20 29	09 28	13 57
20	07 51 48	27 28 25	17 ♍ 30 56	23 ♍ 25 36	23 35	00 19	17 17	25 23	28 25	20 29	09 27	13 55
21	07 55 45	28 25 42	29 ♍ 19 24	05 ♎ 12 53	25 43	00 29	17 56	25 35	28 24	20 30	09 26	13 54
22	07 59 42	29 ♋ 22 59	11 ♎ 06 41	17 ♎ 01 26	27 51	00 37	18 36	25 47	28 22	20 30	09 24	13 52
23	08 03 38	00 ♌ 20 16	22 ♎ 57 48	28 ♎ 56 27	29 ♋ 59	00 42	19 15	26 00	28 22	20 30	09 23	13 51
24	08 07 35	01 17 34	04 ♏ 58 05	11 ♏ 03 21	02 ♌ 06	00 45	19 55	26 12	28 21	20 30	09 22	13 50
25	08 11 31	02 14 52	17 ♏ 12 52	23 ♏ 27 15	04 12	00 R 46	20 34	26 24	28 20	20 30	09 21	13 48
26	08 15 28	03 12 11	29 ♏ 47 03	06 ♐ 12 42	06 17	00 45	21 13	26 37	28 19	20 R 30	09 20	13 47
27	08 19 24	04 09 29	12 ♐ 44 35	19 ♐ 22 57	08 21	00 41	21 52	26 49	28 19	20 30	09 18	13 45
28	08 23 21	05 06 49	26 ♐ 07 55	02 ♑ 59 27	10 24	00 35	22 32	27 02	28 18	20 30	09 17	13 44
29	08 27 17	06 04 09	09 ♑ 57 20	17 ♑ 01 13	12 25	00 26	23 11	27 14	28 18	20 30	09 16	13 43
30	08 31 14	07 01 29	24 ♑ 10 35	01 ♒ 24 44	14 25	00 15	23 50	27 27	28 17	20 30	09 15	13 41
31	08 35 11	07 ♌ 58 51	08 ♒ 42 51	16 ♒ 04 03	16 ♌ 23	00 ♍ 02	24 ♋ 29	27 ♌ 40	28 ♏ 17	20 ♈ 30	09 ♓ 13	13 ♑ 40

Date	Moon True ☊ ° '	Moon Mean ☊ ° '	Moon ☽ Latitude ° '
01	05 ♎ 07	05 ♎ 20	05 N 00
02	04 R 56	05 16	04 56
03	04 45	05 13	04 33
04	04 36	05 10	03 53
05	04 30	05 07	02 58
06	04 27	05 04	01 51
07	04 25	05 01	00 N 38
08	04 D 25	04 57	00 S 37
09	04 R 25	04 54	01 49
10	04 24	04 51	02 54
11	04 20	04 48	03 48
12	04 14	04 45	04 28
13	04 05	04 41	04 53
14	03 54	04 38	05 02
15	03 42	04 35	04 55
16	03 31	04 32	04 34
17	03 20	04 29	03 59
18	03 12	04 26	03 14
19	03 06	04 22	02 21
20	03 02	04 19	01 22
21	03 01	04 16	00 S 20
22	03 D 01	04 13	00 N 43
23	03 02	04 10	01 45
24	03 R 02	04 06	02 42
25	03 01	04 03	03 34
26	02 58	04 00	04 16
27	02 52	03 57	04 47
28	02 45	03 54	05 04
29	02 37	03 51	05 03
30	02 28	03 47	04 44
31	02 ♎ 20	03 ♎ 44	04 N 07

DECLINATIONS

Date	Sun ☉ ° '	Moon ☽ ° '	Mercury ☿ ° '	Venus ♀ ° '	Mars ♂ ° '	Jupiter ♃ ° '	Saturn ♄ ° '	Uranus ♅ ° '	Neptune ♆ ° '	Pluto ♇ ° '
01	23 N 06	18 S 25	20 N 46	14 N 43	24 N 06	15 N 06	17 S 54	07 N 19	08 S 39	20 S 41
02	23 02	17 38	21 04	14 23	24 05	15 02	17 53	07 20	08 40	20 41
03	22 57	15 44	21 22	14 03	24 04	14 59	17 53	07 20	08 40	20 41
04	22 52	12 52	21 40	13 43	24 02	14 55	17 53	07 20	08 40	20 41
05	22 47	09 11	21 57	13 23	24 01	14 51	17 52	07 21	08 40	20 42
06	22 41	04 59	22 13	13 04	23 59	14 47	17 52	07 21	08 41	20 42
07	22 35	00 S 31	22 28	12 44	23 57	14 44	17 52	07 21	08 41	20 42
08	22 28	03 N 57	22 42	12 25	23 55	14 40	17 51	07 22	08 41	20 42
09	22 21	08 09	22 54	12 05	23 52	14 36	17 51	07 22	08 42	20 43
10	22 14	11 51	23 04	11 46	23 49	14 32	17 51	07 22	08 42	20 43
11	22 06	14 51	23 13	11 27	23 46	14 28	17 51	07 23	08 42	20 43
12	21 58	17 00	23 19	11 08	23 43	14 24	17 50	07 23	08 43	20 43
13	21 50	18 11	23 23	10 50	23 40	14 20	17 50	07 23	08 43	20 44
14	21 41	18 22	23 25	10 32	23 37	14 16	17 50	07 23	08 44	20 44
15	21 31	17 35	23 24	10 14	23 33	14 12	17 50	07 23	08 44	20 44
16	21 22	15 56	23 21	09 57	23 29	14 08	17 50	07 24	08 44	20 44
17	21 12	13 33	23 15	09 40	23 25	14 04	17 50	07 24	08 45	20 45
18	21 01	10 37	23 06	09 23	23 21	14 00	17 50	07 24	08 45	20 45
19	20 51	07 16	22 54	09 07	23 17	13 56	17 49	07 24	08 46	20 45
20	20 40	03 N 40	22 40	08 51	23 12	13 52	17 49	07 24	08 46	20 45
21	20 28	00 S 02	22 23	08 36	23 07	13 48	17 49	07 24	08 47	20 46
22	20 17	03 44	22 04	08 21	23 02	13 44	17 49	07 24	08 47	20 46
23	20 04	07 18	21 42	08 07	22 57	13 39	17 49	07 24	08 47	20 46
24	19 52	10 37	21 18	07 54	22 52	13 35	17 49	07 24	08 48	20 47
25	19 39	13 33	20 51	07 41	22 46	13 31	17 49	07 24	08 48	20 47
26	19 26	15 55	20 23	07 29	22 41	13 27	17 50	07 24	08 49	20 47
27	19 13	17 34	19 53	07 18	22 35	13 23	17 50	07 24	08 49	20 47
28	18 59	18 19	19 21	07 07	22 29	13 18	17 50	07 24	08 50	20 48
29	18 45	18 01	18 47	06 58	22 23	13 14	17 50	07 24	08 50	20 48
30	18 31	16 36	18 12	06 49	22 16	13 10	17 50	07 24	08 51	20 48
31	18 N 16	14 S 06	17 N 36	06 N 41	22 N 10	13 N 05	17 S 50	07 N 24	08 S 51	20 S 48

ZODIAC SIGN ENTRIES

Date	h	m	Planets
01	09	11	☽ ♑
03	12	21	☽ ♒
05	14	23	☽ ♓
07	16	37	☽ ♈
08	18	52	☿ ♋
09	19	49	☽ ♉
12	00	16	☽ ♊
14	06	14	☽ ♋
16	14	15	☽ ♌
18	22	38	♀ ♍
19	00	47	☽ ♍
21	13	23	☽ ♎
23	03	30	☉ ♌
23	12	14	☿ ♌
24	02	07	☽ ♏
26	12	24	☽ ♐
28	18	47	☽ ♑
30	21	40	☽ ♒
31	15	27	♀ ♌

LATITUDES

Date	Mercury ☿ ° '	Venus ♀ ° '	Mars ♂ ° '	Jupiter ♃ ° '	Saturn ♄ ° '	Uranus ♅ ° '	Neptune ♆ ° '	Pluto ♇ ° '
01	02 S 11	00 N 30	00 N 45	00 N 52	02 N 06	00 S 38	00 S 47	02 N 00
04	01 34	00 N 06	00 47	00 52	02 05	00 38	00 48	01 59
07	00 57	00 S 20	00 48	00 51	02 05	00 39	00 48	01 59
10	00 S 19	00 49	00 49	00 51	02 04	00 39	00 48	01 59
13	00 N 16	01 19	00 50	00 51	02 03	00 39	00 48	01 58
16	00 46	01 53	00 52	00 51	02 03	00 39	00 48	01 58
19	01 12	02 28	00 53	00 51	02 02	00 39	00 48	01 58
22	01 30	03 06	00 54	00 51	02 01	00 39	00 48	01 57
25	01 42	03 45	00 55	00 51	02 01	00 39	00 48	01 57
28	01 47	04 25	00 56	00 51	02 00	00 39	00 48	01 57
31	01 N 45	05 S 06	00 N 58	00 N 51	01 N 59	00 S 39	00 S 48	01 N 56

DATA

Julian Date	2457205
Delta T	+69 seconds
Ayanamsa	24° 04' 27"
Synetic vernal point	05° ♓ 02' 32"
True obliquity of ecliptic	23° 26' 04"

LONGITUDES

Date	Chiron ⚷ ° '	Ceres ⚳ ° '	Pallas ⚴ ° '	Juno ⚵ ° '	Vesta ⚶ ° '	Black Moon Lilith ⚸ ° '
01	21 ♓ 32	06 ♒ 51	15 ♐ 31	00 ♍ 12	07 ♈ 33	23 ♍ 57
11	21 ♓ 26	05 ♒ 02	13 ♐ 32	03 ♍ 54	09 ♈ 54	25 ♍ 03
21	21 ♓ 14	02 ♒ 56	12 ♐ 15	07 ♍ 41	11 ♈ 42	26 ♍ 09
31	20 ♓ 58	00 ♒ 44	11 ♐ 41	11 ♍ 30	12 ♈ 54	27 ♍ 16

MOON'S PHASES, APSIDES AND POSITIONS ☽

Date	h	m	Phase	Longitude ° '	Eclipse Indicator
02	02	20	○	09 ♑ 55	
08	20	24	☾	16 ♈ 22	
16	01	24	●	23 ♋ 14	
24	04	04	☽	00 ♏ 59	
31	10	43	○	07 ♒ 56	

Day	h	m		
05	19	03	Perigee	
21	11	05	Apogee	
01	06	48	Max dec	18° S 27'
07	14	43	0N	
14	04	20	Max dec	18° N 25'
21	11	47	0S	
28	17	29	Max dec	18° S 21'

ASPECTARIAN

h m	Aspects	h m	Aspects	h m	Aspects
01 Wednesday		07 21	☽ ⊻ ♅	11 07	☉ △ ♄
05 13	☽ Q ♆	08 35	☽ ∥ ♃	13 20	☽ Q ♂
05 35	☿ # ♇	09 27	☽ Q ♆	14 24	☽ ⊻ ♀
07 34	☽ ⊻ ♄	12 56	☽ □ ♃	16 41	☽ ⊥ ♃
07 51	♀ ☌ ♃	17 56	☽ ⊥ ♅	**22 Wednesday**	
17 32	☽ ☍ ♂	18 05	☽ ∠ ♂	02 45	☽ ⊥ ♀
17 57	☽ ⊥ ♄	19 57	☽ □ ♀	08 33	☽ ⊼ ♆
20 50	☽ ᵽ ♃	21 52	☽ ☍ ♄	08 52	☽ Q ☿
21 10	☉ △ ♆	22 44	☽ ᵽ ♇	11 20	☽ ∠ ♃
21 18	☽ ᵽ ♀	**12 Sunday**		12 36	☽ Q ☉
02 Thursday		00 19	☽ ⊥ ☿	16 36	☽ ∠ ♄
01 30	♂ ± ♄	08 37	☽ ∠ ☉	17 36	☽ □ ♇
01 58	☽ ✶ ♆	09 54	☽ ∠ ♅	17 51	☿ △ ♄
02 20	☽ ☍ ☉	10 52	☽ ⊥ ♂	20 43	☽ ± ♆
06 51	☽ ∥ ♄	12 47	☽ ⊻ ☿	21 13	☽ ∠ ♀
09 20	☽ ∠ ♄	14 42	☽ ± ♇	**23 Thursday**	
09 56	☽ ☌ ♇	17 19	☽ □ ♆	04 04	☽ □ ♂
12 27	☽ ± ♃	21 22	☽ Q ♃	07 02	☽ ☍ ♅
13 27	☽ ± ♀	22 08	☽ ⊻ ♂	10 47	☽ ⊥ ♄
18 50	☿ ✶ ♅	**13 Monday**		12 41	☽ # ♅
19 58	☽ □ ♅	01 25	☽ ⊼ ♇	14 52	☽ ᵽ ♆
20 05	☽ ⊼ ☿	01 57	☽ ⊥ ☉	17 22	☽ # ♀
22 45	☽ ⊼ ♃	02 31	☽ # ♄	18 12	☽ ✶ ♃
03 Friday		02 53	☉ □ ♅	19 24	☉ ☌ ☿
00 06	☽ ⊼ ♀	05 06	☽ Q ♀	20 28	☿ ⊻ ♀
03 26	☽ ∠ ♆	12 09	☿ ∠ ♃	21 53	☉ ⊻ ♀
07 22	☽ ± ☿	12 50	☽ ✶ ♅	22 30	☽ ∥ ♆
10 38	☽ ✶ ♄	13 36	☽ ⊻ ☉	22 49	☽ ⊻ ♄
18 32	☽ ⊥ ♆	19 14	☿ △ ♆	**24 Friday**	
19 20	☽ # ♃	19 21	☽ ✶ ♃	03 36	☽ ✶ ♀
22 54	☽ ⊼ ♂	**14 Tuesday**		04 04	☽ □ ☉
04 Saturday		03 31	☽ ✶ ♀	05 05	☽ □ ☿
00 13	☽ ᵽ ☿	03 37	☽ ⊼ ♄	05 46	☽ Q ♇
00 53	☿ ✶ ♃	06 48	♀ □ ♄	18 30	☽ Q ♃
02 14	☽ Q ♅	10 43	☽ Q ♅	20 40	☽ △ ♆
04 34	☽ ⊻ ♆	14 36	☽ ± ♄	**25 Saturday**	
04 46	☽ # ♀	20 27	♂ ᵽ ♄	02 23	☿ ± ♆
06 38	☽ Q ♄	23 17	☽ ∠ ♃	03 22	☽ Q ♀
08 32	☽ ⊼ ☉	23 48	☽ △ ♆	03 30	♂ ⊥ ♃
09 24	☽ ± ♂	**15 Wednesday**		05 23	☽ ✶ ♇
12 16	☽ ⊻ ♇	05 10	☽ ☌ ☿	09 29	♀ St R
19 14	☽ ± ☉	06 44	☽ # ♄	09 43	♂ □ ♅
22 15	☽ ⊥ ♇	07 13	☽ ᵽ ♄	11 44	☽ # ♃
22 16	☽ ✶ ♅	07 50	☽ ☌ ♂	15 56	☿ # ♇
05 Sunday		07 59	☽ ∠ ♀	18 21	☽ ⊼ ♅
01 07	☽ ᵽ ♂	08 10	☽ ☍ ♇	18 49	☽ △ ♂
01 37	☽ ☍ ♃	13 55	♀ ∠ ♂	**26 Sunday**	
04 14	☽ △ ☿	14 12	♂ ☍ ♇	05 48	☽ ± ♅
04 32	☽ ☍ ♀	14 28	♀ ᵽ ♇	05 55	☽ □ ♃
08 19	☿ ✶ ♀	16 19	☽ ⊥ ♃	09 14	☽ ☌ ♄
11 17	☽ ᵽ ☉	17 49	☿ ᵽ ♄	10 07	☽ ∠ ♇
12 31	☽ □ ♄	20 10	☽ □ ♅	10 40	♅ St R
13 12	☽ ∠ ♇	23 40	☿ ☍ ♇	13 48	☽ □ ♀
15 05	☽ ∥ ♆	23 47	☉ ⊥ ♀	15 05	☉ ± ♆
15 45	♂ ∠ ♃	**16 Thursday**		18 55	☽ △ ☉
22 47	☽ # ♅	01 19	☽ ⊥ ♀	22 30	♀ ∥ ♅
23 16	☽ ∠ ♅	01 24	☽ ☌ ☉	22 41	☽ ᵽ ♅
06 Monday		01 32	☿ ∠ ♀	**27 Monday**	
03 19	☽ △ ♂	03 48	☽ ⊻ ♃	00 39	☽ ᵽ ♂
05 23	☉ ᵽ ♄	03 50	☽ ᵽ ♆	02 26	☽ △ ☿
06 30	☽ ☌ ♆	04 14	☿ ☌ ♂	02 52	☽ ⊥ ♇
14 05	☽ △ ☉	11 24	☽ △ ♄	05 43	☽ □ ♆
14 11	☽ ✶ ♇	13 00	☽ ⊻ ♀	13 51	☽ ⊻ ♇
14 19	☽ ⊥ ♅	20 58	☽ ± ♆	17 33	☽ ∥ ♄
15 38	☉ ☍ ♇	**17 Friday**		17 59	☽ ± ♂
07 Tuesday		07 07	☽ ∥ ♃	23 06	☿ ⊼ ♆
00 23	☽ ⊻ ♅	08 31	☽ ⊼ ♆	**28 Tuesday**	
04 23	☽ ⊼ ♃	09 17	☉ ᵽ ♆	00 29	☽ ᵽ ☉
08 45	☽ ⊼ ♀	17 03	☉ ⊻ ♃	02 00	☽ △ ♅
10 17	☽ Q ♇	17 13	☽ ⊼ ♇	05 18	☽ ⊼ ♂
12 56	☽ □ ☿	20 14	☽ ⊻ ♂	10 28	☽ ᵽ ☿
14 36	☽ △ ♄	**18 Saturday**		13 37	☽ △ ♃
14 37	☽ ± ♃	01 54	☽ ⊻ ☿	14 02	☽ Q ♆
19 17	☽ ± ♀	04 57	☽ ± ♇	15 49	☽ ⊻ ♄
19 50	☉ ∥ ☿	05 55	☽ △ ♅	17 38	☽ ± ☉
08 Wednesday		07 39	☿ ⊥ ♃	19 44	☽ △ ♀
02 19	☿ ⊼ ♄	08 40	☽ ⊥ ♂	**29 Wednesday**	
06 04	☽ ᵽ ♃	14 50	☽ ☌ ♃	02 16	☽ ⊥ ♄
08 16	☽ □ ♂	16 13	☽ ⊥ ☿	04 50	☽ ⊼ ☉
08 57	☽ ⊻ ♆	16 18	☽ ⊻ ☉	04 54	☽ ± ☿
11 08	☽ ᵽ ♀	21 41	☽ □ ♄	10 49	☽ ✶ ♆
15 58	☽ ᵽ ♄	21 53	☽ ∥ ♀	14 26	☉ ∥ ☿
16 44	☽ □ ♇	22 42	☽ ᵽ ♇	15 58	☽ ᵽ ♃
19 10	☽ ⊥ ♆	**19 Sunday**		16 39	☽ ∥ ♄
20 24	☽ □ ☉	00 49	☽ ☌ ♀	16 54	☽ ⊼ ☿
22 42	♂ △ ♆	01 07	☿ □ ♅	17 41	☽ ∠ ♄
23 50	☽ Q ☿	01 35	☽ # ♆	18 23	☽ ☌ ♇
09 Thursday		03 33	☽ ∠ ♂	21 13	☽ ᵽ ♀
03 17	☽ ☌ ♅	05 16	☽ ⊥ ☉	**30 Thursday**	
07 19	☽ ± ♄	11 05	☽ ∥ ♅	03 21	☿ ⊼ ♇
07 20	☽ ∥ ♅	11 44	☽ ᵽ ♅	05 51	☽ □ ♅
08 02	☽ △ ♃	14 02	☽ ∠ ☿	07 23	☽ ± ♃
10 34	☽ ∠ ♆	19 43	☽ ☍ ♆	11 24	☽ ☍ ♂
13 47	☽ △ ♀	23 50	☉ # ♇	12 07	☽ ∠ ♆
15 19	☽ # ♆	**20 Monday**		12 08	☽ ± ♀
16 36	☽ Q ♂	00 55	☽ ∠ ☉	17 32	☽ ⊼ ♃
17 37	☽ ⊼ ♄	04 44	☽ △ ♇	18 50	☽ ✶ ♄
18 26	☉ ⊥ ♃	05 52	☽ ± ♅	21 57	☽ ⊼ ♀
23 35	☽ ✶ ☿	09 46	☽ Q ♄	23 44	♀ ⊥ ♂
10 Friday		11 30	☽ ✶ ♂	**31 Friday**	
02 25	☿ Q ♅	18 02	☽ ⊼ ♅	02 29	♂ ᵽ ♆
05 39	☽ Q ☉	19 42	♀ # ♆	02 36	☿ # ♄
11 27	☽ ∥ ♀	21 04	☿ ⊥ ♀	03 00	☽ ⊥ ♆
12 29	☽ ✶ ♆	21 38	☿ ᵽ ♆	10 43	☽ ☍ ☉
14 26	☽ ✶ ♂	**21 Tuesday**		11 38	☽ Q ♅
20 24	☽ △ ♇	03 02	☽ ✶ ☿	12 50	☽ ⊻ ♆
11 Saturday		04 15	☽ ⊻ ♃	14 34	☽ Q ♄
04 08	☽ ✶ ☉	10 01	☽ ✶ ☉	19 54	☽ # ♃
05 50	☽ ∠ ☿	10 06	☽ ✶ ♄	20 05	☽ ⊻ ♇
07 21	☿ ± ♄	10 18	☿ ⊻ ♃		

All ephemeris data is given at 12.00 UT and the Moon's longitude is additionally given for 24.00 UT

AUGUST 2015

LONGITUDES

Date	Sidereal time h m s	Sun ☉ ° ' "	Moon ☽ ° ' "	Moon ☽ 24.00 ° ' "	Mercury ☿ ° '	Venus ♀ ° '	Mars ♂ ° '	Jupiter ♃ ° '	Saturn ♄ ° '	Uranus ♅ ° '	Neptune ♆ ° '	Pluto ♇ ° '
01	08 39 07	08 ♌ 56 13	23 ♒ 27 22	00 ♓ 51 46	18 ♌ 20	29 ♌ 46	25 ♋ 08	27 ♌ 52	28 ♏ 17	20 ♈ 29	09 ♓ 12	13 ♑ 39
02	08 43 04	09 53 35	08 ♓ 16 18	15 ♓ 40 03	20 15	29 R 28	25 47	28 05	28 D 17	20 R 29	09 R 10	13 R 37
03	08 47 00	10 50 59	23 ♓ 02 11	00 ♈ 21 58	22 09	29 08	26 26	28 18	28 17	20 29	09 09	13 36
04	08 50 57	11 48 24	07 ♈ 38 48	14 ♈ 52 13	24 01	28 45	27 06	28 31	28 17	20 28	09 08	13 35
05	08 54 53	12 45 51	22 ♈ 01 51	29 ♈ 07 29	25 52	28 21	27 45	28 43	28 17	20 28	09 06	13 34
06	08 58 50	13 43 18	06 ♉ 08 58	13 ♉ 06 15	27 41	27 54	28 23	28 56	28 18	20 27	09 05	13 32
07	09 02 46	14 40 47	19 ♉ 59 21	26 ♉ 48 21	29 ♌ 28	27 26	29 02	29 09	28 18	20 27	09 03	13 31
08	09 06 43	15 38 17	03 ♊ 33 19	10 ♊ 14 23	01 ♍ 14	26 56	29 ♋ 41	29 22	28 19	20 26	09 02	13 30
09	09 10 40	16 35 49	16 ♊ 51 42	23 ♊ 25 22	02 58	26 24	00 ♌ 20	29 35	28 20	20 25	09 00	13 29
10	09 14 36	17 33 21	29 ♊ 55 31	06 ♋ 22 16	04 40	25 51	00 59	29 ♌ 48	28 20	20 25	08 59	13 27
11	09 18 33	18 30 56	12 ♋ 45 44	19 ♋ 06 00	06 22	25 16	01 38	00 ♍ 00	28 21	20 24	08 57	13 26
12	09 22 29	19 28 31	25 ♋ 23 10	01 ♌ 37 21	08 01	24 41	02 17	00 13	28 22	20 23	08 56	13 25
13	09 26 26	20 26 08	07 ♌ 48 38	13 ♌ 57 08	09 39	24 04	02 55	00 26	28 23	20 22	08 54	13 24
14	09 30 22	21 23 46	20 ♌ 02 59	26 ♌ 06 20	11 16	23 28	03 34	00 39	28 24	20 21	08 53	13 23
15	09 34 19	22 21 25	02 ♍ 07 22	08 ♍ 06 19	12 50	22 51	04 13	00 52	28 25	20 21	08 51	13 22
16	09 38 15	23 19 05	14 ♍ 03 24	19 ♍ 58 56	14 24	22 13	04 51	01 05	28 27	20 20	08 50	13 21
17	09 42 12	24 16 47	25 ♍ 53 15	01 ♎ 46 44	15 56	21 36	05 30	01 18	28 28	20 19	08 48	13 20
18	09 46 09	25 14 29	07 ♎ 39 46	13 ♎ 32 50	17 26	20 59	06 09	01 31	28 30	20 17	08 47	13 19
19	09 50 05	26 12 13	19 ♎ 26 25	25 ♎ 21 03	18 55	20 23	06 47	01 44	28 31	20 16	08 45	13 18
20	09 54 02	27 09 58	01 ♏ 17 18	07 ♏ 15 45	20 22	19 48	07 26	01 57	28 33	20 15	08 43	13 17
21	09 57 58	28 07 44	13 ♏ 17 00	19 ♏ 21 40	21 48	19 14	08 04	02 10	28 35	20 14	08 42	13 16
22	10 01 55	29 ♌ 05 31	25 ♏ 30 21	01 ♐ 43 41	23 12	18 40	08 43	02 23	28 37	20 13	08 40	13 15
23	10 05 51	00 ♍ 03 19	08 ♐ 02 12	14 ♐ 26 28	24 35	18 09	09 21	02 36	28 39	20 12	08 39	13 14
24	10 09 48	01 01 09	20 ♐ 56 55	27 ♐ 33 57	25 56	17 39	10 00	02 49	28 41	20 10	08 37	13 13
25	10 13 44	01 59 00	04 ♑ 17 49	11 ♑ 08 41	27 15	17 10	10 38	03 03	28 43	20 09	08 35	13 12
26	10 17 41	02 56 52	18 ♑ 06 32	25 ♑ 11 12	28 33	16 44	11 17	03 16	28 45	20 07	08 34	13 11
27	10 21 38	03 54 45	02 ♒ 22 20	09 ♒ 39 24	29 ♍ 48	16 20	11 55	03 29	28 48	20 06	08 32	13 11
28	10 25 34	04 52 40	17 ♒ 01 42	24 ♒ 28 23	01 ♎ 02	15 57	12 33	03 42	28 50	20 05	08 30	13 10
29	10 29 31	05 50 36	01 ♓ 58 26	09 ♓ 30 46	02 14	15 38	13 11	03 55	28 53	20 03	08 29	13 09
30	10 33 27	06 48 33	17 ♓ 04 14	24 ♓ 37 39	03 24	15 20	13 50	04 08	28 55	20 02	08 27	13 08
31	10 37 24	07 ♍ 46 32	02 ♈ 09 55	09 ♈ 39 58	04 ♎ 32	15 ♌ 05	14 ♌ 28	04 ♍ 21	28 ♏ 58	20 ♈ 00	08 ♓ 25	13 ♑ 07

Date	Moon True ☊ ° '	Moon Mean ☊ ° '	Moon ☽ Latitude ° '
01	02 ♎ 13	03 ♎ 41	03 N 13
02	02 R 09	03 38	02 05
03	02 07	03 35	00 N 49
04	02 D 07	03 32	00 S 30
05	02 08	03 28	01 45
06	02 09	03 25	02 53
07	02 R 09	03 22	03 50
08	02 07	03 19	04 32
09	02 04	03 16	04 58
10	01 59	03 12	05 09
11	01 52	03 09	05 04
12	01 44	03 06	04 43
13	01 37	03 03	04 10
14	01 30	03 00	03 26
15	01 25	02 57	02 32
16	01 22	02 53	01 33
17	01 20	02 50	00 S 30
18	01 D 20	02 47	00 N 35
19	01 21	02 44	01 37
20	01 23	02 41	02 37
21	01 25	02 38	03 30
22	01 26	02 34	04 14
23	01 R 25	02 31	04 48
24	01 24	02 28	05 09
25	01 21	02 25	05 13
26	01 17	02 22	05 01
27	01 14	02 18	04 29
28	01 10	02 15	03 39
29	01 07	02 12	02 34
30	01 05	02 09	01 17
31	01 ♎ 05	02 ♎ 06	00 S 06

DECLINATIONS

Date	Sun ☉ ° '	Moon ☽ ° '	Mercury ☿ ° '	Venus ♀ ° '	Mars ♂ ° '	Jupiter ♃ ° '	Saturn ♄ ° '	Uranus ♅ ° '	Neptune ♆ ° '	Pluto ♇ ° '
01	18 N 01	10 S 40	16 N 58	06 N 33	22 N 03	13 N 01	17 S 51	07 N 24	08 S 52	20 S 49
02	17 46	06 32	16 20	06 27	21 56	12 57	17 51	07 24	08 53	20 49
03	17 30	02 S 01	15 41	06 22	21 49	12 52	17 51	07 23	08 53	20 49
04	17 15	02 N 35	15 01	06 17	21 42	12 48	17 51	07 23	08 54	20 49
05	16 59	06 57	14 20	06 14	21 35	12 43	17 52	07 23	08 54	20 50
06	16 42	10 50	13 38	06 11	21 27	12 39	17 52	07 23	08 55	20 50
07	16 26	14 03	12 57	06 10	21 20	12 34	17 52	07 23	08 55	20 50
08	16 09	16 25	12 14	06 09	21 12	12 30	17 53	07 22	08 56	20 50
09	15 52	17 50	11 32	06 10	21 04	12 25	17 53	07 22	08 56	20 51
10	15 34	18 17	10 49	06 11	20 56	12 21	17 53	07 22	08 57	20 51
11	15 16	17 47	10 06	06 13	20 48	12 16	17 54	07 22	08 58	20 51
12	14 59	16 25	09 23	06 16	20 39	12 12	17 54	07 21	08 58	20 51
13	14 40	14 17	08 39	06 20	20 31	12 07	17 55	07 21	08 59	20 52
14	14 22	11 32	07 56	06 25	20 22	12 03	17 55	07 20	08 59	20 52
15	14 03	08 21	07 13	06 30	20 13	11 58	17 56	07 20	09 00	20 52
16	13 45	04 51	06 30	06 37	20 04	11 54	17 56	07 20	09 01	20 52
17	13 26	01 N 11	05 47	06 44	19 55	11 49	17 57	07 19	09 01	20 53
18	13 06	02 S 31	05 05	06 51	19 46	11 44	17 57	07 19	09 02	20 53
19	12 47	06 06	04 22	06 59	19 37	11 40	17 58	07 18	09 02	20 53
20	12 27	09 28	03 40	07 08	19 27	11 35	17 59	07 18	09 03	20 53
21	12 07	12 29	02 59	07 17	19 17	11 31	17 59	07 18	09 04	20 54
22	11 47	15 01	02 17	07 26	19 08	11 26	18 00	07 17	09 04	20 54
23	11 27	16 54	01 37	07 36	18 58	11 21	18 01	07 17	09 05	20 54
24	11 07	18 00	00 56	07 46	18 48	11 17	18 01	07 16	09 06	20 54
25	10 46	18 09	00 N 16	07 56	18 38	11 12	18 02	07 16	09 06	20 54
26	10 25	17 15	00 S 23	08 06	18 27	11 07	18 03	07 15	09 07	20 55
27	10 04	15 15	01 01	08 16	18 17	11 03	18 03	07 14	09 07	20 55
28	09 43	12 15	01 39	08 26	18 07	10 58	18 04	07 14	09 08	20 55
29	09 22	08 23	02 16	08 37	17 56	10 53	18 05	07 13	09 09	20 55
30	09 01	03 56	02 53	08 47	17 45	10 49	18 06	07 13	09 09	20 56
31	08 N 39	00 N 46	03 S 28	08 N 57	17 N 35	10 N 44	18 S 07	07 N 12	09 S 10	20 S 56

ZODIAC SIGN ENTRIES

Date	h m	Planets
01	22 36	☽ ♓
03	23 24	☽ ♈
06	01 29	☽ ♉
07	19 15	☿ ♍
08	05 40	☽ ♊
08	23 32	♂ ♌
10	12 08	☽ ♋
11	11 11	♃ ♍
12	20 52	☽ ♌
15	07 45	☽ ♍
17	20 22	☽ ♎
20	09 24	☽ ♏
22	20 41	☽ ♐
23	10 37	☉ ♍
25	04 22	☽ ♑
27	08 03	☽ ♒
27	15 44	☿ ♎
29	08 51	☽ ♓
31	08 33	☽ ♈

LATITUDES

Date	Mercury ☿ ° '	Venus ♀ ° '	Mars ♂ ° '	Jupiter ♃ ° '	Saturn ♄ ° '	Uranus ♅ ° '	Neptune ♆ ° '	Pluto ♇ ° '
01	01 N 44	05 S 19	00 N 58	00 N 51	01 N 59	00 S 39	00 S 48	01 N 56
04	01 35	05 59	00 59	00 52	01 58	00 39	00 48	01 56
07	01 22	06 35	01 00	00 52	01 57	00 39	00 49	01 55
10	01 06	07 08	01 01	00 52	01 57	00 39	00 49	01 55
13	00 46	07 35	01 02	00 52	01 56	00 40	00 49	01 55
16	00 N 24	07 54	01 03	00 52	01 55	00 40	00 49	01 54
19	00 S 01	08 06	01 04	00 52	01 54	00 40	00 49	01 54
22	00 26	08 11	01 05	00 52	01 54	00 40	00 49	01 53
25	00 53	08 07	01 06	00 52	01 53	00 40	00 49	01 53
28	01 21	07 57	01 07	00 52	01 52	00 40	00 49	01 52
31	01 S 49	07 S 42	01 N 08	00 N 53	01 N 52	00 S 40	00 S 49	01 N 52

DATA

Julian Date	2457236
Delta T	+69 seconds
Ayanamsa	24° 04' 32"
Synetic vernal point	05° ♓ 02' 27"
True obliquity of ecliptic	23° 26' 05"

MOON'S PHASES, APSIDES AND POSITIONS ☽

Date	h m	Phase	Longitude ° '	Eclipse Indicator
07	02 03	☾	14 ♉ 17	
14	14 53	●	21 ♌ 31	
22	19 31	☽	29 ♏ 24	
29	18 35	○	06 ♓ 06	

Day	h m		
02	10 12	Perigee	
18	02 40	Apogee	
30	15 29	Perigee	
03	22 29	0N	
10	11 09	Max dec	18° N 17'
17	19 39	0S	
25	03 39	Max dec	18° S 12'
31	08 05	0N	

LONGITUDES

Date	Chiron ⚷ ° '	Ceres ⚳ ° '	Pallas ⚴ ° '	Juno ⚵ ° '	Vesta ⚶ ° '	Black Moon Lilith ⚸ ° '
01	20 ♓ 56	00 ♒ 31	11 ♐ 40	11 ♍ 53	12 ♈ 59	27 ♍ 23
11	20 ♓ 35	28 ♑ 30	11 ♐ 52	15 ♍ 43	13 ♈ 25	28 ♍ 29
21	20 ♓ 11	26 ♑ 51	12 ♐ 41	19 ♍ 36	13 ♈ 05	29 ♍ 35
31	19 ♓ 45	25 ♑ 43	14 ♐ 03	23 ♍ 28	11 ♈ 59	00 ♎ 42

ASPECTARIAN

01 Saturday
- 02 26 ☽ ☍ ☿
- 05 49 ☽ ⊥ ♇
- 07 11 ☽ ✶ ♅
- 14 51 ☽ ⊼ ♂
- 18 24 ☉ ⊼ ♆
- 19 16 ☽ ☍ ♃
- 19 49 ☽ □ ♄
- 20 24 ☽ ∠ ♇
- 22 02 ☽ ☍ ♀
- 22 50 ☽ ∥ ♆

02 Sunday
- 01 01 ☽ ± ♂
- 04 09 ☿ ± ♇
- 04 38 ☉ # ♄
- 05 53 ♄ St D
- 07 15 ☽ # ♅
- 07 29 ☽ ∠ ♅
- 12 28 ☽ # ♀
- 13 28 ☽ ☌ ♆
- 14 49 ☽ ⊼ ☉
- 14 50 ☿ △ ♅
- 16 16 ☽ ⚼ ♂
- 20 40 ☽ ✶ ♇
- 22 04 ☽ ⊥ ♅

03 Monday
- 01 13 ☽ ± ☉
- 07 50 ☽ ⊻ ♅
- 10 21 ☽ ⊼ ☿
- 10 36 ♃ □ ♄
- 16 11 ☽ Q ♇
- 16 55 ☽ ⚼ ☉
- 17 49 ☽ △ ♂
- 20 35 ☽ △ ♄
- 20 44 ☽ ⊼ ♃
- 21 36 ☽ ± ☿
- 21 44 ☽ ⊼ ♀

04 Tuesday
- 06 44 ☽ ± ♃
- 07 21 ☽ ± ♀
- 14 27 ☽ ⊻ ♆
- 14 36 ☽ ⚼ ☿
- 19 22 ♃ ⚼ ♇
- 19 23 ☽ △ ☉
- 21 22 ☽ ⚼ ♄
- 21 47 ♀ ☌ ♃
- 21 50 ☽ □ ♇
- 21 52 ☽ ⚼ ♀
- 21 53 ☽ ⚼ ♃
- 23 14 ♀ ⚼ ♇

05 Wednesday
- 00 25 ☽ ⊥ ♆
- 07 58 ☽ ∥ ♀
- 09 22 ☽ ☌ ♅
- 12 26 ☽ ± ♄
- 14 31 ☽ ∥ ♅
- 15 08 ♀ □ ♄
- 15 29 ☽ ∠ ♆
- 19 25 ☽ △ ☿
- 22 07 ☽ □ ♂
- 22 22 ☽ △ ♀
- 22 35 ☽ ⊼ ♄
- 23 29 ☽ △ ♃
- 23 38 ☽ # ♆

06 Thursday
- 01 24 ♀ ⊻ ♂
- 07 30 ☉ ⊼ ♇
- 08 29 ♂ △ ♄
- 14 25 ☿ ☌ ♀
- 17 02 ☽ ✶ ♆
- 20 19 ☿ □ ♄
- 23 23 ☿ ⚼ ♇

07 Friday
- 00 31 ☽ ∥ ♃
- 00 44 ☽ △ ♇
- 02 03 ☽ □ ☉
- 02 59 ☿ ⊻ ♂
- 04 42 ☽ ∥ ☿
- 06 35 ☽ Q ♂
- 07 08 ☿ ☌ ♃
- 12 48 ☽ ⊻ ♅
- 13 52 ☽ Q ♆
- 17 59 ♂ ⊻ ♃
- 23 21 ☽ ⊥ ♅

08 Saturday
- 00 38 ☽ □ ♀
- 02 03 ☿ ∥ ♃
- 02 40 ☽ ☍ ♄
- 03 01 ☽ ⚼ ♇
- 04 25 ☽ □ ♃
- 04 46 ☽ ✶ ♂
- 07 14 ☽ □ ☿
- 09 09 ☽ ∥ ☉
- 12 10 ☽ Q ☉
- 15 22 ☽ ∠ ♅
- 19 03 ☽ ± ♇
- 21 48 ☽ □ ♆

09 Sunday
- 05 52 ☽ ⊼ ♇
- 07 42 ☽ Q ♀
- 09 05 ☽ ∠ ♂
- 11 29 ☽ ✶ ☉
- 13 13 ☽ # ♄
- 13 20 ☽ Q ♃
- 18 30 ☽ ✶ ♅
- 20 37 ☽ Q ☿

10 Monday
- 02 24 ☽ ⊥ ♂
- 04 46 ☽ ✶ ♀
- 09 03 ☽ ⊼ ♄
- 11 45 ☽ ✶ ♃
- 14 04 ☽ ⊻ ♂
- 16 37 ☽ Q ♅
- 17 17 ☽ ∠ ☉
- 20 13 ☽ ± ♄
- 22 10 ☽ ✶ ☿
- 22 23 ☿ ⚼ ♅

11 Tuesday
- 02 04 ♂ # ♇
- 04 51 ☽ △ ♆
- 07 31 ☽ ∠ ♀
- 09 04 ☽ # ♄
- 11 30 ☽ ⊥ ☉
- 13 07 ☽ ⚼ ♄
- 13 16 ☽ ☍ ♇
- 16 19 ☽ ∠ ♃
- 23 46 ☽ ⊥ ♀
- 23 47 ☽ ⊻ ☉

12 Wednesday
- 02 27 ☽ □ ♅
- 06 47 ☽ ∠ ☿
- 09 13 ☽ ⚼ ♆
- 09 44 ☽ ⊥ ♃
- 10 38 ☉ ± ♇
- 10 42 ☽ ⊻ ♀
- 17 44 ☽ △ ♄
- 18 16 ☿ ⊥ ♂
- 21 28 ☽ ⊻ ♃

13 Thursday
- 01 11 ☿ ☍ ♆
- 01 23 ☿ # ♆
- 02 00 ☽ ☌ ♂
- 02 30 ☽ ± ♆
- 02 42 ☽ ⊥ ☿
- 07 30 ☽ ∥ ☉
- 10 27 ☉ △ ♅
- 11 24 ♂ ± ♆
- 14 08 ☽ ⊼ ♆
- 16 08 ☽ ⊻ ☿
- 22 17 ♄ ∠ ♇
- 22 54 ☽ ⊼ ♇

14 Friday
- 07 47 ☽ ∥ ♃
- 10 41 ☽ ± ♇
- 12 36 ☽ △ ♅
- 14 53 ☽ ☌ ☉
- 18 25 ☽ ☌ ♀

15 Saturday
- 04 30 ☽ ⚼ ♇
- 04 36 ☽ □ ♄
- 07 18 ☽ # ♆
- 08 06 ☿ ∥ ♅
- 09 27 ☽ ☌ ♃
- 16 25 ☽ ⊻ ♂
- 18 26 ☽ ⚼ ♅
- 19 06 ☽ ∥ ♅
- 19 22 ☉ ☌ ♀
- 19 55 ☿ △ ♇
- 21 56 ☽ ∥ ☿

16 Sunday
- 00 26 ☽ ∥ ♀
- 01 29 ☽ ☍ ♆
- 05 10 ☽ ⊥ ♂
- 08 52 ☿ ∥ ♀
- 10 34 ☽ △ ♇
- 10 53 ☿ ± ♅
- 12 33 ☽ ± ♅
- 12 48 ☽ ☌ ☿
- 16 51 ☽ Q ♄

17 Monday
- 00 25 ☽ ∠ ♂
- 00 41 ☽ ⊼ ♅
- 03 43 ☽ ⊻ ♀
- 08 26 ☽ ⊻ ☉
- 15 19 ☽ ⊥ ♀
- 17 16 ☽ ✶ ♄
- 20 42 ☿ Q ♄
- 21 44 ☽ ⊥ ☉
- 23 14 ☽ ⊻ ♃

18 Tuesday
- 08 44 ☽ ✶ ♂
- 08 45 ☽ ∠ ♀
- 11 42 ☽ ⊥ ♃
- 14 16 ☽ ⊼ ♆
- 17 44 ☽ ∠ ☉
- 23 30 ☽ □ ♇
- 23 55 ☽ ∠ ♄

19 Wednesday
- 02 14 ☽ # ☿
- 02 28 ☽ ± ♆
- 06 24 ☽ ∠ ♃
- 10 36 ☽ Q ♂
- 10 47 ☽ ⊻ ☿
- 13 41 ☽ ☍ ♅
- 13 50 ☽ ✶ ♀
- 16 44 ♀ △ ♅
- 18 17 ☽ ⊥ ♄
- 18 23 ☽ # ♀
- 20 21 ☽ # ♅
- 20 44 ☽ ⚼ ♆

20 Thursday
- 00 41 ☽ ⊥ ☿
- 02 56 ☽ ✶ ☉
- 05 11 ☿ ⊻ ♀
- 06 28 ☽ ⊻ ♄
- 08 53 ☽ ∥ ♆
- 10 02 ☿ ⊼ ♅
- 11 59 ☽ Q ♇
- 12 59 ☽ Q ♀
- 13 22 ☽ ✶ ♃
- 21 20 ☽ ∠ ☿

21 Friday
- 01 02 ☽ □ ♂
- 02 53 ☽ △ ♆
- 04 04 ☽ # ♃
- 05 11 ☽ Q ☉
- 09 14 ☽ # ☉
- 10 21 ♀ ± ♇
- 11 58 ☽ ✶ ♇
- 13 47 ♀ ∥ ♅
- 13 48 ☽ Q ♃
- 15 17 ☉ ⚼ ♇
- 23 13 ☽ □ ♀
- 23 40 ☉ □ ♄

22 Saturday
- 01 41 ☽ ⊼ ♅
- 06 57 ☽ ✶ ☿
- 10 22 ♂ ⊼ ♆
- 13 22 ☽ ± ♅
- 17 18 ☽ ∠ ♇
- 18 02 ☽ ☌ ♄
- 19 31 ☽ □ ☉

23 Sunday
- 01 30 ☽ □ ♃
- 04 29 ☿ ∠ ♂
- 06 27 ☿ ⊥ ♀
- 06 37 ☽ ⚼ ♅
- 08 55 ☽ Q ☿
- 10 29 ☽ ⊥ ♇
- 13 08 ☽ □ ♆
- 14 37 ☽ △ ♂
- 20 45 ☉ ∥ ♃
- 21 44 ☽ ⊻ ♇

24 Monday
- 06 09 ☽ △ ♀
- 10 35 ☽ △ ♅
- 13 08 ☽ ∥ ♄
- 19 44 ☽ ⚼ ♂
- 22 04 ☽ □ ☿
- 22 16 ☽ Q ♆

25 Tuesday
- 02 03 ☽ ⊻ ♄
- 07 35 ☽ △ ☉
- 08 22 ☽ ⚼ ♀
- 09 45 ☽ △ ♃
- 12 38 ☽ ± ♂
- 12 45 ☽ ⊥ ♄
- 17 23 ☽ ∥ ♄
- 19 32 ☽ ✶ ♆
- 23 39 ☽ ⊼ ♂
- 23 40 ☽ ± ♀

26 Wednesday
- 03 33 ☽ ☌ ♇
- 04 30 ☽ ∠ ♄
- 09 43 ☽ ⊼ ♀
- 11 42 ☽ ⚼ ☉
- 12 16 ☽ ⚼ ♃
- 15 26 ☽ □ ♅
- 16 07 ☿ ✶ ♄
- 21 14 ☽ ∠ ♆
- 22 02 ☉ ☌ ♃

27 Thursday
- 03 43 ☽ ± ♃
- 04 02 ☽ ± ☉
- 06 02 ☽ ✶ ♄
- 07 20 ☽ △ ☿
- 12 16 ☽ ⊥ ♆
- 13 52 ☽ ⊼ ♃
- 14 44 ☽ ⊼ ☉
- 21 26 ☽ Q ♅
- 22 08 ☽ ⊻ ♆

28 Friday
- 01 54 ☽ Q ♄
- 04 24 ☽ ☍ ♂
- 05 44 ☽ ⊻ ♇
- 10 15 ☽ ⚼ ☿
- 10 18 ☽ ☍ ♀
- 10 47 ☿ ∠ ♀
- 15 27 ☽ ⊥ ♇
- 16 49 ☉ ⚼ ♅
- 16 55 ☽ ✶ ♅
- 17 11 ♂ # ♄
- 20 38 ☽ # ♃

29 Saturday
- 02 02 ☽ ± ☿
- 05 43 ☽ # ☉
- 05 54 ☽ ∠ ♇
- 07 03 ☽ □ ♄
- 07 35 ☽ ∥ ♆
- 10 28 ♂ ⊼ ♇
- 10 43 ☽ # ♀
- 12 27 ☽ ⊼ ☿
- 15 08 ☽ ☍ ♃
- 16 54 ☽ ∠ ♅
- 18 30 ☽ # ♅
- 18 35 ☽ ☍ ☉
- 22 20 ☽ ☌ ♆

30 Sunday
- 02 33 ☉ # ♆
- 05 46 ☽ ✶ ♇
- 06 38 ☽ ⊼ ♂
- 07 10 ☽ ⊥ ♅
- 09 17 ☽ ⊼ ♀
- 16 34 ☽ ± ♂
- 16 41 ☽ ⊻ ♅
- 16 49 ☽ ∥ ☿
- 18 39 ☽ ± ♀
- 22 39 ☉ ∥ ♀

31 Monday
- 00 48 ☽ Q ♇
- 06 53 ☽ △ ♄
- 06 59 ☿ ⊻ ♃
- 07 30 ☽ ⚼ ♂
- 08 43 ☽ ⚼ ♀
- 15 32 ☽ ⊼ ♃
- 16 05 ☽ ☍ ☿
- 21 35 ☽ ⊼ ☉
- 21 59 ☽ ⊻ ♆

All ephemeris data is given at 12.00 UT and the Moon's longitude is additionally given for 24.00 UT

SEPTEMBER 2015

LONGITUDES

Date	Sidereal time h m s	Sun ☉ ° ' "	Moon ☽ ° ' "	Moon ☽ 24.00 ° ' "	Mercury ☿ ° '	Venus ♀ ° '	Mars ♂ ° '	Jupiter ♃ ° '	Saturn ♄ ° '	Uranus ♅ ° '	Neptune ♆ ° '	Pluto ♇ ° '
01	10 41 20	08 ♍ 44 33	17 ♈ 06 52	24 ♈ 29 48	05 ♎ 38	14 ♌ 52	15 ♌ 06	04 ♍ 34	29 ♏ 01	19 ♈ 58	08 ♓ 24	13 ♑ 07
02	10 45 17	09 42 36	01 ♉ 48 06	09 ♉ 01 16	06 41	14 R 41	15 44	04 47	29 04	19 R 57	08 R 22	13 R 06
03	10 49 13	10 40 41	16 ♉ 08 56	23 ♉ 10 54	07 42	14 33	16 22	05 00	29 07	19 55	08 20	13 05
04	10 53 10	11 38 47	00 ♊ 07 04	06 ♊ 57 29	08 41	14 27	17 01	05 13	29 10	19 53	08 19	13 05
05	10 57 07	12 36 56	13 ♊ 42 16	20 ♊ 21 36	09 37	14 24	17 39	05 26	29 13	19 51	08 17	13 04
06	11 01 03	13 35 07	26 ♊ 55 46	03 ♋ 25 01	10 30	14 D 23	18 17	05 39	29 16	19 50	08 16	13 04
07	11 05 00	14 33 20	09 ♋ 49 43	16 ♋ 10 10	11 20	14 25	18 55	05 52	29 20	19 48	08 14	13 03
08	11 08 56	15 31 34	22 ♋ 26 43	28 ♋ 39 43	12 06	14 29	19 33	06 05	29 23	19 46	08 12	13 03
09	11 12 53	16 29 51	04 ♌ 49 29	10 ♌ 56 20	12 49	14 35	20 11	06 18	29 27	19 44	08 11	13 02
10	11 16 49	17 28 10	17 ♌ 00 35	23 ♌ 02 32	13 29	14 43	20 49	06 31	29 30	19 42	08 09	13 02
11	11 20 46	18 26 30	29 ♌ 02 26	05 ♍ 00 36	14 04	14 54	21 27	06 44	29 34	19 40	08 07	13 01
12	11 24 42	19 24 53	10 ♍ 57 16	16 ♍ 52 42	14 35	15 06	22 05	06 57	29 37	19 38	08 06	13 01
13	11 28 39	20 23 17	22 ♍ 47 11	28 ♍ 40 58	15 02	15 21	22 43	07 10	29 41	19 36	08 04	13 01
14	11 32 36	21 21 43	04 ♎ 34 21	10 ♎ 27 38	15 23	15 37	23 20	07 22	29 45	19 34	08 03	13 00
15	11 36 32	22 20 11	16 ♎ 21 06	22 ♎ 15 06	15 40	15 56	23 58	07 35	29 49	19 32	08 01	13 00
16	11 40 29	23 18 41	28 ♎ 09 59	04 ♏ 06 09	15 50	16 16	24 36	07 48	29 53	19 30	07 59	13 00
17	11 44 25	24 17 12	10 ♏ 03 59	16 ♏ 03 56	15 55	16 38	25 14	08 01	29 ♏ 57	19 28	07 58	12 59
18	11 48 22	25 15 45	22 ♏ 06 26	28 ♏ 11 58	15 R 53	17 02	25 52	08 14	00 ♐ 02	19 26	07 56	12 59
19	11 52 18	26 14 20	04 ♐ 21 02	10 ♐ 34 06	15 45	17 28	26 29	08 26	00 06	19 24	07 55	12 59
20	11 56 15	27 12 57	16 ♐ 51 41	23 ♐ 14 15	15 30	17 55	27 07	08 39	00 10	19 21	07 53	12 59
21	12 00 11	28 11 35	29 ♐ 42 16	06 ♑ 16 09	15 08	18 24	27 45	08 52	00 15	19 19	07 52	12 59
22	12 04 08	29 ♍ 10 15	12 ♑ 56 14	19 ♑ 42 49	14 39	18 54	28 22	09 04	00 19	19 17	07 50	12 59
23	12 08 05	00 ♎ 08 57	26 ♑ 36 04	03 ♒ 36 02	14 03	19 26	29 00	09 17	00 24	19 15	07 49	12 59
24	12 12 01	01 07 40	10 ♒ 42 38	17 ♒ 55 36	13 19	19 59	29 ♌ 38	09 30	00 28	19 12	07 47	12 59
25	12 15 58	02 06 25	25 ♒ 14 32	02 ♓ 38 50	12 30	20 34	00 ♍ 15	09 42	00 33	19 10	07 46	12 D 58
26	12 19 54	03 05 12	10 ♓ 07 45	17 ♓ 40 19	11 34	21 10	00 53	09 55	00 38	19 08	07 44	12 59
27	12 23 51	04 04 00	25 ♓ 15 31	02 ♈ 52 11	10 33	21 47	01 30	10 07	00 43	19 05	07 43	12 59
28	12 27 47	05 02 51	10 ♈ 29 05	18 ♈ 04 59	09 29	22 26	02 08	10 20	00 48	19 03	07 41	12 59
29	12 31 44	06 01 43	25 ♈ 38 43	03 ♉ 09 09	08 22	23 05	02 45	10 32	00 53	19 01	07 40	12 59
30	12 35 40	07 ♎ 00 38	10 ♉ 35 16	17 ♉ 56 15	07 ♎ 15	23 ♌ 46	03 ♍ 23	10 ♍ 44	00 ♐ 58	18 ♈ 58	07 ♓ 39	12 ♑ 59

Date	Moon True ☊ ° '	Moon Mean ☊ ° '	Moon ☽ Latitude ° '
01	01 ♎ 05	02 ♎ 03	01 S 27
02	01 D 07	01 59	02 42
03	01 08	01 56	03 44
04	01 09	01 53	04 31
05	01 R 09	01 50	05 02
06	01 09	01 47	05 15
07	01 07	01 44	05 13
08	01 06	01 40	04 55
09	01 03	01 37	04 23
10	01 01	01 34	03 40
11	01 00	01 31	02 48
12	00 58	01 28	01 49
13	00 58	01 24	00 S 45
14	00 D 57	01 21	00 N 20
15	00 58	01 18	01 25
16	00 58	01 15	02 26
17	00 59	01 12	03 21
18	01 00	01 09	04 08
19	01 00	01 05	04 44
20	01 01	01 02	05 09
21	01 R 01	00 59	05 18
22	01 01	00 56	05 11
23	01 01	00 53	04 47
24	01 D 01	00 50	04 05
25	01 01	00 46	03 06
26	01 01	00 43	01 54
27	01 01	00 40	00 N 32
28	01 R 01	00 37	00 S 52
29	01 01	00 34	02 13
30	01 ♎ 00	00 ♎ 30	03 S 23

DECLINATIONS

Date	Sun ☉ ° '	Moon ☽ ° '	Mercury ☿ ° '	Venus ♀ ° '	Mars ♂ ° '	Jupiter ♃ ° '	Saturn ♄ ° '	Uranus ♅ ° '	Neptune ♆ ° '	Pluto ♇ ° '
01	08 N 17	05 N 22	04 S 03	09 N 06	17 N 24	10 N 39	18 S 07	07 N 11	09 S 11	20 S 56
02	07 56	09 34	04 36	09 16	17 13	10 34	18 08	07 11	09 11	20 56
03	07 34	13 05	05 05	09 25	17 01	10 30	18 09	07 10	09 12	20 56
04	07 12	15 45	05 40	09 34	16 50	10 25	18 10	07 09	09 12	20 57
05	06 49	17 27	06 10	09 42	16 39	10 20	18 11	07 09	09 13	20 57
06	06 27	18 09	06 39	09 50	16 27	10 16	18 12	07 08	09 13	20 57
07	06 05	17 52	07 07	09 58	16 16	10 11	18 13	07 07	09 14	20 57
08	05 42	16 43	07 33	10 05	16 04	10 06	18 14	07 07	09 15	20 57
09	05 20	14 48	07 57	10 12	15 52	10 01	18 15	07 06	09 16	20 58
10	04 57	12 14	08 19	10 18	15 41	09 57	18 16	07 05	09 16	20 58
11	04 34	09 11	08 40	10 24	15 29	09 52	18 17	07 04	09 17	20 58
12	04 11	05 47	08 58	10 29	15 17	09 47	18 18	07 03	09 17	20 58
13	03 48	02 N 10	09 15	10 34	15 05	09 43	18 19	07 03	09 18	20 58
14	03 25	01 S 31	09 28	10 38	14 52	09 38	18 20	07 02	09 19	20 59
15	03 02	05 08	09 39	10 42	14 40	09 33	18 21	07 01	09 19	20 59
16	02 39	08 33	09 47	10 45	14 28	09 29	18 22	07 00	09 20	20 59
17	02 16	11 39	09 53	10 48	14 15	09 24	18 23	07 00	09 20	20 59
18	01 53	14 18	09 54	10 50	14 03	09 19	18 24	06 59	09 21	20 59
19	01 30	16 21	09 52	10 52	13 50	09 14	18 25	06 58	09 22	20 59
20	01 06	17 40	09 47	10 53	13 37	09 10	18 26	06 57	09 22	21 00
21	00 43	18 08	09 37	10 53	13 25	09 05	18 27	06 56	09 23	21 00
22	00 N 20	17 38	09 23	10 53	13 12	09 00	18 28	06 56	09 23	21 00
23	00 S 04	16 08	09 05	10 52	12 59	08 56	18 30	06 55	09 24	21 00
24	00 27	13 37	08 43	10 51	12 46	08 51	18 31	06 54	09 24	21 00
25	00 50	10 11	08 16	10 49	12 33	08 47	18 32	06 53	09 25	21 00
26	01 14	06 01	07 45	10 47	12 20	08 42	18 33	06 52	09 25	21 01
27	01 37	01 S 24	07 10	10 44	12 07	08 37	18 34	06 51	09 26	21 01
28	02 00	03 N 21	06 33	10 40	11 54	08 33	18 35	06 50	09 27	21 01
29	02 24	07 51	05 52	10 36	11 40	08 28	18 37	06 49	09 27	21 01
30	02 S 47	11 N 47	05 S 10	10 N 32	11 N 27	08 N 24	18 S 38	06 N 49	09 S 28	21 S 01

ZODIAC SIGN ENTRIES

Date	h	m	Planets
02	09	02	☽ ♉
04	11	48	☽ ♊
06	17	40	☽ ♋
09	02	36	☽ ♌
11	13	55	☽ ♍
14	02	41	☽ ♎
16	15	43	☽ ♏
18	02	49	♄ ♐
19	03	32	☽ ♐
21	12	33	☽ ♑
23	08	21	☉ ♎
23	17	51	☽ ♒
25	02	18	♂ ♍
25	19	44	☽ ♓
27	19	29	☽ ♈
29	18	57	☽ ♉

LATITUDES

Date	Mercury ☿ ° '	Venus ♀ ° '	Mars ♂ ° '	Jupiter ♃ ° '	Saturn ♄ ° '	Uranus ♅ ° '	Neptune ♆ ° '	Pluto ♇ ° '
01	01 S 58	07 S 36	01 N 08	00 N 53	01 N 51	00 S 40	00 S 49	01 N 52
04	02 26	07 14	01 09	00 53	01 51	00 40	00 49	01 51
07	02 52	06 50	01 10	00 53	01 50	00 40	00 49	01 51
10	03 16	06 23	01 11	00 53	01 49	00 40	00 49	01 50
13	03 36	05 55	01 12	00 54	01 49	00 40	00 49	01 50
16	03 52	05 27	01 13	00 54	01 48	00 40	00 49	01 50
19	03 59	04 58	01 14	00 54	01 47	00 40	00 49	01 49
22	03 55	04 29	01 14	00 54	01 47	00 40	00 49	01 49
25	03 37	04 01	01 15	00 55	01 46	00 40	00 49	01 48
28	03 02	03 33	01 16	00 55	01 46	00 40	00 49	01 48
31	02 S 11	03 S 06	01 N 17	00 N 55	01 N 45	00 S 40	00 S 49	01 N 47

DATA

Julian Date	2457267
Delta T	+69 seconds
Ayanamsa	24° 04' 35"
Synetic vernal point	05° ♓ 02' 24"
True obliquity of ecliptic	23° 26' 05"

MOON'S PHASES, APSIDES AND POSITIONS ☽

Date	h	m	Phase	Longitude ° '	Eclipse Indicator
05	09	54	☾	12 ♊ 32	
13	06	41	●	20 ♍ 10	Partial
21	08	59	☽	28 ♐ 04	
28	02	50	○	04 ♈ 40	total

Day	h	m	
14	11	41	Apogee
28	01	52	Perigee

Day	h	m		
06	17	05	Max dec	18° N 10'
14	02	09	0S	
21	12	00	Max dec	18° S 08'
27	19	03	0N	

LONGITUDES

Date	Chiron ⚷ ° '	Ceres ⚳ ° '	Pallas ⚴ ° '	Juno ⚵ ° '	Vesta ⚶ ° '	Black Moon Lilith ⚸ ° '
01	19 ♓ 43	25 ♑ 38	14 ♐ 13	23 ♍ 51	11 ♈ 50	00 ♎ 49
11	19 ♓ 15	25 ♑ 08	16 ♐ 05	27 ♍ 43	09 ♈ 59	01 ♎ 55
21	18 ♓ 47	25 ♑ 14	18 ♐ 21	01 ♎ 34	07 ♈ 38	03 ♎ 01
31	18 ♓ 21	25 ♑ 54	20 ♐ 55	05 ♎ 23	05 ♈ 05	04 ♎ 08

All ephemeris data is given at 12.00 UT and the Moon's longitude is additionally given for 24.00 UT

ASPECTARIAN

h m	Aspects
01 Tuesday	
01 17	☽ ± ♃
03 38	☉ ☍ ♆
03 55	☽ # ☿
05 04	♀ ☌ ♂
05 33	☽ □ ♇
06 59	☽ ⚻ ♄
07 37	☽ ⊥ ♆
07 54	☽ ± ☉
08 24	☽ △ ♀
08 36	☽ △ ♂
16 02	☽ ⚻ ♃
16 37	☽ ☌ ♅
21 37	☽ ± ♄
21 59	☽ ∥ ♅
22 11	☽ ∠ ♆
23 32	☽ ⚻ ☉
23 48	♀ # ♆
02 Wednesday	
03 01	☽ ∥ ☉
07 28	☽ ⊼ ♄
09 40	☽ # ♆
10 03	☽ ∥ ♀
17 01	☽ △ ♃
18 14	☽ ∥ ♃
20 44	☽ ⊼ ☿
22 53	☽ ⚹ ♆
03 Thursday	
02 06	☽ △ ☉
03 52	♃ ⚻ ♅
06 50	☽ △ ♇
07 34	☽ ± ☿
09 19	☽ □ ♀
12 24	☽ □ ♂
18 24	☽ ⊻ ♅
19 07	☽ Q ♆
04 Friday	
00 02	☽ ⚻ ☿
03 02	☿ ⊼ ♆
04 41	☽ ⊥ ♅
08 28	☽ ⚻ ♇
10 20	☽ ☍ ♄
14 26	☉ ∥ ♅
16 04	☽ Q ♀
20 20	☽ ∠ ♅
20 59	☽ Q ♂
21 04	☽ □ ♃
05 Saturday	
00 12	☽ ± ♇
00 15	☽ ∥ ♂
02 22	☽ □ ♆
04 10	☽ △ ☿
09 54	☽ □ ☉
10 52	☽ ⊼ ♇
13 15	☽ ⚹ ♀
19 26	☽ ⚹ ♂
23 04	☽ ⚹ ♅
23 10	☉ △ ♇
06 Sunday	
05 53	☽ Q ♃
06 14	☉ # ☿
08 29	♀ St D
16 20	☽ ⊼ ♄
16 32	☽ ∠ ♀
17 49	☉ ± ♅
21 02	☽ Q ♅
21 18	☽ Q ☉
07 Monday	
00 21	☽ ∠ ♂
03 31	☽ ± ♄
04 26	☽ ⚹ ♃
08 21	☉ ⊻ ♀
09 00	☽ △ ♆
09 20	☽ ⊥ ♀
12 32	☿ # ♅
15 01	☽ □ ☿
17 13	♂ ± ♇
18 05	☽ ☍ ♇
18 08	☽ ⊥ ♂
20 32	☽ ⚻ ♄
20 42	☽ ⊻ ♀
21 41	☽ ⚹ ☉
08 Tuesday	
06 09	☽ ⊻ ♂
06 53	☽ □ ♅
09 20	☽ ∠ ♃
11 00	☿ ⊥ ♃
13 27	☽ ⚻ ♆
14 36	♀ ∥ ♃
19 53	♂ △ ♅
22 11	☽ ∥ ♂
09 Wednesday	
01 28	☽ △ ♄
03 01	☽ ⊥ ♃
03 44	☽ Q ☿
04 57	☽ ∠ ☉
06 51	☽ ± ♆
14 56	☽ ⊻ ♃
18 33	☽ ⊼ ♆
19 27	☿ □ ♇
10 Thursday	
00 06	☽ ⊥ ☉
04 08	☽ ⊼ ♇
04 38	☽ ⚹ ☿
07 24	☽ ☌ ♀
12 51	☉ Q ♄
12 59	☽ ⊻ ☉
16 00	☽ ± ♇
17 20	☽ △ ♅
19 59	☽ ☌ ♂
11 Friday	
03 04	☽ ∥ ♀
06 43	☽ ∥ ♃
09 58	☽ ⚻ ♇
11 16	☽ # ♆
12 04	☽ ∠ ☿
13 03	☽ □ ♄
14 11	☿ ± ♆
15 25	☽ # ☿
23 17	☽ ⚻ ♅
12 Saturday	
03 08	☽ ∥ ♅
03 45	☽ ☌ ♃
06 14	☽ ☍ ♆
07 01	☽ ⊥ ☿
14 07	☿ ∠ ♄
16 10	☽ △ ♇
17 18	☉ ⊼ ♅
17 25	☽ ± ♅
19 40	☽ ⊻ ☿
20 34	☽ ⊻ ♀
23 57	☽ ∥ ☉
13 Sunday	
01 35	☽ Q ♄
05 33	☽ ⊼ ♅
06 41	☽ ☌ ☉
09 00	☽ ⊥ ♀
11 50	☽ ⊻ ♂
17 50	☿ ∥ ♆
14 Monday	
00 44	☽ ⊥ ♂
02 08	☽ ⚹ ♄
03 45	☽ ∠ ♀
17 49	☽ ⊻ ♃
19 03	☽ ⊼ ♆
20 07	☽ ∠ ♂
21 14	☉ ⊥ ♀
23 23	☽ # ☉
15 Tuesday	
02 13	☿ # ♃
05 10	☽ □ ♇
06 16	☽ ⊥ ♃
07 15	☽ ± ♆
08 52	☽ ∠ ♄
10 34	☽ ☌ ☿
11 07	☽ ⚹ ♀
18 27	☽ ☍ ♅
16 Wednesday	
00 55	☽ ∠ ♃
01 00	☽ # ♅
01 16	☽ ⊻ ☉
01 31	☽ ⚻ ♆
03 16	☽ ⊥ ♄
04 22	☽ ⚹ ♂
12 13	☽ Q ♀
14 32	☽ ⊥ ☉
15 30	☽ ⊻ ♄
17 43	☽ Q ♇
17 46	☽ ∥ ♆
18 41	☽ # ♃
21 33	☽ ∥ ☿
17 Thursday	
04 59	☽ # ♀
05 59	☽ Q ♂
06 54	♃ ☍ ♆
07 47	☽ △ ♆
07 48	☽ ⚹ ♃
10 18	☽ ∠ ☉
17 51	☽ ⚹ ♇
18 09	☿ St R
23 42	☽ ⊻ ☿
18 Friday	
01 35	☽ □ ♀
03 49	♃ # ♆
06 42	☽ ⊼ ♅
08 13	☽ Q ♃
09 41	☽ # ♂
11 35	☽ ⊥ ☿
18 32	☽ ± ♅
18 46	☽ ⚹ ☉
19 49	☽ □ ♂
23 35	☽ ∠ ♇
19 Saturday	
03 40	☽ ☌ ♄
05 06	☽ ∠ ☿
12 05	☽ ⚻ ♅
17 06	☽ ⊥ ♇
18 53	☽ □ ♆
20 03	☽ □ ♃
20 10	☽ Q ☉
20 Sunday	
04 37	☽ ⊻ ♇
05 17	☉ ⊻ ♂
09 29	☽ ⚹ ☿
14 05	☽ △ ♀
16 42	☽ △ ♅
21 Monday	
04 54	☽ Q ♆
06 38	☿ ∠ ♄
07 24	☽ Q ☿
08 12	☽ △ ♂
08 59	☽ □ ☉
13 00	☽ ⊻ ♄
19 03	☽ ⚻ ♀
20 51	♂ ⚻ ♇
21 55	☿ ⊥ ♃
22 Tuesday	
00 02	☽ ⊥ ♄
02 51	☽ ⚹ ♆
04 57	☽ △ ♃
11 57	☽ ± ♀
11 57	☿ ∥ ♆
12 04	☽ ☌ ♇
12 49	☽ ⚻ ♂
14 56	☽ □ ☿
15 15	♀ ± ♇
16 16	☽ ∠ ♄
23 00	☽ ⊼ ♀
23 13	☽ □ ♅
23 Wednesday	
04 00	♀ △ ♅
05 26	☽ ∠ ♆
05 27	☽ ± ♂
07 55	☽ ⚻ ♃
12 48	☿ ∠ ♂
16 20	☽ ⊼ ♂
18 34	☽ △ ☉
18 34	☽ ⚹ ♄
18 36	☉ ⚹ ♄
20 29	☿ ± ♆
20 56	☽ ⊥ ♆
23 38	☽ ± ♃
24 Thursday	
01 10	☿ # ♃
06 07	☽ Q ♅
07 06	☽ ⊻ ♆
09 56	☽ ⊼ ♃
14 58	☽ Q ♄
15 47	☽ ⊻ ♇
16 08	☽ △ ☿
18 57	☽ # ♂
21 41	☽ ⚻ ☉
22 26	☿ □ ♇
25 Friday	
01 44	☽ ⊥ ♇
02 04	☽ ⚹ ♅
04 02	☽ ☍ ♀
06 53	♇ St D
07 51	☽ # ♀
13 31	☽ ± ☉
15 27	☽ ⚻ ☿
16 27	☽ ∠ ♇
16 40	☽ ∥ ♆
20 30	☽ ☍ ♂
20 39	☽ # ♃
20 40	☽ □ ♄
23 55	☽ ⊼ ☉
26 Saturday	
01 01	☽ ∥ ☿
01 12	♂ □ ♄
02 25	☽ ∠ ♅
05 08	☽ ± ☿
07 19	☽ # ♅
08 11	☽ ☌ ♆
11 39	☽ ☍ ♃
14 09	☽ ⊼ ☿
16 32	☽ ⚹ ♇
16 46	☽ ⊥ ♅
27 Sunday	
02 16	☽ ⊻ ♅
06 17	☽ ⊼ ♀
10 58	☽ ∥ ☉
11 33	☽ Q ♇
16 10	☽ ± ♀
20 22	☿ ⊻ ♃
20 39	☽ △ ♄
22 16	☽ ⊼ ♂
28 Monday	
00 41	☿ # ♅
02 50	☽ ☍ ☉
04 33	☽ # ☉
06 58	☽ ⚻ ♀
07 36	☽ ⊻ ♆
08 08	☽ ± ♂
10 32	☽ ☍ ☿
11 45	☽ ⊼ ♃
15 56	☽ □ ♇
17 03	☽ ⊥ ♆
20 26	☽ ⚻ ♄
21 21	☽ ± ♃
22 56	☽ ⚻ ♂
29 Tuesday	
01 30	☽ ☌ ♅
02 36	☽ # ☿
06 21	☽ ∥ ♅
06 45	☿ ⊥ ♂
07 16	☽ ∠ ♆
07 45	☽ △ ♀
10 46	☽ ± ♄
11 49	☽ ⚻ ♃
15 28	☽ ∥ ♃
15 48	☿ ∠ ♀
20 24	☽ ⊼ ♄
21 17	☽ # ♆
23 51	☽ △ ♂
30 Wednesday	
03 19	☿ ⊼ ♆
04 01	☽ ∥ ♀
05 48	☽ ⊼ ☉
06 58	☽ ⊼ ☿
07 14	☽ ⚹ ♆
09 55	☽ ∥ ♂
12 15	☽ △ ♃
14 38	☉ ☌ ☿
15 54	☽ △ ♇
16 01	☽ ± ☿
16 13	☽ ± ☉

OCTOBER 2015

LONGITUDES

Date	Sidereal time h m s	Sun ☉ ° ' "	Moon ☽ ° ' "	Moon ☽ 24.00 ° ' "	Mercury ☿ ° '	Venus ♀ ° '	Mars ♂ ° '	Jupiter ♃ ° '	Saturn ♄ ° '	Uranus ♅ ° '	Neptune ♆ ° '	Pluto ♇ ° '
01	12 39 37	07 ♎ 59 35	25 ♉ 11 23	02 ♊ 20 11	06 ♎ 08	24 ♌ 28	04 ♍ 00	10 ♍ 57	01 ♐ 03	18 ♈ 56	07 ♓ 37	12 ♑ 59
02	12 43 34	08 58 34	09 ♊ 22 19	16 ♊ 17 38	05 R 03	25 11	04 37	11 09	01 08	18 R 54	07 R 36	12 59
03	12 47 30	09 57 36	23 ♊ 06 08	29 ♊ 47 56	04 04	25 55	05 15	11 21	01 13	18 51	07 34	13 00
04	12 51 27	10 56 40	06 ♋ 23 18	12 ♋ 52 35	03 10	26 40	05 52	11 33	01 19	18 49	07 33	13 00
05	12 55 23	11 55 46	19 ♋ 16 10	25 ♋ 34 33	02 23	27 26	06 29	11 45	01 24	18 46	07 32	13 00
06	12 59 20	12 54 54	01 ♌ 48 12	07 ♌ 57 39	01 46	28 12	07 07	11 57	01 30	18 44	07 31	13 00
07	13 03 16	13 54 05	14 ♌ 03 26	20 ♌ 06 03	01 18	29 00	07 44	12 09	01 35	18 42	07 29	13 01
08	13 07 13	14 53 18	26 ♌ 06 01	02 ♍ 03 49	01 01	29 ♌ 49	08 21	12 21	01 41	18 39	07 28	13 01
09	13 11 09	15 52 33	07 ♍ 59 55	13 ♍ 54 44	00 54	00 ♍ 38	08 58	12 33	01 46	18 37	07 27	13 02
10	13 15 06	16 51 51	19 ♍ 48 41	25 ♍ 42 09	00 D 58	01 28	09 36	12 45	01 52	18 34	07 26	13 02
11	13 19 03	17 51 10	01 ♎ 35 28	07 ♎ 28 57	01 13	02 19	10 13	12 57	01 58	18 32	07 25	13 02
12	13 22 59	18 50 32	13 ♎ 22 56	19 ♎ 17 39	01 38	03 11	10 50	13 09	02 03	18 29	07 23	13 03
13	13 26 56	19 49 56	25 ♎ 13 23	01 ♏ 10 23	02 12	04 03	11 27	13 20	02 09	18 27	07 22	13 04
14	13 30 52	20 49 22	07 ♏ 08 53	13 ♏ 09 08	02 55	04 56	12 04	13 32	02 15	18 25	07 21	13 04
15	13 34 49	21 48 49	19 ♏ 11 22	25 ♏ 15 49	03 47	05 50	12 41	13 43	02 21	18 22	07 20	13 05
16	13 38 45	22 48 19	01 ♐ 22 45	07 ♐ 32 26	04 46	06 44	13 18	13 55	02 27	18 20	07 19	13 05
17	13 42 42	23 47 51	13 ♐ 45 10	20 ♐ 01 13	05 52	07 39	13 55	14 06	02 33	18 17	07 18	13 06
18	13 46 38	24 47 24	26 ♐ 20 54	02 ♑ 44 33	07 04	08 35	14 32	14 18	02 39	18 15	07 17	13 07
19	13 50 35	25 46 59	09 ♑ 12 28	15 ♑ 45 00	08 21	09 31	15 09	14 29	02 45	18 12	07 16	13 07
20	13 54 32	26 46 36	22 ♑ 22 25	29 ♑ 05 01	09 42	10 28	15 46	14 40	02 51	18 10	07 15	13 08
21	13 58 28	27 46 15	05 ♒ 53 01	12 ♒ 46 36	11 07	11 25	16 23	14 51	02 58	18 08	07 14	13 09
22	14 02 25	28 45 55	19 ♒ 45 52	26 ♒ 50 48	12 35	12 23	16 59	15 02	03 04	18 05	07 13	13 10
23	14 06 21	29 ♎ 45 37	04 ♓ 01 17	11 ♓ 17 04	14 06	13 21	17 36	15 13	03 10	18 03	07 12	13 11
24	14 10 18	00 ♏ 45 21	18 ♓ 37 43	26 ♓ 02 41	15 39	14 20	18 13	15 24	03 16	18 01	07 12	13 11
25	14 14 14	01 45 06	03 ♈ 31 13	11 ♈ 02 27	17 13	15 19	18 50	15 35	03 23	17 58	07 11	13 12
26	14 18 11	02 44 54	18 ♈ 35 21	26 ♈ 08 50	18 49	16 18	19 26	15 46	03 29	17 56	07 10	13 13
27	14 22 07	03 44 43	03 ♉ 41 43	11 ♉ 12 48	20 27	17 19	20 03	15 57	03 36	17 54	07 09	13 14
28	14 26 04	04 44 34	18 ♉ 40 55	26 ♉ 05 01	22 05	18 19	20 40	16 07	03 42	17 51	07 09	13 15
29	14 30 01	05 44 27	03 ♊ 24 07	10 ♊ 37 26	23 43	19 20	21 16	16 18	03 49	17 49	07 08	13 16
30	14 33 57	06 44 22	17 ♊ 44 20	24 ♊ 44 23	25 22	20 21	21 53	16 28	03 55	17 47	07 07	13 17
31	14 37 54	07 ♏ 44 20	01 ♋ 37 22	08 ♋ 23 11	27 ♎ 01	21 ♍ 23	22 ♍ 29	16 ♍ 38	04 ♐ 02	17 ♈ 44	07 ♓ 07	13 ♑ 18

Date	Moon True ☊ ° '	Moon Mean ☊ ° '	Moon ☽ Latitude ° '
01	00 ♎ 59	00 ♎ 27	04 S 18
02	00 R 59	00 24	04 56
03	00 58	00 21	05 15
04	00 57	00 18	05 16
05	00 D 57	00 15	05 02
06	00 58	00 11	04 33
07	00 59	00 08	03 52
08	01 00	00 05	03 02
09	01 02	00 ♎ 02	02 04
10	01 03	29 ♍ 59	01 S 02
11	01 R 03	29 55	00 N 03
12	01 02	29 52	01 08
13	01 01	29 49	02 10
14	00 59	29 46	03 07
15	00 55	29 43	03 55
16	00 52	29 40	04 34
17	00 48	29 36	05 01
18	00 45	29 33	05 14
19	00 43	29 30	05 11
20	00 42	29 27	04 53
21	00 D 42	29 24	04 17
22	00 43	29 21	03 27
23	00 45	29 17	02 22
24	00 46	29 14	01 N 06
25	00 R 46	29 11	00 S 15
26	00 45	29 08	01 36
27	00 43	29 05	02 51
28	00 38	29 01	03 52
29	00 33	28 58	04 38
30	00 28	28 55	05 04
31	00 ♎ 22	28 ♍ 52	05 S 12

DECLINATIONS

Date	Sun ☉ ° '	Moon ☽ ° '	Mercury ☿ ° '	Venus ♀ ° '	Mars ♂ ° '	Jupiter ♃ ° '	Saturn ♄ ° '	Uranus ♅ ° '	Neptune ♆ ° '	Pluto ♇ ° '
01	03 S 10	14 N 53	04 S 26	10 N 26	11 N 14	08 N 19	18 S 39	06 N 48	09 S 28	21 S 01
02	03 33	16 59	03 43	10 20	11 00	08 14	18 40	06 47	09 29	21 01
03	03 57	18 01	03 01	10 14	10 47	08 10	18 41	06 46	09 29	21 02
04	04 20	18 01	02 21	10 07	10 33	08 05	18 43	06 45	09 30	21 02
05	04 43	17 04	01 44	10 00	10 20	08 01	18 44	06 44	09 30	21 02
06	05 06	15 19	01 10	09 52	10 06	07 56	18 45	06 43	09 30	21 02
07	05 29	12 54	00 41	09 43	09 52	07 52	18 47	06 42	09 31	21 02
08	05 52	09 58	00 S 18	09 34	09 39	07 47	18 48	06 41	09 31	21 02
09	06 15	06 39	00 N 01	09 24	09 25	07 43	18 49	06 40	09 32	21 02
10	06 38	03 N 05	00 14	09 14	09 11	07 39	18 50	06 39	09 32	21 02
11	07 00	00 S 35	00 22	09 03	08 57	07 34	18 52	06 38	09 33	21 03
12	07 23	04 14	00 24	08 52	08 43	07 30	18 53	06 37	09 33	21 03
13	07 45	07 45	00 21	08 40	08 29	07 25	18 54	06 37	09 34	21 03
14	08 08	10 57	00 N 13	08 28	08 16	07 21	18 56	06 36	09 34	21 03
15	08 30	13 44	00 00	08 15	08 02	07 17	18 57	06 35	09 34	21 03
16	08 52	15 57	00 S 17	08 02	07 48	07 13	18 58	06 34	09 35	21 03
17	09 14	17 28	00 37	07 48	07 34	07 08	18 59	06 33	09 35	21 03
18	09 36	18 09	01 01	07 34	07 19	07 04	19 01	06 32	09 35	21 03
19	09 58	17 56	01 29	07 19	07 05	07 00	19 02	06 31	09 36	21 03
20	10 19	16 46	01 59	07 04	06 51	06 56	19 03	06 30	09 36	21 03
21	10 41	14 38	02 31	06 49	06 37	06 51	19 05	06 29	09 36	21 03
22	11 02	11 37	03 05	06 33	06 23	06 47	19 06	06 28	09 37	21 03
23	11 23	07 50	03 41	06 17	06 09	06 43	19 07	06 28	09 37	21 04
24	11 44	03 S 29	04 18	06 00	05 55	06 39	19 09	06 27	09 37	21 04
25	12 05	01 N 10	04 56	05 43	05 40	06 35	19 10	06 26	09 38	21 04
26	12 25	05 48	05 35	05 25	05 26	06 31	19 11	06 25	09 38	21 04
27	12 46	10 04	06 15	05 07	05 12	06 27	19 13	06 24	09 38	21 04
28	13 06	13 39	06 55	04 49	04 58	06 23	19 14	06 23	09 38	21 04
29	13 26	16 17	07 35	04 30	04 43	06 19	19 15	06 22	09 39	21 04
30	13 46	17 49	08 16	04 11	04 29	06 15	19 17	06 21	09 39	21 04
31	14 S 05	18 N 14	08 S 56	03 N 52	04 N 15	06 N 12	19 S 18	06 N 21	09 S 39	21 S 04

ZODIAC SIGN ENTRIES

Date	h	m	Planets
01	20	03	☽ ♊
04	00	22	☽ ♋
06	08	31	☽ ♌
08	17	29	♀ ♍
08	19	50	☽ ♍
11	08	45	☽ ♎
13	21	38	☽ ♏
16	09	18	☽ ♐
18	18	52	☽ ♑
21	01	38	☽ ♒
23	05	18	☽ ♓
23	17	47	☉ ♏
25	06	22	☽ ♈
27	06	07	☽ ♉
29	06	24	☽ ♊
31	09	09	☽ ♋

LATITUDES

Date	Mercury ☿ ° '	Venus ♀ ° '	Mars ♂ ° '	Jupiter ♃ ° '	Saturn ♄ ° '	Uranus ♅ ° '	Neptune ♆ ° '	Pluto ♇ ° '
01	02 S 11	03 S 06	01 N 17	00 N 55	01 N 45	00 S 40	00 S 49	01 N 47
04	01 11	02 40	01 17	00 56	01 45	00 40	00 49	01 47
07	00 S 11	02 15	01 18	00 56	01 44	00 40	00 49	01 46
10	00 N 41	01 50	01 19	00 57	01 44	00 40	00 49	01 46
13	01 20	01 27	01 19	00 57	01 43	00 40	00 49	01 45
16	01 46	01 05	01 20	00 57	01 43	00 40	00 49	01 45
19	02 00	00 44	01 21	00 58	01 42	00 40	00 49	01 45
22	02 03	00 24	01 21	00 58	01 42	00 40	00 49	01 44
25	01 59	00 S 05	01 22	00 59	01 41	00 40	00 49	01 44
28	01 49	00 N 13	01 22	00 59	01 41	00 40	00 49	01 43
31	01 N 35	00 N 29	01 N 23	01 N 00	01 N 41	00 S 40	00 S 49	01 N 43

DATA

Julian Date	2457297
Delta T	+69 seconds
Ayanamsa	24° 04' 37"
Synetic vernal point	05° ♓ 02' 22"
True obliquity of ecliptic	23° 26' 05"

MOON'S PHASES, APSIDES AND POSITIONS ☽

Date	h	m	Phase	Longitude ° '	Eclipse Indicator
04	21	06	☾	11 ♋ 19	
13	00	06	●	19 ♎ 20	
20	20	31	☽	27 ♑ 08	
27	12	05	○	03 ♉ 45	

Day	h	m		
11	13	35	Apogee	
26	13	07	Perigee	
03	23	54	Max dec	18° N 08'
11	08	11	0S	
18	18	28	Max dec	18° S 11'
25	06	02	0N	
31	09	00	Max dec	18° N 14'

LONGITUDES

Date	Chiron ⚷ ° '	Ceres ⚳ ° '	Pallas ⚴ ° '	Juno ⚵ ° '	Vesta ⚶ ° '	Black Moon Lilith ⚸ ° '
01	18 ♓ 21	25 ♑ 54	20 ♐ 55	05 ♎ 23	05 ♈ 05	04 ♎ 08
11	17 ♓ 56	27 ♑ 06	23 ♐ 44	09 ♎ 10	02 ♈ 38	05 ♎ 14
21	17 ♓ 35	28 ♑ 46	26 ♐ 47	12 ♎ 53	00 ♈ 37	06 ♎ 21
31	17 ♓ 18	00 ♒ 50	00 ♑ 01	16 ♎ 32	29 ♓ 15	07 ♎ 27

ASPECTARIAN

h m	Aspects
01 Thursday	
01 40	☽ ⊻ ♅
02 47	☽ Q ♆
03 05	☉ ⊼ ♆
05 44	☽ bQ ☿
08 05	☽ bQ ☉
09 37	♂ bQ ♅
10 44	☽ □ ♀
11 35	☽ ⊥ ♅
16 40	☽ bQ ♇
21 53	☽ ☍ ♄
02 Friday	
02 40	☽ ∠ ♅
03 30	☽ □ ♂
05 08	☽ △ ☿
07 55	☽ ± ♇
08 58	☽ □ ♆
11 16	☽ △ ☉
15 06	☽ □ ♃
15 29	☉ ∥ ☿
18 15	☽ ⊼ ♇
18 17	☿ ⊻ ♂
18 56	☽ Q ♀
03 Saturday	
04 31	☽ ✶ ♅
12 16	☽ Q ♂
17 18	☽ ✶ ♀
23 22	☽ Q ♃
04 Sunday	
01 52	☽ Q ♅
02 41	☽ ⊼ ♄
06 28	☽ □ ☿
11 00	☽ ✶ ♂
13 43	☽ ± ♄
14 08	☽ △ ♆
19 36	☿ ⊥ ♀
21 06	☽ □ ☉
21 41	☽ ✶ ♃
22 21	☽ ∠ ♀
05 Monday	
00 14	☽ ☍ ♇
06 33	☽ bQ ♄
06 40	☉ ⊻ ♃
11 04	☽ □ ♅
14 01	☽ Q ☿
16 21	☽ ⊥ ♀
16 26	☽ ∠ ♂
18 11	☽ bQ ♆
06 Tuesday	
02 30	☽ ∠ ♃
04 35	☽ ⊻ ♀
05 48	♀ bQ ♇
10 08	☽ Q ☉
10 35	☽ ⊥ ♂
11 24	☽ △ ♄
11 26	☽ ± ♆
11 56	☽ ✶ ☿
14 14	☉ □ ♇
20 13	☽ ⊥ ♃
22 53	☽ ⊻ ♂
22 58	☿ ✶ ♄
23 06	☽ ⊼ ♆
07 Wednesday	
00 56	☉ ⊥ ♂
02 10	☉ ± ♆
02 53	♂ ☍ ♆
08 11	☽ ⊻ ♃
09 56	☽ ⊼ ♇
11 40	☽ ✶ ☉
16 19	☽ ∠ ☿
21 10	☽ △ ♅
21 50	☽ ± ♇
08 Thursday	
01 35	☉ ∠ ♀
09 51	☽ ⊥ ☿
14 34	☽ ∥ ♂
15 09	☽ ∥ ♀
15 17	☽ ⫲ ♆
15 51	☽ bQ ♇
17 44	♀ ⫲ ♆
20 01	☽ ☌ ♀
20 18	☽ ∠ ☉
21 45	☽ ⊻ ☿
23 18	☽ □ ♄
09 Friday	
00 19	♂ ⫲ ♆
03 09	☽ bQ ♅
04 18	☽ ∥ ♃
10 53	☽ ☍ ♆
11 48	☽ ∥ ♅
14 05	☽ ☌ ♂
14 29	☽ ⫲ ☉
14 57	☿ St D
16 09	☽ ⊥ ☉
17 23	♀ ∥ ♂
17 54	♃ ± ♅
19 42	☿ ⊻ ♀
21 20	☽ ± ♅
21 24	☽ ☌ ♃
22 12	☽ △ ♇
10 Saturday	
05 27	☽ ⊻ ☉
09 29	☽ ⊼ ♅
12 00	☉ ∠ ♄
12 06	☽ Q ♄
13 56	☉ ⫲ ♅
11 Sunday	
00 31	♀ □ ♄
05 58	☽ ∥ ☿
10 31	☽ ⫲ ☿
11 12	☽ ☌ ☿
12 45	☽ ✶ ♄
13 36	☽ ⊻ ♀
23 50	☽ ⊼ ♆
23 51	♃ △ ♇
12 Monday	
02 47	☽ ⊥ ♀
03 49	☉ ☍ ♅
06 32	☽ ⊻ ♂
11 19	☽ □ ♇
11 31	☽ ⊻ ♃
12 01	☽ ± ♆
18 18	☉ ⫲ ♃
19 23	☽ ⊥ ♂
19 31	☽ ∠ ♄
20 09	♀ bQ ♅
21 07	☉ ⊥ ♃
22 20	☽ ☍ ♅
22 31	☽ ∠ ♀
23 54	☽ ⊥ ♃
13 Tuesday	
00 06	☽ ☌ ☉
04 06	☽ ⫲ ♅
06 14	☽ bQ ♆
09 48	☽ ⫲ ♃
09 54	☿ ✶ ♄
12 06	☽ ∥ ☉
13 54	☽ ⊥ ♄
14 37	☽ ∠ ♂
17 01	☽ ⫲ ♂
18 16	☽ ⫲ ♀
18 24	☽ ∠ ♃
23 47	☽ Q ♇
14 Wednesday	
01 16	☽ ∥ ♆
02 05	☽ ⊻ ♄
02 56	☽ ⊻ ☿
07 13	☽ ✶ ♀
12 25	☽ △ ♆
15 49	☽ ⊥ ☿
17 14	☉ ⫲ ♂
22 22	☽ ✶ ♂
23 51	☽ ✶ ♇
15 Thursday	
00 31	♂ ± ♅
00 58	☽ ✶ ♃
01 54	☉ ⫲ ♀
09 06	☽ Q ♀
10 23	☽ ⊼ ♅
11 08	☽ ∠ ☿
17 39	☽ ⊻ ☉
22 12	☽ ± ♅
23 26	☽ Q ♂
16 Friday	
00 24	☉ bQ ♆
01 07	☽ Q ♃
03 36	♂ △ ♇
05 33	☽ ∠ ♇
06 31	☽ ⊥ ☉
14 06	☽ ☌ ♄
15 48	☽ bQ ♅
19 15	☽ ✶ ☿
23 08	☽ ⊥ ♇
23 17	☽ □ ♀
23 33	☽ □ ♆
17 Saturday	
01 36	☽ ∠ ☉
02 53	♀ ☍ ♆
10 45	☽ ⊻ ♇
12 20	☽ □ ♂
12 41	☽ □ ♃
20 40	☽ △ ♅
20 42	☽ Q ☿
22 40	♂ ☌ ♃
18 Sunday	
08 48	☽ ✶ ☉
10 00	☽ Q ♆
11 25	☉ ∥ ♆
16 11	☿ ⊼ ♆
23 55	☽ ⊻ ♄
19 Monday	
08 25	☽ ✶ ♆
09 09	☽ Q ☉
10 13	☽ □ ☿
11 09	☽ ⊥ ♄
12 37	☽ △ ♀
19 12	☽ ☌ ♇
21 50	☽ △ ♃
23 26	☽ △ ♂
20 Tuesday	
01 25	♂ ∥ ♃
03 46	☽ ∠ ♄
04 25	☽ □ ♅
11 47	☽ ∠ ♆
14 08	☉ ⊥ ♄
17 58	☽ bQ ♀
20 31	☽ □ ☉
21 Wednesday	
01 13	☽ bQ ♃
03 41	☽ bQ ♂
03 50	☽ ⊥ ♆
06 27	♀ ∥ ♃
06 49	☽ ✶ ♄
11 07	☽ ± ♀
12 26	☽ Q ♅
14 22	☽ ⊻ ♆
17 16	☽ ± ♃
20 12	☽ ± ♂
22 11	☽ △ ☿
22 22	☽ ⊼ ♀
22 Thursday	
00 39	☽ ⊻ ♇
02 04	♂ ∥ ♅
02 30	☿ ⊻ ♀
03 48	☽ ⊼ ♃
03 53	☽ Q ♄
05 06	♀ ± ♅
07 02	☽ ⊼ ♂
09 09	☽ ✶ ♅
10 58	☽ ⊥ ♇
15 38	☽ ∥ ☉
15 59	♄ bQ ♅
18 55	♀ ∥ ♅
21 22	☿ □ ♇
22 08	☿ ± ♆
23 Friday	
01 13	☽ ∥ ♆
02 14	☽ ∠ ♇
02 48	☽ bQ ☿
04 22	☽ △ ☉
07 43	♀ △ ♇
10 23	☽ ∠ ♅
10 34	☽ □ ♄
17 17	☽ ☌ ♆
18 29	☽ ⫲ ♃
19 33	☽ ± ☿
19 52	☽ ⫲ ♅
21 29	☽ ⫲ ♀
22 09	☽ ⫲ ♂
24 Saturday	
01 13	☽ ⊥ ♅
01 40	☉ ∠ ♃
03 07	☽ ✶ ♇
04 22	♂ ⊼ ♅
04 29	☽ ☍ ♀
06 34	☽ ⊼ ☿
06 41	☽ ☍ ♃
06 59	☽ bQ ☉
07 51	☿ ⊻ ♃
08 12	☽ ∥ ☿
11 00	☽ ⊻ ♅
11 18	☽ ☍ ♂
22 38	☽ ± ☉
22 38	☽ Q ♇
22 38	☉ Q ♇
25 Sunday	
08 58	☽ ⊼ ☉
11 47	☽ △ ♄
17 51	☽ ⊻ ♆
20 03	♀ ☌ ♃
22 59	☿ ☍ ♅
26 Monday	
03 24	☽ ⊥ ♆
03 28	☽ □ ♇
03 31	♀ ∥ ♂
06 39	☿ ∠ ♄
07 28	☽ ⊼ ♃
07 49	☿ ⫲ ♀
08 06	☿ ⫲ ♂
08 07	☽ ⊼ ♀
10 05	☽ ∥ ♀
10 10	☽ ∥ ♂
10 38	☽ ⫲ ☿
10 57	☽ ☌ ♅
11 50	☽ bQ ♄
12 25	☽ ☍ ☿
13 24	☽ ⊼ ♂
15 17	☽ ∥ ♅
15 47	☽ ∥ ♃
17 06	☽ ± ♃
17 41	☽ ∠ ♆
18 19	☽ ± ♀
23 20	☽ ± ♂
27 Tuesday	
02 14	☽ ± ♄
02 44	☿ ⊻ ♂
07 34	☽ bQ ♃
07 58	☉ ⊻ ♄
09 24	☽ ⫲ ♆
09 38	☽ bQ ♀
11 50	☽ ⊼ ♄
12 05	☽ ☍ ☉
14 15	☽ bQ ♂
17 31	☽ ✶ ♆
17 35	☿ ⫲ ♅
18 52	☿ ⫲ ♃
28 Wednesday	
01 23	♀ ⊼ ♅
03 15	☽ △ ♇
07 26	☽ ⫲ ☉
07 49	☽ △ ♃
10 40	☽ ⊻ ♅
11 22	☽ △ ♀
12 01	♃ ∥ ♅
12 42	☿ ⊥ ♃
12 45	☽ Q ♆
13 00	☿ bQ ♆
15 20	☽ △ ♂
18 10	☽ ⊼ ☿
20 21	☽ ⊥ ♅
29 Thursday	
03 34	☽ bQ ♇
05 10	☽ ± ☿
11 02	☽ ∠ ♅
12 41	☽ ☍ ♄
16 09	☽ ⊼ ☉
18 10	☽ □ ♆
18 25	☽ ± ♇
21 58	☽ bQ ☿
30 Friday	
02 55	☽ ± ☉
04 28	☽ ⊼ ♇
09 49	☽ □ ♃
12 04	☽ ✶ ♅
13 59	♂ Q ♄
16 49	☽ □ ♀
19 22	☽ bQ ☉
19 24	☽ □ ♂
20 44	☉ ∠ ♂
21 06	☉ △ ♆
31 Saturday	
02 52	☽ △ ☿
08 42	☽ Q ♅
16 17	☽ ⊼ ♄
17 24	☽ Q ♃
21 43	☽ △ ♆
23 43	☽ △ ☉

All ephemeris data is given at 12.00 UT and the Moon's longitude is additionally given for 24.00 UT

NOVEMBER 2015

LONGITUDES

Date	Sidereal time h m s	Sun ☉ ° ' "	Moon ☽ ° ' "	Moon ☽ 24.00 ° ' "	Mercury ☿ ° '	Venus ♀ ° '	Mars ♂ ° '	Jupiter ♃ ° '	Saturn ♄ ° '	Uranus ♅ ° '	Neptune ♆ ° '	Pluto ♇ ° '
01	14 41 50	08 ♏ 44 19	15 ♋ 01 57	21 ♋ 33 54	28 ♎ 41	22 ♍ 25	23 ♍ 06	16 ♍ 49	04 ♐ 09	17 ♈ 42	07 ♓ 06	13 ♑ 19
02	14 45 47	09 44 21	27 ♋ 59 26	04 ♌ 18 59	00 ♏ 20	23 28	23 42	16 59	04 15	17 R 40	07 R 06	13 20
03	14 49 43	10 44 24	10 ♌ 33 06	16 ♌ 42 22	02 00	24 31	24 19	17 09	04 22	17 38	07 05	13 22
04	14 53 40	11 44 30	22 ♌ 47 25	28 ♌ 48 54	03 39	25 34	24 55	17 19	04 29	17 36	07 05	13 23
05	14 57 36	12 44 38	04 ♍ 47 27	10 ♍ 43 42	05 18	26 38	25 32	17 28	04 35	17 33	07 04	13 24
06	15 01 33	13 44 48	16 ♍ 38 18	22 ♍ 31 50	06 57	27 42	26 08	17 38	04 42	17 31	07 04	13 25
07	15 05 30	14 45 00	28 ♍ 24 51	04 ♎ 17 54	08 35	28 46	26 44	17 48	04 49	17 29	07 03	13 26
08	15 09 26	15 45 14	10 ♎ 11 26	16 ♎ 05 55	10 14	29 ♍ 51	27 21	17 57	04 56	17 27	07 03	13 28
09	15 13 23	16 45 30	22 ♎ 01 44	27 ♎ 59 12	11 52	00 ♎ 55	27 57	18 07	05 03	17 25	07 03	13 29
10	15 17 19	17 45 48	03 ♏ 58 38	10 ♏ 00 15	13 29	02 01	28 33	18 16	05 10	17 23	07 02	13 30
11	15 21 16	18 46 07	16 ♏ 04 16	22 ♏ 10 50	15 07	03 06	29 09	18 25	05 17	17 21	07 02	13 32
12	15 25 12	19 46 29	28 ♏ 20 04	04 ♐ 32 04	16 44	04 12	29 ♍ 45	18 35	05 23	17 19	07 02	13 33
13	15 29 09	20 46 52	10 ♐ 46 53	17 ♐ 04 35	18 21	05 18	00 ♎ 22	18 44	05 30	17 17	07 02	13 34
14	15 33 05	21 47 16	23 ♐ 25 11	29 ♐ 48 45	19 57	06 24	00 58	18 52	05 37	17 15	07 01	13 36
15	15 37 02	22 47 43	06 ♑ 15 20	12 ♑ 45 00	21 33	07 30	01 34	19 01	05 44	17 14	07 01	13 37
16	15 40 59	23 48 10	19 ♑ 17 49	25 ♑ 53 53	23 09	08 37	02 10	19 10	05 51	17 12	07 01	13 39
17	15 44 55	24 48 39	02 ♒ 33 21	09 ♒ 16 18	24 44	09 44	02 46	19 18	05 58	17 10	07 01	13 40
18	15 48 52	25 49 09	16 ♒ 02 55	22 ♒ 53 18	26 20	10 51	03 21	19 27	06 05	17 08	07 01	13 42
19	15 52 48	26 49 41	29 ♒ 47 35	06 ♓ 45 52	27 55	11 59	03 57	19 35	06 13	17 06	07 D 01	13 43
20	15 56 45	27 50 13	13 ♓ 48 09	20 ♓ 54 24	29 ♏ 30	13 06	04 33	19 43	06 20	17 05	07 01	13 45
21	16 00 41	28 50 47	28 ♓ 04 30	05 ♈ 18 11	01 ♐ 04	14 14	05 09	19 51	06 27	17 03	07 01	13 46
22	16 04 38	29 ♏ 51 22	12 ♈ 35 07	19 ♈ 54 45	02 38	15 22	05 45	19 59	06 34	17 01	07 01	13 48
23	16 08 34	00 ♐ 51 58	27 ♈ 16 29	04 ♉ 39 32	04 13	16 30	06 20	20 07	06 41	17 00	07 02	13 50
24	16 12 31	01 52 35	12 ♉ 03 02	19 ♉ 26 01	05 47	17 39	06 56	20 15	06 48	16 58	07 02	13 51
25	16 16 28	02 53 14	26 ♉ 47 29	04 ♊ 06 25	07 20	18 47	07 31	20 22	06 55	16 57	07 02	13 53
26	16 20 24	03 53 54	11 ♊ 21 51	18 ♊ 32 53	08 54	19 56	08 07	20 29	07 02	16 55	07 02	13 55
27	16 24 21	04 54 36	25 ♊ 38 45	02 ♋ 38 49	10 28	21 05	08 43	20 37	07 09	16 54	07 02	13 56
28	16 28 17	05 55 19	09 ♋ 32 38	16 ♋ 19 54	12 01	22 14	09 18	20 44	07 16	16 53	07 03	13 58
29	16 32 14	06 56 03	23 ♋ 00 29	29 ♋ 34 28	13 35	23 23	09 53	20 51	07 23	16 51	07 03	14 00
30	16 36 10	07 ♐ 56 49	06 ♌ 02 02	12 ♌ 23 29	15 ♐ 08	24 ♎ 33	10 ♎ 29	20 ♍ 58	07 ♐ 31	16 ♈ 50	07 ♓ 04	14 ♑ 01

Date	Moon True ☊ ° '	Moon Mean ☊ ° '	Moon ☽ Latitude ° '
01	00 ♎ 18	28 ♍ 49	05 S 02
02	00 R 16	28 46	04 36
03	00 15	28 42	03 58
04	00 D 16	28 39	03 10
05	00 17	28 36	02 14
06	00 19	28 33	01 14
07	00 20	28 30	00 S 10
08	00 R 19	28 27	00 N 53
09	00 17	28 23	01 55
10	00 12	28 20	02 52
11	00 05	28 17	03 41
12	29 ♍ 56	28 14	04 22
13	29 46	28 11	04 50
14	29 36	28 07	05 04
15	29 27	28 04	05 04
16	29 20	28 01	04 48
17	29 16	27 58	04 16
18	29 13	27 55	03 30
19	29 D 13	27 52	02 31
20	29 13	27 48	01 22
21	29 14	27 45	00 N 06
22	29 R 13	27 42	01 S 11
23	29 10	27 39	02 24
24	29 05	27 36	03 28
25	28 56	27 33	04 17
26	28 46	27 29	04 50
27	28 35	27 26	05 03
28	28 25	27 23	04 58
29	28 16	27 20	04 36
30	28 ♍ 09	27 ♍ 17	04 S 00

DECLINATIONS

Date	Sun ☉ ° '	Moon ☽ ° '	Mercury ☿ ° '	Venus ♀ ° '	Mars ♂ ° '	Jupiter ♃ ° '	Saturn ♄ ° '	Uranus ♅ ° '	Neptune ♆ ° '	Pluto ♇ ° '
01	14 S 25	17 N 36	09 S 36	03 N 32	04 N 01	06 N 08	19 S 19	06 N 20	09 S 39	21 S 04
02	14 44	16 03	10 17	03 12	03 46	06 04	19 21	06 19	09 39	21 04
03	15 03	13 46	10 57	02 51	03 32	06 00	19 22	06 18	09 40	21 04
04	15 21	10 55	11 36	02 31	03 18	05 56	19 23	06 17	09 40	21 04
05	15 40	07 40	12 15	02 10	03 04	05 53	19 25	06 17	09 40	21 04
06	15 58	04 08	12 54	01 49	02 49	05 49	19 26	06 16	09 40	21 04
07	16 16	00 N 28	13 32	01 27	02 35	05 46	19 27	06 15	09 40	21 04
08	16 33	03 S 13	14 09	01 05	02 21	05 42	19 29	06 14	09 40	21 04
09	16 50	06 48	14 46	00 43	02 06	05 39	19 30	06 13	09 40	21 04
10	17 07	10 09	15 22	00 N 21	01 52	05 35	19 31	06 13	09 41	21 04
11	17 24	13 06	15 57	00 S 01	01 38	05 32	19 33	06 12	09 41	21 04
12	17 41	15 32	16 31	00 24	01 24	05 28	19 34	06 11	09 41	21 04
13	17 57	17 16	17 05	00 47	01 10	05 25	19 35	06 11	09 41	21 04
14	18 13	18 12	17 38	01 10	00 55	05 22	19 37	06 10	09 41	21 04
15	18 28	18 13	18 10	01 33	00 41	05 18	19 38	06 09	09 41	21 04
16	18 43	17 18	18 41	01 56	00 27	05 15	19 39	06 08	09 41	21 04
17	18 58	15 25	19 11	02 20	00 N 13	05 12	19 40	06 08	09 41	21 04
18	19 12	12 41	19 40	02 43	00 S 01	05 09	19 42	06 07	09 41	21 04
19	19 27	09 11	20 08	03 07	00 15	05 06	19 43	06 07	09 41	21 04
20	19 40	05 07	20 35	03 30	00 29	05 03	19 44	06 06	09 41	21 04
21	19 54	00 S 40	21 01	03 54	00 43	05 00	19 46	06 05	09 41	21 04
22	20 07	03 N 53	21 27	04 18	00 57	04 57	19 47	06 05	09 41	21 04
23	20 20	08 16	21 51	04 42	01 11	04 54	19 48	06 04	09 41	21 04
24	20 32	12 09	22 14	05 06	01 25	04 51	19 49	06 04	09 41	21 04
25	20 44	15 16	22 36	05 30	01 39	04 49	19 51	06 03	09 40	21 04
26	20 55	17 21	22 56	05 54	01 53	04 46	19 52	06 03	09 40	21 04
27	21 07	18 19	23 16	06 19	02 07	04 43	19 53	06 02	09 40	21 04
28	21 17	18 09	23 34	06 43	02 21	04 41	19 54	06 02	09 40	21 04
29	21 28	16 56	23 52	07 07	02 35	04 38	19 55	06 01	09 40	21 04
30	21 S 38	14 N 53	24 S 08	07 S 31	02 S 48	04 N 36	19 S 57	06 N 01	09 S 40	21 S 04

ZODIAC SIGN ENTRIES

Date	h	m	Planets
02	07	06	☿ ♏
02	15	48	☽ ♌
05	02	22	☽ ♍
07	15	14	☽ ♎
08	15	31	♀ ♎
10	04	02	☽ ♏
12	15	14	☽ ♐
12	21	41	♂ ♎
15	00	21	☽ ♑
17	07	24	☽ ♒
19	12	21	☽ ♓
20	19	43	☿ ♐
21	15	12	☽ ♈
22	15	25	☉ ♐
23	16	26	☽ ♉
25	17	15	☽ ♊
27	19	27	☽ ♋
30	00	47	☽ ♌

LATITUDES

Date	Mercury ☿ ° '	Venus ♀ ° '	Mars ♂ ° '	Jupiter ♃ ° '	Saturn ♄ ° '	Uranus ♅ ° '	Neptune ♆ ° '	Pluto ♇ ° '
01	01 N 30	00 N 34	01 N 23	01 N 00	01 N 41	00 S 40	00 S 49	01 N 43
04	01 12	00 49	01 24	01 01	01 40	00 40	00 49	01 42
07	00 53	01 03	01 24	01 01	01 40	00 40	00 49	01 42
10	00 33	01 15	01 25	01 02	01 40	00 40	00 49	01 41
13	00 N 13	01 27	01 25	01 02	01 39	00 40	00 48	01 41
16	00 S 07	01 37	01 26	01 03	01 39	00 40	00 48	01 41
19	00 27	01 46	01 26	01 04	01 39	00 40	00 48	01 40
22	00 46	01 53	01 27	01 04	01 39	00 39	00 48	01 40
25	01 05	02 00	01 27	01 05	01 38	00 39	00 48	01 39
28	01 21	02 06	01 27	01 06	01 38	00 39	00 48	01 39
31	01 S 37	02 N 10	01 N 28	01 N 06	01 N 38	00 S 39	00 S 48	01 N 39

DATA

Julian Date	2457328
Delta T	+69 seconds
Ayanamsa	24° 04' 40"
Synetic vernal point	05° ♓ 02' 19"
True obliquity of ecliptic	23° 26' 05"

MOON'S PHASES, APSIDES AND POSITIONS ☽

Date	h	m	Phase	Longitude ° '	Eclipse Indicator
03	12	24	☾	10 ♌ 45	
11	17	47	●	19 ♏ 01	
19	06	27	☽	26 ♒ 36	
25	22	44	○	03 ♊ 20	

Day	h	m		
07	22	00	Apogee	
23	20	13	Perigee	
07	15	03	OS	
15	00	35	Max dec	18° S 20'
21	15	32	ON	
27	20	08	Max dec	18° N 23'

LONGITUDES

Date	Chiron ⚷ ° '	Ceres ⚳ ° '	Pallas ⚴ ° '	Juno ⚵ ° '	Vesta ⚶ ° '	Black Moon Lilith ⚸ ° '
01	17 ♓ 16	01 ♒ 04	00 ♑ 21	16 ♎ 54	29 ♓ 09	07 ♎ 34
11	17 ♓ 04	03 ♒ 31	03 ♑ 43	20 ♎ 27	28 ♓ 36	08 ♎ 40
21	16 ♓ 58	06 ♒ 16	07 ♑ 13	23 ♎ 54	28 ♓ 48	09 ♎ 47
31	16 ♓ 57	09 ♒ 17	10 ♑ 47	27 ♎ 13	29 ♓ 43	10 ♎ 53

All ephemeris data is given at 12.00 UT and the Moon's longitude is additionally given for 24.00 UT

ASPECTARIAN

h m	Aspects
01 Sunday	
02 02	☿ ⊥ ♀
02 57	☽ Q ♀
03 05	☽ ± ♄
03 40	☿ ⊥ ♄
04 32	☽ Q ♂
04 45	♀ Q ♄
08 53	☽ ☍ ♇
13 42	☿ ∥ ♆
15 17	☽ ⚹ ♃
16 52	☽ □ ♅
19 36	☽ bQ ♄
21 34	☿ ⊥ ♂
02 Monday	
00 59	☽ bQ ♆
02 47	☽ ⚹ ♀
03 35	☽ ⚹ ♂
17 06	☽ □ ☿
17 52	☽ ± ♆
19 39	☽ ∠ ♃
23 59	☽ △ ♄
03 Tuesday	
00 56	☽ ǂ ☉
01 10	♀ ☌ ♂
02 42	☿ Q ♇
05 19	☽ ⊼ ♆
09 29	☽ ∠ ♂
09 48	☽ ∠ ♀
12 24	☽ □ ☉
13 10	☽ ⊥ ♃
14 25	☿ ∠ ♃
17 28	☽ ⊼ ♇
04 Wednesday	
01 02	☽ ⊻ ♃
01 46	☽ △ ♅
03 57	☽ ⊥ ♂
05 02	☽ ⊥ ♀
05 15	☽ ± ♇
07 35	☽ ǂ ☿
09 23	☽ Q ☿
16 27	☽ ⊻ ♂
18 03	☽ ⊻ ♀
21 34	☽ ǂ ♆
23 09	☽ bQ ♇
05 Thursday	
00 56	☿ ⊻ ♄
03 07	☽ Q ☉
07 31	☽ bQ ♅
11 36	☽ □ ♄
13 12	☽ ⚹ ☿
16 36	☽ ☍ ♆
21 40	☽ ∥ ♅
22 01	♃ ⊼ ♅
06 Friday	
00 33	☽ ∥ ♃
01 38	☽ ± ♅
04 01	☉ ⚹ ♇
05 27	☽ △ ♇
05 35	☽ ⚹ ☉
13 39	☿ △ ♆
13 48	☽ ⊼ ♅
14 04	☽ ☌ ♃
21 18	☽ ∥ ♂
07 Saturday	
00 28	☽ Q ♄
00 34	☽ ∠ ☿
04 57	☽ ∥ ♀
08 24	☽ ☌ ♂
12 47	☽ ☌ ♀
14 59	☽ ∠ ☉
21 54	☽ ⊥ ☿
23 21	☽ ǂ ♀
08 Sunday	
01 11	☽ ⚹ ♄
05 36	☽ ⊼ ♆
06 38	☽ ǂ ♂
11 02	☽ ⊥ ☉
12 05	☽ ⊻ ☿
17 48	☽ ± ♆
18 40	☽ □ ♇
09 Monday	
00 21	☽ ⊻ ☉
02 42	☽ ☍ ♅
03 59	☽ ⊻ ♃
04 14	☽ ǂ ♃
07 57	☽ ∠ ♄
08 03	☽ ǂ ♅
12 02	☽ bQ ♆
16 16	☽ ⊥ ♃
10 Tuesday	
00 33	☽ ⊻ ♂
02 16	☽ ⊥ ♄
03 16	☉ ⊼ ♅
07 03	☽ Q ♇
07 40	☽ ⊻ ♀
08 28	☽ ∥ ♆
10 34	☽ ∠ ♃
12 14	☿ ⚹ ♇
13 13	☽ ⊥ ♂
13 27	☿ ∠ ♂
14 23	☽ ⊻ ♄
18 06	☽ △ ♆
20 50	☽ ⊥ ♀
11 Wednesday	
02 17	☉ ⚹ ♃
06 58	☽ ⚹ ♇
08 01	☽ ∠ ♂
09 49	☽ ☌ ☿
14 31	☽ ⊼ ♅
16 23	☽ ∠ ♀
16 41	☽ ⚹ ♃
17 47	☽ ☌ ☉
12 Thursday	
02 15	☽ ± ♅
12 25	☽ ∠ ♇
14 54	☽ ⚹ ♂
16 24	☽ Q ♃
19 42	☽ bQ ♅
20 36	☿ ⊼ ♅
13 Friday	
00 27	☽ ⚹ ♀
01 47	☽ ☌ ♄
04 48	☽ □ ♆
05 50	☽ ⊥ ♇
06 46	☽ ∥ ☿
15 10	☽ Q ♂
17 12	♀ ⚹ ♄
17 21	☽ ⊻ ♇
18 18	☿ ⚹ ♃
14 Saturday	
00 22	☽ △ ♅
01 36	☽ Q ♀
02 48	♀ ǂ ♂
03 18	☽ □ ♃
04 30	☽ ⊻ ☿
08 39	☽ ⊻ ☉
12 38	☽ ∥ ☉
15 01	☽ Q ♆
17 27	☽ ⊥ ☿
20 55	☽ ⊥ ☉
23 13	☽ ∥ ☉
15 Sunday	
01 33	♀ ⊼ ♆
02 50	☽ □ ♂
11 02	☽ ⊻ ♄
12 38	☽ ∠ ☿
13 25	☽ ⚹ ♆
13 33	☽ ∥ ☿
14 32	☽ □ ♀
15 06	☽ ∠ ☉
21 57	☉ ± ♅
22 14	☽ ⊥ ♄
16 Monday	
01 38	☽ ☌ ♇
08 10	☽ □ ♅
11 45	☽ △ ♃
12 41	☿ ± ♅
14 52	☽ ∠ ♄
15 59	☉ ∥ ☿
16 58	☽ ∠ ♆
19 59	☽ ⚹ ☿
20 53	☽ ⚹ ☉
17 Tuesday	
09 14	☽ ⊥ ♆
11 36	☿ ∠ ♀
12 23	☽ △ ♂
14 53	☉ ☌ ☿
15 10	☽ bQ ♃
16 40	☽ Q ♅
18 10	☽ ⚹ ♄
19 59	☽ ⊻ ♆
20 14	☽ Q ☉
20 30	☽ Q ☿
18 Wednesday	
01 59	☽ △ ♀
04 53	☉ ∠ ♀
07 21	☽ ± ♃
07 50	☽ ⊻ ♇
13 38	☿ ∥ ♄
13 55	☽ ⚹ ♅
15 38	☽ Q ♄
16 15	☽ bQ ♂
16 28	♆ St D
18 02	☽ ⊼ ♃
18 25	☽ ⊥ ♇
19 Thursday	
06 27	☽ □ ☉
06 41	☽ bQ ♀
08 19	☽ □ ☿
08 40	☽ ± ♂
08 51	☽ ∥ ♆
10 08	☽ ∠ ♇
15 59	☽ ∠ ♅
19 30	☽ ⊼ ♂
23 08	☽ □ ♄
23 35	☽ ± ♀
20 Friday	
00 26	☽ ☌ ♆
00 29	☿ ∠ ♇
06 25	☽ ǂ ♅
07 23	☽ ⊥ ♅
10 15	♀ ± ♆
10 42	☽ ⊼ ♀
11 54	☽ ⚹ ♇
12 21	☽ ǂ ♃
17 32	☽ ⊻ ♅
19 31	☉ ∥ ♄
20 06	☽ ∥ ♀
22 06	☽ ☍ ♃
21 Saturday	
02 02	♀ □ ♇
08 09	☽ Q ♇
10 13	☉ ∠ ♇
11 44	☽ ∥ ♂
13 23	☽ △ ☉
14 21	☿ ∥ ♇
17 35	☽ △ ☿
19 44	☽ ǂ ♂
22 Sunday	
00 15	☽ ☍ ♂
01 05	☿ Q ♃
02 00	☽ △ ♄
02 45	☿ bQ ♅
02 50	☽ ⊻ ♆
12 43	☽ ⊥ ♆
14 00	☽ □ ♇
14 28	☽ ǂ ♀
16 00	☽ bQ ☉
16 57	☽ ☍ ♀
17 38	☽ ∥ ♃
19 16	☽ ☌ ♅
21 17	☽ bQ ☿
23 47	☽ ∥ ♅
23 Monday	
00 14	☽ ⊼ ♃
02 49	☽ bQ ♄
03 27	☽ ∠ ♆
07 47	☽ ± ☉
10 06	☽ ± ♃
13 42	☽ ± ☿
17 35	☽ ± ♄
18 16	☽ ⊼ ☉
20 18	☽ ǂ ♆
22 13	♀ ☍ ♅
22 40	♀ ǂ ♃
24 Tuesday	
00 37	☽ ⊼ ☿
00 51	☽ bQ ♃
03 20	☽ ⊼ ♂
03 24	☽ ⊼ ♄
03 51	☽ ⚹ ♆
05 19	♂ ⚹ ♄
13 29	☽ ± ♂
14 15	☉ bQ ♅
14 56	☽ △ ♇
15 58	♂ ⊼ ♆
19 59	☽ ⊻ ♅
21 51	☽ ⊼ ♀
21 55	☉ Q ♃
23 21	☽ Q ♆
25 Wednesday	
01 26	☽ △ ♃
04 44	☽ bQ ♂
04 57	☿ ☌ ♄
05 44	☽ ⊥ ♅
07 15	☿ □ ♆
08 27	☽ ± ♀
15 26	☽ bQ ♇
16 32	☿ ⚹ ♂
20 26	☽ ∠ ♅
20 27	☿ ⊥ ♇
22 44	☽ ☍ ☉
26 Thursday	
00 27	☽ bQ ♀
04 46	☽ ☍ ♄
04 50	☽ □ ♆
06 16	☽ ± ♇
06 23	☽ △ ♂
07 26	☽ ☍ ☿
12 15	♄ □ ♆
16 15	☽ ⊼ ♇
19 52	♀ ǂ ♅
21 16	☽ ⚹ ♅
27 Friday	
01 01	♀ ⊻ ♃
03 24	☽ □ ♃
03 35	☽ △ ♀
05 40	☉ ∥ ♇
17 33	☽ Q ♅
28 Saturday	
05 10	☽ ⊼ ☉
07 38	☽ △ ♆
08 00	☽ ⊼ ♄
08 04	♀ bQ ♆
10 34	☽ Q ♃
11 33	☽ □ ♂
12 52	♀ ∠ ♄
16 31	☽ ± ☉
16 55	☽ ⊼ ☿
18 38	☽ ± ♄
19 49	☽ ☍ ♇
29 Sunday	
00 57	☽ □ ♅
04 59	☽ ± ☿
08 03	☽ ⚹ ♃
09 54	☽ bQ ☉
10 16	☽ bQ ♆
10 52	☽ bQ ♄
12 46	☽ □ ♀
14 49	☉ □ ♆
18 35	☿ ⊻ ♇
21 19	☽ Q ♂
23 32	☽ bQ ☿
30 Monday	
00 16	☉ ☌ ♄
02 44	☽ ± ♆
11 52	☽ ∠ ♃
13 53	☉ ⊥ ♇
13 55	☽ ⊼ ♆
14 48	☽ △ ♄
15 54	☽ △ ☉
20 47	☽ ⚹ ♂

DECEMBER 2015

LONGITUDES

Date	Sidereal time h m s	Sun ☉ ° ' "	Moon ☽ ° ' "	Moon ☽ 24.00 ° ' "	Mercury ☿ ° '	Venus ♀ ° '	Mars ♂ ° '	Jupiter ♃ ° '	Saturn ♄ ° '	Uranus ♅ ° '	Neptune ♆ ° '	Pluto ♇ ° '
01	16 40 07	08 ♐ 57 36	18 ♌ 39 18	24 ♌ 49 59	16 ♐ 41	25 ♎ 43	11 ♎ 04	21 ♍ 04	07 ♐ 38	16 ♈ 49	07 ♓ 04	14 ♑ 03
02	16 44 03	09 58 25	00 ♍ 56 08	06 ♍ 58 23	18 14	26 52	11 39	21 11	07 45	16 R 48	07 04	14 05
03	16 48 00	10 59 15	12 ♍ 57 27	18 ♍ 54 01	19 47	28 02	12 15	21 17	07 52	16 47	07 05	14 07
04	16 51 57	12 00 07	24 ♍ 48 47	00 ♎ 42 27	21 20	29 ♎ 12	12 50	21 23	07 59	16 45	07 05	14 09
05	16 55 53	13 01 00	06 ♎ 35 42	12 ♎ 29 12	22 53	00 ♏ 23	13 25	21 29	08 06	16 44	07 06	14 10
06	16 59 50	14 01 54	18 ♎ 23 34	24 ♎ 19 22	24 26	01 33	14 00	21 35	08 13	16 43	07 07	14 12
07	17 03 46	15 02 49	00 ♏ 17 09	06 ♏ 17 21	25 59	02 44	14 35	21 41	08 20	16 42	07 07	14 14
08	17 07 43	16 03 46	12 ♏ 20 23	18 ♏ 26 35	27 31	03 54	15 10	21 47	08 27	16 41	07 08	14 16
09	17 11 39	17 04 44	24 ♏ 36 13	00 ♐ 49 27	29 ♐ 04	05 05	15 45	21 52	08 34	16 41	07 09	14 18
10	17 15 36	18 05 43	07 ♐ 06 23	13 ♐ 27 04	00 ♑ 36	06 16	16 20	21 57	08 42	16 40	07 09	14 20
11	17 19 32	19 06 43	19 ♐ 51 27	26 ♐ 19 26	02 09	07 27	16 55	22 03	08 49	16 39	07 10	14 22
12	17 23 29	20 07 43	02 ♑ 50 52	09 ♑ 25 34	03 41	08 38	17 29	22 08	08 56	16 38	07 11	14 24
13	17 27 26	21 08 45	16 ♑ 03 18	22 ♑ 43 52	05 12	09 49	18 04	22 12	09 03	16 38	07 12	14 26
14	17 31 22	22 09 47	29 ♑ 27 02	06 ♒ 12 36	06 44	11 01	18 39	22 17	09 10	16 37	07 13	14 27
15	17 35 19	23 10 50	13 ♒ 00 25	19 ♒ 50 18	08 15	12 12	19 13	22 21	09 17	16 36	07 13	14 29
16	17 39 15	24 11 53	26 ♒ 42 11	03 ♓ 36 00	09 45	13 24	19 48	22 26	09 24	16 36	07 14	14 31
17	17 43 12	25 12 57	10 ♓ 31 42	17 ♓ 29 17	11 15	14 35	20 22	22 30	09 31	16 35	07 15	14 33
18	17 47 08	26 14 01	24 ♓ 28 45	01 ♈ 30 04	12 44	15 47	20 56	22 34	09 38	16 35	07 16	14 35
19	17 51 05	27 15 05	08 ♈ 33 13	15 ♈ 38 06	14 13	16 59	21 31	22 37	09 44	16 35	07 17	14 37
20	17 55 01	28 16 10	22 ♈ 44 35	29 ♈ 52 27	15 40	18 11	22 05	22 41	09 51	16 34	07 18	14 39
21	17 58 58	29 ♐ 17 15	07 ♉ 01 23	14 ♉ 10 59	17 06	19 23	22 39	22 44	09 58	16 34	07 20	14 41
22	18 02 55	00 ♑ 18 20	21 ♉ 20 46	28 ♉ 30 09	18 30	20 35	23 13	22 48	10 05	16 34	07 21	14 43
23	18 06 51	01 19 25	05 ♊ 38 29	12 ♊ 45 05	19 53	21 47	23 47	22 51	10 12	16 34	07 22	14 45
24	18 10 48	02 20 31	19 ♊ 49 15	26 ♊ 50 18	21 13	22 59	24 21	22 54	10 19	16 34	07 23	14 47
25	18 14 44	03 21 37	03 ♋ 47 34	10 ♋ 40 28	22 31	24 12	24 55	22 56	10 25	16 34	07 24	14 50
26	18 18 41	04 22 44	17 ♋ 28 33	24 ♋ 11 25	23 46	25 24	25 29	22 59	10 32	16 D 34	07 25	14 52
27	18 22 37	05 23 51	00 ♌ 48 51	07 ♌ 20 44	24 57	26 36	26 03	23 01	10 39	16 34	07 27	14 54
28	18 26 34	06 24 58	13 ♌ 47 05	20 ♌ 08 04	26 04	27 49	26 36	23 03	10 45	16 34	07 28	14 56
29	18 30 30	07 26 05	26 ♌ 23 55	02 ♍ 35 02	27 07	29 ♏ 02	27 10	23 05	10 52	16 34	07 29	14 58
30	18 34 27	08 27 14	08 ♍ 41 50	14 ♍ 44 53	28 04	00 ♐ 14	27 43	23 07	10 59	16 34	07 31	15 00
31	18 38 24	09 ♑ 28 22	20 ♍ 44 44	26 ♍ 42 03	28 ♑ 54	01 ♐ 27	28 ♎ 17	23 ♍ 09	11 ♐ 05	16 ♈ 34	07 ♓ 32	15 ♑ 02

Date	Moon True ☊ ° '	Moon Mean ☊ ° '	Moon ☽ Latitude ° '
01	28 ♍ 05	27 ♍ 13	03 S 13
02	28 R 03	27 10	02 19
03	28 D 02	27 07	01 19
04	28 02	27 04	00 S 17
05	28 R 02	27 01	00 N 45
06	28 00	26 58	01 46
07	27 56	26 54	02 42
08	27 49	26 51	03 32
09	27 39	26 48	04 13
10	27 27	26 45	04 42
11	27 13	26 42	04 58
12	26 59	26 39	04 59
13	26 46	26 35	04 44
14	26 36	26 32	04 13
15	26 28	26 29	03 28
16	26 24	26 26	02 30
17	26 22	26 23	01 23
18	26 D 21	26 19	00 N 10
19	26 R 21	26 16	01 S 04
20	26 20	26 13	02 15
21	26 16	26 10	03 17
22	26 10	26 07	04 07
23	26 00	26 04	04 42
24	25 48	26 00	04 59
25	25 36	25 57	04 57
26	25 23	25 54	04 39
27	25 12	25 51	04 05
28	25 03	25 48	03 19
29	24 57	25 44	02 25
30	24 54	25 41	01 25
31	24 ♍ 53	25 ♍ 38	00 S 22

DECLINATIONS

Date	Sun ☉ ° '	Moon ☽ ° '	Mercury ☿ ° '	Venus ♀ ° '	Mars ♂ ° '	Jupiter ♃ ° '	Saturn ♄ ° '	Uranus ♅ ° '	Neptune ♆ ° '	Pluto ♇ ° '
01	21 S 47	12 N 10	24 S 22	07 S 55	03 S 02	04 N 33	19 S 58	06 N 00	09 S 40	21 S 04
02	21 56	08 58	24 36	08 19	03 16	04 31	19 59	06 00	09 39	21 03
03	22 05	05 28	24 48	08 43	03 29	04 29	20 00	05 59	09 39	21 03
04	22 13	01 N 48	24 59	09 07	03 43	04 26	20 01	05 59	09 39	21 03
05	22 21	01 S 55	25 09	09 30	03 56	04 24	20 03	05 59	09 39	21 03
06	22 29	05 35	25 17	09 54	04 10	04 22	20 04	05 58	09 39	21 03
07	22 36	09 02	25 24	10 17	04 23	04 20	20 05	05 58	09 38	21 03
08	22 42	12 10	25 29	10 41	04 37	04 18	20 06	05 58	09 38	21 03
09	22 48	14 50	25 33	11 04	04 50	04 16	20 07	05 57	09 38	21 03
10	22 54	16 51	25 36	11 27	05 03	04 14	20 08	05 57	09 37	21 03
11	22 59	18 06	25 37	11 50	05 17	04 13	20 09	05 57	09 37	21 03
12	23 04	18 26	25 37	12 13	05 30	04 11	20 10	05 56	09 37	21 03
13	23 08	17 46	25 35	12 35	05 43	04 09	20 11	05 56	09 37	21 03
14	23 12	16 08	25 32	12 57	05 56	04 08	20 13	05 56	09 36	21 03
15	23 16	13 35	25 27	13 19	06 09	04 06	20 14	05 56	09 36	21 02
16	23 18	10 16	25 21	13 41	06 22	04 05	20 15	05 56	09 36	21 02
17	23 21	06 21	25 13	14 03	06 35	04 03	20 16	05 56	09 35	21 02
18	23 23	02 S 02	25 04	14 24	06 48	04 02	20 17	05 55	09 35	21 02
19	23 24	02 N 25	24 53	14 45	07 00	04 01	20 18	05 55	09 34	21 02
20	23 25	06 46	24 41	15 05	07 13	04 00	20 19	05 55	09 34	21 02
21	23 26	10 45	24 28	15 26	07 26	03 58	20 20	05 55	09 34	21 02
22	23 26	14 07	24 13	15 46	07 38	03 57	20 21	05 55	09 33	21 02
23	23 26	16 37	23 57	16 05	07 51	03 57	20 22	05 55	09 33	21 02
24	23 25	18 05	23 39	16 25	08 03	03 56	20 23	05 55	09 32	21 01
25	23 24	18 25	23 21	16 44	08 15	03 55	20 24	05 55	09 32	21 01
26	23 22	17 41	23 01	17 02	08 28	03 54	20 25	05 55	09 31	21 01
27	23 19	15 59	22 41	17 20	08 40	03 53	20 26	05 55	09 31	21 01
28	23 17	13 30	22 20	17 38	08 52	03 53	20 26	05 55	09 30	21 01
29	23 14	10 26	21 58	17 56	09 04	03 52	20 27	05 55	09 30	21 01
30	23 10	07 00	21 36	18 12	09 16	03 52	20 28	05 55	09 29	21 01
31	23 S 06	03 N 20	21 S 14	18 S 29	09 S 28	03 N 51	20 S 29	05 N 56	09 S 29	21 S 01

ZODIAC SIGN ENTRIES

Date	h	m	Planets
02	10	09	☽ ♍
04	22	34	☽ ♎
05	04	15	♀ ♏
07	11	26	☽ ♏
09	22	25	☽ ♐
10	02	34	☿ ♑
12	06	46	☽ ♑
14	12	59	☽ ♒
16	17	45	☽ ♓
18	21	26	☽ ♈
21	00	13	☽ ♉
22	04	48	☉ ♑
23	02	31	☽ ♊
25	05	26	☽ ♋
27	10	31	☽ ♌
29	18	58	☽ ♍
30	07	16	♀ ♐

LATITUDES

Date	Mercury ☿ ° '	Venus ♀ ° '	Mars ♂ ° '	Jupiter ♃ ° '	Saturn ♄ ° '	Uranus ♅ ° '	Neptune ♆ ° '	Pluto ♇ ° '
01	01 S 37	02 N 10	01 N 28	01 N 06	01 N 38	00 S 39	00 S 48	01 N 39
04	01 50	02 13	01 28	01 07	01 38	00 39	00 48	01 38
07	02 01	02 16	01 28	01 08	01 38	00 39	00 48	01 38
10	02 10	02 17	01 29	01 09	01 38	00 39	00 48	01 38
13	02 15	02 17	01 29	01 09	01 38	00 39	00 48	01 37
16	02 16	02 16	01 29	01 10	01 38	00 39	00 48	01 37
19	02 13	02 15	01 29	01 11	01 38	00 39	00 48	01 37
22	02 04	02 12	01 30	01 12	01 38	00 38	00 48	01 36
25	01 49	02 09	01 30	01 13	01 38	00 38	00 48	01 36
28	01 25	02 05	01 30	01 13	01 38	00 38	00 48	01 36
31	00 S 52	02 N 00	01 N 30	01 N 14	01 N 38	00 S 38	00 S 48	01 N 35

DATA

Julian Date	2457358
Delta T	+69 seconds
Ayanamsa	24° 04' 44"
Synetic vernal point	05° ♓ 02' 15"
True obliquity of ecliptic	23° 26' 04"

MOON'S PHASES, APSIDES AND POSITIONS ☽

Date	h	m	Phase	Longitude ° '	Eclipse Indicator
03	07	40	☾	10 ♍ 48	
11	10	29	●	19 ♐ 03	
18	15	14	☽	26 ♓ 22	
25	11	11	○	03 ♋ 20	

Day	h	m		
05	15	01	Apogee	
21	09	06	Perigee	
04	23	35	0S	
12	08	12	Max dec	18° S 26'
18	23	02	0N	
25	07	26	Max dec	18° N 27'

LONGITUDES

Date	Chiron ⚷ ° '	Ceres ⚳ ° '	Pallas ⚴ ° '	Juno ⚵ ° '	Vesta ⚶ ° '	Black Moon Lilith ⚸ ° '
01	16 ♓ 57	09 ♒ 17	10 ♑ 47	27 ♎ 13	29 ♓ 43	10 ♎ 53
11	17 ♓ 01	12 ♒ 30	14 ♑ 25	00 ♏ 22	01 ♈ 14	12 ♎ 00
21	17 ♓ 12	15 ♒ 53	18 ♑ 06	03 ♏ 19	03 ♈ 18	13 ♎ 06
31	17 ♓ 27	19 ♒ 25	21 ♑ 47	06 ♏ 03	05 ♈ 50	14 ♎ 13

ASPECTARIAN

h m Aspects

01 Tuesday
01 33 ☽ Q ♀
03 09 ☽ ⊼ ♇
05 03 ☽ ⊥ ♃
07 40 ☽ △ ☿
08 28 ☽ △ ♅
14 00 ☿ △ ♅
14 43 ☽ ± ♇
16 43 ☽ ⊻ ♃

02 Wednesday
03 09 ☽ ∠ ♂
03 09 ☽ ⚹ ♀
07 04 ☽ ⋕ ♆
08 20 ☽ ⅊ ♇
13 42 ☽ ⅊ ♅
16 09 ☽ ⋕ ♀
18 57 ♀ ⊥ ♃
21 51 ☽ ⊥ ♂

03 Thursday
00 13 ☽ ☍ ♆
01 41 ☽ □ ♄
07 37 ☽ ± ♅
07 40 ☽ □ ☉
08 32 ☽ ∥ ♅
10 29 ☽ ⊻ ♂
12 11 ☽ ∠ ♀
14 20 ☽ △ ♇
18 38 ☽ ∥ ♃
19 41 ☽ ⊼ ♅

04 Friday
00 17 ☽ ⋕ ♂
03 52 ☽ □ ☿
04 59 ☽ ☌ ♃
08 23 ☽ ⊥ ♀
12 53 ☿ □ ♃
14 24 ☽ Q ♄
21 56 ☽ ⊻ ♀
22 52 ♂ ± ♆
23 33 ☽ Q ☉

05 Saturday
13 02 ☽ ⊼ ♆
15 06 ☽ ⚹ ♄
20 28 ♀ ∥ ♆
22 04 ☽ Q ☿

06 Sunday
01 16 ☽ ± ♆
02 01 ☽ ∥ ♂
02 19 ☽ ⚹ ☉
02 37 ☽ ☌ ♂
03 28 ☽ □ ♇
04 03 ☽ ⋕ ♃
07 30 ♄ ⊥ ♇
08 37 ☽ ☍ ♅
10 14 ☉ ⚹ ♂
14 39 ☽ ⋕ ♅
16 12 ☉ ⊻ ♇
18 32 ☽ ⊻ ♃
19 32 ☽ ⅊ ♆
20 52 ♂ □ ♇
21 52 ☽ ∠ ♄
22 36 ☿ Q ♆

07 Monday
01 41 ♀ Q ♇
02 03 ☽ ⚹ ☿
03 11 ♀ ⊥ ♄
06 44 ☽ ⊥ ♃
07 00 ♂ ⋕ ♃
11 29 ☽ ∠ ☉
15 55 ☽ Q ♇
16 09 ☽ ⊥ ♄
16 23 ☽ ∥ ♆
17 25 ☽ ☌ ♀
22 33 ☽ ∥ ♀

08 Tuesday
00 53 ☽ ∠ ♃
01 40 ☽ △ ♆
03 07 ☿ Q ♂
04 14 ☽ ⊻ ♄
07 05 ☽ ⊥ ☉
12 25 ☽ ∠ ☿
15 49 ☽ ⚹ ♇
17 51 ☽ ⊻ ♂
20 00 ☽ ⊻ ☉
20 33 ☽ ⊼ ♅

09 Wednesday
02 38 ☉ △ ♅
06 11 ☽ ⊥ ♂
06 39 ☽ ⚹ ♃
08 16 ☽ ± ♅
08 35 ☽ ⊥ ☿
21 05 ☽ ∠ ♇
21 50 ☽ ⊻ ☿

10 Thursday
00 26 ☽ ∠ ♂
01 37 ☽ ⅊ ♅
05 57 ☽ Q ♃
10 14 ☽ ⊻ ♀
12 06 ☽ □ ♆
14 20 ☽ ⊥ ♇
15 02 ☽ ☌ ♄
22 46 ☽ ⊥ ♀

11 Friday
01 31 ♂ ☍ ♅
01 41 ☽ ⊻ ♇
03 07 ♀ ∠ ♃
06 01 ☽ △ ♅
06 13 ♀ △ ♆
06 14 ☽ ⚹ ♂
10 29 ☽ ☌ ☉
16 06 ☽ □ ♃
17 19 ☽ ∠ ♀
21 52 ☽ Q ♆

12 Saturday
05 33 ☽ Q ♂
13 43 ☽ ☌ ☿
18 33 ♀ ⊻ ♄
19 55 ☽ ⚹ ♆
23 12 ☽ ⊻ ♄
23 36 ☽ ⚹ ♀

13 Sunday
09 03 ☽ ☌ ♇
10 10 ☽ ⊥ ♄
13 02 ☽ □ ♅
15 47 ☽ □ ♂
21 55 ☽ ⊻ ☉
23 03 ☽ ∠ ♆
23 07 ☽ △ ♃
23 23 ☽ Q ♀

14 Monday
02 29 ☽ ∠ ♄
09 31 ☽ ⊥ ☉
12 13 ♂ ⋕ ♅
15 02 ☉ □ ♃
15 08 ☽ ⊥ ♆
19 40 ☿ ⚹ ♆
21 10 ☽ Q ♅

15 Tuesday
01 47 ☽ ⊻ ♆
01 58 ☽ ⅊ ♃
02 33 ☽ ⊻ ☿
02 48 ☽ ∠ ☉
05 22 ☽ ⚹ ♄
10 27 ☽ □ ♀
13 53 ☽ ∥ ♀
14 27 ☽ ⊥ ☿
14 37 ☽ ⊻ ♇
17 55 ☽ ± ♃
18 20 ☽ ⚹ ♅
23 24 ☽ △ ♂

16 Wednesday
01 10 ☽ ⊥ ♇
02 39 ☽ Q ♄
04 30 ☽ ⊼ ♃
05 46 ☿ ⊻ ♄
07 17 ☽ ⚹ ☉
08 11 ☽ ∠ ☿
16 20 ☽ ∥ ♆
16 56 ☽ ∠ ♇
20 31 ☽ ∠ ♅

17 Thursday
02 41 ☽ ⅊ ♂
05 49 ☽ Q ☉
06 20 ☽ ☌ ♆
10 13 ☽ □ ♄
10 42 ☽ ∥ ♂
11 19 ♀ ⚹ ♇
12 06 ☽ ⊥ ♅
12 58 ☉ Q ♆
13 24 ☽ ⚹ ☿
14 24 ☽ ⋕ ♅
18 55 ☽ ± ♂
18 58 ☽ ⚹ ♇
19 40 ☽ △ ♀
22 27 ☽ ⊻ ♅

18 Friday
01 01 ☽ ⋕ ♃
05 40 ☽ ⊼ ♂
08 42 ☽ ☍ ♃
12 30 ☽ Q ☿
15 14 ☽ □ ☉
15 37 ☽ Q ♇
23 47 ☽ ⅊ ♀

19 Saturday
03 56 ♀ ⊼ ♅
09 51 ☽ ⊻ ♆
14 02 ☽ △ ♄
16 30 ☽ ± ♀
18 56 ☿ ☌ ♇
20 02 ☽ ⊥ ♆
20 40 ☽ ∥ ♃
22 19 ☽ □ ♇
22 41 ☽ □ ☿

20 Sunday
01 35 ☽ ☌ ♅
03 36 ☽ ⊼ ♀
07 16 ☽ ∥ ♅
10 50 ☽ ☍ ♂
11 16 ☽ ∠ ♆
11 54 ☽ ⊼ ♃
14 45 ☽ ⋕ ♂
15 27 ☿ ⊥ ♄
15 35 ☽ ⅊ ♄
21 49 ♂ ⅊ ♆
22 01 ☽ △ ☉
22 02 ☽ ± ♃

21 Monday
03 07 ☿ □ ♅
04 32 ☽ ⋕ ♆
06 50 ☽ ± ♄
12 30 ☽ ⚹ ♆
13 13 ☽ ⅊ ♃
16 13 ♂ ⊻ ♃
16 59 ☽ ⊼ ♄

22 Tuesday
00 53 ☽ △ ♇
01 06 ☽ ⅊ ☉
03 59 ☽ ⊻ ♅
06 43 ☽ △ ☿
08 38 ☽ Q ♆
10 36 ☽ ☍ ♀
14 02 ☽ ⊥ ♅
14 26 ☽ △ ♃
15 16 ☽ ⊼ ♂
17 20 ☽ ± ☉

23 Wednesday
01 45 ☽ ± ♂
02 05 ☽ ⅊ ♇
04 11 ☽ ⊼ ☉
05 08 ☽ ∠ ♅
05 09 ☽ ⋕ ♀
10 35 ☽ ⅊ ☿
14 54 ☽ □ ♆
17 16 ☽ ± ♇
17 31 ☽ ⅊ ♂
19 45 ☽ ☍ ♄

24 Thursday
03 22 ☽ ± ☿
03 26 ☽ ⊼ ♇
03 29 ♀ ± ♅
06 27 ☽ ⚹ ♅
10 03 ♀ ⚹ ♃
14 38 ☽ ⊼ ☿
17 16 ☽ □ ♃
17 55 ☽ ⊼ ♀
20 04 ☽ △ ♂

25 Friday
02 57 ☽ Q ♅
05 11 ☽ ± ♀
08 12 ☉ ∥ ☿
09 51 ☿ ∠ ♆
11 11 ☽ ☍ ☉
18 17 ☽ △ ♆
20 19 ☿ △ ♃
22 19 ☽ ⅊ ♀
23 39 ☽ ⊼ ♄

26 Saturday
00 30 ☽ Q ♃
03 55 ♅ St D
07 21 ☽ ☍ ♇
10 19 ☽ ± ♄
10 22 ☽ □ ♅
14 58 ♂ ∠ ♄
14 58 ♀ ⊻ ♂
20 50 ☽ ⅊ ♆
20 58 ☽ ⋕ ♀
21 51 ☽ ⚹ ♃

27 Sunday
00 21 ☽ ☍ ☿
02 33 ☽ ⅊ ♄
02 57 ☽ □ ♂
03 36 ☽ △ ♀
13 09 ☽ ± ♆
21 07 ☽ ⊼ ☉

28 Monday
00 12 ☽ ⊼ ♆
01 17 ☽ ∠ ♃
04 23 ☿ ∠ ♄
06 17 ☽ △ ♄
09 13 ☽ ± ☉
13 36 ☽ Q ♂
14 09 ☽ ⊼ ♇
17 14 ☽ △ ♅
18 11 ☽ ⊥ ♃

29 Tuesday
01 33 ☽ ± ♇
03 43 ☽ ⅊ ☉
05 37 ☽ ⊻ ♃
13 17 ☉ ⚹ ♆
13 30 ☽ ⊼ ☿
13 33 ☽ ⚹ ♂
14 53 ☿ □ ♂
17 38 ☽ □ ♀
18 49 ☽ ⋕ ♆
18 55 ☽ ⅊ ♇
21 18 ☽ ⋕ ♂
22 01 ☽ ⅊ ♅

30 Wednesday
02 09 ☽ ± ☿
07 04 ♀ ∠ ♇
09 39 ☽ ☍ ♆
11 29 ☽ △ ☉
15 42 ☽ ± ♅
16 33 ☽ □ ♄
19 07 ☽ ∥ ♅
20 21 ☽ ∠ ♂
21 19 ☽ ⅊ ☿

31 Thursday
00 32 ☽ △ ♇
03 38 ☽ ⊼ ♅
08 35 ☽ ∥ ♃
09 07 ☽ Q ♀
14 08 ♂ ∥ ♆
14 22 ♀ ⅊ ♅
15 14 ☽ ⊥ ♂
16 50 ☽ ☌ ♃

All ephemeris data is given at 12.00 UT and the Moon's longitude is additionally given for 24.00 UT

JANUARY 2016

LONGITUDES

Date	Sidereal time h m s	Sun ☉ ° ′ ″	Moon ☽ ° ′ ″	Moon ☽ 24.00 ° ′ ″	Mercury ☿ ° ′	Venus ♀ ° ′	Mars ♂ ° ′	Jupiter ♃ ° ′	Saturn ♄ ° ′	Uranus ♅ ° ′	Neptune ♆ ° ′	Pluto ♇ ° ′
01	18 42 20	10 ♑ 29 31	02 ♎ 37 29	08 ♎ 31 45	29 ♑ 38	02 ♐ 40	28 ♎ 50	23 ♍ 10	11 ♐ 12	16 ♈ 35	07 ♓ 33	15 ♑ 04
02	18 46 17	11 30 40	14 ♎ 25 32	20 ♎ 19 32	00 ♒ 13	03 53	29 23	23 11	11 18	16 35	07 35	15 06
03	18 50 13	12 31 50	26 ♎ 14 28	02 ♏ 10 59	00 40	05 06	29 ♎ 56	23 12	11 25	16 35	07 36	15 08
04	18 54 10	13 32 59	08 ♏ 09 43	14 ♏ 11 17	00 57	06 19	00 ♏ 30	23 13	11 31	16 36	07 38	15 10
05	18 58 06	14 34 09	20 ♏ 16 13	26 ♏ 25 00	01 03	07 32	01 03	23 14	11 38	16 36	07 39	15 12
06	19 02 03	15 35 20	02 ♐ 38 01	08 ♐ 55 37	00 R 57	08 45	01 36	23 14	11 44	16 37	07 41	15 14
07	19 05 59	16 36 30	15 ♐ 17 59	21 ♐ 45 15	00 40	09 58	02 08	23 14	11 50	16 37	07 42	15 16
08	19 09 56	17 37 41	28 ♐ 17 54	04 ♑ 54 26	00 ♒ 12	11 11	02 41	23 R 14	11 56	16 38	07 44	15 18
09	19 13 53	18 38 51	11 ♑ 36 02	18 ♑ 21 58	29 ♑ 31	12 24	03 14	23 14	12 03	16 39	07 46	15 21
10	19 17 49	19 40 01	25 ♑ 11 50	02 ♒ 05 12	28 40	13 38	03 46	23 14	12 09	16 40	07 47	15 23
11	19 21 46	20 41 11	09 ♒ 01 36	16 ♒ 00 31	27 39	14 51	04 19	23 13	12 15	16 41	07 49	15 25
12	19 25 42	21 42 21	23 ♒ 01 26	00 ♓ 03 53	26 30	16 04	04 51	23 12	12 21	16 41	07 51	15 27
13	19 29 39	22 43 30	07 ♓ 07 25	14 ♓ 11 37	25 15	17 18	05 23	23 12	12 27	16 42	07 52	15 29
14	19 33 35	23 44 38	21 ♓ 16 09	28 ♓ 20 45	23 57	18 31	05 55	23 10	12 33	16 43	07 54	15 31
15	19 37 32	24 45 46	05 ♈ 25 11	12 ♈ 29 15	22 37	19 45	06 27	23 09	12 39	16 44	07 56	15 33
16	19 41 28	25 46 53	19 ♈ 32 49	26 ♈ 35 46	21 20	20 58	06 59	23 08	12 45	16 45	07 58	15 35
17	19 45 25	26 47 59	03 ♉ 37 59	10 ♉ 39 22	20 06	22 12	07 31	23 06	12 50	16 46	08 00	15 37
18	19 49 22	27 49 05	17 ♉ 39 45	24 ♉ 39 01	18 57	23 25	08 03	23 04	12 56	16 48	08 01	15 39
19	19 53 18	28 50 09	01 ♊ 36 56	08 ♊ 33 17	17 56	24 39	08 34	23 02	13 02	16 49	08 03	15 41
20	19 57 15	29 ♑ 51 13	15 ♊ 27 49	22 ♊ 20 12	17 03	25 53	09 06	23 00	13 07	16 50	08 05	15 43
21	20 01 11	00 ♒ 52 16	29 ♊ 10 08	05 ♋ 57 17	16 19	27 06	09 37	22 57	13 13	16 51	08 07	15 45
22	20 05 08	01 53 19	12 ♋ 41 20	19 ♋ 21 58	15 45	28 20	10 08	22 55	13 19	16 53	08 09	15 47
23	20 09 04	02 54 20	25 ♋ 58 54	02 ♌ 31 57	15 19	29 ♐ 34	10 39	22 52	13 24	16 54	08 11	15 49
24	20 13 01	03 55 21	09 ♌ 00 54	15 ♌ 25 42	15 03	00 ♑ 48	11 10	22 49	13 29	16 56	08 13	15 51
25	20 16 57	04 56 20	21 ♌ 46 18	28 ♌ 02 47	14 55	02 01	11 41	22 45	13 35	16 57	08 15	15 53
26	20 20 54	05 57 20	04 ♍ 15 17	10 ♍ 24 01	14 D 56	03 15	12 12	22 42	13 40	16 59	08 17	15 55
27	20 24 51	06 58 18	16 ♍ 29 17	22 ♍ 31 26	15 04	04 29	12 42	22 39	13 45	17 00	08 19	15 57
28	20 28 47	07 59 16	28 ♍ 30 54	04 ♎ 28 11	15 20	05 43	13 13	22 35	13 50	17 02	08 21	15 59
29	20 32 44	09 00 13	10 ♎ 23 48	16 ♎ 18 19	15 42	06 57	13 43	22 31	13 55	17 04	08 23	16 01
30	20 36 40	10 01 09	22 ♎ 12 22	28 ♎ 06 35	16 10	08 11	14 13	22 27	14 00	17 05	08 25	16 03
31	20 40 37	11 ♒ 02 05	04 ♏ 01 36	09 ♏ 58 04	16 ♑ 44	09 ♑ 25	14 ♏ 43	22 ♍ 23	14 ♐ 05	17 ♈ 07	08 ♓ 27	16 ♑ 05

Date	Moon True ☊ ° ′	Moon Mean ☊ ° ′	Moon ☽ Latitude ° ′
01	24 ♍ 53	25 ♍ 35	00 N 41
02	24 R 53	25 32	01 42
03	24 52	25 29	02 39
04	24 49	25 25	03 29
05	24 44	25 22	04 10
06	24 36	25 19	04 41
07	24 25	25 16	04 59
08	24 13	25 13	05 02
09	24 01	25 10	04 49
10	23 50	25 06	04 19
11	23 41	25 03	03 34
12	23 35	25 00	02 35
13	23 31	24 57	01 26
14	23 30	24 54	00 N 12
15	23 D 30	24 50	01 S 03
16	23 31	24 47	02 15
17	23 R 31	24 44	03 18
18	23 29	24 41	04 09
19	23 25	24 38	04 45
20	23 18	24 35	05 03
21	23 10	24 31	05 05
22	23 01	24 28	04 49
23	22 51	24 25	04 17
24	22 43	24 22	03 33
25	22 37	24 19	02 38
26	22 33	24 16	01 37
27	22 31	24 12	00 S 33
28	22 D 31	24 09	00 N 33
29	22 32	24 06	01 36
30	22 34	24 03	02 34
31	22 ♍ 35	24 ♍ 00	03 N 26

DECLINATIONS

Date	Sun ☉ ° ′	Moon ☽ ° ′	Mercury ☿ ° ′	Venus ♀ ° ′	Mars ♂ ° ′	Jupiter ♃ ° ′	Saturn ♄ ° ′	Uranus ♅ ° ′	Neptune ♆ ° ′	Pluto ♇ ° ′
01	23 S 01	00 S 25	20 S 52	18 S 45	09 S 39	03 N 51	20 S 30	05 N 56	09 S 28	21 S 00
02	22 56	04 07	20 30	19 00	09 51	03 51	20 31	05 56	09 28	21 00
03	22 51	07 40	20 09	19 15	10 03	03 51	20 32	05 56	09 27	21 00
04	22 45	10 56	19 50	19 30	10 14	03 51	20 33	05 56	09 27	21 00
05	22 38	13 47	19 31	19 44	10 26	03 51	20 33	05 57	09 26	21 00
06	22 31	16 05	19 14	19 58	10 37	03 51	20 34	05 57	09 25	21 00
07	22 24	17 40	19 00	20 11	10 48	03 51	20 35	05 57	09 25	21 00
08	22 16	18 23	18 47	20 23	11 00	03 51	20 36	05 57	09 24	20 59
09	22 08	18 08	18 37	20 35	11 11	03 52	20 37	05 58	09 24	20 59
10	22 00	16 50	18 29	20 46	11 22	03 52	20 37	05 58	09 23	20 59
11	21 51	14 33	18 24	20 57	11 33	03 52	20 38	05 58	09 22	20 59
12	21 41	11 24	18 21	21 08	11 43	03 53	20 39	05 59	09 22	20 59
13	21 31	07 33	18 20	21 17	11 54	03 54	20 40	05 59	09 21	20 59
14	21 21	03 S 17	18 21	21 26	12 05	03 54	20 40	05 59	09 20	20 59
15	21 10	01 N 11	18 24	21 35	12 15	03 55	20 41	06 00	09 20	20 58
16	20 59	05 34	18 28	21 43	12 26	03 56	20 42	06 00	09 19	20 58
17	20 48	09 37	18 34	21 50	12 36	03 57	20 42	06 01	09 18	20 58
18	20 36	13 07	18 41	21 56	12 46	03 58	20 43	06 01	09 18	20 58
19	20 23	15 50	18 48	22 02	12 57	03 59	20 44	06 02	09 17	20 58
20	20 11	17 37	18 57	22 08	13 07	04 00	20 44	06 02	09 16	20 58
21	19 58	18 21	19 05	22 12	13 17	04 01	20 45	06 03	09 15	20 58
22	19 44	18 02	19 15	22 17	13 27	04 03	20 46	06 03	09 15	20 57
23	19 30	16 44	19 24	22 20	13 36	04 04	20 46	06 04	09 14	20 57
24	19 16	14 35	19 34	22 23	13 46	04 05	20 47	06 05	09 13	20 57
25	19 02	11 45	19 43	22 25	13 55	04 07	20 47	06 05	09 13	20 57
26	18 47	08 26	19 52	22 26	14 05	04 09	20 48	06 06	09 12	20 57
27	18 32	04 50	20 01	22 27	14 14	04 10	20 48	06 06	09 11	20 57
28	18 16	01 N 05	20 10	22 27	14 24	04 12	20 49	06 07	09 10	20 57
29	18 00	02 S 39	20 18	22 27	14 33	04 14	20 49	06 08	09 10	20 56
30	17 44	06 15	20 25	22 25	14 42	04 15	20 50	06 09	09 09	20 56
31	17 S 27	09 S 37	20 S 32	22 S 23	14 S 51	04 N 17	20 S 50	06 N 09	09 S 08	20 S 56

ZODIAC SIGN ENTRIES

Date	h m	Planets
01	06 41	☽ ♎
02	02 20	♀ ♐
03	14 32	♂ ♏
03	19 36	☽ ♏
06	06 56	☽ ♐
08	15 07	☽ ♑
08	19 36	☿ ♑
10	20 23	☽ ♒
12	23 53	☽ ♓
15	02 48	☽ ♈
17	05 48	☽ ♉
19	09 13	☽ ♊
20	15 27	☉ ♒
21	13 28	☽ ♋
23	19 21	☽ ♌
23	20 31	♀ ♑
26	03 46	☽ ♍
28	14 59	☽ ♎
31	03 50	☽ ♏

LATITUDES

Date	Mercury ☿ ° ′	Venus ♀ ° ′	Mars ♂ ° ′	Jupiter ♃ ° ′	Saturn ♄ ° ′	Uranus ♅ ° ′	Neptune ♆ ° ′	Pluto ♇ ° ′
01	00 S 39	01 N 59	01 N 30	01 N 15	01 N 38	00 S 38	00 S 48	01 N 35
04	00 N 07	01 53	01 30	01 15	01 38	00 38	00 48	01 35
07	01 02	01 47	01 30	01 16	01 38	00 38	00 48	01 35
10	01 59	01 40	01 30	01 17	01 38	00 38	00 48	01 34
13	02 48	01 33	01 30	01 18	01 38	00 38	00 48	01 34
16	03 19	01 25	01 30	01 19	01 38	00 37	00 47	01 34
19	03 27	01 17	01 29	01 20	01 39	00 37	00 47	01 34
22	03 17	01 09	01 29	01 20	01 39	00 37	00 47	01 33
25	02 54	01 00	01 29	01 21	01 39	00 37	00 47	01 33
28	02 24	00 52	01 29	01 22	01 39	00 37	00 47	01 33
31	01 N 52	00 N 43	01 N 28	01 N 23	01 N 39	00 S 37	00 S 47	01 N 33

DATA

Julian Date	2457389
Delta T	+69 seconds
Ayanamsa	24° 04′ 49″
Synetic vernal point	05° ♓ 02′ 10″
True obliquity of ecliptic	23° 26′ 04″

MOON'S PHASES, APSIDES AND POSITIONS ☽

Date	h m	Phase	Longitude ° ′	Eclipse Indicator
02	05 30	☾	11 ♎ 14	
10	01 31	●	19 ♑ 13	
16	23 26	☽	26 ♈ 16	
24	01 46	○	03 ♌ 29	

Day	h m		
02	11 54	Apogee	
15	02 09	Perigee	
30	09 06	Apogee	
01	09 20	0S	
08	17 54	Max dec	18° S 25′
15	05 39	0N	
21	16 44	Max dec	18° N 23′
28	18 57	0S	

LONGITUDES

Date	Chiron ⚷ ° ′	Ceres ⚳ ° ′	Pallas ⚴ ° ′	Juno ⚵ ° ′	Vesta ⚶ ° ′	Black Moon Lilith ⚸ ° ′
01	17 ♓ 29	19 ♒ 47	22 ♑ 09	06 ♏ 19	06 ♈ 06	14 ♎ 19
11	17 ♓ 50	23 ♒ 27	25 ♑ 51	08 ♏ 45	09 ♈ 02	15 ♎ 26
21	18 ♓ 15	27 ♒ 13	29 ♑ 32	10 ♏ 53	12 ♈ 16	16 ♎ 33
31	18 ♓ 44	01 ♓ 03	03 ♒ 10	12 ♏ 39	15 ♈ 46	17 ♎ 39

All ephemeris data is given at 12.00 UT and the Moon's longitude is additionally given for 24.00 UT

ASPECTARIAN

01 Friday
02 26 ☿ ∥ ♇
03 56 ☽ ⚺ ♂
04 59 ☽ Q ♄
05 33 ☽ △ ☿
12 05 ☽ ✶ ♀
22 03 ☽ ⚻ ♆

02 Saturday
02 54 ♂ ⊥ ♃
05 30 ☽ □ ☉
05 35 ☽ ✶ ♄
05 41 ☉ Q ♂
06 35 ☉ ⚺ ♄
10 14 ☽ # ♃
10 17 ☽ ± ♆
11 14 ☿ ∥ ♄
13 23 ☽ □ ♇
16 23 ☽ ☍ ♅
22 06 ☽ ∠ ♀

03 Sunday
00 07 ☽ # ♅
04 37 ☽ ꝑ ♆
05 50 ☽ ⚺ ♃
12 21 ☽ ∠ ♄
14 09 ♀ Q ♃
18 00 ☽ ⊥ ♃
18 26 ☽ ⊥ ♀
19 51 ☽ ☌ ♂
21 12 ☽ □ ☿
21 29 ☽ Q ☉

04 Monday
00 48 ☽ ∥ ♆
01 57 ☽ Q ♇
06 22 ☽ ∥ ♂
06 40 ☽ ⊥ ♄
07 53 ☽ ⚺ ♀
10 56 ☽ △ ♆
12 06 ☽ ∠ ♃
18 32 ♀ ⊥ ♂
18 46 ☽ ⚺ ♄
23 43 ☽ ✶ ☉

05 Tuesday
01 59 ☽ ✶ ♇
02 16 ☿ ∥ ♀
04 47 ☽ ⚻ ♅
09 35 ☽ Q ☿
12 00 ☿ □ ♂
13 05 ☿ St R
14 33 ♀ □ ♆
16 35 ☽ ± ♅
17 47 ☽ ✶ ♃

06 Wednesday
03 28 ☉ ☌ ♇
07 23 ☽ ∠ ♇
07 43 ☽ ∠ ☉
08 49 ☽ ✶ ☿
09 54 ☽ ⚺ ♂
10 02 ☽ ꝑ ♅
16 59 ☽ Q ♃
21 39 ☽ □ ♆
21 54 ☽ ⊥ ♂
21 57 ♀ ⊥ ♇

07 Thursday
00 38 ☽ ⊥ ♇
00 54 ☽ ☌ ♀
02 25 ☽ ⊥ ☉
05 27 ☽ ☌ ♄
11 57 ☽ ⚺ ♇
12 22 ☉ □ ♅
12 41 ☽ ∠ ☿
14 29 ☽ △ ♅
14 39 ☽ ⚺ ☉
15 35 ☽ ∠ ♂

08 Friday
02 44 ☽ □ ♃
04 40 ♃ St R
04 48 ☽ ⊥ ☿
07 19 ☽ Q ♆
15 19 ☽ ⚺ ☿
20 11 ☉ ⊥ ♄
20 20 ☽ ✶ ♂

09 Saturday
04 11 ♀ ☌ ♄
05 07 ☽ ✶ ♆
12 48 ☽ ⚺ ♄
13 35 ☽ ⚺ ♀
15 15 ♀ ∥ ♄
17 24 ♂ Q ♇
18 40 ☽ ☌ ♇
18 43 ☽ Q ♂
20 58 ☽ □ ♅
23 31 ☽ ⊥ ♄

10 Sunday
01 16 ☽ ⊥ ♀
01 31 ☽ ☌ ☉
07 46 ☽ ∠ ♆
08 33 ☽ △ ♃
12 23 ☿ ∠ ♀
15 26 ☽ ∠ ♄
16 25 ☉ ⊥ ♀
17 39 ☽ ☌ ☿
18 34 ☽ ∠ ♀
23 30 ☽ ⊥ ♆

11 Monday
03 31 ☽ □ ♂
04 28 ☽ Q ♅
09 55 ☽ ⚺ ♆
10 37 ☽ ꝑ ♃
15 57 ♀ ∥ ♇
17 35 ☽ ✶ ♄
19 56 ☿ ∠ ♄
22 59 ☽ ✶ ♀
23 00 ☽ ⚺ ♇
23 18 ♀ ⚺ ♇

12 Tuesday
01 09 ☽ ✶ ♅
02 04 ☽ ± ♃
09 18 ☽ ⊥ ♇
09 34 ☽ ⚺ ☉
09 51 ☽ ∥ ♂
12 19 ☽ ⚻ ♃
14 17 ☽ Q ♄
17 27 ☽ ⚺ ☿
20 36 ☽ ⊥ ☉
21 26 ☽ Q ♀

13 Wednesday
00 11 ♀ △ ♅
00 41 ☽ ∠ ♇
01 11 ☽ ∥ ♆
02 47 ☽ ∠ ♅
02 51 ☽ ⊥ ☿
08 56 ☽ △ ♂
13 06 ☽ ∠ ☉
13 17 ☽ ☌ ♆
15 37 ☉ ∠ ♆
16 52 ☽ ∠ ☿
18 05 ☽ ⊥ ♅
21 01 ☽ # ♅
21 06 ☽ □ ♄
22 49 ☉ △ ♃

14 Thursday
02 13 ☽ ✶ ♇
04 17 ☽ ⚺ ♅
05 20 ☉ ∥ ♀
06 34 ☿ ⊥ ♀
06 54 ☽ □ ♀
08 35 ☽ # ♃
11 23 ☽ ꝑ ♂
14 05 ☉ ☌ ☿
15 13 ☽ ☍ ♃
16 09 ☽ ✶ ☿
16 31 ☽ ✶ ☉
22 37 ☽ Q ♇

15 Friday
02 15 ☿ △ ♃
03 15 ☽ ± ♂
06 31 ☿ ∠ ♆
10 46 ☽ Q ☿
13 50 ☽ ⚻ ♂
14 27 ☽ Q ☉
16 16 ☽ ⚺ ♆
22 37 ♂ ⊥ ♄

16 Saturday
00 21 ☽ △ ♄
02 29 ☽ ⊥ ♆
02 54 ☽ ∥ ♃
05 15 ☽ □ ♇
07 15 ☽ ☌ ♅
13 37 ☉ ∥ ♇
14 30 ☽ ∥ ♅
14 39 ☽ △ ♀
14 47 ☽ □ ☿
15 25 ☿ ⚺ ♀
17 49 ☽ ∠ ♆
18 05 ☽ ⚻ ♃
23 26 ☽ □ ☉

17 Sunday
02 03 ☽ ꝑ ♄
04 17 ☽ ± ♃
10 02 ☽ # ♆
17 31 ☽ ± ♄
18 40 ☽ ꝑ ♀
18 53 ☽ ☍ ♂
19 28 ☽ ✶ ♆
19 37 ☽ ꝑ ♃
20 08 ☿ Q ♂
22 03 ☉ ∥ ♄
22 55 ♀ ∠ ♂

18 Monday
03 51 ☽ ⚻ ♄
05 10 ♀ □ ♃
08 33 ☽ △ ♇
09 19 ☽ # ♂
10 30 ☽ ⚺ ♅
11 01 ♂ △ ♆
11 33 ☽ ± ♀
12 20 ☿ ⊥ ♄
12 59 ♂ ∠ ♃
14 03 ☽ △ ☿
15 05 ☉ ∠ ♄
16 04 ☽ Q ♆
20 49 ☽ ⊥ ♅
21 15 ☽ △ ♃
22 51 ☽ ⚻ ♀

19 Tuesday
06 50 ☽ △ ☉
10 24 ☽ ꝑ ♇
12 20 ☽ ∠ ♅
14 08 ☽ ꝑ ☿
23 09 ☽ □ ♆

20 Wednesday
00 30 ☽ ⚻ ♂
02 00 ☽ ± ♇
04 46 ☽ ± ☿
07 54 ☽ ☍ ♄
10 51 ☽ ꝑ ☉
11 20 ☽ ± ♂
12 27 ☽ ⚻ ♇
14 24 ☽ ✶ ♅
14 37 ☽ ⚻ ☿
16 09 ♀ Q ♆
18 26 ☿ □ ♅

21 Thursday
01 07 ☽ □ ♃
03 41 ☽ ꝑ ♂
03 50 ☽ ± ☉
08 01 ☽ ☍ ♀
11 27 ☽ Q ♅
15 15 ☽ ⚻ ☉

22 Friday
03 53 ☽ △ ♆
07 15 ☽ △ ♂
08 50 ☽ Q ♃
09 59 ☿ ☌ ♇
13 07 ☽ ⚻ ♄
17 17 ☽ ☍ ☿
17 34 ☽ ☍ ♇
18 20 ☉ ⊥ ♆
19 32 ☽ □ ♅
23 59 ☽ ± ♄

23 Saturday
06 21 ☽ ✶ ♃
06 54 ☽ ꝑ ♆
16 27 ☽ ꝑ ♄
18 22 ☉ ∥ ☿
19 13 ☽ ⚻ ♀
23 23 ☽ ± ♆

24 Sunday
01 46 ☽ ☍ ☉
07 26 ☽ ± ♀
09 46 ☽ ∠ ♃
10 30 ☽ ⚻ ♆
16 11 ☽ □ ♂
19 00 ☽ # ♂
20 25 ☽ △ ♄
23 09 ☽ ⚻ ☿

25 Monday
00 50 ☽ ⚻ ♇
02 02 ☽ ꝑ ♀
02 33 ☽ ⊥ ♃
02 51 ☽ △ ♅
10 24 ☽ ± ☿
12 13 ☽ ± ♇
12 19 ☉ Q ♅
13 52 ☽ ⚺ ♃
21 51 ☿ St D

26 Tuesday
03 36 ☽ ꝑ ☿
03 48 ☽ Q ♂
05 31 ☽ ꝑ ♇
06 42 ☽ # ♆
07 35 ☽ ꝑ ♅
09 50 ☽ △ ♀
15 36 ☽ ⚻ ☉
19 52 ☽ ☍ ♆

27 Wednesday
01 10 ☽ ± ♅
03 41 ☽ ∥ ♅
04 12 ☽ ✶ ♂
04 25 ☽ ± ☉
06 33 ☽ □ ♄
09 10 ☽ △ ☿
10 56 ☽ △ ♇
13 02 ☽ ⚻ ♅
16 15 ☽ ∥ ♃
23 54 ☽ ꝑ ☉

28 Thursday
00 11 ☽ ☌ ♃
02 57 ☉ ꝑ ♃
11 21 ☽ ∠ ♂
18 44 ☽ Q ♄
20 45 ☉ ⚺ ♆

29 Friday
04 12 ☽ □ ♀
06 19 ☽ ⊥ ♂
07 54 ☽ ⚻ ♆
08 55 ☽ △ ☉
19 02 ☽ ⚺ ♂
19 12 ☽ ✶ ♄
20 07 ☽ ± ♆
22 26 ☽ # ♃
23 11 ☽ □ ☿
23 27 ☽ □ ♇
23 56 ♂ ⚺ ♄

30 Saturday
01 34 ☽ ☍ ♅
05 58 ☿ ☌ ♇
11 12 ☽ # ♅
12 29 ☽ ⚺ ♃
14 28 ☽ ꝑ ♆
16 44 ♀ ✶ ♆
21 01 ☽ Q ♀

31 Sunday
00 37 ☽ ⊥ ♃
01 55 ☽ ∠ ♄
08 24 ☽ ∥ ♆
12 07 ☽ Q ♇
13 30 ☽ Q ☿
18 44 ☽ ∠ ♃
20 16 ☽ ⊥ ♄
20 58 ☽ △ ♆

FEBRUARY 2016

LONGITUDES

Date	Sidereal time h m s	Sun ☉ ° ' "	Moon ☽ ° ' "	Moon ☽ 24.00 ° ' "	Mercury ☿ ° '	Venus ♀ ° '	Mars ♂ ° '	Jupiter ♃ ° '	Saturn ♄ ° '	Uranus ♅ ° '	Neptune ♆ ° '	Pluto ♇ ° '
01	20 44 33	12 ♒ 03 00	15 ♏ 56 41	21 ♏ 58 03	17 ♑ 22	10 ♑ 39	15 ♏ 13	22 ♍ 18	14 ♐ 10	17 ♈ 09	08 ♓ 29	16 ♑ 07
02	20 48 30	13 03 54	28 ♏ 02 50	04 ♐ 11 35	18 05	11 53	15 42	22 R 14	14 15	17 11	08 31	16 09
03	20 52 26	14 04 47	10 ♐ 24 51	16 ♐ 43 06	18 53	13 07	16 12	22 09	14 20	17 13	08 33	16 11
04	20 56 23	15 05 40	23 ♐ 06 45	29 ♐ 36 04	19 44	14 21	16 41	22 04	14 24	17 15	08 35	16 13
05	21 00 20	16 06 32	06 ♑ 11 17	12 ♑ 52 26	20 38	15 35	17 10	21 59	14 29	17 17	08 37	16 14
06	21 04 16	17 07 22	19 ♑ 39 29	26 ♑ 32 15	21 35	16 49	17 40	21 54	14 33	17 19	08 40	16 16
07	21 08 13	18 08 12	03 ♒ 30 23	10 ♒ 33 27	22 36	18 03	18 08	21 48	14 38	17 21	08 42	16 18
08	21 12 09	19 09 01	17 ♒ 40 52	24 ♒ 51 59	23 39	19 17	18 37	21 43	14 42	17 23	08 44	16 20
09	21 16 06	20 09 48	02 ♓ 06 03	09 ♓ 22 19	24 44	20 31	19 06	21 37	14 46	17 25	08 46	16 22
10	21 20 02	21 10 34	16 ♓ 39 57	23 ♓ 58 12	25 52	21 45	19 34	21 31	14 50	17 27	08 48	16 23
11	21 23 59	22 11 18	01 ♈ 16 18	08 ♈ 33 33	27 02	22 59	20 02	21 25	14 54	17 30	08 50	16 25
12	21 27 55	23 12 01	15 ♈ 49 23	23 ♈ 03 14	28 13	24 13	20 30	21 19	14 58	17 32	08 53	16 27
13	21 31 52	24 12 42	00 ♉ 14 40	07 ♉ 23 21	29 ♑ 27	25 27	20 58	21 13	15 02	17 34	08 55	16 29
14	21 35 49	25 13 22	14 ♉ 29 01	21 ♉ 31 26	00 ♒ 42	26 41	21 25	21 07	15 06	17 37	08 57	16 30
15	21 39 45	26 14 00	28 ♉ 30 30	05 ♊ 26 08	01 58	27 56	21 52	21 00	15 10	17 39	08 59	16 32
16	21 43 42	27 14 36	12 ♊ 18 17	19 ♊ 06 55	03 17	29 ♑ 10	22 20	20 54	15 14	17 41	09 02	16 34
17	21 47 38	28 15 10	25 ♊ 52 05	02 ♋ 33 46	04 36	00 ♒ 24	22 46	20 47	15 17	17 44	09 04	16 35
18	21 51 35	29 ♒ 15 43	09 ♋ 12 00	15 ♋ 46 49	05 57	01 38	23 13	20 40	15 21	17 46	09 06	16 37
19	21 55 31	00 ♓ 16 14	22 ♋ 18 16	28 ♋ 46 21	07 19	02 52	23 40	20 33	15 24	17 49	09 08	16 39
20	21 59 28	01 16 43	05 ♌ 11 09	11 ♌ 32 40	08 43	04 06	24 06	20 26	15 27	17 51	09 11	16 40
21	22 03 24	02 17 10	17 ♌ 51 00	24 ♌ 06 12	10 07	05 20	24 32	20 19	15 31	17 54	09 13	16 42
22	22 07 21	03 17 36	00 ♍ 18 22	06 ♍ 27 38	11 33	06 35	24 58	20 12	15 34	17 57	09 15	16 43
23	22 11 18	04 18 00	12 ♍ 34 09	18 ♍ 38 05	13 00	07 49	25 23	20 05	15 37	17 59	09 17	16 45
24	22 15 14	05 18 23	24 ♍ 39 39	00 ♎ 39 08	14 28	09 03	25 48	19 58	15 40	18 02	09 20	16 47
25	22 19 11	06 18 44	06 ♎ 36 48	12 ♎ 32 59	15 57	10 17	26 13	19 50	15 43	18 05	09 22	16 48
26	22 23 07	07 19 03	18 ♎ 28 05	24 ♎ 22 30	17 27	11 31	26 38	19 43	15 46	18 08	09 24	16 50
27	22 27 04	08 19 21	00 ♏ 16 41	06 ♏ 11 07	18 58	12 46	27 03	19 35	15 48	18 10	09 26	16 51
28	22 31 00	09 19 37	12 ♏ 06 19	18 ♏ 02 50	20 31	14 00	27 27	19 28	15 51	18 13	09 29	16 52
29	22 34 57	10 ♓ 19 52	24 ♏ 01 13	00 ♐ 02 03	22 ♒ 04	15 ♒ 14	27 ♏ 51	19 ♍ 20	15 ♐ 53	18 ♈ 16	09 ♓ 31	16 ♑ 54

Moon Nodes and Latitude

Date	Moon True ☊ ° '	Moon Mean ☊ ° '	Moon ☽ Latitude ° '
01	22 ♍ 35	23 ♍ 56	04 N 10
02	22 R 34	23 53	04 43
03	22 31	23 50	05 04
04	22 26	23 47	05 11
05	22 20	23 44	05 02
06	22 14	23 41	04 37
07	22 09	23 37	03 54
08	22 04	23 34	02 57
09	22 01	23 31	01 47
10	21 59	23 28	00 N 29
11	22 D 00	23 25	00 S 51
12	22 01	23 22	02 07
13	22 02	23 18	03 15
14	22 04	23 15	04 10
15	22 R 04	23 12	04 49
16	22 03	23 09	05 10
17	22 01	23 06	05 14
18	21 58	23 02	05 01
19	21 55	22 59	04 32
20	21 51	22 56	03 50
21	21 48	22 53	02 57
22	21 46	22 50	01 56
23	21 45	22 47	00 S 51
24	21 D 45	22 43	00 N 16
25	21 45	22 40	01 21
26	21 47	22 37	02 23
27	21 48	22 34	03 18
28	21 49	22 31	04 04
29	21 ♍ 50	22 ♍ 28	04 N 41

DECLINATIONS

Date	Sun ☉ ° '	Moon ☽ ° '	Mercury ☿ ° '	Venus ♀ ° '	Mars ♂ ° '	Jupiter ♃ ° '	Saturn ♄ ° '	Uranus ♅ ° '	Neptune ♆ ° '	Pluto ♇ ° '
01	17 S 11	12 S 37	20 S 38	22 S 21	15 S 00	04 N 19	20 S 51	06 N 10	09 S 07	20 S 56
02	16 53	15 07	20 44	22 18	15 08	04 21	20 51	06 11	09 06	20 56
03	16 36	16 59	20 48	22 14	15 17	04 23	20 52	06 11	09 06	20 56
04	16 18	18 04	20 51	22 09	15 25	04 26	20 52	06 12	09 05	20 55
05	16 00	18 15	20 54	22 04	15 34	04 28	20 53	06 13	09 04	20 55
06	15 42	17 26	20 55	21 58	15 42	04 30	20 53	06 14	09 03	20 55
07	15 23	15 34	20 56	21 52	15 50	04 32	20 54	06 15	09 02	20 55
08	15 05	12 43	20 55	21 44	15 58	04 35	20 54	06 16	09 02	20 55
09	14 46	09 04	20 53	21 36	16 06	04 37	20 54	06 16	09 01	20 55
10	14 26	04 49	20 50	21 28	16 14	04 40	20 55	06 17	09 00	20 55
11	14 07	00 S 16	20 46	21 19	16 22	04 42	20 55	06 18	08 59	20 54
12	13 47	04 N 16	20 41	21 09	16 30	04 45	20 55	06 19	08 58	20 54
13	13 27	08 31	20 34	20 59	16 37	04 47	20 56	06 20	08 57	20 54
14	13 07	12 12	20 27	20 48	16 45	04 50	20 56	06 21	08 57	20 54
15	12 46	15 08	20 18	20 36	16 52	04 53	20 56	06 22	08 56	20 54
16	12 26	17 08	20 08	20 24	16 59	04 56	20 57	06 23	08 55	20 54
17	12 05	18 08	19 56	20 11	17 06	04 58	20 57	06 24	08 54	20 54
18	11 44	18 07	19 43	19 58	17 13	05 01	20 57	06 25	08 53	20 53
19	11 22	17 07	19 29	19 44	17 20	05 04	20 58	06 26	08 52	20 53
20	11 01	15 15	19 14	19 29	17 27	05 07	20 58	06 27	08 52	20 53
21	10 40	12 41	18 58	19 14	17 34	05 10	20 58	06 28	08 51	20 53
22	10 18	09 33	18 40	18 58	17 40	05 13	20 58	06 29	08 50	20 53
23	09 56	06 04	18 21	18 42	17 47	05 16	20 58	06 30	08 49	20 53
24	09 34	02 N 22	18 01	18 25	17 53	05 19	20 59	06 31	08 48	20 53
25	09 12	01 S 23	17 39	18 08	17 59	05 22	20 59	06 32	08 47	20 53
26	08 49	05 02	17 16	17 50	18 05	05 25	20 59	06 33	08 47	20 52
27	08 27	08 29	16 52	17 31	18 12	05 28	20 59	06 34	08 46	20 52
28	08 04	11 35	16 26	17 13	18 18	05 31	20 59	06 35	08 45	20 52
29	07 S 42	14 S 14	16 S 00	16 S 53	18 S 23	05 N 34	21 S 00	06 N 36	08 S 44	20 S 52

ZODIAC SIGN ENTRIES

Date	h	m	Planets
02	15	50	☽ ♐
05	00	44	☽ ♑
07	05	59	☽ ♒
09	08	31	☽ ♓
11	09	55	☽ ♈
13	11	35	☽ ♉
13	22	43	☿ ♒
15	14	35	☽ ♊
17	04	17	♀ ♒
17	19	24	☽ ♋
19	05	34	☉ ♓
20	02	17	☽ ♌
22	11	24	☽ ♍
24	22	41	☽ ♎
27	11	26	☽ ♏
29	23	56	☽ ♐

LATITUDES

Date	Mercury ☿ ° '	Venus ♀ ° '	Mars ♂ ° '	Jupiter ♃ ° '	Saturn ♄ ° '	Uranus ♅ ° '	Neptune ♆ ° '	Pluto ♇ ° '
01	01 N 41	00 N 40	01 N 28	01 N 23	01 N 40	00 S 37	00 S 47	01 N 33
04	01 08	00 31	01 27	01 24	01 40	00 37	00 47	01 32
07	00 37	00 22	01 27	01 24	01 40	00 37	00 47	01 32
10	00 N 08	00 13	01 26	01 25	01 40	00 37	00 47	01 32
13	00 S 19	00 N 04	01 25	01 25	01 41	00 37	00 47	01 32
16	00 44	00 S 05	01 24	01 26	01 41	00 36	00 47	01 31
19	01 05	00 13	01 23	01 27	01 41	00 36	00 47	01 31
22	01 24	00 21	01 22	01 27	01 42	00 36	00 47	01 31
25	01 40	00 29	01 21	01 27	01 42	00 36	00 47	01 31
28	01 53	00 37	01 20	01 28	01 42	00 36	00 47	01 31
31	02 S 03	00 S 44	01 N 18	01 N 28	01 N 43	00 S 36	00 S 48	01 N 31

DATA

Julian Date	2457420
Delta T	+69 seconds
Ayanamsa	24° 04' 54"
Synetic vernal point	05° ♓ 02' 05"
True obliquity of ecliptic	23° 26' 05"

MOON'S PHASES, APSIDES AND POSITIONS ☽

Date	h	m	Phase	Longitude ° '	Eclipse Indicator
01	03	28	☾	11 ♏ 41	
08	14	39	●	19 ♒ 16	
15	07	46	☽	26 ♉ 03	
22	18	20	○	03 ♍ 34	

Day	h	m		
11	02	35	Perigee	
27	03	19	Apogee	
05	04	32	Max dec	18° S 18'
11	13	25	0N	
17	23	24	Max dec	18° N 15'
25	03	09	0S	

LONGITUDES

Date	Chiron ⚷ ° '	Ceres ⚳ ° '	Pallas ⚴ ° '	Juno ⚵ ° '	Vesta ⚶ ° '	Black Moon Lilith ⚸ ° '
01	18 ♓ 47	01 ♓ 26	03 ♒ 32	12 ♏ 48	16 ♈ 08	17 ♎ 46
11	19 ♓ 20	05 ♓ 20	07 ♒ 06	14 ♏ 05	19 ♈ 51	18 ♎ 52
21	19 ♓ 54	09 ♓ 15	10 ♒ 36	14 ♏ 54	23 ♈ 43	19 ♎ 59
31	20 ♓ 30	13 ♓ 11	14 ♒ 01	15 ♏ 10	27 ♈ 44	21 ♎ 06

All ephemeris data is given at 12.00 UT and the Moon's longitude is additionally given for 24.00 UT

ASPECTARIAN

01 Monday
00 08 ☽ ⚹ ♀
03 28 ☽ □ ☉
03 41 ☿ □ ♅
08 25 ☽ ⊻ ♄
10 28 ☽ ☌ ♂
12 20 ☽ ⚹ ♇
14 25 ☽ ⊼ ♅
15 01 ☽ ⚹ ☿

02 Tuesday
00 35 ☽ ⚹ ♃
02 23 ☽ ± ♅
09 26 ☽ ∠ ♀
12 17 ☽ ∥ ♂
18 05 ☽ ∠ ♇
18 26 ☽ Q ☉
20 06 ☽ bQ ♅
22 30 ☽ ∠ ☿
23 59 ☽ Q ♃

03 Wednesday
04 57 ☽ ⊥ ♀
07 07 ☽ ∥ ☉
08 25 ☽ □ ♆
10 53 ♂ ⚹ ♇
11 33 ☽ ⊥ ♇
17 02 ☽ ⊥ ☿
17 42 ☽ ⊻ ♀
18 17 ☉ ⚹ ♄
19 31 ☽ ☌ ♄
19 37 ☽ ⚹ ☉
23 00 ☽ ⊻ ♇
23 28 ☽ ⊻ ♂

04 Thursday
00 58 ☽ △ ♅
05 12 ☽ ⊻ ☿
10 04 ☽ □ ♃
11 11 ☽ ⊥ ♂
13 15 ♀ ⊻ ♄
18 28 ☽ Q ♆
21 11 ☿ ∥ ♄

05 Friday
01 59 ☽ ∠ ☉
04 26 ☽ ∠ ♂
07 41 ☿ ⊥ ♄
09 13 ☉ ± ♃
15 12 ☉ ⊻ ♇
16 24 ☽ ⚹ ♆
17 31 ♂ ⊼ ♅
19 38 ☽ ⊥ ☉

06 Saturday
01 15 ♀ ☌ ♇
02 56 ☽ ⊻ ♄
06 01 ☽ ☌ ♇
06 29 ☽ ☌ ♀
07 11 ☽ ⊻ ☉
07 52 ☽ □ ♅
07 55 ☿ ∥ ♇
08 21 ☽ ⚹ ♂
11 54 ☉ ∥ ♂
13 35 ☽ ⊥ ♄
15 39 ☽ ☌ ☿
15 54 ☽ △ ♃
16 39 ☉ ⚹ ♅
18 45 ☿ △ ♃
19 01 ☽ ∠ ♆
22 03 ♀ □ ♅

07 Sunday
05 18 ☽ ∠ ♄
06 01 ☽ Q ♂
09 19 ☽ ∥ ♂
10 36 ☽ ⊥ ♆
12 08 ☉ □ ♂
13 59 ☽ ∥ ☉
15 01 ♀ ⚹ ♂
15 09 ☽ Q ♅
17 36 ☽ bQ ♃
20 52 ☽ ⊻ ♆
21 59 ☉ ⊻ ♀

08 Monday
06 58 ☽ ⚹ ♄
08 43 ☽ ± ♃
09 44 ☽ ⊻ ♇
11 30 ☽ ⚹ ♅
13 37 ☽ □ ♂
13 56 ☿ ∠ ♆
14 39 ☽ ☌ ☉
14 56 ☽ ⊻ ♀
16 20 ☿ ∥ ♇
18 42 ☽ ⊼ ♃
19 48 ☽ ⊥ ♇
22 47 ☽ ⊻ ☿

09 Tuesday
01 32 ☿ ∥ ♄
01 52 ☽ ⊥ ♀
03 07 ☽ Q ♄
09 34 ☽ ⊥ ☿
10 46 ☽ ∠ ♇
12 17 ☽ ∥ ♆
12 32 ☽ ∠ ♅
17 16 ♀ ⊥ ♄
18 10 ☽ ∠ ♀
23 02 ☽ ☌ ♆

10 Wednesday
01 39 ☽ ∠ ☿
03 25 ☽ ⊥ ♅
03 59 ☽ # ♅
04 06 ♄ ∥ ♇
07 54 ♀ △ ♃
08 59 ☽ □ ♄
11 33 ☽ ⚹ ♇
12 49 ☽ # ♃
13 18 ☽ ⊻ ♅
16 55 ☽ △ ♂
19 28 ☉ ⊼ ♃
19 55 ☽ ☍ ♃
19 58 ☽ ⊻ ☉
21 07 ☽ ⚹ ♀

11 Thursday
04 25 ☽ ⚹ ☿
06 33 ☽ ⊥ ☉
07 18 ☽ Q ♇
17 41 ☉ ⊥ ♇
18 24 ☽ bQ ♂
18 40 ☽ Q ♀
22 28 ☽ ∠ ☉

12 Friday
00 30 ☽ ⊻ ♆
01 54 ☽ Q ☿
05 09 ♀ ∠ ♆
09 44 ☽ ± ♂
10 26 ☽ ⊥ ♆
10 35 ☽ △ ♄
13 02 ☽ □ ♇
14 37 ☽ ∥ ♃
14 50 ☽ ☌ ♅
20 00 ☽ ⊼ ♂
21 03 ☽ ⊼ ♃
23 19 ☽ ∥ ♅

13 Saturday
01 10 ☽ ⚹ ☉
01 24 ☽ ∠ ♆
03 15 ☽ □ ♀
06 59 ☽ ± ♃
10 32 ☽ □ ☿
11 39 ☽ bQ ♄
14 41 ☽ # ♆
18 18 ♀ ∥ ♄
21 57 ☽ bQ ♃
22 14 ♀ ∥ ♇
22 47 ☽ Q ☉
22 59 ♂ ⚹ ♃

14 Sunday
00 05 ☿ ∠ ♄
02 37 ☽ ⚹ ♆
02 51 ☽ ± ♄
13 03 ☽ ⊼ ♄
15 27 ☽ △ ♇
17 20 ☽ ⊻ ♅
18 06 ☽ # ☉
23 03 ☽ Q ♆
23 13 ☽ △ ♃

15 Monday
00 13 ☽ ☍ ♂
03 37 ☽ ⊥ ♅
07 46 ☽ □ ☉
10 54 ☽ △ ♀
17 15 ☽ bQ ♇
18 36 ☽ △ ☿
19 11 ☽ ∠ ♅

16 Tuesday
06 15 ☽ □ ♆
07 18 ☿ ⊥ ♆
08 57 ☽ ± ♇
09 29 ☽ # ♂
11 35 ☉ Q ♄
15 35 ☽ bQ ♀
17 10 ☽ ☍ ♄
19 30 ☽ ⊼ ♇
21 31 ☽ ⚹ ♅
23 39 ☽ bQ ☿

17 Wednesday
03 02 ☽ □ ♃
06 18 ☽ ⊼ ♂
09 07 ☽ ± ♀
09 45 ♀ ∠ ♄
16 37 ☽ △ ☉
16 38 ☿ Q ♂
17 22 ☽ ± ♂
17 26 ☽ ± ☿
18 56 ☽ Q ♅
20 56 ☽ ⊼ ♀

18 Thursday
05 26 ☽ ⊼ ☿
07 28 ☿ bQ ♃
08 47 ☿ Q ♅
10 09 ☽ bQ ♂
11 03 ☽ Q ♃
11 49 ☽ △ ♆
21 59 ☽ bQ ☉
23 15 ☽ ⊼ ♄

19 Friday
01 34 ☽ ☍ ♇
03 42 ☽ □ ♅
08 31 ☽ # ♂
08 48 ☽ ⚹ ♃
10 20 ☽ ± ♄
14 36 ☽ △ ♂
15 24 ☽ bQ ♆
15 57 ☽ ± ☉
17 24 ♀ ⊥ ♆
21 27 ♂ ± ♅

20 Saturday
03 06 ☽ bQ ♄
04 03 ☽ ⊼ ☉
08 13 ☽ ± ♆
09 45 ☽ ☍ ♀
12 28 ☽ ∠ ♃
19 28 ☽ ☍ ☿
19 32 ☽ ⊼ ♆
20 13 ☿ ⊻ ♆
21 38 ☉ ∠ ♇

21 Sunday
05 20 ☽ ⊥ ♃
07 31 ☽ △ ♄
09 48 ☽ ⊼ ♇
11 40 ♀ bQ ♃
12 06 ☽ △ ♅
16 41 ☽ ⊻ ♃
21 19 ☽ ± ♇
23 16 ♀ Q ♅

22 Monday
01 17 ☽ □ ♂
03 20 ☉ ∠ ♅
05 55 ☽ # ☉
14 46 ☽ bQ ♇
17 09 ☽ bQ ♅
17 10 ☽ # ♆
18 20 ☽ ☍ ☉
23 19 ♀ Q ♂

23 Tuesday
01 36 ☽ ⊼ ♀
05 32 ☽ ☍ ♆
09 08 ☽ ∥ ♅
10 51 ☽ ± ♅
12 57 ☽ ⊼ ☿
13 40 ☽ Q ♂
14 44 ☽ ± ♀
17 12 ☽ ∥ ♃
18 02 ☽ □ ♄
20 17 ☽ △ ♇
22 46 ☽ ⊼ ♅

24 Wednesday
02 28 ☽ ± ☿
02 44 ☽ ☌ ♃
04 27 ☿ ± ♃
10 39 ☽ bQ ♀
14 22 ☽ ⚹ ♂
17 32 ♀ ⊻ ♆
18 33 ☿ ∥ ♂
22 58 ☽ bQ ☿

25 Thursday
06 08 ☽ Q ♄
08 05 ☿ ⚹ ♄
11 20 ☽ ⊼ ☉
17 35 ☽ ⊼ ♆
20 17 ☽ △ ♀
20 19 ♀ ∥ ♂
21 39 ☽ ∠ ♂

26 Friday
00 35 ☽ ± ☉
01 53 ☿ ⊻ ♇
05 46 ☽ ± ♆
06 29 ☽ ⚹ ♄
08 40 ☽ □ ♇
09 38 ☽ △ ☿
11 18 ☽ ☍ ♅
14 30 ☽ ⊻ ♃
14 35 ☽ # ♃
15 06 ☉ ∥ ♆
16 34 ☽ ⊥ ♂
20 33 ☽ bQ ☉
22 23 ☽ # ♅
23 02 ☿ ⚹ ♅

27 Saturday
00 06 ☽ bQ ♆
02 34 ☽ ⊥ ♃
05 12 ☽ ⊻ ♂
11 48 ☽ ∥ ☉
13 04 ☽ ∠ ♄
14 04 ☽ ∥ ♆
20 40 ☽ ∠ ♃
20 56 ☿ ⊼ ♃
21 18 ☽ Q ♇

28 Sunday
02 34 ♀ ± ♃
05 51 ☽ △ ☉
06 40 ☽ △ ♆
07 25 ☽ ⊥ ♄
15 47 ☉ ☌ ♆
16 16 ☽ □ ♀
19 35 ☽ ⊻ ♄
21 39 ☽ ⚹ ♇

29 Monday
00 24 ☽ ⊼ ♅
02 42 ☽ ⚹ ♃
07 30 ☽ □ ☿
12 30 ☽ ± ♅
19 55 ☽ ☌ ♂

MARCH 2016

LONGITUDES

Date	Sidereal time h m s	Sun ☉ ° ' "	Moon ☽ ° ' "	Moon ☽ 24.00 ° ' "	Mercury ☿ ° '	Venus ♀ ° '	Mars ♂ ° '	Jupiter ♃ ° '	Saturn ♄ ° '	Uranus ♅ ° '	Neptune ♆ ° '	Pluto ♇ ° '
01	22 38 53	11 ♓ 20 05	06 ♐ 05 56	12 ♐ 13 26	23 ♒ 38	16 ♒ 28	28 ♏ 15	19 ♍ 12	15 ♐ 56	18 ♈ 19	09 ♓ 33	16 ♑ 55
02	22 42 50	12 20 17	18 ♐ 25 08	24 ♐ 41 34	25 14	17 43	28 38	19 R 05	15 58	18 22	09 36	16 57
03	22 46 47	13 20 27	01 ♑ 03 16	07 ♑ 30 40	26 50	18 57	29 01	18 57	16 00	18 25	09 38	16 58
04	22 50 43	14 20 36	14 ♑ 04 09	20 ♑ 44 01	28 ♒ 28	20 11	29 24	18 49	16 02	18 28	09 40	16 59
05	22 54 40	15 20 44	27 ♑ 30 27	04 ♒ 23 29	00 ♓ 07	21 25	29 ♏ 47	18 41	16 04	18 31	09 42	17 00
06	22 58 36	16 20 49	11 ♒ 23 02	18 ♒ 28 52	01 46	22 40	00 ♐ 09	18 34	16 06	18 34	09 45	17 02
07	23 02 33	17 20 53	25 ♒ 40 35	02 ♓ 57 36	03 27	23 54	00 31	18 26	16 08	18 37	09 47	17 03
08	23 06 29	18 20 55	10 ♓ 19 14	17 ♓ 44 36	05 09	25 08	00 52	18 18	16 10	18 40	09 49	17 04
09	23 10 26	19 20 55	25 ♓ 12 46	02 ♈ 42 41	06 52	26 22	01 13	18 10	16 12	18 43	09 51	17 05
10	23 14 22	20 20 54	10 ♈ 13 17	17 ♈ 43 29	08 36	27 36	01 34	18 02	16 13	18 46	09 54	17 06
11	23 18 19	21 20 50	25 ♈ 12 15	02 ♉ 38 37	10 22	28 ♒ 51	01 55	17 54	16 15	18 49	09 56	17 08
12	23 22 16	22 20 44	10 ♉ 01 43	17 ♉ 20 50	12 08	00 ♓ 05	02 15	17 47	16 16	18 52	09 58	17 09
13	23 26 12	23 20 36	24 ♉ 35 21	01 ♊ 44 50	13 56	01 19	02 34	17 39	16 17	18 55	10 00	17 10
14	23 30 09	24 20 26	08 ♊ 48 59	15 ♊ 47 37	15 45	02 33	02 54	17 31	16 18	18 59	10 03	17 11
15	23 34 05	25 20 13	22 ♊ 40 41	29 ♊ 28 14	17 35	03 48	03 13	17 23	16 19	19 02	10 05	17 12
16	23 38 02	26 19 59	06 ♋ 10 25	12 ♋ 47 26	19 26	05 02	03 31	17 16	16 20	19 05	10 07	17 13
17	23 41 58	27 19 42	19 ♋ 19 33	25 ♋ 47 03	21 18	06 16	03 49	17 08	16 21	19 08	10 09	17 14
18	23 45 55	28 19 22	02 ♌ 10 17	08 ♌ 29 33	23 12	07 30	04 07	17 01	16 22	19 12	10 12	17 15
19	23 49 51	29 ♓ 19 01	14 ♌ 45 12	20 ♌ 57 34	25 07	08 44	04 24	16 53	16 23	19 15	10 14	17 16
20	23 53 48	00 ♈ 18 37	27 ♌ 06 57	03 ♍ 13 39	27 03	09 59	04 41	16 46	16 23	19 18	10 16	17 17
21	23 57 45	01 18 11	09 ♍ 17 59	15 ♍ 20 12	29 ♓ 00	11 13	04 58	16 38	16 24	19 21	10 18	17 17
22	00 01 41	02 17 43	21 ♍ 20 35	27 ♍ 19 22	00 ♈ 58	12 27	05 14	16 31	16 24	19 25	10 20	17 18
23	00 05 38	03 17 13	03 ♎ 16 50	09 ♎ 13 13	02 57	13 41	05 29	16 24	16 24	19 28	10 22	17 19
24	00 09 34	04 16 40	15 ♎ 08 45	21 ♎ 03 42	04 57	14 55	05 44	16 16	16 24	19 31	10 24	17 20
25	00 13 31	05 16 06	26 ♎ 58 21	02 ♏ 52 57	06 58	16 09	05 59	16 09	16 R 24	19 35	10 27	17 20
26	00 17 27	06 15 30	08 ♏ 47 50	14 ♏ 43 19	08 59	17 24	06 13	16 02	16 24	19 38	10 29	17 21
27	00 21 24	07 14 52	20 ♏ 39 45	26 ♏ 37 30	11 01	18 38	06 26	15 55	16 24	19 41	10 31	17 22
28	00 25 20	08 14 12	02 ♐ 36 58	08 ♐ 38 36	13 04	19 52	06 39	15 49	16 24	19 45	10 33	17 23
29	00 29 17	09 13 31	14 ♐ 42 50	20 ♐ 50 09	15 06	21 06	06 52	15 42	16 23	19 48	10 35	17 23
30	00 33 14	10 12 47	27 ♐ 01 02	03 ♑ 16 00	17 09	22 20	07 04	15 35	16 23	19 52	10 37	17 24
31	00 37 10	11 ♈ 12 02	09 ♑ 35 31	16 ♑ 00 05	19 ♈ 11	23 ♓ 34	07 ♐ 15	15 ♍ 29	16 ♐ 22	19 ♈ 55	10 ♓ 39	17 ♑ 24

Date	Moon True ☊ ° '	Moon Mean ☊ ° '	Moon ☽ Latitude ° '
01	21 ♍ 51	22 ♍ 24	05 N 06
02	21 R 51	22 21	05 17
03	21 50	22 18	05 13
04	21 50	22 15	04 54
05	21 49	22 12	04 18
06	21 48	22 08	03 26
07	21 47	22 05	02 20
08	21 47	22 02	01 N 03
09	21 D 47	21 59	00 S 19
10	21 47	21 56	01 41
11	21 47	21 53	02 55
12	21 R 47	21 49	03 57
13	21 47	21 46	04 43
14	21 47	21 43	05 10
15	21 47	21 40	05 18
16	21 D 47	21 37	05 08
17	21 47	21 33	04 42
18	21 48	21 30	04 02
19	21 49	21 27	03 12
20	21 49	21 24	02 13
21	21 50	21 21	01 09
22	21 50	21 18	00 S 03
23	21 R 50	21 14	01 N 03
24	21 49	21 11	02 06
25	21 48	21 08	03 03
26	21 45	21 05	03 52
27	21 43	21 02	04 31
28	21 41	20 59	04 59
29	21 39	20 55	05 14
30	21 37	20 52	05 15
31	21 ♍ 37	20 ♍ 49	05 N 01

DECLINATIONS

Date	Sun ☉ ° '	Moon ☽ ° '	Mercury ☿ ° '	Venus ♀ ° '	Mars ♂ ° '	Jupiter ♃ ° '	Saturn ♄ ° '	Uranus ♅ ° '	Neptune ♆ ° '	Pluto ♇ ° '
01	07 S 19	16 S 18	15 S 32	16 S 33	18 S 29	05 N 37	21 S 00	06 N 37	08 S 43	20 S 52
02	06 56	17 40	15 02	16 13	18 35	05 40	21 00	06 39	08 42	20 52
03	06 33	18 12	14 32	15 52	18 40	05 43	21 00	06 40	08 41	20 52
04	06 10	17 49	14 00	15 31	18 46	05 47	21 00	06 41	08 41	20 52
05	05 46	16 26	13 27	15 09	18 51	05 50	21 00	06 42	08 40	20 52
06	05 23	14 03	12 52	14 47	18 56	05 53	21 00	06 43	08 39	20 51
07	05 00	10 45	12 16	14 25	19 01	05 56	21 00	06 44	08 38	20 51
08	04 36	06 43	11 40	14 02	19 07	05 59	21 00	06 45	08 37	20 51
09	04 13	02 S 12	11 01	13 39	19 11	06 02	21 00	06 47	08 36	20 51
10	03 49	02 N 30	10 22	13 15	19 16	06 05	21 00	06 48	08 36	20 51
11	03 26	07 02	09 41	12 51	19 21	06 08	21 00	06 49	08 35	20 51
12	03 02	11 04	08 59	12 27	19 26	06 11	21 00	06 50	08 34	20 51
13	02 39	14 20	08 16	12 02	19 30	06 14	21 00	06 51	08 33	20 51
14	02 15	16 40	07 32	11 37	19 35	06 18	21 00	06 53	08 32	20 51
15	01 51	17 57	06 46	11 12	19 39	06 21	21 00	06 54	08 31	20 51
16	01 27	18 10	06 00	10 46	19 44	06 24	21 00	06 55	08 31	20 51
17	01 04	17 23	05 12	10 20	19 48	06 27	21 00	06 56	08 30	20 50
18	00 40	15 44	04 23	09 54	19 52	06 29	21 00	06 58	08 29	20 50
19	00 S 16	13 21	03 33	09 28	19 56	06 32	21 00	06 59	08 28	20 50
20	00 N 07	10 23	02 42	09 01	20 00	06 35	21 00	07 00	08 27	20 50
21	00 31	07 01	01 50	08 34	20 04	06 38	21 00	07 01	08 27	20 50
22	00 55	03 N 23	00 57	08 07	20 08	06 41	21 00	07 03	08 26	20 50
23	01 18	00 S 20	00 S 03	07 39	20 11	06 44	21 00	07 04	08 25	20 50
24	01 42	04 02	00 N 51	07 12	20 15	06 46	21 00	07 05	08 24	20 50
25	02 06	07 33	01 47	06 44	20 19	06 49	21 00	07 06	08 23	20 50
26	02 29	10 46	02 42	06 16	20 22	06 52	21 00	07 08	08 23	20 50
27	02 53	13 33	03 39	05 48	20 25	06 54	21 00	07 09	08 22	20 50
28	03 16	15 48	04 35	05 19	20 29	06 57	20 59	07 10	08 21	20 50
29	03 39	17 22	05 32	04 51	20 32	07 00	20 59	07 12	08 20	20 50
30	04 03	18 10	06 29	04 22	20 35	07 02	20 59	07 13	08 20	20 50
31	04 N 26	18 S 05	07 N 25	03 S 53	20 S 38	07 N 05	20 S 59	07 N 14	08 S 19	20 S 50

ZODIAC SIGN ENTRIES

Date	h	m	Planets
03	10	01	☽ ♑
05	10	23	☿ ♓
05	16	22	☽ ♒
06	02	29	♂ ♐
07	19	08	☽ ♓
09	19	40	☽ ♈
11	19	44	☽ ♉
12	10	24	♀ ♓
13	21	03	☽ ♊
16	00	57	☽ ♋
18	07	54	☽ ♌
20	04	30	☉ ♈
20	17	39	☽ ♍
22	00	19	☿ ♈
23	05	23	☽ ♎
25	18	09	☽ ♏
28	06	46	☽ ♐
30	17	45	☽ ♑

LATITUDES

Date	Mercury ☿ ° '	Venus ♀ ° '	Mars ♂ ° '	Jupiter ♃ ° '	Saturn ♄ ° '	Uranus ♅ ° '	Neptune ♆ ° '	Pluto ♇ ° '
01	02 S 00	00 S 41	01 N 19	01 N 28	01 N 42	00 S 36	00 S 48	01 N 31
04	02 07	00 48	01 17	01 28	01 43	00 36	00 48	01 30
07	02 11	00 55	01 15	01 28	01 43	00 36	00 48	01 30
10	02 11	01 01	01 13	01 29	01 43	00 36	00 48	01 30
13	02 07	01 06	01 11	01 29	01 44	00 36	00 48	01 30
16	01 58	01 11	01 09	01 29	01 44	00 36	00 48	01 30
19	01 45	01 16	01 06	01 29	01 44	00 36	00 48	01 30
22	01 27	01 19	01 04	01 29	01 45	00 36	00 48	01 29
25	01 04	01 23	01 00	01 28	01 45	00 36	00 48	01 29
28	00 37	01 25	00 57	01 28	01 46	00 36	00 48	01 29
31	00 S 06	01 S 27	00 N 53	01 N 28	01 N 46	00 S 36	00 S 48	01 N 29

DATA

Julian Date	2457449
Delta T	+69 seconds
Ayanamsa	24° 04' 57"
Synetic vernal point	05° ♓ 02' 02"
True obliquity of ecliptic	23° 26' 05"

MOON'S PHASES, APSIDES AND POSITIONS ☽

Date	h	m	Phase	Longitude ° '	Eclipse Indicator
01	23	11	☾	11 ♐ 48	
09	01	54	●	18 ♓ 56	Total
15	17	03	☽	25 ♊ 33	
23	12	01	○	03 ♎ 17	
31	15	17	☾	11 ♑ 20	

Day	h	m		
10	06	58	Perigee	
25	14	01	Apogee	
03	14	17	Max dec	18° S 13'
09	23	13	0N	
16	05	05	Max dec	18° N 12'
23	09	49	0S	
30	22	12	Max dec	18° S 14'

LONGITUDES

Date	Chiron ⚷ ° '	Ceres ⚳ ° '	Pallas ⚴ ° '	Juno ⚵ ° '	Vesta ⚶ ° '	Black Moon Lilith ⚸ ° '
01	20 ♓ 27	12 ♓ 48	13 ♒ 41	15 ♏ 10	27 ♈ 20	20 ♎ 59
11	21 ♓ 03	16 ♓ 44	16 ♒ 59	14 ♏ 56	01 ♉ 26	22 ♎ 06
21	21 ♓ 40	20 ♓ 40	20 ♒ 09	14 ♏ 07	05 ♉ 38	23 ♎ 12
31	22 ♓ 15	24 ♓ 34	23 ♒ 10	12 ♏ 45	09 ♉ 53	24 ♎ 19

ASPECTARIAN

h m	Aspects
01 Tuesday	
00 53	☿ ⊥ ♇
01 07	♀ ✶ ♄
02 26	☽ Q ♃
03 44	☽ ∠ ♇
04 01	☽ ∥ ☿
06 29	☽ ⚼ ♅
08 26	☽ Q ♀
15 01	☽ ∥ ♀
18 48	☽ □ ♆
20 50	♀ ⊻ ♇
21 28	☽ ⊥ ♇
23 11	☽ □ ☉
02 Wednesday	
00 28	☽ Q ☿
07 15	☽ ☌ ♄
09 09	☽ ⊻ ♇
10 29	☽ ✶ ♀
11 54	☽ △ ♅
12 39	☉ ⊥ ♅
13 15	☽ □ ♃
03 Thursday	
01 13	♀ ✶ ♅
02 55	☽ ✶ ☿
05 12	☉ # ♅
05 33	☽ Q ♆
08 04	☽ ⊻ ♂
12 04	♀ ⚻ ♃
12 35	☽ Q ☉
17 58	☽ ∠ ♀
19 37	☽ ⊥ ♂
04 Friday	
03 57	☽ ✶ ♆
05 37	☿ Q ♄
10 45	☽ ∠ ☿
12 14	☽ ⊥ ♀
12 32	☽ ✶ ☉
12 37	☽ ∠ ♂
15 35	☽ ⊻ ♄
17 17	☽ ☌ ♇
19 58	☽ □ ♅
20 30	☽ △ ♃
05 Saturday	
00 08	☽ ⊻ ♀
02 22	☽ ⊥ ♄
05 11	☽ ⊥ ☿
05 43	☿ □ ♂
07 03	☽ ∠ ♆
09 06	☉ # ♃
16 05	☽ ✶ ♂
17 11	☽ ⊻ ☿
17 22	☽ ∠ ☉
18 15	☽ ∠ ♄
22 41	☽ ⚼ ♃
22 51	☽ ⊥ ♆
06 Sunday	
03 43	☽ Q ♅
04 28	☽ ∥ ♀
06 04	☉ □ ♄
09 12	☽ ⊻ ♆
10 06	☽ ⊥ ☉
11 45	♃ ⚻ ♅
13 20	☽ Q ♂
13 59	☽ ± ♃
15 42	☿ ∠ ♇
19 18	♀ ⊥ ♇
20 01	☽ ✶ ♄
21 03	☽ ⊻ ☉
21 34	☽ ⊻ ♇
23 26	☽ ∥ ☿
07 Monday	
00 01	☽ ⚻ ♃
00 11	☽ ✶ ♅
04 41	☉ ✶ ♇
07 38	☽ ⊥ ♇
08 05	♂ Q ♃
08 46	☽ ☌ ♀
14 19	☿ ∠ ♅
16 05	☽ Q ♄
20 11	☽ □ ♂
22 31	☽ ∠ ♇
08 Tuesday	
01 07	☽ ∠ ♅
01 08	☽ ∥ ♆
02 29	☽ ☌ ☿
10 57	☉ ☍ ♃
11 11	☽ ☌ ♆
11 47	☽ # ♅
15 49	☽ ⊥ ♅
15 59	☽ # ♃
19 58	☉ ⊻ ♅
21 29	☽ □ ♄
22 56	☽ ✶ ♇
09 Wednesday	
00 30	☽ ∥ ☉
00 47	☽ ☍ ♃
01 32	☽ ⊻ ♅
01 54	☽ ☌ ☉
14 01	☽ ⊻ ♀
18 13	☽ Q ♇
21 51	☽ △ ♂
10 Thursday	
00 29	☽ ⊥ ♀
09 05	☽ ⊻ ☿
11 29	☽ ⊻ ♆
16 09	☽ ∠ ♀
18 18	☽ # ☉
19 56	☽ ⊥ ☿
21 06	☽ ⊥ ♆
21 36	☽ △ ♄
22 23	☽ ⚼ ♂
23 02	☽ □ ♇
11 Friday	
00 06	♀ Q ♄
00 24	☽ ⚻ ♃
01 43	☽ ☌ ♅
05 22	☽ ⊻ ☉
06 01	☿ ☌ ♆
07 04	☽ ∥ ♃
09 56	☽ ± ♃
10 50	☽ ∥ ♅
11 34	☽ ∠ ♆
12 17	☽ ∠ ☿
13 10	☽ ± ♂
15 42	☽ ⊥ ☉
18 24	☽ ✶ ♀
20 46	☽ # ♆
21 45	☽ ⚼ ♄
23 04	☽ ⚻ ♂
12 Saturday	
00 19	☽ ⚼ ♃
01 06	☽ # ☿
04 30	♂ ∠ ♇
07 19	☽ ∠ ☉
11 54	☽ ✶ ♆
12 23	☽ ± ♄
15 40	☽ Q ♀
15 55	☽ ✶ ☿
20 23	☽ # ♀
22 08	☿ ⊥ ♅
22 14	☽ ⚻ ♄
23 41	☽ △ ♇
13 Sunday	
00 36	☽ △ ♃
02 30	☿ ∥ ♆
02 34	☽ ⊻ ♅
07 42	☽ Q ♆
09 46	☽ ✶ ☉
12 34	☽ ⊥ ♅
14 34	☽ Q ☿
14 Monday	
00 21	☽ □ ♀
00 43	☽ ⚼ ♇
01 42	☽ ☍ ♂
03 44	☽ ∠ ♅
04 36	♀ ∠ ♇
07 28	☽ Q ☉
14 06	☽ □ ♆
16 03	☽ ± ♇
19 26	☿ □ ♄
20 48	♀ □ ♂
15 Tuesday	
00 54	☽ ☍ ♄
01 44	☽ □ ☿
02 25	☽ ⚻ ♇
02 51	☽ □ ♃
05 36	☽ ✶ ♅
06 58	☿ ✶ ♇
08 07	☿ # ♅
09 42	☿ ☍ ♃
16 48	♀ ∠ ♅
17 03	☽ □ ☉
16 Wednesday	
00 31	☿ # ♃
02 50	☽ Q ♅
07 07	☽ ⚻ ♂
07 22	☿ ⊻ ♅
09 44	☽ △ ♀
10 23	☽ Q ♃
18 11	☽ ± ♂
19 09	☽ △ ♆
20 06	♃ △ ♇
17 Thursday	
06 31	☽ ⚻ ♄
08 00	☽ ✶ ♃
08 08	☽ ☍ ♇
11 03	☽ ⚼ ♂
11 39	☽ □ ♅
15 58	☽ ⚼ ♀
16 17	☽ △ ☿
17 37	☽ ± ♄
22 51	☽ ⚼ ♆
18 Friday	
04 09	☽ △ ☉
10 29	☽ ⚼ ♄
10 36	☽ ± ♀
11 42	☽ ∠ ♃
15 46	☽ △ ♂
15 50	☽ ± ♆
19 43	♂ ⚼ ♅
23 13	☽ ⚻ ♀
19 Saturday	
01 29	☽ ⚼ ☿
03 17	☽ ⚻ ♆
04 39	☽ ⊥ ♃
10 37	☉ Q ♇
11 05	☽ ⚼ ☉
15 08	☽ △ ♄
16 04	☽ ⊻ ♃
16 51	☽ ⚻ ♇
20 43	☽ △ ♅
21 58	☽ ± ☿
20 Sunday	
04 30	☽ ± ♇
06 02	☽ ± ☉
11 50	☽ ⚻ ☿
17 45	♀ ☌ ♆
18 49	☽ ⚻ ☉
22 08	☽ ⚼ ♇
23 36	☽ # ♀
21 Monday	
02 05	☽ # ♆
02 11	☽ ⚼ ♅
03 13	☽ □ ♂
11 56	☽ ∥ ♅
13 59	☽ ☍ ♆
14 32	☽ ∥ ♃
15 40	☿ Q ♇
16 14	☽ ☍ ♀
18 44	♀ ∥ ♆
20 06	☽ ± ♅
22 Tuesday	
02 07	☽ □ ♄
02 27	☽ ☌ ♃
03 55	☽ △ ♇
08 07	☽ ⚻ ♅
12 41	☉ # ☿
15 51	☽ Q ♂
23 Wednesday	
02 26	☽ ∥ ☉
07 32	♀ ⊥ ♅
08 41	☽ # ☿
10 15	♃ □ ♄
10 31	☽ ∥ ☿
11 11	☽ ☍ ☿
12 01	☽ ☍ ☉
14 16	☽ Q ♄
16 33	☽ ✶ ♂
18 58	☽ # ☉
20 11	☉ ☌ ☿
24 Thursday	
02 22	☽ ⚻ ♆
11 30	☽ ⚻ ♀
14 16	☽ ⊻ ♃
14 33	☽ ✶ ♄
14 34	☽ ± ♆
16 26	☽ □ ♇
17 19	♀ # ♅
20 55	☽ ☍ ♅
22 45	☿ △ ♂
23 35	☽ ∠ ♂
25 Friday	
01 05	☽ ± ♀
02 19	☽ ⊥ ♃
06 49	☽ # ♃
06 56	☽ ∥ ♀
07 47	♀ # ♃
08 52	☽ # ♅
08 53	☽ ⚼ ♆
09 59	♄ St R
11 58	♀ ☍ ♃
16 48	♀ □ ♄
17 59	☽ ∥ ♆
18 14	☽ ⊥ ♂
20 25	☽ ∠ ♃
21 00	☽ ∠ ♄
21 29	☽ ⚼ ♀
26 Saturday	
02 07	☉ ∥ ☿
05 00	☽ Q ♇
06 23	☽ ⚻ ☉
06 39	☽ ⊻ ♂
10 33	☉ △ ♂
11 12	♀ ✶ ♇
12 27	☽ ⚻ ☿
15 15	☽ ⊥ ♄
15 25	☽ △ ♆
19 39	☽ ± ☉
27 Sunday	
02 31	☽ ✶ ♃
03 07	☽ ± ☿
03 24	☽ ⊻ ♄
05 20	☽ ✶ ♇
05 55	☿ ⊻ ♆
07 25	☽ △ ♀
10 02	☽ ⚻ ♅
15 29	☽ ⚼ ☉
22 10	☽ ± ♅
28 Monday	
01 00	☽ ⚼ ☿
02 28	☽ Q ♃
09 33	♀ ⊻ ♅
11 31	☽ ∠ ♇
16 16	☽ ⚼ ♅
20 12	☽ ☌ ♂
29 Tuesday	
00 11	☽ △ ☉
00 24	☿ # ♀
03 49	☽ □ ♆
05 26	☽ ⊥ ♇
12 56	☽ △ ☿
13 55	☽ □ ♃
15 18	☽ ☌ ♄
17 15	☽ ⊻ ♇
18 36	☿ ⚻ ♃
22 02	☽ △ ♅
30 Wednesday	
01 55	☽ □ ♀
03 02	☿ △ ♄
05 37	☿ ⊥ ♆
14 53	☿ □ ♇
15 05	☽ Q ♆
21 01	☉ # ♀
22 08	☉ ⊻ ♆
31 Thursday	
02 49	☿ ∥ ♃
07 13	☿ ∥ ♅
07 31	☽ ⊻ ♂
14 00	☽ ✶ ♆
15 17	☽ □ ☉
16 08	☽ Q ♀
18 59	☽ ⊥ ♂
20 49	☿ ☌ ♅
22 57	☽ △ ♃

All ephemeris data is given at 12.00 UT and the Moon's longitude is additionally given for 24.00 UT

LONGITUDES

Date	Sidereal time h m s	Sun ☉ ° ' "	Moon ☽ ° ' "	Moon ☽ 24.00 ° ' "
01	00 41 07	12 ♈ 11 15	22 ♑ 30 10	29 ♑ 06 10
02	00 45 03	13 10 26	05 ♒ 48 26	12 ♒ 37 15
03	00 49 00	14 09 36	19 ♒ 32 46	26 ♒ 35 01
04	00 52 56	15 08 43	03 ♓ 43 53	10 ♓ 59 06
05	00 56 53	16 07 49	18 ♓ 20 11	25 ♓ 46 29
06	01 00 49	17 06 53	03 ♈ 17 10	10 ♈ 51 13
07	01 04 46	18 05 55	18 ♈ 27 30	26 ♈ 04 47
08	01 08 43	19 04 55	03 ♉ 41 46	11 ♉ 17 09
09	01 12 39	20 03 53	18 ♉ 49 44	26 ♉ 18 22
10	01 16 36	21 02 49	03 ♊ 42 05	11 ♊ 00 06
11	01 20 32	22 01 42	18 ♊ 11 48	25 ♊ 16 46
12	01 24 29	23 00 34	02 ♋ 14 49	09 ♋ 05 54
13	01 28 25	23 59 23	15 ♋ 50 08	22 ♋ 27 46
14	01 32 22	24 58 09	28 ♋ 59 10	05 ♌ 24 44
15	01 36 18	25 56 54	11 ♌ 44 59	18 ♌ 00 25
16	01 40 15	26 55 36	24 ♌ 11 35	00 ♍ 19 00
17	01 44 12	27 54 16	06 ♍ 23 14	12 ♍ 24 46
18	01 48 08	28 52 53	18 ♍ 24 05	24 ♍ 21 40
19	01 52 05	29 ♈ 51 29	00 ♎ 17 54	06 ♎ 13 11
20	01 56 01	00 ♉ 50 02	12 ♎ 07 52	18 ♎ 02 17
21	01 59 58	01 48 34	23 ♎ 56 42	29 ♎ 51 24
22	02 03 54	02 47 04	05 ♏ 46 36	11 ♏ 42 33
23	02 07 51	03 45 31	17 ♏ 39 26	23 ♏ 37 29
24	02 11 47	04 43 57	29 ♏ 36 55	05 ♐ 37 55
25	02 15 44	05 42 22	11 ♐ 40 46	17 ♐ 45 41
26	02 19 41	06 40 44	23 ♐ 52 58	00 ♑ 02 54
27	02 23 37	07 39 05	06 ♑ 15 50	12 ♑ 32 06
28	02 27 34	08 37 24	18 ♑ 52 05	25 ♑ 16 10
29	02 31 30	09 35 42	01 ♒ 44 45	08 ♒ 18 15
30	02 35 27	10 ♉ 33 59	14 ♒ 57 02	21 ♒ 41 26

Date	Mercury ☿ ° '	Venus ♀ ° '	Mars ♂ ° '	Jupiter ♃ ° '	Saturn ♄ ° '	Uranus ♅ ° '	Neptune ♆ ° '	Pluto ♇ ° '
01	21 ♈ 13	24 ♓ 49	07 ♐ 26	15 ♍ 23	16 ♐ 22	19 ♈ 58	10 ♓ 41	17 ♑ 25
02	23 14	26 03	07 36	15 R 16	16 R 21	20 02	10 43	17 25
03	25 13	27 17	07 46	15 10	16 20	20 05	10 45	17 26
04	27 11	28 31	07 55	15 04	16 19	20 09	10 47	17 26
05	29 ♈ 07	29 ♓ 45	08 04	14 58	16 18	20 12	10 49	17 27
06	01 ♉ 00	00 ♈ 59	08 12	14 53	16 17	20 15	10 51	17 27
07	02 51	02 13	08 19	14 47	16 16	20 19	10 53	17 27
08	04 39	03 27	08 25	14 42	16 15	20 22	10 55	17 28
09	06 23	04 41	08 31	14 36	16 13	20 26	10 57	17 28
10	08 03	05 56	08 37	14 31	16 12	20 29	10 58	17 28
11	09 39	07 10	08 41	14 26	16 10	20 33	11 00	17 28
12	11 11	08 24	08 45	14 21	16 08	20 36	11 02	17 29
13	12 38	09 38	08 48	14 17	16 06	20 39	11 04	17 29
14	14 00	10 52	08 51	14 12	16 05	20 43	11 06	17 29
15	15 17	12 06	08 53	14 08	16 03	20 46	11 07	17 29
16	16 29	13 20	08 54	14 04	16 01	20 50	11 09	17 29
17	17 36	14 34	08 54	13 59	15 58	20 53	11 11	17 29
18	18 37	15 48	08 R 54	13 56	15 56	20 57	11 12	17 R 29
19	19 33	17 02	08 53	13 52	15 54	21 00	11 14	17 29
20	20 23	18 16	08 51	13 48	15 51	21 03	11 16	17 29
21	21 08	19 30	08 48	13 45	15 49	21 07	11 17	17 29
22	21 46	20 44	08 45	13 42	15 46	21 10	11 19	17 29
23	22 19	21 58	08 41	13 39	15 44	21 14	11 21	17 29
24	22 46	23 12	08 36	13 36	15 41	21 17	11 22	17 29
25	23 07	24 26	08 30	13 33	15 38	21 20	11 24	17 28
26	23 22	25 39	08 24	13 31	15 35	21 24	11 25	17 28
27	23 32	26 53	08 17	13 29	15 32	21 27	11 27	17 28
28	23 36	28 07	08 09	13 27	15 29	21 30	11 28	17 28
29	23 R 34	29 ♈ 21	08 00	13 25	15 26	21 34	11 29	17 27
30	23 ♉ 28	00 ♉ 35	07 ♐ 51	13 ♍ 23	15 ♐ 23	21 ♈ 37	11 ♓ 31	17 ♑ 27

Date	Moon True ☊ ° '	Moon Mean ☊ ° '	Moon ☽ Latitude ° '
01	21 ♍ 37	20 ♍ 46	04 N 31
02	21 D 38	20 43	03 47
03	21 39	20 39	02 48
04	21 41	20 36	01 37
05	21 41	20 33	00 N 18
06	21 R 41	20 30	01 S 03
07	21 39	20 27	02 22
08	21 37	20 24	03 31
09	21 33	20 20	04 24
10	21 29	20 17	04 59
11	21 25	20 14	05 13
12	21 22	20 11	05 08
13	21 20	20 08	04 45
14	21 D 20	20 05	04 08
15	21 21	20 01	03 20
16	21 23	19 58	02 23
17	21 24	19 55	01 21
18	21 25	19 52	00 S 16
19	21 R 25	19 49	00 N 48
20	21 23	19 45	01 50
21	21 19	19 42	02 48
22	21 13	19 39	03 37
23	21 05	19 36	04 18
24	20 57	19 33	04 47
25	20 49	19 30	05 04
26	20 41	19 26	05 08
27	20 36	19 23	04 57
28	20 32	19 20	04 32
29	20 30	19 17	03 52
30	20 ♍ 29	19 ♍ 14	03 N 00

DECLINATIONS

Date	Sun ☉ ° '	Moon ☽ ° '	Mercury ☿ ° '	Venus ♀ ° '	Mars ♂ ° '	Jupiter ♃ ° '	Saturn ♄ ° '	Uranus ♅ ° '	Neptune ♆ ° '	Pluto ♇ ° '
01	04 N 49	17 S 06	08 N 21	03 S 25	20 S 41	07 N 07	20 S 59	07 N 15	08 S 18	20 S 50
02	05 12	15 09	09 17	02 56	20 44	07 09	20 59	07 17	08 18	20 50
03	05 35	12 17	10 11	02 26	20 47	07 12	20 58	07 18	08 17	20 50
04	05 58	08 37	11 05	01 57	20 50	07 14	20 58	07 19	08 16	20 50
05	06 21	04 S 20	11 57	01 28	20 53	07 16	20 58	07 21	08 15	20 50
06	06 43	00 N 20	12 48	00 59	20 56	07 18	20 58	07 22	08 14	20 50
07	07 06	05 03	13 37	00 S 30	20 58	07 20	20 58	07 23	08 14	20 50
08	07 28	09 26	14 24	00 00	21 01	07 22	20 57	07 25	08 13	20 50
09	07 51	13 11	15 09	00 N 29	21 03	07 24	20 57	07 26	08 12	20 50
10	08 13	16 00	15 52	00 59	21 06	07 26	20 57	07 27	08 12	20 50
11	08 35	17 43	16 32	01 28	21 08	07 28	20 57	07 28	08 11	20 50
12	08 57	18 17	17 10	01 57	21 11	07 29	20 56	07 30	08 10	20 50
13	09 18	17 46	17 45	02 27	21 13	07 31	20 56	07 31	08 10	20 50
14	09 40	16 18	18 18	02 56	21 15	07 33	20 56	07 32	08 09	20 50
15	10 01	14 03	18 47	03 25	21 17	07 34	20 55	07 34	08 08	20 50
16	10 23	11 05	19 15	03 54	21 19	07 36	20 55	07 35	08 08	20 50
17	10 44	07 54	19 39	04 23	21 21	07 37	20 55	07 36	08 07	20 50
18	11 04	04 20	20 01	04 52	21 23	07 38	20 55	07 37	08 06	20 50
19	11 25	00 N 37	20 19	05 21	21 25	07 40	20 54	07 39	08 06	20 50
20	11 46	03 S 06	20 35	05 50	21 27	07 41	20 54	07 40	08 05	20 50
21	12 06	06 41	20 48	06 19	21 29	07 42	20 54	07 41	08 05	20 50
22	12 26	10 01	20 59	06 47	21 30	07 43	20 53	07 43	08 04	20 50
23	12 46	12 58	21 07	07 16	21 32	07 44	20 53	07 44	08 04	20 50
24	13 06	15 23	21 11	07 44	21 33	07 45	20 53	07 45	08 03	20 51
25	13 25	17 09	21 13	08 12	21 35	07 46	20 52	07 46	08 02	20 51
26	13 45	18 10	21 13	08 40	21 36	07 47	20 52	07 48	08 02	20 51
27	14 04	18 21	21 09	09 08	21 38	07 47	20 51	07 49	08 01	20 51
28	14 22	17 37	21 03	09 35	21 39	07 48	20 51	07 50	08 01	20 51
29	14 41	15 59	20 55	10 03	21 40	07 49	20 51	07 51	08 00	20 51
30	14 N 59	13 S 29	20 N 44	10 N 30	21 S 41	07 N 49	20 S 50	07 N 53	08 S 00	20 S 51

ZODIAC SIGN ENTRIES

Date	h	m	Planets	
02	01	37	☽	♒
04	05	45	☽	♓
05	16	50	♀	♈
05	23	09	☿	♉
06	06	46	☽	♈
08	06	10	☽	♉
10	05	59	☽	♊
12	08	06	☽	♋
14	13	53	☽	♌
16	23	23	☽	♍
19	11	24	☽	♎
19	15	29	☉	♉
22	00	17	☽	♏
24	12	46	☽	♐
26	23	54	☽	♑
29	08	47	☽	♒
30	00	36	♀	♉

LATITUDES

Date	Mercury ☿ ° '	Venus ♀ ° '	Mars ♂ ° '	Jupiter ♃ ° '	Saturn ♄ ° '	Uranus ♅ ° '	Neptune ♆ ° '	Pluto ♇ ° '
01	00 N 05	01 S 28	00 N 52	01 N 28	01 N 46	00 S 36	00 S 48	01 N 29
04	00 39	01 29	00 48	01 28	01 46	00 36	00 48	01 29
07	01 14	01 30	00 44	01 27	01 47	00 36	00 48	01 29
10	01 46	01 30	00 39	01 27	01 47	00 36	00 48	01 29
13	02 14	01 29	00 33	01 27	01 47	00 35	00 48	01 28
16	02 35	01 28	00 28	01 26	01 47	00 35	00 48	01 28
19	02 49	01 27	00 22	01 26	01 48	00 35	00 48	01 28
22	02 53	01 24	00 15	01 25	01 48	00 35	00 49	01 28
25	02 46	01 21	00 08	01 25	01 48	00 35	00 49	01 28
28	02 28	01 18	00 N 01	01 24	01 48	00 35	00 49	01 28
31	01 N 59	01 S 14	00 S 07	01 N 23	01 N 48	00 S 36	00 S 49	01 N 27

DATA

Julian Date	2457480
Delta T	+69 seconds
Ayanamsa	24° 05' 00"
Synetic vernal point	05° ♓ 02' 00"
True obliquity of ecliptic	23° 26' 05"

MOON'S PHASES, APSIDES AND POSITIONS ☽

Date	h	m	Phase	Longitude ° '	Eclipse Indicator
07	11	24	●	18 ♈ 04	
14	03	59	☽	24 ♋ 39	
22	05	24	○	02 ♏ 31	
30	03	29	☾	10 ♒ 13	

Day	h	m		
07	17	29	Perigee	
21	15	49	Apogee	
06	10	18	0N	
12	12	15	Max dec	18° N 17'
19	15	59	0S	
27	04	47	Max dec	18° S 23'

LONGITUDES

Date	Chiron ⚷ ° '	Ceres ⚳ ° '	Pallas ⚴ ° '	Juno ⚵ ° '	Vesta ⚶ ° '	Black Moon Lilith ⚸ ° '
01	22 ♓ 19	24 ♓ 57	23 ♒ 28	12 ♏ 36	10 ♉ 19	24 ♎ 26
11	22 ♓ 53	28 ♓ 49	26 ♒ 16	10 ♏ 44	14 ♉ 37	25 ♎ 32
21	23 ♓ 24	02 ♈ 36	28 ♒ 51	08 ♏ 33	18 ♉ 58	26 ♎ 39
31	23 ♓ 53	06 ♈ 19	01 ♓ 10	06 ♏ 17	23 ♉ 19	27 ♎ 46

ASPECTARIAN

01 Friday
00 41 ☽ ⊻ ♄
02 37 ☽ ☌ ♇
07 20 ☽ □ ♅
09 13 ☽ □ ☿
10 38 ☿ # ♆
11 45 ☽ ⊥ ♄
11 53 ☽ ∠ ♂
13 47 ☿ ± ♃
16 39 ☽ ✶ ♀
17 49 ☽ ∠ ♆

02 Saturday
02 11 ☽ bQ ♃
03 04 ☽ Q ☉
03 52 ☿ bQ ♂
04 03 ☽ ∠ ♄
10 03 ☽ ⊥ ♆
15 14 ☽ ✶ ♂
15 57 ☽ Q ♅
18 05 ☽ ± ♃
20 41 ☽ ⊻ ♆
22 09 ☽ ∠ ♀
23 12 ☽ Q ☿

03 Sunday
01 58 ☽ ✶ ☉
04 30 ☽ ⊼ ♃
06 28 ☽ ✶ ♄
08 21 ☽ ⊻ ♇
12 23 ☽ Q ♂
12 56 ☽ ✶ ♅
15 16 ☽ ⊥ ♀
18 31 ☿ ∠ ♆
18 39 ☽ ⊥ ♇
23 16 ☽ ✶ ☿
23 32 ☽ # ☿

04 Monday
02 26 ☽ ⊻ ♀
02 57 ☽ Q ♄
05 34 ☽ ∠ ☉
08 59 ♂ ∥ ♇
09 50 ☽ ∠ ♇
10 19 ☉ ⊼ ♃
14 08 ☽ ∥ ♆
14 21 ☽ ∠ ♅
19 01 ☽ □ ♂
19 33 ☽ # ♅
20 03 ☽ # ♃
21 37 ☽ ⊥ ☉
23 42 ☽ ☌ ♆

05 Tuesday
01 58 ☽ # ☉
04 06 ☽ ∠ ☿
05 14 ☽ ⊥ ♅
06 00 ♀ Q ♇
06 34 ☽ ☍ ♃
08 09 ☽ ⊻ ☉
08 42 ☽ □ ♄
10 33 ☽ ✶ ♇
15 02 ☽ ⊻ ♅
16 09 ☉ △ ♄
20 51 ☽ ⊥ ☿
22 16 ☿ bQ ♃

06 Wednesday
04 33 ☽ ∥ ♀
05 15 ☉ ⊥ ♆
05 53 ☽ Q ♇
07 51 ☽ ⊻ ☿
08 01 ☽ ☌ ♀
11 14 ☿ ⊻ ♀
14 57 ☽ # ♀
15 33 ☿ bQ ♄
19 51 ☽ △ ♂
20 13 ☉ □ ♇

07 Thursday
00 01 ☽ ⊻ ♆
04 28 ☿ ± ♂
05 28 ♂ ∥ ♄
06 14 ☽ ⊼ ♃
08 33 ☽ △ ♄
09 30 ☽ ⊥ ♆
10 25 ☽ □ ♇
11 24 ☽ ☌ ☉
14 56 ☽ ☌ ♅
15 38 ☽ ± ♃
19 42 ☽ bQ ♂
23 43 ☽ ∠ ♆
23 53 ☽ ∥ ☉

08 Friday
00 16 ☽ ∥ ♃
00 31 ☽ ∥ ♅
04 45 ☉ ∥ ♃
05 03 ☽ # ♆
05 44 ☽ bQ ♃
07 49 ☉ ∥ ♅
08 08 ☽ bQ ♄
09 59 ☽ ± ♂
11 35 ☽ ⊻ ♀
13 41 ☽ ☌ ☿
19 31 ☽ ⊼ ♂
21 55 ☽ ⊥ ♀
22 20 ☽ ± ♄
23 26 ☽ ✶ ♆

09 Saturday
05 18 ☽ △ ♃
07 51 ☽ ⊼ ♄
09 49 ☽ △ ♇
13 30 ☽ ∠ ♀
14 07 ☽ ⊻ ☉
14 34 ☽ ⊻ ♅
18 36 ☽ Q ♆
21 27 ☉ ☌ ♅

10 Sunday
00 07 ☉ ± ♃
00 15 ☽ ⊥ ♅
00 26 ☽ ⊥ ☉
09 56 ☽ ∥ ☿
09 59 ☽ bQ ♇
10 53 ☉ # ♆
14 56 ☽ ∠ ♅
15 58 ☽ ✶ ♀
16 07 ☽ ∠ ☉
17 29 ♃ ± ♅
20 01 ☽ ⊻ ☿
20 06 ☽ ☍ ♂
20 46 ☿ ⊼ ♂
23 59 ☽ □ ♆

11 Monday
00 47 ☽ ± ♇
05 44 ☽ □ ♃
07 12 ☽ ⊥ ☿
08 36 ☽ ☍ ♄
10 47 ☽ ⊼ ♇
13 46 ☽ Q ♀
15 58 ☽ ✶ ♅
18 57 ☽ ✶ ☉
19 52 ☿ ± ♄

12 Tuesday
00 16 ☽ ∠ ☿
09 40 ☿ ✶ ♆
12 11 ☽ Q ♃
12 37 ☽ Q ♅
17 11 ☽ Q ☉
19 16 ♀ △ ♂
23 26 ☽ ⊼ ♂
23 50 ☽ □ ♀

13 Wednesday
03 28 ☽ △ ♆
05 35 ☽ ✶ ☿
07 14 ☉ bQ ♂
07 58 ♃ ∥ ♅
09 13 ☽ ✶ ♃
10 09 ☽ ± ♂
12 21 ☽ ∥ ☿
12 29 ☽ ⊼ ♄
14 58 ☽ ☍ ♇
20 45 ☽ □ ♅
23 19 ☽ ± ♄

14 Thursday
02 30 ☽ bQ ♂
03 59 ☽ □ ☉
05 51 ☽ Q ☿
06 39 ☽ bQ ♆
12 24 ☽ ∠ ♃
15 27 ☿ △ ♃
15 52 ☽ bQ ♄
16 36 ♀ ⊻ ♆
23 26 ☽ ± ♆

15 Friday
05 10 ☽ ⊥ ♃
06 32 ☽ △ ♂
10 48 ☽ ⊼ ♆
12 44 ☽ △ ♀
16 25 ☉ ∠ ♆
16 31 ☽ ⊻ ♃
19 30 ☽ □ ☿
20 12 ☽ △ ♄
22 59 ☽ ⊼ ♇

16 Saturday
02 28 ☿ ⊼ ♄
05 26 ☽ △ ♅
10 37 ☽ ± ♇
17 39 ☽ ∥ ☉
17 48 ☽ △ ☉
21 00 ☽ bQ ♀

17 Sunday
01 26 ♀ ⊼ ♃
04 16 ☽ bQ ♇
09 23 ☿ △ ♇
10 32 ☽ # ♆
11 00 ☽ bQ ♅
12 14 ♂ St R
13 59 ☽ ∥ ♃
14 06 ☽ ∥ ♅
16 49 ☽ ± ♀
17 00 ☽ □ ♂
21 34 ☽ ☍ ♆

18 Monday
02 08 ☽ bQ ☉
03 04 ☽ ☌ ♃
05 02 ☽ ± ♅
06 10 ☽ ⊼ ♀
07 04 ☽ □ ♄
07 26 ♇ St R
08 52 ☽ ∥ ♀
10 10 ☽ △ ♇
12 29 ☽ △ ☿
13 03 ☉ bQ ♃
14 37 ♀ △ ♄
17 08 ☽ ⊼ ♅
21 49 ☽ ± ☉

19 Tuesday
05 06 ☽ Q ♂
11 02 ☽ ⊼ ☉
16 07 ♀ ⊥ ♆
19 16 ☽ Q ♄
20 51 ♀ □ ♇
21 18 ☽ bQ ☿

20 Wednesday
05 21 ☽ ✶ ♂
10 14 ☽ ⊼ ♆
12 32 ☉ bQ ♄
15 23 ☽ ⊻ ♃
16 54 ☽ ± ☿
19 32 ☽ ✶ ♄
22 27 ☽ ± ♆
22 52 ☽ □ ♇

21 Thursday
01 54 ☽ ☍ ♀
03 31 ☽ ⊥ ♃
05 55 ☽ ⊼ ☿
06 13 ☽ ☍ ♅
09 00 ☽ # ♀
11 27 ☿ ⊻ ♅
11 43 ☽ ∠ ♂
15 52 ☿ # ♇
16 46 ♀ ± ♃
16 46 ☽ bQ ♆
18 59 ☽ # ♅
19 05 ☽ # ♃
21 41 ☽ ∥ ♆
21 43 ☽ ∠ ♃
22 36 ☿ # ♄

22 Friday
01 54 ☽ ∠ ♄
05 24 ☽ ☍ ☉
05 54 ☽ ⊥ ♂
11 09 ☉ ± ♂
11 24 ☽ Q ♇
17 59 ☽ ⊻ ♂
20 03 ☽ ⊥ ♄
21 00 ♀ ☌ ♅
23 14 ☽ △ ♆

23 Saturday
03 57 ☽ ✶ ♃
08 08 ☽ ⊻ ♄
10 02 ☽ # ☉
11 38 ☽ ✶ ♇
19 13 ☽ ⊼ ♅
21 39 ☽ ⊼ ♀
21 46 ☽ ☍ ☿
23 18 ☿ ⊻ ♀

24 Sunday
04 00 ☽ Q ♃
07 19 ☽ ± ♅
11 04 ☽ ± ♀
12 53 ♀ ∥ ♅
12 53 ♀ ∥ ♃
13 03 ♃ ∥ ♅
17 43 ☽ ∠ ♇
19 22 ♀ bQ ♂
23 06 ☽ ⊼ ☉

25 Monday
01 21 ☽ bQ ♅
03 49 ♀ # ♆
05 46 ☽ ☌ ♂
07 02 ☽ bQ ♀
11 26 ☽ □ ♆
11 35 ☽ ⊥ ♇
12 03 ☽ ± ☉
15 42 ☽ □ ♃
19 47 ☽ ☌ ♄
23 26 ☽ ⊻ ♇

26 Tuesday
07 06 ☽ △ ♅
07 19 ☽ bQ ☉
10 59 ☽ ⊼ ☿
15 51 ☽ △ ♀
18 24 ☿ Q ♆
22 48 ☽ Q ♆
22 51 ☽ ± ☿

27 Wednesday
03 08 ♀ ∠ ♆
14 53 ☽ △ ☉
15 50 ☽ ⊻ ♂
16 24 ☽ bQ ☿
21 56 ☽ ✶ ♆

28 Thursday
01 42 ☉ ⊼ ♂
01 45 ☽ △ ♃
03 10 ☽ ⊥ ♂
05 38 ☽ ⊻ ♄
09 21 ☽ ☌ ♇
16 54 ☽ ⊥ ♄
16 59 ☽ □ ♅
17 20 ☿ St R
18 04 ♀ bQ ♃
19 57 ☽ ∠ ♂
20 28 ♄ ∥ ♇
20 53 ☽ △ ☿

29 Friday
02 15 ☽ ∠ ♆
05 51 ☽ bQ ♃
07 07 ☽ □ ♀
08 16 ☉ ± ♄
09 36 ☽ ∠ ♄
18 53 ☽ ⊥ ♆
20 36 ☿ # ♇
21 42 ☿ # ♄
22 21 ☽ ± ♃
23 19 ☽ ✶ ♂

30 Saturday
00 03 ☽ # ☉
02 20 ☽ Q ♅
03 29 ☽ □ ☉
04 24 ☿ Q ♆
05 48 ☽ ⊻ ♆
08 13 ♀ bQ ♄
09 11 ☽ ⊼ ♃
12 46 ☽ ✶ ♄
16 28 ☽ ⊻ ♇
19 09 ☽ Q ♀
20 38 ☽ Q ♂
23 55 ☽ ✶ ♅

All ephemeris data is given at 12.00 UT and the Moon's longitude is additionally given for 24.00 UT

MAY 2016

LONGITUDES

Date	Sidereal time h m s	Sun ☉ ° ′ ″	Moon ☽ ° ′ ″	Moon ☽ 24.00 ° ′ ″	Mercury ☿ ° ′	Venus ♀ ° ′	Mars ♂ ° ′	Jupiter ♃ ° ′	Saturn ♄ ° ′	Uranus ♅ ° ′	Neptune ♆ ° ′	Pluto ♇ ° ′
01	02 39 23	11 ♉ 32 13	28 ♒ 31 46	05 ♓ 28 12	23 ♉ 16	01 ♉ 49	07 ♐ 41	13 ♍ 21	15 ♐ 20	21 ♈ 40	11 ♓ 32	17 ♑ 27
02	02 43 20	12 30 27	12 ♓ 30 52	19 ♓ 39 43	22 R 59	03 03	07 R 30	13 R 20	15 R 16	21 43	11 34	17 R 26
03	02 47 16	13 28 38	26 ♓ 54 35	04 ♈ 15 05	22 39	04 17	07 18	13 19	15 13	21 47	11 35	17 26
04	02 51 13	14 26 49	11 ♈ 40 40	19 ♈ 10 36	22 14	05 31	07 06	13 18	15 10	21 50	11 36	17 25
05	02 55 10	15 24 57	26 ♈ 43 55	04 ♉ 19 31	21 46	06 45	06 53	13 17	15 06	21 53	11 37	17 25
06	02 59 06	16 23 05	11 ♉ 56 11	19 ♉ 32 34	21 15	07 58	06 39	13 16	15 03	21 56	11 39	17 24
07	03 03 03	17 21 10	27 ♉ 07 20	04 ♊ 39 12	20 41	09 12	06 24	13 16	14 59	22 00	11 40	17 24
08	03 06 59	18 19 14	12 ♊ 06 56	19 ♊ 29 28	20 06	10 26	06 09	13 15	14 55	22 03	11 41	17 23
09	03 10 56	19 17 17	26 ♊ 45 58	03 ♋ 55 44	19 30	11 40	05 53	13 15	14 51	22 06	11 42	17 23
10	03 14 52	20 15 17	10 ♋ 58 20	17 ♋ 53 32	18 53	12 54	05 37	13 D 15	14 48	22 09	11 43	17 22
11	03 18 49	21 13 16	24 ♋ 41 18	01 ♌ 21 48	18 17	14 08	05 20	13 16	14 44	22 12	11 44	17 21
12	03 22 45	22 11 13	07 ♌ 55 18	14 ♌ 22 40	17 41	15 22	05 03	13 16	14 40	22 15	11 45	17 21
13	03 26 42	23 09 08	20 ♌ 43 07	26 ♌ 58 31	17 07	16 35	04 45	13 17	14 36	22 18	11 46	17 20
14	03 30 39	24 07 01	03 ♍ 09 03	09 ♍ 15 22	16 35	17 49	04 27	13 18	14 32	22 21	11 47	17 19
15	03 34 35	25 04 52	15 ♍ 18 07	21 ♍ 17 57	16 06	19 03	04 08	13 19	14 28	22 24	11 48	17 18
16	03 38 32	26 02 42	27 ♍ 15 28	03 ♎ 11 18	15 39	20 17	03 48	13 20	14 24	22 27	11 49	17 18
17	03 42 28	27 00 29	09 ♎ 05 59	15 ♎ 00 03	15 16	21 31	03 29	13 21	14 20	22 30	11 50	17 17
18	03 46 25	27 58 16	20 ♎ 53 58	26 ♎ 48 09	14 57	22 44	03 09	13 23	14 16	22 33	11 51	17 16
19	03 50 21	28 56 01	02 ♏ 43 00	08 ♏ 38 51	14 41	23 58	02 49	13 24	14 11	22 36	11 52	17 15
20	03 54 18	29 ♉ 53 44	14 ♏ 35 57	20 ♏ 34 35	14 30	25 12	02 28	13 26	14 07	22 39	11 53	17 14
21	03 58 14	00 ♊ 51 26	26 ♏ 34 54	02 ♐ 37 07	14 23	26 26	02 08	13 28	14 03	22 42	11 53	17 13
22	04 02 11	01 49 06	08 ♐ 41 20	14 ♐ 47 41	14 20	27 39	01 47	13 31	13 59	22 45	11 54	17 13
23	04 06 08	02 46 46	20 ♐ 56 17	27 ♐ 07 13	14 D 22	28 ♉ 53	01 26	13 33	13 54	22 48	11 55	17 12
24	04 10 04	03 44 24	03 ♑ 20 35	09 ♑ 36 32	14 29	00 ♊ 07	01 05	13 36	13 50	22 50	11 56	17 11
25	04 14 01	04 42 01	15 ♑ 55 10	22 ♑ 16 41	14 40	01 21	00 44	13 38	13 46	22 53	11 56	17 10
26	04 17 57	05 39 37	28 ♑ 41 14	05 ♒ 09 03	14 55	02 34	00 23	13 41	13 41	22 56	11 57	17 09
27	04 21 54	06 37 12	11 ♒ 40 22	18 ♒ 15 27	15 15	03 48	00 ♐ 02	13 44	13 37	22 59	11 57	17 08
28	04 25 50	07 34 46	24 ♒ 54 34	01 ♓ 37 59	15 40	05 02	29 ♏ 41	13 48	13 33	23 01	11 58	17 07
29	04 29 47	08 32 19	08 ♓ 25 58	15 ♓ 18 44	16 08	06 16	29 20	13 51	13 28	23 04	11 59	17 06
30	04 33 43	09 29 52	22 ♓ 16 26	29 ♓ 19 09	16 41	07 29	28 59	13 55	13 24	23 07	11 59	17 04
31	04 37 40	10 ♊ 27 23	06 ♈ 26 52	13 ♈ 39 24	17 ♉ 18	08 ♊ 43	28 ♏ 39	13 ♍ 58	13 ♐ 19	23 ♈ 09	11 ♓ 59	17 ♑ 03

Moon Nodes and Latitude

Date	Moon True ☊ ° ′	Moon Mean ☊ ° ′	Moon ☽ Latitude ° ′
01	20 ♍ 30	19 ♍ 11	01 N 56
02	20 D 31	19 07	00 N 43
03	20 R 31	19 04	00 S 34
04	20 29	19 01	01 51
05	20 25	18 58	03 02
06	20 19	18 55	04 00
07	20 11	18 51	04 41
08	20 02	18 48	05 02
09	19 54	18 45	05 03
10	19 47	18 42	04 45
11	19 41	18 39	04 11
12	19 38	18 36	03 24
13	19 37	18 32	02 28
14	19 D 38	18 29	01 27
15	19 38	18 26	00 S 23
16	19 R 38	18 23	00 N 41
17	19 36	18 20	01 42
18	19 32	18 17	02 38
19	19 25	18 13	03 28
20	19 16	18 10	04 09
21	19 04	18 07	04 39
22	18 52	18 04	04 57
23	18 39	18 01	05 01
24	18 27	17 57	04 51
25	18 17	17 54	04 27
26	18 10	17 51	03 50
27	18 06	17 48	03 00
28	18 04	17 45	01 59
29	18 D 03	17 42	00 N 51
30	18 R 03	17 38	00 S 22
31	18 ♍ 02	17 ♍ 35	01 S 36

DECLINATIONS

Date	Sun ☉ ° ′	Moon ☽ ° ′	Mercury ☿ ° ′	Venus ♀ ° ′	Mars ♂ ° ′	Jupiter ♃ ° ′	Saturn ♄ ° ′	Uranus ♅ ° ′	Neptune ♆ ° ′	Pluto ♇ ° ′
01	15 N 17	10 S 10	20 N 30	10 N 56	21 S 42	07 N 50	20 S 50	07 N 54	07 S 59	20 S 51
02	15 35	06 12	20 14	11 23	21 43	07 50	20 49	07 55	07 59	20 51
03	15 53	01 S 45	19 56	11 49	21 44	07 50	20 49	07 56	07 58	20 51
04	16 10	02 N 55	19 36	12 15	21 44	07 50	20 49	07 57	07 58	20 52
05	16 27	07 29	19 14	12 41	21 45	07 51	20 48	07 59	07 57	20 52
06	16 44	11 36	18 50	13 07	21 45	07 51	20 48	08 00	07 57	20 52
07	17 01	14 57	18 25	13 32	21 46	07 51	20 48	08 01	07 57	20 52
08	17 17	17 15	18 00	13 57	21 46	07 51	20 47	08 02	07 56	20 52
09	17 33	18 21	17 33	14 21	21 46	07 50	20 47	08 03	07 56	20 52
10	17 48	18 15	17 06	14 45	21 46	07 50	20 46	08 04	07 55	20 52
11	18 04	17 04	16 40	15 09	21 46	07 50	20 46	08 06	07 55	20 52
12	18 19	15 00	16 13	15 32	21 46	07 50	20 45	08 07	07 55	20 53
13	18 33	12 14	15 47	15 55	21 46	07 49	20 45	08 08	07 54	20 53
14	18 48	09 00	15 22	16 18	21 46	07 49	20 44	08 09	07 54	20 53
15	19 02	05 26	14 59	16 40	21 45	07 48	20 44	08 10	07 54	20 53
16	19 16	01 N 43	14 37	17 02	21 45	07 47	20 43	08 11	07 53	20 53
17	19 29	02 S 05	14 16	17 23	21 44	07 47	20 43	08 12	07 53	20 53
18	19 42	05 43	13 58	17 44	21 43	07 46	20 42	08 13	07 53	20 54
19	19 55	09 09	13 42	18 05	21 42	07 45	20 42	08 14	07 52	20 54
20	20 07	12 15	13 27	18 24	21 41	07 44	20 41	08 15	07 52	20 54
21	20 20	14 52	13 16	18 44	21 40	07 43	20 41	08 17	07 52	20 54
22	20 31	16 52	13 06	19 03	21 39	07 42	20 40	08 18	07 52	20 54
23	20 43	18 07	12 59	19 21	21 38	07 41	20 40	08 19	07 51	20 54
24	20 54	18 32	12 54	19 39	21 36	07 40	20 40	08 20	07 51	20 55
25	21 04	18 04	12 52	19 57	21 35	07 39	20 39	08 21	07 51	20 55
26	21 15	16 40	12 51	20 14	21 34	07 37	20 39	08 22	07 51	20 55
27	21 25	14 24	12 53	20 30	21 32	07 36	20 38	08 23	07 51	20 55
28	21 34	11 20	12 58	20 46	21 31	07 34	20 38	08 24	07 50	20 55
29	21 43	07 37	13 04	21 01	21 29	07 33	20 37	08 25	07 50	20 56
30	21 52	03 24	13 12	21 16	21 27	07 31	20 37	08 26	07 50	20 56
31	22 N 01	01 N 06	13 N 22	21 N 30	21 S 26	07 N 30	20 S 36	08 N 26	07 S 50	20 S 56

ZODIAC SIGN ENTRIES

Date	h	m	Planets	
01	14	33	☽	♓
03	17	04	☽	♈
05	17	10	☽	♉
07	16	34	☽	♊
09	17	24	☽	♋
11	21	32	☽	♌
14	05	52	☽	♍
16	17	33	☽	♎
19	06	29	☽	♏
20	14	36	☉	♊
21	18	48	☽	♐
24	05	34	☽	♑
24	09	45	♀	♊
26	14	27	☽	♒
27	13	51	♂	♏
28	21	06	☽	♓
31	01	09	☽	♈

LATITUDES

Date	Mercury ☿ ° ′	Venus ♀ ° ′	Mars ♂ ° ′	Jupiter ♃ ° ′	Saturn ♄ ° ′	Uranus ♅ ° ′	Neptune ♆ ° ′	Pluto ♇ ° ′
01	01 N 59	01 S 14	00 S 07	01 N 23	01 N 48	00 S 36	00 S 49	01 N 27
04	01 19	01 10	00 15	01 23	01 49	00 36	00 49	01 27
07	00 N 31	01 05	00 24	01 22	01 49	00 36	00 49	01 27
10	00 S 21	01 00	00 33	01 22	01 49	00 36	00 49	01 27
13	01 12	00 55	00 42	01 21	01 49	00 36	00 49	01 27
16	02 00	00 49	00 51	01 20	01 49	00 36	00 49	01 26
19	02 40	00 42	01 00	01 20	01 49	00 36	00 50	01 26
22	03 11	00 36	01 10	01 19	01 49	00 36	00 50	01 26
25	03 32	00 29	01 19	01 19	01 49	00 36	00 50	01 26
28	03 43	00 23	01 28	01 18	01 49	00 36	00 50	01 26
31	03 S 46	00 S 16	01 S 37	01 N 17	01 N 48	00 S 36	00 S 50	01 N 25

DATA

Julian Date	2457510
Delta T	+69 seconds
Ayanamsa	24° 05′ 02″
Synetic vernal point	05° ♓ 01′ 57″
True obliquity of ecliptic	23° 26′ 05″

MOON'S PHASES, APSIDES AND POSITIONS ☽

Date	h	m	Phase	Longitude ° ′	Eclipse Indicator
06	19	29	●	16 ♉ 41	
13	17	02	☽	23 ♌ 21	
21	21	14	○	01 ♐ 14	
29	12	12	☾	08 ♓ 33	

Day	h	m		
06	04	06	Perigee	
18	21	58	Apogee	
03	21	04	ON	
09	21	57	Max dec	18° N 27′
16	22	54	OS	
24	11	20	Max dec	18° S 32′
31	06	14	ON	

LONGITUDES

Date	Chiron ⚷ ° ′	Ceres ⚳ ° ′	Pallas ⚴ ° ′	Juno ⚵ ° ′	Vesta ⚶ ° ′	Black Moon Lilith ⚸ ° ′
01	23 ♓ 53	06 ♈ 19	01 ♓ 10	06 ♏ 17	23 ♉ 19	27 ♎ 46
11	24 ♓ 18	09 ♈ 57	03 ♓ 12	04 ♏ 06	27 ♉ 42	28 ♎ 52
21	24 ♓ 39	13 ♈ 28	04 ♓ 52	02 ♏ 13	02 ♊ 04	29 ♎ 59
31	24 ♓ 56	16 ♈ 51	06 ♓ 10	00 ♏ 45	06 ♊ 26	01 ♏ 06

All ephemeris data is given at 12.00 UT and the Moon's longitude is additionally given for 24.00 UT

ASPECTARIAN

h m Aspects

01 Sunday
02 56 ☽ □ ☿
03 06 ☽ ⊥ ♇
07 27 ☽ # ♀
09 36 ♀ ± ♂
09 55 ☽ Q ♄
12 00 ☉ ⚹ ♆
13 53 ☽ Q ☉
18 16 ☽ ⚹ ♀
18 47 ☽ ∠ ♇

02 Monday
01 39 ☽ ∥ ♆
02 07 ☽ ∠ ♅
02 07 ☽ # ♅
02 34 ☽ # ♃
03 35 ☽ □ ♂
09 28 ☽ Q ☿
10 23 ☽ ☌ ♆
11 59 ☽ ⚹ ☉
13 23 ☽ ☍ ♃
16 38 ☽ □ ♄
17 26 ☽ ⊥ ♅
20 17 ☽ ⚹ ♇
22 11 ☽ ∠ ♀

03 Tuesday
03 29 ☽ ⊻ ♅
05 08 ☽ ⚹ ☿
07 58 ☉ △ ♃
14 28 ☽ ⊥ ♀
14 45 ☽ ∠ ☉
16 08 ☽ Q ♇

04 Wednesday
01 09 ☽ ⊻ ♀
04 43 ☽ △ ♂
05 02 ☽ ∠ ☿
06 26 ☽ ⊥ ☉
11 53 ☽ ⊻ ♆
14 36 ☽ ⊼ ♃
16 45 ☽ ⊻ ☉
17 34 ☽ △ ♄
19 05 ☽ ⊥ ☿
20 21 ♅ # ♆
21 12 ☽ □ ♇
21 30 ☽ ⊥ ♆

05 Thursday
00 11 ☽ ± ♃
04 17 ☽ ☌ ♅
04 22 ☽ ⊻ ☿
04 24 ☽ bQ ♂
04 39 ☉ ⊼ ♄
06 31 ☿ ⊻ ♅
11 10 ♂ bQ ♅
11 50 ☽ ∠ ♆
13 59 ☽ ∥ ♃
14 11 ♀ ⊼ ♂
14 27 ☽ bQ ♃
14 37 ☽ # ♆
14 44 ☽ ∥ ♅
17 19 ☽ bQ ♄
18 27 ☽ ± ♂

06 Friday
03 47 ☽ ⊼ ♂
05 12 ☽ ☌ ♀
07 27 ☽ ± ♄
11 32 ☽ ⚹ ♆
14 06 ☽ △ ♃
16 53 ☽ ⊼ ♄
19 29 ☽ ☌ ☉
20 37 ☽ △ ♇
23 25 ☽ ∥ ♀

07 Saturday
02 10 ☽ ☌ ☿
03 51 ☽ ⊻ ♅
06 31 ☽ Q ♆
07 50 ♀ ± ♄
13 03 ☉ △ ♇
13 23 ☽ ⊥ ♅
20 23 ☽ bQ ♇

08 Sunday
02 34 ☽ ☍ ♂
03 48 ☽ ∠ ♅
09 03 ☽ ⊻ ♀
10 49 ☽ ± ♇
11 18 ☽ □ ♆
12 35 ☽ ∥ ☉
13 51 ☽ □ ♃
16 32 ☽ ☍ ♄
19 39 ☽ ⊥ ♀
20 33 ☽ ⊼ ♇
21 24 ☽ ∥ ☿
22 48 ☽ ⊻ ☉

09 Monday
00 29 ☽ ⊻ ☿
04 15 ☽ ⚹ ♅
09 22 ☽ ⊥ ☉
09 59 ☽ ⊥ ☿
11 49 ☽ ∠ ♀
12 14 ♃ St D
12 16 ☉ ∥ ☿
12 41 ♀ ⚹ ♆
15 12 ☉ ☌ ☿
19 30 ☽ Q ♃

10 Tuesday
00 20 ☽ Q ♅
00 26 ☽ ∠ ☿
01 32 ☽ ∠ ☉
03 02 ☽ ⊼ ♂
13 05 ☽ ± ♂
13 17 ☽ △ ♆
15 39 ☽ ⚹ ♀
15 56 ☽ ⚹ ♃
18 34 ☽ ⊼ ♄
19 00 ♀ △ ♃
21 27 ☽ ∥ ☉
23 04 ☽ ☍ ♇

11 Wednesday
01 10 ☽ ⚹ ☿
04 27 ☽ bQ ♂
05 01 ☽ ± ♄
05 23 ☽ ⚹ ☉
07 34 ☽ □ ♅
14 50 ☽ Q ♀
15 40 ☽ bQ ♆
18 24 ☽ ∠ ♃
19 21 ☽ ∥ ☿
21 00 ☽ bQ ♄
21 36 ☽ Q ☿
23 08 ♀ ⊼ ♄

12 Thursday
04 36 ☽ Q ☉
06 50 ☽ △ ♂
07 20 ☽ ∥ ♀
08 01 ☽ ± ♆
10 48 ☽ ⊥ ♃
13 44 ☉ ⊻ ♅
19 07 ☽ ⊼ ♆
21 57 ☽ ⊻ ♃

13 Friday
00 29 ☽ △ ♄
02 45 ☿ △ ♇
03 20 ☽ □ ♀
05 28 ☽ □ ☿
05 35 ☽ ⊼ ♇
08 00 ☿ ∥ ♀
15 02 ☽ △ ♅
16 59 ☽ ± ♇
17 02 ☽ □ ☉
19 10 ☿ ☌ ♀

14 Saturday
02 20 ♀ △ ♇
03 43 ☉ Q ♆
10 23 ☽ bQ ♇
14 28 ☽ □ ♂
17 50 ☽ ∥ ♅
19 35 ☽ # ♆
20 12 ☽ ∥ ♃
20 17 ☽ bQ ♅

15 Sunday
05 02 ☽ ☍ ♆
08 02 ☽ ☌ ♃
10 21 ☽ □ ♄
13 31 ☽ △ ☿
14 13 ☽ ± ♅
16 00 ☽ △ ♇
20 21 ☽ △ ♀

16 Monday
01 19 ☽ Q ♂
02 17 ☽ ⊼ ♅
09 20 ☽ △ ☉
18 38 ☽ bQ ☿
22 20 ☽ Q ♄

17 Tuesday
00 54 ☽ ⚹ ♂
06 08 ☽ bQ ♀
12 20 ☽ ± ☿
17 34 ☽ ⊼ ♆
18 26 ☽ bQ ☉
20 40 ☽ ⊻ ♃
22 34 ☽ ⚹ ♄

18 Wednesday
00 11 ☽ ⊼ ☿
02 33 ☽ ± ♀
04 37 ☽ □ ♇
05 47 ☽ ± ♆
06 34 ☽ ∠ ♂
08 14 ♀ ⊻ ♅
08 54 ☽ ⊥ ♃
14 22 ☽ ± ☉
15 23 ☽ ☍ ♅
16 11 ☽ ⊼ ♀

19 Thursday
00 07 ☽ bQ ♆
00 21 ☽ ⊥ ♂
02 01 ☽ # ♃
02 52 ☽ ∥ ♆
03 14 ☽ ∠ ♃
03 19 ☉ ⊥ ♅
03 39 ☽ ⊼ ☉
04 53 ☽ ∠ ♄
05 24 ☽ # ♅
09 56 ♀ Q ♆
12 11 ☽ ⊻ ♂
17 08 ☽ Q ♇
23 01 ☽ ⊥ ♄

20 Friday
06 31 ☽ △ ♆
09 39 ☽ ⚹ ♃
11 03 ☽ ⊻ ♄
11 48 ☽ ☍ ☿
17 18 ☽ ⚹ ♇
21 42 ☽ # ☿

21 Saturday
04 13 ☽ ⊼ ♅
04 53 ♂ ∠ ♇
09 47 ☽ Q ♃
11 40 ☽ ☍ ♀
16 14 ☽ ± ♅
21 14 ☽ ☍ ☉
22 43 ☽ ☌ ♂
23 12 ☽ ∠ ♇

22 Sunday
10 08 ☽ bQ ♅
11 17 ☉ ☍ ♂
13 19 ☿ St D
16 57 ☽ ⊥ ♇
18 20 ☽ □ ♆
21 31 ☽ □ ♃
21 36 ☉ bQ ♇
22 20 ☽ ☌ ♄
23 07 ☽ ⊼ ☿

23 Monday
04 42 ☽ ⊻ ♇
06 32 ☉ # ♄
10 07 ♀ ⊥ ♅
10 53 ☽ ± ☿
15 37 ☽ △ ♅

24 Tuesday
04 29 ☽ bQ ☿
05 06 ☽ ⊼ ♀
05 25 ☽ Q ♆
07 46 ☽ ⊻ ♂
12 50 ☽ ⊼ ☉
14 00 ☉ # ♇
17 54 ☽ ± ♀
18 58 ☽ ⊥ ♂

25 Wednesday
01 16 ☽ ± ☉
02 38 ♀ ☍ ♂
04 26 ☽ ⚹ ♆
07 39 ☽ △ ♃
07 56 ☽ ⊻ ♄
09 35 ☽ △ ☿
11 39 ☽ ∠ ♂
12 54 ☽ bQ ♀
14 21 ☽ ☌ ♇
19 13 ☽ ⊥ ♄
19 44 ☽ bQ ☉

26 Thursday
01 11 ☽ □ ♅
03 44 ♀ bQ ♇
08 45 ☽ ∠ ♆
12 00 ☽ bQ ♃
12 00 ☽ ∠ ♄
12 28 ♃ □ ♄
15 04 ☽ ⚹ ♂
19 59 ☽ △ ♀

27 Friday
01 29 ☽ ⊥ ♆
01 58 ☽ △ ☉
04 45 ☽ ± ♃
10 43 ☽ Q ♅
12 31 ☽ ⊻ ♆
12 38 ☽ Q ♂
15 32 ☽ ⚹ ♄
15 48 ☽ ⊼ ♃
18 44 ☽ □ ☿
21 56 ☽ ⊻ ♇
23 43 ♀ # ♄

28 Saturday
00 19 ☽ # ☿
04 03 ☉ # ♂
08 36 ☽ ⚹ ♅
08 46 ☽ ⊥ ♇
13 08 ☽ Q ♄
20 19 ☽ □ ♂
23 36 ☉ ∠ ♅

29 Sunday
00 50 ☽ ∠ ♇
03 08 ♀ # ♇
04 09 ☽ Q ☿
07 14 ☽ # ♅
07 48 ☽ □ ♀
10 42 ☽ ∥ ♆
11 21 ☽ ∠ ♅
12 12 ☽ □ ☉
12 27 ☽ # ♃
18 12 ☽ ☌ ♆
20 45 ☽ □ ♄
21 30 ☽ ☍ ♃

30 Monday
01 58 ☽ ⚹ ☿
03 04 ☽ ⚹ ♇
03 05 ☽ ⊥ ♅
04 31 ♂ ± ♅
13 26 ☽ ⊻ ♅
18 01 ☽ Q ♀
21 34 ☽ Q ☉
23 10 ☽ △ ♂
23 34 ☽ Q ♇

31 Tuesday
00 32 ♀ ∠ ♅
03 09 ☿ △ ♇
04 42 ☽ ∠ ☿
05 32 ♀ # ♂
16 09 ☽ ⚹ ♀
19 10 ☽ ⚹ ☉
20 28 ☽ ⊥ ☿
21 15 ☽ ⊻ ♆
23 23 ☽ △ ♄
23 43 ☽ bQ ♂

JUNE 2016

LONGITUDES

Date	Sidereal time h m s	Sun ☉ ° ' "	Moon ☽ ° ' "	Moon ☽ 24.00 ° ' "	Mercury ☿ ° '	Venus ♀ ° '	Mars ♂ ° '	Jupiter ♃ ° '	Saturn ♄ ° '	Uranus ♅ ° '	Neptune ♆ ° '	Pluto ♇ ° '
01	04 41 37	11 ♊ 24 54	20 ♈ 56 27	28 ♈ 17 32	17 ♉ 59	09 ♊ 57	28 ♏ 19	14 ♍ 02	13 ♐ 15	23 ♈ 12	12 ♓ 00	17 ♑ 02
02	04 45 33	12 22 24	05 ♉ 42 01	13 ♉ 09 04	18 43	11 11	27 R 59	14 06	13 R 10	23 14	12 00	17 R 01
03	04 49 30	13 19 53	20 ♉ 37 44	28 ♉ 06 56	19 31	12 24	27 40	14 11	13 06	23 17	12 01	17 00
04	04 53 26	14 17 22	05 ♊ 35 30	13 ♊ 02 15	20 23	13 38	27 21	14 15	13 02	23 19	12 01	16 59
05	04 57 23	15 14 49	20 ♊ 26 02	27 ♊ 45 47	21 19	14 52	27 02	14 20	12 57	23 22	12 01	16 58
06	05 01 19	16 12 16	05 ♋ 00 33	12 ♋ 09 32	22 18	16 06	26 44	14 24	12 53	23 24	12 02	16 56
07	05 05 16	17 09 42	19 ♋ 12 08	26 ♋ 07 59	23 21	17 19	26 27	14 29	12 48	23 26	12 02	16 55
08	05 09 12	18 07 07	02 ♌ 56 50	09 ♌ 38 42	24 26	18 33	26 10	14 34	12 44	23 29	12 02	16 54
09	05 13 09	19 04 30	16 ♌ 13 43	22 ♌ 42 09	25 35	19 47	25 54	14 39	12 39	23 31	12 02	16 53
10	05 17 06	20 01 53	29 ♌ 04 27	05 ♍ 21 06	26 48	21 01	25 38	14 45	12 35	23 33	12 02	16 51
11	05 21 02	20 59 14	11 ♍ 32 42	17 ♍ 39 51	28 03	22 14	25 23	14 50	12 31	23 35	12 02	16 50
12	05 24 59	21 56 35	23 ♍ 43 13	29 ♍ 43 30	29 ♉ 22	23 28	25 09	14 56	12 26	23 38	12 02	16 49
13	05 28 55	22 53 54	05 ♎ 41 21	11 ♎ 37 27	00 ♊ 43	24 42	24 55	15 02	12 22	23 40	12 02	16 47
14	05 32 52	23 51 12	17 ♎ 32 27	23 ♎ 26 57	02 08	25 55	24 42	15 08	12 18	23 42	12 R 02	16 46
15	05 36 48	24 48 30	29 ♎ 21 31	05 ♏ 16 43	03 36	27 09	24 30	15 14	12 13	23 44	12 02	16 45
16	05 40 45	25 45 47	11 ♏ 13 01	17 ♏ 10 51	05 07	28 23	24 18	15 20	12 09	23 46	12 02	16 43
17	05 44 41	26 43 03	23 ♏ 10 36	29 ♏ 12 33	06 40	29 ♊ 37	24 08	15 26	12 05	23 48	12 02	16 42
18	05 48 38	27 40 19	05 ♐ 17 00	11 ♐ 24 07	08 17	00 ♋ 50	23 58	15 33	12 01	23 50	12 02	16 41
19	05 52 35	28 37 33	17 ♐ 34 03	23 ♐ 46 55	09 57	02 04	23 49	15 39	11 56	23 52	12 02	16 39
20	05 56 31	29 ♊ 34 48	00 ♑ 02 44	06 ♑ 21 31	11 39	03 18	23 41	15 46	11 52	23 53	12 02	16 38
21	06 00 28	00 ♋ 32 02	12 ♑ 44 35	19 ♑ 07 56	13 25	04 31	23 33	15 53	11 48	23 55	12 01	16 36
22	06 04 24	01 29 15	25 ♑ 35 29	02 ♒ 05 54	15 13	05 45	23 27	16 00	11 44	23 57	12 01	16 35
23	06 08 21	02 26 28	08 ♒ 39 07	15 ♒ 15 09	17 04	06 59	23 21	16 07	11 40	23 59	12 01	16 34
24	06 12 17	03 23 41	21 ♒ 54 01	28 ♒ 35 44	18 57	08 12	23 16	16 14	11 36	24 00	12 01	16 32
25	06 16 14	04 20 54	05 ♓ 20 22	12 ♓ 08 01	20 53	09 26	23 12	16 22	11 32	24 02	12 00	16 31
26	06 20 10	05 18 07	18 ♓ 58 44	25 ♓ 52 37	22 52	10 40	23 09	16 29	11 28	24 04	12 00	16 29
27	06 24 07	06 15 20	02 ♈ 49 43	09 ♈ 50 04	24 53	11 54	23 06	16 37	11 25	24 05	11 59	16 28
28	06 28 04	07 12 33	16 ♈ 53 38	24 ♈ 00 18	26 55	13 07	23 04	16 45	11 21	24 07	11 59	16 26
29	06 32 00	08 09 46	01 ♉ 09 53	08 ♉ 22 04	29 ♊ 00	14 21	23 04	16 52	11 17	24 08	11 59	16 25
30	06 35 57	09 ♋ 06 59	15 ♉ 36 26	22 ♉ 52 28	01 ♋ 07	15 ♋ 35	23 ♏ 04	17 ♍ 00	11 ♐ 13	24 ♈ 10	11 ♓ 58	16 ♑ 23

Date	Moon True ☊ ° '	Moon Mean ☊ ° '	Moon ☽ Latitude ° '
01	18 ♍ 00	17 ♍ 32	02 S 45
02	17 R 54	17 29	03 44
03	17 46	17 26	04 28
04	17 36	17 22	04 54
05	17 25	17 19	05 00
06	17 14	17 16	04 46
07	17 04	17 13	04 15
08	16 57	17 10	03 30
09	16 52	17 07	02 34
10	16 50	17 03	01 33
11	16 D 49	17 00	00 S 28
12	16 R 49	16 57	00 N 37
13	16 49	16 54	01 38
14	16 47	16 51	02 35
15	16 43	16 48	03 25
16	16 36	16 44	04 06
17	16 27	16 41	04 37
18	16 15	16 38	04 55
19	16 03	16 35	05 01
20	15 50	16 32	04 52
21	15 38	16 28	04 28
22	15 29	16 25	03 51
23	15 21	16 22	03 01
24	15 17	16 19	02 01
25	15 15	16 16	00 N 52
26	15 D 15	16 13	00 S 20
27	15 15	16 09	01 32
28	15 R 15	16 06	02 40
29	15 13	16 03	03 39
30	15 ♍ 09	16 ♍ 00	04 S 24

DECLINATIONS

Date	Sun ☉ ° '	Moon ☽ ° '	Mercury ☿ ° '	Venus ♀ ° '	Mars ♂ ° '	Jupiter ♃ ° '	Saturn ♄ ° '	Uranus ♅ ° '	Neptune ♆ ° '	Pluto ♇ ° '
01	22 N 09	05 N 38	13 N 34	21 N 43	21 S 24	07 N 28	20 S 36	08 N 27	07 S 50	20 S 56
02	22 16	09 54	13 48	21 56	21 22	07 26	20 35	08 28	07 50	20 56
03	22 24	13 36	14 03	22 08	21 20	07 24	20 35	08 29	07 50	20 57
04	22 31	16 24	14 20	22 20	21 19	07 23	20 34	08 30	07 50	20 57
05	22 37	18 06	14 38	22 31	21 17	07 21	20 34	08 31	07 49	20 57
06	22 43	18 34	14 58	22 41	21 15	07 19	20 33	08 32	07 49	20 57
07	22 49	17 51	15 19	22 51	21 14	07 17	20 33	08 33	07 49	20 58
08	22 54	16 05	15 40	23 00	21 12	07 14	20 32	08 34	07 49	20 58
09	22 59	13 31	16 03	23 08	21 10	07 12	20 32	08 34	07 49	20 58
10	23 04	10 21	16 27	23 16	21 09	07 10	20 31	08 35	07 49	20 58
11	23 08	06 48	16 51	23 23	21 08	07 08	20 31	08 36	07 49	20 58
12	23 11	03 N 03	17 16	23 29	21 06	07 05	20 30	08 37	07 49	20 59
13	23 15	00 S 45	17 42	23 35	21 05	07 03	20 30	08 38	07 49	20 59
14	23 18	04 30	18 08	23 40	21 04	07 00	20 29	08 38	07 49	20 59
15	23 20	08 03	18 34	23 44	21 03	06 58	20 29	08 39	07 49	20 59
16	23 22	11 17	19 01	23 48	21 02	06 55	20 29	08 40	07 49	21 00
17	23 24	14 06	19 27	23 51	21 01	06 53	20 28	08 40	07 49	21 00
18	23 25	16 20	19 53	23 53	21 00	06 50	20 28	08 41	07 49	21 00
19	23 26	17 52	20 19	23 54	21 00	06 47	20 27	08 42	07 50	21 00
20	23 26	18 34	20 45	23 55	20 59	06 45	20 27	08 42	07 50	21 01
21	23 26	18 22	21 09	23 55	20 59	06 42	20 26	08 43	07 50	21 01
22	23 26	17 14	21 33	23 54	20 59	06 39	20 26	08 44	07 50	21 01
23	23 25	15 11	21 56	23 53	20 59	06 36	20 26	08 44	07 50	21 01
24	23 23	12 18	22 18	23 51	20 59	06 33	20 25	08 45	07 50	21 02
25	23 22	08 44	22 39	23 48	21 00	06 30	20 25	08 46	07 51	21 02
26	23 20	04 40	22 57	23 45	21 00	06 27	20 24	08 46	07 51	21 02
27	23 17	00 S 17	23 14	23 40	21 01	06 24	20 24	08 47	07 51	21 03
28	23 14	04 N 10	23 29	23 35	21 02	06 21	20 24	08 47	07 51	21 03
29	23 11	08 27	23 42	23 30	21 03	06 17	20 23	08 48	07 51	21 03
30	23 N 07	12 N 17	23 N 53	23 N 24	21 S 04	06 N 14	20 S 23	08 N 48	07 S 52	21 S 03

ZODIAC SIGN ENTRIES

Date	h	m	Planets
02	02	46	☽ ♉
04	03	01	☽ ♊
06	03	41	☽ ♋
08	06	47	☽ ♌
10	13	46	☽ ♍
12	23	22	☿ ♊
13	00	33	☽ ♎
15	13	18	☽ ♏
17	19	39	♀ ♋
18	01	34	☽ ♐
20	11	55	☽ ♑
20	22	34	☉ ♋
22	20	08	☽ ♒
25	02	30	☽ ♓
27	07	08	☽ ♈
29	10	03	☽ ♉
29	23	24	☿ ♋

LATITUDES

Date	Mercury ☿ ° '	Venus ♀ ° '	Mars ♂ ° '	Jupiter ♃ ° '	Saturn ♄ ° '	Uranus ♅ ° '	Neptune ♆ ° '	Pluto ♇ ° '
01	03 S 46	00 S 13	01 S 40	01 N 17	01 N 48	00 S 36	00 S 50	01 N 25
04	03 38	00 S 06	01 48	01 17	01 48	00 36	00 50	01 25
07	03 24	00 N 01	01 55	01 16	01 48	00 36	00 50	01 25
10	03 04	00 08	02 03	01 15	01 48	00 36	00 50	01 25
13	02 39	00 15	02 09	01 15	01 47	00 36	00 50	01 24
16	02 10	00 22	02 15	01 14	01 47	00 36	00 51	01 24
19	01 38	00 29	02 21	01 14	01 47	00 36	00 51	01 24
22	01 04	00 36	02 26	01 13	01 46	00 36	00 51	01 24
25	00 S 29	00 42	02 30	01 13	01 46	00 36	00 51	01 23
28	00 N 05	00 48	02 34	01 12	01 45	00 36	00 51	01 23
31	00 N 37	00 N 54	02 S 38	01 N 12	01 N 45	00 S 36	00 S 51	01 N 23

DATA

Julian Date	2457541
Delta T	+69 seconds
Ayanamsa	24° 05' 06"
Synetic vernal point	05° ♓ 01' 53"
True obliquity of ecliptic	23° 26' 04"

MOON'S PHASES, APSIDES AND POSITIONS ☽

Date	h	m	Phase	Longitude ° '	Eclipse Indicator
05	03	00	●	14 ♊ 53	
12	08	10	☽	21 ♍ 47	
20	11	02	○	29 ♐ 33	
27	18	19	☾	06 ♈ 30	

Day	h	m		
03	10	46	Perigee	
15	11	57	Apogee	
06	09	10	Max dec	18° N 35'
13	07	14	0S	
20	18	53	Max dec	18° S 37'
27	13	32	0N	

LONGITUDES

Date	Chiron ⚷ ° '	Ceres ⚳ ° '	Pallas ⚴ ° '	Juno ⚵ ° '	Vesta ⚶ ° '	Black Moon Lilith ⚸ ° '
01	24 ♓ 57	17 ♈ 11	06 ♓ 16	00 ♏ 38	06 ♊ 52	01 ♏ 13
11	25 ♓ 08	20 ♈ 23	07 ♓ 03	29 ♎ 45	11 ♊ 12	02 ♏ 20
21	25 ♓ 14	23 ♈ 24	07 ♓ 19	29 ♎ 25	15 ♊ 31	03 ♏ 26
31	25 ♓ 15	26 ♈ 11	07 ♓ 02	29 ♎ 36	19 ♊ 48	04 ♏ 33

All ephemeris data is given at 12.00 UT and the Moon's longitude is additionally given for 24.00 UT

ASPECTARIAN

h m Aspects

01 Wednesday
00 35 ☽ ⊼ ♃
02 43 ☉ ± ♇
05 36 ☽ □ ♇
06 53 ☽ ⊻ ☿
07 10 ☽ ⊥ ♆
10 31 ☽ ± ♃
14 12 ☽ ± ♂
15 42 ☽ ☌ ♅
19 09 ☽ ∠ ♀
21 34 ☽ ∠ ☉
21 54 ☽ ∠ ♆
21 59 ☽ ∥ ♃
23 46 ☽ ⊼ ♂
23 52 ☽ ⚼ ♄

02 Thursday
00 04 ☽ # ♆
01 16 ☽ ⚼ ♃
02 42 ☉ □ ♆
03 40 ☽ ∥ ♅
08 56 ♀ ± ♇
11 05 ☽ ⊥ ♀
13 10 ☽ ⊥ ☉
14 22 ☽ ± ♄
21 37 ☽ ⊻ ♀
22 10 ☽ ⚹ ♆
23 29 ☽ ⊻ ☉
23 59 ☽ ⊼ ♄

03 Friday
01 36 ☽ △ ♃
04 14 ♀ □ ♆
06 11 ☽ △ ♇
06 37 ☉ ☍ ♄
10 07 ☽ ☌ ☿
15 44 ☽ ∥ ☿
16 16 ☽ ⊻ ♅
17 25 ☽ Q ♆
23 02 ☽ ☍ ♂

04 Saturday
00 47 ♀ ☍ ♄
01 54 ☽ ⊥ ♅
06 12 ☽ ⚼ ♇
10 57 ☉ □ ♃
16 24 ☽ ∠ ♅
20 40 ☽ ± ♇
22 21 ☽ □ ♆
23 55 ☽ ☍ ♄

05 Sunday
00 49 ♀ □ ♃
02 02 ☽ □ ♃
02 08 ☽ ☌ ♀
03 00 ☽ ☌ ☉
06 21 ☽ ⊼ ♇
13 32 ☽ ⊻ ☿
16 47 ☽ ⚹ ♅
22 35 ☽ ⊼ ♂

06 Monday
00 04 ☽ ⊥ ☿
07 39 ☽ Q ♃
08 18 ☽ ± ♂
12 39 ☽ Q ♅
16 07 ☽ ∠ ☿
21 49 ☉ ☌ ♀
23 04 ☽ ⚼ ♂
23 47 ☽ △ ♆
23 48 ☉ ∥ ♀

07 Tuesday
01 09 ☽ ⊼ ♄
03 54 ☽ ⚹ ♃
04 15 ♀ ⊼ ♇
06 02 ☉ ⊼ ♇
08 06 ☽ ☍ ♇
08 15 ☽ ⊻ ☉
08 28 ☽ ⊻ ♀
09 24 ♂ Q ♃
11 19 ☽ ± ♄
14 14 ☿ ⊻ ♅
19 20 ☽ □ ♅
19 20 ☽ ⊥ ☉
19 45 ☽ ⚹ ☿
19 48 ☽ ⊥ ♀

08 Wednesday
00 18 ☽ △ ♂
01 34 ☽ ⚼ ♆
02 51 ☽ ⚼ ♄
03 12 ☿ Q ♆
05 59 ☽ ∠ ♃
12 20 ☽ ∠ ☉
13 11 ☽ ∠ ♀
15 47 ☽ ∥ ☿
17 30 ☽ ± ♆
18 48 ☽ Q ☿
22 08 ☽ ⊥ ♃

09 Thursday
04 20 ☽ ⊼ ♆
05 30 ☽ △ ♄
09 06 ☽ ⊻ ♃
13 11 ☽ ⊼ ♇
17 01 ☿ ☍ ♂
17 40 ☽ ⚹ ☉
19 15 ☽ ⚹ ♀

10 Friday
00 18 ☽ ± ♇
01 34 ☽ △ ♅
05 37 ☽ □ ♂
07 14 ☽ □ ☿
17 17 ☽ ⚼ ♇
18 06 ☽ Q ☉
20 19 ☽ Q ♀

11 Saturday
00 07 ☽ ∥ ♅
05 18 ☽ # ♆
06 14 ☽ ⚼ ♅
09 50 ☽ ∥ ♃
12 58 ☽ ☍ ♆
13 52 ☽ □ ♄
15 31 ☽ Q ♂
18 29 ☽ ☌ ♃
22 21 ☽ △ ♇
23 53 ☽ ± ♅

12 Sunday
08 10 ☽ □ ☉
11 26 ☽ □ ♀
11 49 ☽ ⊼ ♅
14 47 ☽ ⚹ ♂
15 13 ♀ ⚹ ♅
16 51 ☿ ⊥ ♅

13 Monday
00 42 ☽ △ ☿
01 21 ☽ Q ♄
15 39 ♀ ⊼ ♂
20 23 ☽ ∠ ♂
20 41 ♆ St R

14 Tuesday
00 51 ☽ ⊼ ♆
01 25 ☽ ⚹ ♄
05 57 ☿ ⚼ ♇
07 03 ☽ ⊻ ♃
07 53 ☉ ⚹ ♅
10 26 ☽ □ ♇
11 04 ☽ ⚼ ☿
13 01 ☽ ± ♆
14 19 ☽ ⊥ ♂
19 21 ☽ ⊥ ♃

15 Wednesday
00 32 ☽ ☍ ♅
01 57 ☽ △ ☉
02 18 ☽ ⊻ ♂
04 36 ☽ # ♃
05 31 ☉ ⊼ ♂
07 00 ☽ △ ♀
07 18 ☽ ⚼ ♆
07 41 ☽ ∠ ♄
07 55 ☽ ± ☿
10 26 ☽ ∥ ♆
13 47 ☽ ∠ ♃
16 18 ☽ # ♅
21 50 ☽ ⊼ ☿
22 54 ☽ Q ♇

16 Thursday
01 50 ☽ ⊥ ♄
11 00 ☽ ⚼ ☉
13 39 ☽ △ ♆
13 52 ☽ ⊻ ♄
16 52 ☽ ⚼ ♀
20 21 ☽ ⚹ ♃
23 04 ☽ ⚹ ♇

17 Friday
06 40 ☽ ± ☉
12 58 ☽ ± ♀
13 14 ☽ ⊼ ♅
13 52 ☽ ☌ ♂
19 39 ☽ ⊼ ☉
20 33 ☽ Q ♃
20 57 ♀ ± ♂

18 Saturday
01 12 ☽ ± ♅
02 14 ☽ ⊼ ♀
03 29 ♄ □ ♆
04 54 ☽ ∠ ♇
18 38 ♂ ∥ ♇
18 48 ☽ ☍ ☿
18 59 ☽ ⚼ ♅
20 07 ☿ ∠ ♅
22 34 ☽ ⊥ ♇

19 Sunday
01 07 ☽ ☌ ♄
01 14 ☽ □ ♆
05 55 ♂ ⊼ ♅
08 15 ☽ □ ♃
10 14 ☽ ⊻ ♇
19 30 ☿ # ♄
21 56 ☿ ± ♇
23 56 ☽ ⊻ ♂

20 Monday
00 11 ☽ △ ♅
11 02 ☽ ☍ ☉
11 18 ☽ ⊥ ♂
11 58 ☽ Q ♆
14 10 ☉ ± ♂
14 55 ☿ ☍ ♄
17 11 ☿ □ ♆
18 51 ☽ ☍ ♀
22 10 ♀ Q ♃

21 Tuesday
02 08 ☿ # ♂
03 39 ☿ # ♇
04 13 ☽ ∠ ♂
10 17 ☽ ⊻ ♄
10 41 ☽ ⚹ ♆
13 30 ☽ ⊼ ☿
17 59 ☽ △ ♃
19 16 ☽ ☌ ♇
21 28 ☽ ⊥ ♄

22 Wednesday
02 32 ☽ ± ☿
08 03 ☽ ⚹ ♂
08 57 ☽ □ ♅
14 06 ☽ ∠ ♄
14 38 ☽ ∠ ♆
16 00 ♀ Q ♅
21 56 ☽ ⚼ ☿
22 04 ☽ ⚼ ♃
22 59 ☿ □ ♃
23 44 ☽ ⊼ ☉

23 Thursday
05 39 ☿ ⊼ ♇
06 00 ☽ Q ♂
07 11 ☽ ⊥ ♆
08 38 ☽ ⊼ ♀
11 35 ☽ ± ☉
14 42 ☽ ± ♃
17 28 ☽ ⚹ ♄
18 04 ☽ Q ♅
18 07 ☽ ⊻ ♆
20 41 ☽ ± ♀

24 Friday
01 41 ☽ ⊼ ♃
02 21 ☽ ⊻ ♇
05 12 ☽ ⚼ ☉
05 48 ☽ △ ☿
13 04 ♀ ⚼ ♂
13 09 ☽ ⊥ ♇
14 26 ☽ □ ♂
14 35 ☽ ⚼ ♀
15 03 ☽ Q ♄
15 48 ☽ ⚹ ♅

25 Saturday
05 13 ☽ ∠ ♇
10 07 ☽ △ ☉
11 52 ☽ # ♅
12 21 ☉ Q ♃
17 30 ☽ ∥ ♆
18 33 ☽ ∠ ♅
19 58 ☽ △ ♀
22 54 ☽ □ ♄
23 46 ☽ ☌ ♆

26 Sunday
01 39 ☽ # ♃
07 36 ☽ ☍ ♃
07 39 ☽ ⚹ ♇
10 24 ☽ ⊥ ♅
12 30 ♃ △ ♇
15 16 ☿ ⊼ ♂
19 14 ☽ △ ♂
19 55 ☽ □ ☿
20 52 ☽ ⊻ ♅

27 Monday
02 31 ☿ ⚹ ♅
03 00 ♀ ⊼ ♄
04 29 ☽ Q ♇
07 39 ☉ Q ♅
13 53 ♀ △ ♆
15 40 ☉ ∥ ☿
18 19 ☽ □ ☉
21 01 ☽ ⚼ ♂

28 Tuesday
02 37 ☽ △ ♄
03 40 ☽ ⊻ ♆
04 59 ☽ □ ♀
08 05 ☽ Q ☿
11 14 ☽ □ ♇
11 44 ☽ ⊼ ♃
12 18 ☽ ± ♂
13 51 ☽ ⊥ ♆
19 41 ☿ ∥ ♀
21 58 ☽ ± ♃
22 25 ☽ ⊼ ♂
23 48 ☽ ∥ ♃

29 Wednesday
00 12 ☽ ☌ ♅
03 02 ☽ Q ☉
03 52 ☽ ⚼ ♄
04 59 ☽ ∠ ♆
07 46 ☽ ⚹ ☿
08 31 ☽ # ♆
09 25 ☉ ⚼ ♂
12 39 ☿ ± ♂
13 12 ☽ ⚼ ♃
14 00 ☽ ∥ ♅
14 10 ☽ Q ♀
18 50 ☽ ± ♄
23 38 ♂ St D

30 Thursday
00 29 ☽ ⚹ ☉
03 42 ♂ ∥ ♇
04 46 ☽ ⊼ ♄
05 59 ☽ ⚹ ♆
11 57 ☽ ⚹ ♀
12 58 ☽ ∠ ☿
13 18 ☽ △ ♇
14 20 ☽ △ ♃

JULY 2016

LONGITUDES

Date	Sidereal time h m s	Sun ☉ ° ' "	Moon ☽ ° ' "	Moon ☽ 24.00 ° ' "	Mercury ☿ ° '	Venus ♀ ° '	Mars ♂ ° '	Jupiter ♃ ° '	Saturn ♄ ° '	Uranus ♅ ° '	Neptune ♆ ° '	Pluto ♇ ° '
01	06 39 53	10 ♋ 04 12	00 ♊ 09 31	07 ♊ 26 50	03 ♋ 14	16 ♋ 49	23 ♏ 04	17 ♍ 08	11 ♐ 10	24 ♈ 11	11 ♓ 57	16 ♑ 22
02	06 43 50	11 01 26	14 ♊ 43 37	21 ♊ 59 00	05 23	18 02	23 D 06	17 17	11 R 06	24 12	11 R 57	16 R 21
03	06 47 46	11 58 39	29 ♊ 12 07	06 ♋ 22 09	07 33	19 16	23 09	17 25	11 03	24 14	11 56	16 19
04	06 51 43	12 55 53	13 ♋ 28 18	20 ♋ 29 54	09 43	20 30	23 12	17 34	10 59	24 15	11 56	16 18
05	06 55 39	13 53 06	27 ♋ 26 23	04 ♌ 17 19	11 53	21 44	23 16	17 42	10 56	24 16	11 55	16 16
06	06 59 36	14 50 20	11 ♌ 02 27	17 ♌ 41 37	14 04	22 57	23 21	17 51	10 53	24 17	11 54	16 15
07	07 03 33	15 47 33	24 ♌ 14 52	00 ♍ 42 17	16 14	24 11	23 27	18 00	10 49	24 18	11 54	16 13
08	07 07 29	16 44 46	07 ♍ 04 11	13 ♍ 20 54	18 23	25 25	23 33	18 08	10 46	24 19	11 53	16 12
09	07 11 26	17 42 00	19 ♍ 32 53	25 ♍ 40 38	20 32	26 39	23 40	18 17	10 43	24 20	11 52	16 10
10	07 15 22	18 39 13	01 ♎ 44 36	07 ♎ 45 51	22 39	27 52	23 48	18 27	10 40	24 21	11 51	16 09
11	07 19 19	19 36 26	13 ♎ 44 32	19 ♎ 41 28	24 46	29 ♋ 06	23 57	18 36	10 37	24 22	11 50	16 07
12	07 23 15	20 33 38	25 ♎ 37 18	01 ♏ 32 40	26 51	00 ♌ 20	24 07	18 45	10 34	24 22	11 50	16 06
13	07 27 12	21 30 51	07 ♏ 28 13	13 ♏ 24 33	28 ♋ 55	01 34	24 17	18 54	10 31	24 24	11 49	16 04
14	07 31 08	22 28 04	19 ♏ 22 13	25 ♏ 21 45	00 ♌ 57	02 47	24 28	19 04	10 29	24 25	11 48	16 03
15	07 35 05	23 25 17	01 ♐ 23 39	07 ♐ 28 19	02 58	04 01	24 39	19 14	10 26	24 25	11 47	16 01
16	07 39 02	24 22 31	13 ♐ 36 08	19 ♐ 47 24	04 56	05 15	24 52	19 23	10 23	24 26	11 46	16 00
17	07 42 58	25 19 44	26 ♐ 02 19	02 ♑ 21 05	06 54	06 28	25 05	19 33	10 21	24 27	11 45	15 58
18	07 46 55	26 16 58	08 ♑ 43 46	15 ♑ 10 23	08 49	07 42	25 18	19 43	10 18	24 27	11 44	15 57
19	07 50 51	27 14 12	21 ♑ 40 54	28 ♑ 15 11	10 43	08 56	25 33	19 53	10 16	24 28	11 43	15 55
20	07 54 48	28 11 27	04 ♒ 53 07	11 ♒ 34 29	12 35	10 10	25 47	20 03	10 14	24 28	11 42	15 54
21	07 58 44	29 ♋ 08 42	18 ♒ 19 05	25 ♒ 06 39	14 25	11 23	26 03	20 13	10 12	24 29	11 41	15 53
22	08 02 41	00 ♌ 05 57	01 ♓ 56 57	08 ♓ 49 45	16 13	12 37	26 19	20 23	10 10	24 29	11 40	15 51
23	08 06 37	01 03 14	15 ♓ 44 49	22 ♓ 41 56	17 59	13 51	26 36	20 34	10 08	24 29	11 38	15 50
24	08 10 34	02 00 31	29 ♓ 40 54	06 ♈ 41 32	19 44	15 05	26 53	20 44	10 06	24 30	11 37	15 48
25	08 14 31	02 57 49	13 ♈ 43 38	20 ♈ 47 03	21 27	16 18	27 11	20 54	10 04	24 30	11 36	15 47
26	08 18 27	03 55 08	27 ♈ 51 36	04 ♉ 57 05	23 08	17 32	27 30	21 05	10 02	24 30	11 35	15 45
27	08 22 24	04 52 28	12 ♉ 03 17	19 ♉ 09 56	24 48	18 46	27 49	21 16	10 00	24 30	11 34	15 44
28	08 26 20	05 49 49	26 ♉ 16 46	03 ♊ 23 25	26 25	20 00	28 08	21 26	09 59	24 30	11 32	15 43
29	08 30 17	06 47 11	10 ♊ 29 32	17 ♊ 34 41	28 01	21 13	28 29	21 37	09 57	24 30	11 31	15 41
30	08 34 13	07 44 35	24 ♊ 38 25	01 ♋ 40 15	29 ♌ 36	22 27	28 49	21 48	09 56	24 R 30	11 30	15 40
31	08 38 10	08 ♌ 41 59	08 ♋ 39 43	15 ♋ 36 21	01 ♍ 08	23 ♌ 41	29 ♏ 11	21 ♍ 59	09 ♐ 55	24 ♈ 30	11 ♓ 29	15 ♑ 39

Date	Moon True ☊ ° '	Moon Mean ☊ ° '	Moon ☽ Latitude ° '
01	15 ♍ 02	15 ♍ 57	04 S 53
02	14 R 54	15 54	05 03
03	14 44	15 50	04 54
04	14 34	15 47	04 26
05	14 26	15 44	03 43
06	14 20	15 41	02 48
07	14 16	15 38	01 45
08	14 15	15 34	00 S 38
09	14 D 14	15 31	00 N 29
10	14 14	15 28	01 33
11	14 15	15 25	02 32
12	14 R 15	15 22	03 24
13	14 14	15 19	04 07
14	14 10	15 15	04 39
15	14 05	15 12	05 00
16	13 58	15 09	05 07
17	13 49	15 06	05 00
18	13 41	15 03	04 39
19	13 33	15 00	04 02
20	13 27	14 56	03 13
21	13 23	14 53	02 11
22	13 20	14 50	01 N 01
23	13 D 20	14 47	00 S 13
24	13 20	14 44	01 28
25	13 22	14 40	02 38
26	13 23	14 37	03 38
27	13 R 23	14 34	04 26
28	13 21	14 31	04 57
29	13 18	14 28	05 11
30	13 13	14 25	05 05
31	13 ♍ 08	14 ♍ 21	04 S 41

DECLINATIONS

Date	Sun ☉ ° '	Moon ☽ ° '	Mercury ☿ ° '	Venus ♀ ° '	Mars ♂ ° '	Jupiter ♃ ° '	Saturn ♄ ° '	Uranus ♅ ° '	Neptune ♆ ° '	Pluto ♇ ° '
01	23 N 03	15 N 24	24 N 01	23 N 17	21 S 05	06 N 11	20 S 23	08 N 49	07 S 52	21 S 04
02	22 59	17 32	24 06	23 09	21 08	06 07	20 22	08 49	07 52	21 04
03	22 54	18 32	24 09	23 00	21 08	06 04	20 22	08 50	07 52	21 04
04	22 48	18 20	24 09	22 51	21 10	06 01	20 22	08 50	07 53	21 04
05	22 43	17 01	24 06	22 42	21 12	05 57	20 21	08 50	07 53	21 05
06	22 37	14 46	24 00	22 31	21 14	05 54	20 21	08 51	07 53	21 05
07	22 30	11 47	23 52	22 21	21 16	05 50	20 21	08 51	07 53	21 05
08	22 23	08 19	23 41	22 09	21 18	05 46	20 21	08 52	07 54	21 06
09	22 16	04 35	23 27	21 56	21 21	05 43	20 20	08 52	07 54	21 06
10	22 08	00 N 43	23 11	21 43	21 23	05 39	20 20	08 52	07 54	21 06
11	22 00	03 S 05	22 52	21 30	21 26	05 35	20 20	08 53	07 55	21 06
12	21 52	06 44	22 31	21 15	21 29	05 32	20 20	08 53	07 55	21 07
13	21 43	10 07	22 08	21 01	21 32	05 28	20 19	08 53	07 55	21 07
14	21 34	13 05	21 43	20 45	21 36	05 24	20 19	08 53	07 56	21 07
15	21 24	15 33	21 16	20 29	21 39	05 20	20 19	08 54	07 56	21 07
16	21 14	17 21	20 48	20 13	21 42	05 16	20 19	08 54	07 57	21 08
17	21 04	18 22	20 18	19 55	21 46	05 12	20 19	08 54	07 57	21 08
18	20 53	18 31	19 46	19 38	21 50	05 08	20 18	08 54	07 57	21 08
19	20 42	17 42	19 13	19 19	21 54	05 04	20 18	08 54	07 58	21 09
20	20 31	15 55	18 40	19 01	21 57	05 00	20 18	08 55	07 58	21 09
21	20 19	13 15	18 05	18 41	22 02	04 56	20 18	08 55	07 59	21 09
22	20 08	09 49	17 29	18 21	22 06	04 52	20 18	08 55	07 59	21 09
23	19 55	05 49	16 52	18 01	22 10	04 48	20 18	08 55	08 00	21 10
24	19 43	01 S 28	16 15	17 40	22 14	04 44	20 18	08 55	08 00	21 10
25	19 30	03 N 00	15 37	17 19	22 19	04 39	20 18	08 55	08 01	21 10
26	19 16	07 19	14 59	16 57	22 23	04 35	20 18	08 55	08 01	21 11
27	19 03	11 14	14 20	16 35	22 27	04 31	20 18	08 55	08 02	21 11
28	18 49	14 30	13 41	16 12	22 32	04 26	20 18	08 55	08 02	21 11
29	18 34	16 53	13 01	15 49	22 37	04 22	20 18	08 55	08 03	21 11
30	18 20	18 15	12 22	15 25	22 41	04 18	20 18	08 55	08 03	21 12
31	18 N 05	18 N 29	11 N 42	15 N 01	22 S 46	04 N 13	20 S 18	08 N 55	08 S 04	21 S 12

ZODIAC SIGN ENTRIES

Date	h	m	Planets
01	11	44	☽ ♊
03	13	20	☽ ♋
05	16	28	☽ ♌
07	22	41	☽ ♍
10	08	32	☽ ♎
12	05	34	☿ ♌
12	20	52	☽ ♏
14	00	47	♀ ♌
15	09	14	☽ ♐
17	19	33	☽ ♑
20	03	10	☽ ♒
22	08	35	☽ ♓
22	09	30	☉ ♌
24	12	33	☽ ♈
26	15	37	☽ ♉
28	18	17	☽ ♊
30	18	18	☿ ♍
30	21	09	☽ ♋

LATITUDES

Date	Mercury ☿ ° '	Venus ♀ ° '	Mars ♂ ° '	Jupiter ♃ ° '	Saturn ♄ ° '	Uranus ♅ ° '	Neptune ♆ ° '	Pluto ♇ ° '
01	00 N 37	00 N 54	02 S 38	01 N 12	01 N 45	00 S 36	00 S 51	01 N 23
04	01 04	01 00	02 41	01 11	01 44	00 37	00 51	01 23
07	01 25	01 05	02 43	01 11	01 44	00 37	00 51	01 22
10	01 40	01 10	02 46	01 11	01 43	00 37	00 51	01 22
13	01 48	01 14	02 47	01 10	01 43	00 37	00 52	01 22
16	01 49	01 18	02 49	01 10	01 42	00 37	00 52	01 21
19	01 45	01 21	02 50	01 09	01 42	00 37	00 52	01 21
22	01 35	01 24	02 51	01 09	01 41	00 37	00 52	01 21
25	01 21	01 26	02 52	01 09	01 41	00 37	00 52	01 20
28	01 02	01 28	02 52	01 08	01 40	00 37	00 52	01 20
31	00 N 40	01 N 29	02 S 52	01 N 08	01 N 39	00 S 37	00 S 52	01 N 20

DATA

Julian Date	2457571
Delta T	+69 seconds
Ayanamsa	24° 05' 11"
Synetic vernal point	05° ♓ 01' 48"
True obliquity of ecliptic	23° 26' 04"

LONGITUDES

Date	Chiron ⚷ ° '	Ceres ⚳ ° '	Pallas ⚴ ° '	Juno ⚵ ° '	Vesta ⚶ ° '	Black Moon Lilith ⚸ ° '
01	25 ♓ 15	26 ♈ 11	07 ♓ 02	29 ♎ 36	19 ♊ 48	04 ♏ 33
11	25 ♓ 10	28 ♈ 43	06 ♓ 10	00 ♏ 17	24 ♊ 01	05 ♏ 40
21	25 ♓ 00	00 ♉ 55	04 ♓ 44	01 ♏ 25	28 ♊ 12	06 ♏ 47
31	24 ♓ 45	02 ♉ 45	02 ♓ 47	02 ♏ 56	02 ♋ 17	07 ♏ 54

MOON'S PHASES, APSIDES AND POSITIONS ☽

Date	h	m	Phase	Longitude ° '	Eclipse Indicator
04	11	01	●	12 ♋ 54	
12	00	52	☽	20 ♎ 07	
19	22	57	○	27 ♑ 40	
26	23	00	☾	04 ♉ 21	

Day	h	m		
01	06	30	Perigee	
13	05	23	Apogee	
27	11	33	Perigee	
03	20	01	Max dec	18° N 36'
10	16	31	0S	
18	03	38	Max dec	18° S 34'
24	19	53	0N	
31	04	51	Max dec	18° N 32'

ASPECTARIAN

01 Friday
00 19 ☽ ☍ ♂
01 48 ☽ Q ♆
02 09 ☽ ⊻ ♅
03 02 ☽ ∠ ☉
03 32 ♀ ☍ ♇
06 22 ☽ ⊥ ☿
12 02 ☽ ⊥ ♅
13 59 ☽ ⚼ ♇
14 58 ☽ ∠ ♀
17 56 ☽ ⊻ ☿
18 34 ♀ ± ♄
18 53 ☽ ⊥ ☉
19 18 ♀ ✶ ♃

02 Saturday
02 53 ☽ ∠ ♅
04 47 ☽ ± ♇
05 28 ☽ ⊻ ☉
06 03 ☽ ☍ ♄
07 09 ☽ ⊥ ♀
07 25 ☽ □ ♆
10 45 ☿ Q ♃
13 52 ☉ ⊼ ♄
14 40 ☽ ⊼ ♇
16 15 ☽ □ ♃
17 59 ☽ ⊻ ♀
21 14 ☿ Q ♅

03 Sunday
01 53 ☽ ⊼ ♂
03 43 ☽ ✶ ♅
11 01 ☉ △ ♆
11 54 ☽ ± ♂
18 47 ☿ ⚼ ♂
22 30 ☽ Q ♃
23 47 ☽ Q ♅

04 Monday
03 02 ☽ ⚼ ♂
04 29 ☽ ☌ ☿
07 48 ☽ ⊼ ♄
09 23 ☽ △ ♆
11 01 ☽ ☌ ☉
16 48 ☽ ☍ ♇
17 58 ☽ ± ♄
19 02 ☽ ✶ ♃

05 Tuesday
01 09 ☽ ☌ ♀
01 42 ☿ ⊼ ♄
04 44 ☽ △ ♂
06 23 ☉ ∥ ♀
06 29 ☽ □ ♅
09 23 ☽ ⚼ ♄
11 05 ☽ ⚼ ♆
12 19 ☿ △ ♆
21 18 ☽ ∠ ♃

06 Wednesday
02 52 ☽ ± ♆
11 42 ☽ △ ♄
13 28 ☽ ⊥ ♃
13 33 ☽ ⊼ ♆
18 29 ☽ ⊻ ☿
19 21 ☽ ⊻ ☉
20 17 ♀ △ ♂
21 21 ☽ ⊼ ♇

07 Thursday
00 25 ☽ ⊻ ♃
03 24 ☉ ☌ ☿
07 08 ☽ ⊥ ☉
07 33 ☽ ⊥ ☿
08 16 ☽ ± ♇
10 30 ☽ □ ♂
11 52 ☽ ⊻ ♀
11 55 ☿ ☍ ♇
12 06 ☽ △ ♅
14 25 ♀ □ ♅
18 27 ☿ ± ♄
22 27 ☉ ☍ ♇

08 Friday
00 12 ☽ ⊥ ♀
00 56 ☽ ⚼ ♇
01 09 ☽ ∠ ☉
03 37 ☽ ∠ ☿
08 26 ☽ ∥ ♅
09 05 ☿ ✶ ♃
12 32 ☉ ± ♄
14 47 ☽ # ♆
16 18 ☽ ⚼ ♅
18 40 ♀ ⚼ ♄
19 01 ☽ □ ♄
19 04 ☽ ∠ ♀
20 38 ☽ Q ♂
21 10 ☽ ☍ ♆

09 Saturday
04 43 ☽ ∥ ♃
05 27 ☽ △ ♇
08 06 ☽ ✶ ☉
09 32 ☽ ☌ ♃
09 39 ☽ ± ♅
14 19 ☽ ✶ ☿
16 22 ♀ ⚼ ♆
20 09 ☽ ✶ ♂
20 42 ♃ ± ♅
21 23 ☽ ⊼ ♅

10 Sunday
03 28 ☽ ✶ ♀
05 41 ☉ ✶ ♃
05 55 ☽ Q ♄
09 39 ☽ Q ☉
19 02 ☽ Q ☿

11 Monday
02 00 ☿ △ ♂
02 15 ☽ ∠ ♂
05 45 ☽ ✶ ♄
06 05 ☽ Q ♀
07 29 ☿ □ ♅
08 11 ☽ ⊼ ♆
16 47 ☽ □ ♇
16 48 ♀ # ♂
20 15 ☽ ± ♆
20 36 ☽ ⊥ ♂
21 34 ☿ ⚼ ♄
21 55 ☽ ⊻ ♃

12 Tuesday
00 52 ☽ □ ☉
04 00 ☽ # ♃
08 54 ☽ ⊻ ♂
09 30 ☽ ☍ ♅
10 13 ☽ ⊥ ♃
11 44 ☿ ⚼ ♆
11 54 ☽ ∠ ♄
14 26 ☽ ⚼ ♆
15 01 ☽ □ ☿
20 09 ☽ ∥ ♆
22 39 ☽ □ ♀

13 Wednesday
02 10 ♀ # ♇
02 59 ☽ # ♅
04 41 ☽ ∠ ♃
05 08 ☽ Q ♇
06 03 ☽ ⊥ ♄
18 09 ☽ ⊻ ♄
20 46 ☽ △ ♆

14 Thursday
05 13 ♂ ⊼ ♅
05 20 ☽ ✶ ♇
08 29 ☉ # ♂
11 23 ☽ ✶ ♃
18 00 ☿ # ♂
18 45 ☽ △ ☉
22 07 ☽ ⊼ ♅
22 22 ☽ ☌ ♂

15 Friday
01 02 ☉ ∥ ☿
10 04 ☽ ± ♅
11 16 ☽ ∠ ♇
11 40 ☽ Q ♃
15 42 ☽ △ ☿
16 42 ♀ ∠ ♃
17 46 ☽ △ ♀
19 23 ☿ # ♇

16 Saturday
03 02 ☽ ⚼ ☉
03 04 ♀ # ♄
03 51 ☽ ⚼ ♅
04 40 ☿ ∠ ♃
04 58 ☽ ⊥ ♇
05 45 ☽ ☌ ♄
08 25 ☽ □ ♆
13 32 ☉ □ ♅
16 39 ☽ ⊻ ♇
21 59 ☿ ☌ ♀
22 00 ☿ ± ♆
22 01 ♀ ± ♆
22 02 ☽ ± ☉
23 22 ☽ □ ♃

17 Sunday
02 17 ☽ ⚼ ♀
02 35 ☽ ⚼ ☿
03 05 ☉ # ♇
03 45 ☉ △ ♂
08 57 ☽ △ ♅
10 08 ☽ ⊻ ♂
10 32 ☽ ⊼ ☉
11 12 ☿ # ♄
12 27 ☉ ⚼ ♄
19 03 ☽ Q ♆
21 21 ☽ ± ♀
21 46 ☽ ⊥ ♂
22 54 ☽ ± ☿

18 Monday
09 52 ☽ ⊼ ♀
12 12 ☽ ⊼ ☿
14 56 ☽ ⊻ ♄
15 00 ☽ ∠ ♂
17 36 ☽ ✶ ♆
23 06 ☉ ⚼ ♆

19 Tuesday
01 25 ☽ ☌ ♇
02 03 ☽ ⊥ ♄
02 10 ☿ ∥ ♀
06 28 ☿ △ ♄
08 39 ☽ △ ♃
17 06 ☽ □ ♅
18 33 ☽ ∠ ♄
19 12 ☽ ✶ ♂
21 11 ☽ ∠ ♆
22 57 ☽ ☍ ☉

20 Wednesday
00 44 ☿ ⊼ ♆
12 18 ☽ ⚼ ♃
13 19 ♀ △ ♄
13 28 ☽ ⊥ ♆
17 20 ☽ Q ♂
21 34 ☽ ✶ ♄
22 26 ☽ ☍ ♀

21 Thursday
00 12 ☽ ⊻ ♆
01 37 ☽ Q ♅
03 58 ☽ ☍ ☿
04 37 ☽ ± ♃
07 40 ☽ ⊻ ♇
09 10 ☿ ⊥ ♃
14 53 ☉ # ♄
15 24 ☽ ⊼ ♃
17 33 ♀ ⊼ ♆
18 17 ☽ ⊥ ♇
18 50 ☽ Q ♄
22 54 ☽ ✶ ♅

22 Friday
01 56 ☽ □ ♂
07 11 ☿ ⊼ ♇
08 31 ☽ ⊼ ☉
10 05 ☽ ∠ ♇
17 43 ☽ # ♅
19 47 ☽ ± ☉
23 21 ☽ ∥ ♆

23 Saturday
01 09 ☽ ∠ ♅
02 17 ☽ □ ♄
04 54 ☽ ☌ ♆
08 23 ☽ ⊼ ♀
12 08 ☽ ✶ ♇
12 34 ☽ ⚼ ☉
16 27 ☽ ⊼ ☿
16 45 ☽ ⊥ ♅
17 52 ☽ # ♃
19 46 ☽ ± ♀
20 25 ☽ ☍ ♃

24 Sunday
03 05 ☽ ⊻ ♅
04 10 ♀ ⊥ ♃
04 16 ☽ ± ☿
07 06 ☽ △ ♂
08 47 ☽ Q ♇
12 45 ☽ ⚼ ♀
16 17 ☽ △ ☉
21 52 ☽ ⚼ ☿

25 Monday
01 57 ♀ ⊼ ♇
03 26 ☿ ⊻ ♃
05 46 ☽ △ ♄
08 23 ☽ ⊻ ♆
09 19 ☽ ⚼ ♂
15 29 ☽ □ ♇
16 34 ☿ ± ♇
16 48 ☽ △ ♀
18 35 ☽ ⊥ ♆
20 55 ☽ ∥ ♃

26 Tuesday
00 22 ☽ ⊼ ♃
00 58 ☽ ± ♂
02 55 ☽ △ ☿
06 19 ☽ ☌ ♅
07 13 ☽ ⚼ ♄
09 50 ☽ ∠ ♆
10 40 ☽ ± ♃
11 22 ☽ ⊼ ♂
16 08 ☽ # ♆
21 29 ☽ ∥ ♅
22 26 ☽ ± ♄
23 00 ☽ □ ☉

27 Wednesday
02 05 ☽ ⚼ ♃
07 45 ☿ △ ♅
08 33 ☽ ⊼ ♄
11 10 ☽ ✶ ♆
18 12 ☽ △ ♇

28 Thursday
00 23 ☽ □ ♀
03 44 ☽ △ ♃
04 54 ☉ ± ♆
06 36 ☽ ∥ ☿
07 23 ☽ Q ♆
07 34 ☽ Q ☉
09 01 ☽ ⊻ ♅
12 17 ☽ □ ☿
15 13 ☽ ☍ ♂
19 08 ☽ ⊥ ♅
19 28 ☽ ⚼ ♇

29 Friday
01 42 ☽ ∥ ♀
05 17 ☽ ✶ ☉
06 45 ☉ ∠ ♃
09 39 ☽ Q ♀
10 20 ☽ ∠ ♅
10 39 ☽ ± ♇
11 06 ☽ ☍ ♄
13 44 ☽ □ ♆
20 47 ☽ ⊼ ♇
20 49 ☿ □ ♂
20 58 ♀ ± ♇
21 03 ♀ ⊻ ♃
21 06 ♅ St R
22 32 ☽ Q ☿

30 Saturday
07 06 ☽ □ ♃
07 55 ☽ ✶ ♀
08 32 ☽ ∠ ☉
11 46 ☽ ✶ ♅
13 57 ☽ ∥ ☉
19 19 ☽ ⊼ ♂
21 30 ☽ ✶ ☿

31 Sunday
01 01 ☽ ⊥ ☉
04 27 ☿ ⚼ ♇
05 51 ☽ ± ♂
08 18 ☽ Q ♅
12 02 ☽ ∠ ♀
12 04 ☽ ⊻ ☉
14 09 ☽ ⊼ ♄
14 18 ☽ Q ♃
16 51 ☽ △ ♆
21 46 ☽ ⚼ ♂

All ephemeris data is given at 12.00 UT and the Moon's longitude is additionally given for 24.00 UT

LONGITUDES

Date	Sidereal time h m s	Sun ☉ ° ' "	Moon ☽ ° ' "	Moon ☽ 24.00 ° ' "	Mercury ☿ ° '	Venus ♀ ° '	Mars ♂ ° '	Jupiter ♃ ° '	Saturn ♄ ° '	Uranus ♅ ° '	Neptune ♆ ° '	Pluto ♇ ° '
01	08 42 06	09 ♌ 39 24	22 ♋ 29 42	29 ♋ 19 21	02 ♍ 39	24 ♌ 54	29 ♏ 32	22 ♍ 10	09 ♐ 54	24 ♈ 30	11 ♓ 27	15 ♑ 37
02	08 46 03	10 36 51	06 ♌ 04 58	12 ♌ 46 17	04 07	26 08	29 ♏ 55	22 21	09 R 53	24 R 30	11 R 26	15 R 36
03	08 50 00	11 34 18	19 ♌ 23 04	25 ♌ 55 14	05 35	27 22	00 ♐ 17	22 32	09 52	24 30	11 25	15 35
04	08 53 56	12 31 46	02 ♍ 22 44	08 ♍ 45 38	07 00	28 36	00 41	22 43	09 51	24 30	11 23	15 33
05	08 57 53	13 29 14	15 ♍ 04 05	21 ♍ 18 18	08 23	29 ♌ 49	01 04	22 54	09 50	24 29	11 22	15 32
06	09 01 49	14 26 44	27 ♍ 28 35	03 ♎ 35 19	09 45	01 ♍ 03	01 28	23 06	09 49	24 29	11 20	15 31
07	09 05 46	15 24 15	09 ♎ 38 54	15 ♎ 39 49	11 04	02 17	01 53	23 17	09 48	24 29	11 19	15 30
08	09 09 42	16 21 46	21 ♎ 38 35	27 ♎ 35 46	12 22	03 31	02 18	23 29	09 48	24 28	11 18	15 28
09	09 13 39	17 19 18	03 ♏ 31 55	09 ♏ 27 39	13 38	04 44	02 44	23 40	09 47	24 28	11 16	15 27
10	09 17 35	18 16 51	15 ♏ 23 33	21 ♏ 20 14	14 51	05 58	03 10	23 52	09 47	24 27	11 15	15 26
11	09 21 32	19 14 25	27 ♏ 18 18	03 ♐ 18 20	16 03	07 12	03 36	24 03	09 47	24 27	11 13	15 25
12	09 25 29	20 12 00	09 ♐ 20 53	15 ♐ 26 29	17 12	08 25	04 03	24 15	09 47	24 26	11 12	15 24
13	09 29 25	21 09 36	21 ♐ 35 38	27 ♐ 48 44	18 19	09 39	04 30	24 27	09 D 47	24 25	11 10	15 23
14	09 33 22	22 07 12	04 ♑ 06 12	10 ♑ 28 18	19 23	10 53	04 58	24 39	09 47	24 25	11 09	15 21
15	09 37 18	23 04 50	16 ♑ 55 17	23 ♑ 27 16	20 25	12 06	05 26	24 50	09 47	24 24	11 07	15 20
16	09 41 15	24 02 29	00 ♒ 04 18	06 ♒ 46 21	21 25	13 20	05 54	25 02	09 47	24 23	11 05	15 19
17	09 45 11	25 00 09	13 ♒ 33 16	20 ♒ 24 49	22 21	14 34	06 23	25 14	09 48	24 22	11 04	15 18
18	09 49 08	25 57 50	27 ♒ 20 39	04 ♓ 20 24	23 15	15 47	06 52	25 26	09 48	24 21	11 02	15 17
19	09 53 04	26 55 33	11 ♓ 23 36	18 ♓ 29 42	24 06	17 01	07 21	25 38	09 49	24 20	11 01	15 16
20	09 57 01	27 53 17	25 ♓ 38 12	02 ♈ 48 29	24 53	18 15	07 51	25 50	09 49	24 19	10 59	15 15
21	10 00 58	28 51 02	10 ♈ 00 00	17 ♈ 12 11	25 38	19 28	08 21	26 03	09 50	24 18	10 58	15 14
22	10 04 54	29 ♌ 48 49	24 ♈ 24 30	01 ♉ 36 26	26 18	20 42	08 52	26 15	09 51	24 17	10 56	15 13
23	10 08 51	00 ♍ 46 38	08 ♉ 47 33	15 ♉ 57 25	26 55	21 55	09 22	26 27	09 52	24 16	10 54	15 12
24	10 12 47	01 44 29	23 ♉ 05 40	00 ♊ 11 58	27 27	23 09	09 53	26 39	09 53	24 15	10 53	15 11
25	10 16 44	02 42 21	07 ♊ 16 04	14 ♊ 17 43	27 56	24 22	10 25	26 52	09 54	24 14	10 51	15 10
26	10 20 40	03 40 16	21 ♊ 16 43	28 ♊ 12 53	28 20	25 36	10 57	27 04	09 55	24 12	10 49	15 09
27	10 24 37	04 38 12	05 ♋ 06 03	11 ♋ 56 08	28 39	26 50	11 29	27 16	09 56	24 11	10 48	15 09
28	10 28 33	05 36 10	18 ♋ 42 59	25 ♋ 26 31	28 53	28 03	12 01	27 29	09 58	24 10	10 46	15 08
29	10 32 30	06 34 10	02 ♌ 06 40	08 ♌ 43 21	29 02	29 ♍ 17	12 34	27 41	09 59	24 08	10 45	15 07
30	10 36 27	07 32 11	15 ♌ 16 32	21 ♌ 46 10	29 05	00 ♎ 30	13 06	27 54	10 01	24 07	10 43	15 06
31	10 40 23	08 ♍ 30 14	28 ♌ 12 17	04 ♍ 34 53	29 ♍ 02	01 ♎ 44	13 ♐ 40	28 ♍ 06	10 ♐ 03	24 ♈ 05	10 ♓ 41	15 ♑ 06

Date	Moon True ☊ ° '	Moon Mean ☊ ° '	Moon ☽ Latitude ° '
01	13 ♍ 03	14 ♍ 18	04 S 01
02	12 R 59	14 15	03 08
03	12 55	14 12	02 05
04	12 53	14 09	00 S 57
05	12 D 53	14 06	00 N 12
06	12 54	14 02	01 19
07	12 55	13 59	02 22
08	12 57	13 56	03 17
09	12 58	13 53	04 03
10	12 59	13 50	04 39
11	12 R 59	13 46	05 03
12	12 57	13 43	05 14
13	12 55	13 40	05 11
14	12 53	13 37	04 54
15	12 50	13 34	04 21
16	12 47	13 31	03 34
17	12 45	13 27	02 34
18	12 43	13 24	01 24
19	12 43	13 21	00 N 07
20	12 D 43	13 18	01 S 11
21	12 44	13 15	02 25
22	12 45	13 11	03 31
23	12 46	13 08	04 23
24	12 46	13 05	04 58
25	12 R 46	13 02	05 15
26	12 46	12 59	05 13
27	12 45	12 56	04 53
28	12 45	12 52	04 17
29	12 44	12 49	03 27
30	12 43	12 46	02 26
31	12 ♍ 43	12 ♍ 43	01 S 20

DECLINATIONS

Date	Sun ☉ ° '	Moon ☽ ° '	Mercury ☿ ° '	Venus ♀ ° '	Mars ♂ ° '	Jupiter ♃ ° '	Saturn ♄ ° '	Uranus ♅ ° '	Neptune ♆ ° '	Pluto ♇ ° '
01	17 N 50	17 N 36	11 N 02	14 N 37	22 S 51	04 N 09	20 S 18	08 N 55	08 S 04	21 S 12
02	17 34	15 43	10 22	14 12	22 56	04 04	20 18	08 55	08 05	21 12
03	17 19	13 01	09 43	13 47	23 01	04 00	20 18	08 55	08 05	21 13
04	17 03	09 44	09 03	13 21	23 05	03 55	20 18	08 55	08 06	21 13
05	16 46	06 04	08 24	12 55	23 10	03 51	20 18	08 55	08 06	21 13
06	16 30	02 N 13	07 45	12 29	23 15	03 46	20 18	08 54	08 07	21 13
07	16 13	01 S 39	07 06	12 03	23 20	03 42	20 18	08 54	08 07	21 14
08	15 56	05 23	06 27	11 36	23 25	03 37	20 18	08 54	08 08	21 14
09	15 38	08 52	05 49	11 08	23 30	03 33	20 19	08 54	08 09	21 14
10	15 21	12 00	05 12	10 41	23 35	03 28	20 19	08 54	08 09	21 14
11	15 03	14 38	04 35	10 13	23 40	03 23	20 19	08 53	08 10	21 15
12	14 45	16 40	03 59	09 45	23 45	03 18	20 19	08 53	08 10	21 15
13	14 27	18 00	03 23	09 17	23 50	03 14	20 19	08 53	08 11	21 15
14	14 08	18 29	02 48	08 48	23 55	03 09	20 20	08 53	08 12	21 15
15	13 49	18 03	02 14	08 20	24 00	03 04	20 20	08 52	08 12	21 16
16	13 30	16 39	01 41	07 51	24 04	02 59	20 20	08 52	08 13	21 16
17	13 11	14 17	01 09	07 21	24 09	02 55	20 20	08 52	08 13	21 16
18	12 52	11 05	00 38	06 52	24 14	02 50	20 21	08 51	08 14	21 16
19	12 32	07 11	00 N 09	06 22	24 19	02 45	20 21	08 51	08 15	21 17
20	12 12	02 S 49	00 S 20	05 53	24 23	02 40	20 21	08 51	08 15	21 17
21	11 52	01 N 44	00 47	05 23	24 28	02 35	20 22	08 50	08 16	21 17
22	11 32	06 11	01 12	04 53	24 32	02 30	20 22	08 50	08 17	21 17
23	11 12	10 16	01 36	04 22	24 37	02 25	20 22	08 49	08 17	21 18
24	10 51	13 44	01 57	03 52	24 41	02 21	20 23	08 49	08 18	21 18
25	10 30	16 20	02 17	03 22	24 45	02 16	20 23	08 48	08 18	21 18
26	10 10	17 56	02 35	02 51	24 49	02 11	20 23	08 48	08 19	21 18
27	09 49	18 27	02 51	02 21	24 54	02 06	20 24	08 47	08 20	21 19
28	09 27	17 53	03 04	01 50	24 58	02 01	20 24	08 47	08 20	21 19
29	09 06	16 20	03 14	01 19	25 01	01 56	20 25	08 46	08 21	21 19
30	08 44	13 55	03 21	00 48	25 05	01 51	20 25	08 46	08 22	21 19
31	08 N 23	10 N 51	03 S 26	00 N 17	25 S 09	01 N 46	20 S 26	08 N 45	08 S 22	21 S 19

ZODIAC SIGN ENTRIES

Date	h	m	Planets
02	01	12	☽ ♌
02	17	49	♂ ♐
04	07	34	☽ ♍
05	15	27	♀ ♍
06	16	56	☽ ♎
09	04	51	☽ ♏
11	17	24	☽ ♐
14	04	11	☽ ♑
16	11	52	☽ ♒
18	16	34	☽ ♓
20	19	18	☽ ♈
22	16	38	☉ ♍
22	21	19	☽ ♉
24	23	40	☽ ♊
27	03	06	☽ ♋
29	08	11	☽ ♌
30	02	07	♀ ♎
31	15	22	☽ ♍

LATITUDES

Date	Mercury ☿ ° '	Venus ♀ ° '	Mars ♂ ° '	Jupiter ♃ ° '	Saturn ♄ ° '	Uranus ♅ ° '	Neptune ♆ ° '	Pluto ♇ ° '
01	00 N 33	01 N 29	02 S 52	01 N 08	01 N 39	00 S 37	00 S 52	01 N 20
04	00 N 07	01 29	02 52	01 08	01 38	00 37	00 52	01 19
07	00 S 20	01 29	02 52	01 07	01 38	00 37	00 52	01 19
10	00 50	01 28	02 51	01 07	01 37	00 37	00 52	01 19
13	01 21	01 26	02 51	01 07	01 36	00 38	00 52	01 18
16	01 52	01 24	02 50	01 07	01 36	00 38	00 52	01 18
19	02 24	01 21	02 49	01 06	01 35	00 38	00 52	01 17
22	02 54	01 18	02 48	01 06	01 34	00 38	00 53	01 17
25	03 24	01 14	02 47	01 06	01 34	00 38	00 53	01 17
28	03 49	01 09	02 45	01 06	01 33	00 38	00 53	01 16
31	04 S 09	01 N 04	02 S 44	01 N 06	01 N 33	00 S 38	00 S 53	01 N 16

DATA

Julian Date	2457602
Delta T	+69 seconds
Ayanamsa	24° 05' 16"
Synetic vernal point	05° ♓ 01' 43"
True obliquity of ecliptic	23° 26' 05"

MOON'S PHASES, APSIDES AND POSITIONS ☽

Date	h	m	Phase	Longitude ° '	Eclipse Indicator
02	20	45	●	10 ♌ 58	
10	18	21	☽	18 ♏ 32	
18	09	27	○	25 ♒ 52	
25	03	41	☾	02 ♊ 22	

Day	h	m		
10	00	05	Apogee	
22	01	30	Perigee	
07	01	42	0S	
14	13	00	Max dec	18° S 29'
21	02	53	0N	
27	11	19	Max dec	18° N 27'

LONGITUDES

Date	Chiron ⚷ ° '	Ceres ⚳ ° '	Pallas ⚴ ° '	Juno ⚵ ° '	Vesta ⚶ ° '	Black Moon Lilith ⚸ ° '
01	24 ♓ 43	02 ♉ 54	02 ♓ 34	03 ♏ 06	02 ♋ 42	08 ♏ 01
11	24 ♓ 24	04 ♉ 16	00 ♓ 12	04 ♏ 59	06 ♋ 42	09 ♏ 08
21	24 ♓ 01	05 ♉ 07	27 ♒ 39	07 ♏ 10	10 ♋ 36	10 ♏ 15
31	23 ♓ 36	05 ♉ 24	25 ♒ 08	09 ♏ 36	14 ♋ 22	11 ♏ 22

All ephemeris data is given at 12.00 UT and the Moon's longitude is additionally given for 24.00 UT

ASPECTARIAN

h m Aspects

01 Monday
00 03 ☽ ☍ ♇
00 31 ☽ ± ♄
02 30 ☽ ∠ ☿
04 08 ♀ △ ♅
05 08 ☽ ⊥ ♀
06 38 ☽ ∥ ☉
11 25 ☽ ⚹ ♃
15 31 ☽ □ ♅
16 12 ☽ bQ ♄
16 39 ☽ ⚺ ♀
17 49 ☉ △ ♄
18 56 ☽ bQ ♆
20 10 ☽ ⊥ ☿

02 Tuesday
00 44 ☽ △ ♂
08 05 ☽ ⚺ ☿
10 51 ☽ ± ♆
14 17 ☽ ∠ ♃
18 47 ☽ △ ♄
20 45 ☽ ☌ ☉
21 34 ☽ ⚻ ♆

03 Wednesday
04 47 ☽ ∥ ♀
05 05 ☽ ⚻ ♇
06 44 ☽ ⊥ ♃
08 02 ☉ ⚻ ♆
16 00 ☽ ± ♇
17 51 ☽ ⚺ ♃
21 23 ☽ △ ♅

04 Thursday
01 01 ♂ ± ♅
04 13 ☽ ☌ ♀
05 05 ♂ ∠ ♇
08 36 ☽ bQ ♇
08 43 ☽ □ ♂
17 06 ☿ ∥ ♅
17 33 ☽ ∥ ♅
17 39 ☽ ∥ ☿
21 44 ☽ ☌ ☿
22 55 ☽ # ♆

05 Friday
01 23 ☽ bQ ♅
02 02 ☽ □ ♄
04 57 ☽ ☍ ♆
08 44 ☽ ⚺ ☉
12 54 ☽ △ ♇
18 34 ☽ ± ♅
19 56 ☽ Q ♂
21 12 ☽ ⊥ ☉
22 28 ☿ # ♆

06 Saturday
01 40 ♀ bQ ♇
02 10 ☽ ∥ ♃
03 20 ☽ ☌ ♃
06 10 ☽ ⚻ ♅
07 22 ☿ bQ ♅
12 40 ☽ Q ♄
13 18 ☿ □ ♄
16 11 ☽ ∠ ☉
19 47 ☽ ⚺ ♀
20 06 ☽ ⚹ ♂

07 Sunday
00 20 ♀ □ ♂
08 59 ☽ ⊥ ♀
12 19 ☽ ⚹ ♄
14 12 ☉ ⚻ ♇
15 11 ☽ ⚺ ☿
15 19 ☽ ⚻ ♆
16 24 ☿ ☍ ♆
23 38 ☽ □ ♇

08 Monday
00 29 ☽ ⚹ ☉
00 43 ☽ # ♃
02 58 ☽ ∠ ♂
03 17 ☽ ± ♆
04 37 ☽ ⊥ ☿
04 59 ☽ ∠ ♀
15 45 ☽ ⚺ ♃
17 41 ☽ ☍ ♅
18 03 ☽ # ☿
18 21 ☽ ∠ ♄
21 21 ☽ bQ ♆
21 44 ☽ ⊥ ♂

09 Tuesday
00 55 ☽ ∠ ☿
02 44 ☽ Q ☉
04 04 ☽ ⊥ ♃
06 47 ☽ ∥ ♆
10 19 ☽ ⚺ ♂
11 50 ☽ Q ♇
12 11 ☽ # ♅
12 31 ☽ ⊥ ♄
14 43 ☽ ⚹ ♀
22 34 ☽ ∠ ♃
22 53 ☉ ⊥ ♃

10 Wednesday
00 40 ☽ ⚺ ♄
02 53 ☽ # ♀
03 37 ☽ △ ♆
10 47 ☽ ⚹ ☿
12 05 ☽ ⚹ ♇
17 48 ☽ Q ♀
18 21 ☽ □ ☉
23 25 ☿ △ ♇

11 Thursday
05 22 ☽ ⚹ ♃
06 15 ☽ ⚻ ♅
13 38 ☽ Q ☿
15 51 ☽ # ☉
18 13 ☽ ∠ ♇
18 17 ☽ ± ♅

12 Friday
01 04 ☽ ☌ ♂
05 46 ☽ Q ♃
09 58 ☽ □ ♀
12 05 ☽ ⊥ ♇
12 10 ☽ bQ ♅
12 51 ☽ ☌ ♄
15 38 ☽ □ ♆
23 53 ☽ ⚺ ♇

13 Saturday
04 59 ☽ □ ☿
07 33 ♀ bQ ♅
09 02 ♃ ⚻ ♅
09 51 ♄ St D
11 05 ☽ △ ☉
14 21 ☿ ± ♅
14 31 ♀ □ ♄
17 16 ☉ ± ♇
17 28 ☽ △ ♅
17 37 ☽ □ ♃
19 22 ☿ ∥ ♃

14 Sunday
02 34 ☽ Q ♆
08 21 ♀ ∥ ♅
13 41 ☽ ⚺ ♂
17 03 ♀ ☍ ♆
18 10 ☽ bQ ☉
22 42 ☽ ⚺ ♄

15 Monday
01 14 ☽ ⚹ ♆
01 24 ☽ ⊥ ♂
02 07 ☽ △ ♀
09 04 ☽ ☌ ♇
09 54 ☽ ⊥ ♄
12 19 ☽ ± ☉
17 59 ♀ # ♆
18 42 ☽ ∠ ♂
18 59 ☽ △ ☿

16 Tuesday
00 13 ☽ ⚻ ☉
01 42 ☽ □ ♅
02 26 ☽ ∠ ♄
02 45 ☽ △ ♃
04 49 ☽ ∠ ♆
08 33 ☽ bQ ♀
20 23 ☉ △ ♅
20 59 ☽ ⊥ ♆
22 50 ☽ ⚹ ♂

17 Wednesday
00 13 ☽ bQ ☿
02 18 ☽ ± ♀
05 21 ☽ ⚹ ♄
06 04 ☽ bQ ♃
07 37 ☽ ⚺ ♆
09 55 ☽ Q ♅
13 57 ☽ ⚻ ♀
15 04 ☽ ⚺ ♇
17 16 ☽ ± ☿
19 26 ☉ ⚺ ♃
20 46 ☽ Q ♂
21 58 ☽ # ☉
22 06 ☽ ± ♃

18 Thursday
01 32 ☽ ⊥ ♇
02 17 ♀ △ ♇
02 24 ☽ Q ♄
04 27 ☽ ⚻ ☿
06 50 ☽ ⚹ ♅
08 40 ☽ ⚻ ♃
09 27 ☽ ☍ ☉
17 03 ☽ ∠ ♇

19 Friday
02 08 ☽ # ♅
04 53 ☽ □ ♂
05 47 ☽ ∥ ♆
08 31 ☽ ∠ ♅
09 19 ☽ □ ♄
11 21 ☽ ☌ ♆
17 11 ☽ # ♀
18 33 ☽ ⚹ ♇
18 55 ☿ ⚻ ♅
22 24 ☽ ☍ ♀
23 43 ☽ ⊥ ♅

20 Saturday
09 48 ☽ ⚺ ♅
10 41 ☽ ☍ ☿
11 18 ♂ Q ♃
12 21 ☽ ☍ ♃
12 49 ☽ # ♃
13 30 ♀ ± ♅
14 42 ☽ Q ♇
16 03 ☽ ⚻ ☉
23 59 ☽ ∥ ☿

21 Sunday
02 48 ☽ ± ☉
06 26 ☽ # ☿
09 09 ☽ △ ♂
11 43 ☽ △ ♄
13 36 ☽ ⚺ ♆
16 27 ☽ ∥ ♃
18 53 ☽ bQ ☉
20 43 ☽ □ ♇
23 34 ☽ ⊥ ♆

22 Monday
05 14 ☽ ⚻ ♀
05 32 ☽ ∥ ♀
09 05 ☿ ☌ ♃
11 03 ☽ bQ ♂
11 48 ☽ ☌ ♅
12 44 ☽ bQ ♄
14 32 ☽ ∠ ♆
15 06 ☽ ⚻ ♃
15 18 ☽ ⚻ ☿
16 10 ☽ ± ♀
21 39 ☽ △ ☉
21 55 ☉ bQ ♇
23 55 ☽ # ♆

23 Tuesday
01 15 ☽ ± ♃
01 45 ☽ ± ☿
02 36 ☽ ± ♂
03 08 ☽ ∥ ♅
03 45 ☽ ± ♄
04 22 ♂ ⊥ ♇
07 14 ♂ bQ ♅
08 35 ☽ bQ ♀
13 00 ☽ ⚻ ♂
13 47 ☽ ⚻ ♄
15 32 ☽ ⚹ ♆
16 31 ☽ bQ ♃
17 26 ☽ ∥ ☉
17 26 ☽ bQ ☿
22 43 ☽ △ ♇

24 Wednesday
11 26 ♂ ☌ ♄
11 38 ☽ Q ♆
12 06 ☽ △ ♀
13 56 ☽ ⚺ ♅
18 06 ☽ △ ♃
19 38 ☽ △ ☿
23 58 ☽ bQ ♇

25 Thursday
00 04 ☽ ⊥ ♅
03 41 ☽ □ ☉
09 08 ♀ ⚻ ♅
10 05 ☿ Q ♄
10 12 ☿ # ♃
15 15 ☽ ± ♇
15 20 ☽ ∠ ♅
16 29 ☽ ☍ ♄
17 34 ☽ ☍ ♂
18 06 ☽ □ ♆

26 Friday
01 29 ☽ ⚻ ♇
06 56 ♂ □ ♆
12 44 ☽ Q ☉
17 03 ☽ ⚹ ♅
20 12 ☽ □ ♀
20 13 ☿ # ♀
22 09 ☽ □ ♃

27 Saturday
00 30 ☽ □ ☿
11 08 ☽ ⚹ ☉
13 53 ☽ Q ♅
20 30 ☽ ⚻ ♄
21 58 ☽ △ ♆
22 29 ♀ ☌ ♃
23 39 ☽ ⚻ ♂

28 Sunday
01 54 ♀ ∥ ♃
05 39 ☽ ☍ ♇
06 10 ☽ Q ♃
06 48 ☽ Q ♀
07 06 ☽ ± ♄
08 42 ☽ Q ☿
10 11 ♀ Q ♄
10 42 ☽ ± ♂
15 37 ☽ ∠ ☉
21 41 ☽ □ ♅
23 10 ☽ bQ ♄

29 Monday
00 34 ☽ bQ ♆
03 27 ☽ bQ ♂
03 54 ☽ ⚹ ♃
06 22 ☽ ⚹ ♀
06 23 ☽ ⚹ ☿
06 32 ☿ ☌ ♀
09 00 ☽ ⊥ ☉
16 45 ☽ ± ♆
20 43 ☽ ⚺ ☉

30 Tuesday
02 20 ☽ △ ♄
03 39 ☽ ⚻ ♆
07 34 ☽ ∠ ♃
07 51 ☽ △ ♂
09 48 ☽ ∠ ☿
10 24 ☉ ∥ ♅
11 41 ☽ ⚻ ♇
12 28 ☽ ∠ ♀
13 03 ☿ St R
22 45 ☽ ± ♇

31 Wednesday
00 26 ☽ ⊥ ♃
02 24 ☽ ⊥ ☿
03 43 ♃ Q ♄
04 20 ☽ △ ♅
06 53 ☽ ⊥ ♀
09 29 ♀ Q ♂
11 49 ☽ ⚺ ♃
12 38 ☉ # ♆
13 32 ☽ ⚺ ☿
15 32 ☽ bQ ♇
19 20 ☽ ⚺ ♀

SEPTEMBER 2016

LONGITUDES

Date	Sidereal time h m s	Sun ☉ ° ' "	Moon ☽ ° ' "	Moon ☽ 24.00 ° ' "	Mercury ☿ ° '	Venus ♀ ° '	Mars ♂ ° '	Jupiter ♃ ° '	Saturn ♄ ° '	Uranus ♅ ° '	Neptune ♆ ° '	Pluto ♇ ° '
01	10 44 20	09 ♍ 28 19	10 ♍ 54 01	17 ♍ 09 47	28 ♍ 53	02 ♎ 57	14 ♐ 13	28 ♍ 19	10 ♐ 04	24 ♈ 04	10 ♓ 40	15 ♑ 05
02	10 48 16	10 26 25	23 ♍ 22 18	29 ♍ 31 43	28 R 38	04 11	14 47	28 31	10 06	24 R 02	10 R 38	15 R 04
03	10 52 13	11 24 33	05 ♎ 38 15	11 ♎ 42 08	28 17	05 24	15 21	28 44	10 08	24 01	10 36	15 03
04	10 56 09	12 22 43	17 ♎ 43 40	23 ♎ 43 09	27 50	06 38	15 55	28 57	10 10	23 59	10 35	15 03
05	11 00 06	13 20 54	29 ♎ 40 59	05 ♏ 37 33	27 16	07 51	16 30	29 09	10 12	23 57	10 33	15 02
06	11 04 02	14 19 07	11 ♏ 33 17	17 ♏ 28 42	26 36	09 05	17 05	29 22	10 15	23 56	10 31	15 02
07	11 07 59	15 17 21	23 ♏ 24 16	29 ♏ 20 32	25 51	10 18	17 40	29 35	10 17	23 54	10 30	15 01
08	11 11 56	16 15 37	05 ♐ 17 22	11 ♐ 17 22	25 01	11 32	18 15	29 ♍ 48	10 19	23 52	10 28	15 00
09	11 15 52	17 13 54	17 ♐ 19 02	23 ♐ 23 44	24 06	12 45	18 51	00 ♎ 00	10 22	23 50	10 26	15 00
10	11 19 49	18 12 13	29 ♐ 31 55	05 ♑ 44 10	23 09	13 59	19 26	00 13	10 25	23 49	10 25	14 59
11	11 23 45	19 10 34	12 ♑ 00 59	18 ♑ 22 51	22 08	15 12	20 02	00 26	10 27	23 47	10 23	14 59
12	11 27 42	20 08 56	24 ♑ 50 09	01 ♒ 23 13	21 07	16 25	20 39	00 39	10 30	23 45	10 22	14 59
13	11 31 38	21 07 19	08 ♒ 02 19	14 ♒ 47 33	20 06	17 39	21 15	00 52	10 33	23 43	10 20	14 58
14	11 35 35	22 05 45	21 ♒ 38 58	28 ♒ 36 27	19 07	18 52	21 52	01 05	10 36	23 41	10 18	14 58
15	11 39 31	23 04 12	05 ♓ 39 44	12 ♓ 48 25	18 11	20 05	22 29	01 17	10 39	23 39	10 17	14 57
16	11 43 28	24 02 40	20 ♓ 01 58	27 ♓ 19 42	17 20	21 19	23 06	01 30	10 42	23 37	10 15	14 57
17	11 47 25	25 01 11	04 ♈ 40 50	12 ♈ 04 29	16 34	22 32	23 43	01 43	10 45	23 35	10 14	14 57
18	11 51 21	25 59 43	19 ♈ 29 43	26 ♈ 55 32	15 56	23 45	24 21	01 56	10 49	23 33	10 12	14 57
19	11 55 18	26 58 18	04 ♉ 20 58	11 ♉ 45 06	15 25	24 58	24 58	02 09	10 52	23 31	10 10	14 56
20	11 59 14	27 56 55	19 ♉ 07 04	26 ♉ 26 07	15 04	26 12	25 36	02 22	10 56	23 29	10 09	14 56
21	12 03 11	28 55 34	03 ♊ 41 35	10 ♊ 52 59	14 52	27 25	26 14	02 35	10 59	23 27	10 07	14 56
22	12 07 07	29 ♍ 54 15	17 ♊ 59 54	25 ♊ 02 05	14 D 50	28 38	26 52	02 48	11 03	23 24	10 06	14 56
23	12 11 04	00 ♎ 52 58	01 ♋ 59 25	08 ♋ 51 50	14 57	29 ♎ 51	27 31	03 01	11 07	23 22	10 04	14 56
24	12 15 00	01 51 44	15 ♋ 39 25	22 ♋ 22 16	15 15	01 ♏ 05	28 09	03 14	11 10	23 20	10 03	14 56
25	12 18 57	02 50 32	29 ♋ 00 34	05 ♌ 34 32	15 42	02 18	28 48	03 27	11 14	23 18	10 01	14 56
26	12 22 54	03 49 23	12 ♌ 04 25	18 ♌ 30 27	16 18	03 31	29 ♐ 27	03 40	11 18	23 16	10 00	14 56
27	12 26 50	04 48 15	24 ♌ 52 54	01 ♍ 12 01	17 03	04 44	00 ♑ 06	03 53	11 22	23 13	09 58	14 D 56
28	12 30 47	05 47 10	07 ♍ 28 03	13 ♍ 41 12	17 57	05 57	00 46	04 06	11 26	23 11	09 57	14 56
29	12 34 43	06 46 07	19 ♍ 51 43	25 ♍ 59 47	18 58	07 10	01 25	04 19	11 31	23 09	09 55	14 56
30	12 38 40	07 ♎ 45 06	02 ♎ 05 35	08 ♎ 09 20	20 ♍ 06	08 ♏ 23	02 ♑ 05	04 ♎ 32	11 ♐ 35	23 ♈ 06	09 ♓ 54	14 ♑ 56

DECLINATIONS

Date	Moon True ☊ ° '	Moon Mean ☊ ° '	Moon ☽ Latitude ° '	Sun ☉ ° '	Moon ☽ ° '	Mercury ☿ ° '	Venus ♀ ° '	Mars ♂ ° '	Jupiter ♃ ° '	Saturn ♄ ° '	Uranus ♅ ° '	Neptune ♆ ° '	Pluto ♇ ° '
01	12 ♍ 43	12 ♍ 40	00 S 10	08 N 01	07 N 19	03 S 27	00 S 14	25 S 13	01 N 41	20 S 26	08 N 45	08 S 23	21 S 20
02	12 D 43	12 37	00 N 59	07 39	03 N 32	03 24	00 44	25 16	01 36	20 27	08 44	08 23	21 20
03	12 43	12 33	02 04	07 17	00 S 21	03 18	01 15	25 19	01 31	20 27	08 44	08 24	21 20
04	12 43	12 30	03 02	06 55	04 09	03 09	01 46	25 23	01 25	20 28	08 43	08 25	21 20
05	12 R 43	12 27	03 52	06 33	07 44	02 55	02 17	25 26	01 20	20 28	08 42	08 25	21 20
06	12 43	12 24	04 32	06 10	10 59	02 38	02 48	25 29	01 15	20 29	08 42	08 26	21 21
07	12 42	12 21	04 59	05 48	13 47	02 17	03 19	25 31	01 10	20 29	08 41	08 27	21 21
08	12 42	12 17	05 15	05 25	16 01	01 52	03 50	25 34	01 05	20 30	08 41	08 27	21 21
09	12 D 42	12 14	05 16	05 03	17 35	01 24	04 20	25 37	01 00	20 30	08 40	08 28	21 21
10	12 42	12 11	05 04	04 40	18 22	00 52	04 51	25 39	00 55	20 31	08 39	08 29	21 21
11	12 43	12 08	04 37	04 17	18 18	00 S 18	05 22	25 41	00 50	20 31	08 38	08 29	21 22
12	12 43	12 05	03 56	03 54	17 18	00 N 18	05 52	25 43	00 45	20 32	08 38	08 30	21 22
13	12 44	12 02	03 01	03 31	15 20	00 55	06 22	25 45	00 40	20 33	08 37	08 30	21 22
14	12 45	11 58	01 54	03 08	12 29	01 33	06 53	25 47	00 34	20 33	08 36	08 31	21 22
15	12 45	11 55	00 N 39	02 45	08 50	02 11	07 23	25 48	00 29	20 34	08 35	08 32	21 22
16	12 R 45	11 52	00 S 40	02 22	04 S 34	02 48	07 53	25 50	00 24	20 35	08 35	08 32	21 22
17	12 45	11 49	01 58	01 59	00 N 03	03 24	08 22	25 51	00 19	20 35	08 34	08 33	21 23
18	12 43	11 46	03 10	01 35	04 42	03 57	08 52	25 52	00 14	20 36	08 33	08 33	21 23
19	12 42	11 43	04 08	01 12	09 04	04 27	09 21	25 53	00 09	20 37	08 32	08 34	21 23
20	12 40	11 39	04 50	00 49	12 51	04 54	09 50	25 54	00 N 04	20 37	08 32	08 35	21 23
21	12 38	11 36	05 12	00 26	15 47	05 16	10 19	25 54	00 S 02	20 38	08 31	08 35	21 23
22	12 37	11 33	05 14	00 N 02	17 40	05 34	10 48	25 55	00 07	20 39	08 30	08 36	21 23
23	12 D 37	11 30	04 58	00 S 21	18 27	05 48	11 17	25 55	00 12	20 39	08 29	08 36	21 24
24	12 37	11 27	04 25	00 44	18 08	05 57	11 45	25 55	00 17	20 40	08 28	08 37	21 24
25	12 39	11 23	03 38	01 08	16 48	06 01	12 13	25 54	00 22	20 41	08 27	08 37	21 24
26	12 40	11 20	02 41	01 31	14 36	06 00	12 41	25 54	00 27	20 41	08 27	08 38	21 24
27	12 41	11 17	01 37	01 55	11 42	05 55	13 08	25 53	00 32	20 42	08 26	08 39	21 24
28	12 42	11 14	00 S 29	02 18	08 19	05 45	13 35	25 52	00 37	20 43	08 25	08 39	21 24
29	12 R 42	11 11	00 N 39	02 41	04 37	05 31	14 02	25 51	00 43	20 44	08 24	08 40	21 24
30	12 ♍ 41	11 ♍ 08	01 N 45	03 S 04	00 N 46	05 N 13	14 S 29	25 S 50	00 S 48	20 S 44	08 N 23	08 S 40	21 S 24

ZODIAC SIGN ENTRIES

Date	h	m	Planets
03	00	55	☽ ♎
05	12	38	☽ ♏
08	01	20	☽ ♐
09	11	18	♃ ♎
10	12	55	☽ ♑
12	21	28	☽ ♒
15	02	23	☽ ♓
17	04	22	☽ ♈
19	04	58	☽ ♉
21	05	53	☽ ♊
22	14	21	☉ ♎
23	08	33	☽ ♋
23	14	51	♀ ♏
25	13	48	☽ ♌
27	08	07	♂ ♑
27	21	43	☽ ♍
30	07	52	☽ ♎

LATITUDES

Date	Mercury ☿ ° '	Venus ♀ ° '	Mars ♂ ° '	Jupiter ♃ ° '	Saturn ♄ ° '	Uranus ♅ ° '	Neptune ♆ ° '	Pluto ♇ ° '
01	04 S 14	01 N 02	02 S 44	01 N 06	01 N 32	00 S 38	00 S 53	01 N 16
04	04 22	00 56	02 42	01 06	01 32	00 38	00 53	01 15
07	04 17	00 50	02 40	01 06	01 31	00 38	00 53	01 15
10	03 55	00 43	02 39	01 06	01 30	00 38	00 53	01 15
13	03 16	00 36	02 37	01 06	01 30	00 38	00 53	01 14
16	02 23	00 29	02 35	01 06	01 29	00 38	00 53	01 14
19	01 24	00 20	02 33	01 06	01 29	00 38	00 53	01 13
22	00 S 26	00 12	02 31	01 06	01 28	00 38	00 53	01 13
25	00 N 25	00 N 03	02 28	01 06	01 27	00 38	00 53	01 13
28	01 04	00 S 06	02 26	01 06	01 27	00 38	00 53	01 12
31	01 N 33	00 S 15	02 S 24	01 N 06	01 N 26	00 S 38	00 S 53	01 N 12

DATA

Julian Date	2457633
Delta T	+69 seconds
Ayanamsa	24° 05' 19"
Synetic vernal point	05° ♓ 01' 40"
True obliquity of ecliptic	23° 26' 05"

MOON'S PHASES, APSIDES AND POSITIONS ☽

Date	h	m	Phase	Longitude ° '	Eclipse Indicator
01	09	03	●	09 ♍ 21	Annular
09	11	49	☽	17 ♐ 13	
16	19	05	○	24 ♓ 20	
23	09	56	☾	00 ♋ 48	

Day	h	m		
06	18	48	Apogee	
18	17	09	Perigee	
03	09	52	0S	
10	22	02	Max dec	18° S 27'
17	11	46	0N	
23	16	46	Max dec	18° N 28'
30	16	46	0S	

LONGITUDES

Date	Chiron ⚷ ° '	Ceres ⚳ ° '	Pallas ⚴ ° '	Juno ⚵ ° '	Vesta ⚶ ° '	Black Moon Lilith ⚸ ° '
01	23 ♓ 33	05 ♉ 24	24 ♒ 54	09 ♏ 52	14 ♋ 44	11 ♏ 28
11	23 ♓ 06	05 ♉ 19	22 ♒ 40	12 ♏ 32	18 ♋ 19	12 ♏ 35
21	22 ♓ 39	04 ♉ 01	20 ♒ 53	15 ♏ 22	21 ♋ 43	13 ♏ 43
31	22 ♓ 12	02 ♉ 26	19 ♒ 38	18 ♏ 22	24 ♋ 54	14 ♏ 50

ASPECTARIAN

h m	Aspects
01 Thursday	
02 09	☉ Ꝓ ♅
02 35	☽ ∥ ♅
05 04	☽ # ♆
06 57	☽ ∥ ☉
08 31	☽ Ꝓ ♅
09 03	☽ ☌ ☉
10 25	☽ □ ♄
11 33	☽ ☍ ♆
18 39	☽ □ ♂
19 59	☽ △ ♇
02 Friday	
01 42	☽ ± ♅
03 22	☉ □ ♄
12 47	☽ # ☿
13 18	☽ ⊼ ♅
16 38	☉ ☍ ♆
17 18	☿ ☌ ♃
21 14	☽ Q ♄
22 00	☽ ☌ ☿
22 13	☽ ☌ ♃
23 54	♂ ⊻ ♇
03 Saturday	
00 16	☽ ∥ ♃
03 15	☽ # ♀
07 16	☽ Q ♂
11 30	☽ ☌ ♀
18 34	☽ ∥ ♀
19 07	☽ # ♃
19 54	☿ Q ♄
20 55	☽ ✶ ♄
21 48	☽ ⊼ ♆
22 08	♀ # ♃
04 Sunday	
00 25	☽ ⊻ ☉
05 54	☽ ∥ ☿
06 39	☽ □ ♇
08 13	☽ ✶ ♂
09 43	☽ ± ♆
13 25	☽ ⊥ ☉
05 Monday	
00 30	☽ ☍ ♅
02 58	☽ ∠ ♄
03 42	☽ Ꝓ ♆
04 36	☽ # ☉
07 22	☽ ⊻ ☿
09 04	☽ ∠ ☉
10 55	☽ ⊻ ♃
15 51	☽ ∠ ♂
16 52	☽ ∥ ♆
18 46	☽ Q ♇
18 51	☽ # ♅
18 52	☽ ⊥ ☿
21 10	☽ ⊥ ♄
23 15	☽ ⊥ ♃
06 Tuesday	
06 25	☽ ⊻ ♀
07 07	☿ ∥ ♀
09 20	☽ ⊻ ♄
09 55	☽ △ ♆
10 59	☽ ⊥ ♂
12 06	☽ ∠ ☿
17 48	☽ ∠ ♃
18 06	☽ ✶ ☉
19 02	☽ ✶ ♇
19 58	☽ ⊥ ♀
23 46	☽ ⊻ ♂
07 Wednesday	
05 19	☉ △ ♇
11 34	♀ ✶ ♄
13 00	☽ ⊼ ♅
15 39	♀ ⊼ ♆
16 17	☽ ∠ ♀
16 38	☽ ✶ ☿
20 33	☽ Q ☉
08 Thursday	
00 43	☽ ✶ ♃
01 06	☽ ± ♅
01 21	☽ ∠ ♇
15 13	☽ Q ☿
19 09	☽ Ꝓ ♅
19 26	☽ ⊥ ♇
22 06	☽ ☌ ♄
22 20	☽ □ ♆
09 Friday	
01 14	☽ Q ♃
01 54	☽ ✶ ♀
07 24	☽ ⊻ ♇
11 49	☽ □ ☉
15 11	☽ ☌ ♂
18 58	☿ ⊼ ♅
10 Saturday	
00 26	☽ □ ☿
00 51	☽ △ ♅
02 36	☉ ± ♅
04 18	☽ Q ♀
06 58	☉ # ♀
09 45	☿ # ♃
09 50	☽ Q ♆
13 04	♄ □ ♆
13 22	☽ □ ♃
11 Sunday	
07 47	♀ □ ♇
08 54	☽ ✶ ♆
09 01	☽ ⊻ ♄
17 37	☽ ☌ ♇
18 40	☽ □ ♀
20 25	☽ ⊥ ♄
12 Monday	
02 36	☽ △ ☉
03 51	☽ ⊻ ♂
05 37	☽ △ ☿
10 00	☽ □ ♅
10 47	♀ ± ♆
12 58	☽ ∠ ♆
13 14	☽ ∠ ♄
15 30	☽ ⊥ ♂
18 58	☿ □ ♂
22 50	☽ △ ♃
23 40	☉ ☌ ☿
13 Tuesday	
03 21	☿ ∥ ♃
05 21	☽ ⊥ ♆
07 06	☽ Ꝓ ☿
08 17	☽ Ꝓ ☉
08 39	☽ ∠ ♂
16 05	☽ ⊻ ♆
16 30	☽ ✶ ♄
18 33	☽ Q ♅
20 39	☉ □ ♂
22 03	☽ ± ☿
14 Wednesday	
00 18	☽ ⊻ ♇
01 33	☽ ± ☉
02 06	☽ Ꝓ ♃
06 41	☽ △ ♀
07 53	☽ ⊼ ☿
10 48	☽ ⊥ ♇
12 23	☽ ✶ ♂
12 50	☽ ⊼ ☉
13 39	☽ Q ♄
14 44	☿ ⊻ ♀
15 31	☽ ✶ ♅
18 01	☽ ± ♃
15 Thursday	
02 19	☽ ∠ ♇
04 28	☽ ⊼ ♃
09 54	☽ Q ♂
10 56	☽ Ꝓ ♀
13 25	☽ # ♅
13 47	☽ ∥ ♆
17 02	☽ ∠ ♅
19 35	☽ ∥ ♀
19 45	☽ ☌ ♆
20 25	☽ □ ♄
16 Friday	
01 26	☉ ∥ ☿
01 51	☉ ⊼ ♅
03 11	☿ ± ♅
03 27	☽ ± ♀
03 35	☽ ✶ ♇
07 46	☽ ☍ ☿
08 01	☽ ⊥ ♅
14 18	☽ ⊼ ♀
17 17	☽ □ ♂
17 54	☽ ⊻ ♅
19 05	☽ ☍ ☉
20 13	☽ # ☿
23 23	☽ Q ♇
17 Saturday	
00 38	☽ # ☉
07 04	♂ △ ♅
07 06	☽ ☍ ♃
10 06	☽ # ♃
12 27	☿ ⊥ ♀
13 22	☽ ∥ ♃
20 42	♀ ∥ ♆
20 59	☽ ⊻ ♆
21 08	☽ ∥ ☉
21 10	♀ # ♅
21 54	☽ △ ♄
18 Sunday	
04 39	☽ □ ♇
06 27	☽ ⊼ ☿
06 41	☽ ⊥ ♆
07 03	♅ # ♆
07 33	☽ ∥ ☿
08 06	♀ ☍ ♅
15 47	☽ ± ☿
18 32	☽ ☌ ♅
19 30	☽ ☍ ♀
20 10	☽ △ ♂
21 12	☽ ∠ ♆
22 14	☽ Ꝓ ♄
23 14	☽ ⊼ ☉
19 Monday	
05 50	☽ Ꝓ ☿
08 24	☽ ⊼ ♃
08 57	☽ ∥ ♅
09 05	☽ # ♆
09 37	☽ ± ☉
11 53	♀ ✶ ♂
12 51	☽ ± ♄
13 53	☽ # ♀
15 49	♀ Ꝓ ♆
18 15	☽ ± ♃
21 25	☽ ✶ ♆
21 31	☽ Ꝓ ♂
22 36	☽ ⊼ ♄
20 Tuesday	
01 14	☽ Ꝓ ☉
05 11	☽ △ ♇
05 31	☽ △ ☿
06 27	♀ ∠ ♄
09 06	☽ Ꝓ ♃
12 50	☽ ± ♂
16 57	☽ Q ♆
19 07	☽ ⊻ ♅
23 07	☽ ⊼ ♂
21 Wednesday	
00 40	☽ ⊼ ♀
01 26	☿ △ ♇
03 32	☽ △ ☉
04 59	☽ ⊥ ♅
05 46	☽ Ꝓ ♇
10 08	☽ △ ♃
11 30	☽ ± ♀
13 34	☉ Q ♄
19 54	☽ ∠ ♅
20 44	☽ ± ♇
22 42	☽ □ ♆
22 Thursday	
00 13	☽ ☍ ♄
03 56	☽ Ꝓ ♀
05 30	☿ St D
06 38	☽ □ ☿
06 49	☽ ⊼ ♇
08 20	☉ # ♃
21 11	☽ ✶ ♅
23 46	☉ ∥ ♃
23 Friday	
03 53	☽ ☍ ♂
07 57	☽ △ ♀
08 51	☿ △ ♇
09 56	☽ □ ☉
13 42	☽ Q ☿
13 48	☽ □ ♃
14 24	☿ ∠ ♀
17 52	☽ Q ♅
24 Saturday	
02 06	☽ △ ♆
04 01	☽ ⊼ ♄
07 58	♂ Q ♆
10 42	☽ ☍ ♇
11 15	☽ ✶ ☿
14 43	☽ ± ♄
20 05	☽ Q ☉
22 07	☽ Q ♃
25 Sunday	
01 42	☽ □ ♅
04 47	☽ Ꝓ ♆
06 57	☽ Ꝓ ♄
11 36	☽ ⊼ ♂
15 12	☽ ∠ ☿
18 36	☽ □ ♀
19 33	☽ ✶ ☉
20 14	☽ ✶ ♃
21 08	☽ ± ♆
23 08	☽ ± ♂
26 Monday	
00 27	♀ Q ♇
07 00	☉ ☌ ♃
08 09	☽ ⊼ ♆
08 33	☽ ⊥ ☿
10 34	☽ △ ♄
15 04	♇ St D
15 35	♀ ⊻ ♃
16 40	☽ Ꝓ ♂
17 18	☽ ⊼ ♇
20 20	☽ ⊻ ☿
27 Tuesday	
00 30	☽ ∠ ♃
01 38	☽ ∠ ☉
02 14	☽ # ♀
04 33	☽ ± ♇
07 31	☽ Q ♀
08 52	☽ △ ♅
16 30	☿ ± ♅
17 47	☽ ⊥ ♃
19 14	☉ ⊻ ♀
20 04	☽ ⊥ ☉
21 34	☽ Ꝓ ♇
22 27	☽ △ ♂
28 Wednesday	
01 18	♀ ⊥ ♄
05 25	☽ ⊻ ♃
08 30	☽ ⊻ ☉
08 46	☽ ✶ ♀
09 46	☽ # ♆
11 21	☽ ∥ ♅
13 22	☽ Ꝓ ♅
16 46	☽ ☍ ♆
19 42	☽ □ ♄
29 Thursday	
02 24	☽ △ ♇
05 51	☽ ∥ ☿
06 44	☽ ± ♅
10 05	☽ ☌ ☿
17 00	☽ ∠ ♀
18 24	☽ ⊼ ♅
22 59	☽ # ☉
30 Friday	
07 01	☽ Q ♄
11 50	☽ # ♃
11 58	☽ □ ♂
12 39	☽ ⊥ ♀
16 54	☽ ☌ ♃
21 56	☽ ∥ ♃

All ephemeris data is given at 12.00 UT and the Moon's longitude is additionally given for 24.00 UT

OCTOBER 2016

LONGITUDES

Date	Sidereal time h m s	Sun ☉ ° ' "	Moon ☽ ° ' "	Moon ☽ 24.00 ° ' "	Mercury ☿ ° '	Venus ♀ ° '	Mars ♂ ° '	Jupiter ♃ ° '	Saturn ♄ ° '	Uranus ♅ ° '	Neptune ♆ ° '	Pluto ⯓ ° '
01	12 42 36	08 ♎ 44 07	14 ♎ 11 13	20 ♎ 11 25	21 ♍ 20	09 ♏ 36	02 ♑ 45	04 ♎ 45	11 ♐ 39	23 ♈ 04	09 ♓ 53	14 ♑ 56
02	12 46 33	09 43 09	26 ♎ 10 09	02 ♏ 07 38	22 39	10 49	03 24	04 58	11 44	23 R 02	09 R 51	14 56
03	12 50 29	10 42 14	08 ♏ 04 06	13 ♏ 59 50	24 04	12 02	04 05	05 11	11 48	22 59	09 50	14 56
04	12 54 26	11 41 21	19 ♏ 55 06	25 ♏ 50 15	25 32	13 15	04 45	05 24	11 53	22 57	09 48	14 56
05	12 58 23	12 40 30	01 ♐ 45 38	07 ♐ 41 40	27 04	14 28	05 25	05 36	11 57	22 55	09 47	14 57
06	13 02 19	13 39 41	13 ♐ 37 45	19 ♐ 37 23	28 ♍ 39	15 41	06 06	05 49	12 02	22 52	09 46	14 57
07	13 06 16	14 38 53	25 ♐ 38 03	01 ♑ 41 17	00 ♎ 17	16 54	06 47	06 02	12 07	22 50	09 45	14 57
08	13 10 12	15 38 07	07 ♑ 47 37	13 ♑ 57 37	01 56	18 07	07 27	06 15	12 12	22 47	09 43	14 58
09	13 14 09	16 37 23	20 ♑ 11 52	26 ♑ 30 55	03 37	19 20	08 08	06 28	12 17	22 45	09 42	14 58
10	13 18 05	17 36 41	02 ♒ 55 19	09 ♒ 25 33	05 19	20 33	08 49	06 41	12 22	22 43	09 41	14 59
11	13 22 02	18 36 01	16 ♒ 02 05	22 ♒ 45 15	07 02	21 46	09 31	06 54	12 27	22 40	09 40	14 59
12	13 25 58	19 35 22	29 ♒ 35 19	06 ♓ 32 26	08 46	22 59	10 12	07 07	12 32	22 38	09 38	14 59
13	13 29 55	20 34 45	13 ♓ 36 33	20 ♓ 47 28	10 30	24 12	10 53	07 19	12 37	22 35	09 37	15 00
14	13 33 52	21 34 10	28 ♓ 04 48	05 ♈ 27 56	12 14	25 24	11 35	07 32	12 42	22 33	09 36	15 00
15	13 37 48	22 33 37	12 ♈ 56 04	20 ♈ 28 13	13 58	26 37	12 17	07 45	12 47	22 30	09 35	15 01
16	13 41 45	23 33 06	28 ♈ 03 15	05 ♉ 39 52	15 43	27 50	12 59	07 58	12 53	22 28	09 34	15 02
17	13 45 41	24 32 36	13 ♉ 16 46	20 ♉ 52 36	17 27	29 ♏ 02	13 40	08 10	12 58	22 25	09 33	15 02
18	13 49 38	25 32 09	28 ♉ 26 06	05 ♊ 56 04	19 10	00 ♐ 15	14 23	08 23	13 04	22 23	09 32	15 03
19	13 53 34	26 31 45	13 ♊ 21 30	20 ♊ 41 34	20 54	01 28	15 05	08 36	13 09	22 21	09 31	15 04
20	13 57 31	27 31 22	27 ♊ 55 37	05 ♋ 03 15	22 37	02 40	15 47	08 48	13 15	22 18	09 30	15 04
21	14 01 27	28 31 02	12 ♋ 04 13	18 ♋ 58 30	24 19	03 53	16 29	09 01	13 21	22 16	09 29	15 05
22	14 05 24	29 ♎ 30 44	25 ♋ 46 12	02 ♌ 27 32	26 01	05 05	17 12	09 14	13 26	22 13	09 28	15 06
23	14 09 21	00 ♏ 30 28	09 ♌ 02 53	15 ♌ 32 38	27 42	06 18	17 54	09 26	13 32	22 11	09 27	15 07
24	14 13 17	01 30 15	21 ♌ 57 15	28 ♌ 17 13	29 ♎ 23	07 30	18 37	09 39	13 38	22 08	09 26	15 07
25	14 17 14	02 30 04	04 ♍ 33 02	10 ♍ 45 12	01 ♏ 04	08 43	19 20	09 51	13 44	22 06	09 25	15 08
26	14 21 10	03 29 55	16 ♍ 54 10	23 ♍ 00 22	02 44	09 55	20 02	10 04	13 50	22 04	09 25	15 09
27	14 25 07	04 29 48	29 ♍ 04 15	05 ♎ 06 08	04 23	11 08	20 45	10 16	13 56	22 01	09 24	15 10
28	14 29 03	05 29 43	11 ♎ 06 24	17 ♎ 05 19	06 02	12 20	21 28	10 28	14 02	21 59	09 23	15 11
29	14 33 00	06 29 40	23 ♎ 03 10	29 ♎ 00 10	07 40	13 33	22 12	10 41	14 08	21 57	09 22	15 12
30	14 36 56	07 29 39	04 ♏ 56 32	10 ♏ 52 28	09 18	14 45	22 55	10 53	14 14	21 54	09 22	15 13
31	14 40 53	08 ♏ 29 41	16 ♏ 48 08	22 ♏ 43 44	10 ♏ 55	15 ♐ 57	23 ♑ 38	11 ♎ 05	14 ♐ 20	21 ♈ 52	09 ♓ 21	15 ♑ 14

Date	Moon True ☊ ° '	Moon Mean ☊ ° '	Moon ☽ Latitude ° '
01	12 ♍ 38	11 ♍ 04	02 N 44
02	12 R 34	11 01	03 36
03	12 29	10 58	04 18
04	12 24	10 55	04 49
05	12 19	10 52	05 07
06	12 15	10 49	05 12
07	12 12	10 45	05 04
08	12 10	10 42	04 42
09	12 D 10	10 39	04 06
10	12 11	10 36	03 18
11	12 12	10 33	02 18
12	12 14	10 29	01 N 08
13	12 R 14	10 26	00 S 07
14	12 13	10 23	01 25
15	12 10	10 20	02 39
16	12 05	10 17	03 43
17	11 59	10 14	04 31
18	11 53	10 10	05 00
19	11 47	10 07	05 09
20	11 42	10 04	04 57
21	11 38	10 01	04 27
22	11 37	09 58	03 42
23	11 D 37	09 55	02 47
24	11 38	09 51	01 44
25	11 39	09 48	00 S 38
26	11 R 39	09 45	00 N 28
27	11 38	09 42	01 32
28	11 34	09 39	02 31
29	11 27	09 35	03 23
30	11 18	09 32	04 05
31	11 ♍ 07	09 ♍ 29	04 N 37

DECLINATIONS

Date	Sun ☉ ° '	Moon ☽ ° '	Mercury ☿ ° '	Venus ♀ ° '	Mars ♂ ° '	Jupiter ♃ ° '	Saturn ♄ ° '	Uranus ♅ ° '	Neptune ♆ ° '	Pluto ⯓ ° '
01	03 S 28	03 S 04	04 N 51	14 S 55	25 S 48	00 S 53	20 S 45	08 N 22	08 S 41	21 S 25
02	03 51	06 45	04 26	15 21	25 46	00 58	20 46	08 22	08 41	21 25
03	04 14	10 07	03 58	15 46	25 44	01 03	20 47	08 21	08 42	21 25
04	04 37	13 05	03 27	16 11	25 42	01 08	20 48	08 20	08 42	21 25
05	05 00	15 30	02 53	16 36	25 40	01 13	20 48	08 19	08 43	21 25
06	05 23	17 16	02 17	17 01	25 37	01 18	20 49	08 18	08 43	21 25
07	05 46	18 18	01 40	17 24	25 34	01 23	20 50	08 17	08 44	21 25
08	06 09	18 31	01 01	17 48	25 31	01 28	20 51	08 16	08 44	21 25
09	06 32	17 51	00 N 20	18 11	25 28	01 34	20 52	08 15	08 45	21 25
10	06 55	16 17	00 S 21	18 34	25 25	01 39	20 52	08 14	08 45	21 25
11	07 17	13 50	01 03	18 56	25 21	01 44	20 53	08 14	08 46	21 26
12	07 40	10 33	01 47	19 17	25 17	01 49	20 54	08 13	08 46	21 26
13	08 02	06 34	02 30	19 39	25 13	01 54	20 55	08 12	08 46	21 26
14	08 24	02 S 04	03 14	19 59	25 08	01 59	20 56	08 11	08 47	21 26
15	08 47	02 N 40	03 58	20 19	25 03	02 04	20 56	08 10	08 47	21 26
16	09 09	07 19	04 42	20 39	24 59	02 09	20 57	08 09	08 48	21 26
17	09 31	11 30	05 26	20 58	24 53	02 14	20 58	08 08	08 48	21 26
18	09 52	14 55	06 10	21 17	24 48	02 19	20 59	08 07	08 48	21 26
19	10 14	17 18	06 54	21 35	24 42	02 23	21 00	08 06	08 49	21 26
20	10 35	18 29	07 37	21 52	24 37	02 28	21 01	08 05	08 49	21 26
21	10 57	18 28	08 20	22 09	24 31	02 33	21 01	08 05	08 49	21 26
22	11 18	17 21	09 03	22 25	24 24	02 38	21 02	08 04	08 50	21 26
23	11 39	15 19	09 45	22 41	24 18	02 43	21 03	08 03	08 50	21 26
24	12 00	12 33	10 27	22 56	24 11	02 48	21 04	08 02	08 50	21 26
25	12 20	09 15	11 08	23 10	24 04	02 53	21 05	08 01	08 51	21 26
26	12 41	05 36	11 49	23 24	23 57	02 58	21 06	08 00	08 51	21 26
27	13 01	01 N 47	12 29	23 37	23 49	03 02	21 07	07 59	08 51	21 26
28	13 21	02 S 04	13 08	23 50	23 42	03 07	21 07	07 58	08 52	21 26
29	13 41	05 49	13 46	24 01	23 34	03 12	21 08	07 57	08 52	21 26
30	14 01	09 18	14 24	24 13	23 26	03 17	21 09	07 57	08 52	21 26
31	14 S 20	12 S 25	15 S 01	24 S 23	23 S 17	03 S 21	21 S 10	07 N 56	08 S 52	21 S 26

ZODIAC SIGN ENTRIES

Date	h	m	Planets
02	19	43	☽ ♏
05	08	26	☽ ♐
07	07	56	☿ ♎
07	20	40	☽ ♑
10	06	33	☽ ♒
12	12	43	☽ ♓
14	15	08	☽ ♈
16	15	04	☽ ♉
18	07	01	♀ ♐
18	14	30	☽ ♊
20	15	28	☽ ♋
22	19	34	☽ ♌
22	23	46	☉ ♏
24	20	46	☿ ♏
25	03	16	☽ ♍
27	13	51	☽ ♎
30	02	01	☽ ♏

LATITUDES

Date	Mercury ☿ ° '	Venus ♀ ° '	Mars ♂ ° '	Jupiter ♃ ° '	Saturn ♄ ° '	Uranus ♅ ° '	Neptune ♆ ° '	Pluto ⯓ ° '
01	01 N 33	00 S 15	02 S 24	01 N 06	01 N 26	00 S 38	00 S 53	01 N 12
04	01 49	00 24	02 21	01 06	01 26	00 38	00 53	01 11
07	01 56	00 33	02 19	01 06	01 25	00 38	00 53	01 11
10	01 55	00 42	02 16	01 06	01 25	00 38	00 53	01 11
13	01 48	00 51	02 14	01 06	01 24	00 38	00 53	01 10
16	01 36	01 00	02 11	01 06	01 24	00 38	00 53	01 10
19	01 21	01 09	02 08	01 06	01 23	00 38	00 53	01 09
22	01 04	01 18	02 06	01 07	01 23	00 38	00 53	01 09
25	00 45	01 26	02 03	01 07	01 22	00 38	00 52	01 09
28	00 25	01 34	02 00	01 07	01 22	00 38	00 52	01 08
31	00 N 05	01 S 42	01 S 57	01 N 07	01 N 22	00 S 38	00 S 52	01 N 08

DATA

Julian Date	2457663
Delta T	+69 seconds
Ayanamsa	24° 05' 22"
Synetic vernal point	05° ♓ 01' 38"
True obliquity of ecliptic	23° 26' 05"

MOON'S PHASES, APSIDES AND POSITIONS ☽

Date	h	m	Phase	Longitude ° '	Eclipse Indicator
01	00	11	●	08 ♎ 15	
09	04	33	☽	16 ♑ 59	
16	04	23	○	23 ♈ 14	
22	19	14	☾	29 ♋ 49	
30	17	38	●	07 ♏ 44	

Day	h	m		
04	11	10	Apogee	
16	23	41	Perigee	
31	19	43	Apogee	
08	06	03	Max dec	18° S 33'
14	22	32	ON	
20	23	38	Max dec	18° N 37'
27	23	03	OS	

LONGITUDES

Date	Chiron ⚷ ° '	Ceres ⚳ ° '	Pallas ⚴ ° '	Juno ⚵ ° '	Vesta ⚶ ° '	Black Moon Lilith ⚸ ° '
01	22 ♓ 12	02 ♉ 26	19 ♒ 38	18 ♏ 22	24 ♋ 54	14 ♏ 50
11	21 ♓ 47	00 ♉ 26	18 ♒ 58	21 ♏ 29	27 ♋ 48	15 ♏ 57
21	21 ♓ 25	28 ♈ 10	18 ♒ 54	24 ♏ 35	00 ♌ 21	17 ♏ 04
31	21 ♓ 06	25 ♈ 55	19 ♒ 22	28 ♏ 01	02 ♌ 30	18 ♏ 11

All ephemeris data is given at 12.00 UT and the Moon's longitude is additionally given for 24.00 UT

ASPECTARIAN

h m	Aspects
01 Saturday	
00 11	☽ ☌ ☉
01 51	☽ ⊻ ♀
03 26	☽ ⊼ ♆
06 55	☽ ⚹ ♄
13 29	☽ □ ⯓
14 48	☽ ∥ ☉
15 22	☽ ± ♆
17 13	♀ △ ♆
22 21	☽ # ☿
02 Sunday	
01 53	☽ Q ♂
04 02	☽ ⊻ ☿
05 43	☽ ☍ ♅
09 22	☽ ♇ ♆
13 08	☽ ∠ ♄
15 11	☉ ⊼ ♆
15 17	♀ ⊥ ♃
17 39	☽ ⊥ ☿
18 19	☿ ⊼ ♅
23 07	☽ # ♅
03 Monday	
01 32	☽ ∥ ♆
01 38	☽ Q ⯓
03 27	☽ ⚹ ♂
04 29	☉ # ☿
06 03	☽ ⊻ ♃
06 59	♀ ⊻ ♄
07 23	☽ ⊥ ♄
14 17	☽ ∠ ☿
15 33	☽ △ ♆
17 49	☽ ⊻ ☉
18 24	☽ ⊥ ♃
19 36	☽ ⊻ ♄
20 57	☽ ☌ ♀
04 Tuesday	
01 55	☽ ⚹ ⯓
07 04	☽ ⊥ ☉
11 38	☽ ∠ ♂
12 59	☽ ∠ ♃
16 59	☉ ⚹ ♄
18 07	☽ ⊼ ♅
05 Wednesday	
01 04	☽ ⚹ ☿
02 58	☽ ∠ ☉
06 15	☽ ± ♅
06 59	☽ ⊥ ♂
08 20	☽ ∠ ⯓
19 51	☽ ⊻ ♂
19 56	☽ ⚹ ♃
21 22	♀ ⚹ ⯓
21 45	♂ □ ♃
06 Thursday	
00 24	☽ ♇ ♅
02 32	☽ ⊥ ⯓
04 11	☽ □ ♆
05 02	☽ Q ☿
06 20	☽ ∥ ♀
08 44	☽ ☌ ♄
12 02	☽ ⚹ ☉
14 38	☽ ⊻ ⯓
16 35	☽ ⊻ ♀
20 33	☽ Q ♃
07 Friday	
05 57	☽ ⊥ ♀
06 26	☽ △ ♅
09 28	☿ Q ♄
14 12	☽ Q ☉
16 11	☽ Q ♆
19 33	☉ □ ⯓
21 03	☿ # ♃
22 40	☽ □ ☿
08 Saturday	
01 49	☽ ∠ ♀
08 56	☽ □ ♃
11 18	☽ ☌ ♂
14 02	☉ ± ♆
15 45	☽ ⚹ ♆
20 38	☽ ⊻ ♄
09 Sunday	
01 57	☽ ☌ ⯓
04 33	☽ □ ☉
06 21	☽ ∥ ♀
08 18	☽ ⊥ ♄
10 10	☽ ⚹ ♀
16 51	☽ □ ♅
20 33	☽ ∠ ♆
10 Monday	
01 31	☽ ∠ ♄
11 14	☽ Q ♀
13 24	☽ ⊥ ♆
17 07	☽ △ ☿
19 05	☽ △ ♃
23 05	☿ ∠ ♀
23 30	☽ ⊻ ♂
11 Tuesday	
00 27	☽ ⊻ ♆
02 18	☽ Q ♅
05 28	☽ ⚹ ♄
09 45	☿ ☌ ♃
10 06	☽ ⊻ ⯓
11 00	☽ ⊥ ♂
15 09	♀ ∠ ♃
16 58	☽ △ ☉
17 03	♂ ⚹ ♆
20 52	☽ ⊥ ⯓
22 39	☽ ♇ ♃
23 15	☽ □ ♀
23 49	☽ ⚹ ♅
12 Wednesday	
00 18	☽ ♇ ☿
03 05	☽ Q ♄
03 55	☽ ∠ ♂
05 16	♀ ⊼ ♅
12 42	☽ ∠ ⯓
13 20	☿ ∥ ♃
14 41	☽ ± ♃
18 18	☽ ± ☿
21 19	☽ ♇ ☉
23 10	☽ ∥ ♆
23 59	☿ ⊼ ♆
13 Thursday	
01 11	☽ ⊼ ♃
01 49	☽ ∠ ♅
02 33	☽ # ♅
04 14	☽ ∥ ☉
05 16	☽ ☌ ♆
06 01	☽ ⊼ ☿
07 10	☽ ⚹ ♂
10 19	☽ □ ♄
13 45	☽ ± ☉
14 20	☽ ⚹ ⯓
16 59	☽ ⊥ ♅
21 00	☿ □ ♂
21 52	☉ # ♅
14 Friday	
00 30	☽ ⊼ ☉
02 56	☽ ⊻ ♅
04 15	☽ Q ♂
06 47	☽ ∥ ☿
07 13	☽ △ ♀
10 15	☽ Q ⯓
12 26	☽ ∥ ♃
18 47	☿ ⚹ ♄
15 Saturday	
03 34	☽ ☍ ♃
06 38	☽ ⊻ ♆
08 53	☽ # ♃
09 42	☽ ♇ ♀
10 43	☉ ☍ ♅
10 54	☽ □ ♂
11 46	☽ △ ♄
12 41	☉ ∥ ♆
13 53	☽ ☍ ☿
15 20	☽ □ ⯓
16 13	☽ ⊥ ♆
19 48	☽ # ☿
16 Sunday	
01 18	☽ ± ♀
02 29	☿ □ ⯓
03 11	☽ ☌ ♅
04 23	☽ ☍ ☉
06 30	☽ ∠ ♆
08 14	♂ ⊻ ♄
10 01	☿ ± ♆
11 37	☽ ⊼ ♀
11 43	☽ ♇ ♄
16 31	☽ ∥ ♅
18 51	♀ ∠ ♂
20 06	☽ # ♆
23 00	☽ # ☉
17 Monday	
00 10	♀ ± ♅
02 00	☽ ± ♄
03 50	☽ ⊼ ♃
06 08	☽ ⚹ ♆
11 31	☽ ⊼ ♄
11 54	♀ ∥ ♄
12 07	☉ ♇ ♆
12 39	☽ △ ♂
13 26	☽ ± ♃
14 46	☽ △ ⯓
19 25	☽ ⊼ ☿
18 Tuesday	
01 03	☽ Q ♆
02 25	☽ ⊻ ♅
03 52	☽ ♇ ♃
06 08	☽ ± ☿
07 04	☽ ⊼ ☉
07 56	♀ ∠ ⯓
11 55	☽ ⊥ ♅
13 34	☽ ♇ ♂
14 34	☽ ♇ ⯓
15 09	☽ ☍ ♀
17 18	☽ ± ☉
22 22	☽ ♇ ☿
19 Wednesday	
00 24	♀ ∥ ⯓
02 18	☽ ∠ ♅
04 11	☽ △ ♃
04 43	☽ ± ♂
05 02	☽ ± ⯓
05 47	☽ □ ♆
08 49	☽ ♇ ☉
11 22	♂ ☌ ⯓
11 40	☽ ☍ ♄
14 46	☽ ⊼ ⯓
14 56	☽ ⊼ ♂
20 Thursday	
01 59	☽ △ ☿
02 41	☽ ⚹ ♅
07 48	☿ ☍ ♅
11 16	☽ △ ☉
20 43	☽ ⊼ ♀
22 42	☽ Q ♅
21 Friday	
03 16	☿ # ♅
06 41	☽ □ ♃
07 20	☉ ∠ ♄
07 33	☽ △ ♆
07 53	☽ ± ♀
14 13	☽ ⊼ ♄
14 19	☿ ♇ ♆
17 13	☽ ☍ ⯓
20 04	☽ ☍ ♂
22 Saturday	
00 44	☽ ± ♄
00 59	☽ ♇ ♀
04 24	☿ ∥ ♆
05 44	☽ □ ♅
09 41	☽ ♇ ♆
12 30	☽ □ ☿
14 38	☽ Q ♃
16 48	☽ ♇ ♄
19 14	☽ □ ☉
23 Sunday	
01 48	☽ ± ♆
06 28	☽ △ ♀
12 44	☽ ⚹ ♃
12 44	☽ ⊼ ♆
13 33	♃ ⊼ ♆
20 20	☽ △ ♄
23 12	☽ ⊼ ⯓
24 Monday	
00 31	☿ ∠ ♄
02 09	☽ Q ☿
04 56	♀ ♇ ♅
05 22	☽ ⊼ ♂
07 00	☽ Q ☉
10 26	☽ ± ⯓
11 34	☉ ⊥ ♀
12 21	☽ △ ♅
15 49	☽ # ☉
17 10	☽ ∠ ♃
17 19	☽ ± ♂
25 Tuesday	
00 56	☽ # ☿
03 31	☽ ♇ ⯓
04 16	☽ ⚹ ☿
07 43	☽ ⚹ ☉
10 38	☽ ⊥ ♃
11 33	☽ ♇ ♂
14 44	☽ # ♆
16 54	☽ ♇ ♅
20 20	☽ ∥ ♅
20 29	♀ ⊥ ⯓
20 55	☽ □ ♀
21 24	☽ ☍ ♆
22 26	☽ ⊻ ♃
26 Wednesday	
01 54	♀ □ ♆
03 32	☉ Q ⯓
03 39	♂ ⊥ ♄
05 56	☽ □ ♄
08 34	☽ △ ⯓
10 21	☽ ± ♅
13 52	☽ ∠ ☿
15 19	♀ ⚹ ♃
15 24	☽ ∠ ☉
18 14	☿ Q ⯓
18 33	☽ △ ♂
22 06	☽ ⊼ ♅
27 Thursday	
04 20	☽ # ♃
10 25	☽ ⊥ ☿
10 46	☽ ⊥ ☉
12 08	☽ Q ♀
16 16	☉ ☌ ☿
17 43	☽ Q ♄
23 46	☽ ⊻ ☉
28 Friday	
00 14	☽ ⊻ ☿
02 19	♀ ∥ ♂
08 33	☽ ⊼ ♆
10 43	☽ ☌ ♃
14 44	☽ ⚹ ♀
17 54	☽ ⚹ ♄
18 44	☽ ∥ ♃
20 11	☽ □ ⯓
20 34	☽ ± ♆
29 Saturday	
04 06	♂ □ ♅
05 06	☿ ⊥ ♀
05 06	☉ ∥ ☿
09 46	☽ ☍ ♅
10 09	☽ □ ♂
14 39	☽ ♇ ♆
19 15	☿ ⊥ ♄
30 Sunday	
00 19	☽ ∠ ♀
00 22	☽ ∠ ♄
00 45	♀ ☌ ♄
02 24	☽ # ♅
08 30	☽ Q ⯓
08 51	☽ ∥ ♆
12 57	☿ △ ♆
17 38	☽ ☌ ☉
18 42	☽ ⊥ ♄
20 34	☽ ⊥ ♀
20 56	☽ △ ♆
21 27	♀ ⊻ ⯓
22 12	☽ ☌ ☿
31 Monday	
00 14	☽ ⊻ ♃
00 52	☽ Q ♂
06 57	☽ ⊻ ♄
07 40	☉ ⊥ ♄
08 49	☽ ⚹ ⯓
10 05	☽ ⊻ ♀
12 35	☽ ⊥ ♃
14 54	☿ ⊻ ♃
22 13	☽ ⊼ ♅

NOVEMBER 2016

LONGITUDES

Date	Sidereal time h m s	Sun ☉ ° ' "	Moon ☽ ° ' "	Moon ☽ 24.00 ° ' "	Mercury ☿ ° '	Venus ♀ ° '	Mars ♂ ° '	Jupiter ♃ ° '	Saturn ♄ ° '	Uranus ♅ ° '	Neptune ♆ ° '	Pluto ♇ ° '
01	14 44 50	09 ♏ 29 44	28 ♏ 39 25	04 ♐ 35 25	12 ♏ 32	17 ♐ 09	24 ♑ 21	11 ♎ 17	14 ♐ 26	21 ♈ 50	09 ♓ 20	15 ♑ 15
02	14 48 46	10 29 48	10 ♐ 31 56	16 ♐ 29 13	14 08	18 22	25 05	11 30	14 32	21 R 47	09 R 20	15 16
03	14 52 43	11 29 55	22 ♐ 27 33	28 ♐ 27 15	15 44	19 34	25 48	11 42	14 39	21 45	09 19	15 17
04	14 56 39	12 30 03	04 ♑ 28 40	10 ♑ 32 12	17 20	20 46	26 32	11 54	14 45	21 43	09 19	15 18
05	15 00 36	13 30 13	16 ♑ 38 16	22 ♑ 47 23	18 55	21 58	27 16	12 06	14 51	21 41	09 18	15 19
06	15 04 32	14 30 25	29 ♑ 00 02	05 ♒ 16 44	20 29	23 10	28 00	12 18	14 58	21 38	09 18	15 21
07	15 08 29	15 30 38	11 ♒ 38 04	18 ♒ 04 33	22 04	24 22	28 43	12 30	15 04	21 36	09 17	15 22
08	15 12 25	16 30 52	24 ♒ 36 45	01 ♓ 15 07	23 37	25 34	29 ♑ 27	12 41	15 11	21 34	09 17	15 23
09	15 16 22	17 31 08	08 ♓ 00 06	14 ♓ 52 02	25 11	26 46	00 ♒ 11	12 53	15 17	21 32	09 16	15 24
10	15 20 19	18 31 25	21 ♓ 51 09	28 ♓ 57 29	26 44	27 58	00 55	13 05	15 24	21 30	09 16	15 26
11	15 24 15	19 31 44	06 ♈ 10 56	13 ♈ 31 09	28 17	29 ♐ 09	01 39	13 17	15 30	21 28	09 16	15 27
12	15 28 12	20 32 04	20 ♈ 57 33	28 ♈ 29 21	29 ♏ 50	00 ♑ 21	02 24	13 28	15 37	21 26	09 15	15 28
13	15 32 08	21 32 25	06 ♉ 05 29	13 ♉ 44 45	01 ♐ 22	01 33	03 08	13 40	15 44	21 24	09 15	15 30
14	15 36 05	22 32 49	21 ♉ 25 43	29 ♉ 06 55	02 54	02 44	03 52	13 51	15 50	21 22	09 15	15 31
15	15 40 01	23 33 14	06 ♊ 46 50	14 ♊ 23 59	04 26	03 56	04 36	14 02	15 57	21 20	09 15	15 32
16	15 43 58	24 33 40	21 ♊ 57 03	29 ♊ 24 51	05 57	05 08	05 21	14 14	16 04	21 18	09 15	15 34
17	15 47 54	25 34 09	06 ♋ 46 25	14 ♋ 01 04	07 28	06 19	06 05	14 25	16 11	21 16	09 15	15 35
18	15 51 51	26 34 39	21 ♋ 08 20	28 ♋ 08 00	08 59	07 30	06 50	14 36	16 18	21 14	09 15	15 37
19	15 55 48	27 35 11	05 ♌ 00 02	11 ♌ 44 37	10 30	08 42	07 34	14 47	16 24	21 12	09 14	15 38
20	15 59 44	28 35 44	18 ♌ 22 05	24 ♌ 52 52	12 00	09 53	08 19	14 58	16 31	21 10	09 D 14	15 40
21	16 03 41	29 ♏ 36 20	01 ♍ 17 31	07 ♍ 36 36	13 30	11 04	09 03	15 09	16 38	21 09	09 14	15 41
22	16 07 37	00 ♐ 36 57	13 ♍ 50 44	20 ♍ 00 34	15 00	12 15	09 48	15 20	16 45	21 07	09 15	15 43
23	16 11 34	01 37 36	26 ♍ 06 43	02 ♎ 09 48	16 29	13 26	10 33	15 31	16 52	21 05	09 15	15 44
24	16 15 30	02 38 16	08 ♎ 10 23	14 ♎ 09 00	17 58	14 37	11 17	15 42	16 59	21 03	09 15	15 46
25	16 19 27	03 38 58	20 ♎ 06 08	26 ♎ 02 15	19 26	15 48	12 02	15 52	17 06	21 02	09 15	15 48
26	16 23 23	04 39 42	01 ♏ 57 45	07 ♏ 52 57	20 54	16 59	12 47	16 03	17 13	21 00	09 15	15 49
27	16 27 20	05 40 27	13 ♏ 48 10	19 ♏ 43 40	22 22	18 10	13 32	16 13	17 20	20 59	09 15	15 51
28	16 31 17	06 41 13	25 ♏ 39 38	01 ♐ 36 16	23 49	19 20	14 17	16 24	17 27	20 57	09 16	15 53
29	16 35 13	07 42 01	07 ♐ 33 44	13 ♐ 32 10	25 15	20 31	15 01	16 34	17 34	20 56	09 16	15 54
30	16 39 10	08 ♐ 42 50	19 ♐ 31 41	25 ♐ 32 26	26 ♐ 40	21 ♑ 41	15 ♒ 46	16 ♎ 44	17 ♐ 41	20 ♈ 54	09 ♓ 16	15 ♑ 56

Date	Moon True ☊ ° '	Moon Mean ☊ ° '	Moon ☽ Latitude ° '
01	10 ♍ 55	09 ♍ 26	04 N 57
02	10 R 43	09 23	05 04
03	10 33	09 20	04 57
04	10 24	09 16	04 38
05	10 18	09 13	04 05
06	10 15	09 10	03 21
07	10 13	09 07	02 26
08	10 D 13	09 04	01 22
09	10 14	09 00	00 N 12
10	10 R 13	08 57	01 S 01
11	10 11	08 54	02 13
12	10 06	08 51	03 18
13	09 58	08 48	04 11
14	09 48	08 45	04 46
15	09 37	08 41	05 01
16	09 27	08 38	04 54
17	09 18	08 35	04 28
18	09 11	08 32	03 45
19	09 07	08 29	02 50
20	09 05	08 26	01 47
21	09 D 05	08 22	00 S 41
22	09 R 05	08 19	00 N 25
23	09 04	08 16	01 29
24	09 01	08 13	02 27
25	08 55	08 10	03 18
26	08 46	08 06	04 01
27	08 35	08 03	04 32
28	08 21	08 00	04 52
29	08 05	07 57	04 59
30	07 ♍ 50	07 ♍ 54	04 N 53

DECLINATIONS

Date	Sun ☉ ° '	Moon ☽ ° '	Mercury ☿ ° '	Venus ♀ ° '	Mars ♂ ° '	Jupiter ♃ ° '	Saturn ♄ ° '	Uranus ♅ ° '	Neptune ♆ ° '	Pluto ♇ ° '
01	14 S 39	15 S 02	15 S 37	24 S 33	23 S 09	03 S 26	21 S 11	07 N 55	08 S 53	21 S 26
02	14 58	17 01	16 13	24 42	23 00	03 31	21 12	07 54	08 53	21 26
03	15 17	18 17	16 47	24 50	22 51	03 35	21 12	07 53	08 53	21 26
04	15 35	18 44	17 21	24 58	22 41	03 40	21 13	07 52	08 53	21 26
05	15 53	18 21	17 54	25 05	22 32	03 44	21 14	07 52	08 53	21 26
06	16 11	17 05	18 26	25 11	22 22	03 49	21 15	07 51	08 53	21 26
07	16 29	14 57	18 57	25 17	22 12	03 54	21 16	07 50	08 54	21 26
08	16 46	12 01	19 27	25 22	22 02	03 58	21 17	07 49	08 54	21 26
09	17 03	08 23	19 56	25 26	21 52	04 02	21 17	07 48	08 54	21 26
10	17 20	04 S 10	20 24	25 29	21 41	04 07	21 18	07 48	08 54	21 26
11	17 37	00 N 25	20 51	25 32	21 30	04 11	21 19	07 47	08 54	21 26
12	17 53	05 07	21 17	25 34	21 19	04 16	21 20	07 46	08 54	21 26
13	18 09	09 36	21 42	25 35	21 08	04 20	21 21	07 45	08 54	21 26
14	18 24	13 30	22 06	25 35	20 57	04 24	21 21	07 45	08 54	21 26
15	18 39	16 29	22 29	25 35	20 45	04 29	21 22	07 44	08 54	21 26
16	18 54	18 18	22 51	25 34	20 33	04 33	21 23	07 43	08 54	21 26
17	19 09	18 48	23 12	25 32	20 21	04 37	21 24	07 43	08 54	21 26
18	19 23	18 04	23 31	25 30	20 09	04 41	21 25	07 42	08 54	21 26
19	19 37	16 16	23 50	25 27	19 57	04 46	21 25	07 41	08 54	21 26
20	19 51	13 37	24 07	25 23	19 44	04 50	21 26	07 41	08 54	21 26
21	20 04	10 22	24 23	25 18	19 31	04 54	21 27	07 40	08 54	21 26
22	20 17	06 44	24 37	25 13	19 18	04 58	21 28	07 39	08 54	21 26
23	20 29	02 N 54	24 51	25 07	19 05	05 02	21 29	07 39	08 54	21 26
24	20 41	00 S 59	25 03	25 00	18 52	05 06	21 29	07 38	08 54	21 26
25	20 53	04 47	25 13	24 52	18 38	05 10	21 30	07 38	08 54	21 26
26	21 04	08 23	25 23	24 44	18 25	05 14	21 31	07 37	08 54	21 26
27	21 15	11 39	25 31	24 35	18 11	05 18	21 32	07 36	08 54	21 25
28	21 25	14 26	25 38	24 26	17 57	05 22	21 32	07 36	08 54	21 25
29	21 35	16 39	25 43	24 15	17 43	05 25	21 33	07 35	08 54	21 25
30	21 S 45	18 S 09	25 S 47	24 S 04	17 S 28	05 S 29	21 S 34	07 N 35	08 S 54	21 S 25

ZODIAC SIGN ENTRIES

Date	h	m	Planets
01	14	43	☽ ♐
04	03	05	☽ ♑
06	13	55	☽ ♒
08	21	45	☽ ♓
09	05	51	♂ ♒
11	01	45	☽ ♈
12	04	54	♀ ♑
12	14	40	☿ ♐
13	02	24	☽ ♉
15	01	23	☽ ♊
17	00	57	☽ ♋
19	03	14	☽ ♌
21	09	34	☽ ♍
21	21	22	☉ ♐
23	19	42	☽ ♎
26	08	01	☽ ♏
28	20	46	☽ ♐

LATITUDES

Date	Mercury ☿ ° '	Venus ♀ ° '	Mars ♂ ° '	Jupiter ♃ ° '	Saturn ♄ ° '	Uranus ♅ ° '	Neptune ♆ ° '	Pluto ♇ ° '
01	00 S 02	01 S 44	01 S 56	01 N 07	01 N 21	00 S 38	00 S 52	01 N 08
04	00 22	01 52	01 53	01 08	01 21	00 38	00 52	01 07
07	00 41	01 58	01 50	01 08	01 21	00 38	00 52	01 07
10	01 00	02 04	01 47	01 08	01 20	00 38	00 52	01 07
13	01 18	02 09	01 44	01 09	01 20	00 38	00 52	01 06
16	01 35	02 14	01 41	01 09	01 20	00 38	00 52	01 06
19	01 50	02 18	01 37	01 09	01 19	00 38	00 52	01 06
22	02 03	02 21	01 34	01 10	01 19	00 38	00 52	01 05
25	02 13	02 23	01 31	01 10	01 19	00 37	00 52	01 05
28	02 20	02 25	01 28	01 11	01 19	00 37	00 52	01 04
31	02 S 24	02 S 25	01 S 25	01 N 11	01 N 18	00 S 37	00 S 52	01 N 04

DATA

Julian Date	2457694
Delta T	+69 seconds
Ayanamsa	24° 05' 24"
Synetic vernal point	05° ♓ 01' 35"
True obliquity of ecliptic	23° 26' 05"

MOON'S PHASES, APSIDES AND POSITIONS ☽

Date	h	m	Phase	Longitude ° '	Eclipse Indicator
07	19	51	☽	15 ♒ 50	
14	13	52	○	22 ♉ 38	
21	08	33	☾	29 ♌ 28	
29	12	18	●	07 ♐ 43	

Day	h	m		
14	11	27	Perigee	
27	20	25	Apogee	
04	13	05	Max dec	18° S 44'
11	09	52	0N	
17	09	30	Max dec	18° N 49'
24	05	54	0S	

LONGITUDES

Date	Chiron ⚷ ° '	Ceres ⚳ ° '	Pallas ⚴ ° '	Juno ⚵ ° '	Vesta ⚶ ° '	Black Moon Lilith ⚸ ° '
01	21 ♓ 05	25 ♈ 42	19 ♒ 27	28 ♏ 21	02 ♌ 42	18 ♏ 18
11	20 ♓ 51	23 ♈ 45	20 ♒ 28	01 ♐ 44	04 ♌ 18	19 ♏ 25
21	20 ♓ 43	22 ♈ 15	21 ♒ 54	05 ♐ 09	05 ♌ 20	20 ♏ 32
31	20 ♓ 40	21 ♈ 22	23 ♒ 44	08 ♐ 36	05 ♌ 43	21 ♏ 39

All ephemeris data is given at 12.00 UT and the Moon's longitude is additionally given for 24.00 UT

ASPECTARIAN

h m	Aspects
01 Tuesday	
02 44	☽ ✶ ♂
07 08	☽ ∠ ♃
07 19	☿ Q ♂
07 36	☽ ∥ ☉
08 17	☉ △ ♆
10 20	☽ ± ♅
11 23	♂ ∠ ♆
15 14	☽ ∠ ♇
20 51	☽ ∥ ☿
02 Wednesday	
04 28	☽ bQ ♅
09 27	☽ ⊥ ♇
09 34	☽ □ ♆
11 02	☽ ∠ ♂
11 55	☽ ⊻ ☉
13 58	☽ ✶ ♃
18 29	☿ ⊻ ♄
20 09	☽ ☌ ♄
20 24	☽ ⊻ ☿
21 34	☽ ⊻ ♇
03 Thursday	
01 07	☽ ⊥ ☉
05 10	☿ ✶ ♇
05 32	☽ ☌ ♀
06 20	☽ ⊥ ♂
10 19	☽ ⊥ ☿
10 35	☽ △ ♅
14 31	☽ Q ♃
17 54	☉ ⊻ ♃
19 08	☽ ⊻ ♂
20 50	☽ ∠ ☉
21 43	☽ Q ♆
04 Friday	
00 23	♀ ⊥ ♂
07 04	☽ ∠ ☿
21 34	☽ ✶ ♆
21 52	☿ ⊥ ♃
05 Saturday	
02 56	☽ □ ♃
05 18	☽ ✶ ☉
06 23	♀ △ ♅
08 29	☽ ⊻ ♄
09 25	☽ ☌ ♇
17 06	☽ ✶ ☿
19 09	☽ ∥ ☿
20 19	☽ ⊥ ♄
21 48	☽ □ ♅
23 31	☽ ⊻ ♀
06 Sunday	
02 55	☽ ∠ ♆
06 47	☽ Q ☉
09 56	☽ ☌ ♂
12 21	☽ ⊥ ♀
13 52	☽ ∠ ♄
19 38	☽ Q ☿
20 13	☽ ⊥ ♆
21 49	☽ ∥ ☉
07 Monday	
00 15	☉ ⊻ ♄
05 11	☿ ⚻ ♅
07 18	☽ ∠ ♀
07 35	☽ ⊻ ♆
08 12	☽ Q ♅
08 25	☉ ✶ ♇
13 38	☽ △ ♃
15 02	♀ Q ♃
18 29	☽ ✶ ♄
18 59	☽ ⊻ ♇
19 51	☽ □ ☉
08 Tuesday	
06 05	☽ ⊥ ♇
06 27	☽ ✶ ♅
09 57	☽ □ ☿
13 54	☽ ✶ ♀
16 42	☽ Q ♄
17 40	☽ bQ ♃
21 17	☽ ⊻ ♂
22 28	☽ ∠ ♇
09 Wednesday	
08 37	☽ ⊥ ♂
08 51	☽ ∥ ♆
09 25	☽ ∠ ♅
10 00	☽ ± ♃
13 28	☽ Q ♀
14 14	☽ ☌ ♆
15 28	☽ # ♅
15 53	♂ ∠ ♄
20 41	☽ ⚻ ♃
22 09	♀ Q ♆
10 Thursday	
00 50	☽ □ ♄
00 57	☽ ✶ ♇
01 07	☽ ⊥ ♅
01 15	☽ ∠ ♂
05 52	☽ △ ☉
11 24	☽ ⊻ ♅
12 18	☽ ∥ ♃
19 39	♄ ⊻ ♇
21 17	☽ △ ☿
21 27	☽ Q ♇
23 16	☽ □ ♀
23 30	☿ ± ♅
11 Friday	
04 06	☽ ✶ ♂
04 31	☉ ⊥ ♃
09 04	☽ bQ ☉
11 48	☿ ∠ ♃
17 03	☽ ⊻ ♆
20 44	♂ ∥ ♇
23 20	☉ Q ♂
23 46	☽ ☍ ♃
12 Saturday	
00 52	☽ Q ♂
00 54	☽ ± ☉
00 59	☽ bQ ☿
02 49	☽ ⊥ ♆
03 09	☽ □ ♇
03 20	☽ △ ♄
07 33	☽ # ♃
10 59	♂ ∥ ♄
11 16	☽ ⚻ ☉
12 45	☽ ☌ ♅
13 17	☿ ∥ ♂
14 24	☿ ∥ ♄
17 07	☽ ± ☿
17 16	☽ ∠ ♆
20 27	☿ ∥ ♇
22 08	☿ ∠ ♇
13 Sunday	
01 55	☽ ∥ ♅
03 29	☽ bQ ♄
03 43	☽ ⚻ ☿
04 14	☽ △ ♀
07 06	☽ □ ♂
08 07	☽ # ♆
08 38	☉ ⚻ ♅
16 03	♂ ⊥ ♆
16 58	☽ ✶ ♆
17 45	☽ ± ♄
14 Monday	
00 01	☽ ⚻ ♃
00 43	☿ ⊻ ♀
02 45	☽ △ ♇
03 13	☽ ⚻ ♄
05 46	☽ bQ ♀
09 30	☽ ± ♃
11 43	☽ Q ♆
11 54	☽ ⊻ ♅
13 52	☽ ☍ ☉
20 59	☽ ± ♀
21 14	☽ ⊥ ♅
23 44	☽ bQ ♃
15 Tuesday	
02 13	☽ bQ ♇
07 10	☽ ⚻ ♀
07 54	☽ ☍ ☿
08 25	☽ △ ♂
11 18	☽ ∠ ♅
15 52	☽ □ ♆
16 20	☽ ± ♇
17 22	☿ ✶ ♂
23 35	☽ △ ♃
16 Wednesday	
01 49	☽ ⚻ ♇
02 34	☽ ☍ ♄
09 18	☽ bQ ♂
10 57	☽ ✶ ♅
16 29	☽ ⚻ ☉
17 19	☿ bQ ♅
23 38	♀ ⊻ ♂
17 Thursday	
00 31	☽ ± ♂
02 52	☽ ± ☉
06 16	☽ Q ♅
10 48	☽ ⚻ ♂
11 11	☽ ☍ ♀
13 17	☽ ⚻ ☿
16 04	☽ △ ♆
18 44	☽ bQ ☉
18 Friday	
00 24	☽ ± ☿
00 50	☽ □ ♃
02 39	☽ ☍ ♇
03 45	☽ ⚻ ♄
12 10	☽ □ ♅
13 59	☽ ± ♄
16 04	☿ □ ♆
17 18	☽ bQ ♆
17 27	☽ bQ ☿
22 02	☽ △ ☉
22 06	☿ ⊥ ♇
19 Saturday	
03 09	☉ ± ♅
05 38	☽ bQ ♄
08 04	☽ Q ♃
08 54	☽ ± ♆
16 48	☽ ☍ ♂
19 11	☽ ⚻ ♀
19 31	☽ ⚻ ♆
23 00	☽ △ ☿
23 03	♀ ✶ ♆
20 Sunday	
04 38	♆ St D
05 05	♄ ∥ ♇
05 44	☽ ✶ ♃
06 03	☉ ∥ ♂
07 01	☽ ± ♀
07 04	☽ ⚻ ♇
08 36	☽ △ ♄
17 08	☽ △ ♅
18 03	☽ ± ♇
21 Monday	
01 13	☽ bQ ♀
08 33	☽ □ ☉
09 50	☽ ∠ ♃
10 51	☽ bQ ♇
14 48	♂ Q ♅
18 06	♂ ⊻ ♆
21 11	☽ bQ ♅
21 56	☽ # ♆
23 19	♃ ± ♆
22 Tuesday	
03 08	☽ ☍ ♆
03 11	☽ ⊥ ♃
03 42	☽ ⚻ ♂
03 57	☉ ∠ ♃
06 06	☽ ∥ ♅
08 36	☽ △ ♀
14 22	☉ ∠ ♇
14 27	☽ ± ♅
14 32	☽ □ ☿
14 56	☽ ⊻ ♃
15 38	☽ △ ♇
16 02	☽ ± ♂
17 41	☽ □ ♄
18 19	☿ ✶ ♃
22 06	☽ Q ☉
23 01	☽ # ♃
23 48	☿ ⊻ ♇
23 Wednesday	
02 08	☽ ⚻ ♅
10 48	☽ bQ ♂
18 44	☿ ☌ ♄
23 56	☽ ✶ ☉
24 Thursday	
05 33	☽ Q ♄
06 57	☽ Q ☿
08 18	☿ ∥ ♀
14 09	☽ ⚻ ♆
18 40	☽ △ ♂
23 00	♃ □ ♇
25 Friday	
02 13	☽ ± ♆
02 22	☽ □ ♀
03 17	☽ □ ♇
03 21	☽ ☌ ♃
05 52	☽ ✶ ♄
08 48	☽ ∠ ☉
10 28	☽ ✶ ☿
11 50	♀ ☌ ♇
13 45	♀ □ ♃
13 52	☽ ☍ ♅
14 28	☽ ∥ ♃
20 23	☽ bQ ♆
26 Saturday	
04 41	☽ ⊥ ☉
06 43	☽ # ♅
12 31	☽ ∠ ♄
13 38	☿ △ ♅
15 36	☽ ∥ ♆
15 46	☽ Q ♇
17 13	♀ ⊻ ♄
17 59	☽ ⊻ ☉
18 48	☽ Q ♀
21 07	☽ ∠ ☿
27 Sunday	
02 47	☽ △ ♆
06 56	☽ ⊥ ♄
11 24	☽ □ ♂
16 09	☽ ✶ ♇
16 59	☽ ⊻ ♃
17 54	☽ ⊥ ☿
19 03	☉ bQ ♅
19 13	☽ ⊻ ♄
21 48	☽ ✶ ♀
28 Monday	
02 30	☽ ⚻ ♅
05 18	☽ ⊥ ♃
07 44	☽ ⊻ ☿
12 15	☉ ∥ ♇
14 36	☽ ± ♅
22 33	☽ ∠ ♇
23 45	☽ ∠ ♃
29 Tuesday	
02 15	☽ Q ♂
05 46	☉ ∥ ♄
07 26	☽ ∠ ♀
08 43	☽ bQ ♅
12 18	☽ ☌ ☉
15 26	☽ □ ♆
16 43	☽ ⊥ ♇
20 20	♀ □ ♅
30 Wednesday	
01 21	☽ ∥ ♂
03 29	☽ ⊥ ♀
03 59	☽ ✶ ♂
04 47	☽ ⊻ ♇
06 20	☽ ✶ ♃
08 16	☽ ☌ ♄
14 45	☽ △ ♅
16 47	☽ ⊻ ♀
17 15	♂ ⊻ ♇
22 18	☿ Q ♆

DECEMBER 2016

LONGITUDES

Date	Sidereal time h m s	Sun ☉ ° ' "	Moon ☽ ° ' "	Moon ☽ 24.00 ° ' "	Mercury ☿ ° '	Venus ♀ ° '	Mars ♂ ° '	Jupiter ♃ ° '	Saturn ♄ ° '	Uranus ♅ ° '	Neptune ♆ ° '	Pluto ♇ ° '
01	16 43 06	09 ♐ 43 41	01 ♑ 34 34	07 ♑ 38 14	28 ♐ 05	22 ♑ 52	16 ♒ 31	16 ♎ 54	17 ♐ 48	20 ♈ 53	09 ♓ 17	15 ♑ 58
02	16 47 03	10 44 32	13 ♑ 43 38	19 ♑ 50 59	29 ♐ 28	24 02	17 16	17 04	17 55	20 R 52	09 17	15 59
03	16 50 59	11 45 24	26 ♑ 00 33	02 ♒ 12 39	00 ♑ 50	25 12	18 02	17 14	18 02	20 50	09 18	16 01
04	16 54 56	12 46 18	08 ♒ 27 36	14 ♒ 45 48	02 11	26 22	18 47	17 24	18 09	20 49	09 18	16 03
05	16 58 52	13 47 12	21 ♒ 07 40	27 ♒ 33 38	03 30	27 32	19 32	17 34	18 16	20 48	09 19	16 05
06	17 02 49	14 48 06	04 ♓ 04 49	10 ♓ 39 41	04 48	28 42	20 17	17 44	18 23	20 47	09 19	16 07
07	17 06 46	15 49 02	17 ♓ 20 38	24 ♓ 07 25	06 03	29 ♑ 52	21 02	17 53	18 30	20 46	09 20	16 08
08	17 10 42	16 49 58	01 ♈ 00 19	07 ♈ 59 32	07 15	01 ♒ 01	21 47	18 02	18 37	20 45	09 20	16 10
09	17 14 39	17 50 55	15 ♈ 05 10	22 ♈ 17 05	08 25	02 11	22 32	18 12	18 44	20 44	09 21	16 12
10	17 18 35	18 51 52	29 ♈ 35 01	06 ♉ 58 29	09 31	03 20	23 18	18 21	18 52	20 43	09 22	16 14
11	17 22 32	19 52 50	14 ♉ 26 46	21 ♉ 58 55	10 33	04 29	24 03	18 30	18 59	20 42	09 22	16 16
12	17 26 28	20 53 49	29 ♉ 33 50	07 ♊ 10 15	11 31	05 38	24 48	18 39	19 06	20 41	09 23	16 18
13	17 30 25	21 54 49	14 ♊ 46 47	22 ♊ 22 03	12 24	06 47	25 33	18 48	19 13	20 40	09 24	16 20
14	17 34 21	22 55 49	29 ♊ 54 40	07 ♋ 23 21	13 11	07 56	26 19	18 57	19 20	20 39	09 25	16 22
15	17 38 18	23 56 50	14 ♋ 47 00	22 ♋ 04 40	13 51	09 05	27 04	19 05	19 27	20 38	09 26	16 23
16	17 42 14	24 57 52	29 ♋ 15 39	06 ♌ 19 27	14 24	10 13	27 49	19 14	19 34	20 38	09 26	16 25
17	17 46 11	25 58 54	13 ♌ 15 51	20 ♌ 04 47	14 48	11 21	28 34	19 22	19 41	20 37	09 27	16 27
18	17 50 08	26 59 58	26 ♌ 46 22	03 ♍ 20 55	15 03	12 29	29 ♒ 20	19 30	19 48	20 37	09 28	16 29
19	17 54 04	28 01 02	09 ♍ 48 50	16 ♍ 10 39	15 R 08	13 37	00 ♓ 05	19 38	19 55	20 36	09 29	16 31
20	17 58 01	29 ♐ 02 07	22 ♍ 26 55	28 ♍ 38 18	15 02	14 45	00 50	19 46	20 02	20 36	09 30	16 33
21	18 01 57	00 ♑ 03 13	04 ♎ 45 27	10 ♎ 49 00	14 44	15 53	01 36	19 54	20 09	20 35	09 31	16 35
22	18 05 54	01 04 20	16 ♎ 49 38	22 ♎ 48 00	14 15	17 00	02 21	20 02	20 16	20 35	09 32	16 37
23	18 09 50	02 05 27	28 ♎ 44 41	04 ♏ 40 16	13 34	18 07	03 06	20 10	20 23	20 34	09 33	16 39
24	18 13 47	03 06 35	10 ♏ 35 17	16 ♏ 30 14	12 42	19 14	03 51	20 17	20 30	20 34	09 35	16 41
25	18 17 43	04 07 44	22 ♏ 25 31	28 ♏ 21 33	11 40	20 21	04 37	20 24	20 37	20 34	09 36	16 43
26	18 21 40	05 08 53	04 ♐ 18 40	10 ♐ 17 07	10 30	21 28	05 22	20 32	20 44	20 34	09 37	16 45
27	18 25 37	06 10 03	16 ♐ 17 10	22 ♐ 18 59	09 13	22 34	06 07	20 39	20 51	20 34	09 38	16 47
28	18 29 33	07 11 13	28 ♐ 22 43	04 ♑ 28 31	07 52	23 40	06 53	20 46	20 58	20 33	09 39	16 49
29	18 33 30	08 12 23	10 ♑ 36 24	16 ♑ 46 31	06 30	24 46	07 38	20 52	21 05	20 D 33	09 41	16 51
30	18 37 26	09 13 33	22 ♑ 58 55	29 ♑ 13 40	05 09	25 52	08 23	20 59	21 12	20 33	09 42	16 53
31	18 41 23	10 ♑ 14 44	05 ♒ 30 53	11 ♒ 50 37	03 ♑ 53	26 ♒ 57	09 ♓ 09	21 ♎ 06	21 ♐ 19	20 ♈ 34	09 ♓ 43	16 ♑ 56

Date	Moon True ☊ ° '	Moon Mean ☊ ° '	Moon ☽ Latitude ° '
01	07 ♍ 37	07 ♍ 51	04 N 34
02	07 R 25	07 47	04 02
03	07 16	07 44	03 19
04	07 11	07 41	02 25
05	07 08	07 38	01 24
06	07 D 08	07 35	00 N 16
07	07 R 08	07 32	00 S 54
08	07 07	07 28	02 03
09	07 05	07 25	03 06
10	07 00	07 22	04 00
11	06 52	07 19	04 39
12	06 42	07 16	04 59
13	06 31	07 12	04 58
14	06 20	07 09	04 36
15	06 10	07 06	03 56
16	06 03	07 03	03 01
17	05 58	07 00	01 57
18	05 56	06 57	00 S 49
19	05 D 56	06 53	00 N 21
20	05 56	06 50	01 27
21	05 R 56	06 47	02 27
22	05 55	06 44	03 20
23	05 51	06 41	04 03
24	05 45	06 38	04 35
25	05 36	06 34	04 56
26	05 25	06 31	05 03
27	05 14	06 28	04 58
28	05 02	06 25	04 39
29	04 51	06 22	04 07
30	04 41	06 18	03 23
31	04 ♍ 35	06 ♍ 15	02 N 29

DECLINATIONS

Date	Sun ☉ ° '	Moon ☽ ° '	Mercury ☿ ° '	Venus ♀ ° '	Mars ♂ ° '	Jupiter ♃ ° '	Saturn ♄ ° '	Uranus ♅ ° '	Neptune ♆ ° '	Pluto ♇ ° '
01	21 S 54	18 S 52	25 S 49	23 S 53	17 S 14	05 S 33	21 S 34	07 N 34	08 S 53	21 S 25
02	22 03	18 43	25 50	23 41	16 59	05 36	21 35	07 34	08 53	21 25
03	22 12	17 41	25 50	23 28	16 44	05 40	21 36	07 34	08 53	21 25
04	22 19	15 48	25 48	23 14	16 29	05 44	21 37	07 33	08 53	21 25
05	22 27	13 08	25 45	23 00	16 14	05 47	21 37	07 33	08 53	21 25
06	22 34	09 46	25 40	22 45	15 59	05 51	21 38	07 32	08 52	21 25
07	22 41	05 49	25 34	22 30	15 43	05 54	21 39	07 32	08 52	21 25
08	22 47	01 S 29	25 27	22 14	15 28	05 58	21 39	07 32	08 52	21 25
09	22 53	03 N 05	25 18	21 58	15 12	06 01	21 40	07 31	08 52	21 24
10	22 58	07 35	25 08	21 41	14 56	06 04	21 41	07 31	08 52	21 24
11	23 03	11 44	24 57	21 23	14 40	06 08	21 41	07 30	08 51	21 24
12	23 07	15 11	24 45	21 05	14 24	06 11	21 42	07 30	08 51	21 24
13	23 11	17 38	24 32	20 46	14 08	06 14	21 42	07 30	08 50	21 24
14	23 15	18 50	24 18	20 27	13 51	06 17	21 43	07 30	08 50	21 24
15	23 18	18 42	24 03	20 07	13 35	06 20	21 44	07 29	08 50	21 24
16	23 20	17 20	23 48	19 47	13 18	06 23	21 44	07 29	08 49	21 24
17	23 22	14 58	23 32	19 27	13 02	06 26	21 45	07 29	08 49	21 23
18	23 24	11 50	23 16	19 05	12 45	06 29	21 45	07 29	08 49	21 23
19	23 25	08 12	22 59	18 44	12 28	06 32	21 46	07 29	08 48	21 23
20	23 26	04 19	22 43	18 22	12 11	06 35	21 47	07 28	08 48	21 23
21	23 26	00 N 22	22 27	17 59	11 54	06 38	21 47	07 28	08 48	21 23
22	23 26	03 S 32	22 11	17 37	11 37	06 40	21 48	07 28	08 47	21 23
23	23 25	07 14	21 56	17 13	11 19	06 43	21 48	07 28	08 47	21 22
24	23 24	10 38	21 41	16 50	11 02	06 46	21 49	07 28	08 46	21 22
25	23 22	13 36	21 27	16 26	10 44	06 48	21 49	07 28	08 46	21 22
26	23 20	16 02	21 14	16 01	10 27	06 51	21 50	07 28	08 45	21 22
27	23 17	17 47	21 02	15 37	10 09	06 53	21 50	07 28	08 45	21 22
28	23 14	18 46	20 51	15 12	09 51	06 56	21 51	07 28	08 44	21 22
29	23 11	18 54	20 40	14 46	09 33	06 58	21 51	07 28	08 44	21 22
30	23 07	18 08	20 32	14 21	09 16	07 00	21 52	07 28	08 44	21 22
31	23 S 02	16 S 29	20 S 24	13 S 55	08 S 58	07 S 03	21 S 52	07 N 28	08 S 43	21 S 21

ZODIAC SIGN ENTRIES

Date	h	m	Planets
01	08	52	☽ ♑
02	21	18	☿ ♑
03	19	44	☽ ♒
06	04	31	☽ ♓
07	14	51	♀ ♒
08	10	16	☽ ♈
10	12	41	☽ ♉
12	12	41	☽ ♊
14	12	09	☽ ♋
16	13	15	☽ ♌
18	17	52	☽ ♍
19	09	23	♂ ♓
21	02	40	☽ ♎
21	10	44	☉ ♑
23	14	32	☽ ♏
26	03	19	☽ ♐
28	15	12	☽ ♑
31	01	29	☽ ♒

LATITUDES

Date	Mercury ☿ ° '	Venus ♀ ° '	Mars ♂ ° '	Jupiter ♃ ° '	Saturn ♄ ° '	Uranus ♅ ° '	Neptune ♆ ° '	Pluto ♇ ° '
01	02 S 24	02 S 25	01 S 25	01 N 11	01 N 18	00 S 37	00 S 52	01 N 04
04	02 23	02 24	01 22	01 12	01 18	00 37	00 52	01 04
07	02 16	02 23	01 18	01 12	01 18	00 37	00 52	01 03
10	02 03	02 20	01 15	01 13	01 18	00 37	00 52	01 03
13	01 41	02 16	01 12	01 13	01 18	00 37	00 52	01 03
16	01 09	02 11	01 09	01 14	01 17	00 37	00 52	01 02
19	00 S 25	02 05	01 06	01 14	01 17	00 37	00 51	01 02
22	00 N 29	01 58	01 03	01 15	01 17	00 37	00 51	01 02
25	01 28	01 49	00 59	01 15	01 17	00 36	00 51	01 02
28	02 22	01 40	00 56	01 16	01 17	00 36	00 51	01 01
31	02 N 58	01 S 29	00 S 53	01 N 17	01 N 17	00 S 36	00 S 51	01 N 01

DATA

Julian Date	2457724
Delta T	+69 seconds
Ayanamsa	24° 05' 29"
Synetic vernal point	05° ♓ 01' 30"
True obliquity of ecliptic	23° 26' 04"

MOON'S PHASES, APSIDES AND POSITIONS ☽

Date	h	m	Phase	Longitude ° '	Eclipse Indicator
07	09	03	☽	15 ♓ 42	
14	00	06	○	22 ♊ 26	
21	01	56	☾	29 ♍ 38	
29	06	53	●	07 ♑ 59	

Day	h	m		
12	23	35	Perigee	
25	06	05	Apogee	
01	19	57	Max dec	18° S 54'
08	19	51	0N	
14	21	38	Max dec	18° N 56'
21	14	11	0S	
29	03	32	Max dec	18° S 58'

LONGITUDES

Date	Chiron ⚷ ° '	Ceres ⚳ ° '	Pallas ⚴ ° '	Juno ⚵ ° '	Vesta ⚶ ° '	Black Moon Lilith ⚸ ° '
01	20 ♓ 40	21 ♈ 22	23 ♒ 44	08 ♐ 36	05 ♌ 43	21 ♏ 39
11	20 ♓ 43	21 ♈ 08	25 ♒ 53	12 ♐ 04	05 ♌ 21	22 ♏ 46
21	20 ♓ 52	21 ♈ 33	28 ♒ 18	15 ♐ 31	04 ♌ 14	23 ♏ 53
31	21 ♓ 05	22 ♈ 34	00 ♓ 57	18 ♐ 56	02 ♌ 25	25 ♏ 01

ASPECTARIAN

h m	Aspects
01 Thursday	
01 17	☉ □ ♆
03 27	☽ Q ♆
04 08	☽ ☌ ☿
06 38	☽ Q ♃
11 53	☽ ∠ ♂
17 41	☉ ⊥ ♇
02 Friday	
03 15	☽ ⚹ ♆
03 46	♂ △ ♃
04 16	☿ Q ♃
05 36	☽ ⊻ ☉
06 51	☽ ⊥ ♂
09 19	♀ ⊥ ♄
16 27	☽ ☌ ♇
17 13	♀ ∠ ♆
18 27	☽ ⊥ ☉
18 40	☽ □ ♃
19 25	☽ ⊻ ♂
20 18	☽ ⊻ ♄
03 Saturday	
01 57	☽ □ ♅
08 07	☽ ⊥ ♄
08 40	☽ ∠ ♆
10 16	☽ ☌ ♀
12 15	♂ ⚹ ♄
13 35	☽ ∠ ☉
22 30	☽ ⊻ ☿
04 Sunday	
01 43	☽ ∠ ♄
02 06	☽ ⊥ ♆
03 24	☽ ∥ ♂
11 25	☽ ⊥ ☿
12 41	☽ Q ♅
13 36	☽ ⊻ ♆
20 56	☽ ⚹ ☉
05 Monday	
02 28	☽ ⊻ ♇
05 13	☽ △ ♃
06 30	☽ ∠ ☿
06 34	☽ ⚹ ♄
08 49	☽ ☌ ♂
11 23	☽ ⚹ ♅
13 47	☽ ⊥ ♇
21 27	☽ Q ☉
06 Tuesday	
01 08	☽ ⊻ ♀
05 10	☽ Q ♄
06 33	☽ ∠ ♇
09 30	☽ bQ ♃
13 16	☽ ⊥ ♀
13 28	☽ ⚹ ☿
15 07	☽ ∠ ♅
17 43	☽ ∥ ♆
21 34	☽ ☌ ♆
07 Wednesday	
00 21	☉ ∥ ♀
01 57	☽ # ♅
02 05	☽ ± ♃
03 30	♂ ⚹ ♅
07 08	☽ ∠ ♀
07 23	☽ ⊥ ♅
09 03	☽ □ ☉
09 51	☽ ⚹ ♇
11 30	☿ ∠ ♂
11 33	☽ ∥ ♃
12 58	☽ ⊼ ♃
13 22	☽ Q ☿
14 05	☽ □ ♄
18 04	☽ ⊻ ♅
18 56	☽ ⊻ ♂
19 51	☉ ⊻ ♇
08 Thursday	
06 05	☽ ⊥ ♂
07 04	☽ Q ♇
12 02	☽ ⚹ ♀
22 30	☽ ∠ ♂
23 43	☽ □ ☿
09 Friday	
00 45	♂ ⊥ ♇
02 18	☽ ⊻ ♆
10 21	☽ Q ♀
12 26	☽ ⊥ ♆
13 52	☽ □ ♇
16 59	☽ △ ☉
17 16	☽ ☍ ♃
18 10	☽ △ ♄
21 24	☽ ☌ ♅
21 39	☉ ⚹ ♃
10 Saturday	
01 06	☽ ⚹ ♂
03 25	☽ ∠ ♆
03 44	☽ # ♃
08 25	☿ ⚹ ♆
11 38	☽ ∥ ♅
11 51	☉ ☌ ♄
12 11	♀ ∥ ♄
12 30	♀ ⊥ ♆
18 38	☽ □ ♀
19 01	☽ bQ ♄
19 04	☽ # ♆
19 29	☽ bQ ☉
21 47	☽ Q ♂
11 Sunday	
00 08	♀ ∠ ♄
03 51	☽ ⚹ ♆
05 19	☽ △ ☿
09 38	☽ ± ♄
10 36	♀ ∥ ♇
11 02	☽ ± ☉
14 55	☽ △ ♇
18 32	☽ ⊼ ♃
19 17	☽ ⊼ ♄
21 17	☽ ⊼ ☉
21 57	☽ ⊻ ♅
23 02	☽ Q ♆
12 Monday	
04 04	☽ □ ♂
04 09	☽ ± ♃
06 25	☽ # ♂
06 53	☽ bQ ☿
06 56	☉ △ ♅
07 27	☽ ⊥ ♅
14 44	☽ bQ ♇
18 31	☽ bQ ♃
21 38	☽ ∠ ♅
22 00	☽ ± ☿
22 22	☽ △ ♀
13 Tuesday	
03 30	☽ □ ♆
04 58	☽ ± ♇
08 02	☽ ⊼ ☿
14 27	☽ ⊼ ♇
18 24	☽ △ ♃
19 04	☽ ☍ ♄
21 18	☽ ⚹ ♅
23 59	☽ bQ ♀
14 Wednesday	
00 06	☽ ☍ ☉
05 57	☽ △ ♂
11 13	☉ ∠ ♀
15 30	☽ ± ♀
16 23	☽ Q ♅
15 Thursday	
01 57	☽ ⊼ ♀
02 55	♀ Q ♅
03 17	☽ △ ♆
07 20	☽ bQ ♂
10 25	☽ ☍ ☿
14 38	☽ ☍ ♇
19 08	☽ □ ♃
19 24	♀ ⊻ ♆
19 43	☽ ⊼ ♄
21 37	☽ □ ♅
22 53	☽ ± ♂
16 Friday	
03 55	☽ bQ ♆
04 15	☽ ⊼ ☉
05 45	☽ ± ♄
09 26	☽ ⊼ ♂
15 06	☽ ± ☉
19 05	☽ ± ♆
21 04	☽ bQ ♄
17 Saturday	
01 41	☽ Q ♃
05 23	☽ ⊼ ♆
07 43	☽ bQ ☉
08 23	☽ ☍ ♀
14 45	☽ ⊼ ☿
17 36	☽ ⊼ ♇
22 51	☽ ⚹ ♃
23 24	☽ △ ♄
18 Sunday	
00 41	☉ ∥ ☿
00 57	☽ △ ♅
01 34	☽ ± ☿
04 17	☽ ± ♇
04 47	☽ # ♂
12 27	☽ △ ☉
16 55	☽ ☍ ♂
18 01	☽ bQ ☿
20 36	☽ bQ ♇
23 18	☉ Q ♆
19 Monday	
02 17	☽ ∠ ♃
04 10	☽ bQ ♅
08 11	☽ # ♆
10 54	☿ St R
11 23	☽ ☍ ♆
13 31	☿ ∠ ♂
16 36	☽ ∥ ♅
19 16	☽ ⊥ ♃
19 52	☽ ⊼ ♀
21 00	☽ ± ♅
21 59	☽ △ ☿
22 20	☽ # ♃
20 Tuesday	
00 41	☽ △ ♇
06 48	☽ ⊻ ♃
07 19	☽ □ ♄
08 25	☽ ± ♀
08 26	☽ ⊼ ♅
16 58	☿ ⊻ ♀
21 Wednesday	
01 56	☽ □ ☉
03 37	☽ bQ ♀
05 22	☽ ⊼ ♂
11 50	♂ ∠ ♇
17 58	☽ ± ♂
18 47	☽ Q ♄
21 26	☽ ⊼ ♆
22 Thursday	
03 36	♀ ⊻ ♇
07 05	☽ □ ☿
09 10	♂ Q ♄
09 25	☽ ± ♆
11 35	☽ □ ♇
12 23	☽ △ ♀
13 07	☽ bQ ♂
16 55	☽ Q ☉
18 30	☽ ☌ ♃
18 59	☽ ⚹ ♄
19 31	☽ ☍ ♅
23 Friday	
03 32	☽ bQ ♆
08 28	☽ ∥ ♃
13 32	☽ # ♅
13 53	☉ Q ♃
17 21	☽ Q ☿
19 25	☽ ⚹ ☉
21 25	☽ △ ♂
22 31	☽ ∥ ♆
24 Saturday	
00 00	☽ Q ♇
00 25	☿ ∥ ♄
01 35	☽ ∠ ♄
05 44	☿ ⊥ ♀
09 57	☽ △ ♆
14 43	☽ ∥ ♂
15 58	☽ ⚹ ☿
20 02	☽ ⊥ ♄
25 Sunday	
00 21	♄ △ ♅
00 25	☽ ⚹ ♇
04 42	☽ ∠ ☉
07 22	☽ □ ♀
07 52	☽ ⊻ ♃
08 14	☽ ⊼ ♅
08 19	☽ ⊻ ♄
13 23	♀ △ ♃
16 35	♀ ⚹ ♅
18 33	♀ ⚹ ♄
19 50	☽ ∠ ☿
20 08	☽ ⊥ ♃
20 22	☽ ± ♅
21 26	☿ ∥ ♇
26 Monday	
00 37	☽ ⊥ ☉
06 50	☽ ∠ ♇
11 57	☽ ∥ ♀
12 20	☽ ⊥ ☿
13 50	☽ ⊻ ☉
14 16	☽ □ ♂
14 28	☽ ∠ ♃
14 31	☽ bQ ♅
17 59	♂ bQ ♃
18 04	♂ ∠ ♅
18 35	♃ ☍ ♅
22 41	☽ □ ♆
23 14	☽ ⊻ ☿
23 24	☽ Q ♀
27 Tuesday	
00 59	☽ ⊥ ♇
04 19	☿ ⚹ ♆
08 07	☉ ⚹ ♂
13 00	☽ ⊻ ♇
17 01	♀ ⊥ ♇
20 30	☽ △ ♅
20 46	☽ ⚹ ♃
21 11	☽ ☌ ♄
28 Wednesday	
01 45	☽ ⚹ ♀
04 09	☿ ∠ ♀
04 37	☽ Q ♂
10 34	☽ Q ♆
18 47	☉ ☌ ☿
20 43	☽ Q ♃
23 06	☿ ⚹ ♂
29 Thursday	
04 46	☽ ☌ ☿
05 49	☽ ⚹ ♂
06 53	☽ ☌ ☉
09 29	♅ St D
10 11	☽ ⚹ ♆
10 12	☽ ∠ ♀
30 Friday	
00 12	☽ ☌ ♇
05 23	☽ ⊥ ♀
07 19	☽ □ ♅
08 07	☽ □ ♃
08 32	☽ ⊻ ♄
12 50	☽ ∠ ♂
15 19	☽ ∠ ♆
18 04	☽ ⊻ ♀
20 11	☽ ⊥ ♄
23 25	☉ ⚹ ♆
31 Saturday	
07 12	☽ ⊥ ♂
08 35	☽ ⊥ ♆
09 10	☽ ⊻ ☿
13 32	☽ ∠ ♄
17 47	☽ Q ♅
19 20	☽ ⊻ ♂
19 33	☽ ⊥ ☿
20 00	☽ ⊻ ♆
21 46	☽ ⊻ ☉

All ephemeris data is given at 12.00 UT and the Moon's longitude is additionally given for 24.00 UT

JANUARY 2017

LONGITUDES

Date	Sidereal time h m s	Sun ☉ ° ′ ″	Moon ☽ ° ′ ″	Moon ☽ 24.00 ° ′ ″	Mercury ☿ ° ′	Venus ♀ ° ′	Mars ♂ ° ′	Jupiter ♃ ° ′	Saturn ♄ ° ′	Uranus ♅ ° ′	Neptune ♆ ° ′	Pluto ♇ ° ′
01	18 45 19	11 ♑ 15 54	18 ♒ 13 02	24 ♒ 38 16	02 ♑ 42	28 ♒ 02	09 ♓ 54	21 ♎ 12	21 ♐ 26	20 ♈ 34	09 ♓ 45	16 ♑ 58
02	18 49 16	12 17 05	01 ♓ 06 31	07 ♓ 37 58	01 R 40	29 ♒ 07	10 39	21 18	21 32	20 34	09 46	17 00
03	18 53 12	13 18 15	14 ♓ 12 51	20 ♓ 51 25	00 47	00 ♓ 11	11 25	21 24	21 39	20 34	09 48	17 02
04	18 57 09	14 19 25	27 ♓ 33 54	04 ♈ 20 31	00 ♑ 04	01 16	12 10	21 30	21 46	20 34	09 49	17 04
05	19 01 06	15 20 34	11 ♈ 11 29	18 ♈ 06 55	29 ♐ 31	02 19	12 55	21 36	21 53	20 35	09 51	17 06
06	19 05 02	16 21 43	25 ♈ 06 54	02 ♉ 11 25	29 08	03 23	13 41	21 41	21 59	20 35	09 52	17 08
07	19 08 59	17 22 52	09 ♉ 20 18	16 ♉ 33 17	28 55	04 26	14 26	21 47	22 06	20 36	09 54	17 10
08	19 12 55	18 24 00	23 ♉ 49 58	01 ♊ 09 44	28 D 51	05 29	15 11	21 52	22 13	20 36	09 55	17 12
09	19 16 52	19 25 08	08 ♊ 31 55	15 ♊ 55 40	28 56	06 32	15 56	21 57	22 19	20 37	09 57	17 14
10	19 20 48	20 26 15	23 ♊ 20 00	00 ♋ 43 57	29 09	07 34	16 42	22 02	22 26	20 37	09 58	17 16
11	19 24 45	21 27 22	08 ♋ 06 27	15 ♋ 26 30	29 30	08 36	17 27	22 07	22 32	20 38	10 00	17 18
12	19 28 41	22 28 29	22 ♋ 43 07	29 ♋ 55 26	29 ♐ 57	09 37	18 12	22 11	22 39	20 39	10 02	17 20
13	19 32 38	23 29 35	07 ♌ 02 43	14 ♌ 04 23	00 ♑ 31	10 38	18 57	22 16	22 45	20 39	10 03	17 22
14	19 36 35	24 30 41	21 ♌ 00 01	27 ♌ 49 21	01 10	11 39	19 42	22 20	22 52	20 40	10 05	17 24
15	19 40 31	25 31 47	04 ♍ 32 17	11 ♍ 08 53	01 54	12 39	20 27	22 24	22 58	20 41	10 07	17 26
16	19 44 28	26 32 52	17 ♍ 39 20	24 ♍ 03 55	02 42	13 38	21 12	22 28	23 04	20 42	10 08	17 28
17	19 48 24	27 33 57	00 ♎ 23 03	06 ♎ 37 12	03 35	14 38	21 57	22 32	23 11	20 43	10 10	17 30
18	19 52 21	28 35 02	12 ♎ 46 54	18 ♎ 52 44	04 31	15 36	22 43	22 35	23 17	20 44	10 12	17 32
19	19 56 17	29 ♑ 36 06	24 ♎ 55 19	00 ♏ 55 16	05 30	16 34	23 28	22 39	23 23	20 45	10 14	17 35
20	20 00 14	00 ♒ 37 10	06 ♏ 53 12	12 ♏ 49 46	06 32	17 32	24 13	22 42	23 29	20 46	10 16	17 37
21	20 04 10	01 38 14	18 ♏ 45 33	24 ♏ 41 08	07 37	18 29	24 58	22 45	23 35	20 47	10 18	17 39
22	20 08 07	02 39 17	00 ♐ 37 06	06 ♐ 33 56	08 45	19 26	25 43	22 48	23 41	20 48	10 19	17 41
23	20 12 04	03 40 20	12 ♐ 32 09	18 ♐ 32 09	09 54	20 22	26 27	22 51	23 47	20 50	10 21	17 43
24	20 16 00	04 41 22	24 ♐ 34 20	00 ♑ 39 01	11 06	21 17	27 12	22 53	23 53	20 51	10 23	17 45
25	20 19 57	05 42 24	06 ♑ 46 29	12 ♑ 56 57	12 19	22 12	27 57	22 55	23 59	20 52	10 25	17 47
26	20 23 53	06 43 25	19 ♑ 10 36	25 ♑ 27 32	13 34	23 06	28 42	22 57	24 05	20 54	10 27	17 49
27	20 27 50	07 44 25	01 ♒ 47 49	08 ♒ 11 28	14 50	23 59	29 ♓ 27	22 59	24 11	20 55	10 29	17 51
28	20 31 46	08 45 25	14 ♒ 38 29	21 ♒ 08 49	16 08	24 52	00 ♈ 12	23 01	24 17	20 57	10 31	17 52
29	20 35 43	09 46 23	27 ♒ 42 25	04 ♓ 19 10	17 27	25 44	00 57	23 03	24 22	20 58	10 33	17 54
30	20 39 39	10 47 20	10 ♓ 59 00	17 ♓ 41 48	18 48	26 36	01 41	23 04	24 28	21 00	10 35	17 56
31	20 43 36	11 ♒ 48 17	24 ♓ 27 30	01 ♈ 15 58	20 ♑ 09	27 ♓ 26	02 ♈ 26	23 ♎ 05	24 ♐ 34	21 ♈ 02	10 ♓ 37	17 ♑ 58

Date	Moon True ☊ ° ′	Moon Mean ☊ ° ′	Moon ☽ Latitude ° ′
01	04 ♍ 31	06 ♍ 12	01 N 26
02	04 R 30	06 09	00 N 18
03	04 D 30	06 06	00 S 52
04	04 31	06 03	02 01
05	04 32	05 59	03 05
06	04 R 32	05 56	03 59
07	04 29	05 53	04 39
08	04 25	05 50	05 03
09	04 19	05 47	05 07
10	04 12	05 44	04 51
11	04 05	05 40	04 16
12	03 59	05 37	03 24
13	03 54	05 34	02 20
14	03 52	05 31	01 S 09
15	03 D 51	05 28	00 N 04
16	03 52	05 24	01 14
17	03 53	05 21	02 19
18	03 55	05 18	03 16
19	03 56	05 15	04 03
20	03 R 56	05 12	04 38
21	03 54	05 09	05 01
22	03 51	05 05	05 12
23	03 47	05 02	05 09
24	03 42	04 59	04 52
25	03 37	04 56	04 22
26	03 32	04 53	03 39
27	03 28	04 49	02 45
28	03 25	04 46	01 41
29	03 24	04 43	00 N 31
30	03 D 24	04 40	00 S 42
31	03 ♍ 25	04 ♍ 37	01 S 53

DECLINATIONS

Date	Sun ☉ ° ′	Moon ☽ ° ′	Mercury ☿ ° ′	Venus ♀ ° ′	Mars ♂ ° ′	Jupiter ♃ ° ′	Saturn ♄ ° ′	Uranus ♅ ° ′	Neptune ♆ ° ′	Pluto ♇ ° ′
01	22 S 57	14 S 00	20 S 19	13 S 29	08 S 40	07 S 05	21 S 53	07 N 28	08 S 42	21 S 21
02	22 52	10 48	20 15	13 02	08 21	07 07	21 53	07 28	08 42	21 21
03	22 46	07 01	20 13	12 36	08 03	07 09	21 54	07 28	08 41	21 21
04	22 40	02 S 49	20 12	12 09	07 45	07 11	21 54	07 29	08 41	21 21
05	22 33	01 N 36	20 14	11 42	07 27	07 13	21 54	07 29	08 40	21 20
06	22 26	06 01	20 17	11 14	07 09	07 15	21 55	07 29	08 39	21 20
07	22 18	10 11	20 21	10 47	06 50	07 16	21 55	07 29	08 39	21 20
08	22 10	13 50	20 27	10 19	06 32	07 18	21 56	07 29	08 38	21 20
09	22 02	16 40	20 34	09 51	06 13	07 20	21 56	07 30	08 38	21 20
10	21 53	18 25	20 42	09 23	05 55	07 21	21 56	07 30	08 37	21 20
11	21 43	18 56	20 50	08 55	05 36	07 23	21 57	07 30	08 36	21 19
12	21 34	18 10	20 59	08 27	05 18	07 24	21 57	07 30	08 36	21 19
13	21 23	16 15	21 08	07 59	04 59	07 26	21 57	07 31	08 35	21 19
14	21 13	13 24	21 17	07 30	04 41	07 27	21 58	07 31	08 34	21 19
15	21 02	09 54	21 26	07 02	04 22	07 29	21 58	07 31	08 34	21 19
16	20 50	06 01	21 35	06 33	04 03	07 30	21 58	07 32	08 33	21 19
17	20 39	01 N 59	21 43	06 05	03 45	07 31	21 59	07 32	08 32	21 18
18	20 26	02 S 02	21 52	05 36	03 26	07 32	21 59	07 33	08 32	21 18
19	20 14	05 53	21 59	05 07	03 07	07 33	21 59	07 33	08 31	21 18
20	20 01	09 25	22 06	04 39	02 48	07 34	22 00	07 34	08 30	21 18
21	19 47	12 34	22 12	04 10	02 30	07 35	22 00	07 34	08 30	21 18
22	19 34	15 11	22 17	03 41	02 11	07 36	22 00	07 34	08 29	21 18
23	19 20	17 12	22 22	03 13	01 52	07 36	22 01	07 35	08 28	21 17
24	19 05	18 28	22 26	02 44	01 34	07 37	22 01	07 35	08 27	21 17
25	18 50	18 54	22 28	02 16	01 15	07 38	22 01	07 36	08 27	21 17
26	18 35	18 27	22 30	01 48	00 56	07 38	22 01	07 37	08 26	21 17
27	18 20	17 05	22 30	01 19	00 37	07 39	22 01	07 37	08 25	21 17
28	18 04	14 49	22 30	00 51	00 S 19	07 39	22 02	07 38	08 24	21 17
29	17 48	11 47	22 28	00 S 23	00 00	07 40	22 02	07 38	08 24	21 16
30	17 31	08 05	22 25	00 N 05	00 N 19	07 40	22 02	07 39	08 23	21 16
31	17 S 15	03 S 56	22 S 21	00 N 32	00 N 37	07 S 40	22 S 02	07 N 40	08 S 22	21 S 16

ZODIAC SIGN ENTRIES

Date	h	m	Planets
02	09	57	☽ ♓
03	07	47	♀ ♓
04	14	17	☿ ♐
04	16	20	☽ ♈
06	20	18	☽ ♉
08	22	06	☽ ♊
10	22	49	☽ ♋
12	14	03	☿ ♑
13	00	08	☽ ♌
15	03	52	☽ ♍
17	11	16	☽ ♎
19	21	24	☉ ♒
19	22	09	☽ ♏
22	10	45	☽ ♐
24	22	43	☽ ♑
27	08	37	☽ ♒
28	05	39	♂ ♈
29	16	10	☽ ♓
31	21	46	☽ ♈

LATITUDES

Date	Mercury ☿ ° ′	Venus ♀ ° ′	Mars ♂ ° ′	Jupiter ♃ ° ′	Saturn ♄ ° ′	Uranus ♅ ° ′	Neptune ♆ ° ′	Pluto ♇ ° ′
01	03 N 06	01 S 25	00 S 52	01 N 17	01 N 17	00 S 36	00 S 51	01 N 01
04	03 14	01 12	00 49	01 18	01 17	00 36	00 51	01 01
07	03 04	00 58	00 46	01 18	01 17	00 36	00 51	01 00
10	02 44	00 43	00 43	01 19	01 17	00 36	00 51	01 00
13	02 18	00 26	00 40	01 20	01 17	00 36	00 51	01 00
16	01 50	00 S 08	00 37	01 21	01 17	00 36	00 51	01 00
19	01 20	00 N 12	00 34	01 21	01 17	00 36	00 51	00 59
22	00 51	00 32	00 31	01 22	01 17	00 35	00 51	00 59
25	00 N 24	00 54	00 28	01 23	01 17	00 35	00 51	00 59
28	00 S 02	01 17	00 26	01 23	01 17	00 35	00 51	00 58
31	00 S 26	01 N 42	00 S 23	01 N 24	01 N 17	00 S 35	00 S 51	00 N 58

DATA

Julian Date	2457755
Delta T	+70 seconds
Ayanamsa	24° 05′ 34″
Synetic vernal point	05° ♓ 01′ 25″
True obliquity of ecliptic	23° 26′ 04″

MOON'S PHASES, APSIDES AND POSITIONS ☽

Date	h	m	Phase	Longitude ° ′	Eclipse Indicator
05	19	47	☽	15 ♈ 40	
12	11	34	○	22 ♋ 27	
19	22	13	☾	00 ♏ 02	
28	00	07	●	08 ♒ 15	

Day	h	m		
10	06	07	Perigee	
22	00	19	Apogee	
05	03	23	0N	
11	09	30	Max dec	18° N 56′
17	23	46	0S	
25	12	01	Max dec	18° S 54′

LONGITUDES

Date	Chiron ⚷ ° ′	Ceres ⚳ ° ′	Pallas ⚴ ° ′	Juno ⚵ ° ′	Vesta ⚶ ° ′	Black Moon Lilith ⚸ ° ′
01	21 ♓ 07	22 ♈ 41	01 ♓ 14	19 ♐ 16	02 ♌ 12	25 ♏ 07
11	21 ♓ 26	24 ♈ 17	04 ♓ 06	22 ♐ 38	29 ♋ 49	26 ♏ 15
21	21 ♓ 50	26 ♈ 20	07 ♓ 07	25 ♐ 56	27 ♋ 11	27 ♏ 22
31	22 ♓ 18	28 ♈ 47	10 ♓ 17	29 ♐ 08	24 ♋ 38	28 ♏ 29

ASPECTARIAN

01 Sunday
02 28 ☿ Q ♃
06 53 ♂ ☌ ♆
08 19 ♂ ∥ ♆
09 38 ☽ ⊻ ♇
10 03 ☽ ⊥ ☉
11 07 ☽ ∠ ☿
16 24 ☽ ⚹ ♅
16 59 ☽ ∥ ♀
17 38 ☽ △ ♃
18 04 ☽ ⚹ ♄
20 54 ☽ ⊥ ♇

02 Monday
04 19 ☽ ∠ ☉
07 59 ☽ ☌ ♀
12 58 ☽ ⚹ ☿
13 38 ☽ ∠ ♇
16 31 ☽ Q ♄
20 13 ☽ ∠ ♅
21 38 ☽ ꝑ ♃

03 Tuesday
01 44 ☽ ∥ ♆
03 56 ☽ ☌ ♆
05 09 ☽ ∥ ♂
06 36 ☽ ☌ ♂
09 14 ☽ ⋕ ♅
09 33 ☽ Q ☿
10 13 ☽ ⚹ ☉
11 12 ☽ ∥ ♃
12 39 ☽ ⊥ ♅
14 10 ☽ ± ♃
17 07 ☽ ⚹ ♇
19 41 ☿ ⚹ ♀
23 29 ☽ ⊻ ♅

04 Wednesday
01 05 ☽ ⚻ ♃
01 33 ☽ □ ♄
09 36 ☽ Q ☉
10 09 ☿ Q ♂
14 40 ☽ Q ♇
16 14 ☽ □ ☿
19 07 ☽ ⊻ ♀

05 Thursday
06 34 ☽ ⊥ ♀
06 41 ♀ ∠ ♇
09 30 ♂ ⋕ ♅
09 38 ☽ ⊻ ♆
15 11 ☽ ⊻ ♂
19 47 ☽ □ ☉
20 05 ☽ ⊥ ♆
22 16 ☽ □ ♇
23 31 ☽ ∠ ♀

06 Friday
02 09 ☽ ⊥ ♂
04 15 ☽ ☌ ♅
04 46 ♂ ∥ ♃
06 06 ☽ ☍ ♃
06 37 ☽ △ ♄
11 35 ☽ ∠ ♆
17 52 ☽ ⋕ ♂
18 24 ☽ ∠ ♂
18 41 ☽ △ ☿
18 55 ☽ ⋕ ♃
20 13 ☽ ∥ ♅

07 Saturday
02 55 ☽ ⋕ ♆
03 08 ☽ ⚹ ♀
03 23 ♀ Q ♄
06 45 ☉ ☌ ♇
08 14 ☽ ꝑ ♄
12 56 ☽ ⚹ ♆
15 15 ☽ ⋕ ♀
17 14 ♂ ⊥ ♅
19 33 ☽ ꝑ ☿
20 57 ☽ ⚹ ♂
23 20 ☽ ± ♄

08 Sunday
00 44 ☽ Q ♀
01 02 ☽ △ ♇
02 23 ☽ △ ☉
06 41 ☽ ⊻ ♅
08 45 ☽ ⚻ ♃
08 51 ☽ Q ♆
09 19 ☽ ⚻ ♄
09 44 ☿ St D
10 23 ☽ ± ☿
14 39 ♀ ∠ ♅
16 33 ☽ ⊥ ♅
17 48 ☽ Q ♂
18 39 ☽ ± ♃
20 15 ☽ ⚻ ☿

09 Monday
01 43 ☽ ꝑ ♇
04 49 ☽ ꝑ ☉
07 15 ☽ ∠ ♅
08 30 ☽ □ ♀
09 25 ☽ ꝑ ♃
12 25 ♂ ± ♃
14 18 ☽ □ ♆
16 24 ☽ ± ♇
20 31 ☽ ± ☉
22 34 ♀ ꝑ ♃

10 Tuesday
00 40 ☽ □ ♂
02 09 ☽ ⚻ ♇
02 57 ☉ ∥ ♄
06 58 ☽ ⚻ ☉
07 36 ☽ ⚹ ♅
09 53 ☽ △ ♃
10 32 ☽ ☍ ♄
16 22 ☉ □ ♅
21 38 ☽ ☍ ☿

11 Wednesday
03 05 ☽ Q ♅
07 15 ♂ ⚹ ♇
12 51 ☽ △ ♀
15 06 ☽ △ ♆

12 Thursday
03 06 ☽ ☍ ♇
04 08 ☽ △ ♂
04 20 ♀ ∥ ♆
04 43 ☉ □ ♃
08 34 ☽ □ ♅
11 07 ☽ □ ♃
11 34 ☽ ☍ ☉
11 53 ☽ ⚻ ♄
15 23 ☽ ꝑ ♀
15 50 ☽ ꝑ ♆
16 32 ☉ ⊻ ♄
21 54 ♀ ☌ ♆
21 56 ☽ ± ♄

13 Friday
00 31 ☽ ⚻ ☿
06 28 ☽ ꝑ ♂
06 56 ☽ ± ♆
07 36 ☽ ± ♀
11 04 ☽ ± ☿
13 13 ☽ ꝑ ♄
17 08 ☽ ⚻ ♆
17 30 ☽ Q ♃
18 35 ☽ ⚻ ♀
22 10 ☉ ∥ ♇
22 39 ☽ ± ♂

14 Saturday
03 11 ☽ ꝑ ☿
05 44 ☽ ⚻ ♇
07 03 ☉ ∥ ☿
09 37 ☽ ⚻ ♂
11 14 ♀ ⋕ ♅
11 25 ☽ △ ♅
14 21 ☽ ⚹ ♃
14 23 ♀ ∥ ♃
15 17 ☽ △ ♄
16 13 ☽ ± ♇
17 04 ☿ ∥ ♇
18 39 ☽ ⚻ ☉

15 Sunday
01 52 ☉ ∠ ♆
06 09 ☽ ± ☉
06 58 ☽ △ ☿
08 13 ☽ ꝑ ♇
14 04 ☽ ꝑ ♅
17 12 ☽ ∠ ♃
19 29 ♂ ⊻ ♅
20 30 ☽ ⋕ ♆
22 08 ☽ ☍ ♆
23 47 ☽ ꝑ ☉

16 Monday
02 52 ☽ ∥ ♅
03 06 ☽ ⋕ ♃
03 58 ☽ ☍ ♀
06 31 ☽ ± ♅
08 21 ☽ ⋕ ♀
09 47 ☽ ⊥ ♃
11 40 ☽ △ ♇
17 41 ☽ ⚻ ♅
19 02 ☽ ☍ ♂
21 02 ☽ ⊻ ♃
22 13 ☽ □ ♄

17 Tuesday
00 42 ☽ ⋕ ♂
06 09 ☽ △ ☉
14 14 ♀ ⊥ ♅
18 37 ☽ □ ☿

18 Wednesday
06 57 ☽ ⚻ ♆
07 51 ♂ ⚻ ♃
09 02 ☽ Q ♄
13 56 ☿ Q ♃
18 01 ☽ ⚻ ♀
18 44 ☽ ± ♆
19 55 ☽ ∥ ♂
21 23 ☽ □ ♇

19 Thursday
03 42 ☽ ☍ ♅
06 19 ☉ ⊥ ♄
06 55 ☽ ± ♀
07 27 ☽ ☌ ♃
07 43 ☽ ∥ ♀
08 54 ☽ ⚻ ♂
08 55 ☽ ⚹ ♄
08 55 ☽ Q ☿
09 16 ♂ □ ♄
12 37 ☽ ꝑ ♆
13 08 ♃ ⋕ ♅
13 13 ☿ ∥ ♄
13 54 ♀ ± ♃
21 40 ☽ ± ♂
22 13 ☽ □ ☉
23 04 ☽ ⋕ ♅
23 05 ☽ ∥ ♃

20 Friday
02 29 ☽ ꝑ ♀
05 34 ☽ ∥ ♆
09 25 ☽ Q ♇
11 14 ☽ ⚹ ☿
13 57 ♀ ⚹ ♇
15 15 ☽ ∠ ♄
17 00 ☽ ꝑ ♂
18 50 ☽ △ ♆

21 Saturday
09 37 ☽ ⊥ ♄
09 44 ☽ ⚹ ♇
11 24 ☽ △ ♀
13 57 ☽ Q ☉
16 07 ☽ ⚻ ♅
20 07 ☽ ⊻ ♃
20 38 ☽ ∠ ☿
21 52 ☽ ⊻ ♄

22 Sunday
01 24 ☽ △ ♂
04 17 ☽ ± ♅
08 18 ☽ ⊥ ♃
16 10 ☽ ∠ ♇
16 30 ☽ ⚹ ☉
16 45 ☽ ⊥ ☿
22 30 ☽ ꝑ ♅

23 Monday
02 32 ☽ ∠ ♃
06 08 ☽ ⊻ ☿
07 37 ☽ □ ♆
10 20 ☽ ⊥ ♇
21 29 ☿ ⚹ ♆
22 23 ☽ ⊻ ♇

24 Tuesday
00 24 ♀ ⊻ ♅
01 24 ☽ ∠ ☉
04 36 ☽ △ ♅
04 39 ☉ ⊥ ♆
04 56 ☽ □ ♀
08 38 ☽ ⚹ ♃
10 38 ☽ ☌ ♄
17 33 ☽ □ ♂
19 34 ☽ Q ♆
20 53 ☽ ⊥ ☉

25 Wednesday
07 37 ☽ ∥ ☉
08 22 ☽ Q ♃
09 44 ☽ ⊻ ☉
19 07 ☽ ⚹ ♆
19 12 ☽ Q ♀
23 58 ☽ ☌ ☿

26 Thursday
05 42 ☽ ∥ ☉
06 57 ☽ Q ♂
08 02 ♀ ⚻ ♃
09 22 ☽ ☌ ♇
15 18 ☽ □ ♅
19 15 ☽ □ ♃
20 05 ☽ ⚹ ♀
21 28 ☽ ⊻ ♄

27 Friday
00 01 ☽ ∠ ♆
07 18 ☽ ⚹ ♂
08 56 ☽ ⊥ ♄
17 04 ☽ ⊥ ♆
17 50 ♀ □ ♄

28 Saturday
00 07 ☽ ☌ ☉
01 08 ♂ Q ♇
01 23 ☽ Q ♅
01 57 ☽ ∠ ♄
02 29 ☽ ∠ ♀
04 19 ☽ ⊻ ♆
13 06 ☽ ∠ ♂
15 04 ☽ ⊻ ☿
16 35 ☉ Q ♅
17 59 ☽ ⊻ ♇
20 22 ☽ ⊥ ♀
23 39 ☽ ⚹ ♅

29 Sunday
01 35 ☉ ∠ ♄
03 21 ☽ ⊥ ☿
03 28 ☽ △ ♃
05 03 ☽ ⊥ ♇
05 52 ☽ ⚹ ♄
06 39 ☽ ⊥ ♂
08 09 ☽ ⊻ ♀
18 14 ☽ ⊻ ♂
20 21 ☿ ☌ ♇
21 28 ☽ ∠ ♇
21 35 ☽ ∠ ☿

30 Monday
00 03 ♀ ⋕ ♂
03 01 ☽ ∠ ♅
03 49 ☽ Q ♄
06 45 ☽ ꝑ ♃
07 03 ☉ ⊻ ♆
10 12 ☽ ∥ ♆
11 17 ☽ ☌ ♆
11 37 ☽ ⊻ ☉
14 33 ☽ ∥ ♃
14 38 ☽ ⋕ ♅
19 12 ☽ ⊥ ♅
22 54 ☽ ± ♃
23 14 ☽ ⊥ ☉

31 Tuesday
00 28 ☽ ⚹ ♇
03 31 ☽ ⚹ ☿
05 54 ☽ ⊻ ♅
09 34 ☽ ⚻ ♃
12 11 ☽ □ ♄
16 29 ☽ ∠ ☉
17 36 ☽ ☌ ♀
21 45 ☽ Q ♇
22 47 ☿ Q ♂

All ephemeris data is given at 12.00 UT and the Moon's longitude is additionally given for 24.00 UT

FEBRUARY 2017

Date	Sidereal time h m s	Sun ☉ ° ' "	Moon ☽ ° ' "	Moon ☽ 24.00 ° ' "	Mercury ☿ ° '	Venus ♀ ° '	Mars ♂ ° '	Jupiter ♃ ° '	Saturn ♄ ° '	Uranus ♅ ° '	Neptune ♆ ° '	Pluto ♇ ° '
01	20 47 33	12 ♒ 49 11	08 ♈ 07 09	15 ♈ 00 55	21 ♑ 32	28 ♓ 16	03 ♈ 11	23 ♎ 06	24 ♐ 39	21 ♈ 03	10 ♓ 39	18 ♑ 00
02	20 51 29	13 50 05	21 ♈ 57 10	28 ♈ 55 50	22 56	29 04	03 56	23 07	24 45	21 05	10 41	18 02
03	20 55 26	14 50 57	05 ♉ 56 45	12 ♉ 59 46	24 20	29 ♓ 52	04 40	23 08	24 50	21 07	10 43	18 04
04	20 59 22	15 51 48	20 ♉ 04 41	27 ♉ 11 17	25 46	00 ♈ 39	05 25	23 08	24 55	21 09	10 46	18 06
05	21 03 19	16 52 38	04 ♊ 19 16	11 ♊ 28 18	27 13	01 25	06 09	23 08	25 01	21 10	10 48	18 08
06	21 07 15	17 53 26	18 ♊ 37 58	25 ♊ 47 50	28 ♑ 40	02 10	06 54	23 R 08	25 06	21 12	10 50	18 10
07	21 11 12	18 54 12	02 ♋ 57 23	10 ♋ 06 05	00 ♒ 09	02 54	07 39	23 08	25 11	21 14	10 52	18 11
08	21 15 08	19 54 57	17 ♋ 13 22	24 ♋ 18 40	01 38	03 37	08 23	23 08	25 16	21 16	10 54	18 13
09	21 19 05	20 55 41	01 ♌ 21 26	08 ♌ 21 08	03 09	04 19	09 08	23 07	25 21	21 18	10 56	18 15
10	21 23 01	21 56 23	15 ♌ 17 16	22 ♌ 09 24	04 40	05 00	09 52	23 07	25 26	21 20	10 58	18 17
11	21 26 58	22 57 04	28 ♌ 57 12	05 ♍ 40 22	06 12	05 39	10 36	23 06	25 31	21 23	11 01	18 19
12	21 30 55	23 57 43	12 ♍ 18 46	18 ♍ 52 16	07 44	06 17	11 21	23 05	25 36	21 25	11 03	18 20
13	21 34 51	24 58 21	25 ♍ 20 53	01 ♎ 44 43	09 18	06 54	12 05	23 03	25 40	21 27	11 05	18 22
14	21 38 48	25 58 58	08 ♎ 03 57	14 ♎ 18 50	10 52	07 29	12 49	23 02	25 45	21 29	11 07	18 24
15	21 42 44	26 59 33	20 ♎ 29 43	26 ♎ 36 58	12 28	08 03	13 33	23 00	25 49	21 31	11 09	18 26
16	21 46 41	28 00 07	02 ♏ 41 04	08 ♏ 42 28	14 04	08 36	14 18	22 59	25 54	21 34	11 12	18 27
17	21 50 37	29 ♒ 00 41	14 ♏ 41 44	20 ♏ 39 24	15 41	09 07	15 02	22 57	25 58	21 36	11 14	18 29
18	21 54 34	00 ♓ 01 12	26 ♏ 36 03	02 ♐ 32 17	17 19	09 36	15 46	22 55	26 03	21 38	11 16	18 31
19	21 58 31	01 01 43	08 ♐ 28 42	14 ♐ 25 52	18 58	10 04	16 30	22 52	26 07	21 41	11 18	18 32
20	22 02 27	02 02 12	20 ♐ 24 22	26 ♐ 24 46	20 38	10 30	17 14	22 49	26 11	21 43	11 21	18 34
21	22 06 24	03 02 40	02 ♑ 27 36	08 ♑ 33 22	22 19	10 55	17 58	22 47	26 15	21 46	11 23	18 35
22	22 10 20	04 03 07	14 ♑ 42 30	20 ♑ 55 25	24 00	11 17	18 42	22 44	26 19	21 48	11 25	18 37
23	22 14 17	05 03 32	27 ♑ 12 28	03 ♒ 33 54	25 43	11 38	19 26	22 41	26 23	21 51	11 27	18 38
24	22 18 13	06 03 56	09 ♒ 59 56	16 ♒ 30 41	27 27	11 57	20 10	22 37	26 27	21 54	11 30	18 40
25	22 22 10	07 04 18	23 ♒ 06 11	29 ♒ 46 24	29 ♒ 11	12 13	20 54	22 34	26 31	21 56	11 32	18 42
26	22 26 06	08 04 38	06 ♓ 31 10	13 ♓ 20 16	00 ♓ 57	12 28	21 38	22 30	26 35	21 59	11 34	18 43
27	22 30 03	09 04 57	20 ♓ 13 25	27 ♓ 10 15	02 43	12 40	22 22	22 26	26 38	22 02	11 36	18 44
28	22 34 00	10 ♓ 05 14	04 ♈ 10 19	11 ♈ 13 09	04 ♓ 31	12 ♈ 51	23 ♈ 05	22 ♎ 22	26 ♐ 42	22 ♈ 04	11 ♓ 39	18 ♑ 46

Date	Moon True ☊ ° '	Moon Mean ☊ ° '	Moon ☽ Latitude ° '
01	03 ♍ 27	04 ♍ 34	03 S 00
02	03 28	04 30	03 57
03	03 29	04 27	04 40
04	03 R 29	04 24	05 07
05	03 29	04 21	05 16
06	03 27	04 18	05 05
07	03 25	04 15	04 35
08	03 23	04 11	03 48
09	03 22	04 08	02 48
10	03 21	04 05	01 38
11	03 20	04 02	00 S 24
12	03 D 20	03 59	00 N 50
13	03 21	03 55	01 59
14	03 22	03 52	03 01
15	03 23	03 49	03 53
16	03 23	03 46	04 32
17	03 23	03 43	05 01
18	03 24	03 40	05 15
19	03 R 24	03 36	05 16
20	03 23	03 33	05 03
21	03 23	03 30	04 37
22	03 D 23	03 27	03 59
23	03 24	03 24	03 08
24	03 24	03 21	02 06
25	03 24	03 17	00 N 57
26	03 R 24	03 14	00 S 17
27	03 24	03 11	01 32
28	03 ♍ 23	03 ♍ 08	02 S 42

DECLINATIONS

Date	Sun ☉ ° '	Moon ☽ ° '	Mercury ☿ ° '	Venus ♀ ° '	Mars ♂ ° '	Jupiter ♃ ° '	Saturn ♄ ° '	Uranus ♅ ° '	Neptune ♆ ° '	Pluto ♇ ° '
01	16 S 58	00 N 28	22 S 16	01 N 00	00 N 56	07 S 40	22 S 03	07 N 40	08 S 21	21 S 16
02	16 40	04 53	22 10	01 27	01 14	07 40	22 03	07 41	08 21	21 16
03	16 23	09 05	22 02	01 54	01 33	07 40	22 03	07 42	08 20	21 15
04	16 05	12 49	21 53	02 21	01 51	07 40	22 03	07 42	08 19	21 15
05	15 47	15 50	21 43	02 48	02 10	07 40	22 03	07 43	08 18	21 15
06	15 28	17 53	21 32	03 14	02 28	07 40	22 03	07 44	08 17	21 15
07	15 09	18 49	21 19	03 40	02 47	07 40	22 04	07 45	08 17	21 15
08	14 50	18 33	21 05	04 05	03 05	07 39	22 04	07 45	08 16	21 15
09	14 31	17 07	20 50	04 31	03 23	07 39	22 04	07 46	08 15	21 14
10	14 12	14 41	20 34	04 56	03 42	07 38	22 04	07 47	08 14	21 14
11	13 52	11 27	20 16	05 20	04 00	07 38	22 04	07 48	08 13	21 14
12	13 32	07 42	19 57	05 45	04 18	07 37	22 04	07 49	08 12	21 14
13	13 12	03 N 40	19 36	06 08	04 36	07 36	22 04	07 50	08 12	21 14
14	12 51	00 S 26	19 14	06 32	04 54	07 36	22 04	07 50	08 11	21 14
15	12 31	04 25	18 51	06 54	05 12	07 35	22 05	07 51	08 10	21 14
16	12 10	08 07	18 26	07 17	05 30	07 34	22 05	07 52	08 09	21 13
17	11 49	11 27	18 00	07 38	05 48	07 33	22 05	07 53	08 08	21 13
18	11 28	14 17	17 33	08 00	06 06	07 32	22 05	07 54	08 07	21 13
19	11 06	16 31	17 05	08 20	06 23	07 31	22 05	07 55	08 06	21 13
20	10 45	18 03	16 35	08 40	06 41	07 30	22 05	07 56	08 06	21 13
21	10 23	18 47	16 03	08 59	06 59	07 28	22 05	07 57	08 05	21 13
22	10 01	18 40	15 31	09 18	07 16	07 27	22 05	07 58	08 04	21 13
23	09 39	17 39	14 57	09 36	07 34	07 26	22 05	07 59	08 03	21 12
24	09 17	15 43	14 21	09 53	07 51	07 24	22 05	08 00	08 02	21 12
25	08 55	12 55	13 45	10 09	08 08	07 23	22 05	08 01	08 01	21 12
26	08 32	09 23	13 07	10 24	08 25	07 21	22 05	08 02	08 00	21 12
27	08 10	05 17	12 27	10 39	08 42	07 20	22 05	08 03	08 00	21 12
28	07 S 47	00 S 50	11 S 47	10 N 52	08 N 59	07 S 18	22 S 05	08 N 04	07 S 59	21 S 12

ZODIAC SIGN ENTRIES

Date	h	m	Planets
03	01	50	☽ ♉
03	15	51	♀ ♈
05	04	44	☽ ♊
07	07	03	☽ ♋
07	09	35	☿ ♒
09	09	41	☽ ♌
11	13	52	☽ ♍
13	20	43	☽ ♎
16	06	41	☽ ♏
18	11	31	☉ ♓
18	18	52	☽ ♐
21	07	08	☽ ♑
23	17	17	☽ ♒
25	23	07	☿ ♓
26	00	24	☽ ♓
28	04	52	☽ ♈

LATITUDES

Date	Mercury ☿ ° '	Venus ♀ ° '	Mars ♂ ° '	Jupiter ♃ ° '	Saturn ♄ ° '	Uranus ♅ ° '	Neptune ♆ ° '	Pluto ♇ ° '
01	00 S 34	01 N 50	00 S 22	01 N 24	01 N 17	00 S 35	00 S 51	00 N 58
04	00 55	02 17	00 19	01 25	01 17	00 35	00 51	00 58
07	01 14	02 44	00 17	01 26	01 17	00 35	00 51	00 58
10	01 31	03 13	00 14	01 27	01 17	00 35	00 51	00 57
13	01 44	03 42	00 11	01 27	01 18	00 35	00 51	00 57
16	01 55	04 13	00 09	01 28	01 18	00 35	00 51	00 57
19	02 03	04 44	00 06	01 29	01 18	00 35	00 51	00 57
22	02 07	05 15	00 04	01 30	01 18	00 34	00 51	00 57
25	02 07	05 47	00 S 01	01 30	01 18	00 34	00 51	00 56
28	02 04	06 18	00 N 01	01 31	01 18	00 34	00 51	00 56
31	01 S 56	06 N 49	00 N 03	01 N 31	01 N 18	00 S 34	00 S 51	00 N 56

DATA

Julian Date	2457786
Delta T	+70 seconds
Ayanamsa	24° 05' 39"
Synetic vernal point	05° ♓ 01' 20"
True obliquity of ecliptic	23° 26' 05"

MOON'S PHASES, APSIDES AND POSITIONS ☽

Date	h	m	Phase	Longitude ° '	Eclipse Indicator
04	04	19	☽	15 ♉ 32	
11	00	33	○	22 ♌ 28	
18	19	33	☾	00 ♐ 20	
26	14	58	●	08 ♓ 12	Annular

Day	h	m		
06	14	08	Perigee	
18	21	14	Apogee	
01	09	29	0N	
07	18	39	Max dec	18° N 52'
14	09	29	0S	
21	20	52	Max dec	18° S 51'
28	16	22	0N	

LONGITUDES

Date	Chiron ⚷ ° '	Ceres ⚳ ° '	Pallas ⚴ ° '	Juno ⚵ ° '	Vesta ⚶ ° '	Black Moon Lilith ⚸ ° '
01	22 ♓ 21	29 ♈ 03	10 ♓ 36	29 ♐ 27	24 ♋ 24	28 ♏ 36
11	22 ♓ 52	01 ♉ 51	13 ♓ 53	02 ♑ 30	22 ♋ 19	29 ♏ 43
21	23 ♓ 25	04 ♉ 56	17 ♓ 25	05 ♑ 05	20 ♋ 52	00 ♐ 50
31	24 ♓ 00	08 ♉ 15	20 ♓ 40	08 ♑ 09	20 ♋ 11	01 ♐ 58

All ephemeris data is given at 12.00 UT and the Moon's longitude is additionally given for 24.00 UT

ASPECTARIAN

h m	Aspects
01 Wednesday	
01 14	♀ ∥ ♂
02 52	☽ ☌ ♂
03 05	☽ Q ☿
03 32	☿ □ ♅
04 51	☽ # ♀
04 58	☽ # ♂
07 27	♃ # ♅
14 41	☽ ∥ ♂
15 11	☽ ∥ ♀
16 26	☽ ⊻ ♆
20 50	☽ ⚹ ☉
02 Thursday	
02 53	☽ ⊥ ♆
05 13	☽ □ ♇
10 30	☽ ☌ ♅
13 52	☽ □ ☿
14 01	☽ ☍ ♃
15 15	☿ □ ♃
16 50	☽ △ ♄
18 27	☽ ∠ ♆
19 12	☽ Q ☉
03 Friday	
00 59	☽ ⊻ ♀
03 40	☽ # ♃
03 46	☽ ∥ ♅
07 30	☽ # ♆
09 42	☽ ⊻ ♂
10 17	☿ ∥ ♄
11 52	☽ ⊥ ♀
14 34	☉ ∠ ♀
18 07	♀ Q ♇
18 40	☽ Ꝓ ♄
20 10	☽ ⚹ ♆
20 30	☽ ⊥ ♂
20 50	☿ ⊻ ♄
04 Saturday	
04 05	☽ ∠ ♀
04 19	☽ □ ☉
08 38	☽ △ ♇
10 02	☽ ± ♄
11 49	☿ ∠ ♆
12 36	☽ ∠ ♂
13 48	☽ ⊻ ♅
16 32	☽ Q ♆
17 10	☽ ⊼ ♃
20 14	☽ ⊼ ♄
22 42	☽ △ ☿
23 57	☽ ⊥ ♅
05 Sunday	
03 17	☽ ± ♃
06 51	☽ ⚹ ♀
09 59	☽ Ꝓ ♇
11 33	☽ # ☉
15 07	☽ ∠ ♅
15 15	☽ ⚹ ♂
18 25	☽ Ꝓ ♃
22 53	☽ □ ♆
06 Monday	
01 08	☽ ± ♇
02 45	☽ Ꝓ ☿
04 07	☽ Q ♀
06 52	♃ St R
10 40	☽ △ ☉
11 12	☽ ⊼ ♇
12 28	☽ Q ♂
16 19	☽ ⚹ ♅
18 34	☉ ⊻ ♇
19 33	☽ ± ☿
19 33	☽ △ ♃
22 53	☽ ☍ ♄
07 Tuesday	
06 45	☽ ⊼ ☿
11 55	☽ □ ♀
12 28	☽ Q ♅
13 43	☽ Ꝓ ☉
20 09	☿ ∥ ♇
20 18	☽ □ ♂
08 Wednesday	
01 19	☽ △ ♆
05 40	☿ ⊥ ♄
06 00	☽ ± ☉
13 41	☽ ☍ ♇
16 54	☽ ⊼ ☉
18 52	☽ □ ♅
22 00	☽ □ ♃
09 Thursday	
01 42	☽ ⊼ ♄
02 44	☽ Ꝓ ♆
11 59	☽ ± ♄
15 25	☽ ☍ ☿
17 20	☽ △ ♀
18 09	☽ ± ♆
21 16	☉ ⚹ ♅
10 Friday	
02 05	☽ △ ♂
03 32	☽ Ꝓ ♄
04 30	☽ ⊼ ♆
04 46	☽ Q ♃
16 29	☽ # ☉
17 03	☿ ⊥ ♆
17 13	☽ ⊼ ♇
20 38	☽ Ꝓ ♀
21 18	☿ ⚹ ♀
22 36	☽ △ ♅
11 Saturday	
00 33	☽ ☍ ☉
01 40	☽ ⚹ ♃
03 46	☽ ± ♇
05 44	☽ Ꝓ ♂
05 52	☽ △ ♄
13 18	☽ ± ♀
15 25	☉ △ ♃
19 47	☽ Ꝓ ♇
22 40	☽ ± ♂
12 Sunday	
00 34	☽ ⊼ ♀
01 18	☽ Ꝓ ♅
01 49	♂ ⊻ ♆
02 38	☽ ⊼ ☿
04 21	☽ ∠ ♃
08 54	☽ # ♆
09 42	☽ ☍ ♆
10 08	☽ ⊼ ♂
11 21	☽ ∥ ♅
12 31	☽ # ♃
14 57	☽ ± ☿
17 40	☽ ± ♅
20 42	☽ ⊥ ♃
21 13	☉ ⊥ ♇
22 44	☽ ∥ ♀
23 03	☽ △ ♇
13 Monday	
04 44	☽ ⊼ ♅
06 54	☽ ∥ ♂
07 45	☽ ⊻ ♃
09 47	☽ Ꝓ ☿
11 14	☽ ⊼ ☉
12 36	☽ □ ♄
14 21	☿ Q ♅
23 27	☽ ± ☉
14 Tuesday	
05 58	☉ ⚹ ♄
10 00	☿ ∠ ♄
10 51	☽ ☍ ♀
15 50	☿ ⊻ ♆
17 52	☽ ⊼ ♆
18 05	☽ Ꝓ ☉
18 09	☽ △ ☿
21 42	☽ ☍ ♂
22 59	☽ Q ♄
15 Wednesday	
05 29	☽ ± ♆
07 57	☽ □ ♇
14 01	☽ ☍ ♅
16 54	☽ ☌ ♃
17 23	☽ # ♂
22 30	☽ ⚹ ♄
23 08	☽ Ꝓ ♆
16 Thursday	
01 54	☽ △ ☉
05 41	☽ # ♀
08 16	☽ ∥ ♃
10 16	☽ # ♅
12 11	☽ ∥ ♆
18 15	☿ ⚹ ♂
19 31	☽ Q ♇
17 Friday	
00 19	☽ ⊼ ♀
04 29	☽ ∠ ♄
05 02	☽ △ ♆
06 17	♀ # ♃
12 43	☽ ⊼ ♂
12 53	☽ ± ♀
14 18	☽ □ ☿
14 33	☽ ∥ ☉
19 38	☽ ⚹ ♇
22 41	☽ ⊥ ♄
18 Saturday	
01 36	☽ ± ♂
01 57	☽ ⊼ ♅
04 34	☽ ⊻ ♃
05 21	♀ ∥ ♅
07 48	☽ Ꝓ ♀
10 52	☽ ⊻ ♄
14 06	☽ ± ♅
16 39	☽ ⊥ ♃
19 33	☽ □ ☉
20 40	♀ # ♆
20 58	☽ Ꝓ ♂
19 Sunday	
02 00	☽ ∠ ♇
05 38	☿ ⊻ ♇
08 22	☽ Ꝓ ♅
08 27	☽ Q ☿
10 46	☽ ∠ ♃
15 20	☽ △ ♀
17 43	☽ □ ♆
17 54	☽ ∥ ☿
20 12	☽ ⊥ ♇
20 Monday	
05 13	☽ △ ♂
08 18	☽ ⊻ ♇
11 12	☽ Q ☉
12 31	☽ ⚹ ☿
14 39	☽ △ ♅
15 44	♂ ⊥ ♆
16 49	☽ ⚹ ♃
23 37	☽ ☌ ♄
21 Tuesday	
04 01	☿ ⚹ ♅
05 24	☉ ∠ ♂
05 53	☽ Q ♆
13 16	☽ ⚹ ☉
16 33	☽ Q ♃
18 27	☿ △ ♃
23 05	☽ ∠ ☿
22 Wednesday	
01 19	☉ ∠ ♇
05 08	☽ □ ♀
05 35	☽ ⚹ ♆
09 03	♂ □ ♇
19 24	☽ ⊥ ☿
19 34	☽ ☌ ♇
20 13	☽ □ ♂
20 43	☿ ⊥ ♇
21 08	☽ ∠ ☉
21 58	♀ ⊻ ♆
23 Thursday	
01 44	☽ □ ♅
02 10	♂ # ♃
03 24	☽ □ ♃
08 43	☽ ⊻ ☿
10 26	☽ ⊻ ♄
10 34	☽ ∠ ♆
14 08	☉ # ♀
15 49	☽ ⊥ ☉
16 43	☽ Q ♀
21 45	☿ ⚹ ♄
21 50	☽ ⊥ ♄
24 Friday	
03 35	☽ ⊥ ♆
03 38	☿ ∠ ♀
04 04	☽ ⊻ ☉
07 53	☉ ⊥ ♀
08 24	☽ Q ♂
11 48	☽ Q ♅
14 42	☽ ∠ ♄
14 47	☽ ⊻ ♆
15 41	☽ ⚹ ♀
25 Saturday	
01 10	♂ ∥ ♅
03 03	♂ # ♆
03 52	☽ ∥ ☿
03 58	☽ ⊻ ♇
07 46	☽ ⚹ ♂
08 37	☉ ∠ ♅
09 53	☽ ⚹ ♅
11 02	☽ △ ♃
14 53	☽ ⊥ ♇
18 11	☽ ⚹ ♄
19 28	♅ # ♆
19 34	☽ ∠ ♀
23 04	☉ Ꝓ ♃
26 Sunday	
00 36	☽ ☌ ☿
05 55	☽ # ♀
07 01	☽ ∠ ♇
11 54	☽ ⊥ ♀
12 12	☽ ∠ ♂
12 49	☽ ∠ ♅
13 44	☽ Ꝓ ♃
14 58	☽ ☌ ☉
15 39	☽ Q ♄
16 19	☉ # ♂
17 31	☽ # ♂
17 44	☽ ∥ ☉
20 13	☽ # ♅
20 24	☽ ∥ ♆
20 56	☽ ☌ ♆
22 39	☽ ⊻ ♀
27 Monday	
00 17	☽ ∥ ♃
00 19	♂ ☌ ♅
00 44	☉ Q ♄
04 40	☽ ⊥ ♅
04 54	☽ ⊥ ♂
05 27	☽ ± ♃
09 25	☽ ⚹ ♇
14 24	♂ ☍ ♃
15 08	☽ ⊻ ♅
15 49	☽ ⊼ ♃
15 54	☽ ⊻ ♂
19 05	☉ # ♅
23 08	☽ □ ♄
23 12	☉ ∥ ♆
28 Tuesday	
01 50	☿ ∠ ♇
06 10	☽ Q ♇
12 41	☽ ⊻ ☿
22 51	☽ ⊻ ☉

MARCH 2017

LONGITUDES

Date	Sidereal time h m s	Sun ☉ ° ' "	Moon ☽ ° ' "	Moon ☽ 24.00 ° ' "	Mercury ☿ ° '	Venus ♀ ° '	Mars ♂ ° '	Jupiter ♃ ° '	Saturn ♄ ° '	Uranus ♅ ° '	Neptune ♆ ° '	Pluto ♇ ° '
01	22 37 56	11 ♓ 05 29	18 ♈ 18 16	25 ♈ 25 07	06 ♓ 20	12 ♈ 59	23 ♈ 49	22 ♎ 18	26 ♐ 45	22 ♈ 07	11 ♓ 41	18 ♑ 47
02	22 41 53	12 05 42	02 ♉ 33 12	09 ♉ 42 01	08 09	13 05	24 33	22 R 14	26 49	22 10	11 43	18 49
03	22 45 49	13 05 53	16 ♉ 51 03	23 ♉ 59 54	10 00	13 08	25 16	22 09	26 52	22 13	11 46	18 50
04	22 49 46	14 06 02	01 ♊ 08 07	08 ♊ 15 22	11 52	13 R 09	26 00	22 05	26 55	22 15	11 48	18 51
05	22 53 42	15 06 09	15 ♊ 21 20	22 ♊ 25 43	13 44	13 07	26 44	22 00	26 58	22 18	11 50	18 53
06	22 57 39	16 06 14	29 ♊ 28 18	06 ♋ 28 53	15 38	13 03	27 27	21 55	27 01	22 21	11 52	18 54
07	23 01 35	17 06 16	13 ♋ 27 16	20 ♋ 23 17	17 33	12 57	28 10	21 50	27 04	22 24	11 55	18 55
08	23 05 32	18 06 17	27 ♋ 16 49	04 ♌ 07 43	19 28	12 48	28 54	21 45	27 07	22 27	11 57	18 57
09	23 09 29	19 06 15	10 ♌ 55 51	17 ♌ 41 05	21 24	12 36	29 ♈ 37	21 39	27 10	22 30	11 59	18 58
10	23 13 25	20 06 11	24 ♌ 23 17	01 ♍ 02 22	23 21	12 22	00 ♉ 21	21 34	27 12	22 33	12 02	18 59
11	23 17 22	21 06 05	07 ♍ 38 12	14 ♍ 10 42	25 19	12 05	01 04	21 28	27 15	22 36	12 04	19 00
12	23 21 18	22 05 57	20 ♍ 39 47	27 ♍ 05 26	27 17	11 46	01 47	21 22	27 17	22 39	12 06	19 01
13	23 25 15	23 05 47	03 ♎ 27 36	09 ♎ 46 20	29 ♓ 15	11 25	02 30	21 16	27 20	22 42	12 08	19 02
14	23 29 11	24 05 35	16 ♎ 01 42	22 ♎ 13 48	01 ♈ 13	11 01	03 14	21 10	27 22	22 45	12 11	19 04
15	23 33 08	25 05 22	28 ♎ 22 49	04 ♏ 28 57	03 12	10 36	03 57	21 04	27 24	22 49	12 13	19 05
16	23 37 04	26 05 06	10 ♏ 32 28	16 ♏ 33 43	05 10	10 08	04 40	20 58	27 26	22 52	12 15	19 06
17	23 41 01	27 04 49	22 ♏ 33 02	28 ♏ 30 52	07 07	09 38	05 23	20 51	27 28	22 55	12 17	19 07
18	23 44 58	28 04 30	04 ♐ 27 39	10 ♐ 23 54	09 04	09 06	06 06	20 45	27 30	22 58	12 19	19 08
19	23 48 54	29 ♓ 04 09	16 ♐ 20 08	22 ♐ 16 56	10 59	08 33	06 49	20 38	27 32	23 01	12 22	19 09
20	23 52 51	00 ♈ 03 47	28 ♐ 14 53	04 ♑ 14 33	12 52	07 59	07 32	20 31	27 34	23 04	12 24	19 10
21	23 56 47	01 03 23	10 ♑ 16 35	16 ♑ 21 34	14 43	07 23	08 15	20 25	27 35	23 08	12 26	19 10
22	00 00 44	02 02 57	22 ♑ 30 04	28 ♑ 42 41	16 32	06 47	08 57	20 18	27 37	23 11	12 28	19 11
23	00 04 40	03 02 29	04 ♒ 59 56	11 ♒ 22 16	18 18	06 10	09 40	20 11	27 38	23 14	12 30	19 12
24	00 08 37	04 02 00	17 ♒ 50 07	24 ♒ 23 47	20 00	05 32	10 23	20 03	27 40	23 17	12 33	19 13
25	00 12 33	05 01 28	01 ♓ 03 30	07 ♓ 49 20	21 39	04 55	11 06	19 56	27 41	23 21	12 35	19 14
26	00 16 30	06 00 55	14 ♓ 41 17	21 ♓ 39 09	23 13	04 17	11 48	19 49	27 42	23 24	12 37	19 15
27	00 20 27	07 00 20	28 ♓ 42 36	05 ♈ 51 09	24 42	03 40	12 31	19 42	27 43	23 27	12 39	19 15
28	00 24 23	07 59 43	13 ♈ 04 11	20 ♈ 20 56	26 06	03 03	13 14	19 34	27 44	23 31	12 41	19 16
29	00 28 20	08 59 04	27 ♈ 40 32	05 ♉ 02 03	27 25	02 27	13 56	19 27	27 45	23 34	12 43	19 17
30	00 32 16	09 58 22	12 ♉ 24 33	19 ♉ 47 02	28 38	01 52	14 39	19 19	27 45	23 37	12 45	19 17
31	00 36 13	10 ♈ 57 39	27 ♉ 08 35	04 ♊ 28 22	29 ♈ 45	01 ♈ 18	15 ♉ 21	19 ♎ 12	27 ♐ 46	23 ♈ 41	12 ♓ 47	19 ♑ 18

MOON

Date	Moon True ☊ ° '	Moon Mean ☊ ° '	Moon ☽ Latitude ° '
01	03 ♍ 22	03 ♍ 05	03 S 44
02	03 R 21	03 01	04 32
03	03 20	02 58	05 04
04	03 20	02 55	05 16
05	03 D 19	02 52	05 10
06	03 20	02 49	04 44
07	03 21	02 46	04 02
08	03 22	02 42	03 07
09	03 23	02 39	02 01
10	03 24	02 36	00 S 50
11	03 R 24	02 33	00 N 23
12	03 23	02 30	01 34
13	03 21	02 27	02 38
14	03 18	02 23	03 34
15	03 15	02 20	04 18
16	03 11	02 17	04 50
17	03 08	02 14	05 09
18	03 05	02 11	05 14
19	03 03	02 07	05 06
20	03 02	02 04	04 44
21	03 D 02	02 01	04 10
22	03 04	01 58	03 24
23	03 05	01 55	02 28
24	03 07	01 52	01 23
25	03 08	01 48	00 N 11
26	03 R 07	01 45	01 S 03
27	03 05	01 42	02 15
28	03 01	01 39	03 20
29	02 56	01 36	04 14
30	02 50	01 33	04 51
31	02 ♍ 45	01 ♍ 29	05 S 09

DECLINATIONS

Date	Sun ☉ ° '	Moon ☽ ° '	Mercury ☿ ° '	Venus ♀ ° '	Mars ♂ ° '	Jupiter ♃ ° '	Saturn ♄ ° '	Uranus ♅ ° '	Neptune ♆ ° '	Pluto ♇ ° '
01	07 S 24	03 N 43	11 S 05	11 N 05	09 N 16	07 S 16	22 S 05	08 N 05	07 S 58	21 S 12
02	07 01	08 06	10 21	11 17	09 33	07 15	22 05	08 06	07 57	21 11
03	06 38	12 01	09 37	11 27	09 50	07 13	22 05	08 07	07 56	21 11
04	06 15	15 13	08 51	11 36	10 06	07 11	22 05	08 08	07 55	21 11
05	05 52	17 30	08 04	11 44	10 23	07 09	22 05	08 09	07 54	21 11
06	05 29	18 42	07 16	11 51	10 39	07 07	22 05	08 10	07 54	21 11
07	05 06	18 44	06 26	11 57	10 55	07 05	22 05	08 11	07 53	21 11
08	04 42	17 39	05 36	12 01	11 11	07 03	22 05	08 13	07 52	21 11
09	04 19	15 33	04 44	12 04	11 28	07 01	22 05	08 14	07 51	21 11
10	03 55	12 37	03 51	12 05	11 43	06 58	22 05	08 15	07 50	21 11
11	03 32	09 04	02 58	12 05	11 59	06 56	22 05	08 16	07 49	21 11
12	03 08	05 08	02 04	12 04	12 15	06 54	22 05	08 17	07 48	21 10
13	02 44	01 N 03	01 09	12 01	12 31	06 51	22 05	08 18	07 48	21 10
14	02 21	03 S 01	00 S 15	11 56	12 46	06 49	22 05	08 19	07 47	21 10
15	01 57	06 53	00 N 43	11 50	13 01	06 47	22 05	08 21	07 46	21 10
16	01 33	10 23	01 39	11 42	13 16	06 44	22 05	08 22	07 45	21 10
17	01 10	13 26	02 35	11 33	13 32	06 42	22 05	08 23	07 44	21 10
18	00 46	15 53	03 31	11 23	13 46	06 39	22 05	08 24	07 43	21 10
19	00 S 22	17 40	04 27	11 11	14 01	06 36	22 05	08 25	07 43	21 10
20	00 N 02	18 41	05 22	10 58	14 16	06 34	22 05	08 27	07 42	21 10
21	00 25	18 53	06 16	10 43	14 31	06 31	22 05	08 28	07 41	21 10
22	00 49	18 12	07 09	10 27	14 45	06 28	22 05	08 29	07 40	21 10
23	01 13	16 37	08 01	10 10	14 59	06 26	22 05	08 30	07 39	21 10
24	01 36	14 10	08 52	09 52	15 13	06 23	22 05	08 31	07 39	21 10
25	02 00	10 55	09 40	09 33	15 27	06 20	22 05	08 33	07 38	21 10
26	02 23	07 00	10 26	09 13	15 41	06 17	22 05	08 34	07 37	21 10
27	02 47	02 S 35	11 10	08 53	15 55	06 15	22 05	08 35	07 36	21 10
28	03 10	02 N 05	11 52	08 32	16 08	06 12	22 05	08 36	07 35	21 09
29	03 34	06 41	12 31	08 10	16 22	06 09	22 05	08 38	07 35	21 09
30	03 57	10 56	13 07	07 49	16 35	06 06	22 05	08 39	07 34	21 09
31	04 N 20	14 N 30	13 N 40	07 N 27	16 N 48	06 S 03	22 S 05	08 N 40	07 S 33	21 S 09

ZODIAC SIGN ENTRIES

Date	h	m	Planets
02	07	43	☽ ♉
04	10	05	☽ ♊
06	12	54	☽ ♋
08	16	45	☽ ♌
10	00	34	♂ ♉
10	22	07	☽ ♍
13	05	28	☽ ♎
13	21	07	☿ ♈
15	15	11	☽ ♏
18	03	00	☽ ♐
20	10	29	☉ ♈
20	15	31	☽ ♑
23	02	28	☽ ♒
25	10	06	☽ ♓
27	14	11	☽ ♈
29	15	48	☽ ♉
31	16	40	☽ ♊
31	17	30	☿ ♉

LATITUDES

Date	Mercury ☿ ° '	Venus ♀ ° '	Mars ♂ ° '	Jupiter ♃ ° '	Saturn ♄ ° '	Uranus ♅ ° '	Neptune ♆ ° '	Pluto ♇ ° '
01	02 S 02	06 N 29	00 N 02	01 N 31	01 N 18	00 S 34	00 S 51	00 N 56
04	01 53	06 58	00 04	01 32	01 19	00 34	00 51	00 56
07	01 39	07 26	00 06	01 32	01 19	00 34	00 51	00 56
10	01 20	07 50	00 09	01 33	01 19	00 34	00 51	00 55
13	00 56	08 09	00 11	01 33	01 19	00 34	00 51	00 55
16	00 S 26	08 23	00 13	01 34	01 19	00 34	00 51	00 55
19	00 N 07	08 29	00 15	01 34	01 20	00 34	00 51	00 55
22	00 43	08 27	00 17	01 34	01 20	00 34	00 51	00 55
25	01 20	08 17	00 19	01 34	01 20	00 34	00 51	00 54
28	01 55	07 59	00 21	01 35	01 20	00 34	00 51	00 54
31	02 N 26	07 N 33	00 N 23	01 N 35	01 N 20	00 S 34	00 S 51	00 N 54

DATA

Julian Date	2457814
Delta T	+70 seconds
Ayanamsa	24° 05' 42"
Synetic vernal point	05° ♓ 01' 18"
True obliquity of ecliptic	23° 26' 06"

MOON'S PHASES, APSIDES AND POSITIONS ☽

Date	h	m	Phase	Longitude ° '	Eclipse Indicator
05	11	32	☽	15 ♊ 05	
12	14	54	○	22 ♍ 13	
20	15	58	☾	00 ♑ 14	
28	02	57	●	07 ♈ 37	

Day	h	m		
03	07	28	Perigee	
18	17	21	Apogee	
30	12	26	Perigee	
07	00	49	Max dec	18° N 52'
13	18	07	0S	
21	05	23	Max dec	18° S 55'
28	01	21	0N	

LONGITUDES

Date	Chiron ⚷ ° '	Ceres ⚳ ° '	Pallas ⚴ ° '	Juno ⚵ ° '	Vesta ⚶ ° '	Black Moon Lilith ⚸ ° '
01	23 ♓ 53	07 ♉ 34	19 ♓ 58	07 ♑ 37	20 ♋ 15	01 ♐ 44
11	24 ♓ 29	11 ♉ 03	23 ♓ 26	10 ♑ 10	20 ♋ 11	02 ♐ 51
21	25 ♓ 05	14 ♉ 41	26 ♓ 56	12 ♑ 28	20 ♋ 51	03 ♐ 59
31	25 ♓ 41	18 ♉ 27	00 ♈ 28	14 ♑ 27	22 ♋ 10	05 ♐ 06

ASPECTARIAN

h m	Aspects
01 Wednesday	
00 24	☽ ⊥ ☿
00 46	☽ ⊻ ♆
02 55	☽ ☌ ♀
09 48	☽ ⊥ ☉
10 57	☽ ⊥ ♆
11 49	☿ # ♀
12 49	☽ □ ♇
17 51	☽ ∠ ☿
18 27	☽ ☌ ♅
18 43	☽ ☍ ♃
21 02	☉ ∥ ♃
21 08	☿ ⊥ ♀
21 48	☽ ☌ ♂
22 40	☿ ∠ ♅
02 Thursday	
00 21	☿ ⚼ ♃
02 08	☽ ∠ ☉
02 10	☽ ∠ ♆
02 18	☽ △ ♄
02 44	☉ ☌ ♆
06 27	☽ # ☉
07 13	☽ # ♃
11 12	☽ # ♆
12 02	☽ ∥ ♅
20 49	☿ Q ♄
21 09	☽ ∥ ♂
22 48	☽ ⚹ ☿
23 17	☽ # ☿
03 Friday	
01 15	♃ ☍ ♅
03 26	☽ ⚹ ♆
03 36	☽ ⚼ ♄
05 14	☽ ⚹ ☉
05 44	☽ ⊻ ♀
06 59	☿ # ♂
08 08	☽ ∥ ♀
12 49	☉ ⊻ ♀
15 20	☽ △ ♇
15 50	☽ ⊥ ♀
17 47	☿ ∠ ♂
18 46	☽ ± ♄
20 51	☽ ⊼ ♃
21 02	☽ ⊻ ♅
21 56	☽ Q ☿
23 38	☽ Q ♆
04 Saturday	
02 54	☽ Q ☉
02 54	☽ ⊻ ♂
04 53	☽ ⊼ ♄
06 53	☽ ± ♃
06 58	☽ ∠ ♀
07 09	☽ ⊥ ♅
09 09	♀ St R
11 10	☿ ☌ ♆
13 32	☽ ⊥ ♂
16 35	☽ ⚼ ♇
21 57	☽ ⚼ ♃
22 21	☽ ∠ ♅
05 Sunday	
04 17	☿ ⊻ ♀
05 32	☽ ∠ ♂
06 02	☽ □ ♆
07 48	☽ ± ♇
08 14	☽ ⚹ ♀
08 51	☽ □ ☿
09 21	☿ # ♅
11 32	☽ □ ☉
15 53	♂ ∠ ♆
16 47	☿ ∥ ♆
17 59	☽ ⊼ ♇
20 46	♂ △ ♄
23 12	☽ △ ♃
23 50	☽ ⚹ ♅
06 Monday	
04 31	☽ Q ♀
07 48	☽ ☍ ♄
07 50	☉ ± ♃
08 22	☽ ⚹ ♂
15 25	☿ ± ♃
16 27	☿ ∥ ♃
18 19	☉ ⊥ ♅
20 23	☽ Q ♅
21 19	☿ ⊥ ♅
07 Tuesday	
00 29	☉ ☌ ☿
06 02	☽ Q ♂
09 20	☽ △ ♆
11 08	☽ □ ♀
18 48	☽ △ ☉
20 12	☽ △ ☿
21 28	☽ ☍ ♇
08 Wednesday	
02 25	☽ □ ♃
03 33	☽ □ ♅
05 25	☿ ⚹ ♇
11 25	☽ ⚼ ♆
11 43	☽ ⊼ ♄
14 59	☽ □ ♂
22 15	☽ ± ♄
23 00	☽ ⚼ ☉
09 Thursday	
02 41	☽ ⚼ ☿
03 15	☽ ± ♆
08 33	☉ ⚹ ♇
09 45	☽ Q ♃
13 53	☽ ⊼ ♆
14 11	☽ ⚼ ♄
14 55	☽ △ ♀
14 57	☿ ⊼ ♃
16 10	☽ ± ☉
21 16	☽ ± ☿
10 Friday	
01 55	☿ ⊻ ♅
02 18	☽ ⊼ ♇
03 42	☽ ⊼ ☉
06 58	☽ ⚹ ♃
08 42	☽ △ ♅
08 59	☉ ∥ ☿
09 49	☽ ⊼ ☿
13 04	☽ ± ♇
15 47	☽ ∥ ♀
17 05	☽ △ ♄
17 15	☽ ⚼ ♀
17 53	☽ ∥ ♂
23 22	☽ △ ♂
11 Saturday	
05 22	☽ ⚼ ♇
07 16	☿ ⊥ ♂
09 14	☽ ± ♀
09 53	☽ ∠ ♃
11 56	☽ ⚼ ♅
13 53	♀ ⊻ ♆
17 01	☽ ∥ ♅
19 49	☽ # ♆
19 58	☽ ⊼ ♀
20 00	☉ ⊼ ♃
20 08	☽ ☍ ♆
20 20	♀ ∥ ♂
12 Sunday	
01 21	☽ # ♃
02 16	☽ ⊥ ♃
04 24	☽ ⚼ ♂
04 32	☽ ± ♅
08 57	☽ △ ♇
12 10	☿ □ ♄
13 18	☽ ⊻ ♃
14 54	☽ ☍ ☉
15 43	☽ ⊼ ♅
22 08	☽ ± ♂
13 Monday	
00 25	☽ □ ♄
01 03	☽ # ☉
02 04	☉ ⊻ ♅
02 36	☽ ☍ ☿
10 05	☽ ⊼ ♂
11 14	☽ # ☿
22 29	☽ ∥ ☿
14 Tuesday	
02 41	☽ ☍ ♀
04 34	☽ ⊼ ♆
08 20	☽ ∥ ☉
09 58	☿ Q ♇
10 44	☽ Q ♄
16 09	☽ ± ♆
17 52	☽ □ ♇
21 52	☽ ☌ ♃
15 Wednesday	
01 05	☽ ☍ ♅
05 00	☽ ⊼ ☉
09 42	☽ ⚼ ♆
10 05	☽ ⚹ ♄
11 22	☽ ∥ ♃
17 47	☽ ± ☉
17 49	☽ ∥ ♆
21 45	☽ # ♅
23 17	☽ ⊼ ☿
23 38	☽ ☍ ♂
16 Thursday	
02 21	☿ ⊻ ♂
05 09	☽ Q ♇
10 23	☉ # ☿
11 13	☽ ⊼ ♀
13 11	☽ ⚼ ☉
13 29	☽ ± ☿
15 25	☽ △ ♆
15 47	☽ ∠ ♄
21 29	☽ # ♀
22 42	☽ ± ♀
17 Friday	
05 05	☽ ⚹ ♇
08 38	☽ ⊻ ♃
09 50	☽ ⊥ ♄
10 58	☽ ⚼ ☿
12 44	☽ ⊼ ♅
12 57	☽ # ♂
16 01	☽ ⚼ ♀
20 35	☽ ⊥ ♃
21 48	☉ □ ♄
21 56	☽ ⊻ ♄
21 56	☽ △ ☉
18 Saturday	
00 52	☽ ± ♅
11 20	☽ ∠ ♇
12 27	☿ ☌ ♀
14 34	☽ ∠ ♃
15 31	☽ ⊼ ♂
19 07	☽ ⚼ ♅
20 59	☽ △ ♀
23 06	☽ △ ☿
19 Sunday	
03 57	☽ □ ♆
04 25	☽ ± ♂
05 32	☽ ⊥ ♇
17 41	☽ ⊻ ♇
20 36	☽ ⚹ ♃
23 46	☽ ⚼ ♂
20 Monday	
01 33	☽ △ ♅
05 51	☿ ⊻ ♆
10 37	☽ ☌ ♄
15 58	☽ □ ☉
16 19	☽ Q ♆
20 22	♀ ⊻ ♂
20 29	☽ Q ♃
21 Tuesday	
06 32	☽ □ ♀
07 43	☽ △ ♂
14 53	☉ Q ♇
16 17	☽ ⚹ ♆
18 20	☿ # ♃
22 20	☽ □ ☿
22 Wednesday	
05 32	☽ ☌ ♇
06 48	☽ Q ☉
07 44	☽ □ ♃
13 20	☽ □ ♅
16 13	☽ Q ♀
21 38	☽ ∠ ♆
21 55	☽ ⊻ ♄
23 Thursday	
01 54	☿ # ♆
07 58	☽ ⚹ ☉
09 25	☽ ⊥ ♄
14 06	☽ ⚹ ♀
14 51	☽ Q ☿
14 52	☽ ⊥ ♆
14 57	☿ ⊥ ♆
21 20	☽ □ ♂
23 48	☽ Q ♅
24 Friday	
00 45	☿ □ ♇
02 02	☿ ∥ ♅
02 10	☽ ⊻ ♆
02 24	☽ ∠ ♄
03 28	☽ # ♂
12 45	☿ ☍ ♃
14 23	☽ ∠ ☉
14 33	☽ ⊻ ♇
16 03	☽ △ ♃
16 34	☽ ⚹ ☿
16 44	☽ ∠ ♀
22 02	☽ ⚹ ♅
25 Saturday	
01 30	☽ ⊥ ♇
05 56	☽ ⚹ ♄
08 04	☽ ⊥ ☉
08 17	☽ Q ♂
08 19	☽ ⊥ ♀
08 41	♀ ⊥ ♂
09 31	☿ ∥ ♀
10 17	☉ ☌ ♀
17 39	☽ ∠ ♇
17 58	☉ ⊥ ♂
18 33	☽ ⊻ ♀
18 42	☽ # ☿
18 51	☽ ⚼ ♃
19 37	☽ ⊻ ☉
21 36	☽ # ♀
23 14	☽ ∠ ☿
26 Sunday	
00 58	☽ ∠ ♅
02 52	☽ # ♅
03 17	☽ Q ♄
06 43	☽ ⚹ ♂
08 23	☽ ☌ ♆
08 25	☽ ∥ ♆
10 30	☽ ± ♃
15 06	☿ ☌ ♅
16 01	☽ ∥ ♃
16 43	☽ ⊥ ♅
16 54	☽ ⊥ ☿
19 52	☽ ⚹ ♇
20 47	☽ ⊼ ♃
27 Monday	
03 03	☽ ⊻ ♅
04 25	☽ ⊻ ☿
09 52	☽ ∠ ♂
10 19	☽ □ ♄
11 02	☽ # ☉
16 18	☽ Q ♇
16 44	♂ ⚹ ♆
18 57	♂ ⚼ ♄
19 59	☽ ☌ ♀
28 Tuesday	
01 47	☽ ⊥ ♂
02 57	☽ ☌ ☉
07 07	♀ ∥ ♅
11 22	☽ ⊻ ♆
12 16	☽ ⊻ ♂
18 06	☽ ∥ ☉
21 17	☽ ⊥ ♆
22 14	☽ □ ♇
22 38	☽ ☍ ♃
29 Wednesday	
05 15	☽ ☌ ♅
09 08	☽ # ♃
11 33	☽ ☌ ☿
12 04	☽ ∠ ♆
12 07	☽ △ ♄
16 46	☽ # ♆
17 52	☿ ∠ ♆
18 16	☿ △ ♄
19 26	☽ ∥ ♀
19 29	☽ ⊻ ♀
22 38	☽ ∥ ♅
30 Thursday	
04 53	☽ ⊥ ♀
07 45	☽ ⊻ ☉
12 34	☽ ⚹ ♆
12 34	☽ ⚼ ♄
15 49	☽ ☌ ♂
18 13	☽ ⊥ ☉
18 19	♃ □ ♇
18 59	☽ ∠ ♀
23 09	☽ ⊼ ♃
23 12	☽ △ ♇
31 Friday	
03 14	☽ ± ♄
04 41	☽ ∥ ☿
04 42	♀ # ♆
06 20	☽ ⊻ ♅
08 09	☽ Q ♆
08 51	☽ ± ♃
09 56	☽ ∠ ☉
12 10	♀ Q ♇
13 01	☽ ⊼ ♄
16 10	☽ ⊥ ♅
16 36	☽ ⊻ ☿
18 33	☽ ⚹ ♀
23 27	☽ ⚼ ♃
23 43	☽ ⚼ ♇

All ephemeris data is given at 12.00 UT and the Moon's longitude is additionally given for 24.00 UT

APRIL 2017

Date	Sidereal time h m s	Sun ☉ ° ' "	Moon ☽ ° ' "	Moon ☽ 24.00 ° ' "	Mercury ☿ ° '	Venus ♀ ° '	Mars ♂ ° '	Jupiter ♃ ° '	Saturn ♄ ° '	Uranus ♅ ° '	Neptune ♆ ° '	Pluto ♇ ° '
01	00 40 09	11 ♈ 56 53	11 ♊ 45 39	18 ♊ 59 47	00 ♉ 46	00 ♈ 46	16 ♉ 03	19 ♎ 04	27 ♐ 47	23 ♈ 44	12 ♓ 49	19 ♑ 18
02	00 44 06	12 56 05	26 ♊ 10 19	03 ♋ 16 52	01 41	00 ♈ 15	16 46	18 R 57	27 47	23 48	12 51	19 19
03	00 48 02	13 55 15	10 ♋ 19 13	17 ♋ 17 15	02 28	29 ♓ 46	17 28	18 49	27 47	23 51	12 53	19 20
04	00 51 59	14 54 22	24 ♋ 10 58	01 ♌ 00 26	03 09	29 R 19	18 10	18 41	27 48	23 54	12 55	19 20
05	00 55 56	15 53 27	07 ♌ 45 46	14 ♌ 27 08	03 43	28 54	18 53	18 34	27 48	23 58	12 57	19 21
06	00 59 52	16 52 29	21 ♌ 04 44	27 ♌ 38 46	04 10	28 32	19 35	18 26	27 R 48	24 01	12 59	19 21
07	01 03 49	17 51 30	04 ♍ 09 26	10 ♍ 36 56	04 30	28 11	20 17	18 18	27 48	24 05	13 01	19 21
08	01 07 45	18 50 28	17 ♍ 01 27	23 ♍ 23 06	04 44	27 53	20 59	18 11	27 47	24 08	13 03	19 22
09	01 11 42	19 49 23	29 ♍ 42 03	05 ♎ 58 25	04 50	27 37	21 41	18 03	27 47	24 11	13 05	19 22
10	01 15 38	20 48 17	12 ♎ 12 17	18 ♎ 23 45	04 R 50	27 24	22 23	17 55	27 47	24 15	13 07	19 22
11	01 19 35	21 47 09	24 ♎ 32 55	00 ♏ 39 53	04 43	27 13	23 05	17 48	27 46	24 18	13 09	19 23
12	01 23 31	22 45 58	06 ♏ 44 45	12 ♏ 47 39	04 31	27 05	23 47	17 40	27 46	24 22	13 11	19 23
13	01 27 28	23 44 46	18 ♏ 48 44	24 ♏ 48 13	04 12	26 59	24 29	17 32	27 45	24 25	13 13	19 23
14	01 31 25	24 43 32	00 ♐ 46 20	06 ♐ 43 20	03 48	26 56	25 11	17 25	27 44	24 29	13 15	19 23
15	01 35 21	25 42 16	12 ♐ 39 34	18 ♐ 35 23	03 20	26 D 55	25 53	17 17	27 43	24 32	13 16	19 24
16	01 39 18	26 40 59	24 ♐ 31 12	00 ♑ 27 30	02 47	26 56	26 34	17 09	27 43	24 35	13 18	19 24
17	01 43 14	27 39 39	06 ♑ 24 46	12 ♑ 23 34	02 11	27 00	27 16	17 02	27 41	24 39	13 20	19 24
18	01 47 11	28 38 18	18 ♑ 24 28	24 ♑ 28 04	01 33	27 06	27 58	16 55	27 40	24 42	13 22	19 24
19	01 51 07	29 ♈ 36 56	00 ♒ 35 00	06 ♒ 45 53	00 52	27 14	28 39	16 47	27 39	24 46	13 23	19 24
20	01 55 04	00 ♉ 35 31	13 ♒ 01 22	19 ♒ 22 03	00 ♉ 10	27 24	29 ♉ 21	16 40	27 38	24 49	13 25	19 24
21	01 59 00	01 34 05	25 ♒ 48 28	02 ♓ 21 09	29 ♈ 28	27 37	00 ♊ 03	16 33	27 36	24 53	13 27	19 R 24
22	02 02 57	02 32 37	09 ♓ 00 30	15 ♓ 46 50	28 46	27 51	00 44	16 25	27 35	24 56	13 28	19 24
23	02 06 54	03 31 08	22 ♓ 40 17	29 ♓ 40 53	28 05	28 08	01 26	16 18	27 33	24 59	13 30	19 24
24	02 10 50	04 29 37	06 ♈ 48 26	14 ♈ 02 32	27 26	28 27	02 07	16 11	27 31	25 03	13 32	19 24
25	02 14 47	05 28 04	21 ♈ 22 34	28 ♈ 47 45	26 49	28 47	02 48	16 04	27 30	25 06	13 33	19 24
26	02 18 43	06 26 29	06 ♉ 17 03	13 ♉ 49 19	26 16	29 09	03 30	15 57	27 28	25 10	13 35	19 23
27	02 22 40	07 24 53	21 ♉ 23 16	28 ♉ 57 37	25 46	29 33	04 11	15 51	27 26	25 13	13 36	19 23
28	02 26 36	08 23 15	06 ♊ 31 01	14 ♊ 02 15	25 20	29 ♓ 59	04 52	15 44	27 24	25 16	13 38	19 23
29	02 30 33	09 21 35	21 ♊ 30 12	28 ♊ 53 54	24 58	00 ♈ 26	05 34	15 37	27 22	25 20	13 39	19 23
30	02 34 29	10 ♉ 19 53	06 ♋ 12 37	13 ♋ 25 45	24 ♈ 40	00 ♈ 55	06 ♊ 15	15 ♎ 31	27 ♐ 19	25 ♈ 23	13 ♓ 41	19 ♑ 22

Date	Moon True ☊ ° '	Moon Mean ☊ ° '	Moon ☽ Latitude ° '	DECLINATIONS Sun ☉ ° '	Moon ☽ ° '	Mercury ☿ ° '	Venus ♀ ° '	Mars ♂ ° '	Jupiter ♃ ° '	Saturn ♄ ° '	Uranus ♅ ° '	Neptune ♆ ° '	Pluto ♇ ° '
01	02 ♍ 41	01 ♍ 26	05 S 06	04 N 43	17 N 08	14 N 10	07 N 05	17 N 01	06 S 00	22 S 05	08 N 41	07 S 32	21 S 09
02	02 R 38	01 23	04 45	05 06	18 38	14 37	06 43	17 14	05 57	22 05	08 43	07 31	21 09
03	02 36	01 20	04 06	05 29	18 57	15 00	06 21	17 26	05 54	22 04	08 44	07 31	21 09
04	02 D 37	01 17	03 13	05 52	18 06	15 20	05 59	17 39	05 51	22 04	08 45	07 30	21 09
05	02 38	01 13	02 11	06 15	16 13	15 36	05 38	17 51	05 49	22 04	08 46	07 29	21 09
06	02 39	01 10	01 S 02	06 38	13 29	15 49	05 18	18 03	05 46	22 04	08 48	07 29	21 09
07	02 R 40	01 07	00 N 08	07 00	10 07	15 59	04 58	18 15	05 43	22 04	08 49	07 28	21 09
08	02 39	01 04	01 17	07 23	06 18	16 05	04 39	18 27	05 40	22 04	08 50	07 27	21 09
09	02 35	01 01	02 21	07 45	02 N 16	16 07	04 21	18 38	05 37	22 04	08 51	07 26	21 09
10	02 30	00 58	03 17	08 07	01 S 48	16 05	04 03	18 50	05 34	22 04	08 53	07 26	21 09
11	02 22	00 54	04 02	08 29	05 45	16 00	03 46	19 01	05 31	22 04	08 54	07 25	21 09
12	02 13	00 51	04 36	08 51	09 25	15 52	03 30	19 12	05 28	22 04	08 55	07 24	21 09
13	02 03	00 48	04 58	09 13	12 38	15 40	03 16	19 23	05 25	22 04	08 57	07 24	21 10
14	01 54	00 45	05 06	09 35	15 19	15 25	03 02	19 33	05 23	22 04	08 58	07 23	21 10
15	01 45	00 42	05 00	09 56	17 21	15 07	02 49	19 44	05 20	22 04	08 59	07 22	21 10
16	01 38	00 38	04 42	10 17	18 38	14 46	02 37	19 54	05 17	22 04	09 00	07 22	21 10
17	01 33	00 35	04 11	10 38	19 06	14 22	02 26	20 04	05 14	22 04	09 02	07 21	21 10
18	01 31	00 32	03 30	10 59	18 43	13 57	02 16	20 14	05 11	22 03	09 03	07 20	21 10
19	01 D 30	00 29	02 38	11 20	17 27	13 29	02 08	20 24	05 09	22 03	09 04	07 20	21 10
20	01 31	00 26	01 37	11 41	15 21	13 00	02 00	20 34	05 06	22 03	09 05	07 19	21 10
21	01 32	00 23	00 N 31	12 01	12 26	12 31	01 53	20 43	05 03	22 03	09 07	07 18	21 10
22	01 R 31	00 19	00 S 40	12 21	08 48	12 01	01 48	20 53	05 00	22 03	09 08	07 18	21 10
23	01 30	00 16	01 50	12 41	04 S 36	11 30	01 43	21 02	04 58	22 03	09 09	07 17	21 10
24	01 25	00 13	02 56	13 01	00 00	11 00	01 39	21 10	04 55	22 03	09 10	07 17	21 10
25	01 19	00 10	03 53	13 21	04 N 44	10 31	01 37	21 19	04 53	22 03	09 12	07 16	21 10
26	01 10	00 07	04 35	13 40	09 17	10 03	01 35	21 28	04 50	22 03	09 13	07 15	21 10
27	01 00	00 04	04 58	13 59	13 18	09 37	01 34	21 36	04 48	22 03	09 14	07 15	21 10
28	00 50	00 ♍ 00	05 01	14 18	16 27	09 12	01 34	21 44	04 45	22 03	09 15	07 14	21 11
29	00 42	29 ♌ 57	04 43	14 37	18 27	08 49	01 35	21 52	04 43	22 03	09 17	07 14	21 11
30	00 ♍ 36	29 ♌ 54	04 S 07	14 N 55	19 N 11	08 N 29	01 N 37	21 N 59	04 S 40	22 S 03	09 N 18	07 S 13	21 S 11

ZODIAC SIGN ENTRIES

Date	h m	Planets
02	18 27	☽ ♋
03	00 25	♀ ♓
04	22 13	☽ ♌
07	04 20	☽ ♍
09	12 34	☽ ♎
11	22 42	☽ ♏
14	10 27	☽ ♐
16	23 04	☽ ♑
19	10 52	☽ ♒
19	21 27	☉ ♉
20	17 37	☿ ♈
21	10 32	♂ ♊
21	19 43	☽ ♓
24	00 32	☽ ♈
26	01 56	☽ ♉
28	01 39	☽ ♊
28	13 13	♀ ♈
30	01 48	☽ ♋

LATITUDES

Date	Mercury ☿ ° '	Venus ♀ ° '	Mars ♂ ° '	Jupiter ♃ ° '	Saturn ♄ ° '	Uranus ♅ ° '	Neptune ♆ ° '	Pluto ♇ ° '
01	02 N 35	07 N 23	00 N 23	01 N 35	01 N 20	00 S 34	00 S 51	00 N 54
04	02 57	06 50	00 25	01 35	01 21	00 34	00 52	00 54
07	03 09	06 12	00 27	01 35	01 21	00 34	00 52	00 53
10	03 09	05 32	00 29	01 35	01 21	00 34	00 52	00 53
13	02 55	04 51	00 30	01 35	01 21	00 34	00 52	00 53
16	02 28	04 11	00 32	01 35	01 21	00 34	00 52	00 53
19	01 50	03 31	00 34	01 34	01 21	00 34	00 52	00 53
22	01 03	02 53	00 35	01 34	01 22	00 34	00 52	00 52
25	00 N 12	02 17	00 37	01 34	01 22	00 34	00 52	00 52
28	00 S 38	01 43	00 38	01 33	01 22	00 34	00 52	00 52
31	01 S 24	01 N 11	00 N 40	01 N 33	01 N 22	00 S 34	00 S 52	00 N 52

DATA

Julian Date	2457845
Delta T	+70 seconds
Ayanamsa	24° 05' 44"
Synetic vernal point	05° ♓ 01' 15"
True obliquity of ecliptic	23° 26' 06"

MOON'S PHASES, APSIDES AND POSITIONS ☽

Date	h m	Phase	Longitude ° '	Eclipse Indicator
03	18 39	☽	14 ♋ 12	
11	06 08	○	21 ♎ 33	
19	09 57	☾	29 ♑ 32	
26	12 16	●	06 ♉ 27	

Day	h m	
15	09 55	Apogee
27	16 09	Perigee
03	06 15	Max dec 18° N 59'
10	01 20	0S
17	13 16	Max dec 19° S 06'
24	12 00	0N
30	13 35	Max dec 19° N 11'

LONGITUDES

Date	Chiron ⚷ ° '	Ceres ⚳ ° '	Pallas ⚴ ° '	Juno ⚵ ° '	Vesta ⚶ ° '	Black Moon Lilith ⚸ ° '
01	25 ♓ 44	18 ♉ 50	00 ♈ 49	14 ♑ 38	22 ♋ 20	05 ♐ 12
11	26 ♓ 19	22 ♉ 43	04 ♈ 20	16 ♑ 14	24 ♋ 17	06 ♐ 20
21	26 ♓ 51	26 ♉ 42	07 ♈ 52	17 ♑ 26	26 ♋ 42	07 ♐ 27
31	27 ♓ 20	00 ♊ 45	11 ♈ 22	18 ♑ 09	29 ♋ 33	08 ♐ 34

ASPECTARIAN

h m	Aspects
01 Saturday	
03 11	☽ ⊥ ☿
06 18	♀ ∠ ♂
06 59	☽ ∠ ♅
10 31	☽ ∥ ♂
11 53	☿ ⊻ ♀
12 20	☽ ✶ ☉
13 36	☽ Q ♀
13 46	☽ □ ♆
14 34	☽ ± ♇
19 06	☽ ∠ ☿
19 29	☽ ⊻ ♂
02 Sunday	
00 01	☽ △ ♃
00 32	☽ ⊼ ♇
05 59	☽ ⊥ ♂
08 00	☽ ✶ ♅
09 46	☽ Q ☉
10 02	☉ ⊻ ♆
14 43	☽ ☍ ♄
18 39	☽ □ ♀
21 52	☽ ✶ ☿
21 55	☽ ∠ ♂
03 Monday	
01 11	♃ ± ♆
04 20	☽ Q ♅
16 25	☽ △ ♆
18 39	☽ □ ☉
19 32	☽ Q ☿
04 Tuesday	
00 58	☽ ✶ ♂
02 31	☽ □ ♃
03 33	☽ ☍ ♇
11 15	☉ # ♃
11 31	☽ □ ♅
15 55	☉ ∥ ♀
18 20	☽ ⊼ ♄
18 20	☽ ∥ ♂
18 35	☽ ⚼ ♆
20 45	☽ △ ♀
22 33	♀ # ♃
23 06	☽ Q ♂
05 Wednesday	
02 53	♂ ⊼ ♃
04 31	☽ □ ☿
04 56	☽ ± ♄
09 53	☽ Q ♃
10 34	☽ ± ♆
17 28	☽ ∥ ☿
21 01	☽ ⚼ ♄
21 20	☽ ⊼ ♆
22 42	☽ ⚼ ♀
06 Thursday	
03 46	☽ △ ☉
04 01	♂ △ ♇
05 06	♄ St R
07 15	☽ ✶ ♃
08 51	☽ ⊼ ♇
09 08	☽ □ ♂
14 34	☽ ± ♀
17 23	☽ △ ♅
19 48	☽ ± ♇
23 59	☿ ⊥ ♀
07 Friday	
00 16	☽ △ ♄
01 16	☽ ⊼ ♀
09 24	☽ ⚼ ☉
10 26	☽ ∠ ♃
12 22	☽ ⚼ ♇
12 40	☽ △ ☿
20 21	☽ ∥ ♅
21 10	☽ ⚼ ♅
21 39	☉ ☍ ♃
08 Saturday	
03 00	☽ ⊥ ♃
03 30	☽ ± ☉
04 32	☽ ☍ ♆
04 57	☽ # ♆
05 59	☽ ∥ ☉
14 06	☽ ± ♅
14 09	☽ ⊻ ♃
15 42	☽ ⊼ ☉
15 55	☽ # ♃
16 24	☽ △ ♇
16 29	☉ # ♆
17 09	☽ ⚼ ☿
17 24	☉ ⊥ ♆
19 54	☽ △ ♂
20 28	♀ □ ♄
22 46	☽ ∥ ♀
09 Sunday	
00 49	☉ □ ♇
01 29	☽ ⊼ ♅
08 07	☽ ☍ ♀
08 21	☽ □ ♄
10 21	☽ ± ☿
15 24	♂ ± ♄
21 50	☽ ⊼ ☿
23 15	☿ St R
10 Monday	
02 10	☽ ⚼ ♂
13 46	☽ ⊼ ♆
18 55	☽ Q ♄
20 35	☽ ± ♂
22 58	☽ ☌ ♃
11 Tuesday	
00 35	☽ # ♀
01 26	☽ ± ♆
01 54	☽ □ ♇
06 08	☽ ☍ ☉
08 58	☽ ⊼ ♂
10 34	☽ ∥ ♃
11 31	☽ ☍ ♅
17 10	☽ ⊼ ♀
18 19	☽ ✶ ♄
19 05	☽ ⚼ ♆
22 36	☽ ∥ ♆
12 Wednesday	
04 50	☽ ± ♀
07 40	☽ ☍ ☿
07 45	☽ # ☉
08 30	♂ ± ♃
08 38	☽ # ♅
13 16	☽ Q ♇
16 46	☉ ∥ ♅
22 29	☽ ⚼ ♀
23 56	☽ ∠ ♄
13 Thursday	
00 48	☽ △ ♆
09 29	☽ ⊻ ♃
09 37	♂ ⊻ ♅
13 09	☽ ✶ ♇
17 52	☽ ⊥ ♄
21 22	☽ ⊥ ♃
22 46	☽ ⊼ ☉
23 17	☽ ⊼ ♅
14 Friday	
00 04	☽ ☍ ♂
04 18	☽ △ ♀
05 30	☉ ☌ ♅
05 54	☽ ⊻ ♄
11 24	☽ ± ♅
11 54	☽ ± ☉
12 48	☽ # ☿
14 15	♂ Q ♆
15 16	☽ ∠ ♃
17 54	☽ ⊼ ☿
19 18	☽ ∠ ♇
15 Saturday	
05 33	☽ ± ☿
05 39	☽ ⚼ ♅
07 41	☽ ⚼ ☉
10 18	♀ St D
13 15	☽ □ ♆
13 29	☽ ⊥ ♇
21 15	☽ ✶ ♃
23 00	☽ ⚼ ☿
16 Sunday	
01 38	☽ ⊻ ♇
02 37	☉ ⊻ ♂
06 23	☿ ⊥ ♀
12 09	☽ △ ♅
16 25	☽ ⊼ ♂
16 46	☽ △ ☉
16 54	☽ □ ♀
18 26	☉ ⊻ ♀
18 26	☽ ☌ ♄
21 17	☽ Q ♃
17 Monday	
01 26	♀ ✶ ♂
01 44	☽ Q ♆
03 55	☽ △ ☿
05 16	☽ ± ♂
12 43	☉ △ ♄
18 Tuesday	
00 28	☽ ⚼ ♂
01 55	☽ ✶ ♆
02 14	♂ ⊼ ♄
04 58	☉ ∠ ♆
05 20	☽ Q ♀
09 03	☽ □ ♃
13 58	☽ ☌ ♇
19 Wednesday	
00 32	☽ □ ♅
05 21	☽ ✶ ♀
06 16	☽ ⊻ ♄
07 42	☽ ∠ ♆
08 00	☽ △ ♂
09 57	☽ □ ☉
12 31	☽ □ ☿
17 58	☽ ⊥ ♄
20 Thursday	
01 14	☽ ⊥ ♆
05 54	☉ ☌ ☿
10 48	☽ ∠ ♀
11 15	☽ ∠ ♄
11 37	☽ Q ♅
12 45	☽ ⊻ ♆
12 48	♇ St R
18 50	☽ △ ♃
21 14	☽ Q ☿
23 25	☽ Q ☉
21 Friday	
00 04	☽ ⊻ ♇
02 00	☿ ⊻ ♂
04 04	☽ ⊥ ♀
10 16	☽ ✶ ♅
11 09	♀ □ ♄
11 15	☽ ⊥ ♇
11 20	☽ # ☿
14 44	☽ # ☉
15 19	☽ ✶ ♄
15 23	☽ ⊻ ♀
18 23	☽ ✶ ☿
20 13	☽ □ ♂
22 26	☽ ⚼ ♃
23 25	☽ ✶ ☉
22 Saturday	
02 10	☉ ∥ ☿
03 43	☽ ∠ ♇
10 02	☽ # ♅
13 01	☽ Q ♄
13 39	☽ ∠ ♅
14 30	☽ ± ♃
19 31	♂ ⊥ ♅
19 57	☽ ☌ ♆
20 02	☽ ∠ ☿
20 58	☽ ∥ ♆
21 38	☿ ∠ ♆
23 Sunday	
01 01	☽ ⊼ ♃
04 15	☽ ∠ ☉
05 35	☽ ⊥ ♅
06 05	☽ Q ♂
06 19	☽ ✶ ♇
08 22	♂ ⚼ ♃
09 59	☽ ∥ ♃
10 39	☿ ⊻ ♀
11 02	☽ ⊥ ☿
16 00	☽ ⊻ ♅
20 22	☽ □ ♄
20 52	☽ ⊻ ☿
20 56	☽ ⊥ ☉
21 34	☽ ☌ ♀
24 Monday	
02 54	☽ Q ♇
03 24	☽ # ♀
03 43	☽ ✶ ♂
07 50	☽ ⊻ ☉
08 15	☿ △ ♄
10 08	☉ ⊥ ♀
11 26	♂ # ♇
20 18	☽ ∥ ♀
23 10	☽ ⊻ ♆
25 Tuesday	
03 24	☽ ☍ ♃
05 53	☽ ∠ ♂
08 46	☽ □ ♇
09 02	☽ ⊥ ♆
12 17	☿ ⊥ ♂
12 44	☽ # ♃
18 04	☽ ☌ ♅
20 29	☽ ☌ ☿
21 13	☽ ⊥ ♂
21 53	☽ △ ♄
23 38	☽ ∠ ♆
26 Wednesday	
00 17	☽ ⊻ ♀
01 06	☽ # ♆
07 20	☽ ⊻ ♂
10 09	☽ ⊥ ♀
11 37	☽ ∥ ♅
12 16	☽ ☌ ☉
15 55	☽ ∥ ☿
21 49	☽ ⚼ ♄
23 38	☽ ✶ ♆
27 Thursday	
00 52	☽ ∠ ♀
03 16	☽ ⊼ ♃
08 50	☽ △ ♇
12 04	☽ ± ♄
12 43	☽ ± ♃
17 06	☽ ∥ ☉
18 05	☽ ⊻ ♅
18 42	☽ Q ♆
18 44	☽ ⊻ ☿
19 01	♂ ⚼ ♇
21 33	☽ ⊼ ♄
28 Friday	
01 18	☽ ✶ ♀
02 52	☽ ⚼ ♃
03 38	☽ ⊥ ♅
03 58	☽ ⊥ ☿
08 36	☽ ⚼ ♇
09 00	☿ ∥ ♅
09 16	☽ ☌ ♂
14 49	☿ ☌ ♅
15 11	☽ ⊻ ☉
17 55	☽ ∠ ☿
18 00	☽ ∠ ♅
20 58	☽ Q ♀
22 57	☽ ± ♇
23 22	☽ □ ♆
29 Saturday	
01 26	☽ ⊥ ☉
02 36	☽ △ ♃
08 35	☽ ⊼ ♇
16 57	☽ ∠ ☉
17 29	☽ ✶ ☿
18 13	☽ ✶ ♅
21 28	☽ ☍ ♄
30 Sunday	
02 59	☽ □ ♀
12 04	☽ ⊻ ♂
12 45	☽ Q ☿
13 57	☽ Q ♅
19 19	☽ ✶ ☉
21 44	♂ # ♄
22 32	☽ ⊥ ♂

All ephemeris data is given at 12.00 UT and the Moon's longitude is additionally given for 24.00 UT

MAY 2017

Date	Sidereal time h m s	LONGITUDES Sun ☉ ° ' "	Moon ☽ ° ' "	Moon ☽ 24.00 ° ' "	Mercury ☿ ° '	Venus ♀ ° '	Mars ♂ ° '	Jupiter ♃ ° '	Saturn ♄ ° '	Uranus ♅ ° '	Neptune ♆ ° '	Pluto ♇ ° '
01	02 38 26	11 ♉ 18 09	20 ♋ 32 59	27 ♋ 34 08	24 ♈ 27	01 ♈ 25	06 ♊ 56	15 ♎ 24	27 ♐ 17	25 ♈ 26	13 ♓ 42	19 ♑ 22
02	02 42 23	12 16 22	04 ♌ 29 10	11 ♌ 18 15	24 R 19	01 56	07 37	15 R 18	27 R 15	25 30	13 44	19 R 22
03	02 46 19	13 14 34	18 ♌ 01 37	24 ♌ 39 34	24 16	02 29	08 18	15 12	27 12	25 33	13 45	19 21
04	02 50 16	14 12 44	01 ♍ 12 30	07 ♍ 40 49	24 D 17	03 04	08 59	15 06	27 10	25 36	13 46	19 21
05	02 54 12	15 10 51	14 ♍ 04 57	20 ♍ 25 17	24 24	03 39	09 40	15 00	27 07	25 40	13 48	19 21
06	02 58 09	16 08 57	26 ♍ 42 15	02 ♎ 56 12	24 35	04 16	10 21	14 54	27 04	25 43	13 49	19 20
07	03 02 05	17 07 01	09 ♎ 07 29	15 ♎ 16 24	24 50	04 54	11 02	14 49	27 01	25 46	13 50	19 20
08	03 06 02	18 05 03	21 ♎ 23 13	27 ♎ 28 10	25 10	05 33	11 43	14 43	26 59	25 49	13 51	19 19
09	03 09 58	19 03 03	03 ♏ 31 28	09 ♏ 33 15	25 35	06 14	12 24	14 38	26 56	25 53	13 53	19 19
10	03 13 55	20 01 02	15 ♏ 33 44	21 ♏ 33 01	26 03	06 55	13 05	14 33	26 53	25 56	13 54	19 18
11	03 17 52	20 58 59	27 ♏ 31 17	03 ♐ 28 40	26 36	07 37	13 45	14 28	26 49	25 59	13 55	19 18
12	03 21 48	21 56 54	09 ♐ 25 20	15 ♐ 21 30	27 13	08 21	14 26	14 23	26 46	26 02	13 56	19 17
13	03 25 45	22 54 48	21 ♐ 17 22	27 ♐ 13 11	27 54	09 05	15 07	14 18	26 43	26 05	13 57	19 16
14	03 29 41	23 52 41	03 ♑ 09 16	09 ♑ 05 58	28 38	09 50	15 47	14 13	26 40	26 09	13 58	19 16
15	03 33 38	24 50 32	15 ♑ 03 39	21 ♑ 02 45	29 ♈ 26	10 36	16 28	14 09	26 36	26 12	13 59	19 15
16	03 37 34	25 48 22	27 ♑ 03 46	03 ♒ 07 13	00 ♉ 17	11 23	17 09	14 05	26 33	26 15	14 00	19 14
17	03 41 31	26 46 11	09 ♒ 13 39	15 ♒ 23 39	01 12	12 10	17 49	14 01	26 30	26 18	14 01	19 13
18	03 45 27	27 43 59	21 ♒ 37 50	27 ♒ 56 48	02 10	12 59	18 30	13 57	26 26	26 21	14 02	19 13
19	03 49 24	28 41 45	04 ♓ 21 10	10 ♓ 51 30	03 11	13 48	19 10	13 53	26 22	26 24	14 03	19 12
20	03 53 21	29 ♉ 39 31	17 ♓ 28 18	24 ♓ 12 01	04 15	14 38	19 50	13 49	26 19	26 27	14 04	19 11
21	03 57 17	00 ♊ 37 15	01 ♈ 02 57	08 ♈ 01 19	05 22	15 28	20 31	13 46	26 15	26 30	14 05	19 10
22	04 01 14	01 34 58	15 ♈ 07 05	22 ♈ 20 05	06 32	16 20	21 11	13 42	26 11	26 33	14 06	19 09
23	04 05 10	02 32 40	29 ♈ 39 52	07 ♉ 05 46	07 44	17 11	21 52	13 39	26 07	26 36	14 07	19 09
24	04 09 07	03 30 22	14 ♉ 36 55	22 ♉ 12 11	09 00	18 04	22 32	13 36	26 04	26 39	14 07	19 08
25	04 13 03	04 28 02	29 ♉ 50 17	07 ♊ 29 48	10 18	18 57	23 12	13 33	26 00	26 42	14 08	19 07
26	04 17 00	05 25 41	15 ♊ 09 17	22 ♊ 47 17	11 39	19 50	23 52	13 31	25 56	26 45	14 09	19 06
27	04 20 56	06 23 18	00 ♋ 22 25	07 ♋ 53 27	13 02	20 44	24 33	13 28	25 52	26 47	14 09	19 05
28	04 24 53	07 20 55	15 ♋ 19 23	22 ♋ 39 23	14 28	21 39	25 13	13 26	25 48	26 50	14 10	19 04
29	04 28 50	08 18 30	29 ♋ 52 51	06 ♌ 59 27	15 56	22 34	25 53	13 24	25 43	26 53	14 11	19 03
30	04 32 46	09 16 03	13 ♌ 59 02	20 ♌ 51 37	17 27	23 30	26 33	13 22	25 39	26 56	14 11	19 02
31	04 36 43	10 ♊ 13 35	27 ♌ 37 25	04 ♍ 16 44	19 ♉ 01	24 ♈ 26	27 ♊ 13	13 ♎ 20	25 ♐ 35	26 ♈ 58	14 ♓ 12	19 ♑ 01

Date	Moon True ☊ ° '	Moon Mean ☊ ° '	Moon ☽ Latitude ° '
01	00 ♍ 32	29 ♌ 51	03 S 15
02	00 R 30	29 48	02 13
03	00 D 30	29 44	01 S 06
04	00 R 30	29 41	00 N 04
05	00 30	29 38	01 11
06	00 27	29 35	02 14
07	00 22	29 32	03 09
08	00 14	29 29	03 55
09	00 03	29 25	04 29
10	29 ♌ 51	29 22	04 51
11	29 37	29 19	05 00
12	29 23	29 16	04 55
13	29 11	29 13	04 38
14	29 00	29 10	04 09
15	28 53	29 06	03 29
16	28 48	29 03	02 39
17	28 45	29 00	01 41
18	28 D 45	28 57	00 N 38
19	28 R 45	28 54	00 S 30
20	28 44	28 50	01 37
21	28 42	28 47	02 42
22	28 37	28 44	03 39
23	28 30	28 41	04 24
24	28 21	28 38	04 52
25	28 10	28 35	05 00
26	27 59	28 31	04 47
27	27 50	28 28	04 14
28	27 43	28 25	03 24
29	27 38	28 22	02 21
30	27 36	28 19	01 11
31	27 ♌ 35	28 ♌ 16	00 00

DECLINATIONS

Date	Sun ☉ ° '	Moon ☽ ° '	Mercury ☿ ° '	Venus ♀ ° '	Mars ♂ ° '	Jupiter ♃ ° '	Saturn ♄ ° '	Uranus ♅ ° '	Neptune ♆ ° '	Pluto ♇ ° '
01	15 N 13	18 N 39	08 N 11	01 N 39	22 N 07	04 S 38	22 S 02	09 N 19	07 S 13	21 S 11
02	15 31	16 58	07 55	01 43	22 14	04 36	22 02	09 20	07 12	21 11
03	15 49	14 23	07 41	01 47	22 21	04 34	22 02	09 21	07 12	21 11
04	16 06	11 06	07 31	01 52	22 28	04 32	22 02	09 23	07 11	21 11
05	16 23	07 22	07 22	01 57	22 35	04 29	22 02	09 24	07 11	21 11
06	16 40	03 N 22	07 17	02 04	22 41	04 27	22 02	09 25	07 10	21 12
07	16 57	00 S 43	07 14	02 11	22 48	04 25	22 02	09 26	07 10	21 12
08	17 13	04 43	07 13	02 18	22 54	04 23	22 02	09 27	07 09	21 12
09	17 29	08 28	07 15	02 27	23 00	04 22	22 02	09 28	07 09	21 12
10	17 45	11 51	07 19	02 36	23 05	04 20	22 02	09 30	07 09	21 12
11	18 00	14 44	07 25	02 45	23 11	04 18	22 02	09 31	07 08	21 12
12	18 15	16 59	07 33	02 56	23 16	04 16	22 02	09 32	07 08	21 12
13	18 30	18 31	07 44	03 06	23 21	04 15	22 01	09 33	07 07	21 13
14	18 44	19 15	07 56	03 18	23 26	04 13	22 01	09 34	07 07	21 13
15	18 58	19 07	08 11	03 29	23 31	04 12	22 01	09 35	07 07	21 13
16	19 12	18 08	08 27	03 42	23 35	04 10	22 01	09 36	07 06	21 13
17	19 26	16 19	08 45	03 54	23 39	04 09	22 01	09 38	07 06	21 13
18	19 39	13 42	09 04	04 07	23 43	04 07	22 01	09 39	07 06	21 13
19	19 52	10 22	09 25	04 21	23 47	04 06	22 01	09 40	07 05	21 14
20	20 04	06 27	09 48	04 35	23 51	04 05	22 01	09 41	07 05	21 14
21	20 17	02 S 03	10 12	04 49	23 54	04 04	22 01	09 42	07 05	21 14
22	20 28	02 N 35	10 37	05 04	23 57	04 03	22 01	09 43	07 04	21 14
23	20 40	07 14	11 03	05 19	24 00	04 02	22 01	09 44	07 04	21 14
24	20 51	11 34	11 31	05 35	24 03	04 01	22 00	09 45	07 04	21 15
25	21 02	15 13	11 59	05 50	24 05	04 00	22 00	09 46	07 04	21 15
26	21 12	17 51	12 29	06 07	24 08	03 59	22 00	09 47	07 03	21 15
27	21 22	19 12	12 59	06 23	24 10	03 58	22 00	09 48	07 03	21 15
28	21 32	19 11	13 30	06 39	24 12	03 58	22 00	09 49	07 03	21 16
29	21 41	17 52	14 02	06 56	24 14	03 57	22 00	09 50	07 03	21 16
30	21 50	15 29	14 35	07 13	24 15	03 57	22 00	09 51	07 03	21 16
31	21 N 59	12 N 18	15 N 08	07 N 30	24 N 16	03 S 56	22 S 00	09 N 52	07 S 02	21 S 16

ZODIAC SIGN ENTRIES

Date	h	m	Planets
02	04	12	☽ ♌
04	09	46	☽ ♍
06	18	20	☽ ♎
09	05	01	☽ ♏
11	16	59	☽ ♐
14	05	37	☽ ♑
16	04	07	☿ ♉
16	17	50	☽ ♒
19	03	52	☽ ♓
20	20	31	☉ ♊
21	10	11	☽ ♈
23	12	33	☽ ♉
25	12	15	☽ ♊
27	11	24	☽ ♋
29	12	12	☽ ♌
31	16	16	☽ ♍

LATITUDES

Date	Mercury ☿ ° '	Venus ♀ ° '	Mars ♂ ° '	Jupiter ♃ ° '	Saturn ♄ ° '	Uranus ♅ ° '	Neptune ♆ ° '	Pluto ♇ ° '
01	01 S 24	01 N 11	00 N 40	01 N 33	01 N 22	00 S 34	00 S 52	00 N 52
04	02 03	00 42	00 41	01 32	01 22	00 34	00 52	00 52
07	02 34	00 N 15	00 42	01 32	01 22	00 34	00 53	00 51
10	02 57	00 S 10	00 44	01 31	01 22	00 34	00 53	00 51
13	03 12	00 32	00 45	01 31	01 22	00 34	00 53	00 51
16	03 20	00 53	00 46	01 30	01 22	00 34	00 53	00 51
19	03 21	01 11	00 48	01 29	01 22	00 34	00 53	00 50
22	03 15	01 28	00 49	01 29	01 22	00 34	00 53	00 50
25	03 04	01 42	00 50	01 28	01 22	00 34	00 53	00 50
28	02 47	01 55	00 51	01 27	01 22	00 34	00 53	00 50
31	02 S 26	02 S 06	00 N 52	01 N 26	01 N 22	00 S 34	00 S 53	00 N 50

DATA

Julian Date	2457875
Delta T	+70 seconds
Ayanamsa	24° 05' 47"
Synetic vernal point	05° ♓ 01' 12"
True obliquity of ecliptic	23° 26' 05"

MOON'S PHASES, APSIDES AND POSITIONS ☽

Date	h	m	Phase	Longitude ° '	Eclipse Indicator
03	02	47	☽	12 ♌ 52	
10	21	42	○	20 ♏ 24	
19	00	33	☾	28 ♒ 14	
25	19	44	●	04 ♊ 47	

Day	h	m		
12	19	36	Apogee	
26	01	15	Perigee	
07	07	46	OS	
14	20	36	Max dec	19° S 18'
21	22	43	ON	
27	23	38	Max dec	19° N 22'

LONGITUDES

Date	Chiron ⚷ ° '	Ceres ⚳ ° '	Pallas ⚴ ° '	Juno ⚵ ° '	Vesta ⚶ ° '	Black Moon Lilith ⚸ ° '
01	27 ♓ 20	00 ♊ 45	11 ♈ 22	18 ♑ 09	29 ♋ 33	08 ♐ 34
11	27 ♓ 47	04 ♊ 52	14 ♈ 49	18 ♑ 20	02 ♌ 44	09 ♐ 41
21	28 ♓ 09	09 ♊ 01	18 ♈ 14	17 ♑ 58	06 ♌ 12	10 ♐ 49
31	28 ♓ 27	13 ♊ 12	21 ♈ 35	17 ♑ 00	09 ♌ 55	11 ♐ 56

All ephemeris data is given at 12.00 UT and the Moon's longitude is additionally given for 24.00 UT

ASPECTARIAN

h m	Aspects
01 Monday	
00 26	☽ △ ♆
03 23	☽ □ ♃
09 58	♀ Q ♇
10 00	☽ ☍ ♇
14 28	☽ ∠ ♂
17 02	☽ Q ☉
18 35	☽ □ ☿
20 23	☽ □ ♅
23 29	☽ ⊼ ♄
02 Tuesday	
01 59	☽ bQ ♆
07 23	☽ △ ♀
09 50	☽ ± ♄
09 57	☽ Q ♃
11 19	☉ bQ ♄
17 41	☽ ± ♆
17 47	☽ ⚹ ♂
03 Wednesday	
00 57	☽ ∥ ☉
01 38	☽ bQ ♄
02 47	☽ □ ☉
04 20	☽ ⊼ ♆
06 58	☽ ⚹ ♃
11 00	☽ bQ ♀
11 14	♃ Q ♄
14 24	☽ ⊼ ♇
16 19	☽ Q ♂
16 33	☿ St D
23 17	☽ △ ☿
04 Thursday	
00 51	☉ ⚹ ♆
01 16	☽ ± ♇
01 41	☽ △ ♅
04 02	☽ ± ♀
04 35	☽ △ ♄
09 59	☽ ∠ ♃
15 35	☽ ⊼ ♀
17 48	☽ bQ ♇
23 17	☽ ∥ ♅
05 Friday	
00 09	☿ ∠ ♂
02 32	☽ ⊥ ♃
03 07	☽ bQ ☿
03 15	☽ □ ♂
05 33	☽ bQ ♅
07 59	☉ ⊼ ♃
11 27	☽ ☍ ♆
11 54	☽ ∥ ☿
13 06	☽ # ♆
13 43	☽ ⊻ ♃
14 15	☽ △ ☉
20 15	☽ ± ☿
21 57	☽ △ ♇
22 36	☽ ± ♅
06 Saturday	
05 27	☽ # ♃
07 51	☽ ⊼ ☿
10 06	☽ ⊼ ♅
12 42	☽ □ ♄
19 26	☽ ∥ ♀
21 16	☽ bQ ☉
07 Sunday	
01 54	♂ ∠ ♅
03 22	☽ ☍ ♀
15 56	☽ △ ♂
16 13	☽ ± ☉
20 58	☽ # ♀
21 12	☽ ⊼ ♆
23 01	☽ ☌ ♃
23 28	☽ Q ♄
08 Monday	
04 57	☽ ⊼ ☉
07 56	☽ □ ♇
08 59	☽ ± ♆
10 04	☽ ∥ ♃
19 42	☽ ☍ ☿
20 47	☽ ☍ ♅
22 59	☽ ⚹ ♄
23 08	☽ bQ ♂
09 Tuesday	
02 46	☽ bQ ♆
03 22	☽ ∥ ♆
03 52	☽ # ☿
17 42	☽ ⊼ ♀
18 03	☽ ± ♂
18 24	☉ △ ♇
18 51	☽ # ♅
19 31	☽ Q ♇
10 Wednesday	
04 40	☽ ∠ ♄
05 20	☿ ☌ ♅
06 23	☽ ± ♀
06 44	☽ ⊼ ♂
08 40	☽ △ ♆
09 59	☽ ⊻ ♃
19 29	☽ ⚹ ♇
19 52	♂ ± ♇
21 42	☽ ☍ ☉
21 55	☽ ⊥ ♃
22 36	☽ ⊥ ♄
11 Thursday	
00 05	☉ ± ♃
01 32	☽ bQ ♀
08 16	☉ ± ♄
08 54	☽ ⊼ ♅
10 04	☽ ⊼ ☿
10 36	☽ ⊻ ♄
15 53	☽ ∠ ♃
17 52	♂ □ ♆
20 14	☿ △ ♄
21 02	☽ ± ♅
22 47	☽ ± ☿
12 Friday	
01 38	☽ ∠ ♇
09 41	☽ △ ♀
10 19	♂ △ ♃
15 17	☽ bQ ♅
17 59	☽ bQ ☿
19 48	☽ ⊥ ♇
21 08	☽ □ ♆
21 57	☽ ⚹ ♃
22 45	☽ ☍ ♂
13 Saturday	
07 55	☽ ⊻ ♇
11 22	☽ # ☉
15 35	☽ ⊼ ☉
21 46	☽ △ ♅
22 05	☽ Q ♃
22 56	☽ ☌ ♄
14 Sunday	
02 14	☽ △ ☿
04 47	☽ ± ☉
09 36	☽ Q ♆
22 37	☿ ∠ ♆
15 Monday	
00 34	☽ bQ ☉
02 24	☽ □ ♀
09 51	☽ ⚹ ♆
10 11	☽ □ ♃
14 59	☽ ⊼ ♂
16 11	☽ # ☉
20 24	☽ ☌ ♇
16 Tuesday	
03 43	☽ ± ♂
09 17	☽ △ ☉
10 22	☽ □ ♅
10 59	☽ ⊻ ♄
15 52	☽ ∠ ♆
16 55	☽ Q ♀
17 03	☉ Q ♆
18 54	☽ □ ☿
22 40	☽ bQ ♂
22 49	☽ ⊥ ♄
23 35	☉ ⊻ ♅
17 Wednesday	
05 29	☉ ⊼ ♄
08 37	♃ ⊼ ♆
09 38	☽ ⊥ ♆
16 24	☽ ∠ ♄
18 09	☽ ⚹ ♀
21 16	☽ △ ♃
21 21	☽ ⊻ ♆
21 55	☽ Q ♅
18 Thursday	
05 38	☽ △ ♂
07 22	☽ ⊻ ♇
08 57	☽ Q ☿
11 56	♀ # ♃
18 49	☽ ⊥ ♇
21 01	☽ ⚹ ♅
21 06	☽ ⚹ ♄
19 Friday	
00 33	☽ □ ☉
00 53	☽ ∠ ♀
01 49	☽ bQ ♃
06 14	♄ △ ♅
09 38	☽ ⚹ ☿
11 43	☽ ∠ ♇
13 06	♂ ⊼ ♇
14 11	♀ ☍ ♃
16 19	☉ bQ ♃
16 34	☽ # ♅
17 34	☽ # ☿
18 30	☽ ± ♃
18 49	☽ ⊥ ♀
19 24	☽ Q ♄
19 30	♀ ⊻ ♆
20 Saturday	
01 02	☽ ∠ ♅
04 12	☿ ∥ ♅
05 26	☽ ⊼ ♃
05 51	☽ ☌ ♆
06 31	☽ ⊻ ♀
06 37	☉ ∠ ♀
08 17	☽ ∥ ♆
12 22	☽ Q ☉
15 04	☽ ⚹ ♇
15 28	☽ ∠ ☿
16 28	☽ □ ♂
17 21	☽ ⊥ ♅
21 53	☽ # ♀
21 Sunday	
01 13	☽ ∥ ♃
03 39	☽ □ ♄
04 01	☽ ⊻ ♅
08 48	☽ ⊥ ☿
11 12	☽ ⚹ ☉
12 13	☽ Q ♇
19 32	☿ ∠ ♂
20 07	☽ ⊻ ☿
22 Monday	
01 29	☽ Q ♂
09 38	☽ ☍ ♃
10 17	☽ ⊻ ♆
14 09	☽ ☌ ♀
14 38	☽ ∠ ☉
18 44	☽ □ ♇
19 26	☽ # ♃
20 18	☽ ⊥ ♆
22 36	☽ ⚹ ♂
23 Tuesday	
01 27	☽ ∥ ♀
06 15	☽ △ ♄
06 33	☽ ⊥ ☉
06 59	☽ ☌ ♅
11 06	☽ ∠ ♆
11 08	☽ # ♆
13 25	☉ ⊥ ♅
17 00	☽ ⊻ ☉
24 Wednesday	
00 10	☽ ∠ ♂
01 34	☽ ∥ ♅
02 13	☽ ☌ ☿
06 22	☽ bQ ♄
10 24	☽ ⊼ ♃
11 13	☽ ⚹ ♆
11 41	☽ ∥ ☿
15 11	☽ ⊥ ♂
17 48	☽ ⊻ ♀
19 08	☽ △ ♇
19 53	☽ ± ♃
20 35	☽ ± ♄
25 Thursday	
01 06	☽ ⊻ ♂
03 16	☉ bQ ♇
03 51	☽ ⊥ ♀
05 59	☽ ⊼ ♄
06 11	☽ Q ♆
07 03	☽ ⊻ ♅
10 00	☽ bQ ♃
16 22	♀ □ ♇
16 30	☽ ⊥ ♅
18 41	☽ bQ ♇
18 50	☽ ∠ ♀
19 44	☽ ☌ ☉
23 55	☿ bQ ♄
26 Friday	
05 57	☽ ⊻ ☿
06 38	☽ ∠ ♅
08 46	☽ ± ♇
09 26	☽ △ ♃
10 25	☽ □ ♆
16 17	☽ ⊥ ☿
18 11	☽ ⊼ ♇
18 55	☉ # ♇
19 49	☽ ⚹ ♀
20 15	♀ ⊥ ♆
27 Saturday	
02 21	☽ ☌ ♂
04 53	☽ ☍ ♄
06 18	☽ ⚹ ♅
07 54	☽ ∠ ☿
16 01	☽ Q ♀
19 17	☿ ⊼ ♃
22 15	☽ ⊻ ☉
28 Sunday	
01 29	☽ Q ♅
07 04	☿ ⚹ ♆
08 35	☽ ⊥ ☉
08 57	☽ □ ♃
10 07	☽ △ ♆
10 27	☽ ⚹ ☿
18 05	☽ ☍ ♇
23 02	☽ □ ♀
29 Monday	
00 18	☽ ∠ ☉
05 01	☽ ⊻ ♂
05 06	☽ ⊼ ♄
06 55	♂ ☍ ♄
06 59	☽ □ ♅
08 23	☽ Q ☿
10 49	☽ bQ ♆
14 33	☽ Q ♃
15 05	☽ ± ♄
15 31	☽ ⊥ ♂
21 08	♀ # ♆
30 Tuesday	
02 02	☽ ± ♆
03 17	☽ ⚹ ☉
06 18	☽ bQ ♄
07 35	☽ ∠ ♂
10 56	☽ ⚹ ♃
12 21	☽ ⊼ ♆
18 16	☽ ∥ ☿
18 47	☽ □ ☿
20 46	☽ ⊼ ♇
31 Wednesday	
01 41	☽ Q ☉
02 39	♂ ⚹ ♅
05 53	☽ △ ♀
07 21	☽ ± ♇
08 23	☽ △ ♄
10 50	☽ △ ♅
11 14	☽ ⚹ ♂
12 03	☿ △ ♇
13 17	☽ ∠ ♃
14 54	☉ # ♄
16 57	☿ ± ♃
20 23	☿ ± ♄
23 30	☽ bQ ♇

JUNE 2017

Date	Sidereal time h m s	LONGITUDES Sun ☉ ° ' "	Moon ☽ ° ' "	Moon ☽ 24.00 ° ' "	Mercury ☿ ° '	Venus ♀ ° '	Mars ♂ ° '	Jupiter ♃ ° '	Saturn ♄ ° '	Uranus ♅ ° '	Neptune ♆ ° '	Pluto ♇ ° '
01	04 40 39	11 ♊ 11 06	10 ♍ 49 58	17 ♍ 17 36	20 ♉ 36	25 ♈ 22	27 ♊ 53	13 ♎ 19	25 ♐ 31	27 ♈ 01	14 ♓ 12	19 ♑ 00
02	04 44 36	12 08 35	23 ♍ 40 09	29 ♍ 58 07	22 15	26 19	28 33	13 R 17	25 R 27	27 04	14 13	18 R 59
03	04 48 32	13 06 03	06 ♎ 12 03	12 ♎ 22 27	23 56	27 17	29 13	13 16	25 23	27 06	14 13	18 57
04	04 52 29	14 03 30	18 ♎ 29 49	24 ♎ 34 37	25 39	28 14	29 ♊ 53	13 15	25 18	27 09	14 13	18 56
05	04 56 25	15 00 56	00 ♏ 37 15	06 ♏ 38 06	27 25	29 ♈ 13	00 ♋ 33	13 14	25 14	27 12	14 14	18 55
06	05 00 22	15 58 21	12 ♏ 37 32	18 ♏ 35 50	29 ♉ 13	00 ♉ 11	01 13	13 14	25 10	27 14	14 14	18 54
07	05 04 19	16 55 45	24 ♏ 33 16	00 ♐ 30 05	01 ♊ 04	01 10	01 52	13 13	25 05	27 17	14 15	18 53
08	05 08 15	17 53 07	06 ♐ 26 29	12 ♐ 22 41	02 57	02 09	02 32	13 13	25 01	27 19	14 15	18 52
09	05 12 12	18 50 29	18 ♐ 18 50	24 ♐ 15 09	04 52	03 09	03 12	13 13	24 57	27 22	14 15	18 50
10	05 16 08	19 47 50	00 ♑ 11 47	06 ♑ 08 58	06 50	04 09	03 52	13 D 13	24 52	27 24	14 15	18 49
11	05 20 05	20 45 10	12 ♑ 06 53	18 ♑ 05 49	08 49	05 09	04 31	13 13	24 48	27 26	14 15	18 48
12	05 24 01	21 42 30	24 ♑ 06 00	00 ♒ 07 46	10 51	06 10	05 11	13 14	24 43	27 29	14 16	18 47
13	05 27 58	22 39 49	06 ♒ 11 28	12 ♒ 17 28	12 55	07 11	05 51	13 14	24 39	27 31	14 16	18 45
14	05 31 54	23 37 08	18 ♒ 26 13	24 ♒ 38 09	15 01	08 12	06 30	13 15	24 34	27 33	14 16	18 44
15	05 35 51	24 34 26	00 ♓ 53 46	07 ♓ 13 33	17 08	09 13	07 10	13 16	24 30	27 35	14 16	18 43
16	05 39 48	25 31 43	13 ♓ 38 02	20 ♓ 07 42	19 16	10 15	07 50	13 17	24 26	27 38	14 R 16	18 41
17	05 43 44	26 29 01	26 ♓ 43 02	03 ♈ 24 26	21 26	11 17	08 29	13 19	24 21	27 40	14 16	18 40
18	05 47 41	27 26 18	10 ♈ 12 16	17 ♈ 06 46	23 37	12 20	09 09	13 20	24 17	27 42	14 16	18 39
19	05 51 37	28 23 35	24 ♈ 08 03	01 ♉ 16 05	25 48	13 22	09 48	13 22	24 12	27 44	14 16	18 37
20	05 55 34	29 ♊ 20 51	08 ♉ 30 35	15 ♉ 51 09	27 ♊ 59	14 25	10 27	13 24	24 08	27 46	14 16	18 36
21	05 59 30	00 ♋ 18 08	23 ♉ 17 07	00 ♊ 47 37	00 ♋ 11	15 28	11 07	13 26	24 04	27 48	14 15	18 35
22	06 03 27	01 15 24	08 ♊ 21 36	15 ♊ 57 51	02 23	16 31	11 46	13 28	23 59	27 50	14 15	18 33
23	06 07 23	02 12 40	23 ♊ 35 04	01 ♋ 11 52	04 34	17 35	12 26	13 30	23 55	27 52	14 15	18 32
24	06 11 20	03 09 56	08 ♋ 46 56	16 ♋ 18 57	06 44	18 39	13 05	13 33	23 50	27 54	14 15	18 30
25	06 15 17	04 07 12	23 ♋ 46 48	01 ♌ 09 30	08 54	19 43	13 44	13 35	23 46	27 56	14 15	18 29
26	06 19 13	05 04 27	08 ♌ 26 17	15 ♌ 36 34	11 02	20 47	14 24	13 38	23 42	27 57	14 14	18 28
27	06 23 10	06 01 41	22 ♌ 39 59	29 ♌ 36 23	13 09	21 51	15 03	13 41	23 38	27 59	14 14	18 26
28	06 27 06	06 58 55	06 ♍ 25 47	13 ♍ 08 21	15 14	22 56	15 42	13 45	23 33	28 01	14 14	18 25
29	06 31 03	07 56 09	19 ♍ 44 11	26 ♍ 14 12	17 18	24 00	16 21	13 48	23 29	28 02	14 14	18 23
30	06 34 59	08 ♋ 53 22	02 ♎ 38 20	08 ♎ 57 18	19 ♋ 20	25 ♉ 05	17 ♋ 01	13 ♎ 52	23 ♐ 25	28 ♈ 04	14 ♓ 13	18 ♑ 22

Date	Moon True ☊ ° '	Moon Mean ☊ ° '	Moon ☽ Latitude ° '
01	27 ♌ 35	28 ♌ 12	01 N 10
02	27 R 35	28 09	02 14
03	27 33	28 06	03 09
04	27 28	28 03	03 55
05	27 21	28 00	04 30
06	27 12	27 56	04 52
07	27 00	27 53	05 01
08	26 48	27 50	04 57
09	26 35	27 47	04 40
10	26 24	27 44	04 11
11	26 15	27 41	03 31
12	26 08	27 37	02 41
13	26 03	27 34	01 44
14	26 01	27 31	00 N 40
15	26 D 01	27 28	00 S 26
16	26 02	27 25	01 33
17	26 R 02	27 22	02 37
18	26 02	27 18	03 34
19	25 59	27 15	04 20
20	25 54	27 12	04 52
21	25 48	27 09	05 05
22	25 40	27 06	04 58
23	25 32	27 02	04 30
24	25 25	26 59	03 43
25	25 20	26 56	02 41
26	25 16	26 53	01 29
27	25 15	26 50	00 S 14
28	25 D 16	26 47	01 N 00
29	25 17	26 43	02 08
30	25 ♌ 18	26 ♌ 40	03 N 08

DECLINATIONS

Date	Sun ☉ ° '	Moon ☽ ° '	Mercury ☿ ° '	Venus ♀ ° '	Mars ♂ ° '	Jupiter ♃ ° '	Saturn ♄ ° '	Uranus ♅ ° '	Neptune ♆ ° '	Pluto ♇ ° '
01	22 N 07	08 N 35	15 N 41	07 N 48	24 N 17	03 S 56	22 S 00	09 N 53	07 S 02	21 S 16
02	22 15	04 33	16 15	08 06	24 18	03 56	21 59	09 54	07 02	21 17
03	22 22	00 N 26	16 48	08 23	24 19	03 55	21 59	09 55	07 02	21 17
04	22 29	03 S 37	17 22	08 41	24 19	03 55	21 59	09 56	07 02	21 17
05	22 36	07 28	17 56	08 59	24 20	03 55	21 59	09 57	07 02	21 17
06	22 42	10 59	18 30	09 17	24 20	03 55	21 59	09 58	07 02	21 18
07	22 48	14 02	19 03	09 35	24 20	03 55	21 59	09 59	07 02	21 18
08	22 53	16 30	19 36	09 54	24 19	03 55	21 59	09 59	07 02	21 18
09	22 58	18 16	20 08	10 12	24 19	03 56	21 59	10 00	07 01	21 18
10	23 03	19 15	20 39	10 31	24 18	03 56	21 59	10 01	07 01	21 19
11	23 07	19 23	21 09	10 49	24 17	03 56	21 59	10 02	07 01	21 19
12	23 11	18 38	21 38	11 07	24 16	03 57	21 58	10 03	07 01	21 19
13	23 14	17 03	22 05	11 26	24 15	03 57	21 58	10 04	07 01	21 19
14	23 17	14 39	22 31	11 44	24 13	03 58	21 58	10 04	07 01	21 20
15	23 19	11 34	22 55	12 03	24 11	03 58	21 58	10 05	07 01	21 20
16	23 22	07 52	23 17	12 21	24 09	03 59	21 58	10 06	07 01	21 20
17	23 23	03 S 42	23 37	12 39	24 07	04 00	21 58	10 07	07 01	21 20
18	23 25	00 N 46	23 54	12 58	24 05	04 00	21 58	10 07	07 01	21 21
19	23 25	05 20	24 09	13 16	24 02	04 01	21 58	10 08	07 02	21 21
20	23 26	09 44	24 21	13 34	23 59	04 02	21 58	10 09	07 02	21 21
21	23 26	13 40	24 31	13 52	23 56	04 03	21 57	10 09	07 02	21 22
22	23 26	16 47	24 37	14 10	23 53	04 04	21 57	10 10	07 02	21 22
23	23 25	18 47	24 41	14 27	23 50	04 06	21 57	10 11	07 02	21 22
24	23 24	19 26	24 42	14 45	23 47	04 07	21 57	10 11	07 02	21 22
25	23 22	18 42	24 40	15 02	23 43	04 08	21 57	10 12	07 02	21 23
26	23 20	16 43	24 36	15 19	23 39	04 10	21 57	10 13	07 02	21 23
27	23 18	13 44	24 29	15 36	23 35	04 11	21 57	10 13	07 03	21 23
28	23 15	10 05	24 19	15 53	23 31	04 12	21 57	10 14	07 03	21 24
29	23 12	06 02	24 07	16 10	23 26	04 14	21 57	10 15	07 03	21 24
30	23 N 08	01 N 49	23 N 52	16 N 26	23 N 21	04 S 16	21 S 57	10 N 15	07 S 03	21 S 24

ZODIAC SIGN ENTRIES

Date	h m	Planets
03	00 04	☽ ♎
04	16 16	♂ ♋
05	10 46	☽ ♏
06	07 26	♀ ♉
06	22 15	☿ ♊
07	22 59	☽ ♐
10	11 36	☽ ♑
12	23 45	☽ ♒
15	10 17	☽ ♓
17	17 55	☽ ♈
19	21 53	☽ ♉
21	04 24	☉ ♋
21	09 57	☿ ♋
21	22 44	☽ ♊
23	22 07	☽ ♋
25	22 06	☽ ♌
28	00 41	☽ ♍
30	07 02	☽ ♎

LATITUDES

Date	Mercury ☿ ° '	Venus ♀ ° '	Mars ♂ ° '	Jupiter ♃ ° '	Saturn ♄ ° '	Uranus ♅ ° '	Neptune ♆ ° '	Pluto ♇ ° '
01	02 S 18	02 S 10	00 N 52	01 N 26	01 N 22	00 S 34	00 S 53	00 N 49
04	01 51	02 19	00 53	01 25	01 22	00 34	00 54	00 49
07	01 21	02 26	00 54	01 25	01 22	00 34	00 54	00 49
10	00 48	02 32	00 55	01 24	01 22	00 34	00 54	00 49
13	00 S 15	02 37	00 56	01 23	01 21	00 34	00 54	00 48
16	00 N 17	02 40	00 57	01 22	01 21	00 34	00 54	00 48
19	00 47	02 43	00 58	01 21	01 21	00 34	00 54	00 48
22	01 13	02 43	00 59	01 21	01 21	00 34	00 54	00 48
25	01 33	02 43	01 00	01 20	01 20	00 34	00 54	00 47
28	01 46	02 42	01 00	01 19	01 20	00 34	00 55	00 47
31	01 N 53	02 S 40	01 N 01	01 N 18	01 N 20	00 S 34	00 S 55	00 N 47

DATA

Julian Date	2457906
Delta T	+70 seconds
Ayanamsa	24° 05' 52"
Synetic vernal point	05° ♓ 01' 07"
True obliquity of ecliptic	23° 26' 05"

MOON'S PHASES, APSIDES AND POSITIONS ☽

Date	h m	Phase	Longitude ° '	Eclipse Indicator
01	12 42	☽	11 ♍ 13	
09	13 10	○	18 ♐ 53	
17	11 33	☾	26 ♓ 28	
24	02 31	●	02 ♋ 47	

Day	h m		
08	22 06	Apogee	
23	10 45	Perigee	
03	14 33	0S	
11	03 40	Max dec	19° S 26'
18	07 57	0N	
24	11 10	Max dec	19° N 26'
30	22 26	0S	

LONGITUDES

Date	Chiron ⚷ ° '	Ceres ⚳ ° '	Pallas ⚴ ° '	Juno ⚵ ° '	Vesta ⚶ ° '	Black Moon Lilith ⚸ ° '
01	28 ♓ 29	13 ♊ 37	21 ♈ 55	16 ♑ 53	10 ♌ 18	12 ♐ 03
11	28 ♓ 42	17 ♊ 49	25 ♈ 10	15 ♑ 20	14 ♌ 16	13 ♐ 10
21	28 ♓ 49	22 ♊ 01	28 ♈ 21	13 ♑ 21	18 ♌ 16	14 ♐ 17
31	28 ♓ 52	26 ♊ 14	01 ♉ 18	11 ♑ 05	22 ♌ 40	15 ♐ 24

All ephemeris data is given at 12.00 UT and the Moon's longitude is additionally given for 24.00 UT

ASPECTARIAN

h m	Aspects
01 Thursday	
03 53	☽ ∥ ♅
05 33	☽ ⊥ ♃
10 09	☽ Q ♂
11 05	☽ ⚼ ♀
12 42	☽ □ ☉
14 12	☽ ⚼ ♅
15 23	♀ △ ♄
16 25	☽ ∥ ♀
16 35	☽ ⊻ ♃
18 15	☽ ☍ ♆
21 20	☽ # ♆
02 Friday	
03 10	☽ △ ♇
05 10	☽ ± ♀
07 04	☽ ± ♅
08 54	☽ △ ☿
09 55	☉ ∠ ♅
15 21	☽ □ ♄
15 41	☽ # ♃
17 27	☽ ⊼ ♀
18 28	☽ ⊼ ♅
19 17	☿ ⊥ ♂
21 48	☽ □ ♂
03 Saturday	
07 32	♀ ☌ ♅
08 27	☉ ± ♇
16 12	☉ △ ♃
18 08	☽ ⚼ ☿
04 Sunday	
01 44	☽ ☌ ♃
01 53	☽ Q ♄
02 33	☽ △ ☉
03 37	☽ ⊼ ♆
07 24	☿ ⊼ ♄
12 52	☽ □ ♇
13 49	☽ ∥ ♃
14 39	☽ ± ☿
15 24	☽ ± ♆
16 13	☉ □ ♆
19 55	☿ Q ♆
05 Monday	
01 22	☽ ✶ ♄
04 32	☽ ⊼ ☿
05 10	☽ ☍ ♅
08 21	♃ Q ♄
08 57	☽ ☍ ♀
08 58	☿ ⊻ ♅
09 09	☽ ∥ ♆
09 14	☽ ⚼ ♆
10 42	☽ ⚼ ☉
11 51	☽ △ ♂
12 32	♀ ∠ ♆
22 57	☽ # ♀
23 01	☿ ⚼ ♃
06 Tuesday	
00 33	☽ Q ♇
04 41	☽ # ♅
06 13	☽ ± ☉
07 05	☽ ∠ ♄
13 13	☽ ⊻ ♃
15 14	☽ △ ♆
19 18	☽ ⊼ ☉
19 37	☽ ⚼ ♂
07 Wednesday	
00 35	☽ ✶ ♇
01 03	☽ ⊥ ♄
01 16	☽ ⊥ ♃
13 04	☽ ⊻ ♄
14 49	☽ ± ♂
14 54	☿ ⊻ ♀
17 31	☽ ⊼ ♅
19 24	☽ ∠ ♃
08 Thursday	
02 33	☽ ⊼ ♀
03 36	☽ ☍ ☿
03 39	☽ ⊼ ♂
04 05	☿ ⊻ ♂
05 40	☽ ± ♅
06 47	☽ ∠ ♇
15 47	☽ ± ♀
16 49	☿ ⊥ ♅
19 42	♀ ∥ ♅
23 22	☿ ⚼ ♇
23 55	☽ ⚼ ♅
09 Friday	
00 57	☽ ⊥ ♇
01 42	☽ ✶ ♃
03 47	☽ □ ♆
11 38	☽ ⚼ ♀
11 57	☉ ⊼ ♇
13 04	☽ ⊻ ♇
13 10	☽ ☍ ☉
14 02	♃ St D
15 41	♀ ✶ ♂
10 Saturday	
01 19	☽ ☌ ♄
01 57	☽ Q ♃
06 20	☽ △ ♅
16 09	☽ Q ♆
19 50	☽ ☍ ♂
20 42	☽ △ ♀
11 Sunday	
04 02	☽ ⊼ ☿
14 13	☽ □ ♃
16 18	☽ ✶ ♆
18 32	☽ ± ☿
20 04	☿ # ♇
12 Monday	
01 23	☽ ☌ ♇
06 49	☽ ⊼ ☉
13 14	☽ ⊻ ♄
16 13	☽ ⚼ ☿
18 45	☽ □ ♅
19 48	☽ ± ☉
22 17	☽ ∠ ♆
13 Tuesday	
01 06	☽ ⊥ ♄
05 39	☿ # ♄
07 17	☿ ∠ ♅
10 10	☿ ± ♇
11 17	☽ ⊼ ♂
14 08	☽ □ ♀
15 09	☽ ⚼ ☉
15 45	☿ △ ♃
16 05	☽ ⊥ ♆
17 55	☿ ⊥ ♀
18 46	☽ ∠ ♄
23 46	☽ ± ♂
14 Wednesday	
01 52	☽ △ ♃
03 29	☿ □ ♆
03 52	☽ ⊻ ♆
03 56	☽ △ ☿
06 22	☽ Q ♅
12 35	☽ ⊻ ♇
18 17	☽ ⚼ ♂
22 52	☽ △ ☉
23 49	☽ ✶ ♄
15 Thursday	
00 10	☽ ⊥ ♇
04 21	☽ Q ♀
05 40	☽ ✶ ♅
06 58	☽ ⚼ ♃
08 50	☽ # ♀
10 18	☉ ☍ ♄
17 21	☽ ∠ ♇
18 02	♀ ⚼ ♄
21 58	☽ # ♅
22 34	☽ Q ♄
16 Friday	
00 06	☽ ± ♃
00 32	☽ △ ♂
05 08	☽ ✶ ♀
05 35	☿ ⊼ ♇
10 07	☽ ∠ ♅
11 07	♆ St R
11 21	☽ ⊼ ♃
13 10	☽ ☌ ♆
17 02	☽ ∥ ♆
17 30	☉ ∥ ☿
21 20	☽ ✶ ♇
17 Saturday	
00 29	☽ □ ☿
02 47	☽ ⊥ ♅
07 44	☽ □ ♄
10 22	☽ ∥ ♃
11 09	☽ ∠ ♀
11 33	☽ □ ☉
13 43	☽ ⊻ ♅
19 06	☽ Q ♇
18 Sunday	
04 37	☽ ⊥ ♀
10 03	☽ □ ♂
14 55	☽ Q ☿
16 01	☽ ⊻ ♀
17 28	☽ ☍ ♃
18 47	☉ ✶ ♅
19 04	☽ ⊻ ♆
19 07	☿ ☍ ♄
21 47	☽ Q ☉
19 Monday	
01 54	☿ ∥ ♂
02 37	☽ □ ♇
05 06	☽ # ♃
05 24	☽ ⊥ ♆
09 24	♂ Q ♅
11 49	♀ ⊼ ♃
12 07	☽ △ ♄
15 20	☽ ✶ ☿
18 05	☽ ☌ ♅
18 08	☉ ∠ ♀
18 29	☽ Q ♂
19 42	☽ ✶ ☉
20 39	☽ ∠ ♆
21 05	☽ # ♆
20 Tuesday	
08 25	♀ ✶ ♆
09 30	☿ ✶ ♅
13 01	☽ ⚼ ♄
14 24	☽ ∥ ♅
15 21	☽ ✶ ♂
20 01	☽ ⊼ ♃
20 38	☽ ∠ ☿
21 25	☽ ✶ ♆
22 13	☽ ∠ ☉
22 24	☽ ☌ ♀
21 Wednesday	
03 37	☽ ± ♄
04 26	☽ △ ♇
05 46	☽ ± ♃
13 14	☽ ⊼ ♄
13 30	☽ ∥ ♀
13 42	☽ ⊥ ☿
13 45	☽ ⊥ ☉
14 14	☉ ☌ ☿
16 45	☽ ∠ ♂
16 46	☽ Q ♆
17 55	☿ ∠ ♀
19 15	☽ ⊻ ♅
20 15	☽ ⚼ ♃
23 59	☽ ⊻ ☉
22 Thursday	
00 55	☽ ⊻ ☿
04 24	☽ ⚼ ♇
04 49	☽ ⊥ ♅
07 43	☽ ⊥ ♂
17 38	☽ ⊻ ♂
18 37	☽ ± ♇
19 05	☽ ∠ ♅
20 05	☽ △ ♃
21 18	☽ □ ♆
23 Friday	
01 50	☽ ⊻ ♀
04 03	☽ ⊼ ♇
12 00	☽ ⊥ ♀
12 31	☽ ☍ ♄
18 45	☽ ✶ ♅
19 01	♀ ± ♄
24 Saturday	
02 31	☽ ☌ ☉
03 15	☽ ∠ ♀
08 13	☽ ☌ ☿
08 59	♀ △ ♇
13 46	☽ Q ♅
19 09	☽ ☌ ♂
19 36	☽ □ ♃
20 42	☽ △ ♆
25 Sunday	
03 29	☽ ☍ ♇
04 56	☽ ✶ ♀
06 07	♂ □ ♃
09 11	♀ ± ♃
11 59	☽ ⊼ ♄
18 44	☽ □ ♅
20 52	☽ ⚼ ♆
21 41	☽ ± ♄
23 44	☿ Q ♅
26 Monday	
00 45	☽ Q ♃
01 55	☽ Q ♀
06 03	☽ ⊻ ☉
06 18	♂ △ ♆
11 40	☽ ± ♆
12 26	☽ ⚼ ♄
16 42	☽ ⊥ ☉
17 04	☽ ⊻ ☿
20 43	☽ ✶ ♃
21 41	☽ ⊼ ♆
22 26	☽ ⊻ ♂
23 00	☽ ∥ ♀
27 Tuesday	
04 48	☽ ⊼ ♇
04 56	☽ ⊥ ☿
09 00	☽ ∠ ☉
09 06	☽ ⊥ ♂
10 29	☽ □ ♀
13 38	☽ △ ♄
15 02	☽ ± ♇
18 21	☿ □ ♃
21 12	☽ △ ♅
22 27	☽ ∠ ♃
23 09	☽ ∠ ☿
28 Wednesday	
00 22	☿ △ ♆
01 25	☽ ∠ ♂
06 41	☽ ⚼ ♇
11 03	☽ ∥ ♅
13 03	☽ ✶ ☉
14 21	☽ ⊥ ♃
19 50	☿ ☌ ♂
23 48	☽ ⚼ ♅
29 Thursday	
01 07	♀ ⊼ ♄
01 09	☽ ⊻ ♃
01 57	☽ ☍ ♆
05 30	☽ ✶ ♂
06 06	☽ # ♆
06 44	☽ ✶ ☿
09 32	☽ △ ♇
12 23	☽ Q ☉
16 14	☽ ± ♅
18 51	☽ □ ♄
20 34	☽ △ ♀
22 13	☽ # ♃
30 Friday	
00 36	☿ ☍ ♇
03 24	☽ ⊼ ♅
04 49	☽ Q ♂
09 05	☽ Q ☿

JULY 2017

LONGITUDES

Date	Sidereal time h m s	Sun ☉ ° ' "	Moon ☽ ° ' "	Moon ☽ 24.00 ° ' "	Mercury ☿ ° '	Venus ♀ ° '	Mars ♂ ° '	Jupiter ♃ ° '	Saturn ♄ ° '	Uranus ♅ ° '	Neptune ♆ ° '	Pluto ⯓ ° '
01	06 38 56	09 ♋ 50 34	15 ♎ 11 37	21 ♎ 21 51	21 ♋ 21	26 ♉ 10	17 ♋ 40	13 ♎ 55	23 ♐ 21	28 ♈ 06	14 ♓ 12	18 ♑ 20
02	06 42 52	10 47 46	27 ♎ 28 33	03 ♏ 32 18	23 19	27 16	18 19	13 59	23 R 17	28 07	14 R 12	18 R 19
03	06 46 49	11 44 58	09 ♏ 33 35	15 ♏ 32 56	25 16	28 21	18 58	14 03	23 12	28 09	14 11	18 17
04	06 50 46	12 42 10	21 ♏ 30 48	27 ♏ 27 38	27 10	29 ♉ 27	19 37	14 07	23 08	28 10	14 11	18 16
05	06 54 42	13 39 21	03 ♐ 23 50	09 ♐ 19 44	29 ♋ 03	00 ♊ 32	20 16	14 12	23 04	28 12	14 10	18 15
06	06 58 39	14 36 33	15 ♐ 15 42	21 ♐ 12 01	00 ♌ 54	01 38	20 55	14 16	23 00	28 13	14 09	18 13
07	07 02 35	15 33 44	27 ♐ 08 56	03 ♑ 06 42	02 42	02 44	21 34	14 21	22 57	28 14	14 09	18 12
08	07 06 32	16 30 55	09 ♑ 05 33	15 ♑ 05 39	04 29	03 51	22 13	14 26	22 53	28 15	14 08	18 10
09	07 10 28	17 28 06	21 ♑ 07 15	27 ♑ 10 29	06 13	04 57	22 52	14 31	22 49	28 17	14 07	18 09
10	07 14 25	18 25 18	03 ♒ 15 36	09 ♒ 22 47	07 56	06 04	23 31	14 36	22 45	28 18	14 07	18 07
11	07 18 21	19 22 30	15 ♒ 32 14	21 ♒ 44 13	09 36	07 10	24 10	14 41	22 41	28 19	14 06	18 06
12	07 22 18	20 19 42	27 ♒ 58 58	04 ♓ 16 45	11 15	08 17	24 49	14 47	22 38	28 20	14 05	18 04
13	07 26 15	21 16 54	10 ♓ 37 53	17 ♓ 02 39	12 51	09 24	25 28	14 52	22 34	28 21	14 04	18 03
14	07 30 11	22 14 07	23 ♓ 31 22	00 ♈ 04 20	14 26	10 31	26 07	14 58	22 31	28 22	14 04	18 01
15	07 34 08	23 11 20	06 ♈ 41 51	13 ♈ 24 11	15 58	11 39	26 46	15 04	22 27	28 23	14 03	18 00
16	07 38 04	24 08 34	20 ♈ 11 33	27 ♈ 04 08	17 29	12 46	27 24	15 10	22 24	28 24	14 02	17 58
17	07 42 01	25 05 49	04 ♉ 01 59	11 ♉ 05 05	18 57	13 54	28 03	15 16	22 20	28 25	14 01	17 57
18	07 45 57	26 03 04	18 ♉ 13 17	25 ♉ 26 18	20 23	15 01	28 42	15 22	22 17	28 25	14 00	17 55
19	07 49 54	27 00 20	02 ♊ 43 42	10 ♊ 04 54	21 48	16 09	29 21	15 28	22 14	28 26	13 59	17 54
20	07 53 50	27 57 37	17 ♊ 29 12	24 ♊ 55 43	23 10	17 17	29 ♋ 59	15 35	22 10	28 27	13 58	17 53
21	07 57 47	28 54 55	02 ♋ 23 30	09 ♋ 51 32	24 30	18 25	00 ♌ 38	15 42	22 07	28 28	13 57	17 51
22	08 01 44	29 ♋ 52 14	17 ♋ 18 45	24 ♋ 44 05	25 47	19 33	01 17	15 49	22 04	28 28	13 56	17 50
23	08 05 40	00 ♌ 49 33	02 ♌ 06 33	09 ♌ 25 13	27 03	20 41	01 56	15 56	22 01	28 29	13 55	17 48
24	08 09 37	01 46 52	16 ♌ 39 18	23 ♌ 48 07	28 16	21 50	02 34	16 03	21 59	28 29	13 54	17 47
25	08 13 33	02 44 12	00 ♍ 51 12	07 ♍ 48 10	29 ♌ 26	22 58	03 13	16 10	21 56	28 30	13 53	17 45
26	08 17 30	03 41 33	14 ♍ 38 51	21 ♍ 23 11	00 ♍ 35	24 07	03 52	16 17	21 53	28 30	13 51	17 44
27	08 21 26	04 38 54	28 ♍ 01 17	04 ♎ 33 19	01 40	25 16	04 30	16 25	21 50	28 30	13 50	17 43
28	08 25 23	05 36 15	10 ♎ 59 37	17 ♎ 20 34	02 43	26 25	05 09	16 32	21 48	28 31	13 49	17 41
29	08 29 19	06 33 37	23 ♎ 36 36	29 ♎ 48 14	03 44	27 33	05 47	16 40	21 45	28 31	13 48	17 40
30	08 33 16	07 30 59	05 ♏ 55 59	12 ♏ 00 25	04 41	28 43	06 26	16 48	21 43	28 31	13 47	17 38
31	08 37 13	08 ♌ 28 22	18 ♏ 02 05	24 ♏ 01 33	05 ♍ 36	29 ♊ 52	07 ♌ 05	16 ♎ 56	21 ♐ 40	28 ♈ 31	13 ♓ 45	17 ♑ 37

Date	Moon True ☊ ° '	Moon Mean ☊ ° '	Moon ☽ Latitude ° '
01	25 ♌ 18	26 ♌ 37	03 N 57
02	25 R 16	26 34	04 34
03	25 13	26 31	04 58
04	25 08	26 27	05 08
05	25 01	26 24	05 06
06	24 54	26 21	04 50
07	24 47	26 18	04 22
08	24 40	26 15	03 42
09	24 35	26 12	02 51
10	24 31	26 08	01 53
11	24 29	26 05	00 N 49
12	24 D 29	26 02	00 S 19
13	24 30	25 59	01 27
14	24 31	25 56	02 32
15	24 33	25 53	03 31
16	24 34	25 49	04 19
17	24 R 33	25 46	04 54
18	24 32	25 43	05 11
19	24 29	25 40	05 10
20	24 26	25 37	04 49
21	24 22	25 33	04 07
22	24 19	25 30	03 09
23	24 17	25 27	01 59
24	24 16	25 24	00 S 42
25	24 D 16	25 21	00 N 36
26	24 17	25 18	01 50
27	24 18	25 14	02 56
28	24 19	25 11	03 50
29	24 20	25 08	04 32
30	24 21	25 05	05 00
31	24 ♌ 21	25 ♌ 02	05 N 14

DECLINATIONS

Date	Sun ☉ ° '	Moon ☽ ° '	Mercury ☿ ° '	Venus ♀ ° '	Mars ♂ ° '	Jupiter ♃ ° '	Saturn ♄ ° '	Uranus ♅ ° '	Neptune ♆ ° '	Pluto ⯓ ° '
01	23 N 04	02 S 20	23 N 36	16 N 42	23 N 17	04 S 17	21 S 57	10 N 16	07 S 03	21 S 24
02	23 00	06 19	23 17	16 58	23 12	04 19	21 56	10 16	07 04	21 25
03	22 55	09 58	22 56	17 13	23 06	04 21	21 56	10 17	07 04	21 25
04	22 50	13 10	22 34	17 29	23 01	04 23	21 56	10 17	07 04	21 25
05	22 44	15 50	22 10	17 44	22 56	04 25	21 56	10 18	07 04	21 26
06	22 38	17 49	21 44	17 58	22 50	04 27	21 56	10 18	07 05	21 26
07	22 32	19 03	21 17	18 13	22 44	04 29	21 56	10 19	07 05	21 26
08	22 25	19 26	20 49	18 27	22 38	04 31	21 56	10 19	07 05	21 26
09	22 18	18 57	20 20	18 40	22 32	04 33	21 56	10 19	07 05	21 27
10	22 10	17 35	19 49	18 53	22 25	04 35	21 56	10 20	07 06	21 27
11	22 02	15 24	19 18	19 06	22 19	04 37	21 56	10 20	07 06	21 27
12	21 54	12 28	18 46	19 19	22 12	04 40	21 56	10 21	07 06	21 28
13	21 45	08 55	18 13	19 31	22 05	04 42	21 56	10 21	07 07	21 28
14	21 36	04 54	17 40	19 43	21 58	04 44	21 56	10 21	07 07	21 28
15	21 27	00 S 34	17 06	19 54	21 51	04 47	21 56	10 21	07 07	21 28
16	21 17	03 N 53	16 31	20 05	21 43	04 49	21 55	10 22	07 08	21 29
17	21 07	08 15	15 57	20 15	21 36	04 52	21 55	10 22	07 08	21 29
18	20 56	12 16	15 22	20 25	21 28	04 55	21 55	10 22	07 09	21 29
19	20 45	15 38	14 47	20 35	21 20	04 57	21 55	10 23	07 09	21 30
20	20 34	18 03	14 12	20 44	21 12	05 00	21 55	10 23	07 09	21 30
21	20 22	19 17	13 37	20 52	21 04	05 03	21 55	10 23	07 10	21 30
22	20 10	19 11	13 02	21 01	20 56	05 06	21 55	10 23	07 10	21 31
23	19 58	17 45	12 27	21 08	20 47	05 09	21 55	10 23	07 11	21 31
24	19 46	15 11	11 52	21 15	20 39	05 11	21 55	10 24	07 11	21 31
25	19 33	11 44	11 17	21 22	20 30	05 14	21 55	10 24	07 12	21 31
26	19 19	07 44	10 43	21 28	20 21	05 17	21 55	10 24	07 12	21 32
27	19 06	03 N 28	10 10	21 33	20 12	05 21	21 55	10 24	07 13	21 32
28	18 52	00 S 49	09 37	21 38	20 03	05 24	21 55	10 24	07 13	21 32
29	18 38	04 57	09 05	21 43	19 54	05 27	21 55	10 24	07 14	21 33
30	18 23	08 47	08 33	21 47	19 44	05 30	21 55	10 24	07 14	21 33
31	18 N 08	12 S 10	08 N 02	21 N 50	19 N 35	05 S 33	21 S 55	10 N 24	07 S 15	21 S 33

ZODIAC SIGN ENTRIES

Date	h	m	Planets
02	16	59	☽ ♏
05	00	11	♀ ♊
05	05	08	☽ ♐
06	00	20	☿ ♌
07	17	45	☽ ♑
10	05	35	☽ ♒
12	15	51	☽ ♓
14	23	52	☽ ♈
17	05	04	☽ ♉
19	07	31	☽ ♊
20	12	19	♂ ♌
21	08	09	☽ ♋
22	15	15	☉ ♌
23	08	34	☽ ♌
25	10	32	☽ ♍
25	23	41	☿ ♍
27	15	37	☽ ♎
30	00	23	☽ ♏
31	14	54	♀ ♋

LATITUDES

Date	Mercury ☿ ° '	Venus ♀ ° '	Mars ♂ ° '	Jupiter ♃ ° '	Saturn ♄ ° '	Uranus ♅ ° '	Neptune ♆ ° '	Pluto ⯓ ° '
01	01 N 53	02 S 40	01 N 01	01 N 18	01 N 20	00 S 34	00 S 55	00 N 47
04	01 53	02 37	01 02	01 18	01 19	00 34	00 55	00 46
07	01 47	02 33	01 02	01 17	01 19	00 34	00 55	00 46
10	01 36	02 28	01 03	01 16	01 18	00 35	00 55	00 46
13	01 19	02 22	01 04	01 15	01 18	00 35	00 55	00 46
16	00 59	02 16	01 04	01 15	01 18	00 35	00 55	00 45
19	00 34	02 09	01 05	01 14	01 17	00 35	00 55	00 45
22	00 N 07	02 01	01 05	01 13	01 17	00 35	00 55	00 45
25	00 S 24	01 53	01 06	01 13	01 16	00 35	00 55	00 44
28	00 57	01 45	01 06	01 12	01 16	00 35	00 56	00 44
31	01 S 32	01 S 36	01 N 07	01 N 11	01 N 15	00 S 35	00 S 56	00 N 44

DATA

Julian Date	2457936
Delta T	+70 seconds
Ayanamsa	24° 05' 56"
Synetic vernal point	05° ♓ 01' 03"
True obliquity of ecliptic	23° 26' 05"

MOON'S PHASES, APSIDES AND POSITIONS ☽

Date	h	m	Phase	Longitude ° '	Eclipse Indicator
01	00	51	☽	09 ♎ 24	
09	04	07	○	17 ♑ 09	
16	19	26	☾	24 ♈ 26	
23	09	46	●	00 ♌ 44	
30	15	23	☽	07 ♏ 39	

Day	h	m		
06	04	20	Apogee	
21	17	04	Perigee	
08	10	48	Max dec	19° S 26'
15	15	04	0N	
21	22	07	Max dec	19° N 25'
28	07	22	0S	

LONGITUDES

Date	Chiron ⚷ ° '	Ceres ⚳ ° '	Pallas ⚴ ° '	Juno ⚵ ° '	Vesta ⚶ ° '	Black Moon Lilith ⚸ ° '
01	28 ♓ 52	26 ♊ 14	01 ♉ 18	11 ♑ 05	22 ♌ 40	15 ♐ 24
11	28 ♓ 49	00 ♋ 25	04 ♉ 07	08 ♑ 46	27 ♌ 05	16 ♐ 31
21	28 ♓ 41	04 ♋ 36	06 ♉ 44	06 ♑ 36	01 ♍ 37	17 ♐ 38
31	28 ♓ 29	08 ♋ 44	09 ♉ 04	04 ♑ 49	06 ♍ 15	18 ♐ 45

ASPECTARIAN

01 Saturday
00 51 ☽ □ ☉
03 31 ☽ ♇ ♀
04 37 ☽ Q ♄
09 32 ☽ ☌ ♃
10 05 ☽ ⊼ ♆
12 42 ♀ Q ♆
17 03 ☽ □ ♂
18 06 ☽ □ ⯓
18 31 ☉ Q ♅
21 44 ☽ ± ♆
22 37 ☽ ± ♀
23 39 ☽ ∥ ♃

02 Sunday
02 16 ☽ □ ☿
03 47 ☽ ⚹ ♄
11 29 ☿ ⊼ ♄
11 32 ☽ ⊼ ♀
12 02 ♂ ☍ ⯓
13 16 ☽ ☍ ♅
15 24 ☽ ♇ ♆
16 43 ☽ ∥ ♆
20 21 ☿ ∥ ♂

03 Monday
05 29 ☽ Q ⯓
07 23 ♀ ⊻ ♅
09 19 ☽ ∠ ♄
13 51 ☉ ∥ ☿
14 12 ☽ # ♅
16 46 ☽ △ ☉
21 03 ☽ ⊻ ♃
21 16 ☽ △ ♆

04 Tuesday
03 15 ☽ ⊥ ♄
04 28 ♀ ♇ ♃
05 29 ☽ ⚹ ⯓
07 58 ☽ △ ♂
09 11 ☽ ⊥ ♃
15 16 ☽ ⊻ ♄

05 Wednesday
00 51 ☿ □ ♅
01 27 ☽ ⊼ ♅
01 34 ☽ △ ☿
01 35 ☽ ♇ ☉
03 27 ☽ ∠ ♃
04 19 ♃ ⊼ ♆
05 38 ☽ ☍ ♀
11 41 ☽ ∠ ⯓
12 19 ☿ ± ♄
13 32 ☿ ♇ ♆
13 37 ☽ ± ♅
16 00 ☽ ♇ ♂
21 22 ☽ ± ☉

06 Thursday
00 46 ☉ △ ♆
00 53 ☿ # ♄
02 44 ☉ □ ♃
05 51 ☽ ⊥ ⯓
07 51 ☽ ♇ ♅
09 46 ☽ □ ♆
09 59 ☽ ⚹ ♃
10 34 ☽ ⊼ ☉
11 16 ☽ ± ♂
13 30 ☽ ♇ ☿
14 45 ☽ # ♀
17 58 ☽ ⊻ ⯓

07 Friday
00 06 ☽ ⊼ ♂
03 34 ☽ ☌ ♄
04 11 ☿ # ⯓
07 03 ☿ Q ♃
10 23 ☽ Q ♃
10 56 ☽ ± ☿
13 19 ☿ ⚹ ♀
14 12 ☽ △ ♅
21 39 ♀ ♇ ⯓
22 03 ☽ Q ♆

08 Saturday
00 24 ☽ ⊼ ♀
01 08 ☽ ⊼ ☿
13 40 ☽ ± ♀
21 08 ♀ ⊥ ♅
22 05 ☽ ⚹ ♆
22 45 ☽ □ ♃

09 Sunday
04 07 ☽ ☍ ☉
06 06 ☽ ☌ ⯓
09 26 ☽ ♇ ♀
10 09 ♂ ⊼ ♄
15 21 ☽ ⊻ ♄
15 40 ☽ ☍ ♂
17 25 ☽ # ♀

10 Monday
02 12 ☽ □ ♅
03 10 ☽ ⊥ ♄
03 50 ☽ ∠ ♆
04 35 ☉ ☍ ⯓
09 33 ☿ ♇ ♄
14 33 ☿ ± ♆
18 03 ☽ △ ♀
20 46 ☽ ∠ ♄
21 31 ☽ ⊥ ♆
22 38 ☽ ☍ ☿

11 Tuesday
09 13 ☽ ⊻ ♆
10 20 ☽ △ ♃
13 31 ☽ Q ♅
16 57 ☽ ⊻ ⯓
18 15 ☿ ∥ ♀
20 03 ☽ ⊼ ☉

12 Wednesday
01 46 ☽ ⚹ ♄
04 31 ☽ ⊥ ⯓
05 36 ☽ ⊼ ♂
06 38 ☉ # ♄
08 34 ☽ ± ☉
12 40 ☽ ⚹ ♅
15 27 ☽ ♇ ♃
17 42 ☽ ± ♂
21 41 ☽ ∠ ⯓

13 Thursday
00 36 ☽ Q ♄
02 49 ☽ # ♅
03 08 ☽ ♇ ☉
08 39 ☽ ± ♃
09 28 ☽ □ ♀
11 40 ☽ ♇ ♂
16 46 ☽ ⊼ ☿
17 07 ☽ ∠ ♅
18 27 ☽ ☌ ♆
20 00 ☽ ⊼ ♃
23 05 ☽ ∥ ♆

14 Friday
01 50 ☽ ⚹ ⯓
05 30 ☽ ± ☿
06 19 ☿ ⊼ ♆
09 26 ☽ △ ☉
09 52 ☽ ⊥ ♅
10 08 ☽ □ ♄
12 54 ☽ ∥ ♃
17 00 ☽ △ ♂
18 29 ☉ ⊼ ♄
20 32 ♂ # ♄
20 46 ☿ ⚹ ♃
20 54 ☽ ⊻ ♅
22 02 ☽ Q ♀
23 53 ☽ Q ⯓

15 Saturday
00 17 ☽ ♇ ☿
07 19 ☉ # ⯓
17 50 ♀ ∠ ♂
19 24 ♀ ± ⯓
21 40 ☽ ⚹ ♀

16 Sunday
01 08 ☽ ⊻ ♆
03 03 ☽ ☍ ♃
06 38 ☽ △ ☿
08 06 ☽ □ ⯓
11 43 ☽ ⊥ ♆
15 50 ☽ △ ♄
17 06 ☽ # ♃
19 26 ☽ □ ☉
19 50 ☿ ⊼ ⯓

17 Monday
01 12 ☽ □ ♂
01 38 ♀ ∠ ♅
02 19 ☽ ☌ ♅
02 23 ☽ ∠ ♀
03 23 ☽ ∠ ♆
05 44 ☽ # ♆
14 33 ♀ □ ♆
17 37 ☽ ♇ ♄
19 09 ☽ ⊥ ♀
21 43 ♂ ± ♄

18 Tuesday
00 19 ☽ ∥ ♅
01 37 ♂ □ ♅
04 30 ☽ Q ☉
04 55 ☽ ⚹ ♆
06 11 ☽ ⊻ ♀
07 11 ☽ ⊼ ♃
08 40 ♂ # ⯓
08 46 ☽ ± ♄
09 20 ☽ Q ♂
11 30 ☽ △ ⯓
16 01 ☽ □ ☿
17 17 ☽ ± ♃
18 45 ☽ ⊼ ♄
20 09 ♀ △ ♃
22 51 ♂ ♇ ♆

19 Wednesday
00 55 ☽ Q ♆
01 56 ☽ ⚹ ☉
04 57 ☽ ⊻ ♅
06 11 ☽ ⚹ ♂
06 25 ☽ ∥ ☿
08 17 ☽ ♇ ♃
12 17 ☽ ♇ ⯓
14 48 ☽ ⊥ ♅
19 16 ☿ △ ♄

20 Thursday
00 16 ☉ ∥ ♀
00 43 ☽ Q ☿
02 56 ☽ ± ⯓
04 10 ☽ ∠ ☉
05 28 ☽ ∠ ♅
06 19 ☽ □ ♆
07 47 ☽ ∠ ♂
08 54 ☽ △ ♃
11 39 ☽ ☌ ♀
12 38 ☽ ⊼ ⯓
17 07 ☉ ± ♄
19 32 ☽ ☍ ♄
19 43 ☽ ⊥ ☉
22 04 ☽ ⚹ ☿
22 58 ☽ ⊥ ♂

21 Friday
00 17 ♀ ⊼ ⯓
00 26 ☉ □ ♅
00 33 ☿ ± ⯓
05 41 ☽ ⚹ ♅
06 02 ☽ ⊻ ☉
09 03 ☽ ⊻ ♂
12 49 ☉ ♇ ♆

22 Saturday
00 30 ☽ ∠ ☿
00 59 ☽ Q ♅
05 13 ♀ ∥ ♂
06 33 ☽ △ ♆
09 34 ☽ □ ♃
12 50 ☽ ☍ ⯓
15 55 ☽ ⊻ ♀
16 22 ☽ ⊥ ☿
19 40 ☽ ⊼ ♄

23 Sunday
02 26 ☽ ⊥ ♀
02 59 ☽ ⊻ ☿
05 22 ☽ ± ♄
06 05 ☽ □ ♅
06 48 ☽ ♇ ♆
09 46 ☽ ☌ ☉
11 41 ☽ ☌ ♂
15 00 ☽ Q ♃
18 22 ☽ ∠ ♀
20 02 ☽ ♇ ♄
21 30 ☽ ± ♆

24 Monday
07 25 ☽ ⊼ ♆
10 58 ☽ ⚹ ♃
13 53 ☽ ⊼ ⯓
14 54 ♀ ☍ ♄
16 33 ☿ △ ♅
20 53 ☽ △ ♄
21 26 ☽ ⚹ ♀
23 57 ☽ ± ⯓

25 Tuesday
07 58 ☽ △ ♅
09 22 ☽ ☌ ☿
12 32 ☽ ∠ ♃
15 16 ☽ ♇ ⯓
15 17 ☽ ∥ ☿
15 28 ☽ ⊻ ☉
16 15 ☽ ⊻ ♂
19 44 ☽ Q ♀
20 19 ☽ ∥ ♅

26 Wednesday
02 39 ☽ ⊥ ☉
03 10 ☽ ⊥ ♂
04 16 ☽ ⊥ ♃
09 59 ☽ ♇ ♅
10 36 ☽ ☍ ♆
14 56 ☽ ⊻ ♃
15 03 ☽ # ♆
17 28 ☽ △ ⯓
19 44 ☽ ∠ ☉
19 51 ☽ ∠ ♂

27 Thursday
00 51 ☽ □ ♄
00 57 ☉ ☌ ♂
01 41 ☽ # ♃
02 00 ☿ ∥ ♅
02 01 ☽ ± ♅
05 10 ☉ Q ♃
05 18 ♀ # ⯓
05 26 ☿ ∠ ♃
06 31 ☽ □ ♀
07 43 ♂ Q ♃
12 53 ☽ ⊼ ♅
19 17 ☽ ⊻ ☿

28 Friday
00 32 ☽ ⚹ ♂
01 08 ☽ ⚹ ☉
07 23 ☽ ⊥ ☿
09 46 ☽ Q ♄
11 08 ☿ ♇ ⯓
17 19 ☽ ⊼ ♆
22 35 ☽ ☌ ♃

29 Saturday
00 15 ☽ Q ♂
00 38 ☽ □ ⯓
01 31 ☽ Q ☉
01 50 ☽ ∠ ☿
04 42 ☽ ± ♆
08 26 ☽ ⚹ ♄
15 00 ☽ ∥ ♃
16 39 ☉ ♇ ♄
20 25 ☽ △ ♀
21 30 ☽ ☍ ♅
22 01 ☽ ♇ ♆

30 Sunday
02 00 ☽ ∥ ♆
08 04 ♀ ⚹ ♅
09 21 ☽ ⚹ ☿
10 41 ☽ # ☿
11 25 ☽ Q ⯓
13 02 ☽ □ ♂
13 32 ☽ ∠ ♄
15 23 ☽ □ ☉
18 24 ☉ ± ♆
21 49 ♂ ♇ ♄
23 04 ☽ # ♅

31 Monday
03 29 ☽ △ ♆
05 00 ☽ ♇ ♀
07 18 ☽ ⊥ ♄
09 46 ☽ ⊻ ♃
11 03 ☽ Q ☿
11 10 ☽ ⚹ ⯓
19 15 ☽ ⊻ ♄
21 55 ☽ ⊥ ♃

All ephemeris data is given at 12.00 UT and the Moon's longitude is additionally given for 24.00 UT

AUGUST 2017

Date	Sidereal time h m s	Sun ☉ ° ' "	Moon ☽ ° ' "	Moon ☽ 24.00 ° ' "	Mercury ☿ ° '	Venus ♀ ° '	Mars ♂ ° '	Jupiter ♃ ° '	Saturn ♄ ° '	Uranus ♅ ° '	Neptune ♆ ° '	Pluto ♇ ° '
01	08 41 09	09 ♌ 25 46	29 ♏ 59 23	05 ♐ 56 06	06 ♍ 27	01 ♋ 01	07 ♌ 43	17 ♎ 04	21 ♐ 38	28 ♈ 31	13 ♓ 44	17 ♑ 36
02	08 45 06	10 23 10	11 ♐ 52 14	17 ♐ 48 16	07 15	02 10	08 22	17 12	21 R 36	28 32	13 R 43	17 R 34
03	08 49 02	11 20 35	23 ♐ 44 39	29 ♐ 41 50	08 00	03 20	09 00	17 21	21 34	28 R 32	13 41	17 33
04	08 52 59	12 18 01	05 ♑ 40 12	11 ♑ 40 07	08 41	04 29	09 39	17 29	21 32	28 32	13 40	17 32
05	08 56 55	13 15 27	17 ♑ 41 53	23 ♑ 45 48	09 18	05 39	10 17	17 37	21 30	28 31	13 39	17 30
06	09 00 52	14 12 55	29 ♑ 52 06	06 ♒ 01 01	09 51	06 49	10 56	17 46	21 28	28 31	13 37	17 29
07	09 04 48	15 10 23	12 ♒ 12 43	18 ♒ 27 21	10 21	07 58	11 34	17 55	21 26	28 31	13 36	17 28
08	09 08 45	16 07 52	24 ♒ 45 03	01 ♓ 05 56	10 46	09 08	12 12	18 04	21 25	28 31	13 35	17 27
09	09 12 42	17 05 22	07 ♓ 30 04	13 ♓ 57 32	11 06	10 18	12 51	18 13	21 23	28 31	13 33	17 25
10	09 16 38	18 02 54	20 ♓ 28 22	27 ♓ 02 38	11 21	11 28	13 29	18 22	21 22	28 30	13 32	17 24
11	09 20 35	19 00 26	03 ♈ 40 22	10 ♈ 21 35	11 32	12 39	14 08	18 31	21 20	28 30	13 30	17 23
12	09 24 31	19 58 00	17 ♈ 06 17	23 ♈ 54 27	11 38	13 49	14 46	18 40	21 19	28 29	13 29	17 22
13	09 28 28	20 55 36	00 ♉ 46 05	07 ♉ 41 07	11 R 38	14 59	15 24	18 50	21 18	28 29	13 27	17 21
14	09 32 24	21 53 13	14 ♉ 39 26	21 ♉ 40 56	11 33	16 10	16 03	18 59	21 17	28 28	13 26	17 19
15	09 36 21	22 50 52	28 ♉ 45 25	05 ♊ 52 39	11 22	17 20	16 41	19 09	21 16	28 28	13 24	17 18
16	09 40 17	23 48 32	13 ♊ 02 19	20 ♊ 14 04	11 06	18 31	17 19	19 18	21 15	28 27	13 23	17 17
17	09 44 14	24 46 14	27 ♊ 27 27	04 ♋ 41 59	10 44	19 42	17 58	19 28	21 14	28 27	13 21	17 16
18	09 48 11	25 43 57	11 ♋ 57 05	19 ♋ 12 09	10 16	20 53	18 36	19 38	21 13	28 26	13 20	17 15
19	09 52 07	26 41 42	26 ♋ 26 32	03 ♌ 39 35	09 44	22 04	19 14	19 48	21 13	28 25	13 18	17 14
20	09 56 04	27 39 28	10 ♌ 50 38	17 ♌ 59 01	09 06	23 15	19 53	19 58	21 12	28 24	13 17	17 13
21	10 00 00	28 37 16	25 ♌ 04 08	02 ♍ 05 26	08 24	24 26	20 31	20 08	21 12	28 23	13 15	17 12
22	10 03 57	29 ♌ 35 05	09 ♍ 02 26	15 ♍ 54 45	07 37	25 37	21 09	20 19	21 11	28 23	13 13	17 11
23	10 07 53	00 ♍ 32 56	22 ♍ 42 05	29 ♍ 24 15	06 48	26 48	21 47	20 29	21 11	28 22	13 12	17 10
24	10 11 50	01 30 48	06 ♎ 01 07	12 ♎ 32 44	05 55	27 59	22 26	20 39	21 11	28 21	13 10	17 09
25	10 15 46	02 28 41	18 ♎ 59 12	25 ♎ 20 42	05 02	29 ♋ 11	23 04	20 50	21 11	28 20	13 09	17 08
26	10 19 43	03 26 36	01 ♏ 37 32	07 ♏ 50 03	04 07	00 ♌ 22	23 42	21 00	21 D 11	28 19	13 07	17 07
27	10 23 40	04 24 31	13 ♏ 58 40	20 ♏ 03 51	03 13	01 34	24 20	21 11	21 11	28 17	13 05	17 06
28	10 27 36	05 22 29	26 ♏ 06 06	02 ♐ 05 59	02 21	02 46	24 59	21 22	21 11	28 16	13 04	17 05
29	10 31 33	06 20 27	08 ♐ 04 03	14 ♐ 00 54	01 32	03 57	25 37	21 33	21 12	28 15	13 02	17 04
30	10 35 29	07 18 27	19 ♐ 57 05	25 ♐ 53 14	00 46	05 09	26 15	21 43	21 12	28 14	13 01	17 04
31	10 39 26	08 ♍ 16 28	01 ♑ 49 53	07 ♑ 47 37	00 ♍ 05	06 ♌ 21	26 ♌ 53	21 ♎ 54	21 ♐ 13	28 ♈ 13	12 ♓ 59	17 ♑ 03

Date	Moon True ☊ ° '	Moon Mean ☊ ° '	Moon ☽ Latitude ° '
01	24 ♌ 20	24 ♌ 59	05 N 14
02	24 R 18	24 55	05 01
03	24 16	24 52	04 35
04	24 15	24 49	03 57
05	24 13	24 46	03 09
06	24 12	24 43	02 11
07	24 11	24 39	01 N 06
08	24 D 11	24 36	00 S 03
09	24 11	24 33	01 13
10	24 12	24 30	02 21
11	24 12	24 27	03 22
12	24 13	24 24	04 13
13	24 13	24 20	04 51
14	24 13	24 17	05 13
15	24 R 13	24 14	05 16
16	24 13	24 11	05 00
17	24 D 13	24 08	04 25
18	24 13	24 05	03 34
19	24 13	24 01	02 28
20	24 14	23 58	01 S 14
21	24 R 14	23 55	00 N 05
22	24 14	23 52	01 21
23	24 13	23 49	02 31
24	24 12	23 45	03 31
25	24 11	23 42	04 19
26	24 10	23 39	04 52
27	24 09	23 36	05 12
28	24 09	23 33	05 16
29	24 D 08	23 30	05 07
30	24 09	23 26	04 45
31	24 ♌ 10	23 ♌ 23	04 N 11

DECLINATIONS

Date	Sun ☉ ° '	Moon ☽ ° '	Mercury ☿ ° '	Venus ♀ ° '	Mars ♂ ° '	Jupiter ♃ ° '	Saturn ♄ ° '	Uranus ♅ ° '	Neptune ♆ ° '	Pluto ♇ ° '
01	17 N 53	15 S 01	07 N 32	21 N 53	19 N 25	05 S 36	21 S 55	10 N 24	07 S 15	21 S 33
02	17 38	17 14	07 03	21 55	19 15	05 40	21 55	10 24	07 16	21 34
03	17 22	18 42	06 36	21 57	19 05	05 43	21 55	10 24	07 16	21 34
04	17 06	19 22	06 09	21 58	18 55	05 47	21 55	10 24	07 17	21 34
05	16 50	19 09	05 44	21 58	18 45	05 50	21 56	10 24	07 17	21 35
06	16 34	18 03	05 21	21 58	18 34	05 54	21 56	10 24	07 18	21 35
07	16 17	16 05	04 59	21 58	18 24	05 57	21 56	10 24	07 18	21 35
08	16 00	13 19	04 38	21 56	18 13	06 01	21 56	10 24	07 19	21 35
09	15 43	09 53	04 20	21 54	18 02	06 04	21 56	10 24	07 20	21 36
10	15 25	05 56	04 04	21 52	17 51	06 08	21 56	10 24	07 20	21 36
11	15 07	01 S 38	03 50	21 49	17 40	06 11	21 56	10 23	07 21	21 36
12	14 49	02 N 49	03 38	21 45	17 29	06 15	21 56	10 23	07 21	21 36
13	14 31	07 11	03 29	21 41	17 18	06 19	21 56	10 23	07 22	21 37
14	14 13	11 15	03 23	21 36	17 07	06 23	21 56	10 23	07 22	21 37
15	13 54	14 44	03 19	21 31	16 55	06 26	21 56	10 23	07 23	21 37
16	13 35	17 24	03 19	21 25	16 44	06 30	21 57	10 22	07 24	21 37
17	13 16	18 59	03 21	21 18	16 32	06 34	21 57	10 22	07 24	21 38
18	12 56	19 21	03 27	21 11	16 20	06 38	21 57	10 22	07 25	21 38
19	12 37	18 26	03 36	21 03	16 08	06 42	21 57	10 22	07 25	21 38
20	12 17	16 20	03 48	20 55	15 57	06 46	21 57	10 21	07 26	21 38
21	11 57	13 14	04 03	20 46	15 44	06 50	21 57	10 21	07 27	21 39
22	11 37	09 26	04 22	20 36	15 32	06 54	21 57	10 21	07 27	21 39
23	11 17	05 13	04 43	20 26	15 20	06 58	21 58	10 20	07 28	21 39
24	10 56	00 N 51	05 07	20 15	15 08	07 02	21 58	10 20	07 29	21 39
25	10 35	03 S 26	05 33	20 04	14 55	07 06	21 58	10 19	07 29	21 40
26	10 15	07 28	06 00	19 52	14 43	07 10	21 58	10 19	07 30	21 40
27	09 54	11 04	06 30	19 39	14 30	07 14	21 58	10 19	07 31	21 40
28	09 32	14 09	07 00	19 26	14 17	07 18	21 59	10 18	07 31	21 40
29	09 11	16 36	07 31	19 13	14 04	07 23	21 59	10 18	07 32	21 40
30	08 50	18 19	08 01	18 59	13 52	07 27	21 59	10 17	07 32	21 41
31	08 N 28	19 S 15	08 N 31	18 N 44	13 N 39	07 S 31	21 S 59	10 N 17	07 S 33	21 S 41

ZODIAC SIGN ENTRIES

Date	h	m	Planets
01	12	01	☽ ♐
04	00	37	☽ ♑
06	12	15	☽ ♒
08	21	56	☽ ♓
11	05	22	☽ ♈
13	10	40	☽ ♉
15	14	06	☽ ♊
17	16	13	☽ ♋
19	17	55	☽ ♌
21	20	25	☽ ♍
22	22	20	☉ ♍
24	01	04	☽ ♎
26	04	30	♀ ♌
26	08	53	☽ ♏
28	19	47	☽ ♐
31	08	18	☽ ♑
31	15	28	☿ ♌

LATITUDES

Date	Mercury ☿ ° '	Venus ♀ ° '	Mars ♂ ° '	Jupiter ♃ ° '	Saturn ♄ ° '	Uranus ♅ ° '	Neptune ♆ ° '	Pluto ♇ ° '
01	01 S 44	01 S 33	01 N 07	01 N 11	01 N 15	00 S 35	00 S 56	00 N 44
04	02 20	01 24	01 07	01 10	01 15	00 35	00 56	00 43
07	02 56	01 14	01 08	01 10	01 14	00 35	00 56	00 43
10	03 30	01 05	01 08	01 09	01 13	00 35	00 56	00 43
13	04 01	00 55	01 08	01 09	01 13	00 35	00 56	00 42
16	04 25	00 45	01 09	01 08	01 12	00 35	00 56	00 42
19	04 40	00 35	01 09	01 08	01 12	00 35	00 56	00 42
22	04 41	00 25	01 09	01 07	01 11	00 35	00 56	00 41
25	04 26	00 16	01 09	01 07	01 11	00 36	00 56	00 41
28	03 54	00 S 06	01 10	01 06	01 10	00 36	00 56	00 41
31	03 S 07	00 N 03	01 N 10	01 N 06	01 N 10	00 S 36	00 S 56	00 N 40

DATA

Julian Date	2457967
Delta T	+70 seconds
Ayanamsa	24° 06' 01"
Synetic vernal point	05° ♓ 00' 58"
True obliquity of ecliptic	23° 26' 06"

MOON'S PHASES, APSIDES AND POSITIONS ☽

Date	h	m	Phase	Longitude ° '	Eclipse Indicator
07	18	11	○	15 ♒ 25	partial
15	01	15	☾	22 ♉ 25	
21	18	30	●	28 ♌ 53	Total
29	08	13	☽	06 ♐ 11	

Day	h	m		
02	17	51	Apogee	
18	13	08	Perigee	
30	11	24	Apogee	
04	18	14	Max dec	19° S 24'
11	20	49	0N	
18	06	50	Max dec	19° N 23'
24	16	40	0S	

LONGITUDES

Date	Chiron ⚷ ° '	Ceres ⚳ ° '	Pallas ⚴ ° '	Juno ⚵ ° '	Vesta ⚶ ° '	Black Moon Lilith ⚸ ° '
01	28 ♓ 27	09 ♋ 08	09 ♉ 17	04 ♑ 40	06 ♍ 43	18 ♐ 52
11	28 ♓ 09	13 ♋ 38	11 ♉ 15	03 ♑ 26	11 ♍ 27	19 ♐ 59
21	27 ♓ 48	17 ♋ 14	12 ♉ 48	02 ♑ 48	16 ♍ 18	21 ♐ 06
31	27 ♓ 24	21 ♋ 09	13 ♉ 50	02 ♑ 47	21 ♍ 09	22 ♐ 13

All ephemeris data is given at 12.00 UT and the Moon's longitude is additionally given for 24.00 UT

ASPECTARIAN

h m	Aspects
01 Tuesday	
00 55	☽ ± ♀
09 03	☽ ⊼ ♅
12 34	♂ ± ♆
14 17	☽ ⊼ ♀
16 14	☽ ∠ ♃
17 15	☽ ∠ ♇
21 09	☽ ± ♅
02 Wednesday	
01 57	☿ # ♆
02 00	☽ □ ☿
04 30	☽ △ ♂
08 44	☽ △ ☉
11 24	☽ ⊥ ♇
15 21	☽ bQ ♅
15 30	♀ # ♄
15 43	☽ □ ♆
16 44	☽ # ☉
20 53	♀ ⊥ ♂
22 55	☽ ⚹ ♃
23 31	☽ ⊻ ♇
03 Thursday	
05 28	♅ St R
07 37	☽ ☌ ♄
12 33	☽ bQ ♂
17 42	☽ bQ ☉
20 12	☽ # ♂
21 38	☽ △ ♅
23 25	☽ Q ♃
04 Friday	
03 59	☽ Q ♆
07 42	☽ ± ♂
09 22	☽ ☍ ♀
13 22	☽ ± ☉
18 22	☽ △ ☿
18 48	♃ □ ♇
20 24	☽ ⊼ ♂
05 Saturday	
02 24	☽ ⊼ ☉
03 58	☽ ⚹ ♆
07 02	☿ # ♃
11 37	☽ ☌ ♇
11 51	☽ □ ♃
19 31	☽ ⊻ ♄
21 31	☉ ⊼ ♆
06 Sunday	
01 18	☽ # ♂
01 42	☽ bQ ☿
07 18	☽ ⊥ ♄
09 22	☽ □ ♅
09 34	☽ ∠ ♆
20 08	☽ ± ☿
07 Monday	
00 51	☽ ∠ ♄
02 57	☽ ⊼ ♀
03 06	☽ ⊥ ♆
08 15	☽ ⊼ ☿
09 37	☽ # ☉
10 41	☽ ☍ ♂
14 40	☽ ⊻ ♆
15 45	☽ ± ♀
18 11	☽ ☍ ☉
20 17	☽ Q ♅
22 05	☽ ⊻ ♇
23 06	☽ △ ♃
08 Tuesday	
05 40	☽ ⚹ ♄
09 31	☽ ⊥ ♇
10 43	☽ bQ ♀
19 07	☽ ⚹ ♅
20 06	♀ # ♄
09 Wednesday	
02 30	☽ ∠ ♇
03 53	☽ bQ ♃
04 19	☽ Q ♄
08 41	☽ # ♅
16 11	♀ Q ♅
17 45	☽ △ ♀
18 51	☽ ☍ ☿
20 10	☉ ⊼ ♇
20 52	☽ ± ♃
22 28	☽ ⊼ ♂
23 10	☽ ∠ ♅
23 14	☽ ☌ ♆
10 Thursday	
03 47	☽ ∥ ♆
06 22	☽ ⚹ ♇
07 12	☽ ⊼ ☉
08 05	☽ ⊼ ♃
09 03	☿ ⚹ ♀
10 06	☽ ± ♂
10 52	☽ ∥ ♃
13 31	♂ ⊼ ♆
13 38	☽ □ ♄
15 43	☽ ⊥ ♅
19 03	☽ ± ☉
21 24	☉ ⚹ ♃
23 12	☽ # ☿
11 Friday	
02 39	☽ ⊻ ♅
03 22	☽ bQ ♂
04 15	☽ Q ♇
12 39	☽ bQ ☉
21 49	♅ ∠ ♆
12 Saturday	
02 13	☽ ⊼ ☿
05 17	♀ △ ♆
05 35	☽ ⊻ ♆
05 36	☽ □ ♀
07 39	☽ △ ♂
12 27	☽ □ ♇
12 56	☽ ± ☿
14 48	☽ ☍ ♃
16 12	☽ ⊥ ♆
16 17	☽ ∥ ☿
17 27	☽ △ ☉
19 26	☽ △ ♄
13 Sunday	
01 00	☿ St R
04 47	☽ bQ ☿
05 02	☉ ⊥ ♀
07 02	☽ # ♃
07 58	☽ ∠ ♆
08 01	☽ ☌ ♅
13 00	☽ # ♆
16 13	☽ Q ♀
21 06	☉ △ ♄
21 35	☽ bQ ♄
14 Monday	
06 39	☽ ∥ ♅
06 42	☽ △ ☿
06 43	♀ ⊻ ♂
09 08	♀ # ♇
09 54	☽ ⚹ ♆
13 04	☽ ± ♄
14 30	☽ □ ♂
14 49	☽ ⚹ ♀
16 34	☽ △ ♇
19 29	☽ ⊼ ♃
23 18	☽ ⊼ ♄
15 Tuesday	
01 15	☽ □ ☉
05 49	☽ ± ♃
06 18	☽ ∥ ☉
06 20	☽ Q ♆
11 17	♀ ☍ ♇
11 30	☽ ⊻ ♅
17 59	☽ bQ ♇
18 36	☽ ∠ ♀
21 12	☽ bQ ♃
21 37	☽ ⊥ ♅
22 28	☽ Q ♂
23 11	☉ ± ♇
16 Wednesday	
05 37	☽ ∥ ♂
08 49	☽ □ ☿
09 04	☽ ± ♇
09 48	☽ Q ☉
10 37	♂ ⊼ ♇
11 03	☽ ⊥ ♀
12 34	☽ □ ♆
12 42	☽ ∠ ♅
19 05	☽ ⊼ ♇
19 29	☽ ⚹ ♂
21 16	♃ ± ♆
21 57	☽ ⊻ ♀
22 35	☽ △ ♃
17 Thursday	
01 40	☽ ☍ ♄
06 40	♀ □ ♃
07 13	☽ ⚹ ☉
13 38	☽ ⚹ ♅
14 03	☽ Q ☿
21 33	☽ ∠ ♂
18 Friday	
09 19	☽ ⚹ ☿
09 29	☽ Q ♅
09 50	☽ ∠ ☉
13 07	☽ ⊥ ♂
14 17	☽ △ ♆
18 55	♀ ⊼ ♄
20 45	☽ ☍ ♇
23 31	☽ ⊻ ♂
19 Saturday	
00 52	☽ □ ♃
01 48	☽ ⊥ ☉
03 20	☽ ⊼ ♄
04 05	☽ ☌ ♀
09 16	☽ ∠ ☿
12 27	☽ ⊻ ☉
13 17	☽ ± ♄
15 05	☽ bQ ♆
15 17	☽ □ ♅
23 37	☽ ⊥ ☿
20 Sunday	
04 15	☽ bQ ♄
06 03	☽ ± ♆
07 08	☽ Q ♃
09 12	☽ ⊻ ☿
15 42	☽ ∥ ♂
16 04	☽ ⊼ ♆
16 42	♂ ⚹ ♃
22 41	☽ ⊼ ♇
22 58	☿ ∠ ♀
21 Monday	
03 32	☽ ⚹ ♃
03 55	☽ ☌ ♂
05 26	☽ △ ♄
06 22	☉ △ ♅
08 49	☽ ± ♇
10 49	☽ ⊻ ♀
17 39	☽ △ ♅
18 30	☽ ☌ ☉
21 22	☽ ∥ ☉
22 00	☽ ⊥ ♀
22 Tuesday	
00 10	☽ bQ ♇
05 28	☽ ∠ ♃
06 33	☽ ∥ ♅
09 40	☽ ☌ ☿
13 21	♂ △ ♄
15 00	☽ ∠ ♀
19 16	☽ ☍ ♆
19 33	☽ bQ ♅
21 18	☽ ⊥ ♃
23 25	☽ # ♆
23 Wednesday	
02 13	☽ △ ♇
02 23	☽ # ♃
08 01	☽ ⊻ ♃
09 18	☽ □ ♄
10 18	☽ ⊻ ♂
11 24	☽ ± ♅
14 32	☽ ∥ ☿
19 43	♀ ± ♄
20 02	☽ ⚹ ♀
21 33	☽ ⊥ ♂
22 07	☽ ⊼ ♅
24 Thursday	
03 10	☽ ⊻ ☉
11 50	☽ ⊻ ☿
14 43	☽ ∠ ♂
14 57	☽ ⊥ ☉
15 33	♀ bQ ♆
17 48	☽ Q ♄
18 05	☿ ∠ ♃
19 01	♀ □ ♅
20 01	☽ Q ♀
22 09	☽ ⊥ ☿
25 Friday	
01 08	☽ ⊼ ♆
03 32	☉ bQ ♇
08 32	☽ □ ♇
08 57	☽ ∠ ☉
10 15	☿ ⊥ ♀
12 08	♄ St D
12 18	☽ ± ♆
13 50	☽ ∠ ☿
14 26	♂ ± ♇
15 31	☽ ☌ ♃
16 08	☽ ⚹ ♄
20 05	☽ ⚹ ♂
26 Saturday	
01 54	☽ # ☿
05 18	☽ bQ ♆
05 39	☽ ☍ ♅
06 44	☉ ∥ ♅
09 20	☽ □ ♀
10 09	☽ ∥ ♃
12 15	☽ ∥ ♆
15 48	☽ ⚹ ☉
16 29	☽ ⚹ ☿
18 43	☽ Q ♇
20 18	☽ Q ♂
20 42	☉ ☌ ☿
20 48	☽ ∠ ♄
27 Sunday	
04 33	☽ # ☉
06 41	☽ # ♅
10 16	☽ △ ♆
12 15	♃ ⚹ ♄
14 17	☽ Q ☿
14 22	☽ ⊥ ♄
17 12	☽ Q ☉
18 08	☽ ⚹ ♇
28 Monday	
02 14	☽ ⊻ ♄
02 26	☽ ⊻ ♃
07 14	☿ ⊻ ♀
09 38	☽ □ ♂
13 08	☽ # ♂
14 33	☽ ⊥ ♃
16 19	☽ ⊼ ♅
19 38	☿ bQ ♇
23 41	☽ □ ☿
23 58	☽ ∠ ♇
29 Tuesday	
02 48	☽ △ ♀
04 20	☽ ± ♅
04 43	☿ # ♃
08 13	☽ □ ☉
08 53	☽ ∠ ♃
13 01	☿ # ♆
18 03	☽ ⊥ ♇
18 10	☉ ∠ ♃
22 00	☽ □ ♆
22 26	☽ bQ ♅
30 Wednesday	
06 10	☽ ⊻ ♇
12 27	☽ bQ ♀
14 32	☽ ☌ ♄
15 38	☽ ⚹ ♃
23 30	☽ # ♀
31 Thursday	
01 27	☽ △ ♂
04 42	☽ △ ♅
08 40	☽ △ ☿
08 40	☽ ± ♀
09 15	♀ bQ ♄
10 17	☽ Q ♆
10 41	☉ ∥ ☿
16 15	☽ Q ♃
22 07	☽ ⊼ ♀

SEPTEMBER 2017

LONGITUDES

Date	Sidereal time h m s	Sun ☉ ° ' "	Moon ☽ ° ' "	Moon ☽ 24.00 ° ' "	Mercury ☿ ° '	Venus ♀ ° '	Mars ♂ ° '	Jupiter ♃ ° '	Saturn ♄ ° '	Uranus ♅ ° '	Neptune ♆ ° '	Pluto ♇ ° '
01	10 43 22	09 ♍ 14 31	13 ♑ 46 57	19 ♑ 48 24	29 ♌ 31	07 ♌ 33	27 ♌ 31	22 ♎ 05	21 ♐ 13	28 ♈ 11	12 ♓ 57	17 ♑ 02
02	10 47 19	10 12 35	25 ♑ 52 26	01 ♒ 59 29	29 R 03	08 45	28 09	22 17	21 14	28 R 10	12 R 56	17 R 01
03	10 51 15	11 10 40	08 ♒ 09 54	14 ♒ 24 02	28 42	09 57	28 48	22 28	21 15	28 08	12 54	17 01
04	10 55 12	12 08 47	20 ♒ 42 06	27 ♒ 04 19	28 30	11 09	29 ♌ 26	22 39	21 16	28 07	12 52	17 00
05	10 59 09	13 06 55	03 ♓ 30 49	10 ♓ 01 36	28 D 25	12 21	00 ♍ 04	22 50	21 17	28 05	12 51	16 59
06	11 03 05	14 05 05	16 ♓ 36 41	23 ♓ 15 58	28 30	13 34	00 42	23 02	21 18	28 04	12 49	16 59
07	11 07 02	15 03 18	29 ♓ 59 16	06 ♈ 46 22	28 43	14 46	01 20	23 13	21 19	28 02	12 47	16 58
08	11 10 58	16 01 31	13 ♈ 36 59	20 ♈ 30 48	29 06	15 58	01 58	23 25	21 20	28 01	12 46	16 57
09	11 14 55	16 59 47	27 ♈ 27 27	04 ♉ 26 34	29 ♌ 37	17 11	02 36	23 36	21 21	27 59	12 44	16 57
10	11 18 51	17 58 04	11 ♉ 27 45	18 ♉ 30 37	00 ♍ 16	18 23	03 14	23 48	21 23	27 57	12 42	16 56
11	11 22 48	18 56 24	25 ♉ 34 48	02 ♊ 39 56	01 04	19 36	03 52	24 00	21 25	27 56	12 41	16 56
12	11 26 44	19 54 46	09 ♊ 45 41	16 ♊ 51 44	01 59	20 49	04 31	24 11	21 27	27 54	12 39	16 55
13	11 30 41	20 53 10	23 ♊ 57 49	01 ♋ 03 40	03 02	22 02	05 09	24 23	21 28	27 52	12 37	16 55
14	11 34 38	21 51 37	08 ♋ 09 00	15 ♋ 13 38	04 12	23 14	05 47	24 35	21 30	27 50	12 36	16 54
15	11 38 34	22 50 05	22 ♋ 17 17	29 ♋ 19 43	05 27	24 27	06 25	24 47	21 32	27 49	12 34	16 54
16	11 42 31	23 48 36	06 ♌ 20 42	13 ♌ 19 59	06 49	25 40	07 03	24 59	21 34	27 47	12 33	16 54
17	11 46 27	24 47 08	20 ♌ 17 15	27 ♌ 12 15	08 16	26 53	07 41	25 11	21 37	27 45	12 31	16 53
18	11 50 24	25 45 43	04 ♍ 04 41	10 ♍ 54 15	09 47	28 06	08 19	25 23	21 39	27 43	12 29	16 53
19	11 54 20	26 44 20	17 ♍ 40 40	24 ♍ 23 40	11 22	29 ♌ 20	08 57	25 35	21 41	27 41	12 28	16 53
20	11 58 17	27 42 58	01 ♎ 03 00	07 ♎ 38 28	13 01	00 ♍ 33	09 35	25 47	21 44	27 39	12 26	16 52
21	12 02 13	28 41 39	14 ♎ 09 56	20 ♎ 37 17	14 42	01 46	10 13	26 00	21 46	27 37	12 25	16 52
22	12 06 10	29 ♍ 40 21	27 ♎ 00 30	03 ♏ 19 38	16 26	02 59	10 51	26 12	21 49	27 35	12 23	16 52
23	12 10 07	00 ♎ 39 05	09 ♏ 34 47	15 ♏ 46 10	18 11	04 13	11 29	26 24	21 51	27 33	12 22	16 52
24	12 14 03	01 37 52	21 ♏ 54 01	27 ♏ 58 40	19 58	05 26	12 07	26 37	21 54	27 30	12 20	16 52
25	12 18 00	02 36 39	04 ♐ 00 31	10 ♐ 00 02	21 46	06 40	12 45	26 49	21 57	27 28	12 18	16 51
26	12 21 56	03 35 29	15 ♐ 57 41	21 ♐ 54 02	23 34	07 53	13 23	27 01	22 00	27 26	12 17	16 51
27	12 25 53	04 34 21	27 ♐ 49 39	03 ♑ 45 10	25 24	09 07	14 01	27 14	22 03	27 24	12 15	16 51
28	12 29 49	05 33 14	09 ♑ 41 10	15 ♑ 38 19	27 13	10 21	14 39	27 26	22 06	27 22	12 14	16 51
29	12 33 46	06 32 09	21 ♑ 37 16	27 ♑ 38 37	29 ♍ 02	11 34	15 17	27 39	22 10	27 20	12 12	16 D 51
30	12 37 42	07 ♎ 31 06	03 ♒ 43 00	09 ♒ 51 00	00 ♎ 51	12 ♍ 48	15 ♍ 55	27 ♎ 52	22 ♐ 13	27 ♈ 17	12 ♓ 11	16 ♑ 51

DECLINATIONS

Date	Moon True ☊ ° '	Moon Mean ☊ ° '	Moon ☽ Latitude ° '	Sun ☉ ° '	Moon ☽ ° '	Mercury ☿ ° '	Venus ♀ ° '	Mars ♂ ° '	Jupiter ♃ ° '	Saturn ♄ ° '	Uranus ♅ ° '	Neptune ♆ ° '	Pluto ♇ ° '
01	24 ♌ 11	23 ♌ 20	03 N 25	08 N 06	19 S 19	08 N 59	18 N 29	13 N 25	07 S 35	21 S 59	10 N 16	07 S 34	21 S 41
02	24 D 12	23 17	02 30	07 44	18 30	09 26	18 13	13 12	07 39	22 00	10 16	07 34	21 41
03	24 13	23 14	01 28	07 22	16 49	09 51	17 57	12 59	07 44	22 00	10 15	07 35	21 42
04	24 14	23 11	00 N 19	07 00	14 17	10 13	17 40	12 46	07 48	22 00	10 15	07 36	21 42
05	24 R 14	23 07	00 S 51	06 38	11 01	10 32	17 23	12 32	07 52	22 00	10 14	07 36	21 42
06	24 13	23 04	02 00	06 16	07 08	10 49	17 05	12 19	07 57	22 01	10 14	07 37	21 42
07	24 11	23 01	03 04	05 53	02 S 49	11 01	16 47	12 05	08 01	22 01	10 13	07 38	21 42
08	24 08	22 58	03 59	05 31	01 N 42	11 10	16 28	11 52	08 05	22 01	10 13	07 38	21 42
09	24 04	22 55	04 41	05 08	06 12	11 16	16 09	11 38	08 10	22 01	10 12	07 39	21 43
10	24 01	22 51	05 06	04 45	10 24	11 17	15 49	11 24	08 14	22 02	10 11	07 39	21 43
11	23 59	22 48	05 14	04 23	14 04	11 15	15 29	11 11	08 19	22 02	10 11	07 40	21 43
12	23 57	22 45	05 02	04 00	16 56	11 09	15 09	10 57	08 23	22 02	10 10	07 41	21 43
13	23 D 57	22 42	04 32	03 37	18 46	10 59	14 48	10 43	08 27	22 03	10 09	07 41	21 43
14	23 58	22 39	03 45	03 14	19 26	10 45	14 26	10 29	08 32	22 03	10 09	07 42	21 44
15	23 59	22 36	02 45	02 51	18 53	10 27	14 05	10 15	08 36	22 03	10 08	07 42	21 44
16	24 01	22 32	01 35	02 27	17 09	10 06	13 42	10 01	08 41	22 04	10 07	07 43	21 44
17	24 02	22 29	00 S 20	02 04	14 24	09 42	13 20	09 47	08 45	22 04	10 07	07 44	21 44
18	24 R 01	22 26	00 N 55	01 41	10 52	09 15	12 57	09 32	08 50	22 04	10 06	07 44	21 44
19	23 59	22 23	02 06	01 18	06 48	08 45	12 33	09 18	08 54	22 05	10 05	07 45	21 44
20	23 56	22 20	03 08	00 54	02 N 28	08 12	12 10	09 04	08 59	22 05	10 05	07 46	21 44
21	23 51	22 16	04 00	00 31	01 S 54	07 37	11 46	08 49	09 03	22 05	10 04	07 46	21 45
22	23 45	22 13	04 37	00 N 08	06 06	07 00	11 21	08 35	09 08	22 05	10 03	07 47	21 45
23	23 38	22 10	05 01	00 S 16	09 55	06 21	10 56	08 20	09 12	22 06	10 02	07 48	21 45
24	23 32	22 07	05 10	00 39	13 14	05 41	10 31	08 06	09 17	22 06	10 02	07 48	21 45
25	23 27	22 04	05 05	01 02	15 57	04 59	10 06	07 51	09 22	22 07	10 01	07 49	21 45
26	23 24	22 01	04 47	01 26	17 57	04 16	09 40	07 37	09 26	22 07	10 00	07 49	21 45
27	23 22	21 57	04 16	01 49	19 09	03 32	09 14	07 22	09 31	22 07	09 59	07 50	21 45
28	23 D 22	21 54	03 35	02 12	19 31	02 47	08 48	07 08	09 35	22 08	09 58	07 50	21 46
29	23 23	21 51	02 44	02 36	19 00	02 01	08 22	06 53	09 40	22 08	09 58	07 51	21 46
30	23 ♌ 24	21 ♌ 48	01 N 45	02 S 59	17 S 37	01 N 16	07 N 55	06 N 38	09 S 44	22 S 08	09 N 57	07 S 52	21 S 46

ZODIAC SIGN ENTRIES

Date	h	m	Planets
02	20	06	☽ ♒
05	05	28	☽ ♓
05	09	35	♂ ♍
07	12	01	☽ ♈
09	16	23	☽ ♉
10	02	52	☿ ♍
11	19	29	☽ ♊
13	22	12	☽ ♋
16	01	09	☽ ♌
18	04	52	☽ ♍
20	01	15	♀ ♍
20	10	06	☽ ♎
22	17	40	☽ ♏
22	20	02	☉ ♎
25	04	01	☽ ♐
27	16	24	☽ ♑
30	00	42	☿ ♎
30	04	40	☽ ♒

LATITUDES

Date	Mercury ☿ ° '	Venus ♀ ° '	Mars ♂ ° '	Jupiter ♃ ° '	Saturn ♄ ° '	Uranus ♅ ° '	Neptune ♆ ° '	Pluto ♇ ° '
01	02 S 50	00 N 06	01 N 10	01 N 06	01 N 09	00 S 36	00 S 56	00 N 40
04	01 54	00 15	01 10	01 05	01 09	00 36	00 56	00 40
07	00 57	00 24	01 10	01 05	01 08	00 36	00 56	00 40
10	00 S 06	00 32	01 10	01 04	01 08	00 36	00 56	00 39
13	00 N 38	00 40	01 10	01 04	01 07	00 36	00 56	00 39
16	01 11	00 48	01 10	01 04	01 07	00 36	00 56	00 39
19	01 34	00 54	01 10	01 03	01 06	00 36	00 56	00 38
22	01 47	01 01	01 10	01 03	01 06	00 36	00 56	00 38
25	01 52	01 07	01 10	01 03	01 05	00 36	00 56	00 37
28	01 49	01 12	01 10	01 03	01 05	00 36	00 56	00 37
31	01 N 41	01 N 17	01 N 10	01 N 02	01 N 04	00 S 36	00 S 56	00 N 37

DATA

Julian Date	2457998
Delta T	+70 seconds
Ayanamsa	24° 06' 04"
Synetic vernal point	05° ♓ 00' 55"
True obliquity of ecliptic	23° 26' 06"

MOON'S PHASES, APSIDES AND POSITIONS ☽

Date	h	m	Phase	Longitude ° '	Eclipse Indicator
06	07	03	○	13 ♓ 53	
13	06	25	☾	20 ♊ 40	
20	05	30	●	27 ♍ 27	
28	02	54	☽	05 ♑ 11	

Day	h	m		
13	16	04	Perigee	
27	06	51	Apogee	
01	02	02	Max dec	19° S 24'
08	03	03	0N	
14	13	02	Max dec	19° N 26'
21	01	29	0S	
28	10	06	Max dec	19° S 31'

LONGITUDES

Date	Chiron ⚷ ° '	Ceres ⚳ ° '	Pallas ⚴ ° '	Juno ⚵ ° '	Vesta ⚶ ° '	Black Moon Lilith ⚸ ° '
01	27 ♓ 21	21 ♋ 33	13 ♉ 55	02 ♑ 49	21 ♍ 38	22 ♐ 20
11	26 ♓ 55	25 ♋ 21	14 ♉ 17	03 ♑ 27	26 ♍ 35	23 ♐ 27
21	26 ♓ 27	29 ♋ 01	13 ♉ 55	04 ♑ 38	01 ♎ 36	24 ♐ 34
31	26 ♓ 00	02 ♌ 32	12 ♉ 45	06 ♑ 19	06 ♎ 38	25 ♐ 41

All ephemeris data is given at 12.00 UT and the Moon's longitude is additionally given for 24.00 UT

ASPECTARIAN

01 Friday
00 24 ♀ ± ♆
02 06 ☽ △ ☉
02 27 ♃ ∥ ♆
09 20 ☽ bQ ♂
10 21 ☽ ⚹ ♆
13 24 ☽ bQ ☿
18 29 ☽ ☌ ♇

02 Saturday
02 49 ☽ ⊻ ♄
04 15 ☽ ± ♂
04 47 ☽ □ ♃
06 36 ☽ ± ☿
10 35 ☽ bQ ☉
12 13 ♂ △ ♅
14 41 ☽ ⊥ ♄
16 02 ☽ ∠ ♆
16 30 ☽ □ ♅
16 30 ☉ # ♃
16 44 ☽ ⊼ ♂
18 02 ☽ ⊼ ☿
18 11 ☽ # ♀
22 35 ☉ # ♆

03 Sunday
05 43 ☽ ± ☉
08 17 ☽ ∠ ♄
09 33 ☽ ⊥ ♆
09 38 ☿ ☌ ♂
15 49 ☽ ☍ ♀
18 18 ☽ ⊼ ☉
21 06 ☽ ⊻ ♆

04 Monday
00 10 ♀ Q ♃
03 18 ☽ Q ♅
04 58 ☽ ⊻ ♇
13 04 ☽ ⚹ ♄
13 54 ☿ ∥ ♅
15 44 ☽ △ ♃
16 20 ☽ ⊥ ♇

05 Tuesday
00 37 ☽ # ♂
01 56 ☽ ⚹ ♅
02 33 ☽ ☍ ☿
05 15 ☽ ☍ ♂
05 28 ☉ ☍ ♆
09 10 ☽ ∠ ♇
11 25 ☉ bQ ♅
11 29 ☿ St D
11 34 ☽ Q ♄
14 52 ☽ # ☿
17 04 ☽ # ♅
20 06 ☽ bQ ♃
21 33 ♀ ⊼ ♆

06 Wednesday
05 07 ☽ ☌ ♆
05 34 ☽ ∠ ♅
05 54 ☽ ⊼ ♀
07 03 ☽ ☍ ☉
07 18 ☽ ∥ ♃
09 10 ☽ ∥ ♆
12 40 ☽ ⚹ ♇
12 46 ☽ ± ♃
17 30 ☽ # ☉
17 52 ☽ ± ♀
20 29 ☽ □ ♄
21 49 ☽ ⊥ ♅
23 44 ☽ ⊼ ♃

07 Thursday
08 32 ☽ ⊻ ♅
09 42 ☽ ⊼ ☿
10 11 ☽ Q ♇
11 34 ☽ bQ ♀
14 30 ☽ ⊼ ♂
20 35 ☽ ± ☿

08 Friday
01 37 ☽ ± ♂
10 31 ☽ ⊻ ♆
11 28 ♂ bQ ♇
12 52 ☽ bQ ☿
16 30 ☽ △ ♀
16 31 ☽ ⊼ ☉
17 26 ☉ ⊻ ♀
17 49 ☽ □ ♇
18 08 ☽ bQ ♂
20 57 ☽ ⊥ ♆

09 Saturday
01 27 ☽ △ ♄
03 43 ☽ ± ☉
05 15 ☽ ☍ ♃
06 43 ☽ ∥ ☉
07 23 ♀ ⊼ ♇
10 45 ☉ △ ♇
12 28 ☽ ∠ ♆
12 54 ☽ ☌ ♅
15 52 ☽ △ ☿
20 03 ☽ # ♆
20 23 ☽ bQ ☉
21 16 ☽ △ ♂
23 07 ☽ # ♃

10 Sunday
03 19 ☽ bQ ♄
06 44 ☉ ⊥ ♃
10 42 ☽ ∥ ♅
14 07 ☽ ⚹ ♆
17 23 ☽ ∥ ☿
17 49 ☽ ∥ ♂
18 42 ☽ ± ♄
21 19 ☽ △ ♇
23 54 ☽ △ ☉

11 Monday
00 54 ☽ □ ♀
02 05 ☿ ∥ ♂
04 55 ☽ ⊼ ♄
09 16 ☽ ⊼ ♃
10 29 ☽ Q ♆
15 58 ☽ ⊻ ♅
19 35 ☽ ± ♃
21 49 ☽ ∥ ♀
21 54 ☽ □ ☿
22 45 ☽ bQ ♇

12 Tuesday
02 06 ☽ ⊥ ♅
02 42 ☽ □ ♂
10 15 ☽ Q ♀
10 26 ☿ bQ ♇
11 01 ☽ bQ ♃
13 57 ☽ ± ♇
16 53 ☽ □ ♆
17 18 ☽ ∠ ♅

13 Wednesday
00 05 ☽ ⊼ ♇
00 50 ♀ △ ♄
06 25 ☽ □ ☉
06 38 ☽ Q ☿
07 47 ☽ ☍ ♄
08 25 ☽ ⚹ ♀
10 33 ☽ Q ♂
12 43 ☽ △ ♃
18 35 ☽ ⚹ ♅

14 Thursday
02 58 ☉ □ ♄
04 41 ☽ ⚹ ☿
05 26 ♀ ± ♇
07 48 ☽ ⚹ ♂
11 30 ☉ ± ♅
12 10 ☽ ∠ ♀
14 51 ☽ Q ♅
15 07 ☽ Q ☉
19 31 ☽ △ ♆

15 Friday
02 50 ☽ ☍ ♇
04 52 ☽ ⊥ ♀
08 34 ☽ ∠ ☿
10 26 ☽ ∠ ♂
10 43 ☽ ⊼ ♄
13 00 ☽ ⚹ ☉
16 02 ☽ ⊻ ♀
16 18 ☽ □ ♃
19 44 ♀ ⚹ ♃
20 58 ☽ ± ♄
20 59 ☽ bQ ♆
21 23 ☽ □ ♅

16 Saturday
00 11 ♂ ∥ ♅
01 30 ☽ ⊥ ☿
02 30 ☽ ⊥ ♂
11 02 ☿ ∥ ♅
12 20 ☽ ± ♆
12 23 ☽ bQ ♄
12 54 ☽ ⊻ ☿
13 16 ☽ ⊻ ♂
16 33 ☽ ∠ ☉
19 01 ☿ ☌ ♂
22 37 ☽ ⊼ ♆
23 34 ☽ Q ♃

17 Sunday
02 37 ☿ ∥ ♂
06 08 ☽ ⊼ ♇
09 12 ☽ ⊥ ☉
14 18 ☽ △ ♄
16 30 ☽ ± ♇
20 23 ☽ ⊻ ☉
20 37 ☽ ⚹ ♃
20 43 ☽ ∥ ♀

18 Monday
00 17 ☉ ⊻ ♃
00 34 ☽ ☌ ♀
00 55 ☽ △ ♅
04 27 ♀ △ ♅
08 09 ☽ bQ ♇
16 43 ☽ ∥ ♅
19 48 ☽ ☌ ♂
20 35 ☽ ∥ ♂
22 32 ☿ ∠ ♃
23 09 ☽ ∥ ☿
23 15 ☽ ∠ ♃
23 20 ☽ ☌ ☿

19 Tuesday
00 04 ☽ # ♃
02 46 ☽ ☍ ♆
03 09 ☽ bQ ♅
05 41 ☿ # ♃
06 34 ☽ # ♆
10 34 ☽ △ ♇
15 27 ☽ ⊥ ♃
19 07 ☽ ± ♅
19 10 ☽ □ ♄

20 Wednesday
02 22 ☽ ⊻ ♃
03 49 ☿ ☍ ♆
05 30 ☽ ☌ ☉
05 52 ☽ ⊼ ♅
06 50 ☿ bQ ♅
10 21 ☉ ⊼ ♅
11 00 ☽ ⊻ ♀
18 18 ♂ # ♃
21 19 ☽ ∥ ☉
23 01 ☽ ⊥ ♀

21 Thursday
03 52 ☽ Q ♄
04 21 ☽ ⊻ ♂
04 58 ☽ # ☉
05 59 ☿ # ♆
08 46 ☽ ⊼ ♆
13 08 ☽ ⊻ ☿
13 57 ♀ bQ ♇
16 00 ☽ ⊥ ♂
17 00 ☽ □ ♇
17 20 ☽ ∠ ♀
19 52 ☽ ± ♆

22 Friday
02 01 ☽ ⊥ ☿
02 12 ☽ ⚹ ♄
09 42 ☽ ∠ ♂
10 27 ☽ ☌ ♃
12 43 ☽ bQ ♆
13 04 ☽ ☍ ♅
16 44 ☽ # ☿
17 28 ☽ ⊻ ☉
18 01 ☿ △ ♇
21 43 ☽ ∠ ☿
22 19 ☽ ∥ ♆

23 Saturday
00 35 ☽ ⚹ ♀
02 25 ☽ # ♂
02 56 ☽ Q ♇
05 53 ☽ ⊥ ☉
06 44 ☽ ∠ ♄
07 15 ☽ ∥ ♃
07 25 ♂ ∠ ♃
12 49 ☽ # ♅
15 53 ☽ ⚹ ♂
17 22 ☽ △ ♆
18 14 ☽ # ♀

24 Sunday
00 13 ☽ ⊥ ♄
00 47 ☽ ∠ ☉
02 08 ☽ ⚹ ♇
02 17 ☽ Q ♀
07 33 ☽ ⚹ ☿
12 00 ☽ ⊻ ♄
16 36 ☽ Q ♂
19 49 ♂ ☍ ♆
21 27 ☽ ⊻ ♃
21 46 ☿ ⊥ ♃
23 02 ☽ ⊼ ♅

25 Monday
01 58 ♂ bQ ♅
07 43 ☽ ∠ ♇
08 13 ☿ ± ♅
08 58 ☽ ⚹ ☉
09 35 ☽ ⊥ ♃
10 56 ☽ ± ♅
11 25 ☽ Q ☿
14 36 ☿ □ ♄
16 20 ♂ # ♆
17 05 ♀ ∥ ♅
17 55 ☽ □ ♀

26 Tuesday
01 43 ☽ ⊥ ♇
03 56 ☽ ∠ ♃
04 36 ☽ □ ♆
04 55 ☽ bQ ♅
06 31 ☽ □ ♂
11 11 ☽ Q ☉
13 48 ☽ ⊻ ♇
23 14 ♀ # ♃

27 Wednesday
00 15 ☽ ☌ ♄
06 10 ☽ □ ☿
10 46 ☽ ⚹ ♃
11 08 ☽ △ ♅
14 37 ♃ bQ ♆
16 55 ☽ Q ♆

28 Thursday
02 54 ☽ □ ☉
04 25 ♃ ☍ ♅
11 30 ☽ Q ♃
13 29 ☽ △ ♀
13 55 ☿ ⊼ ♅
15 22 ☿ ⊻ ♃
17 08 ☽ ⚹ ♆
19 38 ♇ St D
22 35 ☽ △ ♂

29 Friday
00 02 ☉ # ☿
02 27 ☽ ☌ ♇
13 05 ☽ ⊻ ♄
22 59 ☽ bQ ♀
23 07 ☽ ∠ ♆
23 20 ☽ □ ♅

30 Saturday
00 12 ♀ ☍ ♆
00 13 ☽ □ ♃
01 05 ☽ ⊥ ♄
02 18 ♀ bQ ♅
05 22 ☽ △ ☿
06 11 ☽ bQ ♂
13 24 ♀ ∠ ♃
14 58 ♀ # ♆
16 50 ☽ ⊥ ♆
18 43 ☽ ± ♀
18 53 ☽ ∠ ♄
20 06 ☽ △ ☉

OCTOBER 2017

LONGITUDES

Date	Sidereal time h m s	Sun ☉ ° ' "	Moon ☽ ° ' "	Moon ☽ 24.00 ° ' "	Mercury ☿ ° '	Venus ♀ ° '	Mars ♂ ° '	Jupiter ♃ ° '	Saturn ♄ ° '	Uranus ♅ ° '	Neptune ♆ ° '	Pluto ⯓ ° '
01	12 41 39	08 ♎ 30 04	16 ♒ 03 10	22 ♒ 19 57	02 ♎ 40	14 ♍ 02	16 ♍ 33	28 ♎ 04	22 ♐ 16	27 ♈ 15	12 ♓ 10	16 ♑ 51
02	12 45 36	09 29 04	28 ♒ 41 48	05 ♓ 09 02	04 29	15 16	17 11	28 17	22 20	27 R 13	12 R 08	16 51
03	12 49 32	10 28 06	11 ♓ 41 53	18 ♓ 20 28	06 17	16 30	17 49	28 30	22 23	27 10	12 07	16 52
04	12 53 29	11 27 10	25 ♓ 04 46	01 ♈ 54 39	08 04	17 44	18 27	28 42	22 27	27 08	12 05	16 52
05	12 57 25	12 26 16	08 ♈ 49 54	15 ♈ 49 54	09 51	18 58	19 05	28 55	22 31	27 06	12 04	16 52
06	13 01 22	13 25 24	22 ♈ 54 20	00 ♉ 02 28	11 37	20 12	19 43	29 08	22 35	27 03	12 03	16 52
07	13 05 18	14 24 34	07 ♉ 13 35	14 ♉ 26 54	13 22	21 26	20 21	29 21	22 38	27 01	12 01	16 52
08	13 09 15	15 23 46	21 ♉ 41 35	28 ♉ 56 53	15 07	22 40	20 59	29 34	22 42	26 59	12 00	16 53
09	13 13 11	16 23 00	06 ♊ 12 00	13 ♊ 26 15	16 51	23 54	21 37	29 46	22 46	26 56	11 59	16 53
10	13 17 08	17 22 17	20 ♊ 39 02	27 ♊ 49 52	18 34	25 08	22 15	29 ♎ 59	22 51	26 54	11 57	16 53
11	13 21 05	18 21 36	04 ♋ 58 20	12 ♋ 04 09	20 17	26 23	22 53	00 ♏ 12	22 55	26 51	11 56	16 54
12	13 25 01	19 20 58	19 ♋ 07 08	26 ♋ 07 10	21 58	27 37	23 30	00 25	22 59	26 49	11 55	16 54
13	13 28 58	20 20 22	03 ♌ 04 13	09 ♌ 58 17	23 39	28 ♍ 51	24 08	00 38	23 03	26 47	11 54	16 54
14	13 32 54	21 19 48	16 ♌ 49 25	23 ♌ 37 41	25 20	00 ♎ 06	24 46	00 51	23 08	26 44	11 52	16 55
15	13 36 51	22 19 16	00 ♍ 23 07	07 ♍ 05 48	26 59	01 20	25 24	01 04	23 12	26 42	11 51	16 55
16	13 40 47	23 18 47	13 ♍ 45 45	20 ♍ 22 58	28 ♎ 38	02 35	26 02	01 17	23 17	26 39	11 50	16 56
17	13 44 44	24 18 19	26 ♍ 57 27	03 ♎ 29 10	00 ♏ 16	03 49	26 40	01 30	23 22	26 37	11 49	16 56
18	13 48 40	25 17 54	09 ♎ 58 02	16 ♎ 24 01	01 54	05 04	27 18	01 43	23 26	26 34	11 48	16 57
19	13 52 37	26 17 31	22 ♎ 47 02	29 ♎ 07 02	03 31	06 18	27 56	01 56	23 31	26 32	11 47	16 58
20	13 56 34	27 17 10	05 ♏ 23 58	11 ♏ 37 51	05 07	07 33	28 34	02 09	23 36	26 30	11 46	16 58
21	14 00 30	28 16 52	17 ♏ 48 42	23 ♏ 56 36	06 43	08 48	29 12	02 22	23 41	26 27	11 45	16 59
22	14 04 27	29 ♎ 16 35	00 ♐ 01 42	06 ♐ 04 11	08 18	10 02	29 ♍ 50	02 35	23 46	26 25	11 44	17 00
23	14 08 23	00 ♏ 16 19	12 ♐ 04 18	18 ♐ 02 24	09 52	11 17	00 ♎ 28	02 48	23 51	26 22	11 43	17 00
24	14 12 20	01 16 06	23 ♐ 58 49	29 ♐ 54 02	11 26	12 32	01 06	03 02	23 56	26 20	11 42	17 01
25	14 16 16	02 15 55	05 ♑ 48 32	11 ♑ 42 51	13 00	13 47	01 43	03 15	24 01	26 17	11 41	17 02
26	14 20 13	03 15 45	17 ♑ 37 34	23 ♑ 33 19	14 33	15 01	02 21	03 28	24 07	26 15	11 40	17 03
27	14 24 09	04 15 37	29 ♑ 30 46	05 ♒ 30 36	16 05	16 16	02 59	03 41	24 12	26 12	11 39	17 04
28	14 28 06	05 15 30	11 ♒ 33 28	17 ♒ 40 06	17 37	17 31	03 37	03 54	24 17	26 10	11 38	17 04
29	14 32 03	06 15 25	23 ♒ 51 08	00 ♓ 07 13	19 08	18 46	04 15	04 07	24 23	26 08	11 38	17 05
30	14 35 59	07 15 22	06 ♓ 28 56	12 ♓ 56 49	20 39	20 01	04 53	04 20	24 28	26 05	11 37	17 06
31	14 39 56	08 ♏ 15 21	19 ♓ 31 15	26 ♓ 12 34	22 ♏ 09	21 ♎ 16	05 ♎ 31	04 ♏ 33	24 ♐ 34	26 ♈ 03	11 ♓ 36	17 ♑ 07

Date	Moon True ☊ ° '	Moon Mean ☊ ° '	Moon ☽ Latitude ° '
01	23 ♌ 25	21 ♌ 45	00 N 40
02	23 R 25	21 42	00 S 28
03	23 24	21 38	01 37
04	23 20	21 35	02 42
05	23 14	21 32	03 40
06	23 07	21 29	04 25
07	22 58	21 26	04 55
08	22 50	21 22	05 06
09	22 43	21 19	04 58
10	22 38	21 16	04 31
11	22 36	21 13	03 47
12	22 35	21 10	02 49
13	22 D 35	21 07	01 43
14	22 36	21 03	00 S 31
15	22 R 36	21 00	00 N 42
16	22 34	20 57	01 51
17	22 29	20 54	02 52
18	22 22	20 51	03 44
19	22 12	20 48	04 24
20	22 00	20 44	04 50
21	21 48	20 41	05 01
22	21 36	20 38	04 59
23	21 25	20 35	04 43
24	21 16	20 32	04 15
25	21 11	20 28	03 36
26	21 07	20 25	02 48
27	21 06	20 22	01 52
28	21 D 06	20 19	00 N 50
29	21 R 06	20 16	00 S 15
30	21 05	20 13	01 20
31	21 ♌ 02	20 ♌ 09	02 S 24

DECLINATIONS

Date	Sun ☉ ° '	Moon ☽ ° '	Mercury ☿ ° '	Venus ♀ ° '	Mars ♂ ° '	Jupiter ♃ ° '	Saturn ♄ ° '	Uranus ♅ ° '	Neptune ♆ ° '	Pluto ⯓ ° '
01	03 S 22	15 S 23	00 N 29	07 N 28	06 N 23	09 S 49	22 S 09	09 N 56	07 S 52	21 S 46
02	03 45	12 22	00 S 17	07 01	06 08	09 54	22 09	09 55	07 53	21 46
03	04 09	08 40	01 03	06 33	05 54	09 58	22 09	09 54	07 53	21 46
04	04 32	04 S 26	01 50	06 06	05 39	10 03	22 10	09 53	07 54	21 46
05	04 55	00 N 02	02 36	05 38	05 24	10 07	22 10	09 53	07 54	21 46
06	05 18	04 48	03 22	05 10	05 09	10 12	22 10	09 52	07 55	21 46
07	05 41	09 16	04 08	04 42	04 54	10 16	22 11	09 51	07 55	21 46
08	06 04	13 15	04 54	04 13	04 39	10 21	22 11	09 50	07 56	21 46
09	06 26	16 27	05 39	03 45	04 24	10 26	22 12	09 49	07 56	21 47
10	06 49	18 36	06 24	03 16	04 09	10 30	22 12	09 48	07 57	21 47
11	07 12	19 34	07 09	02 47	03 54	10 35	22 12	09 47	07 57	21 47
12	07 34	19 17	07 52	02 19	03 39	10 39	22 13	09 47	07 58	21 47
13	07 57	17 48	08 36	01 50	03 24	10 44	22 13	09 46	07 58	21 47
14	08 19	15 18	09 19	01 21	03 09	10 48	22 14	09 45	07 58	21 47
15	08 41	11 59	10 01	00 51	02 54	10 53	22 14	09 44	07 59	21 47
16	09 03	08 05	10 43	00 N 22	02 38	10 58	22 14	09 43	07 59	21 47
17	09 25	03 N 51	11 23	00 S 07	02 23	11 02	22 15	09 42	08 00	21 47
18	09 47	00 S 30	12 04	00 36	02 08	11 07	22 15	09 41	08 00	21 47
19	10 09	04 46	12 43	01 05	01 53	11 11	22 15	09 40	08 01	21 47
20	10 30	08 46	13 22	01 35	01 38	11 16	22 16	09 40	08 01	21 47
21	10 52	12 19	14 00	02 04	01 23	11 20	22 16	09 39	08 01	21 47
22	11 13	15 17	14 38	02 33	01 08	11 25	22 17	09 38	08 02	21 47
23	11 34	17 33	15 14	03 03	00 52	11 29	22 17	09 37	08 02	21 47
24	11 55	19 03	15 50	03 32	00 37	11 34	22 17	09 36	08 02	21 47
25	12 15	19 43	16 25	04 01	00 22	11 38	22 18	09 35	08 03	21 47
26	12 36	19 30	17 00	04 30	00 N 07	11 43	22 18	09 34	08 03	21 47
27	12 56	18 25	17 33	04 59	00 S 08	11 47	22 19	09 33	08 03	21 47
28	13 16	16 30	18 05	05 28	00 23	11 52	22 19	09 33	08 04	21 47
29	13 36	13 48	18 37	05 57	00 38	11 56	22 19	09 32	08 04	21 47
30	13 56	10 23	19 07	06 26	00 54	12 01	22 20	09 31	08 04	21 47
31	14 S 15	06 S 22	19 S 37	06 S 54	01 S 09	12 S 05	22 S 20	09 N 30	08 S 04	21 S 47

ZODIAC SIGN ENTRIES

Date	h	m	Planets
02	14	26	☽ ♓
04	20	40	☽ ♈
06	23	56	☽ ♉
09	01	44	☽ ♊
10	13	20	♃ ♏
11	03	38	☽ ♋
13	06	41	☽ ♌
14	10	11	♀ ♎
15	11	19	☽ ♍
17	07	58	☿ ♏
17	17	35	☽ ♎
20	01	41	☽ ♏
22	11	57	☽ ♐
22	18	29	♂ ♎
23	05	27	☉ ♏
25	00	12	☽ ♑
27	12	59	☽ ♒
29	23	46	☽ ♓

LATITUDES

Date	Mercury ☿ ° '	Venus ♀ ° '	Mars ♂ ° '	Jupiter ♃ ° '	Saturn ♄ ° '	Uranus ♅ ° '	Neptune ♆ ° '	Pluto ⯓ ° '
01	01 N 41	01 N 17	01 N 10	01 N 02	01 N 04	00 S 36	00 S 56	00 N 37
04	01 29	01 21	01 10	01 02	01 04	00 36	00 56	00 36
07	01 14	01 25	01 10	01 02	01 03	00 36	00 56	00 36
10	00 57	01 27	01 10	01 02	01 03	00 36	00 56	00 36
13	00 38	01 30	01 10	01 02	01 02	00 36	00 56	00 36
16	00 N 18	01 31	01 10	01 01	01 02	00 36	00 56	00 35
19	00 S 02	01 32	01 09	01 01	01 01	00 36	00 56	00 35
22	00 23	01 33	01 09	01 01	01 01	00 36	00 56	00 34
25	00 43	01 32	01 09	01 01	01 00	00 36	00 56	00 34
28	01 03	01 31	01 09	01 01	01 00	00 36	00 56	00 34
31	01 S 22	01 N 30	01 N 08	01 N 01	00 N 59	00 S 36	00 S 56	00 N 33

DATA

Julian Date	2458028
Delta T	+70 seconds
Ayanamsa	24° 06' 07"
Synetic vernal point	05° ♓ 00' 52"
True obliquity of ecliptic	23° 26' 06"

MOON'S PHASES, APSIDES AND POSITIONS ☽

Date	h	m	Phase	Longitude ° '	Eclipse Indicator
05	18	40	○	12 ♈ 43	
12	12	25	☾	19 ♋ 22	
19	19	12	●	26 ♎ 35	
27	22	22	☽	04 ♒ 41	

Day	h	m		
09	06	06	Perigee	
25	02	29	Apogee	
05	11	18	ON	
11	18	23	Max dec	19° N 37'
18	09	12	OS	
25	18	14	Max dec	19° S 44'

LONGITUDES

Date	Chiron ⚷ ° '	Ceres ⚳ ° '	Pallas ⚴ ° '	Juno ⚵ ° '	Vesta ⚶ ° '	Black Moon Lilith ⚸ ° '
01	26 ♓ 00	02 ♌ 32	12 ♉ 45	06 ♑ 19	06 ♎ 38	25 ♐ 41
11	25 ♓ 35	05 ♌ 49	10 ♉ 46	08 ♑ 25	11 ♎ 43	26 ♐ 48
21	25 ♓ 11	08 ♌ 52	08 ♉ 05	10 ♑ 54	16 ♎ 50	27 ♐ 55
31	24 ♓ 52	11 ♌ 36	04 ♉ 56	13 ♑ 43	21 ♎ 57	29 ♐ 02

ASPECTARIAN

h m	Aspects
01 Sunday	
00 47	☽ ± ♂
04 30	☽ ⊻ ♆
07 41	☽ ⊼ ♀
10 28	☽ Q ♅
13 00	☽ ⊼ ♂
13 33	☽ ⊻ ⯓
15 38	☽ ♇ ☿
23 36	♂ △ ⯓
23 56	☽ ✶ ♄
02 Monday	
01 00	☽ ⊥ ⯓
03 25	☽ ♇ ☉
09 13	☽ ✶ ♅
11 13	☽ △ ♃
11 32	☽ ± ☿
17 54	☽ ∠ ⯓
19 02	♃ # ♅
21 39	☽ ± ☉
22 32	☽ Q ♄
03 Tuesday	
00 30	☽ ⊼ ☿
04 10	☽ ∥ ♃
04 23	☽ # ♅
09 35	☽ ⊼ ☉
09 57	☉ Q ♄
12 45	☽ ☌ ♆
12 52	☽ ∠ ♅
15 19	☽ ♇ ♃
16 38	☽ ∥ ♆
19 09	♀ △ ⯓
21 20	☽ ✶ ⯓
21 34	☽ ☍ ♀
23 36	☽ ☍ ♂
04 Wednesday	
01 47	☽ # ♀
04 59	☽ # ♂
05 01	☽ ⊥ ♅
07 19	☽ □ ♄
07 43	☽ ± ♃
11 32	☽ ∥ ☉
15 37	☽ ⊻ ♅
18 29	☽ ⊼ ♃
18 40	☽ Q ⯓
23 49	☽ ∥ ☿
05 Thursday	
03 09	☉ ⊼ ♆
14 01	☽ ☍ ☿
16 53	♀ ☌ ♂
17 33	☽ ⊻ ♆
18 40	☽ ☍ ☉
21 19	☿ Q ♄
06 Friday	
01 46	☽ □ ⯓
03 11	☽ # ☿
03 47	☽ ⊥ ♆
06 20	☉ # ♂
06 21	☽ ⊼ ♂
06 58	☽ ⊼ ♀
08 13	☉ # ♀
11 26	☽ △ ♄
13 41	♀ ∥ ♂
13 43	☽ ∥ ♀
13 43	☽ ∥ ♂
14 50	☽ # ☉
16 57	☽ ± ♂
17 44	☿ ⊼ ♆
18 04	☽ ± ♀
18 58	☽ ∠ ♆
18 58	☽ ☌ ♅
22 38	☽ ☍ ♃
07 Saturday	
04 17	♀ ± ♅
04 31	☽ # ♆
07 14	♅ ∠ ♆
08 43	☽ ♇ ♂
10 33	☽ ♇ ♀
12 42	☽ ♇ ♄
15 14	☽ ∥ ♅
17 48	☽ # ♃
19 58	☽ ✶ ♆
22 48	☿ # ♀
23 37	☽ ⊼ ☿
08 Sunday	
00 49	☽ ⊼ ☉
03 42	☽ ± ♄
04 01	☽ △ ⯓
06 02	☿ # ♂
10 46	☽ △ ♂
10 55	☽ ± ☿
11 28	☽ ± ☉
11 59	♂ ± ♅
12 54	♀ □ ♄
13 41	☽ ⊼ ♄
13 45	☽ △ ♀
15 49	☽ Q ♆
20 43	☽ ⊻ ♅
20 54	☉ ☌ ☿
09 Monday	
01 12	☽ ⊼ ♃
03 27	☽ ♇ ☉
03 50	☽ ♇ ☿
04 51	☽ ♇ ⯓
06 37	☽ ⊥ ♅
09 01	♀ ⊥ ♃
11 17	☽ ± ♃
12 26	☿ □ ⯓
19 46	☽ ± ⯓
21 29	☽ ∠ ♅
21 34	☽ □ ♆
10 Tuesday	
00 11	☉ □ ⯓
02 26	☽ ♇ ♃
03 30	☿ ± ♆
05 44	☽ ⊼ ⯓
06 08	☽ △ ☉
08 04	☽ △ ☿
14 47	☽ □ ♂
15 41	☽ ☍ ♄
20 12	☽ □ ♀
22 25	☽ ✶ ♅
11 Wednesday	
01 54	☉ ± ♆
03 51	☽ △ ♃
13 37	♂ □ ♄
15 40	☉ ∥ ☿
18 33	☽ Q ♅
21 04	♀ ⊼ ♅
22 27	☽ Q ♂
23 45	☽ △ ♆
12 Thursday	
05 27	☽ Q ♀
08 13	☽ ☍ ⯓
12 25	☽ □ ☉
14 52	☿ ∥ ♆
17 33	☽ □ ☿
18 39	☽ ⊼ ♄
19 52	☽ ✶ ♂
13 Friday	
01 10	☽ □ ♅
01 21	☽ ♇ ♆
03 02	☿ ✶ ♄
04 00	☽ ✶ ♀
05 02	☽ ± ♄
07 43	☽ □ ♃
13 25	☉ ∥ ♆
16 54	☽ ± ♆
20 43	☽ ♇ ♄
21 52	☽ Q ☉
23 04	☽ ∠ ♂
23 07	☿ ⊻ ♂
14 Saturday	
03 20	☽ ⊼ ♆
05 01	☽ Q ☿
08 40	☽ ∠ ♀
12 10	☽ ⊼ ⯓
15 36	☽ ⊥ ♂
15 37	☽ Q ♃
16 31	♂ ⊥ ♃
20 34	☽ ✶ ☉
22 45	☽ ± ⯓
23 11	☽ △ ♄
15 Sunday	
02 07	☽ ⊥ ♀
02 33	☿ # ♅
02 43	☽ ⊻ ♂
05 07	☽ ✶ ☿
05 27	☽ △ ♅
05 45	♀ ⊻ ♃
07 52	☿ ☍ ♅
10 05	☿ ♇ ♆
13 14	☽ ✶ ♃
13 52	☽ ⊻ ♀
14 45	☽ ♇ ⯓
18 56	☽ # ♃
22 35	☽ # ☿
16 Monday	
01 24	☽ ∠ ☉
02 14	☽ ∥ ♅
06 45	☽ # ☉
08 13	☽ ♇ ♅
08 32	☽ ☍ ♆
11 13	☉ ✶ ♄
11 44	☽ ∠ ☿
12 34	☽ # ♆
16 38	☽ ∠ ♃
17 44	☽ △ ⯓
18 57	☽ ⊥ ☉
21 53	☿ ∥ ♃
17 Tuesday	
00 27	☽ ± ♅
05 23	☽ □ ♄
06 24	☽ ⊥ ☿
06 45	☽ ⊻ ☉
09 17	☽ ⊥ ♃
10 02	♂ ⊼ ♅
11 22	☽ ⊼ ♅
11 27	☽ ☌ ♂
18 57	☽ ⊻ ☿
20 29	☽ ⊻ ♃
20 33	☽ ∥ ♂
18 Wednesday	
01 57	☽ ☌ ♀
05 53	☉ # ♅
06 29	☽ # ♀
08 54	☿ ☌ ♃
12 36	☽ ∥ ♀
14 45	☽ Q ♄
15 24	☽ ⊼ ♆
20 33	☽ # ♂
19 Thursday	
01 02	☽ □ ⯓
02 36	☽ ± ♆
13 24	☽ ✶ ♄
17 35	☉ ☍ ♅
19 04	☽ ☍ ♅
19 12	☽ ☌ ☉
19 33	☽ ♇ ♆
22 16	☽ ⊻ ♂
22 18	☿ ⊥ ♂
23 36	☉ ♇ ♆
20 Friday	
05 41	☽ ☌ ♃
07 20	☽ ∥ ♆
09 45	☿ Q ⯓
10 19	☽ ⊥ ♂
11 11	☽ Q ⯓
11 23	☽ ☌ ☿
13 39	♀ # ♂
16 35	☽ ⊻ ♀
17 45	☽ # ♅
18 12	☽ ∠ ♄
21 Saturday	
00 14	☽ △ ♆
00 38	☽ ∥ ☉
04 36	☽ ∠ ♂
04 54	☽ ∥ ♃
05 28	☽ ⊥ ♀
10 23	☽ ✶ ⯓
11 45	☽ ⊥ ♄
23 34	☽ ⊻ ♄
22 Sunday	
01 02	☽ ∠ ♀
04 40	☽ ∥ ☿
04 53	☽ ⊼ ♅
10 23	☽ ⊻ ☉
11 35	☽ ✶ ♂
15 54	☽ ∠ ⯓
16 42	☽ ± ♅
17 10	☽ ⊻ ♃
19 29	☿ ∠ ♄
23 21	☽ ⊥ ☉
23 Monday	
05 15	☉ ∥ ♃
05 21	☽ ⊥ ♃
06 56	☽ ⊻ ☿
09 52	☽ ⊥ ⯓
10 14	☽ ✶ ♀
10 36	☽ ♇ ♅
11 17	☽ □ ♆
12 49	☽ Q ♂
19 01	☽ ∠ ☉
20 11	♀ ⊼ ♆
20 48	☽ ⊥ ☿
21 56	☽ ⊻ ⯓
23 43	♀ Q ♄
23 45	☽ ∠ ♃
24 Tuesday	
00 27	☉ ⊻ ♂
11 55	☽ ☌ ♄
13 15	☽ Q ♀
15 55	☿ △ ♆
16 44	☽ △ ♅
17 44	☽ ∠ ☿
23 34	☽ Q ♆
25 Wednesday	
03 14	☽ □ ♂
04 08	☽ ✶ ☉
06 41	☽ ✶ ♃
23 55	☽ ✶ ♆
26 Thursday	
04 49	☽ ✶ ☿
06 06	☽ □ ♀
06 46	☽ Q ☉
07 32	☽ Q ♃
10 49	☽ ☌ ⯓
18 09	☉ ☌ ♃
27 Friday	
01 13	☽ ⊻ ♄
05 22	☽ □ ♅
06 15	☽ ∠ ♆
08 42	☽ Q ☿
13 23	☽ ⊥ ♄
19 21	☽ △ ♂
20 30	☽ □ ♃
21 30	☽ ∥ ☿
22 22	☽ □ ☉
28 Saturday	
00 16	☽ ⊥ ♆
03 21	♀ □ ⯓
03 25	☿ ✶ ⯓
03 42	☿ ⊻ ♀
07 29	☽ ∠ ♄
07 29	☉ Q ⯓
12 10	☽ ⊻ ♆
14 19	♀ ± ♆
17 07	☽ Q ♅
22 51	☽ ⊻ ⯓
23 16	☿ ⊥ ♄
29 Sunday	
01 02	☽ △ ♀
01 35	☽ □ ☿
02 36	☽ ♇ ♂
04 16	♂ ⊻ ♃
10 31	☽ ⊥ ⯓
13 01	☽ ✶ ♄
13 21	☽ ∥ ☉
15 04	☿ ∠ ♂
16 22	☽ ✶ ♅
20 53	☽ ± ♂
30 Monday	
01 22	☽ ∥ ♃
03 45	☽ ∠ ⯓
07 54	☽ △ ♃
08 51	☽ ⊼ ♂
08 57	☽ ♇ ♀
11 59	☽ Q ♄
13 34	☽ △ ☉
17 28	☽ # ♅
20 32	☽ ∠ ♅
21 32	☽ ☌ ♆
31 Tuesday	
02 10	☽ ∥ ♆
03 26	☽ ± ♀
07 38	☽ ✶ ⯓
09 14	☽ ∥ ♀
12 04	☽ ♇ ♃
12 57	☽ ⊥ ♅
15 28	☽ ⊼ ♀
17 21	☽ △ ☿
19 16	☽ ♇ ☉
21 08	☽ □ ♄
23 41	☽ ⊻ ♅

All ephemeris data is given at 12.00 UT and the Moon's longitude is additionally given for 24.00 UT

NOVEMBER 2017

LONGITUDES

Date	Sidereal time h m s	Sun ☉ ° ′ ″	Moon ☽ ° ′ ″	Moon ☽ 24.00 ° ′ ″	Mercury ☿ ° ′	Venus ♀ ° ′	Mars ♂ ° ′	Jupiter ♃ ° ′	Saturn ♄ ° ′	Uranus ♅ ° ′	Neptune ♆ ° ′	Pluto ♇ ° ′
01	14 43 52	09 ♏ 15 21	03 ♈ 00 53	09 ♈ 56 13	23 ♏ 39	22 ♎ 31	06 ♎ 09	04 ♏ 46	24 ♐ 39	26 ♈ 00	11 ♓ 35	17 ♑ 08
02	14 47 49	10 15 22	16 ♈ 58 20	24 ♈ 06 51	25 09	23 46	06 46	04 59	24 45	25 R 58	11 R 35	17 09
03	14 51 45	11 15 26	01 ♉ 21 11	08 ♉ 40 32	26 37	25 01	07 24	05 12	24 51	25 56	11 34	17 10
04	14 55 42	12 15 31	16 ♉ 03 57	23 ♉ 30 21	28 06	26 16	08 02	05 26	24 57	25 53	11 33	17 11
05	14 59 38	13 15 38	00 ♊ 58 34	08 ♊ 27 25	29 ♏ 33	27 31	08 40	05 39	25 02	25 51	11 33	17 12
06	15 03 35	14 15 48	15 ♊ 55 44	23 ♊ 22 24	01 ♐ 01	28 ♎ 46	09 18	05 52	25 08	25 49	11 32	17 13
07	15 07 32	15 15 59	00 ♋ 46 28	08 ♋ 07 06	02 27	00 ♏ 01	09 56	06 05	25 14	25 47	11 32	17 15
08	15 11 28	16 16 12	15 ♋ 23 03	22 ♋ 35 41	03 53	01 16	10 34	06 18	25 20	25 44	11 31	17 16
09	15 15 25	17 16 27	29 ♋ 42 51	06 ♌ 45 02	05 19	02 31	11 11	06 31	25 26	25 42	11 31	17 17
10	15 19 21	18 16 44	13 ♌ 42 12	20 ♌ 34 28	06 43	03 47	11 49	06 44	25 32	25 40	11 30	17 18
11	15 23 18	19 17 03	27 ♌ 22 01	04 ♍ 05 04	08 07	05 02	12 27	06 57	25 38	25 38	11 30	17 19
12	15 27 14	20 17 25	10 ♍ 43 55	17 ♍ 18 51	09 30	06 17	13 05	07 10	25 45	25 35	11 30	17 21
13	15 31 11	21 17 48	23 ♍ 50 09	00 ♎ 18 05	10 53	07 32	13 43	07 23	25 51	25 33	11 29	17 22
14	15 35 07	22 18 13	06 ♎ 42 54	13 ♎ 04 48	12 14	08 47	14 21	07 35	25 57	25 31	11 29	17 23
15	15 39 04	23 18 39	19 ♎ 23 59	25 ♎ 40 36	13 34	10 03	14 59	07 48	26 03	25 29	11 29	17 25
16	15 43 01	24 19 08	01 ♏ 54 45	08 ♏ 06 32	14 53	11 18	15 36	08 01	26 10	25 27	11 29	17 26
17	15 46 57	25 19 38	14 ♏ 16 02	20 ♏ 23 19	16 10	12 33	16 14	08 14	26 16	25 25	11 28	17 27
18	15 50 54	26 20 10	26 ♏ 28 28	02 ♐ 31 33	17 26	13 49	16 52	08 27	26 23	25 23	11 28	17 29
19	15 54 50	27 20 44	08 ♐ 32 41	14 ♐ 31 59	18 40	15 04	17 30	08 40	26 29	25 21	11 28	17 30
20	15 58 47	28 21 19	20 ♐ 29 38	26 ♐ 25 52	19 53	16 19	18 08	08 52	26 36	25 19	11 28	17 32
21	16 02 43	29 ♏ 21 55	02 ♑ 20 55	08 ♑ 15 06	21 02	17 35	18 45	09 05	26 42	25 17	11 28	17 33
22	16 06 40	00 ♐ 22 33	14 ♑ 08 48	20 ♑ 02 26	22 10	18 50	19 23	09 18	26 49	25 15	11 28	17 35
23	16 10 36	01 23 12	25 ♑ 56 29	01 ♒ 51 27	23 14	20 05	20 01	09 30	26 55	25 13	11 D 28	17 36
24	16 14 33	02 23 52	07 ♒ 47 55	13 ♒ 46 29	24 15	21 21	20 39	09 43	27 02	25 12	11 28	17 38
25	16 18 30	03 24 33	19 ♒ 47 50	25 ♒ 52 35	25 12	22 36	21 17	09 56	27 09	25 10	11 28	17 39
26	16 22 26	04 25 15	02 ♓ 01 27	08 ♓ 15 06	26 05	23 52	21 54	10 08	27 15	25 08	11 28	17 41
27	16 26 23	05 25 59	14 ♓ 34 12	20 ♓ 59 22	26 53	25 07	22 32	10 21	27 22	25 06	11 28	17 42
28	16 30 19	06 26 43	27 ♓ 31 09	04 ♈ 10 02	27 36	26 22	23 10	10 33	27 29	25 05	11 29	17 44
29	16 34 16	07 27 28	10 ♈ 56 21	17 ♈ 50 19	28 12	27 38	23 48	10 46	27 35	25 03	11 29	17 46
30	16 38 12	08 ♐ 28 14	24 ♈ 51 56	02 ♉ 01 01	28 ♐ 41	28 ♏ 53	24 ♎ 25	10 ♏ 58	27 ♐ 42	25 ♈ 01	11 ♓ 29	17 ♑ 47

Date	Moon True ☊ ° ′	Moon Mean ☊ ° ′	Moon ☽ Latitude ° ′
01	20 ♌ 57	20 ♌ 06	03 S 22
02	20 R 49	20 03	04 10
03	20 38	20 00	04 44
04	20 27	19 57	05 00
05	20 15	19 54	04 55
06	20 05	19 50	04 31
07	19 57	19 47	03 48
08	19 52	19 44	02 50
09	19 50	19 41	01 44
10	19 D 49	19 38	00 S 32
11	19 R 49	19 34	00 N 40
12	19 48	19 31	01 48
13	19 45	19 28	02 49
14	19 40	19 25	03 41
15	19 31	19 22	04 20
16	19 19	19 19	04 46
17	19 05	19 15	04 59
18	18 51	19 12	04 57
19	18 36	19 09	04 42
20	18 24	19 06	04 15
21	18 13	19 03	03 37
22	18 05	19 00	02 49
23	18 01	18 56	01 54
24	17 59	18 53	00 N 54
25	17 D 58	18 50	00 S 10
26	17 59	18 47	01 14
27	17 R 58	18 44	02 16
28	17 56	18 40	03 14
29	17 52	18 37	04 03
30	17 ♌ 45	18 ♌ 34	04 S 39

DECLINATIONS

Date	Sun ☉ ° ′	Moon ☽ ° ′	Mercury ☿ ° ′	Venus ♀ ° ′	Mars ♂ ° ′	Jupiter ♃ ° ′	Saturn ♄ ° ′	Uranus ♅ ° ′	Neptune ♆ ° ′	Pluto ♇ ° ′
01	14 S 35	01 S 54	20 S 06	07 S 23	01 S 24	12 S 09	22 S 20	09 N 29	08 S 05	21 S 47
02	14 54	02 N 49	20 34	07 51	01 39	12 14	22 21	09 28	08 05	21 47
03	15 12	07 30	21 00	08 19	01 54	12 18	22 21	09 28	08 05	21 47
04	15 31	11 51	21 26	08 47	02 09	12 22	22 21	09 27	08 05	21 47
05	15 49	15 32	21 51	09 15	02 24	12 27	22 22	09 26	08 06	21 47
06	16 07	18 12	22 14	09 43	02 39	12 31	22 22	09 25	08 06	21 47
07	16 25	19 38	22 37	10 10	02 54	12 35	22 22	09 24	08 06	21 47
08	16 42	19 43	22 58	10 37	03 09	12 40	22 23	09 23	08 06	21 47
09	16 59	18 31	23 18	11 04	03 24	12 44	22 23	09 23	08 06	21 47
10	17 16	16 12	23 37	11 31	03 39	12 48	22 23	09 22	08 06	21 47
11	17 33	13 00	23 55	11 57	03 54	12 52	22 24	09 21	08 07	21 47
12	17 49	09 12	24 11	12 24	04 08	12 57	22 24	09 20	08 07	21 47
13	18 05	05 02	24 27	12 49	04 23	13 01	22 24	09 20	08 07	21 47
14	18 21	00 N 43	24 41	13 15	04 38	13 05	22 25	09 19	08 07	21 47
15	18 36	03 S 35	24 53	13 40	04 53	13 09	22 25	09 18	08 07	21 47
16	18 51	07 39	25 05	14 05	05 08	13 13	22 25	09 17	08 07	21 47
17	19 05	11 22	25 15	14 29	05 22	13 17	22 26	09 17	08 07	21 46
18	19 20	14 33	25 24	14 54	05 37	13 22	22 26	09 16	08 07	21 46
19	19 34	17 05	25 31	15 17	05 51	13 26	22 26	09 15	08 07	21 46
20	19 47	18 51	25 37	15 41	06 06	13 30	22 26	09 14	08 07	21 46
21	20 01	19 48	25 41	16 04	06 21	13 34	22 27	09 14	08 07	21 46
22	20 14	19 53	25 44	16 26	06 35	13 38	22 27	09 13	08 07	21 46
23	20 26	19 05	25 46	16 48	06 49	13 42	22 27	09 12	08 07	21 46
24	20 38	17 27	25 46	17 10	07 04	13 46	22 27	09 12	08 07	21 46
25	20 50	15 02	25 44	17 31	07 18	13 50	22 28	09 11	08 07	21 46
26	21 01	11 54	25 41	17 52	07 32	13 54	22 28	09 11	08 07	21 46
27	21 12	08 10	25 37	18 12	07 47	13 57	22 28	09 10	08 07	21 46
28	21 23	03 S 57	25 31	18 32	08 01	14 01	22 28	09 09	08 07	21 45
29	21 33	00 N 36	25 24	18 51	08 15	14 05	22 29	09 09	08 07	21 45
30	21 S 43	05 N 17	25 S 15	19 S 10	08 S 29	14 S 09	22 S 29	09 N 08	08 S 07	21 S 45

ZODIAC SIGN ENTRIES

Date	h	m	Planets	
01	06	43	☽	♈
03	09	46	☽	♉
05	10	26	☽	♊
05	19	19	☿	♐
07	10	44	☽	♋
07	11	38	♀	♏
09	12	29	☽	♌
11	16	41	☽	♍
13	23	26	☽	♎
16	08	19	☽	♏
18	18	59	☽	♐
21	07	14	☽	♑
22	03	05	☉	♐
23	20	14	☽	♒
26	08	04	☽	♓
28	16	30	☽	♈
30	20	38	☽	♉

LATITUDES

Date	Mercury ☿ ° ′	Venus ♀ ° ′	Mars ♂ ° ′	Jupiter ♃ ° ′	Saturn ♄ ° ′	Uranus ♅ ° ′	Neptune ♆ ° ′	Pluto ♇ ° ′
01	01 S 28	01 N 29	01 N 08	01 N 01	00 N 59	00 S 36	00 S 56	00 N 33
04	01 45	01 27	01 08	01 01	00 59	00 36	00 56	00 33
07	02 00	01 24	01 07	01 01	00 59	00 36	00 56	00 33
10	02 13	01 20	01 07	01 01	00 58	00 36	00 56	00 32
13	02 24	01 16	01 07	01 01	00 58	00 36	00 56	00 32
16	02 31	01 12	01 06	01 01	00 58	00 36	00 56	00 32
19	02 34	01 07	01 06	01 01	00 57	00 36	00 56	00 31
22	02 32	01 01	01 05	01 01	00 57	00 35	00 56	00 31
25	02 23	00 56	01 04	01 01	00 57	00 35	00 56	00 31
28	02 06	00 50	01 04	01 01	00 56	00 35	00 55	00 30
31	01 S 38	00 N 43	01 N 03	01 N 01	00 N 56	00 S 35	00 S 55	00 N 30

DATA

Julian Date	2458059
Delta T	+70 seconds
Ayanamsa	24° 06′ 10″
Synetic vernal point	05° ♓ 00′ 49″
True obliquity of ecliptic	23° 26′ 06″

MOON'S PHASES, APSIDES AND POSITIONS ☽

Date	h	m	Phase	Longitude ° ′	Eclipse Indicator
04	05	23	○	11 ♉ 59	
10	20	36	☾	18 ♌ 38	
18	11	42	●	26 ♏ 19	
26	17	03	☽	04 ♓ 38	

Day	h	m		
06	00	18	Perigee	
21	19	01	Apogee	
01	21	45	ON	
08	01	29	Max dec	19° N 51′
14	15	56	0S	
22	02	06	Max dec	19° S 57′
29	08	53	0N	

LONGITUDES

Date	Chiron ⚷ ° ′	Ceres ⚳ ° ′	Pallas ⚴ ° ′	Juno ⚵ ° ′	Vesta ⚶ ° ′	Black Moon Lilith ⚸ ° ′
01	24 ♓ 50	11 ♌ 51	04 ♉ 36	14 ♑ 01	22 ♎ 28	29 ♐ 09
11	24 ♓ 35	14 ♌ 11	01 ♉ 23	17 ♑ 09	27 ♎ 36	00 ♑ 16
21	24 ♓ 24	16 ♌ 03	28 ♈ 33	20 ♑ 32	02 ♏ 44	01 ♑ 23
31	24 ♓ 19	17 ♌ 23	26 ♈ 27	23 ♑ 48	07 ♏ 50	02 ♑ 29

ASPECTARIAN

h m	Aspects
01 Wednesday	
04 25	☽ ± ♃
05 11	☽ Q ♇
12 27	☽ ± ☉
14 28	☽ ∥ ♂
15 07	☽ ⊼ ♃
17 43	☽ ☍ ♂
22 37	☉ ∠ ♄
22 58	☽ ⚼ ☿
23 40	☽ ⊼ ☉
02 Thursday	
02 50	☽ ⊻ ♆
05 16	☿ ⊻ ♄
05 47	☽ # ♂
12 18	☽ □ ♇
13 02	☽ ⊥ ♆
16 05	☽ ± ☿
23 44	♀ ∥ ♆
03 Friday	
00 30	☽ ☍ ♀
01 02	☿ ⊼ ♅
01 09	☽ △ ♄
03 03	☽ ☌ ♅
03 17	☽ ⊼ ☿
04 05	☽ ∠ ♆
07 29	♃ Q ♇
08 31	♀ ✶ ♄
15 05	☽ # ♆
16 51	☽ # ♀
18 26	☽ ☍ ♃
19 23	☉ △ ♆
22 22	☽ ⊼ ♂
22 25	☽ ∥ ♅
04 Saturday	
02 00	☽ ⚼ ♄
04 42	☽ ✶ ♆
05 02	♀ ☍ ♅
05 23	☽ ☍ ☉
08 34	☽ ± ♂
13 49	☽ △ ♇
15 08	☽ # ♃
16 41	☽ ± ♄
17 35	♀ ⚼ ♆
23 44	☽ ⚼ ♂
05 Sunday	
00 05	☽ Q ♆
02 24	☽ ⊼ ♄
03 48	☽ ⊻ ♅
05 56	☽ ⊼ ♀
08 35	☿ ∥ ♇
09 28	☽ ☍ ☿
13 24	☽ ⊥ ♅
13 58	☽ ⚼ ♇
14 23	☽ # ☉
16 27	☽ ± ♀
19 36	☽ ⊼ ♃
20 55	♀ # ♅
06 Monday	
00 53	☽ △ ♂
03 48	☽ ∠ ♅
04 26	☽ ± ♇
04 57	☽ □ ♆
05 22	☽ ± ♃
08 12	☽ ⚼ ♀
09 08	☽ ⊼ ☉
14 05	☽ ⊼ ♇
19 29	☽ ± ☉
20 04	☽ ⚼ ♃
20 32	☿ ∥ ♄
07 Tuesday	
01 00	☿ ± ♅
02 57	☽ ☍ ♄
03 54	☽ ✶ ♅
08 27	☿ ∠ ♇
10 39	☽ △ ♀
11 07	☽ ⚼ ☉
15 02	☽ ⊼ ☿
20 47	☽ △ ♃
23 25	☽ Q ♅
08 Wednesday	
01 55	☽ ± ☿
03 39	☽ □ ♂
05 36	☽ △ ♆
13 34	☽ △ ☉
15 06	☽ ☍ ♇
18 27	☽ ⚼ ☿
09 Thursday	
04 43	☽ ⊼ ♄
05 14	☽ □ ♅
06 36	☽ ⚼ ♆
06 39	☉ ⊥ ♂
11 04	☽ Q ♂
12 12	☉ ✶ ♇
14 57	☽ ± ♄
17 14	☽ □ ♀
21 53	☽ ± ♆
22 36	☽ △ ☿
23 46	☽ □ ♃
10 Friday	
00 10	♂ ⊼ ♆
03 08	☽ # ☉
06 29	☽ ⚼ ♄
08 12	☽ ⊼ ♆
08 35	☽ ✶ ♂
12 07	☿ ⊻ ♃
18 17	☽ ⊼ ♇
20 36	☽ □ ☉
11 Saturday	
03 33	☽ Q ♀
04 50	☽ ± ♇
07 38	☽ Q ♃
08 55	☽ △ ♄
08 55	☽ △ ♅
09 45	♄ △ ♅
12 10	☽ ∠ ♂
12 53	☽ # ♃
17 43	♀ Q ♇
18 14	☽ # ♀
20 51	☽ ⚼ ♇
21 28	☉ ⊥ ♄
12 Sunday	
03 07	☽ ✶ ♀
05 04	☽ ⊥ ♂
05 26	☽ ✶ ♃
07 13	☽ Q ☉
09 31	☽ □ ☿
11 13	☽ ∥ ♅
11 45	☽ ⚼ ♅
13 23	☽ ☍ ♆
16 29	☽ ⊻ ♂
18 28	☽ # ♆
13 Monday	
00 05	☽ △ ♇
04 08	☽ ± ♅
06 30	☿ ⚼ ♅
06 55	☽ ✶ ☉
08 15	♀ ☌ ♃
09 15	☽ ∠ ♃
09 21	☽ ∠ ♀
15 10	☽ ⊼ ♅
15 26	☽ # ♂
15 45	☽ □ ♄
18 04	♂ Q ♄
20 47	☿ ⊥ ♇
22 27	☽ Q ☿
22 47	☿ □ ♆
14 Tuesday	
00 55	♀ ∥ ♃
02 14	☽ ⊥ ♃
03 51	☽ ⊥ ♀
13 12	☽ ∠ ☉
13 40	☽ ⊻ ♃
16 20	☽ ⊻ ♀
20 59	☽ ⊼ ♆
23 37	☽ ✶ ☿
15 Wednesday	
01 46	☽ Q ♄
03 09	☽ ☌ ♂
07 41	☽ ⊥ ☉
08 12	☽ □ ♇
08 21	☽ ± ♆
17 12	☿ ⊥ ♃
19 58	☽ ∥ ♂
20 07	☽ ⊻ ☉
23 36	☽ ☍ ♅
16 Thursday	
00 50	☽ ✶ ♄
01 32	☽ ⚼ ♆
07 37	☽ ∠ ☿
09 06	♀ ∠ ♄
14 51	☽ ∥ ♆
15 21	♀ △ ♆
18 49	☽ Q ♇
22 12	☽ # ♅
17 Friday	
00 02	☽ ☌ ♃
03 05	☽ ⊥ ☿
06 06	☽ ∠ ♄
06 33	☽ △ ♆
08 17	☽ ☌ ♀
14 04	☉ ⊼ ♅
14 24	☿ ✶ ♂
16 04	☽ ⊻ ♂
16 09	☽ ⊻ ☿
18 15	☽ ✶ ♇
23 52	☽ ⊥ ♄
18 Saturday	
02 20	☽ ∥ ♃
04 29	☽ ⊥ ♂
09 51	☽ ⊼ ♅
11 42	☽ ☌ ☉
11 48	☽ ⊻ ♄
12 50	☿ ⊻ ♇
13 03	☉ ⊻ ♄
15 26	☽ ∥ ♀
21 42	☽ ± ♅
23 17	☽ ∠ ♂
23 56	☽ ∠ ♇
19 Sunday	
10 55	♂ ± ♆
12 14	☽ ⊻ ♃
12 15	♂ □ ♇
15 36	☽ ⚼ ♅
17 51	☽ □ ♆
17 56	☽ ⊥ ♇
20 Monday	
00 29	☽ ⊥ ♃
02 37	☽ ⊻ ♀
06 01	☽ ⊻ ♇
06 58	☽ ✶ ♂
10 37	☽ ☌ ☿
16 08	☽ ⊥ ♀
18 57	☽ ∠ ♃
21 43	☽ △ ♅
21 Tuesday	
00 26	☽ ☌ ♄
05 23	☽ ⊻ ☉
06 09	☽ Q ♆
08 35	☽ Q ♂
11 29	♀ ✶ ♇
12 31	☽ ∠ ♀
18 42	☽ ⊥ ☉
22 Wednesday	
01 57	☽ ✶ ♃
06 32	☽ ✶ ♆
14 19	♆ St D
14 44	☽ ∠ ☉
19 00	☽ ☌ ♇
22 41	☽ ✶ ♀
23 16	☽ □ ♂
23 Thursday	
02 49	☽ Q ♃
05 58	☽ ⊻ ☿
08 15	☉ ± ♅
09 10	♀ ⊻ ♂
10 33	☽ □ ♅
13 04	☽ ∠ ♆
14 00	☽ ⊻ ♄
19 20	☽ ⊥ ☿
24 Friday	
00 05	☽ ✶ ☉
01 57	☽ Q ♀
02 17	☽ ⊥ ♄
05 23	♀ ⊥ ♄
07 18	☽ ⊥ ♆
14 46	☽ ∥ ♀
15 11	☽ ∠ ☿
15 56	☽ □ ♃
17 37	☉ ∠ ♇
19 23	☽ ⊻ ♆
20 35	☽ ∠ ♄
22 49	☽ Q ♅
25 Saturday	
02 28	☽ Q ☉
02 46	☿ ∠ ♃
07 44	☽ ⊻ ♇
10 56	☿ △ ♅
15 05	☽ △ ♂
18 12	☽ □ ♀
19 39	☽ ⊥ ♇
21 39	☽ ∥ ♃
22 34	☽ ✶ ♅
23 32	☽ ✶ ☿
26 Sunday	
02 37	☽ ✶ ♄
13 17	☽ ∠ ♇
17 03	☽ □ ☉
21 55	☽ ⚼ ♂
27 Monday	
00 30	☽ Q ☿
02 02	☽ Q ♄
03 34	☽ ∠ ♅
03 52	☽ △ ♃
05 55	☽ # ♅
06 08	☽ ☌ ♆
11 48	♀ ⊼ ♅
12 20	☽ ∥ ♆
14 13	☽ ∥ ♂
15 53	☽ ± ♂
17 54	☽ ✶ ♇
20 29	☽ ⊥ ♅
28 Tuesday	
03 37	☽ ⊼ ♂
06 58	☿ ☌ ♄
07 33	☽ ⊻ ♅
08 21	☽ ⚼ ♃
09 41	☽ △ ♀
11 56	☽ □ ♄
12 09	☽ □ ☿
16 02	☽ Q ♇
22 12	♂ ∥ ♆
29 Wednesday	
00 53	☽ ± ♃
05 22	☽ △ ☉
11 11	♀ ⊻ ♄
11 41	☽ ⊼ ♃
12 57	☽ ⊻ ♆
15 16	☽ ⚼ ♀
23 23	☽ ⊥ ♆
23 53	☽ □ ♇
30 Thursday	
06 00	☿ ⊻ ♀
08 19	☽ ± ♀
09 27	☽ ⚼ ☉
11 13	☽ ☍ ♂
12 16	☽ ☌ ♅
14 44	☽ ∠ ♆
16 50	☽ △ ♄
17 59	☉ ∥ ♇
18 37	☽ △ ☿
19 26	☽ ⊼ ♀

All ephemeris data is given at 12.00 UT and the Moon's longitude is additionally given for 24.00 UT

DECEMBER 2017

Date	Sidereal time h m s	Sun ☉ ° ′ ″	Moon ☽ ° ′ ″	Moon ☽ 24.00 ° ′ ″	Mercury ☿ ° ′	Venus ♀ ° ′	Mars ♂ ° ′	Jupiter ♃ ° ′	Saturn ♄ ° ′	Uranus ♅ ° ′	Neptune ♆ ° ′	Pluto ⯓ ° ′
01	16 42 09	09 ♐ 29 02	09 ♉ 17 10	16 ♉ 39 43	29 ♐ 02	00 ♐ 09	25 ♎ 03	11 ♏ 10	27 ♐ 49	25 ♈ 00	11 ♓ 29	17 ♑ 49
02	16 46 05	10 29 50	24 ♉ 07 49	01 ♊ 40 22	29 15	01 24	25 41	11 23	27 56	24 R 58	11 30	17 51
03	16 50 02	11 30 39	09 ♊ 16 08	16 ♊ 53 45	29 R 18	02 40	26 19	11 35	28 03	24 57	11 30	17 52
04	16 53 59	12 31 30	24 ♊ 31 48	02 ♋ 08 55	29 11	03 55	26 56	11 47	28 10	24 56	11 30	17 54
05	16 57 55	13 32 22	09 ♋ 43 46	17 ♋ 15 10	28 53	05 10	27 34	11 59	28 17	24 54	11 31	17 56
06	17 01 52	14 33 15	24 ♋ 42 07	02 ♌ 03 50	28 24	06 26	28 12	12 11	28 24	24 53	11 31	17 58
07	17 05 48	15 34 09	09 ♌ 19 44	16 ♌ 29 25	27 43	07 41	28 49	12 23	28 31	24 51	11 32	17 59
08	17 09 45	16 35 04	23 ♌ 32 43	00 ♍ 29 36	26 52	08 57	29 ♎ 27	12 35	28 38	24 50	11 32	18 01
09	17 13 41	17 36 01	07 ♍ 20 12	14 ♍ 04 45	25 50	10 12	00 ♏ 05	12 47	28 45	24 49	11 33	18 03
10	17 17 38	18 36 59	20 ♍ 43 34	27 ♍ 17 03	24 40	11 28	00 42	12 59	28 52	24 48	11 33	18 05
11	17 21 34	19 37 58	03 ♎ 45 35	10 ♎ 09 37	23 23	12 43	01 20	13 11	28 59	24 47	11 34	18 07
12	17 25 31	20 38 58	16 ♎ 29 34	22 ♎ 45 51	22 02	13 59	01 58	13 22	29 06	24 46	11 35	18 09
13	17 29 28	21 39 59	28 ♎ 58 51	05 ♏ 08 58	20 39	15 14	02 35	13 34	29 13	24 45	11 35	18 10
14	17 33 24	22 41 01	11 ♏ 16 29	17 ♏ 21 44	19 17	16 30	03 13	13 46	29 20	24 44	11 36	18 12
15	17 37 21	23 42 04	23 ♏ 24 57	29 ♏ 26 25	17 59	17 45	03 51	13 57	29 27	24 43	11 37	18 14
16	17 41 17	24 43 08	05 ♐ 26 18	11 ♐ 24 49	16 47	19 01	04 28	14 09	29 34	24 42	11 38	18 16
17	17 45 14	25 44 13	17 ♐ 22 09	23 ♐ 18 28	15 44	20 16	05 06	14 20	29 41	24 41	11 39	18 18
18	17 49 10	26 45 18	29 ♐ 13 58	05 ♑ 08 49	14 50	21 32	05 44	14 31	29 48	24 40	11 39	18 20
19	17 53 07	27 46 24	11 ♑ 03 14	16 ♑ 57 27	14 06	22 48	06 21	14 43	29 ♐ 55	24 39	11 40	18 22
20	17 57 03	28 47 31	22 ♑ 51 45	28 ♑ 46 24	13 34	24 03	06 59	14 54	00 ♑ 02	24 39	11 41	18 24
21	18 01 00	29 ♐ 48 38	04 ♒ 41 45	10 ♒ 38 10	13 13	25 19	07 36	15 05	00 09	24 38	11 42	18 26
22	18 04 57	00 ♑ 49 45	16 ♒ 36 05	22 ♒ 35 56	13 02	26 34	08 14	15 16	00 16	24 37	11 43	18 28
23	18 08 53	01 50 53	28 ♒ 38 15	04 ♓ 43 31	13 D 01	27 50	08 51	15 27	00 23	24 37	11 44	18 30
24	18 12 50	02 52 00	10 ♓ 52 20	17 ♓ 05 14	13 10	29 ♐ 05	09 29	15 38	00 30	24 36	11 45	18 32
25	18 16 46	03 53 08	23 ♓ 22 50	29 ♓ 45 40	13 27	00 ♑ 21	10 07	15 48	00 38	24 36	11 46	18 34
26	18 20 43	04 54 16	06 ♈ 14 19	12 ♈ 49 15	13 52	01 36	10 44	15 59	00 45	24 36	11 48	18 36
27	18 24 39	05 55 24	19 ♈ 30 54	26 ♈ 19 35	14 24	02 52	11 22	16 10	00 52	24 35	11 49	18 38
28	18 28 36	06 56 31	03 ♉ 15 30	10 ♉ 18 41	15 02	04 07	11 59	16 20	00 59	24 35	11 50	18 40
29	18 32 32	07 57 39	17 ♉ 29 00	24 ♉ 46 06	15 46	05 23	12 36	16 31	01 06	24 35	11 51	18 42
30	18 36 29	08 58 47	02 ♊ 09 25	09 ♊ 38 12	16 35	06 38	13 14	16 41	01 13	24 34	11 52	18 44
31	18 40 26	09 ♑ 59 55	17 ♊ 11 26	24 ♊ 47 59	17 ♐ 29	07 ♑ 54	13 ♏ 51	16 ♏ 51	01 ♑ 20	24 ♈ 34	11 ♓ 54	18 ♑ 46

Date	Moon True ☊ ° ′	Moon Mean ☊ ° ′	Moon ☽ Latitude ° ′
01	17 ♌ 36	18 ♌ 31	05 S 00
02	17 R 26	18 28	05 01
03	17 15	18 25	04 41
04	17 06	18 21	04 01
05	16 59	18 18	03 04
06	16 55	18 15	01 55
07	16 53	18 12	00 S 40
08	16 D 53	18 09	00 N 36
09	16 53	18 05	01 47
10	16 R 54	18 02	02 51
11	16 52	17 59	03 44
12	16 49	17 56	04 24
13	16 43	17 53	04 51
14	16 35	17 50	05 04
15	16 25	17 46	05 03
16	16 15	17 43	04 49
17	16 04	17 40	04 22
18	15 55	17 37	03 44
19	15 47	17 34	02 56
20	15 42	17 31	02 00
21	15 39	17 27	00 N 59
22	15 D 39	17 24	00 S 05
23	15 39	17 21	01 10
24	15 41	17 18	02 13
25	15 42	17 15	03 11
26	15 R 43	17 11	04 01
27	15 42	17 08	04 40
28	15 39	17 05	05 04
29	15 35	17 02	05 11
30	15 29	16 59	04 58
31	15 ♌ 24	16 ♌ 56	04 S 25

DECLINATIONS

Date	Sun ☉ ° ′	Moon ☽ ° ′	Mercury ☿ ° ′	Venus ♀ ° ′	Mars ♂ ° ′	Jupiter ♃ ° ′	Saturn ♄ ° ′	Uranus ♅ ° ′	Neptune ♆ ° ′	Pluto ⯓ ° ′
01	21 S 52	09 N 51	25 S 04	19 S 28	08 S 43	14 S 13	22 S 29	09 N 08	08 S 06	21 S 45
02	22 01	13 56	24 53	19 46	08 57	14 16	22 29	09 07	08 06	21 45
03	22 10	17 12	24 39	20 03	09 11	14 18	22 29	09 07	08 06	21 45
04	22 18	19 18	24 24	20 20	09 25	14 24	22 30	09 06	08 06	21 45
05	22 25	20 01	24 08	20 36	09 38	14 27	22 30	09 06	08 06	21 45
06	22 32	19 17	23 50	20 51	09 52	14 31	22 30	09 05	08 06	21 44
07	22 39	17 16	23 31	21 06	10 06	14 35	22 30	09 05	08 05	21 44
08	22 46	14 13	23 11	21 20	10 19	14 38	22 30	09 04	08 05	21 44
09	22 51	10 28	22 49	21 34	10 33	14 42	22 30	09 04	08 05	21 44
10	22 57	06 18	22 26	21 47	10 46	14 45	22 31	09 04	08 05	21 44
11	23 02	01 N 56	22 03	21 59	10 59	14 49	22 31	09 03	08 04	21 44
12	23 06	02 S 25	21 39	22 11	11 13	14 52	22 31	09 03	08 04	21 44
13	23 10	06 34	21 16	22 22	11 26	14 56	22 31	09 03	08 04	21 43
14	23 14	10 23	20 54	22 32	11 39	14 59	22 31	09 02	08 04	21 43
15	23 17	13 43	20 33	22 41	11 52	15 02	22 31	09 02	08 03	21 43
16	23 20	16 28	20 14	22 50	12 05	15 05	22 31	09 02	08 03	21 43
17	23 22	18 29	19 58	22 59	12 18	15 09	22 31	09 01	08 03	21 43
18	23 24	19 42	19 46	23 06	12 30	15 12	22 31	09 01	08 02	21 43
19	23 25	20 03	19 36	23 13	12 43	15 16	22 32	09 01	08 02	21 42
20	23 26	19 31	19 29	23 19	12 56	15 19	22 32	09 01	08 02	21 42
21	23 26	18 08	19 26	23 25	13 08	15 22	22 32	09 00	08 01	21 42
22	23 26	15 56	19 25	23 29	13 21	15 25	22 32	09 00	08 01	21 42
23	23 25	13 02	19 28	23 33	13 33	15 28	22 32	09 00	08 00	21 42
24	23 24	09 32	19 32	23 37	13 45	15 31	22 32	09 00	08 00	21 42
25	23 23	05 33	19 39	23 39	13 57	15 34	22 32	09 00	07 59	21 41
26	23 21	01 S 12	19 48	23 41	14 09	15 37	22 32	09 00	07 59	21 41
27	23 18	03 N 19	19 58	23 42	14 21	15 40	22 32	09 00	07 59	21 41
28	23 15	07 50	20 09	23 42	14 33	15 43	22 32	08 59	07 58	21 41
29	23 12	12 04	20 21	23 42	14 45	15 46	22 32	08 59	07 58	21 41
30	23 08	15 43	20 34	23 41	14 56	15 49	22 32	08 59	07 57	21 41
31	23 S 03	18 N 25	20 S 47	23 S 39	15 S 08	15 S 52	22 S 32	08 N 59	07 S 57	21 S 40

ZODIAC SIGN ENTRIES

Date	h m	Planets
01	09 14	♀ ♐
02	21 21	☽ ♊
04	20 37	☽ ♋
06	20 37	☽ ♌
08	23 09	☽ ♍
09	08 59	♂ ♏
11	05 01	☽ ♎
13	13 59	☽ ♏
16	01 07	☽ ♐
18	13 33	☽ ♑
20	04 49	♄ ♑
21	02 29	☽ ♒
21	16 28	☉ ♑
23	14 42	☽ ♓
25	05 26	♀ ♑
26	00 27	☽ ♈
28	06 23	☽ ♉
30	08 31	☽ ♊

LATITUDES

Date	Mercury ☿ ° ′	Venus ♀ ° ′	Mars ♂ ° ′	Jupiter ♃ ° ′	Saturn ♄ ° ′	Uranus ♅ ° ′	Neptune ♆ ° ′	Pluto ⯓ ° ′
01	01 S 38	00 N 43	01 N 03	01 N 01	00 N 56	00 S 35	00 S 55	00 N 30
04	00 58	00 37	01 02	01 01	00 56	00 35	00 55	00 30
07	00 S 06	00 30	01 02	01 01	00 55	00 35	00 55	00 30
10	00 N 54	00 23	01 01	01 01	00 55	00 35	00 55	00 29
13	01 51	00 16	01 00	01 02	00 55	00 35	00 55	00 29
16	02 33	00 08	00 59	01 02	00 55	00 35	00 55	00 29
19	02 55	00 N 01	00 58	01 02	00 55	00 35	00 55	00 28
22	02 57	00 S 06	00 57	01 02	00 55	00 35	00 55	00 28
25	02 47	00 13	00 56	01 02	00 54	00 34	00 55	00 28
28	02 28	00 20	00 55	01 03	00 54	00 34	00 55	00 27
31	02 N 04	00 S 27	00 N 54	01 N 03	00 N 54	00 S 34	00 S 55	00 N 27

DATA

Julian Date	2458089
Delta T	+70 seconds
Ayanamsa	24° 06′ 14″
Synetic vernal point	05° ♓ 00′ 45″
True obliquity of ecliptic	23° 26′ 06″

MOON'S PHASES, APSIDES AND POSITIONS ☽

Date	h m	Phase	Longitude ° ′	Eclipse Indicator
03	15 47	○	11 ♊ 40	
10	07 51	☾	18 ♍ 26	
18	06 30	●	26 ♐ 31	
26	09 20	☽	04 ♈ 47	

Day	h m		
04	08 53	Perigee	
19	01 42	Apogee	
05	11 42	Max dec	20° N 01′
11	22 35	0S	
19	09 31	Max dec	20° S 04′
26	18 27	0N	

LONGITUDES

Date	Chiron ⚷ ° ′	Ceres ⚳ ° ′	Pallas ⚴ ° ′	Juno ⚵ ° ′	Vesta ⚶ ° ′	Black Moon Lilith ⚸ ° ′
01	24 ♓ 19	17 ♌ 23	26 ♈ 27	24 ♑ 08	07 ♏ 50	02 ♑ 29
11	24 ♓ 20	18 ♌ 06	25 ♈ 18	27 ♑ 55	12 ♏ 54	03 ♑ 36
21	24 ♓ 26	18 ♌ 20	25 ♈ 09	01 ♒ 01	17 ♏ 56	04 ♑ 43
31	24 ♓ 38	17 ♌ 26	25 ♈ 58	05 ♒ 58	22 ♏ 53	05 ♑ 50

ASPECTARIAN

01 Friday
01 43 ☽ ± ☉
02 40 ☽ # ♆
05 33 ☽ # ♂
08 07 ☽ ∥ ♅
10 05 ♂ ☍ ♅
12 21 ☽ ⊼ ☉
15 08 ☽ ☍ ♃
15 36 ☽ ✱ ♆
17 49 ☽ ♇ ♄
19 53 ☽ ♇ ☿
23 54 ☉ ♇ ♅

02 Saturday
01 53 ☽ △ ⯓
03 59 ♀ ± ♅
08 27 ☽ ± ♄
10 35 ☽ ± ☿
10 59 ☽ Q ♆
13 21 ☽ ⊻ ♅
14 12 ☽ # ♃
14 35 ☽ ⊼ ♂
18 07 ☽ ⊼ ♄
20 13 ☽ ⊼ ☿
22 36 ♀ ⊥ ♂
22 52 ☽ ⊥ ♅
23 22 ☉ ∠ ♂

03 Sunday
00 32 ☽ ± ♂
00 37 ☽ ☍ ♀
01 53 ☽ ♇ ⯓
02 19 ♃ △ ♆
05 18 ♂ # ♅
07 33 ☿ St R
11 44 ☉ □ ♆
13 04 ☽ ∠ ♅
14 01 ☉ ⊻ ♃
15 21 ☽ ♇ ♂
15 31 ☽ □ ♆
15 41 ☽ ⊼ ♃
15 47 ☽ ☍ ☉
16 07 ☽ ± ⯓
16 11 ♀ ∠ ⯓
19 23 ♂ ♇ ♆
20 50 ☉ ⊥ ⯓

04 Monday
01 15 ☽ ± ♃
01 34 ☽ ⊼ ⯓
12 37 ☽ ✱ ♅
15 35 ☽ ♇ ♃
15 57 ☽ △ ♂
17 46 ☽ ☍ ♄
19 13 ☽ ☍ ☿

05 Tuesday
04 08 ☽ ⊼ ♀
07 31 ☽ Q ♅
14 30 ☽ ± ♀
14 50 ☽ △ ♆
15 38 ☽ △ ♃
18 30 ☽ ⊼ ☉

06 Wednesday
01 07 ☽ ☍ ⯓
03 20 ☉ ∥ ♄
04 49 ☽ ± ☉
06 14 ☽ ♇ ♀
12 05 ☿ ☌ ♄
12 17 ☽ □ ♅
14 57 ☽ ♇ ♆
15 58 ☿ ✱ ♂
17 46 ☽ ⊼ ☿
17 56 ☽ □ ♂
18 03 ☽ ⊼ ♄
20 29 ☽ ♇ ☉
21 21 ♂ ✱ ♄

07 Thursday
03 10 ☽ ± ☿
03 57 ☽ ± ♄
05 42 ☽ ± ♆
09 01 ☽ △ ♀
15 40 ☽ ⊼ ♆
17 10 ☽ □ ♃
17 22 ☽ ♇ ☿
19 02 ☽ ♇ ♄
20 05 ☿ ∠ ♃
23 15 ☽ △ ☉

08 Friday
01 09 ☽ Q ♂
02 34 ☽ ⊼ ⯓
09 10 ☽ # ♃
12 49 ☽ ± ⯓
14 13 ☽ △ ♅
17 20 ☽ △ ☿
20 50 ☽ △ ♄
22 40 ☽ ✱ ♂

09 Saturday
00 20 ☽ Q ♃
04 27 ☽ ♇ ⯓
04 41 ♀ ♇ ♅
09 46 ☉ ∥ ☿
11 36 ☽ # ♂
16 23 ☽ ♇ ♅
17 37 ☽ □ ♀
19 29 ☽ ☍ ♆
20 17 ☽ ∥ ♅
21 50 ☽ ✱ ♃
22 58 ☉ ⊻ ⯓

10 Sunday
01 56 ☽ # ♆
02 28 ☽ ∠ ♂
06 54 ♀ ∥ ⯓
07 12 ☽ △ ⯓
07 18 ☿ ∥ ♄
07 51 ☽ □ ☉
08 30 ☽ ± ♅
09 29 ☿ △ ♅
13 47 ♀ □ ♆
18 34 ☽ □ ☿
19 25 ☽ ⊼ ♅
19 38 ☽ ⊥ ♂
22 41 ☉ ⊥ ♃

11 Monday
00 03 ♀ ⊥ ⯓
01 29 ☽ ∠ ♃
03 02 ☽ □ ♄
05 45 ☽ Q ♀
07 16 ☽ ⊻ ♂
14 31 ☿ ∥ ♀
18 29 ☽ ⊥ ♃
19 52 ☽ Q ☉
22 16 ♀ ⊻ ♃

12 Tuesday
00 57 ☽ Q ☿
02 40 ☽ ⊼ ♆
05 59 ☽ ⊻ ♃
06 42 ☽ ✱ ♀
07 19 ☿ ∥ ⯓
13 09 ☽ Q ♄
14 04 ☽ ± ♆
15 09 ☽ □ ⯓
20 39 ☽ ✱ ☉
21 32 ☽ ✱ ☿

13 Wednesday
01 49 ☉ ☌ ☿
03 49 ☽ ☍ ♅
07 22 ☽ ♇ ♆
12 27 ☽ ✱ ♄
14 43 ☽ ∠ ♀
19 23 ☽ ☌ ♂
21 07 ☽ ∥ ♆
23 40 ☽ ∠ ☿

14 Thursday
02 02 ☽ Q ⯓
03 15 ☽ # ♅
04 19 ☽ ∠ ☉
04 33 ☿ ⊥ ♃
10 12 ♀ ∥ ♄
10 18 ☽ ⊥ ♀
12 39 ☽ △ ♆
15 33 ☽ ⊥ ☿
16 58 ☽ ☌ ♃
18 04 ☽ ∠ ♄
21 14 ☽ ∥ ♂
23 29 ☽ ⊻ ♀
23 38 ☽ ⊥ ☉

15 Friday
01 06 ☿ ∠ ♂
01 42 ☽ ✱ ⯓
02 15 ☽ ⊻ ☿
07 16 ☿ ⊻ ⯓
12 04 ☽ ⊥ ♄
12 37 ☽ ⊻ ☉
14 08 ☿ ☌ ♀
14 34 ☽ ⊼ ♅
21 23 ♀ ⊻ ⯓
23 03 ☽ ∥ ♃

16 Saturday
00 08 ☽ ⊻ ♄
02 31 ☽ ± ♅
07 38 ☽ ∠ ⯓
09 57 ☽ ⊻ ♂
11 28 ☉ △ ♅
20 32 ☽ ♇ ♅
22 40 ☽ ⊥ ♂

17 Sunday
00 27 ☽ □ ♆
01 45 ☽ ⊥ ⯓
03 11 ♄ Q ♆
05 22 ♀ ∠ ♂
05 47 ☽ ⊻ ♃
08 56 ☽ ☌ ☿
13 20 ♀ ⊥ ♃
13 53 ☽ ⊻ ⯓
17 49 ☽ ∠ ♂
18 05 ☽ ⊥ ♃
18 34 ☽ ☌ ♀

18 Monday
02 46 ☽ △ ♅
06 30 ☽ ☌ ☉
12 36 ☽ ∠ ♃
12 52 ☽ Q ♆
13 10 ☽ ☌ ♄
13 19 ☽ ∥ ☿
19 34 ☿ ⊻ ♃

19 Tuesday
01 55 ☽ ✱ ♂
12 27 ♂ Q ⯓
13 16 ☽ ✱ ♆
17 54 ☽ ⊻ ☿
19 33 ☽ ✱ ♃

20 Wednesday
02 54 ☽ ☌ ⯓
03 40 ☽ Q ♂
05 34 ☽ ⊥ ☿
12 59 ☽ ∥ ☿
14 42 ☽ ⊻ ♀
15 37 ☽ □ ♅
19 47 ☽ ∠ ♆
20 19 ☽ Q ♃
23 12 ☽ ∠ ☿
23 13 ♀ △ ♅

21 Thursday
01 10 ☽ ⊻ ☉
01 48 ☿ ⊥ ♂
02 42 ☽ ⊻ ♄
04 20 ☽ ⊥ ♀
09 27 ☉ Q ♆
14 03 ☽ ⊥ ♆
14 28 ☽ ⊥ ☉
14 59 ☽ ⊥ ♄
18 13 ☽ □ ♂
18 39 ☉ ∥ ♀
19 45 ☉ ∠ ♃
21 08 ☉ ☌ ♄

22 Friday
00 41 ☽ ∠ ♀
02 10 ☽ ⊻ ♆
04 01 ☽ Q ♅
04 54 ☽ ✱ ☿
09 16 ☽ □ ♃
09 18 ☽ ∠ ♄
10 18 ☽ ∠ ☉
14 54 ♃ ∠ ♄
15 44 ☽ ⊻ ⯓
16 40 ☽ ∥ ♃

23 Saturday
01 52 ☿ St D
03 45 ☽ ⊥ ⯓
04 01 ☽ ✱ ♅
04 47 ☽ Q ☿
08 23 ☽ ∥ ♂
10 13 ☽ ✱ ♀
15 30 ☽ ✱ ♄
18 55 ☽ ✱ ☉
21 37 ☽ ∠ ⯓

24 Sunday
09 09 ☽ △ ♂
09 32 ☽ ∠ ♅
12 28 ☽ Q ♀
13 43 ☽ ☌ ♆
15 12 ☽ Q ♄
15 23 ☽ # ♅
16 31 ☽ □ ☿
20 25 ☽ Q ☉
21 20 ☽ △ ♃
21 34 ☽ ∥ ♆

25 Monday
00 57 ♀ Q ♆
02 48 ☽ ✱ ⯓
02 54 ☽ ⊥ ♅
14 18 ☽ ⊻ ♅
15 26 ☽ ♇ ♂
17 55 ♀ ☌ ♄
22 15 ♀ ∠ ♃

26 Tuesday
01 32 ☽ Q ⯓
01 44 ☽ □ ♄
02 09 ☽ ♇ ♃
02 30 ☽ □ ♀
09 05 ☽ ± ♂
09 20 ☽ □ ☉
18 57 ☽ ± ♃
20 37 ☽ ⊼ ♂
22 09 ☽ ⊻ ♆

27 Wednesday
02 26 ☽ △ ☿
05 56 ☽ ⊼ ♃
08 58 ☽ ⊥ ♆
10 25 ☽ □ ⯓
20 57 ☽ ☌ ♅

28 Thursday
00 52 ☽ ∠ ♆
05 59 ♂ △ ♆
06 10 ☽ ♇ ☿
08 03 ☽ △ ♄
12 45 ☽ # ♆
13 37 ☽ △ ♀
18 22 ☽ ∥ ♅
18 47 ☽ △ ☉
22 22 ☽ ± ☿

29 Friday
02 35 ☽ ✱ ♆
03 29 ☽ ☍ ♂
09 00 ☽ ⊼ ☿
09 41 ☽ ♇ ♄
10 22 ☽ ☍ ♃
14 01 ☽ △ ⯓
17 15 ☽ ♇ ♀
21 43 ☽ ♇ ☉
22 31 ☽ Q ♆
23 41 ☽ ⊻ ♅

30 Saturday
00 38 ☽ ± ♄
06 07 ☽ # ♂
09 19 ☽ ± ♀
09 27 ☽ ⊥ ♅
10 28 ☽ ⊼ ♄
12 46 ☽ # ♃
13 25 ☽ ± ☉
14 32 ☽ ♇ ⯓
15 09 ☿ ⊻ ♃
19 52 ☽ ⊼ ♀
23 45 ☽ ⊼ ☉
23 54 ☽ ∠ ♅

31 Sunday
03 35 ☽ □ ♆
04 58 ☽ ± ⯓
06 29 ☽ ⊼ ♂
11 27 ☽ ⊼ ♃
12 29 ☽ ☍ ☿
14 29 ☽ ⊼ ⯓
16 23 ☽ ± ♂
21 02 ☽ ± ♃
23 38 ☽ ✱ ♅

All ephemeris data is given at 12.00 UT and the Moon's longitude is additionally given for 24.00 UT

JANUARY 2018

Date	Sidereal time h m s	Sun ☉ ° ' "	Moon ☽ ° ' "	Moon ☽ 24.00 ° ' "	Mercury ☿ ° '	Venus ♀ ° '	Mars ♂ ° '	Jupiter ♃ ° '	Saturn ♄ ° '	Uranus ♅ ° '	Neptune ♆ ° '	Pluto ⯓ ° '
01	18 44 22	11 ♑ 01 03	02 ♋ 26 35	10 ♋ 05 52	18 ♐ 26	09 ♑ 09	14 ♏ 29	17 ♏ 01	01 ♑ 27	24 ♈ 34	11 ♓ 55	18 ♑ 48
02	18 48 19	12 02 11	17 ♋ 44 27	25 ♋ 21 00	19 27	10 25	15 06	17 11	01 34	24 R 34	11 56	18 50
03	18 52 15	13 03 18	02 ♌ 54 18	10 ♌ 23 15	20 32	11 40	15 44	17 21	01 41	24 D 34	11 58	18 52
04	18 56 12	14 04 26	17 ♌ 46 56	25 ♌ 04 37	21 38	12 56	16 21	17 31	01 48	24 34	11 59	18 54
05	19 00 08	15 05 34	02 ♍ 15 49	09 ♍ 20 13	22 48	14 11	16 58	17 41	01 55	24 34	12 00	18 56
06	19 04 05	16 06 43	16 ♍ 17 39	23 ♍ 08 11	23 59	15 27	17 36	17 50	02 02	24 35	12 02	18 58
07	19 08 01	17 07 51	29 ♍ 51 57	06 ♎ 29 15	25 13	16 42	18 13	18 00	02 09	24 35	12 03	19 00
08	19 11 58	18 08 59	13 ♎ 00 27	19 ♎ 25 58	26 28	17 58	18 50	18 09	02 16	24 35	12 05	19 02
09	19 15 55	19 10 08	25 ♎ 46 18	02 ♏ 01 57	27 45	19 13	19 28	18 19	02 23	24 35	12 06	19 04
10	19 19 51	20 11 17	08 ♏ 13 25	14 ♏ 21 12	29 ♐ 03	20 29	20 05	18 28	02 29	24 36	12 08	19 06
11	19 23 48	21 12 25	20 ♏ 25 51	26 ♏ 27 48	00 ♑ 23	21 44	20 42	18 37	02 36	24 36	12 09	19 08
12	19 27 44	22 13 34	02 ♐ 27 32	08 ♐ 25 29	01 44	23 00	21 19	18 46	02 43	24 37	12 11	19 10
13	19 31 41	23 14 42	14 ♐ 22 02	20 ♐ 17 33	03 06	24 15	21 57	18 55	02 50	24 37	12 13	19 12
14	19 35 37	24 15 51	26 ♐ 12 24	02 ♑ 06 53	04 28	25 30	22 34	19 03	02 57	24 38	12 14	19 14
15	19 39 34	25 16 59	08 ♑ 01 17	13 ♑ 55 52	05 52	26 46	23 11	19 12	03 04	24 39	12 16	19 16
16	19 43 30	26 18 06	19 ♑ 50 54	25 ♑ 46 36	07 17	28 01	23 48	19 20	03 10	24 39	12 18	19 18
17	19 47 27	27 19 14	01 ♒ 43 13	07 ♒ 40 59	08 43	29 ♑ 17	24 25	19 29	03 17	24 40	12 19	19 20
18	19 51 24	28 20 20	13 ♒ 40 09	19 ♒ 40 56	10 09	00 ♒ 32	25 03	19 37	03 24	24 41	12 21	19 22
19	19 55 20	29 ♑ 21 26	25 ♒ 43 37	01 ♓ 48 29	11 36	01 48	25 40	19 45	03 30	24 42	12 23	19 24
20	19 59 17	00 ♒ 22 31	07 ♓ 55 49	14 ♓ 05 57	13 04	03 03	26 17	19 53	03 37	24 43	12 25	19 26
21	20 03 13	01 23 36	20 ♓ 19 12	26 ♓ 35 56	14 32	04 19	26 54	20 01	03 43	24 43	12 27	19 29
22	20 07 10	02 24 39	02 ♈ 56 30	09 ♈ 21 17	16 01	05 34	27 31	20 09	03 50	24 44	12 28	19 31
23	20 11 06	03 25 42	15 ♈ 50 38	22 ♈ 24 54	17 31	06 49	28 08	20 16	03 56	24 46	12 30	19 33
24	20 15 03	04 26 43	29 ♈ 03 24	05 ♉ 49 24	19 02	08 05	28 45	20 24	04 03	24 47	12 32	19 34
25	20 18 59	05 27 44	12 ♉ 40 05	19 ♉ 36 34	20 33	09 20	29 22	20 31	04 09	24 48	12 34	19 36
26	20 22 56	06 28 43	26 ♉ 38 51	03 ♊ 46 50	22 05	10 36	29 ♏ 59	20 38	04 16	24 49	12 36	19 38
27	20 26 53	07 29 42	11 ♊ 00 14	18 ♊ 18 39	23 37	11 51	00 ♐ 35	20 45	04 22	24 50	12 38	19 40
28	20 30 49	08 30 39	25 ♊ 41 30	03 ♋ 08 04	25 10	13 06	01 12	20 52	04 28	24 52	12 40	19 42
29	20 34 46	09 31 35	10 ♋ 37 30	18 ♋ 08 47	26 44	14 22	01 49	20 59	04 35	24 53	12 42	19 44
30	20 38 42	10 32 30	25 ♋ 40 52	03 ♌ 12 39	28 18	15 37	02 26	21 05	04 41	24 54	12 44	19 46
31	20 42 39	11 ♒ 33 24	10 ♌ 42 59	18 ♌ 10 46	29 ♑ 53	16 ♒ 52	03 ♐ 03	21 ♏ 12	04 ♑ 47	24 ♈ 56	12 ♓ 46	19 ♑ 48

Date	Moon True ☊ ° '	Moon Mean ☊ ° '	Moon ☽ Latitude ° '
01	15 ♌ 19	16 ♌ 52	03 S 33
02	15 R 15	16 49	02 25
03	15 13	16 46	01 S 07
04	15 D 12	16 43	00 N 14
05	15 13	16 40	01 32
06	15 15	16 37	02 42
07	15 16	16 33	03 41
08	15 17	16 30	04 26
09	15 R 17	16 27	04 56
10	15 16	16 24	05 12
11	15 13	16 21	05 13
12	15 10	16 17	05 00
13	15 06	16 14	04 35
14	15 02	16 11	03 57
15	14 59	16 08	03 10
16	14 56	16 05	02 14
17	14 55	16 02	01 12
18	14 54	15 58	00 N 07
19	14 D 54	15 55	01 S 00
20	14 55	15 52	02 04
21	14 57	15 49	03 04
22	14 58	15 46	03 56
23	14 59	15 43	04 38
24	14 59	15 39	05 05
25	14 R 59	15 36	05 17
26	14 59	15 33	05 11
27	14 58	15 30	04 45
28	14 57	15 27	04 01
29	14 57	15 23	02 59
30	14 56	15 20	01 45
31	14 ♌ 56	15 ♌ 17	00 S 23

DECLINATIONS

Date	Sun ☉ ° '	Moon ☽ ° '	Mercury ☿ ° '	Venus ♀ ° '	Mars ♂ ° '	Jupiter ♃ ° '	Saturn ♄ ° '	Uranus ♅ ° '	Neptune ♆ ° '	Pluto ⯓ ° '
01	22 S 59	19 N 52	21 S 00	23 S 36	15 S 19	15 S 54	22 S 32	08 N 59	07 S 56	21 S 40
02	22 53	19 52	21 14	23 33	15 31	15 57	22 32	08 59	07 56	21 40
03	22 48	18 25	21 27	23 29	15 42	16 00	22 32	08 59	07 55	21 40
04	22 41	15 43	21 40	23 24	15 53	16 03	22 32	08 59	07 55	21 40
05	22 35	12 06	21 53	23 19	16 04	16 05	22 32	09 00	07 54	21 39
06	22 28	07 54	22 05	23 13	16 15	16 08	22 32	09 00	07 53	21 39
07	22 20	03 N 26	22 17	23 06	16 26	16 10	22 32	09 00	07 53	21 39
08	22 12	01 S 03	22 28	22 58	16 36	16 13	22 32	09 00	07 52	21 39
09	22 04	05 21	22 39	22 50	16 47	16 15	22 31	09 00	07 52	21 39
10	21 55	09 19	22 48	22 40	16 57	16 18	22 31	09 00	07 51	21 38
11	21 46	12 49	22 57	22 31	17 08	16 20	22 31	09 00	07 50	21 38
12	21 36	15 44	23 04	22 20	17 18	16 22	22 31	09 01	07 50	21 38
13	21 26	17 58	23 11	22 09	17 28	16 25	22 31	09 01	07 49	21 38
14	21 15	19 26	23 17	21 57	17 38	16 27	22 31	09 01	07 48	21 38
15	21 05	20 02	23 21	21 45	17 48	16 29	22 31	09 01	07 48	21 37
16	20 53	19 45	23 25	21 32	17 57	16 31	22 31	09 02	07 47	21 37
17	20 42	18 36	23 27	21 18	18 07	16 34	22 31	09 02	07 47	21 37
18	20 29	16 37	23 28	21 04	18 17	16 36	22 30	09 02	07 46	21 37
19	20 17	13 53	23 28	20 49	18 26	16 38	22 30	09 03	07 45	21 37
20	20 04	10 31	23 27	20 33	18 35	16 40	22 30	09 03	07 44	21 36
21	19 51	06 39	23 25	20 17	18 44	16 42	22 30	09 03	07 44	21 36
22	19 37	02 S 27	23 21	20 00	18 53	16 44	22 30	09 04	07 43	21 36
23	19 23	01 N 58	23 16	19 42	19 02	16 45	22 30	09 04	07 42	21 36
24	19 09	06 23	23 10	19 24	19 11	16 47	22 30	09 05	07 42	21 36
25	18 54	10 36	23 02	19 06	19 19	16 49	22 29	09 05	07 41	21 36
26	18 39	14 21	22 53	18 47	19 28	16 51	22 29	09 06	07 40	21 35
27	18 24	17 23	22 43	18 27	19 36	16 53	22 29	09 06	07 39	21 35
28	18 08	19 22	22 32	18 07	19 44	16 54	22 29	09 07	07 39	21 35
29	17 52	20 02	22 19	17 46	19 52	16 56	22 29	09 07	07 38	21 35
30	17 36	19 17	22 04	17 25	20 00	16 58	22 28	09 08	07 37	21 35
31	17 S 19	17 N 10	21 S 49	17 S 03	20 S 08	16 S 59	22 S 28	09 N 08	07 S 36	21 S 34

ZODIAC SIGN ENTRIES

Date	h	m	Planets
01	08	10	☽ ♋
03	07	22	☽ ♌
05	08	12	☽ ♍
07	12	14	☽ ♎
09	20	05	☽ ♏
11	05	09	☿ ♑
12	07	04	☽ ♐
14	19	42	☽ ♑
17	08	32	☽ ♒
18	01	44	♀ ♒
19	20	26	☽ ♓
20	03	09	☉ ♒
22	06	27	☽ ♈
24	13	39	☽ ♉
26	12	56	♂ ♐
26	17	40	☽ ♊
28	18	57	☽ ♋
30	18	53	☽ ♌
31	13	39	☿ ♒

LATITUDES

Date	Mercury ☿ ° '	Venus ♀ ° '	Mars ♂ ° '	Jupiter ♃ ° '	Saturn ♄ ° '	Uranus ♅ ° '	Neptune ♆ ° '	Pluto ⯓ ° '
01	01 N 56	00 S 29	00 N 54	01 N 03	00 N 54	00 S 34	00 S 55	00 N 27
04	01 30	00 36	00 53	01 03	00 54	00 34	00 55	00 27
07	01 04	00 42	00 51	01 04	00 53	00 34	00 55	00 27
10	00 38	00 49	00 50	01 04	00 53	00 34	00 55	00 26
13	00 N 13	00 54	00 49	01 04	00 53	00 34	00 55	00 26
16	00 S 11	01 00	00 47	01 05	00 53	00 34	00 55	00 26
19	00 33	01 05	00 46	01 05	00 53	00 34	00 54	00 25
22	00 53	01 09	00 44	01 05	00 53	00 33	00 54	00 25
25	01 11	01 13	00 43	01 06	00 53	00 33	00 54	00 25
28	01 27	01 17	00 41	01 06	00 53	00 33	00 54	00 25
31	01 S 41	01 S 20	00 N 39	01 N 07	00 N 53	00 S 33	00 S 54	00 N 24

DATA

Julian Date	2458120
Delta T	+70 seconds
Ayanamsa	24° 06' 19"
Synetic vernal point	05° ♓ 00' 40"
True obliquity of ecliptic	23° 26' 06"

MOON'S PHASES, APSIDES AND POSITIONS ☽

Date	h	m	Phase	Longitude ° '	Eclipse Indicator
02	02	24	○	11 ♋ 38	
08	22	25	☾	18 ♎ 36	
17	02	17	●	26 ♑ 54	
24	22	20	☽	04 ♉ 53	
31	13	27	○	11 ♌ 37	total

Day	h	m		
01	21	55	Perigee	
15	02	26	Apogee	
30	10	03	Perigee	
01	23	58	Max dec	20° N 03'
08	06	20	OS	
15	16	30	Max dec	20° S 03'
23	01	23	ON	
29	11	32	Max dec	20° N 02'

LONGITUDES

Date	Chiron ⚷ ° '	Ceres ⚳ ° '	Pallas ⚴ ° '	Juno ⚵ ° '	Vesta ⚶ ° '	Black Moon Lilith ⚸ ° '
01	24 ♓ 39	17 ♌ 20	26 ♈ 06	06 ♒ 23	23 ♏ 23	05 ♑ 57
11	24 ♓ 56	15 ♌ 53	27 ♈ 52	10 ♒ 37	28 ♏ 14	07 ♑ 03
21	25 ♓ 18	13 ♌ 54	00 ♉ 24	14 ♒ 57	02 ♐ 58	08 ♑ 10
31	25 ♓ 44	11 ♌ 36	03 ♉ 33	19 ♒ 23	07 ♐ 34	09 ♑ 17

All ephemeris data is given at 12.00 UT and the Moon's longitude is additionally given for 24.00 UT

ASPECTARIAN

h m	Aspects
01 Monday	
07 09	☽ ♇ ♂
10 26	☽ ☍ ♄
11 20	☽ ♇ ♃
18 28	☽ Q ♅
20 51	☿ ⊻ ⯓
23 28	☽ ☍ ♀
02 Tuesday	
02 24	☽ ☍ ☉
02 52	☽ △ ♆
07 41	☽ △ ♂
09 37	☉ ⚹ ♆
11 07	☽ △ ♃
13 43	☽ ☍ ⯓
14 11	♅ St D
14 54	☽ ⊼ ☿
22 46	☽ □ ♅
03 Wednesday	
01 05	☽ ± ☿
02 32	☽ ♇ ♆
10 02	☽ ⊼ ♄
16 31	☽ ♇ ☿
16 53	☽ ± ♆
17 38	♀ ⚹ ♆
19 42	☽ ± ♄
04 Thursday	
02 34	☽ ⊼ ♆
03 22	☽ ⊼ ♀
05 31	☽ ⊼ ☉
09 34	☽ □ ♂
09 38	☽ ⋕ ♃
10 23	☽ ♇ ♄
10 37	☿ ∥ ⯓
10 52	☽ ⋕ ♂
11 34	☽ □ ♃
13 50	☽ ⊼ ⯓
14 03	☽ ± ♀
16 02	☽ ± ☉
18 52	☽ △ ☿
23 10	☽ △ ♅
23 44	☽ ± ⯓
05 Friday	
06 21	☽ ♇ ♀
08 05	☽ ♇ ☉
09 19	♂ ∠ ♄
11 25	☽ △ ♄
14 49	☽ ♇ ⯓
15 06	♂ ∥ ♃
16 47	☽ Q ♂
17 50	☽ Q ♃
19 34	☿ ⊥ ♂
22 58	☉ ∥ ♄
06 Saturday	
00 24	☽ ♇ ♅
04 37	☽ ☍ ♆
05 58	☽ ∥ ♅
08 34	☿ ⊥ ♃
10 23	☽ △ ♀
11 39	☽ △ ☉
12 04	☽ ⋕ ♆
14 22	☽ ⚹ ♂
14 43	☽ ⚹ ♃
15 59	☽ ± ♅
16 40	☽ △ ⯓
23 39	☿ △ ♅
07 Sunday	
00 39	♂ ☌ ♃
02 33	☽ ⊼ ♅
02 51	☽ □ ☿
15 49	☉ ∥ ☿
16 09	☽ □ ♄
17 43	☽ ∠ ♃
18 21	☽ ∠ ♂
08 Monday	
10 17	☽ ⊼ ♆
10 24	☽ ⊥ ♃
11 40	☽ ⊥ ♂
12 07	☉ ⚹ ♃
15 01	☽ Q ☿
16 13	♀ ⚹ ♃
19 18	☿ ∥ ♄
19 59	♂ ⚹ ⯓
21 29	☽ ± ♆
21 43	☽ ⊻ ♃
22 15	☽ □ ♀
22 25	☽ □ ☉
23 17	☽ □ ⯓
23 26	☽ ⊻ ♂
09 Tuesday	
01 41	☽ Q ♄
07 02	☉ ☌ ♀
09 03	♀ ☌ ⯓
09 33	☉ ☌ ⯓
09 45	☽ ☍ ♅
14 33	☽ ♇ ♆
16 13	☽ ⚹ ☿
21 08	♀ ⚹ ♂
10 Wednesday	
00 47	☽ ⚹ ♄
02 09	☿ ∥ ♀
02 52	☽ ∥ ♆
05 36	☉ ⚹ ♂
09 49	☽ Q ⯓
10 01	☽ ⋕ ♅
11 55	☽ Q ☉
12 33	☽ Q ♀
19 39	☽ △ ♆
11 Thursday	
00 48	☽ ∠ ☿
06 21	☽ ∠ ♄
07 54	☿ Q ♆
08 21	☽ ☌ ♃
09 26	☽ ⚹ ⯓
10 40	♀ ∥ ♄
12 34	☽ ☌ ♂
13 41	☽ ⚹ ☉
14 53	☽ ⚹ ♀
20 18	☽ ⊼ ♅
20 50	☽ ⊥ ☿
12 Friday	
00 24	☽ ⊥ ♄
07 08	☉ ∥ ⯓
08 18	☽ ± ♅
10 21	☽ ⊻ ☿
12 32	☽ ⊻ ♄
15 27	☽ ∠ ⯓
18 11	☽ ∥ ♃
22 29	☽ ∠ ☉
13 Saturday	
00 27	☽ ∠ ♀
02 24	☽ ♇ ♅
05 16	☽ ∥ ♂
07 03	☿ ☌ ♄
07 38	☽ □ ♆
09 39	☽ ⊥ ⯓
18 22	☽ ⊥ ☉
19 09	♀ □ ♅
20 48	☽ ⊥ ♀
21 19	☽ ⊻ ♃
21 49	☽ ⊻ ⯓
14 Sunday	
03 55	☿ ∠ ♃
04 12	☽ ⊻ ♂
07 41	☽ ⊻ ☉
08 48	☽ △ ♅
09 38	☽ ⊥ ♃
10 25	☽ ⊻ ♀
17 03	☽ ⊥ ♂
20 13	☽ Q ♆
20 45	☉ □ ♅
15 Monday	
01 49	☽ ☌ ♄
04 08	☽ ∠ ♃
07 03	☽ ☌ ☿
12 21	☽ ∠ ♂
20 39	☽ ⚹ ♆
21 47	♀ ∠ ♆
16 Tuesday	
01 50	♀ ∥ ⯓
04 13	♃ ⚹ ⯓
10 54	☽ ☌ ⯓
10 58	☽ ⚹ ♃
20 27	☽ ⚹ ♂
21 44	☽ □ ♅
17 Wednesday	
02 17	☽ ☌ ☉
03 06	☽ ∠ ♆
06 30	☽ ☌ ♀
11 31	☽ Q ♃
12 05	☉ ∠ ♆
15 11	☽ ⊻ ♄
18 14	☽ ∥ ♂
21 17	☽ ⊥ ♆
21 37	♂ ⊼ ♅
21 59	☽ Q ♂
18 Thursday	
03 21	☽ ⊥ ♄
03 59	☽ ⊻ ☿
08 55	☿ ∠ ♂
09 22	☽ ⊻ ♆
10 01	☽ Q ♅
12 10	☽ ∥ ♃
17 38	☽ ⊥ ☿
21 31	☽ ∠ ♄
23 25	☽ ⊻ ⯓
19 Friday	
00 00	☽ □ ♃
09 57	☽ ⚹ ♅
11 05	♀ Q ♃
11 22	☽ ⊥ ⯓
11 52	☽ □ ♂
13 58	☽ ∠ ☿
19 50	☽ ⊻ ☉
20 Saturday	
01 08	☿ ⚹ ♆
01 21	☽ ⊻ ♀
03 28	☽ ⚹ ♄
05 09	☽ ∠ ⯓
08 41	☽ ⊥ ☉
14 26	☽ ⊥ ♀
15 28	☽ ∠ ♅
20 45	☽ ☌ ♆
21 25	☽ ⋕ ♅
23 20	☽ ⚹ ☿
23 45	♀ ⊻ ♄
21 Sunday	
03 04	☽ Q ♄
03 46	☽ ∠ ☉
05 33	☽ ∥ ♆
08 56	☽ ⊥ ♅
09 50	☽ ∠ ♀
10 22	☽ ⚹ ⯓
11 25	☽ △ ♃
20 26	☽ ⊻ ♅
22 Monday	
01 13	☽ △ ♂
01 28	☽ Q ☿
04 50	☉ Q ♃
09 18	☽ Q ⯓
10 55	☽ ⚹ ☉
13 41	☽ □ ♄
16 11	☽ ♇ ♃
17 28	☽ ⚹ ♀
23 Tuesday	
03 06	♂ ⊥ ♄
05 45	♀ ⊥ ♆
05 50	☽ ⊻ ♆
06 45	☽ ♇ ♂
09 05	☽ ± ♃
11 10	☽ Q ☉
15 29	☽ □ ☿
16 53	☽ ⊥ ♆
18 02	☽ Q ♀
18 47	☽ □ ⯓
20 11	☽ ⊼ ♃
24 Wednesday	
00 03	☽ ± ♂
01 31	☉ ⊻ ♄
04 16	☽ ☌ ♅
09 14	☽ ∠ ♆
10 09	☉ ∥ ♂
11 23	☽ ⊼ ♂
19 17	☽ ⋕ ♆
20 50	☿ ☌ ⯓
20 56	☽ △ ♄
22 20	☽ □ ☉
25 Thursday	
00 14	♀ ∥ ♂
03 10	☽ ∥ ♅
05 35	☽ □ ♀
11 28	☿ ⚹ ♃
11 49	☽ ⚹ ♆
23 18	☽ ♇ ♄
26 Friday	
00 02	☽ △ ⯓
01 40	☽ ☍ ♃
03 17	☽ △ ☿
05 06	♀ ⊥ ♄
08 31	☽ Q ♆
08 53	☽ ⊻ ♅
14 45	☽ ± ♄
14 55	☉ ⊥ ♆
17 52	☽ ☍ ♂
19 03	☽ ⊥ ♅
27 Saturday	
00 54	☽ ⊼ ♄
01 28	☽ ♇ ⯓
05 45	☽ △ ☉
07 25	☽ ⋕ ♃
07 35	☽ ♇ ☿
10 04	☽ ∠ ♅
13 31	☽ △ ♀
14 41	☽ □ ♆
16 25	☽ ± ⯓
21 25	☽ ⋕ ☉
21 41	☽ ⋕ ♀
22 00	♂ ± ♅
28 Sunday	
00 09	☽ ± ☿
02 15	☽ ⊼ ⯓
03 22	♀ ⊻ ♆
04 07	☽ ⊼ ♃
06 03	☉ ∥ ♀
07 08	☿ □ ♅
07 15	♀ Q ♅
08 12	☽ ♇ ☉
10 39	☽ ⚹ ♅
11 03	☽ ⊼ ☿
13 55	☽ ± ♃
15 48	♀ Q ♂
16 15	☽ ♇ ♀
17 23	☿ ∥ ♄
21 17	☽ ⊼ ♂
21 30	☽ ⋕ ♂
23 48	☽ ± ☉
29 Monday	
02 15	☽ ☍ ♄
04 31	☽ ♇ ♃
06 00	☽ Q ♅
07 19	☽ ± ♂
08 03	☽ ± ♀
10 07	☽ ⊼ ☉
15 19	☽ △ ♆
18 30	☽ ⊼ ♀
21 03	☽ ⋕ ♂
22 18	☽ ♇ ♂
30 Tuesday	
02 34	☽ ☍ ⯓
03 02	☿ ∠ ♆
04 38	☽ △ ♃
10 46	☽ □ ♅
15 16	☽ ♇ ♆
15 38	☉ ⊥ ♄
16 40	☽ ☍ ☿
23 13	☽ △ ♂
31 Wednesday	
02 27	☽ ⊼ ♄
05 39	☽ ± ♆
10 33	☽ ⋕ ☉
12 06	☽ ± ♄
13 09	☽ ⋕ ♀
13 27	☽ ☍ ☉
13 35	☽ ⋕ ♃
15 17	☽ ⊼ ♆
16 13	♀ ∥ ♃
22 48	☽ ☍ ♀

FEBRUARY 2018

Date	Sidereal time h m s	Sun ☉ ° ' "	Moon ☽ ° ' "	Moon ☽ 24.00 ° ' "	Mercury ☿ ° '	Venus ♀ ° '	Mars ♂ ° '	Jupiter ♃ ° '	Saturn ♄ ° '	Uranus ♅ ° '	Neptune ♆ ° '	Pluto ♇ ° '
01	20 46 35	12 ♒ 34 17	25 ♌ 35 01	02 ♍ 54 47	01 ♒ 29	18 ♒ 07	03 ♐ 39	21 ♏ 18	04 ♑ 53	24 ♈ 57	12 ♓ 48	19 ♑ 50
02	20 50 32	13 35 08	10 ♍ 09 20	17 ♍ 18 03	03 06	19 23	04 16	21 24	04 59	24 59	12 50	19 52
03	20 54 28	14 35 59	24 ♍ 20 27	01 ♎ 16 16	04 43	20 38	04 53	21 30	05 05	25 00	12 52	19 54
04	20 58 25	15 36 49	08 ♎ 05 23	14 ♎ 47 49	06 21	21 53	05 30	21 36	05 11	25 02	12 54	19 56
05	21 02 22	16 37 38	21 ♎ 23 44	27 ♎ 53 23	07 59	23 08	06 06	21 42	05 17	25 04	12 56	19 58
06	21 06 18	17 38 26	04 ♏ 17 08	10 ♏ 35 26	09 39	24 24	06 43	21 47	05 23	25 05	12 58	20 00
07	21 10 15	18 39 13	16 ♏ 49 47	22 ♏ 57 42	11 19	25 39	07 19	21 53	05 29	25 07	13 00	20 01
08	21 14 11	19 39 59	29 ♏ 02 47	05 ♐ 04 36	13 00	26 54	07 56	21 58	05 34	25 09	13 02	20 03
09	21 18 08	20 40 45	11 ♐ 03 43	17 ♐ 00 44	14 42	28 09	08 32	22 03	05 40	25 11	13 04	20 05
10	21 22 04	21 41 29	22 ♐ 56 13	28 ♐ 50 41	16 24	29 ♒ 25	09 09	22 08	05 46	25 13	13 07	20 07
11	21 26 01	22 42 12	04 ♑ 44 40	10 ♑ 38 39	18 07	00 ♓ 40	09 45	22 13	05 51	25 15	13 09	20 09
12	21 29 57	23 42 54	16 ♑ 33 05	22 ♑ 28 24	19 52	01 55	10 22	22 17	05 57	25 17	13 11	20 10
13	21 33 54	24 43 35	28 ♑ 24 58	04 ♒ 23 07	21 37	03 10	10 58	22 22	06 02	25 19	13 13	20 12
14	21 37 51	25 44 15	10 ♒ 23 10	16 ♒ 25 23	23 22	04 25	11 34	22 26	06 08	25 21	13 15	20 14
15	21 41 47	26 44 53	22 ♒ 30 01	28 ♒ 37 14	25 09	05 40	12 10	22 30	06 13	25 23	13 17	20 16
16	21 45 44	27 45 29	04 ♓ 47 14	11 ♓ 00 09	26 57	06 55	12 47	22 34	06 18	25 25	13 20	20 17
17	21 49 40	28 46 04	17 ♓ 16 07	23 ♓ 35 13	28 ♒ 45	08 10	13 23	22 37	06 24	25 27	13 22	20 19
18	21 53 37	29 ♒ 46 38	29 ♓ 57 34	06 ♈ 23 14	00 ♓ 34	09 25	13 59	22 41	06 29	25 29	13 24	20 21
19	21 57 33	00 ♓ 47 10	12 ♈ 52 17	19 ♈ 24 49	02 24	10 41	14 35	22 44	06 34	25 32	13 26	20 22
20	22 01 30	01 47 40	26 ♈ 00 51	02 ♉ 40 29	04 15	11 56	15 11	22 47	06 39	25 34	13 29	20 24
21	22 05 26	02 48 08	09 ♉ 23 45	16 ♉ 10 42	06 07	13 11	15 47	22 50	06 44	25 36	13 31	20 26
22	22 09 23	03 48 34	23 ♉ 01 21	29 ♉ 55 42	07 59	14 26	16 23	22 53	06 49	25 39	13 33	20 27
23	22 13 20	04 48 59	06 ♊ 53 44	13 ♊ 55 22	09 51	15 40	16 59	22 56	06 54	25 41	13 35	20 29
24	22 17 16	05 49 22	21 ♊ 00 28	28 ♊ 08 51	11 44	16 55	17 35	22 58	06 58	25 44	13 38	20 30
25	22 21 13	06 49 42	05 ♋ 20 13	12 ♋ 34 15	13 38	18 10	18 10	23 01	07 03	25 46	13 40	20 32
26	22 25 09	07 50 01	19 ♋ 50 28	27 ♋ 08 22	15 32	19 25	18 46	23 03	07 08	25 49	13 42	20 33
27	22 29 06	08 50 18	04 ♌ 27 20	11 ♌ 46 39	17 25	20 40	19 22	23 05	07 12	25 51	13 44	20 35
28	22 33 02	09 ♓ 50 33	19 ♌ 05 36	26 ♌ 23 23	19 ♓ 19	21 ♓ 55	19 ♐ 57	23 ♏ 06	07 ♑ 17	25 ♈ 54	13 ♓ 47	20 ♑ 36

Date	Moon True ☊ ° '	Moon Mean ☊ ° '	Moon ☽ Latitude ° '
01	14 ♌ 56	15 ♌ 14	00 N 59
02	14 D 56	15 11	02 16
03	14 56	15 08	03 22
04	14 R 56	15 04	04 15
05	14 56	15 01	04 52
06	14 56	14 58	05 12
07	14 56	14 55	05 18
08	14 D 56	14 52	05 08
09	14 56	14 49	04 46
10	14 57	14 45	04 11
11	14 58	14 42	03 26
12	14 59	14 39	02 32
13	14 59	14 36	01 31
14	15 00	14 33	00 N 26
15	15 R 00	14 29	00 S 42
16	14 59	14 26	01 48
17	14 58	14 23	02 49
18	14 56	14 20	03 44
19	14 53	14 17	04 28
20	14 51	14 14	04 59
21	14 49	14 10	05 14
22	14 48	14 07	05 13
23	14 D 47	14 04	04 53
24	14 48	14 01	04 15
25	14 49	13 58	03 21
26	14 51	13 54	02 14
27	14 52	13 51	00 57
28	14 ♌ 52	13 ♌ 48	00 N 23

DECLINATIONS

Date	Sun ☉ ° '	Moon ☽ ° '	Mercury ☿ ° '	Venus ♀ ° '	Mars ♂ ° '	Jupiter ♃ ° '	Saturn ♄ ° '	Uranus ♅ ° '	Neptune ♆ ° '	Pluto ♇ ° '
01	17 S 02	13 N 55	21 S 32	16 S 41	20 S 15	17 S 01	22 S 28	09 N 09	07 S 36	21 S 34
02	16 45	09 51	21 13	16 18	20 23	17 02	22 28	09 09	07 35	21 34
03	16 27	05 20	20 53	15 55	20 30	17 03	22 28	09 10	07 34	21 34
04	16 09	00 N 42	20 32	15 32	20 37	17 05	22 27	09 11	07 33	21 34
05	15 51	03 S 50	20 09	15 08	20 44	17 06	22 27	09 11	07 32	21 33
06	15 33	08 02	19 45	14 43	20 51	17 07	22 27	09 12	07 32	21 33
07	15 14	11 46	19 20	14 19	20 58	17 09	22 27	09 13	07 31	21 33
08	14 55	14 55	18 53	13 53	21 04	17 10	22 26	09 13	07 30	21 33
09	14 36	17 23	18 24	13 28	21 11	17 11	22 26	09 14	07 29	21 33
10	14 16	19 04	17 54	13 02	21 17	17 12	22 26	09 14	07 28	21 33
11	13 57	19 56	17 23	12 36	21 23	17 13	22 26	09 15	07 27	21 32
12	13 37	19 54	16 50	12 09	21 29	17 14	22 26	09 16	07 27	21 32
13	13 17	19 00	16 16	11 42	21 35	17 15	22 25	09 17	07 26	21 32
14	12 56	17 13	15 41	11 15	21 41	17 16	22 25	09 17	07 25	21 32
15	12 36	14 40	15 04	10 47	21 47	17 17	22 25	09 19	07 24	21 32
16	12 15	11 25	14 25	10 20	21 52	17 18	22 25	09 19	07 23	21 31
17	11 54	07 38	13 46	09 52	21 57	17 19	22 24	09 20	07 22	21 31
18	11 33	03 S 26	13 04	09 23	22 02	17 19	22 24	09 21	07 22	21 31
19	11 12	00 N 58	12 22	08 55	22 07	17 20	22 24	09 22	07 21	21 31
20	10 50	05 24	11 38	08 26	22 12	17 21	22 24	09 23	07 20	21 31
21	10 28	09 39	10 53	07 57	22 17	17 21	22 23	09 24	07 19	21 31
22	10 07	13 29	10 06	07 27	22 22	17 22	22 23	09 24	07 18	21 30
23	09 45	16 39	09 19	06 58	22 26	17 22	22 23	09 25	07 17	21 30
24	09 22	18 53	08 30	06 28	22 30	17 23	22 23	09 26	07 16	21 30
25	09 00	19 59	07 40	05 59	22 34	17 23	22 22	09 27	07 16	21 30
26	08 38	19 46	06 49	05 29	22 38	17 24	22 22	09 28	07 15	21 30
27	08 15	18 13	05 58	04 58	22 42	17 24	22 22	09 29	07 14	21 30
28	07 S 53	15 N 28	05 S 05	04 S 28	22 S 46	17 S 24	22 S 22	09 N 30	07 S 13	21 S 30

ZODIAC SIGN ENTRIES

Date	h	m	Planets
01	19	13	☽ ♍
03	21	47	☽ ♎
06	03	56	☽ ♏
08	13	53	☽ ♐
10	23	19	♀ ♓
11	02	21	☽ ♑
13	15	11	☽ ♒
16	02	42	☽ ♓
18	04	28	☿ ♓
18	12	05	☽ ♈
18	17	18	☉ ♓
20	19	12	☽ ♉
23	00	07	☽ ♊
25	03	06	☽ ♋
27	04	42	☽ ♌

LATITUDES

Date	Mercury ☿ ° '	Venus ♀ ° '	Mars ♂ ° '	Jupiter ♃ ° '	Saturn ♄ ° '	Uranus ♅ ° '	Neptune ♆ ° '	Pluto ♇ ° '
01	01 S 45	01 S 21	00 N 38	01 N 07	00 N 53	00 S 33	00 S 54	00 N 24
04	01 55	01 24	00 36	01 07	00 53	00 33	00 54	00 24
07	02 02	01 26	00 34	01 08	00 53	00 33	00 54	00 24
10	02 05	01 27	00 32	01 08	00 53	00 33	00 54	00 24
13	02 05	01 28	00 30	01 09	00 53	00 33	00 54	00 23
16	02 01	01 28	00 28	01 09	00 53	00 33	00 54	00 23
19	01 52	01 27	00 25	01 09	00 53	00 33	00 54	00 23
22	01 39	01 26	00 23	01 10	00 53	00 32	00 54	00 22
25	01 20	01 25	00 20	01 10	00 53	00 32	00 54	00 22
28	00 56	01 23	00 17	01 11	00 53	00 32	00 54	00 22
31	00 S 26	01 S 20	00 N 15	01 N 11	00 N 53	00 S 32	00 S 54	00 N 22

DATA

Julian Date	2458151
Delta T	+70 seconds
Ayanamsa	24° 06' 24"
Synetic vernal point	05° ♓ 00' 35"
True obliquity of ecliptic	23° 26' 06"

MOON'S PHASES, APSIDES AND POSITIONS ☽

Date	h	m	Phase	Longitude ° '	Eclipse Indicator
07	15	54	☾	18 ♏ 49	
15	21	05	●	27 ♒ 08	Partial
23	08	09	☽	04 ♊ 39	

Day	h	m		
11	14	26	Apogee	
27	14	45	Perigee	
04	15	36	0S	
11	23	22	Max dec	20° S 02'
19	06	48	0N	
25	20	11	Max dec	20° N 03'

LONGITUDES

Date	Chiron ⚷ ° '	Ceres ⚳ ° '	Pallas ⚴ ° '	Juno ⚵ ° '	Vesta ⚶ ° '	Black Moon Lilith ⚸ ° '
01	25 ♓ 47	11 ♌ 22	03 ♉ 54	19 ♒ 50	08 ♐ 01	09 ♑ 23
11	26 ♓ 17	09 ♌ 04	07 ♉ 39	24 ♒ 20	12 ♐ 25	10 ♑ 30
21	26 ♓ 49	07 ♌ 04	11 ♉ 52	28 ♒ 55	16 ♐ 36	11 ♑ 37
31	27 ♓ 23	05 ♌ 37	16 ♉ 27	03 ♓ 33	20 ♐ 31	12 ♑ 44

All ephemeris data is given at 12.00 UT and the Moon's longitude is additionally given for 24.00 UT

ASPECTARIAN

h m	Aspects
01 Thursday	
02 39	☽ ⊼ ♇
02 41	☽ P ♄
05 00	☽ □ ♃
08 28	☿ ∥ ♇
10 58	☽ △ ♅
12 25	☽ ± ♇
13 40	☉ ∥ ♃
17 30	☉ ⊻ ♆
21 18	☉ Q ♅
22 51	☽ ⊼ ☿
02 Friday	
01 48	☽ □ ♂
03 12	☽ P ♇
03 21	☽ △ ♄
10 01	☽ ± ☿
10 44	☽ Q ♃
11 42	☽ P ♅
15 49	☽ ∥ ♅
16 29	☽ ☍ ♆
16 57	☿ Q ♃
18 11	☽ ⊼ ☉
21 36	♀ ⊻ ♇
03 Saturday	
00 19	☽ # ♆
00 37	♀ ∠ ♄
02 52	☽ ± ♅
03 04	☽ P ☿
04 23	☽ △ ♇
05 02	☽ ⊼ ♀
05 06	☽ ± ☉
07 07	☽ ✶ ♃
09 23	☽ Q ♂
12 46	♂ ∠ ♇
13 09	☽ ⊼ ♅
15 57	☿ ✶ ♂
16 20	☽ ± ♀
17 51	☿ ⊻ ♄
21 36	♂ ⊻ ♄
21 49	☽ P ☉
04 Sunday	
06 07	♀ □ ♃
06 50	☽ □ ♄
07 11	☽ ✶ ♂
07 47	☿ ∥ ♂
08 29	☽ △ ☿
09 21	☽ ∠ ♃
09 39	☽ P ♀
20 17	☿ ⊥ ♆
20 36	☽ ⊼ ♆
05 Monday	
01 33	☽ ⊥ ♃
02 35	☽ △ ☉
07 29	☽ ± ♆
09 22	☽ □ ♇
11 26	☽ ∠ ♂
12 34	☽ ⊻ ♃
15 30	☽ Q ♄
15 33	☽ △ ♀
18 46	☽ ☍ ♅
06 Tuesday	
00 07	☽ P ♆
04 56	☽ ⊥ ♂
08 57	☽ ∥ ♆
14 05	☽ ✶ ♄
16 50	☽ ⊻ ♂
19 03	☽ Q ♇
19 07	☽ # ♅
23 45	☽ □ ☿
07 Wednesday	
01 37	♀ ✶ ♅
04 37	☽ △ ♆
14 29	☿ ⊥ ♄
15 54	☽ □ ☉
18 16	☽ ✶ ♇
19 12	☽ ∠ ♄
19 22	♀ ⊥ ♇
21 57	☽ ☌ ♃
08 Thursday	
04 17	☽ ⊼ ♅
04 36	☽ ∥ ♀
07 16	☽ □ ♀
11 57	☽ ∥ ☉
12 37	☿ ⊻ ♆
13 03	☽ ⊥ ♄
14 13	☿ Q ♅
16 11	☽ ± ♅
16 30	☽ Q ☿
21 29	☉ ⊻ ♇
23 59	☽ ∠ ♇
09 Friday	
01 06	☽ ⊻ ♄
03 40	☉ Q ♂
06 40	☽ ☌ ♂
06 46	☽ Q ☉
09 44	☽ ∥ ♃
10 13	☽ P ♅
11 45	☉ ∠ ♄
16 04	☽ □ ♆
18 06	☽ ⊥ ♇
20 32	☽ ✶ ☿
22 09	☽ ∥ ☿
23 29	☽ Q ♀
10 Saturday	
06 16	☽ ⊻ ♇
09 14	☽ ✶ ☉
10 21	☽ ⊻ ♃
16 38	☽ △ ♅
22 37	☽ ⊥ ♃
23 21	☉ □ ♃
11 Sunday	
02 42	☽ ✶ ♀
04 39	☽ Q ♆
08 08	☽ ∠ ☿
14 17	☽ ☌ ♄
17 03	☽ ∠ ♃
18 35	☽ ∠ ☉
19 11	☿ ∥ ♃
22 44	☽ ⊻ ♂
12 Monday	
05 08	☽ ✶ ♆
05 35	☽ ⊥ ☿
08 34	♂ P ♅
11 35	☽ ⊥ ♂
12 49	☽ ∠ ♀
14 35	☽ ⊥ ☉
16 24	☿ ⊻ ♇
19 22	☽ ☌ ♇
19 52	☽ ⊻ ☿
22 56	♂ ∥ ♇
23 42	☽ ✶ ♃
13 Tuesday	
03 47	☿ ∠ ♄
03 52	☽ ⊻ ☉
05 43	☽ □ ♅
06 48	☽ ∠ ♂
09 11	☽ ⊥ ♀
11 36	☽ ∠ ♆
22 39	☿ □ ♃
22 40	☽ ⊻ ♀
14 Wednesday	
00 01	☽ Q ♃
02 23	☉ ✶ ♅
03 26	☽ ⊻ ♄
05 44	☽ ⊥ ♆
11 31	☽ ∥ ♃
14 29	☽ ✶ ♂
15 30	☽ ⊥ ♄
16 01	☿ Q ♂
17 44	☽ ⊻ ♆
17 55	☽ Q ♅
15 Thursday	
00 07	☉ ⊥ ♇
03 58	♀ ∠ ♇
07 35	☽ ⊻ ♇
07 44	☽ ∥ ☿
09 28	☽ ∠ ♄
12 00	☽ □ ♃
15 07	☿ ✶ ♅
15 28	☽ Q ♂
17 41	☽ ✶ ♅
18 06	☽ ☌ ☿
19 24	☽ ⊥ ♇
21 05	☽ ☌ ☉
23 19	♀ ✶ ♄
16 Friday	
03 06	☿ ⊥ ♇
05 38	☽ ∥ ☉
12 59	☽ ∠ ♇
14 58	☽ ✶ ♄
16 36	☽ ☌ ♀
20 18	☽ ∥ ♀
22 54	☽ ∠ ♅
17 Saturday	
01 36	☽ # ♅
04 11	☽ □ ♂
04 31	☽ ☌ ♆
11 20	♂ □ ♆
12 27	☉ ☌ ☿
13 32	☽ ∥ ♆
14 10	☽ Q ♄
16 10	☽ ⊥ ♅
17 49	☽ ✶ ♇
22 13	☽ △ ♃
18 Sunday	
03 34	☽ ⊻ ♅
11 38	☽ ⊻ ☉
13 20	☽ ⊻ ☿
13 49	♀ # ♅
16 29	☽ Q ♇
23 47	☽ ⊥ ☉
19 Monday	
00 15	☽ □ ♄
02 23	☽ ⊥ ☿
02 28	☽ P ♃
03 13	♂ ⊥ ♇
07 31	☽ ⊻ ♀
09 06	♀ ∠ ♅
13 03	☽ ⊻ ♆
15 18	☽ △ ♂
17 48	☽ ∠ ☉
19 08	☽ ± ♃
19 44	☽ ⊥ ♀
21 41	☽ ∠ ☿
20 Tuesday	
00 05	☽ ⊥ ♆
01 47	☽ □ ♇
06 08	☽ ⊼ ♃
11 11	☽ ☌ ♅
13 49	☽ ∠ ♀
16 28	☽ ∠ ♆
19 53	☽ P ♂
22 41	☽ # ♆
23 16	☽ ✶ ☉
21 Wednesday	
03 05	☿ ∠ ♇
03 12	☽ # ♀
05 12	☽ ✶ ☿
07 14	☽ △ ♄
10 31	☽ ∥ ♅
12 43	☽ ± ♂
16 31	☽ # ☉
18 11	☽ # ☿
18 42	♀ ☌ ♆
19 19	☽ ✶ ♆
19 23	☽ ✶ ♀
20 23	☿ ✶ ♄
22 20	☽ Q ☉
23 49	☽ ⊼ ♂
22 Thursday	
05 50	☽ Q ☿
07 30	☽ △ ♇
09 53	☽ P ♄
11 46	☽ ☍ ♃
11 46	☉ ∥ ☿
16 25	☽ Q ♆
16 35	☽ ⊻ ♅
18 31	☽ Q ♀
19 47	♂ ∥ ♄
19 52	♀ ∥ ♆
23 Friday	
01 37	☽ ± ♄
03 01	☽ ⊥ ♅
08 09	☽ □ ☉
08 44	☿ # ♅
09 34	☽ P ♇
12 00	☽ ⊼ ♄
17 51	☽ □ ☿
18 30	☽ ∠ ♅
18 42	☽ # ♃
22 50	☿ ∠ ♅
23 28	☽ □ ♆
24 Saturday	
00 58	☽ ± ♇
04 17	☉ ∠ ♇
04 26	☽ □ ♀
05 57	☽ ☍ ♂
08 02	☉ # ♅
11 09	☽ ⊼ ♇
15 19	☽ ⊼ ♃
19 58	☽ ✶ ♅
25 Sunday	
01 25	☽ ± ♃
12 01	♀ □ ♂
12 26	☿ ☌ ♆
14 40	☽ △ ☉
14 52	☽ ☍ ♄
16 03	☽ Q ♅
16 27	☽ P ♃
17 46	☉ ✶ ♄
23 53	☿ ∥ ♆
26 Monday	
01 51	☽ △ ♆
03 49	☽ △ ☿
06 01	♀ Q ♄
10 09	☽ ⊼ ♂
11 15	☽ △ ♀
13 11	☽ ☍ ♇
17 17	☽ △ ♃
17 17	☽ P ☉
19 47	♀ ⊥ ♅
20 27	☽ ± ♂
21 51	☽ □ ♅
27 Tuesday	
02 36	☽ P ♆
08 10	☽ P ☿
09 09	☽ ± ☉
10 20	♀ ✶ ♇
11 50	☽ P ♂
14 10	☽ P ♀
16 32	☽ ⊼ ♄
17 24	☽ ± ♆
19 43	☽ ⊼ ☉
20 14	☽ # ♃
28 Wednesday	
01 07	☽ ± ☿
02 25	☽ ± ♄
03 15	☽ ⊼ ♆
06 18	☽ ± ♀
11 33	☿ Q ♄
12 25	☽ ⊼ ☿
13 29	☽ △ ♂
14 30	☽ ⊼ ♇
17 04	☽ ⊼ ♀
17 16	☽ P ♄
18 36	☽ □ ♃
19 37	☿ ⊥ ♅
23 13	☽ △ ♅
23 56	☿ □ ♂

MARCH 2018

LONGITUDES

Date	Sidereal time h m s	Sun ☉ ° ' "	Moon ☽ ° ' "	Moon ☽ 24.00 ° ' "	Mercury ☿ ° '	Venus ♀ ° '	Mars ♂ ° '	Jupiter ♃ ° '	Saturn ♄ ° '	Uranus ♅ ° '	Neptune ♆ ° '	Pluto ♇ ° '
01	22 36 59	10 ♓ 50 45	03 ♍ 39 12	10 ♍ 52 15	21 ♓ 12	23 ♓ 10	20 ♐ 33	23 ♏ 08	07 ♑ 21	25 ♈ 56	13 ♓ 49	20 ♑ 38
02	22 40 55	11 50 56	18 ♍ 01 48	25 ♍ 07 10	23 04	24 25	21 08	23 09	07 25	25 59	13 51	20 39
03	22 44 52	12 51 06	02 ♎ 07 45	09 ♎ 03 05	24 55	25 39	21 44	23 10	07 30	26 02	13 54	20 41
04	22 48 49	13 51 13	15 ♎ 52 47	22 ♎ 36 39	26 45	26 54	22 19	23 11	07 34	26 05	13 56	20 42
05	22 52 45	14 51 19	29 ♎ 14 34	05 ♏ 46 34	28 ♓ 33	28 09	22 55	23 12	07 38	26 07	13 58	20 44
06	22 56 42	15 51 23	12 ♏ 12 49	18 ♏ 33 34	00 ♈ 19	29 ♓ 23	23 30	23 13	07 42	26 10	14 00	20 45
07	23 00 38	16 51 26	24 ♏ 49 10	01 ♐ 00 04	02 02	00 ♈ 38	24 05	23 13	07 46	26 13	14 03	20 46
08	23 04 35	17 51 27	07 ♐ 06 46	13 ♐ 09 49	03 42	01 53	24 40	23 13	07 50	26 16	14 05	20 47
09	23 08 31	18 51 27	19 ♐ 09 49	25 ♐ 07 23	05 19	03 07	25 15	23 R 13	07 53	26 19	14 07	20 49
10	23 12 28	19 51 24	01 ♑ 03 10	06 ♑ 57 48	06 50	04 22	25 50	23 13	07 57	26 21	14 09	20 50
11	23 16 24	20 51 21	12 ♑ 51 57	18 ♑ 46 12	08 18	05 37	26 25	23 13	08 01	26 24	14 12	20 51
12	23 20 21	21 51 15	24 ♑ 41 11	00 ♒ 37 29	09 39	06 51	27 00	23 12	08 04	26 27	14 14	20 52
13	23 24 18	22 51 08	06 ♒ 35 37	12 ♒ 36 06	10 55	08 06	27 35	23 12	08 08	26 30	14 16	20 54
14	23 28 14	23 50 59	18 ♒ 39 22	24 ♒ 45 49	12 05	09 20	28 10	23 11	08 11	26 33	14 19	20 55
15	23 32 11	24 50 48	00 ♓ 55 46	07 ♓ 09 29	13 08	10 35	28 44	23 10	08 14	26 36	14 21	20 56
16	23 36 07	25 50 36	13 ♓ 27 09	19 ♓ 48 54	14 03	11 49	29 19	23 08	08 17	26 39	14 23	20 57
17	23 40 04	26 50 21	26 ♓ 14 45	02 ♈ 44 41	14 52	13 03	29 ♐ 53	23 07	08 20	26 43	14 25	20 58
18	23 44 00	27 50 05	09 ♈ 18 36	15 ♈ 56 20	15 32	14 18	00 ♑ 28	23 05	08 23	26 46	14 27	20 59
19	23 47 57	28 49 46	22 ♈ 37 41	29 ♈ 22 24	16 05	15 32	01 02	23 03	08 26	26 49	14 30	21 00
20	23 51 53	29 ♓ 49 25	06 ♉ 10 11	13 ♉ 00 46	16 29	16 47	01 36	23 01	08 29	26 52	14 32	21 01
21	23 55 50	00 ♈ 49 03	19 ♉ 53 51	26 ♉ 49 07	16 45	18 01	02 10	22 59	08 32	26 55	14 34	21 02
22	23 59 47	01 48 37	03 ♊ 46 20	10 ♊ 45 15	16 53	19 15	02 44	22 57	08 34	26 58	14 36	21 03
23	00 03 43	02 48 10	17 ♊ 45 37	24 ♊ 47 16	16 R 54	20 29	03 18	22 54	08 37	27 01	14 39	21 04
24	00 07 40	03 47 41	01 ♋ 50 02	08 ♋ 53 45	16 46	21 44	03 52	22 51	08 39	27 05	14 41	21 05
25	00 11 36	04 47 09	15 ♋ 58 17	23 ♋ 03 27	16 31	22 58	04 26	22 48	08 42	27 08	14 43	21 05
26	00 15 33	05 46 34	00 ♌ 09 05	07 ♌ 14 58	16 09	24 12	04 59	22 45	08 44	27 11	14 45	21 06
27	00 19 29	06 45 58	14 ♌ 20 53	21 ♌ 26 30	15 40	25 26	05 33	22 42	08 46	27 14	14 47	21 07
28	00 23 26	07 45 19	28 ♌ 31 30	05 ♍ 35 28	15 06	26 40	06 06	22 39	08 48	27 18	14 49	21 08
29	00 27 22	08 44 38	12 ♍ 37 59	19 ♍ 38 33	14 27	27 54	06 40	22 35	08 50	27 21	14 51	21 09
30	00 31 19	09 43 54	26 ♍ 36 41	03 ♎ 31 54	13 44	29 ♈ 08	07 13	22 32	08 52	27 24	14 54	21 09
31	00 35 16	10 ♈ 43 09	10 ♎ 23 42	17 ♎ 11 39	12 ♈ 57	00 ♉ 22	07 ♑ 46	22 ♏ 27	08 ♑ 54	27 ♈ 28	14 ♓ 56	21 ♑ 10

Date	Moon True ☊ ° '	Moon Mean ☊ ° '	Moon ☽ Latitude ° '
01	14 ♌ 51	13 ♌ 45	01 N 42
02	14 R 49	13 42	02 52
03	14 45	13 39	03 51
04	14 40	13 35	04 35
05	14 36	13 32	05 02
06	14 31	13 29	05 13
07	14 28	13 26	05 08
08	14 26	13 23	04 49
09	14 D 25	13 20	04 17
10	14 26	13 16	03 35
11	14 27	13 13	02 44
12	14 29	13 10	01 46
13	14 30	13 07	00 N 43
14	14 R 30	13 04	00 S 23
15	14 29	13 00	01 28
16	14 26	12 57	02 31
17	14 20	12 54	03 27
18	14 13	12 51	04 13
19	14 05	12 48	04 47
20	13 58	12 45	05 05
21	13 51	12 41	05 06
22	13 46	12 38	04 50
23	13 43	12 35	04 16
24	13 41	12 32	03 26
25	13 D 42	12 29	02 24
26	13 43	12 26	01 S 12
27	13 R 43	12 22	00 N 03
28	13 42	12 19	01 19
29	13 39	12 16	02 29
30	13 34	12 13	03 29
31	13 ♌ 26	12 ♌ 10	04 N 16

DECLINATIONS

Date	Sun ☉ ° '	Moon ☽ ° '	Mercury ☿ ° '	Venus ♀ ° '	Mars ♂ ° '	Jupiter ♃ ° '	Saturn ♄ ° '	Uranus ♅ ° '	Neptune ♆ ° '	Pluto ♇ ° '
01	07 S 30	11 N 45	04 S 12	03 S 58	22 S 49	17 S 24	22 S 21	09 N 31	07 S 12	21 S 29
02	07 07	07 22	03 19	03 27	22 53	17 25	22 21	09 32	07 11	21 29
03	06 44	02 N 41	02 25	02 57	22 56	17 25	22 21	09 33	07 10	21 29
04	06 21	02 S 01	01 31	02 26	22 59	17 25	22 21	09 34	07 09	21 29
05	05 58	06 29	00 S 38	01 55	23 02	17 25	22 20	09 35	07 09	21 29
06	05 35	10 32	00 N 16	01 25	23 05	17 25	22 20	09 36	07 08	21 29
07	05 11	13 59	01 08	00 54	23 08	17 25	22 20	09 37	07 07	21 29
08	04 48	16 45	01 59	00 S 23	23 10	17 25	22 20	09 38	07 06	21 29
09	04 24	18 43	02 50	00 N 08	23 13	17 25	22 19	09 39	07 05	21 28
10	04 01	19 51	03 38	00 39	23 15	17 24	22 19	09 40	07 04	21 28
11	03 37	20 05	04 25	01 10	23 17	17 24	22 19	09 41	07 03	21 28
12	03 14	19 27	05 10	01 40	23 19	17 24	22 19	09 43	07 02	21 28
13	02 50	17 56	05 52	02 11	23 21	17 24	22 19	09 44	07 02	21 28
14	02 27	15 35	06 31	02 42	23 23	17 23	22 18	09 45	07 01	21 28
15	02 03	12 31	07 07	03 13	23 24	17 23	22 18	09 46	07 00	21 28
16	01 39	08 49	07 40	03 44	23 26	17 22	22 18	09 47	06 59	21 28
17	01 15	04 39	08 10	04 14	23 27	17 22	22 18	09 48	06 58	21 28
18	00 52	00 S 12	08 35	04 45	23 28	17 21	22 18	09 49	06 57	21 28
19	00 28	04 N 21	08 57	05 15	23 29	17 21	22 17	09 50	06 57	21 28
20	00 S 04	08 46	09 15	05 45	23 30	17 20	22 17	09 51	06 56	21 28
21	00 N 19	12 47	09 29	06 16	23 31	17 19	22 17	09 53	06 55	21 27
22	00 43	16 09	09 38	06 46	23 32	17 19	22 17	09 54	06 55	21 27
23	01 07	18 38	09 43	07 16	23 32	17 18	22 17	09 55	06 54	21 27
24	01 30	19 59	09 43	07 45	23 33	17 17	22 16	09 56	06 53	21 27
25	01 54	20 06	09 39	08 15	23 33	17 16	22 16	09 57	06 52	21 27
26	02 18	18 56	09 31	08 44	23 33	17 15	22 16	09 58	06 51	21 27
27	02 41	16 35	09 19	09 13	23 33	17 14	22 16	10 00	06 50	21 27
28	03 05	13 13	09 03	09 42	23 33	17 13	22 16	10 01	06 49	21 27
29	03 28	09 06	08 44	10 11	23 33	17 12	22 16	10 02	06 48	21 27
30	03 51	04 33	08 21	10 40	23 33	17 11	22 16	10 03	06 47	21 27
31	04 N 15	00 S 11	07 N 55	11 N 08	23 S 32	17 S 10	22 S 15	10 N 04	06 S 47	21 S 27

ZODIAC SIGN ENTRIES

Date	h	m	Planets
01	05	57	☽ ♍
03	08	20	☽ ♎
05	13	23	☽ ♏
06	07	34	☿ ♈
06	23	45	♀ ♈
07	22	03	☽ ♐
10	09	52	☽ ♑
12	22	44	☽ ♒
15	10	12	☽ ♓
17	16	40	♂ ♑
17	18	57	☽ ♈
20	01	07	☽ ♉
20	16	15	☉ ♈
22	05	30	☽ ♊
24	08	53	☽ ♋
26	11	45	☽ ♌
28	14	30	☽ ♍
30	17	52	☽ ♎
31	04	54	♀ ♉

LATITUDES

Date	Mercury ☿ ° '	Venus ♀ ° '	Mars ♂ ° '	Jupiter ♃ ° '	Saturn ♄ ° '	Uranus ♅ ° '	Neptune ♆ ° '	Pluto ♇ ° '
01	00 S 47	01 S 22	00 N 17	01 N 11	00 N 53	00 S 32	00 S 54	00 N 22
04	00 S 15	01 19	00 14	01 12	00 53	00 32	00 54	00 22
07	00 N 21	01 15	00 10	01 12	00 53	00 32	00 54	00 21
10	01 00	01 11	00 07	01 13	00 53	00 32	00 54	00 21
13	01 40	01 07	00 N 04	01 13	00 53	00 32	00 54	00 21
16	02 18	01 02	00 00	01 13	00 53	00 32	00 55	00 21
19	02 51	00 56	00 S 03	01 14	00 53	00 32	00 55	00 20
22	03 15	00 50	00 07	01 14	00 53	00 32	00 55	00 20
25	03 26	00 44	00 11	01 14	00 53	00 32	00 55	00 20
28	03 22	00 37	00 15	01 15	00 53	00 32	00 55	00 20
31	03 N 03	00 S 30	00 S 20	01 N 15	00 N 53	00 S 32	00 S 55	00 N 19

DATA

Julian Date	2458179
Delta T	+70 seconds
Ayanamsa	24° 06' 27"
Synetic vernal point	05° ♓ 00' 32"
True obliquity of ecliptic	23° 26' 07"

MOON'S PHASES, APSIDES AND POSITIONS ☽

Date	h	m	Phase	Longitude ° '	Eclipse Indicator
02	00	51	○	11 ♍ 23	
09	11	20	☾	18 ♐ 50	
17	13	12	●	26 ♓ 53	
24	15	35	☽	03 ♋ 57	
31	12	37	○	10 ♎ 45	

Day	h	m		
11	09	18	Apogee	
26	17	23	Perigee	
04	01	37	0S	
11	06	37	Max dec	20° S 07'
18	13	02	0N	
25	02	07	Max dec	20° N 13'
31	11	02	0S	

LONGITUDES

Date	Chiron ⚷ ° '	Ceres ⚳ ° '	Pallas ⚴ ° '	Juno ⚵ ° '	Vesta ⚶ ° '	Black Moon Lilith ⚸ ° '
01	27 ♓ 16	05 ♌ 51	15 ♉ 30	02 ♓ 37	19 ♐ 46	12 ♑ 30
11	27 ♓ 51	04 ♌ 56	20 ♉ 20	07 ♓ 17	23 ♐ 26	13 ♑ 37
21	28 ♓ 27	04 ♌ 45	25 ♉ 27	12 ♓ 00	26 ♐ 44	14 ♑ 43
31	29 ♓ 03	05 ♌ 15	00 ♊ 47	16 ♓ 44	29 ♐ 36	15 ♑ 50

All ephemeris data is given at 12.00 UT and the Moon's longitude is additionally given for 24.00 UT

ASPECTARIAN

h m Aspects

01 Thursday
00 23 ☽ ± ♇
04 42 ☿ ✶ ♇
11 22 ♀ △ ♃
14 23 ☉ ∠ ♅
15 17 ☽ bQ ♇
15 32 ♂ ⊻ ♇
18 10 ☽ △ ♄

02 Friday
00 09 ☽ bQ ♅
00 27 ☽ Q ♃
00 30 ☽ ∥ ♅
00 51 ☽ ☍ ☉
02 58 ☿ ∥ ♀
04 58 ☽ ☍ ♆
07 29 ☉ ∥ ♆
12 59 ☽ # ♆
13 05 ☿ △ ♃
13 28 ☽ # ☉
15 18 ☽ ± ♅
16 26 ☽ △ ♇
17 29 ☽ □ ♂
20 40 ☽ ✶ ♃
21 48 ☽ ☍ ☿
23 50 ☽ ☍ ♀

03 Saturday
01 31 ☽ ⊼ ♅
10 31 ☽ # ♀
13 41 ☽ # ☿
19 31 ♀ ⊻ ♅
21 20 ☽ □ ♄
22 29 ☽ ∠ ♃

04 Sunday
01 47 ☽ Q ♂
02 49 ☿ ⊻ ♅
08 08 ☽ ⊼ ☉
08 33 ☽ ⊼ ♆
09 50 ☽ ∥ ☿
13 54 ☉ ☌ ♆
13 57 ☽ ∥ ♀
14 19 ☽ ⊥ ♃
18 05 ☿ ☌ ♀
19 13 ☽ ± ♆
19 38 ☽ ± ☉
20 36 ☽ □ ♇

05 Monday
00 01 ☽ ✶ ♂
01 03 ☽ ⊻ ♃
05 25 ☽ Q ♄
06 19 ☽ ☍ ♅
09 18 ☽ ∥ ☉
09 48 ☽ ⊼ ♀
10 33 ☽ ⊼ ☿
11 30 ☽ bQ ♆
13 13 ☽ bQ ☉
15 40 ☽ ∥ ♆
21 56 ☽ ± ♀
23 17 ☽ ± ☿

06 Tuesday
00 07 ♂ ⊻ ♃
03 31 ☽ ✶ ♄
04 43 ☽ ∠ ♂
05 30 ☽ Q ♇
06 10 ☽ # ♅
15 23 ☽ △ ♆
16 32 ☽ bQ ♀
18 47 ☽ bQ ☿
19 27 ☽ △ ☉
22 28 ☽ ⊥ ♂

07 Wednesday
04 12 ☽ ✶ ♇
07 55 ☿ # ♀
08 01 ☽ ∠ ♄
08 55 ☽ ☌ ♃
10 31 ☽ ⊻ ♂
14 42 ☽ ⊼ ♅
22 32 ☿ Q ♇

08 Thursday
00 33 ☽ △ ♀
01 34 ☽ ⊥ ♄
02 26 ☽ ± ♅
04 15 ☽ △ ☿
09 23 ☽ ∠ ♇
13 25 ☽ ⊻ ♄
19 06 ☽ ∥ ♃
20 15 ☽ bQ ♅

09 Friday
01 52 ☽ □ ♆
03 16 ☽ ⊥ ♇
04 45 ♃ St R
05 54 ♀ Q ♇
11 20 ☽ □ ☉
15 19 ☽ ⊻ ♇
20 10 ☽ ⊻ ♃

10 Saturday
00 54 ☽ ☌ ♂
02 27 ☽ △ ♅
08 17 ☽ ⊥ ♃
14 15 ☽ Q ♆
14 24 ☉ Q ♄
19 31 ☽ □ ♀
19 39 ☉ # ☿

11 Sunday
00 40 ☉ ⊥ ♅
01 25 ☽ □ ☿
02 05 ☽ ☌ ♄
02 33 ☽ ∠ ♃
03 06 ☽ Q ☉
07 00 ☿ □ ♄
10 40 ☿ bQ ♃
11 23 ♂ △ ♅
11 56 ☉ ✶ ♇
14 43 ☽ ✶ ♆

12 Monday
04 15 ☽ ☌ ♇
05 44 ☽ ✶ ☉
09 00 ☽ ✶ ♃
12 22 ☽ Q ♀
15 36 ☽ □ ♅
16 56 ☽ ⊻ ♂
18 45 ☽ Q ☿
21 13 ☽ ∠ ♆

13 Tuesday
05 39 ☽ ⊥ ♂
09 12 ☽ Q ♃
12 39 ♀ □ ♄
13 55 ♀ bQ ♃
14 45 ☽ ∠ ☉
15 05 ☽ ⊻ ♄
15 21 ☽ ✶ ♀
15 22 ☽ ⊥ ♆
18 17 ☽ ∥ ♃
20 06 ☉ △ ♃
21 36 ☽ ✶ ☿

14 Wednesday
00 34 ☽ ∠ ♂
03 06 ☽ ⊥ ♄
03 22 ☽ ⊻ ♆
03 51 ☽ Q ♅
05 07 ☉ # ♀
10 16 ☽ ⊥ ☉
11 03 ♃ ∠ ♄
16 27 ☽ ⊻ ♇
20 53 ☽ □ ♃
20 56 ☽ ∠ ♄
23 07 ☽ ⊻ ☉

15 Thursday
00 25 ☽ ∠ ♀
03 34 ☽ ✶ ♅
04 13 ☽ ⊥ ♇
06 05 ☽ ∠ ☿
07 06 ☿ # ♆
07 32 ☽ ✶ ♂
19 49 ☽ ⊥ ♀
21 40 ☽ ∠ ♇

16 Friday
00 56 ☽ ⊥ ☿
02 07 ☽ ✶ ♄
04 58 ♂ ⊥ ♃
06 07 ☽ # ♅
07 45 ☽ Q ♂
08 33 ☽ ⊻ ♀
08 35 ☽ ∠ ♅
13 14 ☽ ⊻ ☿
13 46 ☽ ☌ ♆
18 06 ☽ # ☿
21 44 ☿ ⊻ ♆
22 54 ☽ ∥ ♆

17 Saturday
00 56 ☽ Q ♄
01 38 ☽ ⊥ ♅
02 09 ☽ ✶ ♇
06 11 ☽ △ ♃
08 41 ☉ ⊻ ♅
12 52 ☽ ⊻ ♅
13 12 ☽ ☌ ☉
14 04 ☽ # ♀
19 03 ☽ □ ♂

18 Sunday
00 25 ☽ Q ♇
08 08 ☽ ∥ ☉
09 47 ☽ bQ ♃
10 19 ☽ □ ♄
15 13 ♀ ⊻ ♆
17 07 ☽ # ☉
21 21 ☽ ⊻ ♆
21 58 ☽ ☌ ♀
23 47 ☽ ☌ ☿

19 Monday
02 02 ☽ ± ♃
08 10 ☽ ⊥ ♆
09 05 ☽ □ ♇
12 46 ☽ ⊼ ♃
17 23 ☽ ∥ ♀
19 29 ☽ ☌ ♅
23 55 ☽ ⊻ ☉

20 Tuesday
00 15 ☽ ∠ ♆
01 51 ☽ # ♆
03 36 ☽ △ ♂
04 03 ☿ ☌ ♀
11 21 ☽ ⊥ ☉
14 55 ☽ ∥ ☿
16 05 ☽ △ ♄
16 39 ♀ ± ♃
18 15 ☽ ∥ ♅

21 Wednesday
02 42 ☽ ✶ ♆
04 21 ☽ ∠ ☉
06 27 ☽ ⊻ ☿
07 03 ☽ bQ ♂
08 24 ☽ ⊻ ♀
13 58 ☽ △ ♇
17 02 ☽ ⊥ ☿
17 21 ☽ ☍ ♃
18 19 ☽ bQ ♄
19 51 ☽ ⊥ ♀
23 21 ☽ ± ♂
23 36 ☽ Q ♆

22 Thursday
00 13 ☽ ⊻ ♅
05 57 ♂ Q ♆
08 22 ☽ ✶ ☉
08 44 ☽ ∠ ☿
09 56 ☽ ± ♄
10 09 ☽ ⊼ ♂
10 37 ☽ ⊥ ♅
12 54 ☽ ∠ ♀
15 55 ☽ bQ ♇
18 29 ♀ # ♆
20 17 ☽ ⊼ ♄
21 58 ☽ # ♃

23 Friday
00 19 ☿ St R
02 08 ☽ ∠ ♅
06 33 ☽ Q ☉
06 39 ☽ □ ♆
07 23 ☽ ± ♇
10 31 ☽ ✶ ☿
15 04 ♀ ⊥ ♆
17 07 ☽ ✶ ♀
17 39 ☽ ⊼ ♇
18 23 ☉ Q ♇
20 45 ☽ ⊼ ♃
23 17 ♀ □ ♇

24 Saturday
03 52 ☽ ✶ ♅
06 50 ☽ Q ☿
06 57 ☽ ± ♃
15 31 ☽ Q ♀
15 35 ☽ □ ☉
15 36 ☽ ☍ ♂
16 08 ☉ □ ♂
22 12 ☽ bQ ♃
23 37 ☽ ☍ ♄

25 Sunday
00 21 ☽ Q ♅
09 08 ♀ ⊼ ♃
09 52 ☽ △ ♆
12 54 ☽ □ ☿
20 41 ☽ ☍ ♇
23 32 ☽ △ ♃

26 Monday
00 58 ☽ □ ♀
06 58 ☽ □ ♅
11 19 ☽ bQ ♆
20 31 ☽ ⊼ ♂
22 13 ☽ △ ☉

27 Tuesday
02 33 ☽ ⊼ ♄
02 34 ☽ ± ♆
06 14 ☽ # ♃
07 05 ☽ ± ♂
12 43 ☽ ± ♄
12 45 ☽ ⊼ ♆
14 09 ☽ △ ☿
15 11 ☿ ∥ ♀
22 55 ☽ bQ ♂
23 28 ☽ ⊼ ♇

28 Wednesday
01 29 ☽ bQ ☉
02 04 ☽ □ ♃
03 59 ☽ bQ ♄
08 33 ☽ △ ♀
09 26 ☉ bQ ♃
09 38 ☽ ± ♇
09 54 ☽ △ ♅
14 34 ☽ bQ ☿
17 54 ☽ ± ☉
22 01 ☿ ⊻ ♆

29 Thursday
00 47 ♀ ☌ ♅
00 56 ☽ bQ ♇
01 24 ☽ △ ♂
04 06 ♀ ∥ ♅
04 52 ☽ ⊼ ☉
05 11 ☽ ± ☿
05 30 ☽ △ ♄
06 37 ☽ ∥ ♀
06 54 ☽ ∥ ♅
08 31 ☽ Q ♃
11 31 ☽ bQ ♅
12 30 ☽ bQ ♀
14 15 ☽ ∥ ☿
14 16 ☉ □ ♄
14 57 ☽ ⊼ ☿
15 49 ☽ ☍ ♆

30 Friday
00 21 ☽ # ♆
02 35 ☽ △ ♇
02 59 ☽ ± ♅
04 59 ☽ ✶ ♃
05 25 ☽ ± ♀
13 23 ☽ ⊼ ♅
15 15 ☽ ∥ ☉
16 47 ☽ ⊼ ♀
23 45 ♂ ∠ ♃

31 Saturday
03 15 ♀ ∠ ♆
06 52 ☽ ∠ ♃
07 12 ☽ □ ♂
09 22 ☽ □ ♄
12 37 ☽ ☍ ☉
16 15 ☽ ☍ ☿
20 00 ☽ ⊼ ♆
22 38 ☽ ⊥ ♃

APRIL 2018

Date	Sidereal time h m s	Sun ☉ ° ' "	Moon ☽ ° ' "	Moon ☽ 24.00 ° ' "	Mercury ☿ ° '	Venus ♀ ° '	Mars ♂ ° '	Jupiter ♃ ° '	Saturn ♄ ° '	Uranus ♅ ° '	Neptune ♆ ° '	Pluto ♇ ° '
01	00 39 12	11 ♈ 42 21	23 ♎ 55 23	00 ♏ 34 35	12 ♈ 09	01 ♉ 36	08 ♑ 19	22 ♏ 23	08 ♑ 55	27 ♈ 31	14 ♓ 58	21 ♑ 11
02	00 43 09	12 41 31	07 ♏ 09 02	13 ♏ 38 36	11 R 19	02 50	08 52	22 R 19	08 57	27 34	15 00	21 11
03	00 47 05	13 40 40	20 ♏ 03 18	26 ♏ 23 11	10 30	04 04	09 25	22 14	08 58	27 38	15 02	21 11
04	00 51 02	14 39 47	02 ♐ 38 27	08 ♐ 49 23	09 41	05 17	09 58	22 10	09 00	27 41	15 04	21 12
05	00 54 58	15 38 52	14 ♐ 56 24	20 ♐ 59 45	08 54	06 31	10 30	22 05	09 01	27 44	15 06	21 13
06	00 58 55	16 37 55	27 ♐ 00 09	02 ♑ 58 05	08 09	07 45	11 03	22 00	09 02	27 48	15 08	21 13
07	01 02 51	17 36 56	08 ♑ 54 10	14 ♑ 49 02	07 28	08 59	11 35	21 55	09 03	27 51	15 10	21 14
08	01 06 48	18 35 56	20 ♑ 43 21	26 ♑ 37 48	06 51	10 12	12 07	21 50	09 04	27 55	15 12	21 14
09	01 10 45	19 34 54	02 ♒ 33 05	08 ♒ 29 51	06 18	11 26	12 39	21 45	09 05	27 58	15 14	21 15
10	01 14 41	20 33 50	14 ♒ 28 47	20 ♒ 30 30	05 50	12 39	13 11	21 39	09 06	28 01	15 16	21 15
11	01 18 38	21 32 44	26 ♒ 35 36	02 ♓ 44 38	05 27	13 53	13 43	21 33	09 07	28 05	15 18	21 15
12	01 22 34	22 31 37	08 ♓ 58 02	15 ♓ 16 14	05 09	15 07	14 14	21 28	09 07	28 08	15 20	21 16
13	01 26 31	23 30 27	21 ♓ 39 31	28 ♓ 08 06	04 56	16 20	14 46	21 22	09 08	28 12	15 22	21 16
14	01 30 27	24 29 16	04 ♈ 42 03	11 ♈ 21 21	04 49	17 33	15 17	21 16	09 08	28 15	15 24	21 16
15	01 34 24	25 28 03	18 ♈ 05 49	24 ♈ 55 11	04 D 47	18 47	15 48	21 10	09 09	28 18	15 25	21 16
16	01 38 20	26 26 48	01 ♉ 49 03	08 ♉ 46 57	04 50	20 00	16 19	21 03	09 09	28 22	15 27	21 17
17	01 42 17	27 25 31	15 ♉ 48 16	22 ♉ 52 25	04 58	21 14	16 50	20 57	09 09	28 25	15 29	21 17
18	01 46 14	28 24 12	29 ♉ 58 43	07 ♊ 06 30	05 11	22 27	17 20	20 50	09 R 09	28 29	15 31	21 17
19	01 50 10	29 ♈ 22 51	14 ♊ 15 11	21 ♊ 24 09	05 29	23 40	17 51	20 44	09 09	28 32	15 33	21 17
20	01 54 07	00 ♉ 21 28	28 ♊ 32 54	05 ♋ 41 01	05 52	24 53	18 21	20 37	09 09	28 36	15 34	21 17
21	01 58 03	01 20 03	12 ♋ 48 08	19 ♋ 53 59	06 19	26 07	18 51	20 30	09 08	28 39	15 36	21 17
22	02 02 00	02 18 35	26 ♋ 58 23	04 ♌ 01 12	06 50	27 20	19 21	20 24	09 08	28 43	15 38	21 17
23	02 05 56	03 17 06	11 ♌ 02 21	18 ♌ 01 46	07 26	28 33	19 50	20 17	09 07	28 46	15 40	21 R 17
24	02 09 53	04 15 34	24 ♌ 59 26	01 ♍ 55 16	08 05	29 ♉ 46	20 20	20 10	09 07	28 49	15 41	21 17
25	02 13 49	05 13 59	08 ♍ 49 15	15 ♍ 41 17	08 48	00 ♊ 59	20 49	20 03	09 06	28 53	15 43	21 17
26	02 17 46	06 12 23	22 ♍ 31 14	29 ♍ 18 58	09 35	02 12	21 18	19 55	09 05	28 56	15 44	21 17
27	02 21 43	07 10 45	06 ♎ 04 19	12 ♎ 47 03	10 25	03 25	21 47	19 48	09 05	29 00	15 46	21 17
28	02 25 39	08 09 04	19 ♎ 26 57	26 ♎ 03 48	11 18	04 37	22 16	19 41	09 04	29 03	15 48	21 17
29	02 29 36	09 07 22	02 ♏ 37 23	09 ♏ 07 30	12 14	05 50	22 44	19 33	09 03	29 07	15 49	21 17
30	02 33 32	10 ♉ 05 38	15 ♏ 33 59	21 ♏ 56 43	13 ♈ 13	07 ♊ 03	23 ♑ 12	19 ♏ 26	09 ♑ 01	29 ♈ 10	15 ♓ 51	21 ♑ 16

DECLINATIONS

Date	Moon True ☊ ° '	Moon Mean ☊ ° '	Moon ☽ Latitude ° '	Sun ☉ ° '	Moon ☽ ° '	Mercury ☿ ° '	Venus ♀ ° '	Mars ♂ ° '	Jupiter ♃ ° '	Saturn ♄ ° '	Uranus ♅ ° '	Neptune ♆ ° '	Pluto ♇ ° '
01	13 ♌ 16	12 ♌ 06	04 N 48	04 N 38	04 S 49	07 N 27	11 N 36	23 S 32	17 S 09	22 S 15	10 N 06	06 S 46	21 S 27
02	13 R 05	12 03	05 03	05 01	09 07	06 58	12 04	23 31	17 08	22 15	10 07	06 45	21 27
03	12 55	12 00	05 02	05 24	12 54	06 27	12 31	23 31	17 07	22 15	10 08	06 44	21 27
04	12 47	11 57	04 47	05 47	16 00	05 55	12 58	23 30	17 05	22 15	10 09	06 44	21 27
05	12 40	11 54	04 18	06 09	18 19	05 23	13 25	23 29	17 04	22 15	10 11	06 43	21 27
06	12 36	11 51	03 38	06 32	19 46	04 51	13 51	23 28	17 03	22 15	10 12	06 42	21 27
07	12 34	11 47	02 50	06 55	20 19	04 21	14 18	23 27	17 01	22 15	10 13	06 41	21 27
08	12 D 33	11 44	01 54	07 17	19 58	03 51	14 43	23 26	17 00	22 15	10 14	06 41	21 27
09	12 34	11 41	00 N 53	07 40	18 43	03 23	15 09	23 24	16 58	22 15	10 15	06 40	21 27
10	12 R 34	11 38	00 S 10	08 02	16 39	02 57	15 34	23 23	16 57	22 14	10 17	06 39	21 27
11	12 33	11 35	01 14	08 24	13 49	02 33	15 59	23 22	16 55	22 14	10 18	06 38	21 27
12	12 30	11 32	02 16	08 46	10 18	02 11	16 23	23 22	16 54	22 14	10 19	06 38	21 27
13	12 24	11 28	03 12	09 08	06 15	01 52	16 47	23 18	16 52	22 14	10 20	06 37	21 27
14	12 16	11 25	04 00	09 29	01 S 48	01 35	17 10	23 17	16 51	22 14	10 22	06 36	21 27
15	12 06	11 22	04 36	09 51	02 N 51	01 21	17 33	23 15	16 49	22 14	10 23	06 36	21 27
16	11 54	11 19	04 57	10 12	07 27	01 09	17 56	23 13	16 47	22 14	10 24	06 35	21 27
17	11 43	11 16	05 01	10 33	11 46	01 00	18 18	23 11	16 46	22 14	10 25	06 34	21 27
18	11 32	11 12	04 46	10 54	15 29	00 54	18 39	23 09	16 44	22 14	10 26	06 34	21 27
19	11 24	11 09	04 14	11 15	18 18	00 50	19 01	23 07	16 42	22 14	10 28	06 33	21 28
20	11 18	11 06	03 26	11 36	20 00	00 49	19 21	23 05	16 40	22 14	10 29	06 32	21 28
21	11 15	11 03	02 24	11 56	20 25	00 50	19 41	23 03	16 39	22 14	10 30	06 32	21 28
22	11 14	11 00	01 15	12 16	19 32	00 53	20 01	23 01	16 37	22 14	10 31	06 31	21 28
23	11 D 14	10 57	00 S 01	12 36	17 26	00 59	20 20	22 59	16 35	22 14	10 33	06 30	21 28
24	11 R 14	10 53	01 N 12	12 56	14 19	01 07	20 39	22 57	16 33	22 14	10 34	06 30	21 28
25	11 12	10 50	02 20	13 16	10 26	01 17	20 57	22 55	16 31	22 14	10 35	06 29	21 28
26	11 07	10 47	03 20	13 35	06 01	01 29	21 14	22 52	16 29	22 14	10 36	06 28	21 28
27	11 00	10 44	04 07	13 54	01 N 22	01 43	21 31	22 50	16 27	22 14	10 38	06 28	21 28
28	10 50	10 41	04 40	14 13	03 S 17	01 59	21 47	22 47	16 25	22 14	10 39	06 27	21 28
29	10 37	10 38	04 57	14 32	07 43	02 17	22 03	22 45	16 23	22 15	10 40	06 27	21 29
30	10 ♌ 24	10 ♌ 34	04 N 59	14 N 50	11 S 43	02 N 36	22 N 18	22 S 43	16 S 21	22 S 15	10 N 41	06 S 26	21 S 29

ZODIAC SIGN ENTRIES

Date	h	m	Planets
01	22	57	☽ ♏
04	06	55	☽ ♐
06	18	01	☽ ♑
09	06	50	☽ ♒
11	18	40	☽ ♓
14	03	26	☽ ♈
16	08	51	☽ ♉
18	12	02	☽ ♊
20	03	13	☉ ♉
20	14	26	☽ ♋
22	17	09	☽ ♌
24	16	40	♀ ♊
24	20	40	☽ ♍
27	01	13	☽ ♎
29	07	11	☽ ♏

LATITUDES

Date	Mercury ☿ ° '	Venus ♀ ° '	Mars ♂ ° '	Jupiter ♃ ° '	Saturn ♄ ° '	Uranus ♅ ° '	Neptune ♆ ° '	Pluto ♇ ° '
01	02 N 53	00 S 28	00 S 21	01 N 15	00 N 53	00 S 32	00 S 55	00 N 19
04	02 16	00 20	00 26	01 16	00 53	00 32	00 55	00 19
07	01 30	00 12	00 31	01 16	00 53	00 32	00 55	00 19
10	00 N 41	00 S 04	00 36	01 16	00 53	00 32	00 55	00 19
13	00 S 07	00 N 04	00 42	01 16	00 53	00 32	00 55	00 18
16	00 50	00 12	00 47	01 16	00 53	00 32	00 55	00 18
19	01 28	00 20	00 53	01 16	00 53	00 32	00 55	00 18
22	01 59	00 28	00 59	01 17	00 53	00 31	00 55	00 17
25	02 24	00 36	01 06	01 17	00 53	00 31	00 55	00 17
28	02 42	00 45	01 13	01 17	00 53	00 31	00 56	00 17
31	02 S 54	00 N 53	01 S 20	01 N 16	00 N 53	00 S 31	00 S 56	00 N 17

DATA

Julian Date	2458210
Delta T	+70 seconds
Ayanamsa	24° 06' 30"
Synetic vernal point	05° ♓ 00' 29"
True obliquity of ecliptic	23° 26' 07"

MOON'S PHASES, APSIDES AND POSITIONS ☽

Date	h	m	Phase	Longitude ° '	Eclipse Indicator
08	07	18	☾	18 ♑ 24	
16	01	57	●	26 ♈ 02	
22	21	46	☽	02 ♌ 42	
30	00	58	○	09 ♏ 39	

Day	h	m		
08	05	31	Apogee	
20	14	36	Perigee	
07	14	35	Max dec	20° S 19'
14	21	22	0N	
21	07	39	Max dec	20° N 27'
27	19	00	0S	

LONGITUDES

Date	Chiron ⚷ ° '	Ceres ⚳ ° '	Pallas ⚴ ° '	Juno ⚵ ° '	Vesta ⚶ ° '	Black Moon Lilith ⚸ ° '
01	29 ♓ 06	05 ♌ 21	01 ♊ 19	17 ♓ 13	29 ♐ 51	15 ♑ 57
11	29 ♓ 41	06 ♌ 34	06 ♊ 51	21 ♓ 58	02 ♑ 08	17 ♑ 03
21	00 ♈ 13	08 ♌ 21	12 ♊ 30	26 ♓ 45	03 ♑ 47	18 ♑ 10
31	00 ♈ 44	10 ♌ 38	18 ♊ 16	01 ♈ 31	04 ♑ 43	19 ♑ 16

ASPECTARIAN

h m	Aspects
01 Sunday	
06 41	☽ ± ♆
07 05	☽ □ ♇
09 16	☽ ⊻ ♃
10 53	☽ # ☉
16 29	☽ Q ♂
17 24	☽ Q ♄
17 53	☉ ☌ ☿
18 29	☽ ☍ ♅
22 31	☽ ∥ ♆
22 55	☽ bQ ♆
02 Monday	
00 56	☽ # ☿
03 17	☽ ☍ ♀
15 18	☽ ⚹ ♂
15 19	☽ ⚹ ♄
15 44	♂ ☌ ♄
15 45	☽ Q ♇
18 00	☽ # ♅
19 14	☽ ⊼ ☿
22 01	☿ # ♆
23 05	☽ ⊼ ☉
03 Tuesday	
02 34	☽ △ ♆
05 43	☽ ± ☿
09 00	☽ # ♀
11 14	☽ ± ☉
14 09	☽ ⚹ ♇
16 06	☽ ☌ ♃
19 25	☽ ∠ ♄
20 37	☽ ∠ ♂
21 40	☽ bQ ☿
04 Wednesday	
02 26	☽ ⊼ ♅
05 47	☽ bQ ☉
07 05	☿ □ ♂
12 41	☽ ⊥ ♄
14 01	☽ ± ♅
14 40	☽ ⊥ ♂
15 36	☉ ∥ ☿
17 41	☽ ⊼ ♀
18 54	☽ ∠ ♇
22 10	☉ ⊻ ♆
22 11	☽ ∥ ♃
05 Thursday	
00 22	☽ ⊻ ♄
00 51	☽ △ ☿
02 53	☽ ⊻ ♂
06 42	☽ ± ♀
07 39	☽ bQ ♅
08 22	☿ □ ♄
12 19	☽ □ ♆
12 33	☽ ⊥ ♇
13 31	☽ △ ☉
21 49	☉ ± ♃
06 Friday	
00 27	☽ ⊻ ♇
02 04	☽ ⊻ ♃
02 31	☽ bQ ♀
13 36	☽ △ ♅
13 59	☽ ⊥ ♃
17 05	☿ ⊻ ♀
22 08	☉ # ♆
07 Saturday	
00 22	☽ Q ♆
08 01	☽ ∠ ♃
09 15	☽ □ ☿
12 10	☽ △ ♀
12 19	☽ ☌ ♄
13 36	♀ △ ♄
17 41	☽ ☌ ♂
08 Sunday	
00 45	☽ ⚹ ♆
07 18	☽ □ ☉
12 56	☿ bQ ♃
13 03	☽ ☌ ♇
14 14	☽ ⚹ ♃
20 00	☽ Q ☿
09 Monday	
02 40	☽ □ ♅
07 18	☽ ∠ ♆
14 23	☽ Q ♃
19 16	☽ ⚹ ☿
23 04	☽ Q ☉
10 Tuesday	
00 10	☿ ⊥ ♀
01 12	☽ ⊻ ♄
01 31	☽ ⊥ ♆
07 56	☽ □ ♀
08 59	☽ ∥ ♃
09 17	☽ ⊻ ♂
13 15	☽ ⊥ ♄
13 34	☽ ⊻ ♆
15 06	☽ Q ♅
20 37	☽ # ♀
21 48	☽ ⊥ ♂
11 Wednesday	
00 14	☽ ∠ ☿
01 11	☽ ⚹ ☉
01 29	☽ ⊻ ♇
02 09	☽ □ ♃
04 54	☉ □ ♇
05 44	☉ ⊥ ♆
06 03	♀ △ ♂
07 07	☽ ∠ ♄
12 15	☉ ⊼ ♃
13 18	☽ ⊥ ♇
14 56	☽ ⚹ ♅
16 20	☽ ∠ ♂
17 26	☽ ⊥ ☿
23 28	☽ Q ♀
12 Thursday	
04 48	☽ ⊻ ☿
06 48	☽ ∠ ♇
09 00	☽ ∠ ☉
11 54	☽ # ♅
12 18	☽ ⚹ ♄
16 26	♀ ⚹ ♆
19 59	☽ ∠ ♅
20 39	☽ # ☉
22 29	☽ ⚹ ♂
13 Friday	
00 08	☽ ☌ ♆
00 57	☽ ⚹ ♀
03 34	☽ ⊥ ☉
09 55	☽ ∥ ♆
11 01	☽ Q ♄
11 16	☽ ⚹ ♇
11 27	☽ △ ♃
13 00	☽ ⊥ ♅
15 44	☽ ⊻ ☉
17 21	♀ # ♃
21 52	☽ Q ♂
14 Saturday	
00 10	☽ ⊻ ♅
07 42	☽ ∠ ♀
09 24	☽ Q ♇
09 58	♃ ⚹ ♇
12 12	☽ ☌ ☿
13 13	☽ # ☿
14 48	☽ bQ ♃
17 27	♂ ⚹ ♆
20 01	☽ □ ♄
15 Sunday	
01 36	☽ ⊥ ♀
04 39	☽ ∥ ☿
06 50	☽ ± ♃
07 15	☽ ⊻ ♆
07 46	☽ □ ♂
09 21	☿ St D
13 20	☽ ⊻ ♀
17 22	☽ ⊼ ♃
17 36	☽ □ ♇
17 53	☽ ⊥ ♆
16 Monday	
01 57	☽ ☌ ☉
05 59	☽ ☌ ♅
07 23	☽ # ♆
08 20	☿ ∠ ♀
09 38	☽ ∠ ♆
17 15	☽ ⊻ ☿
17 Tuesday	
00 38	☽ △ ♄
02 18	☉ ∥ ♅
03 39	☽ ⊥ ☿
04 11	☽ ∥ ♅
04 21	☽ ∥ ☉
07 00	♀ ☍ ♃
11 27	☽ ⚹ ♆
13 04	♀ △ ♇
13 48	☽ △ ♂
19 10	☽ ∠ ☿
20 40	☽ ☍ ♃
21 18	☽ △ ♇
22 05	☽ ☌ ♀
18 Wednesday	
01 46	♄ St R
02 09	☽ bQ ♄
07 50	☽ Q ♆
09 09	☽ ⊻ ☉
09 28	☽ ⊻ ♅
14 00	☉ ☌ ♅
16 07	☽ bQ ♂
17 20	☽ ± ♄
19 37	☽ ⊥ ♅
20 00	☽ ⊥ ☉
20 57	☽ ⚹ ☿
21 31	☽ # ♃
22 37	☽ bQ ♇
19 Thursday	
03 26	☽ ⊼ ♄
07 49	☽ ± ♂
10 48	☽ ∠ ♅
12 14	☽ ∠ ☉
13 44	☽ ± ♇
14 10	☽ □ ♆
17 34	☽ Q ☿
18 15	☽ ⊼ ♂
21 23	♀ bQ ♄
21 56	☽ ∥ ♀
22 47	☽ ⊼ ♃
23 48	☽ ⊼ ♇
20 Friday	
00 24	☿ bQ ♃
05 17	☽ ⊻ ♀
08 47	☽ ± ♃
12 05	☽ ⚹ ♅
15 16	☽ ⚹ ☉
16 18	☽ ⊥ ♀
17 27	☉ ∠ ♆
23 48	☽ bQ ♃
21 Saturday	
00 42	☽ □ ☿
05 49	☽ ☍ ♄
08 21	☽ Q ♅
08 53	☽ ∠ ♀
12 58	☽ Q ☉
16 44	☽ △ ♆
22 36	☽ ☍ ♂
22 Sunday	
00 56	☽ △ ♃
02 21	☽ ☍ ♇
05 02	☽ ∥ ♀
12 40	☽ ⚹ ♀
14 58	☽ □ ♅
15 26	♇ St R
18 06	♀ Q ♆
18 14	☽ bQ ♆
21 46	☽ □ ☉
23 Monday	
05 32	☽ △ ☿
08 43	☽ ⊼ ♄
09 38	☽ ± ♆
11 05	☽ Q ♀
16 34	♀ ⊻ ♅
19 00	☽ ± ♄
19 26	☽ # ♃
19 56	☽ ⊼ ♆
24 Tuesday	
03 40	☽ ⊼ ♂
03 44	☽ □ ♃
05 13	♂ ⚹ ♃
05 37	☽ ⊼ ♇
08 32	☽ bQ ☿
10 29	☽ bQ ♄
14 24	☽ ± ♂
15 58	☽ ± ♇
18 40	☽ △ ♅
20 18	☽ ∥ ☉
21 03	☽ □ ♀
25 Wednesday	
00 56	☽ ± ☿
05 17	☽ △ ☉
06 35	☽ bQ ♂
07 35	☽ bQ ♇
10 39	☽ Q ♃
11 07	☽ ∥ ♅
11 58	☽ ⊼ ☿
12 30	☽ △ ♄
20 52	☽ bQ ♅
21 28	☿ □ ♄
26 Thursday	
00 04	☽ ☍ ♆
07 28	☽ ⚹ ♃
09 31	☽ bQ ☉
09 38	☽ # ♆
09 47	☽ △ ♂
09 49	☽ △ ♇
11 00	♂ ☌ ♇
12 44	☽ ± ♅
23 23	☽ ⊼ ♅
27 Friday	
02 38	☽ ± ☉
05 29	♀ ± ♄
06 48	☽ △ ♀
07 50	♀ # ♇
09 45	☽ ∠ ♃
10 19	☽ ∥ ☿
14 08	☽ ⊼ ☉
17 21	☽ □ ♄
20 17	☽ ☍ ☿
28 Saturday	
01 42	☽ ⊥ ♃
04 47	☽ # ☿
05 24	☽ ⊼ ♆
12 21	☽ bQ ♀
12 25	☽ ⊻ ♃
15 19	☽ □ ♇
16 15	☽ ± ♆
17 17	☽ □ ♂
20 52	♀ ⊥ ♅
29 Sunday	
01 48	☽ Q ♄
04 55	☽ ∥ ♆
05 32	☽ ☍ ♅
06 22	☽ ± ♀
08 41	☽ bQ ♆
10 04	☉ △ ♄
18 32	☽ ⊼ ♀
20 38	♀ bQ ♇
23 50	☽ ⚹ ♄
30 Monday	
00 17	☽ Q ♇
00 58	☽ ☍ ☉
03 33	☽ Q ♂
05 25	☽ # ♅
06 04	♀ # ♄
07 15	☽ ⊼ ☿
12 31	☽ △ ♆
16 24	☿ ± ♃
19 11	☽ ☌ ♃
19 28	☽ ± ☿
22 43	☽ ⚹ ♇

All ephemeris data is given at 12.00 UT and the Moon's longitude is additionally given for 24.00 UT

MAY 2018

LONGITUDES

Date	Sidereal time h m s	Sun ☉ ° ′ ″	Moon ☽ ° ′ ″	Moon ☽ 24.00 ° ′ ″	Mercury ☿ ° ′	Venus ♀ ° ′	Mars ♂ ° ′	Jupiter ♃ ° ′	Saturn ♄ ° ′	Uranus ♅ ° ′	Neptune ♆ ° ′	Pluto ♇ ° ′
01	02 37 29	11 ♉ 03 52	28 ♏ 15 41	04 ♐ 30 53	14 ♈ 16	08 ♊ 16	23 ♑ 40	19 ♏ 19	09 ♑ 00	29 ♈ 13	15 ♓ 52	21 ♑ 16
02	02 41 25	12 02 05	10 ♐ 42 24	16 ♐ 50 24	15 21	09 28	24 08	19 R 11	08 R 59	29 17	15 54	21 R 16
03	02 45 22	13 00 16	22 ♐ 55 09	28 ♐ 56 56	16 28	10 41	24 35	19 03	08 57	29 20	15 55	21 16
04	02 49 18	13 58 25	04 ♑ 56 10	10 ♑ 53 16	17 39	11 54	25 03	18 56	08 56	29 23	15 56	21 15
05	02 53 15	14 56 33	16 ♑ 48 46	22 ♑ 43 13	18 51	13 06	25 30	18 48	08 54	29 27	15 58	21 15
06	02 57 12	15 54 40	28 ♑ 37 13	04 ♒ 31 25	20 07	14 19	25 56	18 41	08 53	29 30	15 59	21 14
07	03 01 08	16 52 45	10 ♒ 26 28	16 ♒ 23 04	21 24	15 31	26 23	18 33	08 51	29 33	16 01	21 14
08	03 05 05	17 50 49	22 ♒ 21 54	28 ♒ 23 39	22 44	16 43	26 49	18 25	08 49	29 37	16 02	21 14
09	03 09 01	18 48 51	04 ♓ 28 59	10 ♓ 38 34	24 07	17 56	27 15	18 18	08 47	29 40	16 03	21 13
10	03 12 58	19 46 52	16 ♓ 52 58	23 ♓ 12 43	25 31	19 08	27 40	18 10	08 45	29 43	16 04	21 13
11	03 16 54	20 44 51	29 ♓ 38 17	06 ♈ 09 59	26 58	20 20	28 06	18 02	08 43	29 47	16 06	21 12
12	03 20 51	21 42 49	12 ♈ 48 03	19 ♈ 32 33	28 26	21 33	28 31	17 55	08 40	29 50	16 07	21 12
13	03 24 47	22 40 46	26 ♈ 23 26	03 ♉ 20 27	29 ♈ 57	22 45	28 55	17 47	08 38	29 53	16 08	21 11
14	03 28 44	23 38 42	10 ♉ 23 12	17 ♉ 31 07	01 ♉ 31	23 57	29 19	17 40	08 36	29 ♈ 56	16 09	21 10
15	03 32 41	24 36 36	24 ♉ 43 29	01 ♊ 59 30	03 06	25 09	29 ♑ 43	17 32	08 33	00 ♉ 00	16 10	21 10
16	03 36 37	25 34 29	09 ♊ 18 14	16 ♊ 38 45	04 43	26 21	00 ♒ 07	17 25	08 31	00 03	16 11	21 09
17	03 40 34	26 32 20	24 ♊ 00 04	01 ♋ 21 17	06 23	27 33	00 30	17 17	08 28	00 06	16 12	21 08
18	03 44 30	27 30 09	08 ♋ 41 33	16 ♋ 00 06	08 04	28 45	00 53	17 10	08 25	00 09	16 13	21 08
19	03 48 27	28 27 57	23 ♋ 16 19	00 ♌ 29 41	09 48	29 ♊ 56	01 15	17 02	08 23	00 12	16 14	21 07
20	03 52 23	29 ♉ 25 44	07 ♌ 39 49	14 ♌ 46 29	11 34	01 ♋ 08	01 38	16 55	08 20	00 15	16 15	21 06
21	03 56 20	00 ♊ 23 28	21 ♌ 49 31	28 ♌ 48 50	13 22	02 20	01 59	16 48	08 17	00 18	16 16	21 05
22	04 00 16	01 21 11	05 ♍ 44 28	12 ♍ 36 28	15 12	03 32	02 21	16 40	08 14	00 22	16 17	21 05
23	04 04 13	02 18 52	19 ♍ 24 55	26 ♍ 09 55	17 04	04 43	02 42	16 33	08 11	00 25	16 18	21 04
24	04 08 10	03 16 32	02 ♎ 51 34	09 ♎ 29 59	18 58	05 55	03 02	16 26	08 08	00 28	16 19	21 03
25	04 12 06	04 14 10	16 ♎ 05 14	22 ♎ 37 24	20 54	07 06	03 22	16 19	08 04	00 31	16 20	21 02
26	04 16 03	05 11 47	29 ♎ 06 30	05 ♏ 32 36	22 52	08 18	03 42	16 12	08 01	00 34	16 21	21 01
27	04 19 59	06 09 22	11 ♏ 55 41	18 ♏ 15 47	24 52	09 29	04 01	16 05.	07 58	00 37	16 21	21 00
28	04 23 56	07 06 56	24 ♏ 32 56	00 ♐ 47 07	26 55	10 40	04 20	15 59	07 54	00 39	16 22	20 59
29	04 27 52	08 04 29	06 ♐ 58 26	13 ♐ 06 55	28 ♉ 58	11 51	04 38	15 52	07 51	00 42	16 23	20 58
30	04 31 49	09 02 01	19 ♐ 12 42	25 ♐ 15 55	01 ♊ 04	13 03	04 56	15 45	07 47	00 45	16 23	20 57
31	04 35 45	09 ♊ 59 32	01 ♑ 16 46	07 ♑ 15 30	03 ♊ 11	14 ♋ 14	05 ♒ 13	15 ♏ 39	07 ♑ 44	00 ♉ 48	16 ♓ 24	20 ♑ 56

Date	Moon True ☊ ° ′	Moon Mean ☊ ° ′	Moon ☽ Latitude ° ′
01	10 ♌ 12	10 ♌ 31	04 N 46
02	10 R 00	10 28	04 19
03	09 51	10 25	03 41
04	09 44	10 22	02 53
05	09 41	10 18	01 58
06	09 39	10 15	00 N 58
07	09 D 39	10 12	00 S 04
08	09 R 39	10 09	01 07
09	09 38	10 06	02 08
10	09 35	10 03	03 04
11	09 30	09 59	03 52
12	09 23	09 56	04 30
13	09 13	09 53	04 54
14	09 02	09 50	05 02
15	08 51	09 47	04 50
16	08 41	09 43	04 20
17	08 33	09 40	03 33
18	08 27	09 37	02 31
19	08 25	09 34	01 20
20	08 24	09 31	00 S 04
21	08 D 24	09 28	01 N 11
22	08 R 24	09 24	02 20
23	08 23	09 21	03 20
24	08 20	09 18	04 08
25	08 14	09 15	04 42
26	08 06	09 12	05 00
27	07 56	09 09	05 03
28	07 45	09 05	04 51
29	07 34	09 02	04 25
30	07 25	08 59	03 48
31	07 ♌ 17	08 ♌ 56	03 N 00

DECLINATIONS

Date	Sun ☉ ° ′	Moon ☽ ° ′	Mercury ☿ ° ′	Venus ♀ ° ′	Mars ♂ ° ′	Jupiter ♃ ° ′	Saturn ♄ ° ′	Uranus ♅ ° ′	Neptune ♆ ° ′	Pluto ♇ ° ′
01	15 N 09	15 S 07	02 N 57	22 N 33	22 S 40	16 S 19	22 S 15	10 N 42	06 S 26	21 S 29
02	15 27	17 47	03 20	22 47	22 38	16 17	22 15	10 44	06 25	21 29
03	15 44	19 35	03 44	23 00	22 35	16 15	22 15	10 45	06 25	21 29
04	16 02	20 28	04 09	23 12	22 33	16 13	22 15	10 46	06 24	21 29
05	16 19	20 26	04 36	23 24	22 30	16 11	22 15	10 47	06 24	21 29
06	16 36	19 29	05 05	23 36	22 28	16 09	22 15	10 48	06 23	21 29
07	16 53	17 41	05 34	23 46	22 26	16 07	22 15	10 49	06 23	21 30
08	17 09	15 07	06 05	23 56	22 23	16 05	22 15	10 51	06 22	21 30
09	17 25	11 51	06 37	24 06	22 21	16 03	22 16	10 52	06 22	21 30
10	17 41	08 00	07 10	24 14	22 18	16 01	22 16	10 53	06 21	21 30
11	17 56	03 S 42	07 43	24 22	22 16	15 59	22 16	10 54	06 21	21 30
12	18 11	00 N 54	08 18	24 29	22 14	15 57	22 16	10 55	06 20	21 30
13	18 26	05 36	08 54	24 36	22 12	15 55	22 16	10 56	06 20	21 31
14	18 41	10 09	09 31	24 42	22 09	15 53	22 16	10 57	06 19	21 31
15	18 55	14 15	10 08	24 47	22 07	15 51	22 17	10 59	06 19	21 31
16	19 09	17 33	10 46	24 51	22 05	15 49	22 17	11 00	06 19	21 31
17	19 23	19 46	11 25	24 55	22 03	15 47	22 17	11 01	06 18	21 31
18	19 36	20 38	12 04	24 58	22 01	15 45	22 17	11 02	06 18	21 32
19	19 49	20 07	12 43	25 01	21 59	15 43	22 17	11 03	06 18	21 32
20	20 01	18 17	13 23	25 02	21 57	15 41	22 17	11 04	06 17	21 32
21	20 14	15 21	14 04	25 03	21 55	15 39	22 18	11 05	06 17	21 32
22	20 26	11 35	14 44	25 03	21 54	15 37	22 18	11 06	06 17	21 32
23	20 37	07 15	15 25	25 03	21 52	15 36	22 18	11 07	06 16	21 33
24	20 48	02 N 39	16 05	25 01	21 51	15 34	22 18	11 08	06 16	21 33
25	20 59	01 S 59	16 45	24 59	21 49	15 32	22 19	11 09	06 16	21 33
26	21 10	06 28	17 25	24 57	21 48	15 30	22 19	11 10	06 15	21 33
27	21 20	10 36	18 05	24 53	21 47	15 28	22 19	11 11	06 15	21 33
28	21 30	14 12	18 44	24 49	21 46	15 26	22 19	11 12	06 15	21 34
29	21 39	17 07	19 21	24 45	21 45	15 25	22 19	11 13	06 15	21 34
30	21 48	19 13	19 58	24 39	21 44	15 23	22 20	11 14	06 15	21 34
31	21 N 57	20 S 26	20 N 34	24 N 33	21 S 43	15 S 21	22 S 20	11 N 15	06 S 14	21 S 34

ZODIAC SIGN ENTRIES

Date	h	m	Planets
01	15	19	☽ ♐
04	02	06	☽ ♑
06	14	48	☽ ♒
09	03	11	☽ ♓
11	12	40	☽ ♈
13	12	40	☿ ♉
13	18	15	☽ ♉
15	15	16	♅ ♉
15	20	43	☽ ♊
16	04	55	♂ ♒
17	21	47	☽ ♋
19	13	11	♀ ♋
19	23	11	☽ ♌
21	02	15	☉ ♊
22	02	03	☽ ♍
24	06	52	☽ ♎
26	13	39	☽ ♏
28	22	29	☽ ♐
29	23	49	☿ ♊
31	09	26	☽ ♑

LATITUDES

Date	Mercury ☿ ° ′	Venus ♀ ° ′	Mars ♂ ° ′	Jupiter ♃ ° ′	Saturn ♄ ° ′	Uranus ♅ ° ′	Neptune ♆ ° ′	Pluto ♇ ° ′
01	02 S 54	00 N 53	01 S 20	01 N 16	00 N 53	00 S 31	00 S 56	00 N 17
04	03 00	01 00	01 27	01 16	00 53	00 31	00 56	00 16
07	03 00	01 08	01 35	01 16	00 53	00 31	00 56	00 16
10	02 54	01 15	01 43	01 16	00 53	00 31	00 56	00 16
13	02 44	01 22	01 52	01 16	00 53	00 31	00 56	00 16
16	02 28	01 28	02 00	01 15	00 53	00 31	00 56	00 15
19	02 08	01 34	02 10	01 15	00 53	00 32	00 56	00 15
22	01 44	01 40	02 19	01 15	00 53	00 32	00 56	00 15
25	01 16	01 45	02 29	01 14	00 53	00 32	00 57	00 15
28	00 46	01 49	02 40	01 14	00 53	00 32	00 57	00 14
31	00 S 14	01 N 53	02 S 51	01 N 13	00 N 53	00 S 32	00 S 57	00 N 14

DATA

Julian Date	2458240
Delta T	+70 seconds
Ayanamsa	24° 06′ 33″
Synetic vernal point	05° ♓ 00′ 26″
True obliquity of ecliptic	23° 26′ 07″

MOON'S PHASES, APSIDES AND POSITIONS ☽

Date	h	m	Phase	Longitude ° ′	Eclipse Indicator
08	02	09	☾	17 ♒ 27	
15	11	48	●	24 ♉ 36	
22	03	49	☽	01 ♍ 02	
29	14	20	○	08 ♐ 10	

Day	h	m		
06	00	30	Apogee	
17	20	59	Perigee	
04	23	00	Max dec	20° S 34′
12	07	22	0N	
18	15	04	Max dec	20° N 39′
25	01	41	0S	

LONGITUDES

Date	Chiron ⚷ ° ′	Ceres ⚳ ° ′	Pallas ⚴ ° ′	Juno ⚵ ° ′	Vesta ⚶ ° ′	Black Moon Lilith ⚸ ° ′
01	00 ♈ 44	10 ♌ 38	18 ♊ 16	01 ♈ 31	04 ♑ 43	19 ♑ 16
11	01 ♈ 11	13 ♌ 18	24 ♊ 06	06 ♈ 17	04 ♑ 52	20 ♑ 23
21	01 ♈ 35	16 ♌ 20	29 ♊ 58	11 ♈ 02	04 ♑ 12	21 ♑ 30
31	01 ♈ 54	19 ♌ 38	05 ♋ 52	15 ♈ 46	02 ♑ 45	22 ♑ 36

All ephemeris data is given at 12.00 UT and the Moon's longitude is additionally given for 24.00 UT

ASPECTARIAN

h m	Aspects
01 Tuesday	
02 56	☽ ⚹ ♂
03 55	☽ ∠ ♄
12 11	☽ ⋕ ☉
13 51	☽ ⊼ ♅
14 05	☽ ⚼ ☿
21 04	☽ ⊥ ♄
21 52	☽ ∥ ♃
22 57	♀ ⋕ ♂
02 Wednesday	
01 06	♀ ⚼ ♂
01 26	☽ ± ♅
02 24	♀ ⊼ ♄
03 23	☽ ∠ ♇
08 39	☽ ⊻ ♄
08 49	☽ ∠ ♂
09 20	☽ ☍ ♀
14 49	☽ ⊼ ☉
19 00	☽ ⚼ ♅
20 54	☽ ⊥ ♇
21 58	☽ △ ☿
22 10	☽ □ ♆
03 Thursday	
00 03	☿ ⊻ ♆
03 07	☽ ⊥ ♂
03 35	☽ ± ☉
04 27	☽ ⊻ ♃
08 43	☽ ⊻ ♇
15 27	☽ ⊻ ♂
16 12	☽ ⊥ ♃
23 00	☽ ⚼ ☉
04 Friday	
00 50	☽ △ ♅
10 00	☽ Q ♆
10 00	☽ ∠ ♃
20 02	☽ ☌ ♄
05 Saturday	
02 31	☉ ⋕ ♃
03 38	☽ ⊼ ♀
07 52	☽ △ ☉
10 17	☽ ⚹ ♆
11 05	☿ ⊼ ♃
16 00	☽ ⚹ ♃
16 38	☽ □ ☿
17 11	☽ ± ♀
21 00	☽ ☌ ♇
06 Sunday	
06 20	☽ ☌ ♂
13 34	☽ ⚼ ♀
13 48	☽ □ ♅
13 56	☉ ⚹ ♆
15 59	♀ ∠ ♅
16 08	☽ Q ♃
16 49	☽ ∠ ♆
07 Monday	
06 23	♀ ± ♇
08 47	☽ ⊻ ♄
08 52	☿ □ ♇
09 39	☽ Q ☿
11 08	☽ ⊥ ♆
19 30	☽ ⋕ ☉
20 53	☽ ⊥ ♄
21 58	♀ □ ♆
23 09	☿ ⊥ ♆
23 16	☽ ⊻ ♆
23 25	☽ △ ♀
08 Tuesday	
02 09	☽ □ ☉
02 26	☽ Q ♅
03 33	☽ ∥ ♃
04 11	☽ □ ♃
09 43	☽ ⊻ ♇
12 50	☽ ⚹ ☿
14 53	☽ ∠ ♄
21 12	☽ ⊻ ♂
21 41	☽ ⊥ ♇
09 Wednesday	
00 39	☉ ☍ ♃
00 57	☿ ⋕ ♆
02 29	☽ ⚹ ♅
09 29	☽ ⊥ ♂
15 24	☽ ∠ ♇
16 57	☽ Q ☉
18 26	☽ ⋕ ♅
18 36	♀ ⊼ ♃
20 22	☽ ⚹ ♄
22 10	☽ ∠ ☿
10 Thursday	
03 38	☽ ∠ ♂
07 51	☽ ∠ ♅
10 27	☽ ☌ ♆
14 26	☽ △ ♃
16 18	☽ ⋕ ☿
16 45	☽ □ ♀
17 39	☽ ⊥ ☿
17 58	☽ ⚹ ☉
19 20	☽ Q ♄
20 13	☽ ⚹ ♇
21 28	☽ ∥ ♆
11 Friday	
01 01	☽ ⊥ ♅
06 23	☽ ⊻ ☿
09 02	☽ ⚹ ♂
12 16	☽ ⊻ ♅
14 17	♂ ∥ ♄
18 13	☽ ⚼ ♃
18 34	☽ Q ♇
23 10	☉ △ ♇
12 Saturday	
00 08	☽ ∠ ☉
04 35	☽ □ ♄
05 05	♀ ⊼ ♇
05 33	☽ Q ♀
07 45	☽ Q ♂
10 25	☽ ± ♃
13 30	☿ □ ♂
17 37	☽ ⊥ ☉
17 56	☽ ⊻ ♆
21 02	☽ ⊼ ♃
13 Sunday	
02 54	☽ □ ♇
04 33	☽ ⊥ ♆
05 01	☽ ⚹ ♀
05 02	☽ ⊻ ☉
05 25	☉ ⊻ ♀
10 50	☿ ☌ ♅
15 44	☽ ⋕ ♆
16 31	☽ □ ♂
17 16	♀ ± ♂
18 05	☽ ☌ ♅
18 57	☽ ☌ ☿
20 13	☽ ∠ ♆
14 Monday	
06 28	☿ ∠ ♆
06 50	♀ ± ♃
07 55	☽ ∥ ☿
08 58	☽ △ ♄
09 20	☽ ∠ ♀
10 50	☉ ⚼ ♄
16 28	☽ ∥ ♅
21 44	☽ ⚹ ♆
15 Tuesday	
00 08	☽ ☍ ♃
01 53	☽ ⊥ ♀
03 22	♂ Q ♃
06 05	☽ △ ♇
10 04	☽ ⚼ ♄
11 48	☽ ☌ ☉
12 46	☽ ⊻ ♀
17 43	☽ Q ♆
20 30	☽ △ ♂
20 44	☽ ⊻ ♅
22 42	☽ ⋕ ♃
16 Wednesday	
00 53	☽ ± ♄
03 32	☽ ⊻ ☿
06 38	☽ ⊥ ♅
06 50	☽ ⚼ ♇
07 04	♂ □ ♅
10 42	☽ ⊼ ♄
14 37	☽ ⊥ ☿
20 51	☿ ∥ ♅
21 25	☽ ∠ ♅
21 33	☽ ± ♇
21 46	☽ ⚼ ♂
23 16	☽ □ ♆
17 Thursday	
01 08	☽ ⊼ ♃
05 57	☽ ∥ ☉
07 10	☽ ∠ ☿
07 20	☽ ⊼ ♇
10 50	☽ ± ♃
12 50	☽ ± ♂
16 26	☽ ⊻ ☉
18 18	☽ ☌ ♀
21 59	☽ ⚹ ♅
22 53	☽ ⊼ ♂
18 Friday	
01 24	☽ ⚼ ♃
02 55	☽ ⊥ ☉
10 51	☽ ⚹ ☿
11 34	☽ ☍ ♄
16 49	☿ △ ♄
17 42	☽ Q ♅
18 41	☽ ∠ ☉
19 Saturday	
00 23	☽ △ ♆
01 47	☽ △ ♃
06 17	☉ Q ♆
08 26	☽ ☍ ♇
09 14	☽ Q ☿
10 50	♂ ∠ ♆
16 46	☽ ∥ ☉
17 30	♀ ⚹ ♅
21 14	☽ ⚹ ☉
23 33	☽ □ ♅
20 Sunday	
00 05	☽ ⊻ ♀
01 16	☽ ⚼ ♆
01 37	☽ ☍ ♂
11 02	☽ ⊥ ♀
13 07	☽ ⊼ ♄
16 22	☽ ± ♆
18 48	☽ Q ☉
19 31	☽ □ ☿
23 12	☽ ± ♄
21 Monday	
02 06	♀ ⊼ ♂
02 08	♀ ⚼ ♃
02 32	☽ ⊼ ♆
03 30	☽ □ ♃
03 38	☽ ∠ ♀
09 47	☽ ⋕ ♃
09 47	☉ ⊻ ♅
10 45	☽ ⊼ ♇
14 29	☽ ⚼ ♄
19 21	☽ ∥ ☿
21 01	☽ ± ♇
22 Tuesday	
02 38	☽ △ ♅
03 49	☽ □ ☉
05 57	☽ ⊼ ♂
07 47	☽ ⚹ ♀
10 09	☽ Q ♃
12 35	☽ ⚼ ♇
14 45	☽ ∥ ♅
16 19	☽ △ ♄
16 39	☽ ± ♂
23 Wednesday	
02 13	☿ ⚹ ♆
04 54	☽ ⚼ ♅
05 54	☿ ☍ ♃
06 30	☽ ☍ ♆
06 47	☽ Q ♀
06 59	☽ ⚹ ♃
07 10	☽ △ ☿
08 46	☉ ± ♄
08 52	☽ ⚼ ♂
14 55	☽ △ ♇
17 14	☽ ⋕ ♆
18 13	☿ ⋕ ♃
20 54	☽ ± ♅
24 Thursday	
02 39	☉ △ ♂
07 40	☽ ⊼ ♅
09 28	☽ ∠ ♃
12 19	☽ △ ♂
12 48	☽ △ ☉
14 19	☽ ⚼ ☿
18 03	☽ □ ♀
21 28	☽ □ ♄
25 Friday	
01 35	☽ ⊥ ♃
09 27	☽ ± ☿
09 52	♃ △ ♆
12 25	☽ ⊻ ♃
12 27	☽ ⊼ ♆
13 38	☿ △ ♇
18 14	☽ ⚼ ☉
21 04	☽ □ ♇
22 23	☽ ⊼ ☿
23 28	☽ ± ♆
26 Saturday	
06 18	☽ Q ♄
06 40	♀ ☍ ♄
10 49	☽ ∥ ♆
12 11	☽ ± ☉
13 43	☿ ⚼ ♄
14 42	☽ ☍ ♅
16 10	☽ ⚼ ♆
20 46	☽ □ ♂
27 Sunday	
00 16	☽ ⊼ ☉
00 35	☿ ∠ ♀
04 34	☽ ⚹ ♄
06 30	☽ Q ♇
06 55	☽ △ ♀
08 16	☉ ⚼ ♇
15 42	☽ ⋕ ♅
19 48	☽ ☌ ♃
20 23	☽ △ ♆
23 55	☉ ⊥ ♅
28 Monday	
05 12	☽ ⚹ ♇
07 39	☽ Q ♂
08 52	☽ ∠ ♄
14 23	☽ ⚼ ♀
17 25	☽ ☍ ☿
21 26	☽ ∥ ♃
22 29	☉ ⋕ ♇
23 48	☽ ⊼ ♅
29 Tuesday	
02 06	☽ ⊥ ♄
05 05	☿ Q ♆
06 36	☉ ⊼ ♄
07 20	☽ ⚹ ♂
09 36	☽ ± ♀
10 03	☽ ∠ ♇
11 29	☽ ± ♅
13 41	☽ ⊻ ♄
14 20	☽ ☍ ☉
22 33	☽ ⊼ ♀
30 Wednesday	
01 49	☉ ⋕ ♂
03 38	☽ ⊥ ♇
05 09	☽ ⚼ ♅
05 15	☽ ⊻ ♃
05 55	♀ Q ♅
06 26	☽ □ ♆
08 22	☿ ⊻ ♅
13 27	☽ ∠ ♂
15 27	☽ ⊻ ♇
16 59	☽ ⊥ ♃
19 59	☿ ± ♄
31 Thursday	
07 47	☽ ⊥ ♂
10 45	☽ ∠ ♃
11 02	☽ △ ♅
16 39	☽ ⊼ ☿
18 15	☽ Q ♆
20 05	☽ ⊻ ♂

JUNE 2018

Date	Sidereal time h m s	LONGITUDES Sun ☉	Moon ☽	Moon ☽ 24.00	Mercury ☿	Venus ♀	Mars ♂	Jupiter ♃	Saturn ♄	Uranus ♅	Neptune ♆	Pluto ♇
01	04 39 42	10 ♊ 57 02	13 ♑ 12 25	19 ♑ 07 50	05 ♊ 20	15 ♋ 25	05 ♒ 30	15 ♏ 33	07 ♑ 40	00 ♉ 51	16 ♓ 25	20 ♑ 55
02	04 43 39	11 54 31	25 ♑ 02 10	00 ♒ 55 52	07 29	16 36	05 46	15 R 26	07 R 36	00 54	16 25	20 R 54
03	04 47 35	12 51 59	06 ♒ 49 24	12 ♒ 43 20	09 40	17 46	06 02	15 20	07 33	00 56	16 26	20 53
04	04 51 32	13 49 26	18 ♒ 38 13	24 ♒ 34 39	11 51	18 57	06 17	15 14	07 29	00 59	16 26	20 52
05	04 55 28	14 46 52	00 ♓ 33 17	06 ♓ 34 45	14 03	20 08	06 32	15 08	07 25	01 02	16 27	20 51
06	04 59 25	15 44 18	12 ♓ 39 42	18 ♓ 48 47	16 15	21 19	06 46	15 03	07 21	01 05	16 27	20 50
07	05 03 21	16 41 43	25 ♓ 02 37	01 ♈ 21 47	18 27	22 29	07 00	14 57	07 17	01 07	16 27	20 49
08	05 07 18	17 39 08	07 ♈ 46 51	14 ♈ 18 14	20 39	23 40	07 13	14 52	07 13	01 10	16 28	20 48
09	05 11 14	18 36 32	20 ♈ 56 19	27 ♈ 41 19	22 50	24 50	07 25	14 46	07 09	01 12	16 28	20 46
10	05 15 11	19 33 55	04 ♉ 33 21	11 ♉ 32 18	25 01	26 00	07 37	14 41	07 05	01 15	16 28	20 45
11	05 19 08	20 31 18	18 ♉ 37 57	25 ♉ 49 50	27 10	27 11	07 48	14 36	07 01	01 18	16 29	20 44
12	05 23 04	21 28 40	03 ♊ 07 20	10 ♊ 29 40	29 ♊ 18	28 21	07 58	14 31	06 57	01 20	16 29	20 43
13	05 27 01	22 26 02	17 ♊ 55 52	25 ♊ 24 53	01 ♋ 24	29 ♋ 31	08 08	14 26	06 52	01 22	16 29	20 42
14	05 30 57	23 23 23	02 ♋ 55 35	10 ♋ 26 50	03 29	00 ♌ 41	08 17	14 22	06 48	01 25	16 29	20 40
15	05 34 54	24 20 44	17 ♋ 57 29	25 ♋ 26 30	05 32	01 51	08 26	14 17	06 44	01 27	16 29	20 39
16	05 38 50	25 18 04	02 ♌ 52 55	10 ♌ 15 55	07 33	03 01	08 34	14 13	06 40	01 30	16 29	20 38
17	05 42 47	26 15 22	17 ♌ 34 50	24 ♌ 49 09	09 32	04 11	08 41	14 09	06 35	01 32	16 30	20 37
18	05 46 43	27 12 40	01 ♍ 58 30	09 ♍ 02 39	11 29	05 20	08 47	14 05	06 31	01 34	16 30	20 35
19	05 50 40	28 09 57	16 ♍ 01 31	22 ♍ 55 06	13 23	06 30	08 53	14 01	06 27	01 37	16 R 30	20 34
20	05 54 37	29 ♊ 07 14	29 ♍ 43 29	06 ♎ 26 50	15 16	07 40	08 58	13 57	06 22	01 39	16 30	20 33
21	05 58 33	00 ♋ 04 29	13 ♎ 05 22	19 ♎ 39 19	17 06	08 49	09 02	13 54	06 18	01 41	16 29	20 31
22	06 02 30	01 01 44	26 ♎ 08 57	02 ♏ 34 33	18 54	09 58	09 06	13 50	06 14	01 43	16 29	20 30
23	06 06 26	01 58 58	08 ♏ 56 21	15 ♏ 14 39	20 40	11 08	09 09	13 47	06 09	01 45	16 29	20 28
24	06 10 23	02 56 11	21 ♏ 29 41	27 ♏ 41 41	22 23	12 17	09 11	13 44	06 05	01 47	16 29	20 27
25	06 14 19	03 53 24	03 ♐ 50 55	09 ♐ 57 33	24 04	13 26	09 12	13 42	06 00	01 49	16 29	20 26
26	06 18 16	04 50 37	16 ♐ 01 50	22 ♐ 03 58	25 43	14 35	09 13	13 39	05 56	01 51	16 29	20 24
27	06 22 12	05 47 49	28 ♐ 04 10	04 ♑ 02 38	27 20	15 43	09 R 13	13 36	05 52	01 53	16 28	20 23
28	06 26 09	06 45 01	09 ♑ 59 37	15 ♑ 55 20	28 ♋ 54	16 52	09 12	13 34	05 47	01 55	16 28	20 22
29	06 30 06	07 42 13	21 ♑ 50 05	27 ♑ 44 07	00 ♌ 25	18 01	09 10	13 32	05 43	01 57	16 28	20 20
30	06 34 02	08 ♋ 39 24	03 ♒ 37 46	09 ♒ 31 23	01 ♌ 55	19 ♌ 09	09 ♒ 08	13 ♏ 30	05 ♑ 38	01 ♉ 59	16 ♓ 27	20 ♑ 19

Date	Moon True ☊	Moon Mean ☊	Moon ☽ Latitude
01	07 ♌ 12	08 ♌ 53	02 N 05
02	07 R 09	08 49	01 05
03	07 08	08 46	00 N 02
04	07 D 09	08 43	01 S 02
05	07 10	08 40	02 03
06	07 11	08 37	03 00
07	07 R 10	08 34	03 49
08	07 08	08 30	04 29
09	07 04	08 27	04 57
10	06 59	08 24	05 08
11	06 52	08 21	05 02
12	06 45	08 18	04 37
13	06 39	08 15	03 53
14	06 34	08 11	02 52
15	06 30	08 08	01 39
16	06 29	08 05	00 S 20
17	06 D 29	08 02	01 N 00
18	06 31	07 59	02 14
19	06 32	07 55	03 19
20	06 R 32	07 52	04 10
21	06 32	07 49	04 46
22	06 29	07 46	05 07
23	06 25	07 43	05 11
24	06 21	07 40	05 01
25	06 15	07 36	04 37
26	06 10	07 33	04 00
27	06 05	07 30	03 13
28	06 01	07 27	02 18
29	05 59	07 24	01 17
30	05 ♌ 58	07 ♌ 21	00 N 13

DECLINATIONS

Date	Sun ☉	Moon ☽	Mercury ☿	Venus ♀	Mars ♂	Jupiter ♃	Saturn ♄	Uranus ♅	Neptune ♆	Pluto ♇
01	22 N 05	20 S 42	21 N 08	24 N 26	21 S 43	15 S 20	22 S 20	11 N 16	06 S 14	21 S 35
02	22 13	20 04	21 41	24 19	21 42	15 18	22 20	11 17	06 14	21 35
03	22 20	18 32	22 12	24 10	21 42	15 16	22 21	11 18	06 14	21 35
04	22 27	16 13	22 41	24 02	21 42	15 15	22 21	11 19	06 14	21 35
05	22 34	13 11	23 07	23 52	21 42	15 13	22 21	11 20	06 14	21 36
06	22 40	09 34	23 31	23 42	21 42	15 12	22 21	11 21	06 13	21 36
07	22 46	05 29	23 53	23 31	21 42	15 11	22 22	11 22	06 13	21 36
08	22 52	01 S 02	24 12	23 20	21 43	15 09	22 22	11 23	06 13	21 36
09	22 57	03 N 35	24 29	23 08	21 44	15 08	22 22	11 24	06 13	21 36
10	23 01	08 12	24 42	22 55	21 45	15 06	22 23	11 25	06 13	21 37
11	23 06	12 31	24 53	22 42	21 46	15 05	22 23	11 26	06 13	21 37
12	23 10	16 15	25 01	22 28	21 47	15 04	22 23	11 26	06 13	21 38
13	23 13	19 02	25 06	22 13	21 49	15 03	22 23	11 27	06 13	21 38
14	23 16	20 32	25 09	21 58	21 50	15 02	22 24	11 28	06 13	21 38
15	23 18	20 36	25 09	21 43	21 52	15 01	22 24	11 29	06 13	21 38
16	23 21	19 12	25 06	21 26	21 54	14 59	22 24	11 30	06 13	21 39
17	23 23	16 31	25 01	21 10	21 57	14 58	22 24	11 30	06 13	21 39
18	23 24	12 52	24 53	20 52	21 59	14 58	22 25	11 31	06 13	21 39
19	23 25	08 34	24 44	20 35	22 02	14 57	22 25	11 32	06 13	21 40
20	23 26	03 N 56	24 32	20 16	22 05	14 56	22 25	11 33	06 13	21 40
21	23 26	00 S 46	24 18	19 58	22 08	14 55	22 26	11 33	06 13	21 40
22	23 26	05 20	24 02	19 38	22 11	14 54	22 26	11 34	06 13	21 40
23	23 25	09 33	23 45	19 18	22 15	14 54	22 26	11 35	06 13	21 41
24	23 24	13 17	23 26	18 58	22 19	14 53	22 26	11 36	06 13	21 41
25	23 23	16 23	23 06	18 37	22 22	14 52	22 27	11 36	06 13	21 41
26	23 21	18 43	22 44	18 16	22 27	14 52	22 27	11 37	06 14	21 42
27	23 18	20 12	22 21	17 55	22 31	14 51	22 27	11 38	06 14	21 42
28	23 16	20 46	21 57	17 33	22 35	14 51	22 28	11 38	06 14	21 42
29	23 13	20 24	21 33	17 10	22 40	14 50	22 28	11 39	06 14	21 43
30	23 N 09	19 S 08	21 N 07	16 N 47	22 S 45	14 S 50	22 S 28	11 N 39	06 S 14	21 S 43

ZODIAC SIGN ENTRIES

Date	h	m	Planets
02	22	06	☽ ♒
05	10	53	☽ ♓
07	21	26	☽ ♈
10	04	04	☽ ♉
12	06	53	☽ ♊
12	20	00	☿ ♋
13	21	54	♀ ♌
14	07	20	☽ ♋
16	07	21	☽ ♌
18	08	40	☽ ♍
20	12	29	☽ ♎
21	10	07	☉ ♋
22	19	11	☽ ♏
25	04	29	☽ ♐
27	15	52	☽ ♑
29	05	16	☿ ♌
30	04	37	☽ ♒

LATITUDES

Date	Mercury ☿	Venus ♀	Mars ♂	Jupiter ♃	Saturn ♄	Uranus ♅	Neptune ♆	Pluto ♇
01	00 S 03	01 N 54	02 S 55	01 N 13	00 N 53	00 S 32	00 S 57	00 N 14
04	00 N 29	01 57	03 06	01 12	00 53	00 32	00 57	00 14
07	00 57	01 59	03 18	01 12	00 52	00 32	00 57	00 13
10	01 22	02 01	03 30	01 11	00 52	00 32	00 57	00 13
13	01 41	02 01	03 43	01 10	00 52	00 32	00 57	00 13
16	01 53	02 01	03 56	01 10	00 52	00 32	00 57	00 13
19	01 58	02 00	04 09	01 09	00 52	00 32	00 58	00 12
22	01 57	01 58	04 22	01 08	00 52	00 32	00 58	00 12
25	01 50	01 55	04 36	01 08	00 52	00 32	00 58	00 12
28	01 37	01 51	04 49	01 07	00 51	00 32	00 58	00 12
31	01 N 18	01 N 46	05 S 02	01 N 06	00 N 51	00 S 32	00 S 58	00 N 11

DATA

Julian Date	2458271
Delta T	+70 seconds
Ayanamsa	24° 06' 37"
Synetic vernal point	05° ♓ 00' 22"
True obliquity of ecliptic	23° 26' 07"

MOON'S PHASES, APSIDES AND POSITIONS ☽

Date	h	m	Phase	Longitude	Eclipse Indicator
06	18	32	☾	16 ♓ 00	
13	19	43	●	22 ♊ 44	
20	10	51	☽	29 ♍ 04	
28	04	53	○	06 ♑ 28	

Day	h	m		
02	16	25	Apogee	
14	23	47	Perigee	
30	02	28	Apogee	
01	07	09	Max dec	20° S 44'
08	17	26	ON	
15	00	55	Max dec	20° N 45'
21	08	02	OS	
28	14	27	Max dec	20° S 46'

LONGITUDES

Date	Chiron ⚷	Ceres ⚳	Pallas ⚴	Juno ⚵	Vesta ⚶	Black Moon Lilith ⚸
01	01 ♈ 56	19 ♌ 59	06 ♋ 27	16 ♈ 14	02 ♑ 34	22 ♑ 43
11	02 ♈ 11	23 ♌ 33	12 ♋ 20	21 ♈ 55	00 ♑ 29	23 ♑ 49
21	02 ♈ 20	27 ♌ 19	18 ♋ 12	25 ♈ 32	28 ♐ 06	24 ♑ 56
31	02 ♈ 25	01 ♍ 15	24 ♋ 01	00 ♉ 04	25 ♐ 47	26 ♑ 02

ASPECTARIAN

h m Aspects

01 Friday
00 53 ☽ ☌ ♄
07 03 ☽ ⚻ ☉
07 22 ☽ ± ☿
14 13 ☿ △ ♂
14 29 ♀ △ ♃
16 41 ☽ ✶ ♃
16 57 ☽ ☍ ♀
18 29 ☽ ✶ ♆
18 36 ☿ ⚼ ♇
20 15 ☽ ± ☉

02 Saturday
03 37 ☽ ☌ ♇
05 18 ☿ ⊥ ♅
05 39 ☽ ⚼ ☿
07 23 ☿ # ♇
08 26 ♀ △ ♆
12 57 ☿ # ♂
13 16 ☿ ⚻ ♄
16 09 ☽ ⚼ ☉
16 51 ☽ Q ♃
23 58 ☽ □ ♅

03 Sunday
01 00 ☽ ∠ ♆
10 22 ☽ ☌ ♂
13 27 ☽ ⚺ ♄
13 30 ☉ # ♄
19 06 ☽ △ ☿
19 14 ☿ # ♄
19 20 ☽ ⊥ ♆
21 03 ☉ ∥ ☿

04 Monday
01 23 ☽ △ ☉
01 36 ☽ ⊥ ♄
05 10 ☽ □ ♃
07 32 ☽ ⚺ ♆
12 43 ☽ Q ♅
12 43 ☽ ⚻ ♀
16 31 ☽ ⚺ ♇
19 44 ☽ ∠ ♄
20 19 ☽ ∥ ♃

05 Tuesday
02 09 ☽ ± ♀
04 35 ☽ ⊥ ♇
12 58 ☽ ✶ ♅
13 43 ☉ ± ♇
13 51 ☿ ⊥ ♀
20 10 ☉ ⚻ ♃
20 37 ☿ ± ♇
22 07 ☽ ⚼ ♀
22 32 ☽ ∠ ♇
23 21 ☿ ⚻ ♃

06 Wednesday
00 09 ☽ ⚺ ♂
00 41 ☽ # ♅
01 35 ☽ ✶ ♄
02 02 ☉ ☌ ☿
02 25 ♀ ☍ ♇
10 01 ☿ ∠ ♅
12 13 ☽ ⊥ ♂
14 07 ☿ □ ♆
16 38 ☽ △ ♃
18 32 ☽ □ ☉
18 42 ☽ ∠ ♅
19 25 ☽ ☌ ♆
19 35 ☿ ∥ ♀
20 34 ☽ □ ☿
20 54 ☉ ∠ ♅

07 Thursday
00 58 ☽ Q ♄
03 53 ☽ ✶ ♇
05 58 ☉ □ ♆
06 03 ☽ ∠ ♂
06 35 ☽ △ ♀
07 48 ☽ ∥ ♆
12 09 ☽ ⊥ ♅
21 16 ☽ ⚼ ♃
23 35 ☽ ⚺ ♅

08 Friday
02 42 ☽ Q ♇
07 43 ☽ Q ☉
10 56 ☽ ✶ ♂
10 57 ☽ □ ♄
11 01 ☉ ⊥ ♀
12 22 ♂ ⚺ ♄
13 32 ☿ ⚻ ♇
13 57 ☽ Q ☿
13 59 ☽ ± ♃
14 10 ☿ ± ♃

09 Saturday
00 55 ☽ ⚻ ♃
03 56 ☽ ⚺ ♆
06 54 ☿ ⚼ ♂
07 29 ☽ ✶ ☉
09 14 ☽ Q ♂
11 42 ☽ □ ♇
14 44 ☽ ⊥ ♆
16 03 ☽ ✶ ☿
19 37 ☽ □ ♀

10 Sunday
01 37 ☽ # ♆
03 05 ☉ ∥ ♀
06 14 ☽ ☌ ♅
06 38 ☽ ∠ ♆
12 01 ☽ ∠ ☉
16 21 ☽ △ ♄
17 21 ☽ □ ♂
23 06 ☽ ∠ ☿

11 Monday
04 34 ☽ ⊥ ☉
05 14 ☽ ☍ ♃
05 39 ☽ Q ♀
05 40 ☽ ∥ ♅
08 22 ☽ ✶ ♆
12 23 ☿ ⚺ ♀
13 47 ☉ ± ♃
15 24 ☽ ⚺ ☉
15 31 ☽ △ ♇
16 59 ☽ ⊥ ☿
17 14 ☉ ⚻ ♇
17 38 ☽ ⚼ ♄

12 Tuesday
03 29 ☽ ✶ ♀
03 51 ☽ # ♃
04 23 ☽ Q ♆
04 39 ☽ ⚺ ☿
08 27 ☽ ± ♄
09 04 ☽ ⚺ ♅
14 25 ☿ ⚼ ♃
16 14 ☽ ⚼ ♇
18 12 ☽ ⚻ ♄
18 53 ☽ ⊥ ♅
19 51 ♀ # ♄
20 00 ☽ △ ♂

13 Wednesday
06 02 ☽ ∠ ♀
06 24 ☽ ⚻ ♃
06 48 ☽ ± ♇
09 29 ☽ ∠ ♅
09 40 ☽ □ ♆
11 41 ☿ ✶ ♅
16 00 ☽ ± ♃
16 26 ☽ ⚻ ♇
19 43 ☽ ☌ ☉
20 27 ☽ ⚼ ♂
21 09 ☿ ± ♂
21 43 ☽ ⊥ ♀

14 Thursday
06 20 ☽ ⚼ ♃
08 07 ☽ ⚺ ♀
09 04 ☉ ⚼ ♂
09 35 ☽ ✶ ♅
10 59 ☽ ± ♂
13 02 ☽ ☌ ☿
18 09 ☽ ☍ ♄
20 39 ☽ ⚻ ♂
22 59 ♀ # ♂

15 Friday
03 33 ♀ □ ♅
04 31 ♀ ⚼ ♆
04 47 ☽ Q ♅
06 09 ☽ △ ♃
09 39 ☽ △ ♆
16 18 ☽ ☍ ♇
18 15 ♀ # ♇
22 56 ☽ ⚺ ☉

16 Saturday
01 47 ☿ ☍ ♄
09 16 ☽ ⊥ ☉
09 41 ♅ ∠ ♆
09 45 ☽ ⚼ ♆
09 45 ☽ □ ♅
12 14 ☽ ☌ ♀
18 06 ☽ ⚻ ♄
20 45 ☽ ⚺ ☿
21 18 ☽ ☍ ♂

17 Sunday
00 22 ☽ ± ♆
00 54 ☽ ∠ ☉
01 03 ☿ ⚻ ♂
03 50 ☽ ± ♄
06 23 ☽ □ ♃
08 06 ☽ ⊥ ☿
10 12 ☽ ⚻ ♆
17 00 ☽ ⚻ ♇
18 36 ☽ ⚼ ♄
22 49 ☽ # ♃

18 Monday
01 20 ☽ ∠ ☿
02 58 ☽ ± ♇
03 25 ☽ ✶ ☉
11 19 ☽ △ ♅
12 10 ☽ Q ♃
18 06 ☽ ⚼ ♇
18 12 ☽ ⚺ ♀
19 39 ☽ △ ♄
19 45 ☽ ∥ ♅
23 29 ♆ St R
23 39 ☽ ⚻ ♂

19 Tuesday
01 11 ☽ Q ☉
05 22 ☽ ⊥ ♀
06 44 ☽ ✶ ☿
08 33 ☽ ✶ ♃
10 01 ☽ ± ♂
10 54 ♀ ⚻ ♄
12 49 ☽ ☍ ♆
13 01 ☽ ⚼ ♅
14 50 ☿ Q ♅
19 43 ☿ △ ♃
19 52 ☽ △ ♇
22 24 ☽ ∠ ♀

20 Wednesday
00 17 ☽ # ♆
01 47 ☽ ⚼ ♂
04 46 ☽ ± ♅
06 57 ☽ Q ☿
08 04 ☉ ⚼ ♃
10 38 ☽ ∠ ♃
10 51 ☽ □ ☉
15 25 ☽ ⚻ ♅
23 48 ☽ □ ♄

21 Thursday
02 39 ☽ ⊥ ♃
03 32 ☽ ✶ ♀
03 58 ☿ △ ♆
04 38 ☽ △ ♂
13 28 ☽ ⚺ ♃
16 54 ♀ ☍ ♂
18 12 ☽ ⚻ ♆
20 29 ☽ □ ☿

22 Friday
01 34 ☽ □ ♇
03 31 ☽ Q ♀
05 13 ☽ ± ♆
08 27 ☽ Q ♄
16 53 ☽ ∥ ♆
21 50 ☽ △ ☉
21 58 ☽ ⚼ ♆
22 25 ☽ ☍ ♅
22 44 ♀ ± ♆

23 Saturday
05 58 ☉ ✶ ♅
06 46 ☽ ✶ ♄
09 26 ☿ ☍ ♇
11 07 ☽ Q ♇
12 24 ☽ □ ♂
16 34 ☽ □ ♀
21 11 ☽ ☌ ♃

24 Sunday
00 38 ☽ # ♅
02 22 ☽ △ ♆
04 36 ☽ ⚼ ☉
08 08 ♀ ± ♄
10 00 ☽ ✶ ♇
11 12 ☽ ∠ ♄
14 00 ☽ △ ☿
14 45 ☉ ∥ ☿
18 23 ☉ ± ♂
23 02 ☽ Q ♂
23 24 ☽ ± ☉
23 39 ☽ ∥ ♃

25 Monday
04 33 ☽ ⊥ ♄
08 01 ☽ ⚻ ♅
12 05 ☽ ⚻ ☉
15 05 ☽ ∠ ♇
16 12 ☽ ⚺ ♄
17 19 ♀ □ ♃
19 49 ☽ ± ♅
22 32 ☽ ✶ ♂
23 52 ☽ ⚼ ☿

26 Tuesday
07 18 ☽ ⚺ ♃
07 24 ☽ # ♀
08 47 ☽ ⊥ ♇
08 49 ☽ △ ♀
12 53 ☽ □ ♆
13 38 ☽ ⚼ ♅
14 09 ♂ ∥ ♄
19 10 ☽ ⊥ ♃
20 28 ☽ ± ☿
20 40 ☽ ⚺ ♇
21 04 ♂ St R

27 Wednesday
03 45 ☿ # ♂
04 18 ☽ ∠ ♂
06 07 ☿ # ♄
10 17 ☽ ⚻ ☿
13 04 ☽ ∠ ♃
13 28 ☉ ☍ ♄
17 53 ☽ ⚼ ♀
19 41 ☽ △ ♅
22 19 ☽ ⊥ ♂

28 Thursday
00 52 ☽ Q ♆
03 34 ☽ ☌ ♄
03 41 ♀ ⚻ ♆
04 53 ☽ ☍ ☉
10 24 ☽ ⚺ ♂
13 57 ☽ ± ♀
19 13 ☽ ✶ ♃

29 Friday
01 06 ☽ ✶ ♆
02 35 ☿ # ♇
03 24 ☽ ⚻ ♀
08 58 ☽ ☌ ♇
19 30 ☽ Q ♃

30 Saturday
04 35 ☿ ⚼ ♆
07 35 ☽ ∠ ♆
08 01 ☽ ☍ ☿
08 38 ☽ □ ♅
13 01 ☿ □ ♅
16 04 ☽ ⚺ ♄
23 08 ☽ ⚻ ☉
23 10 ☽ ☌ ♂
23 29 ☉ ⚻ ♂

All ephemeris data is given at 12.00 UT and the Moon's longitude is additionally given for 24.00 UT

JULY 2018

LONGITUDES

Date	Sidereal time h m s	Sun ☉ ° ' "	Moon ☽ ° ' "	Moon ☽ 24.00 ° ' "	Mercury ☿ ° '	Venus ♀ ° '	Mars ♂ ° '	Jupiter ♃ ° '	Saturn ♄ ° '	Uranus ♅ ° '	Neptune ♆ ° '	Pluto ♇ ° '
01	06 37 59	09 ♋ 36 36	15 ♒ 25 19	21 ♒ 20 00	03 ♌ 22	20 ♌ 17	09 ♒ 05	13 ♏ 28	05 ♑ 34	02 ♉ 00	16 ♓ 27	20 ♑ 17
02	06 41 55	10 33 48	27 ♒ 15 53	03 ♓ 13 24	04 47	21 25	09 R 01	13 R 27	05 R 30	02 02	16 R 27	20 R 16
03	06 45 52	11 30 59	09 ♓ 13 04	15 ♓ 15 25	06 09	22 34	08 57	13 25	05 25	02 04	16 26	20 14
04	06 49 48	12 28 11	21 ♓ 20 58	27 ♓ 30 17	07 29	23 42	08 51	13 24	05 21	02 05	16 26	20 13
05	06 53 45	13 25 23	03 ♈ 43 55	10 ♈ 02 23	08 46	24 49	08 45	13 23	05 16	02 07	16 25	20 11
06	06 57 41	14 22 35	16 ♈ 26 14	22 ♈ 55 54	10 00	25 57	08 38	13 22	05 12	02 09	16 25	20 10
07	07 01 38	15 19 48	29 ♈ 31 49	06 ♉ 14 19	11 12	27 05	08 31	13 22	05 08	02 10	16 24	20 09
08	07 05 35	16 17 00	13 ♉ 03 36	19 ♉ 59 46	12 22	28 12	08 22	13 21	05 03	02 12	16 24	20 07
09	07 09 31	17 14 14	27 ♉ 02 47	04 ♊ 12 24	13 28	29 ♌ 19	08 13	13 21	04 59	02 13	16 23	20 06
10	07 13 28	18 11 28	11 ♊ 28 16	18 ♊ 49 46	14 32	00 ♍ 26	08 04	13 21	04 55	02 14	16 22	20 04
11	07 17 24	19 08 42	26 ♊ 16 10	03 ♋ 46 33	15 33	01 34	07 53	13 D 21	04 51	02 16	16 22	20 03
12	07 21 21	20 05 56	11 ♋ 19 52	18 ♋ 55 00	16 31	02 40	07 43	13 21	04 46	02 17	16 21	20 01
13	07 25 17	21 03 11	26 ♋ 30 44	04 ♌ 05 51	17 25	03 47	07 31	13 21	04 42	02 18	16 20	20 00
14	07 29 14	22 00 26	11 ♌ 39 13	19 ♌ 09 43	18 17	04 54	07 19	13 22	04 38	02 19	16 19	19 58
15	07 33 10	22 57 41	26 ♌ 36 23	03 ♍ 58 25	19 05	06 00	07 06	13 23	04 34	02 21	16 19	19 57
16	07 37 07	23 54 57	11 ♍ 15 07	18 ♍ 26 00	19 49	07 06	06 53	13 24	04 30	02 22	16 18	19 55
17	07 41 04	24 52 12	25 ♍ 30 44	02 ♎ 29 08	20 30	08 12	06 39	13 25	04 26	02 23	16 17	19 54
18	07 45 00	25 49 27	09 ♎ 21 10	16 ♎ 06 55	21 07	09 18	06 25	13 26	04 22	02 24	16 16	19 52
19	07 48 57	26 46 43	22 ♎ 46 34	29 ♎ 20 23	21 40	10 24	06 10	13 28	04 18	02 25	16 15	19 51
20	07 52 53	27 43 58	05 ♏ 48 43	12 ♏ 11 58	22 09	11 30	05 56	13 29	04 14	02 26	16 14	19 50
21	07 56 50	28 41 14	18 ♏ 30 30	24 ♏ 44 48	22 33	12 35	05 40	13 31	04 10	02 26	16 13	19 48
22	08 00 46	29 ♋ 38 31	00 ♐ 55 16	07 ♐ 02 21	22 54	13 40	05 25	13 33	04 07	02 27	16 12	19 47
23	08 04 43	00 ♌ 35 47	13 ♐ 06 29	19 ♐ 08 04	23 09	14 45	05 09	13 35	04 03	02 28	16 11	19 45
24	08 08 39	01 33 04	25 ♐ 07 31	01 ♑ 05 11	23 20	15 50	04 53	13 38	03 59	02 29	16 10	19 44
25	08 12 36	02 30 22	07 ♑ 01 27	12 ♑ 56 38	23 26	16 54	04 37	13 40	03 56	02 29	16 09	19 42
26	08 16 33	03 27 40	18 ♑ 51 05	24 ♑ 45 04	23 R 27	17 59	04 21	13 43	03 52	02 30	16 08	19 41
27	08 20 29	04 24 58	00 ♒ 38 55	06 ♒ 32 55	23 23	19 03	04 04	13 46	03 49	02 31	16 07	19 40
28	08 24 26	05 22 17	12 ♒ 27 20	18 ♒ 22 28	23 14	20 07	03 48	13 49	03 45	02 31	16 06	19 38
29	08 28 22	06 19 37	24 ♒ 18 36	00 ♓ 16 02	23 00	21 10	03 31	13 52	03 42	02 32	16 05	19 37
30	08 32 19	07 16 58	06 ♓ 15 03	12 ♓ 16 00	22 41	22 14	03 15	13 56	03 39	02 32	16 03	19 35
31	08 36 15	08 ♌ 14 20	18 ♓ 19 12	24 ♓ 25 00	22 ♌ 17	23 ♍ 17	02 ♒ 59	13 ♏ 59	03 ♑ 35	02 ♉ 32	16 ♓ 02	19 ♑ 34

Date	Moon True ☊ ° '	Moon Mean ☊ ° '	Moon ☽ Latitude ° '
01	05 ♌ 58	07 ♌ 17	00 S 52
02	06 D 00	07 14	01 55
03	06 01	07 11	02 53
04	06 03	07 08	03 45
05	06 04	07 05	04 27
06	06 R 04	07 01	04 58
07	06 04	06 58	05 14
08	06 02	06 55	05 14
09	06 00	06 52	04 55
10	05 58	06 49	04 18
11	05 56	06 46	03 22
12	05 54	06 42	02 12
13	05 53	06 39	00 S 52
14	05 D 53	06 36	00 N 32
15	05 54	06 33	01 52
16	05 55	06 30	03 04
17	05 55	06 27	04 02
18	05 56	06 23	04 44
19	05 57	06 20	05 09
20	05 R 56	06 17	05 17
21	05 56	06 14	05 10
22	05 55	06 11	04 48
23	05 55	06 07	04 13
24	05 54	06 04	03 28
25	05 54	06 01	02 34
26	05 53	05 58	01 33
27	05 53	05 55	00 N 29
28	05 D 53	05 52	00 S 36
29	05 R 53	05 48	01 41
30	05 53	05 45	02 41
31	05 ♌ 53	05 ♌ 42	03 S 35

DECLINATIONS

Date	Sun ☉ ° '	Moon ☽ ° '	Mercury ☿ ° '	Venus ♀ ° '	Mars ♂ ° '	Jupiter ♃ ° '	Saturn ♄ ° '	Uranus ♅ ° '	Neptune ♆ ° '	Pluto ♇ ° '
01	23 N 05	17 S 02	20 N 40	16 N 24	22 S 50	14 S 50	22 S 28	11 N 40	06 S 14	21 S 43
02	23 01	14 13	20 13	16 01	22 56	14 50	22 29	11 41	06 15	21 44
03	22 56	10 47	19 45	15 37	23 01	14 49	22 29	11 41	06 15	21 44
04	22 51	06 52	19 17	15 12	23 07	14 49	22 29	11 42	06 15	21 44
05	22 45	02 S 36	18 48	14 48	23 12	14 49	22 29	11 42	06 15	21 44
06	22 40	01 N 53	18 20	14 23	23 18	14 49	22 30	11 43	06 15	21 45
07	22 33	06 24	17 50	13 57	23 24	14 49	22 30	11 43	06 16	21 45
08	22 27	10 46	17 21	13 32	23 31	14 49	22 30	11 44	06 16	21 45
09	22 19	14 42	16 52	13 06	23 37	14 50	22 30	11 44	06 16	21 46
10	22 12	17 54	16 23	12 40	23 43	14 50	22 31	11 45	06 17	21 46
11	22 04	20 01	15 55	12 13	23 50	14 50	22 31	11 45	06 17	21 46
12	21 56	20 46	15 26	11 47	23 56	14 50	22 31	11 46	06 17	21 47
13	21 47	20 00	14 58	11 20	24 03	14 51	22 32	11 46	06 17	21 47
14	21 38	17 48	14 30	10 52	24 10	14 51	22 32	11 46	06 18	21 47
15	21 29	14 24	14 04	10 25	24 16	14 52	22 32	11 47	06 18	21 48
16	21 19	10 11	13 37	09 57	24 23	14 52	22 32	11 47	06 18	21 48
17	21 09	05 29	13 12	09 30	24 29	14 53	22 33	11 47	06 19	21 48
18	20 59	00 N 39	12 47	09 02	24 36	14 53	22 33	11 48	06 19	21 48
19	20 48	04 S 04	12 24	08 33	24 43	14 54	22 33	11 48	06 20	21 49
20	20 37	08 28	12 02	08 05	24 49	14 55	22 33	11 48	06 20	21 49
21	20 25	12 22	11 41	07 36	24 56	14 56	22 33	11 49	06 20	21 49
22	20 13	15 38	11 21	07 08	25 02	14 56	22 34	11 49	06 21	21 50
23	20 01	18 11	11 04	06 39	25 08	14 57	22 34	11 49	06 21	21 50
24	19 49	19 53	10 47	06 10	25 14	14 58	22 34	11 49	06 22	21 50
25	19 36	20 41	10 33	05 41	25 20	14 59	22 34	11 50	06 22	21 51
26	19 23	20 34	10 20	05 12	25 26	15 00	22 35	11 50	06 23	21 51
27	19 09	19 32	10 10	04 43	25 31	15 02	22 35	11 50	06 23	21 51
28	18 55	17 39	10 02	04 14	25 37	15 03	22 35	11 50	06 24	21 52
29	18 41	15 00	09 56	03 44	25 42	15 04	22 35	11 50	06 24	21 52
30	18 27	11 43	09 52	03 15	25 47	15 05	22 36	11 50	06 24	21 52
31	18 N 12	07 S 55	09 N 51	02 N 45	25 S 52	15 S 06	22 S 36	11 N 50	06 S 25	21 S 52

ZODIAC SIGN ENTRIES

Date	h	m	Planets
02	17	31	☽ ♓
05	04	50	☽ ♈
07	12	51	☽ ♉
09	16	58	☽ ♊
10	02	32	♀ ♍
11	17	59	☽ ♋
13	17	31	☽ ♌
15	17	31	☽ ♍
17	19	42	☽ ♎
20	01	13	☽ ♏
22	10	12	☽ ♐
22	21	00	☉ ♌
24	21	49	☽ ♑
27	10	41	☽ ♒
29	23	28	☽ ♓

LATITUDES

Date	Mercury ☿ ° '	Venus ♀ ° '	Mars ♂ ° '	Jupiter ♃ ° '	Saturn ♄ ° '	Uranus ♅ ° '	Neptune ♆ ° '	Pluto ♇ ° '
01	01 N 18	01 N 46	05 S 02	01 N 06	00 N 51	00 S 32	00 S 58	00 N 11
04	00 55	01 41	05 16	01 05	00 51	00 32	00 58	00 11
07	00 N 27	01 34	05 28	01 04	00 50	00 32	00 58	00 11
10	00 S 05	01 26	05 40	01 04	00 50	00 32	00 58	00 10
13	00 41	01 18	05 52	01 03	00 50	00 32	00 59	00 10
16	01 19	01 08	06 02	01 02	00 49	00 32	00 59	00 10
19	01 59	00 58	06 11	01 01	00 49	00 32	00 59	00 09
22	02 40	00 46	06 19	01 01	00 49	00 32	00 59	00 09
25	03 20	00 34	06 25	01 00	00 48	00 33	00 59	00 09
28	03 57	00 20	06 30	00 59	00 48	00 33	00 59	00 08
31	04 S 28	00 N 06	06 S 33	00 N 58	00 N 47	00 S 33	00 S 59	00 N 08

DATA

Julian Date	2458301
Delta T	+70 seconds
Ayanamsa	24° 06' 42"
Synetic vernal point	05° ♓ 00' 17"
True obliquity of ecliptic	23° 26' 07"

MOON'S PHASES, APSIDES AND POSITIONS ☽

Date	h	m	Phase	Longitude ° '	Eclipse Indicator
06	07	51	☾	14 ♈ 13	
13	02	48	●	20 ♋ 41	Partial
19	19	52	☽	27 ♎ 05	
27	20	20	○	04 ♒ 45	total

Day	h	m		
13	08	18	Perigee	
27	05	29	Apogee	
06	02	01	0N	
12	12	00	Max dec	20° N 46'
18	15	14	0S	
25	20	50	Max dec	20° S 45'

LONGITUDES

Date	Chiron ⚷ ° '	Ceres ⚳ ° '	Pallas ⚴ ° '	Juno ⚵ ° '	Vesta ⚶ ° '	Black Moon Lilith ⚸ ° '
01	02 ♈ 25	01 ♍ 15	24 ♋ 01	00 ♉ 04	25 ♐ 47	26 ♑ 02
11	02 ♈ 24	05 ♍ 18	29 ♋ 46	04 ♉ 29	23 ♐ 52	27 ♑ 09
21	02 ♈ 18	09 ♍ 29	05 ♌ 28	08 ♉ 47	22 ♐ 37	28 ♑ 15
31	02 ♈ 08	13 ♍ 45	11 ♌ 05	12 ♉ 53	22 ♐ 09	29 ♑ 22

ASPECTARIAN

01 Sunday
01 54 ☽ ⊥ ♆
04 12 ☽ ⊥ ♄
08 03 ☽ □ ♃
11 59 ♀ ⊼ ♇
12 25 ☽ ± ☉
14 05 ☽ ⊻ ♆
17 30 ♀ ⚼ ♄
18 56 ☽ # ♀
21 20 ☽ Q ♅
21 52 ☽ ⊻ ♇
22 23 ☽ ∠ ♄
22 56 ☽ ☍ ♀

02 Monday
07 14 ☽ ∥ ♃
08 16 ☽ ⚼ ☉
09 59 ☽ ⊥ ♇
21 38 ☽ ✱ ♅
23 49 ☿ ⊼ ♄

03 Tuesday
00 44 ☉ # ♂
04 04 ☽ ∠ ♇
04 27 ☽ ✱ ♄
05 05 ☽ ⊼ ☿
06 03 ☽ # ♅
11 27 ☽ ⊻ ♂
16 58 ☽ △ ☉
18 34 ☽ ± ☿
20 21 ☽ △ ♃
23 18 ☽ ⊥ ♂

04 Wednesday
02 19 ☽ ☌ ♆
03 36 ☽ ∠ ♅
04 10 ☽ Q ♄
09 47 ☽ ✱ ♇
14 28 ☽ ⚼ ☿
15 36 ☽ ∥ ♆
16 52 ☽ ∠ ♂
17 03 ☽ ⊼ ♀
21 16 ☽ ⊥ ♅

05 Thursday
01 43 ☽ ⚼ ♃
05 51 ☽ ± ♀
08 54 ☽ ⊻ ♅
09 03 ☽ Q ♇
10 25 ♀ # ♃
11 04 ☉ △ ♃
11 48 ☿ ☍ ♂
14 56 ☽ □ ♄
18 58 ☽ ± ♃
21 29 ☽ ✱ ♂
22 38 ☽ △ ☿

06 Friday
00 43 ☽ ⚼ ♀
05 58 ☉ Q ♅
06 17 ☽ ⊼ ♃
07 51 ☽ □ ☉
11 57 ☽ ⊻ ♆
16 31 ♀ ± ♇
18 54 ☽ □ ♇
19 43 ☽ Q ♂
19 55 ☿ ± ♆
23 02 ☽ ⊥ ♆

07 Saturday
07 09 ☽ △ ♀
10 30 ☿ ± ♄
11 14 ☽ # ♆
15 22 ☽ ∠ ♆
16 45 ☽ ☌ ♅
19 21 ☽ Q ☉
21 59 ☽ △ ♄
23 34 ☉ # ♄

08 Sunday
03 52 ☽ □ ♂
10 40 ☽ □ ☿
12 31 ☽ ☍ ♃
14 42 ☉ △ ♆
17 34 ☽ ∥ ♅
17 47 ☽ ✱ ♆
18 01 ☽ ✱ ☉

09 Monday
00 03 ☽ ⚼ ♄
00 11 ☽ △ ♇
02 47 ☽ ∥ ♀
09 14 ☿ □ ♃
12 48 ☽ # ♃
14 15 ☽ Q ♆
15 15 ☽ ± ♄
16 10 ☽ □ ♀
20 03 ☽ Q ☿
20 42 ☽ ⊻ ♅
21 20 ☽ ∠ ☉

10 Tuesday
01 14 ☽ ⊼ ♄
01 23 ☽ ∥ ☿
01 27 ☽ ⚼ ♇
06 27 ☽ △ ♂
06 40 ☽ ⊥ ♅
13 16 ☽ ⊥ ☉
15 04 ☽ ⊼ ♃
16 15 ☽ ± ♇
17 02 ♃ St D
17 24 ☽ ✱ ☿
20 00 ☽ □ ♆
21 26 ☽ ∠ ♅
23 43 ☽ ⊻ ☉

11 Wednesday
00 18 ☽ Q ♀
00 50 ☽ ± ♃
01 59 ☽ ⊼ ♇
06 38 ☽ ⚼ ♂
07 25 ♀ Q ♃
15 20 ☽ ⚼ ♃
19 20 ☽ ∠ ☿
20 53 ☽ ± ♂
21 09 ☽ ✱ ♀
21 36 ☽ ✱ ♅

12 Thursday
01 38 ☽ ☍ ♄
03 26 ♀ △ ♅
06 19 ☽ ⊼ ♂
07 52 ☿ ⊼ ♆
10 04 ☉ ☍ ♇
10 37 ☽ ⊥ ☿
12 52 ♀ ∥ ♅
15 12 ☽ △ ♃
16 41 ☽ Q ♅
19 56 ☽ △ ♆
20 44 ☽ ⊻ ☿
22 50 ☽ ∠ ♀

13 Friday
01 43 ☽ ☍ ♇
02 48 ☽ ☌ ☉
12 51 ☉ # ♇
14 10 ☽ ⊥ ♀
18 13 ☿ # ♃
19 37 ☽ ⚼ ♆
21 10 ☽ □ ♅

14 Saturday
00 25 ☽ ⊻ ♀
00 54 ☽ ⊼ ♄
05 12 ☽ ☍ ♂
06 44 ♀ △ ♄
09 53 ☽ ± ♆
10 23 ☽ ± ♄
13 37 ♀ ⚼ ♇
14 44 ☽ □ ♃
19 27 ☽ ⊼ ♆
23 12 ☽ ☌ ☿

15 Sunday
00 42 ☽ ⚼ ♄
01 17 ☽ ⊼ ♇
05 42 ☽ ⊻ ☉
09 09 ☽ # ♃
10 56 ☽ ± ♇
14 24 ☽ ∥ ☿
16 05 ☽ ⊥ ☉
19 46 ☽ Q ♃
21 21 ☽ △ ♅

16 Monday
00 55 ☽ △ ♄
01 34 ☽ ⚼ ♇
02 21 ☿ ⚼ ♄
03 18 ☽ ∥ ♅
04 35 ☽ ☌ ♀
04 53 ☽ ⊼ ♂
07 52 ☽ ∠ ☉
07 56 ♀ ⊼ ♂
13 18 ☽ ∥ ♀
14 40 ☽ ± ♂
15 24 ☿ ⊼ ♇
15 34 ☽ ✱ ♃
20 24 ☽ ☍ ♆
22 13 ☽ ⚼ ♅

17 Tuesday
02 29 ☽ △ ♇
03 04 ☽ ⊻ ☿
05 32 ☽ ⚼ ♂
07 51 ☽ # ♆
10 49 ☽ ✱ ☉
13 29 ☽ ± ♅
13 46 ☽ ⊥ ☿
16 59 ☽ ∠ ♃
23 50 ☽ ⊼ ♅

18 Wednesday
03 19 ☽ □ ♄
06 04 ☽ ∠ ☿
06 56 ☽ △ ♂
08 38 ☽ ⊥ ♃
09 07 ☽ Q ☉
11 55 ☽ ⊻ ♀
19 14 ☽ ⊻ ♃
23 30 ☽ ⊥ ♀

19 Thursday
00 15 ☽ ⊼ ♆
06 43 ☽ □ ♇
09 54 ☽ ✱ ☿
11 03 ☽ ± ♆
11 09 ☽ Q ♄
17 13 ☽ ∠ ♀
19 52 ☽ □ ☉

20 Friday
00 05 ☽ ∥ ♆
03 30 ☽ ⚼ ♆
05 42 ☽ ☍ ♅
08 47 ☽ Q ☿
09 05 ☽ ✱ ♄
10 04 ☽ # ♀
12 12 ☽ □ ♂
13 36 ♄ Q ♆
15 46 ☽ Q ♇
19 42 ♀ ± ♂
23 40 ☽ ✱ ♀

21 Saturday
02 29 ☽ ☌ ♃
03 08 ☿ ∥ ♅
07 38 ☽ △ ♆
07 55 ☽ # ☿
08 22 ☽ # ♅
13 16 ☽ ∠ ♄
14 28 ☽ ✱ ♇
20 00 ☽ □ ☿
21 43 ☽ Q ♂

22 Sunday
00 48 ☽ Q ♀
06 24 ☽ ∥ ♃
06 33 ☽ ⊥ ♄
09 18 ☽ △ ☉
09 21 ♀ ✱ ♃
15 00 ☽ ⊼ ♅
18 13 ☽ ⊻ ♄
19 32 ☽ ∠ ♇
20 37 ☽ ✱ ♂

23 Monday
02 48 ☽ ± ♅
12 58 ☽ ⊻ ♃
13 17 ☽ ⊥ ♇
15 35 ☽ □ ♀
17 22 ☽ ⚼ ☉
18 07 ☽ □ ♆
20 41 ☽ ⚼ ♅

24 Tuesday
00 57 ☽ ⊥ ♃
01 13 ☽ ⊻ ♇
01 43 ☽ ∠ ♂
02 33 ☉ ⚼ ♆
02 35 ♀ # ♆
08 22 ☽ △ ☿
10 49 ☽ # ☉
12 56 ☽ ± ☉
19 04 ☽ ∠ ♃
19 22 ♀ ☍ ♆
19 23 ☽ ⊥ ♂

25 Wednesday
02 04 ☽ ⊼ ☉
02 49 ☽ △ ♅
05 46 ☽ ☌ ♄
06 12 ☽ Q ♆
07 14 ☽ ⊻ ♂
11 36 ☉ □ ♅
14 52 ☽ ⚼ ☿

26 Thursday
01 09 ♀ ⚼ ♅
01 32 ☽ ✱ ♃
05 02 ☿ St R
06 29 ☽ ✱ ♆
09 10 ☽ ± ☿
10 03 ☽ △ ♀
13 41 ☽ ☌ ♇
21 20 ☽ ⊼ ☿
21 40 ☉ ⊼ ♄

27 Friday
02 01 ☽ Q ♃
05 13 ☉ ☍ ♂
12 22 ♀ ⚼ ♂
12 57 ☽ ∠ ♆
15 47 ☽ □ ♅
18 24 ☽ ⊻ ♄
18 43 ☽ # ☉
18 48 ☽ ☌ ♂
19 36 ☽ ⚼ ♀
20 20 ☽ ☍ ☉

28 Saturday
01 26 ♀ △ ♇
06 32 ☽ ⊥ ♄
07 13 ☽ ⊥ ♆
14 47 ☽ □ ♃
15 42 ☽ ± ♀
16 40 ♂ ⊻ ♄
19 22 ☽ ⊻ ♆

29 Sunday
00 42 ☽ ∠ ♄
02 31 ☽ ⊻ ♇
04 21 ☽ Q ♅
05 03 ☽ ⊼ ♀
09 25 ☽ ☍ ☿
11 28 ☽ ∥ ♃
14 37 ☽ ⊥ ♇

30 Monday
04 33 ☽ ✱ ♅
06 07 ☽ ⊻ ♂
06 48 ☽ ✱ ♄
08 41 ☽ ∠ ♇
11 07 ☽ # ♅
14 15 ☽ ⊼ ☉
17 52 ☽ ⊥ ♂
19 35 ☿ ⊻ ♀

31 Tuesday
00 02 ☽ # ☿
00 29 ☉ ∠ ♀
03 13 ☽ ± ☉
03 23 ☽ △ ♃
06 37 ☽ Q ♄
07 29 ☽ ☌ ♆
10 27 ☽ ∠ ♅
11 20 ☽ ∠ ♂
14 27 ☽ ✱ ♇
19 32 ☽ ⊼ ☿
20 48 ☽ ∥ ♆
22 31 ☽ ⚼ ☉
22 42 ☽ ☍ ♀

All ephemeris data is given at 12.00 UT and the Moon's longitude is additionally given for 24.00 UT

AUGUST 2018

LONGITUDES

Date	Sidereal time h m s	Sun ☉ ° ' "	Moon ☽ ° ' "	Moon ☽ 24.00 ° ' "	Mercury ☿ ° '	Venus ♀ ° '	Mars ♂ ° '	Jupiter ♃ ° '	Saturn ♄ ° '	Uranus ♅ ° '	Neptune ♆ ° '	Pluto ♇ ° '
01	08 40 12	09 ♌ 11 42	00 ♈ 33 45	06 ♈ 45 51	21 ♌ 48	24 ♍ 20	02 ♒ 43	14 ♏ 03	03 ♑ 32	02 ♉ 33	16 ♓ 01	19 ♑ 33
02	08 44 08	10 09 06	13 ♈ 01 40	19 ♈ 21 35	21 R 16	25 23	02 R 27	14 07	03 R 29	02 33	16 R 00	19 R 31
03	08 48 05	11 06 31	25 ♈ 46 00	02 ♉ 15 16	20 39	26 25	02 11	14 11	03 26	02 33	15 58	19 30
04	08 52 01	12 03 57	08 ♉ 49 45	15 ♉ 29 44	19 58	27 27	01 56	14 15	03 23	02 33	15 57	19 29
05	08 55 58	13 01 24	22 ♉ 15 30	29 ♉ 07 11	19 15	28 29	01 40	14 20	03 20	02 34	15 56	19 27
06	08 59 55	13 58 53	06 ♊ 04 54	13 ♊ 08 37	18 30	29 ♍ 31	01 26	14 24	03 17	02 34	15 55	19 26
07	09 03 51	14 56 23	20 ♊ 18 11	27 ♊ 33 18	17 43	00 ♎ 32	01 11	14 29	03 15	02 34	15 53	19 25
08	09 07 48	15 53 54	04 ♋ 53 31	12 ♋ 18 14	16 56	01 33	00 57	14 34	03 12	02 R 34	15 52	19 23
09	09 11 44	16 51 27	19 ♋ 46 43	27 ♋ 18 02	16 08	02 34	00 44	14 39	03 09	02 34	15 50	19 22
10	09 15 41	17 49 00	04 ♌ 51 13	12 ♌ 25 08	15 22	03 34	00 31	14 44	03 07	02 33	15 49	19 21
11	09 19 37	18 46 35	19 ♌ 58 38	27 ♌ 30 36	14 38	04 34	00 19	14 49	03 04	02 33	15 48	19 20
12	09 23 34	19 44 11	04 ♍ 59 54	12 ♍ 25 30	13 57	05 34	00 ♒ 07	14 55	03 02	02 33	15 46	19 18
13	09 27 30	20 41 48	19 ♍ 46 28	27 ♍ 02 02	13 20	06 33	29 ♑ 56	15 00	03 00	02 33	15 45	19 17
14	09 31 27	21 39 26	04 ♎ 11 35	11 ♎ 14 41	12 47	07 32	29 45	15 06	02 58	02 33	15 43	19 16
15	09 35 24	22 37 05	18 ♎ 11 03	25 ♎ 00 35	12 19	08 31	29 35	15 12	02 55	02 32	15 42	19 15
16	09 39 20	23 34 45	01 ♏ 43 20	08 ♏ 19 29	11 57	09 29	29 26	15 18	02 53	02 32	15 40	19 14
17	09 43 17	24 32 26	14 ♏ 49 19	21 ♏ 13 13	11 42	10 27	29 18	15 24	02 52	02 31	15 39	19 13
18	09 47 13	25 30 08	27 ♏ 31 39	03 ♐ 45 08	11 34	11 24	29 10	15 31	02 50	02 31	15 37	19 11
19	09 51 10	26 27 51	09 ♐ 54 11	15 ♐ 59 23	11 D 32	12 21	29 03	15 37	02 48	02 30	15 36	19 10
20	09 55 06	27 25 36	22 ♐ 01 19	28 ♐ 00 33	11 38	13 18	28 57	15 44	02 46	02 30	15 34	19 09
21	09 59 03	28 23 21	03 ♑ 57 37	09 ♑ 53 05	11 52	14 14	28 52	15 51	02 45	02 29	15 33	19 08
22	10 02 59	29 ♌ 21 08	15 ♑ 47 27	21 ♑ 41 13	12 13	15 09	28 47	15 58	02 43	02 28	15 31	19 07
23	10 06 56	00 ♍ 18 56	27 ♑ 34 49	03 ♒ 28 41	12 42	16 05	28 43	16 05	02 42	02 28	15 30	19 06
24	10 10 53	01 16 45	09 ♒ 23 12	15 ♒ 18 43	13 19	16 59	28 40	16 12	02 41	02 27	15 28	19 05
25	10 14 49	02 14 35	21 ♒ 15 32	27 ♒ 13 57	14 03	17 53	28 38	16 19	02 39	02 26	15 26	19 04
26	10 18 46	03 12 27	03 ♓ 14 13	09 ♓ 16 34	14 54	18 47	28 37	16 27	02 38	02 25	15 25	19 03
27	10 22 42	04 10 20	15 ♓ 21 11	21 ♓ 28 16	15 53	19 39	28 37	16 34	02 37	02 24	15 23	19 02
28	10 26 39	05 08 15	27 ♓ 37 59	03 ♈ 50 29	16 58	20 32	28 D 37	16 42	02 36	02 23	15 22	19 01
29	10 30 35	06 06 12	10 ♈ 05 56	16 ♈ 24 29	18 10	21 23	28 38	16 50	02 36	02 22	15 20	19 00
30	10 34 32	07 04 10	22 ♈ 46 17	29 ♈ 11 29	19 27	22 15	28 40	16 58	02 35	02 21	15 18	19 00
31	10 38 28	08 ♍ 02 10	05 ♉ 40 17	12 ♉ 12 48	20 ♌ 51	23 ♎ 05	28 ♑ 43	17 ♏ 06	02 ♑ 34	02 ♉ 20	15 ♓ 17	18 ♑ 59

Date	Moon True ☊ ° '	Moon Mean ☊ ° '	Moon ☽ Latitude ° '
01	05 ♌ 53	05 ♌ 39	04 S 19
02	05 R 52	05 36	04 53
03	05 52	05 32	05 13
04	05 52	05 29	05 17
05	05 D 52	05 26	05 05
06	05 52	05 23	04 35
07	05 53	05 20	03 47
08	05 54	05 17	02 44
09	05 54	05 13	01 28
10	05 55	05 10	00 S 06
11	05 R 54	05 07	01 N 17
12	05 53	05 04	02 34
13	05 52	05 01	03 40
14	05 50	04 58	04 30
15	05 48	04 54	05 02
16	05 46	04 51	05 15
17	05 45	04 48	05 12
18	05 D 45	04 45	04 54
19	05 45	04 42	04 22
20	05 47	04 38	03 39
21	05 48	04 35	02 47
22	05 50	04 32	01 48
23	05 51	04 29	00 N 45
24	05 R 51	04 26	00 S 19
25	05 50	04 23	01 24
26	05 48	04 19	02 25
27	05 44	04 16	03 20
28	05 40	04 13	04 07
29	05 35	04 10	04 42
30	05 30	04 07	05 05
31	05 ♌ 25	04 ♌ 04	05 S 12

DECLINATIONS

Date	Sun ☉ ° '	Moon ☽ ° '	Mercury ☿ ° '	Venus ♀ ° '	Mars ♂ ° '	Jupiter ♃ ° '	Saturn ♄ ° '	Uranus ♅ ° '	Neptune ♆ ° '	Pluto ♇ ° '
01	17 N 57	03 S 45	09 N 53	02 N 16	25 S 56	15 S 08	22 S 36	11 N 51	06 S 25	21 S 53
02	17 42	00 N 39	09 56	01 46	26 00	15 09	22 36	11 51	06 26	21 53
03	17 26	05 06	10 03	01 17	26 04	15 11	22 36	11 51	06 27	21 53
04	17 10	09 26	10 12	00 47	26 08	15 12	22 37	11 51	06 27	21 54
05	16 54	13 25	10 23	00 N 18	26 12	15 14	22 37	11 51	06 28	21 54
06	16 38	16 48	10 36	00 S 12	26 15	15 15	22 37	11 51	06 28	21 54
07	16 21	19 18	10 51	00 41	26 18	15 17	22 37	11 51	06 29	21 54
08	16 04	20 37	11 08	01 11	26 20	15 18	22 37	11 51	06 29	21 55
09	15 47	20 31	11 27	01 40	26 22	15 20	22 38	11 51	06 30	21 55
10	15 29	18 57	11 46	02 09	26 24	15 22	22 38	11 51	06 30	21 55
11	15 12	16 02	12 07	02 39	26 26	15 24	22 38	11 51	06 31	21 56
12	14 54	12 04	12 29	03 08	26 27	15 26	22 38	11 50	06 31	21 56
13	14 35	07 25	12 50	03 37	26 29	15 27	22 38	11 50	06 32	21 56
14	14 17	02 N 27	13 12	04 06	26 29	15 29	22 39	11 50	06 33	21 56
15	13 58	02 S 29	13 33	04 35	26 30	15 31	22 39	11 50	06 33	21 57
16	13 39	07 08	13 54	05 04	26 30	15 33	22 39	11 50	06 34	21 57
17	13 20	11 18	14 14	05 32	26 30	15 35	22 39	11 50	06 34	21 57
18	13 01	14 50	14 32	06 01	26 29	15 37	22 39	11 50	06 35	21 57
19	12 42	17 37	14 49	06 29	26 29	15 39	22 40	11 49	06 36	21 58
20	12 22	19 33	15 05	06 57	26 28	15 42	22 40	11 49	06 36	21 58
21	12 02	20 36	15 18	07 25	26 26	15 44	22 40	11 49	06 37	21 58
22	11 42	20 42	15 29	07 53	26 25	15 46	22 40	11 49	06 38	21 58
23	11 22	19 54	15 38	08 21	26 23	15 48	22 40	11 48	06 38	21 59
24	11 01	18 13	15 45	08 48	26 21	15 50	22 40	11 48	06 39	21 59
25	10 40	15 44	15 48	09 15	26 18	15 53	22 41	11 48	06 39	21 59
26	10 20	12 34	15 49	09 42	26 16	15 55	22 41	11 47	06 40	21 59
27	09 59	08 51	15 47	10 09	26 13	15 57	22 41	11 47	06 41	22 00
28	09 38	04 43	15 42	10 36	26 10	16 00	22 41	11 47	06 41	22 00
29	09 16	00 S 20	15 33	11 02	26 06	16 02	22 41	11 46	06 42	22 00
30	08 55	04 N 08	15 22	11 28	26 03	16 05	22 41	11 46	06 43	22 00
31	08 N 33	08 N 30	15 N 07	11 S 54	25 S 59	16 S 07	22 S 42	11 N 46	06 S 43	22 S 00

ZODIAC SIGN ENTRIES

Date	h m	Planets
01	10 54	☽ ♈
03	19 51	☽ ♉
06	01 32	☽ ♊
06	23 27	♀ ♎
08	04 01	☽ ♋
10	04 18	☽ ♌
12	03 59	☽ ♍
13	02 14	♂ ♑
14	04 57	☽ ♎
16	08 54	☽ ♏
18	16 45	☽ ♐
21	04 00	☽ ♑
23	04 09	☉ ♍
23	16 56	☽ ♒
26	05 32	☽ ♓
28	16 35	☽ ♈
31	01 30	☽ ♉

LATITUDES

Date	Mercury ☿ ° '	Venus ♀ ° '	Mars ♂ ° '	Jupiter ♃ ° '	Saturn ♄ ° '	Uranus ♅ ° '	Neptune ♆ ° '	Pluto ♇ ° '
01	04 S 36	00 N 01	06 S 34	00 N 58	00 N 47	00 S 33	00 S 59	00 N 08
04	04 52	00 S 15	06 35	00 57	00 47	00 33	00 59	00 08
07	04 54	00 31	06 34	00 57	00 47	00 33	00 59	00 08
10	04 39	00 48	06 32	00 56	00 46	00 33	00 59	00 07
13	04 09	01 06	06 28	00 55	00 46	00 33	01 00	00 07
16	03 26	01 25	06 23	00 54	00 45	00 33	01 00	00 07
19	02 36	01 45	06 16	00 54	00 45	00 33	01 00	00 06
22	01 42	02 05	06 09	00 53	00 44	00 33	01 00	00 06
25	00 51	02 26	06 00	00 52	00 44	00 33	01 00	00 06
28	00 S 03	02 47	05 51	00 52	00 44	00 33	01 00	00 05
31	00 N 37	03 S 09	05 S 42	00 N 51	00 N 43	00 S 33	01 S 00	00 N 05

DATA

Julian Date	2458332
Delta T	+70 seconds
Ayanamsa	24° 06' 47"
Synetic vernal point	05° ♓ 00' 12"
True obliquity of ecliptic	23° 26' 07"

MOON'S PHASES, APSIDES AND POSITIONS ☽

Date	h m	Phase	Longitude ° '	Eclipse Indicator
04	18 18	☾	12 ♉ 19	
11	09 58	●	18 ♌ 42	Partial
18	07 49	☽	25 ♏ 20	
26	11 56	○	03 ♓ 12	

Day	h m		
10	18 00	Perigee	
23	11 15	Apogee	
02	08 31	0N	
08	22 29	Max dec	20° N 45'
14	23 51	0S	
22	02 55	Max dec	20° S 46'
29	13 46	0N	

LONGITUDES

Date	Chiron ⚷ ° '	Ceres ⚳ ° '	Pallas ⚴ ° '	Juno ⚵ ° '	Vesta ⚶ ° '	Black Moon Lilith ⚸ ° '
01	02 ♈ 06	14 ♍ 11	11 ♌ 38	13 ♉ 17	22 ♐ 08	29 ♑ 28
11	01 ♈ 50	18 ♍ 32	17 ♌ 09	17 ♉ 08	22 ♐ 32	00 ♒ 35
21	01 ♈ 30	22 ♍ 57	22 ♌ 35	20 ♉ 41	23 ♐ 41	01 ♒ 41
31	01 ♈ 07	27 ♍ 25	27 ♌ 55	23 ♉ 51	25 ♐ 29	02 ♒ 48

All ephemeris data is given at 12.00 UT and the Moon's longitude is additionally given for 24.00 UT

ASPECTARIAN

h m	Aspects
01 Wednesday	
04 10	☽ ⊥ ♅
06 51	☽ ± ☿
09 02	☽ ꝑ ♃
13 54	☽ Q ♇
15 51	☽ ⊻ ♅
16 05	☽ ⚹ ♂
17 45	☽ □ ♄
20 06	☉ ± ♄
21 13	☽ ⋕ ♀
23 35	☽ ꝑ ☿
02 Thursday	
02 33	☽ ± ♃
02 39	♂ □ ♅
06 03	☽ △ ☉
08 10	☉ ± ♆
14 05	☽ ⊼ ♃
14 39	☽ Q ♂
17 29	☽ ∥ ♀
17 38	☽ ⊻ ♆
03 Friday	
00 17	☽ □ ♇
02 52	☽ △ ☿
04 56	☽ ⊥ ♆
13 19	☽ ⊼ ♀
15 09	♀ ± ♅
15 13	☿ ⊥ ♀
19 21	☽ ⋕ ♆
21 38	☽ ∠ ♆
23 38	☽ □ ♂
04 Saturday	
00 33	☽ ☌ ♅
01 22	☽ ± ♀
02 07	☽ △ ♄
16 36	☽ ∥ ☿
18 18	☽ □ ☉
19 06	☽ ꝑ ♀
21 50	☽ ☍ ♃
05 Sunday	
00 48	☽ ⚹ ♆
02 13	☽ ∥ ♅
05 05	☽ ꝑ ♄
05 15	☿ ⊼ ♇
06 58	☽ □ ☿
07 03	☽ △ ♇
20 52	☽ ± ♄
21 55	☽ Q ♆
23 46	☽ △ ♀
06 Monday	
00 15	☽ ⋕ ♃
04 08	☽ △ ♂
04 26	☽ Q ☉
05 57	☽ ⊻ ♅
07 13	☽ ⊼ ♄
09 11	☽ ꝑ ♇
09 13	♀ ∠ ♃
10 42	☽ ∥ ☉
12 40	☽ Q ☿
16 14	☽ ⊥ ♅
18 53	☿ ꝑ ♄
23 27	☉ □ ♃
07 Tuesday	
00 28	☽ ± ♇
02 12	☽ ⊼ ♃
02 23	☽ ⚹ ☉
04 38	☽ □ ♆
05 14	☽ ꝑ ♂
07 25	☽ ∠ ♅
07 54	☽ ⚹ ☿
10 31	☽ ⊼ ♇
12 18	☽ ± ♃
16 50	♅ St R
19 58	☽ ± ♂
08 Wednesday	
00 33	♀ △ ♂
03 15	☽ ꝑ ♃
05 02	☽ ∠ ☉
05 41	☽ ⊼ ♂
06 09	☽ □ ♀
07 25	☽ ∠ ☿
08 12	☽ ⚹ ♅
09 15	☽ ☍ ♄
11 10	☉ ⊼ ♆
17 00	☿ ∠ ♀
20 41	☽ ⊥ ☉
21 17	☽ ⊥ ☿
22 45	♂ ∠ ♆
09 Thursday	
02 06	☉ ☌ ☿
03 38	☽ Q ♅
03 43	☽ △ ♃
05 42	☽ △ ♆
06 27	☽ ⊻ ☿
07 00	☽ ⊻ ☉
11 21	☽ ☍ ♇
11 57	♀ ⊼ ♅
13 21	☽ Q ♀
21 29	☿ ⊼ ♆
10 Friday	
01 34	♀ □ ♄
05 12	☽ ☍ ♂
05 36	☽ ꝑ ♆
08 21	☽ □ ♅
09 15	☽ ⊼ ♄
09 49	☽ ⚹ ♀
16 55	☿ ∥ ♅
18 44	☽ ± ♄
19 07	☉ ꝑ ♄
19 52	☽ ± ♆
21 15	☉ ⋕ ♃
11 Saturday	
03 46	☽ □ ♃
03 54	☽ ☌ ☿
05 22	☽ ⊼ ♆
06 31	☿ □ ♃
08 59	☽ ꝑ ♄
09 58	☽ ☌ ☉
10 58	☽ ⊼ ♇
11 18	☽ ∠ ♀
16 17	☽ ⋕ ♃
18 06	☽ ∥ ☉
20 30	☽ ± ♇
12 Sunday	
01 28	☉ ⊼ ♇
02 40	☽ ⊥ ♀
04 16	☽ ⊼ ♂
08 04	☽ △ ♅
08 38	☽ Q ♃
08 51	☽ △ ♄
09 56	☽ ∥ ☿
10 53	☽ ꝑ ♇
12 59	☽ ⊻ ♀
13 15	☽ ∥ ♅
13 46	☽ ± ♂
13 Monday	
01 53	☽ ⊻ ☿
04 09	☽ ⚹ ♃
04 10	☽ ꝑ ♂
05 25	☽ ☍ ♆
08 21	☽ ꝑ ♅
11 12	☽ △ ♇
11 18	☽ ⊥ ☿
13 37	☽ ⊻ ☉
16 19	☽ ⋕ ♆
23 11	☽ ± ♅
14 Tuesday	
00 15	☽ ⊥ ☉
01 37	☽ ∠ ☿
04 37	☽ △ ♂
04 49	☽ ⋕ ♀
05 04	☽ ∠ ♃
09 13	☽ ⊼ ♅
09 56	☽ □ ♄
16 29	☽ ∠ ☉
18 05	☽ ☌ ♀
20 24	☽ ⊥ ♃
15 Wednesday	
02 09	☽ ⚹ ☿
06 47	☽ ⊻ ♃
07 41	☽ ⊼ ♆
13 51	☽ □ ♇
16 47	☽ Q ♄
18 08	☽ ± ♆
20 22	☽ ⚹ ☉
22 28	☽ Q ☿
23 48	☽ ∥ ♀
16 Thursday	
03 07	☉ ∥ ☿
06 56	♀ ⊥ ♃
07 56	☽ □ ♂
08 57	☽ ∥ ♆
10 07	☽ ꝑ ♆
13 27	☽ ☍ ♅
14 06	☽ ⚹ ♄
19 32	☽ Q ☉
21 59	☽ Q ♇
17 Friday	
03 15	☽ ⊻ ♀
06 18	☽ □ ☿
13 06	☽ ☌ ♃
13 32	☽ △ ♆
15 17	☽ ⊥ ♀
15 19	☽ ⋕ ♅
16 34	☽ Q ♂
17 40	☽ ∠ ♄
20 12	☽ ⚹ ♇
18 Saturday	
00 10	☽ ⋕ ☉
04 21	☉ ± ♇
07 49	☽ □ ☉
09 36	☽ ⋕ ☿
09 41	☽ ∠ ♀
10 40	☽ ⊥ ♄
15 07	☽ ⚹ ♂
15 35	☿ ⚹ ♀
18 14	☽ ∥ ♃
21 36	☽ ⊼ ♅
22 11	☽ ⊻ ♄
19 Sunday	
00 50	☽ ∠ ♇
04 24	☿ St D
07 44	♃ △ ♆
09 16	☽ ± ♅
15 13	☽ △ ☿
17 14	☽ ⚹ ♀
17 41	♀ ∥ ♆
18 25	☽ ⊥ ♇
20 06	☽ ∠ ♂
23 12	☽ □ ♆
23 22	☽ ⊻ ♃
20 Monday	
02 59	☽ ꝑ ♅
06 17	☽ ⊻ ♇
11 25	☽ ⊥ ♃
13 50	☽ ⊥ ♂
19 07	☽ Q ♀
21 24	☽ ꝑ ☿
23 47	☽ △ ☉
21 Tuesday	
01 47	☽ ⊻ ♂
05 39	☽ ∠ ♃
09 01	☽ △ ♅
09 33	☽ ☌ ♄
11 10	☽ Q ♆
15 57	☽ ± ☿
22 52	☉ ⊼ ♂
22 Wednesday	
03 47	☉ ∥ ♅
04 29	☽ ⊼ ☿
08 49	☽ ꝑ ☉
10 36	☽ □ ♀
11 27	☽ ⚹ ♆
12 21	☽ ⚹ ♃
18 46	☽ ☌ ♇
21 10	♀ ⊼ ♆
23 Thursday	
04 46	☽ ± ☉
12 04	♀ ⊻ ♃
13 01	☽ Q ♃
14 19	☽ ☌ ♂
17 55	☽ ∠ ♆
18 04	☽ ⊼ ☉
21 55	☽ □ ♅
22 24	☽ ⊻ ♄
24 Friday	
10 34	☽ ⊥ ♄
12 10	☽ ⊥ ♆
20 27	☽ ☍ ☿
25 Saturday	
00 17	☽ ⊻ ♆
01 56	☽ □ ♃
04 39	☽ △ ♀
04 45	☽ ∠ ♄
07 35	☽ ⊻ ♇
10 20	☽ Q ♅
10 47	☽ ∥ ♃
11 26	☽ ⋕ ☿
16 38	☉ △ ♅
19 39	☽ ⊥ ♇
22 07	☉ △ ♄
26 Sunday	
02 47	☽ ⊻ ♂
10 22	☽ ⚹ ♅
10 49	☽ ⚹ ♄
11 56	☽ ☍ ☉
13 10	☽ ꝑ ♀
13 37	☽ ∠ ♇
14 45	☽ ⊥ ♂
17 17	☽ ⋕ ♅
19 21	♀ □ ♇
27 Monday	
00 29	☿ ⊼ ♆
04 18	☽ ⋕ ☉
04 45	☽ ∥ ♀
06 41	☉ ⋕ ♀
08 24	☽ ± ♀
08 34	☽ ∠ ♂
08 41	☉ ꝑ ♇
10 34	☽ Q ♄
12 04	☽ ☌ ♆
13 08	☽ ⊼ ☿
14 05	♂ St D
14 25	☽ △ ♃
16 01	☽ ∠ ♅
19 14	☽ ⚹ ♇
19 24	♂ Q ♃
21 06	☽ ⊼ ♀
22 54	☉ ± ♂
23 26	☉ Q ♃
28 Tuesday	
00 47	☽ ∥ ♆
02 01	☽ ± ☿
05 32	☿ □ ♃
09 35	☽ ⊥ ♅
13 54	☽ ⚹ ♂
14 07	♀ Q ♄
18 33	☽ Q ♇
19 57	☽ ꝑ ♃
21 11	☽ ⊻ ♅
21 15	☽ ꝑ ☿
21 37	☽ □ ♄
29 Wednesday	
00 57	☿ ꝑ ♄
03 42	☽ ⊼ ☉
10 24	♀ ± ♆
13 02	☽ Q ♂
13 25	☽ ± ♃
16 08	☽ ± ☉
21 56	☽ ⊻ ♆
30 Thursday	
00 56	☽ ⊼ ♃
03 40	☿ ⊼ ♇
04 54	☽ □ ♇
05 02	☽ △ ☿
09 15	☽ ⊥ ♆
10 34	☽ ꝑ ☉
10 56	☽ ☍ ♀
23 04	☽ □ ♂
31 Friday	
02 02	☽ ∠ ♆
02 04	☽ ⋕ ♆
04 38	♀ ⋕ ♅
05 50	☽ ☌ ♅
06 17	☽ △ ♄
12 19	☽ ∥ ☉
16 42	☽ △ ☉
20 09	☉ ∠ ♀

SEPTEMBER 2018

LONGITUDES

Date	Sidereal time h m s	Sun ☉ ° ' "	Moon ☽ ° ' "	Moon ☽ 24.00 ° ' "	Mercury ☿ ° '	Venus ♀ ° '	Mars ♂ ° '	Jupiter ♃ ° '	Saturn ♄ ° '	Uranus ♅ ° '	Neptune ♆ ° '	Pluto ♇ ° '
01	10 42 25	09 ♍ 00 12	18 ♉ 49 15	25 ♉ 29 46	22 ♌ 19	23 ♎ 55	28 ♑ 46	17 ♏ 14	02 ♑ 34	02 ♉ 19	15 ♓ 15	18 ♑ 58
02	10 46 22	09 58 16	02 ♊ 14 31	09 ♊ 03 38	23 52	24 44	28 51	17 22	02 R 33	02 R 18	15 R 13	18 R 57
03	10 50 18	10 56 22	15 ♊ 57 13	22 ♊ 55 19	25 30	25 32	28 56	17 31	02 33	02 16	15 12	18 56
04	10 54 15	11 54 30	29 ♊ 57 56	07 ♋ 04 59	27 11	26 19	29 02	17 39	02 33	02 15	15 10	18 56
05	10 58 11	12 52 40	14 ♋ 16 15	21 ♋ 31 27	28 ♌ 55	27 06	29 09	17 48	02 33	02 14	15 08	18 55
06	11 02 08	13 50 52	28 ♋ 50 11	06 ♌ 11 54	00 ♍ 42	27 52	29 16	17 57	02 D 33	02 12	15 07	18 54
07	11 06 04	14 49 06	13 ♌ 35 55	21 ♌ 01 28	02 31	28 37	29 24	18 06	02 33	02 11	15 05	18 53
08	11 10 01	15 47 21	28 ♌ 27 38	05 ♍ 53 28	04 22	29 ♎ 21	29 34	18 15	02 33	02 10	15 04	18 53
09	11 13 57	16 45 39	13 ♍ 17 58	20 ♍ 40 07	06 14	00 ♏ 05	29 43	18 24	02 33	02 08	15 02	18 52
10	11 17 54	17 43 58	27 ♍ 58 56	05 ♎ 13 33	08 07	00 47	29 ♑ 54	18 33	02 33	02 07	15 00	18 52
11	11 21 51	18 42 19	12 ♎ 23 09	19 ♎ 27 06	10 01	01 28	00 ♒ 05	18 43	02 34	02 05	14 59	18 51
12	11 25 47	19 40 42	26 ♎ 24 55	03 ♏ 16 16	11 56	02 09	00 17	18 52	02 34	02 03	14 57	18 50
13	11 29 44	20 39 06	10 ♏ 00 59	16 ♏ 39 05	13 50	02 48	00 30	19 02	02 35	02 02	14 55	18 50
14	11 33 40	21 37 32	23 ♏ 10 42	29 ♏ 36 07	15 44	03 26	00 44	19 11	02 36	02 00	14 54	18 49
15	11 37 37	22 36 00	05 ♐ 55 42	12 ♐ 09 57	17 38	04 03	00 58	19 21	02 37	01 58	14 52	18 49
16	11 41 33	23 34 30	18 ♐ 19 24	24 ♐ 24 39	19 32	04 38	01 13	19 31	02 37	01 57	14 50	18 48
17	11 45 30	24 33 01	00 ♑ 26 20	06 ♑ 25 06	21 25	05 13	01 28	19 41	02 39	01 55	14 49	18 48
18	11 49 26	25 31 33	12 ♑ 21 37	18 ♑ 16 32	23 17	05 46	01 45	19 51	02 40	01 53	14 47	18 48
19	11 53 23	26 30 08	24 ♑ 10 30	00 ♒ 04 08	25 08	06 18	02 01	20 01	02 41	01 51	14 46	18 47
20	11 57 20	27 28 44	05 ♒ 58 01	11 ♒ 52 43	26 59	06 48	02 19	20 11	02 42	01 49	14 44	18 47
21	12 01 16	28 27 21	17 ♒ 48 44	23 ♒ 46 32	28 ♍ 49	07 17	02 37	20 22	02 44	01 47	14 42	18 47
22	12 05 13	29 ♍ 26 01	29 ♒ 46 32	05 ♓ 49 03	00 ♎ 38	07 44	02 56	20 32	02 45	01 45	14 41	18 46
23	12 09 09	00 ♎ 24 42	11 ♓ 54 25	18 ♓ 02 51	02 26	08 09	03 15	20 43	02 47	01 44	14 39	18 46
24	12 13 06	01 23 25	24 ♓ 14 31	00 ♈ 29 32	04 13	08 33	03 35	20 53	02 48	01 42	14 38	18 46
25	12 17 02	02 22 10	06 ♈ 47 56	13 ♈ 09 45	05 59	08 56	03 55	21 04	02 50	01 39	14 36	18 46
26	12 20 59	03 20 57	19 ♈ 34 57	26 ♈ 03 26	07 44	09 16	04 16	21 15	02 52	01 37	14 34	18 46
27	12 24 55	04 19 46	02 ♉ 35 08	09 ♉ 09 55	09 28	09 35	04 38	21 26	02 54	01 35	14 33	18 46
28	12 28 52	05 18 37	15 ♉ 47 41	22 ♉ 28 19	11 12	09 51	05 00	21 37	02 56	01 33	14 31	18 45
29	12 32 49	06 17 31	29 ♉ 11 43	05 ♊ 57 49	12 54	10 06	05 22	21 48	02 58	01 31	14 30	18 45
30	12 36 45	07 ♎ 16 27	12 ♊ 46 34	19 ♊ 37 54	14 ♎ 35	10 ♏ 19	05 ♒ 45	21 ♏ 59	03 ♑ 01	01 ♉ 29	14 ♓ 28	18 ♑ 45

Moon nodes and latitude

Date	Moon True ☊ ° '	Moon Mean ☊ ° '	Moon ☽ Latitude ° '
01	05 ♌ 22	04 ♌ 00	05 S 04
02	05 R 21	03 57	04 39
03	05 D 21	03 54	03 57
04	05 21	03 51	03 01
05	05 23	03 48	01 53
06	05 24	03 44	00 S 36
07	05 R 24	03 41	00 N 45
08	05 22	03 38	02 02
09	05 18	03 35	03 11
10	05 13	03 32	04 07
11	05 06	03 29	04 45
12	04 59	03 25	05 06
13	04 52	03 22	05 08
14	04 47	03 19	04 53
15	04 44	03 16	04 25
16	04 42	03 13	03 44
17	04 D 42	03 10	02 54
18	04 43	03 06	01 58
19	04 44	03 03	00 N 57
20	04 R 45	03 00	00 S 07
21	04 44	02 57	01 10
22	04 41	02 54	02 10
23	04 35	02 50	03 06
24	04 27	02 47	03 53
25	04 18	02 44	04 30
26	04 07	02 41	04 55
27	03 57	02 38	05 04
28	03 47	02 35	04 58
29	03 38	02 31	04 35
30	03 ♌ 35	02 ♌ 28	03 S 56

DECLINATIONS

Date	Sun ☉ ° '	Moon ☽ ° '	Mercury ☿ ° '	Venus ♀ ° '	Mars ♂ ° '	Jupiter ♃ ° '	Saturn ♄ ° '	Uranus ♅ ° '	Neptune ♆ ° '	Pluto ♇ ° '
01	08 N 12	12 N 33	14 N 50	12 S 19	25 S 55	16 S 10	22 S 42	11 N 45	06 S 44	22 S 01
02	07 50	16 03	14 29	12 44	25 51	16 12	22 42	11 45	06 45	22 01
03	07 28	18 46	14 05	13 09	25 46	16 15	22 42	11 44	06 45	22 01
04	07 06	20 25	13 39	13 33	25 42	16 17	22 42	11 44	06 46	22 01
05	06 43	20 48	13 09	13 57	25 37	16 20	22 42	11 43	06 47	22 01
06	06 21	19 48	12 38	14 21	25 32	16 23	22 42	11 43	06 47	22 02
07	05 59	17 27	12 03	14 44	25 27	16 25	22 43	11 42	06 48	22 02
08	05 36	13 55	11 27	15 07	25 21	16 28	22 43	11 42	06 48	22 02
09	05 14	09 30	10 49	15 30	25 15	16 31	22 43	11 41	06 49	22 02
10	04 51	04 N 34	10 09	15 52	25 10	16 33	22 43	11 41	06 50	22 02
11	04 28	00 S 31	09 28	16 14	25 04	16 36	22 43	11 40	06 50	22 02
12	04 05	05 26	08 45	16 35	24 57	16 39	22 43	11 40	06 51	22 02
13	03 42	09 57	08 02	16 56	24 51	16 42	22 43	11 39	06 52	22 02
14	03 19	13 50	07 17	17 16	24 44	16 44	22 44	11 39	06 52	22 03
15	02 56	16 57	06 31	17 36	24 38	16 47	22 44	11 38	06 53	22 03
16	02 33	19 12	05 45	17 56	24 31	16 50	22 44	11 37	06 54	22 03
17	02 10	20 32	04 58	18 15	24 24	16 53	22 44	11 37	06 54	22 04
18	01 47	20 54	04 11	18 33	24 17	16 56	22 44	11 36	06 55	22 04
19	01 24	20 21	03 24	18 50	24 09	16 59	22 44	11 36	06 55	22 04
20	01 00	18 53	02 37	19 08	24 02	17 02	22 44	11 35	06 56	22 04
21	00 37	16 36	01 49	19 24	23 54	17 05	22 44	11 35	06 57	22 04
22	00 N 14	13 35	01 02	19 40	23 46	17 08	22 45	11 34	06 57	22 04
23	00 S 10	09 57	00 N 14	19 55	23 38	17 10	22 45	11 33	06 58	22 04
24	00 33	05 51	00 S 33	20 10	23 30	17 13	22 45	11 32	06 59	22 04
25	00 57	01 S 26	01 20	20 23	23 21	17 16	22 45	11 31	06 59	22 05
26	01 20	03 N 07	02 07	20 36	23 13	17 19	22 45	11 31	07 00	22 05
27	01 43	07 36	02 53	20 48	23 04	17 22	22 45	11 30	07 00	22 05
28	02 07	11 49	03 39	21 00	22 55	17 25	22 45	11 29	07 01	22 05
29	02 30	15 30	04 25	21 10	22 46	17 28	22 45	11 28	07 01	22 05
30	02 S 53	18 N 25	05 S 10	21 S 20	22 S 37	17 S 31	22 S 45	11 N 28	07 S 02	22 S 05

ZODIAC SIGN ENTRIES

Date	h	m	Planets
02	08	02	☽ ♊
04	12	03	☽ ♋
06	02	39	☿ ♍
06	13	54	☽ ♌
08	14	29	☽ ♍
09	09	25	♀ ♏
10	15	20	☽ ♎
11	00	56	♂ ♒
12	18	15	☽ ♏
15	00	45	☽ ♐
17	11	07	☽ ♑
19	23	52	☽ ♒
22	03	39	☿ ♎
22	12	27	☽ ♓
23	01	54	☉ ♎
24	23	04	☽ ♈
27	07	16	☽ ♉
29	13	26	☽ ♊

LATITUDES

Date	Mercury ☿ ° '	Venus ♀ ° '	Mars ♂ ° '	Jupiter ♃ ° '	Saturn ♄ ° '	Uranus ♅ ° '	Neptune ♆ ° '	Pluto ♇ ° '
01	00 N 48	03 S 16	05 S 38	00 N 51	00 N 43	00 S 33	01 S 00	00 N 05
04	01 16	03 39	05 28	00 50	00 43	00 33	01 00	00 05
07	01 35	04 02	05 17	00 50	00 42	00 33	01 00	00 04
10	01 46	04 24	05 07	00 49	00 42	00 33	01 00	00 04
13	01 48	04 47	04 56	00 49	00 41	00 33	01 00	00 04
16	01 45	05 10	04 45	00 48	00 41	00 33	01 00	00 03
19	01 36	05 33	04 34	00 48	00 40	00 34	01 00	00 03
22	01 24	05 54	04 23	00 47	00 40	00 34	01 00	00 03
25	01 08	06 15	04 12	00 47	00 39	00 34	01 00	00 02
28	00 50	06 35	04 02	00 46	00 39	00 34	01 00	00 02
31	00 N 31	06 S 52	03 S 51	00 N 46	00 N 39	00 S 34	01 S 00	00 N 02

DATA

Julian Date	2458363
Delta T	+70 seconds
Ayanamsa	24° 06' 50"
Synetic vernal point	05° ♓ 00' 09"
True obliquity of ecliptic	23° 26' 08"

MOON'S PHASES, APSIDES AND POSITIONS ☽

Date	h	m	Phase	Longitude ° '	Eclipse Indicator
03	02	37	☾	10 ♊ 34	
09	18	01	●	17 ♍ 09	
16	23	15	☽	24 ♐ 02	
25	02	52	○	02 ♈ 00	

Day	h	m		
08	01	12	Perigee	
20	00	50	Apogee	
05	06	53	Max dec	20° N 50'
11	09	34	0S	
18	09	33	Max dec	20° S 55'
25	19	37	0N	

LONGITUDES

Date	Chiron ⚷ ° '	Ceres ⚳ ° '	Pallas ⚴ ° '	Juno ⚵ ° '	Vesta ⚶ ° '	Black Moon Lilith ⚸ ° '
01	01 ♈ 05	27 ♍ 52	28 ♌ 27	24 ♉ 08	25 ♐ 42	02 ♒ 54
11	00 ♈ 39	02 ♎ 21	03 ♍ 40	26 ♉ 47	28 ♐ 07	04 ♒ 01
21	00 ♈ 12	06 ♎ 53	08 ♍ 47	28 ♉ 49	01 ♑ 05	05 ♒ 07
31	29 ♓ 45	11 ♎ 24	13 ♍ 46	00 ♊ 09	04 ♑ 18	06 ♒ 13

All ephemeris data is given at 12.00 UT and the Moon's longitude is additionally given for 24.00 UT

ASPECTARIAN

h m Aspects

01 Saturday
05 33 ☽ ⚹ ♆
07 07 ☽ ∥ ♅
09 06 ☽ ☍ ♃
09 44 ☽ ᵽ ♄
10 24 ☽ # ♀
12 16 ☽ △ ♇
19 07 ☽ □ ☿
21 46 ☽ ⊼ ♀

02 Sunday
01 45 ☽ ∥ ☿
01 54 ☽ ± ♄
03 06 ☽ Q ♆
05 56 ☽ △ ♂
09 09 ☽ ± ♀
12 05 ☽ ⊻ ♅
12 33 ☽ ⊼ ♄
13 12 ☽ # ♃
15 01 ☽ ᵽ ♇
22 38 ☽ ⊥ ♅

03 Monday
01 59 ☽ ᵽ ♀
02 37 ☽ □ ☉
03 54 ☿ ± ♇
06 46 ☽ ± ♇
07 10 ☽ Q ☿
08 28 ☽ ᵽ ♂
10 41 ☽ □ ♆
12 58 ☿ ⚹ ♀
14 17 ☽ ∠ ♅
14 44 ☽ ⊼ ♃
17 09 ☽ ⊼ ♇
17 42 ♃ ∠ ♄

04 Tuesday
00 06 ☽ ± ♂
01 09 ☽ ± ♃
05 27 ☽ △ ♀
06 37 ☽ ⚹ ☿
10 24 ☽ ⊼ ♂
11 54 ☽ Q ☉
14 26 ☿ # ♀
15 52 ☽ ⚹ ♅
16 22 ☽ ☍ ♄
16 36 ☽ ᵽ ♃

05 Wednesday
08 49 ☉ # ♆
09 31 ☽ ⚹ ☉
11 20 ☽ ∠ ☿
11 56 ☽ Q ♅
13 27 ☽ △ ♆
15 18 ☿ ⊼ ♂
17 55 ☽ △ ♃
19 41 ☽ ☍ ♇

06 Thursday
04 16 ☽ ⊥ ☿
10 20 ☽ □ ♀
11 09 ♄ St D
12 01 ☽ ∠ ☉
12 43 ☽ ☍ ♂
14 05 ☽ ᵽ ♆
15 28 ☽ ⊻ ☿
17 30 ☽ □ ♅
18 03 ☽ ⊼ ♄

07 Friday
00 07 ☉ ᵽ ♂
03 43 ☽ ⊥ ☉
03 48 ☽ ± ♄
04 42 ☽ ± ♆
07 41 ☿ △ ♅
12 20 ☿ △ ♄
14 07 ☽ ⊻ ☉
14 24 ☽ ⊼ ♆
17 08 ☽ Q ♀
18 23 ☽ ᵽ ♄
18 27 ☉ ☍ ♆
19 21 ☽ □ ♃
19 40 ☽ # ♃
20 33 ☽ ⊼ ♇

08 Saturday
02 17 ☿ ∥ ♅
05 14 ☽ # ♀
05 45 ☿ ᵽ ♇
06 14 ☽ ± ♇
13 31 ☽ ⚹ ♀
13 48 ☽ ⊼ ♂
17 58 ☽ △ ♅
18 36 ☽ △ ♄
20 38 ♀ □ ♂
20 45 ☽ ᵽ ♇
22 54 ☽ ☌ ☿
23 35 ☽ ± ♂

09 Sunday
00 33 ☽ ∥ ♅
00 42 ☽ Q ♃
04 06 ☽ ∥ ☿
04 47 ☿ ± ♂
10 31 ♀ ᵽ ♆
14 17 ☿ Q ♃
14 20 ☽ ᵽ ♂
14 48 ☽ ☍ ♆
15 02 ☽ ∠ ♀
18 01 ☽ ☌ ☉
18 14 ☽ ᵽ ♅
20 23 ☽ ⚹ ♃
21 00 ☉ ᵽ ♅
21 03 ☽ △ ♇

10 Monday
01 15 ☽ # ♆
06 28 ☽ ⊥ ♀
08 55 ☽ ± ♅
10 36 ☽ ∥ ☉
15 12 ☽ △ ♂
16 51 ☽ ⊻ ♀
18 49 ☽ ⊼ ♅
19 34 ☽ □ ♄
21 19 ☽ ∠ ♃
23 41 ♂ ∠ ♆

11 Tuesday
07 25 ☽ ⊻ ☿
12 10 ☉ ⚹ ♃
12 33 ☽ ⊥ ♃
15 31 ☉ △ ♇
16 22 ☽ ⊼ ♆
19 07 ☽ ⊥ ☿
22 51 ☽ ⊻ ♃
22 58 ☽ □ ♇
23 31 ☽ ⊻ ☉

12 Wednesday
01 55 ☽ Q ♄
02 35 ☽ ± ♆
05 45 ☽ # ☉
07 55 ♃ ⚹ ♇
09 02 ♀ ☍ ♅
10 37 ☽ ⊥ ☉
13 02 ☽ ∠ ☿
16 52 ♀ ∥ ♃
18 09 ☽ ᵽ ♆
18 52 ☽ □ ♂
19 15 ☽ ∥ ♆
21 50 ☽ ☍ ♅
22 31 ☽ ☌ ♀
22 47 ☽ ⚹ ♄

13 Thursday
02 55 ☽ # ☿
03 37 ☽ ∠ ☉
03 58 ♀ ⚹ ♄
06 19 ☽ Q ♇
20 02 ☽ ⚹ ☿
20 50 ☽ △ ♆
22 01 ☽ # ♅

14 Friday
01 31 ☿ ☍ ♆
01 43 ☽ ∠ ♄
03 39 ☽ Q ♂
03 59 ☽ ⚹ ♇
04 33 ☽ ☌ ♃
08 54 ☽ ⚹ ☉
11 53 ☿ ᵽ ♂
18 22 ☽ ⊥ ♄
21 59 ☽ Q ☿

15 Saturday
00 47 ☿ # ♆
02 23 ☽ ⚹ ♂
03 45 ☿ ᵽ ♅
04 30 ☽ ⊼ ♅
05 40 ☽ ⊻ ♄
07 59 ☽ ∠ ♇
08 14 ☽ ⊻ ♀
09 15 ☽ Q ☉
10 35 ☽ ∥ ♃
15 54 ☽ ± ♅
19 02 ☽ ∥ ♀
20 18 ☽ ⊥ ♀

16 Sunday
01 15 ☽ ⊥ ♇
02 54 ☿ △ ♇
05 12 ☽ □ ♆
07 47 ☽ ∠ ♂
09 18 ☽ ᵽ ♅
11 51 ☿ ⚹ ♃
12 57 ☽ ⊻ ♇
14 04 ☿ ∠ ♀
14 22 ☽ ⊻ ♃
14 43 ☽ ∠ ♀
14 48 ☽ □ ☿
23 15 ☽ □ ☉

17 Monday
01 53 ☽ ⊥ ♂
02 23 ☽ ⊥ ♃
14 07 ☽ ⊻ ♂
14 57 ☽ △ ♅
16 25 ☽ ☌ ♄
16 45 ☽ Q ♆
20 38 ☽ ∠ ♃
22 03 ☽ ⚹ ♀

18 Tuesday
16 54 ☽ ⚹ ♆
20 33 ☉ ± ♅
23 01 ♂ □ ♅
23 29 ☽ Q ♀

19 Wednesday
01 03 ☽ ☌ ♇
03 25 ☽ ⚹ ♃
11 22 ♂ Q ♃
14 20 ☽ △ ☿
17 10 ☽ △ ☉
21 07 ☿ ± ♅
23 21 ☽ ∠ ♆

20 Thursday
03 35 ☽ □ ♅
04 12 ☽ Q ♃
04 23 ☽ ☌ ♂
05 21 ☽ ⊻ ♄
09 15 ☽ ∥ ♀
11 20 ♀ Q ♇
13 46 ☽ □ ♀
17 34 ☽ ⊥ ♄
17 36 ☽ ⊥ ♆

21 Friday
01 43 ♄ Q ♆
01 52 ☉ ☌ ☿
02 24 ☽ ᵽ ☉
02 27 ☽ ᵽ ☿
05 44 ☽ ⊻ ♆
07 40 ☽ ∥ ♃
11 50 ☽ ∠ ♄
13 57 ☽ ⊻ ♇
15 59 ☽ Q ♅
17 13 ☽ □ ♃
21 20 ♂ ⊻ ♄
22 11 ☽ ± ☉
23 53 ☽ ± ☿

22 Saturday
02 00 ☽ ⊥ ♇
11 15 ☽ ⊼ ☉
14 00 ☽ ⊼ ☿
15 56 ☽ ⚹ ♅
17 56 ☽ ⚹ ♄
18 26 ☽ ⊻ ♂
19 57 ☽ ∠ ♇

23 Sunday
01 53 ☽ # ♅
02 46 ☿ ⊼ ♅
04 22 ☽ △ ♀
06 38 ☽ ⊥ ♂
07 17 ☿ ⊥ ♀
13 30 ☉ # ☿
16 46 ☿ □ ♄
17 22 ☽ ☌ ♆
17 38 ☽ Q ♄
21 24 ☽ ∠ ♅

24 Monday
00 44 ☽ ∠ ♂
01 24 ☽ ⚹ ♇
01 29 ☿ △ ♂
05 26 ☽ △ ♃
05 41 ☽ ∥ ♆
10 38 ☽ ᵽ ♀
12 10 ☉ ∥ ☿
14 47 ☽ ⊥ ♅
19 09 ☉ ⊼ ♅

25 Tuesday
00 31 ☽ Q ♇
02 15 ☽ ⊻ ♅
02 52 ☽ ☍ ☉
04 26 ☽ ± ♀
04 27 ☽ □ ♄
06 23 ☽ ⚹ ♂
10 12 ☽ ☍ ☿
10 36 ☽ ᵽ ♃
12 29 ☽ ∥ ☿
13 19 ☿ ∠ ♃
14 26 ☽ ∥ ☉
16 08 ☽ ⊼ ♀
23 50 ☉ □ ♄

26 Wednesday
01 46 ☽ # ☉
02 40 ☽ ⊻ ♆
03 48 ☽ ± ♃
05 39 ☽ Q ♂
05 40 ☽ # ☿
09 02 ☉ ⊥ ♀
10 28 ☽ □ ♇
13 50 ☽ ⊥ ♆
15 08 ☽ ⊼ ♃

27 Thursday
06 26 ☽ ∠ ♆
08 44 ☽ # ♆
10 11 ☽ ☌ ♅
12 35 ☽ △ ♄
13 48 ☿ ⊻ ♀
15 27 ☽ ⊼ ☉
15 50 ☽ □ ♂
23 34 ☉ △ ♂

28 Friday
01 02 ☽ ☍ ♀
02 26 ☽ ⊼ ☿
03 15 ☽ ± ☉
09 43 ☽ ⚹ ♆
10 03 ☽ ∥ ♅
14 53 ☽ ± ☿
15 52 ☽ ᵽ ♄
17 20 ☽ △ ♇
20 46 ☽ ᵽ ☉
22 36 ☽ ☍ ♃

29 Saturday
07 12 ☽ Q ♆
08 02 ☽ ± ♄
09 22 ☽ ᵽ ☿
15 12 ♂ ∥ ♄
16 07 ☽ ⊻ ♅
18 44 ☽ ⊼ ♄
20 06 ☽ ᵽ ♇
23 16 ☽ △ ♂

30 Sunday
01 34 ☽ △ ☉
02 42 ☽ ⊥ ♅
03 11 ☉ ∠ ♃
03 39 ☽ # ♃
07 37 ☽ ⊼ ♀
10 22 ☿ ⊼ ♆
11 58 ☽ ± ♇
14 58 ☽ □ ♆
15 38 ☽ △ ☿
18 17 ☽ ± ♀
18 29 ☽ ∠ ♅
22 28 ☽ ⊼ ♇

OCTOBER 2018

Date	Sidereal time h m s	Sun ☉ ° ′ ″	Moon ☽ ° ′ ″	Moon ☽ 24.00 ° ′ ″	Mercury ☿ ° ′	Venus ♀ ° ′	Mars ♂ ° ′	Jupiter ♃ ° ′	Saturn ♄ ° ′	Uranus ♅ ° ′	Neptune ♆ ° ′	Pluto ⯓ ° ′
01	12 40 42	08 ♎ 15 25	26 ♊ 31 51	03 ♋ 28 22	16 ♎ 16	10 ♏ 29	06 ♒ 09	22 ♏ 10	03 ♑ 03	01 ♉ 27	14 ♓ 27	18 ♑ 45
02	12 44 38	09 14 25	10 ♋ 27 27	17 ♋ 29 04	17 56	10 38	06 33	22 21	03 05	01 R 25	14 R 25	18 D 45
03	12 48 35	10 13 28	24 ♋ 33 10	01 ♌ 39 39	19 35	10 44	06 57	22 33	03 08	01 22	14 24	18 45
04	12 52 31	11 12 33	08 ♌ 48 20	15 ♌ 58 58	21 13	10 48	07 22	22 44	03 11	01 20	14 23	18 46
05	12 56 28	12 11 41	23 ♌ 11 13	00 ♍ 24 38	22 50	10 50	07 47	22 56	03 13	01 18	14 21	18 46
06	13 00 24	13 10 50	07 ♍ 38 42	14 ♍ 52 46	24 26	10 R 50	08 13	23 07	03 16	01 16	14 20	18 46
07	13 04 21	14 10 02	22 ♍ 06 09	29 ♍ 18 05	26 02	10 47	08 39	23 19	03 19	01 13	14 18	18 46
08	13 08 18	15 09 16	06 ♎ 27 48	13 ♎ 34 32	27 37	10 42	09 06	23 30	03 22	01 11	14 17	18 46
09	13 12 14	16 08 32	20 ♎ 37 33	27 ♎ 36 11	29 ♎ 11	10 34	09 33	23 42	03 25	01 09	14 16	18 46
10	13 16 11	17 07 50	04 ♏ 29 53	11 ♏ 18 13	00 ♏ 44	10 24	10 00	23 54	03 28	01 06	14 14	18 47
11	13 20 07	18 07 11	18 ♏ 00 52	24 ♏ 37 40	02 17	10 12	10 28	24 06	03 32	01 04	14 13	18 47
12	13 24 04	19 06 33	01 ♐ 08 37	07 ♐ 33 48	03 48	09 57	10 56	24 18	03 35	01 01	14 12	18 47
13	13 28 00	20 05 57	13 ♐ 53 28	20 ♐ 07 58	05 20	09 40	11 25	24 30	03 38	00 59	14 10	18 48
14	13 31 57	21 05 22	26 ♐ 17 44	02 ♑ 23 18	06 50	09 20	11 54	24 42	03 42	00 57	14 09	18 48
15	13 35 53	22 04 50	08 ♑ 25 14	14 ♑ 24 11	08 20	08 59	12 23	24 54	03 46	00 54	14 08	18 48
16	13 39 50	23 04 19	20 ♑ 20 48	26 ♑ 15 47	09 49	08 35	12 53	25 06	03 49	00 52	14 07	18 49
17	13 43 47	24 03 50	02 ♒ 09 51	08 ♒ 03 40	11 17	08 09	13 23	25 19	03 53	00 49	14 06	18 49
18	13 47 43	25 03 23	13 ♒ 57 57	19 ♒ 53 20	12 45	07 41	13 53	25 31	03 57	00 47	14 04	18 50
19	13 51 40	26 02 58	25 ♒ 50 28	01 ♓ 49 57	14 11	07 12	14 24	25 43	04 01	00 44	14 03	18 50
20	13 55 36	27 02 34	07 ♓ 51 46	13 ♓ 58 01	15 37	06 40	14 55	25 56	04 05	00 42	14 02	18 51
21	13 59 33	28 02 12	20 ♓ 07 29	26 ♓ 21 03	17 03	06 08	15 26	26 08	04 09	00 40	14 01	18 52
22	14 03 29	29 ♎ 01 52	02 ♈ 38 56	09 ♈ 01 17	18 27	05 34	15 58	26 21	04 13	00 37	14 00	18 52
23	14 07 26	00 ♏ 01 34	15 ♈ 28 10	21 ♈ 59 32	19 51	04 59	16 29	26 33	04 17	00 35	13 59	18 53
24	14 11 22	01 01 17	28 ♈ 35 14	05 ♉ 15 04	21 14	04 23	17 01	26 46	04 22	00 32	13 58	18 54
25	14 15 19	02 01 03	11 ♉ 58 44	18 ♉ 45 51	22 35	03 47	17 34	26 59	04 26	00 30	13 57	18 54
26	14 19 16	03 00 51	25 ♉ 36 04	02 ♊ 28 57	23 56	03 10	18 06	27 11	04 30	00 27	13 56	18 55
27	14 23 12	04 00 41	09 ♊ 24 05	16 ♊ 21 05	25 16	02 33	18 39	27 24	04 35	00 25	13 55	18 56
28	14 27 09	05 00 33	23 ♊ 19 34	00 ♋ 19 14	26 35	01 57	19 12	27 37	04 40	00 22	13 54	18 57
29	14 31 05	06 00 27	07 ♋ 19 48	14 ♋ 21 04	27 52	01 21	19 46	27 50	04 44	00 20	13 53	18 57
30	14 35 02	07 00 23	21 ♋ 22 51	28 ♋ 25 02	29 ♏ 08	00 45	20 19	28 02	04 49	00 18	13 53	18 58
31	14 38 58	08 ♏ 00 22	05 ♌ 27 30	12 ♌ 30 12	00 ♐ 23	00 ♏ 11	20 ♒ 53	28 ♏ 15	04 ♑ 54	00 ♉ 15	13 ♓ 52	18 ♑ 59

DECLINATIONS

Date	Moon True ☊ ° ′	Moon Mean ☊ ° ′	Moon ☽ Latitude ° ′	Sun ☉ ° ′	Moon ☽ ° ′	Mercury ☿ ° ′	Venus ♀ ° ′	Mars ♂ ° ′	Jupiter ♃ ° ′	Saturn ♄ ° ′	Uranus ♅ ° ′	Neptune ♆ ° ′	Pluto ⯓ ° ′
01	03 ♌ 33	02 ♌ 25	03 S 04	03 S 16	20 N 20	05 S 55	21 S 28	22 S 28	17 S 34	22 S 45	11 N 27	07 S 03	22 S 05
02	03 D 32	02 22	02 00	03 40	21 02	06 40	21 36	22 19	17 37	22 45	11 26	07 03	22 05
03	03 32	02 19	00 S 48	04 03	20 25	07 23	21 43	22 09	17 40	22 46	11 25	07 04	22 05
04	03 R 32	02 16	00 N 28	04 26	18 30	08 07	21 49	21 59	17 43	22 46	11 25	07 04	22 06
05	03 31	02 12	01 43	04 49	15 24	08 49	21 53	21 50	17 46	22 46	11 24	07 05	22 06
06	03 27	02 09	02 51	05 12	11 20	09 31	21 57	21 40	17 50	22 46	11 23	07 05	22 06
07	03 21	02 06	03 48	05 35	06 37	10 13	21 59	21 29	17 53	22 46	11 22	07 06	22 06
08	03 11	02 03	04 30	05 58	01 N 34	10 54	22 00	21 19	17 56	22 46	11 21	07 06	22 06
09	03 00	02 00	04 55	06 21	03 S 30	11 34	22 00	21 09	17 59	22 46	11 21	07 07	22 06
10	02 48	01 56	05 02	06 44	08 17	12 13	21 58	20 58	18 02	22 46	11 20	07 07	22 06
11	02 37	01 53	04 51	07 06	12 32	12 52	21 55	20 47	18 05	22 46	11 19	07 08	22 06
12	02 27	01 50	04 25	07 29	16 03	13 30	21 51	20 36	18 08	22 46	11 18	07 08	22 06
13	02 20	01 47	03 47	07 51	18 43	14 07	21 45	20 25	18 11	22 46	11 17	07 09	22 06
14	02 15	01 44	02 58	08 14	20 25	14 44	21 38	20 14	18 14	22 46	11 17	07 09	22 06
15	02 13	01 41	02 03	08 36	21 08	15 19	21 30	20 03	18 17	22 46	11 16	07 10	22 06
16	02 12	01 37	01 N 03	08 58	20 52	15 54	21 20	19 52	18 20	22 46	11 15	07 10	22 06
17	02 D 12	01 34	00 00	09 20	19 40	16 28	21 09	19 40	18 23	22 46	11 14	07 11	22 06
18	02 R 12	01 31	01 S 02	09 42	17 37	17 02	20 56	19 28	18 26	22 46	11 13	07 11	22 06
19	02 10	01 28	02 02	10 04	14 49	17 34	20 42	19 16	18 29	22 46	11 12	07 12	22 06
20	02 06	01 25	02 57	10 25	11 21	18 05	20 26	19 04	18 32	22 46	11 11	07 12	22 06
21	01 59	01 21	03 45	10 46	07 21	18 36	20 09	18 52	18 35	22 46	11 11	07 12	22 06
22	01 49	01 18	04 23	11 08	02 S 58	19 06	19 51	18 40	18 38	22 46	11 10	07 13	22 06
23	01 37	01 15	04 49	11 29	01 N 39	19 34	19 32	18 28	18 41	22 46	11 09	07 13	22 06
24	01 24	01 12	05 00	11 50	06 18	20 02	19 11	18 15	18 44	22 46	11 08	07 14	22 06
25	01 11	01 09	04 55	12 10	10 45	20 28	18 50	18 02	18 47	22 46	11 07	07 14	22 06
26	00 59	01 06	04 33	12 31	14 44	20 54	18 27	17 50	18 50	22 46	11 06	07 14	22 06
27	00 50	01 02	03 56	12 51	17 58	21 18	18 04	17 37	18 53	22 46	11 05	07 15	22 06
28	00 43	00 59	03 03	13 11	20 13	21 42	17 40	17 24	18 56	22 46	11 05	07 15	22 06
29	00 39	00 56	02 00	13 31	21 14	22 04	17 16	17 11	18 59	22 46	11 04	07 15	22 06
30	00 38	00 53	00 49	13 51	20 56	22 25	16 52	16 57	19 02	22 46	11 03	07 16	22 06
31	00 ♌ 38	00 ♌ 50	00 N 26	14 S 11	19 N 19	22 S 45	16 S 27	16 S 44	19 S 05	22 S 46	11 N 02	07 S 16	22 S 06

ZODIAC SIGN ENTRIES

Date	h	m	Planets
01	18	00	☽ ♋
03	21	12	☽ ♌
05	23	19	☽ ♍
08	01	10	☽ ♎
10	00	40	☿ ♏
10	04	09	☽ ♏
12	09	53	☽ ♐
14	19	17	☽ ♑
17	07	36	☽ ♒
19	20	20	☽ ♓
22	06	58	☽ ♈
23	11	22	☉ ♏
24	14	33	☽ ♉
26	19	41	☽ ♊
28	23	27	☽ ♋
31	02	42	☽ ♌
31	04	38	☿ ♐
31	19	42	♀ ♎

LATITUDES

Date	Mercury ☿ ° ′	Venus ♀ ° ′	Mars ♂ ° ′	Jupiter ♃ ° ′	Saturn ♄ ° ′	Uranus ♅ ° ′	Neptune ♆ ° ′	Pluto ⯓ ° ′
01	00 N 31	06 S 52	03 S 51	00 N 46	00 N 39	00 S 34	01 S 00	00 N 02
04	00 N 11	07 07	03 41	00 45	00 38	00 34	01 00	00 02
07	00 S 10	07 18	03 31	00 45	00 38	00 34	01 00	00 02
10	00 31	07 25	03 21	00 44	00 37	00 34	01 00	00 01
13	00 52	07 27	03 12	00 44	00 37	00 34	01 00	00 01
16	01 13	07 23	03 03	00 44	00 37	00 34	01 00	00 N 01
19	01 33	07 11	02 53	00 43	00 36	00 34	01 00	00 00
22	01 51	06 53	02 45	00 43	00 36	00 34	01 00	00 00
25	02 08	06 27	02 36	00 42	00 35	00 34	01 00	00 00
28	02 22	05 54	02 28	00 42	00 35	00 34	01 00	00 00
31	02 S 34	05 S 15	02 S 19	00 N 42	00 N 35	00 S 34	01 S 00	00 S 01

DATA

Julian Date	2458393
Delta T	+70 seconds
Ayanamsa	24° 06′ 53″
Synetic vernal point	05° ♓ 00′ 06″
True obliquity of ecliptic	23° 26′ 08″

MOON'S PHASES, APSIDES AND POSITIONS ☽

Date	h	m	Phase	Longitude ° ′	Eclipse Indicator
02	09	45	☾	09 ♋ 09	
09	03	47	●	15 ♎ 48	
16	18	02	◑	23 ♑ 19	
24	16	45	○	01 ♉ 13	
31	16	40	☾	08 ♌ 12	

Day	h	m		
05	22	16	Perigee	
17	19	15	Apogee	
31	20	22	Perigee	
02	13	03	Max dec	21° N 02′
08	19	22	0S	
15	17	25	Max dec	21° S 09′
23	03	30	0N	
29	18	36	Max dec	21° N 17′

LONGITUDES

Date	Chiron ⚷ ° ′	Ceres ⚳ ° ′	Pallas ⚴ ° ′	Juno ⚵ ° ′	Vesta ⚶ ° ′	Black Moon Lilith ⚸ ° ′
01	29 ♓ 45	11 ♎ 24	13 ♍ 46	00 ♊ 09	04 ♑ 18	06 ♒ 13
11	29 ♓ 19	15 ♎ 56	18 ♍ 39	00 ♊ 39	07 ♑ 56	07 ♒ 20
21	28 ♓ 55	20 ♎ 27	23 ♍ 23	00 ♊ 16	11 ♑ 50	08 ♒ 26
31	28 ♓ 34	24 ♎ 56	27 ♍ 58	29 ♉ 02	15 ♑ 58	09 ♒ 33

All ephemeris data is given at 12.00 UT and the Moon's longitude is additionally given for 24.00 UT

ASPECTARIAN

h m Aspects

01 Monday
02 04 ⯓ St D
02 22 ☽ ♇ ♂
04 19 ☽ ⊼ ♃
10 11 ☽ ♇ ♀
10 25 ☿ ⊥ ♃
14 53 ☽ ± ♃
18 26 ☽ ± ♂
20 29 ☽ ⚹ ♅
23 18 ☽ ☍ ♄

02 Tuesday
05 05 ☽ ⊼ ♂
06 36 ☽ ♇ ♃
09 45 ☽ □ ☉
12 18 ☽ △ ♀
17 02 ☽ Q ♅
18 46 ☽ △ ♆

03 Wednesday
00 03 ☿ □ ⯓
01 06 ☿ ∥ ♆
02 10 ☽ ☍ ⯓
02 27 ☽ □ ☿
08 33 ☽ △ ♃
18 40 ☽ Q ☉
20 11 ☽ ♇ ♆
21 07 ♂ ∥ ⯓
23 29 ☽ □ ♅
23 55 ☿ ± ♆

04 Thursday
01 36 ☉ ⊻ ♀
02 31 ☽ ⊼ ♄
09 31 ☽ ☍ ♂
11 17 ☽ ± ♆
11 32 ☿ Q ♄
12 38 ☽ ± ♄
12 46 ☽ Q ☿
15 22 ☽ □ ♀
16 19 ☽ ⚹ ☉
18 48 ☽ ⋕ ♃
21 18 ☽ ⊼ ♆

05 Friday
03 43 ☽ ♇ ♄
04 38 ☽ ⊼ ⯓
05 58 ♀ ∥ ♂
11 20 ☽ ⚹ ☿
11 34 ☽ □ ♃
13 39 ☿ ⊻ ♃
14 37 ☽ ± ⯓
19 04 ♀ St R
19 09 ☽ ∠ ☉
21 24 ☽ Q ♀

06 Saturday
01 26 ☽ △ ♅
04 43 ☽ △ ♄
05 34 ☽ ♇ ⯓
11 10 ☽ ⊥ ☉
11 46 ☽ ∥ ♅
12 59 ☽ ⊼ ♂
15 20 ☽ ∠ ☿
17 16 ☽ ⚹ ♀
17 50 ☽ Q ♃
17 56 ♂ ⊥ ♆
20 19 ☽ ⋕ ☿
21 51 ☽ ⊻ ☉
23 04 ☽ ☍ ♆
23 14 ☽ ± ♂

07 Sunday
02 15 ☽ ♇ ♅
06 27 ☽ △ ⯓
08 08 ☽ ⊥ ☿
09 40 ☽ ⋕ ♆
14 03 ☽ ⚹ ♃
14 40 ☽ ♇ ♂
15 17 ☉ ⊼ ♆
16 38 ☽ ⋕ ☉
17 11 ☽ ± ♅
18 06 ☽ ∠ ♀
19 21 ☽ ⊻ ☿

08 Monday
03 10 ☽ ⊼ ♅
06 47 ☽ □ ♄
09 03 ☽ ⊥ ♀
15 29 ☽ ∠ ♃
16 34 ☽ △ ♂
19 04 ☽ ⊻ ♀

09 Tuesday
01 11 ☽ ⊼ ♆
03 47 ☽ ☌ ☉
04 19 ☿ ⋕ ♅
04 24 ♂ ⊥ ♄
06 57 ☽ ⊥ ♃
08 50 ☽ □ ⯓
11 23 ☽ ± ♆
13 15 ☿ ♇ ♆
13 22 ☽ Q ♄
17 21 ☽ ⊻ ♃

10 Wednesday
02 51 ☽ ♇ ♆
03 16 ☽ ∥ ☉
04 36 ☽ ☌ ☿
05 59 ☽ ∥ ♆
06 05 ☽ ☍ ♅
10 12 ☽ ⚹ ♄
16 00 ☽ Q ⯓
17 36 ☿ ☍ ♅
22 02 ☽ □ ♂
22 15 ☽ ☌ ♀

11 Thursday
02 29 ♀ □ ♂
04 46 ☽ ⋕ ♅
05 12 ☽ △ ♆
11 24 ☉ ⊥ ♃
12 12 ☽ ⊻ ☉
12 56 ☽ ∠ ♄
13 23 ☽ ⚹ ⯓
13 47 ☉ ∥ ♆
14 28 ☽ ∥ ☿
23 12 ☽ ☌ ♃
23 58 ☽ ⊥ ☉

12 Friday
04 11 ☉ □ ⯓
05 23 ☽ ⊥ ♄
07 46 ☽ Q ♂
08 20 ☿ ⚹ ♄
11 47 ☽ ⊼ ♅
16 34 ☽ ⊻ ♄
16 55 ☽ ∠ ⯓
17 37 ☽ ⊻ ☿
17 59 ☽ ∠ ☉
22 57 ☽ ± ♅

13 Saturday
04 09 ☽ ⊻ ♀
06 21 ☽ ∥ ♃
06 27 ☽ ⊥ ☿
07 06 ☽ ⚹ ♂
09 54 ☽ ⊥ ⯓
12 32 ☽ □ ♆
13 46 ☉ ± ♆
15 19 ☽ ⊥ ♀
16 00 ☽ ♇ ♅
21 25 ☽ ⊻ ⯓

14 Sunday
00 58 ☽ ⚹ ☉
02 06 ☽ ∠ ☿
08 17 ☽ ∠ ♀
08 50 ☽ ⊻ ♃
09 06 ☽ ∥ ♂
11 28 ☿ Q ⯓
13 14 ☽ ∠ ♂
20 49 ☽ ⊥ ♃
21 07 ☽ △ ♅
23 31 ☽ Q ♆

15 Monday
02 35 ☽ Q ☉
02 40 ☽ ☌ ♄
03 42 ☉ Q ♄
07 46 ☽ ⊥ ♂
11 47 ☽ ⚹ ☿
13 05 ☽ ⚹ ♀
15 01 ☽ ∠ ♃
20 17 ☽ ⊻ ♂
20 21 ☿ ☌ ♀
23 26 ☽ ⚹ ♆

16 Tuesday
08 54 ☽ ☌ ⯓
12 28 ☽ Q ♀
15 23 ☽ Q ☿
18 02 ☽ □ ☉
21 49 ☽ ⚹ ♃

17 Wednesday
05 46 ☽ ∠ ♆
09 17 ☽ □ ♅
12 05 ☽ ∥ ♂
15 31 ☽ ⊻ ♄
22 40 ☽ Q ♃
23 44 ☽ □ ♀

18 Thursday
00 03 ☽ ⊥ ♆
03 45 ☽ ∥ ♃
03 47 ☽ ⊥ ♄
09 10 ☽ □ ☿
11 50 ☽ ☌ ♂
12 13 ☽ ⊻ ♆
16 40 ☽ ∥ ☿
20 30 ♂ ⊻ ♆
21 44 ☽ Q ♅
21 52 ☽ ⊻ ⯓
22 09 ☽ ∠ ♄

19 Friday
02 01 ☉ ⊻ ♃
09 47 ☿ △ ♆
09 59 ☽ ⊥ ⯓
11 45 ☽ □ ♃
12 27 ☽ △ ☉
17 23 ☿ □ ♂
21 47 ☽ ⚹ ♅

20 Saturday
04 01 ☽ ∠ ⯓
04 07 ♀ Q ⯓
04 26 ☽ ⚹ ♄
09 44 ☽ △ ♀
13 01 ☽ ⋕ ♅
17 22 ☽ ∥ ☉
20 57 ☽ ♇ ☉

21 Sunday
00 07 ☽ ☌ ♆
02 28 ☽ ⊻ ♂
03 20 ☽ ∠ ♅
04 13 ☽ Q ♄
05 14 ☽ △ ☿
09 33 ☽ ⚹ ⯓
11 25 ☿ ∥ ♃
12 50 ☽ ∥ ♆
13 52 ☽ ♇ ♀
14 39 ☽ ⊥ ♂
16 01 ☽ ± ☉
20 44 ☽ ⊥ ♅
21 17 ☿ ∥ ♂
23 47 ☽ △ ♃

22 Monday
04 32 ☽ ⊼ ☉
06 24 ☽ ± ♀
08 10 ☽ ⊻ ♅
08 37 ☽ Q ⯓
08 39 ☽ ∠ ♂
11 16 ☉ ♇ ♆
13 43 ☽ ♇ ☿
14 13 ☉ ⋕ ♅
14 35 ♂ ∥ ♃
14 59 ☽ □ ♄
17 16 ☽ ⊼ ♀
19 14 ☿ ⚹ ⯓

23 Tuesday
01 49 ☿ ∠ ♄
04 37 ☽ ♇ ♃
08 38 ☽ ± ☿
09 15 ☽ ⊻ ♆
10 39 ☿ ∥ ♀
13 58 ☽ ⚹ ♂
18 18 ☽ □ ⯓
20 18 ☽ ⊥ ♆
21 01 ☽ ⊼ ☿
21 31 ☽ ± ♃

24 Wednesday
00 47 ☉ ☍ ♅
08 39 ☽ ⊼ ♃
12 41 ☽ ∠ ♆
12 49 ☽ Q ♂
12 52 ♀ ⚹ ♄
15 31 ☽ ☌ ♅
16 45 ☽ ☍ ☉
16 54 ☽ ⋕ ♆
22 00 ☽ ☍ ♀
22 27 ☽ △ ♄

25 Thursday
14 02 ♀ ∥ ♃
14 07 ☽ ∥ ♅
15 30 ☽ ⚹ ♆
20 58 ☽ ⋕ ☉
22 18 ☽ □ ♂

26 Friday
00 15 ☽ △ ⯓
01 15 ☽ ♇ ♄
08 46 ☽ ☍ ☿
12 35 ☽ Q ♆
14 16 ☉ ☌ ♀
14 49 ☽ ☍ ♃
17 06 ☽ ± ♄
20 27 ☽ ⊻ ♅

27 Saturday
00 38 ☽ ⊼ ♀
01 56 ☽ ⊼ ☉
02 15 ♂ Q ♅
02 30 ☽ ♇ ⯓
03 36 ☽ ⊼ ♄
06 50 ☽ ⊥ ♅
09 10 ☽ ⋕ ♂
10 36 ☽ ± ♀
12 43 ☽ ⋕ ♀
13 08 ☽ ± ☉
18 06 ☽ ± ⯓
19 48 ☽ □ ♆
20 39 ☽ ⋕ ♃
22 21 ☽ ∠ ♅

28 Sunday
00 19 ♂ ⊻ ⯓
01 29 ☽ ♇ ♀
02 52 ☉ ⚹ ♄
04 27 ☽ ⊼ ⯓
04 37 ☽ △ ♂
05 51 ☽ ♇ ☉
18 09 ☽ ⊼ ☿
19 28 ☽ ⊼ ♃

29 Monday
00 03 ☽ ⚹ ♅
02 10 ☽ △ ♀
05 29 ☽ ± ☿
05 55 ☽ ± ♃
07 25 ☽ ♇ ♂
07 32 ☽ ☍ ♄
09 34 ☽ △ ☉
10 47 ♂ ∠ ♄
11 05 ☿ ☌ ♃
14 34 ☿ ∥ ⯓
20 32 ☽ Q ♅
21 32 ☽ ♇ ♃
22 25 ☽ ♇ ☿
23 12 ☽ △ ♆
23 27 ☽ ± ♂
23 41 ♀ ∥ ♂

30 Tuesday
05 31 ☿ ⊥ ♄
07 53 ☽ ☍ ⯓
10 07 ☽ ⊼ ♂
11 08 ☉ Q ⯓
23 32 ☽ △ ♃

31 Wednesday
00 46 ☽ ♇ ♆
02 31 ☽ △ ☿
03 09 ☽ □ ♅
03 21 ☽ □ ♀
08 45 ♀ ☍ ♅
09 21 ☿ ⊻ ♀
09 36 ☿ ⊼ ♅
11 02 ☽ ⊼ ♄
13 41 ☿ ∥ ♄
14 19 ☽ ⋕ ♃
16 05 ☽ ± ♆
16 40 ☽ □ ☉
21 19 ☽ ± ♄

NOVEMBER 2018

LONGITUDES

Date	Sidereal time h m s	Sun ☉ ° ' "	Moon ☽ ° ' "	Moon ☽ 24.00 ° ' "	Mercury ☿ ° '	Venus ♀ ° '	Mars ♂ ° '	Jupiter ♃ ° '	Saturn ♄ ° '	Uranus ♅ ° '	Neptune ♆ ° '	Pluto ♇ ° '
01	14 42 55	09 ♏ 00 23	19 ♌ 33 02	26 ♌ 35 54	01 ♐ 36	29 ♎ 38	21 ♒ 27	28 ♏ 28	04 ♑ 59	00 ♉ 13	13 ♓ 51	19 ♑ 00
02	14 46 51	10 00 26	03 ♍ 38 40	10 ♍ 41 09	02 47	29 R 06	22 01	28 41	05 04	00 R 10	13 R 50	19 01
03	14 50 48	11 00 31	17 ♍ 43 08	24 ♍ 44 19	03 56	28 35	22 36	28 54	05 09	00 08	13 49	19 02
04	14 54 45	12 00 38	01 ♎ 44 21	08 ♎ 42 51	05 04	28 06	23 10	29 07	05 14	00 05	13 49	19 03
05	14 58 41	13 00 47	15 ♎ 39 22	22 ♎ 33 26	06 08	27 40	23 45	29 20	05 19	00 03	13 48	19 04
06	15 02 38	14 00 58	29 ♎ 24 35	06 ♏ 12 23	07 10	27 15	24 20	29 33	05 24	00 ♉ 01	13 47	19 05
07	15 06 34	15 01 11	12 ♏ 56 24	19 ♏ 36 18	08 09	26 52	24 56	29 ♏ 46	05 29	29 ♈ 58	13 47	19 06
08	15 10 31	16 01 26	26 ♏ 11 47	02 ♐ 42 39	09 05	26 32	25 31	00 ♐ 00	05 35	29 56	13 46	19 07
09	15 14 27	17 01 42	09 ♐ 08 49	15 ♐ 30 16	09 57	26 14	26 07	00 13	05 40	29 54	13 46	19 08
10	15 18 24	18 02 01	21 ♐ 47 08	27 ♐ 59 35	10 45	25 58	26 43	00 26	05 46	29 51	13 45	19 09
11	15 22 20	19 02 21	04 ♑ 07 56	10 ♑ 12 33	11 27	25 45	27 19	00 39	05 51	29 49	13 45	19 11
12	15 26 17	20 02 42	16 ♑ 13 54	22 ♑ 12 29	12 05	25 34	27 55	00 52	05 57	29 47	13 44	19 12
13	15 30 14	21 03 05	28 ♑ 08 53	04 ♒ 03 42	12 36	25 25	28 31	01 06	06 02	29 45	13 44	19 13
14	15 34 10	22 03 29	09 ♒ 57 37	15 ♒ 51 17	13 01	25 19	29 08	01 19	06 08	29 42	13 44	19 14
15	15 38 07	23 03 55	21 ♒ 45 25	27 ♒ 40 41	13 19	25 16	29 ♒ 44	01 32	06 14	29 40	13 43	19 16
16	15 42 03	24 04 22	03 ♓ 37 48	09 ♓ 37 25	13 28	25 D 15	00 ♓ 21	01 46	06 20	29 38	13 43	19 17
17	15 46 00	25 04 50	15 ♓ 40 12	21 ♓ 46 43	13 R 29	25 16	00 58	01 59	06 26	29 36	13 43	19 18
18	15 49 56	26 05 19	27 ♓ 57 33	04 ♈ 13 09	13 20	25 20	01 35	02 12	06 31	29 34	13 42	19 20
19	15 53 53	27 05 50	10 ♈ 33 56	17 ♈ 00 09	13 01	25 26	02 12	02 26	06 37	29 32	13 42	19 21
20	15 57 49	28 06 22	23 ♈ 32 02	00 ♉ 09 36	12 31	25 34	02 50	02 39	06 43	29 29	13 42	19 22
21	16 01 46	29 ♏ 06 56	06 ♉ 52 48	13 ♉ 41 26	11 51	25 45	03 27	02 52	06 49	29 27	13 42	19 24
22	16 05 43	00 ♐ 07 31	20 ♉ 35 10	27 ♉ 33 32	11 01	25 57	04 05	03 06	06 56	29 25	13 42	19 25
23	16 09 39	01 08 07	04 ♊ 36 01	11 ♊ 41 57	10 01	26 12	04 42	03 19	07 02	29 23	13 42	19 27
24	16 13 36	02 08 45	18 ♊ 50 40	26 ♊ 01 26	08 52	26 29	05 20	03 33	07 08	29 21	13 42	19 28
25	16 17 32	03 09 24	03 ♋ 13 33	10 ♋ 26 20	07 37	26 48	05 58	03 46	07 14	29 20	13 D 42	19 30
26	16 21 29	04 10 05	17 ♋ 39 08	24 ♋ 51 24	06 17	27 09	06 36	03 59	07 20	29 18	13 42	19 31
27	16 25 25	05 10 48	02 ♌ 02 40	09 ♌ 12 30	04 54	27 32	07 14	04 13	07 27	29 16	13 42	19 33
28	16 29 22	06 11 32	16 ♌ 20 37	23 ♌ 26 46	03 33	27 57	07 52	04 26	07 33	29 14	13 42	19 34
29	16 33 18	07 12 17	00 ♍ 30 45	07 ♍ 32 29	02 15	28 24	08 31	04 40	07 40	29 12	13 42	19 36
30	16 37 15	08 ♐ 13 04	14 ♍ 31 51	21 ♍ 28 48	01 ♐ 02	28 ♎ 52	09 ♓ 09	04 ♐ 53	07 ♑ 46	29 ♈ 10	13 ♓ 42	19 ♑ 37

Date	Moon True ☊ ° '	Moon Mean ☊ ° '	Moon ☽ Latitude ° '
01	00 ♌ 37	00 ♌ 47	01 N 38
02	00 R 36	00 43	02 45
03	00 32	00 40	03 42
04	00 25	00 37	04 25
05	00 15	00 34	04 52
06	00 03	00 31	05 01
07	29 ♋ 51	00 27	04 53
08	29 38	00 24	04 30
09	29 28	00 21	03 53
10	29 19	00 18	03 05
11	29 14	00 15	02 09
12	29 11	00 12	01 08
13	29 10	00 08	00 N 05
14	29 D 10	00 05	00 S 57
15	29 11	00 ♌ 02	01 58
16	29 R 10	29 ♋ 59	02 53
17	29 08	29 56	03 42
18	29 03	29 53	04 21
19	28 56	29 49	04 49
20	28 47	29 46	05 03
21	28 37	29 43	05 01
22	28 27	29 40	04 42
23	28 17	29 37	04 06
24	28 10	29 33	03 14
25	28 05	29 30	02 09
26	28 02	29 27	00 S 56
27	28 D 02	29 24	00 N 22
28	28 02	29 21	01 37
29	28 03	29 18	02 46
30	28 ♋ 03	29 ♋ 14	03 N 44

DECLINATIONS

Date	Sun ☉ ° '	Moon ☽ ° '	Mercury ☿ ° '	Venus ♀ ° '	Mars ♂ ° '	Jupiter ♃ ° '	Saturn ♄ ° '	Uranus ♅ ° '	Neptune ♆ ° '	Pluto ♇ ° '
01	14 S 30	16 N 31	23 S 03	16 S 02	16 S 30	19 S 08	22 S 46	11 N 01	07 S 16	22 S 06
02	14 49	12 44	23 20	15 37	16 17	19 11	22 46	11 00	07 16	22 06
03	15 08	08 15	23 36	15 13	16 03	19 14	22 46	11 00	07 17	22 06
04	15 26	03 N 21	23 51	14 49	15 49	19 17	22 46	10 59	07 17	22 06
05	15 45	01 S 40	24 04	14 25	15 35	19 20	22 46	10 58	07 17	22 06
06	16 03	06 34	24 15	14 02	15 21	19 23	22 45	10 57	07 17	22 06
07	16 20	11 03	24 25	13 39	15 07	19 26	22 45	10 56	07 18	22 06
08	16 38	14 55	24 33	13 18	14 53	19 29	22 45	10 56	07 18	22 06
09	16 55	17 59	24 40	12 57	14 38	19 32	22 45	10 55	07 18	22 06
10	17 12	20 06	24 45	12 37	14 24	19 34	22 45	10 54	07 18	22 06
11	17 29	21 13	24 48	12 18	14 09	19 37	22 45	10 53	07 18	22 06
12	17 45	21 19	24 49	12 00	13 54	19 40	22 45	10 52	07 18	22 05
13	18 01	20 26	24 48	11 44	13 39	19 43	22 45	10 52	07 19	22 05
14	18 17	18 40	24 45	11 28	13 24	19 46	22 44	10 51	07 19	22 05
15	18 32	16 06	24 40	11 14	13 09	19 48	22 44	10 50	07 19	22 05
16	18 47	12 52	24 32	11 00	12 54	19 51	22 44	10 49	07 19	22 05
17	19 02	09 03	24 22	10 48	12 39	19 54	22 44	10 49	07 19	22 05
18	19 16	04 48	24 09	10 37	12 24	19 57	22 44	10 48	07 19	22 05
19	19 30	00 S 15	23 53	10 28	12 08	19 59	22 44	10 47	07 19	22 05
20	19 44	04 N 26	23 35	10 19	11 53	20 02	22 43	10 46	07 19	22 05
21	19 57	09 03	23 13	10 11	11 37	20 05	22 43	10 46	07 19	22 05
22	20 10	13 21	22 48	10 05	11 21	20 07	22 43	10 45	07 19	22 04
23	20 23	17 01	22 21	10 00	11 06	20 10	22 43	10 44	07 19	22 04
24	20 35	19 45	21 51	09 56	10 50	20 13	22 43	10 44	07 19	22 04
25	20 47	21 15	21 18	09 53	10 34	20 15	22 42	10 43	07 19	22 04
26	20 59	21 21	20 45	09 51	10 18	20 18	22 42	10 42	07 19	22 04
27	21 10	20 03	20 11	09 49	10 02	20 20	22 42	10 42	07 19	22 04
28	21 20	17 29	19 37	09 49	09 46	20 23	22 41	10 41	07 19	22 04
29	21 31	13 52	19 04	09 50	09 29	20 25	22 41	10 40	07 19	22 04
30	21 S 40	09 N 32	18 S 34	09 S 52	09 S 13	20 S 28	22 S 41	10 N 40	07 S 19	22 S 03

ZODIAC SIGN ENTRIES

Date	h m	Planets
02	05 48	☽ ♍
04	09 01	☽ ♎
06	13 02	☽ ♏
06	19 00	♅ ♈
08	12 38	♃ ♐
08	18 59	☽ ♐
11	03 55	☽ ♑
13	15 45	☽ ♒
15	22 21	♂ ♓
16	04 41	☽ ♓
18	15 56	☽ ♈
20	23 43	☽ ♉
22	09 01	☉ ♐
23	04 10	☽ ♊
25	06 38	☽ ♋
27	08 35	☽ ♌
29	11 08	☽ ♍

LATITUDES

Date	Mercury ☿ ° '	Venus ♀ ° '	Mars ♂ ° '	Jupiter ♃ ° '	Saturn ♄ ° '	Uranus ♅ ° '	Neptune ♆ ° '	Pluto ♇ ° '
01	02 S 38	05 S 01	02 S 17	00 N 42	00 N 35	00 S 34	01 S 00	00 S 01
04	02 45	04 18	02 09	00 41	00 34	00 33	01 00	00 01
07	02 47	03 32	02 01	00 41	00 34	00 33	00 59	00 01
10	02 43	02 47	01 54	00 41	00 34	00 33	00 59	00 02
13	02 31	02 02	01 47	00 41	00 33	00 33	00 59	00 02
16	02 08	01 20	01 40	00 40	00 33	00 33	00 59	00 02
19	01 33	00 40	01 33	00 40	00 33	00 33	00 59	00 02
22	00 S 43	00 S 04	01 26	00 40	00 32	00 33	00 59	00 03
25	00 N 16	00 N 29	01 20	00 40	00 32	00 33	00 59	00 03
28	01 16	00 59	01 14	00 39	00 32	00 33	00 59	00 03
31	02 N 04	01 N 26	01 S 08	00 N 39	00 N 31	00 S 33	00 S 59	00 S 03

DATA

Julian Date	2458424
Delta T	+70 seconds
Ayanamsa	24° 06' 56"
Synetic vernal point	05° ♓ 00' 03"
True obliquity of ecliptic	23° 26' 08"

MOON'S PHASES, APSIDES AND POSITIONS ☽

Date	h m	Phase	Longitude ° '	Eclipse Indicator
07	16 02	●	15 ♏ 11	
15	14 54	☽	23 ♒ 11	
23	05 39	○	00 ♊ 52	
30	00 19	☾	07 ♍ 43	

Day	h m		
14	15 57	Apogee	
26	12 23	Perigee	
05	04 00	OS	
12	02 18	Max dec	21° S 24'
19	13 19	ON	
26	01 50	Max dec	21° N 29'

LONGITUDES

Date	Chiron ⚷ ° '	Ceres ⚳ ° '	Pallas ⚴ ° '	Juno ⚵ ° '	Vesta ⚶ ° '	Black Moon Lilith ⚸ ° '
01	28 ♓ 32	25 ♎ 23	28 ♍ 25	28 ♉ 52	16 ♑ 23	09 ♒ 39
11	28 ♓ 15	29 ♎ 49	02 ♎ 49	26 ♉ 55	20 ♑ 43	10 ♒ 46
21	28 ♓ 03	04 ♏ 12	07 ♎ 02	24 ♉ 41	25 ♑ 13	11 ♒ 52
31	27 ♓ 56	08 ♏ 31	11 ♎ 02	22 ♉ 36	29 ♑ 50	12 ♒ 59

All ephemeris data is given at 12.00 UT and the Moon's longitude is additionally given for 24.00 UT

ASPECTARIAN

h m	Aspects	h m	Aspects	h m	Aspects
01 Thursday		05 04	☽ ⊻ ♃	09 59	☽ △ ♇
02 18	☽ ⊼ ♆	07 19	☽ Q ♆	13 13	☿ ∠ ♀
08 51	☽ Q ♀	11 48	☽ ∠ ☉	14 20	☽ ⚼ ♄
11 04	☽ ⊼ ♇	15 22	☉ ⚹ ♇	16 47	☿ ∥ ♄
12 01	☽ # ♂	15 25	☽ ☌ ♄	20 48	☽ Q ♆
12 44	☽ ⚼ ♄	17 03	☽ ⊥ ♃	21 25	☽ ⊼ ♀
15 22	☽ ☍ ♂	19 00	☽ Q ♀	**23 Friday**	
15 50	☽ # ♀	**12 Monday**		01 39	♂ ∠ ♇
21 17	☽ ± ♇	03 17	☽ ⊻ ☿	03 09	☽ ⊻ ♅
02 Friday		05 01	☽ ∠ ♂	05 39	☽ ☍ ☉
00 25	☽ # ☉	07 01	☽ ⚹ ♆	05 53	☽ ± ♄
01 40	☽ Q ☉	11 16	☽ ∠ ♃	07 51	☽ ± ♀
03 25	☽ □ ♃	15 53	☽ ⊥ ☿	09 11	☉ ⊥ ♄
04 32	☽ ⚹ ♀	17 57	☽ ☌ ♇	09 47	☽ ☍ ♃
06 06	☽ △ ♅	20 21	☽ ⚹ ☉	11 44	☽ ⚼ ♇
10 24	☽ □ ☿	**13 Tuesday**		12 11	☽ □ ♂
12 38	☽ ⚼ ♇	00 01	☽ ⊥ ♂	13 20	☽ ⊥ ♅
14 26	☽ △ ♄	06 32	☽ □ ♀	16 09	☽ ⊼ ♄
21 39	☽ ∥ ♅	10 51	☽ ∠ ☿	20 30	☽ ☍ ☿
23 40	☽ ⚹ ☉	11 42	☉ ∠ ♄	23 23	☽ ⚼ ♀
03 Saturday		12 47	☽ ⊻ ♂	**24 Saturday**	
00 17	♀ ⚼ ♆	13 11	☽ ∠ ♆	01 20	☿ ∥ ♇
01 22	♀ ⊻ ♃	15 13	☽ □ ♅	02 58	☽ ± ♇
05 11	☽ ∠ ♀	18 05	☽ ⚹ ♃	03 21	☽ □ ♆
05 21	☽ ☍ ♆	22 52	☽ Q ☉	04 29	☽ ∠ ♅
07 36	☽ ⚼ ♅	22 58	☽ ∥ ♃	13 03	☽ ⊼ ♇
10 35	☽ Q ♃	**14 Wednesday**		17 44	☽ # ♃
13 59	☿ ∠ ♇	04 09	☽ ⊻ ♄	21 34	♂ # ♅
14 15	☽ △ ♇	07 27	☽ ⊥ ♆	23 44	☉ ⊥ ♀
14 49	☉ ∥ ♀	12 11	♀ ⊥ ♃	**25 Sunday**	
16 54	☽ # ♆	15 44	☽ ∥ ☉	00 30	☽ # ☉
19 51	☽ Q ☿	16 28	☽ ⊥ ♄	01 03	☽ △ ♀
20 02	☽ ⊥ ♀	18 25	☽ ⚹ ☿	01 10	♆ St D
20 41	☽ ⊼ ♂	18 58	☽ Q ♃	05 31	☽ ⚹ ♅
21 35	☿ ⊥ ♀	19 40	☽ ⊻ ♆	11 53	☽ ⊼ ☉
22 56	☽ ± ♅	**15 Thursday**		12 55	☽ ⊼ ♃
04 Sunday		03 43	☽ Q ♅	13 07	☽ # ☿
03 16	☽ ∠ ☉	06 25	☿ ⊥ ♇	16 46	☽ △ ♂
05 58	☽ ⊻ ♀	06 55	☽ ⊻ ♇	18 22	☿ ⊻ ♄
07 24	☽ ± ♂	09 31	♂ ⚹ ♅	18 41	☽ ⊼ ☿
07 26	☽ ⚹ ♃	10 55	☽ ∠ ♄	18 43	☽ ☍ ♄
09 11	☽ ⊼ ♅	14 54	☽ □ ☉	22 37	☽ ± ☉
16 01	☿ ⊻ ♄	19 05	☽ △ ♀	23 03	☽ ± ♃
18 02	☽ □ ♄	19 07	☽ ⊥ ♇	**26 Monday**	
18 12	☽ ⚹ ☿	19 20	☽ Q ☿	01 27	☽ Q ♅
19 55	☽ ⊥ ☉	**16 Friday**		03 49	☽ ± ☿
23 33	☽ ⚼ ♂	03 58	☽ ⚹ ♅	04 44	☉ ∥ ☿
05 Monday		05 02	☽ ☌ ♂	05 25	☽ △ ♆
05 04	☉ ∥ ♂	08 10	☽ □ ♃	06 33	☉ ☌ ♃
07 04	☽ ⊻ ☉	10 51	♀ St D	08 06	☿ □ ♂
07 47	♃ ⊥ ♄	11 43	☽ ∥ ♂	14 16	☽ ⚼ ♃
08 47	☽ ⊼ ♆	13 19	☽ ∠ ♇	14 43	☽ ⚼ ☉
09 41	☽ ∠ ♃	17 27	☽ ⚹ ♄	15 07	☽ ☍ ♇
10 05	☿ ± ♅	**17 Saturday**		17 31	☽ ⚼ ☿
17 56	☽ □ ♇	00 48	☽ ∥ ♀	18 53	☽ ⚼ ♂
19 12	☽ ± ♆	01 15	☽ ⚼ ♀	20 33	☉ ∠ ♇
22 19	☽ ∠ ☿	01 18	☽ # ♅	20 39	☽ # ☉
06 Tuesday		01 32	☿ St R	**27 Tuesday**	
01 24	☽ Q ♄	07 41	☽ □ ☿	04 16	☽ □ ♀
01 35	☽ ⊥ ♃	08 08	☽ ☌ ♆	05 38	☿ ± ♅
02 43	☽ △ ♂	09 53	☽ ∠ ♅	05 42	☿ ∥ ♃
06 40	☉ △ ♆	11 26	♀ # ♅	06 24	☽ ⚼ ♆
08 19	☽ ☌ ♀	16 35	☉ ⊻ ♀	07 22	☽ □ ♅
10 55	☽ ⚼ ♆	17 28	☽ Q ♄	08 25	☽ # ♃
12 16	☽ ⊻ ♃	19 06	☽ ± ♀	09 15	☉ ☌ ☿
13 03	☽ ☍ ♅	19 10	☽ ⚹ ♇	09 45	☽ # ☿
15 21	☽ ⊥ ☿	22 05	☽ ∥ ♆	10 35	☽ ± ♂
15 43	☽ ∥ ♆	**18 Sunday**		13 55	☉ ± ♅
22 39	☽ ⚹ ♄	03 30	☽ ⊥ ♅	15 41	☽ △ ♃
07 Wednesday		06 52	☽ ⊼ ♀	16 22	☽ △ ☿
01 34	☽ Q ♇	08 04	☽ △ ☉	17 39	☽ △ ☉
02 48	☽ ⊻ ☿	12 12	☿ ⊥ ♇	18 12	☿ ∠ ♇
11 21	☽ # ♅	15 05	☽ ⊻ ♅	21 06	☽ ⊼ ♂
13 30	☽ △ ♆	18 29	☽ Q ♇	21 07	☽ ⊼ ♄
16 02	☽ ☌ ☉	19 20	☽ ⊻ ♂	21 28	☽ ± ♆
19 21	♂ ⊥ ♇	20 18	☽ △ ♃	21 31	♂ ⚹ ♄
23 06	☽ ⚹ ♇	**19 Monday**		22 27	☿ ☌ ♃
08 Thursday		04 30	☽ □ ♄	**28 Wednesday**	
01 42	☽ ∠ ♄	07 19	☽ ⊥ ♂	06 30	☿ ⊥ ♀
02 16	☽ ∥ ♀	15 07	☽ ⚼ ☉	06 36	♀ ∥ ♂
06 20	♃ ⊼ ♅	16 26	☽ △ ☿	07 16	☽ ± ♄
10 42	☽ □ ♂	17 52	☽ ⊻ ♆	07 33	☽ ⊼ ♆
11 43	☽ ∥ ♂	**20 Tuesday**		11 19	☽ Q ♀
12 36	☽ ⊻ ♀	01 00	☽ ∠ ♂	17 27	☽ ⊼ ♇
18 15	☽ ⊥ ♄	01 01	☽ ⚼ ♃	22 34	☽ ⚼ ♄
18 51	☽ ⊼ ♅	01 30	♂ □ ♃	**29 Thursday**	
19 06	☽ ☌ ♃	04 22	☽ □ ♇	03 38	☽ ± ♇
23 23	☽ ⊥ ♀	04 59	☽ ⊥ ♆	04 28	♃ ∠ ♇
09 Friday		09 10	☽ ± ☉	08 17	☽ ⚹ ♀
01 52	☽ ∥ ☉	15 45	☽ ☍ ♀	09 47	☽ △ ♅
02 38	☽ ∠ ♇	17 46	☽ ± ♃	14 42	☽ □ ☿
05 27	☽ ⊻ ♄	18 56	☽ ⚼ ☿	18 59	☽ ⚼ ♇
05 56	☽ ± ♅	20 59	☽ ⊼ ☉	19 11	☽ □ ♃
13 37	☽ ☌ ☿	21 22	☽ ∠ ♆	22 24	☿ ⊥ ♄
15 12	♀ △ ♂	22 46	☽ ☌ ♅	**30 Friday**	
15 50	☽ ∠ ♀	**21 Wednesday**		00 03	☉ ⊻ ♄
19 32	☽ ⊥ ♇	02 51	☽ # ♆	00 18	☽ △ ♄
20 41	☽ □ ♆	04 45	☽ ⊼ ♃	00 19	☽ □ ☉
21 49	☽ Q ♂	05 36	☽ ⚹ ♂	02 19	☽ ☍ ♂
22 49	☽ ⚼ ♅	10 16	☽ ± ☿	03 51	♀ ⚼ ♆
10 Saturday		11 54	☽ △ ♄	06 01	☽ ∥ ♅
04 11	☽ ⊻ ☉	17 57	☽ # ♀	10 17	☽ # ♀
04 39	☽ ∥ ♃	19 51	☉ ⊼ ♅	10 34	☽ ☍ ♆
06 57	☽ ⊻ ♇	20 18	☽ ⊼ ☿	10 49	☽ ∠ ♀
16 43	☽ ⊥ ☉	21 12	☽ ∥ ♅	11 23	☽ ⚼ ♅
17 26	☿ ∠ ♀	**22 Thursday**		13 42	♀ ⊥ ♃
19 55	☽ ⚹ ♀	00 01	☽ ⚹ ♆	13 43	☽ # ♂
21 59	☽ ⚹ ♂	01 10	☽ # ♂	19 12	☽ Q ☿
11 Sunday		03 48	☽ Q ♂	20 48	☽ △ ♇
03 35	☽ △ ♅	04 49	☉ ∥ ♃	23 18	☽ # ♆

DECEMBER 2018

LONGITUDES

Date	Sidereal time h m s	Sun ☉ ° ' "	Moon ☽ ° ' "	Moon ☽ 24.00 ° ' "	Mercury ☿ ° '	Venus ♀ ° '	Mars ♂ ° '	Jupiter ♃ ° '	Saturn ♄ ° '	Uranus ♅ ° '	Neptune ♆ ° '	Pluto ♇ ° '
01	16 41 12	09 ♐ 13 53	28 ♍ 23 17	05 ♎ 15 14	29 ♏ 58	29 ♎ 22	09 ♓ 48	05 ♐ 06	07 ♑ 53	29 ♈ 09	13 ♓ 42	19 ♑ 39
02	16 45 08	10 14 43	12 ♎ 04 36	18 ♎ 51 18	29 R 03	29 ♎ 53	10 26	05 20	07 59	29 R 07	13 43	19 41
03	16 49 05	11 15 34	25 ♎ 35 14	02 ♏ 16 17	28 20	00 ♏ 26	11 05	05 33	08 06	29 05	13 43	19 42
04	16 53 01	12 16 27	08 ♏ 54 22	15 ♏ 29 20	27 47	01 01	11 44	05 47	08 12	29 04	13 43	19 44
05	16 56 58	13 17 21	22 ♏ 01 03	28 ♏ 29 26	27 27	01 37	12 23	06 00	08 19	29 02	13 43	19 46
06	17 00 54	14 18 17	04 ♐ 54 23	11 ♐ 15 50	27 17	02 14	13 02	06 13	08 25	29 01	13 44	19 47
07	17 04 51	15 19 13	17 ♐ 33 46	23 ♐ 48 13	27 D 18	02 52	13 41	06 27	08 32	28 59	13 44	19 49
08	17 08 47	16 20 11	29 ♐ 59 16	06 ♑ 07 02	27 29	03 32	14 20	06 40	08 39	28 58	13 45	19 51
09	17 12 44	17 21 09	12 ♑ 11 44	18 ♑ 13 37	27 50	04 13	14 59	06 53	08 46	28 56	13 45	19 53
10	17 16 41	18 22 08	24 ♑ 12 58	00 ♒ 10 11	28 18	04 55	15 38	07 07	08 52	28 55	13 46	19 54
11	17 20 37	19 23 08	06 ♒ 05 41	11 ♒ 59 56	28 54	05 38	16 18	07 20	08 59	28 54	13 46	19 56
12	17 24 34	20 24 09	17 ♒ 53 26	23 ♒ 46 46	29 ♏ 37	06 23	16 57	07 33	09 06	28 52	13 47	19 58
13	17 28 30	21 25 10	29 ♒ 40 31	05 ♓ 35 17	00 ♐ 26	07 08	17 37	07 46	09 13	28 51	13 47	20 00
14	17 32 27	22 26 11	11 ♓ 31 44	17 ♓ 30 29	01 19	07 54	18 16	08 00	09 20	28 50	13 48	20 02
15	17 36 23	23 27 14	23 ♓ 32 12	29 ♓ 37 32	02 18	08 41	18 56	08 13	09 27	28 49	13 49	20 04
16	17 40 20	24 28 16	05 ♈ 47 04	12 ♈ 01 25	03 20	09 29	19 36	08 26	09 34	28 48	13 50	20 05
17	17 44 16	25 29 19	18 ♈ 21 05	24 ♈ 46 33	04 26	10 18	20 15	08 39	09 41	28 47	13 50	20 07
18	17 48 13	26 30 23	01 ♉ 18 10	07 ♉ 56 13	05 35	11 08	20 55	08 52	09 48	28 46	13 51	20 09
19	17 52 10	27 31 26	14 ♉ 40 51	21 ♉ 32 02	06 47	11 59	21 35	09 05	09 54	28 45	13 52	20 11
20	17 56 06	28 32 31	28 ♉ 29 39	05 ♊ 33 22	08 01	12 50	22 15	09 18	10 01	28 44	13 53	20 13
21	18 00 03	29 ♐ 33 35	12 ♊ 42 44	19 ♊ 57 08	09 16	13 42	22 55	09 31	10 08	28 43	13 54	20 15
22	18 03 59	00 ♑ 34 40	27 ♊ 15 48	04 ♋ 37 53	10 34	14 35	23 35	09 45	10 15	28 42	13 55	20 17
23	18 07 56	01 35 46	12 ♋ 02 26	19 ♋ 28 28	11 54	15 29	24 15	09 57	10 23	28 41	13 56	20 19
24	18 11 52	02 36 52	26 ♋ 55 00	04 ♌ 21 02	13 14	16 23	24 55	10 10	10 30	28 41	13 56	20 21
25	18 15 49	03 37 58	11 ♌ 45 42	19 ♌ 08 10	14 36	17 18	25 35	10 23	10 37	28 40	13 58	20 23
26	18 19 45	04 39 06	26 ♌ 27 44	03 ♍ 43 49	15 59	18 14	26 15	10 36	10 44	28 39	13 59	20 25
27	18 23 42	05 40 13	10 ♍ 55 57	18 ♍ 03 46	17 23	19 10	26 55	10 49	10 51	28 39	14 00	20 27
28	18 27 39	06 41 21	25 ♍ 07 05	02 ♎ 05 44	18 48	20 07	27 35	11 02	10 58	28 38	14 01	20 29
29	18 31 35	07 42 30	08 ♎ 59 43	15 ♎ 49 02	20 14	21 04	28 15	11 15	11 05	28 38	14 02	20 31
30	18 35 32	08 43 39	22 ♎ 33 48	29 ♎ 14 08	21 40	22 02	28 56	11 27	11 12	28 37	14 03	20 33
31	18 39 28	09 ♑ 44 49	05 ♏ 50 13	12 ♏ 22 13	23 ♐ 07	23 ♏ 00	29 ♓ 36	11 ♐ 40	11 ♑ 19	28 ♈ 37	14 ♓ 04	20 ♑ 35

MOON

Date	Moon True ☊ ° '	Moon Mean ☊ ° '	Moon ☽ Latitude ° '
01	28 ♋ 02	29 ♋ 11	04 N 28
02	27 R 58	29 08	04 57
03	27 52	29 05	05 08
04	27 44	29 02	05 02
05	27 36	28 59	04 41
06	27 28	28 55	04 05
07	27 21	28 52	03 18
08	27 15	28 49	02 22
09	27 12	28 46	01 21
10	27 11	28 43	00 N 16
11	27 D 11	28 39	00 S 49
12	27 12	28 36	01 51
13	27 14	28 33	02 48
14	27 16	28 30	03 39
15	27 R 16	28 27	04 21
16	27 16	28 24	04 52
17	27 14	28 20	05 10
18	27 10	28 17	05 13
19	27 06	28 14	04 59
20	27 01	28 11	04 28
21	26 57	28 08	03 39
22	26 53	28 05	02 36
23	26 51	28 01	01 S 21
24	26 D 51	27 58	00 00
25	26 51	27 55	01 N 21
26	26 53	27 52	02 36
27	26 54	27 49	03 40
28	26 55	27 45	04 29
29	26 R 55	27 42	05 01
30	26 55	27 39	05 15
31	26 ♋ 53	27 ♋ 36	05 N 12

DECLINATIONS

Date	Sun ☉ ° '	Moon ☽ ° '	Mercury ☿ ° '	Venus ♀ ° '	Mars ♂ ° '	Jupiter ♃ ° '	Saturn ♄ ° '	Uranus ♅ ° '	Neptune ♆ ° '	Pluto ♇ ° '
01	21 S 50	04 N 44	18 S 07	09 S 54	08 S 57	20 S 30	22 S 41	10 N 39	07 S 19	22 S 03
02	21 59	00 S 13	17 44	09 58	08 40	20 33	22 40	10 39	07 19	22 03
03	22 07	05 06	17 26	10 02	08 24	20 35	22 40	10 38	07 19	22 03
04	22 16	09 41	17 11	10 07	08 07	20 38	22 40	10 38	07 18	22 03
05	22 23	13 44	17 01	10 13	07 51	20 40	22 39	10 37	07 18	22 03
06	22 31	17 05	16 56	10 19	07 34	20 42	22 39	10 37	07 18	22 02
07	22 38	19 34	16 55	10 26	07 17	20 45	22 39	10 36	07 18	22 02
08	22 44	21 04	16 57	10 34	07 01	20 47	22 38	10 36	07 18	22 02
09	22 50	21 32	17 03	10 42	06 44	20 49	22 38	10 35	07 18	22 02
10	22 56	21 00	17 11	10 50	06 27	20 52	22 38	10 35	07 17	22 02
11	23 01	19 32	17 23	11 00	06 10	20 54	22 37	10 34	07 17	22 02
12	23 05	17 13	17 36	11 10	05 53	20 56	22 37	10 34	07 17	22 01
13	23 09	14 13	17 51	11 20	05 36	20 58	22 37	10 33	07 17	22 01
14	23 13	10 37	18 08	11 30	05 19	21 00	22 36	10 33	07 16	22 01
15	23 16	06 33	18 25	11 41	05 02	21 03	22 36	10 33	07 16	22 01
16	23 19	02 S 10	18 44	11 53	04 45	21 05	22 35	10 32	07 16	22 01
17	23 22	02 N 25	19 03	12 05	04 28	21 07	22 35	10 32	07 15	22 01
18	23 23	07 02	19 22	12 17	04 11	21 09	22 35	10 32	07 15	22 00
19	23 25	11 29	19 42	12 29	03 54	21 11	22 34	10 31	07 15	22 00
20	23 26	15 28	20 01	12 42	03 37	21 13	22 34	10 31	07 14	22 00
21	23 26	18 42	20 21	12 55	03 19	21 15	22 33	10 31	07 14	22 00
22	23 26	20 49	20 40	13 08	03 02	21 17	22 33	10 30	07 14	22 00
23	23 26	21 33	20 59	13 21	02 45	21 19	22 32	10 30	07 13	21 59
24	23 25	20 47	21 17	13 35	02 28	21 21	22 32	10 30	07 13	21 59
25	23 23	18 34	21 35	13 48	02 10	21 23	22 31	10 30	07 12	21 59
26	23 21	15 08	21 52	14 02	01 53	21 24	22 31	10 30	07 12	21 59
27	23 19	10 51	22 08	14 16	01 36	21 26	22 30	10 29	07 12	21 59
28	23 16	06 03	22 23	14 30	01 18	21 28	22 30	10 29	07 11	21 58
29	23 13	01 N 03	22 38	14 44	01 01	21 30	22 29	10 29	07 11	21 58
30	23 09	03 S 54	22 52	14 58	00 44	21 32	22 29	10 29	07 10	21 58
31	23 S 05	08 S 33	23 S 04	15 S 12	00 S 27	21 S 33	22 S 28	10 N 29	07 S 10	21 S 58

ZODIAC SIGN ENTRIES

Date	h m	Planets
01	11 12	☿ ♏
01	14 49	☽ ♎
02	17 02	♀ ♏
03	19 55	☽ ♏
06	02 49	☽ ♐
08	12 01	☽ ♑
10	23 39	☽ ♒
12	23 43	☿ ♐
13	12 40	☽ ♓
16	00 44	☽ ♈
18	09 37	☽ ♉
20	14 34	☽ ♊
21	22 23	☉ ♑
22	16 28	☽ ♋
24	16 58	☽ ♌
26	17 50	☽ ♍
28	20 23	☽ ♎
31	01 23	☽ ♏

LATITUDES

Date	Mercury ☿ ° '	Venus ♀ ° '	Mars ♂ ° '	Jupiter ♃ ° '	Saturn ♄ ° '	Uranus ♅ ° '	Neptune ♆ ° '	Pluto ♇ ° '
01	02 N 04	01 N 26	01 S 08	00 N 39	00 N 31	00 S 33	00 S 59	00 S 03
04	02 32	01 49	01 02	00 39	00 31	00 33	00 59	00 04
07	02 43	02 10	00 57	00 39	00 31	00 33	00 59	00 04
10	02 39	02 27	00 51	00 39	00 31	00 33	00 59	00 04
13	02 26	02 42	00 46	00 38	00 30	00 33	00 59	00 05
16	02 08	02 55	00 41	00 38	00 30	00 33	00 59	00 05
19	01 46	03 05	00 36	00 38	00 30	00 32	00 58	00 05
22	01 22	03 13	00 32	00 38	00 30	00 32	00 58	00 05
25	00 58	03 19	00 27	00 38	00 29	00 32	00 58	00 06
28	00 34	03 24	00 23	00 38	00 29	00 32	00 58	00 06
31	00 N 11	03 N 26	00 S 19	00 N 38	00 N 29	00 S 32	00 S 58	00 S 06

DATA

Julian Date	2458454
Delta T	+70 seconds
Ayanamsa	24° 07' 01"
Synetic vernal point	04° ♓ 59' 58"
True obliquity of ecliptic	23° 26' 08"

MOON'S PHASES, APSIDES AND POSITIONS ☽

Date	h m	Phase	Longitude ° '	Eclipse Indicator
07	07 20	●	15 ♐ 07	
15	11 49	☽	23 ♓ 27	
22	17 49	○	00 ♋ 49	
29	09 34	☾	07 ♎ 36	

Day	h m		
12	12 28	Apogee	
24	09 57	Perigee	
02	10 56	0S	
09	11 07	Max dec	21° S 32'
16	23 26	0N	
23	11 48	Max dec	21° N 33'
29	17 01	0S	

LONGITUDES

Date	Chiron ⚷ ° '	Ceres ⚳ ° '	Pallas ⚴ ° '	Juno ⚵ ° '	Vesta ⚶ ° '	Black Moon Lilith ⚸ ° '
01	27 ♓ 56	08 ♏ 31	11 ♎ 02	22 ♉ 36	29 ♑ 50	12 ♒ 59
11	27 ♓ 54	12 ♏ 44	14 ♎ 47	21 ♉ 04	04 ♒ 33	14 ♒ 05
21	27 ♓ 58	16 ♏ 50	18 ♎ 15	20 ♉ 21	09 ♒ 21	15 ♒ 12
31	28 ♓ 07	20 ♏ 48	21 ♎ 23	20 ♉ 35	14 ♒ 13	16 ♒ 18

All ephemeris data is given at 12.00 UT and the Moon's longitude is additionally given for 24.00 UT

ASPECTARIAN

h m Aspects

01 Saturday
02 13 ♀ ☍ ♅
02 40 ☽ Q ♃
02 54 ☽ ± ♅
02 56 ☽ ⊥ ♀
09 50 ☽ Q ☉
13 19 ☽ ⊼ ♅
13 46 ☽ ⊻ ♀
14 34 ☽ ✶ ☿
15 40 ♃ ± ♅
21 49 ☿ ⊻ ♀
23 56 ☽ ✶ ♃

02 Sunday
04 44 ☽ □ ♄
08 31 ☽ ✶ ☉
08 58 ☽ ⊼ ♂
10 15 ☿ ⊼ ♅
14 53 ☽ ⊼ ♆
15 18 ☽ ∠ ☿
20 06 ☽ ± ♂
23 25 ☉ ∥ ♇

03 Monday
00 34 ☉ □ ♂
01 29 ☽ □ ♇
01 31 ☽ ± ♆
02 52 ☽ ∠ ♃
06 27 ☽ ⊥ ☿
12 55 ☽ Q ♄
12 56 ☽ ⚼ ♂
13 18 ☽ ∠ ☉
16 42 ☽ ⊻ ☿
17 36 ☽ ⚼ ♆
18 16 ☽ ☍ ♅
19 14 ☽ ⊥ ♃
21 05 ☽ ☌ ♀
23 16 ☽ ∥ ♆

04 Tuesday
04 03 ☽ ∥ ♂
06 14 ☽ ⊻ ♃
06 50 ☽ ⊥ ☉
09 52 ☽ Q ♇
10 43 ☽ ✶ ♄
14 26 ☽ ∥ ♀
17 16 ☽ # ♅
17 24 ☽ △ ♂
18 38 ☽ ⊻ ☉
20 46 ☽ △ ♆

05 Wednesday
07 50 ☽ ✶ ♇
14 25 ☽ ∠ ♄
21 53 ☽ ☌ ☿
22 22 ☉ □ ♆
23 31 ☉ ⊥ ♇

06 Thursday
00 59 ☽ ⊼ ♅
05 13 ☉ ⚼ ♅
06 43 ☽ ⊻ ♀
07 18 ☽ ⊥ ♄
10 47 ☽ ∥ ☿
11 47 ☽ ∠ ♇
12 12 ☽ ± ♅
14 31 ☽ ☌ ♃
18 35 ☽ ⊥ ♀
18 41 ☽ ⊻ ♄
21 23 ☿ St D

07 Friday
04 11 ☽ □ ♂
04 42 ☽ □ ♆
04 50 ☽ ⊥ ♇
05 11 ☽ ⚼ ♅
07 20 ☽ ☌ ☉
11 19 ♂ ∥ ♆
12 37 ☽ ∠ ♀
14 11 ♂ ☌ ♆
15 58 ☉ ∥ ♄
16 20 ☽ ⊻ ♇
22 52 ♂ ∠ ♅

08 Saturday
05 50 ☽ ∥ ♃
07 02 ☽ ⊻ ☿
10 00 ☽ △ ♅
15 26 ☽ Q ♆
16 50 ☽ Q ♂
17 50 ♀ # ♅
19 00 ☽ ⊥ ☿
19 20 ☽ ✶ ♀

09 Sunday
01 19 ☽ ⊻ ♃
05 09 ☽ ☌ ♄
13 18 ☽ ∠ ☿
13 24 ☽ ⊥ ♃
15 06 ☽ ✶ ♆
17 51 ☽ ✶ ♂
20 29 ☽ Q ♀
23 12 ☽ ⊻ ☉

10 Monday
03 20 ☽ ☌ ♇
07 41 ☽ ∠ ♃
12 20 ☽ ⊥ ☉
15 04 ☽ ∥ ♃
20 38 ☽ ✶ ☿
21 10 ☽ ∠ ♆
21 27 ☽ □ ♅

11 Tuesday
01 43 ☽ ∠ ♂
08 12 ☽ ∠ ☉
11 01 ☽ □ ♀
11 38 ☿ ⊼ ♅
14 34 ☽ ✶ ♃
15 24 ☽ ⊥ ♆
17 56 ☽ ⊻ ♄
21 02 ☽ ⊥ ♂
22 22 ☽ Q ☿

12 Wednesday
01 26 ☉ ⊻ ♇
03 37 ☽ ⊻ ♆
06 16 ☽ ⊥ ♄
08 52 ☽ ∥ ☿
09 56 ☽ Q ♅
09 59 ☽ ⊻ ♂
15 27 ☽ Q ♃
16 15 ☽ ⊻ ♇
17 36 ☽ ✶ ☉

13 Thursday
00 47 ☽ ∠ ♄
04 30 ☽ ⊥ ♇
10 20 ☽ ✶ ♅
13 39 ☽ □ ☿
20 19 ☽ Q ☉
22 50 ☽ ∠ ♇

14 Friday
04 10 ☽ △ ♀
04 44 ☽ □ ♃
06 36 ☽ ∥ ♀
07 31 ☽ ✶ ♄
12 24 ☽ # ♅
15 59 ♀ ⊻ ♃
16 07 ♀ Q ♇
16 35 ☽ ☌ ♆
16 37 ☽ ∠ ♅

15 Saturday
02 19 ☽ ☌ ♂
05 05 ☽ ✶ ♇
07 49 ☽ Q ♄
07 57 ☽ ∥ ♆
10 34 ☽ ⊥ ♅
11 12 ♅ ∠ ♆
11 49 ☽ □ ☉
12 19 ☽ ⚼ ♀
21 05 ☽ ∥ ♂
22 23 ☽ ⊻ ♅

16 Sunday
04 48 ☽ Q ♇
06 48 ☽ △ ☿
07 14 ☽ ± ♀
14 08 ☉ ∠ ♀
14 26 ♀ ✶ ♄
17 12 ☽ △ ♃
17 35 ☿ ⊥ ♄
19 21 ☽ □ ♄
19 39 ☽ ⊼ ♀

17 Monday
03 27 ☽ ⊻ ♆
06 57 ♂ ✶ ♇
14 14 ☽ ⚼ ☿
14 48 ☽ ⊥ ♆
15 20 ☽ □ ♇
15 46 ☽ ⊻ ♂
19 10 ☿ ± ♅
22 00 ☽ # ♂
22 05 ☽ ⚼ ♃

18 Tuesday
02 27 ☽ △ ☉
02 54 ☿ ∠ ♇
03 32 ☽ ⊥ ♂
07 21 ☽ ☌ ♅
07 31 ☽ ∠ ♆
08 33 ☽ ± ☿
13 08 ☽ # ♆
14 54 ☽ ± ♃
20 31 ☽ ⊼ ☿
20 48 ☽ ∠ ♂

19 Wednesday
01 54 ☽ ⊼ ♃
03 27 ☽ △ ♄
06 42 ☽ ∥ ♅
06 54 ☽ ☍ ♀
07 52 ☽ ⚼ ☉
10 33 ☽ ✶ ♆
18 04 ☽ # ♀
21 41 ☽ △ ♇

20 Thursday
00 42 ☽ ✶ ♂
00 56 ☽ ± ☉
02 17 ♂ Q ♄
05 59 ☽ ⚼ ♄
07 30 ☽ Q ♆
12 05 ☽ ⊼ ☉
12 24 ☽ ⊻ ♅
16 22 ☉ △ ♅
21 29 ☽ ± ♄
22 16 ☽ Q ♂
22 36 ☽ ⊥ ♅
23 27 ☽ ⚼ ♇

21 Friday
05 03 ♂ ⊥ ♅
05 41 ☽ ☍ ☿
06 36 ☽ ☍ ♃
07 40 ☽ ⊼ ♄
13 40 ☽ ∠ ♅
13 46 ☽ ⊼ ♀
13 58 ☽ □ ♆
14 34 ☽ ± ♇
17 11 ♀ △ ♆
17 37 ☿ ☌ ♃

22 Saturday
00 21 ☽ ± ♀
00 31 ☽ ⊼ ♇
05 41 ☽ □ ♂
05 41 ☿ ⊻ ♄
09 11 ☽ # ☿
14 21 ☽ ✶ ♅
16 02 ☽ ⚼ ♀
17 49 ☽ ☍ ☉
21 51 ☽ # ♃

23 Sunday
08 35 ☽ ⊼ ♃
09 17 ☽ ☍ ♄
09 49 ☽ Q ♅
11 44 ☽ ⊼ ☿
15 03 ☽ △ ♆
17 55 ☽ △ ♀
18 25 ☽ ± ♃
19 53 ☉ Q ♆
22 23 ☽ ± ☿

24 Monday
00 41 ☽ # ♃
01 23 ☽ ☍ ♇
04 19 ☽ # ☿
08 37 ☽ △ ♂
09 09 ☽ ⚼ ♃
14 21 ☽ ⚼ ☿
14 50 ☽ □ ♅
15 16 ☽ ⚼ ♆
17 21 ☿ ∥ ♃
19 40 ☿ ⚼ ♅
21 52 ☽ ⊼ ☉

25 Tuesday
00 32 ☿ □ ♆
05 50 ☽ ± ♆
07 57 ☿ ⊥ ♇
08 18 ☽ ± ☉
09 44 ☽ △ ♃
10 00 ☽ ⚼ ♂
10 07 ☽ ⊼ ♄
15 34 ☽ ⊼ ♆
17 06 ☽ △ ☿
19 57 ☽ ± ♄
21 37 ☽ □ ♀

26 Wednesday
00 01 ☽ ⚼ ☉
01 20 ☽ ± ♂
02 04 ☽ ⊼ ♇
10 47 ☽ ⚼ ♄
11 38 ☽ ⊼ ♂
11 55 ☽ ± ♇
15 37 ☽ △ ♅
18 15 ☽ # ♀
22 02 ☿ ∥ ♇

27 Thursday
02 33 ☽ △ ☉
02 49 ☽ ⚼ ♇
05 16 ☽ Q ♀
11 48 ☽ □ ♃
11 51 ☽ △ ♄
13 54 ☽ ∥ ♅
16 33 ☽ ⚼ ♅
17 09 ☽ ☍ ♆
19 18 ♃ ⊻ ♄

28 Friday
00 04 ☽ □ ☿
02 52 ☽ ✶ ♀
04 04 ☽ △ ♇
06 28 ☽ # ♆
07 46 ☽ ± ♅
16 27 ☽ ☍ ♂
18 02 ☽ ⊼ ♅
18 49 ☽ Q ♃
21 31 ♀ ✶ ♇
22 00 ☿ ∥ ♄

29 Saturday
06 31 ☽ ∠ ♀
09 34 ☽ □ ☉
10 31 ☽ Q ☿
12 08 ☽ # ♂
15 41 ☽ □ ♄
16 00 ☽ ✶ ♃
16 44 ☿ ⊻ ♇
20 51 ☽ ⊼ ♆
21 23 ☽ ∥ ♂
23 29 ☽ ⊥ ♀

30 Sunday
01 14 ♂ ⊻ ♅
07 31 ☽ ± ♆
08 23 ☽ □ ♇
10 13 ☽ ✶ ☿
10 59 ☽ ⊻ ♀
19 06 ☽ ∠ ♃
20 06 ☽ Q ☉
22 53 ☽ ☍ ♅
23 41 ☽ ⚼ ♆

31 Monday
00 03 ☽ Q ♄
00 03 ☽ ⊼ ♂
04 38 ☽ ∥ ♆
06 00 ☿ ⊻ ♀
11 33 ☽ ± ♂
11 41 ☽ ⊥ ♃
12 35 ☉ ∥ ☿
16 43 ☽ ∠ ☿
17 02 ☽ Q ♇
19 46 ☽ ✶ ☉
22 09 ☽ ✶ ♄
22 43 ☽ # ♅
22 52 ☽ ⊻ ♃

JANUARY 2019

LONGITUDES

Date	Sidereal time h m s	Sun ☉ ° ' "	Moon ☽ ° ' "	Moon ☽ 24.00 ° '	Mercury ☿ ° '	Venus ♀ ° '	Mars ♂ ° '	Jupiter ♃ ° '	Saturn ♄ ° '	Uranus ♅ ° '	Neptune ♆ ° '	Pluto ♇ ° '
01	18 43 25	10 ♑ 45 59	18 ♏ 50 21	25 ♏ 14 49	24 ♐ 35	23 ♏ 59	00 ♈ 16	11 ♐ 52	11 ♑ 26	28 ♈ 37	14 ♓ 05	20 ♑ 37
02	18 47 21	11 47 09	01 ♐ 35 47	07 ♐ 53 29	26 03	24 59	00 57	12 05	11 33	28 R 37	14 07	20 39
03	18 51 18	12 48 20	14 ♐ 08 04	20 ♐ 19 44	27 32	25 58	01 37	12 18	11 40	28 36	14 08	20 41
04	18 55 14	13 49 31	26 ♐ 28 40	02 ♑ 35 03	29 ♐ 01	26 59	02 17	12 30	11 47	28 36	14 09	20 43
05	18 59 11	14 50 42	08 ♑ 39 03	14 ♑ 40 52	00 ♑ 31	27 59	02 58	12 42	11 55	28 36	14 11	20 45
06	19 03 08	15 51 53	20 ♑ 40 43	26 ♑ 38 48	02 01	29 ♏ 00	03 38	12 55	12 02	28 36	14 12	20 47
07	19 07 04	16 53 04	02 ♒ 35 23	08 ♒ 30 43	03 32	00 ♐ 02	04 19	13 07	12 09	28 D 36	14 13	20 49
08	19 11 01	17 54 14	14 ♒ 25 06	20 ♒ 18 52	05 03	01 04	04 59	13 19	12 16	28 36	14 15	20 51
09	19 14 57	18 55 24	26 ♒ 12 21	02 ♓ 05 58	06 35	02 06	05 40	13 31	12 23	28 36	14 16	20 53
10	19 18 54	19 56 34	08 ♓ 00 07	13 ♓ 55 17	08 07	03 08	06 20	13 44	12 30	28 36	14 18	20 55
11	19 22 50	20 57 43	19 ♓ 51 56	25 ♓ 50 35	09 39	04 11	07 01	13 56	12 37	28 37	14 19	20 57
12	19 26 47	21 58 52	01 ♈ 51 47	07 ♈ 56 03	11 12	05 15	07 41	14 08	12 44	28 37	14 21	20 59
13	19 30 43	23 00 00	14 ♈ 03 59	20 ♈ 16 08	12 46	06 18	08 22	14 19	12 51	28 37	14 22	21 01
14	19 34 40	24 01 08	26 ♈ 33 02	02 ♉ 55 14	14 19	07 22	09 03	14 31	12 58	28 38	14 24	21 03
15	19 38 37	25 02 15	09 ♉ 23 12	15 ♉ 57 22	15 54	08 26	09 43	14 43	13 05	28 38	14 26	21 05
16	19 42 33	26 03 21	22 ♉ 38 05	29 ♉ 25 35	17 28	09 30	10 24	14 55	13 12	28 38	14 27	21 07
17	19 46 30	27 04 27	06 ♊ 19 59	13 ♊ 21 18	19 04	10 35	11 04	15 06	13 19	28 39	14 29	21 09
18	19 50 26	28 05 32	20 ♊ 29 20	27 ♊ 43 45	20 39	11 40	11 45	15 18	13 26	28 40	14 31	21 11
19	19 54 23	29 ♑ 06 36	05 ♋ 04 00	12 ♋ 29 24	22 16	12 45	12 26	15 29	13 33	28 40	14 32	21 13
20	19 58 19	00 ♒ 07 39	19 ♋ 59 04	27 ♋ 32 00	23 53	13 51	13 06	15 41	13 40	28 41	14 34	21 15
21	20 02 16	01 08 42	05 ♌ 07 04	12 ♌ 43 04	25 30	14 57	13 47	15 52	13 47	28 42	14 36	21 17
22	20 06 12	02 09 44	20 ♌ 18 46	27 ♌ 52 58	27 08	16 03	14 28	16 03	13 54	28 42	14 38	21 19
23	20 10 09	03 10 45	05 ♍ 24 32	12 ♍ 52 26	28 ♑ 46	17 09	15 08	16 15	14 00	28 43	14 39	21 21
24	20 14 06	04 11 46	20 ♍ 15 46	27 ♍ 33 50	00 ♒ 26	18 15	15 49	16 26	14 07	28 44	14 41	21 23
25	20 18 02	05 12 46	04 ♎ 46 03	11 ♎ 52 03	02 05	19 22	16 29	16 37	14 14	28 45	14 43	21 25
26	20 21 59	06 13 46	18 ♎ 51 37	25 ♎ 44 33	03 46	20 29	17 10	16 48	14 21	28 46	14 45	21 27
27	20 25 55	07 14 45	02 ♏ 31 24	09 ♏ 11 53	05 27	21 36	17 51	16 59	14 27	28 47	14 47	21 29
28	20 29 52	08 15 43	15 ♏ 46 26	22 ♏ 15 25	07 08	22 43	18 31	17 09	14 34	28 48	14 49	21 31
29	20 33 48	09 16 42	28 ♏ 39 13	04 ♐ 58 17	08 51	23 50	19 12	17 20	14 41	28 49	14 51	21 33
30	20 37 45	10 17 39	11 ♐ 13 04	17 ♐ 24 02	10 34	24 58	19 53	17 30	14 48	28 51	14 53	21 35
31	20 41 41	11 ♒ 18 36	23 ♐ 31 37	29 ♐ 36 15	12 ♒ 17	26 ♐ 06	20 ♈ 33	17 ♐ 41	14 ♑ 54	28 ♈ 52	14 ♓ 55	21 ♑ 37

DECLINATIONS

Date	Moon True ☊ ° '	Moon Mean ☊ ° '	Moon ☽ Latitude ° '	Sun ☉ ° '	Moon ☽ ° '	Mercury ☿ ° '	Venus ♀ ° '	Mars ♂ ° '	Jupiter ♃ ° '	Saturn ♄ ° '	Uranus ♅ ° '	Neptune ♆ ° '	Pluto ♇ ° '
01	26 ♋ 51	27 ♋ 33	04 N 53	23 S 00	12 S 43	23 S 16	15 S 26	00 S 09	21 S 35	22 S 28	10 N 29	07 S 09	21 S 58
02	26 R 49	27 30	04 20	22 55	16 14	23 26	15 40	00 N 08	21 37	22 27	10 29	07 09	21 57
03	26 47	27 26	03 35	22 49	18 56	23 36	15 53	00 25	21 39	22 27	10 29	07 08	21 57
04	26 45	27 23	02 40	22 43	20 43	23 44	16 07	00 43	21 40	22 26	10 29	07 08	21 57
05	26 44	27 20	01 39	22 37	21 31	23 51	16 21	01 00	21 42	22 26	10 29	07 07	21 57
06	26 43	27 17	00 N 33	22 30	21 18	23 57	16 35	01 17	21 43	22 25	10 29	07 07	21 56
07	26 D 43	27 14	00 S 33	22 22	20 06	24 02	16 48	01 34	21 45	22 24	10 29	07 06	21 56
08	26 43	27 10	01 37	22 14	18 03	24 06	17 01	01 51	21 46	22 24	10 29	07 05	21 56
09	26 44	27 07	02 37	22 06	15 14	24 08	17 14	02 09	21 48	22 23	10 29	07 05	21 56
10	26 45	27 04	03 30	21 57	11 49	24 09	17 27	02 26	21 49	22 23	10 29	07 04	21 56
11	26 46	27 01	04 15	21 48	07 55	24 08	17 40	02 43	21 51	22 22	10 29	07 04	21 55
12	26 46	26 58	04 49	21 39	03 S 40	24 07	17 53	03 00	21 52	22 21	10 29	07 03	21 55
13	26 47	26 55	05 10	21 29	00 N 47	24 04	18 05	03 17	21 54	22 21	10 29	07 02	21 55
14	26 R 47	26 51	05 18	21 18	05 18	24 00	18 17	03 34	21 55	22 20	10 30	07 02	21 55
15	26 47	26 48	05 10	21 07	09 42	23 54	18 28	03 51	21 56	22 20	10 30	07 01	21 54
16	26 46	26 45	04 46	20 56	13 48	23 47	18 40	04 08	21 58	22 19	10 30	07 01	21 54
17	26 D 46	26 42	04 06	20 44	17 20	23 38	18 51	04 25	21 59	22 18	10 30	07 00	21 54
18	26 46	26 39	03 09	20 32	19 58	23 28	19 02	04 42	22 00	22 18	10 30	06 59	21 54
19	26 47	26 36	01 58	20 20	21 22	23 17	19 12	04 59	22 02	22 17	10 31	06 59	21 54
20	26 47	26 32	00 S 38	20 07	21 20	23 04	19 22	05 16	22 03	22 16	10 31	06 58	21 53
21	26 R 47	26 29	00 N 46	19 54	19 44	22 50	19 32	05 33	22 04	22 16	10 31	06 57	21 53
22	26 47	26 26	02 07	19 40	16 43	22 34	19 41	05 49	22 05	22 15	10 32	06 56	21 53
23	26 46	26 23	03 19	19 27	12 36	22 17	19 50	06 06	22 06	22 15	10 32	06 56	21 53
24	26 45	26 20	04 16	19 12	07 46	21 59	19 59	06 23	22 08	22 14	10 33	06 55	21 52
25	26 44	26 16	04 55	18 58	02 N 37	21 38	20 07	06 39	22 09	22 13	10 33	06 54	21 52
26	26 43	26 13	05 15	18 43	02 S 32	21 17	20 15	06 56	22 10	22 13	10 33	06 54	21 52
27	26 43	26 10	05 16	18 27	07 24	20 54	20 22	07 12	22 11	22 12	10 33	06 53	21 52
28	26 D 43	26 07	05 00	18 12	11 46	20 29	20 29	07 29	22 12	22 11	10 34	06 52	21 52
29	26 44	26 04	04 30	17 56	15 28	20 03	20 35	07 45	22 13	22 11	10 34	06 51	21 51
30	26 45	26 01	03 47	17 40	18 22	19 35	20 41	08 01	22 14	22 10	10 35	06 51	21 51
31	26 ♋ 46	25 ♋ 57	02 N 54	17 S 23	20 S 23	19 S 06	20 S 46	08 N 17	22 S 15	22 S 09	10 N 35	06 S 50	21 S 51

ZODIAC SIGN ENTRIES

Date	h	m	Planets
01	02	20	♂ ♈
02	08	58	☽ ♐
04	18	55	☽ ♑
05	03	40	☿ ♑
07	06	46	☽ ♒
07	11	18	♀ ♐
09	19	44	☽ ♓
12	08	18	☽ ♈
14	18	31	☽ ♉
17	01	00	☽ ♊
19	03	44	☽ ♋
20	09	00	☉ ♒
21	03	54	☽ ♌
23	03	22	☽ ♍
24	05	49	☿ ♒
25	04	02	☽ ♎
27	07	31	☽ ♏
29	14	33	☽ ♐

LATITUDES

Date	Mercury ☿ ° '	Venus ♀ ° '	Mars ♂ ° '	Jupiter ♃ ° '	Saturn ♄ ° '	Uranus ♅ ° '	Neptune ♆ ° '	Pluto ♇ ° '
01	00 N 04	03 N 27	00 S 17	00 N 38	00 N 29	00 S 32	00 S 58	00 S 06
04	00 S 18	03 27	00 13	00 38	00 29	00 32	00 58	00 07
07	00 39	03 26	00 09	00 38	00 29	00 32	00 58	00 07
10	00 58	03 23	00 06	00 37	00 28	00 32	00 58	00 07
13	01 15	03 20	00 S 02	00 37	00 28	00 32	00 58	00 07
16	01 30	03 15	00 N 01	00 37	00 28	00 31	00 58	00 08
19	01 43	03 09	00 05	00 37	00 28	00 31	00 58	00 08
22	01 53	03 02	00 08	00 37	00 28	00 31	00 58	00 08
25	02 00	02 54	00 11	00 37	00 27	00 31	00 58	00 08
28	02 04	02 46	00 14	00 37	00 27	00 31	00 58	00 09
31	02 S 05	02 N 36	00 N 17	00 N 37	00 N 27	00 S 31	00 S 58	00 S 09

DATA

Julian Date	2458485
Delta T	+70 seconds
Ayanamsa	24° 07' 06"
Synetic vernal point	04° ♓ 59' 53"
True obliquity of ecliptic	23° 26' 08"

MOON'S PHASES, APSIDES AND POSITIONS ☽

Date	h	m	Phase	Longitude ° '	Eclipse Indicator
06	01	28	●	15 ♑ 25	Partial
14	06	46	☽	23 ♈ 48	
21	05	16	○	00 ♌ 52	total
27	21	10	☾	07 ♏ 38	

Day	h	m		
09	04	37	Apogee	
21	20	06	Perigee	
05	18	43	Max dec	21° S 33'
13	07	52	0N	
19	23	18	Max dec	21° N 33'
26	00	06	0S	

LONGITUDES

Date	Chiron ⚷ ° '	Ceres ⚳ ° '	Pallas ⚴ ° '	Juno ⚵ ° '	Vesta ⚶ ° '	Black Moon Lilith ⚸ ° '
01	28 ♓ 09	21 ♏ 12	21 ♎ 40	20 ♉ 39	14 ♒ 42	16 ♒ 25
11	28 ♓ 24	24 ♏ 58	24 ♎ 22	21 ♉ 51	19 ♒ 37	17 ♒ 32
21	28 ♓ 44	28 ♏ 33	26 ♎ 35	23 ♉ 52	24 ♒ 34	18 ♒ 38
31	29 ♓ 08	01 ♐ 53	28 ♎ 14	26 ♉ 32	29 ♒ 31	19 ♒ 45

ASPECTARIAN

h m Aspects

01 Tuesday
03 10 ☽ △ ♆
05 00 ☽ ⚼ ♂
11 28 ☽ ⊥ ☿
15 19 ☽ ⚹ ♇
22 26 ☽ ☌ ♀

02 Wednesday
00 09 ☽ ⊻ ☿
02 06 ☽ ∠ ☉
02 22 ☽ ∠ ♄
05 49 ☉ ☌ ♄
06 20 ☽ ⊼ ♅
07 24 ☽ ∥ ♀
10 41 ☽ △ ♂
17 44 ☽ ± ♅
19 36 ☽ ⊥ ♄
19 43 ☽ ∠ ♇
20 41 ☽ ⊥ ☉
20 49 ☉ ⊻ ♃

03 Thursday
07 13 ☽ ⊻ ♄
08 23 ☽ ☌ ♃
09 13 ☽ ⊻ ☉
10 59 ☽ ⚼ ♅
12 00 ☽ □ ♆
13 03 ☽ ⊥ ♇

04 Friday
00 43 ☽ ⊻ ♇
05 13 ☿ △ ♅
06 59 ♀ ∠ ♄
13 04 ☽ ⊻ ♀
16 10 ☽ △ ♅
17 41 ☽ ☌ ☿
19 57 ☉ ⚹ ♆
23 11 ☽ Q ♆

05 Saturday
00 05 ☽ □ ♂
01 56 ☽ ⊥ ♀
03 49 ♂ Q ♇
18 32 ☽ ☌ ♄
20 12 ☽ ⊻ ♃
21 25 ☽ ∠ ♀
23 01 ☽ ⚹ ♆

06 Sunday
01 28 ☽ ☌ ☉
02 28 ♀ ⊼ ♅
08 24 ☽ ⊥ ♃
12 12 ☽ ☌ ♇
14 02 ☽ Q ♂
14 51 ☿ Q ♆
20 26 ♅ St D

07 Monday
02 48 ☽ ∠ ♃
03 56 ☽ □ ♅
04 10 ☉ ∥ ♄
05 11 ☽ ∠ ♆
06 20 ☽ ⚹ ♀
14 12 ☽ ⊻ ☿
15 42 ☽ ⚹ ♂
23 26 ☽ ⊥ ♆

08 Tuesday
04 09 ☽ ⊥ ☿
07 34 ☽ ⊻ ♄
08 58 ☽ Q ♀
09 44 ☽ ⚹ ♃
10 05 ☿ □ ♂
11 39 ☽ ⊻ ♆
16 26 ☽ Q ♅
19 46 ☽ ⊻ ☉
19 54 ☽ ⊥ ♄
20 45 ☽ ∥ ♀

09 Wednesday
00 01 ☽ ∠ ♂
01 07 ☽ ⊻ ♇
01 10 ☽ ∠ ☿
09 08 ☽ ⊥ ☉
10 35 ☽ Q ♃
13 23 ☽ ⊥ ♇
14 25 ☽ ∠ ♄
16 53 ☽ ⚹ ♅
19 28 ☽ ⊥ ♂
21 33 ♃ ⚼ ♅

10 Thursday
01 09 ☽ □ ♀
05 12 ☽ ∠ ☉
05 37 ☉ ⊥ ♃
07 45 ☽ ∠ ♇
08 25 ☽ ⊻ ♂
12 16 ☽ ⚹ ☿
16 39 ☉ ∥ ♇
20 28 ☽ # ♅
21 13 ☽ ⚹ ♄
23 22 ☽ ∠ ♅
23 48 ☽ □ ♃

11 Friday
00 47 ☽ ☌ ♆
05 43 ☉ ∥ ♃
11 38 ☉ ☌ ♇
14 11 ☽ ⚹ ♇
14 25 ☽ ⚹ ☉
16 08 ☽ Q ☿
16 58 ☽ ∥ ♆
17 31 ☽ ⊥ ♅
21 38 ☽ Q ♄
21 39 ♀ ± ♅

12 Saturday
05 32 ☽ ⊻ ♅
13 53 ☿ ⊥ ♀
14 13 ☽ Q ♇
15 27 ☽ # ♂
16 35 ☽ Q ☉
19 20 ☽ △ ♀

13 Sunday
00 12 ☽ ☌ ♂
05 19 ♀ ∠ ♇
09 05 ☽ □ ☿
09 36 ☽ □ ♄
12 31 ☽ △ ♃
12 36 ☽ ⊻ ♆
13 31 ☿ ☌ ♄
18 58 ♃ □ ♆

14 Monday
00 14 ☽ ⊥ ♆
01 28 ☽ □ ♇
01 56 ♀ ⊥ ♄
02 14 ☽ ∥ ♂
03 17 ☽ ⚼ ♀
05 06 ♃ ∥ ♇
06 46 ☽ □ ☉
13 13 ☿ ⚹ ♆
15 30 ☿ ⊻ ♃
15 56 ☽ ☌ ♅
17 24 ☽ ∠ ♆
17 42 ☽ ⚼ ♃
21 19 ☽ # ♆
21 55 ☽ ± ♀

15 Tuesday
10 05 ☽ ⊼ ♀
10 45 ☽ ± ♃
12 39 ☽ ⊻ ♂
16 27 ☽ ∥ ♅
18 50 ☽ △ ♄
21 15 ☽ ⚹ ♆
21 54 ☽ ⊼ ♃

16 Wednesday
00 12 ☽ ⊥ ♂
01 30 ☽ △ ☿
09 17 ☽ △ ♇
17 10 ☽ ∠ ♂
18 34 ☽ △ ☉
18 47 ☽ Q ♆
21 56 ☽ ⚼ ♄
22 38 ☽ ⊻ ♅

17 Thursday
07 34 ☽ ⚼ ☿
09 05 ☽ ⊥ ♅
11 41 ☽ ⚼ ♇
13 42 ☽ ± ♄
18 24 ♃ ⊥ ♇
19 54 ☽ ☍ ♀
20 32 ☽ ⚹ ♂
22 35 ☽ ⚼ ☉

18 Friday
00 02 ☽ ⊼ ♄
00 30 ☽ ∠ ♅
00 58 ☽ ± ☿
01 37 ☽ # ♀
01 56 ☽ □ ♆
03 04 ☽ ± ♇
03 10 ☽ ☍ ♃
12 19 ☽ ⊼ ☿
13 10 ☽ ⊼ ♇
14 52 ☽ ± ☉
16 49 ♀ △ ♂
17 42 ☽ Q ♂
18 38 ☽ # ☉
20 03 ☿ ☌ ♇
22 56 ☿ ⊥ ♃

19 Saturday
01 31 ☉ □ ♅
01 32 ☽ ⚹ ♅
01 32 ☽ ⊼ ☉
21 05 ☽ Q ♅
22 26 ☉ ∠ ♆

20 Sunday
00 28 ☽ □ ♂
01 24 ☽ ⊼ ♀
01 48 ☽ ☍ ♄
03 19 ☽ △ ♆
05 02 ☽ ⊼ ♃
07 29 ♀ ⊻ ♄
08 19 ♀ ⚼ ♅
11 46 ☽ ± ♀
14 01 ☽ ☍ ♇
14 44 ☽ ± ♃
18 56 ☽ ☍ ☿

21 Monday
01 50 ☽ □ ♅
03 11 ☽ ⚼ ♀
03 15 ☽ ⚼ ♆
04 03 ☉ ∠ ♃
04 15 ♀ □ ♆
05 12 ☽ ⚼ ♃
05 16 ☽ ☍ ☉
10 03 ☽ # ☉
11 48 ♂ □ ♄
13 52 ☽ # ♀
17 30 ☽ ± ♆
19 42 ♀ ⊥ ♇

22 Tuesday
01 47 ☽ ⊼ ♄
02 19 ☽ △ ♂
03 00 ☽ ⊼ ♆
04 43 ☽ △ ♀
05 11 ☽ △ ♃
11 05 ☉ ∥ ♀
11 20 ☽ ± ♄
12 26 ♀ ☌ ♃
13 36 ☽ ⊼ ♇
18 13 ♂ ⊻ ♆
23 08 ☽ ± ♇

23 Wednesday
00 07 ☽ ⊼ ☿
01 19 ☽ △ ♅
01 42 ☽ ⚼ ♄
03 11 ☽ ⚼ ♂
08 11 ☽ ⊼ ☉
09 00 ☉ Q ♂
10 52 ☽ ± ☿
11 13 ☿ □ ♅
13 31 ☽ ⚼ ♇
15 40 ☿ ∥ ♄
18 16 ☽ ± ♂
18 29 ☽ ± ☉
22 37 ☽ ∥ ♅

24 Thursday
01 05 ☿ ∠ ♆
01 18 ☿ ∥ ♃
01 23 ☽ ⚼ ♅
01 56 ☽ △ ♄
02 55 ☽ ☍ ♆
03 08 ☽ ⚼ ☿
04 25 ☽ ⊼ ♂
05 41 ☽ □ ♃
08 27 ☽ □ ♀
10 08 ☽ ⚼ ☉
13 50 ☽ △ ♇
16 03 ☽ # ♆
16 03 ☽ ± ♅
18 14 ☽ ∥ ♂
19 35 ☿ ∥ ♇

25 Friday
01 57 ☽ ⊼ ♅
04 18 ☿ ∠ ♃
06 56 ☽ △ ☿
11 44 ☽ Q ♃
12 48 ☽ △ ☉
16 44 ☽ Q ♀
17 53 ♂ △ ♃

26 Saturday
04 10 ☽ □ ♄
04 54 ☽ ⊼ ♆
08 23 ☽ ⚹ ♃
08 56 ☽ ☍ ♂
08 59 ♂ # ♆
15 03 ☽ ⚹ ♀
15 17 ☽ ± ♆
16 30 ☽ □ ♇

27 Sunday
05 21 ☽ ☍ ♅
07 07 ☽ ⚼ ♆
09 22 ☽ ∥ ♆
09 32 ♀ ⊻ ♇
10 57 ☽ # ♂
11 01 ☽ ∠ ♃
11 53 ☽ Q ♄
17 59 ☽ □ ☿
19 58 ☽ ∠ ♀
21 10 ☽ □ ☉
21 29 ☿ Q ♂

28 Monday
00 33 ☽ Q ♇
03 06 ♃ ∥ ♄
03 26 ☽ ⊥ ♃
05 04 ☽ # ♅
09 46 ☽ ⚹ ♄
10 14 ☽ △ ♆
12 17 ☿ ∥ ♀
13 54 ☽ ⊥ ♀
14 35 ☽ ⊻ ♃
17 21 ☽ ⊼ ♂
22 39 ☽ ⚹ ♇

29 Tuesday
01 25 ☉ ⊥ ♆
02 05 ☽ ⊻ ♀
05 08 ☽ ± ♂
08 04 ☽ Q ☿
09 11 ☽ Q ☉
11 44 ☿ ∠ ♀
11 59 ☿ ⊥ ♆
12 19 ☽ ⊼ ♅
13 58 ☽ ∠ ♄
23 08 ☽ ⚼ ♂
23 44 ☽ ± ♅

30 Wednesday
02 52 ☉ ☌ ☿
03 03 ☽ ∠ ♇
06 00 ☽ ∥ ☉
07 17 ☽ ⊥ ♄
10 04 ☽ ⚹ ☉
10 32 ☽ ⚹ ☿
17 05 ☽ ⚼ ♅
18 59 ☽ ⊻ ♄
19 06 ☽ □ ♆
20 29 ☽ ⊥ ♇
22 39 ☽ ∥ ☿

31 Thursday
00 23 ☽ ☌ ♃
05 50 ☽ △ ♂
08 14 ☽ ⊻ ♇
14 15 ♄ ⚹ ♆
17 35 ☽ ☌ ♀
17 59 ☽ ∠ ☉
19 25 ☽ ∥ ♀
20 39 ☽ ∠ ☿
22 33 ☽ △ ♅

All ephemeris data is given at 12.00 UT and the Moon's longitude is additionally given for 24.00 UT

FEBRUARY 2019

LONGITUDES

Date	Sidereal time h m s	Sun ☉ ° ' "	Moon ☽ ° ' "	Moon ☽ 24.00 ° ' "	Mercury ☿ ° '	Venus ♀ ° '	Mars ♂ ° '	Jupiter ♃ ° '	Saturn ♄ ° '	Uranus ♅ ° '	Neptune ♆ ° '	Pluto ♇ ° '
01	20 45 38	12 ♒ 19 31	05 ♑ 38 21	11 ♑ 38 19	14 ♒ 02	27 ♐ 14	21 ♈ 14	17 ♐ 51	15 ♑ 01	28 ♈ 53	14 ♓ 57	21 ♑ 39
02	20 49 35	13 20 26	17 ♑ 36 30	23 ♑ 33 15	15 47	28 22	21 55	18 02	15 07	28 54	14 59	21 41
03	20 53 31	14 21 21	29 ♑ 28 53	05 ♒ 23 42	17 32	29 ♐ 30	22 35	18 12	15 14	28 56	15 01	21 43
04	20 57 28	15 22 13	11 ♒ 17 57	17 ♒ 11 56	19 18	00 ♑ 39	23 16	18 22	15 20	28 57	15 03	21 44
05	21 01 24	16 23 05	23 ♒ 05 52	29 ♒ 00 01	21 05	01 47	23 57	18 32	15 27	28 59	15 05	21 46
06	21 05 21	17 23 56	04 ♓ 54 37	10 ♓ 49 56	22 52	02 56	24 37	18 42	15 33	29 00	15 07	21 48
07	21 09 17	18 24 45	16 ♓ 46 14	22 ♓ 43 48	24 40	04 05	25 18	18 51	15 39	29 02	15 09	21 50
08	21 13 14	19 25 33	28 ♓ 42 55	04 ♈ 43 55	26 28	05 14	25 58	19 01	15 46	29 03	15 11	21 52
09	21 17 10	20 26 20	10 ♈ 47 09	16 ♈ 52 59	28 ♒ 17	06 23	26 39	19 11	15 52	29 05	15 13	21 54
10	21 21 07	21 27 05	23 ♈ 01 49	29 ♈ 14 04	00 ♓ 05	07 32	27 20	19 20	15 58	29 07	15 15	21 56
11	21 25 04	22 27 48	05 ♉ 30 10	11 ♉ 50 34	01 54	08 41	28 00	19 29	16 04	29 09	15 17	21 57
12	21 29 00	23 28 30	18 ♉ 15 43	24 ♉ 46 02	03 42	09 51	28 41	19 39	16 11	29 10	15 19	21 59
13	21 32 57	24 29 10	01 ♊ 21 56	08 ♊ 03 47	05 31	11 00	29 ♈ 21	19 48	16 17	29 12	15 21	22 01
14	21 36 53	25 29 49	14 ♊ 51 53	21 ♊ 46 27	07 18	12 10	00 ♉ 02	19 57	16 23	29 14	15 24	22 03
15	21 40 50	26 30 26	28 ♊ 47 35	05 ♋ 55 16	09 05	13 20	00 42	20 05	16 29	29 16	15 26	22 04
16	21 44 46	27 31 01	13 ♋ 09 18	20 ♋ 29 19	10 50	14 29	01 23	20 14	16 35	29 18	15 28	22 06
17	21 48 43	28 31 34	27 ♋ 54 48	05 ♌ 24 58	12 34	15 39	02 04	20 23	16 41	29 20	15 30	22 08
18	21 52 39	29 ♒ 32 06	12 ♌ 58 56	20 ♌ 35 35	14 17	16 49	02 44	20 31	16 46	29 22	15 32	22 10
19	21 56 36	00 ♓ 32 36	28 ♌ 13 42	05 ♍ 51 59	15 56	18 00	03 25	20 40	16 52	29 24	15 35	22 11
20	22 00 33	01 33 05	13 ♍ 29 06	21 ♍ 03 42	17 33	19 10	04 05	20 48	16 58	29 26	15 37	22 13
21	22 04 29	02 33 32	28 ♍ 33 36	06 ♎ 00 40	19 06	20 20	04 45	20 56	17 04	29 29	15 39	22 15
22	22 08 26	03 33 57	13 ♎ 20 58	20 ♎ 34 48	20 36	21 31	05 26	21 04	17 09	29 31	15 41	22 16
23	22 12 22	04 34 21	27 ♎ 41 38	04 ♏ 41 10	22 00	22 41	06 06	21 12	17 15	29 33	15 44	22 18
24	22 16 19	05 34 44	11 ♏ 33 16	18 ♏ 18 03	23 20	23 52	06 47	21 19	17 20	29 35	15 46	22 19
25	22 20 15	06 35 05	24 ♏ 55 42	01 ♐ 26 36	24 33	25 02	07 27	21 27	17 26	29 38	15 48	22 21
26	22 24 12	07 35 25	07 ♐ 51 11	14 ♐ 10 00	25 40	26 13	08 07	21 34	17 31	29 40	15 50	22 23
27	22 28 08	08 35 44	20 ♐ 23 37	26 ♐ 32 39	26 39	27 24	08 48	21 42	17 37	29 43	15 53	22 24
28	22 32 05	09 ♓ 36 01	02 ♑ 37 42	08 ♑ 39 24	27 ♓ 31	28 ♑ 35	09 ♉ 28	21 ♐ 49	17 ♑ 42	29 ♈ 45	15 ♓ 55	22 ♑ 26

MOON

Date	Moon True ☊ ° '	Moon Mean ☊ ° '	Moon ☽ Latitude ° '
01	26 ♋ 47	25 ♋ 54	01 N 55
02	26 D 48	25 51	00 N 51
03	26 R 49	25 48	00 S 15
04	26 48	25 45	01 19
05	26 46	25 42	02 20
06	26 43	25 38	03 15
07	26 40	25 35	04 02
08	26 36	25 32	04 38
09	26 32	25 29	05 03
10	26 28	25 26	05 14
11	26 26	25 22	05 10
12	26 25	25 19	04 52
13	26 D 25	25 16	04 18
14	26 26	25 13	03 29
15	26 28	25 10	02 26
16	26 29	25 07	01 S 13
17	26 R 30	25 03	00 N 08
18	26 29	25 00	01 29
19	26 26	24 57	02 45
20	26 22	24 54	03 49
21	26 16	24 51	04 36
22	26 11	24 48	05 04
23	26 05	24 44	05 11
24	26 01	24 41	05 00
25	25 59	24 38	04 33
26	25 D 58	24 35	03 52
27	25 59	24 32	03 01
28	26 ♋ 00	24 ♋ 28	02 N 04

DECLINATIONS

Date	Sun ☉ ° '	Moon ☽ ° '	Mercury ☿ ° '	Venus ♀ ° '	Mars ♂ ° '	Jupiter ♃ ° '	Saturn ♄ ° '	Uranus ♅ ° '	Neptune ♆ ° '	Pluto ♇ ° '
01	17 S 06	21 S 24	18 S 36	20 S 51	08 N 34	22 S 16	22 S 09	10 N 36	06 S 49	21 S 51
02	16 49	21 26	18 04	20 56	08 50	22 17	22 08	10 36	06 48	21 51
03	16 31	20 30	17 30	21 00	09 06	22 18	22 07	10 37	06 48	21 50
04	16 13	18 39	16 55	21 03	09 21	22 18	22 06	10 37	06 47	21 50
05	15 55	16 01	16 18	21 06	09 37	22 19	22 06	10 38	06 46	21 50
06	15 37	12 44	15 40	21 08	09 53	22 20	22 05	10 38	06 45	21 50
07	15 18	08 56	15 01	21 10	10 09	22 21	22 04	10 39	06 44	21 49
08	15 00	04 46	14 20	21 11	10 24	22 22	22 04	10 39	06 43	21 49
09	14 40	00 S 22	13 38	21 12	10 40	22 23	22 03	10 40	06 43	21 49
10	14 21	04 N 05	12 55	21 12	10 55	22 23	22 02	10 41	06 42	21 49
11	14 01	08 28	12 11	21 11	11 11	22 24	22 02	10 41	06 41	21 49
12	13 42	12 35	11 25	21 10	11 26	22 25	22 01	10 42	06 40	21 48
13	13 21	16 13	10 38	21 09	11 41	22 25	22 00	10 43	06 39	21 48
14	13 01	19 07	09 51	21 06	11 56	22 26	22 00	10 43	06 39	21 48
15	12 41	21 00	09 03	21 04	12 11	22 27	21 59	10 44	06 38	21 48
16	12 20	21 35	08 14	21 00	12 26	22 27	21 58	10 45	06 37	21 48
17	11 59	20 42	07 25	20 56	12 41	22 28	21 58	10 46	06 36	21 48
18	11 38	18 20	06 35	20 52	12 55	22 29	21 57	10 46	06 35	21 47
19	11 17	14 40	05 46	20 47	13 10	22 29	21 56	10 47	06 34	21 47
20	10 55	10 01	04 57	20 41	13 24	22 30	21 56	10 48	06 33	21 47
21	10 34	04 N 47	04 09	20 35	13 39	22 30	21 55	10 49	06 33	21 47
22	10 12	00 S 36	03 21	20 28	13 53	22 31	21 54	10 50	06 32	21 47
23	09 50	05 48	02 35	20 21	14 07	22 31	21 54	10 50	06 31	21 47
24	09 28	10 32	01 51	20 13	14 21	22 32	21 53	10 51	06 30	21 46
25	09 06	14 35	01 09	20 04	14 35	22 32	21 52	10 52	06 29	21 46
26	08 43	17 48	00 S 29	19 55	14 48	22 33	21 52	10 53	06 28	21 46
27	08 21	20 04	00 N 08	19 46	15 02	22 33	21 51	10 54	06 27	21 46
28	07 S 58	21 S 21	00 N 42	19 S 35	15 N 16	22 S 34	21 S 50	10 N 55	06 S 26	21 S 46

ZODIAC SIGN ENTRIES

Date	h	m	Planets
01	00	47	☽ ♑
03	13	03	☽ ♒
03	22	29	♀ ♑
06	02	02	☽ ♓
08	14	34	☽ ♈
10	10	51	☿ ♓
11	01	28	☽ ♉
13	09	32	☽ ♊
14	10	51	♂ ♉
15	14	03	☽ ♋
17	15	21	☽ ♌
18	23	04	☉ ♓
19	14	47	☽ ♍
21	14	17	☽ ♎
23	15	56	☽ ♏
25	21	19	☽ ♐
28	06	48	☽ ♑

LATITUDES

Date	Mercury ☿ ° '	Venus ♀ ° '	Mars ♂ ° '	Jupiter ♃ ° '	Saturn ♄ ° '	Uranus ♅ ° '	Neptune ♆ ° '	Pluto ♇ ° '
01	02 S 04	02 N 33	00 N 18	00 N 37	00 N 27	00 S 31	00 S 58	00 S 09
04	01 59	02 23	00 21	00 37	00 27	00 31	00 58	00 09
07	01 50	02 13	00 23	00 37	00 27	00 31	00 58	00 10
10	01 35	02 02	00 26	00 37	00 27	00 31	00 58	00 10
13	01 14	01 51	00 28	00 37	00 26	00 31	00 58	00 10
16	00 47	01 39	00 31	00 37	00 26	00 30	00 58	00 11
19	00 S 14	01 27	00 33	00 37	00 26	00 30	00 58	00 11
22	00 N 24	01 16	00 35	00 37	00 26	00 30	00 58	00 11
25	01 07	01 04	00 37	00 37	00 26	00 30	00 58	00 11
28	01 51	00 52	00 39	00 37	00 26	00 30	00 58	00 12
31	02 N 33	00 N 40	00 N 41	00 N 37	00 N 26	00 S 30	00 S 58	00 S 12

DATA

Julian Date	2458516
Delta T	+70 seconds
Ayanamsa	24° 07' 11"
Synetic vernal point	04° ♓ 59' 48"
True obliquity of ecliptic	23° 26' 08"

MOON'S PHASES, APSIDES AND POSITIONS ☽

Date	h	m	Phase	Longitude ° '	Eclipse Indicator
04	21	04	●	15 ♒ 45	
12	22	26	☽	23 ♉ 55	
19	15	54	○	00 ♍ 42	
26	11	28	☾	07 ♐ 34	

Day	h	m		
05	09	45	Apogee	
19	09	09	Perigee	
02	00	46	Max dec	21° S 33'
09	14	01	ON	
16	09	55	Max dec	21° N 35'
22	09	17	0S	

LONGITUDES

Date	Chiron ⚷ ° '	Ceres ⚳ ° '	Pallas ⚴ ° '	Juno ⚵ ° '	Vesta ⚶ ° '	Black Moon Lilith ⚸ ° '
01	29 ♓ 11	02 ♐ 12	28 ♎ 22	26 ♉ 50	00 ♓ 01	19 ♒ 51
11	29 ♓ 39	05 ♐ 14	29 ♎ 17	00 ♊ 07	04 ♓ 59	20 ♒ 58
21	00 ♈ 10	07 ♐ 55	29 ♎ 27	03 ♊ 49	09 ♓ 56	22 ♒ 05
31	00 ♈ 43	10 ♐ 13	28 ♎ 48	07 ♊ 51	14 ♓ 52	23 ♒ 11

ASPECTARIAN

h m	Aspects
01 Friday	
01 11	♂ ⊥ ♆
06 37	☽ Q ♆
13 30	☽ ⊥ ☉
17 35	☽ ⊥ ☿
02 Saturday	
00 49	☿ ⊻ ♆
02 25	☿ ⊻ ♄
02 37	☽ ⊻ ☉
03 20	♂ □ ♇
06 41	☽ ⚹ ♆
06 57	☽ ☌ ♄
07 22	☉ ∠ ♀
07 41	☽ ⊻ ☿
12 51	☽ ⊻ ♃
20 14	☽ ☌ ♇
21 12	☽ □ ♂
23 41	♀ △ ♅
03 Sunday	
01 09	☽ ⊥ ♃
02 48	☽ ∥ ♀
03 38	☿ Q ♅
10 53	☽ □ ♅
12 03	☽ ⊻ ♀
13 04	☽ ∠ ♆
19 39	☽ ∠ ♃
21 54	☿ ⚹ ♃
04 Monday	
01 31	☽ ⊥ ♀
04 00	☉ ⊻ ♆
07 24	☽ ⊥ ♆
11 09	☉ ⊻ ♄
11 56	☽ Q ♂
19 38	☽ ⊻ ♆
20 17	☽ ⊻ ♄
21 04	☽ ☌ ☉
21 47	☽ ∠ ♀
23 32	☽ Q ♅
05 Tuesday	
02 35	☽ ⚹ ♃
07 11	☽ ☌ ☿
08 36	☽ ⊥ ♄
09 05	☽ ∥ ☿
09 18	☽ ⊻ ♇
12 54	☽ ∥ ☉
13 49	☽ ⚹ ♂
17 09	☿ ⊥ ♄
21 24	☿ ⊻ ♇
21 32	☽ ⊥ ♇
23 59	☽ ⚹ ♅
06 Wednesday	
02 25	☉ Q ♅
03 04	☽ ∠ ♄
03 19	☽ Q ♃
07 33	☽ ⚹ ♀
15 51	☽ ∠ ♇
15 55	♀ Q ♆
16 02	☉ ∥ ☿
22 07	☽ ∠ ♂
07 Thursday	
01 32	☽ # ♅
05 06	☽ # ♂
06 27	☽ ∠ ♅
08 43	☽ ☌ ♆
09 44	☽ ⚹ ♄
10 27	☽ Q ♀
15 37	☽ ⊻ ☉
16 16	☽ □ ♃
17 24	☽ ⊥ ♂
22 14	☽ ⚹ ♇
08 Friday	
00 32	☉ ⚹ ♃
00 38	☽ ⊥ ♅
00 53	☽ ∥ ♆
01 24	☿ ⚹ ♂
04 48	☽ ⊥ ☉
06 11	☽ ⊻ ♂
06 43	☽ ⊻ ☿
10 05	☽ Q ♄
12 41	☽ ⊻ ♅
20 49	☽ ⊥ ☿
22 18	☽ Q ♇
09 Saturday	
00 26	☽ ∠ ☉
02 21	☽ □ ♀
06 51	☿ ⊥ ♇
12 28	♂ ∥ ♅
17 46	☽ ∠ ☿
20 45	☽ ⊻ ♆
22 06	☽ □ ♄
22 53	☿ ⚹ ♅
10 Sunday	
04 42	☽ △ ♃
08 32	☽ ⊥ ♆
08 39	☽ ⚹ ☉
09 51	☽ □ ♇
20 48	☽ ☌ ♂
23 36	☉ ⊻ ♇
23 48	☽ ☌ ♅
11 Monday	
00 26	☿ ∠ ♄
01 45	☉ ⊥ ♄
02 00	☽ ∠ ♆
02 08	☽ # ♆
03 57	☽ ⚹ ☿
06 05	☿ Q ♃
09 51	☽ Q ☉
10 03	☽ ⚼ ♃
18 39	☽ △ ♀
12 Tuesday	
00 43	☽ ∥ ♅
03 17	☽ ± ♃
04 35	☽ ∥ ♂
06 05	☽ # ☿
06 28	☽ Q ☿
06 31	☽ ⚹ ♆
08 05	☽ △ ♄
11 38	☿ # ♂
14 36	☽ ⊼ ♃
18 22	☽ # ☉
18 54	☽ △ ♇
22 26	☽ □ ☉
13 Wednesday	
01 19	☽ ⚼ ♀
04 43	☽ Q ♆
06 21	♂ ☌ ♅
08 05	☽ ⊻ ♅
08 10	☽ ⊻ ♂
09 45	☿ # ♅
11 51	☽ ⚼ ♄
18 55	☽ ⊥ ♅
19 10	☽ ± ♀
19 33	☽ ⊥ ♂
20 36	☽ □ ☿
22 09	☽ ⚼ ♇
14 Thursday	
04 03	☽ ± ♄
06 49	☽ ⊼ ♀
08 30	☿ ∠ ♇
10 54	☽ ∠ ♅
12 19	☽ ∠ ♂
12 56	☽ □ ♆
14 04	☽ ± ♇
14 40	☽ ⊼ ♄
20 56	☽ ☍ ♃
15 Friday	
00 30	☽ ⊼ ♇
01 35	♂ ∠ ♆
07 49	☽ △ ☉
12 48	☽ ⚹ ♅
13 16	☽ # ♀
15 24	☽ ⚹ ♂
16 Saturday	
07 39	☽ △ ☿
08 02	☉ # ♂
08 56	☽ Q ♅
10 52	☽ ⚼ ☉
12 24	☽ Q ♂
14 23	☽ ☍ ♀
15 49	☽ △ ♆
17 40	☽ ☍ ♄
23 42	☽ ⊼ ♃
17 Sunday	
02 21	☉ ⊥ ♇
02 39	☽ ☍ ♇
02 40	☽ ± ☉
08 02	☽ # ♀
08 44	♀ ⚹ ♆
09 31	☽ ± ♃
11 23	☽ ⚼ ☿
13 03	☽ ⊼ ☉
14 17	☽ □ ♅
16 10	☽ ⚼ ♆
18 57	☽ □ ♂
18 Monday	
00 03	☽ ⚼ ♃
03 37	☽ ± ☿
06 32	☽ ± ♆
07 55	☉ ⚹ ♅
10 52	♀ ☌ ♄
12 05	☿ ∥ ♆
13 21	☿ ∠ ♅
14 18	☽ ⊼ ☿
16 03	☽ ⊼ ♆
18 01	☽ ⊼ ♄
18 34	☽ ⊼ ♀
19 Tuesday	
00 00	☽ △ ♃
02 30	☽ ⊼ ♇
03 32	☽ ± ♄
04 48	☽ ± ♀
06 37	☿ ☌ ♆
11 56	☽ ± ♇
13 51	☽ △ ♅
15 54	☽ ☍ ☉
17 45	☽ ⚼ ♄
19 47	☽ ∥ ♂
20 06	☽ ⚼ ♀
20 31	☽ △ ♂
20 Wednesday	
02 06	☽ ⚼ ♇
02 39	☿ ⚹ ♄
07 14	☽ # ☉
08 12	☽ ∥ ♅
13 31	☽ ⚼ ♅
15 22	☽ ☍ ♆
17 32	☽ △ ♄
19 11	☽ ☍ ☿
19 50	☉ # ♅
21 16	☽ ⚼ ♂
21 45	☽ △ ♀
22 55	☉ ∠ ♄
23 41	☽ □ ♃
21 Thursday	
01 52	☽ △ ♇
03 49	☽ ± ♅
04 05	☽ # ♆
12 18	☽ ± ♂
13 27	☽ ⊼ ♅
15 20	☽ # ☿
18 53	☽ ⊼ ☉
22 15	☉ Q ♃
22 26	☽ ⊼ ♂
22 Friday	
01 44	♀ ⊻ ♃
04 55	☽ Q ♃
05 20	☽ ± ☉
07 02	☿ ∠ ♂
15 52	☽ ⊼ ♆
18 20	☽ □ ♄
20 40	☿ □ ♃
21 17	☽ ⚼ ☉
22 53	☽ ∥ ☿
23 Saturday	
00 56	☽ ⚹ ♃
01 21	☽ ⊼ ☿
01 54	☽ ± ♆
02 46	☽ □ ♀
02 52	☽ □ ♇
03 53	♀ ☌ ♇
12 35	☽ ± ☿
15 11	☽ ☍ ♅
15 24	☽ ∥ ♆
15 59	♂ ⚼ ♃
17 11	☽ ⚼ ♆
17 18	☿ ⚹ ♇
24 Sunday	
00 44	☽ △ ☉
01 04	☽ Q ♄
02 45	☽ ∠ ♃
03 12	☽ ☍ ♂
05 45	☽ ⚼ ☿
06 42	☽ ∥ ☉
09 50	☽ Q ♇
12 36	☽ Q ♀
13 44	☽ # ♅
17 11	☿ ⊥ ♅
18 44	☽ ⊥ ♃
19 29	☽ △ ♆
22 21	☽ ⚹ ♄
25 Monday	
05 37	☽ ⊻ ♃
07 18	☽ ⚹ ♇
11 14	☽ △ ☿
11 58	☽ # ♂
12 14	☽ ⚹ ♀
20 40	☽ ⊼ ♅
26 Tuesday	
01 56	☽ ∠ ♄
06 45	☉ ∠ ♇
07 53	☽ ± ♅
11 06	☽ ∠ ♇
11 28	☽ □ ☉
12 32	☽ ⊼ ♂
19 00	☽ ⊥ ♄
19 02	☽ ∠ ♀
27 Wednesday	
00 36	☽ ± ♂
01 00	☽ ⚼ ♅
03 15	☽ □ ♆
04 16	☽ ⊥ ♇
06 35	☽ ⊻ ♄
08 20	☽ ∥ ♀
14 10	☽ ⊥ ♀
14 33	☽ ☌ ♃
15 55	☽ ⊻ ♇
18 39	♀ ⊥ ♃
19 00	☽ ⚼ ♂
28 Thursday	
01 11	☽ □ ☿
01 11	☽ Q ☉
02 33	☉ ⚹ ♂
03 09	☽ ⊻ ♀
06 17	☽ △ ♅
14 33	☽ Q ♆

All ephemeris data is given at 12.00 UT and the Moon's longitude is additionally given for 24.00 UT

MARCH 2019

LONGITUDES

Date	Sidereal time h m s	Sun ☉ ° ' "	Moon ☽ ° ' "	Moon ☽ 24.00 ° ' "	Mercury ☿ ° '	Venus ♀ ° '	Mars ♂ ° '	Jupiter ♃ ° '	Saturn ♄ ° '	Uranus ♅ ° '	Neptune ♆ ° '	Pluto ♇ ° '
01	22 36 02	10 ♓ 36 16	14 ♑ 38 22	20 ♑ 35 09	28 ♓ 14	29 ♑ 46	10 ♉ 08	21 ♐ 56	17 ♑ 47	29 ♈ 47	15 ♓ 57	22 ♑ 27
02	22 39 58	11 36 31	26 ♑ 30 19	02 ♒ 24 23	28 49	00 ♒ 57	10 49	22 03	17 52	29 50	15 59	22 29
03	22 43 55	12 36 43	08 ♒ 17 48	14 ♒ 11 01	29 15	02 08	11 29	22 09	17 57	29 53	16 02	22 30
04	22 47 51	13 36 54	20 ♒ 04 25	25 ♒ 58 20	29 32	03 19	12 09	22 16	18 02	29 55	16 04	22 31
05	22 51 48	14 37 03	01 ♓ 53 04	07 ♓ 48 53	29 39	04 30	12 50	22 22	18 07	29 ♈ 58	16 06	22 33
06	22 55 44	15 37 10	13 ♓ 46 01	19 ♓ 44 39	29 R 36	05 42	13 30	22 29	18 12	00 ♉ 00	16 08	22 34
07	22 59 41	16 37 15	25 ♓ 44 58	01 ♈ 47 08	29 25	06 53	14 10	22 35	18 17	00 03	16 11	22 36
08	23 03 37	17 37 19	07 ♈ 51 17	13 ♈ 57 36	29 05	08 04	14 50	22 41	18 22	00 06	16 13	22 37
09	23 07 34	18 37 20	20 ♈ 06 12	26 ♈ 17 15	28 37	09 16	15 31	22 46	18 26	00 09	16 15	22 38
10	23 11 31	19 37 20	02 ♉ 30 58	08 ♉ 47 32	28 01	10 27	16 11	22 52	18 31	00 11	16 18	22 39
11	23 15 27	20 37 17	15 ♉ 07 11	21 ♉ 30 10	27 19	11 39	16 51	22 57	18 35	00 14	16 20	22 41
12	23 19 24	21 37 12	27 ♉ 56 47	04 ♊ 27 18	26 31	12 51	17 31	23 03	18 40	00 17	16 22	22 42
13	23 23 20	22 37 05	11 ♊ 02 04	17 ♊ 41 22	25 39	14 02	18 11	23 08	18 44	00 20	16 24	22 43
14	23 27 17	23 36 56	24 ♊ 25 31	01 ♋ 14 46	24 44	15 14	18 51	23 13	18 48	00 23	16 27	22 44
15	23 31 13	24 36 45	08 ♋ 09 21	15 ♋ 09 23	23 47	16 26	19 31	23 18	18 53	00 26	16 29	22 46
16	23 35 10	25 36 31	22 ♋ 14 55	29 ♋ 25 53	22 49	17 37	20 11	23 22	18 57	00 29	16 31	22 47
17	23 39 06	26 36 15	06 ♌ 42 01	14 ♌ 02 57	21 53	18 49	20 51	23 27	19 01	00 32	16 33	22 48
18	23 43 03	27 35 57	21 ♌ 28 05	28 ♌ 56 40	20 58	20 01	21 31	23 31	19 05	00 35	16 36	22 49
19	23 47 00	28 35 37	06 ♍ 27 46	14 ♍ 00 19	20 06	21 13	22 11	23 35	19 09	00 38	16 38	22 50
20	23 50 56	29 ♓ 35 14	21 ♍ 33 06	29 ♍ 04 54	19 17	22 25	22 51	23 39	19 12	00 41	16 40	22 51
21	23 54 53	00 ♈ 34 49	06 ♎ 34 28	14 ♎ 00 34	18 34	23 37	23 31	23 43	19 16	00 44	16 42	22 52
22	23 58 49	01 34 23	21 ♎ 22 07	28 ♎ 38 10	17 55	24 49	24 11	23 47	19 20	00 47	16 45	22 53
23	00 02 46	02 33 54	05 ♏ 47 57	12 ♏ 50 54	17 22	26 01	24 51	23 50	19 23	00 50	16 47	22 54
24	00 06 42	03 33 24	19 ♏ 46 41	26 ♏ 35 08	16 55	27 13	25 31	23 53	19 27	00 53	16 49	22 55
25	00 10 39	04 32 52	03 ♐ 16 17	09 ♐ 50 22	16 34	28 25	26 11	23 56	19 30	00 56	16 51	22 56
26	00 14 35	05 32 18	16 ♐ 17 42	22 ♐ 38 46	16 18	29 ♒ 37	26 51	23 59	19 34	00 59	16 53	22 57
27	00 18 32	06 31 42	28 ♐ 54 07	05 ♑ 04 22	16 09	00 ♓ 49	27 31	24 02	19 37	01 03	16 56	22 57
28	00 22 29	07 31 05	11 ♑ 10 10	17 ♑ 12 13	16 06	02 01	28 10	24 05	19 40	01 06	16 58	22 58
29	00 26 25	08 30 26	23 ♑ 09 46	29 ♑ 07 47	16 D 08	03 13	28 50	24 07	19 43	01 09	17 00	22 59
30	00 30 22	09 29 45	05 ♒ 02 39	10 ♒ 56 27	16 16	04 26	29 ♉ 30	24 09	19 46	01 12	17 02	23 00
31	00 34 18	10 ♈ 29 02	16 ♒ 49 47	22 ♒ 43 12	16 ♓ 29	05 ♓ 38	00 ♊ 10	24 ♐ 11	19 ♑ 49	01 ♉ 16	17 ♓ 04	23 ♑ 00

Moon nodes and latitude

Date	Moon True ☊ ° '	Mean ☊ ° '	Moon ☽ Latitude ° '
01	26 ♋ 02	24 ♋ 25	01 N 02
02	26 R 02	24 22	00 S 03
03	26 01	24 19	01 06
04	25 58	24 16	02 06
05	25 53	24 13	03 01
06	25 45	24 09	03 48
07	25 36	24 06	04 26
08	25 26	24 03	04 52
09	25 15	24 00	05 04
10	25 06	23 57	05 03
11	24 59	23 53	04 47
12	24 54	23 50	04 17
13	24 51	23 47	03 32
14	24 D 50	23 44	02 35
15	24 51	23 41	01 28
16	24 51	23 38	00 S 14
17	24 R 51	23 34	01 N 03
18	24 48	23 31	02 17
19	24 43	23 28	03 23
20	24 35	23 25	04 15
21	24 26	23 22	04 49
22	24 15	23 19	05 03
23	24 05	23 15	04 57
24	23 57	23 12	04 33
25	23 51	23 09	03 55
26	23 47	23 06	03 05
27	23 45	23 03	02 08
28	23 D 45	22 59	01 06
29	23 45	22 56	00 N 03
30	23 R 45	22 53	01 S 00
31	23 ♋ 42	22 ♋ 50	01 S 59

DECLINATIONS

Date	Sun ☉ ° '	Moon ☽ ° '	Mercury ☿ ° '	Venus ♀ ° '	Mars ♂ ° '	Jupiter ♃ ° '	Saturn ♄ ° '	Uranus ♅ ° '	Neptune ♆ ° '	Pluto ♇ ° '
01	07 S 35	21 S 37	01 N 13	19 S 25	15 N 29	22 S 34	21 S 50	10 N 56	06 S 26	21 S 46
02	07 13	20 53	01 40	19 13	15 42	22 34	21 49	10 57	06 25	21 45
03	06 50	19 15	02 02	19 01	15 55	22 35	21 48	10 57	06 24	21 45
04	06 27	16 47	02 20	18 49	16 08	22 35	21 48	10 58	06 23	21 45
05	06 03	13 37	02 34	18 36	16 21	22 36	21 47	10 59	06 22	21 45
06	05 40	09 54	02 43	18 23	16 34	22 36	21 47	11 00	06 21	21 45
07	05 17	05 45	02 47	18 09	16 47	22 36	21 46	11 01	06 20	21 45
08	04 53	01 S 21	02 46	17 54	16 59	22 37	21 45	11 02	06 19	21 45
09	04 30	03 N 09	02 40	17 39	17 11	22 37	21 45	11 03	06 19	21 45
10	04 07	07 36	02 30	17 23	17 24	22 37	21 44	11 04	06 18	21 44
11	03 43	11 47	02 16	17 07	17 36	22 37	21 44	11 05	06 17	21 44
12	03 19	15 32	01 57	16 51	17 48	22 38	21 43	11 06	06 16	21 44
13	02 56	18 35	01 35	16 34	17 59	22 38	21 43	11 07	06 15	21 44
14	02 32	20 44	01 10	16 16	18 11	22 38	21 42	11 08	06 14	21 44
15	02 08	21 43	00 42	15 58	18 22	22 38	21 41	11 09	06 13	21 44
16	01 45	21 22	00 N 13	15 40	18 34	22 39	21 41	11 10	06 12	21 44
17	01 21	19 37	00 S 18	15 21	18 45	22 39	21 40	11 11	06 12	21 44
18	00 57	16 30	00 49	15 02	18 56	22 39	21 40	11 13	06 11	21 44
19	00 34	12 17	01 21	14 42	19 07	22 39	21 39	11 14	06 10	21 44
20	00 S 10	07 15	01 52	14 22	19 18	22 39	21 39	11 15	06 09	21 43
21	00 N 14	01 N 48	02 21	14 02	19 28	22 39	21 38	11 16	06 08	21 43
22	00 38	03 S 39	02 50	13 41	19 38	22 40	21 38	11 17	06 07	21 43
23	01 01	08 47	03 17	13 19	19 49	22 40	21 37	11 18	06 06	21 43
24	01 25	13 18	03 41	12 58	19 59	22 40	21 37	11 19	06 06	21 43
25	01 48	16 58	04 04	12 36	20 09	22 40	21 36	11 20	06 05	21 43
26	02 12	19 40	04 24	12 13	20 18	22 40	21 36	11 21	06 04	21 43
27	02 36	21 18	04 41	11 50	20 28	22 40	21 36	11 22	06 03	21 43
28	02 59	21 52	04 57	11 27	20 37	22 40	21 35	11 24	06 02	21 43
29	03 22	21 24	05 09	11 04	20 47	22 40	21 35	11 25	06 01	21 43
30	03 46	19 58	05 19	10 40	20 56	22 41	21 34	11 26	06 01	21 43
31	04 N 09	17 S 41	05 S 27	10 S 16	21 N 05	22 S 41	21 S 34	11 N 27	06 S 00	21 S 43

ZODIAC SIGN ENTRIES

Date	h	m	Planets
01	16	45	♀ ♒
02	19	06	☽ ♒
05	08	11	☽ ♓
06	08	26	♅ ♉
07	20	27	☽ ♈
10	07	10	☽ ♉
12	15	48	☽ ♊
14	21	49	☽ ♋
17	00	57	☽ ♌
19	01	41	☽ ♍
20	21	58	☉ ♈
21	01	28	☽ ♎
23	02	16	☽ ♏
25	06	06	☽ ♐
26	19	43	♀ ♓
27	14	07	☽ ♑
30	01	46	☽ ♒
31	06	12	♂ ♊

LATITUDES

Date	Mercury ☿ ° '	Venus ♀ ° '	Mars ♂ ° '	Jupiter ♃ ° '	Saturn ♄ ° '	Uranus ♅ ° '	Neptune ♆ ° '	Pluto ♇ ° '
01	02 N 05	00 N 48	00 N 40	00 N 37	00 N 26	00 S 30	00 S 58	00 S 12
04	02 45	00 37	00 41	00 37	00 26	00 30	00 58	00 12
07	03 17	00 25	00 43	00 38	00 26	00 30	00 58	00 12
10	03 35	00 14	00 45	00 38	00 25	00 30	00 58	00 13
13	03 36	00 N 03	00 47	00 38	00 25	00 30	00 58	00 13
16	03 20	00 S 07	00 48	00 38	00 25	00 30	00 58	00 13
19	02 48	00 17	00 50	00 38	00 25	00 30	00 58	00 13
22	02 07	00 27	00 51	00 38	00 25	00 30	00 58	00 14
25	01 21	00 36	00 52	00 38	00 25	00 30	00 58	00 14
28	00 N 35	00 45	00 54	00 38	00 25	00 30	00 58	00 14
31	00 S 07	00 S 53	00 N 55	00 N 38	00 N 25	00 S 30	00 S 58	00 S 15

DATA

Julian Date	2458544
Delta T	+70 seconds
Ayanamsa	24° 07' 14"
Synetic vernal point	04° ♓ 59' 45"
True obliquity of ecliptic	23° 26' 09"

MOON'S PHASES, APSIDES AND POSITIONS ☽

Date	h	m	Phase	Longitude ° '	Eclipse Indicator
06	16	04	●	15 ♓ 47	
14	10	27	☽	23 ♊ 33	
21	01	43	○	00 ♎ 09	
28	04	10	☾	07 ♑ 12	

Day	h	m		
04	11	42	Apogee	
19	19	54	Perigee	
01	06	18	Max dec	21° S 38'
08	19	13	0N	
15	18	01	Max dec	21° N 46'
21	19	53	0S	
28	12	56	Max dec	21° S 52'

LONGITUDES

Date	Chiron ⚷ ° '	Ceres ⚳ ° '	Pallas ⚴ ° '	Juno ⚵ ° '	Vesta ⚶ ° '	Black Moon Lilith ⚸ ° '
01	00 ♈ 36	09 ♐ 48	29 ♎ 00	07 ♊ 01	13 ♓ 53	22 ♒ 58
11	01 ♈ 11	11 ♐ 44	27 ♎ 41	11 ♊ 17	18 ♓ 41	24 ♒ 05
21	01 ♈ 46	13 ♐ 10	25 ♎ 34	15 ♊ 45	23 ♓ 40	25 ♒ 11
31	02 ♈ 22	14 ♐ 02	22 ♎ 50	20 ♊ 22	28 ♓ 30	26 ♒ 18

All ephemeris data is given at 12.00 UT and the Moon's longitude is additionally given for 24.00 UT

ASPECTARIAN

h m Aspects

01 Friday
02 26 ☽ △ ♂
03 09 ☽ ✶ ☉
12 32 ♀ □ ♅
14 39 ☽ ✶ ♆
15 25 ☽ Q ☿
18 23 ☽ ☌ ♄

02 Saturday
02 52 ☽ ⊻ ♃
03 49 ☽ ☌ ♇
12 14 ☽ ∠ ☉
12 50 ♀ ∠ ♆
15 09 ☽ ⊥ ♃
16 55 ☽ ✶ ☿
18 47 ☽ □ ♅
21 09 ☽ ∠ ♆
22 03 ☽ ☌ ♀

03 Sunday
08 15 ☽ ⊥ ☉
09 39 ☽ ∠ ♃
14 49 ☽ ∥ ♀
15 32 ☽ ⊥ ♆
18 54 ☽ □ ♂
21 37 ☽ ⊻ ☉

04 Monday
00 28 ☽ ∠ ☿
03 49 ☽ ⊻ ♆
07 36 ☽ Q ♅
07 49 ☽ ⊻ ♄
15 51 ☉ ∥ ♆
16 30 ☽ ✶ ♃
16 58 ☽ # ♂
17 00 ☽ ⊻ ♇
19 08 ☽ ⊥ ☿
20 07 ☽ ⊥ ♄

05 Tuesday
05 13 ☽ ⊥ ♇
07 26 ☽ ⊻ ☿
08 05 ☽ ✶ ♅
09 44 ☽ Q ♂
14 31 ☽ ∠ ♄
17 05 ☽ Q ♃
17 54 ☽ ⊻ ♀
18 19 ☿ St R
20 38 ☉ ∠ ♅
23 29 ☽ ∠ ♇

06 Wednesday
05 12 ☽ # ♅
07 22 ☽ ⊥ ♀
11 26 ☽ ✶ ♂
14 30 ☽ ∠ ♅
16 04 ☽ ☌ ☉
16 47 ☽ ☌ ♆
20 58 ☽ ✶ ♄

07 Thursday
01 00 ☉ ☌ ♆
03 26 ☽ ∠ ♀
05 37 ☽ □ ♃
05 41 ☽ ✶ ♇
08 36 ☽ ⊥ ♅
08 43 ☽ ∥ ♆
14 54 ☽ ∥ ☉
16 33 ♃ ⊻ ♇
18 18 ☿ ∠ ♂
19 08 ☽ ☌ ☿
19 12 ☽ ∠ ♂
20 35 ☽ ⊻ ♅
21 04 ☽ Q ♄

08 Friday
03 15 ♀ ∠ ♃
04 19 ☽ # ☿
05 35 ☽ Q ♇
12 29 ☽ ✶ ♀
14 03 ☽ ⊥ ♂

09 Saturday
02 31 ☽ ⊻ ♂
04 28 ☽ ⊻ ♆
07 10 ☉ ✶ ♄
08 44 ☽ □ ♄
08 52 ☽ ⊻ ☉
09 31 ☽ ∥ ☿
14 30 ☽ Q ♀
16 12 ☽ ⊥ ♆
16 56 ☽ □ ♇
17 14 ☽ △ ♃
18 37 ☽ # ☉
21 33 ☽ ⊥ ☉

10 Sunday
02 59 ♄ ∥ ♇
03 46 ☽ ⊻ ☿
04 54 ☽ # ♆
07 31 ☽ ☌ ♅
08 36 ♀ ⊥ ♆
09 39 ☽ ∠ ♆
11 49 ♀ # ♂
14 44 ☽ ⊥ ☿
16 20 ♂ ✶ ♆
16 23 ☽ ∠ ☉
22 19 ☽ ⚼ ♃

11 Monday
04 45 ☽ □ ♀
06 59 ☽ ∠ ☿
07 50 ☽ ∥ ♅
14 18 ☽ ✶ ♆
15 26 ☽ ☌ ♂
15 29 ☽ ± ♃
16 31 ♂ ± ♃
18 34 ☽ △ ♄
20 10 ☿ ∠ ♀
23 13 ☽ ✶ ☉

12 Tuesday
02 13 ☽ △ ♇
02 49 ☽ ⊼ ♃
09 31 ☽ ✶ ☿
12 47 ☽ Q ♆
16 20 ☽ ⊻ ♅
20 49 ☽ # ♀
22 36 ☽ ⚼ ♄
23 20 ☽ Q ☉

13 Wednesday
03 24 ☽ ⊥ ♅
05 58 ☽ ⚼ ♇
06 14 ☽ Q ☿
06 21 ☽ ∥ ♂
14 29 ☉ ✶ ♇
15 06 ☽ ± ♄
17 58 ☽ △ ♀
19 47 ☽ ∠ ♅
21 44 ☽ □ ♆
22 16 ☽ ± ♇

14 Thursday
01 29 ☉ □ ♃
01 34 ☽ ⊻ ♂
01 57 ☽ ⊼ ♄
09 00 ☽ ⊼ ♇
09 50 ☽ ☍ ♃
10 02 ♂ △ ♄
10 27 ☽ □ ☉
12 30 ☽ □ ☿
12 48 ☽ ⊥ ♂
20 32 ☿ ⊥ ♅
22 31 ☽ ✶ ♅
23 12 ☽ ⚼ ♀

15 Friday
01 48 ☉ ☌ ☿
05 23 ☽ ∠ ♂
07 20 ☉ ⊥ ♅
10 18 ☽ # ♄
12 58 ☽ # ♇
13 11 ♀ ⊻ ♆
16 16 ☽ ± ♀
19 22 ☽ Q ♅
23 08 ☽ # ♇
23 16 ☿ □ ♃

16 Saturday
02 01 ☽ # ♄
02 18 ☽ △ ♆
03 07 ☿ ⊥ ♀
03 28 ☽ ⊼ ♀
06 24 ☽ ☍ ♄
08 22 ☽ ✶ ♂
12 53 ☽ ☍ ♇
12 54 ☽ △ ☿
13 08 ☿ ✶ ♇
13 54 ☽ ⊼ ♃
18 03 ☽ △ ☉
23 58 ☽ ± ♃

17 Sunday
01 47 ☽ □ ♅
03 30 ☽ ⚼ ♆
05 22 ☽ Q ♂
05 55 ♀ Q ♅
12 16 ☽ ⚼ ☿
14 53 ☽ ⚼ ♃
16 09 ♀ ⊻ ♄
18 20 ☽ ± ♆
19 16 ☽ ∥ ♂
20 36 ☽ ⚼ ☉

18 Monday
02 05 ☽ ± ☿
03 23 ☿ ✶ ♂
04 07 ☽ ⊼ ♆
08 08 ☽ ⊼ ♄
09 27 ☽ ☍ ♀
11 14 ☽ ⊼ ☿
12 06 ☽ □ ♂
12 14 ☽ ± ☉
14 10 ☽ ⊼ ♇
15 19 ☽ △ ♃
15 24 ☉ ∥ ☿
17 50 ☽ ± ♄
21 49 ☽ # ♀
22 33 ☽ ⊼ ☉
22 56 ☿ ⊻ ♀
23 48 ☽ ± ♇

19 Tuesday
02 40 ☽ △ ♅
08 17 ☽ ⚼ ♄
14 11 ☽ ⚼ ♇
17 16 ☽ ∥ ♅

20 Wednesday
02 38 ☽ ⚼ ♅
04 13 ☽ ☍ ♆
08 15 ☽ △ ♄
08 35 ☽ ☍ ☿
11 41 ♂ △ ♇
13 29 ☽ ⊼ ♀
14 04 ☽ △ ♇
14 10 ☽ △ ♂
14 27 ☿ ✶ ♄
15 22 ☽ □ ♃
16 57 ☽ # ♆
17 00 ☽ ± ♅
20 54 ♀ ⊻ ♇
23 52 ☽ ± ♀

21 Thursday
01 43 ☽ ☍ ☉
02 36 ☽ ⊼ ♅
08 07 ♀ □ ♂
09 48 ☽ # ☿
14 16 ♀ ✶ ♃
15 17 ☽ ⚼ ♂
15 34 ☽ ⚼ ♀
15 48 ☉ ⊻ ♅
18 25 ☽ ∥ ☉
19 43 ♂ ⊼ ♃
20 19 ☽ Q ♃
21 34 ☽ # ☉

22 Friday
04 25 ☽ ⊼ ♆
05 44 ☉ Q ♄
06 33 ☽ ± ♂
06 35 ☽ ⊼ ☿
08 00 ☽ ∥ ☿
08 39 ☽ □ ♄
14 16 ☽ ± ♆
14 29 ☽ □ ♇
15 59 ☽ ✶ ♃
16 02 ☽ ± ☿
16 52 ☽ ⊼ ♂
18 10 ☽ △ ♀
22 58 ♀ ⊥ ♄
23 14 ☽ ∥ ♆

23 Saturday
03 38 ☽ ☍ ♅
05 14 ☽ ⚼ ♆
06 09 ☽ ⊼ ☉
06 26 ☽ ⚼ ☿
14 42 ☽ Q ♄
17 02 ☽ ± ☉
17 10 ☽ ∠ ♃
20 40 ☽ Q ♇

24 Sunday
00 57 ☽ # ♅
06 50 ☽ △ ♆
07 10 ☽ △ ☿
08 42 ☽ ⊥ ♃
09 42 ☽ ⚼ ☉
10 14 ☽ ∥ ♀
11 25 ☽ ✶ ♄
17 28 ☿ ☌ ♆
17 30 ☽ ✶ ♇
19 15 ☽ ⊻ ♃
22 38 ☽ ☍ ♂

25 Monday
02 24 ☽ □ ♀
07 46 ☽ ⊼ ♅
14 15 ☽ ∠ ♄
14 30 ☽ △ ☉
18 42 ☽ ± ♅
20 30 ☽ ∠ ♇
21 21 ☉ Q ♇
22 27 ♀ ⊥ ♇

26 Tuesday
03 30 ☉ ⊥ ♀
06 52 ☽ ⊥ ♄
11 26 ☽ ⚼ ♅
12 01 ☽ □ ☿
13 07 ☽ □ ♆
13 13 ☽ ⊥ ♇
14 44 ☽ Q ♀
18 10 ☽ ⊻ ♄
20 32 ☽ # ♂

27 Wednesday
00 35 ☽ ⊻ ♇
02 37 ☽ ☌ ♃
09 10 ☽ ⊼ ♂
16 06 ☽ ✶ ♀
16 10 ☽ △ ♅
16 45 ♀ ✶ ♅
19 37 ☽ ∥ ♄
21 27 ☽ ± ♂
22 08 ☽ Q ☿
23 45 ☽ Q ♆

28 Thursday
00 13 ☽ ∥ ♇
04 10 ☽ □ ☉
09 18 ☿ Q ♂
11 39 ☿ ∠ ♅
13 59 ☿ St D
15 42 ♀ # ♅
16 12 ☽ ⚼ ♂
21 48 ☽ ✶ ☿
23 33 ☽ ✶ ♆

29 Friday
00 56 ☽ ∠ ♀
01 47 ☽ ∥ ♇
05 00 ☽ ☌ ♄
06 47 ☽ ∥ ♄
11 35 ☽ ☌ ♇
13 53 ☽ ⊻ ♃
18 16 ♂ Q ♆
19 18 ☽ Q ☉
19 50 ☿ ∠ ♅
21 04 ☽ ⊥ ♀
23 07 ☽ # ♂

30 Saturday
00 05 ☽ △ ♂
02 03 ☽ ⊥ ♃
04 10 ☽ □ ♅
04 14 ☽ ∠ ☿
05 52 ☽ ∠ ♆
10 36 ☽ ⊻ ♀
19 02 ♀ ∠ ♄
20 23 ☽ ∠ ♃
21 53 ☽ ✶ ☉
22 49 ☽ ⊥ ☿

31 Sunday
00 14 ☽ ⊥ ♆
11 17 ☽ ⊻ ☿
12 30 ☽ ⊻ ♆
16 58 ☽ Q ♅
18 06 ☽ ⊻ ♄
23 22 ♀ Q ♃

APRIL 2019

LONGITUDES

Date	Sidereal time h m s	Sun ☉ ° ' "	Moon ☽ ° ' "	Moon ☽ 24.00 ° ' "	Mercury ☿ ° '	Venus ♀ ° '	Mars ♂ ° '	Jupiter ♃ ° '	Saturn ♄ ° '	Uranus ♅ ° '	Neptune ♆ ° '	Pluto ♇ ° '
01	00 38 15	11 ♈ 28 17	28 ♒ 37 13	04 ♓ 32 19	16 ♓ 48	06 ♓ 50	00 ♊ 49	24 ♐ 13	19 ♑ 52	01 ♉ 19	17 ♓ 06	23 ♑ 01
02	00 42 11	12 27 31	10 ♓ 28 54	16 ♓ 27 18	17 11	08 02	01 29	24 15	19 54	01 22	17 08	23 02
03	00 46 08	13 26 42	22 ♓ 27 50	28 ♓ 30 44	17 39	09 15	02 09	24 16	19 57	01 25	17 11	23 02
04	00 50 04	14 25 52	04 ♈ 36 09	10 ♈ 44 15	18 11	10 27	02 48	24 17	19 59	01 29	17 13	23 03
05	00 54 01	15 24 59	16 ♈ 55 06	23 ♈ 08 45	18 47	11 40	03 28	24 18	20 02	01 32	17 15	23 04
06	00 57 58	16 24 05	29 ♈ 25 13	05 ♉ 44 30	19 27	12 52	04 08	24 19	20 04	01 35	17 17	23 04
07	01 01 54	17 23 08	12 ♉ 06 36	18 ♉ 31 29	20 11	14 04	04 47	24 20	20 06	01 39	17 19	23 05
08	01 05 51	18 22 09	24 ♉ 59 11	01 ♊ 29 42	20 59	15 17	05 27	24 21	20 08	01 42	17 21	23 05
09	01 09 47	19 21 09	08 ♊ 03 05	14 ♊ 39 24	21 50	16 29	06 06	24 21	20 11	01 46	17 23	23 06
10	01 13 44	20 20 06	21 ♊ 18 46	28 ♊ 01 18	22 43	17 42	06 46	24 21	20 12	01 49	17 25	23 06
11	01 17 40	21 19 00	04 ♋ 47 09	11 ♋ 36 26	23 40	18 54	07 25	24 R 21	20 14	01 52	17 27	23 07
12	01 21 37	22 17 53	18 ♋ 29 20	25 ♋ 25 56	24 40	20 07	08 05	24 21	20 16	01 56	17 29	23 07
13	01 25 33	23 16 43	02 ♌ 26 18	09 ♌ 30 28	25 43	21 19	08 44	24 20	20 18	01 59	17 31	23 07
14	01 29 30	24 15 31	16 ♌ 38 20	23 ♌ 49 42	26 48	22 32	09 24	24 20	20 19	02 03	17 33	23 08
15	01 33 27	25 14 16	01 ♍ 04 16	08 ♍ 21 33	27 56	23 44	10 03	24 19	20 21	02 06	17 35	23 08
16	01 37 23	26 12 59	15 ♍ 40 59	23 ♍ 01 50	29 ♓ 06	24 57	10 42	24 18	20 22	02 09	17 36	23 08
17	01 41 20	27 11 40	00 ♎ 23 16	07 ♎ 44 20	00 ♈ 18	26 10	11 22	24 17	20 23	02 13	17 38	23 08
18	01 45 16	28 10 19	15 ♎ 04 06	22 ♎ 21 32	01 33	27 22	12 01	24 15	20 25	02 16	17 40	23 09
19	01 49 13	29 ♈ 08 56	29 ♎ 35 43	06 ♏ 45 45	02 50	28 35	12 40	24 14	20 26	02 20	17 42	23 09
20	01 53 09	00 ♉ 07 31	13 ♏ 50 52	20 ♏ 50 26	04 09	29 ♓ 47	13 20	24 12	20 27	02 23	17 44	23 09
21	01 57 06	01 06 04	27 ♏ 44 00	04 ♐ 31 15	05 30	01 ♈ 00	13 59	24 10	20 28	02 27	17 46	23 09
22	02 01 02	02 04 35	11 ♐ 12 04	17 ♐ 46 29	06 53	02 13	14 38	24 08	20 28	02 30	17 47	23 09
23	02 04 59	03 03 05	24 ♐ 14 39	00 ♑ 36 53	08 18	03 25	15 17	24 06	20 29	02 34	17 49	23 09
24	02 08 56	04 01 33	06 ♑ 53 35	13 ♑ 05 16	09 44	04 38	15 57	24 03	20 30	02 37	17 51	23 09
25	02 12 52	05 00 00	19 ♑ 12 31	25 ♑ 15 55	11 13	05 51	16 36	24 01	20 30	02 40	17 52	23 R 09
26	02 16 49	05 58 25	01 ♒ 16 09	07 ♒ 13 54	12 44	07 03	17 15	23 58	20 31	02 44	17 54	23 09
27	02 20 45	06 56 48	13 ♒ 09 50	19 ♒ 04 38	14 16	08 16	17 54	23 55	20 31	02 47	17 56	23 09
28	02 24 42	07 55 09	24 ♒ 59 00	00 ♓ 53 32	15 50	09 29	18 33	23 52	20 31	02 51	17 57	23 09
29	02 28 38	08 53 29	06 ♓ 48 52	12 ♓ 45 33	17 27	10 42	19 12	23 48	20 31	02 54	17 59	23 09
30	02 32 35	09 ♉ 51 48	18 ♓ 44 08	24 ♓ 45 04	19 ♈ 05	11 ♈ 54	19 ♊ 51	23 ♐ 45	20 ♑ 31	02 ♉ 58	18 ♓ 01	23 ♑ 09

MOON NODES AND LATITUDE

Date	Moon True ☊ ° '	Moon Mean ☊ ° '	Moon ☽ Latitude ° '
01	23 ♋ 38	22 ♋ 47	02 S 54
02	23 R 30	22 44	03 41
03	23 20	22 40	04 18
04	23 07	22 37	04 45
05	22 54	22 34	04 58
06	22 40	22 31	04 58
07	22 27	22 28	04 43
08	22 17	22 25	04 13
09	22 09	22 21	03 30
10	22 05	22 18	02 35
11	22 03	22 15	01 30
12	22 D 02	22 12	00 S 19
13	22 R 02	22 09	00 N 55
14	22 01	22 05	02 06
15	21 58	22 02	03 11
16	21 52	21 59	04 03
17	21 44	21 56	04 40
18	21 33	21 53	04 59
19	21 21	21 50	04 57
20	21 10	21 46	04 38
21	21 00	21 43	04 02
22	20 53	21 40	03 13
23	20 48	21 37	02 16
24	20 45	21 34	01 13
25	20 44	21 31	00 N 08
26	20 D 45	21 27	00 S 56
27	20 R 45	21 24	01 56
28	20 43	21 21	02 51
29	20 40	21 18	03 39
30	20 ♋ 34	21 ♋ 15	04 S 18

DECLINATIONS

Date	Sun ☉ ° '	Moon ☽ ° '	Mercury ☿ ° '	Venus ♀ ° '	Mars ♂ ° '	Jupiter ♃ ° '	Saturn ♄ ° '	Uranus ♅ ° '	Neptune ♆ ° '	Pluto ♇ ° '
01	04 N 32	14 S 40	05 S 32	09 S 52	21 N 13	22 S 41	21 S 34	11 N 28	05 S 59	21 S 43
02	04 55	11 02	05 35	09 27	21 21	22 41	21 33	11 29	05 58	21 43
03	05 18	06 57	05 35	09 02	21 30	22 41	21 33	11 30	05 57	21 43
04	05 41	02 S 32	05 33	08 37	21 38	22 41	21 33	11 32	05 57	21 43
05	06 04	02 N 03	05 29	08 12	21 46	22 41	21 32	11 33	05 56	21 43
06	06 27	06 37	05 23	07 46	21 54	22 41	21 32	11 34	05 55	21 43
07	06 50	10 58	05 14	07 20	22 02	22 41	21 32	11 35	05 54	21 43
08	07 12	14 55	05 04	06 54	22 09	22 41	21 31	11 36	05 54	21 43
09	07 34	18 12	04 52	06 28	22 17	22 41	21 31	11 37	05 53	21 43
10	07 57	20 35	04 38	06 02	22 24	22 41	21 31	11 39	05 52	21 43
11	08 19	21 51	04 22	05 35	22 31	22 41	21 31	11 40	05 51	21 43
12	08 41	21 51	04 05	05 08	22 38	22 41	21 30	11 41	05 51	21 43
13	09 03	20 30	03 45	04 41	22 44	22 41	21 30	11 42	05 50	21 43
14	09 24	17 51	03 25	04 14	22 51	22 41	21 30	11 43	05 49	21 43
15	09 46	14 04	03 02	03 47	22 57	22 41	21 30	11 45	05 48	21 43
16	10 07	09 23	02 38	03 19	23 03	22 41	21 30	11 46	05 48	21 43
17	10 28	04 N 08	02 13	02 52	23 09	22 41	21 30	11 47	05 47	21 43
18	10 49	01 S 20	01 46	02 24	23 14	22 41	21 29	11 48	05 46	21 43
19	11 10	06 41	01 18	01 57	23 20	22 41	21 29	11 49	05 46	21 43
20	11 31	11 34	00 49	01 29	23 25	22 41	21 29	11 51	05 45	21 44
21	11 51	15 43	00 S 18	01 01	23 30	22 40	21 29	11 52	05 44	21 44
22	12 12	18 56	00 N 14	00 33	23 35	22 40	21 29	11 53	05 44	21 44
23	12 32	21 03	00 47	00 S 05	23 40	22 40	21 29	11 54	05 43	21 44
24	12 52	22 03	01 21	00 N 23	23 44	22 40	21 29	11 55	05 42	21 44
25	13 11	21 56	01 56	00 51	23 48	22 40	21 29	11 56	05 42	21 44
26	13 31	20 47	02 33	01 19	23 53	22 40	21 29	11 58	05 41	21 44
27	13 50	18 43	03 10	01 47	23 56	22 40	21 29	11 59	05 40	21 44
28	14 09	15 53	03 48	02 15	24 00	22 40	21 29	12 00	05 40	21 44
29	14 28	12 24	04 28	02 43	24 04	22 40	21 29	12 01	05 39	21 44
30	14 N 46	08 S 24	05 N 08	03 N 11	24 N 07	22 S 39	21 S 29	12 N 02	05 S 39	21 S 45

ZODIAC SIGN ENTRIES

Date	h	m	Planets
01	14	48	☽ ♓
04	02	56	☽ ♈
06	13	06	☽ ♉
08	21	15	☽ ♊
11	03	31	☽ ♋
13	07	50	☽ ♌
15	10	14	☽ ♍
17	06	01	☿ ♈
17	11	22	☽ ♎
19	12	40	☽ ♏
20	08	55	☉ ♉
20	16	11	♀ ♈
21	15	59	☽ ♐
23	22	50	☽ ♑
26	09	27	☽ ♒
28	22	11	☽ ♓

LATITUDES

Date	Mercury ☿ ° '	Venus ♀ ° '	Mars ♂ ° '	Jupiter ♃ ° '	Saturn ♄ ° '	Uranus ♅ ° '	Neptune ♆ ° '	Pluto ♇ ° '
01	00 S 21	00 S 56	00 N 55	00 N 38	00 N 25	00 S 29	00 S 58	00 S 15
04	00 57	01 03	00 56	00 38	00 25	00 29	00 58	00 15
07	01 28	01 10	00 58	00 38	00 24	00 29	00 58	00 15
10	01 54	01 16	00 59	00 38	00 24	00 29	00 58	00 16
13	02 14	01 21	01 00	00 38	00 24	00 29	00 58	00 16
16	02 29	01 26	01 01	00 38	00 24	00 29	00 58	00 16
19	02 39	01 30	01 01	00 38	00 24	00 29	00 59	00 17
22	02 43	01 33	01 02	00 38	00 24	00 29	00 59	00 17
25	02 43	01 36	01 03	00 38	00 24	00 29	00 59	00 17
28	02 38	01 38	01 04	00 38	00 24	00 29	00 59	00 18
31	02 S 28	01 S 39	01 N 05	00 N 38	00 N 23	00 S 29	00 S 59	00 S 18

DATA

Julian Date	2458575
Delta T	+70 seconds
Ayanamsa	24° 07' 17"
Synetic vernal point	04° ♓ 59' 42"
True obliquity of ecliptic	23° 26' 09"

MOON'S PHASES, APSIDES AND POSITIONS ☽

Date	h	m	Phase	Longitude ° '	Eclipse Indicator
05	08	50	●	15 ♈ 17	
12	19	06	☽	22 ♋ 35	
19	11	12	○	29 ♎ 07	
26	22	18	☾	06 ♒ 23	

Day	h	m		
01	00	23	Apogee	
16	22	11	Perigee	
28	18	24	Apogee	
05	01	18	0N	
12	00	00	Max dec	22° N 01'
18	06	07	0S	
24	21	19	Max dec	22° S 08'

LONGITUDES

Date	Chiron ⚷ ° '	Ceres ⚳ ° '	Pallas ⚴ ° '	Juno ⚵ ° '	Vesta ⚶ ° '	Black Moon Lilith ⚸ ° '
01	02 ♈ 25	14 ♐ 05	22 ♎ 32	20 ♊ 50	28 ♓ 58	26 ♒ 25
11	03 ♈ 00	14 ♐ 15	19 ♎ 27	25 ♊ 34	03 ♈ 44	27 ♒ 31
21	03 ♈ 33	13 ♐ 47	16 ♎ 26	00 ♋ 23	08 ♈ 26	28 ♒ 38
31	04 ♈ 04	12 ♐ 40	13 ♎ 50	05 ♋ 15	13 ♈ 02	29 ♒ 45

All ephemeris data is given at 12.00 UT and the Moon's longitude is additionally given for 24.00 UT

ASPECTARIAN

h m Aspects

01 Monday
00 36 ☽ ⊻ ♇
03 02 ☽ ⚹ ♃
06 22 ☽ ⊥ ♄
07 14 ☽ ∠ ☉
12 49 ☽ ⊥ ♇
16 44 ☽ □ ♂
17 30 ☽ ⚹ ♅

02 Tuesday
00 42 ☽ ∠ ♄
03 09 ☽ ⊥ ☉
03 26 ☽ Q ♃
06 31 ☽ ☌ ♀
07 03 ☽ ∠ ♇
07 29 ♂ ⊻ ♅
09 13 ☽ # ♅
09 36 ☿ ☌ ♆
11 47 ♀ ∠ ♇
16 20 ☽ ⊻ ☉
22 39 ☽ ∥ ♀
23 53 ☽ ∠ ♅

03 Wednesday
01 25 ☽ ☌ ♆
01 58 ☽ ☌ ☿
06 58 ☽ ⚹ ♄
07 06 ☽ Q ♂
13 09 ☽ ⚹ ♇
15 36 ☽ □ ♃
17 31 ☽ ∥ ♆
17 55 ☽ ⊥ ♅
19 35 ☽ ∥ ☿
19 36 ♂ # ♄
20 20 ☽ # ☉

04 Thursday
04 15 ☉ # ☿
05 50 ☽ ⊻ ♅
06 51 ☽ Q ♄
08 16 ☽ ⚹ ♂
12 53 ☽ Q ♇

05 Friday
00 41 ☽ ⊻ ♀
01 14 ♂ # ♇
03 34 ☉ # ♆
08 50 ☽ ☌ ☉
12 38 ☽ ⊻ ♆
13 35 ☽ ⊥ ♀
15 09 ☽ ∠ ♂
15 48 ☽ ⊻ ☿
18 02 ☽ □ ♄
23 51 ☽ □ ♇

06 Saturday
00 13 ☽ ⊥ ♆
02 15 ☽ △ ♃
03 59 ☽ ⊥ ☿
05 36 ☽ # ☿
08 18 ☽ # ♆
08 43 ☽ ∠ ♀
09 24 ☽ ⊥ ♂
11 02 ☽ ∥ ☉
16 09 ☽ ☌ ♅
17 27 ☽ ∠ ♆
17 39 ☽ # ♀
21 26 ☽ ⊻ ♂
22 08 ☽ ∠ ☿

07 Sunday
06 47 ☽ ꝗ ♃
09 17 ☿ ⚹ ♄
10 10 ☉ ⊻ ♆
15 33 ☽ ∥ ♅
16 04 ☽ ⚹ ♀
21 46 ☽ ⚹ ♆
22 42 ☽ ⊻ ☉
23 39 ☽ ± ♃

08 Monday
00 18 ♂ ꝗ ♄
02 59 ☽ △ ♄
03 14 ☉ # ♀
04 04 ☽ ⚹ ☿
08 29 ☽ △ ♇
10 46 ☽ ⊥ ☉
10 48 ☽ ⊼ ♃
16 41 ☽ Q ♀
20 04 ☽ Q ♆

09 Tuesday
00 26 ☽ ⊻ ♅
03 44 ☽ Q ☿
04 42 ☽ ∠ ☉
06 44 ☽ ꝗ ♄
08 15 ☽ ☌ ♂
11 28 ☽ ⊥ ♅
12 05 ☽ ꝗ ♇
17 38 ♀ ∠ ♅
23 09 ☽ ± ♄

10 Wednesday
03 52 ☽ ∠ ♅
04 25 ☽ ± ♇
04 51 ☽ □ ♀
04 58 ☽ □ ♆
06 13 ♀ ☌ ♆
08 47 ☉ □ ♄
10 01 ☽ ⊼ ♄
10 06 ☽ ⚹ ☉
14 43 ☽ □ ☿
15 13 ☽ ⊼ ♇
17 00 ♃ St R
17 27 ☽ ☍ ♃
20 51 ♀ ∥ ♆
21 45 ☿ ⚹ ♇

11 Thursday
02 54 ☽ # ♄
06 49 ☽ ⚹ ♅
07 45 ☽ # ♇
09 12 ☽ Q ☉
16 53 ☽ ⊻ ♂

12 Friday
03 56 ☽ ⊥ ♂
04 02 ☽ Q ♅
04 18 ☿ □ ♃
06 00 ♂ ⊥ ♅
10 15 ☽ △ ♆
13 18 ♂ ꝗ ♇
15 05 ☽ △ ♀
15 05 ☽ ☍ ♄
15 08 ♀ ⚹ ♄
16 03 ☽ # ♇
19 06 ☽ □ ☉
20 01 ☽ ☍ ♇
20 20 ☽ ∠ ♂
20 55 ☽ # ♄
22 07 ☽ ⊼ ♃
23 33 ☽ △ ☿
23 38 ♂ # ♃

13 Saturday
08 07 ☉ □ ♇
08 25 ☽ ± ♃
11 14 ☽ □ ♅
12 08 ☽ ꝗ ♆
17 54 ☉ ⊥ ♆
18 25 ☿ ⊥ ♅
19 13 ☽ ꝗ ♀
23 13 ☽ ⚹ ♂
23 42 ☽ ꝗ ♃

14 Sunday
03 11 ☽ ꝗ ☿
03 25 ☽ ± ♆
11 48 ☽ ± ♀
13 31 ☽ ⊼ ♆
13 41 ☉ △ ♃
18 10 ☽ ⊼ ♄
19 33 ☽ ± ☿
20 20 ☽ Q ♂
22 07 ☉ ∠ ♂
22 45 ☽ ⊼ ♀
22 50 ☽ ⊼ ♇
23 52 ♀ ⚹ ♇

15 Monday
00 49 ☽ △ ♃
01 38 ☽ △ ☉
04 10 ☽ ± ♄
06 22 ☽ ⊼ ☿
08 48 ☽ ± ♇
13 42 ☽ △ ♅
17 52 ☿ Q ♂
19 03 ☽ ꝗ ♄
23 15 ♀ □ ♃
23 38 ☽ ꝗ ♇

16 Tuesday
00 18 ☽ ∥ ♅
03 28 ☽ □ ♂
04 10 ☽ ꝗ ☉
08 40 ☽ ∥ ☉
14 25 ☽ ꝗ ♅
15 09 ☽ ☍ ♆
19 40 ☽ △ ♄
19 56 ☽ ± ☉

17 Wednesday
00 10 ☽ △ ♇
02 03 ☽ □ ♃
04 29 ☽ ☍ ♀
04 37 ☽ # ♆
05 10 ☽ ± ♅
06 25 ☽ ⊼ ☉
11 51 ☽ ☍ ☿
13 09 ♀ ⊥ ♅
15 00 ☽ ⊼ ♅
18 06 ☽ # ♀
21 09 ☽ # ☿

18 Thursday
06 46 ☽ △ ♂
07 24 ☽ Q ♃
13 45 ☽ ∥ ☿
16 17 ☽ ⊼ ♆
16 19 ☽ ∥ ♀
20 48 ☽ □ ♄

19 Friday
01 18 ☽ □ ♇
02 12 ☽ ± ♆
02 17 ☿ ⊻ ♅
03 07 ☽ ⚹ ♃
04 27 ☿ Q ♄
07 45 ☽ ∥ ♆
08 39 ☽ ꝗ ♂
10 09 ☽ ⊼ ♀
11 12 ☽ ☍ ☉
16 35 ☽ ☍ ♅
17 12 ☽ ꝗ ♆
17 56 ☽ ⊼ ☿
21 06 ☽ ± ♀

20 Saturday
00 25 ☽ ± ♂
02 49 ☽ Q ♄
04 08 ☽ ∠ ♃
05 04 ☽ ± ☿
07 25 ☽ Q ♇
11 04 ☽ ⊼ ♂
11 41 ☽ # ☉
13 26 ☽ # ♅
13 45 ☽ ꝗ ♀
18 39 ☽ △ ♆
19 26 ☽ ⊥ ♃
22 02 ☽ ꝗ ☿
23 20 ☽ ⚹ ♄

21 Sunday
04 00 ☽ ⚹ ♇
05 47 ☽ ⊻ ♃
05 55 ☿ Q ♇
12 28 ☉ ∥ ♅
18 19 ☽ △ ♀
18 23 ☽ ⊼ ☉
20 21 ☽ ⊼ ♅
22 19 ☉ ⊻ ♀

22 Monday
01 41 ☽ ∠ ♄
03 18 ☽ △ ☿
05 48 ♂ ± ♄
05 55 ☽ ± ☉
06 30 ☽ ∠ ♇
07 06 ☽ ± ♅
17 14 ♀ Q ♄
17 57 ☽ ⊥ ♄
18 03 ♀ ⊻ ♅
18 35 ☽ ☍ ♂
19 40 ☿ # ♀
22 51 ☽ ⊥ ♇
23 07 ☉ ☌ ♅
23 33 ☽ ꝗ ♅
23 35 ☽ ꝗ ☉

23 Tuesday
00 03 ☽ □ ♆
05 00 ☽ ⊻ ♄
06 04 ☉ ∠ ♆
06 22 ♀ Q ♂
09 58 ☽ ⊻ ♇
11 43 ☽ ☌ ♃
19 26 ☽ ∥ ♄

24 Wednesday
00 59 ☽ ∥ ♇
03 46 ☽ △ ♅
06 02 ☽ △ ☉
07 12 ☽ □ ♀
09 59 ☽ Q ♆
18 14 ☽ □ ☿
18 47 ♇ St R
22 16 ♀ Q ♇

25 Thursday
06 35 ☽ ⊼ ♂
09 22 ☽ ⚹ ♆
14 33 ☽ ☌ ♄
17 57 ☽ ∥ ♇
19 05 ☽ ± ♂
19 48 ☽ ☌ ♇
21 28 ☽ ⊻ ♃
22 12 ☽ Q ♀
23 51 ☽ ∥ ♄

26 Friday
08 21 ♂ ± ♇
09 24 ☽ ⊥ ♃
10 45 ☽ Q ☿
14 05 ☽ ꝗ ♂
14 57 ☽ □ ♅
15 17 ☽ ∠ ♆
22 18 ☽ □ ☉

27 Saturday
00 58 ☽ ⚹ ♀
03 26 ☽ ∠ ♃
07 23 ♂ ∠ ♅
09 30 ☽ ⊥ ♆
13 03 ♂ □ ♆
14 35 ☽ ⚹ ☿
21 42 ☽ ⊻ ♆
22 11 ☽ △ ♂

28 Sunday
02 55 ☽ ⊻ ♄
03 33 ☽ Q ♅
08 16 ☽ ⊻ ♇
09 44 ☽ ⚹ ♃
10 52 ☽ ∠ ♀
14 04 ☽ Q ☉
15 07 ☽ ⊥ ♄
20 28 ☽ ⊥ ♇
23 24 ☽ # ☉

29 Monday
01 45 ☽ ∠ ☿
04 02 ☽ ⚹ ♅
07 13 ☽ ⊥ ♀
09 23 ☽ ∠ ♄
09 58 ☽ Q ♃
09 59 ☉ ꝗ ♃
14 23 ☽ # ♅
14 42 ☽ ∠ ♇
16 34 ☽ ⚹ ☉
20 08 ☿ ⊻ ♆
20 44 ☽ ⊻ ♀
22 49 ☽ ⊥ ☿

30 Tuesday
00 54 ♄ St R
10 26 ☽ ∠ ♅
10 33 ☽ ☌ ♆
12 47 ☽ ⊻ ☿
14 22 ☽ □ ♂
15 34 ☽ ⚹ ♄
20 48 ☽ ⚹ ♇
21 57 ☽ □ ♃

MAY 2019

LONGITUDES

Date	Sidereal time h m s	Sun ☉ ° ' "	Moon ☽ ° ' "	Moon ☽ 24.00 ° ' "	Mercury ☿ ° '	Venus ♀ ° '	Mars ♂ ° '	Jupiter ♃ ° '	Saturn ♄ ° '	Uranus ♅ ° '	Neptune ♆ ° '	Pluto ♇ ° '
01	02 36 31	10 ♉ 50 05	00 ♈ 48 46	06 ♈ 55 33	20 ♈ 44	13 ♈ 07	20 ♊ 31	23 ♐ 41	20 ♑ 31	03 ♉ 01	18 ♓ 02	23 ♑ 08
02	02 40 28	11 48 20	13 ♈ 05 42	19 ♈ 19 24	22 26	14 20	21 10	23 R 37	20 31	03 04	18 04	23 R 08
03	02 44 25	12 46 33	25 ♈ 36 47	01 ♉ 57 54	24 09	15 33	21 49	23 33	20 31	03 08	18 05	23 08
04	02 48 21	13 44 45	08 ♉ 22 44	14 ♉ 51 12	25 55	16 46	22 28	23 29	20 30	03 11	18 07	23 08
05	02 52 18	14 42 56	21 ♉ 23 12	27 ♉ 58 33	27 42	17 58	23 07	23 25	20 30	03 15	18 08	23 07
06	02 56 14	15 41 04	04 ♊ 37 04	11 ♊ 18 34	29 ♈ 31	19 11	23 46	23 20	20 29	03 18	18 10	23 07
07	03 00 11	16 39 11	18 ♊ 02 49	24 ♊ 49 39	01 ♉ 21	20 24	24 25	23 15	20 28	03 21	18 11	23 07
08	03 04 07	17 37 16	01 ♋ 38 52	08 ♋ 30 20	03 14	21 37	25 04	23 11	20 28	03 25	18 12	23 06
09	03 08 04	18 35 20	15 ♋ 23 56	22 ♋ 19 33	05 08	22 50	25 42	23 06	20 27	03 28	18 14	23 06
10	03 12 00	19 33 21	29 ♋ 17 06	06 ♌ 16 32	07 05	24 03	26 21	23 01	20 26	03 32	18 15	23 06
11	03 15 57	20 31 21	13 ♌ 17 46	20 ♌ 20 43	09 03	25 15	27 00	22 55	20 25	03 35	18 16	23 05
12	03 19 54	21 29 18	27 ♌ 25 16	04 ♍ 31 15	11 02	26 28	27 39	22 50	20 24	03 38	18 18	23 05
13	03 23 50	22 27 14	11 ♍ 38 28	18 ♍ 46 37	13 04	27 41	28 18	22 44	20 22	03 42	18 19	23 04
14	03 27 47	23 25 08	25 ♍ 55 21	03 ♎ 04 15	15 07	28 ♈ 54	28 57	22 39	20 21	03 45	18 20	23 04
15	03 31 43	24 23 00	10 ♎ 14 22	17 ♎ 20 27	17 12	00 ♉ 07	29 ♊ 36	22 33	20 20	03 48	18 21	23 03
16	03 35 40	25 20 50	24 ♎ 26 37	01 ♏ 30 41	19 18	01 20	00 ♋ 14	22 27	20 18	03 51	18 22	23 03
17	03 39 36	26 18 39	08 ♏ 32 01	15 ♏ 30 04	21 26	02 33	00 53	22 21	20 16	03 55	18 24	23 02
18	03 43 33	27 16 26	22 ♏ 24 16	29 ♏ 14 10	23 35	03 45	01 32	22 15	20 15	03 58	18 25	23 01
19	03 47 29	28 14 12	05 ♐ 59 23	12 ♐ 39 40	25 44	04 58	02 10	22 08	20 13	04 01	18 26	23 00
20	03 51 26	29 ♉ 11 56	19 ♐ 14 51	25 ♐ 44 53	27 ♉ 55	06 11	02 49	22 02	20 11	04 04	18 27	23 00
21	03 55 23	00 ♊ 09 39	02 ♑ 09 48	08 ♑ 29 48	00 ♊ 06	07 24	03 28	21 55	20 09	04 07	18 28	22 59
22	03 59 19	01 07 21	14 ♑ 45 08	20 ♑ 56 08	02 18	08 37	04 07	21 49	20 07	04 11	18 29	22 58
23	04 03 16	02 05 02	27 ♑ 03 14	03 ♒ 06 55	04 29	09 50	04 45	21 42	20 05	04 14	18 30	22 57
24	04 07 12	03 02 42	09 ♒ 07 42	15 ♒ 06 11	06 41	11 03	05 24	21 35	20 03	04 17	18 31	22 57
25	04 11 09	04 00 21	21 ♒ 02 57	26 ♒ 58 39	08 52	12 16	06 02	21 28	20 00	04 20	18 32	22 56
26	04 15 05	04 57 59	02 ♓ 53 55	08 ♓ 49 23	11 02	13 29	06 41	21 21	19 58	04 23	18 32	22 55
27	04 19 02	05 55 35	14 ♓ 45 41	20 ♓ 43 26	13 11	14 42	07 20	21 14	19 56	04 26	18 33	22 54
28	04 22 58	06 53 11	26 ♓ 43 14	02 ♈ 45 37	15 19	15 55	07 58	21 07	19 53	04 29	18 34	22 53
29	04 26 55	07 50 46	08 ♈ 51 07	15 ♈ 00 11	17 26	17 08	08 37	21 00	19 50	04 32	18 35	22 52
30	04 30 52	08 48 20	21 ♈ 13 12	27 ♈ 30 31	19 30	18 21	09 15	20 53	19 48	04 35	18 35	22 51
31	04 34 48	09 ♊ 45 53	03 ♉ 52 21	10 ♉ 18 52	21 ♊ 33	19 ♉ 34	09 ♋ 54	20 ♐ 45	19 ♑ 45	04 ♉ 38	18 ♓ 36	22 ♑ 50

DECLINATIONS

Date	Moon True ☊ ° '	Moon Mean ☊ ° '	Moon ☽ Latitude ° '	Sun ☉ ° '	Moon ☽ ° '	Mercury ☿ ° '	Venus ♀ ° '	Mars ♂ ° '	Jupiter ♃ ° '	Saturn ♄ ° '	Uranus ♅ ° '	Neptune ♆ ° '	Pluto ♇ ° '
01	20 ♋ 25	21 ♋ 11	04 S 45	15 N 04	04 S 02	05 N 49	03 N 39	24 N 10	22 S 39	21 S 29	12 N 04	05 S 38	21 S 45
02	20 R 15	21 08	05 00	15 22	00 N 34	06 31	04 07	24 13	22 39	21 29	12 05	05 37	21 45
03	20 03	21 05	05 01	15 40	05 14	07 13	04 35	24 16	22 39	21 29	12 06	05 37	21 45
04	19 51	21 02	04 47	15 58	09 46	07 57	05 03	24 18	22 39	21 29	12 07	05 36	21 45
05	19 41	20 59	04 18	16 15	13 57	08 41	05 30	24 21	22 39	21 29	12 08	05 36	21 45
06	19 32	20 56	03 35	16 32	17 32	09 25	05 58	24 23	22 38	21 29	12 09	05 35	21 45
07	19 25	20 52	02 39	16 49	20 16	10 10	06 26	24 25	22 38	21 30	12 10	05 35	21 46
08	19 22	20 49	01 33	17 05	21 53	10 56	06 53	24 27	22 38	21 30	12 12	05 34	21 46
09	19 20	20 46	00 S 21	17 21	22 12	11 41	07 20	24 28	22 38	21 30	12 13	05 34	21 46
10	19 D 21	20 43	00 N 53	17 37	21 10	12 27	07 47	24 30	22 38	21 30	12 14	05 33	21 46
11	19 21	20 40	02 05	17 53	18 49	13 14	08 14	24 31	22 37	21 30	12 15	05 33	21 46
12	19 R 21	20 37	03 09	18 08	15 19	14 00	08 41	24 32	22 37	21 31	12 16	05 32	21 46
13	19 20	20 33	04 02	18 23	10 56	14 46	09 08	24 32	22 37	21 31	12 17	05 32	21 47
14	19 16	20 30	04 41	18 38	05 55	15 31	09 34	24 33	22 37	21 31	12 18	05 31	21 47
15	19 10	20 27	05 02	18 52	00 N 35	16 17	10 00	24 33	22 36	21 31	12 20	05 31	21 47
16	19 03	20 24	05 04	19 06	04 S 45	17 01	10 26	24 34	22 36	21 31	12 21	05 31	21 47
17	18 54	20 21	04 48	19 20	09 48	17 45	10 52	24 34	22 36	21 32	12 22	05 30	21 47
18	18 46	20 17	04 15	19 33	14 15	18 28	11 17	24 33	22 35	21 32	12 23	05 30	21 48
19	18 39	20 14	03 28	19 46	17 53	19 10	11 43	24 33	22 35	21 33	12 24	05 29	21 48
20	18 33	20 11	02 31	19 59	20 30	19 50	12 08	24 32	22 35	21 33	12 25	05 29	21 48
21	18 30	20 08	01 27	20 11	21 58	20 29	12 33	24 32	22 35	21 33	12 26	05 29	21 48
22	18 28	20 05	00 N 20	20 23	22 17	21 06	12 57	24 31	22 34	21 34	12 27	05 28	21 49
23	18 D 29	20 02	00 S 46	20 35	21 30	21 41	13 21	24 29	22 34	21 34	12 28	05 28	21 49
24	18 30	19 58	01 49	20 46	19 44	22 14	13 45	24 28	22 34	21 34	12 29	05 28	21 49
25	18 31	19 55	02 47	20 57	17 07	22 45	14 09	24 26	22 33	21 35	12 30	05 27	21 49
26	18 32	19 52	03 37	21 07	13 49	23 13	14 32	24 25	22 33	21 35	12 31	05 27	21 49
27	18 R 31	19 49	04 18	21 18	09 58	23 39	14 55	24 23	22 32	21 35	12 32	05 27	21 50
28	18 29	19 46	04 48	21 27	05 43	24 02	15 17	24 21	22 32	21 36	12 33	05 27	21 50
29	18 26	19 42	05 06	21 37	01 S 11	24 22	15 40	24 18	22 32	21 36	12 34	05 26	21 50
30	18 20	19 39	05 10	21 46	03 N 29	24 40	16 01	24 16	22 31	21 37	12 35	05 26	21 50
31	18 ♋ 14	19 ♋ 36	04 S 59	21 N 55	08 N 07	24 N 55	16 N 23	24 N 13	22 S 31	21 S 37	12 N 36	05 S 26	21 S 51

ZODIAC SIGN ENTRIES

Date	h	m	Planets
01	10	24	☽ ♈
03	20	18	☽ ♉
06	03	40	☽ ♊
06	18	25	☿ ♉
08	09	06	☽ ♋
10	13	14	☽ ♌
12	16	22	☽ ♍
14	18	51	☽ ♎
15	09	46	♀ ♉
16	03	09	♂ ♋
16	21	26	☽ ♏
19	01	21	☽ ♐
21	07	56	☽ ♑
21	07	59	☉ ♊
21	10	52	☿ ♊
23	17	49	☽ ♒
26	06	07	☽ ♓
28	18	32	☽ ♈
31	04	43	☽ ♉

LATITUDES

Date	Mercury ☿ ° '	Venus ♀ ° '	Mars ♂ ° '	Jupiter ♃ ° '	Saturn ♄ ° '	Uranus ♅ ° '	Neptune ♆ ° '	Pluto ♇ ° '
01	02 S 28	01 S 39	01 N 05	00 N 38	00 N 23	00 S 29	00 S 59	00 S 18
04	02 13	01 40	01 05	00 38	00 23	00 29	00 59	00 18
07	01 53	01 40	01 06	00 38	00 23	00 29	00 59	00 18
10	01 30	01 39	01 06	00 38	00 23	00 29	00 59	00 19
13	01 03	01 38	01 07	00 37	00 23	00 29	00 59	00 19
16	00 33	01 36	01 07	00 37	00 23	00 29	01 00	00 19
19	00 S 02	01 33	01 08	00 37	00 23	00 29	01 00	00 20
22	00 N 30	01 30	01 08	00 37	00 22	00 29	01 00	00 20
25	00 59	01 26	01 09	00 37	00 22	00 29	01 00	00 20
28	01 25	01 22	01 09	00 37	00 22	00 29	01 00	00 21
31	01 N 45	01 S 17	01 N 09	00 N 36	00 N 22	00 S 29	01 S 00	00 S 21

DATA

Julian Date	2458605
Delta T	+70 seconds
Ayanamsa	24° 07' 20"
Synetic vernal point	04° ♓ 59' 39"
True obliquity of ecliptic	23° 26' 09"

MOON'S PHASES, APSIDES AND POSITIONS ☽

Date	h	m	Phase	Longitude ° '	Eclipse Indicator
04	22	45	●	14 ♉ 11	
12	01	12	☽	21 ♌ 03	
18	21	11	○	27 ♏ 39	
26	16	34	☾	05 ♓ 09	

Day	h	m		
13	21	58	Perigee	
26	13	27	Apogee	
02	09	06	0N	
09	05	45	Max dec	22° N 15'
15	14	37	0S	
22	06	39	Max dec	22° S 19'
29	18	05	0N	

LONGITUDES

Date	Chiron ⚷ ° '	Ceres ⚳ ° '	Pallas ⚴ ° '	Juno ⚵ ° '	Vesta ⚶ ° '	Black Moon Lilith ⚸ ° '
01	04 ♈ 04	12 ♐ 40	13 ♎ 50	05 ♋ 15	13 ♈ 02	29 ♒ 45
11	04 ♈ 33	11 ♐ 00	11 ♎ 53	10 ♋ 09	17 ♈ 34	00 ♓ 51
21	04 ♈ 58	08 ♐ 58	10 ♎ 44	15 ♋ 03	21 ♈ 59	01 ♓ 58
31	05 ♈ 19	06 ♐ 44	10 ♎ 23	19 ♋ 57	26 ♈ 17	03 ♓ 05

All ephemeris data is given at 12.00 UT and the Moon's longitude is additionally given for 24.00 UT

ASPECTARIAN

h m	Aspects
01 Wednesday	
01 17	☽ ∠ ☉
03 23	☽ ∥ ♆
03 42	☽ ⋕ ☿
04 28	☽ ⊥ ♅
05 43	☿ ⋕ ♆
06 37	☿ ⚹ ♂
08 50	☿ □ ♄
12 17	♂ ⊼ ♄
13 50	☽ ⋕ ♀
15 21	☽ Q ♄
16 22	☽ ⊻ ♅
20 30	☽ Q ♇
20 35	☽ ⊥ ☉
02 Thursday	
02 49	♅ ∠ ♆
03 56	☽ Q ♂
09 17	☽ ⊻ ☉
14 39	☽ ☌ ♀
21 36	☽ ⊻ ♆
21 51	☿ □ ♇
03 Friday	
02 17	☽ □ ♄
03 59	☿ △ ♃
04 22	☽ ⚹ ♂
07 17	☽ □ ♇
08 06	☽ △ ♃
08 19	☽ ∥ ♀
08 47	☽ ☌ ☿
09 06	☽ ⊥ ♆
11 03	☿ ⊥ ♆
14 01	☽ ⋕ ♆
04 Saturday	
00 23	☽ ∥ ☿
02 08	☽ ∠ ♆
02 15	☽ ☌ ♅
10 12	☽ ∠ ♂
12 12	☽ ⚼ ♃
22 45	☽ ☌ ☉
05 Sunday	
01 12	☽ ∥ ♅
03 45	☽ ⊥ ♂
04 45	☽ ± ♃
05 06	☽ ⊻ ♀
06 02	☽ ⚹ ♆
10 22	☽ △ ♄
12 29	♂ ⊼ ♇
15 10	☽ △ ♇
15 18	♀ ⊻ ♆
15 19	☽ ⊻ ♂
15 41	☽ ⊼ ♃
16 33	♀ ⋕ ♆
17 12	☽ ⊥ ♀
21 57	♂ ☍ ♃
06 Monday	
01 18	☽ ⊻ ☿
03 57	☽ Q ♆
04 05	☽ ∥ ☉
09 37	☽ ⊻ ♅
11 09	☽ ∠ ♀
13 33	☽ ⚼ ♄
13 52	☽ ⊥ ☿
18 17	☽ ⚼ ♇
20 27	☽ ⊥ ♅
07 Tuesday	
05 39	☽ ± ♄
08 31	☽ ∠ ☿
09 20	☽ ⊻ ☉
10 20	☽ ± ♇
12 15	☽ □ ♆
12 33	☽ ∠ ♅
13 26	♀ □ ♄
16 18	☽ ⊼ ♄
16 35	☽ ⚹ ♀
20 47	☽ ⊥ ☉
20 58	☽ ⊼ ♇
21 10	☽ ☍ ♃
23 50	☽ ☌ ♂
08 Wednesday	
01 51	☉ ± ♃
04 22	☽ ⋕ ♄
09 26	☽ ⋕ ♇
11 41	☿ ∠ ♆
13 50	☽ ∠ ☉
14 23	☿ ☌ ♅
15 07	☽ ⚹ ♅
15 13	☽ ⚹ ☿
15 47	☽ Q ♀
09 Thursday	
02 52	☉ ⚹ ♆
10 30	♃ ⊻ ♇
12 07	☽ Q ♅
15 30	☽ Q ☿
16 55	☽ △ ♆
16 56	♀ △ ♃
17 20	♀ □ ♇
17 57	☽ ⚹ ☉
20 44	☽ ☍ ♄
10 Friday	
01 15	☽ ⊼ ♃
01 20	☽ ☍ ♇
01 49	☽ ⋕ ♇
02 06	☽ □ ♀
04 47	☿ ∥ ♅
06 43	☽ ⊻ ♂
06 46	☽ ⋕ ♄
11 32	☽ ± ♃
16 12	☽ Q ☉
16 13	♀ ⊥ ♆
17 32	☽ ⊥ ♂
18 50	☽ ⚼ ♆
19 19	☽ □ ♅
22 56	☿ ⚼ ♃
11 Saturday	
02 52	☽ ⚼ ♃
03 33	☽ □ ☿
09 19	☉ △ ♄
09 41	☽ ∠ ♂
10 15	☽ ± ♆
18 41	☽ ∥ ☉
20 29	☽ ⊼ ♆
12 Sunday	
00 06	☽ ⊼ ♄
01 12	☽ □ ☉
04 16	☽ △ ♃
04 38	☽ ⊼ ♇
10 14	☽ △ ♀
10 16	☽ ± ♄
12 24	☽ ⚹ ♂
14 48	☽ ± ♇
18 33	☽ ∥ ☿
22 33	☽ △ ♅
13 Monday	
00 17	☉ ⊥ ♂
01 27	☽ ⚼ ♄
05 02	☽ ∥ ♅
05 59	☽ ⚼ ♇
09 38	☽ Q ♂
13 55	☽ ⚼ ♀
14 48	☽ △ ☿
15 59	☿ ∠ ♂
18 27	☉ ⊼ ♃
20 11	☽ ∥ ♀
23 14	☽ ☍ ♆
23 54	☽ ⚼ ♅
14 Tuesday	
02 39	☽ △ ♄
03 07	☉ △ ♇
06 27	☽ ± ♀
06 32	☽ □ ♃
07 12	☽ △ ♇
07 30	☽ △ ☉
13 48	☽ ⋕ ♆
13 58	♀ ⚹ ♂
15 04	☽ ± ♅
17 19	☽ □ ♂
17 28	☽ ⊼ ♀
20 14	☽ ⚼ ☿
15 Wednesday	
01 11	☽ ⊼ ♅
04 51	☿ ± ♃
10 30	☽ ⚼ ☉
12 33	☽ Q ♃
13 57	☽ ± ☿
16 Thursday	
01 20	☿ ⚹ ♆
01 44	☽ ⊼ ♆
01 48	☽ ⊼ ☿
02 46	☽ ± ☉
05 00	☽ □ ♄
08 39	☽ ⚹ ♃
09 37	☽ □ ♇
11 53	☽ ± ♆
13 39	☽ ⊼ ☉
15 30	☽ ∥ ♆
22 18	☽ △ ♂
23 09	☿ △ ♄
17 Friday	
00 47	☽ ☍ ♀
03 12	☽ ⚼ ♆
04 04	☽ ☍ ♅
09 59	☽ ∠ ♃
11 33	☽ Q ♄
16 17	☽ Q ♇
18 00	☽ ⋕ ♀
21 49	☿ ⊼ ♃
18 Saturday	
01 17	☽ ⚼ ♂
01 22	☽ ⊥ ♃
01 24	☽ ⋕ ♅
05 02	☽ △ ♆
05 04	♀ ∠ ♆
05 48	☿ △ ♇
08 15	☽ ⚹ ♄
11 43	☽ ⊻ ♃
13 04	☽ ⚹ ♇
14 26	☽ ☍ ☿
16 17	♀ ☌ ♅
17 45	☽ ± ♂
21 11	☽ ☍ ☉
19 Sunday	
04 52	☽ ⊼ ♂
08 28	☽ ⊼ ♅
10 00	☽ ⊼ ♀
10 37	☽ ∠ ♄
15 37	☽ ∠ ♇
18 50	☿ ⊥ ♂
19 15	☽ ± ♅
21 51	☽ ± ♀
20 Monday	
02 17	☽ ⋕ ☿
02 47	☽ ⊥ ♄
05 45	☽ ⋕ ☉
07 53	☽ ⊥ ♇
10 32	☽ □ ♆
11 41	☽ ⚼ ♅
13 43	☽ ⊻ ♄
15 56	☽ ⚼ ♀
17 05	☽ ☌ ♃
18 54	☽ ⊻ ♇
19 15	☉ ∥ ☿
21 Tuesday	
02 57	☽ ∥ ♄
03 20	♀ ⚼ ♃
05 21	♀ ∥ ♅
07 20	☽ ⊼ ☿
07 56	☽ ⊼ ☉
07 59	☽ ∥ ♇
13 07	☉ ☌ ☿
14 35	☽ ☍ ♂
15 43	☽ △ ♅
15 58	☿ Q ♆
19 40	☉ Q ♆
20 08	☽ Q ♆
20 11	☽ ± ☉
21 01	☽ ± ☿
22 58	☽ △ ♀
22 Wednesday	
14 46	♂ ⚹ ♅
14 52	☽ ⚼ ☉
17 59	☽ ⚼ ☿
19 14	☽ ⚹ ♆
22 22	☽ ☌ ♄
23 Thursday	
01 36	☽ ⊻ ♃
03 58	☽ ☌ ♇
05 38	☽ ∥ ♇
06 41	☿ ⋕ ♄
09 05	☿ ⊻ ♅
09 31	☽ ⋕ ☿
10 49	☽ ∥ ♄
13 16	☽ ⊥ ♃
16 04	☿ ⊻ ♂
17 15	☿ ⋕ ♇
18 22	☿ ⚼ ♄
22 49	☽ △ ☉
24 Friday	
00 44	☽ ⋕ ☉
00 46	☽ ∠ ♆
02 17	☽ □ ♅
04 07	☽ ⊼ ♂
06 01	☽ △ ☿
06 58	☽ ∠ ♃
16 17	☽ □ ♀
16 48	☽ ± ♂
18 48	☽ ⊥ ♆
25 Saturday	
01 48	☿ ⚼ ♇
02 36	☿ ⋕ ♃
06 54	☽ ⊻ ♆
09 54	☽ ⊻ ♄
11 59	☽ ⚼ ♂
12 51	☽ ⚹ ♃
14 37	☽ Q ♅
15 48	☽ ⊻ ♇
20 40	☉ ⊻ ♅
22 00	☽ ⊥ ♄
26 Sunday	
03 56	☽ ⊥ ♇
04 38	☿ ⊥ ♅
07 37	☽ ⋕ ♀
08 48	☽ Q ♀
12 00	☉ ⚼ ♄
12 55	☽ Q ♃
15 02	☽ ⚹ ♅
16 11	☽ ∠ ♄
16 34	☽ □ ☉
20 07	☽ △ ♂
20 22	☽ ⋕ ♅
22 09	☽ ∠ ♇
27 Monday	
08 07	☽ □ ☿
11 52	☽ ⚹ ♀
19 39	☽ ☌ ♆
20 08	☿ ± ♄
21 27	☽ ∠ ♅
21 44	♀ ± ♃
22 22	☽ ⚹ ♄
28 Tuesday	
00 54	☽ □ ♃
04 21	☽ ⚹ ♇
08 01	☽ Q ☉
13 27	☽ ∥ ♆
15 32	☽ ⊥ ♅
21 17	☽ ∠ ♀
22 14	☽ Q ♄
29 Wednesday	
03 29	☽ ⊻ ♅
03 52	☽ Q ☿
04 01	☿ ⊻ ♀
04 11	☽ Q ♇
05 43	☿ ± ♇
07 48	☿ ∥ ♂
09 51	☽ ⚹ ☉
10 40	☉ ⋕ ♄
11 30	☽ □ ♂
12 39	☉ ⚼ ♇
16 57	☽ ⊥ ♀
30 Thursday	
01 22	☿ □ ♆
05 53	☽ ⊻ ♀
06 56	☽ ⊻ ♆
08 03	☽ ⚹ ☿
09 16	☽ □ ♄
11 21	☽ △ ♃
13 00	☿ ∠ ♅
15 08	☽ □ ♇
15 19	☿ ⊼ ♄
16 50	♀ ⚹ ♆
17 22	☽ ∠ ☉
18 27	☽ ⊥ ♆
22 00	☽ ⋕ ♆
31 Friday	
00 08	☽ Q ♂
00 41	☉ ⋕ ♇
03 12	☿ ☍ ♃
11 30	☽ ∠ ♆
11 47	☽ ⊥ ☉
13 26	☽ ☌ ♅
15 26	♀ △ ♄
15 30	☽ ⚼ ♃
17 56	☽ ∠ ☿
22 02	☉ ⊻ ♂
23 49	☽ ⚹ ♂
23 52	☽ ⊻ ☉

JUNE 2019

LONGITUDES

Date	Sidereal time h m s	Sun ☉ ° ′ ″	Moon ☽ ° ′ ″	Moon ☽ 24.00 ° ′ ″	Mercury ☿ ° ′	Venus ♀ ° ′	Mars ♂ ° ′	Jupiter ♃ ° ′	Saturn ♄ ° ′	Uranus ♅ ° ′	Neptune ♆ ° ′	Pluto ♇ ° ′
01	04 38 45	10 ♊ 43 25	16 ♉ 50 09	23 ♉ 26 10	23 ♊ 34	20 ♉ 47	10 ♋ 32	20 ♐ 38	19 ♑ 42	04 ♉ 41	18 ♓ 37	22 ♑ 49
02	04 42 41	11 40 56	00 ♊ 06 49	06 ♊ 51 53	25 32	22 00	11 11	20 R 31	19 R 39	04 44	18 38	22 R 48
03	04 46 38	12 38 27	13 ♊ 41 07	20 ♊ 34 09	27 28	23 13	11 49	20 23	19 36	04 47	18 38	22 47
04	04 50 34	13 35 56	27 ♊ 30 37	04 ♋ 30 04	29 ♊ 22	24 26	12 28	20 16	19 33	04 50	18 39	22 46
05	04 54 31	14 33 25	11 ♋ 32 03	18 ♋ 36 07	01 ♋ 14	25 39	13 06	20 08	19 30	04 53	18 39	22 45
06	04 58 27	15 30 52	25 ♋ 41 48	02 ♌ 48 40	03 02	26 52	13 45	20 00	19 27	04 56	18 40	22 44
07	05 02 24	16 28 18	09 ♌ 56 18	17 ♌ 04 20	04 49	28 05	14 23	19 53	19 23	04 58	18 40	22 43
08	05 06 21	17 25 43	24 ♌ 12 24	01 ♍ 20 11	06 32	29 ♉ 19	15 02	19 45	19 20	05 01	18 41	22 42
09	05 10 17	18 23 07	08 ♍ 27 23	15 ♍ 33 45	08 13	00 ♊ 32	15 40	19 38	19 17	05 04	18 41	22 41
10	05 14 14	19 20 29	22 ♍ 39 01	29 ♍ 42 57	09 52	01 45	16 18	19 30	19 13	05 06	18 41	22 40
11	05 18 10	20 17 51	06 ♎ 45 18	13 ♎ 45 49	11 27	02 58	16 57	19 22	19 10	05 09	18 42	22 39
12	05 22 07	21 15 11	20 ♎ 44 17	27 ♎ 40 26	13 00	04 11	17 35	19 15	19 06	05 12	18 42	22 37
13	05 26 03	22 12 31	04 ♏ 34 02	11 ♏ 24 51	14 30	05 24	18 14	19 07	19 02	05 14	18 42	22 36
14	05 30 00	23 09 49	18 ♏ 12 38	24 ♏ 57 10	15 58	06 37	18 52	18 59	18 59	05 17	18 43	22 35
15	05 33 56	24 07 07	01 ♐ 38 14	08 ♐ 15 41	17 22	07 50	19 30	18 52	18 55	05 20	18 43	22 34
16	05 37 53	25 04 24	14 ♐ 49 23	21 ♐ 19 13	18 44	09 04	20 08	18 44	18 51	05 22	18 43	22 33
17	05 41 50	26 01 41	27 ♐ 45 10	04 ♑ 07 14	20 03	10 17	20 47	18 37	18 47	05 24	18 43	22 31
18	05 45 46	26 58 57	10 ♑ 25 28	16 ♑ 40 00	21 19	11 30	21 25	18 29	18 43	05 27	18 43	22 30
19	05 49 43	27 56 12	22 ♑ 51 01	28 ♑ 58 45	22 32	12 43	22 03	18 22	18 39	05 29	18 43	22 29
20	05 53 39	28 53 27	05 ♒ 03 29	11 ♒ 05 34	23 42	13 56	22 42	18 14	18 35	05 32	18 43	22 27
21	05 57 36	29 ♊ 50 41	17 ♒ 05 23	23 ♒ 03 22	24 49	15 10	23 20	18 07	18 31	05 34	18 43	22 26
22	06 01 32	00 ♋ 47 56	28 ♒ 59 59	04 ♓ 55 46	25 53	16 23	23 58	17 59	18 27	05 36	18 R 43	22 25
23	06 05 29	01 45 10	10 ♓ 51 14	16 ♓ 46 56	26 54	17 36	24 36	17 52	18 23	05 39	18 43	22 24
24	06 09 25	02 42 24	22 ♓ 43 27	28 ♓ 41 22	27 51	18 50	25 15	17 45	18 19	05 41	18 43	22 22
25	06 13 22	03 39 37	04 ♈ 41 17	10 ♈ 43 45	28 45	20 03	25 53	17 38	18 15	05 43	18 43	22 21
26	06 17 19	04 36 51	16 ♈ 49 22	22 ♈ 58 38	29 ♋ 35	21 16	26 31	17 31	18 11	05 45	18 43	22 19
27	06 21 15	05 34 05	29 ♈ 12 05	05 ♉ 30 10	00 ♌ 22	22 29	27 09	17 24	18 06	05 47	18 43	22 18
28	06 25 12	06 31 18	11 ♉ 53 17	18 ♉ 21 44	01 05	23 43	27 47	17 17	18 02	05 49	18 43	22 17
29	06 29 08	07 28 32	24 ♉ 55 47	01 ♊ 35 33	01 45	24 56	28 26	17 10	17 58	05 51	18 42	22 15
30	06 33 05	08 ♋ 25 46	08 ♊ 21 04	15 ♊ 12 15	02 ♌ 20	26 ♊ 10	29 ♋ 04	17 ♐ 03	17 ♑ 53	05 ♉ 53	18 ♓ 42	22 ♑ 14

Date	Moon True ☊ ° ′	Moon Mean ☊ ° ′	Moon ☽ Latitude ° ′
01	18 ♋ 08	19 ♋ 33	04 S 32
02	18 R 02	19 30	03 51
03	17 57	19 27	02 56
04	17 54	19 23	01 49
05	17 53	19 20	00 S 35
06	17 D 53	19 17	00 N 43
07	17 54	19 14	01 58
08	17 55	19 11	03 06
09	17 56	19 08	04 02
10	17 R 57	19 04	04 44
11	17 56	19 01	05 08
12	17 54	18 58	05 13
13	17 51	18 55	05 00
14	17 47	18 52	04 31
15	17 44	18 48	03 46
16	17 41	18 45	02 51
17	17 38	18 42	01 47
18	17 37	18 39	00 N 40
19	17 D 37	18 36	00 S 29
20	17 38	18 33	01 35
21	17 39	18 29	02 36
22	17 41	18 26	03 29
23	17 42	18 23	04 13
24	17 43	18 20	04 47
25	17 R 43	18 17	05 08
26	17 43	18 14	05 16
27	17 42	18 10	05 10
28	17 41	18 07	04 49
29	17 39	18 04	04 12
30	17 ♋ 38	18 ♋ 01	03 S 21

DECLINATIONS

Date	Sun ☉ ° ′	Moon ☽ ° ′	Mercury ☿ ° ′	Venus ♀ ° ′	Mars ♂ ° ′	Jupiter ♃ ° ′	Saturn ♄ ° ′	Uranus ♅ ° ′	Neptune ♆ ° ′	Pluto ♇ ° ′
01	22 N 03	12 N 30	25 N 07	16 N 44	24 N 07	22 S 30	21 S 38	12 N 37	05 S 26	21 S 51
02	22 11	16 24	25 16	17 04	24 07	22 30	21 38	12 38	05 25	21 51
03	22 19	19 32	25 23	17 25	24 04	22 30	21 39	12 39	05 25	21 52
04	22 26	21 36	25 28	17 44	24 01	22 29	21 39	12 40	05 25	21 52
05	22 33	22 22	25 30	18 04	23 57	22 29	21 40	12 41	05 25	21 52
06	22 39	21 42	25 29	18 22	23 53	22 28	21 40	12 42	05 25	21 52
07	22 45	19 39	25 27	18 41	23 49	22 28	21 41	12 43	05 24	21 53
08	22 51	16 22	25 22	18 59	23 45	22 27	21 41	12 44	05 24	21 53
09	22 56	12 08	25 16	19 16	23 41	22 27	21 42	12 45	05 24	21 53
10	23 00	07 15	25 08	19 33	23 36	22 26	21 42	12 46	05 24	21 54
11	23 05	02 N 02	24 58	19 49	23 31	22 26	21 43	12 47	05 24	21 54
12	23 09	03 S 15	24 46	20 05	23 26	22 25	21 44	12 47	05 24	21 54
13	23 12	08 19	24 33	20 21	23 21	22 25	21 44	12 48	05 24	21 54
14	23 16	12 55	24 18	20 35	23 16	22 24	21 45	12 49	05 24	21 55
15	23 18	16 47	24 03	20 50	23 11	22 24	21 45	12 50	05 24	21 55
16	23 21	19 45	23 46	21 03	23 05	22 24	21 46	12 51	05 24	21 55
17	23 23	21 38	23 28	21 16	22 59	22 23	21 47	12 52	05 24	21 56
18	23 24	22 22	23 09	21 29	22 54	22 22	21 47	12 52	05 24	21 56
19	23 25	21 58	22 50	21 41	22 47	22 22	21 48	12 53	05 24	21 56
20	23 26	20 32	22 29	21 52	22 41	22 21	21 48	12 54	05 24	21 57
21	23 26	18 11	22 08	22 03	22 35	22 21	21 49	12 55	05 24	21 57
22	23 26	15 05	21 47	22 13	22 28	22 20	21 50	12 56	05 24	21 57
23	23 25	11 24	21 25	22 22	22 21	22 20	21 50	12 56	05 24	21 57
24	23 24	07 17	21 03	22 31	22 15	22 19	21 51	12 57	05 24	21 58
25	23 23	02 S 51	20 41	22 39	22 08	22 19	21 52	12 58	05 24	21 58
26	23 21	01 N 44	20 19	22 47	22 00	22 18	21 52	12 58	05 24	21 58
27	23 19	06 21	19 57	22 54	21 53	22 18	21 53	12 59	05 24	21 59
28	23 16	10 49	19 35	23 00	21 45	22 17	21 53	13 00	05 24	21 59
29	23 13	14 55	19 13	23 06	21 38	22 17	21 54	13 00	05 24	21 59
30	23 N 10	18 N 23	18 N 52	23 N 10	21 N 30	22 S 16	21 S 55	13 N 01	05 S 25	22 S 00

ZODIAC SIGN ENTRIES

Date	h	m	Planets
02	11	48	☽ ♊
04	16	17	☽ ♋
04	20	04	☿ ♋
06	19	16	☽ ♌
08	21	45	☽ ♍
09	01	37	♀ ♊
11	00	29	☽ ♎
13	04	02	☽ ♏
15	09	03	☽ ♐
17	16	13	☽ ♑
20	02	01	☽ ♒
21	15	54	☉ ♋
22	14	01	☽ ♓
25	02	38	☽ ♈
27	00	19	☿ ♌
27	13	32	☽ ♉
29	21	09	☽ ♊

LATITUDES

Date	Mercury ☿ ° ′	Venus ♀ ° ′	Mars ♂ ° ′	Jupiter ♃ ° ′	Saturn ♄ ° ′	Uranus ♅ ° ′	Neptune ♆ ° ′	Pluto ♇ ° ′
01	01 N 50	01 S 16	01 N 09	00 N 36	00 N 22	00 S 29	01 S 00	00 S 21
04	02 02	01 10	01 10	00 36	00 22	00 29	01 00	00 21
07	02 06	01 05	01 10	00 35	00 22	00 29	01 00	00 22
10	02 04	00 59	01 10	00 35	00 21	00 29	01 01	00 22
13	01 55	00 52	01 10	00 35	00 21	00 29	01 01	00 22
16	01 39	00 46	01 10	00 34	00 21	00 29	01 01	00 23
19	01 17	00 39	01 11	00 34	00 21	00 29	01 01	00 23
22	00 50	00 32	01 11	00 33	00 20	00 29	01 01	00 23
25	00 N 17	00 25	01 11	00 33	00 20	00 30	01 01	00 23
28	00 S 20	00 17	01 11	00 32	00 20	00 30	01 01	00 24
31	01 S 02	00 S 10	01 N 11	00 N 32	00 N 20	00 S 30	01 S 01	00 S 24

DATA

Julian Date	2458636
Delta T	+70 seconds
Ayanamsa	24° 07′ 25″
Synetic vernal point	04° ♓ 59′ 35″
True obliquity of ecliptic	23° 26′ 09″

MOON'S PHASES, APSIDES AND POSITIONS ☽

Date	h	m	Phase	Longitude ° ′	Eclipse Indicator
03	10	02	●	12 ♊ 34	
10	05	59	☽	19 ♍ 06	
17	08	31	○	25 ♐ 53	
25	09	46	☾	03 ♈ 34	

Day	h	m		
07	23	10	Perigee	
23	07	45	Apogee	
05	12	59	Max dec	22° N 22′
11	21	10	0S	
18	15	29	Max dec	22° S 23′
26	02	59	0N	

LONGITUDES

Date	Chiron ⚷ ° ′	Ceres ⚳ ° ′	Pallas ⚴ ° ′	Juno ⚵ ° ′	Vesta ⚶ ° ′	Black Moon Lilith ⚸ ° ′
01	05 ♈ 21	06 ♐ 31	10 ♎ 23	20 ♋ 27	26 ♈ 42	03 ♓ 12
11	05 ♈ 37	04 ♐ 24	10 ♎ 51	25 ♋ 20	00 ♉ 51	04 ♓ 19
21	05 ♈ 48	02 ♐ 37	11 ♎ 58	00 ♌ 11	04 ♉ 51	05 ♓ 25
31	05 ♈ 55	01 ♐ 20	13 ♎ 40	05 ♌ 01	08 ♉ 40	06 ♓ 32

All ephemeris data is given at 12.00 UT and the Moon's longitude is additionally given for 24.00 UT

ASPECTARIAN

h m Aspects

01 Saturday
03 14 ☿ ⊼ ♇
08 00 ☽ ± ♃
09 20 ♀ ⊼ ♃
11 00 ☉ ⊥ ♅
12 39 ☽ ∥ ♅
13 34 ☽ ⊥ ☿
15 15 ☽ ⚹ ♆
17 12 ☽ △ ♄
18 52 ☽ ⊼ ♃
19 55 ☽ ☌ ♀
22 53 ☽ △ ♇

02 Sunday
02 22 ☽ ⊻ ☿
04 36 ☽ ∠ ♂
12 55 ☽ Q ♆
17 05 ☽ ∥ ♀
20 03 ☽ bQ ♄
20 15 ☽ ⊻ ♅
21 28 ☽ ⊥ ♂

03 Monday
01 39 ☽ bQ ♇
03 42 ♀ △ ♇
06 53 ☽ ⊥ ♅
08 34 ☽ ⊻ ♂
10 02 ☽ ☌ ☉
11 51 ☽ ± ♄
17 25 ☽ ± ♇
20 39 ☽ □ ♆
22 17 ☽ ⊼ ♄
22 40 ☽ ∠ ♅
23 35 ☽ ☍ ♃

04 Tuesday
03 50 ☽ ⊼ ♇
06 11 ☽ ⊻ ♀
10 49 ☉ ± ♄
12 57 ☽ # ♄
15 42 ☽ ☌ ☿
16 53 ☽ # ♇
17 31 ☽ ⊥ ♀
23 01 ☉ # ♃

05 Wednesday
00 36 ☽ ⚹ ♅
10 22 ☽ ∠ ♀
14 48 ☽ ☌ ♂
17 31 ☽ ⊻ ☉
21 07 ☽ Q ♅

06 Thursday
00 06 ☽ △ ♆
01 28 ☽ ☍ ♄
02 28 ☽ ⊼ ♃
04 25 ☽ ⊥ ☉
05 03 ☿ ⊥ ♀
07 00 ☽ ☍ ♇
08 46 ☽ # ♇
12 30 ☽ # ♄
12 31 ☽ ± ♃
14 10 ☽ ⚹ ♀
20 43 ☽ ∠ ☉
23 19 ♃ bQ ♅

07 Friday
01 26 ☽ bQ ♆
02 09 ☽ ⊻ ☿
03 34 ☽ bQ ♃
03 37 ☽ □ ♅
12 17 ☽ Q ♀
13 40 ☽ ⊥ ☿
14 16 ☿ ⚹ ♅
16 36 ☽ ± ♆
18 07 ☉ ± ♇
19 16 ☽ ∥ ♀
19 50 ☽ ⊻ ♂
23 47 ☽ ⚹ ☉

08 Saturday
02 42 ☽ ⊼ ♆
03 50 ☽ ⊼ ♄
04 34 ☽ △ ♃
06 24 ☽ ⊥ ♂
06 54 ☽ ∠ ☿
09 28 ☽ ⊼ ♇
13 53 ☽ ± ♄
19 33 ☽ ± ♇
21 23 ☽ □ ♀
21 25 ☽ Q ☉
22 15 ☽ ∠ ♂

09 Sunday
04 59 ☽ bQ ♄
06 16 ☽ △ ♅
08 49 ☽ ∥ ♅
10 42 ☽ bQ ♇
11 33 ☽ ⚹ ☿
15 07 ♀ Q ♆
17 48 ♀ ∠ ♂
19 34 ☉ □ ♆

10 Monday
00 45 ☽ ⚹ ♂
05 17 ☽ ☍ ♆
05 59 ☽ □ ☉
06 12 ☽ △ ♄
06 42 ☽ □ ♃
07 41 ☽ bQ ♅
09 04 ☉ ⊼ ♄
10 29 ☽ Q ☿
12 01 ☽ △ ♇
15 28 ☉ ☍ ♃
20 38 ☽ # ♆
22 04 ☽ Q ♂
23 00 ☽ ± ♅

11 Tuesday
04 55 ☽ △ ♀
08 11 ☉ ∠ ♅
09 15 ☽ ⊼ ♅
13 03 ☽ Q ♃
20 18 ♂ Q ♅
21 03 ☽ □ ☿

12 Wednesday
06 18 ☽ □ ♂
08 29 ☽ ⊼ ♆
09 04 ☽ bQ ♀
09 11 ☽ □ ♄
09 27 ☽ ⚹ ♃
10 25 ♀ bQ ♄
12 57 ☽ △ ☉
15 15 ☽ □ ♇
18 51 ☽ ± ♆
21 57 ☽ ∥ ♆

13 Thursday
02 08 ☽ ± ♀
08 41 ♀ ⊻ ♅
10 30 ☽ bQ ♆
11 13 ☽ ∠ ♃
13 11 ☽ ☍ ♅
13 36 ☽ ⊼ ♀
16 18 ☽ Q ♄
16 58 ☽ bQ ☉
21 45 ☉ ⊼ ♇
22 34 ☽ Q ♇

14 Friday
02 51 ☽ ⊥ ♃
06 11 ♂ △ ♆
07 33 ☽ △ ☿
10 00 ☽ ± ☉
11 29 ☽ # ♅
12 53 ☽ △ ♆
13 13 ☽ △ ♂
13 21 ☽ ⚹ ♄
13 22 ☽ ⊻ ♃
13 51 ☉ ∥ ♂
15 50 ♂ ☍ ♄
15 53 ♂ ⊼ ♃
16 28 ♃ ⊻ ♄
19 46 ☽ ⚹ ♇
21 29 ☽ ⊼ ☉

15 Saturday
06 39 ♀ bQ ♇
11 12 ☿ Q ♅
13 29 ☽ bQ ☿
16 05 ☽ ∠ ♄
17 26 ☽ bQ ♂
18 41 ☽ ⊼ ♅
22 43 ☽ ∠ ♇

16 Sunday
00 23 ☽ ☍ ♀
05 39 ☽ ± ♅
07 44 ☽ ± ☿
08 24 ☽ ⊥ ♄
10 41 ☽ ± ♂
11 43 ☿ △ ♆
12 02 ☿ ⊼ ♃
14 00 ☿ ☍ ♄
15 10 ☽ ⊥ ♇
15 22 ♃ □ ♆
19 08 ☽ ☌ ♃
19 11 ☽ □ ♆
19 23 ☽ ⊻ ♄
20 02 ☽ ⊼ ☿
22 16 ☽ bQ ♅
22 19 ☽ ⊼ ♂

17 Monday
02 15 ☽ ⊻ ♇
05 13 ☽ # ♀
08 31 ☽ ☍ ☉
14 50 ☽ ∥ ♄
18 08 ☽ ∥ ♇
18 23 ☉ ∥ ☿

18 Tuesday
02 29 ☽ △ ♅
04 56 ☽ Q ♆
10 58 ♀ ⊥ ♅
11 47 ♄ ⚹ ♆
12 44 ☽ ∥ ♃
14 17 ☽ ⊼ ♀
16 04 ☿ ☌ ♂
18 33 ☽ ∥ ♃

19 Wednesday
03 06 ☽ ± ♀
03 22 ☽ ⊻ ♃
03 53 ☽ ☌ ♄
03 59 ☽ ⚹ ♆
10 22 ☽ ☍ ♂
10 47 ♀ ± ♄
10 56 ☿ ☍ ♇
11 17 ☽ ☌ ♇
11 19 ☽ ☍ ☿
12 56 ☽ ∥ ♇
14 55 ☽ ⊥ ♃
15 39 ☿ ∥ ♂
16 08 ☽ ∥ ♄
17 40 ☽ # ♀
22 35 ☽ bQ ♀
22 48 ☽ ⊼ ☉

20 Thursday
03 26 ♂ ☍ ♇
03 41 ♀ # ♄
08 26 ☽ ∠ ♃
09 21 ☽ ∠ ♆
11 38 ☽ ± ☉
12 56 ☽ □ ♅
21 25 ☿ # ♃
22 12 ☿ ± ♃
22 18 ♀ # ♇

21 Friday
03 15 ☽ ⊥ ♆
07 07 ☽ bQ ☉
07 42 ☽ △ ♀
14 02 ☽ ⚹ ♃
14 35 ♆ St R
14 52 ☽ ⊻ ♄
15 17 ☽ ⊻ ♆
16 25 ☿ ∥ ♀
22 44 ☽ ⊻ ♇

22 Saturday
00 58 ☿ # ♇
01 04 ☽ Q ♅
01 16 ☽ ⊼ ♂
02 52 ☽ ⊥ ♄
05 06 ☽ ⊼ ☿
09 21 ☿ # ♄
10 49 ☽ ⊥ ♇
12 37 ♀ ± ♇
12 41 ♂ ± ♃
13 59 ☽ Q ♃
14 04 ☽ ± ♂
15 57 ☽ △ ☉
18 23 ☽ ± ☿
20 58 ☽ ∠ ♄

23 Sunday
01 25 ☽ ⚹ ♅
02 25 ☽ # ♅
05 00 ☽ ∠ ♇
06 14 ♀ # ♃
09 20 ☽ bQ ♂
10 53 ♀ ∥ ♂
14 18 ☽ bQ ☿
16 45 ♀ ☍ ♃
18 01 ♂ # ♃

24 Monday
02 03 ☽ □ ♃
02 32 ♀ ⊼ ♄
03 09 ☽ ⚹ ♄
03 14 ☽ □ ♀
03 55 ☽ ☌ ♆
07 52 ☽ ∠ ♅
09 58 ♀ □ ♆
11 17 ☽ ⚹ ♇
17 22 ☽ △ ♂
22 21 ☽ ∥ ♆
23 10 ☽ △ ☿

25 Tuesday
02 02 ☽ ⊥ ♅
03 10 ☽ Q ♄
05 09 ♀ ⊥ ♂
09 46 ☽ □ ☉
11 19 ☽ Q ♇
14 03 ☽ ⊻ ♅
19 26 ☽ Q ♀

26 Wednesday
01 34 ♀ ∠ ♅
13 20 ☽ △ ♃
14 38 ☽ □ ♄
15 42 ☽ ⊻ ♆
17 51 ♂ # ♇
21 38 ☽ ⚹ ♀
22 43 ☽ □ ♇

27 Thursday
00 14 ☽ Q ☉
03 22 ☽ ⊥ ♆
07 03 ☽ # ♆
07 51 ☽ □ ♂
08 20 ♀ ⊼ ♇
12 21 ♂ # ♄
14 23 ☽ □ ☿
17 45 ☉ ⚹ ♅
18 03 ☽ bQ ♃
20 36 ☽ ∠ ♆

28 Friday
00 34 ☽ ☌ ♅
01 06 ☽ ⚹ ☉
05 26 ☽ ∠ ♀
10 53 ☽ ± ♃
19 38 ☽ Q ♂
21 55 ☽ ⊼ ♃
23 20 ☽ △ ♄
23 56 ☽ ⊥ ♀

29 Saturday
00 26 ☽ ∥ ♅
00 38 ☽ ⚹ ♆
02 04 ☽ Q ☿
07 09 ☽ △ ♇
07 11 ☽ ∠ ☉
12 01 ☽ ⊻ ♀
18 38 ☽ ⚹ ♂
22 25 ☽ Q ♆

30 Sunday
00 42 ☽ ⊥ ☉
00 51 ☽ ⚹ ☿
02 06 ☿ bQ ♃
02 22 ☽ bQ ♄
07 38 ☽ ⊻ ♅
10 02 ☽ bQ ♇
10 51 ☉ ∥ ♀
12 09 ☽ ⊻ ☉
15 23 ☽ ∥ ☿
18 11 ☽ ± ♄
18 14 ☽ ⊥ ♅
18 15 ☿ ⊥ ♀
22 30 ☽ ∠ ♂

JULY 2019

LONGITUDES

Date	Sidereal time h m s	Sun ☉ ° ' "	Moon ☽ ° ' "	Moon ☽ 24.00 ° ' "	Mercury ☿ ° '	Venus ♀ ° '	Mars ♂ ° '	Jupiter ♃ ° '	Saturn ♄ ° '	Uranus ♅ ° '	Neptune ♆ ° '	Pluto ♇ ° '
01	06 37 01	09 ♋ 23 00	22 ♊ 08 54	29 ♊ 10 41	02 ♌ 51	27 ♊ 23	29 ♋ 42	16 ♐ 57	17 ♑ 49	05 ♉ 55	18 ♓ 42	22 ♑ 12
02	06 40 58	10 20 14	06 ♋ 17 09	13 ♋ 27 45	03 19	28 36	00 ♌ 20	16 R 50	17 R 45	05 57	18 R 42	22 R 11
03	06 44 54	11 17 27	20 ♋ 41 50	27 ♋ 58 42	03 41	29 ♊ 50	00 58	16 44	17 40	05 59	18 41	22 10
04	06 48 51	12 14 41	05 ♌ 17 34	12 ♌ 37 38	04 00	01 ♋ 03	01 37	16 38	17 36	06 01	18 41	22 08
05	06 52 48	13 11 54	19 ♌ 58 05	27 ♌ 18 10	04 14	02 17	02 15	16 31	17 32	06 03	18 40	22 07
06	06 56 44	14 09 07	04 ♍ 37 08	11 ♍ 54 20	04 23	03 30	02 53	16 25	17 27	06 05	18 40	22 05
07	07 00 41	15 06 20	19 ♍ 09 10	26 ♍ 21 07	04 27	04 44	03 31	16 19	17 23	06 06	18 39	22 04
08	07 04 37	16 03 33	03 ♎ 29 48	10 ♎ 34 53	04 R 27	05 57	04 09	16 14	17 18	06 08	18 39	22 02
09	07 08 34	17 00 45	17 ♎ 36 08	24 ♎ 33 23	04 22	07 11	04 47	16 08	17 14	06 10	18 38	22 01
10	07 12 30	17 57 57	01 ♏ 26 33	08 ♏ 15 37	04 13	08 24	05 25	16 02	17 09	06 11	18 38	22 00
11	07 16 27	18 55 09	15 ♏ 00 37	21 ♏ 41 35	03 59	09 38	06 03	15 57	17 05	06 13	18 37	21 58
12	07 20 23	19 52 22	28 ♏ 18 36	04 ♐ 51 49	03 40	10 52	06 42	15 52	17 01	06 14	18 36	21 57
13	07 24 20	20 49 34	11 ♐ 21 19	17 ♐ 47 16	03 17	12 05	07 20	15 47	16 56	06 16	18 36	21 55
14	07 28 17	21 46 46	24 ♐ 09 47	00 ♑ 29 01	02 51	13 19	07 58	15 42	16 52	06 17	18 35	21 54
15	07 32 13	22 43 58	06 ♑ 45 09	12 ♑ 58 18	02 20	14 32	08 36	15 37	16 47	06 18	18 34	21 52
16	07 36 10	23 41 11	19 ♑ 08 39	25 ♑ 16 22	01 47	15 46	09 14	15 32	16 43	06 20	18 34	21 51
17	07 40 06	24 38 24	01 ♒ 21 40	07 ♒ 24 44	01 10	17 00	09 52	15 28	16 39	06 21	18 33	21 49
18	07 44 03	25 35 37	13 ♒ 25 49	19 ♒ 25 08	00 ♌ 32	18 13	10 30	15 23	16 34	06 22	18 32	21 48
19	07 47 59	26 32 51	25 ♒ 23 00	01 ♓ 19 41	29 ♋ 52	19 27	11 08	15 19	16 30	06 23	18 31	21 46
20	07 51 56	27 30 05	07 ♓ 15 32	13 ♓ 10 56	29 11	20 41	11 46	15 15	16 26	06 24	18 30	21 45
21	07 55 52	28 27 20	19 ♓ 06 15	25 ♓ 01 56	28 30	21 54	12 24	15 11	16 21	06 25	18 29	21 44
22	07 59 49	29 ♋ 24 35	00 ♈ 58 25	06 ♈ 56 13	27 49	23 08	13 02	15 07	16 17	06 26	18 29	21 42
23	08 03 46	00 ♌ 21 52	12 ♈ 55 49	18 ♈ 57 45	27 10	24 22	13 40	15 04	16 13	06 27	18 28	21 41
24	08 07 42	01 19 09	25 ♈ 02 33	01 ♉ 10 47	26 32	25 36	14 19	15 00	16 09	06 28	18 27	21 39
25	08 11 39	02 16 27	07 ♉ 22 58	13 ♉ 39 39	25 58	26 50	14 57	14 57	16 05	06 29	18 26	21 38
26	08 15 35	03 13 46	20 ♉ 01 19	26 ♉ 28 27	25 26	28 03	15 35	14 54	16 00	06 30	18 25	21 36
27	08 19 32	04 11 06	03 ♊ 01 27	09 ♊ 40 38	24 59	29 ♋ 17	16 13	14 51	15 56	06 31	18 24	21 35
28	08 23 28	05 08 27	16 ♊ 26 15	23 ♊ 18 25	24 36	00 ♌ 31	16 51	14 49	15 52	06 32	18 22	21 34
29	08 27 25	06 05 49	00 ♋ 17 08	07 ♋ 22 14	24 18	01 45	17 29	14 46	15 48	06 32	18 21	21 32
30	08 31 21	07 03 12	14 ♋ 33 23	21 ♋ 50 08	24 05	02 59	18 07	14 44	15 44	06 33	18 20	21 31
31	08 35 18	08 ♌ 00 36	29 ♋ 11 47	06 ♌ 37 33	23 ♋ 58	04 ♌ 13	18 ♌ 45	14 ♐ 42	15 ♑ 40	06 ♉ 34	18 ♓ 19	21 ♑ 29

Date	Moon True ☊ ° '	Moon Mean ☊ ° '	Moon ☽ Latitude ° '
01	17 ♋ 37	17 ♋ 58	02 S 17
02	17 R 37	17 54	01 S 02
03	17 D 37	17 51	00 N 17
04	17 37	17 48	01 37
05	17 37	17 45	02 50
06	17 38	17 42	03 52
07	17 38	17 39	04 39
08	17 38	17 35	05 08
09	17 R 38	17 32	05 17
10	17 D 38	17 29	05 08
11	17 38	17 26	04 42
12	17 38	17 23	04 01
13	17 39	17 20	03 08
14	17 39	17 16	02 07
15	17 39	17 13	01 N 00
16	17 R 39	17 10	00 S 08
17	17 39	17 07	01 15
18	17 38	17 04	02 18
19	17 37	17 00	03 14
20	17 36	16 57	04 01
21	17 35	16 54	04 38
22	17 33	16 51	05 03
23	17 32	16 48	05 15
24	17 32	16 45	05 14
25	17 D 31	16 41	04 58
26	17 32	16 38	04 27
27	17 33	16 35	03 42
28	17 34	16 32	02 44
29	17 36	16 29	01 34
30	17 36	16 26	00 17
31	17 ♋ 36	16 ♋ 22	01 N 04

DECLINATIONS

Date	Sun ☉ ° '	Moon ☽ ° '	Mercury ☿ ° '	Venus ♀ ° '	Mars ♂ ° '	Jupiter ♃ ° '	Saturn ♄ ° '	Uranus ♅ ° '	Neptune ♆ ° '	Pluto ♇ ° '
01	23 N 06	20 N 56	18 N 31	23 N 15	21 N 22	22 S 16	21 S 55	13 N 02	05 S 25	22 S 00
02	23 02	22 15	18 10	23 18	21 14	22 16	21 56	13 02	05 25	22 00
03	22 57	22 07	17 51	23 21	21 05	22 15	21 57	13 03	05 25	22 01
04	22 52	20 30	17 32	23 23	20 57	22 15	21 57	13 03	05 25	22 01
05	22 47	17 31	17 14	23 25	20 48	22 14	21 58	13 04	05 26	22 01
06	22 41	13 25	16 57	23 26	20 40	22 14	21 59	13 05	05 26	22 02
07	22 35	08 34	16 41	23 26	20 31	22 13	21 59	13 05	05 26	22 02
08	22 28	03 N 19	16 26	23 25	20 22	22 13	22 00	13 06	05 26	22 02
09	22 21	02 S 01	16 13	23 24	20 12	22 12	22 01	13 06	05 26	22 03
10	22 14	07 09	16 01	23 22	20 03	22 12	22 01	13 07	05 27	22 03
11	22 06	11 50	15 50	23 20	19 54	22 12	22 02	13 07	05 27	22 03
12	21 58	15 52	15 41	23 16	19 44	22 11	22 03	13 08	05 27	22 04
13	21 49	19 02	15 34	23 12	19 34	22 11	22 03	13 08	05 28	22 04
14	21 40	21 12	15 28	23 08	19 25	22 11	22 04	13 09	05 28	22 04
15	21 31	22 16	15 24	23 02	19 15	22 10	22 04	13 09	05 28	22 05
16	21 22	22 12	15 22	22 56	19 04	22 10	22 05	13 09	05 29	22 05
17	21 12	21 05	15 21	22 49	18 54	22 10	22 06	13 10	05 29	22 05
18	21 01	19 00	15 23	22 42	18 44	22 09	22 06	13 10	05 29	22 06
19	20 50	16 06	15 26	22 34	18 33	22 09	22 07	13 11	05 30	22 06
20	20 39	12 35	15 30	22 25	18 23	22 09	22 08	13 11	05 30	22 06
21	20 28	08 35	15 36	22 16	18 12	22 09	22 08	13 11	05 30	22 07
22	20 16	04 S 15	15 43	22 06	18 01	22 08	22 09	13 12	05 31	22 07
23	20 04	00 N 16	15 52	21 55	17 50	22 08	22 09	13 12	05 31	22 07
24	19 52	04 49	16 02	21 44	17 39	22 08	22 10	13 12	05 32	22 08
25	19 39	09 17	16 13	21 32	17 28	22 08	22 11	13 12	05 32	22 08
26	19 26	13 27	16 25	21 19	17 16	22 08	22 11	13 13	05 32	22 08
27	19 13	17 08	16 37	21 06	17 05	22 08	22 12	13 13	05 33	22 09
28	18 59	20 02	16 50	20 52	16 53	22 08	22 12	13 13	05 33	22 09
29	18 45	21 52	17 04	20 38	16 42	22 07	22 13	13 13	05 34	22 09
30	18 30	22 22	17 17	20 23	16 30	22 07	22 13	13 14	05 34	22 09
31	18 N 16	21 N 21	17 N 31	20 N 07	16 N 18	22 S 07	22 S 14	13 N 14	05 S 35	22 S 10

ZODIAC SIGN ENTRIES

Date	h	m	Planets
01	23	19	♂ ♌
02	01	24	☽ ♋
03	15	18	♀ ♋
04	03	19	☽ ♌
06	04	25	☽ ♍
08	06	07	☽ ♎
10	09	29	☽ ♏
12	15	05	☽ ♐
14	23	05	☽ ♑
17	09	19	☽ ♒
19	07	06	☿ ♋
19	21	19	☽ ♓
22	10	02	☽ ♈
23	02	50	☉ ♌
24	21	42	☽ ♉
27	06	29	☽ ♊
28	01	54	♀ ♌
29	11	31	☽ ♋
31	13	18	☽ ♌

LATITUDES

Date	Mercury ☿ ° '	Venus ♀ ° '	Mars ♂ ° '	Jupiter ♃ ° '	Saturn ♄ ° '	Uranus ♅ ° '	Neptune ♆ ° '	Pluto ♇ ° '
01	01 S 02	00 S 10	01 N 11	00 N 32	00 N 20	00 S 30	01 S 01	00 S 24
04	01 46	00 S 03	01 11	00 31	00 20	00 30	01 02	00 24
07	02 32	00 N 05	01 11	00 31	00 19	00 30	01 02	00 25
10	03 17	00 12	01 11	00 30	00 19	00 30	01 02	00 25
13	03 58	00 19	01 10	00 30	00 19	00 30	01 02	00 25
16	04 30	00 26	01 10	00 29	00 18	00 30	01 02	00 25
19	04 51	00 33	01 10	00 29	00 18	00 30	01 02	00 26
22	04 58	00 39	01 10	00 28	00 18	00 30	01 02	00 26
25	04 49	00 45	01 10	00 28	00 18	00 30	01 02	00 26
28	04 26	00 51	01 10	00 27	00 17	00 30	01 03	00 27
31	03 S 51	00 N 57	01 N 09	00 N 26	00 N 17	00 S 30	01 S 03	00 S 27

DATA

Julian Date	2458666
Delta T	+70 seconds
Ayanamsa	24° 07' 30"
Synetic vernal point	04° ♓ 59' 29"
True obliquity of ecliptic	23° 26' 09"

MOON'S PHASES, APSIDES AND POSITIONS ☽

Date	h	m	Phase	Longitude ° '	Eclipse Indicator
02	19	16	●	10 ♋ 38	Total
09	10	55	☽	16 ♎ 58	
16	21	38	○	24 ♑ 04	partial
25	01	18	☾	01 ♉ 51	

Day	h	m		
05	04	54	Perigee	
20	23	49	Apogee	
02	22	04	Max dec	22° N 23'
09	02	53	0S	
15	22	44	Max dec	22° S 22'
23	10	36	0N	
30	08	08	Max dec	22° N 23'

LONGITUDES

Date	Chiron ⚷ ° '	Ceres ⚳ ° '	Pallas ⚴ ° '	Juno ⚵ ° '	Vesta ⚶ ° '	Black Moon Lilith ⚸ ° '
01	05 ♈ 55	01 ♐ 20	13 ♎ 40	05 ♌ 01	08 ♉ 40	06 ♓ 32
11	05 ♈ 56	00 ♐ 38	15 ♎ 50	09 ♌ 48	12 ♉ 16	07 ♓ 39
21	05 ♈ 52	00 ♐ 33	18 ♎ 24	14 ♌ 32	15 ♉ 37	08 ♓ 46
31	05 ♈ 43	01 ♐ 04	21 ♎ 18	19 ♌ 13	18 ♉ 41	09 ♓ 53

All ephemeris data is given at 12.00 UT and the Moon's longitude is additionally given for 24.00 UT

ASPECTARIAN

h m	Aspects
01 Monday	
01 46	☽ ± ♇
03 06	☽ ☍ ♃
04 20	☽ ∠ ☿
04 34	☽ ⊼ ♄
06 04	☽ □ ♆
09 53	☽ ∠ ♅
12 06	☽ ⊼ ♇
14 47	☽ ⊥ ♂
17 16	☽ ∥ ♂
20 20	☽ ⊥ ☿
21 48	☽ ☌ ♀
02 Tuesday	
01 30	☽ ⊻ ♂
03 20	☽ # ♄
04 59	☽ # ♇
06 50	☽ ⊻ ☿
11 27	☽ ⚹ ♅
12 26	☽ # ♃
19 16	☽ ☌ ☉
03 Wednesday	
05 29	☽ ⊼ ♃
07 01	☽ ☍ ♄
07 30	☽ Q ♅
07 53	☽ # ♃
08 40	☽ △ ♆
11 46	☿ bQ ♆
14 25	☽ ☍ ♇
14 47	☽ # ♇
15 20	☽ ± ♃
16 13	☽ # ♄
04 Thursday	
04 25	☽ ⊻ ♀
05 41	☽ ☌ ♂
06 02	☽ bQ ♃
06 47	☽ ∥ ♂
09 21	☽ bQ ♆
09 50	☽ ☌ ☿
12 32	♂ bQ ♃
13 11	☽ □ ♅
15 09	☽ ⊥ ♀
05 Friday	
00 05	☽ ± ♆
00 10	☽ ⊻ ☉
06 24	☽ △ ♃
07 12	☽ ∠ ♀
08 02	☽ ⊼ ♄
09 53	☽ ⊼ ♆
10 32	♀ ⊻ ♂
10 39	☽ ⊥ ☉
14 02	☽ ∥ ☿
15 30	☽ ⊼ ♇
17 47	☽ ± ♄
06 Saturday	
01 18	☽ ± ♇
02 24	☽ ∠ ☉
08 28	☽ bQ ♄
09 01	☽ ⊻ ♂
10 00	☽ ⚹ ♀
11 36	☽ ⊻ ☿
13 47	☽ ∥ ♅
14 24	☽ △ ♅
16 03	☽ bQ ♇
19 20	☽ ⊥ ♂
21 33	☽ ⊥ ☿
07 Sunday	
04 49	☽ ⚹ ☉
06 26	☿ ⊻ ♀
07 20	☽ □ ♃
07 37	☽ Q ♀
09 04	☽ △ ♄
10 54	☽ ∠ ♂
11 11	☽ ☍ ♆
12 30	☽ ∠ ☿
15 15	☽ bQ ♅
16 50	☽ △ ♇
17 15	♂ bQ ♆
23 14	☿ St R
08 Monday	
02 12	☽ Q ☉
02 26	☽ # ♆
06 20	☽ ± ♅
13 09	☽ ⚹ ♂
13 13	☽ Q ♃
13 37	☽ ⚹ ☿
15 32	♀ ⚹ ♅
15 48	☉ ⊼ ♃
16 28	☽ ⊼ ♅
16 33	☽ □ ♀
22 27	☿ ☌ ♂
09 Tuesday	
09 30	☽ ⚹ ♃
09 54	☽ Q ☿
10 32	☽ Q ♂
10 55	☽ □ ☉
11 22	☽ □ ♄
13 47	☽ ⊼ ♆
17 07	☉ ☍ ♄
19 35	☽ □ ♇
10 Wednesday	
00 08	☽ ± ♆
03 50	☽ ∥ ♆
11 18	☽ ∠ ♃
15 50	☽ bQ ♆
16 47	☽ □ ☿
17 40	☉ Q ♅
17 41	☉ # ♃
18 29	☽ Q ♄
19 20	☽ □ ♂
20 21	☽ ☍ ♅
11 Thursday	
01 29	☽ △ ♀
03 02	☽ Q ♇
03 03	☽ ⊥ ♃
04 31	☉ △ ♆
04 34	♃ ⊥ ♇
13 40	☽ ⊻ ♃
15 41	☽ ⚹ ♄
18 01	♂ □ ♅
18 28	☽ △ ♆
19 11	☽ # ♅
19 33	☽ △ ☉
19 48	☉ # ♇
23 28	☉ # ♄
12 Friday	
00 28	☽ ⚹ ♇
07 05	☽ bQ ♀
10 52	☽ # ☿
16 19	☿ bQ ♆
18 43	☽ ∠ ♄
21 33	☽ △ ☿
13 Saturday	
00 58	☽ bQ ☉
01 14	☽ ± ♀
02 33	☽ ⊼ ♅
03 48	☽ ∠ ♇
04 10	☽ △ ♂
11 14	☽ ⊥ ♄
13 30	☽ ⊼ ♀
13 41	☽ ± ♅
16 43	☽ # ♂
18 59	☽ ± ☉
20 11	☽ ☌ ♃
20 29	☽ ⊥ ♇
22 21	☽ ⊻ ♄
14 Sunday	
00 31	☽ bQ ☿
01 30	☽ □ ♆
06 33	☽ bQ ♅
07 08	☽ ⊼ ☉
07 44	☽ ⊻ ♇
09 37	☽ bQ ♂
09 59	☉ ± ♃
14 51	☉ ☍ ♇
16 53	☽ ± ☿
19 05	☽ # ☉
15 Monday	
03 37	☽ ± ♂
03 53	☽ ⊼ ☿
05 04	☽ ∥ ♄
05 13	☽ ∥ ♇
08 19	☽ ∥ ♃
11 08	☽ △ ♅
11 39	☽ Q ♆
15 44	☽ ⊼ ♂
16 Tuesday	
03 29	♄ ∥ ♇
04 42	☽ ☍ ♀
05 01	☽ ⊻ ♃
07 18	☽ ☌ ♄
07 44	♀ ⊼ ♃
10 52	☽ ⚹ ♆
13 28	☽ ∥ ♃
16 06	☽ ∥ ♄
16 10	☽ ∥ ♇
16 38	☽ ⊥ ♃
17 16	☽ ☌ ♇
21 38	☽ ☍ ☉
17 Wednesday	
05 34	♀ ☍ ♄
10 03	☽ # ☉
10 14	☽ ∠ ♃
11 39	☽ ☍ ☿
16 19	☽ ∠ ♆
21 54	☽ □ ♅
18 Thursday	
05 50	☽ ☍ ♂
10 13	☽ ⊥ ♆
14 38	☽ # ♂
14 55	♀ Q ♅
15 53	☽ ⚹ ♃
17 51	☿ bQ ♃
18 03	♀ △ ♆
18 15	☽ ⊻ ♄
22 13	☽ ⊻ ♆
22 42	☽ ⊼ ♀
19 Friday	
04 45	☽ ⊻ ♇
06 14	☽ ⊥ ♄
09 59	☽ Q ♅
12 09	☽ ± ♀
14 33	☽ ⊼ ☉
15 53	☽ Q ♃
16 49	☽ ⊥ ♇
16 50	☽ # ☿
20 33	☽ ⊼ ☿
20 Saturday	
00 16	☽ ∠ ♄
03 44	☽ ± ☉
08 01	☽ ± ☿
08 08	☽ # ♅
08 26	☽ bQ ♀
10 16	☽ ⚹ ♅
10 58	☽ ∠ ♇
21 39	☽ ⊼ ♂
22 35	♀ ± ♃
23 33	☽ bQ ☉
21 Sunday	
01 15	☽ bQ ☿
04 06	☽ □ ♃
06 28	☽ ⚹ ♄
08 32	♀ ☍ ♇
10 30	☽ ± ♂
10 46	☽ ☌ ♆
12 34	☉ ☌ ☿
15 10	♂ ± ♆
16 42	☽ ∠ ♅
17 18	☽ ⚹ ♇
18 20	☽ △ ♀
22 Monday	
02 09	♃ ∥ ♄
05 08	☽ ∥ ♆
05 27	♀ # ♄
05 44	♀ # ♃
05 45	☽ bQ ♂
05 58	☽ △ ☿
06 37	☽ Q ♄
08 34	☽ △ ☉
09 32	♀ # ♇
10 56	☽ ⊥ ♅
17 29	☽ Q ♇
23 01	☽ ⊻ ♅
23 Tuesday	
04 52	☉ bQ ♃
13 34	☽ △ ♂
16 14	☽ △ ♃
18 31	☽ □ ♄
22 59	☽ ⊻ ♆
24 Wednesday	
05 20	☽ □ ♇
10 49	☽ ⊥ ♆
13 12	☽ □ ♀
14 48	☽ □ ☿
15 44	☽ # ♆
21 40	☽ bQ ♃
25 Thursday	
00 26	☿ ☌ ♀
01 18	☽ □ ☉
04 22	☽ ∠ ♆
10 17	☽ ☌ ♅
12 22	♂ △ ♃
12 54	♃ ∥ ♇
15 00	☽ ± ♃
26 Friday	
00 03	☽ Q ☿
02 24	☽ ⊼ ♃
03 12	☽ □ ♂
03 44	☽ Q ♀
04 29	☽ △ ♄
08 59	☽ ⚹ ♆
10 32	☽ ∥ ♅
14 26	☽ Q ☉
14 57	☽ △ ♇
16 27	☉ bQ ♆
21 44	☽ ⚹ ☿
27 Saturday	
02 41	♂ ⊼ ♄
04 28	☽ ⚹ ♀
07 12	☽ Q ♆
08 10	☽ ∥ ☿
08 13	☽ bQ ♄
11 42	☽ ∥ ♂
14 16	☽ Q ♂
14 16	☽ ⚹ ☉
18 20	☽ ⊻ ♅
18 26	☽ bQ ♇
22 41	♀ bQ ♃
28 Sunday	
00 11	☽ ∠ ☿
00 24	☽ ± ♄
03 07	☽ ∥ ☉
05 05	☽ ⊥ ♅
09 08	☽ ☍ ♃
10 13	☽ ∠ ♀
10 27	☽ ± ♇
11 00	☽ ⊼ ♄
12 45	☽ ⚹ ♂
15 08	☿ ∥ ♂
15 24	☽ □ ♆
15 42	☽ ⊥ ☿
18 59	☽ ∠ ☉
20 10	☽ ∥ ♀
20 55	☽ ∠ ♅
20 57	☽ ⊼ ♇
29 Monday	
01 55	☽ ⊻ ☿
03 28	☽ ⊥ ♀
11 39	☽ ⊥ ☉
14 44	☽ ⊻ ♀
15 55	☽ ∠ ♂
17 57	☽ # ♃
18 50	☽ # ♇
20 53	☽ # ♄
22 34	☽ ⊻ ☉
22 37	☽ ⚹ ♅
23 14	☉ □ ♅
30 Tuesday	
07 45	☽ ⊥ ♂
12 17	☽ ⊼ ♃
13 57	☽ ☍ ♄
18 09	☽ ⊻ ♂
18 15	☽ △ ♆
18 36	☽ Q ♅
18 50	♀ bQ ♆
19 00	☽ # ♄
20 09	♂ ⊼ ♆
21 08	☽ # ♇
22 10	☽ ± ♃
22 11	☽ # ♃
23 27	☽ ☍ ♇
31 Wednesday	
03 32	☽ ☌ ☿
12 48	☽ bQ ♃
18 40	☽ bQ ♆
20 51	☽ ☌ ♀
23 54	☽ □ ♅

AUGUST 2019

LONGITUDES

Date	Sidereal time h m s	Sun ☉ ° ' "	Moon ☽ ° ' "	Moon ☽ 24.00 ° ' "	Mercury ☿ ° '	Venus ♀ ° '	Mars ♂ ° '	Jupiter ♃ ° '	Saturn ♄ ° '	Uranus ♅ ° '	Neptune ♆ ° '	Pluto ♇ ° '
01	08 39 15	08 ♌ 58 00	14 ♌ 06 29	21 ♌ 37 31	23 ♋ 57	05 ♌ 27	19 ♌ 23	14 ♐ 40	15 ♑ 36	06 ♉ 34	18 ♓ 18	21 ♑ 28
02	08 43 11	09 55 26	29 ♌ 09 32	06 ♍ 41 23	24 D 02	06 41	20 01	14 R 38	15 R 33	06 35	18 R 17	21 R 27
03	08 47 08	10 52 52	14 ♍ 11 54	21 ♍ 40 03	24 14	07 55	20 39	14 36	15 29	06 35	18 16	21 25
04	08 51 04	11 50 18	29 ♍ 04 52	06 ♎ 25 30	24 32	09 09	21 17	14 35	15 25	06 36	18 14	21 24
05	08 55 01	12 47 46	13 ♎ 41 17	20 ♎ 51 42	24 56	10 23	21 55	14 34	15 21	06 36	18 13	21 23
06	08 58 57	13 45 14	27 ♎ 56 25	04 ♏ 55 15	25 27	11 37	22 34	14 33	15 18	06 36	18 12	21 21
07	09 02 54	14 42 43	11 ♏ 48 07	18 ♏ 35 09	26 05	12 51	23 12	14 32	15 14	06 36	18 10	21 20
08	09 06 50	15 40 12	25 ♏ 16 29	01 ♐ 52 24	26 49	14 05	23 50	14 31	15 11	06 37	18 09	21 19
09	09 10 47	16 37 43	08 ♐ 23 14	14 ♐ 49 20	27 39	15 19	24 28	14 31	15 07	06 37	18 08	21 17
10	09 14 44	17 35 14	21 ♐ 11 04	27 ♐ 28 51	28 36	16 33	25 06	14 30	15 04	06 37	18 06	21 16
11	09 18 40	18 32 46	03 ♑ 43 04	09 ♑ 54 06	29 ♋ 38	17 47	25 44	14 30	15 01	06 37	18 05	21 15
12	09 22 37	19 30 19	16 ♑ 02 19	22 ♑ 08 02	00 ♌ 47	19 01	26 22	14 D 30	14 57	06 R 37	18 04	21 14
13	09 26 33	20 27 53	28 ♑ 11 36	04 ♒ 13 17	02 01	20 15	27 00	14 31	14 54	06 37	18 02	21 12
14	09 30 30	21 25 28	10 ♒ 13 22	16 ♒ 12 07	03 21	21 30	27 38	14 31	14 51	06 37	18 01	21 11
15	09 34 26	22 23 05	22 ♒ 09 45	28 ♒ 06 30	04 47	22 44	28 16	14 32	14 48	06 37	17 59	21 10
16	09 38 23	23 20 42	04 ♓ 02 36	09 ♓ 58 16	06 17	23 58	28 54	14 33	14 45	06 36	17 58	21 09
17	09 42 19	24 18 21	15 ♓ 53 45	21 ♓ 49 16	07 51	25 12	29 ♌ 33	14 34	14 42	06 36	17 57	21 08
18	09 46 16	25 16 01	27 ♓ 45 06	03 ♈ 41 32	09 30	26 26	00 ♍ 11	14 35	14 39	06 36	17 55	21 06
19	09 50 13	26 13 42	09 ♈ 38 53	15 ♈ 37 29	11 13	27 40	00 49	14 36	14 37	06 36	17 54	21 05
20	09 54 09	27 11 25	21 ♈ 37 44	27 ♈ 40 00	12 59	28 ♌ 55	01 27	14 38	14 34	06 35	17 52	21 04
21	09 58 06	28 09 10	03 ♉ 44 45	09 ♉ 52 28	14 48	00 ♍ 09	02 05	14 39	14 31	06 35	17 51	21 03
22	10 02 02	29 ♌ 06 56	16 ♉ 03 36	22 ♉ 18 41	16 40	01 23	02 43	14 41	14 29	06 35	17 49	21 02
23	10 05 59	00 ♍ 04 44	28 ♉ 38 15	05 ♊ 02 47	18 34	02 38	03 21	14 43	14 26	06 34	17 47	21 01
24	10 09 55	01 02 34	11 ♊ 32 47	18 ♊ 08 43	20 29	03 52	03 59	14 46	14 24	06 33	17 46	21 00
25	10 13 52	02 00 26	24 ♊ 50 58	01 ♋ 39 53	22 26	05 06	04 38	14 48	14 22	06 33	17 44	20 59
26	10 17 48	02 58 19	08 ♋ 35 39	15 ♋ 38 21	24 24	06 20	05 16	14 51	14 20	06 32	17 43	20 58
27	10 21 45	03 56 14	22 ♋ 47 54	00 ♌ 04 01	26 23	07 35	05 54	14 54	14 17	06 31	17 41	20 57
28	10 25 42	04 54 11	07 ♌ 26 15	14 ♌ 53 55	28 ♌ 22	08 49	06 32	14 57	14 15	06 30	17 40	20 56
29	10 29 38	05 52 09	22 ♌ 26 06	00 ♍ 01 44	00 ♍ 21	10 04	07 10	15 00	14 14	06 30	17 38	20 55
30	10 33 35	06 50 10	07 ♍ 39 35	15 ♍ 18 19	02 20	11 18	07 49	15 03	14 12	06 29	17 36	20 54
31	10 37 31	07 ♍ 48 11	22 ♍ 56 31	00 ♎ 32 49	04 ♍ 18	12 ♍ 32	08 ♍ 27	15 ♐ 07	14 ♑ 10	06 ♉ 28	17 ♓ 35	20 ♑ 53

Date	Moon True ☊ ° '	Moon Mean ☊ ° '	Moon ☽ Latitude ° '
01	17 ♋ 35	16 ♋ 19	02 N 21
02	17 R 32	16 16	03 29
03	17 29	16 13	04 23
04	17 26	16 10	04 58
05	17 23	16 06	05 13
06	17 22	16 03	05 09
07	17 21	16 00	04 46
08	17 D 21	15 57	04 08
09	17 22	15 54	03 18
10	17 24	15 51	02 19
11	17 25	15 47	01 14
12	17 26	15 44	00 N 08
13	17 R 25	15 41	00 S 59
14	17 23	15 38	02 01
15	17 19	15 35	02 58
16	17 14	15 31	03 47
17	17 08	15 28	04 25
18	17 01	15 25	04 52
19	16 54	15 22	05 07
20	16 48	15 19	05 08
21	16 44	15 16	04 56
22	16 41	15 12	04 30
23	16 40	15 09	03 51
24	16 D 40	15 06	02 59
25	16 42	15 03	01 55
26	16 43	15 00	00 S 44
27	16 R 43	14 57	00 N 33
28	16 41	14 53	01 49
29	16 37	14 50	03 00
30	16 31	14 47	03 59
31	16 ♋ 24	14 ♋ 44	04 N 41

DECLINATIONS

Date	Sun ☉ ° '	Moon ☽ ° '	Mercury ☿ ° '	Venus ♀ ° '	Mars ♂ ° '	Jupiter ♃ ° '	Saturn ♄ ° '	Uranus ♅ ° '	Neptune ♆ ° '	Pluto ♇ ° '
01	18 N 01	18 N 50	17 N 45	19 N 51	16 N 06	22 S 07	22 S 15	13 N 14	05 S 35	22 S 10
02	17 46	15 02	17 58	19 34	15 54	22 07	22 15	13 14	05 36	22 10
03	17 30	10 15	18 11	19 17	15 42	22 07	22 16	13 14	05 36	22 11
04	17 14	04 N 55	18 23	18 59	15 29	22 07	22 16	13 14	05 37	22 11
05	16 58	00 S 35	18 34	18 41	15 17	22 07	22 17	13 14	05 37	22 11
06	16 42	05 56	18 44	18 22	15 05	22 07	22 17	13 14	05 38	22 11
07	16 25	10 50	18 54	18 02	14 52	22 08	22 18	13 15	05 38	22 12
08	16 08	15 04	19 01	17 42	14 39	22 08	22 18	13 15	05 39	22 12
09	15 51	18 27	19 08	17 22	14 26	22 08	22 19	13 15	05 39	22 12
10	15 34	20 50	19 12	17 01	14 14	22 08	22 19	13 15	05 40	22 13
11	15 16	22 09	19 15	16 40	14 01	22 08	22 20	13 15	05 41	22 13
12	14 58	22 21	19 15	16 18	13 48	22 08	22 20	13 15	05 41	22 13
13	14 40	21 29	19 13	15 55	13 34	22 08	22 21	13 15	05 42	22 14
14	14 22	19 37	19 09	15 33	13 21	22 09	22 21	13 15	05 42	22 14
15	14 03	16 56	19 03	15 09	13 08	22 09	22 21	13 15	05 43	22 14
16	13 44	13 32	18 53	14 46	12 54	22 09	22 22	13 14	05 43	22 14
17	13 25	09 38	18 41	14 22	12 41	22 10	22 22	13 14	05 44	22 15
18	13 06	05 22	18 27	13 57	12 27	22 10	22 23	13 14	05 45	22 15
19	12 46	00 S 53	18 09	13 33	12 14	22 10	22 23	13 14	05 45	22 15
20	12 27	03 N 39	17 49	13 08	12 00	22 11	22 23	13 14	05 46	22 15
21	12 07	08 07	17 26	12 42	11 46	22 11	22 24	13 14	05 46	22 16
22	11 47	12 20	17 01	12 16	11 32	22 11	22 24	13 14	05 47	22 16
23	11 27	16 06	16 33	11 50	11 19	22 12	22 25	13 13	05 48	22 16
24	11 06	19 13	16 02	11 23	11 05	22 12	22 25	13 13	05 48	22 16
25	10 46	21 25	15 29	10 57	10 50	22 13	22 25	13 13	05 49	22 17
26	10 25	22 26	14 54	10 30	10 36	22 13	22 26	13 13	05 50	22 17
27	10 04	22 03	14 18	10 02	10 22	22 14	22 26	13 13	05 50	22 17
28	09 43	20 10	13 39	09 34	10 08	22 14	22 26	13 12	05 51	22 17
29	09 21	16 52	12 59	09 06	09 53	22 15	22 27	13 12	05 52	22 18
30	09 00	12 24	12 18	08 38	09 39	22 15	22 27	13 12	05 52	22 18
31	08 N 39	07 N 06	11 N 35	08 N 10	09 N 25	22 S 16	22 S 27	13 N 11	05 S 53	22 S 18

ZODIAC SIGN ENTRIES

Date	h	m	Planets
02	13	20	☽ ♍
04	13	30	☽ ♎
06	15	31	☽ ♏
08	20	35	☽ ♐
11	04	50	☽ ♑
11	19	46	☿ ♌
13	15	35	☽ ♒
16	03	49	☽ ♓
18	05	18	♂ ♍
18	16	33	☽ ♈
21	04	37	☽ ♉
21	09	06	♀ ♍
23	10	02	☉ ♍
23	14	34	☽ ♊
25	21	05	☽ ♋
27	23	53	☽ ♌
29	07	48	☿ ♍
29	23	57	☽ ♍
31	23	08	☽ ♎

LATITUDES

Date	Mercury ☿ ° '	Venus ♀ ° '	Mars ♂ ° '	Jupiter ♃ ° '	Saturn ♄ ° '	Uranus ♅ ° '	Neptune ♆ ° '	Pluto ♇ ° '
01	03 S 37	00 N 58	01 N 09	00 N 26	00 N 17	00 S 30	01 S 03	00 S 27
04	02 53	01 03	01 09	00 26	00 17	00 30	01 03	00 27
07	02 04	01 08	01 09	00 25	00 16	00 30	01 03	00 28
10	01 16	01 12	01 08	00 24	00 16	00 30	01 03	00 28
13	00 S 30	01 15	01 08	00 24	00 16	00 30	01 03	00 28
16	00 N 12	01 18	01 08	00 23	00 16	00 30	01 03	00 28
19	00 47	01 21	01 07	00 23	00 15	00 30	01 03	00 29
22	01 14	01 23	01 07	00 22	00 15	00 31	01 03	00 29
25	01 32	01 24	01 06	00 22	00 15	00 31	01 03	00 29
28	01 43	01 25	01 06	00 21	00 14	00 31	01 03	00 29
31	01 N 46	01 N 25	01 N 05	00 N 21	00 N 14	00 S 31	01 S 03	00 S 30

DATA

Julian Date	2458697
Delta T	+70 seconds
Ayanamsa	24° 07' 35"
Synetic vernal point	04° ♓ 59' 24"
True obliquity of ecliptic	23° 26' 10"

MOON'S PHASES, APSIDES AND POSITIONS ☽

Date	h	m	Phase	Longitude ° '	Eclipse Indicator
01	03	12	●	08 ♌ 37	
07	17	31	☽	14 ♏ 56	
15	12	29	○	22 ♒ 24	
23	14	56	☾	00 ♊ 12	
30	10	37	●	06 ♍ 47	

Day	h	m		
02	07	05	Perigee	
17	10	34	Apogee	
30	15	47	Perigee	
05	09	27	0S	
12	04	24	Max dec	22° S 24'
19	16	41	0N	
26	17	48	Max dec	22° N 28'

LONGITUDES

Date	Chiron ⚷ ° '	Ceres ⚳ ° '	Pallas ⚴ ° '	Juno ⚵ ° '	Vesta ⚶ ° '	Black Moon Lilith ⚸ ° '
01	05 ♈ 42	01 ♐ 09	21 ♎ 36	19 ♌ 41	18 ♉ 59	10 ♓ 00
11	05 ♈ 28	02 ♐ 16	24 ♎ 49	24 ♌ 19	21 ♉ 40	11 ♓ 07
21	05 ♈ 10	03 ♐ 51	28 ♎ 15	28 ♌ 52	23 ♉ 57	12 ♓ 14
31	04 ♈ 48	05 ♐ 52	01 ♏ 53	03 ♍ 22	25 ♉ 45	13 ♓ 21

All ephemeris data is given at 12.00 UT and the Moon's longitude is additionally given for 24.00 UT

ASPECTARIAN

h m Aspects

01 Thursday
02 58 ☽ ∥ ♀
03 12 ☽ ☌ ☉
03 57 ☿ St D
09 07 ☽ ± ♆
12 53 ☽ △ ♃
14 23 ☽ ⊼ ♄
18 21 ☽ ∥ ☉
18 41 ☽ ⊼ ♆
19 16 ☽ ∥ ☿
20 48 ☽ ☌ ♂
23 44 ☽ ⊼ ♇
23 55 ☽ ± ♄

02 Friday
01 40 ☉ ∥ ☿
03 47 ☽ ⊻ ☿
06 50 ☽ ∥ ♂
09 16 ☽ ± ♇
10 00 ♀ □ ♅
13 25 ☽ ⊥ ☿
14 12 ☽ bQ ♄
21 29 ☽ ∥ ♅
23 35 ☽ bQ ♇
23 50 ☽ △ ♅

03 Saturday
01 03 ☽ ⊻ ♀
03 56 ☽ ∠ ☿
06 20 ☽ ⊻ ☉
11 30 ☽ ⊥ ♀
12 39 ☽ □ ♃
14 03 ☽ △ ♄
16 36 ☽ ⊥ ☉
18 30 ☽ ☍ ♆
22 50 ☽ ⊻ ♂
23 35 ☽ △ ♇
23 52 ☽ bQ ♅

04 Sunday
03 17 ☽ ∠ ♀
04 27 ☽ ✱ ☿
08 06 ☽ ∠ ☉
08 58 ☽ ⊥ ♂
08 59 ☽ # ♆
14 28 ☽ ± ♅
15 59 ♂ ⊼ ♇
16 28 ♂ ± ♄
17 42 ☽ Q ♃
21 49 ☉ ± ♆

05 Monday
00 17 ☽ ⊼ ♅
00 19 ☽ ∠ ♂
00 30 ☽ Q ☿
06 00 ☽ ✱ ♀
10 25 ☽ ✱ ☉
13 27 ☽ ✱ ♃
14 46 ☽ □ ♄
17 20 ☿ ∥ ♀
19 33 ☽ ⊼ ♆

06 Tuesday
00 51 ☽ □ ♇
02 26 ☽ ✱ ♂
03 57 ☽ Q ♀
05 39 ☽ ± ♆
07 36 ☽ □ ☿
08 00 ☽ Q ☉
10 36 ☽ ∥ ♆
14 44 ☽ ∠ ♃
21 00 ☽ bQ ♆
21 09 ☽ Q ♄
23 09 ♀ ± ♆
23 55 ☽ Q ♂

07 Wednesday
02 55 ☽ ☍ ♅
06 17 ☽ ⊥ ♃
07 31 ☉ △ ♃
07 41 ☽ Q ♇
14 01 ☽ □ ♀
16 48 ☽ ⊻ ♃
17 31 ☽ □ ☉
18 02 ☽ ✱ ♄
23 15 ☽ △ ♆

08 Thursday
00 24 ☉ ⊼ ♄
01 08 ☽ # ♅
04 53 ☽ ✱ ♇
09 16 ☽ □ ♂
09 35 ☽ # ♂
14 58 ☽ △ ☿
18 26 ☽ # ☉
20 28 ♀ △ ♃
20 52 ☽ ∠ ♄

09 Friday
04 23 ☽ # ♀
08 07 ☽ ∠ ♇
08 24 ♀ ⊼ ♄
08 43 ☽ ⊼ ♅
13 21 ☽ ⊥ ♄
18 05 ☽ # ☿
19 52 ☽ ± ♅
20 33 ☽ bQ ☿
23 25 ☽ ☌ ♃

10 Saturday
00 30 ☽ ⊻ ♄
00 51 ☽ ⊥ ♇
02 19 ☽ △ ♀
04 39 ☽ △ ☉
06 12 ☽ □ ♆
12 09 ☽ ⊻ ♇
12 49 ☽ bQ ♅
14 55 ☽ ± ☿
19 50 ☽ △ ♂

11 Sunday
00 44 ☉ ⊼ ♆
03 24 ☽ ⊼ ☿
09 01 ☿ bQ ♃
10 00 ☽ bQ ♀
11 38 ☽ bQ ☉
11 42 ☽ ∥ ♃
13 38 ♃ St D
14 32 ☽ ∥ ♇
16 34 ☽ Q ♆
17 37 ☽ △ ♅
17 43 ♀ ⊼ ♆
19 43 ☽ ∥ ♄

12 Monday
02 22 ☽ bQ ♂
02 27 ♅ St R
05 26 ☽ ± ♀
06 37 ☽ ± ☉
09 00 ☽ ⊻ ♃
09 53 ☽ ☌ ♄
12 46 ☽ ∥ ♄
15 58 ☽ ✱ ♆
18 09 ☽ ∥ ♇
18 31 ☽ ⊼ ♀
19 24 ☽ ⊼ ☉
20 47 ☽ ⊥ ♃
20 59 ☽ ± ♂
21 05 ☽ ∥ ♃
22 11 ☽ ☌ ♇

13 Tuesday
09 30 ☽ ⊼ ♂
14 37 ☽ ∠ ♃
19 27 ♂ ± ♇
20 33 ☽ ☍ ☿
21 37 ☽ ∠ ♆
22 24 ☉ ± ♄

14 Wednesday
00 03 ♀ ± ♄
04 47 ☽ □ ♅
06 06 ☿ bQ ♆
06 07 ☉ ☌ ♀
06 08 ♀ ⊼ ♇
06 08 ☉ ⊼ ♇
15 35 ☽ ⊥ ♆
16 55 ☽ # ☿
20 37 ☽ ✱ ♃
21 15 ☽ ⊻ ♄
23 58 ♂ ∥ ♅

15 Thursday
03 37 ☽ ⊻ ♆
09 16 ☽ ⊥ ♄
10 00 ☽ ⊻ ♇
12 29 ☽ ☍ ☉
13 16 ☽ ☍ ♀
16 56 ☽ Q ♅
20 49 ☽ Q ♃
22 05 ☽ ⊥ ♇

16 Friday
01 02 ☽ ☍ ♂
02 45 ☽ # ♀
03 21 ☽ ∠ ♄
10 36 ☽ # ☉
13 57 ☽ # ♅
16 15 ☽ ∠ ♇
16 21 ☽ # ♂
17 07 ☿ □ ♅
17 11 ☽ ✱ ♅
17 12 ☽ ⊼ ☿

17 Saturday
07 13 ☽ ± ☿
09 17 ☽ □ ♃
09 36 ☽ ✱ ♄
16 08 ☽ ☌ ♆
17 39 ♂ bQ ♄
22 34 ☽ ✱ ♇
23 33 ☽ ∠ ♅

18 Sunday
01 30 ☉ ∥ ♅
04 21 ☽ bQ ☿
06 32 ☽ ⊼ ☉
09 02 ☽ ⊼ ♀
09 48 ☽ Q ♄
09 56 ☽ ∥ ♆
17 11 ☽ ⊼ ♂
17 45 ☽ ⊥ ♅
19 44 ☽ ± ☉
22 34 ☽ ± ♀
22 48 ☽ Q ♇

19 Monday
00 47 ♀ ± ♇
05 51 ☽ ⊻ ♅
05 58 ☽ ± ♂
14 47 ♃ ⊻ ♄
15 27 ☽ bQ ☉
15 41 ☽ △ ☿
18 47 ☽ bQ ♀
21 11 ☿ ± ♆
21 56 ☽ □ ♄
21 58 ☽ △ ♃

20 Tuesday
01 04 ☽ bQ ♂
04 30 ☽ ⊻ ♆
05 51 ♀ ∥ ♅
09 02 ☉ ± ♇
10 53 ☽ □ ♇
16 27 ☽ ⊥ ♆
23 15 ☽ # ♆

21 Wednesday
00 01 ☽ △ ☉
00 15 ♀ bQ ♄
03 55 ☽ bQ ♃
04 06 ☽ △ ♀
08 26 ☿ ⊼ ♄
08 33 ☽ △ ♂
10 05 ☿ △ ♃
10 14 ☽ ∠ ♆
17 33 ☽ ☌ ♅
21 39 ☽ ± ♃

22 Thursday
07 36 ☽ ∥ ♂
08 57 ☽ △ ♄
09 00 ☽ ∥ ☉
09 21 ☽ ⊼ ♃
11 41 ☽ ∥ ♀
13 22 ☽ □ ☿
15 23 ☽ ✱ ♆
17 25 ☽ ∥ ♅
20 43 ☉ bQ ♄
21 33 ☽ △ ♇

23 Friday
02 26 ☿ ⊼ ♆
13 30 ☽ bQ ♄
14 10 ☽ Q ♆
14 41 ☽ ∥ ☿
14 56 ☽ □ ☉
20 17 ☽ □ ♀
21 19 ☽ □ ♂

24 Saturday
01 47 ☽ bQ ♇
02 48 ☽ ⊻ ♅
05 24 ☽ Q ☿
06 14 ☽ ± ♄
10 56 ☿ ± ♄
13 50 ☽ ⊥ ♅
17 05 ♀ ☌ ♂
17 12 ☽ ⊼ ♄
17 53 ☽ ☍ ♃
17 57 ☉ ∥ ♂
18 14 ☿ ⊼ ♇
18 17 ☽ ± ♇
23 17 ☽ □ ♆

25 Sunday
02 41 ☽ Q ☉
05 06 ☽ ⊼ ♇
06 07 ☽ ∠ ♅
06 58 ☽ ✱ ☿
07 51 ☽ Q ♂
08 35 ☽ Q ♀
23 28 ♀ ∥ ♂

26 Monday
01 33 ☽ ✱ ☉
03 24 ☽ # ♃
04 47 ♀ bQ ♇
05 23 ☽ # ♇
05 59 ☽ ✱ ♂
07 44 ☽ ✱ ♀
08 27 ☽ ✱ ♅
11 49 ☽ # ♄
13 37 ☽ ∠ ☿
15 38 ♀ △ ♅
21 45 ☽ ☍ ♄
22 42 ☽ ⊼ ♃
23 35 ☽ # ♄

27 Tuesday
03 28 ☽ △ ♆
04 51 ☽ Q ♅
05 05 ☽ ∠ ☉
05 40 ☉ ∥ ♀
05 56 ☽ # ♇
07 19 ☽ ⊥ ☿
07 42 ☽ # ♃
08 41 ☽ ∠ ♂
08 49 ☽ ± ♃
08 55 ☽ ☍ ♇
11 36 ☽ ∠ ♀
13 46 ♂ bQ ♇
18 48 ☿ ± ♇
18 52 ☽ ⊻ ☿
21 06 ☽ ⊥ ☉
23 45 ☽ bQ ♃

28 Wednesday
00 16 ☽ ⊥ ♂
03 49 ☽ ⊥ ♀
04 15 ☽ bQ ♆
07 36 ☽ ⊻ ☉
07 58 ♃ ⊥ ♇
10 29 ☽ ⊻ ♂
10 30 ☽ □ ♅
10 53 ♂ △ ♅
14 26 ☽ ⊻ ♀
18 48 ☽ ± ♆
22 38 ☿ bQ ♄
22 57 ☽ ⊼ ♄

29 Thursday
00 07 ☽ △ ♃
04 19 ☿ ∥ ♅
04 23 ☽ ⊼ ♆
08 30 ☽ ± ♄
09 36 ☽ ⊼ ♇
13 08 ☉ bQ ♇
19 05 ☽ ± ♇
22 43 ☽ bQ ♄

30 Friday
02 22 ☽ ☌ ☿
03 14 ☉ △ ♅
08 03 ☽ ∥ ♅
09 14 ☽ bQ ♇
10 09 ☽ △ ♅
10 37 ☽ ☌ ☉
12 15 ☽ ☌ ♂
12 33 ☽ ∥ ☿
18 13 ☽ ☌ ♀
22 14 ☽ △ ♄
23 39 ☽ □ ♃

31 Saturday
01 23 ☽ ∥ ♂
03 35 ☽ ☍ ♆
04 48 ☽ ∥ ☉
06 57 ☽ ∥ ♀
08 46 ☽ △ ♇
09 41 ☽ bQ ♅
17 15 ☽ # ♆
23 51 ☽ ± ♅

SEPTEMBER 2019

Date	Sidereal time h m s	Sun ☉ ° ' "	Moon ☽ ° ' "	Moon ☽ 24.00 ° ' "	Mercury ☿ ° '	Venus ♀ ° '	Mars ♂ ° '	Jupiter ♃ ° '	Saturn ♄ ° '	Uranus ♅ ° '	Neptune ♆ ° '	Pluto ♇ ° '
01	10 41 28	08 ♍ 46 14	08 ♎ 05 54	15 ♎ 34 37	06 ♍ 16	13 ♍ 47	09 ♍ 05	15 ♐ 10	14 ♑ 08	06 ♉ 27	17 ♓ 33	20 ♑ 52
02	10 45 24	09 44 19	22 ♎ 57 58	00 ♏ 15 10	08 13	15 01	09 43	15 14	14 R 07	06 R 26	17 R 31	20 R 51
03	10 49 21	10 42 25	07 ♏ 25 38	14 ♏ 29 03	10 10	16 16	10 21	15 18	14 05	06 25	17 30	20 51
04	10 53 17	11 40 32	21 ♏ 25 16	28 ♏ 14 19	12 05	17 30	11 00	15 22	14 04	06 24	17 28	20 50
05	10 57 14	12 38 41	04 ♐ 56 26	11 ♐ 31 56	14 00	18 44	11 38	15 27	14 03	06 23	17 27	20 49
06	11 01 11	13 36 52	18 ♐ 01 14	24 ♐ 24 51	15 53	19 59	12 16	15 31	14 01	06 21	17 25	20 48
07	11 05 07	14 35 04	00 ♑ 43 19	06 ♑ 57 12	17 45	21 13	12 55	15 36	14 00	06 20	17 23	20 48
08	11 09 04	15 33 17	13 ♑ 07 05	19 ♑ 13 32	19 37	22 28	13 33	15 41	13 59	06 19	17 22	20 47
09	11 13 00	16 31 32	25 ♑ 17 04	01 ♒ 18 13	21 27	23 42	14 11	15 46	13 58	06 18	17 20	20 46
10	11 16 57	17 29 48	07 ♒ 17 27	13 ♒ 15 13	23 16	24 57	14 49	15 51	13 58	06 16	17 18	20 46
11	11 20 53	18 28 06	19 ♒ 11 54	25 ♒ 07 51	25 04	26 11	15 28	15 57	13 57	06 15	17 17	20 45
12	11 24 50	19 26 26	01 ♓ 03 24	06 ♓ 58 48	26 51	27 26	16 06	16 02	13 56	06 13	17 15	20 44
13	11 28 46	20 24 48	12 ♓ 54 18	18 ♓ 50 08	28 ♍ 36	28 40	16 44	16 08	13 56	06 12	17 13	20 44
14	11 32 43	21 23 11	24 ♓ 46 27	00 ♈ 43 28	00 ♎ 21	29 ♍ 55	17 23	16 14	13 55	06 10	17 12	20 43
15	11 36 40	22 21 36	06 ♈ 41 21	12 ♈ 40 15	02 04	01 ♎ 09	18 01	16 20	13 55	06 09	17 10	20 43
16	11 40 36	23 20 03	18 ♈ 40 22	24 ♈ 41 54	03 46	02 24	18 39	16 26	13 55	06 07	17 08	20 42
17	11 44 33	24 18 32	00 ♉ 45 04	06 ♉ 50 08	05 28	03 38	19 18	16 32	13 55	06 06	17 07	20 42
18	11 48 29	25 17 03	12 ♉ 57 22	19 ♉ 07 06	07 08	04 53	19 56	16 38	13 D 55	06 04	17 05	20 41
19	11 52 26	26 15 36	25 ♉ 19 41	01 ♊ 35 31	08 47	06 07	20 35	16 45	13 55	06 02	17 04	20 41
20	11 56 22	27 14 11	07 ♊ 55 02	14 ♊ 18 39	10 25	07 22	21 13	16 52	13 55	06 01	17 02	20 40
21	12 00 19	28 12 49	20 ♊ 46 52	27 ♊ 20 06	12 02	08 36	21 51	16 59	13 55	05 59	17 00	20 40
22	12 04 15	29 ♍ 11 29	03 ♋ 58 48	10 ♋ 43 23	13 38	09 51	22 30	17 06	13 55	05 57	16 59	20 40
23	12 08 12	00 ♎ 10 11	17 ♋ 34 09	24 ♋ 31 22	15 14	11 05	23 08	17 13	13 56	05 55	16 57	20 39
24	12 12 09	01 08 56	01 ♌ 35 08	08 ♌ 45 24	16 48	12 20	23 47	17 20	13 56	05 53	16 55	20 39
25	12 16 05	02 07 42	16 ♌ 01 57	23 ♌ 24 22	18 21	13 35	24 25	17 28	13 57	05 51	16 54	20 39
26	12 20 02	03 06 31	00 ♍ 51 59	08 ♍ 23 58	19 53	14 49	25 04	17 35	13 58	05 49	16 52	20 39
27	12 23 58	04 05 22	15 ♍ 59 13	23 ♍ 36 32	21 25	16 04	25 42	17 43	13 59	05 47	16 51	20 39
28	12 27 55	05 04 15	01 ♎ 14 31	08 ♎ 51 47	22 55	17 18	26 21	17 51	14 00	05 45	16 49	20 38
29	12 31 51	06 03 10	16 ♎ 26 53	23 ♎ 58 30	24 25	18 33	26 59	17 59	14 01	05 43	16 48	20 38
30	12 35 48	07 ♎ 02 08	01 ♏ 25 25	08 ♏ 46 38	25 ♎ 53	19 ♎ 47	27 ♍ 38	18 ♐ 07	14 ♑ 02	05 ♉ 41	16 ♓ 46	20 ♑ 38

Date	Moon True ☊ ° '	Moon Mean ☊ ° '	Moon ☽ Latitude ° '
01	16 ♋ 16	14 ♋ 41	05 N 03
02	16 R 09	14 37	05 04
03	16 03	14 34	04 45
04	15 59	14 31	04 10
05	15 57	14 28	03 22
06	15 D 57	14 25	02 24
07	15 58	14 22	01 21
08	15 58	14 18	00 N 15
09	15 R 58	14 15	00 S 50
10	15 56	14 12	01 51
11	15 51	14 09	02 47
12	15 43	14 06	03 36
13	15 34	14 03	04 15
14	15 22	13 59	04 43
15	15 09	13 56	04 58
16	14 57	13 53	05 01
17	14 46	13 50	04 50
18	14 37	13 47	04 26
19	14 30	13 43	03 49
20	14 26	13 40	03 01
21	14 25	13 37	02 02
22	14 D 25	13 34	00 S 55
23	14 R 25	13 31	00 N 17
24	14 24	13 28	01 30
25	14 21	13 24	02 39
26	14 15	13 21	03 39
27	14 07	13 18	04 25
28	13 57	13 15	04 53
29	13 46	13 12	05 00
30	13 ♋ 35	13 ♋ 09	04 N 46

DECLINATIONS

Date	Sun ☉ ° '	Moon ☽ ° '	Mercury ☿ ° '	Venus ♀ ° '	Mars ♂ ° '	Jupiter ♃ ° '	Saturn ♄ ° '	Uranus ♅ ° '	Neptune ♆ ° '	Pluto ♇ ° '
01	08 N 17	01 N 26	10 N 51	07 N 41	09 N 10	22 S 16	22 S 27	13 N 11	05 S 53	22 S 18
02	07 55	04 S 13	10 07	07 12	08 56	22 17	22 28	13 11	05 54	22 19
03	07 33	09 29	09 21	06 43	08 41	22 18	22 28	13 10	05 55	22 19
04	07 11	14 05	08 35	06 14	08 26	22 18	22 28	13 10	05 55	22 19
05	06 49	17 49	07 49	05 45	08 12	22 19	22 28	13 10	05 56	22 19
06	06 27	20 30	07 02	05 15	07 57	22 20	22 29	13 09	05 57	22 19
07	06 04	22 05	06 15	04 45	07 42	22 20	22 29	13 09	05 57	22 20
08	05 42	22 32	05 28	04 15	07 27	22 21	22 29	13 08	05 58	22 20
09	05 19	21 53	04 40	03 45	07 12	22 22	22 29	13 08	05 59	22 20
10	04 56	20 14	03 53	03 15	06 57	22 22	22 29	13 07	05 59	22 20
11	04 34	17 43	03 05	02 45	06 42	22 23	22 30	13 07	06 00	22 20
12	04 11	14 28	02 18	02 15	06 27	22 24	22 30	13 06	06 01	22 20
13	03 48	10 38	01 31	01 44	06 12	22 25	22 30	13 06	06 01	22 21
14	03 25	06 24	00 N 44	01 14	05 57	22 25	22 30	13 05	06 02	22 21
15	03 02	01 S 55	00 S 03	00 43	05 42	22 26	22 30	13 05	06 03	22 21
16	02 39	02 N 40	00 49	00 N 13	05 26	22 27	22 30	13 04	06 03	22 21
17	02 16	07 12	01 36	00 S 18	05 11	22 28	22 31	13 04	06 04	22 21
18	01 52	11 30	02 21	00 48	04 56	22 28	22 31	13 03	06 05	22 21
19	01 29	15 23	03 07	01 19	04 41	22 29	22 31	13 03	06 05	22 21
20	01 06	18 39	03 52	01 50	04 25	22 30	22 31	13 02	06 06	22 22
21	00 43	21 05	04 36	02 20	04 10	22 31	22 31	13 02	06 06	22 22
22	00 N 19	22 28	05 20	02 51	03 55	22 32	22 31	13 01	06 07	22 22
23	00 S 04	22 34	06 04	03 21	03 39	22 33	22 31	13 00	06 08	22 22
24	00 27	21 16	06 47	03 52	03 24	22 33	22 31	13 00	06 08	22 22
25	00 51	18 33	07 29	04 22	03 08	22 34	22 31	12 59	06 09	22 22
26	01 14	14 35	08 11	04 52	02 53	22 35	22 31	12 58	06 10	22 22
27	01 38	09 36	08 52	05 23	02 37	22 36	22 31	12 58	06 10	22 22
28	02 01	03 N 59	09 33	05 53	02 22	22 37	22 31	12 57	06 11	22 23
29	02 24	01 S 51	10 13	06 23	02 06	22 38	22 31	12 56	06 11	22 23
30	02 S 48	07 S 29	10 S 52	06 S 53	01 N 51	22 S 39	22 S 31	12 N 56	06 S 12	22 S 23

ZODIAC SIGN ENTRIES

Date	h	m	Planets
02	23	35	☽ ♏
05	03	08	☽ ♐
07	10	37	☽ ♑
09	21	24	☽ ♒
12	09	52	☽ ♓
14	07	14	☿ ♎
14	13	43	♀ ♎
14	22	32	☽ ♈
17	10	31	☽ ♉
19	20	58	☽ ♊
22	04	50	☽ ♋
23	07	50	☉ ♎
24	09	19	☽ ♌
26	10	37	☽ ♍
28	10	03	☽ ♎
30	09	42	☽ ♏

LATITUDES

Date	Mercury ☿ ° '	Venus ♀ ° '	Mars ♂ ° '	Jupiter ♃ ° '	Saturn ♄ ° '	Uranus ♅ ° '	Neptune ♆ ° '	Pluto ♇ ° '
01	01 N 46	01 N 25	01 N 05	00 N 20	00 N 14	00 S 31	01 S 03	00 S 30
04	01 41	01 24	01 05	00 20	00 13	00 31	01 03	00 30
07	01 32	01 23	01 04	00 19	00 13	00 31	01 03	00 30
10	01 19	01 21	01 03	00 19	00 13	00 31	01 03	00 30
13	01 03	01 19	01 03	00 18	00 13	00 31	01 04	00 31
16	00 44	01 16	01 02	00 18	00 12	00 31	01 04	00 31
19	00 24	01 13	01 02	00 17	00 12	00 31	01 04	00 31
22	00 N 03	01 09	01 01	00 17	00 12	00 31	01 04	00 31
25	00 S 19	01 04	01 00	00 17	00 11	00 31	01 04	00 32
28	00 41	01 00	00 59	00 16	00 11	00 31	01 04	00 32
31	01 S 03	00 N 54	00 N 59	00 N 16	00 N 11	00 S 31	01 S 04	00 S 32

DATA

Julian Date	2458728
Delta T	+70 seconds
Ayanamsa	24° 07' 39"
Synetic vernal point	04° ♓ 59' 21"
True obliquity of ecliptic	23° 26' 11"

MOON'S PHASES, APSIDES AND POSITIONS ☽

Date	h	m	Phase	Longitude ° '	Eclipse Indicator
06	03	10	☽	13 ♐ 15	
14	04	33	○	21 ♓ 05	
22	02	41	☾	28 ♊ 49	
28	18	26	●	05 ♎ 20	

Day	h	m		
13	13	17	Apogee	
28	02	17	Perigee	
01	18	01	0S	
08	09	37	Max dec	22° S 32'
15	22	02	0N	
23	01	50	Max dec	22° N 41'
29	04	24	0S	

LONGITUDES

Date	Chiron ⚷ ° '	Ceres ⚳ ° '	Pallas ⚴ ° '	Juno ⚵ ° '	Vesta ⚶ ° '	Black Moon Lilith ⚸ ° '
01	04 ♈ 46	06 ♐ 05	02 ♏ 15	03 ♍ 48	25 ♉ 54	13 ♓ 27
11	04 ♈ 21	08 ♐ 29	06 ♏ 04	08 ♍ 13	27 ♉ 05	14 ♓ 34
21	03 ♈ 54	11 ♐ 12	10 ♏ 00	12 ♍ 31	27 ♉ 38	15 ♓ 42
31	03 ♈ 27	14 ♐ 09	14 ♏ 02	16 ♍ 44	27 ♉ 27	16 ♓ 49

All ephemeris data is given at 12.00 UT and the Moon's longitude is additionally given for 24.00 UT

ASPECTARIAN

01 Sunday
- 04 08 ☽ Q ♃
- 07 11 ☿ ⅊ ♇
- 08 39 ☽ ⊻ ☿
- 09 22 ☽ ⊼ ♅
- 13 09 ☽ ⊻ ☉
- 13 39 ☽ ⊻ ♂
- 14 11 ☿ △ ♅
- 18 49 ♀ △ ♄
- 19 41 ☽ ⊥ ☿
- 21 40 ☽ □ ♄
- 21 56 ☽ ⊻ ♀
- 23 24 ☽ ✶ ♃
- 23 27 ☽ ⊥ ☉
- 23 42 ☽ ⊥ ♂

02 Monday
- 03 10 ☽ ⊼ ♆
- 08 32 ☽ ⊥ ♀
- 08 34 ☽ □ ♇
- 10 42 ☉ ☌ ♂
- 12 29 ☽ ∠ ☿
- 12 55 ☽ ± ♆
- 15 00 ☽ ∠ ♂
- 15 07 ☽ ∠ ☉
- 16 26 ♀ □ ♃
- 19 26 ☽ ∥ ♆

03 Tuesday
- 00 02 ☽ ∠ ♃
- 00 11 ☽ # ♀
- 00 42 ☽ ∠ ♀
- 03 04 ☽ Q ♄
- 03 30 ☽ # ☉
- 03 45 ☽ ⅊ ♆
- 08 21 ☽ # ♂
- 10 18 ☽ ☍ ♅
- 11 27 ☽ # ☿
- 14 23 ☽ Q ♇
- 15 11 ☽ ⊥ ♃
- 15 40 ☿ ☌ ♂
- 17 12 ☽ ✶ ♂
- 17 21 ☽ ✶ ☿
- 17 58 ☽ ✶ ☉
- 23 18 ☽ ✶ ♄

04 Wednesday
- 01 28 ☽ ⊻ ♃
- 01 40 ☉ ☌ ☿
- 04 31 ☽ ✶ ♀
- 05 09 ☽ △ ♆
- 06 50 ☽ # ♅
- 10 58 ☽ ✶ ♇
- 11 26 ♀ ☍ ♆
- 14 53 ☽ Q ♂
- 16 15 ☽ Q ☉
- 17 26 ☽ Q ☿
- 18 56 ☿ ∥ ♂

05 Thursday
- 01 27 ☽ ∠ ♄
- 02 54 ♀ # ♆
- 03 41 ☽ Q ♀
- 12 37 ☿ △ ♄
- 13 35 ☽ ∠ ♇
- 14 35 ☽ ⊼ ♅
- 17 37 ☽ ⊥ ♄

06 Friday
- 00 49 ☽ □ ♂
- 00 55 ♃ ∥ ♇
- 01 32 ☽ ± ♅
- 03 10 ☽ □ ☉
- 04 36 ☽ ⊻ ♄
- 06 02 ☽ ⊥ ♇
- 07 11 ☿ □ ♃
- 07 20 ☽ ☌ ♃
- 07 21 ☽ □ ☿
- 10 53 ☽ □ ♆
- 16 03 ☽ □ ♀
- 17 12 ☽ ⊻ ♇
- 18 14 ☽ ⅊ ♅
- 21 56 ☉ △ ♄

07 Saturday
- 03 46 ♀ △ ♇
- 07 18 ☿ ☍ ♆
- 14 09 ♀ ⅊ ♅
- 18 41 ☽ ∥ ♇
- 19 02 ☉ # ♆
- 19 11 ☽ ∥ ♃
- 20 48 ☿ # ♆
- 20 57 ☽ Q ♆
- 22 27 ☉ ∥ ☿
- 22 47 ☽ △ ♅

08 Sunday
- 01 51 ☽ ∥ ♄
- 12 53 ☽ △ ♂
- 13 42 ☽ ☌ ♄
- 15 27 ☉ □ ♃
- 17 04 ☽ ⊻ ♃
- 17 11 ☽ △ ☉
- 17 15 ☽ ∥ ♄
- 20 18 ☽ ✶ ♆
- 23 41 ☽ ∥ ♃

09 Monday
- 00 34 ☽ ∥ ♇
- 03 02 ☽ △ ☿
- 03 03 ☽ ☌ ♇
- 03 09 ☿ △ ♇
- 04 14 ♂ △ ♄
- 04 58 ☽ ⊥ ♃
- 08 30 ☽ △ ♀
- 09 59 ☿ ⅊ ♅
- 20 12 ☽ ⅊ ♂
- 23 00 ☽ ∠ ♃

10 Tuesday
- 01 32 ☽ ⅊ ☉
- 02 02 ☽ ∠ ♆
- 07 24 ☉ ☍ ♆
- 09 57 ☽ □ ♅
- 14 18 ☽ ⅊ ☿
- 15 15 ☽ ± ♂
- 17 57 ☽ ⅊ ♀
- 20 03 ☽ ⊥ ♆
- 21 13 ☽ ± ☉

11 Wednesday
- 01 25 ☽ ⊻ ♄
- 04 02 ☽ ⊼ ♂
- 05 22 ☽ ✶ ♃
- 08 08 ☽ ⊻ ♆
- 10 24 ☽ ⊼ ☉
- 11 41 ☽ ± ☿
- 13 31 ☽ ⊥ ♄
- 14 14 ☽ ± ♀
- 15 08 ☽ ⊻ ♇
- 22 11 ☽ Q ♅

12 Thursday
- 01 58 ☽ ⊼ ☿
- 03 16 ☽ ⊥ ♇
- 03 47 ☽ ⊼ ♀
- 05 50 ☽ Q ♃
- 07 43 ☽ ∠ ♄
- 09 06 ♂ □ ♃
- 16 27 ☿ ∥ ♀
- 20 52 ☽ # ♅
- 21 29 ☽ ∠ ♇
- 22 27 ☽ ✶ ♅

13 Friday
- 14 04 ☽ ✶ ♄
- 15 11 ☿ ☌ ♀
- 18 35 ☽ □ ♃
- 19 42 ☉ △ ♇
- 20 12 ☽ ☍ ♂
- 20 43 ☽ ☌ ♆

14 Saturday
- 03 49 ☽ ✶ ♇
- 04 17 ♂ # ♆
- 04 33 ☽ ☍ ☉
- 04 45 ☽ ∠ ♅
- 05 25 ♂ ☍ ♆
- 06 53 ☉ ⅊ ♅
- 09 40 ☿ ± ♅
- 14 01 ☽ ∥ ♆
- 14 19 ☽ Q ♄
- 14 38 ☽ # ♂
- 16 58 ♀ ± ♅
- 22 52 ☽ ⊥ ♅
- 23 34 ☽ ☍ ♀

15 Sunday
- 01 08 ☽ ☍ ☿
- 04 00 ☽ Q ♇
- 05 34 ☽ # ☉
- 10 55 ☽ ⊻ ♅
- 19 02 ☽ # ♀
- 20 22 ☽ ∥ ☿

16 Monday
- 00 24 ☽ # ☿
- 00 26 ☽ ∥ ♀
- 00 35 ☿ # ♀
- 02 29 ☽ □ ♄
- 07 29 ☽ △ ♃
- 08 57 ☽ ⊻ ♆
- 11 53 ☽ ∥ ☉
- 11 58 ☽ ⊼ ♂
- 16 03 ☽ □ ♇
- 20 53 ☽ ⊥ ♆
- 21 54 ☿ Q ♃
- 22 06 ☽ ⊼ ☉

17 Tuesday
- 00 35 ☽ ± ♂
- 01 49 ☽ ∥ ♂
- 05 54 ☽ # ♆
- 11 03 ☽ ± ☉
- 13 33 ☽ ⅊ ♃
- 14 41 ☽ ∠ ♆
- 18 21 ☽ ⊼ ♀
- 19 23 ☽ ⅊ ♂
- 20 57 ☿ ⊼ ♅
- 22 31 ☽ ☌ ♅
- 22 46 ☽ ⊼ ☿

18 Wednesday
- 01 54 ☉ # ☿
- 06 19 ☽ ⅊ ☉
- 06 57 ♀ Q ♃
- 07 26 ☽ ± ♃
- 07 29 ☽ ± ♀
- 08 47 ♄ St D
- 12 24 ☽ ± ☿
- 13 52 ☽ △ ♄
- 19 15 ☽ ⊼ ♃
- 20 02 ☽ ✶ ♆
- 21 15 ☽ ∥ ♅

19 Thursday
- 02 19 ☽ △ ♂
- 02 59 ☽ ⅊ ♀
- 03 02 ☽ △ ♇
- 08 34 ☽ ⅊ ☿
- 10 27 ♀ ⊼ ♅
- 13 57 ☽ △ ☉
- 15 53 ♂ △ ♇
- 16 35 ☉ # ♀
- 18 53 ☽ ⅊ ♄
- 19 09 ☽ Q ♆

20 Friday
- 04 36 ♂ ⅊ ♅
- 07 46 ☽ ⅊ ♇
- 08 24 ☽ ⊻ ♅
- 10 50 ☽ △ ♀
- 12 00 ☽ ± ♄
- 17 24 ☽ △ ☿
- 19 41 ☽ ⊥ ♅
- 23 16 ☽ ⊼ ♄

21 Saturday
- 00 40 ☽ ± ♇
- 01 27 ☿ # ♂
- 04 54 ☽ ☍ ♃
- 05 02 ☽ □ ♆
- 11 32 ♃ ∥ ♄
- 11 47 ☽ ⊼ ♇
- 12 22 ☽ ∠ ♅
- 14 05 ☽ □ ♂
- 16 44 ♃ □ ♆

22 Sunday
- 02 41 ☽ □ ☉
- 09 16 ☽ # ♇
- 13 53 ☽ # ♄
- 14 25 ☽ # ♃
- 15 31 ☽ ✶ ♅
- 16 19 ☿ □ ♄
- 23 30 ☽ □ ♀

23 Monday
- 00 10 ☽ Q ♂
- 05 39 ☽ ☍ ♄
- 06 02 ☉ ± ♅
- 07 23 ☽ □ ☿
- 10 56 ☽ △ ♆
- 11 22 ☽ ⊼ ♃
- 12 36 ☽ Q ♅
- 12 38 ☽ # ♃
- 13 07 ☽ Q ☉
- 13 37 ☽ # ♄
- 14 12 ☿ ∥ ♆
- 17 21 ☽ ☍ ♇
- 18 01 ☽ # ♇
- 21 25 ♀ # ♂
- 21 50 ☽ ± ♃
- 22 05 ☽ ✶ ♂

24 Tuesday
- 09 41 ☽ Q ♀
- 11 13 ☽ ✶ ☉
- 12 34 ☽ ⅊ ♆
- 13 16 ☽ ⅊ ♃
- 13 57 ☿ ⊼ ♆
- 18 03 ☽ Q ☿
- 19 12 ☽ □ ♅
- 21 01 ☿ ✶ ♃

25 Wednesday
- 00 36 ☽ ∠ ♂
- 03 34 ☽ ± ♆
- 07 35 ☽ ✶ ♀
- 08 35 ☽ ⊼ ♄
- 13 25 ☽ ⊼ ♆
- 13 55 ☽ ∠ ☉
- 14 21 ☽ △ ♃
- 16 05 ☽ ⊥ ♂
- 16 14 ☽ ✶ ☿
- 18 24 ☽ ± ♄
- 19 20 ♀ □ ♄
- 19 32 ☽ ⊼ ♇

26 Thursday
- 02 15 ☽ ⊻ ♂
- 05 14 ☽ ± ♇
- 05 33 ☽ ⊥ ☉
- 08 57 ☽ ⅊ ♄
- 10 10 ☽ ∠ ♀
- 15 50 ☽ ⊻ ☉
- 19 09 ☽ ∠ ☿
- 19 37 ☽ ⅊ ♇
- 19 54 ☽ △ ♅
- 20 13 ☽ ∥ ♅
- 23 52 ☿ □ ♇

27 Friday
- 01 48 ☽ ⊥ ♀
- 08 50 ☽ △ ♄
- 11 00 ☽ ⊥ ☿
- 12 08 ☽ ⊻ ♀
- 13 21 ☽ ☍ ♆
- 14 45 ☽ □ ♃
- 14 53 ☽ # ☿
- 19 20 ☽ △ ♇
- 19 33 ☽ ⅊ ♅
- 21 29 ☽ ⊻ ☿

28 Saturday
- 02 50 ♀ ⊼ ♆
- 02 52 ☽ # ♆
- 03 58 ☽ ☌ ♂
- 04 45 ☽ # ♀
- 09 40 ☽ ± ♅
- 10 26 ☿ ± ♆
- 18 26 ☽ ☌ ☉
- 19 01 ☽ ∥ ♂
- 19 05 ☽ ⊼ ♅
- 19 18 ☽ Q ♃
- 19 38 ☽ # ☉
- 23 41 ♀ ✶ ♃

29 Sunday
- 00 53 ☉ # ♂
- 02 44 ♀ ∥ ♆
- 04 14 ☉ ⊼ ♅
- 08 08 ☽ □ ♄
- 09 53 ☉ Q ♃
- 12 33 ☽ ⊼ ♆
- 13 02 ☽ # ♂
- 14 27 ☽ ✶ ♃
- 14 29 ☽ ∥ ☉
- 15 38 ☽ ☌ ♀
- 18 40 ☽ □ ♇
- 22 06 ☽ ± ♆

30 Monday
- 02 06 ☽ ☌ ☿
- 05 36 ☽ ⊻ ♂
- 06 21 ☽ ∥ ♆
- 09 02 ☽ ∥ ♀
- 12 33 ☽ ⅊ ♆
- 12 59 ☽ Q ♄
- 14 46 ☽ ∠ ♃
- 15 45 ☽ ⊥ ♂
- 18 56 ☽ ☍ ♅
- 21 48 ☽ ⊻ ☉
- 23 46 ☽ Q ♇

OCTOBER 2019

LONGITUDES

Date	Sidereal time h m s	Sun ☉ ° ' "	Moon ☽ ° ' "	Moon ☽ 24.00 ° ' "	Mercury ☿ ° '	Venus ♀ ° '	Mars ♂ ° '	Jupiter ♃ ° '	Saturn ♄ ° '	Uranus ♅ ° '	Neptune ♆ ° '	Pluto ♇ ° '
01	12 39 44	08 ♎ 01 07	16 ♏ 01 19	23 ♏ 08 54	27 ♎ 21	21 ♎ 02	28 ♍ 16	18 ♐ 15	14 ♑ 03	05 ♉ 39	16 ♓ 45	20 ♑ 38
02	12 43 41	09 00 07	00 ♐ 47 20	07 ♐ 01 41	28 ♎ 48	22 17	28 55	18 23	14 04	05 R 37	16 R 43	20 R 38
03	12 47 38	09 59 10	13 ♐ 46 50	20 ♐ 24 46	00 ♏ 13	23 31	29 ♍ 34	18 32	14 06	05 35	16 42	20 D 38
04	12 51 34	10 58 15	26 ♐ 55 50	03 ♑ 20 34	01 38	24 46	00 ♎ 12	18 41	14 07	05 33	16 40	20 38
05	12 55 31	11 57 21	09 ♑ 39 31	15 ♑ 53 17	03 02	26 00	00 51	18 49	14 09	05 31	16 39	20 38
06	12 59 27	12 56 29	22 ♑ 02 33	28 ♑ 07 57	04 25	27 15	01 30	18 58	14 11	05 29	16 37	20 38
07	13 03 24	13 55 39	04 ♒ 10 09	10 ♒ 09 47	05 46	28 30	02 08	19 07	14 13	05 26	16 36	20 38
08	13 07 20	14 54 50	16 ♒ 07 27	22 ♒ 03 42	07 07	29 ♎ 44	02 47	19 16	14 15	05 24	16 34	20 38
09	13 11 17	15 54 03	27 ♒ 59 06	03 ♓ 54 06	08 26	00 ♏ 59	03 26	19 26	14 17	05 22	16 33	20 39
10	13 15 13	16 53 19	09 ♓ 49 08	15 ♓ 44 34	09 44	02 13	04 04	19 35	14 19	05 20	16 32	20 39
11	13 19 10	17 52 36	21 ♓ 40 45	27 ♓ 37 56	11 01	03 28	04 43	19 44	14 21	05 17	16 30	20 39
12	13 23 07	18 51 55	03 ♈ 36 22	09 ♈ 36 14	12 16	04 42	05 22	19 54	14 23	05 15	16 29	20 39
13	13 27 03	19 51 15	15 ♈ 37 41	21 ♈ 40 50	13 30	05 57	06 00	20 04	14 26	05 13	16 28	20 40
14	13 31 00	20 50 38	27 ♈ 45 48	03 ♉ 52 42	14 43	07 12	06 39	20 14	14 28	05 10	16 26	20 40
15	13 34 56	21 50 03	10 ♉ 01 35	16 ♉ 12 36	15 53	08 26	07 18	20 23	14 31	05 08	16 25	20 40
16	13 38 53	22 49 31	22 ♉ 25 50	28 ♉ 41 26	17 02	09 41	07 57	20 33	14 33	05 05	16 24	20 41
17	13 42 49	23 49 00	04 ♊ 59 34	11 ♊ 20 28	18 09	10 55	08 36	20 44	14 36	05 03	16 22	20 41
18	13 46 46	24 48 32	17 ♊ 44 20	24 ♊ 11 27	19 14	12 10	09 14	20 54	14 39	05 01	16 21	20 41
19	13 50 42	25 48 06	00 ♋ 42 07	07 ♋ 16 39	20 16	13 24	09 53	21 04	14 42	04 58	16 20	20 42
20	13 54 39	26 47 42	13 ♋ 55 21	20 ♋ 38 34	21 16	14 39	10 32	21 15	14 45	04 56	16 19	20 42
21	13 58 36	27 47 20	27 ♋ 26 34	04 ♌ 19 35	22 13	15 54	11 11	21 25	14 48	04 53	16 18	20 43
22	14 02 32	28 47 01	11 ♌ 17 47	18 ♌ 21 13	23 06	17 08	11 50	21 36	14 51	04 51	16 17	20 43
23	14 06 29	29 ♎ 46 44	25 ♌ 29 51	02 ♍ 43 27	23 56	18 23	12 29	21 46	14 55	04 48	16 16	20 44
24	14 10 25	00 ♏ 46 30	10 ♍ 01 37	17 ♍ 23 50	24 43	19 37	13 08	21 57	14 58	04 46	16 14	20 45
25	14 14 22	01 46 17	24 ♍ 49 19	02 ♎ 17 11	25 25	20 52	13 47	22 08	15 01	04 44	16 13	20 45
26	14 18 18	02 46 07	09 ♎ 46 22	17 ♎ 15 45	26 02	22 06	14 26	22 19	15 05	04 41	16 12	20 46
27	14 22 15	03 45 58	24 ♎ 44 07	02 ♏ 10 15	26 34	23 21	15 05	22 30	15 09	04 39	16 11	20 47
28	14 26 11	04 45 52	09 ♏ 33 03	16 ♏ 51 27	27 00	24 36	15 44	22 41	15 12	04 36	16 10	20 47
29	14 30 08	05 45 48	24 ♏ 04 34	01 ♐ 11 42	27 20	25 50	16 23	22 53	15 16	04 34	16 10	20 48
30	14 34 05	06 45 46	08 ♐ 12 20	15 ♐ 06 10	27 33	27 05	17 02	23 04	15 20	04 31	16 09	20 49
31	14 38 01	07 ♏ 45 45	21 ♐ 53 02	28 ♐ 33 01	27 ♏ 38	28 ♏ 19	17 ♎ 41	23 ♐ 15	15 ♑ 24	04 ♉ 29	16 ♓ 08	20 ♑ 50

Date	Moon True ☊ ° '	Moon Mean ☊ ° '	Moon ☽ Latitude ° '
01	13 ♋ 26	13 ♋ 05	04 N 14
02	13 R 19	13 02	03 27
03	13 15	12 59	02 29
04	13 15	12 56	01 25
05	13 D 13	12 53	00 N 19
06	13 R 13	12 49	00 S 47
07	13 12	12 46	01 48
08	13 09	12 43	02 44
09	13 03	12 40	03 33
10	12 55	12 37	04 12
11	12 44	12 34	04 40
12	12 30	12 30	04 56
13	12 16	12 27	04 59
14	12 02	12 24	04 49
15	11 49	12 21	04 25
16	11 39	12 18	03 48
17	11 31	12 15	03 00
18	11 27	12 11	02 02
19	11 25	12 08	00 S 56
20	11 D 24	12 05	00 N 13
21	11 25	12 02	01 24
22	11 R 24	11 59	02 31
23	11 22	11 55	03 31
24	11 17	11 52	04 19
25	11 09	11 49	04 50
26	11 00	11 46	05 02
27	10 49	11 43	04 53
28	10 39	11 40	04 25
29	10 30	11 36	03 40
30	10 24	11 33	02 42
31	10 ♋ 20	11 ♋ 30	01 N 37

DECLINATIONS

Date	Sun ☉ ° '	Moon ☽ ° '	Mercury ☿ ° '	Venus ♀ ° '	Mars ♂ ° '	Jupiter ♃ ° '	Saturn ♄ ° '	Uranus ♅ ° '	Neptune ♆ ° '	Pluto ♇ ° '
01	03 S 11	12 S 35	11 S 31	07 S 22	01 N 35	22 S 39	22 S 31	12 N 55	06 S 13	22 S 23
02	03 34	16 49	12 09	07 52	01 19	22 40	22 31	12 54	06 13	22 23
03	03 57	19 59	12 46	08 21	01 04	22 41	22 31	12 54	06 14	22 23
04	04 20	21 59	13 22	08 50	00 48	22 42	22 31	12 53	06 14	22 23
05	04 44	22 46	13 58	09 19	00 33	22 43	22 31	12 52	06 15	22 23
06	05 07	22 24	14 33	09 48	00 17	22 44	22 31	12 51	06 15	22 23
07	05 30	20 58	15 06	10 17	00 N 01	22 45	22 31	12 51	06 16	22 23
08	05 53	18 37	15 40	10 45	00 S 14	22 46	22 30	12 50	06 16	22 23
09	06 15	15 30	16 12	11 13	00 30	22 46	22 30	12 49	06 17	22 23
10	06 38	11 46	16 43	11 41	00 45	22 47	22 30	12 48	06 18	22 23
11	07 01	07 35	17 13	12 09	01 01	22 48	22 30	12 48	06 18	22 23
12	07 23	03 S 06	17 42	12 36	01 17	22 49	22 30	12 47	06 19	22 23
13	07 46	01 N 33	18 10	13 03	01 32	22 50	22 30	12 46	06 19	22 23
14	08 08	06 11	18 37	13 30	01 48	22 51	22 30	12 45	06 20	22 24
15	08 30	10 38	19 03	13 56	02 03	22 52	22 29	12 45	06 20	22 24
16	08 53	14 42	19 28	14 22	02 19	22 52	22 29	12 44	06 21	22 24
17	09 15	18 11	19 51	14 48	02 34	22 53	22 29	12 43	06 21	22 24
18	09 36	20 51	20 14	15 13	02 50	22 54	22 29	12 42	06 21	22 24
19	09 58	22 30	20 34	15 38	03 06	22 55	22 29	12 41	06 22	22 24
20	10 20	22 56	20 54	16 03	03 21	22 56	22 28	12 41	06 22	22 24
21	10 41	22 02	21 11	16 27	03 37	22 57	22 28	12 40	06 23	22 24
22	11 02	19 49	21 27	16 51	03 52	22 57	22 28	12 39	06 23	22 24
23	11 24	16 20	21 42	17 14	04 08	22 58	22 28	12 38	06 24	22 24
24	11 45	11 48	21 54	17 37	04 23	22 59	22 27	12 37	06 24	22 24
25	12 05	06 30	22 05	17 59	04 38	23 00	22 27	12 36	06 24	22 24
26	12 26	00 N 46	22 14	18 21	04 54	23 01	22 27	12 36	06 25	22 23
27	12 46	05 S 01	22 20	18 43	05 09	23 01	22 26	12 35	06 25	22 23
28	13 06	10 29	22 24	19 03	05 24	23 02	22 26	12 34	06 25	22 23
29	13 26	15 14	22 25	19 24	05 40	23 03	22 26	12 33	06 26	22 23
30	13 46	19 00	22 24	19 44	05 55	23 03	22 25	12 32	06 26	22 23
31	14 S 06	21 S 35	22 S 19	20 S 03	06 S 10	23 S 04	22 S 25	12 N 32	06 S 26	22 S 23

ZODIAC SIGN ENTRIES

Date	h	m	Planets
02	11	44	☽ ♐
03	08	14	☿ ♏
04	04	22	♂ ♎
04	17	43	☽ ♑
07	03	42	☽ ♒
08	17	06	♀ ♏
09	16	05	☽ ♓
12	04	46	☽ ♈
14	16	24	☽ ♉
17	02	30	☽ ♊
19	10	43	☽ ♋
21	16	29	☽ ♌
23	17	20	☉ ♏
23	19	30	☽ ♍
25	20	20	☽ ♎
27	20	29	☽ ♏
29	21	58	☽ ♐

LATITUDES

Date	Mercury ☿ ° '	Venus ♀ ° '	Mars ♂ ° '	Jupiter ♃ ° '	Saturn ♄ ° '	Uranus ♅ ° '	Neptune ♆ ° '	Pluto ♇ ° '
01	01 S 03	00 N 54	00 N 59	00 N 16	00 N 11	00 S 31	01 S 04	00 S 32
04	01 25	00 48	00 58	00 15	00 10	00 31	01 04	00 32
07	01 46	00 42	00 57	00 15	00 10	00 31	01 03	00 33
10	02 06	00 36	00 56	00 14	00 10	00 31	01 03	00 33
13	02 24	00 29	00 56	00 14	00 10	00 31	01 03	00 33
16	02 39	00 22	00 55	00 14	00 09	00 31	01 03	00 33
19	02 52	00 14	00 54	00 13	00 09	00 31	01 03	00 33
22	03 01	00 N 07	00 53	00 13	00 09	00 31	01 03	00 34
25	03 04	00 S 01	00 52	00 12	00 09	00 31	01 03	00 34
28	03 00	00 09	00 51	00 12	00 08	00 31	01 03	00 34
31	02 S 46	00 S 17	00 N 50	00 N 12	00 N 08	00 S 31	01 S 03	00 S 34

DATA

Julian Date	2458758
Delta T	+70 seconds
Ayanamsa	24° 07' 41"
Synetic vernal point	04° ♓ 59' 18"
True obliquity of ecliptic	23° 26' 11"

MOON'S PHASES, APSIDES AND POSITIONS ☽

Date	h	m	Phase	Longitude ° '	Eclipse Indicator
05	16	47	☽	12 ♑ 09	
13	21	08	○	20 ♈ 14	
21	12	39	☾	27 ♋ 49	
28	03	38	●	04 ♏ 25	

Day	h	m	
10	18	21	Apogee
26	10	31	Perigee
05	16	02	Max dec 22° S 47'
13	04	02	0N
20	08	06	Max dec 22° N 57'
26	15	08	0S

LONGITUDES

Date	Chiron ⚷ ° '	Ceres ⚳ ° '	Pallas ⚴ ° '	Juno ⚵ ° '	Vesta ⚶ ° '	Black Moon Lilith ⚸ ° '
01	03 ♈ 27	14 ♐ 09	14 ♏ 02	16 ♍ 44	27 ♉ 27	16 ♓ 49
11	03 ♈ 01	17 ♐ 20	18 ♏ 10	20 ♍ 51	26 ♉ 32	17 ♓ 56
21	02 ♈ 36	20 ♐ 42	22 ♏ 22	24 ♍ 49	24 ♉ 55	19 ♓ 03
31	02 ♈ 13	24 ♐ 14	26 ♏ 37	28 ♍ 40	22 ♉ 42	20 ♓ 10

ASPECTARIAN

h m	Aspects
01 Tuesday	
04 18	♀ □ ♇
05 40	☽ ⊥ ♃
05 53	☽ ∥ ☿
07 13	☽ ∠ ♂
08 25	☽ ⊥ ☉
08 43	☽ ⚹ ♄
13 12	☽ △ ♆
13 45	☽ # ♅
15 46	☽ ⊻ ♃
19 45	☽ ⚹ ♇
21 14	☽ ⊻ ♀
02 Wednesday	
00 40	☽ ∠ ☉
08 27	☽ ⊥ ♀
09 23	☽ ⊻ ☿
09 46	☽ ⚹ ♂
10 08	☽ ∠ ♄
15 45	☿ ⊻ ♂
20 21	♀ ± ♆
21 01	☽ ⊥ ☿
21 30	☽ ⊼ ♅
21 33	☽ ∠ ♇
03 Thursday	
01 41	☽ ∠ ♀
01 52	☽ ⊥ ♄
04 42	☽ ⚹ ☉
06 40	♇ St D
07 50	☽ Q ♂
08 05	☽ ± ♅
12 34	☽ ⊻ ♄
12 50	♂ ± ♅
13 32	☽ ⊥ ♇
14 54	☽ ∠ ☿
17 01	☿ # ♅
17 14	☽ □ ♆
20 40	☽ ☌ ♃
04 Friday	
00 17	☽ Ᵽ ♅
00 24	☽ ⊻ ♇
04 05	☽ Q ☉
07 34	☽ ⚹ ♀
12 33	☿ Ᵽ ♆
18 26	☽ □ ♂
20 06	☽ ∥ ♇
20 32	☿ Q ♄
21 52	☽ ⚹ ☿
23 34	☽ ∥ ♄
05 Saturday	
02 29	☽ Q ♆
04 08	☽ △ ♅
07 21	☽ ∥ ♃
08 30	☽ Q ♀
16 47	☽ □ ☉
20 39	☽ ☌ ♄
23 38	☽ Q ☿
06 Sunday	
00 08	☽ ∥ ♃
01 26	☽ ⚹ ♆
03 24	☿ ∠ ♃
05 55	☽ ⊻ ♃
08 49	☽ ∥ ♄
09 15	☽ ☌ ♇
12 19	☽ ∥ ♇
17 50	☽ ⊥ ♃
23 25	☽ □ ♀
07 Monday	
06 17	☿ ☍ ♅
06 53	☽ ∠ ♆
07 43	☽ △ ♂
11 54	☽ ∠ ♃
14 32	☽ □ ♅
15 36	☽ □ ☿
19 07	☉ □ ♄
08 Tuesday	
00 51	☽ ⊥ ♆
08 12	☽ ⊻ ♄
09 20	☽ △ ☉
12 54	☽ ⊻ ♆
15 32	☽ Ᵽ ♂
18 27	☽ ⚹ ♃
20 21	☽ ⊥ ♄
21 08	☽ ⊻ ♇
09 Wednesday	
02 40	☽ Q ♅
07 45	☽ ∥ ☿
09 17	☽ ⊥ ♇
10 48	☽ ± ♂
13 44	☉ ∥ ♆
14 38	☽ ∠ ♄
15 51	☿ Q ♇
18 27	☽ Ᵽ ☉
18 47	☽ △ ♀
19 04	☽ Q ♃
22 49	♀ Ᵽ ♆
23 40	☽ ⊼ ♂
10 Thursday	
02 55	☽ ⚹ ♅
03 24	☉ ⊼ ♆
03 32	☽ ∠ ♇
05 38	☽ # ♅
11 49	☽ △ ☿
12 27	☽ ∥ ♀
13 47	♀ Q ♄
14 22	☽ ± ☉
21 08	☽ ⚹ ♄
11 Friday	
00 35	☿ ⊥ ♂
01 34	☽ ☌ ♆
03 37	☽ ⊼ ☉
04 45	☽ Ᵽ ♀
08 02	☽ □ ♃
09 12	☽ ∠ ♅
09 55	☽ ⚹ ♇
14 54	☽ ∥ ☉
19 00	☽ ∥ ♆
21 27	☽ Q ♄
21 47	☽ Ᵽ ☿
12 Saturday	
01 01	☽ ± ♀
03 17	☽ ⊥ ♅
08 00	♂ ⊼ ♅
10 05	☽ Q ♇
14 28	☽ ⊼ ♀
15 17	☽ ⊻ ♅
15 43	☽ ☍ ♂
16 16	♀ ∠ ♃
17 57	☽ ± ☿
20 57	☽ ∥ ♂
21 17	♀ # ♅
22 07	♀ ☍ ♅
13 Sunday	
07 18	☽ ⊼ ☿
09 36	☽ □ ♄
11 56	☽ # ♂
13 39	☽ ⊻ ♆
14 17	♀ ⊻ ♂
18 02	☉ ⚹ ♃
20 55	☽ △ ♃
21 08	☽ ☍ ☉
21 59	☽ □ ♇
14 Monday	
00 44	☿ ⊥ ♃
01 31	☽ ⊥ ♆
05 26	♃ Ᵽ ♅
06 56	☿ ⚹ ♄
07 38	☉ □ ♇
12 46	☽ # ♆
19 12	☽ ∠ ♆
23 18	☽ # ☉
15 Tuesday	
02 29	☽ ☌ ♅
02 50	☽ Ᵽ ♃
06 24	☽ ⊼ ♂
08 33	☽ ☍ ♀
16 33	♀ Q ♇
18 43	☽ ± ♂
20 35	☽ ± ♃
20 44	☽ △ ♄
22 45	☿ △ ♆
16 Wednesday	
00 04	☽ ∥ ♅
00 23	☽ ⚹ ♆
00 33	☽ ☍ ☿
01 48	☉ ± ♆
08 21	☽ ⊼ ♃
08 37	☽ △ ♇
09 41	☽ # ♀
12 49	☽ ⊼ ☉
13 03	☽ Ᵽ ♂
23 25	☽ Q ♆
17 Thursday	
01 18	☽ ± ☉
01 42	☽ Ᵽ ♄
05 45	♃ ⊻ ♇
12 07	☽ ⊻ ♅
13 19	☽ Ᵽ ♇
18 39	♂ Q ♃
18 51	☽ ± ♄
19 11	☽ △ ♂
19 51	☽ Ᵽ ☉
23 25	☽ ⊥ ♅
18 Friday	
00 25	☽ ⊼ ♀
04 28	☽ # ☿
06 12	☽ ⊼ ♄
06 18	☽ ± ♇
09 25	☽ □ ♆
12 53	☽ ± ♀
15 02	☽ ⊼ ☿
16 13	☽ ∠ ♅
17 30	☽ ⊼ ♇
17 58	☽ ☍ ♃
19 Saturday	
02 14	☽ △ ☉
03 08	☽ ± ☿
07 20	☽ Ᵽ ♀
09 50	☽ # ♇
11 36	☽ # ♄
19 47	☽ ⚹ ♅
21 02	☽ Ᵽ ☿
22 21	☿ ⚹ ♇
20 Sunday	
03 45	☽ # ♃
05 35	☽ □ ♂
11 23	☿ ⊻ ♃
11 59	☽ # ♃
13 26	☽ △ ♀
13 29	☽ ☍ ♄
13 58	♀ ⚹ ♄
16 17	☽ △ ♆
17 22	☽ Q ♅
21 Monday	
00 07	☽ ☍ ♇
01 14	☽ ⊼ ♃
01 18	♀ ⊥ ♃
02 06	☽ △ ☿
04 24	☽ # ♄
05 57	☽ # ♇
11 57	☽ ± ♃
12 39	☽ □ ☉
15 12	☽ Q ♂
18 44	☽ Ᵽ ♆
19 40	♀ △ ♆
21 41	☽ # ☿
22 Tuesday	
00 56	☽ □ ♅
03 49	☽ Ᵽ ♃
10 15	☽ ± ♆
12 58	☽ ⚹ ♂
18 05	☽ ⊼ ♄
20 28	☽ ⊼ ♆
22 03	☽ Q ☉
22 54	☽ □ ♀
23 Wednesday	
04 01	☽ ⊼ ♇
04 17	☽ ± ♄
05 41	☽ △ ♃
07 00	☽ # ♀
09 14	☽ □ ☿
14 04	☽ ± ♇
15 28	☽ ∠ ♂
16 10	♀ ⊥ ♂
19 22	☽ Ᵽ ♄
19 39	☽ ⚹ ☉
24 Thursday	
03 24	☽ △ ♅
04 58	☽ Ᵽ ♇
07 02	☽ ⊥ ♂
07 42	☽ Q ♀
07 58	☽ ∥ ♅
12 16	☽ # ☉
16 37	☽ Q ☿
17 18	☽ ⊻ ♂
20 05	☽ △ ♄
22 03	☽ ∠ ☉
22 07	☽ ☍ ♆
23 03	☉ Ᵽ ♆
25 Friday	
03 48	☽ Ᵽ ♅
05 02	☽ ⚹ ♀
05 26	☽ △ ♇
07 37	☽ □ ♃
09 52	♀ ⚹ ♇
12 23	☽ # ♆
13 00	☽ ⚹ ☿
13 38	☽ ⊥ ☉
18 16	☽ ± ♅
19 33	☽ # ♂
23 58	☽ ⊻ ☉
26 Saturday	
03 52	☽ ⊼ ♅
07 21	☽ ∠ ♀
12 53	☽ Q ♃
14 06	☽ ∠ ☿
16 47	♀ ⊻ ♃
19 48	☽ ☌ ♂
20 01	☉ Q ♄
20 32	☽ □ ♄
22 18	☽ ⊼ ♆
23 01	☉ # ♅
23 04	☽ ⊥ ♀
27 Sunday	
05 04	☽ ⊥ ☿
05 38	☽ □ ♇
07 55	☽ ± ♆
08 22	☽ ⚹ ♃
09 34	☽ ⊻ ♀
12 34	☽ ∥ ♂
14 31	♂ □ ♄
15 03	☽ ⊻ ☿
17 57	☽ ∥ ♆
22 24	☽ Ᵽ ♆
28 Monday	
01 38	☽ Q ♄
03 38	☽ ☌ ☉
03 58	☽ ☍ ♅
08 15	☉ ☍ ♅
08 29	☿ ∥ ♇
08 55	☽ ∠ ♃
10 45	☽ Q ♇
21 19	☽ ⚹ ♄
22 00	☽ # ♅
22 37	☽ ⊻ ♂
22 52	☽ △ ♆
23 53	☽ ⊥ ♃
29 Tuesday	
01 43	☽ ∥ ☉
03 57	♂ ⊼ ♆
06 32	☽ ⚹ ♇
09 02	☽ ⊥ ♂
09 58	☽ ⊻ ♃
15 14	☽ ☌ ♀
17 34	☽ ☌ ☿
22 28	☽ ∠ ♄
30 Wednesday	
00 55	☽ ∠ ♂
05 41	☽ ⊼ ♅
07 53	☽ ∠ ♇
09 19	☽ ⊻ ☉
13 57	☽ ⊥ ♄
14 00	☿ ∥ ♇
16 00	☽ ± ♅
18 27	☽ ∥ ♀
20 32	☽ ⊥ ☉
22 05	☿ ☌ ♀
23 31	☽ ⊥ ♇
31 Thursday	
00 28	☽ ⊻ ♄
01 49	☽ □ ♆
04 10	☽ ⚹ ♂
07 44	☽ Ᵽ ♅
10 07	☽ ⊻ ♇
13 42	☽ ∠ ☉
14 29	☽ ☌ ♃
15 40	☿ St R
21 53	☽ ∥ ☿
22 20	☽ ⊻ ☿
23 58	☽ ∥ ♇

All ephemeris data is given at 12.00 UT and the Moon's longitude is additionally given for 24.00 UT

NOVEMBER 2019

LONGITUDES

Date	Sidereal time h	m	s	Sun ☉ °	'	"	Moon ☽ °	'	"	Moon ☽ 24.00 °	'	"	Mercury ☿ °	'	Venus ♀ °	'	Mars ♂ °	'	Jupiter ♃ °	'	Saturn ♄ °	'	Uranus ♅ °	'	Neptune ♆ °	'	Pluto ♇ °	'
01	14	41	58	08 ♏	45	46	05 ♑	06	19	11 ♑	33	15	27 ♏	35	29 ♏	34	18 ♎	20	23 ♐	27	15 ♑	28	04 ♉	26	16 ♓	07	20 ♑	51
02	14	45	54	09	45	49	17 ♑	54	17	24 ♑	09	56	27 R	24	00 ♐	48	18	59	23	39	15	32	04 R	24	16 R	06	20	52
03	14	49	51	10	45	54	00 ♒	20	48	06 ♒	27	29	27	03	02	03	19	38	23	50	15	36	04	21	16	05	20	52
04	14	53	47	11	46	00	12 ♒	30	40	18 ♒	30	59	26	32	03	18	20	18	24	02	15	41	04	19	16	04	20	53
05	14	57	44	12	46	07	24 ♒	29	07	00 ♓	25	42	25	52	04	32	20	57	24	14	15	45	04	17	16	04	20	54
06	15	01	40	13	46	16	06 ♓	21	20	12 ♓	16	38	25	02	05	47	21	36	24	26	15	49	04	14	16	03	20	55
07	15	05	37	14	46	27	18 ♓	12	08	24 ♓	08	19	24	04	07	01	22	15	24	38	15	54	04	12	16	02	20	56
08	15	09	34	15	46	39	00 ♈	05	39	06 ♈	04	33	22	57	08	16	22	54	24	50	15	58	04	09	16	02	20	57
09	15	13	30	16	46	52	12 ♈	05	19	18 ♈	08	17	21	44	09	30	23	34	25	02	16	03	04	07	16	01	20	58
10	15	17	27	17	47	08	24 ♈	13	39	00 ♉	21	37	20	27	10	45	24	13	25	14	16	08	04	04	16	00	20	59
11	15	21	23	18	47	25	06 ♉	32	18	12 ♉	45	47	19	07	11	59	24	52	25	26	16	13	04	02	16	00	21	01
12	15	25	20	19	47	43	19 ♉	02	08	25 ♉	21	20	17	48	13	14	25	31	25	39	16	18	04	00	15	59	21	02
13	15	29	16	20	48	04	01 ♊	43	24	08 ♊	08	19	16	31	14	28	26	11	25	51	16	22	03	57	15	59	21	03
14	15	33	13	21	48	26	14 ♊	36	04	21 ♊	06	37	15	20	15	43	26	50	26	03	16	27	03	55	15	58	21	04
15	15	37	09	22	48	50	27 ♊	39	59	04 ♋	16	09	14	16	16	57	27	29	26	16	16	33	03	53	15	58	21	05
16	15	41	06	23	49	15	10 ♋	55	11	17 ♋	37	06	13	21	18	11	28	09	26	28	16	38	03	51	15	58	21	07
17	15	45	03	24	49	43	24 ♋	21	59	01 ♌	09	53	12	38	19	26	28	48	26	41	16	43	03	48	15	57	21	08
18	15	48	59	25	50	12	08 ♌	00	54	14 ♌	55	04	12	05	20	40	29 ♎	28	26	54	16	48	03	46	15	57	21	09
19	15	52	56	26	50	44	21 ♌	52	24	28 ♌	52	55	11	45	21	55	00 ♏	07	27	06	16	53	03	44	15	57	21	10
20	15	56	52	27	51	17	05 ♍	56	32	13 ♍	03	04	11	36	23	09	00	47	27	19	16	59	03	42	15	56	21	12
21	16	00	49	28	51	52	20 ♍	12	19	27 ♍	23	55	11 D	38	24	24	01	26	27	32	17	04	03	39	15	56	21	13
22	16	04	45	29 ♏	52	28	04 ♎	37	25	11 ♎	52	18	11	51	25	38	02	06	27	45	17	10	03	37	15	56	21	14
23	16	08	42	00 ♐	53	06	19 ♎	07	54	26 ♎	23	30	12	13	26	53	02	45	27	58	17	15	03	35	15	56	21	16
24	16	12	38	01	53	46	03 ♏	38	22	10 ♏	51	39	12	44	28	07	03	25	28	11	17	21	03	33	15	56	21	17
25	16	16	35	02	54	28	18 ♏	02	35	25 ♏	10	25	13	24	29 ♐	21	04	04	28	24	17	27	03	31	15	56	21	19
26	16	20	32	03	55	11	02 ♐	14	25	09 ♐	14	01	14	10	00 ♑	36	04	44	28	37	17	32	03	29	15	56	21	20
27	16	24	28	04	55	56	16 ♐	08	41	22 ♐	58	03	15	03	01	50	05	24	28	50	17	38	03	27	15	56	21	22
28	16	28	25	05	56	41	29 ♐	41	53	06 ♑	20	04	16	01	03	05	06	03	29	03	17	44	03	25	15 D	56	21	23
29	16	32	21	06	57	28	12 ♑	52	36	19 ♑	19	37	17	04	04	19	06	43	29	17	17	50	03	23	15	56	21	25
30	16	36	18	07 ♐	58	16	25 ♑	41	21	01 ♒	58	08	18 ♏	11	05 ♑	33	07 ♏	23	29 ♐	30	17 ♑	56	03 ♉	21	15 ♓	56	21 ♑	26

Moon True ☊, Mean ☊ and Latitude

Date	Moon True ☊ °	'	Moon Mean ☊ °	'	Moon ☽ Latitude °	'
01	10 ♋	18	11 ♋	27	00 N	28
02	10 D	18	11	24	00 S	41
03	10	19	11	20	01	45
04	10 R	20	11	17	02	43
05	10	19	11	14	03	34
06	10	16	11	11	04	14
07	10	11	11	08	04	44
08	10	03	11	05	05	01
09	09	54	11	01	05	05
10	09	44	10	58	04	56
11	09	34	10	55	04	33
12	09	25	10	52	03	56
13	09	17	10	49	03	08
14	09	12	10	46	02	08
15	09	10	10	42	01 S	02
16	09 D	09	10	39	00 N	10
17	09	10	10	36	01	21
18	09	11	10	33	02	30
19	09	12	10	30	03	31
20	09 R	12	10	26	04	20
21	09	10	10	23	04	53
22	09	06	10	20	05	09
23	09	01	10	17	05	05
24	08	55	10	14	04	42
25	08	49	10	11	04	01
26	08	44	10	07	03	05
27	08	41	10	04	02	00
28	08	39	10	01	00 N	49
29	08 D	38	09	58	00 S	23
30	08 ♋	39	09 ♋	55	01 S	32

DECLINATIONS

Date	Sun ☉ °	'	Moon ☽ °	'	Mercury ☿ °	'	Venus ♀ °	'	Mars ♂ °	'	Jupiter ♃ °	'	Saturn ♄ °	'	Uranus ♅ °	'	Neptune ♆ °	'	Pluto ♇ °	'
01	14 S	25	22 S	53	22 S	11	20 S	22	06 S	25	23 S	05	22 S	25	12 N	31	06 S	27	22 S	23
02	14	44	22	55	22	00	20	40	06	41	23	06	22	24	12	30	06	27	22	23
03	15	03	21	47	21	45	20	58	06	56	23	06	22	24	12	29	06	27	22	23
04	15	22	19	40	21	26	21	15	07	11	23	07	22	23	12	28	06	28	22	23
05	15	40	16	43	21	03	21	32	07	26	23	08	22	23	12	27	06	28	22	23
06	15	58	13	07	20	36	21	48	07	41	23	08	22	23	12	27	06	28	22	23
07	16	16	09	01	20	05	22	03	07	56	23	09	22	22	12	26	06	28	22	23
08	16	34	04 S	34	19	31	22	18	08	11	23	09	22	22	12	25	06	29	22	23
09	16	51	00 N	06	18	53	22	32	08	25	23	10	22	21	12	24	06	29	22	23
10	17	08	04	48	18	13	22	45	08	40	23	11	22	21	12	23	06	29	22	23
11	17	25	09	24	17	32	22	58	08	55	23	11	22	20	12	23	06	29	22	22
12	17	41	13	41	16	50	23	10	09	10	23	12	22	20	12	22	06	29	22	22
13	17	57	17	26	16	09	23	22	09	24	23	12	22	19	12	21	06	30	22	22
14	18	13	20	25	15	30	23	32	09	39	23	13	22	19	12	20	06	30	22	22
15	18	28	22	23	14	55	23	42	09	53	23	13	22	18	12	20	06	30	22	22
16	18	44	23	09	14	24	23	52	10	08	23	13	22	18	12	19	06	30	22	22
17	18	58	22	35	13	57	24	00	10	22	23	14	22	17	12	18	06	30	22	22
18	19	13	20	40	13	36	24	08	10	36	23	14	22	16	12	17	06	30	22	22
19	19	27	17	32	13	20	24	16	10	51	23	15	22	16	12	17	06	30	22	22
20	19	41	13	21	13	10	24	22	11	05	23	15	22	15	12	16	06	30	22	21
21	19	54	08	22	13	05	24	28	11	19	23	15	22	15	12	15	06	31	22	21
22	20	07	02 N	54	13	05	24	33	11	33	23	16	22	14	12	14	06	31	22	21
23	20	20	02 S	47	13	09	24	37	11	47	23	16	22	13	12	14	06	31	22	21
24	20	32	08	18	13	17	24	41	12	01	23	16	22	13	12	13	06	31	22	21
25	20	44	13	21	13	29	24	43	12	15	23	17	22	12	12	12	06	31	22	21
26	20	56	17	35	13	44	24	45	12	28	23	17	22	11	12	12	06	31	22	21
27	21	07	20	44	14	02	24	47	12	42	23	17	22	11	12	11	06	31	22	20
28	21	18	22	37	14	21	24	47	12	55	23	17	22	10	12	10	06	31	22	20
29	21	28	23	12	14	43	24	47	13	09	23	18	22	09	12	10	06	31	22	20
30	21 S	38	22 S	31	15 S	06	24 S	46	13 S	22	23 S	18	22 S	09	12 N	09	06 S	30	22 S	20

ZODIAC SIGN ENTRIES

Date	h	m	Planets
01	02	38	☽ ♑
01	20	25	♀ ♐
03	11	19	☽ ♒
05	23	08	☽ ♓
08	11	49	☽ ♈
10	23	18	☽ ♉
13	08	46	☽ ♊
15	16	15	☽ ♋
17	21	57	☽ ♌
19	07	40	♂ ♏
20	01	54	☽ ♍
22	04	20	☽ ♎
22	14	59	☉ ♐
24	05	58	☽ ♏
26	00	28	♀ ♑
26	08	11	☽ ♐
28	12	33	☽ ♑
30	20	13	☽ ♒

LATITUDES

Date	Mercury ☿ °	'	Venus ♀ °	'	Mars ♂ °	'	Jupiter ♃ °	'	Saturn ♄ °	'	Uranus ♅ °	'	Neptune ♆ °	'	Pluto ♇ °	'
01	02 S	38	00 S	19	00 N	50	00 N	12	00 N	08	00 S	31	01 S	03	00 S	34
04	02	07	00	27	00	48	00	11	00	08	00	31	01	03	00	35
07	01	21	00	35	00	47	00	11	00	07	00	31	01	03	00	35
10	00 S	22	00	43	00	46	00	11	00	07	00	31	01	03	00	35
13	00 N	39	00	50	00	45	00	10	00	07	00	31	01	03	00	35
16	01	31	00	57	00	44	00	10	00	07	00	31	01	03	00	36
19	02	07	01	04	00	43	00	10	00	06	00	31	01	03	00	36
22	02	25	01	11	00	42	00	09	00	06	00	31	01	03	00	36
25	02	29	01	17	00	40	00	09	00	06	00	31	01	03	00	36
28	02	22	01	23	00	39	00	09	00	06	00	31	01	02	00	36
31	02 N	09	01 S	29	00 N	38	00 N	08	00 N	05	00 S	30	01 S	02	00 S	37

DATA

Julian Date	2458789
Delta T	+70 seconds
Ayanamsa	24° 07' 45"
Synetic vernal point	04° ♓ 59' 14"
True obliquity of ecliptic	23° 26' 10"

MOON'S PHASES, APSIDES AND POSITIONS ☽

Date	h	m	Phase	Longitude °	'	Eclipse Indicator
04	10	23	☽	11 ♒	42	
12	13	34	○	19 ♉	52	
19	21	11	☾	27 ♌	14	
26	15	06	●	04 ♐	03	

Day	h	m		
07	08	33	Apogee	
23	07	30	Perigee	
02	00	35	Max dec	23° S 03'
09	11	32	ON	
16	13	53	Max dec	23° N 09'
23	00	15	OS	
29	10	35	Max dec	23° S 12'

LONGITUDES

Date	Chiron ⚷ °	'	Ceres ⚳ °	'	Pallas ⚴ °	'	Juno ⚵ °	'	Vesta ⚶ °	'	Black Moon Lilith ⚸ °	'
01	02 ♈	11	24 ♐	35	27 ♏	02	29 ♍	02	22 ♉	28	20 ♓	17
11	01 ♈	53	28 ♐	15	01 ♐	19	02 ♎	41	19 ♉	54	21 ♓	24
21	01 ♈	39	02 ♑	01	05 ♐	37	06 ♎	08	17 ♉	20	22 ♓	31
31	01 ♈	30	05 ♑	51	09 ♐	55	09 ♎	22	15 ♉	04	23 ♓	38

All ephemeris data is given at 12.00 UT and the Moon's longitude is additionally given for 24.00 UT

ASPECTARIAN

h m	Aspects
01 Friday	
00 26	☽ ∥ ♄
00 47	☽ ⊻ ♀
02 40	☉ ∠ ♃
02 48	☽ Q ♂
09 14	☽ ⊥ ☿
10 11	☽ Q ♆
10 46	☽ △ ♅
12 56	☽ ⊥ ♀
13 58	☉ Q ♇
14 10	♂ ∥ ♆
19 21	☽ ⚹ ☉
02 Saturday	
01 46	☽ ∠ ☿
06 26	♀ ∠ ♄
07 29	☽ ☌ ♄
07 35	☽ ∠ ♀
08 35	☽ ⚹ ♆
14 11	☽ □ ♂
17 39	☽ ☌ ♇
20 01	☽ Q ☉
23 10	☽ ⊻ ♃
03 Sunday	
01 42	☽ ∥ ♄
01 58	☽ ∥ ♇
05 46	☽ ⚹ ☿
10 59	☽ ⊥ ♃
12 38	☽ ∥ ☿
13 27	☽ ∠ ♆
15 42	☽ ⚹ ♀
19 50	☽ □ ♅
21 12	☽ ∥ ♀
04 Monday	
04 27	☽ Q ☿
04 59	☽ ∠ ♃
07 10	☽ ⊥ ♆
10 23	☽ □ ☉
16 37	☿ ⊥ ♂
18 11	☽ Q ♀
18 21	☽ ⊻ ♄
18 42	☿ ∥ ♀
19 06	☽ ⊻ ♆
05 Tuesday	
04 28	☽ △ ♂
04 47	☽ ⊻ ♇
06 27	☽ ⊥ ♄
07 09	♀ ⊼ ♅
07 34	☽ Q ♅
10 10	♄ ∥ ♇
10 28	♂ □ ♇
11 29	☽ ⚹ ♃
14 37	☽ □ ☿
16 53	☽ ⊥ ♇
18 47	☽ ∥ ☉
06 Wednesday	
00 44	☽ ∠ ♄
07 43	☽ ⚹ ♅
10 41	☽ □ ♀
11 07	☽ ∠ ♇
12 09	☽ Q ♃
12 31	☽ bQ ♂
14 50	♀ ∠ ♇
16 06	☽ # ♅
07 Thursday	
00 48	☿ ⊻ ♃
04 19	♂ ± ♆
04 25	☽ △ ☉
07 18	☽ ⚹ ♄
07 38	☽ ☌ ♆
07 49	☽ ± ♂
14 00	☽ ∠ ♅
17 33	☽ ⚹ ♇
17 40	☽ ∥ ♂
20 40	☽ ⊼ ♂
21 32	♀ ∠ ♂
22 52	☽ △ ☿
08 Friday	
01 13	☽ □ ♃
01 52	☽ ∥ ♆
07 42	☽ Q ♄
08 07	☽ ⊥ ♅
12 38	☿ ⊻ ♂
13 30	☽ bQ ☉
17 06	☉ ⚹ ♄
17 45	☽ Q ♇
17 56	☉ △ ♆
18 08	♀ ∥ ♄
20 07	☽ ⊻ ♅
20 11	♀ ∥ ♇
09 Saturday	
02 20	☽ bQ ☿
02 45	♄ ⚹ ♆
06 15	☽ △ ♀
09 10	☽ ± ☉
18 34	☽ ± ☿
19 48	☽ ⊻ ♆
19 55	☽ □ ♄
22 09	☽ ⊼ ☉
23 21	♀ ⊥ ♄
23 28	♀ ± ♅
10 Sunday	
02 09	☿ ⚹ ♇
05 17	☽ ⊼ ☿
05 37	☽ □ ♇
07 38	☽ ⊥ ♆
11 58	☽ ☍ ♂
14 00	☽ △ ♃
15 19	☽ bQ ♀
20 40	☽ # ♆
11 Monday	
01 15	☽ ∠ ♆
07 01	☿ ⊥ ♃
07 10	☽ ☌ ♅
09 16	☽ # ♂
10 49	☽ ± ♀
14 51	☉ ∥ ☿
15 22	☉ ☌ ☿
19 39	☽ bQ ♃
23 40	☽ ⊼ ♀
12 Tuesday	
04 22	☽ ∥ ♅
06 11	☽ ⚹ ♆
06 44	☽ △ ♄
07 25	☉ ⊥ ♃
09 51	☽ ☍ ☿
13 11	☽ ± ♃
13 34	☽ ☍ ☉
15 00	♀ ∥ ♃
15 48	☽ △ ♇
18 21	♂ ⚹ ♃
13 Wednesday	
00 45	☽ ⊼ ♃
00 59	☽ ⊼ ♂
04 37	☽ # ☿
04 58	☽ Q ♆
11 20	☽ bQ ♄
12 54	☽ ± ♂
14 35	☿ ⚹ ♄
15 59	☽ # ☉
16 11	☽ ⊻ ♅
18 00	☉ ⚹ ♇
20 07	☽ bQ ♇
22 34	☿ △ ♆
23 24	♀ ⊥ ♇
14 Thursday	
03 20	☽ ⊥ ♅
04 16	☽ ± ♄
06 36	☽ bQ ♂
08 09	☿ ⊻ ♀
12 52	☽ ± ♇
13 14	☽ ⊼ ☿
14 16	☽ ☍ ♀
14 32	☽ □ ♆
15 27	☽ ⊼ ♄
17 06	♀ □ ♆
19 57	☽ ∠ ♅
23 26	☽ ± ☿
23 56	☽ ⊼ ♇
15 Friday	
02 23	☽ ⊼ ☉
03 32	♀ ⊻ ♄
09 24	☽ ☍ ♃
10 33	☽ # ♄
11 38	☽ # ♇
11 40	☽ △ ♂
14 16	☽ ± ☉
14 43	☽ bQ ☿
23 16	☽ ⚹ ♅
16 Saturday	
07 55	☽ bQ ☉
16 08	☽ △ ☿
20 48	☽ Q ♅
21 02	☽ △ ♆
22 18	☽ ☍ ♄
17 Sunday	
00 14	♀ bQ ♅
01 56	☿ ⊥ ♀
02 21	☽ ⊼ ♀
06 15	☽ ☍ ♇
12 53	☽ △ ☉
14 05	☽ ± ♀
15 49	☽ # ♇
16 10	☽ ⊼ ♃
17 08	☽ # ♄
20 14	☽ □ ♂
23 38	☽ bQ ♆
18 Monday	
02 54	☽ ± ♃
04 35	☽ □ ♅
07 30	☽ bQ ♀
15 22	☽ ± ♆
18 52	☽ bQ ♃
18 53	☽ □ ☿
19 28	☿ ∠ ♃
21 25	♀ ⊻ ♇
23 24	☽ # ☉
19 Tuesday	
01 47	☽ ⊼ ♆
03 21	☽ ⊼ ♄
05 13	☽ Q ♂
10 48	☽ ⊼ ♇
12 05	☽ △ ♀
13 46	☽ ± ♄
19 54	☉ ⊻ ♃
21 06	☽ ± ♇
21 06	☽ △ ♃
21 11	☽ □ ☉
20 Wednesday	
01 17	☽ Q ☿
02 48	☽ ⚹ ♂
05 14	☽ bQ ♄
08 12	☽ △ ♅
12 26	☽ bQ ♇
12 57	☽ # ☿
17 32	☽ ∥ ♅
17 57	♂ bQ ♆
19 12	☿ St D
21 32	☽ ⚹ ☿
22 49	☽ # ♂
21 Thursday	
04 51	☽ ☍ ♆
05 23	☽ ∠ ♂
05 59	☽ Q ☉
06 43	☽ △ ♄
09 25	☽ bQ ♅
13 42	☽ △ ♇
19 39	☽ □ ♀
20 21	☽ # ♆
21 09	☽ ⊥ ♂
22 51	☽ ∠ ☿
22 Friday	
00 24	☽ ± ♅
00 25	☽ □ ♃
03 31	☽ ⚹ ☉
07 36	☽ ⊻ ♂
10 21	☽ ⊼ ♅
14 04	☽ ⊥ ☿
23 Saturday	
00 14	☽ ⊻ ☿
04 19	☽ Q ♀
06 14	☽ ∠ ☉
06 41	☽ Q ♃
06 43	☽ ⊼ ♆
08 53	☽ □ ♄
15 32	☽ □ ♇
16 37	☽ ± ♆
22 13	☽ ⊥ ☉
22 51	☿ ∠ ♀
24 Sunday	
02 00	☽ ⚹ ♀
02 49	☽ ⚹ ♃
04 03	☽ ∥ ♆
07 31	☽ bQ ♆
08 54	☽ ⊻ ☉
11 36	☽ ☌ ♂
11 51	☽ ☍ ♅
13 33	♀ ☌ ♃
14 51	☽ Q ♄
16 51	♂ ☍ ♅
21 24	☽ Q ♇
23 53	☉ ∠ ♄
25 Monday	
03 50	☽ ☌ ☿
04 07	☽ ∠ ♃
05 15	☽ ∠ ♀
06 12	☽ ∥ ♂
06 18	☽ # ♅
08 25	♂ # ♅
08 27	☽ △ ♆
10 59	☽ ⚹ ♄
12 12	☿ ∠ ♃
12 46	☽ ∥ ☿
17 30	☽ ⚹ ♇
19 26	☽ ⊥ ♃
21 47	☽ ⊥ ♀
26 Tuesday	
02 00	☉ ⊼ ♅
05 44	☽ ⊻ ♃
08 56	☽ ⊻ ♀
12 31	☽ ∠ ♄
14 07	☽ ⊼ ♅
15 06	☽ ☌ ☉
16 28	☽ ⊻ ♂
19 01	☽ ∠ ♇
27 Wednesday	
00 24	☽ ± ♅
03 19	☽ ⊥ ♂
04 06	☽ ⊥ ♄
09 57	☽ ⊻ ☿
10 38	☽ ⊥ ♇
11 37	☽ □ ♆
12 31	♆ St D
14 38	☽ ⊻ ♄
16 02	☽ bQ ♅
16 06	☽ ∥ ☉
19 50	☽ ∠ ♂
21 11	☽ ⊻ ♇
21 15	☽ ⊥ ☿
22 27	♂ Q ♄
28 Thursday	
04 28	☽ ∥ ♄
06 59	☽ ∥ ♇
09 51	☿ △ ♆
10 50	☽ ☌ ♃
14 34	☽ ∠ ☿
18 27	♀ △ ♅
18 41	☽ △ ♅
18 43	☽ ☌ ♀
19 21	☉ ⊻ ♂
19 38	☽ Q ♆
22 44	☉ ∠ ♇
29 Friday	
00 06	☽ ⚹ ♂
00 14	☽ ⊻ ☉
04 30	♀ Q ♆
12 10	☽ ⊥ ☉
17 39	☽ ⚹ ♆
20 30	☽ ⚹ ☿
21 17	☽ ☌ ♄
23 27	☽ Q ♂
30 Saturday	
03 57	☽ ☌ ♇
06 13	☿ ⚹ ♄
06 24	☽ ∠ ☉
15 11	☽ ∥ ♇
18 14	☽ ∥ ♄
19 23	☽ ⊻ ♃
21 27	☽ Q ☿
22 00	☽ ∠ ♆

DECEMBER 2019

Date	Sidereal time h m s	Sun ☉ ° ' "	Moon ☽ ° ' "	Moon ☽ 24.00 ° ' "	Mercury ☿ ° '	Venus ♀ ° '	Mars ♂ ° '	Jupiter ♃ ° '	Saturn ♄ ° '	Uranus ♅ ° '	Neptune ♆ ° '	Pluto ♇ ° '
01	16 40 14	08 ♐ 59 05	08 ♒ 10 23	14 ♒ 18 34	19 ♏ 22	06 ♑ 48	08 ♏ 02	29 ♐ 43	18 ♑ 02	03 ♉ 19	15 ♓ 56	21 ♑ 28
02	16 44 11	09 59 55	20 ♒ 23 13	26 ♒ 24 54	20 36	08 02	08 42	29 ♐ 56	18 08	03 R 17	15 56	21 29
03	16 48 07	11 00 46	02 ♓ 24 12	08 ♓ 21 45	21 52	09 16	09 22	00 ♑ 10	18 14	03 16	15 56	21 31
04	16 52 04	12 01 38	14 ♓ 18 10	20 ♓ 14 04	23 11	10 30	10 01	00 23	18 20	03 14	15 56	21 33
05	16 56 01	13 02 30	26 ♓ 10 03	02 ♈ 06 43	24 31	11 45	10 41	00 37	18 26	03 12	15 57	21 34
06	16 59 57	14 03 24	08 ♈ 04 37	14 ♈ 04 19	25 54	12 59	11 21	00 50	18 33	03 10	15 57	21 36
07	17 03 54	15 04 18	20 ♈ 06 16	26 ♈ 10 55	27 17	14 13	12 01	01 04	18 39	03 09	15 57	21 38
08	17 07 50	16 05 13	02 ♉ 18 40	08 ♉ 29 51	28 ♏ 42	15 27	12 41	01 17	18 45	03 07	15 58	21 39
09	17 11 47	17 06 08	14 ♉ 44 44	21 ♉ 03 31	00 ♐ 08	16 42	13 20	01 31	18 52	03 06	15 58	21 41
10	17 15 43	18 07 05	27 ♉ 26 19	03 ♊ 53 13	01 35	17 56	14 00	01 44	18 58	03 04	15 58	21 43
11	17 19 40	19 08 02	10 ♊ 24 13	16 ♊ 59 13	03 03	19 10	14 40	01 58	19 05	03 03	15 59	21 45
12	17 23 36	20 09 01	23 ♊ 38 07	00 ♋ 20 44	04 31	20 24	15 20	02 12	19 11	03 01	15 59	21 46
13	17 27 33	21 10 00	07 ♋ 06 50	13 ♋ 56 11	06 00	21 38	16 00	02 25	19 18	03 00	16 00	21 48
14	17 31 30	22 11 00	20 ♋ 48 28	27 ♋ 43 25	07 30	22 52	16 40	02 39	19 24	02 58	16 01	21 50
15	17 35 26	23 12 01	04 ♌ 40 42	11 ♌ 40 02	08 59	24 06	17 20	02 53	19 31	02 57	16 01	21 52
16	17 39 23	24 13 03	18 ♌ 41 06	25 ♌ 43 38	10 30	25 20	18 00	03 06	19 37	02 56	16 02	21 54
17	17 43 19	25 14 06	02 ♍ 47 19	09 ♍ 51 54	12 00	26 34	18 40	03 20	19 44	02 54	16 02	21 55
18	17 47 16	26 15 09	16 ♍ 57 06	24 ♍ 02 40	13 31	27 48	19 20	03 34	19 51	02 53	16 03	21 57
19	17 51 12	27 16 14	01 ♎ 08 20	08 ♎ 13 49	15 02	29 ♑ 02	20 00	03 48	19 57	02 52	16 04	21 59
20	17 55 09	28 17 19	15 ♎ 18 51	22 ♎ 23 09	16 34	00 ♒ 16	20 40	04 01	20 04	02 51	16 05	22 01
21	17 59 05	29 ♐ 18 26	29 ♎ 26 24	06 ♏ 28 17	18 05	01 30	21 20	04 15	20 11	02 50	16 05	22 03
22	18 03 02	00 ♑ 19 33	13 ♏ 28 28	20 ♏ 26 36	19 37	02 44	22 01	04 29	20 18	02 49	16 06	22 05
23	18 06 59	01 20 41	27 ♏ 22 22	04 ♐ 15 25	21 10	03 58	22 41	04 43	20 25	02 48	16 07	22 07
24	18 10 55	02 21 50	11 ♐ 05 26	17 ♐ 52 07	22 42	05 12	23 21	04 56	20 32	02 47	16 08	22 09
25	18 14 52	03 22 59	24 ♐ 35 12	01 ♑ 14 28	24 15	06 26	24 01	05 10	20 38	02 46	16 09	22 10
26	18 18 48	04 24 09	07 ♑ 49 45	14 ♑ 20 55	25 48	07 39	24 41	05 24	20 45	02 45	16 10	22 12
27	18 22 45	05 25 19	20 ♑ 47 57	27 ♑ 10 51	27 21	08 53	25 22	05 38	20 52	02 45	16 11	22 14
28	18 26 41	06 26 29	03 ♒ 29 41	09 ♒ 44 37	28 ♐ 54	10 07	26 02	05 52	20 59	02 44	16 12	22 16
29	18 30 38	07 27 39	15 ♒ 55 51	22 ♒ 03 41	00 ♑ 28	11 21	26 42	06 06	21 06	02 43	16 13	22 18
30	18 34 34	08 28 49	28 ♒ 08 25	04 ♓ 10 28	02 02	12 34	27 23	06 19	21 13	02 42	16 14	22 20
31	18 38 31	09 ♑ 29 59	10 ♓ 10 17	16 ♓ 08 19	03 ♑ 36	13 ♒ 48	28 ♏ 03	06 ♑ 33	21 ♑ 20	02 ♉ 42	16 ♓ 15	22 ♑ 22

Date	Moon True ☊ ° '	Moon Mean ☊ ° '	Moon ☽ Latitude ° '
01	08 ♋ 41	09 ♋ 52	02 S 35
02	08 D 43	09 48	03 29
03	08 44	09 45	04 13
04	08 R 45	09 42	04 46
05	08 44	09 39	05 07
06	08 42	09 36	05 14
07	08 39	09 32	05 08
08	08 35	09 29	04 48
09	08 31	09 26	04 14
10	08 28	09 23	03 27
11	08 25	09 20	02 28
12	08 24	09 17	01 21
13	08 23	09 13	00 S 07
14	08 D 23	09 10	01 N 08
15	08 24	09 07	02 20
16	08 26	09 04	03 25
17	08 27	09 01	04 17
18	08 27	08 58	04 55
19	08 R 28	08 54	05 14
20	08 27	08 51	05 14
21	08 26	08 48	04 56
22	08 25	08 45	04 19
23	08 24	08 42	03 28
24	08 24	08 38	02 26
25	08 23	08 35	01 16
26	08 23	08 32	00 N 03
27	08 D 23	08 29	01 S 08
28	08 23	08 26	02 15
29	08 23	08 23	03 14
30	08 R 23	08 19	04 02
31	08 ♋ 23	08 ♋ 16	04 S 40

DECLINATIONS

Date	Sun ☉ ° '	Moon ☽ ° '	Mercury ☿ ° '	Venus ♀ ° '	Mars ♂ ° '	Jupiter ♃ ° '	Saturn ♄ ° '	Uranus ♅ ° '	Neptune ♆ ° '	Pluto ♇ ° '
01	21 S 48	20 S 43	15 S 30	24 S 44	13 S 36	23 S 18	22 S 08	12 N 09	06 S 30	22 S 20
02	21 57	17 59	15 55	24 42	13 49	23 18	22 07	12 08	06 30	22 20
03	22 06	14 33	16 20	24 39	14 02	23 18	22 06	12 07	06 30	22 19
04	22 14	10 34	16 46	24 35	14 15	23 18	22 06	12 07	06 30	22 19
05	22 22	06 13	17 12	24 30	14 28	23 18	22 05	12 06	06 30	22 19
06	22 29	01 S 36	17 38	24 24	14 40	23 18	22 04	12 06	06 30	22 19
07	22 36	03 N 06	18 05	24 18	14 53	23 18	22 03	12 05	06 30	22 19
08	22 43	07 46	18 30	24 11	15 06	23 18	22 03	12 05	06 30	22 18
09	22 49	12 13	18 56	24 04	15 18	23 18	22 02	12 04	06 29	22 18
10	22 54	16 14	19 21	23 55	15 31	23 18	22 01	12 04	06 29	22 18
11	22 59	19 34	19 45	23 46	15 43	23 18	22 00	12 03	06 29	22 18
12	23 04	21 57	20 09	23 36	15 55	23 18	21 59	12 03	06 29	22 18
13	23 09	23 08	20 32	23 26	16 07	23 18	21 58	12 02	06 28	22 17
14	23 12	22 57	20 55	23 15	16 19	23 18	21 58	12 02	06 28	22 17
15	23 16	21 22	21 16	23 03	16 31	23 17	21 57	12 01	06 28	22 17
16	23 19	18 28	21 37	22 50	16 42	23 17	21 56	12 01	06 28	22 17
17	23 21	14 29	21 57	22 37	16 54	23 17	21 55	12 00	06 27	22 17
18	23 23	09 40	22 15	22 23	17 05	23 17	21 54	12 00	06 27	22 16
19	23 24	04 N 21	22 33	22 08	17 17	23 16	21 53	12 00	06 27	22 16
20	23 25	01 S 12	22 50	21 53	17 28	23 16	21 52	11 59	06 27	22 16
21	23 26	06 40	23 06	21 37	17 39	23 16	21 51	11 59	06 26	22 16
22	23 26	11 45	23 20	21 21	17 50	23 15	21 50	11 59	06 26	22 15
23	23 26	16 11	23 34	21 04	18 01	23 15	21 49	11 58	06 25	22 15
24	23 25	19 42	23 46	20 46	18 11	23 15	21 48	11 58	06 25	22 15
25	23 24	22 04	23 57	20 28	18 22	23 14	21 48	11 58	06 25	22 15
26	23 22	23 09	24 07	20 09	18 32	23 14	21 47	11 58	06 24	22 15
27	23 20	22 57	24 15	19 50	18 42	23 13	21 46	11 57	06 24	22 14
28	23 17	21 33	24 23	19 30	18 53	23 13	21 45	11 57	06 23	22 14
29	23 14	19 08	24 29	19 09	19 02	23 12	21 44	11 57	06 23	22 14
30	23 10	15 54	24 34	18 48	19 12	23 12	21 43	11 57	06 23	22 14
31	23 S 06	12 S 04	24 S 37	18 S 27	19 S 22	23 S 11	21 S 42	11 N 57	06 S 22	22 S 13

ZODIAC SIGN ENTRIES

Date	h	m	Planets
02	18	20	♃ ♑
03	07	11	☽ ♓
05	19	44	☽ ♈
08	07	29	☽ ♉
09	09	42	☿ ♐
10	16	47	☽ ♊
12	23	23	☽ ♋
15	03	56	☽ ♌
17	07	16	☽ ♍
19	10	04	☽ ♎
20	06	41	♀ ♒
21	12	57	☽ ♏
22	04	19	☉ ♑
23	16	34	☽ ♐
25	21	45	☽ ♑
28	05	21	☽ ♒
29	04	55	☿ ♑
30	15	41	☽ ♓

LATITUDES

Date	Mercury ☿ ° '	Venus ♀ ° '	Mars ♂ ° '	Jupiter ♃ ° '	Saturn ♄ ° '	Uranus ♅ ° '	Neptune ♆ ° '	Pluto ♇ ° '
01	02 N 09	01 S 29	00 N 38	00 N 08	00 N 05	00 S 30	01 S 02	00 S 37
04	01 51	01 34	00 36	00 08	00 05	00 30	01 02	00 37
07	01 31	01 38	00 35	00 08	00 05	00 30	01 02	00 37
10	01 09	01 42	00 33	00 07	00 05	00 30	01 02	00 37
13	00 47	01 45	00 32	00 07	00 04	00 30	01 02	00 38
16	00 24	01 48	00 30	00 07	00 04	00 30	01 02	00 38
19	00 N 03	01 50	00 29	00 06	00 04	00 30	01 02	00 38
22	00 S 18	01 51	00 27	00 06	00 04	00 30	01 02	00 38
25	00 38	01 52	00 25	00 06	00 04	00 30	01 02	00 39
28	00 57	01 52	00 24	00 06	00 03	00 30	01 02	00 39
31	01 S 14	01 S 51	00 N 22	00 N 05	00 N 03	00 S 30	01 S 02	00 S 39

DATA

Julian Date	2458819
Delta T	+70 seconds
Ayanamsa	24° 07' 49"
Synetic vernal point	04° ♓ 59' 10"
True obliquity of ecliptic	23° 26' 10"

MOON'S PHASES, APSIDES AND POSITIONS ☽

Date	h	m	Phase	Longitude ° '	Eclipse Indicator
04	06	58	☽	11 ♓ 49	
12	05	12	○	19 ♊ 52	
19	04	57	☾	26 ♍ 58	
26	05	13	●	04 ♑ 07	Annular

Day	h	m		
05	04	07	Apogee	
18	20	28	Perigee	
06	20	12	0N	
13	20	57	Max dec	23° N 14'
20	06	51	0S	
26	20	09	Max dec	23° S 14'

LONGITUDES

Date	Chiron ⚷ ° '	Ceres ⚳ ° '	Pallas ⚴ ° '	Juno ⚵ ° '	Vesta ⚶ ° '	Black Moon Lilith ⚸ ° '
01	01 ♈ 30	05 ♑ 51	09 ♐ 55	09 ♎ 22	15 ♉ 04	23 ♓ 38
11	01 ♈ 26	09 ♑ 46	14 ♐ 12	12 ♎ 19	13 ♉ 22	24 ♓ 45
21	01 ♈ 28	13 ♑ 43	18 ♐ 27	14 ♎ 58	12 ♉ 22	25 ♓ 53
31	01 ♈ 35	17 ♑ 42	22 ♐ 39	17 ♎ 16	12 ♉ 06	27 ♓ 00

All ephemeris data is given at 12.00 UT and the Moon's longitude is additionally given for 24.00 UT

ASPECTARIAN

h m Aspects

01 Sunday
00 16 ☽ ∥ ☉
02 38 ☽ □ ♅
07 09 ☽ ⊥ ♃
09 01 ☽ ⊻ ♀
11 43 ☽ □ ♂
13 43 ☽ ⚹ ☉
15 25 ☽ ⊥ ♆
19 45 ☉ ± ♅
22 02 ☽ ⊥ ♀

02 Monday
01 03 ☽ ∠ ♃
03 12 ☽ ⊻ ♆
07 30 ☽ ⊻ ♄
12 27 ☽ □ ☿
13 47 ☽ Q ♅
14 12 ☽ ⊻ ♇
15 30 ☽ Q ☉
17 51 ☽ ∠ ♀
19 31 ☽ ⊥ ♄

03 Tuesday
01 24 ☽ ∥ ☿
02 11 ☽ ⊥ ♇
05 23 ☿ ⚹ ♇
07 25 ☽ ⚹ ♃
13 41 ☽ ∠ ♄
13 43 ☽ ⚹ ♅
14 32 ☉ ∥ ♄
15 08 ☽ ∥ ♂
15 47 ♀ ⚹ ♂
17 55 ♂ Q ♇
20 18 ☽ ∠ ♇

04 Wednesday
02 51 ☽ △ ♂
03 00 ☽ # ♅
03 26 ☽ ⚹ ♀
06 58 ☽ □ ☉
08 03 ☽ Q ♃
15 19 ☽ ☌ ♆
19 56 ☽ ∠ ♅
20 12 ☉ ⊥ ♄
20 14 ☽ ⚹ ♄

05 Thursday
02 41 ☽ ⚹ ♇
03 57 ☉ ∥ ♇
06 32 ☽ Q ♀
08 15 ☽ △ ☿
10 27 ☽ ∥ ♆
10 58 ☽ bQ ♂
13 54 ☿ ⊥ ♃
14 05 ☽ ⊥ ♅
20 42 ☽ Q ♄
21 09 ☽ □ ♃

06 Friday
02 10 ☽ ⊻ ♅
02 59 ☽ Q ♇
06 12 ☽ ± ♂
18 23 ☽ bQ ☿
18 57 ☽ ⊼ ♂
22 57 ☽ □ ♀

07 Saturday
01 04 ☽ △ ☉
03 45 ☽ ⊻ ♆
09 05 ☽ □ ♄
14 39 ☽ ± ☿
15 01 ☽ □ ♇
15 40 ☽ ⊥ ♆

08 Sunday
01 32 ☉ ⊥ ♇
04 02 ☽ ⊼ ☿
05 22 ☽ # ♆
09 00 ☉ □ ♆
09 22 ☽ ∠ ♆
09 24 ☽ bQ ☉
09 58 ☽ △ ♃
13 34 ☽ ☌ ♅
21 48 ♀ ⚹ ♆

09 Monday
04 24 ☽ ± ☉
09 10 ☽ ☍ ♂
11 11 ☽ ∥ ♅
14 20 ☽ ⚹ ♆
15 26 ☽ bQ ♃
16 07 ☽ △ ♀
16 53 ☽ ⊼ ☉
19 54 ☽ △ ♄

10 Tuesday
01 13 ☽ △ ♇
07 11 ☽ # ♂
08 45 ☽ ± ♃
10 50 ☉ bQ ♅
13 00 ☽ Q ♆
14 58 ☿ ⊻ ♃
20 10 ☽ ⊼ ♃
20 43 ☽ ☍ ☿
22 28 ☽ ⊻ ♅
23 18 ☽ bQ ♀

11 Wednesday
00 15 ☽ bQ ♄
05 16 ☽ bQ ♇
08 26 ☉ ⊻ ♀
09 30 ☽ ⊥ ♅
10 05 ♀ ☌ ♄
10 29 ☉ ⊻ ♄
11 54 ☿ ⊼ ♅
13 55 ☽ # ☿
16 56 ☽ ± ♄
17 34 ☽ ± ♀
20 12 ☽ ⊼ ♂
21 46 ☽ ± ♇
22 11 ☽ □ ♆

12 Thursday
01 53 ☽ ∠ ♅
03 55 ☽ ⊼ ♄
05 12 ☽ ☍ ☉
05 35 ☽ ⊼ ♀
06 07 ☿ ∠ ♄
07 39 ☽ ± ♂
08 38 ☽ ⊼ ♇
12 37 ☽ # ♄
16 58 ☽ # ♇

13 Friday
00 36 ☽ bQ ♂
03 33 ☽ ☍ ♃
04 43 ☽ ⚹ ♅
09 48 ☽ ⊼ ☿
11 55 ♂ △ ♆
12 46 ☽ # ☉
15 16 ♀ ☌ ♇
21 39 ☽ ± ☿

14 Saturday
01 10 ☿ ∠ ♇
01 50 ☽ Q ♅
02 56 ☽ # ☉
03 27 ☉ ⊻ ♇
03 37 ☽ △ ♆
04 25 ☽ △ ♂
05 41 ♀ ∥ ♃
09 32 ☽ ☍ ♄
13 47 ☽ ☍ ♇
14 35 ☽ ⊼ ☉
15 17 ☽ bQ ☿
15 38 ☉ ∥ ♀
15 57 ☽ ☍ ♀

15 Sunday
00 37 ☽ # ♇
01 48 ☽ ± ☉
05 12 ☽ # ♄
05 41 ☽ bQ ♆
08 51 ☽ ⊼ ♃
09 02 ☽ □ ♅
11 22 ☿ ± ♅
12 49 ☽ # ☿
18 32 ☽ bQ ☉
19 01 ♃ △ ♅
19 20 ☽ ± ♃
20 18 ☽ △ ☿
21 11 ☽ ± ♆
21 12 ☉ ⊥ ♂
22 29 ☿ ∠ ♀

16 Monday
00 53 ☉ ∥ ♃
07 28 ☽ ⊼ ♆
10 46 ☽ □ ♂
11 00 ☽ bQ ♃
13 37 ☽ ⊼ ♄
17 29 ☽ ⊼ ♇
17 38 ♂ ∠ ♃
22 10 ☽ △ ☉
22 47 ☽ # ♂
23 55 ☽ ± ♄

17 Tuesday
00 26 ☽ ⊼ ♀
03 43 ☽ ± ♇
10 01 ☿ ∥ ♄
11 36 ☽ ± ♀
12 12 ☽ △ ♅
12 56 ☽ △ ♃
15 20 ☽ bQ ♄
18 55 ☽ Q ♂
19 02 ☽ bQ ♇

18 Wednesday
00 46 ☽ ∥ ♅
04 19 ☽ bQ ♀
05 29 ☽ □ ☿
10 29 ☽ ☍ ♆
13 14 ☿ ∥ ♇
13 35 ☽ bQ ♅
16 14 ☽ ⚹ ♂
16 56 ☽ △ ♄
17 36 ☿ ⊥ ♄
17 36 ☿ ∥ ♀
20 29 ☽ △ ♇
23 14 ♀ ∥ ♇

19 Thursday
02 43 ☽ # ♆
04 47 ☽ ± ♅
04 57 ☽ □ ☉
08 07 ☽ △ ♀
10 00 ♂ ⚹ ♄
14 55 ☽ ⊼ ♅
15 36 ☽ Q ☿
16 34 ☽ □ ♃
18 52 ☽ ∠ ♂

20 Friday
03 14 ☿ ⊥ ♇
04 19 ☿ □ ♆
10 52 ☽ ⊥ ♂
13 18 ☽ ⊼ ♆
13 38 ♀ ∥ ♄
13 47 ☽ Q ☉
14 22 ☽ ⚹ ☿
18 14 ♃ Q ♆
20 08 ☽ □ ♄
21 33 ☽ ⊻ ♂
23 24 ☽ □ ♇
23 29 ☽ ± ♆
23 34 ☽ Q ♃

21 Saturday
03 51 ♀ ∠ ♆
07 59 ☿ bQ ♅
11 00 ☽ ∥ ♆
11 45 ☽ ⚹ ☉
14 49 ☽ bQ ♆
15 51 ☽ □ ♀
17 47 ☽ ☍ ♅
18 59 ☽ ∠ ☿
20 21 ☽ ⚹ ♃

22 Sunday
03 03 ☽ Q ♄
04 09 ☿ ∥ ♃
06 10 ☽ Q ♇
12 17 ☽ ⊥ ☿
13 08 ☽ # ♅
13 30 ♀ □ ♅
14 32 ♂ ⚹ ♇
15 26 ☽ ∠ ☉
16 32 ☽ △ ♆
22 14 ☉ ∥ ☿
22 31 ☽ ∠ ♃
23 24 ☿ ⊻ ♄
23 51 ☽ ⚹ ♄
23 54 ☽ ⊻ ☿

23 Monday
01 43 ☽ Q ♀
02 52 ☽ ⚹ ♇
03 27 ☽ ☌ ♂
08 12 ☽ ⊥ ☉
14 22 ☽ ⊥ ♃
19 28 ☽ ⊻ ☉
21 26 ☽ ⊼ ♅

24 Tuesday
00 11 ☽ ∥ ♂
00 38 ☽ ⚹ ♀
01 01 ☽ ⊻ ♃
02 08 ☽ ∠ ♄
03 08 ☿ ⊻ ♇
05 02 ☽ ∠ ♇
05 51 ♀ ⊻ ♃
07 57 ☽ ± ♅
18 07 ☽ ⊥ ♄
18 32 ♀ Q ♂
20 24 ☽ ∥ ♀
20 56 ☽ □ ♆
20 57 ☽ ⊥ ♇
21 44 ☉ △ ♅
23 50 ☽ bQ ♅

25 Wednesday
04 53 ☽ ⊻ ♄
05 46 ☽ ∠ ♀
05 53 ☿ ⊻ ♂
07 40 ☽ ⊻ ♇
08 29 ☽ ∥ ♄
10 56 ☽ ⊻ ♂
11 18 ☽ ☌ ☿
14 38 ☽ ∥ ♇
22 19 ☽ ⊥ ♂
23 36 ☽ ⊥ ♀

26 Thursday
02 45 ☽ △ ♅
05 13 ☽ ☌ ☉
05 18 ☽ Q ♆
06 21 ☉ Q ♆
07 29 ☽ ☌ ♃
11 39 ☽ ⊻ ♀
15 36 ☽ ∠ ♂
19 02 ☽ ∥ ♃
21 32 ☽ ∥ ♃

27 Friday
03 23 ☽ ⚹ ♆
12 08 ☽ ☌ ♄
14 42 ☽ ☌ ♇
18 25 ☉ ☌ ♃
21 03 ☽ ⚹ ♂

28 Saturday
02 02 ☽ ⊻ ☿
02 35 ☽ ∥ ♇
07 37 ☽ ∠ ♆
09 36 ☽ ∥ ♄
10 32 ☽ □ ♅
13 41 ♀ ⊥ ♆
15 04 ☽ ⊥ ☿
16 37 ☽ ⊻ ♃
18 09 ☽ ⊻ ☉
21 12 ☽ Q ♂

29 Sunday
00 54 ☽ ⊥ ♆
02 07 ☽ ☌ ♀
04 24 ☽ ⊥ ♃
06 46 ☽ ⊥ ☉
10 57 ☽ ∠ ☿
11 49 ☽ ∥ ♀
12 34 ☽ ⊻ ♆
12 45 ☽ ∥ ♂
17 24 ♀ ∥ ♂
21 21 ☽ Q ♅
22 13 ☽ ⊻ ♄
22 18 ☽ ∠ ♃
22 53 ☉ ∥ ♃

30 Monday
00 31 ☽ ⊻ ♇
01 57 ☽ ∠ ☉
06 05 ♀ ⊥ ♃
10 10 ☽ ⊥ ♄
10 24 ☽ □ ♂
12 23 ☽ ⊥ ♇
20 52 ☽ ⚹ ☿
21 04 ☽ ⚹ ♅
22 22 ☿ △ ♅

31 Tuesday
04 15 ☽ ∠ ♄
04 37 ☽ ⚹ ♃
06 22 ☽ ∠ ♇
10 32 ☽ ⚹ ☉
12 43 ☽ # ♅
20 07 ☽ ⊻ ♀
22 10 ☿ Q ♆

JANUARY 2020

LONGITUDES

Date	Sidereal time h m s	Sun ☉ ° ' "	Moon ☽ ° ' "	Moon ☽ 24.00 ° ' "	Mercury ☿ ° '	Venus ♀ ° '	Mars ♂ ° '	Jupiter ♃ ° '	Saturn ♄ ° '	Uranus ♅ ° '	Neptune ♆ ° '	Pluto ♇ ° '
01	18 42 28	10 ♑ 31 09	22 ♓ 05 06	28 ♓ 01 11	05 ♑ 10	15 ♒ 01	28 ♏ 43	06 ♑ 47	21 ♑ 27	02 ♉ 41	16 ♓ 16	22 ♑ 24
02	18 46 24	11 32 19	03 ♈ 57 07	09 ♈ 53 30	06 45	16 15	29 ♏ 24	07 01	21 34	02 R 41	16 18	22 26
03	18 50 21	12 33 29	15 ♈ 50 57	21 ♈ 50 02	08 20	17 28	00 ♐ 04	07 15	21 41	02 40	16 19	22 28
04	18 54 17	13 34 38	27 ♈ 51 20	03 ♉ 55 27	09 56	18 42	00 44	07 29	21 48	02 40	16 20	22 30
05	18 58 14	14 35 48	10 ♉ 02 56	16 ♉ 14 16	11 31	19 55	01 25	07 42	21 55	02 40	16 21	22 32
06	19 02 10	15 36 56	22 ♉ 29 55	28 ♉ 50 19	13 08	21 08	02 05	07 56	22 02	02 40	16 23	22 34
07	19 06 07	16 38 05	05 ♊ 15 46	11 ♊ 46 33	14 44	22 22	02 46	08 10	22 10	02 39	16 24	22 36
08	19 10 03	17 39 14	18 ♊ 22 48	25 ♊ 04 35	16 21	23 35	03 26	08 24	22 17	02 39	16 25	22 38
09	19 14 00	18 40 22	01 ♋ 51 51	08 ♋ 44 25	17 59	24 48	04 07	08 37	22 24	02 39	16 27	22 40
10	19 17 57	19 41 29	15 ♋ 42 00	22 ♋ 44 10	19 36	26 01	04 47	08 51	22 31	02 39	16 28	22 42
11	19 21 53	20 42 37	29 ♋ 50 27	07 ♌ 00 13	21 15	27 14	05 28	09 05	22 38	02 D 39	16 30	22 44
12	19 25 50	21 43 44	14 ♌ 12 47	21 ♌ 27 26	22 53	28 27	06 08	09 19	22 45	02 39	16 31	22 46
13	19 29 46	22 44 51	28 ♌ 43 25	05 ♍ 59 57	24 32	29 ♒ 40	06 49	09 32	22 52	02 39	16 33	22 48
14	19 33 43	23 45 58	13 ♍ 16 19	20 ♍ 31 48	26 12	00 ♓ 53	07 30	09 46	22 59	02 39	16 34	22 50
15	19 37 39	24 47 05	27 ♍ 45 49	04 ♎ 57 49	27 52	02 05	08 10	10 00	23 06	02 39	16 36	22 52
16	19 41 36	25 48 11	12 ♎ 07 19	19 ♎ 14 00	29 ♑ 33	03 18	08 51	10 13	23 14	02 40	16 37	22 54
17	19 45 32	26 49 18	26 ♎ 17 34	03 ♏ 17 50	01 ♒ 14	04 31	09 32	10 27	23 21	02 40	16 39	22 56
18	19 49 29	27 50 24	10 ♏ 14 42	17 ♏ 08 05	02 55	05 43	10 12	10 40	23 28	02 40	16 40	22 58
19	19 53 26	28 51 30	23 ♏ 58 00	00 ♐ 44 29	04 37	06 56	10 53	10 54	23 35	02 41	16 42	23 00
20	19 57 22	29 ♑ 52 36	07 ♐ 27 33	14 ♐ 07 18	06 19	08 08	11 34	11 07	23 42	02 41	16 44	23 02
21	20 01 19	00 ♒ 53 41	20 ♐ 43 48	27 ♐ 17 06	08 02	09 21	12 15	11 21	23 49	02 42	16 46	23 04
22	20 05 15	01 54 46	03 ♑ 47 17	10 ♑ 14 24	09 45	10 33	12 55	11 34	23 56	02 42	16 47	23 06
23	20 09 12	02 55 50	16 ♑ 38 31	22 ♑ 59 40	11 28	11 45	13 36	11 48	24 03	02 43	16 49	23 08
24	20 13 08	03 56 54	29 ♑ 17 55	05 ♒ 33 20	13 11	12 57	14 17	12 01	24 10	02 44	16 51	23 10
25	20 17 05	04 57 57	11 ♒ 45 58	17 ♒ 55 55	14 55	14 09	14 58	12 14	24 17	02 44	16 53	23 12
26	20 21 01	05 58 59	24 ♒ 03 18	00 ♓ 08 16	16 38	15 21	15 39	12 28	24 24	02 45	16 54	23 14
27	20 24 58	07 00 00	06 ♓ 11 00	12 ♓ 11 44	18 21	16 33	16 20	12 41	24 31	02 46	16 56	23 16
28	20 28 55	08 01 00	18 ♓ 10 42	24 ♓ 08 14	20 04	17 45	17 00	12 54	24 38	02 47	16 58	23 18
29	20 32 51	09 01 59	00 ♈ 04 41	06 ♈ 00 27	21 47	18 56	17 41	13 07	24 45	02 48	17 00	23 20
30	20 36 48	10 02 57	11 ♈ 55 59	17 ♈ 51 45	23 29	20 08	18 22	13 21	24 52	02 49	17 02	23 22
31	20 40 44	11 ♒ 03 54	23 ♈ 48 17	29 ♈ 46 09	25 ♒ 09	21 ♓ 19	19 ♐ 03	13 ♑ 34	24 ♑ 59	02 ♉ 50	17 ♓ 04	23 ♑ 24

Date	Moon True ☊ ° '	Moon Mean ☊ ° '	Moon ☽ Latitude ° '
01	08 ♋ 23	08 ♋ 13	05 S 04
02	08 R 23	08 10	05 16
03	08 D 23	08 07	05 14
04	08 23	08 04	04 59
05	08 24	08 00	04 30
06	08 24	07 57	03 48
07	08 25	07 54	02 54
08	08 26	07 51	01 49
09	08 26	07 48	00 S 36
10	08 R 26	07 44	00 N 40
11	08 26	07 41	01 56
12	08 24	07 38	03 05
13	08 23	07 35	04 03
14	08 21	07 32	04 46
15	08 19	07 29	05 10
16	08 18	07 25	05 15
17	08 17	07 22	05 00
18	08 D 18	07 19	04 28
19	08 19	07 16	03 41
20	08 20	07 13	02 42
21	08 22	07 09	01 36
22	08 23	07 06	00 N 25
23	08 R 22	07 03	00 S 45
24	08 21	07 00	01 52
25	08 18	06 57	02 53
26	08 14	06 54	03 45
27	08 08	06 50	04 25
28	08 03	06 47	04 54
29	07 58	06 44	05 09
30	07 53	06 41	05 11
31	07 ♋ 50	06 ♋ 38	05 S 00

DECLINATIONS

Date	Sun ☉ ° '	Moon ☽ ° '	Mercury ☿ ° '	Venus ♀ ° '	Mars ♂ ° '	Jupiter ♃ ° '	Saturn ♄ ° '	Uranus ♅ ° '	Neptune ♆ ° '	Pluto ♇ ° '
01	23 S 01	07 S 48	24 S 39	18 S 05	19 S 31	23 S 10	21 S 41	11 N 56	06 S 22	22 S 13
02	22 56	03 S 16	24 40	17 42	19 41	23 10	21 40	11 56	06 21	22 13
03	22 51	01 N 24	24 39	17 19	19 50	23 09	21 38	11 56	06 21	22 13
04	22 45	06 03	24 37	16 56	19 59	23 09	21 37	11 56	06 20	22 12
05	22 38	10 33	24 33	16 32	20 08	23 08	21 36	11 56	06 20	22 12
06	22 31	14 43	24 29	16 08	20 17	23 07	21 35	11 56	06 19	22 12
07	22 24	18 20	24 22	15 43	20 25	23 06	21 34	11 56	06 19	22 12
08	22 16	21 07	24 14	15 18	20 34	23 06	21 33	11 56	06 18	22 11
09	22 08	22 49	24 05	14 52	20 42	23 05	21 32	11 56	06 17	22 11
10	21 59	23 11	23 54	14 26	20 50	23 04	21 31	11 56	06 17	22 11
11	21 50	22 04	23 42	14 00	20 58	23 03	21 30	11 56	06 16	22 11
12	21 41	19 31	23 28	13 33	21 05	23 02	21 29	11 56	06 16	22 10
13	21 31	15 43	23 13	13 06	21 13	23 01	21 28	11 56	06 15	22 10
14	21 21	10 58	22 56	12 39	21 20	23 01	21 27	11 56	06 15	22 10
15	21 10	05 38	22 37	12 11	21 28	23 00	21 25	11 56	06 14	22 10
16	20 59	00 N 03	22 17	11 43	21 35	22 59	21 24	11 56	06 13	22 09
17	20 47	05 S 29	21 56	11 15	21 42	22 58	21 23	11 56	06 13	22 09
18	20 35	10 39	21 33	10 47	21 48	22 57	21 22	11 57	06 12	22 09
19	20 23	15 12	21 08	10 18	21 55	22 56	21 21	11 57	06 11	22 09
20	20 10	18 53	20 42	09 49	22 01	22 55	21 20	11 57	06 11	22 09
21	19 57	21 31	20 14	09 19	22 07	22 54	21 19	11 57	06 10	22 08
22	19 44	22 58	19 45	08 50	22 13	22 53	21 17	11 57	06 09	22 08
23	19 30	23 09	19 14	08 20	22 19	22 52	21 16	11 58	06 09	22 08
24	19 16	22 08	18 42	07 50	22 24	22 50	21 15	11 58	06 08	22 08
25	19 01	20 01	18 09	07 20	22 30	22 49	21 14	11 58	06 07	22 07
26	18 46	17 02	17 34	06 50	22 35	22 48	21 13	11 59	06 06	22 07
27	18 31	13 21	16 57	06 19	22 40	22 47	21 12	11 59	06 06	22 07
28	18 16	09 10	16 20	05 49	22 45	22 46	21 11	11 59	06 05	22 07
29	18 00	04 42	15 41	05 18	22 49	22 45	21 09	11 59	06 04	22 06
30	17 44	00 03	15 01	04 47	22 54	22 43	21 08	12 00	06 03	22 06
31	17 S 27	04 N 35	14 S 20	04 S 16	22 S 58	22 S 42	21 S 07	12 N 00	06 S 03	22 S 06

ZODIAC SIGN ENTRIES

Date	h	m	Planets
02	04	00	☽ ♈
03	09	37	♂ ♐
04	16	15	☽ ♉
07	02	11	☽ ♊
09	08	43	☽ ♋
11	12	16	☽ ♌
13	14	06	☽ ♍
13	18	39	♀ ♓
15	15	43	☽ ♎
16	18	31	☿ ♒
17	18	20	☽ ♏
19	22	41	☽ ♐
20	14	55	☉ ♒
22	05	00	☽ ♑
24	13	20	☽ ♒
26	23	44	☽ ♓
29	11	51	☽ ♈

LATITUDES

Date	Mercury ☿ ° '	Venus ♀ ° '	Mars ♂ ° '	Jupiter ♃ ° '	Saturn ♄ ° '	Uranus ♅ ° '	Neptune ♆ ° '	Pluto ♇ ° '
01	01 S 19	01 S 50	00 N 21	00 N 05	00 N 03	00 S 30	01 S 02	00 S 39
04	01 33	01 48	00 20	00 05	00 03	00 30	01 02	00 39
07	01 46	01 45	00 18	00 05	00 03	00 29	01 01	00 40
10	01 55	01 42	00 16	00 04	00 02	00 29	01 01	00 40
13	02 02	01 37	00 14	00 04	00 02	00 29	01 01	00 40
16	02 06	01 32	00 12	00 04	00 02	00 29	01 01	00 41
19	02 06	01 26	00 10	00 04	00 02	00 29	01 01	00 41
22	02 01	01 19	00 08	00 03	00 02	00 29	01 01	00 41
25	01 52	01 12	00 05	00 03	00 01	00 29	01 01	00 41
28	01 37	01 03	00 03	00 03	00 01	00 29	01 01	00 42
31	01 S 16	00 S 54	00 01	00 N 02	00 N 01	00 S 29	01 S 01	00 S 42

DATA

Julian Date	2458850
Delta T	+71 seconds
Ayanamsa	24° 07' 55"
Synetic vernal point	04° ♓ 59' 04"
True obliquity of ecliptic	23° 26' 10"

MOON'S PHASES, APSIDES AND POSITIONS ☽

Date	h	m	Phase	Longitude ° '	Eclipse Indicator
03	04	45	☽	12 ♈ 15	
10	19	21	○	20 ♋ 00	
17	12	58	☾	26 ♎ 52	
24	21	42	●	04 ♒ 22	

Day	h	m		
02	01	31	Apogee	
13	20	32	Perigee	
29	21	31	Apogee	
03	04	51	ON	
10	06	02	Max dec	23° N 13'
16	12	11	OS	
23	03	35	Max dec	23° S 14'
30	12	18	ON	

LONGITUDES

Date	Chiron ⚷ ° '	Ceres ⚳ ° '	Pallas ⚴ ° '	Juno ⚵ ° '	Vesta ⚶ ° '	Black Moon Lilith ⚸ ° '
01	01 ♈ 36	18 ♑ 06	23 ♐ 04	17 ♎ 28	12 ♉ 07	27 ♓ 06
11	01 ♈ 49	22 ♑ 06	27 ♐ 11	19 ♎ 18	12 ♉ 39	28 ♓ 14
21	02 ♈ 08	26 ♑ 06	01 ♑ 13	20 ♎ 39	13 ♉ 50	29 ♓ 21
31	02 ♈ 30	00 ♒ 04	05 ♑ 09	21 ♎ 27	15 ♉ 35	00 ♈ 28

All ephemeris data is given at 12.00 UT and the Moon's longitude is additionally given for 24.00 UT

ASPECTARIAN

h m	Aspects
01 Wednesday	
00 02	☿ ⊥ ♂
00 15	☽ ☌ ♆
00 34	☽ Q ☿
03 08	☽ ∠ ♅
05 12	☽ Q ♃
09 36	☽ ⊥ ♀
10 43	☽ ⚹ ♄
12 39	☽ ⚹ ♇
12 58	☽ Q ☉
19 45	☽ ∥ ♆
21 18	☽ ⊥ ♅
02 Thursday	
02 14	☽ △ ♂
05 54	☽ ∠ ♀
09 26	☽ ⊻ ♅
11 13	☽ Q ♄
12 56	♀ ⊻ ♆
12 59	☽ Q ♇
16 42	☿ ☌ ♃
18 19	☽ □ ♃
18 32	☽ □ ☿
03 Friday	
04 45	☽ □ ☉
10 20	☽ bQ ♂
12 56	☽ ⊻ ♆
15 38	☽ ⚹ ♀
23 49	☽ □ ♄
04 Saturday	
00 59	☽ ⊥ ♆
01 18	☽ □ ♇
05 26	☽ ± ♂
13 29	☽ # ♆
18 03	☽ ⊼ ♂
18 15	☽ Q ♀
18 54	☽ ∠ ♆
21 31	☽ ☌ ♅
05 Sunday	
07 20	☽ △ ♃
15 18	☽ △ ☿
19 42	☽ ∥ ♅
21 37	☽ △ ☉
06 Monday	
00 15	☽ ⚹ ♆
02 36	♀ Q ♅
03 48	♂ ⊥ ♃
09 08	☽ □ ♀
11 07	☽ △ ♄
12 08	☽ △ ♇
12 51	☽ bQ ♃
19 59	☽ # ♀
23 09	☽ Q ♆
07 Tuesday	
00 12	☽ bQ ☿
04 40	☽ bQ ☉
06 08	☽ ± ♃
06 21	☉ ⚹ ♆
07 05	☽ ☍ ♂
07 09	☽ ⊻ ♅
07 38	♀ ⊻ ♄
08 13	♂ ⊼ ♅
15 33	☽ bQ ♄
16 20	☽ bQ ♇
16 53	♀ ⊻ ♇
17 28	☽ ⊼ ♃
18 16	☽ ⊥ ♅
19 20	☽ ± ☿
22 45	☽ ± ☉
08 Wednesday	
06 07	☽ # ♂
07 32	♀ ∠ ♃
07 49	☽ ⊼ ☿
08 10	☽ ± ♄
08 27	☽ □ ♆
08 50	☽ ± ♇
10 35	☽ ⊼ ☉
10 41	☽ ∠ ♅
13 03	☿ ⚹ ♆
16 43	☽ # ♄
19 04	☽ ⊼ ♄
19 40	☽ ⊼ ♇
20 54	☉ ⊥ ♀
22 16	☽ △ ♀
09 Thursday	
00 57	☽ # ♇
01 05	☽ # ☉
02 46	☉ ∥ ♇
13 23	☽ ⚹ ♅
16 09	☽ ⊼ ♂
19 17	☽ # ♃
10 Friday	
00 00	☽ ☍ ♃
03 07	☽ ± ♂
03 10	☽ bQ ♀
10 12	☽ Q ♅
13 19	☽ △ ♆
15 19	☉ ☌ ☿
16 32	☿ ∠ ♂
17 08	☽ # ♃
18 43	☉ ∠ ♂
19 20	☽ bQ ♂
19 21	☽ ☍ ☉
19 33	☽ ☍ ☿
20 04	☽ ± ♀
23 43	☽ ☍ ♄
23 58	☽ ☍ ♇
11 Saturday	
01 48	♅ St D
07 12	☽ ⊼ ♀
10 30	☽ # ♇
11 24	☿ ⊥ ♀
14 47	☽ bQ ♆
15 02	☽ # ☉
16 43	☽ □ ♅
18 42	☽ # ♄
21 54	☽ △ ♂
23 25	☽ # ♂
12 Sunday	
03 43	☽ ⊼ ♃
05 51	☽ ± ♆
09 51	☿ ☌ ♄
10 14	☿ ☌ ♇
13 51	☽ ± ♃
15 50	☽ ⊼ ♆
16 59	♄ ☌ ♇
18 30	♀ ⊥ ♇
18 37	♀ ⊥ ♄
13 Monday	
01 23	☽ ⊼ ☉
02 12	☽ ⊼ ♇
02 15	☽ ⊼ ♄
04 12	☽ ⊼ ☿
04 59	☽ bQ ♃
12 03	☽ ± ☉
12 08	☽ ± ♇
12 15	☽ ± ♄
13 21	☉ ☌ ♇
13 42	☽ ☍ ♀
15 16	☉ ☌ ♄
15 23	☽ ± ☿
18 29	☽ △ ♅
20 57	☉ ∥ ♄
14 Tuesday	
02 00	☽ □ ♂
03 01	☽ bQ ♇
03 06	☽ # ♀
03 13	☽ bQ ♄
04 00	☽ bQ ☉
05 08	☿ ∥ ♃
06 07	☽ △ ♃
07 25	☽ ∥ ♅
08 08	☽ bQ ☿
12 26	☉ ∥ ♂
17 27	☽ ☍ ♆
19 15	☽ bQ ♅
15 Wednesday	
00 48	♂ ∠ ♇
03 51	☽ △ ♇
04 12	☽ △ ♄
05 38	♂ ∥ ♄
06 41	☽ △ ☉
09 13	☽ Q ♂
09 17	♂ ∠ ♄
09 23	☽ # ♆
10 10	☽ ± ♅
12 12	☽ △ ☿
19 52	☽ ⊼ ♀
20 09	☽ ⊼ ♅
23 18	♀ ⚹ ♅
16 Thursday	
00 57	♀ # ♅
05 24	♂ ± ♅
06 14	☽ ⚹ ♂
06 49	☽ ± ♀
08 45	☽ □ ♃
19 36	☽ ⊼ ♆
21 09	☿ ∥ ♇
23 24	☽ bQ ♀
17 Friday	
05 47	☽ ± ♆
06 16	☽ □ ♇
06 56	☽ □ ♄
08 50	☽ ∠ ♂
12 58	☽ □ ☉
15 16	☽ ∥ ♆
15 44	☽ Q ♃
18 05	☿ ∠ ♆
21 11	☽ bQ ♆
21 36	☽ □ ☿
22 55	☽ ☍ ♅
23 43	☿ ∥ ♂
18 Saturday	
01 02	☽ ⊥ ♂
03 26	☽ △ ♀
08 32	☿ □ ♅
11 56	☽ ⊻ ♂
12 35	☽ ∥ ♀
12 45	☽ ⚹ ♃
13 16	☽ Q ♇
14 08	☽ Q ♄
18 29	☽ # ♅
22 31	☽ Q ☉
23 12	☿ ∥ ♄
23 13	☽ △ ♆
19 Sunday	
09 17	☽ Q ☿
10 18	☽ ⚹ ♇
11 19	☽ ⚹ ♄
12 46	♂ ⊻ ♃
15 28	☽ ∠ ♃
21 22	☽ ⚹ ☉
20 Monday	
03 28	☽ ⊼ ♅
07 45	☽ ⊥ ♃
09 40	☽ ⚹ ☿
09 58	♀ ∠ ♇
13 02	☽ ∠ ♇
13 20	☽ □ ♀
14 12	☽ ± ♅
14 15	☽ ∠ ♄
18 42	☽ ⊻ ♃
19 47	☽ ☌ ♂
21 41	☽ ∥ ☉
21 Tuesday	
00 25	♀ ∠ ♄
00 59	☽ ∥ ☿
02 29	☽ ∠ ☉
04 46	☽ □ ♆
05 20	☽ ⊥ ♇
06 29	☽ bQ ♅
06 39	☽ ⊥ ♄
09 37	☽ ∥ ♄
16 08	♂ ∥ ♇
16 17	☽ ⊻ ♇
16 50	☽ ∠ ☿
17 41	☽ ⊻ ♄
19 57	☽ ∥ ♇
20 13	☽ ∥ ♂
20 15	☽ ⊥ ☉
22 Wednesday	
01 20	☽ Q ♀
08 14	☽ ⊻ ☉
08 58	☉ ∠ ♆
09 44	☽ ∥ ♃
10 00	☽ △ ♅
11 55	☽ ⊥ ☿
13 50	☉ ∥ ☿
13 51	☽ Q ♆
23 Thursday	
00 47	☽ ⊻ ☿
01 52	☽ ⚹ ♀
02 45	☽ ☌ ♃
02 48	☿ ⊥ ♆
05 58	☽ ⊻ ♂
06 54	☉ □ ♅
12 20	☽ ⚹ ♆
13 07	♀ ⚹ ♃
17 17	☿ ⊻ ♃
17 54	☽ ⊥ ♂
22 20	☽ ∥ ♃
24 Friday	
00 18	☽ ☌ ♇
01 07	☿ ⊻ ♀
02 08	☽ ☌ ♄
07 45	☽ ∥ ♂
09 09	☽ ∠ ♀
11 58	☽ ∠ ♂
12 02	☽ ∥ ♇
16 53	☽ ∠ ♆
18 34	☽ □ ♅
21 42	☽ ☌ ☉
23 30	☽ ∥ ♄
25 Saturday	
04 16	☽ ⊥ ♀
10 16	☽ ⊥ ♆
12 56	☽ ⊻ ♃
13 09	☿ ⚹ ♂
17 08	☽ ⊻ ♀
18 34	☽ ⚹ ♂
19 06	☽ ☌ ☿
21 35	☽ ∥ ☉
21 58	☽ ⊻ ♆
26 Sunday	
00 50	☽ ⊥ ♃
05 31	☽ Q ♅
07 18	☽ ∥ ☿
10 23	☽ ⊻ ♇
12 42	☽ ⊻ ♄
15 50	☿ ⊻ ♆
18 50	☽ ∠ ♃
19 29	☽ Q ♂
22 15	☽ ⊥ ♇
27 Monday	
00 39	☽ ⊥ ♄
01 37	♀ □ ♂
05 12	☽ ⚹ ♅
13 47	☽ ⊻ ☉
16 10	☽ ∠ ♇
17 13	☿ ⊥ ♃
18 43	☽ ∠ ♄
20 00	♀ ☌ ♆
20 06	☽ # ♅
23 01	♀ ∥ ♆
28 Tuesday	
01 13	☽ ⚹ ♃
02 52	☽ ⊥ ☉
09 30	☽ □ ♂
09 34	☽ ☌ ♆
10 34	♂ □ ♆
11 02	☽ ☌ ♀
11 12	☽ ∠ ♅
12 46	♀ ∠ ♅
16 09	♂ ∥ ♃
16 27	☽ ⊻ ☿
22 02	☿ Q ♅
22 21	☽ ⚹ ♇
22 39	☽ ∠ ☉
22 55	♂ ⊥ ♇
29 Wednesday	
01 08	☽ ⚹ ♄
01 48	☽ Q ♃
04 43	☽ ∥ ♆
05 22	☽ ⊥ ♅
06 35	☽ ⊥ ☿
08 24	☽ ∥ ♀
15 55	♂ bQ ♅
17 31	☽ ⊻ ♅
22 40	☽ Q ♇
30 Thursday	
01 39	☽ Q ♄
03 50	☽ ∠ ☿
07 50	☽ ⚹ ☉
10 26	☿ ⊻ ♇
14 54	☽ □ ♃
22 21	☽ ⊻ ♆
31 Friday	
01 49	☽ △ ♂
06 26	☽ ⊻ ♀
09 10	♂ ⊥ ♄
09 25	☿ ⊻ ♄
10 22	☽ Q ☉
10 30	☽ ⊥ ♆
10 31	☽ # ♀
11 11	☽ □ ♇
11 57	☉ ⊥ ♆
14 24	☽ □ ♄
15 09	☽ ⚹ ☿
19 37	☽ # ♆
19 52	☽ ⊥ ♀

FEBRUARY 2020

LONGITUDES

Date	Sidereal time h m s	Sun ☉ ° ' "	Moon ☽ ° ' "	Moon ☽ 24.00 ° ' "	Mercury ☿ ° '	Venus ♀ ° '	Mars ♂ ° '	Jupiter ♃ ° '	Saturn ♄ ° '	Uranus ♅ ° '	Neptune ♆ ° '	Pluto ♇ ° '
01	20 44 41	12 ♒ 04 49	05 ♉ 45 55	11 ♉ 48 12	26 ♒ 48	22 ♓ 31	19 ♐ 44	13 ♑ 47	25 ♑ 06	02 ♉ 51	17 ♓ 06	23 ♑ 26
02	20 48 37	13 05 44	17 ♉ 53 36	24 ♉ 02 43	28 ♒ 26	23 42	20 25	14 00	25 13	02 52	17 08	23 28
03	20 52 34	14 06 36	00 ♊ 16 11	06 ♊ 34 34	00 ♓ 01	24 53	21 06	14 13	25 20	02 53	17 10	23 30
04	20 56 30	15 07 28	12 ♊ 58 24	19 ♊ 28 10	01 34	26 04	21 47	14 26	25 27	02 54	17 12	23 32
05	21 00 27	16 08 18	26 ♊ 04 16	02 ♋ 46 59	03 04	27 15	22 28	14 38	25 34	02 56	17 14	23 34
06	21 04 24	17 09 07	09 ♋ 36 29	16 ♋ 32 49	04 30	28 26	23 09	14 51	25 40	02 57	17 16	23 35
07	21 08 20	18 09 55	23 ♋ 35 49	00 ♌ 45 10	05 52	29 ♓ 36	23 50	15 04	25 47	02 58	17 18	23 37
08	21 12 17	19 10 41	08 ♌ 00 22	15 ♌ 20 43	07 08	00 ♈ 47	24 31	15 17	25 54	03 00	17 20	23 39
09	21 16 13	20 11 25	22 ♌ 45 20	00 ♍ 13 13	08 19	01 57	25 12	15 29	26 01	03 01	17 22	23 41
10	21 20 10	21 12 09	07 ♍ 43 15	15 ♍ 14 13	09 23	03 08	25 54	15 42	26 07	03 03	17 24	23 43
11	21 24 06	22 12 51	22 ♍ 44 55	00 ♎ 14 12	10 20	04 18	26 35	15 54	26 14	03 04	17 26	23 45
12	21 28 03	23 13 32	07 ♎ 40 58	15 ♎ 04 18	11 09	05 28	27 16	16 07	26 21	03 06	17 28	23 46
13	21 31 59	24 14 12	22 ♎ 23 23	29 ♎ 37 38	11 49	06 37	27 57	16 19	26 27	03 08	17 30	23 48
14	21 35 56	25 14 50	06 ♏ 46 35	13 ♏ 50 01	12 20	07 47	28 38	16 31	26 34	03 09	17 32	23 50
15	21 39 53	26 15 28	20 ♏ 47 47	27 ♏ 39 58	12 41	08 57	29 ♐ 20	16 44	26 40	03 11	17 34	23 52
16	21 43 49	27 16 05	04 ♐ 26 42	11 ♐ 08 13	12 52	10 06	00 ♑ 01	16 56	26 47	03 13	17 37	23 54
17	21 47 46	28 16 40	17 ♐ 44 50	24 ♐ 16 52	12 R 52	11 15	00 42	17 08	26 53	03 15	17 39	23 55
18	21 51 42	29 ♒ 17 14	00 ♑ 44 43	07 ♑ 08 43	12 42	12 24	01 23	17 20	26 59	03 17	17 41	23 57
19	21 55 39	00 ♓ 17 47	13 ♑ 29 14	19 ♑ 46 37	12 22	13 33	02 05	17 32	27 06	03 19	17 43	23 59
20	21 59 35	01 18 19	26 ♑ 01 09	02 ♒ 13 08	11 52	14 42	02 46	17 44	27 12	03 21	17 45	24 00
21	22 03 32	02 18 49	08 ♒ 22 48	14 ♒ 30 22	11 13	15 51	03 27	17 56	27 18	03 23	17 48	24 02
22	22 07 28	03 19 17	20 ♒ 36 02	26 ♒ 39 57	10 27	16 59	04 09	18 07	27 25	03 25	17 50	24 04
23	22 11 25	04 19 44	02 ♓ 42 15	08 ♓ 43 06	09 33	18 07	04 50	18 19	27 31	03 27	17 52	24 05
24	22 15 22	05 20 10	14 ♓ 42 38	20 ♓ 40 58	08 34	19 15	05 31	18 31	27 37	03 29	17 54	24 07
25	22 19 18	06 20 33	26 ♓ 38 19	02 ♈ 34 49	07 32	20 23	06 13	18 42	27 43	03 31	17 56	24 09
26	22 23 15	07 20 55	08 ♈ 30 43	14 ♈ 26 16	06 27	21 31	06 54	18 53	27 49	03 33	17 59	24 10
27	22 27 11	08 21 15	20 ♈ 21 45	26 ♈ 17 31	05 22	22 38	07 36	19 05	27 55	03 36	18 01	24 12
28	22 31 08	09 21 33	02 ♉ 13 57	08 ♉ 11 29	04 19	23 46	08 17	19 16	28 01	03 38	18 03	24 13
29	22 35 04	10 ♓ 21 49	14 ♉ 10 35	20 ♉ 11 48	03 ♓ 17	24 ♈ 53	08 ♑ 58	19 ♑ 27	28 ♑ 07	03 ♉ 40	18 ♓ 05	24 ♑ 15

MOON NODES, LATITUDE AND DECLINATIONS

Date	Moon True ☊ ° '	Moon Mean ☊ ° '	Moon ☽ Latitude ° '	Sun ☉ ° '	Moon ☽ ° '	Mercury ☿ ° '	Venus ♀ ° '	Mars ♂ ° '	Jupiter ♃ ° '	Saturn ♄ ° '	Uranus ♅ ° '	Neptune ♆ ° '	Pluto ♇ ° '
01	07 ♋ 48	06 ♋ 35	04 S 36	17 S 10	09 N 06	13 S 38	03 S 45	23 S 02	22 S 41	21 S 06	12 N 01	06 S 02	22 S 06
02	07 D 48	06 31	03 59	16 53	13 20	12 56	03 14	23 06	22 40	21 05	12 01	06 01	22 05
03	07 49	06 28	03 10	16 36	17 06	12 13	02 43	23 10	22 39	21 03	12 01	06 00	22 05
04	07 51	06 25	02 11	16 18	20 11	11 30	02 11	23 13	22 37	21 02	12 02	06 00	22 05
05	07 52	06 22	01 S 04	16 00	22 19	10 47	01 40	23 16	22 36	21 01	12 02	05 59	22 05
06	07 R 53	06 19	00 N 09	15 42	23 15	10 04	01 09	23 19	22 35	21 00	12 03	05 58	22 04
07	07 52	06 15	01 24	15 23	22 46	09 22	00 37	23 22	22 33	20 59	12 03	05 57	22 04
08	07 49	06 12	02 36	15 04	20 46	08 41	00 S 06	23 25	22 32	20 58	12 04	05 56	22 04
09	07 44	06 09	03 39	14 45	17 22	08 01	00 N 26	23 27	22 31	20 56	12 04	05 56	22 04
10	07 37	06 06	04 28	14 26	12 48	07 23	00 57	23 29	22 29	20 55	12 05	05 55	22 04
11	07 30	06 03	04 58	14 06	07 26	06 48	01 29	23 31	22 28	20 54	12 06	05 54	22 03
12	07 24	06 00	05 08	13 46	01 N 40	06 14	02 00	23 33	22 27	20 53	12 06	05 53	22 03
13	07 18	05 56	04 58	13 26	04 S 06	05 44	02 31	23 35	22 25	20 52	12 07	05 52	22 03
14	07 14	05 53	04 28	13 06	09 33	05 18	03 03	23 36	22 24	20 50	12 07	05 51	22 03
15	07 13	05 50	03 44	12 46	14 21	04 55	03 34	23 37	22 22	20 49	12 08	05 51	22 03
16	07 D 13	05 47	02 47	12 25	18 17	04 36	04 05	23 38	22 21	20 48	12 09	05 50	22 02
17	07 14	05 44	01 43	12 04	21 10	04 22	04 36	23 39	22 19	20 47	12 09	05 49	22 02
18	07 15	05 41	00 N 35	11 43	22 51	04 13	05 07	23 40	22 18	20 46	12 10	05 48	22 02
19	07 R 15	05 37	00 S 33	11 22	23 18	04 08	05 38	23 40	22 17	20 45	12 11	05 47	22 02
20	07 13	05 34	01 39	11 01	22 34	04 09	06 08	23 40	22 15	20 44	12 11	05 46	22 02
21	07 09	05 31	02 39	10 39	20 43	04 14	06 39	23 40	22 14	20 42	12 12	05 46	22 01
22	07 01	05 28	03 30	10 17	17 57	04 24	07 10	23 40	22 12	20 41	12 13	05 45	22 01
23	06 52	05 25	04 12	09 55	14 25	04 38	07 40	23 39	22 11	20 40	12 13	05 44	22 01
24	06 41	05 21	04 42	09 33	10 21	04 56	08 10	23 39	22 09	20 39	12 14	05 43	22 01
25	06 29	05 18	04 59	09 11	05 54	05 17	08 40	23 38	22 08	20 38	12 15	05 42	22 01
26	06 17	05 15	05 03	08 49	01 S 16	05 41	09 10	23 37	22 06	20 37	12 16	05 41	22 00
27	06 06	05 12	04 54	08 26	03 N 25	06 07	09 39	23 35	22 05	20 36	12 17	05 40	22 00
28	05 58	05 09	04 32	08 04	07 59	06 35	10 09	23 34	22 03	20 34	12 17	05 39	22 00
29	05 ♋ 52	05 ♋ 06	03 S 59	07 S 41	12 N 17	07 S 03	10 N 38	23 S 32	22 S 02	20 S 33	12 N 18	05 S 39	22 S 00

ZODIAC SIGN ENTRIES

Date	h m	Planets
01	00 28	☽ ♉
03	11 29	☽ ♊
03	11 37	☿ ♓
05	19 03	☽ ♋
07	20 02	♀ ♈
07	22 45	☽ ♌
09	23 39	☽ ♍
11	23 37	☽ ♎
14	00 37	☽ ♏
16	04 07	☽ ♐
16	11 33	♂ ♑
18	10 37	☽ ♑
19	04 57	☉ ♓
20	19 42	☽ ♒
23	06 37	☽ ♓
25	18 47	☽ ♈
28	07 30	☽ ♉

LATITUDES

Date	Mercury ☿ ° '	Venus ♀ ° '	Mars ♂ ° '	Jupiter ♃ ° '	Saturn ♄ ° '	Uranus ♅ ° '	Neptune ♆ ° '	Pluto ♇ ° '
01	01 S 08	00 S 51	00 00	00 N 02	00 N 01	00 S 29	01 S 01	00 S 42
04	00 38	00 41	00 S 02	00 02	00 N 01	00 29	01 01	00 42
07	00 S 01	00 30	00 05	00 02	00 00	00 28	01 01	00 43
10	00 N 43	00 19	00 07	00 02	00 00	00 28	01 01	00 43
13	01 30	00 S 07	00 10	00 01	00 00	00 28	01 01	00 43
16	02 18	00 N 05	00 12	00 01	00 00	00 28	01 01	00 44
19	03 01	00 18	00 15	00 N 01	00 00	00 28	01 01	00 44
22	03 31	00 32	00 18	00 00	00 S 01	00 28	01 01	00 44
25	03 43	00 45	00 20	00 00	00 01	00 28	01 01	00 44
28	03 36	01 00	00 23	00 00	00 01	00 28	01 01	00 45
31	03 N 11	01 N 14	00 S 26	00 01	00 S 01	00 S 28	01 S 01	00 S 45

DATA

Julian Date	2458881
Delta T	+71 seconds
Ayanamsa	24° 08' 00"
Synetic vernal point	04° ♓ 58' 59"
True obliquity of ecliptic	23° 26' 11"

MOON'S PHASES, APSIDES AND POSITIONS ☽

Date	h m	Phase	Longitude ° '	Eclipse Indicator
02	01 42	☽	12 ♉ 40	
09	07 33	○	20 ♌ 00	
15	22 17	☾	26 ♏ 41	
23	15 32	●	04 ♓ 29	

Day	h m	
10	20 36	Perigee
26	11 43	Apogee
06	16 08	Max dec 23° N 16'
12	18 52	0S
19	08 55	Max dec 23° S 19'
26	18 29	0N

LONGITUDES

Date	Chiron ⚷ ° '	Ceres ⚳ ° '	Pallas ⚴ ° '	Juno ⚵ ° '	Vesta ⚶ ° '	Black Moon Lilith ☾ ° '
01	02 ♈ 33	00 ♒ 28	05 ♑ 32	21 ♎ 29	15 ♉ 47	00 ♈ 35
11	02 ♈ 59	04 ♒ 23	09 ♑ 18	21 ♎ 38	18 ♉ 03	01 ♈ 42
21	03 ♈ 29	08 ♒ 16	12 ♑ 54	21 ♎ 08	20 ♉ 43	02 ♈ 49
31	04 ♈ 02	12 ♒ 04	16 ♑ 18	20 ♎ 01	23 ♉ 42	03 ♈ 57

ASPECTARIAN

h m Aspects

01 Saturday
04 39 ☽ ∠ ♆
06 10 ☽ ☌ ♅
09 49 ☽ ⅊ ♂
15 52 ☽ ∠ ♀
19 00 ☽ Q ☿

02 Sunday
01 42 ☽ □ ☉
04 11 ☽ △ ♃
04 17 ☽ ∥ ♅
04 46 ☽ ± ♂
07 07 ♀ ✶ ♇
09 59 ☽ # ☿
10 30 ☽ ✶ ♆
17 14 ☽ ⊼ ♂
21 42 ☿ ∠ ♃
22 54 ☽ △ ♇

03 Monday
00 32 ☽ ✶ ♀
02 24 ☽ △ ♄
03 47 ☿ ⊥ ♇
08 48 ☽ # ☉
09 52 ☽ Q ♆
09 56 ☽ ⅊ ♃
11 28 ☽ □ ☿
15 00 ☉ ⊻ ♃
17 01 ☽ ⊻ ♅
18 28 ☿ # ♅
22 01 ♀ ✶ ♄

04 Tuesday
01 52 ☽ Q ♀
03 20 ☽ ± ♃
03 39 ☽ ⅊ ♇
04 23 ☽ ⊥ ♅
07 14 ☽ ⅊ ♄
09 50 ☿ ⊥ ♄
14 45 ☽ ⊼ ♃
16 20 ☽ △ ☉
19 50 ☽ □ ♆
20 13 ☽ # ♄
20 27 ☽ ± ♇
20 52 ♀ Q ♃
21 09 ☽ ∠ ♅

05 Wednesday
00 04 ☽ ± ♄
05 07 ☽ ☍ ♂
05 23 ♀ ⊥ ♅
07 27 ☽ ⊼ ♇
08 35 ☽ # ♇
09 43 ☿ ✶ ♅
11 04 ☽ ⊼ ♄
14 20 ☽ □ ♀
16 37 ☽ # ♃
21 49 ☽ ⅊ ☉

06 Thursday
00 17 ☽ ✶ ♅
02 00 ☽ △ ☿
14 39 ☉ ⊻ ♆
14 54 ☽ ± ☉
21 14 ☽ ☍ ♃
21 16 ☽ Q ♅

07 Friday
01 15 ☽ △ ♆
02 03 ☽ ⊼ ☉
04 03 ♂ ⊻ ♇
06 53 ☽ ⅊ ☿
11 10 ☿ Q ♂
12 02 ☽ ☍ ♇
12 26 ☽ ⊼ ♂
15 41 ☽ # ♃
15 43 ☽ ☍ ♄
22 34 ☽ # ♇
22 59 ☽ △ ♀
23 00 ☽ ± ♂
23 33 ☽ ± ☿

08 Saturday
02 36 ☽ ⅊ ♆
03 43 ☽ □ ♅
10 19 ☽ # ♄
10 26 ☽ ⊼ ☿
14 37 ☽ ⅊ ♂
17 28 ☽ ± ♆

09 Sunday
00 04 ☽ ⊼ ♃
01 48 ☽ ⅊ ♀
03 15 ☽ ⊼ ♆
07 33 ☽ ☍ ☉
09 55 ☽ ± ♃
13 30 ☽ ⊼ ♇
16 08 ☽ △ ♂
17 17 ☽ ⊼ ♄
17 35 ☽ ± ♀
20 12 ☿ ∠ ♇
23 10 ☽ ± ♇

10 Monday
00 36 ☽ ⅊ ♃
02 59 ☽ ± ♄
03 27 ☽ # ☉
04 02 ☽ ⊼ ♀
04 31 ☽ △ ♅
08 15 ☉ Q ♅
10 23 ♀ ⊻ ♅
13 35 ☽ ⅊ ♇
14 51 ☽ ☍ ☿
15 24 ☽ ∥ ♅
17 28 ☽ ⅊ ♄
21 30 ♂ ⊻ ♄

11 Tuesday
00 55 ☽ △ ♃
02 44 ☉ ⊥ ♃
03 29 ☽ ☍ ♆
04 31 ☽ ⅊ ♅
11 05 ☽ ⊼ ☉
13 36 ☽ △ ♇
15 03 ☽ # ☿
17 37 ☽ △ ♄
18 26 ☽ □ ♂
18 31 ☽ # ♆
18 56 ☽ ± ♅
21 23 ☽ ± ☉

12 Wednesday
04 36 ☽ ⊼ ♅
08 06 ☽ ☍ ♀
10 46 ☽ ∥ ♀
12 57 ☽ ⅊ ☉
17 54 ☽ ⊼ ☿
18 39 ♀ Q ♇
19 33 ☿ ∠ ♄

13 Thursday
00 55 ☽ Q ♂
01 25 ☉ ⊻ ♇
01 54 ☽ □ ♃
03 57 ☽ ⊼ ♆
04 10 ☽ ± ☿
04 40 ☽ # ♀
05 05 ☿ ∥ ♆
13 50 ☽ ± ♆
14 20 ☽ □ ♇
15 17 ☽ △ ☉
18 26 ☽ ∥ ☿
18 46 ☽ □ ♄
19 33 ☽ ∥ ♆
19 38 ☽ ⅊ ☿
21 40 ☽ ✶ ♂

14 Friday
04 51 ☽ ⅊ ♆
05 54 ☽ ☍ ♅
08 09 ☽ Q ♃
13 51 ☽ ⊼ ♀
20 36 ☽ Q ♇
21 43 ☽ △ ☿

15 Saturday
00 16 ☽ ∠ ♂
00 25 ☽ # ♅
01 00 ☽ ± ♀
01 21 ☽ Q ♄
04 11 ☽ ∥ ☉
04 52 ☽ ✶ ♃
05 43 ♀ Q ♄
06 25 ☽ △ ♆
16 38 ☽ ⊥ ♂
17 21 ☽ ✶ ♇
17 59 ☽ ⅊ ♀
22 17 ☽ □ ☉
22 20 ☽ ✶ ♄
22 57 ☉ ⊻ ♄

16 Sunday
03 43 ☽ ⊻ ♂
07 28 ☽ ∠ ♃
09 49 ☽ ⊼ ♅
19 59 ☽ ∠ ♇
20 34 ☽ ± ♅
23 06 ☽ △ ♀
23 48 ☽ ⊥ ♃

17 Monday
00 54 ☿ St R
01 16 ☽ ∠ ♄
03 10 ☽ □ ☿
04 20 ☿ # ♀
06 28 ☉ # ♅
08 13 ☽ ∥ ♄
09 06 ☽ Q ☉
10 52 ☽ ⊻ ♃
11 49 ☽ □ ♆
12 19 ☽ ⊥ ♇
12 55 ☽ ⅊ ♅
17 12 ☿ Q ♂
17 48 ☽ ⊥ ♄
22 15 ☽ ∥ ♇
23 22 ☽ ⊻ ♇

18 Tuesday
02 13 ☽ ∥ ♃
04 57 ☽ ⊻ ♄
09 03 ☽ ✶ ☉
11 55 ☽ Q ☿
13 16 ☽ ☌ ♂
16 45 ☽ △ ♅
17 02 ☿ ⊻ ♀
21 16 ☽ Q ♆

19 Wednesday
04 13 ☉ ⊥ ♇
09 56 ☽ ✶ ☿
12 08 ☽ □ ♀
15 44 ☽ ∠ ☉
19 13 ♀ # ♆
19 50 ☽ ☌ ♃
20 05 ☽ ✶ ♆
23 31 ☿ ∠ ♄

20 Thursday
08 07 ☽ ☌ ♇
10 30 ☽ ⊥ ☉
13 34 ☽ ∠ ☿
14 18 ☽ ☌ ♄
15 56 ♃ ✶ ♆
17 16 ☽ ∥ ♃
20 33 ☽ ∥ ♇
23 08 ☽ ⊻ ☉

21 Friday
01 05 ☽ ∠ ♆
01 50 ☽ ⊻ ♂
02 13 ☽ □ ♅
02 15 ☽ Q ♀
06 10 ☽ ⊥ ☿
09 10 ♂ △ ♅
12 09 ☽ ∥ ♄
14 13 ☽ ⊥ ♂
17 15 ☽ ⊻ ☿
18 42 ☽ ⊥ ♆

22 Saturday
04 08 ☽ ✶ ♀
05 23 ☿ ⊥ ♀
06 07 ☉ ∠ ♃
06 31 ☽ ⊻ ♆
07 02 ☽ ⊻ ♃
08 57 ☽ ∠ ♂
13 36 ☽ Q ♅
14 13 ☉ ✶ ♅
14 23 ☉ ⊥ ♄
18 51 ☽ ⊻ ♇
19 04 ☽ ⊥ ♃

23 Sunday
01 36 ☽ ⊻ ♄
06 27 ♀ ⊻ ♆
06 47 ☽ ⊥ ♇
12 55 ☽ ∠ ♀
13 14 ☽ ∠ ♃
13 29 ☽ ✶ ♅
13 38 ☽ ⊥ ♄
15 32 ☽ ☌ ☉
16 30 ☽ ✶ ♂
16 59 ♀ □ ♃
23 15 ☿ ∠ ♇

24 Monday
00 39 ☽ ☌ ☿
00 46 ☽ ∠ ♇
01 13 ☽ # ♅
07 46 ☽ ∠ ♄
08 47 ☽ ⊥ ♀
16 48 ☽ ∥ ☉
17 59 ☽ Q ♂
18 26 ☽ ☌ ♆
19 36 ☽ ∠ ♅
19 45 ☽ ✶ ♃
22 05 ☽ ⊻ ♀
22 45 ☽ # ♀

25 Tuesday
02 03 ♂ Q ♆
02 06 ☉ ✶ ♂
06 57 ☽ ✶ ♇
13 04 ☽ ∥ ♆
13 47 ☽ ⊥ ♅
14 12 ☽ ✶ ♄
14 59 ☽ ∥ ☿
20 20 ☽ Q ♃

26 Wednesday
01 45 ☉ ☌ ☿
01 56 ☽ ⊻ ♅
02 20 ☉ # ♀
05 59 ☿ ✶ ♂
07 15 ☽ Q ♇
08 11 ☽ ⊻ ☿
08 32 ☽ □ ♂
09 26 ☽ ⊻ ☉
11 23 ☿ ∠ ♀
11 53 ☿ ∥ ♆
14 40 ☽ Q ♄
19 19 ☽ ⊥ ☿
22 42 ☽ ⊥ ☉

27 Thursday
07 14 ☽ ⊻ ♆
09 22 ☽ □ ♃
12 01 ☽ ∠ ☿
17 05 ☽ ☌ ♀
18 37 ☽ ∠ ☉
19 25 ☽ ⊥ ♆
19 47 ☽ □ ♇
23 42 ☽ # ♆

28 Friday
03 25 ☽ □ ♄
03 40 ☽ # ☿
08 36 ☉ ∠ ♇
12 23 ☽ # ☉
12 50 ☿ ∠ ♃
13 40 ☽ ∠ ♆
14 50 ☽ ☌ ♅
15 51 ☽ ✶ ☿
18 09 ☿ ⊥ ♄
18 33 ♀ ⊥ ♆
22 08 ♀ □ ♇

29 Saturday
00 56 ☽ △ ♂
01 18 ☽ ∥ ♀
03 13 ☿ ✶ ♅
03 40 ☽ ✶ ☉
12 04 ☽ ∥ ♅
14 03 ☽ Q ☿
19 50 ☽ ✶ ♆
22 41 ☽ △ ♃

All ephemeris data is given at 12.00 UT and the Moon's longitude is additionally given for 24.00 UT

MARCH 2020

LONGITUDES

Date	Sidereal time h m s	Sun ☉ ° ' "	Moon ☽ ° ' "	Moon ☽ 24.00 ° ' "	Mercury ☿ ° '	Venus ♀ ° '	Mars ♂ ° '	Jupiter ♃ ° '	Saturn ♄ ° '	Uranus ♅ ° '	Neptune ♆ ° '	Pluto ♇ ° '
01	22 39 01	11 ♓ 22 04	26 ♉ 15 41	02 ♊ 22 49	02 ♓ 20	25 ♈ 59	09 ♑ 40	19 ♑ 38	28 ♑ 13	03 ♉ 43	18 ♓ 08	24 ♑ 16
02	22 42 57	12 22 16	08 ♊ 33 48	14 ♊ 49 18	01 R 27	27 06	10 21	19 49	28 19	03 45	18 10	24 18
03	22 46 54	13 22 26	21 ♊ 09 54	27 ♊ 36 12	00 ♓ 40	28 12	11 03	20 00	28 24	03 47	18 12	24 19
04	22 50 51	14 22 34	04 ♋ 08 46	10 ♋ 48 06	29 ♒ 59	29 ♈ 18	11 44	20 11	28 30	03 50	18 15	24 21
05	22 54 47	15 22 40	17 ♋ 34 34	24 ♋ 28 26	29 24	00 ♉ 24	12 26	20 22	28 36	03 52	18 17	24 22
06	22 58 44	16 22 44	01 ♌ 29 50	08 ♌ 38 39	28 56	01 30	13 07	20 32	28 41	03 55	18 19	24 23
07	23 02 40	17 22 45	15 ♌ 54 36	23 ♌ 17 09	28 36	02 35	13 49	20 43	28 47	03 58	18 21	24 25
08	23 06 37	18 22 45	00 ♍ 45 33	08 ♍ 18 46	28 22	03 40	14 30	20 53	28 52	04 00	18 24	24 26
09	23 10 33	19 22 42	15 ♍ 55 38	23 ♍ 34 47	28 14	04 45	15 12	21 03	28 57	04 03	18 26	24 27
10	23 14 30	20 22 38	01 ♎ 14 46	08 ♎ 54 05	28 D 13	05 50	15 53	21 13	29 03	04 06	18 28	24 29
11	23 18 26	21 22 32	16 ♎ 31 18	24 ♎ 05 05	28 18	06 54	16 35	21 23	29 08	04 08	18 30	24 30
12	23 22 23	22 22 24	01 ♏ 34 17	08 ♏ 57 57	28 29	07 58	17 17	21 33	29 13	04 11	18 33	24 31
13	23 26 20	23 22 14	16 ♏ 15 21	23 ♏ 26 00	28 46	09 01	17 58	21 43	29 18	04 14	18 35	24 33
14	23 30 16	24 22 02	00 ♐ 29 38	07 ♐ 26 12	29 08	10 05	18 40	21 53	29 23	04 17	18 37	24 34
15	23 34 13	25 21 49	14 ♐ 15 49	20 ♐ 58 45	29 ♒ 35	11 08	19 21	22 03	29 28	04 19	18 40	24 35
16	23 38 09	26 21 35	27 ♐ 35 22	04 ♑ 06 08	00 ♓ 06	12 10	20 03	22 12	29 33	04 22	18 42	24 36
17	23 42 06	27 21 18	10 ♑ 31 31	16 ♑ 52 05	00 42	13 13	20 45	22 22	29 38	04 25	18 44	24 37
18	23 46 02	28 21 00	23 ♑ 08 21	29 ♑ 20 52	01 22	14 15	21 26	22 31	29 43	04 28	18 46	24 38
19	23 49 59	29 ♓ 20 40	05 ♒ 30 07	11 ♒ 36 34	02 05	15 16	22 08	22 40	29 48	04 31	18 49	24 39
20	23 53 55	00 ♈ 20 18	17 ♒ 40 41	23 ♒ 42 50	02 53	16 18	22 50	22 49	29 52	04 34	18 51	24 40
21	23 57 52	01 19 55	29 ♒ 43 22	05 ♓ 42 36	03 43	17 18	23 31	22 58	29 ♑ 57	04 37	18 53	24 41
22	00 01 48	02 19 29	11 ♓ 40 48	17 ♓ 38 10	04 37	18 19	24 13	23 07	00 ♒ 01	04 40	18 55	24 42
23	00 05 45	03 19 02	23 ♓ 34 56	29 ♓ 31 16	05 34	19 19	24 55	23 15	00 06	04 43	18 58	24 43
24	00 09 42	04 18 32	05 ♈ 27 20	11 ♈ 23 17	06 34	20 19	25 36	23 24	00 10	04 46	19 00	24 44
25	00 13 38	05 18 01	17 ♈ 19 16	23 ♈ 15 29	07 36	21 18	26 18	23 32	00 15	04 49	19 02	24 45
26	00 17 35	06 17 27	29 ♈ 12 06	05 ♉ 09 22	08 41	22 16	27 00	23 41	00 19	04 52	19 04	24 46
27	00 21 31	07 16 51	11 ♉ 07 30	17 ♉ 06 50	09 48	23 15	27 41	23 49	00 23	04 55	19 06	24 47
28	00 25 28	08 16 14	23 ♉ 07 40	29 ♉ 10 25	10 57	24 13	28 23	23 57	00 27	04 59	19 09	24 48
29	00 29 24	09 15 33	05 ♊ 15 29	11 ♊ 23 21	12 09	25 10	29 05	24 05	00 31	05 02	19 11	24 49
30	00 33 21	10 14 51	17 ♊ 34 30	23 ♊ 49 30	13 23	26 07	29 ♑ 47	24 12	00 35	05 05	19 13	24 49
31	00 37 17	11 ♈ 14 07	00 ♋ 08 54	06 ♋ 33 15	14 ♓ 38	27 ♉ 03	00 ♒ 28	24 ♑ 20	00 ♒ 39	05 ♉ 08	19 ♓ 15	24 ♑ 50

MOON NODES AND LATITUDE / DECLINATIONS

Date	Moon True ☊ ° '	Moon Mean ☊ ° '	Moon ☽ Latitude ° '	Sun ☉ ° '	Moon ☽ ° '	Mercury ☿ ° '	Venus ♀ ° '	Mars ♂ ° '	Jupiter ♃ ° '	Saturn ♄ ° '	Uranus ♅ ° '	Neptune ♆ ° '	Pluto ♇ ° '
01	05 ♋ 48	05 ♋ 02	03 S 14	07 S 18	16 N 10	07 S 31	11 N 07	23 S 30	22 S 00	20 S 32	12 N 19	05 S 38	22 S 00
02	05 R 47	04 59	02 19	06 55	19 26	07 59	11 35	23 28	21 59	20 31	12 20	05 37	22 00
03	05 D 47	04 56	01 17	06 32	21 52	08 26	12 04	23 26	21 57	20 30	12 21	05 36	21 59
04	05 47	04 53	00 S 09	06 09	23 14	08 52	12 32	23 23	21 56	20 29	12 22	05 35	21 59
05	05 R 47	04 50	01 N 02	05 46	23 19	09 16	13 00	23 20	21 54	20 28	12 22	05 34	21 59
06	05 44	04 47	02 12	05 22	21 58	09 38	13 27	23 17	21 53	20 27	12 23	05 33	21 59
07	05 39	04 43	03 16	04 59	19 11	09 58	13 55	23 14	21 52	20 26	12 24	05 32	21 59
08	05 32	04 40	04 08	04 36	15 04	10 17	14 22	23 11	21 50	20 25	12 25	05 31	21 59
09	05 22	04 37	04 44	04 12	09 55	10 33	14 48	23 07	21 49	20 24	12 26	05 31	21 59
10	05 11	04 34	05 00	03 49	04 N 06	10 46	15 15	23 03	21 47	20 23	12 27	05 30	21 59
11	05 00	04 31	04 55	03 25	01 S 57	10 58	15 41	22 59	21 46	20 22	12 28	05 29	21 58
12	04 51	04 27	04 29	03 02	07 48	11 07	16 06	22 55	21 44	20 21	12 29	05 28	21 58
13	04 45	04 24	03 46	02 38	13 05	11 14	16 32	22 51	21 43	20 20	12 30	05 27	21 58
14	04 40	04 21	02 50	02 14	17 29	11 19	16 56	22 46	21 41	20 19	12 31	05 26	21 58
15	04 39	04 18	01 46	01 51	20 46	11 22	17 21	22 41	21 40	20 18	12 32	05 25	21 58
16	04 D 38	04 15	00 N 37	01 27	22 48	11 22	17 45	22 37	21 38	20 17	12 33	05 24	21 58
17	04 R 38	04 12	00 S 31	01 03	23 32	11 21	18 09	22 31	21 37	20 16	12 34	05 24	21 58
18	04 37	04 08	01 36	00 39	23 02	11 18	18 32	22 26	21 36	20 15	12 35	05 23	21 58
19	04 34	04 05	02 35	00 S 16	21 24	11 13	18 55	22 21	21 34	20 14	12 36	05 22	21 58
20	04 28	04 02	03 26	00 N 08	18 48	11 06	19 18	22 15	21 33	20 13	12 37	05 21	21 58
21	04 19	03 59	04 07	00 32	15 25	10 57	19 40	22 09	21 31	20 12	12 38	05 20	21 57
22	04 08	03 56	04 37	00 55	11 27	10 46	20 02	22 03	21 30	20 11	12 39	05 19	21 57
23	03 54	03 53	04 54	01 19	07 03	10 34	20 23	21 56	21 29	20 11	12 40	05 18	21 57
24	03 39	03 49	04 59	01 43	02 S 24	10 20	20 44	21 50	21 27	20 10	12 41	05 17	21 57
25	03 25	03 46	04 51	02 06	02 N 19	10 04	21 05	21 43	21 26	20 09	12 42	05 17	21 57
26	03 11	03 43	04 29	02 30	06 59	09 47	21 25	21 36	21 25	20 08	12 43	05 16	21 57
27	03 00	03 40	03 56	02 53	11 25	09 29	21 44	21 29	21 24	20 07	12 44	05 15	21 57
28	02 52	03 37	03 13	03 17	15 27	09 08	22 03	21 22	21 22	20 06	12 45	05 14	21 57
29	02 47	03 33	02 20	03 40	18 53	08 47	22 22	21 15	21 21	20 06	12 46	05 13	21 57
30	02 45	03 30	01 19	04 03	21 32	08 24	22 40	21 07	21 20	20 05	12 47	05 12	21 57
31	02 ♋ 44	03 ♋ 27	00 S 14	04 N 27	23 N 13	07 S 59	22 N 57	20 S 59	21 S 19	20 S 04	12 N 48	05 S 12	21 S 57

ZODIAC SIGN ENTRIES

Date	h m	Planets
01	19 21	☽ ♊
04	04 25	☽ ♋
04	11 08	☿ ♒
05	03 07	♀ ♉
06	09 28	☽ ♌
08	10 47	☽ ♍
10	10 03	☽ ♎
12	09 28	☽ ♏
14	11 09	☽ ♐
16	07 42	☿ ♓
16	16 25	☽ ♑
19	01 16	☽ ♒
20	03 50	☉ ♈
21	12 33	☽ ♓
22	03 58	♄ ♒
24	00 58	☽ ♈
26	13 37	☽ ♉
29	01 38	☽ ♊
30	19 43	♂ ♒
31	11 43	☽ ♋

LATITUDES

Date	Mercury ☿ ° '	Venus ♀ ° '	Mars ♂ ° '	Jupiter ♃ ° '	Saturn ♄ ° '	Uranus ♅ ° '	Neptune ♆ ° '	Pluto ♇ ° '
01	03 N 21	01 N 09	00 S 25	00 00	00 S 01	00 S 28	01 S 01	00 S 45
04	02 48	01 24	00 28	00 S 01	00 01	00 28	01 01	00 45
07	02 07	01 38	00 31	00 01	00 02	00 28	01 01	00 46
10	01 24	01 53	00 34	00 01	00 02	00 28	01 01	00 46
13	00 43	02 08	00 37	00 02	00 02	00 27	01 01	00 46
16	00 N 04	02 23	00 41	00 02	00 03	00 27	01 01	00 47
19	00 S 31	02 38	00 44	00 02	00 03	00 27	01 01	00 47
22	01 01	02 52	00 47	00 03	00 03	00 27	01 01	00 47
25	01 28	03 06	00 51	00 03	00 03	00 27	01 01	00 48
28	01 49	03 20	00 54	00 03	00 03	00 27	01 01	00 48
31	02 S 06	03 N 33	00 S 58	00 S 04	00 S 04	00 S 27	01 S 01	00 S 48

DATA

Julian Date	2458910
Delta T	+71 seconds
Ayanamsa	24° 08' 03"
Synetic vernal point	04° ♓ 58' 56"
True obliquity of ecliptic	23° 26' 12"

MOON'S PHASES, APSIDES AND POSITIONS ☽

Date	h m	Phase	Longitude ° '	Eclipse Indicator
02	19 57	☽	12 ♊ 42	
09	17 48	○	19 ♍ 37	
16	09 34	☾	26 ♐ 16	
24	09 28	●	04 ♈ 12	

Day	h m		
10	06 37	Perigee	
24	15 39	Apogee	
05	01 33	Max dec	23° N 27'
11	04 15	0S	
17	14 02	Max dec	23° S 32'
25	00 13	0N	

LONGITUDES

Date	Chiron ⚷ ° '	Ceres ⚳ ° '	Pallas ⚴ ° '	Juno ⚵ ° '	Vesta ⚶ ° '	Black Moon Lilith ⚸ ° '
01	03 ♈ 58	11 ♒ 42	15 ♑ 58	20 ♎ 09	23 ♉ 24	03 ♈ 50
11	04 ♈ 32	15 ♒ 25	19 ♑ 10	18 ♎ 31	26 ♉ 38	04 ♈ 57
21	05 ♈ 07	19 ♒ 02	22 ♑ 05	16 ♎ 26	00 ♊ 06	06 ♈ 04
31	05 ♈ 43	22 ♒ 31	24 ♑ 42	14 ♎ 06	03 ♊ 46	07 ♈ 12

All ephemeris data is given at 12.00 UT and the Moon's longitude is additionally given for 24.00 UT

ASPECTARIAN

h m	Aspects
01 Sunday	
05 46	☽ Q ☉
05 48	☉ ∥ ☿
08 04	☽ △ ♇
08 39	☽ ⚼ ♂
11 25	☽ ⊻ ♀
15 52	☽ △ ♄
19 37	☽ Q ♆
23 05	☽ □ ☿
23 44	♃ ∥ ♇
02 Monday	
00 21	☽ ⊥ ♀
02 38	☽ ⊻ ♅
03 22	☽ ± ♂
04 39	☽ ⚼ ♃
13 25	☽ ⚼ ♇
14 18	☽ ⊥ ♅
15 39	☽ ⊼ ♂
19 28	☽ ∠ ♀
19 57	☽ □ ☉
21 11	☽ ⚼ ♄
21 33	☽ # ♄
22 14	☽ ± ♃
03 Tuesday	
06 24	☽ □ ♆
06 38	☽ ± ♇
07 31	☽ ∠ ♅
09 47	☽ ⊼ ♃
12 50	☉ ∠ ♄
13 11	☽ # ♃
13 38	☽ # ♇
14 21	☽ ± ♄
16 44	♀ □ ♄
17 55	☽ ⊼ ♇
23 07	☿ ⊥ ♇
04 Wednesday	
01 35	☽ ⊼ ♄
02 20	☽ ⚹ ♀
02 55	♀ ∥ ♅
04 44	☽ △ ☿
11 26	☽ ⚹ ♅
17 57	☽ # ♂
21 24	☿ ⚹ ♀
05 Thursday	
02 03	☽ Q ♀
02 25	☽ ☍ ♂
04 55	☉ ∠ ♀
06 37	☽ ⚼ ☿
07 49	☽ △ ☉
09 00	☽ Q ♅
10 42	☽ # ♂
13 14	☽ △ ♆
16 56	☽ ☍ ♃
21 48	☽ ± ☿
23 50	☽ ☍ ♇
06 Friday	
00 28	☉ ∥ ♆
07 11	☽ ☍ ♄
07 47	☽ ⊼ ☿
11 47	☽ ⚼ ☉
11 48	☽ # ♇
12 00	☽ □ ♀
12 58	☽ # ♃
15 05	☽ ⚼ ♆
16 06	☽ □ ♅
07 Saturday	
01 08	☿ ⊻ ♄
02 36	☽ # ♄
03 59	☽ ± ☉
06 08	☽ ± ♆
06 43	☿ ∠ ♂
08 23	☽ ⊼ ♂
14 35	☽ ⊼ ☉
16 01	☽ ⊼ ♆
18 41	☽ ± ♂
19 55	☽ ⊼ ♃
08 Sunday	
01 50	☽ ⊼ ♇
05 36	♀ ∠ ♆
05 43	☽ ± ♃
08 12	☽ ☍ ☿
08 58	☽ ⊼ ♄
09 54	☽ ⚼ ♂
11 29	☽ ± ♇
12 23	☉ ☌ ♆
15 15	☽ ∥ ♀
17 00	☽ △ ♀
17 11	☽ △ ♅
18 35	☽ ± ♄
19 38	♀ ☌ ♅
20 15	☽ ⚼ ♃
09 Monday	
00 47	☽ ∥ ♅
01 48	☽ ⚼ ♇
03 40	☉ ∠ ♅
08 53	☽ ⚼ ♄
09 22	☽ # ☿
10 48	☽ △ ♂
15 57	☽ ☍ ♆
16 55	☽ ⚼ ♅
17 48	☽ ☍ ☉
18 28	☽ ⚼ ♀
20 08	☽ △ ♃
10 Tuesday	
01 24	☽ △ ♇
03 49	☿ St D
06 22	☽ # ♆
07 03	☽ ± ♅
07 15	☽ ⊼ ☿
08 32	☽ △ ♄
09 37	☽ ± ♀
13 13	☽ # ☉
16 28	☽ ⊼ ♅
16 40	☽ ± ☿
19 43	☽ ⊼ ♀
11 Wednesday	
06 53	☽ ⚼ ☿
12 06	☽ □ ♂
12 27	☉ ⚹ ♃
15 09	☽ ⊼ ♆
17 31	☽ ∥ ☉
19 48	☽ □ ♃
20 14	☽ ⊼ ☉
12 Thursday	
00 41	☽ □ ♇
00 42	☽ ± ♆
02 14	☽ ∥ ♆
06 29	☽ ± ☉
06 58	☽ △ ☿
08 12	☽ □ ♄
15 12	☽ ⚼ ♆
16 14	☽ ☍ ♅
18 17	☽ Q ♂
22 05	☽ ⚼ ☉
23 10	☽ ☍ ♀
13 Friday	
01 07	☽ Q ♃
03 00	☽ ∥ ☿
05 51	☽ Q ♇
09 08	☽ # ♅
13 45	☽ Q ♄
14 59	☽ ⚹ ♂
15 53	☽ △ ♆
21 14	☽ ⚹ ♃
14 Saturday	
00 47	☽ △ ☉
01 53	☽ ⚹ ♇
08 21	☽ # ♀
09 36	☽ □ ☿
10 06	☽ ⚹ ♄
10 32	♂ ⚹ ♆
16 47	☉ ⚹ ♇
17 44	☽ ∠ ♂
18 32	☽ ⊼ ♅
23 10	☽ ∠ ♃
15 Sunday	
03 44	☽ ∠ ♇
05 01	☽ ± ♅
05 50	☿ ⊻ ♄
06 00	☽ ⊼ ♀
08 00	☽ ∥ ♄
10 18	☽ ⊥ ♂
12 23	☽ ∠ ♄
15 12	☽ ⊥ ♃
17 32	☽ ± ♀
18 07	☽ Q ☿
19 42	☽ ⊥ ♇
19 51	☽ □ ♆
20 48	☽ ∥ ♃
21 03	☽ ⚼ ♅
21 35	☽ ⊻ ♂
16 Monday	
00 19	☽ ∥ ♇
02 05	☽ ⊻ ♃
04 37	☽ ⊥ ♄
06 33	☽ ⊻ ♇
08 21	☿ Q ♀
09 07	☽ ∥ ♂
09 34	☽ □ ☉
11 10	☽ ⚼ ♀
15 38	☽ ⊻ ♄
16 49	☽ ⚹ ☿
17 Tuesday	
00 33	☽ △ ♅
04 53	☽ Q ♆
09 00	☿ ⊥ ♇
17 31	☽ △ ♀
21 54	☽ Q ☉
22 18	☽ ∠ ☿
18 Wednesday	
03 36	☽ ⚹ ♆
08 32	☽ ☌ ♂
10 47	☽ ☌ ♃
14 53	☽ ☌ ♇
14 59	☉ ⊥ ♅
16 32	☽ ⊥ ☿
22 57	☽ ⚹ ☉
23 23	☽ ∥ ♂
19 Thursday	
00 48	☽ ☌ ♄
04 54	☽ ⊻ ☿
05 18	☽ ∥ ♇
08 41	☽ ∠ ♆
10 02	☽ ∥ ♃
10 04	☽ □ ♅
23 50	☉ ⚹ ♄
23 51	☽ ∥ ♄
20 Friday	
02 25	☽ ⊥ ♆
06 57	☽ ∠ ☉
08 25	☽ # ♀
09 00	☽ □ ♀
11 35	♂ ☌ ♃
14 20	☽ ⊻ ♆
21 45	☽ Q ♅
22 21	☽ ⊻ ♃
22 52	☽ ⊻ ♂
21 Saturday	
01 56	☽ ⊻ ♇
02 26	☽ ⊥ ☉
10 28	☽ ⊥ ♃
11 34	☽ ⊥ ♂
12 27	☽ ⊻ ♄
13 56	☽ ⊥ ♇
15 31	☽ ⊻ ☉
20 39	☽ ☌ ☿
21 51	☽ ⚹ ♅
22 Sunday	
00 13	☽ Q ♀
00 34	☽ ⊥ ♄
04 44	☽ ∠ ♃
05 05	☽ # ♅
06 44	☽ ∠ ♂
08 02	☽ ∠ ♇
13 19	☿ ⚹ ♅
16 00	☽ ∥ ☿
18 47	☽ ∠ ♄
22 10	♀ # ♄
23 Monday	
02 36	☽ ⚹ ♀
02 38	☽ ☌ ♆
03 08	♀ ⚹ ♆
04 10	☽ ∠ ♅
05 20	♂ ☌ ♇
07 58	♂ ∥ ♇
11 20	☽ ⚹ ♃
14 18	☽ ⚹ ♇
14 51	☽ ⚹ ♂
21 09	☽ ∥ ♆
22 25	☽ ⊥ ♅
24 Tuesday	
01 15	☽ ⚹ ♄
02 02	☿ ⊥ ♄
09 28	☽ ☌ ☉
10 36	☽ ⊻ ♅
11 41	☽ ∠ ♀
11 53	☽ Q ♃
14 27	☽ ⊻ ☿
14 36	☽ Q ♇
15 16	☽ # ☉
16 37	☽ Q ♂
23 44	☉ ⊻ ♅
25 Wednesday	
01 40	☽ Q ♄
03 44	☽ ⊥ ☿
07 32	☽ ⊥ ♀
10 47	☽ ∥ ☉
15 28	☽ ⊻ ♆
18 46	☉ Q ♃
20 45	☽ ⊻ ♀
23 43	☽ ∠ ☿
26 Thursday	
00 43	☽ □ ♃
03 02	☽ □ ♇
03 04	☽ # ♆
03 38	☽ ⊥ ♆
07 16	☽ □ ♂
12 01	☿ ∠ ♃
12 23	♀ # ♃
14 15	☽ □ ♄
21 51	☽ ∠ ♆
22 33	♀ # ♂
23 29	☽ ☌ ♅
23 46	☉ Q ♇
27 Friday	
02 00	☽ # ☿
03 35	☽ ⊻ ☉
09 03	☽ ⚹ ☿
11 42	☿ ∠ ♇
16 43	☽ ⊥ ☉
19 35	☽ ∥ ♅
28 Saturday	
04 02	☽ ⚹ ♆
04 24	♀ △ ♃
04 27	♀ # ♇
10 32	♂ ∥ ♃
11 37	☽ Q ☿
12 19	☽ ∠ ☉
13 39	☽ △ ♃
14 20	☽ ☌ ♀
15 19	☽ △ ♇
23 05	☽ △ ♂
29 Sunday	
02 36	☽ △ ♄
02 57	♀ △ ♇
03 56	☽ Q ♆
11 33	☽ ⊻ ♅
19 34	☽ ⚼ ♃
20 32	☽ ⚹ ☉
20 56	☽ ⚼ ♇
21 55	☽ # ♄
23 21	☽ ⊥ ♅
30 Monday	
02 58	☽ □ ☿
06 16	☽ ⚼ ♂
07 46	☽ # ♂
08 07	☽ ⚼ ♄
09 46	☽ # ♃
13 14	☽ ± ♃
14 25	☽ ± ♇
15 10	☽ □ ♆
16 41	☽ # ♇
16 51	☽ ∠ ♅
21 45	☽ Q ☉
31 Tuesday	
00 36	☽ ± ♂
00 52	☽ ⊼ ♃
01 31	☽ ± ♄
01 55	☽ ⊼ ♇
05 40	☽ ⊻ ♀
05 42	☽ ∥ ♀
12 39	☽ ⊼ ♂
12 56	☽ ⊼ ♄
17 53	☽ ⊥ ♀
18 31	♂ ☌ ♄
21 24	☽ ⚹ ♅

APRIL 2020

LONGITUDES

Date	Sidereal time h m s	Sun ☉ ° ' "	Moon ☽ ° ' "	Moon ☽ 24.00 ° ' "	Mercury ☿ ° '	Venus ♀ ° '	Mars ♂ ° '	Jupiter ♃ ° '	Saturn ♄ ° '	Uranus ♅ ° '	Neptune ♆ ° '	Pluto ♇ ° '
01	00 41 14	12 ♈ 13 20	13 ♋ 03 06	19 ♋ 38 58	15 ♓ 56	27 ♉ 59	01 ♒ 10	24 ♑ 28	00 ♒ 42	05 ♉ 11	19 ♓ 17	24 ♑ 51
02	00 45 11	13 12 30	26 ♋ 21 20	03 ♌ 10 34	17 16	28 54	01 52	24 35	00 46	05 15	19 19	24 52
03	00 49 07	14 11 39	10 ♌ 06 54	17 ♌ 10 29	18 37	29 ♉ 48	02 33	24 42	00 50	05 18	19 21	24 52
04	00 53 04	15 10 45	24 ♌ 21 14	01 ♍ 38 50	20 00	00 ♊ 42	03 15	24 49	00 53	05 21	19 24	24 53
05	00 57 00	16 09 48	09 ♍ 02 49	16 ♍ 32 25	21 25	01 36	03 57	24 56	00 56	05 25	19 26	24 54
06	01 00 57	17 08 50	24 ♍ 06 39	01 ♎ 44 20	22 51	02 28	04 39	25 03	01 00	05 28	19 28	24 54
07	01 04 53	18 07 49	09 ♎ 24 07	17 ♎ 04 33	24 20	03 20	05 20	25 09	01 03	05 31	19 30	24 55
08	01 08 50	19 06 46	24 ♎ 44 08	02 ♏ 21 26	25 49	04 11	06 02	25 16	01 06	05 34	19 32	24 55
09	01 12 46	20 05 41	09 ♏ 55 06	17 ♏ 23 56	27 21	05 02	06 44	25 22	01 09	05 38	19 34	24 56
10	01 16 43	21 04 34	24 ♏ 46 59	02 ♐ 03 28	28 ♓ 54	05 51	07 25	25 28	01 12	05 41	19 36	24 56
11	01 20 40	22 03 26	09 ♐ 12 54	16 ♐ 15 00	00 ♈ 29	06 40	08 07	25 34	01 15	05 45	19 38	24 57
12	01 24 36	23 02 16	23 ♐ 09 40	29 ♐ 57 01	02 05	07 28	08 49	25 40	01 18	05 48	19 40	24 57
13	01 28 33	24 01 04	06 ♑ 37 20	13 ♑ 11 00	03 43	08 16	09 30	25 45	01 20	05 51	19 42	24 57
14	01 32 29	24 59 50	19 ♑ 38 28	26 ♑ 00 16	05 22	09 02	10 12	25 51	01 23	05 55	19 44	24 58
15	01 36 26	25 58 35	02 ♒ 17 00	08 ♒ 29 16	07 03	09 48	10 54	25 56	01 26	05 58	19 46	24 58
16	01 40 22	26 57 18	14 ♒ 37 38	20 ♒ 42 41	08 45	10 32	11 35	26 01	01 28	06 02	19 48	24 58
17	01 44 19	27 55 59	26 ♒ 45 00	02 ♓ 45 06	10 30	11 16	12 17	26 06	01 30	06 05	19 50	24 59
18	01 48 15	28 54 38	08 ♓ 43 28	14 ♓ 40 33	12 15	11 58	12 59	26 11	01 32	06 08	19 51	24 59
19	01 52 12	29 ♈ 53 16	20 ♓ 36 46	26 ♓ 32 28	14 03	12 40	13 40	26 16	01 35	06 12	19 53	24 59
20	01 56 09	00 ♉ 51 52	02 ♈ 27 59	08 ♈ 23 36	15 52	13 20	14 22	26 20	01 37	06 15	19 55	24 59
21	02 00 05	01 50 26	14 ♈ 19 32	20 ♈ 16 02	17 42	14 00	15 04	26 25	01 39	06 19	19 57	24 59
22	02 04 02	02 48 58	26 ♈ 13 18	02 ♉ 11 29	19 35	14 38	15 45	26 29	01 40	06 22	19 59	24 59
23	02 07 58	03 47 29	08 ♉ 10 46	14 ♉ 11 19	21 29	15 15	16 27	26 33	01 42	06 26	20 00	24 59
24	02 11 55	04 45 57	20 ♉ 13 18	26 ♉ 16 56	23 24	15 50	17 08	26 37	01 44	06 29	20 02	25 00
25	02 15 51	05 44 24	02 ♊ 22 24	08 ♊ 29 58	25 21	16 25	17 50	26 40	01 45	06 33	20 04	25 00
26	02 19 48	06 42 48	14 ♊ 39 52	20 ♊ 52 25	27 20	16 57	18 31	26 44	01 47	06 36	20 06	25 R 00
27	02 23 44	07 41 11	27 ♊ 07 57	03 ♋ 26 48	29 ♈ 20	17 29	19 13	26 47	01 48	06 39	20 07	24 59
28	02 27 41	08 39 32	09 ♋ 49 23	16 ♋ 16 04	01 ♉ 22	17 59	19 54	26 50	01 50	06 43	20 09	24 59
29	02 31 38	09 37 50	22 ♋ 47 16	29 ♋ 23 22	03 25	18 27	20 36	26 53	01 51	06 46	20 11	24 59
30	02 35 34	10 ♉ 36 07	06 ♌ 04 43	12 ♌ 51 39	05 ♉ 30	18 ♊ 54	21 ♒ 17	26 ♑ 56	01 ♒ 52	06 ♉ 50	20 ♓ 12	24 ♑ 59

Date	Moon True ☊ ° '	Moon Mean ☊ ° '	Moon ☽ Latitude ° '
01	02 ♋ 44	03 ♋ 24	00 N 54
02	02 R 43	03 21	02 02
03	02 41	03 18	03 04
04	02 37	03 14	03 57
05	02 30	03 11	04 37
06	02 20	03 08	04 58
07	02 10	03 05	04 59
08	02 00	03 02	04 38
09	01 50	02 58	03 58
10	01 44	02 55	03 02
11	01 39	02 52	01 56
12	01 37	02 49	00 N 45
13	01 D 37	02 46	00 S 27
14	01 38	02 43	01 34
15	01 R 38	02 39	02 35
16	01 36	02 36	03 28
17	01 32	02 33	04 10
18	01 26	02 30	04 40
19	01 17	02 27	04 58
20	01 07	02 24	05 03
21	00 55	02 20	04 55
22	00 44	02 17	04 34
23	00 33	02 14	04 01
24	00 25	02 11	03 17
25	00 19	02 08	02 24
26	00 15	02 04	01 23
27	00 14	02 01	00 S 17
28	00 D 14	01 58	00 N 51
29	00 15	01 55	01 58
30	00 ♋ 16	01 ♋ 52	03 N 01

DECLINATIONS

Date	Sun ☉ ° '	Moon ☽ ° '	Mercury ☿ ° '	Venus ♀ ° '	Mars ♂ ° '	Jupiter ♃ ° '	Saturn ♄ ° '	Uranus ♅ ° '	Neptune ♆ ° '	Pluto ♇ ° '
01	04 N 50	23 N 42	07 S 34	23 N 14	20 S 51	21 S 17	20 S 03	12 N 49	05 S 11	21 S 57
02	05 13	22 52	07 06	23 31	20 43	21 16	20 03	12 51	05 10	21 57
03	05 36	20 40	06 38	23 47	20 35	21 15	20 02	12 52	05 09	21 57
04	05 59	17 08	06 08	24 03	20 26	21 14	20 01	12 53	05 08	21 57
05	06 21	12 27	05 37	24 18	20 18	21 13	20 01	12 54	05 08	21 57
06	06 44	06 54	05 05	24 32	20 09	21 12	20 00	12 55	05 07	21 57
07	07 07	00 N 51	04 32	24 46	20 00	21 11	19 59	12 56	05 06	21 57
08	07 29	05 S 16	03 57	25 00	19 51	21 10	19 59	12 57	05 05	21 57
09	07 51	11 01	03 22	25 13	19 42	21 09	19 58	12 58	05 04	21 57
10	08 13	16 01	02 45	25 25	19 32	21 08	19 58	12 59	05 04	21 57
11	08 35	19 55	02 07	25 37	19 23	21 07	19 57	13 01	05 03	21 57
12	08 57	22 31	01 28	25 49	19 13	21 06	19 57	13 02	05 02	21 57
13	09 19	23 43	00 48	26 00	19 03	21 05	19 56	13 03	05 01	21 57
14	09 41	23 33	00 S 07	26 10	18 53	21 04	19 56	13 04	05 01	21 57
15	10 02	22 10	00 N 35	26 20	18 43	21 03	19 55	13 05	05 00	21 57
16	10 23	19 45	01 18	26 29	18 33	21 02	19 55	13 06	04 59	21 57
17	10 44	16 30	02 02	26 38	18 23	21 01	19 54	13 07	04 58	21 57
18	11 05	12 37	02 46	26 46	18 12	21 01	19 54	13 09	04 58	21 58
19	11 26	08 17	03 32	26 54	18 01	21 00	19 53	13 10	04 57	21 58
20	11 46	03 S 39	04 19	27 01	17 50	20 59	19 53	13 11	04 56	21 58
21	12 07	01 N 07	05 06	27 08	17 40	20 58	19 53	13 12	04 56	21 58
22	12 27	05 51	05 54	27 14	17 28	20 58	19 52	13 13	04 55	21 58
23	12 47	10 25	06 42	27 20	17 17	20 57	19 52	13 14	04 54	21 58
24	13 07	14 38	07 31	27 25	17 06	20 57	19 52	13 15	04 54	21 58
25	13 26	18 17	08 21	27 30	16 54	20 56	19 52	13 17	04 53	21 58
26	13 45	21 11	09 11	27 34	16 43	20 56	19 51	13 18	04 52	21 58
27	14 04	23 08	10 01	27 38	16 31	20 55	19 51	13 19	04 52	21 58
28	14 23	23 56	10 52	27 41	16 19	20 55	19 51	13 20	04 51	21 58
29	14 42	23 27	11 42	27 44	16 08	20 54	19 51	13 21	04 51	21 59
30	15 N 00	21 N 40	12 N 33	27 N 46	15 S 56	20 S 54	19 S 51	13 N 22	04 S 50	21 S 59

ZODIAC SIGN ENTRIES

Date	h	m	Planets
02	18	26	☽ ♌
03	17	11	♀ ♊
04	21	18	☽ ♍
06	21	16	☽ ♎
08	20	17	☽ ♏
10	20	35	☽ ♐
11	04	48	☿ ♈
13	00	05	☽ ♑
15	07	37	☽ ♒
17	18	29	☽ ♓
19	14	45	☉ ♉
20	07	00	☽ ♈
22	19	36	☽ ♉
25	07	20	☽ ♊
27	17	28	☽ ♋
27	19	53	☿ ♉
30	01	06	☽ ♌

LATITUDES

Date	Mercury ☿ ° '	Venus ♀ ° '	Mars ♂ ° '	Jupiter ♃ ° '	Saturn ♄ ° '	Uranus ♅ ° '	Neptune ♆ ° '	Pluto ♇ ° '
01	02 S 11	03 N 38	00 S 59	00 S 04	00 S 04	00 S 27	01 S 01	00 S 48
04	02 22	03 50	01 03	00 04	00 04	00 27	01 01	00 49
07	02 29	04 02	01 06	00 05	00 04	00 27	01 01	00 49
10	02 31	04 12	01 10	00 05	00 04	00 27	01 02	00 49
13	02 29	04 22	01 14	00 05	00 05	00 27	01 02	00 50
16	02 22	04 31	01 18	00 06	00 05	00 27	01 02	00 50
19	02 10	04 38	01 22	00 06	00 05	00 27	01 02	00 50
22	01 54	04 43	01 26	00 07	00 06	00 27	01 02	00 51
25	01 34	04 47	01 30	00 07	00 06	00 27	01 02	00 51
28	01 09	04 49	01 34	00 08	00 06	00 27	01 02	00 52
31	00 S 41	04 N 48	01 S 38	00 S 08	00 S 06	00 S 27	01 S 02	00 S 52

DATA

Julian Date	2458941
Delta T	+71 seconds
Ayanamsa	24° 08' 06"
Synetic vernal point	04° ♓ 58' 53"
True obliquity of ecliptic	23° 26' 12"

MOON'S PHASES, APSIDES AND POSITIONS ☽

Date	h	m	Phase	Longitude ° '	Eclipse Indicator
01	10	21	☽	12 ♋ 09	
08	02	35	○	18 ♎ 44	
14	22	56	☾	25 ♑ 27	
23	02	26	●	03 ♉ 24	
30	20	38	☽	10 ♌ 57	

Day	h	m		
07	18	15	Perigee	
20	19	15	Apogee	
01	09	12	Max dec	23° N 42'
07	15	20	0S	
13	21	01	Max dec	23° S 48'
21	06	26	0N	
28	15	21	Max dec	23° N 56'

LONGITUDES

Date	Chiron ⚷ ° '	Ceres ⚳ ° '	Pallas ⚴ ° '	Juno ⚵ ° '	Vesta ⚶ ° '	Black Moon Lilith ⚸ ° '
01	05 ♈ 46	22 ♒ 51	24 ♑ 56	13 ♎ 51	04 ♊ 08	07 ♈ 18
11	06 ♈ 21	26 ♒ 10	27 ♑ 08	11 ♎ 31	07 ♊ 58	08 ♈ 25
21	06 ♈ 54	29 ♒ 18	28 ♑ 54	09 ♎ 24	11 ♊ 55	09 ♈ 33
31	07 ♈ 26	02 ♓ 12	00 ♒ 09	07 ♎ 41	15 ♊ 58	10 ♈ 40

All ephemeris data is given at 12.00 UT and the Moon's longitude is additionally given for 24.00 UT

ASPECTARIAN

h m	Aspects
01 Wednesday	
07 35	☿ ∠ ♄
10 21	☽ □ ☉
11 51	☽ ∠ ♀
14 38	☿ Q ♀
17 51	☽ △ ☿
19 35	☽ Q ♅
20 57	☿ ∠ ♂
23 23	☽ △ ♆
02 Thursday	
00 32	☉ Q ♄
01 34	☽ ∥ ♀
08 49	☽ ☍ ♃
09 07	☉ # ♆
09 20	☽ ☍ ♇
16 49	☽ ⚹ ♀
19 49	☽ ☍ ♄
22 13	☽ ☍ ♂
23 32	☽ ⚼ ☿
23 50	☽ # ♇
03 Friday	
02 02	☽ ⚼ ♆
03 40	☽ □ ♅
06 47	☽ # ♃
12 40	☽ # ♂
15 05	☽ Q ♀
16 44	☽ ± ☿
16 58	☽ # ♄
17 33	☽ ± ♆
19 29	☽ △ ☉
04 Saturday	
01 15	☿ ☌ ♆
03 42	☽ ⊼ ♆
03 58	☽ ⊼ ☿
12 46	☽ ⊼ ♃
12 53	☽ ⊼ ♇
16 23	☉ # ☿
17 09	♀ △ ♄
18 01	☉ Q ♂
18 16	☿ ∠ ♅
22 17	☽ ⚼ ☉
22 44	☽ ± ♃
22 45	☽ ± ♇
22 48	☽ ⊼ ♄
23 08	☽ □ ♀
05 Sunday	
02 45	♃ ☌ ♇
03 20	☽ ⊼ ♂
06 06	☽ △ ♅
07 21	♀ Q ♆
08 35	☽ ± ♄
09 54	☽ ∥ ♅
13 22	☽ ⚼ ♇
13 26	☽ ⚼ ♃
13 31	☽ ± ♂
13 55	☽ ± ☉
23 05	☽ ⚼ ♄
06 Monday	
00 12	☽ ⊼ ☉
04 35	☽ ⚼ ♂
04 38	☽ ☍ ♆
05 28	♂ ∠ ♆
06 13	☽ ⚼ ♅
09 49	☽ ☍ ☿
10 54	☿ ∥ ♆
12 38	☽ ∥ ☉
13 15	☽ △ ♇
13 29	☽ △ ♃
19 14	☽ # ♆
20 03	☽ # ☿
20 28	☽ ± ♅
22 52	☽ △ ♄
07 Tuesday	
01 56	☽ △ ♀
05 20	☽ △ ♂
05 54	☽ ⊼ ♅
14 01	♂ ∥ ♄
18 50	♂ □ ♅
21 28	☿ ⚹ ♇
08 Wednesday	
02 20	☿ ⚹ ♃
02 35	☽ ☍ ☉
02 48	☽ ⚼ ♀
03 49	☽ ⊼ ♆
07 16	☽ ∥ ☿
11 18	☽ ∥ ♆
12 17	☽ □ ♇
12 50	☽ □ ♃
13 15	☽ ± ♆
13 54	☽ ⊼ ☿
17 45	☽ ± ♀
21 35	☽ # ☉
22 03	☽ □ ♄
22 34	☉ ⊻ ♆
09 Thursday	
00 24	☽ ± ☿
00 58	☉ ∠ ♀
03 28	☽ ⚼ ♆
03 46	☽ ⊼ ♀
05 09	☽ ☍ ♅
06 41	☽ □ ♂
16 20	☽ ⚼ ☿
16 49	☽ Q ♇
17 33	☽ Q ♃
20 53	☽ # ♅
10 Friday	
02 53	☽ Q ♄
03 32	☽ △ ♆
05 31	☽ ⊼ ☉
06 40	♀ ⊻ ♅
12 15	☽ ⚹ ♇
13 06	☽ Q ♂
13 08	☽ ⚹ ♃
16 02	☽ ± ☉
19 35	☽ △ ☿
22 37	☽ ⚹ ♄
11 Saturday	
00 29	☿ ⊥ ♅
06 08	☽ ⊼ ♅
07 27	☽ ☍ ♀
08 06	☽ ⚼ ☉
08 21	☽ ∥ ♂
10 03	☽ ⚹ ♂
12 15	☽ ∥ ♄
13 14	☽ ∠ ♇
14 18	☽ ∠ ♃
16 18	☽ ± ♅
21 27	☽ ∥ ♃
23 58	☿ ⚹ ♄
12 Sunday	
00 02	☽ ∠ ♄
04 39	☽ ⊥ ♇
05 33	☽ ∥ ♇
05 51	☽ ⊥ ♃
05 53	☽ □ ♆
07 51	☽ ⚼ ♅
11 46	☽ △ ☉
13 12	☽ ∠ ♂
15 08	☽ ⊻ ♇
15 46	☽ ⊥ ♄
16 26	☽ ⊻ ♃
13 Monday	
02 27	☽ ⊻ ♄
06 00	☽ □ ☿
06 04	☽ ⊥ ♂
10 36	☽ △ ♅
13 57	☽ Q ♆
15 10	☽ ⊼ ♀
17 33	☽ ⊻ ♂
14 Tuesday	
02 53	☽ ± ♀
11 07	☉ □ ♇
12 10	☽ ⚹ ♆
20 04	☽ Q ☿
20 06	☿ ⊻ ♅
20 48	☽ ⚼ ♀
22 02	☽ ☌ ♇
22 56	☽ □ ☉
23 47	☽ ☌ ♃
15 Wednesday	
06 34	☉ ⊥ ♆
10 21	☽ ☌ ♄
10 50	☿ Q ♇
10 59	☉ □ ♃
14 37	☽ ∥ ♇
16 47	☽ ∠ ♆
17 35	♀ ⚼ ♇
19 09	☽ □ ♅
22 40	☽ ⚹ ☿
16 Thursday	
00 21	☽ ∥ ♃
01 12	☿ Q ♃
03 29	☽ △ ♀
05 42	☽ ☌ ♂
10 22	☽ ⊥ ♆
10 40	☽ ∥ ♄
12 42	☽ Q ☉
22 06	☽ ∥ ♂
22 13	☽ ⊻ ♆
17 Friday	
06 11	♀ ⚼ ♃
06 40	☽ Q ♅
08 28	☽ ⊻ ♇
09 04	☽ ∠ ☿
10 43	☽ ⊻ ♃
14 34	☽ ⚹ ☉
20 27	☽ ⊥ ♇
21 32	☽ ⊻ ♄
22 47	☽ ⊥ ♃
18 Saturday	
05 36	☿ ⚹ ♀
06 10	☽ ⊥ ☿
06 47	☽ ⚹ ♅
08 58	☽ # ♅
09 37	☽ ⊥ ♄
14 32	☽ ∠ ♇
17 00	☽ ∠ ♃
18 18	♀ ⊥ ♅
18 57	☽ □ ♀
20 05	☽ # ☉
20 22	☽ ⊻ ☿
21 07	☽ ⊻ ♂
23 23	☽ ∠ ☉
19 Sunday	
03 49	☽ ∠ ♄
03 55	☿ ⚹ ♂
05 37	☿ Q ♄
09 59	☽ ⊥ ♂
10 32	☽ ☌ ♆
13 11	☽ ∠ ♅
19 13	☽ ⊥ ☉
19 43	♂ ⊥ ♆
20 51	☽ ⚹ ♇
23 31	☽ ⚹ ♃
20 Monday	
05 20	☽ ∠ ♂
05 26	☽ ∥ ♆
07 30	☽ ⊥ ♅
08 28	☽ ⊻ ☉
09 09	☽ # ☿
09 35	☽ Q ♀
10 16	☽ ⚹ ♄
19 42	☽ ⊻ ♅
21 09	☽ Q ♇
23 58	☽ Q ♃
21 Tuesday	
07 00	☉ □ ♄
07 00	☿ # ♆
10 37	☽ Q ♄
11 17	☽ ⚹ ♀
13 35	☽ ⚹ ♂
20 06	☽ ☌ ☿
23 23	☽ ⊻ ♆
22 Wednesday	
07 13	☽ # ♆
09 31	☽ □ ♇
11 31	☽ ⊥ ♆
12 13	☽ ∥ ☿
12 32	☽ □ ♃
15 17	☽ Q ♂
17 11	☿ ⊻ ♆
19 14	☽ ∠ ♀
22 59	☽ □ ♄
23 Thursday	
02 26	☽ ☌ ☉
05 38	☽ ∠ ♆
08 29	☽ ☌ ♅
14 14	☽ ⊥ ♀
24 Friday	
02 15	☽ ∥ ☉
02 50	☽ ⊻ ♀
03 50	☽ ∥ ♅
05 30	☽ □ ♂
11 38	☽ ⚹ ♆
18 54	☉ ∠ ♆
19 30	☽ ⊻ ☿
21 27	☽ △ ♇
23 36	☉ ∥ ♅
25 Saturday	
00 43	☽ △ ♃
02 55	☽ # ♂
07 36	☿ □ ♇
09 36	☽ ⊥ ☿
10 47	☽ △ ♄
11 24	☽ Q ♆
18 54	♇ St R
19 10	☽ ⊻ ☉
20 13	☽ ⊻ ♅
20 50	☿ ⊥ ♆
26 Sunday	
00 07	☽ # ♄
02 55	☽ ⚼ ♇
03 54	♀ ⚼ ♄
04 31	☿ □ ♃
06 16	☽ ⚼ ♃
06 36	☽ ∠ ☿
07 53	☽ ⊥ ☉
07 58	☽ ⊥ ♅
09 01	☉ ☌ ♅
09 30	☽ # ♃
16 07	☽ ⚼ ♄
16 39	☽ ☌ ♀
19 55	☽ △ ♂
20 15	☽ # ♇
20 22	☽ ± ♇
22 31	☽ □ ♆
23 46	☽ ± ♃
27 Monday	
01 28	☽ ∠ ♅
02 46	☽ ∠ ☉
07 54	☽ ⊼ ♇
09 28	☽ ± ♄
11 20	☽ ⊼ ♃
17 00	☽ ⚹ ☿
20 54	☽ ⊼ ♄
28 Tuesday	
02 14	☽ ⚼ ♂
06 08	☽ ⚹ ♅
09 38	☽ ⚹ ☉
17 28	☿ □ ♄
19 52	☽ Q ☿
20 03	☽ ± ♂
20 48	♂ ⊻ ♆
21 35	☿ Q ♂
29 Wednesday	
03 45	☽ ⊻ ♀
04 36	☽ Q ♅
07 12	☽ △ ♆
07 46	☽ ⊼ ♂
09 43	☽ Q ☉
12 33	☿ ∠ ♀
15 10	☽ ⊥ ♀
16 01	☽ ☍ ♇
19 29	☽ ☍ ♃
30 Thursday	
04 27	☽ ☍ ♄
07 59	☽ ∠ ♀
08 35	☿ ∠ ♆
08 49	☽ # ♇
10 26	☽ ⚼ ♆
10 47	☽ □ ☿
13 21	☽ □ ♅
16 31	♀ ± ♇
19 03	☽ # ♃
20 38	☽ □ ☉

MAY 2020

LONGITUDES

Date	Sidereal time h m s	Sun ☉ ° ' "	Moon ☽ ° ' "	Moon ☽ 24.00 ° ' "	Mercury ☿ ° '	Venus ♀ ° '	Mars ♂ ° '	Jupiter ♃ ° '	Saturn ♄ ° '	Uranus ♅ ° '	Neptune ♆ ° '	Pluto ♇ ° '
01	02 39 31	11 ♉ 34 22	19 ♌ 44 22	26 ♌ 43 02	07 ♉ 36	19 ♊ 20	21 ♒ 59	26 ♑ 58	01 ♒ 53	06 ♉ 53	20 ♓ 14	24 ♑ 59
02	02 43 27	12 32 34	03 ♍ 47 38	10 ♍ 58 04	09 43	19 43	22 40	27 01	01 54	06 57	20 15	24 R 59
03	02 47 24	13 30 44	18 ♍ 14 00	25 ♍ 34 56	11 51	20 05	23 21	27 03	01 55	07 00	20 17	24 59
04	02 51 20	14 28 52	03 ♎ 00 13	10 ♎ 29 00	14 00	20 25	24 03	27 05	01 55	07 04	20 19	24 58
05	02 55 17	15 26 59	18 ♎ 00 15	25 ♎ 32 52	16 10	20 42	24 44	27 07	01 56	07 07	20 20	24 58
06	02 59 13	16 25 03	03 ♏ 05 37	10 ♏ 37 16	18 20	20 58	25 25	27 08	01 56	07 10	20 22	24 58
07	03 03 10	17 23 06	18 ♏ 06 38	25 ♏ 32 33	20 30	21 12	26 06	27 10	01 57	07 14	20 23	24 58
08	03 07 07	18 21 07	02 ♐ 54 03	10 ♐ 10 15	22 40	21 24	26 48	27 11	01 57	07 17	20 24	24 57
09	03 11 03	19 19 07	17 ♐ 20 31	24 ♐ 24 20	24 50	21 34	27 29	27 12	01 57	07 21	20 26	24 57
10	03 15 00	20 17 05	01 ♑ 21 26	08 ♑ 11 42	26 59	21 41	28 10	27 13	01 57	07 24	20 27	24 56
11	03 18 56	21 15 02	14 ♑ 55 11	21 ♑ 32 03	29 ♉ 07	21 47	28 51	27 14	01 R 57	07 27	20 28	24 56
12	03 22 53	22 12 58	28 ♑ 02 38	04 ♒ 27 20	01 ♊ 14	21 50	29 ♒ 32	27 14	01 57	07 31	20 30	24 56
13	03 26 49	23 10 52	10 ♒ 46 36	17 ♒ 00 59	03 20	21 R 50	00 ♓ 13	27 14	01 57	07 34	20 31	24 55
14	03 30 46	24 08 45	23 ♒ 11 01	29 ♒ 17 19	05 23	21 49	00 54	27 14	01 57	07 37	20 32	24 55
15	03 34 42	25 06 37	05 ♓ 20 27	11 ♓ 21 00	07 25	21 45	01 35	27 R 14	01 56	07 41	20 34	24 54
16	03 38 39	26 04 27	17 ♓ 19 33	23 ♓ 16 37	09 24	21 38	02 16	27 14	01 56	07 44	20 35	24 53
17	03 42 36	27 02 17	29 ♓ 12 45	05 ♈ 08 25	11 21	21 29	02 57	27 14	01 55	07 47	20 36	24 53
18	03 46 32	28 00 05	11 ♈ 04 05	17 ♈ 00 10	13 16	21 18	03 38	27 13	01 55	07 51	20 37	24 52
19	03 50 29	28 57 52	22 ♈ 57 01	28 ♈ 55 00	15 08	21 04	04 18	27 12	01 54	07 54	20 38	24 52
20	03 54 25	29 ♉ 55 37	04 ♉ 54 24	10 ♉ 55 28	16 57	20 48	04 59	27 11	01 53	07 57	20 39	24 51
21	03 58 22	00 ♊ 53 22	16 ♉ 58 28	23 ♉ 03 33	18 43	20 30	05 40	27 10	01 52	08 01	20 40	24 50
22	04 02 18	01 51 05	29 ♉ 10 56	05 ♊ 20 45	20 26	20 09	06 20	27 08	01 51	08 04	20 41	24 50
23	04 06 15	02 48 47	11 ♊ 33 10	17 ♊ 48 17	22 06	19 46	07 01	27 07	01 50	08 07	20 42	24 49
24	04 10 11	03 46 28	24 ♊ 06 14	00 ♋ 27 10	23 43	19 21	07 41	27 05	01 49	08 10	20 43	24 48
25	04 14 08	04 44 07	06 ♋ 51 12	13 ♋ 18 28	25 17	18 54	08 21	27 03	01 47	08 13	20 44	24 47
26	04 18 05	05 41 45	19 ♋ 49 07	26 ♋ 23 17	26 47	18 25	09 02	27 01	01 46	08 17	20 45	24 46
27	04 22 01	06 39 22	03 ♌ 01 07	09 ♌ 42 44	28 15	17 54	09 42	26 59	01 45	08 20	20 46	24 46
28	04 25 58	07 36 57	16 ♌ 28 17	23 ♌ 17 51	29 ♊ 39	17 21	10 22	26 56	01 43	08 23	20 47	24 45
29	04 29 54	08 34 30	00 ♍ 11 28	07 ♍ 09 09	01 ♋ 00	16 48	11 02	26 53	01 41	08 26	20 48	24 44
30	04 33 51	09 32 03	14 ♍ 10 50	21 ♍ 16 22	02 17	16 13	11 42	26 51	01 39	08 29	20 48	24 43
31	04 37 47	10 ♊ 29 33	28 ♍ 25 31	05 ♎ 37 57	03 ♋ 31	15 ♊ 37	12 ♓ 22	26 ♑ 47	01 ♒ 38	08 ♉ 32	20 ♓ 49	24 ♑ 42

Date	Moon True ☊ ° '	Moon Mean ☊ ° '	Moon ☽ Latitude ° '
01	00 ♋ 16	01 ♋ 49	03 N 55
02	00 R 14	01 45	04 36
03	00 10	01 42	05 01
04	00 05	01 39	05 08
05	29 ♊ 58	01 36	04 53
06	29 52	01 33	04 18
07	29 46	01 30	03 26
08	29 42	01 26	02 20
09	29 39	01 23	01 N 06
10	29 D 39	01 20	00 S 09
11	29 40	01 17	01 22
12	29 41	01 14	02 28
13	29 42	01 10	03 25
14	29 43	01 07	04 11
15	29 R 43	01 04	04 44
16	29 40	01 01	05 04
17	29 37	00 58	05 11
18	29 32	00 55	05 05
19	29 27	00 51	04 46
20	29 22	00 48	04 14
21	29 17	00 45	03 31
22	29 13	00 42	02 37
23	29 11	00 39	01 35
24	29 10	00 36	00 S 28
25	29 D 10	00 32	00 N 42
26	29 11	00 29	01 51
27	29 12	00 26	02 56
28	29 14	00 23	03 52
29	29 15	00 20	04 36
30	29 R 15	00 16	05 04
31	29 ♊ 14	00 ♋ 13	05 N 15

DECLINATIONS

Date	Sun ☉ ° '	Moon ☽ ° '	Mercury ☿ ° '	Venus ♀ ° '	Mars ♂ ° '	Jupiter ♃ ° '	Saturn ♄ ° '	Uranus ♅ ° '	Neptune ♆ ° '	Pluto ♇ ° '
01	15 N 18	18 N 36	13 N 24	27 N 47	15 S 44	20 S 53	19 S 50	13 N 23	04 S 49	21 S 59
02	15 36	14 24	14 14	27 48	15 31	20 53	19 50	13 25	04 49	21 59
03	15 54	09 16	15 03	27 49	15 19	20 53	19 50	13 26	04 48	21 59
04	16 11	03 N 31	15 52	27 49	15 07	20 53	19 50	13 27	04 48	21 59
05	16 28	02 S 33	16 40	27 48	14 54	20 52	19 50	13 28	04 47	22 00
06	16 45	08 30	17 27	27 47	14 42	20 52	19 50	13 29	04 46	22 00
07	17 01	13 56	18 13	27 46	14 29	20 52	19 50	13 30	04 46	22 00
08	17 17	18 27	18 57	27 44	14 16	20 52	19 50	13 31	04 45	22 00
09	17 33	21 44	19 39	27 41	14 03	20 52	19 50	13 32	04 45	22 00
10	17 49	23 35	20 19	27 37	13 50	20 52	19 50	13 33	04 44	22 00
11	18 04	23 58	20 58	27 33	13 37	20 52	19 50	13 35	04 44	22 01
12	18 19	22 58	21 34	27 29	13 24	20 52	19 50	13 36	04 43	22 01
13	18 34	20 49	22 08	27 24	13 11	20 52	19 51	13 37	04 43	22 01
14	18 48	17 44	22 39	27 18	12 58	20 52	19 51	13 38	04 43	22 01
15	19 02	13 57	23 08	27 12	12 45	20 52	19 51	13 39	04 42	22 01
16	19 16	09 40	23 34	27 05	12 31	20 53	19 51	13 40	04 42	22 02
17	19 30	05 05	23 58	26 57	12 18	20 53	19 51	13 41	04 41	22 02
18	19 43	00 S 18	24 19	26 48	12 05	20 53	19 52	13 42	04 41	22 02
19	19 56	04 N 30	24 37	26 39	11 51	20 53	19 52	13 43	04 40	22 02
20	20 08	09 10	24 53	26 29	11 38	20 54	19 52	13 44	04 40	22 03
21	20 20	13 32	25 06	26 19	11 24	20 54	19 52	13 45	04 40	22 03
22	20 32	17 25	25 17	26 07	11 10	20 55	19 53	13 46	04 39	22 03
23	20 43	20 36	25 26	25 55	10 57	20 55	19 53	13 47	04 39	22 03
24	20 54	22 51	25 33	25 42	10 43	20 56	19 53	13 48	04 39	22 04
25	21 05	23 58	25 37	25 29	10 29	20 56	19 54	13 49	04 38	22 04
26	21 15	23 48	25 39	25 15	10 15	20 57	19 54	13 50	04 38	22 04
27	21 25	22 20	25 40	25 00	10 02	20 57	19 55	13 51	04 38	22 04
28	21 35	19 35	25 38	24 44	09 48	20 58	19 55	13 52	04 37	22 05
29	21 44	15 42	25 35	24 28	09 34	20 59	19 56	13 53	04 37	22 05
30	21 53	10 54	25 30	24 11	09 20	20 59	19 56	13 54	04 37	22 05
31	22 N 01	05 N 27	25 N 24	23 N 54	09 S 06	21 S 00	19 S 57	13 N 55	04 S 37	22 S 05

ZODIAC SIGN ENTRIES

Date	h	m	Planets
02	05	35	☽ ♍
04	07	09	☽ ♎
06	07	05	☽ ♏
08	07	15	☽ ♐
10	09	38	☽ ♑
11	21	58	☿ ♊
12	15	39	☽ ♒
13	04	17	♂ ♓
15	01	24	☽ ♓
17	13	36	☽ ♈
20	02	10	☽ ♉
20	13	49	☉ ♊
22	13	36	☽ ♊
24	23	09	☽ ♋
27	06	33	☽ ♌
28	18	09	☿ ♋
29	11	40	☽ ♍
31	14	38	☽ ♎

LATITUDES

Date	Mercury ☿ ° '	Venus ♀ ° '	Mars ♂ ° '	Jupiter ♃ ° '	Saturn ♄ ° '	Uranus ♅ ° '	Neptune ♆ ° '	Pluto ♇ ° '
01	00 S 41	04 N 48	01 S 38	00 S 08	00 S 06	00 S 27	01 S 02	00 S 52
04	00 S 11	04 44	01 42	00 08	00 07	00 27	01 02	00 52
07	00 N 21	04 38	01 46	00 09	00 07	00 27	01 02	00 53
10	00 52	04 27	01 51	00 09	00 07	00 27	01 03	00 53
13	01 20	04 13	01 55	00 10	00 07	00 27	01 03	00 53
16	01 44	03 55	01 59	00 10	00 08	00 27	01 03	00 54
19	02 01	03 32	02 04	00 11	00 08	00 27	01 03	00 54
22	02 12	03 04	02 08	00 11	00 08	00 27	01 03	00 54
25	02 16	02 31	02 13	00 12	00 09	00 27	01 03	00 55
28	02 12	01 54	02 17	00 12	00 09	00 27	01 03	00 55
31	02 N 01	01 N 15	02 S 22	00 S 13	00 S 09	00 S 27	01 S 03	00 S 55

DATA

Julian Date	2458971
Delta T	+71 seconds
Ayanamsa	24° 08' 10"
Synetic vernal point	04° ♓ 58' 49"
True obliquity of ecliptic	23° 26' 12"

MOON'S PHASES, APSIDES AND POSITIONS ☽

Date	h	m	Phase	Longitude ° '	Eclipse Indicator
07	10	45	○	17 ♏ 20	
14	14	03	☾	24 ♒ 14	
22	17	39	●	02 ♊ 05	
30	03	30	☽	09 ♍ 12	

Day	h	m		
06	03	09	Perigee	
18	07	53	Apogee	
05	01	58	0S	
11	06	16	Max dec	24° S 00'
18	13	31	0N	
25	21	12	Max dec	24° N 03'

LONGITUDES

Date	Chiron ⚷ ° '	Ceres ⚳ ° '	Pallas ⚴ ° '	Juno ⚵ ° '	Vesta ⚶ ° '	Black Moon Lilith ⚸ ° '
01	07 ♈ 26	02 ♓ 12	00 ♒ 09	07 ♎ 41	15 ♊ 58	10 ♈ 40
11	07 ♈ 55	04 ♓ 52	00 ♒ 50	06 ♎ 30	20 ♊ 07	11 ♈ 47
21	08 ♈ 21	07 ♓ 14	00 ♒ 54	05 ♎ 54	24 ♊ 19	12 ♈ 54
31	08 ♈ 43	09 ♓ 15	00 ♒ 16	05 ♎ 52	28 ♊ 35	14 ♈ 02

All ephemeris data is given at 12.00 UT and the Moon's longitude is additionally given for 24.00 UT

ASPECTARIAN

h m	Aspects
01 Friday	
02 23	☽ ± ♆
03 23	☽ # ♄
03 41	☿ ☌ ♅
11 16	☽ ✶ ♀
11 55	☿ ∥ ♅
12 51	☽ ⊼ ♆
16 04	☽ ☍ ♂
21 02	☽ ⊼ ♇
02 Saturday	
00 28	☽ ⊼ ♃
05 51	☽ # ♂
06 06	☽ ∥ ☉
07 15	☽ ± ♇
08 14	☉ # ♂
08 24	☽ Q ♀
08 48	☽ ⊼ ♄
10 41	☽ ± ♃
12 45	☽ ∥ ☿
16 56	☽ ∥ ♅
17 19	☽ △ ♅
18 53	☽ ± ♄
22 21	☽ ⚼ ♇
23 38	☽ △ ☿
03 Sunday	
01 46	☽ ⚼ ♃
03 40	☽ △ ☉
09 49	☽ ⚼ ♄
15 06	☽ □ ♀
15 22	☽ ☍ ♆
18 10	☿ # ♂
18 12	☽ ⚼ ♅
20 47	☽ ⊼ ♂
23 01	☽ △ ♇
04 Monday	
02 25	☽ △ ♃
03 52	♀ □ ♆
04 27	☽ ⚼ ☿
05 55	☽ ⚼ ☉
06 48	☽ # ♆
07 00	☽ ± ♂
08 51	☽ ± ♅
09 21	☉ ⊥ ♀
10 15	☽ △ ♄
17 19	☿ ⊥ ♀
18 32	☽ ⊼ ♅
21 23	☽ ± ☿
21 24	☽ ± ☉
21 41	☉ ☌ ☿
22 10	☽ ⚼ ♂
05 Tuesday	
02 30	☉ ∥ ☿
07 39	☽ ⊼ ☉
08 34	☽ ⊼ ☿
15 43	☽ ⊼ ♆
16 23	☽ △ ♀
20 11	♂ ⊻ ♇
20 53	☽ ∥ ♆
23 05	☽ □ ♇
23 13	☽ △ ♂
06 Wednesday	
01 16	☽ ± ♆
02 31	☽ □ ♃
02 34	♂ Q ♅
10 10	☽ □ ♄
15 37	☽ ⚼ ♆
16 40	☽ ⚼ ♀
18 31	☽ ☍ ♅
07 Thursday	
03 44	☽ Q ♇
06 24	♀ ± ♃
07 15	☽ Q ♃
07 16	☽ ± ♀
09 58	☽ # ♅
10 42	☿ ✶ ♆
10 45	☽ ☍ ☉
14 33	☽ ∥ ♂
14 57	☽ Q ♄
15 40	☽ △ ♆
16 30	☽ ☍ ☿
17 03	☽ ⊼ ♀
20 39	☿ ⊻ ♀
23 03	☽ ✶ ♇
08 Friday	
01 33	☽ □ ♂
02 39	☽ ✶ ♃
04 47	☽ # ☉
10 27	☽ ✶ ♄
15 47	☽ # ☿
19 15	☽ ⊼ ♅
20 58	☽ ∥ ♄
23 38	☽ ∠ ♇
09 Saturday	
01 53	♂ ⊻ ♃
03 22	☽ ∠ ♃
04 35	☽ ∥ ♃
05 16	☽ ± ♅
08 43	☽ Q ♂
11 21	☽ ∠ ♄
13 17	☿ △ ♇
14 35	☽ ∥ ♇
14 43	☽ ⊥ ♇
15 35	☽ ⊼ ☉
17 14	☽ □ ♆
18 32	☽ ⊥ ♃
18 35	☿ # ♄
19 13	☽ ☍ ♀
20 31	☽ ⚼ ♅
10 Sunday	
00 55	☽ ⊻ ♇
02 34	☽ ± ☉
02 40	☽ ⊥ ♄
03 03	☽ ⊼ ☿
04 49	☽ ⊻ ♃
06 11	☽ ✶ ♂
13 03	☽ ⊻ ♄
14 36	☿ △ ♃
15 22	☽ ± ☿
16 16	☉ ✶ ♆
19 23	☽ ⚼ ☉
22 38	☽ △ ♅
11 Monday	
00 29	☽ Q ♆
04 09	♄ St R
07 33	☿ □ ♂
08 18	☿ # ♃
09 59	☽ ∠ ♂
10 17	☽ ⚼ ☿
16 56	♂ # ♅
22 05	☽ ✶ ♆
12 Tuesday	
00 24	☽ △ ☉
00 30	☽ ⊼ ♀
01 57	☉ ⊻ ♀
03 13	☽ ⊥ ♂
06 14	☽ ☌ ♇
10 30	☽ ☌ ♃
11 36	☽ ± ♀
14 56	☽ ⊻ ♂
19 07	☽ △ ☿
19 18	☽ ☌ ♄
20 14	☿ △ ♄
13 Wednesday	
00 05	☽ ∥ ♇
01 30	☽ # ☿
01 59	☽ ∠ ♆
02 36	☿ Q ♆
04 30	☽ ⚼ ♀
05 52	☽ □ ♅
06 45	♀ St R
07 08	☿ # ♇
11 30	☽ ∥ ♃
19 11	☽ ⊥ ♆
20 16	☽ ∥ ♄
14 Thursday	
04 50	☽ # ☉
06 50	☽ ⊻ ♆
09 20	☽ △ ♀
12 14	♂ ⊥ ♇
14 03	☽ □ ☉
14 32	♃ St R
15 23	☽ ⊻ ♇
16 48	☽ Q ♅
19 58	☽ ⊻ ♃
15 Friday	
03 12	☽ ⊥ ♇
04 06	☽ ☌ ♂
05 15	☽ ⊻ ♄
06 49	☉ △ ♇
07 49	☽ ⊥ ♃
13 46	☽ # ♅
15 16	☿ ⊻ ♅
16 41	☽ ✶ ♅
16 57	☽ □ ☿
17 11	☽ ⊥ ♄
19 23	☽ ∥ ♂
21 05	☽ ∠ ♇
16 Saturday	
00 25	♂ ⊻ ♄
01 47	☽ ∠ ♃
04 53	☽ Q ☉
11 13	☽ ∠ ♄
17 54	☿ ⚼ ♇
18 34	☽ ☌ ♆
20 35	☽ □ ♀
22 57	☽ ∠ ♅
17 Sunday	
03 15	☽ ✶ ♇
07 13	☽ ✶ ☉
07 59	☽ ✶ ♃
12 21	☽ Q ☿
13 59	☽ ∥ ♆
16 40	☉ △ ♃
17 15	☽ ⊥ ♅
17 29	☽ ✶ ♄
20 01	☽ ⊻ ♂
21 44	♂ ⊥ ♃
22 50	☿ ⚼ ♃
18 Monday	
03 31	☽ Q ♇
05 27	☽ ⊻ ♅
08 15	☽ Q ♃
08 29	☽ Q ♀
08 54	☽ ⊥ ♂
16 15	☽ ∠ ☉
17 17	☽ ✶ ☿
17 45	☽ Q ♄
19 39	☿ ⊥ ♅
21 32	♀ ± ♃
19 Tuesday	
04 12	☽ ∠ ♂
04 55	☉ # ♄
07 20	☽ ⊻ ♆
08 17	☽ ✶ ♀
12 02	☽ ⊥ ☉
12 55	☽ # ♆
15 51	☽ □ ♇
19 26	☽ ⊥ ♆
20 33	☽ □ ♃
20 Wednesday	
01 09	☽ ⊻ ☉
05 01	☽ ∠ ☿
05 58	☽ □ ♄
11 14	☿ ⚼ ♄
12 10	☽ ✶ ♂
13 30	☽ ∠ ♆
13 45	☽ ∠ ♀
18 07	☽ ☌ ♅
23 03	♀ □ ♆
21 Thursday	
00 36	☽ # ♂
02 07	☽ ⊥ ☿
07 13	☽ ⊥ ♀
13 16	☽ ∥ ♅
13 26	☽ Q ♂
13 43	☿ ± ♇
16 00	☽ ⊻ ☿
18 46	☽ ⊻ ♀
19 19	☽ ✶ ♆
22 Friday	
03 29	☽ △ ♇
08 01	☽ △ ♃
08 41	☿ ☌ ♀
12 02	☉ △ ♄
15 43	☿ □ ♆
17 12	☽ △ ♄
17 39	☽ ☌ ☉
18 51	☽ Q ♆
21 59	☿ ± ♃
23 Saturday	
02 43	☽ □ ♂
05 20	☽ ⊻ ♅
06 02	☽ # ♄
08 39	☽ ⚼ ♇
09 17	☉ Q ♆
13 05	☽ ⚼ ♃
13 10	☽ ∥ ☉
14 54	☽ # ♃
16 57	☽ ⊥ ♅
22 08	☽ ⚼ ♄
24 Sunday	
01 55	☽ ± ♇
02 16	☽ # ♇
03 15	☽ ☌ ♀
03 33	☿ ∠ ♅
05 34	☽ □ ♆
06 16	☽ ± ♃
10 13	☽ ∠ ♅
11 09	☽ ☌ ☿
13 19	☽ ⊼ ♇
15 14	☽ ± ♄
15 23	☉ # ♃
16 26	♂ ⊥ ♄
17 38	☽ ⊼ ♃
25 Monday	
01 06	☿ ∥ ♀
02 32	☽ ⊼ ♄
04 26	☿ ⊼ ♇
06 48	♂ ✶ ♅
07 43	☽ ⊻ ☉
14 34	☽ ✶ ♅
14 58	☽ △ ♂
17 35	♀ ± ♇
19 49	☽ ⊥ ☉
19 55	☿ ± ♄
26 Tuesday	
09 31	☽ ⊻ ♀
12 51	☽ Q ♅
13 43	☽ △ ♆
13 44	☽ ∠ ☉
15 35	☿ ⊼ ♃
20 05	☽ ⊥ ♀
20 07	☽ ⚼ ♂
21 03	☽ ☍ ♇
27 Wednesday	
01 06	☽ ☍ ♃
02 19	☽ ⊻ ☿
09 42	☽ ☍ ♄
11 47	☽ ∠ ♀
13 17	☽ ± ♂
14 09	♂ ∠ ♇
14 28	☽ ⊥ ☿
14 46	☽ # ♇
16 57	☽ ⚼ ♆
19 02	☽ ✶ ☉
20 39	☽ ∥ ☉
21 34	☽ □ ♅
28 Thursday	
00 36	☽ ⊼ ♂
01 13	☽ # ♃
08 25	☽ ∠ ☿
09 00	☽ ± ♆
09 30	☽ # ♄
13 30	☽ ✶ ♀
17 58	☽ Q ☉
19 36	☽ ⊼ ♆
29 Friday	
02 31	☽ ⊼ ♇
06 17	☽ ⊼ ♃
08 13	☉ ⊻ ♅
09 40	☽ Q ♀
12 56	☽ ± ♇
13 32	☽ ✶ ☿
14 35	☽ ⊼ ♄
16 39	☽ ± ♃
16 43	♀ ⚼ ♄
21 30	☽ ∥ ♅
30 Saturday	
00 25	☿ ⊼ ♄
00 53	☽ ± ♄
02 15	☽ △ ♅
03 30	☽ □ ☉
04 24	☽ ⚼ ♇
07 34	☽ ☍ ♂
08 02	☽ ⚼ ♃
12 12	☽ Q ☿
15 18	☽ □ ♀
16 12	☽ ⚼ ♄
16 29	☉ ⚼ ♇
16 44	♂ ∠ ♃
19 28	☽ # ♂
23 14	☽ ☍ ♆
31 Sunday	
03 47	☽ ⚼ ♅
05 46	☽ △ ♇
09 17	☽ △ ♃
15 30	☽ # ♆
17 20	☽ △ ♄
18 53	☽ ± ♅
21 16	☽ □ ☿

JUNE 2020

LONGITUDES

Date	Sidereal time h m s	Sun ☉ ° ' "	Moon ☽ ° ' "	Moon ☽ 24.00 ° ' "	Mercury ☿ ° '	Venus ♀ ° '	Mars ♂ ° '	Jupiter ♃ ° '	Saturn ♄ ° '	Uranus ♅ ° '	Neptune ♆ ° '	Pluto ♇ ° '
01	04 41 44	11 ♊ 27 03	12 ♎ 53 13	20 ♎ 10 46	04 ♋ 42	15 ♊ 00	13 ♓ 02	26 ♑ 44	01 ♒ 36	08 ♉ 35	20 ♓ 50	24 ♑ 41
02	04 45 40	12 24 31	27 ♎ 29 58	04 ♏ 50 05	05 49	14 R 22	13 42	26 R 41	01 R 34	08 38	20 51	24 R 40
03	04 49 37	13 21 58	12 ♏ 10 20	19 ♏ 29 53	06 53	13 45	14 21	26 37	01 31	08 41	20 51	24 39
04	04 53 34	14 19 24	26 ♏ 47 55	04 ♐ 03 35	07 53	13 07	15 01	26 34	01 29	08 44	20 52	24 38
05	04 57 30	15 16 49	11 ♐ 16 07	18 ♐ 24 49	08 49	12 29	15 40	26 30	01 27	08 47	20 53	24 37
06	05 01 27	16 14 13	25 ♐ 29 04	02 ♑ 28 22	09 41	11 52	16 20	26 25	01 25	08 50	20 53	24 36
07	05 05 23	17 11 36	09 ♑ 22 21	16 ♑ 10 46	10 30	11 16	16 59	26 21	01 22	08 53	20 54	24 35
08	05 09 20	18 08 58	22 ♑ 53 28	29 ♑ 30 26	11 15	10 40	17 38	26 17	01 20	08 56	20 54	24 34
09	05 13 16	19 06 20	06 ♒ 01 48	12 ♒ 27 44	11 55	10 06	18 17	26 12	01 17	08 59	20 55	24 33
10	05 17 13	20 03 41	18 ♒ 48 33	25 ♒ 04 37	12 32	09 33	18 56	26 08	01 14	09 01	20 55	24 32
11	05 21 09	21 01 01	01 ♓ 16 20	07 ♓ 24 12	13 04	09 01	19 35	26 03	01 11	09 04	20 55	24 31
12	05 25 06	21 58 21	13 ♓ 28 44	19 ♓ 30 29	13 32	08 31	20 14	25 58	01 09	09 07	20 56	24 29
13	05 29 03	22 55 41	25 ♓ 30 00	01 ♈ 27 52	13 55	08 03	20 53	25 52	01 06	09 10	20 56	24 28
14	05 32 59	23 53 00	07 ♈ 24 38	13 ♈ 20 54	14 15	07 37	21 31	25 47	01 03	09 12	20 56	24 27
15	05 36 56	24 50 19	19 ♈ 17 11	25 ♈ 14 02	14 29	07 13	22 09	25 42	01 00	09 15	20 57	24 26
16	05 40 52	25 47 37	01 ♉ 11 57	07 ♉ 11 24	14 39	06 51	22 48	25 36	00 56	09 18	20 57	24 25
17	05 44 49	26 44 55	13 ♉ 12 50	19 ♉ 16 38	14 45	06 32	23 26	25 30	00 53	09 20	20 57	24 23
18	05 48 45	27 42 13	25 ♉ 23 10	01 ♊ 32 44	14 R 46	06 14	24 04	25 24	00 50	09 23	20 57	24 22
19	05 52 42	28 39 30	07 ♊ 45 36	14 ♊ 01 58	14 42	06 00	24 42	25 18	00 47	09 25	20 57	24 21
20	05 56 38	29 ♊ 36 47	20 ♊ 21 59	26 ♊ 45 46	14 34	05 47	25 20	25 12	00 43	09 28	20 57	24 20
21	06 00 35	00 ♋ 34 04	03 ♋ 13 22	09 ♋ 44 46	14 22	05 37	25 57	25 06	00 40	09 30	20 58	24 18
22	06 04 32	01 31 20	16 ♋ 19 58	22 ♋ 58 51	14 06	05 29	26 35	24 59	00 36	09 33	20 58	24 17
23	06 08 28	02 28 36	29 ♋ 41 19	06 ♌ 27 12	13 45	05 24	27 12	24 53	00 33	09 35	20 R 58	24 16
24	06 12 25	03 25 51	13 ♌ 16 20	20 ♌ 08 30	13 22	05 21	27 49	24 46	00 29	09 37	20 58	24 14
25	06 16 21	04 23 05	27 ♌ 03 31	04 ♍ 01 06	12 54	05 D 20	28 26	24 40	00 25	09 40	20 58	24 13
26	06 20 18	05 20 19	11 ♍ 01 02	18 ♍ 03 03	12 25	05 22	29 03	24 33	00 21	09 42	20 57	24 12
27	06 24 14	06 17 33	25 ♍ 06 54	02 ♎ 12 16	11 52	05 26	29 ♓ 39	24 26	00 18	09 44	20 57	24 10
28	06 28 11	07 14 46	09 ♎ 18 54	16 ♎ 26 27	11 18	05 32	00 ♈ 16	24 19	00 14	09 46	20 57	24 09
29	06 32 07	08 11 58	23 ♎ 34 38	00 ♏ 43 06	10 43	05 40	00 52	24 12	00 10	09 49	20 57	24 08
30	06 36 04	09 ♋ 09 10	07 ♏ 51 28	14 ♏ 59 23	10 ♋ 07	05 ♊ 51	01 ♈ 28	24 ♑ 05	00 ♒ 06	09 ♉ 51	20 ♓ 57	24 ♑ 06

Date	Moon True ☊ ° '	Moon Mean ☊ ° '	Moon ☽ Latitude ° '
01	29 ♊ 13	00 ♋ 10	05 N 06
02	29 R 11	00 07	04 38
03	29 09	00 04	03 51
04	29 07	00 ♋ 01	02 49
05	29 06	29 ♊ 57	01 37
06	29 05	29 54	00 N 20
07	29 D 05	29 51	00 S 57
08	29 06	29 48	02 08
09	29 07	29 45	03 10
10	29 08	29 42	04 02
11	29 08	29 38	04 40
12	29 09	29 35	05 05
13	29 R 09	29 32	05 16
14	29 09	29 29	05 13
15	29 08	29 26	04 57
16	29 08	29 22	04 29
17	29 08	29 19	03 48
18	29 07	29 16	02 56
19	29 07	29 13	01 56
20	29 D 07	29 10	00 S 48
21	29 R 07	29 07	00 N 23
22	29 07	29 03	01 34
23	29 07	29 00	02 41
24	29 07	28 57	03 41
25	29 06	28 54	04 29
26	29 05	28 51	05 01
27	29 05	28 47	05 16
28	29 D 05	28 44	05 12
29	29 05	28 41	04 48
30	29 ♊ 06	28 ♊ 38	04 N 07

DECLINATIONS

Date	Sun ☉ ° '	Moon ☽ ° '	Mercury ☿ ° '	Venus ♀ ° '	Mars ♂ ° '	Jupiter ♃ ° '	Saturn ♄ ° '	Uranus ♅ ° '	Neptune ♆ ° '	Pluto ♇ ° '
01	22 N 09	00 S 23	25 N 16	23 N 36	08 S 52	21 S 01	19 S 57	13 N 56	04 S 36	22 S 06
02	22 17	06 16	25 07	23 18	08 38	21 02	19 58	13 57	04 36	22 06
03	22 24	11 49	24 57	22 59	08 24	21 03	19 58	13 58	04 36	22 06
04	22 31	16 42	24 46	22 40	08 11	21 03	19 59	13 59	04 36	22 07
05	22 37	20 32	24 34	22 22	07 57	21 04	19 59	14 00	04 35	22 07
06	22 44	23 02	24 21	22 03	07 43	21 05	20 00	14 01	04 35	22 07
07	22 49	24 03	24 07	21 44	07 29	21 06	20 01	14 02	04 35	22 07
08	22 54	23 36	23 52	21 25	07 15	21 07	20 01	14 03	04 35	22 08
09	22 59	21 50	23 37	21 07	07 01	21 08	20 02	14 04	04 35	22 08
10	23 04	19 01	23 22	20 48	06 47	21 09	20 03	14 05	04 35	22 08
11	23 08	15 23	23 05	20 31	06 33	21 10	20 03	14 06	04 35	22 09
12	23 12	11 11	22 49	20 13	06 19	21 12	20 04	14 06	04 34	22 09
13	23 15	06 37	22 32	19 57	06 05	21 13	20 05	14 07	04 34	22 09
14	23 18	01 S 52	22 16	19 41	05 52	21 14	20 06	14 08	04 34	22 10
15	23 20	02 N 57	21 59	19 25	05 38	21 15	20 06	14 09	04 34	22 10
16	23 22	07 41	21 42	19 11	05 24	21 16	20 07	14 10	04 34	22 10
17	23 24	12 11	21 26	18 57	05 10	21 17	20 08	14 11	04 34	22 11
18	23 25	16 15	21 09	18 44	04 57	21 19	20 09	14 11	04 34	22 11
19	23 26	19 42	20 53	18 32	04 43	21 20	20 10	14 12	04 34	22 11
20	23 26	22 17	20 38	18 21	04 29	21 21	20 10	14 13	04 34	22 12
21	23 26	23 47	20 22	18 10	04 16	21 23	20 11	14 14	04 34	22 12
22	23 26	24 00	20 08	18 01	04 02	21 24	20 12	14 15	04 34	22 12
23	23 25	22 51	19 54	17 53	03 49	21 25	20 13	14 15	04 34	22 13
24	23 24	20 22	19 41	17 45	03 36	21 27	20 14	14 16	04 34	22 13
25	23 22	16 42	19 29	17 38	03 22	21 28	20 15	14 17	04 34	22 13
26	23 20	12 04	19 17	17 32	03 09	21 29	20 16	14 17	04 34	22 14
27	23 17	06 46	19 07	17 27	02 56	21 31	20 17	14 18	04 34	22 14
28	23 14	01 N 05	18 57	17 23	02 43	21 32	20 18	14 19	04 35	22 14
29	23 11	04 S 41	18 49	17 19	02 30	21 33	20 18	14 19	04 35	22 15
30	23 N 07	10 S 14	18 N 42	17 N 17	02 S 17	21 S 35	20 S 19	14 N 20	04 S 35	22 S 15

ZODIAC SIGN ENTRIES

Date	h m	Planets
02	16 06	☽ ♏
04	17 17	☽ ♐
06	19 44	☽ ♑
09	00 54	☽ ♒
11	09 32	☽ ♓
13	21 03	☽ ♈
16	09 35	☽ ♉
18	21 00	☽ ♊
20	21 44	☉ ♋
21	06 02	☽ ♋
23	12 33	☽ ♌
25	17 05	☽ ♍
27	20 16	☽ ♎
28	01 45	♂ ♈
29	22 48	☽ ♏

LATITUDES

Date	Mercury ☿ ° '	Venus ♀ ° '	Mars ♂ ° '	Jupiter ♃ ° '	Saturn ♄ ° '	Uranus ♅ ° '	Neptune ♆ ° '	Pluto ♇ ° '
01	01 N 55	01 N 01	02 S 23	00 S 13	00 S 09	00 S 27	01 S 03	00 S 55
04	01 34	00 N 18	02 28	00 13	00 10	00 27	01 04	00 56
07	01 06	00 S 24	02 32	00 14	00 10	00 27	01 04	00 56
10	00 N 31	01 05	02 37	00 14	00 10	00 27	01 04	00 56
13	00 S 10	01 43	02 42	00 15	00 10	00 27	01 04	00 57
16	00 56	02 18	02 46	00 15	00 11	00 27	01 04	00 57
19	01 45	02 49	02 51	00 16	00 11	00 27	01 04	00 57
22	02 34	03 15	02 55	00 16	00 11	00 27	01 04	00 58
25	03 21	03 37	03 00	00 17	00 12	00 27	01 04	00 58
28	04 01	03 54	03 04	00 17	00 12	00 27	01 05	00 58
31	04 S 30	04 S 08	03 S 09	00 S 18	00 S 12	00 S 27	01 S 05	00 S 59

DATA

Julian Date	2459002
Delta T	+71 seconds
Ayanamsa	24° 08' 14"
Synetic vernal point	04° ♓ 58' 45"
True obliquity of ecliptic	23° 26' 12"

MOON'S PHASES, APSIDES AND POSITIONS ☽

Date	h m	Phase	Longitude ° '	Eclipse Indicator
05	19 12	○	15 ♐ 34	
13	06 24	☾	22 ♓ 42	
21	06 41	●	00 ♋ 21	Annular
28	08 16	☽	07 ♎ 06	

Day	h m		
03	03 45	Perigee	
15	01 01	Apogee	
30	02 17	Perigee	
01	10 26	0S	
07	16 24	Max dec	24° S 04'
14	21 16	0N	
22	03 54	Max dec	24° N 04'
28	16 30	0S	

LONGITUDES

Date	Chiron ⚷ ° '	Ceres ⚳ ° '	Pallas ⚴ ° '	Juno ⚵ ° '	Vesta ⚶ ° '	Black Moon Lilith ⚸ ° '
01	08 ♈ 45	09 ♓ 26	00 ♒ 10	05 ♎ 53	29 ♊ 01	14 ♈ 08
11	09 ♈ 03	11 ♓ 01	28 ♑ 48	06 ♎ 26	03 ♋ 20	15 ♈ 15
21	09 ♈ 16	12 ♓ 08	26 ♑ 48	07 ♎ 27	07 ♋ 40	16 ♈ 23
31	09 ♈ 23	12 ♓ 44	24 ♑ 19	08 ♎ 54	12 ♋ 02	17 ♈ 30

All ephemeris data is given at 12.00 UT and the Moon's longitude is additionally given for 24.00 UT

ASPECTARIAN

h m Aspects

01 Monday
01 15 ☉ # ♇
04 52 ☽ ⊼ ♅
09 28 ☽ △ ☉
12 15 ☽ ⊼ ♂
15 20 ☽ △ ♀
18 47 ☉ ⚼ ♃
22 36 ☽ ± ♂

02 Tuesday
01 05 ☽ ⊼ ♆
02 41 ♀ ⊥ ♅
05 10 ☽ ∥ ♆
07 22 ☽ □ ♇
10 40 ☽ □ ♃
10 56 ☽ ± ♆
11 50 ☽ ⚼ ☉
14 03 ☽ ⚼ ♂
14 56 ☽ ⚼ ♀
18 38 ☽ □ ♄
21 37 ☽ ∥ ♂

03 Wednesday
00 41 ♀ □ ♂
01 40 ☽ ⚼ ♆
02 41 ☽ △ ☿
03 35 ☽ ± ☉
05 03 ☽ ± ♀
06 17 ☽ ☍ ♅
12 47 ☽ Q ♇
14 05 ☽ ⊼ ☉
14 28 ☽ ⊼ ♀
15 44 ☽ △ ♂
15 59 ☽ Q ♃
17 44 ☉ ☌ ♀
22 06 ☽ # ♅

04 Thursday
00 01 ☽ Q ♄
02 14 ☽ △ ♆
05 05 ☽ ⚼ ☿
08 27 ☽ ⚹ ♇
11 36 ☽ ⚹ ♃
19 43 ☽ ⚹ ♄
20 56 ☉ ∥ ♀
20 59 ☽ ± ☿
22 52 ☉ ⊥ ♅

05 Friday
07 38 ☽ ⊼ ☿
07 50 ☽ ⊼ ♅
08 06 ☽ ∥ ♄
09 15 ☽ ∠ ♇
11 05 ☿ ⚹ ♅
12 22 ☽ ∠ ♃
13 57 ☽ ☍ ♀
16 16 ☽ ∥ ♃
17 55 ☽ ± ♅
19 12 ☽ ☍ ☉
19 44 ☽ □ ♂
20 40 ☽ ∠ ♄

06 Saturday
00 20 ☽ ⊥ ♇
01 34 ☽ ∥ ♇
02 07 ☽ # ♀
03 27 ☽ ⊥ ♃
04 10 ☽ □ ♆
06 19 ♀ # ♇
07 58 ☽ # ☉
09 10 ☽ ⚼ ♅
10 30 ☽ ⊻ ♇
11 52 ☽ ⊥ ♄
13 36 ☽ ⊻ ♃
14 50 ♂ ∠ ♄
16 09 ☉ ⚼ ♄
19 11 ☉ □ ♂
22 08 ☽ ⊻ ♄

07 Sunday
03 58 ☽ Q ♂
07 50 ♀ ⚼ ♃
11 08 ☽ △ ♅
11 10 ☽ Q ♆
14 06 ☽ ☍ ☿
15 11 ☽ ⊼ ♀
16 24 ☽ # ☿

08 Monday
00 22 ☽ # ☿
01 20 ☽ ± ♀
01 29 ☿ ⊻ ♀
02 07 ☽ ⚹ ♂
02 51 ☽ ⊼ ☉
08 25 ☽ ⚹ ♆
14 27 ☽ ± ☉
15 01 ☽ ☌ ♇
16 48 ☽ ⚼ ♀
18 05 ☽ ☌ ♃
22 14 ☉ ± ♇
22 54 ☽ # ☉

09 Tuesday
03 17 ☽ ☌ ♄
06 40 ☽ ∠ ♂
08 10 ☽ ⚼ ☉
08 51 ☽ ∥ ♇
09 44 ♀ # ♃
11 47 ☽ ∠ ♆
17 30 ☽ □ ♅
18 42 ☽ ∥ ♃
19 15 ☽ △ ♀
19 56 ☽ # ♀
23 33 ☽ ⊼ ☿

10 Wednesday
00 18 ☽ ⊥ ♂
04 06 ☽ ∥ ♄
04 37 ☽ ⊥ ♆
11 27 ☽ ± ☿
12 15 ☽ ⊻ ♂
12 44 ♀ ⚼ ♇
13 29 ☉ ± ♃
14 35 ☽ △ ☉
16 01 ☽ ⊻ ♆
22 56 ☽ ⊻ ♇

11 Thursday
01 56 ☽ ⊻ ♃
03 49 ☽ Q ♅
05 30 ☽ ⚼ ☿
09 04 ☉ ∥ ☿
09 37 ☉ □ ♆
09 50 ♀ ⊻ ♅
10 31 ☽ ⊥ ♇
11 51 ☽ ⊻ ♄
13 30 ☽ ⊥ ♃
19 38 ☽ # ♅
23 32 ☽ ⊥ ♄

12 Friday
02 34 ☽ □ ♀
03 20 ☽ ⚹ ♅
04 07 ☽ ∠ ♇
07 03 ☽ ∠ ♃
12 07 ☽ △ ☿
17 16 ☽ ∠ ♄

13 Saturday
00 34 ♀ # ♄
02 12 ☽ ☌ ♂
02 51 ☽ ☌ ♆
06 24 ☽ □ ☉
09 18 ☽ ∠ ♅
09 56 ☽ ⚹ ♇
12 45 ☽ ⚹ ♃
13 04 ☽ Q ♀
14 13 ♂ ☌ ♆
14 51 ☽ ∥ ♂
15 49 ☿ ⊥ ♀
22 24 ☽ ∥ ♆
23 12 ☽ ⚹ ♄

14 Sunday
03 29 ☽ ⊥ ♅
10 04 ☽ Q ♇
12 24 ☽ ⚹ ♀
12 45 ☽ Q ♃
15 38 ☽ ⊻ ♅
20 27 ☉ ∠ ♅
20 35 ☿ # ♇
21 50 ☽ Q ☉
23 20 ☽ Q ♄

15 Monday
01 58 ☉ ⊼ ♇
02 07 ☽ □ ☿
15 21 ☽ ⊻ ♆
15 41 ☉ ± ♄
17 44 ☽ ∠ ♀
18 08 ☽ ⊻ ♂
20 06 ☽ # ♆
22 22 ☽ □ ♇

16 Tuesday
00 11 ☽ ⚹ ☉
00 49 ☽ □ ♃
00 50 ☽ # ♂
03 27 ☽ ⊥ ♆
06 54 ☽ ⊥ ♂
07 33 ☉ ⊼ ♃
11 20 ☽ ⊥ ♀
11 29 ☽ □ ♄
14 57 ☽ Q ☿
21 31 ☽ ∠ ♆
23 01 ☽ ⊻ ♀

17 Wednesday
01 57 ☽ ∠ ♂
04 15 ☽ ☌ ♅
08 50 ☽ ∠ ☉
15 03 ☽ ⚹ ☿
23 07 ☿ # ♃
23 26 ☽ ∥ ♅

18 Thursday
03 18 ☽ ⚹ ♆
04 10 ☽ ⊥ ☉
04 59 ☿ St R
09 16 ☽ ⚹ ♂
10 01 ☽ △ ♇
12 02 ☽ △ ♃
16 41 ♀ Q ♂
16 54 ☽ ⊻ ☉
20 31 ☽ ∠ ☿
22 34 ☽ △ ♄
23 08 ♂ ⚹ ♇

19 Friday
00 43 ♂ ∠ ♅
02 44 ☽ Q ♆
03 46 ☽ ∥ ♀
08 40 ☽ ☌ ♀
09 51 ☽ Q ♂
13 48 ☽ ⊥ ☿
15 03 ☽ ⚼ ♇
15 12 ☽ ⊻ ♅
15 44 ☽ # ♄
16 51 ☽ ⚼ ♃
21 04 ☽ ∥ ☿

20 Saturday
01 09 ☽ ⊻ ☿
02 11 ☽ # ♃
02 41 ☽ ⊥ ♅
03 15 ☽ ⚼ ♄
03 49 ♂ ∥ ♆
07 56 ♂ ⚹ ♃
08 09 ☽ ± ♇
09 49 ☽ ± ♃
10 58 ☽ # ♇
13 07 ☽ □ ♆
19 26 ☽ ⊼ ♇
19 43 ☽ ∠ ♅
20 09 ☽ ± ♄
21 01 ☽ ⊼ ♃
21 48 ☽ □ ♂

21 Sunday
04 27 ☽ ∥ ☉
06 41 ☽ ☌ ☉
07 17 ☽ ⊼ ♄
14 14 ☉ ⊼ ♄
16 22 ☽ ⊻ ♀
23 35 ☽ ⚹ ♅

22 Monday
03 16 ☽ ⊥ ♀
05 17 ☿ # ♄
08 01 ☽ ☌ ☿
19 27 ☽ ∠ ♀
20 22 ☽ △ ♆
21 27 ☽ Q ♅

23 Tuesday
02 19 ☽ ☍ ♇
03 17 ☽ ∥ ☉
03 29 ☽ ☍ ♃
04 31 ♆ St R
07 20 ☽ △ ♂
13 31 ☽ ☍ ♄
17 20 ☽ ⊻ ☉
19 26 ☽ # ♇
22 05 ☽ ⚹ ♀
23 08 ☽ ⚼ ♆

24 Wednesday
03 03 ☽ # ♃
04 45 ☽ ⊥ ☉
05 34 ☽ □ ♅
11 10 ☽ ⚼ ♂
12 09 ☽ ⊻ ☿
12 58 ☽ # ♄
14 57 ☽ ± ♆
17 23 ☽ ∥ ☿
19 07 ☽ Q ♀
21 41 ☽ ∠ ☉
22 19 ☽ ⊥ ☿

25 Thursday
01 25 ☽ ⊼ ♆
03 36 ☽ ± ♂
06 18 ☽ ∥ ♀
06 48 ♀ St D
07 05 ☽ ⊼ ♇
07 53 ☽ ⊼ ♃
13 25 ☽ ∠ ☿
14 29 ☽ ⊼ ♂
17 27 ☽ ± ♇
17 47 ☽ ⊼ ♄
18 10 ☽ ± ♃

26 Friday
01 02 ☽ ∥ ♅
01 33 ☽ ⚹ ☉
02 17 ☽ □ ♀
04 03 ☽ ± ♄
08 53 ☽ ⚼ ♇
09 30 ☽ ⚼ ♃
09 44 ☽ △ ♅
12 41 ☉ ⊻ ♀
14 18 ☽ ⚹ ☿
19 23 ☽ ⚼ ♄
23 34 ☽ Q ☉

27 Saturday
04 56 ☽ ☍ ♆
09 58 ☽ Q ☿
10 24 ☽ △ ♇
10 51 ☽ △ ♃
11 21 ☽ ⚼ ♅
20 02 ☽ ☍ ♂
20 44 ☽ △ ♄
21 23 ☽ # ♆

28 Sunday
02 38 ☽ ± ♅
04 11 ☿ ⊥ ♀
04 57 ☽ # ♂
05 34 ☽ △ ♀
08 16 ☽ □ ☉
10 59 ♂ ⚹ ♄
12 47 ☽ ⊼ ♅
15 13 ☽ □ ☿

29 Monday
03 11 ☽ ∥ ♂
07 04 ☽ ⚼ ♀
07 35 ☽ ⊼ ♆
11 33 ☽ ∥ ♆
12 55 ☽ □ ♇
13 02 ☽ □ ♃
17 40 ☽ ± ♆
22 21 ☽ ± ♀
23 01 ☽ □ ♄

30 Tuesday
00 47 ☽ ⊼ ♂
05 46 ♃ ☌ ♇
08 34 ☽ ⊼ ♀
08 47 ☽ ⚼ ♆
11 18 ☽ ± ♂
14 20 ☽ △ ☉
15 21 ☽ ☍ ♅
15 39 ☽ △ ☿
19 02 ☽ Q ♃
19 08 ☽ Q ♇
22 12 ☿ ⚹ ♅

JULY 2020

LONGITUDES

Date	Sidereal time h m s	Sun ☉ ° ' "	Moon ☽ ° ' "	Moon ☽ 24.00 ° ' "	Mercury ☿ ° '	Venus ♀ ° '	Mars ♂ ° '	Jupiter ♃ ° '	Saturn ♄ ° '	Uranus ♅ ° '	Neptune ♆ ° '	Pluto ♇ ° '
01	06 40 01	10 ♋ 06 22	22 ♏ 06 27	29 ♏ 12 14	09 ♋ 31	06 ♊ 03	02 ♈ 04	23 ♑ 57	00 ♒ 02	09 ♉ 53	20 ♓ 56	24 ♑ 05
02	06 43 57	11 03 33	06 ♐ 16 20	13 ♐ 18 19	08 R 55	06 17	02 39	23 R 50	29 ♑ 58	09 55	20 R 56	24 R 03
03	06 47 54	12 00 44	20 ♐ 17 47	27 ♐ 14 18	08 21	06 34	03 15	23 43	29 R 54	09 57	20 56	24 02
04	06 51 50	12 57 55	04 ♑ 07 31	10 ♑ 57 06	07 48	06 52	03 50	23 35	29 50	09 59	20 56	24 01
05	06 55 47	13 55 06	17 ♑ 42 44	24 ♑ 24 12	07 18	07 12	04 25	23 28	29 46	10 01	20 55	23 59
06	06 59 43	14 52 17	01 ♒ 01 19	07 ♒ 33 59	06 50	07 34	05 00	23 20	29 41	10 03	20 55	23 58
07	07 03 40	15 49 28	14 ♒ 02 10	20 ♒ 25 55	06 26	07 58	05 34	23 13	29 37	10 05	20 54	23 56
08	07 07 36	16 46 39	26 ♒ 45 21	03 ♓ 00 40	06 06	08 23	06 09	23 05	29 33	10 06	20 54	23 55
09	07 11 33	17 43 50	09 ♓ 12 07	15 ♓ 20 02	05 50	08 50	06 43	22 58	29 29	10 08	20 53	23 53
10	07 15 30	18 41 02	21 ♓ 24 49	27 ♓ 26 52	05 38	09 18	07 17	22 50	29 24	10 10	20 53	23 52
11	07 19 26	19 38 14	03 ♈ 26 42	09 ♈ 24 50	05 31	09 48	07 51	22 42	29 20	10 12	20 52	23 51
12	07 23 23	20 35 27	15 ♈ 21 48	21 ♈ 18 11	05 D 30	10 19	08 24	22 34	29 16	10 13	20 52	23 49
13	07 27 19	21 32 40	27 ♈ 14 34	03 ♉ 11 32	05 33	10 52	08 57	22 27	29 11	10 15	20 51	23 48
14	07 31 16	22 29 54	09 ♉ 09 42	15 ♉ 09 39	05 42	11 26	09 30	22 19	29 07	10 16	20 50	23 46
15	07 35 12	23 27 08	21 ♉ 11 56	27 ♉ 17 06	05 56	12 01	10 03	22 11	29 03	10 18	20 50	23 45
16	07 39 09	24 24 23	03 ♊ 25 40	09 ♊ 38 04	06 15	12 38	10 35	22 04	28 58	10 19	20 49	23 43
17	07 43 05	25 21 39	15 ♊ 54 45	22 ♊ 16 00	06 41	13 15	11 07	21 56	28 54	10 21	20 48	23 42
18	07 47 02	26 18 55	28 ♊ 42 08	05 ♋ 13 17	07 11	13 54	11 39	21 48	28 49	10 22	20 47	23 40
19	07 50 59	27 16 12	11 ♋ 49 33	18 ♋ 30 54	07 47	14 34	12 10	21 40	28 45	10 24	20 47	23 39
20	07 54 55	28 13 29	25 ♋ 17 14	02 ♌ 08 17	08 29	15 15	12 41	21 33	28 40	10 25	20 46	23 38
21	07 58 52	29 ♋ 10 47	09 ♌ 03 45	16 ♌ 03 12	09 16	15 57	13 12	21 25	28 36	10 26	20 45	23 36
22	08 02 48	00 ♌ 08 05	23 ♌ 06 08	00 ♍ 11 57	10 08	16 40	13 42	21 18	28 32	10 27	20 44	23 35
23	08 06 45	01 05 24	07 ♍ 20 05	14 ♍ 29 53	11 06	17 24	14 12	21 10	28 27	10 28	20 43	23 33
24	08 10 41	02 02 43	21 ♍ 40 42	28 ♍ 51 57	12 09	18 08	14 42	21 03	28 23	10 30	20 42	23 32
25	08 14 38	03 00 02	06 ♎ 03 03	13 ♎ 13 30	13 17	18 54	15 11	20 55	28 18	10 31	20 41	23 30
26	08 18 34	03 57 22	20 ♎ 22 51	27 ♎ 30 43	14 30	19 40	15 40	20 48	28 14	10 32	20 40	23 29
27	08 22 31	04 54 42	04 ♏ 36 48	11 ♏ 40 50	15 47	20 28	16 09	20 40	28 09	10 33	20 39	23 27
28	08 26 28	05 52 03	18 ♏ 42 39	25 ♏ 42 07	17 10	21 16	16 37	20 33	28 05	10 33	20 38	23 26
29	08 30 24	06 49 24	02 ♐ 39 07	09 ♐ 33 34	18 37	22 04	17 05	20 26	28 01	10 34	20 37	23 25
30	08 34 21	07 46 45	16 ♐ 25 25	23 ♐ 14 36	20 09	22 54	17 33	20 19	27 56	10 35	20 36	23 23
31	08 38 17	08 ♌ 44 08	00 ♑ 01 02	06 ♑ 44 40	21 ♋ 45	23 ♊ 44	18 ♈ 00	20 ♑ 12	27 ♑ 52	10 ♉ 36	20 ♓ 35	23 ♑ 22

Date	Moon True ☊ ° '	Moon Mean ☊ ° '	Moon ☽ Latitude ° '
01	29 ♊ 07	28 ♊ 35	03 N 11
02	29 D 08	28 32	02 03
03	29 08	28 28	00 N 49
04	29 R 08	28 25	00 S 28
05	29 08	28 22	01 41
06	29 06	28 19	02 47
07	29 04	28 16	03 43
08	29 02	28 13	04 26
09	28 59	28 09	04 56
10	28 57	28 06	05 12
11	28 55	28 03	05 14
12	28 54	28 00	05 02
13	28 D 54	27 57	04 37
14	28 55	27 53	04 00
15	28 56	27 50	03 13
16	28 58	27 47	02 16
17	28 59	27 44	01 11
18	29 00	27 41	00 S 02
19	28 R 59	27 38	01 N 10
20	28 57	27 34	02 19
21	28 54	27 31	03 22
22	28 50	27 28	04 13
23	28 45	27 25	04 50
24	28 41	27 22	05 09
25	28 37	27 19	05 09
26	28 35	27 15	04 49
27	28 34	27 12	04 12
28	28 D 35	27 09	03 20
29	28 36	27 06	02 16
30	28 37	27 03	01 06
31	28 ♊ 38	26 ♊ 59	00 S 08

DECLINATIONS

Date	Sun ☉ ° '	Moon ☽ ° '	Mercury ☿ ° '	Venus ♀ ° '	Mars ♂ ° '	Jupiter ♃ ° '	Saturn ♄ ° '	Uranus ♅ ° '	Neptune ♆ ° '	Pluto ♇ ° '
01	23 N 03	15 S 13	18 N 36	17 N 15	02 S 04	21 S 36	20 S 20	14 N 21	04 S 35	22 S 15
02	22 59	19 20	18 32	17 13	01 51	21 38	20 21	14 21	04 35	22 16
03	22 54	22 16	18 29	17 13	01 38	21 39	20 22	14 22	04 35	22 16
04	22 48	23 50	18 27	17 13	01 26	21 41	20 23	14 23	04 35	22 16
05	22 43	23 56	18 26	17 13	01 13	21 42	20 24	14 23	04 36	22 17
06	22 36	22 39	18 27	17 14	01 01	21 43	20 25	14 24	04 36	22 17
07	22 30	20 10	18 29	17 16	00 49	21 45	20 26	14 24	04 36	22 17
08	22 23	16 46	18 33	17 18	00 36	21 46	20 27	14 25	04 36	22 18
09	22 16	12 41	18 37	17 21	00 24	21 48	20 28	14 26	04 37	22 18
10	22 08	08 11	18 43	17 24	00 S 12	21 49	20 29	14 26	04 37	22 18
11	22 00	03 S 26	18 50	17 28	00 00	21 51	20 30	14 27	04 37	22 19
12	21 52	01 N 24	18 57	17 32	00 N 12	21 52	20 31	14 27	04 37	22 19
13	21 43	06 11	19 06	17 36	00 23	21 53	20 32	14 28	04 38	22 19
14	21 34	10 45	19 15	17 40	00 35	21 55	20 33	14 28	04 38	22 20
15	21 24	14 57	19 25	17 45	00 46	21 56	20 34	14 29	04 38	22 20
16	21 14	18 37	19 36	17 50	00 58	21 58	20 35	14 29	04 38	22 21
17	21 04	21 31	19 47	17 55	01 09	21 59	20 36	14 29	04 39	22 21
18	20 53	23 24	19 58	18 01	01 20	22 00	20 37	14 30	04 39	22 21
19	20 42	24 04	20 09	18 06	01 31	22 02	20 38	14 30	04 40	22 22
20	20 31	23 21	20 21	18 12	01 42	22 03	20 39	14 31	04 40	22 22
21	20 19	21 14	20 32	18 18	01 52	22 04	20 40	14 31	04 40	22 22
22	20 07	17 48	20 42	18 23	02 03	22 06	20 41	14 31	04 41	22 23
23	19 55	13 18	20 52	18 29	02 13	22 07	20 42	14 32	04 41	22 23
24	19 42	08 02	21 01	18 35	02 24	22 08	20 43	14 32	04 41	22 23
25	19 29	02 N 19	21 10	18 41	02 34	22 09	20 44	14 32	04 42	22 24
26	19 16	03 S 30	21 17	18 47	02 44	22 11	20 45	14 33	04 42	22 24
27	19 02	09 06	21 22	18 52	02 53	22 12	20 46	14 33	04 43	22 24
28	18 48	14 11	21 27	18 58	03 03	22 13	20 47	14 33	04 43	22 25
29	18 34	18 27	21 29	19 04	03 12	22 14	20 48	14 33	04 44	22 25
30	18 19	21 39	21 30	19 09	03 22	22 16	20 49	14 34	04 44	22 25
31	18 N 04	23 S 34	21 N 28	19 N 14	03 N 31	22 S 17	20 S 50	14 N 34	04 S 44	22 S 26

ZODIAC SIGN ENTRIES

Date	h	m	Planets
01	23	37	♄ ♑
02	01	21	☽ ♐
04	04	48	☽ ♑
06	10	08	☽ ♒
08	18	12	☽ ♓
11	05	06	☽ ♈
13	17	34	☽ ♉
16	05	19	☽ ♊
18	14	24	☽ ♋
20	20	16	☽ ♌
22	08	37	☉ ♌
22	23	40	☽ ♍
25	01	54	☽ ♎
27	04	12	☽ ♏
29	07	25	☽ ♐
31	11	58	☽ ♑

LATITUDES

Date	Mercury ☿ ° '	Venus ♀ ° '	Mars ♂ ° '	Jupiter ♃ ° '	Saturn ♄ ° '	Uranus ♅ ° '	Neptune ♆ ° '	Pluto ♇ ° '
01	04 S 30	04 S 08	03 S 09	00 S 18	00 S 12	00 S 27	01 S 05	00 S 59
04	04 46	04 18	03 13	00 18	00 12	00 27	01 05	00 59
07	04 48	04 25	03 18	00 19	00 13	00 27	01 05	00 59
10	04 36	04 29	03 22	00 19	00 13	00 27	01 05	01 00
13	04 14	04 31	03 26	00 20	00 13	00 27	01 05	01 00
16	03 42	04 31	03 30	00 20	00 14	00 27	01 05	01 00
19	03 03	04 28	03 34	00 20	00 14	00 27	01 06	01 00
22	02 21	04 24	03 38	00 21	00 14	00 27	01 06	01 01
25	01 37	04 18	03 42	00 21	00 14	00 27	01 06	01 01
28	00 54	04 11	03 46	00 22	00 15	00 27	01 06	01 01
31	00 13	04 S 03	03 S 50	00 S 22	00 S 15	00 S 28	01 S 06	01 S 02

DATA

Julian Date	2459032
Delta T	+71 seconds
Ayanamsa	24° 08' 19"
Synetic vernal point	04° ♓ 58' 40"
True obliquity of ecliptic	23° 26' 12"

MOON'S PHASES, APSIDES AND POSITIONS ☽

Date	h	m	Phase	Longitude ° '	Eclipse Indicator
05	04	44	○	13 ♑ 38	
12	23	29	☾	21 ♈ 03	
20	17	33	●	28 ♋ 27	
27	12	33	☽	04 ♏ 56	

Day	h	m		
12	19	26	Apogee	
25	04	56	Perigee	
05	01	36	Max dec	24° S 04'
12	05	02	0N	
19	11	50	Max dec	24° N 04'
25	21	34	0S	

LONGITUDES

Date	Chiron ⚷ ° '	Ceres ⚳ ° '	Pallas ⚴ ° '	Juno ⚵ ° '	Vesta ⚶ ° '	Black Moon Lilith ⚸ ° '
01	09 ♈ 23	12 ♓ 44	24 ♑ 19	08 ♎ 54	12 ♋ 02	17 ♈ 30
11	09 ♈ 26	12 ♓ 46	21 ♑ 36	10 ♎ 42	16 ♋ 25	18 ♈ 37
21	09 ♈ 24	12 ♓ 13	18 ♑ 53	12 ♎ 48	20 ♋ 49	19 ♈ 44
31	09 ♈ 16	11 ♓ 05	16 ♑ 26	15 ♎ 10	25 ♋ 12	20 ♈ 51

All ephemeris data is given at 12.00 UT and the Moon's longitude is additionally given for 24.00 UT

ASPECTARIAN

h m Aspects

01 Wednesday
02 53 ☉ ☌ ☿
03 07 ☽ ⚼ ♂
05 09 ☽ Q ♄
06 07 ☉ ⚹ ♅
07 31 ☽ # ♅
10 02 ☽ △ ♆
15 06 ☽ ⚹ ♃
15 20 ☽ ⚹ ♇
15 54 ☽ ⚼ ☿
17 26 ☽ ⚼ ☉
23 00 ☽ # ♀

02 Thursday
01 20 ☽ ⚹ ♄
05 35 ☽ △ ♂
06 32 ☽ ± ☿
06 54 ☽ # ☿
09 47 ☽ ± ☉
12 02 ☽ ☍ ♀
16 20 ☽ ∠ ♃
16 20 ☽ ⚻ ☿
16 44 ☽ ∠ ♇
18 13 ☽ ⚻ ♅
19 17 ☽ ∥ ♄
20 45 ☽ ⚻ ☉

03 Friday
02 46 ☽ ∠ ♄
04 31 ☽ ± ♅
05 52 ☽ ∥ ♃
07 36 ☽ ⊥ ♃
08 07 ☽ ⊥ ♇
11 57 ☽ ∥ ♇
13 06 ☽ □ ♆
17 51 ☽ ⊻ ♃
18 11 ☽ ⊥ ♄
18 26 ☽ ⊻ ♇
18 55 ☽ # ☉
20 03 ☽ ⚼ ♅

04 Saturday
04 32 ☽ ⊻ ♄
08 18 ☉ ⊥ ♀
11 28 ☽ □ ♂
16 55 ☽ ⚻ ♀
18 13 ☽ ☍ ☿
18 29 ♂ ⊥ ♅
20 25 ☽ Q ♆
22 19 ☽ △ ♅

05 Sunday
03 46 ☽ ± ♀
04 44 ☽ ☍ ☉
14 45 ☿ ⊻ ♀
17 44 ☽ ⚹ ♆
20 16 ☽ ⚼ ♀
20 48 ☽ Q ♂
22 13 ☽ ☌ ♃
23 14 ☽ ☌ ♇

06 Monday
09 35 ☽ ☌ ♄
12 29 ☽ # ☉
16 15 ☽ ∥ ♇
19 37 ☽ ⚹ ♂
20 57 ☽ ∠ ♆
22 06 ☽ ∥ ♃
22 20 ☽ ⚻ ☿
23 40 ♂ Q ♃

07 Tuesday
00 22 ☽ △ ♀
04 37 ☽ □ ♅
09 06 ☽ ± ☿
09 51 ☽ ∥ ♄
13 37 ☽ ⊥ ♆
15 37 ☽ ⚻ ☉
23 14 ♀ ⚼ ♃

08 Wednesday
00 23 ☽ # ☿
00 51 ☽ ∠ ♂
00 53 ☽ ⊻ ♆
01 31 ☽ ⚼ ☿
02 38 ♂ Q ♇
03 49 ☽ ± ☉
05 05 ☽ ⊻ ♃
06 36 ☽ ⊻ ♇
08 33 ☽ # ♀
10 42 ☿ □ ♂
14 35 ☽ Q ♅
16 25 ☽ ⊥ ♃
17 19 ☽ ⊻ ♄
18 02 ☽ ⊥ ♇
18 48 ☽ ⊥ ♂
22 25 ☽ ⚼ ☉

09 Thursday
02 11 ☽ # ♅
04 36 ☉ # ♇
04 48 ☽ ⊥ ♄
05 35 ☽ △ ☿
06 56 ☽ ⊻ ♂
09 36 ☽ ∠ ♃
11 15 ☽ □ ♀
11 24 ☽ ∠ ♇
13 50 ☽ ⚹ ♅
15 03 ♀ ⚼ ♇
22 15 ☽ ∠ ♄

10 Friday
06 08 ☽ △ ☉
10 57 ☽ ☌ ♆
14 47 ☽ ⚹ ♃
16 52 ☽ ⚹ ♇
19 28 ☽ ∠ ♅

11 Saturday
00 13 ☽ Q ♀
03 49 ☽ ⚹ ♄
06 05 ☽ ∥ ♆
13 30 ☽ ⊥ ♅
14 30 ☽ Q ♃
16 09 ☽ □ ☿
16 48 ☽ Q ♇
21 16 ☽ ☌ ♂

12 Sunday
01 21 ☽ ⚹ ♀
01 36 ☽ ⊻ ♅
03 46 ☽ Q ♄
04 22 ☽ # ♂
05 44 ☽ ∥ ♂
07 12 ♀ ⊻ ♅
08 26 ☿ St D
11 01 ☉ # ♃
18 43 ☉ △ ♆
23 06 ☽ ⊻ ♆
23 29 ☽ □ ☉

13 Monday
02 25 ☽ □ ♃
04 07 ☽ # ♆
04 29 ☽ Q ☿
05 03 ☽ □ ♇
09 05 ☽ ∠ ♀
11 12 ☽ ⊥ ♆
15 54 ☽ □ ♄

14 Tuesday
04 08 ☽ ⊥ ♀
04 56 ☽ ⚹ ☿
05 20 ☽ ∠ ♆
06 12 ☉ Q ♅
07 58 ☉ ☍ ♃
12 43 ☽ ⊻ ♂
14 14 ☽ ☌ ♅
14 55 ☽ Q ☉
16 47 ☽ ⊻ ♀

15 Wednesday
01 17 ☽ ⊥ ♂
09 08 ☽ ∥ ♅
11 16 ☽ ⚹ ♆
11 27 ☽ ∠ ☿
13 56 ☽ △ ♃
16 50 ☽ ⚹ ☉
17 01 ☽ △ ♇
19 12 ☉ ☍ ♇
19 57 ☽ ∠ ♂
23 54 ♂ ⊻ ♅

16 Thursday
03 21 ☽ △ ♄
05 38 ☽ ⊥ ☿
06 21 ☽ ∥ ♀
10 49 ☽ Q ♆
17 40 ☽ ⊻ ☿
18 58 ☽ ⚼ ♃
19 51 ☽ ∥ ☿
22 13 ☽ ⚼ ♇

17 Friday
00 31 ☽ ∠ ☉
01 21 ☽ ⊻ ♅
02 26 ☽ ⚹ ♂
03 16 ♂ Q ♄
03 34 ☽ # ♄
06 40 ☽ ☌ ♀
08 00 ☽ ∥ ☉
08 11 ☽ ⚼ ♄
12 02 ☽ ± ♃
12 50 ☽ ⊥ ♅
15 23 ☽ ± ♇
16 54 ☽ # ♃
19 04 ☽ ⊥ ☉
20 59 ☽ # ♇
21 14 ☽ □ ♆
23 15 ☽ ⚻ ♃

18 Saturday
01 06 ☽ ± ♄
02 11 ☽ Q ♂
02 39 ☽ ⚻ ♇
05 48 ☽ ∠ ♅
07 13 ☽ ⊻ ☉
09 16 ♀ ⚼ ♄
12 13 ☽ ⚻ ♄

19 Sunday
04 19 ☽ ☌ ☿
09 24 ☽ ⚹ ♅
12 38 ☽ □ ♂
17 12 ☽ ⊻ ♀
20 11 ☉ # ♄

20 Monday
04 01 ☽ △ ♆
04 29 ☽ ⊥ ♀
05 27 ☽ ☍ ♃
06 55 ☽ Q ♅
09 04 ☽ ☍ ♇
17 33 ☽ ☌ ☉
17 55 ☽ ☍ ♄
20 37 ♀ ± ♃
21 10 ☽ ∠ ♀
22 28 ☉ ☍ ♄
22 52 ☉ ∥ ☿

21 Tuesday
01 03 ☽ # ♇
04 15 ☽ # ♃
06 17 ☽ ⚼ ♆
12 22 ☽ ⊻ ☿
14 22 ☽ □ ♅
16 34 ☽ # ♄
17 22 ☽ ∥ ☿
19 23 ☽ △ ♂
19 44 ☽ ∥ ☉
21 46 ☽ ± ♆
23 20 ☽ ⊥ ☿

22 Wednesday
00 27 ☽ ⚹ ♀
04 48 ♀ ⊥ ♅
07 59 ☽ ⚻ ♆
08 29 ☽ ∥ ♀
08 57 ☽ ⚻ ♃
09 03 ☿ # ♄
12 48 ☽ ⚻ ♇
15 41 ☽ ∠ ☿
19 02 ☽ ± ♃
20 25 ☿ ⚹ ♅
21 08 ☽ ⚻ ♄
21 50 ☽ ⚼ ♂
21 55 ☽ Q ♀
22 56 ☽ ± ♇

23 Thursday
00 45 ☽ ⊻ ☉
05 55 ☽ ∥ ♅
07 11 ☽ ± ♄
10 03 ☽ ⚼ ♃
11 34 ☽ ⊥ ☉
13 31 ☽ ± ♂
14 02 ☽ ⚼ ♇
16 59 ♀ ± ♇
17 16 ☽ △ ♅
18 47 ☽ ⚹ ☿
22 12 ☽ ⚼ ♄
23 55 ☽ ⚻ ♂

24 Friday
03 42 ☽ ∠ ☉
05 46 ☽ □ ♀
10 23 ☽ ☍ ♆
10 57 ☽ △ ♃
15 05 ☽ △ ♇
16 27 ☽ Q ☿
18 23 ☽ ⚼ ♅
23 08 ☽ △ ♄

25 Saturday
02 10 ☽ # ♆
06 32 ☽ ⚹ ☉
09 25 ☽ ± ♅
11 03 ☽ ∥ ♂
19 28 ☽ ⚻ ♅

26 Sunday
01 11 ☽ □ ☿
03 50 ☽ ☍ ♂
04 03 ☽ Q ☉
08 43 ☽ # ♂
10 45 ☽ △ ♀
12 29 ☽ ⚻ ♆
12 41 ☽ □ ♃
17 04 ☽ ∥ ♆
17 12 ☽ □ ♇
22 34 ☽ ± ♆

27 Monday
01 09 ☽ □ ♄
12 33 ☽ □ ☉
13 31 ☽ ⚼ ♀
13 46 ☽ ⚼ ♆
16 07 ♃ ⚹ ♆
17 36 ♀ ⚻ ♃
17 48 ♀ □ ♆
18 50 ☽ Q ♃
21 46 ☿ □ ♂
22 04 ☽ ☍ ♅
23 36 ☽ Q ♇
23 51 ☉ ∥ ♀

28 Tuesday
05 44 ☽ ± ♀
06 21 ☉ ⚼ ♆
07 32 ☽ Q ♄
08 18 ☽ ⚻ ♂
09 04 ☽ △ ☿
13 54 ☽ # ♅
15 08 ☽ ⚹ ♃
15 18 ☽ △ ♆
16 38 ☽ ⚻ ♀
18 56 ☽ ± ♂
20 05 ☽ ⚹ ♇

29 Wednesday
04 01 ☽ ⚹ ♄
10 24 ♀ ± ♄
10 59 ☽ ⚼ ♂
12 39 ☽ # ☉
13 53 ☽ ⚼ ☿
16 03 ☽ # ♀
16 47 ☽ ∠ ♃
19 47 ☽ △ ☉
21 59 ☽ ∠ ♇

30 Thursday
01 47 ☽ ⚻ ♅
04 42 ☽ ∥ ♄
05 56 ☽ ∠ ♄
07 31 ☽ ± ☿
08 20 ☽ ⊥ ♃
10 34 ☽ # ☿
12 17 ☽ ± ♅
13 41 ☽ ⊥ ♇
14 02 ☽ △ ♂
14 17 ☿ ☍ ♃
18 04 ☽ ∥ ♃
18 45 ☿ △ ♆
18 47 ☽ ⊻ ♃
19 20 ☽ □ ♆
19 25 ☽ ⚻ ☿
19 48 ☽ ∥ ♇
21 39 ☽ ⊥ ♄

31 Friday
00 02 ☽ ⚼ ☉
00 08 ☽ ☍ ♀
00 14 ☽ ⊻ ♇
01 51 ♀ ⚻ ♇
04 09 ☽ ⚼ ♅
08 12 ☽ ⊻ ♄
10 47 ☉ ∠ ♀
17 13 ☽ ± ☉

AUGUST 2020

LONGITUDES

Date	Sidereal time h m s	Sun ☉ ° ' "	Moon ☽ ° ' "	Moon ☽ 24.00 ° ' "	Mercury ☿ ° '	Venus ♀ ° '	Mars ♂ ° '	Jupiter ♃ ° '	Saturn ♄ ° '	Uranus ♅ ° '	Neptune ♆ ° '	Pluto ♇ ° '
01	08 42 14	09 ♌ 41 31	13 ♑ 25 25	20 ♑ 03 12	23 ♋ 25	24 ♊ 35	18 ♈ 26	20 ♑ 05	27 ♑ 48	10 ♉ 37	20 ♓ 34	23 ♑ 21
02	08 46 10	10 38 54	26 ♑ 37 55	03 ♒ 09 28	25 09	25 26	18 52	19 R 58	27 R 43	10 37	20 R 33	23 R 19
03	08 50 07	11 36 19	09 ♒ 37 48	16 ♒ 02 49	26 57	26 18	19 18	19 52	27 39	10 38	20 32	23 18
04	08 54 03	12 33 44	22 ♒ 24 31	28 ♒ 42 52	28 ♋ 47	27 11	19 43	19 45	27 35	10 38	20 30	23 16
05	08 58 00	13 31 10	04 ♓ 57 55	11 ♓ 09 45	00 ♌ 40	28 04	20 08	19 38	27 31	10 39	20 29	23 15
06	09 01 57	14 28 38	17 ♓ 18 30	23 ♓ 24 21	02 36	28 58	20 33	19 32	27 27	10 39	20 28	23 14
07	09 05 53	15 26 06	29 ♓ 27 33	05 ♈ 28 25	04 34	29 ♊ 52	20 57	19 26	27 23	10 40	20 27	23 12
08	09 09 50	16 23 36	11 ♈ 27 19	17 ♈ 24 39	06 34	00 ♋ 47	21 20	19 20	27 18	10 40	20 25	23 11
09	09 13 46	17 21 07	23 ♈ 20 54	29 ♈ 16 34	08 34	01 43	21 43	19 14	27 14	10 41	20 24	23 10
10	09 17 43	18 18 39	05 ♉ 12 13	11 ♉ 08 25	10 36	02 39	22 05	19 08	27 10	10 41	20 23	23 09
11	09 21 39	19 16 13	17 ♉ 05 47	23 ♉ 04 58	12 39	03 35	22 27	19 02	27 06	10 41	20 21	23 07
12	09 25 36	20 13 48	29 ♉ 06 36	05 ♊ 11 19	14 42	04 32	22 48	18 56	27 03	10 41	20 20	23 06
13	09 29 32	21 11 24	11 ♊ 19 45	17 ♊ 32 31	16 45	05 29	23 09	18 51	26 59	10 41	20 19	23 05
14	09 33 29	22 09 02	23 ♊ 50 09	00 ♋ 13 11	18 48	06 27	23 29	18 45	26 55	10 41	20 17	23 04
15	09 37 26	23 06 42	06 ♋ 42 01	13 ♋ 17 01	20 51	07 26	23 49	18 40	26 51	10 42	20 16	23 02
16	09 41 22	24 04 23	19 ♋ 58 22	26 ♋ 46 09	22 53	08 24	24 08	18 35	26 47	10 R 41	20 14	23 01
17	09 45 19	25 02 06	03 ♌ 40 19	10 ♌ 40 36	24 54	09 23	24 26	18 30	26 44	10 41	20 13	23 00
18	09 49 15	25 59 50	17 ♌ 46 36	24 ♌ 57 46	26 54	10 23	24 44	18 26	26 40	10 41	20 12	22 59
19	09 53 12	26 57 35	02 ♍ 13 20	09 ♍ 32 27	28 ♌ 53	11 23	25 01	18 21	26 37	10 41	20 10	22 58
20	09 57 08	27 55 21	16 ♍ 54 10	24 ♍ 17 27	00 ♍ 52	12 23	25 17	18 17	26 33	10 41	20 09	22 56
21	10 01 05	28 53 09	01 ♎ 41 16	09 ♎ 04 37	02 49	13 23	25 33	18 12	26 30	10 41	20 07	22 55
22	10 05 01	29 ♌ 50 58	16 ♎ 26 33	23 ♎ 46 14	04 44	14 24	25 48	18 08	26 26	10 40	20 06	22 54
23	10 08 58	00 ♍ 48 48	01 ♏ 02 59	08 ♏ 16 13	06 39	15 25	26 02	18 04	26 23	10 40	20 04	22 53
24	10 12 55	01 46 40	15 ♏ 25 33	22 ♏ 30 41	08 32	16 27	26 16	18 01	26 20	10 40	20 02	22 52
25	10 16 51	02 44 32	29 ♏ 31 31	06 ♐ 28 00	10 24	17 29	26 29	17 57	26 17	10 39	20 01	22 51
26	10 20 48	03 42 26	13 ♐ 20 12	20 ♐ 08 15	12 14	18 31	26 41	17 54	26 14	10 39	19 59	22 50
27	10 24 44	04 40 21	26 ♐ 52 19	03 ♑ 32 37	14 03	19 33	26 52	17 50	26 11	10 38	19 58	22 49
28	10 28 41	05 38 18	10 ♑ 09 21	16 ♑ 42 44	15 51	20 36	27 03	17 47	26 08	10 37	19 56	22 48
29	10 32 37	06 36 16	23 ♑ 12 57	29 ♑ 40 10	17 38	21 39	27 13	17 45	26 05	10 37	19 55	22 47
30	10 36 34	07 34 15	06 ♒ 04 32	12 ♒ 26 10	19 23	22 42	27 22	17 42	26 02	10 36	19 53	22 46
31	10 40 30	08 ♍ 32 15	18 ♒ 45 10	25 ♒ 01 35	21 ♍ 07	23 ♋ 46	27 ♈ 30	17 ♑ 39	26 ♑ 00	10 ♉ 35	19 ♓ 51	22 ♑ 45

Moon

Date	Moon True ☊ ° '	Moon Mean ☊ ° '	Moon ☽ Latitude ° '
01	28 ♊ 37	26 ♊ 56	01 S 19
02	28 R 34	26 53	02 26
03	28 29	26 50	03 23
04	28 22	26 47	04 09
05	28 15	26 44	04 43
06	28 06	26 40	05 02
07	27 59	26 37	05 07
08	27 52	26 34	04 59
09	27 47	26 31	04 38
10	27 44	26 28	04 04
11	27 43	26 25	03 21
12	27 D 43	26 21	02 28
13	27 44	26 18	01 27
14	27 45	26 15	00 S 21
15	27 R 45	26 12	00 N 48
16	27 43	26 09	01 56
17	27 38	26 05	03 00
18	27 32	26 02	03 54
19	27 23	25 59	04 35
20	27 14	25 56	04 59
21	27 05	25 53	05 03
22	26 57	25 50	04 47
23	26 52	25 46	04 12
24	26 49	25 43	03 21
25	26 47	25 40	02 20
26	26 D 48	25 37	01 N 11
27	26 R 48	25 34	00 00
28	26 47	25 31	01 S 10
29	26 45	25 27	02 15
30	26 39	25 24	03 12
31	26 ♊ 31	25 ♊ 21	03 S 59

DECLINATIONS

Date	Sun ☉	Moon ☽	Mercury ☿	Venus ♀	Mars ♂	Jupiter ♃	Saturn ♄	Uranus ♅	Neptune ♆	Pluto ♇
01	17 N 49	24 S 05	21 N 24	19 N 20	03 N 40	22 S 18	20 S 51	14 N 34	04 S 45	22 S 26
02	17 34	23 13	21 18	19 25	03 48	22 19	20 52	14 34	04 45	22 26
03	17 18	21 06	21 10	19 29	03 57	22 20	20 52	14 34	04 46	22 26
04	17 02	17 58	20 59	19 34	04 05	22 21	20 53	14 35	04 46	22 27
05	16 46	14 04	20 45	19 38	04 14	22 22	20 54	14 35	04 47	22 27
06	16 29	09 39	20 28	19 42	04 22	22 23	20 55	14 35	04 47	22 27
07	16 12	04 55	20 09	19 46	04 30	22 24	20 56	14 35	04 48	22 28
08	15 55	00 S 03	19 47	19 50	04 37	22 25	20 57	14 35	04 49	22 28
09	15 38	04 N 46	19 23	19 53	04 45	22 26	20 58	14 35	04 49	22 28
10	15 20	09 25	18 56	19 56	04 52	22 27	20 59	14 35	04 50	22 29
11	15 02	13 44	18 27	19 59	04 59	22 28	21 00	14 35	04 50	22 29
12	14 44	17 33	17 56	20 01	05 06	22 29	21 00	14 35	04 51	22 29
13	14 26	20 46	17 23	20 03	05 13	22 30	21 01	14 35	04 51	22 29
14	14 08	22 57	16 48	20 04	05 19	22 30	21 02	14 35	04 52	22 30
15	13 49	24 04	16 11	20 05	05 26	22 31	21 03	14 35	04 52	22 30
16	13 30	23 52	15 33	20 06	05 32	22 32	21 04	14 35	04 53	22 30
17	13 11	22 15	14 53	20 07	05 38	22 33	21 04	14 35	04 54	22 31
18	12 51	19 13	14 12	20 07	05 43	22 34	21 05	14 35	04 54	22 31
19	12 31	14 58	13 30	20 06	05 49	22 34	21 06	14 35	04 55	22 31
20	12 12	09 45	12 47	20 05	05 54	22 35	21 07	14 35	04 55	22 31
21	11 52	03 N 57	12 03	20 04	05 59	22 36	21 07	14 35	04 56	22 32
22	11 31	02 S 03	11 19	20 02	06 04	22 36	21 08	14 35	04 57	22 32
23	11 11	07 54	10 34	20 00	06 08	22 37	21 09	14 35	04 57	22 32
24	10 50	13 15	09 48	19 57	06 13	22 37	21 09	14 35	04 58	22 32
25	10 30	17 46	09 02	19 54	06 17	22 38	21 10	14 35	04 59	22 33
26	10 09	21 13	08 16	19 51	06 21	22 38	21 11	14 34	04 59	22 33
27	09 48	23 24	07 30	19 47	06 24	22 39	21 11	14 34	05 00	22 33
28	09 27	24 13	06 43	19 42	06 28	22 39	21 12	14 34	05 00	22 33
29	09 05	23 40	05 57	19 37	06 31	22 40	21 12	14 34	05 01	22 34
30	08 44	21 51	05 10	19 32	06 34	22 40	21 13	14 33	05 02	22 34
31	08 N 22	18 S 59	04 N 24	19 N 25	06 N 37	22 S 41	21 S 14	14 N 33	05 S 02	22 S 34

ZODIAC SIGN ENTRIES

Date	h	m	Planets
02	18	11	☽ ♒
05	02	28	☽ ♓
05	03	32	☿ ♌
07	13	05	☽ ♈
07	15	21	♀ ♋
10	01	28	☽ ♉
12	13	46	☽ ♊
14	23	35	☽ ♋
17	05	38	☽ ♌
19	08	20	☽ ♍
20	01	30	☿ ♍
21	09	16	☽ ♎
22	15	45	☉ ♍
23	10	16	☽ ♏
25	12	49	☽ ♐
27	17	37	☽ ♑
30	00	37	☽ ♒

LATITUDES

Date	Mercury ☿	Venus ♀	Mars ♂	Jupiter ♃	Saturn ♄	Uranus ♅	Neptune ♆	Pluto ♇
01	00 00	04 S 00	03 S 51	00 S 22	00 S 15	00 S 28	01 S 06	01 S 02
04	00 N 35	03 50	03 55	00 22	00 15	00 28	01 06	01 02
07	01 04	03 40	03 58	00 23	00 15	00 28	01 06	01 02
10	01 25	03 29	04 01	00 23	00 16	00 28	01 06	01 02
13	01 39	03 17	04 04	00 23	00 16	00 28	01 06	01 03
16	01 45	03 04	04 07	00 23	00 16	00 28	01 06	01 03
19	01 45	02 51	04 09	00 24	00 16	00 28	01 07	01 03
22	01 39	02 38	04 11	00 24	00 17	00 28	01 07	01 03
25	01 29	02 25	04 13	00 24	00 17	00 28	01 07	01 03
28	01 15	02 11	04 14	00 24	00 17	00 28	01 07	01 04
31	00 N 57	01 S 57	04 S 15	00 S 25	00 S 17	00 S 28	01 S 07	01 S 04

DATA

Julian Date	2459063
Delta T	+71 seconds
Ayanamsa	24° 08' 25"
Synetic vernal point	04° ♓ 58' 34"
True obliquity of ecliptic	23° 26' 12"

MOON'S PHASES, APSIDES AND POSITIONS ☽

Date	h	m	Phase	Longitude ° '	Eclipse Indicator
03	15	59	○	11 ♒ 46	
11	16	45	☾	19 ♉ 28	
19	02	42	●	26 ♌ 35	
25	17	58	☽	02 ♐ 59	

Day	h	m		
09	13	45	Apogee	
21	10	51	Perigee	
01	08	45	Max dec	24° S 05'
08	12	16	0N	
15	20	38	Max dec	24° N 09'
22	03	48	0S	
28	14	05	Max dec	24° S 13'

LONGITUDES

Date	Chiron ⚷ ° '	Ceres ⚳ ° '	Pallas ⚴ ° '	Juno ⚵ ° '	Vesta ⚶ ° '	Black Moon Lilith ⚸ ° '
01	09 ♈ 15	10 ♓ 56	16 ♑ 12	15 ♎ 25	25 ♋ 38	20 ♈ 58
11	09 ♈ 03	09 ♓ 15	14 ♑ 17	18 ♎ 02	00 ♌ 01	22 ♈ 05
21	08 ♈ 45	07 ♓ 11	12 ♑ 58	20 ♎ 49	04 ♌ 22	23 ♈ 12
31	08 ♈ 24	04 ♓ 59	12 ♑ 18	23 ♎ 46	08 ♌ 42	24 ♈ 19

ASPECTARIAN

h m	Aspects
01 Saturday	
00 20	☿ Q ♅
03 16	☽ Q ♆
04 46	☽ ⊼ ☉
06 56	☽ △ ♅
10 52	☿ ☍ ♇
21 23	☽ □ ♂
23 57	☽ ☌ ♃
02 Sunday	
00 55	☽ ✶ ♆
05 57	☽ ☌ ♇
08 53	☽ ☍ ☿
09 39	☽ ⊼ ♀
11 19	☉ □ ♅
13 59	☽ ☌ ♄
17 17	♀ ∠ ♅
19 26	☿ ⊻ ♀
21 27	☽ ± ♀
22 23	☽ ∥ ♇
23 43	☽ ∥ ♃
03 Monday	
04 23	☽ ∠ ♆
07 31	☽ Q ♂
11 24	☽ # ☿
13 52	☽ □ ♅
14 00	☽ ∥ ♄
15 21	☽ bQ ♀
15 59	☽ ☍ ☉
21 00	☿ ☍ ♄
21 08	☽ ⊥ ♆
04 Tuesday	
00 53	☽ # ♀
06 45	☽ ✶ ♂
07 01	☽ ⊻ ♃
08 24	☽ ⊻ ♆
13 07	♂ □ ♃
13 38	☽ ⊻ ♇
18 17	☽ ⊥ ♃
18 37	☽ # ☉
20 57	☿ # ♄
21 45	☽ △ ♀
21 47	☽ ⊻ ♄
22 07	♀ ⊼ ♄
23 52	☽ Q ♅
05 Wednesday	
01 03	☽ ⊥ ♇
02 16	☽ ⊼ ☿
09 04	☽ # ♅
09 13	☽ ⊥ ♄
11 23	☽ ∠ ♃
12 21	☽ ∠ ♂
15 54	☽ ± ☿
18 20	☽ ∠ ♇
23 01	☽ ✶ ♅
06 Thursday	
02 33	☽ ∠ ♄
05 59	☽ ⊼ ☉
06 24	☽ ⊥ ♂
07 30	♂ ⊻ ♆
11 43	☉ ± ♆
12 41	☽ bQ ☿
16 20	☽ ✶ ♃
18 11	☽ ☌ ♆
18 34	☽ ⊻ ♂
18 45	☽ ± ☉
23 38	☽ ✶ ♇
07 Friday	
04 28	☽ ∠ ♅
07 53	☽ ✶ ♄
12 33	☽ ∥ ♆
12 53	☽ □ ♀
14 01	☽ # ♂
14 07	☽ bQ ☉
15 53	☽ Q ♃
22 23	☽ ⊥ ♅
22 31	☿ bQ ♆
23 27	☽ Q ♇
08 Saturday	
00 12	☽ △ ☿
07 42	☽ Q ♄
09 52	☿ ∥ ♀
10 25	☽ ⊻ ♅
17 04	☿ ⊥ ♀
22 49	☽ △ ☉
09 Sunday	
03 44	☽ □ ♃
04 02	☽ Q ♀
06 03	☽ ⊻ ♆
08 35	☽ ☌ ♂
11 38	☽ □ ♇
11 53	☽ ∥ ♂
12 15	☽ # ♆
18 10	☽ ⊥ ♆
19 50	☽ □ ♄
10 Monday	
03 06	♂ # ♆
06 23	☽ ✶ ♀
12 21	☽ ∠ ♆
12 52	☿ □ ♅
23 05	☽ ☌ ♅
11 Tuesday	
01 12	☽ □ ☿
06 35	☉ ⊼ ♃
15 15	☽ ∠ ♀
15 51	☽ △ ♃
16 45	☽ □ ☉
17 06	☽ ∥ ♅
18 32	☽ ✶ ♆
19 17	☽ ∥ ☉
23 04	☽ ⊻ ♂
12 Wednesday	
00 03	☽ △ ♇
07 44	☿ ± ♆
07 55	☽ △ ♄
10 46	☽ ⊥ ♀
11 23	☽ ⊥ ♂
14 16	☽ ∥ ☿
14 32	☉ ⊼ ♆
18 22	☽ Q ♆
20 33	☽ Q ☿
21 28	☽ bQ ♃
22 43	♀ Q ♂
23 37	☽ ⊻ ♀
23 50	☉ ∥ ♅
13 Thursday	
05 21	♃ ∥ ♇
05 38	☽ ∠ ♂
05 41	☽ bQ ♇
06 25	☽ ∥ ♀
07 14	♂ □ ♇
07 29	☽ Q ☉
10 45	☽ ⊻ ♅
13 15	☽ bQ ♄
14 53	☽ # ♄
14 55	☽ ± ♃
22 22	☽ ⊥ ♅
23 06	☽ ± ♇
14 Friday	
00 33	☽ ✶ ☿
02 24	☽ ⊼ ♃
05 16	☽ □ ♆
06 12	☽ # ♇
06 19	☽ # ♃
06 29	☽ ± ♄
08 32	☽ ✶ ☉
10 32	☽ ⊼ ♇
11 19	☽ ✶ ♂
11 28	☿ ⊼ ♃
15 30	☽ ∠ ♅
17 47	☽ ⊼ ♄
15 Saturday	
05 13	☿ ⊼ ♆
10 08	☽ ∠ ☿
10 13	☉ ⊼ ♇
10 20	☽ Q ♂
13 26	☽ ☌ ♀
14 25	♅ St R
14 48	☽ ∠ ☉
19 18	☽ ✶ ♅
16 Sunday	
05 29	☽ ⊥ ☿
08 21	☽ ⊥ ☉
09 33	☽ ☍ ♃
12 29	☽ △ ♆
13 38	☿ ⊼ ♇
14 02	☉ △ ♂
16 49	☽ Q ♅
17 24	☽ ☍ ♇
18 03	☽ ⊻ ☿
19 32	☽ □ ♂
19 49	☽ ⊻ ☉
23 48	☉ ± ♃
23 59	☽ ☍ ♄
17 Monday	
00 06	☿ ∠ ♀
05 29	☿ △ ♂
07 29	☿ ± ♃
08 45	☽ # ♃
09 09	☽ # ♇
14 39	☽ bQ ♆
15 07	☉ ☌ ☿
22 19	☿ ∥ ♅
22 33	☽ ⊻ ♀
22 33	☽ # ♄
18 Tuesday	
00 01	☽ □ ♅
05 55	☽ ∥ ♀
05 58	☽ ± ♆
09 17	☿ ⊼ ♄
09 28	☽ ⊥ ♀
13 05	☽ ⊼ ♃
16 02	☽ ⊼ ♆
19 28	♀ ✶ ♅
20 41	☽ ⊼ ♇
23 03	☽ ± ♃
23 51	☽ △ ♂
19 Wednesday	
01 38	☽ ∠ ♀
02 42	☽ ☌ ☉
02 46	☽ ⊼ ♄
03 48	☉ ⊼ ♄
05 38	☽ ☌ ☿
06 38	☽ ± ♇
12 38	☽ ± ♄
12 50	☿ ± ♇
13 51	☽ bQ ♃
13 52	☽ ∥ ♅
20 18	☽ ∥ ☿
21 24	☽ bQ ♇
20 Thursday	
00 28	☽ ∥ ☉
01 01	☽ bQ ♂
01 52	☽ △ ♅
03 19	☽ bQ ♄
04 05	☽ ✶ ♀
14 13	☽ △ ♃
15 57	☽ ± ♂
17 15	☽ ☍ ♆
21 48	☽ △ ♇
21 Friday	
01 02	☽ Q ♀
01 52	☽ ⊼ ♂
02 15	☽ bQ ♅
03 37	☽ △ ♄
03 55	☽ ∥ ♂
07 08	☽ ⊻ ☉
08 04	☽ # ♆
08 15	☿ ± ♄
12 54	☉ ± ♇
14 06	☽ ⊻ ☿
16 45	☿ bQ ♃
16 51	☽ ± ♅
17 33	☽ ⊥ ☉
23 20	☉ ∥ ☿
22 Saturday	
01 18	☽ ⊥ ☿
02 36	☽ ⊼ ♅
08 25	☽ □ ♀
09 13	☽ ∠ ☉
14 45	☽ □ ♃
17 57	☽ ⊼ ♆
18 12	☽ ∠ ☿
22 34	☽ □ ♇
23 43	☽ ∥ ♆
23 Sunday	
03 35	☽ ☍ ♂
03 47	☽ ± ♆
04 20	☽ □ ♄
04 31	☽ # ♂
11 35	☽ ✶ ☉
14 59	♂ ⊥ ♆
18 39	☽ bQ ♆
20 18	☽ Q ♃
22 09	☽ # ☿
22 42	☽ ✶ ☿
24 Monday	
01 30	☽ # ☉
03 37	☿ bQ ♇
04 00	☽ ☍ ♅
04 21	☽ Q ♇
09 02	☽ Q ☉
10 10	☽ Q ♄
13 51	☽ △ ♀
16 21	☽ ✶ ♃
18 19	♂ □ ♄
18 34	☽ # ♅
19 47	☽ △ ♆
21 57	☽ Q ☿
25 Tuesday	
00 36	☽ ✶ ♇
01 06	☉ ± ♄
06 27	☽ ✶ ♄
06 41	☽ ⊼ ♂
15 18	☿ △ ♅
16 53	☉ bQ ♃
17 10	☽ ± ♂
17 30	☽ bQ ♀
17 53	☽ ∠ ♃
17 58	☽ □ ☉
22 26	♀ ☍ ♃
23 11	☿ bQ ♄
26 Wednesday	
01 35	☽ # ♀
02 23	☽ ∠ ♇
03 48	☿ bQ ♂
07 17	☽ ⊼ ♅
08 19	☽ ∠ ♄
09 03	☽ bQ ♂
09 29	☽ ⊥ ♃
09 46	☽ □ ☿
10 26	☽ ± ♀
11 37	☽ ∥ ♄
17 49	☽ ± ♅
18 09	☽ ⊥ ♇
20 00	☽ ⊻ ♃
21 53	☽ ⊼ ♀
23 43	☽ □ ♆
27 Thursday	
00 07	☽ ⊥ ♄
00 46	☽ ∥ ♇
01 50	☽ ∥ ♃
04 46	☽ ⊻ ♇
09 47	☽ bQ ♅
10 46	☽ ⊻ ♄
12 00	☽ △ ♂
21 12	♀ △ ♆
28 Friday	
03 09	☽ △ ☉
07 58	☽ Q ♆
12 51	☽ △ ♅
19 34	☿ ∥ ♂
23 46	☉ ∠ ♀
29 Saturday	
00 04	☽ △ ☿
01 56	☽ ☌ ♃
05 54	☽ ✶ ♆
08 47	☽ bQ ☉
08 51	☽ ☍ ♀
11 12	☽ ☌ ♇
13 28	☿ △ ♃
17 18	☽ ☌ ♄
19 31	☽ □ ♂
30 Sunday	
02 52	☽ ± ☉
03 00	☽ ∥ ♃
04 16	☽ ∥ ♇
08 20	☽ bQ ☿
09 42	♀ Q ♅
09 46	☽ ∠ ♆
13 31	♀ ☍ ♇
15 03	☽ ⊼ ☉
16 25	☿ # ♆
16 53	☉ bQ ♇
18 04	☽ ∥ ♄
18 44	☿ ☍ ♆
20 31	☽ □ ♅
31 Monday	
02 43	☽ ⊥ ♆
04 00	☽ ± ☿
05 45	☽ Q ♂
08 38	☽ # ♀
09 55	☽ ⊻ ♃
14 06	☽ ⊻ ♆
17 14	☽ ⊼ ☿
17 48	☿ ± ♂
19 38	☽ ⊻ ♇
21 21	☽ ⊥ ♃
22 28	☽ ⊼ ♀

All ephemeris data is given at 12.00 UT and the Moon's longitude is additionally given for 24.00 UT

SEPTEMBER 2020

Date	Sidereal time h m s	LONGITUDES Sun ☉ ° ' "	Moon ☽ ° ' "	Moon ☽ 24.00 ° ' "	Mercury ☿ ° '	Venus ♀ ° '	Mars ♂ ° '	Jupiter ♃ ° '	Saturn ♄ ° '	Uranus ♅ ° '	Neptune ♆ ° '	Pluto ♇ ° '
01	10 44 27	09 ♍ 30 17	01 ♓ 15 29	07 ♓ 26 56	22 ♍ 50	24 ♋ 50	27 ♈ 38	17 ♑ 37	25 ♑ 57	10 ♉ 34	19 ♓ 50	22 ♑ 44
02	10 48 24	10 28 21	13 ♓ 35 59	19 ♓ 42 42	24 31	25 54	27 45	17 R 35	25 R 54	10 R 34	19 R 48	22 R 44
03	10 52 20	11 26 26	25 ♓ 47 11	01 ♈ 49 33	26 12	26 58	27 51	17 33	25 52	10 33	19 47	22 43
04	10 56 17	12 24 33	07 ♈ 49 58	13 ♈ 48 38	27 51	28 03	27 56	17 31	25 50	10 32	19 45	22 42
05	11 00 13	13 22 42	19 ♈ 45 48	25 ♈ 41 47	29 ♍ 29	29 ♋ 08	28 00	17 30	25 47	10 31	19 43	22 41
06	11 04 10	14 20 52	01 ♉ 36 54	07 ♉ 31 36	01 ♎ 05	00 ♌ 13	28 03	17 29	25 45	10 30	19 42	22 40
07	11 08 06	15 19 05	13 ♉ 26 20	19 ♉ 21 36	02 41	01 18	28 06	17 27	25 43	10 29	19 40	22 40
08	11 12 03	16 17 20	25 ♉ 17 57	01 ♊ 16 00	04 15	02 24	28 08	17 26	25 41	10 28	19 38	22 39
09	11 15 59	17 15 36	07 ♊ 16 22	13 ♊ 19 41	05 48	03 29	28 08	17 26	25 39	10 26	19 37	22 38
10	11 19 56	18 13 55	19 ♊ 26 39	25 ♊ 37 55	07 20	04 35	28 R 08	17 25	25 37	10 25	19 35	22 37
11	11 23 53	19 12 16	01 ♋ 54 09	08 ♋ 15 57	08 51	05 42	28 07	17 25	25 36	10 24	19 33	22 37
12	11 27 49	20 10 39	14 ♋ 43 55	21 ♋ 18 32	10 21	06 48	28 06	17 24	25 34	10 23	19 32	22 36
13	11 31 46	21 09 04	28 ♋ 00 10	04 ♌ 49 06	11 49	07 55	28 03	17 D 24	25 32	10 21	19 30	22 36
14	11 35 42	22 07 32	11 ♌ 45 25	18 ♌ 49 01	13 16	09 02	27 59	17 25	25 31	10 20	19 28	22 35
15	11 39 39	23 06 01	25 ♌ 59 36	03 ♍ 16 40	14 43	10 09	27 55	17 25	25 29	10 19	19 27	22 34
16	11 43 35	24 04 32	10 ♍ 39 27	18 ♍ 07 01	16 07	11 16	27 49	17 26	25 28	10 17	19 25	22 34
17	11 47 32	25 03 05	25 ♍ 38 15	03 ♎ 11 53	17 31	12 23	27 43	17 26	25 27	10 16	19 23	22 33
18	11 51 28	26 01 40	10 ♎ 46 36	18 ♎ 21 04	18 54	13 31	27 36	17 27	25 26	10 14	19 22	22 33
19	11 55 25	27 00 17	25 ♎ 53 58	03 ♏ 24 09	20 15	14 39	27 28	17 28	25 25	10 12	19 20	22 32
20	11 59 22	27 58 56	10 ♏ 50 35	18 ♏ 12 25	21 35	15 47	27 19	17 30	25 24	10 11	19 19	22 32
21	12 03 18	28 57 36	25 ♏ 29 02	02 ♐ 39 59	22 53	16 55	27 10	17 31	25 23	10 09	19 17	22 32
22	12 07 15	29 ♍ 56 18	09 ♐ 45 01	16 ♐ 44 04	24 11	18 03	26 59	17 33	25 22	10 08	19 15	22 31
23	12 11 11	00 ♎ 55 02	23 ♐ 37 13	00 ♑ 24 38	25 26	19 11	26 48	17 35	25 22	10 06	19 14	22 31
24	12 15 08	01 53 48	07 ♑ 06 36	13 ♑ 43 28	26 40	20 20	26 36	17 37	25 21	10 04	19 12	22 31
25	12 19 04	02 52 35	20 ♑ 15 36	26 ♑ 43 23	27 53	21 29	26 24	17 39	25 21	10 02	19 10	22 30
26	12 23 01	03 51 24	03 ♒ 07 13	09 ♒ 27 29	29 ♎ 03	22 38	26 10	17 42	25 21	10 00	19 09	22 30
27	12 26 57	04 50 15	15 ♒ 44 31	21 ♒ 58 39	00 ♏ 12	23 47	25 56	17 45	25 20	09 59	19 07	22 30
28	12 30 54	05 49 07	28 ♒ 10 10	04 ♓ 19 19	01 19	24 56	25 42	17 47	25 20	09 57	19 06	22 30
29	12 34 51	06 48 01	10 ♓ 26 20	16 ♓ 31 23	02 24	26 05	25 27	17 51	25 D 20	09 55	19 04	22 30
30	12 38 47	07 ♎ 46 57	22 ♓ 34 40	28 ♓ 36 18	03 ♏ 27	27 ♌ 15	25 ♈ 11	17 ♑ 54	25 ♑ 20	09 ♉ 53	19 ♓ 03	22 ♑ 29

Date	Moon True ☊ ° '	Moon Mean ☊ ° '	Moon ☽ Latitude ° '
01	26 ♊ 20	25 ♊ 18	04 S 33
02	26 R 08	25 15	04 53
03	25 55	25 11	05 00
04	25 42	25 08	04 54
05	25 31	25 05	04 34
06	25 22	25 02	04 03
07	25 15	24 59	03 22
08	25 12	24 56	02 31
09	25 10	24 52	01 33
10	25 D 10	24 49	00 S 30
11	25 R 10	24 46	00 N 36
12	25 09	24 43	01 42
13	25 06	24 40	02 44
14	25 00	24 36	03 40
15	24 52	24 33	04 23
16	24 42	24 30	04 51
17	24 31	24 27	05 00
18	24 20	24 24	04 48
19	24 10	24 21	04 16
20	24 03	24 17	03 26
21	23 58	24 14	02 24
22	23 56	24 11	01 14
23	23 D 56	24 08	00 N 02
24	23 R 56	24 05	01 S 09
25	23 55	24 02	02 14
26	23 52	23 58	03 11
27	23 46	23 55	03 58
28	23 38	23 52	04 32
29	23 27	23 49	04 53
30	23 ♊ 14	23 ♊ 46	05 S 00

DECLINATIONS

Date	Sun ☉ ° '	Moon ☽ ° '	Mercury ☿ ° '	Venus ♀ ° '	Mars ♂ ° '	Jupiter ♃ ° '	Saturn ♄ ° '	Uranus ♅ ° '	Neptune ♆ ° '	Pluto ♇ ° '
01	08 N 00	15 S 16	03 N 38	19 N 19	06 N 39	22 S 41	21 S 14	14 N 33	05 S 03	22 S 34
02	07 38	10 58	02 52	19 12	06 41	22 41	21 15	14 33	05 04	22 35
03	07 16	06 16	02 06	19 05	06 44	22 42	21 15	14 32	05 04	22 35
04	06 54	01 S 24	01 20	18 57	06 45	22 42	21 16	14 32	05 05	22 35
05	06 32	03 N 29	00 N 35	18 48	06 47	22 42	21 16	14 32	05 06	22 35
06	06 10	08 14	00 S 10	18 39	06 48	22 42	21 17	14 31	05 06	22 35
07	05 47	12 40	00 55	18 30	06 49	22 43	21 17	14 31	05 07	22 36
08	05 25	16 39	01 39	18 20	06 50	22 43	21 18	14 31	05 08	22 36
09	05 02	19 59	02 23	18 09	06 51	22 43	21 18	14 30	05 08	22 36
10	04 39	22 31	03 07	17 58	06 51	22 43	21 18	14 30	05 09	22 36
11	04 16	24 01	03 50	17 47	06 52	22 43	21 19	14 30	05 10	22 36
12	03 53	24 18	04 32	17 35	06 51	22 43	21 19	14 29	05 10	22 36
13	03 30	23 14	05 14	17 22	06 51	22 43	21 19	14 29	05 11	22 37
14	03 07	20 46	05 55	17 09	06 51	22 43	21 20	14 28	05 12	22 37
15	02 44	16 59	06 36	16 56	06 50	22 43	21 20	14 28	05 12	22 37
16	02 21	12 04	07 16	16 42	06 49	22 43	21 20	14 27	05 13	22 37
17	01 58	06 19	07 56	16 27	06 48	22 43	21 21	14 27	05 14	22 37
18	01 35	00 N 09	08 35	16 12	06 46	22 43	21 21	14 26	05 14	22 37
19	01 11	06 S 02	09 13	15 57	06 45	22 43	21 21	14 26	05 15	22 37
20	00 48	11 48	09 50	15 41	06 43	22 43	21 21	14 25	05 16	22 38
21	00 25	16 48	10 27	15 25	06 41	22 43	21 22	14 25	05 16	22 38
22	00 N 01	20 41	11 02	15 08	06 39	22 43	21 22	14 24	05 17	22 38
23	00 S 22	23 15	11 37	14 51	06 37	22 42	21 22	14 24	05 17	22 38
24	00 45	24 24	12 11	14 33	06 34	22 42	21 22	14 23	05 18	22 38
25	01 09	24 07	12 44	14 15	06 31	22 42	21 22	14 23	05 19	22 38
26	01 32	22 34	13 16	13 57	06 29	22 42	21 22	14 22	05 19	22 38
27	01 55	19 54	13 47	13 38	06 26	22 42	21 22	14 22	05 20	22 38
28	02 19	16 21	14 17	13 18	06 22	22 41	21 22	14 21	05 21	22 38
29	02 42	12 10	14 46	12 59	06 19	22 41	21 23	14 20	05 21	22 39
30	03 S 05	07 S 33	15 S 13	12 N 38	06 N 16	22 S 40	21 S 23	14 N 20	05 S 22	22 S 39

ZODIAC SIGN ENTRIES

Date	h	m	Planets
01	09	34	☽ ♓
03	20	22	☽ ♈
05	19	46	☿ ♎
06	07	22	♀ ♌
06	08	43	☽ ♉
08	21	28	☽ ♊
11	08	23	☽ ♋
13	15	32	☽ ♌
15	18	37	☽ ♍
17	18	56	☽ ♎
19	18	33	☽ ♏
21	19	32	☽ ♐
22	13	31	☉ ♎
23	23	16	☽ ♑
26	06	08	☽ ♒
27	07	41	☿ ♏
28	15	34	☽ ♓

LATITUDES

Date	Mercury ☿ ° '	Venus ♀ ° '	Mars ♂ ° '	Jupiter ♃ ° '	Saturn ♄ ° '	Uranus ♅ ° '	Neptune ♆ ° '	Pluto ♇ ° '
01	00 N 51	01 S 52	04 S 15	00 S 25	00 S 17	00 S 28	01 S 07	01 S 04
04	00 31	01 38	04 16	00 25	00 17	00 28	01 07	01 04
07	00 N 10	01 25	04 15	00 25	00 18	00 28	01 07	01 04
10	00 S 13	01 11	04 14	00 25	00 18	00 28	01 07	01 05
13	00 37	00 57	04 12	00 25	00 18	00 28	01 07	01 05
16	01 01	00 44	04 09	00 25	00 18	00 28	01 07	01 05
19	01 24	00 31	04 06	00 25	00 18	00 28	01 07	01 05
22	01 48	00 18	04 01	00 26	00 18	00 28	01 07	01 06
25	02 10	00 S 06	03 56	00 26	00 19	00 28	01 07	01 06
28	02 31	00 N 06	03 49	00 26	00 19	00 28	01 07	01 06
31	02 S 50	00 N 17	03 S 41	00 S 26	00 S 19	00 S 28	01 S 07	01 S 06

DATA

Julian Date	2459094
Delta T	+71 seconds
Ayanamsa	24° 08' 29"
Synetic vernal point	04° ♓ 58' 30"
True obliquity of ecliptic	23° 26' 13"

MOON'S PHASES, APSIDES AND POSITIONS ☽

Date	h	m	Phase	Longitude ° '	Eclipse Indicator
02	05	22	○	10 ♓ 12	
10	09	26	☾	18 ♊ 08	
17	11	00	●	25 ♍ 01	
24	01	55	☽	01 ♑ 29	

Day	h	m		
06	06	19	Apogee	
18	13	42	Perigee	
04	18	49	ON	
12	05	21	Max dec	24° N 21'
18	12	35	OS	
24	19	12	Max dec	24° S 28'

LONGITUDES

Date	Chiron ⚷ ° '	Ceres ⚳ ° '	Pallas ⚴ ° '	Juno ⚵ ° '	Vesta ⚶ ° '	Black Moon Lilith ⚸ ° '
01	08 ♈ 22	04 ♓ 45	12 ♑ 16	24 ♎ 04	09 ♌ 07	24 ♈ 26
11	07 ♈ 58	02 ♓ 38	12 ♑ 18	27 ♎ 09	13 ♌ 24	25 ♈ 33
21	07 ♈ 32	00 ♓ 50	12 ♑ 54	00 ♏ 20	17 ♌ 37	26 ♈ 40
31	07 ♈ 05	29 ♒ 31	14 ♑ 01	03 ♏ 36	21 ♌ 45	27 ♈ 47

All ephemeris data is given at 12.00 UT and the Moon's longitude is additionally given for 24.00 UT

ASPECTARIAN

h m	Aspects
01 Tuesday	
01 49	☽ ⊻ ♄
04 56	☽ ⚹ ♂
06 50	☽ Q ♅
07 09	☽ ⊥ ♇
10 42	☿ △ ♇
11 05	☽ ± ♀
13 20	☽ ⊥ ♄
14 37	☽ ∠ ♃
16 13	☽ # ♅
02 Wednesday	
00 33	☽ ∠ ♇
05 22	☽ ☍ ☉
06 04	☽ ⚹ ♅
06 13	☽ Ᵽ ♀
06 45	☽ ∠ ♄
10 19	☽ ∠ ♂
12 18	♀ ☍ ♄
14 09	☉ △ ♅
19 48	☽ ⚹ ♃
22 22	☉ Ᵽ ♄
03 Thursday	
00 09	☽ ☌ ♆
02 43	☿ Ᵽ ♅
04 08	☽ ⊥ ♂
05 55	☽ ⚹ ♇
06 33	☽ # ☉
07 22	☿ △ ♄
09 44	☽ # ♂
11 31	☽ ∠ ♅
12 10	☽ ⚹ ♄
12 56	☽ ☍ ☿
14 34	☽ △ ♀
16 06	☽ ⊻ ♂
17 56	☽ ∥ ♆
19 27	☽ Q ♃
04 Friday	
05 24	☽ ⊥ ♅
05 44	☽ Q ♇
09 12	♀ □ ♂
11 59	☽ Q ♄
12 21	☽ # ☿
13 15	☿ ⊼ ♂
17 24	☽ ⊻ ♅
20 32	☿ ⚹ ♀
20 55	☉ ∥ ♂
22 00	☽ ⊼ ☉
23 34	☽ ∥ ☿
05 Saturday	
01 54	☉ Ᵽ ♂
07 26	☽ □ ♃
11 09	☽ ± ☉
11 55	☽ ⊻ ♆
17 54	☽ □ ♇
20 01	☽ # ♆
06 Sunday	
00 01	☽ ⊥ ♆
00 09	☽ □ ♄
02 10	☽ ∥ ☉
04 38	☽ ∥ ♂
04 45	☽ ☌ ♂
06 59	☽ Ᵽ ☉
08 52	☽ □ ♀
10 46	☽ ⊼ ☿
18 14	☽ ∠ ♆
07 Monday	
00 50	☽ ± ☿
06 00	☽ ☌ ♅
16 09	☽ △ ☉
20 08	☽ △ ♃
21 55	☽ Ᵽ ☿
22 46	☽ ∥ ♅
08 Tuesday	
00 35	☽ ⚹ ♆
01 05	☽ Q ♀
06 39	☽ △ ♇
12 46	☽ △ ♄
15 12	☿ ± ♅
17 42	☽ ⊻ ♂
22 52	☽ ∥ ♀
09 Wednesday	
00 43	☽ Q ♆
02 20	☽ Ᵽ ♃
03 41	☽ ⚹ ♀
05 26	☉ # ♆
05 45	☽ ⊥ ♂
08 38	☽ △ ☿
12 43	☽ Ᵽ ♇
16 04	☉ △ ♃
18 17	☽ ⊻ ♅
18 42	☽ Ᵽ ♄
20 14	☽ ± ♃
22 22	♂ St R
23 23	☽ # ♄
23 38	☽ ∠ ♂
10 Thursday	
06 06	☽ ⊥ ♅
06 29	☽ ± ♇
08 02	☽ ⊼ ♃
09 26	☽ □ ☉
11 53	♀ Ᵽ ♆
12 16	☽ □ ♆
12 19	☽ ∠ ♀
12 21	☽ ± ♄
13 01	☽ # ♇
14 27	☽ # ♃
18 11	☽ ⊼ ♇
23 34	☽ ∠ ♅
23 57	☽ ⊼ ♄
11 Friday	
04 48	☽ ⚹ ♂
07 23	☽ ⊥ ♀
19 51	☽ ⊻ ♀
20 26	☉ ☍ ♆
21 45	☉ # ☿
22 50	☽ Q ☉
12 Saturday	
02 49	☽ □ ☿
03 26	☽ Q ♂
03 57	☽ ⚹ ♅
12 35	☿ ⊼ ♅
16 54	☽ ☍ ♃
20 45	☽ △ ♆
22 44	☽ ⚹ ☉
13 Sunday	
00 41	♃ St D
01 55	☽ Q ♅
02 19	☽ ☍ ♇
07 37	☽ ☍ ♄
10 08	☿ ∥ ♆
12 05	☽ □ ♂
15 36	☽ Q ☿
18 22	☽ # ♃
19 36	☽ # ♇
23 25	☽ Ᵽ ♆
14 Monday	
03 25	☽ ∠ ☉
06 54	☽ ☌ ♀
07 36	☽ # ♄
08 49	☉ ± ♂
09 34	☽ □ ♅
14 53	☽ ⚹ ☿
14 56	☽ ± ♆
20 00	☽ ⊥ ☉
21 38	☽ ⊼ ♃
23 09	☉ △ ♇
15 Tuesday	
01 05	☽ ⊼ ♆
06 19	☽ ⊼ ♇
06 50	☽ ⊻ ☉
07 43	☽ ± ♃
11 10	☽ ⊼ ♄
12 19	☽ ∥ ♀
15 10	☽ △ ♂
15 29	♀ □ ♅
16 16	☽ ± ♇
18 48	☽ ∠ ☿
19 58	☿ # ♂
21 04	☽ ± ♄
22 36	☽ Ᵽ ♃
16 Wednesday	
00 53	☽ ∥ ♅
06 59	☽ Ᵽ ♇
11 03	☽ ⊥ ☿
11 24	☽ △ ♅
11 42	☽ Ᵽ ♄
13 04	☽ ⊻ ♀
15 28	☽ Ᵽ ♂
21 43	☽ ⊻ ☿
22 54	☽ △ ♃
23 30	☽ ⊥ ♀
17 Thursday	
02 03	☽ ☍ ♆
05 48	☽ ± ♂
06 10	☽ # ☿
07 06	☽ △ ♇
10 06	☽ ∥ ♂
10 34	☿ □ ♃
11 00	☽ ☌ ☉
11 24	☽ Ᵽ ♅
11 42	☽ △ ♄
15 00	☽ ∠ ♀
15 17	☽ ⊼ ♂
16 20	☽ # ♆
17 00	☉ Ᵽ ♅
21 36	☉ △ ♄
18 Friday	
01 40	☽ ± ♅
06 09	☽ ∥ ☉
08 52	♀ ± ♆
11 09	☽ ⊼ ♅
16 41	☽ ⚹ ♀
18 16	☽ # ☉
20 05	☿ ⊼ ♆
22 35	☽ □ ♃
19 Saturday	
01 35	☽ ⊼ ♆
02 08	☽ ☌ ☿
06 39	☽ □ ♇
08 55	☽ ∥ ♆
11 06	☽ ± ♆
11 14	☽ □ ♄
13 17	☽ Q ♀
13 53	☽ ⊻ ☉
14 29	☽ ☍ ♂
14 51	☽ # ♂
21 55	☉ ⊼ ♂
20 Sunday	
00 09	☽ ⊥ ☉
01 28	☽ Ᵽ ♆
02 30	☽ ∥ ☿
03 21	☽ Q ♃
10 56	☽ ☍ ♅
11 30	☽ Q ♇
15 43	☽ ∠ ☉
16 09	☽ Q ♄
20 42	☽ □ ♀
22 51	☽ ⚹ ♃
21 Monday	
00 00	☽ # ♅
01 47	☽ △ ♆
05 14	☽ # ♀
05 21	☿ □ ♇
07 07	☽ ⚹ ♇
07 17	☽ ⊻ ☿
11 50	☽ ⚹ ♄
14 45	☽ ⊼ ♂
18 13	☽ ⚹ ☉
18 14	☽ ⊥ ☿
23 47	☽ ∠ ♃
22 Tuesday	
00 41	☽ ± ♂
01 12	♀ ⊼ ♃
08 12	☽ ∠ ♇
10 55	☽ ∠ ☿
12 38	☽ ⊼ ♅
13 04	☽ ∠ ♄
15 05	☽ ⊥ ♃
15 47	☽ Ᵽ ♂
16 01	☽ Q ☉
17 13	☽ ∥ ♄
22 55	☽ ± ♅
23 38	☽ ⊥ ♇
23 Wednesday	
01 27	☽ ⊻ ♃
03 34	☽ △ ♀
04 20	☽ □ ♆
04 34	☽ ⊥ ♄
04 47	☽ ∥ ♇
05 35	☽ ∥ ♃
08 05	☿ ± ♆
10 04	☽ ⊻ ♇
10 38	☿ □ ♄
12 45	♀ ⊼ ♆
14 35	☽ Ᵽ ♅
15 04	☽ ⊻ ♄
15 31	☽ ⚹ ☿
17 31	☽ △ ♂
24 Thursday	
01 55	☽ □ ☉
08 30	☽ Ᵽ ♀
10 53	☿ ☍ ♂
12 10	☽ Q ♆
15 06	☽ Q ☿
17 20	☽ △ ♅
25 Friday	
01 40	♀ ∥ ♅
02 22	☽ ± ♀
07 11	☽ ☌ ♃
10 00	☽ ⚹ ♆
14 28	☽ ⊼ ♀
16 09	☽ ☌ ♇
21 26	☽ ☌ ♄
23 12	☽ □ ♂
26 Saturday	
03 36	☽ □ ☿
09 22	♀ ⊼ ♇
10 29	☽ ∥ ♃
11 07	☽ ∥ ♇
13 30	☽ △ ☉
13 56	☽ ∠ ♆
15 34	☉ ± ♅
23 48	☽ ∥ ♄
27 Sunday	
01 01	☽ □ ♅
07 00	☽ ⊥ ♆
07 18	☿ # ♀
08 37	☽ Q ♂
11 12	♀ ± ♃
15 51	☽ ⊻ ♃
18 29	☽ ⊻ ♆
20 33	☽ Ᵽ ☉
28 Monday	
01 00	☽ ⊻ ♇
03 28	☽ ⊥ ♃
05 04	☽ ☍ ♀
06 30	☽ ⊻ ♄
07 18	☽ ⚹ ♂
11 34	☽ Q ♅
12 38	☽ ⊥ ♇
14 54	☿ # ♅
15 29	☽ ± ☉
18 10	☽ ⊥ ♄
18 44	☽ △ ☿
20 23	♀ ⊼ ♄
21 03	☽ ∠ ♃
22 58	☽ ∥ ☿
23 55	☽ # ♅
29 Tuesday	
01 01	♀ △ ♂
04 14	☽ ⊼ ☉
05 11	♄ St D
06 13	☽ ∠ ♇
07 13	☽ # ♀
10 58	☽ ⚹ ♅
11 48	☽ ∠ ♄
12 01	☽ ∠ ♂
21 49	♂ □ ♄
30 Wednesday	
02 40	☽ ⚹ ♃
03 02	☽ Ᵽ ☿
05 00	☽ ☌ ♆
05 24	☽ ⊥ ♂
11 50	☽ ⚹ ♇
16 34	☽ ∠ ♅
17 04	☽ ⊻ ♂
17 29	☽ ⚹ ♄
18 27	☽ # ♂
22 17	☽ ⊼ ♀
22 34	☽ ± ☿
22 49	☽ ∥ ♆

OCTOBER 2020

LONGITUDES

Date	Sidereal time h m s	Sun ☉ ° ' "	Moon ☽ ° ' "	Moon ☽ 24.00 ° '	Mercury ☿ ° '	Venus ♀ ° '	Mars ♂ ° '	Jupiter ♃ ° '	Saturn ♄ ° '	Uranus ♅ ° '	Neptune ♆ ° '	Pluto ♇ ° '
01	12 42 44	08 ♎ 45 55	04 ♈ 36 26	10 ♈ 35 14	04 ♏ 27	28 ♌ 25	24 ♈ 55	17 ♑ 57	25 ♑ 20	09 ♉ 51	19 ♓ 01	22 ♑ 29
02	12 46 40	09 44 55	16 ♈ 32 49	22 ♈ 29 22	05 24	29 ♌ 34	24 R 38	18 01	25 21	09 R 49	19 R 00	22 R 29
03	12 50 37	10 43 57	28 ♈ 25 04	04 ♉ 20 09	06 19	00 ♍ 44	24 21	18 05	25 21	09 47	18 58	22 29
04	12 54 33	11 43 02	10 ♉ 14 51	16 ♉ 09 29	07 10	01 54	24 03	18 08	25 22	09 45	18 57	22 29
05	12 58 30	12 42 08	22 ♉ 04 24	27 ♉ 59 58	07 59	03 05	23 45	18 13	25 22	09 43	18 55	22 D 29
06	13 02 26	13 41 17	03 ♊ 56 38	09 ♊ 54 53	08 43	04 15	23 27	18 17	25 23	09 40	18 54	22 29
07	13 06 23	14 40 28	15 ♊ 55 14	21 ♊ 58 15	09 24	05 25	23 08	18 21	25 24	09 38	18 52	22 29
08	13 10 20	15 39 41	28 ♊ 04 32	04 ♋ 14 43	10 00	06 36	22 49	18 26	25 25	09 36	18 51	22 29
09	13 14 16	16 38 57	10 ♋ 29 24	16 ♋ 49 13	10 31	07 47	22 30	18 31	25 25	09 34	18 49	22 30
10	13 18 13	17 38 14	23 ♋ 14 46	29 ♋ 46 37	10 57	08 58	22 11	18 36	25 27	09 32	18 48	22 30
11	13 22 09	18 37 35	06 ♌ 25 14	13 ♌ 11 02	11 17	10 09	21 52	18 41	25 28	09 29	18 46	22 30
12	13 26 06	19 36 57	20 ♌ 04 15	27 ♌ 05 00	11 32	11 20	21 33	18 46	25 29	09 27	18 45	22 30
13	13 30 02	20 36 22	04 ♍ 13 12	11 ♍ 28 31	11 39	12 31	21 14	18 52	25 30	09 25	18 44	22 30
14	13 33 59	21 35 49	18 ♍ 50 26	26 ♍ 18 11	11 R 39	13 42	20 55	18 58	25 32	09 23	18 42	22 31
15	13 37 55	22 35 18	03 ♎ 50 46	11 ♎ 27 02	11 32	14 54	20 36	19 04	25 33	09 20	18 41	22 31
16	13 41 52	23 34 49	19 ♎ 05 38	26 ♎ 45 10	11 17	16 05	20 17	19 10	25 35	09 18	18 40	22 31
17	13 45 49	24 34 23	04 ♏ 24 10	12 ♏ 01 17	10 53	17 17	19 58	19 16	25 37	09 16	18 39	22 32
18	13 49 45	25 33 58	19 ♏ 35 12	27 ♏ 04 49	10 20	18 29	19 40	19 22	25 39	09 13	18 37	22 32
19	13 53 42	26 33 36	04 ♐ 29 10	11 ♐ 47 34	09 39	19 41	19 22	19 29	25 41	09 11	18 36	22 33
20	13 57 38	27 33 15	18 ♐ 59 30	26 ♐ 04 41	08 50	20 53	19 05	19 35	25 43	09 08	18 35	22 33
21	14 01 35	28 32 56	03 ♑ 03 01	09 ♑ 54 33	07 52	22 05	18 47	19 42	25 45	09 06	18 34	22 33
22	14 05 31	29 ♎ 32 39	16 ♑ 39 30	23 ♑ 18 10	06 48	23 17	18 31	19 49	25 47	09 04	18 33	22 34
23	14 09 28	00 ♏ 32 23	29 ♑ 50 57	06 ♒ 18 18	05 38	24 29	18 15	19 56	25 49	09 01	18 31	22 35
24	14 13 24	01 32 09	12 ♒ 40 40	18 ♒ 58 34	04 24	25 41	17 59	20 03	25 52	08 59	18 30	22 35
25	14 17 21	02 31 57	25 ♒ 12 29	01 ♓ 22 54	03 08	26 54	17 44	20 11	25 54	08 56	18 29	22 36
26	14 21 18	03 31 46	07 ♓ 30 16	13 ♓ 35 01	01 52	28 06	17 29	20 18	25 57	08 54	18 28	22 36
27	14 25 14	04 31 37	19 ♓ 37 32	25 ♓ 38 12	00 ♏ 39	29 ♍ 19	17 16	20 26	26 00	08 51	18 27	22 37
28	14 29 11	05 31 30	01 ♈ 37 19	07 ♈ 35 13	29 ♎ 31	00 ♎ 31	17 02	20 34	26 03	08 49	18 26	22 38
29	14 33 07	06 31 24	13 ♈ 32 09	19 ♈ 28 22	28 30	01 44	16 50	20 42	26 05	08 46	18 25	22 38
30	14 37 04	07 31 21	25 ♈ 24 05	01 ♉ 19 31	27 37	02 57	16 38	20 50	26 08	08 44	18 24	22 39
31	14 41 00	08 ♏ 31 19	07 ♉ 14 53	13 ♉ 10 22	26 ♎ 54	04 ♎ 10	16 ♈ 27	20 ♑ 58	26 ♑ 11	08 ♉ 41	18 ♓ 23	22 ♑ 40

Date	Moon True ☊ ° '	Moon Mean ☊ ° '	Moon ☽ Latitude ° '
01	23 ♊ 00	23 ♊ 42	04 S 54
02	22 R 47	23 39	04 36
03	22 35	23 36	04 05
04	22 26	23 33	03 23
05	22 19	23 30	02 33
06	22 15	23 27	01 36
07	22 13	23 23	00 S 34
08	22 D 13	23 20	00 N 31
09	22 14	23 17	01 36
10	22 R 14	23 14	02 38
11	22 12	23 11	03 33
12	22 09	23 08	04 18
13	22 03	23 04	04 50
14	21 55	23 01	05 04
15	21 46	22 58	04 58
16	21 37	22 55	04 31
17	21 29	22 52	03 44
18	21 23	22 48	02 41
19	21 19	22 45	01 29
20	21 18	22 42	00 N 12
21	21 D 19	22 39	01 S 03
22	21 20	22 36	02 12
23	21 20	22 33	03 12
24	21 R 20	22 29	04 01
25	21 17	22 26	04 37
26	21 12	22 23	04 59
27	21 05	22 20	05 08
28	20 57	22 17	05 02
29	20 48	22 14	04 44
30	20 40	22 10	04 14
31	20 ♊ 32	22 ♊ 07	03 S 32

DECLINATIONS

Date	Sun ☉ ° '	Moon ☽ ° '	Mercury ☿ ° '	Venus ♀ ° '	Mars ♂ ° '	Jupiter ♃ ° '	Saturn ♄ ° '	Uranus ♅ ° '	Neptune ♆ ° '	Pluto ♇ ° '
01	03 S 29	02 S 40	15 S 40	12 N 18	06 N 13	22 S 40	21 S 23	14 N 19	05 S 22	22 S 39
02	03 52	02 N 16	16 04	11 57	06 09	22 39	21 23	14 18	05 23	22 39
03	04 15	07 06	16 28	11 36	06 05	22 39	21 23	14 18	05 24	22 39
04	04 38	11 40	16 50	11 14	06 02	22 38	21 23	14 17	05 24	22 39
05	05 01	15 49	17 10	10 52	05 58	22 38	21 22	14 16	05 25	22 39
06	05 24	19 22	17 28	10 30	05 54	22 37	21 22	14 16	05 25	22 39
07	05 47	22 08	17 44	10 07	05 51	22 37	21 22	14 15	05 26	22 39
08	06 10	23 57	17 58	09 44	05 47	22 36	21 22	14 14	05 26	22 39
09	06 33	24 37	18 10	09 21	05 43	22 35	21 22	14 14	05 27	22 39
10	06 55	24 01	18 19	08 57	05 39	22 35	21 22	14 13	05 27	22 39
11	07 18	22 06	18 26	08 33	05 36	22 34	21 22	14 12	05 28	22 39
12	07 40	18 52	18 30	08 09	05 32	22 33	21 22	14 11	05 29	22 39
13	08 03	14 28	18 31	07 44	05 28	22 32	21 22	14 11	05 29	22 39
14	08 25	09 04	18 28	07 20	05 25	22 32	21 21	14 10	05 30	22 39
15	08 47	03 N 02	18 21	06 55	05 21	22 31	21 21	14 09	05 30	22 39
16	09 09	03 S 18	18 11	06 29	05 18	22 30	21 21	14 09	05 31	22 39
17	09 31	09 28	17 57	06 04	05 15	22 29	21 20	14 08	05 31	22 39
18	09 53	15 02	17 38	05 38	05 12	22 29	21 20	14 07	05 31	22 39
19	10 15	19 35	17 14	05 12	05 09	22 28	21 20	14 06	05 32	22 39
20	10 36	22 46	16 46	04 46	05 06	22 27	21 19	14 06	05 32	22 39
21	10 57	24 27	16 14	04 20	05 04	22 26	21 19	14 05	05 33	22 39
22	11 19	24 35	15 38	03 53	05 01	22 25	21 19	14 04	05 33	22 39
23	11 40	23 19	14 58	03 27	04 59	22 24	21 18	14 03	05 34	22 39
24	12 00	20 51	14 15	03 00	04 57	22 23	21 18	14 02	05 34	22 39
25	12 21	17 27	13 31	02 33	04 55	22 22	21 17	14 02	05 35	22 39
26	12 41	13 22	12 45	02 06	04 54	22 21	21 17	14 01	05 35	22 39
27	13 02	08 49	12 00	01 38	04 52	22 20	21 17	14 00	05 35	22 39
28	13 22	03 S 59	11 17	01 11	04 51	22 19	21 16	13 59	05 36	22 39
29	13 42	00 N 59	10 37	00 44	04 50	22 18	21 16	13 59	05 36	22 39
30	14 01	05 53	10 01	00 16	04 50	22 16	21 15	13 58	05 36	22 39
31	14 S 21	10 N 35	09 S 29	00 S 11	04 N 49	22 S 15	21 S 15	13 N 57	05 S 37	22 S 39

ZODIAC SIGN ENTRIES

Date	h	m	Planets
01	02	47	☽ ♈
02	20	48	♀ ♍
03	15	12	☽ ♉
06	04	03	☽ ♊
08	15	45	☽ ♋
11	00	24	☽ ♌
13	04	56	☽ ♍
15	05	54	☽ ♎
17	05	05	☽ ♏
19	04	43	☽ ♐
21	06	44	☽ ♑
22	22	59	☉ ♏
23	12	17	☽ ♒
25	21	18	☽ ♓
28	01	33	☿ ♎
28	01	41	♀ ♎
28	08	45	☽ ♈
30	21	19	☽ ♉

LATITUDES

Date	Mercury ☿ ° '	Venus ♀ ° '	Mars ♂ ° '	Jupiter ♃ ° '	Saturn ♄ ° '	Uranus ♅ ° '	Neptune ♆ ° '	Pluto ♇ ° '
01	02 S 50	00 N 17	03 S 41	00 S 26	00 S 19	00 S 28	01 S 07	01 S 06
04	03 06	00 28	03 33	00 26	00 19	00 28	01 07	01 06
07	03 17	00 38	03 23	00 26	00 19	00 28	01 07	01 06
10	03 23	00 48	03 13	00 26	00 19	00 28	01 07	01 06
13	03 21	00 57	03 02	00 26	00 19	00 28	01 07	01 07
16	03 08	01 05	02 50	00 26	00 20	00 28	01 07	01 07
19	02 41	01 13	02 38	00 26	00 20	00 28	01 07	01 07
22	01 57	01 20	02 25	00 27	00 20	00 28	01 07	01 07
25	01 S 01	01 26	02 12	00 27	00 20	00 28	01 07	01 07
28	00 N 01	01 31	02 00	00 27	00 20	00 28	01 07	01 07
31	00 N 57	01 N 36	01 S 47	00 S 27	00 S 20	00 S 28	01 S 07	01 S 08

DATA

Julian Date	2459124
Delta T	+71 seconds
Ayanamsa	24° 08' 31"
Synetic vernal point	04° ♓ 58' 28"
True obliquity of ecliptic	23° 26' 13"

MOON'S PHASES, APSIDES AND POSITIONS ☽

Date	h	m	Phase	Longitude ° '	Eclipse Indicator
01	21	05	○	09 ♈ 08	
10	00	40	☾	17 ♋ 10	
16	19	31	●	23 ♎ 53	
23	13	23	☽	00 ♒ 36	
31	14	49	○	08 ♉ 38	

Day	h	m		
03	17	06	Apogee	
16	23	40	Perigee	
30	18	30	Apogee	
02	01	00	0N	
09	13	04	Max dec	24° N 37'
15	23	31	0S	
22	02	06	Max dec	24° S 42'
29	07	17	0N	

LONGITUDES

Date	Chiron ⚷ ° '	Ceres ⚳ ° '	Pallas ⚴ ° '	Juno ⚵ ° '	Vesta ⚶ ° '	Black Moon Lilith ⚸ ° '
01	07 ♈ 05	29 ♒ 31	14 ♑ 01	03 ♏ 36	21 ♌ 45	27 ♈ 47
11	06 ♈ 38	28 ♒ 46	15 ♑ 34	06 ♏ 56	25 ♌ 47	28 ♈ 54
21	06 ♈ 13	28 ♒ 38	17 ♑ 29	10 ♏ 19	29 ♌ 40	00 ♉ 01
31	05 ♈ 50	29 ♒ 05	19 ♑ 44	13 ♏ 43	03 ♍ 26	01 ♉ 08

All ephemeris data is given at 12.00 UT and the Moon's longitude is additionally given for 24.00 UT

ASPECTARIAN

h m Aspects

01 Thursday
01 38 ♂ ⊥ ♆
01 55 ☿ P ♆
02 39 ☽ Q ♃
08 23 ☽ ∥ ☉
10 29 ☽ ⊥ ♅
11 34 ☽ ± ♀
11 39 ☽ ⊼ ☿
11 46 ☽ Q ♇
13 38 ♀ ± ♇
17 29 ☽ Q ♄
21 05 ☽ ☍ ☉
22 29 ☽ ⊻ ♅

02 Friday
07 35 ☽ P ♀
13 31 ☉ ⊼ ♅
14 58 ☽ □ ♃
16 55 ☽ ⊻ ♆
20 32 ☽ # ☉

03 Saturday
00 00 ☽ □ ♇
03 25 ☽ # ♆
03 57 ☽ ☌ ♂
05 02 ☽ ⊥ ♆
05 07 ☿ Q ♃
05 47 ☽ □ ♄
06 59 ☽ ∥ ♂
11 02 ♃ ∥ ♇
17 13 ☽ △ ♀
23 14 ☽ ∠ ♆

04 Sunday
00 42 ♀ ± ♄
05 17 ☽ ☍ ☿
09 48 ☽ ∥ ♀
10 59 ☽ ☌ ♅
13 34 ♇ St D
15 15 ☽ ⊼ ☉

05 Monday
02 43 ☽ ∥ ♅
04 07 ☽ △ ♃
04 33 ☽ ± ☉
05 37 ☽ ✶ ♆
12 50 ☽ △ ♇
14 54 ♀ P ♃
15 19 ☽ ⊻ ♂
18 41 ☽ △ ♄
21 20 ☽ # ☿

06 Tuesday
00 26 ☽ P ☉
03 09 ☽ ⊥ ♂
05 52 ☽ Q ♆
10 40 ☽ P ♃
12 41 ☽ □ ♀
13 17 ☉ ∥ ♆
19 08 ☽ P ♇
20 50 ☽ ∠ ♂
22 11 ☽ ⊼ ☿
23 29 ☽ ⊻ ♅

07 Wednesday
00 57 ☽ P ♄
04 33 ☽ # ♄
04 51 ☽ ± ♃
09 18 ☽ △ ☉
10 53 ☽ ± ☿
11 26 ☽ ⊥ ♅
13 08 ☽ ± ♇
15 19 ☉ # ♂
16 53 ☽ ⊼ ♃
17 07 ☽ # ♃
17 35 ☽ # ♇
17 51 ☽ □ ♆
18 55 ☽ ± ♄
20 55 ☿ ☍ ♅

08 Thursday
01 01 ☽ ⊼ ♇
01 57 ☽ ✶ ♂
04 27 ☽ Q ♀
05 12 ☽ ∠ ♅
05 39 ☽ P ☿
06 46 ☽ ⊼ ♄

09 Friday
00 48 ☽ Q ♂
06 08 ♀ P ♇
06 16 ☽ ✶ ♀
07 38 ♀ P ♂
10 14 ☽ ✶ ♅
10 53 ☿ Q ♇
12 03 ☽ △ ☿
13 09 ♂ □ ♇

10 Saturday
00 40 ☽ □ ☉
03 17 ☽ ☍ ♃
03 44 ☽ △ ♆
08 49 ☽ Q ♅
10 05 ☽ □ ♂
10 36 ☽ ☍ ♇
13 27 ☽ ∠ ♀
16 04 ☽ ☍ ♄
23 08 ♀ △ ♅

11 Sunday
06 35 ☽ # ♇
07 15 ☽ P ♆
07 26 ☽ # ♃
07 31 ☽ ⊥ ♀
12 24 ☽ Q ☉
13 34 ☉ □ ♃
15 30 ☉ ⊼ ♆
17 27 ☽ □ ♅
18 26 ☽ # ♄
18 33 ♀ P ♄
19 16 ☽ ⊻ ♀
20 50 ☽ □ ☿
23 16 ☽ ± ♆

12 Monday
07 06 ♃ ✶ ♆
09 43 ☽ ⊼ ♆
09 45 ☽ ⊼ ♃
11 09 ☽ ✶ ☉
14 16 ☽ # ☿
14 29 ☽ △ ♂
16 11 ☽ ⊼ ♇
16 39 ☿ ✶ ♀
20 07 ☽ ± ♃
21 18 ☽ ⊼ ♄

13 Tuesday
02 24 ☽ ± ♇
02 38 ☉ # ♀
04 18 ☽ Q ☿
07 27 ☽ ± ♄
08 25 ♂ # ♆
11 24 ☽ P ♃
13 21 ☽ ∥ ♅
14 29 ☽ ∠ ☉
15 16 ☽ P ♂
17 28 ☽ P ♇
20 35 ☽ △ ♅
22 26 ☽ P ♄
23 26 ☉ ☍ ♂

14 Wednesday
00 19 ☽ ✶ ☿
01 04 ☿ St R
02 55 ☽ ☌ ♀
05 45 ☽ ± ♂
06 21 ☽ ⊥ ☉
11 47 ☽ ☍ ♆
12 12 ☽ △ ♃
14 33 ☽ # ☉
15 16 ☽ ⊼ ♂
16 46 ☽ ⊻ ☉
17 55 ☽ △ ♇
19 40 ☽ ∥ ♀
20 54 ☽ P ♅
22 47 ☽ △ ♄

15 Thursday
00 29 ☽ ∠ ☿
02 25 ☽ # ♆
02 53 ☽ ∥ ♂
07 12 ♀ ± ♂
10 15 ☉ □ ♇
11 12 ☽ ± ♅
14 38 ☽ ⊥ ☿
20 39 ☽ ⊼ ♅
23 57 ☽ ⊻ ☿

16 Friday
06 53 ☽ ⊻ ♀
11 20 ☽ ⊼ ♆
12 06 ☽ □ ♃
13 49 ☽ ☍ ♂
17 05 ☽ ⊥ ♀
17 22 ☽ □ ♇
19 31 ☽ ☌ ☉
19 34 ☽ # ♂
20 26 ☽ ∥ ♆
20 43 ☽ ± ♆
22 11 ☽ □ ♄
23 26 ☽ # ♀

17 Saturday
08 23 ☽ ∠ ♀
10 48 ☽ P ♆
12 13 ☽ ∥ ☉
13 38 ☉ ± ♆
16 32 ☽ Q ♃
19 37 ☽ ☍ ♅
21 39 ☽ Q ♇
21 52 ☽ ☌ ☿

18 Sunday
02 33 ☽ Q ♄
04 02 ☿ Q ♇
07 47 ☽ # ♅
10 05 ☽ ✶ ♀
10 28 ☽ △ ♆
11 39 ☽ ✶ ♃
12 08 ☽ ⊼ ♂
13 58 ☉ □ ♄
14 49 ♀ ☍ ♆
16 42 ☽ ✶ ♇
18 08 ♀ # ♆
21 32 ☽ ± ♂
21 43 ☽ ✶ ♄
22 15 ☽ ⊻ ☉
23 56 ☽ ∥ ☿

19 Monday
05 37 ♂ □ ♃
07 02 ☽ Q ♀
07 03 ♀ ⊼ ♂
07 35 ♀ △ ♃
08 38 ☽ ⊥ ☉
11 49 ☽ P ♂
11 59 ☽ ∠ ♃
15 31 ♀ ∥ ♂
17 00 ☽ ∠ ♇
19 40 ☽ ⊼ ♅
20 02 ☽ ⊻ ☿
22 11 ☽ ∠ ♄
23 45 ☽ ∥ ♄

20 Tuesday
00 28 ☽ ∠ ☉
02 54 ☿ ☍ ♅
02 54 ☽ ⊥ ♃
05 26 ☽ ⊥ ☿
05 34 ☽ ± ♅
07 54 ☽ ⊥ ♇
08 55 ☽ ∥ ♃
10 50 ☽ ∥ ♇
11 19 ☽ □ ♆
12 08 ☽ △ ♂
13 00 ☽ ⊻ ♃
13 13 ☽ ⊥ ♄
15 28 ☽ □ ♀
18 00 ☽ ⊻ ♇
19 40 ☽ ∠ ☿
20 40 ☽ P ♅
23 24 ☽ ⊻ ♄

21 Wednesday
03 38 ☽ ✶ ☉
15 33 ☿ Q ♃
18 07 ☽ Q ♆
19 49 ☽ ✶ ☿
20 30 ☿ ∠ ♀
21 42 ♀ △ ♇
22 33 ☽ △ ♅

22 Thursday
02 10 ☽ Q ☉
09 10 ♂ ⊻ ♆
15 16 ☽ □ ♂
15 23 ☽ ✶ ♆
15 33 ☽ Q ☿
17 44 ☽ ☌ ♃
22 40 ☽ ☌ ♇

23 Friday
01 09 ☽ △ ♀
03 05 ♀ P ♅
04 35 ☽ ☌ ♄
13 23 ☽ □ ☉
18 44 ☉ ⊥ ♀
18 48 ☽ ∠ ♆
19 33 ☽ ∥ ♇
21 49 ☽ □ ☿
22 15 ☽ ∥ ♃
23 39 ☽ Q ♂

24 Saturday
05 02 ☽ □ ♅
07 50 ☽ P ♀
08 16 ☽ ∥ ♄
11 40 ☽ ⊥ ♆
15 41 ♀ △ ♄
19 03 ☿ # ♅
21 54 ☽ ✶ ♂
23 05 ☽ ⊻ ♆

25 Sunday
02 13 ☽ ⊻ ♃
02 47 ☽ ± ♀
05 16 ☿ P ♆
06 57 ☽ ⊻ ♇
13 21 ☽ ⊻ ♄
13 54 ☽ ⊥ ♃
14 21 ☿ ⊥ ♀
15 20 ☽ Q ♅
15 37 ☽ ⊼ ♀
18 23 ☉ ☌ ☿
18 35 ☽ ⊥ ♇

26 Monday
01 04 ☽ ⊥ ♄
01 59 ☽ △ ☿
02 21 ☽ ∠ ♂
03 30 ☽ △ ☉
07 38 ☽ ∠ ♃
08 25 ☽ # ♅
10 34 ☉ P ♆
12 12 ☽ ∠ ♇
13 22 ☉ ∥ ☿
14 44 ☽ ✶ ♅
15 28 ☽ ∥ ☉
16 05 ☽ ∥ ☿
18 49 ☽ ∠ ♄
19 43 ☽ ⊥ ♂

27 Tuesday
04 49 ☽ P ☿
07 23 ☽ ⊻ ♂
09 40 ☽ ☌ ♆
11 47 ☽ P ☉
13 38 ☽ ✶ ♃
17 58 ☽ ✶ ♇
20 25 ☽ ∠ ♅
21 09 ☽ ± ☿

28 Wednesday
00 46 ☽ ✶ ♄
01 37 ☿ ⊻ ♀
04 06 ☽ ∥ ♆
07 24 ☽ ± ☉
07 43 ☽ # ♂
08 08 ☽ ⊼ ☿
09 32 ☽ ☍ ♀
13 55 ☽ Q ♃
14 23 ☽ ⊥ ♅
18 03 ☽ Q ♇
20 34 ☽ ⊼ ☉

29 Thursday
00 58 ☽ Q ♄
02 25 ☽ ⊻ ♅
02 55 ☽ # ♀
10 54 ☽ ∥ ♀
18 33 ☽ ☌ ♂
21 51 ☽ ⊻ ♆

30 Friday
02 39 ☽ □ ♃
06 26 ☽ □ ♇
06 47 ☽ ∥ ♂
07 54 ♀ ± ♅
07 59 ☉ # ♅
09 59 ☽ ⊥ ♆
10 37 ☽ # ♆
13 30 ☽ □ ♄
16 12 ☽ ☍ ☿

31 Saturday
04 11 ☽ ∠ ♆
05 02 ☽ ⊼ ♀
06 48 ☽ # ☿
14 49 ☽ ☍ ☉
14 55 ☽ ☌ ♅
15 53 ☉ ☍ ♅
18 34 ☽ ± ♀

NOVEMBER 2020

LONGITUDES

Date	Sidereal time h m s	Sun ☉ ° ' "	Moon ☽ ° ' "	Moon ☽ 24.00 ° ' "	Mercury ☿ ° '	Venus ♀ ° '	Mars ♂ ° '	Jupiter ♃ ° '	Saturn ♄ ° '	Uranus ♅ ° '	Neptune ♆ ° '	Pluto ♇ ° '
01	14 44 57	09 ♏ 31 19	19 ♉ 06 12	25 ♉ 02 36	26 ♎ 23	05 ♎ 23	16 ♈ 17	21 ♑ 07	26 ♑ 15	08 ♉ 39	18 ♓ 22	22 ♑ 41
02	14 48 53	10 31 21	00 ♊ 59 48	06 ♊ 58 06	26 R 03	06 36	16 R 07	21 15	26 18	08 R 36	18 R 21	22 42
03	14 52 50	11 31 25	12 ♊ 57 48	18 ♊ 59 12	25 54	07 49	15 59	21 24	26 21	08 34	18 21	22 42
04	14 56 47	12 31 32	25 ♊ 02 41	01 ♋ 08 38	25 D 57	09 02	15 51	21 33	26 25	08 32	18 20	22 43
05	15 00 43	13 31 40	07 ♋ 17 30	13 ♋ 29 44	26 11	10 15	15 43	21 42	26 28	08 29	18 19	22 44
06	15 04 40	14 31 50	19 ♋ 45 47	26 ♋ 06 09	26 35	11 29	15 37	21 51	26 32	08 27	18 19	22 45
07	15 08 36	15 32 02	02 ♌ 31 19	09 ♌ 01 44	27 08	12 42	15 31	22 00	26 35	08 24	18 18	22 46
08	15 12 33	16 32 16	15 ♌ 37 51	22 ♌ 20 02	27 50	13 55	15 26	22 09	26 39	08 22	18 17	22 47
09	15 16 29	17 32 33	29 ♌ 08 35	06 ♍ 03 41	28 39	15 09	15 22	22 19	26 43	08 19	18 16	22 48
10	15 20 26	18 32 51	13 ♍ 05 24	20 ♍ 13 39	29 ♎ 35	16 23	15 19	22 28	26 47	08 17	18 16	22 49
11	15 24 22	19 33 11	27 ♍ 28 09	04 ♎ 48 26	00 ♏ 37	17 36	15 17	22 38	26 51	08 14	18 15	22 50
12	15 28 19	20 33 33	12 ♎ 13 52	19 ♎ 43 36	01 44	18 50	15 15	22 48	26 55	08 12	18 14	22 51
13	15 32 16	21 33 57	27 ♎ 16 37	04 ♏ 51 47	02 56	20 04	15 14	22 58	26 59	08 10	18 14	22 53
14	15 36 12	22 34 23	12 ♏ 27 51	20 ♏ 03 32	04 11	21 18	15 D 14	23 08	27 04	08 07	18 13	22 54
15	15 40 09	23 34 51	27 ♏ 37 35	05 ♐ 08 46	05 29	22 31	15 15	23 18	27 08	08 05	18 13	22 55
16	15 44 05	24 35 20	12 ♐ 36 01	19 ♐ 58 22	06 51	23 45	15 16	23 28	27 12	08 03	18 12	22 56
17	15 48 02	25 35 51	27 ♐ 15 04	04 ♑ 25 32	08 14	24 59	15 19	23 39	27 17	08 00	18 12	22 57
18	15 51 58	26 36 23	11 ♑ 29 22	18 ♑ 26 23	09 40	26 13	15 22	23 49	27 21	07 58	18 12	22 59
19	15 55 55	27 36 57	25 ♑ 16 30	01 ♒ 59 51	11 07	27 28	15 26	24 00	27 26	07 56	18 11	23 00
20	15 59 51	28 37 32	08 ♒ 36 40	15 ♒ 07 16	12 35	28 42	15 31	24 10	27 31	07 53	18 11	23 01
21	16 03 48	29 ♏ 38 08	21 ♒ 32 05	27 ♒ 51 35	14 05	29 ♎ 56	15 36	24 21	27 35	07 51	18 11	23 02
22	16 07 45	00 ♐ 38 45	04 ♓ 06 16	10 ♓ 16 41	15 35	01 ♏ 10	15 42	24 32	27 40	07 49	18 10	23 04
23	16 11 41	01 39 23	16 ♓ 23 23	22 ♓ 26 55	17 07	02 24	15 49	24 43	27 45	07 47	18 10	23 05
24	16 15 38	02 40 02	28 ♓ 27 48	04 ♈ 26 35	18 38	03 39	15 57	24 54	27 50	07 45	18 10	23 07
25	16 19 34	03 40 42	10 ♈ 23 44	16 ♈ 19 45	20 11	04 53	16 05	25 06	27 55	07 42	18 10	23 08
26	16 23 31	04 41 24	22 ♈ 15 02	28 ♈ 10 01	21 43	06 07	16 14	25 17	28 00	07 40	18 10	23 09
27	16 27 27	05 42 07	04 ♉ 05 04	10 ♉ 00 30	23 16	07 22	16 24	25 28	28 06	07 38	18 10	23 11
28	16 31 24	06 42 51	15 ♉ 56 40	21 ♉ 53 49	24 49	08 36	16 34	25 40	28 11	07 36	18 10	23 12
29	16 35 20	07 43 36	27 ♉ 52 12	03 ♊ 52 05	26 22	09 51	16 45	25 51	28 16	07 34	18 D 10	23 14
30	16 39 17	08 ♐ 44 22	09 ♊ 53 41	15 ♊ 57 11	27 ♏ 56	11 ♏ 05	16 ♈ 57	26 ♑ 03	28 ♑ 22	07 ♉ 32	18 ♓ 10	23 ♑ 15

Moon Nodes and Latitude

Date	Moon True ☊ ° '	Moon Mean ☊ ° '	Moon ☽ Latitude ° '
01	20 ♊ 26	22 ♊ 04	02 S 41
02	20 R 22	22 01	01 43
03	20 20	21 58	00 S 40
04	20 D 20	21 54	00 N 26
05	20 21	21 51	01 31
06	20 23	21 48	02 34
07	20 24	21 45	03 30
08	20 25	21 42	04 17
09	20 R 24	21 39	04 52
10	20 22	21 35	05 10
11	20 19	21 32	05 10
12	20 15	21 29	04 50
13	20 10	21 26	04 10
14	20 07	21 23	03 12
15	20 04	21 20	02 00
16	20 03	21 16	00 N 41
17	20 D 03	21 13	00 S 40
18	20 04	21 10	01 55
19	20 05	21 07	03 02
20	20 06	21 04	03 57
21	20 07	21 00	04 37
22	20 R 08	20 57	05 03
23	20 07	20 54	05 15
24	20 05	20 51	05 12
25	20 03	20 48	04 56
26	20 01	20 45	04 27
27	19 58	20 41	03 47
28	19 56	20 38	02 57
29	19 55	20 35	01 59
30	19 ♊ 54	20 ♊ 32	00 S 55

DECLINATIONS

Date	Sun ☉ ° '	Moon ☽ ° '	Mercury ☿ ° '	Venus ♀ ° '	Mars ♂ ° '	Jupiter ♃ ° '	Saturn ♄ ° '	Uranus ♅ ° '	Neptune ♆ ° '	Pluto ♇ ° '
01	14 S 40	14 N 55	09 S 03	00 S 39	04 N 49	22 S 13	21 S 14	13 N 56	05 S 37	22 S 39
02	14 59	18 40	08 43	01 07	04 49	22 12	21 13	13 55	05 37	22 38
03	15 17	21 41	08 28	01 35	04 50	22 11	21 13	13 55	05 38	22 38
04	15 36	23 46	08 20	02 02	04 51	22 09	21 12	13 54	05 38	22 38
05	15 54	24 45	08 17	02 30	04 51	22 08	21 12	13 53	05 38	22 38
06	16 12	24 31	08 19	02 58	04 53	22 07	21 11	13 52	05 39	22 38
07	16 29	23 00	08 26	03 26	04 54	22 05	21 10	13 52	05 39	22 38
08	16 47	20 14	08 37	03 54	04 56	22 04	21 10	13 51	05 39	22 38
09	17 04	16 19	08 52	04 21	04 58	22 02	21 09	13 50	05 39	22 38
10	17 21	11 25	09 11	04 49	05 00	22 01	21 08	13 49	05 40	22 38
11	17 37	05 N 45	09 33	05 17	05 02	21 59	21 07	13 48	05 40	22 38
12	17 53	00 S 23	09 57	05 44	05 05	21 57	21 07	13 48	05 40	22 37
13	18 09	06 37	10 23	06 12	05 08	21 56	21 06	13 47	05 40	22 37
14	18 25	12 32	10 51	06 39	05 11	21 54	21 05	13 46	05 40	22 37
15	18 40	17 41	11 20	07 07	05 15	21 52	21 04	13 45	05 40	22 37
16	18 55	21 38	11 50	07 34	05 19	21 51	21 04	13 45	05 41	22 37
17	19 09	24 04	12 21	08 01	05 23	21 49	21 03	13 44	05 41	22 37
18	19 24	24 51	12 53	08 28	05 27	21 47	21 02	13 43	05 41	22 37
19	19 38	24 04	13 25	08 55	05 32	21 45	21 01	13 43	05 41	22 36
20	19 51	21 55	13 58	09 21	05 36	21 44	21 00	13 42	05 41	22 36
21	20 04	18 42	14 30	09 48	05 41	21 42	20 59	13 41	05 41	22 36
22	20 17	14 43	15 03	10 14	05 46	21 40	20 58	13 40	05 41	22 36
23	20 29	10 12	15 35	10 40	05 52	21 38	20 57	13 40	05 41	22 36
24	20 41	05 23	16 07	11 06	05 58	21 36	20 56	13 39	05 41	22 36
25	20 53	00 S 26	16 38	11 31	06 03	21 34	20 55	13 38	05 41	22 35
26	21 04	04 N 31	17 09	11 56	06 10	21 32	20 55	13 38	05 41	22 35
27	21 15	09 19	17 39	12 21	06 16	21 30	20 54	13 37	05 41	22 35
28	21 26	13 47	18 09	12 46	06 22	21 28	20 52	13 36	05 41	22 35
29	21 36	17 45	18 38	13 11	06 29	21 26	20 51	13 36	05 41	22 35
30	21 S 45	21 N 01	19 S 06	13 S 35	06 N 36	21 S 24	20 S 50	13 N 35	05 S 41	22 S 34

ZODIAC SIGN ENTRIES

Date	h	m	Planets
02	10	00	☽ ♊
04	21	45	☽ ♋
07	07	18	☽ ♌
09	13	30	☽ ♍
10	21	55	☿ ♏
11	16	09	☽ ♎
13	16	19	☽ ♏
15	15	47	☽ ♐
17	16	35	☽ ♑
19	20	25	☽ ♒
21	13	22	♀ ♏
21	20	40	☉ ♐
22	04	06	☽ ♓
24	15	05	☽ ♈
27	03	43	☽ ♉
29	16	16	☽ ♊

LATITUDES

Date	Mercury ☿ ° '	Venus ♀ ° '	Mars ♂ ° '	Jupiter ♃ ° '	Saturn ♄ ° '	Uranus ♅ ° '	Neptune ♆ ° '	Pluto ♇ ° '
01	01 N 13	01 N 37	01 S 43	00 S 27	00 S 20	00 S 28	01 S 07	01 S 08
04	01 49	01 40	01 30	00 27	00 20	00 28	01 06	01 08
07	02 10	01 43	01 19	00 27	00 21	00 28	01 06	01 08
10	02 17	01 45	01 07	00 27	00 21	00 28	01 06	01 08
13	02 14	01 46	00 56	00 27	00 21	00 28	01 06	01 08
16	02 04	01 47	00 46	00 27	00 21	00 28	01 06	01 09
19	01 50	01 46	00 36	00 27	00 21	00 28	01 06	01 09
22	01 32	01 45	00 26	00 28	00 21	00 28	01 06	01 09
25	01 12	01 44	00 18	00 28	00 21	00 28	01 06	01 09
28	00 51	01 41	00 09	00 28	00 22	00 28	01 06	01 09
31	00 N 30	01 N 38	00 S 02	00 S 28	00 S 22	00 S 28	01 S 06	01 S 10

DATA

Julian Date	2459155
Delta T	+71 seconds
Ayanamsa	24° 08' 35"
Synetic vernal point	04° ♓ 58' 24"
True obliquity of ecliptic	23° 26' 13"

MOON'S PHASES, APSIDES AND POSITIONS ☽

Date	h	m	Phase	Longitude ° '	Eclipse Indicator
08	13	46	☾	16 ♌ 37	
15	05	07	●	23 ♏ 18	
22	04	45	☽	00 ♓ 20	
30	09	30	○	08 ♊ 38	

Day	h	m		
14	11	36	Perigee	
27	00	21	Apogee	
05	19	29	Max dec	24° N 49'
12	10	33	0S	
18	11	35	Max dec	24° S 51'
25	14	03	0N	

LONGITUDES

Date	Chiron ⚷ ° '	Ceres ⚳ ° '	Pallas ⚴ ° '	Juno ⚵ ° '	Vesta ⚶ ° '	Black Moon Lilith ⚸ ° '
01	05 ♈ 47	29 ♒ 10	19 ♑ 58	14 ♏ 04	03 ♍ 48	01 ♉ 15
11	05 ♈ 28	00 ♓ 15	22 ♑ 30	17 ♏ 29	07 ♍ 21	02 ♉ 21
21	05 ♈ 13	01 ♓ 49	25 ♑ 15	20 ♏ 54	10 ♍ 38	03 ♉ 28
31	05 ♈ 02	03 ♓ 49	28 ♑ 10	24 ♏ 17	13 ♍ 38	04 ♉ 35

All ephemeris data is given at 12.00 UT and the Moon's longitude is additionally given for 24.00 UT

ASPECTARIAN

01 Sunday
00 34 ☉ Q ♃
06 22 ☽ ⊻ ♂
06 23 ☽ ∥ ♅
10 26 ☽ # ☉
10 31 ☽ ⚹ ♆
14 52 ☽ bQ ♀
16 07 ☽ △ ♃
18 20 ☽ ⊥ ♂
19 06 ☿ □ ♄
19 14 ☽ △ ♇

02 Monday
02 15 ☽ ⊼ ☿
02 29 ☽ △ ♄
10 43 ☽ Q ♆
12 15 ☽ ∠ ♂
14 04 ☽ ± ☿
16 09 ☉ Q ♇
22 42 ☽ bQ ♃

03 Tuesday
00 31 ☽ △ ♀
01 28 ☽ bQ ♇
03 14 ☽ ⊻ ♅
07 43 ☽ # ♄
07 54 ☽ bQ ☿
08 46 ☽ bQ ♄
08 52 ☽ ⊼ ☉
15 11 ☽ ⊥ ♅
16 41 ☽ # ♃
16 55 ☽ ± ♃
17 50 ☿ St D
17 56 ☽ ⚹ ♂
19 29 ☽ ± ♇
21 32 ☽ # ♇
21 55 ☽ ± ☉
22 43 ☽ □ ♆

04 Wednesday
02 22 ♀ ⊼ ♅
02 47 ☽ ± ♄
05 00 ☽ ⊼ ♃
07 24 ☽ ⊼ ♇
09 01 ☽ ∠ ♅
13 49 ☽ △ ☿
14 43 ☽ ⊼ ♄
17 20 ☽ bQ ☉
17 28 ☽ Q ♂

05 Thursday
14 19 ☽ ⚹ ♅
18 22 ☽ □ ♀

06 Friday
01 07 ☽ △ ☉
04 08 ☽ □ ♂
09 12 ☿ □ ♄
09 13 ☽ △ ♆
11 59 ☉ Q ♄
13 18 ☽ Q ♅
16 01 ☽ ☍ ♃
17 41 ☽ ☍ ♇

07 Saturday
00 52 ☽ ☍ ♄
01 27 ☽ □ ☿
08 15 ☽ Q ♀
11 41 ☉ ⊼ ♂
13 26 ☽ bQ ♆
15 57 ☽ # ♇
21 19 ☽ # ♃
22 49 ☽ □ ♅

08 Sunday
05 04 ☽ # ♄
05 56 ☽ ± ♆
08 36 ☽ ⚹ ♀
11 39 ☽ △ ♂
12 23 ☽ Q ☿
13 46 ☽ □ ☉
16 46 ☽ ⊼ ♆
23 49 ☽ ⊼ ♃

09 Monday
00 49 ☽ ⊼ ♇
07 44 ☽ ⊼ ♄
08 10 ☽ # ☉
10 32 ☽ ± ♃
11 05 ☽ ⚹ ☿
11 24 ☽ ± ♇
13 56 ☽ ∠ ♀
14 08 ☽ bQ ♂
16 08 ♀ ☍ ♂
18 15 ☽ ± ♄
23 58 ☽ Q ☉

10 Tuesday
00 45 ☽ ∥ ♅
02 19 ☽ bQ ♃
03 01 ☽ bQ ♇
03 50 ☽ △ ♅
05 11 ☉ △ ♆
05 36 ☽ ± ♂
06 57 ☽ ⊥ ♀
09 46 ☽ bQ ♄
14 43 ☽ ∠ ☿
15 45 ☽ ⊼ ♂
18 04 ☽ ⊻ ♀
20 42 ☽ ☍ ♆
21 13 ☽ # ☿
21 53 ☽ ⚹ ☉
22 14 ♀ # ♂

11 Wednesday
03 55 ☽ △ ♃
04 20 ☽ △ ♇
05 02 ☽ bQ ♅
06 55 ☽ ⊥ ☿
10 59 ☽ △ ♄
12 22 ☽ # ♆
13 45 ☽ # ♀
14 50 ☽ ∥ ♂
17 35 ☽ ⊻ ☿
19 48 ☽ ± ♅

12 Thursday
00 26 ☽ ∠ ☉
00 31 ♀ ⊼ ♆
05 31 ☽ ⊼ ♅
08 02 ♀ ∥ ♆
16 00 ☽ ⊥ ☉
16 50 ☽ ☍ ♂
21 37 ☽ ⊼ ♆
21 39 ♃ ☌ ♇
23 31 ☽ ☌ ♀

13 Friday
02 17 ☽ ⊻ ☉
04 11 ☉ ± ♂
05 00 ☽ □ ♇
05 05 ☽ □ ♃
06 14 ☽ # ♂
07 10 ☽ ± ♆
08 20 ☽ ∥ ♆
10 16 ☽ ∥ ♀
11 32 ☽ □ ♄
17 53 ☿ bQ ♆
21 25 ☽ bQ ♆
21 44 ☽ ☌ ☿

14 Saturday
00 36 ♂ St D
04 18 ☽ ∥ ☿
05 10 ☽ ☍ ♅
09 31 ☽ Q ♇
09 52 ☽ Q ♃
16 07 ☽ Q ♄
16 23 ☽ ⊼ ♂
17 23 ☽ # ♅
19 48 ☉ ⚹ ♇
21 05 ☽ △ ♆

15 Sunday
01 52 ☽ ± ♂
03 11 ☽ ⊻ ♀
03 57 ☉ ⚹ ♃
04 31 ☽ ⚹ ♇
05 03 ☽ ⚹ ♃
05 07 ☽ ☌ ☉
11 13 ☽ ⚹ ♄
13 33 ☽ ⊥ ♀
16 11 ☽ bQ ♂
17 37 ☽ ∥ ☉
19 43 ♀ □ ♇

16 Monday
01 48 ☽ ⊻ ☿
04 28 ☽ ∠ ♇
04 40 ☽ ⊼ ♅
05 14 ☽ ∠ ♀
05 16 ☽ ∠ ♃
05 33 ♀ □ ♃
07 56 ☽ ∥ ♄
11 21 ☽ ∠ ♄
12 26 ☽ ⊥ ☿
13 37 ☽ ∥ ♃
14 20 ☽ ± ♅
16 21 ☽ △ ♂
19 03 ☽ ⊥ ♇
19 55 ☽ ∥ ♇
20 00 ☽ ⊥ ♃
20 43 ♀ ± ♆
21 07 ☽ □ ♆

17 Tuesday
02 06 ☽ ⊥ ♄
04 39 ☽ ∠ ☿
04 53 ☽ ⊻ ♇
05 00 ☽ bQ ♅
05 58 ☽ ⊻ ♃
07 54 ☽ ⚹ ♀
08 07 ☿ ☍ ♅
09 03 ☽ ⊻ ☉
12 03 ☽ ⊻ ♄
19 48 ☽ ⊥ ☉

18 Wednesday
02 59 ☽ Q ♆
05 54 ☽ Q ♀
06 00 ☽ △ ♅
08 31 ☽ ⚹ ☿
12 13 ☽ ∠ ☉
18 42 ☽ □ ♂
23 34 ☽ ⚹ ♆

19 Thursday
07 18 ☉ ⚹ ♄
07 44 ☽ Q ☿
07 58 ☽ ☌ ♇
09 43 ☽ ☌ ♃
10 02 ☿ Q ♇
11 29 ♀ □ ♄
15 51 ☽ ☌ ♄
16 16 ☽ □ ♀
16 30 ☽ ⚹ ☉

20 Friday
00 23 ☿ # ♅
02 08 ☽ ∠ ♆
02 40 ☽ Q ♂
04 21 ☿ Q ♃
04 44 ☉ ⊻ ♀
05 34 ☽ ∥ ♇
10 41 ☽ □ ♅
13 37 ☽ ∥ ♃
16 00 ☽ Q ☉
18 34 ☽ ⊥ ♆
19 34 ☽ ∥ ♄
20 15 ☽ □ ☿

21 Saturday
00 49 ☽ ⚹ ♂
03 07 ☽ ∥ ☉
05 42 ☽ ⊻ ♆
11 31 ♂ # ♆
14 51 ☽ ⊻ ♇
17 24 ☽ ⊻ ♃
20 09 ☽ Q ♅
23 34 ☽ ⊻ ♄

22 Sunday
02 17 ☽ ⊥ ♇
04 45 ☽ □ ☉
05 01 ☽ ⊥ ♃
05 24 ☽ ∠ ♂
05 43 ☽ △ ♀
10 20 ☽ ∥ ☿
11 10 ☽ ⊥ ♄
13 24 ☿ Q ♄
13 36 ☉ bQ ♂
14 00 ☿ ⊼ ♂
17 44 ☽ # ♅
19 11 ☽ ⚹ ♅
19 42 ☽ ∠ ♇
22 43 ☽ ∠ ♃
22 59 ☽ ⊥ ♂

23 Monday
04 48 ☽ ∠ ♄
09 50 ☽ ∥ ♀
10 52 ☽ ⊻ ♂
13 37 ☽ △ ☿
14 14 ☽ bQ ♀
15 31 ☽ ☌ ♆

24 Tuesday
00 37 ☽ ∠ ♅
01 18 ☽ ⚹ ♇
02 50 ♀ bQ ♆
04 40 ☿ △ ♆
04 47 ☽ ⚹ ♃
09 15 ☽ # ♂
10 10 ☽ ± ♀
10 31 ☽ ∥ ♆
10 44 ☽ ⚹ ♄
18 33 ☽ ⊥ ♅
21 13 ☽ △ ☉
23 36 ☽ ⊼ ♀
23 55 ☽ bQ ☿

25 Wednesday
01 22 ☽ Q ♇
05 14 ☽ Q ♃
06 35 ☽ ⊻ ♅
11 02 ☽ Q ♄
16 32 ☉ ∥ ♄
20 47 ☽ ± ☿
23 39 ☽ ☌ ♂

26 Thursday
03 43 ☽ ⊻ ♆
06 19 ☽ bQ ☉
10 46 ☽ ⊼ ☿
13 50 ☽ □ ♇
15 53 ☽ ⊥ ♆
17 44 ☽ # ♆
18 15 ☽ □ ♃
20 14 ☽ ∥ ♂
21 00 ☿ ± ♂
23 46 ☽ □ ♄

27 Friday
02 17 ☽ ± ☉
10 08 ☽ ∠ ♆
10 38 ☿ ⚹ ♇
15 35 ☽ ⊼ ☉
17 11 ♀ ☍ ♅
19 10 ☽ ☌ ♅
19 25 ☽ ☍ ♀

28 Saturday
05 46 ☽ # ♀
11 01 ☽ ∥ ♅
13 17 ☽ ⊻ ♂
16 04 ☉ ∥ ♃
16 29 ☽ ⚹ ♆

29 Sunday
00 34 ♆ St D
01 34 ☽ ⊥ ♂
02 40 ☽ △ ♇
02 51 ☿ ⚹ ♃
07 53 ☽ △ ♃
08 20 ☉ ⊼ ♅
08 33 ☽ ☍ ☿
12 48 ☽ △ ♄
16 36 ☽ Q ♆
18 51 ☽ # ☿
19 55 ☽ ∠ ♂

30 Monday
00 14 ☉ ∠ ♇
07 19 ☽ ⊻ ♅
08 44 ☽ bQ ♇
09 30 ☽ ☍ ☉
10 30 ☽ # ♄
12 20 ♀ # ♅
14 20 ☽ bQ ♃
14 38 ☽ ⊼ ♀
15 08 ☽ # ♃
15 23 ♀ Q ♇
18 52 ☽ # ☉
18 55 ☽ bQ ♄
19 01 ☿ ⚹ ♄
19 12 ☽ ⊥ ♅

DECEMBER 2020

LONGITUDES

Date	Sidereal time h m s	Sun ☉ ° ' "	Moon ☽ ° ' "	Moon ☽ 24.00 ° ' "	Mercury ☿ ° '	Venus ♀ ° '	Mars ♂ ° '	Jupiter ♃ ° '	Saturn ♄ ° '	Uranus ♅ ° '	Neptune ♆ ° '	Pluto ♇ ° '
01	16 43 14	09 ♐ 45 10	22 ♊ 02 47	28 ♊ 10 42	29 ♏ 29	12 ♏ 20	17 ♈ 09	26 ♑ 15	28 ♑ 27	07 ♉ 30	18 ♓ 10	23 ♑ 17
02	16 47 10	10 45 59	04 ♋ 21 07	10 ♋ 34 14	01 ♐ 03	13 34	17 22	26 27	28 33	07 R 28	18 10	23 18
03	16 51 07	11 46 49	16 ♋ 50 16	23 ♋ 09 24	02 37	14 49	17 36	26 38	28 38	07 26	18 10	23 20
04	16 55 03	12 47 40	29 ♋ 31 54	05 ♌ 57 57	04 10	16 03	17 50	26 50	28 44	07 24	18 10	23 22
05	16 59 00	13 48 33	12 ♌ 27 48	19 ♌ 01 41	05 44	17 18	18 04	27 03	28 49	07 22	18 10	23 23
06	17 02 56	14 49 27	25 ♌ 39 48	02 ♍ 22 21	07 18	18 33	18 20	27 15	28 55	07 21	18 11	23 25
07	17 06 53	15 50 22	09 ♍ 09 30	16 ♍ 01 20	08 51	19 48	18 35	27 27	29 01	07 19	18 11	23 26
08	17 10 49	16 51 19	22 ♍ 57 55	29 ♍ 59 14	10 25	21 02	18 52	27 39	29 07	07 17	18 11	23 28
09	17 14 46	17 52 16	07 ♎ 05 08	14 ♎ 15 25	11 59	22 17	19 08	27 52	29 13	07 15	18 12	23 30
10	17 18 43	18 53 15	21 ♎ 29 44	28 ♎ 47 37	13 33	23 32	19 26	28 04	29 19	07 14	18 12	23 32
11	17 22 39	19 54 15	06 ♏ 08 29	13 ♏ 31 39	15 07	24 47	19 44	28 17	29 25	07 12	18 12	23 33
12	17 26 36	20 55 16	20 ♏ 56 18	28 ♏ 21 34	16 41	26 02	20 02	28 29	29 31	07 10	18 13	23 35
13	17 30 32	21 56 19	05 ♐ 46 31	13 ♐ 10 11	18 15	27 17	20 21	28 42	29 37	07 09	18 13	23 37
14	17 34 29	22 57 22	20 ♐ 31 39	27 ♐ 50 00	19 49	28 31	20 40	28 55	29 43	07 07	18 14	23 38
15	17 38 25	23 58 26	05 ♑ 04 24	12 ♑ 14 10	21 23	29 ♏ 46	21 00	29 08	29 49	07 06	18 14	23 40
16	17 42 22	24 59 31	19 ♑ 18 39	26 ♑ 17 26	22 57	01 ♐ 01	21 20	29 20	29 ♑ 56	07 04	18 15	23 42
17	17 46 18	26 00 36	03 ♒ 10 10	09 ♒ 56 41	24 32	02 16	21 40	29 33	00 ♒ 02	07 03	18 16	23 44
18	17 50 15	27 01 41	16 ♒ 36 57	23 ♒ 11 04	26 06	03 31	22 01	29 46	00 08	07 02	18 16	23 46
19	17 54 12	28 02 47	29 ♒ 39 14	06 ♓ 01 45	27 41	04 46	22 23	29 ♑ 59	00 15	07 00	18 17	23 48
20	17 58 08	29 ♐ 03 53	12 ♓ 19 01	18 ♓ 31 30	29 ♐ 16	06 01	22 45	00 ♒ 13	00 21	06 59	18 18	23 49
21	18 02 05	00 ♑ 05 00	24 ♓ 39 44	00 ♈ 44 14	00 ♑ 51	07 16	23 07	00 26	00 27	06 58	18 18	23 51
22	18 06 01	01 06 06	06 ♈ 45 37	12 ♈ 44 27	02 26	08 31	23 29	00 39	00 34	06 57	18 19	23 53
23	18 09 58	02 07 13	18 ♈ 41 23	24 ♈ 36 58	04 02	09 46	23 52	00 52	00 40	06 56	18 20	23 55
24	18 13 54	03 08 20	00 ♉ 31 49	06 ♉ 26 29	05 37	11 01	24 16	01 06	00 47	06 54	18 21	23 57
25	18 17 51	04 09 27	12 ♉ 21 31	18 ♉ 17 25	07 13	12 17	24 39	01 19	00 54	06 53	18 22	23 59
26	18 21 47	05 10 34	24 ♉ 14 41	00 ♊ 13 43	08 50	13 32	25 04	01 32	01 00	06 52	18 23	24 01
27	18 25 44	06 11 42	06 ♊ 14 56	12 ♊ 18 40	10 26	14 47	25 28	01 46	01 07	06 52	18 24	24 03
28	18 29 41	07 12 49	18 ♊ 25 12	24 ♊ 34 49	12 03	16 02	25 53	01 59	01 14	06 51	18 25	24 04
29	18 33 37	08 13 57	00 ♋ 47 40	07 ♋ 03 55	13 39	17 17	26 18	02 13	01 20	06 50	18 26	24 06
30	18 37 34	09 15 05	13 ♋ 23 40	19 ♋ 46 57	15 17	18 32	26 43	02 26	01 27	06 49	18 27	24 08
31	18 41 30	10 ♑ 16 13	26 ♋ 13 47	02 ♌ 44 07	16 ♑ 54	19 ♐ 47	27 ♈ 09	02 ♒ 40	01 ♒ 34	06 ♉ 48	18 ♓ 28	24 ♑ 10

Date	Moon True ☊ ° '	Moon Mean ☊ ° '	Moon ☽ Latitude ° '
01	19 ♊ 54	20 ♊ 29	00 N 12
02	19 D 54	20 25	01 19
03	19 55	20 22	02 24
04	19 55	20 19	03 23
05	19 56	20 16	04 12
06	19 57	20 13	04 50
07	19 57	20 10	05 12
08	19 R 57	20 06	05 17
09	19 57	20 03	05 04
10	19 57	20 00	04 31
11	19 56	19 57	03 40
12	19 D 57	19 54	02 34
13	19 57	19 51	01 N 18
14	19 R 57	19 47	00 S 03
15	19 57	19 44	01 23
16	19 56	19 41	02 36
17	19 56	19 38	03 38
18	19 55	19 35	04 25
19	19 54	19 31	04 58
20	19 53	19 28	05 14
21	19 52	19 25	05 16
22	19 D 53	19 22	05 04
23	19 53	19 19	04 38
24	19 54	19 16	04 01
25	19 55	19 12	03 14
26	19 57	19 09	02 18
27	19 58	19 06	01 15
28	19 59	19 03	00 S 09
29	19 R 58	19 00	00 N 59
30	19 57	18 57	02 06
31	19 ♊ 54	18 ♊ 53	03 N 07

DECLINATIONS

Date	Sun ☉ ° '	Moon ☽ ° '	Mercury ☿ ° '	Venus ♀ ° '	Mars ♂ ° '	Jupiter ♃ ° '	Saturn ♄ ° '	Uranus ♅ ° '	Neptune ♆ ° '	Pluto ♇ ° '
01	21 S 55	23 N 24	19 S 33	13 S 59	06 N 43	21 S 21	20 S 49	13 N 34	05 S 41	22 S 34
02	22 03	24 41	20 00	14 22	06 50	21 19	20 48	13 34	05 41	22 34
03	22 12	24 45	20 25	14 45	06 57	21 17	20 47	13 33	05 41	22 34
04	22 20	23 33	20 50	15 08	07 05	21 15	20 46	13 33	05 41	22 34
05	22 27	21 06	21 14	15 30	07 13	21 12	20 45	13 32	05 41	22 33
06	22 34	17 30	21 37	15 52	07 20	21 10	20 44	13 31	05 41	22 33
07	22 41	12 57	21 58	16 14	07 29	21 08	20 43	13 31	05 41	22 33
08	22 47	07 39	22 19	16 35	07 37	21 05	20 42	13 30	05 41	22 33
09	22 53	01 N 50	22 39	16 56	07 45	21 03	20 40	13 30	05 40	22 33
10	22 58	04 S 11	22 57	17 16	07 53	21 00	20 39	13 29	05 40	22 32
11	23 03	10 06	23 14	17 36	08 02	20 58	20 38	13 29	05 40	22 32
12	23 08	15 30	23 31	17 56	08 11	20 55	20 37	13 28	05 40	22 32
13	23 11	19 59	23 45	18 15	08 20	20 53	20 36	13 28	05 40	22 32
14	23 15	23 09	23 59	18 33	08 28	20 50	20 34	13 27	05 39	22 31
15	23 18	24 43	24 12	18 51	08 38	20 48	20 33	13 27	05 39	22 31
16	23 21	24 37	24 23	19 09	08 47	20 45	20 32	13 26	05 39	22 31
17	23 23	22 58	24 33	19 26	08 56	20 42	20 30	13 26	05 39	22 31
18	23 24	20 04	24 42	19 42	09 05	20 40	20 29	13 26	05 38	22 30
19	23 25	16 14	24 49	19 58	09 15	20 37	20 28	13 25	05 38	22 30
20	23 26	11 47	24 55	20 13	09 25	20 34	20 27	13 25	05 38	22 30
21	23 26	06 57	25 00	20 28	09 34	20 32	20 25	13 24	05 37	22 30
22	23 26	01 S 58	25 03	20 42	09 44	20 29	20 24	13 24	05 37	22 29
23	23 25	03 N 02	25 05	20 56	09 54	20 26	20 23	13 24	05 37	22 29
24	23 24	07 54	25 05	21 09	10 04	20 23	20 21	13 23	05 36	22 29
25	23 22	12 28	25 04	21 21	10 14	20 20	20 20	13 23	05 36	22 29
26	23 20	16 36	25 01	21 33	10 24	20 17	20 18	13 23	05 36	22 28
27	23 18	20 07	24 58	21 44	10 34	20 14	20 17	13 23	05 35	22 28
28	23 14	22 47	24 52	21 54	10 44	20 11	20 16	13 22	05 35	22 28
29	23 11	24 26	24 45	22 04	10 55	20 08	20 14	13 22	05 34	22 28
30	23 07	24 51	24 37	22 13	11 05	20 05	20 13	13 22	05 34	22 27
31	23 S 02	23 N 58	24 S 27	22 S 22	11 N 15	20 S 02	20 S 11	13 N 22	05 S 33	22 S 27

ZODIAC SIGN ENTRIES

Date	h	m	Planets
01	19	51	☿ ♐
02	03	33	☽ ♋
04	12	53	☽ ♌
06	19	46	☽ ♍
09	00	01	☽ ♎
11	01	59	☽ ♏
13	02	39	☽ ♐
15	03	35	☽ ♑
15	16	21	♀ ♐
17	05	04	♄ ♒
17	06	27	☽ ♒
19	12	39	☽ ♓
19	13	07	♃ ♒
20	23	07	☿ ♑
21	10	02	☉ ♑
21	22	32	☽ ♈
24	10	55	☽ ♉
26	23	33	☽ ♊
29	10	28	☽ ♋
31	18	58	☽ ♌

LATITUDES

Date	Mercury ☿ ° '	Venus ♀ ° '	Mars ♂ ° '	Jupiter ♃ ° '	Saturn ♄ ° '	Uranus ♅ ° '	Neptune ♆ ° '	Pluto ♇ ° '
01	00 N 30	01 N 38	00 S 02	00 S 28	00 S 22	00 S 28	01 S 06	01 S 10
04	00 N 09	01 35	00 N 06	00 28	00 22	00 28	01 06	01 10
07	00 S 12	01 30	00 12	00 28	00 22	00 28	01 06	01 10
10	00 32	01 26	00 19	00 28	00 22	00 28	01 05	01 10
13	00 51	01 20	00 24	00 28	00 22	00 28	01 05	01 10
16	01 08	01 15	00 30	00 29	00 22	00 27	01 05	01 11
19	01 24	01 08	00 35	00 29	00 23	00 27	01 05	01 11
22	01 38	01 02	00 40	00 29	00 23	00 27	01 05	01 11
25	01 50	00 55	00 44	00 29	00 23	00 27	01 05	01 11
28	01 59	00 48	00 48	00 29	00 23	00 27	01 05	01 11
31	02 S 05	00 N 41	00 N 52	00 S 29	00 S 23	00 S 27	01 S 05	01 S 12

DATA

Julian Date	2459185
Delta T	+71 seconds
Ayanamsa	24° 08' 39"
Synetic vernal point	04° ♓ 58' 20"
True obliquity of ecliptic	23° 26' 13"

LONGITUDES

Date	Chiron ⚷ ° '	Ceres ⚳ ° '	Pallas ⚴ ° '	Juno ⚵ ° '	Vesta ⚶ ° '	Black Moon Lilith ⚸ ° '
01	05 ♈ 02	03 ♓ 49	28 ♑ 10	24 ♏ 17	13 ♍ 38	04 ♉ 35
11	04 ♈ 57	06 ♓ 11	01 ♒ 14	27 ♏ 37	16 ♍ 15	05 ♉ 42
21	04 ♈ 57	08 ♓ 53	04 ♒ 25	00 ♐ 54	18 ♍ 26	06 ♉ 49
31	05 ♈ 03	11 ♓ 50	07 ♒ 41	04 ♐ 06	20 ♍ 05	07 ♉ 56

MOON'S PHASES, APSIDES AND POSITIONS ☽

Date	h	m	Phase	Longitude ° '	Eclipse Indicator
08	00	37	☾	16 ♍ 22	
14	16	17	●	23 ♐ 08	Total
21	23	41	☽	00 ♈ 35	
30	03	28	○	08 ♋ 53	

Day	h	m		
12	20	34	Perigee	
24	16	28	Apogee	
03	01	22	Max dec	24° N 53'
09	19	21	0S	
15	22	26	Max dec	24° S 53'
22	21	24	0N	
30	07	53	Max dec	24° N 52'

ASPECTARIAN

h m	Aspects
01 Tuesday	
02 13	☽ ⚹ ♂
02 27	☽ # ♇
02 36	☽ ± ♇
03 51	☽ ± ♀
04 22	☽ □ ♆
08 24	☽ ± ♃
12 48	☽ ± ♄
12 53	☽ ∠ ♅
14 26	☽ ⊼ ♇
20 22	☽ ⊼ ♃
23 30	☽ ⚼ ♀
02 Wednesday	
00 38	☽ ⊼ ♄
02 09	☽ Q ♂
04 40	☽ ⊼ ☿
17 58	☽ ± ☿
18 00	☽ ⚹ ♅
03 Thursday	
01 28	☽ ⊼ ☉
07 43	☽ △ ♀
07 54	☉ ∠ ♃
08 03	♀ Q ♃
11 45	☿ ⚼ ♂
13 28	☽ □ ♂
13 41	☽ ⚼ ☿
13 57	☽ ± ☉
14 32	☽ △ ♆
16 56	☽ Q ♅
04 Friday	
00 22	☽ ☍ ♇
06 52	☽ ☍ ♃
08 09	☿ ∥ ♄
08 28	☽ ⚼ ☉
10 29	☽ ☍ ♄
18 48	☽ ⚼ ♆
21 52	☽ △ ☿
23 09	☽ # ♇
05 Saturday	
00 44	☽ # ☉
01 59	☉ ± ♅
02 02	♀ Q ♄
02 38	☽ □ ♅
10 32	☿ ∥ ♃
11 02	☽ # ☿
11 06	☽ # ♃
11 28	☽ ± ♆
12 23	☉ ∠ ♄
14 40	☽ # ♄
14 41	☽ △ ☉
21 44	♂ ⊻ ♆
21 47	☽ □ ♀
22 27	☽ ⊼ ♆
22 28	☽ △ ♂
06 Sunday	
04 53	♀ △ ♆
06 41	♀ ⊼ ♂
07 56	☽ ⊼ ♇
08 03	☉ ∥ ♇
12 43	☿ ⊼ ♅
14 53	☽ ⊼ ♃
17 53	☽ ⊼ ♄
18 44	☽ ± ♇
20 27	☽ # ♀
07 Monday	
01 45	☽ ± ♃
01 58	☽ ⚼ ♂
04 38	☽ ± ♄
05 29	☿ ∠ ♇
08 45	☽ △ ♅
09 14	☽ ∥ ♅
09 21	☽ Q ♀
10 44	☽ ⚼ ♇
11 24	☽ □ ☿
17 51	☽ ⚼ ♃
18 08	☽ ± ♂
20 34	☽ ⚼ ♄
08 Tuesday	
00 37	☽ □ ☉
03 45	☽ ☍ ♆
04 47	☽ ⊼ ♂
08 21	☽ ⚹ ♀
10 50	☽ ⚼ ♅
12 08	☽ ∥ ♂
12 52	☽ △ ♇
20 09	☽ △ ♃
20 19	☽ # ♆
22 30	☽ Q ☿
22 35	☽ △ ♄
09 Wednesday	
02 10	☽ ± ♅
02 55	☉ ⊥ ♇
04 31	☿ ∥ ♇
09 48	☽ Q ☉
12 17	☽ ⊼ ♅
12 22	☽ ∠ ♀
19 41	☉ □ ♆
21 13	☽ ⚹ ☿
10 Thursday	
03 30	☿ ∠ ♃
04 49	☽ ⊥ ♀
06 33	☽ ⊼ ♆
07 09	☿ ± ♅
07 22	☽ ⚹ ☉
08 31	☽ ☍ ♂
11 52	♀ ⚹ ♇
14 19	☉ ∥ ☿
15 21	☽ □ ♇
15 40	☽ ⊻ ♀
16 28	☽ ± ♆
17 54	☽ ∥ ♆
22 58	☽ □ ♃
11 Friday	
00 30	☿ ∠ ♄
00 56	☽ □ ♄
00 59	☽ ∠ ☿
03 14	☽ # ♂
06 01	☉ △ ♂
07 13	☽ ⚼ ♆
09 50	☽ ∠ ☉
13 43	☽ ☍ ♅
17 24	☽ ⊥ ☿
20 49	☽ Q ♇
12 Saturday	
01 32	☽ ⊥ ☉
02 36	☽ # ♅
04 18	☽ ⊻ ☿
04 42	☽ Q ♃
06 25	☽ Q ♄
07 35	☽ △ ♆
10 30	☽ ⊼ ♂
11 58	☽ ⊻ ☉
12 04	♀ ± ♂
16 17	☽ ⚹ ♇
20 25	☽ ± ♂
20 59	☽ ☌ ♀
13 Sunday	
00 23	☽ ⚹ ♃
01 09	☽ ∥ ♀
01 58	☽ ⚹ ♄
02 06	☿ ⊥ ♇
11 17	☽ ⚼ ♂
11 38	☿ □ ♆
14 13	☽ ⊼ ♅
15 49	☽ ∥ ♄
16 36	☽ ∠ ♇
16 48	☉ ⚼ ♅
17 42	☽ ∥ ♃
23 57	☽ ± ♅
14 Monday	
01 03	☽ ∠ ♃
02 27	☽ ∠ ♄
06 11	☽ ∥ ♇
07 16	☽ ⊥ ♇
08 15	☽ □ ♆
10 40	☉ ⊥ ♃
10 42	☽ ☌ ☿
12 14	☽ △ ♂
13 01	☽ ∥ ☉
14 36	☽ ⚼ ♅
15 58	☽ ⊥ ♃
16 17	☽ ☌ ☉
17 07	☽ ⊻ ♇
17 16	☽ ⊥ ♄
20 58	♀ ⚹ ♃
23 05	☽ ∥ ☿
15 Tuesday	
01 59	☽ ⊻ ♃
02 23	☽ ⊻ ♀
03 13	☽ ⊻ ♄
04 24	☿ △ ♂
04 39	☉ ⊻ ♇
07 59	☉ ⊥ ♄
13 00	♀ ⚹ ♄
13 17	☽ ⊥ ♀
13 57	☽ Q ♆
15 22	☽ △ ♅
22 43	☿ ⚼ ♅
16 Wednesday	
05 52	☽ ∠ ♀
10 11	☽ ⚹ ♆
15 32	☽ □ ♂
16 36	☽ ∥ ☿
18 47	☿ ⊥ ♃
19 02	☽ ⊻ ☿
19 33	☽ ☌ ♇
22 32	☽ ⊻ ☉
23 35	☿ ⊻ ♇
17 Thursday	
03 50	☿ ⊥ ♄
05 34	☽ ☌ ♃
06 28	☽ ☌ ♄
06 47	☽ ⊥ ☿
07 42	☽ ∥ ☉
09 48	☽ ⊥ ☉
10 16	☽ ⚹ ♀
12 10	☽ ∠ ♆
16 33	☽ ∥ ♇
18 50	☽ □ ♅
23 49	☽ Q ♂
18 Friday	
00 45	☽ ∠ ☿
03 03	☽ ∠ ☉
04 10	☽ ⊥ ♆
07 40	☽ ∥ ♃
09 00	☽ ∥ ♄
09 49	☽ Q ♀
14 21	☽ ∥ ♀
15 01	☽ ⊻ ♆
22 08	☽ ⚹ ♂
19 Saturday	
01 06	☽ ⊻ ♇
03 22	☽ Q ♅
07 49	☽ ⚹ ☿
08 45	☽ ⚹ ☉
12 16	☽ ⊥ ♇
12 38	☽ ⊻ ♃
13 07	☽ ⊻ ♄
22 40	☽ □ ♀
20 Sunday	
00 08	☽ ⊥ ♃
00 31	☽ ⊥ ♄
01 50	☽ ⚹ ♅
03 00	☽ ∠ ♂
03 26	☉ ☌ ☿
03 28	☽ # ♅
05 18	☽ ∠ ♇
09 23	☽ Q ☉
09 42	☽ Q ☿
17 40	☽ ∠ ♃
17 54	☽ ∠ ♄
20 48	☽ ⊥ ♂
23 33	☽ # ♂
23 34	☽ ☌ ♆
21 Monday	
04 34	☿ ⊻ ♃
05 37	☿ ⊻ ♄
06 12	♀ ⊼ ♅
06 43	☽ ∠ ♅
07 43	♀ ∥ ♄
08 52	☽ ⊻ ♂
10 24	☽ ⚹ ♇
16 55	♀ ∥ ♃
18 21	♃ ☌ ♄
18 27	☽ ∥ ♆
21 52	☉ ⊻ ♄
22 22	☉ ⊻ ♃
23 33	☽ ⚹ ♄
23 36	☽ ⚹ ♃
23 41	☽ □ ☉
22 Tuesday	
00 26	☽ ⊥ ♅
02 04	☽ □ ☿
10 15	☽ Q ♇
11 07	♀ ⚼ ♂
12 22	☽ ⊻ ♅
15 56	☽ △ ♀
19 08	♀ ∠ ♇
23 45	☽ Q ♄
23 Wednesday	
00 02	☽ Q ♃
11 17	☽ ⊻ ♆
14 53	♂ □ ♇
22 37	☽ □ ♇
22 51	☽ ☌ ♂
23 27	☽ ⊥ ♆
24 Thursday	
00 35	☽ # ♆
01 46	☽ ⚼ ♀
12 31	☽ □ ♄
13 10	☽ □ ♃
17 30	♂ ⊥ ♆
17 44	☽ ∠ ♆
17 48	☽ △ ☉
22 12	☽ ± ♀
23 00	☿ Q ♆
23 35	☽ ∥ ♂
23 57	☽ △ ☿
25 Friday	
00 56	☽ ☌ ♅
07 05	☿ △ ♅
11 49	☽ ⊼ ♀
17 03	☽ ∥ ♅
18 51	♃ ∥ ♄
23 40	♀ ± ♅
26 Saturday	
00 10	☽ ⚹ ♆
03 02	☽ ⚼ ☉
11 02	☽ ⚼ ☿
11 32	☽ △ ♇
13 42	☽ ⊻ ♂
22 49	☽ ± ☉
27 Sunday	
00 19	☽ Q ♆
01 41	☽ △ ♄
02 08	☽ ⊥ ♂
02 54	☽ △ ♃
07 50	☽ ± ☿
11 53	☽ ⊼ ☉
12 57	☽ # ♃
13 13	☽ ⊻ ♅
13 18	☽ # ♄
16 48	☉ Q ♆
17 33	☽ ⚼ ♇
20 39	☽ ∠ ♂
21 33	☽ ⊼ ☿
28 Monday	
01 04	☽ ⊥ ♅
02 28	☽ # ♀
03 25	☉ △ ♅
06 47	☽ ☍ ♀
07 40	☽ ⚼ ♄
08 34	☽ # ♇
09 09	☽ ⚼ ♃
11 19	☽ ± ♇
11 59	☽ □ ♆
16 12	♀ ∠ ♄
17 04	☽ # ☉
18 41	☽ ∠ ♅
23 03	☽ ⊼ ♇
29 Tuesday	
01 23	☽ ± ♄
03 00	☽ ± ♃
03 01	☽ ⚹ ♂
10 27	♀ ∠ ♃
13 04	☽ ⊼ ♄
14 47	☽ ⊼ ♃
19 58	☽ # ☿
23 32	☽ ⚹ ♅
30 Wednesday	
02 50	☽ Q ♂
03 28	☽ ☍ ☉
04 15	♀ ⊥ ♇
10 19	♀ □ ♆
16 04	☽ ☍ ☿
21 31	☽ △ ♆
22 11	☽ Q ♅
22 42	☽ ⊼ ♀
31 Thursday	
01 18	☽ # ☿
08 10	☽ ☍ ♇
11 05	☽ ± ♀
13 45	☽ □ ♂
21 56	☽ ☍ ♄

All ephemeris data is given at 12.00 UT and the Moon's longitude is additionally given for 24.00 UT

JANUARY 2021

LONGITUDES

Date	Sidereal time h m s	Sun ☉ ° ' "	Moon ☽ ° ' "	Moon ☽ 24.00 ° ' "	Mercury ☿ ° '	Venus ♀ ° '	Mars ♂ ° '	Jupiter ♃ ° '	Saturn ♄ ° '	Uranus ♅ ° '	Neptune ♆ ° '	Pluto ♇ ° '
01	18 45 27	11 ♑ 17 21	09 ♌ 17 55	15 ♌ 55 05	18 ♑ 32	21 ♐ 02	27 ♈ 34	02 ♒ 54	01 ♒ 41	06 ♉ 48	18 ♓ 29	24 ♑ 12
02	18 49 23	12 18 30	22 ♌ 35 31	29 ♌ 19 04	20 10	22 17	28 01	03 08	01 48	06 R 47	18 30	24 14
03	18 53 20	13 19 39	06 ♍ 05 38	12 ♍ 55 05	21 48	23 33	28 27	03 21	01 55	06 46	18 31	24 16
04	18 57 16	14 20 47	19 ♍ 47 17	26 ♍ 42 06	23 26	24 48	28 54	03 35	02 02	06 46	18 32	24 18
05	19 01 13	15 21 56	03 ♎ 39 23	10 ♎ 39 01	25 04	26 03	29 21	03 49	02 08	06 45	18 34	24 20
06	19 05 10	16 23 06	17 ♎ 40 52	24 ♎ 44 44	26 43	27 18	29 ♈ 48	04 02	02 15	06 45	18 35	24 22
07	19 09 06	17 24 15	01 ♏ 50 28	08 ♏ 57 50	28 ♑ 22	28 33	00 ♉ 16	04 17	02 22	06 45	18 36	24 24
08	19 13 03	18 25 25	16 ♏ 06 34	23 ♏ 16 22	00 ♒ 00	29 ♐ 48	00 43	04 31	02 29	06 44	18 38	24 26
09	19 16 59	19 26 35	00 ♐ 26 51	07 ♐ 37 38	01 38	01 ♑ 04	01 11	04 45	02 36	06 44	18 39	24 28
10	19 20 56	20 27 45	14 ♐ 48 13	21 ♐ 58 05	03 16	02 19	01 40	04 59	02 43	06 44	18 40	24 30
11	19 24 52	21 28 54	29 ♐ 06 40	06 ♑ 13 23	04 54	03 34	02 08	05 13	02 50	06 44	18 42	24 32
12	19 28 49	22 30 04	13 ♑ 17 38	20 ♑ 18 50	06 31	04 49	02 37	05 27	02 58	06 43	18 43	24 34
13	19 32 45	23 31 13	27 ♑ 16 27	04 ♒ 09 58	08 08	06 05	03 06	05 41	03 05	06 43	18 45	24 36
14	19 36 42	24 32 22	10 ♒ 58 58	17 ♒ 43 06	09 43	07 20	03 35	05 55	03 12	06 D 43	18 46	24 38
15	19 40 39	25 33 31	24 ♒ 22 10	00 ♓ 55 59	11 17	08 35	04 04	06 09	03 19	06 43	18 48	24 40
16	19 44 35	26 34 38	07 ♓ 24 35	13 ♓ 48 01	12 50	09 50	04 33	06 23	03 26	06 43	18 49	24 42
17	19 48 32	27 35 45	20 ♓ 06 29	26 ♓ 20 16	14 20	11 05	05 03	06 37	03 33	06 44	18 51	24 44
18	19 52 28	28 36 52	02 ♈ 29 45	08 ♈ 35 21	15 49	12 21	05 33	06 51	03 40	06 44	18 52	24 46
19	19 56 25	29 ♑ 37 57	14 ♈ 37 37	20 ♈ 37 04	17 14	13 36	06 03	07 06	03 47	06 44	18 54	24 48
20	20 00 21	00 ♒ 39 02	26 ♈ 34 20	02 ♉ 30 01	18 36	14 51	06 33	07 20	03 54	06 44	18 56	24 50
21	20 04 18	01 40 05	08 ♉ 24 48	14 ♉ 19 19	19 54	16 06	07 04	07 34	04 02	06 45	18 57	24 52
22	20 08 14	02 41 08	20 ♉ 14 15	26 ♉ 10 13	21 07	17 22	07 35	07 48	04 09	06 45	18 59	24 54
23	20 12 11	03 42 10	02 ♊ 07 52	08 ♊ 07 47	22 15	18 37	08 05	08 02	04 16	06 45	19 01	24 56
24	20 16 08	04 43 11	14 ♊ 10 33	20 ♊ 16 41	23 16	19 52	08 36	08 17	04 23	06 46	19 02	24 58
25	20 20 04	05 44 11	26 ♊ 26 38	02 ♋ 40 47	24 11	21 07	09 07	08 31	04 30	06 47	19 04	25 00
26	20 24 01	06 45 10	08 ♋ 59 26	15 ♋ 22 50	24 58	22 22	09 39	08 45	04 37	06 47	19 06	25 02
27	20 27 57	07 46 08	21 ♋ 51 05	28 ♋ 24 13	25 35	23 38	10 10	08 59	04 44	06 48	19 08	25 04
28	20 31 54	08 47 05	05 ♌ 02 10	11 ♌ 44 45	26 04	24 53	10 42	09 14	04 52	06 49	19 10	25 06
29	20 35 50	09 48 01	18 ♌ 31 41	25 ♌ 22 37	26 22	26 08	11 13	09 28	04 59	06 49	19 11	25 08
30	20 39 47	10 48 56	02 ♍ 17 06	09 ♍ 14 40	26 29	27 23	11 45	09 42	05 06	06 50	19 13	25 10
31	20 43 43	11 ♒ 49 50	16 ♍ 14 46	23 ♍ 16 53	26 ♒ 25	28 ♑ 38	12 ♉ 17	09 ♒ 56	05 ♒ 13	06 ♉ 51	19 ♓ 15	25 ♑ 12

MOON NODES AND LATITUDE / DECLINATIONS

Date	Moon True ☊ ° '	Moon Mean ☊ ° '	Moon ☽ Latitude ° '	Sun ☉ ° '	Moon ☽ ° '	Mercury ☿ ° '	Venus ♀ ° '	Mars ♂ ° '	Jupiter ♃ ° '	Saturn ♄ ° '	Uranus ♅ ° '	Neptune ♆ ° '	Pluto ♇ ° '
01	19 ♊ 51	18 ♊ 50	03 N 59	22 S 57	21 N 46	24 S 15	22 S 30	11 N 26	19 S 59	20 S 10	13 N 21	05 S 33	22 S 27
02	19 R 48	18 47	04 40	22 52	18 23	24 02	22 37	11 36	19 56	20 08	13 21	05 33	22 27
03	19 44	18 44	05 05	22 46	14 00	23 47	22 44	11 47	19 53	20 07	13 21	05 32	22 26
04	19 42	18 41	05 14	22 40	08 51	23 31	22 49	11 57	19 50	20 05	13 21	05 32	22 26
05	19 40	18 37	05 05	22 33	03 N 13	23 13	22 55	12 08	19 47	20 04	13 21	05 31	22 26
06	19 D 39	18 34	04 38	22 26	02 S 40	22 54	22 59	12 19	19 43	20 02	13 21	05 31	22 26
07	19 40	18 31	03 53	22 18	08 28	22 33	23 03	12 29	19 40	20 01	13 21	05 30	22 25
08	19 41	18 28	02 54	22 10	13 53	22 11	23 06	12 40	19 37	19 59	13 20	05 29	22 25
09	19 43	18 25	01 44	22 02	18 33	21 47	23 08	12 51	19 34	19 58	13 20	05 29	22 25
10	19 44	18 22	00 N 27	21 53	22 07	21 21	23 10	13 01	19 30	19 56	13 20	05 28	22 24
11	19 R 44	18 18	00 S 51	21 43	24 17	20 55	23 11	13 12	19 27	19 55	13 20	05 28	22 24
12	19 42	18 15	02 05	21 34	24 51	20 26	23 11	13 23	19 24	19 53	13 20	05 27	22 24
13	19 38	18 12	03 11	21 23	23 49	19 57	23 10	13 33	19 20	19 51	13 20	05 27	22 23
14	19 33	18 09	04 04	21 13	21 12	19 26	23 09	13 44	19 17	19 50	13 20	05 26	22 23
15	19 27	18 06	04 42	21 02	17 49	18 54	23 07	13 55	19 13	19 48	13 20	05 25	22 23
16	19 20	18 03	05 04	20 50	13 29	18 22	23 04	14 06	19 10	19 47	13 20	05 25	22 23
17	19 14	17 59	05 11	20 38	08 40	17 48	23 01	14 16	19 07	19 45	13 20	05 24	22 22
18	19 09	17 56	05 02	20 26	03 S 38	17 13	22 57	14 27	19 03	19 44	13 21	05 23	22 22
19	19 06	17 53	04 41	20 14	01 N 27	16 38	22 52	14 38	18 59	19 42	13 21	05 23	22 22
20	19 05	17 50	04 07	20 01	06 25	16 03	22 47	14 49	18 56	19 40	13 21	05 22	22 21
21	19 D 05	17 47	03 23	19 47	11 06	15 28	22 40	14 59	18 52	19 39	13 21	05 21	22 21
22	19 06	17 43	02 30	19 33	15 23	14 53	22 34	15 10	18 49	19 37	13 21	05 21	22 21
23	19 08	17 40	01 31	19 19	19 06	14 18	22 26	15 21	18 45	19 35	13 21	05 20	22 21
24	19 09	17 37	00 S 27	19 05	22 03	13 45	22 18	15 31	18 42	19 34	13 21	05 19	22 20
25	19 R 09	17 34	00 N 39	18 50	24 03	13 12	22 09	15 42	18 38	19 32	13 22	05 19	22 20
26	19 06	17 31	01 45	18 35	24 53	12 42	21 59	15 52	18 34	19 31	13 22	05 18	22 20
27	19 02	17 28	02 47	18 20	24 25	12 14	21 49	16 03	18 31	19 29	13 22	05 17	22 20
28	18 56	17 24	03 41	18 04	22 35	11 48	21 38	16 13	18 27	19 27	13 22	05 17	22 19
29	18 47	17 21	04 25	17 48	19 28	11 26	21 26	16 24	18 23	19 26	13 23	05 16	22 19
30	18 38	17 18	04 54	17 31	15 13	11 07	21 14	16 34	18 20	19 24	13 23	05 15	22 19
31	18 ♊ 29	17 ♊ 15	05 N 06	17 S 14	10 N 07	10 S 52	21 S 01	16 N 44	18 S 16	19 S 22	13 N 23	05 S 14	22 S 19

ZODIAC SIGN ENTRIES

Date	h m	Planets
03	01 13	☽ ♍
05	05 42	☽ ♎
06	22 27	♂ ♉
07	08 53	☽ ♏
08	12 00	☿ ♒
08	15 41	♀ ♑
09	11 15	☽ ♐
11	13 30	☽ ♑
13	16 44	☽ ♒
15	22 17	☽ ♓
18	07 07	☽ ♈
19	20 40	☉ ♒
20	18 56	☽ ♉
23	07 43	☽ ♊
25	18 52	☽ ♋
28	02 54	☽ ♌
30	08 02	☽ ♍

LATITUDES

Date	Mercury ☿ ° '	Venus ♀ ° '	Mars ♂ ° '	Jupiter ♃ ° '	Saturn ♄ ° '	Uranus ♅ ° '	Neptune ♆ ° '	Pluto ♇ ° '
01	02 S 07	00 N 38	00 N 53	00 S 29	00 S 23	00 S 27	01 S 05	01 S 12
04	02 09	00 31	00 56	00 30	00 23	00 27	01 05	01 12
07	02 06	00 23	00 59	00 30	00 24	00 27	01 05	01 12
10	01 59	00 15	01 02	00 30	00 24	00 27	01 05	01 12
13	01 47	00 N 08	01 05	00 30	00 24	00 27	01 05	01 13
16	01 28	00 00	01 07	00 30	00 24	00 27	01 05	01 13
19	01 01	00 S 08	01 10	00 31	00 24	00 26	01 04	01 13
22	00 S 26	00 15	01 12	00 31	00 25	00 26	01 04	01 14
25	00 N 16	00 22	01 14	00 31	00 25	00 26	01 04	01 14
28	01 05	00 29	01 15	00 31	00 25	00 26	01 04	01 14
31	01 N 58	00 S 36	01 N 17	00 S 32	00 S 25	00 S 26	01 S 04	01 S 14

DATA

Julian Date	2459216
Delta T	+71 seconds
Ayanamsa	24° 08' 46"
Synetic vernal point	04° ♓ 58' 13"
True obliquity of ecliptic	23° 26' 13"

MOON'S PHASES, APSIDES AND POSITIONS ☽

Date	h m	Phase	Longitude ° '	Eclipse Indicator
06	09 37	☾	16 ♎ 17	
13	05 00	●	23 ♑ 13	
20	21 02	☽	01 ♉ 02	
28	19 16	○	09 ♌ 06	

Day	h m		
09	15 25	Perigee	
21	13 10	Apogee	
06	01 10	0S	
12	08 18	Max dec	24° S 52'
19	05 07	0N	
26	15 38	Max dec	24° N 54'

LONGITUDES

Date	Chiron ⚷ ° '	Ceres ⚳ ° '	Pallas ⚴ ° '	Juno ⚵ ° '	Vesta ⚶ ° '	Black Moon Lilith ⚸ ° '
01	05 ♈ 04	12 ♓ 09	08 ♒ 01	04 ♐ 24	20 ♍ 13	08 ♉ 02
11	05 ♈ 16	15 ♓ 21	11 ♒ 21	07 ♐ 28	21 ♍ 09	09 ♉ 09
21	05 ♈ 32	18 ♓ 45	14 ♒ 44	10 ♐ 24	21 ♍ 23	10 ♉ 16
31	05 ♈ 54	22 ♓ 17	18 ♒ 07	13 ♐ 09	20 ♍ 49	11 ♉ 23

ASPECTARIAN

h m	Aspects
01 Friday	
00 05	☽ ☍ ♃
00 16	☽ ♯ ☉
01 21	☽ bQ ♆
02 55	♀ ∥ ♇
05 25	☽ bQ ♀
05 42	☽ ♯ ♀
05 51	☽ ♯ ♇
07 26	☽ □ ♅
11 18	☿ ⚹ ♆
15 55	☽ ⊼ ☉
17 48	☽ ± ♆
16 56	☽ ∠ ♄
17 56	☿ ∠ ♆
18 12	☽ ⊥ ♇
18 29	☽ □ ♆
18 33	☽ ∠ ☿
20 39	♀ ⊻ ♄
20 48	☽ ∠ ♃
21 33	☽ ∥ ♀
22 12	☽ ⊻ ☉
23 36	☽ bQ ♅
02 Saturday	
00 28	☽ ♯ ♄
01 48	☽ ♯ ♃
02 22	♀ bQ ♅
03 40	☽ ± ☉
04 39	☽ ⊼ ♆
07 02	☽ ⊼ ☿
11 24	☽ △ ♀
14 57	☽ ⊼ ♇
19 16	☽ ± ☿
21 07	☽ bQ ☉
22 00	☽ △ ♂
03 Sunday	
01 40	☽ ± ♇
04 32	☽ ⊼ ♄
07 05	☽ ⊼ ♃
13 12	☽ △ ♅
13 24	☽ bQ ☿
15 13	☽ ∥ ♅
15 14	☽ ± ♄
17 08	☉ ∥ ♀
17 37	☽ bQ ♇
17 51	☽ ± ♃
22 22	☽ ∥ ♂
04 Monday	
01 22	☽ bQ ♂
01 44	☽ △ ☉
02 19	♀ ⊻ ♇
06 58	♃ ∠ ♆
07 09	☽ bQ ♄
09 49	☽ ☍ ♆
09 52	☽ bQ ♃
15 26	☽ bQ ♅
17 35	☽ ± ♂
19 11	☽ △ ☿
19 52	☽ △ ♇
21 34	☽ □ ♀
05 Tuesday	
00 58	☿ ☌ ♇
02 22	☽ ♯ ♆
04 19	☽ ⊼ ♂
07 00	☽ ± ♅
09 22	☽ △ ♄
12 17	☽ △ ♃
13 58	♀ ⊥ ♄
17 19	☽ ⊼ ♅
06 Wednesday	
07 00	☿ ∥ ♀
07 33	☽ Q ♀
09 37	☽ □ ☉
13 16	☉ ∥ ♇
13 32	☽ ⊼ ♆
23 23	☽ □ ♇
23 39	☽ ∥ ♆
23 45	☽ ± ♆
07 Thursday	
05 21	☽ □ ☿
05 32	♀ ⊥ ♃
05 55	☽ ⚹ ♀
09 14	☽ ☍ ♂
12 54	☽ □ ♄
14 59	☽ bQ ♆
16 11	☽ □ ♃
18 28	☽ Q ☉
20 16	☽ ☍ ♅
20 39	☿ ∥ ♇
08 Friday	
00 04	☿ ⊻ ♀
05 49	☽ Q ♇
06 12	☽ ♯ ♂
09 30	☽ ♯ ♅
09 36	☽ ∠ ♀
12 47	☉ ∥ ☿
15 35	☽ Q ☿
16 11	☽ ⚹ ☉
16 14	☽ △ ♆
16 53	☉ ⚹ ♆
19 24	☽ Q ♄
22 54	☽ Q ♃
09 Saturday	
01 59	☽ ⚹ ♇
02 08	☽ ⊥ ♀
02 44	☿ □ ♂
13 07	☽ ⊻ ♀
13 17	☽ ⊼ ♂
14 15	☽ ⚹ ☿
15 38	☽ ⚹ ♄
15 53	♀ △ ♂
17 55	☽ ∥ ♃
19 11	☽ ∠ ☉
19 18	☽ ⚹ ♃
20 27	☽ ∥ ♄
22 30	☽ ⊼ ♅
23 39	☽ ± ♂
10 Sunday	
03 07	☽ ∠ ♇
03 17	☿ ☌ ♄
06 44	☽ ∥ ☿
08 32	☽ ± ♅
10 07	☽ ∥ ☉
11 23	☽ ⊥ ☉
14 23	☽ ∥ ♇
15 13	☽ bQ ♂
11 Monday	
04 17	☽ ⊻ ♇
08 09	☽ ⊥ ♄
11 36	☽ ⊥ ☿
12 10	☽ ⊥ ♃
17 16	☽ △ ♂
17 19	☿ ☌ ♃
18 20	☽ ⊻ ♄
20 14	☽ ☌ ♀
22 28	☽ ⊻ ♃
23 02	☽ ⊻ ☿
12 Tuesday	
00 49	☽ Q ♆
00 51	☽ △ ♅
06 30	♂ ∥ ♅
15 00	☿ □ ♅
21 17	☽ ⚹ ♆
13 Wednesday	
02 40	♀ ⊻ ♃
05 00	☽ ☌ ☉
07 22	☽ ☌ ♇
11 02	♂ □ ♄
16 36	☿ ∥ ♄
19 49	☽ ∥ ♀
22 11	☽ ☌ ♄
22 30	☽ □ ♂
23 17	☽ ∠ ♆
14 Thursday	
00 22	♀ △ ♅
01 03	♀ Q ♆
02 55	☽ ☌ ♃
03 26	☽ ∥ ♇
04 29	☽ □ ♅
04 54	☽ ⊻ ♀
08 36	♅ St D
09 28	☽ ☌ ☿
13 20	☽ ∥ ☉
14 19	☉ ☌ ♇
15 10	☽ ⊥ ♆
16 35	☽ ⊥ ♀
20 03	☿ ∥ ♃
21 55	♂ ∠ ♆
23 18	☽ ∥ ♄
15 Friday	
01 55	☽ ⊻ ♆
03 06	☽ ∥ ♃
04 10	☽ ∥ ☿
07 40	☽ Q ♂
10 25	☽ ∠ ♀
12 33	☽ ⊻ ♇
12 38	☽ Q ♅
14 21	☽ ⊻ ☉
23 32	☽ ⊥ ♇
16 Saturday	
02 16	☽ ⊥ ☉
04 33	☽ ⊻ ♄
06 29	☽ ⚹ ♂
08 56	☽ ♯ ♂
10 03	☽ ⊻ ♃
10 43	☽ ⚹ ♅
11 51	☿ ⊥ ♆
12 45	☽ ♯ ♅
15 49	☽ ⊥ ♄
16 18	☽ ∠ ♇
17 02	☽ ⚹ ♀
20 29	☽ ∠ ☉
21 30	☽ ⊥ ♃
23 33	☽ ⊻ ☿
17 Sunday	
09 00	☽ ∠ ♄
09 35	☽ ☌ ♆
11 53	☽ ∠ ♂
12 30	☽ ⊥ ☿
14 57	☽ ∠ ♃
15 06	☽ ∠ ♅
18 22	☽ Q ♀
20 55	☽ ⚹ ♇
22 50	♃ □ ♅
18 Monday	
03 40	☽ ∥ ♆
03 44	☽ ⚹ ☉
06 00	☽ ⊥ ♂
08 16	☽ ∠ ☿
08 33	☽ ⊥ ♅
14 20	☽ ⚹ ♄
18 16	☽ ⊻ ♂
20 20	☽ ⊻ ♅
20 25	☽ Q ♇
20 45	☽ ⚹ ♃
19 Tuesday	
05 29	☽ Q ☉
09 43	☽ □ ♀
14 21	☽ Q ♄
17 53	☽ ⚹ ☿
20 34	☽ ⊻ ♆
21 07	☽ Q ♃
20 Wednesday	
06 53	☽ ♯ ♆
08 29	☽ □ ♇
08 40	☽ ⊥ ♆
16 57	♄ ∠ ♆
18 04	☿ ⊻ ♆
20 38	♂ ☌ ♅
21 02	☽ □ ☉
21 10	☽ Q ☿
21 Thursday	
02 56	☽ ∠ ♆
03 00	☽ □ ♄
08 37	☽ ☌ ♅
09 08	☽ ☌ ♂
10 15	☽ □ ♃
22 Friday	
00 15	☽ ∥ ♅
02 51	☿ ♯ ♂
04 54	☉ ∥ ♄
05 28	☽ △ ♀
09 21	☽ ♯ ☿
09 27	☽ ⚹ ♆
10 37	☽ ∥ ♂
13 59	☽ □ ☿
21 28	☽ △ ♇
23 Saturday	
07 49	♂ □ ♃
09 36	☽ ♯ ♃
09 45	☽ Q ♆
13 29	☽ ♯ ☉
15 19	☽ bQ ♀
15 27	☽ △ ☉
15 32	☽ ♯ ♄
16 19	☽ △ ♄
19 29	☉ ∠ ♆
19 49	♀ ⚹ ♆
21 16	☽ ⊻ ♅
24 Sunday	
00 04	☽ △ ♃
00 27	☽ ⊻ ♂
03 01	☉ ☌ ♄
03 27	♀ ∥ ♇
03 38	☽ bQ ♇
09 13	☽ ⊥ ♅
11 19	☽ ± ♀
12 53	☽ ⊥ ♂
14 14	☽ ♯ ♀
14 51	☽ ♯ ♇
21 27	☽ ± ♇
21 36	☽ □ ♆
22 21	☽ bQ ♄
23 54	☽ bQ ☉
25 Monday	
00 28	☽ ⊼ ♀
02 55	☽ ∠ ♅
04 56	☿ ♯ ♅
06 12	☽ bQ ♃
07 17	☽ △ ☿
07 18	☽ ∠ ♂
09 12	☽ ⊼ ♇
16 01	☽ ± ♄
18 55	☽ ± ☉
23 55	☽ ± ♃
26 Tuesday	
03 38	☽ ⊼ ♄
06 09	☿ Q ♅
07 23	☽ ⊼ ☉
07 49	☽ ⚹ ♅
11 32	☽ ⊼ ♃
12 48	☉ □ ♅
13 17	☽ ⚹ ♂
13 21	☉ ∥ ♃
13 56	☽ bQ ☿
14 35	☿ ⊻ ♇
27 Wednesday	
06 21	☽ Q ♅
06 57	☽ △ ♆
07 38	☽ ± ☿
12 36	☽ Q ♂
15 37	☽ ☍ ♀
17 55	☽ ☍ ♇
19 09	☽ ⊼ ☿
28 Thursday	
10 25	☽ bQ ♆
11 41	☽ ☍ ♄
14 25	☽ ♯ ♇
15 11	☽ □ ♅
16 18	♀ ☌ ♇
19 16	☽ ☍ ☉
19 39	☽ ☍ ♃
20 58	☽ ♯ ♀
22 32	☽ □ ♂
29 Friday	
01 40	☉ ☌ ♃
02 33	☽ ± ♆
12 14	☽ ♯ ♄
13 10	☽ ⊼ ♆
17 19	☿ ⊻ ♀
18 46	☽ ♯ ♃
22 52	☽ ♯ ☉
23 36	☽ ⊼ ♇
30 Saturday	
01 53	☽ ☍ ☿
02 39	☽ ⊼ ♀
05 13	☽ ∥ ♂
10 03	☽ ± ♇
14 05	☽ ± ♀
15 51	☿ St R
16 54	☽ ⊼ ♄
19 52	☽ △ ♅
21 02	☽ ∥ ♅
31 Sunday	
01 01	☽ ⊼ ♃
01 37	☽ bQ ♇
03 19	☽ ± ♄
03 51	☽ ⊼ ☉
04 57	☽ △ ♂
07 06	☽ bQ ♀
08 33	☽ ♯ ☿
11 28	☽ ± ♃
14 55	☽ ± ☉
17 09	☽ ☍ ♆
18 50	☽ bQ ♄
21 34	☽ bQ ♅

All ephemeris data is given at 12.00 UT and the Moon's longitude is additionally given for 24.00 UT

FEBRUARY 2021

Date	Sidereal time h m s	Sun ☉ ° ' "	Moon ☽ ° ' "	Moon ☽ 24.00 ° ' "	Mercury ☿ ° '	Venus ♀ ° '	Mars ♂ ° '	Jupiter ♃ ° '	Saturn ♄ ° '	Uranus ♅ ° '	Neptune ♆ ° '	Pluto ♇ ° '
01	20 47 40	12 ♒ 50 43	00 ♎ 20 29	07 ♎ 25 05	26 ♒ 10	29 ♑ 53	12 ♉ 49	10 ♒ 11	05 ♒ 20	06 ♉ 52	19 ♓ 17	25 ♑ 14
02	20 51 37	13 51 35	14 ♎ 30 14	21 ♎ 35 32	25 R 44	01 ♒ 09	13 21	10 25	05 27	06 53	19 19	25 16
03	20 55 33	14 52 27	28 ♎ 40 38	05 ♏ 45 18	25 07	02 24	13 53	10 39	05 34	06 54	19 21	25 17
04	20 59 30	15 53 18	12 ♏ 49 18	19 ♏ 52 30	24 21	03 39	14 26	10 53	05 41	06 55	19 23	25 19
05	21 03 26	16 54 08	26 ♏ 54 45	03 ♐ 55 58	23 26	04 54	14 58	11 08	05 48	06 56	19 25	25 21
06	21 07 23	17 54 57	10 ♐ 56 04	17 ♐ 54 57	22 24	06 09	15 31	11 22	05 55	06 57	19 27	25 23
07	21 11 19	18 55 45	24 ♐ 52 31	01 ♑ 48 39	21 17	07 25	16 04	11 36	06 03	06 58	19 29	25 25
08	21 15 16	19 56 33	08 ♑ 43 08	15 ♑ 35 48	20 06	08 40	16 37	11 50	06 10	07 00	19 31	25 27
09	21 19 12	20 57 19	22 ♑ 26 23	29 ♑ 14 38	18 55	09 55	17 10	12 04	06 17	07 01	19 33	25 29
10	21 23 09	21 58 04	06 ♒ 00 13	12 ♒ 42 52	17 44	11 10	17 43	12 19	06 24	07 02	19 35	25 31
11	21 27 06	22 58 48	19 ♒ 22 16	25 ♒ 58 09	16 36	12 25	18 16	12 33	06 31	07 04	19 37	25 32
12	21 31 02	23 59 30	02 ♓ 30 17	08 ♓ 58 29	15 31	13 40	18 49	12 47	06 37	07 05	19 39	25 34
13	21 34 59	25 00 11	15 ♓ 22 37	21 ♓ 42 40	14 32	14 55	19 23	13 01	06 44	07 07	19 41	25 36
14	21 38 55	26 00 51	27 ♓ 58 39	04 ♈ 10 42	13 40	16 11	19 56	13 15	06 51	07 08	19 43	25 38
15	21 42 52	27 01 29	10 ♈ 19 01	16 ♈ 23 53	12 55	17 26	20 30	13 29	06 58	07 10	19 46	25 39
16	21 46 48	28 02 05	22 ♈ 25 40	28 ♈ 24 48	12 17	18 41	21 03	13 43	07 05	07 12	19 48	25 41
17	21 50 45	29 ♒ 02 39	04 ♉ 21 47	10 ♉ 17 10	11 47	19 56	21 37	13 57	07 12	07 13	19 50	25 43
18	21 54 41	00 ♓ 03 12	16 ♉ 11 34	22 ♉ 05 37	11 24	21 11	22 11	14 11	07 19	07 15	19 52	25 45
19	21 58 38	01 03 43	27 ♉ 59 58	03 ♊ 55 20	11 10	22 26	22 45	14 25	07 25	07 17	19 54	25 46
20	22 02 35	02 04 12	09 ♊ 52 24	15 ♊ 51 52	11 02	23 41	23 19	14 39	07 32	07 19	19 56	25 48
21	22 06 31	03 04 39	21 ♊ 54 25	28 ♊ 00 42	11 D 02	24 56	23 53	14 53	07 39	07 21	19 59	25 50
22	22 10 28	04 05 05	04 ♋ 11 21	10 ♋ 26 56	11 09	26 11	24 27	15 07	07 46	07 23	20 01	25 51
23	22 14 24	05 05 29	16 ♋ 47 54	23 ♋ 14 40	11 21	27 26	25 01	15 21	07 52	07 24	20 03	25 53
24	22 18 21	06 05 50	29 ♋ 47 30	06 ♌ 26 34	11 40	28 41	25 36	15 35	07 59	07 26	20 05	25 55
25	22 22 17	07 06 10	13 ♌ 11 51	20 ♌ 03 14	12 04	29 ♒ 56	26 10	15 49	08 05	07 29	20 07	25 56
26	22 26 14	08 06 28	27 ♌ 00 23	04 ♍ 02 50	12 33	01 ♓ 11	26 45	16 02	08 12	07 31	20 10	25 58
27	22 30 10	09 06 44	11 ♍ 10 00	18 ♍ 21 07	13 07	02 26	27 19	16 16	08 18	07 33	20 12	25 59
28	22 34 07	10 ♓ 06 59	25 ♍ 35 22	02 ♎ 51 51	13 ♒ 46	03 ♓ 41	27 ♉ 54	16 ♒ 30	08 ♒ 25	07 ♉ 35	20 ♓ 14	26 ♑ 01

DECLINATIONS

Date	Moon True ☊ ° '	Moon Mean ☊ ° '	Moon ☽ Latitude ° '	Sun ☉ ° '	Moon ☽ ° '	Mercury ☿ ° '	Venus ♀ ° '	Mars ♂ ° '	Jupiter ♃ ° '	Saturn ♄ ° '	Uranus ♅ ° '	Neptune ♆ ° '	Pluto ♇ ° '
01	18 ♊ 21	17 ♊ 12	04 N 59	16 S 57	04 N 26	10 S 41	20 S 48	16 N 55	18 S 12	19 S 21	13 N 24	05 S 14	22 S 19
02	18 R 15	17 09	04 35	16 40	01 S 30	10 34	20 34	17 05	18 08	19 19	13 24	05 13	22 18
03	18 12	17 05	03 53	16 22	07 22	10 32	20 19	17 15	18 05	19 17	13 24	05 12	22 18
04	18 11	17 02	02 58	16 04	12 52	10 34	20 04	17 25	18 01	19 16	13 25	05 11	22 18
05	18 D 11	16 59	01 51	15 46	17 40	10 40	19 48	17 35	17 57	19 14	13 25	05 10	22 18
06	18 12	16 56	00 N 39	15 28	21 26	10 50	19 31	17 45	17 53	19 12	13 26	05 10	22 17
07	18 R 12	16 53	00 S 36	15 09	23 56	11 04	19 14	17 55	17 49	19 11	13 26	05 09	22 17
08	18 10	16 49	01 47	14 50	24 56	11 20	18 57	18 05	17 45	19 09	13 26	05 08	22 17
09	18 05	16 46	02 52	14 31	24 24	11 39	18 38	18 15	17 41	19 07	13 27	05 07	22 17
10	17 58	16 43	03 46	14 11	22 25	12 00	18 20	18 25	17 38	19 06	13 27	05 06	22 16
11	17 48	16 40	04 26	13 51	19 13	12 21	18 00	18 34	17 34	19 04	13 28	05 06	22 16
12	17 36	16 37	04 52	13 31	15 07	12 44	17 41	18 44	17 30	19 02	13 28	05 05	22 16
13	17 24	16 34	05 02	13 11	10 24	13 06	17 20	18 54	17 26	19 01	13 29	05 04	22 16
14	17 12	16 30	04 57	12 51	05 21	13 28	17 00	19 03	17 22	18 59	13 29	05 03	22 15
15	17 02	16 27	04 38	12 30	00 S 11	13 49	16 38	19 12	17 18	18 57	13 30	05 02	22 15
16	16 53	16 24	04 07	12 09	04 N 55	14 09	16 17	19 22	17 14	18 56	13 31	05 01	22 15
17	16 48	16 21	03 25	11 48	09 45	14 28	15 55	19 31	17 10	18 54	13 31	05 01	22 15
18	16 45	16 18	02 35	11 27	14 13	14 45	15 32	19 40	17 06	18 52	13 32	05 00	22 15
19	16 44	16 14	01 38	11 06	18 07	15 01	15 09	19 49	17 02	18 51	13 32	04 59	22 14
20	16 D 44	16 11	00 S 36	10 44	21 20	15 15	14 46	19 58	16 58	18 49	13 33	04 58	22 14
21	16 R 44	16 08	00 N 27	10 23	23 39	15 27	14 22	20 07	16 54	18 48	13 34	04 57	22 14
22	16 43	16 05	01 31	10 01	24 53	15 38	13 57	20 15	16 50	18 46	13 34	04 56	22 14
23	16 40	16 02	02 32	09 39	24 54	15 47	13 33	20 24	16 46	18 44	13 35	04 55	22 14
24	16 34	15 59	03 27	09 16	23 34	15 54	13 08	20 32	16 42	18 43	13 36	04 55	22 13
25	16 25	15 55	04 12	08 54	20 53	16 00	12 43	20 41	16 38	18 41	13 36	04 54	22 13
26	16 14	15 52	04 44	08 32	16 57	16 04	12 17	20 49	16 34	18 39	13 37	04 53	22 13
27	16 02	15 49	04 59	08 09	11 59	16 06	11 51	20 57	16 30	18 38	13 38	04 52	22 13
28	15 ♊ 50	15 ♊ 46	04 N 56	07 S 46	06 N 17	16 S 07	11 S 25	21 N 05	16 S 26	18 S 36	13 N 38	04 S 51	22 S 13

ZODIAC SIGN ENTRIES

Date	h m	Planets
01	11 25	☽ ♎
01	14 05	♀ ♒
03	14 14	☽ ♏
05	17 16	☽ ♐
07	20 52	☽ ♑
10	01 20	☽ ♒
12	07 23	☽ ♓
14	15 54	☽ ♈
17	03 12	☽ ♉
18	10 44	☉ ♓
19	16 04	☽ ♊
22	03 53	☽ ♋
24	12 23	☽ ♌
25	13 11	♀ ♓
26	17 07	☽ ♍
28	19 17	☽ ♎

LATITUDES

Date	Mercury ☿ ° '	Venus ♀ ° '	Mars ♂ ° '	Jupiter ♃ ° '	Saturn ♄ ° '	Uranus ♅ ° '	Neptune ♆ ° '	Pluto ♇ ° '
01	02 N 15	00 S 38	01 N 17	00 S 32	00 S 25	00 S 26	01 S 04	01 S 15
04	03 01	00 45	01 19	00 32	00 26	00 26	01 04	01 15
07	03 31	00 51	01 20	00 32	00 26	00 26	01 04	01 15
10	03 42	00 57	01 21	00 33	00 26	00 26	01 04	01 15
13	03 31	01 02	01 22	00 33	00 26	00 26	01 04	01 16
16	03 05	01 07	01 23	00 33	00 27	00 26	01 04	01 16
19	02 31	01 11	01 24	00 34	00 27	00 26	01 04	01 16
22	01 52	01 15	01 25	00 34	00 27	00 25	01 04	01 17
25	01 13	01 18	01 26	00 34	00 27	00 25	01 04	01 17
28	00 36	01 21	01 26	00 35	00 28	00 25	01 04	01 17
31	00 N 02	01 S 23	01 N 27	00 S 35	00 S 28	00 S 25	01 S 04	01 S 18

DATA

Julian Date	2459247
Delta T	+71 seconds
Ayanamsa	24° 08' 51"
Synetic vernal point	04° ♓ 58' 08"
True obliquity of ecliptic	23° 26' 14"

MOON'S PHASES, APSIDES AND POSITIONS ☽

Date	h m	Phase	Longitude ° '	Eclipse Indicator
04	17 37	☾	16 ♏ 08	
11	19 06	●	23 ♒ 17	
19	18 47	☽	01 ♊ 21	
27	08 17	○	08 ♍ 57	

Day	h m		
03	19 10	Perigee	
18	10 23	Apogee	
02	05 59	0S	
08	15 32	Max dec	24° S 57'
15	12 50	0N	
23	00 11	Max dec	25° N 03'

LONGITUDES

Date	Chiron ⚷ ° '	Ceres ⚳ ° '	Pallas ⚴ ° '	Juno ⚵ ° '	Vesta ⚶ ° '	Black Moon Lilith ⚸ ° '
01	05 ♈ 56	22 ♓ 39	18 ♒ 28	13 ♐ 25	20 ♍ 43	11 ♉ 29
11	06 ♈ 22	26 ♓ 20	21 ♒ 51	15 ♐ 56	19 ♍ 18	12 ♉ 36
21	06 ♈ 51	00 ♈ 06	25 ♒ 13	18 ♐ 12	17 ♍ 13	13 ♉ 43
31	07 ♈ 23	03 ♈ 57	28 ♒ 33	20 ♐ 10	14 ♍ 43	14 ♉ 49

All ephemeris data is given at 12.00 UT and the Moon's longitude is additionally given for 24.00 UT

ASPECTARIAN

Day	h m	Aspects
01 Monday	03 05	☽ ⚼ ♃
	03 17	☽ △ ♇
	05 05	☽ ⚻ ☿
	07 26	☽ ⚼ ☉
	07 33	☽ ⚼ ♂
	08 45	☽ ⋕ ♆
	10 34	☉ □ ♂
	11 10	☽ △ ♀
	12 53	☽ ± ♅
	14 12	☉ ⋕ ♂
	15 02	☽ ± ☿
	20 32	☽ △ ♄
	22 44	☉ ⊥ ♆
	23 05	☽ ⚻ ♅
	23 25	☽ ± ♂
02 Tuesday	04 57	☽ △ ♃
	05 50	☽ ⚼ ☿
	09 58	☽ ⚻ ♂
	10 50	☽ △ ☉
	20 10	☽ ⚻ ♆
03 Wednesday	03 03	☽ ∥ ♆
	06 15	☽ □ ♇
	06 15	☽ △ ☿
	06 21	☽ ± ♆
	06 22	☿ ⚺ ♇
	18 55	☽ □ ♀
	19 19	☿ Q ♅
	21 38	☽ ⚼ ♆
	23 47	☽ □ ♄
04 Thursday	01 34	☽ ∥ ☿
	01 57	☽ ☍ ♅
	08 40	☽ □ ♃
	12 51	☽ Q ♇
	14 33	☽ ⋕ ♅
	14 51	☽ ☍ ♂
	17 37	☽ □ ☉
	23 11	☽ △ ♆
05 Friday	02 23	♀ ∠ ♆
	02 37	☽ ∥ ☉
	04 29	☽ Q ♀
	06 27	☽ □ ☿
	06 39	☽ Q ♄
	09 20	☽ ⚹ ♇
	11 36	☽ ⋕ ♂
	13 35	☽ ∥ ♃
	15 51	☽ Q ♃
	21 04	☽ ∥ ♄
	23 47	☽ ∥ ♀
06 Saturday	02 43	☽ Q ☉
	03 00	☽ ⚹ ♀
	03 20	☽ ⚹ ♄
	05 10	☽ ⚻ ♅
	07 07	♀ ☌ ♄
	11 03	☽ ∠ ♇
	11 09	☽ Q ☿
	12 45	☽ ⚹ ♃
	15 28	☽ ± ♅
	18 47	☽ ∥ ♇
	20 12	☽ ⚻ ♂
07 Sunday	00 56	☽ ⚹ ☉
	01 19	♂ ⋕ ♃
	02 34	☽ ⊥ ♇
	02 40	☽ □ ♆
	03 33	♀ □ ♅
	05 20	☽ ∠ ♄
	06 16	☽ ⚹ ☿
	06 57	☽ ± ♂
	06 59	☽ ⚼ ♅
	07 19	☽ ∠ ♀
	12 56	☽ ⚺ ♇
	15 02	☽ ∠ ♃
	17 22	♀ ∥ ♄
	21 01	☽ ⊥ ♄
	23 09	☽ ⚼ ♂
08 Monday	00 26	☽ ⊥ ♀
	01 32	☉ ⚺ ♆
	04 55	☽ ∠ ☉
	06 13	☽ ∠ ☿
	06 54	☽ ⊥ ♃
	07 31	☽ ⚺ ♄
	09 00	☽ △ ♅
	09 54	☽ Q ♆
	11 53	☽ ⚺ ♀
	13 48	☉ ☌ ☿
	17 32	☽ ⚺ ♃
	20 39	☽ ⊥ ☿
	21 50	☽ ⊥ ☉
	23 35	☿ ⚺ ♆
09 Tuesday	02 21	☽ △ ♂
	06 18	☽ ⚺ ☿
	06 55	☽ ⚹ ♆
	09 11	☽ ⚺ ☉
	17 22	☽ ☌ ♇
10 Wednesday	07 44	♀ ⋕ ♂
	09 28	☽ ∠ ♆
	12 16	☿ □ ♂
	12 42	☽ ☌ ♄
	13 16	☽ ∥ ♇
	13 51	☽ □ ♅
	22 11	☽ ☌ ♀
	23 29	☽ ☌ ♃
11 Thursday	01 36	☽ ⊥ ♆
	07 22	☽ ☌ ☿
	09 55	☽ □ ♂
	12 27	☽ ⚺ ♆
	13 01	☽ ∥ ♄
	15 00	♀ ☌ ♃
	15 57	☽ ⋕ ♂
	19 06	☽ ☌ ☉
	20 16	☽ ∥ ♀
	22 22	☽ Q ♅
	22 25	☽ ∥ ♃
	23 14	☽ ⚺ ♇
12 Friday	10 16	☽ ⊥ ♇
	11 38	♀ ⊥ ♆
	15 32	☉ ⋕ ♅
	19 42	☽ ⚺ ♄
	20 21	☽ Q ♂
	20 30	☽ ⚹ ♅
	20 38	☽ ⋕ ♅
	21 02	☽ ∥ ☉
	23 33	☽ ∥ ☿
13 Saturday	03 01	☽ ∠ ♇
	04 02	♀ ∥ ♃
	07 00	☽ ⊥ ♄
	07 29	☽ ⚺ ♃
	07 48	☿ ☌ ♀
	10 32	☽ ⚺ ☿
	11 03	☽ ⚺ ♀
	14 42	☉ Q ♅
	15 02	☉ ∥ ☿
	19 01	☽ ⊥ ♃
	19 55	☽ ⚹ ♂
	20 11	☽ ☌ ♆
	21 07	☽ ⊥ ☿
	23 40	☽ ⊥ ♀
14 Sunday	00 10	☽ ∠ ♄
	00 48	☽ ∠ ♅
	02 13	♂ ⚹ ♆
	02 34	☉ ⚺ ♇
	03 25	♂ ⋕ ♄
	07 29	☽ ⚹ ♇
	07 54	☽ ⚺ ☉
	10 24	☿ ⊥ ♆
	12 32	☽ ∠ ♃
	13 15	☽ ∠ ☿
	13 22	☽ ∥ ♆
	14 00	☿ ⋕ ♅
	18 07	☽ ⊥ ♅
	18 52	☽ ∠ ♀
	20 29	☽ ⊥ ☉
	21 40	☿ ☌ ♃
15 Monday	02 07	☽ ∠ ♂
	05 23	☽ ⚹ ♄
	05 49	☽ ⚺ ♅
	06 47	☽ Q ♇
	15 40	☽ ∠ ☉
	16 50	☽ ⚹ ☿
	18 22	☽ ⚹ ♃
	20 38	☽ ⊥ ♂
16 Tuesday	03 40	☽ ⚹ ♀
	05 16	☽ Q ♄
	06 44	☽ ⚺ ♆
	09 08	☽ ⚺ ♂
	12 33	☽ ⋕ ♆
	15 32	☽ Q ☿
	18 32	☽ □ ♇
	18 44	☽ Q ♃
	18 46	☽ ⊥ ♆
	20 49	♃ ⊥ ♆
17 Wednesday	00 17	☽ ⚹ ☉
	06 31	☽ Q ♀
	10 00	♀ ⚺ ♆
	12 57	☽ ∠ ♆
	17 48	☽ □ ♄
	17 48	☽ ☌ ♅
	19 08	♄ □ ♅
	21 58	☽ ⋕ ☉
18 Thursday	02 32	☽ □ ☿
	02 48	☽ Q ☉
	07 51	☽ □ ♃
	08 09	☽ ∥ ♅
	15 18	☽ ⋕ ☿
	19 03	☽ ⋕ ♀
	19 30	☽ ⚹ ♆
	23 21	☽ □ ♀
19 Friday	00 48	☽ ☌ ♂
	05 04	☽ ⋕ ♃
	07 28	☽ △ ♇
	16 54	☽ ⋕ ♄
	17 18	☿ ∥ ♀
	18 47	☽ □ ☉
	19 56	☽ Q ♆
	23 04	♀ □ ♂
20 Saturday	00 30	☽ ∥ ♂
	05 25	☉ ⊥ ♇
	06 50	☽ ⚺ ♅
	07 15	☽ △ ♄
	13 52	☽ ⚼ ♇
	14 20	☽ △ ☿
	18 55	☽ ⊥ ♅
	20 15	☽ ⋕ ♇
	21 46	☽ △ ♃
21 Sunday	00 53	☿ St D
	07 53	☽ ± ♇
	08 10	☽ □ ♆
	12 52	☽ ∠ ♅
	13 29	☽ ⚼ ♄
	16 06	☽ ⚺ ♂
	18 39	☽ △ ♀
	19 45	☽ ⚻ ♇
	20 01	♀ Q ♅
	20 11	☽ ⚼ ☿
22 Monday	03 58	☽ ⚼ ♃
	04 25	☽ ⊥ ♂
	05 31	♀ ⚺ ♇
	07 15	☽ ± ♄
	11 47	☽ △ ☉
	13 52	☽ ± ☿
	18 09	☽ ⚹ ♅
	18 55	☽ ⚻ ♄
	21 38	☽ ± ♃
	22 35	☽ ∠ ♂
23 Tuesday	01 31	☽ ⚻ ☿
	02 53	☽ ⚼ ♀
	09 14	☽ ⚻ ♃
	10 06	♀ ⋕ ♅
	16 54	☽ Q ♅
	18 06	☽ △ ♆
	18 40	☽ ⚼ ☉
	21 35	☽ ± ♀
24 Wednesday	03 59	☽ ⚹ ♂
	04 54	☽ ☍ ♇
	09 47	☽ ⚻ ♀
	12 36	☽ ± ☉
	21 36	☽ ⚼ ♆
25 Thursday	00 18	☽ ⚻ ☉
	01 26	☽ ⋕ ♇
	01 49	☽ □ ♅
	01 52	♂ △ ♇
	02 41	☽ Q ♂
	02 52	☽ ☍ ♄
	09 56	☽ ☍ ☿
	13 21	☽ ∥ ♂
	13 38	☽ ± ♆
	16 40	☽ ☍ ♃
	21 13	☉ ⚹ ♅
26 Friday	00 09	☽ ⚻ ♆
	02 21	☽ ⋕ ♄
	10 13	☽ ⚻ ♇
	11 32	☽ □ ♂
	14 01	☽ ⋕ ♃
	14 25	☉ ⚺ ♄
	16 35	☽ ⋕ ☿
	19 50	☽ ☍ ♀
	20 29	☽ ± ♇
27 Saturday	03 14	♀ ⊥ ♇
	04 32	☽ ∥ ♅
	05 54	☽ △ ♅
	07 09	☽ ⚻ ♄
	08 17	☽ ☍ ☉
	11 42	☽ ⚼ ♇
	12 40	☽ ⋕ ♀
	15 25	☽ ⚻ ☿
	17 18	☽ ± ♄
	20 40	☽ ⚻ ♃
28 Sunday	01 53	☽ ± ☿
	03 06	☽ ☍ ♆
	05 31	☽ ⋕ ☉
	06 48	☽ ± ♃
	07 01	☽ ⚼ ♅
	08 22	☽ ⚼ ♄
	12 42	☽ △ ♇
	15 58	☽ △ ♂
	17 30	☽ ⚼ ☿
	17 43	☽ ⋕ ♆
	21 54	☽ ⚼ ♃
	21 55	☽ ± ♅

MARCH 2021

LONGITUDES

Date	Sidereal time h m s	Sun ☉ ° ' "	Moon ☽ ° ' "	Moon ☽ 24.00 ° ' "	Mercury ☿ ° '	Venus ♀ ° '	Mars ♂ ° '	Jupiter ♃ ° '	Saturn ♄ ° '	Uranus ♅ ° '	Neptune ♆ ° '	Pluto ♇ ° '
01	22 38 04	11 ♓ 07 11	10 ♎ 09 38	17 ♎ 27 50	14 ♒ 28	04 ♓ 56	28 ♉ 28	16 ♒ 43	08 ♒ 31	07 ♉ 37	20 ♓ 16	26 ♑ 03
02	22 42 00	12 07 22	24 ♎ 45 37	02 ♏ 02 12	15 14	06 11	29 03	16 57	08 38	07 39	20 19	26 04
03	22 45 57	13 07 32	09 ♏ 16 58	16 ♏ 29 22	16 04	07 26	29 ♉ 38	17 11	08 44	07 42	20 21	26 05
04	22 49 53	14 07 40	23 ♏ 39 01	00 ♐ 45 37	16 57	08 41	00 ♊ 12	17 24	08 50	07 44	20 23	26 07
05	22 53 50	15 07 47	07 ♐ 49 02	14 ♐ 49 11	17 53	09 56	00 47	17 38	08 56	07 46	20 25	26 08
06	22 57 46	16 07 52	21 ♐ 46 04	28 ♐ 39 45	18 52	11 11	01 22	17 51	09 03	07 49	20 28	26 10
07	23 01 43	17 07 55	05 ♑ 30 18	12 ♑ 17 51	19 54	12 26	01 57	18 04	09 09	07 51	20 30	26 11
08	23 05 39	18 07 57	19 ♑ 02 28	25 ♑ 44 16	20 58	13 41	02 32	18 18	09 15	07 54	20 32	26 13
09	23 09 36	19 07 58	02 ♒ 23 17	08 ♒ 59 36	22 04	14 56	03 07	18 31	09 21	07 56	20 35	26 14
10	23 13 33	20 07 56	15 ♒ 33 10	22 ♒ 04 00	23 13	16 10	03 42	18 44	09 27	07 59	20 37	26 15
11	23 17 29	21 07 53	28 ♒ 32 02	04 ♓ 57 13	24 24	17 25	04 17	18 58	09 33	08 01	20 39	26 17
12	23 21 26	22 07 48	11 ♓ 19 30	17 ♓ 38 47	25 36	18 40	04 52	19 11	09 39	08 04	20 41	26 18
13	23 25 22	23 07 41	23 ♓ 55 04	00 ♈ 08 18	26 51	19 55	05 28	19 24	09 45	08 07	20 44	26 19
14	23 29 19	24 07 32	06 ♈ 18 33	12 ♈ 25 51	28 07	21 10	06 03	19 37	09 51	08 09	20 46	26 20
15	23 33 15	25 07 21	18 ♈ 30 21	24 ♈ 32 14	29 ♒ 25	22 25	06 38	19 50	09 56	08 12	20 48	26 22
16	23 37 12	26 07 08	00 ♉ 31 44	06 ♉ 29 09	00 ♓ 45	23 39	07 14	20 03	10 02	08 15	20 50	26 23
17	23 41 08	27 06 53	12 ♉ 24 52	18 ♉ 19 18	02 07	24 54	07 49	20 15	10 08	08 17	20 53	26 24
18	23 45 05	28 06 35	24 ♉ 12 57	00 ♊ 06 21	03 30	26 09	08 25	20 28	10 13	08 20	20 55	26 25
19	23 49 02	29 ♓ 06 16	06 ♊ 00 05	11 ♊ 54 45	04 54	27 24	09 00	20 41	10 19	08 23	20 57	26 26
20	23 52 58	00 ♈ 05 54	17 ♊ 51 02	23 ♊ 49 36	06 20	28 38	09 36	20 54	10 24	08 26	21 00	26 27
21	23 56 55	01 05 30	29 ♊ 51 08	05 ♋ 56 20	07 48	29 ♓ 53	10 11	21 06	10 30	08 29	21 02	26 28
22	00 00 51	02 05 04	12 ♋ 05 53	18 ♋ 20 27	09 16	01 ♈ 08	10 47	21 19	10 35	08 32	21 04	26 29
23	00 04 48	03 04 35	24 ♋ 40 36	01 ♌ 06 54	10 47	02 22	11 23	21 31	10 40	08 35	21 06	26 30
24	00 08 44	04 04 05	07 ♌ 39 48	14 ♌ 19 36	12 18	03 37	11 58	21 43	10 45	08 38	21 09	26 31
25	00 12 41	05 03 31	21 ♌ 06 31	28 ♌ 00 34	13 51	04 52	12 34	21 56	10 51	08 41	21 11	26 32
26	00 16 37	06 02 56	05 ♍ 01 35	12 ♍ 09 14	15 26	06 06	13 10	22 08	10 56	08 44	21 13	26 33
27	00 20 34	07 02 18	19 ♍ 22 57	26 ♍ 42 00	17 01	07 21	13 45	22 20	11 01	08 47	21 15	26 34
28	00 24 31	08 01 38	04 ♎ 05 29	11 ♎ 32 20	18 39	08 35	14 21	22 32	11 06	08 50	21 17	26 35
29	00 28 27	09 00 56	19 ♎ 01 26	26 ♎ 31 35	20 17	09 50	14 57	22 44	11 11	08 53	21 20	26 36
30	00 32 24	10 00 12	04 ♏ 01 37	11 ♏ 30 26	21 57	11 04	15 33	22 56	11 15	08 56	21 22	26 37
31	00 36 20	10 ♈ 59 27	18 ♏ 57 01	26 ♏ 20 30	23 ♓ 38	12 ♈ 19	16 ♊ 09	23 ♒ 08	11 ♒ 20	08 ♉ 59	21 ♓ 24	26 ♑ 38

Date	Moon True ☊ ° '	Moon Mean ☊ ° '	Moon ☽ Latitude ° '
01	15 ♊ 39	15 ♊ 43	04 N 34
02	15 R 31	15 40	03 53
03	15 25	15 36	02 58
04	15 22	15 33	01 52
05	15 D 22	15 30	00 N 40
06	15 R 21	15 27	00 S 34
07	15 21	15 24	01 44
08	15 18	15 20	02 48
09	15 13	15 17	03 41
10	15 05	15 14	04 22
11	14 54	15 11	04 48
12	14 41	15 08	04 59
13	14 27	15 05	04 56
14	14 13	15 01	04 38
15	14 01	14 58	04 08
16	13 51	14 55	03 27
17	13 45	14 52	02 38
18	13 41	14 49	01 41
19	13 39	14 46	00 S 41
20	13 D 39	14 42	00 N 22
21	13 39	14 39	01 25
22	13 R 39	14 36	02 25
23	13 36	14 33	03 20
24	13 32	14 30	04 06
25	13 25	14 26	04 41
26	13 16	14 23	05 00
27	13 06	14 20	05 01
28	12 56	14 17	04 43
29	12 47	14 14	04 05
30	12 40	14 11	03 10
31	12 ♊ 35	14 ♊ 07	02 N 03

DECLINATIONS

Date	Sun ☉ ° '	Moon ☽ ° '	Mercury ☿ ° '	Venus ♀ ° '	Mars ♂ ° '	Jupiter ♃ ° '	Saturn ♄ ° '	Uranus ♅ ° '	Neptune ♆ ° '	Pluto ♇ ° '
01	07 S 24	00 N 10	16 S 06	10 S 58	21 N 13	16 S 22	18 S 35	13 N 39	04 S 50	22 S 12
02	07 01	05 S 58	16 03	10 31	21 21	16 18	18 33	13 40	04 49	22 12
03	06 38	11 46	15 59	10 04	21 29	16 14	18 32	13 41	04 48	22 12
04	06 15	16 53	15 53	09 36	21 37	16 10	18 30	13 41	04 47	22 12
05	05 52	20 57	15 46	09 09	21 44	16 06	18 28	13 42	04 47	22 12
06	05 28	23 45	15 37	08 41	21 51	16 02	18 27	13 43	04 46	22 12
07	05 05	25 04	15 27	08 12	21 59	15 58	18 25	13 44	04 45	22 11
08	04 42	24 51	15 15	07 44	22 06	15 54	18 24	13 45	04 44	22 11
09	04 18	23 13	15 02	07 15	22 13	15 50	18 22	13 45	04 43	22 11
10	03 55	20 20	14 48	06 46	22 20	15 47	18 21	13 46	04 42	22 11
11	03 31	16 29	14 31	06 17	22 26	15 43	18 19	13 47	04 41	22 11
12	03 07	11 56	14 14	05 48	22 33	15 39	18 18	13 48	04 40	22 11
13	02 44	06 57	13 55	05 19	22 39	15 35	18 16	13 49	04 39	22 11
14	02 20	01 S 45	13 35	04 49	22 46	15 31	18 15	13 50	04 39	22 11
15	01 56	03 N 25	13 14	04 20	22 52	15 27	18 13	13 51	04 38	22 10
16	01 33	08 25	12 51	03 50	22 58	15 23	18 12	13 52	04 37	22 10
17	01 09	13 04	12 27	03 20	23 04	15 19	18 11	13 53	04 36	22 10
18	00 45	17 11	12 02	02 50	23 10	15 15	18 09	13 53	04 35	22 10
19	00 S 21	20 38	11 35	02 20	23 15	15 11	18 08	13 54	04 34	22 10
20	00 N 02	23 15	11 07	01 50	23 21	15 07	18 06	13 55	04 33	22 10
21	00 26	24 51	10 38	01 20	23 26	15 03	18 05	13 56	04 32	22 10
22	00 50	25 18	10 07	00 50	23 31	14 59	18 04	13 57	04 32	22 10
23	01 13	24 28	09 36	00 S 20	23 37	14 56	18 02	13 58	04 31	22 10
24	01 37	22 19	09 03	00 N 11	23 41	14 52	18 01	13 59	04 30	22 10
25	02 01	18 53	08 29	00 41	23 46	14 48	18 00	14 00	04 29	22 10
26	02 24	14 19	07 54	01 11	23 51	14 44	17 59	14 01	04 28	22 10
27	02 48	08 49	07 17	01 41	23 55	14 40	17 57	14 02	04 27	22 10
28	03 11	02 N 42	06 40	02 11	24 00	14 36	17 56	14 03	04 26	22 09
29	03 34	03 S 40	06 01	02 42	24 04	14 33	17 55	14 04	04 25	22 09
30	03 58	09 52	05 21	03 12	24 08	14 29	17 54	14 05	04 25	22 09
31	04 N 21	15 S 29	04 S 40	03 N 42	24 N 11	14 S 25	17 S 52	14 N 06	04 S 24	22 S 09

ZODIAC SIGN ENTRIES

Date	h m	Planets
02	20 38	☽ ♏
04	03 29	♂ ♊
04	22 43	☽ ♐
07	02 20	☽ ♑
09	07 41	☽ ♒
11	14 44	☽ ♓
13	23 44	☽ ♈
15	22 26	☿ ♓
16	10 56	☽ ♉
18	23 47	☽ ♊
20	09 37	☉ ♈
21	12 18	☽ ♋
21	14 16	♀ ♈
23	21 56	☽ ♌
26	03 25	☽ ♍
28	05 22	☽ ♎
30	05 33	☽ ♏

LATITUDES

Date	Mercury ☿ ° '	Venus ♀ ° '	Mars ♂ ° '	Jupiter ♃ ° '	Saturn ♄ ° '	Uranus ♅ ° '	Neptune ♆ ° '	Pluto ♇ ° '
01	00 N 25	01 S 22	01 N 27	00 S 35	00 S 28	00 S 25	01 S 04	01 S 18
04	00 S 08	01 24	01 27	00 35	00 28	00 25	01 04	01 18
07	00 38	01 25	01 27	00 35	00 28	00 25	01 04	01 18
10	01 04	01 26	01 28	00 36	00 29	00 25	01 04	01 19
13	01 27	01 26	01 28	00 36	00 29	00 25	01 04	01 19
16	01 46	01 26	01 28	00 37	00 29	00 25	01 04	01 19
19	02 00	01 25	01 29	00 37	00 30	00 25	01 04	01 20
22	02 12	01 24	01 29	00 38	00 30	00 25	01 04	01 20
25	02 19	01 22	01 29	00 38	00 30	00 25	01 04	01 20
28	02 22	01 19	01 29	00 39	00 31	00 25	01 04	01 21
31	02 S 20	01 S 16	01 N 29	00 S 39	00 S 31	00 S 25	01 S 04	01 S 21

DATA

Julian Date	2459275
Delta T	+71 seconds
Ayanamsa	24° 08' 54"
Synetic vernal point	04° ♓ 58' 05"
True obliquity of ecliptic	23° 26' 14"

MOON'S PHASES, APSIDES AND POSITIONS ☽

Date	h m	Phase	Longitude ° '	Eclipse Indicator
06	01 30	☾	15 ♐ 42	
13	10 21	●	23 ♓ 04	
21	14 40	☽	01 ♋ 12	
28	18 48	○	08 ♎ 18	

Day	h m		
02	05 28	Perigee	
18	05 07	Apogee	
30	06 24	Perigee	
01	12 40	0S	
07	20 41	Max dec	25° S 10'
14	20 06	0N	
22	08 34	Max dec	25° N 19'
28	22 13	0S	

LONGITUDES

Date	Chiron ⚷ ° '	Ceres ⚳ ° '	Pallas ⚴ ° '	Juno ⚵ ° '	Vesta ⚶ ° '	Black Moon Lilith ⚸ ° '
01	07 ♈ 16	03 ♈ 11	27 ♒ 54	19 ♐ 48	15 ♍ 14	14 ♉ 36
11	07 ♈ 49	07 ♈ 05	01 ♓ 11	21 ♐ 29	12 ♍ 37	15 ♉ 43
21	08 ♈ 24	11 ♈ 01	04 ♓ 24	22 ♐ 48	10 ♍ 12	16 ♉ 49
31	08 ♈ 59	14 ♈ 59	07 ♓ 32	23 ♐ 39	08 ♍ 17	17 ♉ 56

All ephemeris data is given at 12.00 UT and the Moon's longitude is additionally given for 24.00 UT

ASPECTARIAN

h m	Aspects
01 Monday	
02 36	☽ ⚻ ♀
05 18	☿ ⊥ ♆
07 49	☽ ⚻ ♅
09 17	☽ △ ♄
10 05	☉ ∠ ♇
13 24	☽ ± ♀
13 42	☽ ⚻ ☉
17 40	☽ ⚼ ♂
19 27	☽ △ ☿
22 57	☽ △ ♃
02 Tuesday	
00 17	☽ ± ☉
04 40	☽ ⚻ ♆
05 34	☽ ⚼ ♀
07 28	☽ ∥ ♆
09 04	☽ ± ♂
14 09	☽ □ ♇
14 34	☽ ± ♆
15 56	☽ ∥ ☉
16 11	☽ ⚼ ☉
19 21	☽ ⚻ ♂
03 Wednesday	
05 14	☽ ∥ ♀
05 28	☽ ⚼ ♆
08 39	☽ △ ♀
09 21	☽ ☍ ♅
11 05	☽ □ ♄
17 09	♀ ✶ ♅
18 52	☽ △ ☉
20 01	☽ Q ♇
20 32	☽ # ♅
04 Thursday	
00 01	☽ □ ☿
01 22	☽ □ ♃
06 30	☽ △ ♆
07 07	☽ ∥ ☿
08 29	☽ ∥ ♃
15 14	♀ ⚺ ♄
16 10	☽ ✶ ♇
17 24	☽ Q ♄
20 42	☽ ∥ ♄
23 32	☽ ☍ ♂
05 Friday	
03 27	☿ ☌ ♃
06 58	☉ ⊥ ♄
08 12	☽ Q ♃
08 28	☽ Q ☿
11 55	☽ ⚻ ♅
13 56	☽ ✶ ♄
15 58	☽ □ ♀
17 42	☽ ∠ ♇
17 51	☽ # ♂
21 14	☽ ∥ ♇
22 14	☽ ± ♅
06 Saturday	
01 30	☽ □ ☉
05 07	☽ ✶ ♃
06 35	☽ ✶ ☿
09 13	☽ ⊥ ♇
09 44	☽ □ ♆
11 40	♀ ∠ ♇
13 49	☽ ⚼ ♅
15 59	☽ ∠ ♄
19 39	☽ ⚺ ♇
07 Sunday	
02 12	☽ Q ♀
05 29	☽ ⚻ ♂
07 39	☽ ∠ ♃
07 49	☽ ⊥ ♄
10 50	☽ ∠ ☿
11 17	☽ Q ☉
16 09	☽ △ ♅
16 30	☽ ± ♂
17 18	☽ Q ♆
18 28	☽ ⚺ ♄
23 48	☽ ⊥ ♃
08 Monday	
01 29	☽ ✶ ♀
02 12	☿ ⚺ ♆
04 06	☽ ⊥ ☿
09 11	☽ ⚼ ♂
09 28	☉ ∥ ♆
10 15	☽ ✶ ☉
10 39	☽ ⚺ ♃
12 12	♂ Q ♆
14 41	☽ ✶ ♆
15 44	☽ ⚺ ☿
17 04	☉ ⚺ ♃
09 Tuesday	
00 52	☽ ☌ ♇
06 33	♂ # ♇
07 05	☽ ∠ ♀
13 23	☽ △ ♂
15 25	☽ ∠ ☉
17 48	☽ ∠ ♆
20 54	♀ ⊥ ♄
21 04	☽ # ♂
21 41	☽ ∥ ♇
22 06	☽ □ ♅
10 Wednesday	
00 45	☽ ☌ ♄
01 07	☽ ⊥ ♀
09 11	☽ ⊥ ☉
10 16	☽ ⊥ ♆
13 16	☽ ⚺ ♀
17 58	☽ ☌ ♃
21 08	☽ ⚺ ☉
21 21	☽ ⚺ ♆
11 Thursday	
00 01	☉ ☌ ♆
01 06	☽ ∥ ♄
03 32	☽ ☌ ☿
07 19	☽ Q ♅
07 48	☽ ⚺ ♇
16 22	☽ ∥ ♃
19 00	☽ ⊥ ♇
23 16	☽ □ ♂
23 24	☽ ∥ ☿
12 Friday	
02 31	☽ # ♅
05 50	☽ ✶ ♅
08 49	☽ ⚺ ♄
11 57	☽ ∠ ♇
20 16	☽ ⊥ ♄
21 16	☿ Q ♅
23 53	♀ ⚺ ♃
13 Saturday	
01 41	☿ ⚺ ♇
03 11	☽ ⚺ ♃
03 29	☽ ☌ ♀
05 52	☽ ☌ ♆
10 21	☽ ☌ ☉
10 27	☽ ∠ ♅
11 05	☽ Q ♂
11 32	☉ ∠ ♅
13 36	☽ ∠ ♄
14 54	☽ ⊥ ♃
16 38	☽ ✶ ♇
18 17	☽ ⚺ ☿
19 30	☿ # ♅
20 23	☽ ∥ ♀
22 40	☽ ∥ ♆
14 Sunday	
03 53	☽ ⊥ ♅
04 08	♀ ☌ ♆
07 14	☽ ⊥ ☿
07 35	☉ Q ♂
08 38	☽ ∠ ♃
09 08	☽ ∥ ☉
11 28	☽ ✶ ♂
15 37	☽ ⚺ ♅
15 59	☽ Q ♇
18 58	☽ ✶ ♄
21 07	♀ ∥ ♆
15 Monday	
02 57	☽ ∠ ☿
05 33	☽ # ☉
07 07	☉ ∠ ♄
14 40	☽ ✶ ♃
15 53	☽ # ♀
16 35	☽ ⚺ ♆
17 40	☽ # ♆
18 33	☽ ∠ ♂
18 52	☽ Q ♄
20 39	☽ ⚺ ♀
16 Tuesday	
02 22	☽ ⚺ ☉
03 40	☽ □ ♇
03 48	♀ ∠ ♅
04 35	☽ ⊥ ♆
09 41	☉ ⊥ ♃
10 02	☽ ⊥ ♀
12 31	☽ ✶ ☿
13 29	☽ ⊥ ♂
15 06	☽ Q ♃
15 29	☽ ⊥ ☉
18 26	☉ ✶ ♇
22 44	☽ ∠ ♆
17 Wednesday	
02 12	☽ ⚺ ♂
03 37	☽ ☌ ♅
06 18	☽ ∠ ♀
07 20	☽ □ ♄
08 59	☽ # ☿
11 20	☽ ∠ ☉
15 54	☽ Q ☿
16 30	☽ ∥ ♅
16 43	♀ ∠ ♄
17 07	☿ ⊥ ♇
18 Thursday	
00 29	☽ # ♃
04 14	☽ □ ♃
05 16	☽ ✶ ♆
08 48	♂ ⚺ ♅
16 24	☽ ✶ ♀
16 30	☽ △ ♇
17 20	♀ ✶ ♇
18 10	☽ # ♄
19 31	♀ ⊥ ♃
20 40	☽ ✶ ☉
21 41	♀ Q ♂
19 Friday	
05 47	☽ Q ♆
09 28	☽ □ ☿
16 52	☽ ⚺ ♅
18 25	☽ ☌ ♂
19 42	☽ Q ♀
20 49	☽ △ ♄
23 03	☽ ⚼ ♇
23 19	☽ Q ☉
20 Saturday	
00 57	☽ # ♇
05 04	☽ ⊥ ♅
13 08	☽ ∥ ♂
17 15	☽ ± ♇
18 14	☽ △ ♃
18 20	☽ □ ♆
23 15	☽ ∠ ♅
21 Sunday	
01 51	♃ ⚺ ♆
03 16	☽ ⚼ ♄
05 17	☽ ⚻ ♇
12 04	☽ □ ♀
14 40	☽ □ ☉
21 14	☽ ± ♄
23 35	☿ ✶ ♅
22 Monday	
00 32	☽ ⚼ ♃
02 35	♂ △ ♄
05 02	☽ ✶ ♅
05 46	☽ △ ☿
09 02	☽ ⚻ ♄
09 19	☽ ⚺ ♂
12 05	☉ # ♀
18 18	☽ ± ♃
21 28	☽ ⊥ ♂
23 19	☉ ⊥ ♅
23 Tuesday	
04 14	☽ Q ♅
05 14	☽ △ ♆
05 57	☽ ⚻ ♃
10 12	☿ ⚺ ♄
14 21	☽ ⚼ ☿
15 20	☽ ∠ ♂
15 26	☽ ☍ ♇
16 10	♀ ⊥ ♅
17 32	♂ ⚼ ♇
22 58	☽ ∥ ♂
23 39	☿ ∠ ♇
24 Wednesday	
03 26	☿ □ ♂
03 50	☽ △ ♀
04 54	☽ △ ☉
09 12	☽ ± ☿
09 13	☽ ⚼ ♆
13 19	☽ # ♇
13 45	☽ □ ♅
17 38	☽ ☍ ♄
20 08	☽ ✶ ♂
21 28	☽ ⚻ ☿
25 Thursday	
01 29	☽ ± ♆
09 35	☽ ⚼ ♀
10 01	☽ ⚼ ☉
12 07	☽ ⚻ ♆
13 28	☽ ☍ ♃
17 10	☽ # ♄
18 18	☽ Q ♂
21 28	☽ ⚻ ♇
26 Friday	
02 47	☽ ± ♀
02 52	☽ ± ☉
06 58	☉ ☌ ♀
07 48	☽ ± ♇
09 58	☽ # ♃
13 23	☽ ∥ ♅
13 52	☽ ⚻ ☉
14 00	☽ ⚻ ♀
18 17	☽ △ ♅
22 00	☽ ⚻ ♄
23 01	☽ ⚼ ♇
27 Saturday	
02 16	☽ □ ♂
07 37	☽ ☍ ☿
08 04	☽ ± ♄
11 46	♀ ∠ ♃
11 48	☿ ⊥ ♄
15 05	☽ ☍ ♆
16 56	☽ ⚻ ♃
18 53	☽ # ☿
19 15	☽ ⚼ ♅
21 02	☉ ∠ ♃
22 56	☽ ⚼ ♄
23 48	☽ △ ♇
28 Sunday	
02 52	☽ ± ♃
05 20	☽ # ♆
09 57	☽ ± ♅
10 16	☽ ∥ ☉
11 58	♀ Q ♇
13 48	☽ ∥ ♀
16 53	♀ ⚺ ♅
17 38	☽ ⚼ ♃
18 48	☽ ☍ ☉
19 40	☽ ⚻ ♅
19 55	☽ ☍ ♀
23 21	☽ △ ♄
29 Monday	
01 46	☉ Q ♇
05 12	☽ △ ♂
08 01	☽ # ♀
08 34	☉ ⚺ ♅
09 00	♂ ⊥ ♅
11 38	☽ # ☉
14 16	☽ ⚻ ☿
14 52	☽ ∥ ♆
15 42	☽ ⚻ ♆
18 01	☽ △ ♃
20 05	☽ ∥ ☿
30 Tuesday	
00 08	☽ □ ♇
01 03	☽ ± ☿
01 19	☽ ± ♆
03 24	☿ ☌ ♆
06 12	☽ ⚼ ♂
15 45	☽ ⚼ ♆
15 47	♀ ✶ ♄
17 16	☽ ⚼ ☿
19 54	☽ ☍ ♅
21 13	☽ ± ♂
22 16	☽ ⚻ ☉
23 39	☽ □ ♄
31 Wednesday	
00 20	☽ ⚻ ♀
03 54	☿ ⚺ ♃
05 01	☽ Q ♇
05 42	☽ # ♅
07 12	☽ ∥ ♃
07 17	☽ ⚻ ♂
08 36	☽ ± ☉
10 53	☽ ± ♀
14 51	☉ # ♆
15 58	☽ △ ♆
17 05	☿ ∠ ♅
18 52	☽ □ ♃
19 13	☉ # ☿
20 35	☽ △ ☿
21 04	☉ ✶ ♄
21 46	☿ ∥ ♆
23 34	☽ ∥ ♄

APRIL 2021

LONGITUDES

Date	Sidereal time h m s	Sun ☉ ° ' "	Moon ☽ ° ' "	Moon ☽ 24.00 ° ' "	Mercury ☿ ° '	Venus ♀ ° '	Mars ♂ ° '	Jupiter ♃ ° '	Saturn ♄ ° '	Uranus ♅ ° '	Neptune ♆ ° '	Pluto ♇ ° '
01	00 40 17	11 ♈ 58 39	03 ♐ 40 09	10 ♐ 55 27	25 ♓ 21	13 ♈ 33	16 ♊ 45	23 ♒ 20	11 ♒ 25	09 ♉ 02	21 ♓ 26	26 ♑ 38
02	00 44 13	12 57 50	18 ♐ 06 00	25 ♐ 11 34	27 05	14 48	17 21	23 31	11 29	09 05	21 28	26 39
03	00 48 10	13 56 59	02 ♑ 12 03	09 ♑ 07 29	28 ♓ 50	16 02	17 56	23 43	11 34	09 09	21 30	26 40
04	00 52 06	14 56 06	15 ♑ 57 58	22 ♑ 43 40	00 ♈ 37	17 17	18 32	23 54	11 38	09 12	21 33	26 41
05	00 56 03	15 55 11	29 ♑ 24 49	06 ♒ 01 39	02 25	18 31	19 08	24 06	11 43	09 15	21 35	26 41
06	01 00 00	16 54 15	12 ♒ 34 25	19 ♒ 03 24	04 15	19 46	19 44	24 17	11 47	09 18	21 37	26 42
07	01 03 56	17 53 17	25 ♒ 28 50	01 ♓ 50 56	06 06	21 00	20 21	24 28	11 51	09 22	21 39	26 42
08	01 07 53	18 52 17	08 ♓ 09 55	14 ♓ 25 58	07 59	22 14	20 57	24 39	11 55	09 25	21 41	26 43
09	01 11 49	19 51 15	20 ♓ 39 15	26 ♓ 49 56	09 53	23 29	21 33	24 50	11 59	09 28	21 43	26 44
10	01 15 46	20 50 11	02 ♈ 58 08	09 ♈ 03 59	11 49	24 43	22 09	25 01	12 03	09 31	21 45	26 44
11	01 19 42	21 49 06	15 ♈ 07 37	21 ♈ 09 12	13 46	25 57	22 45	25 12	12 07	09 35	21 47	26 45
12	01 23 39	22 47 58	27 ♈ 08 53	03 ♉ 06 51	15 44	27 12	23 21	25 23	12 11	09 38	21 49	26 45
13	01 27 35	23 46 48	09 ♉ 03 18	14 ♉ 58 29	17 44	28 26	23 57	25 33	12 15	09 41	21 51	26 46
14	01 31 32	24 45 37	20 ♉ 52 41	26 ♉ 46 14	19 45	29 ♈ 40	24 34	25 44	12 19	09 45	21 53	26 46
15	01 35 29	25 44 23	02 ♊ 39 29	08 ♊ 32 52	21 47	00 ♉ 55	25 10	25 54	12 22	09 48	21 55	26 46
16	01 39 25	26 43 07	14 ♊ 26 48	20 ♊ 21 48	23 51	02 09	25 46	26 05	12 26	09 52	21 57	26 47
17	01 43 22	27 41 49	26 ♊ 18 24	02 ♋ 17 10	25 56	03 23	26 22	26 15	12 29	09 55	21 59	26 47
18	01 47 18	28 40 29	08 ♋ 18 40	14 ♋ 23 32	28 ♈ 01	04 37	26 59	26 25	12 32	09 58	22 01	26 47
19	01 51 15	29 ♈ 39 07	20 ♋ 32 23	26 ♋ 45 49	00 ♉ 08	05 51	27 35	26 35	12 36	10 02	22 03	26 47
20	01 55 11	00 ♉ 37 42	03 ♌ 04 28	09 ♌ 28 52	02 15	07 06	28 11	26 45	12 39	10 05	22 05	26 48
21	01 59 08	01 36 15	15 ♌ 59 32	22 ♌ 36 55	04 23	08 20	28 48	26 55	12 42	10 09	22 07	26 48
22	02 03 04	02 34 46	29 ♌ 21 19	06 ♍ 12 57	06 31	09 34	29 ♊ 24	27 04	12 45	10 12	22 08	26 48
23	02 07 01	03 33 15	13 ♍ 11 51	20 ♍ 17 54	08 39	10 48	00 ♋ 00	27 14	12 48	10 16	22 10	26 48
24	02 10 58	04 31 42	27 ♍ 30 45	04 ♎ 49 53	10 46	12 02	00 37	27 23	12 51	10 19	22 12	26 48
25	02 14 54	05 30 06	12 ♎ 14 35	19 ♎ 43 56	12 53	13 16	01 13	27 32	12 53	10 22	22 14	26 48
26	02 18 51	06 28 29	27 ♎ 16 51	04 ♏ 52 11	15 00	14 30	01 49	27 42	12 56	10 26	22 16	26 48
27	02 22 47	07 26 49	12 ♏ 28 39	20 ♏ 04 59	17 05	15 44	02 26	27 51	12 58	10 29	22 17	26 48
28	02 26 44	08 25 08	27 ♏ 39 57	05 ♐ 12 24	19 08	16 58	03 02	28 00	13 01	10 33	22 19	26 R 48
29	02 30 40	09 23 26	12 ♐ 41 17	20 ♐ 05 45	21 10	18 12	03 39	28 08	13 03	10 36	22 21	26 48
30	02 34 37	10 ♉ 21 41	27 ♐ 25 05	04 ♑ 38 46	23 ♉ 09	19 ♉ 26	04 ♋ 15	28 ♒ 17	13 ♒ 05	10 ♉ 40	22 ♓ 23	26 ♑ 48

Date	Moon True ☊ ° '	Moon Mean ☊ ° '	Moon ☽ Latitude ° '
01	12 ♊ 33	14 ♊ 04	00 N 47
02	12 D 33	14 01	00 S 30
03	12 33	13 58	01 43
04	12 R 34	13 55	02 49
05	12 33	13 52	03 44
06	12 30	13 48	04 26
07	12 25	13 45	04 53
08	12 17	13 42	05 05
09	12 08	13 39	05 03
10	11 58	13 36	04 46
11	11 49	13 32	04 16
12	11 40	13 29	03 36
13	11 33	13 26	02 46
14	11 29	13 23	01 49
15	11 27	13 20	00 S 47
16	11 D 26	13 17	00 N 16
17	11 27	13 13	01 20
18	11 29	13 10	02 21
19	11 30	13 07	03 16
20	11 R 30	13 04	04 04
21	11 29	13 01	04 41
22	11 26	12 58	05 04
23	11 22	12 54	05 11
24	11 17	12 51	04 59
25	11 12	12 48	04 27
26	11 07	12 45	03 36
27	11 03	12 42	02 30
28	11 01	12 38	01 N 12
29	11 D 00	12 35	00 S 09
30	11 ♊ 01	12 ♊ 32	01 S 29

DECLINATIONS

Date	Sun ☉ ° '	Moon ☽ ° '	Mercury ☿ ° '	Venus ♀ ° '	Mars ♂ ° '	Jupiter ♃ ° '	Saturn ♄ ° '	Uranus ♅ ° '	Neptune ♆ ° '	Pluto ♇ ° '
01	04 N 44	20 S 07	03 S 58	04 N 12	24 N 15	14 S 22	17 S 51	14 N 07	04 S 23	22 S 09
02	05 07	23 24	03 15	04 42	24 19	14 18	17 50	14 08	04 22	22 09
03	05 30	25 08	02 31	05 11	24 22	14 14	17 49	14 09	04 21	22 09
04	05 53	25 17	01 46	05 41	24 25	14 11	17 48	14 10	04 20	22 09
05	06 16	23 55	01 00	06 11	24 28	14 07	17 47	14 11	04 20	22 09
06	06 38	21 16	00 S 13	06 40	24 31	14 03	17 46	14 12	04 19	22 09
07	07 01	17 37	00 N 35	07 09	24 34	14 00	17 45	14 13	04 18	22 09
08	07 23	13 13	01 24	07 39	24 36	13 56	17 43	14 14	04 17	22 09
09	07 46	08 20	02 14	08 07	24 39	13 53	17 42	14 15	04 16	22 09
10	08 08	03 S 12	03 04	08 36	24 41	13 49	17 41	14 17	04 16	22 09
11	08 30	02 N 01	03 56	09 05	24 43	13 46	17 40	14 18	04 15	22 09
12	08 52	07 06	04 48	09 33	24 45	13 42	17 40	14 19	04 14	22 09
13	09 14	11 54	05 40	10 01	24 46	13 39	17 39	14 20	04 13	22 10
14	09 35	16 13	06 33	10 29	24 48	13 36	17 38	14 21	04 13	22 10
15	09 57	19 55	07 27	10 57	24 49	13 32	17 37	14 22	04 12	22 10
16	10 18	22 48	08 21	11 24	24 50	13 29	17 36	14 23	04 11	22 10
17	10 39	24 43	09 15	11 51	24 51	13 26	17 35	14 24	04 10	22 10
18	11 00	25 31	10 09	12 18	24 52	13 22	17 34	14 25	04 10	22 10
19	11 21	25 06	11 03	12 45	24 53	13 19	17 34	14 26	04 09	22 10
20	11 41	23 26	11 57	13 11	24 53	13 16	17 33	14 27	04 08	22 10
21	12 02	20 31	12 50	13 37	24 54	13 13	17 32	14 28	04 07	22 10
22	12 22	16 27	13 42	14 03	24 54	13 10	17 31	14 29	04 07	22 10
23	12 42	11 23	14 34	14 28	24 54	13 07	17 31	14 31	04 06	22 10
24	13 02	05 N 34	15 25	14 53	24 54	13 04	17 30	14 32	04 05	22 10
25	13 21	00 S 44	16 14	15 17	24 53	13 01	17 29	14 33	04 05	22 10
26	13 41	07 08	17 02	15 41	24 53	12 58	17 29	14 34	04 04	22 11
27	14 00	13 12	17 48	16 05	24 52	12 55	17 28	14 35	04 03	22 11
28	14 19	18 28	18 33	16 29	24 51	12 52	17 27	14 36	04 03	22 11
29	14 37	22 28	19 15	16 51	24 50	12 49	17 27	14 37	04 02	22 11
30	14 N 56	24 S 53	19 N 55	17 N 14	24 N 49	12 S 46	17 S 26	14 N 38	04 S 01	22 S 11

ZODIAC SIGN ENTRIES

Date	h	m	Planets
01	05	59	☽ ♐
03	08	13	☽ ♑
04	03	41	☿ ♈
05	13	04	☽ ♒
07	20	30	☽ ♓
10	06	11	☽ ♈
12	17	44	☽ ♉
14	18	22	♀ ♉
15	06	35	☽ ♊
17	19	25	☽ ♋
19	10	29	☿ ♉
19	20	33	☉ ♉
20	06	11	☽ ♌
22	13	08	☽ ♍
23	11	49	♂ ♋
24	16	06	☽ ♎
26	16	18	☽ ♏
28	15	42	☽ ♐
30	16	16	☽ ♑

LATITUDES

Date	Mercury ☿ ° '	Venus ♀ ° '	Mars ♂ ° '	Jupiter ♃ ° '	Saturn ♄ ° '	Uranus ♅ ° '	Neptune ♆ ° '	Pluto ♇ ° '
01	02 S 19	01 S 15	01 N 29	00 S 39	00 S 31	00 S 25	01 S 04	01 S 21
04	02 12	01 11	01 29	00 40	00 31	00 25	01 04	01 22
07	02 00	01 07	01 29	00 40	00 32	00 25	01 05	01 22
10	01 44	01 03	01 29	00 41	00 32	00 25	01 05	01 22
13	01 23	00 57	01 28	00 41	00 33	00 24	01 05	01 23
16	00 59	00 52	01 28	00 42	00 33	00 24	01 05	01 23
19	00 S 30	00 46	01 28	00 42	00 33	00 24	01 05	01 24
22	00 N 01	00 40	01 28	00 43	00 34	00 24	01 05	01 24
25	00 33	00 33	01 27	00 44	00 34	00 24	01 05	01 24
28	01 05	00 27	01 27	00 44	00 34	00 24	01 05	01 25
31	01 N 34	00 S 20	01 N 27	00 S 45	00 S 35	00 S 24	01 S 05	01 S 25

DATA

Julian Date	2459306
Delta T	+71 seconds
Ayanamsa	24° 08' 57"
Synetic vernal point	04° ♓ 58' 02"
True obliquity of ecliptic	23° 26' 15"

MOON'S PHASES, APSIDES AND POSITIONS ☽

Date	h	m	Phase	Longitude ° '	Eclipse Indicator
04	10	02	☾	14 ♑ 51	
12	02	31	●	22 ♈ 25	
20	06	59	☽	00 ♌ 25	
27	03	32	○	07 ♏ 06	

Day	h	m		
14	17	55	Apogee	
27	15	29	Perigee	
04	02	05	Max dec	25° S 25'
11	02	42	0N	
18	16	01	Max dec	25° N 32'
25	09	14	0S	

LONGITUDES

Date	Chiron ⚷ ° '	Ceres ⚳ ° '	Pallas ⚴ ° '	Juno ⚵ ° '	Vesta ⚶ ° '	Black Moon Lilith ⚸ ° '
01	09 ♈ 03	15 ♈ 23	07 ♓ 50	23 ♐ 43	08 ♍ 08	18 ♉ 03
11	09 ♈ 38	19 ♈ 22	10 ♓ 50	24 ♐ 02	07 ♍ 01	19 ♉ 09
21	10 ♈ 12	23 ♈ 20	13 ♓ 42	23 ♐ 48	06 ♍ 42	20 ♉ 16
31	10 ♈ 44	27 ♈ 18	16 ♓ 25	23 ♐ 02	07 ♍ 11	21 ♉ 22

All ephemeris data is given at 12.00 UT and the Moon's longitude is additionally given for 24.00 UT

ASPECTARIAN

h m	Aspects
01 Thursday	
00 15	☽ ⚼ ☉
00 29	☽ ⚹ ♇
02 50	☽ ⚼ ♀
04 59	☽ Q ♄
07 35	☿ # ♀
20 45	♀ # ♆
20 54	☽ ⚻ ♅
02 Friday	
00 51	☽ Q ♃
00 53	☽ ⚹ ♄
01 12	☽ ∠ ♇
01 37	☽ ∥ ♇
02 46	☽ △ ☉
03 28	☿ ∠ ♄
05 56	☽ △ ♀
06 04	☿ ⚹ ♇
06 56	☽ ± ♅
10 40	☽ ☍ ♂
16 18	☽ ⊥ ♇
17 42	☽ □ ♆
21 17	☽ ⚹ ♃
22 10	☽ ⚼ ♅
22 12	☽ # ♂
03 Saturday	
02 17	☽ ∠ ♄
02 30	☽ ⚺ ♇
05 24	☽ □ ☿
17 51	☽ ⊥ ♄
23 27	☽ ∠ ♃
04 Sunday	
00 05	☽ △ ♅
00 42	☽ Q ♆
01 16	☿ ⊥ ♃
04 21	☽ ⚺ ♄
10 02	☽ □ ☉
13 44	♃ # ♅
14 33	☽ □ ♀
15 29	☽ ⊥ ♃
16 46	☽ ⚻ ♂
17 25	☽ Q ☿
21 55	☽ ⚹ ♆
05 Monday	
02 18	☽ ⚺ ♃
03 57	☽ ± ♂
05 04	☽ # ♂
07 05	☽ ☌ ♇
18 19	☽ ⚹ ☿
20 49	☽ Q ☉
20 58	☽ ⚼ ♂
23 13	☿ ⊥ ♅
06 Tuesday	
01 02	☽ ∠ ♆
02 15	☽ Q ♀
05 04	☽ ∥ ♇
05 58	☽ □ ♅
06 07	☉ ∥ ♀
10 32	☽ ☌ ♄
11 18	♀ ⚹ ♂
17 38	☽ ⊥ ♆
20 40	☽ ⚹ ☉
07 Wednesday	
01 55	☽ △ ♂
02 26	☽ ∠ ☿
02 44	☽ ⚹ ♀
04 49	☽ ⚺ ♆
10 05	☽ ☌ ♃
11 15	☽ ∥ ♄
14 18	☽ ⚺ ♇
15 33	☽ Q ♅
22 13	☽ ⊥ ☿
08 Thursday	
00 55	♀ ⚺ ♆
01 38	☽ ⊥ ♇
02 52	♂ ± ♇
03 09	☽ ∠ ☉
06 43	☽ # ♅
08 14	☽ ∥ ♃
10 03	☽ ∠ ♀
11 36	☽ ⚺ ☿
14 24	☽ ⚹ ♅
18 48	☽ ∠ ♇
19 13	☽ ⚺ ♄
21 20	☿ Q ♇
21 46	☽ ⊥ ☉
09 Friday	
05 11	☽ ⊥ ♀
05 45	☿ Q ♂
06 37	☿ ⚺ ♅
06 49	☽ ⊥ ♄
10 19	☽ ⚺ ☉
11 21	☿ ∠ ♃
12 55	☽ # ♀
13 49	☽ □ ♂
14 04	☽ ☌ ♆
14 32	☽ # ☉
18 05	☽ ⚺ ♀
19 18	♂ □ ♆
19 26	☽ ∠ ♅
20 15	☽ ⚺ ♃
22 28	♀ Q ♄
23 48	☽ ⚹ ♇
10 Saturday	
00 23	☽ ∠ ♄
07 03	☽ ∥ ♆
08 08	☽ ⊥ ♃
12 28	☽ # ☿
13 06	☽ ⊥ ♅
15 09	☿ ⚹ ♄
18 53	♀ ⚹ ♃
23 21	☽ Q ♇
11 Sunday	
00 58	☽ ⚺ ♅
02 06	☽ ∠ ♃
02 52	☽ Q ♂
06 01	☽ ⚹ ♄
08 46	☽ ☌ ☿
11 10	☉ ⚺ ♆
20 46	☿ # ♆
22 24	☽ # ♆
22 44	☽ ∥ ☿
12 Monday	
01 18	☽ ⚺ ♆
02 31	☽ ☌ ☉
03 20	♀ □ ♇
03 59	☽ ⚹ ♂
06 02	☽ Q ♄
08 24	☽ ⚹ ♃
11 12	☽ □ ♇
12 06	☽ ☌ ♀
13 21	☽ ⊥ ♆
21 20	☽ ∥ ☉
13 Tuesday	
00 25	♀ ⊥ ♆
01 21	☽ ∥ ♀
07 32	☽ ∠ ♆
08 56	☽ Q ♃
11 47	☽ ∠ ♂
13 18	☽ ☌ ♅
18 30	☽ □ ♄
21 15	☽ # ♃
23 09	☉ ⚹ ♂
14 Wednesday	
00 14	☉ Q ♄
01 11	☽ ∥ ♅
07 02	☽ ⊥ ♂
09 14	☽ ⚺ ☿
14 03	☽ ⚹ ♆
19 54	☽ ⚺ ♂
20 17	♂ ∠ ♅
20 34	☽ # ♄
20 37	☽ ⚺ ☉
22 02	☽ □ ♃
15 Thursday	
00 00	☽ ⊥ ☿
00 00	☽ △ ♇
08 01	☽ ⚺ ♀
09 57	☽ ⊥ ☉
13 32	☿ ⚺ ♆
14 35	☽ Q ♆
16 59	☉ ⚹ ♃
21 41	☽ ⊥ ♀
22 12	☽ ∠ ☿
16 Friday	
02 38	☽ ⚺ ♅
05 57	☽ ∠ ☉
05 58	☽ # ♇
06 34	☽ ⚼ ♇
07 53	☽ △ ♄
13 27	☉ □ ♇
14 53	☽ ⊥ ♅
18 07	☽ ∠ ♀
18 53	☿ Q ♄
17 Saturday	
00 51	☽ ± ♇
03 15	☽ □ ♆
05 14	♂ △ ♃
09 11	☽ ∠ ♅
11 05	☽ ⚹ ☿
11 53	☽ △ ♃
12 08	☽ ☌ ♂
12 57	☽ ⚻ ♇
14 23	☽ ⚼ ♄
14 35	☽ ∥ ♂
15 03	☽ ⚹ ☉
16 00	☿ ⚹ ♃
19 09	☿ ⚹ ♂
19 17	☉ ⊥ ♆
21 49	☿ □ ♇
18 Sunday	
03 49	☽ ⚹ ♀
04 26	♂ ⚻ ♇
08 28	☽ ± ♄
11 54	☿ ⊥ ♆
15 18	☽ ⚹ ♅
16 06	☽ Q ☿
17 05	☽ Q ☉
18 14	☽ ⚼ ♃
20 24	☽ ⚻ ♄
19 Monday	
01 49	☉ ☌ ☿
06 12	☽ Q ♀
12 05	☽ ± ♃
12 33	♂ ⚼ ♄
14 54	☽ Q ♅
14 56	☽ △ ♆
16 31	☽ ∥ ♂
23 49	☽ ⚻ ♃
20 Tuesday	
00 03	☽ ☍ ♇
01 01	☉ ∥ ☿
02 15	☽ ⚺ ♂
06 59	☽ □ ☉
10 08	☽ □ ☿
11 44	♀ ∠ ♆
14 12	☽ ⊥ ♂
15 55	♀ # ♃
19 09	♃ ⚺ ♇
19 33	☽ ⚼ ♆
20 21	☽ □ ♀
23 36	☽ # ♇
21 Wednesday	
01 11	☽ □ ♅
05 56	☽ ☍ ♄
07 46	☽ ∠ ♂
12 13	☽ ± ♆
21 50	☿ # ♃
23 07	☽ ⚻ ♆
22 Thursday	
01 01	♀ Q ♃
06 12	☽ # ♄
07 29	☽ ⚻ ♇
07 54	☽ ☍ ♃
12 05	☽ ⚹ ♂
18 03	☽ ± ♇
18 06	☽ △ ☉
19 10	☿ ∠ ♆
21 44	☽ ∥ ♅
22 59	☽ ∥ ♀
23 32	☽ ∥ ☿
23 Friday	
01 01	♀ ☌ ♅
02 47	☽ △ ☿
04 11	☽ # ♃
06 17	☿ ∥ ♀
06 29	☽ ∥ ☉
06 57	☽ △ ♅
07 30	☽ △ ♀
09 37	☽ ⚼ ♇
09 52	☽ Q ♂
10 16	☿ ∥ ♅
11 19	☽ ⚻ ♄
14 37	♀ ∥ ♅
19 06	☿ Q ♃
21 31	☽ ± ♄
21 44	☽ ⚼ ☉
24 Saturday	
03 10	☽ ☍ ♆
06 42	☿ ☌ ♅
08 21	☽ ⚼ ♅
08 38	☽ ⚼ ☿
10 50	☽ △ ♇
11 08	☽ ⚼ ♀
11 47	☽ ⚻ ♃
12 33	☽ ⚼ ♄
13 48	☽ ± ☉
13 55	☉ # ♃
17 19	☽ □ ♂
17 43	☽ # ♆
21 45	☽ ± ♃
23 12	☽ ± ♅
25 Sunday	
00 19	☽ ⚻ ☉
01 54	☽ ± ☿
03 14	☽ ± ♀
04 22	♀ □ ♄
08 59	☽ ⚻ ♅
11 58	☿ □ ♄
12 29	☽ ⚼ ♃
13 02	☽ △ ♄
13 13	☽ ⚻ ☿
13 48	☽ ⚻ ♀
22 19	☿ ☌ ♀
26 Monday	
00 27	☽ ∥ ♆
04 01	☽ ⚻ ♆
11 15	☽ □ ♇
12 40	☽ △ ♃
13 33	☽ ± ♆
19 29	☽ △ ♂
27 Tuesday	
01 28	☿ # ♄
03 32	☽ ☍ ☉
03 48	☽ ⚼ ♆
08 00	☉ ∠ ♆
08 51	☽ ☍ ♅
10 48	☽ ∥ ♃
12 47	☽ □ ♄
15 33	☽ # ☉
15 41	☽ Q ♇
17 35	☽ ☍ ♀
17 50	☿ ∠ ♂
17 53	☽ # ♅
20 06	♇ St R
20 08	☽ ⚼ ♂
20 24	☽ ☍ ☿
28 Wednesday	
01 35	☽ # ♀
03 31	☽ △ ♆
07 01	☽ ∥ ♄
10 38	☽ ⚹ ♇
10 58	☽ ± ♂
12 31	☽ # ☿
12 31	☽ □ ♃
17 20	☽ Q ♄
20 54	☽ ⚻ ♂
29 Thursday	
06 20	☽ ⚻ ☉
08 38	☽ ⚻ ♅
09 56	☽ ∥ ♇
10 35	☽ ∠ ♇
11 49	☉ ∥ ♅
12 35	☽ ⚹ ♄
16 40	☽ ± ☉
17 38	☽ Q ♃
18 21	☽ ± ♅
21 44	☽ ⚻ ♀
30 Friday	
01 10	☽ ⊥ ♇
02 27	☿ ⚹ ♆
03 42	☽ □ ♆
03 54	☽ ⚻ ☿
05 09	♀ ∠ ♂
08 22	☽ ⚼ ☉
08 26	☽ ± ♀
09 06	☽ ⚼ ♅
09 43	☉ Q ♃
10 53	☽ # ♂
11 00	☽ ⚺ ♇
13 07	☽ ∠ ♄
13 26	☽ ⚹ ♃
15 19	☽ ± ☿
19 54	☉ ☌ ♅
23 51	☽ ☍ ♂

MAY 2021

LONGITUDES

Date	Sidereal time h m s	Sun ☉ ° ' "	Moon ☽ ° ' "	Moon ☽ 24.00 ° ' "	Mercury ☿ ° '	Venus ♀ ° '	Mars ♂ ° '	Jupiter ♃ ° '	Saturn ♄ ° '	Uranus ♅ ° '	Neptune ♆ ° '	Pluto ♇ ° '
01	02 38 33	11 ♉ 19 56	11 ♑ 46 26	18 ♑ 47 54	25 ♉ 06	20 ♉ 40	04 ♋ 52	28 ♒ 26	13 ♒ 08	10 ♉ 43	22 ♓ 24	26 ♑ 48
02	02 42 30	12 18 08	25 ♑ 43 08	02 ♒ 32 11	27 01	21 54	05 28	28 34	13 10	10 47	22 26	26 R 48
03	02 46 27	13 16 20	09 ♒ 15 16	15 ♒ 52 37	28 ♉ 53	23 08	06 05	28 42	13 12	10 50	22 27	26 48
04	02 50 23	14 14 30	22 ♒ 24 34	28 ♒ 51 27	00 ♊ 41	24 22	06 41	28 50	13 14	10 54	22 29	26 48
05	02 54 20	15 12 38	05 ♓ 13 41	11 ♓ 31 39	02 27	25 35	07 18	28 58	13 15	10 57	22 31	26 48
06	02 58 16	16 10 45	17 ♓ 45 44	23 ♓ 56 21	04 09	26 49	07 54	29 06	13 17	11 00	22 32	26 47
07	03 02 13	17 08 50	00 ♈ 03 51	06 ♈ 08 37	05 47	28 03	08 31	29 14	13 19	11 04	22 34	26 47
08	03 06 09	18 06 54	12 ♈ 10 58	18 ♈ 11 15	07 22	29 ♉ 17	09 07	29 21	13 20	11 07	22 35	26 47
09	03 10 06	19 04 56	24 ♈ 09 45	00 ♉ 06 45	08 53	00 ♊ 31	09 44	29 29	13 22	11 11	22 37	26 47
10	03 14 02	20 02 57	06 ♉ 02 33	11 ♉ 57 24	10 21	01 44	10 21	29 36	13 23	11 14	22 38	26 46
11	03 17 59	21 00 57	17 ♉ 51 34	23 ♉ 45 20	11 45	02 58	10 57	29 43	13 24	11 18	22 39	26 46
12	03 21 56	21 58 55	29 ♉ 38 56	05 ♊ 32 40	13 04	04 12	11 34	29 50	13 25	11 21	22 41	26 45
13	03 25 52	22 56 51	11 ♊ 26 49	17 ♊ 21 42	14 20	05 26	12 10	29 ♒ 57	13 26	11 24	22 42	26 45
14	03 29 49	23 54 46	23 ♊ 17 40	29 ♊ 15 02	15 32	06 39	12 47	00 ♓ 04	13 27	11 28	22 43	26 45
15	03 33 45	24 52 39	05 ♋ 14 12	11 ♋ 15 35	16 40	07 53	13 24	00 10	13 28	11 31	22 45	26 44
16	03 37 42	25 50 31	17 ♋ 19 35	23 ♋ 26 40	17 43	09 07	14 00	00 17	13 29	11 35	22 46	26 44
17	03 41 38	26 48 21	29 ♋ 37 17	05 ♌ 51 55	18 42	10 20	14 37	00 23	13 29	11 38	22 47	26 43
18	03 45 35	27 46 09	12 ♌ 10 42	18 ♌ 35 06	19 37	11 34	15 14	00 29	13 30	11 41	22 48	26 42
19	03 49 31	28 43 56	25 ♌ 04 32	01 ♍ 39 45	20 28	12 48	15 51	00 35	13 30	11 45	22 50	26 42
20	03 53 28	29 ♉ 41 41	08 ♍ 21 05	15 ♍ 08 47	21 14	14 01	16 27	00 40	13 31	11 48	22 51	26 41
21	03 57 25	00 ♊ 39 24	22 ♍ 03 00	29 ♍ 03 47	21 56	15 15	17 04	00 46	13 31	11 51	22 52	26 41
22	04 01 21	01 37 05	06 ♎ 11 01	13 ♎ 24 26	22 33	16 28	17 41	00 51	13 31	11 55	22 53	26 40
23	04 05 18	02 34 45	20 ♎ 43 37	28 ♎ 07 56	23 06	17 42	18 17	00 57	13 R 31	11 58	22 54	26 39
24	04 09 14	03 32 24	05 ♏ 36 38	13 ♏ 08 47	23 34	18 55	18 54	01 02	13 31	12 01	22 55	26 39
25	04 13 11	04 30 01	20 ♏ 43 19	28 ♏ 19 07	23 57	20 09	19 31	01 07	13 31	12 05	22 56	26 38
26	04 17 07	05 27 37	05 ♐ 54 58	13 ♐ 29 40	24 15	21 22	20 08	01 11	13 31	12 08	22 57	26 37
27	04 21 04	06 25 11	21 ♐ 02 04	28 ♐ 31 05	24 29	22 36	20 45	01 16	13 30	12 11	22 58	26 36
28	04 25 00	07 22 45	05 ♑ 55 46	13 ♑ 15 17	24 38	23 49	21 21	01 20	13 30	12 14	22 59	26 36
29	04 28 57	08 20 18	20 ♑ 29 01	27 ♑ 36 27	24 43	25 03	21 58	01 25	13 29	12 17	23 00	26 35
30	04 32 54	09 17 49	04 ♒ 37 18	11 ♒ 31 25	24 R 42	26 16	22 35	01 29	13 29	12 21	23 01	26 34
31	04 36 50	10 ♊ 15 20	18 ♒ 18 48	24 ♒ 59 36	24 ♊ 38	27 ♊ 29	23 ♋ 12	01 ♓ 33	13 ♒ 28	12 ♉ 24	23 ♓ 02	26 ♑ 33

Date	Moon True ☊ ° '	Moon Mean ☊ ° '	Moon ☽ Latitude ° '
01	11 ♊ 03	12 ♊ 29	02 S 41
02	11 D 04	12 26	03 41
03	11 05	12 23	04 27
04	11 R 05	12 19	04 58
05	11 03	12 16	05 13
06	11 01	12 13	05 12
07	10 57	12 10	04 57
08	10 54	12 07	04 29
09	10 50	12 03	03 49
10	10 47	12 00	03 00
11	10 44	11 57	02 03
12	10 43	11 54	01 S 01
13	10 D 43	11 51	00 N 04
14	10 43	11 48	01 09
15	10 44	11 44	02 12
16	10 45	11 41	03 09
17	10 47	11 38	03 59
18	10 47	11 35	04 39
19	10 48	11 32	05 05
20	10 R 48	11 29	05 17
21	10 47	11 25	05 11
22	10 46	11 22	04 47
23	10 45	11 19	04 03
24	10 45	11 16	03 03
25	10 44	11 13	01 49
26	10 44	11 09	00 N 27
27	10 D 44	11 06	00 S 57
28	10 44	11 03	02 16
29	10 44	11 00	03 24
30	10 44	10 57	04 17
31	10 ♊ 45	10 ♊ 54	04 S 54

DECLINATIONS

Date	Sun ☉ ° '	Moon ☽ ° '	Mercury ☿ ° '	Venus ♀ ° '	Mars ♂ ° '	Jupiter ♃ ° '	Saturn ♄ ° '	Uranus ♅ ° '	Neptune ♆ ° '	Pluto ♇ ° '
01	15 N 14	25 S 35	20 N 33	17 N 36	24 N 48	12 S 43	17 S 26	14 N 39	04 S 01	22 S 11
02	15 32	24 37	21 09	17 58	24 46	12 41	17 25	14 40	04 00	22 11
03	15 49	22 14	21 42	18 19	24 44	12 38	17 25	14 41	04 00	22 12
04	16 07	18 44	22 12	18 39	24 42	12 35	17 25	14 43	03 59	22 12
05	16 24	14 26	22 40	19 00	24 40	12 33	17 24	14 44	03 58	22 12
06	16 41	09 37	23 06	19 19	24 38	12 30	17 24	14 45	03 58	22 12
07	16 57	04 S 31	23 29	19 38	24 36	12 28	17 24	14 46	03 57	22 12
08	17 13	00 N 41	23 50	19 57	24 33	12 25	17 23	14 47	03 57	22 12
09	17 29	05 49	24 08	20 15	24 30	12 23	17 23	14 48	03 56	22 13
10	17 45	10 42	24 24	20 33	24 27	12 21	17 23	14 49	03 56	22 13
11	18 01	15 11	24 37	20 50	24 24	12 18	17 23	14 50	03 55	22 13
12	18 16	19 05	24 49	21 06	24 21	12 16	17 22	14 51	03 55	22 13
13	18 30	22 13	24 58	21 22	24 17	12 14	17 22	14 52	03 54	22 13
14	18 45	24 25	25 05	21 37	24 14	12 12	17 22	14 53	03 54	22 14
15	18 59	25 32	25 10	21 52	24 10	12 10	17 22	14 54	03 53	22 14
16	19 13	25 27	25 14	22 06	24 06	12 08	17 22	14 55	03 53	22 14
17	19 26	24 07	25 15	22 19	24 02	12 06	17 22	14 56	03 52	22 14
18	19 40	21 36	25 15	22 32	23 58	12 04	17 22	14 57	03 52	22 15
19	19 53	17 57	25 13	22 44	23 53	12 02	17 22	14 58	03 51	22 15
20	20 05	13 20	25 09	22 55	23 49	12 00	17 22	14 59	03 51	22 15
21	20 17	07 55	25 04	23 06	23 44	11 58	17 22	15 00	03 51	22 15
22	20 29	01 N 56	24 58	23 16	23 39	11 57	17 22	15 01	03 50	22 16
23	20 40	04 S 14	24 50	23 26	23 34	11 55	17 22	15 02	03 50	22 16
24	20 52	10 31	24 41	23 35	23 29	11 54	17 22	15 03	03 49	22 16
25	21 02	16 11	24 30	23 43	23 23	11 52	17 23	15 04	03 49	22 16
26	21 13	20 51	24 19	23 50	23 18	11 51	17 23	15 05	03 49	22 17
27	21 23	24 05	24 06	23 57	23 12	11 49	17 23	15 06	03 48	22 17
28	21 32	25 34	23 53	24 03	23 06	11 48	17 23	15 07	03 48	22 17
29	21 42	25 14	23 38	24 08	23 00	11 47	17 24	15 08	03 48	22 17
30	21 51	23 16	23 23	24 13	22 54	11 45	17 24	15 09	03 47	22 18
31	21 N 59	20 S 00	23 N 06	24 N 17	22 N 47	11 S 44	17 S 24	15 N 10	03 S 47	22 S 18

ZODIAC SIGN ENTRIES

Date	h	m	Planets
02	19	31	☽ ♒
04	02	49	☿ ♊
05	02	08	☽ ♓
07	11	52	☽ ♈
09	02	01	♀ ♊
09	23	46	☽ ♉
12	12	43	☽ ♊
13	22	36	♃ ♓
15	01	30	☽ ♋
17	12	44	☽ ♌
19	20	59	☽ ♍
20	19	37	☉ ♊
22	01	35	☽ ♎
24	03	00	☽ ♏
26	02	39	☽ ♐
28	02	23	☽ ♑
30	04	04	☽ ♒

LATITUDES

Date	Mercury ☿ ° '	Venus ♀ ° '	Mars ♂ ° '	Jupiter ♃ ° '	Saturn ♄ ° '	Uranus ♅ ° '	Neptune ♆ ° '	Pluto ♇ ° '
01	01 N 34	00 S 20	01 N 27	00 S 45	00 S 35	00 S 24	01 S 05	01 S 25
04	01 57	00 13	01 26	00 46	00 35	00 24	01 05	01 26
07	02 15	00 S 05	01 26	00 46	00 36	00 24	01 05	01 26
10	02 25	00 N 02	01 25	00 47	00 36	00 24	01 06	01 26
13	02 28	00 10	01 25	00 48	00 37	00 24	01 06	01 27
16	02 22	00 17	01 25	00 48	00 37	00 24	01 06	01 27
19	02 07	00 24	01 24	00 49	00 37	00 24	01 06	01 28
22	01 44	00 32	01 23	00 50	00 38	00 24	01 06	01 28
25	01 13	00 39	01 23	00 51	00 38	00 24	01 06	01 28
28	00 N 33	00 46	01 22	00 51	00 39	00 24	01 06	01 29
31	00 S 13	00 N 52	01 N 22	00 S 52	00 S 39	00 S 24	01 S 06	01 S 29

DATA

Julian Date	2459336
Delta T	+71 seconds
Ayanamsa	24° 09' 01"
Synetic vernal point	04° ♓ 57' 58"
True obliquity of ecliptic	23° 26' 14"

MOON'S PHASES, APSIDES AND POSITIONS ☽

Date	h	m	Phase	Longitude ° '	Eclipse Indicator
03	19	50	☾	13 ♒ 35	
11	19	00	●	21 ♉ 18	
19	11	13	☽	29 ♌ 01	
26	11	14	○	05 ♐ 26	total

Day	h	m		
11	22	09	Apogee	
26	01	56	Perigee	
01	09	40	Max dec	25° S 36'
08	08	49	0N	
15	22	22	Max dec	25° N 38'
22	19	28	0S	
28	19	26	Max dec	25° S 39'

LONGITUDES

Date	Chiron ⚷ ° '	Ceres ⚳ ° '	Pallas ⚴ ° '	Juno ⚵ ° '	Vesta ⚶ ° '	Black Moon Lilith ⚸ ° '
01	10 ♈ 44	27 ♈ 18	16 ♓ 25	23 ♐ 02	07 ♍ 11	21 ♉ 22
11	11 ♈ 14	01 ♉ 14	18 ♓ 55	21 ♐ 43	08 ♍ 24	22 ♉ 29
21	11 ♈ 41	05 ♉ 07	21 ♓ 13	19 ♐ 57	10 ♍ 16	23 ♉ 35
31	12 ♈ 05	08 ♉ 57	23 ♓ 14	17 ♐ 50	12 ♍ 41	24 ♉ 42

All ephemeris data is given at 12.00 UT and the Moon's longitude is additionally given for 24.00 UT

ASPECTARIAN

h m	Aspects
01 Saturday	
00 44	☽ ⚼ ♀
01 12	♀ # ♄
04 08	☽ ⊥ ♄
08 44	☽ ⚼ ☿
09 40	☽ Q ♆
10 12	☽ △ ♅
11 12	☽ △ ☉
14 18	☽ ⊻ ♄
14 50	☽ ∠ ♃
02 Sunday	
04 42	☽ △ ♀
06 16	☽ ⚹ ♆
06 27	☽ ⊥ ♃
09 19	☿ △ ♇
09 49	☽ # ♂
13 54	☽ ☌ ♇
14 38	☽ △ ☿
17 02	☽ ⊻ ♃
22 38	♀ ⚹ ♆
03 Monday	
06 02	☽ ⊼ ♂
08 46	☽ ∠ ♆
09 33	☿ □ ♃
10 02	☉ □ ♄
12 16	☽ ∥ ♇
14 52	☽ □ ♅
15 32	☽ # ☿
17 20	☽ ± ♂
19 08	☽ ☌ ♄
19 50	☽ □ ☉
04 Tuesday	
01 05	☽ ⊥ ♆
10 36	☽ ⚼ ♂
11 29	☿ # ♇
12 00	☿ ⊥ ♂
12 08	☽ ⊻ ♆
12 24	☽ # ♀
16 00	☽ □ ♀
19 47	☽ ∥ ♄
20 09	☽ ⊻ ♇
05 Wednesday	
00 05	☽ ☌ ♃
00 07	☽ Q ♅
02 07	☽ # ☉
05 14	♃ Q ♅
05 54	☽ □ ☿
07 24	☽ ⊥ ♇
07 52	☽ Q ☉
10 23	♂ ± ♄
10 28	☽ # ♅
16 07	☽ △ ♂
21 44	☽ ∥ ♃
22 57	☽ ⚹ ♅
06 Thursday	
00 30	☽ ∠ ♇
03 21	☽ ⊻ ♄
05 42	☽ Q ♀
08 41	☽ ⚹ ☉
11 25	♀ △ ♇
14 57	☽ ⊥ ♄
17 43	☿ Q ♆
21 17	☽ ☌ ♆
21 50	☽ Q ☿
07 Friday	
04 07	☽ ∠ ♅
05 34	☽ ⚹ ♇
07 36	☽ ⚹ ♀
08 33	☽ ∠ ♄
10 21	☽ ⊻ ♃
14 36	☽ ∥ ♆
16 27	☽ ∠ ☉
21 55	☽ ⊥ ♅
22 18	☽ ⊥ ♃
08 Saturday	
01 01	☽ ⚹ ☿
05 14	☽ Q ♇
05 35	☽ □ ♂
09 53	☽ ⊻ ♅
11 51	☽ ⊥ ☉
13 38	♀ □ ♃
14 18	☽ ⚹ ♄
16 23	☽ ∠ ♃
16 40	☽ ∠ ♀
09 Sunday	
00 54	☽ ⊻ ☉
02 27	☉ # ♄
03 08	☽ # ♆
08 52	☽ ⊻ ♆
11 23	☽ ∠ ☿
12 47	☽ ⊥ ♀
14 25	☽ Q ♄
17 16	☽ □ ♇
19 35	☽ Q ♂
20 59	☽ ⊥ ♆
22 50	☽ ⚹ ♃
10 Monday	
02 17	☽ ⊻ ♀
08 06	☽ ⊥ ☿
11 48	☿ ⊻ ♂
15 14	☽ ∠ ♆
16 48	☿ ∥ ♂
20 25	☽ # ♃
21 12	☽ ⚹ ♂
21 55	☽ ⊻ ☿
22 35	☽ ☌ ♅
23 24	☽ Q ♃
11 Tuesday	
02 55	☽ □ ♄
03 48	☿ ⊻ ♅
10 00	☽ ∥ ♅
12 21	☿ ⚼ ♇
19 00	☽ ☌ ☉
21 47	☽ ⚹ ♆
12 Wednesday	
00 57	☽ # ♄
02 48	♂ ⚹ ♅
05 22	☽ ∠ ♂
06 07	☽ △ ♇
06 10	☽ ∥ ☉
12 23	☽ □ ♃
18 34	☿ △ ♄
21 33	♀ Q ♆
22 16	☽ Q ♆
22 20	☽ ☌ ♀
13 Thursday	
00 42	☽ ⊥ ♂
04 03	☽ ∥ ♀
05 45	☉ ⚹ ♆
11 55	☽ ⊻ ♅
12 02	☽ # ♇
12 37	☽ ⚼ ♇
13 33	☽ ⊻ ♂
16 03	☽ △ ♄
18 32	☽ ☌ ☿
14 Friday	
00 09	☽ ⊥ ♅
06 51	☽ ± ♇
09 30	☽ ∥ ♂
10 51	☽ □ ♆
13 22	☽ ⊻ ☉
16 59	♀ ⊥ ♂
18 25	☽ ∠ ♅
18 57	☽ ⊼ ♇
22 24	☽ ⚼ ♄
15 Saturday	
00 22	☽ ∥ ☿
01 45	☽ △ ♃
02 30	☽ ⊥ ☉
14 49	♂ ⊼ ♄
16 27	☽ ± ♄
17 53	☽ ⊻ ♀
22 03	☽ ∠ ☉
16 Sunday	
00 35	☽ ⚹ ♅
04 24	☽ ⊼ ♄
05 06	☽ ☌ ♂
07 08	☽ ⊥ ♀
07 55	☽ ⚼ ♃
08 32	☿ ⊥ ♅
12 51	☽ ⊻ ☿
17 50	☽ ∥ ☿
22 42	☽ △ ♆
17 Monday	
00 19	☽ Q ♅
01 39	☽ ⊥ ☿
01 44	☽ ± ♃
02 47	☽ ∠ ♀
03 26	♀ # ♇
06 05	☽ ⚹ ☉
06 23	☽ ☍ ♇
09 49	☉ △ ♇
13 08	☽ ∥ ♂
13 29	☽ ⊼ ♃
20 30	☽ ∠ ☿
18 Tuesday	
03 42	☽ ⚼ ♆
04 52	☽ ∥ ♀
06 47	☽ # ♇
07 03	☽ Q ☉
10 43	☽ ⚹ ♀
11 04	☽ □ ♅
14 29	☽ ☍ ♄
14 31	♀ ⊻ ♅
14 43	♀ ⚼ ♇
18 01	☽ ⊻ ♂
20 42	☽ ± ♆
23 43	♂ ⚼ ♃
19 Wednesday	
00 50	☽ ∥ ☉
02 55	☽ ⚹ ☿
05 45	☽ ⊥ ♂
07 51	☽ ⊼ ♆
11 26	☽ Q ♀
14 58	☽ ⊼ ♇
15 20	☽ # ♄
18 53	☿ ± ♇
19 13	☽ □ ☉
22 07	☽ ☍ ♃
23 02	☽ ∠ ♂
20 Thursday	
01 52	☽ ± ♇
01 58	♀ △ ♄
02 18	☽ Q ☿
03 55	☽ ∥ ♅
17 55	☽ ⚼ ♇
18 09	☽ △ ♅
18 13	☽ # ♃
21 08	☽ ⊼ ♄
23 01	☽ □ ♀
21 Friday	
02 57	☽ ⚹ ♂
07 37	☽ ± ♄
11 47	☽ □ ☿
13 24	☽ ☍ ♆
15 03	☉ □ ♃
19 56	☽ △ ♇
20 17	☽ ⚼ ♅
23 04	☽ ⚼ ♄
22 Saturday	
00 33	☽ Q ♂
02 59	☽ ⊼ ♃
03 46	☽ △ ☉
04 32	☽ # ♆
11 33	☽ ± ♅
13 08	☽ ± ♃
21 34	☽ ⊼ ♅
23 Sunday	
00 11	☽ △ ♄
02 43	☿ □ ♆
04 08	☽ ⚼ ♃
06 30	☽ ⚼ ☉
06 36	☽ △ ♀
07 51	☽ □ ♂
09 18	♄ St R
10 06	☽ ∥ ♆
15 33	☽ ⊼ ♆
15 59	☽ △ ☿
17 31	♀ ⊥ ♅
21 36	☽ □ ♇
22 09	☽ ± ☉
24 Monday	
01 15	☽ ± ♆
01 27	♀ ∥ ♂
04 37	☽ △ ♃
08 28	☽ ⊼ ☉
09 04	☽ ⚼ ♀
11 15	♀ ⊻ ♂
15 41	☽ ⚼ ♆
16 51	☽ ⚼ ☿
17 33	☽ ∥ ♃
22 15	☽ ☍ ♅
25 Tuesday	
00 35	☽ □ ♄
00 40	☽ ± ♀
02 22	☽ Q ♇
07 01	☽ # ♅
07 30	☽ ± ☿
10 01	☽ △ ♂
11 01	☽ ⊼ ♀
13 07	☉ ∠ ♂
15 30	☽ △ ♆
17 13	☽ ⊼ ☿
17 36	☽ ∥ ♄
21 20	☽ ⚹ ♇
21 23	♀ ± ♇
23 06	☉ Q ♆
26 Wednesday	
04 30	☽ □ ♃
05 02	☽ Q ♄
10 42	☽ ⚼ ♂
11 14	☽ ☍ ☉
14 17	☽ # ☉
21 01	☽ ∠ ♇
21 11	☽ ∥ ♇
21 52	☽ ⊼ ♅
27 Thursday	
00 01	☽ ⚹ ♄
01 33	☽ ± ♂
04 19	☽ # ♂
07 26	☽ ± ♅
09 10	☽ Q ♃
10 39	☽ # ♀
11 19	☽ ⊥ ♇
11 31	☽ ⊼ ♂
12 13	☽ # ☿
14 43	☽ ☍ ♀
15 06	☽ □ ♆
17 35	☽ ☍ ☿
19 25	♀ □ ♆
20 55	☽ ⊻ ♇
21 53	☽ ⚼ ♅
23 18	☿ ∥ ♀
23 58	☽ ∠ ♄
28 Friday	
04 31	☽ ⚹ ♃
14 32	☽ ⊼ ☉
14 33	☽ ⊥ ♄
20 16	☽ Q ♆
22 22	☽ △ ♅
29 Saturday	
00 23	☽ ⊻ ♄
01 04	☽ ± ☉
05 11	☽ ∠ ♃
05 13	☿ ☌ ♀
14 36	☽ ☍ ♂
16 13	☽ ⚹ ♆
17 08	☽ ⚼ ☉
19 06	☽ ⊼ ☿
20 19	☽ ⊥ ♃
20 23	☽ ⊼ ♀
22 15	☽ ☌ ♇
22 34	☿ St R
30 Sunday	
02 47	☽ # ♀
05 17	☽ ± ☿
06 34	☽ ⊻ ♃
07 34	☽ ± ♀
10 49	☽ # ☿
15 18	☽ # ♂
17 47	♀ ⊼ ♇
17 53	☽ ∠ ♆
20 00	☽ ∥ ♇
20 43	☽ △ ☉
20 47	☽ ⚼ ☿
22 57	☽ # ☉
31 Monday	
00 41	☽ ⚼ ♀
01 30	☽ □ ♅
03 25	☽ ☌ ♄
05 15	♂ △ ♆
09 43	☽ ⊥ ♆
10 07	♀ ∠ ♅
20 27	☽ ⊻ ♆
21 11	☽ ⊼ ♂
23 13	☽ △ ☿

JUNE 2021

LONGITUDES

Date	Sidereal time h m s	Sun ☉ ° ' "	Moon ☽ ° ' "	Moon ☽ 24.00 ° ' "	Mercury ☿ ° '	Venus ♀ ° '	Mars ♂ ° '	Jupiter ♃ ° '	Saturn ♄ ° '	Uranus ♅ ° '	Neptune ♆ ° '	Pluto ⯓ ° '
01	04 40 47	11 ♊ 12 50	01 ♓ 34 01	08 ♓ 02 26	24 ♊ 28	28 ♊ 43	23 ♋ 49	01 ♓ 36	13 ♒ 27	12 ♉ 27	23 ♓ 02	26 ♑ 32
02	04 44 43	12 10 20	14 ♓ 25 14	20 ♓ 42 52	24 R 15	29 ♊ 56	24 25	01 40	13 R 26	12 30	23 03	26 R 31
03	04 48 40	13 07 48	26 ♓ 55 49	03 ♈ 04 38	23 58	01 ♋ 09	25 02	01 43	13 25	12 33	23 04	26 30
04	04 52 36	14 05 16	09 ♈ 09 49	15 ♈ 11 53	23 37	02 23	25 39	01 46	13 24	12 36	23 05	26 29
05	04 56 33	15 02 43	21 ♈ 11 22	27 ♈ 08 44	23 13	03 36	26 16	01 49	13 23	12 39	23 05	26 28
06	05 00 29	16 00 09	03 ♉ 04 29	08 ♉ 59 03	22 46	04 49	26 53	01 52	13 21	12 42	23 06	26 27
07	05 04 26	16 57 35	14 ♉ 52 52	20 ♉ 46 20	22 17	06 02	27 30	01 55	13 20	12 45	23 07	26 26
08	05 08 23	17 55 00	26 ♉ 39 49	02 ♊ 33 41	21 46	07 16	28 07	01 57	13 19	12 48	23 07	26 25
09	05 12 19	18 52 24	08 ♊ 28 23	14 ♊ 28 13	21 13	08 29	28 44	01 59	13 17	12 51	23 08	26 24
10	05 16 16	19 49 47	20 ♊ 20 33	26 ♊ 18 54	20 40	09 42	29 21	02 01	13 15	12 54	23 08	26 23
11	05 20 12	20 47 10	02 ♋ 19 02	08 ♋ 21 13	20 06	10 55	29 ♋ 58	02 03	13 14	12 57	23 09	26 22
12	05 24 09	21 44 32	14 ♋ 25 41	20 ♋ 32 40	19 33	12 08	00 ♌ 35	02 05	13 12	13 00	23 09	26 21
13	05 28 05	22 41 53	26 ♋ 42 25	02 ♌ 55 11	19 01	13 22	01 12	02 06	13 10	13 03	23 10	26 20
14	05 32 02	23 39 14	09 ♌ 11 12	15 ♌ 30 45	18 30	14 35	01 49	02 07	13 08	13 06	23 10	26 19
15	05 35 58	24 36 33	21 ♌ 54 05	28 ♌ 21 26	18 01	15 48	02 26	02 09	13 06	13 09	23 10	26 17
16	05 39 55	25 33 51	04 ♍ 53 05	11 ♍ 29 15	17 34	17 01	03 03	02 09	13 03	13 11	23 11	26 16
17	05 43 52	26 31 09	18 ♍ 10 09	24 ♍ 55 59	17 10	18 14	03 40	02 10	13 01	13 14	23 11	26 15
18	05 47 48	27 28 26	01 ♎ 46 52	08 ♎ 42 54	16 50	19 27	04 17	02 11	12 59	13 17	23 11	26 14
19	05 51 45	28 25 42	15 ♎ 44 03	22 ♎ 50 15	16 33	20 40	04 54	02 11	12 56	13 19	23 11	26 13
20	05 55 41	29 ♊ 22 57	00 ♏ 01 17	07 ♏ 16 50	16 21	21 53	05 31	02 11	12 54	13 22	23 12	26 11
21	05 59 38	00 ♋ 20 11	14 ♏ 36 29	21 ♏ 59 37	16 12	23 06	06 08	02 R 11	12 51	13 25	23 12	26 10
22	06 03 34	01 17 25	29 ♏ 25 34	06 ♐ 53 30	16 08	24 19	06 45	02 11	12 49	13 27	23 12	26 09
23	06 07 31	02 14 38	14 ♐ 22 31	21 ♐ 51 36	16 D 08	25 32	07 22	02 10	12 46	13 30	23 12	26 08
24	06 11 27	03 11 51	29 ♐ 19 45	06 ♑ 45 56	16 13	26 44	07 59	02 10	12 43	13 32	23 12	26 06
25	06 15 24	04 09 04	14 ♑ 09 08	21 ♑ 28 27	16 23	27 57	08 36	02 09	12 40	13 35	23 12	26 05
26	06 19 21	05 06 16	28 ♑ 43 02	05 ♒ 52 13	16 38	29 ♋ 10	09 13	02 08	12 37	13 37	23 R 12	26 04
27	06 23 17	06 03 28	12 ♒ 55 25	19 ♒ 52 17	16 57	00 ♌ 23	09 50	02 07	12 34	13 40	23 12	26 02
28	06 27 14	07 00 40	26 ♒ 42 35	03 ♓ 26 13	17 21	01 36	10 28	02 05	12 31	13 42	23 12	26 01
29	06 31 10	07 57 52	10 ♓ 03 17	16 ♓ 33 59	17 50	02 48	11 05	02 04	12 28	13 44	23 12	26 00
30	06 35 07	08 ♋ 55 04	22 ♓ 58 37	29 ♓ 17 36	18 ♊ 24	04 ♌ 01	11 ♌ 42	02 ♓ 02	12 ♒ 24	13 ♉ 47	23 ♓ 12	25 ♑ 58

Date	Moon True ☊ ° '	Moon Mean ☊ ° '	Moon ☽ Latitude ° '
01	10 ♊ 45	10 ♊ 50	05 S 14
02	10 D 45	10 47	05 17
03	10 45	10 44	05 05
04	10 45	10 41	04 40
05	10 45	10 38	04 02
06	10 46	10 35	03 15
07	10 46	10 31	02 19
08	10 47	10 28	01 18
09	10 47	10 25	00 S 13
10	10 R 47	10 22	00 N 53
11	10 46	10 19	01 57
12	10 45	10 15	02 56
13	10 44	10 12	03 48
14	10 42	10 09	04 30
15	10 40	10 06	05 00
16	10 39	10 03	05 15
17	10 38	10 00	05 14
18	10 D 38	09 56	04 55
19	10 38	09 53	04 19
20	10 40	09 50	03 26
21	10 41	09 47	02 19
22	10 42	09 44	01 N 02
23	10 R 42	09 41	00 S 20
24	10 41	09 37	01 41
25	10 39	09 34	02 54
26	10 36	09 31	03 54
27	10 33	09 28	04 39
28	10 29	09 25	05 05
29	10 26	09 21	05 14
30	10 ♊ 24	09 ♊ 18	05 S 07

DECLINATIONS

Date	Sun ☉ ° '	Moon ☽ ° '	Mercury ☿ ° '	Venus ♀ ° '	Mars ♂ ° '	Jupiter ♃ ° '	Saturn ♄ ° '	Uranus ♅ ° '	Neptune ♆ ° '	Pluto ⯓ ° '
01	22 N 07	15 S 48	22 N 50	24 N 20	22 N 41	11 S 43	17 S 25	15 N 11	03 S 47	22 S 18
02	22 15	11 00	22 32	24 23	22 34	11 42	17 25	15 12	03 47	22 19
03	22 22	05 53	22 14	24 25	22 27	11 41	17 25	15 13	03 46	22 19
04	22 29	00 S 39	21 56	24 26	22 20	11 40	17 26	15 14	03 46	22 19
05	22 36	04 N 31	21 37	24 26	22 13	11 40	17 26	15 15	03 46	22 20
06	22 42	09 29	21 19	24 26	22 06	11 39	17 27	15 16	03 46	22 20
07	22 48	14 05	21 00	24 25	21 58	11 38	17 27	15 17	03 45	22 20
08	22 53	18 09	20 42	24 23	21 51	11 38	17 28	15 18	03 45	22 21
09	22 58	21 30	20 24	24 20	21 43	11 37	17 29	15 19	03 45	22 21
10	23 03	23 58	20 07	24 17	21 35	11 37	17 29	15 20	03 45	22 21
11	23 07	25 22	19 50	24 13	21 27	11 36	17 30	15 20	03 45	22 21
12	23 11	25 34	19 34	24 09	21 19	11 36	17 30	15 21	03 45	22 22
13	23 14	24 32	19 20	24 03	21 10	11 36	17 31	15 22	03 45	22 22
14	23 17	22 17	19 07	23 57	21 02	11 35	17 32	15 23	03 44	22 22
15	23 20	18 56	18 55	23 50	20 53	11 35	17 33	15 24	03 44	22 23
16	23 22	14 36	18 44	23 43	20 45	11 35	17 33	15 25	03 44	22 23
17	23 24	09 29	18 35	23 35	20 36	11 35	17 34	15 25	03 44	22 23
18	23 25	03 N 48	18 28	23 26	20 27	11 35	17 35	15 26	03 44	22 24
19	23 26	02 S 12	18 23	23 16	20 18	11 35	17 36	15 27	03 44	22 24
20	23 26	08 15	18 18	23 06	20 08	11 36	17 37	15 28	03 44	22 25
21	23 26	14 00	18 16	22 55	19 59	11 36	17 37	15 29	03 44	22 25
22	23 26	19 01	18 16	22 44	19 49	11 36	17 38	15 29	03 44	22 25
23	23 25	22 51	18 17	22 32	19 40	11 37	17 39	15 30	03 44	22 26
24	23 24	25 07	18 20	22 19	19 30	11 37	17 40	15 31	03 44	22 26
25	23 22	25 34	18 25	22 06	19 20	11 38	17 41	15 32	03 44	22 26
26	23 20	24 14	18 31	21 52	19 10	11 38	17 42	15 32	03 44	22 27
27	23 18	21 23	18 38	21 37	19 00	11 39	17 43	15 33	03 44	22 27
28	23 15	17 23	18 47	21 22	18 49	11 40	17 44	15 34	03 44	22 27
29	23 12	12 38	18 57	21 06	18 39	11 41	17 45	15 34	03 44	22 28
30	23 N 08	07 S 29	19 N 08	20 N 50	18 N 28	11 S 42	17 S 46	15 N 35	03 S 44	22 S 28

ZODIAC SIGN ENTRIES

Date	h	m	Planets
01	09	07	☽ ♓
02	13	19	♀ ♋
03	17	59	☽ ♈
06	05	46	☽ ♉
08	18	47	☽ ♊
11	07	22	☽ ♋
11	13	34	♂ ♌
13	18	22	☽ ♌
16	03	02	☽ ♍
18	08	54	☽ ♎
20	11	58	☽ ♏
21	03	32	☉ ♋
22	12	55	☽ ♐
24	13	05	☽ ♑
26	14	08	☽ ♒
27	04	27	♀ ♌
28	17	51	☽ ♓

LATITUDES

Date	Mercury ☿ ° '	Venus ♀ ° '	Mars ♂ ° '	Jupiter ♃ ° '	Saturn ♄ ° '	Uranus ♅ ° '	Neptune ♆ ° '	Pluto ⯓ ° '
01	00 S 30	00 N 54	01 N 22	00 S 53	00 S 39	00 S 24	01 S 07	01 S 29
04	01 21	01 01	01 21	00 53	00 40	00 24	01 07	01 30
07	02 13	01 07	01 21	00 54	00 40	00 24	01 07	01 30
10	03 00	01 12	01 20	00 55	00 41	00 24	01 07	01 30
13	03 40	01 18	01 19	00 56	00 41	00 24	01 07	01 31
16	04 08	01 22	01 18	00 57	00 41	00 24	01 07	01 31
19	04 25	01 27	01 17	00 57	00 42	00 24	01 07	01 31
22	04 29	01 30	01 17	00 58	00 42	00 24	01 08	01 32
25	04 21	01 33	01 16	00 59	00 43	00 24	01 08	01 32
28	04 04	01 36	01 15	01 00	00 43	00 24	01 08	01 32
31	03 S 39	01 N 38	01 N 14	01 S 01	00 S 43	00 S 24	01 S 08	01 S 33

DATA

Julian Date	2459367
Delta T	+71 seconds
Ayanamsa	24° 09' 06"
Synetic vernal point	04° ♓ 57' 53"
True obliquity of ecliptic	23° 26' 14"

MOON'S PHASES, APSIDES AND POSITIONS ☽

Date	h	m	Phase	Longitude ° '	Eclipse Indicator
02	07	24	☾	11 ♓ 59	
10	10	53	●	19 ♊ 47	Annular
18	03	54	☽	27 ♍ 09	
24	18	40	○	03 ♑ 28	

Day	h	m		
08	02	41	Apogee	
23	10	01	Perigee	
04	15	00	0N	
12	04	07	Max dec	25° N 38'
19	03	17	0S	
25	05	52	Max dec	25° S 38'

LONGITUDES

Date	Chiron ⚷ ° '	Ceres ⚳ ° '	Pallas ⚴ ° '	Juno ⚵ ° '	Vesta ⚶ ° '	Black Moon Lilith ⚸ ° '
01	12 ♈ 07	09 ♉ 20	23 ♓ 25	17 ♐ 37	12 ♍ 57	24 ♉ 48
11	12 ♈ 26	13 ♉ 05	25 ♓ 06	15 ♐ 21	15 ♍ 53	25 ♉ 55
21	12 ♈ 41	16 ♉ 45	26 ♓ 24	13 ♐ 10	19 ♍ 12	27 ♉ 01
31	12 ♈ 51	20 ♉ 18	27 ♓ 18	11 ♐ 15	22 ♍ 51	28 ♉ 08

All ephemeris data is given at 12.00 UT and the Moon's longitude is additionally given for 24.00 UT

ASPECTARIAN

h m Aspects

01 Tuesday
02 48 ☽ ⊻ ⯓
03 16 ☽ ∥ ♄
06 14 ☽ △ ♀
06 57 ♀ ♇ ♄
08 37 ☽ ± ♂
09 56 ☽ Q ♅
12 04 ☽ ☌ ♃
13 47 ☽ ⊥ ⯓
15 10 ☽ # ♅
19 54 ☉ ♇ ⯓

02 Wednesday
02 07 ☽ ♇ ♂
06 32 ☽ ∠ ⯓
07 14 ☿ ⊻ ♂
07 24 ☽ □ ☉
07 44 ☿ ∥ ♂
08 21 ☽ ⚹ ♅
08 37 ☽ ∥ ♃
10 08 ☽ ⊻ ♄
15 18 ♂ Q ♅
20 43 ☉ ⊻ ♅
21 32 ☽ ⊥ ♄

03 Thursday
00 18 ☉ # ⯓
04 10 ☉ ∥ ☿
04 31 ☽ ☌ ♆
05 40 ☿ # ⯓
06 24 ☽ □ ☿
08 08 ☽ △ ♂
11 10 ☽ ⚹ ⯓
13 13 ☽ ∠ ♅
14 53 ☽ ∠ ♄
19 05 ☉ △ ♄
20 11 ☉ ∥ ♂
20 53 ☽ Q ☉
21 09 ☽ □ ♀
21 23 ☽ ⊻ ♃
21 43 ☽ ∥ ♆
23 33 ♀ △ ♃

04 Friday
06 55 ☽ ⊥ ♅
09 14 ☽ ⊥ ♃
10 40 ☽ Q ⯓
14 54 ♂ # ⯓
16 43 ☽ Q ☿
17 00 ♂ ± ♃
18 52 ☽ ⊻ ♅
20 24 ☽ ⚹ ♄
22 38 ☽ ⚹ ☉

05 Saturday
03 12 ☽ ∠ ♃
08 27 ☽ # ♆
12 55 ☽ Q ♀
15 49 ☽ ⊻ ♆
15 56 ☽ ⚹ ☿
19 05 ☿ □ ♆
19 45 ♂ ☍ ⯓
20 25 ☽ Q ♄
22 37 ☽ □ ⯓
22 47 ☽ □ ♂

06 Sunday
03 57 ☽ ⊥ ♆
07 26 ☽ ∠ ☉
09 32 ☽ ⚹ ♃
15 57 ☽ ⚹ ♀
21 10 ☽ ∠ ☿
22 13 ☽ ∠ ♆
22 58 ☽ # ♃

07 Monday
03 19 ☽ ⊥ ☉
07 39 ☽ ☌ ♅
08 51 ☽ □ ♄
10 01 ☽ Q ♃
13 19 ☽ Q ♂
14 44 ☽ ⊥ ☿
16 36 ☽ ⊻ ☉
18 44 ☽ ∥ ♅

08 Tuesday
02 00 ☽ ∠ ♀
02 27 ☽ ⊻ ☿
04 41 ☿ ⊥ ♂
04 47 ☽ ⚹ ♆
07 41 ☽ # ♄
11 30 ☽ △ ⯓
12 57 ♀ ± ♄
15 07 ☽ ⚹ ♂
22 26 ☽ ⊥ ♀
22 47 ☽ □ ♃

09 Wednesday
04 09 ☽ ∥ ☿
05 13 ☽ Q ♆
11 31 ☉ ⊥ ♅
12 01 ☽ ⊻ ♀
13 40 ☽ ∥ ♂
17 56 ☽ ♇ ⯓
19 15 ☽ # ⯓
20 55 ☽ ⊻ ♅
21 44 ☽ △ ♄
23 14 ☽ ∠ ♂

10 Thursday
01 34 ☽ ∥ ☉
09 05 ☽ ⊥ ♅
10 53 ☽ ☌ ☉
12 05 ☽ ± ⯓
12 37 ☽ ☌ ☿
16 01 ☽ ∥ ♀
17 37 ☽ □ ♆
18 22 ☽ ⊥ ♂

11 Friday
00 07 ☽ ⚻ ⯓
00 24 ☿ ± ⯓
01 13 ☉ ☌ ☿
01 40 ☉ ± ⯓
03 15 ☽ ∠ ♅
03 51 ☽ ♇ ♄
07 02 ☽ ⊻ ♂
11 28 ☽ △ ♃
21 44 ☽ ± ♄

12 Saturday
06 59 ☽ ☌ ♀
09 11 ☽ ⚹ ♅
09 35 ☽ ⚻ ♄
17 13 ☽ ♇ ♃
21 38 ☽ ⊻ ☿

13 Sunday
03 33 ☽ ⊻ ☉
05 06 ☽ △ ♆
05 38 ♀ ⚹ ♅
08 15 ♀ ⚻ ♄
08 46 ☽ Q ♅
08 50 ☽ ⊥ ☿
10 24 ☿ ⊥ ♅
10 50 ☽ ± ♃
11 16 ☽ ☍ ⯓
16 11 ☽ ⊥ ☉
18 46 ☽ ∥ ♀
21 08 ☽ ☌ ♂
22 27 ☽ ⚻ ♃
23 39 ☉ □ ♆

14 Monday
01 32 ☽ ∠ ☿
03 02 ☽ ∥ ☉
10 03 ☽ ♇ ♆
10 54 ☽ ∠ ☉
11 17 ☽ # ⯓
19 27 ☽ □ ♅
19 28 ☽ ☍ ♄
22 01 ♄ □ ♅
22 20 ☽ ∥ ♂
23 19 ☽ ⊻ ♀

15 Tuesday
00 39 ♂ ⚻ ♃
03 07 ☽ ± ♆
04 58 ☽ ⚹ ☿
11 47 ☽ ⊥ ♀
12 06 ☽ ∥ ☿
14 22 ☽ ⚻ ♆
17 27 ☽ ⚹ ☉
20 10 ☽ ⚻ ⯓
20 12 ☽ # ♄

16 Wednesday
01 07 ☿ ∠ ♂
02 33 ☽ Q ☿
06 12 ☽ ∠ ♀
07 00 ☽ ☍ ♃
07 13 ☽ ± ⯓
07 50 ☽ ∥ ♅
08 28 ☽ ⊻ ♂
14 52 ♀ ♇ ♃
17 16 ☽ Q ☉
19 57 ☽ ⊥ ♂
20 11 ☿ ⊻ ♀
23 35 ☽ ♇ ⯓

17 Thursday
02 31 ☽ # ♃
02 47 ☽ ⚻ ♄
03 07 ☽ △ ♅
05 24 ☉ ⚻ ⯓
10 16 ☽ □ ☿
12 07 ☽ ⚹ ♀
12 55 ☽ ∠ ♂
13 31 ☽ ± ♄
20 55 ☽ ☍ ♆

18 Friday
02 18 ☽ △ ⯓
03 54 ☽ □ ☉
05 23 ☽ ♇ ♄
05 52 ☽ ♇ ♅
11 22 ☽ Q ♀
12 17 ☽ # ♆
12 41 ☽ ⚻ ♃
14 46 ☉ ∥ ♀
16 32 ☽ ⚹ ♂
21 33 ☽ ± ♅
23 05 ☽ ± ♃

19 Saturday
00 12 ☉ ♇ ♄
07 15 ☽ △ ♄
07 53 ☽ ⚻ ♅
09 14 ☉ ∠ ♅
13 22 ☽ △ ☿
14 04 ☽ Q ♂
14 28 ☽ ♇ ♃
18 02 ☽ ∥ ♆
21 07 ☽ □ ♀

20 Sunday
00 36 ☽ ⚻ ♆
05 38 ☽ □ ⯓
10 37 ☽ ± ♆
10 52 ☽ △ ☉
14 10 ☽ ♇ ☿
15 05 ♃ St R
15 35 ☽ △ ♃
18 10 ♃ ⊥ ⯓
20 05 ☿ ⊥ ♀
21 15 ☉ ⊥ ♂
21 29 ☽ □ ♂

21 Monday
01 30 ☽ ♇ ♆
01 41 ☽ ∥ ♃
04 51 ☽ ± ☿
09 09 ☽ □ ♄
10 02 ☽ ☍ ♅
11 17 ☽ Q ⯓
13 16 ☽ ♇ ☉
13 57 ♀ △ ♆
14 35 ☽ ⚻ ☿
18 38 ☽ # ♅

22 Tuesday
01 57 ☽ △ ♆
03 01 ☽ △ ♀
04 52 ☽ ± ☉
04 53 ☽ ∥ ♄
06 43 ☽ ⚹ ⯓
08 03 ☽ # ☿
14 13 ☽ Q ♄
15 12 ☽ ⚻ ☉
16 14 ☽ # ♂
16 26 ☽ □ ♃
22 00 ☿ St D

23 Wednesday
00 17 ☽ △ ♂
05 17 ☽ ♇ ♀
06 48 ☽ ∠ ⯓
08 46 ☽ ∥ ⯓
09 25 ☽ ⚹ ♄
09 41 ☽ # ♀
10 11 ☉ △ ♃
10 35 ☽ ⚻ ♅
11 24 ♀ Q ♅
14 50 ☽ ☍ ☿
16 31 ☽ # ☉
20 14 ☽ ± ♅
20 59 ☽ ± ♀
21 12 ☽ ⊥ ⯓
21 17 ☽ Q ♃
23 39 ♀ ☍ ⯓
23 40 ♀ # ⯓

24 Thursday
00 39 ♀ ± ♃
01 22 ☽ ♇ ♂
02 09 ☽ □ ♆
06 49 ☽ ⊻ ⯓
07 28 ☽ ⚻ ♀
09 25 ☽ ∠ ♄
10 43 ☽ ♇ ♅
16 28 ☽ ± ♂
16 33 ☽ ⚹ ♃
18 40 ☽ ☍ ☉
20 21 ♂ ♇ ♆
23 53 ☽ ⊥ ♄

25 Friday
02 35 ☽ ⚻ ♂
07 12 ☽ Q ♆
09 35 ☽ ⊻ ♄
11 04 ☽ △ ♅
15 42 ☽ ⚻ ☿
16 53 ☽ ∠ ♃
19 18 ♆ St R

26 Saturday
01 43 ☽ ± ☿
02 51 ☽ ⚹ ♆
07 35 ☽ ☌ ⯓
07 42 ☽ ⊥ ♃
12 49 ☽ ☍ ♀
16 58 ☽ ♇ ☿
17 42 ☽ ⊻ ♃
21 00 ☽ # ☉
23 29 ☽ ⚻ ☉

27 Sunday
03 57 ☽ ∠ ♆
04 14 ☽ ∥ ⯓
06 29 ☽ ☍ ♂
10 13 ☽ # ♀
10 25 ☽ ± ☉
11 23 ☽ ☌ ♄
13 16 ☽ □ ♅
19 08 ☽ △ ☿
19 22 ☽ ⊥ ♆

28 Monday
00 07 ☉ ± ♄
03 07 ☽ ♇ ☉
03 38 ☽ # ♂
04 28 ☽ # ☿
05 49 ☽ ⊻ ♆
10 05 ☽ ∥ ♄
10 47 ☽ ⊻ ⯓
14 44 ☿ ∥ ♂
20 55 ☽ Q ♅
21 26 ☽ ⊥ ⯓
21 31 ♀ ⚻ ♃
21 32 ☽ # ♅
21 33 ☽ ☌ ♃
21 34 ☽ ⚻ ♀

29 Tuesday
07 54 ☽ △ ☉
09 29 ☽ ± ♀
12 56 ☿ ∠ ♀
13 43 ☽ ∠ ⯓
13 58 ☽ ⚻ ♂
16 24 ☽ ⊻ ♄
16 34 ☽ ∥ ♃
18 47 ☽ ⚹ ♅

30 Wednesday
01 36 ☽ ± ♂
03 00 ☽ □ ☿
03 28 ☽ ⊥ ♄
03 48 ☽ ♇ ♀
12 25 ☽ ☌ ♆
17 39 ☽ ⚹ ⯓
19 25 ☽ ♇ ♂
20 22 ☽ ∠ ♄
23 03 ☽ ∠ ♅

JULY 2021

LONGITUDES

Date	Sidereal time h m s	Sun ☉ ° ' "	Moon ☽ ° ' "	Moon ☽ 24.00 ° ' "	Mercury ☿ ° '	Venus ♀ ° '	Mars ♂ ° '	Jupiter ♃ ° '	Saturn ♄ ° '	Uranus ♅ ° '	Neptune ♆ ° '	Pluto ♇ ° '
01	06 39 03	09 ♋ 52 16	05 ♈ 31 26	11 ♈ 40 37	19 ♊ 02	05 ♌ 14	12 ♌ 19	02 ♓ 00	12 ♒ 21	13 ♉ 49	23 ♓ 11	25 ♑ 57
02	06 43 00	10 49 28	17 ♈ 45 44	23 ♈ 47 25	19 45	06 27	12 56	01 R 58	12 R 18	13 51	23 R 11	25 R 56
03	06 46 56	11 46 41	29 ♈ 46 15	05 ♉ 42 51	20 32	07 39	13 33	01 55	12 14	13 53	23 11	25 54
04	06 50 53	12 43 53	11 ♉ 37 50	17 ♉ 31 48	21 24	08 52	14 11	01 53	12 11	13 56	23 11	25 53
05	06 54 50	13 41 06	23 ♉ 25 19	29 ♉ 18 54	22 21	10 04	14 48	01 50	12 07	13 58	23 10	25 51
06	06 58 46	14 38 20	05 ♊ 13 04	11 ♊ 08 17	23 22	11 17	15 25	01 47	12 03	14 00	23 10	25 50
07	07 02 43	15 35 33	17 ♊ 04 59	23 ♊ 03 32	24 27	12 30	16 03	01 44	12 00	14 02	23 10	25 49
08	07 06 39	16 32 47	29 ♊ 04 16	05 ♋ 07 29	25 36	13 42	16 40	01 41	11 56	14 04	23 09	25 47
09	07 10 36	17 30 00	11 ♋ 13 24	17 ♋ 22 15	26 50	14 55	17 17	01 37	11 52	14 06	23 09	25 46
10	07 14 32	18 27 14	23 ♋ 34 10	29 ♋ 49 16	28 08	16 07	17 54	01 33	11 48	14 08	23 08	25 44
11	07 18 29	19 24 29	06 ♌ 07 37	12 ♌ 29 17	29 ♊ 30	17 19	18 32	01 30	11 44	14 10	23 08	25 43
12	07 22 25	20 21 43	18 ♌ 54 17	25 ♌ 22 38	00 ♋ 56	18 32	19 09	01 26	11 40	14 11	23 07	25 41
13	07 26 22	21 18 57	01 ♍ 54 20	08 ♍ 29 21	02 26	19 44	19 47	01 21	11 36	14 13	23 07	25 40
14	07 30 19	22 16 12	15 ♍ 07 42	21 ♍ 49 22	03 59	20 57	20 24	01 17	11 32	14 15	23 06	25 39
15	07 34 15	23 13 26	28 ♍ 34 21	05 ♎ 22 38	05 37	22 09	21 01	01 13	11 28	14 17	23 06	25 37
16	07 38 12	24 10 40	12 ♎ 14 14	19 ♎ 09 08	07 18	23 21	21 39	01 08	11 24	14 18	23 05	25 36
17	07 42 08	25 07 55	26 ♎ 07 18	03 ♏ 08 41	09 03	24 33	22 16	01 03	11 20	14 20	23 04	25 34
18	07 46 05	26 05 10	10 ♏ 13 12	17 ♏ 20 43	10 51	25 46	22 54	00 58	11 16	14 22	23 04	25 33
19	07 50 01	27 02 24	24 ♏ 31 00	01 ♐ 43 46	12 42	26 58	23 31	00 53	11 11	14 23	23 03	25 31
20	07 53 58	27 59 40	08 ♐ 58 40	16 ♐ 15 13	14 36	28 10	24 09	00 48	11 07	14 25	23 02	25 30
21	07 57 54	28 56 55	23 ♐ 32 51	00 ♑ 50 56	16 33	29 ♌ 22	24 46	00 42	11 03	14 26	23 01	25 28
22	08 01 51	29 ♋ 54 11	08 ♑ 08 44	15 ♑ 25 29	18 31	00 ♍ 34	25 24	00 37	10 59	14 28	23 01	25 27
23	08 05 48	00 ♌ 51 27	22 ♑ 40 22	29 ♑ 52 35	20 32	01 46	26 01	00 31	10 54	14 29	23 00	25 26
24	08 09 44	01 48 44	07 ♒ 01 21	14 ♒ 05 58	22 35	02 58	26 39	00 25	10 50	14 30	22 59	25 24
25	08 13 41	02 46 01	21 ♒ 05 47	28 ♒ 00 19	24 39	04 10	27 16	00 19	10 46	14 31	22 58	25 23
26	08 17 37	03 43 19	04 ♓ 49 10	11 ♓ 32 07	26 45	05 22	27 54	00 13	10 41	14 33	22 57	25 21
27	08 21 34	04 40 38	18 ♓ 09 02	24 ♓ 39 58	28 ♋ 51	06 34	28 31	00 07	10 37	14 34	22 56	25 20
28	08 25 30	05 37 57	01 ♈ 05 05	07 ♈ 24 39	00 ♌ 57	07 46	29 09	00 ♓ 00	10 32	14 35	22 55	25 19
29	08 29 27	06 35 18	13 ♈ 39 03	19 ♈ 48 46	03 04	08 58	29 ♌ 47	29 ♒ 54	10 28	14 36	22 54	25 17
30	08 33 23	07 32 39	25 ♈ 54 19	01 ♉ 56 18	05 10	10 09	00 ♍ 24	29 47	10 23	14 37	22 53	25 16
31	08 37 20	08 ♌ 30 02	07 ♉ 55 19	13 ♉ 52 04	07 ♌ 16	11 ♍ 21	01 ♍ 02	29 ♒ 40	10 ♒ 19	14 ♉ 38	22 ♓ 52	25 ♑ 14

Date	Moon True ☊ ° '	Moon Mean ☊ ° '	Moon ☽ Latitude ° '
01	10 ♊ 23	09 ♊ 15	04 S 44
02	10 D 23	09 12	04 10
03	10 24	09 09	03 25
04	10 26	09 06	02 31
05	10 27	09 02	01 32
06	10 29	08 59	00 S 29
07	10 R 28	08 56	00 N 36
08	10 27	08 53	01 40
09	10 23	08 50	02 40
10	10 18	08 47	03 33
11	10 12	08 43	04 17
12	10 05	08 40	04 49
13	09 59	08 37	05 06
14	09 53	08 34	05 08
15	09 49	08 31	04 53
16	09 46	08 27	04 21
17	09 D 46	08 24	03 34
18	09 46	08 21	02 33
19	09 48	08 18	01 22
20	09 48	08 15	00 N 04
21	09 R 47	08 12	01 S 14
22	09 45	08 08	02 27
23	09 40	08 05	03 30
24	09 33	08 02	04 19
25	09 25	07 59	04 51
26	09 16	07 56	05 05
27	09 08	07 52	05 02
28	09 01	07 49	04 44
29	08 56	07 46	04 12
30	08 54	07 43	03 29
31	08 ♊ 53	07 ♊ 40	02 S 38

DECLINATIONS

Date	Sun ☉ ° '	Moon ☽ ° '	Mercury ☿ ° '	Venus ♀ ° '	Mars ♂ ° '	Jupiter ♃ ° '	Saturn ♄ ° '	Uranus ♅ ° '	Neptune ♆ ° '	Pluto ♇ ° '
01	23 N 04	02 S 10	19 N 20	20 N 33	18 N 18	11 S 43	17 S 47	15 N 36	03 S 45	22 S 28
02	23 00	03 N 07	19 33	20 15	18 07	11 44	17 48	15 36	03 45	22 29
03	22 55	08 12	19 47	19 57	17 56	11 45	17 49	15 37	03 45	22 29
04	22 50	12 55	20 02	19 38	17 45	11 46	17 51	15 38	03 45	22 30
05	22 44	17 08	20 16	19 19	17 34	11 47	17 52	15 38	03 45	22 30
06	22 38	20 42	20 31	18 59	17 22	11 48	17 53	15 39	03 45	22 30
07	22 32	23 24	20 46	18 39	17 11	11 50	17 54	15 40	03 45	22 31
08	22 25	25 06	21 01	18 18	17 00	11 51	17 55	15 40	03 46	22 31
09	22 18	25 37	21 16	17 57	16 48	11 53	17 56	15 41	03 46	22 31
10	22 10	24 53	21 31	17 36	16 36	11 54	17 57	15 41	03 46	22 32
11	22 02	22 53	21 45	17 13	16 24	11 56	17 59	15 42	03 46	22 32
12	21 54	19 43	21 58	16 51	16 13	11 57	18 00	15 42	03 47	22 32
13	21 45	15 34	22 10	16 28	16 00	11 59	18 01	15 43	03 47	22 33
14	21 36	10 35	22 21	16 04	15 48	12 01	18 02	15 43	03 47	22 33
15	21 26	05 N 03	22 30	15 41	15 36	12 03	18 04	15 44	03 47	22 34
16	21 17	00 S 49	22 39	15 16	15 24	12 05	18 05	15 44	03 48	22 34
17	21 06	06 46	22 45	14 52	15 11	12 07	18 06	15 45	03 48	22 34
18	20 56	12 28	22 49	14 27	14 59	12 09	18 07	15 45	03 48	22 35
19	20 45	17 35	22 51	14 01	14 46	12 11	18 09	15 46	03 49	22 35
20	20 34	21 43	22 51	13 36	14 33	12 13	18 10	15 46	03 49	22 35
21	20 22	24 30	22 49	13 09	14 21	12 15	18 11	15 47	03 49	22 36
22	20 10	25 38	22 44	12 43	14 08	12 17	18 12	15 47	03 50	22 36
23	19 58	24 59	22 37	12 16	13 55	12 19	18 14	15 47	03 50	22 36
24	19 45	22 41	22 26	11 49	13 42	12 22	18 15	15 48	03 50	22 37
25	19 32	19 03	22 13	11 22	13 29	12 24	18 16	15 48	03 51	22 37
26	19 19	14 28	21 58	10 54	13 15	12 26	18 18	15 48	03 51	22 37
27	19 06	09 19	21 40	10 26	13 02	12 29	18 19	15 49	03 52	22 38
28	18 52	03 S 55	21 19	09 58	12 48	12 31	18 20	15 49	03 52	22 38
29	18 37	01 N 31	20 56	09 30	12 35	12 34	18 22	15 49	03 52	22 38
30	18 23	06 45	20 30	09 01	12 21	12 36	18 23	15 50	03 53	22 39
31	18 N 08	11 N 39	20 N 02	08 N 32	12 N 08	12 S 39	18 S 24	15 N 50	03 S 53	22 S 39

ZODIAC SIGN ENTRIES

Date	h	m	Planets
01	01	21	☽ ♈
03	12	28	☽ ♉
06	01	24	☽ ♊
08	13	51	☽ ♋
11	00	21	☽ ♌
11	20	35	☿ ♋
13	08	30	☽ ♍
15	14	32	☽ ♎
17	18	38	☽ ♏
19	21	08	☽ ♐
21	22	36	☽ ♑
22	00	37	♀ ♍
22	14	26	☉ ♌
24	00	12	☽ ♒
26	03	30	☽ ♓
28	01	12	☿ ♌
28	09	58	☽ ♈
28	12	43	♃ ♒
29	20	32	♂ ♍
30	20	08	☽ ♉

LATITUDES

Date	Mercury ☿ ° '	Venus ♀ ° '	Mars ♂ ° '	Jupiter ♃ ° '	Saturn ♄ ° '	Uranus ♅ ° '	Neptune ♆ ° '	Pluto ♇ ° '
01	03 S 39	01 N 38	01 N 14	01 S 01	00 S 43	00 S 24	01 S 08	01 S 33
04	03 08	01 39	01 14	01 02	00 44	00 24	01 08	01 33
07	02 33	01 40	01 13	01 02	00 44	00 24	01 08	01 33
10	01 55	01 40	01 12	01 03	00 44	00 24	01 08	01 34
13	01 15	01 39	01 11	01 04	00 45	00 25	01 08	01 34
16	00 S 36	01 38	01 10	01 05	00 45	00 25	01 09	01 34
19	00 N 02	01 36	01 09	01 05	00 45	00 25	01 09	01 35
22	00 35	01 33	01 08	01 06	00 46	00 25	01 09	01 35
25	01 03	01 29	01 07	01 07	00 46	00 25	01 09	01 35
28	01 24	01 25	01 07	01 07	00 46	00 25	01 09	01 35
31	01 N 38	01 N 20	01 N 06	01 S 08	00 S 47	00 S 25	01 S 09	01 S 36

DATA

Julian Date	2459397
Delta T	+71 seconds
Ayanamsa	24° 09' 11"
Synetic vernal point	04° ♓ 57' 48"
True obliquity of ecliptic	23° 26' 14"

MOON'S PHASES, APSIDES AND POSITIONS ☽

Date	h	m	Phase	Longitude ° '	Eclipse Indicator
01	21	11	☾	10 ♈ 14	
10	01	17	●	18 ♋ 02	
17	10	11	☽	25 ♎ 04	
24	02	37	○	01 ♒ 26	
31	13	16	☾	08 ♉ 33	

Day	h	m		
05	14	55	Apogee	
21	10	31	Perigee	
01	21	45	0N	
09	10	03	Max dec	25° N 37'
16	08	41	0S	
22	15	10	Max dec	25° S 39'
29	05	16	0N	

LONGITUDES

Date	Chiron ⚷ ° '	Ceres ⚳ ° '	Pallas ⚴ ° '	Juno ⚵ ° '	Vesta ⚶ ° '	Black Moon Lilith ⚸ ° '
01	12 ♈ 51	20 ♉ 18	27 ♓ 18	11 ♐ 15	22 ♍ 51	28 ♉ 08
11	12 ♈ 55	23 ♉ 43	27 ♓ 42	09 ♐ 45	26 ♍ 47	29 ♉ 14
21	12 ♈ 55	26 ♉ 59	27 ♓ 35	08 ♐ 47	00 ♎ 57	00 ♊ 21
31	12 ♈ 49	00 ♊ 03	26 ♓ 54	08 ♐ 21	05 ♎ 19	01 ♊ 27

All ephemeris data is given at 12.00 UT and the Moon's longitude is additionally given for 24.00 UT

ASPECTARIAN

01 Thursday
04 53 ☽ ∥ ♆
05 12 ☽ ⚺ ♃
11 22 ☽ △ ♀
13 08 ♂ ☍ ♄
15 06 ☽ Q ☿
16 28 ☽ ⊥ ♅
16 42 ☽ Q ♇
16 47 ☽ ⊥ ♃
21 11 ☽ □ ☉

02 Friday
01 16 ☽ ✶ ♄
01 58 ☽ △ ♂
04 15 ☽ ⚺ ♅
10 25 ☽ ∠ ♃
14 54 ☽ # ♆
15 35 ☿ ⊥ ♅
16 12 ☽ ✶ ☿
17 30 ☿ ± ♇
22 48 ☽ ⚺ ♆

03 Saturday
00 57 ☽ Q ♄
04 15 ☽ □ ♇
10 49 ☽ ⊥ ♆
12 01 ☽ Q ☉
16 19 ☽ ✶ ♃
19 09 ☿ ∥ ♀
21 06 ♃ Q ♅
22 29 ♀ bQ ♆
22 51 ☉ ⚻ ♄

04 Sunday
00 32 ☽ ∠ ☿
00 50 ♂ # ♄
01 40 ♂ □ ♅
05 00 ☽ ∠ ♆
05 44 ☽ □ ♀
05 54 ☽ # ♃
09 19 ♃ ⊥ ♇
13 06 ☽ □ ♄
14 26 ☽ ✶ ☉
16 33 ☽ Q ♃
16 41 ☽ ☌ ♅
17 28 ☽ □ ♂
20 19 ☽ ⊥ ☿

05 Monday
03 03 ☽ ∥ ♅
09 36 ☽ ⚺ ☿
11 30 ☽ ✶ ♆
14 29 ☽ ∥ ♂
16 30 ☽ # ♄
16 57 ☽ △ ♇
19 14 ☉ ✶ ♅
22 33 ☽ Q ♀
23 40 ☽ ∠ ☉

06 Tuesday
00 53 ☽ ∥ ♀
05 03 ☽ □ ♃
07 39 ☿ □ ♆
08 09 ☽ Q ♂
10 34 ☽ ∥ ☿
11 54 ☽ Q ♆
19 33 ☽ ⊥ ☉
23 22 ☽ bQ ♇

07 Wednesday
01 41 ☽ ✶ ♀
01 47 ☽ △ ♄
02 36 ♀ ☍ ♄
03 03 ☽ # ♇
03 35 ☽ ∥ ☉
05 50 ☽ ⚺ ♅
08 44 ☽ ⚺ ☉
09 47 ☽ ✶ ♂
15 18 ☉ # ♇
17 28 ☽ ± ♇
17 57 ☽ ⊥ ♅

08 Thursday
00 12 ☽ □ ♆
04 20 ☽ ☌ ☿
05 28 ☽ ⚻ ♇
07 46 ☽ bQ ♄
11 11 ☽ ∠ ♀
11 59 ☽ ∠ ♅
15 06 ☉ bQ ♃
15 36 ☿ ⚻ ♇
17 09 ☽ △ ♃
17 26 ☽ ∠ ♂
19 25 ♀ □ ♅
20 29 ☉ ⚺ ♂

09 Friday
01 31 ☽ ± ♄
06 48 ♂ ± ♆
06 58 ☽ ⊥ ♀
12 08 ☽ ⊥ ♂
12 40 ☿ bQ ♄
13 02 ♀ # ♄
13 15 ☽ ⚻ ♄
17 38 ☽ ✶ ♅
19 59 ☽ ⚺ ♀
22 29 ☽ bQ ♃

10 Saturday
00 28 ☽ ⚺ ♂
01 17 ☽ ☌ ☉
11 11 ☽ △ ♆
15 49 ☽ ± ♃
16 10 ☽ ☍ ♇
16 56 ☽ Q ♅
21 49 ☽ ⚺ ☿

11 Sunday
03 14 ☽ ⚻ ♃
06 06 ☿ ∠ ♅
08 13 ♀ ± ♆
10 39 ☽ ⊥ ☿
15 08 ☽ # ♇
15 48 ☽ bQ ♆
19 43 ☽ ∥ ☉
21 01 ☽ ∥ ☿
22 32 ☽ ☍ ♄

12 Monday
03 10 ☽ □ ♅
05 44 ☽ ∠ ☿
07 24 ☉ ∥ ☿
08 41 ☽ ± ♆
11 14 ☽ ☌ ♀
12 29 ☽ ☌ ♂
14 56 ☽ ⚺ ☉
19 45 ☿ △ ♃
19 50 ☽ ⚻ ♆
22 33 ☽ # ♄

13 Tuesday
00 33 ☽ ⚻ ♇
02 55 ☽ ⊥ ☉
06 44 ☽ ∥ ♀
09 32 ☽ ∥ ♂
11 00 ☽ ☍ ♃
11 11 ☽ ∥ ♅
11 34 ☽ ± ♇
13 05 ☽ ✶ ☿
13 33 ♀ ☌ ♂
20 41 ☽ ∠ ☉

14 Wednesday
03 55 ☽ bQ ♇
05 29 ☽ # ♃
05 33 ☽ ⚻ ♄
10 25 ☽ △ ♅
13 46 ☽ Q ☿
16 19 ☽ ± ♄
21 20 ♂ ∥ ♅
21 55 ☽ ⚺ ♂
23 28 ☽ ⚺ ♀

15 Thursday
01 46 ☽ ✶ ☉
02 17 ☽ ☍ ♆
06 46 ☽ △ ♇
08 17 ☽ bQ ♄
08 46 ♀ ∥ ♅
08 49 ☉ △ ♆
09 07 ☽ ⊥ ♂
09 57 ☿ ± ♄
11 11 ☽ ⊥ ♀
13 15 ☽ bQ ♅
16 38 ☽ ⚻ ♃
17 13 ☽ # ♆
20 55 ☿ # ♇
21 05 ♀ ∥ ♂
21 19 ☿ ∠ ♂

16 Friday
00 37 ☽ Q ☉
01 46 ☽ ∠ ♂
02 09 ☽ □ ☿
03 08 ☽ ± ♃
04 34 ☽ ∠ ♀
05 07 ☽ ± ♅
06 41 ♀ ⚻ ♆
10 33 ☽ △ ♄
15 37 ☽ ⚻ ♅
18 44 ☽ bQ ♃
23 59 ☽ ∥ ♆

17 Saturday
05 04 ☽ ✶ ♂
06 46 ☽ ⚻ ♆
09 04 ☽ ✶ ♀
10 08 ☉ ± ♃
10 11 ☽ □ ☉
11 03 ☽ □ ♇
17 03 ☽ ± ♆
20 23 ☽ △ ♃
22 46 ☉ ☍ ♇

18 Sunday
02 33 ☽ Q ♂
07 27 ☽ Q ♀
07 48 ♀ ⚻ ♇
08 21 ☽ bQ ♆
08 41 ☿ ∠ ♀
10 37 ☽ ∥ ♃
13 13 ☽ △ ☿
13 45 ☽ □ ♄
17 13 ☿ ⚻ ♄
17 36 ☽ Q ♇
18 22 ♂ ⚻ ♆
19 00 ☽ ☍ ♅
19 06 ☉ Q ♅
20 14 ☽ # ♀
22 56 ☽ # ♂

19 Monday
03 03 ☽ # ♅
09 33 ☽ △ ♆
10 16 ☽ □ ♂
13 41 ☽ ✶ ♇
14 57 ☽ ∥ ♄
16 27 ☽ □ ♀
16 30 ☽ △ ☉
18 06 ☽ bQ ☿
19 17 ☉ ⚺ ♀
19 45 ☽ Q ♄
22 32 ☽ □ ♃

20 Tuesday
04 53 ☽ # ☉
09 38 ☿ ✶ ♅
11 17 ☽ ± ☿
14 30 ☽ ∠ ♇
15 31 ☽ ✶ ♄
18 15 ☽ ∥ ♇
19 06 ☽ bQ ☉
20 15 ☽ # ☿
20 59 ☽ ⚻ ♅
22 42 ☽ ⚻ ☿

21 Wednesday
02 10 ☿ bQ ♃
04 05 ☽ Q ♃
05 19 ☽ ⊥ ♇
06 52 ☽ ± ♅
10 57 ☽ ± ☉
11 09 ☽ □ ♆
14 06 ☽ △ ♂
15 10 ☽ ⚺ ♇
16 05 ☽ ∠ ♄
21 30 ☽ ⚻ ☉
21 42 ☽ bQ ♅
22 26 ☽ △ ♀
23 41 ☽ ✶ ♃

22 Thursday
06 49 ☽ ⊥ ♄
12 45 ♀ ☍ ♃
14 09 ♂ ⚻ ♇
15 52 ☽ bQ ♂
16 38 ☽ ⚺ ♄
16 43 ☽ Q ♆
22 25 ☽ △ ♅

23 Friday
00 14 ☽ ∠ ♃
01 21 ☽ bQ ♀
03 02 ☿ ⊥ ♂
04 10 ☉ ⚻ ♃
05 16 ♀ ± ♇
07 24 ☽ ± ♂
07 53 ☽ ☍ ☿
09 23 ♀ # ♃
12 26 ☿ # ♇
12 32 ☽ ✶ ♆
15 02 ☽ ⊥ ♃
16 34 ☽ ☌ ♇
17 37 ☽ ± ♀
17 49 ☽ ⚻ ♂

24 Saturday
00 59 ☽ ⚺ ♃
02 37 ☽ ☍ ☉
04 33 ☽ ⚻ ♀
12 34 ☽ ∥ ♇
13 37 ☽ ∠ ♆
14 02 ☽ # ☿
16 35 ☿ △ ♆
18 25 ☽ ☌ ♄

25 Sunday
00 43 ☽ □ ♅
04 55 ☽ ⊥ ♆
08 16 ☿ ± ♃
08 59 ☽ # ☉
15 14 ☽ ⚺ ♆
16 22 ☽ ∥ ♄
19 15 ☽ ⚻ ☿
19 24 ☽ ⚺ ♇
20 15 ☿ ☍ ♇
23 14 ☽ ☍ ♂

26 Monday
03 56 ☽ ☌ ♃
05 23 ☽ # ♅
05 53 ☽ ⊥ ♇
07 39 ☽ ± ☿
07 58 ☽ Q ♅
09 43 ☿ Q ♅
09 54 ☽ ⚻ ☉
13 04 ☽ ☍ ♀
18 08 ☽ # ♂
21 25 ☽ ± ☉
21 37 ☽ ∥ ♃
21 52 ☽ ∠ ♇
22 25 ☽ ⚺ ♄

27 Tuesday
02 41 ☽ bQ ☿
05 28 ☽ ✶ ♅
06 26 ☽ # ♀
06 48 ☿ ⚺ ♂
09 12 ☽ ⊥ ♄
15 01 ☽ bQ ☉
20 47 ☽ ☌ ♆

28 Wednesday
01 13 ☽ ✶ ♇
01 41 ☽ ∠ ♄
01 45 ☿ ⚻ ♃
08 11 ☽ ⚻ ♂
09 10 ☽ ∠ ♅
09 59 ☽ ⚺ ♃
11 42 ☽ △ ☿
12 11 ☽ ∥ ♆
20 06 ☽ ± ♂
21 14 ☽ ⊥ ♃
21 19 ☽ △ ☉
23 47 ☽ Q ♇

29 Thursday
02 00 ☽ ⚻ ♀
02 16 ☽ ⊥ ♅
05 53 ☽ ✶ ♄
09 22 ☿ ⊥ ♀
13 42 ♂ # ♃
13 51 ☽ ⚺ ♅
14 18 ☽ bQ ♂
14 23 ☽ ∠ ♃
14 48 ☽ ± ♀
15 50 ♂ ☍ ♃
22 41 ☽ # ♆

30 Friday
05 06 ☽ Q ♄
06 03 ☽ ⚺ ♆
10 21 ☽ bQ ♀
10 44 ☽ □ ♇
12 11 ☉ # ♄
14 07 ♀ bQ ♇
16 27 ♀ ⚻ ♄
17 55 ☽ ⊥ ♆
19 38 ☽ ✶ ♃
20 28 ☉ bQ ♆
21 26 ☽ △ ♂
21 54 ☽ ∥ ♀

31 Saturday
10 24 ☽ □ ☿
11 54 ☽ ∠ ♆
13 16 ☽ □ ☉
14 20 ☽ ∥ ♂
16 48 ☽ □ ♄
17 12 ☽ # ♃
18 51 ☿ bQ ♆
19 29 ☽ Q ♃
19 36 ♂ ± ♇
19 41 ☽ △ ♀

AUGUST 2021

Date	Sidereal time h m s	Sun ☉ ° ' "	Moon ☽ ° ' "	Moon ☽ 24.00 ° ' "	Mercury ☿ ° '	Venus ♀ ° '	Mars ♂ ° '	Jupiter ♃ ° '	Saturn ♄ ° '	Uranus ♅ ° '	Neptune ♆ ° '	Pluto ♇ ° '
01	08 41 17	09 ♌ 27 26	19 ♉ 47 11	25 ♉ 41 21	09 ♌ 21	12 ♍ 33	01 ♍ 40	29 ♒ 33	10 ♒ 14	14 ♉ 39	22 ♓ 51	25 ♑ 13
02	08 45 13	10 24 51	01 ♊ 35 15	07 ♊ 29 30	11 26	13 44	02 17	29 R 27	10 R 10	14 40	22 R 50	25 R 12
03	08 49 10	11 22 17	13 ♊ 24 45	19 ♊ 21 34	13 30	14 56	02 55	29 19	10 06	14 41	22 49	25 10
04	08 53 06	12 19 45	25 ♊ 20 31	01 ♋ 22 05	15 32	16 07	03 33	29 12	10 01	14 42	22 48	25 09
05	08 57 03	13 17 13	07 ♋ 26 43	13 ♋ 34 47	17 34	17 19	04 11	29 05	09 57	14 42	22 47	25 07
06	09 00 59	14 14 43	19 ♋ 46 35	26 ♋ 02 21	19 34	18 30	04 48	28 58	09 52	14 43	22 45	25 06
07	09 04 56	15 12 14	02 ♌ 22 13	08 ♌ 46 15	21 33	19 42	05 26	28 50	09 48	14 44	22 44	25 05
08	09 08 52	16 09 45	15 ♌ 14 26	21 ♌ 46 41	23 30	20 53	06 04	28 43	09 43	14 44	22 43	25 03
09	09 12 49	17 07 18	28 ♌ 22 49	05 ♍ 02 38	25 26	22 04	06 42	28 35	09 39	14 45	22 42	25 02
10	09 16 46	18 04 52	11 ♍ 45 52	18 ♍ 32 14	27 21	23 16	07 20	28 28	09 34	14 45	22 40	25 01
11	09 20 42	19 02 27	25 ♍ 21 24	02 ♎ 13 05	29 ♌ 14	24 27	07 58	28 20	09 30	14 46	22 39	24 59
12	09 24 39	20 00 03	09 ♎ 06 57	16 ♎ 02 46	01 ♍ 05	25 38	08 36	28 12	09 26	14 46	22 38	24 58
13	09 28 35	20 57 39	23 ♎ 00 16	29 ♎ 59 16	02 55	26 49	09 14	28 05	09 21	14 47	22 37	24 57
14	09 32 32	21 55 17	06 ♏ 59 34	14 ♏ 01 04	04 44	28 00	09 52	27 57	09 17	14 47	22 35	24 56
15	09 36 28	22 52 56	21 ♏ 03 37	28 ♏ 07 08	06 31	29 ♍ 11	10 30	27 49	09 13	14 47	22 34	24 54
16	09 40 25	23 50 35	05 ♐ 11 30	12 ♐ 16 35	08 16	00 ♎ 22	11 08	27 41	09 09	14 47	22 32	24 53
17	09 44 21	24 48 16	19 ♐ 22 13	26 ♐ 28 13	10 00	01 33	11 46	27 33	09 04	14 47	22 31	24 52
18	09 48 18	25 45 58	03 ♑ 34 18	10 ♑ 40 08	11 43	02 44	12 24	27 26	09 00	14 48	22 30	24 51
19	09 52 15	26 43 40	17 ♑ 45 21	24 ♑ 49 30	13 24	03 55	13 02	27 18	08 56	14 48	22 28	24 50
20	09 56 11	27 41 24	01 ♒ 52 04	08 ♒ 52 31	15 04	05 06	13 40	27 10	08 52	14 R 48	22 27	24 48
21	10 00 08	28 39 09	15 ♒ 50 20	22 ♒ 44 58	16 42	06 16	14 18	27 02	08 48	14 48	22 25	24 47
22	10 04 04	29 ♌ 36 56	29 ♒ 35 54	06 ♓ 22 43	18 19	07 27	14 56	26 54	08 44	14 47	22 24	24 46
23	10 08 01	00 ♍ 34 43	13 ♓ 05 01	19 ♓ 42 31	19 54	08 37	15 34	26 46	08 40	14 47	22 22	24 45
24	10 11 57	01 32 33	26 ♓ 15 03	02 ♈ 42 33	21 29	09 48	16 12	26 39	08 36	14 47	22 21	24 44
25	10 15 54	02 30 23	09 ♈ 05 02	15 ♈ 22 40	23 01	10 58	16 51	26 31	08 32	14 47	22 19	24 43
26	10 19 50	03 28 16	21 ♈ 35 42	27 ♈ 44 27	24 33	12 09	17 29	26 23	08 28	14 47	22 18	24 42
27	10 23 47	04 26 10	03 ♉ 49 22	09 ♉ 50 55	26 03	13 19	18 07	26 15	08 24	14 46	22 16	24 41
28	10 27 44	05 24 06	15 ♉ 49 42	21 ♉ 46 17	27 31	14 29	18 45	26 08	08 21	14 46	22 15	24 40
29	10 31 40	06 22 04	27 ♉ 41 21	03 ♊ 35 32	28 ♍ 59	15 39	19 24	26 00	08 17	14 45	22 13	24 39
30	10 35 37	07 20 04	09 ♊ 29 32	15 ♊ 24 04	00 ♎ 24	16 49	20 02	25 52	08 13	14 45	22 11	24 38
31	10 39 33	08 ♍ 18 05	21 ♊ 19 47	27 ♊ 17 23	01 ♎ 49	17 ♎ 59	20 ♍ 40	25 ♒ 45	08 ♒ 10	14 ♉ 44	22 ♓ 10	24 ♑ 37

Date	Moon True ☊ ° '	Moon Mean ☊ ° '	Moon ☽ Latitude ° '
01	08 ♊ 53	07 ♊ 37	01 S 41
02	08 D 54	07 33	00 S 39
03	08 R 54	07 30	00 N 24
04	08 53	07 27	01 27
05	08 49	07 24	02 26
06	08 43	07 21	03 20
07	08 35	07 18	04 05
08	08 24	07 14	04 38
09	08 13	07 11	04 58
10	08 02	07 08	05 02
11	07 52	07 05	04 48
12	07 44	07 02	04 19
13	07 39	06 58	03 33
14	07 36	06 55	02 35
15	07 36	06 52	01 27
16	07 D 36	06 49	00 N 13
17	07 R 35	06 46	01 S 02
18	07 33	06 43	02 13
19	07 29	06 39	03 15
20	07 22	06 36	04 06
21	07 12	06 33	04 40
22	07 00	06 30	04 58
23	06 48	06 27	04 59
24	06 36	06 24	04 43
25	06 26	06 20	04 14
26	06 18	06 17	03 33
27	06 13	06 14	02 42
28	06 11	06 11	01 46
29	06 D 10	06 08	00 S 45
30	06 R 10	06 04	00 N 18
31	06 ♊ 09	06 ♊ 01	01 N 20

DECLINATIONS

Date	Sun ☉ ° '	Moon ☽ ° '	Mercury ☿ ° '	Venus ♀ ° '	Mars ♂ ° '	Jupiter ♃ ° '	Saturn ♄ ° '	Uranus ♅ ° '	Neptune ♆ ° '	Pluto ♇ ° '
01	17 N 53	16 N 04	19 N 32	08 N 03	11 N 54	12 S 42	18 S 25	15 N 50	03 S 54	22 S 39
02	17 38	19 50	19 00	07 34	11 40	12 44	18 27	15 51	03 54	22 40
03	17 22	22 48	18 27	07 05	11 26	12 47	18 28	15 51	03 55	22 40
04	17 06	24 48	17 52	06 35	11 12	12 49	18 29	15 51	03 55	22 40
05	16 50	25 40	17 15	06 05	10 58	12 52	18 31	15 51	03 55	22 41
06	16 33	25 16	16 37	05 35	10 44	12 55	18 32	15 51	03 56	22 41
07	16 17	23 36	15 58	05 05	10 30	12 58	18 33	15 52	03 57	22 41
08	15 59	20 41	15 18	04 35	10 16	13 00	18 34	15 52	03 57	22 42
09	15 42	16 41	14 37	04 05	10 01	13 03	18 36	15 52	03 58	22 42
10	15 25	11 48	13 55	03 34	09 47	13 06	18 37	15 52	03 58	22 42
11	15 07	06 15	13 13	03 04	09 32	13 09	18 38	15 52	03 59	22 43
12	14 49	00 N 21	12 30	02 33	09 18	13 12	18 39	15 52	03 59	22 43
13	14 31	05 S 38	11 46	02 02	09 03	13 14	18 41	15 52	04 00	22 43
14	14 12	11 25	11 03	01 31	08 49	13 17	18 42	15 52	04 00	22 44
15	13 53	16 38	10 18	01 00	08 34	13 20	18 43	15 52	04 01	22 44
16	13 34	20 57	09 34	00 N 30	08 19	13 23	18 44	15 53	04 02	22 44
17	13 15	24 03	08 49	00 S 01	08 04	13 26	18 45	15 53	04 02	22 44
18	12 56	25 36	08 05	00 32	07 50	13 28	18 47	15 53	04 03	22 45
19	12 36	25 29	07 20	01 03	07 35	13 31	18 48	15 53	04 03	22 45
20	12 16	23 44	06 36	01 34	07 20	13 34	18 49	15 53	04 04	22 45
21	11 57	20 33	05 51	02 05	07 05	13 37	18 50	15 53	04 04	22 45
22	11 36	16 16	05 07	02 36	06 50	13 40	18 51	15 52	04 05	22 46
23	11 16	11 15	04 22	03 07	06 35	13 42	18 52	15 52	04 06	22 46
24	10 56	05 49	03 38	03 38	06 19	13 45	18 53	15 52	04 06	22 46
25	10 35	00 S 17	02 55	04 08	06 04	13 48	18 54	15 52	04 07	22 47
26	10 14	05 N 08	02 11	04 39	05 49	13 51	18 55	15 52	04 08	22 47
27	09 53	10 15	01 28	05 10	05 34	13 53	18 56	15 52	04 08	22 47
28	09 32	14 53	00 45	05 40	05 18	13 56	18 57	15 52	04 09	22 47
29	09 10	18 55	00 N 03	06 11	05 03	13 59	18 58	15 52	04 10	22 48
30	08 49	22 10	00 S 39	06 41	04 48	14 01	18 59	15 52	04 10	22 48
31	08 N 27	24 N 29	01 S 20	07 S 11	04 N 32	14 S 04	19 S 00	15 N 51	04 S 11	22 S 48

ZODIAC SIGN ENTRIES

Date	h	m	Planets
02	08	46	☽ ♊
04	21	17	☽ ♋
07	07	31	☽ ♌
09	14	56	☽ ♍
11	20	08	☽ ♎
11	21	57	☿ ♍
14	00	01	☽ ♏
16	03	12	☽ ♐
16	04	27	♀ ♎
18	05	58	☽ ♑
20	08	49	☽ ♒
22	12	42	☽ ♓
22	21	35	☉ ♍
24	18	57	☽ ♈
27	04	27	☽ ♉
29	16	42	☽ ♊
30	05	10	☿ ♎

LATITUDES

Date	Mercury ☿ ° '	Venus ♀ ° '	Mars ♂ ° '	Jupiter ♃ ° '	Saturn ♄ ° '	Uranus ♅ ° '	Neptune ♆ ° '	Pluto ♇ ° '
01	01 N 41	01 N 18	01 N 05	01 S 08	00 S 47	00 S 25	01 S 09	01 S 36
04	01 46	01 12	01 04	01 09	00 47	00 25	01 09	01 36
07	01 45	01 06	01 03	01 09	00 47	00 25	01 09	01 36
10	01 38	00 58	01 02	01 10	00 47	00 25	01 10	01 36
13	01 26	00 50	01 01	01 10	00 48	00 25	01 10	01 37
16	01 11	00 42	01 00	01 11	00 48	00 25	01 10	01 37
19	00 53	00 33	00 59	01 11	00 48	00 25	01 10	01 37
22	00 32	00 23	00 58	01 11	00 48	00 25	01 10	01 37
25	00 N 09	00 13	00 57	01 12	00 48	00 25	01 10	01 37
28	00 S 15	00 N 03	00 56	01 12	00 48	00 25	01 10	01 38
31	00 S 40	00 S 09	00 N 55	01 S 12	00 S 49	00 S 25	01 S 10	01 S 38

DATA

Julian Date	2459428
Delta T	+71 seconds
Ayanamsa	24° 09' 16"
Synetic vernal point	04° ♓ 57' 43"
True obliquity of ecliptic	23° 26' 15"

MOON'S PHASES, APSIDES AND POSITIONS ☽

Date	h	m	Phase	Longitude ° '	Eclipse Indicator
08	13	50	●	16 ♌ 14	
15	15	20	☽	23 ♏ 01	
22	12	02	○	29 ♒ 37	
30	07	13	☾	07 ♊ 09	

Day	h	m		
02	07	39	Apogee	
17	09	20	Perigee	
30	02	22	Apogee	
05	16	46	Max dec	25° N 41'
12	13	24	0S	
18	22	21	Max dec	25° S 46'
25	13	16	0N	

LONGITUDES

Date	Chiron ⚷ ° '	Ceres ⚳ ° '	Pallas ⚴ ° '	Juno ⚵ ° '	Vesta ⚶ ° '	Black Moon Lilith ⚸ ° '
01	12 ♈ 49	00 ♊ 21	26 ♓ 48	08 ♐ 21	05 ♎ 46	01 ♊ 34
11	12 ♈ 38	03 ♊ 10	25 ♓ 28	08 ♐ 32	10 ♎ 20	02 ♊ 40
21	12 ♈ 22	05 ♊ 42	23 ♓ 35	09 ♐ 15	15 ♎ 03	03 ♊ 47
31	12 ♈ 03	07 ♊ 55	21 ♓ 17	10 ♐ 27	19 ♎ 53	04 ♊ 53

ASPECTARIAN

h m	Aspects
01 Sunday	
01 34	☽ ☌ ♅
10 40	☽ ∥ ♅
14 07	☉ ☌ ☿
18 13	☽ ✶ ♆
21 50	☿ ☍ ♄
22 19	☽ ∥ ☉
23 01	☽ △ ♇
02 Monday	
02 32	☽ # ♄
04 58	☽ Q ☉
06 14	☉ ☍ ♄
06 41	☽ Q ☿
07 03	☽ ∥ ☿
07 41	☽ □ ♃
13 30	☽ □ ♂
18 35	☽ Q ♆
03 Tuesday	
05 19	☽ △ ♄
05 27	☽ ⊑ ♇
06 53	♀ △ ♅
07 30	☽ ✶ ☉
10 42	☽ # ♇
11 17	☿ # ♄
12 12	☽ ✶ ☿
14 34	☽ ⊻ ♅
15 25	☽ □ ♀
23 36	☽ ± ♇
04 Wednesday	
01 57	☿ □ ♅
02 40	☽ ⊥ ♅
03 58	☽ Q ♂
06 55	☽ □ ♆
10 00	♀ ± ♄
11 21	☽ ⊑ ♄
11 37	☽ ⊼ ♇
16 19	☽ ∠ ☉
19 38	☽ △ ♃
20 41	☽ ∠ ♅
05 Thursday	
00 27	☽ ∠ ☿
02 43	☿ ± ♆
04 43	☿ ⊻ ♀
05 08	☽ ± ♄
05 12	☽ ✶ ♂
07 21	☽ Q ♀
11 40	☽ ⊥ ☉
16 52	☽ ⊼ ♄
21 39	☽ ⊥ ☿
06 Friday	
00 24	☽ ⊻ ☉
00 51	☽ ⊑ ♃
02 12	☽ ✶ ♅
09 17	☽ ✶ ♀
11 31	☽ ⊻ ☿
12 04	☽ ∠ ♂
16 20	☉ ∥ ☿
17 43	☽ △ ♆
18 04	☽ ± ♃
22 12	☽ ☍ ♇
23 57	☉ □ ♅
07 Saturday	
01 18	☽ Q ♅
05 23	☽ ⊼ ♃
06 10	☽ ⊥ ♂
15 59	☿ ∥ ♅
16 50	☽ ∠ ♀
18 04	☽ ⊻ ♂
20 37	☽ # ♇
22 03	☽ ⊑ ♆
08 Sunday	
01 50	☽ ☍ ♄
02 24	☿ ⊼ ♆
11 04	☽ □ ♅
11 17	☽ ⊥ ♀
13 50	☽ ☌ ☉
14 43	☽ ± ♆
22 41	☉ ∥ ♅
23 24	☽ ⊻ ♀
09 Monday	
01 22	☽ # ♄
01 34	☉ ± ♆
01 42	☽ ⊼ ♆
05 45	☽ ☌ ☿
05 57	☽ ⊼ ♇
07 03	☿ ⊼ ♇
12 23	☽ ☍ ♃
16 19	☽ ∥ ♅
16 47	☽ ± ♇
17 16	♀ # ♆
17 30	☽ ∥ ☉
10 Tuesday	
00 20	♀ ☍ ♆
00 22	☽ ∥ ☿
03 42	☽ ☌ ♂
06 00	☽ # ♃
08 07	☽ ⊼ ♄
08 53	☽ ⊑ ♇
17 19	☽ △ ♅
18 44	☽ ± ♄
21 23	☽ ∥ ♂
11 Wednesday	
00 03	☽ ⊻ ☉
01 20	☿ ☍ ♃
07 16	☽ ☍ ♆
10 15	☽ ☌ ♀
10 30	☽ ⊑ ♄
11 22	☽ △ ♇
11 24	☽ ⊥ ☉
12 59	♀ ⊑ ♄
14 11	☿ # ♃
17 10	☽ ⊼ ♃
19 43	☽ ⊑ ♅
19 51	☽ ⊻ ☿
21 21	☽ # ♆
22 46	♀ △ ♇
12 Thursday	
02 20	☽ ∥ ♀
03 33	☽ ± ♃
04 19	☽ ∠ ☉
07 56	☽ ⊥ ☿
10 31	☿ ± ♇
11 03	☽ ⊻ ♂
11 24	☽ ± ♅
12 32	☽ △ ♄
19 02	☽ ⊑ ♃
21 48	☽ ⊼ ♅
21 57	☽ ⊥ ♂
22 40	☽ # ♀
13 Friday	
01 54	☽ ∠ ☿
05 23	☽ ∥ ♆
08 13	☽ ✶ ☉
11 19	☽ ⊼ ♆
14 12	☽ ∠ ♂
15 20	☽ □ ♇
16 27	♂ ⊼ ♄
19 10	☽ ⊻ ♀
20 39	☽ △ ♃
21 37	☽ ± ♆
14 Saturday	
01 27	☽ # ♂
02 55	☉ ⊥ ♀
06 21	☽ Q ☉
06 25	☽ ⊥ ♀
07 33	☽ ✶ ☿
10 36	☽ # ☿
10 58	♀ ⊼ ♃
13 01	☽ ⊑ ♆
14 32	♂ ⊑ ♇
15 54	☽ □ ♄
17 08	☽ ✶ ♂
20 20	☽ ∥ ♃
22 08	☽ Q ♇
23 13	☽ ∠ ♀
23 43	☽ # ☉
15 Sunday	
01 18	☽ ☍ ♅
04 13	☉ ⊼ ♆
07 02	☽ Q ☿
08 18	☽ # ♅
14 33	☽ Q ♂
14 33	☽ △ ♆
15 20	☽ □ ☉
18 32	☽ ✶ ♇
22 25	☽ Q ♄
22 54	☽ ∥ ♄
23 23	☽ □ ♃
16 Monday	
00 06	♀ ⊑ ♅
03 05	☽ ✶ ♀
17 57	☽ □ ☿
18 40	☽ ✶ ♄
19 57	☽ ∠ ♇
22 31	☽ □ ♂
23 35	☿ ⊼ ♄
17 Tuesday	
00 30	☽ ∥ ♇
00 37	☉ # ♃
01 16	☽ Q ♀
04 15	☽ ⊼ ♅
05 37	☽ Q ♃
10 08	☿ ⊑ ♇
11 09	☽ ⊥ ♇
13 31	☉ ⊼ ♇
14 24	☽ ± ♅
17 19	☽ □ ♆
19 55	☽ ∠ ♄
21 17	☽ ⊻ ♇
21 51	☽ △ ☉
18 Wednesday	
01 43	☽ ✶ ♃
05 37	☽ ⊑ ♅
10 27	☽ □ ♀
11 03	☽ ⊥ ♄
21 08	☽ ⊻ ♄
23 41	☽ Q ♆
19 Thursday	
00 17	☿ ∥ ♂
00 41	♀ ± ♃
01 03	☽ ⊑ ☉
02 50	☽ ∠ ♃
03 28	☿ ☌ ♂
03 37	☽ △ ♂
03 38	☽ △ ☿
06 59	☽ △ ♅
17 25	☽ ± ☉
17 57	☽ ⊥ ♃
19 59	☽ ✶ ♆
23 59	☽ ☌ ♇
20 Friday	
00 28	☉ ☍ ♃
01 41	♅ St R
04 03	☽ ⊻ ♃
04 21	☽ ⊼ ☉
06 16	☽ ⊑ ♂
08 06	☿ △ ♅
08 30	☽ ⊑ ☿
09 16	☿ ± ♄
18 01	☽ △ ♀
20 26	☽ ∥ ♇
21 32	☽ ∠ ♆
22 23	☽ ± ♂
23 55	☽ ☌ ♄
21 Saturday	
01 58	☽ ± ☿
09 13	☽ ⊼ ♂
10 11	☽ □ ♅
13 00	☽ ⊥ ♆
13 41	☽ ⊼ ☿
22 11	☽ ∥ ♄
22 18	☽ ⊑ ♀
23 24	☽ ⊻ ♆
22 Sunday	
03 32	☽ ⊻ ♇
05 04	♂ ± ♄
06 38	♂ △ ♅
07 19	☽ ☌ ♃
12 02	☽ ☍ ☉
13 57	☽ # ♅
14 04	☽ ⊥ ♇
15 34	☽ ± ♀
17 38	☽ Q ♅
23 Monday	
00 39	☽ ∥ ♃
03 14	☽ ⊼ ♀
04 07	☽ ⊻ ♄
06 02	☽ ∠ ♇
11 53	☽ # ☉
12 48	♀ △ ♄
14 50	☽ ⊥ ♄
15 04	☽ ✶ ♅
15 23	♀ ± ♅
16 13	☉ ± ♇
16 43	☽ ☍ ♂
20 51	☿ # ♆
24 Tuesday	
02 03	☽ ☍ ☿
04 50	☽ ☌ ♆
07 08	☽ ∠ ♄
09 12	☽ ✶ ♇
09 43	☽ # ♂
12 13	☿ # ♀
12 43	☽ ⊻ ♃
18 33	☽ ∠ ♅
19 26	☽ ∥ ♆
20 43	☽ ∥ ♀
22 37	☽ ⊼ ☉
22 55	☽ # ☿
23 45	☽ ⊥ ♃
25 Wednesday	
01 14	☿ ☍ ♆
07 32	☽ Q ♇
10 49	☽ ± ☉
10 56	♀ ∥ ♆
10 58	☽ ✶ ♄
11 26	☽ ⊥ ♅
15 57	☽ ☍ ♀
16 34	☽ ∠ ♃
19 41	☿ ⊑ ♄
21 59	♀ ⊑ ♃
22 51	☽ ⊻ ♅
26 Thursday	
00 23	☽ ∥ ☿
03 37	☽ ⊼ ♂
05 27	☽ ⊑ ☉
07 29	☽ # ♆
09 37	☽ # ♀
09 50	☽ Q ♄
13 21	☽ ⊻ ♆
14 23	☿ △ ♇
14 59	☽ ∥ ♂
15 52	☽ ± ♂
18 02	☽ □ ♇
18 33	☽ ⊼ ☿
21 14	☽ ✶ ♃
27 Friday	
01 04	☽ ⊥ ♆
07 59	☽ ± ☿
10 21	☽ ∥ ☉
10 31	☽ ⊑ ♂
13 19	☽ △ ☉
15 05	☿ ⊼ ♃
18 50	☽ ∠ ♆
20 43	☽ Q ♃
21 04	☽ □ ♄
28 Saturday	
00 31	♀ # ♂
04 25	☽ ⊑ ☿
06 45	☽ # ♃
09 00	☽ ⊼ ♀
09 52	☽ ☌ ♅
17 27	☽ ∥ ♅
17 43	♀ ⊼ ♅
18 14	☽ △ ♂
22 25	☽ ± ♀
29 Sunday	
00 56	☽ ✶ ♆
05 50	☽ △ ♇
08 36	☽ □ ♃
12 23	☽ # ♄
14 59	☽ △ ☿
18 41	☽ ⊑ ♀
30 Monday	
00 59	☿ ⊑ ♅
01 14	☽ Q ♆
07 13	☽ □ ☉
09 26	☽ △ ♄
12 17	☽ ⊑ ♇
17 39	☽ # ♇
22 40	☽ ⊻ ♅
31 Tuesday	
04 30	☽ △ ♀
06 31	☽ ± ♇
08 46	☉ ⊼ ♄
10 36	☽ □ ♂
10 48	☽ ⊥ ♅
10 59	☿ ± ♃
13 41	☽ □ ♆
15 41	☽ ⊑ ♄
18 37	☽ ⊼ ♇
20 48	☽ △ ♃
22 54	☽ Q ☉

All ephemeris data is given at 12.00 UT and the Moon's longitude is additionally given for 24.00 UT

SEPTEMBER 2021

LONGITUDES

Date	Sidereal time h m s	Sun ☉ ° ' "	Moon ☽ ° ' "	Moon ☽ 24.00 ° ' "	Mercury ☿ ° '	Venus ♀ ° '	Mars ♂ ° '	Jupiter ♃ ° '	Saturn ♄ ° '	Uranus ♅ ° '	Neptune ♆ ° '	Pluto ♇ ° '
01	10 43 30	09 ♍ 16 09	03 ♋ 17 31	09 ♋ 20 47	03 ♎ 11	19 ♎ 09	21 ♍ 19	25 ♒ 37	08 ♒ 06	14 ♉ 44	22 ♓ 08	24 ♑ 36
02	10 47 26	10 14 14	15 ♋ 27 44	21 ♋ 38 53	04 33	20 19	21 57	25 R 30	08 R 03	14 R 43	22 R 07	24 R 35
03	10 51 23	11 12 21	27 ♋ 54 38	04 ♌ 15 20	05 53	21 29	22 36	25 23	07 59	14 42	22 05	24 34
04	10 55 19	12 10 30	10 ♌ 41 13	17 ♌ 12 24	07 11	22 38	23 14	25 15	07 56	14 42	22 03	24 33
05	10 59 16	13 08 41	23 ♌ 48 53	00 ♍ 30 33	08 28	23 48	23 53	25 08	07 53	14 41	22 02	24 32
06	11 03 13	14 06 54	07 ♍ 17 10	14 ♍ 08 24	09 43	24 58	24 31	25 01	07 50	14 40	22 00	24 32
07	11 07 09	15 05 09	21 ♍ 03 47	28 ♍ 02 47	10 56	26 07	25 10	24 54	07 47	14 39	21 59	24 31
08	11 11 06	16 03 25	05 ♎ 04 51	12 ♎ 09 21	12 07	27 17	25 48	24 47	07 44	14 38	21 57	24 30
09	11 15 02	17 01 42	19 ♎ 15 40	26 ♎ 23 11	13 17	28 26	26 27	24 41	07 41	14 37	21 55	24 29
10	11 18 59	18 00 02	03 ♏ 31 22	10 ♏ 39 41	14 24	29 ♎ 35	27 05	24 34	07 38	14 36	21 54	24 29
11	11 22 55	18 58 23	17 ♏ 47 45	24 ♏ 55 10	15 30	00 ♏ 44	27 44	24 27	07 35	14 35	21 52	24 28
12	11 26 52	19 56 46	02 ♐ 01 41	09 ♐ 07 05	16 33	01 53	28 23	24 21	07 32	14 34	21 50	24 27
13	11 30 48	20 55 10	16 ♐ 11 11	23 ♐ 13 54	17 34	03 02	29 02	24 15	07 30	14 33	21 49	24 26
14	11 34 45	21 53 36	00 ♑ 15 08	07 ♑ 14 48	18 32	04 11	29 ♍ 40	24 09	07 27	14 32	21 47	24 26
15	11 38 42	22 52 03	14 ♑ 12 49	21 ♑ 09 06	19 27	05 20	00 ♎ 19	24 03	07 25	14 31	21 45	24 25
16	11 42 38	23 50 32	28 ♑ 03 30	04 ♒ 55 54	20 20	06 28	00 58	23 57	07 22	14 29	21 44	24 25
17	11 46 35	24 49 03	11 ♒ 46 06	18 ♒ 33 53	21 09	07 37	01 37	23 51	07 20	14 28	21 42	24 24
18	11 50 31	25 47 35	25 ♒ 19 01	02 ♓ 01 16	21 55	08 45	02 15	23 45	07 18	14 27	21 40	24 24
19	11 54 28	26 46 09	08 ♓ 40 23	15 ♓ 16 08	22 38	09 54	02 54	23 40	07 16	14 25	21 39	24 23
20	11 58 24	27 44 45	21 ♓ 48 17	28 ♓ 16 41	23 16	11 02	03 33	23 35	07 14	14 24	21 37	24 23
21	12 02 21	28 43 22	04 ♈ 41 14	11 ♈ 01 51	23 51	12 10	04 12	23 30	07 12	14 22	21 35	24 22
22	12 06 17	29 ♍ 42 02	17 ♈ 18 33	23 ♈ 31 25	24 20	13 18	04 51	23 25	07 10	14 21	21 34	24 22
23	12 10 14	00 ♎ 40 43	29 ♈ 40 37	05 ♉ 46 22	24 45	14 26	05 30	23 20	07 08	14 19	21 32	24 21
24	12 14 11	01 39 27	11 ♉ 48 59	17 ♉ 48 51	25 05	15 34	06 09	23 15	07 06	14 18	21 31	24 21
25	12 18 07	02 38 13	23 ♉ 46 22	29 ♉ 42 04	25 19	16 41	06 48	23 11	07 05	14 16	21 29	24 21
26	12 22 04	03 37 01	05 ♊ 36 28	11 ♊ 30 09	25 27	17 49	07 27	23 06	07 03	14 14	21 27	24 20
27	12 26 00	04 35 52	17 ♊ 23 45	23 ♊ 17 56	25 R 28	18 56	08 06	23 02	07 02	14 13	21 26	24 20
28	12 29 57	05 34 44	29 ♊ 13 20	05 ♋ 10 38	25 23	20 03	08 46	22 58	07 01	14 11	21 24	24 20
29	12 33 53	06 33 39	11 ♋ 10 32	17 ♋ 13 40	25 10	21 10	09 25	22 55	06 59	14 09	21 22	24 20
30	12 37 50	07 ♎ 32 36	23 ♋ 20 42	29 ♋ 32 13	24 ♎ 50	22 ♏ 17	10 ♎ 04	22 ♒ 51	06 ♒ 58	14 ♉ 07	21 ♓ 21	24 ♑ 19

Date	Moon True ☊ ° '	Moon Mean ☊ ° '	Moon ☽ Latitude ° '
01	06 ♊ 08	05 ♊ 58	02 N 18
02	06 R 04	05 55	03 12
03	05 57	05 52	03 58
04	05 48	05 49	04 33
05	05 37	05 45	04 55
06	05 25	05 42	05 01
07	05 13	05 39	04 50
08	05 02	05 36	04 21
09	04 54	05 33	03 36
10	04 48	05 30	02 37
11	04 45	05 26	01 29
12	04 44	05 23	00 N 14
13	04 D 44	05 20	01 S 00
14	04 R 44	05 17	02 11
15	04 43	05 14	03 13
16	04 39	05 10	04 04
17	04 32	05 07	04 39
18	04 23	05 04	04 59
19	04 12	05 01	05 02
20	04 01	04 58	04 48
21	03 50	04 55	04 20
22	03 40	04 51	03 40
23	03 33	04 48	02 50
24	03 28	04 45	01 53
25	03 25	04 42	00 S 51
26	03 D 25	04 39	00 N 12
27	03 26	04 36	01 14
28	03 26	04 32	02 14
29	03 R 26	04 29	03 08
30	03 ♊ 25	04 ♊ 26	03 N 55

DECLINATIONS

Date	Sun ☉ ° '	Moon ☽ ° '	Mercury ☿ ° '	Venus ♀ ° '	Mars ♂ ° '	Jupiter ♃ ° '	Saturn ♄ ° '	Uranus ♅ ° '	Neptune ♆ ° '	Pluto ♇ ° '
01	08 N 06	25 N 42	02 S 01	07 S 41	04 N 17	14 S 06	19 S 01	15 N 51	04 S 12	22 S 48
02	07 44	25 43	02 41	08 11	04 01	14 09	19 02	15 51	04 12	22 48
03	07 22	24 28	03 21	08 41	03 46	14 11	19 03	15 51	04 13	22 49
04	07 00	21 55	04 00	09 11	03 30	14 14	19 04	15 51	04 13	22 49
05	06 37	18 13	04 38	09 40	03 14	14 16	19 05	15 50	04 14	22 49
06	06 15	13 29	05 16	10 09	02 59	14 19	19 06	15 50	04 15	22 49
07	05 53	07 59	05 53	10 38	02 43	14 21	19 07	15 50	04 15	22 50
08	05 30	01 N 59	06 29	11 07	02 27	14 23	19 07	15 50	04 16	22 50
09	05 07	04 S 12	07 04	11 36	02 12	14 26	19 08	15 49	04 17	22 50
10	04 45	10 13	07 38	12 04	01 56	14 28	19 09	15 49	04 17	22 50
11	04 22	15 43	08 11	12 32	01 40	14 30	19 10	15 49	04 18	22 50
12	03 59	20 20	08 43	13 00	01 25	14 32	19 10	15 48	04 19	22 50
13	03 36	23 43	09 14	13 28	01 09	14 34	19 11	15 48	04 19	22 51
14	03 13	25 37	09 43	13 55	00 53	14 36	19 12	15 48	04 20	22 51
15	02 50	25 53	10 12	14 22	00 37	14 38	19 12	15 47	04 21	22 51
16	02 27	24 31	10 38	14 49	00 21	14 40	19 13	15 47	04 21	22 51
17	02 04	21 43	11 04	15 15	00 N 06	14 42	19 14	15 46	04 22	22 51
18	01 40	17 46	11 27	15 41	00 S 10	14 44	19 14	15 46	04 23	22 51
19	01 17	12 58	11 49	16 07	00 26	14 45	19 15	15 46	04 23	22 51
20	00 54	07 40	12 09	16 33	00 42	14 47	19 15	15 45	04 24	22 52
21	00 30	02 S 07	12 27	16 58	00 58	14 49	19 16	15 45	04 25	22 52
22	00 N 07	03 N 24	12 43	17 22	01 14	14 50	19 16	15 44	04 25	22 52
23	00 S 16	08 42	12 56	17 47	01 29	14 52	19 17	15 44	04 26	22 52
24	00 40	13 35	13 07	18 11	01 45	14 53	19 17	15 43	04 27	22 52
25	01 03	17 53	13 15	18 34	02 01	14 55	19 18	15 43	04 27	22 52
26	01 26	21 26	13 20	18 57	02 17	14 56	19 18	15 42	04 28	22 52
27	01 50	24 04	13 21	19 20	02 33	14 57	19 18	15 42	04 29	22 52
28	02 13	25 40	13 19	19 42	02 49	14 59	19 19	15 41	04 29	22 52
29	02 36	26 06	13 13	20 04	03 04	15 00	19 19	15 41	04 30	22 53
30	03 S 00	25 N 17	13 S 03	20 S 26	03 S 20	15 S 01	19 S 19	15 N 40	04 S 30	22 S 53

ZODIAC SIGN ENTRIES

Date	h m	Planets
01	05 26	☽ ♋
03	15 58	☽ ♌
05	23 06	☽ ♍
08	03 20	☽ ♎
10	06 05	☽ ♏
10	20 39	♀ ♏
12	08 34	☽ ♐
14	11 34	☽ ♑
15	00 14	♂ ♎
16	15 23	☽ ♒
18	20 22	☽ ♓
21	03 13	☽ ♈
22	19 21	☉ ♎
23	12 38	☽ ♉
26	00 36	☽ ♊
28	13 34	☽ ♋

LATITUDES

Date	Mercury ☿ ° '	Venus ♀ ° '	Mars ♂ ° '	Jupiter ♃ ° '	Saturn ♄ ° '	Uranus ♅ ° '	Neptune ♆ ° '	Pluto ♇ ° '
01	00 S 49	00 S 12	00 N 54	01 S 12	00 S 49	00 S 25	01 S 10	01 S 38
04	01 15	00 24	00 53	01 12	00 49	00 25	01 10	01 38
07	01 41	00 36	00 52	01 12	00 49	00 25	01 10	01 38
10	02 07	00 48	00 51	01 12	00 49	00 25	01 10	01 38
13	02 32	01 00	00 50	01 12	00 49	00 25	01 10	01 38
16	02 55	01 12	00 48	01 12	00 49	00 25	01 10	01 39
19	03 15	01 25	00 47	01 12	00 49	00 25	01 10	01 39
22	03 32	01 37	00 46	01 11	00 49	00 25	01 10	01 39
25	03 43	01 49	00 45	01 11	00 49	00 25	01 10	01 39
28	03 46	02 02	00 43	01 11	00 49	00 25	01 10	01 39
31	03 S 37	02 S 14	00 N 42	01 S 11	00 S 49	00 S 25	01 S 10	01 S 39

DATA

Julian Date	2459459
Delta T	+71 seconds
Ayanamsa	24° 09' 20"
Synetic vernal point	04° ♓ 57' 39"
True obliquity of ecliptic	23° 26' 15"

MOON'S PHASES, APSIDES AND POSITIONS ☽

Date	h m	Phase	Longitude ° '	Eclipse Indicator
07	00 52	●	14 ♍ 38	
13	20 39	☽	21 ♐ 16	
20	23 55	○	28 ♓ 14	
29	01 57	☾	06 ♋ 09	

Day	h m		
11	09 58	Perigee	
26	21 39	Apogee	
02	00 23	Max dec	25° N 52'
08	19 41	0S	
15	03 46	Max dec	25° S 59'
21	21 09	0N	
29	08 26	Max dec	26° N 07'

LONGITUDES

Date	Chiron ⚷ ° '	Ceres ⚳ ° '	Pallas ⚴ ° '	Juno ⚵ ° '	Vesta ⚶ ° '	Black Moon Lilith ⚸ ° '
01	12 ♈ 00	08 ♊ 07	21 ♓ 02	10 ♐ 36	20 ♎ 23	05 ♊ 00
11	11 ♈ 37	09 ♊ 53	18 ♓ 27	12 ♐ 15	25 ♎ 21	06 ♊ 06
21	11 ♈ 12	11 ♊ 11	15 ♓ 51	14 ♐ 17	00 ♏ 24	07 ♊ 13
31	10 ♈ 45	11 ♊ 57	13 ♓ 28	16 ♐ 38	05 ♏ 33	08 ♊ 19

All ephemeris data is given at 12.00 UT and the Moon's longitude is additionally given for 24.00 UT

ASPECTARIAN

h m	Aspects
01 Wednesday	
04 54	☽ ∠ ♅
09 39	☽ ± ♄
11 46	☽ □ ☿
19 37	♂ # ♆
20 03	☉ ⚼ ♇
21 30	☽ ⚻ ♄
23 19	☉ # ♀
02 Thursday	
00 36	☽ Q ♂
00 52	☽ ✶ ☉
02 22	☽ ⚼ ♃
10 33	☽ ✶ ♅
17 43	♂ ☍ ♆
19 46	☽ ± ♃
22 24	☽ □ ♀
03 Friday	
00 52	☽ △ ♆
01 16	☽ ✶ ♂
03 23	☽ Q ☿
05 37	☽ ☍ ♇
07 12	☽ ⚻ ♃
08 29	☽ ∠ ☉
09 43	☽ Q ♅
22 47	☿ # ♂
04 Saturday	
00 13	♀ ⚻ ♆
01 45	♂ ⚼ ♄
02 55	☽ ⊥ ☉
04 45	☽ ✶ ☿
04 48	☽ # ♇
05 16	☽ ⚼ ♆
06 55	☽ ☍ ♄
07 12	☽ ∠ ♂
11 54	☽ Q ♀
14 59	☽ ⚺ ☉
19 23	☽ □ ♅
20 32	☿ ‖ ♆
21 53	☽ ± ♆
05 Sunday	
00 40	☽ ⊥ ♂
01 30	☿ △ ♄
06 58	☽ # ♄
08 47	☽ ⚻ ♆
11 18	☽ ∠ ☿
11 58	☽ ✶ ♀
12 07	☽ ⚺ ♂
13 18	☽ ⚻ ♇
14 22	☽ ☍ ♃
15 26	♀ ⚺ ♂
16 09	☿ ± ♅
06 Monday	
00 03	☽ ± ♇
00 36	☽ ‖ ♅
03 07	♀ □ ♇
05 03	☽ ⊥ ☿
05 18	☉ ± ♄
08 07	☽ # ♃
12 20	♂ △ ♇
12 57	☽ ⚻ ♄
13 05	♀ △ ♃
15 56	☽ ⚼ ♇
16 41	☽ ⚺ ☿
17 08	☽ ∠ ♀
17 28	☿ ⚼ ♃
23 25	☽ ± ♄
07 Tuesday	
00 52	☽ ☌ ☉
00 54	☽ △ ♅
01 29	☉ △ ♅
01 39	☽ # ♀
03 50	♂ ⚻ ♃
10 13	☽ ⊥ ♀
11 58	☉ # ☿
13 34	☽ ☍ ♆
14 57	☽ ⚼ ♄
17 56	☽ △ ♇
18 33	☽ ⚻ ♃
19 24	☽ ☌ ♂
19 46	☽ # ☿
21 08	☽ ‖ ☉
21 29	☽ ⚺ ♀
08 Wednesday	
02 44	☽ ⚼ ♅
03 00	☽ # ♆
04 45	☽ ± ♃
10 02	☽ ‖ ♂
16 29	☽ △ ♄
18 02	☽ ± ♅
19 56	☽ ⚼ ♃
09 Thursday	
01 01	☽ ☌ ☿
01 39	♀ ± ♆
04 11	☽ ⚻ ♅
04 31	☽ # ♂
07 57	☽ ⚺ ☉
12 18	☽ ‖ ♆
15 23	☽ # ☉
16 28	☽ ⚻ ♆
18 49	☽ ⊥ ☉
20 48	☽ □ ♇
21 03	☽ △ ♃
10 Friday	
00 24	☽ ‖ ☿
00 40	☽ ⚺ ♂
02 33	☽ ± ♆
04 48	☽ ☌ ♀
11 03	☽ ∠ ☉
11 14	☽ ⊥ ♂
16 17	☿ ⚻ ♅
17 39	☽ ⚼ ♆
18 53	☽ □ ♄
20 27	☽ ‖ ♀
11 Saturday	
03 02	☽ Q ♇
03 05	☽ ∠ ♂
06 21	☽ ‖ ♃
06 37	☽ ☍ ♅
07 49	☽ ⚺ ☿
10 06	♃ ⚺ ♇
12 26	☽ # ♅
14 08	☽ ✶ ☉
15 46	☉ # ♆
18 44	☽ ⊥ ☿
18 50	☽ △ ♆
23 08	☽ □ ♃
23 13	☽ ✶ ♇
12 Sunday	
01 05	☽ Q ♄
05 24	☽ ‖ ♄
05 33	☽ ✶ ♂
11 07	☽ ∠ ☿
11 45	☽ ⚺ ♀
11 51	☽ Q ☉
21 18	☽ ✶ ♄
22 47	☽ ⊥ ♀
13 Monday	
00 33	☽ ∠ ♇
02 49	☽ Q ♂
04 44	☽ ‖ ♇
05 21	☽ Q ♃
09 13	☽ ⚻ ♅
14 31	☽ ✶ ☿
15 26	☽ ∠ ♀
15 50	☽ ⊥ ♇
19 25	☽ ± ♅
20 39	☽ □ ☉
21 33	☽ □ ♆
22 43	☽ ∠ ♄
14 Tuesday	
01 38	☽ ✶ ♃
02 03	☽ ⚺ ♇
06 57	♂ ⚼ ♅
09 21	☉ ☍ ♆
10 46	☽ ⚼ ♅
10 57	☽ □ ♂
12 30	☽ Q ☿
14 03	☽ ⊥ ♄
19 21	☽ ✶ ♀
15 Wednesday	
00 19	☽ ⚺ ♄
01 14	☉ ⚼ ♄
03 09	☽ ∠ ♃
03 10	♂ ± ♃
04 20	☽ Q ♆
12 31	☽ △ ♅
17 52	☽ Q ♀
18 34	☽ ⊥ ♃
21 41	☽ □ ☿
16 Thursday	
01 01	☽ ✶ ♆
03 32	♀ ‖ ♃
04 07	☽ △ ☉
04 54	☽ ⚺ ♃
05 40	☽ ☌ ♇
14 18	☉ ⚻ ♃
17 11	♀ ⚼ ♆
17 19	☽ △ ♂
17 Friday	
01 53	☉ △ ♇
03 07	☽ ∠ ♆
03 36	☽ ‖ ♇
04 03	☽ □ ♀
04 14	☽ ☌ ♄
06 15	♀ □ ♄
08 18	☽ ⚼ ☉
11 41	♀ ⊥ ♂
16 45	☽ □ ♅
18 55	☽ ⊥ ♆
20 58	☽ ⚼ ♂
18 Saturday	
01 25	☽ ± ☉
03 46	☽ ‖ ♄
04 18	☿ ⚻ ♆
05 32	☽ ⚺ ♆
05 36	☽ △ ☿
09 14	☽ ☌ ♃
10 21	☽ ⚺ ♇
12 55	☽ ⚻ ☉
13 46	☽ ± ♂
16 18	♀ # ♅
21 04	☽ ⊥ ♇
21 55	☽ ‖ ♀
22 26	☽ # ♅
19 Sunday	
00 44	☽ Q ♅
01 04	☽ ⚻ ♂
03 29	☽ ‖ ♃
09 27	☽ ⚺ ♄
10 00	☽ ⚼ ☿
13 17	☽ ∠ ♇
14 26	☽ △ ♀
17 01	☽ ‖ ☿
20 19	☽ ⊥ ♄
22 26	☽ ✶ ♅
20 Monday	
03 15	☽ ± ☿
11 39	☽ ☌ ♆
12 47	☽ ∠ ♄
14 50	☽ ⚻ ☿
15 15	☽ ⚺ ♃
16 45	☽ ✶ ♇
19 19	☉ # ♂
20 35	☽ ⚼ ♀
22 53	☿ △ ♃
23 55	☽ ☍ ☉
21 Tuesday	
02 04	☽ ∠ ♅
02 09	☽ ‖ ♆
02 20	☽ ⊥ ♃
11 02	☽ ☍ ♂
15 04	☽ ± ♀
15 10	☽ Q ♇
16 14	♀ Q ♇
16 43	☽ ✶ ♄
16 46	☽ ‖ ♂
18 56	☽ ⊥ ♅
19 08	☽ ∠ ♃
19 29	☽ # ☉
22 36	☽ ‖ ☉
22 Wednesday	
01 59	☽ # ♂
03 32	☉ ⚼ ♅
03 34	☽ ⚻ ♀
05 26	☉ ± ♃
06 20	☽ ⚺ ♅
13 12	☿ □ ♇
15 34	☽ Q ♄
16 31	☽ # ♆
20 11	☽ ⚺ ♆
23 42	☽ ✶ ♃
23 Thursday	
01 37	☽ □ ♇
02 05	☽ ☍ ☿
07 49	☽ ⊥ ♆
09 41	♀ ☍ ♅
14 08	☽ ⚻ ☉
23 03	☽ Q ♃
24 Friday	
00 07	☽ ⚻ ♂
01 29	☽ ∠ ♆
02 40	☽ □ ♄
03 00	☽ ± ☉
09 29	☽ # ☿
12 43	☽ ± ♂
16 56	☽ ☌ ♅
18 56	☽ # ♃
20 16	☽ ☍ ♀
22 33	☽ ⚼ ☉
23 26	☽ ‖ ♅
25 Saturday	
07 23	☽ ✶ ♆
07 48	☽ ⚼ ♂
10 49	☽ □ ♃
13 09	☽ △ ♇
15 10	☽ ⚻ ☿
16 43	☽ # ♀
20 54	☽ # ♄
21 50	♂ △ ♄
26 Sunday	
03 28	☽ ± ☿
07 35	☽ △ ☉
07 38	☽ Q ♆
14 57	☽ △ ♄
15 59	☽ △ ♂
19 36	☽ ⚼ ♇
21 54	☽ ⚼ ☿
23 58	☽ # ♇
27 Monday	
05 09	☿ St R
05 32	☽ ⚺ ♅
09 44	♂ ⚼ ♃
10 05	♀ ‖ ♄
13 55	☽ ± ♇
15 28	☽ ⚻ ♀
15 36	♂ ± ♅
17 43	☽ ⊥ ♅
20 11	☽ □ ♆
21 25	☽ ⚼ ♄
23 25	☽ △ ♃
28 Tuesday	
02 06	☽ ⚻ ♇
04 18	☽ △ ☿
04 55	☽ ± ♀
11 55	☽ ∠ ♅
15 36	☽ ± ♄
29 Wednesday	
00 58	☽ ⚼ ♀
01 57	☽ □ ☉
03 39	☽ ⚻ ♄
05 31	☽ ⚼ ♃
08 17	☽ □ ♂
16 14	♀ △ ♆
17 54	☽ ✶ ♅
22 19	☉ △ ♄
23 19	☽ ± ♃
30 Thursday	
08 06	☽ △ ♆
09 44	☽ △ ♀
11 03	☽ ⚻ ♃
13 54	☽ ☍ ♇
14 49	☽ □ ☿
16 39	☽ Q ☉
17 23	☽ Q ♅
19 06	☉ ⚼ ♃
21 40	☽ Q ♂
23 31	♀ □ ♃

OCTOBER 2021

LONGITUDES

Date	Sidereal time h m s	Sun ☉ ° ' "	Moon ☽ ° ' "	Moon ☽ 24.00 ° ' "	Mercury ☿ ° '	Venus ♀ ° '	Mars ♂ ° '	Jupiter ♃ ° '	Saturn ♄ ° '	Uranus ♅ ° '	Neptune ♆ ° '	Pluto ♇ ° '
01	12 41 46	08 ♎ 31 36	05 ♌ 48 45	12 ♌ 10 48	24 ♎ 22	23 ♏ 24	10 ♎ 43	22 ♒ 48	06 ♒ 57	14 ♉ 05	21 ♓ 19	24 ♑ 19
02	12 45 43	09 30 38	18 ♌ 38 44	25 ♌ 12 49	23 R 47	24 31	11 23	22 R 44	06 R 56	14 R 04	21 R 18	24 R 19
03	12 49 40	10 29 42	01 ♍ 53 12	08 ♍ 39 54	23 04	25 37	12 02	22 41	06 56	14 02	21 16	24 19
04	12 53 36	11 28 48	15 ♍ 32 46	22 ♍ 31 31	22 14	26 44	12 41	22 39	06 55	14 00	21 15	24 19
05	12 57 33	12 27 56	29 ♍ 35 41	06 ♎ 44 42	21 18	27 50	13 21	22 36	06 54	13 58	21 13	24 19
06	13 01 29	13 27 06	13 ♎ 57 51	21 ♎ 14 19	20 16	28 ♏ 56	14 00	22 34	06 54	13 56	21 12	24 19
07	13 05 26	14 26 19	28 ♎ 33 19	05 ♏ 53 40	19 09	00 ♐ 02	14 39	22 31	06 53	13 54	21 10	24 19
08	13 09 22	15 25 33	13 ♏ 14 46	20 ♏ 35 40	17 59	01 07	15 19	22 29	06 53	13 52	21 09	24 D 19
09	13 13 19	16 24 50	27 ♏ 55 34	05 ♐ 13 47	16 48	02 13	15 58	22 27	06 53	13 49	21 07	24 19
10	13 17 15	17 24 08	12 ♐ 29 43	19 ♐ 42 54	15 38	03 18	16 38	22 26	06 53	13 47	21 06	24 19
11	13 21 12	18 23 28	26 ♐ 52 58	03 ♑ 59 38	14 30	04 23	17 18	22 24	06 D 53	13 45	21 04	24 19
12	13 25 09	19 22 50	11 ♑ 02 44	18 ♑ 02 10	13 27	05 28	17 57	22 23	06 53	13 43	21 03	24 19
13	13 29 05	20 22 13	24 ♑ 57 54	01 ♒ 49 56	12 30	06 33	18 37	22 22	06 53	13 41	21 02	24 20
14	13 33 02	21 21 38	08 ♒ 38 19	15 ♒ 23 06	11 42	07 38	19 17	22 21	06 53	13 39	21 00	24 20
15	13 36 58	22 21 05	22 ♒ 04 21	28 ♒ 42 08	11 02	08 42	19 56	22 21	06 54	13 36	20 59	24 20
16	13 40 55	23 20 34	05 ♓ 16 32	11 ♓ 47 35	10 33	09 46	20 36	22 20	06 54	13 34	20 57	24 20
17	13 44 51	24 20 04	18 ♓ 15 21	24 ♓ 39 53	10 15	10 50	21 16	22 20	06 55	13 32	20 56	24 21
18	13 48 48	25 19 36	01 ♈ 01 14	07 ♈ 19 27	10 08	11 53	21 56	22 D 20	06 55	13 29	20 55	24 21
19	13 52 44	26 19 11	13 ♈ 34 34	19 ♈ 46 42	10 D 12	12 56	22 35	22 20	06 56	13 27	20 54	24 21
20	13 56 41	27 18 47	25 ♈ 55 56	02 ♉ 02 23	10 26	13 59	23 15	22 20	06 57	13 25	20 52	24 22
21	14 00 38	28 18 25	08 ♉ 06 12	14 ♉ 07 37	10 52	15 02	23 55	22 21	06 58	13 22	20 51	24 22
22	14 04 34	29 ♎ 18 05	20 ♉ 06 49	26 ♉ 04 07	11 26	16 05	24 35	22 22	06 59	13 20	20 50	24 23
23	14 08 31	00 ♏ 17 47	01 ♊ 59 49	07 ♊ 54 17	12 10	17 07	25 15	22 23	07 00	13 18	20 49	24 23
24	14 12 27	01 17 32	13 ♊ 47 55	19 ♊ 41 11	13 01	18 09	25 55	22 24	07 02	13 15	20 47	24 24
25	14 16 24	02 17 18	25 ♊ 34 32	01 ♋ 28 32	14 00	19 10	26 35	22 25	07 03	13 13	20 46	24 24
26	14 20 20	03 17 07	07 ♋ 23 43	13 ♋ 20 39	15 05	20 11	27 15	22 27	07 05	13 10	20 45	24 25
27	14 24 17	04 16 58	19 ♋ 19 58	25 ♋ 22 15	16 16	21 12	27 55	22 28	07 06	13 08	20 44	24 25
28	14 28 13	05 16 51	01 ♌ 28 09	07 ♌ 38 15	17 31	22 13	28 36	22 30	07 08	13 06	20 43	24 26
29	14 32 10	06 16 46	13 ♌ 53 09	20 ♌ 13 24	18 50	23 13	29 16	22 33	07 10	13 03	20 42	24 27
30	14 36 07	07 16 44	26 ♌ 39 30	03 ♍ 11 52	20 13	24 13	29 ♎ 56	22 35	07 12	13 01	20 41	24 27
31	14 40 03	08 ♏ 16 44	09 ♍ 50 50	16 ♍ 36 37	21 ♎ 39	25 ♐ 12	00 ♏ 36	22 ♒ 37	07 ♒ 14	12 ♉ 58	20 ♓ 40	24 ♑ 28

Date	Moon True ☊ ° '	Moon Mean ☊ ° '	Moon ☽ Latitude ° '
01	03 ♊ 22	04 ♊ 23	04 N 32
02	03 R 16	04 20	04 57
03	03 09	04 16	05 07
04	03 00	04 13	05 00
05	02 51	04 10	04 34
06	02 44	04 07	03 52
07	02 38	04 04	02 53
08	02 34	04 01	01 42
09	02 33	03 57	00 N 25
10	02 D 33	03 54	00 S 54
11	02 34	03 51	02 08
12	02 35	03 48	03 13
13	02 R 35	03 45	04 06
14	02 34	03 41	04 44
15	02 31	03 38	05 05
16	02 26	03 35	05 10
17	02 20	03 32	04 59
18	02 13	03 29	04 32
19	02 06	03 26	03 53
20	02 01	03 22	03 04
21	01 57	03 19	02 07
22	01 55	03 16	01 S 04
23	01 D 54	03 13	00 N 01
24	01 54	03 10	01 05
25	01 56	03 07	02 06
26	01 58	03 03	03 03
27	01 59	03 00	03 52
28	02 00	02 57	04 32
29	02 R 00	02 54	05 00
30	01 59	02 51	05 14
31	01 ♊ 56	02 ♊ 47	05 N 12

DECLINATIONS

Date	Sun ☉ ° '	Moon ☽ ° '	Mercury ☿ ° '	Venus ♀ ° '	Mars ♂ ° '	Jupiter ♃ ° '	Saturn ♄ ° '	Uranus ♅ ° '	Neptune ♆ ° '	Pluto ♇ ° '
01	03 S 23	23 N 13	12 S 48	20 S 47	03 S 36	15 S 02	19 S 19	15 N 40	04 S 31	22 S 53
02	03 46	19 56	12 29	21 07	03 52	15 03	19 20	15 39	04 32	22 53
03	04 09	15 34	12 06	21 27	04 07	15 04	19 20	15 39	04 32	22 53
04	04 32	10 18	11 38	21 47	04 23	15 05	19 20	15 38	04 33	22 53
05	04 56	04 N 21	11 06	22 06	04 39	15 05	19 20	15 37	04 33	22 53
06	05 19	01 S 57	10 30	22 24	04 55	15 06	19 20	15 37	04 34	22 53
07	05 42	08 16	09 51	22 42	05 10	15 07	19 21	15 36	04 35	22 53
08	06 04	14 11	09 09	23 00	05 26	15 07	19 21	15 36	04 35	22 53
09	06 27	19 17	08 24	23 17	05 42	15 08	19 21	15 35	04 35	22 53
10	06 50	23 11	07 39	23 33	05 57	15 08	19 21	15 34	04 36	22 53
11	07 13	25 32	06 55	23 49	06 13	15 08	19 21	15 34	04 37	22 53
12	07 35	26 11	06 11	24 04	06 28	15 09	19 21	15 33	04 37	22 53
13	07 58	25 10	05 30	24 19	06 44	15 09	19 21	15 32	04 38	22 53
14	08 20	22 40	04 52	24 33	06 59	15 09	19 21	15 32	04 38	22 53
15	08 42	18 57	04 19	24 47	07 15	15 09	19 20	15 31	04 39	22 53
16	09 04	14 23	03 51	25 00	07 30	15 09	19 20	15 30	04 40	22 53
17	09 26	09 13	03 28	25 12	07 45	15 09	19 20	15 30	04 40	22 53
18	09 48	03 S 46	03 10	25 24	08 01	15 09	19 20	15 29	04 41	22 53
19	10 09	01 N 46	02 59	25 36	08 16	15 09	19 20	15 28	04 41	22 53
20	10 31	07 10	02 53	25 46	08 31	15 09	19 20	15 28	04 41	22 53
21	10 52	12 13	02 53	25 56	08 46	15 08	19 19	15 27	04 42	22 53
22	11 14	16 44	02 58	26 06	09 01	15 08	19 19	15 26	04 42	22 53
23	11 35	20 34	03 08	26 15	09 16	15 08	19 19	15 25	04 43	22 53
24	11 55	23 32	03 22	26 23	09 32	15 07	19 18	15 25	04 43	22 53
25	12 16	25 28	03 40	26 30	09 46	15 06	19 18	15 24	04 44	22 53
26	12 37	26 16	04 02	26 37	10 01	15 06	19 18	15 23	04 44	22 53
27	12 57	25 52	04 27	26 44	10 16	15 05	19 17	15 22	04 45	22 53
28	13 17	24 15	04 55	26 50	10 31	15 04	19 17	15 22	04 45	22 53
29	13 37	21 26	05 25	26 55	10 46	15 04	19 16	15 21	04 45	22 52
30	13 56	17 32	05 57	26 59	11 00	15 03	19 16	15 20	04 46	22 52
31	14 S 16	12 N 41	06 S 30	27 S 03	11 S 15	15 S 02	19 S 15	15 N 20	04 S 46	22 S 52

ZODIAC SIGN ENTRIES

Date	h	m	Planets
01	00	53	☽ ♌
03	08	38	☽ ♍
05	12	41	☽ ♎
07	11	21	♀ ♐
07	14	22	☽ ♏
09	15	24	☽ ♐
11	17	15	☽ ♑
13	20	47	☽ ♒
16	02	22	☽ ♓
18	10	04	☽ ♈
20	19	59	☽ ♉
23	04	51	☉ ♏
23	07	57	☽ ♊
25	21	00	☽ ♋
28	09	07	☽ ♌
30	14	21	♂ ♏
30	18	09	☽ ♍

LATITUDES

Date	Mercury ☿ ° '	Venus ♀ ° '	Mars ♂ ° '	Jupiter ♃ ° '	Saturn ♄ ° '	Uranus ♅ ° '	Neptune ♆ ° '	Pluto ♇ ° '
01	03 S 37	02 S 14	00 N 42	01 S 11	00 S 49	00 S 25	01 S 10	01 S 39
04	03 13	02 25	00 41	01 10	00 49	00 25	01 10	01 39
07	02 32	02 37	00 39	01 10	00 49	00 25	01 10	01 39
10	01 38	02 47	00 38	01 10	00 49	00 25	01 10	01 40
13	00 S 36	02 58	00 37	01 09	00 49	00 25	01 10	01 40
16	00 N 22	03 07	00 35	01 09	00 49	00 25	01 10	01 40
19	01 09	03 16	00 34	01 09	00 49	00 25	01 10	01 40
22	01 41	03 24	00 32	01 08	00 49	00 25	01 10	01 40
25	02 00	03 31	00 31	01 08	00 49	00 25	01 10	01 40
28	02 07	03 37	00 29	01 07	00 49	00 25	01 10	01 40
31	02 N 05	03 S 42	00 N 28	01 S 07	00 S 49	00 S 25	01 S 10	01 S 40

DATA

Julian Date	2459489
Delta T	+71 seconds
Ayanamsa	24° 09' 24"
Synetic vernal point	04° ♓ 57' 35"
True obliquity of ecliptic	23° 26' 16"

MOON'S PHASES, APSIDES AND POSITIONS ☽

Date	h	m	Phase	Longitude ° '	Eclipse Indicator
06	11	05	●	13 ♎ 25	
13	03	25	☽	20 ♑ 01	
20	14	57	○	27 ♈ 26	
28	20	05	☾	05 ♌ 37	

Day	h	m		
08	17	22	Perigee	
24	15	18	Apogee	
06	04	39	0S	
12	09	10	Max dec	26° S 12'
19	04	17	0N	
26	16	05	Max dec	26° N 17'

LONGITUDES

Date	Chiron ⚷ ° '	Ceres ⚳ ° '	Pallas ⚴ ° '	Juno ⚵ ° '	Vesta ⚶ ° '	Black Moon Lilith ⚸ ° '
01	10 ♈ 45	11 ♊ 57	13 ♓ 28	16 ♐ 38	05 ♏ 33	08 ♊ 19
11	10 ♈ 18	12 ♊ 07	11 ♓ 29	19 ♐ 17	10 ♏ 46	09 ♊ 26
21	09 ♈ 52	11 ♊ 38	10 ♓ 04	22 ♐ 10	16 ♏ 02	10 ♊ 32
31	09 ♈ 28	10 ♊ 30	09 ♓ 15	25 ♐ 15	21 ♏ 22	11 ♊ 39

ASPECTARIAN

h m	Aspects
01 Friday	
01 40	☉ ± ♅
12 58	☽ bQ ♆
14 10	☽ ☍ ♄
14 26	☿ □ ♇
14 55	☽ # ♇
17 34	☽ ⚹ ☉
21 46	☽ ⚹ ♂
23 51	☽ Q ☿
02 Saturday	
01 56	☿ ⊻ ♀
03 32	☽ □ ♅
04 57	☽ # ♀
05 49	☽ ± ♆
07 48	♀ ⚹ ♇
11 23	☉ ∠ ♀
15 42	☽ # ♄
16 51	☽ ⊼ ♆
19 29	☽ ☍ ♃
20 57	☽ ⚹ ☿
21 08	♀ Q ♄
22 22	☽ ⊼ ♇
23 35	☽ ∠ ☉
23 43	☽ □ ♀
03 Sunday	
02 50	☽ ∠ ♂
06 03	☉ ∥ ♂
09 12	☽ ± ♇
11 39	☽ ∥ ♅
14 29	☽ # ♃
17 00	☽ ⊥ ☉
19 44	☽ ⊥ ♂
20 56	☽ ⊼ ♄
22 21	☽ ∠ ☿
04 Monday	
00 05	☿ △ ♃
01 09	☽ bQ ♇
04 23	☽ ⊻ ☉
05 39	☽ # ☿
06 47	☽ ⊻ ♂
07 26	☽ ± ♄
09 19	☽ △ ♅
10 28	☽ Q ♀
12 26	☉ ∥ ♆
13 08	☽ ⊥ ☿
21 48	☽ ☍ ♆
22 48	☽ ⊻ ☿
22 57	☽ bQ ♄
05 Tuesday	
00 10	☽ ⊼ ♃
03 03	☽ △ ♇
03 18	♂ ∥ ♆
05 50	☿ ⊥ ♀
08 46	☽ ⚹ ♀
09 55	☽ # ☉
10 20	☽ ± ♃
10 55	☽ # ♂
10 56	☽ bQ ♅
11 13	☽ # ♆
13 53	☿ ⊼ ♆
06 Wednesday	
00 16	☽ △ ♄
01 23	☽ bQ ♃
02 00	☽ ± ♅
09 27	♂ ⊼ ♅
11 05	☽ ☌ ☉
11 56	☽ ⊼ ♅
11 57	☽ ∠ ♀
12 04	☽ ☌ ♂
15 44	♀ ∠ ♂
18 29	♇ St D
21 40	☽ ☌ ☿
21 53	☽ ∥ ♆
23 11	☉ ⊼ ♅
23 39	☽ ∥ ♂
23 54	☽ ⊼ ♆
07 Thursday	
01 30	☽ ∥ ☉
02 08	☽ △ ♃
03 59	☽ ⊥ ♀
05 03	☽ □ ♇
09 44	☽ ± ♆
14 37	☽ ⊻ ♀
17 34	☽ ∥ ☿
08 Friday	
00 26	☽ bQ ♆
01 37	☽ □ ♄
02 42	♀ ∥ ♇
04 01	☉ ☌ ♂
10 29	☽ Q ♇
13 00	☽ ☍ ♅
15 32	☽ ⊻ ♂
15 49	☽ ⊻ ☉
16 04	☽ ∥ ♃
18 09	☽ # ♅
19 10	☽ ⊻ ☿
09 Saturday	
00 53	☽ △ ♆
01 48	☽ ⊥ ♂
02 19	☽ ⊥ ☉
03 04	☽ □ ♃
04 14	☽ ⊥ ☿
06 05	☽ ⚹ ♇
07 01	☽ Q ♄
07 38	☿ ∠ ♀
12 17	☽ ∥ ♄
16 18	☉ ☌ ☿
17 14	☽ ∠ ♂
17 53	☽ ∠ ☿
18 08	☽ ∠ ☉
19 37	☽ ☌ ♀
22 48	☿ ☌ ♂
10 Sunday	
02 43	☽ ⚹ ♄
06 45	☽ ∠ ♇
08 35	☽ Q ♃
09 49	☽ ∥ ♇
14 08	☽ ⊼ ♅
15 12	☽ ∥ ♀
16 49	☽ ⚹ ☿
19 12	☽ ⚹ ♂
20 45	☽ ⚹ ☉
21 40	☽ ⊥ ♇
11 Monday	
00 06	☽ ± ♅
02 17	♄ St D
02 17	☽ □ ♆
03 37	☽ ∠ ♄
04 30	☽ ⚹ ♃
05 33	☉ ∥ ☿
07 42	☽ ⊻ ♇
11 24	☽ Q ☿
15 08	☽ bQ ♅
16 15	☽ Q ♂
18 21	☽ Q ☉
18 44	☽ ⊥ ♄
12 Tuesday	
01 43	☽ ⊻ ♀
04 54	☽ ⊻ ♄
04 54	☿ ∥ ♂
05 29	☿ ⊼ ♅
05 46	☽ ∠ ♃
08 36	☽ Q ♆
12 47	☽ ⊥ ♀
15 51	☽ □ ☿
16 34	☽ △ ♅
21 09	☽ ⊥ ♃
13 Wednesday	
00 27	☽ □ ♂
03 25	☽ □ ☉
05 11	☽ ⚹ ♆
05 35	☽ ∠ ♀
07 30	☽ ⊻ ♃
10 53	☽ ☌ ♇
19 26	♀ ⚹ ♄
20 50	☽ ∥ ♀
14 Thursday	
03 32	☉ ⊼ ♆
07 21	☽ ∠ ♆
08 54	☽ ☌ ♄
10 03	☽ ⚹ ♀
10 17	☽ ∥ ♇
17 09	☽ △ ☿
20 52	☽ □ ♅
21 28	☿ ∥ ♆
23 18	☽ ⊥ ♆
15 Friday	
07 57	☽ △ ♂
09 18	☽ Q ♀
09 47	☽ ∥ ♄
10 02	☽ ⊻ ♆
11 46	☉ △ ♃
12 29	☽ ☌ ♃
12 33	☽ △ ☉
16 05	☽ ⊻ ♇
18 53	☽ bQ ☿
16 Saturday	
02 22	♀ ∠ ♇
02 58	☽ ⊥ ♇
05 14	☽ Q ♅
06 24	☽ # ♅
08 10	☽ ∥ ♃
10 43	☽ ± ☿
12 38	☽ bQ ♂
14 59	☽ ⊻ ♄
18 06	☽ bQ ☉
19 28	☽ ∠ ♇
20 59	☽ □ ♀
21 27	☽ ⊼ ☿
17 Sunday	
00 32	♂ ⊼ ♆
00 48	♀ Q ♃
01 23	☿ ⚹ ♀
02 04	☽ ⊥ ♄
03 14	☽ ⚹ ♅
06 08	☽ ± ♂
11 06	☽ ∥ ☉
12 10	☽ ± ☉
12 12	☉ □ ♇
17 00	☽ ☌ ♆
17 55	☽ ⊼ ♂
18 14	☽ ∥ ♂
18 50	☽ ∠ ♄
19 37	☽ ⊻ ♃
23 24	☽ ⚹ ♇
18 Monday	
00 20	☽ ⊼ ☉
05 30	♃ St D
06 54	☽ ⊥ ♃
07 14	☽ ∠ ♅
08 02	☽ ∥ ♆
14 39	☽ ∥ ☿
15 17	☿ St D
22 08	☽ Q ♇
23 15	☽ ⚹ ♄
19 Tuesday	
00 01	☽ ∠ ♃
00 17	☽ ⊥ ♅
02 36	♂ △ ♃
05 26	☽ ☍ ☿
10 40	☽ △ ♀
11 46	☽ ⊻ ♅
17 11	☽ # ☿
22 23	☽ Q ♄
23 15	♀ ⊼ ♅
20 Wednesday	
00 51	☽ # ♆
01 34	☉ ± ♆
02 09	☽ ⊻ ♆
04 59	☽ ⚹ ♃
06 28	☽ ☍ ♂
08 56	☽ □ ♇
13 50	☽ ⊥ ♆
14 57	☽ ☍ ☉
18 34	☽ bQ ♀
18 35	☽ # ♂
21 Thursday	
04 33	☽ Q ♃
04 55	☽ # ☉
07 32	☽ ∠ ♆
09 45	☽ □ ♄
14 02	☽ ± ♀
17 44	☽ ⊼ ☿
22 28	☽ ☌ ♅
22 Friday	
03 07	☽ # ♃
03 08	☽ ⊼ ♀
04 19	♂ □ ♇
04 45	☽ ∥ ♅
06 19	☽ ± ☿
13 26	☽ ⚹ ♆
16 32	☽ □ ♃
20 35	☽ △ ♇
21 32	☽ ⊼ ♂
23 Saturday	
01 32	☽ bQ ☿
03 34	☽ # ♄
08 14	☽ ⊼ ☉
10 24	☽ ± ♂
13 39	☽ Q ♆
21 32	☽ ± ☉
22 12	☽ △ ♄
24 Sunday	
03 01	☽ bQ ♇
05 47	☽ bQ ♂
06 01	☽ # ♇
10 17	☽ △ ☿
10 54	☽ ⊻ ♅
17 33	☽ bQ ☉
17 45	☿ ⊼ ♅
17 51	♀ ⊥ ♇
21 22	☽ ± ♇
21 42	☽ ☍ ♀
23 05	☽ ⊥ ♅
25 Monday	
02 14	☽ □ ♆
04 48	☽ bQ ♄
05 33	☽ △ ♃
09 36	☽ ⊼ ♇
13 01	♀ ± ♅
14 11	☽ △ ♂
17 21	☽ ∠ ♅
18 28	♂ ± ♆
23 10	☽ ± ♄
26 Tuesday	
02 54	☽ △ ☉
11 21	☽ ⊼ ♄
12 06	☽ bQ ♃
23 37	☽ ⚹ ♅
27 Wednesday	
01 06	♀ □ ♆
05 10	☽ □ ☿
06 16	☽ ± ♃
14 48	☽ △ ♆
16 04	☽ ⊼ ♀
18 16	☽ ⊼ ♃
22 08	☽ ☍ ♇
23 29	☽ Q ♅
28 Thursday	
03 39	☿ ∥ ♆
05 02	☽ ± ♀
06 02	☽ □ ♂
10 02	♀ ∠ ♄
19 15	♀ ⚹ ♃
20 05	☽ □ ☉
20 16	☽ bQ ♆
20 49	☽ Q ☿
22 19	☉ bQ ♆
23 03	☽ ☍ ♄
29 Friday	
00 10	☽ bQ ♀
00 53	☽ # ♇
10 25	☽ □ ♅
13 33	☽ ± ♆
18 47	☽ Q ♂
22 32	☽ ⚹ ☿
30 Saturday	
00 53	☽ ⊼ ♆
02 06	☽ # ♄
04 24	☽ ☍ ♃
07 05	☽ △ ♀
07 54	☽ ⊼ ♇
09 14	☽ Q ☉
09 53	☉ □ ♄
17 54	♀ ⊻ ♇
18 22	☽ ⚹ ♂
19 00	☽ ± ♇
19 40	☿ ⊼ ♆
23 27	☽ ∥ ♅
31 Sunday	
00 55	☽ # ♃
05 05	☽ # ☉
05 34	☽ ∠ ☿
07 17	☽ ⊼ ♄
08 58	☽ ⚹ ☉
11 19	☽ bQ ♇
17 33	☽ △ ♅
18 02	☽ ± ♄
18 10	☽ # ♂
22 46	☽ ∠ ♂
23 34	☽ ⊥ ☿

All ephemeris data is given at 12.00 UT and the Moon's longitude is additionally given for 24.00 UT

NOVEMBER 2021

LONGITUDES

Date	Sidereal time h m s	Sun ☉ ° ' "	Moon ☽ ° ' "	Moon ☽ 24.00 ° ' "	Mercury ☿ ° '	Venus ♀ ° '	Mars ♂ ° '	Jupiter ♃ ° '	Saturn ♄ ° '	Uranus ♅ ° '	Neptune ♆ ° '	Pluto ♇ ° '
01	14 44 00	09 ♏ 16 45	23 ♍ 29 17	00 ♎ 28 47	23 ♎ 08	26 ♐ 11	01 ♏ 17	22 ♒ 40	07 ♒ 16	12 ♉ 56	20 ♓ 39	24 ♑ 29
02	14 47 56	10 16 49	07 ♎ 34 51	14 ♎ 47 06	24 38	27 10	01 57	22 43	07 18	12 R 53	20 R 38	24 29
03	14 51 53	11 16 55	22 ♎ 04 55	29 ♎ 27 35	26 10	28 08	02 37	22 46	07 20	12 51	20 37	24 30
04	14 55 49	12 17 03	06 ♏ 54 10	14 ♏ 23 41	27 44	29 ♐ 06	03 18	22 50	07 23	12 48	20 36	24 31
05	14 59 46	13 17 12	21 ♏ 55 03	29 ♏ 27 07	29 ♎ 18	00 ♑ 03	03 58	22 53	07 25	12 46	20 35	24 32
06	15 03 42	14 17 24	06 ♐ 58 47	14 ♐ 28 59	00 ♏ 53	01 00	04 39	22 57	07 28	12 43	20 35	24 33
07	15 07 39	15 17 37	21 ♐ 56 42	29 ♐ 21 06	02 29	01 56	05 19	23 01	07 30	12 41	20 34	24 34
08	15 11 35	16 17 52	06 ♑ 41 25	13 ♑ 57 05	04 06	02 52	06 00	23 05	07 33	12 38	20 33	24 35
09	15 15 32	17 18 09	21 ♑ 07 38	28 ♑ 12 47	05 42	03 47	06 41	23 09	07 36	12 36	20 32	24 36
10	15 19 29	18 18 26	05 ♒ 12 22	12 ♒ 06 21	07 19	04 42	07 21	23 13	07 39	12 33	20 32	24 37
11	15 23 25	19 18 46	18 ♒ 54 46	25 ♒ 37 47	08 56	05 36	08 02	23 18	07 42	12 31	20 31	24 38
12	15 27 22	20 19 06	02 ♓ 15 36	08 ♓ 48 29	10 33	06 29	08 43	23 23	07 45	12 29	20 30	24 39
13	15 31 18	21 19 28	15 ♓ 16 42	21 ♓ 40 35	12 10	07 22	09 23	23 28	07 48	12 26	20 30	24 40
14	15 35 15	22 19 51	28 ♓ 00 27	04 ♈ 16 37	13 47	08 14	10 04	23 33	07 51	12 24	20 29	24 41
15	15 39 11	23 20 16	10 ♈ 29 24	16 ♈ 39 08	15 23	09 06	10 45	23 39	07 55	12 21	20 29	24 42
16	15 43 08	24 20 42	22 ♈ 46 04	28 ♈ 50 31	17 00	09 56	11 26	23 44	07 58	12 19	20 28	24 43
17	15 47 04	25 21 09	04 ♉ 52 45	10 ♉ 53 02	18 36	10 46	12 07	23 50	08 02	12 16	20 27	24 44
18	15 51 01	26 21 38	16 ♉ 51 36	22 ♉ 48 43	20 12	11 36	12 48	23 56	08 06	12 14	20 27	24 45
19	15 54 58	27 22 09	28 ♉ 44 37	04 ♊ 39 34	21 48	12 24	13 29	24 02	08 09	12 12	20 27	24 47
20	15 58 54	28 22 41	10 ♊ 33 50	16 ♊ 27 40	23 24	13 11	14 10	24 08	08 13	12 09	20 26	24 48
21	16 02 51	29 ♏ 23 15	22 ♊ 21 23	28 ♊ 15 17	25 00	13 58	14 51	24 14	08 17	12 07	20 26	24 49
22	16 06 47	00 ♐ 23 50	04 ♋ 09 42	10 ♋ 04 59	26 35	14 44	15 32	24 21	08 21	12 05	20 26	24 50
23	16 10 44	01 24 27	16 ♋ 01 32	21 ♋ 59 46	28 11	15 28	16 13	24 28	08 25	12 02	20 25	24 52
24	16 14 40	02 25 05	28 ♋ 00 06	04 ♌ 03 02	29 ♏ 46	16 12	16 54	24 34	08 29	12 00	20 25	24 53
25	16 18 37	03 25 45	10 ♌ 09 00	16 ♌ 18 32	01 ♐ 21	16 55	17 35	24 41	08 33	11 58	20 25	24 54
26	16 22 33	04 26 27	22 ♌ 32 08	28 ♌ 50 17	02 55	17 36	18 16	24 49	08 38	11 56	20 25	24 56
27	16 26 30	05 27 10	05 ♍ 13 31	11 ♍ 42 15	04 30	18 17	18 58	24 56	08 42	11 53	20 24	24 57
28	16 30 27	06 27 55	18 ♍ 16 56	24 ♍ 57 55	06 05	18 56	19 39	25 04	08 47	11 51	20 24	24 59
29	16 34 23	07 28 41	01 ♎ 45 28	08 ♎ 39 46	07 39	19 34	20 20	25 11	08 51	11 49	20 24	25 00
30	16 38 20	08 ♐ 29 29	15 ♎ 40 51	22 ♎ 48 37	09 ♐ 13	20 ♑ 11	21 ♏ 02	25 ♒ 19	08 ♒ 56	11 ♉ 47	20 ♓ 24	25 ♑ 01

Date	Moon True ☊ ° '	Moon Mean ☊ ° '	Moon ☽ Latitude ° '
01	01 ♊ 53	02 ♊ 44	04 N 53
02	01 R 50	02 41	04 16
03	01 47	02 38	03 22
04	01 45	02 35	02 13
05	01 44	02 32	00 N 54
06	01 D 44	02 28	00 S 29
07	01 44	02 25	01 50
08	01 46	02 22	03 02
09	01 47	02 19	04 01
10	01 48	02 16	04 44
11	01 48	02 13	05 09
12	01 R 48	02 09	05 17
13	01 47	02 06	05 08
14	01 46	02 03	04 45
15	01 44	02 00	04 08
16	01 43	01 57	03 20
17	01 42	01 53	02 24
18	01 42	01 50	01 22
19	01 42	01 47	00 S 16
20	01 D 42	01 44	00 N 49
21	01 42	01 41	01 52
22	01 42	01 38	02 51
23	01 42	01 34	03 42
24	01 R 42	01 31	04 25
25	01 42	01 28	04 56
26	01 42	01 25	05 14
27	01 42	01 22	05 17
28	01 D 42	01 19	05 05
29	01 42	01 15	04 35
30	01 ♊ 43	01 ♊ 12	03 N 49

DECLINATIONS

Date	Sun ☉ ° '	Moon ☽ ° '	Mercury ☿ ° '	Venus ♀ ° '	Mars ♂ ° '	Jupiter ♃ ° '	Saturn ♄ ° '	Uranus ♅ ° '	Neptune ♆ ° '	Pluto ♇ ° '
01	14 S 35	07 N 04	07 S 05	27 S 07	11 S 29	15 S 01	19 S 15	15 N 19	04 S 47	22 S 52
02	14 54	00 N 55	07 41	27 09	11 44	15 00	19 14	15 18	04 47	22 52
03	15 13	05 S 29	08 17	27 12	11 58	14 58	19 14	15 17	04 47	22 52
04	15 31	11 43	08 55	27 13	12 12	14 57	19 13	15 17	04 48	22 52
05	15 50	17 22	09 32	27 14	12 27	14 56	19 12	15 16	04 48	22 52
06	16 08	21 57	10 10	27 15	12 41	14 55	19 12	15 15	04 48	22 52
07	16 25	25 01	10 48	27 14	12 55	14 53	19 11	15 14	04 48	22 52
08	16 43	26 18	11 26	27 14	13 09	14 52	19 10	15 14	04 49	22 51
09	17 00	25 45	12 03	27 12	13 23	14 50	19 10	15 13	04 49	22 51
10	17 17	23 33	12 41	27 10	13 36	14 48	19 09	15 12	04 49	22 51
11	17 33	20 03	13 18	27 08	13 50	14 47	19 08	15 11	04 49	22 51
12	17 50	15 36	13 54	27 05	14 04	14 45	19 07	15 11	04 50	22 51
13	18 05	10 32	14 30	27 01	14 17	14 43	19 06	15 10	04 50	22 51
14	18 21	05 S 09	15 06	26 57	14 31	14 41	19 05	15 09	04 50	22 51
15	18 36	00 N 21	15 40	26 53	14 44	14 40	19 05	15 09	04 50	22 50
16	18 51	05 46	16 15	26 48	14 57	14 38	19 04	15 08	04 50	22 50
17	19 06	10 53	16 48	26 42	15 10	14 36	19 03	15 07	04 51	22 50
18	19 20	15 34	17 21	26 36	15 23	14 33	19 02	15 06	04 51	22 50
19	19 34	19 37	17 52	26 30	15 36	14 31	19 01	15 06	04 51	22 50
20	19 48	22 50	18 23	26 23	15 49	14 29	19 00	15 05	04 51	22 50
21	20 01	25 05	18 54	26 16	16 01	14 27	18 59	15 04	04 51	22 49
22	20 14	26 13	19 23	26 08	16 14	14 25	18 58	15 04	04 51	22 49
23	20 26	26 09	19 51	26 00	16 26	14 22	18 57	15 03	04 51	22 49
24	20 39	24 53	20 18	25 51	16 39	14 20	18 56	15 02	04 51	22 49
25	20 50	22 27	20 45	25 42	16 51	14 17	18 55	15 02	04 52	22 49
26	21 02	18 56	21 10	25 33	17 03	14 15	18 53	15 01	04 52	22 48
27	21 13	14 31	21 34	25 23	17 15	14 12	18 52	15 00	04 52	22 48
28	21 23	09 18	21 57	25 13	17 26	14 10	18 51	15 00	04 52	22 48
29	21 33	03 31	22 19	25 03	17 38	14 07	18 50	14 59	04 52	22 48
30	21 S 43	02 S 39	22 S 40	24 S 53	17 S 50	14 S 04	18 S 49	14 N 58	04 S 52	22 S 48

ZODIAC SIGN ENTRIES

Date	h	m	Planets
01	23	11	☽ ♎
04	00	52	☽ ♏
05	10	44	♀ ♑
05	22	35	☿ ♏
06	00	52	☽ ♐
08	01	03	☽ ♑
10	03	03	☽ ♒
12	07	54	☽ ♓
14	15	48	☽ ♈
17	02	18	☽ ♉
19	14	33	☽ ♊
22	02	34	☉ ♐
22	03	33	☽ ♋
24	15	36	☿ ♐
24	15	58	☽ ♌
27	02	12	☽ ♍
29	08	55	☽ ♎

LATITUDES

Date	Mercury ☿ ° '	Venus ♀ ° '	Mars ♂ ° '	Jupiter ♃ ° '	Saturn ♄ ° '	Uranus ♅ ° '	Neptune ♆ ° '	Pluto ♇ ° '
01	02 N 03	03 S 44	00 N 27	01 S 07	00 S 49	00 S 25	01 S 10	01 S 40
04	01 53	03 47	00 26	01 06	00 49	00 25	01 10	01 41
07	01 38	03 49	00 24	01 06	00 49	00 25	01 10	01 41
10	01 21	03 49	00 23	01 06	00 49	00 25	01 10	01 41
13	01 02	03 48	00 21	01 05	00 49	00 25	01 10	01 41
16	00 42	03 45	00 19	01 05	00 49	00 25	01 09	01 41
19	00 21	03 39	00 18	01 04	00 49	00 25	01 09	01 41
22	00 N 01	03 32	00 16	01 04	00 49	00 25	01 09	01 41
25	00 S 19	03 22	00 14	01 04	00 49	00 25	01 09	01 41
28	00 39	03 09	00 13	01 03	00 49	00 25	01 09	01 42
31	00 S 57	02 S 54	00 N 11	01 S 03	00 S 49	00 S 25	01 S 09	01 S 42

DATA

Julian Date	2459520
Delta T	+71 seconds
Ayanamsa	24° 09' 27"
Synetic vernal point	04° ♓ 57' 32"
True obliquity of ecliptic	23° 26' 16"

MOON'S PHASES, APSIDES AND POSITIONS ☽

Date	h	m	Phase	Longitude ° '	Eclipse Indicator
04	21	15	●	12 ♏ 40	
11	12	46	☽	19 ♒ 21	
19	08	57	○	27 ♉ 14	partial
27	12	28	☾	05 ♍ 28	

Day	h	m		
05	22	12	Perigee	
21	01	56	Apogee	
02	15	28	OS	
08	16	34	Max dec	26° S 20'
15	10	27	ON	
22	22	45	Max dec	26° N 21'
30	01	48	OS	

LONGITUDES

Date	Chiron ⚷ ° '	Ceres ⚳ ° '	Pallas ⚴ ° '	Juno ⚵ ° '	Vesta ⚶ ° '	Black Moon Lilith ⚸ ° '
01	09 ♈ 25	10 ♊ 22	09 ♓ 12	25 ♐ 34	21 ♏ 54	11 ♊ 45
11	09 ♈ 05	08 ♊ 36	09 ♓ 05	28 ♐ 52	27 ♏ 16	12 ♊ 52
21	08 ♈ 48	06 ♊ 26	09 ♓ 34	02 ♑ 18	02 ♐ 39	13 ♊ 58
31	08 ♈ 35	04 ♊ 05	10 ♓ 35	05 ♑ 53	08 ♐ 03	15 ♊ 05

All ephemeris data is given at 12.00 UT and the Moon's longitude is additionally given for 24.00 UT

ASPECTARIAN

h m	Aspects
01 Monday	
04 18	☿ △ ♃
07 05	☽ ☍ ♆
09 52	☽ ⚼ ♄
10 35	☽ ⊼ ♃
11 18	☽ ⊻ ☿
11 57	☽ # ☿
13 28	☽ ∠ ☉
13 43	☽ △ ♇
15 15	☽ ⊥ ♂
17 00	☽ □ ♀
19 37	☽ ⚼ ♅
20 56	☽ ± ♃
21 08	☽ # ♆
02 Tuesday	
02 02	☽ ⊻ ♂
06 01	☽ ⊥ ☉
09 39	☿ □ ♇
10 50	☽ ± ♅
11 31	☽ △ ♄
12 14	☽ ⚼ ♃
16 51	☽ ⊻ ☉
18 33	☉ ∥ ♃
20 50	☽ ⊼ ♅
03 Wednesday	
01 32	☽ Q ♀
05 10	♀ ⚼ ♅
09 25	☽ ∥ ♆
09 36	☽ ⊼ ♆
13 08	☽ △ ♃
15 57	☽ □ ♇
17 38	☉ # ♅
18 52	☿ ± ♆
19 23	☽ ± ♆
19 27	☽ ☌ ☿
22 32	☽ ✶ ♀
23 48	☽ ∥ ☿
04 Thursday	
05 55	☽ ☌ ♂
09 55	☽ ⚼ ♆
12 46	☽ □ ♄
14 01	☽ ∥ ♂
17 39	☉ Q ♇
21 00	☽ Q ♇
21 15	☽ ☌ ☉
21 26	☽ ☍ ♅
23 58	☉ ☍ ♅
05 Friday	
00 18	☽ ∠ ♀
01 13	☽ ∥ ♃
02 38	☽ # ♅
04 38	☽ ∥ ☉
09 53	☽ △ ♆
13 33	☽ □ ♃
15 38	☽ ⊥ ♀
16 10	☽ ✶ ♇
17 35	☽ Q ♄
20 48	☽ ∥ ♄
06 Saturday	
01 09	☽ ⊻ ☿
01 49	☽ ⊻ ♀
08 06	☽ ⊻ ♂
11 50	☽ ⊥ ☿
12 46	☽ ✶ ♄
15 59	☿ ✶ ♀
16 06	☽ ∠ ♇
17 55	☽ ∥ ♇
18 08	☽ ⊥ ♂
18 22	☽ Q ♃
21 09	☽ ⊼ ♅
07 Sunday	
00 23	♀ ⊥ ♄
00 32	☽ ⊻ ☉
03 58	☽ ∠ ☿
06 33	☽ ⊥ ♇
06 45	☽ ± ♅
09 16	☽ ∠ ♂
09 46	☽ □ ♆
10 52	☽ ⊥ ☉
12 54	☽ ∠ ♄
13 44	☽ ✶ ♃
16 14	☽ ⊻ ♇
20 23	♂ ⚼ ♆
21 15	☽ ⚼ ♅
08 Monday	
02 31	☽ ∠ ☉
03 33	☽ ⊥ ♄
05 19	☽ ☌ ♀
07 13	☽ ✶ ☿
10 49	☽ ✶ ♂
13 25	☽ ⊻ ♄
14 18	☽ ∠ ♃
15 04	☽ Q ♆
21 48	☽ △ ♅
09 Tuesday	
05 06	☽ ✶ ☉
05 18	☽ ⊥ ♃
05 32	☽ Q ☿
07 41	☽ Q ♂
09 34	☿ ⚼ ♆
11 01	☽ ✶ ♆
15 26	☽ ⊻ ♃
17 51	☽ ☌ ♇
10 Wednesday	
02 56	☽ Q ☉
11 04	☽ ⊻ ♀
12 33	☽ ∠ ♆
12 56	☿ ☌ ♂
15 54	☽ □ ♂
16 08	☽ □ ☿
16 14	☽ ☌ ♄
17 04	☿ □ ♄
17 28	☽ ∥ ♇
22 13	☽ ⊥ ♀
23 14	♂ □ ♄
11 Thursday	
00 45	☽ □ ♅
04 14	☽ ⊥ ♆
12 46	☽ □ ☉
14 51	☽ ⊻ ♆
15 13	☽ ∠ ♀
17 19	☽ ∥ ♄
19 52	☽ ☌ ♃
22 13	☽ ⊻ ♇
12 Friday	
01 09	☽ ∥ ☉
08 46	☽ Q ♅
09 03	☽ ⊥ ♇
14 05	☽ # ♅
16 12	☽ ∥ ♃
16 23	☉ △ ♆
19 12	☽ ∥ ♂
19 24	☽ ∥ ☿
20 18	☽ ✶ ♀
22 05	☽ ⊻ ♄
22 08	☿ ∥ ♂
13 Saturday	
00 28	☽ △ ♂
01 34	☽ ∠ ♇
05 23	☽ △ ☿
06 44	☽ ✶ ♅
09 14	☽ ⊥ ♄
15 57	☿ ☍ ♅
19 30	☿ Q ♇
20 13	☽ Q ♀
20 26	☿ ∥ ♃
21 46	☽ ☌ ♆
14 Sunday	
00 19	☽ △ ☉
00 43	♀ ⊻ ♄
02 11	☽ ∠ ♄
03 29	☽ ⊻ ♃
05 40	☽ ✶ ♇
06 06	☽ ⚼ ♂
10 50	☽ ∠ ♅
13 21	☽ ∥ ♆
13 41	☽ ⚼ ☿
14 27	☿ # ♅
14 58	☽ ⊥ ♃
18 45	♀ Q ♆
21 44	♀ ∠ ♃
15 Monday	
00 16	☽ ± ♂
04 01	☽ ⊥ ♅
04 39	☽ Q ♇
05 12	♂ ∥ ♃
06 59	☽ ✶ ♄
07 28	☽ ⚼ ☉
08 24	☽ ∠ ♃
09 06	☽ □ ♀
09 33	☽ ± ☿
12 32	☽ ⊼ ♂
15 36	☽ ⊻ ♅
19 58	☉ □ ♃
22 58	☽ ⊼ ☿
16 Tuesday	
02 32	☽ ± ☉
06 29	☽ Q ♄
07 29	☽ ⊻ ♆
07 51	☽ # ♆
13 55	☽ ✶ ♃
15 23	☽ ⊼ ☉
15 51	☽ □ ♇
19 18	☽ ⊥ ♆
21 01	☉ ✶ ♇
17 Wednesday	
06 42	♂ # ♅
06 57	☉ ∥ ♄
13 09	☽ ∠ ♆
13 55	☽ Q ♃
17 23	♂ ☍ ♅
18 19	☽ □ ♄
18 Thursday	
00 39	☽ △ ♀
02 44	☽ ☌ ♅
03 20	☽ ☍ ♂
05 12	☉ Q ♄
06 36	☽ # ♃
09 30	☽ ∥ ♅
10 38	♂ Q ♇
10 57	☽ # ♂
15 38	☿ △ ♆
19 14	☽ ✶ ♆
19 48	☽ ☍ ☿
23 31	☽ # ☿
19 Friday	
02 23	☽ □ ♃
03 57	☽ △ ♇
06 08	♀ △ ♅
08 10	☽ # ♄
08 57	☽ ☍ ☉
09 05	☽ ⚼ ♀
11 43	☽ # ☉
15 25	☉ ∠ ♀
19 30	☽ Q ♆
20 Saturday	
04 39	☽ ± ♀
07 12	☽ △ ♄
10 26	☽ ⚼ ♇
11 53	☽ # ♇
15 14	☽ ⊻ ♅
17 43	☽ ⊼ ♀
19 46	☽ ⊼ ♂
23 43	☿ □ ♃
21 Sunday	
03 24	☽ ⊥ ♅
04 47	☽ ± ♇
08 05	☽ □ ♆
08 44	☽ ± ♂
09 14	☿ ✶ ♇
13 54	☽ ⚼ ♄
15 52	☽ △ ♃
16 09	☿ ∥ ♄
17 01	☽ ⊼ ♇
18 13	☽ ⊼ ☿
21 39	☽ ∠ ♅
22 Monday	
03 38	☽ ⊼ ☉
04 10	☽ ⚼ ♂
08 13	☿ Q ♄
08 18	☽ ± ♄
08 19	☽ ± ☿
09 09	☽ # ♀
16 57	☽ ± ☉
20 32	☽ ⊼ ♄
22 37	☽ ⚼ ♃
23 Tuesday	
03 59	☽ ✶ ♅
05 22	☽ ⚼ ☿
10 49	☽ ☍ ♀
12 24	☽ △ ♂
12 50	☽ ⚼ ☉
16 57	☽ ± ♃
17 42	☽ # ♀
20 50	☽ △ ♆
24 Wednesday	
04 02	☽ Q ♅
05 06	☽ ⊼ ♃
05 46	☽ ☍ ♇
16 02	☽ △ ☿
21 34	☽ △ ☉
25 Thursday	
02 42	☽ ⚼ ♆
08 52	☽ ☍ ♄
08 58	☽ # ♇
15 32	☽ □ ♅
20 19	☽ ± ♆
21 44	☉ ∥ ☿
23 09	☽ # ☿
23 14	☽ # ☉
26 Friday	
01 57	☽ ⊼ ♀
03 19	☽ □ ♂
03 28	☿ ∠ ♀
07 55	☽ ⊼ ♆
12 18	☽ # ♄
14 10	☽ ± ♀
16 24	☽ ☍ ♃
16 35	☽ ⊼ ♇
22 22	☽ # ♂
27 Saturday	
03 59	☽ ± ♇
08 10	☽ ⚼ ♀
09 32	☽ ∥ ♅
10 28	☽ □ ☿
12 28	☽ □ ☉
13 31	☽ # ♃
15 25	☽ Q ♂
16 16	♃ ⊻ ♇
18 30	☽ ⊼ ♄
20 47	☽ ⚼ ♇
28 Sunday	
00 18	☽ △ ♅
05 36	☽ ± ♄
13 15	☽ △ ♀
14 37	☽ ✶ ♂
15 50	☽ ☍ ♆
17 45	♀ ⊥ ♃
21 56	☽ ⚼ ♄
23 48	☽ Q ☿
29 Monday	
00 01	☽ Q ☉
00 02	☽ △ ♇
00 17	☽ ⊼ ♃
03 19	☽ ⚼ ♅
04 39	☉ ☌ ☿
06 35	☽ # ♆
10 59	☽ ± ♃
14 10	♂ △ ♆
18 35	☽ ∠ ♂
19 03	☽ ± ♅
22 44	☽ ✶ ☉
23 33	☽ ✶ ☿
30 Tuesday	
00 24	☽ △ ♄
02 46	☽ ⚼ ♃
05 22	☽ ⊼ ♅
07 19	☿ ✶ ♄
10 50	☽ ⊥ ♂
19 56	☽ □ ♀
19 58	☽ ⊼ ♆
20 23	☿ ∥ ♇
20 30	☽ ∥ ♆
20 46	♀ ✶ ♆
21 29	☽ ⊻ ♂
23 14	☉ ✶ ♄

DECEMBER 2021

LONGITUDES

Date	Sidereal time h m s	Sun ☉ ° ' "	Moon ☽ ° ' "	Moon ☽ 24.00 ° ' "	Mercury ☿ ° '	Venus ♀ ° '	Mars ♂ ° '	Jupiter ♃ ° '	Saturn ♄ ° '	Uranus ♅ ° '	Neptune ♆ ° '	Pluto ♇ ° '
01	16 42 16	09 ♐ 30 18	00 ♏ 02 48	07 ♏ 22 57	10 ♐ 47	20 ♑ 47	21 ♏ 43	25 ♒ 27	09 ♒ 00	11 ♉ 45	20 ♓ 24	25 ♑ 03
02	16 46 13	10 31 09	14 ♏ 48 26	22 ♏ 18 27	12 22	21 21	22 25	25 35	09 05	11 R 43	20 D 24	25 04
03	16 50 09	11 32 01	29 ♏ 52 01	07 ♐ 28 03	13 56	21 53	23 06	25 43	09 10	11 41	20 24	25 06
04	16 54 06	12 32 54	15 ♐ 05 19	22 ♐ 42 35	15 30	22 25	23 48	25 52	09 15	11 38	20 24	25 07
05	16 58 02	13 33 49	00 ♑ 18 35	07 ♑ 52 05	17 04	22 54	24 30	26 00	09 20	11 36	20 24	25 09
06	17 01 59	14 34 44	15 ♑ 21 57	22 ♑ 47 12	18 38	23 22	25 11	26 09	09 25	11 35	20 25	25 11
07	17 05 56	15 35 41	00 ♒ 07 01	07 ♒ 20 43	20 12	23 49	25 53	26 18	09 30	11 33	20 25	25 12
08	17 09 52	16 36 38	14 ♒ 27 52	21 ♒ 28 11	21 46	24 13	26 35	26 27	09 35	11 31	20 25	25 14
09	17 13 49	17 37 35	28 ♒ 21 34	05 ♓ 08 05	23 20	24 36	27 16	26 36	09 41	11 29	20 25	25 16
10	17 17 45	18 38 33	11 ♓ 47 54	18 ♓ 21 21	24 54	24 57	27 58	26 45	09 46	11 27	20 26	25 17
11	17 21 42	19 39 32	24 ♓ 48 47	01 ♈ 10 41	26 28	25 16	28 40	26 55	09 51	11 25	20 26	25 19
12	17 25 38	20 40 31	07 ♈ 27 32	13 ♈ 39 51	28 03	25 33	29 ♏ 22	27 04	09 57	11 23	20 26	25 21
13	17 29 35	21 41 31	19 ♈ 48 11	25 ♈ 53 03	29 ♐ 37	25 48	00 ♐ 04	27 14	10 02	11 22	20 27	25 22
14	17 33 31	22 42 32	01 ♉ 54 59	07 ♉ 54 28	01 ♑ 11	26 00	00 46	27 24	10 08	11 20	20 27	25 24
15	17 37 28	23 43 33	13 ♉ 51 59	19 ♉ 47 59	02 46	26 11	01 28	27 33	10 14	11 18	20 28	25 26
16	17 41 25	24 44 34	25 ♉ 42 52	01 ♊ 37 02	04 20	26 19	02 10	27 43	10 19	11 17	20 28	25 27
17	17 45 21	25 45 36	07 ♊ 30 50	13 ♊ 24 34	05 54	26 25	02 52	27 54	10 25	11 15	20 29	25 29
18	17 49 18	26 46 39	19 ♊ 18 33	25 ♊ 13 02	07 29	26 28	03 34	28 04	10 31	11 14	20 29	25 31
19	17 53 14	27 47 43	01 ♋ 08 17	07 ♋ 04 31	09 03	26 R 29	04 16	28 14	10 37	11 12	20 30	25 33
20	17 57 11	28 48 47	13 ♋ 01 58	19 ♋ 00 50	10 38	26 28	04 58	28 25	10 43	11 11	20 30	25 35
21	18 01 07	29 ♐ 49 51	25 ♋ 01 22	01 ♌ 03 46	12 12	26 24	05 40	28 35	10 49	11 09	20 31	25 36
22	18 05 04	00 ♑ 50 56	07 ♌ 08 18	13 ♌ 15 11	13 46	26 18	06 23	28 46	10 55	11 08	20 32	25 38
23	18 09 00	01 52 02	19 ♌ 24 44	25 ♌ 37 12	15 20	26 09	07 05	28 57	11 01	11 07	20 33	25 40
24	18 12 57	02 53 08	01 ♍ 52 55	08 ♍ 12 12	16 53	25 58	07 47	29 08	11 07	11 05	20 33	25 42
25	18 16 54	03 54 15	14 ♍ 35 25	21 ♍ 02 55	18 26	25 44	08 30	29 19	11 13	11 04	20 34	25 44
26	18 20 50	04 55 23	27 ♍ 35 03	04 ♎ 12 09	19 59	25 27	09 12	29 30	11 19	11 03	20 35	25 46
27	18 24 47	05 56 31	10 ♎ 54 33	17 ♎ 42 32	21 31	25 09	09 54	29 41	11 25	11 02	20 36	25 47
28	18 28 43	06 57 40	24 ♎ 36 17	01 ♏ 35 57	23 02	24 48	10 37	29 ♒ 52	11 32	11 01	20 37	25 49
29	18 32 40	07 58 49	08 ♏ 41 32	15 ♏ 52 56	24 32	24 25	11 19	00 ♓ 04	11 38	11 00	20 38	25 51
30	18 36 36	08 59 59	23 ♏ 09 53	00 ♐ 31 57	26 01	23 59	12 02	00 15	11 44	10 59	20 39	25 53
31	18 40 33	10 ♑ 01 09	07 ♐ 58 31	15 ♐ 28 49	27 ♑ 28	23 ♑ 32	12 ♐ 45	00 ♓ 27	11 ♒ 51	10 ♉ 58	20 ♓ 40	25 ♑ 55

Date	Moon True ☊ ° '	Moon Mean ☊ ° '	Moon ☽ Latitude ° '
01	01 ♊ 44	01 ♊ 09	02 N 47
02	01 D 44	01 06	01 33
03	01 44	01 03	00 N 10
04	01 R 44	00 59	01 S 13
05	01 43	00 56	02 32
06	01 42	00 53	03 39
07	01 40	00 50	04 30
08	01 38	00 47	05 03
09	01 37	00 44	05 16
10	01 36	00 40	05 12
11	01 D 36	00 37	04 51
12	01 37	00 34	04 17
13	01 38	00 31	03 31
14	01 40	00 28	02 37
15	01 42	00 25	01 37
16	01 43	00 21	00 S 33
17	01 R 43	00 18	00 N 32
18	01 41	00 15	01 35
19	01 39	00 12	02 35
20	01 35	00 09	03 28
21	01 30	00 05	04 12
22	01 24	00 ♊ 02	04 45
23	01 19	29 ♉ 59	05 06
24	01 15	29 56	05 12
25	01 11	29 53	05 04
26	01 10	29 50	04 40
27	01 D 10	29 46	04 01
28	01 11	29 43	03 07
29	01 12	29 40	02 00
30	01 13	29 37	00 44
31	01 ♊ 13	29 ♉ 34	00 S 37

DECLINATIONS

Date	Sun ☉ ° '	Moon ☽ ° '	Mercury ☿ ° '	Venus ♀ ° '	Mars ♂ ° '	Jupiter ♃ ° '	Saturn ♄ ° '	Uranus ♅ ° '	Neptune ♆ ° '	Pluto ♇ ° '
01	21 S 52	08 S 53	23 S 00	24 S 42	18 S 01	14 S 01	18 S 47	14 N 58	04 S 52	22 S 47
02	22 01	14 48	23 19	24 31	18 12	13 58	18 46	14 57	04 52	22 47
03	22 10	19 57	23 36	24 20	18 23	13 56	18 45	14 56	04 52	22 47
04	22 18	23 49	23 53	24 08	18 34	13 53	18 44	14 56	04 51	22 47
05	22 26	25 58	24 08	23 56	18 45	13 50	18 42	14 55	04 51	22 46
06	22 33	26 11	24 22	23 44	18 56	13 47	18 41	14 55	04 51	22 46
07	22 40	24 31	24 34	23 32	19 06	13 43	18 40	14 54	04 51	22 46
08	22 46	21 19	24 46	23 20	19 17	13 40	18 38	14 53	04 51	22 46
09	22 52	16 59	24 55	23 08	19 27	13 37	18 37	14 53	04 51	22 45
10	22 57	11 56	25 04	22 55	19 37	13 34	18 35	14 52	04 51	22 45
11	23 02	06 31	25 11	22 43	19 47	13 30	18 34	14 52	04 51	22 45
12	23 07	00 S 59	25 17	22 30	19 56	13 27	18 32	14 51	04 50	22 45
13	23 11	04 N 29	25 22	22 18	20 06	13 24	18 31	14 51	04 50	22 44
14	23 14	09 41	25 25	22 05	20 15	13 20	18 30	14 50	04 50	22 44
15	23 17	14 27	25 26	21 52	20 25	13 17	18 28	14 50	04 50	22 44
16	23 20	18 39	25 26	21 40	20 34	13 13	18 26	14 49	04 50	22 44
17	23 22	22 05	25 25	21 27	20 42	13 10	18 25	14 49	04 49	22 43
18	23 24	24 36	25 22	21 14	20 51	13 06	18 23	14 48	04 49	22 43
19	23 25	26 01	25 18	21 02	21 00	13 02	18 22	14 48	04 49	22 43
20	23 26	26 15	25 12	20 49	21 08	12 59	18 20	14 47	04 49	22 43
21	23 26	25 15	25 05	20 37	21 16	12 55	18 19	14 47	04 48	22 42
22	23 26	23 05	24 56	20 25	21 24	12 51	18 17	14 47	04 48	22 42
23	23 25	19 50	24 46	20 12	21 32	12 47	18 15	14 46	04 48	22 42
24	23 24	15 40	24 34	20 00	21 39	12 43	18 14	14 46	04 47	22 42
25	23 23	10 44	24 20	19 48	21 47	12 39	18 12	14 46	04 47	22 41
26	23 21	05 N 15	24 05	19 37	21 54	12 35	18 10	14 45	04 47	22 41
27	23 18	00 S 38	23 49	19 25	22 01	12 31	18 09	14 45	04 46	22 41
28	23 15	06 38	23 31	19 13	22 08	12 27	18 07	14 45	04 46	22 40
29	23 12	12 30	23 12	19 02	22 14	12 23	18 05	14 44	04 45	22 40
30	23 08	17 51	22 52	18 51	22 21	12 19	18 03	14 44	04 45	22 40
31	23 S 04	22 S 14	22 S 30	18 S 40	22 S 27	12 S 15	18 S 02	14 N 44	04 S 45	22 S 40

ZODIAC SIGN ENTRIES

Date	h	m	Planets
01	11	55	☽ ♏
03	12	13	☽ ♐
05	11	31	☽ ♑
07	11	48	☽ ♒
09	14	53	☽ ♓
11	21	46	☽ ♈
13	09	53	♂ ♐
13	17	52	☿ ♑
14	08	11	☽ ♉
16	20	43	☽ ♊
19	09	42	☽ ♋
21	15	59	☉ ♑
21	21	54	☽ ♌
24	08	24	☽ ♍
26	16	24	☽ ♎
28	21	16	☽ ♏
29	04	09	♃ ♓
30	23	08	☽ ♐

LATITUDES

Date	Mercury ☿ ° '	Venus ♀ ° '	Mars ♂ ° '	Jupiter ♃ ° '	Saturn ♄ ° '	Uranus ♅ ° '	Neptune ♆ ° '	Pluto ♇ ° '
01	00 S 57	02 S 54	00 N 11	01 S 03	00 S 49	00 S 25	01 S 09	01 S 42
04	01 14	02 36	00 09	01 02	00 49	00 25	01 09	01 42
07	01 30	02 14	00 07	01 02	00 49	00 25	01 09	01 42
10	01 44	01 49	00 06	01 02	00 49	00 25	01 09	01 42
13	01 55	01 20	00 04	01 01	00 49	00 25	01 09	01 42
16	02 04	00 47	00 N 02	01 01	00 49	00 25	01 09	01 43
19	02 11	00 S 11	00 00	01 01	00 49	00 25	01 08	01 43
22	02 13	00 N 29	00 S 02	01 01	00 49	00 25	01 08	01 43
25	02 11	01 13	00 04	01 00	00 49	00 24	01 08	01 43
28	02 05	01 58	00 06	01 00	00 49	00 24	01 08	01 43
31	01 S 52	02 N 45	00 S 08	01 S 00	00 S 49	00 S 24	01 S 08	01 S 44

DATA

Julian Date	2459550
Delta T	+71 seconds
Ayanamsa	24° 09' 32"
Synetic vernal point	04° ♓ 57' 27"
True obliquity of ecliptic	23° 26' 15"

MOON'S PHASES, APSIDES AND POSITIONS ☽

Date	h	m	Phase	Longitude ° '	Eclipse Indicator
04	07	43	●	12 ♐ 22	Total
11	01	36	☽	19 ♓ 13	
19	04	35	○	27 ♊ 29	
27	02	24	☾	05 ♎ 32	

Day	h	m		
04	09	58	Perigee	
18	02	00	Apogee	
06	02	32	Max dec	26° S 20'
12	16	14	0N	
20	04	34	Max dec	26° N 18'
27	09	29	0S	

LONGITUDES

Date	Chiron ⚷ ° '	Ceres ⚳ ° '	Pallas ⚴ ° '	Juno ⚵ ° '	Vesta ⚶ ° '	Black Moon Lilith ⚸ ° '
01	08 ♈ 35	04 ♊ 05	10 ♓ 35	05 ♑ 53	08 ♐ 03	15 ♊ 05
11	08 ♈ 28	01 ♊ 51	12 ♓ 06	09 ♑ 34	13 ♐ 28	16 ♊ 11
21	08 ♈ 26	00 ♊ 00	14 ♓ 02	13 ♑ 20	18 ♐ 52	17 ♊ 18
31	08 ♈ 30	28 ♉ 41	16 ♓ 20	17 ♑ 11	24 ♐ 16	18 ♊ 24

All ephemeris data is given at 12.00 UT and the Moon's longitude is additionally given for 24.00 UT

ASPECTARIAN

h m	Aspects
01 Wednesday	
00 28	☿ ∠ ♇
02 08	☽ ∠ ☉
03 43	☽ □ ♇
04 06	☽ ∠ ☿
04 20	☽ △ ♃
05 59	☽ ± ♆
13 24	♆ St D
18 06	☽ ⊥ ☉
20 43	☽ ⊥ ☿
20 47	☽ ₽ ♆
02 Thursday	
01 12	☉ ∠ ♇
02 15	☿ ⊼ ♅
02 43	☽ □ ♄
02 50	☽ Q ♀
04 35	☽ ⊻ ☉
07 01	☽ ☍ ♅
07 36	☽ ⊻ ☿
08 32	☽ ∥ ♃
09 12	☽ Q ♇
12 38	☽ # ♅
20 58	☽ △ ♆
22 52	☽ ⚹ ♀
03 Friday	
00 45	☽ ☌ ♂
03 53	☽ ∥ ♂
04 26	☽ ⚹ ♇
05 22	☽ □ ♃
05 57	☽ ∥ ♄
07 42	☽ Q ♄
08 32	☿ Q ♃
15 14	☉ ⊼ ♅
23 29	☽ ∠ ♀
04 Saturday	
01 00	☽ ∥ ☉
02 46	☽ ⚹ ♄
04 10	☽ ∠ ♇
04 30	☽ ∥ ♇
06 35	☽ ⊼ ♅
07 43	☽ ☌ ☉
10 03	☽ Q ♃
12 31	☽ ∥ ☿
12 43	☽ ☌ ☿
14 09	☽ ⊥ ♀
14 22	☽ ∥ ♀
16 01	☽ ± ♅
18 22	☽ ⊥ ♇
20 22	☽ □ ♆
23 55	☽ ⊻ ♀
05 Sunday	
01 35	☿ ∥ ♀
02 22	☽ ⊻ ♂
02 30	☽ ∠ ♄
03 50	☽ ⊻ ♇
05 08	☽ ⚹ ♃
06 10	☽ ₽ ♅
06 22	♂ ∥ ♄
08 25	☿ ⊥ ♀
12 18	☽ ⊥ ♂
16 49	☽ ⊥ ♄
20 08	☿ ± ♅
06 Monday	
00 12	☉ Q ♃
00 52	☽ Q ♆
02 25	☽ ⊻ ♄
03 18	☽ ∠ ♂
05 10	☽ ∠ ♃
05 56	☽ △ ♅
10 39	☽ ⊻ ☉
11 41	♂ ⚹ ♇
17 53	☽ ⊻ ☿
19 48	☽ ⊥ ♃
20 09	☽ ⚹ ♆
20 29	☿ ⊥ ♇
21 02	☽ ⊥ ☉
07 Tuesday	
01 21	☽ ☌ ♀
03 56	☽ ☌ ♇
04 42	☽ ⚹ ♂
04 48	☽ ⊥ ☿
05 40	☽ ⊻ ♃
11 33	☽ ∥ ☿
12 51	☽ ∠ ☉
15 16	☿ □ ♆
20 47	☽ ∠ ♆
21 06	☽ ∥ ♀
21 27	☽ ∠ ☿
08 Wednesday	
01 33	☽ Q ♂
02 20	☽ ∥ ♇
02 37	☽ ∥ ☉
03 43	☽ ☌ ♄
06 21	♂ □ ♃
07 01	☽ □ ♅
11 30	☉ ∥ ♇
11 55	☽ ⊥ ♆
15 56	☽ ⚹ ☉
22 11	☽ ⊻ ♆
23 27	☽ ∥ ♂
09 Thursday	
02 07	☽ ⚹ ☿
03 28	☽ ∥ ♄
05 15	☽ ⊻ ♀
06 34	☽ ⊻ ♇
08 38	☉ ± ♅
08 53	☽ ☌ ♃
09 59	☽ □ ♂
13 58	☽ Q ♅
14 24	☽ Q ☉
16 03	☽ ⊥ ♀
17 07	☽ ⊥ ♇
22 19	☽ # ♅
10 Friday	
02 00	☽ Q ☿
03 59	♂ Q ♄
04 26	☽ ∥ ♃
08 18	☽ ⊻ ♄
08 34	☽ ∠ ♀
09 15	☽ ∠ ♇
09 45	☿ ∠ ♄
09 45	☉ ∥ ♀
11 22	☽ ⚹ ♅
12 50	☿ ⊻ ♀
17 56	☿ ⊻ ♇
19 17	☽ ⊥ ♄
22 36	☉ ⊥ ♀
11 Saturday	
01 36	☽ □ ☉
03 37	☉ ⊥ ♇
03 50	☽ ☌ ♆
08 03	♀ ∥ ♇
11 10	☿ ₽ ♅
12 05	☽ ∠ ♄
12 52	☽ ⚹ ♀
12 56	☽ ⚹ ♇
15 00	☽ ∠ ♅
15 33	☽ □ ☿
15 59	☽ ⊻ ♃
16 29	♀ ☌ ♇
19 15	☽ ∥ ♆
19 24	☿ ⚹ ♃
19 40	☽ △ ♂
12 Sunday	
03 29	☽ ⊥ ♃
06 21	☉ □ ♆
08 02	☽ ⊥ ♅
11 46	☽ Q ♇
12 10	☽ Q ♀
16 50	☽ ⚹ ♄
19 34	☽ ⊻ ♅
21 01	☽ ∠ ♃
13 Monday	
02 10	☽ ₽ ♂
13 16	☽ ⊻ ♆
13 36	☽ # ♆
16 03	☽ △ ☉
16 26	☽ Q ♄
20 54	☽ ± ♂
23 00	☽ □ ♇
14 Tuesday	
00 02	☽ □ ♀
00 16	☿ ⊻ ♂
01 07	☽ ⊥ ♆
02 52	☽ ⚹ ♃
09 33	☽ ⊼ ♂
10 20	☽ △ ☿
19 05	☽ ∠ ♆
15 Wednesday	
00 41	☽ ₽ ☉
03 12	☽ Q ♃
04 36	☽ □ ♄
05 55	☽ # ♃
06 51	☽ ☌ ♅
13 59	☽ ∥ ♅
20 32	☽ ± ☉
21 04	☽ ₽ ☿
16 Thursday	
01 21	☽ ⚹ ♆
09 50	☽ ⊼ ☉
10 42	☽ # ♄
11 29	☽ △ ♇
11 47	☿ ⊥ ♄
13 14	☽ △ ♀
16 08	☽ □ ♃
18 08	☽ ± ☿
17 Friday	
01 10	☽ # ♂
01 44	☽ Q ♆
01 56	☽ ☍ ♂
03 02	☉ ∠ ♄
05 21	☉ ⊻ ♇
07 22	☽ # ♀
08 14	☽ ⊼ ☿
17 16	☽ # ♇
17 57	☽ △ ♄
18 04	☽ ₽ ♇
19 36	☽ ⊻ ♅
19 59	☽ ₽ ♀
23 13	☽ # ☉
23 19	☉ ₽ ♅
18 Saturday	
04 28	☉ ⊻ ♀
07 46	☽ ⊥ ♅
12 25	☽ ± ♇
14 22	☽ ± ♀
14 24	☽ □ ♆
22 18	☽ # ☿
19 Sunday	
00 38	☽ ⊼ ♇
00 42	☽ ₽ ♄
02 01	☽ ∠ ♅
02 35	☽ ⊼ ♀
03 26	☿ Q ♆
04 35	☽ ☍ ☉
06 02	☽ △ ♃
10 36	♀ St R
14 33	♀ ∥ ♂
18 43	☽ ⊼ ♂
19 05	☽ ± ♄
20 Monday	
00 31	☉ ⚹ ♃
06 25	☽ ☍ ☿
07 17	☽ ⊼ ♄
07 35	☽ ± ♂
08 17	☽ ⚹ ♅
12 46	☽ ₽ ♃
13 22	☿ ⊻ ♄
20 18	☿ △ ♅
21 28	☿ ⊥ ♂
21 Tuesday	
02 47	☽ ₽ ♂
03 00	☽ △ ♆
07 04	☽ ± ♃
08 17	☽ Q ♅
13 10	☽ ☍ ♇
14 44	☽ ☍ ♀
14 44	☽ # ☿
19 12	☽ ⊼ ♃
22 26	☽ ⊼ ☉
22 Wednesday	
08 48	☽ # ☉
08 50	☽ ₽ ♆
10 24	☽ △ ♂
11 23	☽ ± ☉
12 01	☿ ∠ ♃
15 15	☽ # ♇
19 28	☽ ☍ ♄
19 50	☽ □ ♅
23 Thursday	
00 49	☽ # ♂
02 31	☽ ± ♆
02 54	☽ ⊼ ☿
06 36	☽ ₽ ☉
09 26	☽ # ♀
14 12	☽ ⊼ ♆
16 15	☽ ± ☿
21 44	☽ # ♄
24 Friday	
00 07	☽ ⊼ ♇
00 50	☽ ⊼ ♀
06 39	☽ ☍ ♃
07 17	♄ □ ♅
07 25	♃ Q ♅
11 39	☽ ± ♇
12 01	☽ ₽ ☿
12 09	☽ ± ♀
14 05	☽ △ ☉
16 38	☽ ∥ ♅
23 52	☽ □ ♂
25 Saturday	
02 55	☽ # ♃
04 45	☽ ₽ ♇
04 54	☽ ₽ ♀
05 24	☽ △ ♅
05 37	☽ ⊼ ♄
12 02	♀ ☌ ♇
16 56	☽ ± ♄
20 08	☽ △ ☿
23 07	☽ ☍ ♆
26 Sunday	
08 12	☽ △ ♀
08 39	☽ △ ♇
09 12	☽ ₽ ♅
09 40	☽ ₽ ♄
11 15	☽ Q ♂
13 57	☽ # ♆
15 32	☽ ⊼ ♃
21 29	☿ ⚹ ♆
22 23	☉ ⊥ ♄
27 Monday	
01 30	☽ ± ♅
02 24	☽ □ ☉
02 32	☽ ± ♃
10 07	☽ ⚹ ♂
12 13	☽ ⊼ ♅
12 55	☽ △ ♄
17 32	♀ ∠ ♂
18 46	☽ ₽ ♃
28 Tuesday	
04 32	☽ ∥ ♆
05 04	☽ ⊼ ♆
08 57	☽ □ ☿
12 19	☽ □ ♀
12 40	☽ Q ☉
13 50	☽ ∠ ♂
14 06	☽ □ ♇
15 28	☽ ± ♆
19 22	♂ ∠ ♇
21 11	☽ △ ♃
29 Wednesday	
01 04	♂ ⊼ ♅
03 20	☿ ⊥ ♃
06 01	☽ ⊥ ♂
06 50	☽ ₽ ♆
10 27	☿ ☌ ♀
10 43	☽ ⚹ ☉
11 30	☽ ∥ ♃
12 20	☉ ∥ ☿
15 51	☽ ☍ ♅
16 38	☽ ⊻ ♂
16 58	☽ □ ♄
18 03	☽ Q ♀
19 10	☽ Q ☿
20 39	☽ Q ♇
21 37	☽ # ♅
30 Thursday	
00 22	♂ ⚹ ♄
01 48	♀ ⊥ ♃
03 29	☉ Q ♆
07 51	☽ △ ♆
09 54	☿ ☌ ♇
12 58	☽ ∥ ♄
13 19	☽ ⚹ ♀
13 28	☽ ∠ ☉
16 27	☽ ⚹ ♇
16 45	☽ ∥ ♀
17 10	☽ ⚹ ☿
22 48	☽ Q ♄
23 42	☽ □ ♃
31 Friday	
01 18	☿ ∥ ♇
05 10	☽ ⊥ ☉
12 52	☽ ∠ ♀
13 24	☽ ∥ ♂
13 30	☽ ∥ ☿
14 21	☿ ∥ ♂
14 44	☽ ∥ ♇
15 31	☽ ⊻ ☉
16 44	☽ ∠ ♇
16 47	☽ ⊼ ♅
17 23	☽ ∥ ☉
18 15	☽ ⚹ ♄
19 57	☽ ∠ ☿
20 01	☽ ☌ ♂
21 11	☿ ∠ ♂

JANUARY 2022

LONGITUDES

Date	Sidereal time h m s	Sun ☉ ° ' "	Moon ☽ ° ' "	Moon ☽ 24.00 ° ' "	Mercury ☿ ° '	Venus ♀ ° '	Mars ♂ ° '	Jupiter ♃ ° '	Saturn ♄ ° '	Uranus ♅ ° '	Neptune ♆ ° '	Pluto ⯓ ° '
01	18 44 29	11 ♑ 02 20	23 ♐ 01 52	00 ♑ 36 34	28 ♑ 53	23 ♑ 03	13 ♐ 27	00 ♓ 39	11 ♒ 57	10 ♉ 57	20 ♓ 41	25 ♑ 57
02	18 48 26	12 03 31	08 ♑ 11 44	15 ♑ 46 06	00 ♒ 17	22 R 32	14 10	00 50	12 04	10 R 56	20 42	25 59
03	18 52 23	13 04 42	23 ♑ 18 23	00 ♒ 47 22	01 37	22 00	14 53	01 02	12 11	10 55	20 43	26 01
04	18 56 19	14 05 52	08 ♒ 11 56	15 ♒ 31 07	02 54	21 26	15 35	01 14	12 17	10 54	20 44	26 03
05	19 00 16	15 07 03	22 ♒ 44 06	29 ♒ 50 18	04 08	20 51	16 18	01 26	12 24	10 54	20 45	26 05
06	19 04 12	16 08 13	06 ♓ 49 21	13 ♓ 41 02	05 17	20 16	17 01	01 39	12 30	10 53	20 46	26 07
07	19 08 09	17 09 23	20 ♓ 25 22	27 ♓ 02 32	06 22	19 39	17 44	01 51	12 37	10 52	20 48	26 09
08	19 12 05	18 10 33	03 ♈ 32 51	09 ♈ 56 43	07 20	19 03	18 27	02 03	12 44	10 52	20 49	26 11
09	19 16 02	19 11 42	16 ♈ 14 41	22 ♈ 27 20	08 12	18 26	19 10	02 16	12 51	10 51	20 50	26 13
10	19 19 58	20 12 50	28 ♈ 35 17	04 ♉ 39 12	08 56	17 49	19 53	02 28	12 57	10 51	20 51	26 15
11	19 23 55	21 13 58	10 ♉ 39 44	16 ♉ 37 34	09 32	17 13	20 36	02 41	13 04	10 50	20 53	26 17
12	19 27 52	22 15 06	22 ♉ 33 19	28 ♉ 27 37	09 58	16 37	21 19	02 53	13 11	10 50	20 54	26 19
13	19 31 48	23 16 13	04 ♊ 21 02	10 ♊ 14 07	10 15	16 03	22 02	03 06	13 18	10 50	20 55	26 21
14	19 35 45	24 17 20	16 ♊ 07 23	22 ♊ 01 15	10 R 20	15 29	22 45	03 19	13 25	10 50	20 57	26 23
15	19 39 41	25 18 25	27 ♊ 56 09	03 ♋ 52 26	10 14	14 57	23 28	03 32	13 32	10 49	20 58	26 25
16	19 43 38	26 19 31	09 ♋ 50 22	15 ♋ 50 14	09 57	14 26	24 11	03 45	13 39	10 49	21 00	26 27
17	19 47 34	27 20 36	21 ♋ 52 14	27 ♋ 56 31	09 27	13 57	24 55	03 58	13 46	10 49	21 01	26 29
18	19 51 31	28 21 40	04 ♌ 03 13	10 ♌ 12 26	08 47	13 30	25 38	04 11	13 53	10 49	21 03	26 30
19	19 55 27	29 ♑ 22 44	16 ♌ 24 13	22 ♌ 38 39	07 55	13 05	26 21	04 24	14 00	10 D 49	21 04	26 32
20	19 59 24	00 ♒ 23 47	28 ♌ 55 46	05 ♍ 15 38	06 55	12 42	27 05	04 37	14 07	10 49	21 06	26 34
21	20 03 21	01 24 50	11 ♍ 38 18	18 ♍ 03 52	05 47	12 21	27 48	04 50	14 14	10 49	21 08	26 36
22	20 07 17	02 25 52	24 ♍ 32 26	01 ♎ 04 07	04 34	12 03	28 31	05 04	14 21	10 49	21 09	26 38
23	20 11 14	03 26 53	07 ♎ 39 04	14 ♎ 17 30	03 18	11 47	29 15	05 17	14 28	10 50	21 11	26 40
24	20 15 10	04 27 54	20 ♎ 59 34	27 ♎ 45 28	02 01	11 34	29 ♐ 58	05 30	14 35	10 50	21 13	26 42
25	20 19 07	05 28 55	04 ♏ 35 25	11 ♏ 29 34	00 ♒ 46	11 23	00 ♑ 42	05 44	14 42	10 50	21 14	26 44
26	20 23 03	06 29 56	18 ♏ 28 01	25 ♏ 30 52	29 ♑ 34	11 15	01 26	05 57	14 50	10 51	21 16	26 46
27	20 27 00	07 30 55	02 ♐ 38 02	09 ♐ 49 24	28 27	11 09	02 09	06 11	14 57	10 51	21 18	26 48
28	20 30 56	08 31 55	17 ♐ 04 41	24 ♐ 23 28	27 28	11 06	02 53	06 25	15 04	10 52	21 19	26 50
29	20 34 53	09 32 54	01 ♑ 45 11	09 ♑ 09 05	26 36	11 D 05	03 36	06 38	15 11	10 52	21 21	26 52
30	20 38 50	10 33 52	16 ♑ 34 20	23 ♑ 59 56	25 52	11 06	04 20	06 52	15 18	10 53	21 23	26 54
31	20 42 46	11 ♒ 34 49	01 ♒ 24 50	08 ♒ 47 57	25 ♑ 17	11 ♑ 10	05 ♑ 04	07 ♓ 06	15 ♒ 25	10 ♉ 53	21 ♓ 25	26 ♑ 56

Date	Moon True ☊ ° '	Moon Mean ☊ ° '	Moon ☽ Latitude ° '
01	01 ♊ 12	29 ♉ 30	01 S 56
02	01 R 08	29 27	03 07
03	01 02	29 24	04 05
04	00 55	29 21	04 45
05	00 48	29 18	05 06
06	00 41	29 15	05 07
07	00 36	29 11	04 51
08	00 33	29 08	04 20
09	00 32	29 05	03 36
10	00 D 32	29 02	02 44
11	00 33	28 59	01 45
12	00 34	28 56	00 S 43
13	00 R 35	28 52	00 N 20
14	00 33	28 49	01 23
15	00 29	28 46	02 22
16	00 23	28 43	03 14
17	00 14	28 40	03 59
18	00 03	28 36	04 33
19	29 ♉ 52	28 33	04 55
20	29 40	28 30	05 04
21	29 30	28 27	04 57
22	29 22	28 24	04 35
23	29 16	28 21	03 59
24	29 13	28 17	03 09
25	29 12	28 14	02 07
26	29 D 12	28 11	00 N 57
27	29 R 13	28 08	00 S 18
28	29 11	28 05	01 34
29	29 08	28 02	02 44
30	29 01	27 58	03 43
31	28 ♉ 52	27 ♉ 55	04 S 28

DECLINATIONS

Date	Sun ☉ ° '	Moon ☽ ° '	Mercury ☿ ° '	Venus ♀ ° '	Mars ♂ ° '	Jupiter ♃ ° '	Saturn ♄ ° '	Uranus ♅ ° '	Neptune ♆ ° '	Pluto ⯓ ° '
01	22 S 59	25 S 11	22 S 07	18 S 30	22 S 33	12 S 11	18 S 00	14 N 43	04 S 44	22 S 39
02	22 53	26 18	21 43	18 19	22 39	12 06	17 58	14 43	04 44	22 39
03	22 48	25 27	21 18	18 09	22 44	12 02	17 56	14 43	04 43	22 39
04	22 42	22 48	20 52	18 00	22 50	11 58	17 54	14 43	04 43	22 38
05	22 35	18 45	20 26	17 50	22 55	11 53	17 52	14 43	04 42	22 38
06	22 28	13 45	19 59	17 41	23 00	11 49	17 51	14 42	04 42	22 38
07	22 20	08 15	19 32	17 32	23 05	11 44	17 49	14 42	04 41	22 37
08	22 12	02 S 34	19 05	17 24	23 09	11 40	17 47	14 42	04 41	22 37
09	22 04	03 N 04	18 39	17 16	23 13	11 35	17 45	14 42	04 40	22 37
10	21 55	08 25	18 13	17 08	23 17	11 31	17 43	14 42	04 40	22 37
11	21 46	13 21	17 49	17 01	23 21	11 26	17 41	14 42	04 39	22 36
12	21 36	17 43	17 26	16 54	23 25	11 22	17 39	14 42	04 39	22 36
13	21 26	21 21	17 05	16 48	23 28	11 17	17 37	14 42	04 38	22 36
14	21 15	24 05	16 46	16 42	23 31	11 12	17 35	14 42	04 38	22 35
15	21 04	25 47	16 30	16 37	23 34	11 08	17 33	14 42	04 37	22 35
16	20 53	26 18	16 17	16 32	23 37	11 03	17 31	14 42	04 36	22 35
17	20 41	25 36	16 07	16 27	23 39	10 58	17 30	14 42	04 36	22 34
18	20 29	23 40	16 00	16 24	23 42	10 53	17 28	14 42	04 35	22 34
19	20 17	20 37	15 56	16 20	23 44	10 49	17 26	14 42	04 34	22 34
20	20 04	16 35	15 56	16 17	23 46	10 44	17 24	14 42	04 34	22 34
21	19 51	11 46	15 59	16 15	23 47	10 39	17 22	14 42	04 33	22 33
22	19 37	06 23	16 04	16 13	23 48	10 34	17 20	14 42	04 32	22 33
23	19 23	00 N 37	16 12	16 12	23 49	10 29	17 18	14 42	04 32	22 33
24	19 09	05 S 17	16 22	16 11	23 50	10 24	17 15	14 42	04 31	22 32
25	18 54	11 03	16 33	16 10	23 51	10 19	17 13	14 42	04 30	22 32
26	18 39	16 25	16 45	16 10	23 51	10 14	17 11	14 42	04 30	22 32
27	18 23	20 59	16 59	16 10	23 51	10 09	17 09	14 42	04 29	22 32
28	18 08	24 22	17 12	16 11	23 51	10 04	17 07	14 43	04 28	22 31
29	17 52	26 09	17 26	16 12	23 51	09 59	17 05	14 43	04 28	22 31
30	17 35	26 06	17 40	16 13	23 50	09 54	17 03	14 43	04 27	22 31
31	17 S 19	24 S 12	17 S 53	16 S 14	23 S 49	09 S 48	17 S 01	14 N 43	04 S 26	22 S 30

ZODIAC SIGN ENTRIES

Date	h	m	Planets
01	23	02	☽ ♑
02	07	10	☿ ♒
03	22	44	☽ ♒
06	00	17	☽ ♓
08	05	26	☽ ♈
10	14	47	☽ ♉
13	03	08	☽ ♊
15	16	11	☽ ♋
18	04	03	☽ ♌
20	02	39	☉ ♒
20	14	02	☽ ♍
22	22	03	☽ ♎
24	12	53	♂ ♑
25	03	57	☽ ♏
26	03	05	☿ ♑
27	07	34	☽ ♐
29	09	09	☽ ♑
31	09	43	☽ ♒

LATITUDES

Date	Mercury ☿ ° '	Venus ♀ ° '	Mars ♂ ° '	Jupiter ♃ ° '	Saturn ♄ ° '	Uranus ♅ ° '	Neptune ♆ ° '	Pluto ⯓ ° '
01	01 S 46	03 N 01	00 S 08	01 S 00	00 S 49	00 S 24	01 S 08	01 S 44
04	01 24	03 47	00 10	01 00	00 50	00 24	01 08	01 44
07	00 53	04 30	00 12	00 59	00 50	00 24	01 08	01 44
10	00 S 12	05 10	00 14	00 59	00 50	00 24	01 08	01 44
13	00 N 37	05 43	00 16	00 59	00 50	00 24	01 08	01 45
16	01 32	06 10	00 18	00 59	00 50	00 24	01 08	01 45
19	02 26	06 29	00 20	00 59	00 50	00 24	01 08	01 45
22	03 08	06 42	00 23	00 59	00 50	00 24	01 08	01 45
25	03 31	06 48	00 25	00 59	00 50	00 24	01 07	01 46
28	03 32	06 49	00 27	00 58	00 51	00 24	01 07	01 46
31	03 N 15	06 N 45	00 S 29	00 S 58	00 S 51	00 S 23	01 S 07	01 S 46

DATA

Julian Date	2459581
Delta T	+71 seconds
Ayanamsa	24° 09' 38"
Synetic vernal point	04° ♓ 57' 21"
True obliquity of ecliptic	23° 26' 15"

MOON'S PHASES, APSIDES AND POSITIONS ☽

Date	h	m	Phase	Longitude ° '	Eclipse Indicator
02	18	33	●	12 ♑ 20	
09	18	11	☽	19 ♈ 27	
17	23	48	○	27 ♋ 51	
25	13	41	☾	05 ♏ 33	

Day	h	m		
01	22	48	Perigee	
14	09	19	Apogee	
30	07	02	Perigee	
02	13	35	Max dec	26° S 18'
08	22	51	0N	
16	10	18	Max dec	26° N 18'
23	14	32	0S	
29	23	23	Max dec	26° S 22'

LONGITUDES

Date	Chiron ⚷ ° '	Ceres ⚳ ° '	Pallas ⚴ ° '	Juno ⚵ ° '	Vesta ⚶ ° '	Black Moon Lilith ⚸ ° '
01	08 ♈ 31	28 ♉ 35	16 ♓ 35	17 ♑ 35	24 ♐ 48	18 ♊ 31
11	08 ♈ 40	28 ♉ 00	19 ♓ 14	21 ♑ 29	00 ♑ 09	19 ♊ 38
21	08 ♈ 55	28 ♉ 07	22 ♓ 10	25 ♑ 26	05 ♑ 27	20 ♊ 44
31	09 ♈ 15	28 ♉ 52	25 ♓ 20	29 ♑ 25	10 ♑ 43	21 ♊ 51

All ephemeris data is given at 12.00 UT and the Moon's longitude is additionally given for 24.00 UT

ASPECTARIAN

h m Aspects

01 Saturday
02 21 ☽ ± ♅
02 48 ☽ ⊥ ♀
04 57 ☽ Q ♃
07 06 ☽ ⊥ ⯓
08 16 ☽ □ ♆
09 50 ☉ △ ♅
11 45 ☽ ⊥ ☿
12 02 ☽ ⊻ ♀
16 37 ☽ ♇ ♅
16 38 ☽ ⊻ ⯓
18 16 ☽ ∠ ♄
22 13 ☽ ⊻ ☿

02 Sunday
00 13 ☽ ✶ ♃
08 36 ☽ ⊥ ♄
12 12 ☉ ⊻ ♄
12 40 ♂ ∥ ⯓
12 48 ☽ Q ♆
16 20 ☽ △ ♅
18 10 ☽ ⊻ ♄
18 33 ☽ ☌ ☉
21 55 ☽ ⊻ ♂
23 44 ☿ ⊻ ♃

03 Monday
00 17 ☽ ∠ ♃
07 52 ☽ ✶ ♆
07 56 ☽ ⊥ ♂
09 59 ☽ ☌ ♀
14 48 ☽ ⊥ ♃
16 20 ☽ ☌ ⯓
19 03 ☉ ∥ ♂
23 04 ☽ ∠ ♂

04 Tuesday
00 34 ☽ ⊻ ♃
02 37 ☽ ☌ ☿
07 59 ☽ ∠ ♆
09 04 ♀ ⊥ ♂
11 50 ☽ ∥ ♂
12 48 ☽ ∥ ☉
13 09 ☽ ∥ ⯓
16 25 ☽ □ ♅
18 44 ☽ ☌ ♄
22 23 ☽ ⊻ ☉
22 43 ☽ ⊥ ♆

05 Wednesday
00 12 ☉ ∥ ⯓
00 45 ☽ ✶ ♂
01 47 ☽ ∥ ☿
04 27 ♀ ∥ ♄
08 41 ☽ ⊻ ♆
08 59 ☽ ⊻ ♀
09 05 ☽ ⊥ ☉
16 03 ♀ ✶ ♆
16 31 ☽ ∥ ♄
16 51 ☽ ∥ ♀
17 38 ☽ ⊻ ⯓
18 40 ☽ ⊥ ♀
21 53 ☽ Q ♂
22 23 ☽ Q ♅

06 Thursday
01 27 ☽ ∠ ☉
02 57 ☽ ☌ ♃
03 52 ☽ ⊥ ⯓
07 30 ♂ ± ♅
07 40 ☽ # ♅
09 07 ☽ ⊻ ☿
09 25 ☽ ∠ ♀
19 04 ☽ ✶ ♅
19 30 ☽ ∠ ⯓
20 28 ☽ ⊥ ☿
20 46 ☽ ∥ ♃
22 01 ☽ ⊻ ♄
22 48 ☿ ∠ ♆

07 Friday
02 55 ☉ ∠ ♃
05 41 ☽ ✶ ☉
06 55 ☽ □ ♂
08 44 ☽ ⊥ ♄
10 41 ☽ ✶ ♀
12 40 ☽ ☌ ♆
13 49 ☽ ∠ ☿
21 51 ☽ ∠ ♅
22 23 ☽ ✶ ⯓

08 Saturday
01 10 ☽ ∠ ♄
03 03 ☽ ∥ ♆
05 13 ☽ Q ☉
07 34 ☽ Q ♀
09 11 ☽ ⊻ ♃
14 27 ☽ ⊥ ♅
19 37 ☽ ✶ ☿
20 34 ☽ ⊥ ♃
20 41 ☽ Q ⯓
22 48 ♀ ⊻ ♂

09 Sunday
00 48 ☉ ☌ ♀
01 44 ☽ ⊻ ♅
05 14 ♃ ⊥ ⯓
05 27 ☽ ✶ ♄
09 32 ☉ ⊻ ♂
13 59 ☽ ∠ ♃
16 01 ☽ □ ♀
17 58 ☽ △ ♂
18 11 ☽ □ ☉
19 04 ☽ # ♆
20 08 ☽ Q ☿
20 52 ☽ ⊻ ♆

10 Monday
04 49 ☽ Q ♄
07 23 ☽ □ ⯓
08 35 ☽ ⊥ ♆
19 48 ☽ ✶ ♃
22 24 ♀ ∠ ♃

11 Tuesday
00 46 ♂ ⊥ ⯓
01 14 ☽ ♇ ♂
02 25 ☽ ∠ ♆
02 33 ☽ # ♃
03 28 ☉ ✶ ♆
09 38 ☽ □ ☿
12 21 ☽ ☌ ♅
15 50 ♂ Q ♃
16 53 ☽ □ ♄
19 02 ☽ ∥ ♅
20 13 ☽ Q ♃
20 25 ☽ ± ♂
20 33 ☿ ∥ ♄
21 43 ♂ □ ♆

12 Wednesday
00 34 ☽ △ ♀
07 22 ☽ # ♀
08 38 ☽ ✶ ♆
09 19 ☽ ⊼ ♂
10 28 ☽ # ☿
11 20 ☽ △ ☉
11 39 ☽ # ♄
19 39 ☽ △ ⯓

13 Thursday
05 35 ☽ ♇ ♀
09 05 ☽ Q ♆
09 24 ☽ □ ♃
12 37 ☽ # ☉
20 45 ☽ ♇ ☉
21 54 ☽ # ⯓
23 05 ☽ ± ♀

14 Friday
00 10 ☽ △ ☿
01 12 ☽ ⊻ ♅
02 18 ☽ ♇ ⯓
06 10 ☽ # ♂
06 26 ☽ △ ♄
10 46 ☽ ⊼ ♀
11 41 ☿ St R
13 26 ☽ ⊥ ♅
16 50 ☽ ± ☉
20 39 ☿ ∥ ♀
20 41 ☽ ± ⯓
21 50 ☽ □ ♆

15 Saturday
02 21 ☽ ☍ ♂
06 10 ☽ ⊼ ☉
06 37 ☽ ♇ ☿
07 43 ☽ ∠ ♅
08 54 ☽ ⊼ ⯓
13 13 ☽ ♇ ♄
23 31 ☽ △ ♃

16 Sunday
00 29 ☽ ± ☿
07 33 ☽ ± ♄
12 12 ☽ ⊼ ☿
13 58 ☽ ✶ ♅
14 51 ☉ ☌ ⯓
19 42 ☽ ⊼ ♄
20 50 ☽ ☍ ♀

17 Monday
03 22 ☿ ∠ ♂
06 07 ☽ ♇ ♃
10 19 ☽ △ ♆
13 53 ☽ Q ♅
18 24 ☽ ⊼ ♂
19 44 ♀ ⊻ ♄
21 08 ☽ ☍ ⯓
23 48 ☽ ☍ ☉

18 Tuesday
00 15 ☽ ± ♃
06 31 ☉ ⊥ ♃
06 58 ☽ ± ♂
11 44 ☽ # ♂
12 15 ☽ ⊼ ♃
15 27 ♅ St D
15 54 ☽ ♇ ♆
18 11 ♂ ♇ ♅
20 39 ☽ ☍ ☿
21 38 ☽ # ⯓

19 Wednesday
01 11 ☽ □ ♅
01 37 ☽ ♇ ♂
05 47 ☽ ⊼ ♀
07 19 ☽ ☍ ♄
09 26 ☽ ± ♆
14 20 ☽ # ☉
17 00 ☽ ± ♀
18 30 ♂ ⊻ ⯓
21 00 ☽ ⊼ ♆

20 Thursday
07 30 ☽ ⊼ ⯓
07 32 ☽ # ♄
08 15 ☽ △ ♂
09 44 ☽ ♇ ♀
13 36 ☽ # ♀
15 02 ☽ ⊼ ☉
15 27 ☽ # ☿
18 56 ☽ ± ⯓
21 50 ☽ ∥ ♅
22 58 ☽ ☍ ♃

21 Friday
01 55 ☽ ⊼ ☿
03 22 ☽ ± ☉
05 14 ☿ ∠ ♆
10 28 ☽ △ ♅
11 56 ☽ ♇ ⯓
12 16 ☽ ± ☿
13 19 ☽ △ ♀
16 54 ☽ ⊼ ♄
17 16 ☽ # ♃
21 42 ☽ ♇ ☉

22 Saturday
03 36 ☽ ♇ ☿
04 00 ☿ ⊻ ♃
04 11 ☽ ± ♄
05 43 ☽ ☍ ♆
12 35 ☿ ⊥ ♂
14 22 ☽ ♇ ♅
15 53 ☽ △ ⯓
19 46 ☽ □ ♂
19 48 ☽ # ♆
20 56 ☽ ♇ ♄

23 Sunday
03 42 ☽ △ ☉
04 47 ☽ △ ☿
06 52 ☽ ± ♅
07 37 ☽ ⊼ ♃
10 28 ☉ ☌ ☿
11 26 ☿ ∥ ♀
17 45 ☽ ⊼ ♅
18 41 ☽ ± ♃
19 22 ☽ □ ♀
20 48 ♂ ∠ ♄

24 Monday
00 26 ☽ △ ♄
06 18 ☽ Q ♂
08 56 ☽ ∥ ♆
11 07 ☽ ♇ ♃
12 23 ☽ ⊼ ♆
22 10 ☽ □ ⯓
23 03 ☽ ± ♆

25 Tuesday
02 59 ☽ Q ♀
04 48 ☽ ✶ ♂
05 51 ☽ □ ☿
08 55 ☽ ∥ ♃
12 46 ☿ ⊻ ♂
13 41 ☽ □ ☉
14 01 ☽ △ ♃
14 53 ☽ ♇ ♆
19 32 ☉ ⊻ ♃
22 52 ☽ ☍ ♅
23 41 ☽ ✶ ♀

26 Wednesday
04 03 ☽ # ♅
05 15 ☿ ⊥ ♃
05 38 ☽ Q ⯓
05 42 ☽ □ ♄
06 21 ☉ ∠ ♆
08 18 ☽ ∠ ♂
10 34 ☽ Q ☿
10 50 ☽ ∥ ♀
13 44 ☽ ∥ ☿
15 44 ☽ ∥ ♄
16 47 ☽ △ ♆
22 31 ☽ ∥ ☉
23 04 ☽ Q ☉

27 Thursday
00 29 ☽ ⊥ ♂
01 08 ☽ ∠ ♀
02 10 ☽ ✶ ⯓
05 28 ☽ ✶ ☿
11 09 ☽ ⊻ ♂
12 32 ☽ Q ♄
16 11 ☽ ⊥ ♀
18 02 ☽ □ ♃
20 47 ☽ ✶ ☉
21 45 ☽ ∥ ⯓

28 Friday
01 43 ☽ ⊼ ♅
02 08 ☽ ⊻ ♀
03 19 ☽ ∠ ⯓
04 29 ☿ ∥ ♄
04 51 ☽ ∠ ☿
07 38 ☽ ∥ ♂
08 39 ☽ ✶ ♄
11 39 ☽ ± ♅
18 11 ☽ ⊥ ⯓
18 47 ☽ ⊥ ☿
19 00 ☽ □ ♆
23 23 ☽ ∠ ☉

29 Saturday
00 13 ☽ Q ♃
02 25 ☽ ♇ ♅
04 02 ☽ ⊻ ⯓
04 03 ☽ ⊻ ☿
04 16 ☿ ☌ ⯓
08 46 ♀ St D
09 26 ☽ ∠ ♄
15 08 ☽ ⊥ ☉
15 10 ☽ ☌ ♂
17 00 ☉ ⊥ ♂
20 03 ☽ ✶ ♃

30 Sunday
00 09 ☽ ⊥ ♄
00 21 ☽ Q ♆
01 34 ☽ ⊻ ☉
02 47 ☽ △ ♅
03 08 ☽ ☌ ♀
08 27 ☉ ∥ ☿
09 56 ☽ ⊻ ♄
19 32 ☉ □ ♅
19 48 ☽ ✶ ♆
20 41 ☽ ∠ ♃

31 Monday
01 29 ☉ ⊻ ♀
02 26 ☽ ☌ ☿
04 44 ☽ ☌ ⯓
09 37 ♄ ⊥ ♆
11 29 ☽ ⊥ ♃
15 07 ☽ ∥ ♂
18 14 ☽ ⊻ ♂
20 08 ☽ ∠ ♆
21 22 ☽ ⊻ ♃

FEBRUARY 2022

LONGITUDES

Date	Sidereal time h m s	Sun ☉ ° ' "	Moon ☽ ° ' "	Moon ☽ 24.00 ° ' "	Mercury ☿ ° '	Venus ♀ ° '	Mars ♂ ° '	Jupiter ♃ ° '	Saturn ♄ ° '	Uranus ♅ ° '	Neptune ♆ ° '	Pluto ♇ ° '
01	20 46 43	12 ♒ 35 45	16 ♒ 08 11	23 ♒ 24 32	24 ♑ 51	11 ♑ 16	05 ♑ 48	07 ♓ 20	15 ♒ 33	10 ♉ 54	21 ♓ 27	26 ♑ 58
02	20 50 39	13 36 40	00 ♓ 36 05	07 ♓ 42 03	24 R 34	11 25	06 32	07 33	15 40	10 55	21 29	27 00
03	20 54 36	14 37 34	14 ♓ 41 51	21 ♓ 35 05	24 24	11 36	07 15	07 47	15 47	10 56	21 31	27 02
04	20 58 32	15 38 26	28 ♓ 21 29	05 ♈ 01 01	24 D 23	11 49	07 59	08 01	15 54	10 57	21 32	27 04
05	21 02 29	16 39 17	11 ♈ 33 49	18 ♈ 00 08	24 29	12 04	08 43	08 15	16 01	10 57	21 34	27 06
06	21 06 25	17 40 07	24 ♈ 20 21	00 ♉ 34 59	24 42	12 21	09 27	08 29	16 09	10 58	21 36	27 07
07	21 10 22	18 40 55	06 ♉ 44 36	12 ♉ 49 50	25 01	12 40	10 11	08 43	16 16	10 59	21 38	27 09
08	21 14 19	19 41 42	18 ♉ 51 21	24 ♉ 49 52	25 27	13 01	10 55	08 58	16 23	11 00	21 40	27 11
09	21 18 15	20 42 28	00 ♊ 46 04	06 ♊ 40 41	25 57	13 23	11 39	09 12	16 30	11 02	21 42	27 13
10	21 22 12	21 43 12	12 ♊ 34 23	18 ♊ 27 50	26 33	13 48	12 23	09 26	16 37	11 03	21 44	27 15
11	21 26 08	22 43 54	24 ♊ 21 40	00 ♋ 16 28	27 13	14 14	13 07	09 40	16 45	11 04	21 46	27 17
12	21 30 05	23 44 35	06 ♋ 12 48	12 ♋ 11 06	27 58	14 42	13 52	09 54	16 52	11 05	21 48	27 19
13	21 34 01	24 45 14	18 ♋ 11 50	24 ♋ 15 20	28 46	15 12	14 36	10 08	16 59	11 07	21 51	27 20
14	21 37 58	25 45 52	00 ♌ 21 54	06 ♌ 31 43	29 ♑ 38	15 43	15 20	10 23	17 06	11 08	21 53	27 22
15	21 41 54	26 46 28	12 ♌ 44 58	19 ♌ 01 42	00 ♒ 33	16 15	16 04	10 37	17 13	11 09	21 55	27 24
16	21 45 51	27 47 02	25 ♌ 21 55	01 ♍ 45 36	01 31	16 49	16 48	10 51	17 20	11 11	21 57	27 26
17	21 49 48	28 47 35	08 ♍ 12 37	14 ♍ 42 52	02 31	17 25	17 33	11 06	17 27	11 12	21 59	27 27
18	21 53 44	29 ♒ 48 07	21 ♍ 16 09	27 ♍ 52 21	03 35	18 02	18 17	11 20	17 35	11 14	22 01	27 29
19	21 57 41	00 ♓ 48 37	04 ♎ 31 16	11 ♎ 12 45	04 40	18 40	19 01	11 34	17 42	11 15	22 03	27 31
20	22 01 37	01 49 05	17 ♎ 56 40	24 ♎ 42 57	05 48	19 19	19 46	11 49	17 49	11 17	22 05	27 33
21	22 05 34	02 49 33	01 ♏ 31 30	08 ♏ 22 18	06 58	19 59	20 30	12 03	17 56	11 19	22 08	27 34
22	22 09 30	03 49 59	15 ♏ 15 20	22 ♏ 10 39	08 10	20 41	21 15	12 18	18 03	11 20	22 10	27 36
23	22 13 27	04 50 23	29 ♏ 08 14	06 ♐ 08 07	09 24	21 23	21 59	12 32	18 10	11 22	22 12	27 38
24	22 17 23	05 50 47	13 ♐ 10 17	20 ♐ 14 40	10 39	22 07	22 44	12 47	18 17	11 24	22 14	27 39
25	22 21 20	06 51 09	27 ♐ 21 09	04 ♑ 29 32	11 56	22 52	23 28	13 01	18 24	11 26	22 16	27 41
26	22 25 17	07 51 30	11 ♑ 39 31	18 ♑ 50 42	13 15	23 37	24 13	13 16	18 31	11 28	22 19	27 43
27	22 29 13	08 51 49	26 ♑ 02 37	03 ♒ 14 39	14 35	24 24	24 57	13 30	18 38	11 30	22 21	27 44
28	22 33 10	09 ♓ 52 07	10 ♒ 26 08	17 ♒ 36 21	15 ♒ 56	25 ♑ 11	25 ♑ 42	13 ♓ 45	18 ♒ 44	11 ♉ 32	22 ♓ 23	27 ♑ 46

Date	Moon True ☊ ° '	Moon Mean ☊ ° '	Moon ☽ Latitude ° '
01	28 ♉ 41	27 ♉ 52	04 S 54
02	28 R 29	27 49	05 01
03	28 18	27 46	04 49
04	28 08	27 42	04 21
05	28 01	27 39	03 39
06	27 57	27 36	02 48
07	27 55	27 33	01 50
08	27 D 54	27 30	00 S 48
09	27 R 54	27 27	00 N 15
10	27 53	27 23	01 17
11	27 51	27 20	02 15
12	27 46	27 17	03 08
13	27 38	27 14	03 53
14	27 28	27 11	04 27
15	27 15	27 08	04 50
16	27 01	27 04	05 00
17	26 47	27 01	04 54
18	26 34	26 58	04 33
19	26 24	26 55	03 57
20	26 17	26 52	03 07
21	26 13	26 48	02 06
22	26 11	26 45	00 N 57
23	26 D 11	26 42	00 S 16
24	26 R 10	26 39	01 29
25	26 09	26 36	02 37
26	26 06	26 33	03 36
27	25 59	26 29	04 21
28	25 ♉ 50	26 ♉ 26	04 S 50

DECLINATIONS

Date	Sun ☉ ° '	Moon ☽ ° '	Mercury ☿ ° '	Venus ♀ ° '	Mars ♂ ° '	Jupiter ♃ ° '	Saturn ♄ ° '	Uranus ♅ ° '	Neptune ♆ ° '	Pluto ♇ ° '
01	17 S 02	20 S 40	18 S 06	16 S 16	23 S 48	09 S 43	16 S 59	14 N 44	04 S 25	22 S 30
02	16 44	15 57	18 18	16 18	23 47	09 38	16 57	14 44	04 25	22 30
03	16 27	10 28	18 30	16 20	23 46	09 33	16 55	14 44	04 24	22 30
04	16 09	04 S 39	18 41	16 22	23 44	09 28	16 53	14 44	04 23	22 29
05	15 51	01 N 12	18 51	16 24	23 42	09 22	16 51	14 45	04 22	22 29
06	15 32	06 50	19 00	16 27	23 39	09 17	16 49	14 45	04 22	22 29
07	15 14	12 02	19 08	16 29	23 37	09 12	16 47	14 45	04 21	22 29
08	14 55	16 40	19 15	16 32	23 34	09 07	16 44	14 46	04 20	22 28
09	14 35	20 33	19 21	16 35	23 31	09 01	16 42	14 46	04 19	22 28
10	14 16	23 34	19 25	16 37	23 28	08 56	16 40	14 46	04 18	22 28
11	13 56	25 34	19 29	16 40	23 24	08 51	16 38	14 47	04 18	22 28
12	13 36	26 25	19 31	16 42	23 21	08 45	16 36	14 47	04 17	22 27
13	13 16	26 03	19 33	16 45	23 17	08 40	16 34	14 48	04 16	22 27
14	12 56	24 25	19 33	16 47	23 12	08 34	16 32	14 48	04 15	22 27
15	12 35	21 37	19 31	16 49	23 08	08 29	16 30	14 49	04 14	22 27
16	12 15	17 46	19 29	16 51	23 03	08 24	16 28	14 49	04 13	22 26
17	11 54	13 02	19 25	16 53	22 58	08 18	16 26	14 50	04 13	22 26
18	11 32	07 38	19 20	16 54	22 53	08 13	16 24	14 50	04 12	22 26
19	11 11	01 N 50	19 14	16 55	22 48	08 07	16 21	14 51	04 11	22 26
20	10 50	04 S 09	19 07	16 56	22 42	08 02	16 19	14 51	04 10	22 25
21	10 28	10 02	18 58	16 57	22 36	07 56	16 17	14 52	04 09	22 25
22	10 06	15 30	18 48	16 58	22 30	07 51	16 15	14 52	04 08	22 25
23	09 44	20 13	18 36	16 58	22 24	07 45	16 13	14 53	04 07	22 25
24	09 22	23 51	18 24	16 58	22 17	07 40	16 11	14 53	04 07	22 25
25	09 00	26 02	18 10	16 57	22 11	07 34	16 09	14 54	04 06	22 24
26	08 37	26 31	17 54	16 56	22 04	07 29	16 07	14 55	04 05	22 24
27	08 15	25 13	17 38	16 55	21 56	07 23	16 05	14 55	04 04	22 24
28	07 S 52	22 S 17	17 S 20	16 S 54	21 S 49	07 S 18	16 S 03	14 N 56	04 S 03	22 S 24

ZODIAC SIGN ENTRIES

Date	h	m	Planets
02	10	59	☽ ♓
04	14	56	☽ ♈
06	22	52	☽ ♉
09	10	27	☽ ♊
11	23	27	☽ ♋
14	11	17	☽ ♌
14	21	54	☿ ♒
16	20	42	☽ ♍
18	16	43	☉ ♓
19	03	51	☽ ♎
21	09	19	☽ ♏
23	13	29	☽ ♐
25	16	27	☽ ♑
27	18	36	☽ ♒

LATITUDES

Date	Mercury ☿ ° '	Venus ♀ ° '	Mars ♂ ° '	Jupiter ♃ ° '	Saturn ♄ ° '	Uranus ♅ ° '	Neptune ♆ ° '	Pluto ♇ ° '
01	03 N 06	06 N 43	00 S 30	00 S 58	00 S 51	00 S 23	01 S 07	01 S 46
04	02 36	06 34	00 32	00 58	00 51	00 23	01 07	01 47
07	02 02	06 22	00 34	00 58	00 51	00 23	01 07	01 47
10	01 27	06 08	00 36	00 58	00 51	00 23	01 07	01 47
13	00 53	05 52	00 39	00 58	00 52	00 23	01 07	01 48
16	00 N 21	05 34	00 41	00 58	00 52	00 23	01 07	01 48
19	00 S 09	05 16	00 43	00 58	00 52	00 23	01 07	01 48
22	00 36	04 56	00 45	00 58	00 52	00 23	01 07	01 49
25	00 59	04 37	00 48	00 58	00 53	00 23	01 07	01 49
28	01 20	04 16	00 50	00 59	00 53	00 23	01 07	01 49
31	01 S 38	03 N 56	00 S 52	00 S 59	00 S 53	00 S 23	01 S 07	01 S 50

DATA

Julian Date	2459612
Delta T	+71 seconds
Ayanamsa	24° 09' 44"
Synetic vernal point	04° ♓ 57' 15"
True obliquity of ecliptic	23° 26' 16"

MOON'S PHASES, APSIDES AND POSITIONS ☽

Date	h	m	Phase	Longitude ° '	Eclipse Indicator
01	05	46	●	12 ♒ 20	
08	13	50	☽	19 ♉ 46	
16	16	56	○	28 ♌ 00	
23	22	32	☾	05 ♐ 17	

Day	h	m		
11	02	34	Apogee	
26	22	13	Perigee	
05	07	00	ON	
12	16	45	Max dec	26° N 27'
19	19	21	0S	
26	06	32	Max dec	26° S 33'

LONGITUDES

Date	Chiron ⚷ ° '	Ceres ⚳ ° '	Pallas ⚴ ° '	Juno ⚵ ° '	Vesta ⚶ ° '	Black Moon Lilith ⚸ ° '
01	09 ♈ 17	28 ♉ 59	25 ♓ 39	29 ♑ 49	11 ♑ 14	21 ♊ 57
11	09 ♈ 41	00 ♊ 23	29 ♓ 02	03 ♒ 48	16 ♑ 24	23 ♊ 04
21	10 ♈ 09	02 ♊ 19	02 ♈ 36	07 ♒ 47	21 ♑ 29	24 ♊ 10
31	10 ♈ 40	04 ♊ 40	06 ♈ 19	11 ♒ 45	26 ♑ 28	25 ♊ 17

ASPECTARIAN

h m	Aspects	h m	Aspects	h m	Aspects
01 Tuesday		17 59	☽ Q ♆	00 06	☽ ⊼ ♅
00 44	☽ ∥ ♇	22 35	☽ ± ♂	00 53	☽ ⊼ ♃
03 25	☽ □ ♅	**10 Thursday**		09 50	☽ bQ ☉
03 58	☽ ⊻ ♀	01 55	☽ ± ♀	11 46	☽ △ ♄
04 31	☽ ⊥ ♂	02 15	☽ # ♇	11 46	☽ ± ♃
05 46	☽ ☌ ☉	05 28	☽ □ ♃	12 03	☽ ∥ ♆
10 52	☽ ⊥ ♆	08 53	☽ ⊻ ♅	12 13	☿ ⊥ ♃
11 01	☽ ☌ ♄	09 48	☽ bQ ☿	14 33	☽ □ ♀
13 53	☽ ⊥ ♀	10 59	☽ # ♂	15 25	☽ □ ♂
16 01	☉ ∥ ♄	11 20	☽ bQ ♇	19 22	☽ ⊼ ♆
20 05	☽ ∠ ♂	11 36	☽ ⊼ ♂	**21 Monday**	
20 46	☽ ⊻ ♆	12 29	☉ ⊻ ♆	03 28	☽ ∥ ♃
02 Wednesday		14 35	☽ ⊼ ♀	03 59	☽ bQ ♃
01 06	☽ ∥ ☿	20 20	☽ △ ♄	05 02	☽ □ ♇
02 05	☽ ⊻ ☿	21 08	☽ ⊥ ♅	06 00	☽ ± ♆
04 55	☽ ∠ ♀	**11 Friday**		13 44	☽ ∥ ☉
05 57	☽ ⊻ ♇	03 59	♀ ∥ ♄	14 28	☽ △ ☉
07 15	☽ ∥ ♄	05 13	☽ ± ☿	15 13	☿ ∠ ♆
08 04	☽ ∥ ☉	05 43	☽ ± ♇	21 51	☽ bQ ♆
09 10	☽ Q ♅	06 43	☽ □ ♆	22 27	☽ □ ☿
10 23	☽ ∥ ♀	08 23	☽ △ ☉	23 55	☽ Q ♀
11 56	☽ ⊥ ☿	14 04	☿ ☌ ♇	**22 Tuesday**	
16 02	☽ ⊥ ♇	15 28	☽ ∠ ♅	00 56	☽ Q ♂
17 34	☽ # ♅	17 56	☽ ⊼ ♇	05 10	☽ ☍ ♅
22 33	☽ ⚹ ♂	18 11	☽ ⊼ ☿	06 17	☉ ⊥ ♇
23 57	☽ ☌ ♃	**12 Saturday**		06 45	☽ △ ♃
03 Thursday		03 08	☽ bQ ♄	09 07	☽ # ♅
02 59	☽ ∠ ☿	17 34	☽ bQ ☉	12 36	☽ Q ♇
05 30	☽ ⚹ ♅	19 35	☽ △ ♃	15 33	☽ ∥ ♄
06 35	☽ ⚹ ♀	21 27	☽ ± ♄	16 53	☽ □ ♄
07 24	☽ ∠ ♇	21 49	☽ ⚹ ♅	18 59	☽ ∥ ♀
11 52	☽ ⊻ ☉	**13 Sunday**		21 55	☽ ⚹ ♀
13 54	☽ ⊻ ♄	04 21	☽ ☍ ♂	22 58	☽ ⚹ ♂
15 55	☽ ∥ ♃	05 45	☽ ☍ ♀	**23 Wednesday**	
20 12	☉ ∥ ♀	09 33	☽ ⊼ ♄	00 00	☽ △ ♆
20 22	☽ Q ♂	13 12	☽ ± ☉	03 37	☽ ∥ ☿
23 09	☽ ⊥ ☉	19 15	☽ △ ♆	08 43	☽ Q ☿
23 54	☽ ☌ ♆	21 45	☽ Q ♅	09 15	♂ ∥ ♇
04 Friday		**14 Monday**		09 24	☽ ⚹ ♇
00 28	☽ ⊥ ♄	02 01	☽ bQ ♃	19 12	♂ ⚹ ♆
03 47	☽ Q ♀	02 09	☽ ⊼ ☉	22 16	♃ ∠ ♇
04 14	☿ St D	06 07	☽ ☍ ♇	22 32	☽ □ ☉
04 55	☽ ⚹ ☿	10 27	☽ ☍ ☿	**24 Thursday**	
07 42	☽ ∠ ♅	19 59	☽ ± ♃	00 09	☽ Q ♄
09 35	☉ ⊥ ♆	23 55	☽ # ♂	01 00	☽ ∥ ♂
09 41	☽ ⚹ ♇	**15 Tuesday**		01 07	☽ ∠ ♀
13 03	☽ ∥ ♆	00 43	☽ bQ ♆	01 27	☽ ∥ ♇
13 38	♂ ⚹ ♃	05 52	☽ # ♇	02 12	☽ ∠ ♂
16 26	☽ ∠ ☉	07 49	☽ ⊼ ♃	07 17	☽ ⚹ ☿
16 36	☽ ∠ ♄	08 56	☽ □ ♅	08 59	☽ ⊼ ♅
19 05	☉ ☌ ♄	18 04	☽ ± ♆	11 07	☽ ∠ ♇
23 53	☽ ⊥ ♅	18 45	☽ ⊼ ♂	11 19	☽ □ ♃
05 Saturday		19 02	☽ ⊼ ♀	16 04	♀ ⚹ ♆
02 34	☽ Q ☿	20 38	☽ ☍ ♄	17 17	☽ ⊥ ♀
05 48	☽ ⊻ ♃	**16 Wednesday**		18 23	☽ ⊥ ♂
06 27	☽ □ ♂	01 54	☽ # ☿	19 12	☽ ± ♅
07 26	☽ Q ♇	03 17	☉ ⊻ ♇	20 45	☽ ⚹ ♄
10 53	☽ ⊻ ♅	05 31	☽ ⊼ ♆	**25 Friday**	
12 56	☽ □ ♀	06 52	☽ ± ♂	02 22	☿ □ ♅
17 05	☽ ⊥ ♃	06 58	☽ ± ♀	02 25	☽ ⊥ ♇
20 22	☽ ⚹ ♄	14 29	♀ ☌ ♂	03 24	☽ □ ♆
22 17	☽ ⚹ ☉	15 54	☽ ⊼ ♇	04 00	☽ ⊻ ♀
06 Sunday		16 56	☽ # ♀	05 06	☽ ⊻ ♂
01 18	☽ # ♆	16 56	☽ ☍ ☉	07 28	☽ Q ☉
06 47	☽ ⊻ ♆	19 03	☽ # ♄	10 27	☽ bQ ♅
10 21	☽ ∠ ♃	**17 Thursday**		11 14	☽ ∠ ☿
12 42	☽ □ ☿	00 31	☽ ⊼ ☿	12 33	☽ ⊻ ♇
17 15	♂ Q ♆	00 42	☽ bQ ♀	18 16	☽ Q ♃
17 21	☽ □ ♇	00 50	☽ bQ ♂	22 15	☽ ∠ ♄
18 16	☽ ⊥ ♆	03 09	☽ ± ♇	**26 Saturday**	
19 22	☽ Q ♄	03 21	☽ ∥ ♅	03 52	☽ ⊥ ☿
22 51	☽ # ♃	08 34	♂ ⊻ ♄	05 10	☽ ⚹ ☉
23 08	☽ Q ☉	12 38	☽ ± ☿	09 44	☽ Q ♆
07 Monday		14 09	♀ ⊻ ♄	11 40	☽ △ ♅
11 14	☉ ⊥ ♀	17 26	☽ ☍ ♃	12 16	☿ ⊻ ♃
11 48	☽ ∠ ♆	17 33	☽ △ ♅	13 26	☽ ⊥ ♄
15 02	♂ ⊥ ♄	17 39	☽ # ☉	14 43	☽ ⚹ ♃
15 58	☽ ⚹ ♃	19 52	☽ bQ ♇	14 56	☽ ⊻ ☿
19 12	☽ △ ♂	22 02	☉ Q ♅	23 32	☽ ⊻ ♄
20 22	☽ ☌ ♅	**18 Friday**		**27 Sunday**	
08 Tuesday		00 13	♃ ⚹ ♅	05 49	☽ ⚹ ♆
00 00	☽ △ ♀	05 11	☽ ⊼ ♄	08 06	☽ ∠ ☉
01 42	☽ ∥ ♅	05 47	☽ △ ♀	09 06	☽ ☌ ♀
03 06	☽ # ☉	06 13	☽ △ ♂	10 06	☽ ☌ ♂
07 01	☽ □ ♄	06 39	☽ bQ ☿	14 49	☽ ☌ ♇
11 16	☽ # ♀	09 30	☽ # ♃	16 10	☽ ∠ ♃
12 26	☽ # ♄	13 22	☽ ☍ ♆	**28 Monday**	
13 50	☽ □ ☉	16 14	☽ ± ♄	00 13	☽ ⊥ ☉
14 57	♂ △ ♅	21 02	☽ bQ ♅	06 53	☽ ∠ ♆
16 18	☽ Q ♃	23 20	☽ △ ♇	07 26	☽ ⊥ ♃
17 40	☽ ⚹ ♆	**19 Saturday**		10 59	☽ ⊻ ☉
22 57	☉ # ♅	02 24	☽ # ♆	11 14	☽ ∥ ♇
09 Wednesday		04 46	☽ ⊼ ☉	13 50	☽ □ ♅
01 48	☽ △ ☿	08 41	☽ bQ ♄	15 03	☽ ∥ ♂
03 08	☽ bQ ♂	12 18	☽ △ ☿	17 37	☽ ⊻ ♃
03 45	☽ # ☿	13 20	☽ ± ♅	19 58	☿ ⊥ ♆
04 48	☽ △ ♇	16 27	☽ ± ☉	21 59	☽ ⊥ ♆
07 01	☽ bQ ♀	**20 Sunday**		22 11	☽ ☌ ☿

All ephemeris data is given at 12.00 UT and the Moon's longitude is additionally given for 24.00 UT

LONGITUDES

Date	Sidereal time h m s	Sun ☉ ° ' "	Moon ☽ ° ' "	Moon ☽ 24.00 ° ' "	Mercury ☿ ° '	Venus ♀ ° '	Mars ♂ ° '	Jupiter ♃ ° '	Saturn ♄ ° '	Uranus ♅ ° '	Neptune ♆ ° '	Pluto ♇ ° '
01	22 37 06	10 ♓ 52 23	24 ♒ 44 32	01 ♓ 49 57	17 ♒ 19	25 ♑ 59	26 ♑ 27	13 ♓ 59	18 ♒ 51	11 ♉ 34	22 ♓ 25	27 ♑ 47
02	22 41 03	11 52 38	08 ♓ 51 51	15 ♓ 49 36	18 43	26 48	27 11	14 14	18 58	11 36	22 27	27 49
03	22 44 59	12 52 50	22 ♓ 42 38	29 ♓ 30 30	20 09	27 38	27 56	14 28	19 05	11 38	22 30	27 50
04	22 48 56	13 53 01	06 ♈ 12 54	12 ♈ 49 39	21 36	28 29	28 41	14 43	19 12	11 40	22 32	27 52
05	22 52 52	14 53 10	19 ♈ 20 42	25 ♈ 46 09	23 04	29 ♑ 20	29 ♑ 26	14 57	19 19	11 42	22 34	27 53
06	22 56 49	15 53 17	02 ♉ 06 11	08 ♉ 21 09	24 33	00 ♒ 12	00 ♒ 10	15 12	19 25	11 45	22 36	27 55
07	23 00 46	16 53 22	14 ♉ 31 28	20 ♉ 37 36	26 03	01 05	00 55	15 26	19 32	11 47	22 39	27 56
08	23 04 42	17 53 24	26 ♉ 40 09	02 ♊ 39 42	27 35	01 58	01 40	15 41	19 39	11 49	22 41	27 58
09	23 08 39	18 53 25	08 ♊ 36 55	14 ♊ 32 27	29 ♒ 07	02 52	02 25	15 55	19 45	11 52	22 43	27 59
10	23 12 35	19 53 23	20 ♊ 27 00	26 ♊ 21 16	00 ♓ 41	03 46	03 10	16 10	19 52	11 54	22 46	28 00
11	23 16 32	20 53 20	02 ♋ 15 54	08 ♋ 11 35	02 16	04 41	03 55	16 24	19 58	11 56	22 48	28 02
12	23 20 28	21 53 14	14 ♋ 08 57	20 ♋ 08 34	03 52	05 37	04 40	16 39	20 05	11 59	22 50	28 03
13	23 24 25	22 53 06	26 ♋ 11 01	02 ♌ 16 46	05 30	06 33	05 25	16 53	20 11	12 01	22 52	28 05
14	23 28 21	23 52 56	08 ♌ 26 14	14 ♌ 39 47	07 08	07 29	06 10	17 08	20 18	12 04	22 55	28 06
15	23 32 18	24 52 43	20 ♌ 57 39	27 ♌ 20 03	08 48	08 26	06 55	17 22	20 24	12 06	22 57	28 07
16	23 36 15	25 52 29	03 ♍ 47 02	10 ♍ 18 35	10 29	09 24	07 40	17 37	20 30	12 09	22 59	28 08
17	23 40 11	26 52 12	16 ♍ 54 37	23 ♍ 34 56	12 11	10 22	08 25	17 51	20 37	12 12	23 02	28 09
18	23 44 08	27 51 53	00 ♎ 19 16	07 ♎ 07 17	13 54	11 20	09 10	18 05	20 43	12 14	23 04	28 11
19	23 48 04	28 51 33	13 ♎ 58 37	20 ♎ 52 52	15 39	12 19	09 55	18 20	20 49	12 17	23 06	28 12
20	23 52 01	29 ♓ 51 10	27 ♎ 49 36	04 ♏ 48 26	17 24	13 18	10 40	18 34	20 55	12 20	23 08	28 13
21	23 55 57	00 ♈ 50 46	11 ♏ 48 56	18 ♏ 50 47	19 11	14 18	11 25	18 49	21 01	12 23	23 11	28 14
22	23 59 54	01 50 19	25 ♏ 53 38	02 ♐ 57 12	21 00	15 18	12 10	19 03	21 07	12 25	23 13	28 15
23	00 03 50	02 49 52	10 ♐ 01 15	17 ♐ 05 35	22 49	16 18	12 55	19 17	21 13	12 28	23 15	28 16
24	00 07 47	03 49 22	24 ♐ 10 00	01 ♑ 14 21	24 40	17 19	13 40	19 32	21 19	12 31	23 17	28 17
25	00 11 44	04 48 51	08 ♑ 18 28	15 ♑ 22 10	26 32	18 20	14 25	19 46	21 25	12 34	23 19	28 18
26	00 15 40	05 48 18	22 ♑ 25 18	29 ♑ 27 38	28 ♓ 26	19 21	15 11	20 00	21 31	12 37	23 22	28 19
27	00 19 37	06 47 43	06 ♒ 28 55	13 ♒ 28 53	00 ♈ 20	20 23	15 56	20 14	21 37	12 40	23 24	28 20
28	00 23 33	07 47 06	20 ♒ 27 12	27 ♒ 23 33	02 16	21 25	16 41	20 28	21 43	12 43	23 26	28 21
29	00 27 30	08 46 28	04 ♓ 17 34	11 ♓ 08 52	04 14	22 27	17 26	20 43	21 48	12 46	23 29	28 22
30	00 31 26	09 45 48	17 ♓ 57 06	24 ♓ 41 56	06 12	23 30	18 12	20 57	21 54	12 49	23 31	28 23
31	00 35 23	10 ♈ 45 05	01 ♈ 23 03	08 ♈ 00 12	08 ♈ 12	24 ♒ 33	18 ♒ 57	21 ♓ 11	22 ♒ 00	12 ♉ 52	23 ♓ 33	28 ♑ 24

Date	Moon True ☊ ° '	Moon Mean ☊ ° '	Moon ☽ Latitude ° '
01	25 ♉ 39	26 ♉ 23	05 S 01
02	25 R 27	26 20	04 53
03	25 16	26 17	04 28
04	25 06	26 14	03 48
05	24 58	26 10	02 57
06	24 53	26 07	01 58
07	24 51	26 04	00 S 55
08	24 D 50	26 01	00 N 10
09	24 51	25 58	01 13
10	24 R 51	25 54	02 12
11	24 50	25 51	03 06
12	24 48	25 48	03 52
13	24 42	25 45	04 28
14	24 35	25 42	04 52
15	24 26	25 39	05 04
16	24 15	25 35	05 00
17	24 05	25 32	04 41
18	23 56	25 29	04 06
19	23 48	25 26	03 16
20	23 43	25 23	02 14
21	23 40	25 19	01 N 03
22	23 D 39	25 16	00 S 12
23	23 40	25 13	01 27
24	23 41	25 10	02 37
25	23 R 41	25 07	03 37
26	23 40	25 04	04 24
27	23 37	25 00	04 55
28	23 32	24 57	05 08
29	23 25	24 54	05 03
30	23 17	24 51	04 40
31	23 ♉ 10	24 ♉ 48	04 S 03

DECLINATIONS

Date	Sun ☉ ° '	Moon ☽ ° '	Mercury ☿ ° '	Venus ♀ ° '	Mars ♂ ° '	Jupiter ♃ ° '	Saturn ♄ ° '	Uranus ♅ ° '	Neptune ♆ ° '	Pluto ♇ ° '
01	07 S 29	18 S 00	17 S 01	16 S 52	21 S 41	07 S 12	16 S 01	14 N 56	04 S 02	22 S 24
02	07 06	12 46	16 41	16 49	21 33	07 07	15 59	14 57	04 01	22 23
03	06 43	06 59	16 19	16 47	21 25	07 01	15 57	14 58	04 00	22 23
04	06 20	01 S 01	15 56	16 44	21 17	06 55	15 55	14 59	03 59	22 23
05	05 57	04 N 51	15 32	16 40	21 08	06 50	15 53	14 59	03 59	22 23
06	05 34	10 22	15 06	16 36	21 00	06 44	15 51	15 00	03 58	22 23
07	05 11	15 19	14 40	16 32	20 51	06 39	15 49	15 01	03 57	22 22
08	04 47	19 34	14 12	16 27	20 42	06 33	15 47	15 01	03 56	22 22
09	04 24	22 56	13 42	16 21	20 32	06 27	15 45	15 02	03 55	22 22
10	04 00	25 18	13 12	16 16	20 23	06 22	15 43	15 03	03 54	22 22
11	03 37	26 31	12 40	16 10	20 13	06 16	15 41	15 04	03 53	22 22
12	03 13	26 32	12 07	16 03	20 03	06 11	15 39	15 04	03 52	22 22
13	02 49	25 18	11 33	15 56	19 53	06 05	15 37	15 05	03 51	22 22
14	02 26	22 51	10 57	15 48	19 42	06 00	15 35	15 06	03 51	22 21
15	02 02	19 18	10 21	15 40	19 32	05 54	15 33	15 07	03 50	22 21
16	01 38	14 47	09 43	15 32	19 21	05 48	15 31	15 08	03 49	22 21
17	01 15	09 29	09 04	15 23	19 10	05 43	15 29	15 08	03 48	22 21
18	00 51	03 N 38	08 24	15 13	18 59	05 37	15 27	15 09	03 47	22 21
19	00 27	02 S 30	07 42	15 03	18 47	05 32	15 25	15 10	03 46	22 21
20	00 S 04	08 37	06 59	14 53	18 36	05 26	15 24	15 11	03 45	22 21
21	00 N 20	14 23	06 16	14 42	18 24	05 21	15 22	15 12	03 44	22 21
22	00 44	19 25	05 31	14 31	18 12	05 15	15 20	15 13	03 43	22 21
23	01 08	23 23	04 45	14 19	18 00	05 09	15 18	15 14	03 43	22 21
24	01 31	25 55	03 58	14 07	17 48	05 04	15 16	15 15	03 42	22 20
25	01 55	26 47	03 09	13 54	17 35	04 58	15 15	15 16	03 41	22 20
26	02 18	25 54	02 20	13 41	17 23	04 53	15 13	15 16	03 40	22 20
27	02 42	23 24	01 30	13 27	17 10	04 47	15 11	15 17	03 39	22 20
28	03 05	19 32	00 S 39	13 13	16 57	04 42	15 09	15 18	03 38	22 20
29	03 29	14 38	00 N 13	12 59	16 44	04 36	15 08	15 19	03 37	22 20
30	03 52	09 04	01 06	12 44	16 31	04 31	15 06	15 20	03 36	22 20
31	04 N 15	03 S 10	02 N 00	12 S 28	16 S 17	04 S 25	15 S 04	15 N 21	03 S 36	22 S 20

ZODIAC SIGN ENTRIES

Date	h	m	Planets
01	20	53	☽ ♓
04	00	52	☽ ♈
06	06	23	♂ ♒
06	06	30	♀ ♒
06	08	00	☽ ♉
08	18	40	☽ ♊
10	01	32	☿ ♓
11	07	24	☽ ♋
13	19	32	☽ ♌
16	04	59	☽ ♍
18	11	26	☽ ♎
20	15	33	☉ ♈
20	15	45	☽ ♏
22	18	59	☽ ♐
24	21	54	☽ ♑
27	00	55	☽ ♒
27	07	44	☿ ♈
29	04	32	☽ ♓
31	09	30	☽ ♈

LATITUDES

Date	Mercury ☿ ° '	Venus ♀ ° '	Mars ♂ ° '	Jupiter ♃ ° '	Saturn ♄ ° '	Uranus ♅ ° '	Neptune ♆ ° '	Pluto ♇ ° '
01	01 S 27	04 N 09	00 S 50	00 S 59	00 S 53	00 S 23	01 S 07	01 S 50
04	01 43	03 49	00 53	00 59	00 53	00 23	01 07	01 50
07	01 56	03 28	00 55	00 59	00 54	00 23	01 07	01 50
10	02 06	03 08	00 57	00 59	00 54	00 22	01 07	01 51
13	02 12	02 48	00 59	00 59	00 54	00 22	01 07	01 51
16	02 15	02 28	01 01	00 59	00 55	00 22	01 07	01 51
19	02 13	02 08	01 04	00 59	00 55	00 22	01 07	01 52
22	02 07	01 49	01 06	01 00	00 55	00 22	01 07	01 52
25	01 56	01 30	01 08	01 00	00 56	00 22	01 07	01 53
28	01 41	01 12	01 10	01 00	00 56	00 22	01 07	01 53
31	01 S 22	00 N 55	01 S 12	01 S 00	00 S 57	00 S 22	01 S 07	01 S 54

DATA

Julian Date	2459640
Delta T	+71 seconds
Ayanamsa	24° 09' 47"
Synetic vernal point	04° ♓ 57' 12"
True obliquity of ecliptic	23° 26' 17"

MOON'S PHASES, APSIDES AND POSITIONS ☽

Date	h	m	Phase	Longitude ° '	Eclipse Indicator
02	17	35	●	12 ♓ 07	
10	10	45	☽	19 ♊ 50	
18	07	18	○	27 ♍ 40	
25	05	37	☾	04 ♑ 33	

Day	h	m		
10	23	03	Apogee	
23	23	45	Perigee	
04	16	07	0N	
12	00	13	Max dec	26° N 40'
19	02	16	0S	
25	11	53	Max dec	26° S 47'

LONGITUDES

Date	Chiron ⚷ ° '	Ceres ⚳ ° '	Pallas ⚴ ° '	Juno ⚵ ° '	Vesta ⚶ ° '	Black Moon Lilith ⚸ ° '
01	10 ♈ 34	04 ♊ 10	05 ♈ 33	10 ♒ 57	25 ♑ 29	25 ♊ 04
11	11 ♈ 06	06 ♊ 50	09 ♈ 22	14 ♒ 53	00 ♒ 21	26 ♊ 10
21	11 ♈ 40	09 ♊ 11	13 ♈ 19	18 ♒ 46	05 ♒ 05	27 ♊ 17
31	12 ♈ 16	13 ♊ 04	17 ♈ 22	22 ♒ 35	09 ♒ 40	28 ♊ 24

All ephemeris data is given at 12.00 UT and the Moon's longitude is additionally given for 24.00 UT

ASPECTARIAN

h m Aspects

01 Tuesday
02 01 ☽ ☌ ♄
08 05 ☽ ⊻ ♆
14 14 ☽ ⊻ ♀
15 02 ☽ ⊻ ♂
17 07 ☽ ∥ ☿
17 09 ☽ ⊻ ♇
17 34 ☽ ∥ ♀
20 10 ☽ Q ♅
21 35 ☽ ∥ ♄

02 Wednesday
00 24 ☿ ∥ ♀
01 01 ☽ ⊥ ♀
01 46 ☽ ⊥ ♂
02 24 ☽ # ♅
02 49 ☉ ∠ ♀
03 21 ☽ ⊥ ♇
05 05 ☉ ⚹ ♅
11 54 ☉ ∥ ♃
16 33 ☿ ☌ ♄
16 43 ☽ ⚹ ♅
17 23 ☽ ∠ ♀
17 35 ☽ ☌ ☉
18 03 ☽ ∠ ♂
18 48 ☽ ∠ ♇
21 24 ☽ ☌ ♃

03 Thursday
05 36 ☽ ⊻ ♄
07 00 ☽ ⊻ ☿
08 43 ♂ ☌ ♇
10 58 ☉ ∠ ♇
11 37 ☽ ☌ ♆
11 54 ☽ ∥ ♃
13 09 ☽ ∥ ☉
16 12 ☽ ⊥ ♄
17 14 ☉ ∠ ♂
17 56 ♀ ☌ ♇
18 46 ☽ ⊥ ☿
18 55 ☽ ∠ ♅
21 03 ☽ ⚹ ♇
21 15 ☽ ⚹ ♀
21 45 ☽ ⚹ ♂

04 Friday
00 03 ☽ ∥ ♆
08 20 ☽ ∠ ♄
11 01 ☽ ⊥ ♅
12 46 ☽ ∠ ☿
13 33 ☿ ∥ ♄
18 37 ☽ Q ♇
20 15 ☽ Q ♀
20 34 ☽ Q ♂
21 55 ☽ ⊻ ♅

05 Saturday
03 05 ☽ ⊻ ☉
03 44 ☽ ⊻ ♃
03 50 ☿ ⊻ ♆
08 23 ☽ # ♆
11 56 ☽ ⚹ ♄
14 06 ☉ ☌ ♃
15 03 ☽ ⊥ ♃
15 06 ☽ ⊥ ☉
16 23 ☽ # ☉
18 01 ☽ ⊻ ♆
19 49 ☽ ⚹ ☿
20 17 ☽ # ♃

06 Sunday
04 02 ☽ □ ♇
05 20 ☽ ⊥ ♆
07 12 ♀ ☌ ♂
08 06 ☽ □ ♂
08 06 ☽ □ ♀
08 18 ☽ ∠ ♃
09 29 ☽ ∠ ☉
10 41 ☽ Q ♄
11 49 ♀ ∠ ♃
12 57 ♂ ∠ ♃
17 41 ☿ # ♅
21 41 ☽ Q ☿
22 36 ☽ ∠ ♆

07 Monday
06 38 ☽ ☌ ♅
08 53 ☽ # ☿
10 23 ☽ ∥ ♅
13 49 ☽ ⚹ ♃
14 31 ☽ # ♄
17 03 ☽ ⚹ ☉
18 15 ☽ # ♀
21 56 ☽ □ ♄

08 Tuesday
04 03 ☽ ⚹ ♆
14 03 ☽ Q ♃
14 04 ☽ □ ☿
14 35 ☽ △ ♇
18 05 ☿ ⊻ ♇
19 00 ☽ # ♂
19 02 ☽ Q ☉
22 40 ☽ △ ♂
23 27 ☽ △ ♀

09 Wednesday
04 07 ☽ Q ♆
07 25 ☽ # ♇
18 35 ☽ ⊻ ♅
20 51 ☽ bQ ♇
23 41 ☿ Q ♅

10 Thursday
03 07 ☽ □ ♃
06 48 ☽ ⊥ ♅
07 03 ☽ bQ ♂
08 18 ☽ bQ ♀
10 45 ☽ □ ☉
10 48 ☽ △ ♄
11 17 ☉ ⊻ ♄
15 10 ☽ ± ♇
16 43 ☽ □ ♆
18 30 ☉ ∥ ♆

11 Friday
01 09 ☽ ∠ ♅
02 34 ☽ ± ♂
03 23 ☽ ⊼ ♇
04 07 ☽ ± ♀
12 00 ☽ △ ☿
15 34 ☽ ⊼ ♂
17 19 ☽ ⊼ ♀
17 32 ☽ bQ ♄

12 Saturday
07 37 ☽ ⚹ ♅
11 52 ☽ ± ♄
14 42 ☿ ⊥ ♇
17 07 ☽ △ ♃
22 56 ☽ bQ ☿
23 59 ☽ ⊼ ♄

13 Sunday
04 52 ☽ △ ☉
05 25 ☽ △ ♆
07 42 ☽ Q ♅
09 44 ☿ ⊻ ♂
11 43 ☉ ☌ ♆
15 44 ☽ ☍ ♇
19 32 ☽ ± ☿
23 27 ☽ bQ ♃

14 Monday
07 17 ☽ ☍ ♂
09 05 ☽ ⊼ ☿
10 00 ☽ ☍ ♀
10 59 ☽ bQ ♆
12 56 ☽ bQ ☉
15 50 ☽ # ♇
17 18 ☽ ± ♃
19 02 ☽ □ ♅
23 15 ♀ ∠ ♆
23 51 ☿ ⊻ ♀

15 Tuesday
04 21 ☽ ± ♆
05 02 ☽ ⊼ ♃
07 43 ☽ ± ☉
10 35 ☽ # ♂
10 56 ☽ ☍ ♄
15 46 ☽ ⊼ ♆
20 01 ☽ ⊼ ☉

16 Wednesday
01 29 ☽ ⊼ ♇
08 12 ☽ # ♀
08 20 ☽ # ♄
10 17 ☽ ∥ ♅
12 39 ☽ ± ♇
13 46 ♀ ∥ ♄
19 35 ☽ ⊼ ♂
23 04 ♂ ∠ ♆
23 09 ☽ ⊼ ♀

17 Thursday
02 08 ☽ ☍ ☿
03 25 ☽ △ ♅
05 02 ☉ ⊥ ♄
05 11 ☽ bQ ♇
07 12 ☽ ± ♂
10 56 ☽ ± ♀
12 13 ☿ ⚹ ♅
13 44 ☽ ☍ ♃
14 00 ☽ # ☿
18 44 ☽ ⊼ ♄
20 13 ☉ ∠ ♅
23 02 ☽ ☍ ♆

18 Friday
00 23 ☽ bQ ♂
01 47 ☿ ∠ ♇
03 53 ☽ # ♃
04 22 ☽ bQ ♀
05 33 ☽ ± ♄
06 31 ☽ bQ ♅
07 18 ☽ ☍ ☉
08 11 ☽ △ ♇
11 24 ☽ # ♆
19 37 ☉ ⚹ ♇
20 37 ♀ # ♅
21 36 ☽ bQ ♄
22 29 ☽ ± ♅
23 43 ☽ # ☉

19 Saturday
04 29 ☽ △ ♂
04 31 ☽ ∥ ☉
08 52 ☽ △ ♀
09 02 ☽ ⊼ ♅
11 16 ♀ □ ♅
12 33 ♀ ⊥ ♃
15 20 ☽ ⊼ ☿
16 55 ☽ ∥ ♆
18 24 ☿ ⊥ ♂
19 43 ☽ ⊼ ♃
23 37 ☽ ∥ ♃
23 59 ☽ △ ♄

20 Sunday
03 15 ☽ ± ☿
03 53 ☽ ⊼ ♆
06 14 ☽ ∥ ☿
06 17 ☽ ± ♃
12 40 ☽ □ ♇
14 16 ☽ ± ♆
15 45 ☽ ⊼ ☉
21 02 ☽ bQ ☿
22 03 ☽ bQ ♃

21 Monday
02 51 ☽ ± ☉
05 45 ☽ bQ ♆
06 06 ☿ ☌ ♃
11 17 ☽ □ ♂
12 58 ☽ ☍ ♅
13 23 ☽ ∥ ♀
15 39 ☽ # ♅
16 21 ☽ ∥ ♄
16 33 ☽ □ ♀
19 24 ☽ bQ ☉
19 33 ☽ Q ♇

22 Tuesday
00 09 ☽ △ ♃
02 26 ☽ △ ☿
03 49 ☽ □ ♄
06 01 ☽ ∥ ♂
07 26 ☽ △ ♆
13 48 ☿ ⊻ ♄
16 01 ☽ ⚹ ♇
19 40 ☽ Q ♂
20 44 ♂ □ ♅
20 49 ☿ ⊥ ♀
21 27 ☿ ∥ ♃
22 52 ☽ △ ☉

23 Wednesday
01 33 ☽ Q ♀
04 52 ☽ ∥ ♇
10 38 ☽ Q ♄
16 10 ☽ ⊼ ♅
17 12 ☽ ⚹ ♂
17 31 ☽ ∠ ♇
17 44 ☿ ☌ ♆
23 29 ☽ ⚹ ♀

24 Thursday
02 23 ☽ ± ♅
03 59 ☽ □ ♃
05 10 ♂ ⊥ ♃
07 09 ☽ ⚹ ♄
08 48 ☽ ⊥ ♇
10 30 ☽ □ ♆
11 24 ♀ ⊥ ♆
12 59 ☽ □ ☿
17 42 ☽ bQ ♅
19 00 ☽ ⊻ ♇
20 04 ☽ ∠ ♂
20 10 ☿ ∥ ♆

25 Friday
02 54 ☽ ∠ ♀
04 03 ♄ # ♅
05 37 ☽ □ ☉
08 47 ☽ ∠ ♄
11 04 ☽ Q ♃
12 13 ☽ ⊥ ♂
17 09 ☽ Q ♆
19 16 ☽ △ ♅
19 22 ☽ ⊥ ♀
22 59 ☽ ⊻ ♂
23 52 ☿ ⊥ ♄

26 Saturday
00 10 ☽ ⊥ ♄
00 13 ☽ Q ☿
01 25 ☿ ∠ ♅
06 22 ☽ ⊻ ♀
07 48 ☽ ⚹ ♃
10 27 ☽ ⊻ ♄
10 35 ☿ ⚹ ♇
12 37 ☉ # ☿
13 37 ☽ ⚹ ♆
14 32 ☽ Q ☉
22 04 ☽ ☌ ♇
23 51 ☽ ⚹ ☿

27 Sunday
07 11 ☉ ∠ ♄
07 34 ♀ ⊻ ♃
08 38 ☉ ⊥ ♅
09 50 ☽ ∠ ♃
12 35 ☽ ⚹ ☉
15 18 ☽ ∠ ♆
19 24 ☽ ∥ ♇
22 38 ☽ □ ♅

28 Monday
00 05 ☿ ∠ ♂
01 32 ☽ ⊥ ♃
05 08 ☽ ☌ ♂
05 38 ☽ ∠ ☿
06 47 ☽ ⊥ ♆
12 02 ☽ ⊻ ♃
13 48 ☽ ☌ ♀
14 11 ☽ ☌ ♄
16 20 ☽ ∠ ☉
17 10 ☽ ⊻ ♆
19 27 ♀ ☌ ♄
23 43 ☽ ⊥ ☿

29 Tuesday
01 40 ☽ ⊻ ♇
01 43 ☽ ∥ ♂
05 50 ☽ Q ♅
08 51 ☽ # ♅
09 09 ☽ ⊥ ☉
09 43 ☽ ∥ ♄
11 52 ☽ ⊻ ☿
12 07 ☽ ⊥ ♇
13 11 ♂ ⊥ ♆
19 41 ☽ ∥ ♀
20 27 ☽ ⊻ ☉
20 31 ☉ # ♆

30 Wednesday
02 54 ☽ ⚹ ♅
03 55 ☽ ∠ ♇
12 15 ♀ ⊻ ♆
12 27 ☽ ⊻ ♂
17 24 ☽ ☌ ♃
19 03 ☽ ⊻ ♄
19 35 ☿ ⊥ ♅
20 51 ☿ ∠ ♄
21 55 ☽ ☌ ♆
22 42 ☽ ⊻ ♀
23 46 ☽ ⊥ ♂

31 Thursday
03 10 ☉ Q ♇
05 39 ☽ ∠ ♅
05 51 ☽ ⊥ ♄
06 37 ☽ ⚹ ♇
06 51 ☽ ∥ ♃
07 52 ☽ # ☉
10 16 ☽ ∥ ♆
10 22 ☽ ⊥ ♀
16 04 ☽ # ☿
16 55 ☽ ∠ ♂
20 23 ☉ # ♃
21 58 ☽ ⊥ ♅
22 14 ☽ ∠ ♄

APRIL 2022

LONGITUDES

Date	Sidereal time h m s	Sun ☉ ° ′ ″	Moon ☽ ° ′ ″	Moon ☽ 24.00 ° ′ ″	Mercury ☿ ° ′	Venus ♀ ° ′	Mars ♂ ° ′	Jupiter ♃ ° ′	Saturn ♄ ° ′	Uranus ♅ ° ′	Neptune ♆ ° ′	Pluto ♇ ° ′
01	00 39 19	11 ♈ 44 21	14 ♈ 33 12	21 ♈ 01 54	10 ♈ 12	25 ♒ 36	19 ♒ 42	21 ♓ 25	22 ♒ 05	12 ♉ 55	23 ♓ 35	28 ♑ 24
02	00 43 16	12 43 34	27 ♈ 26 17	03 ♉ 46 23	12 14	26 40	20 28	21 39	22 11	12 58	23 37	28 25
03	00 47 13	13 42 46	10 ♉ 02 17	16 ♉ 14 12	14 17	27 43	21 13	21 53	22 16	13 01	23 39	28 26
04	00 51 09	14 41 55	22 ♉ 22 24	28 ♉ 27 13	16 20	28 47	21 58	22 07	22 21	13 04	23 42	28 27
05	00 55 06	15 41 03	04 ♊ 29 03	10 ♊ 28 22	18 24	29 ♒ 51	22 43	22 21	22 26	13 07	23 44	28 27
06	00 59 02	16 40 08	16 ♊ 25 40	22 ♊ 21 30	20 29	00 ♓ 56	23 29	22 35	22 32	13 11	23 46	28 28
07	01 02 59	17 39 11	28 ♊ 16 27	04 ♋ 11 08	22 34	02 00	24 14	22 49	22 37	13 14	23 48	28 29
08	01 06 55	18 38 11	10 ♋ 06 09	16 ♋ 02 09	24 39	03 05	24 59	23 02	22 42	13 17	23 50	28 29
09	01 10 52	19 37 09	21 ♋ 59 46	27 ♋ 59 37	26 43	04 10	25 45	23 16	22 47	13 20	23 52	28 30
10	01 14 48	20 36 05	04 ♌ 02 19	10 ♌ 08 26	28 ♈ 47	05 15	26 30	23 30	22 52	13 23	23 54	28 31
11	01 18 45	21 34 59	16 ♌ 18 30	22 ♌ 33 00	00 ♉ 50	06 20	27 15	23 44	22 57	13 27	23 56	28 31
12	01 22 42	22 33 50	28 ♌ 52 21	05 ♍ 16 53	02 52	07 26	28 01	23 57	23 01	13 30	23 59	28 32
13	01 26 38	23 32 39	11 ♍ 46 51	18 ♍ 22 24	04 53	08 31	28 46	24 11	23 06	13 33	24 01	28 32
14	01 30 35	24 31 26	25 ♍ 03 33	01 ♎ 50 14	06 52	09 37	29 ♒ 31	24 24	23 11	13 37	24 03	28 33
15	01 34 31	25 30 11	08 ♎ 42 14	15 ♎ 39 16	08 48	10 43	00 ♓ 17	24 38	23 15	13 40	24 05	28 33
16	01 38 28	26 28 53	22 ♎ 40 53	29 ♎ 46 35	10 42	11 50	01 02	24 51	23 20	13 43	24 07	28 33
17	01 42 24	27 27 34	06 ♏ 55 45	14 ♏ 07 44	12 33	12 56	01 48	25 05	23 24	13 47	24 09	28 34
18	01 46 21	28 26 13	21 ♏ 21 51	28 ♏ 37 23	14 21	14 02	02 33	25 18	23 29	13 50	24 11	28 34
19	01 50 17	29 ♈ 24 50	05 ♐ 53 38	13 ♐ 09 56	16 05	15 09	03 18	25 31	23 33	13 53	24 13	28 34
20	01 54 14	00 ♉ 23 25	20 ♐ 25 39	27 ♐ 40 14	17 46	16 16	04 04	25 44	23 37	13 57	24 14	28 35
21	01 58 11	01 21 59	04 ♑ 53 11	12 ♑ 04 03	19 22	17 23	04 49	25 58	23 41	14 00	24 16	28 35
22	02 02 07	02 20 31	19 ♑ 12 31	26 ♑ 18 15	20 55	18 30	05 34	26 11	23 45	14 04	24 18	28 35
23	02 06 04	03 19 02	03 ♒ 21 02	10 ♒ 20 41	22 23	19 37	06 20	26 24	23 49	14 07	24 20	28 35
24	02 10 00	04 17 31	17 ♒ 17 06	24 ♒ 10 09	23 47	20 45	07 05	26 37	23 53	14 10	24 22	28 36
25	02 13 57	05 15 58	00 ♓ 59 47	07 ♓ 45 58	25 06	21 52	07 51	26 50	23 57	14 14	24 24	28 36
26	02 17 53	06 14 24	14 ♓ 28 39	21 ♓ 07 50	26 21	23 00	08 36	27 02	24 01	14 17	24 26	28 36
27	02 21 50	07 12 47	27 ♓ 43 31	04 ♈ 15 43	27 31	24 08	09 21	27 15	24 04	14 21	24 27	28 36
28	02 25 46	08 11 10	10 ♈ 44 26	17 ♈ 09 42	28 36	25 16	10 07	27 28	24 08	14 24	24 29	28 36
29	02 29 43	09 09 30	23 ♈ 31 35	29 ♈ 50 08	29 ♉ 36	26 24	10 52	27 40	24 11	14 28	24 31	28 36
30	02 33 40	10 ♉ 07 49	06 ♉ 05 26	12 ♉ 17 38	00 ♊ 31	27 ♓ 32	11 ♓ 37	27 ♓ 53	24 ♒ 15	14 ♉ 31	24 ♓ 33	28 ♑ 36

Date	Moon True ☊ ° ′	Moon Mean ☊ ° ′	Moon ☽ Latitude ° ′
01	23 ♉ 04	24 ♉ 45	03 S 13
02	22 R 59	24 41	02 14
03	22 56	24 38	01 10
04	22 55	24 35	00 S 03
05	22 D 56	24 32	01 N 03
06	22 57	24 29	02 05
07	22 59	24 25	03 01
08	23 00	24 22	03 49
09	23 R 01	24 19	04 28
10	23 00	24 16	04 56
11	22 57	24 13	05 11
12	22 54	24 10	05 11
13	22 50	24 06	04 56
14	22 45	24 03	04 25
15	22 41	24 00	03 38
16	22 38	23 57	02 37
17	22 36	23 54	01 25
18	22 35	23 51	00 N 07
19	22 D 35	23 47	01 S 13
20	22 36	23 44	02 27
21	22 38	23 41	03 32
22	22 39	23 38	04 23
23	22 39	23 35	04 58
24	22 R 39	23 31	05 14
25	22 38	23 28	05 12
26	22 36	23 25	04 53
27	22 34	23 22	04 18
28	22 31	23 19	03 31
29	22 30	23 16	02 33
30	22 ♉ 29	23 ♉ 12	01 S 29

DECLINATIONS

Date	Sun ☉ ° ′	Moon ☽ ° ′	Mercury ☿ ° ′	Venus ♀ ° ′	Mars ♂ ° ′	Jupiter ♃ ° ′	Saturn ♄ ° ′	Uranus ♅ ° ′	Neptune ♆ ° ′	Pluto ♇ ° ′
01	04 N 39	02 N 46	02 N 54	12 S 13	16 S 04	04 S 20	15 S 03	15 N 22	03 S 35	22 S 20
02	05 02	08 29	03 49	11 57	15 50	04 14	15 01	15 23	03 34	22 20
03	05 25	13 44	04 45	11 40	15 36	04 09	14 59	15 24	03 33	22 20
04	05 48	18 19	05 41	11 23	15 22	04 04	14 58	15 25	03 32	22 20
05	06 10	22 04	06 37	11 06	15 08	03 58	14 56	15 26	03 31	22 20
06	06 33	24 49	07 33	10 48	14 54	03 53	14 55	15 27	03 31	22 20
07	06 56	26 27	08 29	10 30	14 39	03 47	14 53	15 28	03 30	22 20
08	07 18	26 52	09 25	10 11	14 25	03 42	14 51	15 29	03 29	22 20
09	07 41	26 03	10 20	09 53	14 10	03 37	14 50	15 30	03 28	22 20
10	08 03	24 02	11 15	09 33	13 55	03 31	14 48	15 31	03 27	22 20
11	08 25	20 53	12 09	09 14	13 40	03 26	14 47	15 32	03 26	22 20
12	08 47	16 43	13 02	08 54	13 25	03 21	14 46	15 33	03 26	22 20
13	09 09	11 42	13 53	08 34	13 10	03 15	14 44	15 34	03 25	22 20
14	09 30	06 N 11	14 43	08 13	12 55	03 10	14 43	15 35	03 24	22 20
15	09 52	00 S 07	15 31	07 53	12 40	03 05	14 41	15 36	03 23	22 20
16	10 13	06 24	16 18	07 31	12 24	03 00	14 40	15 37	03 23	22 20
17	10 34	12 29	17 02	07 10	12 08	02 55	14 39	15 38	03 22	22 20
18	10 55	18 00	17 44	06 48	11 53	02 49	14 38	15 39	03 21	22 20
19	11 16	22 29	18 24	06 26	11 37	02 44	14 36	15 40	03 20	22 21
20	11 36	25 33	19 01	06 04	11 21	02 39	14 35	15 41	03 20	22 21
21	11 57	26 53	19 36	05 42	11 05	02 34	14 34	15 42	03 19	22 21
22	12 17	26 24	20 08	05 19	10 49	02 29	14 33	15 43	03 18	22 21
23	12 37	24 14	20 38	04 56	10 33	02 24	14 31	15 44	03 17	22 21
24	12 57	20 38	21 05	04 33	10 16	02 19	14 30	15 45	03 17	22 21
25	13 17	15 59	21 30	04 09	10 00	02 14	14 29	15 46	03 16	22 21
26	13 36	10 37	21 52	03 46	09 44	02 09	14 28	15 47	03 15	22 21
27	13 55	04 S 51	22 11	03 22	09 27	02 04	14 27	15 48	03 15	22 21
28	14 14	01 N 01	22 28	02 58	09 10	01 59	14 26	15 49	03 14	22 21
29	14 33	06 46	22 43	02 34	08 54	01 54	14 25	15 50	03 13	22 22
30	14 N 51	12 N 09	22 N 55	02 S 10	08 S 37	01 S 49	14 S 24	15 N 51	03 S 13	22 S 22

ZODIAC SIGN ENTRIES

Date	h m	Planets
02	16 50	☽ ♉
05	03 04	☽ ♊
05	15 18	♀ ♓
07	15 30	☽ ♋
10	04 00	☽ ♌
11	02 09	☿ ♉
12	14 07	☽ ♍
14	20 46	☽ ♎
15	03 06	♂ ♓
17	00 23	☽ ♏
19	02 16	☽ ♐
20	02 24	☉ ♉
21	03 52	☽ ♑
23	06 17	☽ ♒
25	10 15	☽ ♓
27	16 10	☽ ♈
29	22 23	☿ ♊
30	00 19	☽ ♉

LATITUDES

Date	Mercury ☿ ° ′	Venus ♀ ° ′	Mars ♂ ° ′	Jupiter ♃ ° ′	Saturn ♄ ° ′	Uranus ♅ ° ′	Neptune ♆ ° ′	Pluto ♇ ° ′
01	01 S 14	00 N 49	01 S 13	01 S 01	00 S 57	00 S 22	01 S 07	01 S 54
04	00 48	00 33	01 15	01 01	00 57	00 22	01 07	01 54
07	00 S 19	00 17	01 17	01 01	00 58	00 22	01 07	01 55
10	00 N 13	00 N 02	01 19	01 01	00 58	00 22	01 08	01 55
13	00 47	00 S 12	01 21	01 02	00 58	00 22	01 08	01 55
16	01 20	00 26	01 23	01 02	00 59	00 22	01 08	01 56
19	01 49	00 38	01 25	01 03	00 59	00 22	01 08	01 56
22	02 14	00 50	01 27	01 03	01 00	00 22	01 08	01 57
25	02 32	01 01	01 29	01 03	01 00	00 22	01 08	01 57
28	02 41	01 11	01 30	01 04	01 01	00 22	01 08	01 58
31	02 N 42	01 S 20	01 S 32	01 S 04	01 S 01	00 S 22	01 S 08	01 S 58

DATA

Julian Date	2459671
Delta T	+71 seconds
Ayanamsa	24° 09' 50"
Synetic vernal point	04° ♓ 57' 09"
True obliquity of ecliptic	23° 26' 17"

MOON'S PHASES, APSIDES AND POSITIONS ☽

Date	h m	Phase	Longitude ° ′	Eclipse Indicator
01	06 24	●	11 ♈ 31	
09	06 48	☽	19 ♋ 24	
16	18 55	○	26 ♎ 46	
23	11 56	☾	03 ♒ 19	
30	20 28	●	10 ♉ 28	Partial

Day	h m		
07	19 12	Apogee	
19	15 23	Perigee	
01	00 44	ON	
08	08 14	Max dec	26° N 53'
15	11 34	OS	
21	17 38	Max dec	26° S 56'
28	07 50	ON	

LONGITUDES

Date	Chiron ⚷ ° ′	Ceres ⚳ ° ′	Pallas ⚴ ° ′	Juno ⚵ ° ′	Vesta ⚶ ° ′	Black Moon Lilith ⚸ ° ′
01	12 ♈ 19	13 ♊ 25	17 ♈ 46	22 ♒ 58	10 ♒ 07	28 ♊ 30
11	12 ♈ 54	16 ♊ 55	21 ♈ 56	26 ♒ 41	14 ♒ 28	29 ♊ 37
21	13 ♈ 29	20 ♊ 37	26 ♈ 10	00 ♓ 18	18 ♒ 35	00 ♋ 44
31	14 ♈ 02	24 ♊ 28	00 ♉ 30	03 ♓ 46	22 ♒ 27	01 ♋ 51

All ephemeris data is given at 12.00 UT and the Moon's longitude is additionally given for 24.00 UT

ASPECTARIAN

h m Aspects

01 Friday
02 34 ☽ ☌ ☿
04 07 ☽ ∠ ♀
04 23 ☽ Q ♇
06 24 ☽ ☌ ☉
08 58 ☽ ⊻ ♅
12 39 ☽ ∥ ☿
14 23 ☿ Q ♇
15 19 ☽ # ♆
18 20 ☽ # ♃
20 16 ☽ ∥ ☉
21 48 ☿ ∠ ♀
22 07 ☽ ⚹ ♂

02 Saturday
00 57 ☽ ⊻ ♃
02 04 ☽ ⚹ ♄
04 49 ☽ ⊻ ♆
05 24 ☿ # ♆
10 24 ☽ ⚹ ♀
12 24 ☽ ⊥ ♃
13 51 ☽ □ ♇
16 08 ☽ ⊥ ♆
18 10 ☉ ⊻ ♅
20 49 ☿ ⊻ ♅
21 54 ☿ # ♃
22 06 ☽ Q ♂
23 11 ☉ ☌ ☿

03 Sunday
00 52 ☽ Q ♄
02 46 ☽ # ♀
05 50 ☽ ∠ ♃
09 20 ☽ ∠ ♆
11 20 ☽ Q ♀
17 47 ☽ ☌ ♅
18 12 ☽ # ♄
19 43 ☽ ⊻ ☉
20 20 ☽ ∥ ♅
20 55 ☽ # ♂
21 50 ☽ ⊻ ☿

04 Monday
04 16 ♀ ⊻ ♇
08 03 ♂ # ♅
08 26 ☽ ⊥ ☉
11 09 ☽ □ ♂
11 29 ☽ ⚹ ♃
11 55 ☽ ⊥ ☿
11 58 ☽ □ ♄
14 36 ☽ ⚹ ♆
16 57 ☉ ∥ ☿
18 47 ♂ ⊻ ♃

05 Tuesday
00 00 ☽ △ ♇
01 51 ♂ ☌ ♄
01 53 ☽ □ ♀
03 45 ☽ ∠ ☉
09 24 ☽ ∠ ☿
11 43 ☽ Q ♃
14 01 ☽ # ♇
14 30 ☽ Q ♆

06 Wednesday
03 24 ♃ ⊻ ♄
05 25 ☽ ⊻ ♅
06 01 ☽ Ᵽ ♇
10 43 ♂ ∥ ♄
12 32 ☽ ⚹ ☉
17 35 ☽ ⊥ ♅
17 53 ♀ Q ♅
21 34 ♂ ⊻ ♆
21 57 ☽ ⚹ ☿

07 Thursday
00 14 ☽ ± ♇
00 26 ☽ △ ♄
00 42 ☽ □ ♃
02 54 ☽ □ ♆
03 15 ☽ △ ♂
11 54 ☽ ∠ ♅
12 25 ☽ ⊼ ♇
12 37 ☿ ⚹ ♄
15 03 ☽ Q ☉
15 14 ☿ ⊻ ♃
20 19 ☽ △ ♀

08 Friday
02 33 ☿ ⊻ ♆
03 29 ☽ Q ☿
07 05 ☽ Ᵽ ♄
11 45 ☽ Ᵽ ♂
18 19 ☿ ⚹ ♂
18 28 ☽ ⚹ ♅

09 Saturday
01 26 ☽ ± ♄
02 59 ☿ # ♀
05 44 ☽ Ᵽ ♀
06 48 ☽ □ ☉
07 10 ☽ ± ♂
13 35 ☽ ⊼ ♄
14 36 ☽ △ ♃
15 46 ☽ △ ♆
18 43 ☽ Q ♅
19 31 ♀ ⊥ ♇
20 01 ☽ ⊼ ♂
23 25 ☽ □ ☿

10 Sunday
01 01 ☽ ☍ ♇
01 33 ☽ ± ♀
08 45 ☿ □ ♇
14 37 ☽ ⊼ ♀
20 57 ☽ Ᵽ ♃
21 20 ☿ ⊥ ♃
21 37 ☽ Ᵽ ♆

11 Monday
01 17 ☿ ⊥ ♆
01 56 ☽ # ♇
06 25 ☽ □ ♅
09 28 ♃ ∥ ♆
14 47 ☽ ± ♃
15 10 ☽ ± ♆
23 01 ☽ △ ☉

12 Tuesday
00 50 ☽ ☍ ♄
02 30 ☽ ⊼ ♃
02 42 ☽ ⊼ ♆
10 17 ☽ ☍ ♂
11 21 ☽ ⊼ ♇
14 42 ♃ ☌ ♆
16 09 ☉ # ♀
17 57 ☽ ∥ ♅
20 27 ☿ # ♂
20 55 ☽ △ ☿
21 52 ☽ # ♄
22 36 ☽ ± ♇

13 Wednesday
00 14 ☉ ⚹ ♄
03 25 ☽ ∥ ☿
04 30 ♂ ⊻ ♇
04 59 ☽ # ♂
05 27 ☽ ☍ ♀
05 34 ☽ Ᵽ ☉
14 45 ☿ Q ♄
15 13 ☽ Ᵽ ♇
15 16 ☽ △ ♅
16 07 ☉ ∠ ♀
22 25 ☽ ∥ ☉
23 20 ☽ ± ☉
23 48 ☉ ⊻ ♆

14 Thursday
02 23 ☽ # ♀
05 18 ☽ Ᵽ ☿
08 14 ☉ ⊻ ♃
08 37 ☽ ⊼ ♄
10 11 ☽ ☍ ♆
10 49 ☽ ☍ ♃
10 58 ☽ ⊼ ☉
11 50 ☿ # ♄
18 11 ☽ △ ♇
18 20 ☽ Ᵽ ♅
19 21 ☽ ± ♄
20 23 ☽ ⊼ ♂
22 24 ☽ # ♆
23 27 ☽ # ♃

15 Friday
00 00 ☽ ± ☿
07 32 ☽ ± ♂
10 11 ☽ ± ♅
11 13 ☽ Ᵽ ♄
12 12 ☽ ⊼ ☿
14 11 ☿ ∥ ♅
15 31 ☿ ∠ ♆
15 48 ☽ ⊼ ♀
20 37 ☽ ⊼ ♅
23 12 ☽ ∥ ♃
23 49 ☿ ∠ ♃

16 Saturday
00 00 ☽ Ᵽ ♂
00 30 ☽ ∥ ♆
03 01 ☽ ± ♀
13 07 ☽ △ ♄
14 26 ☽ ⊼ ♆
15 45 ☽ ⊼ ♃
16 07 ☽ ∥ ♀
18 55 ☽ ☍ ☉
19 37 ☽ Ᵽ ♀
21 57 ☽ □ ♇

17 Sunday
00 35 ☽ ± ♆
02 02 ☽ ± ♃
02 55 ☽ △ ♂
03 48 ☽ # ☉
10 39 ☽ ∥ ♂
11 27 ♂ Q ♅
15 42 ☽ Ᵽ ♆
17 20 ☽ Ᵽ ♃
20 59 ☽ ∥ ♄
22 43 ☽ ☍ ☿
22 51 ☽ △ ♀
23 28 ☽ ☍ ♅

18 Monday
01 09 ☿ ⚹ ♀
01 17 ☽ # ♅
01 43 ♀ ∠ ♇
04 03 ☽ Q ♇
04 51 ☿ ☌ ♅
07 15 ♀ ⚹ ♅
10 35 ☽ # ☿
15 14 ☉ □ ♇
15 31 ☽ □ ♄
16 40 ☽ △ ♆
16 51 ☿ Q ♂
18 37 ☽ △ ♃
23 55 ☽ ⚹ ♇

19 Tuesday
00 32 ☽ ⊼ ☉
07 30 ☽ □ ♂
11 08 ☽ ∥ ♇
11 09 ☽ ± ☉
21 23 ☽ Q ♄

20 Wednesday
00 41 ☽ ∠ ♇
01 15 ☽ ⊼ ♅
01 48 ☉ # ♂
03 04 ☽ Ᵽ ☉
04 33 ☽ □ ♀
07 02 ☽ ⊼ ☿
08 12 ☉ ⊥ ♆
11 12 ☽ ± ♅
14 51 ☽ Q ♂
15 34 ☽ ⊥ ♇
17 18 ☽ ⚹ ♄
18 13 ☽ ± ☿
18 20 ☽ □ ♆
20 56 ☽ □ ♃

21 Thursday
01 31 ☽ ⊻ ♇
02 10 ☽ Ᵽ ♅
04 30 ♂ ⊥ ♇
05 43 ☽ △ ☉
11 02 ☽ Ᵽ ☿
11 53 ☽ ⚹ ♂
12 54 ☽ Q ♀
18 22 ☽ ∠ ♄

22 Friday
00 22 ☽ Q ♆
03 18 ☽ △ ♅
03 24 ☽ Q ♃
06 46 ☉ ⊥ ♃
09 32 ☽ ⊥ ♄
10 42 ☽ ⚹ ♀
14 26 ☽ ∠ ♂
15 13 ☽ △ ☿
19 43 ☽ ⊻ ♄
20 38 ☽ ⚹ ♆
23 58 ☽ ⚹ ♃

23 Saturday
03 53 ☽ ☌ ♇
06 33 ☽ ⊥ ♂
11 56 ☽ □ ☉
14 22 ☽ ∠ ♀
17 24 ☽ ⊻ ♂
22 17 ☽ ∠ ♆

24 Sunday
01 34 ☽ ∥ ♇
02 02 ☽ ∠ ♃
06 35 ☽ □ ♅
07 13 ☽ ⊥ ♀
09 38 ☽ # ☿
13 50 ☿ □ ♄
13 53 ☽ ⊥ ♆
17 53 ☽ ⊥ ♃
18 33 ☽ ⊻ ♀
21 23 ☽ Q ☉
22 37 ☿ ⚹ ♆
23 33 ☽ ☌ ♄

25 Monday
00 22 ☽ ⊻ ♆
00 33 ☽ □ ☿
04 32 ☽ ⊻ ♃
07 46 ☽ ⊻ ♇
13 01 ☽ # ♅
14 11 ☽ Q ♅
18 22 ☽ ⊥ ♇
18 57 ☽ ∥ ♄
20 09 ☽ ⚹ ☉
23 41 ☽ # ☉

26 Tuesday
00 51 ☽ ☌ ♂
05 58 ☉ Q ♄
10 25 ☽ ∠ ♇
11 39 ☽ ⚹ ♅
11 45 ☽ Q ☿
15 58 ☽ ∥ ♂

27 Wednesday
01 10 ☽ ∠ ☉
04 50 ☽ ☌ ♀
05 11 ☿ ⚹ ♃
05 18 ☽ ⊻ ♄
06 02 ☽ ☌ ♆
10 43 ♀ ⊻ ♄
11 07 ☽ ☌ ♃
11 35 ☽ ⚹ ☿
13 36 ☽ ⚹ ♇
14 59 ☽ ∠ ♅
16 19 ☽ ⊥ ♄
18 31 ☽ ∥ ♀
18 36 ☽ ∥ ♆
18 54 ☽ ⊥ ☉
19 11 ♀ ☌ ♆
19 35 ♀ ∥ ♆
23 33 ☽ ∥ ♃

28 Thursday
02 08 ☿ # ♇
06 52 ☽ ⊻ ☉
07 38 ☽ ⊥ ♅
09 00 ☽ ∠ ♄
10 45 ☽ ⊻ ♂
11 44 ☽ Q ♇
12 05 ☿ △ ♇
15 55 ☽ # ♃
17 47 ☽ ∠ ☿
18 52 ☽ ⊻ ♅
19 31 ☽ # ♀
21 07 ☽ # ♆
22 39 ☽ ⊥ ♂

29 Friday
02 32 ☉ # ♄
12 08 ☽ ⊥ ☿
13 16 ☽ ⚹ ♄
13 53 ☽ ⊻ ♆
16 43 ☽ ∠ ♂
17 59 ☽ ⊻ ♀
18 37 ♇ St R
20 01 ☽ ⊻ ♃
20 50 ☽ # ♂
21 07 ☉ ∠ ♆
21 38 ☽ □ ♇

30 Saturday
00 28 ☽ ⊻ ☿
01 20 ☽ ⊥ ♆
06 35 ☽ ⊥ ♀
07 41 ☽ ⊥ ♃
12 18 ☽ Q ♄
18 41 ☽ ∠ ♆
20 28 ☽ ☌ ☉
21 14 ♀ ☌ ♃
22 48 ☽ # ♄
23 23 ☽ ⚹ ♂

MAY 2022

LONGITUDES

Date	Sidereal time h m s	Sun ☉ ° ' "	Moon ☽ ° ' "	Moon ☽ 24.00 ° ' "	Mercury ☿ ° '	Venus ♀ ° '	Mars ♂ ° '	Jupiter ♃ ° '	Saturn ♄ ° '	Uranus ♅ ° '	Neptune ♆ ° '	Pluto ♇ ° '
01	02 37 36	11 ♉ 06 06	18 ♉ 26 50	24 ♉ 33 14	01 ♊ 20	28 ♓ 40	12 ♓ 23	28 ♓ 06	24 ♒ 18	14 ♉ 35	24 ♓ 34	28 ♑ 36
02	02 41 33	12 04 22	00 ♊ 37 02	06 ♊ 38 29	02 05	29 ♓ 48	13 08	28 18	24 21	14 38	24 36	28 R 36
03	02 45 29	13 02 36	12 ♊ 37 52	18 ♊ 35 32	02 44	00 ♈ 57	13 53	28 30	24 24	14 41	24 38	28 36
04	02 49 26	14 00 47	24 ♊ 31 48	00 ♋ 27 05	03 18	02 05	14 38	28 43	24 27	14 45	24 39	28 36
05	02 53 22	14 58 57	06 ♋ 21 50	12 ♋ 16 30	03 47	03 14	15 24	28 55	24 30	14 48	24 41	28 35
06	02 57 19	15 57 05	18 ♋ 11 35	24 ♋ 07 37	04 10	04 22	16 09	29 07	24 33	14 52	24 43	28 35
07	03 01 15	16 55 11	00 ♌ 05 08	06 ♌ 04 41	04 29	05 31	16 54	29 19	24 36	14 55	24 44	28 35
08	03 05 12	17 53 16	12 ♌ 06 52	18 ♌ 12 15	04 41	06 40	17 39	29 31	24 39	14 59	24 46	28 35
09	03 09 09	18 51 18	24 ♌ 21 22	00 ♍ 34 47	04 49	07 49	18 24	29 43	24 41	15 02	24 47	28 35
10	03 13 05	19 49 18	06 ♍ 53 01	13 ♍ 16 31	04 R 52	08 58	19 09	29 ♓ 54	24 44	15 06	24 49	28 34
11	03 17 02	20 47 16	19 ♍ 45 43	26 ♍ 20 55	04 49	10 07	19 55	00 ♈ 06	24 46	15 09	24 50	28 34
12	03 20 58	21 45 13	03 ♎ 02 23	09 ♎ 50 13	04 42	11 16	20 40	00 18	24 48	15 13	24 52	28 34
13	03 24 55	22 43 08	16 ♎ 44 26	23 ♎ 44 55	04 30	12 25	21 25	00 29	24 51	15 16	24 53	28 33
14	03 28 51	23 41 00	00 ♏ 51 21	08 ♏ 03 19	04 13	13 34	22 10	00 41	24 53	15 19	24 55	28 33
15	03 32 48	24 38 52	15 ♏ 20 15	22 ♏ 41 26	03 53	14 44	22 55	00 52	24 55	15 23	24 56	28 32
16	03 36 44	25 36 42	00 ♐ 06 00	07 ♐ 33 02	03 29	15 53	23 40	01 03	24 57	15 26	24 57	28 32
17	03 40 41	26 34 31	15 ♐ 01 32	22 ♐ 30 28	03 02	17 03	24 25	01 14	24 59	15 30	24 59	28 32
18	03 44 38	27 32 18	29 ♐ 57 50	07 ♑ 25 37	02 33	18 12	25 10	01 25	25 00	15 33	25 00	28 31
19	03 48 34	28 30 05	14 ♑ 49 57	22 ♑ 11 00	02 01	19 22	25 55	01 36	25 02	15 37	25 01	28 31
20	03 52 31	29 ♉ 27 49	29 ♑ 28 07	06 ♒ 40 43	01 28	20 32	26 40	01 47	25 03	15 40	25 02	28 30
21	03 56 27	00 ♊ 25 33	13 ♒ 48 26	20 ♒ 50 59	00 54	21 42	27 25	01 58	25 05	15 43	25 04	28 29
22	04 00 24	01 23 16	27 ♒ 48 11	04 ♓ 40 03	00 ♊ 19	22 51	28 09	02 09	25 06	15 47	25 05	28 29
23	04 04 20	02 20 58	11 ♓ 26 36	18 ♓ 07 59	29 ♉ 45	24 01	28 54	02 19	25 08	15 50	25 06	28 28
24	04 08 17	03 18 38	24 ♓ 44 25	01 ♈ 16 08	29 11	25 11	29 ♓ 39	02 30	25 09	15 53	25 07	28 28
25	04 12 13	04 16 18	07 ♈ 43 26	14 ♈ 06 35	28 39	26 21	00 ♈ 24	02 40	25 10	15 57	25 08	28 27
26	04 16 10	05 13 56	20 ♈ 25 55	26 ♈ 41 44	28 09	27 32	01 08	02 50	25 11	16 00	25 09	28 26
27	04 20 07	06 11 34	02 ♉ 54 19	09 ♉ 04 00	27 41	28 42	01 53	03 00	25 12	16 03	25 10	28 25
28	04 24 03	07 09 10	15 ♉ 11 01	21 ♉ 15 41	27 16	29 ♈ 52	02 38	03 10	25 12	16 07	25 11	28 25
29	04 28 00	08 06 46	27 ♉ 18 13	03 ♊ 18 54	26 55	01 ♉ 02	03 22	03 20	25 13	16 10	25 12	28 24
30	04 31 56	09 04 21	09 ♊ 17 57	15 ♊ 15 38	26 37	02 13	04 07	03 30	25 14	16 13	25 13	28 23
31	04 35 53	10 ♊ 01 54	21 ♊ 12 11	27 ♊ 07 52	26 ♉ 23	03 ♉ 23	04 ♈ 51	03 ♈ 40	25 ♒ 14	16 ♉ 17	25 ♓ 14	28 ♑ 22

DECLINATIONS

Date	Moon True ☊ ° '	Moon Mean ☊ ° '	Moon ☽ Latitude ° '	Sun ☉ ° '	Moon ☽ ° '	Mercury ☿ ° '	Venus ♀ ° '	Mars ♂ ° '	Jupiter ♃ ° '	Saturn ♄ ° '	Uranus ♅ ° '	Neptune ♆ ° '	Pluto ♇ ° '
01	22 ♉ 28	23 ♉ 09	00 S 22	15 N 09	16 N 58	23 N 04	01 S 45	08 S 20	01 S 45	14 S 23	15 N 52	03 S 12	22 S 22
02	22 D 28	23 06	00 N 45	15 27	21 01	23 12	01 20	08 03	01 40	14 22	15 53	03 11	22 22
03	22 29	23 03	01 49	15 45	24 07	23 16	00 56	07 46	01 35	14 21	15 54	03 11	22 22
04	22 29	23 00	02 49	16 03	26 08	23 19	00 31	07 29	01 30	14 21	15 55	03 10	22 22
05	22 30	22 57	03 40	16 20	26 57	23 19	00 S 06	07 12	01 26	14 20	15 56	03 09	22 23
06	22 31	22 53	04 22	16 37	26 32	23 18	00 N 19	06 55	01 21	14 19	15 57	03 09	22 23
07	22 32	22 50	04 53	16 53	24 54	23 13	00 44	06 38	01 16	14 18	15 58	03 08	22 23
08	22 32	22 47	05 12	17 10	22 09	23 07	01 09	06 21	01 12	14 17	15 59	03 08	22 23
09	22 R 32	22 44	05 17	17 26	18 23	22 59	01 35	06 04	01 07	14 17	16 00	03 07	22 23
10	22 32	22 41	05 07	17 42	13 44	22 49	02 00	05 46	01 03	14 16	16 01	03 07	22 24
11	22 32	22 37	04 42	17 57	08 23	22 36	02 25	05 29	00 58	14 15	16 02	03 06	22 24
12	22 31	22 34	04 02	18 12	02 N 29	22 22	02 51	05 12	00 54	14 15	16 03	03 06	22 24
13	22 D 31	22 31	03 06	18 27	03 S 43	22 06	03 16	04 55	00 49	14 14	16 04	03 05	22 24
14	22 31	22 28	01 57	18 42	09 56	21 48	03 42	04 37	00 45	14 14	16 05	03 04	22 24
15	22 31	22 25	00 N 40	18 56	15 48	21 29	04 07	04 20	00 40	14 13	16 06	03 04	22 25
16	22 R 31	22 22	00 S 42	19 10	20 51	21 08	04 33	04 02	00 36	14 13	16 07	03 03	22 25
17	22 31	22 18	02 02	19 23	24 37	20 47	04 58	03 45	00 32	14 12	16 08	03 03	22 25
18	22 31	22 15	03 13	19 37	26 40	20 24	05 23	03 27	00 28	14 12	16 09	03 02	22 25
19	22 30	22 12	04 11	19 49	26 46	20 01	05 49	03 10	00 24	14 11	16 10	03 02	22 25
20	22 30	22 09	04 52	20 02	25 01	19 37	06 14	02 53	00 19	14 11	16 11	03 02	22 26
21	22 29	22 06	05 13	20 14	21 41	19 13	06 39	02 35	00 15	14 10	16 12	03 01	22 26
22	22 29	22 03	05 16	20 26	17 10	18 49	07 04	02 18	00 11	14 10	16 13	03 01	22 26
23	22 D 29	21 59	05 00	20 38	11 53	18 25	07 29	02 00	00 07	14 10	16 14	03 00	22 27
24	22 29	21 56	04 28	20 49	06 12	18 02	07 54	01 43	00 S 03	14 10	16 15	03 00	22 27
25	22 30	21 53	03 44	21 00	00 S 22	17 40	08 19	01 25	00 N 01	14 10	16 16	03 00	22 27
26	22 31	21 50	02 49	21 10	05 N 23	17 19	08 44	01 08	00 04	14 10	16 17	02 59	22 27
27	22 32	21 47	01 47	21 20	10 48	16 59	09 09	00 50	00 08	14 10	16 18	02 59	22 28
28	22 33	21 43	00 S 41	21 30	15 44	16 41	09 33	00 33	00 12	14 09	16 19	02 58	22 28
29	22 R 33	21 40	00 N 26	21 40	19 59	16 24	09 57	00 S 15	00 16	14 09	16 20	02 58	22 28
30	22 32	21 37	01 35	21 49	23 21	16 10	10 22	00 N 02	00 19	14 09	16 21	02 58	22 29
31	22 ♉ 31	21 ♉ 34	02 N 32	21 N 57	25 N 40	15 N 58	10 N 46	00 N 19	00 N 23	14 S 09	16 N 22	02 S 57	22 S 29

ZODIAC SIGN ENTRIES

Date	h	m	Planets
02	10	46	☽ ♊
02	16	10	♀ ♈
04	23	05	☽ ♋
07	11	50	☽ ♌
09	22	53	☽ ♍
10	23	22	♃ ♈
12	06	34	☽ ♎
14	10	34	☽ ♏
16	11	50	☽ ♐
18	12	02	☽ ♑
20	12	53	☽ ♒
21	01	23	☉ ♊
22	15	49	☽ ♓
23	01	15	☿ ♉
24	21	39	☽ ♈
24	23	17	♂ ♈
27	06	22	☽ ♉
28	14	46	♀ ♉
29	17	23	☽ ♊

LATITUDES

Date	Mercury ☿ ° '	Venus ♀ ° '	Mars ♂ ° '	Jupiter ♃ ° '	Saturn ♄ ° '	Uranus ♅ ° '	Neptune ♆ ° '	Pluto ♇ ° '
01	02 N 42	01 S 20	01 S 32	01 S 04	01 S 01	00 S 22	01 S 08	01 S 58
04	02 33	01 28	01 34	01 05	01 02	00 22	01 08	01 58
07	02 14	01 35	01 35	01 05	01 02	00 22	01 08	01 59
10	01 44	01 41	01 37	01 06	01 03	00 22	01 09	01 59
13	01 05	01 47	01 38	01 06	01 03	00 22	01 09	02 00
16	00 N 18	01 51	01 40	01 07	01 04	00 22	01 09	02 00
19	00 S 34	01 55	01 41	01 08	01 05	00 22	01 09	02 01
22	01 26	01 57	01 42	01 08	01 05	00 22	01 09	02 01
25	02 15	01 59	01 43	01 09	01 06	00 21	01 09	02 01
28	02 57	02 00	01 44	01 09	01 06	00 21	01 09	02 02
31	03 S 29	02 S 00	01 S 45	01 S 10	01 S 07	00 S 21	01 S 09	02 S 02

DATA

Julian Date	2459701
Delta T	+71 seconds
Ayanamsa	24° 09' 54"
Synetic vernal point	04° ♓ 57' 05"
True obliquity of ecliptic	23° 26' 17"

MOON'S PHASES, APSIDES AND POSITIONS ☽

Date	h	m	Phase	Longitude ° '	Eclipse Indicator
09	00	21	☽	18 ♌ 23	
16	04	14	○	25 ♏ 18	total
22	18	43	☾	01 ♓ 39	
30	11	30	●	09 ♊ 03	

Day	h	m		
05	12	50	Apogee	
17	15	35	Perigee	
05	15	54	Max dec	26° N 58'
12	21	42	0S	
19	01	23	Max dec	26° S 58'
25	13	29	0N	

LONGITUDES

Date	Chiron ⚷ ° '	Ceres ⚳ ° '	Pallas ⚴ ° '	Juno ⚵ ° '	Vesta ⚶ ° '	Black Moon Lilith ⚸ ° '
01	14 ♈ 02	24 ♊ 28	00 ♉ 30	03 ♓ 46	22 ♒ 27	01 ♋ 51
11	14 ♈ 33	28 ♊ 27	04 ♉ 53	07 ♓ 05	25 ♒ 59	02 ♋ 57
21	15 ♈ 01	02 ♋ 33	09 ♉ 21	10 ♓ 11	29 ♒ 09	04 ♋ 04
31	15 ♈ 26	06 ♋ 44	13 ♉ 53	13 ♓ 02	01 ♓ 54	05 ♋ 11

ASPECTARIAN

Day	h m	Aspects
01 Sunday	01 22	☽ ∠ ♃
	01 43	☽ ∠ ♀
	01 59	☽ ∥ ☉
	04 24	☽ ☌ ♅
	06 13	☽ ∥ ♅
	10 37	♀ ⚹ ♇
	12 43	♀ ∥ ♃
	23 33	☽ □ ♄
02 Monday	00 04	☽ ⚹ ♆
	00 25	☽ Q ♂
	07 19	☽ ⚹ ♃
	08 00	☽ △ ♇
	08 15	♀ ∠ ♅
	10 13	☽ ⚹ ♀
	15 05	☽ ☌ ☿
	21 32	☽ ⋕ ♇
	23 57	☽ Q ♆
03 Tuesday	00 10	♀ ⊥ ♄
	02 50	♂ ∠ ♇
	04 26	☽ ∥ ☿
	07 39	☽ Q ♃
	12 41	☽ Q ♀
	12 54	☽ ⊻ ☉
	13 56	☽ ⚼ ♇
	14 41	☽ □ ♂
	16 10	☽ ⊻ ♅
	22 33	♃ ⚹ ♇
04 Wednesday	00 51	☉ ∥ ♅
	02 03	☽ ⊥ ☉
	02 29	☉ ∠ ♃
	04 19	☽ ⊥ ♅
	08 05	☽ ± ♇
	11 51	☽ △ ♄
	12 16	☽ □ ♆
	15 47	♂ ⚹ ♅
	20 14	☽ ⊼ ♇
	20 37	☽ □ ♃
	21 54	☽ ∠ ☉
	22 38	☽ ∠ ♅
05 Thursday	04 57	☽ □ ♀
	06 33	☽ ⊻ ☿
	07 21	☉ ☌ ♅
	18 24	☽ ⚼ ♄
	19 12	☽ ⊥ ☿
06 Friday	05 13	☽ ⚹ ♅
	06 01	☿ ⚹ ♀
	07 03	☽ ⚹ ☉
	07 34	☽ △ ♂
	12 44	☽ ± ♄
	14 03	☽ ∠ ☿
07 Saturday	00 55	☽ ⊼ ♄
	01 13	☽ △ ♆
	05 36	☽ Q ♅
	08 59	☽ ☍ ♇
	09 27	☽ Q ☉
	09 48	☉ ⚹ ♂
	10 25	☽ △ ♃
	15 53	☽ ⚼ ♂
	20 59	☽ ⚹ ☿
08 Sunday	00 02	☽ △ ♀
	04 15	☽ ∥ ☿
	07 20	☽ ⚼ ♆
	10 16	☽ ⋕ ♇
	11 02	☽ ± ♂
	13 46	♀ ⋕ ♃
	16 49	☽ ⚼ ♃
	17 41	☽ □ ♅
	21 09	☽ Q ☿
	23 38	☽ ⊼ ♂
09 Monday	00 21	☽ □ ☉
	01 07	☽ ± ♆
	08 42	☽ ⚼ ♀
	10 44	☽ ± ♃
	12 39	☽ ☍ ♄
	12 50	☽ ⊼ ♆
	16 59	☽ ∥ ☉
	20 09	☽ ⊼ ♇
	22 30	☽ ⊼ ♃
10 Tuesday	00 46	☽ ∥ ♅
	03 48	☽ ± ♀
	07 37	☽ ± ♇
	08 10	☽ □ ☿
	09 27	☽ ⋕ ♄
	11 48	☿ St R
	14 56	♀ ⊥ ♅
	16 19	☽ ⊼ ♀
11 Wednesday	00 33	☽ ⚼ ♇
	03 27	☽ △ ♅
	04 33	♀ ∠ ♄
	12 17	☽ ☍ ♂
	14 02	☽ △ ☉
	20 52	♃ ∠ ♅
	21 10	☽ ⊼ ♄
	21 17	☽ ☍ ♆
	21 25	♀ Q ♇
12 Thursday	00 37	☽ ⋕ ♂
	04 00	☽ △ ♇
	06 56	☽ ⚼ ♅
	07 02	☽ ☍ ♃
	08 00	☽ ± ♄
	08 51	☿ ⋕ ♇
	09 36	☽ ⋕ ♆
	10 40	☽ ∥ ♀
	14 54	☽ △ ☿
	18 19	☽ ⋕ ♃
	19 05	☽ ⚼ ☉
	22 57	☽ ± ♅
	23 59	☽ ⚼ ♄
13 Friday	01 00	☽ ∥ ♃
	01 36	♀ ⋕ ♆
	03 49	☽ ☍ ♀
	09 27	☽ ⊼ ♅
	09 35	☽ ∥ ♆
	10 10	☽ ⋕ ♀
	11 58	☽ ± ☉
	16 22	☽ ∥ ♂
	16 39	☽ ⚼ ☿
	20 28	☽ ⊼ ♂
	23 00	☽ ⊼ ☉
14 Saturday	01 54	☽ △ ♄
	01 57	☽ ⊼ ♆
	07 13	☽ ± ♂
	07 40	☽ ± ☿
	08 07	☽ □ ♇
	11 42	☽ ⊼ ♃
	12 05	☽ ± ♆
	13 23	☿ Q ♂
	17 31	☽ ⊼ ☿
	21 51	☽ ± ♃
	23 06	☽ ⚼ ♂
15 Sunday	03 05	☽ ⚼ ♆
	05 18	☽ ∥ ♄
	10 55	☽ ⊼ ♀
	12 04	☽ ☍ ♅
	12 53	☽ ⚼ ♃
	13 20	☽ ⋕ ♅
	13 58	☽ Q ♇
	18 49	☉ □ ♄
	19 05	♀ ⋕ ♂
	19 05	♃ ⊥ ♄
	19 15	☉ ⚹ ♆
	21 34	☽ ± ♀
16 Monday	01 01	☽ △ ♂
	02 14	♀ ⊻ ♅
	02 57	☽ ⋕ ☉
	03 38	☽ □ ♄
	03 40	☽ △ ♆
	04 14	☽ ☍ ☉
	09 28	☽ ⚹ ♇
	13 23	☽ ⚼ ♀
	13 26	☽ ⋕ ☿
	13 34	☽ △ ♃
	17 19	☽ ☍ ☿
	20 51	☽ ∥ ♇
17 Tuesday	08 42	☽ Q ♄
	09 36	☽ ∠ ♇
	12 45	☽ ⊼ ♅
	15 31	☽ △ ♀
	15 39	♄ ⊻ ♆
	22 25	☽ ± ♅
18 Wednesday	00 01	☽ ⊥ ♇
	02 31	☿ ∠ ♀
	03 51	☽ □ ♂
	03 59	☽ □ ♆
	03 59	☽ ⚹ ♄
	06 33	♂ ☌ ♆
	06 43	♂ ⊻ ♄
	07 48	☽ ⊼ ☉
	09 39	☽ ⊻ ♇
	12 56	☽ ⚼ ♅
	14 21	☽ □ ♃
	16 00	☽ ⊼ ☿
	18 07	☽ ± ☉
19 Thursday	01 21	☽ ± ☿
	04 12	☽ ∠ ♄
	09 03	☽ Q ♆
	09 41	☽ ⚼ ☉
	10 25	☽ Q ♂
	12 11	☉ △ ♇
	13 16	☽ △ ♅
	15 26	☽ ⚼ ☿
	18 51	☽ ⊥ ♄
	19 18	☉ ∥ ☿
	19 53	☽ Q ♃
	20 02	☽ □ ♀
	23 14	♂ ∥ ♆
20 Friday	01 33	☿ ⚹ ♃
	04 41	☽ ⚹ ♆
	04 43	☽ ⊻ ♄
	07 07	☽ ⚹ ♂
	10 24	☽ ☌ ♇
	11 59	☽ △ ☉
	15 11	☽ △ ☿
	15 54	☽ ⚹ ♃
21 Saturday	04 26	☽ Q ♀
	05 40	☽ ∠ ♆
	07 19	☽ ∥ ♇
	09 30	☽ ∠ ♂
	15 16	☽ □ ♅
	17 26	☽ ∠ ♃
	19 18	☉ ☌ ☿
	19 53	☽ ⋕ ☉
	20 57	☽ ⊥ ♆
22 Sunday	01 41	☽ ⊥ ♂
	02 40	☽ ⚹ ♀
	03 04	☽ ⋕ ☿
	07 16	☽ ⊻ ♆
	07 19	☽ ☌ ♄
	09 05	☽ ⊥ ♃
	12 39	☽ ⊻ ♂
	13 10	☽ ⊻ ♇
	16 12	☽ □ ☿
	16 31	☽ ⋕ ♅
	18 43	☽ □ ☉
	19 40	☽ ⊻ ♃
	22 15	♂ ⚹ ♇
	22 29	☽ Q ♅
	23 40	☽ ⊥ ♇
23 Monday	01 57	☽ ∥ ♄
	07 17	☽ ∠ ♀
	08 09	☿ ⊥ ♀
	11 05	☉ ⚹ ♃
	15 37	☽ ∠ ♇
	19 54	☽ ⚹ ♅
	22 50	☽ Q ☿
24 Tuesday	00 56	☽ ⊥ ♀
	03 27	☿ ⚹ ♂
	05 16	☽ Q ☉
	05 22	☽ ⋕ ♀
	10 30	♀ ⊻ ♆
	11 06	♀ ⚹ ♄
	12 41	☽ ☌ ♆
	12 45	☽ ⊻ ♄
	12 54	☽ ⊻ ♀
	18 49	☽ ⚹ ♇
	19 50	☽ ⚹ ☿
	21 33	☽ ☌ ♂
	23 21	☽ ∠ ♅
	23 47	☽ ⊥ ♄
25 Wednesday	01 10	☽ ∥ ♆
	02 27	☽ ☌ ♃
	05 03	☽ ⚹ ☉
	07 24	☽ ∥ ♂
	13 26	☽ ⋕ ♃
	13 33	☽ ∥ ♃
	16 11	☽ ⊥ ♅
	16 35	☽ ∠ ♄
	17 06	☽ Q ♇
	19 00	☽ ⋕ ♂
	21 49	☿ △ ♇
	22 42	☽ ∠ ☿
26 Thursday	01 54	☽ ⋕ ♆
	03 33	☽ ⊻ ♅
	07 13	♂ ∠ ♅
	11 35	☽ ∠ ☉
	13 20	♂ ⊥ ♄
	15 10	☽ ⊥ ☿
	21 03	☽ ⊻ ♆
	21 06	☽ ⚹ ♄
	21 07	☿ ⊻ ♀
27 Friday	02 16	☽ ⊻ ☿
	03 00	☽ ☌ ♀
	03 20	☽ □ ♇
	03 50	☽ ∥ ♀
	06 18	☽ ⊥ ☉
	06 29	♀ □ ♇
	08 38	☽ ⊥ ♆
	09 54	☽ ⊻ ♂
	12 12	☽ ⊻ ♃
	18 56	☽ ⊻ ☉
	20 21	☽ Q ♄
	22 19	☽ ⊥ ♂
28 Saturday	00 02	☽ ⊥ ♃
	02 11	☽ ∠ ♆
	03 58	☽ ⋕ ♄
	12 52	☉ Q ♆
	13 50	☽ ☌ ♅
	15 02	☽ ∥ ♅
	16 40	☽ ∥ ☿
	17 08	☽ ∠ ♂
	17 58	☽ ∠ ♃
29 Sunday	07 49	☽ ⚹ ♆
	07 51	☽ □ ♄
	10 31	♂ ☌ ♃
	11 15	☽ ☌ ☿
	11 36	♂ ⋕ ♃
	14 11	☽ △ ♇
	15 28	♀ ⊥ ♆
	18 51	☿ ∥ ♅
	20 15	☽ ⊻ ♀
	23 38	☽ ∥ ☉
30 Monday	00 12	☽ ⚹ ♃
	00 55	☽ ⚹ ♂
	05 04	☽ ⋕ ♇
	07 49	☽ Q ♆
	09 34	☽ ⊥ ♀
	11 30	☽ ☌ ☉
	20 13	☽ ⚼ ♇
31 Tuesday	00 39	☽ Q ♃
	02 00	☽ ⊻ ♅
	02 38	☽ Q ♂
	05 40	☽ ∠ ♀
	14 11	☽ ⊥ ♅
	14 21	☽ ± ♇
	18 34	♀ ⊻ ♃
	18 37	♂ ∥ ♃
	20 10	☽ □ ♆
	20 10	☽ △ ♄
	22 03	♄ ⊻ ♆
	22 19	☽ ⊻ ☿

All ephemeris data is given at 12.00 UT and the Moon's longitude is additionally given for 24.00 UT

JUNE 2022

LONGITUDES

Date	Sidereal time h m s	Sun ☉ ° ' "	Moon ☽ ° ' "	Moon ☽ 24.00 ° ' "	Mercury ☿ ° '	Venus ♀ ° '	Mars ♂ ° '	Jupiter ♃ ° '	Saturn ♄ ° '	Uranus ♅ ° '	Neptune ♆ ° '	Pluto ♇ ° '
01	04 39 49	10 ♊ 59 26	03 ♋ 02 56	08 ♋ 57 40	26 ♉ 13	04 ♉ 33	05 ♈ 36	03 ♈ 49	25 ♒ 15	16 ♉ 20	25 ♓ 15	28 ♑ 21
02	04 43 46	11 56 57	14 ♋ 52 23	20 ♋ 47 23	26 R 07	05 44	06 20	03 58	25 15	16 23	25 16	28 R 21
03	04 47 42	12 54 27	26 ♋ 43 02	02 ♌ 39 42	26 D 05	06 54	07 04	04 08	25 15	16 26	25 16	28 20
04	04 51 39	13 51 56	08 ♌ 37 48	14 ♌ 37 46	26 08	08 05	07 49	04 17	25 15	16 29	25 17	28 19
05	04 55 36	14 49 24	20 ♌ 40 02	26 ♌ 45 07	26 16	09 15	08 33	04 26	25 R 15	16 33	25 18	28 18
06	04 59 32	15 46 50	02 ♍ 53 30	09 ♍ 05 41	26 28	10 26	09 17	04 35	25 15	16 36	25 19	28 17
07	05 03 29	16 44 15	15 ♍ 22 12	21 ♍ 43 32	26 44	11 37	10 01	04 44	25 15	16 39	25 20	28 16
08	05 07 25	17 41 39	28 ♍ 10 10	04 ♎ 42 34	27 05	12 48	10 45	04 52	25 15	16 42	25 20	28 15
09	05 11 22	18 39 02	11 ♎ 21 08	18 ♎ 06 09	27 30	13 58	11 29	05 01	25 14	16 45	25 21	28 14
10	05 15 18	19 36 24	24 ♎ 57 51	01 ♏ 56 21	28 00	15 09	12 13	05 09	25 14	16 48	25 21	28 13
11	05 19 15	20 33 45	09 ♏ 01 35	16 ♏ 13 22	28 34	16 20	12 57	05 17	25 13	16 51	25 22	28 12
12	05 23 11	21 31 05	23 ♏ 31 19	00 ♐ 54 52	29 12	17 31	13 41	05 25	25 12	16 54	25 22	28 11
13	05 27 08	22 28 24	08 ♐ 23 14	15 ♐ 55 31	29 ♉ 54	18 42	14 25	05 33	25 12	16 57	25 23	28 10
14	05 31 05	23 25 42	23 ♐ 30 36	01 ♑ 07 19	00 ♊ 40	19 53	15 09	05 41	25 11	17 00	25 23	28 09
15	05 35 01	24 23 00	08 ♑ 44 21	16 ♑ 20 26	01 30	21 04	15 52	05 49	25 10	17 03	25 24	28 07
16	05 38 58	25 20 17	23 ♑ 54 18	01 ♒ 24 47	02 24	22 15	16 36	05 57	25 09	17 06	25 24	28 06
17	05 42 54	26 17 33	08 ♒ 50 50	16 ♒ 11 35	03 21	23 26	17 20	06 04	25 07	17 09	25 25	28 05
18	05 46 51	27 14 50	23 ♒ 26 22	00 ♓ 34 42	04 23	24 37	18 03	06 11	25 06	17 12	25 25	28 04
19	05 50 47	28 12 06	07 ♓ 36 16	14 ♓ 31 00	05 28	25 49	18 46	06 18	25 05	17 14	25 25	28 03
20	05 54 44	29 ♊ 09 21	21 ♓ 18 57	28 ♓ 00 17	06 37	27 00	19 30	06 25	25 03	17 17	25 26	28 02
21	05 58 40	00 ♋ 06 37	04 ♈ 35 19	11 ♈ 04 28	07 49	28 11	20 13	06 32	25 02	17 20	25 26	28 00
22	06 02 37	01 03 52	17 ♈ 28 08	23 ♈ 46 52	09 04	29 ♉ 23	20 56	06 39	25 00	17 23	25 26	27 59
23	06 06 34	02 01 07	00 ♉ 01 08	06 ♉ 11 28	10 24	00 ♊ 34	21 39	06 45	24 59	17 25	25 26	27 58
24	06 10 30	02 58 22	12 ♉ 18 22	18 ♉ 22 20	11 46	01 45	22 22	06 52	24 57	17 28	25 26	27 57
25	06 14 27	03 55 37	24 ♉ 23 51	00 ♊ 23 20	13 12	02 57	23 05	06 58	24 55	17 31	25 26	27 55
26	06 18 23	04 52 51	06 ♊ 21 11	12 ♊ 17 47	14 42	04 08	23 48	07 04	24 53	17 33	25 27	27 54
27	06 22 20	05 50 06	18 ♊ 13 28	24 ♊ 08 32	16 14	05 20	24 31	07 10	24 51	17 36	25 27	27 53
28	06 26 16	06 47 21	00 ♋ 03 17	05 ♋ 57 57	17 50	06 32	25 14	07 15	24 49	17 38	25 R 27	27 51
29	06 30 13	07 44 35	11 ♋ 52 46	17 ♋ 47 58	19 29	07 43	25 57	07 21	24 46	17 41	25 27	27 50
30	06 34 09	08 ♋ 41 49	23 ♋ 43 46	29 ♋ 40 23	21 ♊ 12	08 ♊ 55	26 ♈ 39	07 ♈ 26	24 ♒ 44	17 ♉ 43	25 ♓ 26	27 ♑ 49

Moon Nodes and Latitude

Date	Moon True ☊ ° '	Moon Mean ☊ ° '	Moon ☽ Latitude ° '
01	22 ♉ 28	21 ♉ 31	03 N 25
02	22 R 26	21 28	04 10
03	22 22	21 24	04 44
04	22 19	21 21	05 06
05	22 17	21 18	05 14
06	22 15	21 15	05 09
07	22 14	21 12	04 49
08	22 D 15	21 08	04 15
09	22 16	21 05	03 26
10	22 17	21 02	02 25
11	22 19	20 59	01 N 12
12	22 R 19	20 56	00 S 07
13	22 18	20 53	01 27
14	22 16	20 49	02 42
15	22 12	20 46	03 47
16	22 08	20 43	04 35
17	22 03	20 40	05 03
18	21 59	20 37	05 12
19	21 56	20 34	05 00
20	21 54	20 30	04 32
21	21 D 54	20 27	03 49
22	21 55	20 24	02 57
23	21 57	20 21	01 57
24	21 58	20 18	00 S 52
25	21 R 58	20 14	00 N 13
26	21 57	20 11	01 17
27	21 54	20 08	02 17
28	21 49	20 05	03 11
29	21 42	20 02	03 56
30	21 ♉ 34	19 ♉ 59	04 N 32

DECLINATIONS

Date	Sun ☉ ° '	Moon ☽ ° '	Mercury ☿ ° '	Venus ♀ ° '	Mars ♂ ° '	Jupiter ♃ ° '	Saturn ♄ ° '	Uranus ♅ ° '	Neptune ♆ ° '	Pluto ♇ ° '
01	22 N 05	26 N 49	15 N 47	11 N 09	00 N 37	00 N 27	14 S 09	16 N 23	02 S 57	22 S 29
02	22 13	26 45	15 39	11 33	00 54	00 30	14 09	16 24	02 57	22 30
03	22 21	25 27	15 33	11 56	01 11	00 34	14 10	16 24	02 57	22 30
04	22 28	23 01	15 29	12 19	01 28	00 37	14 10	16 25	02 56	22 30
05	22 34	19 34	15 27	12 42	01 46	00 40	14 10	16 26	02 56	22 30
06	22 41	15 15	15 28	13 05	02 03	00 44	14 10	16 27	02 56	22 31
07	22 47	10 13	15 30	13 27	02 20	00 47	14 10	16 28	02 56	22 31
08	22 52	04 N 38	15 34	13 50	02 37	00 50	14 11	16 29	02 55	22 31
09	22 57	01 S 20	15 41	14 11	02 54	00 53	14 11	16 30	02 55	22 32
10	23 02	07 25	15 49	14 33	03 11	00 57	14 11	16 31	02 55	22 32
11	23 06	13 22	15 59	14 54	03 28	01 00	14 12	16 32	02 55	22 32
12	23 10	18 46	16 10	15 15	03 45	01 03	14 12	16 32	02 55	22 33
13	23 13	23 08	16 23	15 36	04 01	01 05	14 13	16 33	02 54	22 33
14	23 16	25 59	16 38	15 56	04 18	01 08	14 13	16 34	02 54	22 34
15	23 19	26 55	16 54	16 16	04 35	01 11	14 14	16 35	02 54	22 34
16	23 21	25 50	17 10	16 35	04 51	01 14	14 14	16 36	02 54	22 34
17	23 23	22 55	17 29	16 55	05 08	01 17	14 15	16 37	02 54	22 35
18	23 25	18 36	17 47	17 13	05 24	01 19	14 15	16 37	02 54	22 35
19	23 26	13 21	18 07	17 32	05 41	01 22	14 16	16 38	02 54	22 35
20	23 26	07 36	18 28	17 50	05 57	01 24	14 16	16 39	02 54	22 36
21	23 26	01 S 41	18 49	18 07	06 13	01 27	14 17	16 40	02 54	22 36
22	23 26	04 N 08	19 10	18 24	06 29	01 29	14 18	16 40	02 54	22 36
23	23 25	09 40	19 32	18 41	06 45	01 31	14 19	16 41	02 54	22 37
24	23 24	14 42	19 54	18 57	07 01	01 34	14 19	16 42	02 54	22 37
25	23 23	19 05	20 16	19 13	07 17	01 36	14 20	16 43	02 54	22 37
26	23 21	22 38	20 38	19 29	07 33	01 38	14 21	16 43	02 54	22 38
27	23 19	25 12	20 59	19 43	07 49	01 40	14 22	16 44	02 54	22 38
28	23 16	26 37	21 20	19 58	08 04	01 42	14 23	16 45	02 54	22 39
29	23 13	26 50	21 41	20 12	08 20	01 44	14 24	16 46	02 54	22 39
30	23 N 09	25 N 49	22 N 00	20 N 25	08 N 35	01 N 46	14 S 25	16 N 46	02 S 54	22 S 39

ZODIAC SIGN ENTRIES

Date	h m	Planets
01	05 49	☽ ♋
03	18 38	☽ ♌
06	06 22	☽ ♍
08	15 23	☽ ♎
10	20 41	☽ ♏
12	22 31	☽ ♐
13	15 27	☿ ♊
14	22 14	☽ ♑
16	21 44	☽ ♒
18	23 01	☽ ♓
21	03 37	☽ ♈
21	09 14	☉ ♋
23	00 34	♀ ♊
23	11 58	☽ ♉
25	23 13	☽ ♊
28	11 53	☽ ♋

LATITUDES

Date	Mercury ☿ ° '	Venus ♀ ° '	Mars ♂ ° '	Jupiter ♃ ° '	Saturn ♄ ° '	Uranus ♅ ° '	Neptune ♆ ° '	Pluto ♇ ° '
01	03 S 37	02 S 00	01 S 45	01 S 10	01 S 07	00 S 21	01 S 10	02 S 02
04	03 55	01 59	01 46	01 11	01 08	00 21	01 10	02 03
07	04 02	01 57	01 47	01 12	01 08	00 21	01 10	02 03
10	04 00	01 55	01 47	01 12	01 09	00 21	01 10	02 04
13	03 49	01 52	01 48	01 13	01 09	00 21	01 10	02 04
16	03 32	01 48	01 48	01 14	01 10	00 22	01 10	02 05
19	03 08	01 44	01 49	01 14	01 10	00 22	01 10	02 05
22	02 40	01 39	01 49	01 15	01 11	00 22	01 11	02 05
25	02 08	01 33	01 49	01 16	01 11	00 22	01 11	02 05
28	01 33	01 27	01 49	01 17	01 12	00 22	01 11	02 06
31	00 S 57	01 S 21	01 S 49	01 S 18	01 S 12	00 S 22	01 S 11	02 S 06

DATA

Julian Date	2459732
Delta T	+71 seconds
Ayanamsa	24° 09' 59"
Synetic vernal point	04° ♓ 57' 00"
True obliquity of ecliptic	23° 26' 16"

MOON'S PHASES, APSIDES AND POSITIONS ☽

Date	h m	Phase	Longitude ° '	Eclipse Indicator
07	14 48	☽	16 ♍ 51	
14	11 52	○	23 ♐ 25	
21	03 11	☾	29 ♓ 46	
29	02 52	●	07 ♋ 23	

Day	h m		
02	01 23	Apogee	
14	23 30	Perigee	
29	06 24	Apogee	
01	22 30	Max dec	26° N 56'
09	06 44	0S	
15	11 03	Max dec	26° S 55'
21	18 52	0N	
29	04 04	Max dec	26° N 54'

LONGITUDES

Date	Chiron ⚷ ° '	Ceres ⚳ ° '	Pallas ⚴ ° '	Juno ⚵ ° '	Vesta ⚶ ° '	Black Moon Lilith ⚸ ° '
01	15 ♈ 28	07 ♋ 10	14 ♉ 21	13 ♓ 18	02 ♓ 08	05 ♋ 18
11	15 ♈ 49	11 ♋ 26	18 ♉ 56	15 ♓ 49	04 ♓ 18	06 ♋ 24
21	16 ♈ 06	15 ♋ 46	23 ♉ 35	17 ♓ 59	05 ♓ 52	07 ♋ 31
31	16 ♈ 17	20 ♋ 09	28 ♉ 17	19 ♓ 42	06 ♓ 44	08 ♋ 38

All ephemeris data is given at 12.00 UT and the Moon's longitude is additionally given for 24.00 UT

ASPECTARIAN

h m	Aspects
01 Wednesday	
02 30	☽ ⊼ ♇
08 30	☽ ∠ ♅
10 19	☽ ⊥ ☿
13 35	☽ □ ♃
15 24	☽ ⚹ ♀
17 31	☽ □ ♂
02 Thursday	
02 37	☽ ⚼ ♄
04 25	☽ ∠ ☿
05 33	☽ ⊻ ☉
15 05	☽ ⚹ ♅
18 26	☽ Q ♀
18 47	☽ ⊥ ☉
20 53	☽ ± ♄
03 Friday	
08 00	☿ St D
09 02	☽ ⊼ ♄
09 05	☽ △ ♆
10 44	☽ ⚹ ☿
12 16	☉ ⊥ ♀
14 37	☽ ∠ ☉
15 15	☽ ☍ ♇
15 30	☽ Q ♅
19 05	♀ Q ♄
21 18	♀ ⊻ ♂
22 21	☉ ⚼ ♇
04 Saturday	
03 09	☽ △ ♃
10 15	☽ △ ♂
10 47	☽ □ ♀
11 00	☽ Q ☿
15 20	☽ ⚼ ♆
16 01	☽ # ♇
16 10	☽ ∥ ☉
20 46	☉ # ♇
21 47	♄ St R
23 23	☽ ⚹ ☉
05 Sunday	
03 47	☽ □ ♅
09 17	☽ ± ♆
09 31	☽ ⚼ ♃
18 04	☽ ⚼ ♂
21 03	☽ ☍ ♄
21 09	☽ ⊼ ♆
23 12	☽ □ ☿
06 Monday	
01 10	☽ Q ☉
03 01	☽ ⊼ ♇
03 29	☽ ± ♃
05 43	☽ ∥ ♅
09 29	♀ ∠ ♆
10 53	☽ ∥ ☿
12 49	☽ ± ♂
14 42	☽ ± ♇
15 19	☽ ⊼ ♃
15 23	♀ ⊥ ♃
17 23	☽ # ♄
21 55	☽ ∥ ♀
07 Tuesday	
01 09	☽ ⊼ ♂
04 05	☽ △ ♀
08 00	☽ ⚼ ♇
09 36	☉ ⊻ ♅
11 41	☉ Q ♃
14 26	☽ △ ♅
14 48	☽ □ ☉
19 19	♂ ∠ ♄
19 44	♂ Q ♇
08 Wednesday	
06 34	☽ ⊼ ♄
06 44	☽ ☍ ♆
09 56	☽ △ ☿
09 57	♂ ⊥ ♅
11 14	☽ ⚼ ♀
12 09	☽ △ ♇
17 40	☽ ± ♄
18 31	☽ ⚼ ♅
18 59	☽ # ♆
19 51	☽ ∥ ♂
09 Thursday	
00 26	☽ ☍ ♃
03 15	☽ ∥ ♃
05 20	☽ ± ♀
10 00	☽ ⚼ ♄
10 15	☽ # ♃
10 55	☽ ± ♅
11 35	♀ # ♄
12 16	☽ ☍ ♂
13 42	♂ # ♆
14 08	☽ ⚼ ☿
17 08	☽ ⊼ ♀
18 17	☽ ∥ ♆
18 30	☽ # ♂
21 39	☽ ⊼ ♅
10 Friday	
01 57	☽ △ ☉
06 38	☽ ± ☿
12 27	☽ △ ♄
12 41	☽ ⊼ ♆
17 27	☽ ⊼ ☿
17 36	☽ □ ♇
21 21	☿ △ ♇
23 01	☽ ± ♆
11 Saturday	
05 38	☽ ⊼ ♃
05 44	☽ ⚼ ☉
14 15	☽ ⚼ ♆
15 31	☽ ∥ ♄
15 50	☽ ± ♃
18 55	☽ ⊼ ♂
18 58	☽ # ♀
21 54	☽ ± ☉
22 58	♀ ☌ ♅
23 37	☽ # ☿
23 56	☽ Q ♇
12 Sunday	
01 05	☽ ☍ ♅
01 16	☽ ☍ ♀
01 41	☽ # ♅
05 23	☽ ± ♂
06 53	☽ ⚼ ♃
08 30	☽ ⊼ ☉
14 45	☽ □ ♄
15 01	☽ △ ♆
19 34	☽ ⚹ ♇
20 50	☽ ⚼ ♂
21 40	☽ ☍ ☿
13 Monday	
04 17	☉ ± ♇
07 26	☽ △ ♃
08 20	☽ ∥ ♇
12 36	☽ # ☉
19 36	☽ ∠ ♇
19 39	☽ Q ♄
22 05	☽ △ ♂
14 Tuesday	
00 39	☉ ⊥ ♅
01 40	☽ ⊼ ♅
05 46	☿ ∥ ♅
05 47	☽ ⊼ ♀
09 51	☽ ⊥ ♇
11 12	☽ ± ♅
11 52	☽ ☍ ☉
14 38	☽ ⚹ ♄
14 58	☽ □ ♆
16 04	☽ ± ♀
19 18	☽ ⊻ ♇
23 55	☽ ⊼ ☿
15 Wednesday	
01 26	☽ ⚼ ♅
06 19	♀ ∠ ♃
07 21	☽ □ ♃
07 26	☽ ⚼ ♀
09 55	☽ ± ☿
14 14	☽ ∠ ♄
19 21	☽ Q ♆
23 50	☽ □ ♂
16 Thursday	
01 00	☽ ⚼ ☿
01 10	☽ △ ♅
04 27	☽ ⊥ ♄
07 13	☉ △ ♄
09 09	☽ △ ♀
12 04	☽ Q ♃
12 27	♀ ∥ ♅
13 41	☉ □ ♆
13 58	☽ ⊻ ♄
14 23	☽ ⚹ ♆
14 26	☽ ⊼ ☉
18 41	☽ ☌ ♇
17 Friday	
00 41	☽ ± ☉
02 30	☽ △ ☿
05 37	♂ ⊻ ♅
06 00	☽ Q ♂
06 10	♀ ⊥ ♂
07 27	☽ ⚹ ♃
08 56	☽ # ☉
14 12	☽ ∥ ♇
14 33	☽ ∠ ♆
16 15	☽ ⚼ ☉
18 Saturday	
01 37	☽ □ ♅
02 35	☽ ⚹ ♂
05 19	☽ ⊥ ♆
08 13	☽ ∠ ♃
14 09	☽ □ ♀
14 47	☽ ☌ ♄
15 18	☽ ⊻ ♆
15 42	☽ # ☿
18 16	☽ # ♀
18 50	☽ △ ☉
19 45	☽ ⊻ ♇
21 25	☽ # ♅
21 32	♀ □ ♄
23 26	☽ ⊥ ♃
19 Sunday	
04 06	♀ ⚹ ♆
05 05	☽ ∠ ♂
05 54	☽ ⊥ ♇
07 56	☽ Q ♅
08 00	☽ □ ☿
08 03	☽ ∥ ♄
08 10	☉ ⊼ ♇
09 45	☽ ⊻ ♃
21 25	☽ ∠ ♇
21 27	☽ ⊥ ♂
23 47	☽ Q ♀
20 Monday	
04 50	☽ ⚹ ♅
07 44	☿ ⚹ ♃
08 35	☽ ⊻ ♂
18 26	☽ # ♂
18 27	☽ Q ☿
18 41	☽ ⊻ ♄
19 21	☽ ☌ ♆
23 11	☽ ⚹ ♀
21 Tuesday	
00 01	☽ ⚹ ♇
03 11	☽ □ ☉
04 29	☿ Q ♆
05 30	☽ ⊥ ♄
07 06	☽ ∥ ♆
07 51	☽ ∠ ♅
08 23	♀ △ ♇
12 58	☽ # ♃
15 37	☽ ☌ ♃
18 34	☽ ⚹ ☿
22 00	☽ Q ♇
22 02	☽ ∠ ♄
22 Wednesday	
00 32	☽ ⊥ ♅
00 54	☽ ∥ ♃
05 35	☽ ∠ ♀
06 48	☽ # ♆
11 50	☽ ⊻ ♅
15 16	☽ Q ☉
18 58	☽ ☌ ♂
22 31	☽ ∥ ♂
23 Thursday	
00 24	☽ ⊥ ♀
02 01	☽ ∠ ☿
02 19	☽ ⚹ ♄
03 10	☽ ⊻ ♆
08 02	☽ □ ♇
13 10	☽ ⊻ ♀
14 45	☽ ⊥ ♆
16 12	☽ ⚹ ☉
21 33	☽ ⊥ ☿
22 40	☉ ∠ ♅
24 Friday	
01 13	☽ ⊻ ♃
01 30	☽ Q ♄
08 19	☽ ∠ ♆
10 06	☽ # ♄
10 48	☽ ⊻ ☿
13 06	☽ ⊥ ♃
22 15	☽ ☌ ♅
22 29	☽ ∥ ♅
25 Saturday	
00 10	☽ ∠ ☉
07 06	☽ ∠ ♃
07 25	☿ ⚼ ♇
09 14	☽ ⊻ ♂
12 54	☽ ∥ ♀
13 02	☽ □ ♄
14 05	☽ ⚹ ♆
19 02	☽ △ ♇
19 40	☽ ⊥ ☉
20 10	☽ ∥ ☿
22 00	☽ ⊥ ♂
26 Sunday	
07 03	☽ ☌ ♀
08 47	☽ ⊻ ☉
11 58	☽ # ♇
13 27	☽ ⚹ ♃
14 12	☽ Q ♆
17 16	☽ ∠ ♂
17 42	☽ ∥ ☉
27 Monday	
01 12	☽ ⚼ ♇
07 22	☽ ☌ ☿
10 44	☽ ⊻ ♅
13 55	☽ Q ♃
19 24	☽ ± ♇
22 29	♂ ⚹ ♄
22 56	☽ ⊥ ♅
28 Tuesday	
01 23	☽ △ ♄
01 35	☽ ⚹ ♂
02 38	☽ □ ♆
07 33	☽ ⊼ ♇
07 57	♆ St R
09 02	☿ ⊻ ♅
17 16	☽ ∠ ♅
19 08	♂ ⊻ ♆
29 Wednesday	
00 59	☉ □ ♃
02 37	☽ ⊻ ♀
02 44	☽ □ ♃
02 52	☽ ☌ ☉
03 30	☽ Q ♂
03 52	♀ ⚹ ♃
06 24	♀ Q ♆
07 44	☽ ⚼ ♄
09 52	☿ Q ♃
14 09	☉ ⊻ ♀
16 09	☽ ⊥ ♀
23 48	☽ ⚹ ♅
30 Thursday	
01 55	☽ ± ♄
06 00	☽ ⊻ ☿
07 31	☉ Q ♂
12 25	☽ ∠ ♀
14 01	☽ ⊼ ♄
15 28	☽ △ ♆
18 17	☽ □ ♂
20 12	☽ ⊥ ☿
20 14	☽ ☍ ♇
20 26	☿ ± ♇

JULY 2022

LONGITUDES

Date	Sidereal time h m s	Sun ☉ ° ' "	Moon ☽ ° ' "	Moon ☽ 24.00 ° ' "	Mercury ☿ ° '	Venus ♀ ° '	Mars ♂ ° '	Jupiter ♃ ° '	Saturn ♄ ° '	Uranus ♅ ° '	Neptune ♆ ° '	Pluto ♇ ° '
01	06 38 06	09 ♋ 39 03	05 ♌ 38 03	11 ♌ 36 59	22 ♊ 57	10 ♊ 07	27 ♈ 22	07 ♈ 31	24 ♒ 42	17 ♉ 46	25 ♓ 26	27 ♑ 47
02	06 42 03	10 36 16	17 ♌ 37 27	23 ♌ 39 43	24 45	11 18	28 04	07 36	24 R 39	17 48	25 R 26	27 R 46
03	06 45 59	11 33 29	29 ♌ 44 07	05 ♍ 50 58	26 37	12 30	28 46	07 41	24 36	17 51	25 26	27 45
04	06 49 56	12 30 42	12 ♍ 00 38	18 ♍ 13 31	28 ♊ 31	13 42	29 ♈ 28	07 46	24 34	17 53	25 26	27 43
05	06 53 52	13 27 55	24 ♍ 30 02	00 ♎ 50 37	00 ♋ 27	14 54	00 ♉ 10	07 50	24 31	17 55	25 26	27 42
06	06 57 49	14 25 07	07 ♎ 15 44	13 ♎ 45 48	02 26	16 06	00 52	07 55	24 28	17 58	25 25	27 41
07	07 01 45	15 22 20	20 ♎ 21 17	27 ♎ 02 33	04 28	17 18	01 34	07 59	24 25	18 00	25 25	27 39
08	07 05 42	16 19 32	03 ♏ 49 57	10 ♏ 43 46	06 31	18 30	02 16	08 03	24 22	18 02	25 25	27 38
09	07 09 38	17 16 43	17 ♏ 44 09	24 ♏ 51 07	08 36	19 42	02 58	08 07	24 19	18 04	25 25	27 36
10	07 13 35	18 13 55	02 ♐ 04 31	09 ♐ 24 03	10 42	20 54	03 39	08 10	24 16	18 06	25 24	27 35
11	07 17 32	19 11 07	16 ♐ 49 10	24 ♐ 19 09	12 50	22 06	04 21	08 14	24 13	18 08	25 24	27 34
12	07 21 28	20 08 19	01 ♑ 53 04	09 ♑ 29 48	14 58	23 18	05 02	08 17	24 10	18 10	25 23	27 32
13	07 25 25	21 05 30	17 ♑ 08 04	24 ♑ 46 31	17 07	24 30	05 43	08 20	24 06	18 12	25 23	27 31
14	07 29 21	22 02 43	02 ♒ 23 45	09 ♒ 58 25	19 16	25 42	06 24	08 23	24 03	18 14	25 22	27 29
15	07 33 18	22 59 55	17 ♒ 29 14	24 ♒ 55 03	21 26	26 55	07 05	08 26	24 00	18 16	25 22	27 28
16	07 37 14	23 57 08	02 ♓ 14 57	09 ♓ 28 11	23 34	28 07	07 46	08 28	23 56	18 18	25 21	27 26
17	07 41 11	24 54 21	16 ♓ 34 17	23 ♓ 32 58	25 43	29 ♊ 19	08 27	08 30	23 52	18 20	25 21	27 25
18	07 45 07	25 51 35	00 ♈ 24 08	07 ♈ 07 56	27 50	00 ♋ 32	09 08	08 32	23 49	18 22	25 20	27 24
19	07 49 04	26 48 49	13 ♈ 44 38	20 ♈ 14 37	29 ♋ 57	01 44	09 48	08 34	23 45	18 23	25 19	27 22
20	07 53 01	27 46 05	26 ♈ 38 23	02 ♉ 56 29	02 ♌ 02	02 56	10 29	08 36	23 41	18 25	25 19	27 21
21	07 56 57	28 43 21	09 ♉ 09 33	15 ♉ 18 11	04 07	04 09	11 09	08 38	23 38	18 27	25 18	27 19
22	08 00 54	29 ♋ 40 38	21 ♉ 23 03	27 ♉ 24 45	06 10	05 21	11 49	08 39	23 34	18 28	25 17	27 18
23	08 04 50	00 ♌ 37 55	03 ♊ 23 54	09 ♊ 21 05	08 11	06 34	12 29	08 40	23 30	18 30	25 17	27 16
24	08 08 47	01 35 14	15 ♊ 16 50	21 ♊ 11 39	10 11	07 47	13 09	08 41	23 26	18 31	25 16	27 15
25	08 12 43	02 32 33	27 ♊ 05 59	03 ♋ 00 15	12 09	08 59	13 49	08 42	23 22	18 33	25 15	27 14
26	08 16 40	03 29 54	08 ♋ 54 48	14 ♋ 49 56	14 06	10 12	14 29	08 43	23 18	18 34	25 14	27 12
27	08 20 36	04 27 14	20 ♋ 45 57	26 ♋ 43 05	16 01	11 25	15 08	08 43	23 14	18 36	25 13	27 11
28	08 24 33	05 24 36	02 ♌ 41 30	08 ♌ 41 23	17 54	12 37	15 48	08 43	23 09	18 37	25 12	27 09
29	08 28 30	06 21 59	14 ♌ 42 53	20 ♌ 46 09	19 46	13 50	16 27	08 R 43	23 05	18 38	25 11	27 08
30	08 32 26	07 19 22	26 ♌ 51 18	02 ♍ 58 28	21 36	15 03	17 06	08 43	23 01	18 40	25 10	27 06
31	08 36 23	08 ♌ 16 45	09 ♍ 07 49	15 ♍ 19 31	23 ♌ 24	16 ♋ 16	17 ♉ 45	08 ♈ 42	22 ♒ 57	18 ♉ 41	25 ♓ 09	27 ♑ 05

Date	Moon True ☊ ° '	Moon Mean ☊ ° '	Moon ☽ Latitude ° '
01	21 ♉ 25	19 ♉ 55	04 N 55
02	21 R 16	19 52	05 06
03	21 08	19 49	05 03
04	21 02	19 46	04 47
05	20 58	19 43	04 16
06	20 56	19 40	03 32
07	20 D 56	19 36	02 37
08	20 57	19 33	01 31
09	20 58	19 30	00 N 17
10	20 R 57	19 27	00 S 59
11	20 55	19 24	02 14
12	20 50	19 20	03 20
13	20 43	19 17	04 14
14	20 35	19 14	04 49
15	20 26	19 11	05 03
16	20 17	19 08	04 57
17	20 10	19 05	04 32
18	20 06	19 01	03 52
19	20 03	18 58	03 01
20	20 02	18 55	02 01
21	20 D 02	18 52	00 S 58
22	20 R 03	18 49	00 N 07
23	20 02	18 46	01 10
24	19 59	18 42	02 10
25	19 54	18 39	03 03
26	19 46	18 36	03 48
27	19 35	18 33	04 24
28	19 23	18 30	04 48
29	19 10	18 26	04 59
30	18 57	18 23	04 57
31	18 ♉ 45	18 ♉ 20	04 N 42

DECLINATIONS

Date	Sun ☉ ° '	Moon ☽ ° '	Mercury ☿ ° '	Venus ♀ ° '	Mars ♂ ° '	Jupiter ♃ ° '	Saturn ♄ ° '	Uranus ♅ ° '	Neptune ♆ ° '	Pluto ♇ ° '
01	23 N 05	23 N 38	22 N 19	20 N 38	08 N 50	01 N 48	14 S 26	16 N 47	02 S 54	22 S 40
02	23 01	20 24	22 36	20 50	09 05	01 49	14 27	16 48	02 54	22 40
03	22 56	16 17	22 51	21 02	09 20	01 51	14 28	16 48	02 54	22 41
04	22 51	11 28	23 05	21 13	09 35	01 53	14 29	16 49	02 54	22 41
05	22 45	06 06	23 18	21 23	09 50	01 54	14 30	16 50	02 54	22 41
06	22 39	00 N 22	23 28	21 33	10 05	01 56	14 31	16 50	02 54	22 42
07	22 33	05 S 32	23 36	21 43	10 19	01 57	14 32	16 51	02 55	22 42
08	22 26	11 22	23 42	21 52	10 34	01 58	14 33	16 51	02 55	22 42
09	22 19	16 51	23 45	22 00	10 48	01 59	14 34	16 52	02 55	22 43
10	22 12	21 33	23 45	22 08	11 03	02 01	14 36	16 53	02 55	22 43
11	22 04	25 00	23 43	22 15	11 17	02 02	14 37	16 53	02 55	22 44
12	21 56	26 46	23 38	22 21	11 31	02 03	14 38	16 54	02 56	22 44
13	21 47	26 32	23 30	22 27	11 44	02 04	14 39	16 54	02 56	22 44
14	21 38	24 19	23 20	22 32	11 58	02 05	14 41	16 55	02 56	22 45
15	21 29	20 24	23 07	22 37	12 12	02 05	14 42	16 55	02 56	22 45
16	21 19	15 18	22 51	22 41	12 25	02 06	14 43	16 56	02 57	22 45
17	21 09	09 29	22 33	22 44	12 39	02 07	14 44	16 56	02 57	22 46
18	20 58	03 S 24	22 12	22 47	12 52	02 07	14 46	16 57	02 57	22 46
19	20 48	02 N 39	21 49	22 49	13 05	02 08	14 47	16 57	02 57	22 47
20	20 36	08 23	21 24	22 51	13 18	02 08	14 49	16 58	02 58	22 47
21	20 25	13 38	20 57	22 51	13 30	02 08	14 50	16 58	02 58	22 47
22	20 13	18 13	20 28	22 51	13 43	02 09	14 51	16 59	02 58	22 48
23	20 01	21 59	19 57	22 51	13 56	02 09	14 53	16 59	02 59	22 48
24	19 48	24 47	19 25	22 50	14 08	02 09	14 54	16 59	02 59	22 48
25	19 35	26 27	18 51	22 48	14 20	02 09	14 56	17 00	02 59	22 49
26	19 22	26 56	18 17	22 45	14 32	02 09	14 57	17 00	03 00	22 49
27	19 09	26 11	17 40	22 42	14 44	02 09	14 59	17 01	03 00	22 50
28	18 55	24 14	17 03	22 38	14 56	02 09	15 00	17 01	03 01	22 50
29	18 41	21 11	16 25	22 34	15 07	02 08	15 02	17 01	03 01	22 50
30	18 26	17 13	15 47	22 29	15 19	02 08	15 03	17 02	03 01	22 51
31	18 N 12	12 N 30	15 N 07	22 N 23	15 N 30	02 N 08	15 S 05	17 N 02	03 S 02	22 S 51

ZODIAC SIGN ENTRIES

Date	h m	Planets
01	00 40	☽ ♌
03	12 31	☽ ♍
05	06 04	♂ ♉
05	06 25	☿ ♋
05	22 25	☽ ♎
08	05 15	☽ ♏
10	08 34	☽ ♐
12	09 01	☽ ♑
14	08 13	☽ ♒
16	08 18	☽ ♓
18	01 32	♀ ♋
18	11 17	☽ ♈
19	12 35	☿ ♌
20	18 23	☽ ♉
22	20 07	☉ ♌
23	05 11	☽ ♊
25	17 54	☽ ♋
28	06 36	☽ ♌
30	18 11	☽ ♍

LATITUDES

Date	Mercury ☿ ° '	Venus ♀ ° '	Mars ♂ ° '	Jupiter ♃ ° '	Saturn ♄ ° '	Uranus ♅ ° '	Neptune ♆ ° '	Pluto ♇ ° '
01	00 S 57	01 S 21	01 S 49	01 S 18	01 S 12	00 S 22	01 S 11	02 S 06
04	00 S 20	01 14	01 49	01 19	01 13	00 22	01 11	02 06
07	00 N 14	01 07	01 48	01 19	01 13	00 22	01 11	02 07
10	00 45	01 00	01 48	01 20	01 14	00 22	01 11	02 07
13	01 11	00 52	01 47	01 21	01 14	00 22	01 12	02 07
16	01 30	00 44	01 47	01 22	01 15	00 22	01 12	02 08
19	01 42	00 36	01 46	01 23	01 15	00 22	01 12	02 08
22	01 48	00 28	01 45	01 24	01 16	00 22	01 12	02 08
25	01 47	00 20	01 44	01 25	01 16	00 22	01 12	02 09
28	01 40	00 12	01 43	01 26	01 16	00 22	01 12	02 09
31	01 N 29	00 S 04	01 S 41	01 S 26	01 S 17	00 S 22	01 S 12	02 S 09

DATA

Julian Date	2459762
Delta T	+71 seconds
Ayanamsa	24° 10' 05"
Synetic vernal point	04° ♓ 56' 54"
True obliquity of ecliptic	23° 26' 16"

MOON'S PHASES, APSIDES AND POSITIONS ☽

Date	h m	Phase	Longitude ° '	Eclipse Indicator
07	02 14	☽	14 ♎ 59	
13	18 38	○	21 ♑ 21	
20	14 19	☾	27 ♈ 52	
28	17 55	●	05 ♌ 39	

Day	h m		
13	09 12	Perigee	
26	10 36	Apogee	
06	13 31	0S	
12	21 17	Max dec	26° S 55'
19	01 23	0N	
26	09 18	Max dec	26° N 57'

LONGITUDES

Date	Chiron ⚷ ° '	Ceres ⚳ ° '	Pallas ⚴ ° '	Juno ⚵ ° '	Vesta ⚶ ° '	Black Moon Lilith ⚸ ° '
01	16 ♈ 17	20 ♋ 09	28 ♉ 17	19 ♓ 42	06 ♓ 44	08 ♋ 38
11	16 ♈ 24	24 ♋ 34	03 ♊ 01	20 ♓ 53	06 ♓ 51	09 ♋ 45
21	16 ♈ 26	29 ♋ 00	07 ♊ 47	21 ♓ 28	06 ♓ 10	10 ♋ 52
31	16 ♈ 22	03 ♌ 28	12 ♊ 35	21 ♓ 23	04 ♓ 43	11 ♋ 59

All ephemeris data is given at 12.00 UT and the Moon's longitude is additionally given for 24.00 UT

ASPECTARIAN

h m Aspects

01 Friday
00 09 ☽ Q ♅
13 01 ☉ ⚼ ♄
15 49 ☽ △ ♃
16 42 ☽ ∥ ☉
17 28 ☽ ∠ ☿
19 58 ☽ # ♇
20 46 ☽ ⊻ ☉
21 39 ☽ ⚼ ♆
21 48 ☽ ∥ ☿
21 59 ☽ ✱ ♀
23 09 ☿ ⊥ ♅

02 Saturday
02 14 ♂ □ ♇
09 18 ☽ ∥ ♀
09 48 ☽ ⊥ ☉
10 39 ☿ △ ♄
12 22 ☽ □ ♅
15 37 ☽ ± ♆
18 43 ☿ # ♇
20 53 ☿ □ ♆
21 58 ☽ ⚼ ♃

03 Sunday
00 32 ☽ Q ♀
01 55 ☽ ☍ ♄
03 31 ☽ ⊼ ♆
04 43 ☽ ✱ ☿
05 12 ☽ ∠ ☉
08 05 ☽ ⊼ ♇
09 14 ☽ ∥ ♅
09 59 ☽ △ ♂
15 52 ☽ ± ♃
16 45 ♀ ⚼ ♇
17 45 ☉ ∥ ☿
19 52 ☽ ± ♇
21 27 ☽ # ♄

04 Monday
02 13 ☿ ⊼ ♇
03 41 ☽ ⊼ ♃
08 33 ☽ Q ☿
13 03 ☽ ✱ ☉
13 23 ☽ ⚼ ♇
15 37 ☽ □ ♀
17 03 ☽ ⚼ ♂
20 15 ☽ ∥ ♂
23 23 ☽ △ ♅

05 Tuesday
06 37 ☿ ✱ ♂
11 20 ☽ ± ♂
12 02 ☽ ⊼ ♄
13 46 ☽ ☍ ♆
13 59 ☽ Q ☉
18 03 ☽ △ ♇
23 21 ☽ ± ♄
23 22 ☽ ⊼ ♂

06 Wednesday
01 13 ♀ ∠ ♂
01 20 ☽ □ ☿
01 31 ☽ # ♆
03 55 ☉ # ♇
03 57 ☽ ⚼ ♅
05 37 ☽ ∥ ♃
13 13 ☽ ☍ ♃
16 05 ☽ ⚼ ♄
18 19 ☿ ∠ ♅
20 43 ☽ ± ♅
21 24 ☽ # ♃

07 Thursday
01 22 ☽ ∥ ♆
02 14 ☽ □ ☉
05 54 ☽ △ ♀
06 50 ♂ ⊥ ♆
07 43 ☽ ⊼ ♅
19 18 ☽ △ ♄
21 06 ☽ ⊼ ♆

08 Friday
01 04 ☽ □ ♇
02 27 ♀ ⊻ ♅
07 45 ☽ ± ♆
08 28 ☽ # ♂
09 06 ☽ ☍ ♂
11 21 ☽ ⚼ ♀
17 31 ☽ △ ☿
19 23 ☽ ⊼ ♃
23 27 ☽ ⚼ ♆

09 Saturday
01 42 ☽ ∥ ♄
04 27 ☽ ± ♀
05 47 ☽ ± ♃
06 14 ☿ □ ♃
08 23 ☽ Q ♇
11 10 ☽ △ ☉
12 06 ☽ # ♅
12 34 ☽ ☍ ♅
15 38 ☽ ⊼ ♀
20 04 ☿ ⚼ ♄
20 45 ♀ Q ♃
21 07 ☽ ⚼ ♃
23 04 ☽ □ ♄
23 36 ☽ ⚼ ☿

10 Sunday
00 56 ☽ △ ♆
04 34 ☽ ✱ ♇
08 39 ☉ ✱ ♅
12 54 ☉ ± ♄
14 02 ☽ ⚼ ☉
14 44 ☽ ⊼ ♂
15 35 ☽ # ♀
15 45 ☽ # ☉
17 04 ☽ ± ☿
18 20 ☉ ∥ ♀
19 08 ☽ ∥ ♇
22 02 ☽ △ ♃

11 Monday
01 01 ☽ ± ♂
01 28 ♀ ± ♇
01 58 ☽ # ☿
04 29 ☽ ⊼ ☿
04 36 ☽ Q ♄
05 08 ☽ ∠ ♇
05 44 ☽ ± ☉
14 07 ☽ ⊼ ♅
16 03 ☽ ⊼ ☉
16 15 ☽ ⚼ ♂
19 35 ☽ ⊥ ♇
21 12 ☽ ☍ ♀
23 44 ☽ ± ♅
23 48 ☽ ✱ ♄

12 Tuesday
01 42 ☽ □ ♆
05 07 ☽ ⊻ ♇
14 02 ☽ ⚼ ♅
17 12 ☽ △ ♂
22 07 ☽ □ ♃
23 26 ☽ ∠ ♄

13 Wednesday
04 28 ♀ △ ♄
05 55 ♀ ⊥ ♅
06 07 ☽ Q ♆
11 59 ☽ ☍ ☿
13 31 ☽ ⊥ ♄
13 41 ☽ △ ♅
18 38 ☽ ☍ ☉
21 50 ☿ Q ♂
22 43 ☿ ± ♄
22 55 ☽ ⊻ ♄

14 Thursday
00 17 ☿ ✱ ♅
00 27 ♂ Q ♄
00 34 ☽ ⊼ ♀
00 57 ☽ ✱ ♆
02 30 ☽ Q ♃
04 17 ☽ ☌ ♇
05 24 ♀ □ ♆
10 49 ☽ ± ♀
18 39 ☽ □ ♂
19 21 ☽ # ☿
21 30 ☽ ✱ ♃
22 42 ☽ ∥ ♇
23 44 ☽ # ♀

15 Friday
00 38 ☽ ∠ ♆
02 18 ☽ ⚼ ♀
05 58 ☽ # ☉
13 16 ☽ □ ♅
15 01 ☽ ⊥ ♆
19 25 ☽ ⊼ ☿
21 30 ☽ ⊼ ☉
21 36 ☽ ∠ ♃
22 27 ☽ ☌ ♄
22 51 ♀ ⊼ ♇

16 Saturday
00 43 ☽ ⊻ ♆
00 53 ☽ Q ♂
04 07 ☽ ⊻ ♇
04 36 ☽ △ ♀
04 47 ☽ # ♅
06 51 ☽ ± ☿
07 57 ☽ ± ☉
11 34 ☉ ⊼ ♄
12 22 ☽ ⊥ ♃
13 58 ☽ ⊥ ♇
14 28 ☽ ∥ ♄
15 55 ☿ ⊼ ♄
18 43 ☽ Q ♅
19 38 ☉ ☌ ☿
19 43 ☿ # ♇
21 37 ☽ ✱ ♂
22 21 ☽ ⊻ ♃
23 28 ☿ ∥ ♀
23 40 ☽ # ♂
23 56 ☽ ⚼ ☉

17 Sunday
00 21 ☽ ⚼ ☿
04 58 ☽ ∠ ♇
07 52 ☿ △ ♆
13 58 ♂ ⊻ ♃
15 01 ☽ ✱ ♅
22 55 ☉ △ ♆

18 Monday
00 27 ☽ ∠ ♂
00 31 ☽ ⊻ ♄
01 38 ♀ # ♇
03 07 ☽ ☌ ♆
03 26 ☽ △ ☉
06 39 ☽ △ ☿
06 43 ☽ ✱ ♇
07 01 ☿ ☍ ♇
10 58 ☽ ⊥ ♄
12 14 ☽ □ ♀
13 44 ☽ ∥ ♆
17 00 ☽ # ♃
17 06 ☽ ⊥ ♂
17 16 ☽ ∠ ♅

19 Tuesday
02 35 ☽ ☌ ♃
02 58 ☽ ∠ ♄
04 03 ☽ Q ♇
04 27 ☽ ⊻ ♂
09 31 ☽ ⊥ ♅
09 54 ☽ ∥ ♃
13 15 ☽ # ♆
17 07 ☿ Q ♅
20 35 ☽ ⊻ ♅

20 Wednesday
00 11 ☽ Q ♀
01 39 ☉ ☍ ♇
06 03 ♂ ∠ ♆
06 28 ☽ ✱ ♄
09 30 ☽ ⊻ ♆
13 20 ☽ □ ♇
14 19 ☽ □ ☉
20 52 ☽ ⊥ ♆
21 44 ♀ ∠ ♅

21 Thursday
00 19 ☽ □ ☿
01 17 ☽ ✱ ♀
05 12 ☽ Q ♄
10 58 ☽ ⊻ ♃
11 17 ♃ ∠ ♄
11 21 ☽ ∥ ♂
13 02 ☿ ⊻ ♀
14 13 ☽ ∠ ♆
16 06 ☽ ☌ ♂
17 56 ☽ # ♄
22 42 ☽ ⊥ ♃

22 Friday
04 03 ☽ Q ☉
05 04 ☽ ∥ ♅
06 14 ☽ ☌ ♅
09 44 ☽ ∠ ♀
16 18 ☽ □ ♄
16 31 ☽ ∠ ♃
18 38 ☽ Q ☿
19 45 ☽ ✱ ♆
23 24 ☽ ∥ ☉
23 45 ☽ △ ♇

23 Saturday
00 00 ☽ ∥ ☿
05 40 ☽ ⊥ ♀
05 58 ☽ ✱ ☉
07 35 ☉ ∥ ☿
08 34 ☉ Q ♅
17 52 ☿ △ ♃
18 11 ☽ # ♇
18 30 ☽ ∥ ♀
19 06 ☽ ⊻ ♀
19 48 ☽ Q ♆
22 38 ☽ ✱ ♃
23 36 ☽ ✱ ☿

24 Sunday
05 52 ☽ ⚼ ♇
07 26 ☽ ⊻ ♂
12 58 ☿ ⚼ ♆
14 53 ☽ ∠ ☉
18 36 ☽ ⊻ ♅
20 20 ☽ ⊥ ♂
22 59 ☽ Q ♃

25 Monday
00 05 ☽ ± ♇
00 16 ♀ ⚼ ♄
02 13 ♀ # ♇
04 27 ☽ △ ♄
06 13 ♀ □ ♃
06 48 ☽ ⊥ ♅
08 14 ☽ □ ♆
10 46 ☽ ⊥ ☉
12 08 ☽ ∠ ☿
12 15 ☽ ⊼ ♇
15 42 ☽ ∠ ♂

26 Tuesday
00 02 ☽ ⊻ ☉
01 08 ☽ ∠ ♅
10 01 ☽ ⊥ ☿
10 45 ☽ ⚼ ♄
11 35 ☽ □ ♃
14 54 ☽ ☌ ♀
19 13 ☿ □ ♂
20 25 ♂ ⊥ ♃
23 57 ☽ ✱ ♂

27 Wednesday
00 33 ☽ ⊻ ☿
04 53 ☽ ± ♄
07 36 ☽ ✱ ♅
16 56 ☽ ⊼ ♄
20 58 ☽ △ ♆

28 Thursday
00 54 ☽ ☍ ♇
01 36 ☽ Q ♂
07 50 ☽ Q ♅
13 34 ☿ ∥ ♅
17 55 ☽ ☌ ☉
20 37 ♃ St R
21 16 ☿ □ ♅
21 59 ♂ # ♄
23 59 ☽ # ♇

29 Friday
00 03 ☽ △ ♃
01 50 ☽ ∥ ♀
03 00 ☽ ⚼ ♆
04 34 ☿ ± ♆
10 04 ☽ ⊻ ♀
14 41 ☿ ⊥ ♀
15 38 ☽ □ ♂
19 48 ☽ □ ♅
20 52 ☽ ± ♆
23 17 ☽ ⊥ ♀
23 49 ☽ ☌ ☿

30 Saturday
04 29 ☽ ☍ ♄
04 41 ☽ ∥ ☉
05 49 ☽ ⚼ ♃
08 42 ☽ ⊼ ♆
12 30 ☽ ⊼ ♇
13 02 ☽ ∥ ♅
18 58 ☽ ∠ ♀
20 57 ☽ ∥ ☿
21 41 ☽ ∥ ♂
23 22 ☽ # ♄
23 29 ☽ ± ♃

31 Sunday
00 14 ☽ ± ♇
01 08 ☿ ∥ ♂
06 05 ☿ ☍ ♄
10 12 ☽ ⊻ ☉
11 11 ☽ ⊼ ♃
13 30 ☿ # ♄
15 59 ☿ ⚼ ♃
17 43 ☽ ⚼ ♇
22 36 ☉ △ ♃
22 49 ☽ ⊥ ☉

AUGUST 2022

Date	Sidereal time h m s	Sun ☉ ° ' "	Moon ☽ ° ' "	Moon ☽ 24.00 ° ' "	Mercury ☿ ° '	Venus ♀ ° '	Mars ♂ ° '	Jupiter ♃ ° '	Saturn ♄ ° '	Uranus ♅ ° '	Neptune ♆ ° '	Pluto ⯓ ° '
01	08 40 19	09 ♌ 14 10	21 ♍ 33 46	27 ♍ 50 47	25 ♌ 11	17 ♋ 29	18 ♉ 23	08 ♈ 42	22 ♒ 53	18 ♉ 42	25 ♓ 08	27 ♑ 04
02	08 44 16	10 11 35	04 ♎ 10 49	10 ♎ 34 10	26 56	18 42	19 02	08 R 41	22 R 48	18 43	25 R 07	27 R 02
03	08 48 12	11 09 01	17 ♎ 01 08	23 ♎ 32 03	28 ♌ 39	19 55	19 40	08 40	22 44	18 44	25 06	27 01
04	08 52 09	12 06 27	00 ♏ 07 17	06 ♏ 47 09	00 ♍ 21	21 08	20 19	08 39	22 40	18 45	25 05	27 00
05	08 56 05	13 03 54	13 ♏ 32 01	20 ♏ 22 07	02 01	22 21	20 57	08 37	22 35	18 46	25 04	26 58
06	09 00 02	14 01 22	27 ♏ 17 43	04 ♐ 18 55	03 40	23 34	21 35	08 36	22 31	18 47	25 03	26 57
07	09 03 59	14 58 50	11 ♐ 25 45	18 ♐ 38 04	05 17	24 47	22 12	08 34	22 26	18 48	25 02	26 55
08	09 07 55	15 56 20	25 ♐ 55 37	03 ♑ 17 52	06 52	26 00	22 50	08 32	22 22	18 49	25 01	26 54
09	09 11 52	16 53 50	10 ♑ 44 12	18 ♑ 13 43	08 25	27 14	23 27	08 30	22 17	18 50	24 59	26 53
10	09 15 48	17 51 21	25 ♑ 45 26	03 ♒ 18 10	09 57	28 27	24 05	08 27	22 13	18 50	24 58	26 51
11	09 19 45	18 48 53	10 ♒ 50 41	18 ♒ 21 42	11 28	29 ♋ 40	24 42	08 25	22 08	18 51	24 57	26 50
12	09 23 41	19 46 26	25 ♒ 49 58	03 ♓ 14 18	12 56	00 ♌ 53	25 18	08 22	22 04	18 52	24 56	26 49
13	09 27 38	20 44 00	10 ♓ 33 40	17 ♓ 47 13	14 23	02 07	25 55	08 19	21 59	18 52	24 54	26 48
14	09 31 34	21 41 35	24 ♓ 54 17	01 ♈ 54 24	15 49	03 20	26 31	08 16	21 55	18 53	24 53	26 46
15	09 35 31	22 39 12	08 ♈ 47 20	15 ♈ 33 02	17 12	04 34	27 08	08 13	21 50	18 53	24 52	26 45
16	09 39 28	23 36 50	22 ♈ 11 38	28 ♈ 43 23	18 34	05 47	27 44	08 09	21 46	18 54	24 50	26 44
17	09 43 24	24 34 30	05 ♉ 08 44	11 ♉ 28 09	19 54	07 01	28 20	08 05	21 41	18 54	24 49	26 43
18	09 47 21	25 32 11	17 ♉ 42 14	23 ♉ 51 36	21 13	08 14	28 55	08 01	21 37	18 54	24 48	26 41
19	09 51 17	26 29 54	29 ♉ 56 54	05 ♊ 58 48	22 29	09 28	29 ♉ 31	07 57	21 33	18 55	24 46	26 40
20	09 55 14	27 27 39	11 ♊ 58 00	17 ♊ 55 08	23 44	10 41	00 ♊ 06	07 53	21 28	18 55	24 45	26 39
21	09 59 10	28 25 25	23 ♊ 50 50	29 ♊ 45 42	24 57	11 55	00 41	07 49	21 24	18 55	24 44	26 38
22	10 03 07	29 ♌ 23 13	05 ♋ 40 18	11 ♋ 35 08	26 07	13 09	01 16	07 44	21 19	18 55	24 42	26 37
23	10 07 03	00 ♍ 21 02	17 ♋ 30 41	23 ♋ 27 21	27 16	14 23	01 50	07 39	21 15	18 55	24 41	26 35
24	10 11 00	01 18 54	29 ♋ 25 31	05 ♌ 25 27	28 22	15 36	02 25	07 35	21 10	18 55	24 39	26 34
25	10 14 57	02 16 46	11 ♌ 27 26	17 ♌ 31 38	29 ♍ 26	16 50	02 59	07 29	21 06	18 R 55	24 38	26 33
26	10 18 53	03 14 41	23 ♌ 38 14	29 ♌ 47 19	00 ♎ 28	18 04	03 32	07 24	21 02	18 55	24 36	26 32
27	10 22 50	04 12 36	05 ♍ 58 58	12 ♍ 13 14	01 27	19 18	04 06	07 19	20 57	18 55	24 35	26 31
28	10 26 46	05 10 33	18 ♍ 30 09	24 ♍ 49 44	02 23	20 32	04 39	07 13	20 53	18 55	24 33	26 30
29	10 30 43	06 08 32	01 ♎ 12 00	07 ♎ 37 00	03 17	21 46	05 12	07 07	20 49	18 55	24 32	26 29
30	10 34 39	07 06 32	14 ♎ 04 46	20 ♎ 35 24	04 07	23 00	05 45	07 02	20 44	18 54	24 30	26 28
31	10 38 36	08 ♍ 04 34	27 ♎ 08 59	03 ♏ 45 38	04 ♎ 54	24 ♌ 14	06 ♊ 17	06 ♈ 56	20 ♒ 40	18 ♉ 54	24 ♓ 28	26 ♑ 27

Date	Moon True ☊ ° '	Moon Mean ☊ ° '	Moon ☽ Latitude ° '
01	18 ♉ 35	18 ♉ 17	04 N 13
02	18 R 28	18 14	03 31
03	18 24	18 11	02 38
04	18 23	18 07	01 35
05	18 D 22	18 04	00 N 26
06	18 R 22	18 01	00 S 47
07	18 21	17 58	01 59
08	18 18	17 55	03 04
09	18 12	17 52	03 59
10	18 04	17 48	04 38
11	17 54	17 45	04 58
12	17 42	17 42	04 57
13	17 32	17 39	04 37
14	17 23	17 36	03 59
15	17 16	17 32	03 08
16	17 12	17 29	02 08
17	17 10	17 26	01 S 03
18	17 D 10	17 23	00 N 03
19	17 R 10	17 20	01 07
20	17 09	17 17	02 07
21	17 07	17 13	03 01
22	17 02	17 10	03 47
23	16 55	17 07	04 23
24	16 44	17 04	04 47
25	16 33	17 01	05 00
26	16 20	16 57	04 58
27	16 07	16 54	04 43
28	15 55	16 51	04 14
29	15 46	16 48	03 32
30	15 40	16 45	02 39
31	15 ♉ 36	16 ♉ 42	01 N 36

DECLINATIONS

Date	Sun ☉ ° '	Moon ☽ ° '	Mercury ☿ ° '	Venus ♀ ° '	Mars ♂ ° '	Jupiter ♃ ° '	Saturn ♄ ° '	Uranus ♅ ° '	Neptune ♆ ° '	Pluto ⯓ ° '
01	17 N 57	07 N 13	14 N 27	22 N 16	15 N 41	02 N 07	15 S 06	17 N 02	03 S 02	22 S 51
02	17 41	01 N 34	13 47	22 09	15 52	02 06	15 08	17 03	03 03	22 52
03	17 26	04 S 16	13 06	22 01	16 03	02 06	15 09	17 03	03 03	22 52
04	17 10	10 02	12 24	21 53	16 14	02 05	15 11	17 03	03 04	22 52
05	16 54	15 30	11 43	21 44	16 24	02 04	15 12	17 03	03 04	22 53
06	16 37	20 19	11 01	21 34	16 35	02 03	15 14	17 04	03 05	22 53
07	16 20	24 06	10 19	21 24	16 45	02 02	15 15	17 04	03 05	22 53
08	16 03	26 27	09 37	21 13	16 55	02 01	15 17	17 04	03 06	22 54
09	15 46	26 58	08 55	21 02	17 05	02 00	15 18	17 04	03 06	22 54
10	15 29	25 33	08 13	20 50	17 15	01 59	15 20	17 04	03 07	22 54
11	15 11	22 17	07 32	20 37	17 24	01 58	15 21	17 05	03 07	22 55
12	14 53	17 34	06 50	20 24	17 34	01 56	15 23	17 05	03 08	22 55
13	14 35	11 52	06 09	20 10	17 43	01 55	15 25	17 05	03 08	22 55
14	14 16	05 S 41	05 28	19 55	17 52	01 54	15 26	17 05	03 09	22 56
15	13 58	00 N 36	04 47	19 40	18 01	01 52	15 28	17 05	03 09	22 56
16	13 39	06 40	04 07	19 24	18 10	01 50	15 29	17 05	03 10	22 56
17	13 20	12 15	03 27	19 08	18 18	01 49	15 31	17 05	03 10	22 57
18	13 00	17 09	02 48	18 51	18 27	01 47	15 32	17 05	03 11	22 57
19	12 41	21 14	02 09	18 34	18 35	01 45	15 34	17 05	03 12	22 57
20	12 21	24 20	01 30	18 16	18 44	01 43	15 35	17 06	03 12	22 57
21	12 01	26 19	00 53	17 58	18 52	01 41	15 37	17 06	03 13	22 58
22	11 41	27 06	00 N 16	17 39	18 59	01 39	15 38	17 06	03 13	22 58
23	11 21	26 38	00 S 20	17 20	19 07	01 37	15 40	17 06	03 14	22 58
24	11 00	24 57	00 56	17 00	19 15	01 35	15 41	17 06	03 15	22 58
25	10 40	22 08	01 30	16 40	19 22	01 33	15 43	17 06	03 15	22 59
26	10 19	18 20	02 04	16 19	19 29	01 30	15 44	17 06	03 16	22 59
27	09 58	13 42	02 36	15 58	19 37	01 28	15 45	17 06	03 16	22 59
28	09 37	08 27	03 07	15 36	19 44	01 26	15 47	17 05	03 17	23 00
29	09 15	02 N 46	03 37	15 14	19 50	01 23	15 48	17 05	03 18	23 00
30	08 54	03 S 07	04 06	14 51	19 57	01 21	15 50	17 05	03 18	23 00
31	08 N 32	08 S 58	04 S 33	14 N 28	20 N 04	01 N 18	15 S 51	17 N 05	03 S 19	23 S 00

ZODIAC SIGN ENTRIES

Date	h	m	Planets
02	04	05	☽ ♎
04	06	58	☿ ♍
04	11	47	☽ ♏
06	16	39	☽ ♐
08	18	39	☽ ♑
10	18	45	☽ ♒
11	18	30	♀ ♌
12	18	44	☽ ♓
14	20	43	☽ ♈
17	02	22	☽ ♉
19	12	06	☽ ♊
20	07	56	♂ ♊
22	00	29	☽ ♋
23	03	16	☉ ♍
24	13	09	☽ ♌
26	01	03	☿ ♎
27	00	25	☽ ♍
29	09	45	☽ ♎
31	17	11	☽ ♏

LATITUDES

Date	Mercury ☿ ° '	Venus ♀ ° '	Mars ♂ ° '	Jupiter ♃ ° '	Saturn ♄ ° '	Uranus ♅ ° '	Neptune ♆ ° '	Pluto ⯓ ° '
01	01 N 24	00 S 01	01 S 41	01 S 27	01 S 17	00 S 22	01 S 12	02 S 09
04	01 08	00 N 06	01 39	01 28	01 17	00 22	01 12	02 09
07	00 48	00 14	01 38	01 29	01 18	00 22	01 13	02 10
10	00 N 25	00 22	01 36	01 29	01 18	00 22	01 13	02 10
13	00 00	00 29	01 34	01 30	01 18	00 22	01 13	02 10
16	00 S 27	00 36	01 32	01 31	01 18	00 22	01 13	02 10
19	00 55	00 43	01 29	01 32	01 19	00 22	01 13	02 10
22	01 24	00 49	01 27	01 32	01 19	00 22	01 13	02 10
25	01 53	00 55	01 25	01 33	01 19	00 22	01 13	02 11
28	02 22	01 00	01 22	01 34	01 19	00 22	01 13	02 11
31	02 S 50	01 N 05	01 S 19	01 S 34	01 S 19	00 S 22	01 S 13	02 S 11

DATA

Julian Date	2459793
Delta T	+71 seconds
Ayanamsa	24° 10' 10"
Synetic vernal point	04° ♓ 56' 49"
True obliquity of ecliptic	23° 26' 17"

MOON'S PHASES, APSIDES AND POSITIONS ☽

Date	h	m	Phase	Longitude ° '	Eclipse Indicator
05	11	07	☽	13 ♏ 02	
12	01	36	○	19 ♒ 21	
19	04	36	☾	26 ♉ 12	
27	08	17	●	04 ♍ 04	

Day	h	m		
10	17	15	Perigee	
22	22	01	Apogee	
02	18	29	0S	
09	06	33	Max dec	27° S 02'
15	09	40	0N	
22	15	08	Max dec	27° N 06'
29	23	21	0S	

LONGITUDES

Date	Chiron ⚷ ° '	Ceres ⚳ ° '	Pallas ⚴ ° '	Juno ⚵ ° '	Vesta ⚶ ° '	Black Moon Lilith ⚸ ° '
01	16 ♈ 22	03 ♌ 55	13 ♊ 03	21 ♓ 20	04 ♓ 33	12 ♋ 06
11	16 ♈ 13	08 ♌ 24	17 ♊ 51	20 ♓ 28	02 ♓ 26	13 ♋ 13
21	15 ♈ 59	12 ♌ 52	22 ♊ 37	18 ♓ 54	29 ♒ 59	14 ♋ 20
31	15 ♈ 41	17 ♌ 19	27 ♊ 20	16 ♓ 47	27 ♒ 32	15 ♋ 27

All ephemeris data is given at 12.00 UT and the Moon's longitude is additionally given for 24.00 UT

ASPECTARIAN

h m	Aspects	h m	Aspects	h m	Aspects
01 Monday		02 29	☽ ± ☿	07 04	☽ △ ♄
00 41	♀ ± ♄	08 08	☽ ✶ ♃	07 40	☿ ☍ ♆
03 19	☽ ✶ ♀	08 12	☽ ∥ ⯓	13 47	☽ □ ♆
05 35	☽ △ ♂	10 35	☽ ∠ ♆	14 10	☽ ⊥ ♅
06 30	☽ △ ♅	12 53	☉ □ ♅	14 29	☽ □ ☿
11 23	☿ ⊼ ♆	13 05	☽ ⊼ ☿	17 38	☽ ⊼ ⯓
14 30	☽ ⊼ ♄	15 19	☉ ± ♆	18 57	☽ ∠ ♀
17 32	☽ ∠ ☉	17 58	☿ ♇ ⯓	22 06	☽ ✶ ☉
18 50	☽ ☍ ♆	21 42	☽ # ♀	**22 Monday**	
20 03	☽ ⊻ ☿	21 44	♂ ✶ ♆	02 35	☽ ⊻ ♂
22 29	☽ △ ⯓	**12 Friday**		08 26	☽ ∠ ♅
23 53	♂ ☌ ♅	00 47	☽ □ ♅	13 18	☽ ♇ ♄
02 Tuesday		00 55	☽ ⊥ ♆	15 21	☽ ⊥ ♀
01 53	☽ ± ♄	01 36	☽ ☍ ☉	15 24	☽ ⊥ ♂
04 43	☽ Q ♀	05 58	☽ ☌ ♄	16 10	☽ □ ♃
05 49	☽ # ♆	08 02	☽ ∠ ♃	16 12	♀ Q ♂
09 17	☽ ⊥ ☿	10 33	☽ ⊻ ♆	21 58	☿ △ ⯓
09 44	☽ ∥ ♃	11 07	☽ □ ♂	**23 Tuesday**	
10 16	☉ ♇ ♆	11 23	♀ Q ♅	04 55	☽ ⊻ ♀
11 08	☽ ♇ ♅	12 01	☽ # ♂	06 59	☽ Q ☿
11 43	☽ ♇ ♂	13 35	☽ ⊻ ⯓	07 14	☽ ∠ ☉
12 25	♀ ✶ ♅	14 11	☽ # ♅	07 27	☽ ± ♄
13 24	☿ ⊼ ⯓	20 55	☽ ⊼ ♀	10 34	☽ ∠ ♂
18 47	☽ ♇ ♄	21 32	☽ ∥ ♄	11 37	☿ ± ♄
20 27	☽ ☍ ♃	22 33	☽ ⊥ ♃	14 51	☽ ✶ ♅
03 Wednesday		23 18	☽ ⊥ ⯓	19 30	☽ ⊼ ♄
00 12	☽ ✶ ☉	**13 Saturday**		**24 Wednesday**	
02 00	♀ ✶ ♂	00 20	☽ # ☉	02 26	☽ △ ♆
03 08	☽ # ♃	05 56	☽ Q ♅	03 01	☽ ⊥ ☉
04 02	☽ ± ♅	07 36	☽ ± ♀	05 17	♀ ∥ ♅
04 48	☽ ∠ ☿	08 19	☽ ⊻ ♃	06 17	☽ ☍ ⯓
05 28	☽ ± ♂	14 02	☽ ∠ ⯓	09 40	☽ ✶ ☿
07 02	☽ ∥ ♆	17 48	☽ Q ♂	13 52	♅ St R
15 11	☽ ⊼ ♅	19 02	☽ ☍ ☿	15 00	☽ Q ♅
17 10	☽ ⊼ ♂	23 53	☽ ♇ ♀	16 07	☽ ⊻ ☉
17 54	☽ □ ♀	**14 Sunday**		17 57	☉ ± ♃
22 28	☽ △ ♄	01 49	☽ ✶ ♅	18 16	☽ ✶ ♂
04 Thursday		06 10	☽ ⊼ ☉	**25 Thursday**	
00 11	☽ Q ☉	06 58	☽ ⊻ ♄	04 10	☽ △ ♃
02 51	☽ ⊼ ♆	11 58	☽ ☌ ♆	05 43	☽ # ⯓
06 20	☽ □ ⯓	12 55	☽ # ☿	08 23	☽ ♇ ♆
12 29	☽ ✶ ☿	14 53	☽ ✶ ♂	13 46	☿ # ♃
13 45	☽ ± ♆	15 10	☽ ✶ ⯓	18 27	☽ ∠ ☿
21 03	☽ # ☿	17 06	☽ ± ☉	18 41	☉ ± ⯓
21 52	☉ ∥ ♅	17 07	☽ ⊥ ♄	19 18	☽ Q ♂
05 Friday		17 11	☉ ☍ ♄	23 50	☽ ☌ ♀
03 18	☽ ⊼ ♃	21 28	♂ △ ⯓	**26 Friday**	
05 52	☽ ♇ ♆	21 38	☽ ∥ ♆	02 08	☽ ± ♆
10 39	☽ ∥ ♄	**15 Monday**		02 45	☽ □ ♅
11 07	☽ □ ☉	02 29	☽ # ♃	05 27	☽ ∥ ♂
12 59	☽ Q ☿	03 26	☽ ∠ ♅	06 55	☽ ☍ ♄
13 55	☽ ± ♃	03 53	☽ △ ♀	09 36	☽ ♇ ♃
14 32	☽ Q ⯓	08 36	☽ ∠ ♄	13 46	☽ ⊥ ☿
16 23	♀ ⊼ ♄	09 51	☽ ♇ ☉	13 53	☽ ⊼ ♆
16 24	☽ # ♂	10 59	☽ ☌ ♃	17 39	☽ ⊼ ⯓
18 14	☽ # ☉	11 56	☽ Q ⯓	18 48	☽ ∥ ♅
19 21	☽ # ♅	16 52	☽ ∥ ♃	22 12	♀ ± ♆
20 35	☿ ± ♃	18 11	☽ ∠ ♂	23 49	☽ ∥ ♀
21 13	☽ ☍ ♅	19 15	☽ ⊥ ♅	**27 Saturday**	
06 Saturday		21 57	☽ # ♆	01 51	☽ # ♄
01 38	☿ ± ⯓	**16 Tuesday**		02 28	☽ ⊻ ☿
01 38	☽ ☍ ♂	01 06	☉ ♇ ♃	03 02	☽ ± ♃
03 46	☽ □ ♄	02 45	☽ ∥ ☿	04 34	♀ □ ♅
04 56	☽ △ ♀	04 41	☽ ⊼ ☿	05 18	☽ ± ⯓
05 37	☽ ♇ ♃	06 01	☽ ⊻ ♅	05 27	☉ □ ♂
08 08	☽ △ ♆	11 07	☽ ⊥ ♂	08 11	☽ □ ♂
11 24	☽ ✶ ⯓	11 14	☽ ✶ ♄	08 17	☽ ☌ ☉
14 14	☉ ∥ ♂	14 48	☽ △ ☉	14 33	☽ ⊼ ♃
18 52	☽ # ♀	16 50	☽ ⊻ ♆	22 38	☽ ♇ ⯓
07 Sunday		16 51	☽ ± ☿	**28 Sunday**	
00 18	☽ □ ☿	17 46	☿ △ ♅	00 44	♀ # ♄
03 21	☽ ∥ ⯓	20 18	☽ □ ⯓	06 30	☽ ∥ ☉
07 12	☽ △ ♃	22 39	☽ ⊻ ♂	12 47	☽ △ ♅
08 59	☽ ♇ ♀	**17 Wednesday**		16 17	☽ ⊻ ♀
10 21	☽ Q ♄	03 55	☽ ⊥ ♆	16 30	☽ ⊼ ♄
12 50	☽ ∠ ⯓	09 17	☽ Q ♄	18 27	♀ ☍ ♄
16 43	♀ △ ♆	11 30	☽ ♇ ☿	19 56	☿ ∥ ♆
18 21	☽ △ ☉	15 54	☽ □ ♀	23 27	☽ ☍ ♆
19 57	♂ □ ♄	16 43	☽ ∥ ☉	**29 Monday**	
08 Monday		17 32	☽ ⊻ ♃	03 08	☽ △ ⯓
00 17	☽ ⊼ ♅	17 57	☉ ⊼ ♆	03 47	☽ ± ♄
01 23	☽ ± ♀	20 50	☽ ∠ ♆	04 51	☽ ⊥ ♀
03 46	☽ ⊥ ⯓	21 53	☿ # ♆	08 46	☽ # ☿
06 11	☽ ✶ ♄	**18 Thursday**		09 50	☽ # ♆
06 42	☽ ⊼ ♂	03 37	☽ # ♄	16 10	☽ ☌ ☿
10 11	☽ ± ♅	04 56	☽ ⊥ ♃	17 05	☽ ♇ ♅
10 30	☽ □ ♆	08 03	♀ △ ♃	17 43	☽ ∥ ♃
12 09	☽ ⊼ ♀	11 39	☽ ∥ ♅	18 29	♀ ♇ ♃
13 36	☽ ⊻ ⯓	14 20	☽ ☌ ♅	19 50	☽ △ ♂
16 57	☽ ± ♂	19 06	☿ ⊼ ♄	20 35	☽ ♇ ♄
20 44	☽ ♇ ☉	19 18	☽ ∥ ♂	22 00	☽ ⊻ ☉
09 Tuesday		19 34	☽ □ ♄	23 00	☽ ☍ ♃
00 51	☽ ♇ ♅	19 37	☽ △ ☿	23 31	☽ ∠ ♀
05 18	♀ ☍ ⯓	20 47	☽ ∥ ♀	**30 Tuesday**	
06 29	☽ ∠ ♄	22 18	☽ ∠ ♃	04 53	☽ # ♃
07 51	☽ △ ☿	**19 Friday**		05 52	☿ ♇ ♅
08 10	☽ ♇ ♂	01 49	☽ ✶ ♆	09 50	☽ ± ♅
08 24	☽ □ ♃	04 36	☽ □ ☉	10 04	☽ ⊥ ☉
10 25	♂ ∥ ♅	05 32	☽ △ ⯓	10 09	☉ ⊼ ♃
12 17	☽ ± ☉	06 32	☽ Q ♀	12 48	☽ ∥ ♆
13 06	☿ ⊼ ♃	10 57	♀ ∥ ♂	16 22	☽ ∥ ☿
13 26	♂ ∠ ♃	11 06	☽ ☌ ♂	20 54	☽ ⊼ ♅
15 37	☽ Q ♆	16 10	☉ ⊼ ⯓	**31 Wednesday**	
20 51	☽ ⊥ ♄	17 55	♀ ♇ ♆	00 13	☽ △ ♄
22 33	☽ ⊼ ☉	**20 Saturday**		00 50	☽ ♇ ♂
10 Wednesday		00 20	☽ # ⯓	03 58	☽ ∠ ☉
00 58	☽ △ ♅	01 33	☽ Q ♆	06 07	☽ ✶ ♀
06 23	☽ ⊻ ♄	03 30	☿ ∥ ♃	07 08	☽ ⊼ ♆
09 12	☽ △ ♂	03 52	☽ ✶ ♃	10 22	☽ # ☉
10 35	☽ ♇ ☿	09 09	☽ ✶ ♀	10 43	☽ □ ⯓
10 45	☽ ✶ ♆	11 22	☽ ♇ ⯓	16 36	♀ ⊼ ♆
13 06	☽ Q ♃	19 39	☽ Q ☉	17 57	☽ ± ♂
13 45	☽ ☌ ⯓	**21 Sunday**		18 02	☽ ± ♆
16 39	☽ ☍ ♀	02 01	☽ ⊻ ♅	19 46	♂ Q ♆
23 06	☉ # ♄	03 53	☽ Q ♃		
11 Thursday		05 30	☽ ± ⯓		

SEPTEMBER 2022

LONGITUDES

Date	Sidereal time h m s	Sun ☉ ° ' "	Moon ☽ ° ' "	Moon ☽ 24.00 ° ' "	Mercury ☿ ° '	Venus ♀ ° '	Mars ♂ ° '	Jupiter ♃ ° '	Saturn ♄ ° '	Uranus ♅ ° '	Neptune ♆ ° '	Pluto ♇ ° '
01	10 42 32	09 ♍ 02 37	10 ♏ 25 31	17 ♏ 08 47	05 ♎ 38	25 ♌ 28	06 ♊ 50	06 ♈ 49	20 ♒ 36	18 ♉ 54	24 ♓ 27	26 ♑ 26
02	10 46 29	10 00 42	23 ♏ 55 37	00 ♐ 46 08	06 18	26 42	07 22	06 R 43	20 R 32	18 R 53	24 R 25	26 R 25
03	10 50 26	10 58 47	07 ♐ 40 30	14 ♐ 38 47	06 54	27 56	07 53	06 37	20 28	18 53	24 24	26 24
04	10 54 22	11 56 55	21 ♐ 41 01	28 ♐ 47 07	07 26	29 ♌ 10	08 25	06 30	20 24	18 52	24 22	26 23
05	10 58 19	12 55 03	05 ♑ 56 55	13 ♑ 10 07	07 54	00 ♍ 24	08 56	06 24	20 20	18 52	24 20	26 22
06	11 02 15	13 53 14	20 ♑ 26 17	27 ♑ 44 52	08 17	01 39	09 26	06 17	20 16	18 51	24 19	26 21
07	11 06 12	14 51 25	05 ♒ 05 11	12 ♒ 26 23	08 35	02 53	09 57	06 10	20 12	18 50	24 17	26 21
08	11 10 08	15 49 38	19 ♒ 47 36	27 ♒ 07 53	08 47	04 07	10 27	06 03	20 08	18 50	24 16	26 20
09	11 14 05	16 47 53	04 ♓ 26 15	11 ♓ 41 45	08 54	05 21	10 57	05 56	20 04	18 49	24 14	26 19
10	11 18 01	17 46 09	18 ♓ 53 30	26 ♓ 00 44	08 R 55	06 36	11 26	05 49	20 01	18 48	24 12	26 18
11	11 21 58	18 44 27	03 ♈ 02 46	09 ♈ 59 08	08 50	07 50	11 55	05 41	19 57	18 47	24 11	26 17
12	11 25 55	19 42 47	16 ♈ 49 27	23 ♈ 33 34	08 38	09 04	12 24	05 34	19 53	18 46	24 09	26 17
13	11 29 51	20 41 09	00 ♉ 11 25	06 ♉ 43 07	08 20	10 19	12 53	05 26	19 50	18 45	24 07	26 16
14	11 33 48	21 39 33	13 ♉ 08 55	19 ♉ 29 09	07 54	11 33	13 21	05 19	19 46	18 44	24 06	26 15
15	11 37 44	22 37 59	25 ♉ 44 17	01 ♊ 54 47	07 22	12 48	13 48	05 11	19 43	18 43	24 04	26 15
16	11 41 41	23 36 28	08 ♊ 01 15	14 ♊ 04 17	06 44	14 02	14 16	05 04	19 40	18 42	24 02	26 14
17	11 45 37	24 34 58	20 ♊ 04 30	26 ♊ 02 34	05 59	15 17	14 43	04 56	19 36	18 41	24 01	26 13
18	11 49 34	25 33 31	01 ♋ 59 08	07 ♋ 54 49	05 08	16 31	15 09	04 48	19 33	18 40	23 59	26 13
19	11 53 30	26 32 06	13 ♋ 50 15	19 ♋ 46 02	04 13	17 46	15 35	04 40	19 30	18 39	23 57	26 12
20	11 57 27	27 30 43	25 ♋ 42 43	01 ♌ 40 50	03 13	19 01	16 01	04 32	19 27	18 38	23 56	26 12
21	12 01 24	28 29 22	07 ♌ 40 51	13 ♌ 43 12	02 10	20 15	16 27	04 24	19 24	18 36	23 54	26 11
22	12 05 20	29 ♍ 28 03	19 ♌ 48 15	25 ♌ 56 17	01 05	21 30	16 52	04 16	19 21	18 35	23 52	26 11
23	12 09 17	00 ♎ 26 46	02 ♍ 07 34	08 ♍ 22 15	00 ♎ 00	22 44	17 16	04 08	19 18	18 34	23 51	26 10
24	12 13 13	01 25 32	14 ♍ 40 29	21 ♍ 02 18	28 ♍ 57	23 59	17 40	04 00	19 16	18 32	23 49	26 10
25	12 17 10	02 24 19	27 ♍ 27 41	03 ♎ 56 37	27 56	25 14	18 04	03 52	19 13	18 31	23 48	26 10
26	12 21 06	03 23 08	10 ♎ 29 00	17 ♎ 04 42	27 00	26 29	18 27	03 44	19 10	18 29	23 46	26 09
27	12 25 03	04 22 00	23 ♎ 43 34	00 ♏ 25 27	26 11	27 43	18 49	03 36	19 08	18 28	23 44	26 09
28	12 28 59	05 20 53	07 ♏ 10 12	13 ♏ 57 40	25 28	28 ♍ 58	19 12	03 28	19 05	18 26	23 43	26 09
29	12 32 56	06 19 48	20 ♏ 47 40	27 ♏ 40 05	24 55	00 ♎ 13	19 33	03 20	19 03	18 24	23 41	26 08
30	12 36 53	07 ♎ 18 45	04 ♐ 34 48	11 ♐ 31 42	24 ♍ 30	01 ♎ 28	19 ♊ 54	03 ♈ 12	19 ♒ 01	18 ♉ 23	23 ♓ 39	26 ♑ 08

Date	Moon True ☊ ° '	Moon Mean ☊ ° '	Moon ☽ Latitude ° '
01	15 ♉ 35	16 ♉ 38	00 N 27
02	15 D 35	16 35	00 S 44
03	15 35	16 32	01 55
04	15 R 35	16 29	03 00
05	15 33	16 26	03 55
06	15 28	16 23	04 36
07	15 22	16 19	04 59
08	15 13	16 16	05 03
09	15 04	16 13	04 47
10	14 55	16 10	04 13
11	14 48	16 07	03 24
12	14 42	16 03	02 23
13	14 39	16 00	01 17
14	14 38	15 57	00 S 08
15	14 D 39	15 54	00 N 59
16	14 40	15 51	02 02
17	14 41	15 48	02 59
18	14 R 41	15 44	03 47
19	14 39	15 41	04 25
20	14 35	15 38	04 52
21	14 29	15 35	05 06
22	14 22	15 32	05 07
23	14 14	15 29	04 53
24	14 06	15 25	04 26
25	13 59	15 22	03 44
26	13 54	15 19	02 51
27	13 50	15 16	01 47
28	13 48	15 13	00 N 36
29	13 D 48	15 09	00 S 38
30	13 ♉ 49	15 ♉ 06	01 S 51

DECLINATIONS

Date	Sun ☉ ° '	Moon ☽ ° '	Mercury ☿ ° '	Venus ♀ ° '	Mars ♂ ° '	Jupiter ♃ ° '	Saturn ♄ ° '	Uranus ♅ ° '	Neptune ♆ ° '	Pluto ♇ ° '
01	08 N 11	14 S 31	04 S 59	14 N 05	20 N 10	01 N 16	15 S 52	17 N 05	03 S 20	23 S 00
02	07 49	19 28	05 23	13 41	20 16	01 13	15 54	17 05	03 20	23 01
03	07 27	23 29	05 45	13 17	20 22	01 10	15 55	17 05	03 21	23 01
04	07 05	26 10	06 05	12 52	20 28	01 08	15 56	17 05	03 22	23 01
05	06 43	27 13	06 23	12 28	20 34	01 05	15 58	17 05	03 22	23 01
06	06 20	26 26	06 39	12 02	20 40	01 02	15 59	17 04	03 23	23 02
07	05 58	23 50	06 52	11 37	20 46	00 59	16 00	17 04	03 23	23 02
08	05 35	19 40	07 03	11 11	20 51	00 56	16 01	17 04	03 24	23 02
09	05 13	14 20	07 10	10 45	20 56	00 53	16 03	17 04	03 25	23 02
10	04 50	08 16	07 15	10 18	21 02	00 50	16 04	17 04	03 25	23 02
11	04 27	01 S 54	07 16	09 51	21 07	00 47	16 05	17 03	03 26	23 03
12	04 04	04 N 24	07 13	09 24	21 12	00 44	16 06	17 03	03 27	23 03
13	03 41	10 20	07 07	08 57	21 17	00 41	16 07	17 03	03 27	23 03
14	03 18	15 39	06 57	08 29	21 21	00 38	16 08	17 02	03 28	23 03
15	02 55	20 09	06 43	08 02	21 26	00 35	16 09	17 02	03 29	23 03
16	02 32	23 40	06 24	07 34	21 30	00 32	16 10	17 02	03 29	23 03
17	02 09	26 02	06 02	07 05	21 35	00 29	16 12	17 02	03 30	23 03
18	01 46	27 12	05 35	06 37	21 39	00 25	16 13	17 01	03 31	23 04
19	01 23	27 07	05 05	06 08	21 43	00 22	16 13	17 01	03 31	23 04
20	00 59	25 46	04 31	05 39	21 48	00 19	16 14	17 01	03 32	23 04
21	00 36	23 16	03 54	05 10	21 52	00 16	16 15	17 00	03 33	23 04
22	00 N 13	19 43	03 15	04 41	21 55	00 13	16 16	17 00	03 33	23 04
23	00 S 11	15 17	02 34	04 12	21 59	00 09	16 17	16 59	03 34	23 04
24	00 34	10 07	01 53	03 42	22 03	00 06	16 18	16 59	03 35	23 04
25	00 57	04 N 26	01 12	03 13	22 07	00 N 03	16 19	16 59	03 35	23 04
26	01 21	01 S 32	00 S 31	02 43	22 10	00 00	16 20	16 58	03 36	23 05
27	01 44	07 33	00 N 07	02 13	22 14	00 S 03	16 20	16 58	03 37	23 05
28	02 07	13 20	00 42	01 43	22 17	00 07	16 21	16 57	03 37	23 05
29	02 31	18 34	01 14	01 13	22 21	00 10	16 22	16 57	03 38	23 05
30	02 S 54	22 S 52	01 N 42	00 N 43	22 N 24	00 S 13	16 S 22	16 N 56	03 S 39	23 S 05

ZODIAC SIGN ENTRIES

Date	h	m	Planets
02	22	39	☽ ♐
05	02	03	☽ ♑
05	04	05	♀ ♍
07	03	41	☽ ♒
09	04	42	☽ ♓
11	06	47	☽ ♈
13	11	39	☽ ♉
15	20	16	☽ ♊
18	07	59	☽ ♋
20	20	38	☽ ♌
23	01	04	☉ ♎
23	07	53	☽ ♍
23	12	04	☿ ♍
25	16	43	☽ ♎
27	23	15	☽ ♏
29	07	49	♀ ♎
30	04	03	☽ ♐

LATITUDES

Date	Mercury ☿ ° '	Venus ♀ ° '	Mars ♂ ° '	Jupiter ♃ ° '	Saturn ♄ ° '	Uranus ♅ ° '	Neptune ♆ ° '	Pluto ♇ ° '
01	02 S 59	01 N 07	01 S 18	01 S 35	01 S 19	00 S 22	01 S 13	02 S 11
04	03 25	01 11	01 15	01 35	01 19	00 22	01 13	02 11
07	03 46	01 15	01 12	01 36	01 19	00 22	01 13	02 11
10	04 02	01 18	01 08	01 36	01 19	00 22	01 13	02 11
13	04 09	01 21	01 04	01 36	01 19	00 22	01 13	02 11
16	04 04	01 23	01 01	01 37	01 19	00 22	01 13	02 11
19	03 43	01 25	00 56	01 37	01 19	00 22	01 13	02 12
22	03 04	01 26	00 52	01 37	01 19	00 22	01 13	02 12
25	02 12	01 26	00 48	01 37	01 19	00 22	01 13	02 12
28	01 12	01 26	00 43	01 37	01 19	00 22	01 13	02 12
31	00 S 13	01 N 25	00 S 38	01 S 37	01 S 19	00 S 22	01 S 13	02 S 12

DATA

Julian Date	2459824
Delta T	+71 seconds
Ayanamsa	24° 10' 14"
Synetic vernal point	04° ♓ 56' 45"
True obliquity of ecliptic	23° 26' 18"

MOON'S PHASES, APSIDES AND POSITIONS ☽

Date	h	m	Phase	Longitude ° '	Eclipse Indicator
03	18	08	☽	11 ♐ 14	
10	09	59	○	17 ♓ 41	
17	21	52	☾	24 ♊ 59	
25	21	55	●	02 ♎ 49	

Day	h	m		
07	18	25	Perigee	
19	14	47	Apogee	
05	13	51	Max dec	27° S 13'
11	22	09	0N	
18	22	11	Max dec	27° N 19'
26	05	54	0S	

LONGITUDES

Date	Chiron ⚷ ° '	Ceres ⚳ ° '	Pallas ⚴ ° '	Juno ⚵ ° '	Vesta ⚶ ° '	Black Moon Lilith ⚸ ° '
01	15 ♈ 39	17 ♌ 46	27 ♊ 49	16 ♓ 33	27 ♒ 18	15 ♋ 33
11	15 ♈ 17	22 ♌ 12	02 ♋ 26	14 ♓ 07	25 ♒ 14	16 ♋ 40
21	14 ♈ 52	26 ♌ 35	06 ♋ 56	11 ♓ 43	23 ♒ 46	17 ♋ 48
31	14 ♈ 25	00 ♍ 55	11 ♋ 13	09 ♓ 41	23 ♒ 03	18 ♋ 55

ASPECTARIAN

h m	Aspects
01 Thursday	
02 54	☽ ⊻ ☿
05 16	☽ ⊼ ♂
05 35	☽ ⊼ ♃
06 08	☽ Q ♀
09 20	☽ ✶ ☉
10 11	☽ ♯ ♀
10 15	☽ Ᵽ ♆
11 01	☿ Ᵽ ♄
11 52	♂ ✶ ♃
14 17	☽ ⊥ ☿
16 16	☽ ± ♃
18 16	☽ ∥ ♄
19 10	☽ Q ♇
23 59	☽ ♯ ♅
02 Friday	
03 06	☽ ☍ ♅
06 02	☽ □ ♄
06 33	♀ ⊼ ♇
07 08	☽ ∠ ☿
08 08	☽ Ᵽ ♃
08 22	☽ Q ☉
12 52	☽ △ ♆
16 23	☽ ✶ ♇
16 26	☽ ♯ ♂
17 22	☽ □ ♀
03 Saturday	
01 49	☿ ☍ ♃
08 49	☽ ∥ ♇
10 10	☽ △ ♃
10 37	☽ ✶ ☿
12 23	☽ ☍ ♂
13 22	☽ Q ♄
18 08	☽ □ ☉
18 25	☽ ∠ ♇
22 17	☉ Ᵽ ♇
04 Sunday	
07 13	☽ ⊼ ♅
08 03	☽ Q ☿
09 48	☽ ⊥ ♇
09 50	☽ ✶ ♄
16 32	☽ □ ♆
17 24	☽ ± ♅
19 57	☽ ⊻ ♇
05 Monday	
01 51	☽ △ ♀
08 31	☽ Ᵽ ♅
10 59	☽ ∠ ♄
11 43	♀ ± ♃
12 44	☽ □ ♃
15 21	☽ □ ☿
17 09	☽ ⊼ ♂
22 37	☽ Q ♆
06 Tuesday	
00 01	☉ ♯ ☿
00 25	☽ △ ☉
01 52	☽ ⊥ ♄
03 27	☽ ± ♂
05 10	☽ Ᵽ ♀
09 23	☽ △ ♅
11 43	☽ ⊻ ♄
18 16	☽ Q ♃
18 21	☽ ✶ ♆
18 49	☽ Ᵽ ♂
21 21	☽ ± ♀
21 43	☽ ☌ ♇
07 Wednesday	
01 40	♀ ± ♇
02 51	☽ Ᵽ ☉
04 32	☿ ⊥ ♀
08 04	☽ ⊼ ♀
13 45	☽ ✶ ♃
17 19	☽ ∥ ♇
17 48	☽ △ ☿
18 35	☽ ± ☉
18 51	☽ ∠ ♆
20 13	☽ △ ♂
08 Thursday	
05 04	☽ ⊼ ☉
06 02	☽ ♯ ♂
09 30	☽ ⊥ ♆
10 26	☽ □ ♅
12 34	☽ ☌ ♄
14 02	☽ ∠ ♃
18 36	☽ Ᵽ ☿
19 17	☽ ⊻ ♆
22 41	☽ ⊻ ♇
09 Friday	
00 15	☽ ♯ ♅
04 39	☽ ⊥ ♃
04 46	☽ ∥ ♄
08 31	☽ ⊥ ♇
09 28	☽ ± ☿
13 39	☽ ☍ ♀
14 26	☽ ⊻ ♃
15 55	☽ Q ♅
19 24	☽ ⊼ ☿
22 06	♀ ⊼ ♃
23 08	☽ □ ♂
23 22	☽ ∠ ♇
10 Saturday	
03 34	☽ ♯ ♀
03 37	☿ St R
05 41	♂ Ᵽ ♇
09 59	☽ ☍ ☉
11 51	☽ ✶ ♅
13 52	☽ ⊻ ♄
15 53	☽ ∥ ☿
20 55	☽ ☌ ♆
23 57	☽ ⊥ ♄
11 Sunday	
00 29	☽ ✶ ♇
01 49	☽ ♯ ☉
06 15	☽ ∥ ♆
06 28	☽ Q ♂
13 09	☉ △ ♅
13 16	☽ ∠ ♅
15 16	☽ ∠ ♄
16 13	☽ ♯ ♃
16 30	☽ ☌ ♃
21 03	☽ Q ♇
21 05	☽ ⊼ ♀
21 52	☽ ☍ ☿
22 03	☽ ∥ ♃
12 Monday	
03 56	☽ ✶ ♂
04 49	☿ ⊻ ♀
04 53	☽ ⊥ ♅
08 17	☽ ♯ ♆
08 36	☽ ± ♀
10 48	☽ ∥ ☉
15 27	☽ ⊻ ♅
16 09	☉ ⊼ ♄
17 25	☽ ✶ ♄
17 31	☽ ⊼ ☉
22 59	☽ ♯ ☿
13 Tuesday	
01 02	☽ ⊻ ♆
02 15	☽ Ᵽ ♀
04 53	☽ □ ♇
05 08	☽ ± ☉
06 37	☽ ∥ ♀
07 38	☽ ∠ ♂
11 53	☽ ⊥ ♆
14 59	☽ Q ♄
21 33	☽ ⊻ ♃
22 54	☽ Ᵽ ☉
14 Wednesday	
00 45	☽ ⊥ ♂
02 12	☉ ♯ ♆
02 32	☽ ⊼ ☿
04 26	☽ ∠ ♆
06 15	♀ Ᵽ ♇
08 36	☽ ⊥ ♃
08 41	☽ △ ♀
12 23	☽ ⊻ ♂
13 22	☽ ± ☿
14 22	☽ ♯ ♄
18 53	☽ ∥ ♅
22 34	☽ ☌ ♅
15 Thursday	
00 30	☽ □ ♄
01 27	☽ ∠ ♃
05 31	☽ △ ☉
05 49	☽ Ᵽ ☿
08 47	☽ ✶ ♆
12 59	☽ △ ♇
20 07	☽ ∥ ♂
16 Friday	
06 14	☽ ✶ ♃
07 17	☽ ♯ ♇
08 06	☽ Q ♆
09 36	☽ △ ☿
18 21	☽ Ᵽ ♇
18 49	♀ □ ♂
22 21	☉ ☍ ♆
17 Saturday	
00 52	☽ ☌ ♂
01 18	☽ □ ♀
05 46	☽ Q ♃
09 13	☽ ⊻ ♅
11 04	☽ △ ♄
12 18	☽ ± ♇
19 54	☽ □ ♆
21 15	☽ ⊥ ♅
21 52	☽ □ ☉
18 Sunday	
00 21	☽ ⊼ ♇
11 53	☉ ± ♄
15 24	☽ ∠ ♅
17 10	☽ Ᵽ ♄
17 38	☽ □ ♃
17 44	☽ Q ♀
17 56	☽ □ ☿
22 34	☿ ☍ ♃
19 Monday	
03 58	☉ △ ♇
04 19	☿ Ᵽ ♄
11 19	☽ ± ♄
13 32	☽ Q ☉
15 41	☽ ⊻ ♂
20 53	☽ ✶ ♀
21 43	☽ ✶ ♅
23 25	☽ ⊼ ♄
20 Tuesday	
02 03	☿ Ᵽ ♅
03 38	☽ Q ☿
04 17	☽ ⊥ ♂
04 44	♀ △ ♅
08 25	☽ △ ♆
12 58	☽ ☍ ♇
15 57	☽ ✶ ☉
20 12	♀ ⊼ ♄
21 52	☽ Q ♅
23 04	☽ ∠ ♂
21 Wednesday	
01 52	☽ ✶ ☿
05 32	☽ △ ♃
06 36	☽ ∠ ♀
10 17	♂ Q ♃
13 08	♃ ∠ ♄
13 34	☽ ♯ ♇
14 26	☽ Ᵽ ♆
22 09	☽ ∥ ♂
22 Thursday	
00 33	☽ ∠ ☉
01 01	☿ ∥ ♆
02 33	☽ ⊥ ♀
05 17	☽ ∠ ☿
06 00	☽ ✶ ♂
08 13	☽ ± ♆
09 36	☽ □ ♅
10 58	☽ Ᵽ ♃
11 07	☽ ☍ ♄
12 01	☉ ∥ ♃
15 42	☽ ⊻ ♀
19 48	☽ ⊥ ☉
19 57	☽ ⊼ ♆
21 30	☽ ⊥ ☿
23 Friday	
00 28	☽ ⊼ ♇
03 14	☽ ∥ ♅
04 22	☽ ± ♃
06 17	☽ Q ♂
06 50	☉ ☌ ☿
06 55	☽ ♯ ♄
08 13	☽ ⊻ ☿
08 28	☽ ⊻ ☉
10 58	☉ ♯ ♃
12 05	☽ ± ♇
15 50	☽ ⊼ ♃
24 Saturday	
05 20	☽ Ᵽ ♇
08 51	♀ ☍ ♆
17 51	☽ □ ♂
17 55	♀ ♯ ♆
19 17	☽ △ ♅
20 38	☽ ⊼ ♄
25 Sunday	
05 10	☽ ☍ ♆
07 24	☽ ☌ ♀
07 50	☽ ± ♄
09 35	☽ △ ♇
11 40	♀ ± ♄
12 49	☽ ☌ ☿
15 28	☽ ♯ ♆
17 19	☉ ∥ ☿
17 28	☽ ∥ ♀
21 55	☽ ☌ ☉
23 11	☽ Ᵽ ♅
23 45	☽ ☍ ♃
26 Monday	
00 27	☽ Ᵽ ♄
01 13	☽ ♯ ☉
02 48	☽ ♯ ☿
05 46	♀ △ ♇
05 51	☽ ∥ ♃
05 57	☽ ♯ ♃
08 23	☽ ∥ ☿
11 12	☽ ∥ ☉
14 14	♂ ⊻ ♅
14 23	☉ Ᵽ ♅
15 39	☽ ± ♅
16 20	☽ ♯ ♀
17 59	☿ ☌ ♀
19 33	☉ ☍ ♃
20 14	☽ ∥ ♆
27 Tuesday	
02 31	☽ ⊼ ♅
02 54	☽ △ ♂
03 44	☽ △ ♄
05 58	☿ ∥ ♃
06 26	☉ Ᵽ ♄
09 32	☿ ♯ ♃
12 01	☽ ⊼ ♆
12 56	☿ △ ♇
16 10	☽ ⊻ ☿
16 21	☽ □ ♇
19 54	☽ ⊻ ♀
22 45	☽ ± ♆
28 Wednesday	
01 04	☉ ♯ ♀
02 21	☽ ⊥ ☿
05 29	☽ ⊼ ♃
05 48	♂ △ ♄
06 34	☽ Ᵽ ♂
07 42	☽ ⊥ ♀
08 31	☽ ⊻ ☉
14 43	☽ Ᵽ ♆
16 02	☽ ± ♃
17 35	☽ ∠ ☿
19 58	♃ ∠ ♅
19 58	☽ ⊥ ☉
22 56	☽ ± ♂
29 Thursday	
00 19	☽ Q ♇
01 14	☽ ∠ ♀
01 27	☽ ∥ ♄
04 13	☽ ♯ ♅
04 50	☿ ± ♄
07 44	☽ Ᵽ ♃
07 49	☽ ☍ ♅
08 57	☽ □ ♄
09 46	☽ ⊼ ♂
11 41	☿ ∥ ♀
13 01	☽ ∠ ☉
17 03	☽ △ ♆
18 58	☽ ✶ ☿
21 20	☽ ✶ ♇
30 Friday	
06 04	☽ ✶ ♀
08 59	☽ ♯ ♂
09 38	☽ △ ♃
13 23	☽ ∥ ♇
15 15	☽ Q ☿
16 12	☽ Q ♄
17 05	☽ ✶ ☉
23 19	☽ ∠ ♇

All ephemeris data is given at 12.00 UT and the Moon's longitude is additionally given for 24.00 UT

OCTOBER 2022

LONGITUDES

Date	Sidereal time h m s	Sun ☉ ° ' "	Moon ☽ ° ' "	Moon ☽ 24.00 ° ' "	Mercury ☿ ° '	Venus ♀ ° '	Mars ♂ ° '	Jupiter ♃ ° '	Saturn ♄ ° '	Uranus ♅ ° '	Neptune ♆ ° '	Pluto ♇ ° '
01	12 40 49	08 ♎ 17 44	18 ♐ 30 38	25 ♐ 31 32	24 ♍ 16	02 ♎ 43	20 ♊ 15	03 ♈ 04	18 ♒ 59	18 ♉ 21	23 ♓ 38	26 ♑ 08
02	12 44 46	09 16 45	02 ♑ 34 13	09 ♑ 38 34	24 D 12	03 58	20 35	02 R 56	18 R 57	18 R 19	23 R 36	26 R 08
03	12 48 42	10 15 47	16 ♑ 44 21	23 ♑ 51 20	24 18	05 12	20 55	02 48	18 55	18 17	23 35	26 07
04	12 52 39	11 14 51	00 ♒ 59 15	08 ♒ 07 43	24 35	06 27	21 14	02 40	18 53	18 16	23 33	26 07
05	12 56 35	12 13 57	15 ♒ 16 20	22 ♒ 24 39	25 02	07 42	21 32	02 33	18 51	18 14	23 32	26 07
06	13 00 32	13 13 04	29 ♒ 32 10	06 ♓ 38 19	25 38	08 57	21 50	02 25	18 49	18 12	23 30	26 07
07	13 04 28	14 12 14	13 ♓ 42 34	20 ♓ 44 21	26 23	10 12	22 08	02 17	18 48	18 10	23 29	26 07
08	13 08 25	15 11 25	27 ♓ 43 08	04 ♈ 38 25	27 16	11 27	22 24	02 09	18 46	18 08	23 27	26 07
09	13 12 22	16 10 38	11 ♈ 29 46	18 ♈ 16 49	28 17	12 42	22 40	02 02	18 45	18 06	23 25	26 D 07
10	13 16 18	17 09 53	24 ♈ 59 18	01 ♉ 36 59	29 ♍ 24	13 57	22 56	01 54	18 43	18 04	23 24	26 07
11	13 20 15	18 09 10	08 ♉ 09 49	14 ♉ 37 46	00 ♎ 38	15 12	23 11	01 47	18 42	18 02	23 23	26 07
12	13 24 11	19 08 29	21 ♉ 00 55	27 ♉ 19 29	01 56	16 27	23 25	01 40	18 41	18 00	23 21	26 07
13	13 28 08	20 07 51	03 ♊ 33 43	09 ♊ 43 56	03 20	17 42	23 39	01 32	18 40	17 58	23 20	26 07
14	13 32 04	21 07 14	15 ♊ 50 40	21 ♊ 54 01	04 46	18 57	23 52	01 25	18 39	17 56	23 18	26 07
15	13 36 01	22 06 41	27 ♊ 54 51	03 ♋ 53 34	06 17	20 12	24 04	01 18	18 38	17 54	23 17	26 08
16	13 39 57	23 06 09	09 ♋ 50 45	15 ♋ 47 00	07 50	21 27	24 15	01 11	18 38	17 51	23 15	26 08
17	13 43 54	24 05 40	21 ♋ 42 54	27 ♋ 39 03	09 25	22 42	24 26	01 04	18 37	17 49	23 14	26 08
18	13 47 51	25 05 12	03 ♌ 36 04	09 ♌ 34 32	11 02	23 57	24 36	00 58	18 36	17 47	23 13	26 08
19	13 51 47	26 04 48	15 ♌ 35 01	21 ♌ 38 04	12 40	25 12	24 46	00 51	18 36	17 45	23 11	26 09
20	13 55 44	27 04 25	27 ♌ 44 11	03 ♍ 53 49	14 19	26 28	24 55	00 45	18 36	17 43	23 10	26 09
21	13 59 40	28 04 04	10 ♍ 07 22	16 ♍ 25 10	16 00	27 43	25 02	00 39	18 35	17 40	23 09	26 09
22	14 03 37	29 ♎ 03 46	22 ♍ 47 30	29 ♍ 14 32	17 41	28 ♎ 58	25 09	00 32	18 35	17 38	23 07	26 10
23	14 07 33	00 ♏ 03 30	05 ♎ 46 24	12 ♎ 23 05	19 22	00 ♏ 13	25 16	00 26	18 D 35	17 36	23 06	26 10
24	14 11 30	01 03 16	19 ♎ 04 31	25 ♎ 50 33	21 03	01 28	25 21	00 20	18 35	17 33	23 05	26 11
25	14 15 26	02 03 04	02 ♏ 40 57	09 ♏ 35 22	22 45	02 43	25 26	00 15	18 35	17 31	23 04	26 11
26	14 19 23	03 02 54	16 ♏ 33 25	23 ♏ 34 40	24 26	03 58	25 30	00 09	18 36	17 29	23 03	26 12
27	14 23 20	04 02 46	00 ♐ 38 39	07 ♐ 44 50	26 07	05 14	25 33	00 ♈ 04	18 36	17 26	23 01	26 12
28	14 27 16	05 02 40	14 ♐ 52 42	22 ♐ 01 44	27 48	06 29	25 35	29 ♓ 59	18 37	17 24	23 00	26 13
29	14 31 13	06 02 36	29 ♐ 11 26	06 ♑ 21 20	29 ♎ 29	07 44	25 36	29 53	18 37	17 22	22 59	26 13
30	14 35 09	07 02 33	13 ♑ 30 58	20 ♑ 39 57	01 ♏ 10	08 59	25 37	29 49	18 38	17 19	22 58	26 14
31	14 39 06	08 ♏ 02 32	27 ♑ 47 54	04 ♒ 54 30	02 ♏ 50	10 ♏ 14	25 ♊ 36	29 ♓ 44	18 ♒ 39	17 ♉ 16	22 ♓ 57	26 ♑ 14

Date	Moon True ☊ ° '	Moon Mean ☊ ° '	Moon ☽ Latitude ° '
01	13 ♉ 51	15 ♉ 03	02 S 58
02	13 D 52	15 00	03 55
03	13 R 52	14 57	04 38
04	13 50	14 54	05 04
05	13 47	14 50	05 12
06	13 44	14 47	05 00
07	13 39	14 44	04 30
08	13 35	14 41	03 45
09	13 31	14 38	02 46
10	13 29	14 35	01 40
11	13 27	14 31	00 S 29
12	13 D 28	14 28	00 N 41
13	13 29	14 25	01 48
14	13 30	14 22	02 49
15	13 32	14 19	03 41
16	13 33	14 15	04 23
17	13 34	14 12	04 53
18	13 R 33	14 09	05 11
19	13 32	14 06	05 15
20	13 31	14 03	05 06
21	13 28	14 00	04 42
22	13 26	13 56	04 04
23	13 24	13 53	03 13
24	13 23	13 50	02 10
25	13 22	13 47	00 N 59
26	13 D 22	13 44	00 S 18
27	13 22	13 41	01 34
28	13 23	13 37	02 46
29	13 24	13 34	03 47
30	13 24	13 31	04 35
31	13 ♉ 25	13 ♉ 28	05 S 05

DECLINATIONS

Date	Sun ☉ ° '	Moon ☽ ° '	Mercury ☿ ° '	Venus ♀ ° '	Mars ♂ ° '	Jupiter ♃ ° '	Saturn ♄ ° '	Uranus ♅ ° '	Neptune ♆ ° '	Pluto ♇ ° '
01	03 S 17	25 S 54	02 N 05	00 N 13	22 N 27	00 S 16	16 S 23	16 N 56	03 S 39	23 S 05
02	03 41	27 20	02 02	00 S 17	22 30	00 19	16 24	16 55	03 40	23 05
03	04 04	26 59	02 36	00 47	22 33	00 22	16 24	16 55	03 40	23 05
04	04 27	24 53	02 44	01 17	22 37	00 25	16 25	16 54	03 41	23 05
05	04 50	21 13	02 46	01 47	22 40	00 28	16 25	16 54	03 42	23 05
06	05 13	16 19	02 44	02 17	22 43	00 31	16 26	16 53	03 42	23 05
07	05 36	10 34	02 36	02 47	22 45	00 34	16 26	16 53	03 43	23 05
08	05 59	04 S 20	02 24	03 17	22 48	00 37	16 27	16 52	03 43	23 05
09	06 22	02 N 00	02 08	03 47	22 51	00 40	16 27	16 52	03 44	23 05
10	06 44	08 08	01 48	04 17	22 54	00 43	16 28	16 51	03 45	23 05
11	07 07	13 46	01 24	04 47	22 57	00 46	16 28	16 51	03 45	23 05
12	07 30	18 41	00 57	05 17	23 00	00 49	16 28	16 50	03 46	23 05
13	07 52	22 38	00 N 27	05 47	23 02	00 52	16 28	16 49	03 46	23 05
14	08 14	25 29	00 S 05	06 16	23 05	00 54	16 29	16 49	03 47	23 05
15	08 37	27 06	00 40	06 45	23 08	00 57	16 29	16 48	03 47	23 05
16	08 59	27 26	01 16	07 15	23 11	01 00	16 29	16 48	03 48	23 05
17	09 21	26 30	01 54	07 44	23 13	01 02	16 29	16 47	03 48	23 05
18	09 43	24 23	02 33	08 13	23 16	01 05	16 29	16 46	03 49	23 05
19	10 04	21 10	03 13	08 42	23 19	01 07	16 29	16 46	03 49	23 05
20	10 26	17 02	03 54	09 10	23 21	01 10	16 29	16 45	03 50	23 05
21	10 47	12 07	04 35	09 39	23 24	01 12	16 29	16 45	03 50	23 05
22	11 08	06 36	05 17	10 07	23 27	01 14	16 29	16 44	03 51	23 05
23	11 30	00 N 40	05 59	10 35	23 29	01 17	16 29	16 43	03 52	23 05
24	11 50	05 S 27	06 41	11 02	23 32	01 19	16 29	16 43	03 52	23 05
25	12 11	11 29	07 23	11 30	23 35	01 21	16 29	16 42	03 52	23 05
26	12 32	17 04	08 05	11 57	23 37	01 23	16 29	16 41	03 53	23 05
27	12 52	21 49	08 47	12 24	23 40	01 25	16 29	16 41	03 53	23 05
28	13 12	25 20	09 29	12 50	23 43	01 27	16 29	16 40	03 54	23 05
29	13 32	27 14	10 10	13 17	23 46	01 29	16 28	16 39	03 54	23 05
30	13 52	27 19	10 51	13 43	23 48	01 30	16 28	16 39	03 55	23 04
31	14 S 11	25 S 35	11 S 31	14 S 08	23 N 51	01 S 32	16 S 28	16 N 38	03 S 55	23 S 04

ZODIAC SIGN ENTRIES

Date	h	m	Planets	
02	07	38	☽	♑
04	10	20	☽	♒
06	12	47	☽	♓
08	15	57	☽	♈
10	21	04	☽	♉
10	23	51	☿	♎
13	05	08	☽	♊
15	16	11	☽	♋
18	04	45	☽	♌
20	16	25	☽	♍
23	01	24	☽	♎
23	07	52	♀	♏
23	10	36	☉	♏
25	07	19	☽	♏
27	10	55	☽	♐
28	05	10	♃	♓
29	13	21	☽	♑
29	19	22	☿	♏
31	15	43	☽	♒

LATITUDES

Date	Mercury ☿ ° '	Venus ♀ ° '	Mars ♂ ° '	Jupiter ♃ ° '	Saturn ♄ ° '	Uranus ♅ ° '	Neptune ♆ ° '	Pluto ♇ ° '
01	00 S 13	01 N 25	00 S 38	01 S 37	01 S 19	00 S 22	01 S 13	02 S 12
04	00 N 38	01 24	00 32	01 37	01 19	00 22	01 13	02 12
07	01 16	01 22	00 27	01 37	01 19	00 22	01 13	02 12
10	01 42	01 19	00 21	01 37	01 19	00 22	01 13	02 12
13	01 56	01 16	00 15	01 36	01 19	00 22	01 13	02 12
16	02 00	01 12	00 08	01 36	01 18	00 22	01 13	02 12
19	01 57	01 08	00 S 01	01 36	01 18	00 22	01 13	02 12
22	01 47	01 04	00 N 06	01 35	01 18	00 22	01 13	02 12
25	01 34	00 59	00 13	01 35	01 18	00 22	01 13	02 13
28	01 18	00 53	00 21	01 34	01 18	00 22	01 13	02 13
31	01 N 00	00 N 48	00 N 29	01 S 33	01 S 17	00 S 22	01 S 13	02 S 13

DATA

Julian Date	2459854
Delta T	+71 seconds
Ayanamsa	24° 10' 17"
Synetic vernal point	04° ♓ 56' 42"
True obliquity of ecliptic	23° 26' 18"

MOON'S PHASES, APSIDES AND POSITIONS ☽

Date	h	m	Phase	Longitude ° '	Eclipse Indicator
03	00	14	☽	09 ♑ 47	
09	20	55	○	16 ♈ 33	
17	17	15	☾	24 ♋ 19	
25	10	49	●	02 ♏ 00	Partial

Day	h	m	
04	16	38	Perigee
17	10	20	Apogee
29	14	30	Perigee

Day	h	m		
02	19	29	Max dec	27° S 25'
09	04	25	0N	
16	06	15	Max dec	27° N 28'
23	14	37	0S	
30	01	05	Max dec	27° S 30'

LONGITUDES

Date	Chiron ⚷ ° '	Ceres ⚳ ° '	Pallas ⚴ ° '	Juno ⚵ ° '	Vesta ⚶ ° '	Black Moon Lilith ⚸ ° '
01	14 ♈ 25	00 ♍ 55	11 ♋ 13	09 ♓ 41	23 ♒ 03	18 ♋ 55
11	13 ♈ 58	05 ♍ 10	15 ♋ 13	08 ♓ 18	23 ♒ 06	20 ♋ 02
21	13 ♈ 32	09 ♍ 20	18 ♋ 51	07 ♓ 43	23 ♒ 53	21 ♋ 09
31	13 ♈ 07	13 ♍ 22	21 ♋ 58	07 ♓ 58	25 ♒ 21	22 ♋ 16

ASPECTARIAN

h m Aspects

01 Saturday
03 29 ♂ ± ♇
04 50 ☽ Q ♀
09 58 ♀ # ♃
11 43 ☽ ⊼ ♅
12 48 ☽ ⚹ ♄
14 46 ☽ ⊥ ♇
15 03 ☽ ☍ ♂
15 17 ☽ Q ☉
18 12 ♀ ☍ ♃
20 45 ☽ □ ♆
21 46 ☽ □ ☿
21 58 ☽ ± ♅

02 Sunday
00 00 ♀ ₽ ♅
01 02 ☽ ⊻ ♇
09 07 ☿ St D
11 09 ☉ ∥ ♆
11 41 ♀ ₽ ♄
12 37 ☽ □ ♃
13 16 ☽ ₽ ♅
14 03 ♀ ∥ ♃
14 20 ☽ ∠ ♄
14 35 ☽ □ ♀

03 Monday
00 14 ☽ □ ☉
03 18 ☽ Q ♆
05 33 ☽ ⊥ ♄
14 37 ☽ △ ♅
15 39 ☽ ⊻ ♄
18 48 ☽ Q ♃
19 12 ☽ ⊼ ♂
23 31 ☽ ⚹ ♆

04 Tuesday
00 59 ☽ △ ☿
03 49 ☽ ☌ ♇
05 32 ☽ ± ♂
14 48 ☽ ⚹ ♃
21 00 ☽ ₽ ♂
22 04 ☽ △ ♀

05 Wednesday
00 41 ☽ ∠ ♆
00 55 ☽ ∥ ♇
02 53 ☽ ₽ ☿
03 40 ☿ ± ♄
03 43 ☽ # ♂
06 31 ☽ △ ☉
11 57 ☉ ± ♅
15 47 ☽ ∠ ♃
15 47 ☽ ⊥ ♆
16 58 ☽ □ ♅
18 00 ☽ ☌ ♄
18 34 ☽ ± ☿
22 46 ☽ △ ♂

06 Thursday
01 41 ☽ ₽ ♀
01 51 ☽ ⊻ ♆
05 06 ☽ ⊼ ☿
06 14 ☽ ⊻ ♇
06 47 ☽ ⊥ ♃
09 25 ☽ # ♅
09 37 ☽ ₽ ☉
11 30 ☽ ∥ ♄
16 21 ☽ ⊥ ♇
16 49 ☽ ⊻ ♃
18 19 ☽ ± ♀
23 14 ☽ Q ♅

07 Friday
01 57 ☽ ± ☉
03 56 ☿ △ ♇
05 06 ☿ # ♀
05 28 ☽ ⊼ ♀
07 36 ☽ ∠ ♇
12 54 ☽ ⊼ ☉
19 35 ☽ ⚹ ♅
20 39 ☽ ⊻ ♄

08 Saturday
02 40 ☽ □ ♂
04 40 ☽ ☌ ♆
06 07 ☽ ∥ ☉
06 56 ☽ ⊥ ♄
09 14 ☽ ⚹ ♇
11 10 ☽ ☍ ☿
14 20 ☽ ∥ ♆
15 41 ☽ ∥ ♀
19 37 ☽ ☌ ♃
19 38 ☽ # ☿
21 21 ☽ ∠ ♅
21 57 ♇ St D
22 28 ☽ ∠ ♄

09 Sunday
00 48 ♀ ± ♅
01 57 ☽ ∥ ♃
06 04 ☽ Q ♇
06 56 ☽ # ♃
09 18 ♀ ∥ ♆
10 31 ☽ Q ♂
12 29 ☽ ∥ ☿
13 04 ☽ ⊥ ♅
14 20 ☽ ☍ ♀
18 41 ☽ # ♆
19 29 ☽ # ♀
20 55 ☽ ☍ ☉
23 39 ☽ ⊻ ♅

10 Monday
00 48 ☽ ⚹ ♄
06 04 ☽ # ☉
08 14 ☽ ⚹ ♂
09 09 ☽ ⊻ ♆
14 02 ☽ □ ♇
19 57 ☽ ⊥ ♆
20 47 ☽ ⊼ ☿
22 21 ☽ Q ♄

11 Tuesday
00 25 ☽ ⊻ ♃
08 53 ☽ ± ☿
09 13 ☉ ⊼ ♅
11 18 ☽ ⊥ ♃
12 02 ☽ ∠ ♂
12 23 ☽ ∠ ♆

12 Wednesday
00 40 ☽ # ♄
01 06 ☉ △ ♄
02 29 ☽ ⊼ ♀
02 30 ☽ ∥ ♅
03 26 ☽ ₽ ☿
03 53 ☽ ∠ ♃
05 06 ☽ ⊥ ♂
05 46 ♂ □ ♆
06 20 ☽ ☌ ♅
07 24 ☿ ☍ ♃
07 37 ☽ □ ♄
08 10 ☽ ⊼ ☉
15 01 ☽ ± ♀
16 25 ☽ ⚹ ♆
16 38 ☽ ⊻ ♂
17 59 ☿ # ♃
20 30 ☽ ± ☉
21 42 ☽ △ ♇

13 Thursday
06 01 ☿ ₽ ♅
08 08 ☽ ⚹ ♃
10 09 ☽ ₽ ♀
11 29 ☽ △ ☿
14 55 ☽ ∥ ♂
15 14 ☽ # ♇
15 18 ☽ ₽ ☉
15 25 ☽ Q ♆
16 57 ♀ ⊼ ♅
17 41 ☿ ₽ ♄

14 Friday
02 43 ☽ ₽ ♇
06 21 ♀ △ ♄
07 17 ☽ Q ♃
13 31 ♂ # ♇
16 07 ☽ ⊻ ♅
17 33 ☽ △ ♄
18 46 ♂ ⊥ ♅
18 51 ☽ △ ♀
20 28 ☽ ± ♇
23 23 ☽ △ ☉

15 Saturday
02 46 ☽ □ ♆
04 00 ☽ ⊥ ♅
04 11 ☽ ☌ ♂
08 25 ☽ ⊼ ♇
18 44 ☽ □ ♃
21 58 ☽ ∠ ♅
23 29 ☽ ₽ ♄

16 Sunday
00 36 ☿ ∥ ♃
07 19 ☽ □ ☿
15 39 ☉ ⊼ ♆
17 37 ☽ ± ♄

17 Monday
04 09 ☽ ⚹ ♅
05 44 ☽ ⊼ ♄
14 14 ☽ □ ♀
15 04 ☽ △ ♆
17 15 ☽ □ ☉
17 35 ☽ ⊻ ♂
20 56 ☽ ☍ ♇
22 00 ♀ ⊼ ♆
22 05 ☉ △ ♂

18 Tuesday
01 19 ☽ Q ☿
04 20 ☽ Q ♅
05 53 ☽ ⊥ ♂
06 44 ☽ △ ♃
21 06 ☽ ∥ ♂
21 15 ☽ ₽ ♆
22 36 ☽ # ♇
22 52 ☿ ± ♅

19 Wednesday
00 14 ☽ ∠ ♂
02 20 ♀ △ ♂
05 16 ☽ ⚹ ☿
06 43 ☽ Q ♀
08 44 ☽ Q ☉
12 32 ☽ ₽ ♃
13 33 ☉ □ ♇
15 11 ☽ ± ♆
16 17 ☽ □ ♅
17 59 ☽ ☍ ♄

20 Thursday
03 02 ☽ ⊼ ♆
06 03 ♀ □ ♇
06 11 ☽ ± ♃
06 23 ☽ ⚹ ♂
08 53 ☽ ⊼ ♇
09 13 ☽ ⚹ ♀
09 54 ☿ ∥ ♆
10 35 ☽ ⚹ ☉
13 30 ☽ ∥ ♅
14 51 ☽ # ♄
15 35 ☽ ∠ ☿
17 50 ☽ ⊼ ♃
20 37 ☽ ± ♇

21 Friday
06 01 ☽ Q ♂
11 43 ☽ ⊥ ☿
13 59 ☽ ₽ ♇
17 30 ☽ ∠ ♀
17 39 ☽ # ☉
18 07 ☽ ∠ ☉
22 11 ☽ # ♀

22 Saturday
00 54 ☽ ⊻ ☿
02 19 ☽ △ ♅
04 06 ☽ ⊼ ♄
11 24 ☿ ⊼ ♅
12 21 ☽ ⊥ ♀
12 33 ☽ ⊥ ☉
12 37 ☽ ☍ ♆
13 28 ☉ ± ♆
15 02 ♀ ± ♆
15 21 ☽ ± ♄
16 27 ☽ □ ♂
16 52 ☽ # ☿
18 17 ☽ △ ♇
21 17 ☉ ☌ ♀
23 15 ☽ # ♆

23 Sunday
00 38 ☽ ⊻ ☉
00 43 ☽ ⊻ ♀
01 00 ☿ △ ♄
02 17 ☽ ☍ ♃
04 07 ♄ St D
06 12 ☽ ₽ ♅
08 00 ☽ ₽ ♄
09 35 ☽ # ♃
15 58 ♀ ⊼ ♃
19 41 ☽ ∥ ♃
20 21 ☉ ⊼ ♃
22 32 ☽ ± ♅

24 Monday
05 46 ☽ ∥ ♆
09 18 ☽ ⊼ ♅
11 08 ☽ △ ♄
16 01 ☽ ☌ ☿
17 27 ☽ ∥ ☿
19 07 ☽ ⊼ ♆
23 12 ☽ △ ♂

25 Tuesday
00 36 ☽ □ ♇
05 40 ☽ ± ♆
07 46 ☽ ⊼ ♃
10 49 ☽ ☌ ☉
12 04 ☽ ∥ ♀
12 05 ☽ ☌ ♀
15 04 ☽ ∥ ☉
16 30 ☿ ⊼ ♆
18 10 ☽ ± ♃
21 21 ☽ ₽ ♆

26 Wednesday
01 31 ☽ ₽ ♂
07 56 ☽ Q ♇
09 21 ☽ ∥ ♄
09 36 ☽ ₽ ♃
10 16 ☽ # ♅
13 34 ☽ ☍ ♅
15 30 ☽ □ ♄
17 03 ☽ ± ♂
23 04 ☽ △ ♆

27 Thursday
03 17 ☽ ⊻ ☿
03 19 ☽ ⊼ ♂
03 37 ☿ △ ♂
04 27 ☽ ⚹ ♇
11 01 ☽ △ ♃
13 09 ☿ □ ♇
14 50 ☽ ⊥ ☿
18 11 ☽ ⊻ ☉
19 33 ☽ ∥ ♇
20 30 ☽ ⊻ ♀
22 05 ☽ Q ♄
23 34 ☽ # ♂

28 Friday
02 57 ♀ ± ♃
05 04 ☽ ⊥ ☉
05 50 ☽ ∠ ♇
07 35 ☽ ⊥ ♀
08 03 ☽ ∠ ☿
16 13 ☽ ⊼ ♅
18 16 ☽ ⚹ ♄
20 57 ☽ ⊥ ♇
21 19 ☽ ∠ ☉

29 Saturday
00 09 ☽ ∠ ♀
01 37 ☽ □ ♆
02 15 ☽ ± ♅
04 57 ☿ ± ♆
05 59 ☽ ☍ ♂
07 01 ☽ ⊻ ♇
08 37 ☉ ± ♃
12 34 ☽ ⚹ ☿
13 10 ☽ □ ♃
16 46 ♀ ₽ ♆
17 17 ☽ ₽ ♅
17 32 ☿ ⊼ ♃
19 26 ☽ ∠ ♄

30 Sunday
00 20 ☽ ⚹ ☉
03 41 ☽ ⚹ ♀
07 44 ☽ Q ♆
10 31 ☽ ⊥ ♄
11 19 ☽ Q ☿
13 26 ♂ St R
18 21 ☽ △ ♅
19 10 ☽ Q ♃
20 36 ☽ ⊻ ♄
21 58 ☽ Q ☉

31 Monday
01 45 ☽ Q ♀
03 51 ☽ ⚹ ♆
08 19 ☽ ⊼ ♂
09 23 ☽ ☌ ♇
09 51 ☉ ₽ ♆
15 14 ☽ ⚹ ♃
18 25 ☽ ± ♂
18 55 ♀ ₽ ♂
21 37 ☽ □ ☿

All ephemeris data is given at 12.00 UT and the Moon's longitude is additionally given for 24.00 UT

NOVEMBER 2022

LONGITUDES

Date	Sidereal time h m s	Sun ☉ ° ′ ″	Moon ☽ ° ′ ″	Moon ☽ 24.00 ° ′ ″	Mercury ☿ ° ′	Venus ♀ ° ′	Mars ♂ ° ′	Jupiter ♃ ° ′	Saturn ♄ ° ′	Uranus ♅ ° ′	Neptune ♆ ° ′	Pluto ♇ ° ′
01	14 43 02	09 ♏ 02 32	11 ♒ 59 27	19 ♒ 02 31	04 ♏ 29	11 ♏ 30	25 ♊ 35	29 ♓ 39	18 ♒ 40	17 ♉ 14	22 ♓ 56	26 ♑ 15
02	14 46 59	10 02 34	26 ♒ 03 27	03 ♓ 02 04	06 09	12 45	25 R 33	29 R 35	18 41	17 R 12	22 R 55	26 16
03	14 50 55	11 02 37	09 ♓ 58 11	16 ♓ 51 38	07 48	14 00	25 30	29 31	18 42	17 09	22 54	26 17
04	14 54 52	12 02 42	23 ♓ 42 16	00 ♈ 29 56	09 26	15 15	25 26	29 27	18 43	17 07	22 53	26 17
05	14 58 49	13 02 49	07 ♈ 14 31	13 ♈ 55 54	11 04	16 31	25 21	29 23	18 44	17 04	22 52	26 18
06	15 02 45	14 02 57	20 ♈ 33 58	27 ♈ 08 37	12 42	17 46	25 16	29 19	18 46	17 02	22 51	26 19
07	15 06 42	15 03 07	03 ♉ 39 49	10 ♉ 07 29	14 19	19 01	25 09	29 16	18 47	16 59	22 50	26 20
08	15 10 38	16 03 19	16 ♉ 31 38	22 ♉ 52 17	15 56	20 16	25 02	29 13	18 49	16 57	22 50	26 21
09	15 14 35	17 03 32	29 ♉ 09 29	05 ♊ 23 20	17 33	21 32	24 54	29 10	18 51	16 54	22 49	26 22
10	15 18 31	18 03 47	11 ♊ 33 59	17 ♊ 41 38	19 09	22 47	24 45	29 07	18 53	16 52	22 48	26 22
11	15 22 28	19 04 05	23 ♊ 46 29	29 ♊ 48 52	20 45	24 02	24 35	29 04	18 55	16 49	22 47	26 23
12	15 26 24	20 04 24	05 ♋ 49 04	11 ♋ 47 30	22 20	25 17	24 24	29 02	18 57	16 47	22 46	26 24
13	15 30 21	21 04 45	17 ♋ 44 33	23 ♋ 40 42	23 55	26 33	24 12	28 59	18 59	16 44	22 46	26 25
14	15 34 18	22 05 07	29 ♋ 36 26	05 ♌ 32 17	25 30	27 48	24 00	28 57	19 01	16 42	22 45	26 26
15	15 38 14	23 05 32	11 ♌ 28 46	17 ♌ 26 29	27 05	29 ♏ 03	23 46	28 55	19 03	16 39	22 44	26 27
16	15 42 11	24 05 58	23 ♌ 26 01	29 ♌ 27 56	28 ♏ 39	00 ♐ 18	23 32	28 54	19 06	16 37	22 44	26 29
17	15 46 07	25 06 27	05 ♍ 32 51	11 ♍ 41 19	00 ♐ 13	01 34	23 17	28 52	19 08	16 34	22 43	26 30
18	15 50 04	26 06 57	17 ♍ 53 53	24 ♍ 11 05	01 47	02 49	23 01	28 51	19 11	16 32	22 43	26 31
19	15 54 00	27 07 29	00 ♎ 33 22	07 ♎ 01 08	03 20	04 04	22 45	28 50	19 13	16 30	22 42	26 32
20	15 57 57	28 08 02	13 ♎ 34 43	20 ♎ 14 20	04 54	05 20	22 28	28 49	19 16	16 27	22 42	26 33
21	16 01 53	29 ♏ 08 38	27 ♎ 00 05	03 ♏ 51 58	06 27	06 35	22 10	28 49	19 19	16 25	22 41	26 34
22	16 05 50	00 ♐ 09 15	10 ♏ 49 50	17 ♏ 53 22	08 00	07 50	21 51	28 48	19 22	16 22	22 41	26 36
23	16 09 47	01 09 53	25 ♏ 02 08	02 ♐ 15 33	09 33	09 05	21 32	28 48	19 25	16 20	22 41	26 37
24	16 13 43	02 10 34	09 ♐ 32 53	16 ♐ 53 20	11 05	10 21	21 12	28 D 48	19 28	16 18	22 40	26 38
25	16 17 40	03 11 15	24 ♐ 15 58	01 ♑ 39 51	12 38	11 36	20 52	28 48	19 32	16 15	22 40	26 39
26	16 21 36	04 11 58	09 ♑ 03 59	16 ♑ 27 26	14 10	12 51	20 31	28 49	19 35	16 13	22 40	26 41
27	16 25 33	05 12 42	23 ♑ 49 19	01 ♒ 08 49	15 42	14 07	20 10	28 49	19 38	16 11	22 39	26 42
28	16 29 29	06 13 27	08 ♒ 25 15	15 ♒ 38 03	17 14	15 22	19 48	28 50	19 42	16 08	22 39	26 43
29	16 33 26	07 14 13	22 ♒ 46 47	29 ♒ 51 09	18 46	16 37	19 26	28 51	19 45	16 06	22 39	26 45
30	16 37 22	08 ♐ 15 00	06 ♓ 50 57	13 ♓ 46 08	20 ♐ 17	17 ♐ 53	19 ♊ 04	28 ♓ 52	19 ♒ 49	16 ♉ 04	22 ♓ 39	26 ♑ 46

Date	Moon True ☊ ° ′	Moon Mean ☊ ° ′	Moon ☽ Latitude ° ′
01	13 ♉ 25	13 ♉ 25	05 S 17
02	13 R 24	13 21	05 09
03	13 24	13 18	04 44
04	13 24	13 15	04 02
05	13 24	13 12	03 07
06	13 D 24	13 09	02 03
07	13 24	13 06	00 S 54
08	13 R 24	13 02	00 N 17
09	13 24	12 59	01 26
10	13 24	12 56	02 30
11	13 23	12 53	03 26
12	13 22	12 50	04 11
13	13 21	12 46	04 46
14	13 20	12 43	05 08
15	13 19	12 40	05 16
16	13 19	12 37	05 12
17	13 D 19	12 34	04 53
18	13 20	12 31	04 21
19	13 21	12 27	03 35
20	13 22	12 24	02 37
21	13 24	12 21	01 29
22	13 24	12 18	00 N 14
23	13 R 24	12 15	01 S 04
24	13 23	12 12	02 19
25	13 20	12 08	03 26
26	13 18	12 05	04 20
27	13 15	12 02	04 57
28	13 12	11 59	05 13
29	13 10	11 56	05 10
30	13 ♉ 09	11 ♉ 52	04 S 48

DECLINATIONS

Date	Sun ☉ ° ′	Moon ☽ ° ′	Mercury ☿ ° ′	Venus ♀ ° ′	Mars ♂ ° ′	Jupiter ♃ ° ′	Saturn ♄ ° ′	Uranus ♅ ° ′	Neptune ♆ ° ′	Pluto ♇ ° ′
01	14 S 31	22 S 16	12 S 11	14 S 33	23 N 54	01 S 34	16 S 27	16 N 37	03 S 55	23 S 04
02	14 50	17 41	12 50	14 58	23 57	01 35	16 27	16 37	03 56	23 04
03	15 08	12 12	13 28	15 23	23 59	01 37	16 27	16 36	03 56	23 04
04	15 27	06 S 12	14 06	15 47	24 02	01 38	16 26	16 35	03 56	23 04
05	15 45	00 00	14 43	16 11	24 05	01 39	16 26	16 34	03 57	23 04
06	16 03	06 N 08	15 20	16 34	24 07	01 41	16 25	16 34	03 57	23 04
07	16 21	11 54	15 55	16 57	24 10	01 42	16 25	16 33	03 57	23 04
08	16 39	17 03	16 30	17 19	24 13	01 43	16 24	16 32	03 58	23 03
09	16 56	21 22	17 04	17 41	24 16	01 44	16 24	16 32	03 58	23 03
10	17 13	24 39	17 37	18 03	24 18	01 45	16 23	16 31	03 58	23 03
11	17 29	26 43	18 09	18 24	24 21	01 46	16 22	16 30	03 59	23 03
12	17 46	27 30	18 41	18 44	24 23	01 46	16 21	16 30	03 59	23 03
13	18 02	26 59	19 11	19 04	24 26	01 47	16 21	16 29	03 59	23 03
14	18 17	25 14	19 40	19 24	24 28	01 48	16 20	16 28	03 59	23 02
15	18 33	22 24	20 09	19 42	24 31	01 48	16 19	16 28	04 00	23 02
16	18 48	18 36	20 36	20 01	24 33	01 48	16 18	16 27	04 00	23 02
17	19 02	14 01	21 03	20 19	24 36	01 49	16 18	16 26	04 00	23 02
18	19 17	08 47	21 28	20 36	24 38	01 49	16 17	16 25	04 00	23 02
19	19 31	03 N 04	21 52	20 53	24 40	01 49	16 16	16 25	04 00	23 02
20	19 45	02 S 56	22 16	21 09	24 42	01 49	16 15	16 24	04 01	23 01
21	19 58	09 01	22 38	21 24	24 44	01 49	16 14	16 23	04 01	23 01
22	20 11	14 51	22 59	21 39	24 46	01 49	16 13	16 23	04 01	23 01
23	20 24	20 03	23 19	21 53	24 48	01 49	16 12	16 22	04 01	23 01
24	20 36	24 11	23 37	22 07	24 50	01 49	16 11	16 21	04 01	23 00
25	20 48	26 45	23 55	22 20	24 51	01 48	16 10	16 21	04 01	23 00
26	20 59	27 27	24 11	22 32	24 53	01 48	16 09	16 20	04 01	23 00
27	21 10	26 12	24 26	22 44	24 54	01 48	16 08	16 19	04 01	23 00
28	21 21	23 11	24 40	22 55	24 55	01 47	16 06	16 19	04 01	23 00
29	21 31	18 48	24 53	23 05	24 56	01 46	16 05	16 18	04 01	22 59
30	21 S 41	13 S 27	25 S 04	23 S 15	24 N 57	01 S 45	16 S 04	16 N 18	04 S 01	22 S 59

ZODIAC SIGN ENTRIES

Date	h	m	Planets
02	18	46	☽ ♓
04	23	07	☽ ♈
07	05	15	☽ ♉
09	13	37	☽ ♊
12	00	22	☽ ♋
14	12	48	☽ ♌
16	06	08	♀ ♐
17	01	04	☽ ♍
17	08	42	☿ ♐
19	10	58	☽ ♎
21	17	16	☽ ♏
22	08	20	☉ ♐
23	20	16	☽ ♐
25	21	18	☽ ♑
27	22	07	☽ ♒
30	00	15	☽ ♓

LATITUDES

Date	Mercury ☿ ° ′	Venus ♀ ° ′	Mars ♂ ° ′	Jupiter ♃ ° ′	Saturn ♄ ° ′	Uranus ♅ ° ′	Neptune ♆ ° ′	Pluto ♇ ° ′
01	00 N 53	00 N 46	00 N 32	01 S 33	01 S 17	00 S 22	01 S 13	02 S 13
04	00 34	00 39	00 40	01 33	01 17	00 22	01 13	02 13
07	00 N 13	00 33	00 49	01 32	01 17	00 22	01 13	02 13
10	00 S 07	00 26	00 58	01 31	01 17	00 22	01 13	02 13
13	00 27	00 19	01 07	01 30	01 17	00 22	01 13	02 13
16	00 46	00 12	01 17	01 30	01 17	00 22	01 13	02 13
19	01 04	00 N 05	01 26	01 29	01 16	00 22	01 13	02 13
22	01 22	00 S 02	01 35	01 28	01 16	00 22	01 12	02 13
25	01 37	00 09	01 44	01 27	01 16	00 22	01 12	02 13
28	01 51	00 17	01 53	01 26	01 16	00 22	01 12	02 13
31	02 S 03	00 S 24	02 N 01	01 S 25	01 S 16	00 S 22	01 S 12	02 S 14

DATA

Julian Date	2459885
Delta T	+71 seconds
Ayanamsa	24° 10′ 21″
Synetic vernal point	04° ♓ 56′ 38″
True obliquity of ecliptic	23° 26′ 18″

MOON'S PHASES, APSIDES AND POSITIONS ☽

Date	h	m	Phase	Longitude ° ′	Eclipse Indicator
01	06	37	☽	08 ♒ 49	
08	11	02	○	16 ♉ 01	total
16	13	27	☾	24 ♌ 10	
23	22	57	●	01 ♐ 38	
30	14	37	☽	08 ♓ 22	

Day	h	m		
14	06	35	Apogee	
26	01	25	Perigee	
05	11	59	0N	
12	14	21	Max dec	27° N 30′
20	00	21	0S	
26	08	33	Max dec	27° S 29′

LONGITUDES

Date	Chiron ⚷ ° ′	Ceres ⚳ ° ′	Pallas ⚴ ° ′	Juno ⚵ ° ′	Vesta ⚶ ° ′	Black Moon Lilith ⚸ ° ′
01	13 ♈ 04	13 ♍ 45	22 ♋ 14	08 ♓ 02	25 ♒ 31	22 ♋ 23
11	12 ♈ 42	17 ♍ 37	24 ♋ 36	09 ♓ 11	27 ♒ 36	23 ♋ 30
21	12 ♈ 23	21 ♍ 17	26 ♋ 06	11 ♓ 04	00 ♓ 10	24 ♋ 37
31	12 ♈ 09	24 ♍ 44	26 ♋ 33	13 ♓ 37	03 ♓ 08	25 ♋ 44

All ephemeris data is given at 12.00 UT and the Moon's longitude is additionally given for 24.00 UT

ASPECTARIAN

h m	Aspects
01 Tuesday	
00 39	☉ ∥ ♀
01 36	☽ # ♂
05 08	☽ ∠ ♆
06 37	☽ □ ☉
06 59	☽ ∥ ♇
09 37	☽ ꝗ ♂
11 05	☽ □ ♀
16 30	☽ ∠ ♃
20 24	☽ ⊥ ♆
20 53	☽ □ ♅
23 22	☽ ☌ ♄
02 Wednesday	
04 09	☿ ± ♃
06 38	☽ ⊻ ♆
07 47	☽ ⊥ ♃
11 08	☽ △ ♂
12 21	☽ ⊻ ♇
16 57	☽ # ♅
17 40	☽ ∥ ♄
18 01	☽ ⊻ ♃
22 41	☽ ⊥ ♇
23 24	☽ ∥ ♀
23 40	☉ ꝗ ♂
03 Thursday	
00 11	☽ ∥ ☉
03 41	☽ Q ♅
07 13	☽ ∥ ☿
07 43	☽ △ ☿
13 32	☿ ꝗ ♆
14 01	☽ △ ☉
14 16	☽ ∠ ♇
17 18	♀ Q ♇
19 43	☽ △ ♀
21 15	♀ ꝗ ♃
04 Friday	
00 28	☽ ⚹ ♅
03 14	☽ ⊻ ♄
10 34	☽ ☌ ♆
13 28	☽ ꝗ ☿
13 47	☽ ⊥ ♄
15 02	☽ □ ♂
16 34	☽ ⚹ ♇
18 21	☽ ꝗ ☉
20 46	☽ ∥ ♆
22 05	☽ ☌ ♃
05 Saturday	
00 45	☽ ꝗ ♀
02 02	☿ ꝗ ♂
02 49	☽ ∠ ♅
05 37	☽ ∥ ♃
05 45	☽ ∠ ♄
07 36	☽ ± ☿
11 37	☽ ± ☉
13 54	☽ Q ♇
18 25	☽ # ♃
18 27	☽ ± ♀
18 50	☽ ⊥ ♅
19 49	☽ ⊼ ☿
22 22	♀ ☍ ♅
22 54	☽ Q ♂
23 15	☽ ⊼ ☉
06 Sunday	
03 16	♀ ∥ ♄
03 22	☽ # ♆
05 36	☽ ⊻ ♅
06 23	☽ ⊼ ♀
08 44	☽ ⚹ ♄
11 52	♀ # ♅
16 09	☽ ⊻ ♆
18 09	☉ ꝗ ♃
18 28	☉ Q ♇
20 30	☽ ⚹ ♂
22 30	☽ □ ♇
07 Monday	
03 07	☽ ⊥ ♆
03 56	☽ ⊻ ♃
06 41	☽ Q ♄
07 34	♀ □ ♄
11 11	☿ ꝗ ♃
12 08	☿ Q ♇
14 26	♀ ± ♂
14 57	☽ ⊥ ♃
16 55	☉ ∥ ♄
19 44	☽ ∠ ♆
23 57	☽ ∠ ♂
08 Tuesday	
03 50	☉ # ♅
07 40	☽ ∠ ♃
08 00	☿ ∥ ♄
08 46	☽ # ♄
08 53	☽ # ☿
09 27	☽ ∥ ♅
09 49	☽ # ☉
10 44	☽ ☍ ☿
11 02	☽ ☍ ☉
12 47	☽ ☌ ♅
13 27	☽ # ♀
13 36	☿ # ♅
16 20	☽ □ ♄
16 41	☽ ⊥ ♂
16 43	☉ ☌ ☿
19 51	☽ ☍ ♀
23 54	☽ ⚹ ♆
09 Wednesday	
00 04	☉ ∥ ☿
02 40	☿ ☍ ♅
03 57	☽ ⊻ ♂
06 38	☽ △ ♇
08 26	☉ ☍ ♅
12 00	☽ ⚹ ♃
22 52	☽ Q ♆
23 17	☽ # ♇
10 Thursday	
06 30	☿ ± ♂
07 52	☿ □ ♄
09 00	☽ ∥ ♂
11 07	☽ Q ♃
11 37	☽ ꝗ ♇
12 22	♀ △ ♆
22 20	☽ ⊻ ♅
11 Friday	
01 52	☽ ⊼ ☉
01 59	☉ ± ♂
02 22	☽ △ ♄
05 06	☽ ⊼ ☿
05 18	☽ ± ♇
08 04	☉ □ ♄
10 03	☽ □ ♆
10 07	☽ ⊥ ♅
12 34	☽ ⊼ ♀
13 34	☽ ☌ ♂
14 48	☽ ± ☉
17 12	☽ ⊼ ♇
18 47	☽ ± ☿
21 07	♀ ⊼ ♂
22 28	☽ □ ♃
12 Saturday	
01 53	☽ ± ♀
03 57	☽ ∠ ♅
08 14	☽ ꝗ ♄
10 22	☽ ꝗ ☉
15 30	☽ ꝗ ☿
18 36	☿ △ ♆
19 57	☿ ∥ ♀
22 02	☽ ꝗ ♀
13 Sunday	
02 22	☽ ± ♄
09 41	♀ ⚹ ♇
09 59	☽ ⚹ ♅
14 30	☽ ⊼ ♄
15 47	☿ ⊼ ♂
19 22	☽ △ ☉
22 08	☽ △ ♆
14 Monday	
00 50	☽ ⊻ ♂
02 25	☽ △ ☿
05 35	☽ ☍ ♇
07 54	☽ △ ♀
10 10	☽ Q ♅
10 41	☽ △ ♃
12 46	☽ ⊥ ♂
19 17	☽ ∥ ♂
15 Tuesday	
02 27	☿ ⚹ ♇
03 43	☉ △ ♆
04 28	☽ ꝗ ♆
06 38	☽ ∠ ♂
07 16	☽ # ♇
09 36	♀ △ ♃
16 55	☽ ꝗ ♃
22 23	☽ □ ♅
22 35	☽ ± ♆
16 Wednesday	
01 07	☉ ⊼ ♂
01 18	☽ # ☿
03 17	☽ ☍ ♄
04 18	☽ # ♀
10 36	☽ ⊼ ♆
10 56	☽ ± ♃
10 58	☽ # ☉
12 12	☽ ⚹ ♂
13 27	☽ □ ☉
15 43	☿ △ ♃
18 04	☽ ⊼ ♇
22 51	☽ ⊼ ♃
23 47	☽ ∥ ♅
23 55	☽ □ ☿
17 Thursday	
00 31	☽ # ♄
03 14	☽ □ ♀
05 59	☽ ± ♇
11 30	☽ Q ♂
23 38	☽ ꝗ ♇
18 Friday	
04 04	☽ Q ☉
09 23	☽ △ ♅
14 28	☽ ⊼ ♄
16 07	☽ Q ☿
18 12	☽ Q ♀
21 12	☽ ☍ ♆
21 35	☽ □ ♂
21 38	☉ ⚹ ♇
19 Saturday	
01 56	☽ ± ♄
04 26	☽ △ ♇
05 00	☽ ⚹ ☉
08 10	☽ # ♆
08 47	☽ ☍ ♃
13 45	☽ ꝗ ♅
15 43	♂ □ ♆
17 04	☽ # ♃
17 54	☽ ⚹ ☿
18 51	☽ ꝗ ♄
19 15	☽ ⚹ ♀
20 Sunday	
06 19	☽ ± ♅
07 35	☽ ∥ ♃
11 08	☽ ∠ ☉
12 45	♂ ⊥ ♅
16 13	☽ ∥ ♆
17 11	☽ ⊼ ♅
22 18	☽ △ ♄
21 Monday	
00 52	☽ ∠ ☿
01 24	☽ ∠ ♀
03 37	☽ △ ♂
04 07	☉ △ ♃
04 22	☽ ⊼ ♆
04 38	☽ ⊥ ☉
11 15	☽ □ ♇
14 58	☽ ± ♆
15 11	☽ ⊼ ♃
16 04	☽ ⊻ ☉
18 49	☽ ⊥ ☿
18 55	☽ ⊥ ♀
22 55	☿ ☌ ♀
22 Tuesday	
01 38	☽ ± ♃
01 56	☿ Q ♄
02 41	♀ Q ♄
05 19	☽ ꝗ ♂
06 21	☽ ⊻ ♀
06 32	☽ ⊻ ☿
06 36	☽ ꝗ ♆
14 18	☿ ∥ ♇
17 04	☽ ꝗ ♃
17 56	☽ ∥ ♄
18 25	☽ Q ♇
18 40	☽ # ♅
20 22	☽ ± ♂
21 24	☽ ☍ ♅
23 Wednesday	
02 33	☽ □ ♄
06 16	☽ ⊼ ♂
08 03	☽ △ ♆
13 48	☽ ∥ ☉
14 38	☽ ⚹ ♇
18 16	☽ △ ♃
22 19	☽ ∥ ♀
22 57	☽ ☌ ☉
23 02	♃ St D
24 Thursday	
04 21	☽ ∥ ♇
07 56	☽ ∥ ☿
08 35	☽ Q ♄
13 26	☽ ☌ ♀
14 49	☽ ☌ ☿
15 25	☽ ∠ ♇
16 50	☽ # ♂
20 40	☿ ∠ ♇
23 00	☽ ⊼ ♅
25 Friday	
04 16	☽ ⚹ ♄
06 08	☽ ⊥ ♇
06 36	☽ ☍ ♂
08 44	☽ ± ♅
09 24	☽ □ ♆
13 05	♀ ∠ ♇
15 53	☽ ⊻ ♇
19 22	☽ □ ♃
23 18	☽ ꝗ ♅
26 Saturday	
01 40	♂ ± ♇
03 32	☽ ⊻ ☉
04 42	☽ ∠ ♄
13 58	☽ ⊥ ☉
14 35	☽ Q ♆
18 43	☽ ⊻ ♀
19 21	☽ ⊥ ♄
21 14	☽ ⊻ ☿
23 35	☽ △ ♅
27 Sunday	
00 35	☽ Q ♃
05 09	☽ ⊻ ♄
05 23	☽ ⊥ ♀
05 41	☽ ∠ ☉
06 10	☽ ⊼ ♂
08 08	☽ ⊥ ☿
10 06	☽ ⚹ ♆
15 44	☽ ± ♂
16 43	☽ ☌ ♇
19 16	☿ ⊼ ♅
20 11	☽ ⚹ ♃
21 28	☽ ∠ ♀
23 54	☽ # ♂
28 Monday	
00 35	☽ ∠ ☿
02 27	☽ ∥ ☿
06 10	☽ ꝗ ♂
08 06	☽ ⚹ ☉
10 44	☽ ∠ ♆
13 15	☽ ∥ ♇
13 41	☽ ∥ ♀
17 56	♂ △ ♄
21 00	☽ ∠ ♃
22 27	☽ ∥ ☉
22 56	♀ ∥ ♇
29 Tuesday	
00 40	☽ ⚹ ♀
00 48	☽ □ ♅
01 42	☽ ⊥ ♆
02 19	♀ ⊼ ♅
04 26	☽ ⚹ ☿
05 35	☽ Q ☉
06 31	☽ △ ♂
06 53	☽ ☌ ♄
11 47	☽ ⊻ ♆
12 07	☽ ⊥ ♃
18 44	☽ ⊻ ♇
20 07	☿ # ♂
20 30	☿ ☍ ♂
22 19	☽ ⊻ ♃
22 52	☽ Q ♀
23 38	☽ # ♅
30 Wednesday	
00 37	☽ ∥ ♄
01 09	☉ Q ♄
03 13	☽ Q ☿
04 18	☿ ⚹ ♄
04 59	☽ ⊥ ♇
07 13	☽ Q ♅
14 37	☽ □ ☉
19 41	☿ ⊥ ♇
20 32	☽ ∠ ♇

DECEMBER 2022

LONGITUDES

Date	Sidereal time h m s	Sun ☉ ° ' "	Moon ☽ ° ' "	Moon ☽ 24.00 ° ' "	Mercury ☿ ° '	Venus ♀ ° '	Mars ♂ ° '	Jupiter ♃ ° '	Saturn ♄ ° '	Uranus ♅ ° '	Neptune ♆ ° '	Pluto ♇ ° '
01	16 41 19	09 ♐ 15 48	20 ♓ 36 42	27 ♓ 22 46	21 ♐ 49	19 ♐ 08	18 ♊ 41	28 ♓ 54	19 ♒ 53	16 ♉ 01	22 ♓ 39	26 ♑ 48
02	16 45 16	10 16 36	04 ♈ 04 28	10 ♈ 42 00	23 20	20 23	18 R 18	28 56	19 57	15 R 59	22 R 39	26 49
03	16 49 12	11 17 26	17 ♈ 15 35	23 ♈ 45 27	24 51	21 38	17 55	28 57	20 01	15 57	22 39	26 51
04	16 53 09	12 18 16	00 ♉ 11 49	06 ♉ 34 55	26 22	22 54	17 32	28 59	20 05	15 55	22 D 39	26 52
05	16 57 05	13 19 07	12 ♉ 54 59	19 ♉ 12 10	27 52	24 09	17 09	29 02	20 09	15 53	22 39	26 54
06	17 01 02	14 20 00	25 ♉ 26 41	01 ♊ 38 40	29 ♐ 22	25 24	16 46	29 04	20 13	15 51	22 39	26 55
07	17 04 58	15 20 53	07 ♊ 48 18	13 ♊ 55 42	00 ♑ 52	26 40	16 23	29 07	20 17	15 49	22 39	26 57
08	17 08 55	16 21 47	20 ♊ 01 01	26 ♊ 04 24	02 21	27 55	16 00	29 10	20 22	15 47	22 39	26 58
09	17 12 51	17 22 42	02 ♋ 06 00	08 ♋ 05 59	03 49	29 ♐ 10	15 37	29 13	20 26	15 45	22 39	27 00
10	17 16 48	18 23 39	14 ♋ 04 35	20 ♋ 01 58	05 17	00 ♑ 25	15 14	29 16	20 31	15 43	22 39	27 01
11	17 20 45	19 24 36	25 ♋ 58 27	01 ♌ 54 17	06 44	01 41	14 52	29 20	20 35	15 41	22 40	27 03
12	17 24 41	20 25 34	07 ♌ 49 49	13 ♌ 45 26	08 10	02 56	14 29	29 23	20 40	15 39	22 40	27 05
13	17 28 38	21 26 33	19 ♌ 41 32	25 ♌ 38 36	09 35	04 11	14 07	29 27	20 45	15 37	22 40	27 06
14	17 32 34	22 27 34	01 ♍ 37 06	07 ♍ 37 36	10 59	05 26	13 46	29 31	20 50	15 35	22 41	27 08
15	17 36 31	23 28 35	13 ♍ 40 38	19 ♍ 46 48	12 21	06 42	13 25	29 35	20 55	15 33	22 41	27 10
16	17 40 27	24 29 37	25 ♍ 56 42	02 ♎ 10 56	13 41	07 57	13 04	29 40	21 00	15 31	22 41	27 11
17	17 44 24	25 30 40	08 ♎ 30 06	14 ♎ 54 45	14 58	09 12	12 44	29 44	21 05	15 30	22 42	27 13
18	17 48 20	26 31 44	21 ♎ 25 25	28 ♎ 02 34	16 14	10 27	12 24	29 49	21 10	15 28	22 42	27 15
19	17 52 17	27 32 49	04 ♏ 46 32	11 ♏ 37 35	17 26	11 43	12 05	29 54	21 15	15 26	22 43	27 17
20	17 56 14	28 33 55	18 ♏ 35 49	25 ♏ 41 10	18 35	12 58	11 46	29 ♓ 59	21 20	15 25	22 43	27 18
21	18 00 10	29 ♐ 35 02	02 ♐ 53 21	10 ♐ 11 54	19 40	14 13	11 28	00 ♈ 05	21 25	15 23	22 44	27 20
22	18 04 07	00 ♑ 36 10	17 ♐ 36 09	25 ♐ 05 12	20 40	15 28	11 11	00 10	21 31	15 21	22 45	27 22
23	18 08 03	01 37 18	02 ♑ 37 58	10 ♑ 13 13	21 34	16 44	10 54	00 16	21 36	15 20	22 45	27 24
24	18 12 00	02 38 26	17 ♑ 49 39	25 ♑ 25 53	22 23	17 59	10 38	00 22	21 42	15 18	22 46	27 26
25	18 15 56	03 39 35	03 ♒ 00 34	10 ♒ 32 28	23 04	19 14	10 23	00 28	21 47	15 17	22 47	27 27
26	18 19 53	04 40 44	18 ♒ 00 27	25 ♒ 23 34	23 37	20 29	10 09	00 34	21 53	15 16	22 47	27 29
27	18 23 49	05 41 53	02 ♓ 41 06	09 ♓ 52 30	24 02	21 45	09 55	00 41	21 59	15 14	22 48	27 31
28	18 27 46	06 43 02	16 ♓ 57 27	23 ♓ 55 50	24 17	23 00	09 42	00 47	22 05	15 13	22 49	27 33
29	18 31 43	07 44 11	00 ♈ 47 41	07 ♈ 33 12	24 R 21	24 15	09 30	00 54	22 10	15 12	22 50	27 35
30	18 35 39	08 45 19	14 ♈ 12 40	20 ♈ 46 29	24 14	25 30	09 19	01 01	22 16	15 11	22 51	27 37
31	18 39 36	09 ♑ 46 28	27 ♈ 15 04	03 ♉ 38 54	23 ♑ 56	26 ♑ 45	09 ♊ 09	01 ♈ 08	22 ♒ 22	15 ♉ 09	22 ♓ 52	27 ♑ 39

Date	Moon True ☊ ° '	Moon Mean ☊ ° '	Moon ☽ Latitude ° '
01	13 ♉ 10	11 ♉ 49	04 S 10
02	13 D 11	11 46	03 19
03	13 13	11 43	02 18
04	13 14	11 40	01 11
05	13 15	11 37	00 S 02
06	13 R 14	11 33	01 N 06
07	13 12	11 30	02 10
08	13 08	11 27	03 07
09	13 02	11 24	03 55
10	12 55	11 21	04 32
11	12 48	11 18	04 57
12	12 42	11 14	05 09
13	12 36	11 11	05 07
14	12 32	11 08	04 53
15	12 29	11 05	04 25
16	12 D 29	11 02	03 45
17	12 29	10 58	02 54
18	12 31	10 55	01 52
19	12 32	10 52	00 N 42
20	12 R 32	10 49	00 S 33
21	12 30	10 46	01 48
22	12 26	10 43	02 57
23	12 20	10 39	03 56
24	12 12	10 36	04 39
25	12 04	10 33	05 02
26	11 56	10 30	05 04
27	11 50	10 27	04 46
28	11 46	10 24	04 11
29	11 44	10 20	03 21
30	11 D 44	10 17	02 22
31	11 ♉ 45	10 ♉ 14	01 S 17

DECLINATIONS

Date	Sun ☉ ° '	Moon ☽ ° '	Mercury ☿ ° '	Venus ♀ ° '	Mars ♂ ° '	Jupiter ♃ ° '	Saturn ♄ ° '	Uranus ♅ ° '	Neptune ♆ ° '	Pluto ♇ ° '
01	21 S 50	07 S 33	25 S 14	23 S 23	24 N 58	01 S 45	16 S 03	16 N 17	04 S 01	22 S 59
02	21 59	01 S 25	25 22	23 32	24 59	01 44	16 02	16 16	04 01	22 59
03	22 08	04 N 39	25 29	23 39	24 59	01 43	16 00	16 16	04 01	22 58
04	22 16	10 26	25 35	23 46	24 59	01 42	15 59	16 15	04 01	22 58
05	22 24	15 41	25 39	23 52	25 00	01 40	15 58	16 14	04 01	22 58
06	22 31	20 12	25 42	23 57	25 00	01 39	15 56	16 14	04 01	22 58
07	22 38	23 45	25 44	24 02	25 00	01 38	15 55	16 13	04 01	22 57
08	22 44	26 10	25 44	24 05	25 00	01 37	15 53	16 13	04 01	22 57
09	22 50	27 20	25 42	24 08	24 59	01 35	15 52	16 12	04 01	22 57
10	22 56	27 12	25 39	24 11	24 59	01 33	15 50	16 12	04 01	22 57
11	23 01	25 48	25 35	24 12	24 58	01 32	15 49	16 11	04 01	22 56
12	23 06	23 17	25 29	24 13	24 58	01 30	15 47	16 10	04 01	22 56
13	23 10	19 46	25 22	24 13	24 57	01 28	15 46	16 10	04 00	22 56
14	23 13	15 27	25 13	24 13	24 56	01 26	15 44	16 09	04 00	22 55
15	23 17	10 30	25 03	24 11	24 55	01 25	15 43	16 09	04 00	22 55
16	23 19	05 N 03	24 52	24 09	24 54	01 23	15 41	16 08	04 00	22 55
17	23 22	00 S 43	24 39	24 06	24 53	01 20	15 39	16 08	04 00	22 55
18	23 24	06 38	24 25	24 02	24 52	01 18	15 38	16 07	03 59	22 54
19	23 25	12 27	24 10	23 58	24 51	01 16	15 36	16 07	03 59	22 54
20	23 26	17 53	23 54	23 53	24 50	01 14	15 34	16 07	03 59	22 54
21	23 26	22 30	23 37	23 47	24 48	01 11	15 32	16 06	03 59	22 53
22	23 26	25 48	23 19	23 40	24 47	01 09	15 31	16 06	03 58	22 53
23	23 26	27 21	23 01	23 33	24 46	01 06	15 29	16 05	03 58	22 53
24	23 25	26 52	22 42	23 25	24 44	01 04	15 27	16 05	03 58	22 52
25	23 23	24 23	22 23	23 16	24 43	01 01	15 25	16 04	03 58	22 52
26	23 21	20 15	22 04	23 07	24 42	00 58	15 23	16 04	03 57	22 52
27	23 19	14 58	21 45	22 57	24 41	00 56	15 22	16 04	03 57	22 52
28	23 16	09 00	21 27	22 46	24 39	00 53	15 20	16 03	03 57	22 51
29	23 13	02 S 46	21 09	22 34	24 38	00 50	15 18	16 03	03 56	22 51
30	23 09	03 N 25	20 53	22 22	24 37	00 47	15 16	16 03	03 56	22 51
31	23 S 05	09 N 18	20 S 37	22 S 09	24 N 36	00 S 44	15 S 14	16 N 02	03 S 55	22 S 50

ZODIAC SIGN ENTRIES

Date	h	m	Planets
02	04	41	☽ ♈
04	11	38	☽ ♉
06	20	49	☽ ♊
06	22	08	☿ ♑
09	07	49	☽ ♋
10	03	54	♀ ♑
11	20	09	☽ ♌
14	08	45	☽ ♍
16	19	49	☽ ♎
19	03	31	☽ ♏
20	14	32	♃ ♈
21	07	13	☽ ♐
21	21	48	☉ ♑
23	07	49	☽ ♑
25	07	14	☽ ♒
27	07	34	☽ ♓
29	10	36	☽ ♈
31	17	08	☽ ♉

LATITUDES

Date	Mercury ☿ ° '	Venus ♀ ° '	Mars ♂ ° '	Jupiter ♃ ° '	Saturn ♄ ° '	Uranus ♅ ° '	Neptune ♆ ° '	Pluto ♇ ° '
01	02 S 03	00 S 24	02 N 01	01 S 25	01 S 16	00 S 22	01 S 12	02 S 14
04	02 12	00 31	02 09	01 25	01 16	00 22	01 12	02 14
07	02 18	00 38	02 16	01 24	01 15	00 22	01 12	02 14
10	02 19	00 45	02 23	01 23	01 15	00 22	01 12	02 14
13	02 17	00 51	02 28	01 22	01 15	00 22	01 12	02 14
16	02 08	00 57	02 34	01 21	01 15	00 22	01 12	02 14
19	01 53	01 03	02 38	01 20	01 15	00 22	01 12	02 14
22	01 29	01 08	02 42	01 20	01 15	00 22	01 12	02 15
25	00 56	01 14	02 45	01 19	01 15	00 22	01 11	02 15
28	00 S 11	01 18	02 47	01 18	01 15	00 21	01 11	02 15
31	00 N 42	01 S 22	02 N 49	01 S 17	01 S 15	00 S 21	01 S 11	02 S 15

DATA

Julian Date	2459915
Delta T	+71 seconds
Ayanamsa	24° 10' 26"
Synetic vernal point	04° ♓ 56' 33"
True obliquity of ecliptic	23° 26' 17"

MOON'S PHASES, APSIDES AND POSITIONS ☽

Date	h	m	Phase	Longitude ° '	Eclipse Indicator
08	04	08	○	16 ♊ 02	
16	08	56	☾	24 ♍ 22	
23	10	17	●	01 ♑ 33	
30	01	21	☽	08 ♈ 18	

Day	h	m		
12	00	18	Apogee	
24	08	21	Perigee	
02	17	33	0N	
09	21	27	Max dec	27° N 26'
17	09	05	0S	
23	18	22	Max dec	27° S 25'
29	22	37	0N	

LONGITUDES

Date	Chiron ⚷ ° '	Ceres ⚳ ° '	Pallas ⚴ ° '	Juno ⚵ ° '	Vesta ⚶ ° '	Black Moon Lilith ⚸ ° '
01	12 ♈ 09	24 ♍ 44	26 ♋ 33	13 ♓ 37	03 ♓ 08	25 ♋ 44
11	12 ♈ 00	27 ♍ 52	25 ♋ 48	16 ♓ 44	06 ♓ 26	26 ♋ 51
21	11 ♈ 56	00 ♎ 40	23 ♋ 53	20 ♓ 20	10 ♓ 01	27 ♋ 59
31	11 ♈ 58	03 ♎ 03	20 ♋ 59	24 ♓ 22	13 ♓ 50	29 ♋ 06

ASPECTARIAN

h m	Aspects
01 Thursday	
03 58	☽ ∗ ♅
05 28	♀ ☍ ♂
08 42	☽ □ ♂
09 08	☽ □ ♀
10 43	☽ ⊻ ♄
14 23	☽ □ ☿
15 15	☿ ± ♅
15 36	☽ ☌ ♆
21 23	☽ ⊥ ♄
22 59	☽ ∗ ♇
02 Friday	
01 08	☿ □ ♆
01 50	☽ ∥ ♆
02 44	☽ ☌ ♃
03 09	♀ ∗ ♄
06 28	☽ ∠ ♅
10 47	☽ ∥ ♃
13 35	☽ ∠ ♄
15 55	☽ Q ♂
20 26	♀ ⊥ ♇
20 36	☽ Q ♇
22 41	☽ ⊥ ♅
03 Saturday	
00 10	☽ △ ☉
00 19	☽ # ♃
09 27	☽ # ♆
09 36	☽ ⊻ ♅
13 11	☽ ∗ ♂
17 06	☽ ∗ ♄
17 47	♀ ± ♅
20 57	☽ △ ♀
21 56	☽ ⊻ ♆
04 Sunday	
00 12	♆ St D
01 24	☉ ∠ ♇
03 53	☽ △ ☿
05 46	☽ □ ♇
06 08	☽ bQ ☉
07 12	♀ □ ♆
09 06	☽ ⊥ ♆
09 44	☽ ⊻ ♃
15 33	☽ Q ♄
16 16	☽ ∠ ♂
20 11	☿ ⊻ ♇
21 02	☽ ⊥ ♃
05 Monday	
00 28	☽ ± ☉
02 00	☽ ∠ ♆
04 04	☽ bQ ♀
08 45	☽ ⊥ ♂
11 54	☽ bQ ☿
12 50	☽ ⊼ ☉
13 20	☽ # ♄
14 07	☽ ∠ ♃
14 44	☽ ∥ ♅
17 38	☽ ☌ ♅
19 50	☽ ⊻ ♂
23 06	☽ ± ♀
06 Tuesday	
01 54	☽ □ ♄
06 37	☽ ∗ ♆
07 05	☿ □ ♃
07 27	☽ ± ☿
11 55	☽ ⊼ ♀
14 51	☽ △ ♇
19 02	☽ ∗ ♃
20 38	☽ ⊼ ☿
07 Wednesday	
03 22	☽ # ☉
05 50	☽ Q ♆
05 58	☽ # ♇
11 11	☿ bQ ♅
14 18	☽ # ♀
17 33	♀ ⊻ ♇
18 30	☽ Q ♃
20 07	☽ bQ ♇
22 34	☉ ⊼ ♅
22 52	☽ ∥ ♂
08 Thursday	
03 40	☽ ⊻ ♅
04 08	☽ ☍ ☉
04 19	☽ ☌ ♂
05 42	☉ ☍ ♂
06 34	☽ # ☿
12 41	☽ △ ♄
13 53	☽ ± ♇
15 28	☽ ⊥ ♅
17 13	☽ □ ♆
09 Friday	
01 49	☽ ⊼ ♇
03 05	♂ ⊻ ♅
05 29	☽ ☍ ♀
06 13	☽ □ ♃
09 18	☽ ∠ ♅
12 55	♀ □ ♃
15 55	☽ ☍ ☿
18 43	☽ bQ ♄
10 Saturday	
12 53	☽ ± ♄
14 16	☽ ⊻ ♂
14 53	☉ ∥ ♇
15 17	☽ ∗ ♅
15 58	☿ ∠ ♄
17 20	♀ bQ ♅
21 30	☽ ⊼ ☉
11 Sunday	
01 03	☽ ⊼ ♄
01 59	☽ ⊥ ♂
05 18	☽ △ ♆
10 45	☽ ± ☉
14 11	☽ ☍ ♇
14 44	☽ # ☿
15 26	☽ Q ♅
18 49	☽ △ ♃
19 37	☽ ∠ ♂
21 08	☽ ∥ ♂
12 Monday	
00 54	☽ ⊼ ♀
04 11	☽ # ♀
06 40	☽ bQ ☉
11 40	☽ bQ ♆
12 47	☽ ⊼ ☿
13 24	☽ # ☉
14 30	☽ ± ♀
14 40	☽ # ♇
18 12	☉ ∗ ♄
13 Tuesday	
01 04	☽ ∗ ♂
01 21	☽ bQ ♃
02 35	☽ ± ☿
03 47	☽ □ ♅
03 49	☉ ⊥ ♇
05 54	☽ ± ♆
10 51	☽ bQ ♀
14 09	☽ ☍ ♄
15 52	☽ △ ☉
15 55	☉ ± ♅
18 01	☽ ⊼ ♆
19 38	☽ ± ♃
23 11	☽ bQ ☿
14 Wednesday	
00 35	☽ Q ♂
02 59	☽ ⊼ ♇
06 47	☿ Q ♆
07 46	☽ ⊼ ♃
08 20	☽ ∥ ♅
10 32	☽ # ♄
15 02	☽ ± ♇
17 10	☉ □ ♆
19 55	♀ ∠ ♄
20 32	☽ △ ♀
15 Thursday	
09 00	☽ bQ ♇
09 02	☽ △ ☿
11 29	☽ □ ♂
15 41	☽ △ ♅
16 Friday	
02 18	☽ ⊼ ♄
03 12	☿ ⊼ ♂
05 41	☽ ☍ ♆
06 37	☿ # ♂
08 56	☽ □ ☉
14 02	☽ ± ♄
14 25	☽ △ ♇
16 29	☽ # ♆
19 13	☽ ☍ ♃
20 48	☽ bQ ♅
17 Saturday	
03 31	☽ # ♃
07 23	☽ bQ ♄
13 28	☽ □ ♀
13 52	☽ ± ♅
14 04	☿ ⊥ ♄
14 33	☽ ∥ ♃
19 44	☽ △ ♂
21 37	☿ △ ♅
22 12	☽ Q ☉
18 Sunday	
01 03	☽ ⊼ ♅
01 20	☽ ∥ ♆
01 26	☽ □ ☿
11 31	☽ △ ♄
14 20	☽ ⊼ ♆
16 46	♀ Q ♆
22 03	☽ ∗ ☉
22 20	♂ bQ ♇
22 35	☽ bQ ♂
22 35	☽ □ ♇
19 Monday	
01 12	☽ ± ♆
02 04	☽ Q ♀
03 17	☽ ⊼ ♃
05 26	☉ ⊻ ♇
13 16	☽ Q ☿
14 00	☽ ± ♃
14 15	☽ ± ♂
17 11	☽ bQ ♆
17 37	♀ ⊼ ♂
22 29	☿ Q ♃
22 30	☿ ± ♂
22 32	♂ Q ♃
20 Tuesday	
00 30	☽ ⊼ ♂
01 22	☽ ∗ ♀
01 30	☽ ∥ ♄
02 40	☽ ∠ ☉
03 52	☽ # ♅
05 47	☽ bQ ♃
06 21	☽ Q ♇
06 33	☽ ☍ ♅
11 58	☽ ∗ ☿
14 37	☿ ∥ ♀
16 41	☽ □ ♄
19 01	☽ △ ♆
19 16	☽ ⊥ ☉
21 Wednesday	
02 45	☽ ∗ ♇
05 20	☽ ∠ ♀
06 06	☽ ⊻ ☉
07 19	☽ △ ♃
14 23	☽ ∥ ♇
15 09	☽ ∠ ☿
17 51	☽ ∥ ☉
18 30	☽ ∥ ☿
19 53	☽ ∥ ♀
21 35	☽ ⊥ ♀
22 48	☽ Q ♄
22 Thursday	
00 50	☉ □ ♃
01 48	☽ ☍ ♂
02 43	☉ ∥ ☿
03 22	☽ # ♂
03 31	☽ ∠ ♇
06 22	☉ bQ ♅
06 55	☽ ⊥ ☿
08 15	☽ ⊻ ♀
08 23	☽ ⊼ ♅
09 48	♀ △ ♅
12 49	♀ ⊥ ♄
17 15	☽ ⊻ ☿
18 01	☽ ± ♅
18 04	☽ ⊥ ♇
18 19	☽ ∗ ♄
20 16	☽ □ ♆
23 Friday	
03 40	☽ ⊻ ♇
08 13	☽ □ ♃
08 21	☽ bQ ♅
10 17	☽ ☌ ☉
13 08	☿ ⊻ ♄
14 47	♀ ± ♂
18 20	☽ ∠ ♄
22 32	☿ ∥ ♇
24 Saturday	
00 43	♃ ∠ ♅
00 51	☽ Q ♆
00 51	☽ ⊼ ♂
08 02	☽ △ ♅
08 37	☽ ⊥ ♄
10 10	☽ ± ♂
12 16	☽ ☌ ♀
12 51	☽ Q ♃
13 05	☉ ∥ ♀
18 08	☽ ⊻ ♄
19 32	☽ ☌ ☿
19 48	☽ ∗ ♆
19 58	♀ Q ♃
25 Sunday	
00 08	☽ bQ ♂
01 17	☿ ∗ ♆
03 11	☽ ☌ ♇
07 56	☽ ∗ ♃
09 32	☽ # ♂
13 06	☽ ⊻ ☉
18 41	☽ ∥ ☉
19 36	☽ ∠ ♆
19 40	☽ ∥ ♀
21 49	☽ ∥ ♇
23 22	☽ ⊥ ☉
23 34	☽ △ ♂
26 Monday	
01 41	☽ ∥ ☿
07 35	☽ □ ♅
08 03	☽ ∠ ♃
10 02	☽ ⊥ ♆
14 54	☽ ∠ ☉
16 23	☽ ⊻ ♀
18 19	☽ ☌ ♄
19 46	☽ ⊻ ♆
21 24	☽ ⊻ ☿
22 44	☽ ⊥ ♃
27 Tuesday	
03 05	☽ ⊥ ♀
03 28	☽ ⊻ ♇
07 19	☽ # ♅
07 31	☽ ⊥ ☿
08 39	☽ ⊻ ♃
10 19	☽ ∥ ♄
12 55	☽ Q ♅
13 23	☽ ⊥ ♇
16 53	♀ ⊻ ♄
17 23	☽ ∗ ☉
19 24	☽ ∠ ♀
22 49	☽ ∠ ☿
23 36	♀ ∥ ♇
23 54	☽ □ ♂
28 Wednesday	
04 30	☽ ∠ ♇
08 32	♀ ∗ ♆
09 02	☽ ∗ ♅
15 15	☽ Q ☉
20 51	☽ ⊻ ♄
21 19	☉ ∠ ♄
22 05	☽ ☌ ♆
23 25	☽ ∗ ♀
29 Thursday	
00 43	☽ ∗ ☿
06 19	☽ Q ♂
06 21	☽ ∗ ♇
07 22	☽ ⊥ ♄
07 29	☽ ∥ ♆
09 31	☿ St R
10 57	☽ ∠ ♅
12 11	☽ ☌ ♃
13 58	☿ ☌ ♀
16 16	♀ bQ ♂
19 28	☽ ∥ ♃
21 49	☽ Q ☿
22 40	☽ Q ♀
23 24	☽ ∠ ♄
30 Friday	
01 21	☽ □ ☉
01 44	☽ # ♃
02 55	☽ ⊥ ♅
03 17	☽ ∗ ♂
03 40	☽ Q ♇
13 45	☽ ⊻ ♅
14 01	☽ # ♆
23 21	☉ ⊼ ♂
31 Saturday	
02 52	☽ ∗ ♄
03 51	☽ ⊻ ♆
06 00	☽ □ ☿
06 18	☽ ∠ ♂
10 59	☽ □ ♀
12 44	☽ □ ♇
15 01	☽ ⊥ ♆
19 20	☽ ⊻ ♃
22 55	☽ ⊥ ♂

All ephemeris data is given at 12.00 UT and the Moon's longitude is additionally given for 24.00 UT

JANUARY 2023

LONGITUDES

Date	Sidereal time h m s	Sun ☉ ° ' "	Moon ☽ ° ' "	Moon ☽ 24.00 ° ' "	Mercury ☿ ° '	Venus ♀ ° '	Mars ♂ ° '	Jupiter ♃ ° '	Saturn ♄ ° '	Uranus ♅ ° '	Neptune ♆ ° '	Pluto ♇ ° '
01	18 43 32	10 ♑ 47 36	09 ♉ 58 27	16 ♉ 14 13	23 ♑ 25	28 ♑ 01	08 ♊ 59	01 ♈ 15	22 ♒ 28	15 ♉ 08	22 ♓ 53	27 ♑ 40
02	18 47 29	11 48 45	22 ♉ 26 38	28 ♉ 36 09	22 R 43	29 ♑ 16	08 R 51	01 23	22 34	15 R 07	22 54	27 42
03	18 51 25	12 49 53	04 ♊ 43 10	10 ♊ 48 02	21 50	00 ♒ 31	08 43	01 30	22 40	15 06	22 55	27 44
04	18 55 22	13 51 01	16 ♊ 51 04	22 ♊ 52 34	20 47	01 46	08 35	01 38	22 46	15 05	22 56	27 46
05	18 59 18	14 52 09	28 ♊ 52 45	04 ♋ 51 51	19 37	03 01	08 29	01 46	22 53	15 04	22 57	27 48
06	19 03 15	15 53 17	10 ♋ 50 03	16 ♋ 47 30	18 20	04 16	08 24	01 54	22 59	15 03	22 58	27 50
07	19 07 12	16 54 25	22 ♋ 44 23	28 ♋ 40 51	17 00	05 31	08 19	02 02	23 05	15 03	22 59	27 52
08	19 11 08	17 55 33	04 ♌ 37 02	10 ♌ 33 08	15 39	06 46	08 15	02 10	23 11	15 02	23 00	27 54
09	19 15 05	18 56 40	16 ♌ 29 21	22 ♌ 25 54	14 20	08 01	08 12	02 19	23 18	15 01	23 02	27 56
10	19 19 01	19 57 48	28 ♌ 23 03	04 ♍ 21 06	13 06	09 16	08 10	02 27	23 24	15 00	23 03	27 58
11	19 22 58	20 58 55	10 ♍ 20 24	16 ♍ 21 21	11 57	10 31	08 08	02 36	23 31	15 00	23 04	28 00
12	19 26 54	22 00 02	22 ♍ 24 22	28 ♍ 29 57	10 57	11 46	08 08	02 45	23 37	14 59	23 05	28 02
13	19 30 51	23 01 09	04 ♎ 38 38	10 ♎ 50 57	10 05	13 01	08 D 08	02 54	23 44	14 59	23 07	28 04
14	19 34 47	24 02 16	17 ♎ 07 28	23 ♎ 28 48	09 22	14 16	08 09	03 03	23 50	14 58	23 08	28 06
15	19 38 44	25 03 23	29 ♎ 55 31	06 ♏ 28 11	08 50	15 31	08 10	03 12	23 57	14 58	23 09	28 08
16	19 42 41	26 04 30	13 ♏ 07 17	19 ♏ 53 16	08 27	16 46	08 13	03 22	24 04	14 58	23 11	28 10
17	19 46 37	27 05 37	26 ♏ 46 27	03 ♐ 47 02	08 13	18 01	08 16	03 31	24 10	14 57	23 12	28 12
18	19 50 34	28 06 44	10 ♐ 55 01	18 ♐ 10 13	08 08	19 16	08 20	03 41	24 17	14 57	23 14	28 13
19	19 54 30	29 ♑ 07 50	25 ♐ 32 11	03 ♑ 00 17	08 D 12	20 31	08 24	03 51	24 24	14 57	23 15	28 15
20	19 58 27	00 ♒ 08 56	10 ♑ 33 34	18 ♑ 10 56	08 23	21 46	08 29	04 01	24 31	14 57	23 17	28 17
21	20 02 23	01 10 01	25 ♑ 51 01	03 ♒ 32 22	08 42	23 01	08 35	04 11	24 38	14 56	23 18	28 19
22	20 06 20	02 11 06	11 ♒ 13 28	18 ♒ 52 46	09 07	24 16	08 42	04 21	24 44	14 56	23 20	28 21
23	20 10 16	03 12 10	26 ♒ 28 50	04 ♓ 00 24	09 38	25 30	08 49	04 31	24 51	14 D 56	23 21	28 23
24	20 14 13	04 13 13	11 ♓ 26 22	18 ♓ 45 52	10 15	26 45	08 57	04 42	24 58	14 56	23 23	28 25
25	20 18 10	05 14 15	25 ♓ 58 19	03 ♈ 03 22	10 56	28 00	09 06	04 52	25 05	14 57	23 25	28 27
26	20 22 06	06 15 16	10 ♈ 00 51	16 ♈ 50 53	11 42	29 ♒ 15	09 15	05 03	25 12	14 57	23 26	28 29
27	20 26 03	07 16 16	23 ♈ 33 41	00 ♉ 09 37	12 32	00 ♓ 29	09 25	05 14	25 19	14 57	23 28	28 31
28	20 29 59	08 17 15	06 ♉ 39 11	13 ♉ 02 55	13 26	01 44	09 35	05 24	25 26	14 57	23 30	28 33
29	20 33 56	09 18 12	19 ♉ 21 24	25 ♉ 35 16	14 23	02 59	09 47	05 35	25 33	14 58	23 31	28 35
30	20 37 52	10 19 09	01 ♊ 45 05	07 ♊ 51 29	15 22	04 13	09 58	05 46	25 40	14 58	23 33	28 37
31	20 41 49	11 ♒ 20 04	13 ♊ 55 01	19 ♊ 56 14	16 ♑ 25	05 ♓ 28	10 ♊ 10	05 ♈ 58	25 ♒ 48	14 ♉ 58	23 ♓ 35	28 ♑ 39

Date	Moon True ☊ ° '	Moon Mean ☊ ° '	Moon ☽ Latitude ° '
01	11 ♉ 45	10 ♉ 11	00 S 10
02	11 R 45	10 08	00 N 57
03	11 42	10 04	01 59
04	11 37	10 01	02 55
05	11 28	09 58	03 43
06	11 17	09 55	04 21
07	11 05	09 52	04 46
08	10 51	09 49	05 00
09	10 38	09 45	05 00
10	10 26	09 42	04 47
11	10 17	09 39	04 21
12	10 10	09 36	03 44
13	10 06	09 33	02 56
14	10 05	09 29	01 58
15	10 D 05	09 26	00 N 54
16	10 R 05	09 23	00 S 16
17	10 04	09 20	01 27
18	10 01	09 17	02 35
19	09 55	09 14	03 35
20	09 46	09 10	04 22
21	09 36	09 07	04 52
22	09 24	09 04	05 00
23	09 13	09 01	04 47
24	09 03	08 58	04 14
25	08 56	08 55	03 25
26	08 52	08 51	02 26
27	08 50	08 48	01 20
28	08 D 50	08 45	00 S 12
29	08 R 50	08 42	00 N 55
30	08 48	08 39	01 58
31	08 ♉ 45	08 ♉ 35	02 N 54

DECLINATIONS

Date	Sun ☉ ° '	Moon ☽ ° '	Mercury ☿ ° '	Venus ♀ ° '	Mars ♂ ° '	Jupiter ♃ ° '	Saturn ♄ ° '	Uranus ♅ ° '	Neptune ♆ ° '	Pluto ♇ ° '
01	23 S 00	14 N 39	20 S 24	21 S 55	24 N 35	00 S 41	15 S 12	16 N 02	03 S 55	22 S 50
02	22 55	19 18	20 11	21 41	24 34	00 38	15 10	16 02	03 55	22 50
03	22 49	23 02	20 01	21 26	24 33	00 34	15 08	16 02	03 54	22 49
04	22 43	25 42	19 52	21 11	24 32	00 31	15 06	16 01	03 54	22 49
05	22 37	27 09	19 45	20 55	24 31	00 28	15 04	16 01	03 53	22 49
06	22 30	27 19	19 40	20 38	24 30	00 24	15 02	16 01	03 53	22 48
07	22 22	26 13	19 36	20 20	24 30	00 21	15 00	16 01	03 52	22 48
08	22 14	23 57	19 33	20 02	24 29	00 17	14 58	16 00	03 52	22 48
09	22 06	20 39	19 33	19 44	24 29	00 14	14 56	16 00	03 51	22 47
10	21 57	16 31	19 33	19 25	24 28	00 10	14 53	16 00	03 51	22 47
11	21 48	11 43	19 35	19 05	24 28	00 07	14 51	16 00	03 50	22 47
12	21 38	06 27	19 38	18 45	24 28	00 S 03	14 49	16 00	03 50	22 46
13	21 28	00 N 51	19 42	18 24	24 28	00 N 01	14 47	16 00	03 49	22 46
14	21 18	04 S 54	19 47	18 03	24 28	00 05	14 45	15 59	03 49	22 46
15	21 07	10 37	19 54	17 41	24 28	00 09	14 43	15 59	03 48	22 46
16	20 56	16 02	20 00	17 19	24 28	00 13	14 40	15 59	03 47	22 45
17	20 44	20 51	20 08	16 56	24 28	00 17	14 38	15 59	03 47	22 45
18	20 32	24 38	20 16	16 33	24 28	00 21	14 36	15 59	03 46	22 45
19	20 20	26 57	20 24	16 10	24 29	00 25	14 34	15 59	03 46	22 44
20	20 07	27 23	20 32	15 45	24 29	00 29	14 31	15 59	03 45	22 44
21	19 54	25 45	20 41	15 21	24 30	00 33	14 29	15 59	03 44	22 44
22	19 40	22 12	20 49	14 56	24 30	00 37	14 27	15 59	03 44	22 43
23	19 26	17 11	20 57	14 31	24 31	00 42	14 24	15 59	03 43	22 43
24	19 12	11 11	21 04	14 05	24 32	00 46	14 22	15 59	03 42	22 43
25	18 57	04 S 45	21 12	13 39	24 33	00 50	14 20	15 59	03 42	22 42
26	18 43	01 N 44	21 18	13 12	24 34	00 55	14 18	15 59	03 41	22 42
27	18 27	07 55	21 24	12 46	24 35	00 59	14 15	15 59	03 40	22 42
28	18 12	13 33	21 29	12 18	24 36	01 03	14 13	16 00	03 40	22 41
29	17 56	18 27	21 34	11 51	24 37	01 08	14 10	16 00	03 39	22 41
30	17 39	22 26	21 37	11 23	24 38	01 13	14 08	16 00	03 38	22 41
31	17 S 23	25 N 21	21 S 40	10 S 55	24 N 40	01 N 17	14 S 06	16 N 00	03 S 38	22 S 41

ZODIAC SIGN ENTRIES

Date	h	m	Planets
03	02	09	♀ ♒
03	02	44	☽ ♊
05	14	15	☽ ♋
08	02	40	☽ ♌
10	15	15	☽ ♍
13	02	56	☽ ♎
15	12	08	☽ ♏
17	17	33	☽ ♐
19	19	11	☽ ♑
20	08	29	☉ ♒
21	18	29	☽ ♒
23	17	36	☽ ♓
25	18	48	☽ ♈
27	02	33	♀ ♓
27	23	42	☽ ♉
30	08	35	☽ ♊

LATITUDES

Date	Mercury ☿ ° '	Venus ♀ ° '	Mars ♂ ° '	Jupiter ♃ ° '	Saturn ♄ ° '	Uranus ♅ ° '	Neptune ♆ ° '	Pluto ♇ ° '
01	01 N 02	01 S 23	02 N 49	01 S 17	01 S 15	00 S 21	01 S 11	02 S 15
04	01 59	01 27	02 50	01 16	01 15	00 21	01 11	02 15
07	02 47	01 30	02 50	01 16	01 15	00 21	01 11	02 16
10	03 15	01 32	02 51	01 15	01 15	00 21	01 11	02 16
13	03 22	01 34	02 50	01 14	01 15	00 21	01 11	02 16
16	03 10	01 35	02 50	01 14	01 15	00 21	01 11	02 16
19	02 47	01 36	02 49	01 13	01 15	00 21	01 11	02 17
22	02 19	01 35	02 47	01 12	01 15	00 21	01 11	02 17
25	01 48	01 35	02 46	01 12	01 15	00 21	01 11	02 17
28	01 17	01 33	02 45	01 11	01 15	00 21	01 10	02 17
31	00 N 47	01 S 31	02 N 43	01 S 11	01 S 15	00 S 21	01 S 10	02 S 18

DATA

Julian Date	2459946
Delta T	+72 seconds
Ayanamsa	24° 10' 32"
Synetic vernal point	04° ♓ 56' 27"
True obliquity of ecliptic	23° 26' 17"

MOON'S PHASES, APSIDES AND POSITIONS ☽

Date	h	m	Phase	Longitude ° '	Eclipse Indicator
06	23	08	○	16 ♋ 22	
15	02	10	☾	24 ♎ 38	
21	20	53	●	01 ♒ 33	
28	15	19	☽	08 ♉ 26	

Day	h	m		
08	09	02	Apogee	
21	20	51	Perigee	
06	03	10	Max dec	27° N 25'
13	15	33	0S	
20	05	04	Max dec	27° S 28'
26	05	31	0N	

LONGITUDES

Date	Chiron ⚷ ° '	Ceres ⚳ ° '	Pallas ⚴ ° '	Juno ⚵ ° '	Vesta ⚶ ° '	Black Moon Lilith ⚸ ° '
01	11 ♈ 58	03 ♎ 16	20 ♋ 40	24 ♓ 47	14 ♓ 13	29 ♋ 12
11	12 ♈ 06	05 ♎ 06	17 ♋ 19	29 ♓ 11	18 ♓ 13	00 ♌ 20
21	12 ♈ 19	06 ♎ 21	14 ♋ 12	03 ♈ 53	22 ♓ 23	01 ♌ 27
31	12 ♈ 37	06 ♎ 56	11 ♋ 53	08 ♈ 51	26 ♓ 39	02 ♌ 34

All ephemeris data is given at 12.00 UT and the Moon's longitude is additionally given for 24.00 UT

ASPECTARIAN

h m Aspects

01 Sunday
01 28 ☽ Q ♄
05 25 ♀ ☌ ♇
06 46 ☽ ⊥ ♃
08 00 ☽ ∠ ♆
10 09 ☽ ⊻ ♂
13 42 ☽ △ ☉
14 02 ☉ Q ♆
14 36 ☽ ⧺ ♄
18 43 ☽ ∥ ♅
21 52 ☽ ☌ ♅

02 Monday
00 09 ☽ ∠ ♃
06 44 ☿ ⚹ ♆
12 15 ☽ □ ♄
12 30 ☽ △ ☿
12 53 ☽ ⚹ ♆
16 00 ☿ ⊻ ♄
16 59 ☽ ⧺ ☿
21 16 ☽ ⚼ ☉
22 16 ☽ △ ♇

03 Tuesday
01 39 ☽ ⧺ ♀
02 48 ☽ △ ♀
05 37 ☽ ⚹ ♃
10 27 ☽ ⧺ ♇
10 28 ☽ ⧺ ☉
11 08 ☉ ∥ ♇
12 23 ☽ Q ♆
15 52 ☽ ⚼ ☿
16 32 ☽ ± ☉
19 47 ☽ ☌ ♂

04 Wednesday
00 16 ☽ ∥ ♂
03 53 ☽ ⚼ ♇
05 30 ☽ ⊼ ☉
05 32 ☽ Q ♃
08 15 ☽ ± ☿
08 30 ☽ ⊻ ♅
09 08 ♀ ⚹ ♃
11 49 ☽ ⚼ ♀
19 10 ☽ ⊼ ☿
20 25 ☽ ⊥ ♅
21 49 ☽ ± ♇
23 54 ☽ △ ♄

05 Thursday
00 08 ☽ □ ♆
03 47 ☉ ± ♂
07 50 ☽ ± ♀
09 19 ☿ Q ♃
09 50 ☽ ⊼ ♇
14 23 ☽ ∠ ♅
16 43 ☉ △ ♅
17 50 ☽ □ ♃
21 16 ☽ ⊼ ♀

06 Friday
06 13 ☽ ⚼ ♄
07 08 ☽ ⊻ ♂
07 56 ♄ ⊻ ♆
19 07 ☽ ⊥ ♂
20 30 ☽ ⚹ ♅
23 08 ☽ ☍ ☉

07 Saturday
00 29 ☽ ± ♄
01 36 ☽ ☍ ☿
10 36 ☿ ⊥ ♄
12 30 ☽ △ ♆
12 42 ☽ ⊼ ♄
12 57 ☉ ☌ ☿
13 10 ☽ ∠ ♂
16 42 ☉ ⊥ ♄
20 41 ☽ Q ♅
22 23 ☽ ☍ ♇

08 Sunday
07 00 ☽ △ ♃
07 14 ☽ ∥ ♂
16 52 ☽ ☍ ♀
18 52 ☽ ⚼ ♆
19 19 ☽ ⚹ ♂
21 14 ☽ ⧺ ♇
23 23 ☿ △ ♅

09 Monday
01 52 ☽ ⧺ ☉
08 05 ☽ ⊼ ☿
09 02 ☽ □ ♅
12 04 ♀ ∠ ♆
13 05 ☽ ± ♆
13 41 ☽ ⚼ ♃
14 41 ☿ ± ♂
15 22 ♀ △ ♂
17 26 ☽ ⊼ ☉
18 14 ☽ ⧺ ♀
18 53 ☽ ⧺ ☿
19 02 ☽ ± ☿
19 28 ☽ Q ♂

10 Tuesday
01 13 ☽ ⊼ ♆
01 52 ☽ ☍ ♄
02 10 ☿ ∥ ♀
06 40 ☽ ± ☉
08 04 ☽ ± ♃
11 09 ☽ ⊼ ♇
11 28 ☽ ⚼ ☿
14 44 ☽ ∥ ♅
20 17 ☽ ⊼ ♃
20 32 ☽ ⧺ ♄
23 15 ☽ ± ♇

11 Wednesday
01 30 ☉ Q ♃
02 28 ☽ ⚼ ☉
07 36 ☽ □ ♂
12 24 ☽ ⊼ ♀
14 58 ☽ △ ☿
17 19 ☽ ⚼ ♇
21 17 ☽ △ ♅

12 Thursday
01 45 ☽ ± ♀
03 02 ☿ ⊻ ♀
08 26 ☿ Q ♆
11 07 ☽ △ ☉
13 21 ☽ ☍ ♆
14 25 ☽ ⊼ ♄
20 56 ♂ St D
21 35 ☽ ⚼ ♀
23 06 ☽ △ ♇
23 21 ☽ ⧺ ♆

13 Friday
02 20 ☽ ± ♄
02 54 ☽ ⚼ ♅
08 34 ☽ ☍ ♃
14 11 ☉ ⚹ ♆
14 40 ☉ ⚼ ♂
15 27 ☽ ∥ ♃
15 40 ☽ ⧺ ♃
18 46 ☽ △ ♂
19 59 ☽ ⚼ ♄
20 24 ☽ ± ♅
21 55 ☽ □ ☿

14 Saturday
05 58 ☽ △ ♀
06 46 ☉ ⊻ ♄
07 28 ☽ ∥ ♆
07 54 ☽ ⊼ ♅
23 22 ☽ ⊼ ♆
23 24 ☽ ⚼ ♂

15 Sunday
00 47 ☽ △ ♄
01 22 ♀ □ ♅
02 10 ☽ □ ☉
02 26 ☿ ⊥ ♀
06 28 ☽ Q ☿
06 57 ☿ ∠ ♄
08 40 ☽ □ ♇
10 34 ☽ ± ♆
16 09 ☽ ± ♂
18 07 ☽ ⊼ ♃

16 Monday
03 05 ☽ ⚼ ♆
03 08 ☽ ⊼ ♂
03 47 ☽ ⚹ ☿
05 10 ☽ ± ♃
05 50 ☽ ∥ ♄
11 48 ☽ ⧺ ♅
13 51 ☽ Q ☉
15 17 ☽ ☍ ♅
17 26 ☽ Q ♇
17 38 ☽ ∥ ♀
19 09 ☽ □ ♀
20 00 ♀ ⊥ ♆
21 25 ☽ ⚼ ♃

17 Tuesday
05 48 ☽ △ ♆
05 54 ☽ ∠ ☿
06 42 ☿ ⊼ ♂
07 27 ☽ □ ♄
08 02 ☽ ∥ ☿
11 26 ☽ ∥ ☉
12 36 ☽ ⚹ ☉
14 27 ☽ ⚹ ♇
21 15 ☽ ⊥ ☿
23 02 ☽ ∥ ♇
23 04 ♀ ∠ ♃
23 41 ☽ △ ♃

18 Wednesday
05 18 ☽ Q ♀
07 21 ☽ ⊻ ☿
07 39 ☽ ☍ ♂
10 43 ☽ ⧺ ♂
13 13 ☿ St D
14 18 ☽ Q ♄
14 44 ☉ ☌ ♇
15 51 ☽ ∠ ♇
15 56 ☽ ∠ ☉
18 42 ☽ ⊼ ♅

19 Thursday
03 05 ☽ ⚹ ♀
04 33 ☽ ± ♅
06 40 ☽ ⊥ ♇
07 21 ☉ ∥ ☿
07 49 ☽ ⊥ ☉
08 17 ☽ □ ♆
10 09 ☽ ⚹ ♄
16 24 ☽ ⊻ ♇
18 13 ☽ ⊻ ☉
19 06 ☽ ⚼ ♅
22 26 ♀ ⧺ ♅

20 Friday
01 29 ☽ □ ♃
05 27 ☽ ∠ ♀
08 30 ☽ ☌ ☿
08 42 ☽ ⊼ ♂
10 20 ☽ ∠ ♄
13 08 ☽ Q ♆
18 14 ☽ ± ♂
18 55 ☽ △ ♅
20 56 ☽ ⊥ ♀

21 Saturday
00 37 ☽ ⊥ ♄
01 06 ☿ ⊼ ♂
06 12 ☽ Q ♃
07 10 ☽ ⊻ ♀
08 01 ☽ ⚹ ♆
08 27 ☽ ⚼ ♂
10 04 ☽ ⊻ ♄
15 52 ☽ ☌ ♇
17 43 ♀ ⊻ ♆
20 53 ☽ ☌ ☉
21 53 ☽ ⧺ ♂

22 Sunday
01 09 ☽ ⚹ ♃
07 28 ☽ ∠ ♆
07 41 ☿ ∠ ♀
08 02 ☽ △ ♂
08 36 ☽ ⊻ ☿
09 07 ☽ ∥ ♇
17 49 ☽ □ ♅
18 18 ☽ ⊥ ☿
19 05 ☽ ∥ ☿
21 35 ☽ ⊥ ♆
22 13 ♀ ☌ ♄
23 01 ♅ St D

23 Monday
00 53 ☽ ∠ ♃
01 26 ☽ ∥ ☉
07 03 ☽ ⊻ ♆
08 58 ☽ ∠ ☿
09 24 ☽ ☌ ♄
10 19 ☽ ☌ ♀
15 02 ☽ ⊻ ♇
15 17 ☽ ⊥ ♃
17 01 ☽ ⧺ ♅
18 23 ♀ ∥ ♄
22 18 ☽ Q ♅
23 09 ☿ ∠ ♄
23 30 ☽ ⊻ ☉
23 31 ☽ ∥ ♄
23 52 ☽ ∥ ♀

24 Tuesday
00 38 ☽ ⊥ ♇
00 58 ☽ ⊻ ♃
07 56 ☽ □ ♂
09 52 ☽ ⊥ ☉
09 58 ☽ ⚹ ☿
15 14 ☽ ∠ ♇
17 43 ☽ ⚹ ♅

25 Wednesday
01 30 ☉ ⚹ ♃
01 43 ☽ ∠ ☉
06 39 ☽ Q ☿
07 42 ☽ ☌ ♆
10 30 ☽ ⊻ ♄
13 55 ☽ Q ♂
15 44 ☽ ⊻ ♀
15 52 ☽ ∥ ♆
16 11 ☽ ⚹ ♇
18 42 ☽ ∠ ♅
20 43 ☽ ⊥ ♄
20 59 ♀ ⊻ ♇

26 Thursday
02 15 ☽ ⧺ ♃
02 57 ☽ ⊥ ♀
03 18 ☽ ☌ ♃
03 39 ☿ Q ♆
04 59 ☽ ⚹ ☉
07 33 ♀ ⊥ ♃
08 53 ☽ ∥ ♃
10 09 ☽ ⊥ ♅
10 40 ☽ ⚹ ♂
12 20 ☽ ∠ ♄
12 49 ☽ Q ♇
15 08 ☽ □ ☿
19 24 ☽ ⧺ ♆
20 09 ☽ ∠ ♀
20 38 ☽ ⊻ ♅

27 Friday
03 41 ☽ Q ☉
11 50 ☽ ⊻ ♆
13 34 ☽ ∠ ♂
15 13 ☽ ⚹ ♄
21 01 ☽ □ ♇
22 45 ☽ ⊥ ♆

28 Saturday
01 56 ☽ ⚹ ♀
06 14 ☽ ⊥ ♂
06 52 ☽ ⧺ ♀
09 39 ☽ ⊻ ♃
13 29 ☽ Q ♄
14 59 ☽ ⧺ ♄
15 19 ☽ □ ☉
15 27 ☽ ∠ ♆
17 03 ☉ ∠ ♆
17 34 ☽ ⊻ ♂
21 02 ☽ ⊥ ♃
23 26 ☽ ∥ ♅

29 Sunday
01 44 ☽ △ ☿
02 45 ☽ Q ♀
03 37 ☽ ☌ ♅
09 21 ☽ ⧺ ☉
11 36 ♀ Q ♅
14 24 ☽ ∠ ♃
20 02 ☽ ⚹ ♆

30 Monday
00 03 ☽ □ ♄
01 45 ☉ △ ♂
02 16 ☿ △ ♅
05 52 ☽ △ ♇
06 27 ☽ ⧺ ☿
09 04 ☽ ⚼ ☿
13 43 ☽ ⧺ ♇
17 24 ☽ □ ♀
19 29 ☽ Q ♆
19 47 ♀ ⊥ ♇
20 01 ☽ ⚹ ♃

31 Tuesday
04 24 ☽ ± ☿
04 27 ☽ ☌ ♂
05 09 ☿ ± ♂
05 22 ☽ ∥ ♂
06 24 ☽ △ ☉
11 28 ☽ ⚼ ♇
14 06 ☽ ⊻ ♅
17 27 ☽ ⊼ ☿
20 11 ☽ Q ♃
23 14 ♀ ⊻ ♃

FEBRUARY 2023

LONGITUDES

Date	Sidereal time h m s	Sun ☉ ° ' "	Moon ☽ ° ' "	Moon ☽ 24.00 ° ' "	Mercury ☿ ° '	Venus ♀ ° '	Mars ♂ ° '	Jupiter ♃ ° '	Saturn ♄ ° '	Uranus ♅ ° '	Neptune ♆ ° '	Pluto ♇ ° '
01	20 45 45	12 ♒ 20 58	25 ♊ 55 36	01 ♋ 53 35	17 ♑ 30	06 ♓ 43	10 ♊ 23	06 ♈ 09	25 ♒ 55	14 ♉ 59	23 ♓ 37	28 ♑ 41
02	20 49 42	13 21 51	07 ♋ 50 33	13 ♋ 46 53	18 37	07 57	10 36	06 20	26 02	14 59	23 39	28 43
03	20 53 39	14 22 43	19 ♋ 42 52	25 ♋ 38 46	19 47	09 12	10 50	06 32	26 09	15 00	23 40	28 45
04	20 57 35	15 23 33	01 ♌ 34 47	07 ♌ 31 08	20 58	10 26	11 05	06 43	26 16	15 01	23 42	28 47
05	21 01 32	16 24 22	13 ♌ 27 58	19 ♌ 25 27	22 11	11 41	11 19	06 55	26 23	15 01	23 44	28 48
06	21 05 28	17 25 10	25 ♌ 23 41	01 ♍ 22 51	23 26	12 55	11 35	07 07	26 30	15 02	23 46	28 50
07	21 09 25	18 25 57	07 ♍ 23 06	13 ♍ 24 35	24 43	14 09	11 50	07 19	26 38	15 03	23 48	28 52
08	21 13 21	19 26 43	19 ♍ 27 32	25 ♍ 32 09	26 01	15 24	12 07	07 31	26 45	15 04	23 50	28 54
09	21 17 18	20 27 27	01 ♎ 38 43	07 ♎ 47 33	27 20	16 38	12 23	07 43	26 52	15 05	23 52	28 56
10	21 21 14	21 28 11	13 ♎ 59 00	20 ♎ 13 27	28 ♑ 40	17 52	12 41	07 55	26 59	15 05	23 54	28 58
11	21 25 11	22 28 53	26 ♎ 31 21	02 ♏ 53 09	00 ♒ 02	19 06	12 58	08 07	27 07	15 06	23 56	29 00
12	21 29 08	23 29 34	09 ♏ 19 18	15 ♏ 50 19	01 25	20 21	13 16	08 19	27 14	15 08	23 58	29 01
13	21 33 04	24 30 15	22 ♏ 26 38	29 ♏ 08 42	02 49	21 35	13 35	08 31	27 21	15 09	24 00	29 03
14	21 37 01	25 30 54	05 ♐ 56 52	12 ♐ 51 25	04 14	22 49	13 53	08 44	27 28	15 10	24 02	29 05
15	21 40 57	26 31 32	19 ♐ 52 31	27 ♐ 00 09	05 41	24 03	14 13	08 56	27 36	15 11	24 04	29 07
16	21 44 54	27 32 09	04 ♑ 14 10	11 ♑ 34 11	07 08	25 17	14 32	09 09	27 43	15 12	24 06	29 09
17	21 48 50	28 32 45	18 ♑ 59 36	26 ♑ 29 35	08 36	26 31	14 52	09 22	27 50	15 13	24 08	29 10
18	21 52 47	29 ♒ 33 20	04 ♒ 03 09	11 ♒ 39 06	10 06	27 45	15 13	09 34	27 57	15 15	24 10	29 12
19	21 56 43	00 ♓ 33 53	19 ♒ 16 05	26 ♒ 52 43	11 36	28 ♓ 59	15 33	09 47	28 05	15 16	24 12	29 14
20	22 00 40	01 34 24	04 ♓ 27 37	11 ♓ 59 27	13 07	00 ♈ 13	15 54	10 00	28 12	15 18	24 15	29 16
21	22 04 37	02 34 54	19 ♓ 26 58	26 ♓ 49 10	14 40	01 26	16 16	10 13	28 19	15 19	24 17	29 17
22	22 08 33	03 35 23	04 ♈ 05 12	11 ♈ 14 26	16 13	02 40	16 38	10 26	28 26	15 21	24 19	29 19
23	22 12 30	04 35 49	18 ♈ 16 31	25 ♈ 11 14	17 47	03 54	17 00	10 39	28 34	15 22	24 21	29 21
24	22 16 26	05 36 14	01 ♉ 58 38	08 ♉ 38 54	19 22	05 07	17 22	10 52	28 41	15 24	24 23	29 22
25	22 20 23	06 36 36	15 ♉ 12 21	21 ♉ 39 25	20 58	06 21	17 45	11 05	28 48	15 26	24 25	29 24
26	22 24 19	07 36 57	28 ♉ 00 38	04 ♊ 16 35	22 35	07 35	18 08	11 19	28 55	15 27	24 28	29 26
27	22 28 16	08 37 16	10 ♊ 27 52	16 ♊ 35 07	24 14	08 48	18 31	11 32	29 02	15 29	24 30	29 27
28	22 32 12	09 ♓ 37 33	22 ♊ 38 57	28 ♊ 40 01	25 ♒ 53	10 ♈ 02	18 ♊ 55	11 ♈ 45	29 ♒ 10	15 ♉ 31	24 ♓ 32	29 ♑ 29

Date	Moon True ☊ ° '	Moon Mean ☊ ° '	Moon ☽ Latitude ° '
01	08 ♉ 39	08 ♉ 32	03 N 41
02	08 R 30	08 29	04 18
03	08 18	08 26	04 44
04	08 04	08 23	04 57
05	07 49	08 20	04 58
06	07 34	08 16	04 45
07	07 21	08 13	04 20
08	07 10	08 10	03 43
09	07 02	08 07	02 55
10	06 58	08 04	01 58
11	06 56	08 01	00 N 55
12	06 D 55	07 57	00 S 13
13	06 55	07 54	01 21
14	06 R 55	07 51	02 27
15	06 53	07 48	03 27
16	06 48	07 45	04 15
17	06 41	07 41	04 48
18	06 32	07 38	05 02
19	06 22	07 35	04 55
20	06 12	07 32	04 27
21	06 03	07 29	03 41
22	05 57	07 26	02 41
23	05 53	07 22	01 32
24	05 51	07 19	00 S 21
25	05 D 52	07 16	00 N 50
26	05 52	07 13	01 56
27	05 R 53	07 10	02 54
28	05 ♉ 52	07 ♉ 07	03 N 43

DECLINATIONS

Date	Sun ☉ ° '	Moon ☽ ° '	Mercury ☿ ° '	Venus ♀ ° '	Mars ♂ ° '	Jupiter ♃ ° '	Saturn ♄ ° '	Uranus ♅ ° '	Neptune ♆ ° '	Pluto ♇ ° '
01	17 S 06	27 N 03	21 S 41	10 S 27	24 N 41	01 N 22	14 S 03	16 N 00	03 S 37	22 S 40
02	16 49	27 30	21 42	09 58	24 42	01 26	14 01	16 00	03 36	22 40
03	16 31	26 40	21 41	09 29	24 44	01 31	13 59	16 00	03 35	22 40
04	16 13	24 38	21 40	09 00	24 45	01 36	13 56	16 01	03 35	22 39
05	15 55	21 32	21 37	08 31	24 47	01 41	13 54	16 01	03 34	22 39
06	15 37	17 32	21 33	08 01	24 48	01 45	13 51	16 01	03 33	22 39
07	15 18	12 49	21 28	07 31	24 50	01 50	13 49	16 01	03 32	22 38
08	14 59	07 35	21 22	07 01	24 51	01 55	13 46	16 02	03 31	22 38
09	14 40	02 N 01	21 15	06 31	24 53	02 00	13 44	16 02	03 31	22 38
10	14 21	03 S 42	21 06	06 01	24 55	02 05	13 42	16 02	03 30	22 38
11	14 01	09 23	20 56	05 30	24 56	02 10	13 39	16 03	03 29	22 37
12	13 41	14 48	20 45	05 00	24 58	02 15	13 37	16 03	03 28	22 37
13	13 21	19 41	20 33	04 29	25 00	02 20	13 34	16 03	03 27	22 37
14	13 01	23 43	20 19	03 58	25 01	02 25	13 32	16 04	03 27	22 37
15	12 40	26 29	20 04	03 27	25 03	02 30	13 29	16 04	03 26	22 36
16	12 20	27 37	19 48	02 56	25 04	02 35	13 27	16 04	03 25	22 36
17	11 59	26 51	19 31	02 25	25 06	02 40	13 24	16 05	03 24	22 36
18	11 38	24 08	19 12	01 54	25 08	02 45	13 22	16 05	03 23	22 36
19	11 16	19 42	18 52	01 22	25 09	02 50	13 20	16 06	03 22	22 35
20	10 55	14 01	18 30	00 51	25 11	02 55	13 17	16 06	03 22	22 35
21	10 33	07 33	18 08	00 S 20	25 13	03 01	13 15	16 07	03 21	22 35
22	10 11	00 S 50	17 44	00 N 12	25 14	03 06	13 12	16 07	03 20	22 35
23	09 49	05 N 44	17 19	00 43	25 16	03 11	13 10	16 07	03 19	22 34
24	09 27	11 50	16 52	01 15	25 17	03 16	13 07	16 08	03 18	22 34
25	09 05	17 11	16 24	01 46	25 19	03 22	13 05	16 08	03 17	22 34
26	08 43	21 36	15 55	02 17	25 20	03 27	13 02	16 09	03 16	22 34
27	08 20	24 53	15 25	02 49	25 22	03 32	13 00	16 10	03 16	22 33
28	07 S 58	26 N 57	14 S 53	03 N 20	25 N 23	03 N 38	12 S 58	16 N 10	03 S 15	22 S 33

ZODIAC SIGN ENTRIES

Date	h m	Planets
01	20 11	☽ ♋
04	08 48	☽ ♌
06	21 14	☽ ♍
09	08 47	☽ ♎
11	11 22	☿ ♒
11	18 34	☽ ♏
14	01 31	☽ ♐
16	05 00	☽ ♑
18	05 35	☽ ♒
18	22 34	☉ ♓
20	04 56	☽ ♓
20	07 56	♀ ♈
22	05 14	☽ ♈
24	08 29	☽ ♉
26	15 48	☽ ♊

LATITUDES

Date	Mercury ☿ ° '	Venus ♀ ° '	Mars ♂ ° '	Jupiter ♃ ° '	Saturn ♄ ° '	Uranus ♅ ° '	Neptune ♆ ° '	Pluto ♇ ° '
01	00 N 37	01 S 30	02 N 42	01 S 11	01 S 15	00 S 21	01 S 10	02 S 18
04	00 N 09	01 27	02 41	01 10	01 15	00 21	01 10	02 18
07	00 S 17	01 24	02 39	01 10	01 15	00 20	01 10	02 18
10	00 41	01 19	02 37	01 09	01 15	00 20	01 10	02 19
13	01 03	01 15	02 35	01 09	01 15	00 20	01 10	02 19
16	01 21	01 09	02 33	01 08	01 16	00 20	01 10	02 19
19	01 37	01 03	02 31	01 08	01 16	00 20	01 10	02 20
22	01 50	00 56	02 29	01 07	01 16	00 20	01 10	02 20
25	02 00	00 49	02 27	01 07	01 16	00 20	01 10	02 20
28	02 07	00 42	02 25	01 07	01 16	00 20	01 10	02 21
31	02 S 10	00 S 34	02 N 23	01 S 06	01 S 17	00 S 20	01 S 10	02 S 21

DATA

Julian Date	2459977
Delta T	+72 seconds
Ayanamsa	24° 10' 37"
Synetic vernal point	04° ♓ 56' 22"
True obliquity of ecliptic	23° 26' 18"

LONGITUDES

Date	Chiron ⚷ ° '	Ceres ⚳ ° '	Pallas ⚴ ° '	Juno ⚵ ° '	Vesta ⚶ ° '	Black Moon Lilith ⚸ ° '
01	12 ♈ 39	06 ♎ 57	11 ♋ 42	09 ♈ 21	27 ♓ 05	02 ♌ 41
11	13 ♈ 02	06 ♎ 45	10 ♋ 35	14 ♈ 32	01 ♈ 27	03 ♌ 48
21	13 ♈ 29	05 ♎ 50	10 ♋ 34	19 ♈ 54	05 ♈ 53	04 ♌ 55
31	13 ♈ 59	04 ♎ 16	11 ♋ 33	25 ♈ 24	10 ♈ 23	06 ♌ 03

MOON'S PHASES, APSIDES AND POSITIONS ☽

Date	h m	Phase	Longitude ° '	Eclipse Indicator
05	18 29	○	16 ♌ 41	
13	16 01	☾	24 ♏ 40	
20	07 06	●	01 ♓ 22	
27	08 06	☽	08 ♊ 27	

Day	h m		
04	08 41	Apogee	
19	08 58	Perigee	
02	08 17	Max dec	27° N 31'
09	20 31	0S	
16	14 32	Max dec	27° S 38'
22	14 59	0N	

ASPECTARIAN

h m	Aspects
01 Wednesday	
02 05	☽ ⊥ ♅
05 28	☽ ± ♇
07 21	☽ □ ♆
11 58	☽ △ ♄
15 07	☽ ꝗ ☉
17 33	☽ ⊼ ♇
20 09	☽ ∠ ♅
02 Thursday	
08 55	☽ □ ♃
10 57	☽ ± ☉
12 15	☽ △ ♀
17 42	☽ ⊻ ♂
18 30	☽ ꝗ ♄
03 Friday	
00 12	☽ ⊼ ☉
02 27	☽ ⚹ ♅
06 04	☽ ⊥ ♂
12 09	☽ ☍ ☿
12 53	☽ ± ♄
20 02	☽ △ ♆
20 19	☿ ⊥ ♄
22 07	☽ ꝗ ♀
04 Saturday	
00 38	☽ ∠ ♂
01 09	☽ ⊼ ♄
02 45	☽ Q ♅
02 50	☉ □ ♅
06 19	☽ ☍ ♇
10 54	☽ ∥ ♂
18 27	☽ ± ♀
22 34	☽ △ ♃
05 Sunday	
02 26	☽ ꝗ ♆
03 29	♀ □ ♂
04 09	☽ # ♇
04 27	☉ # ♅
07 35	☽ ⚹ ♂
07 58	☽ ⊼ ♀
11 24	☽ # ☿
15 08	☽ □ ♅
18 29	☽ ☍ ☉
20 38	☽ ± ♆
06 Monday	
05 18	☽ ꝗ ♃
07 37	☽ ⊼ ☿
08 16	☽ Q ♂
08 44	☽ ⊼ ♆
14 15	☽ ☍ ♄
18 26	☿ ⚹ ♆
18 56	☽ ⊼ ♇
20 01	☽ ∥ ♅
20 33	☉ ⊥ ♆
21 04	☽ ± ☿
22 51	☽ # ☉
23 39	☽ ± ♃
07 Tuesday	
06 21	♀ ∠ ♇
06 58	☽ ± ♇
07 06	☽ # ♄
11 51	☽ ⊼ ♃
17 12	☽ ꝗ ☿
21 05	☽ □ ♂
08 Wednesday	
00 57	☽ ꝗ ♇
01 07	☿ Q ♃
03 01	☽ ☍ ♀
03 16	☽ △ ♅
05 29	♀ ⚹ ♅
11 58	☽ ⊼ ☉
14 44	☽ # ♀
20 40	☽ ☍ ♆
09 Thursday	
00 54	☽ ± ☉
02 30	☽ △ ☿
02 32	☽ ⊼ ♄
02 49	☿ ⊻ ♄
05 39	☽ # ♆
06 40	☽ △ ♇
08 55	☽ ꝗ ♅
12 05	☽ ∥ ♃
13 23	☿ ꝗ ♂
14 25	☽ ± ♄
20 07	☽ ꝗ ☉
10 Friday	
00 02	☽ ☍ ♃
02 31	☽ ± ♅
05 08	☽ # ♃
08 06	☽ ꝗ ♄
09 25	☽ △ ♂
11 10	☽ ∥ ♆
14 08	☽ ⊼ ♅
17 16	☿ ☌ ♇
20 18	☽ ⊼ ♀
20 55	☽ ∥ ♀
11 Saturday	
03 38	☽ △ ☉
07 04	☽ ⊼ ♆
09 01	☽ ± ♀
13 07	☽ △ ♄
14 48	☽ ꝗ ♂
16 41	☽ □ ♇
18 28	☽ ± ♆
19 27	☽ □ ☿
12 Sunday	
03 48	☽ ꝗ ♀
06 37	☽ ∥ ♄
06 49	☉ ∠ ♃
07 13	☽ ∥ ☉
08 06	☽ ± ♂
10 07	☽ ⊼ ♃
11 20	☽ ꝗ ♆
17 50	☽ # ♅
18 15	☉ ∥ ♄
19 28	☽ ⊼ ♂
21 22	☽ ± ♃
22 43	☽ ☍ ♅
23 37	☉ ⊻ ♆
13 Monday	
02 12	☽ Q ♇
08 43	☽ Q ☿
10 16	☽ △ ♀
13 59	☽ ꝗ ♃
14 49	☽ △ ♆
16 01	☽ □ ☉
16 24	☽ ∥ ☿
20 53	☽ □ ♄
23 52	☽ ⚹ ♇
14 Tuesday	
04 43	☽ ∥ ♇
08 39	☽ ⚹ ☿
16 56	☽ △ ♃
21 51	☽ # ♂
15 Wednesday	
02 06	☽ ☍ ♂
02 09	☽ ∠ ♇
02 09	☽ Q ☉
03 59	☽ ⊼ ♅
04 10	♂ ꝗ ♇
04 38	☽ Q ♄
12 25	♀ ☌ ♆
12 57	♀ ∥ ♆
13 31	☽ ∠ ☿
14 13	☽ ± ♅
17 29	☽ ⊥ ♇
19 06	☽ □ ♆
19 43	☽ □ ♀
16 Thursday	
00 03	☽ ⚹ ☉
01 06	☽ ⚹ ♄
03 33	☽ ⊻ ♇
05 19	☽ ꝗ ♅
06 18	☽ ⊥ ☿
16 48	☉ ☌ ♄
17 17	☽ ⊻ ☿
18 40	♃ ⊥ ♅
20 11	☽ □ ♃
17 Friday	
00 54	☽ Q ♆
01 54	♀ # ♃
01 58	☽ ∠ ♄
02 34	☽ ∠ ☉
04 07	☽ Q ♀
05 12	☽ ⊼ ♂
05 55	☽ △ ♅
15 05	☽ ± ♂
16 36	☽ ⊥ ♄
18 07	☽ ⊥ ☉
20 16	☽ ⚹ ♆
20 47	☿ ∠ ♆
18 Saturday	
01 06	☽ ⚹ ♀
01 34	☽ Q ♃
02 13	☿ ⚹ ♃
02 15	☽ ⊻ ♄
03 23	☉ ⊻ ♇
04 18	☽ ☌ ♇
04 22	☽ ⊻ ☉
04 56	☽ # ♂
05 46	☽ ꝗ ♂
14 51	♂ ⊻ ♅
16 32	♀ ⊻ ♄
20 07	☽ ∠ ♆
20 51	☽ ⚹ ♃
21 20	☽ ∥ ♇
22 35	☽ ☌ ☿
19 Sunday	
02 56	☽ ∠ ♀
05 42	☽ □ ♅
06 01	☽ △ ♂
10 20	☽ ⊥ ♆
16 07	☽ ∥ ☿
17 05	♀ ⚹ ♇
18 22	☽ ⊥ ♀
19 48	☽ ⊻ ♆
20 49	☽ ∠ ♃
20 Monday	
02 00	☽ ☌ ♄
03 40	☽ # ♅
03 45	☽ ⊻ ♇
04 40	☽ ⊻ ♀
07 06	☽ ☌ ☉
10 09	☽ Q ♅
11 16	☽ ⊥ ♃
13 16	☽ ⊥ ♇
13 42	♀ ∠ ♅
14 49	☽ ∥ ♄
20 57	☽ ⊻ ♃
21 Tuesday	
00 26	☽ ∥ ☉
03 24	☽ ⊻ ☿
03 40	☽ ∠ ♇
05 20	☽ ⚹ ♅
06 44	☽ □ ♂
14 11	☽ ⊥ ☿
19 52	☽ ☌ ♆
22 22	☿ □ ♅
22 Wednesday	
02 35	☽ ⊻ ♄
03 05	☽ ∥ ♆
04 02	☽ # ♃
04 05	☽ ⚹ ♇
05 47	☽ ∠ ♅
06 01	☉ Q ♅
06 40	☽ ∠ ☿
09 26	☽ ☌ ♀
11 07	☽ ⊻ ☉
12 36	☽ ⊥ ♄
12 55	☽ Q ♂
14 07	☽ # ♀
16 01	☽ ∥ ♀
20 14	☿ △ ♂
20 49	☽ ⊥ ♅
21 55	☽ ⊥ ☉
22 48	☽ ☌ ♃
23 Thursday	
00 09	☽ Q ♇
02 24	☽ ∥ ♃
03 01	☽ # ♆
03 52	☽ ∠ ♄
07 01	☽ ⊻ ♅
09 44	☽ ⚹ ♂
11 03	☽ ⚹ ☿
13 42	☉ ⊥ ♃
14 27	☽ ∠ ☉
20 48	☿ ⊥ ♆
22 34	☽ ⊻ ♆
24 Friday	
02 22	♀ ⊥ ♄
02 53	☽ # ☉
06 06	☽ ⚹ ♄
06 21	☉ ⊥ ♇
07 22	☽ □ ♇
09 10	☽ ⊥ ♆
10 46	☽ Q ☿
12 43	☽ ∠ ♂
17 24	☽ # ♄
18 12	☽ ⊻ ♀
18 50	♀ Q ♂
18 56	♃ # ♆
19 02	☽ ⚹ ☉
25 Saturday	
01 23	☽ ∠ ♆
03 51	☽ Q ♄
04 19	☽ ⊻ ♃
05 27	☽ ⊥ ♂
06 13	☽ ⊥ ♀
06 58	☽ ∥ ♅
08 32	☽ # ☿
12 24	☽ ☌ ♅
15 33	☽ ⊥ ♃
16 51	☽ ⊻ ♂
18 51	☽ Q ☉
26 Sunday	
00 16	☽ □ ☿
00 38	☽ ∠ ♀
00 48	☿ # ♅
05 15	☽ ⚹ ♆
08 43	☽ ∠ ♃
11 12	☿ ∠ ♀
13 45	☽ □ ♄
14 42	☽ △ ♇
16 11	☉ ⊻ ♀
18 14	☽ # ♇
27 Monday	
02 14	♃ Q ♇
04 16	☽ Q ♆
08 06	☽ □ ☉
08 24	☽ ⚹ ♀
14 08	☽ ⚹ ♃
16 03	☿ ⊻ ♆
16 26	☽ ∥ ♂
19 50	☽ ꝗ ♇
21 51	☽ ⊻ ♅
28 Tuesday	
01 42	♀ ⊥ ♅
04 21	☽ ☌ ♂
08 04	♀ # ♆
09 44	☽ ⊥ ♅
10 37	☽ Q ♀
13 40	☽ ± ♇
14 15	☽ Q ♃
15 46	☽ □ ♆
19 27	☽ △ ☿

All ephemeris data is given at 12.00 UT and the Moon's longitude is additionally given for 24.00 UT

MARCH 2023

Date	Sidereal time h	m	s	Sun ☉ °	'	"	Moon ☽ °	'	"	Moon ☽ 24.00 °	'	"	Mercury ☿ °	'	Venus ♀ °	'	Mars ♂ °	'	Jupiter ♃ °	'	Saturn ♄ °	'	Uranus ♅ °	'	Neptune ♆ °	'	Pluto ♇ °	'
01	22	36	09	10 ♓	37	48	04 ♋	38	54	10 ♋	36	11	27 ♒	33	11 ♈	15	19 ♊	19	11 ♈	59	29 ♒	17	15 ♉	33	24 ♓	34	29 ♑	30
02	22	40	06	11	38	01	16 ♋	32	22	22 ♋	27	57	29 ♒	14	12	28	19	43	12	12	29	24	15	35	24	36	29	32
03	22	44	02	12	38	12	28 ♋	23	22	04 ♌	19	03	00 ♓	56	13	42	20	07	12	26	29	31	15	37	24	39	29	34
04	22	47	59	13	38	21	10 ♌	15	19	16 ♌	12	30	02	39	14	55	20	32	12	39	29	38	15	39	24	41	29	35
05	22	51	55	14	38	28	22 ♌	10	50	28 ♌	10	33	04	24	16	08	20	57	12	53	29	45	15	41	24	43	29	37
06	22	55	52	15	38	33	04 ♍	11	51	10 ♍	14	53	06	09	17	21	21	22	13	07	29 ♒	52	15	43	24	45	29	38
07	22	59	48	16	38	36	16 ♍	19	47	22 ♍	26	41	07	55	18	34	21	47	13	20	00 ♓	00	15	45	24	48	29	40
08	23	03	45	17	38	37	28 ♍	35	42	04 ♎	46	56	09	43	19	47	22	13	13	34	00	07	15	47	24	50	29	41
09	23	07	41	18	38	36	11 ♎	00	32	17 ♎	16	38	11	32	21	00	22	39	13	48	00	14	15	49	24	52	29	42
10	23	11	38	19	38	33	23 ♎	35	24	29 ♎	56	59	13	21	22	13	23	05	14	02	00	21	15	51	24	54	29	44
11	23	15	35	20	38	29	06 ♏	21	38	12 ♏	49	32	15	12	23	25	23	31	14	16	00	28	15	54	24	57	29	45
12	23	19	31	21	38	23	19 ♏	20	57	25 ♏	56	09	17	04	24	38	23	58	14	30	00	35	15	56	24	59	29	47
13	23	23	28	22	38	16	02 ♐	35	21	09 ♐	18	50	18	57	25	51	24	25	14	44	00	41	15	58	25	01	29	48
14	23	27	24	23	38	07	16 ♐	06	47	22 ♐	59	23	20	52	27	03	24	52	14	58	00	48	16	01	25	04	29	49
15	23	31	21	24	37	56	29 ♐	56	43	06 ♑	58	48	22	47	28	16	25	19	15	12	00	55	16	03	25	06	29	50
16	23	35	17	25	37	44	14 ♑	05	32	21 ♑	15	43	24	43	29 ♈	28	25	46	15	26	01	02	16	06	25	08	29	52
17	23	39	14	26	37	30	28 ♑	31	57	05 ♒	50	45	26	41	00 ♉	40	26	14	15	40	01	09	16	08	25	10	29	53
18	23	43	10	27	37	14	13 ♒	12	27	20 ♒	36	15	28 ♓	39	01	53	26	41	15	54	01	16	16	11	25	13	29	54
19	23	47	07	28	36	57	28 ♒	01	17	05 ♓	26	33	00 ♈	38	03	05	27	09	16	08	01	22	16	13	25	15	29	55
20	23	51	04	29 ♓	36	37	12 ♓	51	00	20 ♓	13	38	02	38	04	17	27	37	16	22	01	29	16	16	25	17	29	57
21	23	55	00	00 ♈	36	16	27 ♓	33	27	04 ♈	49	31	04	38	05	29	28	06	16	37	01	36	16	18	25	20	29	58
22	23	58	57	01	35	52	12 ♈	01	01	19 ♈	07	19	06	39	06	41	28	34	16	51	01	42	16	21	25	22	29 ♑	59
23	00	02	53	02	35	27	26 ♈	07	53	03 ♉	02	22	08	40	07	53	29	03	17	05	01	49	16	24	25	24	00 ♒	00
24	00	06	50	03	34	59	09 ♉	50	34	16 ♉	32	27	10	41	09	05	29 ♊	31	17	19	01	56	16	26	25	26	00	01
25	00	10	46	04	34	30	23 ♉	08	06	29 ♉	37	44	12	41	10	17	00 ♋	00	17	34	02	02	16	29	25	29	00	02
26	00	14	43	05	33	58	06 ♊	01	40	12 ♊	20	20	14	41	11	28	00	29	17	48	02	09	16	32	25	31	00	03
27	00	18	39	06	33	24	18 ♊	34	11	24 ♊	43	45	16	41	12	40	00	59	18	02	02	15	16	35	25	33	00	04
28	00	22	36	07	32	47	00 ♋	49	37	06 ♋	52	21	18	39	13	51	01	28	18	17	02	22	16	38	25	35	00	05
29	00	26	33	08	32	08	12 ♋	52	33	18 ♋	50	50	20	36	15	03	01	58	18	31	02	28	16	41	25	38	00	06
30	00	30	29	09	31	27	24 ♋	47	46	00 ♌	43	56	22	30	16	14	02	27	18	46	02	34	16	43	25	40	00	07
31	00	34	26	10 ♈	30	44	06 ♌	39	53	12 ♌	36	08	24 ♈	23	17 ♉	25	02 ♋	57	19 ♈	00	02 ♓	40	16 ♉	46	25 ♓	42	00 ♒	08

Date	Moon True ☊ °	'	Moon Mean ☊ °	'	Moon ☽ Latitude °	'
01	05 ♉	49	07 ♉	03	04 N	22
02	05 R	43	07	00	04	49
03	05	35	06	57	05	03
04	05	26	06	54	05	04
05	05	16	06	51	04	52
06	05	06	06	47	04	27
07	04	57	06	44	03	50
08	04	50	06	41	03	02
09	04	45	06	38	02	04
10	04	42	06	35	01 N	00
11	04 D	42	06	32	00 S	09
12	04	42	06	28	01	19
13	04	44	06	25	02	25
14	04	45	06	22	03	25
15	04 R	45	06	19	04	15
16	04	44	06	16	04	50
17	04	41	06	13	05	08
18	04	36	06	09	05	07
19	04	31	06	06	04	45
20	04	25	06	03	04	04
21	04	20	06	00	03	07
22	04	17	05	57	01	58
23	04	16	05	53	00 S	44
24	04 D	15	05	50	00 N	31
25	04	16	05	47	01	41
26	04	18	05	44	02	45
27	04	20	05	41	03	39
28	04	21	05	38	04	22
29	04 R	21	05	34	04	52
30	04	20	05	31	05	09
31	04 ♉	18	05 ♉	28	05 N	13

DECLINATIONS

Date	Sun ☉ °	'	Moon ☽ °	'	Mercury ☿ °	'	Venus ♀ °	'	Mars ♂ °	'	Jupiter ♃ °	'	Saturn ♄ °	'	Uranus ♅ °	'	Neptune ♆ °	'	Pluto ♇ °	'
01	07 S	35	27 N	43	14 S	20	03 N	51	25 N	24	03 N	43	12 S	55	16 N	11	03 S	14	22 S	33
02	07	12	27	11	13	45	04	22	25	25	03	48	12	53	16	11	03	13	22	33
03	06	49	25	25	13	10	04	53	25	27	03	54	12	50	16	12	03	12	22	33
04	06	26	22	33	12	33	05	24	25	28	03	59	12	48	16	12	03	11	22	32
05	06	03	18	43	11	54	05	55	25	29	04	04	12	45	16	13	03	10	22	32
06	05	40	14	07	11	15	06	26	25	30	04	10	12	43	16	14	03	09	22	32
07	05	16	08	56	10	34	06	56	25	31	04	15	12	41	16	14	03	08	22	32
08	04	53	03 N	20	09	52	07	26	25	32	04	21	12	38	16	15	03	08	22	32
09	04	30	02 S	27	09	09	07	57	25	33	04	26	12	36	16	16	03	07	22	32
10	04	06	08	14	08	24	08	27	25	33	04	32	12	33	16	16	03	06	22	31
11	03	43	13	47	07	38	08	56	25	34	04	37	12	31	16	17	03	05	22	31
12	03	19	18	49	06	51	09	26	25	35	04	42	12	29	16	18	03	04	22	31
13	02	55	23	03	06	03	09	55	25	35	04	48	12	26	16	18	03	03	22	31
14	02	32	26	07	05	14	10	25	25	36	04	53	12	24	16	19	03	02	22	31
15	02	08	27	41	04	24	10	54	25	36	04	59	12	22	16	20	03	01	22	31
16	01	44	27	30	03	32	11	22	25	36	05	04	12	19	16	20	03	00	22	30
17	01	21	25	29	02	40	11	51	25	36	05	10	12	17	16	21	02	59	22	30
18	00	57	21	45	01	47	12	19	25	37	05	15	12	15	16	22	02	59	22	30
19	00	33	16	37	00 S	52	12	47	25	37	05	21	12	12	16	23	02	58	22	30
20	00 S	09	10	29	00 N	03	13	15	25	36	05	26	12	10	16	23	02	57	22	30
21	00 N	14	03 S	50	00	58	13	42	25	36	05	32	12	08	16	24	02	56	22	30
22	00	38	02 N	56	01	54	14	09	25	36	05	37	12	05	16	25	02	55	22	30
23	01	02	09	24	02	51	14	35	25	36	05	43	12	03	16	26	02	54	22	30
24	01	25	15	15	03	47	15	02	25	35	05	48	12	01	16	27	02	53	22	30
25	01	49	20	12	04	44	15	28	25	35	05	54	11	59	16	27	02	52	22	30
26	02	13	24	01	05	41	15	53	25	34	06	00	11	57	16	28	02	51	22	29
27	02	36	26	35	06	37	16	19	25	33	06	05	11	54	16	29	02	51	22	29
28	03	00	27	48	07	33	16	43	25	32	06	11	11	52	16	30	02	50	22	29
29	03	23	27	40	08	28	17	08	25	31	06	16	11	50	16	31	02	49	22	29
30	03	46	26	14	09	21	17	32	25	30	06	22	11	48	16	32	02	48	22	29
31	04 N	10	23 N	39	10 N	14	17 N	55	25 N	29	06 N	27	11 S	46	16 N	32	02 S	47	22 S	29

ZODIAC SIGN ENTRIES

Date	h	m	Planets
01	02	40	☽ ♋
02	22	52	☿ ♓
03	15	16	☽ ♌
06	03	38	☽ ♍
07	13	35	♄ ♓
08	14	44	☽ ♎
11	00	06	☽ ♏
13	07	21	☽ ♐
15	12	06	☽ ♑
16	22	34	♀ ♉
17	14	25	☽ ♒
19	04	24	☿ ♈
19	15	12	☽ ♓
20	21	24	☉ ♈
21	16	01	☽ ♈
23	12	13	♇ ♒
23	18	42	☽ ♉
25	11	45	♂ ♋
26	00	41	☽ ♊
28	10	22	☽ ♋
30	22	31	☽ ♌

LATITUDES

Date	Mercury ☿ °	'	Venus ♀ °	'	Mars ♂ °	'	Jupiter ♃ °	'	Saturn ♄ °	'	Uranus ♅ °	'	Neptune ♆ °	'	Pluto ♇ °	'
01	02 S	08	00 S	39	02 N	24	01 S	07	01 S	16	00 S	20	01 S	10	02 S	21
04	02	10	00	31	02	22	01	06	01	17	00	20	01	10	02	21
07	02	07	00	22	02	20	01	06	01	17	00	20	01	10	02	22
10	02	01	00	13	02	18	01	06	01	17	00	20	01	10	02	22
13	01	50	00 S	04	02	16	01	05	01	18	00	20	01	10	02	23
16	01	34	00 N	06	02	14	01	05	01	18	00	20	01	10	02	23
19	01	13	00	15	02	12	01	05	01	18	00	20	01	10	02	23
22	00	48	00	25	02	10	01	05	01	19	00	19	01	10	02	24
25	00 S	18	00	36	02	08	01	05	01	19	00	19	01	10	02	24
28	00 N	15	00	46	02	06	01	04	01	19	00	19	01	10	02	25
31	00 N	51	00 N	56	02 N	04	01 S	04	01 S	20	00 S	19	01 S	10	02 S	25

DATA

Julian Date	2460005
Delta T	+72 seconds
Ayanamsa	24° 10' 41"
Synetic vernal point	04° ♓ 56' 18"
True obliquity of ecliptic	23° 26' 18"

MOON'S PHASES, APSIDES AND POSITIONS ☽

Date	h	m	Phase	Longitude °	'	Eclipse Indicator
07	12	40	○	16 ♍	40	
15	02	08	☾	24 ♐	13	
21	17	23	●	00 ♈	50	
29	02	32	☽	08 ♋	09	

Day	h	m		
03	17	54	Apogee	
19	15	03	Perigee	
31	11	14	Apogee	
01	14	06	Max dec	27° N 43'
09	01	54	0S	
15	21	39	Max dec	27° S 50'
22	01	32	0N	
28	21	28	Max dec	27° N 54'

LONGITUDES

Date	Chiron ⚷ °	'	Ceres ⚳ °	'	Pallas ⚴ °	'	Juno ⚵ °	'	Vesta ⚶ °	'	Black Moon Lilith ⚸ °	'
01	13 ♈	52	04 ♎	37	11 ♋	17	24 ♈	18	09 ♈	29	05 ♌	49
11	14 ♈	22	02 ♎	54	12 ♋	57	29 ♈	54	14 ♈	00	06 ♌	56
21	14 ♈	58	00 ♎	23	15 ♋	19	05 ♉	36	18 ♈	32	08 ♌	04
31	15 ♈	33	28 ♍	08	18 ♋	14	11 ♉	23	23 ♈	05	09 ♌	11

ASPECTARIAN

01 Wednesday
01 07 ☽ △ ♄
01 40 ☽ ⊼ ♇
02 39 ☿ ∠ ♃
03 44 ☽ ∠ ♅
04 24 ♀ ∥ ♃
17 11 ♀ Q ♇

02 Thursday
01 10 ☽ △ ☉
02 50 ☽ □ ♀
03 04 ☽ □ ♃
05 36 ♀ ☌ ♃
06 33 ☽ bQ ☿
07 38 ☽ bQ ♄
10 03 ☽ ⚹ ♅
14 34 ☿ ☌ ♄
16 21 ☿ ⊻ ♇
18 39 ☽ ⊻ ♂

03 Friday
02 02 ☽ ± ♄
03 49 ☽ ± ☿
04 23 ☽ △ ♆
05 40 ☉ ⊻ ♃
07 14 ☽ ⊥ ♂
10 20 ☽ bQ ☉
10 25 ☽ Q ♅
11 48 ☽ ∥ ♂
14 19 ☽ ⊼ ♄
14 22 ☽ ☍ ♇
18 01 ☽ ⊼ ☿
22 26 ♄ ⊻ ♇

04 Saturday
01 36 ☿ ∥ ♄
02 07 ☽ ∠ ♂
06 00 ♀ ∠ ♄
06 14 ☽ ± ☉
10 50 ☽ bQ ♆
12 04 ☽ # ♇
16 56 ☽ △ ♃
19 27 ☽ ⊼ ☉
22 28 ☽ △ ♀
22 54 ☽ □ ♅

05 Sunday
01 57 ☿ Q ♅
02 46 ♀ ⊻ ♅
05 02 ☽ ± ♆
09 26 ☽ ⚹ ♂
11 14 ☉ ∠ ♇
15 33 ☉ # ♀
17 06 ☽ ⊼ ♆
23 38 ☽ bQ ♃

06 Monday
01 28 ☽ ∥ ♅
02 54 ☽ ⊼ ♇
03 18 ☽ ☍ ♄
04 55 ☿ ⊥ ♇
07 55 ☽ bQ ♀
10 17 ☽ Q ♂
13 42 ☉ ⚹ ♅
14 52 ☽ ± ♇
16 32 ☽ ☍ ☿
17 54 ☽ ± ♃
18 46 ☽ # ♄

07 Tuesday
03 01 ☿ ⊥ ♃
03 33 ☽ # ☿
03 45 ☽ ± ♀
06 00 ☽ ⊼ ♃
08 42 ☽ bQ ♇
09 18 ♂ ⊥ ♅
10 51 ☽ △ ♅
12 40 ☽ ☍ ☉
16 53 ☽ ⊼ ♀
19 59 ☽ ∥ ♀
23 06 ☽ □ ♂

08 Wednesday
04 39 ☽ ☍ ♆
05 01 ☽ # ☉
07 50 ☽ ∥ ♃
12 54 ☽ # ♆
14 07 ☽ △ ♇
14 58 ☽ ⊼ ♄
16 16 ☽ bQ ♅

09 Thursday
02 42 ☽ ± ♄
09 42 ☽ ± ♅
13 10 ☽ ⊼ ☿
14 43 ☽ ∥ ♆
14 51 ☉ # ♃
17 27 ☽ ☍ ♃
19 53 ☽ ∥ ☉
20 10 ☽ bQ ♄
20 19 ☽ # ♃
21 14 ☽ ⊼ ♅

10 Friday
02 35 ☽ ± ☿
03 52 ☽ ⊼ ☉
09 07 ☽ ☍ ♀
11 00 ☽ △ ♂
11 13 ☿ # ♀
12 37 ☽ ∥ ☿
12 58 ☽ # ♀
14 30 ☽ ⊼ ♆
16 13 ☽ ± ☉
22 04 ☿ ⊻ ♃
22 31 ☽ bQ ☿
23 36 ☽ □ ♇

11 Saturday
00 51 ☽ △ ♄
01 50 ☽ ± ♆
06 06 ☿ ∠ ♇
06 25 ☽ ∥ ♄
10 33 ☽ bQ ☉
15 05 ♀ ⚹ ♂
16 10 ☽ bQ ♂
18 41 ☽ bQ ♆
21 04 ☿ ⚹ ♅
23 32 ☽ # ♅

12 Sunday
01 15 ♂ ± ♇
02 55 ☽ ⊼ ♃
05 43 ☽ ☍ ♅
07 07 ☽ △ ☿
09 07 ☽ Q ♇
09 22 ☽ ± ♂
14 08 ☽ ± ♃
16 32 ☽ △ ☉
19 09 ♀ ⊻ ♆
20 43 ☽ ⊼ ♂
22 18 ☽ △ ♆
22 37 ☽ ⊼ ♀

13 Monday
03 49 ☉ ∥ ♆
06 46 ☽ bQ ♃
06 58 ☽ ⚹ ♇
08 34 ☽ □ ♄
08 36 ☽ ∥ ♇
10 32 ☽ ± ♀

14 Tuesday
04 09 ☽ bQ ♀
07 00 ☽ # ♂
09 43 ☽ ∠ ♇
09 57 ☽ △ ♃
11 49 ☽ ⊼ ♅
16 45 ☽ Q ♄
18 35 ☿ ⊥ ♀
20 54 ☿ # ♃
21 38 ☽ □ ☿
22 20 ☽ ± ♅
23 39 ♂ □ ♆

15 Wednesday
01 28 ☽ ⊥ ♇
02 08 ☽ □ ☉
03 38 ☽ □ ♆
03 45 ☽ ☍ ♂
08 50 ☽ △ ♀
11 49 ☽ ⊻ ♇
13 41 ☽ ⚹ ♄
13 54 ☽ bQ ♅
23 39 ☉ ☌ ♆

16 Thursday
09 20 ☽ Q ☿
10 23 ☽ Q ♆
11 10 ☽ Q ☉
14 17 ☽ □ ♃
15 17 ☽ ∠ ♄
15 22 ☽ △ ♅
17 13 ☿ ☌ ♆
18 10 ☉ □ ♂
19 59 ♀ □ ♇

17 Friday
02 54 ☿ ∥ ♆
04 48 ☿ □ ♂
06 22 ☽ ⊥ ♄
06 27 ☽ ⚹ ♆
08 04 ☽ ⊼ ♂
08 28 ☽ ⚹ ☿
08 37 ☽ ⚹ ☉
10 45 ☉ ☌ ☿
10 58 ☽ # ♂
14 14 ☽ ☌ ♇
15 50 ☽ □ ♀
16 20 ☽ ⊻ ♄
18 16 ☽ ± ♂
20 34 ☽ Q ♃
22 16 ♀ ⊥ ♆
22 25 ♀ ⚹ ♄

18 Saturday
07 07 ☽ ∠ ♆
07 52 ☽ ∥ ♇
08 30 ♀ # ♄
09 27 ☽ bQ ♂
10 59 ☽ ∠ ☉
12 49 ☽ ∠ ☿
16 27 ☽ ⚹ ♃
16 50 ☽ □ ♅
21 46 ☽ ⊥ ♆
23 47 ☽ Q ♀

19 Sunday
02 38 ☽ ⊥ ☉
03 24 ☿ ⚹ ♇
05 40 ☽ ⊥ ☿
07 30 ☽ ⊻ ♆
10 33 ☽ △ ♂
12 58 ☽ # ♅
13 02 ☽ ⊻ ☉
15 05 ☽ ⊻ ♇
16 52 ☽ ⊻ ☿
17 07 ☽ ∠ ♃
17 28 ☽ ☌ ♄
19 15 ☿ ∠ ♅
20 54 ☽ ⚹ ♀
21 30 ☿ ⊻ ♄
22 03 ☽ Q ♅
22 06 ♃ ⊻ ♅

20 Monday
00 48 ☽ ⊥ ♇
02 16 ☽ # ♀
02 56 ☉ ∥ ☿
05 39 ☽ ∥ ♄
07 55 ☽ ⊥ ♃
14 03 ☉ # ☿
15 24 ☽ ∠ ♇
17 33 ☽ ⚹ ♅
17 49 ☽ ⊻ ♃
20 12 ☉ ⚹ ♇
23 24 ☽ ∠ ♀

21 Tuesday
06 01 ☽ # ♃
08 20 ☽ ☌ ♆
09 26 ♃ ∠ ♄
12 55 ☽ □ ♂
15 10 ☽ ∥ ♆
15 27 ☽ ⊥ ♀
15 58 ☽ ⚹ ♇
17 23 ☽ ☌ ☉
18 12 ☽ ∠ ♅
18 42 ☽ ⊻ ♄
20 53 ☽ # ☿
23 59 ☽ # ☉

22 Wednesday
01 34 ☽ ☌ ☿
02 17 ☽ ⊻ ♀
03 17 ☽ ∥ ☉
04 44 ☽ ⊥ ♄
05 44 ☉ ∠ ♅
07 42 ☽ ∥ ☿
09 12 ☽ ⊥ ♅
11 56 ☽ # ♆
11 56 ☽ Q ♇
13 14 ☿ ⊻ ♀
15 00 ☉ ⊻ ♄
19 19 ☽ ⊻ ♅
19 56 ☽ Q ♂
19 58 ☽ ∠ ♄
20 17 ☽ ☌ ♃
21 55 ☽ ∥ ♃

23 Thursday
01 25 ☿ ⊥ ♄
10 44 ☽ ⊻ ♆
13 26 ☿ # ♆
16 18 ♂ Q ♃
17 13 ☽ ⚹ ♂
18 42 ☽ □ ♇
21 10 ☽ ⊥ ♆
21 57 ☽ ⚹ ♄
22 26 ☽ # ♄

24 Friday
00 05 ☽ ⊻ ☉
09 08 ☿ ⊥ ♅
10 31 ☽ ☌ ♀
10 58 ☽ ∥ ♀
11 30 ☽ ⊥ ☉
13 04 ☽ ∠ ♆
13 45 ☽ ⊻ ☿
17 24 ☽ ∥ ♅
19 21 ☽ Q ♄
20 41 ☽ ∠ ♂
23 52 ☽ ☌ ♅

25 Saturday
01 40 ☽ ⊻ ♃
02 26 ☽ ⊥ ☿
04 09 ☿ Q ♇
04 58 ☽ ∠ ☉
12 48 ☽ ⊥ ♃
13 34 ♂ ⊼ ♇
13 40 ☽ ⊥ ♂
16 10 ♀ ∠ ♆
16 19 ☽ ⚹ ♆
21 56 ☽ ∠ ☿

26 Sunday
00 46 ☽ △ ♇
01 12 ☽ ⊻ ♂
01 28 ☽ # ♇
04 38 ☽ □ ♄
05 49 ☽ ∠ ♃
11 03 ☽ ⚹ ☉
14 49 ☽ Q ♆
20 51 ☿ ∥ ♃
23 25 ☽ ⊻ ♀
23 30 ☉ ⊥ ♀

27 Monday
00 50 ☽ ∥ ♂
05 14 ☽ bQ ♇
07 39 ☽ ⚹ ☿
08 08 ☽ ⊻ ♅
10 45 ☿ ⊻ ♅
10 57 ☽ ⚹ ♃
11 58 ☽ Q ☉
12 12 ☽ ⊥ ♀
19 19 ☿ ∠ ♄
19 50 ☽ ⊥ ♅
22 24 ♀ ∥ ♅
22 43 ☽ ± ♇

28 Tuesday
01 39 ☽ □ ♆
02 10 ☉ # ♆
06 50 ☿ ☌ ♃
07 41 ☽ ∠ ♀
10 32 ☽ ⊼ ♇
10 54 ☽ Q ♃
11 35 ☽ Q ☿
13 19 ☽ ☌ ♂
13 35 ☽ ∠ ♅
15 03 ☽ △ ♄
20 41 ♂ ∠ ♅
23 12 ♀ Q ♄

29 Wednesday
01 26 ☿ Q ♂
02 32 ☽ □ ☉
10 04 ☉ ⊥ ♄
16 50 ☽ ⚹ ♀
19 40 ☽ ⚹ ♅
21 18 ☽ bQ ♄
23 35 ☽ □ ♃

30 Thursday
06 30 ☽ □ ☿
13 45 ☽ △ ♆
15 37 ☽ ± ♄
19 03 ♂ △ ♄
19 43 ☽ Q ♀
19 58 ☽ Q ♅
20 04 ☽ ∥ ♂
22 26 ♀ ☌ ♅
22 46 ☽ ☍ ♇

31 Friday
03 51 ☽ ⊼ ♄
04 10 ☽ ⊻ ♂
16 50 ☽ ⊥ ♂
18 41 ☉ ⊥ ♅
20 11 ☽ bQ ♆
20 29 ☽ △ ☉
20 34 ☽ # ♇

All ephemeris data is given at 12.00 UT and the Moon's longitude is additionally given for 24.00 UT

LONGITUDES

Date	Sidereal time h m s	Sun ☉ ° ′ ″	Moon ☽ ° ′ ″	Moon ☽ 24.00 ° ′ ″	Mercury ☿ ° ′	Venus ♀ ° ′	Mars ♂ ° ′	Jupiter ♃ ° ′	Saturn ♄ ° ′	Uranus ♅ ° ′	Neptune ♆ ° ′	Pluto ♇ ° ′
01	00 38 22	11 ♈ 29 58	18 ♌ 33 11	24 ♌ 31 29	26 ♈ 12	18 ♉ 36	03 ♋ 27	19 ♈ 14	02 ♓ 47	16 ♉ 49	25 ♓ 44	00 ♒ 09
02	00 42 19	12 29 10	00 ♍ 31 26	06 ♍ 33 23	27 59	19 47	03 57	19 29	02 53	16 52	25 46	00 10
03	00 46 15	13 28 20	12 ♍ 37 40	18 ♍ 44 33	29 ♈ 42	20 58	04 27	19 43	02 59	16 55	25 49	00 10
04	00 50 12	14 27 27	24 ♍ 54 16	01 ♎ 06 59	01 ♉ 21	22 09	04 58	19 58	03 05	16 58	25 51	00 11
05	00 54 08	15 26 33	07 ♎ 22 51	13 ♎ 41 57	02 55	23 19	05 28	20 12	03 11	17 01	25 53	00 12
06	00 58 05	16 25 36	20 ♎ 04 21	26 ♎ 30 05	04 26	24 30	05 59	20 27	03 17	17 05	25 55	00 13
07	01 02 02	17 24 37	02 ♏ 59 09	09 ♏ 31 33	05 51	25 40	06 29	20 41	03 23	17 08	25 57	00 13
08	01 05 58	18 23 37	16 ♏ 07 14	22 ♏ 46 10	07 11	26 51	07 00	20 56	03 29	17 11	25 59	00 14
09	01 09 55	19 22 34	29 ♏ 28 17	06 ♐ 13 33	08 26	28 01	07 31	21 10	03 35	17 14	26 02	00 15
10	01 13 51	20 21 30	13 ♐ 01 53	19 ♐ 53 12	09 35	29 ♉ 11	08 02	21 25	03 40	17 17	26 04	00 15
11	01 17 48	21 20 24	26 ♐ 47 26	03 ♑ 44 27	10 39	00 ♊ 21	08 33	21 39	03 46	17 20	26 06	00 16
12	01 21 44	22 19 17	10 ♑ 44 09	17 ♑ 46 21	11 36	01 31	09 05	21 54	03 52	17 23	26 08	00 17
13	01 25 41	23 18 07	24 ♑ 50 52	01 ♒ 57 27	12 28	02 41	09 36	22 08	03 57	17 27	26 10	00 17
14	01 29 37	24 16 56	09 ♒ 05 50	16 ♒ 15 38	13 13	03 50	10 07	22 23	04 03	17 30	26 12	00 18
15	01 33 34	25 15 44	23 ♒ 26 29	00 ♓ 37 55	13 53	05 00	10 39	22 37	04 08	17 33	26 14	00 18
16	01 37 31	26 14 29	07 ♓ 49 26	15 ♓ 00 30	14 26	06 09	11 11	22 52	04 14	17 37	26 16	00 19
17	01 41 27	27 13 13	22 ♓ 10 31	29 ♓ 18 56	14 52	07 19	11 42	23 06	04 19	17 40	26 18	00 19
18	01 45 24	28 11 55	06 ♈ 25 09	13 ♈ 28 35	15 13	08 28	12 14	23 20	04 24	17 43	26 20	00 19
19	01 49 20	29 ♈ 10 35	20 ♈ 28 45	27 ♈ 25 08	15 27	09 37	12 46	23 35	04 29	17 46	26 22	00 20
20	01 53 17	00 ♉ 09 13	04 ♉ 17 21	11 ♉ 05 04	15 35	10 46	13 18	23 49	04 35	17 50	26 24	00 20
21	01 57 13	01 07 49	17 ♉ 48 03	24 ♉ 26 08	15 R 37	11 55	13 50	24 04	04 40	17 53	26 26	00 20
22	02 01 10	02 06 24	00 ♊ 59 15	07 ♊ 27 28	15 34	13 03	14 23	24 18	04 45	17 57	26 28	00 21
23	02 05 06	03 04 56	13 ♊ 50 53	20 ♊ 09 42	15 24	14 12	14 55	24 33	04 50	18 00	26 30	00 21
24	02 09 03	04 03 27	26 ♊ 24 13	02 ♋ 34 46	15 10	15 20	15 27	24 47	04 54	18 03	26 32	00 21
25	02 13 00	05 01 55	08 ♋ 41 45	14 ♋ 45 39	14 50	16 28	16 00	25 01	04 59	18 07	26 34	00 21
26	02 16 56	06 00 21	20 ♋ 46 58	26 ♋ 46 12	14 26	17 36	16 32	25 16	05 04	18 10	26 36	00 21
27	02 20 53	06 58 45	02 ♌ 43 56	08 ♌ 40 44	13 58	18 44	17 05	25 30	05 09	18 14	26 38	00 22
28	02 24 49	07 57 07	14 ♌ 37 11	20 ♌ 33 51	13 27	19 52	17 38	25 44	05 13	18 17	26 39	00 22
29	02 28 46	08 55 27	26 ♌ 31 19	02 ♍ 30 09	12 53	20 59	18 10	25 59	05 18	18 20	26 41	00 22
30	02 32 42	09 ♉ 53 44	08 ♍ 30 52	14 ♍ 34 00	12 ♉ 16	22 ♊ 07	18 ♋ 43	26 ♈ 13	05 ♓ 22	18 ♉ 24	26 ♓ 43	00 ♒ 22

Date	Moon True ☊ ° ′	Moon Mean ☊ ° ′	Moon ☽ Latitude ° ′
01	04 ♉ 15	05 ♉ 25	05 N 04
02	04 R 11	05 22	04 41
03	04 08	05 18	04 06
04	04 05	05 15	03 19
05	04 02	05 12	02 21
06	04 01	05 09	01 16
07	04 00	05 06	00 N 06
08	04 D 00	05 03	01 S 06
09	04 01	04 59	02 16
10	04 02	04 56	03 19
11	04 04	04 53	04 12
12	04 04	04 50	04 51
13	04 R 04	04 47	05 12
14	04 04	04 44	05 15
15	04 03	04 40	04 58
16	04 03	04 37	04 23
17	04 02	04 34	03 31
18	04 01	04 31	02 27
19	04 01	04 28	01 S 14
20	04 D 01	04 24	00 N 02
21	04 01	04 21	01 16
22	04 01	04 18	02 24
23	04 01	04 15	03 23
24	04 R 01	04 12	04 11
25	04 01	04 09	04 47
26	04 01	04 05	05 08
27	04 01	04 02	05 17
28	04 D 01	03 59	05 11
29	04 01	03 56	04 53
30	04 ♉ 01	03 ♉ 53	04 N 21

DECLINATIONS

Date	Sun ☉ ° ′	Moon ☽ ° ′	Mercury ☿ ° ′	Venus ♀ ° ′	Mars ♂ ° ′	Jupiter ♃ ° ′	Saturn ♄ ° ′	Uranus ♅ ° ′	Neptune ♆ ° ′	Pluto ♇ ° ′
01	04 N 33	20 N 05	11 N 05	18 N 19	25 N 27	06 N 33	11 S 44	16 N 33	02 S 46	22 S 29
02	04 56	15 40	11 55	18 41	25 26	06 38	11 41	16 34	02 45	22 29
03	05 19	10 36	12 42	19 03	25 24	06 44	11 39	16 35	02 45	22 29
04	05 42	05 N 04	13 27	19 25	25 23	06 49	11 37	16 36	02 44	22 29
05	06 05	00 S 46	14 11	19 47	25 21	06 54	11 35	16 37	02 43	22 29
06	06 28	06 40	14 51	20 07	25 19	07 00	11 33	16 38	02 42	22 29
07	06 50	12 25	15 29	20 28	25 17	07 05	11 31	16 39	02 41	22 29
08	07 13	17 43	16 04	20 47	25 14	07 11	11 29	16 39	02 40	22 29
09	07 35	22 15	16 37	21 07	25 12	07 16	11 27	16 40	02 39	22 29
10	07 57	25 39	17 07	21 25	25 10	07 22	11 25	16 41	02 39	22 29
11	08 19	27 36	17 33	21 44	25 07	07 27	11 23	16 42	02 38	22 29
12	08 41	27 50	17 57	22 01	25 04	07 33	11 22	16 43	02 37	22 29
13	09 03	26 16	18 18	22 18	25 01	07 38	11 20	16 44	02 36	22 29
14	09 25	23 02	18 35	22 35	24 58	07 43	11 18	16 45	02 35	22 29
15	09 46	18 24	18 50	22 51	24 55	07 49	11 16	16 46	02 35	22 29
16	10 08	12 42	19 01	23 06	24 52	07 54	11 14	16 47	02 34	22 29
17	10 29	06 S 20	19 10	23 21	24 49	07 59	11 12	16 48	02 33	22 29
18	10 50	00 N 18	19 15	23 35	24 45	08 05	11 11	16 49	02 32	22 29
19	11 11	06 51	19 17	23 49	24 42	08 10	11 09	16 50	02 32	22 30
20	11 32	12 58	19 16	24 01	24 38	08 15	11 07	16 51	02 31	22 30
21	11 52	18 21	19 12	24 14	24 34	08 21	11 05	16 51	02 30	22 30
22	12 12	22 42	19 05	24 25	24 30	08 26	11 04	16 52	02 29	22 30
23	12 32	25 49	18 55	24 37	24 26	08 31	11 02	16 53	02 29	22 30
24	12 52	27 34	18 42	24 47	24 22	08 37	11 00	16 54	02 28	22 30
25	13 12	27 55	18 26	24 57	24 17	08 42	10 59	16 55	02 27	22 30
26	13 31	26 55	18 08	25 06	24 13	08 47	10 57	16 56	02 26	22 30
27	13 51	24 41	17 48	25 14	24 08	08 52	10 56	16 57	02 26	22 30
28	14 10	21 24	17 25	25 22	24 03	08 57	10 54	16 58	02 25	22 30
29	14 28	17 15	17 00	25 29	23 58	09 03	10 53	16 59	02 24	22 31
30	14 N 47	12 N 25	16 N 34	25 N 36	23 N 53	09 N 08	10 S 51	17 N 00	02 S 24	22 S 31

ZODIAC SIGN ENTRIES

Date	h m	Planets
02	10 57	☽ ♍
03	16 22	☿ ♉
04	21 51	☽ ♎
07	06 29	☽ ♏
09	12 57	☽ ♐
11	04 47	♀ ♊
11	17 33	☽ ♑
13	20 42	☽ ♒
15	22 57	☽ ♓
18	01 09	☽ ♈
20	04 30	☽ ♉
20	08 14	☉ ♉
22	10 11	☽ ♊
24	18 58	☽ ♋
27	06 30	☽ ♌
29	18 59	☽ ♍

LATITUDES

Date	Mercury ☿ ° ′	Venus ♀ ° ′	Mars ♂ ° ′	Jupiter ♃ ° ′	Saturn ♄ ° ′	Uranus ♅ ° ′	Neptune ♆ ° ′	Pluto ♇ ° ′
01	01 N 03	00 N 59	02 N 04	01 S 04	01 S 20	00 S 19	01 S 10	02 S 25
04	01 37	01 09	02 02	01 04	01 20	00 19	01 10	02 26
07	02 08	01 19	02 00	01 04	01 21	00 19	01 10	02 26
10	02 34	01 29	01 58	01 04	01 21	00 19	01 11	02 27
13	02 51	01 38	01 56	01 04	01 22	00 19	01 11	02 27
16	03 00	01 48	01 54	01 04	01 22	00 19	01 11	02 28
19	02 57	01 56	01 53	01 04	01 23	00 19	01 11	02 28
22	02 42	02 05	01 51	01 04	01 23	00 19	01 11	02 29
25	02 15	02 12	01 49	01 04	01 24	00 19	01 11	02 29
28	01 37	02 20	01 47	01 04	01 24	00 19	01 11	02 30
31	00 N 50	02 N 26	01 N 46	01 S 04	01 S 25	00 S 19	01 S 11	02 S 30

DATA

Julian Date	2460036
Delta T	+72 seconds
Ayanamsa	24° 10′ 45″
Synetic vernal point	04° ♓ 56′ 14″
True obliquity of ecliptic	23° 26′ 19″

MOON'S PHASES, APSIDES AND POSITIONS ☽

Date	h m	Phase	Longitude ° ′	Eclipse Indicator
06	04 34	○	16 ♎ 07	
13	09 11	☾	23 ♑ 11	
20	04 12	●	29 ♈ 50	Ann-Total
27	21 20	☽	07 ♌ 21	

Day	h m		
16	02 13	Perigee	
28	06 43	Apogee	
05	08 54	0S	
12	03 10	Max dec	27° S 57′
18	10 55	0N	
25	05 54	Max dec	27° N 58′

LONGITUDES

Date	Chiron ⚷ ° ′	Ceres ⚳ ° ′	Pallas ⚴ ° ′	Juno ⚵ ° ′	Vesta ⚶ ° ′	Black Moon Lilith ⚸ ° ′
01	15 ♈ 36	27 ♍ 55	18 ♋ 33	11 ♉ 58	23 ♈ 33	09 ♌ 18
11	16 ♈ 12	26 ♍ 00	21 ♋ 55	17 ♉ 48	28 ♈ 05	10 ♌ 25
21	16 ♈ 46	24 ♍ 38	25 ♋ 36	23 ♉ 41	02 ♉ 48	11 ♌ 32
31	17 ♈ 20	23 ♍ 55	29 ♋ 31	29 ♉ 35	07 ♉ 08	12 ♌ 39

ASPECTARIAN

01 Saturday
05 39 ☿ ⊻ ♆
06 44 ♀ ∠ ♂
08 30 ☽ □ ♅
11 47 ☽ ∠ ♂
12 06 ☽ □ ♀
13 25 ☽ △ ♃
14 23 ☽ ± ♆
21 19 ☽ ∥ ♀

02 Sunday
02 28 ☽ ⊼ ♆
03 57 ☉ Q ♇
04 18 ♀ ⊻ ♃
05 23 ☽ ⚼ ☉
05 47 ☿ # ♄
06 03 ☽ △ ☿
07 26 ☽ ∥ ♅
11 16 ☽ ⊼ ♇
16 44 ☽ ☍ ♄
19 08 ☽ ⚹ ♂
20 03 ☽ ⚼ ♃
23 14 ☽ ± ♇

03 Monday
00 55 ☽ ± ☉
03 39 ☽ ∥ ☿
07 12 ☽ # ♄
13 49 ☽ ⊼ ☉
14 12 ☽ ± ♃
16 43 ☽ ⚼ ☿
17 01 ☽ ⚼ ♇
18 55 ☿ □ ♇
19 51 ☽ Q ♂
20 28 ☽ △ ♅

04 Tuesday
02 12 ☽ ⊼ ♃
04 43 ☽ ∥ ♃
06 04 ☽ △ ♀
09 30 ☽ ∥ ☉
12 59 ☽ ± ☿
13 50 ☽ ☍ ♆
19 41 ☿ ⊥ ♆
21 45 ☽ # ♆
22 13 ☽ △ ♇

05 Wednesday
01 42 ☽ ⚼ ♅
02 16 ☽ ⊼ ☿
03 54 ☽ ⊼ ♄
08 12 ☽ □ ♂
13 59 ☽ ⚼ ♀
15 28 ☽ ± ♄
16 21 ☿ ⚹ ♄
18 58 ☽ ± ♅
19 54 ☽ ∥ ♆

06 Thursday
04 34 ☽ ☍ ☉
06 21 ☽ ⊼ ♅
08 37 ☽ ⚼ ♄
08 45 ☽ ± ♀
11 04 ☽ # ☉
12 43 ☽ ☍ ♃
13 22 ☽ # ♃
21 06 ☽ ⊼ ♀
22 57 ☽ ⊼ ♆

07 Friday
04 42 ☉ ⊻ ♅
06 54 ☽ □ ♇
08 11 ☽ ∥ ♄
10 06 ☽ ± ♆
12 44 ☽ △ ♄
17 53 ☽ ☍ ☿
17 58 ♀ ⚹ ♆
18 42 ☽ △ ♂

08 Saturday
02 39 ☽ ⚼ ♆
03 17 ☽ # ☿
06 29 ☿ ⚹ ♂
06 56 ☽ # ♅
09 31 ☉ ∥ ♃
13 56 ☽ ☍ ♅
14 09 ♀ ⊥ ♃
14 20 ☉ ∠ ♄
15 50 ☽ Q ♇
16 27 ☽ ⊼ ☉
20 51 ☽ ⊼ ♃
23 03 ☽ ⚼ ♂

09 Sunday
04 06 ☽ ± ☉
04 55 ☽ # ♀
05 50 ☽ △ ♆
07 49 ☽ ± ♃
09 09 ☽ ☍ ♀
13 23 ☽ ⚹ ♇
13 24 ☽ ∥ ♇
14 39 ☿ ∥ ♅
15 47 ☽ ± ♂
19 21 ☽ □ ♄
21 24 ☽ ⚼ ☉

10 Monday
00 07 ☽ ⚼ ♃
02 51 ☽ ⊼ ♂
05 23 ☽ ⊼ ☿
07 53 ☽ # ♂
15 54 ☽ ∠ ♇
16 52 ☽ ± ☿
19 29 ☽ ⊼ ♅

11 Tuesday
01 48 ☽ △ ☉
02 55 ☽ △ ♃
03 13 ☽ Q ♄
05 59 ☽ ± ♅
07 37 ☽ ⊥ ♇
09 51 ☽ ⚼ ☿
10 14 ♀ △ ♇
10 48 ☽ □ ♆
18 01 ☽ ⊻ ♇
18 43 ☽ ⊼ ♀
21 37 ☽ ⚼ ♅
22 07 ☉ ☌ ♃
23 27 ☿ ∠ ♆

12 Wednesday
00 08 ☽ ⚹ ♄
05 59 ☽ ± ♀
09 03 ☽ ☍ ♂
13 35 ☽ △ ☿
17 49 ☽ Q ♆
22 45 ☽ ⚼ ♀
23 24 ☽ △ ♅

13 Thursday
01 57 ☽ ∠ ♄
07 20 ☽ □ ♃
09 11 ☽ □ ☉
14 14 ☽ ⚹ ♆
17 17 ☽ ⊥ ♄
21 11 ☽ ☌ ♇
22 53 ☽ # ♂

14 Friday
02 23 ☽ △ ♀
03 28 ☽ ⊻ ♄
03 49 ♀ # ♇
13 47 ☽ ⊼ ♂
14 11 ☽ Q ♃
14 33 ☽ # ♀
15 15 ☽ ∥ ♇
15 32 ☽ ∠ ♆
16 38 ♀ □ ♄
17 44 ☽ Q ☉
19 16 ☽ □ ☿
22 42 ♀ ⊥ ♂

15 Saturday
00 13 ☽ ± ♂
02 08 ☽ □ ♅
06 38 ☽ ⊥ ♆
10 03 ☽ # ☿
10 36 ☽ ⚹ ♃
15 16 ☽ ⚹ ☉
15 49 ☽ ⚼ ♂
16 41 ☽ ⊻ ♆
19 15 ☽ # ♅
23 27 ☽ ⊻ ♇

16 Sunday
02 39 ☽ Q ☿
05 58 ☽ ☌ ♄
08 17 ☽ Q ♅
08 58 ☽ □ ♀
09 28 ☽ ⊥ ♇
12 04 ☽ ∠ ♃
12 43 ☉ ⊻ ♆
17 44 ☽ ∥ ♄
17 49 ☽ △ ♂
18 07 ☽ ∠ ☉
21 24 ☽ # ☉
23 24 ☽ ⚹ ☿

17 Monday
00 30 ☽ ∠ ♇
03 21 ☽ ⊥ ♃
04 25 ☽ ⚹ ♅
06 01 ☽ # ♃
10 17 ☽ ⊥ ☉
13 35 ☽ ⊻ ♃
17 43 ☽ Q ♀
18 57 ☽ ☌ ♆
21 06 ☽ ⊻ ☉

18 Tuesday
01 17 ☽ ∠ ☿
01 41 ☽ ⚹ ♇
01 46 ☽ ∥ ♆
05 43 ☽ ∠ ♅
08 34 ☽ ⊻ ♄
08 46 ♀ ∠ ♃
09 17 ♀ Q ♆
15 46 ☽ ⚹ ♀
16 50 ☽ ⊥ ☿
18 48 ☽ ⊥ ♄
20 05 ☽ # ♆
21 02 ☽ ⊥ ♅
22 02 ☽ Q ♇
22 16 ☽ □ ♂

19 Wednesday
03 15 ☽ ⊻ ☿
07 20 ☽ ⊻ ♅
09 50 ☉ # ♄
10 17 ☽ ∠ ♄
17 03 ☽ ∥ ♃
17 27 ☽ ☌ ♃
19 47 ☽ ∠ ♀
22 12 ☽ ⊻ ♆

20 Thursday
04 12 ☽ ☌ ☉
04 30 ☽ # ♄
05 04 ☽ □ ♇
05 44 ☽ ∥ ☉
06 33 ☽ Q ♂
08 41 ☽ ⊥ ♆
12 30 ☽ ⚹ ♄
12 54 ☽ ⊥ ♀
16 27 ☉ □ ♇

21 Friday
00 29 ☽ ⊻ ♀
00 36 ☽ ∠ ♆
04 36 ☽ ⚹ ♂
04 54 ☽ ∥ ♅
08 05 ☽ ☌ ☿
08 35 ☿ St R
09 56 ☽ Q ♄
12 09 ☽ ☌ ♅
16 10 ☽ ∥ ☿
23 32 ☽ ⊻ ♃

22 Saturday
03 41 ☽ ⚹ ♆
08 54 ☽ ∠ ♂
10 41 ☽ # ♇
10 43 ☽ ⊥ ♃
10 49 ☽ △ ♇
14 14 ☽ ⊻ ☉
19 00 ☽ □ ♄
19 01 ♀ ∥ ♂
21 11 ☉ ⊥ ♆

23 Sunday
00 17 ☽ ∥ ♂
00 44 ☽ ∥ ♀
01 55 ☽ Q ♆
02 18 ☽ ⊥ ☉
02 19 ☽ ⊥ ♂
02 44 ☉ Q ♂
03 45 ☽ ∠ ♃
12 43 ☽ ☌ ♀
14 06 ☽ ⊻ ♂
14 50 ☽ ⚼ ♇
14 54 ☽ ⊻ ☿
19 54 ☽ ⊻ ♅
20 42 ☽ ∠ ☉

24 Monday
02 07 ☽ ⊥ ☿
03 19 ☿ ⚹ ♂
07 27 ☽ ⊥ ♅
08 02 ☽ ± ♇
08 49 ☽ ⚹ ♃
09 04 ☿ ⊻ ♀
12 15 ☽ □ ♆
12 23 ♀ ⚼ ♇
16 50 ♀ ⊻ ♂
19 07 ☽ ∠ ☿
19 40 ☽ ⊼ ♇

25 Tuesday
00 59 ☽ ∠ ♅
04 11 ☽ ⚹ ☉
04 40 ☽ △ ♄
08 38 ☽ Q ♃
10 48 ☉ ⚹ ♄
23 47 ☽ ⚹ ☿

26 Wednesday
03 08 ☽ ☌ ♂
05 00 ☽ ⊻ ♀
05 58 ☽ Q ☉
06 45 ☽ ⚹ ♅
10 33 ☽ ⚼ ♄
18 14 ☽ ⊥ ♀
21 09 ☽ □ ♃
22 55 ☽ Q ☿
23 41 ☽ △ ♆

27 Thursday
00 35 ♀ ⊻ ♅
04 43 ☽ ± ♄
06 56 ☽ Q ♅
07 13 ☽ ☍ ♇
07 14 ☽ ∥ ♀
14 14 ☽ ∠ ♀
16 43 ☽ ∥ ♂
16 54 ☽ ⊼ ♄
21 20 ☽ □ ☉

28 Friday
04 40 ☽ # ♇
05 59 ☿ ⊥ ♀
06 00 ☽ ⚼ ♆
09 44 ☽ □ ☿
18 22 ☽ ⊻ ♂
19 26 ☽ □ ♅
23 42 ☽ ⚹ ♀

29 Saturday
00 13 ☽ ± ♆
07 03 ☽ ⊥ ♂
10 53 ☽ △ ♃
12 20 ☽ ⊼ ♆
13 07 ☿ ∥ ♅
13 26 ☽ ∥ ♅
13 28 ☽ ∥ ☿
19 43 ☽ ⊼ ♇
20 05 ♂ ⚹ ♅

30 Sunday
01 19 ☽ ∥ ☉
01 59 ☽ ∠ ♂
02 19 ☽ Q ♀
05 41 ☽ ☍ ♄
07 43 ☽ ± ♇
14 59 ☽ △ ☉
17 28 ☽ ⚼ ♃
19 05 ☽ △ ☿
19 12 ☽ # ♄

All ephemeris data is given at 12.00 UT and the Moon's longitude is additionally given for 24.00 UT

MAY 2023

LONGITUDES

Date	Sidereal time h m s	Sun ☉ ° ' "	Moon ☽ ° ' "	Moon ☽ 24.00 ° ' "	Mercury ☿ ° '	Venus ♀ ° '	Mars ♂ ° '	Jupiter ♃ ° '	Saturn ♄ ° '	Uranus ♅ ° '	Neptune ♆ ° '	Pluto ♇ ° '
01	02 36 39	10 ♉ 52 00	20 ♍ 40 01	26 ♍ 49 21	11 ♉ 38	23 ♊ 14	19 ♋ 16	26 ♈ 27	05 ♓ 26	18 ♉ 27	26 ♓ 45	00 ♒ 22
02	02 40 35	11 50 13	03 ♎ 02 22	09 ♎ 19 25	11 R 00	24 21	19 49	26 41	05 31	18 31	26 46	00 R 22
03	02 44 32	12 48 25	15 ♎ 40 44	22 ♎ 06 30	10 21	25 28	20 22	26 55	05 35	18 34	26 48	00 22
04	02 48 29	13 46 35	28 ♎ 36 52	05 ♏ 11 49	09 43	26 35	20 56	27 10	05 39	18 38	26 50	00 22
05	02 52 25	14 44 43	11 ♏ 51 21	18 ♏ 35 18	09 06	27 41	21 29	27 24	05 43	18 41	26 52	00 22
06	02 56 22	15 42 49	25 ♏ 23 29	02 ♐ 15 36	08 31	28 47	22 02	27 38	05 47	18 45	26 53	00 22
07	03 00 18	16 40 54	09 ♐ 11 21	16 ♐ 10 17	07 58	29 ♊ 53	22 35	27 52	05 51	18 48	26 55	00 21
08	03 04 15	17 38 58	23 ♐ 12 00	00 ♑ 16 01	07 28	00 ♋ 59	23 09	28 06	05 55	18 52	26 56	00 21
09	03 08 11	18 37 00	07 ♑ 21 50	14 ♑ 29 00	07 02	02 05	23 42	28 20	05 58	18 55	26 58	00 21
10	03 12 08	19 35 00	21 ♑ 37 01	28 ♑ 45 26	06 40	03 10	24 16	28 34	06 02	18 58	27 00	00 21
11	03 16 04	20 32 59	05 ♒ 53 50	13 ♒ 01 50	06 21	04 16	24 50	28 48	06 06	19 02	27 01	00 21
12	03 20 01	21 30 57	20 ♒ 09 04	27 ♒ 15 15	06 07	05 21	25 23	29 02	06 09	19 05	27 03	00 20
13	03 23 58	22 28 54	04 ♓ 20 04	11 ♓ 23 19	05 57	06 25	25 57	29 16	06 12	19 09	27 04	00 20
14	03 27 54	23 26 49	18 ♓ 24 44	25 ♓ 24 08	05 52	07 30	26 31	29 29	06 16	19 12	27 06	00 20
15	03 31 51	24 24 43	02 ♈ 21 20	09 ♈ 16 09	05 D 51	08 34	27 05	29 43	06 19	19 16	27 07	00 19
16	03 35 47	25 22 36	16 ♈ 08 24	22 ♈ 57 55	05 55	09 38	27 39	29 ♈ 57	06 22	19 19	27 08	00 19
17	03 39 44	26 20 27	29 ♈ 44 31	06 ♉ 28 03	06 04	10 42	28 13	00 ♉ 11	06 25	19 23	27 10	00 18
18	03 43 40	27 18 17	13 ♉ 08 22	19 ♉ 45 19	06 17	11 46	28 47	00 24	06 28	19 26	27 11	00 18
19	03 47 37	28 16 06	26 ♉ 18 47	02 ♊ 48 40	06 35	12 49	29 21	00 38	06 31	19 30	27 13	00 17
20	03 51 33	29 ♉ 13 54	09 ♊ 14 55	15 ♊ 37 31	06 57	13 52	29 ♋ 55	00 52	06 34	19 33	27 14	00 17
21	03 55 30	00 ♊ 11 40	21 ♊ 56 30	28 ♊ 11 57	07 23	14 55	00 ♌ 29	01 05	06 36	19 37	27 15	00 16
22	03 59 27	01 09 25	04 ♋ 23 59	10 ♋ 32 48	07 53	15 57	01 04	01 18	06 39	19 40	27 16	00 16
23	04 03 23	02 07 08	16 ♋ 38 40	22 ♋ 41 51	08 28	17 00	01 38	01 32	06 41	19 43	27 18	00 15
24	04 07 20	03 04 50	28 ♋ 42 45	04 ♌ 41 44	09 06	18 01	02 12	01 45	06 44	19 47	27 19	00 15
25	04 11 16	04 02 30	10 ♌ 39 16	16 ♌ 35 51	09 49	19 03	02 47	01 59	06 46	19 50	27 20	00 14
26	04 15 13	05 00 09	22 ♌ 32 01	28 ♌ 28 20	10 35	20 04	03 21	02 12	06 48	19 54	27 21	00 13
27	04 19 09	05 57 46	04 ♍ 25 21	10 ♍ 23 43	11 24	21 05	03 56	02 25	06 50	19 57	27 22	00 13
28	04 23 06	06 55 22	16 ♍ 23 59	22 ♍ 26 49	12 17	22 06	04 30	02 38	06 53	20 00	27 23	00 12
29	04 27 02	07 52 56	28 ♍ 32 46	04 ♎ 42 26	13 14	23 06	05 05	02 51	06 54	20 04	27 24	00 11
30	04 30 59	08 50 29	10 ♎ 56 20	17 ♎ 14 59	14 14	24 06	05 40	03 04	06 56	20 07	27 25	00 11
31	04 34 56	09 ♊ 48 00	23 ♎ 38 48	00 ♏ 08 09	15 ♉ 17	25 ♋ 05	06 ♌ 14	03 ♉ 17	06 ♓ 58	20 ♉ 10	27 ♓ 26	00 ♒ 10

Date	Moon True ☊ ° '	Moon Mean ☊ ° '	Moon ☽ Latitude ° '
01	04 ♉ 02	03 ♉ 50	03 N 38
02	04 D 03	03 46	02 43
03	04 04	03 43	01 40
04	04 04	03 40	00 N 30
05	04 R 04	03 37	00 S 43
06	04 03	03 34	01 55
07	04 02	03 30	03 02
08	04 01	03 27	03 59
09	03 59	03 24	04 42
10	03 57	03 21	05 08
11	03 56	03 18	05 15
12	03 55	03 15	05 03
13	03 D 56	03 11	04 32
14	03 57	03 08	03 45
15	03 58	03 05	02 46
16	03 59	03 02	01 37
17	04 00	02 59	00 S 23
18	04 R 00	02 56	00 N 50
19	03 58	02 52	01 59
20	03 56	02 49	03 01
21	03 52	02 46	03 53
22	03 48	02 43	04 32
23	03 43	02 40	04 59
24	03 39	02 36	05 11
25	03 35	02 33	05 10
26	03 33	02 30	04 55
27	03 32	02 27	04 28
28	03 D 33	02 24	03 49
29	03 34	02 21	03 00
30	03 36	02 17	02 00
31	03 ♉ 37	02 ♉ 14	00 N 54

DECLINATIONS

Sun ☉ ° '	Moon ☽ ° '	Mercury ☿ ° '	Venus ♀ ° '	Mars ♂ ° '	Jupiter ♃ ° '	Saturn ♄ ° '	Uranus ♅ ° '	Neptune ♆ ° '	Pluto ♇ ° '
15 N 05	07 N 02	16 N 07	25 N 42	23 N 48	09 N 13	10 S 50	17 N 01	02 S 23	22 S 31
15 23	01 N 17	15 39	25 47	23 42	09 18	10 49	17 02	02 22	22 31
15 41	04 S 38	15 11	25 52	23 37	09 23	10 47	17 03	02 22	22 31
15 58	10 31	14 42	25 56	23 31	09 28	10 46	17 04	02 21	22 31
16 16	16 04	14 14	25 59	23 25	09 33	10 44	17 05	02 20	22 32
16 33	20 58	13 46	26 01	23 19	09 38	10 43	17 06	02 20	22 32
16 49	24 50	13 20	26 03	23 13	09 43	10 42	17 07	02 19	22 32
17 06	27 15	12 55	26 04	23 07	09 48	10 41	17 08	02 18	22 32
17 22	27 56	12 31	26 05	23 01	09 53	10 40	17 09	02 18	22 32
17 38	26 46	12 10	26 05	22 54	09 58	10 38	17 10	02 17	22 33
17 53	23 53	11 50	26 04	22 48	10 03	10 37	17 11	02 17	22 33
18 08	19 33	11 33	26 03	22 41	10 08	10 36	17 12	02 16	22 33
18 23	14 08	11 18	26 01	22 34	10 13	10 35	17 13	02 16	22 33
18 38	08 02	11 05	25 58	22 27	10 18	10 34	17 14	02 15	22 33
18 52	01 S 36	10 55	25 55	22 20	10 22	10 33	17 14	02 15	22 34
19 06	04 N 52	10 47	25 51	22 13	10 27	10 32	17 15	02 14	22 34
19 20	11 01	10 42	25 47	22 06	10 32	10 31	17 16	02 14	22 34
19 33	16 35	10 39	25 42	21 58	10 37	10 30	17 17	02 13	22 34
19 46	21 16	10 38	25 36	21 50	10 41	10 29	17 18	02 13	22 35
19 59	24 49	10 40	25 30	21 43	10 46	10 29	17 19	02 12	22 35
20 11	27 04	10 44	25 23	21 35	10 51	10 28	17 20	02 12	22 35
20 23	27 54	10 50	25 16	21 27	10 55	10 27	17 21	02 11	22 35
20 35	27 20	10 59	25 08	21 18	11 00	10 26	17 22	02 11	22 36
20 46	25 29	11 09	24 59	21 10	11 04	10 26	17 23	02 10	22 36
20 57	22 32	11 21	24 50	21 02	11 09	10 25	17 24	02 10	22 36
21 08	18 39	11 35	24 40	20 53	11 13	10 24	17 25	02 09	22 37
21 18	14 03	11 51	24 30	20 44	11 18	10 24	17 26	02 09	22 37
21 28	08 53	12 08	24 20	20 36	11 22	10 23	17 27	02 09	22 37
21 37	03 N 19	12 27	24 08	20 27	11 27	10 23	17 27	02 08	22 37
21 46	02 S 29	12 47	23 57	20 18	11 31	10 22	17 28	02 08	22 38
21 N 55	08 S 21	13 N 09	23 N 45	20 N 09	11 N 35	10 S 22	17 N 29	02 S 08	22 S 38

ZODIAC SIGN ENTRIES

Date	h	m	Planets
02	06	09	☽ ♎
04	14	32	☽ ♏
06	20	04	☽ ♐
07	14	25	♀ ♋
08	23	33	☽ ♑
11	02	05	☽ ♒
13	04	39	☽ ♓
15	07	56	☽ ♈
16	17	20	♃ ♉
17	12	28	☽ ♉
19	18	48	☽ ♊
20	15	31	♂ ♌
21	07	09	☉ ♊
22	03	28	☽ ♋
24	14	35	☽ ♌
27	03	05	☽ ♍
29	14	51	☽ ♎
31	23	45	☽ ♏

LATITUDES

Date	Mercury ☿ ° '	Venus ♀ ° '	Mars ♂ ° '	Jupiter ♃ ° '	Saturn ♄ ° '	Uranus ♅ ° '	Neptune ♆ ° '	Pluto ♇ ° '
01	00 N 50	02 N 26	01 N 46	01 S 04	01 S 25	00 S 19	01 S 11	02 S 30
04	00 S 01	02 32	01 44	01 04	01 25	00 19	01 11	02 31
07	00 53	02 37	01 42	01 04	01 26	00 19	01 11	02 31
10	01 40	02 41	01 40	01 04	01 27	00 19	01 11	02 32
13	02 20	02 44	01 39	01 04	01 27	00 19	01 12	02 32
16	02 52	02 47	01 37	01 04	01 28	00 19	01 12	02 33
19	03 15	02 48	01 35	01 04	01 28	00 19	01 12	02 33
22	03 29	02 48	01 34	01 04	01 29	00 19	01 12	02 33
25	03 35	02 47	01 32	01 05	01 30	00 19	01 12	02 34
28	03 33	02 44	01 30	01 05	01 30	00 19	01 12	02 34
31	03 S 25	02 N 40	01 N 29	01 S 05	01 S 31	00 S 19	01 S 12	02 S 35

DATA

Julian Date	2460066
Delta T	+72 seconds
Ayanamsa	24° 10' 48"
Synetic vernal point	04° ♓ 56' 11"
True obliquity of ecliptic	23° 26' 18"

MOON'S PHASES, APSIDES AND POSITIONS ☽

Date	h	m	Phase	Longitude ° '	Eclipse Indicator
05	17	34	○	14 ♏ 58	
12	14	28	☾	21 ♒ 37	
19	15	53	●	28 ♉ 25	
27	15	22	☽	06 ♍ 06	

Day	h	m		
11	05	14	Perigee	
26	01	40	Apogee	
02	17	16	0S	
09	08	57	Max dec	27° S 57'
15	17	53	0N	
22	14	08	Max dec	27° N 55'
30	01	49	0S	

LONGITUDES

Date	Chiron ⚷ ° '	Ceres ⚳ ° '	Pallas ⚴ ° '	Juno ⚵ ° '	Vesta ⚶ ° '	Black Moon Lilith ⚸ ° '
01	17 ♈ 20	23 ♍ 55	29 ♋ 31	29 ♉ 35	07 ♉ 08	12 ♌ 39
11	17 ♈ 52	23 ♍ 53	03 ♌ 37	05 ♊ 29	11 ♉ 37	13 ♌ 46
21	18 ♈ 22	24 ♍ 32	07 ♌ 50	11 ♊ 23	16 ♉ 03	14 ♌ 54
31	18 ♈ 48	25 ♍ 46	12 ♌ 10	17 ♊ 15	20 ♉ 27	16 ♌ 01

All ephemeris data is given at 12.00 UT and the Moon's longitude is additionally given for 24.00 UT

ASPECTARIAN

h m	Aspects
01 Monday	
01 34	☽ bQ ♇
02 39	☽ ∥ ♃
07 38	☽ △ ♅
08 07	☿ ∠ ♆
09 08	☽ ⚹ ♂
11 34	☽ ± ♃
17 11	♇ St R
17 31	☽ □ ♀
23 01	☽ bQ ☉
23 04	☽ bQ ☿
23 28	☉ ☌ ☿
23 30	☽ ⊼ ♃
23 53	☽ ☍ ♆
02 Tuesday	
06 51	☽ △ ♇
07 34	☽ ⋕ ♆
09 33	☽ Q ♂
10 24	☉ ∠ ♆
12 17	♀ ± ♇
12 55	☽ bQ ♅
15 34	☽ ± ☿
15 39	♀ ⊥ ♅
16 46	☽ ⊼ ♄
17 48	☽ ± ☉
20 17	☉ ∥ ☿
22 03	♃ ⊻ ♆
03 Wednesday	
02 26	☽ ⊼ ☿
02 51	☽ ∥ ♆
04 14	☽ ± ♄
06 07	☽ ± ♅
06 09	☽ ⊼ ☉
10 22	☿ ∠ ♀
17 26	☽ ⊼ ♅
21 10	☽ □ ♂
21 13	☽ bQ ♄
22 19	♂ bQ ♄
04 Thursday	
07 37	☽ ⋕ ♃
07 54	☽ △ ♀
08 43	☽ ⊼ ♆
09 17	☽ ☍ ♃
13 02	☽ ∥ ♄
15 12	☽ □ ♇
17 40	♀ □ ♆
19 43	☽ ± ♆
05 Friday	
00 53	☽ △ ♄
04 01	☿ Q ♂
04 03	♀ ⚹ ♃
04 27	☽ ⋕ ☿
07 15	☽ ☍ ☿
12 00	☽ bQ ♆
12 54	☽ ⋕ ☉
13 37	☽ bQ ♀
16 39	☽ ⋕ ♅
17 34	☽ ☍ ☉
23 36	☽ Q ♇
06 Saturday	
00 13	☽ ☍ ♅
05 51	☽ △ ♂
07 01	☽ ± ♀
14 38	☽ △ ♆
15 59	☽ ⊼ ♃
18 28	☽ ⊼ ♀
20 41	☽ ⚹ ♇
20 48	☽ ∥ ♇
07 Sunday	
01 20	☽ ⋕ ♂
02 38	☽ ± ♃
06 12	☽ □ ♄
09 08	☽ bQ ♂
09 58	☽ ⊼ ☿
18 26	☽ bQ ♃
19 56	☽ ± ☿
22 10	♀ ⊼ ♇
22 20	☽ ⋕ ♀
22 36	☽ ∠ ♇
08 Monday	
01 15	☽ ± ♂
01 50	☽ ⊼ ☉
04 34	☽ ⊼ ♅
10 48	☽ bQ ☿
11 55	☽ ⊼ ♂
12 49	☽ ± ☉
13 13	☽ Q ♄
13 58	☽ ⊥ ♇
14 50	☽ ± ♅
15 15	☉ ∥ ♅
18 22	☽ □ ♆
18 58	☉ Q ♄
20 28	☽ △ ♃
09 Tuesday	
00 09	☽ ⊻ ♇
02 20	☽ ☍ ♀
05 12	☽ bQ ☉
06 09	☽ bQ ♅
09 39	☽ ⚹ ♄
11 28	☽ △ ☿
19 56	☉ ☌ ♅
10 Wednesday	
00 50	☽ Q ♆
07 32	☽ △ ♅
08 20	☽ △ ☉
11 01	☽ ∠ ♄
16 38	☽ ☍ ♂
19 08	☽ ⋕ ♀
21 03	☽ ⚹ ♆
23 52	☽ □ ♃
11 Thursday	
02 12	☽ ⊥ ♄
02 40	☽ ☌ ♇
06 42	♀ ∠ ♅
09 01	☽ ⊼ ♀
12 20	☽ ⊻ ♄
12 45	☽ □ ☿
18 55	☽ ⋕ ♂
19 56	☽ ± ♀
20 12	☽ ∥ ♇
22 19	☽ ∠ ♆
12 Friday	
06 39	☽ Q ♃
08 42	☿ ⚹ ♄
10 12	☽ □ ♅
12 21	☽ bQ ♀
13 31	☽ ⊥ ♆
14 28	☽ □ ☉
18 21	☽ ⋕ ☉
18 36	☽ Q ☿
21 13	☽ ⊼ ♂
22 54	☽ ⋕ ♅
23 40	☽ ⊻ ♆
13 Saturday	
02 44	☿ ⚹ ♀
03 15	☽ ⚹ ♃
05 13	☽ ⊻ ♇
06 57	♀ △ ♄
07 47	☽ ± ♂
14 43	☽ ⚹ ☿
15 12	☽ ☌ ♄
15 24	☽ ⊥ ♇
15 39	♂ ⋕ ♇
15 50	☽ △ ♀
16 48	☽ Q ♅
23 13	☽ Q ☉
23 43	☽ bQ ♂
23 52	☽ ⋕ ☿
14 Sunday	
02 16	☽ ∥ ♄
03 28	☽ ⋕ ♃
05 11	☽ ∠ ♃
06 43	☽ ∠ ♇
13 22	☽ ⚹ ♅
16 12	☽ ∠ ☿
20 51	☽ ⊥ ♃
21 17	☽ ⚹ ☉
15 Monday	
02 30	☽ △ ♂
02 56	☽ ☌ ♆
03 17	☿ St D
07 22	☽ ⊻ ♃
07 40	☽ ⊥ ☿
08 29	☽ ⚹ ♇
09 36	☽ ∥ ♆
13 36	♄ ⊥ ♇
13 44	♂ △ ♆
15 19	☽ ∠ ♅
18 05	☽ ⊻ ☿
18 54	☽ ⊻ ♄
23 41	☽ □ ♀
16 Tuesday	
01 10	☽ ∠ ☉
02 11	☽ ⋕ ♆
05 19	☽ Q ♇
05 23	☽ ⊥ ♄
07 03	☽ ⊥ ♅
17 36	☽ ⊻ ♅
18 07	☽ ⊥ ☉
21 13	☽ ∠ ♄
17 Wednesday	
05 30	☽ ⊻ ☉
07 25	☽ ⊻ ♆
09 02	♃ ⋕ ♄
09 10	☽ □ ♂
10 00	☽ Q ♀
10 00	☽ ⋕ ♄
10 01	☽ ∥ ♃
10 43	☽ ∥ ☿
12 47	☽ ☌ ♃
13 00	☽ □ ♇
18 06	☽ ⊥ ♆
23 27	☽ ☌ ☿
23 57	☽ ⚹ ♄
18 Thursday	
01 11	♃ □ ♇
09 00	☉ ⚹ ♆
09 18	☽ ⚹ ♀
10 17	☽ ∠ ♆
15 21	☽ ∥ ♅
18 53	☽ Q ♂
20 46	☿ ∥ ♃
21 42	☽ Q ♄
23 28	☽ ☌ ♅
19 Friday	
03 22	☽ ∥ ☉
06 37	♀ Q ♃
06 40	☿ ⚹ ♄
13 39	☽ ⚹ ♆
15 01	☽ ∠ ♀
15 17	☽ ∥ ♂
15 53	☽ ☌ ☉
17 51	☽ ⚹ ♂
19 20	☽ △ ♇
19 57	☽ ⋕ ♇
20 06	☽ ⊻ ♃
20 Saturday	
06 58	☽ □ ♄
07 27	☽ ⊥ ♃
07 33	☽ ⊻ ☿
09 11	☽ ⊥ ♀
11 58	☽ Q ♆
17 37	☽ ∥ ♀
19 10	☽ ⊥ ☿
21 28	☽ ⊻ ♀
23 10	☽ ∠ ♂
23 21	☽ bQ ♇
21 Sunday	
00 40	☽ ∠ ♃
03 12	♂ ☍ ♇
07 32	☽ ⊻ ♅
12 52	☽ ∠ ☿
13 58	☉ △ ♇
16 27	☽ ± ♇
17 06	☽ ⊥ ♂
17 26	♂ ± ♄
19 03	☽ ⊥ ♅
22 12	☽ □ ♆
22 Monday	
04 00	☽ ⊼ ♇
05 11	☽ ⊻ ☉
05 12	☽ ⊻ ♂
05 54	☽ ⚹ ♃
05 57	☉ ⚹ ♂
12 31	☽ ∠ ♅
16 24	☽ △ ♄
16 55	☉ ⊻ ♃
17 50	☽ ⊥ ☉
19 07	☽ ⚹ ☿
23 Tuesday	
05 13	♂ □ ♃
05 45	☽ Q ♃
12 45	☽ ☌ ♀
13 01	☽ ∠ ☉
16 20	♂ Q ♅
18 04	☿ ∥ ♃
18 07	☽ ⚹ ♅
19 58	☽ Q ☿
22 02	☽ bQ ♄
24 Wednesday	
09 12	☽ △ ♆
15 04	☽ ☍ ♇
16 03	☽ ± ♄
17 02	☽ ∥ ♀
18 11	☽ Q ♅
18 13	☽ □ ♃
19 21	☽ ☌ ♂
21 31	☽ ⚹ ☉
25 Thursday	
04 09	☽ ⊼ ♄
09 05	☉ ∠ ♀
10 11	☽ □ ☿
11 28	☽ ⋕ ♇
15 23	☽ bQ ♆
17 31	☉ ∥ ♂
21 54	☽ ∥ ☉
22 17	☽ ∥ ♂
23 50	☽ Q ☉
26 Friday	
06 33	☽ ⊻ ♀
06 38	☽ □ ♅
07 37	♀ ⚹ ♅
09 36	☽ ± ♆
18 47	☽ ∥ ♅
19 49	☽ ⊥ ♀
21 45	☽ ⊼ ♆
27 Saturday	
03 31	☽ ⊼ ♇
07 53	☽ △ ♃
10 57	☽ ⊻ ♂
15 22	☽ □ ☉
15 36	☽ ± ♇
15 39	☽ ∠ ♀
16 53	☽ ☍ ♄
21 58	☽ ∥ ☿
23 38	☽ ⊥ ♂
28 Sunday	
00 53	☽ ∥ ♃
03 07	☽ △ ☿
05 14	☽ ⋕ ♄
06 37	♀ bQ ♄
09 37	☽ bQ ♇
10 46	☉ □ ♄
14 30	☽ bQ ♃
14 40	☿ ∠ ♆
18 29	☽ ∠ ♂
19 12	☽ △ ♅
29 Monday	
00 19	☽ ⚹ ♀
08 37	☽ ± ♃
09 46	☽ ☍ ♆
11 20	☽ bQ ☿
15 13	☽ △ ♇
16 58	☽ ⋕ ♆
20 33	☽ ⊼ ♃
30 Tuesday	
00 45	☽ bQ ♅
01 21	☽ ⚹ ♂
01 52	☽ Q ♀
04 18	☽ ⊼ ♄
06 20	☽ ± ☿
07 39	☽ △ ☉
10 34	☽ ∥ ♆
15 50	☽ ± ♄
18 05	☽ ± ♅
18 50	☽ ⊼ ☿
19 28	☉ ⊥ ♃
31 Wednesday	
01 23	☽ Q ♂
02 47	☉ Q ♆
05 29	☽ ⊼ ♅
08 51	☽ bQ ♄
14 19	☽ bQ ☉
14 53	☽ □ ♀
19 02	☽ ⊼ ♆
20 23	☽ ∥ ♄

JUNE 2023

LONGITUDES

Date	Sidereal time h m s	Sun ☉ ° ' "	Moon ☽ ° ' "	Moon ☽ 24.00 ° ' "	Mercury ☿ ° '	Venus ♀ ° '	Mars ♂ ° '	Jupiter ♃ ° '	Saturn ♄ ° '	Uranus ♅ ° '	Neptune ♆ ° '	Pluto ♇ ° '
01	04 38 52	10 ♊ 45 30	06 ♏ 43 18	13 ♏ 24 22	16 ♉ 23	26 ♋ 04	06 ♌ 49	03 ♉ 30	07 ♓ 00	20 ♉ 14	27 ♓ 27	00 ♒ 09
02	04 42 49	11 43 00	20 ♏ 11 25	27 ♏ 04 19	17 32	27 03	07 24	03 43	07 01	20 17	27 28	00 R 08
03	04 46 45	12 40 28	04 ♐ 02 51	11 ♐ 06 35	18 44	28 01	07 59	03 56	07 03	20 20	27 29	00 07
04	04 50 42	13 37 55	18 ♐ 15 01	25 ♐ 27 28	19 59	28 58	08 34	04 09	07 04	20 24	27 30	00 07
05	04 54 38	14 35 21	02 ♑ 43 12	10 ♑ 01 19	21 18	29 ♋ 56	09 09	04 21	07 05	20 27	27 31	00 06
06	04 58 35	15 32 46	17 ♑ 20 58	24 ♑ 41 14	22 39	00 ♌ 53	09 44	04 34	07 06	20 30	27 32	00 05
07	05 02 31	16 30 10	02 ♒ 01 13	09 ♒ 20 06	24 03	01 49	10 19	04 46	07 07	20 33	27 32	00 04
08	05 06 28	17 27 34	16 ♒ 37 09	23 ♒ 51 44	25 29	02 45	10 54	04 59	07 08	20 36	27 33	00 03
09	05 10 25	18 24 57	01 ♓ 03 20	08 ♓ 11 33	26 59	03 40	11 29	05 11	07 09	20 40	27 34	00 02
10	05 14 21	19 22 20	15 ♓ 16 07	22 ♓ 16 51	28 ♉ 31	04 35	12 04	05 23	07 10	20 43	27 34	00 01
11	05 18 18	20 19 42	29 ♓ 13 42	06 ♈ 06 40	00 ♊ 06	05 29	12 40	05 36	07 11	20 46	27 35	00 ♒ 00
12	05 22 14	21 17 03	12 ♈ 55 49	19 ♈ 41 16	01 44	06 23	13 15	05 48	07 11	20 49	27 36	29 ♑ 59
13	05 26 11	22 14 24	26 ♈ 23 08	03 ♉ 01 34	03 25	07 16	13 50	06 00	07 12	20 52	27 36	29 58
14	05 30 07	23 11 45	09 ♉ 36 42	16 ♉ 08 42	05 08	08 09	14 26	06 12	07 12	20 55	27 37	29 57
15	05 34 04	24 09 05	22 ♉ 37 39	29 ♉ 03 39	06 54	09 01	15 01	06 24	07 12	20 58	27 37	29 56
16	05 38 00	25 06 25	05 ♊ 26 49	11 ♊ 47 12	08 42	09 52	15 37	06 36	07 13	21 01	27 38	29 55
17	05 41 57	26 03 44	18 ♊ 04 50	24 ♊ 19 49	10 34	10 43	16 12	06 47	07 13	21 04	27 38	29 53
18	05 45 54	27 01 03	00 ♋ 32 10	06 ♋ 41 58	12 27	11 33	16 48	06 59	07 R 13	21 07	27 39	29 52
19	05 49 50	27 58 21	12 ♋ 49 19	18 ♋ 54 19	14 23	12 22	17 23	07 11	07 12	21 10	27 39	29 51
20	05 53 47	28 55 39	24 ♋ 57 08	00 ♌ 57 57	16 22	13 10	17 59	07 22	07 12	21 13	27 39	29 50
21	05 57 43	29 ♊ 52 56	06 ♌ 57 00	12 ♌ 54 34	18 23	13 58	18 35	07 33	07 12	21 16	27 40	29 49
22	06 01 40	00 ♋ 50 12	18 ♌ 50 59	24 ♌ 46 37	20 25	14 45	19 10	07 45	07 12	21 19	27 40	29 48
23	06 05 36	01 47 28	00 ♍ 41 56	06 ♍ 37 24	22 30	15 31	19 46	07 56	07 11	21 22	27 40	29 46
24	06 09 33	02 44 43	12 ♍ 33 32	18 ♍ 30 53	24 36	16 16	20 22	08 07	07 10	21 25	27 41	29 45
25	06 13 29	03 41 57	24 ♍ 30 04	00 ♎ 31 42	26 44	17 01	20 58	08 18	07 10	21 27	27 41	29 44
26	06 17 26	04 39 11	06 ♎ 36 25	12 ♎ 44 52	28 ♊ 53	17 44	21 34	08 29	07 09	21 30	27 41	29 43
27	06 21 22	05 36 24	18 ♎ 57 41	25 ♎ 15 28	01 ♋ 03	18 26	22 10	08 40	07 08	21 33	27 41	29 41
28	06 25 19	06 33 37	01 ♏ 38 50	08 ♏ 08 15	03 14	19 08	22 46	08 51	07 07	21 36	27 41	29 40
29	06 29 16	07 30 49	14 ♏ 44 12	21 ♏ 26 58	05 24	19 48	23 22	09 01	07 06	21 38	27 41	29 39
30	06 33 12	08 ♋ 28 01	28 ♏ 16 47	05 ♐ 13 39	07 ♋ 36	20 ♌ 27	23 ♌ 58	09 ♉ 12	07 ♓ 05	21 ♉ 41	27 ♓ 41	29 ♑ 38

Date	Moon True ☊ ° '	Moon Mean ☊ ° '	Moon ☽ Latitude ° '
01	03 ♉ 37	02 ♉ 11	00 S 17
02	03 R 36	02 08	01 29
03	03 33	02 05	02 38
04	03 29	02 02	03 38
05	03 23	01 58	04 26
06	03 17	01 55	04 57
07	03 11	01 52	05 08
08	03 06	01 49	05 00
09	03 03	01 46	04 33
10	03 02	01 42	03 49
11	03 D 02	01 39	02 52
12	03 03	01 36	01 46
13	03 04	01 33	00 S 36
14	03 R 04	01 30	00 N 35
15	03 02	01 27	01 43
16	02 58	01 23	02 45
17	02 52	01 20	03 37
18	02 43	01 17	04 18
19	02 33	01 14	04 47
20	02 23	01 11	05 02
21	02 13	01 07	05 03
22	02 04	01 04	04 52
23	01 58	01 01	04 27
24	01 53	00 58	03 52
25	01 51	00 55	03 05
26	01 D 51	00 52	02 10
27	01 51	00 48	01 08
28	01 52	00 45	00 N 01
29	01 R 51	00 42	01 S 08
30	01 ♉ 48	00 ♉ 39	02 S 16

DECLINATIONS

Date	Sun ☉ ° '	Moon ☽ ° '	Mercury ☿ ° '	Venus ♀ ° '	Mars ♂ ° '	Jupiter ♃ ° '	Saturn ♄ ° '	Uranus ♅ ° '	Neptune ♆ ° '	Pluto ♇ ° '
01	22 N 03	14 S 01	13 N 32	23 N 32	19 N 59	11 N 40	10 S 21	17 N 30	02 S 07	22 S 38
02	22 11	19 13	13 56	23 19	19 50	11 44	10 21	17 31	02 07	22 39
03	22 19	23 32	14 21	23 06	19 40	11 48	10 21	17 32	02 07	22 39
04	22 26	26 32	14 47	22 52	19 31	11 52	10 20	17 33	02 06	22 39
05	22 33	27 50	15 13	22 37	19 21	11 57	10 20	17 34	02 06	22 40
06	22 39	27 13	15 41	22 23	19 11	12 01	10 20	17 34	02 06	22 40
07	22 45	24 43	16 09	22 08	19 01	12 05	10 20	17 35	02 05	22 40
08	22 51	20 37	16 38	21 52	18 51	12 09	10 20	17 36	02 05	22 41
09	22 56	15 21	17 07	21 36	18 41	12 13	10 19	17 37	02 05	22 41
10	23 01	09 19	17 36	21 20	18 30	12 17	10 19	17 38	02 05	22 41
11	23 05	02 S 56	18 06	21 04	18 20	12 21	10 19	17 39	02 05	22 42
12	23 09	03 N 28	18 35	20 47	18 09	12 25	10 19	17 40	02 04	22 42
13	23 13	09 37	19 05	20 30	17 58	12 29	10 19	17 40	02 04	22 42
14	23 16	15 15	19 34	20 12	17 48	12 33	10 19	17 41	02 04	22 43
15	23 19	20 05	20 03	19 55	17 37	12 36	10 20	17 42	02 04	22 43
16	23 21	23 55	20 31	19 37	17 26	12 40	10 20	17 43	02 04	22 44
17	23 23	26 30	20 59	19 19	17 14	12 44	10 20	17 44	02 04	22 44
18	23 24	27 44	21 25	19 00	17 03	12 48	10 20	17 44	02 03	22 44
19	23 25	27 35	21 51	18 42	16 52	12 51	10 20	17 45	02 03	22 45
20	23 26	26 05	22 15	18 23	16 40	12 55	10 21	17 46	02 03	22 45
21	23 26	23 25	22 38	18 04	16 29	12 59	10 21	17 47	02 03	22 45
22	23 26	19 47	22 59	17 45	16 17	13 02	10 21	17 47	02 03	22 46
23	23 26	15 23	23 18	17 26	16 05	13 06	10 22	17 48	02 03	22 46
24	23 25	10 25	23 35	17 06	15 53	13 09	10 22	17 49	02 03	22 47
25	23 23	05 N 01	23 49	16 47	15 41	13 12	10 23	17 50	02 03	22 47
26	23 21	00 S 38	24 02	16 28	15 29	13 16	10 23	17 50	02 03	22 47
27	23 19	06 22	24 11	16 08	15 17	13 19	10 24	17 51	02 03	22 48
28	23 17	12 02	24 18	15 49	15 05	13 22	10 24	17 52	02 03	22 48
29	23 14	17 20	24 23	15 29	14 53	13 26	10 25	17 52	02 03	22 49
30	23 N 10	21 S 59	24 N 24	15 N 09	14 N 40	13 N 29	10 S 26	17 N 53	02 S 03	22 S 49

ZODIAC SIGN ENTRIES

Date	h m	Planets
03	05 03	☽ ♐
05	07 31	☽ ♑
05	13 46	♀ ♌
07	08 42	☽ ♒
09	10 14	☽ ♓
11	09 47	♇ ♑
11	10 27	☿ ♊
11	13 20	☽ ♈
13	18 31	☽ ♉
16	01 46	☽ ♊
18	10 58	☽ ♋
20	22 04	☽ ♌
21	14 58	☉ ♋
23	10 35	☽ ♍
25	22 57	☽ ♎
27	00 24	☿ ♋
28	08 55	☽ ♏
30	14 59	☽ ♐

LATITUDES

Date	Mercury ☿ ° '	Venus ♀ ° '	Mars ♂ ° '	Jupiter ♃ ° '	Saturn ♄ ° '	Uranus ♅ ° '	Neptune ♆ ° '	Pluto ♇ ° '
01	03 S 21	02 N 39	01 N 28	01 S 05	01 S 31	00 S 19	01 S 12	02 S 35
04	03 04	02 33	01 26	01 05	01 32	00 19	01 13	02 35
07	02 43	02 26	01 25	01 06	01 33	00 19	01 13	02 36
10	02 17	02 17	01 23	01 06	01 33	00 19	01 13	02 36
13	01 47	02 07	01 21	01 06	01 34	00 19	01 13	02 37
16	01 15	01 54	01 20	01 06	01 35	00 19	01 13	02 37
19	00 41	01 40	01 18	01 07	01 35	00 19	01 13	02 37
22	00 S 07	01 24	01 17	01 07	01 36	00 19	01 13	02 38
25	00 N 26	01 06	01 15	01 07	01 37	00 19	01 14	02 38
28	00 54	00 46	01 13	01 08	01 37	00 19	01 14	02 39
31	01 N 18	00 N 23	01 N 12	01 S 08	01 S 38	00 S 19	01 S 14	02 S 39

DATA

Julian Date	2460097
Delta T	+72 seconds
Ayanamsa	24° 10' 53"
Synetic vernal point	04° ♓ 56' 06"
True obliquity of ecliptic	23° 26' 18"

MOON'S PHASES, APSIDES AND POSITIONS ☽

Date	h m	Phase	Longitude ° '	Eclipse Indicator
04	03 42	○	13 ♐ 18	
10	19 31	☾	19 ♓ 40	
18	04 37	●	26 ♊ 43	
26	07 50	☽	04 ♎ 29	

Day	h m		
06	23 16	Perigee	
22	18 35	Apogee	
05	16 21	Max dec	27° S 52'
11	22 57	0N	
18	21 06	Max dec	27° N 51'
26	09 22	0S	

LONGITUDES

Date	Chiron ⚷ ° '	Ceres ⚳ ° '	Pallas ⚴ ° '	Juno ⚵ ° '	Vesta ⚶ ° '	Black Moon Lilith ⚸ ° '
01	18 ♈ 51	25 ♍ 56	12 ♌ 37	17 ♊ 50	20 ♉ 53	16 ♌ 08
11	19 ♈ 13	27 ♍ 45	17 ♌ 02	23 ♊ 39	25 ♉ 12	17 ♌ 15
21	19 ♈ 31	00 ♎ 00	21 ♌ 30	29 ♊ 26	29 ♉ 27	18 ♌ 22
31	19 ♈ 45	02 ♎ 39	26 ♌ 01	05 ♋ 09	03 ♊ 37	19 ♌ 29

ASPECTARIAN

01 Thursday
00 02 ☽ □ ♇
01 43 ☽ ⋕ ♃
06 04 ☽ ☍ ♃
06 04 ☽ ± ♆
06 04 ♃ ⊥ ♆
08 10 ☽ ± ☉
09 40 ☽ ⋕ ☿
12 11 ☽ □ ♂
12 30 ☽ △ ♄
19 31 ♂ ⊼ ♄
19 50 ☽ ⊼ ☉
22 19 ☽ ⚼ ♆

02 Friday
03 46 ☽ ⋕ ♅
06 53 ☽ ☍ ☿
08 23 ☽ Q ♇
12 10 ☽ ☍ ♅
14 59 ☽ ⋕ ♂
22 42 ♀ △ ♆

03 Saturday
00 42 ☽ △ ♆
00 51 ☽ △ ♀
04 20 ☽ ⋕ ☉
05 16 ☽ ⚹ ♇
06 32 ☽ ∥ ♇
09 22 ☽ ⋕ ♀
11 48 ☽ ⊼ ♃
17 07 ☽ □ ♄
18 04 ☿ Q ♄
18 59 ☽ △ ♂
22 09 ☽ ± ♃

04 Sunday
03 42 ☽ ☍ ☉
04 19 ☽ ⚼ ♀
06 44 ☽ ∠ ♇
13 31 ☽ ⚼ ♃
15 12 ☽ ⊼ ☿
15 35 ☽ ⊼ ♅
19 49 ☿ ☌ ♅
20 26 ☽ ± ♀
21 14 ☽ ⚼ ♂
21 45 ☽ ⊥ ♇
23 22 ☽ Q ♄

05 Monday
01 36 ☽ ± ♅
02 09 ☽ ± ☿
03 24 ☽ □ ♆
07 05 ☽ ⊼ ♀
07 40 ☽ ⊻ ♇
08 09 ♀ ⋕ ♇
12 44 ☽ ± ♂
14 44 ☽ △ ♃
16 05 ♀ ☍ ♇
16 30 ☽ ⚼ ♅
17 03 ☉ ∥ ♀
18 28 ☽ ⚼ ☿
19 12 ☽ ⚹ ♄
23 00 ☽ ⊼ ♂

06 Tuesday
00 28 ☉ ⚼ ♇
08 51 ☽ ⊼ ☉
09 01 ☽ Q ♆
15 14 ☉ ⋕ ♇
17 10 ☽ △ ♅
17 58 ♀ ± ♄
19 21 ☽ ± ☉
19 47 ☽ ∠ ♄
21 34 ☽ △ ☿

07 Wednesday
04 40 ☽ ⚹ ♆
08 48 ☽ ☌ ♇
10 32 ☽ ⊥ ♄
11 06 ☽ ⚼ ☉
11 39 ☽ ☍ ♀
16 34 ☽ □ ♃
20 23 ☽ ⊻ ♄

08 Thursday
00 14 ☽ ⋕ ☉
00 58 ☽ ∥ ♇
02 11 ☽ ☍ ♂
05 04 ☽ ⋕ ♀
05 17 ☽ ∠ ♆
08 11 ♀ Q ♅
13 29 ☽ △ ☉
18 37 ☽ □ ♅
20 10 ☽ ⊥ ♆
20 52 ☽ ⋕ ♂
22 41 ☽ Q ♃

09 Friday
02 12 ☽ ⋕ ♅
04 24 ☽ □ ☿
05 02 ☽ ⋕ ☿
06 10 ☽ ⊻ ♆
10 17 ☽ ⊻ ♇
16 41 ☽ ⊼ ♀
19 02 ☽ ⚹ ♃
20 21 ☽ ⊥ ♇
21 14 ☿ ⚹ ♆
22 16 ☽ ☌ ♄

10 Saturday
00 36 ☽ ⋕ ♃
00 50 ☽ Q ♅
03 30 ☽ ± ♀
06 20 ☽ ⊼ ♂
08 09 ☽ ∥ ♄
11 34 ☽ ∠ ♇
13 23 ☿ ∥ ♅
14 24 ☽ Q ☿
17 00 ☽ ± ♂
19 31 ☽ □ ☉
19 53 ☽ ⚼ ♀
20 53 ☽ ∠ ♃
21 21 ☽ ⚹ ♅

11 Sunday
08 52 ♂ ⚼ ♆
09 09 ☽ ☌ ♆
09 10 ☽ ⚼ ♂
10 26 ☿ △ ♇
12 39 ☽ ⊥ ♃
13 20 ☽ ⚹ ♇
13 43 ☽ ⚹ ☿
15 13 ☽ ∥ ♆
15 39 ♀ □ ♃
20 18 ☿ ∥ ♂
20 28 ☉ ∠ ♃
23 16 ☽ ⊻ ♃
23 26 ☽ ∠ ♅
23 37 ☉ ⊻ ♅
23 40 ☽ △ ♀

12 Monday
00 57 ☿ Q ♂
01 53 ☽ ⊻ ♄
05 05 ☽ Q ☉
06 43 ☽ ⋕ ♆
10 20 ☽ Q ♇
12 27 ☽ ⊥ ♄
12 35 ☽ △ ♂
15 21 ☽ ⊥ ♅
19 41 ☽ ∠ ☿

13 Tuesday
02 04 ☽ ⊻ ♅
03 59 ☽ ⚹ ☉
04 29 ☽ ∠ ♄
09 59 ♀ ⊼ ♄
14 07 ☽ ⊥ ☿
14 12 ☽ ⊻ ♆
14 51 ☽ ⋕ ♄
18 27 ☽ □ ♇
21 09 ☉ ∠ ♀
23 58 ☽ ∥ ♃

14 Wednesday
01 04 ☽ ⊥ ♆
02 35 ☽ ⊻ ☿
05 40 ☽ ☌ ♃
07 36 ☽ ⚹ ♄
09 08 ☽ □ ♀
09 13 ☽ ∠ ☉
17 30 ☽ ∠ ♆
21 15 ☽ □ ♂
23 34 ☽ ∥ ♅
23 37 ☽ ∥ ♂

15 Thursday
01 09 ♂ ∥ ♅
03 03 ☽ ⊥ ☉
04 24 ☿ ⊻ ♃
05 39 ☽ Q ♄
06 30 ☉ ± ♇
07 41 ☿ ∥ ♀
08 55 ☽ ☌ ♅
11 05 ☽ ∥ ♀
11 45 ☽ ∥ ☿
15 04 ☽ ⊻ ☉
16 09 ☿ □ ♄
20 45 ☽ Q ♀
21 19 ☽ ⚹ ♆

16 Friday
01 36 ☽ △ ♇
03 44 ☽ ⋕ ♇
07 54 ☽ ∥ ☉
08 22 ☽ Q ♂
14 12 ☽ ⊻ ♃
15 20 ☽ □ ♄
19 12 ☽ ☌ ☿
19 55 ☽ Q ♆
20 57 ☽ ⚹ ♀

17 Saturday
00 06 ☿ Q ♆
01 45 ☽ ⊥ ♃
05 55 ☽ ⚼ ♇
08 14 ☽ ⚹ ♂
15 29 ☿ ⚹ ♀
17 29 ♄ St R
17 45 ☽ ⊻ ♅
19 13 ☽ ∠ ♃
23 08 ☽ ± ♇

18 Sunday
03 43 ☽ ∠ ♀
04 37 ☽ ☌ ☉
05 22 ☽ ⊥ ♅
06 24 ☽ □ ♆
10 43 ☽ ⊼ ♇
14 34 ☽ ∠ ♂
14 45 ☉ ⊥ ♅
19 21 ☿ ⊥ ♃
22 26 ☽ ⊥ ♀
22 55 ☽ ∠ ♅

19 Monday
00 45 ☽ ⚹ ♃
01 00 ☽ △ ♄
03 54 ☉ □ ♆
09 02 ☽ ⊥ ♂
11 02 ☽ ⊻ ♀
15 41 ☽ ⊻ ☿
15 53 ♃ ⚹ ♄
17 38 ☿ ⚼ ♇
20 36 ♀ ⚼ ♆
21 28 ☽ ⊻ ♂

20 Tuesday
00 44 ☽ Q ♃
04 33 ☽ ⚹ ♅
05 51 ☽ ⊥ ☿
06 32 ☽ ⚼ ♄
17 24 ☽ △ ♆
20 37 ☽ ⊻ ☉
21 43 ☽ ☍ ♇

21 Wednesday
00 28 ☽ ± ♄
03 22 ☽ ∠ ☿
04 35 ☽ Q ♅
09 40 ☽ ⊥ ☉
10 20 ☉ ⊼ ♇
11 51 ☽ ∥ ☉
12 30 ☽ ⊼ ♄
13 15 ☽ □ ♃
15 23 ☿ ⚹ ♂
16 50 ☽ ⋕ ♇
17 12 ☽ ∥ ☿
20 45 ☿ ⋕ ♇
23 31 ☽ ⚼ ♆

22 Thursday
03 08 ☽ ☌ ♀
05 23 ☽ ∠ ☉
09 09 ♀ ∥ ♅
12 41 ☽ ☌ ♂
15 51 ☽ ⚹ ☿
17 01 ☽ □ ♅
17 42 ☽ ± ♆
22 37 ☿ ⊻ ♅
23 19 ☽ ∥ ♅

23 Friday
00 28 ☽ ∥ ♀
05 52 ☽ ⊼ ♆
08 14 ☽ ∥ ♂
10 08 ☽ ⊼ ♇
14 24 ☽ ⚹ ☉
17 28 ☿ ∠ ♃
21 21 ☽ Q ☿
22 12 ☉ ∥ ☿
22 16 ☽ ± ♇
23 14 ☽ ∥ ♃

24 Saturday
01 07 ☽ ☍ ♄
02 26 ☿ ± ♇
02 53 ☽ △ ♃
12 11 ☽ ⋕ ♄
16 25 ☽ ⚼ ♇
16 48 ☽ Q ☉
19 59 ☽ ⊻ ♀

25 Sunday
04 33 ☽ ⊻ ♂
05 53 ☽ △ ♅
08 49 ☽ ⊥ ♀
09 34 ☽ ⚼ ♃
17 10 ☽ ⊥ ♂
17 25 ☽ □ ☿
18 20 ☽ ☍ ♆
20 16 ☿ ⊥ ♅
22 24 ☽ △ ♇
22 36 ☿ □ ♆

26 Monday
00 44 ☽ ⋕ ♆
03 45 ☽ ± ♃
03 53 ☽ ∠ ♀
07 50 ☽ □ ☉
09 23 ♂ □ ♅
11 48 ☽ ⚼ ♅
11 55 ☽ ∠ ♂
13 04 ☽ ⊼ ♄
15 44 ☽ ⊼ ♃
16 48 ♂ ± ♆
17 57 ☽ ∥ ♆
21 07 ☿ ⊼ ♇

27 Tuesday
00 46 ☽ ± ♄
05 24 ☽ ± ♅
10 56 ☽ ⚹ ♀
16 58 ☽ ⊼ ♅
18 03 ☽ ⚼ ♄
18 26 ☽ ⚹ ♂

28 Wednesday
04 35 ☽ ⊼ ♆
05 00 ☽ ∥ ♄
08 19 ☽ □ ♇
10 59 ☽ Q ♀
12 52 ☉ ∠ ♅
15 32 ☽ △ ☿
15 47 ☽ ± ♆
17 57 ☽ ⋕ ♃
18 04 ☽ Q ♂
21 49 ☽ △ ☉
22 07 ☽ △ ♄

29 Thursday
01 01 ☽ ⋕ ♂
01 29 ☽ ☍ ♃
01 43 ☉ △ ♄
02 25 ☿ ∠ ♀
03 51 ☽ ⋕ ♀
08 18 ☽ ⚼ ♆
14 33 ☽ ⋕ ♅
17 13 ☽ Q ♇
21 32 ☽ □ ♀

30 Friday
00 06 ☽ ⚼ ☿
00 22 ☽ ☍ ♅
01 47 ☿ ∠ ♅
02 56 ☽ ⚼ ☉
04 05 ☽ □ ♂
06 24 ☿ △ ♄
10 58 ☽ △ ♆
14 20 ☽ ⚹ ♇
16 58 ☽ ∥ ♇
18 49 ☽ ± ☿
19 01 ☽ ⋕ ☉
19 47 ☽ ± ☉
21 07 ♆ St R

All ephemeris data is given at 12.00 UT and the Moon's longitude is additionally given for 24.00 UT

JULY 2023

LONGITUDES

Date	Sidereal time h m s	Sun ☉ ° ' "	Moon ☽ ° ' "	Moon ☽ 24.00 ° ' "	Mercury ☿ ° '	Venus ♀ ° '	Mars ♂ ° '	Jupiter ♃ ° '	Saturn ♄ ° '	Uranus ♅ ° '	Neptune ♆ ° '	Pluto ♇ ° '
01	06 37 09	09 ♋ 25 12	12 ♐ 17 27	19 ♐ 27 49	09 ♋ 46	21 ♌ 06	24 ♌ 34	09 ♉ 22	07 ♓ 03	21 ♉ 43	27 ♓ 41	29 ♑ 36
02	06 41 05	10 22 23	26 ♐ 44 14	04 ♑ 05 56	11 57	21 42	25 10	09 32	07 R 02	21 46	27 R 41	29 R 35
03	06 45 02	11 19 35	11 ♑ 32 01	19 ♑ 01 21	14 07	22 18	25 46	09 43	07 00	21 49	27 41	29 34
04	06 48 58	12 16 46	26 ♑ 32 47	04 ♒ 05 03	16 15	22 53	26 22	09 53	06 59	21 51	27 41	29 32
05	06 52 55	13 13 57	11 ♒ 36 53	19 ♒ 07 05	18 23	23 26	26 59	10 03	06 57	21 54	27 41	29 31
06	06 56 51	14 11 08	26 ♒ 34 35	03 ♓ 58 25	20 29	23 57	27 35	10 13	06 55	21 56	27 41	29 30
07	07 00 48	15 08 19	11 ♓ 17 49	18 ♓ 32 12	22 34	24 28	28 11	10 22	06 54	21 58	27 41	29 28
08	07 04 45	16 05 30	25 ♓ 41 12	02 ♈ 44 34	24 38	24 57	28 48	10 32	06 52	22 01	27 40	29 27
09	07 08 41	17 02 42	09 ♈ 42 15	16 ♈ 34 20	26 40	25 24	29 ♌ 24	10 42	06 50	22 03	27 40	29 25
10	07 12 38	17 59 55	23 ♈ 21 01	00 ♉ 02 32	28 ♋ 40	25 50	00 ♍ 01	10 51	06 48	22 05	27 40	29 24
11	07 16 34	18 57 08	06 ♉ 39 12	13 ♉ 11 24	00 ♌ 38	26 14	00 37	11 00	06 45	22 08	27 39	29 23
12	07 20 31	19 54 21	19 ♉ 39 27	26 ♉ 03 45	02 35	26 36	01 14	11 09	06 43	22 10	27 39	29 21
13	07 24 27	20 51 35	02 ♊ 24 37	08 ♊ 42 22	04 30	26 57	01 50	11 18	06 41	22 12	27 39	29 20
14	07 28 24	21 48 49	14 ♊ 57 18	21 ♊ 09 39	06 23	27 16	02 27	11 27	06 38	22 14	27 38	29 18
15	07 32 20	22 46 04	27 ♊ 19 40	03 ♋ 27 31	08 14	27 33	03 03	11 36	06 36	22 16	27 38	29 17
16	07 36 17	23 43 19	09 ♋ 33 22	15 ♋ 37 23	10 03	27 48	03 40	11 45	06 33	22 18	27 37	29 16
17	07 40 14	24 40 35	21 ♋ 39 41	27 ♋ 40 24	11 50	28 02	04 17	11 53	06 30	22 20	27 37	29 14
18	07 44 10	25 37 51	03 ♌ 39 42	09 ♌ 37 42	13 35	28 13	04 54	12 02	06 28	22 22	27 36	29 13
19	07 48 07	26 35 07	15 ♌ 34 35	21 ♌ 30 34	15 19	28 22	05 30	12 10	06 25	22 24	27 36	29 11
20	07 52 03	27 32 24	27 ♌ 25 52	03 ♍ 20 47	17 01	28 29	06 07	12 18	06 22	22 26	27 35	29 10
21	07 56 00	28 29 41	09 ♍ 15 39	15 ♍ 10 48	18 40	28 33	06 44	12 26	06 19	22 28	27 34	29 09
22	07 59 56	29 ♋ 26 58	21 ♍ 06 41	27 ♍ 03 46	20 18	28 36	07 21	12 34	06 16	22 30	27 34	29 07
23	08 03 53	00 ♌ 24 15	03 ♎ 02 33	09 ♎ 03 36	21 54	28 R 36	07 58	12 42	06 12	22 32	27 33	29 06
24	08 07 49	01 21 33	15 ♎ 07 30	21 ♎ 14 52	23 28	28 34	08 35	12 49	06 09	22 33	27 32	29 04
25	08 11 46	02 18 51	27 ♎ 26 19	03 ♏ 42 31	25 01	28 29	09 12	12 56	06 06	22 35	27 32	29 03
26	08 15 43	03 16 10	10 ♏ 04 04	16 ♏ 31 35	26 31	28 22	09 49	13 04	06 02	22 37	27 31	29 01
27	08 19 39	04 13 29	23 ♏ 05 35	29 ♏ 46 33	28 00	28 13	10 26	13 11	05 59	22 38	27 30	29 00
28	08 23 36	05 10 49	06 ♐ 34 48	13 ♐ 30 34	29 ♌ 26	28 01	11 04	13 18	05 56	22 40	27 29	28 59
29	08 27 32	06 08 09	20 ♐ 33 51	27 ♐ 44 30	00 ♍ 51	27 47	11 41	13 25	05 52	22 41	27 28	28 57
30	08 31 29	07 05 29	05 ♑ 02 07	12 ♑ 26 04	02 14	27 31	12 18	13 31	05 48	22 43	27 27	28 56
31	08 35 25	08 ♌ 02 50	19 ♑ 55 31	27 ♑ 29 22	03 ♍ 34	27 ♌ 12	12 ♍ 55	13 ♉ 38	05 ♓ 45	22 ♉ 44	27 ♓ 27	28 ♑ 54

Date	Moon True ☊ ° '	Moon Mean ☊ ° '	Moon ☽ Latitude ° '
01	01 ♉ 43	00 ♉ 36	03 S 17
02	01 R 36	00 33	04 08
03	01 27	00 29	04 44
04	01 17	00 26	05 01
05	01 07	00 23	04 57
06	00 59	00 20	04 32
07	00 53	00 17	03 50
08	00 50	00 13	02 54
09	00 48	00 10	01 49
10	00 D 48	00 07	00 S 39
11	00 R 48	00 04	00 N 31
12	00 47	00 ♉ 01	01 38
13	00 44	29 ♈ 58	02 39
14	00 38	29 54	03 31
15	00 29	29 51	04 12
16	00 18	29 48	04 40
17	00 05	29 45	04 56
18	29 ♈ 51	29 42	04 59
19	29 37	29 39	04 48
20	29 25	29 35	04 25
21	29 16	29 32	03 51
22	29 09	29 29	03 06
23	29 05	29 26	02 13
24	29 03	29 23	01 13
25	29 D 03	29 19	00 N 09
26	29 R 03	29 16	00 S 58
27	29 02	29 13	02 03
28	28 59	29 10	03 04
29	28 54	29 07	03 57
30	28 47	29 04	04 36
31	28 ♈ 37	29 ♈ 00	04 S 57

DECLINATIONS

Date	Sun ☉ ° '	Moon ☽ ° '	Mercury ☿ ° '	Venus ♀ ° '	Mars ♂ ° '	Jupiter ♃ ° '	Saturn ♄ ° '	Uranus ♅ ° '	Neptune ♆ ° '	Pluto ♇ ° '
01	23 N 06	25 S 32	24 N 23	14 N 50	14 N 28	13 N 32	10 S 26	17 N 54	02 S 03	22 S 49
02	23 02	27 32	24 19	14 30	14 15	13 35	10 27	17 54	02 03	22 50
03	22 57	27 39	24 12	14 11	14 02	13 38	10 28	17 55	02 03	22 50
04	22 52	25 46	24 02	13 52	13 49	13 41	10 28	17 56	02 03	22 51
05	22 47	22 03	23 50	13 32	13 37	13 44	10 29	17 56	02 03	22 51
06	22 41	16 55	23 35	13 13	13 24	13 47	10 30	17 57	02 03	22 51
07	22 35	10 53	23 18	12 54	13 10	13 50	10 31	17 58	02 04	22 52
08	22 28	04 S 23	22 59	12 36	12 57	13 53	10 32	17 58	02 04	22 52
09	22 21	02 N 10	22 37	12 17	12 44	13 56	10 33	17 59	02 04	22 53
10	22 14	08 28	22 14	11 59	12 31	13 59	10 34	17 59	02 04	22 53
11	22 06	14 13	21 49	11 41	12 17	14 01	10 35	18 00	02 04	22 53
12	21 58	19 13	21 22	11 23	12 04	14 04	10 36	18 00	02 04	22 54
13	21 49	23 14	20 54	11 06	11 50	14 07	10 37	18 01	02 05	22 54
14	21 40	26 05	20 24	10 49	11 37	14 09	10 38	18 02	02 05	22 54
15	21 31	27 36	19 53	10 32	11 23	14 12	10 39	18 02	02 05	22 55
16	21 21	27 45	19 21	10 16	11 09	14 14	10 40	18 03	02 05	22 55
17	21 11	26 34	18 48	10 00	10 55	14 17	10 42	18 03	02 05	22 56
18	21 01	24 10	18 14	09 45	10 41	14 19	10 43	18 04	02 06	22 56
19	20 50	20 45	17 39	09 30	10 27	14 22	10 44	18 04	02 06	22 56
20	20 39	16 31	17 04	09 15	10 13	14 24	10 45	18 05	02 06	22 57
21	20 28	11 39	16 27	09 02	09 59	14 26	10 47	18 05	02 07	22 57
22	20 16	06 22	15 51	08 48	09 45	14 28	10 48	18 06	02 07	22 58
23	20 04	00 N 49	15 14	08 36	09 30	14 31	10 49	18 06	02 07	22 58
24	19 51	04 S 50	14 36	08 24	09 16	14 33	10 51	18 06	02 08	22 58
25	19 39	10 26	13 59	08 13	09 02	14 35	10 52	18 07	02 08	22 59
26	19 25	15 45	13 21	08 02	08 47	14 37	10 53	18 07	02 08	22 59
27	19 12	20 32	12 43	07 53	08 33	14 39	10 55	18 08	02 09	23 00
28	18 58	24 26	12 05	07 44	08 18	14 41	10 56	18 08	02 09	23 00
29	18 44	27 02	11 27	07 36	08 03	14 43	10 58	18 08	02 09	23 00
30	18 30	27 56	10 49	07 29	07 49	14 44	10 59	18 09	02 10	23 01
31	18 N 15	26 S 52	10 N 11	07 N 22	07 N 34	14 N 46	11 S 01	18 N 09	02 S 10	23 S 01

ZODIAC SIGN ENTRIES

Date	h	m	Planets
02	17	20	☽ ♑
04	17	30	☽ ♒
06	17	32	☽ ♓
08	19	19	☽ ♈
10	11	40	♂ ♍
10	23	55	☽ ♉
11	04	11	☿ ♌
13	07	26	☽ ♊
15	17	13	☽ ♋
18	04	39	☽ ♌
20	17	13	☽ ♍
23	01	50	☉ ♌
23	05	54	☽ ♎
25	16	55	☽ ♏
28	00	24	☽ ♐
28	21	31	☿ ♍
30	03	44	☽ ♑

LATITUDES

Date	Mercury ☿ ° '	Venus ♀ ° '	Mars ♂ ° '	Jupiter ♃ ° '	Saturn ♄ ° '	Uranus ♅ ° '	Neptune ♆ ° '	Pluto ♇ ° '
01	01 N 18	00 N 23	01 N 12	01 S 08	01 S 38	00 S 19	01 S 14	02 S 39
04	01 36	00 S 02	01 10	01 09	01 39	00 19	01 14	02 39
07	01 47	00 29	01 08	01 09	01 39	00 19	01 14	02 40
10	01 51	00 59	01 07	01 10	01 40	00 19	01 14	02 40
13	01 49	01 31	01 05	01 10	01 40	00 19	01 14	02 40
16	01 41	02 06	01 04	01 10	01 41	00 19	01 15	02 41
19	01 28	02 43	01 02	01 11	01 42	00 19	01 15	02 41
22	01 11	03 22	01 00	01 11	01 42	00 19	01 15	02 41
25	00 50	04 02	00 59	01 12	01 43	00 19	01 15	02 41
28	00 N 26	04 43	00 57	01 12	01 43	00 19	01 15	02 42
31	00 S 01	05 S 24	00 N 56	01 S 13	01 S 44	00 S 19	01 S 15	02 S 42

DATA

Julian Date	2460127
Delta T	+72 seconds
Ayanamsa	24° 10' 59"
Synetic vernal point	04° ♓ 56' 00"
True obliquity of ecliptic	23° 26' 18"

MOON'S PHASES, APSIDES AND POSITIONS ☽

Date	h	m	Phase	Longitude ° '	Eclipse Indicator
03	11	39	○	11 ♑ 19	
10	01	48	☾	17 ♈ 36	
17	18	32	●	24 ♋ 56	
25	22	07	☽	02 ♏ 43	

Day	h	m		
04	22	33	Perigee	
20	07	06	Apogee	
03	01	26	Max dec	27° S 51'
09	04	00	0N	
16	02	38	Max dec	27° N 52'
23	15	30	0S	
30	11	13	Max dec	27° S 56'

LONGITUDES

Date	Chiron ⚷ ° '	Ceres ⚳ ° '	Pallas ⚴ ° '	Juno ⚵ ° '	Vesta ⚶ ° '	Black Moon Lilith ⚸ ° '
01	19 ♈ 45	02 ♎ 39	26 ♌ 01	05 ♋ 09	03 ♊ 37	19 ♌ 29
11	19 ♈ 54	05 ♎ 36	00 ♍ 33	10 ♋ 47	07 ♊ 41	20 ♌ 36
21	19 ♈ 58	08 ♎ 49	05 ♍ 07	16 ♋ 21	11 ♊ 38	21 ♌ 43
31	19 ♈ 56	12 ♎ 15	09 ♍ 42	21 ♋ 49	15 ♊ 27	22 ♌ 51

All ephemeris data is given at 12.00 UT and the Moon's longitude is additionally given for 24.00 UT

ASPECTARIAN

h m Aspects

01 Saturday
03 08 ☽ □ ♄
03 17 ☽ # ☿
05 06 ☉ ☌ ☿
06 48 ☽ ⚻ ☉
06 58 ☽ ⚻ ☿
07 00 ☽ ⚻ ♃
07 10 ☿ ⚹ ♃
08 50 ☿ ∠ ♂
10 26 ☉ ⚹ ♃
15 53 ☽ ∠ ♇
17 14 ☽ ± ♃
21 53 ☉ ∠ ♂

02 Sunday
03 21 ☽ △ ♀
03 48 ☽ ⚻ ♅
06 49 ☽ ⊥ ♇
08 21 ☽ bQ ♃
09 12 ☽ Q ♄
09 19 ☽ △ ♂
11 10 ♀ ± ♆
13 33 ☽ □ ♆
13 42 ☽ ± ♅
14 33 ♀ □ ♅
16 39 ☽ ⚺ ♇

03 Monday
04 22 ☽ bQ ♅
04 43 ☽ ⚹ ♄
04 54 ☽ bQ ♀
09 02 ☽ △ ♃
10 43 ☽ bQ ♂
11 39 ☽ ☍ ☉
16 50 ☽ ☍ ☿
18 40 ☽ Q ♆
19 58 ☽ ± ♀

04 Tuesday
01 45 ☽ ± ♂
04 30 ☽ △ ♅
04 44 ☽ ∠ ♄
05 55 ☽ ⚻ ♀
11 43 ☽ ⚻ ♂
13 49 ☽ ⚹ ♆
16 45 ☽ ☌ ♇
19 03 ☽ ⊥ ♄
19 22 ☉ # ♇
20 26 ♀ ∥ ♂
21 28 ☿ ⊥ ♀
23 06 ♀ ∥ ♃

05 Wednesday
00 10 ♂ ∥ ♃
01 09 ☽ # ☿
04 35 ☽ ⚺ ♄
07 36 ☽ ∥ ♇
07 53 ☽ # ☉
09 28 ☽ □ ♃
13 42 ☽ ∠ ♆
14 45 ☽ ⚻ ☉

06 Thursday
00 36 ☽ ⚻ ☿
01 01 ☽ ± ☉
04 07 ☽ ⊥ ♆
04 30 ☽ □ ♅
07 35 ☽ # ♅
07 37 ☽ ☍ ♀
11 50 ☽ ± ☿
13 42 ☽ ☍ ♂
13 47 ☽ ⚺ ♆
14 40 ☽ Q ♃
15 48 ♂ ⚻ ♆
16 31 ☽ bQ ☉
16 43 ☽ ⚺ ♇

07 Friday
00 40 ☽ # ♃
02 27 ☽ ⊥ ♇
02 51 ☽ # ♂
03 48 ☽ # ♀
04 14 ☿ bQ ♄
04 47 ☽ ☌ ♄
04 53 ☽ bQ ☿
04 54 ☿ ⚹ ♅
05 41 ☿ ⊥ ♂
09 28 ☿ Q ♃
09 49 ☽ Q ♅
10 28 ☽ ⚹ ♃
13 22 ☽ ∥ ♄
17 14 ☽ ∠ ♇
18 48 ☽ △ ☉

08 Saturday
05 48 ☽ ⚹ ♅
09 55 ☽ △ ☿
10 42 ☽ ⚻ ♀
11 44 ☽ ∠ ♃
15 21 ☽ ☌ ♆
16 45 ☿ ⚺ ♀
17 30 ☽ ⚻ ♂
18 22 ☽ ⚹ ♇
19 38 ☿ # ♇
20 28 ☽ ∥ ♆
21 14 ☽ ± ♀

09 Sunday
03 15 ☽ ⊥ ♃
04 13 ☽ ± ♂
07 02 ☽ ⚺ ♄
07 24 ☽ ∠ ♅
11 36 ☽ # ♆
12 54 ♂ ⚻ ♇
13 15 ☽ bQ ♀
13 44 ☽ ⚺ ♃
14 59 ☽ Q ♇
17 25 ☽ ⊥ ♄
20 34 ☽ bQ ♂
23 07 ☽ ⊥ ♅
23 57 ☿ △ ♆

10 Monday
01 48 ☽ □ ☉
09 14 ☽ ∠ ♄
09 45 ☽ ⚺ ♅
12 38 ☉ ∥ ☿
16 34 ☽ △ ♀
19 43 ☽ ⚺ ♆
20 29 ☽ # ♄
20 48 ☿ ☍ ♇
22 50 ☽ □ ♇
23 11 ☽ □ ☿

11 Tuesday
00 31 ☽ △ ♂
01 37 ☽ ∥ ♀
03 57 ☽ ∥ ♂
06 33 ☽ ⊥ ♆
11 07 ☽ ∥ ♃
11 36 ☿ ⚺ ♂
12 11 ☽ ⚹ ♄
12 35 ☽ Q ☉
13 24 ☿ ± ♄
20 04 ☽ ☌ ♃
23 01 ☽ ∠ ♆

12 Wednesday
05 45 ☽ ∥ ♅
10 15 ☽ Q ♄
12 30 ☽ ⚹ ☉
14 02 ☽ Q ☿
16 42 ☽ ☌ ♅
22 48 ☽ ∥ ☿

13 Thursday
01 24 ☽ □ ♀
02 59 ☽ ⚹ ♆
03 06 ☽ ∥ ☉
06 11 ☽ △ ♇
08 13 ☿ Q ♅
09 43 ☽ # ♇
10 51 ☽ □ ♂
15 35 ☉ ⊥ ♀
16 40 ☽ ⚹ ☿
19 06 ☽ ∠ ☉
20 06 ☽ □ ♄

14 Friday
01 47 ☽ Q ♆
05 11 ☽ ⚺ ♃
07 44 ☉ bQ ♄
10 45 ☽ bQ ♇
12 37 ☽ Q ♀
13 48 ☽ ⊥ ☉
15 17 ☿ ⚻ ♄
16 53 ☽ ⊥ ♃
23 02 ☉ ⚹ ♅
23 10 ☽ Q ♂

15 Saturday
02 08 ☽ ⚺ ♅
02 23 ☽ ⚺ ☉
02 24 ♀ # ♄
02 37 ☽ ∠ ☿
04 08 ☽ ± ♇
10 34 ☽ ∠ ♃
12 27 ☽ ⚹ ♀
12 35 ☽ □ ♆
13 51 ☽ ⊥ ♅
15 49 ☽ ⚻ ♇
18 42 ♀ ⚻ ♆
23 16 ☽ ⊥ ☿
23 48 ☽ ⚹ ♂

16 Sunday
06 06 ☽ △ ♄
07 33 ☽ ∠ ♅
12 43 ☉ Q ♃
13 08 ☽ ⚺ ☿
16 23 ☽ ⚹ ♃
18 33 ☽ ∠ ♀

17 Monday
07 01 ☽ ∠ ♂
11 41 ☽ bQ ♄
12 44 ☽ ⊥ ♀
12 49 ☿ □ ♃
13 21 ☽ ⚹ ♅
16 29 ☽ Q ♃
18 32 ☽ ☌ ☉
22 33 ☿ bQ ♆
23 52 ☽ △ ♆

18 Tuesday
00 55 ☽ ⚺ ♀
01 56 ☽ ⊥ ♂
03 06 ☽ ☍ ♇
05 36 ☽ ± ♄
09 37 ♂ # ♄
13 26 ☽ Q ♅
14 37 ☽ ⚺ ♂
17 36 ☽ ⚻ ♄
19 04 ☿ ∥ ♅
21 26 ☽ # ♇

19 Wednesday
00 01 ♂ ± ♇
05 02 ☽ □ ♃
05 59 ☽ bQ ♆
11 23 ☽ ☌ ☿
11 25 ☽ ∥ ☉

20 Thursday
00 10 ☽ ± ♆
01 51 ☽ □ ♅
03 36 ☽ ∥ ♅
08 39 ☽ ∥ ☿
12 14 ☽ ⚺ ☉
12 19 ☽ ⚻ ♆
13 07 ☉ △ ♆
14 08 ☽ ☌ ♀
15 31 ☽ ⚻ ♇
20 39 ♂ ☍ ♄
22 40 ☽ ∥ ♃

21 Friday
01 29 ☽ ⊥ ☉
03 40 ☽ ± ♇
06 03 ☽ ☍ ♄
06 36 ☽ ☌ ♂
13 39 ☉ ⚺ ♀
16 06 ☽ # ♄
18 30 ☽ △ ♃
20 09 ☽ ∥ ♂
21 20 ☽ ∠ ☉
21 53 ☽ bQ ♇

22 Saturday
00 40 ☽ ∥ ♀
03 53 ☉ ☍ ♇
10 07 ☽ ⚺ ☿
11 46 ♃ ∠ ♆
14 48 ☽ △ ♅

23 Sunday
00 06 ☽ ⊥ ☿
01 00 ☽ ☍ ♆
01 09 ☽ bQ ♃
01 33 ♀ St R
03 06 ☽ ⚺ ♀
04 06 ☽ △ ♇
06 15 ☽ ⚹ ☉
06 27 ☽ # ♆
06 43 ☿ ± ♆
07 18 ☉ ± ♄
15 06 ☽ ⊥ ♀
18 18 ☽ ⚻ ♄
19 22 ☽ ± ♃
20 52 ☽ ∠ ☿
20 58 ☽ bQ ♅
21 39 ☿ □ ♅
22 22 ☽ ⚺ ♂

24 Monday
00 30 ☽ ∥ ♆
06 10 ☽ ± ♄
07 24 ☽ ⚻ ♃
08 13 ☽ Q ☉
08 56 ☽ ∠ ♀
10 53 ☽ ⊥ ♂
14 08 ☿ ∥ ♃
14 49 ☽ ± ♅
23 46 ☽ bQ ♄

25 Tuesday
02 35 ☽ ⚻ ♅
02 42 ☽ # ♀
05 25 ☽ ∠ ♂
06 10 ☽ # ♂
06 39 ☽ ⚹ ☿
12 10 ☽ ⚻ ♆
13 55 ☽ ∥ ♄
14 00 ☽ ⚹ ♀
15 05 ☽ □ ♇
22 07 ☽ □ ☉
23 39 ☽ ± ♆

26 Wednesday
02 06 ☽ # ☿
04 27 ☽ △ ♄
06 40 ☽ # ♃
08 42 ☽ Q ☿
11 31 ☽ ⚹ ♂
12 34 ☽ Q ♀
16 34 ☽ bQ ♆
17 38 ☽ ☍ ♃
23 28 ☽ # ♅

27 Thursday
00 53 ☽ Q ♇
03 59 ☿ ⚻ ♆
05 15 ☽ # ☉
10 46 ☽ Q ♂
11 10 ☽ ☍ ♅
15 16 ☿ ☌ ♀
19 56 ☽ △ ♆
21 06 ☽ □ ♀
21 53 ☽ □ ☿
22 36 ☽ ⚹ ♇
22 41 ☉ Q ♅

28 Friday
02 20 ☽ ∥ ♇
03 29 ☉ ⊥ ♂
04 24 ☿ ⚻ ♇
09 22 ☽ △ ☉
10 52 ☽ □ ♄
20 09 ☽ □ ♂
23 44 ☽ ⚻ ♃

29 Saturday
00 47 ☽ ∠ ♇
05 38 ☉ ⚻ ♄
10 02 ☽ ± ♃
13 02 ☽ bQ ☉
15 35 ☽ ⚻ ♅
16 00 ☽ ⊥ ♇
17 31 ☽ Q ♄
23 32 ☽ □ ♆
23 51 ☽ △ ♀

30 Sunday
01 13 ☽ bQ ♃
01 36 ☽ ± ♅
01 59 ☽ ⚺ ♇
05 05 ☽ ± ☉
05 36 ☿ # ♄
06 55 ☽ △ ☿
13 15 ☽ ⚹ ♄
15 35 ☽ ⚻ ☉
16 22 ☽ bQ ♅
16 25 ♀ ⚻ ♆
23 53 ☽ bQ ♀

31 Monday
00 18 ☽ △ ♂
01 51 ☽ △ ♃
04 50 ☽ Q ♆
09 38 ☽ bQ ☿
13 18 ☽ ∠ ♄
13 59 ☽ ± ♀
16 29 ☽ △ ♅
21 42 ☉ ∥ ♅
23 17 ☽ ⚻ ♀
23 55 ☽ ⚹ ♆

AUGUST 2023

LONGITUDES

Date	Sidereal time h m s	Sun ☉ ° ' "	Moon ☽ ° ' "	Moon ☽ 24.00 ° ' "	Mercury ☿ ° '	Venus ♀ ° '	Mars ♂ ° '	Jupiter ♃ ° '	Saturn ♄ ° '	Uranus ♅ ° '	Neptune ♆ ° '	Pluto ♇ ° '
01	08 39 22	09 ♌ 00 12	05 ♒ 06 23	12 ♒ 45 11	04 ♍ 53	26 ♌ 51	13 ♍ 33	13 ♉ 44	05 ♓ 41	22 ♉ 46	27 ♓ 26	28 ♑ 53
02	08 43 18	09 57 35	20 ♒ 24 20	28 ♒ 02 23	06 09	26 R 27	14 10	13 50	05 R 37	22 47	27 R 25	28 R 52
03	08 47 15	10 54 58	05 ♓ 38 00	13 ♓ 09 56	07 24	26 02	14 48	13 56	05 33	22 48	27 24	28 50
04	08 51 12	11 52 23	20 ♓ 37 10	27 ♓ 58 50	08 36	25 35	15 25	14 02	05 29	22 49	27 23	28 49
05	08 55 08	12 49 48	05 ♈ 14 20	12 ♈ 23 17	09 46	25 05	16 02	14 08	05 25	22 51	27 22	28 47
06	08 59 05	13 47 15	19 ♈ 25 29	26 ♈ 20 55	10 53	24 35	16 40	14 13	05 21	22 52	27 20	28 46
07	09 03 01	14 44 43	03 ♉ 09 44	09 ♉ 52 12	11 58	24 02	17 18	14 19	05 17	22 53	27 19	28 45
08	09 06 58	15 42 13	16 ♉ 28 41	22 ♉ 59 36	13 01	23 28	17 55	14 24	05 13	22 54	27 18	28 43
09	09 10 54	16 39 43	29 ♉ 25 24	05 ♊ 46 36	14 01	22 54	18 33	14 29	05 09	22 55	27 17	28 42
10	09 14 51	17 37 16	12 ♊ 03 39	18 ♊ 17 02	14 58	22 18	19 11	14 34	05 05	22 56	27 16	28 41
11	09 18 47	18 34 49	24 ♊ 27 13	00 ♋ 34 35	15 52	21 41	19 48	14 39	05 01	22 57	27 15	28 39
12	09 22 44	19 32 24	06 ♋ 39 33	12 ♋ 42 26	16 43	21 04	20 26	14 43	04 56	22 58	27 14	28 38
13	09 26 41	20 30 00	18 ♋ 43 35	24 ♋ 43 15	17 31	20 27	21 04	14 47	04 52	22 58	27 12	28 37
14	09 30 37	21 27 38	00 ♌ 41 41	06 ♌ 39 06	18 16	19 50	21 42	14 51	04 48	22 59	27 11	28 35
15	09 34 34	22 25 17	12 ♌ 35 43	18 ♌ 31 43	18 57	19 13	22 20	14 55	04 43	23 00	27 10	28 34
16	09 38 30	23 22 57	24 ♌ 27 18	00 ♍ 22 37	19 34	18 36	22 58	14 59	04 39	23 01	27 09	28 33
17	09 42 27	24 20 38	06 ♍ 17 54	12 ♍ 13 21	20 07	18 00	23 36	15 03	04 34	23 01	27 07	28 32
18	09 46 23	25 18 20	18 ♍ 09 13	24 ♍ 05 46	20 36	17 25	24 14	15 06	04 30	23 02	27 06	28 30
19	09 50 20	26 16 04	00 ♎ 03 19	06 ♎ 02 13	21 01	16 51	24 52	15 09	04 26	23 02	27 05	28 29
20	09 54 16	27 13 49	12 ♎ 02 49	18 ♎ 05 35	21 21	16 19	25 30	15 12	04 21	23 03	27 03	28 28
21	09 58 13	28 11 35	24 ♎ 10 58	00 ♏ 19 27	21 36	15 48	26 08	15 15	04 17	23 03	27 02	28 27
22	10 02 10	29 ♌ 09 22	06 ♏ 31 35	12 ♏ 47 54	21 46	15 18	26 46	15 18	04 12	23 03	27 00	28 25
23	10 06 06	00 ♍ 07 11	19 ♏ 08 57	25 ♏ 35 16	21 51	14 51	27 24	15 20	04 08	23 04	26 59	28 24
24	10 10 03	01 05 00	02 ♐ 07 22	08 ♐ 45 44	21 R 50	14 25	28 03	15 23	04 03	23 04	26 57	28 23
25	10 13 59	02 02 51	15 ♐ 30 45	22 ♐ 22 41	21 43	14 01	28 41	15 25	03 58	23 04	26 56	28 22
26	10 17 56	03 00 43	29 ♐ 21 42	06 ♑ 27 48	21 31	13 40	29 19	15 27	03 54	23 04	26 55	28 21
27	10 21 52	03 58 37	13 ♑ 40 47	21 ♑ 00 14	21 12	13 21	29 ♍ 58	15 28	03 49	23 04	26 53	28 20
28	10 25 49	04 56 31	28 ♑ 25 33	05 ♒ 55 53	20 47	13 04	00 ♎ 36	15 30	03 45	23 05	26 52	28 19
29	10 29 45	05 54 27	13 ♒ 30 12	21 ♒ 07 17	20 17	12 49	01 15	15 31	03 40	23 R 05	26 50	28 18
30	10 33 42	06 52 24	28 ♒ 45 49	06 ♓ 24 26	19 41	12 37	01 53	15 32	03 36	23 04	26 49	28 17
31	10 37 39	07 ♍ 50 23	14 ♓ 01 43	21 ♓ 36 23	18 ♍ 59	12 ♌ 27	02 ♎ 32	15 ♉ 33	03 ♓ 31	23 ♉ 04	26 ♓ 47	28 ♑ 16

Date	Moon True ☊ ° '	Moon Mean ☊ ° '	Moon ☽ Latitude ° '
01	28 ♈ 27	28 ♈ 57	04 S 59
02	28 R 16	28 54	04 39
03	28 07	28 51	03 59
04	28 01	28 48	03 04
05	27 57	28 45	01 57
06	27 55	28 41	00 S 45
07	27 D 55	28 38	00 N 28
08	27 R 56	28 35	01 37
09	27 55	28 32	02 39
10	27 52	28 29	03 31
11	27 47	28 25	04 13
12	27 40	28 22	04 42
13	27 30	28 19	04 59
14	27 18	28 16	05 02
15	27 06	28 13	04 51
16	26 54	28 10	04 28
17	26 44	28 06	03 54
18	26 35	28 03	03 09
19	26 30	28 00	02 16
20	26 26	27 57	01 16
21	26 25	27 54	00 N 12
22	26 D 25	27 51	00 S 54
23	26 26	27 47	01 59
24	26 R 27	27 44	02 59
25	26 26	27 41	03 52
26	26 23	27 38	04 34
27	26 18	27 35	04 59
28	26 12	27 31	05 06
29	26 04	27 28	04 53
30	25 57	27 25	04 18
31	25 ♈ 51	27 ♈ 22	03 S 25

DECLINATIONS

Date	Sun ☉ ° '	Moon ☽ ° '	Mercury ☿ ° '	Venus ♀ ° '	Mars ♂ ° '	Jupiter ♃ ° '	Saturn ♄ ° '	Uranus ♅ ° '	Neptune ♆ ° '	Pluto ♇ ° '
01	18 N 00	23 S 49	09 N 33	07 N 17	07 N 19	14 N 48	11 S 02	18 N 09	02 S 11	23 S 01
02	17 45	19 05	08 56	07 12	07 04	14 49	11 04	18 10	02 11	23 02
03	17 30	13 09	08 19	07 09	06 49	14 51	11 05	18 10	02 11	23 02
04	17 14	06 S 32	07 43	07 06	06 34	14 53	11 07	18 10	02 12	23 02
05	16 58	00 N 17	07 07	07 04	06 19	14 54	11 09	18 11	02 12	23 03
06	16 41	06 55	06 32	07 03	06 04	14 56	11 10	18 11	02 13	23 03
07	16 25	13 00	05 57	07 04	05 49	14 57	11 12	18 11	02 13	23 04
08	16 08	18 18	05 23	07 05	05 34	14 58	11 13	18 12	02 14	23 04
09	15 51	22 37	04 50	07 06	05 18	15 00	11 15	18 12	02 14	23 04
10	15 33	25 44	04 18	07 09	05 03	15 01	11 17	18 12	02 15	23 05
11	15 15	27 32	03 47	07 13	04 48	15 02	11 18	18 12	02 15	23 05
12	14 58	27 58	03 17	07 17	04 33	15 03	11 20	18 12	02 16	23 05
13	14 39	27 03	02 48	07 22	04 17	15 04	11 22	18 13	02 16	23 06
14	14 21	24 54	02 20	07 28	04 02	15 05	11 23	18 13	02 17	23 06
15	14 02	21 41	01 54	07 34	03 46	15 06	11 25	18 13	02 17	23 06
16	13 43	17 35	01 30	07 41	03 31	15 07	11 27	18 13	02 18	23 07
17	13 24	12 49	01 07	07 48	03 15	15 08	11 29	18 13	02 18	23 07
18	13 05	07 35	00 46	07 56	03 00	15 09	11 30	18 13	02 19	23 07
19	12 46	02 N 03	00 27	08 04	02 44	15 10	11 32	18 14	02 19	23 07
20	12 26	03 S 36	00 N 10	08 13	02 28	15 10	11 34	18 14	02 20	23 08
21	12 06	09 12	00 S 05	08 22	02 13	15 11	11 36	18 14	02 21	23 08
22	11 46	14 33	00 17	08 31	01 57	15 12	11 37	18 14	02 21	23 08
23	11 26	19 25	00 27	08 40	01 41	15 12	11 39	18 14	02 22	23 09
24	11 05	23 31	00 33	08 50	01 26	15 13	11 41	18 14	02 22	23 09
25	10 45	26 30	00 37	08 59	01 10	15 13	11 42	18 14	02 23	23 09
26	10 24	28 00	00 37	09 09	00 54	15 13	11 44	18 14	02 24	23 09
27	10 03	27 42	00 34	09 18	00 38	15 14	11 46	18 14	02 24	23 10
28	09 42	25 28	00 28	09 27	00 23	15 14	11 48	18 14	02 25	23 10
29	09 21	21 26	00 18	09 37	00 N 07	15 14	11 49	18 14	02 25	23 10
30	08 59	15 56	00 05	09 46	00 S 09	15 14	11 51	18 14	02 26	23 10
31	08 N 38	09 S 26	00 N 13	09 N 55	00 S 25	15 N 14	11 S 53	18 N 14	02 S 27	23 S 11

ZODIAC SIGN ENTRIES

Date	h	m	Planets
01	03	58	☽ ♒
03	03	05	☽ ♓
05	03	19	☽ ♈
07	06	24	☽ ♉
09	13	05	☽ ♊
11	22	52	☽ ♋
14	10	36	☽ ♌
16	23	14	☽ ♍
19	11	53	☽ ♎
21	23	22	☽ ♏
23	09	01	☉ ♍
24	08	07	☽ ♐
26	13	05	☽ ♑
27	13	20	♂ ♎
28	14	32	☽ ♒
30	13	56	☽ ♓

LATITUDES

Date	Mercury ☿ ° '	Venus ♀ ° '	Mars ♂ ° '	Jupiter ♃ ° '	Saturn ♄ ° '	Uranus ♅ ° '	Neptune ♆ ° '	Pluto ♇ ° '
01	00 S 11	05 S 37	00 N 55	01 S 13	01 S 44	00 S 19	01 S 15	02 S 42
04	00 41	06 15	00 53	01 14	01 44	00 19	01 15	02 42
07	01 12	06 50	00 52	01 14	01 45	00 19	01 16	02 42
10	01 45	07 19	00 50	01 15	01 45	00 19	01 16	02 43
13	02 19	07 42	00 49	01 15	01 46	00 19	01 16	02 43
16	02 52	07 57	00 47	01 16	01 46	00 19	01 16	02 43
19	03 23	08 04	00 45	01 17	01 46	00 19	01 16	02 43
22	03 52	08 04	00 44	01 17	01 46	00 19	01 16	02 43
25	04 14	07 58	00 42	01 18	01 47	00 19	01 16	02 43
28	04 29	07 45	00 40	01 18	01 47	00 19	01 16	02 43
31	04 S 30	07 S 27	00 N 39	01 S 19	01 S 47	00 S 19	01 S 16	02 S 44

DATA

Julian Date	2460158
Delta T	+72 seconds
Ayanamsa	24° 11' 05"
Synetic vernal point	04° ♓ 55' 54"
True obliquity of ecliptic	23° 26' 18"

MOON'S PHASES, APSIDES AND POSITIONS ☽

Date	h	m	Phase	Longitude ° '	Eclipse Indicator
01	18	32	○	09 ♒ 16	
08	10	28	☾	15 ♉ 39	
16	09	38	●	23 ♌ 17	
24	09	57	☽	01 ♐ 00	
31	01	36	○	07 ♓ 25	

Day	h	m		
02	05	59	Perigee	
16	12	10	Apogee	
30	16	00	Perigee	
05	10	58	0N	
12	07	35	Max dec	28° N 00'
19	20	46	0S	
26	20	17	Max dec	28° S 06'

LONGITUDES

Date	Chiron ⚷ ° '	Ceres ⚳ ° '	Pallas ⚴ ° '	Juno ⚵ ° '	Vesta ⚶ ° '	Black Moon Lilith ⚸ ° '
01	19 ♈ 56	12 ♎ 36	10 ♍ 10	22 ♋ 22	15 ♊ 49	22 ♌ 57
11	19 ♈ 49	16 ♎ 14	14 ♍ 45	27 ♋ 43	19 ♊ 27	24 ♌ 04
21	19 ♈ 36	20 ♎ 01	19 ♍ 21	02 ♌ 58	22 ♊ 54	25 ♌ 11
31	19 ♈ 20	23 ♎ 56	23 ♍ 57	08 ♌ 05	26 ♊ 07	26 ♌ 19

All ephemeris data is given at 12.00 UT and the Moon's longitude is additionally given for 24.00 UT

ASPECTARIAN

h m	Aspects
01 Tuesday	
01 14	☽ ⚼ ♂
01 17	☽ ± ☿
02 13	☽ ☌ ♇
03 30	☽ ⊥ ♄
11 37	☽ ⊼ ☿
12 02	☿ ± ♇
12 54	☽ ⊻ ♄
16 00	☽ ± ♂
16 37	☽ ∥ ♇
17 09	♀ ∥ ♂
18 32	☽ ☍ ☉
20 45	♂ △ ♃
23 29	☽ ∠ ♆
02 Wednesday	
00 29	♂ ⚼ ♇
01 38	☽ □ ♃
01 48	☽ ⊼ ♂
02 18	☿ ☍ ♄
13 35	☽ ⊥ ♆
15 44	☽ □ ♅
16 01	☽ # ♅
18 03	☽ # ☉
21 15	☽ ☍ ♀
23 00	☽ ⊻ ♆
03 Thursday	
01 16	☽ ⊻ ♇
05 31	☽ # ♃
06 07	☽ Q ♃
10 44	☽ ⊥ ♇
11 52	☽ ☌ ♄
15 03	☽ ☍ ☿
19 37	☽ ∥ ♄
20 14	☽ Q ♅
20 59	☽ ⊼ ☉
04 Friday	
01 03	☽ ∠ ♇
01 19	☽ ✶ ♃
03 15	☽ ☍ ♂
07 15	☽ ± ☉
07 23	☽ # ☿
09 58	☽ # ♀
11 52	☽ # ♂
15 35	☽ ✶ ♅
19 49	☽ ⊼ ♀
22 54	☽ ⚼ ☉
23 00	☽ ☌ ♆
05 Saturday	
00 24	☉ ⚼ ♆
01 21	☽ ✶ ♇
01 49	☽ ∠ ♃
03 13	☽ ∥ ♆
05 21	☽ ± ♀
12 18	☽ ⊻ ♄
14 02	☿ ∥ ♀
16 21	☽ ∠ ♅
16 52	☽ ⊥ ♃
18 49	☽ # ♆
19 51	☽ ⚼ ♀
20 14	☽ ⊼ ☿
21 17	☽ Q ♇
22 19	☽ ⊥ ♄
06 Sunday	
01 40	☽ △ ☉
03 03	☽ ⊻ ♃
07 03	☽ ⊼ ♂
07 17	☽ ± ☿
07 36	☽ ⊥ ♅
08 59	☽ ∥ ♂
10 43	☽ ∥ ☿
12 34	☽ ∥ ♀
13 36	☽ ∠ ♄
17 52	☽ ± ♂
17 57	☽ ⊻ ♅
20 35	☽ △ ♀
07 Monday	
00 03	☉ □ ♃
00 10	☽ ⚼ ☿
01 43	☽ ⊻ ♆
04 13	☽ □ ♇
04 34	☽ # ♄
10 23	☽ ⚼ ♂
12 17	☽ ⊥ ♆
15 46	☽ ✶ ♄
20 23	☽ ∥ ♃
22 41	☿ ∥ ♂
08 Tuesday	
02 10	☽ ∥ ☉
04 25	☽ ∠ ♆
05 09	☽ △ ☿
08 11	☽ ☌ ♃
10 28	☽ □ ☉
11 25	☽ ∥ ♅
13 21	☽ Q ♄
14 46	☽ △ ♂
23 50	☽ ☌ ♅
09 Wednesday	
00 21	☽ □ ♀
04 29	☿ ⚼ ♇
08 00	☽ ✶ ♆
10 39	☽ △ ♇
11 08	♀ □ ♅
15 01	☽ # ♇
22 42	☽ Q ☉
22 45	☽ □ ♄
10 Thursday	
00 47	☿ △ ♃
06 39	☽ Q ♆
08 46	☽ Q ♀
15 06	☽ ⚼ ♇
16 51	☽ ⊻ ♃
18 02	☽ □ ☿
23 37	☽ ✶ ☉
11 Friday	
02 28	☽ □ ♂
04 32	☽ ⊥ ♃
06 52	☽ ✶ ♀
08 30	☽ ± ♇
09 03	☽ ⊻ ♅
17 27	☽ □ ♆
20 13	☽ ⊼ ♇
20 48	☽ ⊥ ♅
22 14	☽ ∠ ♃
12 Saturday	
04 47	☉ ∥ ♃
05 49	♀ ± ♆
07 27	☽ ∠ ☉
07 53	☽ Q ☿
08 37	☽ △ ♄
10 54	☽ ∠ ♀
14 35	☽ ∠ ♅
15 42	☽ Q ♂
13 Sunday	
00 15	♀ ⊻ ♂
02 50	☽ ⊥ ☉
03 53	☽ ⊥ ♀
04 06	☽ ✶ ♃
09 26	☽ ✶ ☿
11 16	☉ ☌ ♀
14 16	☽ ⚼ ♄
15 17	☽ ⊻ ♀
15 51	☽ ⊻ ☉
16 56	☽ ✶ ♂
20 30	☽ ✶ ♅
14 Monday	
04 15	☽ Q ♃
04 57	☽ △ ♆
05 16	☉ ± ♆
07 46	☽ ☍ ♇
08 12	☽ ± ♄
15 18	☿ # ♆
17 30	☽ ∠ ☿
20 12	☽ ⊼ ♄
20 39	☽ Q ♅
15 Tuesday	
00 46	☽ ∠ ♂
02 17	☽ # ♇
05 10	☉ ⊻ ♂
11 08	☽ ⚼ ♆
12 45	☽ ⊥ ☿
16 44	☽ □ ♃
17 04	☿ ⊻ ♀
19 58	☽ ⊥ ♂
16 Wednesday	
00 44	☽ ☌ ♀
01 34	☽ ⊻ ☿
02 35	☉ □ ♅
05 18	☽ ± ♆
08 33	☽ ∥ ♅
08 48	☽ ⊻ ♂
09 04	☽ □ ♅
09 38	☽ ☌ ☉
13 53	♂ △ ♅
17 26	☽ ⊼ ♆
20 17	☽ ⊼ ♇
17 Thursday	
00 44	☽ ∥ ♃
08 25	☽ ± ♇
08 32	☽ ☍ ♄
09 00	☽ ∥ ☉
18 18	☽ # ♄
20 05	♀ ⊥ ♂
18 Friday	
02 37	☽ ⚼ ♇
05 48	☽ △ ♃
10 16	♄ ⊥ ♇
10 29	☽ ∥ ♀
10 35	☽ ⊻ ♀
17 09	☽ ☌ ☿
21 51	☽ △ ♅
22 09	☽ ⊥ ♀
19 Saturday	
00 57	☽ ☌ ♂
03 42	☽ ⊻ ☉
06 01	☽ ☍ ♆
08 51	☽ △ ♇
08 58	☽ ∥ ♂
10 52	☽ # ♆
12 12	☽ ⚼ ♃
15 28	☽ ∠ ♀
16 50	☽ ⊥ ☉
19 15	☽ ∥ ☿
20 43	☽ ⊼ ♄
22 07	☽ # ☿
20 Sunday	
04 01	☽ ⚼ ♅
06 19	☽ ± ♃
06 39	☽ ∥ ♆
07 27	☽ # ♂
07 40	☉ ⊼ ♆
08 39	☽ ± ♄
12 24	☽ ∠ ☉
18 19	☽ ⊼ ♃
20 07	☽ ✶ ♀
21 56	☽ ± ♅
21 Monday	
00 19	♂ # ♆
02 24	☽ ⚼ ♄
06 50	☽ ⊻ ☿
08 19	☽ # ♀
09 47	☽ ⊼ ♅
16 02	☽ ⊻ ♂
17 34	☽ ⊼ ♆
18 07	☉ ⊼ ♇
18 37	☿ ⊥ ♀
18 48	☽ Q ♀
18 48	☽ ⊥ ☿
20 19	☽ □ ♇
20 31	☽ ✶ ☉
22 38	☽ ∥ ♄
22 Tuesday	
00 06	☽ # ☉
04 21	☽ ⊥ ♂
05 13	☽ ± ♆
07 33	☽ △ ♄
12 16	♀ □ ♃
12 28	☽ ∠ ☿
15 03	☽ # ♃
20 34	♂ ☍ ♆
21 35	☉ # ♄
21 36	☽ Q ☉
22 28	☽ ⚼ ♆
22 35	☽ ∠ ♂
23 Wednesday	
04 10	☽ □ ♀
04 48	☽ ☍ ♃
05 52	☽ # ♅
06 50	☽ Q ♇
17 03	☽ ✶ ☿
19 19	☽ ☍ ♅
19 59	☿ St R
24 Thursday	
02 33	☽ △ ♆
04 09	☽ ✶ ♂
05 10	☽ ✶ ♇
09 35	☽ ∥ ♇
09 57	☽ □ ☉
15 05	☽ Q ☿
15 29	☽ □ ♄
25 Friday	
00 23	♂ △ ♇
03 01	☽ Q ♂
08 13	☽ ∠ ♇
09 27	☽ △ ♀
11 50	☽ ⊼ ♃
20 04	♀ ∠ ♂
22 21	☽ ± ♃
22 43	☽ □ ☿
23 14	☽ Q ♄
23 58	☽ ⊥ ♇
26 Saturday	
01 12	☽ ⊼ ♅
07 49	☽ □ ♆
10 16	☽ ⊻ ♇
10 51	☽ ⚼ ♀
11 30	☽ ± ♅
11 56	☽ □ ♂
13 51	☽ ⚼ ♃
18 39	☽ △ ☉
19 39	☽ ✶ ♄
27 Sunday	
01 42	☽ ± ♀
02 42	☽ ⚼ ♅
08 28	☉ ☍ ♄
11 28	☽ ⊼ ♀
13 59	☽ Q ♆
14 58	☽ △ ♃
20 24	☽ ∠ ♄
20 36	☉ ± ♇
20 59	☿ # ♂
21 18	☽ ⚼ ☉
28 Monday	
00 00	☽ △ ☿
03 22	☽ △ ♅
07 51	♂ ⚼ ♃
09 29	☽ ✶ ♆
10 55	☽ ⊥ ♄
11 49	☽ ☌ ♇
12 53	☽ ± ☉
15 39	☽ △ ♂
20 29	☽ ⊻ ♄
23 08	☽ ⊼ ☉
23 22	☉ ∥ ♀
23 25	☽ ⚼ ☿
29 Tuesday	
02 40	♅ St R
02 52	☽ ∥ ♇
09 22	☽ ∠ ♆
10 56	☽ ☍ ♀
13 11	☽ ± ☿
15 11	☽ □ ♃
16 31	☽ ⚼ ♂
22 18	☽ ⊼ ☿
23 32	☽ ⊥ ♆
30 Wednesday	
02 37	☽ # ♅
03 04	☽ □ ♅
07 17	☽ ± ♂
08 24	☿ ∥ ♂
08 56	☽ ⊻ ♆
11 14	☽ ⊻ ♇
14 44	☽ # ♃
17 07	☽ ⊼ ♂
19 30	☽ Q ♃
19 33	☽ ☌ ♄
20 39	☽ ⊥ ♇
31 Thursday	
01 36	☽ ☍ ☉
03 11	♃ Q ♄
03 20	☽ ∥ ♄
07 20	☽ Q ♅
09 33	☽ ⊼ ♀
10 23	☽ # ♀
10 47	☽ ∠ ♇
14 25	☽ ✶ ♃
15 00	☽ # ☉
18 56	☽ ± ♀
19 28	☽ ☍ ☿

SEPTEMBER 2023

LONGITUDES

Date	Sidereal time h m s	Sun ☉ ° ' "	Moon ☽ ° ' "	Moon ☽ 24.00 ° ' "	Mercury ☿ ° '	Venus ♀ ° '	Mars ♂ ° '	Jupiter ♃ ° '	Saturn ♄ ° '	Uranus ♅ ° '	Neptune ♆ ° '	Pluto ♇ ° '
01	10 41 35	08 ♍ 48 23	29 ♓ 07 12	06 ♈ 33 10	18 ♍ 13	12 ♌ 20	03 ♎ 11	15 ♉ 34	03 ♓ 27	23 ♉ 04	26 ♓ 45	28 ♑ 15
02	10 45 32	09 46 26	13 ♈ 53 25	21 ♈ 07 19	17 R 22	12 R 15	03 49	15 34	03 R 22	23 R 04	26 R 44	28 R 14
03	10 49 28	10 44 30	28 ♈ 14 28	05 ♉ 14 38	16 28	12 13	04 28	15 35	03 18	23 04	26 42	28 13
04	10 53 25	11 42 35	12 ♉ 07 47	18 ♉ 54 01	15 31	12 D 12	05 07	15 35	03 13	23 03	26 41	28 12
05	10 57 21	12 40 43	25 ♉ 33 35	02 ♊ 06 50	14 33	12 15	05 45	15 R 35	03 09	23 03	26 39	28 11
06	11 01 18	13 38 53	08 ♊ 34 13	14 ♊ 56 11	13 35	12 19	06 24	15 35	03 04	23 03	26 38	28 10
07	11 05 14	14 37 05	21 ♊ 13 16	27 ♊ 26 00	12 37	12 26	07 03	15 34	03 00	23 02	26 36	28 09
08	11 09 11	15 35 19	03 ♋ 34 56	09 ♋ 40 35	11 42	12 35	07 42	15 33	02 56	23 02	26 34	28 08
09	11 13 08	16 33 35	15 ♋ 43 28	21 ♋ 44 04	10 51	12 46	08 21	15 32	02 51	23 01	26 33	28 07
10	11 17 04	17 31 54	27 ♋ 42 50	03 ♌ 40 12	10 05	12 59	09 00	15 31	02 47	23 01	26 31	28 06
11	11 21 01	18 30 14	09 ♌ 36 34	15 ♌ 32 16	09 24	13 15	09 39	15 30	02 43	23 00	26 29	28 06
12	11 24 57	19 28 36	21 ♌ 27 39	27 ♌ 22 59	08 51	13 32	10 18	15 29	02 38	22 59	26 28	28 05
13	11 28 54	20 26 59	03 ♍ 18 34	09 ♍ 14 37	08 26	13 51	10 57	15 27	02 34	22 59	26 26	28 04
14	11 32 50	21 25 25	15 ♍ 11 22	21 ♍ 09 03	08 09	14 12	11 37	15 25	02 30	22 58	26 25	28 03
15	11 36 47	22 23 53	27 ♍ 07 52	03 ♎ 08 02	08 01	14 35	12 16	15 23	02 26	22 57	26 23	28 03
16	11 40 43	23 22 22	09 ♎ 09 47	15 ♎ 13 20	08 D 02	14 59	12 55	15 21	02 22	22 56	26 21	28 02
17	11 44 40	24 20 54	21 ♎ 18 57	27 ♎ 26 53	08 13	15 25	13 35	15 18	02 18	22 55	26 20	28 01
18	11 48 37	25 19 27	03 ♏ 37 26	09 ♏ 50 55	08 33	15 53	14 14	15 15	02 14	22 54	26 18	28 01
19	11 52 33	26 18 02	16 ♏ 07 40	22 ♏ 28 01	09 03	16 22	14 53	15 13	02 10	22 53	26 16	28 00
20	11 56 30	27 16 38	28 ♏ 52 21	05 ♐ 21 00	09 41	16 53	15 33	15 10	02 06	22 52	26 15	27 59
21	12 00 26	28 15 17	11 ♐ 54 20	18 ♐ 32 40	10 28	17 25	16 12	15 06	02 02	22 51	26 13	27 59
22	12 04 23	29 ♍ 13 57	25 ♐ 16 18	02 ♑ 05 27	11 23	17 59	16 52	15 03	01 58	22 50	26 11	27 58
23	12 08 19	00 ♎ 12 39	09 ♑ 00 15	16 ♑ 00 46	12 26	18 34	17 32	14 59	01 54	22 49	26 10	27 58
24	12 12 16	01 11 22	23 ♑ 06 54	00 ♒ 18 26	13 35	19 10	18 11	14 55	01 51	22 47	26 08	27 57
25	12 16 12	02 10 07	07 ♒ 34 59	14 ♒ 56 03	14 51	19 48	18 51	14 51	01 47	22 46	26 06	27 57
26	12 20 09	03 08 54	22 ♒ 20 54	29 ♒ 48 44	16 12	20 27	19 31	14 47	01 44	22 45	26 05	27 57
27	12 24 06	04 07 42	07 ♓ 18 33	14 ♓ 49 20	17 38	21 07	20 11	14 43	01 40	22 43	26 03	27 56
28	12 28 02	05 06 33	22 ♓ 19 57	29 ♓ 49 17	19 08	21 48	20 50	14 38	01 37	22 42	26 01	27 56
29	12 31 59	06 05 25	07 ♈ 16 14	14 ♈ 39 48	20 42	22 30	21 30	14 33	01 33	22 41	26 00	27 55
30	12 35 55	07 ♎ 04 19	21 ♈ 59 04	29 ♈ 13 17	22 ♍ 19	23 ♌ 13	22 ♎ 10	14 ♉ 29	01 ♓ 30	22 ♉ 39	25 ♓ 58	27 ♑ 55

Date	Moon True ☊ ° '	Moon Mean ☊ ° '	Moon ☽ Latitude ° '
01	25 ♈ 46	27 ♈ 19	02 S 19
02	25 R 44	27 16	01 S 04
03	25 D 43	27 12	00 N 14
04	25 44	27 09	01 28
05	25 45	27 06	02 34
06	25 46	27 03	03 31
07	25 R 46	27 00	04 16
08	25 45	26 56	04 47
09	25 41	26 53	05 05
10	25 36	26 50	05 10
11	25 30	26 47	05 01
12	25 23	26 44	04 39
13	25 17	26 41	04 05
14	25 11	26 37	03 20
15	25 06	26 34	02 27
16	25 04	26 31	01 26
17	25 02	26 28	00 N 20
18	25 D 03	26 25	00 S 47
19	25 04	26 22	01 53
20	25 06	26 18	02 55
21	25 07	26 15	03 49
22	25 08	26 12	04 33
23	25 R 08	26 09	05 02
24	25 06	26 06	05 14
25	25 04	26 02	05 07
26	25 01	25 59	04 40
27	24 58	25 56	03 53
28	24 56	25 53	02 50
29	24 54	25 50	01 36
30	24 ♈ 54	25 ♈ 47	00 S 16

DECLINATIONS

Date	Sun ☉ ° '	Moon ☽ ° '	Mercury ☿ ° '	Venus ♀ ° '	Mars ♂ ° '	Jupiter ♃ ° '	Saturn ♄ ° '	Uranus ♅ ° '	Neptune ♆ ° '	Pluto ♇ ° '
01	08 N 16	02 S 28	00 N 33	10 N 03	00 S 41	15 N 14	11 S 54	18 N 14	02 S 27	23 S 11
02	07 54	04 N 30	00 57	10 11	00 57	15 14	11 56	18 14	02 28	23 11
03	07 32	11 04	01 24	10 19	01 13	15 14	11 58	18 14	02 29	23 11
04	07 10	16 52	01 54	10 27	01 28	15 14	11 59	18 14	02 29	23 12
05	06 48	21 39	02 26	10 34	01 44	15 14	12 01	18 14	02 30	23 12
06	06 26	25 12	02 59	10 41	02 00	15 14	12 03	18 14	02 31	23 12
07	06 03	27 24	03 34	10 47	02 16	15 13	12 04	18 13	02 31	23 12
08	05 41	28 11	04 09	10 53	02 32	15 13	12 06	18 13	02 32	23 13
09	05 18	27 34	04 44	10 59	02 48	15 12	12 08	18 13	02 33	23 13
10	04 56	25 41	05 18	11 04	03 04	15 12	12 09	18 13	02 33	23 13
11	04 33	22 40	05 51	11 08	03 20	15 11	12 11	18 13	02 34	23 13
12	04 10	18 45	06 21	11 12	03 35	15 11	12 12	18 13	02 35	23 13
13	03 47	14 06	06 49	11 16	03 51	15 10	12 14	18 12	02 35	23 13
14	03 24	08 55	07 13	11 19	04 07	15 09	12 15	18 12	02 36	23 14
15	03 01	03 N 23	07 34	11 21	04 23	15 09	12 17	18 12	02 37	23 14
16	02 38	02 S 19	07 51	11 23	04 39	15 08	12 18	18 12	02 37	23 14
17	02 15	08 00	08 04	11 25	04 55	15 07	12 20	18 12	02 38	23 14
18	01 51	13 28	08 13	11 26	05 10	15 06	12 21	18 11	02 39	23 14
19	01 28	18 28	08 17	11 26	05 26	15 05	12 23	18 11	02 39	23 14
20	01 05	22 45	08 17	11 26	05 42	15 04	12 24	18 11	02 40	23 15
21	00 42	26 00	08 12	11 25	05 58	15 03	12 25	18 10	02 41	23 15
22	00 N 18	27 54	08 03	11 24	06 13	15 02	12 27	18 10	02 41	23 15
23	00 S 05	28 09	07 50	11 22	06 29	15 01	12 28	18 10	02 42	23 15
24	00 28	26 37	07 33	11 20	06 45	14 59	12 29	18 10	02 43	23 15
25	00 52	23 19	07 13	11 17	07 00	14 58	12 31	18 09	02 43	23 15
26	01 15	18 28	06 49	11 13	07 16	14 57	12 32	18 09	02 44	23 15
27	01 38	12 26	06 22	11 09	07 31	14 55	12 33	18 09	02 44	23 15
28	02 02	05 S 39	05 52	11 05	07 47	14 54	12 34	18 08	02 45	23 15
29	02 25	01 N 25	05 19	11 00	08 02	14 52	12 35	18 08	02 46	23 15
30	02 S 48	08 N 19	04 N 44	10 N 54	08 S 18	14 N 51	12 S 36	18 N 07	02 S 46	23 S 16

ZODIAC SIGN ENTRIES

Date	h	m	Planets
01	13	25	☽ ♈
03	15	00	☽ ♉
05	20	07	☽ ♊
08	05	00	☽ ♋
10	16	36	☽ ♌
13	05	18	☽ ♍
15	17	44	☽ ♎
18	04	58	☽ ♏
20	14	06	☽ ♐
22	20	20	☽ ♑
23	06	50	☉ ♎
24	23	29	☽ ♒
27	00	18	☽ ♓
29	00	17	☽ ♈

LATITUDES

Date	Mercury ☿ ° '	Venus ♀ ° '	Mars ♂ ° '	Jupiter ♃ ° '	Saturn ♄ ° '	Uranus ♅ ° '	Neptune ♆ ° '	Pluto ♇ ° '
01	04 S 28	07 S 20	00 N 38	01 S 19	01 S 47	00 S 19	01 S 16	02 S 44
04	04 08	06 57	00 36	01 20	01 47	00 19	01 16	02 44
07	03 31	06 32	00 35	01 20	01 47	00 19	01 16	02 44
10	02 41	06 06	00 33	01 21	01 47	00 19	01 16	02 44
13	01 43	05 38	00 31	01 21	01 47	00 19	01 17	02 44
16	00 S 45	05 10	00 30	01 22	01 47	00 19	01 17	02 44
19	00 N 07	04 42	00 28	01 22	01 47	00 19	01 17	02 44
22	00 49	04 15	00 26	01 23	01 47	00 19	01 17	02 44
25	01 21	03 47	00 25	01 23	01 47	00 19	01 17	02 44
28	01 42	03 20	00 23	01 24	01 47	00 19	01 17	02 44
31	01 N 52	02 S 54	00 N 21	01 S 24	01 S 47	00 S 19	01 S 17	02 S 44

DATA

Julian Date	2460189
Delta T	+72 seconds
Ayanamsa	24° 11' 09"
Synetic vernal point	04° ♓ 55' 50"
True obliquity of ecliptic	23° 26' 19"

MOON'S PHASES, APSIDES AND POSITIONS ☽

Date	h	m	Phase	Longitude ° '	Eclipse Indicator
06	22	21	☾	14 ♊ 04	
15	01	40	●	21 ♍ 59	
22	19	32	☽	29 ♐ 32	
29	09	57	○	06 ♈ 00	

Day	h	m		
12	15	57	Apogee	
28	01	05	Perigee	
01	20	25	0N	
08	13	14	Max dec	28° N 11'
16	02	17	0S	
23	03	35	Max dec	28° S 16'
29	07	11	0N	

LONGITUDES

Date	Chiron ⚷ ° '	Ceres ⚳ ° '	Pallas ⚴ ° '	Juno ⚵ ° '	Vesta ⚶ ° '	Black Moon Lilith ⚸ ° '
01	19 ♈ 18	24 ♎ 20	24 ♍ 24	08 ♌ 35	26 ♊ 25	26 ♌ 25
11	18 ♈ 57	28 ♎ 21	29 ♍ 00	13 ♌ 33	29 ♊ 20	27 ♌ 32
21	18 ♈ 33	02 ♏ 28	03 ♎ 35	18 ♌ 22	01 ♋ 54	28 ♌ 39
31	18 ♈ 07	06 ♏ 38	08 ♎ 09	22 ♌ 59	04 ♋ 05	29 ♌ 46

All ephemeris data is given at 12.00 UT and the Moon's longitude is additionally given for 24.00 UT

ASPECTARIAN

01 Friday
02 20 ☽ ⚹ ♅
07 54 ☿ ⊥ ♀
08 13 ☽ ☌ ♆
09 09 ☽ ⚼ ♀
10 36 ☽ ⚹ ♇
12 02 ☽ ∥ ♆
14 20 ☽ ∠ ♃
17 52 ☽ ∥ ♂
18 12 ☽ ⋕ ☿
18 50 ☽ ☍ ♂
18 56 ☽ ⊻ ♄
21 01 ♂ ⊼ ♄
22 55 ☽ ∥ ☿
23 10 ☽ ⋕ ♂

02 Saturday
02 28 ☽ ∠ ♅
04 38 ☽ ⊥ ♄
04 46 ☽ ⊼ ☉
04 53 ☽ ⋕ ♆
04 55 ☽ ⊥ ♃
06 00 ☽ Q ♇
09 19 ☽ △ ♀
10 55 ☿ ⋕ ♂
14 47 ☽ ⊻ ♃
15 20 ☽ ± ☉
17 15 ☽ ⊥ ♅
17 25 ☽ ⊼ ☿
19 23 ☽ ∠ ♄
23 32 ☽ ∥ ☉

03 Sunday
02 50 ☽ ± ☿
03 15 ☽ ⊻ ♅
07 27 ☽ ⚼ ☉
09 07 ☽ ∥ ♀
09 24 ☽ ⊻ ♆
11 57 ☽ □ ♇
15 31 ☽ ⋕ ♄
17 10 ☽ ⚼ ☿
19 37 ☽ ⊥ ♆
20 36 ☽ ⚹ ♄
23 10 ☽ ⊼ ♂

04 Monday
01 20 ♀ St D
04 52 ☽ ∥ ♃
10 07 ☽ ± ♂
10 29 ☿ △ ♃
11 12 ☽ △ ☉
11 13 ☽ ∠ ♆
12 08 ☽ □ ♀
14 11 ♃ St R
17 25 ☽ Q ♄
17 35 ☽ △ ☿
18 06 ☽ ☌ ♃
18 18 ☽ ∥ ♅

05 Tuesday
00 42 ☉ ⊻ ♀
02 53 ☽ ⚼ ♂
07 28 ☽ ☌ ♅
13 59 ☽ ⚹ ♆
15 12 ☿ ⋕ ♆
16 46 ☽ △ ♇
20 36 ☽ Q ♀
21 25 ☽ ⋕ ♇

06 Wednesday
00 10 ☉ ⚼ ♇
01 50 ☽ □ ♄
07 45 ☽ △ ♂
11 09 ☉ ☌ ☿
12 06 ☽ Q ♆
19 06 ☽ ⚹ ♀
20 38 ☽ ⚼ ♇
20 46 ☽ □ ☿
22 21 ☽ □ ☉
22 33 ☿ ⚼ ♇

07 Thursday
01 12 ☽ ⊻ ♃
12 40 ☽ ⊥ ♃
13 47 ☽ ± ♇
15 30 ☽ ⊻ ♅
16 18 ☿ ⊻ ♀
22 22 ☽ □ ♆

08 Friday
00 08 ☽ ∠ ♀
01 22 ☽ ⊼ ♇
03 07 ☽ ⊥ ♅
04 56 ☽ Q ☿
06 05 ☽ ∠ ♃
10 43 ☽ △ ♄
11 13 ☉ △ ♃
11 54 ♂ ∥ ♆
12 01 ☽ Q ☉
17 59 ☽ ⊥ ♀
20 33 ☽ □ ♂
20 45 ☽ ∠ ♅
23 58 ♂ ⚼ ♅

09 Saturday
02 57 ☽ ⚹ ☿
06 02 ☽ ⊻ ♀
11 38 ☽ ⚹ ♃
13 49 ☽ ⚹ ☉
16 13 ☽ ⚼ ♄

10 Sunday
02 24 ☉ ∥ ☿
02 34 ☽ ⚹ ♅
04 41 ♂ ± ♄
07 01 ☽ ∠ ☿
09 36 ☽ △ ♆
10 08 ☽ ± ♄
10 29 ☽ Q ♂
11 37 ☽ Q ♃
12 47 ☽ ☍ ♇
22 09 ☽ ⊼ ♄
22 34 ☽ ∠ ☉

11 Monday
00 07 ☽ ⊥ ☿
02 42 ☽ Q ♅
06 36 ♂ ± ♃
07 22 ☿ ⊻ ♂
08 10 ☽ ⋕ ♇
11 37 ☽ ⊻ ☿
12 06 ☽ ⚹ ♂
15 48 ☽ ⚼ ♆
18 23 ☽ ⊥ ☉
19 32 ☽ ☌ ♀
23 54 ☽ □ ♃

12 Tuesday
07 37 ☽ ⊻ ☉
09 59 ☽ ± ♆
13 54 ☉ ⊥ ♀
14 56 ☽ ∥ ♅
15 06 ☽ □ ♅
20 15 ☽ ∠ ♂
22 07 ☽ ⊼ ♆

13 Wednesday
01 24 ☽ ⊼ ♇
06 43 ☽ ∥ ♃
09 21 ☉ ⋕ ♂
10 31 ☽ ☍ ♄
13 32 ☽ ± ♇
15 32 ☽ ⊥ ♂
20 50 ☽ ⋕ ♄
22 04 ☽ ☌ ☿

14 Thursday
01 14 ☽ ∥ ♀
04 22 ☽ ⊻ ♂
07 42 ☽ ⚼ ♇
09 56 ☽ ⊻ ♀
12 27 ☽ △ ♃
19 00 ☽ ∥ ☿
22 24 ☽ ⊥ ♀

15 Friday
01 40 ☽ ☌ ☉
03 37 ☽ △ ♅
07 55 ☽ ⋕ ♂
10 30 ☽ ☍ ♆
13 40 ☽ ∥ ☉
13 49 ☽ △ ♇
15 16 ☽ ⋕ ♆
17 04 ☽ ∠ ♀
18 29 ☽ ⚼ ♃
20 21 ☿ St D
22 32 ☽ ⊼ ♄

16 Saturday
01 24 ☉ △ ♅
09 34 ☽ ⚼ ♅
09 45 ☽ ⊻ ☿
10 25 ☽ ± ♄
12 21 ☽ ± ♃
12 40 ☉ ⋕ ♆
13 14 ☽ ⋕ ☉
13 16 ☽ ∥ ♆
19 53 ☽ ☌ ♂
21 46 ☽ ⊥ ☿
22 16 ☽ ∥ ♂
23 57 ☽ ⚹ ♀

17 Sunday
00 12 ☽ ⊼ ♃
03 22 ☽ ± ♅
04 08 ☽ ⚼ ♄
06 10 ♀ □ ♃
12 18 ☽ ⋕ ☿
15 09 ☽ ⊼ ♅
15 49 ☽ ∠ ☿
18 27 ☽ ⊻ ☉
21 47 ☽ ⊼ ♆

18 Monday
00 24 ☽ Q ♀
01 06 ☽ □ ♇
02 53 ☽ ⋕ ♀
06 59 ☽ ∥ ♄
07 10 ☽ ⊥ ☉
09 19 ☽ △ ♄
09 26 ☽ ± ♆
19 33 ☽ ⋕ ♃
21 52 ☽ ⚹ ☿

19 Tuesday
02 00 ☽ ∠ ☉
02 45 ☽ ⚼ ♆
09 31 ☽ ⊻ ♂
10 16 ☽ ☍ ♃
10 33 ☽ ⋕ ♅
11 17 ☉ ☍ ♆
11 45 ☽ Q ♇
12 29 ☽ □ ♀
21 31 ☽ ⊥ ♂
21 47 ☽ Q ☿
22 48 ♂ ⊼ ♃

20 Wednesday
00 46 ☽ ☍ ♅
07 06 ☽ △ ♆
08 47 ☽ ⚹ ☉
10 21 ☽ ⚹ ♇
15 08 ☽ ∥ ♇
15 17 ☽ ∠ ♂
17 58 ☽ □ ♄

21 Thursday
05 21 ☉ △ ♇
08 45 ☽ Q ☉
09 13 ☽ □ ☿
13 57 ☽ ∠ ♇
17 47 ☽ ⊼ ♃
20 12 ☽ ⚹ ♂
22 25 ☽ △ ♀

22 Friday
02 36 ☽ Q ♄
04 31 ☽ ± ♃
06 08 ☽ ⊥ ♇
07 40 ☽ ⊼ ♅
10 43 ♂ ± ♅
13 37 ☽ □ ♆
15 23 ♂ ⚼ ♄
16 46 ☽ ⊻ ♇
18 17 ☽ ± ♅
18 40 ☽ Q ♂
19 32 ☽ □ ☉
20 23 ☽ ⚼ ♃
23 44 ☽ ⚹ ♄

23 Saturday
02 09 ☽ ⚼ ♀
06 49 ☉ ⚼ ♃
09 57 ☽ ⚼ ♅
18 24 ☽ ± ♀
18 24 ☽ △ ☿
20 50 ☽ Q ♆
22 12 ☽ △ ♃
23 16 ☿ ⚼ ♇

24 Sunday
01 28 ☽ ∠ ♄
03 17 ☽ □ ♂
05 04 ☽ ⊼ ♀
11 27 ☽ △ ♅
16 33 ☽ ⊥ ♄
17 03 ☽ ⚹ ♆
20 05 ☽ ☌ ♇
22 00 ☽ ⚼ ☿

25 Monday
02 26 ☽ △ ☉
02 29 ☽ ⊻ ♄
03 10 ☉ ⊼ ♄
12 10 ☿ △ ♃
12 22 ☽ ∥ ♇
14 16 ☽ ± ☿
17 45 ☽ ∠ ♆
19 49 ☿ ⋕ ♂
23 02 ♀ ± ♆
23 49 ☽ □ ♃

26 Tuesday
01 02 ☽ ⊼ ☿
04 44 ☽ ⚼ ☉
07 13 ☽ △ ♂
08 21 ☽ ⊥ ♆
08 47 ☽ ☍ ♀
12 39 ☽ □ ♅
13 22 ☽ ⋕ ♅
18 00 ☽ ⊻ ♆
20 16 ☽ ± ☉
21 00 ☽ ⊻ ♇

27 Wednesday
02 30 ☽ ⋕ ♃
03 01 ☽ ☌ ♄
04 41 ☽ Q ♃
06 33 ☽ ⊼ ☉
06 36 ☽ ⊥ ♇
08 26 ☽ ⚼ ♂
11 33 ☽ ∥ ♄
16 42 ☽ ⋕ ♀
17 27 ☽ Q ♅
20 59 ☽ ∠ ♇
23 29 ☽ ± ♂
23 46 ☽ ⚹ ♃

28 Thursday
04 53 ☽ ∥ ♂
06 18 ☽ ☍ ☿
09 30 ☽ ⊼ ♂
11 06 ☽ ⊼ ♀
11 13 ☽ ⋕ ☿
12 35 ☽ ⚹ ♅
17 54 ☽ ☌ ♆
20 58 ☽ ⚹ ♇
21 10 ☽ ± ♀
21 50 ☽ ∥ ♆
23 39 ☽ ∠ ♃
23 40 ☽ ∥ ☉

29 Friday
02 49 ☽ ⊻ ♄
09 57 ☽ ☍ ☉
12 23 ☽ ⚼ ♀
12 28 ☽ ⊥ ♄
12 39 ☽ ∠ ♅
14 04 ☽ ⊥ ♃
15 36 ☽ ⋕ ☉
16 18 ☽ Q ♇
16 36 ☽ ⋕ ♆
17 53 ♀ □ ♅
23 46 ☽ ⊻ ♃

30 Saturday
00 21 ☽ ∥ ☿
03 02 ☽ ∠ ♄
03 16 ☽ ⊥ ♅
08 35 ☿ ⊻ ♂
09 54 ☉ ∥ ♆
11 56 ☽ ⋕ ♂
12 19 ☽ ☍ ♂
12 36 ☽ ⊼ ☿
13 06 ☽ ⊻ ♅
14 08 ☽ △ ♀
16 55 ☿ △ ♅
18 35 ☽ ⊻ ♆
21 20 ☽ ∥ ♀
21 49 ☽ □ ♇
21 59 ☉ ± ♄
23 50 ☽ ± ☿

OCTOBER 2023

LONGITUDES

Date	Sidereal time h m s	Sun ☉ ° ' "	Moon ☽ ° ' "	Moon ☽ 24.00 ° ' "	Mercury ☿ ° '	Venus ♀ ° '	Mars ♂ ° '	Jupiter ♃ ° '	Saturn ♄ ° '	Uranus ♅ ° '	Neptune ♆ ° '	Pluto ♇ ° '
01	12 39 52	08 ♎ 03 15	06 ♉ 21 48	13 ♉ 24 12	23 ♍ 58	23 ♌ 57	22 ♎ 50	14 ♉ 23	01 ♓ 27	22 ♉ 38	25 ♓ 57	27 ♑ 55
02	12 43 48	09 02 14	20 ♉ 20 10	27 ♉ 09 36	25 39	24 42	23 30	14 R 18	01 R 24	22 R 36	25 R 55	27 R 55
03	12 47 45	10 01 15	03 ♊ 52 31	10 ♊ 29 03	27 23	25 28	24 10	14 13	01 21	22 34	25 53	27 54
04	12 51 41	11 00 18	16 ♊ 59 27	23 ♊ 24 06	29 ♍ 07	26 15	24 51	14 07	01 18	22 33	25 52	27 54
05	12 55 38	11 59 23	29 ♊ 43 24	05 ♋ 57 50	00 ♎ 52	27 03	25 31	14 02	01 15	22 31	25 50	27 54
06	12 59 35	12 58 31	12 ♋ 07 56	18 ♋ 14 15	02 38	27 52	26 11	13 56	01 12	22 29	25 49	27 54
07	13 03 31	13 57 41	24 ♋ 17 20	00 ♌ 17 46	04 24	28 42	26 51	13 50	01 09	22 28	25 47	27 54
08	13 07 28	14 56 54	06 ♌ 16 07	12 ♌ 12 54	06 11	29 ♌ 32	27 32	13 44	01 07	22 26	25 45	27 54
09	13 11 24	15 56 08	18 ♌ 08 41	24 ♌ 03 56	07 57	00 ♍ 23	28 12	13 37	01 04	22 24	25 44	27 54
10	13 15 21	16 55 25	29 ♌ 59 09	05 ♍ 54 47	09 43	01 15	28 52	13 31	01 02	22 22	25 42	27 54
11	13 19 17	17 54 44	11 ♍ 51 12	17 ♍ 48 49	11 29	02 08	29 ♎ 33	13 24	00 59	22 20	25 41	27 D 54
12	13 23 14	18 54 05	23 ♍ 47 57	29 ♍ 48 54	13 15	03 01	00 ♏ 13	13 18	00 57	22 18	25 39	27 54
13	13 27 10	19 53 29	05 ♎ 51 56	11 ♎ 57 17	15 00	03 55	00 54	13 11	00 55	22 16	25 38	27 54
14	13 31 07	20 52 54	18 ♎ 05 09	24 ♎ 15 43	16 45	04 49	01 35	13 04	00 53	22 14	25 36	27 54
15	13 35 04	21 52 22	00 ♏ 29 07	06 ♏ 45 31	18 29	05 44	02 15	12 57	00 51	22 12	25 35	27 54
16	13 39 00	22 51 52	13 ♏ 05 00	19 ♏ 27 40	20 13	06 40	02 56	12 49	00 49	22 10	25 34	27 54
17	13 42 57	23 51 23	25 ♏ 53 38	02 ♐ 22 58	21 56	07 36	03 37	12 42	00 47	22 08	25 32	27 54
18	13 46 53	24 50 57	08 ♐ 55 44	15 ♐ 32 02	23 38	08 33	04 18	12 35	00 45	22 06	25 31	27 54
19	13 50 50	25 50 32	22 ♐ 11 54	28 ♐ 55 24	25 20	09 30	04 59	12 27	00 44	22 04	25 29	27 55
20	13 54 46	26 50 09	05 ♑ 42 34	12 ♑ 33 25	27 01	10 28	05 39	12 20	00 42	22 02	25 28	27 55
21	13 58 43	27 49 48	19 ♑ 27 56	26 ♑ 26 03	28 ♎ 42	11 26	06 20	12 12	00 41	21 59	25 27	27 55
22	14 02 39	28 49 29	03 ♒ 27 40	10 ♒ 32 38	00 ♏ 22	12 25	07 01	12 04	00 39	21 57	25 25	27 55
23	14 06 36	29 ♎ 49 11	17 ♒ 40 43	24 ♒ 51 35	02 01	13 24	07 43	11 57	00 38	21 55	25 24	27 56
24	14 10 33	00 ♏ 48 55	02 ♓ 04 54	09 ♓ 20 10	03 40	14 24	08 24	11 49	00 37	21 53	25 23	27 56
25	14 14 29	01 48 40	16 ♓ 36 51	23 ♓ 54 21	05 18	15 24	09 05	11 41	00 36	21 50	25 21	27 57
26	14 18 26	02 48 28	01 ♈ 12 00	08 ♈ 29 04	06 55	16 25	09 46	11 33	00 35	21 48	25 20	27 57
27	14 22 22	03 48 17	15 ♈ 44 49	22 ♈ 58 31	08 32	17 26	10 27	11 25	00 34	21 46	25 19	27 57
28	14 26 19	04 48 07	00 ♉ 09 27	07 ♉ 16 55	10 09	18 27	11 09	11 17	00 33	21 43	25 18	27 58
29	14 30 15	05 48 00	14 ♉ 20 20	21 ♉ 19 10	11 45	19 29	11 50	11 09	00 33	21 41	25 17	27 59
30	14 34 12	06 47 55	28 ♉ 13 00	05 ♊ 01 29	13 20	20 31	12 31	11 01	00 32	21 39	25 15	27 59
31	14 38 08	07 ♏ 47 52	11 ♊ 44 27	18 ♊ 21 49	14 ♏ 55	21 ♍ 34	13 ♏ 13	10 ♉ 53	00 ♓ 32	21 ♉ 36	25 ♓ 14	28 ♑ 00

MOON NODES, LATITUDE AND DECLINATIONS

Date	Moon True ☊ ° '	Moon Mean ☊ ° '	Moon ☽ Latitude ° '	Sun ☉ ° '	Moon ☽ ° '	Mercury ☿ ° '	Venus ♀ ° '	Mars ♂ ° '	Jupiter ♃ ° '	Saturn ♄ ° '	Uranus ♅ ° '	Neptune ♆ ° '	Pluto ♇ ° '
01	24 ♈ 54	25 ♈ 43	01 N 03	03 S 12	14 N 38	04 N 07	10 N 48	08 S 33	14 N 49	12 S 38	18 N 07	02 S 47	23 S 16
02	24 D 55	25 40	02 16	03 35	20 01	03 28	10 41	08 49	14 47	12 39	18 07	02 48	23 16
03	24 56	25 37	03 19	03 58	24 11	02 47	10 34	09 04	14 46	12 40	18 06	02 48	23 16
04	24 57	25 34	04 10	04 21	26 57	02 06	10 26	09 19	14 44	12 41	18 06	02 49	23 16
05	24 58	25 31	04 46	04 44	28 13	01 23	10 17	09 34	14 42	12 42	18 05	02 50	23 16
06	24 58	25 28	05 09	05 07	28 00	00 N 40	10 08	09 49	14 40	12 43	18 05	02 50	23 16
07	24 R 58	25 24	05 16	05 30	26 27	00 S 04	09 59	10 05	14 39	12 43	18 04	02 51	23 16
08	24 57	25 21	05 10	05 53	23 43	00 49	09 49	10 20	14 37	12 44	18 04	02 51	23 16
09	24 56	25 18	04 51	06 16	20 00	01 34	09 38	10 35	14 35	12 45	18 04	02 52	23 16
10	24 55	25 15	04 19	06 39	15 31	02 19	09 27	10 49	14 33	12 46	18 03	02 53	23 16
11	24 54	25 12	03 37	07 02	10 27	03 05	09 16	11 04	14 31	12 47	18 03	02 53	23 16
12	24 53	25 08	02 44	07 24	04 N 58	03 50	09 04	11 19	14 29	12 48	18 02	02 54	23 16
13	24 53	25 05	01 44	07 47	00 S 45	04 35	08 51	11 34	14 27	12 48	18 02	02 54	23 16
14	24 53	25 02	00 N 38	08 09	06 31	05 20	08 38	11 49	14 24	12 49	18 01	02 55	23 16
15	24 D 52	24 59	00 S 31	08 31	12 08	06 05	08 25	12 03	14 22	12 50	18 01	02 56	23 16
16	24 53	24 56	01 39	08 53	17 21	06 49	08 11	12 18	14 20	12 50	18 00	02 56	23 16
17	24 53	24 53	02 44	09 15	21 53	07 33	07 56	12 32	14 18	12 51	17 59	02 57	23 16
18	24 53	24 49	03 41	09 37	25 25	08 17	07 41	12 46	14 16	12 51	17 59	02 57	23 16
19	24 R 53	24 46	04 27	09 59	27 39	08 59	07 26	13 01	14 13	12 52	17 58	02 58	23 16
20	24 53	24 43	05 00	10 21	28 18	09 42	07 10	13 15	14 11	12 52	17 58	02 58	23 16
21	24 52	24 40	05 16	10 42	27 14	10 24	06 54	13 29	14 09	12 53	17 57	02 59	23 16
22	24 D 52	24 37	05 14	11 03	24 28	11 05	06 38	13 43	14 06	12 53	17 57	02 59	23 16
23	24 53	24 34	04 53	11 24	20 10	11 46	06 21	13 57	14 04	12 53	17 56	03 00	23 16
24	24 53	24 30	04 13	11 45	14 40	12 26	06 03	14 11	14 02	12 54	17 56	03 00	23 15
25	24 54	24 27	03 17	12 06	08 08	13 05	05 46	14 24	13 59	12 54	17 55	03 01	23 15
26	24 54	24 24	02 08	12 27	01 S 29	13 44	05 27	14 38	13 57	12 54	17 54	03 01	23 15
27	24 55	24 21	00 S 51	12 47	05 N 25	14 22	05 09	14 52	13 54	12 54	17 54	03 02	23 15
28	24 R 55	24 18	00 N 29	13 07	11 59	14 59	04 50	15 05	13 52	12 55	17 53	03 02	23 15
29	24 54	24 14	01 46	13 27	17 49	15 35	04 31	15 18	13 49	12 55	17 53	03 03	23 15
30	24 53	24 11	02 54	13 47	22 35	16 11	04 11	15 32	13 47	12 55	17 52	03 03	23 15
31	24 ♈ 52	24 ♈ 08	03 N 51	14 S 07	26 N 00	16 S 45	03 N 51	15 S 45	13 N 45	12 S 55	17 N 51	03 S 03	23 S 15

ZODIAC SIGN ENTRIES

Date	h	m	Planets
01	01	18	☽ ♉
03	05	03	☽ ♊
05	00	09	☿ ♎
05	12	32	☽ ♋
07	23	24	☽ ♌
09	01	11	♀ ♍
10	12	02	☽ ♍
12	04	04	♂ ♏
13	00	22	☽ ♎
15	11	04	☽ ♏
17	19	36	☽ ♐
20	01	55	☽ ♑
22	06	06	☽ ♒
22	06	49	☿ ♏
23	16	21	☉ ♏
24	08	33	☽ ♓
26	10	02	☽ ♈
28	11	44	☽ ♉
30	15	08	☽ ♊

LATITUDES

Date	Mercury ☿ ° '	Venus ♀ ° '	Mars ♂ ° '	Jupiter ♃ ° '	Saturn ♄ ° '	Uranus ♅ ° '	Neptune ♆ ° '	Pluto ♇ ° '
01	01 N 52	02 S 54	00 N 21	01 S 24	01 S 47	00 S 19	01 S 17	02 S 44
04	01 54	02 29	00 19	01 24	01 47	00 19	01 17	02 44
07	01 50	02 04	00 18	01 25	01 47	00 19	01 17	02 44
10	01 40	01 41	00 16	01 25	01 46	00 19	01 17	02 44
13	01 26	01 19	00 14	01 25	01 46	00 19	01 17	02 44
16	01 10	00 57	00 12	01 25	01 46	00 19	01 16	02 44
19	00 52	00 37	00 11	01 25	01 46	00 19	01 16	02 44
22	00 33	00 S 18	00 09	01 25	01 45	00 19	01 16	02 44
25	00 N 13	00 00	00 07	01 25	01 45	00 19	01 16	02 44
28	00 S 08	00 N 17	00 05	01 25	01 45	00 19	01 16	02 44
31	00 S 28	00 N 33	00 N 04	01 S 25	01 S 44	00 S 19	01 S 16	02 S 44

DATA

Julian Date	2460219
Delta T	+72 seconds
Ayanamsa	24° 11' 12"
Synetic vernal point	04° ♓ 55' 47"
True obliquity of ecliptic	23° 26' 19"

MOON'S PHASES, APSIDES AND POSITIONS ☽

Date	h	m	Phase	Longitude ° '	Eclipse Indicator
06	13	48	☾	13 ♋ 03	
14	17	55	●	21 ♎ 08	Annular
22	03	29	☽	28 ♑ 28	
28	20	24	○	05 ♉ 09	partial

Day	h	m		
10	03	50	Apogee	
26	03	08	Perigee	
05	20	31	Max dec	28° N 18'
13	08	54	0S	
20	09	13	Max dec	28° S 19'
26	17	07	0N	

LONGITUDES

Date	Chiron ⚷ ° '	Ceres ⚳ ° '	Pallas ⚴ ° '	Juno ⚵ ° '	Vesta ⚶ ° '	Black Moon Lilith ⚸ ° '
01	18 ♈ 07	06 ♏ 38	08 ♎ 09	22 ♌ 59	04 ♋ 05	29 ♌ 46
11	17 ♈ 40	10 ♏ 51	12 ♎ 41	27 ♌ 25	05 ♋ 48	00 ♍ 53
21	17 ♈ 13	15 ♏ 07	17 ♎ 03	01 ♍ 38	06 ♋ 57	02 ♍ 00
31	16 ♈ 47	19 ♏ 24	21 ♎ 41	05 ♍ 35	07 ♋ 29	03 ♍ 07

All ephemeris data is given at 12.00 UT and the Moon's longitude is additionally given for 24.00 UT

ASPECTARIAN

h m Aspects

01 Sunday
01 48 ☉ ꝑ ♅
03 45 ☽ ⚹ ♄
04 01 ☽ # ♄
04 34 ☽ ⊥ ♆
04 38 ♂ ⊼ ♅
11 38 ☿ ⊻ ♀
12 46 ☽ ∥ ♃
15 05 ☽ ⊼ ☉
17 01 ☽ ꝑ ☿
19 34 ☉ ± ♃
19 46 ☽ ∠ ♆

02 Monday
00 02 ☽ Q ♄
01 37 ☽ ☌ ♃
02 07 ☽ ± ☉
02 57 ☽ ∥ ♅
09 17 ☉ # ☿
15 34 ☿ ☍ ♆
15 57 ☽ ☌ ♅
17 50 ☽ ⊼ ♂
18 59 ☽ ꝑ ☉
20 07 ☽ □ ♀
21 47 ☽ ⚹ ♆
22 41 ☽ △ ☿

03 Tuesday
01 20 ☽ △ ♇
05 01 ☽ ± ♂
05 58 ☽ # ♇
07 29 ☽ □ ♄
11 30 ☿ # ♆
19 15 ☽ Q ♆
19 20 ☿ △ ♇
22 07 ☽ ꝑ ♂

04 Wednesday
00 03 ☽ △ ☉
00 21 ♀ ⊼ ♆
04 26 ☽ ꝑ ♇
06 37 ☽ Q ♀
06 43 ☽ ⊻ ♃
12 06 ☿ ꝑ ♃
17 48 ☽ ⊥ ♃
21 11 ☽ ± ♇
22 22 ☽ ⊻ ♅

05 Thursday
03 32 ☽ △ ♂
04 37 ☽ □ ♆
06 34 ☽ ⚹ ♀
08 31 ☽ ⊼ ♇
09 42 ☽ ⊥ ♅
10 41 ☽ ∠ ♃
14 33 ☽ □ ☿
14 55 ☽ △ ♄
17 03 ☿ ⊼ ♄
20 58 ☉ ∠ ♀
23 09 ♂ ⊼ ♆

06 Friday
02 58 ☽ ∠ ♅
12 50 ♀ ⊼ ♇
13 33 ☽ ∠ ♀
13 48 ☽ □ ☉
15 29 ☽ ⚹ ♃
19 57 ☽ ꝑ ♄

07 Saturday
06 31 ♀ # ♂
07 36 ☽ Q ☿
08 22 ☽ ⚹ ♅
08 36 ☽ ⊥ ♀
09 04 ☉ ⊼ ♃
13 43 ☽ ± ♄
14 58 ☽ △ ♆
15 03 ☽ Q ♃
17 25 ☽ □ ♂
19 12 ☽ ☍ ♇
19 26 ☿ ⊥ ♀
21 27 ☽ ⊻ ♀

08 Sunday
01 40 ☽ ⊼ ♄
04 43 ☽ Q ☉
08 18 ☽ Q ♅
11 47 ☽ ⚹ ☿
15 13 ☽ # ♇
21 02 ☽ ꝑ ♆

09 Monday
00 21 ☿ ± ♄
01 05 ♂ □ ♇
02 55 ☽ □ ♃
04 39 ☿ ꝑ ♅
07 07 ☽ ⚹ ☉
07 47 ☿ ± ♃
07 49 ☽ Q ♂
15 08 ☉ ꝑ ♄
15 13 ☽ ± ♆
20 36 ☽ □ ♅
22 52 ☽ ∥ ♅
22 54 ☉ ± ♅
23 27 ☽ ∠ ☿

10 Tuesday
03 21 ☽ ⊼ ♆
06 11 ♀ ☍ ♄
07 45 ☽ ⊼ ♇
09 37 ☽ ⚹ ♂
14 06 ☽ ☍ ♄
14 46 ☽ ☌ ♀
16 17 ☽ ∠ ☉
16 50 ☽ ∥ ♃
19 55 ☽ ± ♇
20 54 ☽ ⊥ ☿

11 Wednesday
01 10 ♇ St D
01 17 ☽ # ♄
05 56 ☿ ∥ ♆
09 18 ☽ # ♂
11 08 ☽ ⊻ ☿
12 08 ☽ ⊥ ☉
14 06 ☽ ꝑ ♇
15 06 ☽ △ ♃
17 32 ☽ ∥ ♀
17 45 ☽ ∠ ♂

12 Thursday
01 18 ☽ ⊻ ☉
02 11 ☽ # ☉
09 01 ☽ △ ♅
12 32 ☿ ⊼ ♃
12 54 ☽ ⊥ ♂
15 42 ☽ ☍ ♆
16 17 ☽ # ☿
20 10 ☽ △ ♇
20 46 ☽ # ♆
20 53 ☽ ꝑ ♃

13 Friday
01 34 ☽ ⊻ ♂
02 13 ☽ ⊼ ♄
07 49 ☽ ⊻ ♀
11 34 ♀ ± ♇
12 29 ♂ △ ♄
14 04 ☽ ± ♄
14 34 ☽ ± ♃
14 46 ☽ ꝑ ♅
20 37 ☽ ⊥ ♀
20 59 ☽ ∥ ♆

14 Saturday
00 15 ☿ ꝑ ♄
02 16 ☽ ⊼ ♃
05 04 ☿ ± ♅
06 19 ☽ ∥ ☿
07 42 ☽ ꝑ ♄
08 24 ☽ ± ♅
08 58 ☽ ☌ ☿
13 02 ♂ ± ♆
15 39 ☽ ∠ ♀
17 55 ☽ ☌ ☉
19 22 ☽ ∥ ☉
20 03 ☽ ⊼ ♅
20 36 ☽ # ♀

15 Sunday
02 34 ☽ ⊼ ♆
07 01 ☽ □ ♇
07 34 ☉ # ♀
11 39 ☽ ∥ ♂
12 41 ☽ △ ♄
14 06 ☽ ± ♆
15 06 ☽ ∥ ♄
15 35 ☽ ☌ ♂
19 44 ☉ ⊼ ♅
21 58 ☽ # ♃
22 51 ☽ ⚹ ♀

16 Monday
07 14 ☽ ꝑ ♆
11 31 ☽ ☍ ♃
15 11 ♃ Q ♄
15 13 ☽ # ♅
17 19 ☽ Q ♇
23 20 ☽ Q ♀

17 Tuesday
03 29 ☽ ⊻ ☿
05 01 ☽ ☍ ♅
07 54 ☽ ⊻ ☉
11 20 ☽ △ ♆
14 48 ☿ ⊼ ♅
15 44 ☽ ⚹ ♇
16 21 ☽ ⊥ ☿
19 57 ☽ ⊥ ☉
20 29 ☽ ∥ ♇
21 02 ☽ □ ♄
21 36 ☿ # ♀

18 Wednesday
03 03 ☽ ⊻ ♂
09 11 ☿ ∠ ♀
11 15 ☽ □ ♀
11 23 ☽ ∠ ☿
13 49 ☽ ∠ ☉
14 37 ☽ ⊥ ♂
18 35 ☽ ⊼ ♃
19 14 ☽ ∠ ♇
20 27 ♂ ∥ ♄

19 Thursday
03 40 ☉ ⊼ ♆
05 20 ☽ ± ♃
05 46 ☽ Q ♄
07 48 ☽ ∠ ♂
11 29 ☽ ⊥ ♇
11 45 ☽ ⊼ ♅
14 12 ☿ ⊼ ♆
17 53 ☽ □ ♆
18 25 ☽ ⚹ ☿
19 02 ☽ ⚹ ☉
21 18 ☽ ꝑ ♃
22 12 ☽ ⊻ ♇
22 27 ☽ ± ♅

20 Friday
03 10 ☽ ⚹ ♄
05 38 ☉ ☌ ☿
11 54 ☽ ⚹ ♂
14 19 ☽ ꝑ ♅
17 55 ☽ Q ☉
18 37 ☽ Q ☿
20 59 ☽ △ ♀
23 30 ☽ △ ♃

21 Saturday
00 51 ☿ □ ♇
01 34 ☽ Q ♆
05 27 ☽ ∠ ♄
09 57 ☽ Q ♂
14 09 ☉ □ ♇
16 21 ☽ △ ♅
20 58 ☽ ⊥ ♄
22 17 ☽ ⚹ ♆

22 Sunday
00 55 ☽ ꝑ ♀
02 33 ☽ ☌ ♇
03 29 ☽ □ ☉
04 32 ♀ △ ♃
06 00 ☽ □ ☿
07 13 ☽ ⊻ ♄
09 45 ☉ ∥ ☿
16 12 ☿ △ ♄
17 24 ☽ ± ♀
18 21 ☽ □ ♂
19 34 ☽ ∥ ♇
23 47 ☽ ∠ ♆

23 Monday
00 21 ♀ ꝑ ♇
02 27 ☽ □ ♃
03 12 ☿ ± ♆
04 17 ☽ ⊼ ♀
14 53 ☽ ⊥ ♆
19 04 ☽ □ ♅
22 19 ☽ # ♅
22 24 ♂ # ♃

24 Tuesday
00 53 ☽ ⊻ ♆
05 07 ☽ ⊻ ♇
07 14 ☉ △ ♄
08 16 ☽ Q ♃
09 34 ☽ ☌ ♄
09 45 ☽ △ ☉
13 52 ☽ ∥ ♂
14 33 ☽ # ♃
14 57 ☽ △ ☿
15 04 ☽ ⊥ ♇
18 56 ☽ ∥ ♄
19 53 ☽ ∥ ☿
22 42 ☽ ∥ ☉
22 58 ☽ △ ♂

25 Wednesday
00 52 ☽ Q ♅
01 18 ☉ ± ♆
03 57 ☽ ⚹ ♃
05 02 ☿ ∥ ♄
05 57 ☽ ∠ ♇
09 52 ☽ ☍ ♀
12 21 ☽ ꝑ ☉
18 49 ☽ ꝑ ☿
20 35 ☽ ⚹ ♅
21 29 ☽ # ♀

26 Thursday
00 54 ☽ ꝑ ♂
02 22 ☽ ☌ ♆
04 15 ☽ ± ☉
04 25 ☽ ∠ ♃
06 39 ☽ ∥ ♆
06 39 ☽ ⚹ ♇
10 59 ☽ ⊻ ♄
11 29 ☽ ± ☿
14 50 ☽ ⊼ ☉
16 26 ☽ ± ♂
19 06 ☽ ⊥ ♃
19 39 ☿ # ♃
20 51 ☽ ⊥ ♄
21 12 ☽ ∠ ♅
22 36 ☽ ⊼ ☿

27 Friday
02 26 ☽ Q ♇
02 49 ☽ ⊼ ♂
03 37 ☽ # ♆
04 54 ☽ ⊻ ♃
07 21 ♂ ꝑ ♆
11 05 ☽ ∥ ♀
11 42 ☽ ∠ ♄
12 02 ☽ ⊥ ♅
15 00 ☽ ⊼ ♀
20 45 ☉ ∥ ♄
21 57 ☽ ⊻ ♅

28 Saturday
01 44 ☽ ± ♀
03 53 ☽ ⊻ ♆
08 20 ☽ □ ♇
12 40 ☽ ⚹ ♄
13 55 ☽ ⊥ ♆
14 16 ☿ ꝑ ♆
15 36 ☽ # ♄
16 03 ♂ ☍ ♃
16 41 ☽ # ☉
17 58 ☽ ꝑ ♀
18 31 ☿ ∥ ♂
19 19 ☽ ∥ ♃
20 24 ☽ ☍ ☉

29 Sunday
00 47 ☽ # ♂
01 15 ☽ # ☿
03 44 ☿ ☍ ♃
05 05 ☽ ∠ ♆
06 37 ☽ ☌ ♃
07 00 ☽ ☍ ☿
07 30 ☽ ☍ ♂
08 56 ☽ Q ♄
12 15 ☽ ∥ ♅
14 22 ☿ ☌ ♂
21 33 ☽ △ ♀

30 Monday
00 36 ☽ ☌ ♅
06 50 ☽ ⚹ ♆
11 36 ☽ △ ♇
12 00 ☉ # ♃
15 56 ☽ # ♇
16 04 ☽ □ ♄

31 Tuesday
03 57 ☽ Q ♆
04 22 ☽ ⊼ ☉
10 28 ☽ ⊻ ♃
12 51 ♀ △ ♅
14 16 ☽ ꝑ ♇
14 48 ☽ ⊼ ♂
16 01 ☽ ± ☉
18 30 ☽ ⊼ ☿
21 12 ☽ ⊥ ♃

LONGITUDES

Date	Sidereal time h m s	Sun ☉ ° ' "	Moon ☽ ° ' "	Moon ☽ 24.00 ° ' "	Mercury ☿ ° '	Venus ♀ ° '	Mars ♂ ° '	Jupiter ♃ ° '	Saturn ♄ ° '	Uranus ♅ ° '	Neptune ♆ ° '	Pluto ♇ ° '
01	14 42 05	08 ♏ 47 51	24 ♊ 53 35	01 ♋ 19 54	16 ♏ 29	22 ♍ 37	13 ♏ 54	10 ♉ 44	00 ♓ 31	21 ♉ 34	25 ♓ 13	28 ♑ 00
02	14 46 02	09 47 52	07 ♋ 41 01	13 ♋ 57 13	18 03	23 40	14 36	10 R 36	00 R 31	21 R 31	25 R 12	28 01
03	14 49 58	10 47 55	20 ♋ 08 56	26 ♋ 16 36	19 37	24 44	15 17	10 28	00 31	21 29	25 11	28 02
04	14 53 55	11 48 01	02 ♌ 20 45	08 ♌ 21 56	21 10	25 48	15 59	10 20	00 D 31	21 27	25 10	28 02
05	14 57 51	12 48 08	14 ♌ 20 43	20 ♌ 17 42	22 43	26 52	16 41	10 12	00 31	21 24	25 09	28 03
06	15 01 48	13 48 17	26 ♌ 13 30	02 ♍ 08 44	24 15	27 57	17 23	10 04	00 31	21 22	25 08	28 04
07	15 05 44	14 48 29	08 ♍ 03 59	13 ♍ 59 52	25 47	29 ♍ 02	18 04	09 56	00 31	21 19	25 07	28 05
08	15 09 41	15 48 42	19 ♍ 56 33	25 ♍ 55 41	27 19	00 ♎ 07	18 46	09 48	00 32	21 17	25 06	28 05
09	15 13 37	16 48 58	01 ♎ 56 40	08 ♎ 00 20	28 ♏ 50	01 12	19 28	09 39	00 32	21 14	25 05	28 06
10	15 17 34	17 49 15	14 ♎ 07 03	20 ♎ 17 12	00 ♐ 21	02 18	20 10	09 31	00 33	21 12	25 05	28 07
11	15 21 31	18 49 34	26 ♎ 31 04	02 ♏ 48 51	01 52	03 24	20 52	09 23	00 34	21 09	25 04	28 08
12	15 25 27	19 49 55	09 ♏ 10 43	15 ♏ 36 46	03 22	04 30	21 34	09 16	00 34	21 07	25 03	28 09
13	15 29 24	20 50 18	22 ♏ 06 58	28 ♏ 41 18	04 51	05 37	22 16	09 08	00 35	21 04	25 02	28 10
14	15 33 20	21 50 43	05 ♐ 19 36	12 ♐ 01 43	06 21	06 44	22 59	09 00	00 36	21 02	25 01	28 11
15	15 37 17	22 51 09	18 ♐ 47 23	25 ♐ 36 20	07 50	07 50	23 41	08 52	00 37	20 59	25 01	28 12
16	15 41 13	23 51 37	02 ♑ 28 15	09 ♑ 22 48	09 18	08 58	24 23	08 45	00 39	20 57	25 00	28 13
17	15 45 10	24 52 06	16 ♑ 19 40	23 ♑ 18 31	10 46	10 05	25 05	08 37	00 40	20 54	24 59	28 14
18	15 49 06	25 52 37	00 ♒ 19 02	07 ♒ 20 55	12 14	11 13	25 48	08 30	00 41	20 52	24 59	28 15
19	15 53 03	26 53 08	14 ♒ 23 53	21 ♒ 27 41	13 41	12 21	26 30	08 22	00 43	20 49	24 58	28 16
20	15 57 00	27 53 41	28 ♒ 32 07	05 ♓ 36 56	15 07	13 29	27 13	08 15	00 45	20 47	24 58	28 17
21	16 00 56	28 54 15	12 ♓ 41 56	19 ♓ 46 55	16 33	14 37	27 55	08 08	00 46	20 44	24 57	28 18
22	16 04 53	29 ♏ 54 50	26 ♓ 51 41	03 ♈ 55 59	17 58	15 45	28 38	08 01	00 48	20 42	24 57	28 19
23	16 08 49	00 ♐ 55 27	10 ♈ 59 34	18 ♈ 02 09	19 23	16 54	29 ♏ 20	07 54	00 50	20 40	24 56	28 21
24	16 12 46	01 56 04	25 ♈ 03 25	02 ♉ 03 02	20 46	18 03	00 ♐ 03	07 47	00 52	20 37	24 56	28 22
25	16 16 42	02 56 43	09 ♉ 00 38	15 ♉ 55 49	22 09	19 12	00 46	07 40	00 54	20 35	24 55	28 23
26	16 20 39	03 57 23	22 ♉ 48 13	29 ♉ 37 25	23 30	20 21	01 29	07 34	00 56	20 32	24 55	28 24
27	16 24 35	04 58 04	06 ♊ 23 06	13 ♊ 04 56	24 50	21 30	02 11	07 28	00 59	20 30	24 55	28 26
28	16 28 32	05 58 47	19 ♊ 42 38	26 ♊ 16 02	26 08	22 40	02 54	07 21	01 01	20 27	24 54	28 27
29	16 32 29	06 59 31	02 ♋ 44 59	09 ♋ 09 27	27 25	23 49	03 37	07 15	01 04	20 25	24 54	28 28
30	16 36 25	08 ♐ 00 17	15 ♋ 29 28	21 ♋ 45 10	28 ♐ 40	24 ♎ 59	04 ♐ 20	07 ♉ 09	01 ♓ 06	20 ♉ 23	24 ♓ 54	28 ♑ 30

DECLINATIONS

Date	Moon True ☊ ° '	Moon Mean ☊ ° '	Moon ☽ Latitude ° '	Sun ☉ ° '	Moon ☽ ° '	Mercury ☿ ° '	Venus ♀ ° '	Mars ♂ ° '	Jupiter ♃ ° '	Saturn ♄ ° '	Uranus ♅ ° '	Neptune ♆ ° '	Pluto ♇ ° '
01	24 ♈ 50	24 ♈ 05	04 N 34	14 S 26	27 N 54	17 S 19	03 N 31	15 S 58	13 N 42	12 S 55	17 N 51	03 S 04	23 S 15
02	24 R 48	24 02	05 02	14 45	28 14	17 52	03 10	16 11	13 40	12 55	17 50	03 04	23 15
03	24 46	23 59	05 15	15 04	27 07	18 24	02 50	16 23	13 37	12 55	17 49	03 05	23 14
04	24 45	23 55	05 13	15 22	24 43	18 55	02 29	16 36	13 35	12 55	17 49	03 05	23 14
05	24 D 45	23 52	04 58	15 41	21 16	19 25	02 07	16 49	13 32	12 55	17 48	03 05	23 14
06	24 45	23 49	04 30	15 59	17 00	19 55	01 45	17 01	13 30	12 54	17 48	03 06	23 14
07	24 46	23 46	03 50	16 17	12 06	20 23	01 24	17 13	13 27	12 54	17 47	03 06	23 14
08	24 48	23 43	03 01	16 34	06 45	20 50	01 01	17 25	13 25	12 54	17 46	03 06	23 14
09	24 50	23 40	02 03	16 52	01 N 06	21 16	00 39	17 37	13 23	12 54	17 46	03 07	23 14
10	24 51	23 36	00 N 59	17 09	04 S 40	21 42	00 N 16	17 49	13 20	12 53	17 45	03 07	23 13
11	24 R 51	23 33	00 S 09	17 25	10 22	22 06	00 S 06	18 01	13 18	12 53	17 44	03 07	23 13
12	24 51	23 30	01 18	17 42	15 47	22 29	00 29	18 13	13 15	12 53	17 44	03 08	23 13
13	24 48	23 27	02 24	17 58	20 37	22 51	00 52	18 24	13 13	12 52	17 43	03 08	23 13
14	24 45	23 24	03 24	18 14	24 32	23 11	01 16	18 35	13 11	12 52	17 42	03 08	23 13
15	24 41	23 20	04 13	18 29	27 10	23 31	01 39	18 47	13 08	12 51	17 42	03 09	23 12
16	24 36	23 17	04 49	18 44	28 14	23 50	02 03	18 58	13 06	12 51	17 41	03 09	23 12
17	24 32	23 14	05 09	18 59	27 33	24 07	02 27	19 09	13 04	12 50	17 40	03 09	23 12
18	24 28	23 11	05 11	19 14	25 08	24 23	02 50	19 19	13 02	12 49	17 40	03 09	23 12
19	24 26	23 08	04 54	19 28	21 11	24 38	03 14	19 30	13 00	12 49	17 39	03 09	23 12
20	24 D 25	23 05	04 19	19 41	16 01	24 51	03 38	19 40	12 57	12 48	17 39	03 10	23 11
21	24 26	23 01	03 28	19 55	10 00	25 03	04 02	19 50	12 55	12 47	17 38	03 10	23 11
22	24 27	22 58	02 25	20 08	03 S 28	25 14	04 27	20 00	12 53	12 47	17 37	03 10	23 11
23	24 29	22 55	01 S 13	20 21	03 N 14	25 24	04 51	20 10	12 51	12 46	17 37	03 10	23 11
24	24 R 30	22 52	00 N 03	20 33	09 45	25 32	05 15	20 20	12 49	12 45	17 36	03 10	23 11
25	24 29	22 49	01 18	20 45	15 44	25 39	05 39	20 30	12 47	12 44	17 35	03 10	23 10
26	24 26	22 45	02 28	20 56	20 51	25 44	06 04	20 39	12 45	12 43	17 35	03 10	23 10
27	24 21	22 42	03 27	21 07	24 47	25 48	06 28	20 48	12 44	12 42	17 34	03 11	23 10
28	24 14	22 39	04 15	21 18	27 16	25 51	06 52	20 57	12 42	12 41	17 33	03 11	23 10
29	24 06	22 36	04 47	21 29	28 12	25 52	07 17	21 06	12 40	12 40	17 33	03 11	23 09
30	23 ♈ 58	22 ♈ 33	05 N 05	21 S 39	27 N 35	25 S 52	07 S 41	21 S 15	12 N 38	12 S 39	17 N 32	03 S 11	23 S 09

ZODIAC SIGN ENTRIES

Date	h	m	Planets
01	21	30	☽ ♋
04	07	21	☽ ♌
06	19	39	☽ ♍
08	09	30	♀ ♎
09	08	08	☽ ♎
10	06	25	☿ ♐
11	18	39	☽ ♏
14	02	23	☽ ♐
16	07	41	☽ ♑
18	11	27	☽ ♒
20	14	29	☽ ♓
22	14	03	☉ ♐
22	17	19	☽ ♈
24	10	15	♂ ♐
24	20	29	☽ ♉
27	00	40	☽ ♊
29	06	54	☽ ♋

LATITUDES

Date	Mercury ☿ ° '	Venus ♀ ° '	Mars ♂ ° '	Jupiter ♃ ° '	Saturn ♄ ° '	Uranus ♅ ° '	Neptune ♆ ° '	Pluto ♇ ° '
01	00 S 34	00 N 38	00 N 03	01 S 25	01 S 44	00 S 19	01 S 16	02 S 44
04	00 54	00 53	00 N 01	01 25	01 44	00 19	01 16	02 44
07	01 13	01 06	00 00	01 24	01 44	00 19	01 16	02 44
10	01 30	01 18	00 S 02	01 24	01 43	00 19	01 16	02 45
13	01 46	01 29	00 04	01 24	01 43	00 19	01 16	02 45
16	02 00	01 38	00 06	01 23	01 43	00 19	01 16	02 45
19	02 12	01 47	00 07	01 22	01 42	00 19	01 16	02 45
22	02 21	01 54	00 09	01 22	01 42	00 19	01 16	02 45
25	02 27	02 00	00 11	01 21	01 42	00 19	01 16	02 45
28	02 28	02 06	00 13	01 21	01 41	00 19	01 16	02 45
31	02 S 24	02 N 10	00 S 15	01 S 20	01 S 41	00 S 19	01 S 15	02 S 45

DATA

Julian Date	2460250
Delta T	+72 seconds
Ayanamsa	24° 11' 16"
Synetic vernal point	04° ♓ 55' 43"
True obliquity of ecliptic	23° 26' 19"

MOON'S PHASES, APSIDES AND POSITIONS ☽

Date	h	m	Phase	Longitude ° '	Eclipse Indicator
05	08	37	☾	12 ♌ 40	
13	09	27	●	20 ♏ 44	
20	10	50	☽	27 ♒ 51	
27	09	16	○	04 ♊ 51	

Day	h	m		
06	21	52	Apogee	
21	21	05	Perigee	
02	05	16	Max dec	28° N 18'
09	16	37	0S	
16	14	44	Max dec	28° S 15'
23	00	25	0N	
29	14	12	Max dec	28° N 12'

LONGITUDES

Date	Chiron ⚷ ° '	Ceres ⚳ ° '	Pallas ⚴ ° '	Juno ⚵ ° '	Vesta ⚶ ° '	Black Moon Lilith ⚸ ° '
01	16 ♈ 45	19 ♏ 50	22 ♎ 08	05 ♍ 57	07 ♋ 30	03 ♍ 14
11	16 ♈ 21	24 ♏ 07	26 ♎ 34	09 ♍ 35	07 ♋ 15	04 ♍ 21
21	16 ♈ 01	28 ♏ 35	00 ♏ 25	12 ♍ 38	06 ♋ 15	05 ♍ 28
31	15 ♈ 45	02 ♐ 41	05 ♏ 12	15 ♍ 44	04 ♋ 33	06 ♍ 35

All ephemeris data is given at 12.00 UT and the Moon's longitude is additionally given for 24.00 UT

ASPECTARIAN

h m Aspects

01 Wednesday
02 18 ☽ ± ♂
04 32 ☿ Q ♇
05 53 ☽ ⊻ ♅
06 40 ☽ ± ♇
06 57 ☽ ± ☿
07 26 ☽ □ ♀
09 48 ☽ bQ ☉
12 36 ☽ □ ♆
13 33 ☽ ∠ ♃
16 57 ☽ ⊥ ♅
17 47 ☽ ⊼ ♇
19 53 ☽ bQ ♂
22 29 ☽ △ ♄

02 Thursday
02 01 ☽ bQ ☿
09 48 ☽ ∠ ♅
10 37 ☿ # ♅
16 23 ☽ △ ☉
17 31 ☽ ✶ ♃
19 01 ♀ # ♆
20 19 ☽ Q ♀
21 33 ☉ bQ ♆

03 Friday
02 01 ☽ △ ♂
03 01 ☽ bQ ♄
05 02 ☉ ☍ ♃
10 49 ☽ △ ☿
14 36 ☽ ✶ ♅
16 29 ☽ Q ♃
20 32 ☽ ± ♄
21 49 ☽ ✶ ♀
21 50 ☽ △ ♆
22 06 ♀ ☍ ♆

04 Saturday
02 45 ♀ bQ ♃
03 28 ☽ ☍ ♇
07 04 ♄ St D
08 22 ☽ ⊼ ♄
13 50 ♂ Q ♇
14 10 ☽ Q ♅
16 07 ☿ ☍ ♅
23 06 ☽ # ♇

05 Sunday
03 35 ☽ bQ ♆
03 46 ☽ □ ♃
06 32 ☽ ∠ ♀
08 37 ☽ □ ☉
17 00 ☽ □ ♂
21 04 ♃ ∠ ♆
21 41 ☽ ± ♆
21 44 ☽ # ☿

06 Monday
02 11 ☽ □ ♅
02 28 ☽ ⊥ ♀
07 25 ☽ □ ☿
07 48 ☽ ∥ ♅
09 48 ☽ ⊼ ♆
11 54 ☽ # ♂
14 38 ♀ △ ♇
15 44 ☽ ⊼ ♇
15 50 ☽ ⊻ ♀
16 54 ☽ # ☉
20 42 ☽ ☍ ♄

07 Tuesday
00 21 ☽ Q ☉
01 37 ☿ △ ♆
03 54 ☽ ± ♇
05 33 ☽ ∥ ♃
07 43 ☽ Q ♂
08 14 ☽ # ♄
15 43 ☽ △ ♃
22 09 ☽ bQ ♇

08 Wednesday
01 17 ☽ Q ☿
02 54 ☽ ✶ ☉
09 29 ☽ ✶ ♂
14 40 ☽ △ ♅
18 42 ☉ Q ♇
21 13 ♀ ⊼ ♄
21 37 ☽ bQ ♃
22 20 ☽ ☍ ♆

09 Thursday
00 17 ☿ ✶ ♇
03 34 ☽ # ♆
04 20 ☽ △ ♇
04 55 ☽ ✶ ☿
09 12 ☽ ⊼ ♄
10 23 ☽ ☌ ♀
11 43 ☽ ∠ ☉
14 02 ☽ ∥ ♀
15 22 ☽ ± ♃
17 19 ☽ ∠ ♂
18 53 ☽ # ♀
20 29 ☽ bQ ♅
21 07 ☽ ± ♄

10 Friday
03 05 ☽ ⊼ ♃
03 34 ♂ # ♅
05 35 ☽ ∥ ♆
07 06 ☽ ⊥ ☉
12 07 ☽ ⊥ ♂
14 06 ☽ ± ♅
14 45 ☽ ∠ ☿
14 48 ☽ bQ ♄
15 07 ☿ □ ♄
19 51 ☽ ⊻ ☉

11 Saturday
00 29 ☽ ⊻ ♂
01 43 ☽ ⊼ ♅
09 13 ☽ ⊼ ♆
10 34 ☽ ⊥ ☿
11 50 ♀ ± ♃
15 05 ☽ □ ♇
19 43 ☽ △ ♄
20 40 ☽ ± ♆
21 11 ♂ ☍ ♅
22 54 ☽ ∥ ♄
23 34 ☽ ⊻ ☿

12 Sunday
00 38 ☽ # ♃
02 22 ☽ ⊻ ♀
12 09 ☽ ☍ ♃
13 38 ☽ bQ ♆
14 43 ☽ ⊥ ♀
14 48 ☉ # ♅
21 12 ☽ # ♅
21 36 ☽ ∥ ☉

13 Monday
00 05 ☽ ∥ ♂
01 00 ☽ Q ♇
08 59 ☽ ∠ ♀
09 27 ☽ ☌ ☉
10 05 ☽ ☍ ♅
12 18 ☽ ☌ ♂
17 20 ☽ △ ♆
17 21 ☉ ☍ ♅
21 33 ♀ bQ ♅
23 03 ☽ ✶ ♇

14 Tuesday
02 06 ☽ ∥ ☿
03 07 ☽ ∥ ♇
03 28 ☽ □ ♄
09 21 ♀ ± ♄
13 25 ☿ ∥ ♇
14 04 ☽ ☌ ☿
14 45 ☽ ✶ ♀
18 32 ☽ ⊼ ♃

15 Wednesday
02 04 ☽ ∠ ♇
05 07 ☽ ± ♃
11 42 ☽ Q ♄
12 48 ☿ ✶ ♀
14 01 ☽ Q ♀
14 22 ☉ ∠ ♀
15 52 ☽ ⊼ ♅
18 01 ☽ ⊥ ♇
19 44 ☽ ⊻ ☉
20 52 ☽ bQ ♃
21 05 ☽ ⊻ ♂
22 57 ☽ □ ♆

16 Thursday
02 23 ☽ ± ♅
03 35 ☿ ⊼ ♃
04 33 ☽ ⊻ ♇
07 05 ☽ ⊥ ☉
07 48 ♀ ⊼ ♃
08 10 ☽ ⊥ ♂
08 49 ☽ ✶ ♄
18 02 ☽ bQ ♅
22 48 ☽ △ ♃
23 58 ☽ ∠ ☉

17 Friday
00 16 ☽ □ ♀
00 39 ☽ ∠ ♂
01 17 ☽ ⊻ ☿
06 15 ☽ Q ♆
08 36 ♂ △ ♆
10 51 ☽ ∠ ♄
12 20 ♀ ∠ ♂
12 51 ☽ ⊥ ☿
14 52 ☉ △ ♆
19 51 ☽ △ ♅

18 Saturday
02 21 ☽ ⊥ ♄
02 52 ☽ ✶ ♆
03 49 ☽ ✶ ☉
03 51 ☽ ✶ ♂
05 42 ☉ ☌ ♂
06 07 ☽ ∠ ☿
08 27 ☽ ☌ ♇
12 38 ☽ ⊻ ♄
16 59 ☽ ∥ ☿

19 Sunday
00 50 ☽ ∥ ♇
01 26 ☽ Q ♂
01 50 ☽ □ ♃
01 54 ☽ Q ☉
04 28 ☽ ∠ ♆
04 59 ☿ ∠ ♇
07 01 ♀ ∥ ♆
08 12 ☽ △ ♀
10 39 ☽ ✶ ☿
19 46 ☽ ⊥ ♆
20 08 ☽ ∥ ♂
20 13 ☽ ∥ ☉
22 33 ☿ ± ♃
22 53 ☽ □ ♅

20 Monday
03 11 ☉ ∥ ♂
04 57 ☽ # ♅
05 56 ☽ ⊻ ♆
08 09 ☽ Q ♃
09 20 ☽ Q ☿
09 38 ☽ □ ♂
10 50 ☽ □ ☉
11 34 ☽ ⊻ ♇
11 54 ☽ bQ ♀
15 45 ☽ ☌ ♄
21 26 ☉ ✶ ♇
21 45 ☽ ⊥ ♇

21 Tuesday
00 38 ☽ # ♃
01 11 ☽ ∥ ♄
04 20 ☽ ✶ ♃
04 29 ☽ ± ♀
05 19 ☽ Q ♅
13 02 ☽ ∠ ♇
14 34 ♀ ± ♅
15 32 ☽ ⊼ ♀
19 16 ☽ □ ☿

22 Wednesday
01 18 ♂ ✶ ♇
01 35 ☽ ✶ ♅
05 32 ☽ ∠ ♃
08 40 ☽ ∥ ♀
08 45 ☽ ☌ ♆
13 00 ♀ bQ ♄
13 04 ☽ ∥ ♆
14 29 ☽ ✶ ♇
15 10 ☽ △ ♂
17 35 ☽ △ ☉
18 42 ☽ ⊻ ♄
20 40 ☽ ⊥ ♃

23 Thursday
02 26 ☿ Q ♄
02 57 ☽ ∠ ♅
04 55 ☽ ⊥ ♄
06 47 ☽ ⊻ ♃
09 47 ☉ □ ♄
10 54 ☽ Q ♇
11 46 ☽ # ♆
18 00 ☽ bQ ♂
18 13 ☽ # ♀
18 13 ☽ ⊥ ♅
20 16 ☽ ∠ ♄
21 03 ☽ bQ ☉
22 57 ☽ ☍ ♀

24 Friday
03 52 ☽ △ ☿
04 26 ☽ ⊻ ♅
09 27 ☿ ⊼ ♅
10 11 ☽ ± ♂
11 47 ☽ ⊻ ♆
13 37 ☽ ± ☉
17 40 ☽ □ ♇
21 01 ☽ ⊼ ♂
21 59 ☽ ✶ ♄
22 04 ☽ ⊥ ♆
23 40 ☽ # ♄
23 54 ☽ ∥ ♃

25 Saturday
00 43 ☽ ⊼ ☉
08 25 ☽ bQ ☿
09 43 ☽ ☌ ♃
13 35 ☽ ∠ ♆
16 18 ☿ ⊥ ♇
16 57 ♂ □ ♄
18 46 ☽ Q ♄
20 09 ☽ ∥ ♅
20 40 ☿ bQ ♃

26 Sunday
01 43 ☽ ± ☿
07 18 ☽ ⊼ ♀
08 03 ☽ ☌ ♅
10 55 ☽ # ♂
12 29 ☽ # ☉
13 21 ☽ ⊼ ☿
15 42 ☽ ✶ ♆
15 49 ♀ ⊼ ♅
18 48 ☽ ± ♀
21 52 ☽ △ ♇

27 Monday
01 08 ☽ # ♇
02 22 ☽ □ ♄
04 08 ☽ ☍ ♂
09 16 ☽ ☍ ☉
12 14 ☽ bQ ♀
12 56 ☽ Q ♆
13 27 ☿ □ ♆
13 54 ☽ ⊻ ♃
20 23 ☽ # ☿

28 Tuesday
00 35 ☽ ⊥ ♃
00 38 ☽ bQ ♇
13 21 ☽ ⊻ ♅
16 47 ☽ ∠ ♃
17 00 ☽ ± ♇
17 45 ☿ ± ♅
17 54 ☽ △ ♀
21 30 ☽ □ ♆

29 Wednesday
00 04 ♃ # ♄
00 19 ☽ ⊥ ♅
01 03 ☽ ☍ ☿
04 03 ☽ ⊼ ♇
08 51 ☽ △ ♄
13 43 ☽ ⊼ ♂
16 58 ☽ ∠ ♅
17 38 ☉ ⊼ ♃
20 21 ☽ ✶ ♃
20 37 ☽ ⊼ ☉

30 Thursday
01 38 ☽ ± ♂
08 34 ☿ ⊻ ♇
08 55 ☽ ± ☉
10 13 ♀ ⊼ ♆
13 11 ☽ bQ ♄
18 57 ☽ Q ♃
19 48 ☽ bQ ♂
21 20 ☽ ✶ ♅

DECEMBER 2023

Date	Sidereal time h m s	Sun ☉ ° ' "	Moon ☽ ° ' "	Moon ☽ 24.00 ° ' "	Mercury ☿ ° '	Venus ♀ ° '	Mars ♂ ° '	Jupiter ♃ ° '	Saturn ♄ ° '	Uranus ♅ ° '	Neptune ♆ ° '	Pluto ⯓ ° '
01	16 40 22	09 ♐ 01 04	27 ♋ 56 47	04 ♌ 04 35	29 ♐ 53	26 ♎ 09	05 ♐ 03	07 ♉ 03	01 ♓ 09	20 ♉ 20	24 ♓ 54	28 ♑ 31
02	16 44 18	10 01 52	10 ♌ 08 58	16 ♌ 10 21	01 ♑ 03	27 19	05 46	06 R 58	01 12	20 R 18	24 R 54	28 32
03	16 48 15	11 02 42	22 ♌ 09 15	28 ♌ 06 12	02 10	28 29	06 29	06 52	01 15	20 16	24 53	28 34
04	16 52 11	12 03 33	04 ♍ 01 48	09 ♍ 56 41	03 13	29 ♎ 40	07 13	06 47	01 18	20 14	24 53	28 35
05	16 56 08	13 04 25	15 ♍ 51 29	21 ♍ 46 53	04 13	00 ♏ 50	07 56	06 42	01 21	20 11	24 53	28 37
06	17 00 04	14 05 19	27 ♍ 43 32	03 ♎ 42 06	05 08	02 01	08 39	06 37	01 24	20 09	24 53	28 38
07	17 04 01	15 06 13	09 ♎ 43 14	15 ♎ 47 33	05 58	03 12	09 22	06 32	01 27	20 07	24 D 53	28 40
08	17 07 58	16 07 10	21 ♎ 55 38	28 ♎ 08 02	06 43	04 23	10 06	06 27	01 31	20 05	24 53	28 41
09	17 11 54	17 08 07	04 ♏ 25 10	10 ♏ 47 27	07 20	05 34	10 49	06 23	01 34	20 03	24 53	28 43
10	17 15 51	18 09 06	17 ♏ 15 09	23 ♏ 48 27	07 51	06 45	11 33	06 19	01 38	20 00	24 54	28 44
11	17 19 47	19 10 05	00 ♐ 27 23	07 ♐ 11 54	08 13	07 56	12 16	06 15	01 42	19 58	24 54	28 46
12	17 23 44	20 11 06	14 ♐ 01 46	20 ♐ 56 39	08 26	09 08	13 00	06 11	01 45	19 56	24 54	28 47
13	17 27 40	21 12 08	27 ♐ 56 05	04 ♑ 59 29	08 R 29	10 19	13 43	06 07	01 49	19 54	24 54	28 49
14	17 31 37	22 13 10	12 ♑ 06 10	19 ♑ 15 26	08 22	11 31	14 27	06 03	01 53	19 52	24 54	28 51
15	17 35 33	23 14 13	26 ♑ 26 30	03 ♒ 38 37	08 03	12 42	15 11	06 00	01 57	19 50	24 55	28 52
16	17 39 30	24 15 17	10 ♒ 51 04	18 ♒ 03 10	07 32	13 54	15 54	05 57	02 01	19 48	24 55	28 54
17	17 43 27	25 16 21	25 ♒ 14 21	02 ♓ 24 08	06 50	15 06	16 38	05 54	02 05	19 46	24 55	28 56
18	17 47 23	26 17 25	09 ♓ 32 06	16 ♓ 38 00	05 57	16 18	17 22	05 51	02 10	19 45	24 56	28 57
19	17 51 20	27 18 30	23 ♓ 41 36	00 ♈ 42 48	04 54	17 30	18 06	05 49	02 14	19 43	24 56	28 59
20	17 55 16	28 19 35	07 ♈ 41 33	14 ♈ 37 51	03 43	18 42	18 50	05 47	02 18	19 41	24 57	29 01
21	17 59 13	29 ♐ 20 40	21 ♈ 31 43	28 ♈ 23 10	02 25	19 54	19 34	05 44	02 23	19 39	24 57	29 02
22	18 03 09	00 ♑ 21 45	05 ♉ 12 15	11 ♉ 58 59	01 ♑ 03	21 07	20 18	05 43	02 27	19 37	24 58	29 04
23	18 07 06	01 22 51	18 ♉ 43 20	25 ♉ 25 17	29 ♐ 40	22 19	21 02	05 41	02 32	19 36	24 58	29 06
24	18 11 02	02 23 57	02 ♊ 04 44	08 ♊ 41 35	28 19	23 31	21 46	05 39	02 37	19 34	24 59	29 08
25	18 14 59	03 25 03	15 ♊ 15 44	21 ♊ 47 01	27 03	24 44	22 30	05 38	02 42	19 32	25 00	29 10
26	18 18 56	04 26 09	28 ♊ 15 18	04 ♋ 40 27	25 52	25 56	23 15	05 37	02 46	19 31	25 00	29 11
27	18 22 52	05 27 16	11 ♋ 02 22	17 ♋ 20 58	24 50	27 09	23 59	05 36	02 51	19 29	25 01	29 13
28	18 26 49	06 28 23	23 ♋ 36 13	29 ♋ 48 09	23 58	28 22	24 43	05 36	02 56	19 28	25 02	29 15
29	18 30 45	07 29 31	05 ♌ 56 50	12 ♌ 02 27	23 16	29 ♏ 35	25 27	05 35	03 02	19 26	25 02	29 17
30	18 34 42	08 30 38	18 ♌ 05 12	24 ♌ 05 23	22 45	00 ♐ 47	26 12	05 35	03 07	19 25	25 03	29 19
31	18 38 38	09 ♑ 31 46	00 ♍ 03 22	05 ♍ 59 33	22 ♐ 24	02 ♐ 00	26 ♐ 56	05 ♉ 35	03 ♓ 12	19 ♉ 24	25 ♓ 04	29 ♑ 21

Date	Moon True ☊ ° '	Moon Mean ☊ ° '	Moon ☽ Latitude ° '
01	23 ♈ 51	22 ♈ 30	05 N 07
02	23 R 45	22 26	04 56
03	23 41	22 23	04 31
04	23 38	22 20	03 55
05	23 D 38	22 17	03 09
06	23 39	22 14	02 15
07	23 40	22 11	01 14
08	23 41	22 07	00 N 09
09	23 R 40	22 04	00 S 57
10	23 38	22 01	02 03
11	23 32	21 58	03 04
12	23 25	21 55	03 56
13	23 15	21 51	04 35
14	23 05	21 48	04 58
15	22 55	21 45	05 03
16	22 47	21 42	04 49
17	22 41	21 39	04 17
18	22 37	21 36	03 28
19	22 36	21 32	02 28
20	22 D 36	21 29	01 19
21	22 37	21 26	00 S 06
22	22 R 36	21 23	01 N 06
23	22 33	21 20	02 14
24	22 28	21 17	03 13
25	22 19	21 13	04 01
26	22 08	21 10	04 35
27	21 55	21 07	04 55
28	21 41	21 04	05 00
29	21 28	21 01	04 51
30	21 17	20 57	04 29
31	21 ♈ 08	20 ♈ 54	03 N 55

DECLINATIONS

Date	Sun ☉ ° '	Moon ☽ ° '	Mercury ☿ ° '	Venus ♀ ° '	Mars ♂ ° '	Jupiter ♃ ° '	Saturn ♄ ° '	Uranus ♅ ° '	Neptune ♆ ° '	Pluto ⯓ ° '
01	21 S 48	25 N 35	25 S 50	08 S 05	21 S 23	12 N 37	12 S 38	17 N 32	03 S 11	23 S 09
02	21 57	22 27	25 47	08 29	21 31	12 35	12 37	17 31	03 11	23 09
03	22 06	18 24	25 43	08 53	21 39	12 33	12 36	17 30	03 11	23 08
04	22 14	13 41	25 37	09 17	21 47	12 32	12 35	17 30	03 11	23 08
05	22 22	08 29	25 30	09 41	21 55	12 30	12 34	17 29	03 11	23 08
06	22 29	02 N 58	25 22	10 05	22 02	12 29	12 32	17 29	03 11	23 07
07	22 36	02 S 43	25 12	10 28	22 10	12 28	12 31	17 28	03 11	23 07
08	22 43	08 24	25 02	10 52	22 17	12 27	12 30	17 28	03 11	23 07
09	22 49	13 54	24 50	11 15	22 23	12 25	12 28	17 27	03 11	23 07
10	22 55	18 57	24 37	11 38	22 30	12 24	12 27	17 26	03 11	23 06
11	23 00	23 14	24 24	12 01	22 37	12 23	12 26	17 26	03 10	23 06
12	23 05	26 23	24 09	12 24	22 43	12 22	12 24	17 25	03 10	23 06
13	23 09	28 00	23 54	12 46	22 49	12 21	12 23	17 25	03 10	23 05
14	23 13	27 50	23 38	13 08	22 54	12 20	12 21	17 24	03 10	23 05
15	23 16	25 49	23 22	13 30	23 00	12 19	12 20	17 24	03 10	23 05
16	23 19	22 08	23 05	13 52	23 05	12 19	12 18	17 23	03 10	23 04
17	23 21	17 08	22 48	14 13	23 10	12 18	12 17	17 23	03 10	23 04
18	23 23	11 12	22 31	14 35	23 15	12 17	12 15	17 22	03 09	23 04
19	23 25	04 S 46	22 13	14 56	23 20	12 17	12 13	17 22	03 09	23 04
20	23 26	01 N 51	21 56	15 16	23 24	12 16	12 12	17 21	03 09	23 03
21	23 26	08 18	21 39	15 36	23 28	12 16	12 10	17 21	03 09	23 03
22	23 26	14 18	21 22	15 56	23 32	12 16	12 08	17 20	03 09	23 03
23	23 26	19 32	21 06	16 16	23 36	12 15	12 06	17 20	03 08	23 02
24	23 25	23 44	20 52	16 35	23 40	12 15	12 05	17 20	03 08	23 02
25	23 24	26 37	20 39	16 54	23 43	12 15	12 03	17 19	03 08	23 02
26	23 22	28 01	20 28	17 12	23 46	12 15	12 01	17 19	03 08	23 01
27	23 20	27 53	20 20	17 30	23 48	12 15	11 59	17 18	03 07	23 01
28	23 17	26 18	20 13	17 48	23 51	12 15	11 57	17 18	03 07	23 01
29	23 14	23 29	20 09	18 05	23 53	12 15	11 55	17 18	03 06	23 00
30	23 10	19 40	20 08	18 22	23 55	12 15	11 53	17 17	03 06	23 00
31	23 S 06	15 N 07	20 S 08	18 S 38	23 S 57	12 N 16	11 S 51	17 N 17	03 S 06	23 S 00

ZODIAC SIGN ENTRIES

Date	h	m	Planets
01	14	31	☿ ♑
01	16	00	☽ ♌
04	03	50	☽ ♍
04	18	51	♀ ♏
06	16	35	☽ ♎
09	03	35	☽ ♏
11	11	11	☽ ♐
13	15	31	☽ ♑
15	17	56	☽ ♒
17	19	58	☽ ♓
19	22	47	☽ ♈
22	02	50	☽ ♉
22	03	27	☉ ♑
23	06	17	☿ ♐
24	08	15	☽ ♊
26	15	15	☽ ♋
29	00	23	☽ ♌
29	20	24	♀ ♐
31	11	53	☽ ♍

LATITUDES

Date	Mercury ☿ ° '	Venus ♀ ° '	Mars ♂ ° '	Jupiter ♃ ° '	Saturn ♄ ° '	Uranus ♅ ° '	Neptune ♆ ° '	Pluto ⯓ ° '
01	02 S 24	02 N 10	00 S 15	01 S 20	01 S 41	00 S 19	01 S 15	02 S 45
04	02 13	02 13	00 17	01 19	01 41	00 19	01 15	02 45
07	01 54	02 15	00 18	01 18	01 40	00 19	01 15	02 45
10	01 25	02 16	00 20	01 18	01 40	00 19	01 15	02 45
13	00 S 44	02 16	00 22	01 17	01 40	00 19	01 15	02 45
16	00 N 08	02 15	00 24	01 16	01 39	00 19	01 15	02 45
19	01 08	02 13	00 26	01 15	01 39	00 19	01 15	02 45
22	02 04	02 10	00 27	01 14	01 39	00 19	01 15	02 46
25	02 45	02 07	00 29	01 13	01 39	00 19	01 14	02 46
28	03 05	02 03	00 31	01 12	01 38	00 18	01 14	02 46
31	03 N 05	01 N 58	00 S 33	01 S 11	01 S 38	00 S 18	01 S 14	02 S 46

DATA

Julian Date	2460280
Delta T	+72 seconds
Ayanamsa	24° 11' 21"
Synetic vernal point	04° ♓ 55' 38"
True obliquity of ecliptic	23° 26' 18"

MOON'S PHASES, APSIDES AND POSITIONS ☽

Date	h	m	Phase	Longitude ° '	Eclipse Indicator
05	05	49	☾	12 ♍ 49	
12	23	32	●	20 ♐ 40	
19	18	39	☽	27 ♓ 35	
27	00	33	○	04 ♋ 58	

Day	h	m		
04	18	41	Apogee	
16	18	46	Perigee	
07	00	36	0S	
13	21	52	Max dec	28° S 10'
20	05	17	0N	
26	21	52	Max dec	28° N 09'

LONGITUDES

Date	Chiron ⚷ ° '	Ceres ⚳ ° '	Pallas ⚴ ° '	Juno ⚵ ° '	Vesta ⚶ ° '	Black Moon Lilith ⚸ ° '
01	15 ♈ 45	02 ♐ 41	05 ♏ 12	15 ♍ 44	04 ♋ 33	06 ♍ 35
11	15 ♈ 34	06 ♐ 55	09 ♏ 23	18 ♍ 09	02 ♋ 18	07 ♍ 41
21	15 ♈ 28	11 ♐ 07	13 ♏ 27	20 ♍ 01	29 ♊ 43	08 ♍ 48
31	15 ♈ 28	15 ♐ 14	17 ♏ 23	21 ♍ 16	27 ♊ 07	09 ♍ 55

ASPECTARIAN

h m Aspects

01 Friday
03 42 ☽ ♇ ☉
06 04 ☽ △ ♆
06 33 ☽ ± ♄
08 09 ☽ □ ♀
09 38 ☽ # ☿
13 07 ☽ ☍ ⯓
16 10 ☽ ⚻ ☿
18 17 ☽ ⚻ ♄
20 34 ☽ Q ♅

02 Saturday
02 48 ☽ △ ♂
05 12 ☽ ± ☿
05 44 ☽ □ ♃
07 16 ☽ # ⯓
11 29 ☽ ♇ ♆
11 45 ☽ △ ☉
15 04 ☽ # ☉
15 27 ☿ ⚹ ♄
17 42 ☽ # ♂
23 25 ☽ Q ♀

03 Sunday
00 57 ☽ ♇ ☿
05 27 ☽ ± ♆
08 13 ☽ □ ♅
13 29 ♀ □ ⯓
16 47 ☽ ∥ ♅
17 31 ☽ ⚻ ♆
23 20 ♂ ⚻ ♃

04 Monday
00 57 ☽ ⚻ ⯓
02 11 ☽ ⚹ ♀
06 26 ☽ ☍ ♄
10 12 ☽ △ ☿
13 08 ☽ ± ⯓
17 16 ☽ # ♄
17 30 ☽ ∥ ♃
17 32 ☽ △ ♃
18 52 ☽ □ ♂

05 Tuesday
03 46 ☉ ± ♃
05 49 ☽ □ ☉
07 00 ☽ # ♀
07 26 ☽ ♇ ⯓
11 58 ☽ ∠ ♀
12 59 ♀ ± ♆
20 45 ☽ △ ♅
22 51 ♀ △ ♄
23 45 ☽ ♇ ♃

06 Wednesday
01 02 ☉ ∠ ⯓
06 17 ☽ ☍ ♆
08 11 ☽ ⊥ ♀
09 42 ☽ Q ♂
11 06 ☽ # ♆
12 25 ☿ ♇ ♅
13 20 ♆ St D
13 50 ☽ △ ⯓
17 46 ☽ ± ♃
19 26 ☽ ⚻ ♄
21 34 ☽ ⚺ ♀
21 35 ☽ Q ☉

07 Thursday
02 51 ☽ ♇ ♅
04 00 ☽ □ ☿
05 42 ☽ ⚻ ♃
07 29 ☽ ± ♄
11 16 ☽ ⚹ ♂
13 59 ☽ ∥ ♆
20 40 ☽ ± ♅
21 05 ♀ ⊥ ♂
23 37 ☽ ⚹ ☉

08 Friday
01 22 ☽ ♇ ♄
04 09 ☿ △ ♃
08 24 ☽ ⚻ ♅
17 43 ☽ Q ☿
17 45 ☽ ⚻ ♆
18 31 ☽ ∠ ♂
23 24 ☽ ∥ ♀

09 Saturday
01 05 ☽ □ ⯓
05 17 ☽ ± ♆
05 28 ☽ # ♃
05 41 ☽ ∥ ♄
06 33 ☽ △ ♄
07 17 ☽ ∠ ☉
12 48 ☽ ⊥ ♂
14 24 ☽ ☌ ♀
15 41 ☽ ☍ ♃
17 46 ☽ ⚹ ☿
22 19 ☽ ♇ ♆

10 Sunday
00 46 ☽ ⚺ ♂
01 44 ☽ ⊥ ☉
03 35 ♀ ☍ ♃
04 32 ☽ # ♅
11 03 ☽ Q ⯓
13 48 ☽ ⚺ ☉
17 03 ☽ ☍ ♅
22 35 ☽ ∠ ☿

11 Monday
01 58 ☽ △ ♆
07 59 ☽ ∥ ♂
08 57 ☽ ⚹ ⯓
10 27 ☽ ∥ ☉
11 07 ☽ ∥ ⯓
11 11 ♂ ± ♃
14 13 ☽ □ ♄
15 13 ☽ ⊥ ☿
19 12 ☽ ∥ ☿
19 17 ☿ ⚹ ♀
22 15 ☽ ⚻ ♃

12 Tuesday
01 10 ☉ Q ♄
02 03 ☽ ⚺ ☿
02 35 ☽ ⚺ ♀
06 21 ☉ ⚻ ♅
08 47 ☽ ± ♃
10 05 ☽ ☌ ♂
10 34 ♀ # ♃
11 35 ☽ ∠ ⯓
12 40 ♀ ∥ ♄
14 06 ☽ ⊥ ♀
17 55 ☉ ∥ ⯓
21 59 ☽ Q ♄
22 14 ☽ ⚻ ♅
23 32 ☽ ☌ ☉

13 Wednesday
00 21 ☽ ♇ ♃
03 13 ☽ ⊥ ⯓
03 34 ♀ ♇ ♆
06 48 ☽ □ ♆
07 07 ☽ ∠ ♀
07 08 ☿ St R
08 32 ☽ ± ♅
10 04 ☉ ♇ ♃
13 31 ☽ ⚺ ⯓
15 16 ♂ ∠ ⯓
18 39 ☽ ⚹ ♄
23 49 ☽ ♇ ♅

14 Thursday
01 51 ☽ △ ♃
05 47 ☽ ☌ ☿
10 55 ☽ ⚹ ♀
13 21 ☽ Q ♆
16 09 ☽ ⚺ ♂
20 04 ☽ ∠ ♄

15 Friday
01 00 ☽ △ ♅
02 45 ☽ ⊥ ♂
03 09 ☉ ⊥ ⯓
06 15 ☽ ⚺ ☉
08 51 ☽ Q ♀
09 27 ☽ ⚹ ♆
11 11 ☽ ⊥ ♄
16 04 ☽ ☌ ⯓
17 01 ☽ ⊥ ☉
18 34 ☽ ∠ ♂
18 57 ♃ # ♄
19 11 ☉ ∥ ☿
21 13 ☽ ⚺ ♄

16 Saturday
03 52 ☽ □ ♃
05 24 ☽ ∥ ☉
06 17 ☽ ∥ ☿
06 42 ☽ ⚺ ☿
06 44 ☽ ∥ ⯓
06 46 ☽ ∥ ♂
08 20 ♂ ∥ ⯓
09 08 ☽ ∠ ☉
10 27 ☽ ∠ ♆
11 44 ☿ ∥ ♂
12 52 ☿ ∥ ⯓
16 17 ☽ ⊥ ☿
17 33 ☽ □ ♀
20 52 ☽ ⚹ ♂

17 Sunday
01 27 ☽ ⊥ ♆
02 53 ☽ □ ♅
03 43 ☉ □ ♆
06 36 ☽ ∠ ☿
09 46 ☽ Q ♃
10 55 ☽ # ♅
11 28 ☽ ⚺ ♆
12 04 ☽ ⚹ ☉
18 00 ☽ Q ♂
18 11 ☽ ⚺ ⯓
23 27 ☽ ∥ ♀
23 28 ☉ ± ♅
23 32 ☽ ☌ ♄

18 Monday
04 16 ☽ ⊥ ⯓
05 49 ☽ ⚹ ♃
06 21 ☽ ⚹ ☿
07 48 ☽ # ♃
07 57 ☽ ∥ ♄
08 59 ☽ Q ♅
09 44 ☽ Q ☉
14 28 ☿ △ ♃
19 29 ☽ ∠ ⯓

19 Tuesday
00 30 ☽ △ ♀
01 17 ☽ Q ☿
01 25 ♀ Q ⯓
01 58 ☽ □ ♂
05 14 ☽ ⚹ ♅
07 07 ☽ ∠ ♃
14 07 ☽ ☌ ♆
16 06 ☿ ♇ ♅
17 51 ☽ ∥ ♆
18 39 ☽ □ ☉
21 03 ☽ ⚹ ⯓
22 26 ☽ ⊥ ♃

20 Wednesday
02 41 ☽ ⚺ ♄
04 29 ☽ ♇ ♀
05 42 ☽ □ ☿
06 50 ☽ ∠ ♅
08 42 ☽ ⚺ ♃
12 04 ☿ ∠ ♀
13 04 ☽ ⊥ ♄
16 45 ☽ # ♆
17 45 ☽ Q ⯓
18 38 ♀ ⚺ ♂
21 16 ☉ ∥ ♂
21 29 ☽ ± ♀
22 20 ☽ ⊥ ♅

21 Thursday
04 39 ☉ ⚺ ⯓
04 44 ☽ ∠ ♄
07 04 ♀ ☍ ♅
08 23 ☽ △ ♂
08 44 ☽ ⚺ ♅
08 54 ☽ ⚻ ♀
12 33 ☿ ⚹ ♄
14 44 ♂ ⚻ ♅
17 59 ☽ ⚺ ♆

22 Friday
01 11 ☽ □ ⯓
02 47 ☽ △ ☉
03 05 ☽ # ♄
03 34 ☽ ∥ ♃
04 31 ☽ ⊥ ♆
05 21 ☽ △ ☿
07 08 ☽ ⚹ ♄
12 11 ☽ ♇ ♂
12 53 ☽ ☌ ♃
17 44 ♂ Q ♄
18 54 ☉ ☌ ☿
19 32 ☽ # ♀
20 25 ☽ ∠ ♆

23 Saturday
00 54 ♂ ♇ ♃
01 24 ☽ ∥ ♅
04 29 ☽ Q ♄
05 03 ☽ ± ♂
05 27 ☽ ♇ ☿
07 29 ☽ ♇ ☉
13 33 ☽ ☌ ♅
16 22 ☽ ⚻ ♂
19 04 ☽ ☍ ♀
19 47 ☽ # ☿
20 02 ☽ ± ☿
21 51 ☿ ⚺ ⯓
23 12 ☽ ⚹ ♆

24 Sunday
00 40 ☿ ⊥ ♀
00 55 ☽ ± ☉
05 50 ☽ ⚻ ☿
06 40 ☽ △ ⯓
07 29 ☽ # ⯓
09 57 ☽ # ☉
11 32 ☽ # ♂
12 38 ☽ ⚻ ☉
12 58 ☽ □ ♄
17 28 ☉ ⚹ ♄
18 28 ☽ ⚺ ♃
20 54 ☽ Q ♆

25 Monday
05 22 ☽ ⊥ ♃
09 59 ☽ ♇ ⯓
17 15 ♀ △ ♆
19 51 ☽ ⚺ ♅
21 52 ☽ ∠ ♃

26 Tuesday
02 08 ☽ ☍ ♂
02 35 ☽ ± ⯓
05 57 ☽ □ ♆
06 55 ☽ ⊥ ♅
07 15 ☽ ⚻ ♀
07 55 ☽ ☍ ☿
10 13 ♂ ⊥ ⯓
11 18 ☿ ⚺ ♀
13 45 ☽ ⚻ ⯓
13 49 ☉ ♇ ♅
19 36 ☽ ± ♀
20 07 ☿ ± ♅
20 23 ♀ # ♅
20 30 ☽ △ ♄
23 41 ☽ ∠ ♅

27 Wednesday
00 33 ☽ ☍ ☉
01 45 ☽ ⚹ ♃
07 43 ☿ □ ♆
14 20 ☽ ♇ ♀
15 28 ☉ △ ♃

28 Thursday
00 28 ☽ Q ♃
00 31 ☿ ☌ ♂
01 03 ☽ ♇ ♄
04 04 ☽ ⚹ ♅
12 40 ☽ ⚻ ☿
14 17 ☽ ⚻ ♂
14 45 ☽ △ ♆
18 30 ☽ ± ♄
22 12 ☽ △ ♀
22 16 ♂ □ ♆
22 57 ☽ ☍ ⯓
23 37 ☽ ± ☿

29 Friday
02 40 ☽ ± ♂
03 12 ☽ Q ♅
06 01 ♀ ⚹ ⯓
06 15 ☽ ⚻ ♄
09 06 ☽ # ♂
11 17 ☽ □ ♃
11 29 ♂ ± ♅
11 30 ☿ ⊥ ⯓
13 51 ☽ # ☉
15 18 ☽ ⚻ ☉
15 22 ☽ # ⯓
16 21 ☽ ♇ ☿
20 04 ☽ ♇ ♆
21 27 ☽ ♇ ♂

30 Saturday
04 14 ☽ ± ☉
09 22 ☽ # ☿
13 56 ☽ ± ♆
14 39 ☽ □ ♅
18 48 ☽ # ♀
21 00 ☽ △ ☿
23 51 ☽ ♇ ☉

31 Sunday
00 58 ☽ ∥ ♅
01 57 ☽ ⚻ ♆
02 40 ♃ St D
05 18 ☽ △ ♂
10 33 ☽ ⚻ ⯓
16 23 ☽ □ ♀
18 24 ☽ ☍ ♄
22 43 ☽ ± ⯓
23 10 ☽ △ ♃

All ephemeris data is given at 12.00 UT and the Moon's longitude is additionally given for 24.00 UT

JANUARY 2024

LONGITUDES

Date	Sidereal time h m s	Sun ☉ ° ' "	Moon ☽ ° ' "	Moon ☽ 24.00 ° ' "	Mercury ☿ ° '	Venus ♀ ° '	Mars ♂ ° '	Jupiter ♃ ° '	Saturn ♄ ° '	Uranus ♅ ° '	Neptune ♆ ° '	Pluto ♇ ° '
01	18 42 35	10 ♑ 32 55	11 ♍ 54 27	17 ♍ 48 35	22 ♐ 13	03 ♐ 13	27 ♐ 41	05 ♉ 35	03 ♓ 17	19 ♉ 22	25 ♓ 05	29 ♑ 22
02	18 46 31	11 34 03	23 ♍ 42 32	29 ♍ 36 57	22 D 11	04 26	28 25	05 D 35	03 23	19 R 21	25 06	29 24
03	18 50 28	12 35 12	05 ♎ 32 30	11 ♎ 29 51	22 19	05 39	29 10	05 36	03 28	19 20	25 07	29 26
04	18 54 25	13 36 22	17 ♎ 29 43	23 ♎ 32 47	22 35	06 52	29 ♐ 54	05 37	03 34	19 19	25 08	29 28
05	18 58 21	14 37 31	29 ♎ 39 47	05 ♏ 51 21	22 59	08 06	00 ♑ 39	05 38	03 39	19 18	25 09	29 30
06	19 02 18	15 38 41	12 ♏ 08 08	18 ♏ 30 40	23 29	09 19	01 24	05 39	03 45	19 16	25 10	29 32
07	19 06 14	16 39 51	24 ♏ 59 28	01 ♐ 34 51	24 05	10 32	02 09	05 41	03 51	19 15	25 11	29 34
08	19 10 11	17 41 01	08 ♐ 17 05	15 ♐ 06 15	24 47	11 45	02 53	05 42	03 56	19 14	25 12	29 36
09	19 14 07	18 42 11	22 ♐ 02 15	29 ♐ 04 47	25 34	12 59	03 38	05 44	04 02	19 14	25 13	29 38
10	19 18 04	19 43 22	06 ♑ 13 24	13 ♑ 27 26	26 25	14 12	04 23	05 46	04 08	19 13	25 14	29 40
11	19 22 00	20 44 32	20 ♑ 46 02	28 ♑ 08 14	27 20	15 26	05 08	05 48	04 14	19 12	25 16	29 41
12	19 25 57	21 45 41	05 ♒ 32 57	12 ♒ 59 03	28 19	16 39	05 53	05 51	04 20	19 11	25 17	29 43
13	19 29 54	22 46 51	20 ♒ 25 25	27 ♒ 50 58	29 ♐ 20	17 53	06 38	05 53	04 26	19 10	25 18	29 45
14	19 33 50	23 47 59	05 ♓ 14 43	12 ♓ 35 48	00 ♑ 25	19 06	07 23	05 56	04 32	19 10	25 19	29 47
15	19 37 47	24 49 07	19 ♓ 53 31	27 ♓ 07 21	01 32	20 20	08 08	05 59	04 38	19 09	25 21	29 49
16	19 41 43	25 50 15	04 ♈ 16 54	11 ♈ 21 58	02 41	21 34	08 53	06 03	04 44	19 08	25 22	29 51
17	19 45 40	26 51 21	18 ♈ 22 27	25 ♈ 18 23	03 53	22 47	09 38	06 06	04 51	19 08	25 23	29 53
18	19 49 36	27 52 27	02 ♉ 09 53	08 ♉ 57 07	05 06	24 01	10 24	06 10	04 57	19 07	25 25	29 55
19	19 53 33	28 53 32	15 ♉ 40 18	22 ♉ 19 41	06 21	25 15	11 09	06 13	05 03	19 07	25 26	29 57
20	19 57 29	29 ♑ 54 36	28 ♉ 55 30	05 ♊ 27 59	07 37	26 28	11 54	06 17	05 10	19 07	25 28	29 ♑ 59
21	20 01 26	00 ♒ 55 40	11 ♊ 57 20	18 ♊ 23 43	08 55	27 42	12 40	06 22	05 16	19 06	25 29	00 ♒ 01
22	20 05 23	01 56 42	24 ♊ 47 17	01 ♋ 08 10	10 14	28 ♐ 56	13 25	06 26	05 23	19 06	25 31	00 03
23	20 09 19	02 57 44	07 ♋ 26 26	13 ♋ 42 09	11 35	00 ♑ 10	14 10	06 31	05 29	19 06	25 32	00 05
24	20 13 16	03 58 44	19 ♋ 55 22	26 ♋ 06 08	12 56	01 24	14 56	06 35	05 36	19 06	25 34	00 07
25	20 17 12	04 59 44	02 ♌ 14 30	08 ♌ 20 31	14 19	02 37	15 41	06 40	05 42	19 05	25 35	00 09
26	20 21 09	06 00 43	14 ♌ 24 17	20 ♌ 25 53	15 43	03 51	16 27	06 46	05 49	19 05	25 37	00 11
27	20 25 05	07 01 42	26 ♌ 25 30	02 ♍ 23 19	17 07	05 05	17 12	06 51	05 56	19 D 05	25 38	00 13
28	20 29 02	08 02 39	08 ♍ 19 34	14 ♍ 14 35	18 33	06 19	17 58	06 56	06 03	19 05	25 40	00 15
29	20 32 58	09 03 36	20 ♍ 08 41	26 ♍ 02 17	19 59	07 33	18 43	07 02	06 09	19 05	25 42	00 16
30	20 36 55	10 04 32	01 ♎ 55 51	07 ♎ 49 53	21 26	08 47	19 29	07 08	06 16	19 06	25 43	00 18
31	20 40 52	11 ♒ 05 27	13 ♎ 44 57	19 ♎ 41 38	22 ♑ 54	10 ♑ 01	20 ♑ 15	07 ♉ 14	06 ♓ 23	19 ♉ 06	25 ♓ 45	00 ♒ 20

Date	Moon True ☊ ° '	Moon Mean ☊ ° '	Moon ☽ Latitude ° '
01	21 ♈ 02	20 ♈ 51	03 N 11
02	20 R 59	20 48	02 19
03	20 57	20 45	01 21
04	20 D 57	20 42	00 N 18
05	20 R 57	20 38	00 S 46
06	20 56	20 35	01 50
07	20 52	20 32	02 50
08	20 46	20 29	03 42
09	20 37	20 26	04 24
10	20 25	20 23	04 51
11	20 13	20 19	05 00
12	20 01	20 16	04 49
13	19 50	20 13	04 19
14	19 42	20 10	03 31
15	19 37	20 07	02 30
16	19 35	20 03	01 20
17	19 D 34	20 00	00 S 06
18	19 R 34	19 57	01 N 06
19	19 33	19 54	02 13
20	19 30	19 51	03 12
21	19 25	19 48	03 59
22	19 16	19 44	04 34
23	19 05	19 41	04 54
24	18 51	19 38	05 00
25	18 37	19 35	04 52
26	18 24	19 32	04 31
27	18 12	19 29	03 57
28	18 03	19 25	03 14
29	17 56	19 22	02 22
30	17 53	19 19	01 24
31	17 ♈ 51	19 ♈ 16	00 N 22

DECLINATIONS

Date	Sun ☉ ° '	Moon ☽ ° '	Mercury ☿ ° '	Venus ♀ ° '	Mars ♂ ° '	Jupiter ♃ ° '	Saturn ♄ ° '	Uranus ♅ ° '	Neptune ♆ ° '	Pluto ♇ ° '
01	23 S 01	10 N 02	20 S 11	18 S 54	23 S 58	12 N 16	11 S 49	17 N 17	03 S 05	22 S 59
02	22 56	04 N 37	20 15	19 09	24 00	12 16	11 47	17 16	03 05	22 59
03	22 51	00 S 58	20 21	19 24	24 01	12 17	11 45	17 16	03 05	22 59
04	22 45	06 35	20 29	19 39	24 01	12 17	11 43	17 16	03 04	22 58
05	22 38	12 04	20 37	19 52	24 02	12 18	11 41	17 15	03 04	22 58
06	22 31	17 13	20 47	20 06	24 02	12 19	11 39	17 15	03 03	22 58
07	22 24	21 45	20 56	20 19	24 02	12 20	11 37	17 15	03 03	22 57
08	22 16	25 21	21 07	20 31	24 02	12 20	11 35	17 15	03 02	22 57
09	22 08	27 35	21 17	20 43	24 01	12 21	11 33	17 14	03 02	22 57
10	21 59	28 08	21 28	20 54	24 00	12 22	11 30	17 14	03 01	22 56
11	21 50	26 46	21 39	21 04	23 59	12 23	11 28	17 14	03 01	22 56
12	21 41	23 33	21 49	21 14	23 58	12 24	11 26	17 14	03 00	22 55
13	21 31	18 46	21 59	21 24	23 57	12 26	11 24	17 14	03 00	22 55
14	21 21	12 52	22 08	21 32	23 55	12 27	11 21	17 13	02 59	22 55
15	21 10	06 S 18	22 17	21 41	23 53	12 28	11 19	17 13	02 59	22 54
16	20 59	00 N 29	22 25	21 48	23 50	12 29	11 17	17 13	02 58	22 54
17	20 47	07 06	22 32	21 55	23 48	12 31	11 14	17 13	02 58	22 54
18	20 35	13 16	22 39	22 01	23 45	12 32	11 12	17 13	02 57	22 53
19	20 23	18 39	22 45	22 07	23 42	12 34	11 10	17 13	02 56	22 53
20	20 10	23 02	22 49	22 12	23 38	12 35	11 07	17 13	02 56	22 53
21	19 57	26 11	22 53	22 16	23 35	12 37	11 05	17 13	02 55	22 52
22	19 44	27 54	22 56	22 20	23 31	12 39	11 02	17 13	02 55	22 52
23	19 30	28 08	22 57	22 23	23 27	12 41	11 00	17 13	02 54	22 52
24	19 16	26 55	22 58	22 26	23 22	12 42	10 58	17 13	02 53	22 51
25	19 01	24 24	22 57	22 27	23 18	12 44	10 55	17 13	02 53	22 51
26	18 46	20 49	22 55	22 29	23 13	12 46	10 53	17 13	02 52	22 51
27	18 31	16 25	22 53	22 29	23 07	12 48	10 50	17 13	02 51	22 50
28	18 15	11 26	22 48	22 29	23 02	12 50	10 48	17 13	02 51	22 50
29	17 59	06 05	22 43	22 28	22 56	12 52	10 45	17 13	02 50	22 50
30	17 43	00 31	22 36	22 26	22 50	12 55	10 43	17 13	02 49	22 49
31	17 S 27	05 S 05	22 S 28	22 S 24	22 S 44	12 N 57	10 S 40	17 N 13	02 S 49	22 S 49

ZODIAC SIGN ENTRIES

Date	h	m	Planets
03	00	47	☽ ♎
04	14	58	♂ ♑
05	12	39	☽ ♏
07	21	08	☽ ♐
10	01	33	☽ ♑
12	03	01	☽ ♒
14	02	49	☿ ♑
14	03	29	☽ ♓
16	04	49	☽ ♈
18	08	12	☽ ♉
20	13	58	☽ ♊
20	14	07	☉ ♒
21	00	50	♇ ♒
22	21	51	☽ ♋
23	08	50	♀ ♑
25	07	37	☽ ♌
27	19	11	☽ ♍
30	08	04	☽ ♎

LATITUDES

Date	Mercury ☿ ° '	Venus ♀ ° '	Mars ♂ ° '	Jupiter ♃ ° '	Saturn ♄ ° '	Uranus ♅ ° '	Neptune ♆ ° '	Pluto ♇ ° '
01	03 N 02	01 N 56	00 S 33	01 S 11	01 S 38	00 S 18	01 S 14	02 S 46
04	02 45	01 50	00 35	01 10	01 38	00 18	01 14	02 46
07	02 22	01 44	00 37	01 09	01 38	00 18	01 14	02 46
10	01 55	01 37	00 39	01 08	01 37	00 18	01 14	02 47
13	01 27	01 30	00 40	01 07	01 37	00 18	01 14	02 47
16	01 00	01 22	00 42	01 06	01 37	00 18	01 14	02 47
19	00 33	01 14	00 44	01 06	01 37	00 18	01 14	02 47
22	00 N 07	01 06	00 45	01 05	01 37	00 18	01 14	02 48
25	00 S 17	00 57	00 47	01 04	01 37	00 18	01 14	02 48
28	00 39	00 49	00 48	01 03	01 37	00 18	01 13	02 48
31	01 S 00	00 N 40	00 S 50	01 S 02	01 S 37	00 S 18	01 S 13	02 S 48

DATA

Julian Date	2460311
Delta T	+72 seconds
Ayanamsa	24° 11' 27"
Synetic vernal point	04° ♓ 55' 32"
True obliquity of ecliptic	23° 26' 18"

MOON'S PHASES, APSIDES AND POSITIONS ☽

Date	h	m	Phase	Longitude ° '	Eclipse Indicator
04	03	30	☾	13 ♎ 15	
11	11	57	●	20 ♑ 44	
18	03	53	☽	27 ♈ 32	
25	17	54	○	05 ♌ 15	

Day	h	m		
01	15	23	Apogee	
13	10	29	Perigee	
29	08	03	Apogee	
03	07	53	OS	
10	07	03	Max dec	28° S 11'
16	10	18	ON	
23	03	41	Max dec	28° N 13'
30	14	12	OS	

LONGITUDES

Date	Chiron ⚷ ° '	Ceres ⚳ ° '	Pallas ⚴ ° '	Juno ⚵ ° '	Vesta ⚶ ° '	Black Moon Lilith ⚸ ° '
01	15 ♈ 28	15 ♐ 39	17 ♏ 46	21 ♍ 22	26 ♊ 52	10 ♍ 02
11	15 ♈ 34	19 ♐ 41	21 ♏ 29	21 ♍ 52	24 ♊ 37	11 ♍ 09
21	15 ♈ 45	23 ♐ 37	25 ♏ 00	21 ♍ 37	22 ♊ 57	12 ♍ 15
31	16 ♈ 01	27 ♐ 26	28 ♏ 15	20 ♍ 37	21 ♊ 59	13 ♍ 22

ASPECTARIAN

h m	Aspects
01 Monday	
01 42	☽ ∥ ♃
03 43	☽ # ♄
08 59	☽ △ ☉
13 26	♀ □ ♄
17 01	☽ ⊡ ♇
21 57	☉ ∥ ♇
02 Tuesday	
03 09	☿ St D
03 09	☽ △ ♅
05 39	☽ ⊡ ♃
08 54	☽ □ ☿
09 07	☽ Q ♀
14 50	☽ ☍ ♆
18 40	☽ # ♆
22 13	☽ □ ♂
23 36	☽ △ ♇
23 58	☽ ± ♃
03 Wednesday	
07 46	☽ ⊼ ♄
09 33	☽ ⊡ ♅
10 55	♀ ⊼ ♃
12 07	☽ ⊼ ♃
12 15	☽ ✱ ♀
19 59	☽ ± ♄
20 59	☽ ∥ ♆
21 09	♂ ⊻ ♇
21 49	☽ Q ☿
04 Thursday	
00 37	☉ Q ♆
03 30	☽ □ ☉
03 39	☽ ± ♅
12 53	☽ Q ♂
14 08	☽ ⊡ ♄
15 37	☽ ⊼ ♅
21 40	☽ ∠ ♀
22 25	☽ ✱ ☿
05 Friday	
03 08	☽ ⊼ ♆
10 18	☽ ∥ ♄
11 41	☽ □ ♇
13 03	☽ # ♃
14 03	☽ ✱ ♂
14 54	☽ ± ♆
17 15	☽ ⊥ ♀
18 17	☽ Q ☉
19 48	☽ △ ♄
23 35	☽ ☍ ♃
06 Saturday	
04 43	☽ ∠ ☿
06 03	☽ ⊻ ♀
08 15	☽ ⊡ ♆
12 11	☽ # ♅
14 14	☿ ⊥ ♇
19 12	☽ ✱ ☉
20 32	☽ ∠ ♂
22 11	☽ Q ♇
22 33	☽ ⊥ ☿
07 Sunday	
01 24	☽ ☍ ♅
03 30	☽ ∥ ♀
07 11	☽ ∥ ☿
10 15	☽ ⊻ ☿
12 21	☽ △ ♆
14 14	☽ ⊥ ♂
15 42	☽ ∥ ☉
19 13	☽ ∥ ♇
20 22	☽ ✱ ♇
08 Monday	
01 10	☽ ∠ ☉
01 47	☽ ⊻ ♂
02 18	☽ ∥ ♂
03 10	☉ ⊥ ♀
04 11	☽ □ ♄
07 24	☽ ⊼ ♃
10 53	♀ ± ♃
18 03	☽ ± ♃
18 29	☽ ⊥ ☉
18 44	☽ ☌ ♀
23 08	☽ ∠ ♇
09 Tuesday	
01 24	☿ □ ♆
02 03	☿ ± ♅
05 48	☽ ⊻ ☉
07 10	☽ ⊼ ♅
09 45	☽ ⊡ ♃
12 00	☽ Q ♄
14 44	☽ ⊥ ♇
17 27	☽ ± ♅
17 27	☽ □ ♆
18 25	☽ ☌ ☿
20 38	☉ ∠ ♄
10 Wednesday	
00 07	☉ △ ♅
00 57	☽ ⊻ ♇
02 40	♂ ✱ ♄
06 29	♂ ⊡ ♅
08 29	☽ ✱ ♄
08 38	☽ ⊡ ♅
08 46	☽ ☌ ♂
11 14	☽ △ ♃
21 08	♀ ∠ ♇
23 39	☽ Q ♆
11 Thursday	
02 27	☽ ⊻ ♀
09 26	☽ △ ♅
09 28	☽ ∠ ♄
11 57	☽ ☌ ☉
13 11	☽ ⊥ ♀
19 20	☽ ✱ ♆
23 27	☽ ⊻ ☿
12 Friday	
00 14	☽ ⊥ ♄
02 14	☉ ∥ ☿
02 33	☽ ☌ ♇
05 08	☽ ∠ ♀
09 28	☽ ∥ ♂
09 51	☽ ⊥ ☿
10 01	☽ ⊻ ♄
10 42	♂ △ ♃
12 29	☽ □ ♃
12 34	☽ ⊻ ♂
15 36	☽ ∥ ♇
19 39	☽ ∠ ♆
21 13	☽ ∥ ☿
22 36	☽ ∥ ☉
22 46	☽ ⊥ ♂
13 Saturday	
00 04	☽ ∥ ♀
01 27	☽ ∠ ☿
07 32	☽ ✱ ♀
09 59	☽ □ ♅
10 11	☽ ⊥ ♆
14 03	☽ ∠ ♂
16 05	☽ ⊻ ☉
17 37	☽ Q ♃
18 38	☽ # ♅
19 53	☽ ⊻ ♆
21 08	☉ ∥ ♀
21 39	☿ ⊻ ♇
14 Sunday	
02 30	☽ ⊥ ☉
03 07	☽ ⊻ ♇
03 32	☽ ✱ ☿
04 40	☽ Q ♀
10 50	☽ ☌ ♄
12 53	☽ ⊥ ♇
13 03	♀ ⊼ ♅
13 08	☽ ✱ ♃
13 34	☽ # ♃
15 07	☽ Q ♅
15 40	☽ ✱ ♂
17 41	☽ ∥ ♄
18 13	☽ ∠ ☉
15 Monday	
00 40	☽ Q ☿
03 38	☽ ∠ ♇
10 46	☽ ✱ ♅
12 26	☽ Q ♂
12 48	☽ □ ♀
13 49	☽ ∠ ♃
20 47	☽ ✱ ☉
21 03	☽ ☌ ♆
23 47	☽ ∥ ♆
16 Tuesday	
00 38	☉ ✱ ♆
01 25	♀ ⊡ ♃
04 33	☽ ✱ ♇
04 51	☽ ⊥ ♃
09 05	☽ □ ☿
11 46	☽ ∠ ♅
12 47	☽ ⊻ ♄
14 59	☽ ⊻ ♃
18 28	☽ Q ☉
20 14	☽ □ ♂
20 53	☽ # ♆
23 01	☽ ⊥ ♄
17 Wednesday	
00 52	☽ Q ♇
03 01	☽ ⊥ ♅
13 14	♀ Q ♄
13 18	☽ ⊻ ♅
14 33	☽ ∠ ♄
16 57	☿ ⊡ ♅
20 22	☽ △ ♀
18 Thursday	
00 10	☽ ⊻ ♆
03 45	☽ # ♄
03 53	☽ □ ☉
08 03	☽ □ ♇
08 49	☿ ✱ ♄
09 02	☽ ∥ ♃
10 02	♀ ⊥ ♇
10 40	☽ ⊥ ♆
16 57	☽ ✱ ♄
17 42	☽ △ ☿
19 05	☽ ☌ ♃
19 Friday	
01 20	☽ ⊡ ♀
02 37	☽ ∠ ♆
03 26	☽ △ ♂
05 10	☽ ∥ ♅
09 30	♀ ± ♅
09 31	☿ △ ♃
14 30	☽ Q ♄
15 49	♀ □ ♆
16 19	☉ ⊥ ♄
18 11	☽ ☌ ♅
19 05	☽ ± ♀
20 21	☽ # ☉
23 18	☽ ⊡ ☿
20 Saturday	
05 40	☽ ✱ ♆
06 46	☽ # ♀
07 04	☽ ⊼ ♀
08 05	☽ ⊡ ♂
10 37	☽ # ☿
11 00	☽ # ♇
13 46	☉ ☌ ♇
13 56	☽ △ ♇
13 57	☽ △ ☉
15 50	☽ # ♂
17 29	☽ ± ☿
23 32	☽ □ ♄
21 Sunday	
01 36	☽ ± ♂
01 36	☽ ⊻ ♃
03 42	☽ Q ♆
05 45	☽ ⊼ ☿
08 04	☿ ∥ ♇
12 46	☽ ⊥ ♃
13 23	☽ ⊼ ♂
17 42	☽ ⊡ ♇
20 02	☽ ⊡ ☉
22 Monday	
01 19	☽ ⊻ ♅
05 39	☽ ∠ ♃
10 36	☽ ± ♇
12 35	☽ ⊥ ♅
13 22	☽ □ ♆
14 22	☽ ± ☉
15 07	♂ Q ♆
20 40	☽ ☍ ♀
21 58	☽ ⊼ ♇
23 Tuesday	
02 43	☽ ⊼ ☉
05 37	☽ ∠ ♅
08 14	☽ △ ♄
10 13	☽ ✱ ♃
10 21	♀ ⊻ ♇
20 53	☽ ☍ ☿
24 Wednesday	
01 44	☽ ☍ ♂
09 24	☽ Q ♃
10 24	☽ ✱ ♅
13 19	☽ ⊡ ♄
22 58	☽ △ ♆
23 03	☿ Q ♆
25 Thursday	
07 00	☽ ± ♄
07 53	☽ ☍ ♇
09 45	☽ Q ♅
12 50	☽ ⊼ ♀
17 54	☽ ☍ ☉
18 52	☽ ⊼ ♄
20 21	☽ # ♂
20 46	☽ □ ♃
22 33	☽ # ☿
23 12	☽ # ♇
26 Friday	
01 43	☽ # ♀
01 58	☽ ± ♀
04 28	☽ ⊡ ♆
06 52	☉ ⊻ ♄
14 56	☽ ⊼ ☿
16 19	☽ ⊼ ♂
16 32	♀ ⊡ ♅
21 19	☽ □ ♅
21 52	☽ ⊡ ♀
22 23	☽ ± ♆
27 Saturday	
00 23	☽ # ☉
04 29	☽ ± ☿
05 06	☽ ± ♂
07 18	☉ □ ♃
07 36	♅ St D
07 58	☽ ∥ ♅
10 25	☽ ⊼ ♆
14 59	☿ ☌ ♂
19 38	☽ ⊼ ♇
28 Sunday	
00 25	☽ ⊡ ♂
01 01	☽ ⊡ ☿
02 21	☿ ∥ ♇
05 31	☽ ∥ ♃
06 03	♀ ✱ ♄
07 20	☽ ☍ ♄
07 28	☽ △ ♀
07 46	☽ ± ♇
09 10	☽ △ ♃
11 23	☽ ⊼ ☉
14 59	☽ # ♄
21 07	☿ △ ♅
29 Monday	
00 41	☽ ± ☉
01 02	♀ △ ♃
02 04	☽ ⊡ ♇
08 54	☽ △ ♂
09 51	☽ △ ♅
11 38	☽ △ ☿
15 52	☽ ⊡ ♃
20 43	☽ ⊡ ☉
23 20	☽ ☍ ♆
23 41	♂ △ ♅
30 Tuesday	
02 05	☽ # ♆
08 41	☽ △ ♇
08 59	☿ ∠ ♄
10 21	☽ ± ♃
16 14	♂ ∥ ♇
16 24	☽ ⊡ ♅
20 55	☽ ⊼ ♄
22 40	☽ ⊼ ♃
23 06	♄ ⊥ ♇
31 Wednesday	
02 14	☽ ∥ ♆
03 34	☽ □ ♀
03 48	☉ ∠ ♆
06 07	☽ △ ☉
09 12	☽ ± ♄
10 41	☽ ± ♅
22 48	☽ ⊼ ♅

All ephemeris data is given at 12.00 UT and the Moon's longitude is additionally given for 24.00 UT

FEBRUARY 2024

Date	Sidereal time h m s	Sun ☉ ° ' "	Moon ☽ ° ' "	Moon ☽ 24.00 ° ' "	Mercury ☿ ° '	Venus ♀ ° '	Mars ♂ ° '	Jupiter ♃ ° '	Saturn ♄ ° '	Uranus ♅ ° '	Neptune ♆ ° '	Pluto ♇ ° '
01	20 44 48	12 ♒ 06 21	25 ♎ 40 35	01 ♏ 42 27	24 ♑ 23	11 ♑ 15	21 ♑ 00	07 ♉ 20	06 ♓ 30	19 ♉ 06	25 ♓ 47	00 ♒ 22
02	20 48 45	13 07 15	07 ♏ 47 55	13 ♏ 57 39	25 53	12 29	21 46	07 26	06 37	19 06	25 49	00 24
03	20 52 41	14 08 08	20 ♏ 12 19	26 ♏ 32 35	27 23	13 43	22 32	07 33	06 44	19 07	25 51	00 26
04	20 56 38	15 09 00	02 ♐ 59 01	09 ♐ 32 07	28 ♑ 54	14 57	23 18	07 39	06 51	19 07	25 52	00 28
05	21 00 34	16 09 51	16 ♐ 12 20	22 ♐ 59 56	00 ♒ 26	16 11	24 03	07 46	06 58	19 08	25 54	00 30
06	21 04 31	17 10 42	29 ♐ 55 01	06 ♑ 57 34	01 59	17 26	24 49	07 53	07 05	19 08	25 56	00 32
07	21 08 27	18 11 32	14 ♑ 07 17	21 ♑ 23 43	03 32	18 40	25 35	08 00	07 12	19 09	25 58	00 34
08	21 12 24	19 12 20	28 ♑ 46 08	06 ♒ 13 39	05 07	19 54	26 21	08 08	07 19	19 09	26 00	00 36
09	21 16 21	20 13 07	13 ♒ 45 09	21 ♒ 19 25	06 42	21 08	27 07	08 15	07 26	19 10	26 02	00 37
10	21 20 17	21 13 54	28 ♒ 55 09	06 ♓ 31 00	08 18	22 22	27 53	08 23	07 33	19 11	26 04	00 39
11	21 24 14	22 14 38	14 ♓ 05 41	21 ♓ 37 58	09 54	23 36	28 39	08 31	07 40	19 11	26 06	00 41
12	21 28 10	23 15 21	29 ♓ 06 49	06 ♈ 31 18	11 32	24 51	29 ♑ 25	08 38	07 47	19 12	26 08	00 43
13	21 32 07	24 16 03	13 ♈ 50 45	21 ♈ 04 38	13 10	26 05	00 ♒ 11	08 47	07 55	19 13	26 10	00 45
14	21 36 03	25 16 43	28 ♈ 12 38	05 ♉ 14 36	14 49	27 19	00 57	08 55	08 02	19 14	26 12	00 47
15	21 40 00	26 17 21	12 ♉ 10 32	19 ♉ 00 33	16 29	28 33	01 44	09 03	08 09	19 15	26 14	00 48
16	21 43 56	27 17 58	25 ♉ 44 51	02 ♊ 23 46	18 10	29 ♑ 47	02 30	09 12	08 16	19 16	26 16	00 50
17	21 47 53	28 18 33	08 ♊ 57 35	15 ♊ 26 43	19 51	01 ♒ 02	03 16	09 20	08 23	19 17	26 18	00 52
18	21 51 50	29 ♒ 19 06	21 ♊ 51 30	28 ♊ 12 21	21 34	02 16	04 02	09 29	08 31	19 18	26 20	00 54
19	21 55 46	00 ♓ 19 37	04 ♋ 29 35	10 ♋ 43 35	23 17	03 30	04 48	09 38	08 38	19 19	26 22	00 55
20	21 59 43	01 20 07	16 ♋ 54 39	23 ♋ 03 05	25 02	04 44	05 35	09 47	08 45	19 21	26 24	00 57
21	22 03 39	02 20 35	29 ♋ 09 08	05 ♌ 13 04	26 47	05 58	06 21	09 56	08 53	19 22	26 26	00 59
22	22 07 36	03 21 01	11 ♌ 15 05	17 ♌ 15 24	28 ♒ 33	07 13	07 07	10 05	09 00	19 23	26 28	01 01
23	22 11 32	04 21 25	23 ♌ 14 12	29 ♌ 11 41	00 ♓ 20	08 27	07 53	10 15	09 07	19 25	26 30	01 02
24	22 15 29	05 21 47	05 ♍ 08 03	11 ♍ 03 29	02 08	09 41	08 40	10 24	09 15	19 26	26 32	01 04
25	22 19 25	06 22 08	16 ♍ 58 13	22 ♍ 52 30	03 57	10 55	09 26	10 34	09 22	19 28	26 35	01 06
26	22 23 22	07 22 28	28 ♍ 46 34	04 ♎ 40 44	05 47	12 10	10 13	10 44	09 29	19 29	26 37	01 07
27	22 27 19	08 22 45	10 ♎ 35 20	16 ♎ 30 44	07 38	13 24	10 59	10 54	09 36	19 31	26 39	01 09
28	22 31 15	09 23 01	22 ♎ 27 20	28 ♎ 25 35	09 30	14 38	11 45	11 04	09 44	19 32	26 41	01 11
29	22 35 12	10 ♓ 23 16	04 ♏ 25 57	10 ♏ 28 57	11 ♓ 23	15 ♒ 52	12 ♒ 32	11 ♉ 14	09 ♓ 51	19 ♉ 34	26 ♓ 43	01 ♒ 12

Date	Moon True ☊ ° '	Moon Mean ☊ ° '	Moon ☽ Latitude ° '
01	17 ♈ 51	19 ♈ 13	00 S 42
02	17 D 52	19 09	01 44
03	17 R 52	19 06	02 44
04	17 50	19 03	03 37
05	17 46	19 00	04 20
06	17 40	18 57	04 50
07	17 32	18 54	05 04
08	17 22	18 50	04 59
09	17 13	18 47	04 33
10	17 05	18 44	03 48
11	16 59	18 41	02 47
12	16 55	18 38	01 34
13	16 53	18 34	00 S 16
14	16 D 54	18 31	01 N 01
15	16 55	18 28	02 12
16	16 56	18 25	03 14
17	16 R 55	18 22	04 03
18	16 52	18 19	04 39
19	16 48	18 15	05 01
20	16 41	18 12	05 08
21	16 33	18 09	05 01
22	16 24	18 06	04 40
23	16 16	18 03	04 07
24	16 09	18 00	03 23
25	16 03	17 56	02 31
26	16 00	17 53	01 32
27	15 58	17 50	00 N 29
28	15 D 58	17 47	00 S 35
29	15 ♈ 59	17 ♈ 44	01 S 39

DECLINATIONS

Date	Sun ☉ ° '	Moon ☽ ° '	Mercury ☿ ° '	Venus ♀ ° '	Mars ♂ ° '	Jupiter ♃ ° '	Saturn ♄ ° '	Uranus ♅ ° '	Neptune ♆ ° '	Pluto ♇ ° '
01	17 S 10	10 S 34	22 S 19	22 S 21	22 S 38	12 N 59	10 S 37	17 N 13	02 S 48	22 S 49
02	16 53	15 45	22 09	22 17	22 31	13 01	10 35	17 13	02 47	22 48
03	16 35	20 25	21 57	22 13	22 24	13 04	10 32	17 13	02 46	22 48
04	16 17	24 18	21 44	22 08	22 17	13 06	10 30	17 13	02 46	22 48
05	15 59	27 02	21 30	22 03	22 10	13 09	10 27	17 13	02 45	22 47
06	15 41	28 17	21 14	21 56	22 02	13 11	10 24	17 14	02 44	22 47
07	15 23	27 44	20 57	21 50	21 54	13 14	10 22	17 14	02 43	22 47
08	15 04	25 17	20 38	21 42	21 46	13 16	10 19	17 14	02 43	22 47
09	14 45	21 03	20 18	21 34	21 38	13 19	10 17	17 14	02 42	22 46
10	14 25	15 24	19 57	21 25	21 29	13 22	10 14	17 14	02 41	22 46
11	14 06	08 49	19 35	21 15	21 20	13 24	10 11	17 15	02 40	22 46
12	13 46	01 S 47	19 11	21 05	21 11	13 27	10 09	17 15	02 40	22 45
13	13 26	05 N 13	18 46	20 55	21 02	13 30	10 06	17 15	02 39	22 45
14	13 06	11 47	18 19	20 43	20 52	13 33	10 03	17 15	02 38	22 45
15	12 45	17 35	17 51	20 31	20 43	13 36	10 01	17 16	02 37	22 45
16	12 25	22 20	17 21	20 19	20 33	13 39	09 58	17 16	02 36	22 44
17	12 04	25 48	16 51	20 05	20 22	13 42	09 55	17 16	02 35	22 44
18	11 43	27 50	16 18	19 52	20 12	13 45	09 52	17 17	02 35	22 44
19	11 21	28 23	15 45	19 37	20 01	13 48	09 50	17 17	02 34	22 43
20	11 00	27 28	15 10	19 22	19 50	13 51	09 47	17 17	02 33	22 43
21	10 38	25 14	14 34	19 07	19 39	13 54	09 44	17 18	02 32	22 43
22	10 17	21 53	13 56	18 51	19 28	13 57	09 42	17 18	02 31	22 43
23	09 55	17 39	13 17	18 34	19 17	14 00	09 39	17 18	02 30	22 42
24	09 33	12 47	12 36	18 17	19 05	14 03	09 36	17 19	02 30	22 42
25	09 10	07 28	11 55	18 00	18 53	14 06	09 33	17 19	02 29	22 42
26	08 48	01 N 54	11 11	17 41	18 41	14 10	09 31	17 20	02 28	22 42
27	08 26	03 S 45	10 27	17 23	18 29	14 13	09 28	17 20	02 27	22 41
28	08 03	09 17	09 41	17 04	18 16	14 16	09 25	17 21	02 26	22 41
29	07 S 40	14 S 33	08 S 54	16 S 44	18 S 04	14 N 20	09 S 23	17 N 21	02 S 25	22 S 41

ZODIAC SIGN ENTRIES

Date	h	m	Planets
01	20	37	☽ ♏
04	06	28	☽ ♐
05	05	10	☿ ♒
06	12	09	☽ ♑
08	13	59	☽ ♒
10	13	42	☽ ♓
12	13	26	☽ ♈
13	06	05	♂ ♒
14	15	02	☽ ♉
16	16	05	♀ ♒
16	19	39	☽ ♊
19	03	25	☽ ♋
19	04	13	☉ ♓
21	13	40	☽ ♌
23	07	29	☿ ♓
24	01	37	☽ ♍
26	14	29	☽ ♎
29	03	09	☽ ♏

LATITUDES

Date	Mercury ☿ ° '	Venus ♀ ° '	Mars ♂ ° '	Jupiter ♃ ° '	Saturn ♄ ° '	Uranus ♅ ° '	Neptune ♆ ° '	Pluto ♇ ° '
01	01 S 06	00 N 37	00 S 51	01 S 02	01 S 37	00 S 18	01 S 13	02 S 48
04	01 23	00 28	00 52	01 01	01 37	00 18	01 13	02 49
07	01 38	00 19	00 54	01 00	01 37	00 17	01 13	02 49
10	01 50	00 10	00 55	00 59	01 37	00 17	01 13	02 49
13	01 59	00 N 01	00 57	00 59	01 37	00 17	01 13	02 50
16	02 04	00 S 07	00 58	00 58	01 37	00 17	01 13	02 50
19	02 07	00 15	00 59	00 57	01 37	00 17	01 13	02 50
22	02 05	00 24	01 01	00 57	01 37	00 17	01 13	02 51
25	01 59	00 31	01 02	00 56	01 37	00 17	01 13	02 51
28	01 49	00 39	01 03	00 55	01 37	00 17	01 13	02 52
31	01 S 34	00 S 46	01 S 04	00 S 55	01 S 37	00 S 17	01 S 13	02 S 52

DATA

Julian Date	2460342
Delta T	+72 seconds
Ayanamsa	24° 11' 32"
Synetic vernal point	04° ♓ 55' 27"
True obliquity of ecliptic	23° 26' 19"

MOON'S PHASES, APSIDES AND POSITIONS ☽

Date	h	m	Phase	Longitude ° '	Eclipse Indicator
02	23	18	☾	13 ♏ 36	
09	22	59	●	20 ♒ 41	
16	15	01	☽	27 ♉ 26	
24	12	30	○	05 ♍ 23	

Day	h	m		
10	18	47	Perigee	
25	14	41	Apogee	
06	17	01	Max dec	28° S 19'
12	18	04	0N	
19	08	41	Max dec	28° N 23'
26	20	04	0S	

LONGITUDES

Date	Chiron ⚷ ° '	Ceres ⚳ ° '	Pallas ⚴ ° '	Juno ⚵ ° '	Vesta ⚶ ° '	Black Moon Lilith ⚸ ° '
01	16 ♈ 03	27 ♐ 48	28 ♏ 33	20 ♍ 28	21 ♊ 55	13 ♍ 29
11	16 ♈ 24	01 ♑ 27	01 ♐ 47	18 ♍ 44	21 ♊ 47	14 ♍ 35
21	16 ♈ 50	04 ♑ 55	03 ♐ 58	16 ♍ 28	22 ♊ 18	15 ♍ 42
31	17 ♈ 19	08 ♑ 11	06 ♐ 01	13 ♍ 58	23 ♊ 35	16 ♍ 49

ASPECTARIAN

h m	Aspects
01 Thursday	
02 00	☽ □ ♂
03 33	☽ ⅊ ♄
05 21	☿ ∥ ♀
07 46	☉ # ♅
09 03	☽ □ ☿
12 13	☽ ⊼ ♆
12 14	☽ ∥ ♄
19 56	☽ Q ♀
21 22	☽ □ ♇
23 02	☽ # ♃
02 Friday	
00 11	☽ ± ♆
06 15	♂ ∠ ♄
09 39	☽ △ ♄
10 55	☿ ⚹ ♆
11 17	☽ ☍ ♃
16 06	☽ Q ♂
17 11	☽ ∥ ☉
17 54	☽ ⅊ ♆
19 11	☽ # ♅
22 09	☽ ⚹ ♀
23 18	☽ □ ☉
03 Saturday	
01 28	☽ Q ☿
08 36	☽ Q ♇
09 55	☽ ☍ ♅
14 24	♀ Q ♆
16 42	☽ ⚹ ♂
20 21	☽ ∥ ☿
22 13	☽ ∥ ♀
22 42	☽ △ ♆
23 12	☽ ∥ ♂
04 Sunday	
02 00	☽ ∥ ♇
03 24	☽ ⚹ ☿
05 47	☽ ∠ ♀
07 19	☽ ⚹ ♇
12 20	☽ Q ☉
19 10	☽ □ ♄
20 39	☽ ⊼ ♃
22 21	☽ ∠ ♂
05 Monday	
00 05	☽ ⊥ ♀
07 36	☽ ± ♃
09 03	☉ ⊻ ♀
10 27	☽ ∠ ☿
10 44	☽ ∠ ♇
11 55	☽ ⚹ ☉
11 58	☽ ⊻ ♀
12 58	☿ ☌ ♇
15 30	☽ ⊥ ♂
17 11	☽ ⊼ ♅
20 49	☿ ⊥ ♄
23 42	☽ ⅊ ♃
06 Tuesday	
02 39	☽ ⊥ ♇
02 40	☽ ⊻ ♂
03 33	☽ Q ♄
03 43	☽ ± ♅
04 21	☽ ⊥ ☿
05 06	☽ □ ♆
13 03	☽ ⊻ ♇
15 59	☽ ⊻ ☿
16 11	☽ ∠ ☉
19 13	☽ ⅊ ♅
07 Wednesday	
00 18	☽ ⚹ ♄
00 22	♄ Q ♅
01 41	☽ △ ♃
08 33	☽ ⊥ ☉
11 44	☽ Q ♆
19 14	☽ ⊻ ☉
20 12	☽ ☌ ♀
20 19	☽ △ ♅
21 25	♀ △ ♅
08 Thursday	
00 20	♂ ⚹ ♆
01 25	☽ ∠ ♄
07 30	☽ ⚹ ♆
07 52	☽ ☌ ♂
10 46	☉ □ ♅
14 57	☽ ☌ ♇
16 08	☽ ⊥ ♄
23 25	☽ ☌ ☿
09 Friday	
01 51	☽ ⊻ ♄
03 10	☽ □ ♃
03 17	☽ ∥ ♇
07 21	☉ ⊥ ♆
07 39	☽ ∠ ♆
09 07	☽ ∥ ♂
09 26	☽ ∥ ♀
15 44	☽ ∥ ☿
20 35	☽ □ ♅
21 58	☽ ⊥ ♆
22 59	☽ ☌ ☉
23 59	☿ ⊻ ♄
10 Saturday	
00 44	☽ ⊻ ♀
04 44	☽ # ♅
07 29	☽ ⊻ ♆
07 57	☽ Q ♃
10 17	☽ ⊻ ♂
11 03	☽ ⊥ ♀
13 25	☿ □ ♃
14 45	☽ ⊻ ♇
15 53	♀ ∠ ♄
15 58	☽ ∥ ☉
19 41	☽ # ♃
20 16	☽ ⊥ ♂
11 Sunday	
00 15	☽ ⊥ ♇
01 03	☽ Q ♅
01 45	☽ ☌ ♄
02 32	☽ ∠ ♀
03 04	☽ ⚹ ♃
04 34	☽ ⊻ ☿
07 10	☽ ∥ ♄
11 16	☽ ∠ ♂
14 32	☽ ∠ ♇
15 13	☽ ⊥ ☿
20 07	☽ ⚹ ♅
12 Monday	
01 55	☽ ⊻ ☉
03 08	☽ ∠ ♃
04 31	☽ ⚹ ♀
05 59	☿ ∠ ♆
07 11	☽ ☌ ♆
07 20	☽ ∠ ☿
09 03	☽ ∥ ♆
12 15	☽ ⊥ ☉
12 31	☽ ⚹ ♂
14 35	☽ ⚹ ♇
17 45	☽ ⊥ ♃
20 14	☽ ∠ ♅
13 Tuesday	
01 40	☽ Q ♀
02 11	☽ ⊻ ♄
03 07	☽ # ♆
03 36	☽ ⊻ ♃
03 56	☽ ∠ ☉
07 53	☉ # ♃
09 07	☽ Q ♂
10 11	☽ Q ♇
10 44	☽ ⚹ ☿
10 58	☽ ⊥ ♅
12 06	☽ ⊥ ♄
13 36	♀ ⚹ ♆
20 55	☽ ⊻ ♅
14 Wednesday	
03 12	☽ ∠ ♄
05 30	☽ # ♄
06 06	♂ ☌ ♇
06 40	☽ ⚹ ☉
08 35	☽ ⊻ ♆
09 20	☽ Q ☿
10 20	☽ □ ♀
16 22	☽ □ ♇
16 51	☽ # ☉
16 56	☽ □ ♂
18 47	☽ ⊥ ♆
18 59	☽ ∥ ♃
15 Thursday	
04 43	☽ Q ☉
04 57	☽ ⚹ ♄
06 31	☽ ☌ ♃
10 21	☽ ∠ ♆
10 28	☉ ⊻ ♆
10 34	☽ ∥ ♅
13 07	☽ # ☿
20 36	☽ □ ☿
16 Friday	
00 27	☽ ☌ ♅
01 32	☽ # ♀
02 09	☽ Q ♄
02 36	☽ # ♂
03 40	♂ ⊥ ♄
09 04	☉ Q ♃
12 55	☽ ⚹ ♆
14 24	☽ # ♇
15 01	☽ □ ☉
16 20	☿ # ♅
20 01	☽ △ ♀
21 11	☽ △ ♇
17 Saturday	
00 56	☽ △ ♂
03 53	☿ □ ♅
08 48	♀ ☌ ♇
10 46	☽ Q ♆
10 57	☽ □ ♄
12 42	☽ ⊻ ♃
18 20	☿ ⊥ ♆
23 56	☽ ⊥ ♃
18 Sunday	
00 49	☽ ⅊ ♇
02 28	☽ ⅊ ♀
06 22	☽ ⅊ ♂
07 12	☽ ⊻ ♅
11 22	☽ △ ☿
17 00	☽ ∠ ♃
17 21	♀ ⊥ ♄
17 44	☽ ± ♇
18 31	☽ ⊥ ♅
20 28	☽ □ ♆
21 13	☽ ± ♀
19 Monday	
00 26	☽ ± ♂
03 21	☽ △ ☉
05 10	☽ ⊼ ♇
09 53	☽ ⊼ ♀
11 41	☽ ∠ ♅
12 38	☽ ⊼ ♂
20 02	☽ △ ♄
20 28	☽ ⅊ ☿
22 00	☽ ⚹ ♃
20 Tuesday	
02 38	☉ ⊻ ♇
10 47	☽ ⅊ ☉
16 45	☽ ⚹ ♅
16 49	☽ ± ☿
21 38	☽ Q ♃
21 Wednesday	
01 31	☽ ⅊ ♄
05 58	☽ ± ☉
06 32	☽ ⊼ ☿
06 38	☽ △ ♆
07 10	☿ ⊻ ♆
15 37	☽ ☍ ♇
16 23	☽ Q ♅
18 53	☽ ⊼ ☉
19 26	☽ ± ♄
22 Thursday	
03 02	☽ ☍ ♀
03 13	☽ ☍ ♂
05 09	☿ Q ♃
06 39	☽ # ♇
07 14	♀ ☌ ♂
07 28	☽ ⊼ ♄
09 39	☽ □ ♃
11 26	☿ # ♃
12 26	☽ ⅊ ♆
23 Friday	
02 52	☽ # ♂
04 18	☽ □ ♅
06 30	☽ ± ♆
06 49	☽ # ♀
13 49	☽ ∥ ♅
18 36	☽ ⊼ ♆
21 31	☿ ⊻ ♇
24 Saturday	
02 26	♀ ⊻ ♄
03 46	☽ ⊼ ♇
04 52	☽ ☍ ☿
06 02	☽ ∥ ♃
07 41	☉ ∥ ♄
12 30	☽ ☍ ☉
12 56	☽ # ☿
15 55	☽ ± ♇
19 39	☽ ⊼ ♂
20 24	☽ ☍ ♄
22 18	☽ ⊼ ♀
22 49	☽ △ ♃
25 Sunday	
02 39	☽ # ♄
03 53	☽ # ☉
04 01	♀ □ ♃
08 40	☽ ± ♂
09 20	♂ ⊻ ♄
10 13	☽ ⅊ ♇
11 54	☽ ± ♀
17 04	☽ △ ♅
26 Monday	
01 03	♀ ∠ ♆
04 14	☽ ⅊ ♂
05 43	☽ ⅊ ♃
05 49	☉ ⊥ ♇
07 35	☽ ☍ ♆
08 20	☽ ⅊ ♀
09 35	☽ # ♆
14 42	☉ Q ♅
16 47	☽ △ ♇
23 38	☽ ⅊ ♅
27 Tuesday	
00 16	☽ ± ♃
04 54	☽ ⊼ ☿
05 34	☿ ⊥ ♇
06 30	☽ ∥ ♆
07 06	☽ ⊼ ☉
08 30	♂ □ ♃
09 59	☽ ⊼ ♄
10 19	☿ Q ♅
12 38	☽ ⊼ ♃
12 51	☽ △ ♂
15 10	♀ # ♅
17 56	☽ ± ♅
18 22	☽ △ ♀
19 20	☽ ± ☿
20 24	☽ ± ☉
22 17	☽ ± ♄
28 Wednesday	
06 06	☽ ⊼ ♅
06 56	☽ ∥ ☉
08 43	☉ ☌ ☿
09 42	♂ ∠ ♆
12 36	☽ ∥ ♄
13 34	☽ ∥ ☿
15 07	☿ ☌ ♄
16 14	☽ ⅊ ☉
16 38	☽ ⅊ ♄
16 53	☽ ⅊ ☿
20 32	☽ ⊼ ♆
20 50	☿ ∥ ♄
21 25	☉ ☌ ♄
29 Thursday	
05 33	☽ □ ♇
08 35	☽ ± ♆
09 53	☿ ⚹ ♃
10 55	☽ # ♃
21 52	☽ ∥ ♀
22 52	☽ △ ♄

All ephemeris data is given at 12.00 UT and the Moon's longitude is additionally given for 24.00 UT

MARCH 2024

LONGITUDES

Date	Sidereal time h m s	Sun ☉	Moon ☽	Moon ☽ 24.00	Mercury ☿	Venus ♀	Mars ♂	Jupiter ♃	Saturn ♄	Uranus ♅	Neptune ♆	Pluto ♇
01	22 39 08	11 ♓ 23 29	16 ♏ 35 08	22 ♏ 45 01	13 ♓ 16	17 ♒ 07	13 ♒ 18	11 ♉ 24	09 ♓ 58	19 ♉ 36	26 ♓ 46	01 ♒ 14
02	22 43 05	12 23 40	28 ♏ 59 12	05 ♐ 18 14	15 11	18 21	14 05	11 34	10 06	19 37	26 48	01 15
03	22 47 01	13 23 50	11 ♐ 42 40	18 ♐ 12 59	17 06	19 35	14 51	11 45	10 13	19 39	26 50	01 17
04	22 50 58	14 23 59	24 ♐ 49 40	01 ♑ 33 03	19 01	20 50	15 38	11 55	10 20	19 41	26 52	01 18
05	22 54 54	15 24 06	08 ♑ 23 25	15 ♑ 20 53	20 57	22 04	16 24	12 06	10 28	19 43	26 54	01 20
06	22 58 51	16 24 11	22 ♑ 25 26	29 ♑ 36 51	22 54	23 18	17 11	12 17	10 35	19 45	26 57	01 21
07	23 02 48	17 24 15	06 ♒ 54 43	14 ♒ 18 26	24 50	24 32	17 57	12 28	10 42	19 47	26 59	01 23
08	23 06 44	18 24 17	21 ♒ 47 11	29 ♒ 20 00	26 46	25 47	18 44	12 39	10 50	19 49	27 01	01 24
09	23 10 41	19 24 18	06 ♓ 55 44	14 ♓ 33 10	28 ♓ 43	27 01	19 31	12 50	10 57	19 51	27 03	01 26
10	23 14 37	20 24 16	22 ♓ 10 57	29 ♓ 47 50	00 ♈ 38	28 15	20 17	13 01	11 04	19 53	27 06	01 27
11	23 18 34	21 24 13	07 ♈ 22 32	14 ♈ 53 54	02 33	29 ♒ 30	21 04	13 12	11 12	19 55	27 08	01 29
12	23 22 30	22 24 07	22 ♈ 20 56	29 ♈ 42 48	04 26	00 ♓ 44	21 50	13 24	11 19	19 57	27 10	01 30
13	23 26 27	23 24 00	06 ♉ 58 49	14 ♉ 08 32	06 18	01 58	22 37	13 35	11 26	20 00	27 13	01 31
14	23 30 23	24 23 50	21 ♉ 11 39	28 ♉ 08 03	08 08	03 12	23 24	13 47	11 33	20 02	27 15	01 33
15	23 34 20	25 23 38	04 ♊ 57 46	11 ♊ 40 57	09 55	04 27	24 10	13 58	11 40	20 04	27 17	01 34
16	23 38 17	26 23 24	18 ♊ 17 52	24 ♊ 48 52	11 40	05 41	24 57	14 10	11 48	20 06	27 19	01 35
17	23 42 13	27 23 08	01 ♋ 14 20	07 ♋ 34 44	13 21	06 55	25 44	14 22	11 55	20 09	27 22	01 36
18	23 46 10	28 22 49	13 ♋ 50 32	20 ♋ 02 12	14 58	08 09	26 30	14 34	12 02	20 11	27 24	01 38
19	23 50 06	29 ♓ 22 28	26 ♋ 10 15	02 ♌ 15 08	16 31	09 24	27 17	14 46	12 09	20 14	27 26	01 39
20	23 54 03	00 ♈ 22 05	08 ♌ 17 19	14 ♌ 17 14	17 58	10 38	28 04	14 58	12 16	20 16	27 28	01 40
21	23 57 59	01 21 40	20 ♌ 15 18	26 ♌ 11 55	19 21	11 52	28 50	15 10	12 23	20 19	27 31	01 41
22	00 01 56	02 21 12	02 ♍ 07 27	08 ♍ 02 13	20 38	13 06	29 ♒ 37	15 22	12 30	20 21	27 33	01 42
23	00 05 52	03 20 42	13 ♍ 56 33	19 ♍ 50 45	21 49	14 20	00 ♓ 24	15 34	12 37	20 24	27 35	01 44
24	00 09 49	04 20 10	25 ♍ 45 06	01 ♎ 39 52	22 53	15 35	01 10	15 47	12 44	20 26	27 38	01 45
25	00 13 46	05 19 36	07 ♎ 35 18	13 ♎ 31 40	23 51	16 49	01 57	15 59	12 51	20 29	27 40	01 46
26	00 17 42	06 19 00	19 ♎ 29 14	25 ♎ 28 16	24 42	18 03	02 44	16 11	12 58	20 32	27 42	01 47
27	00 21 39	07 18 22	01 ♏ 29 02	07 ♏ 31 49	25 25	19 17	03 31	16 24	13 05	20 35	27 44	01 48
28	00 25 35	08 17 43	13 ♏ 36 55	19 ♏ 44 40	26 01	20 31	04 17	16 37	13 12	20 37	27 47	01 49
29	00 29 32	09 17 01	25 ♏ 55 22	02 ♐ 09 24	26 30	21 46	05 04	16 49	13 19	20 40	27 49	01 50
30	00 33 28	10 16 17	08 ♐ 27 05	14 ♐ 48 49	26 52	23 00	05 51	17 02	13 26	20 43	27 51	01 51
31	00 37 25	11 ♈ 15 32	21 ♐ 14 55	27 ♐ 45 46	27 ♈ 06	24 ♓ 14	06 ♓ 37	17 ♉ 15	13 ♓ 33	20 ♉ 46	27 ♓ 53	01 ♒ 52

Date	Moon True ☊	Moon Mean ☊	Moon ☽ Latitude
01	16 ♈ 01	17 ♈ 40	02 S 40
02	16 D 03	17 37	03 34
03	16 03	17 34	04 19
04	16 R 03	17 31	04 52
05	16 01	17 28	05 11
06	15 58	17 25	05 11
07	15 54	17 21	04 53
08	15 50	17 18	04 14
09	15 46	17 15	03 17
10	15 44	17 12	02 06
11	15 42	17 09	00 S 46
12	15 D 42	17 06	00 N 37
13	15 43	17 02	01 55
14	15 45	16 59	03 03
15	15 46	16 56	03 59
16	15 47	16 53	04 40
17	15 R 47	16 50	05 06
18	15 46	16 46	05 16
19	15 45	16 43	05 11
20	15 43	16 40	04 52
21	15 40	16 37	04 20
22	15 38	16 34	03 38
23	15 36	16 31	02 46
24	15 35	16 27	01 48
25	15 35	16 24	00 N 44
26	15 D 34	16 21	00 S 22
27	15 35	16 18	01 27
28	15 35	16 15	02 30
29	15 36	16 12	03 26
30	15 37	16 08	04 13
31	15 ♈ 37	16 ♈ 05	04 S 49

DECLINATIONS

Date	Sun ☉	Moon ☽	Mercury ☿	Venus ♀	Mars ♂	Jupiter ♃	Saturn ♄	Uranus ♅	Neptune ♆	Pluto ♇
01	07 S 18	19 S 20	08 S 06	16 S 24	17 S 51	14 N 23	09 S 20	17 N 22	02 S 24	22 S 41
02	06 55	23 24	07 17	16 03	17 38	14 26	09 17	17 22	02 23	22 41
03	06 32	26 28	06 27	15 42	17 25	14 30	09 14	17 23	02 23	22 40
04	06 08	28 12	05 35	15 21	17 11	14 33	09 12	17 23	02 22	22 40
05	05 45	28 20	04 43	14 59	16 58	14 37	09 09	17 24	02 21	22 40
06	05 22	26 41	03 50	14 36	16 44	14 40	09 06	17 24	02 20	22 40
07	04 59	23 16	02 56	14 14	16 30	14 44	09 03	17 25	02 19	22 40
08	04 35	18 15	02 01	13 51	16 16	14 47	09 01	17 25	02 18	22 39
09	04 12	12 01	01 06	13 27	16 02	14 51	08 58	17 26	02 17	22 39
10	03 48	05 S 02	00 S 11	13 03	15 47	14 54	08 55	17 26	02 16	22 39
11	03 25	02 N 14	00 N 45	12 39	15 33	14 58	08 53	17 27	02 15	22 39
12	03 01	09 16	01 40	12 14	15 18	15 01	08 50	17 28	02 14	22 39
13	02 37	15 39	02 35	11 50	15 03	15 05	08 47	17 28	02 14	22 39
14	02 14	21 00	03 30	11 24	14 48	15 09	08 45	17 29	02 13	22 38
15	01 50	25 03	04 23	10 59	14 33	15 12	08 42	17 29	02 12	22 38
16	01 26	27 35	05 16	10 33	14 18	15 16	08 39	17 30	02 11	22 38
17	01 02	28 32	06 07	10 07	14 02	15 20	08 37	17 31	02 10	22 38
18	00 39	27 57	06 56	09 40	13 47	15 23	08 34	17 31	02 09	22 38
19	00 S 15	26 00	07 44	09 14	13 31	15 27	08 31	17 32	02 08	22 38
20	00 N 09	22 53	08 29	08 47	13 15	15 31	08 29	17 33	02 07	22 38
21	00 32	18 51	09 12	08 20	12 59	15 34	08 26	17 33	02 06	22 37
22	00 56	14 07	09 52	07 52	12 43	15 38	08 23	17 34	02 05	22 37
23	01 20	08 53	10 30	07 25	12 27	15 42	08 21	17 35	02 05	22 37
24	01 43	03 N 20	11 04	06 57	12 10	15 45	08 18	17 36	02 04	22 37
25	02 07	02 S 20	11 35	06 29	11 54	15 49	08 16	17 36	02 03	22 37
26	02 30	07 58	12 03	06 01	11 37	15 53	08 13	17 37	02 02	22 37
27	02 54	13 21	12 27	05 33	11 21	15 57	08 11	17 38	02 01	22 37
28	03 17	18 18	12 48	05 04	11 04	16 00	08 08	17 39	02 00	22 37
29	03 41	22 34	13 04	04 36	10 47	16 04	08 06	17 39	01 59	22 37
30	04 04	25 53	13 17	04 07	10 30	16 08	08 03	17 40	01 58	22 37
31	04 N 27	27 S 58	13 N 26	03 S 38	10 S 13	16 N 12	08 S 01	17 N 41	01 S 58	22 S 37

ZODIAC SIGN ENTRIES

Date	h	m	Planets
02	13	56	☽ ♐
04	21	15	☽ ♑
07	00	38	☽ ♒
09	01	03	☽ ♓
10	04	03	☿ ♈
11	00	19	☽ ♈
11	21	50	♀ ♓
13	00	28	☽ ♉
15	03	16	☽ ♊
17	09	40	☽ ♋
19	19	33	☽ ♌
20	03	06	☉ ♈
22	07	42	☽ ♍
22	23	47	♂ ♓
24	20	37	☽ ♎
27	09	03	☽ ♏
29	19	52	☽ ♐

LATITUDES

Date	Mercury ☿	Venus ♀	Mars ♂	Jupiter ♃	Saturn ♄	Uranus ♅	Neptune ♆	Pluto ♇
01	01 S 39	00 S 43	01 S 04	00 S 55	01 S 37	00 S 17	01 S 13	02 S 52
04	01 21	00 50	01 05	00 54	01 37	00 17	01 13	02 52
07	00 57	00 56	01 06	00 54	01 38	00 17	01 13	02 53
10	00 S 28	01 02	01 07	00 53	01 38	00 17	01 13	02 53
13	00 N 06	01 08	01 08	00 52	01 38	00 17	01 13	02 54
16	00 42	01 12	01 09	00 52	01 38	00 17	01 13	02 54
19	01 21	01 17	01 10	00 51	01 39	00 17	01 13	02 55
22	01 58	01 20	01 11	00 51	01 39	00 17	01 13	02 55
25	02 31	01 23	01 12	00 50	01 39	00 16	01 13	02 56
28	02 57	01 26	01 13	00 50	01 40	00 16	01 13	02 56
31	03 N 13	01 S 28	01 S 13	00 S 49	01 S 40	00 S 16	01 S 13	02 S 56

DATA

Julian Date	2460371
Delta T	+72 seconds
Ayanamsa	24° 11' 36"
Synetic vernal point	04° ♓ 55' 23"
True obliquity of ecliptic	23° 26' 19"

MOON'S PHASES, APSIDES AND POSITIONS ☽

Date	h	m	Phase	Longitude	Eclipse Indicator
03	15	23	☾	13 ♐ 32	
10	09	00	●	20 ♓ 17	
17	04	11	☽	27 ♊ 04	
25	07	00	○	05 ♎ 07	

Day	h	m		
10	06	58	Perigee	
23	15	32	Apogee	
05	01	57	Max dec	28° S 29'
11	04	38	0N	
17	14	40	Max dec	28° N 32'
25	02	08	0S	

LONGITUDES

Date	Chiron ⚷	Ceres ⚳	Pallas ⚴	Juno ⚵	Vesta ⚶	Black Moon Lilith ⚸
01	17 ♈ 16	07 ♑ 52	05 ♐ 50	14 ♍ 13	23 ♊ 26	16 ♍ 42
11	17 ♈ 47	10 ♑ 54	07 ♐ 25	11 ♍ 44	25 ♊ 10	17 ♍ 49
21	18 ♈ 20	13 ♑ 40	08 ♐ 21	09 ♍ 31	27 ♊ 23	18 ♍ 55
31	18 ♈ 55	16 ♑ 06	08 ♐ 35	07 ♍ 49	00 ♋ 01	20 ♍ 02

All ephemeris data is given at 12.00 UT and the Moon's longitude is additionally given for 24.00 UT

ASPECTARIAN

h m	Aspects
01 Friday	
00 53	☽ △ ☉
01 40	☽ ☍ ♃
01 41	☽ # ♅
02 29	☽ ꝗ ♆
04 18	☽ △ ☿
04 31	☽ ∥ ♂
05 08	☽ □ ♂
12 15	☉ ⚹ ♃
12 41	☿ ⊻ ♂
13 09	☽ □ ♀
17 10	☽ Q ♇
17 53	☽ ☍ ♅
02 Saturday	
07 16	☽ ∥ ♇
07 47	☽ △ ♆
16 20	☽ ⚹ ♇
18 17	☽ Q ♂
03 Sunday	
01 43	☿ ∠ ♇
03 28	☽ Q ♀
07 42	☉ ∥ ☿
09 12	☽ □ ♄
12 04	☽ ⊼ ♃
13 17	♀ □ ♅
15 23	☽ □ ☉
15 39	♂ # ♅
18 11	☽ ⚹ ♂
20 28	☽ ∠ ♇
23 17	☽ ± ♃
23 39	☽ □ ☿
04 Monday	
02 39	☽ ⊼ ♅
02 57	♃ ∠ ♆
04 01	☽ ⚹ ♀
12 52	☽ ⊥ ♇
12 53	♀ ⊥ ♆
13 33	☽ ± ♅
15 41	☽ □ ♆
15 49	☽ ꝗ ♃
18 21	☽ Q ♄
20 24	☿ ⚹ ♅
23 00	☽ ∠ ♂
23 35	☽ ⊻ ♇
05 Tuesday	
02 34	☽ Q ☉
05 34	☽ ꝗ ♅
09 28	☽ ∠ ♀
13 08	☽ Q ☿
15 38	☽ ⚹ ♄
15 42	☽ ⊥ ♂
18 30	☽ △ ♃
23 17	☽ Q ♆
06 Wednesday	
01 01	☽ ⚹ ☉
02 30	☽ ⊥ ♀
02 37	☽ ⊻ ♂
07 28	☽ △ ♅
08 36	♀ # ♃
10 52	☉ ∠ ♇
12 55	☽ ⚹ ☿
13 37	☽ ⊻ ♀
17 21	☽ ∠ ♄
17 58	☿ ⊥ ♂
19 35	☽ ⚹ ♆
07 Thursday	
02 00	☿ ⊻ ♀
02 54	☽ ☌ ♇
04 04	☽ ∠ ☉
08 22	☽ ⊥ ♄
15 18	☽ ∥ ♇
17 29	☽ ∠ ☿
18 14	☽ ⊻ ♄
19 50	☽ ⊥ ☉
20 16	☽ ∠ ♆
21 08	☽ □ ♃
08 Friday	
04 23	☿ ∥ ♆
06 12	☽ ⊻ ☉
06 51	☽ ☌ ♂
08 50	☽ □ ♅
10 09	☽ ⊥ ☿
10 46	☽ ⊥ ♆
15 06	☿ ☌ ♆
15 26	☽ # ♅
18 56	☽ ☌ ♀
20 21	☽ ⊻ ♆
20 24	☽ ∥ ♂
21 07	☽ ⊻ ☿
23 55	☿ ∠ ♃
09 Saturday	
01 41	☽ # ♃
02 15	☽ Q ♃
03 18	☽ ⊻ ♇
06 27	☽ ∥ ♀
12 47	☽ ⊥ ♇
12 49	♀ ⊻ ♆
13 27	☽ Q ♅
18 23	☽ ☌ ♄
21 24	☽ ⚹ ♃
22 46	☽ ∥ ♄
22 55	♂ □ ♅
23 01	☉ ⚹ ♅
23 20	☉ ⊻ ♂
10 Sunday	
02 58	☽ ∠ ♇
08 22	☽ ⚹ ♅
08 51	☽ ⊻ ♂
09 00	☽ ☌ ☉
16 18	☽ ∥ ☉
18 49	☽ ⊥ ♂
19 45	☽ ☌ ♆
21 09	☽ ∥ ♆
21 18	☽ ∠ ♃
22 21	☿ ⚹ ♇
22 25	☽ ⊻ ♀
11 Monday	
02 38	☽ ⚹ ♇
03 15	☽ ☌ ☿
03 16	☽ # ☿
06 23	☽ ∥ ☿
08 06	☽ ∠ ♅
08 45	☽ ⊥ ♀
09 48	☽ ∠ ♂
11 43	☽ ⊥ ♃
12 06	☽ # ♆
14 15	♂ ⊥ ♆
15 44	☽ # ☉
18 07	☽ ⊻ ♄
21 24	☽ ⊻ ♃
21 44	☽ Q ♇
22 27	☽ ⊥ ♅
12 Tuesday	
00 22	☽ ∠ ♀
03 48	☽ ⊥ ♄
08 07	☽ ⊻ ♅
10 29	☽ # ♄
11 08	☽ ⚹ ♂
12 06	☽ ⊻ ☉
18 30	☽ ∠ ♄
18 43	☿ ∠ ♅
19 52	☽ ⊻ ♆
22 09	☽ # ♀
22 34	☽ ⊥ ☉
13 Wednesday	
02 38	☿ # ♆
02 56	☽ ⚹ ♀
02 57	☽ □ ♇
03 11	♀ ⊻ ♇
03 11	♀ Q ♃
05 44	☽ ⊥ ♆
07 52	☽ Q ♂
09 33	♂ # ♃
09 42	☽ # ♂
09 42	☽ ∥ ♃
10 43	☽ ⊻ ☿
12 35	☉ # ☿
14 32	☽ ∠ ☉
19 30	☽ ⚹ ♄
19 36	☽ ∥ ♅
20 46	☽ ∠ ♆
22 13	☽ ⊥ ☿
23 13	☽ ☌ ♃
14 Thursday	
00 49	☽ Q ♀
06 42	☿ ⊥ ♃
10 00	☽ ☌ ♅
12 55	☉ ∥ ♆
15 50	☽ ∠ ☿
16 01	☽ □ ♂
16 06	☽ Q ♄
17 56	☽ ⚹ ☉
18 05	☿ ∠ ♂
20 43	☽ # ♇
22 29	☽ ⚹ ♆
15 Friday	
06 00	☽ △ ♇
10 59	☽ □ ♀
16 40	☽ Q ☉
19 43	☽ Q ♆
22 11	☽ ⚹ ☿
16 Saturday	
00 06	☽ □ ♄
04 22	☽ ⊻ ♃
08 52	☽ ꝗ ♇
12 52	☿ ⊥ ♀
13 57	☿ ⊻ ♄
15 19	☽ ⊻ ♅
15 28	☽ ⊥ ♃
23 21	☽ Q ☿
17 Sunday	
01 02	☽ △ ♂
01 27	☽ ± ♇
02 27	☽ ⊥ ♅
04 11	☽ □ ☉
04 43	☽ □ ♆
08 25	☽ ∠ ♃
11 22	☉ ☌ ♆
12 42	☽ ⊼ ♇
15 51	☿ Q ♇
19 24	☽ ∠ ♅
23 55	☽ △ ♀
18 Monday	
00 02	☿ ⊥ ♅
01 36	♀ ⊥ ♇
05 02	☿ ⊻ ♃
07 13	☽ ꝗ ♂
08 29	☽ △ ♄
12 37	♀ Q ♅
13 24	☽ ⚹ ♃
14 29	☽ □ ☿
19 Tuesday	
00 20	☽ ⚹ ♅
01 47	☽ ± ♂
08 07	☽ ꝗ ♀
13 11	☽ Q ♃
13 57	☽ ꝗ ♄
14 20	☽ ⊼ ♂
14 30	☽ △ ♆
16 56	♂ ⊻ ♆
18 52	☽ △ ☉
22 49	☽ ☍ ♇
23 39	☉ ∠ ♃
20 Wednesday	
00 00	☽ Q ♅
03 53	☽ ± ♀
07 56	☽ ± ♄
11 43	☿ # ♄
13 42	☽ # ♇
17 13	☽ ⊼ ♀
17 29	☿ ⊥ ♄
17 58	☿ # ♀
20 02	☽ ⊼ ♄
20 24	☽ ꝗ ♆
21 Thursday	
01 35	☽ □ ♃
03 27	☽ ꝗ ☉
05 50	♀ ∥ ♄
09 57	☽ △ ☿
12 07	☽ □ ♅
14 33	☽ ± ♆
18 49	☽ ∥ ♅
20 03	☉ ⚹ ♇
23 09	♀ ☌ ♄
23 15	☽ ± ☉
22 Friday	
02 42	☽ ⊼ ♆
04 41	☽ ∥ ♃
06 27	☿ ⊻ ♅
06 34	☽ ☍ ♂
11 09	☽ ⊼ ♇
12 30	☽ ⊼ ☉
18 57	☽ # ♂
19 56	☽ ꝗ ☿
23 21	☽ ± ♇
23 Saturday	
05 29	☽ ∥ ☿
09 17	☽ ☍ ♄
12 54	☽ ☍ ♀
14 21	☽ # ♄
15 22	☽ △ ♃
16 12	☽ ± ☿
17 40	☽ ꝗ ♇
19 01	☽ # ♀
24 Sunday	
01 10	☽ △ ♅
05 37	☽ ⊼ ☿
15 49	☽ ☍ ♆
16 37	♀ ⚹ ♃
17 26	☽ # ♆
18 24	☽ ∥ ☉
22 23	☽ ꝗ ♃
23 47	☽ ⊼ ♂
25 Monday	
00 11	☽ △ ♇
05 58	♂ ⊻ ♇
07 00	☽ ☍ ☉
07 44	☽ ꝗ ♅
07 54	☉ # ♆
10 47	☽ ∥ ♆
10 57	♀ ∠ ♇
11 00	☽ # ☉
12 47	☽ ± ♂
16 01	☉ ∠ ♅
16 56	☽ ± ♃
21 51	☿ # ♂
22 45	☽ ⊼ ♄
26 Tuesday	
01 59	☽ ± ♅
04 17	☽ ∥ ♀
05 15	☽ ⊼ ♃
08 13	☽ ꝗ ♂
08 47	☽ ⊼ ♀
10 57	☽ ± ♄
13 07	☽ ∥ ♄
14 06	☽ ⊼ ♅
22 13	☽ ± ♀
23 09	☽ ☍ ☿
27 Wednesday	
03 20	☽ ∥ ♂
04 31	☽ ⊼ ♆
05 10	☽ ꝗ ♄
07 34	☽ # ☿
12 37	☽ □ ♇
16 18	☽ △ ♂
16 30	☽ ± ♆
17 27	☿ ⊥ ♀
18 12	☽ ꝗ ♀
28 Thursday	
00 23	☽ # ♃
00 35	☽ ⊼ ☉
08 37	☽ # ♅
10 21	☽ ꝗ ♆
11 11	☽ △ ♄
13 27	☽ ± ☉
13 58	♀ ⚹ ♅
17 59	☽ ☍ ♃
29 Friday	
00 09	☽ Q ♇
01 39	♂ Q ♃
01 46	☽ ☍ ♅
03 02	☽ △ ♀
08 33	☽ ꝗ ☉
12 17	☽ ∥ ♇
13 10	☽ ⊼ ☿
15 40	☽ △ ♆
23 23	☽ ⚹ ♇
30 Saturday	
01 04	☽ ± ☿
06 43	☽ □ ♂
13 19	☉ # ♀
15 44	☽ △ ☉
18 36	☽ ꝗ ☿
21 29	☽ □ ♄
31 Sunday	
03 49	☽ ∠ ♇
04 26	☽ ⊼ ♃
11 06	☽ ⊼ ♅
11 36	☉ ⊥ ♃
15 45	☽ ± ♃
18 06	☽ □ ♀
18 38	☽ Q ♂
20 31	☽ ⊥ ♇
22 12	☽ ± ♅
22 54	☽ △ ☿

APRIL 2024

LONGITUDES

Date	Sidereal time h m s	Sun ☉ ° ′ ″	Moon ☽ ° ′ ″	Moon ☽ 24.00 ° ′ ″	Mercury ☿ ° ′	Venus ♀ ° ′	Mars ♂ ° ′	Jupiter ♃ ° ′	Saturn ♄ ° ′	Uranus ♅ ° ′	Neptune ♆ ° ′	Pluto ♇ ° ′
01	00 41 21	12 ♈ 14 45	04 ♑ 21 40	11 ♑ 02 54	27 ♈ 12	25 ♓ 28	07 ♓ 24	17 ♉ 28	13 ♓ 39	20 ♉ 49	27 ♓ 55	01 ♒ 52
02	00 45 18	13 13 56	17 ♑ 49 43	24 ♑ 42 16	27 R 12	26 42	08 11	17 41	13 46	20 51	27 58	01 53
03	00 49 15	14 13 06	01 ♒ 40 36	08 ♒ 44 41	27 05	27 56	08 57	17 54	13 53	20 54	28 00	01 54
04	00 53 11	15 12 13	15 ♒ 54 22	23 ♒ 09 19	26 50	29 ♓ 11	09 44	18 07	13 59	20 57	28 02	01 55
05	00 57 08	16 11 19	00 ♓ 29 05	07 ♓ 53 04	26 30	00 ♈ 25	10 31	18 20	14 06	21 00	28 04	01 56
06	01 01 04	17 10 23	15 ♓ 20 31	22 ♓ 50 33	26 04	01 39	11 17	18 33	14 13	21 03	28 07	01 56
07	01 05 01	18 09 26	00 ♈ 22 09	07 ♈ 54 17	25 33	02 53	12 04	18 46	14 19	21 06	28 09	01 57
08	01 08 57	19 08 26	15 ♈ 25 49	22 ♈ 55 40	24 58	04 07	12 51	18 59	14 26	21 09	28 11	01 58
09	01 12 54	20 07 24	00 ♉ 22 45	07 ♉ 46 05	24 19	05 21	13 37	19 13	14 32	21 13	28 13	01 59
10	01 16 50	21 06 20	15 ♉ 04 48	22 ♉ 18 10	23 37	06 35	14 24	19 26	14 38	21 16	28 15	01 59
11	01 20 47	22 05 14	29 ♉ 25 36	06 ♊ 26 42	22 53	07 49	15 11	19 39	14 45	21 19	28 17	02 00
12	01 24 44	23 04 06	13 ♊ 21 12	20 ♊ 09 00	22 08	09 04	15 57	19 53	14 51	21 22	28 19	02 00
13	01 28 40	24 02 56	26 ♊ 50 10	03 ♋ 24 52	21 23	10 18	16 44	20 06	14 57	21 25	28 21	02 01
14	01 32 37	25 01 44	09 ♋ 53 24	16 ♋ 16 08	20 38	11 32	17 30	20 20	15 04	21 28	28 24	02 02
15	01 36 33	26 00 29	22 ♋ 33 31	28 ♋ 46 04	19 55	12 46	18 17	20 33	15 10	21 32	28 26	02 02
16	01 40 30	26 59 12	04 ♌ 54 18	10 ♌ 58 48	19 13	14 00	19 04	20 47	15 16	21 35	28 28	02 02
17	01 44 26	27 57 53	17 ♌ 00 08	22 ♌ 58 52	18 35	15 14	19 50	21 00	15 22	21 38	28 30	02 03
18	01 48 23	28 56 32	28 ♌ 55 35	04 ♍ 50 50	18 00	16 28	20 37	21 14	15 28	21 41	28 32	02 03
19	01 52 19	29 ♈ 55 08	10 ♍ 45 07	16 ♍ 38 59	17 29	17 42	21 23	21 28	15 34	21 45	28 34	02 04
20	01 56 16	00 ♉ 53 42	22 ♍ 32 52	28 ♍ 27 14	17 02	18 56	22 10	21 42	15 40	21 48	28 36	02 04
21	02 00 13	01 52 14	04 ♎ 22 28	10 ♎ 18 58	16 40	20 10	22 56	21 55	15 46	21 51	28 38	02 04
22	02 04 09	02 50 44	16 ♎ 17 02	22 ♎ 17 00	16 22	21 24	23 42	22 09	15 51	21 54	28 40	02 05
23	02 08 06	03 49 12	28 ♎ 19 06	04 ♏ 23 36	16 09	22 38	24 29	22 23	15 57	21 58	28 42	02 05
24	02 12 02	04 47 39	10 ♏ 30 41	16 ♏ 40 32	16 02	23 52	25 15	22 37	16 03	22 01	28 44	02 05
25	02 15 59	05 46 03	22 ♏ 53 19	29 ♏ 09 09	15 59	25 06	26 02	22 51	16 08	22 05	28 46	02 05
26	02 19 55	06 44 26	05 ♐ 28 10	11 ♐ 50 28	16 D 01	26 20	26 48	23 05	16 14	22 08	28 47	02 06
27	02 23 52	07 42 47	18 ♐ 16 09	24 ♐ 45 19	16 08	27 34	27 34	23 18	16 20	22 11	28 49	02 06
28	02 27 48	08 41 06	01 ♑ 18 03	07 ♑ 54 27	16 20	28 ♈ 48	28 21	23 32	16 25	22 15	28 51	02 06
29	02 31 45	09 39 24	14 ♑ 34 35	21 ♑ 18 31	16 37	00 ♉ 01	29 07	23 46	16 30	22 18	28 53	02 06
30	02 35 42	10 ♉ 37 41	28 ♑ 06 18	04 ♒ 58 00	16 ♈ 58	01 ♉ 15	29 ♓ 53	24 ♉ 00	16 ♓ 36	22 ♉ 22	28 ♓ 55	02 ♒ 06

Date	Moon True ☊ ° ′	Moon Mean ☊ ° ′	Moon ☽ Latitude ° ′
01	15 ♈ 37	16 ♈ 02	05 S 12
02	15 R 37	15 59	05 17
03	15 37	15 56	05 05
04	15 37	15 52	04 35
05	15 D 37	15 49	03 45
06	15 37	15 46	02 41
07	15 38	15 43	01 24
08	15 38	15 40	00 S 01
09	15 R 38	15 37	01 N 21
10	15 37	15 33	02 36
11	15 36	15 30	03 40
12	15 36	15 27	04 29
13	15 35	15 24	05 01
14	15 34	15 21	05 16
15	15 34	15 18	05 15
16	15 D 34	15 14	04 59
17	15 34	15 11	04 31
18	15 35	15 08	03 51
19	15 37	15 05	03 02
20	15 38	15 02	02 05
21	15 39	14 58	01 N 02
22	15 R 39	14 55	00 S 03
23	15 39	14 52	01 09
24	15 37	14 49	02 13
25	15 35	14 46	03 11
26	15 32	14 43	04 01
27	15 29	14 39	04 40
28	15 25	14 36	05 05
29	15 23	14 33	05 14
30	15 ♈ 21	14 ♈ 30	05 S 07

DECLINATIONS

Date	Sun ☉ ° ′	Moon ☽ ° ′	Mercury ☿ ° ′	Venus ♀ ° ′	Mars ♂ ° ′	Jupiter ♃ ° ′	Saturn ♄ ° ′	Uranus ♅ ° ′	Neptune ♆ ° ′	Pluto ♇ ° ′
01	04 N 50	28 S 33	13 N 31	03 S 09	09 S 56	16 N 15	07 S 58	17 N 42	01 S 57	22 S 37
02	05 13	27 30	13 32	02 40	09 38	16 19	07 56	17 42	01 56	22 37
03	05 36	24 45	13 30	02 11	09 21	16 23	07 53	17 43	01 55	22 37
04	05 59	20 26	13 23	01 42	09 04	16 27	07 51	17 44	01 54	22 37
05	06 22	14 49	13 12	01 12	08 46	16 30	07 48	17 45	01 53	22 37
06	06 45	08 15	12 58	00 43	08 29	16 34	07 46	17 46	01 52	22 37
07	07 07	01 S 08	12 41	00 S 14	08 11	16 38	07 43	17 46	01 52	22 37
08	07 30	06 N 04	12 20	00 N 16	07 53	16 42	07 41	17 47	01 51	22 37
09	07 52	12 52	11 57	00 45	07 35	16 45	07 39	17 48	01 50	22 37
10	08 14	18 51	11 31	01 15	07 18	16 49	07 36	17 49	01 49	22 37
11	08 36	23 36	11 03	01 44	07 00	16 53	07 34	17 50	01 48	22 37
12	08 58	26 51	10 34	02 13	06 42	16 57	07 32	17 51	01 47	22 37
13	09 20	28 24	10 03	02 43	06 24	17 00	07 29	17 51	01 47	22 37
14	09 41	28 19	09 32	03 12	06 06	17 04	07 27	17 52	01 46	22 37
15	10 03	26 44	09 01	03 41	05 47	17 08	07 25	17 53	01 45	22 37
16	10 24	23 53	08 30	04 10	05 29	17 11	07 23	17 54	01 44	22 37
17	10 45	20 02	08 00	04 40	05 11	17 15	07 20	17 55	01 43	22 37
18	11 06	15 27	07 31	05 09	04 53	17 19	07 18	17 56	01 43	22 37
19	11 27	10 20	07 04	05 37	04 35	17 23	07 16	17 56	01 42	22 37
20	11 47	04 N 52	06 38	06 06	04 16	17 26	07 14	17 57	01 41	22 37
21	12 07	00 S 47	06 15	06 35	03 58	17 30	07 12	17 58	01 40	22 37
22	12 28	06 27	05 54	07 03	03 40	17 34	07 10	17 59	01 40	22 37
23	12 47	11 57	05 35	07 32	03 21	17 37	07 08	18 00	01 39	22 37
24	13 07	17 05	05 18	08 00	03 03	17 41	07 06	18 01	01 38	22 38
25	13 27	21 35	05 05	08 28	02 44	17 44	07 04	18 02	01 37	22 38
26	13 46	25 10	04 53	08 56	02 26	17 48	07 02	18 03	01 37	22 38
27	14 05	27 34	04 45	09 23	02 08	17 52	07 00	18 03	01 36	22 38
28	14 24	28 31	04 39	09 51	01 49	17 55	06 58	18 04	01 35	22 38
29	14 42	27 51	04 35	10 18	01 31	17 59	06 56	18 05	01 34	22 38
30	15 N 01	25 S 33	04 N 34	10 N 45	01 S 12	18 N 02	06 S 54	18 N 06	01 S 34	22 S 38

ZODIAC SIGN ENTRIES

Date	h	m	Planets
01	04	05	☽ ♑
03	09	08	☽ ♒
05	04	00	♀ ♈
05	11	13	☽ ♓
07	11	25	☽ ♈
09	11	23	☽ ♉
11	12	58	☽ ♊
13	17	45	☽ ♋
16	02	24	☽ ♌
18	14	10	☽ ♍
19	14	00	☉ ♉
21	03	08	☽ ♎
23	15	20	☽ ♏
26	01	37	☽ ♐
28	09	37	☽ ♑
29	11	31	♀ ♉
30	15	20	☽ ♒
30	15	33	♂ ♈

LATITUDES

Date	Mercury ☿ ° ′	Venus ♀ ° ′	Mars ♂ ° ′	Jupiter ♃ ° ′	Saturn ♄ ° ′	Uranus ♅ ° ′	Neptune ♆ ° ′	Pluto ♇ ° ′
01	03 N 16	01 S 28	01 S 13	00 S 49	01 S 40	00 S 16	01 S 13	02 S 57
04	03 15	01 29	01 14	00 49	01 40	00 16	01 13	02 57
07	03 01	01 30	01 14	00 48	01 41	00 16	01 13	02 58
10	02 32	01 30	01 15	00 48	01 41	00 16	01 13	02 58
13	01 51	01 29	01 15	00 47	01 42	00 16	01 13	02 59
16	01 04	01 28	01 16	00 47	01 42	00 16	01 13	02 59
19	00 N 13	01 26	01 16	00 47	01 43	00 16	01 14	03 00
22	00 S 35	01 23	01 16	00 46	01 43	00 16	01 14	03 00
25	01 19	01 20	01 16	00 46	01 44	00 16	01 14	03 01
28	01 55	01 17	01 16	00 46	01 44	00 16	01 14	03 01
31	02 S 25	01 S 13	01 S 16	00 S 45	01 S 45	00 S 16	01 S 14	03 S 02

DATA

Julian Date	2460402
Delta T	+72 seconds
Ayanamsa	24° 11′ 40″
Synetic vernal point	04° ♓ 55′ 19″
True obliquity of ecliptic	23° 26′ 19″

MOON'S PHASES, APSIDES AND POSITIONS ☽

Date	h	m	Phase	Longitude ° ′	Eclipse Indicator
02	03	15	☾	12 ♑ 52	
08	18	21	●	19 ♈ 24	Total
15	19	13	☽	26 ♋ 18	
23	23	49	○	04 ♏ 18	

Day	h	m		
07	17	44	Perigee	
20	02	04	Apogee	
01	08	50	Max dec	28° S 34′
07	15	46	ON	
13	22	33	Max dec	28° N 34′
21	08	40	OS	
28	14	18	Max dec	28° S 31′

LONGITUDES

Date	Chiron ⚷ ° ′	Ceres ⚳ ° ′	Pallas ⚴ ° ′	Juno ⚵ ° ′	Vesta ⚶ ° ′	Black Moon Lilith ⚸ ° ′
01	18 ♈ 59	16 ♑ 19	08 ♐ 34	07 ♍ 41	00 ♋ 18	20 ♍ 09
11	19 ♈ 34	18 ♑ 20	07 ♐ 56	06 ♍ 40	03 ♋ 18	21 ♍ 15
21	20 ♈ 09	19 ♑ 54	06 ♐ 29	06 ♍ 18	06 ♋ 36	22 ♍ 22
31	20 ♈ 44	20 ♑ 59	04 ♐ 18	06 ♍ 34	10 ♋ 08	23 ♍ 28

ASPECTARIAN

01 Monday
00 16 ☽ □ ♆
07 03 ☽ Q ♄
07 29 ☽ ⊻ ♇
08 30 ☽ ⚼ ♃
14 37 ☽ ⚼ ♅
17 49 ☽ ✶ ♂
22 15 ☿ St R

02 Tuesday
02 53 ♂ ⊥ ♇
03 15 ☽ □ ☉
04 46 ☽ ✶ ♄
05 56 ☽ Q ♀
08 42 ☽ Q ♆
11 44 ☽ △ ♃
17 20 ☽ △ ♅
20 58 ☿ ⊻ ♀
21 55 ☽ ∠ ♂

03 Wednesday
02 42 ☉ ⊻ ♄
04 11 ☽ □ ☿
04 11 ☉ Q ♇
04 58 ☽ ✶ ♀
05 40 ☽ ✶ ♆
07 10 ☽ ∠ ♄
10 22 ♂ Q ♅
12 23 ☽ ☌ ♇
13 00 ☽ Q ☉
13 10 ♀ ☌ ♆
14 19 ☽ ⊥ ♂
22 37 ☽ ⊥ ♄

04 Thursday
00 53 ☽ ∥ ♇
01 04 ☽ ⊻ ♂
01 27 ♀ ∥ ♆
05 39 ☉ ⊥ ♅
07 12 ☽ ∠ ♆
08 47 ☽ ⊻ ♄
08 51 ☽ ∠ ♀
10 16 ☽ Q ☿
10 45 ☽ ✶ ☉
15 43 ☽ □ ♃
20 24 ☽ □ ♅
22 11 ☽ ⊥ ♆

05 Friday
00 07 ☽ # ♅
01 09 ☽ ⊥ ♀
05 19 ☽ # ♃
05 40 ☽ ✶ ☿
08 03 ☽ ⊻ ♆
11 52 ☽ ⊻ ♀
13 14 ☽ ∠ ☉
14 21 ☽ ⊻ ♇
18 21 ☽ # ☿
21 38 ☽ Q ♃

06 Saturday
00 05 ☽ ⊥ ♇
01 51 ☽ Q ♅
04 50 ☽ ⊥ ☉
05 08 ☽ ☌ ♂
05 21 ☽ ∠ ☿
07 47 ☿ ∠ ♂
10 10 ☽ ☌ ♄
11 09 ☽ ∥ ♂
13 40 ☽ ∥ ♄
14 34 ☽ ∠ ♇
15 08 ☽ ⊻ ☉
16 54 ☽ # ☉
17 13 ☽ ✶ ♃
17 46 ♀ ✶ ♇
19 20 ☽ ⊥ ☿
21 11 ☽ ✶ ♅

07 Sunday
01 39 ☉ ⊥ ♂
04 36 ☽ ⊻ ☿
08 27 ☽ ☌ ♆
09 35 ☽ ∥ ♆
14 31 ☽ ✶ ♇
15 14 ☽ ∥ ♀
16 14 ☽ # ♀
16 22 ☽ ☌ ♀
17 29 ☽ ∠ ♃
21 10 ☽ ∠ ♅
21 55 ☽ # ♆

08 Monday
07 11 ☉ ⊻ ♃
07 39 ☽ ⊻ ♂
08 03 ☽ ⊥ ♃
08 54 ♀ ∠ ♃
09 40 ☽ Q ♇
10 23 ☽ ⊻ ♄
11 34 ☽ ⊥ ♅
17 10 ☽ ∥ ☉
17 31 ☽ # ♄
17 45 ☽ ⊥ ♂
17 46 ☽ ⊻ ♃
17 59 ☽ # ♂
18 21 ☽ ☌ ☉
20 03 ☽ ⊥ ♄
21 12 ☽ ⊻ ♅
23 03 ☉ # ♄

09 Tuesday
02 06 ☉ # ♂
02 39 ☽ ☌ ☿
06 57 ♂ ∥ ♄
08 30 ☽ ⊻ ♆
08 48 ☽ ∥ ☿
09 00 ☽ ∠ ♂
10 37 ☽ ∠ ♄
14 35 ☽ □ ♇
18 14 ☽ ⊥ ♆
20 48 ☽ ⊻ ♀
23 16 ☉ ⊥ ♄

10 Wednesday
03 16 ☽ ∥ ♃
05 21 ♀ ∠ ♅
07 30 ☽ ∥ ♅
07 31 ☽ ⊥ ♀
08 59 ☽ ∠ ♆
10 49 ☽ ✶ ♂
11 16 ☽ ✶ ♄
16 00 ☉ ⊻ ♅
19 19 ☽ ☌ ♃
20 36 ♂ ☌ ♄
22 18 ☽ ☌ ♅
22 44 ☽ ⊻ ☉
23 49 ☽ ∠ ♀

11 Thursday
01 31 ☽ ⊻ ☿
06 19 ☽ # ♇
07 26 ☽ Q ♄
07 58 ☽ Q ♂
09 34 ☽ ⊥ ☉
10 04 ☽ ✶ ♆
11 07 ☽ ⊥ ☿
15 19 ♀ # ♆
16 23 ☽ △ ♇
23 03 ☉ ☌ ☿

12 Friday
01 45 ☽ ∠ ☿
02 06 ☽ ∠ ☉
03 47 ☽ ✶ ♀
06 42 ☽ Q ♆
14 39 ☽ □ ♄
14 47 ☿ ⊥ ♂
16 51 ☽ □ ♂
18 26 ☽ ⚼ ♇
23 43 ☽ ⊻ ♃

13 Saturday
02 13 ☽ ⊻ ♅
02 43 ☽ ✶ ☿
03 00 ☽ Q ♀
06 35 ☽ ✶ ☉
10 31 ☽ ± ♇
10 39 ☽ ⊥ ♃
10 47 ☿ ⊻ ♅
13 04 ☽ ⊥ ♅
14 46 ☽ □ ♆
20 55 ♂ ∠ ♇
21 26 ☽ ⚻ ♇
23 17 ☽ Q ☿
23 51 ☿ ⊥ ♄

14 Sunday
03 23 ☽ ∠ ♃
05 37 ☽ ∠ ♅
06 14 ☽ Q ☉
07 56 ☉ ∥ ☿
15 23 ☽ □ ♀
19 38 ☿ ⊻ ♃
21 48 ☽ △ ♄

15 Monday
03 17 ☽ △ ♂
07 12 ☽ □ ☿
08 05 ☽ ✶ ♃
10 00 ☽ ✶ ♅
19 13 ☽ □ ☉
23 22 ☽ △ ♆

16 Tuesday
02 50 ☽ ⚼ ♄
06 23 ☽ ☍ ♇
07 45 ☽ Q ♃
09 23 ☽ Q ♅
10 14 ☽ ⚼ ♂
12 53 ♀ Q ♇
14 46 ☿ ⊻ ♂
20 31 ☽ # ♇
20 40 ☽ ± ♄

17 Wednesday
04 59 ☽ ⚼ ♆
05 14 ☽ ± ♂
06 42 ♀ ⊥ ♃
08 03 ☽ △ ♀
08 42 ☽ ⚻ ♄
14 53 ♀ ⊻ ♄
15 01 ☽ △ ☿
18 04 ☽ ⚻ ♂
20 11 ☽ □ ♃
20 12 ♀ ⊥ ♅
21 20 ☽ □ ♅
23 03 ☽ ± ♆
23 30 ☽ ∥ ♅

18 Thursday
01 31 ☉ ⊻ ♆
02 46 ☽ ∥ ♃
04 02 ♀ # ♂
11 12 ☽ ⚻ ♆
12 02 ☽ △ ☉
17 44 ☽ ⚼ ♀
18 20 ☽ ⚻ ♇
19 54 ☽ ⚼ ☿

19 Friday
00 19 ☿ # ♄
06 32 ☽ ± ♇
07 16 ☽ ∥ ☉
08 59 ☿ ☌ ♀
13 25 ☽ ± ☿
14 09 ☽ ± ♀
15 28 ♂ ✶ ♃
21 15 ☽ ⚼ ☉
21 53 ☽ ☍ ♄
23 56 ♂ ✶ ♅

20 Saturday
00 51 ☽ ⚼ ♇
01 10 ☽ ⚻ ☿
01 41 ☽ # ♄
03 42 ☽ ∥ ☿
03 47 ☽ ⚻ ♀
05 40 ☉ ∠ ♄
07 05 ☽ ∥ ♀
10 14 ☽ △ ♃
10 28 ☽ △ ♅
11 09 ☽ ☍ ♂
14 41 ☽ # ♂
17 12 ☽ ± ☉

21 Sunday
00 19 ☽ ☍ ♆
01 35 ☽ # ♆
02 27 ♃ ☌ ♅
02 43 ☿ ∥ ♀
06 28 ☽ ⚻ ☉
07 20 ☽ △ ♇
15 43 ☽ ∥ ♆
17 02 ☉ □ ♇
17 02 ☽ ⚼ ♅
17 15 ☽ ⚼ ♃

22 Monday
00 44 ☽ ∥ ♂
09 44 ☽ # ☿
11 08 ☽ ⚻ ♄
11 15 ☽ ± ♅
11 44 ☽ ± ♃
12 10 ☽ ☍ ☿
14 48 ☽ # ♀
15 00 ☽ ∥ ♄
17 03 ♀ # ♄
21 43 ♀ ⊥ ♄
22 25 ♀ ⊻ ♅
23 06 ☿ ⊥ ♃
23 14 ☽ ± ♄
23 18 ☽ ⚻ ♅
23 24 ☽ ☍ ♀
23 58 ☽ ⚻ ♃

23 Tuesday
03 51 ☽ ⚻ ♂
06 04 ♀ ⊻ ♃
12 45 ☽ ⚻ ♆
16 01 ☽ # ☉
16 34 ☽ ± ♂
17 15 ☽ ⚼ ♄
19 27 ☽ □ ♇
23 49 ☽ ☍ ☉

24 Wednesday
00 38 ☽ ± ♆
09 14 ☿ ⊻ ♄
10 18 ☉ ⊥ ♆
11 28 ☽ ⚼ ♂
13 10 ☿ ⊥ ♅
15 02 ☽ # ♃
16 42 ☽ # ♅
18 17 ☽ ⚼ ♆
22 41 ☽ ⚻ ☿
22 52 ☽ △ ♄

25 Thursday
06 36 ☽ Q ♇
10 15 ☽ ± ☿
10 26 ☽ ☍ ♅
11 55 ☽ ☍ ♃
12 54 ☿ St D
16 42 ☽ ⚻ ♀
18 22 ☽ ∥ ♇
18 25 ☽ △ ♂
23 17 ☽ △ ♆

26 Friday
03 31 ☽ ⚼ ☿
05 24 ☽ ± ♀
05 36 ☽ ✶ ♇
14 36 ☽ ⚻ ☉

27 Saturday
00 13 ☽ ⚼ ♀
02 49 ☽ ± ☉
07 59 ☽ △ ☿
08 21 ☽ □ ♄
09 49 ☽ ∠ ♇
12 36 ♀ ⊻ ♂
19 18 ☽ ⚻ ♅
20 54 ☽ ⚼ ☉
21 30 ☽ ⚻ ♃
21 42 ☿ ⊥ ♅

28 Sunday
02 28 ☽ ⊥ ♇
06 15 ☽ □ ♂
06 23 ☽ ± ♅
06 56 ☽ △ ♀
07 31 ☽ □ ♆
08 43 ☽ ± ♃
13 13 ♀ ⊻ ♆
13 27 ☽ ⊻ ♇
17 43 ☽ Q ♄
22 49 ☿ ⊻ ♄
22 51 ☽ ⚼ ♅

29 Monday
01 23 ☽ ⚼ ♃
02 28 ☽ △ ☉
04 31 ♂ ☌ ♆
07 04 ♂ ∥ ♆
15 28 ☽ ✶ ♄
15 45 ☽ □ ☿
16 08 ☽ Q ♆
16 49 ☽ Q ♂

30 Tuesday
01 49 ☽ △ ♅
04 39 ☽ △ ♃
13 25 ☽ ✶ ♆
15 19 ☽ ✶ ♂
18 04 ☽ □ ♀
18 09 ☽ ∠ ♄
19 00 ☽ ☌ ♇
19 04 ♀ ∠ ♄

All ephemeris data is given at 12.00 UT and the Moon's longitude is additionally given for 24.00 UT

MAY 2024

Date	Sidereal time h m s	Sun ☉ ° ' "	Moon ☽ ° ' "	Moon ☽ 24.00 ° ' "	Mercury ☿ ° '	Venus ♀ ° '	Mars ♂ ° '	Jupiter ♃ ° '	Saturn ♄ ° '	Uranus ♅ ° '	Neptune ♆ ° '	Pluto ♇ ° '
01	02 39 38	11 ♉ 35 55	11 ♒ 53 36	18 ♒ 53 04	17 ♈ 24	02 ♉ 29	00 ♈ 39	24 ♉ 14	16 ♓ 41	22 ♉ 25	28 ♓ 57	02 ♒ 06
02	02 43 35	12 34 09	25 ♒ 56 20	03 ♓ 03 15	17 54	03 43	01 26	24 28	16 46	22 28	28 58	02 06
03	02 47 31	13 32 21	10 ♓ 13 34	17 ♓ 27 01	18 28	04 57	02 12	24 42	16 51	22 32	29 00	02 R 06
04	02 51 28	14 30 31	24 ♓ 43 11	02 ♈ 01 34	19 06	06 11	02 58	24 57	16 56	22 35	29 02	02 06
05	02 55 24	15 28 40	09 ♈ 21 35	16 ♈ 42 33	19 48	07 25	03 44	25 11	17 01	22 39	29 04	02 06
06	02 59 21	16 26 47	24 ♈ 03 43	01 ♉ 24 16	20 33	08 39	04 30	25 25	17 06	22 42	29 05	02 06
07	03 03 17	17 24 53	08 ♉ 43 21	16 ♉ 00 08	21 22	09 53	05 16	25 39	17 11	22 46	29 07	02 06
08	03 07 14	18 22 58	23 ♉ 13 46	00 ♊ 23 31	22 14	11 07	06 02	25 53	17 16	22 49	29 09	02 06
09	03 11 11	19 21 01	07 ♊ 28 40	14 ♊ 28 40	23 10	12 21	06 48	26 07	17 20	22 53	29 10	02 06
10	03 15 07	20 19 02	21 ♊ 23 04	28 ♊ 11 33	24 08	13 34	07 34	26 21	17 25	22 56	29 12	02 05
11	03 19 04	21 17 01	04 ♋ 53 57	11 ♋ 30 13	25 10	14 48	08 20	26 35	17 30	23 00	29 13	02 05
12	03 23 00	22 14 59	18 ♋ 00 29	24 ♋ 24 56	26 14	16 02	09 06	26 49	17 34	23 03	29 15	02 05
13	03 26 57	23 12 55	00 ♌ 43 54	06 ♌ 57 48	27 22	17 16	09 52	27 04	17 38	23 07	29 17	02 05
14	03 30 53	24 10 49	13 ♌ 07 07	19 ♌ 12 24	28 32	18 30	10 38	27 18	17 43	23 10	29 18	02 04
15	03 34 50	25 08 41	25 ♌ 14 14	01 ♍ 13 14	29 ♈ 44	19 44	11 24	27 32	17 47	23 14	29 20	02 04
16	03 38 46	26 06 32	07 ♍ 10 02	13 ♍ 05 16	01 ♉ 00	20 57	12 09	27 46	17 51	23 17	29 21	02 04
17	03 42 43	27 04 21	18 ♍ 59 36	24 ♍ 53 39	02 17	22 11	12 55	28 00	17 55	23 21	29 22	02 03
18	03 46 40	28 02 08	00 ♎ 48 01	06 ♎ 43 16	03 38	23 25	13 41	28 14	17 59	23 24	29 24	02 03
19	03 50 36	28 59 53	12 ♎ 39 59	18 ♎ 38 38	05 00	24 39	14 26	28 29	18 03	23 28	29 25	02 02
20	03 54 33	29 ♉ 57 37	24 ♎ 39 42	00 ♏ 43 34	06 25	25 53	15 12	28 43	18 07	23 31	29 26	02 02
21	03 58 29	00 ♊ 55 19	06 ♏ 50 35	13 ♏ 01 03	07 53	27 06	15 57	28 57	18 11	23 34	29 28	02 01
22	04 02 26	01 53 01	19 ♏ 15 10	25 ♏ 33 04	09 23	28 20	16 43	29 11	18 14	23 38	29 29	02 01
23	04 06 22	02 50 40	01 ♐ 54 52	08 ♐ 20 34	10 55	29 ♉ 34	17 28	29 25	18 18	23 41	29 30	02 00
24	04 10 19	03 48 19	14 ♐ 50 08	21 ♐ 23 27	12 30	00 ♊ 48	18 14	29 39	18 21	23 45	29 32	02 00
25	04 14 15	04 45 56	28 ♐ 00 22	04 ♑ 40 43	14 06	02 01	18 59	29 ♉ 53	18 25	23 48	29 33	01 59
26	04 18 12	05 43 32	11 ♑ 24 17	18 ♑ 10 51	15 46	03 15	19 44	00 ♊ 08	18 28	23 52	29 34	01 59
27	04 22 09	06 41 08	25 ♑ 00 10	01 ♒ 52 02	17 27	04 29	20 30	00 22	18 31	23 55	29 35	01 58
28	04 26 05	07 38 42	08 ♒ 46 13	15 ♒ 42 32	19 11	05 43	21 15	00 36	18 34	23 59	29 36	01 57
29	04 30 02	08 36 15	22 ♒ 40 49	29 ♒ 40 54	20 57	06 56	22 00	00 50	18 38	24 02	29 37	01 56
30	04 33 58	09 33 47	06 ♓ 42 40	13 ♓ 45 58	22 46	08 10	22 45	01 04	18 41	24 05	29 38	01 56
31	04 37 55	10 ♊ 31 19	20 ♓ 50 39	27 ♓ 56 36	24 ♉ 36	09 ♊ 24	23 ♈ 30	01 ♊ 18	18 ♓ 43	24 ♉ 09	29 ♓ 39	01 ♒ 55

Date	Moon True ☊ ° '	Moon Mean ☊ ° '	Moon ☽ Latitude ° '
01	15 ♈ 21	14 ♈ 27	04 S 41
02	15 D 21	14 23	03 59
03	15 22	14 20	03 01
04	15 24	14 17	01 51
05	15 25	14 14	00 S 33
06	15 R 25	14 11	00 N 47
07	15 23	14 08	02 04
08	15 20	14 04	03 12
09	15 15	14 01	04 07
10	15 10	13 58	04 45
11	15 04	13 55	05 06
12	14 59	13 52	05 11
13	14 55	13 49	05 00
14	14 53	13 45	04 35
15	14 D 52	13 42	03 58
16	14 53	13 39	03 11
17	14 54	13 36	02 16
18	14 56	13 33	01 16
19	14 57	13 29	00 N 12
20	14 R 56	13 26	00 S 52
21	14 54	13 23	01 56
22	14 50	13 20	02 55
23	14 43	13 17	03 46
24	14 35	13 14	04 27
25	14 27	13 10	04 54
26	14 19	13 07	05 06
27	14 11	13 04	05 01
28	14 06	13 01	04 39
29	14 03	12 58	04 00
30	14 02	12 55	03 06
31	14 ♈ 02	12 ♈ 51	02 S 01

DECLINATIONS

Date	Sun ☉ ° '	Moon ☽ ° '	Mercury ☿ ° '	Venus ♀ ° '	Mars ♂ ° '	Jupiter ♃ ° '	Saturn ♄ ° '	Uranus ♅ ° '	Neptune ♆ ° '	Pluto ♇ ° '
01	15 N 19	21 S 43	04 N 36	11 N 12	00 S 54	18 N 06	06 S 52	18 N 07	01 S 33	22 S 38
02	15 37	16 37	04 40	11 38	00 36	18 09	06 50	18 08	01 32	22 39
03	15 54	10 32	04 46	12 04	00 S 17	18 13	06 48	18 09	01 32	22 39
04	16 11	03 S 48	04 54	12 30	00 N 01	18 16	06 47	18 10	01 31	22 39
05	16 29	03 N 12	05 04	12 56	00 20	18 20	06 45	18 11	01 30	22 39
06	16 45	10 04	05 17	13 21	00 38	18 23	06 43	18 11	01 30	22 39
07	17 02	16 22	05 31	13 46	00 56	18 27	06 41	18 12	01 29	22 40
08	17 18	21 41	05 47	14 11	01 15	18 30	06 40	18 13	01 29	22 40
09	17 34	25 37	06 05	14 35	01 33	18 34	06 38	18 14	01 28	22 40
10	17 50	27 54	06 25	14 59	01 51	18 37	06 36	18 15	01 27	22 40
11	18 05	28 27	06 46	15 23	02 09	18 40	06 35	18 16	01 27	22 40
12	18 20	27 22	07 09	15 46	02 28	18 44	06 33	18 17	01 26	22 41
13	18 35	24 52	07 33	16 09	02 46	18 47	06 32	18 18	01 26	22 41
14	18 49	21 16	07 59	16 31	03 04	18 50	06 30	18 19	01 25	22 41
15	19 03	16 50	08 25	16 53	03 22	18 54	06 29	18 19	01 24	22 41
16	19 17	11 50	08 53	17 15	03 40	18 57	06 27	18 20	01 24	22 42
17	19 30	06 27	09 23	17 36	03 58	19 00	06 26	18 21	01 23	22 42
18	19 43	00 N 51	09 53	17 56	04 16	19 03	06 25	18 22	01 23	22 42
19	19 56	04 S 49	10 24	18 17	04 34	19 06	06 23	18 23	01 22	22 42
20	20 08	10 22	10 56	18 36	04 51	19 10	06 22	18 24	01 22	22 43
21	20 21	15 37	11 29	18 55	05 09	19 13	06 21	18 25	01 21	22 43
22	20 32	20 20	12 03	19 14	05 27	19 16	06 20	18 26	01 21	22 43
23	20 44	24 14	12 38	19 32	05 44	19 19	06 18	18 26	01 20	22 43
24	20 55	26 59	13 13	19 50	06 02	19 22	06 17	18 27	01 20	22 44
25	21 05	28 20	13 48	20 07	06 19	19 25	06 16	18 28	01 20	22 44
26	21 16	28 02	14 24	20 24	06 37	19 28	06 15	18 29	01 19	22 44
27	21 25	26 04	15 01	20 40	06 54	19 31	06 14	18 30	01 19	22 45
28	21 35	22 33	15 38	20 55	07 11	19 34	06 13	18 31	01 18	22 45
29	21 44	17 44	16 14	21 10	07 28	19 37	06 12	18 32	01 18	22 45
30	21 53	11 56	16 51	21 25	07 45	19 40	06 11	18 32	01 18	22 46
31	22 N 01	05 S 29	17 N 28	21 N 38	08 N 02	19 N 43	06 S 10	18 N 33	01 S 17	22 S 46

ZODIAC SIGN ENTRIES

Date	h	m	Planets
02	18	52	☽ ♓
04	20	41	☽ ♈
06	21	42	☽ ♉
08	23	20	☽ ♊
11	03	13	☽ ♋
13	10	36	☽ ♌
15	17	05	☿ ♉
15	21	33	☽ ♍
18	10	23	☽ ♎
20	12	59	☉ ♊
20	22	34	☽ ♏
23	08	24	☽ ♐
23	20	30	♀ ♊
25	15	36	☽ ♑
25	23	15	♃ ♊
27	20	45	☽ ♒
30	00	33	☽ ♓

LATITUDES

Date	Mercury ☿ ° '	Venus ♀ ° '	Mars ♂ ° '	Jupiter ♃ ° '	Saturn ♄ ° '	Uranus ♅ ° '	Neptune ♆ ° '	Pluto ♇ ° '
01	02 S 25	01 S 13	01 S 16	00 S 45	01 S 45	00 S 16	01 S 14	03 S 02
04	02 47	01 08	01 16	00 45	01 45	00 16	01 14	03 02
07	03 02	01 04	01 15	00 45	01 46	00 16	01 14	03 03
10	03 10	00 58	01 15	00 45	01 47	00 16	01 14	03 03
13	03 12	00 53	01 15	00 44	01 47	00 16	01 14	03 04
16	03 07	00 47	01 14	00 44	01 48	00 16	01 15	03 04
19	02 57	00 41	01 14	00 44	01 49	00 16	01 15	03 05
22	02 42	00 34	01 13	00 44	01 49	00 16	01 15	03 05
25	02 22	00 27	01 12	00 43	01 50	00 16	01 15	03 06
28	01 58	00 20	01 11	00 43	01 51	00 16	01 15	03 06
31	01 S 30	00 S 13	01 S 11	00 S 43	01 S 51	00 S 16	01 S 15	03 S 07

DATA

Julian Date	2460432
Delta T	+72 seconds
Ayanamsa	24° 11' 44"
Synetic vernal point	04° ♓ 55' 15"
True obliquity of ecliptic	23° 26' 19"

MOON'S PHASES, APSIDES AND POSITIONS ☽

Date	h	m	Phase	Longitude ° '	Eclipse Indicator
01	11	27	☾	11 ♒ 35	
08	03	22	●	18 ♉ 02	
15	11	48	☽	25 ♌ 08	
23	13	53	○	02 ♐ 55	
30	17	13	☾	09 ♓ 46	

Day	h	m		
05	21	55	Perigee	
17	18	56	Apogee	
05	01	03	0N	
11	07	42	Max dec	28° N 29'
18	15	35	0S	
25	19	49	Max dec	28° S 25'

LONGITUDES

Date	Chiron ⚷ ° '	Ceres ⚳ ° '	Pallas ⚴ ° '	Juno ⚵ ° '	Vesta ⚶ ° '	Black Moon Lilith ⚸ ° '
01	20 ♈ 44	20 ♑ 59	04 ♐ 18	06 ♍ 34	10 ♋ 08	23 ♍ 28
11	21 ♈ 16	21 ♑ 30	01 ♐ 34	07 ♍ 24	13 ♋ 52	24 ♍ 35
21	21 ♈ 47	21 ♑ 25	28 ♏ 33	08 ♍ 44	17 ♋ 46	25 ♍ 41
31	22 ♈ 14	20 ♑ 44	25 ♏ 38	10 ♍ 29	21 ♋ 48	26 ♍ 48

All ephemeris data is given at 12.00 UT and the Moon's longitude is additionally given for 24.00 UT

ASPECTARIAN

h m	Aspects
01 Wednesday	
00 23	☽ Q ☿
04 30	♀ □ ♇
06 58	☽ ∥ ♇
09 54	☽ ⊥ ♄
11 27	☽ □ ☉
15 32	☽ ∠ ♆
18 51	☽ ∠ ♂
20 17	☽ ⊻ ♄
21 48	☽ ⚹ ☿
22 50	♃ ∥ ♅
02 Thursday	
04 09	☽ Q ♀
05 22	☽ # ♃
05 25	☽ # ♅
06 06	☽ □ ♅
06 58	☽ ⊥ ♆
09 28	☽ □ ♃
11 05	☽ ⊥ ♂
16 01	☽ # ☉
17 08	☽ ⊻ ♆
17 43	♇ St R
20 23	☽ Q ☉
21 48	☽ ⊻ ♂
22 24	☽ ⊻ ♇
03 Friday	
00 13	☽ ∠ ☿
02 21	☽ ⚹ ♀
06 34	☽ # ♀
08 28	☽ ⊥ ♇
09 06	♂ ⚹ ♇
12 31	☽ Q ♅
13 01	♀ ⊥ ♆
15 54	☽ ⊥ ☿
16 12	☽ Q ♃
17 55	☽ ⚹ ☉
23 04	☽ ☌ ♄
23 26	☽ ∠ ♇
23 51	☉ ∠ ♆
04 Saturday	
01 32	☽ ∥ ♄
02 18	☽ ⊻ ☿
02 52	☿ ⊥ ♃
05 38	☽ ∠ ♀
08 15	☽ # ☿
08 29	☽ ⚹ ♅
12 22	☽ ⚹ ♃
19 06	☽ ☌ ♆
19 52	☽ ∥ ♆
20 26	☽ ∠ ☉
21 48	☽ ⊥ ♀
05 Sunday	
00 08	☽ ⚹ ♇
00 26	☽ # ♂
01 44	☽ ∥ ♂
02 17	☽ ☌ ♂
06 13	☽ # ♆
08 32	☽ ⊻ ♀
09 11	☽ ∠ ♅
12 12	☽ ⊥ ☉
13 21	☽ ∠ ♃
18 38	☽ ∥ ☿
19 45	☽ Q ♇
22 42	☽ ⊻ ☉
23 57	☽ ⊥ ♅
06 Monday	
00 12	☽ # ♄
00 35	☽ ⊻ ♄
04 17	☽ ⊥ ♃
05 57	☽ ☌ ☿
09 46	☽ ⊻ ♅
10 25	☽ ⊥ ♄
11 50	♄ ∠ ♇
14 14	☽ ⊻ ♃
20 14	☽ ⊻ ♆
07 Tuesday	
01 01	☽ ∥ ♀
01 08	☽ □ ♇
01 13	☽ ∠ ♄
05 42	☉ ⚹ ♄
06 01	☽ ⊻ ♂
06 04	☽ ⊥ ♆
14 05	☽ ☌ ♀
14 52	☽ ∥ ☉
16 26	☽ ⊥ ♂
19 45	☽ ∥ ♅
20 52	☽ ∥ ♃
20 54	☽ ∠ ♆
08 Wednesday	
02 02	☽ ⚹ ♄
03 22	☽ ☌ ☉
08 09	☽ ∠ ♂
10 14	☽ ⊻ ☿
11 19	☽ ☌ ♅
16 30	☽ ☌ ♃
17 14	☽ # ♇
20 57	☽ ⊥ ☿
21 55	☽ ⚹ ♆
22 09	☽ Q ♄
09 Thursday	
02 52	☽ △ ♇
04 17	☿ ⊻ ♅
05 31	♂ # ♆
10 47	☽ ⚹ ♂
13 15	☽ ∠ ☿
16 50	☿ ⊥ ♄
18 19	☽ Q ♆
21 08	☽ ⊻ ♀
10 Friday	
04 31	☽ ⚼ ♇
05 03	☽ □ ♄
08 32	☽ ⊥ ♀
08 39	☽ Q ♂
10 00	☽ ⊻ ☉
11 56	♀ ⊥ ♂
14 44	☽ ⊻ ♅
17 13	☽ ⚹ ☿
20 16	☽ ± ♇
20 54	☽ ⊻ ♃
21 21	☽ ⊥ ☉
11 Saturday	
00 20	☿ # ♄
00 22	♂ ∠ ♅
00 27	♀ ∠ ♆
01 23	☽ ⊥ ♅
01 49	☽ □ ♆
01 57	☽ ∠ ♀
06 57	☽ ⊼ ♇
07 46	☽ ⊥ ♃
14 42	☽ ∠ ☉
16 27	☽ Q ☿
17 38	☽ ∠ ♅
18 36	☽ □ ♂
12 Sunday	
00 23	☽ ∠ ♃
06 44	☉ ∥ ♅
07 57	☽ ⚹ ♀
11 10	☽ △ ♄
20 34	☽ ⚹ ☉
21 29	☽ ⚹ ♅
13 Monday	
04 00	☿ ⊻ ♃
04 52	☽ ⚹ ♃
04 56	☽ □ ☿
08 54	☽ Q ♀
09 13	☽ △ ♆
09 13	☉ ☌ ♅
14 35	☽ ☍ ♇
15 41	☽ ⚼ ♄
19 45	♀ ⚹ ♄
20 27	☽ Q ♅
21 21	☽ Q ☉
14 Tuesday	
03 15	☽ # ♇
04 23	☽ Q ♃
06 49	☽ △ ♂
09 14	☽ ± ♄
14 19	☽ ⚼ ♆
14 47	☉ ∥ ♃
21 06	☽ ⊼ ♄
23 47	☽ □ ♀
15 Wednesday	
01 01	☽ ∥ ☉
01 26	☽ ∥ ♃
03 45	☿ ⊻ ♆
04 21	☽ ∥ ♅
07 58	☽ □ ♅
08 11	☽ ± ♆
11 46	☽ ∥ ♀
11 48	☽ □ ☉
14 28	☽ ⚼ ♂
16 41	☽ □ ♃
20 12	☽ ⊼ ♆
22 04	☽ △ ☿
16 Thursday	
01 42	☽ ⊼ ♇
09 49	☽ ± ♂
13 48	☽ ± ♇
22 48	☽ ⊼ ♂
17 Friday	
00 10	☽ ∥ ☿
07 44	☿ □ ♇
08 03	☽ ⚼ ♇
08 06	☽ ⚼ ☿
09 48	☽ ☍ ♄
12 03	☽ # ♄
16 00	♂ ∠ ♃
19 15	☽ △ ♀
20 53	☽ △ ♅
22 10	☽ ∥ ♂
18 Saturday	
00 00	☿ ∠ ♄
04 43	☽ ± ☿
05 53	☽ △ ☉
06 42	☽ △ ♃
09 09	☽ ☍ ♆
09 43	☽ # ♆
11 40	♀ ☌ ♅
14 32	☽ △ ♇
18 29	☽ ⊼ ☿
18 45	☉ ☌ ♃
21 25	☽ ∥ ♆
23 32	♂ Q ♇
19 Sunday	
03 28	☽ ⚼ ♅
05 13	☽ ⚼ ♀
10 52	☽ # ♂
13 40	☽ ⚼ ♃
14 55	☽ ⚼ ☉
15 48	☽ ☍ ♂
18 42	☽ ∥ ♄
19 12	☿ ⊥ ♆
20 08	♀ ∥ ♅
21 40	☽ ± ♅
22 45	☉ ⚹ ♆
22 52	☽ ⊼ ♄
20 Monday	
01 23	☽ ± ♀
08 03	☽ ± ♃
09 43	☽ ⊼ ♅
10 29	☽ ± ☉
10 54	☽ ± ♄
14 41	☽ ⊼ ♀
14 49	☽ # ☿
16 07	☉ Q ♄
20 11	☽ ⊼ ♃
21 29	☽ ⊼ ♆
23 23	☽ ⊼ ☉
21 Tuesday	
02 34	☽ □ ♇
04 47	☽ ⚼ ♄
09 18	☽ ± ♆
14 18	☽ ☍ ☿
16 00	☉ ∠ ♂
22 Wednesday	
01 50	☽ # ♅
02 49	☽ ⚼ ♆
05 39	☽ # ♀
06 10	☽ # ♃
06 49	☽ ⊼ ♂
10 03	☽ △ ♄
13 10	☽ # ☉
13 27	☽ Q ♇
14 45	♀ ∥ ♃
15 14	☉ △ ♇
19 02	☽ ± ♂
20 24	☽ ☍ ♅
23 Thursday	
01 58	☽ ∥ ♇
07 07	☽ ☍ ♀
07 13	☽ ☍ ♃
07 28	☽ △ ♆
08 29	♀ ☌ ♃
10 50	♀ ⚹ ♆
12 10	☽ ⚹ ♇
13 07	☽ ⚼ ♂
13 53	☽ ☍ ☉
19 28	♂ ⊥ ♅
21 44	♃ ⚹ ♆
24 Friday	
03 02	♀ Q ♄
07 05	☽ ⊼ ☿
15 58	☽ ∠ ♇
16 23	♂ ⊻ ♄
18 29	☽ □ ♄
18 36	☽ △ ♂
19 38	☽ ± ☿
25 Saturday	
04 21	☽ ⊼ ♅
07 57	♂ # ♄
08 21	☽ ⊥ ♇
11 16	♀ △ ♇
14 16	☽ ⚼ ☿
14 47	☽ □ ♆
15 15	☽ ± ♅
15 28	☽ ⊼ ♃
18 31	☿ ∠ ♆
19 10	☽ ⊻ ♇
19 58	☽ ⊼ ♀
26 Sunday	
01 06	☽ ⊼ ☉
02 25	☽ ± ♃
03 10	☽ Q ♄
07 27	☽ ⚼ ♅
07 47	☽ ± ♀
12 37	☽ ± ☉
18 43	☽ ⚼ ♃
20 49	☽ △ ☿
22 56	☽ Q ♆
27 Monday	
00 33	☽ ⚹ ♄
01 20	☽ ⚼ ♀
03 37	☽ □ ♂
05 44	☽ ⚼ ☉
10 05	☽ △ ♅
20 02	☽ ⚹ ♆
21 32	☽ △ ♃
28 Tuesday	
00 10	☽ ☌ ♇
02 56	☽ ∠ ♄
03 22	☿ ⚹ ♄
06 10	☽ △ ♀
09 22	♃ Q ♄
09 54	☽ △ ☉
10 50	☽ ∥ ♇
12 53	☽ Q ♂
17 08	☽ # ☉
18 37	☽ ⊥ ♄
20 19	☽ # ♀
22 06	☽ ∠ ♆
29 Wednesday	
03 20	☽ # ♃
05 00	☽ ⊻ ♄
08 24	☽ # ♅
08 36	☽ □ ☿
10 46	☽ ⚹ ♂
13 37	☽ ⊥ ♆
14 20	☽ □ ♅
15 05	♀ ∠ ♂
17 53	☽ # ☿
23 55	☽ ⊻ ♆
30 Thursday	
02 12	☽ □ ♃
03 51	☽ ⊻ ♇
11 54	☿ ⊻ ♂
13 53	☽ ∠ ♂
14 04	☽ ⊥ ♇
14 43	☽ □ ♀
17 13	☽ □ ☉
19 55	☽ Q ☿
21 11	☽ Q ♅
31 Friday	
03 06	☽ # ♂
05 21	☽ ∠ ♇
05 54	☿ ☌ ♅
06 02	☽ ⊥ ♂
08 24	☽ ☌ ♄
09 20	☽ Q ♃
09 31	☽ ∥ ♄
16 45	☽ ⊻ ♂
17 36	☽ ⚹ ♅
19 19	☽ ⚹ ☿

JUNE 2024

LONGITUDES

Date	Sidereal time h m s	Sun ☉ ° ' "	Moon ☽ ° ' "	Moon ☽ 24.00 ° ' "	Mercury ☿ ° '	Venus ♀ ° '	Mars ♂ ° '	Jupiter ♃ ° '	Saturn ♄ ° '	Uranus ♅ ° '	Neptune ♆ ° '	Pluto ♇ ° '
01	04 41 51	11 ♊ 28 50	05 ♈ 03 37	12 ♈ 11 28	26 ♉ 29	10 ♊ 38	24 ♈ 15	01 ♊ 32	18 ♓ 46	24 ♉ 12	29 ♓ 40	01 ♒ 54
02	04 45 48	12 26 20	19 ♈ 19 53	26 ♈ 28 32	28 ♉ 24	11 51	25 00	01 46	18 49	24 15	29 41	01 R 53
03	04 49 44	13 23 49	03 ♉ 37 01	10 ♉ 44 53	00 ♊ 22	13 05	25 45	02 00	18 52	24 19	29 42	01 53
04	04 53 41	14 21 18	17 ♉ 51 37	24 ♉ 56 41	02 21	14 19	26 30	02 14	18 54	24 22	29 43	01 52
05	04 57 38	15 18 46	01 ♊ 59 29	08 ♊ 59 27	04 23	15 33	27 15	02 28	18 56	24 25	29 44	01 51
06	05 01 34	16 16 13	15 ♊ 56 01	22 ♊ 48 40	06 26	16 46	28 00	02 42	18 59	24 29	29 45	01 50
07	05 05 31	17 13 39	29 ♊ 36 57	06 ♋ 20 31	08 31	18 00	28 45	02 56	19 01	24 32	29 46	01 49
08	05 09 27	18 11 04	12 ♋ 59 04	19 ♋ 32 28	10 38	19 14	29 ♈ 29	03 10	19 03	24 35	29 47	01 48
09	05 13 24	19 08 29	26 ♋ 00 40	02 ♌ 23 43	12 46	20 28	00 ♉ 14	03 23	19 05	24 39	29 47	01 47
10	05 17 20	20 05 52	08 ♌ 41 48	14 ♌ 55 11	14 56	21 41	00 58	03 37	19 07	24 42	29 48	01 46
11	05 21 17	21 03 14	21 ♌ 04 15	27 ♌ 09 26	17 06	22 55	01 43	03 51	19 09	24 45	29 49	01 45
12	05 25 13	22 00 36	03 ♍ 11 16	09 ♍ 10 18	19 17	24 09	02 27	04 05	19 11	24 48	29 49	01 44
13	05 29 10	22 57 56	15 ♍ 07 08	21 ♍ 02 27	21 29	25 23	03 12	04 19	19 12	24 51	29 50	01 43
14	05 33 07	23 55 15	26 ♍ 56 54	02 ♎ 51 10	23 41	26 36	03 56	04 32	19 14	24 55	29 51	01 42
15	05 37 03	24 52 33	08 ♎ 45 55	14 ♎ 41 50	25 53	27 50	04 40	04 46	19 15	24 58	29 51	01 41
16	05 41 00	25 49 51	20 ♎ 39 34	26 ♎ 39 44	28 ♊ 05	29 ♊ 04	05 24	05 00	19 17	25 01	29 52	01 40
17	05 44 56	26 47 08	02 ♏ 42 55	08 ♏ 49 40	00 ♋ 16	00 ♋ 17	06 08	05 13	19 18	25 04	29 52	01 39
18	05 48 53	27 44 24	15 ♏ 00 24	21 ♏ 15 34	02 26	01 31	06 52	05 27	19 19	25 07	29 53	01 38
19	05 52 49	28 41 39	27 ♏ 35 25	04 ♐ 00 13	04 35	02 45	07 36	05 40	19 20	25 10	29 53	01 37
20	05 56 46	29 ♊ 38 54	10 ♐ 30 02	17 ♐ 04 52	06 43	03 59	08 20	05 54	19 21	25 13	29 54	01 36
21	06 00 42	00 ♋ 36 08	23 ♐ 44 38	00 ♑ 29 06	08 49	05 12	09 04	06 07	19 22	25 16	29 54	01 35
22	06 04 39	01 33 22	07 ♑ 17 57	14 ♑ 10 47	10 54	06 26	09 48	06 21	19 23	25 19	29 54	01 33
23	06 08 36	02 30 35	21 ♑ 07 07	28 ♑ 06 25	12 57	07 40	10 32	06 34	19 24	25 22	29 55	01 32
24	06 12 32	03 27 48	05 ♒ 08 09	12 ♒ 11 45	14 59	08 53	11 15	06 47	19 24	25 25	29 55	01 31
25	06 16 29	04 25 01	19 ♒ 16 42	26 ♒ 22 29	16 58	10 07	11 59	07 01	19 25	25 28	29 55	01 30
26	06 20 25	05 22 14	03 ♓ 28 40	10 ♓ 34 53	18 55	11 21	12 43	07 14	19 25	25 31	29 55	01 29
27	06 24 22	06 19 26	17 ♓ 40 49	24 ♓ 46 13	20 51	12 34	13 26	07 27	19 25	25 33	29 56	01 27
28	06 28 18	07 16 39	01 ♈ 50 53	08 ♈ 54 42	22 44	13 48	14 09	07 40	19 26	25 36	29 56	01 26
29	06 32 15	08 13 52	15 ♈ 57 31	22 ♈ 59 17	24 35	15 02	14 53	07 53	19 26	25 39	29 56	01 25
30	06 36 11	09 ♋ 11 05	29 ♈ 59 52	06 ♉ 59 10	26 ♋ 24	16 ♋ 16	15 ♉ 36	08 ♊ 07	19 ♓ 26	25 ♉ 42	29 ♓ 56	01 ♒ 24

Date	Moon True ☊ ° '	Moon Mean ☊ ° '	Moon ☽ Latitude ° '
01	14 ♈ 03	12 ♈ 48	00 S 48
02	14 R 03	12 45	00 N 28
03	14 01	12 42	01 43
04	13 58	12 39	02 50
05	13 51	12 35	03 47
06	13 43	12 32	04 29
07	13 32	12 29	04 55
08	13 22	12 26	05 03
09	13 12	12 23	04 56
10	13 03	12 20	04 34
11	12 57	12 16	03 59
12	12 53	12 13	03 15
13	12 52	12 10	02 22
14	12 D 52	12 07	01 24
15	12 52	12 04	00 N 22
16	12 R 52	12 01	00 S 41
17	12 50	11 57	01 43
18	12 46	11 54	02 42
19	12 40	11 51	03 34
20	12 31	11 48	04 16
21	12 20	11 45	04 45
22	12 09	11 41	05 00
23	11 57	11 38	04 57
24	11 47	11 35	04 37
25	11 39	11 32	03 59
26	11 34	11 29	03 07
27	11 31	11 26	02 02
28	11 D 31	11 22	00 S 51
29	11 R 31	11 19	00 N 24
30	11 ♈ 30	11 ♈ 16	01 N 36

DECLINATIONS

Date	Sun ☉ ° '	Moon ☽ ° '	Mercury ☿ ° '	Venus ♀ ° '	Mars ♂ ° '	Jupiter ♃ ° '	Saturn ♄ ° '	Uranus ♅ ° '	Neptune ♆ ° '	Pluto ♇ ° '
01	22 N 09	01 N 17	18 N 04	21 N 51	08 N 19	19 N 46	06 S 09	18 N 34	01 S 17	22 S 46
02	22 17	08 00	18 40	22 04	08 36	19 49	06 08	18 35	01 16	22 47
03	22 24	14 20	19 15	22 16	08 52	19 52	06 08	18 36	01 16	22 47
04	22 31	19 53	19 50	22 27	09 09	19 54	06 07	18 37	01 16	22 47
05	22 38	24 16	20 24	22 38	09 25	19 57	06 06	18 37	01 16	22 48
06	22 44	27 09	20 56	22 48	09 42	20 00	06 05	18 38	01 15	22 48
07	22 50	28 21	21 27	22 57	09 58	20 03	06 05	18 39	01 15	22 48
08	22 55	27 50	21 57	23 06	10 14	20 06	06 04	18 40	01 15	22 49
09	23 00	25 47	22 25	23 14	10 30	20 08	06 04	18 41	01 14	22 49
10	23 04	22 29	22 51	23 21	10 46	20 11	06 03	18 41	01 14	22 49
11	23 08	18 15	23 15	23 28	11 02	20 13	06 03	18 42	01 14	22 50
12	23 12	13 22	23 37	23 34	11 17	20 16	06 02	18 43	01 14	22 50
13	23 15	08 03	23 56	23 39	11 33	20 19	06 02	18 44	01 14	22 51
14	23 18	02 N 30	24 12	23 43	11 48	20 21	06 01	18 44	01 13	22 51
15	23 20	03 S 08	24 26	23 47	12 04	20 24	06 01	18 45	01 13	22 51
16	23 22	08 42	24 38	23 50	12 19	20 26	06 01	18 46	01 13	22 52
17	23 24	14 02	24 46	23 53	12 34	20 29	06 00	18 47	01 13	22 52
18	23 25	18 55	24 51	23 55	12 49	20 31	06 00	18 47	01 13	22 52
19	23 26	23 05	24 54	23 56	13 03	20 34	06 00	18 48	01 13	22 53
20	23 26	26 14	24 54	23 56	13 18	20 36	06 00	18 49	01 12	22 53
21	23 26	28 03	24 51	23 56	13 32	20 38	06 00	18 50	01 12	22 54
22	23 26	28 14	24 46	23 55	13 47	20 41	06 00	18 50	01 12	22 54
23	23 25	26 40	24 38	23 53	14 01	20 43	06 00	18 51	01 12	22 54
24	23 24	23 27	24 28	23 50	14 15	20 45	06 00	18 52	01 12	22 55
25	23 22	18 49	24 15	23 47	14 29	20 47	06 00	18 52	01 12	22 55
26	23 20	13 08	24 00	23 43	14 42	20 50	06 00	18 53	01 12	22 56
27	23 17	06 45	23 43	23 39	14 56	20 52	06 00	18 54	01 12	22 56
28	23 14	00 S 03	23 25	23 33	15 09	20 54	06 00	18 54	01 12	22 56
29	23 11	06 N 38	23 04	23 27	15 23	20 56	06 00	18 55	01 12	22 57
30	23 N 07	12 N 58	22 N 42	23 N 21	15 N 36	20 N 58	06 S 00	18 N 56	01 S 12	22 S 57

ZODIAC SIGN ENTRIES

Date	h	m	Planets
01	03	28	☽ ♈
03	05	55	☽ ♉
03	07	37	☿ ♊
05	08	36	☽ ♊
07	12	41	☽ ♋
09	04	35	♂ ♉
09	19	29	☽ ♌
12	05	39	☽ ♍
14	18	12	☽ ♎
17	06	20	♀ ♋
17	06	38	☽ ♏
17	09	07	☿ ♋
19	16	32	☽ ♐
20	20	51	☉ ♋
21	23	08	☽ ♑
24	03	14	☽ ♒
26	06	08	☽ ♓
28	08	52	☽ ♈
30	12	00	☽ ♉

LATITUDES

Date	Mercury ☿ ° '	Venus ♀ ° '	Mars ♂ ° '	Jupiter ♃ ° '	Saturn ♄ ° '	Uranus ♅ ° '	Neptune ♆ ° '	Pluto ♇ ° '
01	01 S 20	00 S 11	01 S 10	00 S 43	01 S 52	00 S 16	01 S 15	03 S 07
04	00 49	00 S 04	01 09	00 43	01 52	00 16	01 15	03 07
07	00 S 16	00 N 03	01 08	00 43	01 53	00 16	01 16	03 08
10	00 N 16	00 10	01 07	00 42	01 54	00 16	01 16	03 08
13	00 46	00 17	01 06	00 42	01 55	00 16	01 16	03 09
16	01 12	00 24	01 04	00 42	01 55	00 16	01 16	03 09
19	01 33	00 31	01 03	00 42	01 56	00 16	01 16	03 10
22	01 47	00 38	01 02	00 42	01 57	00 16	01 16	03 10
25	01 54	00 44	01 00	00 42	01 58	00 16	01 16	03 10
28	01 55	00 50	00 58	00 42	01 58	00 16	01 17	03 11
31	01 N 50	00 N 56	00 S 57	00 S 42	01 S 59	00 S 16	01 S 17	03 S 11

DATA

Julian Date	2460463
Delta T	+72 seconds
Ayanamsa	24° 11' 49"
Synetic vernal point	04° ♓ 55' 10"
True obliquity of ecliptic	23° 26' 19"

MOON'S PHASES, APSIDES AND POSITIONS ☽

Date	h	m	Phase	Longitude ° '	Eclipse Indicator
06	12	38	●	16 ♊ 18	
14	05	18	☽	23 ♍ 39	
22	01	08	○	01 ♑ 07	
28	21	53	☾	07 ♈ 40	

Day	h	m		
02	07	05	Perigee	
14	13	35	Apogee	
27	11	39	Perigee	
01	07	30	0N	
07	16	36	Max dec	28° N 23'
14	22	38	0S	
22	02	37	Max dec	28° S 21'
28	12	09	0N	

LONGITUDES

Date	Chiron ⚷ ° '	Ceres ⚳ ° '	Pallas ⚴ ° '	Juno ⚵ ° '	Vesta ⚶ ° '	Black Moon Lilith ⚸ ° '
01	22 ♈ 17	20 ♑ 38	25 ♏ 22	10 ♍ 40	22 ♋ 13	26 ♍ 55
11	22 ♈ 41	19 ♑ 19	22 ♏ 54	12 ♍ 49	26 ♋ 24	28 ♍ 01
21	23 ♈ 00	17 ♑ 32	21 ♏ 06	15 ♍ 14	00 ♌ 41	29 ♍ 08
31	23 ♈ 16	15 ♑ 25	20 ♏ 03	17 ♍ 55	05 ♌ 03	00 ♎ 14

All ephemeris data is given at 12.00 UT and the Moon's longitude is additionally given for 24.00 UT

ASPECTARIAN

h m Aspects

01 Saturday
00 08 ☽ Q ♀
01 55 ☽ Q ☉
02 55 ☽ ☌ ♆
02 58 ☽ ∥ ♆
05 57 ☽ ⚹ ♃
06 41 ☽ ⚹ ♇
10 07 ♂ ⊻ ♅
12 01 ☽ ⋕ ♆
16 56 ☉ Q ♆
19 00 ☽ ∠ ♅
22 15 ☽ ⚹ ♀
23 35 ☽ ⚹ ☉

02 Sunday
00 29 ☽ ∠ ☿
02 52 ☽ Q ♇
05 17 ☽ ⋕ ♄
05 31 ♂ ⊥ ♄
07 37 ☽ ∠ ♃
08 30 ☿ ∥ ♅
08 43 ♀ Q ♆
10 11 ☽ ⊥ ♅
11 08 ☽ ⊻ ♄
14 16 ☽ ∥ ♂
17 59 ☽ ⊥ ☿
20 18 ☽ ⊻ ♅
21 14 ☽ ⊥ ♄
22 04 ☽ ☌ ♂
22 59 ☽ ⊥ ♃

03 Monday
00 13 ♃ △ ♇
01 50 ☽ ∠ ♀
02 36 ☽ ∠ ☉
03 57 ☿ ⚹ ♆
05 25 ☽ ⊻ ♆
05 39 ☽ ⊻ ☿
09 05 ☽ □ ♇
09 14 ☽ ⊻ ♃
12 24 ☽ ∠ ♄
15 31 ☽ ⊥ ♆
18 11 ☿ Q ♄
18 23 ☽ ⊥ ♀
18 49 ☽ ⊥ ☉
23 22 ♂ ⊥ ♃

04 Tuesday
05 27 ☽ ⊻ ♀
05 39 ☽ ⊻ ☉
06 05 ☽ ∥ ♅
06 12 ☿ △ ♇
06 42 ☽ ∠ ♆
10 23 ☿ ☌ ♃
11 45 ☽ ∥ ☿
12 08 ☽ ∥ ♃
13 46 ☽ ⚹ ♄
14 53 ☿ ⊥ ♂
15 29 ☿ ∥ ♃
15 33 ☉ ☌ ♀
23 04 ☽ ☌ ♅

05 Wednesday
01 44 ☽ ∥ ♀
01 54 ☽ ∥ ☉
03 04 ☽ ⋕ ♇
03 28 ☽ ⊻ ♂
08 09 ☽ ⚹ ♆
10 12 ☽ Q ♄
11 45 ☽ △ ♇
12 01 ☉ ∥ ♀
12 49 ☽ ☌ ♃
14 16 ☽ ⊥ ♂
16 46 ☽ ☌ ☿

06 Thursday
04 45 ☽ Q ♆
06 37 ☽ ∠ ♂
12 38 ☽ ☌ ☉
12 38 ♀ ⋕ ♇
13 11 ♀ ⚼ ♇
13 34 ☽ ⚼ ♇
13 36 ☽ ☌ ♀
17 19 ☽ □ ♄

07 Friday
01 56 ☉ ⚼ ♇
02 59 ☽ ⊻ ♅
05 18 ☽ ± ♇
06 29 ☉ ⋕ ♇
10 22 ☽ ⚹ ♂
12 16 ☽ □ ♆
13 38 ☽ ⊥ ♅
15 55 ☽ ⊼ ♇
18 00 ☽ ⊻ ♃

08 Saturday
04 57 ☽ ⊥ ♃
05 49 ☽ ∠ ♅
06 55 ☽ ⊻ ☿
08 25 ♀ □ ♄
09 07 ☽ Q ♂
19 57 ☽ ⊥ ☿
21 33 ♂ ⊻ ♆
21 38 ☽ ∠ ♃
22 15 ☽ ⊻ ☉
23 08 ☽ △ ♄

09 Sunday
00 37 ☽ ⊻ ♀
00 59 ☿ Q ♆
09 26 ☽ ⚹ ♅
10 15 ☽ ⊥ ☉
10 36 ☉ □ ♄
12 56 ☽ ⊥ ♀
15 57 ☽ ∠ ☿
19 05 ☽ △ ♆
20 24 ☽ □ ♂
22 50 ☽ ☍ ♇

10 Monday
02 09 ☽ ⚹ ♃
03 15 ☽ ⚼ ♄
04 34 ☽ ∠ ☉
06 35 ☽ ∥ ♀
07 45 ☽ ∠ ♀
08 09 ☽ Q ♅
08 21 ☽ ∥ ☉
09 52 ☽ ⋕ ♇
09 55 ☽ ∥ ☿
10 22 ☿ ⋕ ♇
20 32 ☽ ± ♄
23 47 ☽ ⚼ ♆

11 Tuesday
01 28 ☽ ∥ ♃
01 37 ☽ Q ♃
02 34 ☽ ⚹ ☿
03 31 ☉ ∥ ☿
05 32 ☿ ∠ ♂
08 14 ☽ ⊼ ♄
08 14 ☿ ⚼ ♇
09 38 ☽ ∥ ♅
11 58 ☽ ⚹ ☉
13 21 ♂ □ ♇
16 02 ☽ ⚹ ♀
17 24 ☽ ± ♆
19 16 ☽ □ ♅

12 Wednesday
05 17 ☽ ⊼ ♆
07 09 ☿ ∥ ♀
07 22 ☽ Q ☿
09 07 ☽ ⊼ ♇
10 26 ☽ △ ♂
10 47 ☿ □ ♄
13 47 ☽ Q ☉
13 49 ☽ □ ♃
18 36 ☽ Q ♀
21 05 ☽ ∥ ♂
21 07 ☽ ± ♇

13 Thursday
01 24 ♀ ⊻ ♅
15 14 ☽ ⚼ ♇
18 38 ☽ ⚼ ♂
18 41 ♀ ± ♇
20 18 ☽ ☍ ♄
20 49 ☽ ⋕ ♄

14 Friday
03 51 ☽ □ ☿
05 18 ☽ □ ☉
07 50 ☽ △ ♅
11 13 ☽ □ ♀
14 08 ☽ ± ♂
16 32 ☉ ☌ ☿
17 26 ☽ ⋕ ♆
17 54 ☽ ☍ ♆
21 39 ☽ △ ♇
22 10 ♂ ∠ ♄

15 Saturday
01 41 ☿ ⊻ ♅
03 08 ☽ ⊼ ♂
03 44 ☽ △ ♃
03 49 ☽ ∥ ♆
09 52 ☿ ± ♇
14 16 ☉ ⊻ ♅
14 26 ☽ ⚼ ♅
16 37 ♂ ⊻ ♃

16 Sunday
00 18 ☽ ∥ ♄
08 02 ☉ ± ♇
08 41 ☽ ± ♅
09 14 ☽ ⊼ ♄
10 38 ☽ ⚼ ♃
20 45 ☽ ⊼ ♅
21 16 ☽ ± ♄
23 14 ☽ △ ☉

17 Monday
03 06 ♂ ⊥ ♆
03 46 ♀ □ ♆
04 52 ☽ ⋕ ♂
04 57 ☽ ± ♃
06 05 ☽ △ ☿
06 22 ☽ ⊼ ♆
06 40 ☽ △ ♀
07 41 ☿ □ ♆
09 54 ☽ □ ♇
12 43 ☿ ☌ ♀
15 08 ☽ ⚼ ♄
17 02 ☽ ⊼ ♃
18 13 ☽ ± ♆
19 10 ☽ ☍ ♂
21 04 ☿ ⊥ ♅

18 Tuesday
03 13 ☿ ⊼ ♇
03 48 ♀ ⊥ ♅
07 15 ☽ ⚼ ☉
11 21 ☽ ⋕ ♅
11 45 ☽ ⚼ ♆
14 12 ♀ ⊼ ♇
15 14 ☽ ⚼ ♀
17 39 ☽ ⚼ ☿
20 19 ☽ △ ♄
20 45 ☽ ⋕ ♃
20 53 ☽ Q ♇

19 Wednesday
01 58 ☽ ± ☉
07 24 ☽ ☍ ♅
10 14 ☽ ± ♀
10 40 ☽ ∥ ♇
14 15 ☽ ⊼ ☉
14 15 ☽ ± ☿
14 16 ☽ ⋕ ☉
16 19 ☽ △ ♆
17 38 ☽ ⋕ ♀
19 32 ☽ ⚹ ♇
22 41 ☽ ⊼ ♀

20 Thursday
00 43 ☽ ⋕ ☿
01 40 ☿ ⊻ ♃
03 22 ☽ ☍ ♃
03 40 ☽ ⊼ ☿
07 47 ☽ ⊼ ♂
18 12 ☉ □ ♆
19 26 ☽ ± ♂
23 06 ☽ ∠ ♇

21 Friday
04 08 ☽ □ ♄
12 37 ☽ ⚼ ♂
14 44 ☽ ⊼ ♅
15 16 ☽ ⊥ ♇
16 23 ☿ ⚹ ♂
22 58 ☽ □ ♆

22 Saturday
01 08 ☽ ☍ ☉
01 26 ☽ ± ♅
01 55 ☽ ⊻ ♇
05 02 ☿ ∠ ♅
05 39 ☉ ⊥ ♅
09 56 ♀ ⊻ ♃
10 18 ☽ ⊼ ♃
10 20 ☽ ☍ ♀
12 01 ☉ ⊼ ♇
12 09 ☽ Q ♄
16 37 ☽ △ ♂
17 18 ☽ ⚼ ♅
19 25 ☽ ☍ ☿
20 57 ☽ ± ♃

23 Sunday
06 28 ☽ Q ♆
06 54 ☿ ⊥ ♃
09 02 ☽ ⚹ ♄
12 47 ☽ ⚼ ♃
19 20 ☽ △ ♅

24 Monday
03 05 ☽ ⚹ ♆
05 15 ☽ ⋕ ☿
05 50 ☽ ☌ ♇
08 57 ☽ ⊼ ☉
09 35 ☽ ⋕ ♀
10 45 ☽ ∠ ♄
12 21 ☽ ⋕ ☉
14 52 ☽ △ ♃
15 10 ☽ ∥ ♇
19 00 ☽ ⊼ ♀
19 54 ☽ ± ☉
22 58 ☽ □ ♂

25 Tuesday
02 04 ☽ ⊥ ♄
02 37 ☽ ⋕ ♃
04 37 ☽ ∠ ♆
06 08 ☽ ± ♀
07 27 ☽ ⊼ ☿
11 45 ☽ ⋕ ♅
12 14 ☽ ⊻ ♄
12 15 ☽ ⚼ ☉
19 01 ♀ ∠ ♅
19 15 ☽ ± ☿
19 51 ☽ ⊥ ♆
22 30 ☽ □ ♅
22 49 ☽ ⚼ ♀

26 Wednesday
05 55 ☽ ⋕ ♂
06 00 ☽ ⊻ ♆
07 04 ☽ Q ♂
08 37 ☽ ⊻ ♇
12 52 ☽ ⚼ ☿
15 26 ☽ △ ☉
18 10 ☿ △ ♄
18 27 ☽ □ ♃
18 45 ☽ ⊥ ♇

27 Thursday
02 33 ☽ △ ♀
04 26 ☽ ⚹ ♂
05 00 ☽ Q ♅
09 56 ☽ ∠ ♇
14 43 ☽ ∥ ♄
14 57 ☽ ☌ ♄
18 11 ☽ △ ☿
20 45 ☿ ∥ ♀

28 Friday
01 22 ☽ Q ♃
01 23 ☽ ⚹ ♅
07 11 ☽ ∠ ♂
07 53 ☽ ∥ ♆
08 44 ☽ ☌ ♆
08 53 ♀ ⊥ ♃
11 05 ☿ ∠ ♃
11 18 ☽ ⚹ ♇
16 25 ☽ ⋕ ♆
21 53 ☽ □ ☉
22 03 ☽ ⚹ ♃
23 18 ☽ ⊥ ♂

29 Saturday
00 54 ☉ ⊻ ♃
02 56 ☽ ∠ ♅
02 57 ☉ ∥ ☿
04 49 ♀ ⚹ ♂
07 40 ☽ Q ♇
09 40 ☽ ⋕ ♄
10 04 ☽ ⊻ ♂
10 16 ☽ □ ♀
13 38 ♂ ∠ ♆
17 55 ☽ ⊻ ♄
18 19 ☽ ⊥ ♅
19 08 ♄ St R
20 16 ☿ ⋕ ♇

30 Sunday
00 01 ☽ ∠ ♃
02 20 ☿ ⚹ ♅
04 10 ☽ ⊥ ♄
04 37 ☽ ⊻ ♅
04 56 ☽ □ ☿
06 50 ☽ Q ☉
11 53 ☽ ⊻ ♆
14 23 ☽ □ ♇
15 41 ☽ ⊥ ♃
19 36 ☽ ∠ ♄
20 01 ☽ Q ♀
22 11 ☽ ⊥ ♆
23 10 ☽ ∥ ♂

JULY 2024

LONGITUDES

Date	Sidereal time (h m s)	Sun ☉ (° ' ")	Moon ☽ (° ' ")	Moon ☽ 24.00 (° ' ")	Mercury ☿ (° ')	Venus ♀ (° ')	Mars ♂ (° ')	Jupiter ♃ (° ')	Saturn ♄ (° ')	Uranus ♅ (° ')	Neptune ♆ (° ')	Pluto ♇ (° ')
01	06 40 08	10 ♋ 08 18	13 ♉ 57 04	20 ♉ 53 24	28 ♋ 11	17 ♋ 29	16 ♉ 19	08 ♊ 20	19 ♓ 26	25 ♉ 45	29 ♓ 56	01 ♒ 22
02	06 44 05	11 05 31	27 ♉ 47 58	04 ♊ 40 31	29 ♋ 56	18 43	17 02	08 32	19 R 25	25 47	29 R 56	01 R 21
03	06 48 01	12 02 45	11 ♊ 30 47	18 ♊ 18 29	01 ♌ 39	19 57	17 45	08 45	19 25	25 50	29 56	01 20
04	06 51 58	12 59 59	25 ♊ 03 18	01 ♋ 44 55	03 20	21 11	18 28	08 58	19 25	25 53	29 56	01 18
05	06 55 54	13 57 12	08 ♋ 23 04	14 ♋ 57 30	04 58	22 24	19 11	09 11	19 24	25 55	29 56	01 17
06	06 59 51	14 54 26	21 ♋ 27 59	27 ♋ 54 24	06 35	23 38	19 54	09 24	19 23	25 58	29 56	01 16
07	07 03 47	15 51 40	04 ♌ 16 40	10 ♌ 34 47	08 09	24 52	20 37	09 36	19 23	26 00	29 55	01 14
08	07 07 44	16 48 54	16 ♌ 48 51	22 ♌ 59 02	09 41	26 06	21 20	09 49	19 22	26 03	29 55	01 13
09	07 11 40	17 46 07	29 ♌ 05 36	05 ♍ 08 51	11 11	27 19	22 02	10 01	19 21	26 05	29 55	01 12
10	07 15 37	18 43 21	11 ♍ 09 12	17 ♍ 07 08	12 39	28 33	22 45	10 14	19 20	26 08	29 55	01 10
11	07 19 34	19 40 34	23 ♍ 03 09	28 ♍ 57 51	14 04	29 ♋ 47	23 27	10 26	19 19	26 10	29 55	01 09
12	07 23 30	20 37 48	04 ♎ 51 50	10 ♎ 45 45	15 27	01 ♌ 00	24 09	10 39	19 18	26 12	29 54	01 07
13	07 27 27	21 35 01	16 ♎ 40 16	22 ♎ 36 05	16 48	02 14	24 52	10 51	19 16	26 15	29 54	01 06
14	07 31 23	22 32 15	28 ♎ 33 52	04 ♏ 34 19	18 07	03 28	25 34	11 03	19 15	26 17	29 54	01 05
15	07 35 20	23 29 28	10 ♏ 38 04	16 ♏ 45 46	19 23	04 42	26 16	11 15	19 13	26 19	29 53	01 03
16	07 39 16	24 26 42	22 ♏ 58 00	29 ♏ 15 15	20 37	05 55	26 58	11 27	19 12	26 21	29 53	01 02
17	07 43 13	25 23 56	05 ♐ 37 59	12 ♐ 06 31	21 48	07 09	27 40	11 39	19 10	26 24	29 52	01 00
18	07 47 09	26 21 10	18 ♐ 41 05	25 ♐ 21 46	22 57	08 23	28 22	11 51	19 08	26 26	29 52	00 59
19	07 51 06	27 18 24	02 ♑ 08 31	09 ♑ 01 10	24 03	09 37	29 03	12 03	19 07	26 28	29 51	00 58
20	07 55 03	28 15 39	15 ♑ 59 20	23 ♑ 02 32	25 07	10 50	29 ♉ 45	12 15	19 05	26 30	29 51	00 56
21	07 58 59	29 ♋ 12 54	00 ♒ 10 10	07 ♒ 21 31	26 08	12 04	00 ♊ 26	12 27	19 03	26 32	29 50	00 55
22	08 02 56	00 ♌ 10 10	14 ♒ 35 45	21 ♒ 52 04	27 05	13 18	01 08	12 38	19 00	26 34	29 49	00 53
23	08 06 52	01 07 26	29 ♒ 09 35	06 ♓ 27 29	28 00	14 31	01 49	12 50	18 58	26 36	29 49	00 52
24	08 10 49	02 04 43	13 ♓ 45 00	21 ♓ 01 28	28 51	15 45	02 31	13 01	18 56	26 38	29 48	00 50
25	08 14 45	03 02 01	28 ♓ 16 17	05 ♈ 28 57	29 ♌ 40	16 59	03 12	13 12	18 53	26 40	29 47	00 49
26	08 18 42	03 59 19	12 ♈ 39 07	19 ♈ 46 30	00 ♍ 24	18 13	03 53	13 24	18 51	26 41	29 47	00 48
27	08 22 38	04 56 39	26 ♈ 50 54	03 ♉ 52 14	01 06	19 26	04 34	13 35	18 48	26 43	29 46	00 46
28	08 26 35	05 54 00	10 ♉ 50 27	17 ♉ 45 30	01 43	20 40	05 15	13 46	18 46	26 45	29 45	00 45
29	08 30 32	06 51 21	24 ♉ 37 26	01 ♊ 26 16	02 16	21 54	05 56	13 57	18 43	26 47	29 44	00 43
30	08 34 28	07 48 44	08 ♊ 12 01	14 ♊ 54 43	02 46	23 07	06 36	14 08	18 40	26 48	29 43	00 42
31	08 38 25	08 ♌ 46 08	21 ♊ 34 21	28 ♊ 10 55	03 ♍ 11	24 ♌ 21	07 ♊ 17	14 ♊ 19	18 ♓ 37	26 ♉ 50	29 ♓ 43	00 ♒ 41

Date	Moon True ☊ (° ')	Moon Mean ☊ (° ')	Moon ☽ Latitude (° ')
01	11 ♈ 28	11 ♈ 13	02 N 42
02	11 R 23	11 10	03 39
03	11 15	11 07	04 21
04	11 05	11 03	04 49
05	10 52	11 00	05 00
06	10 40	10 57	04 55
07	10 28	10 54	04 35
08	10 17	10 51	04 02
09	10 09	10 47	03 19
10	10 04	10 44	02 27
11	10 01	10 41	01 29
12	10 00	10 38	00 N 27
13	10 D 00	10 35	00 S 35
14	10 R 00	10 32	01 37
15	09 59	10 28	02 35
16	09 56	10 25	03 27
17	09 50	10 22	04 11
18	09 43	10 19	04 42
19	09 33	10 16	05 00
20	09 22	10 12	05 00
21	09 11	10 09	04 42
22	09 02	10 06	04 07
23	08 55	10 03	03 14
24	08 50	10 00	02 09
25	08 48	09 57	00 S 56
26	08 D 48	09 53	00 N 21
27	08 48	09 50	01 35
28	08 R 48	09 47	02 42
29	08 47	09 44	03 39
30	08 43	09 41	04 23
31	08 ♈ 38	09 ♈ 38	04 N 51

DECLINATIONS

Date	Sun ☉ (° ')	Moon ☽	Mercury ☿	Venus ♀	Mars ♂	Jupiter ♃	Saturn ♄	Uranus ♅	Neptune ♆	Pluto ♇
01	23 N 03	18 N 36	22 N 19	23 N 13	15 N 49	21 N 00	06 S 01	18 N 56	01 S 12	22 S 58
02	22 59	23 13	21 54	23 05	16 02	21 02	06 01	18 57	01 12	22 58
03	22 54	26 29	21 28	22 56	16 14	21 04	06 01	18 58	01 12	22 58
04	22 48	28 09	21 01	22 47	16 27	21 06	06 02	18 58	01 12	22 59
05	22 42	28 10	20 32	22 37	16 39	21 08	06 02	18 59	01 12	22 59
06	22 36	26 35	20 03	22 26	16 51	21 10	06 03	19 00	01 12	23 00
07	22 30	23 38	19 33	22 15	17 03	21 12	06 03	19 00	01 12	23 00
08	22 23	19 39	19 03	22 03	17 15	21 14	06 04	19 01	01 13	23 01
09	22 16	14 53	18 32	21 50	17 26	21 16	06 04	19 01	01 13	23 01
10	22 08	09 38	18 00	21 37	17 38	21 18	06 05	19 02	01 13	23 01
11	22 00	04 N 07	17 28	21 23	17 49	21 20	06 06	19 02	01 13	23 02
12	21 51	01 S 31	16 56	21 08	18 00	21 21	06 06	19 03	01 13	23 02
13	21 42	07 06	16 23	20 53	18 11	21 23	06 07	19 03	01 13	23 03
14	21 33	12 28	15 50	20 37	18 22	21 25	06 08	19 04	01 14	23 03
15	21 24	17 28	15 17	20 21	18 32	21 27	06 09	19 05	01 14	23 03
16	21 14	21 51	14 45	20 04	18 43	21 28	06 09	19 05	01 14	23 04
17	21 03	25 21	14 12	19 46	18 53	21 30	06 10	19 06	01 14	23 04
18	20 53	27 39	13 40	19 28	19 03	21 31	06 11	19 06	01 14	23 05
19	20 42	28 25	13 07	19 10	19 13	21 33	06 12	19 07	01 15	23 05
20	20 31	27 27	12 36	18 51	19 23	21 35	06 13	19 07	01 15	23 05
21	20 19	24 43	12 04	18 31	19 32	21 36	06 14	19 07	01 15	23 06
22	20 07	20 23	11 34	18 11	19 41	21 38	06 15	19 08	01 16	23 06
23	19 54	14 48	11 04	17 50	19 50	21 39	06 16	19 08	01 16	23 07
24	19 42	08 23	10 35	17 29	19 59	21 40	06 17	19 09	01 16	23 07
25	19 29	01 S 32	10 06	17 07	20 08	21 42	06 19	19 09	01 16	23 07
26	19 15	05 N 19	09 39	16 45	20 17	21 43	06 20	19 10	01 17	23 08
27	19 02	11 49	09 12	16 23	20 25	21 45	06 21	19 10	01 17	23 08
28	18 48	17 39	08 47	16 00	20 33	21 46	06 22	19 10	01 18	23 09
29	18 34	22 28	08 23	15 36	20 41	21 47	06 23	19 11	01 18	23 09
30	18 19	26 00	08 01	15 13	20 49	21 49	06 25	19 11	01 18	23 09
31	18 N 04	28 N 01	07 N 40	14 N 48	20 N 56	21 N 50	06 S 26	19 N 12	01 S 19	23 S 10

ZODIAC SIGN ENTRIES

Date	h	m	Planets
02	12	50	☿ ♌
02	15	50	☽ ♊
04	20	51	☽ ♋
07	03	56	☽ ♌
09	13	47	☽ ♍
11	16	19	♀ ♌
12	02	06	☽ ♎
14	14	53	☽ ♏
17	01	25	☽ ♐
19	08	14	☽ ♑
20	20	43	♂ ♊
21	11	43	☽ ♒
22	07	44	☉ ♌
23	13	23	☽ ♓
25	14	52	☽ ♈
25	22	42	☿ ♍
27	17	22	☽ ♉
29	21	28	☽ ♊

LATITUDES

Date	Mercury ☿ (° ')	Venus ♀	Mars ♂	Jupiter ♃	Saturn ♄	Uranus ♅	Neptune ♆	Pluto ♇
01	01 N 50	00 N 56	00 S 57	00 S 42	01 S 59	00 S 16	01 S 17	03 S 11
04	01 39	01 01	00 55	00 42	02 00	00 16	01 17	03 12
07	01 23	01 07	00 53	00 42	02 01	00 16	01 17	03 12
10	01 02	01 11	00 51	00 42	02 01	00 16	01 17	03 12
13	00 37	01 15	00 49	00 42	02 02	00 16	01 17	03 13
16	00 N 08	01 19	00 47	00 42	02 03	00 16	01 17	03 13
19	00 S 24	01 22	00 45	00 41	02 04	00 16	01 18	03 13
22	00 59	01 25	00 43	00 41	02 04	00 16	01 18	03 13
25	01 35	01 27	00 40	00 41	02 05	00 16	01 18	03 14
28	02 13	01 28	00 38	00 41	02 06	00 16	01 18	03 14
31	02 S 51	01 N 29	00 S 36	00 S 41	02 S 06	00 S 16	01 S 18	03 S 14

DATA

Julian Date	2460493
Delta T	+72 seconds
Ayanamsa	24° 11' 55"
Synetic vernal point	04° ♓ 55' 04"
True obliquity of ecliptic	23° 26' 18"

LONGITUDES

Date	Chiron ⚷ (° ')	Ceres ⚳	Pallas ⚴	Juno ⚵	Vesta ⚶	Black Moon Lilith ⚸
01	23 ♈ 16	15 ♑ 25	20 ♏ 03	17 ♍ 55	05 ♌ 03	00 ♎ 14
11	23 ♈ 26	13 ♑ 14	19 ♏ 47	20 ♍ 47	09 ♌ 30	01 ♎ 21
21	23 ♈ 31	11 ♑ 10	20 ♏ 14	23 ♍ 49	14 ♌ 02	02 ♎ 27
31	23 ♈ 31	09 ♑ 27	21 ♏ 19	26 ♍ 59	18 ♌ 36	03 ♎ 33

MOON'S PHASES, APSIDES AND POSITIONS ☽

Date	h	m	Phase	Longitude (° ')	Eclipse Indicator
05	22	57	●	14 ♋ 23	
13	22	49	☽	22 ♎ 01	
21	10	17	○	29 ♑ 09	
28	02	52	☾	05 ♉ 32	

Day	h	m		
12	08	13	Apogee	
24	05	50	Perigee	
05	00	03	Max dec	28° N 22'
12	05	33	OS	
19	10	56	Max dec	28° S 25'
25	17	21	ON	

ASPECTARIAN

h m	Aspects
01 Monday	
02 09	☽ ⊻ ♃
04 57	☽ ✶ ☉
13 33	☽ ∥ ♅
13 42	☽ ∠ ♆
14 59	☿ Q ♂
16 19	☽ ☌ ♂
16 26	☽ Q ☿
18 42	☽ ✶ ♀
21 28	☽ ✶ ♄
23 48	☽ ∥ ♃
02 Tuesday	
03 58	☉ ∠ ♅
05 08	☽ ∥ ☿
08 29	☽ ☌ ♅
08 49	☽ ∠ ☉
10 31	☽ # ♇
10 36	☽ ∥ ☉
10 42	♆ St R
11 15	☽ ∥ ♀
11 53	☿ △ ♆
14 06	☉ # ♇
15 43	☽ ✶ ♆
16 16	☽ ✶ ☿
18 11	☽ △ ♇
18 19	☽ Q ♄
23 21	☽ ∠ ♀
03 Wednesday	
01 41	☽ ⊥ ☉
01 41	♀ △ ♄
06 53	♀ # ♇
07 05	☽ ☌ ♃
07 27	☿ ☍ ♇
12 44	☽ Q ♆
13 01	☽ ⊻ ☉
16 43	☽ ⊥ ♀
20 29	☽ ⚼ ♇
22 22	☽ ∠ ☿
23 38	☽ ⊻ ♂
04 Thursday	
01 58	☽ □ ♄
04 24	☽ ⊻ ♀
05 15	☉ ∥ ♀
07 20	☿ ∥ ♃
10 54	☽ ⊥ ♂
12 27	☽ ± ♇
13 28	☽ ⊻ ♅
16 38	☽ ⊥ ☿
20 44	☽ □ ♆
23 11	☽ ⊼ ♇
05 Friday	
00 16	☽ ⊥ ♅
03 38	☿ ⚼ ♄
03 58	☽ ∠ ♂
04 57	☽ ⊻ ☿
13 28	☽ ⊻ ♃
16 38	☽ ∠ ♅
19 02	♂ ✶ ♄
22 57	☽ ☌ ☉
06 Saturday	
00 37	☽ ⊥ ♃
08 10	☽ △ ♄
08 56	☽ ✶ ♂
16 27	☽ ☌ ♀
17 32	☽ ∠ ♃
20 23	☽ ✶ ♅
07 Sunday	
03 46	☉ ⊥ ♃
03 47	☽ △ ♆
05 56	♀ ∠ ♃
06 16	☽ ☍ ♇
08 40	☽ Q ♂
09 43	☿ Q ♅
12 11	☽ ⚼ ♄
16 14	☽ # ♇
19 06	☽ Q ♅
19 41	☽ ∥ ☉
20 23	☽ ☌ ☿
21 30	☽ ∥ ♀
22 19	☽ ✶ ♃
08 Monday	
03 07	☽ ∥ ♃
05 21	☽ ± ♄
08 21	☽ ⚼ ♆
11 04	♀ ✶ ♅
12 00	☽ ⊻ ☉
13 39	☿ ∥ ♅
14 26	☿ ✶ ♃
15 22	☽ ∥ ♅
15 36	☽ ∥ ☿
16 56	☽ ⊼ ♄
21 18	☽ □ ♂
21 53	☽ Q ♃
23 59	☽ ∥ ♂
09 Tuesday	
00 39	☽ ⊥ ☉
01 50	☽ ± ♆
06 04	☽ □ ♅
08 07	☽ ⊻ ♀
13 38	☽ ⊼ ♆
16 08	☽ ⊼ ♇
19 54	☽ ∠ ☉
21 19	☽ ⊥ ♀
10 Wednesday	
04 03	☽ ± ♇
10 07	☽ □ ♃
15 24	☽ ⊻ ☿
17 22	☽ ∠ ♀
22 04	☽ ⚼ ♇
23 23	☿ ± ♄
11 Thursday	
00 14	☿ ∥ ♂
03 05	☉ △ ♄
03 31	☽ # ♄
04 27	☽ ☍ ♄
04 34	☽ ✶ ☉
05 09	☽ ⊥ ☿
12 51	☽ △ ♂
14 32	♀ △ ♆
16 37	♀ ∥ ♃
18 20	☽ △ ♅
12 Friday	
00 21	☽ # ♆
01 51	☽ ∠ ☿
01 55	☽ ☍ ♆
02 25	☿ ⚼ ♆
03 15	☽ ✶ ♀
04 24	☽ △ ♇
07 03	☽ Q ☉
10 44	☽ ∥ ♆
14 12	♀ ☍ ♇
21 17	☽ ⚼ ♂
23 58	☽ △ ♃
13 Saturday	
00 57	☽ ⚼ ♅
06 29	☽ Q ♀
07 44	☽ ∥ ♄
12 18	☽ ✶ ☿
16 43	☽ ± ♂
17 16	☽ ⊼ ♄
19 16	☽ ± ♅
22 49	☽ □ ☉
14 Sunday	
05 21	☽ ± ♄
05 35	☽ ⊼ ♂
06 52	☽ ⚼ ♃
07 24	☽ ⊼ ♅
14 40	☽ ⊼ ♆
15 29	☽ Q ☿
17 01	☽ □ ♇
22 55	☽ □ ♀
23 20	☽ ⚼ ♄
15 Monday	
01 11	☽ ± ♃
02 19	☽ # ☿
02 37	☽ ± ♆
03 00	♀ ⚼ ♄
06 03	☉ ∥ ♃
08 57	☿ ⊼ ♄
13 15	☽ ⊼ ♃
14 05	♂ ☌ ♅
17 45	☽ # ♂
20 20	☽ ⚼ ♆
20 22	☽ # ♅
16 Tuesday	
02 22	☽ # ♀
04 25	☽ Q ♇
04 45	☽ △ ♄
06 59	☽ □ ☿
08 28	☽ # ☉
09 43	☽ # ♃
15 04	☽ △ ☉
18 31	☽ ☍ ♅
19 34	☽ ∥ ♇
20 06	☽ ☍ ♂
17 Wednesday	
01 10	☽ △ ♆
03 20	☽ ✶ ♇
15 08	☽ △ ♀
21 33	☽ ⚼ ☉
23 20	☽ ☍ ♃
18 Thursday	
07 06	☽ ∠ ♇
12 49	☽ □ ♄
12 56	♃ Q ♆
12 59	♀ Q ♅
14 00	☉ ✶ ♅
15 15	☽ ± ☉
19 36	♂ ∥ ♅
20 24	☽ △ ☿
21 19	☽ ⚼ ♀
23 18	☽ ⊥ ♇
19 Friday	
01 56	☽ ⊼ ♅
02 49	☽ ⊼ ☉
03 55	☉ ∠ ♃
06 16	☽ ⊼ ♂
07 32	☿ ± ♆
07 58	☽ □ ♆
09 25	♀ ∥ ♂
09 55	☽ ⊻ ♇
11 50	☿ Q ♃
12 34	☽ ± ♅
14 50	☽ ± ♀
16 05	♀ ∥ ♅
17 22	☽ ± ♂
20 40	☽ Q ♄
20 Saturday	
01 04	☽ ⚼ ☿
02 17	☽ ⊼ ♀
04 16	☽ ⚼ ♅
05 29	☽ ⊼ ♃
09 46	☽ ⚼ ♂
15 10	☽ Q ♆
15 17	♂ ✶ ♆
15 55	☽ ± ♃
17 15	☽ ✶ ♄
17 45	☽ ± ☿
21 Sunday	
04 42	☽ ⊼ ☿
05 53	☽ △ ♅
07 21	☽ ⚼ ♃
10 17	☽ ☍ ☉
11 26	☽ ✶ ♆
12 29	☽ △ ♂
13 14	☽ ☌ ♇
18 28	☽ ∠ ♄
20 43	♀ ✶ ♃
21 54	☽ ∥ ♇
22 21	☿ □ ♅
22 Monday	
03 25	☉ △ ♆
03 48	♂ △ ♇
04 45	♀ Q ♂
05 53	☽ # ♃
06 33	♀ ± ♄
07 52	♂ Q ♄
08 43	☽ △ ♃
09 23	☽ ⊥ ♄
09 39	☽ ☍ ♀
12 23	☽ ∠ ♆
13 19	☽ # ☉
15 10	☽ # ♂
17 47	☽ # ♅
19 16	☽ ⊻ ♄
22 40	☽ # ♀
23 Tuesday	
03 13	☽ ⊥ ♆
05 38	☉ ☍ ♇
07 47	☽ □ ♅
09 58	☽ ☍ ☿
13 04	☽ ⊻ ♆
14 48	☽ ⊻ ♇
15 27	☽ ⊼ ☉
16 33	☉ ∥ ♂
16 36	☽ □ ♂
17 37	♀ ⚼ ♆
24 Wednesday	
00 39	☽ ⊥ ♇
02 01	☽ ± ☉
03 24	☽ # ☿
10 47	☽ □ ♃
13 27	☽ Q ♅
15 26	☽ ∠ ♇
15 36	☽ ⊼ ♀
17 52	☽ ⚼ ☉
19 24	☽ ∥ ♄
20 31	☽ ☌ ♄
23 42	☽ Q ♂
25 Thursday	
02 25	☽ ± ♀
09 19	☽ ✶ ♅
12 55	☽ ∥ ♆
14 26	☽ ⊼ ☿
14 31	☽ ☌ ♆
16 00	☿ ⊼ ♆
16 13	☽ ✶ ♇
16 57	☽ Q ♃
18 44	☽ ⚼ ♀
20 29	☽ △ ☉
20 36	☽ ✶ ♂
21 48	☽ # ♆
26 Friday	
00 59	☽ ± ☿
02 32	☉ ✶ ♂
08 40	☉ ⚼ ♄
10 23	☽ ∠ ♅
12 14	☽ Q ♇
13 16	☽ ✶ ♃
15 39	☽ # ♄
16 52	☽ ⚼ ☿
21 55	☉ ∥ ♅
22 14	☽ △ ♀
22 24	☽ ⊻ ♄
23 01	☽ ∠ ♂
27 Saturday	
00 07	♀ ⊼ ♄
00 49	☿ ⊼ ♇
01 34	☽ ⊥ ♅
02 44	☽ ∥ ☿
08 32	☽ ⊥ ♄
11 47	☽ ⊻ ♅
15 00	☽ ∠ ♃
15 05	☽ ⊥ ♂
16 58	☽ ⊻ ♆
18 41	☽ □ ♇
19 35	☽ △ ☿
23 51	☽ ∠ ♄
28 Sunday	
01 52	☽ ⊻ ♂
02 52	☽ □ ☉
03 14	☽ ⊥ ♆
05 18	☽ ∥ ♀
06 37	☽ ⊥ ♃
17 01	☽ ∥ ☉
17 08	☽ ⊻ ♃
18 46	☽ ∠ ♆
19 02	☽ ∥ ♅
29 Monday	
01 42	☽ ✶ ♄
02 09	☽ ∥ ♂
06 45	☽ □ ♀
08 13	☽ ∥ ♃
12 26	☽ Q ☉
15 47	☽ ☌ ♅
16 00	☽ # ♇
20 59	☽ ✶ ♆
22 41	☽ Q ♄
22 43	☽ △ ♇
30 Tuesday	
02 00	☽ □ ☿
09 01	☽ ☌ ♂
11 15	☽ ✶ ☉
17 44	☽ Q ♀
18 17	☽ Q ♆
22 45	☽ ☌ ♃
23 38	♀ ± ♆
31 Wednesday	
01 23	☽ ⚼ ♇
06 42	☽ □ ♄
11 16	☽ Q ☿
13 33	☉ Q ♅
16 17	☽ ∠ ☉
17 33	☽ ✶ ♀
17 37	☽ ± ♇
21 33	☽ ⊻ ♅

All ephemeris data is given at 12.00 UT and the Moon's longitude is additionally given for 24.00 UT

AUGUST 2024

Date	Sidereal time h m s	Sun ☉ ° ' "	Moon ☽ ° ' "	Moon ☽ 24.00 ° ' "	Mercury ☿ ° '	Venus ♀ ° '	Mars ♂ ° '	Jupiter ♃ ° '	Saturn ♄ ° '	Uranus ♅ ° '	Neptune ♆ ° '	Pluto ♇ ° '
01	08 42 21	09 ♌ 43 33	04 ♋ 44 24	11 ♋ 14 44	03 ♍ 32	25 ♌ 35	07 ♊ 58	14 ♊ 29	18 ♓ 34	26 ♉ 51	29 ♓ 42	00 ♒ 39
02	08 46 18	10 40 59	17 ♋ 41 55	24 ♋ 05 53	03 48	26 48	08 38	14 40	18 R 31	26 53	29 R 41	00 R 38
03	08 50 14	11 38 26	00 ♌ 26 37	06 ♌ 44 06	03 59	28 02	09 18	14 51	18 28	26 54	29 40	00 36
04	08 54 11	12 35 54	12 ♌ 58 21	19 ♌ 09 27	04 05	29 ♌ 16	09 59	15 01	18 25	26 56	29 39	00 35
05	08 58 07	13 33 23	25 ♌ 17 28	01 ♍ 22 35	04 R 06	00 ♍ 30	10 39	15 11	18 22	26 57	29 38	00 34
06	09 02 04	14 30 53	07 ♍ 24 57	13 ♍ 24 52	04 02	01 43	11 19	15 21	18 18	26 58	29 37	00 32
07	09 06 01	15 28 23	19 ♍ 22 37	25 ♍ 18 33	03 52	02 57	11 59	15 32	18 15	27 00	29 36	00 31
08	09 09 57	16 25 55	01 ♎ 13 07	07 ♎ 06 44	03 38	04 11	12 39	15 42	18 11	27 01	29 35	00 29
09	09 13 54	17 23 27	12 ♎ 59 56	18 ♎ 53 16	03 17	05 24	13 18	15 51	18 08	27 02	29 34	00 28
10	09 17 50	18 21 00	24 ♎ 47 18	00 ♏ 42 41	02 52	06 38	13 58	16 01	18 04	27 03	29 32	00 27
11	09 21 47	19 18 35	06 ♏ 40 01	12 ♏ 39 58	02 22	07 52	14 37	16 11	18 01	27 04	29 31	00 25
12	09 25 43	20 16 10	18 ♏ 43 11	24 ♏ 50 19	01 46	09 05	15 17	16 20	17 57	27 05	29 30	00 24
13	09 29 40	21 13 46	01 ♐ 02 00	07 ♐ 18 48	01 07	10 19	15 56	16 30	17 53	27 06	29 29	00 23
14	09 33 36	22 11 23	13 ♐ 41 15	20 ♐ 09 50	00 ♍ 23	11 32	16 35	16 39	17 49	27 07	29 28	00 22
15	09 37 33	23 09 01	26 ♐ 44 54	03 ♑ 26 41	29 ♌ 37	12 46	17 14	16 48	17 45	27 08	29 26	00 20
16	09 41 30	24 06 40	10 ♑ 15 19	17 ♑ 10 45	28 48	14 00	17 53	16 57	17 41	27 09	29 25	00 19
17	09 45 26	25 04 20	24 ♑ 12 47	01 ♒ 21 00	27 57	15 13	18 32	17 06	17 37	27 10	29 24	00 18
18	09 49 23	26 02 01	08 ♒ 34 53	15 ♒ 53 41	27 06	16 27	19 10	17 15	17 33	27 10	29 23	00 16
19	09 53 19	26 59 43	23 ♒ 16 34	00 ♓ 42 32	26 14	17 40	19 49	17 24	17 29	27 11	29 21	00 15
20	09 57 16	27 57 27	08 ♓ 11 56	15 ♓ 39 33	25 24	18 54	20 27	17 33	17 25	27 12	29 20	00 14
21	10 01 12	28 55 12	23 ♓ 08 28	00 ♈ 36 17	24 37	20 08	21 06	17 41	17 21	27 12	29 19	00 13
22	10 05 09	29 ♌ 52 59	08 ♈ 02 05	15 ♈ 25 02	23 52	21 21	21 44	17 49	17 17	27 13	29 17	00 12
23	10 09 05	00 ♍ 50 47	22 ♈ 44 28	29 ♈ 59 51	23 13	22 35	22 22	17 57	17 12	27 13	29 16	00 10
24	10 13 02	01 48 37	07 ♉ 10 45	14 ♉ 16 55	22 38	23 48	23 00	18 06	17 08	27 14	29 14	00 09
25	10 16 59	02 46 29	21 ♉ 18 10	28 ♉ 14 29	22 09	25 02	23 38	18 13	17 04	27 14	29 13	00 08
26	10 20 55	03 44 22	05 ♊ 05 51	11 ♊ 52 24	21 47	26 15	24 15	18 21	16 59	27 14	29 11	00 07
27	10 24 52	04 42 18	18 ♊ 34 16	25 ♊ 11 39	21 32	27 29	24 53	18 29	16 55	27 15	29 10	00 06
28	10 28 48	05 40 15	01 ♋ 44 44	08 ♋ 13 45	21 25	28 42	25 30	18 36	16 50	27 15	29 09	00 05
29	10 32 45	06 38 14	14 ♋ 38 55	21 ♋ 00 27	21 D 26	29 ♍ 56	26 08	18 44	16 46	27 15	29 07	00 04
30	10 36 41	07 36 15	27 ♋ 18 34	03 ♌ 33 29	21 35	01 ♎ 09	26 45	18 51	16 41	27 15	29 06	00 03
31	10 40 38	08 ♍ 34 18	09 ♌ 45 24	15 ♌ 54 30	21 ♌ 53	02 ♎ 23	27 ♊ 22	18 ♊ 58	16 ♓ 37	27 ♉ 15	29 ♓ 04	00 ♒ 02

Date	Moon True ☊ ° '	Moon Mean ☊ ° '	Moon ☽ Latitude ° '
01	08 ♈ 29	09 ♈ 34	05 N 04
02	08 R 20	09 31	05 01
03	08 10	09 28	04 42
04	08 00	09 25	04 10
05	07 52	09 22	03 27
06	07 45	09 18	02 35
07	07 41	09 15	01 37
08	07 39	09 12	00 N 35
09	07 D 39	09 09	00 S 29
10	07 40	09 06	01 31
11	07 42	09 03	02 30
12	07 42	08 59	03 24
13	07 R 42	08 56	04 09
14	07 40	08 53	04 43
15	07 36	08 50	05 04
16	07 30	08 47	05 09
17	07 24	08 44	04 56
18	07 17	08 40	04 25
19	07 12	08 37	03 36
20	07 07	08 34	02 31
21	07 05	08 31	01 S 15
22	07 D 04	08 28	00 N 05
23	07 05	08 24	01 25
24	07 06	08 21	02 37
25	07 07	08 18	03 38
26	07 R 08	08 15	04 25
27	07 07	08 12	04 57
28	07 05	08 09	05 11
29	07 01	08 05	05 10
30	06 57	08 02	04 53
31	06 ♈ 52	07 ♈ 59	04 N 23

DECLINATIONS

Date	Sun ☉	Moon ☽	Mercury ☿	Venus ♀	Mars ♂	Jupiter ♃	Saturn ♄	Uranus ♅	Neptune ♆	Pluto ♇
01	17 N 49	28 N 25	07 N 21	14 N 24	21 N 04	21 N 51	06 S 27	19 N 12	01 S 19	23 S 10
02	17 33	27 14	07 04	13 59	21 11	21 52	06 29	19 12	01 19	23 11
03	17 18	24 39	06 49	13 33	21 18	21 53	06 30	19 13	01 20	23 11
04	17 02	20 55	06 36	13 08	21 25	21 55	06 32	19 13	01 20	23 11
05	16 45	16 21	06 25	12 41	21 31	21 56	06 33	19 13	01 21	23 12
06	16 29	11 11	06 17	12 15	21 38	21 57	06 35	19 14	01 21	23 12
07	16 12	05 41	06 11	11 48	21 44	21 58	06 36	19 14	01 22	23 12
08	15 55	00 N 03	06 08	11 21	21 50	21 59	06 38	19 14	01 22	23 13
09	15 37	05 S 35	06 08	10 54	21 56	22 00	06 39	19 14	01 23	23 13
10	15 20	11 01	06 10	10 26	22 02	22 01	06 41	19 15	01 23	23 13
11	15 02	16 06	06 16	09 58	22 07	22 02	06 42	19 15	01 24	23 14
12	14 44	20 39	06 24	09 30	22 12	22 03	06 44	19 15	01 24	23 14
13	14 25	24 25	06 35	09 02	22 18	22 04	06 46	19 15	01 25	23 14
14	14 07	27 07	06 50	08 33	22 22	22 05	06 47	19 15	01 25	23 15
15	13 48	28 28	07 06	08 04	22 27	22 06	06 49	19 16	01 26	23 15
16	13 29	28 11	07 26	07 35	22 32	22 06	06 51	19 16	01 26	23 15
17	13 10	26 08	07 47	07 06	22 36	22 07	06 52	19 16	01 27	23 16
18	12 50	22 23	08 10	06 36	22 40	22 08	06 54	19 16	01 27	23 16
19	12 31	17 09	08 35	06 06	22 44	22 09	06 56	19 16	01 28	23 16
20	12 11	10 50	09 01	05 36	22 48	22 10	06 58	19 17	01 28	23 17
21	11 51	03 S 52	09 28	05 06	22 52	22 10	06 59	19 17	01 29	23 17
22	11 31	03 N 16	09 55	04 36	22 55	22 11	07 01	19 17	01 30	23 17
23	11 10	10 09	10 22	04 06	22 59	22 12	07 03	19 17	01 30	23 18
24	10 50	16 23	10 48	03 36	23 02	22 13	07 05	19 17	01 31	23 18
25	10 29	21 36	11 13	03 05	23 05	22 13	07 07	19 17	01 31	23 18
26	10 08	25 30	11 37	02 35	23 07	22 14	07 08	19 17	01 32	23 18
27	09 47	27 52	11 58	02 04	23 10	22 15	07 10	19 17	01 32	23 19
28	09 26	28 37	12 18	01 33	23 12	22 15	07 12	19 17	01 33	23 19
29	09 05	27 46	12 35	01 02	23 15	22 16	07 14	19 17	01 33	23 19
30	08 43	25 29	12 50	00 32	23 17	22 16	07 16	19 17	01 34	23 20
31	08 N 21	22 N 02	13 N 02	00 N 01	23 N 19	22 N 17	07 S 18	19 N 17	01 S 35	23 S 20

ZODIAC SIGN ENTRIES

Date	h	m	Planets
01	03	19	☽ ♋
03	11	09	☽ ♌
05	02	23	♀ ♍
05	21	17	☽ ♍
08	09	31	☽ ♎
10	22	34	☽ ♏
13	10	01	☽ ♐
15	00	15	☿ ♌
15	17	51	☽ ♑
17	21	45	☽ ♒
19	22	52	☽ ♓
21	23	02	☽ ♈
22	14	55	☉ ♍
24	00	00	☽ ♉
26	03	04	☽ ♊
28	08	47	☽ ♋
29	13	23	♀ ♎
30	17	09	☽ ♌

LATITUDES

Date	Mercury ☿	Venus ♀	Mars ♂	Jupiter ♃	Saturn ♄	Uranus ♅	Neptune ♆	Pluto ♇
01	03 S 04	01 N 29	00 S 35	00 S 41	02 S 06	00 S 16	01 S 18	03 S 14
04	03 40	01 29	00 32	00 42	02 07	00 16	01 18	03 14
07	04 11	01 29	00 30	00 42	02 08	00 16	01 18	03 15
10	04 35	01 28	00 27	00 42	02 08	00 16	01 19	03 15
13	04 48	01 26	00 24	00 42	02 09	00 16	01 19	03 15
16	04 46	01 23	00 21	00 42	02 09	00 16	01 19	03 15
19	04 27	01 20	00 19	00 42	02 10	00 16	01 19	03 15
22	03 52	01 17	00 16	00 42	02 10	00 16	01 19	03 15
25	03 05	01 13	00 13	00 42	02 10	00 16	01 19	03 16
28	02 10	01 08	00 09	00 42	02 11	00 16	01 19	03 16
31	01 S 15	01 N 03	00 S 06	00 S 42	02 S 11	00 S 16	01 S 19	03 S 16

DATA

Julian Date	2460524
Delta T	+72 seconds
Ayanamsa	24° 12' 00"
Synetic vernal point	04° ♓ 54' 59"
True obliquity of ecliptic	23° 26' 19"

MOON'S PHASES, APSIDES AND POSITIONS ☽

Date	h	m	Phase	Longitude ° '	Eclipse Indicator
04	11	13	●	12 ♌ 34	
12	15	19	☽	20 ♏ 24	
19	18	26	○	27 ♒ 15	
26	09	26	☾	03 ♊ 38	

Day	h	m		
09	01	36	Apogee	
21	05	10	Perigee	
01	05	52	Max dec	28° N 28'
08	12	12	0S	
15	20	04	Max dec	28° S 34'
22	01	00	0N	
28	10	59	Max dec	28° N 37'

LONGITUDES

Date	Chiron ⚷	Ceres ⚳	Pallas ⚴	Juno ⚵	Vesta ⚶	Black Moon Lilith ⚸
01	23 ♈ 31	09 ♑ 19	21 ♏ 27	27 ♍ 19	19 ♌ 04	03 ♎ 40
11	23 ♈ 25	08 ♑ 09	23 ♏ 08	00 ♎ 36	23 ♌ 42	04 ♎ 47
21	23 ♈ 15	07 ♑ 34	25 ♏ 16	03 ♎ 59	28 ♌ 22	05 ♎ 53
31	22 ♈ 59	07 ♑ 34	27 ♏ 47	07 ♎ 26	03 ♍ 04	06 ♎ 59

ASPECTARIAN

h m	Aspects
01 Thursday	
02 46	☽ □ ♆
04 31	☽ ⊼ ♇
08 32	☽ ⊥ ♅
09 43	☽ ⚹ ☿
09 59	☽ ⊥ ☉
18 15	☽ ⊻ ♂
21 55	☽ ⊻ ☉
23 54	☽ ∠ ♀
02 Friday	
01 09	☽ ∠ ♅
05 59	☽ ⊥ ♂
06 16	☽ ⊻ ♃
08 48	♀ Q ♃
13 27	♀ □ ♅
13 32	☽ △ ♄
14 05	☽ ∠ ☿
17 38	☽ ⊥ ♃
18 26	☽ ⊥ ♀
23 45	☽ ∠ ♂
03 Saturday	
05 17	☽ ⚹ ♅
06 57	☽ ⊻ ♀
07 16	☽ ⊥ ☿
10 31	☽ △ ♆
10 51	☽ ∠ ♃
12 18	☽ ☍ ♇
17 44	☽ ⚼ ♄
18 48	☽ ⊻ ☿
22 12	☽ # ♇
04 Sunday	
04 12	☽ Q ♅
05 54	☽ ⚹ ♂
06 14	☽ ∥ ♃
07 41	☉ ± ♄
09 13	☽ ∥ ♂
10 56	☽ ± ♄
11 13	☽ ☌ ☉
15 14	☽ ⚼ ♆
16 01	☽ ⚹ ♃
19 24	♀ ⊼ ♆
19 40	☿ # ♄
21 22	☽ ∥ ♅
22 31	☽ ⊼ ♄
05 Monday	
04 56	☿ St R
06 31	☽ Q ♂
08 45	☽ ± ♆
09 51	☽ ∥ ☉
13 18	♀ ⊼ ♇
15 16	☽ □ ♅
15 47	☽ Q ♃
20 32	☽ ⊼ ♆
22 22	☽ ⊼ ♇
23 24	☽ ☌ ♀
06 Tuesday	
05 19	☽ ☌ ☿
06 47	☽ ∥ ♀
10 15	☽ ± ♇
14 26	☉ ⚼ ♆
20 15	☽ □ ♂
22 32	♂ Q ♆
07 Wednesday	
03 27	☽ ⊻ ☉
04 08	☽ □ ♃
04 14	☽ ⚼ ♇
08 06	☽ # ♄
09 44	☽ ☍ ♄
09 51	☽ ∥ ☿
10 28	♃ ⚼ ♇
13 37	☉ ⚹ ♃
16 36	☽ ⊥ ☉
08 Thursday	
03 12	☿ ☌ ♀
03 27	☽ △ ♅
06 24	☽ # ♆
08 40	☽ ☍ ♆
10 31	☽ △ ♇
12 28	☽ ∠ ☉
16 47	☽ ⊻ ☿
18 01	☽ ∥ ♆
18 43	☽ ⊻ ♀
09 Friday	
04 40	☽ ⊥ ☿
08 22	☽ ⊥ ♀
10 02	☽ ⚼ ♅
12 40	☽ △ ♂
14 24	☽ # ☿
16 42	☽ ∥ ♄
17 55	☽ △ ♃
21 45	☽ ⚹ ☉
22 24	☽ ⊼ ♄
22 26	☽ ∠ ☿
10 Saturday	
04 24	☽ ± ♅
04 50	☽ ∠ ♀
05 26	☉ ⊼ ♄
08 13	♂ ∥ ♃
08 26	♀ ± ♇
09 35	☽ # ♀
10 33	☽ ± ♄
16 36	☽ ⊼ ♅
20 58	☽ ⚼ ♂
21 37	☽ ⊼ ♆
23 27	☽ □ ♇
11 Sunday	
00 16	☽ Q ☉
00 48	☽ ⚼ ♃
03 42	☽ ⚹ ☿
04 41	☽ ⚼ ♄
07 03	☽ # ☉
09 42	☽ ± ♆
14 40	☽ ⚹ ♀
16 09	☽ ± ♂
19 08	☽ ± ♃
12 Monday	
02 40	☽ Q ☿
03 40	☽ ⚼ ♆
04 13	☽ # ♅
04 48	☽ ⊼ ♂
07 14	☽ ⊼ ♃
10 29	☽ △ ♄
11 22	☽ Q ♇
15 19	☽ □ ☉
16 26	♂ ⚼ ♇
17 11	☽ Q ♀
20 17	☽ # ♃
21 26	☽ # ♂
13 Tuesday	
03 49	☽ ∥ ♇
04 24	☽ ☍ ♅
09 01	☽ △ ♆
10 45	☽ ⚹ ♇
12 09	☽ □ ☿
14 Wednesday	
07 33	☽ □ ♀
08 05	☿ # ♄
13 03	☿ ⊼ ♇
15 07	☽ ∠ ♇
15 22	♂ ☌ ♃
17 35	☽ ☍ ♃
17 41	☽ ☍ ♂
19 38	☽ □ ♄
15 Thursday	
04 57	☽ △ ☉
07 38	☽ ⊥ ♇
12 42	☽ ⊼ ♅
16 50	☽ □ ♆
16 52	☽ △ ☿
17 19	☿ ⊼ ♆
18 17	☿ Q ♂
18 27	☽ ⊻ ♇
19 06	☉ ± ♆
23 27	☽ ± ♅
16 Friday	
04 01	☽ Q ♄
05 30	♂ □ ♄
08 03	☿ Q ♃
09 51	☽ ⚼ ☉
15 18	☽ ⚼ ♅
16 28	☿ ∥ ♀
17 49	☽ ⚼ ☿
19 08	☽ △ ♀
23 45	☽ ⊼ ♃
17 Saturday	
00 24	☽ Q ♆
00 49	☽ ⚹ ♄
01 51	☽ ⊼ ♂
02 36	☽ ± ☉
08 23	☽ ± ☿
10 06	☽ ± ♃
12 34	☽ ± ♂
13 24	♀ ⚼ ♇
13 34	☽ ⊼ ☉
16 59	☽ △ ♅
17 57	☽ ⊼ ☿
20 43	☽ ⚹ ♆
22 04	♀ # ♄
22 13	☽ ☌ ♇
23 04	☽ ⚼ ♀
18 Sunday	
01 24	☽ ⚼ ♃
02 03	☽ ∠ ♄
04 22	☽ ⚼ ♂
07 07	☽ ∥ ♇
09 47	☿ □ ♅
10 26	☽ # ♂
13 15	☽ # ♃
15 21	☽ ± ♀
16 52	☽ ⊥ ♄
21 30	☽ ∠ ♆
19 Monday	
01 58	☉ ☌ ☿
02 05	☽ ⊼ ♀
02 22	☽ △ ♃
02 38	☽ ⊻ ♄
02 57	☽ # ♅
05 53	♀ □ ♃
06 08	☽ △ ♂
08 30	♀ ☍ ♄
12 08	☽ ⊥ ♆
16 32	☽ ☍ ☿
16 45	☉ □ ♅
18 19	☽ □ ♅
18 26	☽ ☍ ☉
21 46	♃ □ ♄
21 48	☽ ⊻ ♆
23 15	☽ ⊻ ♇
20 Tuesday	
06 52	☽ # ☉
08 53	☽ ⊥ ♇
18 01	☽ # ☿
23 16	☽ Q ♅
23 18	☽ ∠ ♇
21 Wednesday	
01 28	☽ ∥ ♄
02 45	☽ ☌ ♄
03 10	☽ □ ♃
06 44	☽ ☍ ♀
07 31	☽ # ♀
08 34	☽ □ ♂
14 15	☽ ⊼ ☿
18 32	☽ ⚹ ♅
20 01	☽ ∥ ♆
21 29	☉ ⊼ ♆
21 54	☽ ☌ ♆
21 56	☽ ⊼ ☉
23 21	☽ ⚹ ♇
23 26	☽ ± ☿
22 Thursday	
06 00	☽ # ♆
08 17	☽ ± ☉
08 23	☽ Q ♃
10 13	☉ Q ♃
13 18	☽ ⚼ ☿
14 52	☽ Q ♂
16 15	☽ ∥ ♀
18 44	☽ Q ♇
18 47	☽ ∠ ♅
19 34	☉ ⊼ ♇
23 54	☽ ⚼ ☉
23 Friday	
00 56	☽ # ♄
02 58	☽ ⊻ ♄
03 20	♀ □ ♂
04 05	☽ ⚹ ♃
09 30	☽ ⊥ ♅
09 50	☿ ± ♆
11 21	☽ ⚹ ♂
11 42	☽ ⊼ ♀
12 44	☽ △ ☿
12 46	☽ ⊥ ♄
12 48	☽ ∥ ☿
15 33	☽ ∥ ☉
19 24	☽ ⊻ ♅
20 15	☿ ⊻ ♀
22 32	☽ ± ♀
22 46	☽ ⊻ ♆
24 Saturday	
00 17	☽ □ ♇
02 22	☽ △ ☉
03 36	☽ ∠ ♄
04 31	☿ ⚹ ♂
05 05	☽ ∠ ♃
08 45	☽ ⊥ ♆
13 05	☉ ∥ ☿
13 26	☽ ∠ ♂
14 59	☽ ⚼ ♀
20 22	☽ ⊥ ♃
23 54	☽ ∠ ♆
25 Sunday	
00 41	☽ ∥ ♅
04 46	☽ ⚹ ♄
05 24	☽ ⊥ ♂
06 40	☽ ⊻ ♃
13 25	☽ □ ☿
15 21	☽ ∥ ♃
16 12	☽ ⊻ ♂
19 03	☽ △ ♀
20 14	☽ ∥ ♂
21 28	☽ # ♇
22 15	☽ ☌ ♅
26 Monday	
01 21	☽ Q ♄
01 40	☽ ⚹ ♆
03 17	☽ △ ♇
06 49	♂ ± ♇
09 26	☽ □ ☉
20 07	☽ Q ☿
22 46	☽ Q ♆
27 Tuesday	
05 46	☽ ⚼ ♇
07 24	♀ △ ♅
09 02	☽ □ ♄
11 50	☽ ☌ ♃
12 58	☿ ⊥ ♀
17 17	☽ ⚹ ☿
20 04	☽ Q ☉
21 59	☽ ± ♇
28 Wednesday	
00 00	☽ ☌ ♂
03 45	☽ ⊻ ♅
05 50	☽ □ ♀
07 14	☽ □ ♆
08 56	☽ ⊼ ♇
12 04	♀ # ♆
14 46	☽ ⊥ ♅
19 50	☽ ⚹ ☉
20 25	♀ ☍ ♆
20 37	☽ ∠ ☿
21 14	☿ St D
21 57	☉ ± ♇
29 Thursday	
07 30	☽ ∠ ♅
13 29	☽ ⊥ ☿
14 32	♀ △ ♇
15 57	☽ △ ♄
18 50	☽ Q ♀
19 46	☽ ⊻ ♃
30 Friday	
00 56	☽ ⊻ ☿
02 17	☽ ∠ ☉
07 15	☽ ⊥ ♃
10 52	☽ ⊻ ♂
11 54	☽ ⚹ ♅
15 24	☽ △ ♆
17 14	☽ ☍ ♇
20 10	☽ ⚹ ♀
20 21	☽ ⚼ ♄
20 56	☽ ⊥ ☉
22 59	☽ ⊥ ♂
31 Saturday	
00 41	☽ ∠ ♃
03 47	☽ # ♇
03 59	☽ ∥ ♂
07 52	♂ ⊻ ♅
09 30	☽ ⊻ ☉
10 27	☽ ∥ ♃
11 02	☽ Q ♅
13 39	☽ ± ♄
17 20	☽ ∠ ♂
20 23	☽ ⚼ ♆

All ephemeris data is given at 12.00 UT and the Moon's longitude is additionally given for 24.00 UT

SEPTEMBER 2024

Date	Sidereal time h m s	LONGITUDES Sun ☉ ° ' "	Moon ☽ ° ' "	Moon ☽ 24.00 ° ' "	Mercury ☿ ° '	Venus ♀ ° '	Mars ♂ ° '	Jupiter ♃ ° '	Saturn ♄ ° '	Uranus ♅ ° '	Neptune ♆ ° '	Pluto ♇ ° '
01	10 44 34	09 ♍ 32 22	22 ♌ 00 59	28 ♌ 05 02	22 ♌ 19	03 ♎ 36	27 ♊ 59	19 ♊ 05	16 ♓ 32	27 ♉ 15	29 ♓ 03	00 ♒ 01
02	10 48 31	10 30 28	04 ♍ 06 51	10 ♍ 06 38	22 53	04 50	28 35	19 12	16 R 28	27 R 15	29 R 01	00 ♒ 00
03	10 52 28	11 28 36	16 ♍ 04 37	22 ♍ 01 02	23 35	06 03	29 12	19 18	16 23	27 15	28 59	29 ♑ 59
04	10 56 24	12 26 45	27 ♍ 56 09	03 ♎ 50 15	24 25	07 17	29 ♊ 48	19 25	16 19	27 15	28 58	29 R 58
05	11 00 21	13 24 56	09 ♎ 43 40	15 ♎ 36 44	25 22	08 30	00 ♋ 25	19 31	16 14	27 15	28 56	29 57
06	11 04 17	14 23 09	21 ♎ 29 50	27 ♎ 23 23	26 27	09 43	01 01	19 37	16 09	27 15	28 55	29 56
07	11 08 14	15 21 23	03 ♏ 17 50	09 ♏ 13 40	27 38	10 57	01 37	19 43	16 05	27 15	28 53	29 55
08	11 12 10	16 19 39	15 ♏ 11 24	21 ♏ 11 33	28 ♌ 55	12 10	02 12	19 49	16 00	27 14	28 51	29 54
09	11 16 07	17 17 56	27 ♏ 14 41	03 ♐ 21 22	00 ♍ 18	13 23	02 48	19 55	15 56	27 14	28 50	29 53
10	11 20 03	18 16 15	09 ♐ 32 10	15 ♐ 47 37	01 46	14 37	03 23	20 00	15 51	27 13	28 48	29 52
11	11 24 00	19 14 36	22 ♐ 08 17	28 ♐ 34 40	03 19	15 50	03 59	20 05	15 46	27 13	28 47	29 51
12	11 27 57	20 12 58	05 ♑ 07 12	11 ♑ 46 15	04 56	17 03	04 34	20 10	15 42	27 12	28 45	29 51
13	11 31 53	21 11 22	18 ♑ 32 05	25 ♑ 24 51	06 36	18 17	05 09	20 15	15 37	27 12	28 43	29 50
14	11 35 50	22 09 47	02 ♒ 24 35	09 ♒ 31 07	08 18	19 30	05 44	20 20	15 33	27 11	28 42	29 49
15	11 39 46	23 08 14	16 ♒ 44 08	24 ♒ 03 08	10 04	20 43	06 18	20 25	15 28	27 11	28 40	29 48
16	11 43 43	24 06 43	01 ♓ 27 27	08 ♓ 56 13	11 51	21 57	06 53	20 29	15 24	27 10	28 38	29 48
17	11 47 39	25 05 13	16 ♓ 28 27	24 ♓ 03 03	13 40	23 10	07 27	20 34	15 19	27 09	28 37	29 47
18	11 51 36	26 03 45	01 ♈ 38 51	09 ♈ 14 37	15 30	24 23	08 01	20 38	15 15	27 08	28 35	29 46
19	11 55 32	27 02 19	16 ♈ 49 12	24 ♈ 21 26	17 21	25 36	08 35	20 42	15 10	27 07	28 33	29 46
20	11 59 29	28 00 55	01 ♉ 50 18	09 ♉ 14 55	19 12	26 49	09 09	20 45	15 06	27 07	28 32	29 45
21	12 03 26	28 59 33	16 ♉ 34 31	23 ♉ 48 32	21 03	28 02	09 42	20 49	15 01	27 06	28 30	29 44
22	12 07 22	29 ♍ 58 13	00 ♊ 56 34	07 ♊ 58 20	22 55	29 ♎ 16	10 16	20 52	14 57	27 05	28 28	29 44
23	12 11 19	00 ♎ 56 56	14 ♊ 53 45	21 ♊ 42 51	24 46	00 ♏ 29	10 49	20 56	14 53	27 04	28 27	29 43
24	12 15 15	01 55 41	28 ♊ 25 45	05 ♋ 02 43	26 37	01 42	11 22	20 59	14 48	27 02	28 25	29 43
25	12 19 12	02 54 29	11 ♋ 34 01	18 ♋ 00 03	28 ♍ 28	02 55	11 55	21 01	14 44	27 01	28 23	29 42
26	12 23 08	03 53 18	24 ♋ 21 12	00 ♌ 37 53	00 ♎ 18	04 08	12 27	21 04	14 40	27 00	28 22	29 42
27	12 27 05	04 52 10	06 ♌ 50 31	12 ♌ 59 34	02 07	05 21	12 59	21 06	14 36	26 59	28 20	29 42
28	12 31 01	05 51 04	19 ♌ 05 25	25 ♌ 08 30	03 56	06 34	13 32	21 09	14 32	26 58	28 19	29 41
29	12 34 58	06 50 00	01 ♍ 09 12	07 ♍ 07 54	05 44	07 47	14 04	21 11	14 28	26 56	28 17	29 41
30	12 38 55	07 ♎ 48 58	13 ♍ 04 56	19 ♍ 00 39	07 ♎ 31	09 ♏ 00	14 ♋ 35	21 ♊ 13	14 ♓ 24	26 ♉ 55	28 ♓ 15	29 ♑ 40

Date	Moon True ☊ ° '	Moon Mean ☊ ° '	Moon ☽ Latitude ° '
01	06 ♈ 47	07 ♈ 56	03 N 41
02	06 R 43	07 53	02 49
03	06 41	07 50	01 51
04	06 39	07 46	00 N 48
05	06 D 39	07 43	00 S 17
06	06 40	07 40	01 21
07	06 41	07 37	02 22
08	06 43	07 34	03 17
09	06 44	07 30	04 05
10	06 45	07 27	04 42
11	06 R 45	07 24	05 06
12	06 45	07 21	05 16
13	06 43	07 18	05 10
14	06 42	07 15	04 45
15	06 40	07 11	04 03
16	06 38	07 08	03 03
17	06 37	07 05	01 49
18	06 37	07 02	00 S 28
19	06 D 37	06 59	00 N 56
20	06 37	06 56	02 16
21	06 38	06 52	03 24
22	06 39	06 49	04 18
23	06 39	06 46	04 55
24	06 39	06 43	05 14
25	06 R 39	06 40	05 17
26	06 39	06 36	05 03
27	06 39	06 33	04 35
28	06 39	06 30	03 55
29	06 D 39	06 27	03 05
30	06 ♈ 39	06 ♈ 24	02 N 07

DECLINATIONS

Date	Sun ☉ ° '	Moon ☽ ° '	Mercury ☿ ° '	Venus ♀ ° '	Mars ♂ ° '	Jupiter ♃ ° '	Saturn ♄ ° '	Uranus ♅ ° '	Neptune ♆ ° '	Pluto ♇ ° '
01	08 N 00	17 N 39	13 N 10	00 S 30	23 N 20	22 N 18	07 S 19	19 N 17	01 S 36	23 S 20
02	07 38	12 38	13 16	01 01	23 22	22 18	07 21	19 17	01 36	23 20
03	07 16	07 12	13 18	01 32	23 23	22 19	07 23	19 17	01 37	23 21
04	06 53	01 N 33	13 16	02 03	23 25	22 19	07 25	19 17	01 38	23 21
05	06 31	04 S 07	13 11	02 34	23 26	22 19	07 27	19 17	01 38	23 21
06	06 09	09 38	13 03	03 05	23 27	22 20	07 29	19 17	01 39	23 21
07	05 46	14 50	12 51	03 36	23 27	22 20	07 31	19 17	01 39	23 21
08	05 24	19 32	12 35	04 06	23 28	22 21	07 32	19 17	01 40	23 22
09	05 01	23 30	12 16	04 37	23 28	22 21	07 34	19 17	01 41	23 22
10	04 38	26 31	11 54	05 08	23 29	22 22	07 36	19 17	01 41	23 22
11	04 15	28 18	11 29	05 38	23 29	22 22	07 38	19 17	01 42	23 22
12	03 53	28 37	11 01	06 09	23 29	22 22	07 40	19 17	01 43	23 22
13	03 30	27 16	10 30	06 39	23 29	22 23	07 41	19 16	01 43	23 23
14	03 07	24 15	09 57	07 09	23 29	22 23	07 43	19 16	01 44	23 23
15	02 43	19 40	09 21	07 39	23 28	22 23	07 45	19 16	01 45	23 23
16	02 20	13 48	08 44	08 09	23 28	22 24	07 47	19 16	01 45	23 23
17	01 57	07 S 01	08 05	08 39	23 27	22 24	07 49	19 16	01 46	23 23
18	01 34	00 N 14	07 24	09 08	23 27	22 24	07 50	19 16	01 47	23 23
19	01 11	07 29	06 41	09 37	23 26	22 24	07 52	19 15	01 47	23 23
20	00 47	14 14	05 58	10 06	23 25	22 25	07 54	19 15	01 48	23 24
21	00 24	20 03	05 13	10 35	23 24	22 25	07 56	19 15	01 49	23 24
22	00 N 01	24 33	04 28	11 04	23 23	22 25	07 57	19 15	01 49	23 24
23	00 S 23	27 28	03 42	11 32	23 21	22 25	07 59	19 15	01 50	23 24
24	00 46	28 40	02 56	12 01	23 20	22 25	08 01	19 14	01 51	23 24
25	01 09	28 11	02 09	12 28	23 18	22 25	08 02	19 14	01 51	23 24
26	01 33	26 13	01 22	12 56	23 17	22 26	08 04	19 14	01 52	23 24
27	01 56	22 59	00 N 35	13 23	23 15	22 26	08 05	19 13	01 53	23 24
28	02 19	18 49	00 S 12	13 50	23 13	22 26	08 07	19 13	01 53	23 24
29	02 43	13 57	00 59	14 17	23 11	22 26	08 08	19 13	01 54	23 25
30	03 S 06	08 N 36	01 S 46	14 S 43	23 N 09	22 N 26	08 S 10	19 N 13	01 S 55	23 S 25

ZODIAC SIGN ENTRIES

Date	h	m	Planets
02	00	10	♇ ♑
02	03	48	☽ ♍
04	16	12	☽ ♎
04	19	46	♂ ♋
07	05	18	☽ ♏
09	06	50	☿ ♍
09	17	26	☽ ♐
12	02	38	☽ ♑
14	07	53	☽ ♒
16	09	39	☽ ♓
18	09	24	☽ ♈
20	09	02	☽ ♉
22	10	24	☽ ♊
22	12	44	☉ ♎
23	02	36	♀ ♏
24	14	50	☽ ♋
26	08	09	☿ ♎
26	22	47	☽ ♌
29	09	42	☽ ♍

LATITUDES

Date	Mercury ☿ ° '	Venus ♀ ° '	Mars ♂ ° '	Jupiter ♃ ° '	Saturn ♄ ° '	Uranus ♅ ° '	Neptune ♆ ° '	Pluto ♇ ° '
01	00 S 57	01 N 01	00 S 05	00 S 42	02 S 11	00 S 16	01 S 19	03 S 16
04	00 S 07	00 55	00 S 02	00 42	02 11	00 16	01 19	03 16
07	00 N 35	00 48	00 N 02	00 42	02 11	00 16	01 19	03 16
10	01 08	00 41	00 05	00 42	02 12	00 16	01 19	03 16
13	01 31	00 34	00 09	00 42	02 12	00 16	01 19	03 16
16	01 45	00 26	00 12	00 42	02 12	00 16	01 20	03 16
19	01 50	00 18	00 16	00 43	02 12	00 16	01 20	03 16
22	01 48	00 09	00 20	00 43	02 12	00 16	01 20	03 16
25	01 41	00 N 01	00 24	00 43	02 12	00 16	01 20	03 16
28	01 29	00 S 08	00 28	00 43	02 12	00 16	01 20	03 16
31	01 N 15	00 S 17	00 N 32	00 S 43	02 S 11	00 S 16	01 S 20	03 S 16

DATA

Julian Date	2460555
Delta T	+72 seconds
Ayanamsa	24° 12' 05"
Synetic vernal point	04° ♓ 54' 54"
True obliquity of ecliptic	23° 26' 19"

MOON'S PHASES, APSIDES AND POSITIONS ☽

Date	h	m	Phase	Longitude ° '	Eclipse Indicator
03	01	56	●	11 ♍ 04	
11	06	06	☽	19 ♐ 00	
18	02	34	○	25 ♓ 41	partial
24	18	50	☾	02 ♋ 12	

Day	h	m		
05	15	04	Apogee	
18	13	29	Perigee	
04	18	33	0S	
12	04	42	Max dec	28° S 41'
18	11	14	0N	
24	16	53	Max dec	28° N 42'

LONGITUDES

Date	Chiron ⚷ ° '	Ceres ⚳ ° '	Pallas ⚴ ° '	Juno ⚵ ° '	Vesta ⚶ ° '	Black Moon Lilith ⚸ ° '
01	22 ♈ 58	07 ♑ 36	28 ♏ 03	07 ♎ 46	03 ♍ 33	07 ♎ 06
11	22 ♈ 38	08 ♑ 15	00 ♐ 54	11 ♎ 17	08 ♍ 17	08 ♎ 13
21	22 ♈ 14	09 ♑ 24	04 ♐ 01	14 ♎ 49	13 ♍ 01	09 ♎ 19
31	21 ♈ 49	11 ♑ 02	07 ♐ 20	18 ♎ 22	17 ♍ 47	10 ♎ 25

All ephemeris data is given at 12.00 UT and the Moon's longitude is additionally given for 24.00 UT

ASPECTARIAN

h m Aspects

01 Sunday
01 18 ☽ ⊼ ♄
03 30 ☽ ∥ ♅
04 32 ☽ ∠ ♀
06 10 ☽ ✶ ♃
08 01 ♂ ⋕ ♇
12 36 ☽ ☌ ☿
14 01 ☽ ± ♆
15 19 ♅ St R
22 22 ☽ □ ♅

02 Monday
00 17 ☽ ⊥ ♀
00 25 ☽ ✶ ♂
01 52 ☽ ⊼ ♆
03 48 ☽ ⊼ ♇
06 08 ☽ Q ♃
09 08 ☽ ∥ ☿
13 35 ☽ ⊻ ♀
15 45 ☽ ± ♇
16 52 ☿ ± ♆
17 22 ☉ Q ♂

03 Tuesday
01 39 ☽ Q ♂
01 56 ☽ ☌ ☉
04 10 ♂ □ ♆
04 25 ☉ ⋕ ♄
09 47 ☽ ⚼ ♇
11 11 ☽ ⋕ ♄
11 42 ☽ ∥ ☉
12 37 ☽ ☍ ♄
15 43 ♀ ∥ ♆
18 34 ☽ □ ♃

04 Wednesday
04 17 ☽ ⊻ ☿
10 04 ☽ ⋕ ♀
10 37 ☽ △ ♅
11 42 ☽ ⋕ ♆
14 05 ☽ ☍ ♆
16 00 ☽ □ ♂
16 06 ☽ △ ♇
17 27 ☽ ⊥ ☿
17 59 ♂ ⊼ ♇

05 Thursday
01 27 ☽ ∥ ♆
04 45 ☽ ∥ ♀
09 12 ☽ ☌ ♀
13 26 ☽ ∠ ☿
17 09 ☽ ⚼ ♅
20 12 ☽ ⊻ ☉
21 41 ☽ ⋕ ☉

06 Friday
01 11 ☽ ⊼ ♄
02 27 ☽ ∥ ♄
08 08 ☽ △ ♃
09 32 ☽ ⊥ ☉
11 29 ☽ ± ♅
13 20 ☽ ± ♄
23 11 ☽ ✶ ☿
23 42 ☽ ⊼ ♅

07 Saturday
01 12 ☉ ⚼ ♇
02 58 ☽ ⋕ ☿
03 04 ☽ ⊼ ♆
04 21 ☿ □ ♅
05 08 ☽ □ ♇
05 30 ☽ ∠ ☉
07 32 ☽ ⚼ ♄
08 24 ☽ △ ♂
14 54 ☽ ⚼ ♃
15 13 ☽ ± ♆

08 Sunday
02 21 ☽ Q ☿
04 35 ☉ ☍ ♄
05 14 ☽ ⊻ ♀
09 13 ☽ ± ♃
09 20 ☽ ⚼ ♆
10 39 ☽ ⋕ ♅
10 52 ☿ ⊼ ♆
13 20 ♀ ⚼ ♅
13 37 ☽ △ ♄
14 29 ☽ ✶ ☉
16 15 ☽ ⚼ ♂
17 25 ☽ Q ♇
18 39 ☽ ⊥ ♀
21 20 ☽ ⊼ ♃

09 Monday
04 28 ☽ ⋕ ♃
04 55 ☿ ⊼ ♇
11 01 ☽ ∥ ♇
11 05 ☽ ± ♂
11 46 ☽ ⋕ ♂
11 58 ☽ ☍ ♅
14 31 ☽ ∠ ♀
15 07 ☽ △ ♆
16 24 ☽ Q ☉
17 11 ☽ ✶ ♇
18 50 ☽ □ ☿
22 46 ☉ ⋕ ♀
23 28 ☽ ⊼ ♂

10 Tuesday
05 18 ♂ ⊥ ♅
15 50 ☿ Q ♃
22 14 ☽ ∠ ♇
22 48 ☽ ✶ ♀

11 Wednesday
00 02 ☽ □ ♄
06 06 ☽ □ ☉
08 07 ☽ ☍ ♃
10 52 ♀ ⊼ ♄
15 13 ☽ ⊥ ♇
21 28 ☽ ⊼ ♅
23 44 ☽ Q ♀

12 Thursday
00 21 ☽ □ ♆
02 21 ☽ ⊻ ♇
03 42 ☿ ✶ ♂
08 31 ☽ ± ♅
09 25 ☽ Q ♄
10 53 ☉ □ ♃
10 56 ☽ ☍ ♂
11 36 ☽ △ ☿

13 Friday
00 46 ☽ ⚼ ♅
01 11 ☿ ± ♇
06 53 ☽ ✶ ♄
08 49 ☽ Q ♆
11 30 ☽ □ ♀
15 03 ☽ ⊼ ♃
17 01 ☽ △ ☉
18 07 ☽ ⚼ ☿

14 Saturday
01 32 ☽ ± ♃
03 04 ☽ △ ♅
05 40 ☽ ✶ ♆
07 35 ☽ ☌ ♇
08 50 ☽ ∠ ♄
11 48 ☽ ± ☿
16 39 ☽ ⋕ ♂
17 00 ☽ ⚼ ♃
17 14 ☽ ∥ ♇
17 52 ☽ ⊼ ♂
20 38 ☽ ⚼ ☉
22 44 ☽ ⋕ ♃
23 21 ☽ ⊼ ☿
23 59 ☽ ⊥ ♄

15 Sunday
04 21 ☽ ± ♂
05 34 ♀ △ ♃
06 56 ☽ ∠ ♆
09 55 ☽ ⊻ ♄
12 43 ☽ ± ☉
13 49 ☽ ⋕ ♅
17 10 ♀ ∥ ♄
18 05 ☽ △ ♃
19 09 ☽ △ ♀
19 49 ☽ ⚼ ♂
20 52 ♀ ± ♅
21 43 ☽ ⊥ ♆
23 15 ☽ ⊼ ☉

16 Monday
01 52 ♀ ± ♄
05 04 ☽ □ ♅
07 27 ☽ ⊻ ♆
09 19 ☽ ⊻ ♇
18 58 ☽ ⊥ ♇
21 03 ☽ △ ♂
21 35 ☽ ⚼ ♀

17 Tuesday
00 14 ☿ ⋕ ♀
06 50 ☽ ∥ ♀
06 55 ☽ ☍ ☿
08 02 ☽ ⋕ ☿
09 19 ☽ ∠ ♇
09 20 ☽ ∥ ♄
09 54 ☽ Q ♅
10 11 ☽ ☌ ♄
13 11 ☽ ± ♀
18 30 ☽ □ ♃
21 03 ☿ ⋕ ♄
23 06 ☉ ⋕ ♆
23 31 ☽ ⊼ ♀

18 Wednesday
02 34 ☽ ☍ ☉
02 34 ☿ ⚼ ♇
04 53 ☽ ✶ ♅
05 25 ☽ ∥ ♆
05 47 ☽ ⋕ ☉
07 10 ☽ ☌ ♆
08 50 ☿ ☍ ♄
09 02 ☽ ✶ ♇
16 09 ☽ ∥ ☉
17 04 ☽ ⋕ ♆
22 27 ☽ □ ♂
23 05 ☽ Q ♃

19 Thursday
03 59 ☽ Q ♇
04 34 ☽ ∠ ♅
09 24 ☽ ⊻ ♄
09 35 ☽ ∥ ☿
12 57 ☽ ⊼ ☿
13 20 ☽ ⋕ ♄
14 04 ☉ △ ♅
18 11 ☽ ✶ ♃
18 50 ☽ ⊥ ♅
18 53 ☽ ⊥ ♄
19 55 ☽ ⋕ ♀
23 50 ☽ ± ☿

20 Friday
03 14 ☽ ☍ ♀
04 10 ☽ Q ♂
04 25 ☽ ⊻ ♅
05 26 ☽ ⊼ ☉
06 41 ☽ ⊻ ♆
07 00 ♄ Q ♅
08 39 ☽ □ ♇
09 13 ☽ ∠ ♄
15 46 ☽ ± ☉
16 20 ☽ ⊥ ♆
16 21 ☽ ⚼ ☿
17 36 ♀ ⊼ ♅
18 21 ☽ ∠ ♃

21 Saturday
00 17 ☉ ☍ ♆
00 18 ☽ ✶ ♂
06 57 ☽ ∠ ♆
07 27 ☽ ⚼ ☉
08 23 ☽ ∥ ♅
08 50 ☿ □ ♃
09 06 ☽ ⊥ ♃
09 28 ☽ ✶ ♄
12 19 ♂ ⋕ ♇
19 03 ☽ ⊻ ♃
20 30 ☽ △ ☿
20 53 ♀ ⊼ ♆
23 36 ☽ ∥ ♃

22 Sunday
00 00 ☿ Q ♂
02 03 ☽ ∠ ♂
04 55 ☽ ∥ ♂
05 00 ☽ ⋕ ♇
05 18 ☽ Q ♄
05 29 ☽ ☌ ♅
06 12 ☉ △ ♇
07 50 ☽ ✶ ♆
08 53 ☽ ⊼ ♀
09 57 ☽ △ ♇
10 14 ☽ △ ☉
17 52 ☽ ⊥ ♂
20 03 ☽ ± ♀
21 15 ♀ □ ♇

23 Monday
00 52 ♀ ⚼ ♄
01 03 ☿ ⊥ ♀
04 17 ☽ Q ♆
04 37 ☽ ⊻ ♂
11 42 ☽ ⚼ ♇
11 58 ☽ □ ♄
13 07 ☽ ⚼ ♀
22 39 ☽ ☌ ♃

24 Tuesday
03 34 ☽ ± ♇
08 13 ☽ □ ☿
09 30 ☽ ⊻ ♅
11 59 ☽ □ ♆
14 19 ☽ ⊼ ♇
17 27 ☿ △ ♅
18 30 ☽ △ ♀
18 50 ☽ □ ☉
20 20 ☽ ⊥ ♅

25 Wednesday
11 07 ☿ ☍ ♆
11 27 ☉ ⊻ ♀
12 40 ☽ ☌ ♂
12 51 ☽ ∠ ♅
16 48 ♂ ∠ ♅
17 52 ☽ △ ♄
21 05 ☿ ⋕ ♆
22 38 ☽ Q ☿
23 09 ♄ ∠ ♇

26 Thursday
04 14 ☿ △ ♇
05 45 ☽ ⊻ ♃
06 56 ☽ Q ☉
08 30 ☉ ⋕ ☿
16 29 ♀ ± ♆
17 02 ☽ ✶ ♅
17 11 ☽ ⊥ ♃
19 38 ☽ △ ♆
22 05 ☽ ⚼ ♄
22 12 ☽ ☍ ♇

27 Friday
01 18 ☽ ✶ ☿
07 51 ☽ ✶ ☉
08 27 ☉ ∥ ♆
08 47 ☽ □ ♀
09 18 ☽ ⋕ ♇
10 19 ☽ ∥ ♂
10 34 ☽ ∠ ♃
15 24 ☽ ± ♄
15 32 ☽ ∥ ♃
16 09 ☽ Q ♅

28 Saturday
00 33 ☽ ⊻ ♂
00 39 ☽ ⚼ ♆
03 04 ☽ ⊼ ♄
03 26 ♀ ⚼ ♃
09 51 ☽ ∥ ♅
11 37 ☽ ∠ ☿
12 54 ☽ ⊥ ♂
15 47 ☽ ∠ ☉
16 04 ☽ ✶ ♃
18 21 ☽ ± ♆

29 Sunday
00 04 ☽ Q ♀
03 36 ☽ □ ♅
06 16 ☽ ⊼ ♆
07 37 ☽ ∠ ♂
08 39 ☽ ⊥ ☿
09 03 ☽ ⊼ ♇
10 32 ☽ ⋕ ♀
11 18 ☽ ⊥ ☉
16 04 ☽ Q ♃
21 05 ☽ ± ♇
22 48 ☽ ⊻ ☿

30 Monday
00 25 ☽ ⊻ ☉
02 49 ☽ ✶ ♀
04 06 ♂ △ ♄
13 55 ☽ ⋕ ♄
14 38 ☽ ☍ ♄
15 11 ☽ ✶ ♂
15 13 ☽ ⚼ ♇
16 46 ☿ ∥ ♆
21 09 ☉ ☌ ☿

OCTOBER 2024

LONGITUDES

Date	Sidereal time h m s	Sun ☉ ° ' "	Moon ☽ ° ' "	Moon ☽ 24.00 ° ' "	Mercury ☿ ° '	Venus ♀ ° '	Mars ♂ ° '	Jupiter ♃ ° '	Saturn ♄ ° '	Uranus ♅ ° '	Neptune ♆ ° '	Pluto ♇ ° '
01	12 42 51	08 ♎ 47 59	24 ♍ 55 21	00 ♎ 49 21	09 ♎ 17	10 ♏ 13	15 ♋ 07	21 ♊ 14	14 ♓ 20	26 ♉ 54	28 ♓ 14	29 ♑ 40
02	12 46 48	09 47 01	06 ♎ 42 55	12 ♎ 36 21	11 03	11 26	15 38	21 16	14 R 16	26 R 52	28 R 12	29 R 40
03	12 50 44	10 46 06	18 ♎ 29 56	24 ♎ 23 55	12 47	12 39	16 09	21 17	14 12	26 51	28 10	29 40
04	12 54 41	11 45 12	00 ♏ 18 36	06 ♏ 14 16	14 31	13 52	16 40	21 18	14 08	26 49	28 09	29 39
05	12 58 37	12 44 21	12 ♏ 11 13	18 ♏ 09 46	16 14	15 05	17 10	21 19	14 04	26 48	28 07	29 39
06	13 02 34	13 43 31	24 ♏ 10 16	00 ♐ 13 04	17 57	16 18	17 40	21 19	14 00	26 46	28 06	29 39
07	13 06 30	14 42 44	06 ♐ 18 33	12 ♐ 27 05	19 38	17 30	18 10	21 20	13 57	26 44	28 04	29 39
08	13 10 27	15 41 58	18 ♐ 39 06	24 ♐ 55 00	21 19	18 43	18 40	21 20	13 53	26 43	28 02	29 39
09	13 14 24	16 41 14	01 ♑ 15 13	07 ♑ 40 10	22 58	19 56	19 10	21 R 20	13 50	26 41	28 01	29 39
10	13 18 20	17 40 32	14 ♑ 10 16	20 ♑ 45 51	24 38	21 09	19 39	21 20	13 46	26 39	27 59	29 39
11	13 22 17	18 39 52	27 ♑ 27 17	04 ♒ 14 49	26 16	22 22	20 08	21 20	13 43	26 37	27 58	29 39
12	13 26 13	19 39 13	11 ♒ 08 37	18 ♒ 08 47	27 53	23 34	20 37	21 19	13 40	26 36	27 56	29 D 39
13	13 30 10	20 38 36	25 ♒ 15 14	02 ♓ 27 47	29 ♎ 30	24 47	21 05	21 18	13 37	26 34	27 55	29 39
14	13 34 06	21 38 01	09 ♓ 46 06	17 ♓ 09 39	01 ♏ 06	26 00	21 33	21 18	13 33	26 32	27 53	29 39
15	13 38 03	22 37 27	24 ♓ 37 44	02 ♈ 09 30	02 42	27 12	22 01	21 16	13 30	26 30	27 52	29 39
16	13 41 59	23 36 56	09 ♈ 43 58	17 ♈ 19 59	04 17	28 25	22 28	21 15	13 27	26 28	27 50	29 39
17	13 45 56	24 36 26	24 ♈ 56 22	02 ♉ 31 52	05 51	29 ♏ 37	22 56	21 13	13 25	26 26	27 49	29 39
18	13 49 53	25 35 59	10 ♉ 05 18	17 ♉ 35 27	07 24	00 ♐ 50	23 23	21 12	13 22	26 24	27 47	29 39
19	13 53 49	26 35 33	25 ♉ 01 19	02 ♊ 21 57	08 57	02 02	23 49	21 10	13 19	26 22	27 46	29 39
20	13 57 46	27 35 10	09 ♊ 36 37	16 ♊ 44 45	10 30	03 15	24 16	21 08	13 16	26 20	27 45	29 40
21	14 01 42	28 34 49	23 ♊ 46 00	00 ♋ 40 09	12 01	04 27	24 42	21 05	13 14	26 18	27 43	29 40
22	14 05 39	29 ♎ 34 31	07 ♋ 27 13	14 ♋ 07 19	13 32	05 40	25 07	21 03	13 11	26 16	27 42	29 40
23	14 09 35	00 ♏ 34 14	20 ♋ 40 43	27 ♋ 07 48	15 03	06 52	25 33	21 00	13 09	26 13	27 41	29 40
24	14 13 32	01 34 00	03 ♌ 29 00	09 ♌ 44 51	16 33	08 05	25 57	20 57	13 07	26 11	27 39	29 41
25	14 17 28	02 33 48	15 ♌ 55 55	22 ♌ 02 45	18 02	09 17	26 22	20 54	13 05	26 09	27 38	29 41
26	14 21 25	03 33 38	28 ♌ 05 59	04 ♍ 06 10	19 31	10 29	26 46	20 51	13 03	26 07	27 37	29 42
27	14 25 22	04 33 31	10 ♍ 03 54	15 ♍ 59 43	20 59	11 41	27 10	20 47	13 01	26 04	27 35	29 42
28	14 29 18	05 33 25	21 ♍ 54 10	27 ♍ 47 44	22 27	12 54	27 34	20 43	12 59	26 02	27 34	29 42
29	14 33 15	06 33 22	03 ♎ 40 53	09 ♎ 34 01	23 54	14 06	27 57	20 40	12 57	26 00	27 33	29 43
30	14 37 11	07 33 20	15 ♎ 27 33	21 ♎ 21 48	25 20	15 18	28 19	20 35	12 55	25 58	27 32	29 43
31	14 41 08	08 ♏ 33 21	27 ♎ 17 06	03 ♏ 13 42	26 ♏ 46	16 ♐ 30	28 ♋ 42	20 ♊ 31	12 ♓ 53	25 ♉ 55	27 ♓ 31	29 ♑ 44

Date	Moon True ☊ ° '	Moon Mean ☊ ° '	Moon ☽ Latitude ° '
01	06 ♈ 39	06 ♈ 21	01 N 05
02	06 R 39	06 17	00 00
03	06 39	06 14	01 S 05
04	06 39	06 11	02 08
05	06 38	06 08	03 05
06	06 37	06 05	03 54
07	06 36	06 01	04 34
08	06 35	05 58	05 02
09	06 34	05 55	05 16
10	06 34	05 52	05 14
11	06 D 34	05 49	04 56
12	06 34	05 46	04 21
13	06 35	05 42	03 30
14	06 37	05 39	02 23
15	06 37	05 36	01 S 06
16	06 R 38	05 33	00 N 17
17	06 37	05 30	01 40
18	06 35	05 27	02 55
19	06 33	05 23	03 56
20	06 30	05 20	04 41
21	06 27	05 17	05 08
22	06 25	05 14	05 15
23	06 23	05 11	05 06
24	06 D 23	05 07	04 41
25	06 24	05 04	04 03
26	06 25	05 01	03 16
27	06 27	04 58	02 20
28	06 29	04 55	01 19
29	06 30	04 52	00 N 15
30	06 R 29	04 48	00 S 49
31	06 ♈ 27	04 ♈ 45	01 S 51

DECLINATIONS

Date	Sun ☉ ° '	Moon ☽ ° '	Mercury ☿ ° '	Venus ♀ ° '	Mars ♂ ° '	Jupiter ♃ ° '	Saturn ♄ ° '	Uranus ♅ ° '	Neptune ♆ ° '	Pluto ♇ ° '
01	03 S 29	03 N 00	02 S 32	15 S 09	23 N 07	22 N 26	08 S 11	19 N 12	01 S 55	23 S 25
02	03 53	02 S 40	03 19	15 35	23 05	22 26	08 13	19 12	01 56	23 25
03	04 16	08 15	04 05	16 00	23 03	22 26	08 14	19 12	01 57	23 25
04	04 39	13 34	04 50	16 25	23 01	22 26	08 16	19 11	01 57	23 25
05	05 02	18 25	05 36	16 50	22 58	22 26	08 17	19 11	01 58	23 25
06	05 25	22 36	06 21	17 14	22 56	22 26	08 19	19 10	01 59	23 25
07	05 48	25 52	07 05	17 38	22 53	22 26	08 20	19 10	01 59	23 25
08	06 11	27 58	07 49	18 01	22 51	22 26	08 21	19 10	02 00	23 25
09	06 34	28 42	08 32	18 24	22 48	22 26	08 22	19 09	02 00	23 25
10	06 56	27 54	09 15	18 46	22 45	22 26	08 24	19 09	02 01	23 25
11	07 19	25 31	09 57	19 08	22 43	22 26	08 25	19 08	02 02	23 25
12	07 41	21 37	10 39	19 30	22 40	22 26	08 26	19 08	02 02	23 25
13	08 04	16 24	11 19	19 51	22 37	22 26	08 27	19 08	02 03	23 25
14	08 26	10 07	12 00	20 11	22 34	22 26	08 28	19 07	02 03	23 25
15	08 48	03 S 09	12 39	20 31	22 32	22 26	08 29	19 07	02 04	23 25
16	09 10	04 N 07	13 18	20 50	22 29	22 26	08 30	19 06	02 04	23 25
17	09 32	11 12	13 56	21 09	22 26	22 26	08 31	19 06	02 05	23 25
18	09 54	17 36	14 33	21 27	22 23	22 26	08 32	19 05	02 06	23 25
19	10 15	22 51	15 10	21 45	22 20	22 26	08 33	19 05	02 06	23 25
20	10 37	26 32	15 45	22 02	22 17	22 25	08 34	19 04	02 07	23 25
21	10 58	28 25	16 20	22 19	22 14	22 25	08 35	19 04	02 07	23 25
22	11 19	28 28	16 54	22 35	22 11	22 25	08 36	19 03	02 08	23 24
23	11 40	26 53	17 28	22 50	22 08	22 25	08 37	19 03	02 08	23 24
24	12 01	23 55	18 00	23 05	22 06	22 25	08 38	19 02	02 09	23 24
25	12 22	19 56	18 31	23 19	22 03	22 25	08 38	19 02	02 09	23 24
26	12 42	15 12	19 02	23 32	22 00	22 24	08 39	19 01	02 10	23 24
27	13 02	09 57	19 31	23 45	21 57	22 24	08 40	19 01	02 10	23 24
28	13 22	04 N 26	20 00	23 57	21 54	22 24	08 40	19 00	02 11	23 24
29	13 42	01 S 14	20 28	24 09	21 52	22 24	08 41	19 00	02 11	23 24
30	14 02	06 50	20 54	24 20	21 49	22 23	08 41	18 59	02 12	23 24
31	14 S 21	12 S 14	21 S 20	24 S 30	21 N 46	22 N 23	08 S 42	18 N 59	02 S 12	23 S 24

ZODIAC SIGN ENTRIES

Date	h m	Planets
01	22 20	☽ ♎
04	11 22	☽ ♏
06	23 34	☽ ♐
09	09 38	☽ ♑
11	16 31	☽ ♒
13	19 23	☿ ♏
13	19 55	☽ ♓
15	20 34	☽ ♈
17	19 28	♀ ♐
17	20 00	☽ ♉
19	20 07	☽ ♊
21	22 50	☽ ♋
22	22 15	☉ ♏
24	05 24	☽ ♌
26	15 47	☽ ♍
29	04 30	☽ ♎
31	17 29	☽ ♏

LATITUDES

Date	Mercury ☿ ° '	Venus ♀ ° '	Mars ♂ ° '	Jupiter ♃ ° '	Saturn ♄ ° '	Uranus ♅ ° '	Neptune ♆ ° '	Pluto ♇ ° '
01	01 N 15	00 S 17	00 N 32	00 S 43	02 S 11	00 S 16	01 S 20	03 S 16
04	00 57	00 27	00 37	00 43	02 11	00 16	01 20	03 16
07	00 39	00 36	00 41	00 43	02 11	00 16	01 20	03 16
10	00 N 19	00 45	00 46	00 43	02 11	00 16	01 20	03 16
13	00 S 02	00 54	00 51	00 43	02 11	00 16	01 20	03 16
16	00 22	01 03	00 56	00 43	02 10	00 16	01 19	03 16
19	00 43	01 12	01 01	00 43	02 10	00 16	01 19	03 16
22	01 03	01 21	01 06	00 43	02 10	00 16	01 19	03 16
25	01 22	01 29	01 11	00 43	02 09	00 16	01 19	03 16
28	01 41	01 38	01 17	00 43	02 09	00 16	01 19	03 16
31	01 S 57	01 S 45	01 N 23	00 S 43	02 S 08	00 S 16	01 S 19	03 S 16

DATA

Julian Date	2460585
Delta T	+72 seconds
Ayanamsa	24° 12' 08"
Synetic vernal point	04° ♓ 54' 51"
True obliquity of ecliptic	23° 26' 20"

MOON'S PHASES, APSIDES AND POSITIONS ☽

Date	h m	Phase	Longitude ° '	Eclipse Indicator
02	18 49	●	10 ♎ 04	Annular
10	18 55	☽	17 ♑ 58	
17	11 26	○	24 ♈ 35	
24	08 03	☾	01 ♌ 24	

Day	h m		
02	19 56	Apogee	
17	00 57	Perigee	
29	23 05	Apogee	
02	00 42	0S	
09	11 42	Max dec	28° S 42'
15	22 25	0N	
22	00 45	Max dec	28° N 40'
29	06 48	0S	

LONGITUDES

Date	Chiron ⚷ ° '	Ceres ⚳ ° '	Pallas ⚴ ° '	Juno ⚵ ° '	Vesta ⚶ ° '	Black Moon Lilith ⚸ ° '
01	21 ♈ 49	11 ♑ 02	07 ♐ 20	18 ♎ 22	17 ♍ 47	10 ♎ 25
11	21 ♈ 22	13 ♑ 04	10 ♐ 49	21 ♎ 56	22 ♍ 32	11 ♎ 32
21	20 ♈ 54	15 ♑ 28	14 ♐ 27	25 ♎ 28	27 ♍ 17	12 ♎ 38
31	20 ♈ 28	18 ♑ 09	18 ♐ 11	29 ♎ 00	01 ♎ 59	13 ♎ 45

ASPECTARIAN

01 Tuesday
04 30 ☽ □ ♃
10 06 ☽ # ☉
12 40 ☽ ∠ ♀
13 45 ☽ # ☿
16 00 ☽ △ ♅
16 35 ☽ # ♆
16 39 ☽ Q ♂
18 42 ☽ ☍ ♆
21 39 ☽ △ ♇

02 Wednesday
08 52 ☽ ∥ ♆
09 05 ☽ ⊥ ♀
15 08 ☽ ∥ ☿
17 29 ☽ ∥ ☉
18 49 ☽ ☌ ☉
22 21 ☽ ☌ ☿
22 29 ☽ ⚼ ♅
22 43 ☽ ⊻ ♀
23 09 ☿ ⚼ ♅

03 Thursday
03 17 ☽ ⊼ ♄
05 26 ☿ ⊻ ♀
07 00 ☽ □ ♂
11 55 ☽ ∥ ♄
15 26 ☽ ± ♄
16 46 ☽ ± ♅
17 40 ☽ △ ♃
22 10 ♀ ⚼ ♆
23 37 ☉ ∥ ☿

04 Friday
04 56 ☽ ⊼ ♅
06 46 ☿ ⊼ ♄
07 37 ☽ ⊼ ♆
09 37 ☽ ⚼ ♄
10 40 ☽ □ ♇
13 33 ☉ ⚼ ♅
17 04 ♀ △ ♄
19 45 ☽ ± ♆

05 Saturday
00 08 ☽ ⚼ ♃
03 02 ☽ ∥ ♀
13 13 ☽ ⊻ ☉
13 52 ☽ ⚼ ♆
15 46 ☽ △ ♄
16 02 ☽ # ♅
16 43 ♀ ± ♃
18 18 ☽ ± ♃
18 28 ☽ ☌ ♀
21 30 ☽ ⊻ ☿
22 27 ☽ △ ♂
22 58 ☽ Q ♇

06 Sunday
02 20 ☽ ⊥ ☉
06 19 ☽ ⊼ ♃
06 37 ☿ □ ♂
11 00 ☽ # ♃
11 28 ☽ ⊥ ☿
14 07 ☽ # ♂
17 09 ☽ ☍ ♅
17 22 ☽ ∥ ♇
18 28 ☉ ⊼ ♄
19 47 ☽ △ ♆
21 51 ☽ ∠ ☉
22 52 ☽ ⚹ ♇

07 Monday
05 34 ☽ ⚼ ♂
08 11 ☽ ∠ ☿
14 47 ♀ Q ♇
16 20 ☿ ± ♄
23 57 ☽ ± ♂

08 Tuesday
02 50 ☽ □ ♄
03 32 ☿ ± ♅
04 16 ☽ ∠ ♇
05 48 ☽ ⚹ ☉
10 22 ♀ △ ♂
12 02 ☽ ⊼ ♂
12 09 ☽ ⊻ ♀
12 23 ☿ △ ♃
17 10 ☽ ☍ ♃
17 54 ☽ ⚹ ☿
21 34 ☽ ⊥ ♇

09 Wednesday
00 52 ☽ ⊥ ♀
03 23 ☽ ⊼ ♅
05 54 ☽ □ ♆
06 23 ☿ ∥ ♄
06 45 ☽ Q ☉
07 04 ♃ St R
08 58 ☽ ⊻ ♇
13 05 ☽ Q ♄
14 41 ☽ ± ♅
19 37 ☽ ∠ ♀
20 01 ☽ Q ☿

10 Thursday
07 23 ☽ ⚼ ♅
11 16 ☽ ⚹ ♄
15 19 ☽ Q ♆
15 44 ♀ ⊼ ♃
18 55 ☽ □ ☉
22 22 ☽ ☍ ♂

11 Friday
01 02 ☽ ⊼ ♃
01 58 ☽ ⚹ ♀
09 35 ☽ □ ☿
10 31 ☽ △ ♅
11 47 ☽ ± ♃
12 19 ♀ # ♅
12 54 ☽ ⚹ ♆
14 14 ☽ ∠ ♄
15 53 ☽ ☌ ♇
17 11 ☿ ⊼ ♅

12 Saturday
00 34 ♇ St D
01 23 ☽ Q ♀
01 59 ☽ ∥ ♇
03 38 ☽ ⚼ ♃
05 59 ☽ ⊥ ♄
06 15 ☽ # ♂
07 35 ☽ # ♃
12 13 ☉ ± ♄
12 41 ☿ ⊼ ♆
15 05 ☽ ∠ ♆
16 19 ☽ ⊻ ♄
21 46 ☽ ∥ ♀
23 05 ☿ ⚼ ♄

13 Sunday
00 07 ☽ # ♅
03 39 ☽ △ ☉
04 44 ☽ ⊼ ♂
05 22 ☽ △ ♃
06 23 ☽ ⊥ ♆
10 05 ☉ ± ♅
11 08 ☽ □ ♀
14 03 ☿ □ ♇
14 11 ☽ □ ♅
15 10 ☽ ± ♂
16 26 ☽ ⊻ ♆
19 19 ☽ ⊻ ♇
19 59 ☽ △ ☿
23 08 ♂ ⊻ ♃

14 Monday
03 52 ☉ △ ♃
05 15 ☽ ⊥ ♇
05 48 ☽ ∥ ☿
06 30 ☽ ⚼ ☉
06 34 ☽ ⚼ ♂
08 15 ☉ □ ♂
14 43 ☉ ∥ ♄
17 40 ☽ ∥ ☉
17 49 ☽ ∥ ♄
18 09 ☽ ☌ ♄
19 44 ☽ Q ♅
19 56 ☽ ∠ ♇
22 13 ☽ ± ☉
22 22 ♀ ☍ ♅
23 32 ☽ ⚼ ☿

15 Tuesday
06 38 ☽ □ ♃
07 41 ☽ △ ♂
08 34 ☽ ⊼ ☉
14 59 ☽ ⚹ ♅
15 35 ☽ ∥ ♆
15 42 ☽ ± ☿
16 28 ☽ △ ♀
17 09 ☽ ☌ ♆
20 00 ☽ ⚹ ♇

16 Wednesday
00 49 ♀ △ ♆
02 22 ☽ ⊼ ☿
05 15 ☽ # ♆
05 23 ☿ ± ♆
11 14 ☽ Q ♃
14 44 ☽ ∠ ♅
15 02 ☽ Q ♇
17 52 ☽ ⊻ ♄
18 19 ☽ ⚼ ♀

17 Thursday
02 44 ☽ # ♄
03 18 ☽ ⊥ ♄
04 54 ☽ ⊥ ♅
05 53 ☽ # ☉
06 09 ☽ ⚹ ♃
08 44 ☽ □ ♂
09 45 ☽ ± ♀
11 26 ☽ ☍ ☉
12 02 ♂ ∥ ♃
12 30 ♀ ⚹ ♇
14 21 ☽ ⊻ ♅
16 32 ☽ ⊻ ♆
17 28 ☽ ∠ ♄
17 40 ☿ ⚼ ♃
19 26 ☽ □ ♇
20 02 ☽ ⊼ ♀
22 53 ☽ # ☿

18 Friday
02 01 ☽ ⊥ ♆
05 49 ☽ ∠ ♃
07 15 ☽ ☍ ☿
14 07 ☽ Q ♂
16 18 ☽ ∠ ♆
17 13 ☽ ⚹ ♄
18 12 ☽ ∥ ♅
20 08 ☽ ⊥ ♃

19 Saturday
05 46 ☽ ⊻ ♃
06 04 ☽ # ♀
06 40 ☉ ⊼ ♅
09 22 ☽ ∥ ♂
09 49 ☽ ∥ ♃
10 00 ☽ ⚹ ♂
12 29 ☽ Q ♄
14 11 ☽ ☌ ♅
14 44 ☽ ⊼ ☉
15 05 ☽ # ♇
16 28 ☽ ⚹ ♆
19 33 ☽ △ ♇

20 Sunday
00 30 ☽ ☍ ♀
01 17 ☽ ± ☉
11 24 ☽ ∠ ♂
12 13 ☽ Q ♆
13 39 ☽ ⊼ ☿
15 44 ☉ ⊼ ♆
17 21 ☽ ⚼ ☉
18 07 ☽ □ ♄
20 28 ☽ ⚼ ♇

21 Monday
00 59 ☽ ± ☿
03 02 ☽ ⊥ ♂
03 55 ☉ ⚼ ♄
06 24 ♀ # ♂
07 25 ☽ ☌ ♃
11 49 ☽ ± ♇
13 39 ☽ ⊻ ♂
16 22 ☽ ⊻ ♅
18 20 ☽ ⚼ ☿
18 50 ☽ □ ♆
21 00 ☽ △ ☉
21 40 ♀ # ♃
22 15 ☽ ⊼ ♇
22 51 ☿ ⚼ ♆

22 Tuesday
01 58 ☉ ⊥ ♀
02 49 ☽ ⊥ ♅
06 35 ☿ △ ♄
08 30 ☽ ⊼ ♀
14 15 ☉ □ ♇
18 48 ☽ ∠ ♅
20 18 ☽ ± ♀
22 17 ☽ △ ♄

23 Wednesday
00 22 ☽ △ ☿
11 14 ☿ ± ♃
12 36 ☽ ⊻ ♃
14 26 ☽ ⚼ ♀
21 20 ☽ ☌ ♂
22 16 ☽ ⚹ ♅
23 43 ☽ ⊥ ♃

24 Thursday
01 00 ☽ △ ♆
01 52 ☽ ⚼ ♄
04 47 ☽ ☍ ♇
08 03 ☽ □ ☉
15 27 ☽ # ♇
16 41 ☽ ∠ ♃
17 13 ☽ # ♀
18 55 ☽ ± ♄
20 58 ☽ Q ♅
21 43 ☽ △ ♀
21 44 ☽ ∥ ♃
23 48 ☽ ∥ ♂

25 Friday
00 13 ♂ ⚹ ♅
05 36 ☽ ⚼ ♆
06 17 ☿ Q ♇
06 27 ☽ ⊼ ♄
16 41 ☽ □ ☿
16 49 ☽ ∥ ♅
18 43 ☽ # ☿
21 17 ♀ ∥ ♇
21 42 ☽ ⚹ ♃
21 53 ☽ Q ☉
23 10 ☽ ± ♆

26 Saturday
08 04 ☽ □ ♅
09 16 ☽ ⊻ ♂
11 02 ☽ ⊼ ♆
11 36 ☿ # ♅
13 12 ☉ ± ♆
15 11 ☽ ⊼ ♇
15 46 ♂ ⊥ ♃
21 26 ☽ Q ♃
21 39 ☽ ⊥ ♂
22 56 ☽ # ☉
23 54 ☽ ⚹ ☉

27 Sunday
03 12 ☽ ± ♇
08 50 ☿ ⊼ ♃
09 31 ☽ Q ☿
15 39 ☽ □ ♀
16 24 ☽ ∠ ♂
17 42 ☽ # ♄
17 56 ☽ ☍ ♄
21 23 ☽ ⚼ ♇

28 Monday
02 10 ♀ ⚼ ♂
09 01 ☽ ∠ ☉
09 37 ☽ □ ♃
12 31 ♂ △ ♆
13 16 ☽ ⚹ ☿
13 35 ♀ □ ♄
15 46 ☉ ⚼ ♃
20 23 ☽ △ ♅
21 33 ☽ # ♆
23 31 ☽ ☍ ♆
23 55 ☽ ⚹ ♂

29 Tuesday
03 54 ☽ △ ♇
05 02 ☽ ⊥ ☉
08 24 ☽ Q ♀
12 00 ♂ ⚼ ♄
16 04 ☽ ∥ ♆
18 24 ☽ ⊻ ☉

30 Wednesday
00 07 ☽ ∠ ☿
00 23 ♀ ∠ ♇
01 12 ☽ Q ♂
02 52 ☽ ⚼ ♅
06 50 ☽ ⊼ ♄
11 39 ☽ ⚹ ♀
19 01 ☽ ± ♄
20 05 ☽ ∥ ♄
20 58 ☽ ⊥ ☿
21 07 ☽ ± ♅
22 15 ☿ ☍ ♅
22 22 ☽ △ ♃

31 Thursday
09 15 ☽ ⊼ ♅
10 47 ☽ ⊻ ☿
12 27 ☽ ⊼ ♆
13 13 ☽ ⚼ ♄
14 57 ☽ □ ♂
16 57 ☽ □ ♇
21 29 ☽ ∠ ♀
22 33 ☽ ∥ ☉

NOVEMBER 2024

LONGITUDES

Date	Sidereal time h m s	Sun ☉ ° ' "	Moon ☽ ° ' "	Moon ☽ 24.00 ° ' "	Mercury ☿ ° '	Venus ♀ ° '	Mars ♂ ° '	Jupiter ♃ ° '	Saturn ♄ ° '	Uranus ♅ ° '	Neptune ♆ ° '	Pluto ♇ ° '
01	14 45 04	09 ♏ 33 24	09 ♏ 11 52	15 ♏ 11 49	28 ♏ 10	17 ♐ 42	29 ♋ 04	20 ♊ 27	12 ♓ 52	25 ♉ 53	27 ♓ 30	29 ♑ 45
02	14 49 01	10 33 28	21 ♏ 13 44	27 ♏ 17 47	29 ♏ 35	18 54	29 25	20 R 22	12 R 51	25 R 50	27 R 28	29 45
03	14 52 57	11 33 35	03 ♐ 24 10	09 ♐ 33 01	00 ♐ 58	20 06	29 ♋ 46	20 17	12 49	25 48	27 27	29 46
04	14 56 54	12 33 43	15 ♐ 44 30	21 ♐ 58 48	02 20	21 18	00 ♌ 07	20 12	12 48	25 46	27 26	29 46
05	15 00 51	13 33 53	28 ♐ 16 05	04 ♑ 36 33	03 42	22 30	00 27	20 07	12 47	25 43	27 25	29 47
06	15 04 47	14 34 05	11 ♑ 00 23	17 ♑ 27 50	05 03	23 42	00 46	20 02	12 46	25 41	27 24	29 48
07	15 08 44	15 34 18	23 ♑ 59 06	00 ♒ 34 27	06 22	24 54	01 06	19 56	12 45	25 38	27 23	29 49
08	15 12 40	16 34 32	07 ♒ 14 06	13 ♒ 58 17	07 40	26 06	01 24	19 51	12 44	25 36	27 22	29 49
09	15 16 37	17 34 48	20 ♒ 47 12	27 ♒ 41 02	08 57	27 18	01 43	19 45	12 44	25 33	27 21	29 50
10	15 20 33	18 35 06	04 ♓ 39 51	11 ♓ 43 41	10 12	28 29	02 00	19 39	12 43	25 31	27 20	29 51
11	15 24 30	19 35 24	18 ♓ 52 28	26 ♓ 06 00	11 26	29 ♐ 41	02 18	19 33	12 42	25 28	27 19	29 52
12	15 28 26	20 35 44	03 ♈ 23 57	10 ♈ 45 50	12 38	00 ♑ 52	02 34	19 27	12 42	25 26	27 19	29 53
13	15 32 23	21 36 06	18 ♈ 11 00	25 ♈ 38 42	13 48	02 04	02 50	19 20	12 42	25 24	27 18	29 54
14	15 36 20	22 36 29	03 ♉ 07 59	10 ♉ 37 50	14 55	03 15	03 06	19 14	12 42	25 21	27 17	29 55
15	15 40 16	23 36 54	18 ♉ 07 08	25 ♉ 34 46	16 00	04 27	03 21	19 07	12 42	25 19	27 16	29 56
16	15 44 13	24 37 20	02 ♊ 59 35	10 ♊ 20 32	17 02	05 38	03 36	19 01	12 D 42	25 16	27 16	29 57
17	15 48 09	25 37 48	17 ♊ 36 40	24 ♊ 47 11	18 00	06 49	03 50	18 54	12 42	25 14	27 15	29 58
18	15 52 06	26 38 18	01 ♋ 51 26	08 ♋ 48 58	18 54	08 00	04 03	18 47	12 42	25 11	27 14	29 ♑ 59
19	15 56 02	27 38 49	15 ♋ 39 32	22 ♋ 23 02	19 44	09 11	04 16	18 40	12 42	25 09	27 14	00 ♒ 00
20	15 59 59	28 39 22	28 ♋ 59 34	05 ♌ 29 22	20 30	10 22	04 28	18 32	12 43	25 06	27 13	00 01
21	16 03 55	29 ♏ 39 57	11 ♌ 52 48	18 ♌ 10 21	21 09	11 33	04 40	18 25	12 43	25 04	27 12	00 02
22	16 07 52	00 ♐ 40 34	24 ♌ 22 33	00 ♍ 30 01	21 43	12 44	04 51	18 18	12 44	25 01	27 12	00 03
23	16 11 49	01 41 12	06 ♍ 33 24	12 ♍ 33 24	22 09	13 55	05 01	18 10	12 45	24 59	27 11	00 04
24	16 15 45	02 41 51	18 ♍ 30 40	24 ♍ 25 55	22 28	15 06	05 11	18 02	12 46	24 56	27 11	00 05
25	16 19 42	03 42 33	00 ♎ 19 47	06 ♎ 12 56	22 39	16 16	05 19	17 55	12 47	24 54	27 11	00 06
26	16 23 38	04 43 16	12 ♎ 05 58	17 ♎ 59 27	22 R 40	17 27	05 28	17 47	12 48	24 51	27 10	00 08
27	16 27 35	05 44 00	23 ♎ 53 54	29 ♎ 49 49	22 31	18 37	05 35	17 39	12 49	24 49	27 10	00 09
28	16 31 31	06 44 47	05 ♏ 47 36	11 ♏ 47 37	22 11	19 48	05 42	17 31	12 50	24 46	27 09	00 10
29	16 35 28	07 45 34	17 ♏ 50 09	23 ♏ 55 26	21 41	20 58	05 48	17 23	12 51	24 44	27 09	00 12
30	16 39 24	08 ♐ 46 23	00 ♐ 03 41	06 ♐ 14 59	20 ♐ 59	22 ♑ 08	05 ♌ 54	17 ♊ 15	12 ♓ 53	24 ♉ 41	27 ♓ 09	00 ♒ 13

Date	Moon True ☊ ° '	Moon Mean ☊ ° '	Moon ☽ Latitude ° '
01	06 ♈ 24	04 ♈ 42	02 S 49
02	06 R 18	04 39	03 40
03	06 12	04 36	04 22
04	06 05	04 33	04 51
05	05 59	04 29	05 08
06	05 53	04 26	05 09
07	05 49	04 23	04 55
08	05 47	04 20	04 25
09	05 D 47	04 17	03 40
10	05 48	04 13	02 41
11	05 49	04 10	01 31
12	05 50	04 07	00 S 13
13	05 R 49	04 04	01 N 06
14	05 47	04 01	02 22
15	05 42	03 58	03 28
16	05 35	03 54	04 19
17	05 27	03 51	04 52
18	05 18	03 48	05 06
19	05 11	03 45	05 02
20	05 05	03 42	04 41
21	05 01	03 39	04 06
22	04 59	03 35	03 20
23	04 D 59	03 32	02 27
24	05 00	03 29	01 27
25	05 00	03 26	00 N 25
26	05 R 00	03 23	00 S 38
27	04 58	03 19	01 39
28	04 54	03 16	02 37
29	04 46	03 13	03 28
30	04 ♈ 36	03 ♈ 10	04 S 10

DECLINATIONS

Date	Sun ☉ ° '	Moon ☽ ° '	Mercury ☿ ° '	Venus ♀ ° '	Mars ♂ ° '	Jupiter ♃ ° '	Saturn ♄ ° '	Uranus ♅ ° '	Neptune ♆ ° '	Pluto ♇ ° '
01	14 S 40	17 S 14	21 S 44	24 S 39	21 N 44	22 N 23	08 S 42	18 N 58	02 S 13	23 S 23
02	14 59	21 36	22 08	24 48	21 41	22 22	08 43	18 58	02 13	23 23
03	15 18	25 07	22 30	24 56	21 39	22 22	08 43	18 57	02 13	23 23
04	15 36	27 30	22 51	25 04	21 36	22 22	08 43	18 56	02 14	23 23
05	15 55	28 33	23 11	25 10	21 34	22 21	08 44	18 56	02 14	23 23
06	16 13	28 07	23 30	25 16	21 32	22 21	08 44	18 55	02 14	23 23
07	16 30	26 09	23 48	25 21	21 29	22 21	08 44	18 55	02 15	23 22
08	16 47	22 44	24 04	25 26	21 27	22 20	08 44	18 54	02 15	23 22
09	17 05	18 02	24 19	25 30	21 25	22 20	08 44	18 54	02 16	23 22
10	17 21	12 18	24 33	25 33	21 23	22 20	08 44	18 53	02 16	23 22
11	17 38	05 S 47	24 45	25 35	21 22	22 19	08 44	18 52	02 16	23 22
12	17 54	01 N 09	24 56	25 37	21 20	22 19	08 44	18 52	02 17	23 22
13	18 10	08 09	25 06	25 38	21 18	22 18	08 44	18 51	02 17	23 21
14	18 25	14 47	25 14	25 38	21 17	22 18	08 44	18 51	02 17	23 21
15	18 41	20 33	25 20	25 37	21 16	22 17	08 44	18 50	02 17	23 21
16	18 55	25 00	25 25	25 36	21 15	22 17	08 44	18 49	02 18	23 21
17	19 10	27 43	25 29	25 34	21 14	22 16	08 44	18 49	02 18	23 21
18	19 24	28 32	25 31	25 31	21 13	22 16	08 44	18 48	02 18	23 20
19	19 38	27 31	25 31	25 27	21 12	22 15	08 43	18 48	02 19	23 20
20	19 52	24 56	25 30	25 23	21 11	22 15	08 43	18 47	02 19	23 20
21	20 05	21 10	25 27	25 18	21 11	22 14	08 43	18 46	02 19	23 20
22	20 17	16 33	25 22	25 12	21 11	22 14	08 42	18 46	02 19	23 19
23	20 30	11 22	25 16	25 06	21 10	22 13	08 42	18 45	02 19	23 19
24	20 42	05 53	25 08	24 59	21 11	22 12	08 41	18 45	02 19	23 19
25	20 54	00 N 15	24 57	24 51	21 11	22 12	08 41	18 44	02 19	23 19
26	21 05	05 S 22	24 45	24 43	21 11	22 11	08 40	18 44	02 20	23 18
27	21 16	10 49	24 31	24 33	21 12	22 11	08 39	18 43	02 20	23 18
28	21 26	15 55	24 15	24 24	21 13	22 10	08 39	18 42	02 20	23 18
29	21 36	20 28	23 56	24 13	21 14	22 09	08 38	18 42	02 20	23 18
30	21 S 46	24 S 14	23 S 35	24 S 02	21 N 15	22 N 09	08 S 37	18 N 41	02 S 20	23 S 17

ZODIAC SIGN ENTRIES

Date	h	m	Planets
02	19	18	☿ ♐
03	05	19	☽ ♐
04	04	10	♂ ♌
05	15	17	☽ ♑
07	22	58	☽ ♒
10	04	00	☽ ♓
11	18	26	♀ ♑
12	06	26	☽ ♈
14	06	59	☽ ♉
16	07	09	☽ ♊
18	08	50	☽ ♋
19	20	29	♇ ♒
20	13	51	☽ ♌
21	19	56	☉ ♐
22	23	01	☽ ♍
25	11	20	☽ ♎
28	00	21	☽ ♏
30	11	53	☽ ♐

LATITUDES

Date	Mercury ☿ ° '	Venus ♀ ° '	Mars ♂ ° '	Jupiter ♃ ° '	Saturn ♄ ° '	Uranus ♅ ° '	Neptune ♆ ° '	Pluto ♇ ° '
01	02 S 02	01 S 48	01 N 25	00 S 43	02 S 08	00 S 16	01 S 19	03 S 16
04	02 16	01 55	01 31	00 43	02 08	00 16	01 19	03 16
07	02 28	02 01	01 37	00 43	02 07	00 16	01 19	03 16
10	02 36	02 07	01 43	00 43	02 07	00 16	01 19	03 16
13	02 39	02 12	01 50	00 43	02 06	00 16	01 19	03 16
16	02 38	02 17	01 57	00 42	02 06	00 16	01 19	03 16
19	02 29	02 20	02 04	00 42	02 05	00 16	01 19	03 16
22	02 12	02 23	02 11	00 42	02 05	00 16	01 19	03 16
25	01 44	02 25	02 19	00 42	02 04	00 16	01 18	03 16
28	01 02	02 26	02 27	00 41	02 04	00 16	01 18	03 16
31	00 S 09	02 S 27	02 N 34	00 S 41	02 S 03	00 S 16	01 S 18	03 S 16

DATA

Julian Date	2460616
Delta T	+72 seconds
Ayanamsa	24° 12' 11"
Synetic vernal point	04° ♓ 54' 48"
True obliquity of ecliptic	23° 26' 19"

MOON'S PHASES, APSIDES AND POSITIONS ☽

Date	h	m	Phase	Longitude ° '	Eclipse Indicator
01	12	47	●	09 ♏ 35	
09	05	55	☽	17 ♒ 20	
15	21	28	○	24 ♉ 01	
23	01	28	☾	01 ♍ 15	

Day	h	m		
14	11	22	Perigee	
26	12	04	Apogee	
05	17	09	Max dec	28° S 35'
12	08	05	0N	
18	10	21	Max dec	28° N 32'
25	13	04	0S	

LONGITUDES

Date	Chiron ⚷ ° '	Ceres ⚳ ° '	Pallas ⚴ ° '	Juno ⚵ ° '	Vesta ⚶ ° '	Black Moon Lilith ⚸ ° '
01	20 ♈ 25	18 ♑ 26	18 ♐ 34	29 ♎ 21	02 ♎ 27	13 ♎ 51
11	20 ♈ 01	21 ♑ 24	22 ♐ 24	02 ♏ 50	07 ♎ 07	14 ♎ 58
21	19 ♈ 39	24 ♑ 35	26 ♐ 17	06 ♏ 15	11 ♎ 43	16 ♎ 04
31	19 ♈ 22	27 ♑ 57	00 ♑ 13	09 ♏ 35	16 ♎ 15	17 ♎ 11

ASPECTARIAN

h m	Aspects
01 Friday	
00 33	☽ ± ♆
00 33	☿ △ ♆
04 31	☽ bQ ♃
11 18	☿ # ♂
12 47	☽ ☌ ☉
17 35	☽ ⊥ ♀
18 35	☽ bQ ♆
19 20	☽ △ ♄
21 02	☽ # ♅
22 26	☽ ± ♃
02 Saturday	
05 05	☽ Q ♇
06 53	☽ ⊻ ♀
08 21	☿ △ ♂
10 18	☽ ⊼ ♃
12 28	☽ # ♂
15 03	☿ ⚹ ♇
15 34	☽ ∥ ☿
16 44	☽ # ♃
21 06	☽ ☍ ♅
23 21	☽ ∥ ♇
03 Sunday	
00 20	☽ △ ♆
03 28	☿ # ♃
04 39	☽ △ ♂
04 51	☽ ⚹ ♇
06 37	☽ ☌ ☿
10 33	☽ ∥ ♀
11 37	♂ ☍ ♇
15 25	♀ ☍ ♃
04 Monday	
05 18	☽ ⊻ ☉
06 19	☽ □ ♄
09 04	☉ bQ ♆
10 08	☽ ∠ ♇
10 45	☽ bQ ♂
17 36	☉ △ ♄
17 55	☽ ⊥ ☉
20 32	☽ ☍ ♃
23 51	☽ ☌ ♀
05 Tuesday	
03 27	☽ ⊥ ♇
04 31	☽ ± ♂
07 10	☽ ⊼ ♅
10 23	☽ □ ♆
12 37	☽ ∠ ☉
14 53	☽ ⊻ ♇
16 15	☽ ⊼ ♂
16 46	☽ Q ♄
18 32	☽ ± ♅
23 30	☽ ⊻ ☿
06 Wednesday	
00 14	☉ ± ♃
02 30	☿ ∥ ♇
11 24	☽ bQ ♅
12 04	☽ ⊥ ☿
13 55	♀ ⊥ ♇
15 17	☽ ⚹ ♄
19 11	☽ ⚹ ☉
20 10	☽ Q ♆
07 Thursday	
04 37	☽ ⊼ ♃
06 39	☽ ∠ ☿
13 51	☽ ⊻ ♀
15 01	☽ △ ♅
15 33	☽ ± ♃
17 16	♀ ± ♂
18 12	☽ ⚹ ♆
18 23	☽ ∥ ♀
18 52	☽ ∠ ♄
19 05	☽ Q ☉
22 38	☽ ☌ ♇
08 Friday	
01 15	☽ ☍ ♂
01 50	☽ ⊥ ♀
02 20	♀ ⊼ ♅
04 18	☽ ∥ ☿
07 45	☽ bQ ♃
08 09	☽ ∥ ♇
11 07	☽ ⊥ ♄
12 52	☽ ⚹ ☿
14 17	☽ # ♃
19 13	☽ # ♂
19 34	☽ ∠ ♀
21 09	☽ ∠ ♆
21 48	☽ ⊻ ♄
09 Saturday	
05 55	☽ □ ☉
08 01	☽ # ♅
10 12	☽ △ ♃
12 19	☽ Q ☿
13 00	☽ ⊥ ♆
13 15	♀ □ ♆
16 05	☽ ∥ ☉
18 12	☉ Q ♇
20 17	☽ □ ♅
23 25	☽ ⊻ ♆
10 Sunday	
00 23	☽ ⚹ ♀
03 44	☽ ⊻ ♇
07 21	☽ ⊼ ♂
14 02	☽ ⊥ ♇
17 49	☽ ± ♂
22 20	☽ □ ☿
22 49	☽ Q ♀
11 Monday	
01 23	☽ ∥ ♄
01 39	☽ ☌ ♄
02 58	☽ Q ♅
05 17	☽ ∠ ♇
09 18	☽ bQ ♂
11 08	☉ ⊼ ♃
13 07	☽ □ ♃
13 17	☽ △ ☉
15 44	♀ ⊻ ♇
22 56	☽ ⚹ ♅
12 Tuesday	
00 17	☽ ∥ ♆
02 01	☽ ☌ ♆
06 13	☽ ⚹ ♇
07 30	☽ □ ♀
08 33	♀ Q ♄
10 37	☽ △ ♂
13 22	☿ □ ♄
15 50	☽ # ♆
15 51	☽ bQ ☉
18 34	☽ Q ♃
22 54	♀ ± ♅
23 26	☽ ∠ ♅
13 Wednesday	
01 49	☽ Q ♇
03 08	☽ ⊻ ♄
04 19	☽ △ ☿
07 32	☽ ± ☉
12 50	☽ ⊥ ♄
13 51	☽ ⚹ ♃
13 57	☽ ⊥ ♅
14 03	☽ # ♄
17 54	☽ ⊼ ☉
23 34	☽ ⊻ ♅
14 Thursday	
02 38	☽ ⊻ ♆
03 17	☽ ∠ ♄
06 27	☽ bQ ☿
06 50	☽ □ ♇
08 02	♀ ⊼ ♂
11 47	☿ ∠ ♇
11 57	☽ □ ♂
12 13	☽ △ ♀
12 15	☽ ⊥ ♆
13 45	☽ ∠ ♃
21 59	☽ ± ☿
15 Friday	
02 38	☽ ∠ ♆
03 16	☽ # ☉
03 18	☽ ⚹ ♄
04 03	☽ ⊥ ♃
04 21	☽ ∥ ♅
08 20	☽ ⊼ ☿
13 36	☽ ⊻ ♃
14 19	☽ bQ ♀
14 20	♄ St D
15 20	☽ ∥ ♂
17 17	☽ Q ♂
20 25	☽ ∥ ♃
21 28	☽ ☍ ☉
22 34	☽ Q ♄
23 32	☽ ☌ ♅
16 Saturday	
02 05	☽ # ♇
02 37	☉ # ♅
02 43	☽ ⚹ ♆
06 05	☽ ± ♀
07 03	☽ △ ♇
13 00	☽ ⚹ ♂
15 00	☽ # ☿
16 07	☽ # ♀
16 40	☽ ⊼ ♀
22 13	☽ Q ♆
17 Sunday	
02 45	☉ ☍ ♅
03 52	☽ □ ♄
07 36	☽ bQ ♇
12 41	☽ ☍ ☿
14 03	☽ ∠ ♂
14 07	☽ ☌ ♃
16 38	♂ ∠ ♃
22 37	☽ ± ♇
18 Monday	
00 42	☽ ⊻ ♅
02 27	☽ ⊼ ☉
04 09	☽ □ ♆
05 25	☽ ⊥ ♂
08 47	☽ ⊼ ♇
08 55	☿ ☍ ♃
10 51	☽ ⊥ ♅
11 01	☿ ∥ ♀
13 26	☽ ± ☉
15 49	☽ ⊻ ♂
17 25	☿ bQ ♂
23 35	☽ ☍ ♀
19 Tuesday	
02 07	☉ △ ♆
02 20	☽ ∠ ♅
06 16	☽ bQ ☉
06 48	☽ △ ♄
17 17	☽ ⊻ ♃
19 42	☽ ⊼ ☿
20 Wednesday	
03 58	☽ ⊥ ♃
04 56	☽ ⚹ ♅
06 37	♀ bQ ♅
07 11	☽ ± ☿
07 38	☽ # ☿
08 31	☽ # ♀
08 46	☽ △ ♆
09 40	☽ bQ ♄
11 20	☽ △ ☉
13 52	☽ ☍ ♇
20 18	☽ ∠ ♃
22 15	☽ ☌ ♂
23 02	☽ # ♇
21 Thursday	
00 41	☽ bQ ☿
02 18	☽ ± ♄
02 58	☽ Q ♅
05 46	☽ ∥ ♃
11 20	☽ ⊼ ♀
11 54	☽ ∥ ♂
12 37	☽ bQ ♆
13 36	☽ ⊼ ♄
17 42	☽ # ☉
20 49	☉ ⚹ ♇
23 57	☽ ± ♀
22 Friday	
00 21	☽ ⚹ ♃
00 52	☽ ∥ ♅
05 50	☽ ± ♆
06 36	☽ △ ☿
11 55	♀ ⚹ ♄
13 15	☽ □ ♅
17 31	☽ ⊼ ♆
19 16	☽ bQ ♀
23 08	☽ ⊼ ♇
23 28	☽ Q ♃
23 Saturday	
01 28	☽ □ ☉
08 53	☽ ⊻ ♂
11 02	☽ ± ♇
21 02	☽ ⊥ ♂
23 51	☽ # ♄
24 Sunday	
00 24	☽ ☍ ♄
04 21	☽ △ ♀
05 05	☽ bQ ♇
11 04	☽ □ ♃
13 44	♀ Q ♆
15 25	☽ ∠ ♂
16 50	☽ Q ☉
20 10	☽ □ ☿
25 Monday	
00 59	☽ △ ♅
03 12	☽ # ♆
05 35	☽ ☍ ♆
11 33	☽ △ ♇
19 32	☽ ⚹ ☉
22 18	☽ ⚹ ♂
22 58	☽ ∥ ♆
26 Tuesday	
02 41	☿ St R
07 26	☽ bQ ♅
09 05	☽ Q ☿
13 25	☽ ⊼ ♄
18 08	♀ ⊼ ♃
23 03	☽ Q ♂
23 27	☽ △ ♃
27 Wednesday	
00 06	☽ □ ♀
01 09	☿ ∥ ♀
01 40	☽ ± ♄
01 42	☽ ± ♅
02 24	☽ ∥ ♄
02 56	☉ # ♂
04 58	☽ ∠ ☉
08 06	☉ △ ♂
09 14	☽ ⚹ ☿
09 46	♄ Q ♅
13 51	☽ ⊼ ♅
18 36	☽ ⊼ ♆
19 57	☽ bQ ♄
28 Thursday	
00 40	☽ □ ♇
00 54	☽ ⊥ ☉
05 30	☽ bQ ♃
06 42	☽ ± ♆
11 49	☽ □ ♂
14 05	☽ ⊻ ☉
14 42	☽ ∠ ☿
16 27	☽ Q ♀
23 20	☽ ± ♃
29 Friday	
00 43	☽ bQ ♆
02 07	☽ △ ♄
02 15	☽ # ♅
07 55	☽ ⊥ ☿
11 08	☽ ⊼ ♃
12 42	☽ Q ♇
16 26	☽ # ♂
18 51	☽ ⚹ ♀
18 57	☽ ∥ ☉
19 13	☽ ⊻ ☿
21 26	☿ ⊻ ♀
22 02	☽ # ♃
30 Saturday	
01 32	☽ ☍ ♅
05 24	☽ ∥ ♇
06 19	☽ △ ♆
07 53	☽ ∥ ☿
10 39	☽ ∥ ♀
12 18	☽ ⚹ ♇
14 34	☿ bQ ♂
23 23	☽ △ ♂

All ephemeris data is given at 12.00 UT and the Moon's longitude is additionally given for 24.00 UT

DECEMBER 2024

LONGITUDES

Date	Sidereal time h m s	Sun ☉ ° ' "	Moon ☽ ° ' "	Moon ☽ 24.00 ° ' "	Mercury ☿ ° '	Venus ♀ ° '	Mars ♂ ° '	Jupiter ♃ ° '	Saturn ♄ ° '	Uranus ♅ ° '	Neptune ♆ ° '	Pluto ♇ ° '
01	16 43 21	09 ♐ 47 13	12 ♐ 29 24	18 ♐ 46 58	20 ♐ 07	23 ♑ 18	05 ♌ 58	17 ♊ 07	12 ♓ 55	24 ♉ 39	27 ♓ 09	00 ♒ 14
02	16 47 18	10 48 05	25 ♐ 07 40	01 ♑ 31 27	19 R 05	24 28	06 02	16 R 59	12 57	24 R 37	27 R 08	00 16
03	16 51 14	11 48 57	07 ♑ 58 14	14 ♑ 27 58	17 55	25 38	06 05	16 51	12 58	24 34	27 08	00 17
04	16 55 11	12 49 51	21 ♑ 00 34	27 ♑ 35 58	16 38	26 48	06 08	16 43	13 00	24 32	27 08	00 18
05	16 59 07	13 50 45	04 ♒ 14 09	10 ♒ 55 06	15 16	27 58	06 09	16 35	13 02	24 30	27 08	00 20
06	17 03 04	14 51 40	17 ♒ 38 49	24 ♒ 25 21	13 53	29 ♑ 07	06 10	16 27	13 05	24 27	27 08	00 21
07	17 07 00	15 52 36	01 ♓ 14 46	08 ♓ 07 08	12 32	00 ♒ 17	06 R 10	16 18	13 07	24 25	27 08	00 23
08	17 10 57	16 53 32	15 ♓ 02 32	22 ♓ 01 01	11 14	01 26	06 09	16 10	13 09	24 23	27 D 08	00 24
09	17 14 53	17 54 29	29 ♓ 02 36	06 ♈ 07 16	10 03	02 35	06 08	16 02	13 12	24 20	27 08	00 26
10	17 18 50	18 55 27	13 ♈ 14 55	20 ♈ 25 20	09 00	03 44	06 05	15 54	13 14	24 18	27 08	00 27
11	17 22 47	19 56 25	27 ♈ 38 16	04 ♉ 53 15	08 07	04 53	06 02	15 46	13 17	24 16	27 08	00 29
12	17 26 43	20 57 24	12 ♉ 09 46	19 ♉ 27 10	07 24	06 02	05 58	15 38	13 20	24 14	27 08	00 30
13	17 30 40	21 58 24	26 ♉ 44 41	04 ♊ 01 30	06 53	07 10	05 53	15 30	13 22	24 12	27 08	00 32
14	17 34 36	22 59 24	11 ♊ 16 44	18 ♊ 29 29	06 33	08 19	05 47	15 22	13 25	24 10	27 09	00 34
15	17 38 33	24 00 25	25 ♊ 38 52	02 ♋ 44 07	06 24	09 27	05 41	15 14	13 28	24 08	27 09	00 35
16	17 42 29	25 01 27	09 ♋ 44 31	16 ♋ 39 29	06 D 26	10 35	05 33	15 06	13 31	24 06	27 09	00 37
17	17 46 26	26 02 29	23 ♋ 28 37	00 ♌ 11 37	06 37	11 43	05 25	14 58	13 35	24 04	27 09	00 38
18	17 50 22	27 03 32	06 ♌ 48 23	13 ♌ 18 58	06 56	12 50	05 16	14 50	13 38	24 02	27 10	00 40
19	17 54 19	28 04 36	19 ♌ 43 32	26 ♌ 02 23	07 24	13 58	05 06	14 42	13 41	24 00	27 10	00 42
20	17 58 16	29 ♐ 05 41	02 ♍ 15 57	08 ♍ 24 44	07 58	15 05	04 55	14 34	13 45	23 58	27 11	00 43
21	18 02 12	00 ♑ 06 46	14 ♍ 29 19	20 ♍ 30 19	08 39	16 12	04 43	14 27	13 48	23 56	27 11	00 45
22	18 06 09	01 07 52	26 ♍ 28 26	02 ♎ 24 21	09 26	17 19	04 31	14 19	13 52	23 54	27 12	00 47
23	18 10 05	02 08 59	08 ♎ 18 46	14 ♎ 12 25	10 18	18 26	04 18	14 12	13 56	23 52	27 12	00 49
24	18 14 02	03 10 06	20 ♎ 05 59	26 ♎ 00 09	11 14	19 32	04 04	14 04	14 00	23 50	27 13	00 50
25	18 17 58	04 11 15	01 ♏ 55 33	07 ♏ 52 50	12 14	20 38	03 49	13 57	14 04	23 49	27 13	00 52
26	18 21 55	05 12 23	13 ♏ 52 30	19 ♏ 55 06	13 17	21 44	03 33	13 50	14 08	23 47	27 14	00 54
27	18 25 51	06 13 33	26 ♏ 01 01	02 ♐ 10 39	14 24	22 50	03 17	13 43	14 12	23 45	27 14	00 56
28	18 29 48	07 14 42	08 ♐ 24 14	14 ♐ 41 58	15 33	23 56	03 00	13 36	14 16	23 44	27 15	00 58
29	18 33 45	08 15 53	21 ♐ 03 57	27 ♐ 30 10	16 45	25 01	02 42	13 29	14 20	23 42	27 16	00 59
30	18 37 41	09 17 03	04 ♑ 00 34	10 ♑ 34 59	17 58	26 06	02 24	13 23	14 25	23 40	27 17	01 01
31	18 41 38	10 ♑ 18 14	17 ♑ 13 10	23 ♑ 54 51	19 ♐ 14	27 ♒ 10	02 ♌ 05	13 ♊ 16	14 ♓ 29	23 ♉ 39	27 ♓ 17	01 ♒ 03

Date	Moon True ☊ ° '	Moon Mean ☊ ° '	Moon ☽ Latitude ° '
01	04 ♈ 25	03 ♈ 07	04 S 41
02	04 R 12	03 04	04 58
03	04 00	03 00	05 01
04	03 49	02 57	04 49
05	03 41	02 54	04 21
06	03 35	02 51	03 38
07	03 32	02 48	02 42
08	03 31	02 45	01 36
09	03 D 31	02 41	00 S 24
10	03 R 31	02 38	00 N 51
11	03 29	02 35	02 04
12	03 25	02 32	03 09
13	03 17	02 29	04 02
14	03 07	02 25	04 39
15	02 55	02 22	04 58
16	02 42	02 19	04 58
17	02 31	02 16	04 41
18	02 20	02 13	04 09
19	02 13	02 10	03 24
20	02 08	02 06	02 31
21	02 06	02 03	01 32
22	02 D 06	02 00	00 N 30
23	02 R 06	01 57	00 S 33
24	02 05	01 54	01 34
25	02 02	01 50	02 31
26	01 57	01 47	03 22
27	01 50	01 44	04 04
28	01 39	01 41	04 36
29	01 26	01 38	04 55
30	01 13	01 35	05 00
31	00 ♈ 59	01 ♈ 31	04 S 48

DECLINATIONS

Date	Sun ☉ ° '	Moon ☽ ° '	Mercury ☿ ° '	Venus ♀ ° '	Mars ♂ ° '	Jupiter ♃ ° '	Saturn ♄ ° '	Uranus ♅ ° '	Neptune ♆ ° '	Pluto ♇ ° '
01	21 S 55	26 S 56	23 S 13	23 S 50	21 N 16	22 N 08	08 S 37	18 N 41	02 S 20	23 S 17
02	22 04	28 19	22 48	23 38	21 18	22 07	08 36	18 40	02 20	23 17
03	22 12	28 13	22 22	23 24	21 20	22 07	08 35	18 40	02 20	23 16
04	22 20	26 33	21 54	23 11	21 22	22 06	08 34	18 39	02 20	23 16
05	22 28	23 25	21 26	22 56	21 24	22 05	08 33	18 38	02 20	23 16
06	22 35	19 00	20 58	22 41	21 26	22 05	08 32	18 38	02 20	23 15
07	22 41	13 33	20 30	22 26	21 29	22 04	08 31	18 37	02 20	23 15
08	22 48	07 22	20 05	22 10	21 32	22 03	08 30	18 37	02 20	23 15
09	22 53	00 S 45	19 41	21 53	21 35	22 03	08 29	18 36	02 20	23 15
10	22 59	06 N 01	19 21	21 36	21 38	22 02	08 28	18 36	02 20	23 14
11	23 03	12 34	19 04	21 18	21 42	22 01	08 27	18 35	02 20	23 14
12	23 08	18 29	18 51	21 00	21 46	22 00	08 25	18 35	02 20	23 14
13	23 12	23 21	18 41	20 41	21 50	22 00	08 24	18 34	02 20	23 13
14	23 15	26 44	18 36	20 22	21 54	21 59	08 23	18 34	02 20	23 13
15	23 18	28 20	18 33	20 02	21 58	21 58	08 22	18 33	02 19	23 13
16	23 21	28 02	18 34	19 41	22 03	21 58	08 20	18 33	02 19	23 12
17	23 23	26 00	18 38	19 21	22 07	21 57	08 19	18 32	02 19	23 12
18	23 24	22 35	18 45	18 59	22 12	21 56	08 17	18 32	02 19	23 12
19	23 25	18 08	18 54	18 38	22 17	21 56	08 16	18 31	02 19	23 11
20	23 26	13 01	19 05	18 16	22 22	21 55	08 15	18 31	02 19	23 11
21	23 26	07 31	19 17	17 53	22 28	21 54	08 13	18 30	02 18	23 11
22	23 26	01 N 51	19 31	17 30	22 34	21 53	08 11	18 30	02 18	23 10
23	23 25	03 S 48	19 45	17 07	22 39	21 53	08 10	18 29	02 18	23 10
24	23 24	09 18	20 00	16 43	22 45	21 52	08 08	18 29	02 18	23 10
25	23 22	14 30	20 16	16 19	22 51	21 51	08 07	18 29	02 17	23 09
26	23 20	19 13	20 32	15 55	22 57	21 51	08 05	18 28	02 17	23 09
27	23 18	23 13	20 48	15 30	23 04	21 50	08 03	18 28	02 17	23 08
28	23 14	26 15	21 04	15 05	23 10	21 49	08 01	18 27	02 16	23 08
29	23 11	28 03	21 19	14 40	23 16	21 49	08 00	18 27	02 16	23 08
30	23 07	28 22	21 34	14 14	23 23	21 48	07 58	18 27	02 16	23 07
31	23 S 02	27 S 06	21 S 49	13 S 48	23 N 29	21 N 48	07 S 56	18 N 26	02 S 15	23 S 07

ZODIAC SIGN ENTRIES

Date	h	m	Planets
02	21	09	☽ ♑
05	04	21	☽ ♒
07	06	13	♀ ♒
07	09	49	☽ ♓
09	13	38	☽ ♈
11	15	55	☽ ♉
13	17	22	☽ ♊
15	19	21	☽ ♋
17	23	39	☽ ♌
20	07	37	☽ ♍
21	09	20	☉ ♑
22	19	08	☽ ♎
25	08	06	☽ ♏
27	19	46	☽ ♐
30	04	37	☽ ♑

LATITUDES

Date	Mercury ☿ ° '	Venus ♀ ° '	Mars ♂ ° '	Jupiter ♃ ° '	Saturn ♄ ° '	Uranus ♅ ° '	Neptune ♆ ° '	Pluto ♇ ° '
01	00 S 09	02 S 27	02 N 34	00 S 41	02 S 03	00 S 16	01 S 18	03 S 16
04	00 N 52	02 26	02 43	00 41	02 03	00 16	01 18	03 16
07	01 48	02 24	02 51	00 40	02 02	00 16	01 18	03 16
10	02 29	02 21	02 59	00 40	02 02	00 16	01 18	03 16
13	02 49	02 17	03 07	00 39	02 01	00 15	01 18	03 16
16	02 51	02 11	03 16	00 39	02 01	00 15	01 18	03 16
19	02 41	02 05	03 24	00 38	02 00	00 15	01 18	03 16
22	02 23	01 57	03 32	00 38	02 00	00 15	01 17	03 16
25	02 01	01 48	03 40	00 37	02 00	00 15	01 17	03 16
28	01 36	01 38	03 47	00 37	01 59	00 15	01 17	03 16
31	01 N 11	01 S 26	03 N 54	00 S 36	01 S 59	00 S 15	01 S 17	03 S 16

DATA

Julian Date	2460646
Delta T	+72 seconds
Ayanamsa	24° 12' 17"
Synetic vernal point	04° ♓ 54' 42"
True obliquity of ecliptic	23° 26' 18"

MOON'S PHASES, APSIDES AND POSITIONS ☽

Date	h	m	Phase	Longitude ° '	Eclipse Indicator
01	06	21	●	09 ♐ 33	
08	15	27	☽	17 ♓ 02	
15	09	02	○	23 ♊ 53	
22	22	18	☾	01 ♎ 34	
30	22	27	●	09 ♑ 44	

Day	h	m		
12	13	26	Perigee	
24	07	28	Apogee	
02	22	23	Max dec	28° S 28'
09	14	38	0N	
15	20	09	Max dec	28° N 26'
22	19	51	0S	
30	04	59	Max dec	28° S 26'

LONGITUDES

Date	Chiron ⚷ ° '	Ceres ⚳ ° '	Pallas ⚴ ° '	Juno ⚵ ° '	Vesta ⚶ ° '	Black Moon Lilith ⚸ ° '
01	19 ♈ 22	27 ♑ 57	00 ♑ 13	09 ♏ 35	16 ♎ 15	17 ♎ 11
11	19 ♈ 09	01 ♒ 28	04 ♑ 10	12 ♏ 48	20 ♎ 39	18 ♎ 17
21	19 ♈ 02	05 ♒ 06	08 ♑ 08	15 ♏ 54	24 ♎ 55	19 ♎ 24
31	19 ♈ 00	08 ♒ 49	12 ♑ 04	18 ♏ 51	29 ♎ 01	20 ♎ 30

All ephemeris data is given at 12.00 UT and the Moon's longitude is additionally given for 24.00 UT

ASPECTARIAN

h m Aspects

01 Sunday
03 08 ☽ ∠ ♀
06 21 ☽ ☌ ☉
07 39 ☿ ∥ ♇
08 34 ♀ ± ♃
12 49 ☽ □ ♄
17 16 ☽ ∠ ♇
20 45 ☽ ☍ ♃
22 08 ☽ ⊥ ♀

02 Monday
01 29 ☽ ☌ ☿
04 14 ☽ ⚼ ♂
10 22 ☽ ⊥ ♇
10 39 ☽ ⊻ ♀
11 02 ☽ ⊼ ♅
14 43 ♀ △ ♅
15 47 ☽ □ ♆
18 26 ☿ ⊥ ♀
21 16 ☽ ± ♂
21 23 ☉ ⋕ ♃
21 39 ☽ ⊻ ♇
22 16 ☽ ± ♅
22 56 ☽ Q ♄

03 Tuesday
08 30 ☽ ⊼ ♂
14 57 ☽ ⚼ ♅
18 28 ☉ ∥ ☿
19 43 ☽ ⊻ ☉
21 17 ☽ ✶ ♄

04 Wednesday
01 14 ☽ Q ♆
01 29 ☿ ⋕ ♃
02 38 ♀ ∥ ♇
04 13 ☽ ⊼ ♃
04 43 ☽ ⊻ ☿
07 41 ☽ ⊥ ☉
10 16 ☿ ☍ ♃
14 41 ☽ ⊥ ☿
15 05 ☽ ± ♃
16 18 ☉ □ ♄
18 24 ☽ △ ♅
18 52 ♀ ✶ ♆
23 09 ☽ ✶ ♆
23 34 ☽ ☌ ♀

05 Thursday
00 46 ☽ ∠ ♄
01 27 ☽ ∠ ☉
04 56 ☽ ☌ ♇
05 31 ☽ ∠ ☿
07 15 ☽ ⚼ ♃
11 02 ☿ ∠ ♇
12 58 ☽ ∥ ♇
13 34 ☿ ⋕ ♂
13 40 ♀ ∠ ♄
15 08 ☽ ∥ ♀
15 28 ☽ ☍ ♂
17 04 ☽ ⊥ ♄
17 37 ☽ ∥ ☉
19 56 ☽ ⋕ ♃
23 39 ☽ ⋕ ♂

06 Friday
00 56 ☽ ∥ ☿
02 10 ☽ ∠ ♆
02 18 ☉ ☌ ☿
03 50 ☽ ⊻ ♄
05 56 ☽ ✶ ☿
06 38 ☽ ✶ ☉
09 49 ☿ ∠ ♀
09 53 ☽ △ ♃
13 46 ☽ ⋕ ♅
18 11 ☽ ⊥ ♆
19 16 ☉ ∥ ♀
23 33 ♂ St R
23 57 ☉ ∠ ♇

07 Saturday
00 01 ☽ □ ♅
01 15 ☽ Q ☿
01 53 ☿ □ ♄
04 46 ☽ ⊻ ♆
05 37 ☽ Q ☉
10 09 ☽ ⊻ ♀
10 29 ☽ ⊻ ♇
14 08 ♀ ☌ ♇
20 36 ☽ ⊼ ♂
20 58 ☉ ☍ ♃
20 59 ☽ ⊥ ♇
21 36 ☽ ⊥ ♀
23 42 ♆ St D

08 Sunday
05 58 ☽ □ ☿
07 01 ☽ ± ♂
07 07 ♀ ⚼ ♃
07 24 ☽ Q ♅
07 47 ☽ ∥ ♄
08 43 ☽ ☌ ♄
12 38 ☽ ∠ ♇
13 56 ☽ □ ♃
14 37 ☽ ∠ ♀
15 27 ☽ □ ☉
21 48 ♀ ⋕ ♃
22 30 ☽ ⚼ ♂

09 Monday
04 00 ☽ ✶ ♅
06 20 ☽ ∥ ♆
08 45 ☽ ☌ ♆
14 22 ☽ ✶ ♇
18 33 ☽ ✶ ♀
20 23 ☽ Q ♃
22 54 ☽ ⋕ ♆
23 59 ☽ △ ♂

10 Tuesday
05 19 ☽ △ ☿
05 23 ☽ ∠ ♅
09 07 ♀ ⋕ ♂
10 40 ☽ Q ♇
11 59 ☽ ⊻ ♄
16 24 ☽ ✶ ♃
16 32 ☽ Q ♀
20 27 ☽ ⊥ ♅
20 47 ☽ ⋕ ♄
22 03 ☽ ⊥ ♄
22 13 ☽ △ ☉

11 Wednesday
04 53 ☽ ⚼ ☿
06 25 ☽ ⊻ ♅
11 10 ☽ ⊻ ♆
13 04 ☽ ∠ ♄
16 43 ☽ □ ♇
17 08 ☽ ∠ ♃
19 02 ☽ ± ☿
21 06 ☽ ⊥ ♆
22 45 ☉ ∠ ♀

12 Thursday
01 00 ☽ ⚼ ☉
01 01 ☽ □ ♀
01 50 ☽ □ ♂
04 29 ☽ ⊼ ☿
07 52 ☽ ⊥ ♃
10 46 ♀ ☍ ♂
11 57 ☽ ∠ ♆
12 11 ☉ ⚼ ♂
12 24 ☽ ∥ ♅
13 32 ☽ ⋕ ☿
13 55 ☽ ✶ ♄
16 57 ☽ ± ☉
17 39 ☽ ⊻ ♃
22 54 ☽ ⋕ ♀

13 Friday
03 34 ☽ ⊼ ☉
03 38 ☽ ∥ ♂
04 39 ☽ ∥ ♃
06 04 ♃ ⚼ ♇
07 19 ☽ Q ♂
07 45 ☿ ✶ ♀
07 49 ☽ ☌ ♅
09 44 ☽ Q ♄
11 05 ☽ ⋕ ☉
11 15 ☽ ⋕ ♇
12 39 ☽ ✶ ♆
16 28 ♀ ⊥ ♄
18 15 ☽ △ ♇
21 31 ☉ ∥ ♇

14 Saturday
02 58 ☽ ✶ ♂
04 20 ☽ ☍ ☿
06 40 ☽ △ ♀
08 28 ☽ Q ♆
15 34 ☽ □ ♄
18 43 ☽ ☌ ♃
19 07 ☽ ⚼ ♇

15 Sunday
03 43 ☽ ∠ ♂
09 02 ☽ ☍ ☉
09 27 ☽ ⊻ ♅
09 48 ☽ ⚼ ♀
10 13 ☽ ± ♇
13 43 ♂ ∥ ♃
14 32 ☽ □ ♆
14 44 ☉ ⊼ ♅
18 45 ☽ ⊥ ♂
19 33 ☽ ⊥ ♅
20 22 ☽ ⊼ ♇
20 57 ☿ St D

16 Monday
02 03 ☉ ⊥ ♇
02 22 ☽ ± ♀
04 52 ☽ ⊻ ♂
06 16 ☽ ⊼ ☿
10 53 ☽ ∠ ♅
13 34 ☽ ⊼ ♀
16 41 ☽ ± ☿
18 34 ☽ △ ♄
21 11 ☽ ⊻ ♃

17 Tuesday
07 36 ☽ ⊥ ♃
08 38 ☽ ⚼ ☿
13 02 ☽ ✶ ♅
16 56 ☽ ⊼ ☉
18 33 ☽ △ ♆
21 08 ☽ ⚼ ♄
21 33 ♀ ∠ ♆
23 28 ☽ ∠ ♃

18 Wednesday
00 50 ☽ ☍ ♇
04 37 ☽ ± ☉
06 55 ☽ ⋕ ☉
08 13 ☽ ⋕ ♇
09 13 ☽ ☌ ♂
10 35 ☽ Q ♅
12 14 ☽ △ ☿
13 31 ☽ ± ♄
14 10 ☽ ∥ ♂
14 29 ☉ □ ♆
15 46 ☽ ∥ ♃
21 52 ☽ ⚼ ♆
22 30 ☽ ⚼ ☉
23 30 ☿ ∥ ♀

19 Thursday
00 10 ☽ ☍ ♀
00 39 ☽ ⊼ ♄
02 40 ☽ ✶ ♃
05 52 ♀ ⊻ ♄
08 15 ☽ ⋕ ☿
09 17 ☽ ⋕ ♀
10 02 ☽ ∥ ♅
14 44 ☽ ± ♆
19 14 ♀ ⋕ ♅
20 04 ☽ □ ♅

20 Friday
01 08 ☽ Q ♃
02 10 ☽ ⊼ ♆
02 11 ♀ △ ♃
05 19 ☽ △ ☉
08 29 ☉ ± ♂
09 00 ☽ ⊼ ♇
17 05 ☽ ⊻ ♂
20 43 ☽ ± ♇
23 46 ☽ □ ☿

21 Saturday
04 40 ☽ ⊥ ♂
07 48 ☉ ± ♅
09 00 ☽ ⋕ ♄
10 39 ☽ ☍ ♄
11 55 ☽ □ ♃
14 31 ☽ ⚼ ♇
15 45 ☽ ⊼ ♀
22 16 ☽ ∠ ♂

22 Sunday
03 32 ☉ ⊻ ♇
04 59 ☽ ± ♀
06 49 ☽ △ ♅
10 07 ☽ ⋕ ♆
13 27 ☽ ☍ ♆
14 05 ☽ Q ☿
20 44 ☽ △ ♇
22 18 ☽ □ ☉

23 Monday
01 03 ☽ ⚼ ♀
03 59 ☽ ✶ ♂
05 35 ☽ ∥ ♆
06 32 ☉ Q ♄
13 08 ☽ ⚼ ♅
16 22 ☽ ✶ ☿
23 30 ☽ ⊼ ♄
23 51 ☽ △ ♃

24 Tuesday
03 57 ☽ Q ♂
06 51 ☽ ∥ ♄
07 24 ☽ ± ♅
10 44 ☽ △ ♀
11 47 ☽ ± ♄
14 23 ☽ Q ☉
19 35 ☽ ⊼ ♅
21 59 ♃ □ ♄

25 Wednesday
01 36 ☽ ∠ ☿
02 28 ☽ ⊼ ♆
05 00 ☉ ⊼ ♂
06 03 ☽ ⚼ ♃
06 10 ☽ ⚼ ♄
09 51 ☽ □ ♇
14 37 ☽ ± ♆
15 44 ☽ □ ♂
17 00 ☽ ✶ ☉
20 14 ☽ ∥ ♀
21 31 ☽ ⊥ ☿

26 Thursday
00 02 ☽ ± ♃
00 48 ♀ ⊥ ♆
08 01 ☽ ⋕ ♅
08 43 ☽ ⚼ ♆
10 43 ☽ ⊻ ☿
11 55 ☽ ⊼ ♃
12 31 ☽ △ ♄
19 54 ☽ ∥ ☿
22 00 ☽ Q ♇
22 48 ☿ ☍ ♃

27 Friday
01 43 ☽ ∠ ☉
03 10 ☽ ⋕ ♃
05 08 ☽ □ ♀
07 29 ☿ □ ♄
07 34 ☽ ☍ ♅
10 57 ☽ ⋕ ♂
11 31 ☽ ∥ ♇
12 32 ☽ ∥ ☉
14 24 ☽ △ ♆
20 57 ☽ ⊥ ☉
21 36 ☽ ✶ ♇

28 Saturday
01 50 ☽ △ ♂
05 30 ♂ ⋕ ♇
07 42 ♀ □ ♅
09 35 ☽ ⊻ ☉
19 22 ☽ Q ♀
20 30 ☿ ∠ ♇
21 50 ☽ ☍ ♃
22 56 ☉ ⋕ ♂
23 15 ☽ □ ♄

29 Sunday
02 25 ☽ ∠ ♇
03 01 ☽ ☌ ☿
05 50 ☽ ⚼ ♂
16 55 ☽ ⊼ ♅
19 21 ☽ ⊥ ♇
20 03 ☽ ✶ ♀
21 58 ☉ ⚼ ♅
22 17 ☽ ± ♂
23 34 ☽ □ ♆

30 Monday
03 07 ☿ ⚼ ♂
04 02 ☽ ± ♅
06 29 ☽ ⊻ ♇
08 26 ☉ ∥ ♇
09 03 ☽ Q ♄
09 07 ☽ ⊼ ♂
20 31 ☽ ⚼ ♅
22 27 ☽ ☌ ☉

31 Tuesday
02 05 ☽ ∠ ♀
04 56 ☽ ⊼ ♃
07 03 ☽ ✶ ♄
08 31 ☽ Q ♆
09 13 ☿ ⋕ ♃
14 38 ♀ ⊻ ♆
15 39 ☽ ± ♃
16 00 ☽ ⊻ ☿
19 43 ☽ ⊥ ♀
23 30 ☽ △ ♅

JANUARY 2025

LONGITUDES

Date	Sidereal time h m s	Sun ☉ ° ' "	Moon ☽ ° ' "	Moon ☽ 24.00 ° ' "	Mercury ☿ ° '	Venus ♀ ° '	Mars ♂ ° '	Jupiter ♃ ° '	Saturn ♄ ° '	Uranus ♅ ° '	Neptune ♆ ° '	Pluto ♇ ° '
01	18 45 34	11 ♑ 19 25	00 ♒ 39 42	07 ♒ 27 23	20 ♐ 31	28 ♒ 15	01 ♌ 45	13 ♊ 10	14 ♓ 34	23 ♉ 37	27 ♓ 18	01 ♒ 05
02	18 49 31	12 20 35	14 ♒ 17 32	21 ♒ 09 51	21 50	29 ♒ 19	01 R 25	13 R 04	14 38	23 R 36	27 19	01 07
03	18 53 27	13 21 46	28 ♒ 04 01	04 ♓ 59 45	23 10	00 ♓ 23	01 04	12 57	14 43	23 35	27 20	01 09
04	18 57 24	14 22 56	11 ♓ 56 50	18 ♓ 55 07	24 31	01 26	00 43	12 51	14 48	23 33	27 21	01 10
05	19 01 20	15 24 06	25 ♓ 54 26	02 ♈ 54 43	25 53	02 29	00 ♌ 21	12 46	14 53	23 32	27 22	01 12
06	19 05 17	16 25 15	09 ♈ 55 53	16 ♈ 57 52	27 16	03 32	29 ♋ 59	12 40	14 58	23 31	27 23	01 14
07	19 09 14	17 26 24	24 ♈ 00 37	01 ♉ 04 01	28 ♐ 40	04 35	29 36	12 35	15 03	23 30	27 24	01 16
08	19 13 10	18 27 33	08 ♉ 07 57	15 ♉ 12 14	00 ♑ 05	05 37	29 13	12 29	15 08	23 28	27 25	01 18
09	19 17 07	19 28 41	22 ♉ 16 37	29 ♉ 20 48	01 31	06 38	28 50	12 24	15 13	23 27	27 26	01 20
10	19 21 03	20 29 49	06 ♊ 24 23	13 ♊ 26 57	02 58	07 40	28 27	12 19	15 18	23 26	27 27	01 22
11	19 25 00	21 30 56	20 ♊ 28 00	27 ♊ 26 59	04 25	08 41	28 03	12 14	15 23	23 25	27 28	01 24
12	19 28 56	22 32 03	04 ♋ 23 24	11 ♋ 16 41	05 52	09 41	27 39	12 10	15 29	23 24	27 30	01 26
13	19 32 53	23 33 09	18 ♋ 06 22	24 ♋ 51 59	07 21	10 41	27 15	12 05	15 34	23 23	27 31	01 27
14	19 36 49	24 34 15	01 ♌ 33 11	08 ♌ 09 41	08 50	11 40	26 51	12 01	15 40	23 23	27 32	01 29
15	19 40 46	25 35 21	14 ♌ 41 19	21 ♌ 08 01	10 20	12 39	26 27	11 57	15 45	23 22	27 33	01 31
16	19 44 43	26 36 26	27 ♌ 29 49	03 ♍ 46 51	11 50	13 38	26 03	11 53	15 51	23 21	27 35	01 33
17	19 48 39	27 37 30	09 ♍ 59 24	16 ♍ 07 45	13 20	14 36	25 39	11 49	15 56	23 20	27 36	01 35
18	19 52 36	28 38 35	22 ♍ 12 22	28 ♍ 13 42	14 52	15 34	25 16	11 46	16 02	23 20	27 37	01 37
19	19 56 32	29 ♑ 39 39	04 ♎ 12 19	10 ♎ 08 49	16 24	16 30	24 52	11 43	16 08	23 19	27 39	01 39
20	20 00 29	00 ♒ 40 42	16 ♎ 03 49	21 ♎ 57 59	17 56	17 27	24 28	11 39	16 14	23 18	27 40	01 41
21	20 04 25	01 41 45	27 ♎ 52 00	03 ♏ 46 33	19 29	18 23	24 05	11 37	16 20	23 18	27 41	01 43
22	20 08 22	02 42 48	09 ♏ 42 19	15 ♏ 39 58	21 03	19 18	23 42	11 34	16 26	23 17	27 43	01 45
23	20 12 18	03 43 51	21 ♏ 40 09	27 ♏ 43 29	22 37	20 12	23 20	11 31	16 32	23 17	27 44	01 47
24	20 16 15	04 44 53	03 ♐ 50 31	10 ♐ 01 45	24 12	21 06	22 58	11 29	16 38	23 17	27 46	01 49
25	20 20 12	05 45 54	16 ♐ 17 37	22 ♐ 38 29	25 47	21 59	22 36	11 27	16 44	23 16	27 47	01 51
26	20 24 08	06 46 55	29 ♐ 04 34	05 ♑ 36 01	27 23	22 52	22 15	11 25	16 50	23 16	27 49	01 53
27	20 28 05	07 47 55	12 ♑ 12 51	18 ♑ 55 00	29 ♑ 00	23 44	21 54	11 23	16 57	23 16	27 51	01 55
28	20 32 01	08 48 55	25 ♑ 42 13	02 ♒ 34 12	00 ♒ 37	24 35	21 34	11 22	17 03	23 16	27 52	01 56
29	20 35 58	09 49 53	09 ♒ 30 32	16 ♒ 30 41	02 15	25 25	21 14	11 20	17 09	23 16	27 54	01 58
30	20 39 54	10 50 51	23 ♒ 34 04	00 ♓ 40 07	03 54	26 14	20 55	11 19	17 16	23 16	27 56	02 00
31	20 43 51	11 ♒ 51 47	07 ♓ 48 09	14 ♓ 57 34	05 ♒ 33	27 ♓ 03	20 ♋ 37	11 ♊ 18	17 ♓ 22	23 ♉ 16	27 ♓ 57	02 ♒ 02

Date	Moon True ☊ ° '	Moon Mean ☊ ° '	Moon ☽ Latitude ° '
01	00 ♈ 47	01 ♈ 28	04 S 21
02	00 R 37	01 25	03 38
03	00 31	01 22	02 42
04	00 27	01 19	01 36
05	00 26	01 16	00 S 24
06	00 D 26	01 12	00 N 50
07	00 R 26	01 09	02 02
08	00 25	01 06	03 06
09	00 21	01 03	03 59
10	00 14	01 00	04 37
11	00 05	00 56	04 58
12	29 ♓ 54	00 53	05 01
13	29 42	00 50	04 47
14	29 31	00 47	04 17
15	29 22	00 44	03 34
16	29 14	00 41	02 40
17	29 10	00 37	01 40
18	29 08	00 34	00 N 37
19	29 D 08	00 31	00 S 27
20	29 09	00 28	01 30
21	29 09	00 25	02 28
22	29 R 09	00 22	03 20
23	29 07	00 18	04 04
24	29 03	00 15	04 38
25	28 57	00 12	04 59
26	28 48	00 09	05 06
27	28 39	00 06	04 58
28	28 30	00 ♈ 02	04 33
29	28 22	29 ♓ 59	03 52
30	28 15	29 56	02 56
31	28 ♓ 11	29 ♓ 53	01 S 48

DECLINATIONS

Date	Sun ☉ ° '	Moon ☽ ° '	Mercury ☿ ° '	Venus ♀ ° '	Mars ♂ ° '	Jupiter ♃ ° '	Saturn ♄ ° '	Uranus ♅ ° '	Neptune ♆ ° '	Pluto ♇ ° '
01	22 S 57	24 S 15	22 S 04	13 S 22	23 N 36	21 N 47	07 S 54	18 N 26	02 S 15	23 S 07
02	22 52	20 01	22 17	12 56	23 43	21 46	07 52	18 26	02 15	23 06
03	22 46	14 41	22 30	12 29	23 49	21 46	07 50	18 25	02 14	23 06
04	22 40	08 34	22 42	12 02	23 56	21 45	07 48	18 25	02 14	23 06
05	22 33	02 S 00	22 53	11 35	24 02	21 45	07 46	18 25	02 13	23 05
06	22 26	04 N 42	23 03	11 08	24 09	21 44	07 44	18 24	02 13	23 05
07	22 18	11 12	23 13	10 40	24 16	21 44	07 42	18 24	02 13	23 04
08	22 10	17 09	23 21	10 13	24 22	21 43	07 40	18 24	02 12	23 04
09	22 01	22 11	23 28	09 45	24 28	21 43	07 38	18 24	02 12	23 04
10	21 52	25 55	23 34	09 17	24 35	21 42	07 36	18 23	02 11	23 03
11	21 43	28 03	23 39	08 49	24 41	21 42	07 34	18 23	02 11	23 03
12	21 33	28 23	23 43	08 21	24 47	21 42	07 32	18 23	02 10	23 03
13	21 23	26 57	23 45	07 53	24 52	21 41	07 29	18 23	02 10	23 02
14	21 12	23 59	23 47	07 25	24 58	21 41	07 27	18 23	02 09	23 02
15	21 01	19 50	23 47	06 56	25 04	21 41	07 25	18 22	02 09	23 02
16	20 50	14 51	23 46	06 28	25 09	21 40	07 22	18 22	02 08	23 01
17	20 38	09 22	23 43	06 00	25 14	21 40	07 20	18 22	02 08	23 01
18	20 26	03 N 40	23 40	05 31	25 19	21 40	07 18	18 22	02 07	23 00
19	20 13	02 S 05	23 35	05 03	25 24	21 39	07 16	18 22	02 06	23 00
20	20 00	07 42	23 29	04 35	25 29	21 39	07 13	18 22	02 06	23 00
21	19 47	13 01	23 21	04 06	25 33	21 39	07 11	18 22	02 05	22 59
22	19 33	17 53	23 12	03 38	25 37	21 39	07 08	18 21	02 05	22 59
23	19 19	22 06	23 01	03 10	25 41	21 39	07 06	18 21	02 04	22 59
24	19 05	25 28	22 50	02 42	25 45	21 39	07 03	18 21	02 03	22 58
25	18 50	27 41	22 37	02 13	25 48	21 39	07 01	18 21	02 03	22 58
26	18 35	28 33	22 22	01 45	25 51	21 39	06 59	18 21	02 02	22 58
27	18 19	27 49	22 06	01 18	25 54	21 39	06 56	18 21	02 01	22 57
28	18 03	25 28	21 49	00 50	25 57	21 39	06 53	18 21	02 01	22 57
29	17 47	21 35	21 30	00 S 22	26 00	21 39	06 51	18 21	02 00	22 57
30	17 31	16 25	21 09	00 N 05	26 02	21 39	06 48	18 21	01 59	22 56
31	17 S 14	10 S 19	20 S 48	00 N 32	26 N 04	21 N 39	06 S 46	18 N 21	01 S 59	22 S 56

ZODIAC SIGN ENTRIES

Date	h	m	Planets
01	10	50	☽ ♒
03	03	24	♀ ♓
03	15	21	☽ ♓
05	19	01	☽ ♈
06	10	44	♂ ♋
07	22	11	☽ ♉
08	10	30	☿ ♑
10	01	07	☽ ♊
12	04	24	☽ ♋
14	09	12	☽ ♌
16	16	46	☽ ♍
19	03	33	☽ ♎
19	20	00	☉ ♒
21	16	20	☽ ♏
24	04	29	☽ ♐
26	13	43	☽ ♑
28	02	53	☿ ♒
28	19	31	☽ ♒
30	22	52	☽ ♓

LATITUDES

Date	Mercury ☿ ° '	Venus ♀ ° '	Mars ♂ ° '	Jupiter ♃ ° '	Saturn ♄ ° '	Uranus ♅ ° '	Neptune ♆ ° '	Pluto ♇ ° '
01	01 N 03	01 S 22	03 N 56	00 S 36	01 S 59	00 S 15	01 S 17	03 S 16
04	00 38	01 09	04 02	00 35	01 58	00 15	01 17	03 17
07	00 N 13	00 54	04 07	00 35	01 58	00 15	01 17	03 17
10	00 S 10	00 38	04 12	00 34	01 57	00 15	01 17	03 17
13	00 31	00 21	04 15	00 34	01 57	00 15	01 17	03 17
16	00 51	00 S 02	04 18	00 33	01 57	00 15	01 17	03 18
19	01 09	00 N 18	04 19	00 32	01 57	00 15	01 17	03 18
22	01 25	00 39	04 20	00 32	01 56	00 15	01 16	03 18
25	01 39	01 02	04 19	00 31	01 56	00 15	01 16	03 18
28	01 50	01 26	04 18	00 30	01 56	00 15	01 16	03 18
31	01 S 58	01 N 52	04 N 16	00 S 30	01 S 56	00 S 15	01 S 16	03 S 19

DATA

Julian Date	2460677
Delta T	+72 seconds
Ayanamsa	24° 12' 23"
Synetic vernal point	04° ♓ 54' 36"
True obliquity of ecliptic	23° 26' 18"

MOON'S PHASES, APSIDES AND POSITIONS ☽

Date	h	m	Phase	Longitude ° '	Eclipse Indicator
06	23	56	☽	16 ♈ 56	
13	22	27	○	24 ♋ 00	
21	20	31	☾	02 ♏ 03	
29	12	36	●	09 ♒ 51	

Day	h	m		
08	00	05	Perigee	
21	04	53	Apogee	
05	19	08	0N	
12	04	24	Max dec	28° N 28'
19	03	15	0S	
26	13	19	Max dec	28° S 33'

LONGITUDES

Date	Chiron ⚷ ° '	Ceres ⚳ ° '	Pallas ⚴ ° '	Juno ⚵ ° '	Vesta ⚶ ° '	Black Moon Lilith ⚸ ° '
01	19 ♈ 00	09 ♒ 12	12 ♑ 28	19 ♏ 08	29 ♎ 25	20 ♎ 37
11	19 ♈ 05	13 ♒ 01	16 ♑ 22	21 ♏ 52	03 ♏ 16	21 ♎ 44
21	19 ♈ 14	16 ♒ 53	20 ♑ 14	24 ♏ 22	06 ♏ 51	22 ♎ 50
31	19 ♈ 29	20 ♒ 47	24 ♑ 01	26 ♏ 36	10 ♏ 06	23 ♎ 57

ASPECTARIAN

h m	Aspects
01 Wednesday	
03 51	☽ ⊥ ☿
06 02	☽ ⚹ ♆
07 21	☽ ⊻ ♀
07 36	☽ bQ ♃
10 02	☽ ∠ ♄
12 45	☽ ☌ ♇
13 53	☽ ☍ ♂
16 06	☽ # ♂
19 14	☽ ∥ ♇
20 20	☽ ∥ ☉
21 29	☽ ∠ ☿
02 Thursday	
00 38	☽ ∥ ☿
02 01	☽ ⊥ ♄
02 52	☽ # ♃
08 19	☽ ⊻ ☉
08 32	☽ ∠ ♆
09 51	☽ △ ♃
12 37	☽ ⊻ ♄
19 38	☽ # ♅
19 39	☽ ⊥ ☉
03 Friday	
00 17	☽ ⊥ ♆
02 34	☽ ⚹ ☿
03 18	☉ ⊼ ♃
04 13	☽ □ ♅
07 21	♂ ☍ ♇
10 44	☽ ⊻ ♆
12 33	☽ ∠ ☉
16 21	☽ ☌ ♀
17 04	☽ ⊼ ♂
17 21	☽ ⊻ ♇
19 19	☿ ⊼ ♅
21 37	☽ ∥ ♀
23 41	♀ ⊼ ♂
04 Saturday	
01 36	☽ Q ☿
03 12	☽ ± ♂
03 45	☽ ⊥ ♇
05 47	♀ ⊻ ♇
09 06	☉ ∥ ☿
11 19	☽ Q ♅
13 33	☽ □ ♃
14 47	☿ ± ♂
14 51	☽ ∥ ♄
16 32	☽ ⚹ ☉
16 56	☽ ☌ ♄
18 19	☽ bQ ♂
19 18	☽ ∠ ♇
22 36	☉ ⚹ ♄
23 52	☿ ⊥ ♇
05 Sunday	
07 56	☽ ⚹ ♅
11 08	☉ Q ♆
11 10	☽ ∥ ♆
11 57	☽ □ ☿
14 30	☽ ☌ ♆
14 46	☽ Q ☉
19 25	☽ △ ♂
20 16	☽ Q ♃
21 06	☽ ⚹ ♇
06 Monday	
00 11	☽ ⊻ ♀
03 05	☽ # ♆
09 35	☽ ∠ ♅
11 16	☽ ⊥ ♀
13 05	♂ bQ ♄
13 56	☿ □ ♆
15 41	☿ ∥ ♇
16 38	☽ ⚹ ♃
17 39	☽ Q ♇
20 38	☽ ⊻ ♄
22 59	☽ # ♄
23 56	☽ □ ☉
07 Tuesday	
00 55	☽ ⊥ ♅
03 51	☽ ∠ ♀
06 55	☽ ⊥ ♄
10 08	☽ # ♀
11 07	☽ ⊻ ♅
17 46	☽ ⊻ ♆
18 02	☽ ∠ ♃
20 49	☽ △ ☿
21 16	☽ □ ♂
22 19	☽ ∠ ♄
08 Wednesday	
00 22	☽ □ ♇
00 28	☿ ⊼ ♂
01 44	☿ ± ♅
03 59	☽ ⊥ ♆
05 23	♀ ± ♂
07 23	☽ ⚹ ♀
09 13	☽ ⊥ ♃
12 37	☉ ± ♃
17 31	☽ ∥ ♅
19 17	☽ ∠ ♆
19 21	☽ ⊻ ♃
23 57	☽ ⚹ ♄
09 Thursday	
01 08	☽ bQ ☿
03 01	☽ Q ♂
05 21	☽ Q ♀
06 53	☽ △ ☉
08 48	☿ ⊻ ♇
09 32	☽ ∥ ♃
11 12	☽ # ☉
14 00	☽ ☌ ♅
16 55	☽ # ♇
18 07	☽ ± ☿
19 28	☽ # ☿
20 26	☽ Q ♄
20 46	☽ ⚹ ♆
22 50	☽ ⚹ ♂
10 Friday	
01 56	☽ ∥ ♂
03 24	☽ △ ♇
04 43	♀ ⊥ ♇
05 28	☽ ⊼ ☿
10 20	☽ bQ ☉
14 18	☽ □ ♀
17 12	☽ Q ♆
18 01	☿ Q ♄
22 01	☽ ☌ ♃
23 40	☽ ∠ ♂
11 Saturday	
02 52	☽ ± ☉
03 15	☽ □ ♄
05 01	☽ bQ ♇
13 56	☽ ⊼ ☉
14 38	☽ ⊥ ♂
15 02	☉ # ♃
17 04	☽ ⊻ ♅
20 29	☽ ± ♇
12 Sunday	
00 03	☽ □ ♆
00 40	☽ ⊻ ♂
03 23	☽ ⊥ ♅
06 51	☽ ⊼ ♇
14 53	☽ ☍ ☿
18 59	☽ ∠ ♅
21 16	♂ △ ♆
21 56	☽ △ ♀
13 Monday	
01 28	☽ ⊻ ♃
07 30	☽ △ ♄
08 13	☉ △ ♅
11 58	☽ ⊥ ♃
21 22	☽ ⚹ ♅
22 27	☽ ☍ ☉
14 Tuesday	
00 10	♂ ∠ ♃
02 32	☽ bQ ♀
03 48	☽ ☌ ♂
03 53	☽ ∠ ♃
04 42	☿ bQ ♅
04 45	☽ △ ♆
04 51	♀ Q ♅
05 21	☽ ∥ ♂
09 49	♀ ∥ ♄
10 23	☽ bQ ♄
11 53	☽ ☍ ♇
13 21	☽ # ☿
15 07	♀ bQ ♂
18 05	☽ # ♇
18 55	☽ Q ♅
19 48	♀ □ ♃
20 04	☽ ± ♀
15 Wednesday	
02 02	☽ ∥ ♃
02 51	☽ ± ♄
02 55	☽ ⊼ ☿
05 22	☽ # ☉
06 58	☽ ⚹ ♃
07 57	☽ ⊼ ♀
08 03	☽ bQ ♆
13 59	☽ ⊼ ♄
15 26	☽ ± ☿
19 22	☽ ∥ ♅
16 Thursday	
00 49	☽ ± ♆
02 39	☉ ☍ ♂
04 10	☽ □ ♅
05 12	☽ Q ♃
09 21	☽ ⊻ ♂
10 10	☽ ⊼ ☉
10 34	☽ bQ ☿
12 09	☽ ⊼ ♆
12 51	☿ ⊼ ♃
18 10	☉ bQ ♃
19 45	☽ ⊼ ♇
20 25	☽ ⊥ ♂
22 37	☽ ± ☉
17 Friday	
07 20	☽ ± ♇
11 20	☉ ⚹ ♆
13 15	☽ ∠ ♂
15 33	☽ □ ♃
17 36	☽ bQ ☉
19 28	☽ △ ☿
20 40	☽ # ♄
21 46	☽ ☍ ♀
23 43	☽ ☍ ♄
18 Saturday	
00 56	☽ bQ ♇
03 32	☽ # ♀
14 13	☽ △ ♅
17 53	☽ ⚹ ♂
18 26	☽ # ♆
22 48	☽ ☍ ♆
19 Sunday	
00 03	☿ Q ♆
01 26	♀ ☌ ♄
02 01	☽ △ ☉
06 51	☽ △ ♇
07 38	☿ ⚹ ♄
12 05	☽ ∥ ♆
15 47	♀ ∠ ♇
16 32	☿ ⚹ ♀
17 12	☽ Q ♂
20 17	☽ bQ ♅
23 35	☽ ∥ ♀
20 Monday	
03 06	☽ △ ♃
07 48	☿ ± ♃
09 56	☽ ∥ ♄
12 21	☽ ⊼ ♄
14 31	☽ ± ♅
15 03	☽ ⊼ ♀
16 23	☽ □ ☿
21 Tuesday	
00 39	☽ ± ♄
02 27	☉ ∠ ♄
02 43	☽ ⊼ ♅
04 18	☽ ± ♀
04 34	☽ □ ♂
09 27	☽ bQ ♃
11 38	☽ ⊼ ♆
12 29	☉ ☌ ♇
19 06	☽ bQ ♄
19 50	☽ □ ♇
20 31	☽ □ ☉
23 51	☽ ± ♆
22 Wednesday	
00 08	☽ bQ ♀
03 39	☽ ± ♃
10 28	☽ Q ☿
14 32	☽ # ♅
15 44	☽ ⊼ ♃
18 05	☽ bQ ♆
20 34	☽ ∥ ☉
23 Thursday	
01 39	☽ △ ♄
08 13	☽ Q ♇
08 51	☽ △ ♀
09 11	☽ # ♃
12 08	☽ Q ☉
14 10	☽ ⚹ ☿
15 07	♂ ⚹ ♅
15 12	☽ △ ♂
15 13	☽ ☍ ♅
17 37	☽ ∥ ♇
17 39	☽ ∥ ☿
18 10	☿ ∥ ♇
20 49	☿ ☍ ♂
22 07	☿ △ ♅
24 Friday	
00 03	☽ △ ♆
08 01	☽ ⚹ ♇
13 56	☽ ⚹ ☉
14 32	☽ # ♂
19 47	☽ bQ ♂
23 55	☽ ∠ ☿
25 Saturday	
02 45	☽ ☍ ♃
12 34	☽ ± ♂
12 51	☽ □ ♄
13 03	☽ ∠ ♇
19 34	☽ ⊥ ☿
21 12	☽ ∠ ☉
21 21	♀ ∥ ♆
21 43	☿ bQ ♃
23 34	☽ □ ♀
23 36	☽ ⊼ ♂
23 54	♀ △ ♂
26 Sunday	
01 11	☽ ⊼ ♅
06 02	☽ ⊥ ♇
08 25	☽ ⊻ ☿
09 40	☽ □ ♆
12 21	☽ ± ♅
15 25	☽ ⊥ ☉
17 11	☽ ⊻ ♇
18 33	☿ ⚹ ♆
22 42	☽ Q ♄
23 11	♀ ⚹ ♅
27 Monday	
00 46	♄ ∠ ♇
03 20	☽ ⊻ ☉
04 51	☽ bQ ♅
08 50	☉ # ♅
10 30	☽ ⊼ ♃
11 04	☽ Q ♀
18 32	☽ Q ♆
20 33	☽ ⚹ ♄
21 15	☽ ± ♃
28 Tuesday	
04 52	☽ ☍ ♂
07 42	☽ △ ♅
08 13	☽ # ♂
09 53	☽ ⚹ ♀
13 09	☽ bQ ♃
15 49	☽ ⚹ ♆
21 45	☽ ☌ ☿
22 56	☽ ☌ ♇
23 11	☽ ∠ ♄
29 Wednesday	
01 01	☿ # ♃
04 34	☽ ∥ ♇
07 52	☿ ☌ ♇
10 31	☿ ∠ ♄
11 44	☽ # ♃
12 32	☽ ∥ ☿
12 36	☽ ☌ ☉
13 39	☽ ∠ ♀
14 51	☽ ⊥ ♄
15 08	☽ △ ♃
17 50	☽ ∠ ♆
30 Thursday	
01 12	☽ ⊻ ♄
03 38	☽ # ♅
06 00	☽ ⊥ ♀
07 06	☽ ∥ ☉
07 36	☽ ⊼ ♂
09 13	☽ ⊥ ♆
11 29	☽ □ ♅
16 21	♅ St D
16 48	☽ ⊻ ♀
17 33	☽ ± ♂
19 23	☽ ⊻ ♆
22 56	☉ ⊥ ♄
22 59	☉ △ ♃
31 Friday	
02 17	☽ ⊻ ♇
07 43	☽ ⊻ ☿
08 24	☽ bQ ♂
12 24	☽ ⊥ ♇
17 48	☽ Q ♅
17 52	☽ □ ♃
19 06	☽ ⊥ ☿
19 20	☽ ⊻ ☉

All ephemeris data is given at 12.00 UT and the Moon's longitude is additionally given for 24.00 UT

FEBRUARY 2025

Date	Sidereal time h m s	Sun ☉ ° ′ ″	Moon ☽ ° ′ ″	Moon ☽ 24.00 ° ′ ″	Mercury ☿ ° ′	Venus ♀ ° ′	Mars ♂ ° ′	Jupiter ♃ ° ′	Saturn ♄ ° ′	Uranus ♅ ° ′	Neptune ♆ ° ′	Pluto ♇ ° ′
01	20 47 47	12 ♒ 52 43	22 ♓ 07 47	29 ♓ 18 14	07 ♒ 13	27 ♓ 50	20 ♋ 19	11 ♊ 18	17 ♓ 29	23 ♉ 16	27 ♓ 59	02 ♒ 04
02	20 51 44	13 53 37	06 ♈ 28 27	13 ♈ 37 59	08 53	28 37	20 R 02	11 R 17	17 35	23 D 16	28 01	02 06
03	20 55 41	14 54 29	20 ♈ 46 30	27 ♈ 53 43	10 35	29 ♓ 23	19 46	11 17	17 42	23 16	28 02	02 08
04	20 59 37	15 55 21	04 ♉ 59 22	12 ♉ 03 18	12 17	00 ♈ 07	19 30	11 D 17	17 48	23 16	28 04	02 10
05	21 03 34	16 56 10	19 ♉ 05 22	26 ♉ 05 24	14 00	00 51	19 15	11 17	17 55	23 17	28 06	02 12
06	21 07 30	17 56 59	03 ♊ 03 20	09 ♊ 59 02	15 44	01 34	19 01	11 17	18 02	23 17	28 08	02 14
07	21 11 27	18 57 46	16 ♊ 52 23	23 ♊ 43 17	17 28	02 15	18 48	11 18	18 09	23 17	28 10	02 16
08	21 15 23	19 58 31	00 ♋ 31 33	07 ♋ 17 05	19 13	02 55	18 35	11 18	18 15	23 18	28 12	02 17
09	21 19 20	20 59 15	13 ♋ 59 42	20 ♋ 39 15	20 59	03 34	18 23	11 19	18 22	23 18	28 13	02 19
10	21 23 16	21 59 58	27 ♋ 15 34	03 ♌ 48 32	22 46	04 12	18 12	11 20	18 29	23 19	28 15	02 21
11	21 27 13	23 00 39	10 ♌ 18 01	16 ♌ 43 56	24 33	04 48	18 02	11 22	18 36	23 19	28 17	02 23
12	21 31 09	24 01 18	23 ♌ 06 13	29 ♌ 24 54	26 21	05 23	17 53	11 23	18 43	23 20	28 19	02 25
13	21 35 06	25 01 56	05 ♍ 39 59	11 ♍ 51 37	28 ♒ 10	05 57	17 44	11 25	18 50	23 21	28 21	02 27
14	21 39 03	26 02 33	17 ♍ 59 56	24 ♍ 05 10	00 ♓ 00	06 29	17 36	11 27	18 57	23 22	28 23	02 28
15	21 42 59	27 03 08	00 ♎ 07 34	06 ♎ 07 30	01 50	06 59	17 29	11 29	19 04	23 22	28 25	02 30
16	21 46 56	28 03 42	12 ♎ 05 19	18 ♎ 01 29	03 40	07 28	17 23	11 31	19 11	23 23	28 27	02 32
17	21 50 52	29 ♒ 04 15	23 ♎ 56 28	29 ♎ 50 46	05 31	07 55	17 17	11 34	19 18	23 24	28 29	02 34
18	21 54 49	00 ♓ 04 46	05 ♏ 44 58	11 ♏ 39 39	07 23	08 21	17 13	11 37	19 25	23 25	28 31	02 36
19	21 58 45	01 05 16	17 ♏ 35 24	23 ♏ 32 51	09 14	08 45	17 09	11 40	19 32	23 26	28 33	02 37
20	22 02 42	02 05 45	29 ♏ 32 37	05 ♐ 35 20	11 06	09 06	17 06	11 43	19 40	23 27	28 35	02 39
21	22 06 38	03 06 13	11 ♐ 41 36	17 ♐ 52 01	12 57	09 26	17 03	11 46	19 47	23 28	28 37	02 41
22	22 10 35	04 06 39	24 ♐ 07 06	00 ♑ 27 21	14 48	09 44	17 02	11 49	19 54	23 29	28 40	02 42
23	22 14 32	05 07 04	06 ♑ 53 12	13 ♑ 24 58	16 38	10 00	17 01	11 53	20 01	23 31	28 42	02 44
24	22 18 28	06 07 28	20 ♑ 02 54	26 ♑ 47 06	18 28	10 14	17 D 01	11 57	20 08	23 32	28 44	02 46
25	22 22 25	07 07 50	03 ♒ 37 33	10 ♒ 34 06	20 16	10 26	17 02	12 01	20 16	23 33	28 46	02 47
26	22 26 21	08 08 10	17 ♒ 36 28	24 ♒ 44 11	22 02	10 35	17 03	12 05	20 23	23 35	28 48	02 49
27	22 30 18	09 08 29	01 ♓ 56 40	09 ♓ 13 14	23 46	10 42	17 05	12 10	20 30	23 36	28 50	02 51
28	22 34 14	10 ♓ 08 46	16 ♓ 33 05	23 ♓ 55 19	25 ♓ 27	10 ♈ 47	17 ♋ 08	12 ♊ 14	20 ♓ 38	23 ♉ 37	28 ♓ 52	02 ♒ 52

Date	Moon True ☊ ° ′	Moon Mean ☊ ° ′	Moon ☽ Latitude ° ′
01	28 ♓ 09	29 ♓ 50	00 S 33
02	28 D 10	29 47	00 N 45
03	28 11	29 43	01 59
04	28 12	29 40	03 06
05	28 R 12	29 37	04 01
06	28 11	29 34	04 41
07	28 08	29 31	05 04
08	28 04	29 28	05 10
09	27 58	29 24	04 58
10	27 51	29 21	04 31
11	27 45	29 18	03 49
12	27 40	29 15	02 57
13	27 37	29 12	01 57
14	27 34	29 08	00 N 52
15	27 D 34	29 05	00 S 14
16	27 35	29 02	01 19
17	27 37	28 59	02 20
18	27 38	28 56	03 15
19	27 40	28 53	04 01
20	27 41	28 49	04 38
21	27 R 40	28 46	05 03
22	27 39	28 43	05 14
23	27 37	28 40	05 11
24	27 34	28 37	04 51
25	27 30	28 34	04 15
26	27 28	28 30	03 22
27	27 26	28 27	02 16
28	27 ♓ 24	28 ♓ 24	01 S 00

DECLINATIONS

Date	Sun ☉ ° ′	Moon ☽ ° ′	Mercury ☿ ° ′	Venus ♀ ° ′	Mars ♂ ° ′	Jupiter ♃ ° ′	Saturn ♄ ° ′	Uranus ♅ ° ′	Neptune ♆ ° ′	Pluto ♇ ° ′
01	16 S 57	03 S 37	20 S 25	00 N 59	26 N 06	21 N 39	06 S 43	18 N 21	01 S 58	22 S 56
02	16 39	03 N 15	20 00	01 26	26 08	21 39	06 41	18 21	01 57	22 55
03	16 22	09 57	19 34	01 53	26 09	21 39	06 38	18 21	01 57	22 55
04	16 04	16 06	19 06	02 19	26 11	21 39	06 35	18 21	01 56	22 54
05	15 46	21 21	18 37	02 45	26 12	21 40	06 33	18 22	01 55	22 54
06	15 27	25 22	18 07	03 10	26 12	21 40	06 30	18 22	01 54	22 54
07	15 08	27 50	17 35	03 36	26 13	21 40	06 27	18 22	01 54	22 54
08	14 49	28 36	17 01	04 01	26 14	21 40	06 24	18 22	01 53	22 53
09	14 30	27 39	16 27	04 25	26 14	21 41	06 22	18 22	01 52	22 53
10	14 11	25 08	15 50	04 50	26 14	21 41	06 19	18 22	01 51	22 53
11	13 51	21 20	15 12	05 13	26 14	21 42	06 16	18 22	01 51	22 52
12	13 31	16 36	14 33	05 37	26 14	21 42	06 14	18 23	01 50	22 52
13	13 11	11 15	13 53	05 59	26 13	21 42	06 11	18 23	01 49	22 52
14	12 50	05 N 33	13 11	06 22	26 13	21 43	06 08	18 23	01 48	22 51
15	12 30	00 S 16	12 28	06 43	26 12	21 43	06 05	18 23	01 47	22 51
16	12 09	05 59	11 43	07 05	26 11	21 44	06 02	18 23	01 47	22 51
17	11 48	11 27	10 57	07 25	26 10	21 44	06 00	18 24	01 46	22 50
18	11 27	16 30	10 10	07 45	26 09	21 45	05 57	18 24	01 45	22 50
19	11 05	20 56	09 22	08 04	26 08	21 46	05 54	18 24	01 44	22 50
20	10 44	24 34	08 33	08 23	26 07	21 46	05 51	18 24	01 43	22 50
21	10 22	27 11	07 43	08 41	26 05	21 47	05 48	18 25	01 42	22 49
22	10 00	28 32	06 52	08 58	26 04	21 47	05 45	18 25	01 42	22 49
23	09 38	28 26	06 01	09 14	26 02	21 48	05 43	18 25	01 41	22 49
24	09 16	26 44	05 09	09 29	26 00	21 49	05 40	18 26	01 40	22 48
25	08 54	23 28	04 17	09 44	25 58	21 50	05 37	18 26	01 39	22 48
26	08 31	18 46	03 25	09 57	25 56	21 50	05 34	18 26	01 38	22 48
27	08 09	12 54	02 33	10 10	25 54	21 51	05 31	18 27	01 37	22 48
28	07 S 46	06 S 13	01 S 42	10 N 21	25 N 52	21 N 52	05 S 28	18 N 27	01 S 36	22 S 47

ZODIAC SIGN ENTRIES

Date	h	m	Planets
02	01	10	☽ ♈
04	03	33	☽ ♉
04	07	57	♀ ♈
06	06	44	☽ ♊
08	11	04	☽ ♋
10	17	01	☽ ♌
13	01	07	☽ ♍
14	12	06	☿ ♓
15	11	45	☽ ♎
18	00	19	☽ ♏
18	10	06	☉ ♓
20	12	55	☽ ♐
22	23	09	☽ ♑
25	05	40	☽ ♒
27	08	46	☽ ♓

LATITUDES

Date	Mercury ☿ ° ′	Venus ♀ ° ′	Mars ♂ ° ′	Jupiter ♃ ° ′	Saturn ♄ ° ′	Uranus ♅ ° ′	Neptune ♆ ° ′	Pluto ♇ ° ′
01	02 S 00	02 N 01	04 N 15	00 S 29	01 S 55	00 S 15	01 S 16	03 S 19
04	02 04	02 28	04 12	00 29	01 55	00 14	01 16	03 19
07	02 05	02 57	04 08	00 28	01 55	00 14	01 16	03 19
10	02 01	03 26	04 04	00 27	01 55	00 14	01 16	03 20
13	01 53	03 57	04 00	00 27	01 55	00 14	01 16	03 20
16	01 40	04 29	03 55	00 26	01 55	00 14	01 16	03 20
19	01 22	05 01	03 50	00 26	01 55	00 14	01 16	03 21
22	00 58	05 33	03 44	00 25	01 55	00 14	01 16	03 21
25	00 S 28	06 06	03 39	00 24	01 55	00 14	01 16	03 22
28	00 N 07	06 37	03 34	00 24	01 55	00 14	01 16	03 22
31	00 N 47	07 N 07	03 N 28	00 S 23	01 S 55	00 S 14	01 S 16	03 S 22

DATA

Julian Date	2460708
Delta T	+72 seconds
Ayanamsa	24° 12′ 29″
Synetic vernal point	04° ♓ 54′ 30″
True obliquity of ecliptic	23° 26′ 19″

MOON'S PHASES, APSIDES AND POSITIONS ☽

Date	h	m	Phase	Longitude ° ′	Eclipse Indicator
05	08	02	☽	16 ♉ 46	
12	13	53	○	24 ♌ 06	
20	17	32	☾	02 ♐ 20	
28	00	45	●	09 ♓ 41	

Day	h	m		
02	02	41	Perigee	
18	01	04	Apogee	
02	00	38	0N	
08	10	31	Max dec	28° N 36′
15	10	54	0S	
22	22	22	Max dec	28° S 41′

LONGITUDES

Date	Chiron ⚷ ° ′	Ceres ⚳ ° ′	Pallas ⚴ ° ′	Juno ⚵ ° ′	Vesta ⚶ ° ′	Black Moon Lilith ⚸ ° ′
01	19 ♈ 31	21 ♒ 11	24 ♑ 24	26 ♏ 48	10 ♏ 25	24 ♎ 03
11	19 ♈ 51	25 ♒ 07	28 ♑ 05	28 ♏ 41	13 ♏ 12	25 ♎ 10
21	20 ♈ 16	29 ♒ 02	01 ♒ 40	00 ♐ 11	15 ♏ 30	26 ♎ 17
31	20 ♈ 44	02 ♓ 57	05 ♒ 07	01 ♐ 15	17 ♏ 12	27 ♎ 23

All ephemeris data is given at 12.00 UT and the Moon's longitude is additionally given for 24.00 UT

ASPECTARIAN

h m	Aspects
01 Saturday	
00 59	☽ ∥ ♄
03 31	☽ ∠ ♇
04 09	☽ ☌ ♄
06 09	☽ ⊥ ☉
07 50	☉ ∠ ♀
09 02	☽ △ ♂
12 09	☽ ∠ ☿
13 54	☽ ✶ ♅
14 32	☉ ∠ ♆
16 33	♀ ☌ ♆
17 47	☽ ∥ ♆
20 38	☽ # ♀
21 49	☽ ☌ ♆
22 06	☽ ☌ ♀
22 21	☽ ∠ ☉
23 58	☽ Q ♃
02 Sunday	
04 40	☽ ✶ ♇
05 11	☽ ∥ ♀
07 27	☽ # ♆
15 00	☽ ∠ ♅
16 35	☽ ✶ ☿
20 03	☽ ✶ ♃
03 Monday	
00 00	☽ # ♄
00 49	☽ Q ♇
01 23	☽ ✶ ☉
06 06	☽ ⊥ ♅
06 47	☽ ⊻ ♄
08 51	♀ Q ♃
10 19	☽ □ ♂
15 27	☽ Q ☿
15 32	♀ # ♆
16 12	☽ ⊻ ♅
16 57	☽ ⊥ ♄
21 16	☽ ∠ ♃
21 51	☿ △ ♃
23 08	☽ Q ☉
04 Tuesday	
00 16	☽ ⊻ ♆
03 19	☽ ⊻ ♀
04 51	☿ ⊥ ♄
07 12	☽ □ ♇
08 17	☽ ∠ ♄
09 40	♃ St D
10 26	☽ ⊥ ♆
11 50	☽ # ☉
12 29	☽ ⊥ ♃
14 02	☽ ⊥ ♀
16 11	☽ Q ♂
21 42	☽ ∥ ♅
22 41	☽ ⊻ ♃
23 14	☿ ∠ ♆
05 Wednesday	
00 00	☽ # ☿
01 45	☽ ∠ ♆
02 06	☽ □ ☿
06 10	☽ ∠ ♀
08 02	☽ □ ☉
09 59	☽ ✶ ♄
12 16	☽ ✶ ♂
13 34	☽ ∥ ♃
19 10	☽ ☌ ♅
20 18	☽ # ♇
06 Thursday	
00 29	☿ # ♅
03 29	☽ ✶ ♆
06 44	☽ Q ♄
09 17	☽ ✶ ♀
10 34	☽ △ ♇
13 38	☽ ∠ ♂
14 10	☉ ⊻ ♄
18 38	☽ ∥ ♂
07 Friday	
00 17	☽ Q ♆
02 16	☽ ☌ ♃
05 00	☽ ⊥ ♂
07 11	☿ ∠ ♀
07 11	☽ Q ♀
08 42	☉ ⊼ ♂
12 14	♀ ✶ ♇
12 40	☽ ⚼ ♇
13 11	☽ △ ☿
14 14	☽ □ ♄
15 18	☽ ⊻ ♂
15 57	☽ △ ☉
21 57	☿ ⊻ ♄
23 15	☽ ⊻ ♅
08 Saturday	
04 14	☿ ⊼ ♂
04 30	☽ ± ♇
07 52	☽ □ ♆
09 49	☽ ⊥ ♅
15 08	☽ ⊼ ♇
16 28	☽ □ ♀
19 32	☽ ⚼ ☿
20 32	☽ ⚼ ☉
09 Sunday	
01 49	☽ ∠ ♅
07 12	☽ ⊻ ♃
12 08	☉ ☌ ☿
13 15	♂ △ ♄
13 56	☽ ± ☉
14 03	☽ ± ☿
18 00	☽ ⊥ ♃
19 48	☽ ☌ ♂
19 57	☽ △ ♄
10 Monday	
01 39	☽ ⊼ ☉
02 33	☽ ⊼ ☿
03 05	☽ ∥ ♂
04 49	☽ ✶ ♅
05 05	☿ ⊥ ♆
10 19	☽ ∠ ♃
13 49	☽ △ ♆
18 18	☉ ⊥ ♆
19 28	☿ □ ♅
21 21	☽ ☍ ♇
23 30	☽ ⚼ ♄
11 Tuesday	
01 21	☽ △ ♀
02 47	☽ Q ♅
03 06	☽ # ♇
05 39	☿ ± ♂
10 01	☽ ∥ ♃
13 59	☽ ✶ ♃
16 19	☽ ± ♄
17 34	☽ ⚼ ♆
19 30	☉ □ ♅
12 Wednesday	
02 16	☽ ⊻ ♂
03 28	☽ ∥ ♅
03 39	☽ ⊼ ♄
06 38	☽ ⚼ ♀
09 00	☉ ± ♂
10 31	☽ ± ♆
12 26	☽ □ ♅
12 33	☽ Q ♃
13 27	☽ ⊥ ♂
13 53	☽ ☍ ☉
19 12	☽ ☍ ☿
21 56	☽ ⊼ ♆
22 47	☽ # ☿
13 Thursday	
00 31	☽ ± ♀
03 01	☽ # ☉
05 47	☽ ⊼ ♇
06 25	☽ ∠ ♂
12 34	☽ ⊼ ♀
14 30	☿ ⊻ ♆
17 23	☽ ± ♇
22 53	♀ # ♄
23 10	☽ □ ♃
14 Friday	
08 49	☽ ∥ ♀
09 32	☽ # ♄
10 58	☽ ⚼ ♇
11 14	☽ ✶ ♂
13 53	☽ ☍ ♄
20 57	☿ ⊥ ♀
22 34	☽ △ ♅
15 Saturday	
03 30	☽ # ♆
05 19	☽ ⊼ ☉
08 35	☽ ☍ ♆
09 54	☉ ∥ ☿
10 44	☽ Q ♂
11 51	♂ ⊥ ♃
16 00	☽ ⊼ ☿
16 45	☽ △ ♇
18 19	☽ ∥ ♆
18 23	☽ ± ☉
20 07	☿ ⚼ ♂
20 58	☿ ⊻ ♇
16 Sunday	
02 19	☽ ☍ ♀
04 32	☽ ⚼ ♅
06 14	☽ ± ☿
10 52	☽ △ ♃
12 13	☽ ∥ ♄
14 09	☽ ⚼ ☉
16 58	☽ # ♀
21 38	☉ ⊻ ♆
22 37	☽ □ ♂
22 43	☽ ± ♅
17 Monday	
02 30	☽ ⊼ ♄
03 46	☽ ⚼ ☿
10 02	☽ ∥ ☿
10 54	☽ ⊼ ♅
13 28	☽ ∥ ☉
14 48	☽ ± ♄
17 21	☽ ⚼ ♃
21 16	☽ ⊼ ♆
23 24	☽ △ ☉
18 Tuesday	
05 34	☽ □ ♇
09 16	☽ ⚼ ♄
09 30	☽ ± ♆
11 43	☽ ± ♃
15 55	☽ △ ☿
16 11	♀ ∠ ♅
17 28	☽ ⊼ ♀
21 51	☽ # ♅
23 57	☽ ⊼ ♃
19 Wednesday	
03 49	☽ ⚼ ♆
03 58	☿ ⊥ ♇
04 02	☿ ⊻ ♀
06 03	☽ ± ♀
11 07	☽ △ ♂
15 59	☽ △ ♄
16 59	☽ # ♃
18 08	☽ Q ♇
23 45	☽ ∥ ♇
23 47	☽ ☍ ♅
20 Thursday	
00 48	☽ ⚼ ♀
10 05	☽ △ ♆
12 00	☉ ⚼ ♂
15 38	☿ # ♀
16 41	☿ Q ♅
17 04	☽ ⚼ ♂
17 32	☽ □ ☉
18 12	☽ ✶ ♇
20 13	☿ □ ♃
21 Friday	
00 44	☽ # ♂
01 35	☉ ⊻ ♇
07 28	☽ △ ♀
10 46	☽ ± ♂
12 09	☽ ☍ ♃
14 53	☽ □ ☿
22 25	☽ ⊼ ♂
23 33	☉ ⊥ ♀
23 40	☽ ∠ ♇
22 Saturday	
03 50	☽ □ ♄
07 50	☽ Q ☉
10 48	☽ ⊼ ♅
16 56	☽ ⊥ ♇
20 38	☽ □ ♆
22 12	☽ ± ♅
23 Sunday	
04 15	☽ ⊻ ♇
07 08	☽ Q ☿
08 26	☽ ✶ ☉
14 07	☽ Q ♄
15 00	☽ ⚼ ♅
16 58	☿ △ ♂
17 52	☽ □ ♀
21 05	☿ ∥ ♄
21 15	☽ ⊼ ♃
24 Monday	
02 00	♂ St D
02 40	☿ ∠ ♇
03 19	☉ # ♀
06 00	☽ Q ♆
06 32	☽ ☍ ♂
08 12	☽ ± ♃
08 42	☽ ✶ ☿
12 10	☽ ✶ ♄
14 05	☽ ∠ ☉
18 14	☽ △ ♅
18 34	☽ # ♂
25 Tuesday	
00 21	☽ ⚼ ♃
02 47	☽ Q ♀
03 28	☽ ✶ ♆
07 18	☽ ⊥ ☉
10 33	☽ ☌ ♇
12 02	☿ ☌ ♄
14 52	☽ ∠ ♄
15 16	☽ ∠ ☿
15 51	☽ ∥ ♇
18 33	☽ ⊻ ☉
21 09	☽ # ♃
23 54	☽ ✶ ♀
26 Wednesday	
02 33	☽ △ ♃
05 31	☽ ∠ ♆
06 29	☽ ⊥ ♄
08 57	☽ ⊥ ☿
11 03	☽ ⊼ ♂
13 27	☽ # ♅
16 44	☽ ⊻ ♄
20 30	☽ ⊻ ☿
20 47	☽ ⊥ ♆
21 12	☽ ± ♂
22 04	☽ □ ♅
27 Thursday	
01 33	☽ ∠ ♀
04 46	☉ ⊥ ♇
06 50	☽ ⊻ ♆
09 38	☿ ✶ ♅
12 14	☽ ⚼ ♂
13 30	☽ ⊻ ♇
16 36	☽ ⊥ ♀
21 49	☽ # ♀
23 24	☽ ⊥ ♇
28 Friday	
00 45	☽ ☌ ☉
02 32	☽ ⊻ ♀
03 56	☽ Q ♅
04 54	☽ □ ♃
06 20	☽ ∥ ☉
12 57	☽ △ ♂
14 10	☽ ∠ ♇
14 27	☿ ∥ ♆
14 37	☽ ∥ ♄
18 42	☽ ☌ ♄
23 32	☽ ✶ ♅

MARCH 2025

LONGITUDES

Date	Sidereal time h m s	Sun ☉ ° ' "	Moon ☽ ° ' "	Moon ☽ 24.00 ° ' "	Mercury ☿ ° '	Venus ♀ ° '	Mars ♂ ° '	Jupiter ♃ ° '	Saturn ♄ ° '	Uranus ♅ ° '	Neptune ♆ ° '	Pluto ♇ ° '
01	22 38 11	11 ♓ 09 01	01 ♈ 19 03	08 ♈ 43 21	27 ♓ 06	10 ♈ 50	17 ♋ 12	12 ♊ 19	20 ♓ 45	23 ♉ 39	28 ♓ 55	02 ♒ 54
02	22 42 07	12 09 15	16 ♈ 07 19	23 ♈ 30 05	28 ♓ 40	10 R 50	17 16	12 24	20 52	23 40	28 57	02 56
03	22 46 04	13 09 26	00 ♉ 50 55	08 ♉ 09 06	00 ♈ 11	10 47	17 21	12 29	21 00	23 42	28 59	02 57
04	22 50 01	14 09 36	15 ♉ 24 05	22 ♉ 35 25	01 36	10 43	17 26	12 34	21 07	23 44	29 01	02 59
05	22 53 57	15 09 43	29 ♉ 42 43	06 ♊ 45 46	02 56	10 35	17 33	12 40	21 14	23 45	29 03	03 00
06	22 57 54	16 09 48	13 ♊ 44 23	20 ♊ 38 32	04 11	10 25	17 40	12 45	21 22	23 47	29 06	03 02
07	23 01 50	17 09 52	27 ♊ 28 13	04 ♋ 13 28	05 18	10 13	17 47	12 51	21 29	23 49	29 08	03 03
08	23 05 47	18 09 52	10 ♋ 54 25	17 ♋ 31 11	06 19	09 58	17 55	12 57	21 37	23 51	29 10	03 05
09	23 09 43	19 09 51	24 ♋ 03 56	00 ♌ 32 51	07 12	09 40	18 04	13 03	21 44	23 53	29 12	03 06
10	23 13 40	20 09 48	06 ♌ 58 06	13 ♌ 19 52	07 57	09 21	18 13	13 09	21 51	23 55	29 15	03 08
11	23 17 36	21 09 42	19 ♌ 38 20	25 ♌ 53 40	08 34	08 58	18 23	13 16	21 59	23 57	29 17	03 09
12	23 21 33	22 09 35	02 ♍ 06 05	08 ♍ 15 45	09 02	08 34	18 34	13 22	22 06	23 59	29 19	03 10
13	23 25 30	23 09 25	14 ♍ 22 50	20 ♍ 27 32	09 22	08 07	18 45	13 29	22 14	24 01	29 21	03 12
14	23 29 26	24 09 13	26 ♍ 30 03	02 ♎ 30 37	09 33	07 39	18 57	13 36	22 21	24 03	29 24	03 13
15	23 33 23	25 08 59	08 ♎ 29 26	14 ♎ 26 46	09 R 35	07 08	19 09	13 43	22 28	24 05	29 26	03 15
16	23 37 19	26 08 44	20 ♎ 22 53	26 ♎ 18 07	09 29	06 36	19 21	13 50	22 36	24 07	29 28	03 16
17	23 41 16	27 08 26	02 ♏ 12 45	08 ♏ 07 11	09 15	06 03	19 34	13 57	22 43	24 09	29 31	03 17
18	23 45 12	28 08 07	14 ♏ 01 47	19 ♏ 56 59	08 53	05 28	19 48	14 05	22 51	24 11	29 33	03 18
19	23 49 09	29 ♓ 07 46	25 ♏ 53 15	01 ♐ 51 03	08 24	04 52	20 02	14 12	22 58	24 14	29 35	03 20
20	23 53 05	00 ♈ 07 24	07 ♐ 50 52	13 ♐ 53 16	07 49	04 15	20 17	14 20	23 05	24 16	29 37	03 21
21	23 57 02	01 06 59	19 ♐ 58 46	26 ♐ 07 55	07 08	03 37	20 32	14 28	23 13	24 18	29 40	03 22
22	00 00 59	02 06 33	02 ♑ 21 15	08 ♑ 39 18	06 22	03 00	20 47	14 36	23 20	24 21	29 42	03 23
23	00 04 55	03 06 05	15 ♑ 02 35	21 ♑ 31 32	05 33	02 22	21 03	14 44	23 28	24 23	29 44	03 24
24	00 08 52	04 05 35	28 ♑ 06 35	04 ♒ 48 01	04 42	01 44	21 20	14 52	23 35	24 26	29 46	03 26
25	00 12 48	05 05 04	11 ♒ 36 05	18 ♒ 30 52	03 49	01 07	21 37	15 01	23 42	24 28	29 49	03 27
26	00 16 45	06 04 31	25 ♒ 32 21	02 ♓ 40 20	02 56	00 ♈ 31	21 54	15 09	23 49	24 31	29 51	03 28
27	00 20 41	07 03 55	09 ♓ 54 27	17 ♓ 14 11	02 04	29 ♓ 55	22 12	15 18	23 57	24 33	29 53	03 29
28	00 24 38	08 03 18	24 ♓ 38 49	02 ♈ 07 30	01 14	29 21	22 30	15 27	24 04	24 36	29 55	03 30
29	00 28 34	09 02 39	09 ♈ 39 13	17 ♈ 12 53	00 ♈ 27	28 48	22 48	15 36	24 11	24 39	29 ♓ 58	03 31
30	00 32 31	10 01 58	24 ♈ 47 19	02 ♉ 21 20	29 ♓ 43	28 16	23 07	15 45	24 19	24 41	00 ♈ 00	03 32
31	00 36 28	11 ♈ 01 15	09 ♉ 53 44	17 ♉ 23 27	29 ♓ 03	27 ♓ 46	23 ♋ 26	15 ♊ 54	24 ♓ 26	24 ♉ 44	00 ♈ 02	03 ♒ 33

Date	Moon True ☊ ° '	Moon Mean ☊ ° '	Moon ☽ Latitude ° '
01	27 ♓ 24	28 ♓ 21	00 N 22
02	27 D 25	28 18	01 42
03	27 26	28 14	02 55
04	27 27	28 11	03 56
05	27 28	28 08	04 41
06	27 28	28 05	05 08
07	27 R 28	28 02	05 17
08	27 27	27 59	05 08
09	27 26	27 55	04 44
10	27 25	27 52	04 05
11	27 24	27 49	03 15
12	27 24	27 46	02 16
13	27 23	27 43	01 12
14	27 23	27 39	00 N 05
15	27 D 23	27 36	01 S 01
16	27 23	27 33	02 04
17	27 23	27 30	03 02
18	27 R 23	27 27	03 51
19	27 23	27 24	04 31
20	27 23	27 20	04 59
21	27 23	27 17	05 15
22	27 D 23	27 14	05 16
23	27 23	27 11	05 03
24	27 23	27 08	04 33
25	27 24	27 05	03 48
26	27 25	27 01	02 48
27	27 25	26 58	01 36
28	27 26	26 55	00 S 15
29	27 R 26	26 52	01 N 07
30	27 25	26 49	02 26
31	27 ♓ 23	26 ♓ 45	03 N 34

DECLINATIONS

Date	Sun ☉ ° '	Moon ☽ ° '	Mercury ☿ ° '	Venus ♀ ° '	Mars ♂ ° '	Jupiter ♃ ° '	Saturn ♄ ° '	Uranus ♅ ° '	Neptune ♆ ° '	Pluto ♇ ° '
01	07 S 23	00 N 51	00 S 51	10 N 32	25 N 50	21 N 53	05 S 25	18 N 28	01 S 35	22 S 47
02	07 00	07 54	00 S 01	10 41	25 48	21 54	05 22	18 28	01 35	22 47
03	06 37	14 30	00 N 47	10 49	25 45	21 54	05 20	18 28	01 34	22 47
04	06 14	20 12	01 34	10 56	25 43	21 55	05 17	18 29	01 33	22 47
05	05 51	24 39	02 19	11 01	25 40	21 56	05 14	18 29	01 32	22 46
06	05 28	27 33	03 01	11 05	25 37	21 57	05 11	18 30	01 31	22 46
07	05 04	28 42	03 40	11 08	25 35	21 58	05 08	18 30	01 30	22 46
08	04 41	28 07	04 17	11 10	25 32	21 59	05 05	18 31	01 29	22 46
09	04 17	25 57	04 50	11 09	25 29	22 00	05 02	18 31	01 28	22 46
10	03 54	22 28	05 20	11 08	25 26	22 01	04 59	18 32	01 27	22 45
11	03 30	18 00	05 46	11 05	25 22	22 02	04 56	18 32	01 27	22 45
12	03 07	12 50	06 07	11 00	25 19	22 03	04 54	18 33	01 26	22 45
13	02 43	07 15	06 25	10 54	25 16	22 04	04 51	18 33	01 25	22 45
14	02 19	01 N 28	06 38	10 46	25 13	22 05	04 48	18 34	01 24	22 45
15	01 56	04 S 18	06 46	10 37	25 09	22 06	04 45	18 34	01 23	22 44
16	01 32	09 53	06 50	10 27	25 06	22 07	04 42	18 35	01 22	22 44
17	01 08	15 05	06 49	10 15	25 02	22 08	04 39	18 35	01 21	22 44
18	00 45	19 43	06 44	10 01	24 58	22 09	04 36	18 36	01 20	22 44
19	00 S 21	23 37	06 34	09 47	24 55	22 10	04 33	18 37	01 19	22 44
20	00 N 03	26 32	06 20	09 31	24 51	22 11	04 30	18 37	01 19	22 44
21	00 27	28 18	06 03	09 14	24 47	22 12	04 28	18 38	01 18	22 44
22	00 50	28 41	05 41	08 56	24 43	22 13	04 25	18 38	01 17	22 44
23	01 14	27 36	05 17	08 37	24 39	22 14	04 22	18 39	01 16	22 43
24	01 38	25 00	04 50	08 18	24 34	22 15	04 19	18 40	01 15	22 43
25	02 01	20 57	04 20	07 58	24 30	22 16	04 16	18 40	01 14	22 43
26	02 25	15 38	03 50	07 37	24 26	22 18	04 13	18 41	01 13	22 43
27	02 48	09 20	03 18	07 16	24 21	22 19	04 11	18 41	01 12	22 43
28	03 12	02 S 22	02 46	06 54	24 17	22 20	04 08	18 42	01 11	22 43
29	03 35	04 N 51	02 14	06 33	24 12	22 21	04 05	18 43	01 10	22 43
30	03 58	11 52	01 43	06 11	24 07	22 22	04 02	18 43	01 10	22 43
31	04 N 22	18 N 10	01 N 13	05 N 50	24 N 03	22 N 23	03 S 59	18 N 44	01 S 09	22 S 43

ZODIAC SIGN ENTRIES

Date	h	m	Planets
01	09	52	☽ ♈
03	09	04	☿ ♈
03	10	37	☽ ♉
05	12	29	☽ ♊
07	16	29	☽ ♋
09	22	59	☽ ♌
12	07	56	☽ ♍
14	18	59	☽ ♎
17	07	30	☽ ♏
19	20	17	☽ ♐
20	09	01	☉ ♈
22	07	29	☽ ♑
24	15	25	☽ ♒
26	19	32	☽ ♓
27	08	41	♀ ♓
28	20	36	☽ ♈
30	02	18	☿ ♓
30	12	00	♆ ♈
30	20	16	☽ ♉

LATITUDES

Date	Mercury ☿ ° '	Venus ♀ ° '	Mars ♂ ° '	Jupiter ♃ ° '	Saturn ♄ ° '	Uranus ♅ ° '	Neptune ♆ ° '	Pluto ♇ ° '
01	00 N 20	06 N 48	03 N 32	00 S 24	01 S 55	00 S 14	01 S 16	03 S 22
04	01 01	07 17	03 26	00 23	01 55	00 14	01 16	03 23
07	01 42	07 43	03 21	00 23	01 55	00 14	01 16	03 23
10	02 22	08 05	03 16	00 22	01 55	00 14	01 16	03 23
13	02 57	08 22	03 10	00 21	01 55	00 14	01 16	03 24
16	03 21	08 32	03 05	00 21	01 55	00 14	01 16	03 24
19	03 32	08 34	03 00	00 20	01 55	00 14	01 16	03 25
22	03 26	08 27	02 55	00 20	01 55	00 14	01 16	03 25
25	03 04	08 12	02 50	00 19	01 56	00 13	01 16	03 26
28	02 29	07 49	02 45	00 19	01 56	00 13	01 16	03 26
31	01 N 45	07 N 19	02 N 41	00 S 19	01 S 56	00 S 13	01 S 16	03 S 27

DATA

Julian Date	2460736
Delta T	+72 seconds
Ayanamsa	24° 12' 32"
Synetic vernal point	04° ♓ 54' 27"
True obliquity of ecliptic	23° 26' 19"

LONGITUDES

Date	Chiron ⚷ ° '	Ceres ⚳ ° '	Pallas ⚴ ° '	Juno ⚵ ° '	Vesta ⚶ ° '	Black Moon Lilith ⚸ ° '
01	20 ♈ 38	02 ♓ 10	04 ♒ 26	01 ♐ 05	16 ♏ 55	27 ♎ 10
11	21 ♈ 09	06 ♓ 04	07 ♒ 46	01 ♐ 46	18 ♏ 04	28 ♎ 17
21	21 ♈ 42	09 ♓ 55	10 ♒ 55	01 ♐ 56	18 ♏ 28	29 ♎ 23
31	22 ♈ 16	13 ♓ 43	13 ♒ 52	01 ♐ 34	18 ♏ 03	00 ♏ 30

MOON'S PHASES, APSIDES AND POSITIONS ☽

Date	h	m	Phase	Longitude ° '	Eclipse Indicator
06	16	32	☽	16 ♊ 21	
14	06	55	○	23 ♍ 57	total
22	11	29	☾	02 ♑ 05	
29	10	58	●	09 ♈ 00	Partial

Day	h	m		
01	21	15	Perigee	
17	16	25	Apogee	
30	05	19	Perigee	
01	09	07	0N	
07	15	44	Max dec	28° N 43'
14	18	04	0S	
22	06	36	Max dec	28° S 44'
28	19	53	0N	

ASPECTARIAN

01 Saturday
03 45 ☽ ∥ ♆
04 08 ☉ ⊻ ♀
04 19 ☽ ☌ ☿
05 31 ☽ ∥ ☿
08 05 ☽ ☌ ♆
10 22 ☽ Q ♃
11 58 ☽ # ☿
14 28 ☽ # ♆
14 34 ☽ ⚹ ♇
23 54 ☽ ∠ ♅

02 Sunday
00 14 ☉ Q ♅
00 36 ♀ St R
03 20 ☽ # ♄
03 25 ☽ ☌ ♀
05 06 ☽ ⊻ ☉
05 55 ☽ ⚹ ♃
09 01 ☽ # ☉
10 03 ☽ Q ♇
13 52 ☽ □ ♂
14 32 ☽ ⊥ ♅
15 32 ☽ ⊥ ☉
16 22 ☿ ☌ ♆
18 19 ☉ □ ♃
19 47 ☽ ⊻ ♄
22 02 ☽ ∥ ♀

03 Monday
00 18 ☽ ⊻ ♅
05 39 ☽ ⊥ ♄
06 28 ☽ ∠ ♃
07 16 ☽ ∠ ☉
08 56 ☽ ⊻ ♆
10 47 ☽ ⊻ ☿
15 27 ☽ □ ♇
17 15 ☿ Q ♃
18 48 ☽ ⊥ ♆
19 26 ☽ Q ♂
20 31 ☽ ∠ ♄
21 18 ☽ ⊥ ♃
21 43 ☽ ⊥ ☿

04 Tuesday
04 14 ☽ ∥ ♅
04 16 ☽ ⊻ ♀
07 16 ☽ ⊻ ♃
09 42 ☽ ∠ ♆
09 47 ☽ ⚹ ☉
11 26 ☿ # ♆
14 10 ☽ ⊥ ♀
14 13 ☽ ∠ ☿
15 25 ☽ ⚹ ♂
20 25 ☽ ∥ ♃
21 37 ☽ ⚹ ♄

05 Wednesday
00 54 ☽ # ♇
01 56 ☽ ☌ ♅
05 06 ☽ ∠ ♀
07 22 ☽ Q ☉
10 53 ☽ ⚹ ♆
13 13 ☿ ⚹ ♇
16 51 ☽ ∠ ♂
17 36 ☽ △ ♇
18 01 ☽ ⚹ ☿
18 03 ☽ Q ♄
18 54 ☽ ∥ ♂

06 Thursday
06 21 ☽ ⚹ ♀
07 25 ☽ Q ♆
08 23 ☽ ⊥ ♂
10 17 ☽ ☌ ♃
16 32 ☽ □ ☉
16 37 ☽ Q ☿
18 52 ☽ ⊻ ♂
19 27 ☽ bQ ♇

07 Friday
01 23 ☽ □ ♄
02 54 ☽ Q ♀
05 33 ☽ ⊻ ♅
07 32 ☉ ∥ ♄
11 16 ☽ ± ♇
14 57 ☽ □ ♆
16 10 ☽ ⊥ ♅
21 56 ☽ ⊼ ♇

08 Saturday
03 05 ☽ □ ☿
05 13 ☉ △ ♂
08 17 ☽ ∠ ♅
09 54 ☉ ∠ ♇
10 20 ☽ □ ♀
15 43 ☽ ⊻ ♃
21 55 ☉ # ☿

09 Sunday
00 52 ☽ ☌ ♂
02 16 ☽ △ ☉
02 43 ☽ ⊥ ♃
07 40 ☽ △ ♄
11 39 ☽ ⚹ ♅
15 51 ☽ ∥ ♂
19 25 ☽ ∠ ♃
20 32 ☿ # ♄
21 32 ☽ △ ♆

10 Monday
04 48 ☽ ☍ ♇
08 20 ☽ bQ ☉
10 01 ☽ Q ♅
10 18 ☽ # ♇
11 47 ☽ bQ ♄
13 56 ☽ △ ☿
14 42 ☽ ∥ ♃
16 21 ☽ △ ♀
23 46 ☽ ⚹ ♃

11 Tuesday
01 46 ☽ bQ ♆
02 44 ☽ ± ☉
04 58 ☽ ± ♄
09 22 ☽ ∥ ♅
09 35 ☽ ⊻ ♂
13 43 ♀ ∠ ♅
15 10 ☽ ⊼ ☉
16 32 ☽ ⊼ ♄
19 00 ☽ ± ♆
19 50 ☽ bQ ☿
20 03 ☽ bQ ♀
20 16 ☽ □ ♅
21 14 ☽ ⊥ ♂
22 52 ☽ Q ♃
22 55 ☿ ☌ ♀

12 Wednesday
06 36 ☽ ⊼ ♆
08 25 ☿ ∠ ♅
10 29 ☉ ☌ ♄
12 52 ☽ ± ♀
13 52 ☽ ± ☿
14 05 ☽ ⊼ ♇
14 53 ☽ ∠ ♂
20 11 ☽ ∥ ♀

13 Thursday
00 10 ☽ ⊼ ♀
01 49 ☽ ± ♇
01 55 ☽ ⊼ ☿
10 13 ☽ □ ♃
15 21 ☽ ∥ ☿
19 32 ☽ bQ ♇
20 45 ☽ ⚹ ♂
22 07 ☽ # ♄

14 Friday
03 40 ☽ ☍ ♄
06 55 ☽ ☍ ☉
07 06 ☽ △ ♅
08 12 ☽ # ☉
09 16 ☉ ⚹ ♅
12 17 ☽ # ♆
17 47 ☽ ☍ ♆
21 01 ☽ Q ♂
23 50 ☽ ∥ ♆

15 Saturday
01 27 ☽ △ ♇
02 42 ☽ ∥ ☉
06 46 ☿ St R
09 24 ☽ ☍ ♀
13 11 ☽ bQ ♅
13 51 ☽ ∥ ♄
14 12 ☽ ☍ ☿
22 38 ☽ △ ♃
22 38 ☽ # ☿

16 Sunday
07 24 ☽ ± ♅
09 53 ☽ □ ♂
14 25 ☽ # ♀
16 32 ☽ ⊼ ♄
19 35 ☽ ⊼ ♅
22 20 ☉ ∥ ♆

17 Monday
00 45 ☽ ⊼ ☉
04 50 ☽ ± ♄
05 19 ☽ bQ ♃
06 29 ☽ ⊼ ♆
14 04 ☽ ± ☉
14 11 ☽ □ ♇
18 28 ☿ ∠ ♅
18 43 ☽ ± ♆
19 25 ☽ ⊼ ♀
23 18 ☽ bQ ♄
23 47 ☽ ± ♃

18 Tuesday
01 54 ☽ ⊼ ☿
05 51 ☽ # ♅
07 02 ☽ ± ♀
10 01 ☽ bQ ☉
12 06 ☽ ⊼ ♃
13 03 ☽ bQ ♆
13 40 ☽ ± ☿
23 56 ☽ △ ♂

19 Wednesday
00 25 ☽ bQ ♀
02 23 ☽ # ♃
02 46 ☽ Q ♇
06 03 ☽ △ ♄
06 04 ☽ ∥ ♇
07 12 ☽ bQ ☿
08 38 ☽ ☍ ♅
19 08 ☽ △ ☉
19 28 ☽ △ ♆
21 24 ☽ # ♂
23 25 ☉ ☌ ♆

20 Thursday
02 59 ☽ ⚹ ♇
05 09 ☽ △ ♀
06 46 ☽ bQ ♂
11 56 ☽ △ ☿
23 00 ♂ ⊥ ♃

21 Friday
01 01 ☽ ☍ ♃
01 03 ☽ ± ♂
08 50 ☽ ∠ ♇
13 06 ☽ ⊼ ♂
18 23 ☽ □ ♄
20 29 ☽ ⊼ ♅
21 32 ♀ ⚹ ♇

22 Saturday
02 25 ☽ ⊥ ♇
06 53 ☽ □ ♆
08 08 ☽ ± ♅
11 29 ☽ □ ☉
13 10 ☽ □ ♀
13 59 ☽ ⊻ ♇
19 13 ☽ □ ☿

23 Sunday
00 26 ♀ Q ♃
01 07 ☉ ☌ ♀
01 21 ☽ bQ ♅
01 44 ☉ Q ♃
05 13 ☽ Q ♄
11 25 ☽ ⊼ ♃
13 48 ☉ # ♆
17 01 ☽ Q ♆
19 32 ☉ ⚹ ♇
21 25 ☽ Q ♀
22 40 ☽ ± ♃
23 23 ☽ ☍ ♂

24 Monday
00 09 ☽ Q ☉
02 46 ☽ Q ☿
03 41 ☽ ⚹ ♄
05 18 ☽ △ ♅
15 01 ☽ ⚹ ♆
15 01 ☽ # ♂
15 13 ☽ bQ ♃
18 15 ☽ ⚹ ♀
19 48 ☉ ☌ ☿
21 34 ☽ ☌ ♇
23 06 ☽ ⚹ ☿
23 36 ☽ ⚹ ☉

25 Tuesday
02 28 ☽ ∥ ♇
05 01 ☽ # ♃
06 52 ☽ ∠ ♄
15 39 ☿ # ♄
17 37 ☽ ∠ ♆
18 01 ☽ △ ♃
19 32 ☽ ∠ ♀
22 02 ☿ ⚹ ♇
22 42 ☽ ⊥ ♄
22 56 ☽ # ♅
23 47 ☽ ∠ ☿

26 Wednesday
03 49 ☽ ∠ ☉
05 40 ☽ ⊼ ♂
06 54 ☿ Q ♃
09 04 ☽ ⊻ ♄
09 07 ☽ ⊥ ♆
10 15 ☽ □ ♅
10 20 ☽ ⊥ ♀
14 14 ☽ ⊥ ☿
16 04 ☽ ± ♂
19 18 ☽ ⊻ ♆
20 03 ☽ ⊻ ♀
20 13 ☽ ⊥ ☉
23 44 ☽ ⊻ ☿

27 Thursday
01 20 ☽ ⊻ ♇
06 58 ☽ ⊻ ☉
07 25 ☽ bQ ♂
11 18 ☽ ⊥ ♇
13 13 ♀ ☌ ♆
16 22 ☽ Q ♅
19 42 ☽ # ♀
20 56 ☽ □ ♃

28 Friday
00 55 ☉ ∥ ☿
02 02 ☽ ∠ ♇
06 01 ☽ ∥ ♄
08 27 ☽ △ ♂
09 21 ☽ # ☉
10 32 ☽ # ☿
11 04 ☽ ☌ ♄
11 55 ☽ ⚹ ♅
15 56 ☽ ∥ ♆
19 17 ☽ ☌ ♀
20 30 ☽ ☌ ♆
22 02 ☽ ☌ ☿
23 48 ☽ # ♆

29 Saturday
02 12 ☽ ⚹ ♇
02 15 ☽ Q ♃
03 54 ☽ ∥ ☿
07 32 ☽ ∥ ☉
09 26 ☽ # ♄
10 58 ☽ ☌ ☉
11 59 ☽ ∠ ♅
17 23 ☽ ∥ ♀
21 19 ☽ Q ♇
21 32 ☽ ⚹ ♃

30 Sunday
02 18 ☽ ⊥ ♅
02 47 ☿ ☌ ♆
03 15 ☉ ∠ ♅
09 18 ☽ □ ♂
11 14 ☽ ⊻ ♄
11 50 ☽ ⊻ ♅
15 26 ☉ # ♄
17 20 ☽ ⊻ ♀
19 28 ☽ ⊻ ☿
20 17 ☽ ⊻ ♆
20 49 ☽ ⊥ ♄
21 33 ☽ ∠ ♃

31 Monday
01 53 ☽ □ ♇
02 33 ☽ ⊥ ♀
04 36 ☽ ⊥ ☿
05 50 ☽ ⊥ ♆
11 15 ☽ ∠ ♄
12 00 ☽ ⊥ ♃
13 55 ☽ ⊻ ☉
14 24 ☽ ∥ ♅
14 31 ☽ Q ♂
15 58 ☿ # ♆
16 27 ☽ ∠ ♀
18 23 ☽ ∠ ☿
20 15 ☽ ∠ ♆
21 42 ☽ ⊻ ♃

All ephemeris data is given at 12.00 UT and the Moon's longitude is additionally given for 24.00 UT

APRIL 2025

LONGITUDES

Date	Sidereal time h m s	Sun ☉ ° ' "	Moon ☽ ° ' "	Moon ☽ 24.00 ° ' "	Mercury ☿ ° '	Venus ♀ ° '	Mars ♂ ° '	Jupiter ♃ ° '	Saturn ♄ ° '	Uranus ♅ ° '	Neptune ♆ ° '	Pluto ♇ ° '
01	00 40 24	12 ♈ 00 30	24 ♉ 49 29	02 ♊ 10 59	28 ♓ 28	27 ♓ 18	23 ♋ 46	16 ♊ 03	24 ♓ 33	24 ♉ 47	00 ♈ 04	03 ♒ 34
02	00 44 21	12 59 42	09 ♊ 27 16	16 ♊ 37 47	27 R 58	26 R 52	24 06	16 13	24 40	24 50	00 07	03 35
03	00 48 17	13 58 53	23 ♊ 42 14	00 ♋ 40 24	27 33	26 28	24 26	16 22	24 47	24 52	00 09	03 35
04	00 52 14	14 58 01	07 ♋ 32 15	14 ♋ 17 54	27 14	26 06	24 47	16 32	24 54	24 55	00 11	03 36
05	00 56 10	15 57 06	20 ♋ 57 33	27 ♋ 31 28	27 00	25 47	25 08	16 42	25 02	24 58	00 13	03 37
06	01 00 07	16 56 09	04 ♌ 00 00	10 ♌ 23 35	26 52	25 30	25 29	16 52	25 09	25 01	00 16	03 38
07	01 04 03	17 55 10	16 ♌ 42 36	22 ♌ 57 31	26 D 50	25 15	25 51	17 01	25 16	25 04	00 18	03 39
08	01 08 00	18 54 09	29 ♌ 08 46	05 ♍ 16 45	26 52	25 03	26 13	17 12	25 23	25 07	00 20	03 39
09	01 11 57	19 53 05	11 ♍ 21 56	17 ♍ 24 40	27 01	24 53	26 35	17 22	25 30	25 10	00 22	03 40
10	01 15 53	20 51 59	23 ♍ 25 20	29 ♍ 24 18	27 14	24 45	26 58	17 32	25 37	25 13	00 24	03 41
11	01 19 50	21 50 51	05 ♎ 21 51	11 ♎ 18 19	27 32	24 40	27 20	17 42	25 44	25 16	00 26	03 42
12	01 23 46	22 49 41	17 ♎ 13 57	23 ♎ 09 03	27 55	24 38	27 44	17 53	25 50	25 19	00 29	03 42
13	01 27 43	23 48 29	29 ♎ 03 50	04 ♏ 58 34	28 22	24 D 38	28 07	18 03	25 57	25 22	00 31	03 43
14	01 31 39	24 47 15	10 ♏ 53 29	16 ♏ 48 50	28 54	24 40	28 31	18 14	26 04	25 25	00 33	03 43
15	01 35 36	25 45 59	22 ♏ 44 53	28 ♏ 41 54	29 ♓ 30	24 45	28 55	18 25	26 11	25 28	00 35	03 44
16	01 39 32	26 44 41	04 ♐ 40 10	10 ♐ 40 00	00 ♈ 10	24 51	29 19	18 36	26 18	25 31	00 37	03 44
17	01 43 29	27 43 22	16 ♐ 41 45	22 ♐ 45 46	00 53	25 01	29 ♋ 43	18 47	26 24	25 35	00 39	03 45
18	01 47 26	28 42 01	28 ♐ 52 27	05 ♑ 02 11	01 40	25 12	00 ♌ 08	18 58	26 31	25 38	00 41	03 45
19	01 51 22	29 ♈ 40 38	11 ♑ 15 26	17 ♑ 32 38	02 30	25 25	00 33	19 09	26 38	25 41	00 43	03 46
20	01 55 19	00 ♉ 39 13	23 ♑ 54 15	00 ♒ 20 44	03 24	25 41	00 58	19 20	26 44	25 44	00 45	03 46
21	01 59 15	01 37 47	06 ♒ 52 31	13 ♒ 30 00	04 21	25 58	01 23	19 31	26 51	25 47	00 47	03 47
22	02 03 12	02 36 19	20 ♒ 13 33	27 ♒ 03 27	05 20	26 17	01 49	19 43	26 57	25 51	00 49	03 47
23	02 07 08	03 34 50	03 ♓ 59 53	11 ♓ 02 55	06 23	26 38	02 15	19 54	27 04	25 54	00 51	03 47
24	02 11 05	04 33 19	18 ♓ 12 29	25 ♓ 28 20	07 28	27 01	02 41	20 05	27 10	25 57	00 53	03 48
25	02 15 01	05 31 46	02 ♈ 50 01	10 ♈ 16 56	08 36	27 26	03 07	20 17	27 17	26 01	00 55	03 48
26	02 18 58	06 30 12	17 ♈ 48 16	25 ♈ 23 00	09 46	27 52	03 34	20 29	27 23	26 04	00 57	03 48
27	02 22 55	07 28 35	02 ♉ 59 59	10 ♉ 37 56	10 58	28 20	04 00	20 41	27 29	26 07	00 59	03 48
28	02 26 51	08 26 58	18 ♉ 15 33	25 ♉ 51 28	12 13	28 50	04 27	20 52	27 36	26 11	01 01	03 49
29	02 30 48	09 25 18	03 ♊ 24 24	10 ♊ 53 10	13 31	29 20	04 55	21 04	27 42	26 14	01 03	03 49
30	02 34 44	10 ♉ 23 36	18 ♊ 16 45	25 ♊ 34 19	14 ♈ 50	29 ♓ 53	05 ♌ 22	21 ♊ 16	27 ♓ 48	26 ♉ 17	01 ♈ 05	03 ♒ 49

Date	Moon True ☊ ° '	Moon Mean ☊ ° '	Moon ☽ Latitude ° '
01	27 ♓ 22	26 ♓ 42	04 N 27
02	27 R 20	26 39	05 01
03	27 18	26 36	05 16
04	27 17	26 33	05 11
05	27 D 17	26 30	04 50
06	27 18	26 26	04 14
07	27 19	26 23	03 26
08	27 21	26 20	02 30
09	27 22	26 17	01 27
10	27 23	26 14	00 N 22
11	27 R 23	26 11	00 S 44
12	27 21	26 07	01 47
13	27 19	26 04	02 45
14	27 15	26 01	03 36
15	27 10	25 58	04 18
16	27 05	25 55	04 49
17	27 00	25 51	05 07
18	26 55	25 48	05 12
19	26 52	25 45	05 03
20	26 51	25 42	04 38
21	26 D 51	25 39	03 59
22	26 52	25 36	03 07
23	26 53	25 32	02 02
24	26 54	25 29	00 S 47
25	26 R 54	25 26	00 N 32
26	26 53	25 23	01 51
27	26 50	25 20	03 03
28	26 44	25 17	04 03
29	26 38	25 13	04 44
30	26 ♓ 32	25 ♓ 10	05 N 06

DECLINATIONS

Date	Sun ☉ ° '	Moon ☽ ° '	Mercury ☿ ° '	Venus ♀ ° '	Mars ♂ ° '	Jupiter ♃ ° '	Saturn ♄ ° '	Uranus ♅ ° '	Neptune ♆ ° '	Pluto ♇ ° '
01	04 N 45	23 N 17	00 N 45	05 N 29	23 N 58	22 N 24	03 S 57	18 N 45	01 S 08	22 S 43
02	05 08	26 50	00 N 18	05 08	23 53	22 25	03 54	18 45	01 07	22 43
03	05 31	28 33	00 S 06	04 47	23 47	22 27	03 51	18 46	01 06	22 43
04	05 54	28 24	00 28	04 27	23 42	22 28	03 48	18 47	01 05	22 43
05	06 17	26 34	00 48	04 08	23 37	22 29	03 46	18 48	01 04	22 43
06	06 39	23 22	01 05	03 49	23 32	22 30	03 43	18 48	01 03	22 43
07	07 02	19 06	01 20	03 31	23 26	22 31	03 40	18 49	01 03	22 43
08	07 24	14 06	01 32	03 14	23 21	22 32	03 38	18 50	01 02	22 43
09	07 47	08 39	01 42	02 58	23 15	22 33	03 35	18 50	01 01	22 43
10	08 09	02 N 57	01 48	02 43	23 09	22 34	03 32	18 51	01 00	22 43
11	08 31	02 S 48	01 53	02 29	23 03	22 35	03 30	18 52	00 59	22 43
12	08 53	08 25	01 55	02 15	22 57	22 37	03 27	18 53	00 58	22 43
13	09 14	13 43	01 54	02 03	22 51	22 38	03 24	18 53	00 58	22 43
14	09 36	18 31	01 51	01 51	22 45	22 39	03 22	18 54	00 57	22 43
15	09 57	22 37	01 46	01 41	22 39	22 40	03 19	18 55	00 56	22 43
16	10 19	25 48	01 39	01 32	22 32	22 41	03 17	18 56	00 55	22 43
17	10 40	27 52	01 30	01 23	22 26	22 42	03 14	18 56	00 54	22 43
18	11 01	28 38	01 18	01 16	22 19	22 43	03 11	18 57	00 54	22 43
19	11 21	27 59	01 05	01 10	22 13	22 44	03 09	18 58	00 53	22 43
20	11 42	25 53	00 50	01 04	22 06	22 45	03 06	18 59	00 52	22 43
21	12 02	22 25	00 32	01 00	21 59	22 46	03 04	19 00	00 51	22 43
22	12 23	17 42	00 S 14	00 57	21 52	22 47	03 02	19 00	00 50	22 43
23	12 43	11 56	00 N 07	00 54	21 45	22 48	02 59	19 01	00 50	22 43
24	13 02	05 S 23	00 29	00 53	21 38	22 49	02 57	19 02	00 49	22 43
25	13 22	01 N 37	00 52	00 52	21 30	22 50	02 54	19 03	00 48	22 44
26	13 41	08 42	01 17	00 52	21 23	22 51	02 52	19 04	00 47	22 44
27	14 00	15 23	01 44	00 54	21 16	22 52	02 50	19 04	00 47	22 44
28	14 19	21 09	02 12	00 56	21 08	22 53	02 47	19 05	00 46	22 44
29	14 38	25 29	02 41	00 58	21 00	22 54	02 45	19 06	00 45	22 44
30	14 N 56	28 N 00	03 N 11	01 N 02	20 N 52	22 N 55	02 S 43	19 N 07	00 S 44	22 S 44

ZODIAC SIGN ENTRIES

Date	h m	Planets
01	20 26	☽ ♊
03	22 50	☽ ♋
06	04 34	☽ ♌
08	13 40	☽ ♍
11	01 12	☽ ♎
13	13 54	☽ ♏
16	02 37	☽ ♐
16	06 25	☿ ♈
18	04 21	♂ ♌
18	14 12	☽ ♑
19	19 56	☉ ♉
20	23 22	☽ ♒
23	05 07	☽ ♓
25	07 24	☽ ♈
27	07 17	☽ ♉
29	06 34	☽ ♊
30	17 16	♀ ♈

LATITUDES

Date	Mercury ☿ ° '	Venus ♀ ° '	Mars ♂ ° '	Jupiter ♃ ° '	Saturn ♄ ° '	Uranus ♅ ° '	Neptune ♆ ° '	Pluto ♇ ° '
01	01 N 29	07 N 08	02 N 39	00 S 18	01 S 56	00 S 13	01 S 16	03 S 27
04	00 N 41	06 33	02 35	00 18	01 56	00 13	01 16	03 28
07	00 S 05	05 54	02 30	00 17	01 57	00 13	01 16	03 28
10	00 46	05 14	02 26	00 17	01 57	00 13	01 16	03 29
13	01 22	04 33	02 22	00 17	01 58	00 13	01 16	03 29
16	01 52	03 53	02 18	00 16	01 58	00 13	01 16	03 30
19	02 16	03 15	02 14	00 16	01 58	00 13	01 16	03 30
22	02 33	02 38	02 10	00 15	01 59	00 13	01 16	03 31
25	02 45	02 04	02 06	00 15	01 59	00 13	01 16	03 31
28	02 52	01 31	02 03	00 15	02 00	00 13	01 17	03 32
31	02 S 53	01 N 01	01 N 59	00 S 14	02 S 00	00 S 13	01 S 17	03 S 33

DATA

Julian Date	2460767
Delta T	+72 seconds
Ayanamsa	24° 12' 36"
Synetic vernal point	04° ♓ 54' 23"
True obliquity of ecliptic	23° 26' 19"

MOON'S PHASES, APSIDES AND POSITIONS ☽

Date	h m	Phase	Longitude ° '	Eclipse Indicator
05	02 15	☽	15 ♋ 33	
13	00 22	○	23 ♎ 20	
21	01 36	☾	01 ♒ 12	
27	19 31	●	07 ♉ 47	

Day	h m		
13	22 31	Apogee	
27	16 11	Perigee	
03	22 04	Max dec	28° N 42'
11	00 17	0S	
18	13 11	Max dec	28° S 38'
25	06 32	0N	

LONGITUDES

Date	Chiron ⚷ ° '	Ceres ⚳ ° '	Pallas ⚴ ° '	Juno ⚵ ° '	Vesta ⚶ ° '	Black Moon Lilith ⚸ ° '
01	22 ♈ 20	14 ♓ 05	14 ♒ 08	01 ♐ 30	17 ♏ 57	00 ♏ 37
11	22 ♈ 55	17 ♓ 48	16 ♒ 49	00 ♐ 30	16 ♏ 39	01 ♏ 43
21	23 ♈ 31	21 ♓ 25	19 ♒ 13	29 ♏ 01	14 ♏ 41	02 ♏ 50
31	24 ♈ 06	24 ♓ 55	21 ♒ 17	27 ♏ 06	12 ♏ 18	03 ♏ 57

ASPECTARIAN

h m Aspects

01 Tuesday
00 13 ☽ ⊥ ☉
07 19 ☽ ∥ ♃
08 56 ☽ # ♇
10 15 ☽ ✶ ♂
11 33 ☽ ✶ ♄
11 56 ☽ ☌ ♅
15 44 ☽ ∥ ♂
15 48 ☽ ∠ ☉
15 54 ☽ ✶ ♀
17 43 ☽ ✶ ☿
20 34 ☽ ✶ ♆

02 Wednesday
02 17 ☽ △ ♇
07 21 ☽ Q ♄
11 03 ☽ Q ♀
11 24 ☽ ∠ ♂
11 52 ☉ ∥ ♀
12 49 ☽ Q ☿
16 26 ☽ Q ♆
18 20 ☽ ✶ ☉
23 25 ☽ ☌ ♃

03 Thursday
02 50 ☽ ⊥ ♂
03 18 ☽ bQ ♇
13 17 ☽ ⊻ ♂
13 52 ☽ □ ♄
14 00 ☽ ⊻ ♅
16 12 ☽ Q ☉
16 37 ☽ □ ♀
18 26 ☽ □ ☿
18 41 ☽ ± ♇
23 07 ☽ □ ♆

04 Friday
00 23 ☽ ⊥ ♅
05 06 ☽ ⊼ ♇
16 13 ☽ ∠ ♅
16 21 ♄ ✶ ♅
23 05 ♂ ✶ ♅

05 Saturday
01 08 ♂ △ ♄
02 15 ☽ □ ☉
03 47 ☉ Q ♇
04 12 ☽ ⊻ ♃
15 12 ☽ ⊥ ♃
19 20 ☽ ✶ ♅
19 29 ☽ △ ♄
19 49 ☽ ☌ ♂
20 36 ☽ △ ♀
22 54 ☽ △ ☿

06 Sunday
05 02 ☽ △ ♆
07 58 ☽ ∠ ♃
09 22 ☿ ∥ ♆
09 44 ☉ ✶ ♃
10 55 ☽ ∥ ♂
11 19 ☽ ☍ ♇
12 13 ♀ △ ♂
16 02 ☽ # ♇
17 17 ☽ ∥ ♃
17 40 ☽ Q ♅
22 00 ♀ # ♄
23 38 ☽ bQ ♄
23 57 ☽ bQ ♀

07 Monday
02 43 ☽ bQ ☿
09 18 ☽ bQ ♆
11 01 ♀ ☌ ♄
11 08 ☿ St D
12 37 ☽ ✶ ♃
13 26 ☽ ∥ ♅
14 31 ☽ △ ☉
16 46 ☽ ± ♀
16 56 ☽ ± ♄
19 54 ☽ ± ☿

08 Tuesday
02 38 ☽ ± ♆
04 08 ☽ □ ♅
04 09 ☽ ⊼ ♀
04 37 ☽ ⊼ ♄
04 48 ♀ ✶ ♅
06 07 ☽ ⊻ ♂
07 33 ☽ ⊼ ☿
12 06 ☽ Q ♃
14 19 ☽ ⊼ ♆
17 27 ☉ ⊥ ♅
18 11 ☽ ⊥ ♂
20 50 ☽ ⊼ ♇
22 07 ☽ bQ ☉

09 Wednesday
08 39 ☽ ± ♇
12 27 ☽ ∠ ♂
15 29 ☽ ∥ ☉
17 26 ☽ ± ☉

10 Thursday
00 04 ☽ □ ♃
02 31 ☽ bQ ♇
06 26 ☽ ⊼ ☉
09 31 ☽ # ♄
13 00 ☽ ∥ ♀
14 39 ☽ ☍ ♀
15 36 ☽ △ ♅
16 26 ☽ ☍ ♄
16 40 ☽ # ☿
19 19 ☽ ✶ ♂
19 49 ☽ ☍ ☿
20 07 ☽ # ♆

11 Friday
02 03 ☽ ☍ ♆
04 25 ☽ ∥ ♆
08 06 ☽ ∥ ☿
08 38 ☽ △ ♇
10 41 ☽ # ♀
14 53 ☽ ∥ ♄
20 18 ☽ Q ♂
21 56 ☽ bQ ♅

12 Saturday
13 20 ☽ △ ♃
14 10 ☽ # ☉
16 14 ☽ ± ♅

13 Sunday
00 22 ☽ ☍ ☉
01 03 ♀ St D
02 59 ☽ ⊼ ♀
04 28 ☽ ⊼ ♅
05 38 ☽ ⊼ ♄
10 01 ☽ □ ♂
10 32 ☽ ⊼ ☿
14 57 ☽ ⊼ ♆
15 11 ☽ ± ♀
17 56 ☽ ± ♄
20 14 ☽ bQ ♃
21 27 ☽ □ ♇
23 16 ☽ ± ☿

14 Monday
03 10 ☽ ± ♆
08 52 ☉ ⊻ ♀
09 30 ☽ bQ ♀
11 57 ☿ # ♀
12 22 ☽ bQ ♄
14 05 ☽ # ♅
14 46 ☽ ± ♃
18 25 ☽ bQ ☿
21 11 ♂ # ♇
21 28 ☽ bQ ♆

15 Tuesday
03 06 ☽ ⊼ ♃
04 20 ☉ ⊻ ♅
08 51 ♂ ∥ ♃
09 57 ☽ Q ♇
12 11 ☽ # ♂
12 18 ☽ # ♃
12 37 ☽ ∥ ♇
16 04 ☽ △ ♀
17 31 ☽ ☍ ♅
18 38 ☽ ⊼ ☉
19 00 ☽ △ ♄
23 32 ☉ ⊻ ♄

16 Wednesday
00 52 ☽ △ ♂
02 24 ☽ △ ☿
03 50 ☽ △ ♆
07 48 ☽ ± ☉
10 08 ☽ ✶ ♇

17 Thursday
03 24 ☽ bQ ☉
04 11 ☿ ☌ ♆
07 56 ☽ bQ ♂
08 16 ♃ bQ ♇
16 04 ☽ ∠ ♇
16 11 ☽ ☍ ♃

18 Friday
02 22 ☽ ± ♂
04 40 ☽ □ ♀
05 37 ☽ ⊼ ♅
07 21 ☽ □ ♄
09 49 ☽ ⊥ ♇
10 23 ♃ # ♇
11 38 ☽ △ ☉
14 33 ☽ ⊼ ♂
15 33 ☽ □ ♆
17 24 ☽ ± ♅
17 50 ☽ □ ☿
20 30 ☿ # ♀
21 31 ☽ ⊻ ♇

19 Saturday
10 53 ☽ bQ ♅
16 13 ☽ Q ♀
18 31 ☽ Q ♄
22 53 ♂ △ ♆

20 Sunday
02 16 ☽ Q ♆
03 16 ☽ ⊼ ♃
06 56 ☽ Q ☿
08 07 ☿ ∥ ♆
14 35 ☉ ⊻ ♆
14 43 ☽ ± ♃
15 23 ☽ ✶ ♀
15 27 ☽ △ ♅
17 21 ☽ ✶ ♄
18 21 ♀ ✶ ♅
21 40 ☿ ✶ ♇

21 Monday
00 47 ☽ ✶ ♆
01 34 ☉ □ ♂
01 35 ☽ ☍ ♂
01 36 ☽ □ ☉
06 20 ☽ ☌ ♇
07 00 ☽ ✶ ☿
07 38 ☽ bQ ♃
09 55 ☽ # ♃
10 12 ☽ ∥ ♇
14 31 ☽ # ♂
19 37 ☽ ∠ ♀
21 06 ☽ ∠ ♄

22 Tuesday
00 13 ☉ ⊥ ♀
04 09 ☽ ∠ ♆
05 53 ☽ # ♅
11 04 ☽ △ ♃
12 07 ☽ ⊥ ♀
12 13 ☽ ∠ ☿
12 43 ☽ Q ☉
13 18 ☽ ⊥ ♄
20 07 ☽ ⊥ ♆
21 44 ☉ ⊥ ♄
21 55 ☽ □ ♅
22 56 ☽ ⊻ ♀
23 55 ☽ ⊻ ♄

23 Wednesday
05 15 ☽ ⊥ ☿
06 35 ☽ ⊻ ♆
08 54 ☽ ⊼ ♂
09 07 ☽ # ☉
11 14 ☽ ✶ ☉
11 39 ☽ ⊻ ♇
16 24 ☽ ⊻ ☿
17 10 ☉ □ ♇
19 29 ☽ ± ♂
21 52 ☽ ⊥ ♇

24 Thursday
04 52 ☽ Q ♅
11 06 ☽ bQ ♂
12 59 ☽ ∠ ♇
14 24 ☽ ∠ ☉
15 10 ☽ □ ♃
20 32 ☽ ∥ ♄

25 Friday
00 02 ♀ ☌ ♄
00 50 ☽ ✶ ♅
02 54 ☽ ☌ ♄
02 57 ☽ ☌ ♀
03 34 ☽ # ♀
03 47 ☽ ∥ ♆
04 01 ☽ # ☿
04 12 ☿ Q ♃
04 29 ☉ ∠ ♃
06 15 ☽ ⊥ ☉
07 52 ☿ # ♆
08 53 ☽ ☌ ♆
09 15 ☽ # ♆
09 20 ☽ ∥ ☿
09 28 ☽ ∥ ♀
11 42 ☿ ∥ ♀
12 29 ☽ △ ♂
13 34 ☽ ✶ ♇
16 19 ☽ # ♄
16 40 ☽ ⊻ ☉
20 55 ☽ Q ♃
22 04 ☽ ☌ ☿

26 Saturday
01 13 ☽ ∠ ♅
08 49 ☽ Q ♇
15 36 ☽ ⊥ ♅
16 18 ☽ ✶ ♃
23 29 ☉ ⊥ ♆

27 Sunday
01 05 ♂ ☍ ♇
01 07 ☽ ⊻ ♅
03 16 ☽ ⊻ ♄
04 25 ☽ ⊻ ♀
06 35 ☽ ∥ ☉
08 49 ☽ ⊻ ♆
12 46 ☽ ⊥ ♄
13 16 ☽ □ ♇
13 38 ☽ □ ♂
14 10 ☽ ⊥ ♀
14 58 ☿ ∠ ♅
16 16 ☽ ∠ ♃
18 17 ☽ ⊥ ♆
19 31 ☽ ☌ ☉

28 Monday
01 39 ☽ ⊻ ☿
02 49 ☽ ∥ ♅
03 01 ☽ ∠ ♄
04 47 ☽ ∠ ♀
06 36 ☽ ⊥ ♃
08 28 ☽ ∠ ♆
11 56 ☽ ∥ ♂
11 56 ☽ ⊥ ☿
16 10 ☽ ⊻ ♃
18 49 ☽ Q ♂
19 47 ☽ # ♇
20 36 ☽ ∥ ♃

29 Tuesday
00 33 ☽ ☌ ♅
02 51 ☽ ✶ ♄
03 29 ☽ ∠ ☿
05 18 ☽ ✶ ♀
08 14 ☽ ✶ ♆
12 39 ☽ △ ♇
14 29 ☽ ✶ ♂
14 58 ☿ # ♄
22 09 ☽ Q ♄
22 19 ☽ ⊻ ☉

30 Wednesday
01 12 ☽ Q ♀
03 32 ☽ Q ♆
05 50 ☽ ✶ ☿
08 42 ☽ ⊥ ☉
12 53 ☽ bQ ♇
15 31 ☽ ∠ ♂
16 58 ☽ ☌ ♃

All ephemeris data is given at 12.00 UT and the Moon's longitude is additionally given for 24.00 UT

MAY 2025

LONGITUDES

Date	Sidereal time h m s	Sun ☉ ° ' "	Moon ☽ ° ' "	Moon ☽ 24.00 ° ' "	Mercury ☿ ° '	Venus ♀ ° '	Mars ♂ ° '	Jupiter ♃ ° '	Saturn ♄ ° '	Uranus ♅ ° '	Neptune ♆ ° '	Pluto ♇ ° '
01	02 38 41	11 ♉ 21 53	02 ♋ 45 13	09 ♋ 49 04	16 ♈ 12	00 ♈ 26	05 ♌ 49	21 ♊ 28	27 ♓ 54	26 ♉ 21	01 ♈ 07	03 ♒ 49
02	02 42 37	12 20 08	16 ♋ 45 38	23 ♋ 34 56	17 36	01 01	06 17	21 40	28 00	26 24	01 09	03 49
03	02 46 34	13 18 20	00 ♌ 17 05	06 ♌ 52 23	19 02	01 37	06 45	21 52	28 06	26 28	01 10	03 49
04	02 50 30	14 16 31	13 ♌ 21 14	19 ♌ 44 07	20 29	02 15	07 13	22 05	28 12	26 31	01 12	03 49
05	02 54 27	15 14 39	26 ♌ 01 36	02 ♍ 14 14	21 59	02 53	07 41	22 17	28 18	26 34	01 14	03 R 49
06	02 58 24	16 12 45	08 ♍ 22 37	14 ♍ 27 22	23 31	03 33	08 10	22 29	28 24	26 38	01 16	03 49
07	03 02 20	17 10 50	20 ♍ 29 04	26 ♍ 28 17	25 05	04 14	08 38	22 42	28 29	26 41	01 17	03 49
08	03 06 17	18 08 52	02 ♎ 25 33	08 ♎ 21 23	26 41	04 55	09 07	22 54	28 35	26 45	01 19	03 49
09	03 10 13	19 06 53	14 ♎ 16 15	20 ♎ 10 33	28 19	05 38	09 36	23 07	28 41	26 48	01 21	03 49
10	03 14 10	20 04 52	26 ♎ 04 42	01 ♏ 59 00	29 ♈ 59	06 22	10 05	23 19	28 46	26 52	01 22	03 49
11	03 18 06	21 02 49	07 ♏ 53 47	13 ♏ 49 19	01 ♉ 41	07 07	10 34	23 32	28 52	26 55	01 24	03 48
12	03 22 03	22 00 45	19 ♏ 45 49	25 ♏ 43 30	03 25	07 52	11 04	23 44	28 57	26 59	01 26	03 48
13	03 25 59	22 58 40	01 ♐ 42 33	07 ♐ 43 08	05 10	08 38	11 33	23 57	29 02	27 02	01 27	03 48
14	03 29 56	23 56 32	13 ♐ 45 26	19 ♐ 49 37	06 58	09 26	12 03	24 10	29 08	27 06	01 29	03 48
15	03 33 53	24 54 24	25 ♐ 55 50	02 ♑ 04 18	08 48	10 14	12 33	24 23	29 13	27 09	01 30	03 47
16	03 37 49	25 52 14	08 ♑ 15 13	14 ♑ 28 49	10 40	11 02	13 03	24 36	29 18	27 13	01 32	03 47
17	03 41 46	26 50 03	20 ♑ 45 22	27 ♑ 05 08	12 34	11 52	13 33	24 49	29 23	27 16	01 33	03 47
18	03 45 42	27 47 51	03 ♒ 28 28	09 ♒ 55 40	14 29	12 42	14 03	25 01	29 28	27 20	01 35	03 46
19	03 49 39	28 45 37	16 ♒ 27 06	23 ♒ 03 06	16 27	13 33	14 33	25 14	29 33	27 23	01 36	03 46
20	03 53 35	29 ♉ 43 23	29 ♒ 44 02	06 ♓ 30 13	18 27	14 24	15 04	25 28	29 38	27 27	01 38	03 46
21	03 57 32	00 ♊ 41 07	13 ♓ 21 53	20 ♓ 19 16	20 28	15 16	15 34	25 41	29 43	27 30	01 39	03 45
22	04 01 28	01 38 50	27 ♓ 22 25	04 ♈ 31 20	22 31	16 09	16 05	25 54	29 48	27 34	01 40	03 45
23	04 05 25	02 36 32	11 ♈ 45 50	19 ♈ 05 33	24 36	17 02	16 36	26 07	29 52	27 37	01 42	03 44
24	04 09 22	03 34 13	26 ♈ 29 58	03 ♉ 58 21	26 42	17 56	17 07	26 20	29 ♓ 57	27 41	01 43	03 44
25	04 13 18	04 31 54	11 ♉ 29 46	19 ♉ 03 10	28 ♉ 50	18 51	17 38	26 33	00 ♈ 02	27 44	01 44	03 43
26	04 17 15	05 29 33	26 ♉ 37 21	04 ♊ 11 01	00 ♊ 59	19 45	18 09	26 47	00 06	27 48	01 46	03 43
27	04 21 11	06 27 11	11 ♊ 42 54	19 ♊ 11 43	03 10	20 41	18 41	27 00	00 10	27 51	01 47	03 42
28	04 25 08	07 24 48	26 ♊ 36 19	03 ♋ 55 41	05 21	21 37	19 12	27 13	00 15	27 54	01 48	03 41
29	04 29 04	08 22 23	11 ♋ 08 57	18 ♋ 15 32	07 32	22 33	19 44	27 26	00 19	27 58	01 49	03 41
30	04 33 01	09 19 58	25 ♋ 15 00	02 ♌ 07 08	09 44	23 30	20 16	27 40	00 23	28 01	01 50	03 40
31	04 36 57	10 ♊ 17 31	08 ♌ 51 57	15 ♌ 29 37	11 ♊ 56	24 ♈ 27	20 ♌ 48	27 ♊ 53	00 ♈ 27	28 ♉ 05	01 ♈ 51	03 ♒ 39

Date	Moon True ☊ ° '	Moon Mean ☊ ° '	Moon ☽ Latitude ° '
01	26 ♓ 26	25 ♓ 07	05 N 07
02	26 R 21	25 04	04 50
03	26 19	25 01	04 17
04	26 18	24 57	03 31
05	26 D 18	24 54	02 36
06	26 20	24 51	01 35
07	26 21	24 48	00 N 32
08	26 R 20	24 45	00 S 33
09	26 19	24 42	01 35
10	26 15	24 38	02 33
11	26 08	24 35	03 24
12	26 00	24 32	04 06
13	25 49	24 29	04 38
14	25 39	24 26	04 58
15	25 28	24 23	05 04
16	25 19	24 19	04 56
17	25 11	24 16	04 35
18	25 06	24 13	03 59
19	25 03	24 10	03 10
20	25 03	24 07	02 11
21	25 D 03	24 03	01 S 02
22	25 R 03	24 00	00 N 12
23	25 02	23 57	01 28
24	24 59	23 54	02 39
25	24 54	23 51	03 40
26	24 45	23 48	04 27
27	24 36	23 44	04 54
28	24 25	23 41	05 02
29	24 15	23 38	04 49
30	24 07	23 35	04 19
31	24 ♓ 01	23 ♓ 32	03 N 35

DECLINATIONS

Date	Sun ☉ ° '	Moon ☽ ° '	Mercury ☿ ° '	Venus ♀ ° '	Mars ♂ ° '	Jupiter ♃ ° '	Saturn ♄ ° '	Uranus ♅ ° '	Neptune ♆ ° '	Pluto ♇ ° '
01	15 N 14	28 N 32	03 N 43	01 N 06	20 N 44	22 N 56	02 S 40	19 N 08	00 S 44	22 S 44
02	15 32	27 11	04 16	01 11	20 36	22 57	02 38	19 08	00 43	22 45
03	15 50	24 16	04 49	01 17	20 28	22 57	02 36	19 09	00 42	22 45
04	16 07	20 11	05 24	01 24	20 20	22 58	02 34	19 10	00 42	22 45
05	16 24	15 17	06 00	01 31	20 11	22 59	02 32	19 11	00 41	22 45
06	16 41	09 54	06 37	01 39	20 03	23 00	02 29	19 12	00 40	22 45
07	16 58	04 N 15	07 15	01 47	19 54	23 01	02 27	19 12	00 40	22 45
08	17 14	01 S 28	07 53	01 57	19 46	23 02	02 25	19 13	00 39	22 46
09	17 30	07 05	08 33	02 06	19 37	23 02	02 23	19 14	00 38	22 46
10	17 46	12 26	09 13	02 17	19 28	23 03	02 21	19 15	00 38	22 46
11	18 01	17 21	09 53	02 27	19 19	23 04	02 19	19 16	00 37	22 46
12	18 16	21 37	10 35	02 39	19 10	23 05	02 17	19 16	00 37	22 47
13	18 31	25 02	11 16	02 51	19 01	23 05	02 15	19 17	00 36	22 47
14	18 45	27 22	11 59	03 03	18 51	23 06	02 13	19 18	00 35	22 47
15	19 00	28 26	12 41	03 16	18 42	23 07	02 11	19 19	00 35	22 47
16	19 13	28 07	13 24	03 29	18 32	23 07	02 09	19 20	00 34	22 47
17	19 27	26 21	14 07	03 43	18 23	23 08	02 08	19 20	00 34	22 48
18	19 40	23 15	14 50	03 57	18 13	23 09	02 06	19 21	00 33	22 48
19	19 53	18 56	15 33	04 12	18 03	23 09	02 04	19 22	00 33	22 48
20	20 05	13 36	16 15	04 27	17 53	23 10	02 02	19 23	00 32	22 49
21	20 18	07 29	16 58	04 42	17 43	23 10	02 01	19 24	00 32	22 49
22	20 29	00 S 51	17 39	04 57	17 33	23 11	01 59	19 24	00 31	22 49
23	20 41	06 N 00	18 20	05 13	17 23	23 11	01 57	19 25	00 31	22 49
24	20 52	12 41	19 01	05 30	17 12	23 12	01 56	19 26	00 30	22 50
25	21 03	18 46	19 40	05 46	17 02	23 12	01 54	19 27	00 30	22 50
26	21 13	23 43	20 17	06 03	16 51	23 13	01 52	19 28	00 29	22 50
27	21 23	27 03	20 54	06 20	16 41	23 13	01 51	19 28	00 29	22 51
28	21 33	28 25	21 28	06 37	16 30	23 13	01 49	19 29	00 28	22 51
29	21 42	27 46	22 01	06 55	16 19	23 14	01 48	19 30	00 28	22 51
30	21 51	25 20	22 32	07 12	16 08	23 14	01 46	19 31	00 28	22 52
31	21 N 59	21 N 30	23 N 00	07 N 30	15 N 57	23 N 15	01 S 45	19 N 32	00 S 27	22 S 52

ZODIAC SIGN ENTRIES

Date	h	m	Planets
01	07	23	☽ ♋
03	11	29	☽ ♌
05	19	40	☽ ♍
08	07	06	☽ ♎
10	12	15	☿ ♉
10	19	58	☽ ♏
13	08	35	☽ ♐
15	19	58	☽ ♑
18	05	29	☽ ♒
20	12	28	☽ ♓
20	18	55	☉ ♊
22	16	26	☽ ♈
24	17	38	☽ ♉
25	03	35	♄ ♈
26	00	59	☿ ♊
26	17	21	☽ ♊
28	17	33	☽ ♋
30	20	17	☽ ♌

LATITUDES

Date	Mercury ☿ ° '	Venus ♀ ° '	Mars ♂ ° '	Jupiter ♃ ° '	Saturn ♄ ° '	Uranus ♅ ° '	Neptune ♆ ° '	Pluto ♇ ° '
01	02 S 53	01 N 01	01 N 59	00 S 14	02 S 00	00 S 13	01 S 17	03 S 33
04	02 48	00 33	01 56	00 14	02 01	00 13	01 17	03 33
07	02 39	00 N 07	01 52	00 13	02 01	00 13	01 17	03 34
10	02 24	00 S 16	01 49	00 13	02 02	00 13	01 17	03 34
13	02 06	00 38	01 46	00 13	02 02	00 13	01 17	03 35
16	01 43	00 57	01 43	00 12	02 03	00 13	01 17	03 35
19	01 16	01 15	01 39	00 12	02 04	00 13	01 17	03 36
22	00 46	01 31	01 36	00 12	02 04	00 13	01 17	03 36
25	00 S 15	01 45	01 33	00 11	02 05	00 13	01 18	03 37
28	00 N 17	01 57	01 30	00 11	02 06	00 13	01 18	03 37
31	00 N 47	02 S 07	01 N 28	00 S 11	02 S 06	00 S 13	01 S 18	03 S 38

DATA

Julian Date	2460797
Delta T	+72 seconds
Ayanamsa	24° 12' 40"
Synetic vernal point	04° ♓ 54' 19"
True obliquity of ecliptic	23° 26' 19"

MOON'S PHASES, APSIDES AND POSITIONS ☽

Date	h	m	Phase	Longitude ° '	Eclipse Indicator
04	13	52	☽	14 ♌ 21	
12	16	56	○	22 ♏ 13	
20	11	59	☾	29 ♒ 43	
27	03	02	●	06 ♊ 06	

Day	h	m		
11	00	34	Apogee	
26	01	26	Perigee	
01	06	25	Max dec	28° N 35'
08	05	50	0S	
15	18	29	Max dec	28° S 29'
22	15	01	0N	
28	16	04	Max dec	28° N 27'

LONGITUDES

Date	Chiron ⚷ ° '	Ceres ⚳ ° '	Pallas ⚴ ° '	Juno ⚵ ° '	Vesta ⚶ ° '	Black Moon Lilith ⚸ ° '
01	24 ♈ 06	24 ♓ 55	21 ♒ 17	27 ♏ 06	12 ♏ 18	03 ♏ 57
11	24 ♈ 40	28 ♓ 17	22 ♒ 58	24 ♏ 55	09 ♏ 52	05 ♏ 03
21	25 ♈ 11	01 ♈ 30	24 ♒ 14	22 ♏ 41	07 ♏ 44	06 ♏ 10
31	25 ♈ 41	04 ♈ 31	25 ♒ 00	20 ♏ 34	06 ♏ 12	07 ♏ 17

All ephemeris data is given at 12.00 UT and the Moon's longitude is additionally given for 24.00 UT

ASPECTARIAN

h m Aspects

01 Thursday
00 33 ☽ ∠ ☉
01 15 ☽ ⊻ ♅
03 34 ☽ Q ☿
03 44 ☽ ± ♇
03 49 ☽ □ ♄
05 19 ☿ Q ♇
06 55 ☽ ⊥ ♂
07 57 ☽ □ ♀
09 14 ☽ □ ♆
11 19 ☽ ⊥ ♅
13 48 ☽ ⊼ ♇
17 22 ☽ ⊻ ♂

02 Friday
02 41 ☽ ∠ ♅
03 45 ☽ ⚹ ☉
13 37 ☽ □ ☿
17 07 ♀ ☌ ♆
20 45 ☽ ⊻ ♃

03 Saturday
02 22 ☽ Q ☉
05 06 ☽ ⚹ ♅
06 18 ☉ ∠ ♄
07 36 ☽ ⊥ ♃
08 02 ☽ △ ♄
13 36 ☽ △ ♆
14 33 ☽ △ ♀
18 25 ☽ ☍ ♇
20 19 ☽ ∥ ♃
21 37 ☽ # ♇
23 21 ♂ ∠ ♃

04 Sunday
00 12 ☽ ∠ ♃
00 13 ☽ ☌ ♂
02 59 ☽ Q ♅
11 10 ☽ ∥ ♂
11 42 ☽ ⚼ ♄
12 27 ☿ ⊥ ♅
13 52 ☽ □ ☉
15 25 ♇ St R
17 14 ☽ ∥ ♅
17 21 ☽ ⚼ ♆
19 41 ☽ ⚼ ♀

05 Monday
03 14 ☽ △ ☿
04 43 ☽ ⚹ ♃
04 48 ☽ ± ♄
07 03 ☽ ∥ ☉
10 28 ☽ ± ♆
13 03 ☽ □ ♅
13 45 ☽ ± ♀
16 24 ☽ ⊼ ♄
17 21 ☿ ⚹ ♃
22 04 ☽ ⊼ ♆

06 Tuesday
02 01 ☽ ⊼ ♀
03 05 ☽ ⊼ ♇
04 15 ☽ Q ♃
11 34 ☽ ⊻ ♂
12 20 ☽ ⚼ ☿
13 13 ☉ ∠ ♆
14 50 ☽ ± ♇
20 41 ☉ ⊥ ♃
21 31 ♀ ⚹ ♇
23 53 ☽ ⊥ ♂

07 Wednesday
00 39 ☽ ∥ ☿
04 50 ☽ △ ☉
08 40 ☽ ⚼ ♇
08 47 ☽ ± ☿
14 55 ♂ Q ♅
16 30 ☽ □ ♃
18 34 ☽ ∠ ♂
19 36 ☽ # ♄
22 04 ☽ ∥ ♀
22 38 ☽ ⊼ ☿

08 Thursday
00 30 ☽ △ ♅
03 05 ☽ # ♆
04 11 ☽ ☍ ♄
08 35 ☽ ∥ ♆
09 45 ☽ ☍ ♆
12 56 ☿ ⊻ ♅
13 35 ☽ ⚼ ☉
14 04 ☽ # ♀
14 48 ☽ △ ♇
16 00 ☽ ∥ ♄
17 22 ☽ ☍ ♀

09 Friday
02 07 ☽ ⚹ ♂
06 58 ☽ ⚼ ♅
09 27 ☽ ± ☉
17 31 ☿ ⊻ ♄
19 16 ☽ # ☿
22 43 ☽ ⊼ ☉

10 Saturday
01 21 ☽ ± ♅
03 32 ☽ Q ♂
06 17 ☽ △ ♃
13 36 ☽ ⊼ ♅
17 31 ☽ ⊼ ♄
20 38 ♀ # ♄
21 15 ☽ ☍ ☿
22 47 ☽ ⊼ ♆

11 Sunday
03 42 ☽ □ ♇
05 48 ☽ ± ♄
08 01 ☿ ⊻ ♆
10 18 ☽ ⊼ ♀
10 59 ☽ ± ♆
13 19 ☽ ⚼ ♃
15 42 ☽ # ☉
17 39 ☽ □ ♂
20 07 ♂ ∥ ♅
22 10 ☽ # ♂
22 14 ☽ # ♅
23 17 ☽ ± ♀

12 Monday
00 10 ☽ ⚼ ♄
05 15 ☽ ⚼ ♆
07 51 ☽ ± ♃
16 07 ☽ Q ♇
16 56 ☽ ☍ ☉
17 23 ☿ □ ♇
18 41 ☽ ⚼ ♀
19 23 ☽ ∥ ♇
20 09 ☽ ⊼ ♃
21 26 ☽ # ♃

13 Tuesday
02 35 ☽ ☍ ♅
06 37 ☽ △ ♄
10 06 ☿ ⊥ ♄
11 29 ☽ △ ♆
16 11 ☽ ⚹ ♇
20 08 ☽ ⊼ ☿

14 Wednesday
02 48 ☽ △ ♀
08 28 ☽ △ ♂
10 10 ☽ ± ☿
17 57 ☉ ∥ ♂
18 48 ☿ ⊥ ♆
19 09 ☉ ⊻ ♃
21 58 ☽ ∠ ♇

15 Thursday
07 05 ☽ ⚼ ☿
08 54 ☽ ☍ ♃
09 49 ☽ ⊼ ☉
14 24 ☽ ⊼ ♅
15 17 ☽ ⚼ ♂
15 39 ☽ ⊥ ♇
18 29 ☽ □ ♄
20 28 ☿ ∠ ♃
22 33 ☽ ± ☉
22 55 ☽ □ ♆

16 Friday
02 10 ☽ ± ♅
03 20 ☽ ⊻ ♇
09 33 ☽ ± ♂
17 29 ☽ ⚼ ☉
17 29 ☽ △ ☿
17 45 ☽ □ ♀
19 40 ☽ ⚼ ♅
20 24 ☿ ⊻ ♀
21 37 ☽ ⊼ ♂
23 46 ☉ ∥ ♅

17 Saturday
05 32 ☽ Q ♄
09 42 ☽ Q ♆
17 05 ☉ ∠ ♀
19 50 ☽ ⊼ ♃
23 32 ☉ ☌ ♅

18 Sunday
00 24 ☽ △ ♅
00 28 ☽ △ ☉
00 36 ♀ ∠ ♅
04 27 ☽ ⚹ ♄
04 36 ☿ □ ♂
06 26 ☽ Q ♀
07 20 ☽ ± ♃
08 27 ☽ ⚹ ♆
11 46 ☿ ∠ ♄
12 34 ☽ ☌ ♇
12 41 ☽ # ♃
14 49 ☽ ∥ ♇

19 Monday
00 23 ☽ ⚼ ♃
00 29 ♀ Q ♃
06 18 ☽ ⚹ ♀
07 25 ☽ # ☉
08 23 ☽ ☍ ♂
08 30 ☽ ∠ ♄
09 50 ☽ # ♅
12 00 ☽ □ ☿
12 05 ♂ ⚼ ♄
12 17 ☽ ∠ ♆
13 53 ☿ ∠ ♆
16 25 ☽ # ♂

20 Tuesday
00 59 ☽ ⊥ ♄
01 53 ☽ # ☿
04 13 ☽ △ ♃
04 38 ☽ ⊥ ♆
07 53 ☽ □ ♅
09 38 ☉ ⚹ ♄
11 22 ☽ ∠ ♀
11 49 ☽ ⊻ ♄
11 59 ☽ □ ☉
15 23 ☽ ⊻ ♆
19 09 ☽ ⊻ ♇

21 Wednesday
01 32 ☿ ⊥ ♃
01 57 ☽ Q ☿
04 23 ☽ ⊥ ♀
05 42 ☽ ⊥ ♇
15 32 ☽ ⊻ ♀
15 43 ☽ Q ♅
15 58 ☽ ⊼ ♂
21 18 ☽ ∠ ♇
21 52 ☽ Q ☉
21 53 ☽ # ♀

22 Thursday
01 05 ♀ Q ♇
02 21 ☽ ⚹ ☿
02 41 ☽ ± ♂
04 30 ☿ ⊥ ♀
07 42 ♀ △ ♂
08 00 ☽ ∥ ♄
08 57 ☿ ∥ ♂
09 27 ☽ □ ♃
12 19 ☽ ⚹ ♅
12 40 ☉ ⚹ ♆
13 11 ☽ ∥ ♆
16 06 ☽ ☌ ♄
16 50 ☽ # ♆
18 29 ☽ ⚼ ♂
19 15 ☽ ☌ ♆
19 43 ☽ ⚹ ☉
21 55 ☽ # ♄
22 42 ☽ ⚹ ♇

23 Friday
07 50 ☽ ∠ ☿
09 11 ☽ ∥ ♀
13 25 ☽ ∠ ♅
15 55 ☽ Q ♃
16 40 ♂ ⚼ ♆
18 31 ☽ Q ♇
20 13 ☽ △ ♂
21 13 ☽ ☌ ♀
22 15 ☽ ∠ ☉

24 Saturday
01 04 ☽ ⊥ ☿
04 10 ☽ ⊥ ♅
07 16 ☿ ⊻ ♃
11 44 ☽ ⚹ ♃
12 23 ☽ ⊻ ☿
13 51 ☽ ⊥ ☉
13 54 ☽ ⊻ ♅
15 55 ☉ △ ♇
17 35 ☽ ⊻ ♄
20 24 ☽ ⊻ ♆
23 15 ☿ ☌ ♅
23 36 ☽ □ ♇

25 Sunday
00 08 ☽ ⊻ ☉
03 14 ☽ ⊥ ♄
03 52 ☿ ∥ ♅
04 58 ☽ ∥ ♂
06 00 ☽ ⊥ ♆
12 06 ☽ ∠ ♃
14 57 ☽ ∥ ♅
16 25 ☽ ∥ ☿
17 38 ☽ ∠ ♄
20 20 ☽ ∠ ♆
22 06 ☽ □ ♂
22 35 ☽ ∥ ☉

26 Monday
00 25 ☽ ⊻ ♀
01 45 ☿ ⚹ ♄
02 36 ☽ ⊥ ♃
07 10 ☽ # ♇
09 10 ☽ ∥ ♃
10 32 ☽ ⊥ ♀
12 15 ☽ ⊻ ♃
13 52 ☽ ☌ ♅
17 33 ☽ ⚹ ♄
20 05 ☽ ☌ ☿
20 10 ☽ ⚹ ♆
20 37 ☿ ⚹ ♆
23 14 ☽ △ ♇

27 Tuesday
01 45 ☽ ∠ ♀
03 02 ☽ ☌ ☉
03 41 ☽ Q ♂
12 44 ☽ Q ♄
15 18 ☽ Q ♆
17 56 ☿ △ ♇
23 12 ☽ ⚼ ♇
23 35 ☽ ⚹ ♂

28 Wednesday
00 29 ☉ Q ♂
03 21 ☽ ⚹ ♀
13 01 ☽ ☌ ♃
13 46 ☽ ± ♇
14 08 ☽ ⊻ ♅
16 24 ☉ ∥ ☿
17 58 ☽ □ ♄
20 07 ♀ ⊥ ♅
20 30 ☽ □ ♆
23 36 ☽ ⊼ ♇

29 Thursday
00 01 ☽ ⊥ ♅
00 16 ☽ Q ♀
00 56 ☽ ∠ ♂
04 54 ☽ ⊻ ☿
07 02 ☽ ⊻ ☉
12 16 ☿ ∠ ♀
14 49 ☿ Q ♂
15 04 ☽ ∠ ♅
16 30 ☽ ⊥ ♂
16 44 ☽ ⊥ ☿
17 49 ☽ ⊥ ☉

30 Friday
03 05 ☽ ⊻ ♂
04 13 ☉ ☌ ☿
08 45 ☽ □ ♀
10 18 ☽ ∠ ☉
10 57 ☽ ∠ ☿
12 24 ♃ ± ♇
16 16 ☽ ⊻ ♃
16 50 ☽ ⚹ ♅
21 00 ☽ △ ♄
23 31 ☽ △ ♆

31 Saturday
01 57 ☽ ∥ ♃
02 43 ☽ ☍ ♇
02 59 ☽ ⊥ ♃
04 13 ☽ # ♇
04 16 ☽ ∥ ☿
04 50 ☿ # ♇
09 20 ☽ ∥ ☉
14 12 ☽ Q ♅
14 46 ☽ ⚹ ☉
17 48 ☿ Q ♄
18 38 ☽ ⚹ ☿
19 23 ☽ ∠ ♃
22 18 ☽ ∥ ♅
23 59 ☽ ⚼ ♄

JUNE 2025

LONGITUDES

Date	Sidereal time h m s	Sun ☉ ° ' "	Moon ☽ ° ' "	Moon ☽ 24.00 ° ' "	Mercury ☿ ° '	Venus ♀ ° '	Mars ♂ ° '	Jupiter ♃ ° '	Saturn ♄ ° '	Uranus ♅ ° '	Neptune ♆ ° '	Pluto ♇ ° '
01	04 40 54	11 ♊ 15 02	22 ♌ 00 25	28 ♌ 24 49	14 ♊ 08	25 ♈ 24	21 ♌ 19	28 ♊ 07	00 ♈ 31	28 ♉ 08	01 ♈ 52	03 ♒ 39
02	04 44 51	12 12 33	04 ♍ 43 20	10 ♍ 56 33	16 19	26 22	21 52	28 20	00 35	28 12	01 54	03 R 38
03	04 48 47	13 10 02	17 ♍ 05 07	23 ♍ 09 41	18 30	27 21	22 24	28 34	00 39	28 15	01 55	03 37
04	04 52 44	14 07 29	29 ♍ 10 56	05 ♎ 09 32	20 40	28 20	22 56	28 47	00 43	28 19	01 56	03 36
05	04 56 40	15 04 56	11 ♎ 06 07	17 ♎ 01 19	22 48	29 ♈ 19	23 28	29 01	00 46	28 22	01 56	03 36
06	05 00 37	16 02 21	22 ♎ 55 42	28 ♎ 49 50	24 55	00 ♉ 18	24 01	29 14	00 50	28 25	01 57	03 35
07	05 04 33	16 59 45	04 ♏ 44 11	10 ♏ 39 14	27 01	01 18	24 34	29 28	00 53	28 29	01 58	03 34
08	05 08 30	17 57 08	16 ♏ 35 21	22 ♏ 32 53	29 ♊ 04	02 18	25 06	29 41	00 57	28 32	01 59	03 33
09	05 12 26	18 54 31	28 ♏ 32 08	04 ♐ 33 19	01 ♋ 06	03 18	25 39	29 ♊ 55	01 00	28 35	02 00	03 32
10	05 16 23	19 51 52	10 ♐ 36 39	16 ♐ 42 15	03 05	04 19	26 12	00 ♋ 08	01 03	28 39	02 01	03 31
11	05 20 20	20 49 13	22 ♐ 50 15	29 ♐ 00 43	05 03	05 20	26 45	00 22	01 06	28 42	02 02	03 30
12	05 24 16	21 46 33	05 ♑ 13 43	11 ♑ 29 17	06 58	06 22	27 18	00 36	01 09	28 45	02 02	03 29
13	05 28 13	22 43 52	17 ♑ 47 30	24 ♑ 08 23	08 51	07 23	27 51	00 49	01 12	28 49	02 03	03 28
14	05 32 09	23 41 11	00 ♒ 32 01	06 ♒ 58 29	10 41	08 25	28 24	01 03	01 15	28 52	02 04	03 27
15	05 36 06	24 38 29	13 ♒ 27 53	20 ♒ 00 23	12 29	09 27	28 58	01 17	01 18	28 55	02 04	03 26
16	05 40 02	25 35 47	26 ♒ 36 07	03 ♓ 15 17	14 15	10 30	29 ♌ 31	01 30	01 21	28 58	02 05	03 25
17	05 43 59	26 33 04	09 ♓ 58 05	16 ♓ 44 43	15 58	11 32	00 ♍ 05	01 44	01 23	29 01	02 06	03 24
18	05 47 55	27 30 21	23 ♓ 35 22	00 ♈ 30 11	17 39	12 35	00 38	01 58	01 26	29 05	02 06	03 23
19	05 51 52	28 27 38	07 ♈ 29 17	14 ♈ 32 42	19 18	13 39	01 12	02 11	01 28	29 08	02 07	03 22
20	05 55 49	29 ♊ 24 55	21 ♈ 40 21	28 ♈ 52 04	20 54	14 42	01 46	02 25	01 30	29 11	02 07	03 21
21	05 59 45	00 ♋ 22 11	06 ♉ 07 30	13 ♉ 26 12	22 27	15 46	02 20	02 39	01 33	29 14	02 08	03 20
22	06 03 42	01 19 28	20 ♉ 47 33	28 ♉ 10 46	23 58	16 49	02 54	02 53	01 35	29 17	02 08	03 19
23	06 07 38	02 16 44	05 ♊ 34 58	12 ♊ 59 11	25 26	17 53	03 28	03 06	01 37	29 20	02 08	03 18
24	06 11 35	03 14 00	20 ♊ 22 20	27 ♊ 43 24	26 52	18 58	04 02	03 20	01 39	29 23	02 09	03 16
25	06 15 31	04 11 16	05 ♋ 01 20	12 ♋ 15 11	28 16	20 02	04 36	03 34	01 40	29 26	02 09	03 15
26	06 19 28	05 08 31	19 ♋ 24 09	26 ♋ 27 31	29 ♋ 36	21 07	05 11	03 47	01 42	29 29	02 09	03 14
27	06 23 24	06 05 47	03 ♌ 24 48	10 ♌ 15 40	00 ♌ 55	22 12	05 45	04 01	01 44	29 32	02 10	03 13
28	06 27 21	07 03 01	16 ♌ 59 56	23 ♌ 37 38	02 10	23 17	06 20	04 15	01 45	29 35	02 10	03 12
29	06 31 18	08 00 15	00 ♍ 08 55	06 ♍ 34 04	03 23	24 22	06 54	04 28	01 47	29 38	02 10	03 10
30	06 35 14	08 ♋ 57 29	12 ♍ 53 30	19 ♍ 07 43	04 ♌ 32	25 ♉ 27	07 ♍ 29	04 ♋ 42	01 ♈ 48	29 ♉ 41	02 ♈ 10	03 ♒ 09

DECLINATIONS

Date	Moon True ☊ ° '	Moon Mean ☊ ° '	Moon ☽ Latitude ° '	Sun ☉ ° '	Moon ☽ ° '	Mercury ☿ ° '	Venus ♀ ° '	Mars ♂ ° '	Jupiter ♃ ° '	Saturn ♄ ° '	Uranus ♅ ° '	Neptune ♆ ° '	Pluto ♇ ° '
01	23 ♓ 58	23 ♓ 28	02 N 41	22 N 08	16 N 42	23 N 26	07 N 48	15 N 46	23 N 15	01 S 44	19 N 32	00 S 27	22 S 52
02	23 R 57	23 25	01 40	22 15	11 20	23 49	08 06	15 34	23 15	01 42	19 33	00 26	22 53
03	23 D 57	23 22	00 N 36	22 23	05 N 39	24 10	08 24	15 23	23 15	01 41	19 34	00 26	22 53
04	23 R 57	23 19	00 S 28	22 30	00 S 06	24 28	08 43	15 12	23 16	01 40	19 35	00 26	22 53
05	23 56	23 16	01 30	22 36	05 46	24 43	09 01	15 00	23 16	01 39	19 35	00 25	22 54
06	23 53	23 13	02 27	22 42	11 11	24 56	09 20	14 48	23 16	01 37	19 36	00 25	22 54
07	23 47	23 09	03 18	22 48	16 12	25 06	09 38	14 37	23 16	01 36	19 37	00 25	22 54
08	23 39	23 06	04 00	22 54	20 38	25 13	09 57	14 25	23 16	01 35	19 38	00 24	22 55
09	23 28	23 03	04 32	22 58	24 15	25 17	10 16	14 13	23 16	01 34	19 38	00 24	22 55
10	23 16	23 00	04 53	23 03	26 52	25 19	10 35	14 01	23 17	01 33	19 39	00 24	22 55
11	23 03	22 57	05 00	23 07	28 14	25 18	10 53	13 49	23 17	01 32	19 40	00 24	22 56
12	22 49	22 54	04 53	23 11	28 13	25 15	11 12	13 36	23 17	01 31	19 41	00 23	22 56
13	22 38	22 50	04 32	23 14	26 45	25 09	11 31	13 24	23 17	01 30	19 41	00 23	22 57
14	22 28	22 47	03 57	23 17	23 53	25 02	11 50	13 12	23 17	01 29	19 42	00 23	22 57
15	22 21	22 44	03 09	23 20	19 48	24 52	12 08	12 59	23 17	01 28	19 43	00 23	22 57
16	22 17	22 41	02 11	23 22	14 42	24 41	12 27	12 47	23 17	01 27	19 43	00 22	22 58
17	22 16	22 38	01 S 05	23 24	08 50	24 27	12 46	12 34	23 17	01 27	19 44	00 22	22 58
18	22 D 16	22 34	00 N 07	23 25	02 S 26	24 12	13 04	12 21	23 16	01 26	19 45	00 22	22 59
19	22 R 15	22 31	01 20	23 26	04 N 11	23 56	13 22	12 09	23 16	01 25	19 45	00 22	22 59
20	22 14	22 28	02 29	23 26	10 45	23 38	13 41	11 56	23 16	01 25	19 46	00 22	22 59
21	22 11	22 25	03 30	23 26	16 51	23 19	13 59	11 43	23 16	01 24	19 47	00 22	23 00
22	22 05	22 22	04 18	23 26	22 05	22 58	14 17	11 30	23 16	01 23	19 48	00 22	23 00
23	21 56	22 19	04 49	23 25	25 58	22 37	14 35	11 17	23 16	01 23	19 48	00 21	23 01
24	21 46	22 15	05 01	23 24	28 05	22 14	14 52	11 03	23 15	01 22	19 49	00 21	23 01
25	21 35	22 12	04 53	23 22	28 13	21 51	15 10	10 50	23 15	01 22	19 50	00 21	23 01
26	21 25	22 09	04 26	23 20	26 25	21 27	15 27	10 37	23 15	01 21	19 50	00 21	23 02
27	21 16	22 06	03 44	23 18	23 01	21 02	15 44	10 23	23 15	01 21	19 51	00 21	23 02
28	21 10	22 03	02 50	23 15	18 26	20 37	16 01	10 10	23 14	01 21	19 51	00 21	23 03
29	21 06	22 00	01 49	23 12	13 07	20 12	16 18	09 56	23 14	01 20	19 52	00 21	23 03
30	21 ♓ 04	21 ♓ 56	00 N 43	23 N 08	07 N 23	19 N 46	16 N 34	09 N 43	23 N 13	01 S 20	19 N 53	00 S 21	23 S 04

ZODIAC SIGN ENTRIES

Date	h	m	Planets
02	03	00	☽ ♍
04	13	38	☽ ♎
06	04	43	♀ ♉
07	02	23	☽ ♏
08	22	58	☿ ♋
09	14	56	☽ ♐
09	21	02	♃ ♋
12	01	55	☽ ♑
14	11	00	☽ ♒
16	18	09	☽ ♓
17	08	35	♂ ♍
18	23	08	☽ ♈
21	01	53	☽ ♉
21	02	42	☉ ♋
23	02	57	☽ ♊
25	03	44	☽ ♋
26	19	09	☿ ♌
27	06	05	☽ ♌
29	11	43	☽ ♍

LATITUDES

Date	Mercury ☿ ° '	Venus ♀ ° '	Mars ♂ ° '	Jupiter ♃ ° '	Saturn ♄ ° '	Uranus ♅ ° '	Neptune ♆ ° '	Pluto ♇ ° '
01	00 N 57	02 S 11	01 N 27	00 S 11	02 S 07	00 S 13	01 S 18	03 S 38
04	01 22	02 19	01 24	00 10	02 07	00 13	01 18	03 39
07	01 41	02 26	01 21	00 10	02 08	00 13	01 18	03 39
10	01 55	02 32	01 18	00 10	02 09	00 13	01 18	03 40
13	02 01	02 36	01 16	00 09	02 10	00 13	01 19	03 40
16	02 01	02 39	01 13	00 09	02 10	00 13	01 19	03 41
19	01 54	02 41	01 10	00 09	02 11	00 13	01 19	03 41
22	01 41	02 41	01 08	00 09	02 12	00 13	01 19	03 41
25	01 22	02 41	01 05	00 08	02 13	00 13	01 19	03 42
28	00 58	02 40	01 03	00 08	02 13	00 13	01 19	03 42
31	00 N 29	02 S 37	01 N 00	00 S 08	02 S 14	00 S 13	01 S 19	03 S 43

DATA

Julian Date	2460828
Delta T	+72 seconds
Ayanamsa	24° 12' 45"
Synetic vernal point	04° ♓ 54' 14"
True obliquity of ecliptic	23° 26' 18"

MOON'S PHASES, APSIDES AND POSITIONS ☽

Date	h	m	Phase	Longitude ° '	Eclipse Indicator
03	03	41	☽	12 ♍ 50	
11	07	44	○	20 ♐ 39	
18	19	19	☾	27 ♓ 48	
25	10	32	●	04 ♋ 08	

Day	h	m		
07	10	38	Apogee	
23	04	35	Perigee	
04	11	35	0S	
11	23	42	Max dec	28° S 24'
18	20	52	0N	
25	01	32	Max dec	28° N 24'

LONGITUDES

Date	Chiron ⚷ ° '	Ceres ⚳ ° '	Pallas ⚴ ° '	Juno ⚵ ° '	Vesta ⚶ ° '	Black Moon Lilith ⚸ ° '
01	25 ♈ 43	04 ♈ 49	25 ♒ 03	20 ♏ 22	06 ♏ 05	07 ♏ 24
11	26 ♈ 09	07 ♈ 35	25 ♒ 13	18 ♏ 36	05 ♏ 24	08 ♏ 31
21	26 ♈ 30	10 ♈ 05	24 ♒ 48	17 ♏ 17	05 ♏ 32	09 ♏ 37
31	26 ♈ 48	12 ♈ 17	23 ♒ 45	16 ♏ 30	06 ♏ 28	10 ♏ 44

All ephemeris data is given at 12.00 UT and the Moon's longitude is additionally given for 24.00 UT

ASPECTARIAN

h m	Aspects
01 Sunday	
01 16	☿ ∥ ♃
02 31	☽ ⚼ ♆
09 08	☿ Q ♆
10 41	☽ ☌ ♂
14 30	☽ Q ☉
15 48	♃ ⚺ ♅
16 32	☽ ∥ ♂
16 42	☽ ± ♄
18 52	☽ △ ♀
19 14	☽ ± ♆
21 18	☽ Q ☿
23 32	☽ □ ♅
23 38	☽ ⚹ ♃
02 Monday	
04 04	☽ ⚻ ♄
06 35	☽ ⚻ ♆
09 55	☽ ⚻ ♇
21 27	☽ ± ♇
22 04	☉ Q ♄
23 01	☽ Q ♃
03 Tuesday	
01 03	☽ ∥ ♀
01 56	☽ ⚼ ♀
03 41	☽ □ ☉
13 17	☿ ⚼ ♇
15 01	☽ ⚼ ♇
15 23	☽ □ ☿
21 08	☽ ± ♀
22 58	☽ ⚺ ♂
04 Wednesday	
04 37	☽ # ♄
06 55	☉ Q ♆
09 48	☽ # ♆
10 08	☽ ⚻ ♀
10 15	☽ △ ♅
11 11	☽ □ ♃
11 29	☽ ⊥ ♂
11 33	♀ ⚺ ♅
13 23	☽ ∥ ♆
15 04	☽ ☍ ♄
17 30	☽ ☍ ♆
18 32	☽ ∥ ♄
20 52	☽ △ ♇
05 Thursday	
02 32	♀ ⚹ ♃
06 26	☽ ∠ ♂
16 36	☽ ⚼ ♅
20 46	☽ △ ☉
22 10	☿ ⚹ ♂
06 Friday	
03 06	☽ # ♀
10 58	☽ ± ♅
14 19	☽ ⚹ ♂
16 56	☽ △ ☿
23 13	☽ ⚻ ♅
07 Saturday	
01 04	☽ △ ♃
01 32	♀ ⚺ ♄
04 09	☽ ⚻ ♄
04 22	☽ ☍ ♀
04 24	☽ # ♂
05 57	☽ ⚼ ☉
06 23	☽ ⚻ ♆
09 37	☽ □ ♇
15 53	☽ Q ♂
16 23	☽ ± ♄
18 23	☿ ± ♇
18 34	☽ ± ♆
08 Sunday	
01 48	☽ ± ☉
04 09	♂ ± ♄
04 23	♀ ⚺ ♆
05 32	☿ ⚺ ♅
05 51	☽ ⚼ ☿
06 12	☽ # ♅
08 05	☽ ⚼ ♃
10 41	☽ ⚼ ♄
12 48	☽ ⚼ ♆
14 59	☽ ⚻ ☉
18 04	☉ # ♇
20 12	☿ ☌ ♃
21 59	☽ Q ♇
09 Monday	
02 24	☽ ∥ ♇
02 34	☽ ± ♃
02 35	☽ # ☉
02 46	☉ ⚼ ♇
03 43	☽ ± ☿
04 52	☽ # ♃
05 57	☽ □ ♂
10 48	☿ □ ♄
12 06	☽ ☍ ♅
14 48	☽ ⚻ ♃
16 56	☽ △ ♄
17 20	♀ □ ♇
18 08	☽ ⚻ ☿
18 55	☽ △ ♆
20 25	☽ # ☿
21 58	☽ ⚹ ♇
22 23	☽ ⚻ ♀
22 54	☿ □ ♆
10 Tuesday	
03 41	♂ ± ♆
11 22	☽ ± ♀
17 13	☿ ⚻ ♇
11 Wednesday	
03 32	☽ ∠ ♇
06 41	☽ ⚼ ♀
07 36	☿ ⊥ ♅
07 44	☽ ☍ ☉
19 41	☿ ⚹ ♀
19 58	☽ △ ♂
21 04	☽ ⊥ ♇
23 27	☽ ⚻ ♅
12 Thursday	
02 54	☽ ☍ ♃
04 07	☽ □ ♄
05 51	☽ □ ♆
08 39	☽ ⚺ ♇
11 05	☽ ± ♅
14 22	☽ △ ♀
15 56	☽ ☍ ☿
13 Friday	
02 10	☽ ⚼ ♂
04 24	☽ ⚼ ♅
07 33	♀ ⊥ ♄
14 41	☽ Q ♄
16 17	☽ Q ♆
20 02	☽ ± ♂
22 06	☽ ⚻ ☉
14 Saturday	
03 17	☽ # ☿
03 38	♀ ⊥ ♆
07 10	☉ ∥ ♃
07 50	☽ ⚻ ♂
08 52	☽ △ ♅
10 17	☽ ± ☉
12 59	☽ ⚻ ♃
13 21	☽ ⚹ ♄
14 51	☽ ⚹ ♆
15 58	☽ # ☉
16 04	☽ # ♃
17 27	☽ ☌ ♇
18 09	☽ ∥ ♇
15 Sunday	
00 22	☽ ± ♃
03 57	☽ □ ♀
04 23	☽ ⚼ ☉
09 47	♂ □ ♅
09 55	☽ ⚻ ☿
12 28	☽ # ♅
14 36	♃ □ ♄
17 14	☽ ∠ ♄
17 16	☽ ⚼ ♃
18 38	☽ ∠ ♆
22 40	☽ ± ☿
16 Monday	
08 01	☿ ∠ ♅
09 43	☽ ⊥ ♄
10 02	☽ △ ☉
11 04	☽ ⊥ ♆
15 43	☽ Q ♀
16 18	☽ □ ♅
17 30	☿ ∠ ♂
17 31	☽ ☍ ♂
17 31	☽ ⚼ ☿
20 27	☽ # ♂
20 35	☽ ⚺ ♄
21 01	☽ △ ♃
21 02	☽ # ♀
21 54	☽ ⚺ ♆
17 Tuesday	
00 17	☽ ⚺ ♇
03 13	♀ ∥ ♂
11 00	☽ ⊥ ♇
14 53	☉ ∠ ♀
15 02	☽ ⚹ ♀
18 Wednesday	
00 09	☽ △ ☿
00 32	☽ Q ♅
02 54	☽ ∠ ♇
09 03	☉ ± ♇
15 41	☽ ∥ ♄
19 19	☽ □ ☉
19 32	☽ ∠ ♀
19 32	☽ ∥ ♆
21 34	☽ ⚹ ♅
22 12	☽ # ♆
19 Thursday	
00 45	☽ ⚻ ♂
01 38	☽ ☌ ♄
02 02	☽ # ♄
02 46	☽ □ ♃
02 46	☽ ☌ ♆
03 16	♃ □ ♆
04 57	☽ ⚹ ♇
11 29	☽ ± ♂
12 17	☽ ⊥ ♀
23 19	☽ ⚺ ♀
23 20	☽ ∠ ♅
20 Friday	
00 08	♂ ⚻ ♄
01 22	☽ Q ♇
03 25	☽ ⚼ ♂
04 20	☽ Q ☉
05 47	☉ ⚺ ♅
09 52	☽ Q ♃
10 32	☽ □ ☿
14 32	☽ ⊥ ♅
16 20	☽ ∥ ♂
23 47	☽ ∥ ♀
21 Saturday	
00 34	☽ ⚺ ♅
01 49	☽ ⚹ ☉
02 36	☉ ∥ ☿
03 14	♂ ⚻ ♆
04 25	☽ ⚺ ♄
05 24	☽ ⚺ ♆
05 29	☽ △ ♂
06 10	☽ ⚹ ♃
07 24	☽ □ ♇
14 20	☽ ⊥ ♄
15 09	☿ ∥ ♃
15 18	☽ ⊥ ♆
19 56	☽ Q ☿
22 Sunday	
00 48	☽ ∥ ♅
04 13	☽ ∠ ☉
05 02	☽ ☌ ♀
05 07	☽ ∠ ♄
06 02	☽ ∠ ♆
06 16	♀ ∠ ♄
07 10	☽ ∠ ♃
09 46	☿ # ♇
10 32	♂ ⚹ ♃
16 21	☽ ∥ ☿
16 54	☽ # ♇
17 44	☽ ⚹ ☿
18 20	☽ ∥ ♃
18 36	☉ □ ♄
19 01	♀ ∠ ♆
19 16	☽ ∥ ☉
19 53	☽ ⊥ ☉
22 02	☽ ⊥ ♃
23 Monday	
01 50	☽ ☌ ♅
05 00	♂ ⚻ ♇
05 33	☽ ⚹ ♄
06 17	☽ ⚺ ☉
06 25	☽ ⚹ ♆
07 55	☽ ⚺ ♃
08 18	☽ △ ♇
08 26	☽ □ ♂
08 29	☉ □ ♆
18 05	♀ ∠ ♃
20 43	☽ ∠ ☿
24 Tuesday	
01 02	☽ Q ♄
01 53	☽ Q ♆
06 15	♃ ⚻ ♇
08 35	☽ ⚼ ♇
09 32	☽ ⚺ ♀
12 54	☽ ⊥ ☿
12 59	☉ ⚻ ♇
14 49	☽ Q ♂
15 17	☉ ☌ ♃
20 04	☽ ⊥ ♀
23 15	☽ ± ♇
23 44	☽ ⚺ ☿
25 Wednesday	
02 47	☽ ⚺ ♅
06 28	☽ □ ♄
07 16	☽ □ ♆
09 05	☽ ⚻ ♇
09 33	☽ ☌ ♃
10 32	☽ ☌ ☉
11 17	☽ ⚹ ♂
12 02	☽ ∠ ♀
12 41	☽ ⊥ ♅
22 28	☿ ⊥ ♂
26 Thursday	
03 42	☽ ∠ ♅
09 45	☿ ⚹ ♅
13 22	☽ ∠ ♂
14 11	☉ ⚹ ♂
15 08	☽ ⚹ ♀
21 11	☉ ⊥ ♅
27 Friday	
05 16	☽ ⚹ ♅
05 23	☽ ⊥ ♂
07 13	☽ ☌ ☿
09 04	☽ △ ♄
09 49	☽ △ ♆
10 20	☽ ∥ ☉
10 41	☽ ∥ ♃
11 39	☽ ☍ ♇
11 54	☽ # ♇
13 04	☽ ⚺ ♃
13 29	☽ Q ♀
16 15	☽ ⚺ ♂
17 02	☽ ⚺ ☉
23 46	☽ ⊥ ♃
28 Saturday	
00 06	☽ ∥ ☿
02 19	☽ Q ♅
03 53	☿ △ ♄
04 24	☽ ⊥ ☉
05 05	☽ ∥ ♅
11 34	☽ ⚼ ♄
11 58	☿ △ ♆
12 18	☽ ⚼ ♆
16 07	☽ ∠ ♃
19 10	☉ ∥ ♃
21 50	☽ ∠ ☉
22 41	☽ ∥ ♀
29 Sunday	
00 23	☽ □ ♀
03 55	☽ ± ♄
04 39	☽ ± ♆
07 57	☿ ☍ ♇
11 03	☽ □ ♅
15 02	☽ ⚻ ♄
15 45	☽ ⚻ ♆
17 37	☽ ⚻ ♇
18 37	☽ ⚺ ☿
20 13	☽ ⚹ ♃
30 Monday	
01 14	☽ ☌ ♂
01 59	☽ ∥ ♂
03 54	☽ ⚹ ☉
04 54	☽ ± ♇
05 51	☿ ∥ ♅
07 04	☽ ⊥ ☿
16 16	☿ ⚺ ♃
19 27	☽ Q ♃
22 06	☽ ⚼ ♇

JULY 2025

LONGITUDES

Date	Sidereal time h m s	Sun ☉ ° ' "	Moon ☽ ° ' "	Moon ☽ 24.00 ° ' "	Mercury ☿ ° '	Venus ♀ ° '	Mars ♂ ° '	Jupiter ♃ ° '	Saturn ♄ ° '	Uranus ♅ ° '	Neptune ♆ ° '	Pluto ♇ ° '
01	06 39 11	09 ♋ 54 43	25 ♍ 17 16	01 ♎ 22 46	05 ♌ 39	26 ♉ 33	08 ♍ 03	04 ♋ 56	01 ♈ 49	29 ♉ 44	02 ♈ 10	03 ♒ 08
02	06 43 07	10 51 55	07 ♎ 24 52	13 ♎ 24 13	06 43	27 38	08 38	05 09	01 50	29 47	02 10	03 R 07
03	06 47 04	11 49 08	19 ♎ 21 30	25 ♎ 17 23	07 44	28 44	09 13	05 23	01 51	29 50	02 10	03 05
04	06 51 00	12 46 20	01 ♏ 12 30	07 ♏ 07 29	08 42	29 ♉ 50	09 48	05 37	01 52	29 52	02 11	03 04
05	06 54 57	13 43 32	13 ♏ 02 54	18 ♏ 59 19	09 37	00 ♊ 56	10 23	05 50	01 53	29 55	02 R 11	03 03
06	06 58 53	14 40 44	24 ♏ 57 13	00 ♐ 57 03	10 28	02 03	10 58	06 04	01 54	29 ♉ 58	02 10	03 01
07	07 02 50	15 37 56	06 ♐ 59 12	13 ♐ 03 59	11 16	03 09	11 34	06 18	01 54	00 ♊ 00	02 10	03 00
08	07 06 47	16 35 08	19 ♐ 11 42	25 ♐ 22 31	12 00	04 16	12 09	06 31	01 55	00 03	02 10	02 59
09	07 10 43	17 32 20	01 ♑ 36 35	07 ♑ 53 59	12 40	05 23	12 44	06 45	01 55	00 06	02 10	02 57
10	07 14 40	18 29 31	14 ♑ 14 44	20 ♑ 38 49	13 17	06 30	13 20	06 58	01 56	00 08	02 10	02 56
11	07 18 36	19 26 43	27 ♑ 06 10	03 ♒ 36 42	13 49	07 37	13 55	07 12	01 56	00 11	02 10	02 55
12	07 22 33	20 23 56	10 ♒ 10 19	16 ♒ 46 54	14 18	08 44	14 31	07 25	01 56	00 13	02 10	02 53
13	07 26 29	21 21 08	23 ♒ 26 20	00 ♓ 08 32	14 42	09 51	15 06	07 39	01 R 56	00 16	02 09	02 52
14	07 30 26	22 18 21	06 ♓ 53 23	13 ♓ 40 50	15 02	10 59	15 42	07 52	01 56	00 18	02 09	02 51
15	07 34 22	23 15 34	20 ♓ 30 50	27 ♓ 23 21	15 17	12 06	16 18	08 06	01 56	00 21	02 09	02 49
16	07 38 19	24 12 48	04 ♈ 18 21	11 ♈ 15 50	15 28	13 14	16 53	08 19	01 56	00 23	02 08	02 48
17	07 42 16	25 10 02	18 ♈ 15 45	25 ♈ 18 02	15 33	14 22	17 29	08 32	01 55	00 26	02 08	02 46
18	07 46 12	26 07 18	02 ♉ 22 34	09 ♉ 29 12	15 R 34	15 30	18 05	08 46	01 55	00 28	02 08	02 45
19	07 50 09	27 04 34	16 ♉ 37 42	23 ♉ 47 44	15 30	16 38	18 41	08 59	01 54	00 30	02 07	02 44
20	07 54 05	28 01 50	00 ♊ 58 55	08 ♊ 10 46	15 22	17 46	19 18	09 12	01 53	00 32	02 07	02 42
21	07 58 02	28 59 08	15 ♊ 22 44	22 ♊ 34 09	15 08	18 54	19 54	09 26	01 53	00 35	02 06	02 41
22	08 01 58	29 ♋ 56 27	29 ♊ 44 23	06 ♋ 52 43	14 50	20 03	20 30	09 39	01 52	00 37	02 06	02 39
23	08 05 55	00 ♌ 53 46	13 ♋ 58 28	21 ♋ 00 59	14 27	21 11	21 06	09 52	01 51	00 39	02 05	02 38
24	08 09 51	01 51 05	27 ♋ 59 38	04 ♌ 53 54	13 59	22 20	21 43	10 05	01 50	00 41	02 04	02 36
25	08 13 48	02 48 26	11 ♌ 43 22	18 ♌ 27 43	13 28	23 29	22 19	10 18	01 48	00 43	02 04	02 35
26	08 17 45	03 45 47	25 ♌ 06 45	01 ♍ 40 24	12 53	24 37	22 56	10 31	01 47	00 45	02 03	02 34
27	08 21 41	04 43 08	08 ♍ 08 42	14 ♍ 31 48	12 15	25 46	23 32	10 45	01 46	00 47	02 02	02 32
28	08 25 38	05 40 30	20 ♍ 49 58	27 ♍ 03 32	11 34	26 55	24 09	10 58	01 44	00 49	02 02	02 31
29	08 29 34	06 37 52	03 ♎ 12 55	09 ♎ 18 35	10 51	28 05	24 46	11 11	01 43	00 51	02 01	02 29
30	08 33 31	07 35 15	15 ♎ 21 06	21 ♎ 21 01	10 07	29 ♊ 14	25 23	11 23	01 41	00 53	02 00	02 28
31	08 37 27	08 ♌ 32 39	27 ♎ 18 56	03 ♏ 15 30	09 ♌ 22	00 ♋ 23	26 ♍ 00	11 ♋ 36	01 ♈ 39	00 ♊ 54	01 ♈ 59	02 ♒ 27

Date	Moon True ☊ ° '	Moon Mean ☊ ° '	Moon ☽ Latitude ° '
01	21 ♓ 04	21 ♓ 53	00 S 22
02	21 D 04	21 50	01 26
03	21 R 04	21 47	02 24
04	21 02	21 44	03 16
05	20 58	21 40	04 00
06	20 52	21 37	04 33
07	20 44	21 34	04 54
08	20 34	21 31	05 03
09	20 23	21 28	04 57
10	20 12	21 25	04 37
11	20 02	21 21	04 03
12	19 54	21 18	03 15
13	19 49	21 15	02 16
14	19 46	21 12	01 S 08
15	19 D 46	21 09	00 N 04
16	19 46	21 06	01 17
17	19 47	21 02	02 27
18	19 R 47	20 59	03 28
19	19 45	20 56	04 17
20	19 42	20 53	04 51
21	19 36	20 50	05 06
22	19 29	20 46	05 02
23	19 21	20 43	04 39
24	19 14	20 40	04 00
25	19 08	20 37	03 07
26	19 03	20 34	02 05
27	19 01	20 31	00 N 58
28	19 D 00	20 27	00 S 10
29	19 01	20 24	01 16
30	19 03	20 21	02 18
31	19 ♓ 04	20 ♓ 18	03 S 13

DECLINATIONS

Date	Sun ☉ ° '	Moon ☽ ° '	Mercury ☿ ° '	Venus ♀ ° '	Mars ♂ ° '	Jupiter ♃ ° '	Saturn ♄ ° '	Uranus ♅ ° '	Neptune ♆ ° '	Pluto ♇ ° '
01	23 N 04	01 N 32	19 N 20	16 N 50	09 N 29	23 N 13	01 S 20	19 N 53	00 S 21	23 S 04
02	23 00	04 S 15	18 53	17 06	09 15	23 13	01 20	19 54	00 21	23 04
03	22 55	09 48	18 27	17 21	09 01	23 12	01 19	19 54	00 21	23 05
04	22 49	14 58	18 01	17 37	08 47	23 12	01 19	19 55	00 21	23 05
05	22 44	19 34	17 35	17 52	08 33	23 11	01 19	19 56	00 21	23 06
06	22 38	23 25	17 09	18 06	08 19	23 11	01 19	19 56	00 21	23 06
07	22 31	26 19	16 44	18 20	08 05	23 10	01 19	19 57	00 21	23 07
08	22 25	28 02	16 19	18 34	07 51	23 10	01 19	19 57	00 21	23 07
09	22 17	28 23	15 55	18 48	07 37	23 09	01 19	19 58	00 22	23 07
10	22 10	27 16	15 31	19 01	07 22	23 08	01 19	19 58	00 22	23 08
11	22 02	24 42	15 09	19 14	07 08	23 08	01 20	19 59	00 22	23 08
12	21 53	20 49	14 47	19 26	06 54	23 07	01 20	20 00	00 22	23 09
13	21 45	15 51	14 26	19 38	06 39	23 06	01 20	20 00	00 22	23 09
14	21 36	10 02	14 07	19 50	06 25	23 06	01 20	20 01	00 22	23 10
15	21 26	03 S 42	13 49	20 01	06 10	23 05	01 21	20 01	00 22	23 10
16	21 16	02 N 54	13 32	20 12	05 55	23 04	01 21	20 02	00 23	23 10
17	21 06	09 25	13 17	20 22	05 41	23 04	01 21	20 02	00 23	23 11
18	20 55	15 33	13 03	20 32	05 26	23 03	01 22	20 03	00 23	23 11
19	20 45	20 55	12 51	20 41	05 11	23 02	01 22	20 03	00 23	23 12
20	20 33	25 05	12 41	20 50	04 56	23 01	01 23	20 04	00 23	23 12
21	20 22	27 42	12 33	20 59	04 41	23 00	01 23	20 04	00 24	23 12
22	20 10	28 28	12 27	21 06	04 26	22 59	01 24	20 04	00 24	23 13
23	19 57	27 20	12 23	21 14	04 11	22 59	01 25	20 05	00 24	23 13
24	19 45	24 29	12 22	21 21	03 56	22 58	01 25	20 05	00 24	23 14
25	19 32	20 16	12 22	21 27	03 41	22 57	01 26	20 06	00 25	23 14
26	19 19	15 07	12 25	21 33	03 26	22 56	01 27	20 06	00 25	23 15
27	19 05	09 25	12 29	21 38	03 11	22 55	01 28	20 07	00 25	23 15
28	18 51	03 N 29	12 36	21 43	02 56	22 54	01 28	20 07	00 26	23 15
29	18 37	02 S 27	12 45	21 47	02 40	22 53	01 29	20 07	00 26	23 16
30	18 22	08 10	12 55	21 51	02 25	22 52	01 30	20 08	00 26	23 16
31	18 N 08	13 S 31	13 N 08	21 N 54	02 N 10	22 N 51	01 S 31	20 N 08	00 S 27	23 S 17

ZODIAC SIGN ENTRIES

Date	h	m	Planets
01	21	16	☽ ♎
04	09	33	☽ ♏
04	15	31	♀ ♊
06	22	06	☽ ♐
07	07	45	♅ ♊
09	08	55	☽ ♑
11	17	21	☽ ♒
13	23	45	☽ ♓
16	04	32	☽ ♈
18	07	59	☽ ♉
20	10	22	☽ ♊
22	12	26	☽ ♋
22	13	29	☉ ♌
24	15	28	☽ ♌
26	20	55	☽ ♍
29	05	43	☽ ♎
31	03	57	♀ ♋
31	17	25	☽ ♏

LATITUDES

Date	Mercury ☿ ° '	Venus ♀ ° '	Mars ♂ ° '	Jupiter ♃ ° '	Saturn ♄ ° '	Uranus ♅ ° '	Neptune ♆ ° '	Pluto ♇ ° '
01	00 N 29	02 S 37	01 N 00	00 S 08	02 S 14	00 S 13	01 S 19	03 S 43
04	00 S 04	02 34	00 58	00 07	02 15	00 13	01 20	03 43
07	00 41	02 30	00 56	00 07	02 16	00 13	01 20	03 43
10	01 22	02 25	00 53	00 07	02 17	00 13	01 20	03 44
13	02 04	02 19	00 51	00 07	02 18	00 13	01 20	03 44
16	02 47	02 12	00 49	00 06	02 18	00 13	01 20	03 44
19	03 29	02 05	00 46	00 06	02 19	00 13	01 20	03 45
22	04 07	01 58	00 44	00 06	02 20	00 13	01 21	03 45
25	04 36	01 50	00 42	00 05	02 21	00 13	01 21	03 45
28	04 54	01 41	00 40	00 05	02 22	00 13	01 21	03 46
31	04 S 57	01 S 32	00 N 37	00 S 05	02 S 22	00 S 13	01 S 21	03 S 46

DATA

Julian Date	2460858
Delta T	+72 seconds
Ayanamsa	24° 12' 51"
Synetic vernal point	04° ♓ 54' 08"
True obliquity of ecliptic	23° 26' 18"

MOON'S PHASES, APSIDES AND POSITIONS ☽

Date	h	m	Phase	Longitude ° '	Eclipse Indicator
02	19	30	☽	11 ♎ 10	
10	20	37	○	18 ♑ 50	
18	00	38	☾	25 ♈ 40	
24	19	11	●	02 ♌ 08	

Day	h	m		
05	02	26	Apogee	
20	13	44	Perigee	
01	18	17	0S	
09	05	54	Max dec	28° S 26'
16	01	30	0N	
22	09	33	Max dec	28° N 29'
29	02	02	0S	

LONGITUDES

Date	Chiron ⚷ ° '	Ceres ⚳ ° '	Pallas ⚴ ° '	Juno ⚵ ° '	Vesta ⚶ ° '	Black Moon Lilith ⚸ ° '
01	26 ♈ 48	12 ♈ 17	23 ♒ 45	16 ♏ 30	06 ♏ 28	10 ♏ 44
11	27 ♈ 00	14 ♈ 06	22 ♒ 08	16 ♏ 15	08 ♏ 07	11 ♏ 51
21	27 ♈ 08	15 ♈ 30	20 ♒ 00	16 ♏ 31	10 ♏ 22	12 ♏ 58
31	27 ♈ 10	16 ♈ 25	17 ♒ 32	17 ♏ 17	13 ♏ 08	14 ♏ 05

All ephemeris data is given at 12.00 UT and the Moon's longitude is additionally given for 24.00 UT

ASPECTARIAN

01 Tuesday
02 05 ☽ ∠ ☿
04 51 ☽ Q ☉
12 34 ☉ # ♇
12 49 ☽ # ♄
14 43 ☽ △ ♀
16 50 ☽ # ♆
19 44 ☽ ∥ ♆
20 47 ☽ △ ♅
23 46 ☽ ∥ ♄

02 Wednesday
00 54 ☽ ☍ ♄
01 34 ☽ ☍ ♆
03 27 ☽ △ ♇
07 25 ☽ □ ♃
10 29 ☽ ✱ ☿
14 34 ☽ ⊻ ♂
19 30 ☽ □ ☉
23 32 ☽ ⚼ ♀

03 Thursday
02 49 ☽ ⚼ ♅
03 14 ☽ ⊥ ♂
06 45 ♂ ± ♇
08 40 ☽ # ♂
12 50 ☽ Q ☿
19 32 ☽ ± ♀
21 04 ☽ ± ♅
22 20 ☽ ∠ ♂

04 Friday
05 47 ♀ ⊥ ♃
08 56 ☽ ⚻ ♀
09 17 ☽ ⚻ ♅
12 45 ♀ ☌ ♅
13 21 ☽ ⚻ ♄
13 58 ☽ ⚻ ♆
15 46 ☽ □ ♇
21 06 ☽ △ ♃
21 35 ♆ St R

05 Saturday
01 32 ☽ ± ♄
02 08 ☽ ± ♆
02 11 ☽ # ♀
02 12 ☽ # ☿
02 17 ☿ ∥ ♀
04 29 ☽ □ ☿
06 20 ☽ ✱ ♂
13 29 ☽ △ ☉
14 05 ☽ # ♅
19 46 ☽ ⚼ ♄
20 20 ☽ ⚼ ♆
22 22 ♃ ⊥ ♅

06 Sunday
04 02 ☽ ⚼ ♃
04 07 ☽ Q ♇
06 47 ☽ # ☉
07 49 ☽ Q ♂
08 43 ♀ ✱ ♄
09 49 ☽ ∥ ♇
10 21 ☽ # ♃
14 47 ♀ ✱ ♆
19 30 ☉ ∠ ♅
22 04 ☽ ☍ ♅
22 17 ☽ ⚼ ☉
22 26 ☽ ± ♃

07 Monday
01 54 ☽ △ ♄
02 26 ☽ △ ♆
03 37 ☽ ☍ ♀
04 06 ☽ ✱ ♇
08 44 ♀ △ ♇
10 36 ☽ ⚻ ♃
17 41 ☽ ± ☉
21 00 ☽ △ ☿
21 29 ☽ □ ♂

08 Tuesday
06 28 ☽ ⚻ ☉
09 38 ☽ ∠ ♇
13 58 ☿ Q ♅

09 Wednesday
03 04 ☽ ⊥ ♇
04 01 ☽ ⚼ ☿
09 05 ☽ ⚻ ♅
12 36 ☽ □ ♄
13 04 ☽ □ ♆
14 34 ☽ ⊻ ♇
16 15 ☿ ⊥ ♃
19 54 ☽ ⚻ ♀
20 36 ☽ ± ♅
21 59 ☽ ☍ ♃
22 10 ☽ ± ☿

10 Thursday
08 23 ☽ ± ♀
10 06 ☽ ⚻ ☿
10 11 ☽ △ ♂
13 41 ☽ ⚼ ♅
20 37 ☽ ☍ ☉
22 40 ☽ Q ♄
23 06 ☽ Q ♆

11 Friday
00 53 ♀ ⊻ ♃
01 57 ♃ # ♇
02 52 ☽ ⚼ ♀
15 31 ☽ ⚼ ♂
17 42 ☽ △ ♅
20 55 ☽ ✱ ♄
21 20 ☽ ✱ ♆
22 35 ☽ ∥ ♇
22 41 ☽ # ♃
22 42 ☽ ☌ ♇

12 Saturday
05 52 ☽ # ☉
06 54 ☽ ⚻ ♃
08 49 ☽ ± ♂
09 07 ☽ △ ♀
16 20 ☽ # ♅
18 01 ☽ ± ♃
18 52 ☽ # ♀
19 45 ☽ ☍ ☿
20 15 ☽ ⚻ ♂

13 Sunday
00 17 ☽ ∠ ♄
00 41 ☽ ∠ ♆
04 08 ♄ St R
07 58 ☽ ⚻ ☉
10 33 ☽ ⚼ ♃
16 29 ☽ ⊥ ♄
16 52 ☽ ⊥ ♆
18 28 ☽ # ☿
19 33 ☽ ± ☉

14 Monday
00 15 ☽ □ ♅
03 12 ☽ ⊻ ♄
03 35 ☽ ⊻ ♆
04 49 ☽ ⊻ ♇
12 48 ☽ ⚼ ☉
13 46 ☽ △ ♃
15 27 ☽ ⊥ ♇
19 53 ☽ □ ♀

15 Tuesday
02 28 ☽ # ♂
02 40 ☽ ⚻ ☿
04 16 ☽ ☍ ♂
07 17 ☽ ∠ ♇
08 12 ☽ Q ♅
12 05 ♀ ∥ ♅
13 22 ☽ ± ☿
17 10 ☽ △ ☉
20 36 ☽ ∥ ♄

16 Wednesday
00 09 ☽ ∥ ♆
02 52 ☽ # ♆
05 11 ☽ ✱ ♅
05 16 ☽ ⚼ ☿
06 12 ☽ Q ♀
06 24 ☽ # ♄
07 53 ☽ ☌ ♄
08 15 ☽ ☌ ♆
09 23 ☽ ✱ ♇
19 02 ☽ □ ♃
22 38 ☽ ∥ ♂

17 Thursday
02 41 ♀ Q ♄
04 44 ☽ ✱ ♀
06 02 ☽ Q ♇
07 08 ☽ ∠ ♅
07 12 ♀ Q ♆
07 21 ☽ △ ☿
10 37 ☽ ⚻ ♂
21 19 ☽ ± ♂
22 32 ☽ ⊥ ♅
22 54 ♂ ⚼ ♇

18 Friday
00 38 ☽ □ ☉
02 17 ☽ ∥ ☿
02 20 ☽ Q ♃
04 45 ☿ St R
08 32 ☽ ∠ ♀
08 45 ☽ ⊻ ♅
11 13 ☽ ⊻ ♄
11 35 ☽ ⊻ ♆
12 38 ☽ □ ♇
13 16 ☽ ⚼ ♂
13 37 ☿ ✱ ♀
21 20 ☽ ⊥ ♄
21 42 ☽ ⊥ ♆
22 57 ☽ ✱ ♃

19 Saturday
01 03 ☽ ⊥ ♀
07 49 ☽ ∥ ♅
09 13 ☽ Q ☉
10 08 ☽ □ ☿
10 52 ☽ ∥ ♀
11 12 ☽ ∥ ☉
12 00 ☽ ⊻ ♀
12 27 ☽ ∠ ♄
12 49 ☽ ∠ ♆
15 36 ☽ △ ♂
15 55 ☉ ∥ ♀
23 10 ☽ ∥ ♃

20 Sunday
00 08 ☽ # ♇
00 31 ☽ ∠ ♃
06 43 ☽ ✱ ☉
10 42 ♀ ⚼ ♇
11 16 ☽ ☌ ♅
13 31 ☽ ✱ ♄
13 53 ☽ ✱ ♆
14 52 ☽ △ ♇
15 46 ☽ ⊥ ♃
15 55 ☽ Q ☿
20 42 ☿ ⊥ ♃

21 Monday
01 56 ☽ ⊻ ♃
09 30 ☽ Q ♄
09 31 ☽ ∠ ☉
09 52 ☽ Q ♆
11 36 ☽ ✱ ☿
15 50 ☽ ⚼ ♇
18 23 ☽ ☌ ♀
19 52 ☽ □ ♂

22 Tuesday
01 36 ☽ ⊥ ☉
06 50 ☽ ± ♇
12 09 ☽ ∠ ☿
12 22 ☽ ⊻ ☉
13 28 ☽ ⊻ ♅
15 33 ☽ □ ♄
15 57 ☽ □ ♆
16 53 ☽ ⚻ ♇
20 09 ☿ ⊥ ♂
22 04 ☉ ∥ ♅
23 35 ☽ ⊥ ♅

23 Wednesday
02 54 ☽ ⊥ ☿
03 23 ☽ Q ♂
04 56 ☽ ☌ ♃
05 32 ☉ ✱ ♅
08 23 ♀ □ ♂
12 46 ☽ ⊻ ☿
14 51 ☽ ∠ ♅

24 Thursday
00 42 ☽ ✱ ♂
01 23 ☽ ⊻ ♀
11 23 ☉ △ ♄
12 38 ☽ ⊥ ♀
16 40 ☽ ✱ ♅
17 31 ☉ △ ♆
18 38 ☽ △ ♄
19 04 ☽ △ ♆
19 11 ☽ ☌ ☉
19 50 ☽ # ♇
19 59 ☽ ☍ ♇
21 28 ☽ ∥ ♃

25 Friday
03 53 ☽ ∠ ♂
05 45 ☽ ∠ ♀
06 00 ☽ ∥ ♀
06 33 ☉ ☍ ♇
09 27 ☽ ⊻ ♃
11 08 ♂ Q ♃
12 52 ☽ ∥ ♅
13 46 ☽ Q ♅
14 58 ☽ ☌ ☿
15 51 ☽ ∥ ☉
20 17 ☽ ⊥ ♃
20 33 ☽ ⊥ ♂
21 01 ☽ ⚼ ♄
21 29 ☽ ⚼ ♆

26 Saturday
07 51 ☽ ⊻ ♂
11 02 ☽ ✱ ♀
12 46 ☽ ∠ ♃
13 13 ☽ ± ♄
13 42 ☽ ± ♆
16 54 ☿ Q ♅
22 20 ☽ □ ♅
23 28 ☽ ∥ ☿

27 Sunday
00 11 ☽ ⚻ ♄
00 41 ☽ ⚻ ♆
01 37 ☽ ⚻ ♇
05 07 ☽ ⊻ ☉
11 14 ☽ Q ♀
12 44 ☽ ± ♇
16 57 ☽ ✱ ♃
17 12 ☽ ⊥ ☉
19 18 ☽ ⊻ ☿

28 Monday
03 36 ♀ ± ♇
05 40 ☽ ⚼ ♇
06 05 ☽ ⊥ ☿
07 20 ☿ ∠ ♀
11 40 ☽ ∠ ☉
14 20 ☽ ∥ ♂
16 09 ☽ Q ♃
18 43 ☽ ☌ ♂
20 04 ☽ # ♄
22 27 ☽ ∠ ☿

29 Tuesday
00 17 ☽ # ♆
00 57 ☽ □ ♀
03 43 ☿ ⊻ ♃
03 47 ☽ ∥ ♆
07 21 ☽ △ ♅
08 05 ☽ ∥ ♄
09 04 ☽ ☍ ♄
09 39 ☽ ☍ ♆
10 35 ☽ △ ♇
12 54 ☽ # ♂
19 17 ☽ ✱ ☉

30 Wednesday
02 12 ☽ ✱ ☿
03 59 ☽ □ ♃
07 22 ☿ ∠ ♂
13 03 ☽ ⚼ ♅
21 12 ☽ Q ☉

31 Thursday
00 45 ☽ Q ☿
07 08 ☽ ± ♅
09 11 ☽ ⊻ ♂
10 07 ☽ # ☿
18 52 ☽ △ ♀
19 16 ☽ ⚻ ♅
20 44 ☽ ⚻ ♄
21 25 ☽ ⚻ ♆
21 58 ☽ ⊥ ♂
22 20 ☽ □ ♇
23 03 ♀ ⊻ ♅
23 41 ☉ ☌ ☿

LONGITUDES

Date	Sidereal time h m s	Sun ☉ ° ' "	Moon ☽ ° ' "	Moon ☽ 24.00 ° ' "	Mercury ☿ ° '	Venus ♀ ° '	Mars ♂ ° '	Jupiter ♃ ° '	Saturn ♄ ° '	Uranus ♅ ° '	Neptune ♆ ° '	Pluto ♇ ° '
01	08 41 24	09 ♌ 30 03	09 ♏ 11 19	15 ♏ 07 00	08 ♌ 38	01 ♋ 33	26 ♍ 37	11 ♋ 49	01 ♈ 37	00 ♊ 56	01 ♈ 59	02 ♒ 25
02	08 45 20	10 27 28	21 ♏ 03 13	27 ♏ 00 31	07 R 55	02 42	27 14	12 02	01 R 35	00 58	01 R 58	02 R 24
03	08 49 17	11 24 53	02 ♐ 59 30	09 ♐ 00 42	07 13	03 52	27 51	12 15	01 33	01 00	01 57	02 22
04	08 53 14	12 22 19	15 ♐ 04 37	21 ♐ 11 42	06 34	05 02	28 28	12 27	01 31	01 01	01 56	02 21
05	08 57 10	13 19 46	27 ♐ 22 19	03 ♑ 36 48	05 59	06 11	29 05	12 40	01 29	01 03	01 55	02 20
06	09 01 07	14 17 14	09 ♑ 55 25	16 ♑ 18 19	05 28	07 21	29 ♍ 42	12 53	01 27	01 04	01 54	02 18
07	09 05 03	15 14 43	22 ♑ 45 36	29 ♑ 17 18	05 02	08 31	00 ♎ 20	13 05	01 24	01 06	01 53	02 17
08	09 09 00	16 12 12	05 ♒ 53 21	12 ♒ 33 37	04 41	09 42	00 57	13 18	01 22	01 07	01 52	02 16
09	09 12 56	17 09 42	19 ♒ 17 55	26 ♒ 05 58	04 26	10 52	01 35	13 30	01 19	01 09	01 51	02 14
10	09 16 53	18 07 14	02 ♓ 57 29	09 ♓ 52 07	04 17	12 02	02 12	13 42	01 17	01 10	01 50	02 13
11	09 20 49	19 04 46	16 ♓ 49 31	23 ♓ 49 18	04 D 15	13 12	02 50	13 55	01 14	01 11	01 49	02 11
12	09 24 46	20 02 20	00 ♈ 51 06	07 ♈ 54 34	04 20	14 23	03 27	14 07	01 11	01 12	01 48	02 10
13	09 28 43	20 59 55	14 ♈ 59 20	22 ♈ 05 04	04 32	15 33	04 05	14 19	01 08	01 14	01 47	02 09
14	09 32 39	21 57 32	29 ♈ 11 28	06 ♉ 18 13	04 51	16 44	04 43	14 31	01 05	01 15	01 45	02 08
15	09 36 36	22 55 10	13 ♉ 25 02	20 ♉ 31 39	05 17	17 55	05 21	14 43	01 02	01 16	01 44	02 06
16	09 40 32	23 52 50	27 ♉ 37 47	04 ♊ 43 11	05 50	19 06	05 59	14 55	00 59	01 17	01 43	02 05
17	09 44 29	24 50 31	11 ♊ 47 33	18 ♊ 50 36	06 30	20 17	06 37	15 07	00 56	01 18	01 42	02 04
18	09 48 25	25 48 14	25 ♊ 52 03	02 ♋ 51 36	07 18	21 28	07 15	15 19	00 52	01 19	01 40	02 02
19	09 52 22	26 45 59	09 ♋ 48 55	16 ♋ 43 43	08 12	22 39	07 53	15 31	00 49	01 20	01 39	02 01
20	09 56 18	27 43 45	23 ♋ 35 41	00 ♌ 24 32	09 13	23 50	08 31	15 43	00 46	01 21	01 38	02 00
21	10 00 15	28 41 33	07 ♌ 10 00	13 ♌ 51 51	10 20	25 01	09 09	15 55	00 42	01 22	01 37	01 59
22	10 04 12	29 ♌ 39 22	20 ♌ 29 53	27 ♌ 03 58	11 34	26 12	09 48	16 06	00 38	01 22	01 35	01 57
23	10 08 08	00 ♍ 37 12	03 ♍ 33 59	09 ♍ 59 55	12 53	27 24	10 26	16 18	00 35	01 23	01 34	01 56
24	10 12 05	01 35 04	16 ♍ 21 48	22 ♍ 39 42	14 18	28 35	11 04	16 29	00 31	01 24	01 33	01 55
25	10 16 01	02 32 58	28 ♍ 53 48	05 ♎ 04 18	15 47	29 ♋ 47	11 43	16 40	00 27	01 24	01 31	01 54
26	10 19 58	03 30 52	11 ♎ 11 28	17 ♎ 15 39	17 22	00 ♌ 58	12 21	16 52	00 24	01 25	01 30	01 53
27	10 23 54	04 28 48	23 ♎ 17 13	29 ♎ 16 37	19 00	02 10	13 00	17 03	00 20	01 25	01 28	01 52
28	10 27 51	05 26 46	05 ♏ 14 17	11 ♏ 10 45	20 43	03 22	13 39	17 14	00 16	01 26	01 27	01 50
29	10 31 47	06 24 44	17 ♏ 06 33	23 ♏ 02 14	22 28	04 34	14 18	17 25	00 12	01 26	01 25	01 49
30	10 35 44	07 22 45	28 ♏ 58 23	04 ♐ 55 34	24 16	05 46	14 56	17 36	00 08	01 27	01 24	01 48
31	10 39 41	08 ♍ 20 46	10 ♐ 54 23	16 ♐ 55 25	26 ♌ 07	06 ♌ 58	15 ♎ 35	17 ♋ 47	00 ♈ 03	01 ♊ 27	01 ♈ 22	01 ♒ 47

Date	Moon True ☊ ° '	Moon Mean ☊ ° '	Moon ☽ Latitude ° '
01	19 ♓ 05	20 ♓ 15	03 S 59
02	19 R 04	20 12	04 35
03	19 02	20 08	04 59
04	18 58	20 05	05 10
05	18 53	20 02	05 07
06	18 47	19 59	04 50
07	18 42	19 56	04 18
08	18 37	19 52	03 32
09	18 33	19 49	02 33
10	18 30	19 46	01 24
11	18 30	19 43	00 S 09
12	18 D 30	19 40	01 N 07
13	18 31	19 37	02 20
14	18 33	19 33	03 25
15	18 34	19 30	04 17
16	18 R 34	19 27	04 54
17	18 33	19 24	05 12
18	18 31	19 21	05 12
19	18 29	19 17	04 53
20	18 26	19 14	04 17
21	18 23	19 11	03 28
22	18 21	19 08	02 27
23	18 19	19 05	01 20
24	18 19	19 02	00 N 11
25	18 D 19	18 58	00 S 58
26	18 20	18 55	02 03
27	18 21	18 52	03 02
28	18 22	18 49	03 51
29	18 23	18 46	04 31
30	18 24	18 43	04 59
31	18 ♓ 24	18 ♓ 39	05 S 14

DECLINATIONS

Date	Sun ☉ ° '	Moon ☽ ° '	Mercury ☿ ° '	Venus ♀ ° '	Mars ♂ ° '	Jupiter ♃ ° '	Saturn ♄ ° '	Uranus ♅ ° '	Neptune ♆ ° '	Pluto ♇ ° '
01	17 N 52	18 S 20	13 N 21	21 N 56	01 N 54	22 N 50	01 S 32	20 N 08	00 S 27	23 S 17
02	17 37	22 26	13 37	21 58	01 39	22 49	01 33	20 09	00 28	23 17
03	17 21	25 38	13 53	22 00	01 24	22 48	01 34	20 09	00 28	23 18
04	17 05	27 44	14 10	22 00	01 08	22 47	01 35	20 09	00 28	23 18
05	16 49	28 32	14 27	22 01	00 53	22 46	01 36	20 10	00 29	23 19
06	16 33	27 53	14 45	22 00	00 37	22 45	01 37	20 10	00 29	23 19
07	16 16	25 45	15 03	21 59	00 22	22 43	01 39	20 10	00 30	23 19
08	15 59	22 13	15 21	21 58	00 N 06	22 42	01 40	20 11	00 30	23 20
09	15 41	17 27	15 39	21 55	00 S 09	22 41	01 41	20 11	00 30	23 20
10	15 24	11 43	15 56	21 53	00 25	22 40	01 42	20 11	00 31	23 20
11	15 06	05 S 21	16 12	21 49	00 41	22 39	01 44	20 12	00 31	23 21
12	14 48	01 N 22	16 26	21 45	00 56	22 38	01 45	20 12	00 32	23 21
13	14 30	08 04	16 40	21 41	01 12	22 36	01 46	20 12	00 32	23 21
14	14 11	14 23	16 52	21 36	01 27	22 35	01 48	20 12	00 33	23 22
15	13 53	19 57	17 03	21 30	01 43	22 34	01 49	20 12	00 33	23 22
16	13 34	24 23	17 11	21 23	01 59	22 33	01 51	20 13	00 34	23 22
17	13 14	27 21	17 18	21 17	02 14	22 32	01 52	20 13	00 34	23 23
18	12 55	28 34	17 22	21 09	02 30	22 30	01 53	20 13	00 35	23 23
19	12 35	27 56	17 24	21 01	02 46	22 29	01 55	20 13	00 36	23 23
20	12 16	25 36	17 23	20 52	03 02	22 28	01 57	20 13	00 36	23 24
21	11 56	21 49	17 20	20 43	03 17	22 26	01 58	20 14	00 37	23 24
22	11 36	16 59	17 14	20 33	03 33	22 25	02 00	20 14	00 37	23 24
23	11 15	11 27	17 05	20 22	03 49	22 24	02 01	20 14	00 38	23 25
24	10 55	05 N 33	16 54	20 11	04 04	22 23	02 03	20 14	00 38	23 25
25	10 34	00 S 27	16 39	19 59	04 20	22 21	02 05	20 14	00 39	23 25
26	10 13	06 19	16 21	19 47	04 36	22 20	02 06	20 14	00 39	23 26
27	09 52	11 51	16 01	19 34	04 51	22 19	02 08	20 14	00 40	23 26
28	09 31	16 54	15 37	19 21	05 07	22 17	02 10	20 14	00 41	23 26
29	09 10	21 16	15 11	19 07	05 23	22 16	02 11	20 15	00 41	23 26
30	08 48	24 47	14 42	18 53	05 38	22 15	02 13	20 15	00 42	23 27
31	08 N 26	27 S 15	14 N 11	18 N 38	05 S 54	22 N 13	02 S 15	20 N 15	00 S 42	23 S 27

ZODIAC SIGN ENTRIES

Date	h	m	Planets
03	06	00	☽ ♐
05	17	04	☽ ♑
06	23	23	♂ ♎
08	01	18	☽ ♒
10	06	50	☽ ♓
12	10	33	☽ ♈
14	13	22	☽ ♉
16	16	01	☽ ♊
18	19	05	☽ ♋
20	23	17	☽ ♌
22	20	34	☉ ♍
23	05	24	☽ ♍
25	14	08	☽ ♎
25	16	27	♀ ♌
28	01	27	☽ ♏
30	14	04	☽ ♐

LATITUDES

Date	Mercury ☿ ° '	Venus ♀ ° '	Mars ♂ ° '	Jupiter ♃ ° '	Saturn ♄ ° '	Uranus ♅ ° '	Neptune ♆ ° '	Pluto ♇ ° '
01	04 S 55	01 S 29	00 N 37	00 S 05	02 S 23	00 S 13	01 S 21	03 S 46
04	04 37	01 20	00 34	00 05	02 23	00 13	01 21	03 46
07	04 04	01 11	00 32	00 04	02 24	00 13	01 21	03 46
10	03 21	01 01	00 30	00 04	02 25	00 13	01 21	03 46
13	02 32	00 51	00 28	00 04	02 25	00 13	01 21	03 47
16	01 41	00 42	00 26	00 03	02 26	00 13	01 22	03 47
19	00 51	00 32	00 24	00 03	02 27	00 13	01 22	03 47
22	00 S 05	00 22	00 22	00 03	02 27	00 13	01 22	03 47
25	00 N 34	00 13	00 19	00 03	02 28	00 13	01 22	03 47
28	01 05	00 S 03	00 17	00 02	02 28	00 13	01 22	03 47
31	01 N 28	00 N 06	00 N 15	00 S 02	02 S 29	00 S 13	01 S 22	03 S 47

DATA

Julian Date	2460889
Delta T	+72 seconds
Ayanamsa	24° 12' 56"
Synetic vernal point	04° ♓ 54' 03"
True obliquity of ecliptic	23° 26' 18"

MOON'S PHASES, APSIDES AND POSITIONS ☽

Date	h	m	Phase	Longitude ° '	Eclipse Indicator
01	12	41	☽	09 ♏ 32	
09	07	55	○	17 ♒ 00	
16	05	12	☾	23 ♉ 36	
23	06	07	●	00 ♍ 23	
31	06	25	☽	08 ♐ 07	

Day	h	m		
01	20	36	Apogee	
14	18	08	Perigee	
29	15	35	Apogee	
05	13	30	Max dec	28° S 32'
12	07	09	0N	
18	15	48	Max dec	28° N 35'
25	10	11	0S	

LONGITUDES

Date	Chiron ⚷ ° '	Ceres ⚳ ° '	Pallas ⚴ ° '	Juno ⚵ ° '	Vesta ⚶ ° '	Black Moon Lilith ⚸ ° '
01	27 ♈ 10	16 ♈ 29	17 ♒ 17	17 ♏ 23	13 ♏ 27	14 ♏ 12
11	27 ♈ 06	16 ♈ 48	14 ♒ 40	18 ♏ 38	16 ♏ 42	15 ♏ 19
21	26 ♈ 57	16 ♈ 31	12 ♒ 11	20 ♏ 16	20 ♏ 19	16 ♏ 26
31	26 ♈ 44	15 ♈ 38	10 ♒ 01	22 ♏ 14	24 ♏ 15	17 ♏ 33

ASPECTARIAN

h m	Aspects
01 Friday	
08 50	☽ ± ♄
09 33	☽ ± ♆
09 43	☽ # ☉
10 56	☽ □ ☿
12 41	☽ □ ☉
13 34	♀ □ ♄
17 10	☽ ∠ ♂
17 25	☽ △ ♃
20 49	♀ □ ♆
22 03	☽ # ♅
02 Saturday	
01 49	☿ ⊥ ♀
03 00	☽ ⚼ ♄
03 44	☽ ⚼ ♆
04 30	☽ ⚼ ♀
05 46	♀ ⊼ ♇
09 01	☽ # ♀
10 41	☽ Q ♇
14 30	☽ # ♃
17 45	☽ ∥ ♇
20 42	♂ # ♄
03 Sunday	
00 16	☽ ⚼ ♃
00 37	☽ ± ♀
01 07	☽ ✶ ♂
07 59	☽ ☍ ♅
09 08	☽ △ ♄
09 55	☽ △ ♆
10 46	☽ ✶ ♇
13 56	☽ ⊼ ♀
18 37	☽ ± ♃
20 00	☽ △ ☿
04 Monday	
02 23	☽ Q ♂
06 12	☽ △ ☉
06 44	☽ ⊼ ♃
14 43	☉ ⊻ ♃
16 28	☽ ∠ ♇
05 Tuesday	
00 09	☽ ⚼ ☿
04 41	☉ Q ♅
09 09	☿ ⊻ ♀
09 59	☽ ⊥ ♇
14 00	☽ ⚼ ☉
15 29	☽ □ ♂
16 50	☽ ± ☿
19 06	☽ ⊼ ♅
19 54	☽ □ ♄
20 44	☽ □ ♆
21 31	☽ ⊻ ♇
06 Wednesday	
03 52	☽ ⊼ ☿
05 59	♀ ⊥ ♅
06 35	☽ ± ♅
06 38	☽ ☍ ♀
08 39	☽ ± ☉
17 40	☽ ☍ ♃
20 53	☽ ⊼ ☉
23 35	☽ ⚼ ♅
07 Thursday	
00 09	♂ # ♆
05 48	☽ Q ♄
06 41	☽ Q ♆
17 53	☉ ∠ ♂
08 Friday	
02 35	☽ △ ♂
03 20	☽ △ ♅
03 49	☽ ✶ ♄
04 43	☽ ✶ ♆
05 22	☽ ∥ ♇
05 26	☽ ☌ ♇
09 08	☽ # ♃
09 52	☽ ☍ ☿
13 28	☽ # ♀
15 51	☉ ⚼ ♄
18 45	♂ △ ♅
19 31	☽ ⊼ ♀
22 57	☽ # ♅
09 Saturday	
01 31	☽ ⊼ ♃
02 52	♂ ☍ ♄
04 18	☉ ⚼ ♆
06 44	☽ ∠ ♄
06 56	☽ ⚼ ♂
07 16	☽ ± ♀
07 39	☽ ∠ ♆
07 55	☽ ☍ ☉
12 22	☽ ± ♃
13 56	☉ ∥ ☿
19 34	☽ # ☿
20 11	☽ # ☉
22 13	♂ ☍ ♆
22 36	☽ ⊥ ♄
23 33	☽ ⊥ ♆
23 37	☽ ± ♂
10 Sunday	
00 40	☽ ⚼ ♀
04 28	☽ ⚼ ♃
08 52	☽ □ ♅
09 05	☽ ⊻ ♄
10 02	☽ ⊻ ♆
10 37	☽ ⊼ ♂
10 42	☽ ⊻ ♇
12 29	♂ △ ♇
14 18	☽ ⊼ ☿
21 07	☽ ⊥ ♇
21 31	♂ ∥ ♆
21 49	☉ ⊥ ♀
11 Monday	
00 40	☽ ± ☿
05 12	☽ △ ♀
06 55	☽ △ ♃
07 29	☿ St D
12 38	☽ ∠ ♇
16 04	☽ Q ♅
16 10	☽ ⊼ ☉
16 11	☽ ⚼ ☿
12 Tuesday	
00 57	☽ ∥ ♄
03 11	☽ ± ☉
03 32	♄ ✶ ♅
04 07	☽ ∥ ♂
05 16	☽ ∥ ♆
05 30	♀ ☌ ♃
09 02	☽ # ♆
10 25	☽ # ♂
12 34	☽ ☌ ♄
12 36	☽ ✶ ♅
13 22	☽ # ♄
13 36	☽ ☌ ♆
14 14	☽ ✶ ♇
14 30	☉ ⊥ ♃
16 38	☽ ☍ ♂
17 59	☽ △ ☿
19 39	☽ ⚼ ☉
13 Wednesday	
10 35	☽ Q ♇
10 51	☽ □ ♃
13 03	☽ □ ♀
14 06	☽ ∠ ♅
22 54	☽ △ ☉
14 Thursday	
01 55	♀ ∠ ♅
05 20	☽ ⊥ ♅
11 17	☽ ∥ ☉
15 11	☽ ⊻ ♄
15 28	☽ ⊻ ♅
16 19	☽ ⊻ ♆
16 57	☽ □ ♇
17 42	☽ Q ♃
21 45	☽ ⊼ ♂
21 49	☽ □ ☿
22 12	☽ Q ♀
22 36	☽ ∥ ☿
15 Friday	
01 16	☽ ⊥ ♄
02 05	☿ ✶ ♂
02 26	☽ ⊥ ♆
08 20	☽ ± ♂
13 14	☽ ∥ ♅
14 14	☽ ✶ ♃
16 24	☽ ∠ ♄
17 36	☽ ∠ ♆
19 27	☽ ∥ ♀
20 16	☽ ✶ ♀
22 08	♂ ∥ ♄
16 Saturday	
00 14	☽ ⚼ ♂
01 12	☽ ∥ ♃
05 12	☽ □ ☉
05 18	☽ Q ☿
05 48	☽ # ♇
15 56	☽ ∠ ♃
17 39	☽ ✶ ♄
18 11	☽ ☌ ♅
18 54	☽ ✶ ♆
19 31	☽ △ ♇
23 56	☽ ∠ ♀
17 Sunday	
02 33	☽ ✶ ☿
02 47	☽ △ ♂
07 24	☽ ⊥ ♃
13 55	☽ Q ☉
13 55	☽ Q ♄
14 02	☉ ± ♄
15 14	☽ Q ♆
16 36	☽ ⊥ ♀
17 45	☽ ⊻ ♃
20 57	☽ ⚼ ♇
18 Monday	
03 46	☽ ⊻ ♀
05 31	☽ ∠ ☿
05 33	☿ ✶ ♂
08 50	☉ ± ♆
11 53	☽ ✶ ☉
12 18	☽ ± ♇
20 33	☽ □ ♄
21 21	☽ ⊻ ♅
21 56	☽ ⊥ ☿
21 57	☽ □ ♆
22 34	☽ ⊼ ♇
19 Tuesday	
07 42	☽ ⊥ ♅
08 30	☽ □ ♂
09 00	☽ ⊻ ☿
15 38	☽ ∠ ☉
22 02	☽ ☌ ♃
23 19	☽ ∠ ♅
20 Wednesday	
08 29	☽ ⊥ ☉
12 27	☽ ☌ ♀
17 23	☽ Q ♂
19 49	☽ ⊻ ☉
21 Thursday	
00 34	☽ △ ♄
01 40	☽ ✶ ♅
02 08	☽ △ ♆
02 47	☽ ☍ ♇
02 55	☽ # ♇
08 32	☽ ∥ ♃
15 44	☽ ✶ ♂
18 09	☽ ∥ ♀
18 13	☽ ☌ ☿
20 25	☽ ∥ ♅
23 06	☽ Q ♅
22 Friday	
03 15	☽ ⚼ ♄
03 55	☽ ⊻ ♃
04 55	☽ ⚼ ♆
10 49	☽ ∥ ☿
14 58	☽ ⊥ ♃
19 31	☽ ± ♄
20 14	☽ ∠ ♂
21 16	☽ ± ♆
23 28	☽ ⊻ ♀
23 Saturday	
06 07	☽ ☌ ☉
06 30	☽ ⊼ ♄
07 44	☽ ∠ ♃
07 57	☽ □ ♅
08 18	☽ ⊼ ♆
08 59	☽ ⊼ ♇
11 05	☉ ⊼ ♄
11 39	☽ ⊥ ♀
12 52	☽ ∥ ☉
13 41	☽ ⊥ ♂
20 08	☽ ± ♇
20 45	☿ Q ♅
24 Sunday	
00 01	♃ ∠ ♅
01 29	☽ ⊻ ♂
05 53	♀ ∥ ♅
06 12	☽ ∠ ♀
07 15	☉ □ ♅
07 35	☽ ⊻ ☿
08 55	☉ ∠ ♃
10 58	☉ ⊼ ♆
12 14	☽ ✶ ♃
13 03	☽ ⚼ ♇
17 39	☽ # ♂
20 06	☉ ⊼ ♇
20 28	☽ ⊥ ☿
25 Monday	
01 55	☽ # ♄
06 56	☿ ⚼ ♄
07 36	☽ # ♆
09 13	♀ Q ♂
11 34	☽ Q ♃
12 48	☽ ∥ ♆
13 53	☽ ✶ ♀
15 00	☽ ☍ ♄
16 12	☽ ∠ ☿
16 52	☽ △ ♅
17 04	☽ ☍ ♆
17 48	☽ △ ♇
18 36	☽ ∥ ♄
19 41	☽ ⊻ ☉
23 04	☿ ⚼ ♆
26 Tuesday	
00 56	♀ △ ♄
03 27	☿ ⊻ ♃
04 31	☽ ∥ ♂
08 25	☽ ⊥ ☉
14 26	☽ ☌ ♂
15 54	☽ Q ♀
20 58	♀ ✶ ♅
22 18	♀ △ ♆
22 20	☽ ⚼ ♅
23 23	☽ □ ♃
27 Wednesday	
02 06	☽ ✶ ☿
03 42	☽ # ☉
03 45	☽ ∠ ☉
05 54	♀ ☍ ♇
16 17	☽ ± ♅
28 Thursday	
02 02	☽ ⊼ ♄
04 20	☽ ⊼ ♅
04 23	☽ ⊼ ♆
05 10	☽ □ ♇
06 03	☽ Q ☿
06 08	☽ # ☿
07 48	☽ □ ♀
12 27	☽ ✶ ☉
14 03	☽ ± ♄
16 27	☽ ± ♆
29 Friday	
00 09	♅ ✶ ♆
00 19	☽ # ♀
05 58	☽ ⊻ ♂
05 58	☽ # ♅
08 09	☽ ⚼ ♄
10 37	☽ ⚼ ♆
12 38	☽ △ ♃
14 52	☽ Q ☉
17 29	☽ Q ♇
18 11	☽ # ♃
18 49	☽ ⊥ ♂
30 Saturday	
00 47	☽ □ ☿
02 04	☽ ∥ ♇
02 08	☿ ⊥ ♃
10 08	☿ ± ♄
14 04	☽ ∠ ♂
14 19	☽ △ ♄
16 53	☽ △ ♆
16 59	☽ ☍ ♅
17 42	☽ ✶ ♇
19 26	☽ ⚼ ♃
22 19	☉ ± ♇
31 Sunday	
02 31	☿ ± ♆
03 12	☽ △ ♀
06 25	☽ □ ☉
13 47	☽ ± ♃
21 53	☽ ✶ ♂
23 42	☽ ∠ ♇

All ephemeris data is given at 12.00 UT and the Moon's longitude is additionally given for 24.00 UT

SEPTEMBER 2025

LONGITUDES

Date	Sidereal time h m s	Sun ☉ ° ′ ″	Moon ☽ ° ′ ″	Moon ☽ 24.00 ° ′ ″	Mercury ☿ ° ′	Venus ♀ ° ′	Mars ♂ ° ′	Jupiter ♃ ° ′	Saturn ♄ ° ′	Uranus ♅ ° ′	Neptune ♆ ° ′	Pluto ♇ ° ′
01	10 43 37	09 ♍ 18 49	22 ♐ 59 14	29 ♐ 06 23	27 ♌ 59	08 ♌ 10	16 ♎ 14	17 ♋ 58	29 ♓ 59	01 ♊ 27	01 ♈ 21	01 ♒ 46
02	10 47 34	10 16 53	05 ♑ 17 21	11 ♑ 32 38	29 ♌ 53	09 22	16 53	18 08	29 R 55	01 27	01 R 19	01 R 45
03	10 51 30	11 14 59	17 ♑ 52 38	24 ♑ 17 41	01 ♍ 48	10 34	17 32	18 19	29 51	01 28	01 18	01 44
04	10 55 27	12 13 06	00 ♒ 48 03	07 ♒ 23 56	03 44	11 46	18 12	18 29	29 47	01 28	01 16	01 43
05	10 59 23	13 11 15	14 ♒ 05 22	20 ♒ 52 21	05 40	12 58	18 51	18 40	29 42	01 28	01 15	01 42
06	11 03 20	14 09 25	27 ♒ 44 43	04 ♓ 42 14	07 37	14 11	19 30	18 50	29 38	01 R 28	01 13	01 41
07	11 07 16	15 07 36	11 ♓ 44 30	18 ♓ 51 03	09 33	15 23	20 09	19 00	29 33	01 28	01 12	01 40
08	11 11 13	16 05 50	26 ♓ 01 19	03 ♈ 14 39	11 29	16 36	20 49	19 11	29 29	01 28	01 10	01 39
09	11 15 10	17 04 05	10 ♈ 30 20	17 ♈ 47 37	13 25	17 48	21 28	19 21	29 25	01 28	01 08	01 38
10	11 19 06	18 02 22	25 ♈ 05 47	02 ♉ 24 03	15 20	19 01	22 08	19 30	29 20	01 27	01 07	01 37
11	11 23 03	19 00 41	09 ♉ 41 43	16 ♉ 58 07	17 14	20 14	22 47	19 40	29 16	01 27	01 05	01 36
12	11 26 59	19 59 02	24 ♉ 12 40	01 ♊ 24 51	19 07	21 27	23 27	19 50	29 11	01 27	01 04	01 36
13	11 30 56	20 57 26	08 ♊ 34 14	15 ♊ 40 28	21 00	22 39	24 07	19 59	29 06	01 26	01 02	01 35
14	11 34 52	21 55 51	22 ♊ 43 17	29 ♊ 42 29	22 52	23 52	24 46	20 09	29 02	01 26	01 00	01 34
15	11 38 49	22 54 19	06 ♋ 37 59	13 ♋ 29 40	24 42	25 05	25 26	20 18	28 57	01 26	00 59	01 33
16	11 42 45	23 52 49	20 ♋ 17 34	27 ♋ 01 40	26 32	26 18	26 06	20 28	28 53	01 25	00 57	01 33
17	11 46 42	24 51 21	03 ♌ 42 02	10 ♌ 18 44	28 ♍ 21	27 32	26 46	20 37	28 48	01 25	00 55	01 32
18	11 50 39	25 49 55	16 ♌ 51 50	23 ♌ 21 27	00 ♎ 09	28 45	27 26	20 46	28 43	01 24	00 54	01 31
19	11 54 35	26 48 32	29 ♌ 47 40	06 ♍ 10 35	01 55	29 ♌ 58	28 06	20 55	28 39	01 23	00 52	01 30
20	11 58 32	27 47 10	12 ♍ 30 19	18 ♍ 46 57	03 41	01 ♍ 11	28 46	21 03	28 34	01 23	00 51	01 30
21	12 02 28	28 45 50	25 ♍ 00 39	01 ♎ 11 31	05 25	02 25	29 ♎ 27	21 12	28 29	01 22	00 49	01 29
22	12 06 25	29 ♍ 44 32	07 ♎ 19 43	13 ♎ 25 25	07 09	03 38	00 ♏ 07	21 21	28 25	01 21	00 47	01 29
23	12 10 21	00 ♎ 43 16	19 ♎ 28 48	25 ♎ 30 06	08 52	04 52	00 47	21 29	28 20	01 20	00 46	01 28
24	12 14 18	01 42 02	01 ♏ 29 34	07 ♏ 27 29	10 33	06 05	01 28	21 37	28 15	01 19	00 44	01 27
25	12 18 14	02 40 50	13 ♏ 24 10	19 ♏ 19 58	12 14	07 19	02 08	21 45	28 11	01 19	00 42	01 27
26	12 22 11	03 39 40	25 ♏ 15 18	01 ♐ 10 34	13 54	08 32	02 49	21 53	28 06	01 18	00 41	01 26
27	12 26 08	04 38 31	07 ♐ 06 14	13 ♐ 02 48	15 32	09 46	03 29	22 01	28 01	01 17	00 39	01 26
28	12 30 04	05 37 24	19 ♐ 00 47	25 ♐ 00 43	17 10	11 00	04 10	22 09	27 57	01 15	00 37	01 25
29	12 34 01	06 36 19	01 ♑ 03 10	07 ♑ 08 43	18 47	12 13	04 51	22 17	27 52	01 14	00 36	01 25
30	12 37 57	07 ♎ 35 16	13 ♑ 17 54	19 ♑ 31 19	20 ♎ 23	13 ♍ 27	05 ♏ 31	22 ♋ 24	27 ♓ 48	01 ♊ 13	00 ♈ 34	01 ♒ 25

Date	Moon True ☊ ° ′	Moon Mean ☊ ° ′	Moon ☽ Latitude ° ′
01	18 ♓ 24	18 ♓ 36	05 S 16
02	18 R 23	18 33	05 03
03	18 22	18 30	04 36
04	18 21	18 27	03 54
05	18 21	18 23	02 59
06	18 20	18 20	01 52
07	18 20	18 17	00 S 36
08	18 D 20	18 14	00 N 43
09	18 20	18 11	02 00
10	18 20	18 08	03 10
11	18 R 20	18 04	04 08
12	18 20	18 01	04 49
13	18 20	17 58	05 12
14	18 D 20	17 55	05 16
15	18 20	17 52	05 01
16	18 20	17 49	04 29
17	18 21	17 45	03 43
18	18 22	17 42	02 46
19	18 22	17 39	01 41
20	18 23	17 36	00 N 32
21	18 R 23	17 33	00 S 37
22	18 22	17 29	01 43
23	18 21	17 26	02 44
24	18 19	17 23	03 36
25	18 17	17 20	04 19
26	18 15	17 17	04 50
27	18 12	17 14	05 09
28	18 11	17 10	05 15
29	18 10	17 07	05 07
30	18 ♓ 10	17 ♓ 04	04 S 46

DECLINATIONS

Sun ☉ ° ′	Moon ☽ ° ′	Mercury ☿ ° ′	Venus ♀ ° ′	Mars ♂ ° ′	Jupiter ♃ ° ′	Saturn ♄ ° ′	Uranus ♅ ° ′	Neptune ♆ ° ′	Pluto ♇ ° ′
08 N 05	28 S 30	13 N 38	18 N 22	06 S 10	22 N 12	02 S 17	20 N 15	00 S 43	23 S 27
07 43	28 23	13 02	18 06	06 25	22 11	02 19	20 15	00 44	23 28
07 21	26 48	12 24	17 50	06 41	22 09	02 20	20 15	00 44	23 28
06 59	23 47	11 45	17 33	06 56	22 08	02 22	20 15	00 45	23 28
06 36	19 27	11 04	17 15	07 12	22 07	02 24	20 15	00 46	23 28
06 14	14 00	10 22	16 57	07 27	22 06	02 26	20 15	00 46	23 28
05 52	07 43	09 39	16 38	07 43	22 04	02 28	20 15	00 47	23 29
05 29	00 S 56	08 54	16 19	07 58	22 03	02 30	20 15	00 48	23 29
05 06	06 N 00	08 09	16 00	08 14	22 02	02 31	20 15	00 48	23 29
04 44	12 40	07 23	15 40	08 29	22 00	02 33	20 15	00 49	23 29
04 21	18 38	06 37	15 20	08 44	21 59	02 35	20 15	00 50	23 30
03 58	23 30	05 50	14 59	09 00	21 58	02 37	20 15	00 50	23 30
03 35	26 52	05 03	14 38	09 15	21 56	02 39	20 15	00 51	23 30
03 12	28 30	04 15	14 16	09 30	21 55	02 41	20 14	00 52	23 30
02 49	28 17	03 28	13 54	09 45	21 54	02 43	20 14	00 52	23 30
02 26	26 20	02 40	13 31	10 00	21 52	02 45	20 14	00 53	23 30
02 03	22 56	01 52	13 08	10 15	21 51	02 47	20 14	00 54	23 31
01 39	18 25	01 05	12 45	10 30	21 50	02 49	20 14	00 54	23 31
01 16	13 07	00 N 17	12 22	10 45	21 49	02 50	20 14	00 55	23 31
00 53	07 22	00 S 30	11 58	11 00	21 47	02 52	20 14	00 56	23 31
00 29	01 N 25	01 17	11 33	11 15	21 46	02 54	20 14	00 56	23 31
00 N 06	04 S 29	02 03	11 09	11 30	21 45	02 56	20 13	00 57	23 31
00 S 17	10 09	02 49	10 44	11 45	21 44	02 58	20 13	00 58	23 31
00 41	15 22	03 35	10 18	11 59	21 43	03 00	20 13	00 58	23 32
01 04	19 58	04 21	09 53	12 14	21 41	03 02	20 13	00 59	23 32
01 27	23 46	05 06	09 27	12 29	21 40	03 04	20 13	01 00	23 32
01 51	26 35	05 50	09 01	12 43	21 39	03 05	20 13	01 00	23 32
02 14	28 13	06 34	08 34	12 57	21 38	03 07	20 12	01 01	23 32
02 37	28 33	07 18	08 08	13 12	21 37	03 09	20 12	01 02	23 32
03 S 01	27 S 31	08 S 00	07 N 41	13 S 26	21 N 36	03 S 11	20 N 12	01 S 02	23 S 32

ZODIAC SIGN ENTRIES

Date	h	m	Planets	
01	08	07	♄	♓
02	01	45	☽	♑
02	13	23	☿	♍
04	10	32	☽	♒
06	15	54	☽	♓
08	18	37	☽	♈
10	20	03	☽	♉
12	21	38	☽	♊
15	00	30	☽	♋
17	05	20	☽	♌
18	10	06	☿	♎
19	12	23	☽	♍
19	12	39	♀	♍
21	21	41	☽	♎
22	07	54	♂	♏
22	18	19	☉	♎
24	09	00	☽	♏
26	21	37	☽	♐
29	09	55	☽	♑

LATITUDES

Date	Mercury ☿ ° ′	Venus ♀ ° ′	Mars ♂ ° ′	Jupiter ♃ ° ′	Saturn ♄ ° ′	Uranus ♅ ° ′	Neptune ♆ ° ′	Pluto ♇ ° ′
01	01 N 33	00 N 09	00 N 15	00 S 02	02 S 29	00 S 13	01 S 22	03 S 47
04	01 44	00 18	00 13	00 02	02 29	00 13	01 22	03 47
07	01 47	00 26	00 11	00 01	02 30	00 13	01 22	03 47
10	01 44	00 34	00 09	00 01	02 30	00 13	01 22	03 47
13	01 36	00 42	00 07	00 S 01	02 30	00 13	01 22	03 47
16	01 24	00 49	00 05	00 00	02 30	00 13	01 22	03 47
19	01 09	00 56	00 03	00 00	02 31	00 13	01 22	03 47
22	00 51	01 02	00 N 01	00 00	02 31	00 13	01 22	03 47
25	00 32	01 08	00 S 01	00 N 01	02 31	00 13	01 22	03 47
28	00 N 11	01 13	00 03	00 01	02 31	00 13	01 22	03 47
31	00 S 10	01 N 18	00 S 05	00 N 01	02 S 31	00 S 13	01 S 22	03 S 47

DATA

Julian Date	2460920
Delta T	+72 seconds
Ayanamsa	24° 13′ 00″
Synetic vernal point	04° ♓ 53′ 59″
True obliquity of ecliptic	23° 26′ 19″

LONGITUDES

Date	Chiron ⚷ ° ′	Ceres ⚳ ° ′	Pallas ⚴ ° ′	Juno ⚵ ° ′	Vesta ⚶ ° ′	Black Moon Lilith ⚸ ° ′
01	26 ♈ 42	15 ♈ 31	09 ♒ 49	22 ♏ 27	24 ♏ 39	17 ♏ 40
11	26 ♈ 23	14 ♈ 01	08 ♒ 12	24 ♏ 45	28 ♏ 52	18 ♏ 47
21	26 ♈ 01	12 ♈ 04	07 ♒ 08	27 ♏ 17	03 ♐ 16	19 ♏ 54
31	25 ♈ 36	09 ♈ 51	06 ♒ 41	00 ♐ 03	07 ♐ 52	21 ♏ 01

MOON'S PHASES, APSIDES AND POSITIONS ☽

Date	h	m	Phase	Longitude ° ′	Eclipse Indicator
07	18	09	○	15 ♓ 23	total
14	10	33	☾	21 ♊ 52	
21	19	54	●	29 ♍ 05	Partial
29	23	54	☽	07 ♑ 06	

Day	h	m		
10	12	19	Perigee	
26	09	51	Apogee	
01	21	55	Max dec	28° S 38′
08	15	13	0N	
14	21	08	Max dec	28° N 38′
21	17	44	0S	
29	05	58	Max dec	28° S 36′

ASPECTARIAN

h m Aspects

01 Monday
01 55 ☽ ⊼ ♃
12 23 ☽ ꝗ ♀
17 28 ☽ ⊥ ♇
20 02 ♂ ꝗ ♅
22 53 ☽ Q ♂
23 37 ☽ △ ☿

02 Tuesday
01 39 ☽ □ ♄
04 20 ☽ □ ♆
04 35 ☽ ⊼ ♅
05 09 ☽ ⊻ ♇
07 52 ☽ ± ♀
12 21 ☿ ⊼ ♄
16 11 ☽ ± ♅
20 40 ☽ ⊼ ♀
22 23 ☽ △ ☉

03 Wednesday
05 44 ☿ ⊼ ♆
07 40 ☿ □ ♅
09 20 ☽ ꝗ ♅
09 37 ☽ ꝗ ☿
11 05 ☿ ⊼ ♇
11 20 ☽ □ ♂
11 57 ☽ Q ♄
12 50 ☽ ☍ ♃
14 40 ☽ Q ♆

04 Thursday
01 47 ☿ ∠ ♂
04 53 ☽ ꝗ ☉
05 23 ☽ ± ☿
08 39 ☿ ∠ ♃
10 01 ☉ ⊥ ♂
10 08 ☽ ⚹ ♄
12 52 ☽ ⚹ ♆
13 13 ☽ △ ♅
13 28 ☉ # ♂
13 40 ☽ ☌ ♇
14 01 ☽ ∥ ♇
18 17 ☽ ⊼ ☿
22 00 ☽ # ♃
22 39 ☽ ± ☉

05 Friday
02 59 ♂ □ ♃
08 02 ☽ # ♅
09 49 ☽ ☍ ♀
10 16 ☽ ⊼ ☉
13 05 ☽ ∠ ♄
15 50 ☽ ∠ ♆
20 13 ☽ ⊼ ♃
20 51 ☽ △ ♂
21 45 ♀ Q ♅
22 49 ☽ # ♀

06 Saturday
04 52 ♅ St R
04 52 ☽ ⊥ ♄
06 53 ☽ ± ♃
07 37 ☽ ⊥ ♆
09 38 ☉ ⊻ ♀
12 52 ☿ ± ♇
15 15 ☽ ⊻ ♄
18 00 ☽ ⊻ ♆
18 26 ☽ □ ♅
18 48 ☽ ⊻ ♇
20 27 ♀ ꝗ ♄
22 39 ☽ ꝗ ♃

07 Sunday
00 13 ☽ ꝗ ♂
04 00 ☽ # ☿
05 05 ☽ ⊥ ♇
07 41 ☽ ☍ ☿
12 02 ☽ ∥ ♂
16 17 ☽ ± ♂
18 09 ☽ ☍ ☉
18 44 ☽ ⊼ ♀
19 06 ☽ # ☉
20 19 ☽ ∠ ♇

08 Monday
00 25 ☽ △ ♃
01 02 ☽ Q ♅
02 52 ☽ ⊼ ♂
03 41 ♀ ꝗ ♆
05 45 ☽ ± ♀
06 36 ☽ ∥ ♄
12 29 ☽ ∥ ♆
17 44 ☽ ☌ ♄
17 58 ☽ # ♆
20 33 ☽ ☌ ♆
21 03 ☽ ⚹ ♅
21 21 ☽ ⚹ ♇
22 07 ☽ ꝗ ♀
23 53 ☽ # ♄

09 Tuesday
01 31 ☉ ꝗ ♇
09 03 ☽ ∥ ☉
10 14 ☿ # ♂
17 09 ☽ Q ♇
17 31 ☽ ⊼ ☿
18 49 ☽ ∥ ☿
20 10 ☽ # ♂
21 48 ☽ ∠ ♅
23 35 ☽ ⊼ ☉

10 Wednesday
01 06 ☽ △ ♀
02 43 ☽ □ ♃
04 52 ☽ ± ☿
06 54 ☽ ☍ ♂
10 08 ☽ ± ☉
12 35 ☽ ⊥ ♅
18 56 ☽ ⊻ ♄
21 52 ☽ ⊻ ♆
21 53 ☽ ꝗ ☿
22 27 ☽ ⊻ ♅
22 43 ☽ □ ♇
23 01 ☽ ∥ ♀
23 12 ♀ ⊻ ♃

11 Thursday
01 59 ☽ ꝗ ☉
03 24 ☿ ⊥ ♂
04 11 ☿ ꝗ ♇
04 44 ☽ ⊥ ♄
07 43 ☽ ⊥ ♆
08 38 ☽ Q ♃
19 17 ☽ ∥ ♅
19 29 ☽ ∠ ♄
22 32 ☽ ∠ ♆

12 Friday
02 18 ☽ △ ☿
03 43 ☽ ∥ ♃
04 29 ☽ △ ☉
04 39 ☽ ⚹ ♃
06 59 ☽ □ ♀
07 30 ☉ ⚹ ♃
10 40 ☽ ⊼ ♂
12 00 ☽ # ♇
20 14 ☽ ⚹ ♄
21 08 ☽ ± ♂
21 53 ☿ ⚹ ♃
23 23 ☽ ⚹ ♆

13 Saturday
00 03 ☽ ☌ ♅
00 17 ☽ △ ♇
05 55 ☽ ∠ ♃
10 52 ☉ ☌ ☿
12 57 ☽ ꝗ ♂
15 51 ☽ Q ♀
16 15 ☽ Q ♄
19 31 ☽ Q ♆
20 21 ♀ ± ♄
21 15 ☽ ⊥ ♃

14 Sunday
01 32 ☽ ꝗ ♇
07 34 ☽ ⊻ ♃
10 33 ☽ □ ☉
12 17 ☽ □ ☿
14 10 ☽ ⚹ ♀
15 41 ☽ △ ♂
16 52 ☽ ± ♇
22 46 ☽ □ ♄

15 Monday
02 13 ☽ □ ♆
02 58 ☽ ⊻ ♅
03 12 ☽ ⊼ ♇
09 52 ♀ ± ♆
11 40 ♂ ± ♅
13 23 ☽ ⊥ ♅
17 53 ☉ # ♄
18 37 ☽ ∠ ♀
20 02 ☽ Q ☉

16 Tuesday
00 16 ☽ Q ☿
02 59 ☿ ⊻ ♀
03 01 ☿ ⊻ ♂
03 04 ♀ ⚹ ♂
05 09 ☽ ∠ ♅
09 39 ☿ # ♄
12 02 ☽ ⊥ ♀
12 18 ☽ ☌ ♃
15 26 ♀ ⊥ ♃
18 53 ☽ ⚹ ☉
22 53 ☽ □ ♂
23 47 ☽ ⊻ ♀

17 Wednesday
00 52 ☽ ⚹ ☿
01 48 ☉ ∥ ☿
03 14 ☽ △ ♄
07 00 ☽ △ ♆
07 52 ☽ ⚹ ♅
08 05 ☽ ☍ ♇
08 31 ☽ # ♇
17 47 ☿ ☍ ♄
18 16 ☽ ∥ ♃

18 Thursday
00 04 ☽ ∠ ☉
02 55 ☽ ∥ ♅
05 39 ☽ Q ♅
06 16 ☽ ꝗ ♄
08 20 ☽ ∠ ☿
09 14 ☽ Q ♂
10 14 ☽ ꝗ ♆
11 34 ♀ ⊼ ♄
17 16 ☿ # ♆
17 55 ☽ ⊥ ☉
19 17 ☽ ⊻ ♃
22 01 ☿ ☍ ♆
22 45 ☽ ± ♄

19 Friday
02 50 ☽ ± ♆
03 37 ☽ ⊥ ☿
04 52 ☿ △ ♅
05 58 ☽ ⊻ ☉
06 27 ☿ △ ♇
06 33 ☽ ⊥ ♃
08 40 ☽ ⚹ ♂
09 52 ☽ ⊼ ♄
12 21 ☽ ☌ ♀
14 01 ☽ ⊼ ♆
14 59 ☽ □ ♅
15 12 ☽ ⊼ ♇
15 31 ☽ ∥ ♀
16 37 ☽ ⊻ ☿
21 37 ☽ # ♂
23 38 ☽ ∠ ♃

20 Saturday
02 30 ☽ ± ♇
02 43 ☿ Q ♃
05 20 ♀ ⊼ ♆
05 24 ♂ ⊼ ♄
09 18 ☉ # ♆
14 33 ☽ ∠ ♂
15 41 ♀ □ ♅
17 59 ♀ ⊼ ♇
19 37 ☽ ꝗ ♇
19 52 ☉ # ☿

21 Sunday
01 21 ☿ ∥ ♆
04 34 ☽ ⚹ ♃
05 46 ☉ ☍ ♄
06 04 ☽ # ♄
08 48 ☽ ⊥ ♂
12 31 ☽ # ☿
13 57 ☽ # ♆
16 00 ☽ ∥ ☉
18 42 ☽ ☍ ♄
19 14 ☽ # ☉
19 54 ☽ ☌ ☉
21 05 ☽ ⊻ ♂
21 32 ☽ ∥ ♆
23 02 ♀ # ♂
23 14 ☽ ☍ ♆

22 Monday
00 20 ☽ △ ♅
00 33 ☽ ∥ ☿
00 34 ☽ △ ♇
03 58 ☽ ⊻ ♀
04 06 ☽ Q ♃
05 36 ☽ ∥ ♄
11 36 ☽ ☌ ☿
17 02 ☽ ⊥ ♀

23 Tuesday
05 46 ☽ ꝗ ♅
11 01 ♂ ⊼ ♆
12 50 ☽ ∠ ♀
12 53 ☉ ☍ ♆
14 23 ☽ # ♀
16 02 ☽ □ ♃
16 38 ☿ ∥ ♄
17 10 ☉ ⊻ ♂
19 27 ☽ ∥ ♂
23 40 ☽ ± ♅

24 Wednesday
02 55 ☉ △ ♅
05 33 ☽ ⊼ ♄
06 05 ☉ △ ♇
07 15 ♂ ⊼ ♅
10 28 ☽ ⊼ ♆
11 40 ☽ ⊼ ♅
11 52 ♂ □ ♇
11 56 ☽ □ ♇
11 56 ☽ ☌ ♂
12 27 ☽ ⊻ ☉
17 31 ☽ ± ♄
22 18 ☽ ⚹ ♀
22 31 ☽ ± ♆
23 49 ♀ ∠ ♃

25 Thursday
01 37 ☽ ⊥ ☉
06 37 ☉ ∥ ♆
09 15 ☽ ⊻ ☿
11 33 ☽ ꝗ ♄
13 22 ☽ # ♅
14 40 ♀ ± ♇
16 38 ☽ ꝗ ♆
21 26 ☽ ∠ ☉
22 06 ☽ # ♃
23 22 ☽ ⊥ ☿

26 Friday
00 13 ☽ Q ♇
01 20 ☽ Q ♀
05 06 ☽ △ ♃
10 16 ☽ ∥ ♇
17 44 ☽ △ ♄
18 30 ☉ Q ♃
20 34 ☽ ∠ ☿
22 58 ☽ △ ♆

27 Saturday
00 13 ☽ ☍ ♅
00 32 ☽ ⚹ ♇
04 14 ☽ ⊻ ♂
06 34 ☽ ⚹ ☉
11 50 ☽ ꝗ ♃
17 07 ☽ ⊥ ♂
18 00 ☽ □ ♀
22 39 ☿ ꝗ ♅

28 Sunday
01 15 ☿ ⊥ ♀
05 05 ♂ ± ♄
06 11 ☽ ± ♃
06 48 ☽ ∠ ♇
07 43 ☽ ⚹ ☿
08 58 ☽ Q ☉
12 19 ☽ ∠ ♂
18 21 ☽ ⊼ ♃

29 Monday
00 49 ☽ ⊥ ♇
05 44 ☽ □ ♄
11 05 ☽ □ ♆
11 24 ☽ Q ☿
12 22 ☽ ⊼ ♅
12 43 ☽ ⊻ ♇
19 55 ☽ ⚹ ♂
23 54 ☽ □ ☉

30 Tuesday
00 10 ☽ ± ♅
05 14 ☿ # ♀
12 20 ☽ △ ♀
16 48 ☽ Q ♄
17 38 ☽ ꝗ ♅
20 38 ☽ Q ♂
22 09 ☽ Q ♆
23 24 ☉ ∥ ♄

All ephemeris data is given at 12.00 UT and the Moon's longitude is additionally given for 24.00 UT

OCTOBER 2025

Date	Sidereal time h m s	LONGITUDES Sun ☉ ° ' "	Moon ☽ ° ' "	Moon ☽ 24.00 ° ' "	Mercury ☿ ° '	Venus ♀ ° '	Mars ♂ ° '	Jupiter ♃ ° '	Saturn ♄ ° '	Uranus ♅ ° '	Neptune ♆ ° '	Pluto ♇ ° '
01	12 41 54	08 ♎ 34 15	25 ♑ 49 30	02 ♒ 12 57	21 ♎ 59	14 ♍ 41	06 ♏ 12	22 ♋ 31	27 ♓ 43	01 ♊ 12	00 ♈ 32	01 ♒ 24
02	12 45 50	09 33 15	08 ♒ 42 06	15 ♒ 17 22	23 33	15 55	06 53	22 39	27 R 39	01 R 11	00 R 31	01 R 24
03	12 49 47	10 32 17	21 ♒ 59 02	28 ♒ 47 17	25 07	17 09	07 34	22 46	27 34	01 09	00 29	01 24
04	12 53 43	11 31 21	05 ♓ 42 11	12 ♓ 43 39	26 39	18 23	08 15	22 53	27 30	01 08	00 27	01 23
05	12 57 40	12 30 26	19 ♓ 51 25	27 ♓ 05 04	28 11	19 37	08 56	22 59	27 25	01 07	00 26	01 23
06	13 01 37	13 29 34	04 ♈ 24 01	11 ♈ 47 31	29 ♎ 42	20 51	09 37	23 06	27 21	01 05	00 24	01 23
07	13 05 33	14 28 43	19 ♈ 14 38	26 ♈ 44 20	01 ♏ 13	22 05	10 19	23 12	27 17	01 04	00 23	01 23
08	13 09 30	15 27 55	04 ♉ 15 32	11 ♉ 47 02	02 42	23 20	11 00	23 19	27 12	01 02	00 21	01 22
09	13 13 26	16 27 08	19 ♉ 17 43	26 ♉ 46 27	04 11	24 34	11 41	23 25	27 08	01 01	00 19	01 22
10	13 17 23	17 26 24	04 ♊ 12 14	11 ♊ 34 12	05 39	25 48	12 23	23 31	27 04	00 59	00 18	01 22
11	13 21 19	18 25 43	18 ♊ 51 37	26 ♊ 03 56	07 06	27 02	13 04	23 37	27 00	00 58	00 16	01 22
12	13 25 16	19 25 03	03 ♋ 10 45	10 ♋ 11 50	08 32	28 17	13 46	23 42	26 56	00 56	00 15	01 22
13	13 29 12	20 24 26	17 ♋ 07 08	23 ♋ 56 41	09 57	29 ♍ 31	14 27	23 48	26 51	00 54	00 13	01 22
14	13 33 09	21 23 52	00 ♌ 40 39	07 ♌ 19 16	11 22	00 ♎ 46	15 09	23 53	26 47	00 52	00 12	01 D 22
15	13 37 06	22 23 19	13 ♌ 52 51	20 ♌ 21 43	12 45	02 00	15 51	23 58	26 43	00 51	00 10	01 22
16	13 41 02	23 22 49	26 ♌ 46 17	03 ♍ 06 53	14 08	03 15	16 32	24 03	26 40	00 49	00 09	01 22
17	13 44 59	24 22 21	09 ♍ 23 55	15 ♍ 37 43	15 29	04 29	17 14	24 08	26 36	00 47	00 07	01 22
18	13 48 55	25 21 55	21 ♍ 48 38	27 ♍ 56 59	16 50	05 44	17 56	24 13	26 32	00 45	00 06	01 22
19	13 52 52	26 21 32	04 ♎ 03 03	10 ♎ 07 06	18 10	06 58	18 38	24 17	26 28	00 43	00 04	01 22
20	13 56 48	27 21 10	16 ♎ 09 22	22 ♎ 10 03	19 28	08 13	19 20	24 21	26 25	00 41	00 03	01 23
21	14 00 45	28 20 51	28 ♎ 09 22	04 ♏ 07 31	20 45	09 28	20 02	24 26	26 21	00 39	00 01	01 23
22	14 04 41	29 ♎ 20 33	10 ♏ 04 41	16 ♏ 01 03	22 01	10 43	20 45	24 30	26 17	00 37	00 ♈ 00	01 23
23	14 08 38	00 ♏ 20 18	21 ♏ 56 51	27 ♏ 52 16	23 15	11 57	21 27	24 33	26 14	00 35	29 ♓ 58	01 23
24	14 12 35	01 20 04	03 ♐ 47 36	09 ♐ 43 04	24 28	13 12	22 09	24 37	26 11	00 33	29 57	01 24
25	14 16 31	02 19 53	15 ♐ 39 02	21 ♐ 35 49	25 39	14 27	22 51	24 40	26 07	00 31	29 56	01 24
26	14 20 28	03 19 43	27 ♐ 33 47	03 ♑ 33 24	26 48	15 42	23 34	24 44	26 04	00 29	29 54	01 24
27	14 24 24	04 19 35	09 ♑ 35 06	15 ♑ 39 23	27 55	16 57	24 16	24 47	26 01	00 27	29 53	01 25
28	14 28 21	05 19 28	21 ♑ 46 48	27 ♑ 57 52	29 ♏ 00	18 12	24 59	24 49	25 58	00 25	29 52	01 25
29	14 32 17	06 19 24	04 ♒ 13 10	10 ♒ 33 16	00 ♐ 02	19 26	25 41	24 52	25 55	00 22	29 51	01 25
30	14 36 14	07 19 20	16 ♒ 58 44	23 ♒ 30 05	01 02	20 41	26 24	24 55	25 52	00 20	29 49	01 26
31	14 40 10	08 ♏ 19 19	00 ♓ 07 49	06 ♓ 52 18	01 ♐ 58	21 ♎ 56	27 ♏ 07	24 ♋ 57	25 ♓ 49	00 ♊ 18	29 ♓ 48	01 ♒ 26

Date	Moon True ☊ ° '	Moon Mean ☊ ° '	Moon ☽ Latitude ° '
01	18 ♓ 11	17 ♓ 01	04 S 10
02	18 D 12	16 58	03 21
03	18 14	16 55	02 20
04	18 15	16 51	01 S 09
05	18 R 15	16 48	00 N 09
06	18 15	16 45	01 28
07	18 13	16 42	02 42
08	18 09	16 39	03 46
09	18 05	16 35	04 34
10	18 01	16 32	05 04
11	17 58	16 29	05 13
12	17 56	16 26	05 02
13	17 54	16 23	04 33
14	17 D 55	16 20	03 50
15	17 56	16 16	02 55
16	17 58	16 13	01 53
17	17 59	16 10	00 N 47
18	17 R 59	16 07	00 S 21
19	17 58	16 04	01 26
20	17 54	16 01	02 27
21	17 49	15 57	03 20
22	17 42	15 54	04 05
23	17 34	15 51	04 38
24	17 25	15 48	04 59
25	17 17	15 45	05 07
26	17 10	15 41	05 03
27	17 05	15 38	04 44
28	17 02	15 35	04 13
29	17 01	15 32	03 30
30	17 D 01	15 29	02 34
31	17 ♓ 02	15 ♓ 26	01 S 30

DECLINATIONS

Date	Sun ☉ ° '	Moon ☽ ° '	Mercury ☿ ° '	Venus ♀ ° '	Mars ♂ ° '	Jupiter ♃ ° '	Saturn ♄ ° '	Uranus ♅ ° '	Neptune ♆ ° '	Pluto ♇ ° '
01	03 S 24	25 S 04	08 S 43	07 N 14	13 S 40	21 N 35	03 S 13	20 N 12	01 S 03	23 S 32
02	03 47	21 19	09 24	06 46	13 54	21 34	03 14	20 11	01 03	23 32
03	04 10	16 23	10 06	06 19	14 08	21 33	03 16	20 11	01 04	23 32
04	04 33	10 29	10 46	05 51	14 22	21 32	03 18	20 11	01 05	23 32
05	04 56	03 S 53	11 26	05 23	14 36	21 31	03 20	20 11	01 05	23 32
06	05 20	03 N 05	12 05	04 55	14 49	21 30	03 21	20 10	01 06	23 32
07	05 42	10 02	12 43	04 27	15 03	21 29	03 23	20 10	01 07	23 32
08	06 05	16 29	13 21	03 58	15 17	21 28	03 25	20 10	01 07	23 32
09	06 28	21 57	13 58	03 30	15 30	21 27	03 26	20 09	01 08	23 32
10	06 51	25 57	14 34	03 01	15 43	21 26	03 28	20 09	01 09	23 32
11	07 13	28 10	15 09	02 32	15 56	21 25	03 30	20 09	01 09	23 32
12	07 36	28 26	15 44	02 03	16 10	21 24	03 31	20 08	01 10	23 32
13	07 58	26 51	16 17	01 34	16 22	21 23	03 33	20 08	01 10	23 32
14	08 21	23 45	16 50	01 05	16 35	21 22	03 34	20 08	01 11	23 32
15	08 43	19 27	17 22	00 36	16 48	21 22	03 36	20 07	01 12	23 32
16	09 05	14 22	17 53	00 N 06	17 01	21 21	03 37	20 07	01 12	23 32
17	09 27	08 46	18 23	00 S 23	17 13	21 20	03 39	20 06	01 13	23 32
18	09 49	02 N 56	18 52	00 52	17 26	21 20	03 40	20 06	01 13	23 32
19	10 10	02 S 56	19 20	01 22	17 38	21 19	03 41	20 06	01 14	23 32
20	10 32	08 37	19 47	01 51	17 50	21 18	03 43	20 05	01 14	23 32
21	10 53	13 56	20 13	02 20	18 02	21 18	03 44	20 05	01 15	23 32
22	11 14	18 42	20 38	02 50	18 14	21 17	03 45	20 04	01 16	23 32
23	11 35	22 43	21 01	03 19	18 25	21 17	03 47	20 04	01 16	23 32
24	11 56	25 48	21 24	03 48	18 37	21 16	03 48	20 04	01 17	23 32
25	12 17	27 46	21 45	04 17	18 48	21 16	03 49	20 03	01 17	23 32
26	12 37	28 28	22 05	04 46	19 00	21 15	03 50	20 03	01 18	23 32
27	12 58	27 49	22 24	05 15	19 11	21 15	03 51	20 02	01 18	23 31
28	13 18	25 50	22 41	05 44	19 22	21 14	03 52	20 02	01 19	23 31
29	13 38	22 36	22 56	06 13	19 32	21 14	03 53	20 01	01 19	23 31
30	13 57	18 12	23 11	06 42	19 43	21 14	03 54	20 01	01 20	23 31
31	14 S 17	12 S 49	23 S 23	07 S 10	19 S 53	21 N 13	03 S 55	20 N 00	01 S 20	23 S 31

ZODIAC SIGN ENTRIES

Date	h	m	Planets
01	19	52	☽ ♒
04	02	07	☽ ♓
06	04	48	☽ ♈
06	16	41	☿ ♏
08	05	12	☽ ♉
10	05	12	☽ ♊
12	06	37	☽ ♋
13	21	19	♀ ♎
14	10	47	☽ ♌
16	18	06	☽ ♍
19	04	01	☽ ♎
21	15	42	☽ ♏
22	09	48	♆ ♓
23	03	51	☉ ♏
24	04	19	☽ ♐
26	16	53	☽ ♑
29	03	55	☽ ♒
29	11	02	☿ ♐
31	11	46	☽ ♓

LATITUDES

Date	Mercury ☿ ° '	Venus ♀ ° '	Mars ♂ ° '	Jupiter ♃ ° '	Saturn ♄ ° '	Uranus ♅ ° '	Neptune ♆ ° '	Pluto ♇ ° '
01	00 S 10	01 N 18	00 S 05	00 N 01	02 S 31	00 S 13	01 S 22	03 S 47
04	00 31	01 22	00 07	00 02	02 31	00 13	01 22	03 47
07	00 53	01 25	00 09	00 02	02 30	00 13	01 22	03 47
10	01 14	01 28	00 11	00 02	02 30	00 13	01 22	03 47
13	01 34	01 30	00 13	00 03	02 30	00 13	01 22	03 47
16	01 53	01 31	00 14	00 03	02 30	00 13	01 22	03 47
19	02 11	01 32	00 16	00 03	02 30	00 13	01 22	03 47
22	02 26	01 32	00 18	00 04	02 29	00 13	01 22	03 47
25	02 39	01 32	00 20	00 04	02 29	00 13	01 22	03 46
28	02 49	01 30	00 22	00 05	02 28	00 13	01 22	03 46
31	02 S 54	01 N 29	00 S 23	00 N 05	02 S 28	00 S 13	01 S 22	03 S 46

DATA

Julian Date	2460950
Delta T	+72 seconds
Ayanamsa	24° 13' 04"
Synetic vernal point	04° ♓ 53' 55"
True obliquity of ecliptic	23° 26' 19"

LONGITUDES

Date	Chiron ⚷ ° '	Ceres ⚳ ° '	Pallas ⚴ ° '	Juno ⚵ ° '	Vesta ⚶ ° '	Black Moon Lilith ⚸ ° '
01	25 ♈ 36	09 ♈ 51	06 ♒ 41	00 ♐ 03	07 ♐ 52	21 ♏ 01
11	25 ♈ 09	07 ♈ 36	06 ♒ 48	02 ♐ 59	12 ♐ 36	22 ♏ 08
21	24 ♈ 42	05 ♈ 35	07 ♒ 27	06 ♐ 05	17 ♐ 28	23 ♏ 15
31	24 ♈ 14	03 ♈ 59	08 ♒ 35	09 ♐ 19	22 ♐ 26	24 ♏ 22

MOON'S PHASES, APSIDES AND POSITIONS ☽

Date	h	m	Phase	Longitude ° '	Eclipse Indicator
07	03	48	○	14 ♈ 08	
13	18	13	☾	20 ♋ 40	
21	12	25	●	28 ♎ 22	
29	16	21	☽	06 ♒ 30	

Day	h	m		
08	12	46	Perigee	
23	23	41	Apogee	
06	01	26	0N	
12	03	15	Max dec	28° N 33'
18	23	58	0S	
26	12	39	Max dec	28° S 28'

ASPECTARIAN

h m	Aspects
01 Wednesday	
03 39	☽ □ ☿
05 40	☽ ☍ ♃
15 33	☽ ⚹ ♄
20 03	☽ Ᵽ ♀
20 51	☽ ⚹ ♆
20 59	☿ □ ♃
22 05	☽ △ ♅
22 29	☽ ☌ ♇
22 48	☽ ∥ ♇
23 15	♂ ± ♆
02 Thursday	
08 29	☽ □ ♂
10 37	☽ # ♃
13 41	☽ △ ☉
14 28	☽ ± ♀
17 58	☽ # ♅
19 10	☽ ∠ ♄
21 20	♀ Ᵽ ♇
03 Friday	
00 22	☽ ∠ ♆
02 28	☽ ⊼ ♀
11 16	☽ ⊥ ♄
12 42	☿ ± ♅
13 24	☽ ⊼ ♃
16 25	☽ ⊥ ♆
18 15	☽ △ ☿
18 47	☽ Ᵽ ☉
21 11	☽ ∥ ♂
21 49	☽ ⊻ ♄
04 Saturday	
00 03	☽ ± ♃
02 56	☽ ⊻ ♆
04 06	☽ □ ♅
04 32	☽ ⊻ ♇
11 01	☽ ∥ ☿
11 40	☽ ± ☉
14 54	☽ ⊥ ♇
15 46	☽ Ᵽ ♃
16 36	☽ △ ♂
22 42	☽ ⊼ ☉
23 25	☽ Ᵽ ☿
05 Sunday	
00 32	☿ ⊼ ♄
06 11	☽ ∠ ♇
06 18	☽ # ♀
08 28	☽ ∥ ☉
10 45	☽ Q ♅
11 34	☽ ☍ ♀
13 56	☽ ∥ ♄
16 21	☽ ± ☿
17 15	☽ △ ♃
19 08	☽ Ᵽ ♂
21 41	☽ ∥ ♆
06 Monday	
00 26	☉ # ♀
00 30	☽ ☌ ♄
03 26	☽ ⊼ ☿
05 12	☽ # ♆
05 28	☽ ☌ ♆
06 36	☽ ⚹ ♅
07 04	☽ ⚹ ♇
10 40	☽ ± ♂
12 55	☽ # ♄
17 51	☽ ∥ ♀
20 06	☽ # ☉
20 54	☽ ⊼ ♂
22 53	☿ ⊼ ♆
07 Tuesday	
02 34	☽ Q ♇
03 48	☽ ☍ ☉
06 54	☽ ∠ ♅
09 41	☿ ⊼ ♅
14 41	☿ □ ♇
16 58	☽ ⊼ ♀
18 24	☽ □ ♃
21 18	☽ ⊥ ♅
22 43	☽ # ☿
08 Wednesday	
00 48	☽ ⊻ ♄
03 25	☽ ± ♀
05 46	☽ ⊻ ♆
06 52	☽ ⊻ ♅
07 07	☽ # ♂
07 24	☽ □ ♇
09 15	☽ ☍ ☿
10 20	☽ ⊥ ♄
11 41	♀ ⚹ ♃
15 20	☽ ⊥ ♆
19 04	☽ Ᵽ ♀
19 47	☿ ± ♄
23 16	☽ ☍ ♂
23 19	☽ Q ♃
09 Thursday	
00 37	☽ ∠ ♄
01 35	☉ Ᵽ ♅
03 31	☽ ∥ ♅
05 40	☽ ∠ ♆
07 08	☽ ⊼ ☉
09 34	☽ ∥ ♃
14 31	♀ # ♄
17 25	☽ ± ☉
18 39	☽ ⚹ ♃
20 26	☽ # ♇
21 12	☽ △ ♀
10 Friday	
00 31	☽ ⚹ ♄
02 04	♂ Ᵽ ♄
05 41	☽ ⚹ ♆
06 48	☽ ☌ ♅
07 25	☽ △ ♇
08 56	☽ Ᵽ ☉
14 35	☽ ⊼ ☿
19 03	☽ ∠ ♃
19 52	☽ Q ♄
22 35	☿ ± ♆
11 Saturday	
01 10	☽ Q ♆
01 27	☽ ± ☿
01 59	☽ ⊼ ♂
07 53	☽ Ᵽ ♇
09 55	☽ ⊥ ♃
11 11	♀ ☍ ♄
11 14	☽ △ ☉
12 22	☽ ± ♂
17 58	☽ Ᵽ ☿
19 57	☽ ⊻ ♃
22 50	☽ ± ♇
12 Sunday	
01 30	☽ □ ♄
02 56	☽ □ ♀
04 09	☽ Ᵽ ♂
07 02	☽ □ ♆
08 12	☽ ⊻ ♅
08 56	☽ ⊼ ♇
18 23	☽ ⊥ ♅
22 10	☽ △ ☿
13 Monday	
07 07	☽ △ ♂
09 12	♀ ∠ ♂
09 53	☽ ∠ ♅
12 46	☽ Q ♀
18 13	☽ □ ☉
18 19	☿ ∥ ♂
23 49	☽ ☌ ♃
14 Tuesday	
01 16	♀ ☍ ♆
02 54	♇ St D
05 05	☽ △ ♄
07 03	♀ # ♆
11 08	☽ △ ♆
12 10	☽ ⚹ ♀
12 21	☽ ⚹ ♅
13 14	☽ ☍ ♇
13 17	☽ # ♇
13 27	♂ Ᵽ ♆
14 10	♀ △ ♅
19 03	☿ Ᵽ ♄
23 45	♀ △ ♇
15 Wednesday	
01 58	☽ ∥ ♃
05 04	☽ Q ☉
08 04	☽ Ᵽ ♄
08 36	☽ ∥ ♅
09 41	☽ □ ☿
10 06	☽ Q ♅
14 22	☽ Ᵽ ♆
15 50	☽ □ ♂
18 22	☽ ∠ ♀
21 17	☽ # ☿
16 Thursday	
00 23	☽ # ♂
00 37	☽ ± ♄
05 06	☽ ⚹ ☉
06 52	☽ ⊻ ♃
07 04	☽ ± ♆
11 47	☽ ⊼ ♄
12 59	☽ ⊥ ♀
18 14	☽ ⊥ ♃
18 21	☽ ⊼ ♆
19 37	☽ □ ♅
20 41	☽ ⊼ ♇
23 21	☽ Q ☿
17 Friday	
01 35	☽ ⊻ ♀
03 35	☽ Q ♂
05 31	☿ Ᵽ ♆
05 43	☉ □ ♃
08 07	☽ ± ♇
09 19	☽ # ☉
11 29	☽ ∠ ♃
11 57	☽ ∠ ☉
21 38	☉ ± ♅
18 Saturday	
01 10	☽ ⚹ ☿
01 26	☽ Ᵽ ♇
04 01	☽ ⚹ ♂
06 50	☽ ⊥ ☉
09 00	☽ # ♄
16 43	☽ ⚹ ♃
18 57	☽ # ♆
19 33	☽ ⊻ ☉
19 46	☽ # ♀
21 10	☽ ☍ ♄
21 54	♀ Q ♃
19 Sunday	
04 11	☽ ☍ ♆
04 56	☽ ∥ ♀
05 00	☽ ∥ ♆
05 27	☽ △ ♅
05 38	♀ ∥ ♆
06 43	☽ △ ♇
10 02	☽ ∠ ☿
11 08	☽ ∠ ♂
14 31	☉ ⊼ ♄
15 10	☽ ∥ ♄
16 26	☽ Q ♃
18 26	☽ ☌ ♀
20 Monday	
06 00	☽ ⊥ ☿
06 03	☽ ⊥ ♂
06 52	☿ ☌ ♂
10 21	☿ Q ♇
11 04	☽ Ᵽ ♅
13 21	♂ Q ♇
18 44	☽ ⊻ ♂
19 24	☽ ⊻ ☿
21 01	☽ ∥ ☉
21 Tuesday	
04 29	☽ □ ♃
04 41	☿ # ♅
05 00	☽ ± ♅
08 23	☽ ⊼ ♄
12 25	☽ ☌ ☉
15 44	☽ ⊼ ♆
17 00	☽ ⊼ ♅
18 29	☽ □ ♇
20 23	☽ ± ♄
22 Wednesday	
03 47	☽ ± ♆
09 20	☽ ∥ ♂
13 25	☽ ⊻ ♀
14 26	☽ Ᵽ ♄
19 38	☽ # ♅
21 55	☽ Ᵽ ♆
23 Thursday	
00 03	☽ ∥ ☿
02 47	☽ # ♃
02 58	☽ ⊥ ♀
03 26	☉ ⊼ ♆
06 49	☽ Q ♇
10 55	☽ ☌ ♂
14 57	☽ ☌ ☿
17 18	☽ △ ♃
17 35	☽ ∥ ♇
17 47	☉ ⊼ ♅
20 38	☽ △ ♄
23 20	☽ ∠ ♀
24 Friday	
03 48	☿ # ♃
04 14	☽ △ ♆
05 27	☽ ☍ ♅
06 34	☽ ⊻ ☉
07 08	☽ ⚹ ♇
11 51	♀ ∥ ♄
13 25	☉ □ ♇
15 08	☿ △ ♃
19 50	☽ ⊥ ☉
23 51	☽ Ᵽ ♃
25 Saturday	
07 12	☉ ± ♄
09 17	☽ ⚹ ♀
13 31	☽ ∠ ♇
15 42	☽ ∠ ☉
18 08	☽ ± ♃
21 17	☿ △ ♄
26 Sunday	
03 27	☽ ⊻ ♂
06 16	☽ ⊼ ♃
07 40	☽ ⊥ ♇
07 59	♀ Ᵽ ♅
09 01	☽ □ ♄
10 19	☽ ⊻ ☿
12 18	☽ Q ♀
16 16	☽ ⊥ ♂
16 42	☽ □ ♆
17 50	☽ ⊼ ♅
19 42	☽ ⊻ ♇
23 35	☽ ⊥ ☿
27 Monday	
00 35	☽ ⚹ ☉
05 47	☽ ± ♅
11 20	☽ ∠ ♂
19 16	☽ ∠ ☿
20 44	☽ Q ♄
23 33	☽ Ᵽ ♅
28 Tuesday	
02 30	☽ Q ☉
04 11	☽ □ ♀
04 21	☽ Q ♆
06 19	♂ △ ♃
17 57	☽ ☍ ♃
18 36	☽ ⚹ ♂
20 06	☽ ⚹ ♄
29 Wednesday	
00 42	☉ ± ♆
03 17	☽ ⚹ ☿
03 38	☽ ⚹ ♆
04 39	☽ △ ♅
05 59	☽ ∥ ♇
06 39	☽ ☌ ♇
07 26	☿ △ ♆
09 56	☽ ∥ ☿
16 21	☽ □ ☉
18 59	☽ Q ♂
19 05	♂ △ ♄
19 36	☿ ☍ ♅
20 06	☽ # ♃
23 06	♀ ⊥ ♂
30 Thursday	
00 38	☽ ∠ ♄
02 45	☽ # ♅
04 03	☽ Q ☿
04 37	☽ ∥ ♂
08 00	☽ ∠ ♆
17 19	☽ ⊥ ♄
19 35	☽ △ ♀
22 06	☿ ⚹ ♇
31 Friday	
00 34	☽ ⊥ ♆
02 36	☽ ⊼ ♃
04 15	☽ ⊻ ♄
06 12	☽ ∥ ☉
06 15	☽ □ ♂
11 25	☽ ⊻ ♆
12 18	☽ □ ♅
13 28	☽ ± ♃
14 21	☽ ⊻ ♇
15 32	☽ □ ☿

All ephemeris data is given at 12.00 UT and the Moon's longitude is additionally given for 24.00 UT

NOVEMBER 2025

LONGITUDES

Date	Sidereal time h m s	Sun ☉ ° ′ ″	Moon ☽ ° ′ ″	Moon ☽ 24.00 ° ′ ″	Mercury ☿ ° ′	Venus ♀ ° ′	Mars ♂ ° ′	Jupiter ♃ ° ′	Saturn ♄ ° ′	Uranus ♅ ° ′	Neptune ♆ ° ′	Pluto ♇ ° ′
01	14 44 07	09 ♏ 19 19	13 ♓ 43 52	20 ♓ 42 41	02 ♐ 51	23 ♎ 11	27 ♏ 50	24 ♋ 59	25 ♓ 46	00 ♊ 16	29 ♓ 47	01 ♒ 27
02	14 48 04	10 19 20	27 ♓ 48 45	05 ♈ 01 53	03 40	24 26	28 32	25 01	25 R 44	00 R 13	29 R 46	01 27
03	14 52 00	11 19 23	12 ♈ 21 43	19 ♈ 47 37	04 25	25 41	29 15	25 03	25 41	00 11	29 45	01 28
04	14 55 57	12 19 28	27 ♈ 18 45	04 ♉ 54 03	05 05	26 56	29 ♏ 58	25 04	25 39	00 09	29 43	01 29
05	14 59 53	13 19 35	12 ♉ 32 19	20 ♉ 12 10	05 39	28 12	00 ♐ 41	25 05	25 36	00 06	29 42	01 29
06	15 03 50	14 19 43	27 ♉ 52 08	05 ♊ 30 47	06 08	29 ♎ 27	01 24	25 06	25 34	00 04	29 41	01 30
07	15 07 46	15 19 54	13 ♊ 06 43	20 ♊ 38 40	06 30	00 ♏ 42	02 07	25 07	25 32	00 ♊ 01	29 40	01 31
08	15 11 43	16 20 06	28 ♊ 05 32	05 ♋ 26 26	06 45	01 57	02 51	25 08	25 30	29 ♉ 59	29 39	01 31
09	15 15 39	17 20 21	12 ♋ 40 42	19 ♋ 47 56	06 51	03 12	03 34	25 09	25 28	29 57	29 38	01 32
10	15 19 36	18 20 37	26 ♋ 47 55	03 ♌ 40 39	06 R 49	04 27	04 17	25 09	25 26	29 54	29 37	01 33
11	15 23 33	19 20 55	10 ♌ 26 19	17 ♌ 05 12	06 38	05 42	05 00	25 09	25 24	29 52	29 36	01 34
12	15 27 29	20 21 16	23 ♌ 37 44	00 ♍ 04 23	06 18	06 58	05 44	25 R 09	25 23	29 49	29 35	01 34
13	15 31 26	21 21 38	06 ♍ 25 43	12 ♍ 42 17	05 47	08 13	06 27	25 09	25 21	29 47	29 34	01 35
14	15 35 22	22 22 02	18 ♍ 54 39	25 ♍ 03 23	05 05	09 28	07 11	25 08	25 19	29 44	29 34	01 36
15	15 39 19	23 22 28	01 ♎ 09 02	07 ♎ 12 05	04 14	10 43	07 55	25 08	25 18	29 42	29 33	01 37
16	15 43 15	24 22 56	13 ♎ 13 02	19 ♎ 12 17	03 14	11 59	08 38	25 07	25 17	29 39	29 32	01 38
17	15 47 12	25 23 25	25 ♎ 10 14	01 ♏ 07 12	02 05	13 14	09 22	25 06	25 16	29 37	29 31	01 39
18	15 51 08	26 23 57	07 ♏ 03 30	12 ♏ 59 22	00 ♐ 50	14 29	10 06	25 04	25 14	29 34	29 30	01 40
19	15 55 05	27 24 29	18 ♏ 55 02	24 ♏ 50 42	29 ♏ 31	15 45	10 50	25 03	25 13	29 32	29 30	01 41
20	15 59 02	28 25 04	00 ♐ 46 31	06 ♐ 42 39	28 10	17 00	11 33	25 01	25 13	29 29	29 29	01 42
21	16 02 58	29 ♏ 25 40	12 ♐ 39 16	18 ♐ 36 31	26 49	18 15	12 17	25 00	25 12	29 27	29 28	01 43
22	16 06 55	00 ♐ 26 18	24 ♐ 34 35	00 ♑ 33 40	25 32	19 31	13 01	24 57	25 11	29 24	29 28	01 44
23	16 10 51	01 26 57	06 ♑ 34 00	12 ♑ 35 50	24 21	20 46	13 45	24 55	25 11	29 22	29 27	01 45
24	16 14 48	02 27 37	18 ♑ 39 29	24 ♑ 45 17	23 17	22 02	14 30	24 53	25 10	29 19	29 27	01 46
25	16 18 44	03 28 18	00 ♒ 53 37	07 ♒ 04 55	22 24	23 17	15 14	24 50	25 10	29 17	29 26	01 47
26	16 22 41	04 29 00	13 ♒ 19 40	19 ♒ 38 20	21 41	24 32	15 58	24 47	25 10	29 14	29 26	01 49
27	16 26 37	05 29 44	26 ♒ 01 28	02 ♓ 29 33	21 10	25 48	16 42	24 44	25 09	29 12	29 25	01 50
28	16 30 34	06 30 28	09 ♓ 03 08	15 ♓ 42 41	20 51	27 03	17 27	24 41	25 D 09	29 09	29 25	01 51
29	16 34 31	07 31 14	22 ♓ 28 37	29 ♓ 21 16	20 43	28 19	18 11	24 38	25 10	29 07	29 24	01 52
30	16 38 27	08 ♐ 32 00	06 ♈ 20 52	13 ♈ 27 30	20 ♏ 45	29 ♏ 34	18 ♐ 55	24 ♋ 34	25 ♓ 10	29 ♉ 04	29 ♓ 24	01 ♒ 54

Date	Moon True ☊ ° ′	Moon Mean ☊ ° ′	Moon ☽ Latitude ° ′
01	17 ♓ 03	15 ♓ 22	00 S 18
02	17 R 02	15 19	00 N 57
03	17 00	15 16	02 11
04	16 55	15 13	03 18
05	16 48	15 10	04 12
06	16 39	15 06	04 48
07	16 30	15 03	05 04
08	16 22	15 00	04 58
09	16 15	14 57	04 33
10	16 11	14 54	03 52
11	16 09	14 51	02 58
12	16 D 08	14 47	01 57
13	16 09	14 44	00 N 52
14	16 R 09	14 41	00 S 15
15	16 08	14 38	01 19
16	16 04	14 35	02 19
17	15 58	14 32	03 11
18	15 49	14 28	03 55
19	15 37	14 25	04 29
20	15 23	14 22	04 51
21	15 09	14 19	05 00
22	14 55	14 16	04 56
23	14 43	14 12	04 39
24	14 33	14 09	04 09
25	14 26	14 06	03 28
26	14 22	14 03	02 36
27	14 20	14 00	01 35
28	14 D 20	13 57	00 S 28
29	14 R 20	13 53	00 N 43
30	14 ♓ 19	13 ♓ 50	01 N 53

DECLINATIONS

Date	Sun ☉ ° ′	Moon ☽ ° ′	Mercury ☿ ° ′	Venus ♀ ° ′	Mars ♂ ° ′	Jupiter ♃ ° ′	Saturn ♄ ° ′	Uranus ♅ ° ′	Neptune ♆ ° ′	Pluto ♇ ° ′
01	14 S 36	06 S 40	23 S 34	07 S 39	20 S 04	21 N 13	03 S 56	20 00	01 S 21	23 S 31
02	14 55	00 N 01	23 44	08 07	20 14	21 13	03 57	20 00	01 21	23 31
03	15 14	06 54	23 51	08 35	20 24	21 13	03 58	19 59	01 21	23 30
04	15 32	13 36	23 56	09 03	20 34	21 13	03 59	19 59	01 22	23 30
05	15 50	19 36	24 00	09 31	20 43	21 13	04 00	19 58	01 22	23 30
06	16 08	24 22	24 01	09 58	20 53	21 12	04 01	19 58	01 23	23 30
07	16 26	27 24	23 59	10 26	21 02	21 12	04 01	19 57	01 23	23 30
08	16 43	28 23	23 55	10 53	21 11	21 12	04 02	19 57	01 23	23 30
09	17 00	27 22	23 49	11 20	21 20	21 12	04 03	19 56	01 24	23 29
10	17 17	24 35	23 39	11 46	21 28	21 13	04 03	19 56	01 24	23 29
11	17 34	20 29	23 27	12 12	21 37	21 13	04 04	19 55	01 24	23 29
12	17 50	15 29	23 11	12 38	21 45	21 13	04 04	19 55	01 25	23 29
13	18 06	09 57	22 52	13 04	21 53	21 13	04 05	19 54	01 25	23 29
14	18 22	04 N 10	22 29	13 29	22 01	21 13	04 05	19 53	01 25	23 28
15	18 37	01 S 40	22 03	13 54	22 09	21 14	04 05	19 53	01 26	23 28
16	18 52	07 21	21 33	14 19	22 16	21 14	04 06	19 52	01 26	23 28
17	19 07	12 42	21 01	14 43	22 24	21 14	04 06	19 52	01 26	23 28
18	19 21	17 34	20 26	15 07	22 31	21 15	04 06	19 51	01 27	23 27
19	19 35	21 45	19 49	15 31	22 38	21 15	04 06	19 51	01 27	23 27
20	19 48	25 03	19 11	15 54	22 44	21 15	04 07	19 50	01 27	23 27
21	20 02	27 16	18 34	16 17	22 51	21 16	04 07	19 50	01 27	23 27
22	20 15	28 15	17 57	16 39	22 57	21 16	04 07	19 49	01 28	23 26
23	20 27	27 55	17 24	17 01	23 03	21 17	04 07	19 49	01 28	23 26
24	20 39	26 15	16 53	17 23	23 08	21 18	04 07	19 48	01 28	23 26
25	20 51	23 20	16 27	17 44	23 14	21 18	04 07	19 48	01 28	23 26
26	21 02	19 19	16 05	18 04	23 19	21 19	04 07	19 47	01 28	23 25
27	21 13	14 20	15 48	18 24	23 24	21 20	04 07	19 47	01 28	23 25
28	21 24	08 36	15 36	18 44	23 29	21 20	04 06	19 46	01 29	23 25
29	21 34	02 20	15 29	19 03	23 34	21 21	04 06	19 46	01 29	23 24
30	21 S 44	04 N 15	15 S 27	19 S 21	23 S 38	21 N 22	04 S 06	19 N 45	01 S 29	23 S 24

ZODIAC SIGN ENTRIES

Date	h m	Planets
02	15 39	☽ ♈
04	13 01	♂ ♐
04	16 16	☽ ♉
06	15 20	☽ ♊
06	22 39	♀ ♏
08	02 22	♅ ♉
08	15 06	☽ ♋
10	17 34	☽ ♌
12	23 52	☽ ♍
15	09 44	☽ ♎
17	21 44	☽ ♏
19	03 20	☿ ♏
20	10 26	☽ ♐
22	01 36	☉ ♐
22	22 53	☽ ♑
25	10 16	☽ ♒
27	19 24	☽ ♓
30	01 07	☽ ♈
30	20 14	♀ ♐

LATITUDES

Date	Mercury ☿ ° ′	Venus ♀ ° ′	Mars ♂ ° ′	Jupiter ♃ ° ′	Saturn ♄ ° ′	Uranus ♅ ° ′	Neptune ♆ ° ′	Pluto ♇ ° ′
01	02 S 54	01 N 28	00 S 24	00 N 05	02 S 28	00 S 13	01 S 22	03 S 46
04	02 50	01 26	00 26	00 06	02 27	00 13	01 22	03 46
07	02 38	01 22	00 27	00 06	02 27	00 13	01 22	03 46
10	02 15	01 19	00 29	00 06	02 26	00 12	01 22	03 46
13	01 37	01 15	00 31	00 07	02 26	00 12	01 22	03 46
16	00 S 46	01 10	00 32	00 07	02 25	00 12	01 22	03 46
19	00 N 14	01 05	00 34	00 08	02 24	00 12	01 22	03 46
22	01 13	01 00	00 35	00 08	02 24	00 12	01 21	03 46
25	01 59	00 54	00 37	00 09	02 23	00 12	01 21	03 46
28	02 27	00 48	00 39	00 09	02 23	00 12	01 21	03 46
31	02 N 37	00 N 41	00 S 40	00 N 10	02 S 22	00 S 12	01 S 21	03 S 46

DATA

Julian Date	2460981
Delta T	+72 seconds
Ayanamsa	24° 13′ 08″
Synetic vernal point	04° ♓ 53′ 51″
True obliquity of ecliptic	23° 26′ 18″

MOON'S PHASES, APSIDES AND POSITIONS ☽

Date	h m	Phase	Longitude ° ′	Eclipse Indicator
05	13 19	○	13 ♉ 23	
12	05 28	☾	20 ♌ 05	
20	06 47	●	28 ♏ 12	
28	06 59	☽	06 ♓ 18	

Day	h m		
05	22 34	Perigee	
20	03 06	Apogee	
02	11 58	0N	
08	11 31	Max dec	28° N 23′
15	05 07	0S	
22	18 00	Max dec	28° S 18′
29	20 34	0N	

LONGITUDES

Date	Chiron ⚷ ° ′	Ceres ⚳ ° ′	Pallas ⚴ ° ′	Juno ⚵ ° ′	Vesta ⚶ ° ′	Black Moon Lilith ⚸ ° ′
01	24 ♈ 12	03 ♈ 51	08 ♒ 43	09 ♐ 39	22 ♐ 56	24 ♏ 29
11	23 ♈ 46	02 ♈ 52	10 ♒ 17	13 ♐ 00	28 ♐ 00	25 ♏ 36
21	23 ♈ 23	02 ♈ 31	12 ♒ 13	16 ♐ 26	03 ♑ 07	26 ♏ 43
31	23 ♈ 04	02 ♈ 48	14 ♒ 26	19 ♐ 56	08 ♑ 18	27 ♏ 50

All ephemeris data is given at 12.00 UT and the Moon's longitude is additionally given for 24.00 UT

ASPECTARIAN

h m	Aspects
01 Saturday	
01 01	☽ ⊥ ♇
01 21	☽ bQ ♀
03 37	♂ # ♅
03 42	☽ △ ☉
03 52	☿ ∥ ♇
05 27	☽ bQ ♃
08 36	☽ ∥ ♀
16 42	☽ ∠ ♇
18 33	☽ ± ♀
19 47	☽ Q ♅
21 58	☽ ∥ ♄
02 Sunday	
05 47	☽ ⊼ ♀
07 14	☽ ∥ ♆
07 17	☽ △ ♃
07 30	☽ bQ ☉
07 55	♀ ± ♅
08 31	☽ ☌ ♄
13 17	☽ △ ♂
15 15	☽ ☌ ♆
16 01	☽ ⚹ ♅
16 41	☽ # ♆
18 05	☽ ⚹ ♇
21 22	☉ bQ ♄
22 18	☽ △ ☿
23 17	♀ □ ♃
23 38	☽ ± ☉
03 Monday	
01 47	☽ # ♄
10 11	☽ ⊼ ☉
11 57	♀ ⊼ ♄
13 48	☽ Q ♇
15 14	☽ bQ ♂
16 34	☽ ∠ ♅
18 21	☽ # ♀
23 57	☽ bQ ☿
04 Tuesday	
03 59	♂ △ ♆
06 25	☽ ± ♂
06 58	☽ ⊥ ♅
08 25	☽ □ ♃
09 22	☽ ⊻ ♄
11 21	☽ ☍ ♀
14 55	☽ ± ☿
15 49	☽ ⊻ ♆
16 25	☽ ⊼ ♂
16 29	☽ ⊻ ♅
17 30	♂ ☍ ♅
18 36	☽ □ ♇
18 51	☽ ⊥ ♄
19 44	☽ # ☉
05 Wednesday	
00 47	☽ ⊼ ☿
01 17	☽ ⊥ ♆
08 59	☽ ∠ ♄
12 52	☽ Q ♃
13 19	☽ ☍ ☉
13 38	☽ ∥ ♅
15 24	☽ ∠ ♆
17 12	☽ # ♂
19 19	☽ ∥ ♃
06 Thursday	
07 01	☽ # ♇
07 40	☽ ⚹ ♃
08 24	☽ ⚹ ♄
09 55	☽ # ☿
14 41	☽ ⊼ ♀
14 51	☽ ⚹ ♆
15 11	♂ ⚹ ♇
15 26	☽ ☌ ♅
16 37	♀ ⊼ ♆
17 42	☽ △ ♇
17 49	☽ ☍ ♂
20 27	☉ bQ ♆
23 31	♀ ⊼ ♅
07 Friday	
00 57	☽ ± ♀
01 19	☽ ☍ ☿
03 12	☽ Q ♄
06 54	☿ ⊥ ♀
07 16	☽ ∠ ♃
09 43	☽ Q ♆
15 46	☽ ⊼ ☉
16 28	☽ bQ ♀
17 24	☽ bQ ♇
21 35	☽ ⊥ ♃
08 Saturday	
02 03	☽ ± ☉
03 36	♀ ± ♄
03 44	♀ □ ♇
07 13	☽ ⊻ ♃
07 49	☽ □ ♄
07 50	☽ ± ♇
14 32	☽ □ ♆
15 04	☽ ⊻ ♅
16 07	♂ # ♃
17 35	☽ ⊼ ♇
17 40	☽ bQ ☉
18 52	☽ △ ♀
20 08	☽ ⊼ ♂
09 Sunday	
00 52	☽ ⊥ ♅
02 17	☽ ⊼ ☿
06 32	☽ ± ♂
12 18	☽ ± ☿
15 47	☽ ∠ ♅
19 01	☿ St R
20 26	☽ △ ☉
22 26	☽ bQ ♂
10 Monday	
03 30	☽ bQ ☿
04 20	♀ ⊻ ♂
09 09	☽ ☌ ♃
09 39	☽ △ ♄
16 54	☽ △ ♆
17 22	☽ ⚹ ♅
18 21	☽ # ☿
19 06	☽ # ♇
20 16	☽ ☍ ♇
11 Tuesday	
01 48	☽ △ ♂
02 44	☽ □ ♀
05 21	☽ △ ☿
06 08	☽ # ♂
08 08	☽ ∥ ♃
08 21	☿ ∥ ♇
10 03	♀ ± ♆
11 56	☽ bQ ♄
14 33	☽ Q ♅
14 54	☽ ∥ ♅
16 42	♃ St R
17 06	☉ Q ♇
19 29	☽ bQ ♆
12 Wednesday	
01 40	☽ # ☉
02 16	☿ ⊻ ♀
04 12	☽ ± ♄
05 28	☽ □ ☉
11 56	☽ ± ♆
14 44	☽ Q ♀
14 49	☽ ⊻ ♃
15 14	☽ ⊼ ♄
23 05	☽ ⊼ ♆
23 15	☿ ☌ ♂
23 29	☽ □ ♅
23 39	☽ # ♀
13 Thursday	
02 01	☽ ⊥ ♃
02 50	☽ ⊼ ♇
10 49	☽ □ ☿
12 03	☽ □ ♂
14 12	☽ ± ♇
15 47	☽ ⚹ ♀
18 05	☽ Q ☉
19 05	☽ ∠ ♃
14 Friday	
07 31	☽ bQ ♇
12 19	☽ # ♄
19 20	☽ ⚹ ☉
19 39	☽ Q ☿
23 14	☽ # ♆
15 Saturday	
00 05	☽ ∠ ♀
00 09	☽ ⚹ ♃
00 30	☽ ☍ ♄
01 01	☽ Q ♂
04 01	♀ bQ ♄
07 55	☿ ∥ ♂
08 50	☽ ☍ ♆
09 08	☽ △ ♅
11 01	☽ ∥ ♆
12 55	☽ △ ♇
17 40	☽ ⚹ ☿
19 54	☽ ⊥ ♀
22 08	☽ ∥ ♄
23 50	☽ Q ♃
16 Sunday	
02 16	☽ ⚹ ♂
03 39	☽ ∠ ☉
09 14	☽ ⊻ ♀
14 52	☽ bQ ♅
21 12	☽ ∠ ☿
23 18	☽ ⊥ ☉
17 Monday	
02 30	☿ # ♃
05 08	☉ △ ♃
08 53	☽ ± ♅
08 56	☉ △ ♄
10 16	☽ ∠ ♂
11 51	☽ □ ♃
12 11	☽ ⊼ ♄
12 29	☽ ⊻ ☉
13 41	☽ ⊥ ☿
20 36	☿ ⚹ ♇
20 46	☽ ⊼ ♆
20 56	☽ ⊼ ♅
22 28	☽ ∥ ♀
18 Tuesday	
00 16	☽ ± ♄
00 39	☽ ⊻ ☿
01 05	☽ □ ♇
05 37	☽ ⊥ ♂
08 52	☽ ± ♆
11 23	♂ bQ ♃
12 20	♀ bQ ♆
18 26	☽ bQ ♄
18 33	☽ ⊻ ♂
22 14	☽ ∥ ☉
19 Wednesday	
00 32	☽ # ♅
01 50	☽ ∥ ☿
03 04	☽ bQ ♆
04 49	☽ ☌ ♀
08 51	☽ # ♃
10 42	☿ # ♅
11 45	☿ ☍ ♅
12 22	☿ △ ♆
13 33	☽ Q ♇
17 55	☽ ∥ ♂
18 34	☉ ∥ ☿
23 30	☽ ∥ ♇
20 Thursday	
00 23	☽ △ ♃
00 45	☽ △ ♄
06 47	☽ ☌ ☉
07 15	☽ ☌ ☿
09 23	☉ ☌ ☿
09 24	☽ △ ♆
09 24	☽ ☍ ♅
13 52	☽ ⚹ ♇
14 39	♅ ⚹ ♆
15 15	☉ # ♅
21 Friday	
06 38	☽ bQ ♃
11 13	☽ ☌ ♂
12 25	☉ ☍ ♅
13 05	☉ △ ♆
20 12	☽ ∠ ♇
22 Saturday	
00 37	☽ ⊻ ♀
00 44	☽ ± ♃
12 46	☽ ⊼ ♃
13 13	☽ □ ♄
13 44	☽ ⊻ ☿
14 06	☽ ⊥ ♀
14 20	☽ ⊥ ♇
16 15	♀ Q ♇
18 45	☿ △ ♄
21 39	☽ ⊼ ♅
21 48	☽ □ ♆
23 37	☿ △ ♃
23 Sunday	
00 39	☽ ⊥ ☿
00 50	☽ ⊻ ☉
02 22	☽ ⊻ ♇
09 36	☽ ± ♅
10 14	☽ ∠ ♀
13 55	☽ ⊥ ☉
17 04	☽ ∠ ☿
19 20	☉ ⚹ ♇
22 10	☿ ∥ ♀
24 Monday	
01 09	☽ Q ♄
03 14	☽ ⊻ ♂
03 27	☽ bQ ♅
09 25	☽ ∠ ☉
09 36	☽ Q ♆
15 51	☽ ⊥ ♂
19 24	☽ ⚹ ♀
20 28	☽ ⚹ ☿
25 Tuesday	
00 12	☽ ☍ ♃
00 48	☽ ⚹ ♄
01 52	☿ ☌ ♀
08 52	☽ △ ♅
09 10	☽ ⚹ ♆
10 37	☽ ∠ ♂
11 23	☽ ∥ ♇
12 42	☽ ∥ ♂
13 45	☽ ☌ ♇
17 27	☽ ⚹ ☉
18 24	☽ Q ☿
21 29	☽ Q ♀
26 Wednesday	
00 51	☽ # ♃
02 51	☽ ∥ ☉
05 56	☽ ∠ ♄
09 26	☽ # ♅
14 06	☽ ∠ ♆
16 31	♀ △ ♃
17 21	☽ ⚹ ♂
17 58	☽ ∥ ♀
18 32	☽ Q ☉
23 06	☽ ⊥ ♄
23 48	♀ △ ♄
27 Thursday	
03 12	☽ □ ☿
04 58	☽ ∥ ☿
07 08	☽ ⊥ ♆
09 36	☽ ⊼ ♃
10 23	☽ ⊻ ♄
11 32	☽ □ ♀
15 29	♂ ∥ ♇
16 11	♂ ∠ ♇
17 18	☽ Q ♂
17 53	☽ □ ♅
18 19	☽ ⊻ ♆
20 43	☽ ± ♃
22 48	☽ ⊻ ♇
28 Friday	
03 31	☉ # ♃
03 51	♄ St D
06 59	☽ □ ☉
09 49	☽ ⊥ ♇
13 08	☽ bQ ♃
29 Saturday	
02 03	☽ ∠ ♇
02 32	☽ Q ♅
03 58	☽ □ ♂
05 23	☽ ∥ ♄
08 54	☽ △ ☿
15 09	☽ ∥ ♆
15 45	☽ △ ♃
16 42	☽ ☌ ♄
17 39	☿ St D
23 13	☽ △ ♀
23 33	☽ ⚹ ♅
30 Sunday	
00 05	☽ ☌ ♆
01 17	♂ ± ♃
01 57	☽ # ♆
02 48	♀ ☍ ♅
04 22	☽ ⚹ ♇
08 48	♀ △ ♆
10 59	☽ bQ ☿
11 26	☽ # ♄
16 00	☽ △ ☉

DECEMBER 2025

LONGITUDES

Date	Sidereal time h m s	Sun ☉ ° ' "	Moon ☽ ° ' "	Moon ☽ 24.00 ° ' "	Mercury ☿ ° '	Venus ♀ ° '	Mars ♂ ° '	Jupiter ♃ ° '	Saturn ♄ ° '	Uranus ♅ ° '	Neptune ♆ ° '	Pluto ♇ ° '
01	16 42 24	09 ♐ 32 47	20 ♈ 41 01	28 ♈ 01 07	20 ♏ 58	00 ♐ 50	19 ♐ 40	24 ♋ 30	25 ♓ 10	29 ♉ 02	29 ♓ 24	01 ♒ 55
02	16 46 20	10 33 35	05 ♉ 27 14	12 ♉ 58 35	21 D 21	02 05	20 24	24 R 26	25 10	28 R 59	29 R 23	01 56
03	16 50 17	11 34 24	20 ♉ 34 08	28 ♉ 12 41	21 51	03 20	21 09	24 22	25 11	28 57	29 23	01 57
04	16 54 13	12 35 15	05 ♊ 52 49	13 ♊ 33 05	22 30	04 36	21 54	24 18	25 12	28 55	29 23	01 59
05	16 58 10	13 36 06	21 ♊ 11 59	28 ♊ 48 02	23 15	05 51	22 38	24 13	25 12	28 52	29 23	02 00
06	17 02 06	14 36 59	06 ♋ 19 54	13 ♋ 46 26	24 06	07 07	23 23	24 09	25 13	28 50	29 23	02 02
07	17 06 03	15 37 52	21 ♋ 06 40	28 ♋ 19 55	25 02	08 22	24 08	24 04	25 14	28 47	29 23	02 03
08	17 10 00	16 38 47	05 ♌ 25 44	12 ♌ 23 52	26 03	09 38	24 53	23 59	25 15	28 45	29 22	02 04
09	17 13 56	17 39 43	19 ♌ 14 19	25 ♌ 57 16	27 08	10 53	25 38	23 53	25 16	28 43	29 22	02 06
10	17 17 53	18 40 40	02 ♍ 33 03	09 ♍ 02 07	28 16	12 09	26 23	23 48	25 18	28 40	29 22	02 07
11	17 21 49	19 41 38	15 ♍ 25 02	21 ♍ 42 22	29 ♏ 27	13 24	27 08	23 43	25 19	28 38	29 D 22	02 09
12	17 25 46	20 42 37	27 ♍ 54 46	04 ♎ 02 55	00 ♐ 41	14 40	27 53	23 37	25 21	28 36	29 22	02 10
13	17 29 42	21 43 38	10 ♎ 07 26	16 ♎ 08 59	01 58	15 55	28 38	23 31	25 22	28 33	29 23	02 12
14	17 33 39	22 44 39	22 ♎ 08 10	28 ♎ 05 33	03 16	17 11	29 ♐ 23	23 25	25 24	28 31	29 23	02 13
15	17 37 35	23 45 42	04 ♏ 01 40	09 ♏ 57 00	04 36	18 26	00 ♑ 08	23 19	25 26	28 29	29 23	02 15
16	17 41 32	24 46 45	15 ♏ 51 59	21 ♏ 47 00	05 57	19 42	00 54	23 13	25 28	28 27	29 23	02 17
17	17 45 29	25 47 49	27 ♏ 42 23	03 ♐ 38 24	07 20	20 57	01 39	23 06	25 30	28 25	29 23	02 18
18	17 49 25	26 48 55	09 ♐ 35 19	15 ♐ 33 17	08 43	22 13	02 24	23 00	25 32	28 23	29 23	02 20
19	17 53 22	27 50 00	21 ♐ 32 30	27 ♐ 33 05	10 08	23 28	03 10	22 53	25 34	28 20	29 24	02 21
20	17 57 18	28 51 07	03 ♑ 35 10	09 ♑ 38 51	11 34	24 44	03 55	22 47	25 36	28 18	29 24	02 23
21	18 01 15	29 ♐ 52 14	15 ♑ 44 15	21 ♑ 51 29	13 00	25 59	04 41	22 40	25 39	28 16	29 24	02 25
22	18 05 11	00 ♑ 53 21	28 ♑ 00 41	04 ♒ 12 02	14 27	27 15	05 26	22 33	25 41	28 14	29 25	02 26
23	18 09 08	01 54 29	10 ♒ 25 43	16 ♒ 41 59	15 55	28 31	06 12	22 26	25 44	28 12	29 25	02 28
24	18 13 04	02 55 37	23 ♒ 01 04	29 ♒ 23 19	17 23	29 ♐ 46	06 58	22 18	25 46	28 10	29 26	02 30
25	18 17 01	03 56 45	05 ♓ 49 02	12 ♓ 18 36	18 52	01 ♑ 02	07 43	22 11	25 49	28 08	29 26	02 32
26	18 20 58	04 57 53	18 ♓ 52 22	25 ♓ 30 44	20 21	02 17	08 29	22 04	25 52	28 07	29 27	02 33
27	18 24 54	05 59 01	02 ♈ 14 02	09 ♈ 02 35	21 51	03 33	09 15	21 56	25 55	28 05	29 27	02 35
28	18 28 51	07 00 09	15 ♈ 56 37	22 ♈ 56 17	23 21	04 48	10 01	21 49	25 58	28 03	29 28	02 37
29	18 32 47	08 01 17	00 ♉ 01 37	07 ♉ 12 31	24 51	06 04	10 46	21 41	26 02	28 01	29 29	02 39
30	18 36 44	09 02 25	14 ♉ 28 39	21 ♉ 49 35	26 22	07 19	11 32	21 33	26 05	27 59	29 29	02 40
31	18 40 40	10 ♑ 03 33	29 ♉ 14 39	06 ♊ 42 59	27 ♐ 53	08 ♑ 35	12 ♑ 18	21 ♋ 25	26 ♓ 08	27 ♉ 58	29 ♓ 30	02 ♒ 42

Date	Moon True ☊ ° '	Moon Mean ☊ ° '	Moon ☽ Latitude ° '
01	14 ♓ 15	13 ♓ 47	02 N 59
02	14 R 09	13 44	03 55
03	14 00	13 41	04 35
04	13 49	13 38	04 57
05	13 37	13 34	04 57
06	13 26	13 31	04 37
07	13 17	13 28	03 58
08	13 10	13 25	03 05
09	13 06	13 22	02 02
10	13 05	13 18	00 N 55
11	13 D 04	13 15	00 S 12
12	13 R 04	13 12	01 18
13	13 03	13 09	02 18
14	13 00	13 06	03 11
15	12 54	13 03	03 55
16	12 44	12 59	04 28
17	12 33	12 56	04 50
18	12 19	12 53	05 00
19	12 04	12 50	04 56
20	11 50	12 47	04 39
21	11 38	12 44	04 09
22	11 28	12 40	03 28
23	11 21	12 37	02 36
24	11 17	12 34	01 35
25	11 15	12 31	00 S 29
26	11 D 15	12 28	00 N 40
27	11 16	12 24	01 49
28	11 R 15	12 21	02 54
29	11 13	12 18	03 49
30	11 09	12 15	04 32
31	11 ♓ 02	12 ♓ 12	04 N 57

DECLINATIONS

Date	Sun ☉ ° '	Moon ☽ ° '	Mercury ☿ ° '	Venus ♀ ° '	Mars ♂ ° '	Jupiter ♃ ° '	Saturn ♄ ° '	Uranus ♅ ° '	Neptune ♆ ° '	Pluto ♇ ° '
01	21 S 53	10 N 50	15 S 29	19 S 39	23 S 42	21 N 23	04 S 06	19 N 44	01 S 29	23 S 24
02	22 02	17 02	15 34	19 56	23 46	21 24	04 05	19 44	01 29	23 24
03	22 10	22 19	15 43	20 13	23 50	21 24	04 05	19 43	01 29	23 23
04	22 18	26 09	15 55	20 30	23 53	21 25	04 04	19 43	01 29	23 23
05	22 26	28 05	16 09	20 45	23 56	21 26	04 04	19 42	01 29	23 23
06	22 33	27 53	16 26	21 00	23 59	21 27	04 03	19 42	01 29	23 22
07	22 40	25 42	16 44	21 15	24 01	21 28	04 03	19 41	01 29	23 22
08	22 46	21 54	17 03	21 29	24 04	21 29	04 02	19 41	01 29	23 22
09	22 52	16 59	17 24	21 42	24 06	21 30	04 01	19 40	01 29	23 21
10	22 57	11 26	17 45	21 54	24 08	21 31	04 01	19 40	01 29	23 21
11	23 02	05 N 33	18 07	22 06	24 09	21 32	04 00	19 39	01 29	23 21
12	23 07	00 S 21	18 29	22 18	24 11	21 34	03 59	19 39	01 29	23 20
13	23 11	06 07	18 52	22 28	24 12	21 35	03 58	19 38	01 29	23 20
14	23 14	11 34	19 14	22 38	24 12	21 36	03 57	19 38	01 29	23 20
15	23 17	16 32	19 36	22 47	24 13	21 37	03 57	19 37	01 29	23 19
16	23 20	20 52	19 58	22 56	24 13	21 38	03 56	19 37	01 29	23 19
17	23 22	24 21	20 20	23 04	24 13	21 40	03 55	19 37	01 29	23 19
18	23 24	26 49	20 41	23 11	24 13	21 41	03 54	19 36	01 28	23 18
19	23 25	28 05	21 01	23 18	24 12	21 42	03 53	19 36	01 28	23 18
20	23 26	28 02	21 21	23 23	24 12	21 43	03 51	19 35	01 28	23 18
21	23 26	26 38	21 40	23 28	24 11	21 45	03 50	19 35	01 28	23 17
22	23 26	23 57	21 58	23 33	24 09	21 46	03 49	19 34	01 28	23 17
23	23 25	20 07	22 15	23 36	24 08	21 47	03 48	19 34	01 27	23 16
24	23 24	15 21	22 31	23 39	24 06	21 49	03 47	19 33	01 27	23 16
25	23 23	09 49	22 47	23 41	24 04	21 50	03 45	19 33	01 27	23 16
26	23 21	03 S 47	23 01	23 43	24 01	21 51	03 44	19 33	01 27	23 15
27	23 18	02 N 34	23 14	23 43	23 59	21 53	03 42	19 32	01 26	23 15
28	23 15	08 57	23 26	23 43	23 56	21 54	03 41	19 32	01 26	23 15
29	23 12	15 03	23 37	23 43	23 52	21 55	03 40	19 31	01 26	23 14
30	23 08	20 30	23 47	23 41	23 49	21 57	03 38	19 31	01 26	23 14
31	23 S 03	24 N 49	23 S 56	23 S 39	23 S 45	21 N 58	03 S 37	19 N 31	01 S 25	23 S 13

ZODIAC SIGN ENTRIES

Date	h	m	Planets
02	03	13	☽ ♉
04	02	48	☽ ♊
06	01	54	☽ ♋
08	02	48	☽ ♌
10	07	20	☽ ♍
11	22	40	☿ ♐
12	16	04	☽ ♎
15	03	51	☽ ♏
15	07	34	♂ ♑
17	16	38	☽ ♐
20	04	53	☽ ♑
21	15	03	☉ ♑
22	15	52	☽ ♒
24	16	26	♀ ♑
25	01	09	☽ ♓
27	08	02	☽ ♈
29	11	57	☽ ♉
31	13	13	☽ ♊

LATITUDES

Date	Mercury ☿ ° '	Venus ♀ ° '	Mars ♂ ° '	Jupiter ♃ ° '	Saturn ♄ ° '	Uranus ♅ ° '	Neptune ♆ ° '	Pluto ♇ ° '
01	02 N 37	00 N 41	00 S 40	00 N 10	02 S 22	00 S 12	01 S 21	03 S 46
04	02 34	00 35	00 42	00 10	02 21	00 12	01 21	03 46
07	02 22	00 28	00 43	00 11	02 21	00 12	01 21	03 46
10	02 04	00 21	00 44	00 11	02 20	00 12	01 21	03 46
13	01 43	00 13	00 46	00 12	02 19	00 12	01 21	03 46
16	01 21	00 N 06	00 47	00 12	02 19	00 12	01 21	03 46
19	00 58	00 S 01	00 48	00 12	02 18	00 12	01 20	03 46
22	00 34	00 08	00 50	00 13	02 17	00 12	01 20	03 46
25	00 N 12	00 15	00 51	00 13	02 17	00 12	01 20	03 46
28	00 S 10	00 22	00 52	00 14	02 16	00 12	01 20	03 46
31	00 S 31	00 S 29	00 S 53	00 N 14	02 S 16	00 S 12	01 S 20	03 S 46

DATA

Julian Date	2461011
Delta T	+72 seconds
Ayanamsa	24° 13' 12"
Synetic vernal point	04° ♓ 53' 47"
True obliquity of ecliptic	23° 26' 18"

MOON'S PHASES, APSIDES AND POSITIONS ☽

Date	h	m	Phase	Longitude ° '	Eclipse Indicator
04	23	14	○	13 ♊ 04	
11	20	52	☾	20 ♍ 04	
20	01	43	●	28 ♐ 25	
27	19	10	☽	06 ♈ 17	

Day	h	m		
04	11	14	Perigee	
17	06	23	Apogee	
05	21	44	Max dec	28° N 16'
12	10	32	0S	
19	23	07	Max dec	28° S 14'
27	02	23	0N	

LONGITUDES

Date	Chiron ⚷ ° '	Ceres ⚳ ° '	Pallas ⚴ ° '	Juno ⚵ ° '	Vesta ⚶ ° '	Black Moon Lilith ⚸ ° '
01	23 ♈ 04	02 ♈ 48	14 ♒ 26	19 ♐ 56	08 ♑ 18	27 ♏ 50
11	22 ♈ 50	03 ♈ 40	16 ♒ 54	23 ♐ 29	13 ♑ 31	28 ♏ 58
21	22 ♈ 40	05 ♈ 05	19 ♒ 35	27 ♐ 05	18 ♑ 46	00 ♐ 05
31	22 ♈ 36	06 ♈ 58	22 ♒ 27	00 ♑ 41	24 ♑ 01	01 ♐ 12

All ephemeris data is given at 12.00 UT and the Moon's longitude is additionally given for 24.00 UT

ASPECTARIAN

h m	Aspects
01 Monday	
00 45	☽ Q ♇
00 59	☽ ∠ ♅
02 21	☽ ± ☿
03 11	☽ ⚼ ♀
10 14	☽ △ ♂
11 01	☉ ⚼ ♃
12 29	☽ ⊼ ☿
15 51	☽ ⊥ ♅
18 15	☽ □ ♃
18 49	☽ ⚼ ☉
19 11	♀ # ♅
19 22	☽ ⊻ ♄
19 26	☽ ± ♀
02 Tuesday	
01 36	☽ ⊻ ♅
02 14	☽ ⊻ ♆
05 07	☽ ⊥ ♄
06 00	☽ # ☿
06 05	☽ ⊼ ♀
06 20	☽ □ ♇
09 07	♀ ⚹ ♇
10 28	☽ ± ☉
11 54	☽ ⊥ ♆
11 55	☽ ⚼ ♂
19 33	☽ ∠ ♄
20 45	☽ ⊼ ☉
23 06	☽ Q ♃
23 36	☽ ∥ ♅
03 Wednesday	
01 17	☽ # ♀
02 14	☽ ∠ ♆
03 00	☽ ± ♂
07 27	☽ ∥ ♃
11 15	☽ # ☉
12 58	☽ ⊼ ♂
14 07	☽ ☍ ☿
17 45	☽ # ♇
17 57	☽ ⚹ ♃
19 15	☽ ⚹ ♄
20 22	☽ # ♂
04 Thursday	
01 07	☽ ☌ ♅
01 50	☽ ⚹ ♆
05 53	☽ △ ♇
09 49	☽ ☍ ♀
14 03	☽ Q ♄
17 19	☽ ∠ ♃
20 36	☽ Q ♆
23 14	☽ ☍ ☉
05 Friday	
05 24	☽ ⚼ ♇
07 20	☽ ⊥ ♃
14 23	☽ ☍ ♂
15 24	☽ ⊼ ☿
16 44	☽ ⊻ ♃
18 19	☽ □ ♄
19 35	☽ ± ♇
06 Saturday	
00 05	☽ ⊻ ♅
00 55	☽ □ ♆
01 27	☽ ± ☿
05 07	☽ ⊼ ♇
09 36	☽ ⊥ ♅
13 05	☿ △ ♃
13 22	☽ ⊼ ♀
16 44	☽ ⚼ ☿
23 57	☽ ± ♀
07 Sunday	
00 03	☽ ∠ ♅
02 21	☽ ⊼ ☉
09 55	♂ ⊼ ♃
12 55	☽ ± ☉
16 05	☽ ⚼ ♀
16 51	☽ ☌ ♃
16 58	☿ △ ♄
17 16	☽ ⊼ ♂
18 50	☽ △ ♄
18 59	☽ △ ☿
23 29	☽ # ♂
08 Monday	
00 21	♀ ⚼ ♃
00 44	☽ ⚹ ♅
01 45	☽ △ ♆
03 40	☽ # ♇
03 51	☽ ± ♂
05 05	☽ ⚼ ☉
06 17	☽ ☍ ♇
07 15	☽ # ☉
13 14	♀ # ♃
14 08	☽ # ♀
14 10	☽ ∥ ♃
19 55	☽ △ ♀
20 05	☽ ⚼ ♂
20 18	☽ ⚼ ♄
21 07	☽ Q ♅
22 21	☉ ∠ ♇
23 22	☽ ∥ ♅
09 Tuesday	
00 16	♂ □ ♄
03 27	☽ ⚼ ♆
08 59	☽ △ ☉
10 15	☽ # ☿
12 04	☽ ± ♄
17 00	☉ ± ♃
19 22	☽ ± ♆
20 14	☽ ⊻ ♃
22 47	☽ ⊼ ♄
10 Wednesday	
00 06	☽ △ ♂
03 26	☽ □ ☿
03 31	♂ ⊥ ♇
04 56	☽ □ ♅
06 12	☽ ⊼ ♆
07 01	☽ ⊥ ♃
11 13	☽ ⊼ ♇
12 24	♆ St D
19 59	☿ ☍ ♅
22 19	☽ ± ♇
23 29	☽ ∠ ♃
11 Thursday	
07 47	☽ □ ♀
10 19	☿ △ ♆
15 18	☽ ⚼ ♇
16 17	☽ Q ☿
18 19	☽ # ♄
20 52	☽ □ ☉
12 Friday	
03 44	☽ ⚹ ♃
04 30	☽ # ♆
07 00	☽ ☍ ♄
11 56	☽ □ ♂
13 19	☽ △ ♅
14 51	☽ ☍ ♆
16 37	☽ ∥ ♆
18 02	☽ ⚹ ☿
20 20	☽ △ ♇
22 21	☽ Q ♀
13 Saturday	
02 58	☽ ∥ ♄
02 58	☽ Q ♃
09 44	♂ ⊼ ♅
11 08	☽ Q ☉
16 32	☿ ⚹ ♇
18 48	☽ ⚼ ♅
14 Sunday	
00 54	☽ ⚹ ♀
01 50	☽ Q ♂
03 16	☽ ∠ ☿
11 44	♂ □ ♆
12 46	☽ ± ♅
12 52	♀ ∠ ♇
13 20	☽ ⚹ ☉
14 34	☽ □ ♃
16 14	♀ ± ♃
18 35	☽ ⊼ ♄
23 37	☽ ⊥ ☿
15 Monday	
00 49	☽ ⊼ ♅
02 30	☉ ⊼ ♃
02 36	☽ ⊼ ♆
03 36	☽ ⚹ ♂
06 43	☽ ± ♄
08 24	☽ □ ♇
10 40	☽ ∠ ♀
13 09	☿ # ♅
13 18	☽ ⊻ ☿
14 44	☽ ± ♆
22 30	☽ ∠ ☉
16 Tuesday	
01 00	☽ ⚼ ♄
02 09	☉ ∥ ♇
04 40	☽ # ♅
06 11	☽ ∥ ☿
07 05	☽ ⊥ ♀
08 59	☽ ⚼ ♆
12 04	☽ ∠ ♂
16 52	☽ # ♃
18 28	☽ ⊥ ☉
20 42	☽ ⊻ ♀
20 58	☽ Q ♇
17 Wednesday	
02 00	☽ ∥ ♀
02 46	☽ △ ♃
04 07	☽ ∥ ♇
04 28	☽ ∥ ☉
04 34	☉ □ ♄
07 30	☽ △ ♄
07 33	☽ ⊥ ♂
07 46	☽ ⊻ ☉
10 56	☽ ∥ ♂
13 25	☽ ☍ ♅
15 24	☽ △ ♆
20 31	☽ ⊻ ♂
21 19	☽ ⚹ ♇
18 Thursday	
00 16	☉ ⊥ ♇
00 28	☿ ⚼ ♃
08 50	☽ ⚼ ♃
09 32	♂ ⊻ ♇
10 02	☽ ☌ ☿
19 Friday	
01 45	♀ ⊼ ♃
02 46	☽ ± ♃
03 36	☽ ∠ ♇
12 54	♀ ∥ ♇
14 40	☽ ⊼ ♃
16 19	☽ ☌ ♀
20 04	☽ □ ♄
21 39	☽ ⊥ ♇
23 34	☉ ⊼ ♅
20 Saturday	
01 32	☽ ⊼ ♅
01 43	☽ ☌ ☉
03 41	☽ □ ♆
09 37	☽ ⊻ ♇
12 42	☽ ☌ ♂
13 25	☽ ± ♅
23 41	♂ ± ♅
21 Sunday	
01 00	☉ ∥ ♀
01 02	☉ □ ♆
05 09	♀ □ ♄
05 54	☽ ⊻ ☿
07 10	☽ ⚼ ♅
07 52	☽ Q ♄
15 17	☽ Q ♆
19 01	☿ # ♃
19 16	☽ ⊥ ☿
20 14	♀ ⊥ ♇
22 Monday	
01 26	☽ ☍ ♃
07 27	☽ ⚹ ♄
10 21	☽ ⊻ ♀
10 29	☽ ∥ ♂
12 26	☽ △ ♅
14 44	☽ ⚹ ♆
14 50	☽ ∥ ♀
15 11	☽ ∠ ☿
15 40	☽ ∥ ☉
16 45	☽ ∥ ♇
18 05	☽ ⊻ ☉
20 37	☽ ☌ ♇
23 19	☽ ⊥ ♀
23 Tuesday	
00 21	☽ ∥ ☿
02 24	☽ # ♃
03 20	☽ ⊻ ♂
06 22	♀ ⊼ ♅
06 43	☽ ⊥ ☉
12 35	☽ ∠ ♄
15 03	☽ # ♅
15 37	☽ ⊥ ♂
18 34	☽ ∠ ♀
19 39	☽ ∠ ♆
19 40	☿ ± ♃
23 54	☽ ⚹ ☿
24 Wednesday	
01 29	☽ ∠ ☉
01 37	☉ ⊻ ♇
05 31	♀ □ ♆
05 50	☽ ⊥ ♄
09 52	☽ ∠ ♂
10 40	☽ ⊼ ♃
12 47	☽ ⊥ ♆
13 49	☿ ∠ ♇
17 13	☽ ⊻ ♄
21 42	☽ □ ♅
21 52	☽ ± ♃
25 Thursday	
00 05	☽ ⊻ ♆
01 34	☽ Q ☿
02 06	☽ ⚹ ♀
05 52	☽ ⊻ ♇
08 13	☽ ⚹ ☉
14 31	☽ ⚼ ♃
15 45	☽ ⚹ ♂
16 28	☉ ± ♅
17 02	☽ ⊥ ♇
26 Friday	
02 44	☽ Q ♀
06 59	☽ Q ♅
08 14	☽ Q ☉
09 36	☽ ∠ ♇
12 12	☽ ∥ ♄
15 02	☽ □ ☿
15 06	☽ Q ♂
17 17	♀ ⊻ ♇
17 43	☽ △ ♃
20 56	☽ ∥ ♆
27 Saturday	
00 41	☽ ☌ ♄
04 37	☽ ⚹ ♅
07 03	☽ ☌ ♆
07 49	☽ # ♆
12 37	☽ ⚹ ♇
13 16	☿ ⊼ ♃
13 17	☿ ∥ ♇
14 33	☽ □ ♀
16 17	☽ # ♄
18 09	☉ ∥ ☿
19 10	☽ □ ☉
21 59	♀ ± ♅
28 Sunday	
01 05	☽ □ ♂
07 00	☽ ∠ ♅
09 42	☽ Q ♇
17 18	☉ ∥ ♇
21 59	☽ □ ♃
22 28	☽ ⊥ ♅
29 Monday	
02 13	☽ △ ☿
05 13	☽ ⊻ ♄
08 38	☽ ⊻ ♅
11 04	☽ ⊻ ♆
15 22	☽ ⊥ ♄
16 24	☽ □ ♇
21 08	☽ ⊥ ♆
22 33	☿ ∥ ♀
23 03	☽ △ ♀
30 Tuesday	
02 22	☽ △ ☉
03 57	☽ Q ♃
06 18	☽ ⚼ ☿
06 23	☽ ∠ ♄
06 54	☽ △ ♂
07 15	☿ □ ♄
07 20	☽ ∥ ♅
12 01	☽ ∠ ♆
14 49	☿ ∥ ♂
16 54	☿ ⊥ ♇
19 15	☽ ∥ ♃
22 44	☽ ± ☿
23 27	☽ ⚹ ♃
31 Wednesday	
01 25	☽ # ☉
01 59	☽ ⚼ ♀
02 11	☽ # ♇
04 42	☽ # ♀
04 45	☽ ⚼ ☉
05 22	☽ # ♂
06 09	☽ # ☿
06 58	☽ ⚹ ♄
08 42	☽ ⚼ ♂
09 34	☽ ⊼ ☿
09 56	☽ ☌ ♅
12 25	☽ ⚹ ♆
13 08	☿ ⊼ ♅
17 35	☽ △ ♇
17 51	☽ ± ♀
20 18	☽ ± ☉
23 26	☽ ∠ ♃
23 57	☽ ± ♂

JANUARY 2026

Date	Sidereal time h m s	LONGITUDES Sun ☉ ° ′ ″	Moon ☽ ° ′ ″	Moon ☽ 24.00 ° ′ ″	Mercury ☿ ° ′	Venus ♀ ° ′	Mars ♂ ° ′	Jupiter ♃ ° ′	Saturn ♄ ° ′	Uranus ♅ ° ′	Neptune ♆ ° ′	Pluto ♇ ° ′
01	18 44 37	11 ♑ 04 41	14 ♊ 13 34	21 ♊ 45 15	29 ♐ 25	09 ♑ 50	13 ♑ 04	21 ♋ 18	26 ♓ 12	27 ♉ 56	29 ♓ 31	02 ♒ 44
02	18 48 33	12 05 49	29 ♊ 16 49	06 ♋ 47 00	00 ♑ 57	11 06	13 50	21 R 10	26 15	27 R 55	29 32	02 46
03	18 52 30	13 06 57	14 ♋ 14 34	21 ♋ 38 23	02 29	12 21	14 36	21 02	26 19	27 53	29 32	02 48
04	18 56 27	14 08 05	28 ♋ 57 25	06 ♌ 10 49	04 02	13 37	15 22	20 54	26 23	27 51	29 33	02 50
05	19 00 23	15 09 13	13 ♌ 17 56	20 ♌ 18 19	05 35	14 52	16 09	20 46	26 27	27 50	29 34	02 51
06	19 04 20	16 10 20	27 ♌ 11 43	03 ♍ 58 03	07 08	16 08	16 55	20 38	26 31	27 49	29 35	02 53
07	19 08 16	17 11 28	10 ♍ 37 26	17 ♍ 10 07	08 42	17 23	17 41	20 30	26 35	27 47	29 36	02 55
08	19 12 13	18 12 36	23 ♍ 36 29	29 ♍ 57 00	10 16	18 39	18 27	20 22	26 39	27 46	29 37	02 57
09	19 16 09	19 13 44	06 ♎ 12 12	12 ♎ 22 42	11 50	19 54	19 14	20 13	26 43	27 44	29 38	02 59
10	19 20 06	20 14 52	18 ♎ 29 07	24 ♎ 32 06	13 25	21 09	20 00	20 05	26 47	27 43	29 39	03 01
11	19 24 02	21 16 01	00 ♏ 32 17	06 ♏ 30 20	15 00	22 25	20 46	19 57	26 52	27 42	29 40	03 03
12	19 27 59	22 17 09	12 ♏ 26 51	18 ♏ 22 25	16 36	23 40	21 33	19 49	26 56	27 41	29 41	03 04
13	19 31 56	23 18 17	24 ♏ 17 38	00 ♐ 12 58	18 13	24 56	22 19	19 41	27 01	27 40	29 42	03 06
14	19 35 52	24 19 25	06 ♐ 08 55	12 ♐ 05 54	19 49	26 11	23 06	19 33	27 05	27 39	29 43	03 08
15	19 39 49	25 20 33	18 ♐ 04 17	24 ♐ 04 24	21 27	27 27	23 52	19 25	27 10	27 38	29 45	03 10
16	19 43 45	26 21 40	00 ♑ 06 31	06 ♑ 10 51	23 04	28 42	24 39	19 17	27 15	27 37	29 46	03 12
17	19 47 42	27 22 47	12 ♑ 17 35	18 ♑ 26 51	24 42	29 ♑ 58	25 25	19 09	27 20	27 36	29 47	03 14
18	19 51 38	28 23 54	24 ♑ 38 43	00 ♒ 53 18	26 21	01 ♒ 13	26 12	19 01	27 25	27 35	29 48	03 16
19	19 55 35	29 ♑ 25 00	07 ♒ 10 37	13 ♒ 30 42	28 01	02 29	26 58	18 53	27 30	27 34	29 50	03 18
20	19 59 31	00 ♒ 26 06	19 ♒ 53 35	26 ♒ 19 18	29 ♑ 40	03 44	27 45	18 45	27 35	27 33	29 51	03 20
21	20 03 28	01 27 11	02 ♓ 47 53	09 ♓ 19 23	01 ♒ 21	04 59	28 32	18 38	27 40	27 32	29 52	03 22
22	20 07 25	02 28 14	15 ♓ 53 52	22 ♓ 31 25	03 02	06 15	29 ♑ 19	18 30	27 45	27 32	29 54	03 24
23	20 11 21	03 29 17	29 ♓ 12 08	05 ♈ 56 06	04 43	07 30	00 ♒ 05	18 22	27 51	27 31	29 55	03 26
24	20 15 18	04 30 19	12 ♈ 43 26	19 ♈ 34 12	06 26	08 46	00 52	18 15	27 56	27 31	29 57	03 27
25	20 19 14	05 31 20	26 ♈ 28 29	03 ♉ 26 18	08 08	10 01	01 39	18 07	28 01	27 30	29 ♓ 58	03 29
26	20 23 11	06 32 20	10 ♉ 27 36	17 ♉ 32 18	09 52	11 16	02 26	18 00	28 07	27 30	00 ♈ 00	03 31
27	20 27 07	07 33 19	24 ♉ 40 11	01 ♊ 50 59	11 35	12 32	03 13	17 53	28 12	27 29	00 01	03 33
28	20 31 04	08 34 16	09 ♊ 04 17	16 ♊ 19 36	13 20	13 47	03 59	17 46	28 18	27 29	00 03	03 35
29	20 35 00	09 35 13	23 ♊ 36 18	00 ♋ 53 43	15 05	15 02	04 46	17 39	28 24	27 28	00 04	03 37
30	20 38 57	10 36 08	08 ♋ 11 05	15 ♋ 27 35	16 50	16 18	05 33	17 32	28 30	27 28	00 06	03 39
31	20 42 54	11 ♒ 37 03	22 ♋ 42 23	29 ♋ 54 41	18 ♒ 35	17 ♒ 33	06 ♒ 20	17 ♋ 25	28 ♓ 35	27 ♉ 28	00 ♈ 08	03 ♒ 41

Date	Moon True ☊ ° ′	Moon Mean ☊ ° ′	Moon ☽ Latitude ° ′
01	10 ♓ 53	12 ♓ 09	05 N 03
02	10 R 43	12 05	04 48
03	10 34	12 02	04 13
04	10 26	11 59	03 22
05	10 21	11 56	02 19
06	10 18	11 53	01 N 10
07	10 D 17	11 50	00 S 02
08	10 18	11 46	01 11
09	10 19	11 43	02 15
10	10 20	11 40	03 11
11	10 R 19	11 37	03 57
12	10 17	11 34	04 32
13	10 12	11 30	04 56
14	10 06	11 27	05 07
15	09 58	11 24	05 04
16	09 49	11 21	04 49
17	09 40	11 18	04 20
18	09 33	11 15	03 38
19	09 27	11 11	02 45
20	09 23	11 08	01 44
21	09 21	11 05	00 S 36
22	09 D 21	11 02	00 N 36
23	09 22	10 59	01 46
24	09 24	10 55	02 52
25	09 25	10 52	03 49
26	09 R 25	10 49	04 33
27	09 24	10 46	05 02
28	09 21	10 43	05 12
29	09 18	10 40	05 03
30	09 13	10 36	04 34
31	09 ♓ 09	10 ♓ 33	03 N 47

DECLINATIONS

Date	Sun ☉ ° ′	Moon ☽ ° ′	Mercury ☿ ° ′	Venus ♀ ° ′	Mars ♂ ° ′	Jupiter ♃ ° ′	Saturn ♄ ° ′	Uranus ♅ ° ′	Neptune ♆ ° ′	Pluto ♇ ° ′
01	22 S 59	27 N 31	24 S 04	23 S 36	23 S 41	21 N 59	03 S 35	19 N 30	01 S 25	23 S 13
02	22 53	28 14	24 10	23 32	23 37	22 01	03 33	19 30	01 25	23 13
03	22 47	26 53	24 15	23 28	23 32	22 02	03 32	19 30	01 24	23 12
04	22 41	23 40	24 19	23 22	23 28	22 04	03 30	19 29	01 24	23 12
05	22 35	19 03	24 21	23 16	23 22	22 05	03 28	19 29	01 23	23 11
06	22 28	13 32	24 23	23 10	23 17	22 06	03 27	19 29	01 23	23 11
07	22 20	07 33	24 22	23 02	23 12	22 08	03 25	19 28	01 23	23 11
08	22 12	01 N 27	24 21	22 54	23 06	22 09	03 23	19 28	01 22	23 10
09	22 04	04 S 31	24 18	22 46	23 00	22 10	03 21	19 28	01 22	23 10
10	21 55	10 11	24 14	22 36	22 53	22 12	03 19	19 28	01 21	23 10
11	21 45	15 21	24 08	22 26	22 46	22 13	03 17	19 27	01 21	23 09
12	21 36	19 53	24 01	22 15	22 40	22 14	03 15	19 27	01 20	23 09
13	21 26	23 37	23 53	22 04	22 32	22 16	03 14	19 27	01 20	23 08
14	21 15	26 22	23 43	21 51	22 25	22 17	03 12	19 27	01 19	23 08
15	21 04	27 57	23 31	21 39	22 17	22 18	03 09	19 26	01 19	23 08
16	20 53	28 15	23 18	21 25	22 09	22 19	03 07	19 26	01 18	23 07
17	20 41	27 11	23 04	21 11	22 01	22 21	03 05	19 26	01 18	23 07
18	20 29	24 46	22 48	20 56	21 53	22 22	03 03	19 26	01 17	23 06
19	20 16	21 09	22 31	20 41	21 44	22 23	03 01	19 26	01 17	23 06
20	20 03	16 29	22 12	20 25	21 35	22 24	02 59	19 26	01 16	23 06
21	19 50	11 02	21 51	20 08	21 26	22 26	02 57	19 25	01 16	23 05
22	19 36	05 S 01	21 30	19 51	21 17	22 27	02 54	19 25	01 15	23 05
23	19 22	01 N 18	21 06	19 33	21 07	22 28	02 52	19 25	01 15	23 05
24	19 08	07 40	20 41	19 15	20 57	22 29	02 50	19 25	01 14	23 04
25	18 53	13 46	20 15	18 56	20 47	22 30	02 48	19 25	01 13	23 04
26	18 38	19 17	19 47	18 37	20 37	22 31	02 45	19 25	01 13	23 03
27	18 23	23 49	19 17	18 17	20 26	22 32	02 43	19 25	01 12	23 03
28	18 07	26 57	18 46	17 56	20 15	22 33	02 41	19 25	01 11	23 03
29	17 51	28 19	18 14	17 35	20 04	22 34	02 38	19 25	01 11	23 02
30	17 35	27 44	17 40	17 14	19 53	22 35	02 36	19 25	01 10	23 02
31	17 S 18	25 N 15	17 S 04	16 S 52	19 S 42	22 N 36	02 S 33	19 N 25	01 S 09	23 S 02

ZODIAC SIGN ENTRIES

Date	h	m	Planets
01	21	11	☿ ♑
02	13	09	☽ ♋
04	13	43	☽ ♌
06	16	57	☽ ♍
09	00	06	☽ ♎
11	10	55	☽ ♏
13	23	34	☽ ♐
16	11	47	☽ ♑
17	12	43	♀ ♒
18	22	18	☽ ♒
20	01	45	☉ ♒
20	16	41	☿ ♒
21	06	50	☽ ♓
23	09	17	♂ ♒
23	13	26	☽ ♈
25	18	05	☽ ♉
26	17	37	♆ ♈
27	20	55	☽ ♊
29	22	32	☽ ♋

LATITUDES

Date	Mercury ☿ ° ′	Venus ♀ ° ′	Mars ♂ ° ′	Jupiter ♃ ° ′	Saturn ♄ ° ′	Uranus ♅ ° ′	Neptune ♆ ° ′	Pluto ♇ ° ′
01	00 S 37	00 S 31	00 S 54	00 N 14	02 S 15	00 S 12	01 S 20	03 S 46
04	00 56	00 38	00 55	00 15	02 15	00 12	01 20	03 46
07	01 13	00 44	00 56	00 15	02 14	00 12	01 20	03 47
10	01 29	00 50	00 57	00 16	02 14	00 12	01 20	03 47
13	01 41	00 56	00 58	00 16	02 13	00 12	01 19	03 47
16	01 52	01 01	00 59	00 16	02 13	00 12	01 19	03 47
19	02 00	01 06	01 00	00 17	02 12	00 12	01 19	03 47
22	02 04	01 11	01 00	00 17	02 12	00 11	01 19	03 48
25	02 05	01 15	01 01	00 18	02 11	00 11	01 19	03 48
28	02 02	01 18	01 02	00 18	02 11	00 11	01 19	03 48
31	01 S 55	01 S 21	01 S 02	00 N 18	02 S 10	00 S 11	01 S 19	03 S 48

DATA

Julian Date	2461042
Delta T	+73 seconds
Ayanamsa	24° 13′ 19″
Synetic vernal point	04° ♓ 53′ 40″
True obliquity of ecliptic	23° 26′ 17″

MOON'S PHASES, APSIDES AND POSITIONS ☽

Date	h	m	Phase	Longitude ° ′	Eclipse Indicator
03	10	03	○	13 ♋ 02	
10	15	48	☾	20 ♎ 25	
18	19	52	●	28 ♑ 44	
26	04	47	☽	06 ♉ 14	

Day	h	m		
01	21	50	Perigee	
13	20	55	Apogee	
29	21	52	Perigee	
02	08	11	Max dec	28° N 16′
08	17	45	0S	
16	05	15	Max dec	28° S 18′
23	07	05	0N	
29	16	54	Max dec	28° N 22′

LONGITUDES

Date	Chiron ⚷ ° ′	Ceres ⚳ ° ′	Pallas ⚴ ° ′	Juno ⚵ ° ′	Vesta ⚶ ° ′	Black Moon Lilith ⚸ ° ′
01	22 ♈ 36	07 ♈ 11	22 ♒ 44	01 ♑ 03	24 ♑ 33	01 ♐ 19
11	22 ♈ 38	09 ♈ 30	25 ♒ 45	04 ♑ 38	29 ♑ 49	02 ♐ 26
21	22 ♈ 46	12 ♈ 10	28 ♒ 52	08 ♑ 13	05 ♒ 04	03 ♐ 33
31	22 ♈ 59	15 ♈ 08	02 ♓ 04	11 ♑ 45	10 ♒ 18	04 ♐ 40

All ephemeris data is given at 12.00 UT and the Moon's longitude is additionally given for 24.00 UT

ASPECTARIAN

h m Aspects

01 Thursday
02 20 ☽ Q ♄
04 21 ☽ ⊼ ♀
06 37 ☽ ⊼ ☉
07 40 ☽ Q ♆
07 54 ♂ ⚼ ♅
10 04 ☽ ⊼ ♂
13 33 ☿ □ ♆
13 41 ☽ ⊥ ♃
17 36 ☽ ⚼ ♇
23 10 ☽ ⊻ ♃

02 Friday
07 09 ☽ □ ♄
07 59 ☽ ± ♇
09 49 ☽ ⊻ ♅
12 24 ☽ □ ♆
14 58 ☽ ☍ ☿
17 35 ☽ ⊼ ♇
19 23 ☽ ⊥ ♅

03 Saturday
02 12 ♂ Q ♄
06 39 ☉ ⚼ ♅
08 40 ☽ ☍ ♀
09 49 ☽ ∠ ♅
10 03 ☽ ☍ ☉
12 37 ☽ ☍ ♂
16 56 ☿ ⊻ ♇
21 55 ♀ ⚼ ♅
22 54 ☽ ☌ ♃

04 Sunday
07 44 ☽ △ ♄
08 02 ☽ # ☿
09 25 ☿ ± ♅
10 11 ☽ ⚹ ♅
12 59 ☽ △ ♆
13 15 ☽ # ♂
13 46 ☽ # ♀
14 45 ☽ # ♇
17 47 ☽ # ☉
18 11 ☉ Q ♄
18 25 ☽ ☍ ♇
21 01 ☽ ∥ ♃
21 25 ☽ ⊼ ☿

05 Monday
03 29 ♀ Q ♄
06 09 ☽ Q ♅
08 43 ☽ ± ☿
08 51 ☽ ⚼ ♄
09 58 ☽ ∥ ♅
14 10 ☽ ⚼ ♆
14 56 ☽ ⊼ ♀
15 24 ☽ ⊼ ☉
17 08 ☽ ⊼ ♂

06 Tuesday
00 18 ☽ ± ♄
00 40 ☽ ⊻ ♃
02 02 ☽ ⚼ ☿
02 16 ☽ ± ♀
02 32 ☽ ± ☉
04 05 ☽ ± ♂
05 41 ☽ ± ♆
07 19 ♀ ∥ ♇
10 48 ☽ ⊼ ♄
11 01 ☽ ⊥ ♃
13 05 ☽ □ ♅
16 13 ☽ ⊼ ♆
16 36 ☉ ☌ ♀
19 36 ☽ ⚼ ☉
19 39 ☽ ⚼ ♀
20 51 ☽ ⚼ ♂
22 06 ☽ ⊼ ♇

07 Wednesday
02 50 ☽ ∠ ♃
08 02 ☽ △ ☿
08 54 ☽ ± ♇
09 19 ♂ Q ♆
15 39 ♂ ∥ ♇
16 09 ♀ Q ♆
21 46 ☉ Q ♆

08 Thursday
01 04 ☽ △ ☉
01 25 ☽ ⚼ ♇
01 44 ☽ △ ♀
01 46 ☽ △ ♂
02 43 ♀ ☌ ♂
04 21 ☽ # ♄
05 59 ☽ ⚹ ♃
12 19 ☽ # ♆
17 46 ☽ ☍ ♄
19 33 ☉ # ♃
19 50 ☽ △ ♅
23 12 ☽ ∥ ♆
23 23 ☽ ☍ ♆

09 Friday
04 26 ☽ Q ♃
05 47 ☽ △ ♇
07 14 ☽ ∥ ♄
11 41 ☉ ☌ ♂
17 35 ♀ ☍ ♃

10 Saturday
00 33 ☽ □ ☿
00 41 ☽ ⚼ ♅
01 33 ☿ ⚼ ♅
08 42 ☉ ☍ ♃
14 25 ♂ ☍ ♃
15 08 ☽ □ ♃
15 12 ☽ □ ♂
15 48 ☽ □ ☉
17 54 ☽ □ ♀
18 24 ☽ ± ♅

11 Sunday
04 36 ☽ ⊼ ♄
06 20 ☽ ⊼ ♅
09 41 ☿ Q ♄
10 15 ☽ ⊼ ♆
16 42 ☽ ± ♄
17 03 ☽ □ ♇
17 43 ☽ Q ☿
22 20 ☽ ± ♆

12 Monday
05 43 ☽ Q ♂
07 13 ☽ Q ☉
09 30 ☽ # ♅
10 15 ☽ Q ♀
10 58 ☽ ⚼ ♄
13 43 ♀ # ♃
16 32 ☽ ⚼ ♆
21 44 ☽ ⚹ ☿
21 51 ☽ ∥ ☉

13 Tuesday
01 48 ☽ ∥ ♀
02 30 ☽ # ♃
02 46 ☽ △ ♃
04 22 ☿ Q ♆
04 40 ☽ ∥ ♂
05 31 ☽ Q ♇
07 43 ☽ ⚹ ♂
08 33 ☽ ∥ ♇
09 48 ☽ ⚹ ☉
13 27 ☽ ⚹ ♀
13 49 ☽ ∥ ☿
17 33 ☽ △ ♄
18 49 ☽ ☍ ♅
22 59 ☽ △ ♆

14 Wednesday
05 54 ☽ ⚹ ♇
08 17 ☿ ☍ ♃
08 48 ☽ ⚼ ♃
08 54 ☽ ∠ ☿
16 12 ☽ ∠ ♂
19 01 ☽ ∠ ☉
23 22 ☽ ∠ ♀

15 Thursday
02 45 ☽ ± ♃
05 54 ☽ ⊥ ☿
06 19 ♀ ⚹ ♄
09 46 ♂ # ♃
11 34 ☽ ⊥ ♂
12 12 ☽ ∠ ♇
14 40 ☽ ⊼ ♃
14 47 ☽ ⊥ ☉
15 22 ♀ △ ♅
19 33 ☽ ⊥ ♀
19 48 ☽ ⊻ ☿

16 Friday
00 23 ☽ ⊻ ♂
03 52 ☽ ⊻ ☉
06 13 ☽ ⊥ ♇
06 17 ☽ □ ♄
07 03 ☽ ⊼ ♅
08 53 ☽ ⊻ ♀
11 19 ☽ □ ♆
18 08 ☽ ⊻ ♇
18 55 ☽ ± ♅

17 Saturday
07 14 ☿ ∥ ♇
08 33 ♀ ⚹ ♆
10 41 ☉ ⚹ ♄
12 35 ☽ ⚼ ♅
16 59 ☉ △ ♅
17 58 ☽ Q ♄
22 44 ☽ Q ♆

18 Sunday
01 14 ☽ ☍ ♃
07 40 ☿ ☌ ♂
15 11 ☽ ☌ ♂
15 48 ☽ ☌ ☿
17 22 ☽ ⚹ ♄
17 39 ☽ △ ♅
19 52 ☽ ☌ ☉
21 57 ☽ ⚹ ♆
23 58 ☽ ∥ ♇

19 Monday
02 03 ☽ ☌ ♀
03 05 ☽ ∥ ☿
04 09 ☿ ⚹ ♄
04 35 ☽ ☌ ♇
04 37 ☽ # ♃
05 38 ☿ △ ♅
08 24 ☽ ∥ ♂
14 47 ☽ ∥ ♀
17 08 ☽ ∥ ☉
21 15 ☿ # ♃
21 25 ☽ # ♅
21 55 ☉ ⚹ ♆
22 09 ☽ ∠ ♄

20 Tuesday
02 30 ☽ ∠ ♆
04 04 ♀ ☌ ♇
05 19 ♄ ⚹ ♅
05 56 ♂ △ ♅
06 02 ♂ ⚹ ♄
09 54 ☽ ⊼ ♃
14 34 ☿ ⚹ ♆
15 11 ☽ ⊥ ♄
19 25 ☽ ⊥ ♆
21 00 ☽ ± ♃

21 Wednesday
02 16 ☽ □ ♅
02 26 ☽ ⊻ ♄
03 36 ☽ ⊻ ♂
06 35 ☽ ⊻ ♆
08 55 ☽ ⊻ ☿
09 18 ☽ ⊻ ☉
13 03 ☽ ⊻ ♇
13 31 ☽ ⚼ ♃
15 24 ☽ ⊥ ♂
15 49 ☉ ☌ ☿
16 28 ☽ ⊻ ♀
21 18 ☽ ⊥ ☉
21 37 ☽ ⊥ ☿

22 Thursday
00 06 ☽ ⊥ ♇
04 39 ☽ ⊥ ♀
08 56 ☽ ∠ ♂
11 20 ☽ Q ♅
15 06 ☽ ∠ ☉
16 26 ☽ ∠ ☿
16 32 ☽ ∠ ♇
16 41 ☽ △ ♃
17 15 ☿ ☌ ♇
20 08 ☽ ∥ ♄
22 43 ☽ ∠ ♀

23 Friday
02 23 ☽ ∥ ♆
06 39 ♂ ⚹ ♆
07 21 ☉ # ♅
08 59 ☽ ⚹ ♅
09 33 ☽ ☌ ♄
10 28 ☉ ☌ ♇
10 29 ☿ ∥ ♂
11 46 ☽ # ♆
13 17 ☽ ☌ ♆
13 41 ☽ ⚹ ♂
17 51 ☽ # ♄
19 33 ☽ ⚹ ♇
20 16 ☽ ⚹ ☉
22 51 ♀ # ♅
23 16 ☽ ⚹ ☿

24 Saturday
04 18 ☽ ⚹ ♀
11 37 ☽ ∠ ♅
12 16 ☽ Q ♂
16 49 ☽ Q ♇
19 10 ☽ Q ☉
21 36 ☽ □ ♃
23 25 ☽ Q ☿

25 Sunday
03 22 ☽ ⊥ ♅
03 29 ☽ Q ♀
13 46 ☽ ⊻ ♅
14 42 ☽ ⊻ ♄
18 03 ☽ ⊻ ♆
21 27 ☽ □ ♂

26 Monday
00 07 ☽ □ ♇
01 05 ☽ ⊥ ♄
03 22 ☉ ∥ ♀
04 22 ☽ ⊥ ♆
04 28 ☽ Q ♃
04 47 ☽ □ ☉
09 04 ☽ # ♀
09 09 ☽ # ☉
10 50 ☽ □ ☿
12 38 ☽ ∥ ♅
13 31 ☽ □ ♀
14 10 ☽ # ☿
16 32 ☽ ∠ ♄
18 14 ☽ # ♂
19 43 ☽ ∠ ♆

27 Tuesday
00 41 ☽ ⚹ ♃
04 32 ☽ ∥ ♃
05 57 ☿ # ♅
07 28 ☽ # ♇
16 43 ☽ ☌ ♅
17 58 ☽ ⚹ ♄
20 58 ☽ ⚹ ♆
23 02 ♂ ☌ ♇

28 Wednesday
01 36 ☽ ∠ ♃
02 02 ♀ ∠ ♄
02 52 ☽ △ ♇
03 05 ☽ △ ♂
11 06 ☽ △ ☉
11 37 ☿ ∠ ♄
14 03 ☽ Q ♄
16 25 ☽ ⊥ ♃
16 56 ☽ Q ♆
20 01 ☽ △ ☿
20 32 ☽ △ ♀

29 Thursday
02 15 ☽ ⊻ ♃
03 46 ☽ ⚼ ♇
05 20 ☽ ⚼ ♂
10 17 ☿ ☌ ♀
11 57 ☿ ∠ ♆
12 38 ♀ ∠ ♆
13 44 ☽ ⚼ ☉
18 22 ☽ ⊻ ♅
18 37 ☽ ± ♇
19 57 ☽ □ ♄
20 59 ☽ ± ♂
22 40 ☽ □ ♆
23 35 ☽ ⚼ ♀

30 Friday
00 06 ☽ ⚼ ☿
04 14 ☽ ⊥ ♅
04 31 ☽ ⊼ ♇
05 40 ☽ ± ☉
07 25 ☽ ⊼ ♂
15 48 ☽ ± ♀
16 17 ☽ ⊼ ☉
16 58 ☽ ± ☿
18 33 ☉ ∥ ☿
19 04 ☽ ∠ ♅
20 58 ☿ ⊼ ♃

31 Saturday
02 39 ☽ ⊼ ♀
03 18 ☽ ☌ ♃
04 14 ☽ ⊼ ☿
09 39 ♀ ⊼ ♃
19 55 ☽ ⚹ ♅
21 52 ☽ △ ♄

FEBRUARY 2026

LONGITUDES

Date	Sidereal time h m s	Sun ☉ ° ' "	Moon ☽ ° ' "	Moon ☽ 24.00 ° ' "	Mercury ☿ ° '	Venus ♀ ° '	Mars ♂ ° '	Jupiter ♃ ° '	Saturn ♄ ° '	Uranus ♅ ° '	Neptune ♆ ° '	Pluto ♇ ° '
01	20 46 50	12 ♒ 37 56	07 ♌ 03 44	14 ♌ 08 49	20 ♒ 21	18 ♒ 48	07 ♒ 07	17 ♋ 18	28 ♓ 41	27 ♉ 28	00 ♈ 09	03 ♒ 43
02	20 50 47	13 38 47	21 ♌ 09 21	28 ♌ 04 52	22 08	20 04	07 54	17 R 12	28 47	27 R 28	00 11	03 45
03	20 54 43	14 39 38	04 ♍ 54 59	11 ♍ 39 29	23 54	21 19	08 41	17 05	28 53	27 28	00 13	03 47
04	20 58 40	15 40 28	18 ♍ 18 17	24 ♍ 51 24	25 40	22 34	09 28	16 59	28 59	27 D 28	00 14	03 48
05	21 02 36	16 41 17	01 ♎ 18 58	07 ♎ 41 15	27 27	23 49	10 15	16 53	29 06	27 28	00 16	03 50
06	21 06 33	17 42 04	13 ♎ 58 34	20 ♎ 11 20	29 ♒ 13	25 05	11 02	16 47	29 12	27 28	00 18	03 52
07	21 10 29	18 42 51	26 ♎ 20 01	02 ♏ 25 08	00 ♓ 58	26 20	11 49	16 41	29 18	27 28	00 20	03 54
08	21 14 26	19 43 37	08 ♏ 27 15	14 ♏ 26 57	02 42	27 35	12 36	16 35	29 24	27 28	00 21	03 56
09	21 18 23	20 44 22	20 ♏ 24 50	26 ♏ 21 30	04 26	28 ♒ 50	13 24	16 30	29 31	27 28	00 23	03 58
10	21 22 19	21 45 06	02 ♐ 17 33	08 ♐ 13 35	06 08	00 ♓ 05	14 11	16 24	29 37	27 29	00 25	04 00
11	21 26 16	22 45 49	14 ♐ 10 11	20 ♐ 07 53	07 48	01 20	14 58	16 19	29 44	27 29	00 27	04 02
12	21 30 12	23 46 31	26 ♐ 07 11	02 ♑ 08 36	09 25	02 36	15 45	16 14	29 50	27 29	00 29	04 03
13	21 34 09	24 47 11	08 ♑ 12 32	14 ♑ 19 24	11 00	03 51	16 32	16 09	29 ♓ 57	27 30	00 31	04 05
14	21 38 05	25 47 51	20 ♑ 29 30	26 ♑ 43 07	12 32	05 06	17 19	16 04	00 ♈ 03	27 30	00 33	04 07
15	21 42 02	26 48 29	03 ♒ 00 28	09 ♒ 21 41	13 59	06 21	18 07	15 59	00 10	27 31	00 35	04 09
16	21 45 58	27 49 06	15 ♒ 46 52	22 ♒ 16 02	15 22	07 36	18 54	15 55	00 17	27 32	00 37	04 11
17	21 49 55	28 49 41	28 ♒ 49 08	05 ♓ 26 06	16 39	08 51	19 41	15 51	00 23	27 32	00 39	04 12
18	21 53 52	29 ♒ 50 15	12 ♓ 06 48	18 ♓ 51 01	17 51	10 06	20 28	15 47	00 30	27 33	00 41	04 14
19	21 57 48	00 ♓ 50 47	25 ♓ 38 34	02 ♈ 29 11	18 56	11 21	21 16	15 43	00 37	27 34	00 43	04 16
20	22 01 45	01 51 18	09 ♈ 22 37	16 ♈ 18 36	19 53	12 36	22 03	15 39	00 44	27 35	00 45	04 18
21	22 05 41	02 51 47	23 ♈ 16 50	00 ♉ 17 02	20 43	13 51	22 50	15 36	00 51	27 36	00 47	04 19
22	22 09 38	03 52 13	07 ♉ 18 56	14 ♉ 22 15	21 24	15 06	23 37	15 32	00 58	27 37	00 49	04 21
23	22 13 34	04 52 39	21 ♉ 26 41	28 ♉ 32 00	21 56	16 21	24 25	15 29	01 05	27 38	00 51	04 23
24	22 17 31	05 53 02	05 ♊ 37 53	12 ♊ 44 04	22 18	17 36	25 12	15 26	01 12	27 39	00 53	04 25
25	22 21 27	06 53 23	19 ♊ 50 16	26 ♊ 56 11	22 31	18 51	25 59	15 23	01 19	27 40	00 55	04 26
26	22 25 24	07 53 42	04 ♋ 01 29	11 ♋ 05 51	22 R 34	20 06	26 46	15 21	01 26	27 41	00 57	04 28
27	22 29 21	08 54 00	18 ♋ 08 57	25 ♋ 10 25	22 27	21 21	27 34	15 19	01 33	27 42	00 59	04 30
28	22 33 17	09 ♓ 54 15	02 ♌ 09 55	09 ♌ 07 03	22 ♓ 10	22 ♓ 35	28 ♒ 21	15 ♋ 16	01 ♈ 40	27 ♉ 43	01 ♈ 02	04 ♒ 31

Date	Moon True ☊ ° '	Moon Mean ☊ ° '	Moon ☽ Latitude ° '
01	09 ♓ 05	10 ♓ 30	02 N 46
02	09 R 05	10 27	01 37
03	09 02	10 24	00 N 23
04	09 D 02	10 21	00 S 51
05	09 03	10 17	02 00
06	09 05	10 14	03 01
07	09 07	10 11	03 52
08	09 08	10 08	04 31
09	09 08	10 05	04 59
10	09 R 08	10 01	05 13
11	09 07	09 58	05 14
12	09 05	09 55	05 02
13	09 03	09 52	04 36
14	09 01	09 49	03 57
15	08 59	09 46	03 06
16	08 57	09 42	02 05
17	08 57	09 39	00 S 56
18	08 D 56	09 36	00 N 18
19	08 57	09 33	01 31
20	08 57	09 30	02 41
21	08 58	09 27	03 42
22	08 59	09 23	04 30
23	08 59	09 20	05 02
24	08 59	09 17	05 16
25	08 R 59	09 14	05 11
26	08 59	09 11	04 47
27	08 59	09 07	04 06
28	08 ♓ 59	09 ♓ 04	03 N 10

DECLINATIONS

Date	Sun ☉ ° '	Moon ☽ ° '	Mercury ☿ ° '	Venus ♀ ° '	Mars ♂ ° '	Jupiter ♃ ° '	Saturn ♄ ° '	Uranus ♅ ° '	Neptune ♆ ° '	Pluto ♇ ° '
01	17 S 01	21 N 11	16 S 27	16 S 29	19 S 30	22 N 37	02 S 31	19 N 25	01 S 09	23 S 01
02	16 44	15 58	15 49	16 06	19 18	22 38	02 28	19 25	01 08	23 01
03	16 26	10 04	15 09	15 43	19 06	22 39	02 26	19 25	01 07	23 00
04	16 08	03 N 51	14 29	15 19	18 54	22 40	02 23	19 25	01 07	23 00
05	15 50	02 S 21	13 47	14 55	18 41	22 41	02 21	19 25	01 06	23 00
06	15 32	08 17	13 03	14 30	18 29	22 42	02 18	19 25	01 05	22 59
07	15 13	13 46	12 19	14 05	18 16	22 43	02 16	19 25	01 04	22 59
08	14 54	18 36	11 34	13 40	18 03	22 43	02 13	19 25	01 04	22 59
09	14 35	22 39	10 48	13 14	17 49	22 44	02 10	19 25	01 03	22 58
10	14 15	25 44	10 02	12 48	17 36	22 45	02 08	19 25	01 02	22 58
11	13 56	27 42	09 15	12 22	17 22	22 46	02 05	19 25	01 01	22 58
12	13 36	28 24	08 28	11 55	17 08	22 46	02 02	19 25	01 01	22 57
13	13 16	27 46	07 41	11 28	16 54	22 47	02 00	19 25	01 00	22 57
14	12 55	25 47	06 55	11 00	16 40	22 48	01 57	19 25	00 59	22 57
15	12 35	22 30	06 09	10 33	16 26	22 48	01 54	19 26	00 58	22 56
16	12 14	18 05	05 25	10 05	16 11	22 49	01 51	19 26	00 57	22 56
17	11 53	12 45	04 41	09 36	15 57	22 50	01 49	19 26	00 57	22 56
18	11 32	06 45	04 00	09 08	15 42	22 50	01 46	19 26	00 56	22 55
19	11 10	00 S 20	03 21	08 39	15 27	22 51	01 43	19 26	00 55	22 55
20	10 49	06 N 11	02 44	08 10	15 11	22 51	01 40	19 26	00 54	22 55
21	10 27	12 28	02 11	07 41	14 56	22 52	01 38	19 27	00 53	22 55
22	10 05	18 12	01 41	07 12	14 40	22 52	01 35	19 27	00 53	22 54
23	09 43	22 59	01 14	06 42	14 25	22 53	01 32	19 27	00 52	22 54
24	09 21	26 26	00 52	06 12	14 09	22 53	01 29	19 27	00 51	22 54
25	08 59	28 13	00 34	05 42	13 53	22 53	01 26	19 28	00 50	22 53
26	08 36	28 10	00 21	05 12	13 37	22 54	01 23	19 28	00 49	22 53
27	08 14	26 16	00 13	04 42	13 21	22 54	01 20	19 28	00 48	22 53
28	07 S 51	22 N 46	00 S 09	04 S 12	13 S 04	22 N 54	01 S 18	19 N 29	00 S 47	22 S 53

ZODIAC SIGN ENTRIES

Date	h	m	Planets	
01	00	09	☽	♌
03	03	21	☽	♍
05	09	32	☽	♎
06	22	48	☿	♓
07	19	13	☽	♏
10	07	22	☽	♐
10	10	19	♀	♓
12	19	44	☽	♑
14	00	11	♄	♈
15	06	17	☽	♒
17	14	09	☽	♓
18	15	52	☉	♓
19	19	39	☽	♈
21	23	31	☽	♉
24	02	29	☽	♊
26	05	11	☽	♋
28	08	17	☽	♌

LATITUDES

Date	Mercury ☿ ° '	Venus ♀ ° '	Mars ♂ ° '	Jupiter ♃ ° '	Saturn ♄ ° '	Uranus ♅ ° '	Neptune ♆ ° '	Pluto ♇ ° '
01	01 S 51	01 S 22	01 S 03	00 N 18	02 S 10	00 S 11	01 S 19	03 S 48
04	01 37	01 24	01 03	00 19	02 10	00 11	01 19	03 49
07	01 16	01 26	01 04	00 19	02 10	00 11	01 19	03 49
10	00 49	01 27	01 04	00 19	02 09	00 11	01 19	03 49
13	00 S 16	01 28	01 05	00 20	02 09	00 11	01 19	03 50
16	00 N 23	01 28	01 05	00 20	02 09	00 11	01 19	03 50
19	01 07	01 27	01 05	00 20	02 08	00 11	01 19	03 50
22	01 53	01 26	01 05	00 20	02 08	00 11	01 18	03 51
25	02 37	01 24	01 06	00 20	02 08	00 11	01 18	03 51
28	03 13	01 22	01 06	00 20	02 08	00 11	01 18	03 52
31	03 N 36	01 S 19	01 S 06	00 N 21	02 S 08	00 S 11	01 S 18	03 S 52

DATA

Julian Date	2461073
Delta T	+73 seconds
Ayanamsa	24° 13' 25"
Synetic vernal point	04° ♓ 53' 34"
True obliquity of ecliptic	23° 26' 18"

MOON'S PHASES, APSIDES AND POSITIONS ☽

Date	h	m	Phase	Longitude ° '	Eclipse Indicator
01	22	09	○	13 ♌ 04	
09	12	43	☾	20 ♏ 46	
17	12	01	●	28 ♒ 50	Annular
24	12	28	☽	05 ♊ 54	

Day	h	m		
10	16	54	Apogee	
24	23	16	Perigee	
05	02	49	0S	
12	12	47	Max dec	28° S 24'
19	13	14	0N	
25	23	15	Max dec	28° N 26'

LONGITUDES

Date	Chiron ⚷ ° '	Ceres ⚳ ° '	Pallas ⚴ ° '	Juno ⚵ ° '	Vesta ⚶ ° '	Black Moon Lilith ⚸ ° '
01	23 ♈ 01	15 ♈ 26	02 ♓ 24	12 ♑ 06	10 ♒ 49	04 ♐ 47
11	23 ♈ 19	18 ♈ 39	05 ♓ 40	15 ♑ 34	16 ♒ 01	05 ♐ 54
21	23 ♈ 43	22 ♈ 04	09 ♓ 00	18 ♑ 57	21 ♒ 10	07 ♐ 02
31	24 ♈ 10	25 ♈ 39	12 ♓ 20	22 ♑ 14	26 ♒ 16	08 ♐ 09

ASPECTARIAN

h m	Aspects
01 Sunday	
00 23	☽ △ ♆
02 09	☽ # ♇
04 26	☽ ∥ ♃
06 21	☽ ☍ ♇
08 50	☿ ∥ ♀
12 06	☽ ☍ ♂
16 03	☽ Q ♅
20 34	☽ # ♂
20 41	☽ ∥ ♅
22 09	☽ ☍ ☉
23 17	♂ # ♅
23 18	☽ ⚼ ♄
02 Monday	
01 45	☽ ⚼ ♆
05 15	☽ ⊻ ♃
08 34	☽ # ☉
09 56	☽ ☍ ♀
11 23	☽ # ♀
12 44	☽ # ☿
13 55	☽ ☍ ☿
14 50	☽ ± ♄
15 30	☽ ⊥ ♃
15 45	☉ ∠ ♄
17 14	☽ ± ♆
21 29	☿ ⊥ ♄
22 55	☽ □ ♅
03 Tuesday	
01 20	☽ ⊼ ♄
01 37	☿ ± ♃
03 42	☽ ⊼ ♆
07 03	☽ ∠ ♃
09 59	☽ ⊼ ♇
16 15	☿ ⊥ ♆
19 06	☽ ⊼ ♂
20 39	☽ ± ♇
04 Wednesday	
01 23	☉ ∠ ♆
02 34	♅ St D
06 32	☽ ± ♂
06 50	☽ ⊼ ☉
09 37	☽ ⚹ ♃
12 55	☽ ⚼ ♇
17 38	☽ # ♄
18 40	☽ ± ☉
19 21	♀ ± ♃
20 37	☽ ⊼ ♀
20 49	♀ ⊥ ♄
22 33	☽ # ♆
05 Thursday	
00 01	☽ ⚼ ♂
03 39	☽ ⊼ ☿
04 49	☽ △ ♅
07 06	☽ ∥ ♆
07 29	☽ Q ♃
07 49	☽ ☍ ♄
08 54	☽ ± ♀
10 02	☽ ☍ ♆
11 59	☽ ∥ ♄
12 13	☿ □ ♅
12 45	☽ ⚼ ☉
16 09	☉ ⊼ ♃
16 38	☽ ± ☿
16 45	☽ △ ♇
20 45	♀ ⊥ ♆
06 Friday	
03 43	☽ ⚼ ♀
06 00	☽ △ ♂
09 06	☽ ⚼ ♅
11 50	☿ ⊻ ♄
12 31	☽ ⚼ ☿
17 21	☽ □ ♃
19 49	☽ △ ☉
07 Saturday	
02 29	☽ ± ♅
03 08	☿ ⊻ ♆
06 14	☽ ∥ ☿
11 59	☽ △ ♀
13 25	☽ ∥ ♀
14 13	☽ ⊼ ♅
17 53	☽ ⊼ ♄
18 28	☽ ∥ ☉
19 53	☽ ⊼ ♆
21 21	☿ ⚼ ♃
22 39	☽ △ ☿
08 Sunday	
02 59	☽ □ ♇
05 52	☽ ± ♄
07 49	☽ ± ♆
09 11	☽ ∥ ♂
09 49	♀ □ ♅
16 26	☽ # ♅
20 53	☽ □ ♂
09 Monday	
00 01	☽ ⚼ ♄
01 52	☽ ⚼ ♆
04 10	☽ △ ♃
05 23	☿ ⊻ ♇
12 37	☽ # ♃
12 43	☽ □ ☉
14 13	☽ ∥ ♇
15 08	☽ Q ♇
10 Tuesday	
02 10	♀ ⊻ ♄
02 16	☽ ☍ ♅
06 33	☽ △ ♄
07 01	☽ □ ♀
08 12	☽ △ ♆
10 13	☽ ⚼ ♃
11 45	☽ Q ♂
15 27	☽ ⚹ ♇
18 30	♀ ⊻ ♆
21 02	☽ □ ☿
11 Wednesday	
02 12	☉ ± ♃
03 35	♂ ∠ ♄
04 17	☽ ± ♃
04 29	☽ Q ☉
11 32	♀ ⚼ ♃
13 43	☽ ⚹ ♂
16 18	☽ ⊼ ♃
21 48	☽ ∠ ♇
23 38	☽ Q ♀
12 Thursday	
03 28	♂ ∠ ♆
06 53	☽ ⚹ ☉
13 35	☉ ⊥ ♄
14 44	☽ ⊼ ♅
15 00	☽ Q ☿
15 53	☽ ⊥ ♇
19 29	☽ □ ♄
20 43	☽ □ ♆
21 44	☿ ⊥ ♇
21 53	☽ ∠ ♂
13 Friday	
01 15	♂ ⊼ ♃
02 23	☽ ⚹ ♀
02 41	☽ ± ♅
03 50	☽ ⊻ ♇
05 19	☉ ⊥ ♆
15 23	☽ ∠ ☉
16 44	♀ ⊻ ♇
16 54	☽ ⊥ ♂
18 17	☽ ⚹ ☿
20 26	☽ ⚼ ♅
14 Saturday	
03 28	☽ ☍ ♃
05 26	☽ ⊻ ♂
07 14	☽ Q ♄
08 13	☽ Q ♆
10 32	☽ ⊥ ☉
11 09	☽ ∠ ♀
23 08	☽ ⊻ ☉
15 Sunday	
01 31	☽ △ ♅
03 21	☽ ∠ ☿
06 23	☽ ⊥ ♀
06 33	☽ ⚹ ♄
07 22	☽ ⚹ ♆
09 13	☽ ∥ ♇
10 04	☽ # ♃
14 10	☽ ☌ ♇
19 01	☽ ⊻ ♀
22 34	☽ ⊥ ☿
16 Monday	
05 02	☉ □ ♅
05 16	☽ # ♅
11 03	☽ ∠ ♄
11 08	☽ ⊻ ☿
11 41	☽ ∠ ♆
12 15	☽ ⊼ ♃
14 57	☿ Q ♅
18 09	☽ ☌ ♂
21 25	☽ ∥ ♂
21 31	☿ △ ♃
23 18	☽ ± ♃
17 Tuesday	
03 50	☽ ⊥ ♄
04 21	☽ ⊥ ♆
09 40	☽ □ ♅
12 01	☽ ☌ ☉
14 53	☽ ⊻ ♄
15 20	☽ ⊻ ♆
15 40	☽ ⚼ ♃
15 52	☽ ∥ ☉
21 48	☽ ⊻ ♇
18 Wednesday	
01 56	☽ ∥ ♀
08 02	☽ ☌ ♀
08 38	☽ ⊥ ♇
14 37	♀ ⊥ ♇
18 09	☽ Q ♅
18 31	☽ △ ♃
23 09	☽ ☌ ☿
23 36	☽ ∥ ☿
19 Thursday	
00 43	☽ ∠ ♇
03 47	☽ ⊻ ♂
05 48	☉ ⊻ ♄
06 51	☽ ∥ ♄
08 41	☉ ⊻ ♆
09 03	☉ ⚼ ♃
09 51	☽ ∥ ♆
15 01	☽ ⊥ ♂
15 23	☽ ⚹ ♅
16 36	☽ # ♆
19 30	☽ # ♄
20 14	☿ ∠ ♇
20 48	☽ ☌ ♄
20 55	☽ ☌ ♆
21 51	☽ ⊻ ☉
20 Friday	
00 22	☽ # ☿
00 51	♂ ± ♃
03 08	☽ ⚹ ♇
07 42	☽ ∠ ♂
09 09	☽ ⊥ ☉
16 53	♄ ☌ ♆
17 33	☽ ∠ ♅
18 09	☽ ⊻ ♀
18 55	☽ # ♀
22 49	☽ □ ♃
21 Saturday	
00 00	☽ Q ♇
01 57	☽ ∠ ☉
04 33	☽ # ☉
05 32	☽ ⊥ ♀
07 20	☽ ⊻ ☿
09 06	☽ ⊥ ♅
11 11	☽ ⚹ ♂
18 13	☽ ⊥ ☿
19 24	☽ ⊻ ♅
21 30	☽ # ♂
22 29	☽ ∠ ♀
22 Sunday	
00 53	☽ ⊻ ♆
01 04	☽ ⊻ ♄
05 35	☽ Q ♃
05 40	☽ ⚹ ☉
06 56	☽ □ ♇
08 57	☽ Q ♂
10 22	☽ ∠ ☿
11 09	☽ ⊥ ♆
11 23	☽ ⊥ ♄
17 37	☿ ∥ ♄
17 46	☽ ∥ ♅
20 02	♀ △ ♃
21 52	♀ Q ♅
23 49	☉ ⊻ ♇
23 Monday	
01 56	☽ ⚹ ♃
02 29	☽ ∠ ♆
02 32	☽ ⚹ ♀
02 49	☽ ∠ ♄
03 40	☽ Q ☉
11 24	☽ ∥ ♃
11 32	☽ # ♇
12 51	☽ ⚹ ☿
17 19	☽ □ ♂
22 29	☽ ☌ ♅
24 Tuesday	
00 49	☽ Q ♀
02 00	♂ ⊥ ♆
03 15	☽ ∠ ♃
03 57	☽ ⚹ ♆
04 26	☽ ⚹ ♄
09 42	☽ Q ☿
09 56	☽ △ ♇
11 48	♂ ⊥ ♄
12 28	☽ □ ☉
13 24	☿ ∥ ♆
18 25	☽ ⊥ ♃
25 Wednesday	
00 17	☽ Q ♆
00 53	☽ Q ♄
04 31	☽ ⊻ ♃
10 10	☽ □ ♀
11 19	☽ ⚼ ♇
14 44	♃ # ♇
16 34	☽ □ ☿
23 00	☽ △ ♂
23 35	♀ ∠ ♇
26 Thursday	
01 15	☽ ⊻ ♅
02 34	☽ ± ♇
06 47	☽ □ ♆
06 48	☿ St R
07 34	☽ □ ♄
11 25	☽ ⊥ ♅
12 45	☽ ⊼ ♇
19 04	☽ △ ☉
27 Friday	
01 56	☽ ⚼ ♂
02 43	☽ ∠ ♅
07 10	☽ ☌ ♃
16 21	♂ □ ♅
17 59	☽ △ ♀
18 10	☽ ± ♂
19 13	☽ △ ☿
22 35	☽ ⚼ ☉
28 Saturday	
04 21	☽ ⚹ ♅
05 03	☽ ⊼ ♂
05 35	☿ ☌ ♀
10 02	☽ △ ♆
11 08	☽ △ ♄
11 11	☽ ∥ ♃
11 21	☽ # ♇
15 14	☽ ± ☉
16 04	☽ ☍ ♇
20 24	☽ ⚼ ☿
22 17	☽ ⚼ ♀

All ephemeris data is given at 12.00 UT and the Moon's longitude is additionally given for 24.00 UT

MARCH 2026

LONGITUDES

Date	Sidereal time h m s	Sun ☉ ° ' "	Moon ☽ ° ' "	Moon ☽ 24.00 ° ' "	Mercury ☿ ° '	Venus ♀ ° '	Mars ♂ ° '	Jupiter ♃ ° '	Saturn ♄ ° '	Uranus ♅ ° '	Neptune ♆ ° '	Pluto ♇ ° '
01	22 37 14	10 ♓ 54 28	16 ♌ 01 31	22 ♌ 52 56	21 ♓ 44	23 ♓ 50	29 ♒ 08	15 ♋ 14	01 ♈ 47	27 ♉ 45	01 ♈ 04	04 ♒ 33
02	22 41 10	11 54 40	29 ♌ 41 02	06 ♍ 25 31	21 R 10	25 05	29 ♒ 56	15 R 13	01 54	27 46	01 06	04 34
03	22 45 07	12 54 49	13 ♍ 06 10	19 ♍ 42 48	20 29	26 20	00 ♓ 43	15 11	02 01	27 47	01 08	04 36
04	22 49 03	13 54 57	26 ♍ 15 17	02 ♎ 43 35	19 41	27 34	01 30	15 10	02 09	27 49	01 10	04 38
05	22 53 00	14 55 02	09 ♎ 07 42	15 ♎ 27 42	18 47	28 ♓ 49	02 17	15 08	02 16	27 50	01 12	04 39
06	22 56 56	15 55 06	21 ♎ 43 43	27 ♎ 55 59	17 50	00 ♈ 04	03 05	15 07	02 23	27 52	01 15	04 41
07	23 00 53	16 55 09	04 ♏ 04 46	10 ♏ 10 24	16 50	01 18	03 52	15 07	02 30	27 53	01 17	04 42
08	23 04 50	17 55 10	16 ♏ 13 15	22 ♏ 13 47	15 49	02 33	04 39	15 06	02 38	27 55	01 19	04 44
09	23 08 46	18 55 09	28 ♏ 12 27	04 ♐ 09 48	14 49	03 48	05 26	15 06	02 45	27 57	01 21	04 45
10	23 12 43	19 55 07	10 ♐ 06 20	16 ♐ 02 39	13 50	05 02	06 14	15 06	02 52	27 58	01 24	04 47
11	23 16 39	20 55 03	21 ♐ 59 20	27 ♐ 56 58	12 54	06 17	07 01	15 D 05	03 00	28 00	01 26	04 48
12	23 20 36	21 54 57	03 ♑ 56 08	09 ♑ 57 26	12 02	07 31	07 48	15 05	03 07	28 02	01 28	04 50
13	23 24 32	22 54 50	16 ♑ 01 25	22 ♑ 08 39	11 15	08 46	08 35	15 05	03 15	28 04	01 30	04 51
14	23 28 29	23 54 41	28 ♑ 19 38	04 ♒ 34 49	10 33	10 00	09 23	15 06	03 22	28 06	01 33	04 52
15	23 32 25	24 54 30	10 ♒ 54 37	17 ♒ 19 21	09 57	11 15	10 10	15 07	03 29	28 08	01 35	04 54
16	23 36 22	25 54 17	23 ♒ 49 16	00 ♓ 24 33	09 27	12 29	10 57	15 08	03 37	28 10	01 37	04 55
17	23 40 19	26 54 03	07 ♓ 05 13	13 ♓ 51 15	09 03	13 44	11 44	15 09	03 44	28 12	01 39	04 56
18	23 44 15	27 53 47	20 ♓ 42 28	27 ♓ 38 35	08 46	14 58	12 32	15 11	03 52	28 14	01 42	04 58
19	23 48 12	28 53 28	04 ♈ 39 12	11 ♈ 43 50	08 35	16 12	13 19	15 12	03 59	28 16	01 44	04 59
20	23 52 08	29 ♓ 53 08	18 ♈ 51 53	26 ♈ 02 42	08 30	17 27	14 06	15 14	04 07	28 18	01 46	05 00
21	23 56 05	00 ♈ 52 45	03 ♉ 15 35	10 ♉ 29 48	08 D 31	18 41	14 53	15 16	04 14	28 20	01 48	05 01
22	00 00 01	01 52 21	17 ♉ 44 37	24 ♉ 59 21	08 38	19 55	15 40	15 18	04 22	28 22	01 51	05 03
23	00 03 58	02 51 54	02 ♊ 13 20	09 ♊ 26 00	08 50	21 09	16 27	15 20	04 29	28 25	01 53	05 04
24	00 07 54	03 51 25	16 ♊ 36 50	23 ♊ 45 25	09 08	22 23	17 14	15 23	04 37	28 27	01 55	05 05
25	00 11 51	04 50 54	00 ♋ 51 24	07 ♋ 54 34	09 30	23 38	18 02	15 25	04 44	28 29	01 58	05 06
26	00 15 48	05 50 20	14 ♋ 54 42	21 ♋ 51 43	09 58	24 52	18 49	15 28	04 52	28 32	02 00	05 07
27	00 19 44	06 49 44	28 ♋ 45 33	05 ♌ 36 11	10 30	26 06	19 36	15 31	04 59	28 34	02 02	05 08
28	00 23 41	07 49 06	12 ♌ 23 37	19 ♌ 07 54	11 06	27 20	20 23	15 34	05 07	28 37	02 04	05 09
29	00 27 37	08 48 25	25 ♌ 49 03	02 ♍ 27 06	11 46	28 34	21 10	15 38	05 14	28 39	02 07	05 10
30	00 31 34	09 47 42	09 ♍ 02 06	15 ♍ 34 04	12 30	29 ♈ 48	21 57	15 41	05 22	28 42	02 09	05 12
31	00 35 30	10 ♈ 46 57	22 ♍ 03 00	28 ♍ 28 57	13 ♓ 17	01 ♉ 02	22 ♓ 44	15 ♋ 45	05 ♈ 29	28 ♉ 44	02 ♈ 11	05 ♒ 12

MOON

Date	Moon True ☊ ° '	Moon Mean ☊ ° '	Moon ☽ Latitude ° '
01	08 ♓ 59	09 ♓ 01	02 N 04
02	08 D 59	08 58	00 N 51
03	08 R 59	08 55	00 S 23
04	08 59	08 52	01 34
05	08 59	08 48	02 39
06	08 58	08 45	03 35
07	08 56	08 42	04 20
08	08 55	08 39	04 52
09	08 54	08 36	05 11
10	08 54	08 33	05 16
11	08 D 53	08 29	05 08
12	08 54	08 26	04 47
13	08 55	08 23	04 12
14	08 56	08 20	03 26
15	08 57	08 17	02 29
16	08 58	08 13	01 23
17	08 59	08 10	00 S 10
18	08 R 59	08 07	01 N 04
19	08 57	08 04	02 17
20	08 55	08 01	03 22
21	08 52	07 58	04 16
22	08 49	07 54	04 53
23	08 46	07 51	05 12
24	08 44	07 48	05 11
25	08 43	07 45	04 51
26	08 D 43	07 42	04 14
27	08 44	07 39	03 22
28	08 46	07 35	02 20
29	08 47	07 32	01 N 10
30	08 R 48	07 29	00 S 01
31	08 ♓ 47	07 ♓ 26	01 S 12

DECLINATIONS

Date	Sun ☉ ° '	Moon ☽ ° '	Mercury ☿ ° '	Venus ♀ ° '	Mars ♂ ° '	Jupiter ♃ ° '	Saturn ♄ ° '	Uranus ♅ ° '	Neptune ♆ ° '	Pluto ♇ ° '
01	07 S 29	18 N 00	00 S 11	03 S 41	12 S 48	22 N 55	01 S 15	19 N 29	00 S 47	22 S 52
02	07 06	12 23	00 17	03 11	12 31	22 55	01 12	19 29	00 46	22 52
03	06 43	06 17	00 28	02 40	12 15	22 55	01 09	19 30	00 45	22 52
04	06 20	00 N 03	00 43	02 09	11 58	22 55	01 06	19 30	00 44	22 52
05	05 56	06 S 04	01 03	01 39	11 41	22 56	01 03	19 30	00 43	22 51
06	05 33	11 48	01 25	01 08	11 24	22 56	01 00	19 31	00 42	22 51
07	05 10	16 57	01 51	00 37	11 07	22 56	00 57	19 31	00 41	22 51
08	04 47	21 21	02 19	00 S 06	10 49	22 56	00 54	19 31	00 40	22 51
09	04 23	24 48	02 48	00 N 25	10 32	22 56	00 51	19 32	00 40	22 50
10	04 00	27 11	03 18	00 56	10 14	22 56	00 48	19 32	00 39	22 50
11	03 36	28 19	03 48	01 27	09 57	22 56	00 45	19 33	00 38	22 50
12	03 12	28 09	04 19	01 58	09 39	22 56	00 43	19 33	00 37	22 50
13	02 49	26 39	04 48	02 28	09 21	22 56	00 40	19 33	00 36	22 50
14	02 25	23 51	05 16	02 59	09 04	22 56	00 37	19 34	00 35	22 50
15	02 01	19 53	05 43	03 30	08 46	22 56	00 34	19 34	00 34	22 49
16	01 38	14 53	06 07	04 01	08 28	22 56	00 31	19 35	00 33	22 49
17	01 14	09 04	06 30	04 31	08 10	22 56	00 28	19 35	00 32	22 49
18	00 50	02 S 42	06 50	05 02	07 51	22 56	00 25	19 36	00 32	22 49
19	00 26	03 N 56	07 08	05 32	07 33	22 56	00 22	19 36	00 31	22 49
20	00 S 03	10 30	07 24	06 02	07 15	22 56	00 19	19 37	00 30	22 49
21	00 N 21	16 36	07 37	06 32	06 57	22 56	00 16	19 37	00 29	22 48
22	00 45	21 48	07 48	07 02	06 38	22 55	00 13	19 38	00 28	22 48
23	01 08	25 42	07 56	07 32	06 20	22 55	00 10	19 38	00 27	22 48
24	01 32	27 55	08 02	08 02	06 01	22 55	00 07	19 39	00 26	22 48
25	01 56	28 17	08 06	08 31	05 43	22 55	00 04	19 39	00 25	22 48
26	02 19	26 48	08 07	09 01	05 24	22 54	00 S 01	19 40	00 24	22 48
27	02 43	23 42	08 06	09 30	05 06	22 54	00 N 02	19 40	00 23	22 48
28	03 06	19 19	08 03	09 59	04 47	22 54	00 05	19 41	00 22	22 48
29	03 29	14 01	07 58	10 27	04 28	22 53	00 08	19 41	00 22	22 48
30	03 53	08 10	07 51	10 56	04 10	22 53	00 11	19 42	00 21	22 47
31	04 N 16	02 N 03	07 S 42	11 N 24	03 S 51	22 N 53	00 N 14	19 N 43	00 S 20	22 S 47

ZODIAC SIGN ENTRIES

Date	h	m	Planets
02	12	34	☽ ♍
02	14	16	♂ ♓
04	18	56	☽ ♎
06	10	46	♀ ♈
07	04	01	☽ ♏
09	15	36	☽ ♐
12	04	07	☽ ♑
14	15	13	☽ ♒
16	23	16	☽ ♓
19	04	03	☽ ♈
20	14	46	☉ ♈
21	06	35	☽ ♉
23	08	19	☽ ♊
25	10	33	☽ ♋
27	14	10	☽ ♌
29	19	33	☽ ♍
30	16	01	♀ ♉

LATITUDES

Date	Mercury ☿ ° '	Venus ♀ ° '	Mars ♂ ° '	Jupiter ♃ ° '	Saturn ♄ ° '	Uranus ♅ ° '	Neptune ♆ ° '	Pluto ♇ ° '
01	03 N 22	01 S 21	01 S 06	00 N 21	02 S 08	00 S 11	01 S 18	03 S 52
04	03 39	01 18	01 06	00 21	02 08	00 11	01 18	03 52
07	03 38	01 14	01 06	00 21	02 08	00 11	01 18	03 53
10	03 19	01 10	01 05	00 21	02 07	00 11	01 18	03 53
13	02 45	01 05	01 05	00 21	02 07	00 11	01 18	03 54
16	02 03	01 00	01 05	00 22	02 07	00 11	01 18	03 54
19	01 19	00 54	01 05	00 22	02 07	00 10	01 18	03 55
22	00 N 35	00 48	01 04	00 22	02 07	00 10	01 18	03 55
25	00 S 06	00 42	01 04	00 22	02 07	00 10	01 18	03 56
28	00 42	00 35	01 03	00 22	02 08	00 10	01 18	03 56
31	01 S 13	00 S 28	01 S 03	00 N 22	02 S 08	00 S 10	01 S 18	03 S 57

DATA

Julian Date	2461101
Delta T	+73 seconds
Ayanamsa	24° 13' 29"
Synetic vernal point	04° ♓ 53' 30"
True obliquity of ecliptic	23° 26' 18"

MOON'S PHASES, APSIDES AND POSITIONS ☽

Date	h	m	Phase	Longitude ° '	Eclipse Indicator
03	11	38	○	12 ♍ 54	total
11	09	38	☾	20 ♐ 49	
19	01	23	●	28 ♓ 27	
25	19	18	☽	05 ♋ 09	

Day	h	m		
10	13	41	Apogee	
22	11	34	Perigee	
04	12	11	0S	
11	21	05	Max dec	28° S 25'
18	21	48	0N	
25	04	35	Max dec	28° N 22'
31	20	02	0S	

LONGITUDES

Date	Chiron ⚷ ° '	Ceres ⚳ ° '	Pallas ⚴ ° '	Juno ⚵ ° '	Vesta ⚶ ° '	Black Moon Lilith ⚸ ° '
01	24 ♈ 04	27 ♈ 55	11 ♓ 40	21 ♑ 35	25 ♒ 15	07 ♐ 55
11	24 ♈ 34	28 ♈ 36	15 ♓ 01	24 ♑ 45	00 ♓ 18	09 ♐ 03
21	25 ♈ 06	02 ♉ 24	18 ♓ 22	27 ♑ 46	05 ♓ 17	10 ♐ 10
31	25 ♈ 41	06 ♉ 17	21 ♓ 41	00 ♒ 35	10 ♓ 11	11 ♐ 17

All ephemeris data is given at 12.00 UT and the Moon's longitude is additionally given for 24.00 UT

ASPECTARIAN

h m	Aspects
01 Sunday	
01 04	☽ Q ♅
02 24	☽ ⊼ ☉
03 08	☉ ⊥ ♇
05 05	☽ ∥ ♅
10 38	☽ ⊻ ♃
11 31	☽ ± ☿
12 04	☽ ⚼ ♆
13 20	☽ ⚼ ♄
15 29	☽ ± ♀
21 06	☽ ⊥ ♃
21 38	☽ ⊼ ☿
02 Monday	
03 03	☽ ⊼ ♀
03 53	☽ ± ♆
05 15	☽ ± ♄
08 36	☽ □ ♅
11 25	☽ # ♂
12 27	☽ ☍ ♂
12 56	☽ ∠ ♃
14 31	☽ ⊼ ♆
15 58	☽ ⊼ ♄
20 21	♂ ⚼ ♃
20 43	☽ ⊼ ♇
03 Tuesday	
07 29	☽ ± ♇
10 15	☽ # ☉
11 38	☽ ☍ ☉
15 45	☽ ⚹ ♃
23 49	☽ ⚼ ♇
04 Wednesday	
00 39	☽ ☍ ☿
01 26	♂ ⊻ ♆
03 09	☽ # ♀
07 54	☽ # ♄
09 21	☽ # ♆
09 30	☽ # ☿
12 41	☿ ∥ ♆
13 23	☿ ∠ ♇
13 40	☽ Q ♃
14 42	☽ ☍ ♀
14 53	☽ △ ♅
15 00	☽ ∥ ♆
15 08	☽ ∥ ☿
16 24	☽ ∥ ♄
16 41	♀ ⚹ ♅
19 53	☽ ∥ ♀
21 08	☽ ☍ ♆
22 21	☽ ⊼ ♂
23 01	☽ ☍ ♄
05 Thursday	
03 35	☽ △ ♇
10 19	☽ ± ♂
11 07	♂ ⊻ ♄
11 33	☽ ∥ ☉
12 23	☿ ∥ ♄
17 14	☉ △ ♃
19 02	☽ ⚼ ♅
23 22	☽ □ ♃
23 55	☽ ⊼ ☉
06 Friday	
04 16	☿ ∥ ♀
04 31	☽ ⚼ ♂
05 04	☽ ⊼ ☿
10 21	☽ ∥ ♂
10 38	☉ Q ♅
12 16	☽ ± ♅
12 24	☽ ± ☉
15 46	☽ ± ☿
18 33	♀ ∥ ♄
23 53	☽ ⊼ ♅
07 Saturday	
05 58	☽ ⊼ ♀
06 30	☽ ⊼ ♆
07 24	☽ ⚼ ☉
07 57	☽ ⚼ ☿
08 29	♀ ∥ ♆
08 53	☽ ⊼ ♄
11 02	☉ ☌ ☿
11 27	♀ ☌ ♆
11 33	☽ △ ♂
13 14	☽ □ ♇
18 19	☽ ± ♆
19 04	☽ ± ♀
20 48	☽ ± ♄
08 Sunday	
01 28	☽ # ♅
09 46	☽ △ ♃
09 53	☿ Q ♅
11 16	☽ △ ☿
12 12	☽ ⚼ ♆
13 40	♀ ☌ ♄
14 22	♂ ⊻ ♇
14 50	☽ ⚼ ♄
14 57	☽ ⚼ ♀
15 42	☽ △ ☉
21 33	☽ ∥ ♇
22 10	☽ # ♃
09 Monday	
01 02	☽ Q ♇
05 23	☿ △ ♃
11 28	☽ ☍ ♅
15 47	☽ ⚼ ♃
18 21	☽ △ ♆
21 15	☽ △ ♄
23 06	♀ # ♆
10 Tuesday	
00 34	☽ △ ♀
01 13	☽ ⚹ ♇
03 37	☽ □ ♂
06 48	♀ # ♄
06 52	♀ ⚹ ♇
08 32	☉ ∠ ♇
09 57	☽ ± ♃
18 59	☽ □ ☿
22 04	☽ ⊼ ♃
11 Wednesday	
03 30	♃ St D
06 28	☉ ∥ ☿
07 35	☽ ∠ ♇
09 38	☽ □ ☉
18 32	☽ Q ♂
12 Thursday	
00 08	☽ ⊼ ♅
01 44	☽ ⊥ ♇
04 43	☽ Q ☿
07 03	☽ □ ♆
10 21	☽ □ ♄
12 12	☽ ± ♅
13 47	☽ ⊻ ♇
19 59	☽ □ ♀
20 15	☽ ⚹ ♂
13 Friday	
00 59	☽ Q ☉
02 51	♀ ⊻ ♂
03 08	☽ ⚹ ☿
06 09	☽ ⚼ ♅
10 10	☽ ☍ ♃
18 52	☽ Q ♆
20 59	☉ # ♀
22 21	☽ Q ♄
14 Saturday	
00 58	☿ ⊥ ♇
02 41	☽ ⚹ ☉
03 50	☽ ∠ ♂
06 54	☽ ∠ ☿
11 19	☽ Q ♀
11 33	☽ △ ♅
18 12	☽ # ♃
18 12	☽ ⚹ ♆
18 56	☽ ∥ ♇
18 58	☿ ⊻ ♀
21 47	☽ ⚹ ♄
22 21	☽ ⊥ ♂
23 23	☽ ⊥ ☿
15 Sunday	
00 35	☽ ☌ ♇
05 58	♄ ∥ ♆
08 08	☿ ☌ ♂
09 57	☽ ∠ ☉
10 16	☽ ⊻ ☿
10 30	☽ ⊻ ♂
12 42	☽ ⚹ ♀
13 37	☽ # ♅
19 54	☽ ⊼ ♃
22 39	☽ ∠ ♆
16 Monday	
02 18	☽ ∠ ♄
04 11	☽ ⊥ ☉
07 03	☽ ± ♃
10 54	♂ ⊥ ♇
15 18	☽ ⊥ ♆
16 08	☽ ⊻ ☉
19 00	☽ ⊥ ♄
19 23	☽ ∠ ♀
19 57	☽ □ ♅
23 31	☽ ⚼ ♃
17 Tuesday	
01 26	♀ ∠ ♅
02 13	☽ ⊻ ♆
05 57	☽ ⊻ ♄
08 09	☽ ⊻ ♇
13 15	☽ ⊥ ♀
15 25	☽ ☌ ☿
15 43	☽ ∥ ♂
18 52	☽ ⊥ ♇
20 47	☽ ☌ ♂
21 22	☽ ∥ ☿
18 Wednesday	
00 57	☽ ⊻ ♀
02 18	☽ △ ♃
04 02	☽ # ♀
04 09	☽ Q ♅
08 43	☿ ⊥ ♀
10 42	☽ ∠ ♇
16 09	♀ □ ♃
19 12	☽ ∥ ☉
19 55	☽ ∥ ♆
20 20	☉ ⚹ ♅
20 22	☽ ∥ ♄
23 13	☽ # ♄
23 40	☽ # ♆
19 Thursday	
00 06	☽ # ☉
01 03	☽ ⚹ ♅
01 23	☽ ☌ ☉
07 00	☽ ☌ ♆
07 40	☉ ∥ ♆
10 51	☽ ☌ ♄
12 34	☽ ⚹ ♇
17 21	☉ ∥ ♄
18 13	☽ ∥ ♀
18 37	☽ ⊻ ☿
20 Friday	
00 04	☽ # ☿
00 31	☽ # ♂
02 38	☽ ∠ ♅
03 21	♀ Q ♇
03 32	☽ ⊻ ♂
04 41	☽ ⊥ ☿
05 32	☿ ∥ ♂
05 53	☽ □ ♃
08 52	☽ Q ♇
09 23	☽ ☌ ♀
14 11	☽ ⊥ ♂
17 46	☽ ⊥ ♅
19 33	☿ St D
19 44	☽ ∠ ☿
21 Saturday	
03 48	☽ ⊻ ♅
06 04	☽ ∠ ♂
07 26	☉ # ♄
07 45	☽ ⊻ ☉
09 35	☽ ⊻ ♆
12 00	☽ Q ♃
13 38	☽ ⊻ ♄
14 56	☽ □ ♇
18 27	☽ ⊥ ☉
19 34	☽ ⊥ ♆
19 37	☉ # ♆
20 46	☽ ⚹ ☿
23 40	☽ ⊥ ♄
22 Sunday	
00 00	♀ # ♂
00 01	♂ △ ♃
01 19	☽ ∥ ♅
07 56	☽ ⚹ ♃
08 22	☽ ⚹ ♂
10 27	☽ ∠ ☉
10 30	☽ ∠ ♆
11 18	☉ ☌ ♆
14 42	☽ ∠ ♄
15 56	☽ ⊻ ♀
16 50	☽ Q ☿
17 22	☽ # ♇
18 02	☽ ∥ ♃
23 Monday	
02 48	☽ ⊥ ♀
05 23	☽ Q ♂
05 40	☽ ☌ ♅
08 51	☽ ∠ ♃
10 38	♂ Q ♅
11 26	☽ ⚹ ♆
13 09	☽ ⚹ ☉
15 48	☽ ⚹ ♄
16 44	☽ △ ♇
19 09	☽ ∠ ♀
23 13	☽ □ ☿
23 52	☽ ⊥ ♃
24 Tuesday	
07 29	☽ Q ♆
09 55	☽ ⊻ ♃
10 38	☽ Q ☉
12 00	☽ Q ♄
12 03	☿ # ♀
13 07	☽ □ ♂
13 14	♀ ⊥ ♅
17 50	☽ ⚼ ♇
22 37	☽ ⚹ ♀
25 Wednesday	
07 59	☽ ⊻ ♅
08 55	☉ ☌ ♄
09 01	☽ ± ♇
13 52	☽ □ ♆
18 11	☽ ⊥ ♅
18 16	☉ ⚹ ♇
18 39	☽ □ ♄
19 13	☽ ⊼ ♇
19 18	☽ □ ☉
20 53	☽ Q ♀
26 Thursday	
03 13	☽ △ ☿
09 18	♀ ⊥ ♂
09 37	☽ ∠ ♅
12 58	☽ ☌ ♃
15 20	☿ ∠ ♀
19 07	☽ △ ♂
27 Friday	
06 04	☽ ⚼ ☿
06 54	☽ □ ♀
11 40	☽ ⚹ ♅
16 53	☽ ∥ ♃
17 30	☽ # ♇
17 45	☽ △ ♆
22 30	☽ ± ☿
22 51	☽ ⚼ ♂
23 01	☽ △ ♄
23 12	☽ ☍ ♇
28 Saturday	
03 16	☽ △ ☉
05 05	♂ ∠ ♇
08 50	☽ Q ♅
09 35	☽ ⊼ ☿
10 11	☽ ∥ ♅
14 25	☿ ⊥ ♇
15 44	☽ ± ♂
17 40	☽ ⊻ ♃
20 21	☽ ⚼ ♆
22 13	♄ ⚹ ♇
29 Sunday	
01 53	☽ ⚼ ♄
03 07	☽ ⊼ ♂
04 26	☽ ⊥ ♃
08 06	☽ ⚼ ☉
12 32	☽ ± ♆
13 52	♀ ⊻ ♅
17 08	☽ □ ♅
17 28	☽ △ ♀
18 14	☽ ± ♄
20 44	☽ ∠ ♃
23 25	☽ ⊼ ♆
30 Monday	
01 40	☽ ± ☉
01 42	☽ ∥ ♀
04 58	☽ ⊼ ♇
05 14	☽ ⊼ ♄
13 16	☽ # ☿
13 30	☽ ⊼ ☉
15 57	☽ ± ♇
18 44	☽ ☍ ☿
21 34	☉ # ♂
23 41	☽ ⚼ ♀
31 Tuesday	
00 17	☽ ⚹ ♃
03 52	☽ ∥ ☉
04 37	☽ # ♂
08 35	☽ ⚼ ♇
13 20	☽ ☍ ♂
18 08	☽ ± ♀
18 45	☽ # ♆
19 06	☽ ∥ ♄
20 59	☽ # ♄
21 19	☽ ∥ ♆
22 41	☽ Q ♃

APRIL 2026

Date	Sidereal time h m s	Sun ☉ ° ′ ″	Moon ☽ ° ′ ″	Moon ☽ 24.00 ° ′ ″	Mercury ☿ ° ′	Venus ♀ ° ′	Mars ♂ ° ′	Jupiter ♃ ° ′	Saturn ♄ ° ′	Uranus ♅ ° ′	Neptune ♆ ° ′	Pluto ♇ ° ′
01	00 39 27	11 ♈ 46 10	04 ♎ 51 54	11 ♎ 11 53	14 ♓ 08	02 ♉ 15	23 ♓ 31	15 ♋ 49	05 ♈ 37	28 ♉ 47	02 ♈ 13	05 ♒ 13
02	00 43 23	12 45 20	17 ♎ 28 56	23 ♎ 43 06	15 01	03 29	24 17	15 53	05 44	28 50	02 16	05 14
03	00 47 20	13 44 29	29 ♎ 54 27	06 ♏ 03 06	15 58	04 43	25 04	15 57	05 51	28 52	02 18	05 15
04	00 51 17	14 43 35	12 ♏ 09 12	18 ♏ 12 55	16 58	05 57	25 51	16 02	05 59	28 55	02 20	05 16
05	00 55 13	15 42 40	24 ♏ 14 30	00 ♐ 14 13	18 00	07 11	26 38	16 06	06 06	28 58	02 22	05 17
06	00 59 10	16 41 43	06 ♐ 12 24	12 ♐ 09 25	19 05	08 24	27 25	16 11	06 14	29 00	02 25	05 18
07	01 03 06	17 40 44	18 ♐ 05 41	24 ♐ 01 40	20 13	09 38	28 12	16 16	06 21	29 03	02 27	05 19
08	01 07 03	18 39 44	29 ♐ 57 53	05 ♑ 54 53	21 23	10 51	28 58	16 21	06 29	29 06	02 29	05 20
09	01 10 59	19 38 41	11 ♑ 53 13	17 ♑ 53 30	22 35	12 05	29 ♓ 45	16 26	06 36	29 09	02 31	05 20
10	01 14 56	20 37 37	23 ♑ 56 20	00 ♒ 02 20	23 49	13 19	00 ♈ 32	16 32	06 43	29 12	02 33	05 21
11	01 18 52	21 36 31	06 ♒ 12 08	12 ♒ 26 20	25 05	14 32	01 19	16 37	06 51	29 15	02 36	05 22
12	01 22 49	22 35 24	18 ♒ 45 29	25 ♒ 10 07	26 24	15 46	02 05	16 43	06 58	29 18	02 38	05 22
13	01 26 46	23 34 14	01 ♓ 40 40	08 ♓ 17 31	27 44	16 59	02 52	16 49	07 05	29 21	02 40	05 23
14	01 30 42	24 33 03	15 ♓ 00 55	21 ♓ 50 59	29 ♓ 06	18 12	03 39	16 55	07 13	29 24	02 42	05 24
15	01 34 39	25 31 50	28 ♓ 47 40	05 ♈ 50 47	00 ♈ 31	19 26	04 25	17 01	07 20	29 27	02 44	05 24
16	01 38 35	26 30 35	12 ♈ 59 58	20 ♈ 14 37	01 56	20 39	05 12	17 07	07 27	29 30	02 46	05 25
17	01 42 32	27 29 18	27 ♈ 34 02	04 ♉ 57 18	03 24	21 52	05 58	17 14	07 34	29 33	02 48	05 25
18	01 46 28	28 28 00	12 ♉ 23 24	19 ♉ 51 15	04 54	23 06	06 45	17 20	07 41	29 36	02 51	05 26
19	01 50 25	29 ♈ 26 39	27 ♉ 19 40	04 ♊ 47 32	06 25	24 19	07 31	17 27	07 49	29 39	02 53	05 26
20	01 54 21	00 ♉ 25 17	12 ♊ 13 47	19 ♊ 37 25	07 58	25 32	08 17	17 34	07 56	29 42	02 55	05 27
21	01 58 18	01 23 52	26 ♊ 57 38	04 ♋ 13 45	09 33	26 45	09 04	17 41	08 03	29 45	02 57	05 27
22	02 02 15	02 22 25	11 ♋ 25 14	18 ♋ 31 45	11 09	27 58	09 50	17 48	08 10	29 49	02 59	05 28
23	02 06 11	03 20 56	25 ♋ 33 08	02 ♌ 29 19	12 47	29 ♉ 11	10 36	17 56	08 17	29 52	03 01	05 28
24	02 10 08	04 19 25	09 ♌ 20 22	16 ♌ 06 27	14 27	00 ♊ 24	11 23	18 03	08 24	29 55	03 03	05 29
25	02 14 04	05 17 51	22 ♌ 47 49	29 ♌ 24 43	16 09	01 37	12 09	18 11	08 31	29 ♉ 58	03 05	05 29
26	02 18 01	06 16 15	05 ♍ 57 29	12 ♍ 26 24	17 52	02 50	12 55	18 19	08 38	00 ♊ 02	03 07	05 29
27	02 21 57	07 14 38	18 ♍ 51 47	25 ♍ 13 56	19 37	04 03	13 41	18 26	08 45	00 05	03 09	05 29
28	02 25 54	08 12 58	01 ♎ 33 07	07 ♎ 49 34	21 24	05 16	14 27	18 34	08 52	00 08	03 11	05 30
29	02 29 50	09 11 16	14 ♎ 03 31	20 ♎ 15 07	23 12	06 28	15 13	18 42	08 59	00 11	03 13	05 30
30	02 33 47	10 ♉ 09 32	26 ♎ 24 32	02 ♏ 31 56	25 ♈ 03	07 ♊ 41	15 ♈ 59	18 ♋ 51	09 ♈ 05	00 ♊ 15	03 ♈ 15	05 ♒ 30

Date	Moon True ☊ ° ′	Moon Mean ☊ ° ′	Moon ☽ Latitude ° ′
01	08 ♓ 45	07 ♓ 23	02 S 17
02	08 R 40	07 19	03 15
03	08 35	07 16	04 03
04	08 28	07 13	04 38
05	08 22	07 10	05 01
06	08 15	07 07	05 10
07	08 10	07 04	05 05
08	08 06	07 00	04 48
09	08 04	06 57	04 18
10	08 D 04	06 54	03 36
11	08 05	06 51	02 44
12	08 06	06 48	01 43
13	08 06	06 44	00 S 35
14	08 R 07	06 41	00 N 37
15	08 06	06 38	01 49
16	08 02	06 35	02 56
17	07 56	06 32	03 54
18	07 48	06 29	04 37
19	07 40	06 25	05 01
20	07 33	06 22	05 05
21	07 27	06 19	04 49
22	07 23	06 16	04 14
23	07 21	06 13	03 25
24	07 D 21	06 10	02 24
25	07 21	06 06	01 17
26	07 22	06 03	00 N 08
27	07 R 21	06 00	01 S 01
28	07 19	05 57	02 05
29	07 13	05 54	03 02
30	07 ♓ 06	05 ♓ 50	03 S 50

DECLINATIONS

Date	Sun ☉ ° ′	Moon ☽ ° ′	Mercury ☿ ° ′	Venus ♀ ° ′	Mars ♂ ° ′	Jupiter ♃ ° ′	Saturn ♄ ° ′	Uranus ♅ ° ′	Neptune ♆ ° ′	Pluto ♇ ° ′
01	04 N 39	04 S 02	07 S 31	11 N 52	03 S 32	22 N 52	00 N 16	19 N 43	00 S 19	22 S 47
02	05 02	09 52	07 19	12 19	03 13	22 52	00 19	19 44	00 18	22 47
03	05 25	15 13	07 04	12 46	02 54	22 51	00 22	19 44	00 17	22 47
04	05 48	19 53	06 48	13 13	02 36	22 51	00 25	19 45	00 16	22 47
05	06 11	23 41	06 30	13 40	02 17	22 50	00 28	19 46	00 15	22 47
06	06 34	26 25	06 11	14 06	01 58	22 50	00 31	19 46	00 15	22 47
07	06 56	27 58	05 50	14 32	01 39	22 49	00 34	19 47	00 14	22 47
08	07 19	28 14	05 27	14 58	01 20	22 49	00 37	19 48	00 13	22 47
09	07 41	27 11	05 03	15 23	01 01	22 48	00 39	19 48	00 12	22 47
10	08 03	24 52	04 38	15 48	00 43	22 47	00 42	19 49	00 11	22 47
11	08 25	21 22	04 11	16 13	00 24	22 47	00 45	19 49	00 10	22 47
12	08 47	16 50	03 43	16 37	00 S 05	22 46	00 48	19 50	00 09	22 47
13	09 09	11 25	03 14	17 00	00 N 14	22 45	00 51	19 51	00 09	22 47
14	09 31	05 S 20	02 43	17 24	00 33	22 45	00 54	19 51	00 08	22 47
15	09 52	01 N 11	02 11	17 46	00 51	22 44	00 56	19 52	00 07	22 47
16	10 14	07 50	01 38	18 09	01 10	22 43	00 59	19 53	00 06	22 47
17	10 35	14 15	01 04	18 31	01 29	22 42	01 02	19 53	00 05	22 47
18	10 56	19 57	00 S 28	18 52	01 47	22 42	01 05	19 54	00 05	22 47
19	11 17	24 27	00 N 08	19 13	02 06	22 41	01 07	19 55	00 04	22 47
20	11 37	27 18	00 46	19 33	02 25	22 40	01 10	19 55	00 03	22 47
21	11 58	28 13	01 24	19 53	02 43	22 39	01 13	19 56	00 02	22 48
22	12 18	27 10	02 04	20 13	03 02	22 38	01 15	19 57	00 01	22 48
23	12 38	24 23	02 44	20 31	03 20	22 37	01 18	19 58	00 S 01	22 48
24	12 58	20 14	03 26	20 50	03 39	22 36	01 21	19 58	00 00	22 48
25	13 17	15 08	04 08	21 07	03 57	22 35	01 23	19 59	00 N 01	22 48
26	13 37	09 27	04 51	21 25	04 15	22 34	01 26	20 00	00 02	22 48
27	13 56	03 N 28	05 35	21 41	04 34	22 33	01 29	20 00	00 03	22 48
28	14 15	02 S 32	06 20	21 57	04 52	22 32	01 31	20 01	00 03	22 48
29	14 33	08 21	07 05	22 13	05 10	22 31	01 34	20 02	00 04	22 48
30	14 N 52	13 S 46	07 N 51	22 N 27	05 N 28	22 N 30	01 N 36	20 N 02	00 N 05	22 S 49

ZODIAC SIGN ENTRIES

Date	h	m	Planets
01	02	51	☽ ♎
03	12	11	☽ ♏
05	23	31	☽ ♐
08	12	04	☽ ♑
09	19	36	♂ ♈
10	23	55	☽ ♒
13	08	55	☽ ♓
15	03	21	☿ ♈
15	14	04	☽ ♈
17	15	58	☽ ♉
19	16	18	☽ ♊
20	01	39	☉ ♉
21	17	00	☽ ♋
23	19	41	☽ ♌
24	04	03	♀ ♊
26	00	50	♅ ♊
26	01	04	☽ ♍
28	09	03	☽ ♎
30	19	02	☽ ♏

LATITUDES

Date	Mercury ☿ ° ′	Venus ♀ ° ′	Mars ♂ ° ′	Jupiter ♃ ° ′	Saturn ♄ ° ′	Uranus ♅ ° ′	Neptune ♆ ° ′	Pluto ♇ ° ′
01	01 S 23	00 S 25	01 S 03	00 N 22	02 S 08	00 S 10	01 S 18	03 S 57
04	01 48	00 18	01 02	00 22	02 08	00 10	01 19	03 57
07	02 07	00 10	01 01	00 23	02 08	00 10	01 19	03 58
10	02 22	00 S 02	01 00	00 23	02 08	00 10	01 19	03 59
13	02 32	00 N 06	00 59	00 23	02 08	00 10	01 19	03 59
16	02 37	00 14	00 59	00 23	02 09	00 10	01 19	04 00
19	02 38	00 23	00 58	00 23	02 09	00 10	01 19	04 00
22	02 33	00 31	00 56	00 23	02 09	00 10	01 19	04 01
25	02 24	00 39	00 55	00 23	02 10	00 10	01 19	04 02
28	02 10	00 47	00 54	00 23	02 10	00 10	01 19	04 02
31	01 S 52	00 N 55	00 S 53	00 N 23	02 S 10	00 S 10	01 S 19	04 S 03

DATA

Julian Date	2461132
Delta T	+73 seconds
Ayanamsa	24° 13′ 32″
Synetic vernal point	04° ♓ 53′ 27″
True obliquity of ecliptic	23° 26′ 18″

MOON'S PHASES, APSIDES AND POSITIONS ☽

Date	h	m	Phase	Longitude ° ′	Eclipse Indicator
02	02	12	○	12 ♎ 21	
10	04	52	☾	20 ♑ 20	
17	11	52	●	27 ♈ 29	
24	02	32	☽	03 ♌ 56	

Day	h	m		
07	08	25	Apogee	
19	06	49	Perigee	
08	04	49	Max dec	28° S 18′
15	07	42	0N	
21	11	00	Max dec	28° N 13′
28	01	50	0S	

LONGITUDES

Date	Chiron ⚷ ° ′	Ceres ⚳ ° ′	Pallas ⚴ ° ′	Juno ⚵ ° ′	Vesta ⚶ ° ′	Black Moon Lilith ⚸ ° ′
01	25 ♈ 44	06 ♉ 41	22 ♓ 01	00 ♒ 51	10 ♓ 40	11 ♐ 24
11	26 ♈ 20	10 ♉ 39	25 ♓ 17	03 ♒ 25	15 ♓ 27	12 ♐ 31
21	26 ♈ 56	14 ♉ 40	28 ♓ 30	05 ♒ 42	20 ♓ 07	13 ♐ 38
31	27 ♈ 32	18 ♉ 44	01 ♈ 39	07 ♒ 39	24 ♓ 40	14 ♐ 45

All ephemeris data is given at 12.00 UT and the Moon's longitude is additionally given for 24.00 UT

ASPECTARIAN

h m	Aspects
01 Wednesday	
00 31	☽ △ ♅
06 34	☽ ⊼ ♀
07 00	☽ ☍ ♆
10 06	☽ ∥ ♂
11 19	♀ ⊻ ♆
12 41	☽ △ ♇
13 25	☽ ☍ ♄
14 39	☽ ⋕ ☉
02 Thursday	
01 40	☽ ∥ ☿
02 12	☽ ☍ ☉
04 09	♄ ⋕ ♆
04 59	☽ ⚼ ♅
06 56	☽ ⊼ ☿
08 55	☽ □ ♃
19 21	☽ ± ☿
20 09	♀ Q ♃
22 19	☽ ± ♅
23 40	☽ ⋕ ♀
03 Friday	
01 59	☽ ⊼ ♂
09 59	☽ ⊼ ♅
11 28	☿ △ ♃
14 15	☽ ⚼ ☿
14 25	☽ ± ♂
15 19	☉ ∠ ♅
16 40	☽ ⊼ ♆
22 26	☽ ☍ ♀
22 27	☽ □ ♇
22 38	♀ □ ♇
23 44	☽ ⊼ ♄
04 Saturday	
04 27	☽ ± ♆
09 16	☽ ⚼ ♂
10 44	☿ Q ♅
11 13	☽ ⋕ ♅
11 40	☽ ± ♄
12 46	♀ ⊻ ♄
17 32	☽ ⊼ ☉
19 42	☽ △ ♃
22 17	☽ ⚼ ♆
22 24	☽ △ ☿
05 Sunday	
05 41	☽ ⚼ ♄
05 44	☽ ∥ ♇
06 06	☽ ⋕ ♃
06 30	☽ ± ☉
10 05	☽ Q ♇
17 07	☽ △ ♂
21 29	☽ ☍ ♅
22 23	☉ □ ♃
22 59	☉ ⋕ ☿
06 Monday	
01 50	☽ ⚼ ♃
02 07	☽ ⚼ ☉
04 20	☽ △ ♆
10 10	☽ ⚹ ♇
12 03	☽ △ ♄
12 07	♀ ⊥ ♆
16 56	☽ ⊼ ♀
20 04	☽ ± ♃
07 Tuesday	
02 56	☉ Q ♇
06 27	☽ ± ♀
08 16	☽ ⊼ ♃
11 05	☽ △ ☉
14 04	☿ ∠ ♇
16 29	☽ ∠ ♇
16 45	☽ □ ☿
08 Wednesday	
02 45	☽ ⚼ ♀
09 52	☽ □ ♂
10 15	☽ ⊼ ♅
10 42	☽ ⊥ ♇
16 11	♂ ⚹ ♅
17 06	☽ □ ♆
22 24	☽ ± ♅
22 50	☽ ⊻ ♇
09 Thursday	
01 16	☽ □ ♄
09 05	☽ Q ☿
12 26	☽ △ ♀
16 33	☽ ⚼ ♅
21 10	☽ ☍ ♃
23 12	♀ ⊥ ♄
10 Friday	
00 32	☽ Q ♂
04 52	☽ □ ☉
05 17	☽ Q ♆
11 44	☽ ⚹ ☿
12 22	♂ ⋕ ♄
13 34	☽ Q ♄
22 24	☽ △ ♅
11 Saturday	
01 51	☽ ⚹ ♂
01 53	♃ ⋕ ♇
03 10	☽ ∥ ♇
03 10	☽ ⋕ ♃
04 58	☽ ⚹ ♆
10 22	☽ ☌ ♇
13 15	☽ ⚹ ♄
19 08	☽ Q ☉
20 22	☽ ∠ ☿
20 44	☽ ⋕ ♅
12 Sunday	
05 43	☽ □ ♀
06 02	♂ ∥ ♆
08 06	☽ ⊼ ♃
08 38	☽ ∠ ♂
09 52	☽ ∠ ♆
12 58	☽ ⋕ ♀
15 26	☽ ⊥ ☿
18 05	☽ ∠ ♄
19 29	☽ ± ♃
19 47	☽ ⚹ ☉
13 Monday	
02 35	☽ ⊥ ♂
02 45	☽ ⊥ ♆
03 55	☽ ⊻ ☿
05 30	♂ ☌ ♆
05 40	♂ ⋕ ♆
06 11	☉ ⊥ ♅
07 42	☽ □ ♅
08 21	♀ ⚹ ♃
10 55	☽ ⊥ ♄
12 15	☽ ⚼ ♃
13 48	☽ ⊻ ♆
14 18	☽ ⊻ ♂
18 38	☽ Q ♀
18 45	☽ ⊻ ♇
20 41	☽ ⋕ ☉
21 55	☽ ⊻ ♄
14 Tuesday	
01 29	☽ ∠ ☉
01 46	♀ ∠ ♆
05 34	☽ ⊥ ♇
15 23	☽ △ ♃
16 13	☽ Q ♅
17 10	☿ ⚹ ♅
18 11	☽ ⚹ ♀
18 43	☽ ⊥ ☉
21 28	☽ ∠ ♇
22 37	☽ ∥ ☿
15 Wednesday	
04 20	☽ ⋕ ♄
04 55	☽ ⋕ ♂
05 57	☽ ⊻ ☉
07 16	☽ ∥ ♆
08 08	☽ ⋕ ♆
10 44	☽ ∥ ♂
11 06	☽ ∥ ♄
11 24	♀ ∠ ♂
13 07	☽ ⚹ ♅
15 16	☽ ☌ ☿
15 19	☽ ⋕ ☿
18 45	☽ ☌ ♆
19 32	♂ ∥ ♄
22 08	☽ ☌ ♂
22 31	☽ ∠ ♀
23 16	☽ ⚹ ♇
16 Thursday	
02 38	☽ ☌ ♄
14 30	☽ ∠ ♅
15 00	☽ ⊥ ♀
18 54	☽ □ ♃
18 56	♂ ⚹ ♇
19 20	☽ Q ♇
21 15	☽ ∥ ☉
17 Friday	
00 42	☿ ⋕ ♂
01 50	☽ ⊻ ♀
02 01	☿ ☌ ♆
05 24	☽ ⊥ ♅
11 52	☽ ☌ ☉
13 10	☿ ⋕ ♄
15 14	☽ ⊻ ♅
20 32	☽ ⊻ ♆
22 33	☽ ⊻ ☿
18 Saturday	
00 32	☽ Q ♃
00 46	☽ □ ♇
02 24	☽ ⊻ ♂
03 14	♀ ∠ ♄
04 22	☽ ⊻ ♄
06 16	☽ ⊥ ♆
06 47	☽ ∥ ♀
09 19	☽ ⊥ ☿
11 49	☽ ∥ ♅
12 36	☽ ⊥ ♂
14 07	☽ ⊥ ♄
20 01	☽ ⚹ ♃
20 34	☿ ⚹ ♇
20 47	☽ ∠ ♆
19 Sunday	
01 26	☽ ∠ ☿
01 41	☽ ∥ ♃
02 15	☽ ⋕ ♇
03 51	☽ ∠ ♂
04 07	☿ ∥ ♆
04 41	☽ ∠ ♄
06 44	☽ ☌ ♀
09 12	☿ ⋕ ♆
15 38	☽ ⊻ ☉
15 45	☽ ☌ ♅
17 22	☉ ⊻ ♅
20 18	☽ ∠ ♃
20 56	☽ ⚹ ♆
22 44	♂ ☌ ♄
20 Monday	
01 03	☽ △ ♇
01 58	☽ ⊥ ☉
04 19	☽ ⚹ ☿
05 00	☽ ⚹ ♄
05 17	☽ ⚹ ♂
10 55	☽ ⊥ ♃
11 22	☿ ☌ ♄
16 21	☽ Q ♆
17 32	☽ ∠ ☉
20 43	☽ ⊻ ♃
21 45	☿ ☌ ♂
21 Tuesday	
00 36	☽ Q ♄
01 21	☽ ⚼ ♇
01 49	☽ Q ♂
02 04	☽ Q ☿
04 26	☿ ∥ ♄
11 37	☽ ⊻ ♀
15 43	♀ ∥ ♅
16 07	☽ ± ♇
16 37	☽ ⊻ ♅
19 50	☽ ⚹ ☉
21 54	☽ □ ♆
22 26	☽ ⊥ ♀
22 Wednesday	
02 03	☽ ⊼ ♇
02 35	☽ ⊥ ♅
06 30	☽ □ ♄
09 12	☽ □ ♂
11 30	☽ □ ☿
14 51	☽ ∠ ♀
17 20	☽ Q ☉
17 43	☽ ∠ ♅
22 52	☽ ☌ ♃
23 Thursday	
03 27	☉ ⊻ ♆
18 52	☽ ⚹ ♀
19 28	☽ ⚹ ♅
21 59	☽ ⋕ ♇
23 03	☽ ∥ ♃
24 Friday	
00 57	☽ △ ♆
01 58	♀ ☌ ♅
02 32	☽ □ ☉
05 13	☽ ☍ ♇
09 06	☽ ∥ ♀
10 20	☽ △ ♄
13 20	☽ ∥ ♅
15 49	☽ △ ♂
16 34	☽ Q ♅
17 57	☽ Q ♀
18 49	☿ ∠ ♅
22 21	☽ △ ☿
25 Saturday	
01 12	☿ ∥ ♂
03 30	☽ ⚼ ♆
03 37	☽ ⊻ ♃
13 19	☽ ⚼ ♄
14 31	☽ ⊥ ♃
16 33	☉ □ ♇
19 33	☽ ∥ ☉
19 47	☽ ± ♆
20 22	☽ ⚼ ♂
26 Sunday	
01 04	☽ □ ♅
05 27	☽ ⚼ ☿
05 40	☽ □ ♀
05 50	☽ ± ♄
06 41	☿ Q ♇
06 45	☽ ⊼ ♆
07 05	☽ ∠ ♃
10 20	☿ ∠ ♀
11 08	☽ ⊼ ♇
12 37	☽ △ ☉
13 04	☉ Q ♃
13 53	☽ ± ♂
16 59	☽ ⊼ ♄
17 43	♀ ⚹ ♆
18 33	☿ □ ♃
22 14	☽ ± ♇
22 32	♀ ∠ ♃
27 Monday	
00 39	☽ ± ☿
01 43	☽ ⊼ ♂
04 29	☽ ∥ ☿
07 51	☽ ∥ ♂
11 12	☽ ⚹ ♃
13 39	☽ ⊼ ☿
15 03	☽ ⚼ ♇
18 53	☽ ⚼ ☉
19 52	☽ ∥ ♄
28 Tuesday	
01 38	☽ ∥ ♆
02 01	☽ ⋕ ♆
07 54	☽ ⋕ ♄
09 17	☽ △ ♅
10 07	☽ Q ♃
13 22	☽ ± ☉
15 07	☽ ☍ ♆
16 40	♀ △ ♇
19 32	☽ △ ♇
19 50	☽ △ ♀
22 00	☽ ⋕ ♂
29 Wednesday	
01 49	☽ ⊼ ☉
02 07	☽ ☍ ♄
05 52	☽ ⋕ ☿
06 05	☉ ⊻ ♄
10 53	♂ ∠ ♅
12 36	☉ ⊥ ♆
14 12	☽ ⚼ ♅
14 24	☽ ☍ ♂
21 06	☽ □ ♃
30 Thursday	
01 18	☿ ⊥ ♅
03 56	☽ ⚼ ♀
07 45	☽ ± ♅
08 52	☽ ☍ ☿
16 06	♀ ∥ ♃
17 33	☽ ⋕ ☉
19 33	☽ ⊼ ♅
23 28	☽ ± ♀

MAY 2026

LONGITUDES

Date	Sidereal time h m s	Sun ☉ ° ' "	Moon ☽ ° ' "	Moon ☽ 24.00 ° '	Mercury ☿ ° '	Venus ♀ ° '	Mars ♂ ° '	Jupiter ♃ ° '	Saturn ♄ ° '	Uranus ♅ ° '	Neptune ♆ ° '	Pluto ♇ ° '
01	02 37 44	11 ♉ 07 46	08 ♏ 37 25	14 ♏ 41 06	26 ♈ 55	08 ♊ 54	16 ♈ 45	18 ♋ 59	09 ♈ 12	00 ♊ 18	03 ♈ 16	05 ♒ 30
02	02 41 40	12 05 59	20 ♏ 43 07	26 ♏ 43 37	28 ♈ 48	10 06	17 31	19 08	09 19	00 21	03 18	05 30
03	02 45 37	13 04 10	02 ♐ 42 44	08 ♐ 40 39	00 ♉ 44	11 19	18 17	19 16	09 26	00 25	03 20	05 30
04	02 49 33	14 02 20	14 ♐ 37 35	20 ♐ 33 47	02 41	12 31	19 03	19 25	09 32	00 28	03 22	05 31
05	02 53 30	15 00 28	26 ♐ 29 33	02 ♑ 25 13	04 40	13 44	19 49	19 34	09 39	00 32	03 24	05 31
06	02 57 26	15 58 34	08 ♑ 21 11	14 ♑ 17 53	06 40	14 56	20 35	19 43	09 45	00 35	03 26	05 31
07	03 01 23	16 56 39	20 ♑ 15 47	26 ♑ 15 26	08 43	16 08	21 20	19 52	09 52	00 38	03 27	05 R 31
08	03 05 19	17 54 42	02 ♒ 17 22	08 ♒ 22 13	10 46	17 21	22 06	20 01	09 58	00 42	03 29	05 31
09	03 09 16	18 52 44	14 ♒ 30 34	20 ♒ 43 04	12 52	18 33	22 52	20 10	10 05	00 45	03 31	05 30
10	03 13 13	19 50 45	27 ♒ 00 20	03 ♓ 23 00	14 58	19 45	23 37	20 19	10 11	00 49	03 33	05 30
11	03 17 09	20 48 44	09 ♓ 51 38	16 ♓ 26 43	17 06	20 57	24 23	20 29	10 17	00 52	03 34	05 30
12	03 21 06	21 46 42	23 ♓ 08 43	29 ♓ 57 56	19 15	22 09	25 08	20 39	10 24	00 56	03 36	05 30
13	03 25 02	22 44 39	06 ♈ 54 31	13 ♈ 58 27	21 25	23 21	25 54	20 48	10 30	00 59	03 38	05 30
14	03 28 59	23 42 35	21 ♈ 09 33	28 ♈ 27 22	23 35	24 33	26 39	20 58	10 36	01 03	03 39	05 30
15	03 32 55	24 40 29	05 ♉ 51 13	13 ♉ 20 15	25 46	25 45	27 25	21 08	10 42	01 06	03 41	05 29
16	03 36 52	25 38 22	20 ♉ 53 21	28 ♉ 29 16	27 ♉ 57	26 57	28 10	21 18	10 48	01 10	03 42	05 29
17	03 40 48	26 36 13	06 ♊ 06 39	13 ♊ 44 04	00 ♊ 09	28 09	28 55	21 28	10 54	01 13	03 44	05 29
18	03 44 45	27 34 03	21 ♊ 20 08	28 ♊ 53 32	02 19	29 ♊ 21	29 ♈ 40	21 38	11 00	01 17	03 46	05 29
19	03 48 42	28 31 52	06 ♋ 23 07	13 ♋ 47 53	04 29	00 ♋ 33	00 ♉ 26	21 49	11 06	01 20	03 47	05 28
20	03 52 38	29 ♉ 29 39	21 ♋ 07 04	28 ♋ 20 07	06 39	01 44	01 11	21 59	11 12	01 24	03 49	05 28
21	03 56 35	00 ♊ 27 24	05 ♌ 26 42	12 ♌ 26 40	08 47	02 56	01 56	22 09	11 18	01 27	03 50	05 28
22	04 00 31	01 25 08	19 ♌ 20 05	26 ♌ 07 06	10 53	04 07	02 41	22 20	11 24	01 31	03 51	05 27
23	04 04 28	02 22 50	02 ♍ 48 01	09 ♍ 23 14	12 58	05 19	03 26	22 31	11 29	01 34	03 53	05 27
24	04 08 24	03 20 30	15 ♍ 53 10	22 ♍ 18 16	15 01	06 30	04 11	22 41	11 35	01 38	03 54	05 26
25	04 12 21	04 18 09	28 ♍ 39 02	04 ♎ 55 56	17 02	07 42	04 55	22 52	11 40	01 41	03 56	05 26
26	04 16 17	05 15 46	11 ♎ 09 25	17 ♎ 19 54	19 01	08 53	05 40	23 03	11 46	01 45	03 57	05 25
27	04 20 14	06 13 22	23 ♎ 27 47	29 ♎ 33 26	20 57	10 04	06 25	23 14	11 51	01 48	03 58	05 25
28	04 24 11	07 10 57	05 ♏ 37 09	11 ♏ 39 12	22 51	11 15	07 10	23 25	11 57	01 52	03 59	05 24
29	04 28 07	08 08 30	17 ♏ 39 51	23 ♏ 39 18	24 43	12 26	07 54	23 36	12 02	01 55	04 01	05 23
30	04 32 04	09 06 02	29 ♏ 37 45	05 ♐ 35 22	26 31	13 37	08 39	23 47	12 07	01 59	04 02	05 23
31	04 36 00	10 ♊ 03 33	11 ♐ 32 19	17 ♐ 28 45	28 ♊ 17	14 ♋ 48	09 ♉ 23	23 ♋ 58	12 ♈ 12	02 ♊ 02	04 ♈ 03	05 ♒ 22

DECLINATIONS

Date	Moon True ☊ ° '	Moon Mean ☊ ° '	Moon ☽ Latitude ° '	Sun ☉ ° '	Moon ☽ ° '	Mercury ☿ ° '	Venus ♀ ° '	Mars ♂ ° '	Jupiter ♃ ° '	Saturn ♄ ° '	Uranus ♅ ° '	Neptune ♆ ° '	Pluto ♇ ° '
01	06 ♓ 55	05 ♓ 47	04 S 26	15 N 10	18 S 35	08 N 38	22 N 41	05 N 46	22 N 29	01 N 39	20 N 03	00 N 05	22 S 49
02	06 R 44	05 44	04 50	15 28	22 36	09 25	22 55	06 04	22 28	01 41	20 04	00 06	22 49
03	06 31	05 41	05 01	15 46	25 37	10 12	23 08	06 22	22 27	01 44	20 05	00 07	22 49
04	06 19	05 38	04 59	16 03	27 30	11 00	23 20	06 40	22 25	01 46	20 05	00 08	22 49
05	06 09	05 35	04 43	16 20	28 07	11 48	23 32	06 57	22 24	01 49	20 06	00 08	22 50
06	06 01	05 31	04 16	16 37	27 26	12 36	23 43	07 15	22 23	01 51	20 07	00 09	22 50
07	05 55	05 28	03 37	16 54	25 29	13 24	23 53	07 33	22 22	01 54	20 07	00 10	22 50
08	05 52	05 25	02 48	17 10	22 22	14 12	24 03	07 50	22 20	01 56	20 08	00 10	22 50
09	05 51	05 22	01 51	17 26	18 15	15 00	24 12	08 07	22 19	01 59	20 09	00 11	22 50
10	05 D 51	05 19	00 S 47	17 42	13 15	15 47	24 20	08 25	22 18	02 01	20 10	00 12	22 50
11	05 R 51	05 16	00 N 21	17 57	07 33	16 34	24 28	08 42	22 16	02 03	20 10	00 12	22 51
12	05 50	05 12	01 30	18 13	01 S 20	17 20	24 34	08 59	22 15	02 06	20 11	00 13	22 51
13	05 47	05 09	02 36	18 27	05 N 08	18 04	24 41	09 16	22 13	02 08	20 12	00 13	22 51
14	05 42	05 06	03 35	18 42	11 35	18 48	24 46	09 33	22 12	02 10	20 13	00 14	22 51
15	05 33	05 03	04 22	18 56	17 35	19 30	24 51	09 50	22 10	02 12	20 13	00 15	22 52
16	05 23	05 00	04 51	19 10	22 39	20 10	24 55	10 06	22 09	02 15	20 14	00 15	22 52
17	05 13	04 56	05 00	19 24	26 15	20 49	24 58	10 23	22 07	02 17	20 15	00 16	22 52
18	05 02	04 53	04 49	19 37	27 57	21 26	25 01	10 39	22 06	02 19	20 15	00 16	22 52
19	04 53	04 50	04 17	19 50	27 33	22 00	25 03	10 56	22 04	02 21	20 16	00 17	22 53
20	04 47	04 47	03 28	20 02	25 13	22 32	25 04	11 12	22 03	02 23	20 17	00 18	22 53
21	04 43	04 44	02 28	20 15	21 18	23 02	25 05	11 28	22 01	02 25	20 18	00 18	22 53
22	04 42	04 41	01 20	20 27	16 18	23 29	25 05	11 44	21 59	02 27	20 18	00 19	22 54
23	04 D 42	04 37	00 N 10	20 38	10 38	23 53	25 04	12 00	21 58	02 29	20 19	00 19	22 54
24	04 R 42	04 34	00 S 59	20 49	04 N 40	24 15	25 02	12 15	21 56	02 31	20 20	00 20	22 54
25	04 40	04 31	02 03	21 00	01 S 21	24 34	25 00	12 31	21 54	02 33	20 20	00 21	22 54
26	04 37	04 28	03 00	21 11	07 10	24 50	24 57	12 46	21 52	02 35	20 21	00 21	22 55
27	04 31	04 25	03 47	21 21	12 37	25 04	24 53	13 02	21 50	02 37	20 22	00 21	22 55
28	04 22	04 22	04 23	21 30	17 32	25 15	24 49	13 17	21 49	02 39	20 23	00 22	22 55
29	04 10	04 18	04 47	21 40	21 41	25 24	24 43	13 32	21 47	02 41	20 23	00 22	22 56
30	03 57	04 15	04 59	21 49	24 56	25 30	24 38	13 47	21 45	02 43	20 24	00 22	22 56
31	03 ♓ 43	04 ♓ 12	04 S 57	21 N 57	27 S 04	25 N 34	24 N 31	14 N 01	21 N 43	02 N 45	20 N 25	00 N 23	22 S 56

ZODIAC SIGN ENTRIES

Date	h m	Planets
03	02 57	☿ ♉
03	06 33	☽ ♐
05	19 06	☽ ♑
08	07 27	☽ ♒
10	17 39	☽ ♓
13	00 04	☽ ♈
15	02 31	☽ ♉
17	02 23	☽ ♊
17	10 26	☿ ♊
18	22 25	♂ ♉
19	01 05	♀ ♋
19	01 46	☽ ♋
21	00 37	☉ ♊
21	02 48	☽ ♌
23	06 57	☽ ♍
25	14 34	☽ ♎
28	00 52	☽ ♏
30	12 45	☽ ♐

LATITUDES

Date	Mercury ☿ ° '	Venus ♀ ° '	Mars ♂ ° '	Jupiter ♃ ° '	Saturn ♄ ° '	Uranus ♅ ° '	Neptune ♆ ° '	Pluto ♇ ° '
01	01 S 52	00 N 55	00 S 53	00 N 23	02 S 10	00 S 10	01 S 19	04 S 03
04	01 30	01 03	00 52	00 24	02 11	00 10	01 19	04 04
07	01 03	01 10	00 50	00 24	02 11	00 10	01 19	04 04
10	00 34	01 18	00 49	00 24	02 12	00 10	01 20	04 04
13	00 S 03	01 24	00 48	00 24	02 12	00 10	01 20	04 05
16	00 N 29	01 31	00 46	00 24	02 13	00 10	01 20	04 06
19	00 59	01 37	00 44	00 24	02 13	00 10	01 20	04 06
22	01 25	01 42	00 43	00 24	02 14	00 10	01 20	04 07
25	01 46	01 47	00 41	00 24	02 14	00 10	01 20	04 07
28	02 01	01 51	00 40	00 25	02 15	00 10	01 20	04 08
31	02 N 09	01 N 55	00 S 38	00 N 25	02 S 16	00 S 10	01 S 20	04 S 08

DATA

Julian Date	2461162
Delta T	+73 seconds
Ayanamsa	24° 13' 35"
Synetic vernal point	04° ♓ 53' 24"
True obliquity of ecliptic	23° 26' 18"

MOON'S PHASES, APSIDES AND POSITIONS ☽

Date	h m	Phase	Longitude ° '	Eclipse Indicator
01	17 23	○	11 ♏ 21	
09	21 10	☾	19 ♒ 15	
16	20 01	●	25 ♉ 58	
23	11 11	☽	02 ♍ 21	
31	08 45	○	09 ♐ 56	

Day	h m		
04	22 18	Apogee	
17	13 38	Perigee	
05	11 22	Max dec	28° S 07'
12	17 01	ON	
18	19 26	Max dec	28° N 03'
25	06 36	OS	

LONGITUDES

Date	Chiron ⚷ ° '	Ceres ⚳ ° '	Pallas ⚴ ° '	Juno ⚵ ° '	Vesta ⚶ ° '	Black Moon Lilith ⚸ ° '
01	27 ♈ 32	18 ♉ 44	01 ♈ 39	07 ♒ 39	24 ♓ 40	14 ♐ 45
11	28 ♈ 06	22 ♉ 49	04 ♈ 42	09 ♒ 13	29 ♓ 04	15 ♐ 53
21	28 ♈ 39	26 ♉ 55	07 ♈ 38	10 ♒ 19	03 ♈ 18	17 ♐ 00
31	29 ♈ 10	01 ♊ 02	10 ♈ 25	10 ♒ 55	07 ♈ 20	18 ♐ 07

All ephemeris data is given at 12.00 UT and the Moon's longitude is additionally given for 24.00 UT

ASPECTARIAN

h m Aspects

01 Friday
01 26 ☽ ⊼ ♆
05 51 ☽ □ ♇
12 36 ☽ ⊼ ♀
13 09 ☽ ⊼ ♄
13 17 ☽ ± ♆
17 23 ☽ ☍ ☉
18 45 ♀ ⚹ ♄
20 15 ☽ # ♅

02 Saturday
00 55 ♀ # ♇
01 09 ☽ ± ♄
05 12 ☽ ⊼ ♂
07 11 ☽ bQ ♆
08 47 ☽ △ ♃
11 07 ☽ # ♃
11 29 ♂ Q ♇
13 31 ☽ ∥ ♇
14 22 ☽ # ♀
17 34 ☽ Q ♇
17 58 ☽ ± ♂
19 15 ☽ bQ ♄

03 Sunday
07 15 ☽ ⊼ ☿
07 22 ☽ ☍ ♅
07 58 ☿ ⊻ ♅
13 14 ☽ bQ ♂
13 16 ☽ △ ♆
15 10 ☽ bQ ♃
17 37 ☽ ⚹ ♇
21 39 ☽ ± ☿

04 Monday
01 38 ☽ △ ♄
07 16 ☽ ☍ ♀
09 31 ☽ ± ♃
10 42 ☽ ⊼ ☉
19 24 ☽ bQ ☿
20 28 ☿ ⊻ ♆
21 33 ☽ △ ♂
21 48 ☽ ⊼ ♃
23 53 ☽ ∠ ♇
23 55 ☽ ± ☉

05 Tuesday
02 08 ♂ □ ♃
08 14 ♀ ⊥ ♃
18 06 ☽ ⊥ ♇
19 45 ☽ bQ ☉
20 12 ☽ ⊼ ♅
22 08 ☿ □ ♇

06 Wednesday
02 01 ☽ □ ♆
05 52 ☉ ⊥ ♄
06 15 ☽ ⊻ ♇
07 54 ☽ △ ☿
08 24 ☽ ± ♅
14 52 ☽ □ ♄
15 32 ♇ St R
22 04 ♀ Q ♆

07 Thursday
01 14 ☿ Q ♃
02 40 ☽ bQ ♅
02 47 ☽ ⊼ ♀
04 44 ☽ △ ☉
11 11 ☽ ☍ ♃
14 18 ☽ □ ♂
14 24 ☽ Q ♆
16 11 ☽ ± ♀
20 51 ☿ ⊥ ♆

08 Friday
00 43 ☽ # ♀
02 13 ☿ ⊻ ♄
03 21 ☽ Q ♄
08 50 ☽ △ ♅
08 55 ☽ ∥ ♇
12 07 ☽ bQ ♀
12 15 ☽ # ♃
14 23 ☽ ⚹ ♆
18 22 ☽ ☌ ♇

09 Saturday
01 38 ☽ # ♅
02 43 ☉ ∠ ♆
03 17 ☽ ⚹ ♄
03 38 ☿ ⊥ ♀
04 25 ☽ Q ♂
08 08 ☽ □ ☿
15 56 ☽ # ☉
19 47 ☽ ∠ ♆
20 40 ☽ △ ♀
21 10 ☽ □ ☉
23 05 ☽ ⊼ ♃

10 Sunday
01 50 ☽ # ☿
05 09 ☽ ⚹ ♂
08 31 ☽ ∠ ♄
10 42 ☽ ± ♃
13 01 ☽ ⊥ ♆
19 13 ☽ □ ♅
21 29 ☉ ⊻ ♀

11 Monday
00 20 ☽ ⊻ ♆
01 09 ♀ ⊻ ♃
01 27 ☽ Q ☿
01 37 ☽ ⊥ ♄
02 14 ☉ ⚹ ♃
02 26 ☿ ⊥ ♄
03 01 ♀ bQ ♇
03 49 ☽ bQ ♃
03 57 ☽ ⊻ ♇
07 33 ☽ # ♂
09 55 ☽ Q ☉
11 04 ☽ ∠ ♂
12 48 ☽ ⊻ ♄
15 01 ☽ ⊥ ♇

12 Tuesday
03 43 ☽ ⚹ ☿
04 25 ☽ ⊥ ♂
04 26 ☽ Q ♅
04 41 ☿ ∠ ♆
04 46 ♂ ⊥ ♅
07 17 ☽ ∠ ♇
07 29 ☽ △ ♃
09 11 ☽ # ♄
09 23 ☽ ⚹ ☉
10 04 ☽ □ ♀
15 44 ☽ ⊻ ♂
16 12 ☽ # ♆
17 14 ♀ Q ♄
17 49 ☽ ∥ ♆

13 Wednesday
00 51 ☽ ∥ ♄
01 44 ☽ ⚹ ♅
04 44 ☿ ⚹ ♃
06 21 ☽ ☌ ♆
09 35 ☽ ⚹ ♇
11 00 ☽ ∠ ☿
13 32 ☽ ∠ ☉
18 10 ☽ ☌ ♄
20 17 ☽ Q ♀

14 Thursday
03 27 ☽ ∠ ♅
03 59 ☽ ∥ ♂
05 00 ☽ ⊥ ☿
05 51 ☽ ⊥ ☉
05 54 ☽ Q ♇
07 05 ☉ ∥ ☿
11 41 ☽ □ ♃
14 24 ☉ ☌ ☿
16 31 ☽ ⊻ ☉
16 43 ☽ ⊻ ☿
18 07 ☽ ⚹ ♀
18 26 ☽ ⊥ ♅
21 33 ☽ ☌ ♂

15 Friday
04 17 ☽ ⊻ ♅
08 29 ☽ ⊻ ♆
11 14 ☿ ∠ ♄
11 25 ☽ □ ♇
11 39 ☿ ⊻ ♀
17 20 ☽ Q ♃
18 10 ☽ ⊥ ♆
18 11 ☽ ∥ ☉
19 51 ☽ ⊻ ♄
20 34 ☽ ∠ ♀
21 43 ☽ ∥ ☿
23 49 ☽ ∥ ♅

16 Saturday
05 29 ☽ ⊥ ♄
08 32 ☽ ∠ ♆
09 20 ☽ ∥ ♃
12 07 ☽ ⊥ ♀
12 39 ☽ ⚹ ♃
13 11 ☽ # ♇
14 15 ☿ ∥ ♅
15 29 ☿ ⊻ ♂
16 37 ☉ ∠ ♄
19 49 ☽ ∠ ♄
20 01 ☽ ☌ ☉
22 24 ☽ ⊻ ♀

17 Sunday
00 05 ☽ ⊻ ♂
01 02 ☽ ☌ ☿
01 55 ☽ ∥ ♀
04 16 ☽ ☌ ♅
08 15 ☽ ⚹ ♆
10 02 ☽ ⊥ ♂
11 01 ☽ △ ♇
12 34 ☽ ∠ ♃
19 36 ☽ ⚹ ♄

18 Monday
00 10 ☿ ☌ ♅
00 56 ☽ ∠ ♂
02 54 ☽ ⊥ ♃
03 11 ☽ Q ♆
10 39 ☽ bQ ♇
12 29 ☽ ⊻ ♃
14 35 ♀ ± ♇
14 40 ☽ Q ♄
22 34 ☽ ⊻ ☉

19 Tuesday
00 56 ☽ ± ♇
01 50 ☽ ☌ ♀
01 57 ☽ ⚹ ♂
03 52 ☽ ⊻ ♅
04 05 ☿ ⚹ ♆
05 37 ♀ ⚹ ♂
07 49 ☽ □ ♆
08 26 ☽ ⊻ ☿
08 49 ☽ ⊥ ☉
10 32 ☽ ⊼ ♇
13 32 ☽ ⊥ ♅
14 55 ☿ ∥ ♃
19 40 ☽ □ ♄
19 46 ☽ ⊥ ☿
22 18 ☽ Q ♂
22 52 ☿ △ ♇

20 Wednesday
00 22 ☽ ∠ ☉
04 12 ☽ ∠ ♅
04 44 ♀ ⊻ ♅
13 01 ☽ ∠ ☿
13 01 ☽ ∥ ♀
13 27 ☽ ☌ ♃
16 06 ☿ ∠ ♃
19 29 ♂ ⊻ ♅
21 12 ☿ ⊥ ♂

21 Thursday
02 57 ☽ ⚹ ☉
03 11 ☽ # ♇
03 22 ☽ ∥ ☿
04 55 ☿ # ♇
05 13 ☽ ⚹ ♅
05 43 ☽ □ ♂
07 21 ☽ ⊻ ♀
08 06 ☽ ∥ ♃
09 16 ☽ △ ♆
12 01 ☽ ☍ ♇
17 10 ☽ ∥ ♅
17 12 ☽ ∥ ☉
18 04 ☉ ∥ ♅
18 30 ☽ ⊥ ♀
18 42 ☽ ⚹ ☿
22 05 ☽ △ ♄

22 Friday
00 55 ☽ Q ☉
01 48 ☽ Q ♅
06 32 ♀ □ ♆
11 10 ☽ bQ ♆
11 36 ☽ ∠ ♀
14 26 ☉ ☌ ♅
17 21 ☽ ⊻ ♃
18 02 ☿ ⚹ ♄
19 25 ☽ Q ☿

23 Saturday
00 35 ☽ bQ ♄
03 08 ☽ ± ♆
04 10 ☽ ⊥ ♃
06 38 ☽ ∥ ♂
09 46 ☽ □ ♅
11 11 ☽ □ ☉
13 12 ☽ △ ♂
13 58 ☽ ⊼ ♆
14 36 ♀ ⊼ ♇
16 48 ☽ ⊼ ♇
16 55 ☽ ± ♄
17 01 ☽ ⚹ ♀
20 41 ☽ ∠ ♃

24 Sunday
02 58 ♂ ⊻ ♆
03 46 ☽ ± ♇
03 58 ☽ ⊼ ♄
10 06 ☽ □ ☿
17 23 ☽ Q ♀
18 31 ☽ bQ ♂
20 29 ☽ ∥ ♄
20 29 ☽ bQ ♇
22 33 ☿ Q ♆

25 Monday
00 54 ☽ ⚹ ♃
02 22 ☉ ⚹ ♆
05 16 ☽ ∥ ♆
07 56 ☽ # ♆
09 44 ☿ ⊥ ♃
11 49 ♀ ⊥ ♅
12 33 ☽ ± ♂
16 56 ☽ # ♄
17 49 ☽ △ ♅
22 05 ☽ ☍ ♆
23 41 ☽ △ ☉

26 Tuesday
00 03 ☽ Q ♃
00 45 ☽ ⊼ ♂
00 57 ☽ △ ♇
04 02 ♂ □ ♇
07 08 ☽ □ ♀
13 11 ☽ ☍ ♄
15 53 ☉ △ ♇
20 29 ☿ ∥ ♀
22 54 ☽ bQ ♅

27 Wednesday
05 13 ☿ bQ ♇
06 10 ☽ △ ☿
07 14 ☽ bQ ☉
11 32 ☽ □ ♃
13 57 ☽ # ♂
16 37 ☽ ± ♅
21 26 ☿ ∠ ♂

28 Thursday
02 27 ☽ ± ☉
04 31 ☽ ⊼ ♅
08 46 ☽ ⊼ ♆
09 33 ☉ ⊻ ♂
11 34 ☽ □ ♇
15 16 ☽ ☍ ♂
15 22 ☽ ⊼ ☉
17 15 ☽ bQ ☿
19 58 ☿ ⊻ ♃
20 42 ☽ ± ♆

29 Friday
00 26 ☽ △ ♀
00 40 ☽ ⊼ ♄
02 41 ☿ Q ♄
03 03 ♀ □ ♄
03 55 ☽ # ♅
11 49 ☽ # ☉
12 34 ☽ # ♃
12 44 ☽ ± ♄
14 28 ☽ ± ☿
14 42 ☽ bQ ♆
20 20 ☽ ∥ ♇
23 28 ☽ Q ♇

30 Saturday
00 05 ☽ △ ♃
02 13 ☉ ∠ ♃
03 22 ☉ ∥ ♃
04 39 ☽ ⊼ ☿
06 55 ☽ bQ ♄
09 30 ☽ # ♀
09 45 ☽ bQ ♀
16 45 ☽ ☍ ♅
17 32 ☽ # ☿
20 53 ☽ △ ♆
23 34 ☽ ⚹ ♇

31 Sunday
05 52 ☽ ± ♀
06 44 ☽ bQ ♃
07 22 ☽ ⊼ ♂
08 45 ☽ ☍ ☉
13 21 ☽ △ ♄
19 19 ☽ ⊼ ♀
20 18 ☽ ± ♂

JUNE 2026

LONGITUDES

Date	Sidereal time h m s	Sun ☉	Moon ☽	Moon ☽ 24.00	Mercury ☿	Venus ♀	Mars ♂	Jupiter ♃	Saturn ♄	Uranus ♅	Neptune ♆	Pluto ♇
01	04 39 57	11 ♊ 01 03	23 ♐ 24 52	29 ♐ 20 51	00 ♋ 00	15 ♋ 59	10 ♉ 08	24 ♋ 10	12 ♈ 17	02 ♊ 06	04 ♈ 04	05 ♒ 22
02	04 43 53	11 58 32	05 ♑ 16 53	11 ♑ 13 15	01 41	17 10	10 52	24 21	12 22	02 09	04 05	05 R 21
03	04 47 50	12 56 00	17 ♑ 10 12	23 ♑ 08 05	03 18	18 21	11 37	24 33	12 27	02 13	04 06	05 20
04	04 51 46	13 53 27	29 ♑ 07 14	05 ♒ 08 04	04 53	19 31	12 21	24 44	12 32	02 16	04 07	05 19
05	04 55 43	14 50 54	11 ♒ 11 01	17 ♒ 16 37	06 24	20 42	13 05	24 56	12 37	02 19	04 09	05 19
06	04 59 40	15 48 19	23 ♒ 25 21	29 ♒ 37 47	07 53	21 52	13 49	25 07	12 41	02 23	04 10	05 18
07	05 03 36	16 45 44	05 ♓ 54 29	12 ♓ 16 03	09 19	23 03	14 33	25 19	12 46	02 26	04 11	05 17
08	05 07 33	17 43 09	18 ♓ 43 02	25 ♓ 15 58	10 42	24 13	15 18	25 31	12 51	02 30	04 11	05 16
09	05 11 29	18 40 33	01 ♈ 55 19	08 ♈ 41 31	12 02	25 23	16 02	25 43	12 55	02 33	04 12	05 15
10	05 15 26	19 37 56	15 ♈ 34 50	22 ♈ 35 23	13 18	26 33	16 46	25 55	13 00	02 36	04 13	05 14
11	05 19 22	20 35 19	29 ♈ 43 10	06 ♉ 57 56	14 32	27 44	17 29	26 07	13 04	02 40	04 14	05 14
12	05 23 19	21 32 41	14 ♉ 19 14	21 ♉ 46 21	15 42	28 ♋ 54	18 13	26 19	13 08	02 43	04 15	05 13
13	05 27 15	22 30 03	29 ♉ 18 23	06 ♊ 54 12	16 50	00 ♌ 04	18 57	26 31	13 12	02 47	04 16	05 12
14	05 31 12	23 27 24	14 ♊ 32 30	22 ♊ 11 55	17 54	01 13	19 41	26 43	13 16	02 50	04 16	05 11
15	05 35 09	24 24 45	29 ♊ 50 59	07 ♋ 28 16	18 54	02 23	20 25	26 55	13 20	02 53	04 17	05 10
16	05 39 05	25 22 05	15 ♋ 02 27	22 ♋ 32 20	19 51	03 33	21 08	27 07	13 24	02 57	04 18	05 09
17	05 43 02	26 19 25	29 ♋ 56 54	07 ♌ 15 24	20 45	04 42	21 52	27 20	13 28	03 00	04 19	05 08
18	05 46 58	27 16 43	14 ♌ 27 16	21 ♌ 32 09	21 35	05 52	22 35	27 32	13 32	03 03	04 19	05 07
19	05 50 55	28 14 01	28 ♌ 29 56	05 ♍ 20 39	22 21	07 01	23 19	27 44	13 35	03 06	04 20	05 06
20	05 54 51	29 ♊ 11 18	12 ♍ 04 31	18 ♍ 41 51	23 04	08 11	24 02	27 57	13 39	03 10	04 20	05 05
21	05 58 48	00 ♋ 08 34	25 ♍ 13 03	01 ♎ 38 37	23 42	09 20	24 45	28 09	13 43	03 13	04 21	05 04
22	06 02 44	01 05 50	07 ♎ 59 03	14 ♎ 14 55	24 16	10 29	25 29	28 22	13 46	03 16	04 21	05 03
23	06 06 41	02 03 05	20 ♎ 26 44	26 ♎ 35 03	24 47	11 38	26 12	28 34	13 49	03 19	04 22	05 01
24	06 10 38	03 00 19	02 ♏ 40 24	08 ♏ 43 14	25 13	12 47	26 55	28 47	13 52	03 22	04 22	05 00
25	06 14 34	03 57 32	14 ♏ 44 02	20 ♏ 43 13	25 34	13 56	27 38	29 00	13 56	03 26	04 23	04 59
26	06 18 31	04 54 45	26 ♏ 41 09	02 ♐ 38 11	25 51	15 04	28 21	29 12	13 59	03 29	04 23	04 58
27	06 22 27	05 51 58	08 ♐ 34 37	14 ♐ 30 43	26 04	16 13	29 04	29 25	14 02	03 32	04 23	04 57
28	06 26 24	06 49 11	20 ♐ 26 45	26 ♐ 22 55	26 12	17 21	29 ♉ 47	29 38	14 04	03 35	04 24	04 56
29	06 30 20	07 46 23	02 ♑ 19 27	08 ♑ 16 31	26 15	18 29	00 ♊ 29	29 ♋ 50	14 07	03 38	04 24	04 54
30	06 34 17	08 ♋ 43 35	14 ♑ 14 20	20 ♑ 13 05	26 ♋ 14	19 ♌ 38	01 ♊ 12	00 ♌ 03	14 ♈ 10	03 ♊ 41	04 ♈ 24	04 ♒ 53

Date	Moon True ☊	Moon Mean ☊	Moon ☽ Latitude
01	03 ♓ 30	04 ♓ 09	04 S 42
02	03 R 18	04 06	04 15
03	03 08	04 02	03 37
04	03 02	03 59	02 49
05	02 57	03 56	01 52
06	02 56	03 53	00 S 50
07	02 D 55	03 50	00 N 16
08	02 R 56	03 47	01 23
09	02 55	03 43	02 27
10	02 53	03 40	03 26
11	02 48	03 37	04 14
12	02 41	03 34	04 47
13	02 32	03 31	05 02
14	02 22	03 28	04 55
15	02 13	03 24	04 28
16	02 05	03 21	03 42
17	01 59	03 18	02 41
18	01 56	03 15	01 32
19	01 54	03 12	00 N 18
20	01 D 55	03 08	00 S 54
21	01 55	03 05	02 01
22	01 R 55	03 02	03 00
23	01 53	02 59	03 49
24	01 49	02 56	04 26
25	01 43	02 53	04 51
26	01 35	02 49	05 03
27	01 25	02 46	05 02
28	01 15	02 43	04 48
29	01 05	02 40	04 21
30	00 ♓ 56	02 ♓ 37	03 S 43

DECLINATIONS

Date	Sun ☉	Moon ☽	Mercury ☿	Venus ♀	Mars ♂	Jupiter ♃	Saturn ♄	Uranus ♅	Neptune ♆	Pluto ♇
01	22 N 06	27 S 58	25 N 36	24 N 24	14 N 16	21 N 41	02 N 46	20 N 25	00 N 23	22 S 57
02	22 13	27 35	25 36	24 16	14 30	21 39	02 48	20 26	00 24	22 57
03	22 21	25 55	25 33	24 08	14 45	21 37	02 50	20 27	00 24	22 57
04	22 28	23 05	25 29	23 59	14 59	21 35	02 51	20 27	00 24	22 58
05	22 35	19 13	25 23	23 49	15 13	21 33	02 53	20 28	00 25	22 58
06	22 41	14 30	25 15	23 38	15 27	21 31	02 55	20 29	00 25	22 58
07	22 47	09 06	25 06	23 27	15 40	21 29	02 56	20 30	00 25	22 59
08	22 52	03 S 12	24 55	23 15	15 54	21 27	02 58	20 30	00 26	22 59
09	22 57	03 N 01	24 43	23 03	16 07	21 25	02 59	20 31	00 26	23 00
10	23 02	09 18	24 30	22 50	16 20	21 23	03 01	20 32	00 26	23 00
11	23 06	15 19	24 16	22 36	16 33	21 20	03 02	20 32	00 27	23 00
12	23 10	20 42	24 00	22 22	16 46	21 18	03 04	20 33	00 27	23 01
13	23 14	24 54	23 44	22 08	16 59	21 16	03 05	20 34	00 27	23 01
14	23 17	27 26	23 27	21 52	17 11	21 14	03 07	20 34	00 27	23 02
15	23 19	27 54	23 09	21 36	17 23	21 11	03 08	20 35	00 28	23 02
16	23 21	26 16	22 51	21 20	17 36	21 09	03 09	20 35	00 28	23 02
17	23 23	22 47	22 33	21 03	17 48	21 07	03 10	20 36	00 28	23 03
18	23 25	17 57	22 14	20 45	17 59	21 04	03 12	20 37	00 28	23 03
19	23 26	12 17	21 54	20 27	18 11	21 02	03 13	20 37	00 29	23 04
20	23 26	06 12	21 35	20 09	18 22	20 59	03 14	20 38	00 29	23 04
21	23 26	00 N 03	21 15	19 50	18 34	20 57	03 15	20 39	00 29	23 04
22	23 26	05 S 55	20 56	19 30	18 45	20 55	03 16	20 39	00 29	23 05
23	23 25	11 31	20 37	19 10	18 55	20 52	03 17	20 40	00 29	23 05
24	23 24	16 34	20 18	18 50	19 06	20 50	03 18	20 40	00 29	23 06
25	23 23	20 54	19 59	18 29	19 17	20 47	03 19	20 41	00 29	23 06
26	23 21	24 20	19 41	18 08	19 27	20 44	03 20	20 42	00 30	23 07
27	23 18	26 42	19 23	17 46	19 37	20 42	03 21	20 42	00 30	23 07
28	23 16	27 53	19 06	17 24	19 47	20 39	03 22	20 43	00 30	23 07
29	23 13	27 46	18 50	17 01	19 56	20 37	03 23	20 43	00 30	23 08
30	23 N 09	26 S 22	18 N 34	16 N 38	20 N 06	20 N 34	03 N 23	20 N 44	00 N 30	23 S 08

ZODIAC SIGN ENTRIES

Date	h	m	Planets	
01	11	56	☿	♋
02	01	19	☽	♑
04	13	46	☽	♒
07	00	43	☽	♓
09	08	33	☽	♈
11	12	28	☽	♉
13	10	47	♀	♌
13	13	06	☽	♊
15	12	14	☽	♋
17	12	05	☽	♌
19	14	37	☽	♍
21	08	24	☉	♋
21	20	55	☽	♎
24	06	43	☽	♏
26	18	41	☽	♐
28	19	29	♂	♊
29	07	18	☽	♑
30	05	52	♃	♌

LATITUDES

Date	Mercury ☿	Venus ♀	Mars ♂	Jupiter ♃	Saturn ♄	Uranus ♅	Neptune ♆	Pluto ♇
01	02 N 10	01 N 56	00 S 37	00 N 25	02 S 16	00 S 10	01 S 20	04 S 09
04	02 08	01 59	00 35	00 25	02 16	00 10	01 21	04 09
07	02 00	02 01	00 34	00 25	02 17	00 10	01 21	04 10
10	01 44	02 02	00 32	00 25	02 18	00 09	01 21	04 10
13	01 22	02 02	00 30	00 25	02 18	00 09	01 21	04 11
16	00 54	02 02	00 28	00 25	02 19	00 09	01 21	04 11
19	00 N 19	02 00	00 26	00 26	02 20	00 09	01 21	04 12
22	00 S 20	01 58	00 24	00 26	02 21	00 09	01 22	04 12
25	01 04	01 55	00 22	00 26	02 21	00 09	01 22	04 13
28	01 51	01 51	00 20	00 26	02 22	00 09	01 22	04 13
31	02 S 38	01 N 46	00 S 18	00 N 26	02 S 23	00 S 09	01 S 22	04 S 14

DATA

Julian Date	2461193
Delta T	+73 seconds
Ayanamsa	24° 13' 41"
Synetic vernal point	04° ♓ 53' 18"
True obliquity of ecliptic	23° 26' 17"

MOON'S PHASES, APSIDES AND POSITIONS ☽

Date	h	m	Phase	Longitude	Eclipse Indicator
08	10	00	☾	17 ♓ 38	
15	02	54	●	24 ♊ 03	
21	21	55	☽	00 ♎ 32	
29	23	57	○	08 ♑ 15	

Date	h	m		
01	04	16	Apogee	
14	23	13	Perigee	
28	06	59	Apogee	
01	16	48	Max dec	28° S 00'
09	00	26	0N	
15	05	18	Max dec	27° N 59'
21	12	12	0S	
28	21	58	Max dec	27° S 59'

LONGITUDES

Date	Chiron ⚷	Ceres ⚳	Pallas ⚴	Juno ⚵	Vesta ⚶	Black Moon Lilith ⚸
01	29 ♈ 13	01 ♊ 26	10 ♈ 41	10 ♒ 57	07 ♈ 44	18 ♐ 14
11	29 ♈ 40	05 ♊ 32	13 ♈ 17	10 ♒ 54	11 ♈ 31	19 ♐ 21
21	00 ♉ 04	09 ♊ 37	15 ♈ 40	10 ♒ 15	15 ♈ 03	20 ♐ 28
31	00 ♉ 23	13 ♊ 39	17 ♈ 47	08 ♒ 59	18 ♈ 17	21 ♐ 35

All ephemeris data is given at 12.00 UT and the Moon's longitude is additionally given for 24.00 UT

ASPECTARIAN

h m	Aspects
01 Monday	
01 13	☽ ± ♃
02 58	☿ ± ♇
05 50	☽ ∠ ♇
10 00	♂ ⊥ ♆
13 32	☽ ⊼ ♃
15 42	☽ ⚼ ♂
02 Tuesday	
00 01	☽ ⊥ ♇
03 32	☽ ☍ ☿
05 38	☽ ⊼ ♅
09 35	☽ □ ♆
11 41	♀ ∠ ♅
12 08	☽ ⊻ ♇
17 50	☽ ± ♅
19 11	☿ ⊻ ♅
22 49	☉ ⚹ ♄
03 Wednesday	
00 03	☽ △ ♂
02 25	☽ □ ♄
02 43	☽ ⊼ ☉
12 05	☽ ⚼ ♅
14 37	☽ ☍ ♀
15 46	☽ # ☿
15 52	☽ ± ☉
21 57	☽ Q ♆
04 Thursday	
00 18	☿ □ ♆
03 04	☽ ☍ ♃
05 00	☽ # ♀
11 30	☽ ⚼ ☉
12 50	☽ ∥ ♇
14 51	☽ Q ♄
16 07	☽ # ☉
18 19	☽ △ ♅
18 44	♂ ⊻ ♄
18 51	☿ ⊼ ♇
22 00	☽ ⚹ ♆
22 06	☽ # ♃
05 Friday	
00 22	☽ ☌ ♇
01 10	☽ ⊼ ☿
04 51	☽ # ♅
05 03	♂ Q ♃
14 45	☽ ± ☿
14 50	☽ ⚹ ♄
16 00	☽ □ ♂
19 51	☽ △ ☉
06 Saturday	
03 41	☽ ∠ ♆
04 38	☉ ⊥ ♀
07 42	☽ # ♂
08 40	☽ ⊼ ♀
10 49	☽ ⚼ ☿
15 21	☽ ⊼ ♃
20 19	☽ ∠ ♄
20 32	☿ ⊥ ♅
21 01	☉ Q ♆
21 11	☽ ⊥ ♆
21 31	☽ ± ♀
07 Sunday	
03 06	☽ ± ♃
05 13	☽ Q ♂
05 21	☽ □ ♅
08 42	☽ ⊻ ♆
10 49	☽ ⊻ ♇
13 39	☽ ⊥ ♄
16 28	☽ ⚼ ♀
19 15	☽ △ ☿
20 28	☽ ⚼ ♃
22 09	☽ ⊥ ♇
08 Monday	
01 01	☽ ⊻ ♄
05 16	☽ ⚹ ♂
10 00	☽ □ ☉
12 54	☽ # ♄
14 52	☽ ∠ ♇
15 17	☽ Q ♅
22 46	☽ # ♆
23 05	☽ △ ♀
09 Tuesday	
00 38	☽ △ ♃
02 06	☽ ∥ ♆
10 18	☽ ∠ ♂
11 54	☽ ∥ ♄
13 08	☽ ⚹ ♅
16 05	☽ ☌ ♆
17 56	☽ ⚹ ♇
18 15	♀ # ♇
19 48	☉ ∥ ♀
19 59	♀ ☌ ♃
21 05	☽ Q ☉
10 Wednesday	
00 32	☉ # ♇
03 09	☽ ⊥ ♂
05 39	☿ □ ♄
07 30	☽ ☌ ♄
07 40	☽ □ ☿
14 09	☽ ⊻ ♂
14 52	☽ Q ♇
15 30	☽ ∠ ♅
19 28	☽ ⚹ ☉
20 49	☉ ⊥ ♃
11 Thursday	
03 04	☉ ⚼ ♇
05 52	☽ □ ♃
06 52	☽ ⊥ ♅
08 22	☽ □ ♀
16 55	☽ ⊻ ♅
17 06	☽ Q ☿
17 22	☽ ∥ ♂
19 30	☽ ⊻ ♆
21 08	☽ □ ♇
22 25	☽ ∠ ☉
12 Friday	
05 22	☽ ⊥ ♆
10 04	☽ ⊻ ♄
11 16	☽ ∥ ♅
11 59	☽ Q ♃
14 07	☽ ⊥ ☉
14 26	☽ ⚹ ☿
15 00	☽ ∥ ♃
16 31	☽ Q ♀
18 38	☽ ☌ ♂
19 48	☽ ⊥ ♄
19 58	☽ ∠ ♆
20 11	☽ ∥ ♀
13 Saturday	
00 13	☽ # ♇
00 26	☽ ⊻ ☉
01 16	☽ ∥ ☉
04 55	☽ ∥ ☿
07 30	☽ ⚹ ♃
10 15	☽ ∠ ♄
13 18	☽ ⚹ ♀
16 18	☽ ∠ ☿
17 31	☽ ☌ ♅
19 51	☽ ⚹ ♆
21 09	♂ ⊥ ♄
21 18	☽ △ ♇
22 22	♂ ∠ ♆
14 Sunday	
07 30	☽ ∠ ♃
07 33	☽ ⊥ ☿
10 00	☽ ⚹ ♄
10 29	☿ ∠ ♅
14 43	☽ Q ♆
14 51	☽ ∠ ♀
17 38	☽ ⊻ ☿
20 27	☽ ⊻ ♂
20 50	☽ ⚼ ♇
21 48	☽ ⊥ ♃
15 Monday	
00 27	☉ ∥ ☿
02 54	☽ ☌ ☉
04 53	☽ Q ♄
06 07	☽ ⊥ ♀
06 20	☽ ⊥ ♂
07 20	☽ ⊻ ♃
10 55	☽ ± ♇
13 16	♀ Q ♂
16 19	☽ ⊻ ♀
16 48	☽ ⊻ ♅
18 59	☽ □ ♆
20 21	☽ ⊼ ♇
21 11	☽ ∠ ♂
21 43	☿ # ♇
22 53	♀ ⚹ ♅
16 Tuesday	
02 17	☽ ⊥ ♅
09 23	☽ □ ♄
12 58	☉ Q ♄
16 39	☽ ∠ ♅
20 12	☽ ☌ ☿
22 15	☽ ⚹ ♂
17 Wednesday	
03 41	♀ △ ♆
05 42	☽ ⊻ ☉
05 56	♀ ∥ ♃
07 41	☽ ☌ ♃
08 33	☽ ∥ ☉
10 31	☽ # ♇
13 29	☽ ∥ ☿
16 09	☽ ⊥ ☉
17 00	☽ ⚹ ♅
18 44	☽ Q ♂
19 09	☽ △ ♆
20 28	☽ ☌ ♀
20 29	☽ ☍ ♇
20 38	♀ ☍ ♇
21 03	☽ ∥ ♃
21 53	☽ ∥ ♀
23 30	☽ ∥ ♅
18 Thursday	
08 06	☽ ∠ ☉
10 27	☽ △ ♄
11 51	☽ ∥ ♂
13 01	☽ Q ♅
20 07	☉ ⊻ ♃
20 14	☽ ⚼ ♆
23 11	♀ ∥ ♅
19 Friday	
00 48	☽ ⊻ ☿
02 33	☽ □ ♂
10 40	☽ ⊻ ♃
11 30	☽ ⚹ ☉
11 42	☽ ± ♆
11 44	☽ ⊥ ☿
12 10	☽ ⚼ ♄
20 05	☽ □ ♅
21 18	☽ ⊥ ♃
22 13	☽ ⊼ ♆
23 33	☽ ⊼ ♇
20 Saturday	
04 03	☽ ± ♄
04 22	☽ ⊻ ♀
04 26	☽ ∠ ☿
09 17	☉ ± ♇
10 13	☽ ± ♇
10 17	☽ Q ☉
13 35	☽ ∠ ♃
14 51	☽ ⊼ ♄
16 09	☽ ⊥ ♀
23 31	☽ ∥ ♄
21 Sunday	
02 30	☽ ⚼ ♇
09 03	☽ ⚹ ☿
10 12	☽ ∠ ♀
10 18	☽ ∥ ♆
11 06	☽ △ ♂
14 06	☽ # ♆
17 33	☽ ⚹ ♃
21 55	☽ □ ☉
22 Monday	
01 09	☽ # ♄
03 01	☽ △ ♅
05 07	☽ ☍ ♆
06 25	☽ △ ♇
08 36	☽ Q ☿
13 50	☿ ∥ ♃
16 37	☽ Q ♃
17 03	☽ ⚼ ♂
17 15	☽ ⚹ ♀
23 07	☽ ☍ ♄
23 Tuesday	
07 51	☽ ⚼ ♅
08 00	☿ ∥ ♅
11 29	☽ ± ♂
18 51	☽ Q ♀
20 47	☽ □ ☿
23 28	♀ ∥ ♂
23 56	☽ ⊼ ♂
24 Wednesday	
01 30	☽ ± ♅
02 43	☉ ⊥ ♂
04 11	☽ □ ♃
12 43	☽ △ ☉
13 24	☽ ⊼ ♅
15 22	☽ ⊼ ♆
16 37	☽ □ ♇
21 49	☉ ⊻ ♅
23 07	☽ # ♀
25 Thursday	
02 05	☽ # ♂
03 18	☽ ± ♆
06 55	☽ # ☿
10 13	☽ □ ♀
10 23	☽ ⊼ ♄
10 44	☽ # ♅
11 20	☽ # ♃
12 02	♀ △ ♄
21 12	☽ ⚼ ☉
21 19	☽ ⚼ ♆
22 27	☽ ± ♄
22 38	☉ □ ♆
26 Friday	
02 39	☽ ∥ ♇
04 28	☽ # ☉
04 32	☽ Q ♇
10 17	☽ △ ☿
13 19	☉ ⊼ ♇
15 34	☽ ☍ ♂
16 38	☽ ⚼ ♄
16 53	☽ ± ☉
17 10	☽ △ ♃
21 00	♀ Q ♅
23 52	☿ ∥ ♂
27 Saturday	
01 46	☽ ☍ ♅
03 32	☽ △ ♆
04 41	☽ ⚹ ♇
06 03	☽ ⊼ ☉
09 00	♃ ∥ ♅
10 41	♂ ∠ ♄
17 06	☽ ⚼ ☿
23 04	☽ △ ♄
28 Sunday	
00 01	☽ ⚼ ♃
04 50	♂ ⚹ ♃
05 05	☽ △ ♀
10 57	☽ ∠ ♇
11 30	☽ ± ☿
18 33	☽ ± ♃
23 42	☽ ⊼ ☿
29 Monday	
05 07	☽ ⊥ ♇
06 54	☽ ⊼ ♃
08 04	☽ ⊼ ♂
14 36	☽ ⚼ ♀
14 39	☽ ⊼ ♅
16 11	☽ □ ♆
17 12	☽ ⊻ ♇
17 36	☿ St R
20 56	☽ ± ♂
23 57	☽ ☍ ☉
30 Tuesday	
02 48	☽ ± ♅
07 18	♀ ⚼ ♆
10 38	☽ ± ♀
11 51	☽ □ ♄
16 12	☽ ⚼ ♂
16 16	☉ # ♇
20 58	☽ ⚼ ♅
23 57	☽ ⊼ ♀

JULY 2026

LONGITUDES

Date	Sidereal time h m s	Sun ☉ ° ' "	Moon ☽ ° ' "	Moon ☽ 24.00 ° ' "	Mercury ☿ ° '	Venus ♀ ° '	Mars ♂ ° '	Jupiter ♃ ° '	Saturn ♄ ° '	Uranus ♅ ° '	Neptune ♆ ° '	Pluto ♇ ° '
01	06 38 13	09 ♋ 40 46	26 ♑ 12 59	02 ♒ 14 15	26 ♋ 08	20 ♌ 46	01 ♊ 55	00 ♌ 16	14 ♈ 12	03 ♊ 44	04 ♈ 24	04 ♒ 52
02	06 42 10	10 37 58	08 ♒ 17 10	14 ♒ 21 58	25 R 58	21 53	02 37	00 29	14 15	03 47	04 25	04 R 51
03	06 46 07	11 35 10	20 ♒ 28 59	26 ♒ 38 33	25 43	23 01	03 20	00 42	14 17	03 50	04 25	04 49
04	06 50 03	12 32 21	02 ♓ 51 02	09 ♓ 06 50	25 24	24 09	04 02	00 55	14 20	03 53	04 25	04 48
05	06 54 00	13 29 33	15 ♓ 26 22	21 ♓ 50 04	25 01	25 16	04 45	01 08	14 22	03 56	04 25	04 47
06	06 57 56	14 26 45	28 ♓ 18 21	04 ♈ 51 38	24 35	26 24	05 27	01 21	14 24	03 59	04 25	04 46
07	07 01 53	15 23 58	11 ♈ 30 19	18 ♈ 14 45	24 05	27 31	06 10	01 34	14 26	04 01	04 R 25	04 44
08	07 05 49	16 21 10	25 ♈ 05 11	02 ♉ 01 46	23 32	28 38	06 52	01 47	14 28	04 04	04 25	04 43
09	07 09 46	17 18 24	09 ♉ 04 33	16 ♉ 13 26	22 57	29 ♌ 45	07 34	02 00	14 30	04 07	04 25	04 42
10	07 13 42	18 15 37	23 ♉ 28 08	00 ♊ 48 11	22 21	00 ♍ 52	08 16	02 13	14 31	04 10	04 25	04 40
11	07 17 39	19 12 52	08 ♊ 12 57	15 ♊ 41 35	21 42	01 59	08 58	02 26	14 33	04 12	04 25	04 39
12	07 21 36	20 10 06	23 ♊ 13 05	00 ♋ 46 21	21 04	03 05	09 40	02 39	14 35	04 15	04 25	04 38
13	07 25 32	21 07 21	08 ♋ 20 10	15 ♋ 53 18	20 25	04 11	10 22	02 52	14 36	04 18	04 25	04 36
14	07 29 29	22 04 37	23 ♋ 24 30	00 ♌ 52 37	19 47	05 18	11 04	03 06	14 37	04 20	04 24	04 35
15	07 33 25	23 01 53	08 ♌ 16 36	15 ♌ 35 35	19 11	06 24	11 45	03 19	14 38	04 23	04 24	04 34
16	07 37 22	23 59 08	22 ♌ 48 51	29 ♌ 55 51	18 36	07 29	12 27	03 32	14 40	04 26	04 24	04 32
17	07 41 18	24 56 24	06 ♍ 56 15	13 ♍ 49 53	18 05	08 35	13 09	03 45	14 41	04 28	04 23	04 31
18	07 45 15	25 53 40	20 ♍ 36 46	27 ♍ 17 02	17 36	09 41	13 50	03 58	14 41	04 31	04 23	04 29
19	07 49 11	26 50 57	03 ♎ 50 58	10 ♎ 18 53	17 12	10 46	14 32	04 12	14 42	04 33	04 23	04 28
20	07 53 08	27 48 13	16 ♎ 41 16	22 ♎ 58 35	16 51	11 51	15 13	04 25	14 43	04 35	04 22	04 27
21	07 57 05	28 45 30	29 ♎ 11 23	05 ♏ 20 12	16 36	12 56	15 55	04 38	14 44	04 38	04 22	04 25
22	08 01 01	29 ♋ 42 46	11 ♏ 25 35	17 ♏ 28 07	16 25	14 00	16 36	04 52	14 44	04 40	04 21	04 24
23	08 04 58	00 ♌ 40 04	23 ♏ 28 20	29 ♏ 26 46	16 20	15 05	17 17	05 05	14 44	04 43	04 21	04 22
24	08 08 54	01 37 21	05 ♐ 23 53	11 ♐ 20 12	16 D 20	16 09	17 58	05 18	14 45	04 45	04 20	04 21
25	08 12 51	02 34 39	17 ♐ 16 07	23 ♐ 12 03	16 26	17 13	18 39	05 31	14 45	04 47	04 20	04 20
26	08 16 47	03 31 58	29 ♐ 08 23	05 ♑ 05 25	16 38	18 17	19 20	05 45	14 45	04 49	04 19	04 18
27	08 20 44	04 29 17	11 ♑ 03 29	17 ♑ 02 51	16 56	19 21	20 01	05 58	14 R 45	04 51	04 19	04 17
28	08 24 40	05 26 37	23 ♑ 03 45	29 ♑ 06 25	17 20	20 24	20 42	06 11	14 45	04 53	04 18	04 15
29	08 28 37	06 23 57	05 ♒ 11 02	11 ♒ 17 49	17 50	21 27	21 23	06 24	14 45	04 56	04 17	04 14
30	08 32 34	07 21 18	17 ♒ 26 55	23 ♒ 38 31	18 26	22 30	22 03	06 38	14 44	04 58	04 17	04 13
31	08 36 30	08 ♌ 18 40	29 ♒ 52 48	06 ♓ 09 56	19 ♋ 08	23 ♍ 32	22 ♊ 44	06 ♌ 51	14 ♈ 44	05 ♊ 00	04 ♈ 16	04 ♒ 11

Date	Moon True ☊ ° '	Moon Mean ☊ ° '	Moon ☽ Latitude ° '
01	00 ♓ 49	02 ♓ 33	02 S 54
02	00 R 45	02 30	01 58
03	00 42	02 27	00 S 55
04	00 D 42	02 24	00 N 12
05	00 42	02 21	01 19
06	00 44	02 18	02 24
07	00 45	02 14	03 22
08	00 R 44	02 11	04 12
09	00 42	02 08	04 48
10	00 39	02 05	05 07
11	00 34	02 02	05 07
12	00 28	01 59	04 45
13	00 22	01 55	04 04
14	00 17	01 52	03 06
15	00 14	01 49	01 56
16	00 12	01 46	00 N 40
17	00 D 12	01 43	00 S 37
18	00 13	01 39	01 49
19	00 15	01 36	02 53
20	00 16	01 33	03 47
21	00 R 17	01 30	04 28
22	00 16	01 27	04 56
23	00 14	01 24	05 10
24	00 11	01 20	05 11
25	00 07	01 17	04 59
26	00 02	01 14	04 33
27	29 ♒ 58	01 11	03 56
28	29 54	01 08	03 08
29	29 51	01 05	02 11
30	29 50	01 01	01 08
31	29 ♒ 49	00 ♓ 58	00 00

DECLINATIONS

Date	Sun ☉ ° '	Moon ☽ ° '	Mercury ☿ ° '	Venus ♀ ° '	Mars ♂ ° '	Jupiter ♃ ° '	Saturn ♄ ° '	Uranus ♅ ° '	Neptune ♆ ° '	Pluto ♇ ° '
01	23 N 05	23 S 46	18 N 20	16 N 15	20 N 15	20 N 31	03 N 24	20 N 45	00 N 30	23 S 09
02	23 01	20 05	18 06	15 51	20 24	20 29	03 25	20 45	00 30	23 09
03	22 56	15 32	17 54	15 27	20 33	20 26	03 26	20 46	00 30	23 10
04	22 51	10 17	17 43	15 02	20 42	20 23	03 26	20 46	00 30	23 10
05	22 45	04 S 32	17 33	14 38	20 51	20 20	03 27	20 47	00 30	23 10
06	22 39	01 N 31	17 25	14 13	20 59	20 18	03 27	20 47	00 30	23 11
07	22 33	07 39	17 17	13 47	21 07	20 15	03 28	20 48	00 30	23 11
08	22 26	13 37	17 12	13 21	21 15	20 12	03 28	20 48	00 30	23 12
09	22 19	19 04	17 07	12 56	21 23	20 09	03 29	20 49	00 30	23 12
10	22 12	23 35	17 04	12 29	21 30	20 06	03 29	20 49	00 30	23 13
11	22 04	26 43	17 03	12 03	21 37	20 03	03 30	20 50	00 29	23 13
12	21 55	28 01	17 03	11 36	21 44	20 00	03 30	20 50	00 29	23 14
13	21 47	27 14	17 04	11 09	21 51	19 57	03 30	20 51	00 29	23 14
14	21 38	24 28	17 07	10 42	21 58	19 54	03 30	20 51	00 29	23 14
15	21 28	20 04	17 11	10 14	22 04	19 51	03 31	20 52	00 29	23 15
16	21 19	14 33	17 16	09 47	22 11	19 48	03 31	20 52	00 29	23 15
17	21 08	08 24	17 23	09 19	22 17	19 45	03 31	20 53	00 29	23 16
18	20 58	02 N 03	17 30	08 51	22 23	19 42	03 31	20 53	00 28	23 16
19	20 47	04 S 11	17 39	08 23	22 28	19 39	03 31	20 54	00 28	23 17
20	20 36	10 03	17 48	07 54	22 34	19 36	03 31	20 54	00 28	23 17
21	20 24	15 21	17 58	07 26	22 39	19 33	03 31	20 55	00 28	23 18
22	20 13	19 57	18 09	06 57	22 44	19 30	03 31	20 55	00 28	23 18
23	20 00	23 39	18 20	06 28	22 49	19 27	03 31	20 55	00 27	23 18
24	19 48	26 18	18 31	06 00	22 53	19 24	03 31	20 56	00 27	23 19
25	19 35	27 47	18 43	05 31	22 58	19 21	03 31	20 56	00 27	23 19
26	19 22	28 00	18 55	05 01	23 02	19 17	03 30	20 57	00 27	23 20
27	19 08	26 54	19 06	04 32	23 06	19 14	03 30	20 57	00 26	23 20
28	18 54	24 34	19 18	04 03	23 09	19 11	03 30	20 57	00 26	23 21
29	18 40	21 06	19 29	03 34	23 13	19 08	03 29	20 58	00 26	23 21
30	18 26	16 41	19 39	03 05	23 16	19 04	03 29	20 58	00 25	23 21
31	18 N 11	11 S 31	19 N 49	02 N 35	23 N 19	19 N 01	03 N 29	20 N 59	00 N 25	23 S 22

ZODIAC SIGN ENTRIES

Date	h m	Planets
01	19 33	☽ ♒
04	06 30	☽ ♓
06	15 07	☽ ♈
08	20 31	☽ ♉
09	17 22	♀ ♍
10	22 42	☽ ♊
12	22 46	☽ ♋
14	22 35	☽ ♌
17	00 07	☽ ♍
19	04 56	☽ ♎
21	13 34	☽ ♏
22	19 13	☉ ♌
24	01 07	☽ ♐
26	13 44	☽ ♑
29	01 46	☽ ♒
31	12 14	☽ ♓

LATITUDES

Date	Mercury ☿ ° '	Venus ♀ ° '	Mars ♂ ° '	Jupiter ♃ ° '	Saturn ♄ ° '	Uranus ♅ ° '	Neptune ♆ ° '	Pluto ♇ ° '
01	02 S 38	01 N 46	00 S 18	00 N 26	02 S 23	00 S 09	01 S 22	04 S 14
04	03 24	01 40	00 15	00 27	02 24	00 09	01 22	04 14
07	04 04	01 33	00 13	00 27	02 25	00 09	01 22	04 14
10	04 34	01 25	00 11	00 27	02 26	00 09	01 22	04 15
13	04 52	01 16	00 09	00 27	02 26	00 09	01 23	04 15
16	04 55	01 06	00 07	00 27	02 27	00 09	01 23	04 15
19	04 43	00 55	00 04	00 28	02 28	00 09	01 23	04 16
22	04 19	00 43	00 S 02	00 28	02 29	00 09	01 23	04 16
25	03 44	00 30	00 00	00 28	02 30	00 09	01 23	04 16
28	03 03	00 16	00 N 03	00 28	02 31	00 09	01 23	04 16
31	02 S 17	00 N 01	00 N 05	00 N 29	02 S 31	00 S 09	01 S 24	04 S 17

DATA

Julian Date	2461223
Delta T	+73 seconds
Ayanamsa	24° 13' 46"
Synetic vernal point	04° ♓ 53' 13"
True obliquity of ecliptic	23° 26' 17"

MOON'S PHASES, APSIDES AND POSITIONS ☽

Date	h m	Phase	Longitude ° '	Eclipse Indicator
07	19 29	☾	15 ♈ 42	
14	09 44	●	21 ♋ 59	
21	11 06	☽	28 ♎ 43	
29	14 36	○	06 ♒ 30	

Day	h m		
13	07 49	Perigee	
25	16 39	Apogee	
06	06 02	ON	
12	15 09	Max dec	28° N 02'
18	19 48	OS	
26	03 51	Max dec	28° S 04'

LONGITUDES

Date	Chiron ⚷ ° '	Ceres ⚳ ° '	Pallas ⚴ ° '	Juno ⚵ ° '	Vesta ⚶ ° '	Black Moon Lilith ⚸ ° '
01	00 ♉ 23	13 ♊ 39	17 ♈ 47	08 ♒ 59	18 ♈ 17	21 ♐ 35
11	00 ♉ 38	17 ♊ 39	19 ♈ 37	07 ♒ 11	21 ♈ 10	22 ♐ 43
21	00 ♉ 47	21 ♊ 35	21 ♈ 04	04 ♒ 58	23 ♈ 37	23 ♐ 50
31	00 ♉ 52	25 ♊ 27	22 ♈ 05	02 ♒ 35	25 ♈ 35	24 ♐ 57

All ephemeris data is given at 12.00 UT and the Moon's longitude is additionally given for 24.00 UT

ASPECTARIAN

01 Wednesday
04 23 ☽ Q ♆
11 51 ☽ ☍ ☿
13 24 ☉ ⊥ ♅
16 30 ☽ ∥ ♇
17 02 ☽ # ☉
20 14 ☽ ☍ ♃
23 59 ☽ Q ♄

02 Thursday
00 04 ☽ △ ♂
03 02 ☽ △ ♅
04 19 ☽ ✶ ♆
05 12 ☽ ☌ ♇
08 05 ☽ # ♅
09 41 ☽ # ♃
10 11 ☽ # ♂
17 02 ☽ ⚻ ☉
20 32 ♂ ∥ ♃
23 27 ☽ # ☿
23 49 ☽ ✶ ♄

03 Friday
05 51 ☽ ± ☉
09 54 ☽ ∠ ♆
12 25 ☽ # ♀
17 27 ☽ ☍ ♀
21 58 ☽ ⚻ ☿

04 Saturday
00 53 ☽ ⚼ ☉
03 26 ☽ ⊥ ♆
05 11 ☽ ∠ ♄
06 08 ♂ ☌ ♅
08 12 ☽ ⚻ ♃
09 17 ☽ ± ☿
13 59 ☽ □ ♅
14 26 ☽ □ ♂
15 01 ☽ ⚺ ♆
15 45 ☽ ⚺ ♇
19 56 ☽ ± ♃
22 32 ☽ ⊥ ♄

05 Sunday
00 38 ♂ ∥ ♅
00 44 ♂ ✶ ♆
02 03 ☽ ⚼ ☿
03 11 ☽ ⊥ ♇
08 01 ☽ △ ☉
08 06 ☿ ⚺ ♀
09 58 ☽ ⚺ ♄
13 07 ♂ △ ♇
13 20 ☽ ⚼ ♃
16 21 ☽ # ♄
20 09 ☽ ∠ ♇

06 Monday
00 13 ☽ Q ♅
02 30 ☽ Q ♂
04 04 ☽ # ♆
05 21 ☽ △ ☿
07 59 ☽ ∥ ♆
08 08 ☽ ⚻ ♀
10 47 ☉ □ ♄
17 41 ☽ △ ♃
19 35 ☽ ∥ ♄
20 12 ☽ ± ♀
22 26 ☽ ✶ ♅
23 12 ☽ ☌ ♆
23 48 ☽ ✶ ♇

07 Tuesday
01 49 ☽ ✶ ♂
10 55 ♆ St R
13 59 ☽ ⚼ ♀
17 15 ☽ ☌ ♄
19 29 ☽ □ ☉
21 19 ☽ Q ♇

08 Wednesday
01 25 ☽ ∠ ♅
06 04 ☽ ∠ ♂
07 20 ♀ ± ♆
09 25 ☽ □ ☿
11 02 ☽ ∥ ♀
17 12 ☽ ⊥ ♅
18 42 ☽ △ ♀
22 32 ☽ ⊥ ♂
23 45 ☽ □ ♃

09 Thursday
00 51 ☿ ⊥ ♀
03 13 ☽ ∥ ☿
03 33 ☽ ⚺ ♅
04 05 ☽ ⚺ ♆
04 34 ☽ □ ♇
05 08 ☽ Q ☉
06 20 ♀ ⚼ ♄
09 18 ☽ ⚺ ♂
14 16 ☽ ⊥ ♆
15 03 ☽ Q ☿
17 14 ☽ ∥ ♃
19 11 ☿ ∠ ♂
20 38 ☽ ∥ ♅
21 08 ☽ ⚺ ♄
23 51 ☽ ∥ ♂

10 Friday
02 47 ☽ ✶ ☉
04 09 ☽ ∥ ☉
05 18 ☽ ∠ ♆
06 33 ☽ Q ♃
07 08 ☽ ⊥ ♄
09 44 ☽ # ♇
10 13 ☽ ✶ ☿
21 56 ☽ ∠ ♄

11 Saturday
01 05 ☽ □ ♀
02 31 ☽ ✶ ♃
05 06 ☽ ∠ ☉
05 30 ☽ ☌ ♅
05 52 ☽ ✶ ♆
06 15 ☽ △ ♇
09 40 ☽ ∠ ☿
11 50 ☉ ∠ ♅
13 16 ☽ ☌ ♂
20 35 ☽ ⊥ ☉
22 11 ☽ ✶ ♄
23 32 ☽ ⊥ ☿

12 Sunday
00 26 ♀ ⚺ ♃
01 09 ☽ Q ♆
03 00 ☽ ∠ ♃
06 17 ☽ ⚼ ♇
06 49 ☽ ⚺ ☉
08 20 ☽ Q ♀
08 43 ☽ ⚺ ☿
17 21 ☽ Q ♄
17 33 ☽ ⊥ ♃
20 35 ☽ ± ♇

13 Monday
01 26 ☉ ☌ ☿
03 12 ☽ ⚺ ♃
04 54 ☽ ✶ ♀
04 59 ☉ ∥ ♂
05 34 ☽ ⚺ ♅
05 46 ☽ □ ♆
06 05 ☽ ⚻ ♇
14 27 ♀ □ ♅
15 07 ☽ ⊥ ♅
15 23 ☽ ⚺ ♂
16 45 ♀ ⚻ ♆
20 52 ♀ ⚻ ♇
21 58 ☽ □ ♄

14 Tuesday
01 23 ☽ ⊥ ♂
05 00 ☿ ∠ ♀
05 29 ☽ ∠ ♅
06 27 ☽ ☌ ☿
06 38 ☽ ∠ ♀
09 44 ☽ ☌ ☉
16 28 ☽ ∠ ♂
19 31 ☽ # ♇
22 12 ☽ ⊥ ♀

15 Wednesday
02 12 ☽ ∥ ♂
03 49 ☽ ☌ ♃
04 24 ☿ ∠ ♅
04 52 ☽ ∥ ☉
05 39 ☽ ✶ ♅
05 42 ☽ △ ♆
05 58 ☽ ☍ ♇
08 09 ☽ ∥ ♅
08 41 ☽ ⚺ ♀
13 01 ☽ ∥ ♃
17 59 ☽ ✶ ♂
20 32 ♅ ✶ ♆
22 27 ☽ △ ♄

16 Thursday
00 48 ☽ ∥ ☿
01 21 ☽ Q ♅
05 15 ☽ ⚺ ☿
06 18 ☽ ⚼ ♆
14 06 ☽ ⚺ ☉
14 53 ☽ Q ♂
14 54 ☽ ⊥ ☿
15 00 ☿ ⊥ ♂
21 24 ☽ ± ♆
23 33 ☽ ⚼ ♄

17 Friday
00 58 ☽ ⊥ ☉
05 36 ☽ ∠ ☿
06 26 ☽ ⚺ ♃
07 37 ☽ ⚻ ♆
07 44 ☽ □ ♅
07 50 ☽ ⚻ ♇
08 13 ☽ ∥ ♀
14 03 ♀ ± ♄
15 01 ☽ ± ♄
15 06 ☽ ☌ ♀
16 58 ☽ ⊥ ♃
17 35 ☽ ∠ ☉
18 11 ☽ ± ♇
23 22 ☽ □ ♂

18 Saturday
01 30 ☽ ⚻ ♄
04 45 ♅ △ ♇
06 27 ☽ ∥ ♄
06 50 ☽ ✶ ☿
09 02 ☽ ∠ ♃
10 00 ☽ ⚼ ♇
17 59 ☽ ∥ ♆
20 17 ♀ ⊥ ♃
21 36 ☽ # ♆
22 02 ☉ ∥ ♅
22 13 ☽ ✶ ☉

19 Sunday
03 43 ☽ Q ☿
05 36 ♀ ± ♇
09 24 ☽ # ♄
12 39 ☽ ✶ ♃
12 59 ☽ ☍ ♆
13 08 ☽ △ ♇
13 18 ☽ △ ♅
18 11 ♂ ✶ ♄
22 00 ☽ Q ☉

20 Monday
02 02 ☽ ⚺ ♀
03 41 ☽ # ♀
07 23 ♃ △ ♆
08 16 ☽ ☍ ♄
09 04 ☽ △ ♂
11 29 ☽ Q ♃
12 19 ☽ □ ☿
14 25 ☽ ⊥ ♀
14 45 ♃ ☍ ♇
17 32 ☽ ⚼ ♅

21 Tuesday
09 19 ☽ ∠ ♀
10 55 ☽ ± ♅
11 06 ☽ □ ☉
11 11 ♃ ✶ ♅
15 32 ☽ ⚼ ♂
22 05 ☽ ⚻ ♆
22 11 ☽ □ ♇
22 39 ☽ ⚻ ♅
22 49 ☽ □ ♃

22 Wednesday
01 40 ☽ # ☿
03 46 ♂ Q ♆
06 47 ☿ ⚺ ♂
09 30 ☽ # ♃
09 53 ☽ ± ♆
10 16 ☽ ± ♂
13 28 ☽ # ☉
17 37 ☽ ✶ ♀
17 46 ☽ # ♅
18 34 ☽ ⚻ ♄
21 48 ☽ △ ☿
22 53 ☽ ⚻ ♂

23 Thursday
03 45 ☽ ⚼ ♆
04 18 ♀ ⚻ ♄
05 52 ☽ # ♂
06 32 ☽ ± ♄
09 31 ☽ ∥ ♇
09 48 ☽ Q ♇
19 58 ☽ Q ♀
22 57 ☿ St D

24 Friday
00 36 ☽ ⚼ ♄
03 43 ☽ △ ☉
03 46 ☽ ⚼ ☿
09 52 ☽ △ ♆
09 53 ☽ ✶ ♇
10 41 ☽ ☍ ♅
11 48 ☽ △ ♃
16 13 ☿ ✶ ♀
22 02 ☽ ± ☿

25 Saturday
05 49 ♆ ✶ ♇
06 54 ☽ △ ♄
10 17 ☽ ⚻ ☿
11 54 ☽ □ ♀
12 41 ☽ ⚼ ☉
14 58 ☽ ☍ ♂
16 09 ☽ ∠ ♇
18 42 ☽ ⚼ ♃

26 Sunday
08 28 ☽ ± ☉
10 19 ☽ ⊥ ♇
10 59 ♂ ⚼ ♇
13 15 ☽ ± ♃
19 56 ♄ St R
21 38 ☽ ⚻ ☉
22 24 ☽ ⚺ ♇
22 25 ☉ ∥ ♃
22 27 ☽ □ ♆
23 29 ☽ ⚻ ♅

27 Monday
01 34 ☽ ⚻ ♃
06 55 ☉ ☍ ♇
07 36 ☉ △ ♆
10 36 ♀ ⚼ ♇
11 36 ☽ ± ♅
13 43 ☉ ∥ ☿
19 24 ☽ □ ♄
21 36 ☉ ✶ ♅

28 Tuesday
00 08 ☽ ☍ ☿
00 34 ☿ ∥ ♃
05 40 ☽ ⚼ ♅
06 11 ☽ △ ♀
07 00 ☽ ⚻ ♂
10 29 ☽ Q ♆
19 39 ☽ ± ♂
21 17 ☽ ∥ ♇
22 25 ☽ # ♂
22 59 ☉ ∠ ♀

29 Wednesday
07 10 ♀ □ ♂
07 11 ☽ Q ♄
09 58 ☉ ∠ ♂
10 08 ☽ ☌ ♇
10 14 ☽ ✶ ♆
10 47 ♀ ∠ ♃
11 29 ☽ △ ♅
12 18 ☉ ☌ ♃
12 47 ☽ # ♅
13 42 ♂ ∠ ♃
14 27 ☽ ☍ ♃
14 29 ☽ ⚼ ♂
14 36 ☽ ☍ ☉
14 44 ☽ ⚼ ♀
15 42 ♀ ∥ ♄
20 57 ☽ # ☿
23 24 ☽ # ♃

30 Thursday
02 30 ☽ # ☉
06 44 ☽ ✶ ♄
09 59 ☽ ± ♀
14 01 ☽ ⚻ ☿
15 33 ☽ ∠ ♆
21 27 ☽ △ ♂
22 41 ☽ ⚻ ♀

31 Friday
02 18 ☽ ± ☿
08 54 ☽ ⊥ ♆
11 43 ☽ ∠ ♄
20 13 ☽ ⚺ ♇
20 22 ☽ ⚺ ♆
20 39 ☽ ⚼ ☿
21 48 ☽ □ ♅

LONGITUDES

Date	Sidereal time h m s	Sun ☉ ° ' "	Moon ☽ ° ' "	Moon ☽ 24.00 ° ' "	Mercury ☿ ° '	Venus ♀ ° '	Mars ♂ ° '	Jupiter ♃ ° '	Saturn ♄ ° '	Uranus ♅ ° '	Neptune ♆ ° '	Pluto ♇ ° '
01	08 40 27	09 ♌ 16 02	12 ♓ 30 05	18 ♓ 53 27	19 ♋ 56	24 ♍ 35	23 ♊ 24	07 ♌ 04	14 ♈ 43	05 ♊ 02	04 ♈ 15	04 ♒ 10
02	08 44 23	10 13 26	25 ♓ 20 13	01 ♈ 50 35	20 50	25 37	24 05	07 18	14 R 43	05 03	04 R 14	04 R 08
03	08 48 20	11 10 50	08 ♈ 24 44	15 ♈ 02 52	21 50	26 38	24 45	07 31	14 42	05 05	04 14	04 07
04	08 52 16	12 08 16	21 ♈ 45 08	28 ♈ 31 42	22 55	27 40	25 26	07 44	14 41	05 07	04 13	04 06
05	08 56 13	13 05 43	05 ♉ 22 38	12 ♉ 18 00	24 06	28 41	26 06	07 57	14 40	05 09	04 12	04 04
06	09 00 09	14 03 12	19 ♉ 17 46	26 ♉ 21 50	25 23	29 ♍ 42	26 46	08 11	14 39	05 11	04 11	04 03
07	09 04 06	15 00 42	03 ♊ 30 00	10 ♊ 41 57	26 45	00 ♎ 42	27 26	08 24	14 38	05 12	04 10	04 01
08	09 08 03	15 58 13	17 ♊ 57 16	25 ♊ 15 26	28 11	01 42	28 06	08 37	14 37	05 14	04 09	04 00
09	09 11 59	16 55 45	02 ♋ 35 48	09 ♋ 57 38	29 ♋ 42	02 42	28 46	08 50	14 35	05 16	04 08	03 59
10	09 15 56	17 53 19	17 ♋ 20 06	24 ♋ 42 23	01 ♌ 18	03 42	29 ♊ 26	09 04	14 34	05 17	04 07	03 57
11	09 19 52	18 50 54	02 ♌ 03 33	09 ♌ 22 45	02 58	04 41	00 ♋ 06	09 17	14 33	05 19	04 06	03 56
12	09 23 49	19 48 30	16 ♌ 39 07	23 ♌ 51 53	04 42	05 40	00 46	09 30	14 31	05 20	04 05	03 55
13	09 27 45	20 46 08	01 ♍ 00 22	08 ♍ 03 59	06 29	06 38	01 25	09 43	14 29	05 22	04 04	03 53
14	09 31 42	21 43 46	15 ♍ 02 17	21 ♍ 54 56	08 19	07 36	02 05	09 56	14 27	05 23	04 03	03 52
15	09 35 38	22 41 25	28 ♍ 41 46	05 ♎ 22 42	10 12	08 34	02 44	10 09	14 25	05 24	04 02	03 51
16	09 39 35	23 39 06	11 ♎ 57 48	18 ♎ 27 13	12 07	09 31	03 24	10 22	14 23	05 26	04 01	03 49
17	09 43 32	24 36 47	24 ♎ 51 14	01 ♏ 10 10	14 04	10 28	04 03	10 36	14 21	05 27	03 59	03 48
18	09 47 28	25 34 30	07 ♏ 24 27	13 ♏ 34 33	16 02	11 24	04 42	10 49	14 19	05 28	03 58	03 47
19	09 51 25	26 32 14	19 ♏ 40 59	25 ♏ 44 16	18 02	12 20	05 21	11 02	14 17	05 29	03 57	03 46
20	09 55 21	27 29 59	01 ♐ 44 58	07 ♐ 43 39	20 02	13 15	06 00	11 15	14 14	05 30	03 56	03 44
21	09 59 18	28 27 45	13 ♐ 40 55	19 ♐ 37 17	22 03	14 10	06 39	11 28	14 12	05 31	03 54	03 43
22	10 03 14	29 ♌ 25 32	25 ♐ 33 20	01 ♑ 29 36	24 04	15 05	07 18	11 41	14 09	05 32	03 53	03 42
23	10 07 11	00 ♍ 23 20	07 ♑ 26 34	13 ♑ 24 42	26 04	15 58	07 57	11 54	14 07	05 33	03 52	03 41
24	10 11 07	01 21 10	19 ♑ 24 29	25 ♑ 26 17	28 ♌ 05	16 51	08 36	12 06	14 04	05 34	03 50	03 39
25	10 15 04	02 19 01	01 ♒ 30 28	07 ♒ 37 22	00 ♍ 05	17 44	09 14	12 19	14 01	05 35	03 49	03 38
26	10 19 01	03 16 53	13 ♒ 47 15	20 ♒ 00 21	02 04	18 36	09 53	12 32	13 58	05 36	03 48	03 37
27	10 22 57	04 14 46	26 ♒ 16 50	02 ♓ 36 50	04 02	19 28	10 32	12 45	13 55	05 37	03 46	03 36
28	10 26 54	05 12 41	09 ♓ 00 26	15 ♓ 27 41	06 00	20 18	11 10	12 58	13 52	05 37	03 45	03 35
29	10 30 50	06 10 37	21 ♓ 58 36	28 ♓ 33 07	07 56	21 08	11 48	13 10	13 49	05 38	03 44	03 34
30	10 34 47	07 08 36	05 ♈ 11 11	11 ♈ 52 42	09 51	21 58	12 26	13 23	13 46	05 39	03 42	03 32
31	10 38 43	08 ♍ 06 35	18 ♈ 37 33	25 ♈ 25 35	11 ♍ 45	22 ♎ 46	13 ♋ 05	13 ♌ 36	13 ♈ 43	05 ♊ 39	03 ♈ 41	03 ♒ 31

Date	Moon True ☊ ° '	Moon Mean ☊ ° '	Moon ☽ Latitude ° '
01	29 ♒ 50	00 ♓ 55	01 N 09
02	29 D 51	00 52	02 16
03	29 52	00 49	03 17
04	29 53	00 45	04 09
05	29 54	00 42	04 47
06	29 R 54	00 39	05 10
07	29 54	00 36	05 15
08	29 52	00 33	05 00
09	29 51	00 30	04 26
10	29 49	00 26	03 34
11	29 48	00 23	02 28
12	29 48	00 20	01 N 12
13	29 D 47	00 17	00 S 07
14	29 48	00 14	01 23
15	29 48	00 11	02 33
16	29 49	00 07	03 33
17	29 49	00 04	04 20
18	29 50	00 ♓ 01	04 53
19	29 50	29 ♒ 58	05 12
20	29 R 50	29 55	05 17
21	29 50	29 51	05 08
22	29 50	29 48	04 46
23	29 D 50	29 45	04 11
24	29 50	29 42	03 26
25	29 50	29 39	02 31
26	29 50	29 36	01 28
27	29 50	29 32	00 S 20
28	29 R 50	29 29	00 N 51
29	29 50	29 26	02 00
30	29 49	29 23	03 03
31	29 ♒ 48	29 ♒ 20	03 N 58

DECLINATIONS

Date	Sun ☉ ° '	Moon ☽ ° '	Mercury ☿ ° '	Venus ♀ ° '	Mars ♂ ° '	Jupiter ♃ ° '	Saturn ♄ ° '	Uranus ♅ ° '	Neptune ♆ ° '	Pluto ♇ ° '
01	17 N 56	05 S 48	19 N 57	02 N 06	23 N 22	18 N 58	03 N 28	20 N 59	00 N 25	23 S 22
02	17 41	00 N 14	20 05	01 36	23 25	18 55	03 28	20 59	00 24	23 23
03	17 25	06 21	20 11	01 07	23 28	18 51	03 27	21 00	00 24	23 23
04	17 09	12 19	20 16	00 38	23 30	18 48	03 27	21 00	00 24	23 23
05	16 53	17 49	20 19	00 N 08	23 32	18 45	03 26	21 00	00 23	23 24
06	16 37	22 31	20 20	00 S 21	23 34	18 41	03 25	21 01	00 23	23 24
07	16 20	26 00	20 19	00 50	23 36	18 38	03 25	21 01	00 22	23 25
08	16 03	27 53	20 15	01 20	23 37	18 35	03 24	21 01	00 22	23 25
09	15 46	27 51	20 10	01 49	23 38	18 31	03 23	21 01	00 22	23 25
10	15 28	25 51	20 02	02 18	23 40	18 28	03 22	21 02	00 21	23 26
11	15 10	22 06	19 51	02 47	23 41	18 24	03 22	21 02	00 21	23 26
12	14 52	16 59	19 38	03 16	23 41	18 21	03 21	21 02	00 20	23 26
13	14 34	11 01	19 22	03 45	23 42	18 18	03 20	21 03	00 20	23 27
14	14 16	04 N 37	19 04	04 14	23 42	18 14	03 19	21 03	00 19	23 27
15	13 57	01 S 50	18 42	04 43	23 42	18 11	03 18	21 03	00 19	23 28
16	13 38	08 00	18 18	05 11	23 42	18 07	03 17	21 03	00 18	23 28
17	13 19	13 39	17 52	05 40	23 42	18 04	03 16	21 04	00 18	23 28
18	13 00	18 36	17 23	06 08	23 41	18 00	03 15	21 04	00 17	23 29
19	12 40	22 39	16 52	06 36	23 41	17 57	03 14	21 04	00 17	23 29
20	12 20	25 41	16 19	07 04	23 40	17 53	03 12	21 04	00 16	23 29
21	12 00	27 32	15 43	07 32	23 39	17 50	03 11	21 04	00 16	23 30
22	11 40	28 07	15 06	07 59	23 38	17 46	03 10	21 04	00 15	23 30
23	11 20	27 25	14 27	08 27	23 36	17 43	03 09	21 05	00 15	23 30
24	11 00	25 26	13 47	08 54	23 35	17 39	03 08	21 05	00 14	23 31
25	10 39	22 16	13 06	09 21	23 33	17 36	03 06	21 05	00 13	23 31
26	10 18	18 05	12 23	09 47	23 31	17 32	03 05	21 05	00 13	23 31
27	09 57	13 04	11 40	10 14	23 29	17 29	03 04	21 05	00 12	23 32
28	09 36	07 25	10 55	10 40	23 27	17 25	03 02	21 05	00 12	23 32
29	09 15	01 S 21	10 10	11 06	23 25	17 22	03 01	21 05	00 11	23 32
30	08 53	04 N 52	09 24	11 31	23 22	17 18	02 59	21 06	00 11	23 32
31	08 N 32	10 N 58	08 N 38	11 S 56	23 N 19	17 N 15	02 N 58	21 N 06	00 N 10	23 S 33

ZODIAC SIGN ENTRIES

Date	h	m	Planets
02	20	37	☽ ♈
05	02	35	☽ ♉
06	19	13	♀ ♎
07	06	08	☽ ♊
09	07	46	☽ ♋
09	16	28	☿ ♌
11	08	30	♂ ♋
11	08	38	☽ ♌
13	10	18	☽ ♍
15	14	20	☽ ♎
17	21	46	☽ ♏
20	08	30	☽ ♐
22	20	59	☽ ♑
23	02	19	☉ ♍
25	09	02	☽ ♒
25	11	04	☿ ♍
27	19	04	☽ ♓
30	02	38	☽ ♈

LATITUDES

Date	Mercury ☿ ° '	Venus ♀ ° '	Mars ♂ ° '	Jupiter ♃ ° '	Saturn ♄ ° '	Uranus ♅ ° '	Neptune ♆ ° '	Pluto ♇ ° '
01	02 S 01	00 S 04	00 N 06	00 N 29	02 S 32	00 S 09	01 S 24	04 S 17
04	01 15	00 20	00 08	00 29	02 33	00 09	01 24	04 17
07	00 S 30	00 37	00 11	00 29	02 33	00 09	01 24	04 17
10	00 N 10	00 54	00 13	00 29	02 34	00 09	01 24	04 17
13	00 44	01 13	00 16	00 30	02 35	00 09	01 24	04 17
16	01 12	01 32	00 18	00 30	02 36	00 09	01 24	04 18
19	01 31	01 53	00 21	00 30	02 36	00 09	01 24	04 18
22	01 42	02 13	00 24	00 31	02 37	00 09	01 24	04 18
25	01 46	02 35	00 26	00 31	02 38	00 09	01 25	04 18
28	01 44	02 57	00 29	00 31	02 38	00 09	01 25	04 18
31	01 N 36	03 S 19	00 N 32	00 N 32	02 S 39	00 S 09	01 S 25	04 S 18

DATA

Julian Date	2461254
Delta T	+73 seconds
Ayanamsa	24° 13' 52"
Synetic vernal point	04° ♓ 53' 07"
True obliquity of ecliptic	23° 26' 17"

MOON'S PHASES, APSIDES AND POSITIONS ☽

Date	h	m	Phase	Longitude ° '	Eclipse Indicator
06	02	21	☾	13 ♉ 40	
12	17	37	●	20 ♌ 02	Total
20	02	46	☽	27 ♏ 08	
28	04	18	○	04 ♓ 54	partial

Day	h	m		
10	11	09	Perigee	
22	08	18	Apogee	
02	11	06	0N	
08	23	38	Max dec	28° N 06'
15	05	08	0S	
22	10	54	Max dec	28° S 07'
29	17	14	0N	

LONGITUDES

Date	Chiron ⚷ ° '	Ceres ⚳ ° '	Pallas ⚴ ° '	Juno ⚵ ° '	Vesta ⚶ ° '	Black Moon Lilith ⚸ ° '
01	00 ♉ 52	25 ♊ 50	22 ♈ 10	02 ♒ 20	25 ♈ 44	25 ♐ 03
11	00 ♉ 50	29 ♊ 35	22 ♈ 37	00 ♒ 01	27 ♈ 04	26 ♐ 11
21	00 ♉ 44	03 ♋ 13	22 ♈ 29	28 ♑ 03	27 ♈ 44	27 ♐ 18
31	00 ♉ 32	06 ♋ 43	21 ♈ 43	26 ♑ 37	27 ♈ 41	28 ♐ 25

ASPECTARIAN

h m Aspects

01 Saturday
01 32 ☽ ⊼ ♃
04 52 ☽ ⊥ ♄
05 23 ☽ ⊼ ☉
07 36 ☽ ⊥ ♇
09 49 ♂ # ♇
13 06 ☽ ± ♃
14 41 ☿ ∠ ♅
16 11 ☽ ⊻ ♄
17 38 ☽ ± ☉
21 24 ☽ # ♄

02 Sunday
00 29 ☽ ∠ ♇
02 59 ☽ △ ☿
04 08 ☽ # ♀
06 15 ☽ Ꝑ ♃
07 46 ☽ Q ♅
09 30 ☽ # ♆
09 33 ☽ □ ♂
11 46 ☽ Ꝑ ☉
12 33 ☽ ☍ ♀
12 42 ☽ ∥ ♆
17 00 ☽ ∥ ♀

03 Monday
00 38 ☽ ∥ ♄
04 11 ☽ ⚹ ♇
04 22 ☽ ☌ ♆
05 56 ☽ ⚹ ♅
10 20 ☽ △ ♃
17 25 ☽ △ ☉
20 17 ☽ Q ♂
23 22 ☽ ☌ ♄

04 Tuesday
01 54 ☽ Q ♇
09 05 ☽ ∠ ♅
14 17 ☽ □ ☿
18 52 ☽ ⚹ ♂
23 19 ☽ ⊼ ♀
23 39 ♀ ∥ ♆

05 Wednesday
01 04 ☽ ⊥ ♅
07 54 ☽ ∥ ☉
09 44 ☽ □ ♇
09 57 ☽ ⊻ ♆
10 42 ☽ ± ♀
11 36 ☽ ⊻ ♅
16 17 ☽ ∥ ♃
16 34 ☽ □ ♃
20 22 ☽ ⊥ ♆
22 26 ☽ ∠ ♂

06 Thursday
00 07 ☽ ∥ ☿
00 49 ☽ Q ☿
02 21 ☽ □ ☉
03 31 ☽ Ꝑ ♀
03 40 ☽ ∥ ♅
04 04 ☽ ⊻ ♄
08 02 ♂ Q ♄
11 48 ☽ ∠ ♆
13 23 ♀ # ♆
14 19 ☽ ⊥ ♄
14 38 ☽ ⊥ ♂
17 16 ☽ # ♇
18 19 ☽ ∥ ♂
23 25 ☽ ⚹ ☿
23 52 ☽ Q ♃

07 Friday
01 18 ☽ ⊻ ♂
02 44 ☉ △ ♄
05 31 ☽ ∠ ♄
06 57 ☽ △ ♀
11 07 ☽ Q ☉
12 53 ☽ △ ♇
13 07 ☽ ⚹ ♆
14 52 ☽ ☌ ♅
20 18 ☽ ⚹ ♃

08 Saturday
03 14 ☽ ∠ ☿
06 30 ☽ ⚹ ♄
08 30 ☽ ⚹ ☉
08 31 ♂ ± ♇
09 02 ☽ Q ♆
09 32 ☿ ⊻ ♂
13 43 ☽ Ꝑ ♇
19 45 ☽ ⊥ ☿
21 27 ☽ ∠ ♃

09 Sunday
02 12 ☽ Q ♄
04 28 ☽ ± ♇
05 27 ☽ ☌ ♂
06 44 ☽ ⊻ ☿
10 50 ☽ ∠ ☉
12 11 ☽ □ ♀
12 24 ☽ ⊥ ♃
14 15 ☽ ⊼ ♇
14 30 ☽ □ ♆
16 21 ☽ ⊻ ♅
20 32 ☉ Q ♅
22 20 ☽ ⊻ ♃

10 Monday
02 09 ☽ ⊥ ♅
02 31 ☽ ⊥ ☉
07 30 ☽ □ ♄
12 58 ☽ ⊻ ☉
16 49 ☽ ∠ ♅
18 09 ♀ △ ♇
19 37 ☽ Q ♀
22 02 ♀ ☍ ♆

11 Tuesday
03 06 ☽ ∥ ♂
04 30 ☽ # ♇
08 38 ☽ ⊻ ♂
13 41 ☽ ☌ ☿
15 04 ☽ ☍ ♇
15 20 ☽ △ ♆
16 36 ☽ ⚹ ♀
17 20 ☽ ⚹ ♅
17 28 ☽ ∥ ♅
18 12 ☉ Ꝑ ♆
18 55 ☽ ⊥ ♂
23 40 ☽ ∥ ☿

12 Wednesday
00 01 ☽ ☌ ♃
01 20 ☿ ☍ ♇
03 39 ☿ △ ♆
03 46 ♀ △ ♅
06 01 ☽ ∥ ♃
08 28 ☽ △ ♄
10 27 ☽ ∠ ♂
13 08 ☽ Q ♅
15 35 ♀ # ♄
16 02 ☽ Ꝑ ♆
17 37 ☽ ☌ ☉
19 09 ☽ ∠ ♀
20 48 ☿ ⚹ ♅
21 17 ☽ ∥ ☉

13 Thursday
07 03 ☽ ± ♆
09 26 ☽ Ꝑ ♄
11 20 ☽ ⊥ ♀
12 44 ☽ ⚹ ♂
16 23 ☿ ⚹ ♀
16 53 ☽ ⊼ ♇
17 10 ☽ ⊼ ♆
19 24 ☽ □ ♅
22 16 ☽ ⊻ ♀
22 41 ☽ ⊻ ☿

14 Friday
00 41 ☽ ± ♄
03 04 ☽ ⊻ ♃
03 06 ☽ ± ♇
07 10 ☿ ⊥ ♂
10 15 ☽ Q ♂
10 34 ☽ ⊥ ☿
11 00 ☽ ⊼ ♄
13 18 ☽ # ♀
13 35 ☽ ⊥ ♃
16 49 ☽ ∥ ♄
18 39 ☽ Ꝑ ♇

15 Saturday
00 33 ☽ ⊻ ☉
03 57 ☽ ∥ ♆
04 47 ☽ ∠ ☿
05 37 ☽ ∠ ♃
06 19 ☽ # ♆
11 23 ☿ ☌ ♃
11 59 ☽ ⊥ ☉
17 34 ☽ # ♄
19 37 ☽ □ ♂
21 13 ☽ △ ♇
21 33 ☽ ☍ ♆
23 57 ☽ ∥ ♀

16 Sunday
00 04 ☽ △ ♅
05 28 ☽ ∠ ☉
07 11 ☽ ☌ ♀
09 02 ☽ ⚹ ♃
12 20 ☽ ⚹ ☿
16 27 ☽ ☍ ♄
23 50 ☿ ∥ ♃

17 Monday
03 15 ♂ ⊼ ♇
03 43 ☽ Ꝑ ♅
07 40 ☽ Q ♃
09 52 ♂ □ ♆
10 35 ☽ # ☉
11 31 ☽ ⚹ ☉
14 43 ☽ Q ☿
15 27 ☿ △ ♄
16 16 ♀ ⚹ ♃
20 44 ☽ ± ♅

18 Tuesday
05 01 ☽ □ ♇
05 23 ☽ ⊼ ♆
06 20 ☽ # ☿
06 29 ☽ △ ♂
08 15 ☽ ⊼ ♅
08 57 ☽ # ♃
12 21 ☽ Q ☉
16 58 ☽ ± ♆
17 58 ♂ ⊥ ♃
18 43 ☽ □ ♃
20 24 ☽ ⊻ ♀

19 Wednesday
01 25 ☽ ⊼ ♄
01 54 ☽ # ♅
05 23 ☿ Q ♅
08 06 ☽ □ ☿
09 07 ☽ ⊥ ♀
10 33 ☽ Ꝑ ♆
13 10 ☽ ± ♄
13 24 ☽ Ꝑ ♂
16 06 ☽ Q ♇
17 00 ♂ ⊻ ♅
17 47 ☽ ∥ ♇
19 11 ☽ # ♂
22 52 ☿ Ꝑ ♆

20 Thursday
02 46 ☽ □ ☉
04 26 ☽ ∠ ♀
07 00 ☽ Ꝑ ♄
08 19 ☽ ± ♂
15 59 ☽ ⚹ ♇
16 21 ☽ △ ♆
19 33 ☽ ☍ ♅
21 02 ☽ ⊼ ♂
22 27 ☉ ± ♆

21 Friday
05 06 ☿ ∠ ♂
07 26 ☽ △ ♃
12 43 ♀ ☍ ♄
13 02 ☽ △ ♄
13 04 ☽ ⚹ ♀
22 09 ☽ ∠ ♇

22 Saturday
05 34 ☉ Ꝑ ♄
08 22 ☽ △ ☿
14 19 ☽ Ꝑ ♃
15 19 ☽ Q ♀
16 19 ☽ ⊥ ♇
20 31 ☽ △ ☉

23 Sunday
04 25 ☽ ⊻ ♇
04 48 ☽ □ ♆
08 11 ☽ ⊼ ♅
08 49 ☽ ± ♃
13 05 ☽ ☍ ♂
20 17 ☽ ± ♅
20 47 ☽ Ꝑ ☿
21 07 ☽ ⊼ ♃

24 Monday
01 21 ☽ □ ♄
05 22 ☽ Ꝑ ☉
06 30 ☽ □ ♀
09 10 ☿ ± ♆
14 19 ☽ Ꝑ ♅
16 51 ☽ Q ♆
18 23 ☽ ± ☿
23 34 ☿ Ꝑ ♄

25 Tuesday
00 51 ☽ ± ☉
03 09 ☽ # ♂
03 31 ☽ ∥ ♇
08 38 ☽ ⊼ ☿
13 00 ☽ Q ♄
13 44 ☽ ⊼ ☉
16 11 ☽ ☌ ♇
16 32 ☽ ⚹ ♆
19 24 ☽ # ♅
20 01 ☽ △ ♅

26 Wednesday
04 00 ☽ ⊼ ♂
09 32 ☽ ☍ ♃
12 21 ☽ ⚹ ♄
12 29 ♂ # ♇
14 51 ☽ # ♃
16 17 ☽ ± ♂
20 10 ☉ ⊼ ♇
21 39 ☽ ∠ ♆
21 59 ☽ △ ♀

27 Thursday
00 31 ☉ ⊼ ♆
03 38 ☉ # ♀
06 42 ☿ ⊼ ♇
08 50 ☿ ⊼ ♆
10 29 ☽ Ꝑ ♂
14 50 ☽ ⊥ ♆
17 00 ☽ ∠ ♄
17 04 ☉ ☌ ☿
19 08 ☽ # ☿
21 07 ☿ ∠ ♀
23 26 ☽ ∥ ♀

28 Friday
01 50 ☽ ⊻ ♇
02 10 ☽ ⊻ ♆
02 22 ☽ # ☉
04 18 ☽ ☍ ☉
04 34 ☽ Ꝑ ♀
05 20 ☽ ☍ ☿
05 39 ☽ □ ♅
07 25 ☿ □ ♅
09 53 ☽ ⊥ ♄
13 04 ☽ ⊥ ♇
16 14 ☽ △ ♂
17 12 ☿ # ♀
19 29 ☽ ⊼ ♃
21 01 ☽ ⊻ ♄
21 13 ♀ Ꝑ ♅
22 19 ☉ □ ♅
22 32 ☽ ± ♀

29 Saturday
05 08 ☉ ∠ ♀
05 27 ♂ ⊥ ♅
05 29 ☽ # ♄
05 44 ☽ ∠ ♇
06 46 ☽ ± ♃
10 21 ☽ ⊼ ♀
10 37 ☿ ± ♄
15 02 ☽ Q ♅
16 31 ☽ # ♆
17 56 ☽ ∥ ♆
23 30 ☽ Ꝑ ♃

30 Sunday
04 47 ☽ ∥ ♄
08 06 ☿ ± ♇
09 02 ☽ ⚹ ♇
09 20 ☽ ☌ ♆
12 49 ☽ ⚹ ♅
15 48 ☽ ⊼ ☉
21 46 ☽ ⊼ ☿

31 Monday
01 39 ☽ □ ♂
02 35 ☉ ± ♄
02 49 ☽ ∥ ☉
02 55 ☽ △ ♃
03 18 ☽ ☌ ♄
03 22 ☽ ± ☉
03 44 ☽ ∥ ☿
06 30 ☽ Q ♇
10 13 ☽ ± ☿
15 35 ☽ ∠ ♅
16 15 ☽ # ♀
18 12 ☉ ∥ ☿
19 47 ☽ ☍ ♀
20 32 ☽ Ꝑ ☉
22 17 ♃ △ ♄

All ephemeris data is given at 12.00 UT and the Moon's longitude is additionally given for 24.00 UT

SEPTEMBER 2026

LONGITUDES

Date	Sidereal time h m s	Sun ☉ ° ' "	Moon ☽ ° ' "	Moon ☽ 24.00 ° ' "	Mercury ☿ ° '	Venus ♀ ° '	Mars ♂ ° '	Jupiter ♃ ° '	Saturn ♄ ° '	Uranus ♅ ° '	Neptune ♆ ° '	Pluto ♇ ° '
01	10 42 40	09 ♍ 04 37	02 ♉ 16 38	09 ♉ 10 33	13 ♍ 38	23 ♎ 34	13 ♋ 43	13 ♌ 48	13 ♈ 39	05 ♊ 40	03 ♈ 39	03 ♒ 30
02	10 46 36	10 02 41	16 ♉ 07 08	23 ♉ 06 11	15 30	24 21	14 21	14 01	13 R 36	05 40	03 R 38	03 R 29
03	10 50 33	11 00 46	00 ♊ 07 30	07 ♊ 10 50	17 21	25 08	14 58	14 13	13 32	05 40	03 36	03 28
04	10 54 30	11 58 54	14 ♊ 15 58	21 ♊ 22 37	19 10	25 53	15 36	14 26	13 29	05 41	03 35	03 27
05	10 58 26	12 57 03	28 ♊ 30 30	05 ♋ 39 19	20 58	26 38	16 14	14 38	13 25	05 41	03 33	03 26
06	11 02 23	13 55 15	12 ♋ 48 41	19 ♋ 58 15	22 45	27 21	16 52	14 51	13 21	05 41	03 32	03 25
07	11 06 19	14 53 29	27 ♋ 07 36	04 ♌ 16 16	24 31	28 04	17 29	15 03	13 17	05 42	03 30	03 24
08	11 10 16	15 51 44	11 ♌ 23 47	18 ♌ 29 42	26 16	28 46	18 07	15 15	13 14	05 42	03 29	03 23
09	11 14 12	16 50 02	25 ♌ 33 29	02 ♍ 34 41	27 59	29 ♎ 27	18 44	15 28	13 10	05 42	03 27	03 22
10	11 18 09	17 48 21	09 ♍ 32 49	16 ♍ 27 27	29 ♍ 42	00 ♏ 06	19 21	15 40	13 06	05 42	03 25	03 21
11	11 22 05	18 46 42	23 ♍ 18 11	00 ♎ 04 42	01 ♎ 23	00 45	19 58	15 52	13 02	05 R 42	03 24	03 20
12	11 26 02	19 45 05	06 ♎ 46 44	13 ♎ 24 07	03 03	01 22	20 35	16 04	12 58	05 42	03 22	03 19
13	11 29 59	20 43 30	19 ♎ 56 43	26 ♎ 24 33	04 42	01 59	21 12	16 16	12 53	05 42	03 21	03 19
14	11 33 55	21 41 56	02 ♏ 47 42	09 ♏ 06 18	06 19	02 34	21 49	16 28	12 49	05 41	03 19	03 18
15	11 37 52	22 40 24	15 ♏ 20 37	21 ♏ 30 56	07 56	03 08	22 26	16 40	12 45	05 41	03 17	03 17
16	11 41 48	23 38 54	27 ♏ 37 40	03 ♐ 41 14	09 32	03 40	23 02	16 52	12 41	05 41	03 16	03 16
17	11 45 45	24 37 25	09 ♐ 42 08	15 ♐ 40 54	11 06	04 11	23 39	17 03	12 36	05 41	03 14	03 15
18	11 49 41	25 35 58	21 ♐ 38 06	27 ♐ 34 19	12 40	04 41	24 15	17 15	12 32	05 40	03 12	03 15
19	11 53 38	26 34 33	03 ♑ 30 09	09 ♑ 26 14	14 13	05 09	24 52	17 27	12 28	05 40	03 11	03 14
20	11 57 34	27 33 09	15 ♑ 23 09	21 ♑ 21 30	15 44	05 35	25 28	17 38	12 23	05 39	03 09	03 13
21	12 01 31	28 31 47	27 ♑ 21 54	03 ♒ 24 52	17 14	06 00	26 04	17 50	12 19	05 39	03 08	03 13
22	12 05 28	29 ♍ 30 27	09 ♒ 30 57	15 ♒ 40 37	18 44	06 23	26 40	18 01	12 14	05 38	03 06	03 12
23	12 09 24	00 ♎ 29 08	21 ♒ 54 17	28 ♒ 12 18	20 12	06 45	27 16	18 13	12 10	05 38	03 04	03 11
24	12 13 21	01 27 52	04 ♓ 34 57	11 ♓ 02 26	21 40	07 04	27 52	18 24	12 05	05 37	03 03	03 11
25	12 17 17	02 26 36	17 ♓ 34 51	24 ♓ 12 13	23 06	07 22	28 27	18 35	12 00	05 36	03 01	03 10
26	12 21 14	03 25 23	00 ♈ 54 26	07 ♈ 41 18	24 31	07 38	29 03	18 46	11 56	05 36	02 59	03 10
27	12 25 10	04 24 12	14 ♈ 32 32	21 ♈ 27 46	25 55	07 52	29 ♋ 38	18 57	11 51	05 35	02 58	03 09
28	12 29 07	05 23 03	28 ♈ 26 32	05 ♉ 28 19	27 18	08 03	00 ♌ 14	19 08	11 46	05 34	02 56	03 08
29	12 33 03	06 21 56	12 ♉ 32 34	19 ♉ 38 43	28 ♎ 40	08 13	00 49	19 19	11 42	05 33	02 54	03 08
30	12 37 00	07 ♎ 20 51	26 ♉ 46 11	03 ♊ 54 26	00 ♏ 01	08 ♏ 20	01 ♌ 24	19 ♌ 30	11 ♈ 37	05 ♊ 32	02 ♈ 53	03 ♒ 08

Date	Moon True ☊ ° '	Moon Mean ☊ ° '	Moon ☽ Latitude ° '
01	29 ♒ 47	29 ♒ 16	04 N 41
02	29 R 46	29 13	05 07
03	29 46	29 10	05 16
04	29 D 46	29 07	05 06
05	29 46	29 04	04 37
06	29 47	29 01	03 51
07	29 48	28 57	02 51
08	29 49	28 54	01 40
09	29 50	28 51	00 N 24
10	29 R 49	28 48	00 S 53
11	29 48	28 45	02 06
12	29 46	28 42	03 10
13	29 43	28 38	04 02
14	29 40	28 35	04 40
15	29 37	28 32	05 04
16	29 34	28 29	05 14
17	29 32	28 26	05 09
18	29 31	28 22	04 51
19	29 D 31	28 19	04 21
20	29 32	28 16	03 39
21	29 33	28 13	02 47
22	29 35	28 10	01 48
23	29 36	28 07	00 S 42
24	29 R 37	28 03	00 N 27
25	29 35	28 00	01 37
26	29 32	27 57	02 42
27	29 28	27 54	03 40
28	29 22	27 51	04 26
29	29 17	27 48	04 57
30	29 ♒ 11	27 ♒ 44	05 N 10

DECLINATIONS

Date	Sun ☉ ° '	Moon ☽ ° '	Mercury ☿ ° '	Venus ♀ ° '	Mars ♂ ° '	Jupiter ♃ ° '	Saturn ♄ ° '	Uranus ♅ ° '	Neptune ♆ ° '	Pluto ♇ ° '
01	08 N 10	16 N 39	07 N 52	12 S 21	23 N 16	17 N 11	02 N 56	21 N 06	00 N 09	23 S 33
02	07 48	21 34	07 05	12 46	23 13	17 08	02 55	21 06	00 09	23 33
03	07 26	25 19	06 18	13 10	23 10	17 04	02 53	21 06	00 08	23 33
04	07 04	27 35	05 31	13 34	23 07	17 01	02 52	21 06	00 08	23 34
05	06 42	28 03	04 43	13 58	23 03	16 57	02 50	21 06	00 07	23 34
06	06 19	26 40	03 56	14 21	22 59	16 54	02 49	21 06	00 06	23 34
07	05 57	23 32	03 09	14 44	22 55	16 50	02 47	21 06	00 06	23 34
08	05 35	18 58	02 22	15 06	22 51	16 47	02 45	21 06	00 05	23 35
09	05 12	13 22	01 35	15 28	22 47	16 43	02 44	21 06	00 04	23 35
10	04 49	07 10	00 49	15 50	22 43	16 40	02 42	21 06	00 04	23 35
11	04 26	00 N 44	00 N 03	16 11	22 38	16 36	02 40	21 06	00 03	23 35
12	04 03	05 S 36	00 S 43	16 32	22 34	16 33	02 39	21 06	00 02	23 35
13	03 41	11 32	01 29	16 52	22 29	16 29	02 37	21 06	00 02	23 36
14	03 18	16 50	02 14	17 11	22 24	16 26	02 35	21 06	00 N 01	23 36
15	02 54	21 17	02 59	17 30	22 19	16 22	02 33	21 06	00 00	23 36
16	02 31	24 43	03 43	17 49	22 14	16 19	02 32	21 06	00 00	23 36
17	02 08	27 00	04 27	18 07	22 09	16 16	02 30	21 06	00 S 01	23 36
18	01 45	28 01	05 11	18 24	22 03	16 12	02 28	21 06	00 02	23 37
19	01 22	27 44	05 53	18 41	21 58	16 09	02 26	21 06	00 02	23 37
20	00 58	26 11	06 36	18 57	21 52	16 05	02 24	21 06	00 03	23 37
21	00 35	23 25	07 17	19 12	21 47	16 02	02 23	21 06	00 04	23 37
22	00 N 12	19 36	07 59	19 27	21 41	15 59	02 21	21 06	00 04	23 37
23	00 S 12	14 52	08 39	19 41	21 35	15 55	02 19	21 05	00 05	23 37
24	00 35	09 24	09 19	19 54	21 29	15 52	02 17	21 05	00 06	23 37
25	00 58	03 S 25	09 58	20 06	21 22	15 49	02 15	21 05	00 06	23 37
26	01 22	02 N 51	10 37	20 18	21 16	15 45	02 13	21 05	00 07	23 38
27	01 45	09 07	11 15	20 29	21 10	15 42	02 11	21 05	00 08	23 38
28	02 08	15 04	11 52	20 38	21 03	15 39	02 10	21 05	00 08	23 38
29	02 32	20 19	12 28	20 47	20 56	15 35	02 08	21 05	00 09	23 38
30	02 S 55	24 N 27	13 S 03	20 S 55	20 N 50	15 N 32	02 N 06	21 N 04	00 S 10	23 S 38

ZODIAC SIGN ENTRIES

Date	h	m	Planets
01	08	01	☽ ♉
03	11	47	☽ ♊
05	14	30	☽ ♋
07	16	49	☽ ♌
09	19	35	☽ ♍
10	08	07	♀ ♏
10	16	21	☿ ♎
11	23	52	☽ ♎
14	06	44	☽ ♏
16	16	41	☽ ♐
19	04	55	☽ ♑
21	17	14	☽ ♒
23	00	05	☉ ♎
24	03	24	☽ ♓
26	10	23	☽ ♈
28	02	49	♂ ♌
28	14	40	☽ ♉
30	11	44	☿ ♏
30	17	26	☽ ♊

LATITUDES

Date	Mercury ☿ ° '	Venus ♀ ° '	Mars ♂ ° '	Jupiter ♃ ° '	Saturn ♄ ° '	Uranus ♅ ° '	Neptune ♆ ° '	Pluto ♇ ° '
01	01 N 33	03 S 27	00 N 33	00 N 32	02 S 39	00 S 09	01 S 25	04 S 18
04	01 20	03 50	00 35	00 32	02 40	00 09	01 25	04 18
07	01 04	04 14	00 38	00 33	02 40	00 09	01 25	04 18
10	00 45	04 37	00 41	00 33	02 41	00 09	01 25	04 18
13	00 25	05 01	00 44	00 33	02 41	00 09	01 25	04 18
16	00 N 04	05 24	00 46	00 34	02 42	00 09	01 25	04 18
19	00 S 19	05 47	00 49	00 34	02 42	00 09	01 25	04 18
22	00 41	06 09	00 52	00 35	02 42	00 09	01 25	04 18
25	01 04	06 30	00 55	00 35	02 42	00 09	01 25	04 18
28	01 27	06 49	00 58	00 35	02 43	00 09	01 25	04 18
31	01 S 48	07 S 06	01 N 01	00 N 36	02 S 43	00 S 09	01 S 25	04 S 18

DATA

Julian Date	2461285
Delta T	+73 seconds
Ayanamsa	24° 13' 56"
Synetic vernal point	04° ♓ 53' 03"
True obliquity of ecliptic	23° 26' 17"

MOON'S PHASES, APSIDES AND POSITIONS ☽

Date	h	m	Phase	Longitude ° '	Eclipse Indicator
04	07	51	☾	11 ♊ 49	
11	03	27	●	18 ♍ 26	
18	20	44	☽	25 ♐ 57	
26	16	49	○	03 ♈ 37	

Day	h	m		
06	20	32	Perigee	
19	03	00	Apogee	
05	06	07	Max dec	28° N 07'
11	14	44	0S	
18	18	48	Max dec	28° S 04'
26	01	11	0N	

LONGITUDES

Date	Chiron ⚷ ° '	Ceres ⚳ ° '	Pallas ⚴ ° '	Juno ⚵ ° '	Vesta ⚶ ° '	Black Moon Lilith ⚸ ° '
01	00 ♉ 30	07 ♋ 03	21 ♈ 36	26 ♑ 30	27 ♈ 38	28 ♐ 31
11	00 ♉ 13	10 ♋ 21	20 ♈ 05	25 ♑ 47	26 ♈ 44	29 ♐ 38
21	29 ♈ 52	13 ♋ 25	17 ♈ 55	25 ♑ 44	25 ♈ 07	00 ♑ 45
31	29 ♈ 28	16 ♋ 13	15 ♈ 17	26 ♑ 22	22 ♈ 56	01 ♑ 52

All ephemeris data is given at 12.00 UT and the Moon's longitude is additionally given for 24.00 UT

ASPECTARIAN

h m Aspects

01 Tuesday
04 38 ☽ Ꝑ ☿
07 25 ☽ ⊥ ♅
09 59 ♂ □ ♄
10 58 ☽ Q ♂
12 09 ☿ ⊼ ♄
13 21 ☿ ⚹ ♂
14 08 ☽ □ ♇
14 22 ☿ ⊻ ♃
14 24 ☽ ⊻ ♆
14 25 ☽ ∥ ♃
17 20 ♂ ⊻ ♃
17 54 ☽ ⊻ ♅
22 22 ☉ ± ♇

02 Wednesday
00 43 ☽ △ ☉
00 48 ☽ ⊥ ♆
07 40 ☽ ⊻ ♄
08 19 ☽ □ ♃
08 48 ☽ ⚹ ♂
09 31 ☽ ∥ ♅
10 47 ☽ △ ☿
16 19 ☽ ∠ ♆
17 57 ☽ ⊥ ♄
21 27 ☽ ∥ ♂
23 40 ☽ # ♇

03 Thursday
02 58 ☽ ⊼ ♀
09 18 ☽ ∠ ♄
11 44 ☽ ∠ ♂
13 48 ☽ ± ♀
15 38 ☽ Q ♃
17 41 ☽ △ ♇
17 55 ☽ ⚹ ♆
21 27 ☽ ☌ ♅

04 Friday
02 34 ☿ Ꝑ ♇
03 45 ☽ ⊥ ♂
05 57 ☽ Ꝑ ♀
07 51 ☽ □ ☉
10 40 ☽ ⚹ ♄
12 17 ☽ ⚹ ♃
14 13 ☽ Q ♆
14 22 ☽ ⊻ ♂
19 03 ☽ Ꝑ ♇
21 29 ☽ □ ☿

05 Saturday
04 06 ☿ ⊥ ♀
06 49 ☽ Q ♄
06 54 ☿ ⊥ ♃
08 40 ☽ △ ♀
10 12 ☽ ± ♇
12 31 ♀ Q ♃
13 56 ☽ ∠ ♃
16 24 ☽ Q ☉
20 16 ☽ ⊼ ♇
20 28 ☽ □ ♆
22 48 ☉ ⊼ ♄

06 Sunday
00 03 ☽ ⊻ ♅
05 15 ☽ ⊥ ♃
08 04 ☽ Q ☿
10 07 ☽ ⊥ ♅
12 54 ☽ □ ♄
14 00 ☽ ⚹ ☉
15 27 ☽ ⊻ ♃
19 06 ☽ ☌ ♂

07 Monday
01 12 ☽ ∠ ♅
07 01 ☽ ⚹ ☿
11 43 ☽ # ♇
13 40 ☽ □ ♀
15 40 ☽ ∥ ♂
16 59 ☉ ⊻ ♃
16 59 ☽ ∠ ☉
22 31 ☽ ☍ ♇
22 41 ☽ △ ♆
23 40 ☿ ∥ ♄

08 Tuesday
01 33 ☽ ∥ ♅
02 24 ☽ ⚹ ♅
09 13 ☽ ⊥ ☉
11 45 ☽ ∠ ☿
15 04 ☽ △ ♄
18 37 ☽ ☌ ♃
20 06 ☽ ⊻ ☉
21 32 ☽ Q ♀
21 54 ☽ ∥ ♃
22 39 ☽ Q ♅
23 52 ☽ ⊻ ♂
23 57 ☽ Ꝑ ♆

09 Wednesday
03 53 ☽ # ♀
05 05 ☽ ⊥ ☿
10 32 ☽ ⊥ ♂
15 13 ☽ ± ♆
16 25 ☽ Ꝑ ♄
16 43 ☽ ⊻ ☿
18 58 ☽ ⚹ ♀
21 05 ♀ ± ♅

10 Thursday
01 21 ☽ ⊼ ♇
01 28 ☽ ⊼ ♆
02 38 ☽ ∠ ♂
05 22 ☽ □ ♅
07 47 ☽ ± ♄
11 40 ☽ ± ♇
18 07 ☽ ⊼ ♄
18 27 ♅ St R
21 20 ☽ ∥ ☉
21 30 ☿ ⊻ ♀
22 07 ☽ ∠ ♀
22 46 ☽ ⊻ ♃

11 Friday
01 18 ☉ Ꝑ ♇
03 18 ☽ Ꝑ ♇
03 27 ☽ ☌ ☉
03 38 ☿ ∠ ♃
04 44 ☽ ∥ ♄
05 52 ☽ ⚹ ♂
09 26 ☽ ⊥ ♃
11 48 ☿ ∥ ♆
14 33 ☽ ∥ ♆
14 35 ☽ # ☿
14 41 ☽ ⊥ ♀
14 55 ☽ ∥ ☿
14 55 ☽ # ♆
14 55 ☿ # ♆

12 Saturday
00 44 ☽ # ♄
01 29 ☿ Q ♂
01 36 ☽ ∠ ♃
01 51 ☽ ⊻ ♀
04 07 ☽ Q ♂
04 21 ☽ ☌ ☿
05 48 ☽ △ ♇
05 53 ☽ ☍ ♆
06 25 ☽ # ☉
10 03 ☽ △ ♅
13 20 ♀ # ♃
15 59 ☿ △ ♇
16 11 ♂ ∠ ♅
16 38 ☿ ☍ ♆
23 08 ☽ ☍ ♄

13 Sunday
05 08 ☽ ⚹ ♃
13 23 ☽ Ꝑ ♅
13 33 ☽ ⊻ ☉
14 26 ☽ □ ♂

14 Monday
01 37 ☽ ⊥ ☉
02 40 ☿ △ ♅
03 43 ☽ Q ♃
06 09 ☽ ± ♅
10 07 ☽ # ♃
11 33 ☽ ☌ ♀
12 57 ☽ □ ♇
12 59 ☽ ⊼ ♆
13 55 ☽ ∥ ♀
17 29 ☽ ⊼ ♅
19 41 ☽ ⊻ ☿
19 53 ☉ ⚹ ♂
20 02 ☽ ∠ ☉
22 51 ☿ # ♄

15 Tuesday
00 23 ☽ ± ♆
07 01 ☽ ⊼ ♄
08 53 ☽ ⊥ ☿
10 26 ☉ # ☿
10 55 ☽ # ♅
11 41 ☉ ⊥ ♃
14 36 ☽ □ ♃
17 42 ☽ Ꝑ ♆
18 22 ☽ # ♂
18 34 ☽ ± ♄
18 35 ♀ □ ♇
18 46 ♀ ⊼ ♆
23 32 ☽ Q ♇

16 Wednesday
01 47 ♆ ⚹ ♇
02 31 ☽ △ ♂
03 19 ☽ ∥ ♇
03 30 ☽ ⚹ ☉
05 00 ☽ ∠ ☿
11 39 ☉ ∥ ♄
12 06 ☽ Ꝑ ♄
23 08 ☽ △ ♆
23 09 ☽ ⚹ ♇

17 Thursday
00 30 ☽ ⊻ ♀
03 58 ☽ ☍ ♅
05 18 ☽ Q ☉
09 47 ☽ Ꝑ ♂
13 01 ☽ ⊥ ♀
15 14 ☽ ⚹ ☿
17 47 ☽ △ ♄

18 Friday
03 01 ☽ △ ♃
04 49 ☽ ± ♂
05 10 ☽ ∠ ♇
07 53 ☽ ∠ ♀
10 01 ☿ ☍ ♄
17 35 ☽ ⊼ ♂
19 03 ☽ Q ☿
20 44 ☽ □ ☉
23 19 ☽ ⊥ ♇

19 Saturday
09 50 ☽ Ꝑ ♃
11 21 ☽ □ ♆
11 27 ☽ ⊻ ♇
15 27 ☽ ⚹ ♀
16 22 ☽ ⊼ ♅

20 Sunday
04 19 ☽ ± ♃
04 29 ☽ ± ♅
05 59 ☽ □ ♄
12 48 ☽ □ ☿
14 09 ♃ Q ♅
15 49 ♀ ⊼ ♅
16 36 ☽ Q ♀
16 37 ☽ ⊼ ♃
22 35 ☽ Ꝑ ♅
23 34 ☽ Q ♆

21 Monday
09 17 ☽ ☍ ♂
10 35 ☽ ∥ ♇
14 31 ☽ △ ☉
17 49 ☽ Q ♄
22 51 ☿ ⚹ ♃
23 24 ☽ ⚹ ♆
23 25 ☽ # ♂
23 35 ☽ ☌ ♇

22 Tuesday
03 20 ☽ # ♅
04 24 ☽ △ ♅
05 40 ☽ □ ♀
12 48 ☽ ∥ ♀
17 17 ☽ ⚹ ♄
19 30 ☉ # ♆
20 23 ♃ Ꝑ ♆
22 34 ☽ Ꝑ ☉

23 Wednesday
04 39 ☽ ∠ ♆
04 47 ☽ ☍ ♃
04 57 ☉ ∥ ♆
06 57 ☽ # ♃
08 18 ☽ △ ☿
17 21 ☽ ± ☉
18 54 ☿ Ꝑ ♅
21 43 ☉ ⊥ ♀
21 50 ☽ ⊥ ♆
21 57 ☽ ∠ ♄
22 43 ☽ ⊼ ♂

24 Thursday
05 40 ☽ ⊼ ☉
09 07 ☽ ⊻ ♆
09 22 ☽ ⊻ ♇
10 35 ☽ ± ♂
12 20 ☽ ∥ ☿
13 56 ☽ □ ♅
14 47 ☽ ⊥ ♄
16 22 ☽ Ꝑ ☿
16 46 ☽ △ ♀
20 33 ☽ ⊥ ♇

25 Friday
01 50 ☽ ⊻ ♄
04 05 ☽ Ꝑ ♂
11 01 ☽ ± ☿
13 04 ☽ ∠ ♇
13 51 ☽ ⊼ ♃
16 34 ☽ # ♄
20 52 ☽ Ꝑ ♀
20 55 ☽ ∥ ☉
22 55 ☽ Q ♅
23 13 ☽ ⊼ ☿

26 Saturday
00 46 ☽ ∥ ♆
00 52 ☽ ± ♃
01 36 ☉ ☍ ♆
01 36 ☽ # ♆
05 35 ☉ △ ♇
06 00 ☽ # ☉
08 32 ☽ △ ♂
09 39 ☽ ∥ ♄
13 19 ☽ ± ♀
15 41 ☽ ☌ ♆
16 00 ☽ ⚹ ♇
16 49 ☽ ☍ ☉
17 09 ☽ Ꝑ ♃
20 18 ☽ ⚹ ♅
22 31 ☉ ∠ ♃

27 Sunday
00 07 ☽ ⊼ ♀
07 20 ☽ ☌ ♄
13 03 ☽ Q ♇
19 46 ☽ △ ♃
21 19 ☽ # ☿
22 28 ☽ ∠ ♅

28 Monday
05 37 ♂ ∥ ♅
09 50 ☽ ☍ ☿
13 12 ☉ # ♄
13 55 ☽ ⊥ ♅
14 27 ☽ ∥ ♃
15 11 ☽ □ ♂
16 22 ☉ △ ♅
19 39 ☽ ⊻ ♆
20 02 ☽ □ ♇

29 Tuesday
00 09 ☽ ⊻ ♅
00 44 ☽ ⊼ ☉
04 35 ☽ ☍ ♀
05 51 ☽ ⊥ ♆
10 34 ☽ ⊻ ♄
11 41 ☽ ± ☉
14 29 ☽ # ♀
15 09 ☽ ∥ ♂
15 56 ☽ ∥ ♅
20 40 ☽ ⊥ ♄
21 03 ☽ ∠ ♆
23 03 ☽ Q ♂
23 36 ☽ □ ♃

30 Wednesday
03 27 ♀ # ♂
03 29 ☿ ± ♅
04 00 ☽ Ꝑ ☉
06 34 ☽ # ♇
11 45 ☽ ∠ ♄
18 01 ☽ ⊼ ☿
20 07 ☽ ⚹ ♂
22 15 ☽ ⚹ ♆
22 41 ☽ △ ♇

OCTOBER 2026

LONGITUDES

Date	Sidereal time h m s	Sun ☉ ° ' "	Moon ☽ ° ' "	Moon ☽ 24.00 ° ' "	Mercury ☿ ° '	Venus ♀ ° '	Mars ♂ ° '	Jupiter ♃ ° '	Saturn ♄ ° '	Uranus ♅ ° '	Neptune ♆ ° '	Pluto ♇ ° '
01	12 40 56	08 ♎ 19 49	11 ♊ 02 56	18 ♊ 11 14	01 ♏ 20	08 ♏ 26	01 ♌ 59	19 ♌ 41	11 ♈ 32	05 ♊ 31	02 ♈ 51	03 ♒ 07
02	12 44 53	09 18 49	25 ♊ 18 56	02 ♋ 25 41	02 39	08 29	02 34	19 52	11 R 28	05 R 30	02 R 49	03 R 07
03	12 48 50	10 17 51	09 ♋ 31 12	16 ♋ 35 16	03 55	08 R 29	03 08	20 02	11 23	05 29	02 48	03 06
04	12 52 46	11 16 56	23 ♋ 37 43	00 ♌ 38 25	05 11	08 28	03 43	20 13	11 18	05 28	02 46	03 06
05	12 56 43	12 16 03	07 ♌ 37 14	14 ♌ 34 05	06 25	08 24	04 17	20 23	11 14	05 27	02 44	03 06
06	13 00 39	13 15 12	21 ♌ 28 51	28 ♌ 21 27	07 37	08 17	04 52	20 33	11 09	05 25	02 43	03 05
07	13 04 36	14 14 23	05 ♍ 11 46	11 ♍ 59 38	08 48	08 09	05 26	20 43	11 04	05 24	02 41	03 05
08	13 08 32	15 13 37	18 ♍ 44 55	25 ♍ 27 27	09 57	07 58	06 00	20 54	10 59	05 23	02 39	03 05
09	13 12 29	16 12 52	02 ♎ 07 03	08 ♎ 43 32	11 04	07 44	06 34	21 04	10 55	05 21	02 38	03 05
10	13 16 25	17 12 10	15 ♎ 16 44	21 ♎ 46 31	12 09	07 28	07 08	21 14	10 50	05 20	02 36	03 05
11	13 20 22	18 11 30	28 ♎ 12 44	04 ♏ 35 19	13 11	07 10	07 42	21 23	10 45	05 19	02 35	03 04
12	13 24 19	19 10 52	10 ♏ 54 15	17 ♏ 09 34	14 11	06 50	08 15	21 33	10 41	05 17	02 33	03 04
13	13 28 15	20 10 16	23 ♏ 21 22	29 ♏ 29 47	15 08	06 27	08 48	21 43	10 36	05 16	02 31	03 04
14	13 32 12	21 09 41	05 ♐ 35 05	11 ♐ 37 32	16 03	06 02	09 22	21 52	10 31	05 14	02 30	03 04
15	13 36 08	22 09 09	17 ♐ 37 32	23 ♐ 35 28	16 54	05 35	09 55	22 02	10 27	05 12	02 28	03 04
16	13 40 05	23 08 39	29 ♐ 31 52	05 ♑ 27 13	17 41	05 07	10 28	22 11	10 22	05 11	02 27	03 D 04
17	13 44 01	24 08 10	11 ♑ 22 07	17 ♑ 17 10	18 24	04 36	11 00	22 20	10 18	05 09	02 25	03 04
18	13 47 58	25 07 43	23 ♑ 13 00	29 ♑ 10 18	19 03	04 05	11 33	22 29	10 13	05 07	02 24	03 04
19	13 51 54	26 07 18	05 ♒ 09 42	11 ♒ 11 54	19 38	03 31	12 06	22 38	10 09	05 05	02 22	03 04
20	13 55 51	27 06 54	17 ♒ 17 33	23 ♒ 27 16	20 07	02 57	12 38	22 47	10 04	05 04	02 21	03 04
21	13 59 48	28 06 32	29 ♒ 41 38	06 ♓ 01 12	20 30	02 22	13 10	22 56	10 00	05 02	02 19	03 05
22	14 03 44	29 ♎ 06 12	12 ♓ 26 25	18 ♓ 57 38	20 46	01 46	13 42	23 04	09 56	05 00	02 18	03 05
23	14 07 41	00 ♏ 05 54	25 ♓ 35 08	02 ♈ 19 00	20 56	01 09	14 14	23 13	09 51	04 58	02 16	03 05
24	14 11 37	01 05 37	09 ♈ 09 12	16 ♈ 05 34	20 R 59	00 ♏ 32	14 45	23 21	09 47	04 56	02 15	03 05
25	14 15 34	02 05 22	23 ♈ 07 43	00 ♉ 15 09	20 53	29 ♎ 56	15 17	23 29	09 43	04 54	02 14	03 05
26	14 19 30	03 05 10	07 ♉ 27 09	14 ♉ 42 56	20 39	29 19	15 48	23 37	09 39	04 52	02 12	03 06
27	14 23 27	04 04 59	22 ♉ 01 35	29 ♉ 22 07	20 15	28 43	16 19	23 45	09 35	04 50	02 11	03 06
28	14 27 23	05 04 50	06 ♊ 43 33	14 ♊ 04 53	19 43	28 08	16 50	23 53	09 31	04 48	02 10	03 06
29	14 31 20	06 04 44	21 ♊ 25 15	28 ♊ 43 48	19 02	27 34	17 21	24 01	09 27	04 46	02 08	03 07
30	14 35 17	07 04 39	05 ♋ 59 52	13 ♋ 12 53	18 11	27 01	17 52	24 08	09 23	04 44	02 07	03 07
31	14 39 13	08 ♏ 04 37	20 ♋ 22 26	27 ♋ 28 17	17 ♏ 12	26 ♎ 30	18 ♌ 22	24 ♌ 16	09 ♈ 19	04 ♊ 42	02 ♈ 06	03 ♒ 07

Date	Moon True ☊ ° '	Moon Mean ☊ ° '	Moon ☽ Latitude ° '
01	29 ♒ 07	27 ♒ 41	05 N 03
02	29 R 05	27 38	04 38
03	29 04	27 35	03 56
04	29 D 04	27 32	03 00
05	29 05	27 28	01 54
06	29 07	27 25	00 N 41
07	29 R 07	27 22	00 S 33
08	29 05	27 19	01 44
09	29 01	27 16	02 48
10	28 55	27 13	03 42
11	28 46	27 09	04 24
12	28 37	27 06	04 52
13	28 28	27 03	05 05
14	28 19	27 00	05 03
15	28 11	26 57	04 49
16	28 06	26 54	04 21
17	28 03	26 50	03 43
18	28 02	26 47	02 55
19	28 D 02	26 44	01 59
20	28 03	26 41	00 S 57
21	28 R 04	26 38	00 N 09
22	28 02	26 34	01 16
23	27 59	26 31	02 21
24	27 53	26 28	03 21
25	27 45	26 25	04 09
26	27 35	26 22	04 44
27	27 25	26 19	05 01
28	27 15	26 15	04 58
29	27 06	26 12	04 36
30	27 00	26 09	03 56
31	26 ♒ 57	26 ♒ 06	03 N 01

DECLINATIONS

Date	Sun ☉ ° '	Moon ☽ ° '	Mercury ☿ ° '	Venus ♀ ° '	Mars ♂ ° '	Jupiter ♃ ° '	Saturn ♄ ° '	Uranus ♅ ° '	Neptune ♆ ° '	Pluto ♇ ° '
01	03 S 18	27 N 06	13 S 38	21 S 02	20 N 43	15 N 29	02 N 04	21 N 04	00 S 10	23 S 38
02	03 41	27 59	14 12	21 07	20 36	15 26	02 02	21 04	00 11	23 38
03	04 05	27 01	14 45	21 12	20 29	15 23	02 00	21 04	00 12	23 38
04	04 28	24 20	15 17	21 15	20 22	15 19	01 58	21 04	00 12	23 38
05	04 51	20 12	15 47	21 17	20 15	15 16	01 57	21 04	00 13	23 38
06	05 14	15 00	16 17	21 18	20 08	15 13	01 55	21 03	00 14	23 38
07	05 37	09 06	16 46	21 18	20 00	15 10	01 53	21 03	00 14	23 38
08	06 00	02 N 51	17 14	21 16	19 53	15 07	01 51	21 03	00 15	23 38
09	06 23	03 S 25	17 40	21 13	19 45	15 04	01 49	21 03	00 15	23 38
10	06 45	09 26	18 06	21 09	19 38	15 01	01 47	21 02	00 16	23 38
11	07 08	14 57	18 30	21 03	19 30	14 58	01 46	21 02	00 17	23 38
12	07 31	19 42	18 52	20 56	19 23	14 55	01 44	21 02	00 17	23 38
13	07 53	23 31	19 13	20 47	19 15	14 52	01 42	21 02	00 18	23 38
14	08 15	26 13	19 33	20 37	19 07	14 49	01 40	21 01	00 19	23 38
15	08 38	27 39	19 51	20 25	18 59	14 46	01 39	21 01	00 19	23 38
16	09 00	27 48	20 07	20 12	18 52	14 43	01 37	21 01	00 20	23 38
17	09 22	26 39	20 21	19 58	18 44	14 41	01 35	21 00	00 20	23 38
18	09 43	24 19	20 33	19 42	18 36	14 38	01 33	21 00	00 21	23 38
19	10 05	20 54	20 43	19 25	18 28	14 35	01 32	21 00	00 22	23 38
20	10 27	16 34	20 50	19 06	18 20	14 32	01 30	20 59	00 22	23 38
21	10 48	11 27	20 55	18 47	18 12	14 30	01 28	20 59	00 23	23 38
22	11 09	05 S 43	20 58	18 26	18 04	14 27	01 27	20 59	00 23	23 38
23	11 30	00 N 24	20 57	18 04	17 56	14 24	01 25	20 58	00 24	23 38
24	11 51	06 42	20 53	17 42	17 47	14 22	01 24	20 58	00 24	23 38
25	12 12	12 51	20 45	17 19	17 39	14 19	01 22	20 58	00 25	23 38
26	12 32	18 28	20 34	16 55	17 31	14 17	01 21	20 57	00 25	23 37
27	12 53	23 07	20 19	16 31	17 23	14 14	01 19	20 57	00 26	23 37
28	13 13	26 20	19 59	16 06	17 15	14 12	01 18	20 57	00 26	23 37
29	13 33	27 45	19 36	15 41	17 06	14 10	01 16	20 56	00 27	23 37
30	13 52	27 14	19 08	15 17	16 58	14 07	01 15	20 56	00 27	23 37
31	14 S 12	24 N 52	18 S 36	14 S 52	16 N 50	14 N 05	01 N 13	20 N 56	00 S 28	23 S 37

ZODIAC SIGN ENTRIES

Date	h	m	Planets
02	19	54	☽ ♋
04	22	54	☽ ♌
07	02	53	☽ ♍
09	08	10	☽ ♎
11	15	21	☽ ♏
14	00	59	☽ ♐
16	12	57	☽ ♑
19	01	40	☽ ♒
21	12	35	☽ ♓
23	09	38	☉ ♏
23	19	53	☽ ♈
25	09	10	♀ ♎
25	23	35	☽ ♉
28	01	02	☽ ♊
30	02	05	☽ ♋

LATITUDES

Date	Mercury ☿ ° '	Venus ♀ ° '	Mars ♂ ° '	Jupiter ♃ ° '	Saturn ♄ ° '	Uranus ♅ ° '	Neptune ♆ ° '	Pluto ♇ ° '
01	01 S 48	07 S 06	01 N 01	00 N 36	02 S 43	00 S 09	01 S 25	04 S 18
04	02 09	07 20	01 05	00 36	02 43	00 09	01 25	04 17
07	02 28	07 30	01 08	00 37	02 43	00 09	01 25	04 17
10	02 45	07 34	01 11	00 37	02 43	00 09	01 25	04 17
13	02 59	07 33	01 14	00 38	02 43	00 09	01 25	04 17
16	03 08	07 25	01 18	00 38	02 43	00 09	01 25	04 17
19	03 12	07 10	01 21	00 39	02 42	00 09	01 25	04 17
22	03 08	06 47	01 25	00 40	02 42	00 09	01 25	04 17
25	02 53	06 16	01 28	00 40	02 42	00 09	01 25	04 17
28	02 25	05 40	01 32	00 41	02 42	00 09	01 25	04 17
31	01 S 42	04 S 59	01 N 35	00 N 41	02 S 41	00 S 09	01 S 25	04 S 16

DATA

Julian Date	2461315
Delta T	+73 seconds
Ayanamsa	24° 13' 59"
Synetic vernal point	04° ♓ 53' 00"
True obliquity of ecliptic	23° 26' 17"

MOON'S PHASES, APSIDES AND POSITIONS ☽

Date	h	m	Phase	Longitude ° '	Eclipse Indicator
03	13	25	☾	10 ♋ 21	
10	15	50	●	17 ♎ 22	
18	16	13	☽	25 ♑ 18	
26	04	12	○	02 ♉ 46	

Day	h	m		
01	21	02	Perigee	
16	22	58	Apogee	
28	18	15	Perigee	
02	11	27	Max dec	27° N 59'
08	22	53	0S	
16	02	38	Max dec	27° S 54'
23	10	26	0N	
29	17	31	Max dec	27° N 48'

LONGITUDES

Date	Chiron ⚷ ° '	Ceres ⚳ ° '	Pallas ⚴ ° '	Juno ⚵ ° '	Vesta ⚶ ° '	Black Moon Lilith ⚸ ° '
01	29 ♈ 28	16 ♋ 13	15 ♈ 17	26 ♑ 22	22 ♈ 56	01 ♑ 52
11	29 ♈ 02	18 ♋ 42	12 ♈ 22	27 ♑ 37	20 ♈ 24	02 ♑ 59
21	28 ♈ 34	20 ♋ 47	09 ♈ 29	29 ♑ 26	17 ♈ 51	04 ♑ 06
31	28 ♈ 06	22 ♋ 24	06 ♈ 56	01 ♒ 45	15 ♈ 36	05 ♑ 13

All ephemeris data is given at 12.00 UT and the Moon's longitude is additionally given for 24.00 UT

ASPECTARIAN

01 Thursday
02 43 ☽ ☌ ♅
05 08 ☽ ± ☿
06 16 ☽ Q ♃
07 06 ☽ △ ☉
07 35 ☽ ⊼ ♀
12 49 ☽ ⚹ ♄
14 35 ☉ ⊻ ♀
17 42 ☽ ± ♀
18 22 ☽ Q ♆
19 14 ☿ Q ♃
21 47 ☽ Ᵽ ☿
22 24 ☽ ∠ ♂
22 44 ♀ ⋕ ♅
23 53 ☽ Ᵽ ♇

02 Friday
02 42 ☽ ⚹ ♃
08 54 ☽ Q ♄
08 54 ☽ Ᵽ ♀
09 14 ☿ □ ♂
14 11 ☽ ⊥ ♂
15 02 ☽ ± ♇
15 15 ☿ ⊼ ♆
20 43 ☿ □ ♇
22 18 ♂ △ ♆

03 Saturday
00 38 ☽ □ ♆
00 45 ☽ ⊻ ♂
01 09 ☽ ⊼ ♇
01 36 ☽ △ ☿
04 19 ☽ ∠ ♃
05 10 ☽ ⊻ ♅
07 16 ♀ St R
10 15 ☽ △ ♀
10 39 ♂ ☍ ♇
13 25 ☽ □ ☉
15 08 ☽ □ ♄
15 19 ☽ ⊥ ♅
19 45 ☽ ⊥ ♃

04 Sunday
06 06 ☽ ⊻ ♃
06 36 ☽ ∠ ♅
12 29 ☉ ☍ ♄
13 59 ☿ ⋕ ♃
16 36 ☽ ⋕ ♇
17 21 ☿ ⊼ ♅
22 24 ☽ Q ☉

05 Monday
03 37 ☽ △ ♆
04 13 ☽ ☍ ♇
06 01 ☽ ☌ ♂
06 20 ☽ ⋕ ♀
07 31 ☽ ∥ ♅
08 15 ☽ ⚹ ♅
09 43 ☽ □ ☿
11 44 ☽ ∥ ♂
13 20 ☽ □ ♀
18 11 ☽ △ ♄
20 38 ☽ ⚹ ☉

06 Tuesday
04 58 ☽ Q ♅
05 28 ☽ Ᵽ ♆
06 50 ☽ ⋕ ☿
10 22 ☽ ☌ ♃
11 01 ☽ ∥ ♃
19 54 ☽ Q ☿
20 05 ☽ Ᵽ ♄
20 18 ☽ Q ♀
21 06 ☽ ± ♆

07 Wednesday
00 10 ☿ ☌ ♀
00 44 ☽ ∠ ☉
07 36 ☽ ⊼ ♆
08 18 ☽ ⊼ ♇
09 42 ☿ ± ♆
10 42 ♂ ⚹ ♅
11 47 ☽ ± ♄
12 22 ☽ □ ♅
12 26 ☽ ⊻ ♂
17 08 ☽ ⚹ ♀
17 47 ☽ ⊥ ☉
18 51 ☽ ± ♇
18 57 ☽ ⚹ ☿
22 18 ☽ ⊼ ♄
23 29 ☽ ⊥ ♂

08 Thursday
00 42 ☽ ⋕ ☉
05 14 ☽ ⊻ ☉
10 49 ☽ Ᵽ ♇
15 51 ☽ ∥ ♄
15 52 ☽ ⊻ ♃
16 12 ☽ ∠ ♂
19 24 ☽ ∠ ♀
21 55 ☽ ⋕ ♆
23 50 ☽ ∥ ♆

09 Friday
00 06 ☽ ∠ ☿
02 46 ☽ ⊥ ♃
05 52 ☽ ⋕ ♄
08 55 ☿ ⊼ ♄
11 19 ☽ ⊥ ♀
12 56 ☽ ☍ ♆
13 44 ☽ △ ♇
17 49 ☽ ⊥ ☿
17 52 ☽ △ ♅
19 14 ☽ ∠ ♃
20 26 ☽ ⚹ ♂
22 00 ☽ ⊻ ♀

10 Saturday
00 24 ☽ ∥ ☉
03 54 ☽ ☍ ♄
05 44 ☽ ⊻ ☿
15 50 ☽ ☌ ☉
19 25 ☽ Q ♂
21 19 ☽ Ᵽ ♅
21 31 ♀ □ ♂
23 07 ☽ ⚹ ♃

11 Sunday
12 06 ☽ ⋕ ♃
14 03 ☽ ± ♅
15 26 ♀ ⋕ ♅
20 11 ☽ ⊼ ♆
21 08 ☽ □ ♇
21 52 ☽ Q ♃

12 Monday
01 20 ☽ ⊼ ♅
04 27 ☽ ☌ ♀
06 43 ☽ □ ♂
07 01 ☽ ∥ ☿
07 31 ☽ ± ♆
10 14 ☽ ⋕ ♂
11 34 ☽ ⊼ ♄
18 47 ☽ ∥ ♀
18 48 ☽ ☌ ☿
19 37 ☽ ⋕ ♅
23 00 ☽ ± ♄

13 Tuesday
00 44 ☽ Ᵽ ♆
05 17 ☽ ⊻ ☉
07 34 ☽ Q ♇
08 46 ☽ □ ♃
12 52 ☽ ∥ ♇
13 37 ☿ ⋕ ♂
14 04 ☉ Ᵽ ♅
16 21 ☽ Ᵽ ♄
17 58 ☽ ⊥ ☉

14 Wednesday
05 55 ☽ △ ♆
07 02 ☽ ⚹ ♇
11 18 ☽ ☍ ♅
12 52 ☽ ⊻ ♀
13 15 ☽ ∠ ☉
19 51 ☽ △ ♂
21 45 ☽ △ ♄
22 55 ☉ Q ♂

15 Thursday
00 16 ☿ ± ♄
00 22 ☽ ⊥ ♀
08 22 ☉ ⚹ ♃
10 25 ☽ ⊻ ☿
12 53 ☽ ∠ ♇
17 44 ☽ ∠ ♀
20 58 ☽ △ ♃
21 56 ☽ ⚹ ☉
23 22 ☽ ⊥ ☿

16 Friday
02 40 ♇ St D
03 23 ☽ Ᵽ ♂
04 51 ☿ Ᵽ ♆
07 01 ☽ ⊥ ♇
08 32 ♂ △ ♄
08 39 ♀ ⊼ ♅
16 36 ☿ ∥ ♀
17 54 ☽ □ ♆
18 49 ☽ ∠ ☿
19 10 ☽ ⊻ ♇
22 28 ☽ ± ♂
22 51 ☽ ⚹ ♀
23 25 ☽ ⊼ ♅

17 Saturday
00 25 ☽ Q ☉
03 42 ☽ Ᵽ ♃
09 50 ☽ □ ♄
11 14 ☽ ⊼ ♂
11 33 ☽ ± ♅
21 46 ♀ Q ♃
22 11 ☽ Q ♀
22 12 ☽ ± ♃

18 Sunday
03 07 ☽ ⚹ ☿
05 45 ☽ Ᵽ ♅
06 19 ☽ Q ♆
10 30 ☽ ⊼ ♃
16 13 ☽ □ ☉
17 27 ☽ ∥ ♇
22 02 ☽ Q ♄

19 Monday
04 36 ☽ Q ☿
06 26 ☽ ⚹ ♆
07 49 ☽ ☌ ♇
08 52 ☽ □ ♀
11 26 ☽ ⋕ ♅
11 52 ☽ △ ♅
13 08 ☽ ∥ ☿
21 29 ☽ ∥ ♀
21 51 ☽ ⚹ ♄

20 Tuesday
02 24 ☽ ☍ ♂
02 30 ☽ ⋕ ♂
06 56 ♀ □ ♇
12 06 ☽ ∠ ♆
17 42 ☽ □ ☿
21 58 ☽ ⋕ ♃
22 49 ☽ ☍ ♃

21 Wednesday
03 02 ☽ ∠ ♄
05 33 ☽ ⊥ ♆
08 42 ☽ △ ☉
13 39 ♀ ⊼ ♆
14 38 ☽ ∥ ☉
16 51 ☽ △ ♀
17 00 ☽ ⊻ ♆
18 26 ☽ ⊻ ♇
20 09 ☽ ⊥ ♄
22 07 ☽ □ ♅

22 Thursday
05 44 ☽ ⊥ ♇
07 21 ☽ ⊻ ♄
09 34 ☉ ± ♅
14 26 ☽ ⊼ ♂
15 20 ☽ Ᵽ ☉
19 37 ☽ Ᵽ ♀
22 23 ☽ ∠ ♇

23 Friday
01 55 ☽ ± ♂
03 31 ☽ △ ☿
04 56 ☽ ⋕ ♄
07 18 ☽ Q ♅
07 40 ☽ ⊼ ♃
08 55 ☽ ∥ ♆
09 06 ☽ ± ☉
11 15 ☽ ± ♀
11 58 ☽ ⋕ ♆
15 52 ☽ ∥ ♄
18 33 ☽ ± ♃
18 47 ☽ Ᵽ ♂
20 42 ☽ ⊼ ☉
21 30 ☽ ⊼ ♀
23 54 ☽ ☌ ♆

24 Saturday
01 21 ☽ ⚹ ♇
02 45 ♀ ⋕ ♂
03 44 ☉ ☌ ♀
04 38 ☽ ⚹ ♅
06 27 ☽ Ᵽ ☿
07 12 ☿ St R
10 35 ☽ Ᵽ ♃
13 06 ☽ ☌ ♄
22 02 ☽ ± ☿
22 05 ☽ △ ♂
22 16 ☽ Q ♇

25 Sunday
06 32 ☽ ∠ ♅
08 14 ☽ ⊼ ☿
09 14 ☽ ⋕ ☉
12 37 ☽ △ ♃
15 15 ☉ ⊼ ♆
17 58 ☽ ∥ ♃
21 43 ☽ ⊥ ♅
22 59 ☽ ☍ ♀

26 Monday
03 17 ☽ ⊻ ♆
04 12 ☽ ☍ ☉
04 45 ☽ □ ♇
05 30 ☽ ⋕ ♀
07 43 ☽ ⊻ ♅
07 48 ☽ ∥ ♂
12 12 ☉ □ ♇
13 15 ☽ ⊥ ♆
15 37 ☽ ⊻ ♄
21 37 ☽ ⋕ ☿

27 Tuesday
00 04 ☽ ∥ ♅
01 28 ☽ ⊥ ♄
02 18 ☽ □ ♂
04 04 ☽ ∠ ♆
07 11 ♀ ± ♅
09 11 ☽ ☍ ☿
14 51 ☽ □ ♃
15 07 ☽ ⋕ ♇
16 10 ☽ ∠ ♄
20 39 ♀ Q ♂
22 31 ☽ ⊼ ♀

28 Wednesday
04 34 ☽ ⚹ ♆
05 29 ☉ ⊼ ♅
06 06 ☽ △ ♇
07 56 ☽ ± ♀
08 49 ☽ Q ♂
08 52 ☽ ☌ ♅
09 07 ☽ ⊼ ☉
10 17 ♂ Q ♅
16 32 ☽ ⚹ ♄
19 37 ☽ ± ☉
20 29 ☽ Q ♃
22 04 ☽ Ᵽ ♀

29 Thursday
00 07 ☽ Q ♆
02 20 ♂ Ᵽ ♆
05 06 ☽ ⚹ ♂
06 35 ☽ Ᵽ ♇
08 16 ☽ ⊼ ☿
10 10 ☉ Q ♃
11 24 ☽ Ᵽ ☉
12 03 ☽ Q ♄
16 17 ☽ ⚹ ♃
17 36 ☽ ± ☿
21 20 ☽ ± ♇
21 43 ☽ △ ♀

30 Friday
05 35 ☽ □ ♆
06 38 ☽ ∠ ♂
07 14 ☽ ⊼ ♇
07 37 ☽ Ᵽ ☿
09 54 ☽ ⊻ ♅
13 55 ☽ △ ☉
17 15 ☽ ∠ ♃
17 20 ☿ □ ♂
17 35 ☽ □ ♄
19 50 ☽ ⊥ ♅
22 06 ☽ ⊥ ♂

31 Saturday
04 15 ☉ ⋕ ♃
07 02 ☽ △ ☿
08 25 ☽ ⊥ ♃
08 30 ☽ ⊻ ♂
10 52 ☽ ∠ ♅
12 25 ☉ ± ♆
14 27 ☿ Ᵽ ♆
18 37 ☽ ⊻ ♃
19 13 ♃ Ᵽ ♄
20 48 ☽ ⋕ ♇
22 00 ☽ □ ♀

LONGITUDES

Date	Sidereal time h m s	Sun ☉ ° ' "	Moon ☽ ° ' "	Moon ☽ 24.00 ° ' "	Mercury ☿ ° '	Venus ♀ ° '	Mars ♂ ° '	Jupiter ♃ ° '	Saturn ♄ ° '	Uranus ♅ ° '	Neptune ♆ ° '	Pluto ♇ ° '
01	14 43 10	09 ♏ 04 37	04 ♌ 30 14	11 ♌ 28 16	16 ♏ 06	26 ♎ 00	18 ♌ 52	24 ♌ 23	09 ♈ 15	04 ♊ 39	02 ♈ 04	03 ♒ 08
02	14 47 06	10 04 39	18 ♌ 22 26	25 ♌ 12 49	14 R 53	25 R 32	19 22	24 30	09 R 12	04 R 37	02 R 03	03 08
03	14 51 03	11 04 44	01 ♍ 59 35	08 ♍ 42 54	13 37	25 06	19 52	24 37	09 08	04 35	02 02	03 09
04	14 54 59	12 04 50	15 ♍ 22 56	21 ♍ 59 51	12 19	24 42	20 22	24 44	09 04	04 33	02 01	03 09
05	14 58 56	13 04 58	28 ♍ 33 48	05 ♎ 04 55	11 01	24 20	20 51	24 51	09 01	04 30	02 00	03 10
06	15 02 52	14 05 09	11 ♎ 33 15	17 ♎ 58 54	09 46	24 01	21 20	24 57	08 57	04 28	01 58	03 11
07	15 06 49	15 05 21	24 ♎ 21 52	00 ♏ 42 11	08 37	23 44	21 49	25 04	08 54	04 26	01 57	03 11
08	15 10 46	16 05 36	06 ♏ 59 50	13 ♏ 14 49	07 35	23 29	22 18	25 10	08 51	04 24	01 56	03 12
09	15 14 42	17 05 52	19 ♏ 27 08	25 ♏ 36 49	06 42	23 17	22 47	25 16	08 48	04 21	01 55	03 13
10	15 18 39	18 06 10	01 ♐ 43 55	07 ♐ 48 30	06 00	23 07	23 15	25 22	08 45	04 19	01 54	03 13
11	15 22 35	19 06 29	13 ♐ 50 42	19 ♐ 50 43	05 29	23 00	23 43	25 28	08 42	04 16	01 53	03 14
12	15 26 32	20 06 51	25 ♐ 48 47	01 ♑ 45 12	05 10	22 55	24 11	25 34	08 39	04 14	01 52	03 15
13	15 30 28	21 07 13	07 ♑ 40 19	13 ♑ 34 32	05 02	22 52	24 38	25 39	08 36	04 12	01 51	03 16
14	15 34 25	22 07 38	19 ♑ 28 20	25 ♑ 22 15	05 D 06	22 D 52	25 06	25 44	08 33	04 09	01 50	03 16
15	15 38 21	23 08 03	01 ♒ 16 50	07 ♒ 12 42	05 20	22 54	25 33	25 50	08 30	04 07	01 49	03 17
16	15 42 18	24 08 30	13 ♒ 10 31	19 ♒ 10 56	05 44	22 59	26 00	25 54	08 28	04 04	01 48	03 18
17	15 46 15	25 08 58	25 ♒ 14 40	01 ♓ 22 25	06 18	23 06	26 26	25 59	08 25	04 02	01 48	03 19
18	15 50 11	26 09 28	07 ♓ 34 51	13 ♓ 52 38	06 59	23 15	26 52	26 04	08 23	03 59	01 47	03 20
19	15 54 08	27 09 59	20 ♓ 16 22	26 ♓ 46 37	07 47	23 27	27 18	26 08	08 21	03 57	01 46	03 21
20	15 58 04	28 10 31	03 ♈ 23 48	10 ♈ 08 14	08 42	23 41	27 44	26 13	08 18	03 54	01 45	03 22
21	16 02 01	29 ♏ 11 04	17 ♈ 00 06	23 ♈ 59 22	09 43	23 56	28 10	26 17	08 16	03 52	01 45	03 23
22	16 05 57	00 ♐ 11 39	01 ♉ 05 49	08 ♉ 19 01	10 48	24 14	28 35	26 21	08 14	03 49	01 44	03 24
23	16 09 54	01 12 15	15 ♉ 38 19	23 ♉ 02 51	11 57	24 34	29 00	26 25	08 12	03 47	01 43	03 25
24	16 13 50	02 12 52	00 ♊ 31 33	08 ♊ 03 15	13 10	24 56	29 24	26 28	08 11	03 44	01 43	03 26
25	16 17 47	03 13 31	15 ♊ 36 38	23 ♊ 10 25	14 26	25 19	29 ♌ 48	26 32	08 09	03 42	01 42	03 27
26	16 21 44	04 14 11	00 ♋ 43 17	08 ♋ 14 02	15 45	25 45	00 ♍ 12	26 35	08 07	03 39	01 41	03 28
27	16 25 40	05 14 53	15 ♋ 41 37	23 ♋ 05 09	17 06	26 12	00 36	26 38	08 06	03 37	01 41	03 29
28	16 29 37	06 15 36	00 ♌ 23 54	07 ♌ 37 23	18 29	26 41	00 59	26 41	08 04	03 34	01 40	03 30
29	16 33 33	07 16 21	14 ♌ 45 17	21 ♌ 47 28	19 53	27 11	01 22	26 44	08 03	03 32	01 40	03 32
30	16 37 30	08 ♐ 17 07	28 ♌ 43 56	05 ♍ 34 49	21 ♏ 19	27 ♎ 43	01 ♍ 45	26 ♌ 46	08 ♈ 02	03 ♊ 29	01 ♈ 39	03 ♒ 33

DECLINATIONS

Date	Moon True ☊ ° '	Moon Mean ☊ ° '	Moon ☽ Latitude ° '	Sun ☉ ° '	Moon ☽ ° '	Mercury ☿ ° '	Venus ♀ ° '	Mars ♂ ° '	Jupiter ♃ ° '	Saturn ♄ ° '	Uranus ♅ ° '	Neptune ♆ ° '	Pluto ♇ ° '
01	26 ♒ 56	26 ♒ 03	01 N 56	14 S 31	21 N 01	18 S 00	14 S 28	16 N 42	14 N 03	01 N 12	20 N 55	00 S 28	23 S 37
02	26 D 56	26 00	00 N 45	14 50	16 02	17 20	14 04	16 34	14 01	01 11	20 55	00 29	23 36
03	26 R 56	25 56	00 S 27	15 09	10 21	16 38	13 40	16 25	13 58	01 09	20 54	00 29	23 36
04	26 55	25 53	01 36	15 28	04 N 17	15 55	13 17	16 17	13 56	01 08	20 54	00 30	23 36
05	26 51	25 50	02 39	15 46	01 S 52	15 11	12 55	16 09	13 54	01 07	20 53	00 30	23 36
06	26 45	25 47	03 33	16 04	07 50	14 29	12 34	16 01	13 52	01 06	20 53	00 31	23 36
07	26 35	25 44	04 14	16 22	13 23	13 48	12 14	15 53	13 50	01 05	20 53	00 31	23 36
08	26 23	25 40	04 43	16 39	18 18	13 10	11 54	15 45	13 48	01 03	20 52	00 31	23 35
09	26 09	25 37	04 58	16 56	22 22	12 37	11 36	15 36	13 47	01 02	20 52	00 32	23 35
10	25 55	25 34	04 58	17 13	25 22	12 09	11 18	15 28	13 45	01 01	20 51	00 32	23 35
11	25 41	25 31	04 45	17 30	27 11	11 46	11 02	15 20	13 43	01 00	20 51	00 33	23 35
12	25 29	25 28	04 19	17 46	27 42	11 29	10 47	15 12	13 41	00 59	20 51	00 33	23 34
13	25 20	25 25	03 43	18 02	26 55	11 17	10 33	15 04	13 40	00 58	20 50	00 33	23 34
14	25 14	25 21	02 57	18 18	24 56	11 11	10 20	14 57	13 38	00 57	20 50	00 34	23 34
15	25 10	25 18	02 03	18 33	21 52	11 10	10 08	14 49	13 37	00 57	20 49	00 34	23 34
16	25 09	25 15	01 S 03	18 48	17 52	11 14	09 57	14 41	13 35	00 56	20 49	00 34	23 34
17	25 D 09	25 12	00 N 01	19 03	13 06	11 23	09 48	14 33	13 34	00 55	20 48	00 35	23 33
18	25 R 08	25 09	01 05	19 17	07 43	11 35	09 40	14 26	13 32	00 54	20 48	00 35	23 33
19	25 07	25 05	02 09	19 31	01 S 53	11 50	09 33	14 18	13 31	00 53	20 47	00 35	23 33
20	25 03	25 02	03 07	19 45	04 N 13	12 09	09 27	14 11	13 30	00 53	20 47	00 35	23 33
21	24 57	24 59	03 58	19 58	10 20	12 30	09 22	14 03	13 29	00 52	20 46	00 36	23 32
22	24 48	24 56	04 35	20 11	16 09	12 52	09 18	13 56	13 27	00 52	20 46	00 36	23 32
23	24 37	24 53	04 57	20 24	21 15	13 17	09 15	13 48	13 26	00 51	20 46	00 36	23 32
24	24 26	24 50	04 59	20 36	25 07	13 43	09 13	13 41	13 25	00 51	20 45	00 36	23 32
25	24 14	24 46	04 40	20 48	27 18	14 10	09 12	13 34	13 24	00 50	20 45	00 37	23 31
26	24 05	24 43	04 02	20 59	27 28	14 38	09 13	13 27	13 23	00 50	20 44	00 37	23 31
27	23 57	24 40	03 07	21 10	25 37	15 06	09 14	13 20	13 23	00 49	20 44	00 37	23 31
28	23 53	24 37	02 01	21 21	22 02	15 35	09 16	13 13	13 22	00 49	20 43	00 37	23 30
29	23 52	24 34	00 48	21 31	17 10	16 04	09 19	13 07	13 21	00 49	20 43	00 37	23 30
30	23 ♒ 51	24 ♒ 31	00 S 26	21 S 41	11 N 31	16 S 33	09 S 22	13 N 00	13 N 21	00 N 48	20 N 42	00 S 37	23 S 30

ZODIAC SIGN ENTRIES

Date	h	m	Planets
01	04	18	☽ ♌
03	08	28	☽ ♍
05	14	38	☽ ♎
07	22	40	☽ ♏
10	08	36	☽ ♐
12	20	27	☽ ♑
15	09	24	☽ ♒
17	21	19	☽ ♓
20	05	52	☽ ♈
22	07	23	☉ ♐
22	10	10	☽ ♉
24	11	10	☽ ♊
25	23	37	♂ ♍
26	10	51	☽ ♋
28	11	21	☽ ♌
30	14	13	☽ ♍

LATITUDES

Date	Mercury ☿ ° '	Venus ♀ ° '	Mars ♂ ° '	Jupiter ♃ ° '	Saturn ♄ ° '	Uranus ♅ ° '	Neptune ♆ ° '	Pluto ♇ ° '
01	01 S 24	04 S 45	01 N 37	00 N 42	02 S 41	00 S 09	01 S 25	04 S 16
04	00 S 24	04 00	01 40	00 42	02 41	00 09	01 25	04 16
07	00 N 36	03 15	01 44	00 42	02 40	00 09	01 25	04 16
10	01 27	02 30	01 48	00 43	02 40	00 09	01 25	04 16
13	02 02	01 47	01 53	00 44	02 39	00 09	01 25	04 16
16	02 20	01 06	01 57	00 45	02 38	00 09	01 24	04 16
19	02 24	00 S 28	02 02	00 45	02 38	00 09	01 24	04 16
22	02 18	00 N 06	02 06	00 46	02 37	00 09	01 24	04 16
25	02 06	00 38	02 10	00 47	02 36	00 09	01 24	04 15
28	01 49	01 06	02 15	00 48	02 36	00 09	01 24	04 15
31	01 N 29	01 N 31	02 N 20	00 N 48	02 S 35	00 S 09	01 S 24	04 S 15

DATA

Julian Date	2461346
Delta T	+73 seconds
Ayanamsa	24° 14' 03"
Synetic vernal point	04° ♓ 52' 56"
True obliquity of ecliptic	23° 26' 17"

MOON'S PHASES, APSIDES AND POSITIONS ☽

Date	h	m	Phase	Longitude ° '	Eclipse Indicator
01	20	28	☾	09 ♌ 26	
09	07	02	●	16 ♏ 53	
17	11	48	☽	25 ♒ 08	
24	14	53	○	02 ♊ 20	

Day	h	m		
13	17	55	Apogee	
25	21	09	Perigee	
05	04	42	OS	
12	09	32	Max dec	27° S 42'
19	19	28	ON	
26	01	55	Max dec	27° N 39'

LONGITUDES

Date	Chiron ⚷ ° '	Ceres ⚳ ° '	Pallas ⚴ ° '	Juno ⚵ ° '	Vesta ⚶ ° '	Black Moon Lilith ⚸ ° '
01	28 ♈ 03	22 ♋ 32	06 ♈ 43	02 ♒ 01	15 ♈ 24	05 ♑ 20
11	27 ♈ 36	23 ♋ 33	04 ♈ 49	04 ♒ 48	13 ♈ 47	06 ♑ 27
21	27 ♈ 12	23 ♋ 57	03 ♈ 40	07 ♒ 58	12 ♈ 52	07 ♑ 34
31	26 ♈ 51	23 ♋ 39	03 ♈ 18	11 ♒ 28	12 ♈ 43	08 ♑ 41

All ephemeris data is given at 12.00 UT and the Moon's longitude is additionally given for 24.00 UT

ASPECTARIAN

h m	Aspects
01 Sunday	
07 51	☽ △ ♆
09 39	☽ ☍ ♇
09 58	☉ ∥ ♀
12 16	☽ ⚹ ♅
12 29	☽ ∥ ♅
16 01	☉ ⊼ ♄
20 08	☽ △ ♄
20 28	☽ □ ☉
02 Monday	
03 51	☽ Q ♀
05 15	☽ # ☿
05 48	☿ ± ♄
06 26	☽ □ ☿
08 57	☽ Q ♅
09 36	☽ ∥ ♂
09 42	☽ ⚼ ♆
13 49	☽ ☌ ♂
15 25	♀ # ♃
16 59	☽ # ☉
20 53	☽ ∥ ♃
21 15	☽ # ♀
22 09	☽ ⚼ ♄
22 50	☽ ☌ ♃
03 Tuesday	
00 10	☽ ⚹ ♀
01 28	☽ ± ♆
06 25	☽ Q ☉
11 23	☽ Q ☿
12 04	☽ ⊼ ♆
14 01	☽ ± ♄
14 03	☽ ⊼ ♇
16 36	☽ □ ♅
21 03	☿ # ♂
04 Wednesday	
00 42	☽ ⊼ ♄
00 47	☽ ± ♇
02 04	☽ ∠ ♀
05 34	☽ ⚹ ☉
06 57	☽ ⚹ ☿
10 31	♀ ⚹ ♃
14 24	☉ ☌ ☿
17 02	☽ ⚼ ♇
17 51	☽ ⊥ ♀
21 23	☽ ⊻ ♂
22 39	☉ ∥ ☿
05 Thursday	
00 19	☽ ∥ ♄
02 45	☽ # ♆
04 28	☽ ⊻ ♀
05 08	☽ ⊻ ♃
06 40	☽ ∥ ♆
07 45	☽ ∠ ☿
08 45	☽ ⊥ ♂
09 05	☽ # ♄
11 03	☽ ∠ ☉
16 14	☽ ⊥ ♃
18 18	☽ ☍ ♆
20 28	☽ △ ♇
22 49	☽ ⊥ ☿
22 55	☽ △ ♅
06 Friday	
01 57	☽ ∠ ♂
05 01	☽ ⊥ ☉
07 12	☽ ☍ ♄
08 58	☽ ⊻ ☿
09 00	☽ ∠ ♃
09 09	☉ # ♂
17 07	☽ ⊻ ☉
07 Saturday	
02 45	☽ ⚼ ♅
05 24	☿ ⊼ ♄
07 02	☽ ⚹ ♂
07 06	☽ ∥ ♀
07 45	☉ ± ♄
10 28	☿ # ♃
10 50	☽ ☌ ♀
13 20	☽ ⚹ ♃
13 41	☽ ∥ ☿
14 04	☽ # ♃
19 40	☽ ± ♅
23 24	☽ # ♂
08 Sunday	
02 22	☽ ⊼ ♆
02 59	☽ ∥ ☉
03 04	☿ ± ♆
04 44	☽ □ ♇
06 39	☽ Q ♂
07 02	☽ ⊼ ♅
12 20	☽ Q ♃
13 02	☽ ☌ ☿
13 48	☽ ± ♆
15 32	☽ ⊼ ♄
21 31	☿ Q ♃
09 Monday	
02 29	☽ # ♅
03 01	☽ ± ♄
07 02	☽ ☌ ☉
07 06	☽ ⚼ ♆
07 49	☉ ⚼ ♆
15 25	☽ Q ♇
18 43	☽ □ ♂
19 20	☽ ⊻ ♀
20 25	☽ ⚼ ♄
20 43	☽ ∥ ♇
23 25	☽ □ ♃
10 Tuesday	
06 48	♀ ⚹ ♂
06 55	☽ ⊥ ♀
12 20	☽ △ ♆
14 56	☽ ⚹ ♇
17 04	☽ ☍ ♅
20 02	☽ ⊻ ☿
11 Wednesday	
00 28	☽ ∠ ♀
01 48	☽ △ ♄
05 45	☿ Q ♂
07 27	☽ ⊥ ☿
10 50	♂ ⚼ ♄
20 47	☽ ∠ ♇
23 29	☽ ⊻ ☉
12 Thursday	
00 53	☽ ∠ ☿
06 11	☽ ⚹ ♀
08 35	☽ △ ♂
11 29	☽ △ ♃
12 40	☽ ⊥ ☉
14 54	☽ ⊥ ♇
13 Friday	
00 13	☽ □ ♆
03 02	☽ ⊻ ♇
04 58	☽ ⊼ ♅
06 20	☽ Q ♀
06 40	☽ ⚹ ☿
08 33	☽ ∠ ☉
13 52	☽ □ ♄
15 22	☉ Q ♇
15 55	☿ St D
16 10	☽ ⚼ ♂
17 06	☽ ± ♅
18 06	☽ ⚼ ♃
14 Saturday	
00 28	♀ St D
07 07	☽ Q ☿
11 12	☽ ± ♂
11 21	☽ ⚼ ♅
12 33	☽ ± ♃
12 45	☽ Q ♆
17 54	☽ ⚹ ☉
18 56	☽ □ ♀
23 41	☽ ∥ ♇
23 54	☽ ⊼ ♂
15 Sunday	
00 51	☽ ⊼ ♃
02 21	☽ Q ♄
06 17	☉ ⊻ ♀
11 23	♃ ± ♆
13 06	☽ ⚹ ♆
16 04	☽ ☌ ♇
17 43	☽ △ ♅
18 49	☽ # ♅
20 28	☽ □ ☿
20 29	☉ ⚼ ♄
20 32	☽ Q ☉
16 Monday	
02 20	♂ ± ♆
02 33	☽ ⚹ ♄
06 24	♂ ☌ ♃
07 05	☽ ∥ ☉
19 16	☽ ∠ ♆
17 Tuesday	
04 50	☽ # ♂
07 44	☽ △ ♀
08 25	☽ ∠ ♄
09 48	☽ # ♃
11 48	☽ □ ☉
13 05	☽ ⊥ ♆
13 28	☽ ☍ ♃
14 26	☽ ☍ ♂
19 41	☽ ∥ ☿
18 Wednesday	
00 48	☽ ⊻ ♆
01 59	☽ ⊥ ♄
03 22	☽ ∥ ♀
03 48	☽ ⊻ ♇
05 06	☽ □ ♅
09 39	☉ □ ♃
10 46	☽ △ ☿
13 19	☽ ⚼ ♀
13 32	☽ ⊻ ♄
15 22	☽ ⊥ ♇
19 Thursday	
06 38	☽ ± ♀
08 24	☽ ∠ ♇
11 24	☿ ± ♆
15 06	☽ Q ♅
15 57	☽ # ♄
17 00	☽ ⚼ ☿
17 09	☽ ∥ ♆
17 49	☉ □ ♂
17 59	☽ ⊼ ♀
21 48	☽ # ♆
22 23	☿ Q ♃
22 54	☽ ⊼ ♃
22 58	☽ ∥ ♄
20 Friday	
01 24	☽ ⊼ ♂
01 46	☽ △ ☉
02 16	☿ ⊼ ♄
09 03	☽ ☌ ♆
09 51	☽ ± ♃
10 39	☽ ± ☿
11 56	☽ ⚹ ♇
12 38	☽ ± ♂
12 54	☽ ⚹ ♅
20 44	☽ ☌ ♄
22 12	☽ ⊼ ☿
21 Saturday	
01 58	☽ ⚼ ♃
05 06	☽ ⚼ ♂
06 43	☽ ⚼ ☉
08 11	☽ # ♀
09 11	☽ Q ♇
15 12	☽ ∠ ♅
21 16	☽ # ☿
23 27	☽ ± ☉
22 Sunday	
00 10	☽ ☍ ♀
00 40	☽ ∥ ♃
02 46	☽ ∥ ♂
03 58	☽ △ ♃
04 26	☿ Q ♂
06 30	☽ ⊥ ♅
07 39	☽ △ ♂
10 22	☽ ⊼ ☉
13 04	☽ ⊻ ♆
13 29	☉ ⊥ ♀
15 51	☽ □ ♇
16 32	☽ ⊻ ♅
23 01	☽ ⊥ ♆
23 51	☽ ⊻ ♄
23 Monday	
05 27	☽ ☍ ☿
07 27	☽ # ☉
09 29	☽ ∥ ♅
09 40	☽ ⊥ ♄
13 45	☽ ∠ ♆
20 20	☿ # ♃
24 Tuesday	
00 08	☉ △ ♆
00 14	☽ ∠ ♄
00 59	☽ # ♇
02 48	☽ ⊼ ♀
05 29	☽ □ ♃
10 09	☽ □ ♂
10 42	☿ # ♂
12 40	☽ ± ♀
13 53	☽ ⚹ ♆
14 53	☽ ☍ ☉
16 39	☽ △ ♇
17 07	☽ ☌ ♅
25 Wednesday	
00 10	☽ ⚹ ♄
03 23	☽ ⚼ ♀
05 21	☉ # ♅
06 40	☿ ± ♄
08 58	☽ Q ♆
09 58	☽ ⊼ ☿
10 17	☽ Q ♃
15 35	☽ Q ♂
16 31	☽ ⚼ ♇
17 27	☉ ⚹ ♇
19 11	☽ Q ♄
20 23	☽ ± ☿
22 41	☉ ☍ ♅
26 Thursday	
03 51	☽ △ ♀
05 23	☽ ⚹ ♃
06 49	☽ ± ♇
11 09	☽ ⚹ ♂
12 03	☽ ⚼ ☿
13 33	☽ □ ♆
16 23	☽ ⊼ ♇
16 40	☽ ⊻ ♅
18 01	☽ ⊼ ☉
23 48	☽ □ ♄
27 Friday	
02 06	♂ ∥ ♃
02 14	☽ ⊥ ♅
04 19	☽ ± ☉
04 38	☿ ⚼ ♆
05 26	☽ ∠ ♃
11 51	☽ ∠ ♂
14 30	☽ △ ☿
16 42	☽ ∠ ♅
19 55	☽ ⚼ ☉
20 02	☽ ⊥ ♃
28 Saturday	
02 52	☽ ⊥ ♂
03 15	☽ # ♇
05 40	☽ □ ♀
05 52	☽ ⊻ ♃
12 05	♀ ⚹ ♃
13 00	☽ ⊻ ♂
14 06	☽ △ ♆
15 34	☽ # ☉
17 09	☽ ☍ ♇
17 14	☽ ⚹ ♅
19 01	☽ ∥ ♅
22 28	☽ △ ☉
29 Sunday	
00 44	☽ △ ♄
11 21	♅ △ ♇
12 46	☽ Q ♀
13 18	☽ Q ♅
15 14	☽ ⚼ ♆
16 31	☽ # ☿
21 44	☽ □ ☿
30 Monday	
02 09	☽ ⚼ ♄
02 12	♀ ± ♅
04 29	☽ ∥ ♃
05 48	☽ ∥ ♂
06 06	☉ △ ♄
06 20	♂ ⊼ ♆
06 40	☽ ± ♆
08 35	☽ ☌ ♃
10 10	☽ ⚹ ♀
15 49	☿ Q ♇
17 06	☽ ⊼ ♆
17 25	☽ ☌ ♂
17 45	☽ ± ♄
20 17	☽ □ ♅
20 26	☽ ⊼ ♇
20 29	☽ # ♀

DECEMBER 2026

LONGITUDES

Date	Sidereal time h m s	Sun ☉ ° ' "	Moon ☽ ° ' "	Moon ☽ 24.00 ° ' "	Mercury ☿ ° '	Venus ♀ ° '	Mars ♂ ° '	Jupiter ♃ ° '	Saturn ♄ ° '	Uranus ♅ ° '	Neptune ♆ ° '	Pluto ♇ ° '
01	16 41 26	09 ♐ 17 55	12 ♍ 20 20	19 ♍ 00 49	22 ♏ 46	28 ♎ 17	02 ♍ 07	26 ♌ 48	08 ♈ 01	03 ♊ 27	01 ♈ 39	03 ♒ 34
02	16 45 23	10 18 44	25 ♍ 36 33	02 ♎ 07 55	24 14	28 52	02 29	26 51	08 R 00	03 R 24	01 R 39	03 35
03	16 49 19	11 19 34	08 ♎ 35 17	14 ♎ 58 59	25 43	29 ♎ 28	02 51	26 53	07 59	03 22	01 38	03 37
04	16 53 16	12 20 26	21 ♎ 19 20	27 ♎ 36 39	27 13	00 ♏ 06	03 12	26 54	07 58	03 19	01 38	03 38
05	16 57 13	13 21 20	03 ♏ 51 11	10 ♏ 03 10	28 ♏ 43	00 45	03 33	26 56	07 58	03 17	01 38	03 39
06	17 01 09	14 22 14	16 ♏ 12 48	22 ♏ 20 15	00 ♐ 13	01 25	03 53	26 57	07 57	03 14	01 37	03 41
07	17 05 06	15 23 10	28 ♏ 25 40	04 ♐ 29 10	01 44	02 07	04 13	26 58	07 57	03 12	01 37	03 42
08	17 09 02	16 24 07	10 ♐ 30 54	16 ♐ 30 57	03 15	02 49	04 33	26 59	07 56	03 09	01 37	03 43
09	17 12 59	17 25 05	22 ♐ 29 30	28 ♐ 26 39	04 47	03 33	04 52	27 00	07 56	03 07	01 37	03 45
10	17 16 55	18 26 04	04 ♑ 22 37	10 ♑ 17 36	06 18	04 18	05 10	27 01	07 56	03 04	01 37	03 46
11	17 20 52	19 27 04	16 ♑ 11 50	22 ♑ 05 38	07 50	05 03	05 29	27 01	07 D 56	03 02	01 37	03 48
12	17 24 48	20 28 04	27 ♑ 59 20	03 ♒ 53 19	09 23	05 50	05 47	27 01	07 56	03 00	01 37	03 49
13	17 28 45	21 29 06	09 ♒ 48 01	15 ♒ 43 55	10 55	06 38	06 04	27 R 01	07 56	02 57	01 D 37	03 51
14	17 32 42	22 30 07	21 ♒ 41 33	27 ♒ 41 30	12 27	07 26	06 21	27 01	07 57	02 55	01 37	03 52
15	17 36 38	23 31 09	03 ♓ 44 21	09 ♓ 50 45	14 00	08 15	06 37	27 01	07 57	02 53	01 37	03 54
16	17 40 35	24 32 12	16 ♓ 01 20	22 ♓ 16 44	15 33	09 06	06 53	27 00	07 58	02 50	01 37	03 55
17	17 44 31	25 33 15	28 ♓ 37 36	05 ♈ 04 32	17 06	09 57	07 09	27 00	07 58	02 48	01 37	03 57
18	17 48 28	26 34 18	11 ♈ 38 04	18 ♈ 18 40	18 39	10 48	07 24	26 59	07 59	02 46	01 37	03 58
19	17 52 24	27 35 22	25 ♈ 06 39	02 ♉ 02 13	20 12	11 41	07 38	26 57	08 00	02 44	01 38	04 00
20	17 56 21	28 36 26	09 ♉ 05 23	16 ♉ 16 00	21 45	12 34	07 52	26 56	08 01	02 41	01 38	04 02
21	18 00 17	29 ♐ 37 31	23 ♉ 33 38	00 ♊ 57 40	23 19	13 28	08 06	26 54	08 02	02 39	01 38	04 03
22	18 04 14	00 ♑ 38 36	08 ♊ 27 14	16 ♊ 01 16	24 53	14 22	08 19	26 53	08 03	02 37	01 38	04 05
23	18 08 11	01 39 41	23 ♊ 38 32	01 ♋ 17 41	26 27	15 18	08 31	26 51	08 04	02 35	01 39	04 06
24	18 12 07	02 40 47	08 ♋ 57 17	16 ♋ 35 57	28 01	16 13	08 43	26 49	08 06	02 33	01 39	04 08
25	18 16 04	03 41 53	24 ♋ 12 19	01 ♌ 45 10	29 ♐ 35	17 10	08 54	26 46	08 07	02 31	01 40	04 10
26	18 20 00	04 42 59	09 ♌ 13 29	16 ♌ 36 23	01 ♑ 09	18 07	09 05	26 44	08 09	02 29	01 40	04 12
27	18 23 57	05 44 06	23 ♌ 53 16	01 ♍ 03 40	02 44	19 04	09 15	26 41	08 11	02 27	01 40	04 13
28	18 27 53	06 45 14	08 ♍ 07 23	15 ♍ 04 21	04 19	20 02	09 24	26 38	08 13	02 25	01 41	04 15
29	18 31 50	07 46 22	21 ♍ 54 39	28 ♍ 38 31	05 55	21 01	09 33	26 35	08 15	02 23	01 42	04 17
30	18 35 46	08 47 30	05 ♎ 16 16	11 ♎ 48 17	07 30	22 00	09 41	26 32	08 17	02 21	01 42	04 18
31	18 39 43	09 ♑ 48 39	18 ♎ 15 00	24 ♎ 36 52	09 ♑ 06	22 ♏ 59	09 ♍ 49	26 ♌ 28	08 ♈ 19	02 ♊ 19	01 ♈ 43	04 ♒ 20

Moon Nodes and Latitude

Date	Moon True ☊ ° '	Moon Mean ☊ ° '	Moon ☽ Latitude ° '
01	23 ♒ 51	24 ♒ 27	01 S 36
02	23 R 50	24 24	02 40
03	23 47	24 21	03 34
04	23 42	24 18	04 15
05	23 33	24 15	04 44
06	23 22	24 11	04 59
07	23 09	24 08	05 00
08	22 55	24 05	04 48
09	22 43	24 02	04 23
10	22 32	23 59	03 46
11	22 23	23 56	03 00
12	22 17	23 52	02 06
13	22 14	23 49	01 06
14	22 13	23 46	00 S 03
15	22 D 13	23 43	01 N 01
16	22 14	23 40	02 04
17	22 R 14	23 37	03 02
18	22 13	23 33	03 53
19	22 10	23 30	04 33
20	22 04	23 27	04 59
21	21 57	23 24	05 06
22	21 49	23 21	04 54
23	21 41	23 17	04 21
24	21 34	23 14	03 28
25	21 29	23 11	02 21
26	21 26	23 08	01 N 05
27	21 D 25	23 05	00 S 13
28	21 26	23 02	01 29
29	21 28	22 58	02 37
30	21 29	22 55	03 35
31	21 ♒ 28	22 ♒ 52	04 S 20

DECLINATIONS

Date	Sun ☉ ° '	Moon ☽ ° '	Mercury ☿ ° '	Venus ♀ ° '	Mars ♂ ° '	Jupiter ♃ ° '	Saturn ♄ ° '	Uranus ♅ ° '	Neptune ♆ ° '	Pluto ♇ ° '
01	21 S 51	05 N 27	17 S 01	09 S 27	12 N 53	13 N 20	00 N 48	20 N 42	00 S 38	23 S 29
02	22 00	00 S 42	17 30	09 32	12 47	13 19	00 48	20 41	00 38	23 29
03	22 08	06 40	17 58	09 38	12 41	13 19	00 48	20 41	00 38	23 29
04	22 16	12 16	18 26	09 45	12 35	13 19	00 48	20 40	00 38	23 28
05	22 24	17 15	18 53	09 52	12 29	13 18	00 48	20 40	00 38	23 28
06	22 31	21 28	19 19	10 00	12 23	13 18	00 48	20 40	00 38	23 28
07	22 38	24 41	19 45	10 08	12 17	13 18	00 48	20 39	00 38	23 27
08	22 45	26 46	20 10	10 17	12 12	13 18	00 48	20 39	00 38	23 27
09	22 51	27 36	20 34	10 27	12 06	13 18	00 48	20 38	00 38	23 27
10	22 56	27 08	20 57	10 37	12 01	13 18	00 48	20 38	00 38	23 26
11	23 01	25 26	21 20	10 47	11 56	13 18	00 49	20 37	00 38	23 26
12	23 06	22 37	21 41	10 58	11 51	13 18	00 49	20 37	00 38	23 26
13	23 10	18 51	22 02	11 10	11 46	13 18	00 49	20 36	00 38	23 25
14	23 14	14 19	22 21	11 21	11 41	13 19	00 50	20 36	00 38	23 25
15	23 17	09 11	22 39	11 34	11 37	13 19	00 50	20 35	00 38	23 25
16	23 19	03 S 36	22 57	11 46	11 33	13 19	00 50	20 35	00 38	23 24
17	23 22	02 N 15	23 13	11 59	11 29	13 20	00 51	20 35	00 38	23 24
18	23 24	08 11	23 28	12 12	11 25	13 21	00 51	20 34	00 38	23 23
19	23 25	13 57	23 42	12 25	11 21	13 21	00 52	20 34	00 38	23 23
20	23 26	19 14	23 54	12 38	11 18	13 22	00 53	20 33	00 37	23 23
21	23 26	23 36	24 06	12 52	11 14	13 23	00 53	20 33	00 37	23 22
22	23 26	26 33	24 16	13 06	11 11	13 23	00 54	20 33	00 37	23 22
23	23 26	27 37	24 25	13 20	11 09	13 24	00 55	20 32	00 37	23 22
24	23 25	26 36	24 33	13 34	11 06	13 25	00 56	20 32	00 37	23 21
25	23 23	23 36	24 39	13 48	11 04	13 26	00 56	20 31	00 37	23 21
26	23 21	19 00	24 44	14 03	11 01	13 27	00 57	20 31	00 36	23 20
27	23 19	13 21	24 48	14 17	10 59	13 29	00 58	20 31	00 36	23 20
28	23 16	07 09	24 50	14 32	10 58	13 30	00 59	20 30	00 36	23 20
29	23 13	00 N 48	24 51	14 46	10 56	13 31	01 00	20 30	00 36	23 19
30	23 09	05 S 23	24 50	15 00	10 55	13 32	01 01	20 29	00 35	23 19
31	23 S 04	11 S 09	24 S 48	15 S 15	10 N 54	13 N 34	01 N 02	20 N 29	00 S 35	23 S 18

ZODIAC SIGN ENTRIES

Date	h	m	Planets
02	20	04	☽ ♎
04	08	13	♀ ♏
05	04	35	☽ ♏
06	08	33	☿ ♐
07	15	06	☽ ♐
10	03	09	☽ ♑
12	16	06	☽ ♒
15	04	36	☽ ♓
17	14	34	☽ ♈
19	20	30	☽ ♉
21	20	50	☉ ♑
21	22	27	☽ ♊
23	21	58	☽ ♋
25	18	22	☿ ♑
25	21	12	☽ ♌
27	22	13	☽ ♍
30	02	27	☽ ♎

LATITUDES

Date	Mercury ☿ ° '	Venus ♀ ° '	Mars ♂ ° '	Jupiter ♃ ° '	Saturn ♄ ° '	Uranus ♅ ° '	Neptune ♆ ° '	Pluto ♇ ° '
01	01 N 29	01 N 31	02 N 20	00 N 48	02 S 35	00 S 09	01 S 24	04 S 15
04	01 08	01 53	02 25	00 49	02 34	00 09	01 24	04 15
07	00 47	02 12	02 30	00 50	02 34	00 09	01 24	04 15
10	00 25	02 29	02 35	00 50	02 33	00 09	01 24	04 15
13	00 N 03	02 43	02 40	00 51	02 32	00 09	01 23	04 15
16	00 S 17	02 55	02 46	00 52	02 31	00 09	01 23	04 15
19	00 37	03 04	02 51	00 53	02 31	00 09	01 23	04 15
22	00 56	03 12	02 57	00 53	02 30	00 09	01 23	04 15
25	01 13	03 17	03 03	00 54	02 29	00 09	01 23	04 15
28	01 28	03 21	03 09	00 55	02 28	00 08	01 23	04 15
31	01 S 41	03 N 23	03 N 15	00 N 56	02 S 27	00 S 08	01 S 23	04 S 15

DATA

Julian Date	2461376
Delta T	+73 seconds
Ayanamsa	24° 14' 08"
Synetic vernal point	04° ♓ 52' 51"
True obliquity of ecliptic	23° 26' 16"

MOON'S PHASES, APSIDES AND POSITIONS ☽

Date	h	m	Phase	Longitude ° '	Eclipse Indicator
01	06	09	☾	09 ♍ 03	
09	00	52	●	16 ♐ 57	
17	05	43	☽	25 ♓ 17	
24	01	28	○	02 ♋ 14	
30	18	59	☾	09 ♎ 05	

Day	h	m		
11	06	57	Apogee	
24	08	38	Perigee	
02	09	15	0S	
09	15	19	Max dec	27° S 36'
17	02	53	0N	
23	12	25	Max dec	27° N 37'
29	15	02	0S	

LONGITUDES

Date	Chiron ⚷ ° '	Ceres ⚳ ° '	Pallas ⚴ ° '	Juno ⚵ ° '	Vesta ⚶ ° '	Black Moon Lilith ⚸ ° '
01	26 ♈ 51	23 ♋ 39	03 ♈ 18	11 ♒ 28	12 ♈ 43	08 ♑ 41
11	26 ♈ 35	22 ♋ 39	03 ♈ 42	15 ♒ 14	13 ♈ 17	09 ♑ 48
21	26 ♈ 23	21 ♋ 01	04 ♈ 49	19 ♒ 16	14 ♈ 30	10 ♑ 54
31	26 ♈ 17	18 ♋ 54	06 ♈ 34	23 ♒ 30	16 ♈ 18	12 ♑ 01

ASPECTARIAN

01 Tuesday
04 19 ☽ ⊼ ♄
05 26 ♂ ± ♄
06 09 ☽ □ ☉
07 03 ☽ ± ♇
08 52 ☽ Q ☿
13 46 ☽ ∠ ♀
15 55 ☿ ⚼ ♄
23 13 ☽ ⚼ ♇

02 Wednesday
06 07 ☽ ∥ ♄
06 45 ☽ ⊥ ♀
06 47 ☽ # ♆
09 11 ☽ ⚹ ☿
11 43 ☽ ∥ ♆
12 24 ☽ # ♄
14 16 ☽ ⊻ ♃
17 22 ☽ Q ☉
18 16 ☽ ⊻ ♀
23 06 ☽ ☍ ♆

03 Thursday
01 01 ☽ ⊻ ♂
01 21 ☽ ⊥ ♃
02 18 ☽ △ ♅
02 43 ☽ △ ♇
10 52 ☽ ☍ ♄
12 29 ☽ ⊥ ♂
16 30 ☽ ∠ ☿
17 34 ☽ ⚹ ☉
18 10 ☽ ∠ ♃

04 Friday
00 41 ☽ ∥ ♀
05 54 ☽ ∠ ♂
06 19 ☽ ⚼ ♅
07 03 ☿ □ ♃
11 45 ☽ ⊥ ☿
13 25 ☽ # ♂
16 47 ☽ # ♃
19 28 ♂ □ ♅
22 40 ☽ ⚹ ♃
23 24 ☽ ± ♅

05 Saturday
00 30 ☽ ∠ ☉
00 45 ☽ ⊻ ☿
05 42 ☽ ☌ ♀
07 43 ☽ ⊼ ♆
10 54 ☽ ⊼ ♅
11 23 ☽ ⚹ ♂
11 37 ☽ □ ♇
19 18 ☽ ± ♆
19 22 ☽ ⊥ ☉
19 56 ☽ ⊼ ♄
20 24 ♂ ⊼ ♇
21 41 ☽ ∥ ☿
21 51 ☽ Q ♃

06 Sunday
07 01 ☽ # ♅
07 35 ☽ ± ♄
08 05 ☽ ⊻ ☉
11 20 ☽ Q ♂
12 48 ☽ ⚼ ♆
19 05 ♀ ⊼ ♆
19 21 ☽ ∥ ☉
22 43 ☽ Q ♇

07 Monday
01 12 ☽ ⚼ ♄
01 55 ☽ ∥ ♇
09 08 ☽ □ ♃
10 15 ☿ △ ♆
18 19 ☽ △ ♆
19 28 ☽ ☌ ☿
19 44 ☽ ⊻ ♀
21 24 ☽ ☍ ♅
22 28 ☽ ⚹ ♇
23 12 ☿ ⊻ ♀
23 47 ☽ □ ♂

08 Tuesday
06 52 ☽ △ ♄
08 25 ☽ ⊥ ♀
10 29 ☿ ☍ ♅
19 32 ☿ ⚹ ♇
22 27 ♀ ⊼ ♅

09 Wednesday
00 52 ☽ ☌ ☉
03 34 ☽ ∠ ♀
04 28 ☽ ∠ ♇
13 40 ☿ □ ♂
16 13 ☿ # ♅
18 38 ♀ □ ♇
21 06 ☽ △ ♃
22 37 ☽ ⊥ ♇

10 Thursday
06 25 ☽ □ ♆
09 22 ☽ ⊼ ♅
10 46 ☽ ⊻ ♇
11 49 ☽ ⚹ ♀
13 40 ☽ △ ♂
16 30 ☽ ⊻ ☿
19 12 ☽ □ ♄
20 08 ☉ ∠ ♇
21 29 ☽ ± ♅
23 32 ♄ St D

11 Friday
03 30 ☽ ⚼ ♃
06 29 ☽ ⊥ ☿
13 25 ☿ △ ♄
13 52 ☽ Q ♀
15 43 ☽ ⚼ ♅
18 57 ☽ Q ♆
19 15 ☽ ⊻ ☉
20 57 ☽ ⚼ ♂
21 49 ☽ ± ♃

12 Saturday
03 33 ☽ ∠ ☿
05 53 ☽ ∥ ♇
07 49 ☽ Q ♄
08 31 ☽ ∥ ☉
08 37 ☽ ⊥ ☉
09 15 ♀ ⚹ ♂
10 02 ☽ ⊼ ♃
15 44 ☽ ± ♂
17 54 ☽ ∥ ☿
19 22 ☽ ⚹ ♆
22 09 ☽ △ ♅
22 19 ♆ St D
23 53 ☽ ☌ ♇

13 Sunday
00 57 ♃ St R
01 28 ☽ # ♅
04 14 ☽ ⊼ ♂
04 38 ☽ ∠ ☉
05 06 ☽ □ ♀
08 13 ☽ ⚹ ♄
14 36 ☽ ⚹ ☿

14 Monday
01 47 ☽ ∠ ♆
04 11 ☉ ∠ ♀
13 46 ☽ ⚹ ☉
14 30 ☽ ∠ ♄
16 54 ☽ # ♃
17 14 ♀ ± ♆
18 22 ☽ Q ☿
19 51 ☽ ⊥ ♆
22 39 ☽ ☍ ♃

15 Tuesday
00 46 ☽ # ♂
01 36 ☽ ∥ ♀
02 58 ♀ ⊼ ♄
07 48 ☽ ⊻ ♆
08 27 ☽ ⊥ ♄
10 18 ☽ □ ♅
12 18 ☽ ⊻ ♇
15 50 ☽ Q ☉
16 56 ♀ # ♂
17 49 ☽ ☍ ♂
20 18 ☽ ⊻ ♄
20 38 ☿ ⊥ ♀
21 32 ☽ △ ♀

16 Wednesday
00 07 ☽ ⊥ ♇
09 29 ♀ Q ♃
10 57 ☽ □ ☿
17 35 ☽ ∠ ♇
21 13 ☽ Q ♅
23 26 ☽ # ♄

17 Thursday
00 18 ☽ ∥ ♆
04 33 ☽ ⚼ ♀
05 27 ☽ # ♆
05 43 ☽ □ ☉
06 20 ☽ ∥ ♄
08 56 ☽ ⊼ ♃
17 36 ☽ ☌ ♆
19 46 ☽ ⚹ ♅
20 08 ☽ ± ♃
21 56 ☽ ⚹ ♇
22 37 ☽ ± ♀

18 Friday
04 08 ☽ ⊼ ♂
04 10 ☉ ∥ ☿
05 01 ☿ ∥ ♇
05 20 ☽ ☌ ♄
10 23 ☽ ⊼ ♀
10 24 ☉ ∥ ♇
12 37 ☽ ⚼ ♃
15 15 ☽ ± ♂
17 07 ☿ ∠ ♇
19 50 ☽ Q ♇
21 21 ☉ △ ♃
23 00 ☽ ∠ ♅

19 Saturday
01 09 ☽ ∥ ♂
02 14 ☽ △ ☿
05 13 ☽ # ♀
07 35 ☽ ⚼ ♂
09 26 ☽ ∥ ♃
14 49 ☽ ⊥ ♅
15 13 ☽ △ ♃
16 40 ☽ △ ☉
21 55 ☉ ⊥ ♇
23 18 ☽ ⊻ ♆

20 Sunday
01 09 ☽ ⊻ ♅
03 23 ☽ □ ♇
07 34 ☽ ⚼ ☿
09 32 ☽ ⊥ ♆
09 55 ☽ △ ♂
10 11 ☽ ⊻ ♄
18 14 ☽ ☍ ♀
18 37 ☽ ∥ ♅
20 09 ☽ ⚼ ☉
20 16 ☽ ⊥ ♄

21 Monday
00 30 ☽ ± ☿
00 36 ☽ ∠ ♆
04 13 ♂ ⊼ ♄
10 32 ☽ # ♇
10 57 ☽ # ☉
11 08 ☽ ∠ ♄
11 33 ☽ ⊼ ☿
12 07 ☽ ± ☉
15 26 ☽ # ☿
17 26 ☽ □ ♃
22 34 ☽ ⊼ ☉

22 Tuesday
01 05 ☽ ⚹ ♆
02 41 ☽ ☌ ♅
03 21 ♀ ± ♄
05 00 ☽ △ ♇
11 22 ☽ ⚹ ♄
11 46 ☽ □ ♂
20 14 ☽ Q ♆
22 00 ☽ ⊼ ♀
22 10 ☽ Q ♃

23 Wednesday
04 51 ☽ ⚼ ♇
06 23 ☽ Q ♄
08 04 ☽ ± ♀
11 37 ☉ □ ♆
16 34 ☽ Q ♂
16 54 ☽ ☍ ☿
17 01 ☽ ⚹ ♃
18 03 ☿ △ ♃
19 01 ☽ ± ♇
19 57 ♀ # ♃
23 06 ☽ ⚼ ♀

24 Thursday
00 33 ☽ □ ♆
01 28 ☽ ☍ ☉
01 59 ☽ ⊻ ♅
04 26 ☽ ⊼ ♇
08 58 ☉ ⊼ ♅
10 39 ☽ □ ♄
11 22 ☽ ⊥ ♅
11 37 ☽ ⚹ ♂
13 57 ☿ ⊥ ♇
16 28 ☽ ∠ ♃
23 04 ♀ ⚼ ♆

25 Friday
00 09 ☽ △ ♀
01 28 ☽ ∠ ♅
05 14 ☽ # ☿
06 35 ☽ ⊥ ♃
11 31 ☽ ∠ ♂
13 16 ☽ # ☉
13 30 ☽ # ♇
16 03 ☽ ⊻ ♃
21 32 ☽ ⊼ ☿
23 18 ☉ ⊻ ♇
23 51 ☽ △ ♆

26 Saturday
01 11 ☽ ⚹ ♅
02 00 ☽ ⊥ ♂
03 53 ☽ ☍ ♇
04 13 ☽ ⊼ ☉
04 46 ☽ ∥ ♅
08 16 ☽ ± ☿
10 16 ☽ △ ♄
11 46 ☽ ⊻ ♂
14 35 ☽ ± ☉
19 47 ☿ □ ♆
20 30 ☽ Q ♅
22 53 ☉ ∥ ♇

27 Sunday
00 06 ☽ ⚼ ♆
00 37 ☽ ⚼ ☿
03 29 ☽ □ ♀
06 23 ☽ ⚼ ☉
07 40 ☿ ⊼ ♅
08 22 ☽ # ♀
10 49 ☽ ⚼ ♄
11 29 ☽ ∥ ♃
14 58 ☽ ± ♆
16 38 ☽ ☌ ♃
21 19 ☽ ∥ ♂

28 Monday
01 03 ☽ ⊼ ♆
01 55 ☽ ± ♄
02 18 ☽ □ ♅
04 42 ☽ △ ☿
05 23 ☽ ⊼ ♇
09 29 ☽ △ ☉
10 54 ☿ ⊻ ♇
11 51 ☽ Q ♀
12 09 ☽ ⊼ ♄
14 13 ☽ ☌ ♂
15 40 ☽ ± ♇

29 Tuesday
07 21 ☽ ⚼ ♇
10 18 ☽ ⚹ ♀
11 13 ☽ ∥ ♄
12 46 ☽ # ♆
16 08 ☿ ∠ ♀
17 18 ☽ ∥ ♆
18 53 ☽ # ♄
20 17 ☽ ⊻ ♃
23 28 ☉ □ ♄

30 Wednesday
01 53 ☉ ± ♅
03 14 ♀ Q ♂
05 31 ☽ ☍ ♆
06 42 ☽ △ ♅
07 02 ☽ ⊥ ♃
10 14 ☽ △ ♇
15 25 ☽ ∠ ♀
16 39 ☽ □ ☿
17 31 ☽ ☍ ♄
18 59 ☽ □ ☉
19 43 ♀ Q ♇
20 10 ☽ ⊻ ♂
23 26 ☽ ∠ ♃
23 55 ☿ □ ♄

31 Thursday
00 28 ☿ ± ♅
07 24 ☽ ⊥ ♂
09 27 ☽ ⊥ ♀
10 16 ☽ ⚼ ♅
10 54 ☽ # ♂
12 03 ☉ △ ♂
20 03 ♀ ⚼ ♄
21 41 ☽ ⊻ ♀
22 51 ☽ # ♃
23 30 ☿ △ ♂

All ephemeris data is given at 12.00 UT and the Moon's longitude is additionally given for 24.00 UT

JANUARY 2027

LONGITUDES

Date	Sidereal time h m s	Sun ☉ ° ' "	Moon ☽ ° ' "	Moon ☽ 24.00 ° ' "	Mercury ☿ ° '	Venus ♀ ° '	Mars ♂ ° '	Jupiter ♃ ° '	Saturn ♄ ° '	Uranus ♅ ° '	Neptune ♆ ° '	Pluto ♇ ° '
01	18 43 40	10 ♑ 49 48	00 ♏ 54 22	07 ♏ 07 57	10 ♑ 42	23 ♏ 59	09 ♍ 56	26 ♌ 25	08 ♈ 21	02 ♊ 17	01 ♈ 43	04 ♒ 22
02	18 47 36	11 50 58	13 ♏ 18 05	19 ♏ 25 12	12 19	25 00	10 02	26 R 21	08 23	02 R 15	01 44	04 24
03	18 51 33	12 52 08	25 ♏ 29 41	01 ♐ 31 55	13 56	26 01	10 07	26 17	08 26	02 14	01 45	04 26
04	18 55 29	13 53 18	07 ♐ 32 15	13 ♐ 30 59	15 33	27 02	10 12	26 13	08 28	02 12	01 46	04 27
05	18 59 26	14 54 29	19 ♐ 28 25	25 ♐ 24 48	17 11	28 04	10 16	26 08	08 31	02 11	01 46	04 29
06	19 03 22	15 55 40	01 ♑ 20 22	07 ♑ 15 21	18 49	29 ♏ 06	10 20	26 04	08 34	02 09	01 47	04 31
07	19 07 19	16 56 50	13 ♑ 09 58	19 ♑ 04 26	20 27	00 ♐ 08	10 22	25 59	08 37	02 07	01 48	04 33
08	19 11 15	17 58 01	24 ♑ 58 59	00 ♒ 53 51	22 06	01 11	10 24	25 54	08 40	02 06	01 49	04 35
09	19 15 12	18 59 11	06 ♒ 49 15	12 ♒ 45 30	23 45	02 14	10 25	25 49	08 43	02 04	01 50	04 37
10	19 19 09	20 00 21	18 ♒ 42 51	24 ♒ 41 41	25 24	03 17	10 26	25 44	08 46	02 03	01 51	04 38
11	19 23 05	21 01 30	00 ♓ 42 19	06 ♓ 45 09	27 04	04 21	10 R 25	25 39	08 49	02 01	01 52	04 40
12	19 27 02	22 02 39	12 ♓ 50 37	18 ♓ 59 09	28 ♑ 44	05 25	10 24	25 33	08 53	02 00	01 53	04 42
13	19 30 58	23 03 47	25 ♓ 11 14	01 ♈ 27 21	00 ♒ 25	06 29	10 22	25 28	08 56	01 58	01 54	04 44
14	19 34 55	24 04 55	07 ♈ 48 00	14 ♈ 13 40	02 06	07 34	10 20	25 22	09 00	01 57	01 55	04 46
15	19 38 51	25 06 02	20 ♈ 44 49	27 ♈ 21 51	03 46	08 39	10 16	25 16	09 03	01 55	01 56	04 48
16	19 42 48	26 07 09	04 ♉ 05 08	10 ♉ 54 56	05 28	09 44	10 12	25 10	09 07	01 55	01 57	04 50
17	19 46 44	27 08 15	17 ♉ 51 26	24 ♉ 54 37	07 09	10 49	10 07	25 04	09 11	01 54	01 58	04 52
18	19 50 41	28 09 20	02 ♊ 04 22	09 ♊ 20 23	08 50	11 55	10 01	24 57	09 15	01 52	02 00	04 53
19	19 54 38	29 ♑ 10 24	16 ♊ 42 08	24 ♊ 08 56	10 31	13 01	09 54	24 51	09 19	01 51	02 01	04 55
20	19 58 34	00 ♒ 11 27	01 ♋ 39 55	09 ♋ 14 03	12 12	14 07	09 47	24 44	09 23	01 50	02 02	04 57
21	20 02 31	01 12 30	16 ♋ 50 09	24 ♋ 27 00	13 53	15 13	09 39	24 37	09 27	01 49	02 03	04 59
22	20 06 27	02 13 32	02 ♌ 03 19	09 ♌ 37 50	15 33	16 20	09 30	24 31	09 32	01 48	02 05	05 01
23	20 10 24	03 14 33	17 ♌ 09 22	24 ♌ 36 51	17 12	17 26	09 20	24 24	09 36	01 48	02 06	05 03
24	20 14 20	04 15 34	01 ♍ 59 22	09 ♍ 16 09	18 51	18 33	09 09	24 17	09 40	01 47	02 08	05 05
25	20 18 17	05 16 34	16 ♍ 26 39	23 ♍ 30 29	20 28	19 40	08 58	24 10	09 45	01 46	02 09	05 07
26	20 22 13	06 17 33	00 ♎ 27 26	07 ♎ 17 29	22 03	20 48	08 46	24 02	09 50	01 45	02 10	05 09
27	20 26 10	07 18 31	14 ♎ 00 43	20 ♎ 37 22	23 36	21 55	08 33	23 55	09 54	01 45	02 12	05 11
28	20 30 07	08 19 29	27 ♎ 07 46	03 ♏ 32 20	25 07	23 03	08 19	23 48	09 59	01 44	02 13	05 12
29	20 34 03	09 20 27	09 ♏ 51 32	16 ♏ 05 52	26 34	24 11	08 05	23 40	10 04	01 44	02 15	05 14
30	20 38 00	10 21 24	22 ♏ 15 55	28 ♏ 22 06	27 58	25 19	07 49	23 33	10 09	01 43	02 16	05 16
31	20 41 56	11 ♒ 22 20	04 ♐ 25 05	10 ♐ 25 23	29 ♒ 18	26 ♐ 27	07 ♍ 33	23 ♌ 25	10 ♈ 14	01 ♊ 43	02 ♈ 18	05 ♒ 18

DECLINATIONS

Date	Moon True ☊ ° '	Moon Mean ☊ ° '	Moon ☽ Latitude ° '	Sun ☉ ° '	Moon ☽ ° '	Mercury ☿ ° '	Venus ♀ ° '	Mars ♂ ° '	Jupiter ♃ ° '	Saturn ♄ ° '	Uranus ♅ ° '	Neptune ♆ ° '	Pluto ♇ ° '
01	21 ♒ 26	22 ♒ 49	04 S 51	23 S 00	16 S 19	24 S 45	15 S 29	10 N 54	13 N 35	01 N 03	20 N 29	00 S 35	23 S 18
02	21 R 22	22 46	05 07	22 55	20 42	24 40	15 44	10 53	13 37	01 05	20 28	00 34	23 18
03	21 16	22 43	05 09	22 49	24 08	24 34	15 58	10 53	13 38	01 06	20 28	00 34	23 17
04	21 10	22 39	04 58	22 43	26 27	24 26	16 12	10 53	13 40	01 07	20 28	00 34	23 17
05	21 02	22 36	04 33	22 36	27 33	24 17	16 26	10 53	13 42	01 08	20 27	00 33	23 16
06	20 55	22 33	03 57	22 29	27 23	24 06	16 40	10 54	13 43	01 10	20 27	00 33	23 16
07	20 49	22 30	03 11	22 22	25 57	23 54	16 53	10 55	13 45	01 11	20 27	00 33	23 16
08	20 45	22 27	02 16	22 14	23 22	23 40	17 07	10 56	13 47	01 12	20 27	00 32	23 15
09	20 42	22 23	01 15	22 06	19 47	23 24	17 20	10 58	13 49	01 14	20 26	00 32	23 15
10	20 41	22 20	00 S 11	21 57	15 23	23 08	17 33	11 00	13 51	01 15	20 26	00 31	23 14
11	20 D 41	22 17	00 N 55	21 48	10 22	22 49	17 46	11 02	13 53	01 17	20 26	00 31	23 14
12	20 43	22 14	01 59	21 38	04 S 55	22 29	17 59	11 04	13 55	01 18	20 25	00 31	23 13
13	20 45	22 11	02 58	21 28	00 N 49	22 07	18 11	11 07	13 57	01 20	20 25	00 30	23 13
14	20 46	22 08	03 51	21 18	06 38	21 44	18 23	11 10	13 59	01 21	20 25	00 30	23 13
15	20 47	22 04	04 33	21 07	12 19	21 19	18 35	11 13	14 02	01 23	20 25	00 29	23 12
16	20 R 47	22 01	05 02	20 55	17 36	20 53	18 47	11 16	14 04	01 25	20 25	00 29	23 12
17	20 46	21 58	05 15	20 44	22 11	20 25	18 58	11 20	14 06	01 26	20 24	00 28	23 11
18	20 43	21 55	05 09	20 32	25 37	19 56	19 08	11 24	14 08	01 28	20 24	00 28	23 11
19	20 40	21 52	04 43	20 19	27 28	19 26	19 19	11 29	14 11	01 30	20 24	00 27	23 11
20	20 37	21 49	03 58	20 06	27 24	18 54	19 29	11 33	14 13	01 32	20 24	00 27	23 10
21	20 35	21 45	02 56	19 53	25 17	18 20	19 39	11 38	14 16	01 34	20 24	00 26	23 10
22	20 33	21 42	01 40	19 40	21 20	17 46	19 48	11 43	14 18	01 36	20 23	00 25	23 09
23	20 33	21 39	00 N 19	19 26	15 59	17 10	19 57	11 49	14 21	01 37	20 23	00 25	23 09
24	20 D 33	21 36	01 S 03	19 12	09 47	16 33	20 05	11 54	14 23	01 39	20 23	00 24	23 09
25	20 34	21 33	02 19	18 57	03 N 13	15 55	20 13	12 00	14 26	01 41	20 23	00 24	23 08
26	20 35	21 29	03 24	18 42	03 S 18	15 16	20 21	12 07	14 28	01 43	20 23	00 23	23 08
27	20 36	21 26	04 15	18 27	09 26	14 37	20 28	12 14	14 31	01 45	20 23	00 22	23 07
28	20 37	21 23	04 51	18 11	14 58	13 57	20 35	12 20	14 33	01 47	20 23	00 22	23 07
29	20 R 37	21 20	05 12	17 55	19 41	13 17	20 41	12 27	14 36	01 50	20 22	00 21	23 07
30	20 37	21 17	05 17	17 39	23 26	12 37	20 47	12 34	14 39	01 52	20 22	00 21	23 06
31	20 ♒ 36	21 ♒ 14	05 S 08	17 S 22	26 S 04	11 S 58	20 S 52	12 N 41	14 N 41	01 N 54	20 N 22	00 S 20	23 S 06

ZODIAC SIGN ENTRIES

Date	h	m	Planets
01	10	16	☽ ♏
03	20	57	☽ ♐
06	09	17	☽ ♑
07	08	53	♀ ♐
08	22	11	☽ ♒
11	10	36	☽ ♓
13	06	06	☿ ♒
13	21	14	☽ ♈
16	04	44	☽ ♉
18	08	33	☽ ♊
20	07	30	☉ ♒
20	09	21	☽ ♋
22	08	45	☽ ♌
24	08	45	☽ ♍
26	11	12	☽ ♎
28	17	21	☽ ♏
31	03	14	☽ ♐

LATITUDES

Date	Mercury ☿ ° '	Venus ♀ ° '	Mars ♂ ° '	Jupiter ♃ ° '	Saturn ♄ ° '	Uranus ♅ ° '	Neptune ♆ ° '	Pluto ♇ ° '
01	01 S 45	03 N 23	03 N 17	00 N 56	02 S 27	00 S 08	01 S 23	04 S 15
04	01 55	03 23	03 24	00 57	02 27	00 08	01 23	04 16
07	02 02	03 22	03 30	00 57	02 26	00 08	01 22	04 16
10	02 06	03 19	03 36	00 58	02 25	00 08	01 22	04 16
13	02 07	03 15	03 43	00 59	02 25	00 08	01 22	04 16
16	02 03	03 10	03 49	00 59	02 24	00 08	01 22	04 16
19	01 54	03 04	03 55	01 00	02 23	00 08	01 22	04 16
22	01 40	02 57	04 01	01 01	02 23	00 08	01 22	04 17
25	01 19	02 49	04 06	01 01	02 22	00 08	01 22	04 17
28	00 51	02 41	04 11	01 02	02 21	00 08	01 22	04 17
31	00 S 16	02 N 31	04 N 16	01 N 02	02 S 20	00 S 08	01 S 22	04 S 17

DATA

Julian Date	2461407
Delta T	+73 seconds
Ayanamsa	24° 14' 14"
Synetic vernal point	04° ♓ 52' 45"
True obliquity of ecliptic	23° 26' 16"

MOON'S PHASES, APSIDES AND POSITIONS ☽

Date	h	m	Phase	Longitude ° '	Eclipse Indicator
07	20	24	●	17 ♑ 18	
15	20	34	☽	25 ♈ 28	
22	12	17	○	02 ♌ 14	
29	10	55	☾	09 ♏ 18	

Day	h	m		
07	08	26	Apogee	
21	21	54	Perigee	
05	20	39	Max dec	27° S 38'
13	08	37	0N	
19	23	09	Max dec	27° N 41'
25	23	46	0S	

LONGITUDES

Date	Chiron ⚷ ° '	Ceres ⚳ ° '	Pallas ⚴ ° '	Juno ⚵ ° '	Vesta ⚶ ° '	Black Moon Lilith ⚸ ° '
01	26 ♈ 17	18 ♋ 41	06 ♈ 47	23 ♒ 57	16 ♈ 31	12 ♑ 08
11	26 ♈ 17	16 ♋ 19	09 ♈ 09	28 ♒ 24	18 ♈ 50	13 ♑ 15
21	26 ♈ 22	14 ♋ 03	12 ♈ 00	03 ♓ 01	21 ♈ 34	14 ♑ 21
31	26 ♈ 34	12 ♋ 10	15 ♈ 17	07 ♓ 48	24 ♈ 37	15 ♑ 28

ASPECTARIAN

h m Aspects

01 Friday
00 29 ☽ ∠ ♂
03 12 ☽ ± ♅
03 27 ☽ ⚹ ♃
07 10 ☽ Q ☿
07 40 ☽ Q ☉
07 41 ☽ ∥ ♀
13 34 ☽ ⊼ ♆
14 39 ☽ ⊼ ♅
17 07 ☉ ☌ ☿
18 40 ☽ □ ♇
22 09 ☿ ⚼ ♃

02 Saturday
00 54 ☉ ⚼ ♃
01 09 ☽ ± ♆
02 24 ☽ Q ♃
02 24 ☽ ⊼ ♄
05 34 ☽ ⚹ ♂
08 55 ☽ ⚹ ☉
09 47 ☽ ⚹ ☿
10 37 ☽ # ♅
14 08 ☽ ± ♄
18 44 ☽ ⚼ ♆

03 Sunday
02 10 ☽ ∥ ☉
05 17 ☽ Q ♂
05 21 ☽ ∥ ♇
05 55 ☽ Q ♇
07 54 ☽ ⚼ ♄
13 07 ☽ ☌ ♀
13 33 ☽ □ ♃
15 32 ☽ ∥ ☿
17 09 ☽ ∠ ☉
17 55 ♀ □ ♃
19 52 ☽ ∠ ☿

04 Monday
00 27 ☽ △ ♆
01 21 ☽ ☍ ♅
05 49 ☽ ⚹ ♇
12 46 ☽ ⊥ ☉
13 53 ☽ △ ♄
16 40 ☽ ⊥ ☿
17 22 ☽ □ ♂

05 Tuesday
01 56 ☽ ⊻ ☉
06 38 ☽ ⊻ ☿
11 56 ☿ ⚼ ♅
12 02 ☽ ∠ ♇

06 Wednesday
01 23 ☽ △ ♃
06 16 ☽ ⊥ ♇
07 01 ☽ ⊻ ♀
12 55 ☽ □ ♆
13 38 ☽ ⊼ ♅
18 28 ☽ ⊻ ♇
20 21 ☽ ⊥ ♀

07 Thursday
01 46 ☽ ± ♅
02 27 ☿ Q ♆
02 43 ☽ □ ♄
05 32 ☿ ± ♃
06 18 ☽ △ ♂
07 36 ☽ ⚼ ♃
15 55 ☉ ⚼ ♅
16 23 ☽ ∠ ♀
20 01 ☽ ⚼ ♅
20 24 ☽ ☌ ☉

08 Friday
01 30 ☽ Q ♆
01 45 ☽ ± ♃
05 11 ☽ ☌ ☿
09 25 ☽ ∥ ☿
12 50 ☽ ∥ ♇
12 51 ☽ ⚼ ♂
13 51 ☽ ⊼ ♃
15 26 ☽ Q ♄
20 36 ☽ ∥ ☉

09 Saturday
01 48 ☽ ⚹ ♀
01 53 ☽ ⚹ ♆
02 23 ☽ △ ♅
02 45 ♀ △ ♆
07 08 ☽ ± ♂
07 31 ☽ ☌ ♇
08 00 ☽ # ♅
08 20 ♀ ☍ ♅
09 13 ♃ ± ♆
15 51 ☽ ⚹ ♄
19 17 ☽ ⊼ ♂

10 Sunday
01 10 ☽ ∥ ♀
02 28 ☿ ∥ ♇
04 26 ☽ Q ♀
06 06 ☉ ± ♃
08 13 ☉ Q ♆
08 14 ☽ ∠ ♆
12 21 ☿ ⚼ ♂
12 59 ♂ St R
14 50 ☽ ⊻ ☉
16 31 ☿ ⊼ ♃
19 34 ☽ # ♃
22 11 ☽ ∠ ♄

11 Monday
01 59 ☽ ☍ ♃
02 20 ☽ ⊥ ♆
03 35 ☽ ⊻ ☿
03 59 ☽ ⊥ ☉
08 20 ☿ Q ♄
09 00 ☽ # ♂
14 19 ☽ ⊻ ♆
14 37 ☽ □ ♅
16 14 ☽ ⊥ ♄
17 27 ☽ ⊥ ☿
19 26 ♀ ⚹ ♇
19 54 ☽ ⊻ ♇
19 57 ☽ □ ♀
23 32 ☽ ∠ ☉

12 Tuesday
04 10 ☽ ⊻ ♄
07 13 ☽ ☍ ♂
07 47 ☽ ⊥ ♇
14 02 ☽ ∠ ☿

13 Wednesday
01 26 ☽ ∠ ♇
01 56 ☽ Q ♅
03 07 ☽ # ♄
06 32 ☽ ∥ ♆
07 32 ☽ ⚹ ☉
10 42 ☽ # ♆
12 31 ☽ ⊼ ♃
14 08 ☽ ∥ ♄
23 33 ☽ ⚹ ☿
23 55 ☽ ± ♃

14 Thursday
00 52 ☽ ☌ ♆
00 58 ☽ ⚹ ♅
06 16 ☽ ⚹ ♇
08 29 ☽ Q ☉
09 29 ☿ ⚹ ♆
10 02 ☿ △ ♅
11 31 ☽ △ ♀
14 15 ☽ ☌ ♄
16 43 ☽ ⊼ ♂
16 46 ☽ ⚼ ♃

15 Friday
01 30 ☽ Q ☿
03 49 ☽ ± ♂
04 44 ☽ Q ♇
05 01 ☽ ∠ ♅
07 14 ☽ ∥ ♂
09 31 ♅ ⚹ ♆
15 29 ☉ ⊼ ♃
15 45 ☉ ⚼ ♂
17 45 ☽ ⚼ ♀
18 47 ☿ ± ♂
19 34 ☽ ∥ ♃
20 09 ☽ △ ♃
20 11 ☽ ⚼ ♂
20 34 ☽ □ ☉
21 24 ☽ ⊥ ♅
21 39 ♀ △ ♄

16 Saturday
02 49 ☿ ☌ ♇
08 09 ☽ ⊻ ♅
08 13 ☽ ⊻ ♆
08 34 ☉ ∥ ☿
11 19 ☽ ± ♀
13 19 ☽ □ ♇
14 47 ☽ □ ☿
17 56 ☽ # ♀
18 50 ☽ ⊥ ♆
20 54 ☽ ⊻ ♄
21 38 ♀ □ ♂
22 41 ☽ △ ♂
22 47 ☽ ⊼ ♀

17 Sunday
02 06 ☽ ∥ ♅
03 07 ☽ # ☿
04 11 ☽ # ☉
07 23 ☽ ⊥ ♄
10 29 ☽ ∠ ♆
13 01 ☿ # ♅
13 09 ☉ Q ♄
18 09 ☽ # ♇
22 49 ☽ ∠ ♄

18 Monday
00 10 ☽ □ ♃
04 57 ☽ △ ☉
11 40 ☽ ☌ ♅
11 52 ☽ ⚹ ♆
16 41 ☽ △ ♇
18 06 ☿ ⚹ ♄
23 54 ☽ ⚹ ♄

19 Tuesday
00 38 ☽ △ ☿
01 01 ☽ □ ♂
03 05 ☉ # ♅
03 47 ☿ ⊼ ♂
05 31 ☽ ☍ ♀
05 47 ☽ Q ♃
07 35 ☽ ⚼ ☉
07 38 ☽ Q ♆
15 51 ☿ ∥ ♀
17 13 ☽ ⚼ ♇
19 29 ☽ Q ♄
23 12 ☽ ± ☉

20 Wednesday
01 01 ☽ ⚹ ♃
04 00 ☽ ⚼ ☿
05 52 ☽ Q ♂
07 40 ☽ ± ♇
09 29 ☽ ⊼ ☉
12 17 ☽ ⊻ ♅
12 35 ☽ □ ♆
17 14 ☽ ⊼ ♇
20 06 ☽ ± ☿
21 47 ☽ ⊥ ♅

21 Thursday
00 18 ☽ □ ♄
00 42 ☽ ∠ ♃
00 45 ☽ ⚹ ♂
06 46 ☽ ⊼ ☿
09 15 ☽ ⊼ ♀
11 59 ☽ ∠ ♅
14 48 ☽ ⊥ ♃
19 27 ☽ ± ♀

22 Friday
00 11 ☽ ⊻ ♃
00 12 ☽ ∠ ♂
02 06 ☽ # ♇
02 19 ☉ △ ♅
03 18 ☉ ∥ ♀
08 30 ☉ ⚹ ♆
08 52 ♂ ⊼ ♄
10 01 ♃ ⚼ ♄
10 45 ☽ ⚼ ♀
11 37 ☽ ⚹ ♅
12 02 ☽ △ ♆
12 17 ☽ ☍ ☉
14 15 ☽ ⊥ ♂
16 39 ☽ ∥ ♅
16 42 ☽ ☍ ♇
19 14 ☽ # ♀
20 28 ☽ # ☉
23 40 ☽ ⊻ ♂
23 53 ☽ △ ♄

23 Saturday
06 33 ☽ # ☿
06 38 ☽ Q ♅
10 28 ☿ ∠ ♆
11 55 ☽ ⚼ ♆
12 05 ☽ ☍ ☿
12 29 ☽ △ ♀
13 49 ☉ ± ♂
18 35 ☽ ∥ ♃
22 33 ☿ ⚹ ♀
23 33 ☽ ☌ ♃

24 Sunday
00 02 ☽ ⚼ ♄
02 26 ☽ ± ♆
04 09 ☽ ∥ ♂
11 40 ☽ □ ♅
12 13 ☽ ⊼ ♆
14 46 ☽ ± ♄
16 00 ☽ ⊼ ☉
17 05 ☽ ⊼ ♇
23 40 ☽ ☌ ♂

25 Monday
00 44 ☽ ⊼ ♄
02 41 ☽ ± ☉
03 03 ☽ ± ♇
08 02 ☉ ☌ ♇
17 33 ☽ ∥ ♄
17 56 ☽ □ ♀
18 13 ☽ ⚼ ♇
18 59 ☽ ⚼ ☉
19 40 ☽ ⊼ ☿
21 40 ♀ ∠ ♇
22 20 ☽ # ♆

26 Tuesday
01 00 ☽ ⊻ ♃
01 12 ☽ ∥ ♆
06 06 ☽ # ♄
07 17 ☽ ± ☿
11 17 ☽ ⊥ ♃
14 16 ☽ △ ♅
15 00 ☽ ☍ ♆
17 34 ♀ # ♅
20 14 ☽ △ ♇
23 04 ☽ △ ☉

27 Wednesday
01 04 ☽ ⚼ ☿
02 23 ☽ ⊻ ♂
02 58 ☽ ∠ ♃
04 00 ☽ Q ♀
04 36 ☽ ☍ ♄
12 57 ☽ ⊥ ♂
15 29 ☿ # ♃
16 37 ☿ ☍ ♃
16 56 ☽ ⚼ ♅
23 55 ☽ # ♂

28 Thursday
03 44 ☽ ⚹ ♀
05 04 ☽ ∠ ♂
05 53 ☽ ⚹ ♃
07 46 ☽ △ ☿
07 51 ☽ ∥ ☿
09 25 ☽ ± ♅
09 51 ☿ ∠ ♄
10 05 ☽ # ♃
11 51 ☉ ⊼ ♂
20 36 ☽ ⊼ ♅
21 32 ☽ ⊼ ♆

29 Friday
02 13 ♀ △ ♃
02 58 ☽ ∥ ☉
03 12 ☽ □ ♇
04 06 ☽ Q ♃
06 33 ☿ ⊥ ♆
08 40 ☽ ⚹ ♂
08 55 ☽ ± ♆
10 35 ☽ ∠ ♀
10 55 ☽ □ ☉
12 24 ☽ ⊼ ♄
15 57 ☽ # ♅
17 56 ☽ ∥ ♀

30 Saturday
00 01 ☽ ± ♄
02 15 ☽ ⚼ ♆
05 40 ☽ ⊥ ♀
06 36 ☉ ⚹ ♄
07 20 ☽ Q ♂
09 35 ☽ ∥ ♇
13 40 ☿ # ♂
13 59 ☽ Q ♇
14 29 ☽ □ ♃
17 41 ☽ ⚼ ♄
18 36 ☽ ⊻ ♀
19 59 ☉ ∠ ♀

31 Sunday
00 36 ☽ □ ☿
01 04 ☽ Q ☉
06 37 ☽ ☍ ♅
07 47 ☽ △ ♆
13 46 ☽ ⚹ ♇
18 07 ☽ □ ♂
23 42 ☽ △ ♄

All ephemeris data is given at 12.00 UT and the Moon's longitude is additionally given for 24.00 UT

FEBRUARY 2027

LONGITUDES

Date	Sidereal time h m s	Sun ☉ ° ' "	Moon ☽ ° ' "	Moon ☽ 24.00 ° ' "	Mercury ☿ ° '	Venus ♀ ° '	Mars ♂ ° '	Jupiter ♃ ° '	Saturn ♄ ° '	Uranus ♅ ° '	Neptune ♆ ° '	Pluto ♇ ° '
01	20 45 53	12 ♒ 23 15	16 ♐ 23 29	22 ♐ 19 55	00 ♓ 32	27 ♐ 36	07 ♍ 17	23 ♌ 17	10 ♈ 19	01 ♊ 42	02 ♈ 20	05 ♒ 20
02	20 49 49	13 24 10	28 ♐ 15 07	04 ♑ 09 34	01 41	28 44	06 R 59	23 R 10	10 24	01 R 42	02 21	05 22
03	20 53 46	14 25 04	10 ♑ 03 38	15 ♑ 57 44	02 43	29 ♐ 53	06 41	23 02	10 29	01 42	02 23	05 24
04	20 57 42	15 25 57	21 ♑ 52 11	27 ♑ 47 19	03 39	01 ♑ 02	06 23	22 54	10 35	01 41	02 25	05 26
05	21 01 39	16 26 49	03 ♒ 43 26	09 ♒ 40 47	04 26	02 11	06 03	22 46	10 40	01 41	02 26	05 28
06	21 05 36	17 27 39	15 ♒ 39 37	21 ♒ 40 11	05 04	03 20	05 44	22 38	10 45	01 41	02 28	05 30
07	21 09 32	18 28 29	27 ♒ 42 41	03 ♓ 47 20	05 32	04 29	05 23	22 30	10 51	01 41	02 30	05 31
08	21 13 29	19 29 17	09 ♓ 54 21	16 ♓ 03 56	05 51	05 38	05 02	22 22	10 57	01 41	02 31	05 33
09	21 17 25	20 30 04	22 ♓ 16 18	28 ♓ 31 40	05 59	06 48	04 41	22 15	11 02	01 D 41	02 33	05 35
10	21 21 22	21 30 49	04 ♈ 50 16	11 ♈ 12 19	05 R 56	07 57	04 19	22 07	11 08	01 41	02 35	05 37
11	21 25 18	22 31 33	17 ♈ 38 03	24 ♈ 07 43	05 42	09 07	03 57	21 59	11 14	01 41	02 37	05 39
12	21 29 15	23 32 15	00 ♉ 41 32	07 ♉ 19 44	05 18	10 17	03 34	21 51	11 20	01 41	02 39	05 41
13	21 33 11	24 32 56	14 ♉ 02 29	20 ♉ 49 58	04 43	11 26	03 11	21 43	11 26	01 42	02 41	05 43
14	21 37 08	25 33 35	27 ♉ 42 18	04 ♊ 39 30	04 00	12 36	02 48	21 35	11 32	01 42	02 42	05 44
15	21 41 05	26 34 12	11 ♊ 41 33	18 ♊ 48 19	03 08	13 46	02 25	21 27	11 38	01 42	02 44	05 46
16	21 45 01	27 34 48	25 ♊ 59 35	03 ♋ 14 59	02 10	14 57	02 01	21 19	11 44	01 43	02 46	05 48
17	21 48 58	28 35 22	10 ♋ 34 04	17 ♋ 56 12	01 07	16 07	01 38	21 11	11 50	01 43	02 48	05 50
18	21 52 54	29 ♒ 35 54	25 ♋ 20 43	02 ♌ 46 47	00 ♓ 01	17 17	01 14	21 03	11 56	01 44	02 50	05 52
19	21 56 51	00 ♓ 36 25	10 ♌ 13 30	17 ♌ 39 54	28 ♒ 53	18 28	00 50	20 56	12 02	01 44	02 52	05 53
20	22 00 47	01 36 53	25 ♌ 05 02	02 ♍ 27 55	27 45	19 38	00 26	20 48	12 09	01 45	02 54	05 55
21	22 04 44	02 37 20	09 ♍ 47 37	17 ♍ 03 18	26 39	20 49	00 ♍ 02	20 40	12 15	01 45	02 56	05 57
22	22 08 40	03 37 46	24 ♍ 14 12	01 ♎ 19 43	25 37	21 59	29 ♌ 38	20 33	12 22	01 46	02 58	05 59
23	22 12 37	04 38 10	08 ♎ 19 22	15 ♎ 12 48	24 39	23 10	29 15	20 25	12 28	01 47	03 00	06 00
24	22 16 34	05 38 32	21 ♎ 59 52	28 ♎ 40 31	23 47	24 21	28 51	20 18	12 35	01 48	03 02	06 02
25	22 20 30	06 38 53	05 ♏ 14 50	11 ♏ 43 03	23 02	25 32	28 28	20 10	12 41	01 49	03 04	06 04
26	22 24 27	07 39 13	18 ♏ 05 29	24 ♏ 22 33	22 23	26 43	28 05	20 03	12 48	01 50	03 06	06 05
27	22 28 23	08 39 31	00 ♐ 34 42	06 ♐ 42 30	21 51	27 54	27 42	19 56	12 54	01 51	03 09	06 07
28	22 32 20	09 ♓ 39 48	12 ♐ 46 30	18 ♐ 47 18	21 ♒ 26	29 ♑ 05	27 ♌ 20	19 ♌ 49	13 ♈ 01	01 ♊ 52	03 ♈ 11	06 ♒ 09

Date	Moon True ☊ ° '	Moon Mean ☊ ° '	Moon ☽ Latitude ° '
01	20 ♒ 35	21 ♒ 10	04 S 46
02	20 R 34	21 07	04 11
03	20 33	21 04	03 26
04	20 32	21 01	02 33
05	20 32	20 58	01 32
06	20 32	20 54	00 S 27
07	20 D 32	20 51	00 N 40
08	20 32	20 48	01 46
09	20 32	20 45	02 47
10	20 R 32	20 42	03 42
11	20 31	20 39	04 27
12	20 31	20 35	04 59
13	20 31	20 32	05 16
14	20 D 31	20 29	05 15
15	20 31	20 26	04 56
16	20 32	20 23	04 19
17	20 32	20 20	03 24
18	20 33	20 16	02 15
19	20 34	20 13	00 N 57
20	20 R 34	20 10	00 S 25
21	20 33	20 07	01 45
22	20 32	20 04	02 56
23	20 30	20 00	03 55
24	20 28	19 57	04 38
25	20 26	19 54	05 05
26	20 24	19 51	05 16
27	20 23	19 48	05 11
28	20 ♒ 23	19 ♒ 45	04 S 52

DECLINATIONS

Date	Sun ☉ ° '	Moon ☽ ° '	Mercury ☿ ° '	Venus ♀ ° '	Mars ♂ ° '	Jupiter ♃ ° '	Saturn ♄ ° '	Uranus ♅ ° '	Neptune ♆ ° '	Pluto ♇ ° '
01	17 S 05	27 S 29	11 S 19	20 S 57	12 N 49	14 N 44	01 N 56	20 N 22	00 S 19	23 S 05
02	16 48	27 37	10 41	21 01	12 57	14 47	01 58	20 22	00 19	23 05
03	16 30	26 29	10 05	21 05	13 05	14 49	02 00	20 22	00 18	23 05
04	16 12	24 10	09 31	21 08	13 13	14 52	02 03	20 22	00 17	23 04
05	15 54	20 49	08 59	21 10	13 21	14 55	02 05	20 22	00 17	23 04
06	15 36	16 34	08 29	21 12	13 29	14 57	02 07	20 22	00 16	23 04
07	15 17	11 39	08 03	21 14	13 38	15 00	02 10	20 22	00 15	23 03
08	14 58	06 13	07 41	21 15	13 46	15 03	02 12	20 22	00 14	23 03
09	14 39	00 S 30	07 23	21 15	13 55	15 05	02 14	20 22	00 14	23 02
10	14 20	05 N 19	07 09	21 15	14 03	15 08	02 17	20 22	00 13	23 02
11	14 00	11 02	06 59	21 14	14 12	15 11	02 19	20 22	00 12	23 02
12	13 40	16 22	06 55	21 13	14 21	15 14	02 22	20 22	00 11	23 01
13	13 20	21 04	06 55	21 11	14 30	15 16	02 24	20 22	00 11	23 01
14	13 00	24 46	07 00	21 09	14 38	15 19	02 26	20 22	00 10	23 01
15	12 39	27 05	07 09	21 06	14 47	15 21	02 29	20 22	00 09	23 00
16	12 19	27 41	07 22	21 02	14 56	15 24	02 31	20 23	00 08	23 00
17	11 58	26 25	07 39	20 58	15 04	15 27	02 34	20 23	00 08	23 00
18	11 37	23 17	08 00	20 53	15 13	15 29	02 37	20 23	00 07	22 59
19	11 15	18 36	08 22	20 48	15 21	15 32	02 39	20 23	00 06	22 59
20	10 54	12 46	08 47	20 42	15 29	15 34	02 42	20 23	00 05	22 59
21	10 32	06 N 17	09 12	20 35	15 37	15 37	02 44	20 23	00 04	22 58
22	10 11	00 S 24	09 38	20 28	15 45	15 39	02 47	20 23	00 04	22 58
23	09 49	06 54	10 04	20 20	15 53	15 42	02 50	20 24	00 03	22 58
24	09 27	12 52	10 30	20 12	16 00	15 44	02 52	20 24	00 02	22 58
25	09 04	18 04	10 54	20 03	16 08	15 46	02 55	20 24	00 S 01	22 57
26	08 42	22 16	11 18	19 54	16 15	15 49	02 58	20 24	00 00	22 57
27	08 19	25 20	11 39	19 44	16 22	15 51	03 00	20 24	00 N 01	22 57
28	07 S 57	27 S 10	11 S 59	19 S 33	16 N 28	15 N 53	03 N 03	20 N 25	00 N 02	22 S 56

ZODIAC SIGN ENTRIES

Date	h	m	Planets
01	01	26	☿ ♓
02	15	33	☽ ♑
03	14	30	♀ ♑
05	04	29	☽ ♒
07	16	32	☽ ♓
10	02	49	☽ ♈
12	10	44	☽ ♉
14	15	59	☽ ♊
16	18	38	☽ ♋
18	12	16	☿ ♒
18	19	31	☽ ♌
18	21	33	☉ ♓
20	19	59	☽ ♍
21	14	13	♂ ♌
22	21	44	☽ ♎
25	02	24	☽ ♏
27	10	52	☽ ♐

LATITUDES

Date	Mercury ☿ ° '	Venus ♀ ° '	Mars ♂ ° '	Jupiter ♃ ° '	Saturn ♄ ° '	Uranus ♅ ° '	Neptune ♆ ° '	Pluto ♇ ° '
01	00 S 02	02 N 28	04 N 17	01 N 02	02 S 20	00 S 08	01 S 22	04 S 17
04	00 N 43	02 18	04 21	01 02	02 20	00 08	01 21	04 18
07	01 32	02 08	04 24	01 03	02 19	00 08	01 21	04 18
10	02 21	01 57	04 27	01 03	02 19	00 08	01 21	04 18
13	03 04	01 46	04 28	01 03	02 18	00 08	01 21	04 19
16	03 34	01 34	04 28	01 04	02 18	00 08	01 21	04 19
19	03 43	01 23	04 28	01 04	02 17	00 08	01 21	04 19
22	03 33	01 11	04 26	01 04	02 17	00 08	01 21	04 20
25	03 06	01 00	04 24	01 04	02 16	00 08	01 21	04 20
28	02 30	00 48	04 21	01 04	02 16	00 08	01 21	04 21
31	01 N 49	00 N 37	04 N 17	01 N 04	02 S 16	00 S 07	01 S 21	04 S 21

DATA

Julian Date	2461438
Delta T	+73 seconds
Ayanamsa	24° 14' 20"
Synetic vernal point	04° ♓ 52' 39"
True obliquity of ecliptic	23° 26' 16"

MOON'S PHASES, APSIDES AND POSITIONS ☽

Date	h	m	Phase	Longitude ° '	Eclipse Indicator
06	15	56	●	17 ♒ 38	Annular
14	07	58	☽	25 ♉ 23	
20	23	24	○	02 ♍ 06	
28	05	16	☾	09 ♐ 23	

Day	h	m		
03	13	44	Apogee	
19	07	37	Perigee	
02	02	33	Max dec	27° S 43'
09	14	04	0N	
16	07	57	Max dec	27° N 43'
22	10	32	0S	

LONGITUDES

Date	Chiron ⚷ ° '	Ceres ⚳ ° '	Pallas ⚴ ° '	Juno ⚵ ° '	Vesta ⚶ ° '	Black Moon Lilith ⚸ ° '
01	26 ♈ 35	12 ♋ 00	15 ♈ 38	08 ♓ 17	24 ♈ 57	15 ♑ 35
11	26 ♈ 52	10 ♋ 45	19 ♈ 19	13 ♓ 12	28 ♈ 18	16 ♑ 42
21	27 ♈ 14	10 ♋ 13	23 ♈ 18	18 ♓ 14	01 ♉ 52	17 ♑ 48
31	27 ♈ 40	10 ♋ 24	27 ♈ 35	23 ♓ 22	05 ♉ 37	18 ♑ 55

ASPECTARIAN

01 Monday
03 12 ☽ ⚹ ☉
16 49 ☽ Q ☿
19 59 ☽ ∠ ♇

02 Tuesday
01 47 ☽ △ ♃
12 18 ☿ □ ♅
12 20 ☽ ∠ ☉
13 05 ☽ ☌ ♀
14 16 ☽ ⊥ ♇
19 00 ☽ ⊼ ♅
19 40 ☽ ⚹ ☿
20 21 ☽ □ ♆

03 Wednesday
01 37 ♀ ⊥ ♇
02 30 ☽ ⊻ ♇
03 32 ☿ ⊻ ♆
05 19 ☽ △ ♂
07 11 ☽ ± ♅
07 55 ☽ ꝗ ♃
08 21 ☽ ⊥ ☉
12 53 ☽ □ ♄
21 42 ☽ ⊻ ☉

04 Thursday
01 29 ☽ ꝗ ♅
02 01 ☽ ± ♃
04 55 ☽ ∠ ☿
09 02 ☽ Q ♆
11 02 ☽ ꝗ ♂
14 04 ☽ ⊼ ♃
20 40 ☽ ∥ ♇

05 Friday
00 34 ☽ ⊥ ☿
01 42 ☽ Q ♄
01 48 ♀ ⊼ ♅
04 47 ☽ ± ♂
07 53 ☽ △ ♅
08 32 ☽ ⊻ ♀
09 24 ☽ ⚹ ♆
09 43 ☽ ∥ ♀
13 30 ☽ ⊻ ☿
14 43 ☽ # ♅
15 31 ☽ ☌ ♇
16 35 ☽ ⊼ ♂
17 34 ♀ □ ♆
21 50 ☿ ⊥ ♄
21 56 ☽ ⊥ ♀

06 Saturday
02 06 ☽ ⚹ ♄
12 07 ☉ ∠ ♆
15 37 ☽ ∠ ♆
15 56 ☽ ☌ ☉
17 19 ☽ ∥ ☉
17 54 ☽ ∠ ♀
20 07 ☽ # ♃

07 Sunday
01 47 ☽ ☍ ♃
02 56 ☽ # ♂
03 16 ♂ ⊼ ♇
07 15 ☿ ☍ ♂
08 17 ☽ ∠ ♄
09 35 ☽ ⊥ ♆
11 08 ☿ ⊻ ♇
19 51 ☽ □ ♅
21 28 ☽ ⊻ ♆

08 Monday
02 12 ☽ ⊥ ♄
02 29 ♀ △ ♂
02 44 ☽ ☍ ♂
02 46 ☽ ⚹ ♀
03 27 ☽ ⊻ ♇
03 53 ☽ ☌ ☿
05 16 ☽ ∥ ☿
07 13 ☉ # ♃
10 15 ♀ ⊻ ♇
12 29 ♅ St D
14 03 ☽ ⊻ ♄
15 14 ☽ ⊥ ♇
17 09 ♂ ± ♄
17 14 ☿ ⚹ ♀

09 Tuesday
04 36 ☽ Q ♀
04 51 ☽ # ♄
07 00 ☽ Q ♅
08 17 ☽ ⊻ ☉
08 44 ☽ ∠ ♇
11 57 ☽ ⊼ ♃
13 08 ☽ ∥ ♆
15 00 ☽ # ♆
17 36 ☿ St R
20 19 ♀ ꝗ ♃
20 50 ☽ ⊥ ☉
23 20 ☽ ± ♃
23 22 ☽ ∥ ♄

10 Wednesday
06 01 ☽ ⚹ ♅
06 26 ♀ ± ♅
07 43 ☽ ☌ ♆
11 03 ☽ ⊼ ♂
13 29 ☽ ⚹ ♇
14 02 ☽ ⊻ ☿
15 27 ☽ ∠ ☉
16 15 ☽ ꝗ ♃
18 28 ☽ □ ♀
19 21 ☽ # ☿
22 03 ☽ ± ♂
23 57 ☽ ☌ ♄

11 Thursday
00 29 ☉ ☍ ♃
01 10 ☽ ⊥ ☿
01 55 ☉ # ♂
10 14 ☽ ∠ ♅
12 01 ☽ Q ♇
14 22 ☽ ꝗ ♂
15 29 ☿ ⊻ ♇
17 32 ☽ ∠ ☿
19 57 ☽ △ ♃
21 49 ☽ ⚹ ☉

12 Friday
00 19 ☽ # ☉
02 22 ☽ ∥ ♂
02 52 ☽ ⊥ ♅
06 35 ☽ ∥ ♃
10 40 ☿ ⊥ ♄
13 49 ☽ ⊻ ♅
15 33 ☽ ⊻ ♆
17 05 ☽ △ ♂
20 01 ☽ ⚹ ☿
21 03 ☽ □ ♇
21 30 ☽ Q ☉

13 Saturday
02 24 ☽ ⊥ ♆
06 56 ☽ △ ♀
07 18 ☽ ⊻ ♄
08 09 ☽ ∥ ♅
11 40 ♀ □ ♄
12 41 ☽ # ♀
16 32 ☽ Q ☿
18 03 ☽ ⊥ ♄
18 27 ☽ ∠ ♆
23 42 ☽ # ♇

14 Sunday
01 25 ☽ □ ♃
07 58 ☽ □ ☉
09 56 ☽ ∠ ♄
11 49 ☽ ꝗ ♀
17 36 ♂ ⊼ ♆
18 55 ☽ ☌ ♅
20 35 ☽ □ ♂
20 40 ☽ ⚹ ♆
22 16 ☽ □ ☿

15 Monday
01 53 ☽ △ ♇
04 44 ☽ ± ♀
08 13 ☽ Q ♃
11 53 ☽ ⚹ ♄
13 30 ☉ ∠ ♄
15 51 ☽ ⊼ ♀
16 09 ☉ ⊥ ♆
17 10 ☽ Q ♆
21 55 ☿ ⊻ ♆

16 Tuesday
02 18 ☽ Q ♂
03 19 ☽ ꝗ ♇
04 17 ☽ ⚹ ♃
08 12 ☽ Q ♄
14 50 ☽ △ ☉
17 42 ☿ ☍ ♂
18 19 ☽ ± ♇
18 52 ♀ ± ♃
21 28 ☽ ⊻ ♅
21 33 ☽ △ ☿
21 43 ☽ ⚹ ♂
22 38 ☿ □ ♅
23 14 ☽ □ ♆

17 Wednesday
04 13 ☽ ⊼ ♇
04 53 ☽ ∠ ♃
06 33 ♂ □ ♅
07 20 ☽ ⊥ ♅
12 03 ☿ ∠ ♀
14 05 ☽ □ ♄
17 18 ☽ ꝗ ☉
19 28 ☽ ⊥ ♃
19 51 ♀ ꝗ ♂
20 25 ☽ ꝗ ☿
21 37 ☽ ∠ ♂
21 49 ☽ ☍ ♀
22 02 ☽ ∠ ♅

18 Thursday
00 28 ♀ ꝗ ♅
05 07 ☽ ⊻ ♃
08 58 ☽ ± ☉
10 00 ☽ ± ☿
11 49 ☽ ⊥ ♂
13 46 ☽ # ♇
16 39 ☉ ☌ ☿
19 00 ☽ ⊼ ☿
19 22 ☽ ⊼ ☉
21 15 ☽ ⊻ ♂
22 19 ☽ ⚹ ♅

19 Friday
00 07 ☽ △ ♆
01 24 ☽ # ♀
03 39 ☽ ∥ ♅
05 00 ☽ ☍ ♇
14 57 ☽ △ ♄
15 51 ☉ ☍ ♂
17 40 ☽ Q ♅

20 Saturday
00 21 ☽ ꝗ ♆
00 59 ☽ ∥ ♃
01 30 ☽ ∥ ♂
02 25 ☽ ⊼ ♀
05 07 ☽ ☌ ♃
12 58 ☽ ± ♀
14 57 ☽ ± ♆
15 11 ☉ □ ♅
15 22 ☽ ꝗ ♄
16 02 ☽ ☍ ☿
19 30 ☽ # ☉
20 28 ☽ ☌ ♂
22 50 ☽ □ ♅
23 24 ☽ ☍ ☉

21 Sunday
00 01 ☿ ∠ ♄
00 45 ☽ ⊼ ♆
02 02 ☽ # ☿
04 54 ☽ ꝗ ♀
05 41 ☽ ⊼ ♇
05 58 ☿ ⊥ ♆
06 09 ☽ ± ♄
09 23 ♀ ⊼ ♃
10 21 ☿ ⊥ ♀
10 29 ♂ ∥ ♃
14 36 ♀ Q ♆
15 33 ☽ ± ♇
16 05 ☽ ⊼ ♄
19 44 ☉ ⊻ ♆

22 Monday
00 37 ☽ ∥ ♄
05 52 ☽ ⊻ ♃
06 31 ☽ ꝗ ♇
07 54 ☽ △ ♀
10 19 ☽ # ♆
10 45 ☽ ∥ ♆
14 10 ☽ ⊼ ☿
15 51 ☽ ⊥ ♃
20 42 ☽ # ♄
20 53 ☽ ⊻ ♂
23 41 ☽ ± ☿

23 Tuesday
00 46 ☽ △ ♅
02 43 ♀ # ♅
02 50 ☽ ☍ ♆
04 08 ☉ ∥ ☿
05 10 ☽ ⊼ ☉
06 51 ☽ ⊥ ♂
07 02 ☽ ∠ ♃
08 00 ☽ △ ♇
13 11 ♀ ± ♂
14 10 ☽ ꝗ ☿
16 19 ☽ ± ☉
19 15 ☽ ☍ ♄
22 01 ☽ ∠ ♂
22 44 ☽ ∥ ☉

24 Wednesday
01 24 ☽ ∥ ☿
02 46 ☽ ꝗ ♅
05 18 ☿ ⊻ ♀
09 00 ☽ ⚹ ♃
09 24 ☽ ꝗ ☉
15 01 ☽ △ ☿
16 37 ☽ □ ♀
18 49 ☽ ± ♅
21 35 ☉ ⊻ ♇
23 58 ☽ ⚹ ♂

25 Thursday
00 51 ☽ # ♃
02 21 ☽ # ♂
05 42 ☽ ⊼ ♅
06 25 ☽ Q ♃
08 00 ☽ ⊼ ♆
12 59 ☉ ⊥ ♄
13 30 ☽ □ ♇
14 48 ☽ △ ☉
19 06 ☽ ± ♆
21 23 ☽ Q ♂
22 16 ☽ ∥ ♀

26 Friday
00 35 ☽ # ♅
01 56 ☽ ⊼ ♄
04 58 ☽ Q ♀
12 02 ☽ ꝗ ♆
13 21 ☽ ± ♄
15 41 ☽ □ ♃
16 35 ☽ ∥ ♇
19 49 ☽ □ ☿
23 28 ☽ Q ♇

27 Saturday
02 04 ♂ ꝗ ♄
06 15 ☽ ⚹ ♀
06 35 ☽ □ ♂
06 46 ☽ ꝗ ♄
09 01 ♀ ⊼ ♂
14 28 ☽ ☍ ♅
17 01 ☽ △ ♆
22 52 ☽ ⚹ ♇

28 Sunday
05 16 ☽ □ ☉
05 35 ☽ Q ☿
12 29 ☽ △ ♄
14 53 ☽ ∠ ♀
21 04 ♂ ± ♆

All ephemeris data is given at 12.00 UT and the Moon's longitude is additionally given for 24.00 UT

MARCH 2027

LONGITUDES

Date	Sidereal time h m s	Sun ☉ ° ' "	Moon ☽ ° ' "	Moon ☽ 24.00 ° ' "	Mercury ☿ ° '	Venus ♀ ° '	Mars ♂ ° '	Jupiter ♃ ° '	Saturn ♄ ° '	Uranus ♅ ° '	Neptune ♆ ° '	Pluto ♇ ° '
01	22 36 16	10 ♓ 40 03	24 ♐ 45 30	00 ♑ 41 45	21 ♒ 09	00 ♒ 16	26 ♌ 58	19 ♌ 41	13 ♈ 08	01 ♊ 53	03 ♈ 13	06 ♒ 10
02	22 40 13	11 40 17	06 ♑ 36 38	12 ♑ 30 44	20 R 59	01 27	26 R 36	19 R 35	13 15	01 54	03 15	06 12
03	22 44 09	12 40 29	18 ♑ 24 39	24 ♑ 18 54	20 55	02 39	26 15	19 28	13 22	01 55	03 17	06 14
04	22 48 06	13 40 40	00 ♒ 14 01	06 ♒ 10 29	20 D 58	03 50	25 54	19 21	13 28	01 56	03 19	06 15
05	22 52 03	14 40 49	12 ♒ 08 42	18 ♒ 09 05	21 08	05 02	25 34	19 14	13 35	01 57	03 21	06 17
06	22 55 59	15 40 56	24 ♒ 11 59	00 ♓ 17 39	21 23	06 13	25 15	19 08	13 42	01 59	03 24	06 18
07	22 59 56	16 41 02	06 ♓ 26 22	12 ♓ 38 17	21 43	07 25	24 55	19 02	13 49	02 00	03 26	06 20
08	23 03 52	17 41 05	18 ♓ 53 35	25 ♓ 12 19	22 09	08 36	24 37	18 55	13 56	02 02	03 28	06 21
09	23 07 49	18 41 07	01 ♈ 34 34	08 ♈ 00 20	22 40	09 48	24 19	18 49	14 03	02 03	03 30	06 23
10	23 11 45	19 41 07	14 ♈ 29 35	21 ♈ 02 16	23 15	10 59	24 02	18 43	14 11	02 05	03 32	06 24
11	23 15 42	20 41 05	27 ♈ 38 18	04 ♉ 17 38	23 54	12 11	23 46	18 37	14 18	02 06	03 35	06 26
12	23 19 38	21 41 00	11 ♉ 00 08	17 ♉ 45 43	24 37	13 23	23 30	18 32	14 25	02 08	03 37	06 27
13	23 23 35	22 40 54	24 ♉ 34 17	01 ♊ 25 44	25 24	14 35	23 15	18 26	14 32	02 09	03 39	06 29
14	23 27 32	23 40 46	08 ♊ 19 59	15 ♊ 16 56	26 14	15 47	23 01	18 21	14 39	02 11	03 41	06 30
15	23 31 28	24 40 35	22 ♊ 16 28	29 ♊ 18 29	27 08	16 58	22 47	18 16	14 47	02 13	03 44	06 31
16	23 35 25	25 40 22	06 ♋ 22 49	13 ♋ 29 20	28 04	18 10	22 34	18 11	14 54	02 15	03 46	06 33
17	23 39 21	26 40 07	20 ♋ 37 47	27 ♋ 47 56	29 ♒ 03	19 22	22 22	18 06	15 01	02 17	03 48	06 34
18	23 43 18	27 39 49	04 ♌ 59 25	12 ♌ 11 51	00 ♓ 05	20 34	22 11	18 01	15 08	02 18	03 50	06 36
19	23 47 14	28 39 29	19 ♌ 24 46	26 ♌ 37 39	01 10	21 46	22 01	17 56	15 16	02 20	03 53	06 37
20	23 51 11	29 ♓ 39 07	03 ♍ 49 55	11 ♍ 00 54	02 16	22 58	21 51	17 52	15 23	02 22	03 55	06 38
21	23 55 07	00 ♈ 38 42	18 ♍ 09 59	25 ♍ 16 30	03 25	24 10	21 42	17 48	15 31	02 24	03 57	06 39
22	23 59 04	01 38 16	02 ♎ 19 47	09 ♎ 19 14	04 36	25 22	21 34	17 44	15 38	02 26	03 59	06 41
23	00 03 01	02 37 47	16 ♎ 14 21	23 ♎ 04 40	05 50	26 35	21 27	17 40	15 45	02 28	04 02	06 42
24	00 06 57	03 37 17	29 ♎ 49 49	06 ♏ 29 36	07 05	27 47	21 20	17 36	15 53	02 31	04 04	06 43
25	00 10 54	04 36 45	13 ♏ 03 53	19 ♏ 32 41	08 22	28 ♒ 59	21 14	17 32	16 00	02 33	04 06	06 44
26	00 14 50	05 36 11	25 ♏ 56 07	02 ♐ 14 24	09 41	00 ♓ 11	21 09	17 29	16 08	02 35	04 09	06 45
27	00 18 47	06 35 35	08 ♐ 27 52	14 ♐ 36 55	11 01	01 23	21 05	17 26	16 15	02 37	04 11	06 47
28	00 22 43	07 34 57	20 ♐ 42 03	26 ♐ 43 48	12 24	02 36	21 02	17 23	16 23	02 40	04 13	06 48
29	00 26 40	08 34 18	02 ♑ 42 45	08 ♑ 39 33	13 48	03 48	20 59	17 20	16 30	02 42	04 15	06 49
30	00 30 36	09 33 37	14 ♑ 34 49	20 ♑ 29 14	15 13	05 00	20 57	17 17	16 38	02 44	04 18	06 50
31	00 34 33	10 ♈ 32 54	26 ♑ 23 28	02 ♒ 18 10	16 ♓ 40	06 ♓ 13	20 ♌ 56	17 ♌ 15	16 ♈ 45	02 ♊ 47	04 ♈ 20	06 ♒ 51

Date	Moon True ☊ ° '	Moon Mean ☊ ° '	Moon ☽ Latitude ° '
01	20 ♒ 24	19 ♒ 41	04 S 21
02	20 D 25	19 38	03 39
03	20 27	19 35	02 48
04	20 29	19 32	01 49
05	20 30	19 29	00 S 46
06	20 R 30	19 26	00 N 20
07	20 29	19 22	01 27
08	20 26	19 19	02 30
09	20 23	19 16	03 27
10	20 18	19 13	04 14
11	20 13	19 10	04 49
12	20 08	19 06	05 08
13	20 04	19 03	05 11
14	20 02	19 00	04 57
15	20 01	18 57	04 24
16	20 D 01	18 54	03 36
17	20 02	18 51	02 34
18	20 04	18 47	01 21
19	20 04	18 44	00 N 04
20	20 R 04	18 41	01 S 14
21	20 01	18 38	02 27
22	19 56	18 35	03 29
23	19 50	18 32	04 18
24	19 43	18 28	04 50
25	19 35	18 25	05 07
26	19 28	18 22	05 07
27	19 22	18 19	04 52
28	19 19	18 16	04 24
29	19 17	18 12	03 45
30	19 D 17	18 09	02 56
31	19 ♒ 18	18 ♒ 06	02 S 00

DECLINATIONS

Date	Sun ☉ ° '	Moon ☽ ° '	Mercury ☿ ° '	Venus ♀ ° '	Mars ♂ ° '	Jupiter ♃ ° '	Saturn ♄ ° '	Uranus ♅ ° '	Neptune ♆ ° '	Pluto ♇ ° '
01	07 S 34	27 S 41	12 S 18	19 S 22	16 N 35	15 N 56	03 N 06	20 N 25	00 N 02	22 S 56
02	07 11	26 55	12 34	19 11	16 41	15 58	03 08	20 25	00 03	22 56
03	06 48	24 57	12 48	18 58	16 47	16 00	03 11	20 25	00 04	22 56
04	06 25	21 53	13 00	18 46	16 53	16 02	03 14	20 26	00 05	22 55
05	06 02	17 53	13 10	18 32	16 58	16 04	03 17	20 26	00 06	22 55
06	05 39	13 08	13 18	18 18	17 03	16 06	03 20	20 26	00 07	22 55
07	05 15	07 48	13 24	18 04	17 08	16 08	03 22	20 26	00 08	22 55
08	04 52	02 S 06	13 28	17 49	17 13	16 10	03 25	20 27	00 08	22 54
09	04 29	03 N 47	13 30	17 34	17 17	16 12	03 28	20 27	00 09	22 54
10	04 05	09 37	13 30	17 18	17 21	16 14	03 31	20 27	00 10	22 54
11	03 41	15 07	13 28	17 02	17 24	16 16	03 34	20 28	00 11	22 54
12	03 18	20 00	13 25	16 45	17 28	16 17	03 36	20 28	00 12	22 53
13	02 54	23 56	13 20	16 28	17 31	16 19	03 39	20 28	00 13	22 53
14	02 31	26 34	13 13	16 10	17 33	16 21	03 42	20 29	00 14	22 53
15	02 07	27 37	13 04	15 52	17 36	16 22	03 45	20 29	00 15	22 53
16	01 43	26 53	12 54	15 33	17 38	16 24	03 48	20 29	00 16	22 53
17	01 19	24 23	12 42	15 14	17 40	16 25	03 51	20 30	00 16	22 52
18	00 56	20 20	12 29	14 54	17 41	16 27	03 54	20 30	00 17	22 52
19	00 32	15 03	12 14	14 34	17 42	16 28	03 57	20 31	00 18	22 52
20	00 S 08	08 57	11 57	14 14	17 43	16 29	03 59	20 31	00 19	22 52
21	00 N 15	02 N 26	11 39	13 53	17 44	16 31	04 02	20 31	00 20	22 52
22	00 39	04 S 07	11 20	13 32	17 44	16 32	04 05	20 32	00 21	22 52
23	01 03	10 21	10 59	13 10	17 44	16 33	04 08	20 32	00 22	22 52
24	01 26	15 56	10 37	12 48	17 44	16 34	04 11	20 33	00 23	22 51
25	01 50	20 37	10 13	12 26	17 44	16 35	04 14	20 33	00 24	22 51
26	02 14	24 12	09 48	12 03	17 43	16 36	04 17	20 34	00 25	22 51
27	02 37	26 31	09 22	11 40	17 42	16 37	04 20	20 34	00 25	22 51
28	03 01	27 30	08 55	11 17	17 41	16 38	04 23	20 34	00 26	22 51
29	03 24	27 09	08 26	10 53	17 40	16 39	04 25	20 35	00 27	22 51
30	03 47	25 34	07 56	10 29	17 38	16 39	04 28	20 35	00 28	22 51
31	04 N 11	22 S 51	07 S 24	10 S 05	17 N 36	16 N 40	04 N 31	20 N 36	00 N 29	22 S 51

ZODIAC SIGN ENTRIES

Date	h m	Planets
01	06 32	♀ ♒
01	22 35	☽ ♑
04	11 32	☽ ♒
06	23 25	☽ ♓
09	09 02	☽ ♈
11	16 16	☽ ♉
13	21 30	☽ ♊
16	01 11	☽ ♋
18	03 41	☽ ♌
18	10 02	☿ ♓
20	05 37	☽ ♍
20	20 25	☉ ♈
22	08 02	☽ ♎
24	12 18	☽ ♏
26	08 16	♀ ♓
26	19 43	☽ ♐
29	06 33	☽ ♑
31	19 20	☽ ♒

LATITUDES

Date	Mercury ☿ ° '	Venus ♀ ° '	Mars ♂ ° '	Jupiter ♃ ° '	Saturn ♄ ° '	Uranus ♅ ° '	Neptune ♆ ° '	Pluto ♇ ° '
01	02 N 16	00 N 44	04 N 19	01 N 04	02 S 16	00 S 08	01 S 21	04 S 21
04	01 35	00 33	04 15	01 04	02 16	00 07	01 21	04 21
07	00 55	00 22	04 10	01 04	02 15	00 07	01 21	04 22
10	00 N 17	00 N 11	04 04	01 04	02 15	00 07	01 21	04 22
13	00 S 18	00 00	03 58	01 04	02 15	00 07	01 21	04 23
16	00 48	00 S 10	03 51	01 04	02 15	00 07	01 21	04 23
19	01 15	00 20	03 44	01 04	02 14	00 07	01 21	04 24
22	01 37	00 30	03 37	01 04	02 14	00 07	01 21	04 24
25	01 56	00 39	03 30	01 03	02 14	00 07	01 21	04 25
28	02 10	00 47	03 23	01 03	02 14	00 07	01 21	04 25
31	02 S 20	00 S 55	03 N 15	01 N 03	02 S 14	00 S 07	01 S 21	04 S 26

DATA

Julian Date	2461466
Delta T	+73 seconds
Ayanamsa	24° 14' 23"
Synetic vernal point	04° ♓ 52' 35"
True obliquity of ecliptic	23° 26' 16"

MOON'S PHASES, APSIDES AND POSITIONS ☽

Date	h m	Phase	Longitude ° '	Eclipse Indicator
08	09 29	●	17 ♓ 35	
15	16 25	☽	24 ♊ 52	
22	10 44	○	01 ♎ 35	
30	00 54	☾	09 ♑ 06	

Day	h m		
03	05 48	Apogee	
19	04 34	Perigee	
31	01 35	Apogee	
01	09 37	Max dec	27° S 41'
08	20 36	0N	
15	14 14	Max dec	27° N 37'
21	20 50	0S	
28	17 39	Max dec	27° S 32'

LONGITUDES

Date	Chiron ⚷ ° '	Ceres ⚳ ° '	Pallas ⚴ ° '	Juno ⚵ ° '	Vesta ⚶ ° '	Black Moon Lilith ⚸ ° '
01	27 ♈ 34	10 ♋ 18	26 ♈ 42	22 ♓ 20	04 ♉ 52	18 ♑ 42
11	28 ♈ 03	11 ♋ 02	01 ♉ 11	27 ♓ 32	08 ♉ 44	19 ♑ 48
21	28 ♈ 35	12 ♋ 23	05 ♉ 53	02 ♈ 50	12 ♉ 44	20 ♑ 55
31	29 ♈ 09	14 ♋ 17	10 ♉ 48	08 ♈ 12	16 ♉ 49	22 ♑ 01

All ephemeris data is given at 12.00 UT and the Moon's longitude is additionally given for 24.00 UT

ASPECTARIAN

h m	Aspects
01 Monday	
01 55	☽ △ ♃
04 46	☽ ∠ ♇
04 53	☽ ⚹ ☿
10 54	☽ ⊥ ♀
16 19	☽ △ ♂
20 37	☽ Q ☉
22 58	☽ ⊥ ♇
02 Tuesday	
00 23	☽ ⊻ ♀
02 25	☽ ⊼ ♅
05 09	☽ □ ♆
07 15	♀ Q ♄
07 54	☽ P ♃
10 44	☽ ∠ ☿
11 10	☽ ⊻ ♇
14 37	☽ ± ♅
21 00	♀ △ ♅
21 51	☽ P ♂
23 15	☽ ⚹ ☉
03 Wednesday	
00 58	☉ ⊥ ♇
01 37	☽ □ ♄
02 02	☽ ± ♃
04 55	☽ ⊥ ☿
08 57	☽ P ♅
12 33	☿ St D
14 07	☽ ⊼ ♃
15 38	☽ ± ♂
17 07	☽ ⊻ ☿
17 52	☽ Q ♆
04 Thursday	
01 17	♀ ⚹ ♆
03 28	☽ ⊼ ♂
04 40	☽ ∥ ♇
06 29	☉ ⊻ ♄
08 33	☽ ∠ ☉
14 32	☽ Q ♄
15 27	☽ △ ♅
18 16	☽ ⚹ ♆
20 06	☽ ☌ ♀
21 21	☽ # ♅
05 Friday	
00 11	☽ ☌ ♇
04 24	☽ ⊥ ☉
08 12	☽ ∥ ♀
14 55	☽ ⚹ ♄
16 50	☽ # ♂
17 32	☽ ⊻ ☉
21 31	☽ # ♃
06 Saturday	
00 27	☽ ∠ ♆
02 02	☽ ☍ ♃
06 17	☽ ☌ ☿
11 14	☽ ∥ ☿
13 48	♀ ☌ ♇
14 00	☽ ☍ ♂
18 19	☽ ⊥ ♆
20 58	☽ ∠ ♄
07 Sunday	
03 20	☽ □ ♅
06 07	☽ ⊻ ♆
11 47	☽ ⊻ ♇
14 05	☽ ⊻ ♀
14 43	☽ ⊥ ♄
23 26	☽ ⊥ ♇
23 40	☽ ∥ ☉
08 Monday	
02 25	☽ ⊻ ♄
02 55	☽ ⊥ ♀
06 35	☽ # ♄
09 29	☽ ☌ ☉
12 03	☽ ⊼ ♃
14 10	☽ Q ♅
16 42	☽ ∠ ♇
18 27	☽ ⊻ ☿
20 00	☽ # ♆
21 12	☽ ∥ ♆
21 54	☽ ∠ ♀
22 38	☽ ⊼ ♂
23 22	☽ ± ♃
09 Tuesday	
06 17	☽ ⊥ ☿
09 42	☽ ± ♂
10 42	☽ ∥ ♄
12 53	☽ ⚹ ♅
14 38	☽ # ☉
14 56	☉ ⊼ ♃
15 37	☽ ☌ ♆
16 10	☽ P ♃
21 00	☽ ⚹ ♇
23 53	☽ ∠ ☿
10 Wednesday	
02 08	☽ P ♂
04 53	☽ ⚹ ♀
08 48	♀ # ♂
11 25	☽ ☌ ♄
16 45	☽ ∠ ♅
19 12	☽ Q ♇
19 42	☽ △ ♃
21 38	☉ Q ♅
22 19	☽ ⊻ ☉
11 Thursday	
04 40	☽ # ☿
04 51	☽ ⚹ ☿
05 06	☽ △ ♂
05 07	☽ Q ♀
08 31	☿ ☍ ♂
09 13	☽ ⊥ ♅
10 08	☽ ⊥ ☉
17 20	☽ ∥ ♃
19 07	☉ # ♄
20 04	☽ ⊻ ♅
20 24	♃ P ♆
20 28	☽ # ♀
22 45	☽ ⊻ ♆
22 53	☽ ∥ ♂
12 Friday	
03 40	☽ ∠ ☉
03 43	☽ Q ☿
03 52	☽ □ ♇
06 21	☉ ∠ ♇
09 31	☽ ⊥ ♆
14 31	☽ ∥ ♅
16 39	☽ □ ♀
18 07	☽ ⊻ ♄
13 Saturday	
01 16	☽ □ ♃
01 33	☽ ∠ ♆
04 50	☽ ⊥ ♄
04 51	☽ # ♇
08 25	☽ ⚹ ☉
09 43	☽ □ ♂
11 01	♀ ⚹ ♄
13 33	☽ □ ☿
20 46	☽ ∠ ♄
22 30	♀ # ♃
22 59	☉ ⊼ ♂
14 Sunday	
01 18	☽ ☌ ♅
03 55	☽ ⚹ ♆
07 02	☽ Q ☉
08 35	☽ Q ♃
08 49	☽ △ ♇
16 33	☽ Q ♂
23 01	☽ ⚹ ♄
15 Monday	
00 44	☽ Q ♆
02 03	☽ △ ♀
02 47	☉ ± ♃
05 10	☽ ⚹ ♃
10 43	☽ P ♇
12 52	☽ ⚹ ♂
16 25	☽ □ ☉
19 45	☽ Q ♄
20 52	☽ △ ☿
16 Tuesday	
02 06	☽ ± ♇
04 05	☿ ⊥ ♆
04 59	☽ ⊻ ♅
06 04	☽ P ♀
06 36	☽ ∠ ♃
07 33	☽ □ ♆
12 05	♀ ☍ ♃
12 17	☽ ⊼ ♇
13 59	☽ ∠ ♂
15 10	☽ ⊥ ♅
21 44	☽ ⊥ ♃
22 41	☽ ± ♀
17 Wednesday	
00 07	☽ P ☿
00 15	♀ ∠ ♆
02 30	☽ □ ♄
04 57	☽ ⊥ ♂
06 21	☽ ∠ ♅
07 46	☽ ⊻ ♃
09 42	☽ ⊼ ♀
14 53	☽ ⊻ ♂
16 22	☽ ± ☿
21 58	☽ # ♇
22 52	☽ △ ☉
18 Thursday	
03 11	☽ ⊼ ☿
07 31	☽ ⚹ ♅
10 05	☽ △ ♆
11 08	☽ ∥ ♅
13 25	☿ ∠ ♄
14 40	☽ ☍ ♇
22 42	☉ ± ♂
19 Friday	
00 34	☽ ∥ ♂
01 43	☽ P ☉
03 33	☽ Q ♅
05 03	☽ △ ♄
06 01	☽ ∥ ♃
09 34	☽ ☌ ♃
11 06	☽ P ♆
14 08	☽ # ♀
16 14	♀ ☍ ♂
16 16	☽ ☌ ♂
16 17	☽ ☍ ♀
17 48	☽ ± ☉
23 57	☽ # ☿
20 Saturday	
01 25	☉ # ♆
02 07	☽ ± ♆
04 31	☽ ⊼ ☉
06 13	☽ P ♄
09 11	☽ ☍ ☿
09 34	☽ □ ♅
12 08	☽ ⊼ ♆
14 10	☿ □ ♅
16 41	☽ ⊼ ♇
21 21	☽ ± ♄
21 Sunday	
02 44	☽ ± ♇
06 11	☽ ∥ ♄
07 30	☽ ⊼ ♄
11 23	☽ ⊻ ♃
16 54	☉ ∥ ♆
17 53	☽ P ♇
17 54	☽ ⊻ ♂
19 27	☽ ∥ ☉
19 36	☽ ∥ ♆
21 27	☽ ⊥ ♃
22 05	☽ # ♆
22 24	☽ # ☉
23 04	☽ ⊼ ♀
23 11	☿ ⊻ ♆
22 Monday	
03 58	☽ ⊥ ♂
10 13	☽ ± ♀
10 44	☽ ☍ ☉
11 52	☽ # ♄
12 11	☽ △ ♅
12 40	☽ ∠ ♃
14 51	☽ ☍ ♆
16 16	☽ ⊼ ☿
19 12	☽ ∠ ♂
19 28	☽ △ ♇
23 Tuesday	
03 08	☽ P ♀
03 35	☽ ± ☿
08 06	☉ ⚹ ♅
11 09	☽ ☍ ♄
12 42	☉ P ♃
14 10	☽ P ♅
14 26	☽ ∥ ☿
14 28	☽ ⚹ ♃
18 04	☉ ⊥ ♀
20 51	☽ P ☿
21 03	☽ ⚹ ♂
22 59	☽ ∥ ♀
24 Wednesday	
05 00	☿ ⊻ ♇
06 04	☽ ± ♅
07 59	☽ △ ♀
11 35	☽ Q ♃
14 58	☽ # ♃
16 49	☽ ⊼ ♅
17 55	♀ ⊥ ♆
18 15	☽ Q ♂
19 21	☽ ⊼ ☉
19 38	☽ ⊼ ♆
20 37	☽ # ♂
23 13	☉ ☌ ♆
25 Thursday	
00 26	☽ □ ♇
02 28	☽ △ ☿
06 33	☽ ± ♆
07 08	☽ ± ☉
11 35	☽ # ♅
17 29	☽ ⊼ ♄
20 14	☽ □ ♃
23 13	☽ P ♆
26 Friday	
01 08	☽ P ☉
01 59	☽ ∥ ♇
03 04	☽ □ ♂
04 46	☽ ± ♄
09 46	☽ Q ♇
20 56	☽ □ ♀
20 57	☿ ⊥ ♄
21 59	☽ P ♄
27 Saturday	
00 31	☉ P ♂
00 42	☽ ☍ ♅
03 42	☽ △ ♆
08 04	☽ △ ☉
08 44	☽ ⚹ ♇
08 59	♀ ∠ ♄
16 31	☉ ⚹ ♇
17 36	☽ □ ☿
28 Sunday	
03 23	☽ △ ♄
05 28	☽ △ ♃
11 46	☽ Q ♀
12 39	☽ △ ♂
13 18	♀ □ ♅
14 10	☽ ∠ ♇
18 59	☿ ⊥ ♇
29 Monday	
08 10	☽ ⊥ ♇
09 54	☽ Q ☿
11 14	☽ P ♃
11 58	☽ ⊼ ♅
14 26	☽ ⚹ ♀
15 07	☽ □ ♆
18 35	☽ P ♂
20 17	☽ ⊻ ♇
21 20	♀ ⊻ ♆
30 Tuesday	
00 07	☽ ± ♅
00 54	☽ □ ☉
05 21	☽ ± ♃
12 45	☽ ± ♂
13 29	☽ ⚹ ☿
16 13	☽ □ ♄
17 29	☽ ⊼ ♃
18 26	☽ P ♅
31 Wednesday	
00 17	☽ ∠ ♀
00 55	☽ ⊼ ♂
03 43	☽ Q ♆
12 00	☽ ∥ ♇
13 29	☿ ⊻ ♄
16 47	☽ Q ☉
20 39	☽ ⊥ ♀
21 06	☿ ⊼ ♃

APRIL 2027

LONGITUDES

Date	Sidereal time h m s	Sun ☉ ° ' "	Moon ☽ ° ' "	Moon ☽ 24.00 ° ' "	Mercury ☿ ° '	Venus ♀ ° '	Mars ♂ ° '	Jupiter ♃ ° '	Saturn ♄ ° '	Uranus ♅ ° '	Neptune ♆ ° '	Pluto ♇ ° '
01	00 38 30	11 ♈ 32 09	08 ♒ 14 00	14 ♒ 11 33	18 ♓ 09	07 ♓ 25	20 ♌ 56	17 ♌ 13	16 ♈ 53	02 ♊ 49	04 ♈ 22	06 ♒ 52
02	00 42 26	12 31 22	20 ♒ 11 25	26 ♒ 14 09	19 39	08 38	20 D 56	17 R 10	17 01	02 52	04 24	06 53
03	00 46 23	13 30 34	02 ♓ 20 14	08 ♓ 30 05	21 11	09 50	20 57	17 08	17 08	02 54	04 27	06 54
04	00 50 19	14 29 43	14 ♓ 44 03	21 ♓ 02 25	22 45	11 03	20 59	17 07	17 16	02 57	04 29	06 55
05	00 54 16	15 28 51	27 ♓ 25 21	03 ♈ 52 57	24 19	12 15	21 01	17 05	17 23	02 59	04 31	06 56
06	00 58 12	16 27 56	10 ♈ 25 13	17 ♈ 02 03	25 56	13 28	21 04	17 04	17 31	03 02	04 33	06 57
07	01 02 09	17 27 00	23 ♈ 43 14	00 ♉ 28 31	27 33	14 40	21 08	17 03	17 38	03 05	04 36	06 57
08	01 06 05	18 26 02	07 ♉ 17 32	14 ♉ 09 54	29 ♓ 13	15 53	21 13	17 02	17 46	03 07	04 38	06 58
09	01 10 02	19 25 01	21 ♉ 05 11	28 ♉ 02 54	00 ♈ 53	17 05	21 18	17 01	17 54	03 10	04 40	06 59
10	01 13 59	20 23 59	05 ♊ 02 36	12 ♊ 03 52	02 36	18 18	21 24	17 00	18 01	03 13	04 42	07 00
11	01 17 55	21 22 54	19 ♊ 06 17	26 ♊ 09 29	04 19	19 31	21 30	17 00	18 09	03 16	04 45	07 01
12	01 21 52	22 21 47	03 ♋ 13 09	10 ♋ 17 04	06 05	20 43	21 37	17 00	18 16	03 18	04 47	07 01
13	01 25 48	23 20 38	17 ♋ 20 59	24 ♋ 24 46	07 51	21 56	21 45	17 D 00	18 24	03 21	04 49	07 02
14	01 29 45	24 19 27	01 ♌ 28 16	08 ♌ 31 22	09 40	23 08	21 54	17 00	18 31	03 24	04 51	07 03
15	01 33 41	25 18 13	15 ♌ 33 57	22 ♌ 35 53	11 29	24 21	22 03	17 00	18 39	03 27	04 53	07 03
16	01 37 38	26 16 56	29 ♌ 37 02	06 ♍ 37 13	13 21	25 34	22 12	17 01	18 46	03 30	04 55	07 04
17	01 41 34	27 15 38	13 ♍ 36 11	20 ♍ 33 40	15 14	26 46	22 22	17 02	18 54	03 33	04 58	07 05
18	01 45 31	28 14 17	27 ♍ 29 22	04 ♎ 22 57	17 08	27 59	22 33	17 02	19 01	03 36	05 00	07 05
19	01 49 28	29 ♈ 12 54	11 ♎ 14 02	18 ♎ 02 14	19 04	29 ♓ 12	22 45	17 04	19 09	03 39	05 02	07 06
20	01 53 24	00 ♉ 11 29	24 ♎ 47 12	01 ♏ 28 34	21 02	00 ♈ 24	22 56	17 05	19 16	03 42	05 04	07 06
21	01 57 21	01 10 03	08 ♏ 06 03	14 ♏ 39 24	23 01	01 37	23 09	17 06	19 24	03 45	05 06	07 07
22	02 01 17	02 08 34	21 ♏ 08 27	27 ♏ 33 05	25 01	02 50	23 22	17 08	19 31	03 48	05 08	07 07
23	02 05 14	03 07 04	03 ♐ 53 20	10 ♐ 09 17	27 03	04 03	23 35	17 10	19 39	03 51	05 10	07 08
24	02 09 10	04 05 32	16 ♐ 21 06	22 ♐ 29 03	29 ♈ 07	05 15	23 49	17 12	19 46	03 54	05 12	07 08
25	02 13 07	05 03 58	28 ♐ 33 29	04 ♑ 34 49	01 ♉ 11	06 28	24 03	17 14	19 54	03 58	05 14	07 08
26	02 17 03	06 02 22	10 ♑ 33 34	16 ♑ 30 15	03 17	07 41	24 18	17 16	20 01	04 01	05 16	07 09
27	02 21 00	07 00 45	22 ♑ 25 28	28 ♑ 19 52	05 24	08 54	24 33	17 19	20 08	04 04	05 18	07 09
28	02 24 57	07 59 07	04 ♒ 14 06	10 ♒ 08 51	07 32	10 06	24 49	17 22	20 16	04 07	05 20	07 09
29	02 28 53	08 57 26	16 ♒ 04 48	22 ♒ 02 39	09 40	11 19	25 05	17 25	20 23	04 11	05 22	07 10
30	02 32 50	09 ♉ 55 45	28 ♒ 03 05	04 ♓ 06 44	11 ♉ 49	12 ♈ 32	25 ♌ 22	17 ♌ 28	20 ♈ 30	04 ♊ 14	05 ♈ 24	07 ♒ 10

Date	Moon True ☊ ° '	Moon Mean ☊ ° '	Moon ☽ Latitude ° '
01	19 ♒ 19	18 ♒ 03	00 S 59
02	19 R 20	18 00	00 N 05
03	19 19	17 57	01 09
04	19 16	17 53	02 12
05	19 10	17 50	03 10
06	19 02	17 47	03 58
07	18 53	17 44	04 36
08	18 43	17 41	04 58
09	18 33	17 38	05 04
10	18 24	17 34	04 51
11	18 18	17 31	04 22
12	18 14	17 28	03 36
13	18 13	17 25	02 37
14	18 D 13	17 22	01 28
15	18 13	17 18	00 N 14
16	18 R 13	17 15	01 S 00
17	18 10	17 12	02 11
18	18 05	17 09	03 13
19	17 58	17 06	04 02
20	17 47	17 03	04 38
21	17 36	16 59	04 57
22	17 23	16 56	05 01
23	17 12	16 53	04 49
24	17 02	16 50	04 23
25	16 54	16 47	03 46
26	16 50	16 43	02 59
27	16 47	16 40	02 05
28	16 46	16 37	01 06
29	16 D 46	16 34	00 04
30	16 ♒ 46	16 ♒ 31	00 N 59

DECLINATIONS

Date	Sun ☉ ° '	Moon ☽ ° '	Mercury ☿ ° '	Venus ♀ ° '	Mars ♂ ° '	Jupiter ♃ ° '	Saturn ♄ ° '	Uranus ♅ ° '	Neptune ♆ ° '	Pluto ♇ ° '
01	04 N 34	19 S 10	06 S 52	09 S 41	17 N 34	16 N 41	04 N 34	20 N 36	00 N 30	22 S 51
02	04 57	14 41	06 18	09 16	17 31	16 41	04 37	20 37	00 31	22 50
03	05 20	09 34	05 43	08 51	17 29	16 42	04 40	20 37	00 32	22 50
04	05 43	03 S 59	05 07	08 25	17 26	16 42	04 43	20 38	00 33	22 50
05	06 06	01 N 52	04 30	08 00	17 23	16 43	04 46	20 38	00 33	22 50
06	06 28	07 47	03 52	07 34	17 19	16 43	04 49	20 39	00 34	22 50
07	06 51	13 28	03 12	07 08	17 16	16 43	04 51	20 39	00 35	22 50
08	07 14	18 38	02 32	06 42	17 12	16 43	04 54	20 40	00 36	22 50
09	07 36	22 54	01 51	06 15	17 08	16 43	04 57	20 41	00 37	22 50
10	07 58	25 55	01 08	05 49	17 04	16 44	05 00	20 41	00 38	22 50
11	08 20	27 20	00 S 25	05 22	17 00	16 44	05 03	20 42	00 39	22 50
12	08 42	27 00	00 N 20	04 55	16 55	16 44	05 06	20 42	00 39	22 50
13	09 04	24 54	01 05	04 28	16 51	16 43	05 09	20 43	00 40	22 50
14	09 26	21 16	01 51	04 00	16 46	16 43	05 11	20 43	00 41	22 50
15	09 47	16 24	02 39	03 33	16 41	16 43	05 14	20 44	00 42	22 50
16	10 09	10 40	03 27	03 06	16 35	16 43	05 17	20 44	00 43	22 50
17	10 30	04 N 26	04 15	02 38	16 30	16 42	05 20	20 45	00 44	22 50
18	10 51	01 S 57	05 05	02 10	16 24	16 42	05 23	20 46	00 44	22 50
19	11 12	08 10	05 55	01 42	16 19	16 42	05 26	20 46	00 45	22 50
20	11 32	13 54	06 46	01 15	16 13	16 41	05 28	20 47	00 46	22 51
21	11 53	18 53	07 37	00 47	16 07	16 41	05 31	20 47	00 47	22 51
22	12 13	22 52	08 28	00 S 19	16 00	16 40	05 34	20 48	00 48	22 51
23	12 33	25 39	09 20	00 N 09	15 54	16 39	05 37	20 49	00 48	22 51
24	12 53	27 06	10 12	00 38	15 47	16 39	05 39	20 49	00 49	22 51
25	13 13	27 12	11 05	01 06	15 41	16 38	05 42	20 50	00 50	22 51
26	13 32	26 00	11 57	01 34	15 34	16 37	05 45	20 50	00 51	22 51
27	13 51	23 38	12 49	02 02	15 27	16 36	05 48	20 51	00 52	22 51
28	14 10	20 16	13 40	02 30	15 19	16 35	05 50	20 52	00 52	22 51
29	14 29	16 04	14 31	02 58	15 12	16 34	05 53	20 52	00 53	22 51
30	14 N 47	11 S 13	15 N 22	03 N 26	15 N 05	16 N 33	05 N 56	20 N 53	00 N 54	22 S 52

ZODIAC SIGN ENTRIES

Date	h	m	Planets
03	07	25	☽ ♓
05	16	48	☽ ♈
07	23	10	☽ ♉
08	23	20	☿ ♈
10	03	21	☽ ♊
12	06	32	☽ ♋
14	09	30	☽ ♌
16	12	39	☽ ♍
18	16	22	☽ ♎
20	03	57	♀ ♈
20	07	18	☉ ♉
20	21	21	☽ ♏
23	04	37	☽ ♐
24	22	18	☿ ♉
25	14	52	☽ ♑
28	03	24	☽ ♒
30	15	52	☽ ♓

LATITUDES

Date	Mercury ☿ ° '	Venus ♀ ° '	Mars ♂ ° '	Jupiter ♃ ° '	Saturn ♄ ° '	Uranus ♅ ° '	Neptune ♆ ° '	Pluto ♇ ° '
01	02 S 22	00 S 58	03 N 13	01 N 03	02 S 14	00 S 07	01 S 21	04 S 26
04	02 26	01 05	03 05	01 03	02 14	00 07	01 21	04 27
07	02 26	01 11	02 58	01 02	02 14	00 07	01 21	04 27
10	02 22	01 17	02 51	01 02	02 14	00 07	01 21	04 28
13	02 12	01 22	02 44	01 02	02 14	00 07	01 21	04 28
16	01 59	01 27	02 37	01 02	02 14	00 07	01 21	04 29
19	01 41	01 31	02 31	01 01	02 14	00 07	01 21	04 30
22	01 18	01 34	02 25	01 01	02 14	00 07	01 21	04 30
25	00 52	01 36	02 18	01 01	02 15	00 07	01 21	04 31
28	00 S 22	01 38	02 12	01 00	02 15	00 07	01 22	04 32
31	00 N 09	01 S 39	02 N 06	01 N 00	02 S 15	00 S 07	01 S 22	04 S 32

DATA

Julian Date	2461497
Delta T	+73 seconds
Ayanamsa	24° 14' 27"
Synetic vernal point	04° ♓ 52' 32"
True obliquity of ecliptic	23° 26' 16"

MOON'S PHASES, APSIDES AND POSITIONS ☽

Date	h	m	Phase	Longitude ° '	Eclipse Indicator
06	23	51	●	16 ♈ 57	
13	22	57	☽	23 ♋ 47	
20	22	27	○	00 ♏ 37	
28	20	18	☾	08 ♒ 19	

Day	h	m		
14	00	36	Perigee	
27	21	20	Apogee	
05	04	23	0N	
11	19	25	Max dec	27° N 25'
18	04	40	0S	
25	01	40	Max dec	27° S 19'

LONGITUDES

Date	Chiron ⚷ ° '	Ceres ⚳ ° '	Pallas ⚴ ° '	Juno ⚵ ° '	Vesta ⚶ ° '	Black Moon Lilith ⚸ ° '
01	29 ♈ 13	14 ♋ 30	11 ♉ 18	08 ♈ 44	17 ♉ 14	22 ♑ 08
11	29 ♈ 49	16 ♋ 55	16 ♉ 25	14 ♈ 11	21 ♉ 24	23 ♑ 15
21	00 ♉ 25	19 ♋ 43	21 ♉ 42	19 ♈ 41	25 ♉ 38	24 ♑ 21
31	01 ♉ 02	22 ♋ 51	27 ♉ 07	25 ♈ 14	29 ♉ 55	25 ♑ 28

ASPECTARIAN

(h m Aspects)

01 Thursday: 00 15 ☽ ∠ ☿; 00 46 ♀ ⊻ ♇; 01 01 ☽ △ ♅; 03 16 ☽ # ♅; 04 10 ☽ ⚹ ♆; 05 09 ☽ Q ♄; 09 14 ☽ ☌ ♇; 10 11 ☽ ⊻ ♀; 12 31 ☉ ∥ ♄; 14 08 ♂ St D; 19 16 ☽ ⚹ ☉; 21 02 ☽ ⊥ ☿; 21 06 ☽ # ♂

02 Friday: 01 45 ☽ # ♃; 05 35 ☽ ⚹ ♄; 05 59 ☽ ☍ ♃; 10 26 ☽ ∠ ♆; 10 47 ☽ ⊻ ☿; 13 29 ☽ ☍ ♂

03 Saturday: 03 49 ☽ ∠ ☉; 04 20 ☽ ⊥ ♆; 07 27 ☿ Q ♅; 08 14 ☿ ⊼ ♂; 11 36 ☽ ∠ ♄; 12 59 ♃ △ ♄; 13 07 ☽ □ ♅; 15 27 ☽ ∥ ♀; 16 08 ☽ ⊻ ♆; 20 54 ☽ ⊻ ♇; 21 33 ☉ # ☿; 22 57 ☽ ⊥ ☉; 23 07 ☿ ∠ ♇

04 Sunday: 04 09 ☽ ☌ ♀; 05 10 ☽ # ☉; 05 16 ☽ ⊥ ♄; 06 39 ☽ ∥ ☿; 08 30 ☽ ⊥ ♇; 08 57 ☽ # ♄; 11 30 ☽ ⊻ ☉; 16 32 ☽ ⊼ ♃; 16 48 ♀ ⊥ ♄; 16 52 ☽ ⊻ ♄; 17 34 ☿ ± ♃; 23 52 ☽ Q ♅; 23 55 ☽ ⊼ ♂

05 Monday: 01 40 ☽ ∠ ♇; 02 08 ☽ # ♆; 02 41 ☿ # ♄; 03 53 ☽ ± ♃; 05 22 ☽ ☌ ☿; 06 38 ☽ ∥ ♆; 11 15 ☽ ± ♂; 20 40 ☽ ꝑ ♃; 21 37 ☽ # ☿; 22 23 ☽ ⚹ ♅; 23 48 ☽ ∥ ♄

06 Tuesday: 01 13 ☽ ☌ ♆; 01 34 ♀ ⊥ ♇; 04 00 ☽ ꝑ ♂; 05 38 ☽ ⚹ ♇; 06 17 ☽ ∥ ☉; 11 11 ☽ # ♀; 18 05 ☽ ⊻ ♀; 23 51 ☽ ☌ ☉

07 Wednesday: 00 02 ☽ △ ♃; 00 59 ☽ ☌ ♄; 01 51 ☽ ∠ ♅; 02 18 ☉ △ ♃; 03 27 ☽ Q ♇; 05 35 ☿ ± ♂; 06 00 ☽ ⊥ ♀; 07 21 ☽ △ ♂; 17 18 ☉ ☌ ♄; 18 00 ☽ ⊥ ♅; 19 46 ☽ ⊻ ☿; 20 18 ☉ # ♀; 23 37 ☽ ∠ ♀

08 Thursday: 02 42 ☽ ∥ ♃; 04 02 ☉ ∠ ♅; 04 39 ☽ ⊻ ♅; 05 05 ☽ ∥ ♂; 07 19 ☽ ⊻ ♆; 07 50 ☽ ⊥ ☿; 11 26 ☽ □ ♇; 17 51 ☽ ⊥ ♆; 22 41 ☽ ∥ ♅

09 Friday: 01 17 ☉ Q ♇; 01 45 ☽ ∠ ☿; 04 25 ☽ ⚹ ♀; 04 57 ☽ □ ♃; 06 25 ☽ ⊻ ♄; 08 54 ☽ ⊻ ☉; 09 32 ☽ ∠ ♆; 10 34 ♀ ⊼ ♃; 11 33 ☽ # ♇; 12 22 ☽ □ ♂; 16 53 ☽ ⊥ ♄; 20 02 ☽ ⊥ ☉

10 Saturday: 03 06 ☽ Q ♀; 03 49 ☿ ꝑ ♃; 05 47 ♀ ⊻ ♄; 07 13 ☽ ⚹ ☿; 08 30 ☽ ∠ ♄; 08 51 ☽ ☌ ♅; 11 25 ☽ ⚹ ♆; 11 56 ☽ Q ♃; 12 39 ☽ ∠ ☉; 15 21 ☽ △ ♇; 19 30 ☽ Q ♂; 20 54 ☿ ⚹ ♅

11 Sunday: 04 28 ☿ # ♆; 06 36 ☽ Q ☿; 07 58 ☽ Q ♆; 08 25 ☽ ⚹ ♃; 10 21 ☽ ⚹ ♄; 12 45 ☽ □ ♀; 15 25 ☉ △ ♂; 16 07 ☽ ⚹ ♂; 16 10 ☽ ⚹ ☉; 16 57 ☽ ꝑ ♇; 17 54 ☿ ☌ ♆

12 Monday: 03 10 ♀ # ♄; 06 57 ☽ Q ♄; 08 16 ☽ ± ♇; 09 55 ☽ ∠ ♃; 12 09 ☽ ⊻ ♅; 14 05 ☽ Q ☉; 14 39 ☽ □ ♆; 17 33 ☽ □ ☿; 17 50 ☽ ∠ ♂; 18 28 ☽ ⊼ ♇; 20 01 ☿ ꝑ ♂; 22 23 ☽ ⊥ ♅; 22 38 ☿ ∥ ♆; 23 23 ♄ ∠ ♅

13 Tuesday: 00 10 ♀ Q ♅; 00 54 ☿ ⚹ ♇; 01 12 ☽ ⊥ ♃; 02 12 ♃ St D; 08 07 ♀ ⊼ ♂; 09 16 ☽ ⊥ ♂; 11 24 ☽ ⊻ ♃; 13 43 ☽ ∠ ♅; 13 48 ☽ □ ♄; 14 06 ♀ ∠ ♇; 19 33 ☽ ⊻ ♂; 20 31 ☽ △ ♀; 22 57 ☽ □ ☉

14 Wednesday: 02 42 ☽ # ♇; 09 12 ♀ ± ♃; 14 57 ☽ ∥ ♅; 15 18 ☽ ⚹ ♅; 17 46 ☽ △ ♆; 21 30 ☽ ☍ ♇

15 Thursday: 00 09 ♂ ∥ ♃; 00 25 ☽ ꝑ ♀; 04 01 ☽ △ ☿; 10 32 ☽ ∥ ♃; 10 42 ☽ ∥ ♂; 11 48 ☽ Q ♅; 14 27 ☽ ☌ ♃; 17 12 ☽ ± ♀; 17 18 ☽ △ ♄; 19 23 ☽ ꝑ ♆; 23 11 ☽ ☌ ♂

16 Friday: 04 24 ☽ ⊼ ♀; 05 21 ☿ # ♀; 05 52 ☽ △ ☉; 09 29 ☽ ꝑ ☿; 10 49 ☽ ± ♆; 13 57 ☽ ∥ ☉; 18 40 ☽ □ ♅; 19 11 ☽ ꝑ ♄; 21 07 ☽ ⊼ ♆

17 Saturday: 00 46 ☽ ⊼ ♇; 03 18 ☽ ± ☿; 08 38 ☽ ∥ ♄; 09 31 ☽ ꝑ ☉; 10 47 ☽ ± ♄; 11 06 ☽ ± ♇; 12 36 ☽ ∥ ☿; 15 14 ☽ ⊼ ☿; 17 54 ☽ ⊻ ♃; 19 19 ☽ # ♀; 19 30 ☉ ⊥ ♅; 21 13 ☽ ⊼ ♄

18 Sunday: 01 54 ☽ ∥ ♆; 02 13 ☽ ± ☉; 02 38 ☽ ꝑ ♇; 03 20 ☽ ⊻ ♂; 04 17 ☽ ⊥ ♃; 07 27 ☽ # ♆; 10 48 ☿ △ ♃; 12 47 ☽ ∥ ♀; 12 56 ☽ ☍ ♀; 13 24 ☽ ⊼ ☉; 13 52 ☽ ⊥ ♂; 19 56 ☽ ∠ ♃; 21 11 ☿ ∥ ♄; 22 40 ☽ △ ♅

19 Monday: 00 41 ♄ Q ♇; 01 06 ☽ ☍ ♆; 01 12 ☽ # ♄; 01 21 ♀ ± ♂; 01 47 ☽ # ☿; 04 44 ☽ △ ♇; 05 47 ☽ ∠ ♂; 06 41 ☿ ∠ ♅; 12 19 ☿ Q ♇; 13 03 ☿ ☌ ♄; 14 04 ☉ ⊻ ♀; 22 17 ☽ ⚹ ♃

20 Tuesday: 01 08 ☽ ꝑ ♅; 01 09 ☽ # ☉; 02 06 ☽ ☍ ♄; 04 10 ☽ ☍ ☿; 08 39 ☽ ⚹ ♂; 17 14 ☽ ± ♅; 19 42 ☽ Q ♃; 22 23 ☽ # ♂; 22 27 ☽ ☍ ☉; 23 05 ☽ ⊼ ♀

21 Wednesday: 00 50 ☽ # ♃; 04 05 ☽ ⊼ ♅; 06 32 ☽ ⊼ ♆; 06 33 ☽ Q ♂; 10 12 ☽ □ ♇; 11 02 ☽ ± ♀; 11 44 ♀ # ♆; 13 46 ☿ △ ♂; 17 29 ☽ ± ♆; 21 51 ♀ ꝑ ♃; 22 38 ☽ # ♅

22 Thursday: 04 33 ☽ □ ♃; 05 13 ☽ ꝑ ♀; 08 58 ☽ ⊼ ♄; 10 07 ☽ ꝑ ♆; 11 48 ☽ ∥ ♇; 16 13 ☽ □ ♂; 19 26 ☽ Q ♇; 20 16 ☽ ± ♄; 20 37 ☽ ⊼ ☿

23 Friday: 08 08 ♀ ⚹ ♅; 10 06 ☽ ± ☿; 10 25 ☽ ⊼ ☉; 11 56 ☽ ☍ ♅; 12 19 ☽ △ ♀; 13 28 ☽ ꝑ ♄; 14 27 ☽ △ ♆; 18 11 ☽ ⚹ ♇; 21 37 ☿ ⊥ ♅; 22 51 ☽ ± ☉

24 Saturday: 06 46 ☽ ꝑ ☿; 07 13 ☉ ⊻ ♅; 10 56 ♀ ☌ ♆; 13 39 ☽ △ ♃; 17 49 ☽ ꝑ ☉; 18 45 ☽ △ ♄; 22 19 ♀ ∥ ♆; 23 19 ☽ ∠ ♇

25 Sunday: 02 55 ☽ △ ♂; 16 22 ☉ ⊻ ♆; 17 08 ☽ ⊥ ♇; 18 20 ☽ △ ☿; 19 20 ☽ ꝑ ♃; 22 49 ☽ ⊼ ♅

26 Monday: 01 20 ♀ ⚹ ♇; 01 21 ☽ □ ♆; 02 07 ☽ △ ☉; 05 08 ☽ ⊻ ♇; 05 34 ☽ □ ♀; 09 25 ☽ ꝑ ♂; 10 54 ☽ ± ♅; 13 26 ☽ ± ♃; 20 30 ☿ ⊻ ♅

27 Tuesday: 01 36 ☽ ⊼ ♃; 03 59 ☽ ± ♂; 05 10 ☽ ꝑ ♅; 07 19 ☽ □ ♄; 10 54 ☿ ⊻ ♆; 13 47 ☽ Q ♆; 15 25 ☉ □ ♇; 16 26 ☽ ⊼ ♂; 18 11 ☽ ∥ ♇; 22 07 ☽ Q ♀

28 Wednesday: 04 42 ♀ ꝑ ♂; 07 47 ☿ □ ♇; 08 13 ☽ # ♅; 11 46 ☽ △ ♅; 14 15 ☽ ⚹ ♆; 17 56 ☽ ☌ ♇; 20 10 ☽ □ ☿; 20 16 ☽ Q ♄; 20 18 ☽ □ ☉; 21 22 ☉ ☌ ☿

29 Thursday: 01 17 ☽ ⚹ ♀; 09 21 ☽ # ♃; 10 12 ☉ ∥ ☿; 14 41 ☽ ☍ ♃; 16 40 ☽ # ♂; 18 47 ☽ # ☿; 19 42 ☽ # ☉; 20 40 ☽ ∠ ♆; 20 46 ☽ ⚹ ♄

30 Friday: 04 56 ☿ ∥ ♂; 06 31 ☽ ☍ ♂; 07 15 ☿ ⊥ ♆; 10 52 ☽ ∠ ♀; 11 44 ☽ Q ☉; 14 42 ☽ ⊥ ♆; 16 17 ☽ Q ☿

All ephemeris data is given at 12.00 UT and the Moon's longitude is additionally given for 24.00 UT

MAY 2027

LONGITUDES

Date	Sidereal time h m s	Sun ☉ ° ' "	Moon ☽ ° ' "	Moon ☽ 24.00 ° ' "	Mercury ☿ ° '	Venus ♀ ° '	Mars ♂ ° '	Jupiter ♃ ° '	Saturn ♄ ° '	Uranus ♅ ° '	Neptune ♆ ° '	Pluto ♇ ° '
01	02 36 46	10 ♉ 54 01	10 ♓ 14 14	16 ♓ 26 08	13 ♉ 59	13 ♈ 45	25 ♌ 39	17 ♌ 31	20 ♈ 38	04 ♊ 17	05 ♈ 26	07 ♒ 10
02	02 40 43	11 52 16	22 ♓ 42 56	29 ♓ 05 02	16 08	14 58	25 57	17 34	20 45	04 20	05 28	07 10
03	02 44 39	12 50 30	05 ♈ 32 45	12 ♈ 06 17	18 17	16 11	26 15	17 38	20 52	04 24	05 30	07 10
04	02 48 36	13 48 42	18 ♈ 45 42	25 ♈ 30 56	20 26	17 24	26 33	17 42	20 59	04 27	05 32	07 10
05	02 52 32	14 46 52	02 ♉ 21 45	09 ♉ 17 48	22 34	18 36	26 52	17 45	21 07	04 30	05 34	07 11
06	02 56 29	15 45 01	16 ♉ 18 37	23 ♉ 23 33	24 40	19 49	27 11	17 50	21 14	04 34	05 36	07 11
07	03 00 26	16 43 08	00 ♊ 31 56	07 ♊ 43 00	26 45	21 02	27 30	17 54	21 21	04 37	05 37	07 11
08	03 04 22	17 41 14	14 ♊ 55 57	22 ♊ 09 59	28 ♉ 49	22 15	27 50	17 58	21 28	04 40	05 39	07 11
09	03 08 19	18 39 18	29 ♊ 24 20	06 ♋ 38 21	00 ♊ 51	23 28	28 11	18 03	21 35	04 44	05 41	07 R 11
10	03 12 15	19 37 20	13 ♋ 51 23	21 ♋ 02 57	02 50	24 41	28 31	18 08	21 42	04 47	05 43	07 11
11	03 16 12	20 35 20	28 ♋ 12 38	05 ♌ 20 07	04 47	25 54	28 52	18 12	21 49	04 51	05 45	07 11
12	03 20 08	21 33 18	12 ♌ 25 14	19 ♌ 27 49	06 41	27 07	29 14	18 18	21 56	04 54	05 46	07 10
13	03 24 05	22 31 14	26 ♌ 27 48	03 ♍ 25 11	08 33	28 20	29 35	18 23	22 03	04 58	05 48	07 10
14	03 28 01	23 29 08	10 ♍ 19 58	17 ♍ 12 09	10 21	29 ♈ 33	29 ♌ 57	18 28	22 10	05 01	05 50	07 10
15	03 31 58	24 27 01	24 ♍ 01 46	00 ♎ 48 48	12 07	00 ♉ 46	00 ♍ 20	18 34	22 17	05 05	05 51	07 10
16	03 35 55	25 24 52	07 ♎ 33 15	14 ♎ 15 03	13 49	01 58	00 42	18 39	22 23	05 08	05 53	07 10
17	03 39 51	26 22 41	20 ♎ 54 08	27 ♎ 30 24	15 28	03 11	01 05	18 45	22 30	05 11	05 55	07 10
18	03 43 48	27 20 28	04 ♏ 03 45	10 ♏ 34 03	17 04	04 24	01 29	18 51	22 37	05 15	05 56	07 09
19	03 47 44	28 18 14	17 ♏ 01 12	23 ♏ 25 06	18 36	05 37	01 52	18 57	22 43	05 18	05 58	07 09
20	03 51 41	29 ♉ 15 59	29 ♏ 45 40	06 ♐ 02 52	20 05	06 50	02 16	19 03	22 50	05 22	05 59	07 09
21	03 55 37	00 ♊ 13 42	12 ♐ 16 43	18 ♐ 27 16	21 31	08 03	02 40	19 10	22 57	05 25	06 01	07 08
22	03 59 34	01 11 24	24 ♐ 34 39	00 ♑ 39 03	22 53	09 16	03 05	19 16	23 03	05 29	06 02	07 08
23	04 03 30	02 09 05	06 ♑ 40 43	12 ♑ 39 57	24 11	10 29	03 29	19 23	23 10	05 32	06 04	07 08
24	04 07 27	03 06 44	18 ♑ 37 08	24 ♑ 32 42	25 26	11 42	03 54	19 30	23 16	05 36	06 05	07 07
25	04 11 24	04 04 23	00 ♒ 27 08	06 ♒ 20 58	26 37	12 55	04 19	19 37	23 22	05 40	06 07	07 07
26	04 15 20	05 02 00	12 ♒ 14 47	18 ♒ 09 13	27 44	14 08	04 45	19 44	23 29	05 43	06 08	07 06
27	04 19 17	05 59 37	24 ♒ 04 53	00 ♓ 02 29	28 48	15 21	05 11	19 51	23 35	05 47	06 09	07 06
28	04 23 13	06 57 12	06 ♓ 02 40	12 ♓ 06 07	29 ♊ 48	16 34	05 37	19 58	23 41	05 50	06 11	07 05
29	04 27 10	07 54 47	18 ♓ 13 30	24 ♓ 25 27	00 ♋ 44	17 47	06 03	20 06	23 47	05 54	06 12	07 05
30	04 31 06	08 52 21	00 ♈ 42 33	07 ♈ 05 21	01 35	19 00	06 29	20 13	23 54	05 57	06 13	07 04
31	04 35 03	09 ♊ 49 53	13 ♈ 34 17	20 ♈ 09 40	02 ♋ 23	20 ♉ 13	06 ♍ 56	20 ♌ 21	24 ♈ 00	06 ♊ 01	06 ♈ 14	07 ♒ 04

Date	Moon True ☊ ° '	Moon Mean ☊ ° '	Moon ☽ Latitude ° '
01	16 ♒ 44	16 ♒ 28	02 N 01
02	16 R 40	16 24	02 58
03	16 34	16 21	03 48
04	16 25	16 18	04 27
05	16 14	16 15	04 52
06	16 02	16 12	05 00
07	15 50	16 09	04 51
08	15 39	16 05	04 23
09	15 32	16 02	03 37
10	15 27	15 59	02 38
11	15 24	15 56	01 30
12	15 D 24	15 53	00 N 16
13	15 R 24	15 49	00 S 58
14	15 23	15 46	02 08
15	15 20	15 43	03 09
16	15 15	15 40	03 59
17	15 07	15 37	04 35
18	14 57	15 34	04 56
19	14 45	15 30	05 01
20	14 32	15 27	04 51
21	14 20	15 24	04 27
22	14 10	15 21	03 50
23	14 02	15 18	03 04
24	13 57	15 15	02 10
25	13 54	15 11	01 11
26	13 53	15 08	00 S 09
27	13 D 53	15 05	00 N 54
28	13 54	15 02	01 55
29	13 R 53	14 59	02 52
30	13 51	14 55	03 43
31	13 ♒ 47	14 ♒ 52	04 N 24

DECLINATIONS

Date	Sun ☉ ° '	Moon ☽ ° '	Mercury ☿ ° '	Venus ♀ ° '	Mars ♂ ° '	Jupiter ♃ ° '	Saturn ♄ ° '	Uranus ♅ ° '	Neptune ♆ ° '	Pluto ♇ ° '
01	15 N 06	05 S 52	16 N 11	03 N 54	14 N 57	16 N 32	05 N 58	20 N 53	00 N 55	22 S 52
02	15 24	00 S 10	16 59	04 22	14 49	16 31	06 01	20 54	00 55	22 52
03	15 42	05 N 41	17 46	04 50	14 41	16 30	06 03	20 55	00 56	22 52
04	15 59	11 27	18 31	05 18	14 33	16 29	06 06	20 55	00 57	22 52
05	16 16	16 51	19 14	05 45	14 25	16 27	06 09	20 56	00 57	22 52
06	16 33	21 30	19 56	06 13	14 17	16 26	06 11	20 57	00 58	22 53
07	16 50	25 00	20 35	06 40	14 08	16 24	06 14	20 57	00 59	22 53
08	17 06	26 56	21 12	07 08	14 00	16 23	06 16	20 58	01 00	22 53
09	17 22	27 04	21 47	07 35	13 51	16 22	06 19	20 58	01 00	22 53
10	17 38	25 21	22 19	08 02	13 42	16 20	06 21	20 59	01 01	22 53
11	17 54	21 59	22 49	08 29	13 33	16 18	06 24	21 00	01 02	22 54
12	18 09	17 20	23 16	08 56	13 24	16 17	06 26	21 00	01 02	22 54
13	18 24	11 47	23 41	09 22	13 15	16 15	06 29	21 01	01 03	22 54
14	18 39	05 N 43	24 03	09 49	13 06	16 13	06 31	21 02	01 03	22 54
15	18 53	00 S 31	24 22	10 15	12 56	16 11	06 34	21 02	01 04	22 55
16	19 07	06 39	24 39	10 41	12 47	16 10	06 36	21 03	01 05	22 55
17	19 21	12 24	24 54	11 06	12 37	16 08	06 38	21 03	01 05	22 55
18	19 34	17 30	25 06	11 32	12 27	16 06	06 41	21 04	01 06	22 55
19	19 47	21 43	25 16	11 57	12 17	16 04	06 43	21 05	01 06	22 56
20	19 59	24 50	25 24	12 22	12 07	16 02	06 45	21 05	01 07	22 56
21	20 12	26 40	25 30	12 46	11 57	16 00	06 48	21 06	01 08	22 56
22	20 24	27 10	25 34	13 11	11 47	15 57	06 50	21 07	01 08	22 56
23	20 35	26 20	25 36	13 35	11 36	15 55	06 52	21 07	01 09	22 57
24	20 47	24 18	25 36	13 59	11 26	15 53	06 55	21 08	01 09	22 57
25	20 58	21 12	25 35	14 22	11 15	15 51	06 57	21 08	01 10	22 57
26	21 08	17 16	25 32	14 45	11 04	15 48	06 59	21 09	01 10	22 58
27	21 18	12 39	25 27	15 08	10 54	15 46	07 01	21 10	01 11	22 58
28	21 28	07 30	25 21	15 30	10 43	15 44	07 03	21 10	01 11	22 58
29	21 38	02 S 01	25 14	15 52	10 32	15 41	07 05	21 11	01 12	22 59
30	21 47	03 N 41	25 05	16 14	10 21	15 39	07 07	21 12	01 12	22 59
31	21 N 55	09 N 24	24 N 56	16 N 35	10 N 09	15 N 36	07 N 09	21 N 12	01 N 13	22 S 59

ZODIAC SIGN ENTRIES

Date	h	m	Planets
03	01	43	☽ ♈
05	07	53	☽ ♉
07	11	06	☽ ♊
09	01	58	☿ ♊
09	12	59	☽ ♋
11	15	00	☽ ♌
13	18	05	☽ ♍
14	14	47	♂ ♍
14	21	01	♀ ♉
15	22	33	☽ ♎
18	04	33	☽ ♏
20	12	27	☽ ♐
21	06	18	☉ ♊
22	22	43	☽ ♑
25	11	05	☽ ♒
27	23	55	☽ ♓
28	17	06	☿ ♋
30	10	39	☽ ♈

LATITUDES

Date	Mercury ☿ ° '	Venus ♀ ° '	Mars ♂ ° '	Jupiter ♃ ° '	Saturn ♄ ° '	Uranus ♅ ° '	Neptune ♆ ° '	Pluto ♇ ° '
01	00 N 09	01 S 39	02 N 06	01 N 00	02 S 15	00 S 07	01 S 22	04 S 32
04	00 41	01 40	02 01	01 00	02 15	00 07	01 22	04 33
07	01 11	01 39	01 55	00 59	02 16	00 07	01 22	04 33
10	01 37	01 38	01 50	00 59	02 16	00 07	01 22	04 34
13	01 58	01 37	01 45	00 59	02 16	00 07	01 22	04 35
16	02 13	01 35	01 40	00 58	02 17	00 07	01 22	04 35
19	02 20	01 32	01 35	00 58	02 17	00 06	01 22	04 36
22	02 19	01 29	01 30	00 58	02 17	00 06	01 22	04 36
25	02 11	01 25	01 26	00 58	02 18	00 06	01 23	04 37
28	01 55	01 20	01 21	00 57	02 18	00 06	01 23	04 38
31	01 N 31	01 S 16	01 N 17	00 N 57	02 S 19	00 S 06	01 S 23	04 S 38

DATA

Julian Date	2461527
Delta T	+73 seconds
Ayanamsa	24° 14' 31"
Synetic vernal point	04° ♓ 52' 28"
True obliquity of ecliptic	23° 26' 16"

MOON'S PHASES, APSIDES AND POSITIONS ☽

Date	h	m	Phase	Longitude ° '	Eclipse Indicator
06	10	59	●	15 ♉ 43	
13	04	44	☽	22 ♌ 14	
20	10	59	○	29 ♏ 14	
28	13	58	☾	07 ♓ 02	

Day	h	m		
09	20	03	Perigee	
25	15	07	Apogee	
02	12	41	ON	
09	01	38	Max dec	27° N 14'
15	09	59	OS	
22	08	48	Max dec	27° S 10'
29	20	31	ON	

LONGITUDES

Date	Chiron ⚷ ° '	Ceres ⚳ ° '	Pallas ⚴ ° '	Juno ⚵ ° '	Vesta ⚶ ° '	Black Moon Lilith ⚸ ° '
01	01 ♉ 02	22 ♋ 51	27 ♉ 07	25 ♈ 14	29 ♉ 55	25 ♑ 28
11	01 ♉ 37	26 ♋ 16	02 ♊ 40	00 ♉ 49	04 ♊ 13	26 ♑ 34
21	02 ♉ 12	29 ♋ 54	08 ♊ 21	06 ♉ 27	08 ♊ 33	27 ♑ 41
31	02 ♉ 44	03 ♌ 43	14 ♊ 07	12 ♉ 05	12 ♊ 53	28 ♑ 47

All ephemeris data is given at 12.00 UT and the Moon's longitude is additionally given for 24.00 UT

ASPECTARIAN

h m Aspects

01 Saturday
00 17 ☽ □ ♅
02 35 ☽ ⊻ ♆
02 54 ☽ ∠ ♄
03 58 ☉ ∥ ♂
06 00 ☽ ⊻ ♇
06 11 ☿ ⊻ ♀
06 37 ☽ ⊥ ♀
11 32 ☽ # ♄
13 24 ☽ ✶ ☉
17 42 ☽ ⊥ ♇
19 33 ☽ ⊻ ♀
19 45 ☽ # ♀
20 36 ☽ ⊥ ♄
20 47 ☽ ✶ ☿
22 18 ☿ ∥ ♃

02 Sunday
01 41 ☉ ⊥ ♆
02 08 ☽ ⊼ ♃
08 13 ☽ ⊻ ♄
08 52 ☽ # ♆
10 58 ☽ ∠ ♇
11 17 ☽ Q ♅
13 38 ☽ ± ♃
16 30 ☽ ∥ ♆
18 15 ☽ ⊼ ♂
20 30 ☽ ∠ ☉

03 Monday
04 28 ☿ □ ♃
05 45 ☽ ± ♂
06 35 ☽ ⚼ ♃
07 00 ☽ ∠ ☿
08 12 ☽ ∥ ♀
09 52 ☽ ✶ ♅
11 55 ☽ ☌ ♆
13 32 ☽ ∥ ♄
14 35 ☽ ⊥ ☉
15 00 ☽ ✶ ♇
22 41 ☽ ⚼ ♂

04 Tuesday
02 23 ☽ ⊻ ☉
02 43 ☽ ⊥ ☿
09 18 ☽ ☌ ♀
10 05 ☽ △ ♃
12 44 ☽ Q ♇
13 09 ☿ ∠ ♆
13 14 ☽ ∠ ♅
15 33 ☽ ⊻ ☿
16 01 ☽ ☌ ♄
18 16 ♀ △ ♃
18 40 ☿ ⊻ ♄

05 Wednesday
01 09 ☽ ∥ ♂
02 09 ☽ △ ♂
05 14 ☽ ⊥ ♅
09 07 ☽ ∥ ☉
10 08 ☽ ∥ ♃
15 44 ☽ ⊻ ♅
17 34 ☽ ⊻ ♆
20 21 ☽ □ ♇
23 15 ♀ Q ♇

06 Thursday
01 46 ☽ ∥ ☿
02 22 ☉ ∥ ♃
03 55 ☽ ⊥ ♆
06 37 ♀ ∠ ♅
08 48 ☽ ∥ ♅
10 20 ♀ ∥ ♄
10 59 ☽ ☌ ☉
14 35 ☽ □ ♃
18 31 ☽ ⊻ ♀
19 17 ☽ ∠ ♆
20 23 ☽ # ♇
20 25 ☽ ⊻ ♄

07 Friday
04 35 ☽ ☌ ☿
05 35 ☽ ⊥ ♀
06 37 ☽ ⊥ ♄
06 48 ☽ □ ♂
18 48 ♀ ☌ ♄
18 52 ☽ ☌ ♅
19 15 ☿ ⊥ ♄
19 50 ☿ ⊥ ♀
20 32 ☽ ✶ ♆
21 01 ☽ Q ♃
21 48 ☽ ∠ ♄
22 03 ☽ ∠ ♀
22 21 ☿ □ ♂
23 06 ☽ △ ♇

08 Saturday
02 24 ☿ ∥ ♅
12 55 ♇ St R
13 33 ☽ Q ♂
16 32 ☽ Q ♆
16 54 ☽ ⊻ ☉
17 04 ☽ ✶ ♃
19 38 ☉ □ ♃
22 56 ☽ ✶ ♄

09 Sunday
00 01 ☽ ⚼ ♇
01 15 ☽ ✶ ♀
03 34 ☽ ⊥ ☉
09 55 ☽ ✶ ♂
14 46 ☽ ⊻ ☿
14 56 ☽ ± ♇
18 04 ☽ ∠ ♃
18 59 ☽ Q ♄
19 33 ☽ ∠ ☉
20 52 ☽ ⊻ ♅
22 26 ☽ □ ♆
22 59 ☽ Q ♀

10 Monday
00 54 ☽ ⊼ ♇
02 19 ☽ ⊥ ☿
06 52 ☽ ⊥ ♅
09 06 ☽ ⊥ ♃
11 26 ☽ ∠ ♂
19 09 ☽ ⊻ ♃
19 41 ☽ ∠ ☿
21 56 ☽ ∠ ♅
22 19 ☽ ✶ ☉

11 Tuesday
01 12 ☽ □ ♄
02 49 ☽ ⊥ ♂
06 25 ☽ # ♇
07 30 ☽ ∥ ☿
07 45 ☽ □ ♀
12 49 ☿ ☌ ♅
13 08 ☽ ⊻ ♂
15 55 ☉ ∠ ♆
16 11 ☿ # ♇
17 36 ☽ ∥ ♅
19 54 ☽ Q ☉
23 13 ☽ ✶ ♅

12 Wednesday
00 13 ☿ ✶ ♆
00 43 ☽ △ ♆
00 47 ☽ ✶ ☿
03 07 ☽ ☍ ♇
06 45 ☿ Q ♃
08 18 ☽ ∥ ☉
15 20 ☿ ∠ ♄
16 49 ☽ ∥ ♃
18 14 ☿ △ ♇
19 40 ☽ Q ♅
22 04 ☽ ☌ ♃
22 39 ☉ ⊻ ♄

13 Thursday
00 19 ☽ Q ☿
02 16 ☽ ⚼ ♆
04 21 ☽ △ ♄
04 44 ☽ □ ☉
05 45 ☽ ∥ ♂
15 31 ☽ △ ♀
17 32 ☽ ☌ ♂
17 45 ☽ ± ♆
21 03 ☽ ∥ ♀

14 Friday
01 07 ♀ ⊥ ♅
02 44 ☽ □ ♅
02 52 ♂ ± ♆
04 09 ☽ ⊼ ♆
06 26 ☽ ⚼ ♄
06 30 ☽ ⊼ ♇
08 55 ☽ ∥ ♄
12 03 ☽ □ ☿
16 57 ☽ ± ♇
20 04 ☽ ⚼ ♀
22 16 ☽ ± ♄
23 46 ♀ △ ♂

15 Saturday
02 19 ☽ ⊻ ♃
05 53 ☽ ∥ ♆
08 43 ☽ ⚼ ♇
08 53 ☽ ⊼ ♄
12 48 ☽ △ ☉
12 57 ☽ ⊥ ♃
13 25 ☽ ± ♀
14 06 ☽ # ♆
23 28 ☽ ⊻ ♂

16 Sunday
01 05 ☽ ⊼ ♀
05 00 ☽ ∠ ♃
07 40 ☽ △ ♅
09 01 ☽ ☍ ♆
10 27 ☽ ⊥ ♂
11 18 ☽ △ ♇
11 47 ☽ # ♄
17 31 ☽ ⚼ ☉

17 Monday
00 49 ☽ △ ☿
03 03 ☽ ∠ ♂
05 56 ☽ # ♀
08 05 ☽ ✶ ♃
10 42 ☽ ⚼ ♅
10 59 ☽ ± ☉
12 55 ☽ # ♂
14 55 ☽ ☍ ♄
22 44 ☽ ⊼ ☉

18 Tuesday
03 09 ☽ ± ♅
05 03 ☽ # ♃
06 04 ☽ Q ♃
07 07 ☽ ✶ ♂
07 50 ☽ ⚼ ☿
12 42 ☽ ☍ ♀
14 12 ☽ ⊼ ♅
15 27 ☽ ⊼ ♆
17 41 ☽ □ ♇
23 38 ☽ # ☉

19 Wednesday
01 39 ☿ Q ♆
02 34 ☽ ± ♆
02 41 ☽ ± ☿
05 30 ♀ ⊻ ♅
05 57 ☽ Q ♂
07 57 ☽ # ♅
15 22 ☽ ⊼ ☿
15 39 ☽ □ ♃
17 54 ☿ ✶ ♃
18 52 ♀ ⊻ ♆
19 24 ☽ ⚼ ♆
20 19 ☽ ∥ ♇
22 47 ☽ ⊼ ♄
23 50 ☉ ⊥ ♄

20 Thursday
01 51 ♀ ∥ ♂
03 16 ☽ Q ♇
10 13 ☽ ± ♄
10 59 ☽ ☍ ☉
16 04 ☿ Q ♂
16 56 ☽ □ ♂
18 03 ♀ □ ♇
18 16 ☽ # ☿
22 45 ☽ ☍ ♅
23 54 ☽ △ ♆

21 Friday
02 06 ☽ ✶ ♇
02 58 ☽ ⊼ ♀
03 34 ☽ ⚼ ♄
15 49 ☽ ± ♀
22 48 ☿ ⚼ ♇

22 Saturday
01 30 ☽ △ ♃
07 12 ☽ ∠ ♇
08 15 ☽ ☍ ☿
08 59 ☽ △ ♄
11 20 ☽ ⚼ ♀
15 22 ☿ ✶ ♄

23 Sunday
00 57 ☽ ⊥ ♇
02 12 ☽ ⊼ ☉
05 25 ☽ △ ♂
07 23 ☽ ⚼ ♃
09 43 ☽ ⊼ ♅
10 46 ☽ □ ♆
12 54 ☽ ⊻ ♇
15 12 ☽ ± ☉
20 29 ☽ △ ♀
21 47 ☽ ± ♅
22 13 ☽ # ☿

24 Monday
01 34 ☽ ± ♃
10 53 ☽ ⚼ ☉
12 36 ☽ ⚼ ♂
13 47 ☽ ⊼ ♃
16 02 ☽ ⚼ ♅
19 42 ♀ ⊥ ♆
21 30 ☽ □ ♄
23 05 ☽ Q ♆
23 20 ☽ ∥ ♇

25 Tuesday
03 21 ☽ ⊼ ☿
07 31 ☽ ± ♂
12 27 ☽ # ♅
13 35 ☽ # ☉
16 53 ☽ ± ☿
20 01 ☽ △ ☉
20 10 ☽ ⊼ ♂
22 39 ☽ △ ♅
23 13 ☉ □ ♂
23 32 ☽ ✶ ♆

26 Wednesday
01 33 ☽ ☌ ♇
10 26 ☽ Q ♄
13 07 ☽ ⚼ ☿
14 23 ☉ ∥ ♅
16 17 ☽ □ ♀
19 58 ☽ # ♃

27 Thursday
00 24 ☽ # ♀
03 21 ☽ ☍ ♃
06 04 ☽ ∠ ♆
06 12 ☉ ☌ ♅
10 59 ☽ ✶ ♄
16 08 ☉ ✶ ♆
20 43 ☽ # ♂
22 23 ☽ △ ☿

28 Friday
00 15 ☽ ⊥ ♆
08 44 ☽ Q ♀
11 06 ☽ ☍ ♂
11 35 ☽ □ ♅
12 16 ☽ ⊻ ♆
13 58 ☽ □ ☉
14 01 ☽ # ♄
14 04 ☽ ⊻ ♇
15 20 ☉ △ ♇
17 18 ☽ ∠ ♄

29 Saturday
01 10 ♀ ∥ ♃
01 56 ☽ ⊥ ♇
02 14 ♂ □ ♅
11 03 ☽ ✶ ♀
11 09 ☽ ⊥ ♄
15 27 ☽ # ♆
15 40 ☽ ⊼ ♃
17 14 ☉ Q ♃
19 28 ☽ ∠ ♇
20 44 ♂ ⊼ ♆
21 27 ☿ ± ♇
22 52 ☽ ⊻ ♄
23 02 ☽ Q ♅

30 Sunday
01 34 ☽ ∥ ♆
03 22 ☽ ± ♃
04 05 ☽ Q ☉
12 35 ☉ ∠ ♄
13 47 ☽ □ ☿
18 53 ☽ ∠ ♀
20 36 ☽ ⚼ ♃
21 55 ☽ ✶ ♅
22 24 ☽ ☌ ♆
23 16 ☽ ⊼ ♂
23 57 ☽ ✶ ♇

31 Monday
02 28 ☽ ∥ ♄
04 33 ☽ ✶ ☉
10 47 ☽ ± ♂
13 19 ☽ ⊥ ♀
14 47 ♀ □ ♃
15 07 ☽ ∥ ♂
18 38 ♂ ⊼ ♇
22 00 ☽ Q ♇

JUNE 2027

LONGITUDES

Date	Sidereal time h m s	Sun ☉ ° ' "	Moon ☽ ° ' "	Moon ☽ 24.00 ° ' "	Mercury ☿ ° '	Venus ♀ ° '	Mars ♂ ° '	Jupiter ♃ ° '	Saturn ♄ ° '	Uranus ♅ ° '	Neptune ♆ ° '	Pluto ♇ ° '
01	04 38 59	10 ♊ 47 25	26 ♈ 51 45	03 ♉ 40 35	03 ♋ 07	21 ♉ 27	07 ♍ 23	20 ♌ 29	24 ♈ 06	06 ♊ 04	06 ♈ 16	07 ♒ 03
02	04 42 56	11 44 57	10 ♉ 36 04	17 ♉ 37 56	03 46	22 40	07 50	20 37	24 12	06 08	06 17	07 R 02
03	04 46 53	12 42 27	24 ♉ 45 45	01 ♊ 58 52	04 21	23 53	08 17	20 45	24 18	06 11	06 18	07 02
04	04 50 49	13 39 57	09 ♊ 16 33	16 ♊ 37 52	04 52	25 06	08 45	20 53	24 23	06 15	06 19	07 01
05	04 54 46	14 37 25	24 ♊ 01 50	01 ♋ 27 25	05 18	26 19	09 13	21 01	24 29	06 18	06 20	07 00
06	04 58 42	15 34 53	08 ♋ 53 33	16 ♋ 19 14	05 40	27 32	09 41	21 10	24 35	06 22	06 21	07 00
07	05 02 39	16 32 20	23 ♋ 43 32	01 ♌ 05 36	05 57	28 45	10 09	21 18	24 41	06 25	06 22	06 59
08	05 06 35	17 29 45	08 ♌ 24 45	15 ♌ 40 25	06 10	29 ♉ 58	10 38	21 27	24 46	06 29	06 23	06 58
09	05 10 32	18 27 10	22 ♌ 52 11	29 ♌ 59 44	06 18	01 ♊ 12	11 06	21 36	24 52	06 32	06 24	06 57
10	05 14 28	19 24 33	07 ♍ 02 55	14 ♍ 01 40	06 22	02 25	11 35	21 45	24 57	06 35	06 25	06 56
11	05 18 25	20 21 55	20 ♍ 55 59	27 ♍ 45 56	06 R 20	03 38	12 04	21 54	25 02	06 39	06 26	06 55
12	05 22 22	21 19 16	04 ♎ 31 41	11 ♎ 13 20	06 15	04 51	12 34	22 03	25 08	06 42	06 27	06 55
13	05 26 18	22 16 36	17 ♎ 51 05	24 ♎ 25 06	06 05	06 04	13 03	22 12	25 13	06 46	06 28	06 54
14	05 30 15	23 13 55	00 ♏ 55 32	07 ♏ 22 34	05 51	07 17	13 33	22 21	25 18	06 49	06 29	06 53
15	05 34 11	24 11 13	13 ♏ 46 19	20 ♏ 06 55	05 33	08 31	14 02	22 31	25 23	06 53	06 30	06 52
16	05 38 08	25 08 30	26 ♏ 24 31	02 ♐ 39 11	05 12	09 44	14 32	22 40	25 28	06 56	06 31	06 51
17	05 42 04	26 05 47	08 ♐ 51 04	15 ♐ 00 16	04 47	10 57	15 03	22 50	25 33	06 59	06 31	06 50
18	05 46 01	27 03 03	21 ♐ 06 54	27 ♐ 11 07	04 19	12 10	15 33	22 59	25 38	07 03	06 32	06 49
19	05 49 57	28 00 19	03 ♑ 13 04	09 ♑ 12 57	03 49	13 23	16 03	23 09	25 43	07 06	06 33	06 48
20	05 53 54	28 57 34	15 ♑ 11 00	21 ♑ 07 27	03 17	14 37	16 34	23 19	25 48	07 09	06 33	06 47
21	05 57 51	29 ♊ 54 48	27 ♑ 02 36	02 ♒ 56 47	02 43	15 50	17 05	23 29	25 53	07 13	06 34	06 46
22	06 01 47	00 ♋ 52 02	08 ♒ 50 25	14 ♒ 43 52	02 09	17 03	17 36	23 39	25 57	07 16	06 35	06 45
23	06 05 44	01 49 16	20 ♒ 37 38	26 ♒ 32 12	01 34	18 17	18 07	23 49	26 02	07 19	06 35	06 44
24	06 09 40	02 46 30	02 ♓ 28 06	08 ♓ 25 54	00 59	19 30	18 38	23 59	26 06	07 23	06 36	06 43
25	06 13 37	03 43 43	14 ♓ 26 11	20 ♓ 29 34	00 ♋ 26	20 43	19 10	24 10	26 11	07 26	06 36	06 42
26	06 17 33	04 40 57	26 ♓ 36 40	02 ♈ 48 04	29 ♊ 54	21 57	19 41	24 20	26 15	07 29	06 37	06 40
27	06 21 30	05 38 10	09 ♈ 04 22	15 ♈ 26 08	29 24	23 10	20 13	24 30	26 19	07 32	06 37	06 39
28	06 25 26	06 35 24	21 ♈ 53 52	28 ♈ 28 00	28 56	24 23	20 45	24 41	26 23	07 35	06 37	06 38
29	06 29 23	07 32 37	05 ♉ 08 52	11 ♉ 56 41	28 32	25 37	21 17	24 52	26 27	07 39	06 38	06 37
30	06 33 20	08 ♋ 29 51	18 ♉ 51 32	25 ♉ 53 20	28 ♊ 11	26 ♊ 50	21 ♍ 49	25 ♌ 02	26 ♈ 31	07 ♊ 42	06 ♈ 38	06 ♒ 36

Date	Moon True ☊ ° '	Moon Mean ☊ ° '	Moon ☽ Latitude ° '
01	13 ♒ 40	14 ♒ 49	04 N 52
02	13 R 32	14 46	05 04
03	13 22	14 43	04 58
04	13 13	14 40	04 34
05	13 05	14 36	03 51
06	12 59	14 33	02 52
07	12 55	14 30	01 41
08	12 54	14 27	00 N 24
09	12 D 54	14 24	00 S 53
10	12 55	14 21	02 06
11	12 R 55	14 17	03 10
12	12 54	14 14	04 01
13	12 52	14 11	04 39
14	12 47	14 08	05 01
15	12 40	14 05	05 07
16	12 32	14 01	04 59
17	12 24	13 58	04 36
18	12 16	13 55	04 00
19	12 09	13 52	03 14
20	12 04	13 49	02 20
21	12 01	13 46	01 20
22	12 00	13 42	00 S 17
23	12 D 00	13 39	00 N 47
24	12 01	13 36	01 49
25	12 03	13 33	02 48
26	12 04	13 30	03 39
27	12 R 04	13 27	04 22
28	12 03	13 23	04 53
29	12 01	13 20	05 10
30	11 ♒ 57	13 ♒ 17	05 N 10

DECLINATIONS

Sun ☉ ° '	Moon ☽ ° '	Mercury ☿ ° '	Venus ♀ ° '	Mars ♂ ° '	Jupiter ♃ ° '	Saturn ♄ ° '	Uranus ♅ ° '	Neptune ♆ ° '	Pluto ♇ ° '
22 N 04	14 N 53	24 N 45	16 N 56	09 N 58	15 N 34	07 N 12	21 N 13	01 N 13	23 S 00
22 12	19 48	24 33	17 16	09 47	15 31	07 14	21 13	01 13	23 00
22 19	23 47	24 21	17 36	09 35	15 28	07 16	21 14	01 14	23 00
22 26	26 21	24 08	17 56	09 23	15 26	07 18	21 15	01 14	23 01
22 33	27 09	23 54	18 15	09 12	15 23	07 19	21 15	01 15	23 01
22 39	26 00	23 39	18 33	09 00	15 20	07 21	21 16	01 15	23 01
22 45	23 01	23 24	18 52	08 48	15 17	07 23	21 16	01 15	23 02
22 51	18 33	23 09	19 09	08 36	15 15	07 25	21 17	01 16	23 02
22 56	13 03	22 53	19 26	08 24	15 12	07 27	21 18	01 16	23 03
23 01	06 58	22 37	19 43	08 12	15 09	07 29	21 18	01 17	23 03
23 05	00 N 41	22 20	19 59	07 59	15 06	07 31	21 19	01 17	23 03
23 09	05 S 30	22 04	20 15	07 47	15 03	07 32	21 19	01 17	23 04
23 13	11 18	21 47	20 30	07 35	15 00	07 34	21 20	01 17	23 04
23 16	16 29	21 31	20 44	07 22	14 57	07 36	21 21	01 18	23 05
23 19	20 51	21 15	20 58	07 10	14 53	07 37	21 21	01 18	23 05
23 21	24 11	20 59	21 11	06 57	14 50	07 39	21 22	01 18	23 05
23 23	26 19	20 43	21 24	06 44	14 47	07 41	21 22	01 18	23 06
23 24	27 08	20 28	21 36	06 31	14 44	07 42	21 23	01 19	23 06
23 25	26 38	20 14	21 48	06 18	14 41	07 44	21 23	01 19	23 07
23 26	24 54	20 00	21 59	06 05	14 37	07 45	21 24	01 19	23 07
23 26	22 04	19 47	22 09	05 52	14 34	07 47	21 25	01 19	23 08
23 26	18 19	19 35	22 19	05 39	14 31	07 48	21 25	01 20	23 08
23 25	13 52	19 25	22 28	05 26	14 27	07 50	21 26	01 20	23 08
23 24	08 54	19 15	22 37	05 12	14 24	07 51	21 26	01 20	23 09
23 23	03 S 33	19 06	22 45	04 59	14 20	07 52	21 27	01 20	23 09
23 21	02 N 01	18 59	22 52	04 46	14 17	07 54	21 27	01 20	23 10
23 19	07 37	18 53	22 58	04 32	14 13	07 55	21 28	01 20	23 10
23 16	13 04	18 49	23 04	04 19	14 10	07 56	21 28	01 20	23 11
23 13	18 06	18 46	23 09	04 05	14 06	07 57	21 29	01 20	23 11
23 N 10	22 N 24	18 N 44	23 N 14	03 N 51	14 N 02	07 N 59	21 N 29	01 N 21	23 S 12

ZODIAC SIGN ENTRIES

Date	h m	Planets
01	17 33	☽ ♉
03	20 43	☽ ♊
05	21 39	☽ ♋
07	22 13	☽ ♌
08	12 32	♀ ♊
10	00 00	☽ ♍
12	03 57	☽ ♎
14	10 17	☽ ♏
16	18 53	☽ ♐
19	05 35	☽ ♑
21	14 11	☉ ♋
21	18 00	☽ ♒
24	07 01	☽ ♓
26	07 19	☿ ♊
26	18 35	☽ ♈
29	02 46	☽ ♉

LATITUDES

Date	Mercury ☿ ° '	Venus ♀ ° '	Mars ♂ ° '	Jupiter ♃ ° '	Saturn ♄ ° '	Uranus ♅ ° '	Neptune ♆ ° '	Pluto ♇ ° '
01	01 N 21	01 S 14	01 N 16	00 N 57	02 S 19	00 S 06	01 S 23	04 S 38
04	00 47	01 09	01 11	00 57	02 19	00 06	01 23	04 39
07	00 N 06	01 03	01 07	00 57	02 20	00 06	01 23	04 40
10	00 S 41	00 57	01 03	00 56	02 20	00 06	01 23	04 40
13	01 31	00 50	01 00	00 56	02 21	00 06	01 23	04 41
16	02 21	00 43	00 56	00 56	02 22	00 06	01 24	04 41
19	03 09	00 37	00 52	00 56	02 22	00 06	01 24	04 42
22	03 50	00 29	00 49	00 56	02 23	00 06	01 24	04 42
25	04 20	00 22	00 45	00 55	02 24	00 06	01 24	04 43
28	04 37	00 15	00 42	00 55	02 24	00 06	01 24	04 43
31	04 S 41	00 S 08	00 N 39	00 N 55	02 S 25	00 S 06	01 S 24	04 S 44

DATA

Julian Date	2461558
Delta T	+73 seconds
Ayanamsa	24° 14' 35"
Synetic vernal point	04° ♓ 52' 24"
True obliquity of ecliptic	23° 26' 15"

MOON'S PHASES, APSIDES AND POSITIONS ☽

Date	h m	Phase	Longitude ° '	Eclipse Indicator
04	19 40	●	13 ♊ 58	
11	10 56	☽	20 ♍ 19	
19	00 44	○	27 ♐ 33	
27	04 54	☾	05 ♈ 21	

Day	h m		
06	14 48	Perigee	
22	04 55	Apogee	
05	09 53	Max dec	27° N 09'
11	14 37	0S	
18	14 51	Max dec	27° S 09'
26	03 23	0N	

LONGITUDES

Date	Chiron ⚷ ° '	Ceres ⚳ ° '	Pallas ⚴ ° '	Juno ⚵ ° '	Vesta ⚶ ° '	Black Moon Lilith ⚸ ° '
01	02 ♉ 47	04 ♌ 07	14 ♊ 42	12 ♉ 39	13 ♊ 19	28 ♑ 54
11	03 ♉ 16	08 ♌ 07	20 ♊ 33	18 ♉ 19	17 ♊ 39	00 ♒ 01
21	03 ♉ 41	12 ♌ 15	26 ♊ 28	23 ♉ 59	21 ♊ 58	01 ♒ 07
31	04 ♉ 03	16 ♌ 30	02 ♋ 25	29 ♉ 38	26 ♊ 17	02 ♒ 14

All ephemeris data is given at 12.00 UT and the Moon's longitude is additionally given for 24.00 UT

ASPECTARIAN

h m — Aspects

01 Tuesday
00 28 ☽ △ ♃
01 08 ☽ Q ☿
01 20 ☽ ⊻ ♀
01 35 ☽ ∠ ♅
03 43 ☽ ₽ ♂
07 02 ☽ ☌ ♄
08 23 ♀ ∠ ♆
09 57 ☽ ∠ ☉
15 06 ☽ ∥ ♃
17 42 ☽ ⊥ ♅
22 16 ☽ ∥ ♀
23 35 ☽ ⚹ ☿

02 Wednesday
02 59 ☽ ⊥ ☉
04 14 ☽ ⊻ ♅
04 32 ☽ ⊻ ♆
05 51 ☽ □ ♇
07 04 ☽ △ ♂
14 07 ☽ ⊻ ☉
14 53 ☽ ⊥ ♆
19 49 ☽ ∥ ♅

03 Thursday
02 00 ☽ ∥ ☉
02 32 ☽ ∠ ☿
05 12 ☽ □ ♃
06 11 ☽ ∠ ♆
06 42 ☽ ⋕ ♇
10 23 ☽ ☌ ♀
11 12 ☽ ⊻ ♄
15 59 ☽ ∥ ☿
18 13 ☽ ⊥ ☿
20 52 ♀ ⊻ ♄
21 16 ☽ ⊥ ♄

04 Friday
04 31 ☽ ⊻ ☿
07 00 ☽ ☌ ♅
07 09 ☽ ⚹ ♆
08 18 ☽ △ ♇
11 07 ☽ □ ♂
11 21 ☽ Q ♃
12 11 ☽ ∠ ♄
19 40 ☽ ☌ ☉

05 Saturday
02 46 ☽ Q ♆
07 05 ☽ ⚹ ♃
08 43 ☽ ₽ ♇
12 44 ☽ ⚹ ♄
16 02 ☽ ⊻ ♀
17 19 ☽ Q ♂
23 16 ☽ ± ♇

06 Sunday
02 35 ☽ ⊥ ♀
05 31 ♂ ₽ ♄
06 41 ☽ ☌ ☿
07 33 ☽ ∠ ♃
07 54 ☽ ⊻ ♅
07 54 ☽ □ ♆
08 15 ☽ Q ♄
08 56 ☽ ⊼ ♇
10 08 ♅ ⚹ ♆
13 19 ☽ ⚹ ♂
17 37 ☽ ⊥ ♅
18 24 ☽ ∠ ♀
22 14 ☽ ⊥ ♃
23 33 ☽ ⊻ ☉

07 Monday
08 02 ☽ ⊻ ♃
08 14 ☽ ∠ ♅
09 17 ☽ ∥ ☿
09 56 ☽ ⊥ ☉
11 53 ☽ ⋕ ♇
13 33 ☽ ∥ ☉
13 33 ☽ □ ♄
14 24 ☽ ∠ ♂
20 55 ☽ ⚹ ♀
22 09 ☽ ∥ ♅

08 Tuesday
01 02 ♃ ₽ ♆
01 37 ☽ ∠ ☉
05 35 ☽ ⊥ ♂
08 16 ☽ ⊻ ☿
08 40 ☽ △ ♆
08 48 ☽ ⚹ ♅
09 15 ☽ ∥ ♀
09 37 ☽ ☍ ♇
15 46 ☽ ⊻ ♂
16 26 ☿ ⊥ ♀
18 16 ☽ ⊥ ☿
18 25 ☽ Q ♀
21 26 ☿ ⋕ ♇

09 Wednesday
02 58 ☽ ∥ ♃
04 06 ☽ ⚹ ☉
04 44 ☽ Q ♅
04 57 ♀ ⊥ ♄
08 07 ☉ ∥ ☿
09 21 ☽ ∠ ☿
09 33 ☽ ₽ ♆
09 51 ☽ ☌ ♃
10 51 ☉ Q ♆
15 22 ☽ △ ♄

10 Thursday
00 43 ☽ ± ♆
01 42 ☽ Q ☉
03 21 ☽ □ ♀
07 07 ☽ ∥ ♂
10 03 ☽ ∥ ♄
10 49 ☽ ⚹ ☿
10 56 ☽ ⊼ ♆
11 13 ☽ □ ♅
11 49 ☽ ⊼ ♇
17 01 ☽ ₽ ♄
18 15 ☿ St R
20 04 ☽ ☌ ♂
22 07 ☽ ± ♇

11 Friday
00 42 ☉ ⋕ ♇
07 30 ☽ Q ☿
08 41 ☽ ± ♄
09 43 ☽ ∥ ♆
10 56 ☽ □ ☉
13 42 ☽ ⊻ ♃
13 44 ☽ ₽ ♇
19 15 ☽ ⊼ ♄
19 33 ☽ ⋕ ♆

12 Saturday
00 22 ☽ ⊥ ♃
12 38 ☽ △ ♀
15 03 ☽ □ ☿
15 27 ☽ ☍ ♆
15 54 ☽ △ ♅
16 15 ☽ △ ♇
16 33 ☽ ∠ ♃
20 17 ☽ ⋕ ♄
20 56 ☽ ⋕ ♂

13 Sunday
02 35 ☉ ₽ ♇
02 58 ☽ ⊻ ♂
04 43 ♂ ± ♇
09 38 ☉ ⚹ ♃
12 15 ☿ ⊻ ♀
13 05 ♂ ∥ ♄
14 16 ☽ ⊥ ♂
18 28 ☽ ₽ ♀
19 10 ☽ ₽ ♅
19 57 ♀ ⚹ ♆
20 01 ☽ ⚹ ♃
20 43 ☽ △ ☉

14 Monday
01 34 ☽ ☍ ♄
02 19 ♀ ☌ ♅
04 03 ♀ △ ♇
04 34 ☽ ⋕ ♃
07 25 ☽ ∠ ♂
11 48 ☽ ± ♅
12 45 ☽ ± ♀
18 26 ☽ Q ♃
20 58 ☽ △ ☿
22 21 ☽ ⊼ ♆
23 01 ☽ ⊼ ♅
23 04 ☽ □ ♇

15 Tuesday
01 05 ☽ ⊼ ♀
02 42 ☽ ₽ ☉
02 52 ☿ ∥ ♅
08 12 ♅ △ ♇
09 36 ☽ ± ♆
12 31 ☽ ⚹ ♂
12 46 ☽ ⋕ ♀
14 19 ☽ ⋕ ☿
15 10 ☽ ⋕ ♅
21 01 ☽ ± ☉

16 Wednesday
00 29 ☽ ₽ ☿
01 39 ☿ ∥ ♀
02 38 ☽ ₽ ♆
03 10 ☽ ∥ ♇
04 46 ☽ □ ♃
05 04 ☽ ⋕ ☉
09 01 ☽ Q ♇
09 23 ☽ ⊼ ☉
10 12 ☽ ⊼ ♄
12 16 ☽ Q ♂
17 11 ☽ ± ☿
21 09 ☉ ⚹ ♄
21 48 ☽ ± ♄

17 Thursday
03 43 ♀ ∠ ♄
04 22 ☽ ⊼ ☿
07 28 ☽ △ ♆
08 05 ☽ ⚹ ♇
08 14 ♀ ∥ ♅
08 22 ☽ ☍ ♅
09 13 ♀ Q ♃
15 20 ☽ ₽ ♄
16 32 ☽ ☍ ♀

18 Friday
00 35 ☽ □ ♂
13 23 ☽ ∠ ♇
15 45 ☽ △ ♃
21 00 ☽ △ ♄

19 Saturday
00 44 ☽ ☍ ☉
06 23 ☿ Q ♂
07 11 ☽ ⊥ ♇
13 09 ☽ ☍ ☿
18 39 ☽ □ ♆
19 09 ☽ ⊻ ♇
19 48 ☽ ⊼ ♅
22 00 ☽ ₽ ♃

20 Sunday
07 54 ☽ ± ♅
10 43 ☽ ⊼ ♀
14 55 ☽ △ ♂
16 22 ☽ ± ♃

21 Monday
00 13 ☽ ± ♀
01 24 ☽ ⋕ ☉
02 09 ☽ ₽ ♅
03 57 ☽ ∥ ♇
04 40 ☽ ⊼ ♃
06 58 ☽ Q ♆
09 37 ☽ □ ♄
11 21 ☽ ⋕ ♀
16 35 ☽ ⋕ ♅
18 21 ☽ ⊼ ☉
20 35 ☽ ₽ ♀
22 42 ☽ ₽ ♂
23 00 ☽ ⊼ ☿

22 Tuesday
04 01 ☽ ⋕ ☿
07 23 ☽ ⚹ ♆
07 38 ☽ ± ☉
07 45 ☽ ☌ ♇
08 47 ☽ △ ♅
09 03 ☉ ± ♇
10 39 ☽ ± ☿
17 52 ☽ ± ♂
22 29 ☽ Q ♄

23 Wednesday
03 34 ☽ ₽ ☉
04 07 ☽ ₽ ☿
06 29 ♀ □ ♂
06 39 ☽ ⊼ ♂
06 40 ☽ △ ♀
08 00 ☉ ☌ ☿
09 01 ☽ ⋕ ♃
13 57 ☽ ∠ ♆
18 06 ♀ Q ♆
18 35 ☽ ☍ ♃
23 02 ☽ ⚹ ♄

24 Thursday
08 13 ☽ ⊥ ♆
09 09 ☽ △ ☿
12 40 ☽ △ ☉
16 46 ☽ ⋕ ♄
20 19 ☽ ⊻ ♆
20 32 ☽ ⊻ ♇
21 56 ☽ □ ♅

25 Friday
00 18 ☿ ± ♇
05 24 ☽ ⋕ ♂
05 27 ☽ ∠ ♄
08 32 ☽ ⊥ ♇
21 38 ☽ ⋕ ♆
21 48 ☽ ☍ ♂
23 26 ☽ ⊥ ♄

26 Saturday
01 50 ☽ □ ♀
02 21 ☽ ∠ ♇
06 48 ♀ ₽ ♇
07 29 ☽ ⊼ ♃
09 07 ☽ ∥ ♆
09 47 ☽ Q ♅
11 17 ☽ ⊻ ♄
18 08 ☽ □ ☿
19 20 ☽ ± ♃
23 19 ☽ ∥ ♂

27 Sunday
04 54 ☽ □ ☉
07 19 ☽ ☌ ♆
07 24 ☽ ⚹ ♇
09 04 ☽ ⚹ ♅
12 50 ☽ ₽ ♃
13 19 ☽ ∥ ♄
16 23 ☽ Q ♀
17 12 ♂ ± ♄

28 Monday
03 07 ☽ Q ☿
05 59 ☽ Q ♇
09 47 ☽ ⊼ ♂
12 51 ☉ □ ♆
13 08 ☉ ⊼ ♇
13 17 ☽ ∠ ♅
16 58 ☽ ∥ ♃
17 03 ☽ ⚹ ♀
17 11 ☽ △ ♃
17 20 ☽ Q ☉
18 44 ♀ ⚹ ♃
20 16 ☽ ☌ ♄
21 15 ☽ ± ♂

29 Tuesday
00 09 ♆ ⚹ ♇
00 28 ☽ ⚹ ☿
05 42 ☽ ⊥ ♅
14 06 ☽ ₽ ♂
14 36 ☽ □ ♇
14 38 ☽ ⊻ ♆
14 39 ☉ ⊻ ♅
15 22 ☽ ∥ ☿
16 27 ☽ ⊻ ♅
16 35 ☽ ⚹ ☉
21 38 ♀ ⋕ ♇
22 37 ☽ ∠ ♀
23 57 ☉ ∥ ♀

30 Wednesday
01 12 ☽ ⊥ ♆
02 23 ☉ ⋕ ♇
02 23 ♂ ₽ ♇
02 24 ☽ ∠ ☿
05 26 ♀ ⚹ ♄
06 27 ☽ ∥ ♅
12 35 ☉ Q ♄
15 43 ☽ ⊥ ♀
16 46 ☽ ∠ ♆
16 58 ☽ ∥ ☉
17 15 ☽ ⋕ ♇
17 17 ☽ △ ♂
17 35 ☽ ⊥ ☿
17 37 ☽ ∥ ♀
20 31 ☽ ∠ ☉
22 42 ☽ □ ♃

JULY 2027

LONGITUDES

Date	Sidereal time h m s	Sun ☉ ° ' "	Moon ☽ ° ' "	Moon ☽ 24.00 ° ' "	Mercury ☿ ° '	Venus ♀ ° '	Mars ♂ ° '	Jupiter ♃ ° '	Saturn ♄ ° '	Uranus ♅ ° '	Neptune ♆ ° '	Pluto ♇ ° '
01	06 37 16	09 ♋ 27 05	03 ♊ 01 49	10 ♊ 16 33	27 ♊ 54	28 ♊ 04	22 ♍ 22	25 ♌ 13	26 ♈ 35	07 ♊ 45	06 ♈ 38	06 ♒ 35
02	06 41 13	10 24 18	17 ♊ 36 54	25 ♊ 02 03	27 R 41	29 ♊ 17	22 54	25 24	26 39	07 48	06 39	06 R 33
03	06 45 09	11 21 32	02 ♋ 31 04	10 ♋ 02 52	27 32	00 ♋ 31	23 27	25 35	26 43	07 51	06 39	06 32
04	06 49 06	12 18 46	17 ♋ 36 18	25 ♋ 10 10	27 28	01 44	24 00	25 46	26 46	07 54	06 39	06 31
05	06 53 02	13 16 00	02 ♌ 43 19	10 ♌ 14 37	27 D 29	02 58	24 32	25 57	26 50	07 57	06 39	06 30
06	06 56 59	14 13 14	17 ♌ 43 06	25 ♌ 07 51	27 35	04 11	25 06	26 08	26 53	08 00	06 39	06 28
07	07 00 55	15 10 27	02 ♍ 28 09	09 ♍ 43 26	27 46	05 25	25 39	26 19	26 56	08 03	06 39	06 27
08	07 04 52	16 07 40	16 ♍ 53 16	23 ♍ 57 24	28 02	06 38	26 12	26 31	26 58	08 06	06 40	06 26
09	07 08 49	17 04 53	00 ♎ 55 43	07 ♎ 48 13	28 23	07 52	26 46	26 42	27 03	08 09	06 40	06 25
10	07 12 45	18 02 06	14 ♎ 34 58	21 ♎ 16 11	28 49	09 05	27 19	26 53	27 06	08 12	06 R 40	06 23
11	07 16 42	18 59 19	27 ♎ 52 05	04 ♏ 22 59	29 21	10 19	27 53	27 05	27 09	08 15	06 40	06 22
12	07 20 38	19 56 31	10 ♏ 49 11	17 ♏ 11 02	29 ♊ 57	11 32	28 27	27 16	27 12	08 18	06 39	06 21
13	07 24 35	20 53 44	23 ♏ 28 52	29 ♏ 43 03	00 ♋ 39	12 46	29 01	27 28	27 15	08 20	06 39	06 19
14	07 28 31	21 50 57	05 ♐ 53 54	12 ♐ 01 46	01 25	14 00	29 ♍ 35	27 40	27 17	08 23	06 39	06 18
15	07 32 28	22 48 10	18 ♐ 06 57	24 ♐ 09 46	02 17	15 13	00 ♎ 09	27 51	27 20	08 26	06 39	06 17
16	07 36 24	23 45 23	00 ♑ 10 30	06 ♑ 09 25	03 13	16 27	00 43	28 03	27 22	08 29	06 39	06 15
17	07 40 21	24 42 36	12 ♑ 06 48	18 ♑ 02 55	04 15	17 41	01 18	28 15	27 25	08 31	06 39	06 14
18	07 44 18	25 39 49	23 ♑ 58 01	29 ♑ 52 23	05 21	18 54	01 52	28 27	27 27	08 34	06 38	06 13
19	07 48 14	26 37 03	05 ♒ 46 18	11 ♒ 40 03	06 32	20 08	02 27	28 39	27 29	08 37	06 38	06 11
20	07 52 11	27 34 18	17 ♒ 33 56	23 ♒ 28 16	07 47	21 22	03 02	28 51	27 32	08 39	06 38	06 10
21	07 56 07	28 31 32	29 ♒ 23 03	05 ♓ 19 41	09 07	22 36	03 37	29 03	27 34	08 42	06 37	06 08
22	08 00 04	29 ♋ 28 48	11 ♓ 17 32	17 ♓ 17 22	10 32	23 49	04 12	29 15	27 36	08 44	06 37	06 07
23	08 04 00	00 ♌ 26 04	23 ♓ 19 35	29 ♓ 24 41	12 01	25 03	04 47	29 27	27 37	08 47	06 37	06 06
24	08 07 57	01 23 21	05 ♈ 33 06	11 ♈ 45 21	13 34	26 17	05 22	29 39	27 39	08 49	06 36	06 04
25	08 11 53	02 20 39	18 ♈ 01 54	24 ♈ 23 14	15 11	27 31	05 58	29 ♌ 51	27 41	08 52	06 36	06 03
26	08 15 50	03 17 58	00 ♉ 49 47	07 ♉ 22 00	16 52	28 45	06 33	00 ♍ 04	27 42	08 54	06 35	06 01
27	08 19 47	04 15 17	14 ♉ 00 13	20 ♉ 44 44	18 37	29 ♋ 58	07 09	00 16	27 44	08 56	06 35	06 00
28	08 23 43	05 12 38	27 ♉ 35 45	04 ♊ 33 22	20 25	01 ♌ 12	07 44	00 28	27 45	08 59	06 34	05 59
29	08 27 40	06 10 00	11 ♊ 37 31	18 ♊ 48 00	22 16	02 26	08 20	00 41	27 46	09 01	06 33	05 57
30	08 31 36	07 07 23	26 ♊ 04 27	03 ♋ 26 21	24 10	03 40	08 56	00 53	27 47	09 03	06 33	05 56
31	08 35 33	08 ♌ 04 47	10 ♋ 52 58	18 ♋ 23 28	26 ♋ 06	04 ♌ 54	09 ♎ 32	01 ♍ 06	27 ♈ 48	09 ♊ 05	06 ♈ 32	05 ♒ 54

Date	Moon True ☊ ° '	Moon Mean ☊ ° '	Moon ☽ Latitude ° '
01	11 ♒ 52	13 ♒ 14	04 N 51
02	11 R 48	13 11	04 14
03	11 44	13 07	03 18
04	11 41	13 04	02 08
05	11 39	13 01	00 N 49
06	11 D 39	12 58	00 S 33
07	11 40	12 55	01 52
08	11 41	12 52	03 02
09	11 43	12 48	03 59
10	11 43	12 45	04 41
11	11 R 43	12 42	05 06
12	11 42	12 39	05 15
13	11 40	12 36	05 08
14	11 37	12 32	04 47
15	11 34	12 29	04 14
16	11 31	12 26	03 29
17	11 29	12 23	02 35
18	11 28	12 20	01 35
19	11 27	12 17	00 S 31
20	11 D 27	12 13	00 N 34
21	11 28	12 10	01 38
22	11 29	12 07	02 38
23	11 30	12 04	03 32
24	11 31	12 01	04 17
25	11 31	11 58	04 51
26	11 32	11 54	05 12
27	11 R 31	11 51	05 17
28	11 31	11 48	05 05
29	11 30	11 45	04 35
30	11 30	11 42	03 47
31	11 ♒ 30	11 ♒ 38	02 N 42

DECLINATIONS

Date	Sun ☉ ° '	Moon ☽ ° '	Mercury ☿ ° '	Venus ♀ ° '	Mars ♂ ° '	Jupiter ♃ ° '	Saturn ♄ ° '	Uranus ♅ ° '	Neptune ♆ ° '	Pluto ♇ ° '
01	23 N 06	25 N 32	18 N 44	23 N 18	03 N 37	13 N 59	08 N 00	21 N 30	01 N 21	23 S 12
02	23 02	27 04	18 45	23 21	03 24	13 55	08 01	21 30	01 21	23 12
03	22 57	26 43	18 48	23 23	03 10	13 51	08 02	21 31	01 21	23 13
04	22 52	24 24	18 52	23 25	02 56	13 48	08 03	21 31	01 21	23 13
05	22 47	20 21	18 58	23 26	02 42	13 44	08 04	21 32	01 21	23 14
06	22 41	15 00	19 04	23 27	02 28	13 40	08 05	21 32	01 21	23 14
07	22 34	08 51	19 12	23 27	02 14	13 36	08 06	21 33	01 21	23 15
08	22 28	02 N 23	19 21	23 26	01 59	13 32	08 07	21 33	01 21	23 15
09	22 21	04 S 01	19 31	23 24	01 45	13 28	08 08	21 34	01 21	23 16
10	22 13	10 03	19 41	23 22	01 31	13 24	08 09	21 34	01 21	23 16
11	22 06	15 28	19 52	23 19	01 16	13 20	08 10	21 35	01 21	23 17
12	21 57	20 03	20 04	23 15	01 02	13 16	08 10	21 35	01 20	23 17
13	21 49	23 37	20 16	23 11	00 48	13 12	08 11	21 36	01 20	23 17
14	21 40	26 00	20 28	23 06	00 33	13 08	08 12	21 36	01 20	23 18
15	21 31	27 07	20 41	23 00	00 19	13 04	08 13	21 37	01 20	23 18
16	21 21	26 55	20 53	22 53	00 N 04	13 00	08 13	21 37	01 20	23 19
17	21 11	25 28	21 05	22 46	00 S 10	12 56	08 14	21 37	01 20	23 19
18	21 01	22 53	21 16	22 38	00 25	12 52	08 14	21 38	01 20	23 20
19	20 50	19 20	21 27	22 30	00 40	12 48	08 15	21 38	01 20	23 20
20	20 39	15 02	21 37	22 21	00 54	12 44	08 16	21 39	01 19	23 21
21	20 27	10 09	21 46	22 11	01 09	12 39	08 16	21 39	01 19	23 21
22	20 15	04 S 53	21 54	22 01	01 24	12 35	08 16	21 39	01 19	23 21
23	20 03	00 N 36	22 00	21 50	01 39	12 31	08 17	21 40	01 19	23 22
24	19 51	06 08	22 05	21 38	01 54	12 27	08 17	21 40	01 19	23 22
25	19 38	11 33	22 08	21 26	02 08	12 22	08 18	21 41	01 18	23 23
26	19 25	16 38	22 09	21 13	02 23	12 18	08 18	21 41	01 18	23 23
27	19 12	21 05	22 08	20 59	02 38	12 14	08 18	21 41	01 18	23 24
28	18 58	24 34	22 04	20 45	02 53	12 09	08 18	21 42	01 18	23 24
29	18 44	26 43	21 58	20 30	03 08	12 05	08 18	21 42	01 17	23 25
30	18 29	27 09	21 50	20 15	03 23	12 00	08 19	21 42	01 17	23 25
31	18 N 15	25 N 41	21 N 39	19 N 59	03 S 38	11 N 56	08 N 19	21 N 43	01 N 17	23 S 25

ZODIAC SIGN ENTRIES

Date	h m	Planets
01	06 56	☽ ♊
03	02 01	♀ ♋
03	07 58	☽ ♋
05	07 40	☽ ♌
07	07 57	☽ ♍
09	10 24	☽ ♎
11	15 55	☽ ♏
12	13 48	☿ ♋
14	00 33	☽ ♐
15	05 40	♂ ♎
16	11 39	☽ ♑
19	00 15	☽ ♒
21	13 14	☽ ♓
23	01 05	☉ ♌
24	01 09	☽ ♈
26	04 49	♃ ♍
26	10 28	☽ ♉
27	12 31	♀ ♌
28	16 10	☽ ♊
30	18 25	☽ ♋

LATITUDES

Date	Mercury ☿ ° '	Venus ♀ ° '	Mars ♂ ° '	Jupiter ♃ ° '	Saturn ♄ ° '	Uranus ♅ ° '	Neptune ♆ ° '	Pluto ♇ ° '
01	04 S 41	00 S 08	00 N 39	00 N 55	02 S 25	00 S 06	01 S 24	04 S 44
04	04 32	00 00	00 35	00 55	02 26	00 06	01 25	04 44
07	04 13	00 N 07	00 32	00 55	02 27	00 06	01 25	04 44
10	03 45	00 14	00 29	00 55	02 27	00 06	01 25	04 45
13	03 10	00 21	00 26	00 55	02 28	00 06	01 25	04 45
16	02 31	00 28	00 23	00 55	02 29	00 06	01 25	04 46
19	01 50	00 35	00 20	00 55	02 30	00 06	01 25	04 46
22	01 08	00 41	00 18	00 54	02 30	00 06	01 26	04 46
25	00 S 27	00 47	00 15	00 54	02 31	00 06	01 26	04 46
28	00 N 11	00 53	00 12	00 54	02 32	00 06	01 26	04 47
31	00 N 44	00 N 58	00 N 09	00 N 54	02 S 33	00 S 06	01 S 26	04 S 47

DATA

Julian Date	2461588
Delta T	+73 seconds
Ayanamsa	24° 14' 41"
Synetic vernal point	04° ♓ 52' 18"
True obliquity of ecliptic	23° 26' 14"

MOON'S PHASES, APSIDES AND POSITIONS ☽

Date	h m	Phase	Longitude ° '	Eclipse Indicator
04	03 02	●	11 ♋ 57	
10	18 39	☽	18 ♎ 18	
18	15 45	○	25 ♑ 49	
26	16 55	☾	03 ♉ 30	

Day	h m		
04	20 49	Perigee	
19	11 36	Apogee	
02	19 44	Max dec	27° N 10'
08	20 50	0S	
15	20 20	Max dec	27° S 12'
23	09 25	0N	
30	05 42	Max dec	27° N 13'

LONGITUDES

Date	Chiron ⚷ ° '	Ceres ⚳ ° '	Pallas ⚴ ° '	Juno ⚵ ° '	Vesta ⚶ ° '	Black Moon Lilith ⚸ ° '
01	04 ♉ 03	16 ♌ 30	02 ♋ 25	29 ♉ 38	26 ♊ 17	02 ♒ 14
11	04 ♉ 20	20 ♌ 50	08 ♋ 24	05 ♊ 15	00 ♋ 34	03 ♒ 20
21	04 ♉ 32	25 ♌ 14	14 ♋ 22	10 ♊ 50	04 ♋ 48	04 ♒ 26
31	04 ♉ 38	29 ♌ 43	20 ♋ 20	16 ♊ 21	09 ♋ 00	05 ♒ 33

All ephemeris data is given at 12.00 UT and the Moon's longitude is additionally given for 24.00 UT

ASPECTARIAN

h m Aspects

01 Thursday
01 08 ☽ ⚺ ♄
02 53 ☽ ⚺ ♀
03 33 ☽ ⚺ ☿
09 22 ☿ ☌ ♀
11 15 ☽ ⊥ ♄
12 45 ☽ ⊥ ☉
17 53 ☽ △ ♇
18 00 ☽ ⚹ ♆
19 52 ☽ ☌ ♅
23 23 ☽ ⚺ ☉

02 Friday
02 13 ☽ ∠ ♄
05 02 ☽ Q ♃
11 49 ☉ ∠ ♃
13 40 ☽ Q ♆
18 23 ☽ ⚼ ♇
20 53 ☽ □ ♂

03 Saturday
00 45 ☽ ⚹ ♃
02 39 ☽ ⚹ ♄
04 05 ☽ ☌ ☿
08 30 ☽ ☌ ♀
08 50 ☽ ± ♇
12 32 ♀ ± ♇
17 05 ☉ Q ♂
18 24 ☽ ⊼ ♇
18 35 ☽ □ ♆
20 32 ☽ ⚺ ♅
21 55 ☽ Q ♄

04 Sunday
01 00 ☽ ∠ ♃
02 45 ☽ Q ♂
03 02 ☽ ☌ ☉
06 06 ☽ ⊥ ♅
15 28 ☽ ⊥ ♃
18 41 ☽ ∥ ♀
19 39 ☿ St D
19 58 ☽ ⋕ ♇
20 26 ☽ ∠ ♅
22 25 ☽ ∥ ☉
22 31 ☽ ⚹ ♂

05 Monday
01 06 ☽ ⚺ ♃
02 36 ☽ □ ♄
03 39 ☽ ⚺ ☿
05 49 ☽ ∥ ♅
12 25 ☽ ⚺ ♀
13 13 ☽ ⊥ ☿
18 00 ☽ ☍ ♇
18 16 ☽ △ ♆
18 33 ☽ ∥ ☿
20 22 ☽ ⚹ ♅
22 50 ☽ ⊥ ♀
23 17 ☽ ∠ ♂

06 Tuesday
03 41 ☽ ∠ ☿
05 59 ☽ ⚺ ☉
06 12 ☉ ⊥ ♅
14 18 ☽ ⊥ ♂
14 35 ☽ ∠ ♀
15 42 ☽ Q ♅
16 19 ☽ ⊥ ☉
17 27 ☽ ∥ ♃
18 22 ☽ ⚼ ♆

07 Wednesday
00 24 ☽ ⚺ ♂
01 49 ☽ ☌ ♃
02 55 ☽ △ ♄
04 11 ☽ ⚹ ☿
07 58 ☽ ∠ ☉
09 02 ☽ ± ♆
14 48 ☽ ∥ ♄
17 18 ☽ ⚹ ♀
18 34 ☽ ⊼ ♇
18 55 ☽ ⊼ ♆
21 15 ☽ □ ♅

08 Thursday
00 17 ☽ Q ☿
03 45 ☽ ⚼ ♄
04 32 ☽ ± ♇
08 04 ♀ ⊼ ♇
10 38 ☽ ⚹ ☉
12 28 ♀ □ ♆
13 31 ☽ ∥ ♂
15 14 ☽ Q ♀
15 51 ☽ ∥ ♆
18 59 ☽ ± ♄
19 41 ☽ ⚼ ♇

09 Friday
01 52 ☽ ⋕ ♆
03 42 ☽ ⋕ ♂
04 30 ☽ ☌ ♂
04 36 ☽ ⚺ ♃
05 17 ☽ ⊼ ♄
07 28 ☽ □ ☿
08 11 ♂ ⚺ ♃
08 34 ☽ Q ☉
15 07 ☽ ⊥ ♃
17 53 ♀ ⚺ ♅
21 32 ☽ △ ♇
21 59 ☽ ☍ ♆
22 41 ♆ St R

10 Saturday
00 39 ☽ △ ♅
01 18 ☽ □ ♀
01 40 ♂ ⊼ ♄
04 10 ☽ ⋕ ♄
07 09 ☽ ∠ ♃
12 15 ♀ Q ♄
18 39 ☽ □ ☉

11 Sunday
02 17 ☽ ⋕ ♃
03 33 ☽ ⚼ ♅
05 11 ♂ ∥ ♆
10 32 ☽ ⚹ ♃
10 41 ☽ ☍ ♄
12 02 ☽ ⚺ ♂
14 50 ☽ △ ☿
20 05 ☽ ± ♅
23 16 ♃ △ ♄
23 34 ☽ ⊥ ♂

12 Monday
00 57 ♀ ⋕ ♇
03 39 ☽ □ ♇
04 14 ☽ ⊼ ♆
07 15 ☽ ⊼ ♅
09 04 ☽ Q ♃
12 06 ☽ ⋕ ☿
13 30 ☽ △ ♀
15 27 ☽ ± ♆
17 10 ☽ ∠ ♂
20 12 ☽ ⚼ ☿
21 29 ☽ ⋕ ♅
23 27 ☽ ⋕ ☉

13 Tuesday
01 32 ☿ ± ♇
05 02 ♀ ∠ ♃
06 39 ☽ △ ☉
08 31 ☽ ⚼ ♆
08 40 ☽ ⋕ ♀
09 27 ☽ ∥ ♇
13 36 ☽ Q ♇
14 22 ☽ ± ☿
19 15 ☽ ⊼ ♄
19 47 ☽ □ ♃
21 08 ☽ ⚼ ♀
23 09 ☽ ⚹ ♂

14 Wednesday
02 42 ☽ ⊼ ☿
06 05 ☉ ⊥ ♃
06 54 ☽ ± ♄
12 47 ☽ ⚹ ♇
13 28 ☽ △ ♆
14 01 ☽ ⚼ ☉
16 33 ☽ ± ♀
16 53 ☽ ☍ ♅
19 57 ♀ ⊥ ♅
21 23 ☉ ∥ ♅
23 40 ☽ Q ♂

15 Thursday
00 33 ☽ ⚼ ♄
05 39 ☽ ⊼ ♀
09 11 ☽ ± ☉
18 15 ☽ ∠ ♇
22 05 ☽ ⊼ ☉

16 Friday
04 38 ☉ ∠ ♅
06 23 ☽ △ ♄
07 41 ☽ △ ♃
12 10 ☽ ⊥ ♇
13 09 ☽ □ ♂
18 39 ☽ ☍ ☿

17 Saturday
00 10 ☽ ⚺ ♇
00 59 ☽ □ ♆
04 44 ☽ ⊼ ♅
14 20 ☽ ⚼ ♃
16 53 ☽ ± ♅
18 41 ☉ ∥ ☿

18 Sunday
00 33 ☽ ☍ ♀
08 25 ☽ ∥ ♇
08 52 ☽ ± ♃
11 11 ☽ ⚼ ♅
13 22 ☽ Q ♆
13 53 ☽ ⋕ ♀
15 45 ☽ ☍ ☉
19 06 ☽ □ ♄
21 07 ☽ ⋕ ♅
21 16 ☽ ⊼ ♃
23 02 ☽ ⋕ ☿

19 Monday
02 04 ☽ ⋕ ☉
04 54 ☽ △ ♂
05 21 ☿ ⊼ ♇
12 51 ☽ ☌ ♇
13 43 ☽ ⊼ ☿
13 45 ☽ ⚹ ♆
14 06 ☿ □ ♆
17 48 ☽ △ ♅
23 44 ♀ Q ♂

20 Tuesday
03 22 ☽ ± ☿
07 50 ☽ Q ♄
10 49 ☉ □ ♄
13 00 ☽ ⚼ ♂
16 08 ☿ ∥ ♅
20 15 ☽ ∠ ♆
20 37 ☽ ⊼ ♀
23 49 ☽ ⋕ ♃
23 56 ☽ ⚼ ☿

21 Wednesday
04 14 ☿ ⚺ ♅
08 13 ☽ ± ♂
08 17 ☽ ⚹ ♄
10 06 ☽ ⊼ ☉
10 12 ☽ ± ♀
11 18 ☽ ☍ ♃
14 30 ☽ ⊥ ♆
19 47 ☿ Q ♄
20 46 ☽ ⋕ ♄
20 59 ☽ ⊼ ♂
22 37 ♀ ⊥ ♃
23 17 ☽ ± ☉

22 Thursday
01 37 ☽ ⚺ ♇
02 36 ☽ ⚺ ♆
04 09 ♂ ⋕ ♆
04 37 ☉ ⚺ ♃
06 28 ☽ ⚼ ♀
06 51 ☽ □ ♅
10 16 ☽ △ ☿
10 18 ♀ ∠ ♅
13 39 ☽ ⊥ ♇
14 37 ☽ ∠ ♄
18 56 ☽ ⚼ ☉
21 22 ☿ ∥ ♀

23 Friday
02 41 ☽ ∥ ♂
03 42 ☽ ⋕ ♆
07 35 ☽ ∠ ♇
08 37 ☽ ⊥ ♄
15 06 ☽ ∥ ♆
15 48 ☽ △ ♀
16 45 ☽ ⋕ ♂
18 51 ☽ Q ♅
20 30 ☽ ⚺ ♄

24 Saturday
00 17 ☽ ⊼ ♃
03 12 ☽ △ ☉
07 37 ♀ ∥ ♅
11 38 ☽ ☍ ♂
12 12 ☽ ± ♃
13 00 ☽ ⚹ ♇
14 02 ☽ ☌ ♆
18 22 ☽ ⚹ ♅
21 24 ☽ ∥ ♄

25 Sunday
05 35 ♂ ⊥ ♃
05 45 ☽ □ ☿
05 51 ☽ ⚼ ♃
06 34 ☿ ∠ ♃
07 10 ☿ ⊥ ♅
12 02 ☽ Q ♇
15 19 ♀ □ ♄
15 25 ♂ △ ♇
15 40 ☽ ∥ ♃
23 03 ☽ ∠ ♅

26 Monday
06 11 ☽ ☌ ♄
07 43 ☽ □ ♀
10 33 ☽ △ ♃
13 24 ♂ ☍ ♆
15 50 ☽ ⊥ ♅
16 55 ☽ □ ☉
20 33 ☽ Q ☿
21 32 ☽ □ ♇
22 34 ☽ ⚺ ♆
23 01 ☽ ⊼ ♂

27 Tuesday
01 49 ☽ ∥ ☉
02 50 ☽ ⚺ ♅
09 26 ☽ ⊥ ♆
10 23 ☽ ± ♂
11 28 ☽ ∥ ♀
15 42 ☽ ∥ ♅
18 21 ☽ ∥ ☿
18 51 ♀ ⚺ ♃
19 48 ☽ Q ♀
21 28 ☽ ⚹ ☿

28 Wednesday
01 28 ☽ ∠ ♆
03 00 ☽ ⋕ ♇
03 08 ☽ ⚼ ♂
03 46 ☽ Q ☉
12 16 ☽ ⚺ ♄
17 03 ☽ □ ♃
18 51 ☽ ⚹ ♀
22 38 ☽ ⊥ ♄
22 41 ♃ ± ♆

29 Thursday
02 05 ☽ ⚹ ☉
02 24 ☽ △ ♇
03 25 ☽ ⚹ ♆
03 29 ☽ ∠ ☿
06 12 ☽ △ ♂
06 48 ☉ ☍ ♇
07 35 ☽ ☌ ♅
13 56 ☽ ∠ ♄
20 57 ☽ ⊥ ☿
21 43 ☉ △ ♆
22 39 ☽ ∠ ♀
23 35 ☽ Q ♆
23 58 ☽ Q ♃

30 Friday
03 33 ☽ ⚼ ♇
05 03 ☽ ∠ ☉
08 23 ☽ ⚺ ☿
10 35 ☿ ∠ ♅
14 49 ☽ ⚹ ♄
14 51 ☽ ⊥ ♀
17 00 ♂ △ ♅
18 17 ☽ ± ♇
19 58 ☽ ⚹ ♃
20 49 ☽ ⊥ ☉
22 03 ☿ ⊥ ♃

31 Saturday
01 30 ☽ ⚺ ♀
03 45 ☿ ∥ ♅
04 00 ☽ ⊼ ♇
05 01 ☽ □ ♆
07 11 ☽ ⚺ ☉
09 07 ☽ ⚺ ♅
09 45 ☽ □ ♂
10 16 ☽ Q ♄
18 45 ☽ ⊥ ♅
20 27 ☽ ∠ ♃

AUGUST 2027

LONGITUDES

Date	Sidereal time h m s	Sun ☉ ° ′ ″	Moon ☽ ° ′ ″	Moon ☽ 24.00 ° ′ ″	Mercury ☿ ° ′	Venus ♀ ° ′	Mars ♂ ° ′	Jupiter ♃ ° ′	Saturn ♄ ° ′	Uranus ♅ ° ′	Neptune ♆ ° ′	Pluto ♇ ° ′
01	08 39 29	09 ♌ 02 12	25 ♋ 56 50	03 ♌ 31 59	28 ♋ 05	06 ♌ 08	10 ♎ 08	01 ♍ 18	27 ♈ 49	09 ♊ 08	06 ♈ 32	05 ♒ 53
02	08 43 26	09 59 37	11 ♌ 07 43	18 ♌ 42 51	00 ♌ 06	07 22	10 45	01 31	27 50	09 10	06 R 31	05 R 52
03	08 47 22	10 57 04	26 ♌ 16 12	03 ♍ 46 41	02 08	08 36	11 21	01 43	27 51	09 12	06 30	05 50
04	08 51 19	11 54 31	11 ♍ 13 16	18 ♍ 35 05	04 11	09 50	11 57	01 56	27 51	09 14	06 29	05 49
05	08 55 16	12 51 59	25 ♍ 51 25	03 ♎ 01 43	06 15	11 04	12 34	02 08	27 52	09 16	06 28	05 47
06	08 59 12	13 49 28	10 ♎ 05 37	17 ♎ 02 54	08 19	12 18	13 11	02 21	27 52	09 18	06 28	05 46
07	09 03 09	14 46 57	23 ♎ 53 31	00 ♏ 37 32	10 24	13 32	13 47	02 34	27 53	09 20	06 27	05 45
08	09 07 05	15 44 28	07 ♏ 15 10	13 ♏ 46 42	12 29	14 46	14 24	02 47	27 53	09 21	06 26	05 43
09	09 11 02	16 41 59	20 ♏ 12 30	26 ♏ 33 00	14 33	16 00	15 01	02 59	27 53	09 23	06 25	05 42
10	09 14 58	17 39 31	02 ♐ 48 40	09 ♐ 00 00	16 37	17 14	15 38	03 12	27 R 53	09 25	06 24	05 41
11	09 18 55	18 37 03	15 ♐ 07 30	21 ♐ 11 41	18 40	18 29	16 15	03 25	27 53	09 27	06 23	05 39
12	09 22 51	19 34 37	27 ♐ 13 01	03 ♑ 12 01	20 42	19 43	16 53	03 38	27 52	09 28	06 22	05 38
13	09 26 48	20 32 12	09 ♑ 09 08	15 ♑ 04 49	22 43	20 57	17 30	03 51	27 52	09 30	06 21	05 37
14	09 30 45	21 29 47	20 ♑ 59 28	26 ♑ 53 30	24 43	22 11	18 07	04 03	27 52	09 32	06 20	05 35
15	09 34 41	22 27 24	02 ♒ 47 15	08 ♒ 41 04	26 41	23 25	18 45	04 16	27 51	09 33	06 19	05 34
16	09 38 38	23 25 02	14 ♒ 35 18	20 ♒ 30 12	28 ♌ 39	24 39	19 23	04 29	27 50	09 35	06 18	05 33
17	09 42 34	24 22 41	26 ♒ 26 04	02 ♓ 23 11	00 ♍ 35	25 53	20 00	04 42	27 50	09 36	06 17	05 31
18	09 46 31	25 20 21	08 ♓ 21 47	14 ♓ 22 08	02 30	27 08	20 38	04 55	27 49	09 38	06 16	05 30
19	09 50 27	26 18 02	20 ♓ 24 29	26 ♓ 29 05	04 23	28 22	21 16	05 08	27 48	09 39	06 14	05 29
20	09 54 24	27 15 45	02 ♈ 36 11	08 ♈ 46 04	06 15	29 ♌ 36	21 54	05 21	27 47	09 40	06 13	05 27
21	09 58 20	28 13 30	14 ♈ 58 59	21 ♈ 15 14	08 06	00 ♍ 50	22 32	05 34	27 46	09 42	06 12	05 26
22	10 02 17	29 ♌ 11 16	27 ♈ 35 06	03 ♉ 58 52	09 55	02 05	23 10	05 47	27 44	09 43	06 11	05 25
23	10 06 14	00 ♍ 09 04	10 ♉ 26 51	16 ♉ 59 20	11 43	03 19	23 49	06 00	27 43	09 44	06 10	05 24
24	10 10 10	01 06 53	23 ♉ 36 34	00 ♊ 18 48	13 29	04 33	24 27	06 13	27 42	09 45	06 08	05 22
25	10 14 07	02 04 45	07 ♊ 06 15	13 ♊ 59 04	15 15	05 48	25 06	06 26	27 40	09 46	06 07	05 21
26	10 18 03	03 02 38	20 ♊ 57 19	28 ♊ 00 58	16 58	07 02	25 44	06 39	27 38	09 47	06 06	05 20
27	10 22 00	04 00 33	05 ♋ 09 54	12 ♋ 23 53	18 41	08 16	26 23	06 52	27 37	09 48	06 04	05 19
28	10 25 56	04 58 30	19 ♋ 42 31	27 ♋ 05 17	20 22	09 31	27 01	07 05	27 35	09 49	06 03	05 18
29	10 29 53	05 56 28	04 ♌ 31 30	12 ♌ 00 22	22 02	10 45	27 40	07 18	27 33	09 50	06 02	05 16
30	10 33 49	06 54 29	19 ♌ 30 57	27 ♌ 02 13	23 40	11 59	28 19	07 31	27 31	09 51	06 00	05 15
31	10 37 46	07 ♍ 52 31	04 ♍ 33 06	12 ♍ 02 28	25 ♍ 17	13 ♍ 14	28 ♎ 58	07 ♍ 44	27 ♈ 29	09 ♊ 52	05 ♈ 59	05 ♒ 14

Moon Nodes and Latitude

Date	Moon True ☊ ° ′	Moon Mean ☊ ° ′	Moon ☽ Latitude ° ′
01	11 ♒ 29	11 ♒ 35	01 N 25
02	11 D 29	11 32	00 N 02
03	11 29	11 29	01 S 21
04	11 R 29	11 26	02 38
05	11 29	11 23	03 43
06	11 29	11 19	04 31
07	11 29	11 16	05 03
08	11 29	11 13	05 17
09	11 D 29	11 10	05 14
10	11 29	11 07	04 56
11	11 29	11 04	04 25
12	11 30	11 00	03 42
13	11 31	10 57	02 50
14	11 32	10 54	01 52
15	11 32	10 51	00 S 48
16	11 R 32	10 48	00 N 17
17	11 32	10 44	01 22
18	11 31	10 41	02 23
19	11 29	10 38	03 19
20	11 27	10 35	04 06
21	11 24	10 32	04 43
22	11 22	10 29	05 07
23	11 20	10 25	05 16
24	11 19	10 22	05 08
25	11 D 19	10 19	04 44
26	11 20	10 16	04 03
27	11 21	10 13	03 07
28	11 23	10 10	01 57
29	11 24	10 06	00 N 38
30	11 R 23	10 03	00 S 45
31	11 ♒ 22	10 ♒ 00	02 S 04

DECLINATIONS

Date	Sun ☉ ° ′	Moon ☽ ° ′	Mercury ☿ ° ′	Venus ♀ ° ′	Mars ♂ ° ′	Jupiter ♃ ° ′	Saturn ♄ ° ′	Uranus ♅ ° ′	Neptune ♆ ° ′	Pluto ♇ ° ′
01	18 N 00	22 N 21	21 N 25	19 N 42	03 S 53	11 N 52	08 N 19	21 N 43	01 N 16	23 S 26
02	17 44	17 28	21 08	19 25	04 08	11 47	08 19	21 43	01 16	23 26
03	17 29	11 29	20 49	19 08	04 23	11 43	08 19	21 44	01 16	23 27
04	17 13	04 N 55	20 28	18 49	04 38	11 38	08 19	21 44	01 15	23 27
05	16 57	01 S 45	20 04	18 31	04 53	11 34	08 19	21 44	01 15	23 28
06	16 41	08 09	19 37	18 11	05 08	11 29	08 19	21 45	01 15	23 28
07	16 24	13 57	19 08	17 52	05 23	11 24	08 18	21 45	01 14	23 28
08	16 07	18 55	18 37	17 31	05 38	11 20	08 18	21 45	01 14	23 29
09	15 50	22 50	18 04	17 11	05 53	11 15	08 18	21 45	01 13	23 29
10	15 32	25 33	17 30	16 50	06 08	11 11	08 18	21 46	01 13	23 30
11	15 15	26 59	16 53	16 28	06 24	11 06	08 17	21 46	01 13	23 30
12	14 57	27 07	16 16	16 06	06 39	11 01	08 17	21 46	01 12	23 30
13	14 39	25 57	15 36	15 43	06 54	10 57	08 17	21 47	01 12	23 31
14	14 20	23 38	14 56	15 20	07 09	10 52	08 16	21 47	01 11	23 31
15	14 02	20 19	14 15	14 57	07 24	10 47	08 16	21 47	01 11	23 31
16	13 43	16 11	13 32	14 33	07 39	10 43	08 15	21 47	01 10	23 32
17	13 24	11 25	12 49	14 09	07 54	10 38	08 15	21 47	01 10	23 32
18	13 04	06 13	12 06	13 44	08 09	10 33	08 14	21 48	01 09	23 33
19	12 45	00 S 45	11 21	13 19	08 24	10 29	08 14	21 48	01 09	23 33
20	12 25	04 N 48	10 37	12 54	08 39	10 24	08 13	21 48	01 08	23 33
21	12 05	10 15	09 51	12 28	08 53	10 19	08 13	21 48	01 08	23 34
22	11 45	15 23	09 06	12 02	09 08	10 14	08 12	21 48	01 07	23 34
23	11 25	19 57	08 20	11 35	09 23	10 10	08 11	21 49	01 07	23 34
24	11 05	23 39	07 35	11 09	09 38	10 05	08 10	21 49	01 06	23 35
25	10 44	26 10	06 49	10 42	09 53	10 00	08 10	21 49	01 06	23 35
26	10 23	27 11	06 03	10 14	10 08	09 55	08 09	21 49	01 05	23 35
27	10 02	26 27	05 17	09 47	10 22	09 51	08 08	21 49	01 05	23 36
28	09 41	23 55	04 32	09 19	10 37	09 46	08 07	21 49	01 04	23 36
29	09 20	19 44	03 46	08 51	10 52	09 41	08 06	21 50	01 04	23 36
30	08 58	14 16	03 01	08 23	11 06	09 36	08 05	21 50	01 03	23 36
31	08 N 37	07 N 55	02 N 16	07 N 54	11 S 21	09 N 31	08 N 04	21 N 50	01 N 02	23 S 37

ZODIAC SIGN ENTRIES

Date	h	m	Planets
01	18	25	☽ ♌
02	10	52	☿ ♌
03	17	57	☽ ♍
05	18	55	☽ ♎
07	22	53	☽ ♏
10	06	36	☽ ♐
12	17	34	☽ ♑
15	06	20	☽ ♒
17	04	43	☿ ♍
17	19	12	☽ ♓
20	06	54	☽ ♈
20	19	43	♀ ♍
22	16	33	☽ ♉
23	08	14	☉ ♍
24	23	27	☽ ♊
27	03	21	☽ ♋
29	04	43	☽ ♌
31	04	44	☽ ♍

LATITUDES

Date	Mercury ☿ ° ′	Venus ♀ ° ′	Mars ♂ ° ′	Jupiter ♃ ° ′	Saturn ♄ ° ′	Uranus ♅ ° ′	Neptune ♆ ° ′	Pluto ♇ ° ′
01	00 N 53	01 N 00	00 N 09	00 N 54	02 S 33	00 S 06	01 S 26	04 S 47
04	01 17	01 05	00 06	00 54	02 34	00 06	01 26	04 47
07	01 34	01 09	00 03	00 54	02 35	00 06	01 26	04 47
10	01 43	01 13	00 N 01	00 54	02 36	00 06	01 26	04 48
13	01 46	01 16	00 S 02	00 55	02 36	00 06	01 27	04 48
16	01 42	01 19	00 04	00 55	02 37	00 06	01 27	04 48
19	01 34	01 22	00 06	00 55	02 38	00 06	01 27	04 48
22	01 21	01 23	00 09	00 55	02 39	00 06	01 27	04 48
25	01 05	01 24	00 11	00 55	02 40	00 06	01 27	04 48
28	00 46	01 25	00 13	00 55	02 40	00 06	01 27	04 48
31	00 N 25	01 N 25	00 S 15	00 N 55	02 S 41	00 S 06	01 S 27	04 S 48

DATA

Julian Date	2461619
Delta T	+73 seconds
Ayanamsa	24° 14′ 47″
Synetic vernal point	04° ♓ 52′ 12″
True obliquity of ecliptic	23° 26′ 14″

MOON'S PHASES, APSIDES AND POSITIONS ☽

Date	h	m	Phase	Longitude ° ′	Eclipse Indicator
02	10	05	●	09 ♌ 55	Total
09	04	54	☽	16 ♏ 25	
17	07	29	○	24 ♒ 12	
25	02	27	☾	01 ♊ 42	
31	17	41	●	08 ♍ 06	

Day	h	m		
02	06	18	Perigee	
15	14	11	Apogee	
30	15	34	Perigee	
05	05	38	0S	
12	02	07	Max dec	27° S 13′
19	15	15	0N	
26	14	14	Max dec	27° N 11′

LONGITUDES

Date	Chiron ⚷ ° ′	Ceres ⚳ ° ′	Pallas ⚴ ° ′	Juno ⚵ ° ′	Vesta ⚶ ° ′	Black Moon Lilith ⚸ ° ′
01	04 ♉ 39	00 ♍ 10	20 ♋ 56	16 ♊ 54	09 ♋ 25	05 ♒ 40
11	04 ♉ 40	04 ♍ 41	26 ♋ 50	22 ♊ 19	13 ♋ 33	06 ♒ 46
21	04 ♉ 35	09 ♍ 15	02 ♌ 41	27 ♊ 37	17 ♋ 36	07 ♒ 52
31	04 ♉ 25	13 ♍ 50	08 ♌ 27	02 ♋ 46	21 ♋ 33	08 ♒ 59

ASPECTARIAN

01 Sunday
05 29 ☽ # ♇
07 14 ♀ ☍ ♇
08 48 ☿ □ ♄
09 06 ☽ ∠ ♅
10 58 ☽ ⊥ ♃
12 55 ☿ Q ♂
14 18 ☉ ✶ ♅
14 58 ☽ □ ♄
15 33 ☽ ∥ ♅
15 37 ☽ Q ♂
15 54 ☽ ☌ ☿
17 31 ☽ ∥ ☿
19 33 ♀ △ ♆
20 36 ☽ ⊻ ♃

02 Monday
02 37 ☽ ∥ ♀
03 41 ☽ ☍ ♇
04 43 ☽ △ ♆
05 32 ☽ ☌ ♀
08 53 ☽ ✶ ♅
10 05 ☽ ☌ ☉
10 44 ☽ ∥ ☉
11 22 ☽ ✶ ♂

03 Tuesday
03 55 ☽ Q ♅
04 26 ☽ bQ ♆
06 39 ☿ ⊻ ♃
11 08 ☽ ∥ ♃
12 08 ☽ ∠ ♂
14 31 ☽ △ ♄
18 45 ☽ ± ♆
20 50 ☽ ☌ ♃
22 50 ☽ ⊻ ☿
23 45 ☽ ∥ ♄
23 53 ♀ ✶ ♅

04 Wednesday
03 08 ☽ ⊥ ♂
03 17 ☽ ⊼ ♇
04 22 ☽ ⊼ ♆
08 46 ☽ □ ♅
09 33 ☽ ⊻ ♀
10 03 ☽ ⊥ ☿
12 58 ☽ ± ♇
12 59 ☽ # ♂
13 12 ☽ ⊻ ☉
13 15 ☽ ⊻ ♂
14 39 ☽ bQ ♄
15 16 ☉ ✶ ♂
20 11 ☽ ⊥ ♀
23 39 ☽ ⊥ ☉

05 Thursday
01 08 ☽ ∥ ♆
03 07 ☽ ∠ ☿
03 38 ☽ bQ ♇
05 23 ☽ ± ♄
06 45 ☿ ☍ ♇
10 10 ☽ # ♆
12 23 ☽ ∠ ♀
14 35 ☿ △ ♆
15 21 ☽ ⊼ ♄
15 35 ☽ ∠ ☉
22 40 ☽ ⊻ ♃

06 Friday
00 00 ☽ ∥ ♂
04 39 ☽ △ ♇
05 49 ☽ ☍ ♆
08 27 ☽ ✶ ☿
08 59 ☽ ⊥ ♃
10 38 ☽ △ ♅
12 36 ☽ # ♄
16 09 ☽ ✶ ♀
17 32 ☽ ☌ ♂
18 53 ☽ ✶ ☉
23 22 ☿ ✶ ♅

07 Saturday
00 43 ☽ ∠ ♃
01 13 ☽ # ♃
08 54 ☽ Q ☿
12 46 ☽ bQ ♅
15 12 ☽ Q ♀
17 31 ☽ Q ☉
19 05 ☽ ☍ ♄
21 48 ♀ ✶ ♂
22 38 ☽ # ☉

08 Sunday
03 44 ☽ ✶ ♃
04 54 ☽ ± ♅
05 15 ☽ # ♀
09 13 ☽ □ ♇
10 30 ☽ ⊼ ♆
10 36 ☽ # ☿
15 52 ☽ ⊼ ♅
21 30 ☽ ± ♆
23 25 ☽ □ ☿

09 Monday
01 49 ☽ ⊻ ♂
02 05 ☽ Q ♃
03 18 ☽ □ ♀
04 40 ☽ # ♅
04 54 ☽ □ ☉
13 36 ☽ ⊥ ♂
14 16 ☽ bQ ♆
16 53 ☽ ∥ ♇
18 07 ♄ St R
18 34 ☽ Q ♇
19 48 ☿ ✶ ♂

10 Tuesday
02 32 ☽ ⊼ ♄
07 36 ☽ ∠ ♂
12 46 ☽ □ ♃
14 04 ☽ ± ♄
17 32 ☽ ✶ ♇
18 56 ☽ △ ♆

11 Wednesday
00 51 ☽ ☍ ♅
06 30 ☿ ☌ ♀
07 35 ☽ bQ ♄
11 02 ☉ ☌ ☿
14 21 ☽ ✶ ♂
19 22 ☽ △ ♀
19 29 ☽ △ ☉
20 23 ☽ △ ☿
22 54 ☽ ∠ ♇

12 Thursday
00 21 ☉ ☌ ♀
13 19 ☽ △ ♄
15 30 ☽ Q ♂
16 50 ☽ ⊥ ♇
19 54 ☿ bQ ♆
21 21 ☿ Q ♅

13 Friday
01 06 ☽ △ ♃
02 32 ☿ ∥ ♀
04 04 ☽ bQ ☉
04 47 ☽ bQ ♀
04 52 ☽ ⊻ ♇
06 21 ☽ □ ♆
08 30 ☽ bQ ☿
12 42 ☽ ⊼ ♅
19 45 ♀ bQ ♆
23 01 ♀ Q ♅
23 52 ☽ ± ☉

14 Saturday
00 53 ☽ ± ♅
01 06 ☽ ± ♀
05 51 ☽ □ ♂
06 26 ☽ ± ☿
08 00 ☽ bQ ♃
08 00 ☉ bQ ♆
12 48 ☉ Q ♅
12 59 ☽ ∥ ♇
13 07 ☽ ⊼ ☉
14 42 ☽ ⊼ ♀
18 47 ☽ Q ♆
19 12 ☽ bQ ♅
21 06 ☽ ⊼ ☿

15 Sunday
01 58 ☽ □ ♄
02 13 ☽ # ♅
02 38 ☽ ± ♃
15 04 ☽ ⊼ ♃
17 39 ☽ ☌ ♇
19 10 ☽ ✶ ♆

16 Monday
01 33 ☉ ∥ ☿
01 48 ☽ △ ♅
02 07 ☿ △ ♄
14 32 ☽ Q ♄
18 16 ♂ ∠ ♃
21 22 ☽ # ♀
22 16 ☽ △ ♂

17 Tuesday
01 35 ☽ ∠ ♆
01 42 ☽ # ☉
04 02 ☽ # ☿
07 29 ☽ ☍ ☉
08 14 ☿ ± ♆
10 47 ☽ ☍ ♀
14 49 ☽ ✶ ♄
15 49 ☽ # ♃
19 45 ☽ ⊥ ♆
21 59 ☽ ☍ ☿

18 Wednesday
02 51 ☽ # ♄
03 42 ☽ ∥ ♂
04 58 ☽ ☍ ♃
06 14 ☽ bQ ♂
06 16 ☽ ⊻ ♇
07 47 ☽ ⊻ ♆
14 32 ☽ □ ♅
18 16 ☽ ⊥ ♇
20 51 ♂ # ♄
20 53 ☽ ∠ ♄

19 Thursday
01 09 ♀ △ ♄
01 14 ☽ ± ♂
10 16 ☽ # ♆
12 08 ☽ ∠ ♇
13 48 ☽ ⊼ ♂
14 45 ☽ ⊥ ♄
20 12 ☽ ∥ ♆
22 47 ☿ ☌ ♃

20 Friday
00 38 ☽ ⊼ ☉
01 49 ☿ ⊼ ♇
02 19 ☽ Q ♅
02 34 ☽ ⊻ ♄
05 28 ☽ ⊼ ♀
11 34 ☿ ⊼ ♆
13 24 ☽ ± ☉
17 27 ☽ ⊼ ♃
17 33 ☽ ✶ ♇
18 30 ☽ ± ♀
19 03 ☽ ☌ ♆
19 33 ☿ ∥ ♃
20 23 ☽ ⊼ ☿
22 59 ♃ ⊼ ♇
23 48 ♀ ± ♆

21 Saturday
00 40 ☉ △ ♄
00 47 ☿ ∠ ♂
01 47 ☽ ✶ ♅
02 55 ☽ ∥ ♄
05 18 ☽ ± ♃
05 38 ☽ # ♂
08 20 ☽ bQ ☉
10 00 ☽ ± ☿
10 28 ☽ ∥ ☿
12 19 ☽ ∥ ♃
13 50 ☽ bQ ♀
16 42 ☽ Q ♇
19 53 ☽ ∥ ☉
21 21 ☽ ∥ ♀
22 52 ☽ bQ ♃

22 Sunday
03 12 ☽ ☍ ♂
06 07 ☽ bQ ☿
06 34 ☽ ∠ ♅
09 14 ☿ □ ♅
11 05 ☿ # ♂
12 18 ☽ ☌ ♄
15 16 ☽ △ ☉
21 21 ☽ △ ♀
23 31 ☽ ⊥ ♅

23 Monday
02 39 ☽ □ ♇
03 37 ☽ △ ♃
04 04 ☽ ⊻ ♆
07 43 ☿ ± ♇
10 41 ☽ ⊻ ♅
12 12 ☉ ± ♆
14 43 ☽ △ ☿
15 09 ☽ ⊥ ♆
16 58 ☿ ∥ ♄
23 18 ☽ ∥ ♅

24 Tuesday
01 19 ☿ bQ ♄
04 20 ♃ ⊼ ♆
07 33 ☽ ∠ ♆
11 28 ☽ # ♇
13 35 ☽ ⊼ ♂
19 19 ☽ ⊻ ♄
23 36 ♂ bQ ♅

25 Wednesday
00 51 ☽ ± ♂
02 27 ☽ □ ☉
03 22 ☉ ∥ ♀
03 37 ♀ ⊼ ♇
05 57 ☽ ⊥ ♄
08 55 ☽ △ ♇
09 28 ☽ □ ♀
10 16 ☽ ✶ ♆
10 48 ☽ □ ♃
16 40 ☽ ☌ ♅
17 29 ☽ bQ ♂
18 10 ♀ ⊼ ♆
20 56 ♂ # ♃
21 42 ☽ ∠ ♄

26 Thursday
03 00 ♀ ☌ ♃
04 12 ☽ □ ☿
07 06 ☽ Q ♆
10 56 ☽ bQ ♇
12 10 ☽ Q ☉
15 53 ♀ # ♂
18 24 ☽ Q ♃
19 37 ☽ Q ♀
20 32 ☽ △ ♂
22 32 ☉ # ♂
23 20 ☽ ✶ ♄

27 Friday
02 12 ☽ ± ♇
08 08 ♀ ∥ ♃
09 56 ☽ ✶ ☉
12 15 ☽ ⊼ ♇
13 31 ☽ □ ♆
14 52 ☽ Q ☿
14 53 ☽ ✶ ♃
17 39 ☽ ✶ ♀
19 23 ☽ Q ♄
19 43 ☽ ⊻ ♅

28 Saturday
05 19 ☉ ∥ ♃
05 37 ☽ ⊥ ♅
10 59 ☿ bQ ♇
12 28 ☽ ∠ ☉
13 13 ☽ ✶ ☿
14 11 ☽ # ♇
15 56 ☽ ∠ ♃
18 04 ♀ □ ♅
19 45 ☉ ⊼ ♇
20 20 ☽ ∠ ♅
20 32 ☽ ∠ ♀

29 Sunday
00 26 ☽ □ ♂
00 46 ☽ □ ♄
01 05 ☽ ∥ ♅
03 31 ☿ ⊥ ♂
04 06 ☽ ⊥ ☉
05 09 ☿ ± ♄
06 44 ☽ ⊥ ♃
07 35 ♂ ☍ ♄
12 24 ☽ ⊥ ♀
13 12 ☽ ☍ ♇
14 04 ☉ ⊼ ♆
14 25 ☽ △ ♆
14 26 ☽ ⊻ ☉
16 31 ☽ ∠ ☿
16 31 ☽ ⊻ ♃
20 32 ☽ ✶ ♅
22 00 ♀ ± ♇
22 53 ☽ ⊻ ♀

30 Monday
06 40 ☽ Q ♂
08 42 ☽ ⊥ ☿
14 22 ☽ bQ ♆
15 43 ☽ Q ♅
19 26 ☽ ⊻ ☿
21 51 ♀ bQ ♄
23 47 ☽ # ♂

31 Tuesday
00 44 ☽ △ ♄
02 41 ☽ ✶ ♂
03 10 ♀ ∥ ♄
04 42 ☽ ± ♆
06 03 ☽ ∥ ♃
07 29 ☉ ☌ ♃
09 18 ☽ ∥ ☉
11 27 ☽ ∥ ♄
12 04 ☽ ∥ ♀
13 06 ☽ ⊼ ♇
14 17 ☽ ⊼ ♆
17 10 ☽ ☌ ♃
17 41 ☽ ☌ ☉
20 30 ☽ □ ♅
22 42 ☽ ± ♇

All ephemeris data is given at 12.00 UT and the Moon's longitude is additionally given for 24.00 UT

LONGITUDES

Date	Sidereal time h m s	Sun ☉ ° ' "	Moon ☽ ° ' "	Moon ☽ 24.00 ° ' "	Mercury ☿ ° '	Venus ♀ ° '	Mars ♂ ° '	Jupiter ♃ ° '	Saturn ♄ ° '	Uranus ♅ ° '	Neptune ♆ ° '	Pluto ♇ ° '
01	10 41 43	08 ♍ 50 34	19 ♍ 29 15	26 ♍ 52 24	26 ♍ 53	14 ♍ 28	29 ♎ 37	07 ♍ 57	27 ♈ 26	09 ♊ 52	05 ♈ 57	05 ♒ 13
02	10 45 39	09 48 39	04 ♎ 11 00	11 ♎ 24 15	28 ♍ 28	15 43	00 ♏ 17	08 10	27 R 24	09 53	05 R 56	05 R 12
03	10 49 36	10 46 46	18 ♎ 31 32	25 ♎ 32 22	00 ♎ 01	16 57	00 56	08 23	27 22	09 54	05 54	05 11
04	10 53 32	11 44 54	02 ♏ 26 28	09 ♏ 13 43	01 34	18 11	01 35	08 36	27 19	09 54	05 53	05 10
05	10 57 29	12 43 03	15 ♏ 54 09	22 ♏ 27 56	03 05	19 26	02 15	08 49	27 16	09 55	05 51	05 09
06	11 01 25	13 41 14	28 ♏ 55 24	05 ♐ 16 55	04 34	20 40	02 54	09 02	27 14	09 55	05 50	05 08
07	11 05 22	14 39 27	11 ♐ 33 00	17 ♐ 44 10	06 03	21 55	03 34	09 15	27 11	09 56	05 48	05 07
08	11 09 18	15 37 41	23 ♐ 51 00	29 ♐ 54 08	07 30	23 09	04 14	09 28	27 08	09 56	05 47	05 06
09	11 13 15	16 35 56	05 ♑ 54 09	11 ♑ 51 43	08 56	24 24	04 53	09 41	27 05	09 56	05 45	05 05
10	11 17 12	17 34 13	17 ♑ 47 24	23 ♑ 41 49	10 20	25 38	05 33	09 54	27 02	09 57	05 44	05 04
11	11 21 08	18 32 32	29 ♑ 35 32	05 ♒ 29 04	11 43	26 53	06 13	10 07	26 59	09 57	05 42	05 03
12	11 25 05	19 30 52	11 ♒ 22 55	17 ♒ 17 34	13 05	28 07	06 53	10 20	26 56	09 57	05 41	05 02
13	11 29 01	20 29 14	23 ♒ 13 24	29 ♒ 10 48	14 26	29 ♍ 21	07 33	10 33	26 52	09 57	05 39	05 01
14	11 32 58	21 27 37	05 ♓ 10 05	11 ♓ 11 32	15 45	00 ♎ 36	08 14	10 46	26 49	09 57	05 38	05 00
15	11 36 54	22 26 02	17 ♓ 15 24	23 ♓ 21 50	17 02	01 50	08 54	10 59	26 46	09 R 57	05 36	04 59
16	11 40 51	23 24 30	29 ♓ 31 02	05 ♈ 43 05	18 18	03 05	09 34	11 12	26 42	09 57	05 34	04 58
17	11 44 47	24 22 58	11 ♈ 58 06	18 ♈ 16 08	19 33	04 19	10 15	11 25	26 39	09 57	05 33	04 57
18	11 48 44	25 21 29	24 ♈ 37 14	01 ♉ 01 28	20 45	05 34	10 55	11 38	26 35	09 57	05 31	04 57
19	11 52 41	26 20 02	07 ♉ 28 50	13 ♉ 59 25	21 56	06 48	11 36	11 50	26 31	09 57	05 29	04 56
20	11 56 37	27 18 38	20 ♉ 33 14	27 ♉ 10 22	23 05	08 03	12 17	12 03	26 27	09 57	05 28	04 55
21	12 00 34	28 17 15	03 ♊ 50 52	10 ♊ 34 49	24 13	09 17	12 57	12 16	26 24	09 56	05 26	04 55
22	12 04 30	29 ♍ 15 55	17 ♊ 22 18	24 ♊ 13 23	25 18	10 32	13 38	12 29	26 20	09 56	05 24	04 54
23	12 08 27	00 ♎ 14 37	01 ♋ 08 09	08 ♋ 06 36	26 20	11 47	14 19	12 41	26 16	09 56	05 23	04 53
24	12 12 23	01 13 21	15 ♋ 08 46	22 ♋ 14 33	27 21	13 01	15 00	12 54	26 12	09 55	05 21	04 53
25	12 16 20	02 12 07	29 ♋ 23 50	06 ♌ 36 21	28 19	14 16	15 41	13 07	26 07	09 55	05 19	04 52
26	12 20 16	03 10 56	13 ♌ 51 46	21 ♌ 09 39	29 ♎ 14	15 30	16 23	13 19	26 03	09 54	05 18	04 51
27	12 24 13	04 09 47	28 ♌ 29 23	05 ♍ 50 19	00 ♏ 06	16 45	17 04	13 32	25 59	09 54	05 16	04 51
28	12 28 10	05 08 40	13 ♍ 11 38	20 ♍ 32 29	00 55	17 59	17 45	13 45	25 55	09 53	05 14	04 50
29	12 32 06	06 07 35	27 ♍ 51 57	05 ♎ 09 07	01 40	19 14	18 27	13 57	25 50	09 52	05 13	04 50
30	12 36 03	07 ♎ 06 32	12 ♎ 23 03	19 ♎ 32 56	02 ♏ 22	20 ♎ 28	19 ♏ 08	14 ♍ 10	25 ♈ 46	09 ♊ 52	05 ♈ 11	04 ♒ 49

Date	Moon True ☊ ° '	Moon Mean ☊ ° '	Moon ☽ Latitude ° '
01	11 ♒ 20	09 ♒ 57	03 S 14
02	11 R 16	09 54	04 10
03	11 12	09 50	04 49
04	11 07	09 47	05 10
05	11 04	09 44	05 12
06	11 01	09 41	04 58
07	11 00	09 38	04 30
08	11 D 01	09 35	03 50
09	11 02	09 31	03 00
10	11 04	09 28	02 04
11	11 05	09 25	01 S 02
12	11 R 06	09 22	00 N 02
13	11 05	09 19	01 06
14	11 02	09 16	02 07
15	10 58	09 12	03 04
16	10 51	09 09	03 52
17	10 44	09 06	04 31
18	10 36	09 03	04 56
19	10 28	09 00	05 08
20	10 22	08 56	05 03
21	10 17	08 53	04 43
22	10 15	08 50	04 06
23	10 D 14	08 47	03 15
24	10 14	08 44	02 11
25	10 15	08 41	00 N 58
26	10 R 16	08 37	00 S 19
27	10 14	08 34	01 36
28	10 11	08 31	02 47
29	10 08	08 28	03 47
30	09 ♒ 56	08 ♒ 25	04 S 31

DECLINATIONS

Date	Sun ☉ ° '	Moon ☽ ° '	Mercury ☿ ° '	Venus ♀ ° '	Mars ♂ ° '	Jupiter ♃ ° '	Saturn ♄ ° '	Uranus ♅ ° '	Neptune ♆ ° '	Pluto ♇ ° '
01	08 N 15	01 N 11	01 N 31	07 N 25	11 S 35	09 N 26	08 N 03	21 N 50	01 N 02	23 S 37
02	07 53	05 S 29	00 46	06 56	11 50	09 22	08 02	21 50	01 01	23 37
03	07 31	11 43	00 N 02	06 27	12 04	09 17	08 01	21 50	01 01	23 38
04	07 09	17 09	00 S 42	05 58	12 19	09 12	08 00	21 50	01 00	23 38
05	06 47	21 34	01 25	05 28	12 33	09 07	07 59	21 50	00 59	23 38
06	06 25	24 46	02 08	04 58	12 47	09 02	07 58	21 50	00 59	23 38
07	06 02	26 37	02 50	04 29	13 01	08 57	07 56	21 50	00 58	23 39
08	05 40	27 07	03 32	03 59	13 15	08 53	07 55	21 51	00 57	23 39
09	05 17	26 19	04 14	03 29	13 30	08 48	07 54	21 51	00 57	23 39
10	04 55	24 18	04 55	02 58	13 44	08 43	07 53	21 51	00 56	23 39
11	04 32	21 15	05 35	02 28	13 57	08 38	07 51	21 51	00 56	23 40
12	04 09	17 20	06 14	01 58	14 11	08 33	07 50	21 51	00 55	23 40
13	03 46	12 45	06 53	01 27	14 25	08 28	07 49	21 51	00 54	23 40
14	03 23	07 37	07 31	00 57	14 39	08 24	07 47	21 51	00 54	23 40
15	03 00	02 S 13	08 09	00 N 26	14 52	08 19	07 46	21 51	00 53	23 40
16	02 37	03 N 22	08 45	00 S 04	15 06	08 14	07 45	21 51	00 52	23 41
17	02 14	08 53	09 21	00 35	15 19	08 09	07 43	21 51	00 52	23 41
18	01 51	14 08	09 56	01 06	15 32	08 04	07 42	21 51	00 51	23 41
19	01 27	18 51	10 30	01 36	15 46	07 59	07 40	21 51	00 50	23 41
20	01 04	22 45	11 03	02 07	15 59	07 55	07 39	21 51	00 49	23 41
21	00 41	25 33	11 35	02 37	16 12	07 50	07 37	21 51	00 49	23 41
22	00 N 18	26 55	12 06	03 08	16 25	07 45	07 36	21 51	00 48	23 42
23	00 S 06	26 41	12 35	03 38	16 38	07 40	07 34	21 51	00 48	23 42
24	00 29	24 45	13 04	04 09	16 50	07 35	07 32	21 50	00 47	23 42
25	00 53	21 13	13 31	04 39	17 03	07 31	07 31	21 50	00 46	23 42
26	01 16	16 21	13 57	05 09	17 15	07 26	07 29	21 50	00 46	23 42
27	01 39	10 29	14 21	05 40	17 28	07 21	07 28	21 50	00 45	23 42
28	02 03	04 N 02	14 44	06 10	17 40	07 16	07 26	21 50	00 44	23 42
29	02 26	02 S 37	15 04	06 40	17 52	07 12	07 24	21 50	00 44	23 42
30	02 S 49	09 S 03	15 S 23	07 S 09	18 S 04	07 N 07	07 N 23	21 N 50	00 N 43	23 S 42

ZODIAC SIGN ENTRIES

Date	h	m	Planets	
02	01	52	♂	♏
02	05	07	☽	♎
03	11	37	☿	♎
04	07	44	☽	♏
06	14	01	☽	♐
09	00	12	☽	♑
11	12	50	☽	♒
14	00	25	♀	♎
14	01	39	☽	♓
16	12	56	☽	♈
18	22	05	☽	♉
21	05	05	☽	♊
23	06	02	☉	♎
23	10	02	☽	♋
25	13	00	☽	♌
27	09	10	☿	♏
27	14	28	☽	♍
29	15	30	☽	♎

LATITUDES

Date	Mercury ☿ ° '	Venus ♀ ° '	Mars ♂ ° '	Jupiter ♃ ° '	Saturn ♄ ° '	Uranus ♅ ° '	Neptune ♆ ° '	Pluto ♇ ° '
01	00 N 18	01 N 25	00 S 16	00 N 55	02 S 41	00 S 06	01 S 27	04 S 48
04	00 S 05	01 24	00 18	00 55	02 42	00 06	01 27	04 48
07	00 29	01 23	00 20	00 56	02 43	00 06	01 27	04 48
10	00 53	01 21	00 22	00 56	02 43	00 06	01 28	04 48
13	01 18	01 18	00 24	00 56	02 44	00 06	01 28	04 48
16	01 43	01 15	00 26	00 56	02 44	00 06	01 28	04 48
19	02 06	01 12	00 28	00 56	02 45	00 06	01 28	04 48
22	02 29	01 08	00 30	00 57	02 45	00 06	01 28	04 48
25	02 50	01 03	00 32	00 57	02 46	00 06	01 28	04 48
28	03 08	00 58	00 34	00 57	02 46	00 06	01 28	04 48
31	03 S 22	00 N 53	00 S 36	00 N 58	02 S 46	00 S 06	01 S 28	04 S 47

DATA

Julian Date	2461650
Delta T	+73 seconds
Ayanamsa	24° 14' 51"
Synetic vernal point	04° ♓ 52' 08"
True obliquity of ecliptic	23° 26' 15"

MOON'S PHASES, APSIDES AND POSITIONS ☽

Date	h	m	Phase	Longitude ° '	Eclipse Indicator
07	18	31	☽	14 ♐ 55	
15	23	03	○	22 ♓ 53	
23	10	20	☾	00 ♋ 11	
30	02	36	●	06 ♎ 43	

Day	h	m		
11	23	31	Apogee	
27	20	02	Perigee	
01	16	12	0S	
08	08	56	Max dec	27° S 08'
15	21	34	0N	
22	20	35	Max dec	27° N 02'
29	02	32	0S	

LONGITUDES

Date	Chiron ⚷ ° '	Ceres ⚳ ° '	Pallas ⚴ ° '	Juno ⚵ ° '	Vesta ⚶ ° '	Black Moon Lilith ° '
01	04 ♉ 24	14 ♍ 18	09 ♌ 01	03 ♋ 17	21 ♋ 57	09 ♒ 06
11	04 ♉ 09	18 ♍ 54	14 ♌ 39	08 ♋ 13	25 ♋ 47	10 ♒ 42
21	03 ♉ 49	23 ♍ 30	20 ♌ 10	12 ♋ 55	29 ♋ 28	11 ♒ 18
31	03 ♉ 26	28 ♍ 06	25 ♌ 32	17 ♋ 20	02 ♌ 59	12 ♒ 25

ASPECTARIAN

h m	Aspects
01 Wednesday	
00 40	☽ ⚼ ♄
03 10	☽ ☌ ♀
03 48	☽ ∠ ♂
10 41	☽ ‖ ☿
12 33	☽ ‖ ♆
13 11	☽ ⚼ ♇
15 09	☽ ± ♄
18 18	♀ ∠ ♂
19 01	☽ ⊥ ♂
19 51	☽ # ♆
20 07	☿ ⊼ ♄
20 38	☽ # ☿
02 Thursday	
00 53	☽ ⊼ ♄
01 29	☽ ☌ ☿
01 57	☉ ‖ ♄
03 49	☿ ‖ ♆
05 16	☽ ⊻ ♂
13 41	☽ △ ♇
13 51	☉ □ ♅
14 53	☽ ☍ ♆
17 00	☽ # ♀
18 42	☽ ⊻ ♃
20 29	☽ # ☉
21 28	☽ △ ♅
21 30	☽ # ♄
22 01	☽ ⊻ ☉
03 Friday	
02 27	☽ # ♃
04 54	☽ ⊥ ♃
08 49	☽ ⊥ ☉
09 04	☽ ⊻ ♀
13 32	☽ ‖ ♂
20 17	☽ ⊥ ♀
20 26	☽ ∠ ♃
21 45	☉ ± ♇
22 54	☽ ⚼ ♅
04 Saturday	
01 21	☽ ∠ ☉
03 06	☽ ☍ ♄
10 16	☽ ⊻ ☿
10 26	☽ ☌ ♂
12 43	☿ ⊻ ♂
13 27	☽ ∠ ♀
14 34	☽ ± ♅
16 47	☽ □ ♇
18 03	☽ ⊼ ♆
21 54	☿ # ♆
22 11	☽ ⊥ ☿
23 04	☽ ⚹ ♃
05 Sunday	
01 13	☽ ⊼ ♅
01 30	☉ ⚼ ♄
04 43	☽ ± ♆
05 48	☽ ⚹ ☉
13 43	☽ # ♅
16 28	☽ ∠ ☿
19 06	☽ ⚹ ♀
21 02	☽ ⚼ ♆
21 07	☽ Q ♃
06 Monday	
01 14	☽ Q ♇
01 39	♀ ⚼ ♇
02 24	☽ ‖ ♇
05 28	☽ Q ☉
08 51	☽ ⊼ ♄
19 49	☽ Q ♀
19 54	☽ ⊻ ♂
20 05	☽ ± ♄
20 56	☿ △ ♇
22 24	♀ ± ♄
23 41	☽ ⚹ ♇
07 Tuesday	
00 04	☽ ⚹ ☿
01 01	☽ △ ♆
07 30	☽ □ ♃
07 58	☽ ⊥ ♂
08 12	☿ ☍ ♆
08 53	☽ ☍ ♅
13 13	☽ ⚼ ♄
18 31	☽ □ ☉
08 Wednesday	
01 21	♂ ± ♅
02 18	☽ Q ☿
02 24	☽ ∠ ♂
04 37	☽ ∠ ♇
10 28	☽ □ ♀
18 28	☽ △ ♄
20 48	☿ # ♀
22 23	☽ ⊥ ♇
09 Thursday	
09 51	☽ ⚹ ♂
10 21	☽ ⊻ ♇
11 42	☽ □ ♆
18 39	♂ □ ♇
18 54	☽ □ ☿
19 45	☽ △ ♃
20 08	☽ ⊼ ♅
10 Friday	
03 16	☿ ⊻ ♃
05 18	☿ △ ♅
08 16	☽ ± ♅
11 30	☽ Q ♂
11 31	☽ △ ☉
12 02	☉ # ☿
16 41	♃ □ ♅
17 46	☽ ‖ ♇
18 07	♂ ⊼ ♆
11 Saturday	
00 03	☽ Q ♆
02 33	☽ ⚼ ♅
02 44	☽ ⚼ ♃
05 49	☽ △ ♀
06 43	☽ □ ♄
07 51	☽ # ♅
13 59	♀ ⊼ ♄
20 46	☽ ⚼ ☉
21 24	☽ ± ♃
23 06	☽ ☌ ♇
12 Sunday	
00 25	☽ ⚹ ♆
02 19	☽ □ ♂
09 06	☽ △ ♅
09 50	☽ ⊼ ♃
15 55	☽ △ ☿
15 56	☽ ⚼ ♀
16 43	☽ ± ☉
19 10	☽ Q ♄
13 Monday	
00 34	☉ ⚼ ♇
04 00	☽ ‖ ♂
05 59	☽ ⊼ ☉
06 49	☽ ∠ ♆
12 18	☽ ± ♀
19 20	☽ ⚹ ♄
21 02	☉ ± ♄
14 Tuesday	
00 55	☽ ⊥ ♆
01 47	☽ ⊼ ♀
02 03	☽ ⚼ ☿
08 32	☽ # ♃
11 20	☽ # ♄
11 40	☽ ⊻ ♇
12 29	☽ ‖ ☿
12 55	☽ ⊻ ♆
14 34	♀ ‖ ♆
18 28	☽ △ ♂
21 33	☽ □ ♅
21 51	☿ # ♄
22 13	☽ ± ☿
23 21	☽ ☍ ♃
23 36	☽ ⊥ ♇
15 Wednesday	
01 11	☽ ∠ ♄
08 19	☽ # ☉
09 09	♅ St R
10 46	☿ ⊥ ♃
11 31	☽ ⊼ ☿
12 34	♃ ± ♇
17 22	☽ ∠ ♇
17 46	☽ # ♆
17 47	☿ # ♃
18 52	☽ ⊥ ♄
20 27	☽ # ♀
22 29	☽ ‖ ♀
23 03	☽ ☍ ☉
16 Thursday	
01 20	☽ ‖ ♆
01 48	☽ ⚼ ♂
06 33	☽ ⊻ ♄
08 58	☽ Q ♅
09 01	☽ ‖ ☉
19 41	☽ ☍ ♀
20 18	☽ ± ♂
22 33	☽ ⚹ ♇
23 42	☽ ☌ ♆
17 Friday	
01 39	♂ ⊼ ♅
06 54	☽ ‖ ♄
08 09	☽ ⚹ ♅
08 31	☽ ⊼ ♂
08 41	♀ ⊥ ♂
08 50	☽ ‖ ♃
10 55	☽ ⊼ ♃
14 21	☽ # ☿
21 30	☽ Q ♇
22 33	☽ ± ♃
18 Saturday	
00 10	♀ △ ♇
00 50	♀ # ♆
03 57	☽ ☍ ☿
08 08	♃ ⚼ ♄
11 05	♀ ☍ ♆
12 37	☽ ∠ ♅
13 30	☽ ⊼ ☉
15 40	☽ ☌ ♄
15 50	☽ ⚼ ♃
19 11	☽ # ♂
19 Sunday	
01 40	☽ ± ☉
05 27	☽ ⊥ ♅
07 17	☽ □ ♇
08 07	☉ # ♀
08 19	☽ ⊻ ♆
08 20	♂ ± ♆
10 37	☽ ⊼ ♀
16 17	☉ ⊼ ♄
16 34	☽ ⊻ ♅
19 23	☽ ⊥ ♆
19 41	☽ ⚼ ☉
20 01	☽ ☍ ♂
20 11	☽ △ ♃
22 52	☽ ± ♀
20 Monday	
00 29	♂ ⚹ ♃
05 49	☽ ‖ ♅
09 15	☉ ∠ ♂
11 50	☽ ∠ ♆
17 00	☽ ⚼ ♀
17 02	☽ ⊼ ☿
18 55	☽ # ♇
22 39	☽ ⊻ ♄
21 Tuesday	
01 13	☽ △ ☉
03 24	☉ ‖ ♆
04 53	☽ ± ☿
09 24	☽ ⊥ ♄
13 54	☽ △ ♇
14 50	☽ ⚹ ♆
22 24	☽ ⚼ ☿
22 42	☽ △ ♀
22 52	☽ ☌ ♅
22 Wednesday	
00 29	♀ △ ♅
01 23	☽ ∠ ♄
03 14	☽ □ ♃
04 00	☿ ⚼ ♅
05 04	☽ ⊼ ♂
12 04	☽ Q ♆
16 11	☽ ± ♂
16 26	☽ ⚼ ♇
23 Thursday	
03 01	☽ △ ☿
03 35	☽ ⚹ ♄
08 07	☽ ± ♇
08 42	☽ ⚼ ♂
10 14	☿ ☍ ♄
10 20	☽ □ ☉
11 13	☽ Q ♃
18 28	☽ ⊼ ♇
19 18	☽ □ ♆
24 Friday	
00 12	☽ Q ♄
03 06	☽ ⊻ ♅
08 02	☽ □ ♀
08 08	☽ ⚹ ♃
09 19	♀ ⊻ ♃
11 45	☽ △ ♂
13 19	☽ ⊥ ♅
19 25	☽ Q ☉
20 19	☽ # ♇
25 Saturday	
04 30	☽ ∠ ♅
05 32	☿ ∠ ♃
05 48	☉ # ♆
06 33	☽ □ ♄
08 26	☽ ‖ ♅
09 49	☽ ∠ ♃
10 04	☽ □ ☿
11 26	♃ ‖ ♄
17 01	☽ ⚹ ☉
17 14	☽ Q ♀
21 06	☽ ☍ ♇
21 51	☽ △ ♆
26 Sunday	
01 02	☽ ⊥ ♃
05 28	☽ ⚹ ♅
08 07	☽ # ♂
11 06	☽ ⊻ ♃
14 57	☽ ⚹ ♀
16 21	☽ □ ♂
17 54	☽ Q ☿
19 37	☽ ∠ ☉
21 35	☽ # ☿
22 34	☽ ⚼ ♆
27 Monday	
01 13	☽ Q ♅
07 55	☽ △ ♄
11 26	☽ ⊥ ☉
13 16	☽ ± ♆
14 47	☽ ⚹ ☿
17 49	☽ ∠ ♀
21 56	☽ ⊻ ☉
22 23	☽ ⊼ ♇
23 03	☽ ⊼ ♆
23 16	☽ Q ♂
23 30	☽ ‖ ♄
23 59	☽ ‖ ♃
28 Tuesday	
01 47	♀ ⊻ ♂
04 34	☉ △ ♇
04 47	☽ # ♀
06 36	☽ □ ♅
08 09	☽ ± ♇
08 18	☽ ⚼ ♄
09 51	☽ ⊥ ♀
12 55	☽ ☌ ♃
14 19	☉ ☍ ♆
16 41	☽ ∠ ☿
18 46	☽ # ☉
19 49	☽ ⚹ ♂
20 33	☽ ⊻ ♀
22 51	☽ ⚼ ♇
22 55	☽ ± ♄
23 53	☽ ‖ ♆
29 Wednesday	
05 10	☽ # ♆
08 13	☽ ⊥ ☿
08 42	☽ ⊼ ♄
11 17	☽ ‖ ☉
18 35	☽ ⊻ ☿
21 38	☽ ∠ ♂
23 28	☽ △ ♇
30 Thursday	
00 05	☽ ☍ ♆
02 36	☽ ☌ ☉
04 10	☽ ‖ ♀
04 43	☽ # ♃
04 47	♀ ⊥ ♃
05 38	☽ # ♄
07 48	☽ △ ♅
10 19	♀ # ♃
13 19	☽ ⊥ ♂
15 01	☽ ⊻ ♃
22 05	♀ # ♄
23 53	☽ ⊻ ♂

All ephemeris data is given at 12.00 UT and the Moon's longitude is additionally given for 24.00 UT

LONGITUDES

Date	Sidereal time h m s	Sun ☉ ° ′ ″	Moon ☽ ° ′ ″	Moon ☽ 24.00 ° ′ ″	Mercury ☿ ° ′	Venus ♀ ° ′	Mars ♂ ° ′	Jupiter ♃ ° ′	Saturn ♄ ° ′	Uranus ♅ ° ′	Neptune ♆ ° ′	Pluto ♇ ° ′
01	12 39 59	08 ♎ 05 31	26 ♎ 38 00	03 ♏ 37 39	02 ♏ 59	21 ♎ 43	19 ♏ 50	14 ♍ 22	25 ♈ 42	09 ♊ 51	05 ♈ 09	04 ♒ 49
02	12 43 56	09 04 33	10 ♏ 31 24	17 ♏ 18 57	03 32	22 58	20 31	14 35	25 R 37	09 R 50	05 R 08	04 R 48
03	12 47 52	10 03 36	24 ♏ 00 06	00 ♐ 34 54	04 01	24 12	21 13	14 47	25 33	09 49	05 06	04 48
04	12 51 49	11 02 41	07 ♐ 03 27	13 ♐ 26 03	04 23	25 27	21 55	14 59	25 28	09 48	05 05	04 48
05	12 55 45	12 01 47	19 ♐ 43 04	25 ♐ 54 59	04 40	26 41	22 37	15 12	25 24	09 47	05 03	04 47
06	12 59 42	13 00 56	02 ♑ 02 21	08 ♑ 05 46	04 51	27 56	23 19	15 24	25 19	09 46	05 01	04 47
07	13 03 39	14 00 06	14 ♑ 05 53	20 ♑ 03 23	04 56	29 ♎ 10	24 01	15 36	25 14	09 45	05 00	04 47
08	13 07 35	14 59 18	25 ♑ 58 57	01 ♒ 53 16	04 R 53	00 ♏ 25	24 43	15 48	25 10	09 44	04 58	04 46
09	13 11 32	15 58 32	07 ♒ 47 01	13 ♒ 40 50	04 42	01 39	25 25	16 00	25 05	09 43	04 56	04 46
10	13 15 28	16 57 47	19 ♒ 35 22	25 ♒ 31 11	04 24	02 54	26 08	16 12	25 00	09 42	04 55	04 46
11	13 19 25	17 57 05	01 ♓ 28 51	07 ♓ 28 51	03 58	04 09	26 50	16 25	24 56	09 40	04 53	04 46
12	13 23 21	18 56 24	13 ♓ 31 36	19 ♓ 37 29	03 23	05 23	27 32	16 36	24 51	09 39	04 51	04 45
13	13 27 18	19 55 45	25 ♓ 46 48	01 ♈ 59 45	02 40	06 38	28 15	16 48	24 46	09 38	04 50	04 45
14	13 31 14	20 55 08	08 ♈ 16 29	14 ♈ 37 05	01 50	07 52	28 57	17 00	24 42	09 36	04 48	04 45
15	13 35 11	21 54 33	21 ♈ 01 32	27 ♈ 29 45	00 ♏ 52	09 07	29 ♏ 40	17 12	24 37	09 35	04 47	04 45
16	13 39 08	22 54 00	04 ♉ 01 38	10 ♉ 36 58	29 ♎ 48	10 21	00 ♐ 23	17 24	24 32	09 33	04 45	04 45
17	13 43 04	23 53 29	17 ♉ 15 33	23 ♉ 57 09	28 39	11 36	01 06	17 35	24 27	09 32	04 44	04 45
18	13 47 01	24 53 00	00 ♊ 41 31	07 ♊ 28 24	27 26	12 50	01 48	17 47	24 22	09 30	04 42	04 D 45
19	13 50 57	25 52 34	14 ♊ 17 35	21 ♊ 08 53	26 12	14 05	02 31	17 59	24 18	09 29	04 40	04 45
20	13 54 54	26 52 09	28 ♊ 02 08	04 ♋ 57 12	24 59	15 19	03 14	18 10	24 13	09 27	04 39	04 45
21	13 58 50	27 51 47	11 ♋ 54 00	18 ♋ 52 28	23 48	16 34	03 57	18 21	24 08	09 25	04 37	04 45
22	14 02 47	28 51 28	25 ♋ 52 34	02 ♌ 54 14	22 42	17 48	04 41	18 33	24 03	09 24	04 36	04 45
23	14 06 43	29 ♎ 51 10	09 ♌ 57 25	17 ♌ 02 02	21 44	19 03	05 24	18 44	23 59	09 22	04 34	04 45
24	14 10 40	00 ♏ 50 55	24 ♌ 07 57	01 ♍ 14 58	20 53	20 18	06 07	18 55	23 54	09 20	04 33	04 45
25	14 14 37	01 50 42	08 ♍ 22 49	15 ♍ 31 08	20 13	21 32	06 50	19 06	23 49	09 18	04 32	04 46
26	14 18 33	02 50 32	22 ♍ 39 31	29 ♍ 47 26	19 44	22 47	07 34	19 18	23 44	09 16	04 30	04 46
27	14 22 30	03 50 23	06 ♎ 54 19	13 ♎ 59 31	19 25	24 01	08 17	19 29	23 40	09 15	04 29	04 46
28	14 26 26	04 50 17	21 ♎ 02 25	28 ♎ 02 21	19 19	25 16	09 01	19 39	23 35	09 13	04 27	04 46
29	14 30 23	05 50 12	04 ♏ 58 43	11 ♏ 50 56	19 D 23	26 30	09 45	19 50	23 30	09 11	04 26	04 47
30	14 34 19	06 50 10	18 ♏ 38 32	25 ♏ 21 09	19 39	27 45	10 28	20 01	23 26	09 09	04 25	04 47
31	14 38 16	07 ♏ 50 09	01 ♐ 58 31	08 ♐ 30 30	20 ♎ 04	28 ♏ 59	11 ♐ 12	20 ♍ 12	23 ♈ 21	09 ♊ 07	04 ♈ 23	04 ♒ 47

Date	Moon True ☊ ° ′	Moon Mean ☊ ° ′	Moon ☽ Latitude ° ′
01	09 ♒ 47	08 ♒ 21	04 S 57
02	09 R 37	08 18	05 05
03	09 28	08 15	04 55
04	09 21	08 12	04 30
05	09 17	08 09	03 53
06	09 14	08 06	03 05
07	09 D 14	08 02	02 10
08	09 14	07 59	01 10
09	09 14	07 56	00 S 08
10	09 R 14	07 53	00 N 55
11	09 11	07 50	01 56
12	09 07	07 47	02 52
13	08 59	07 43	03 41
14	08 49	07 40	04 20
15	08 37	07 37	04 47
16	08 24	07 34	05 00
17	08 12	07 31	04 57
18	08 02	07 27	04 38
19	07 53	07 24	04 03
20	07 48	07 21	03 14
21	07 46	07 18	02 12
22	07 D 45	07 15	01 N 02
23	07 R 45	07 12	00 S 12
24	07 44	07 08	01 26
25	07 42	07 05	02 34
26	07 36	07 02	03 34
27	07 28	06 59	04 19
28	07 17	06 56	04 48
29	07 05	06 53	05 00
30	06 52	06 49	04 54
31	06 ♒ 40	06 ♒ 46	04 S 32

DECLINATIONS

Date	Sun ☉ ° ′	Moon ☽ ° ′	Mercury ☿ ° ′	Venus ♀ ° ′	Mars ♂ ° ′	Jupiter ♃ ° ′	Saturn ♄ ° ′	Uranus ♅ ° ′	Neptune ♆ ° ′	Pluto ♇ ° ′
01	03 S 13	14 S 53	15 S 40	07 S 39	18 S 16	07 N 02	07 N 21	21 N 50	00 N 42	23 S 42
02	03 36	19 48	15 55	08 09	18 28	06 58	07 19	21 50	00 42	23 42
03	03 59	23 32	16 08	08 38	18 39	06 53	07 18	21 50	00 41	23 43
04	04 22	25 56	16 17	09 07	18 51	06 48	07 16	21 49	00 40	23 43
05	04 45	26 54	16 25	09 36	19 02	06 44	07 14	21 49	00 40	23 43
06	05 08	26 30	16 29	10 05	19 13	06 39	07 12	21 49	00 39	23 43
07	05 31	24 51	16 29	10 33	19 24	06 34	07 11	21 49	00 38	23 43
08	05 54	22 06	16 26	11 01	19 35	06 30	07 09	21 49	00 38	23 43
09	06 17	18 27	16 20	11 29	19 46	06 25	07 07	21 49	00 37	23 43
10	06 40	14 04	16 09	11 57	19 57	06 21	07 05	21 49	00 36	23 43
11	07 02	09 09	15 54	12 24	20 07	06 16	07 04	21 48	00 36	23 43
12	07 25	03 S 50	15 35	12 52	20 17	06 11	07 02	21 48	00 35	23 43
13	07 48	01 N 42	15 11	13 18	20 27	06 07	07 00	21 48	00 35	23 43
14	08 10	07 16	14 42	13 45	20 37	06 02	06 58	21 48	00 34	23 43
15	08 32	12 38	14 10	14 11	20 47	05 58	06 57	21 48	00 33	23 43
16	08 54	17 34	13 33	14 37	20 57	05 53	06 55	21 47	00 33	23 43
17	09 16	21 44	12 52	15 03	21 06	05 49	06 53	21 47	00 32	23 43
18	09 38	24 50	12 09	15 28	21 15	05 45	06 51	21 47	00 31	23 43
19	10 00	26 32	11 24	15 52	21 24	05 40	06 50	21 47	00 31	23 42
20	10 21	26 39	10 39	16 17	21 33	05 36	06 48	21 46	00 30	23 42
21	10 43	25 06	09 53	16 41	21 42	05 32	06 46	21 46	00 30	23 42
22	11 04	21 59	09 10	17 04	21 50	05 27	06 45	21 46	00 29	23 42
23	11 25	17 34	08 29	17 27	21 59	05 23	06 43	21 46	00 29	23 42
24	11 46	12 08	07 52	17 50	22 07	05 19	06 41	21 45	00 28	23 42
25	12 07	06 N 02	07 20	18 12	22 15	05 15	06 39	21 45	00 27	23 42
26	12 27	00 S 21	06 54	18 34	22 22	05 10	06 38	21 45	00 27	23 42
27	12 48	06 42	06 32	18 55	22 30	05 06	06 36	21 45	00 26	23 42
28	13 08	12 40	06 17	19 16	22 37	05 02	06 35	21 44	00 26	23 42
29	13 28	17 53	06 08	19 36	22 44	04 58	06 33	21 44	00 25	23 41
30	13 48	22 04	06 04	19 55	22 51	04 54	06 31	21 44	00 25	23 41
31	14 S 07	25 S 00	06 S 06	20 S 15	22 S 58	04 N 50	06 N 30	21 N 43	00 N 24	23 S 41

ZODIAC SIGN ENTRIES

Date	h	m	Planets
01	17	45	☽ ♏
03	22	56	☽ ♐
06	07	59	☽ ♑
08	03	59	♀ ♏
08	20	10	☽ ♒
11	09	02	☽ ♓
13	20	10	☽ ♈
15	23	14	♂ ♐
16	04	37	☽ ♉
16	07	36	☿ ♎
18	10	46	☽ ♊
20	15	25	☽ ♋
22	19	03	☽ ♌
23	15	33	☉ ♏
24	21	54	☽ ♍
27	00	21	☽ ♎
29	03	23	☽ ♏
31	08	24	☽ ♐

LATITUDES

Date	Mercury ☿ ° ′	Venus ♀ ° ′	Mars ♂ ° ′	Jupiter ♃ ° ′	Saturn ♄ ° ′	Uranus ♅ ° ′	Neptune ♆ ° ′	Pluto ♇ ° ′
01	03 S 22	00 N 53	00 S 36	00 N 58	02 S 46	00 S 06	01 S 28	04 S 47
04	03 31	00 47	00 37	00 58	02 47	00 06	01 28	04 47
07	03 32	00 40	00 39	00 58	02 47	00 06	01 28	04 47
10	03 22	00 34	00 41	00 59	02 47	00 06	01 28	04 47
13	02 58	00 27	00 42	00 59	02 47	00 06	01 28	04 47
16	02 18	00 19	00 44	00 59	02 47	00 06	01 28	04 47
19	01 23	00 12	00 45	01 00	02 47	00 06	01 28	04 47
22	00 S 21	00 N 04	00 47	01 00	02 47	00 06	01 28	04 46
25	00 N 36	00 S 03	00 48	01 01	02 47	00 06	01 28	04 46
28	01 22	00 11	00 50	01 01	02 47	00 06	01 28	04 46
31	01 N 53	00 S 19	00 S 51	01 N 02	02 S 46	00 S 06	01 S 28	04 S 46

DATA

Julian Date	2461680
Delta T	+73 seconds
Ayanamsa	24° 14′ 54″
Synetic vernal point	04° ♓ 52′ 05″
True obliquity of ecliptic	23° 26′ 15″

MOON'S PHASES, APSIDES AND POSITIONS ☽

Date	h	m	Phase	Longitude ° ′	Eclipse Indicator
07	11	47	☽	14 ♑ 00	
15	13	47	○	21 ♈ 59	
22	17	29	☾	29 ♋ 05	
29	13	37	●	05 ♏ 54	

Day	h	m		
09	15	44	Apogee	
25	05	23	Perigee	
05	16	50	Max dec	26° S 56′
13	04	41	0N	
20	01	39	Max dec	26° N 48′
26	10	40	0S	

LONGITUDES

Date	Chiron ⚷ ° ′	Ceres ⚳ ° ′	Pallas ⚴ ° ′	Juno ⚵ ° ′	Vesta ⚶ ° ′	Black Moon Lilith ⚸ ° ′
01	03 ♉ 26	28 ♍ 06	25 ♌ 32	17 ♋ 20	02 ♌ 59	12 ♒ 25
11	03 ♉ 00	02 ♎ 40	00 ♍ 43	21 ♋ 23	06 ♌ 18	13 ♒ 31
21	02 ♉ 32	07 ♎ 12	05 ♍ 41	25 ♋ 00	09 ♌ 21	14 ♒ 38
31	02 ♉ 04	11 ♎ 42	10 ♍ 26	28 ♋ 06	12 ♌ 05	15 ♒ 44

All ephemeris data is given at 12.00 UT and the Moon's longitude is additionally given for 24.00 UT

ASPECTARIAN

01 Friday
01 14 ☽ ⊥ ♃
02 51 ☽ ☌ ♀
08 58 ☽ ⚼ ♅
10 25 ☽ ☍ ♄
15 45 ☽ ∥ ☿
16 45 ☽ ∠ ♃
22 58 ♂ ⚼ ♆
23 22 ☽ ☌ ☿

02 Saturday
00 22 ☽ ± ♅
02 03 ☽ □ ♇
02 37 ☽ ⊼ ♆
04 39 ☽ ∥ ♂
09 16 ☽ ⊻ ☉
10 47 ☽ ⊼ ♅
13 04 ☽ ± ♆
19 15 ☽ ⚹ ♃
20 39 ☽ ⊥ ☉

03 Sunday
00 01 ☽ # ♅
01 59 ☿ ± ♅
05 00 ☽ ⚼ ♆
06 11 ☉ △ ♅
06 42 ☽ ☌ ♂
09 50 ☽ Q ♇
12 24 ☽ ⊻ ♀
13 23 ☽ ∥ ♇
14 04 ☽ ∠ ☉
14 47 ☽ ⊼ ♄
17 08 ☽ Q ♃
23 45 ♀ ⚼ ♅

04 Monday
00 29 ☽ ⊥ ♀
01 42 ☽ ± ♄
06 54 ☽ ⊻ ☿
07 47 ☽ ⚹ ♇
08 19 ☽ △ ♆
12 29 ♀ ☍ ♄
17 08 ☽ ☍ ♅
18 22 ☽ ⚼ ♄
18 25 ☽ ⊥ ☿
19 02 ☽ ∠ ♀
20 07 ☽ ⚹ ☉

05 Tuesday
03 12 ☽ □ ♃
11 55 ☽ ∠ ☿
12 08 ☽ ∠ ♇
17 50 ♂ Q ♇
17 55 ☽ ⊻ ♂
21 03 ☽ Q ☉
22 55 ☽ △ ♄

06 Wednesday
00 33 ☿ □ ♇
03 01 ☽ ⚹ ♀
05 36 ☽ ⊥ ♇
06 19 ☽ ⊥ ♂
17 25 ☽ ⊻ ♇
17 37 ☽ ⚹ ☿
17 53 ☽ □ ♆

07 Thursday
01 12 ☽ ∠ ♂
03 19 ☽ ⊼ ♅
05 28 ☽ Q ♀
11 47 ☽ □ ☉
14 36 ☿ St R
15 04 ☽ △ ♃
15 19 ☽ ± ♅
17 41 ☽ Q ☿
23 00 ☽ ∥ ♇

08 Friday
05 54 ☽ Q ♆
09 16 ☽ ⚹ ♂
09 28 ☽ ⚼ ♅
10 21 ☽ □ ♄
14 05 ☽ # ♅
21 00 ♀ ∠ ♃
21 58 ☽ ⚼ ♃
22 04 ☽ □ ♀

09 Saturday
01 44 ♂ ⊼ ♄
04 16 ☽ ∥ ♂
05 24 ☿ □ ♇
05 51 ☽ □ ☿
05 52 ☽ ☌ ♇
06 13 ☽ ⚹ ♆
11 13 ☽ Q ♂
12 58 ☉ ⊻ ♃
15 55 ☽ △ ♅
16 36 ☽ ± ♃
19 01 ☉ # ♃
22 43 ☽ Q ♄

10 Sunday
00 30 ☽ ∥ ☿
05 01 ☽ ⊼ ♃
06 11 ☽ △ ☉
12 39 ☽ ∠ ♆
21 43 ☽ ∥ ♀
22 54 ☽ ⚹ ♄

11 Monday
02 03 ☽ □ ♂
03 03 ♀ ± ♅
06 48 ☽ ⊥ ♆
09 32 ☿ ☌ ♀
13 09 ☉ # ♄
15 13 ☽ ⚼ ☉
16 46 ☽ △ ☿
17 57 ☽ △ ♀
18 34 ☽ ⊻ ♇
18 48 ☽ ⊻ ♆
21 02 ☽ ∥ ☉
21 39 ☽ # ♄
23 53 ♀ □ ♇

12 Tuesday
01 22 ☿ ± ♅
01 22 ☽ # ♃
02 01 ♀ ⊼ ♆
04 20 ☽ □ ♅
04 46 ☽ ∠ ♄
06 31 ☽ ⊥ ♇
10 44 ☽ ± ☉
18 11 ☽ ☍ ♃
21 05 ☽ ⚼ ☿
22 25 ☽ ⊥ ♄
23 36 ☽ ⊼ ☉

13 Wednesday
00 15 ☽ ∠ ♇
02 11 ☽ # ♆
03 00 ☽ ⚼ ♀
07 10 ☽ ∥ ♆
10 03 ☽ ⊻ ♄
13 38 ☽ ± ☿
15 34 ☽ Q ♅
17 04 ☽ △ ♂
22 24 ☽ ± ♀

14 Thursday
00 29 ☽ ⊼ ☿
05 17 ☽ ⚹ ♇
05 24 ☽ ☌ ♆
06 46 ☽ ∥ ♃
08 12 ☿ ∠ ♃
10 45 ☽ ∥ ♄
11 09 ☽ ⊼ ♀
14 12 ♂ Q ♃
14 31 ☽ ⚹ ♅
16 13 ☽ # ☉
23 23 ☽ ⚼ ♂

15 Friday
04 00 ☽ Q ♇
04 44 ☽ ⊼ ♃
11 23 ☿ ∥ ♀
13 47 ☽ ☍ ☉
16 06 ☽ ± ♃
17 12 ☽ ± ♂
18 23 ☽ # ☿
18 36 ☽ ∠ ♅
18 38 ☽ ☌ ♄
19 56 ☽ # ♀
20 52 ♀ ⊼ ♅

16 Saturday
02 04 ♄ ∠ ♅
04 18 ☿ ⊻ ♂
04 51 ☽ ☍ ☿
04 55 ☽ ⊼ ♂
08 58 ☽ ⚼ ♃
11 08 ☽ ⊥ ♅
13 19 ☽ □ ♇
13 19 ☽ ⊻ ♆
14 28 ♆ ⚹ ♇
16 40 ♂ ± ♄
19 31 ♀ ± ♆
22 03 ☽ ⊻ ♅

17 Sunday
00 13 ☽ ⊥ ♆
00 43 ☽ ☍ ♀
02 56 ☉ ⊥ ♃
07 49 ☽ # ♂
12 20 ☽ ∥ ♅
12 36 ☽ △ ♃
16 25 ☽ ∠ ♆

18 Monday
00 35 ☉ ☍ ♄
00 49 ☽ ⊻ ♄
00 50 ☽ ⊼ ☉
02 09 ☽ # ♇
03 04 ☉ ⚼ ♅
03 54 ♇ St D
06 42 ☽ ⊼ ☿
11 26 ☽ ⊥ ♄
12 22 ☽ ± ☉
14 05 ☽ ☍ ♂
16 28 ☽ ± ☿
19 05 ☽ ⚹ ♆
19 11 ☽ △ ♇

19 Tuesday
03 16 ☽ ∠ ♄
03 33 ☽ ☌ ♅
05 31 ☽ ⚼ ☉
07 01 ☽ ⚼ ☿
08 04 ☿ ⊥ ♂
11 35 ☽ ⊼ ♀
15 31 ☉ ☌ ☿
16 10 ☽ Q ♆
18 33 ☽ □ ♃
21 33 ☽ ⚼ ♇
23 09 ☽ ± ♀

20 Wednesday
05 23 ☽ ⚹ ♄
07 07 ☽ △ ☿
09 49 ☽ △ ☉
13 14 ☽ ± ♇
16 10 ♃ ± ♄
16 22 ☽ ⚼ ♀
18 08 ☉ ∥ ☿
21 31 ☽ ⊼ ♂
22 51 ☿ ⚼ ♅
23 27 ☽ □ ♆
23 39 ☽ ⊼ ♇

21 Thursday
02 05 ☿ ⊥ ♃
02 06 ☽ Q ♄
02 18 ☽ Q ♃
04 32 ☿ ☍ ♄
07 44 ☽ ⊻ ♅
08 28 ☽ ± ♂
18 03 ☽ ⊥ ♅
20 09 ☉ ⊥ ♂
20 49 ☽ △ ♀
23 16 ☽ ⚹ ♃
23 35 ♂ # ♅

22 Friday
00 11 ☽ # ♇
00 48 ☽ ⚼ ♂
06 57 ☽ □ ☿
08 54 ☽ □ ♄
09 28 ☽ ∠ ♅
09 32 ♂ △ ♆
12 55 ☽ # ♂
13 25 ☽ ∥ ♅
14 34 ♂ ⚹ ♇
17 29 ☽ □ ☉

23 Saturday
01 16 ☽ ∠ ♃
02 52 ☽ △ ♆
03 09 ☽ ☍ ♇
03 50 ☽ △ ♂
04 51 ♀ ⚹ ♃
11 00 ☽ ⚹ ♅
11 38 ☽ Q ☿
12 29 ☽ # ♀
16 47 ☽ ⊥ ♃
21 57 ♀ ⚼ ♆

24 Sunday
02 24 ☽ Q ☉
03 05 ☽ ⊻ ♃
04 16 ☽ ⚼ ♆
04 53 ☽ □ ♀
06 48 ☽ ⚹ ☿
07 17 ☽ Q ♅
08 20 ☿ ∠ ♂
11 36 ☽ △ ♄
13 24 ☽ # ☉
19 16 ☿ ⊻ ♀
19 26 ☽ ± ♆

25 Monday
00 11 ☽ ⚹ ☉
05 32 ☽ ⊼ ♆
05 55 ☽ ⊼ ♇
06 36 ☽ # ☿
06 54 ☽ ∠ ☿
09 16 ☽ □ ♂
09 38 ☽ ∥ ♄
12 44 ☽ ⚼ ♄
13 33 ☽ □ ♅
14 08 ☽ Q ♀
15 04 ☽ ∥ ♃
16 00 ☽ ± ♇
21 27 ☽ ⊥ ☿

26 Tuesday
03 18 ☽ ∠ ☉
03 47 ☽ ± ♄
06 16 ☽ ☌ ♃
07 08 ☽ ⚼ ♇
07 13 ☽ ⊻ ☿
09 00 ☽ ∥ ♆
11 47 ♀ Q ♇
12 13 ☽ ⚹ ♀
12 21 ☽ # ♆
13 49 ☽ ⊼ ♄
17 09 ☽ Q ♂
19 34 ☽ ⊥ ☉
22 04 ☉ ± ♅

27 Wednesday
05 32 ♀ ⊼ ♄
05 55 ☽ # ♃
06 26 ☽ ⊻ ☉
06 50 ☿ # ♄
07 55 ☽ ☍ ♆
08 24 ☽ △ ♇
08 57 ☿ ⊻ ♃
11 24 ☽ ∥ ☿
11 36 ☽ # ♄
14 28 ☽ ⚹ ♂
15 55 ☽ ∠ ♀
15 57 ☽ △ ♅
23 08 ♂ ⚼ ♄

28 Thursday
03 01 ☉ ⊼ ♆
06 42 ☉ ∠ ♃
08 40 ☽ ⊥ ♀
09 03 ☽ ☌ ☿
09 37 ☽ ⊻ ♃
10 27 ☉ □ ♇
13 00 ☿ ⊥ ♀
14 09 ☽ ∥ ☉
14 11 ☿ St D
16 20 ☽ ☍ ♄
17 22 ☽ ∠ ♂
17 25 ☽ ⚼ ♅
18 06 ♂ ☍ ♅
19 56 ☽ ⊻ ♀
20 01 ☽ ⊥ ♃

29 Friday
03 49 ♃ ⚼ ♇
08 53 ☽ ± ♅
09 44 ☽ ⊥ ♂
11 03 ☽ ⊼ ♆
11 39 ☽ □ ♇
11 45 ☽ ∠ ♃
13 37 ☽ ☌ ☉
19 18 ☽ ⊼ ♅
20 47 ☽ ⊻ ♂
21 30 ☽ ± ♆
21 46 ☽ ∥ ♀

30 Saturday
09 43 ☽ # ♅
13 22 ☽ ⚼ ♆
13 50 ☽ ⊻ ☿
14 29 ☽ ⚹ ♃
17 36 ☽ ∥ ♂
19 24 ☽ Q ♇
20 30 ☽ ⊼ ♄
23 52 ☽ ∥ ♇

31 Sunday
00 54 ☽ ⊥ ☿
06 00 ☽ ☌ ♀
07 16 ☽ ± ♄
12 24 ☽ Q ♃
16 24 ☽ △ ♆
17 09 ☽ ⚹ ♇
17 54 ☽ ∠ ☿
18 43 ♀ ± ♄
20 24 ☿ ⊻ ♃
23 39 ☽ ⚼ ♄
23 39 ☽ ⊻ ☉

NOVEMBER 2027

LONGITUDES

Date	Sidereal time h m s	Sun ☉ ° ' "	Moon ☽ ° ' "	Moon ☽ 24.00 ° ' "	Mercury ☿ ° '	Venus ♀ ° '	Mars ♂ ° '	Jupiter ♃ ° '	Saturn ♄ ° '	Uranus ♅ ° '	Neptune ♆ ° '	Pluto ♇ ° '
01	14 42 12	08 ♏ 50 10	14 ♐ 57 05	21 ♐ 18 24	20 ♎ 39	00 ♐ 14	11 ♐ 56	20 ♍ 22	23 ♈ 17	09 ♊ 05	04 ♈ 22	04 ♒ 48
02	14 46 09	09 50 13	27 ♐ 34 40	03 ♑ 46 12	21 23	01 28	12 40	20 33	23 R 12	09 R 02	04 R 21	04 48
03	14 50 06	10 50 18	09 ♑ 53 27	15 ♑ 56 54	22 14	02 43	13 24	20 43	23 08	09 00	04 19	04 49
04	14 54 02	11 50 24	21 ♑ 57 07	27 ♑ 54 44	23 12	03 57	14 08	20 54	23 03	08 58	04 18	04 49
05	14 57 59	12 50 32	03 ♒ 50 23	09 ♒ 44 47	24 16	05 12	14 52	21 04	22 59	08 56	04 17	04 50
06	15 01 55	13 50 41	15 ♒ 38 36	21 ♒ 32 33	25 25	06 26	15 36	21 14	22 55	08 54	04 16	04 50
07	15 05 52	14 50 52	27 ♒ 27 20	03 ♓ 23 39	26 39	07 41	16 21	21 24	22 51	08 52	04 15	04 51
08	15 09 48	15 51 04	09 ♓ 22 07	15 ♓ 23 22	27 56	08 55	17 05	21 34	22 46	08 49	04 13	04 51
09	15 13 45	16 51 18	21 ♓ 27 58	27 ♓ 36 25	29 ♎ 17	10 10	17 49	21 44	22 42	08 47	04 12	04 52
10	15 17 41	17 51 33	03 ♈ 49 09	10 ♈ 06 30	00 ♏ 41	11 24	18 34	21 54	22 38	08 45	04 11	04 53
11	15 21 38	18 51 50	16 ♈ 28 45	22 ♈ 56 00	02 06	12 39	19 18	22 03	22 34	08 42	04 10	04 53
12	15 25 35	19 52 09	29 ♈ 28 18	06 ♉ 05 36	03 34	13 53	20 03	22 13	22 30	08 40	04 09	04 54
13	15 29 31	20 52 29	12 ♉ 47 41	19 ♉ 34 15	05 04	15 07	20 47	22 22	22 26	08 38	04 08	04 55
14	15 33 28	21 52 50	26 ♉ 24 57	03 ♊ 19 19	06 34	16 22	21 32	22 31	22 22	08 35	04 07	04 55
15	15 37 24	22 53 14	10 ♊ 16 50	17 ♊ 16 57	08 06	17 36	22 17	22 41	22 19	08 33	04 06	04 56
16	15 41 21	23 53 39	24 ♊ 19 07	01 ♋ 22 49	09 38	18 51	23 02	22 50	22 15	08 31	04 05	04 57
17	15 45 17	24 54 06	08 ♋ 27 33	15 ♋ 32 51	11 12	20 05	23 46	22 59	22 11	08 28	04 04	04 58
18	15 49 14	25 54 35	22 ♋ 38 21	29 ♋ 43 44	12 45	21 19	24 31	23 08	22 08	08 26	04 03	04 59
19	15 53 10	26 55 05	06 ♌ 48 43	13 ♌ 53 08	14 19	22 34	25 16	23 17	22 04	08 23	04 02	05 00
20	15 57 07	27 55 38	20 ♌ 56 50	27 ♌ 59 41	15 54	23 48	26 01	23 25	22 01	08 21	04 02	05 01
21	16 01 04	28 56 12	05 ♍ 01 37	12 ♍ 02 31	17 28	25 03	26 46	23 34	21 57	08 18	04 01	05 02
22	16 05 00	29 ♏ 56 48	19 ♍ 02 18	26 ♍ 00 51	19 03	26 17	27 32	23 42	21 54	08 16	04 00	05 02
23	16 08 57	00 ♐ 57 25	02 ♎ 57 59	09 ♎ 53 33	20 38	27 31	28 17	23 50	21 51	08 13	03 59	05 03
24	16 12 53	01 58 05	16 ♎ 47 17	23 ♎ 38 57	22 13	28 ♐ 46	29 02	23 59	21 48	08 11	03 59	05 04
25	16 16 50	02 58 46	00 ♏ 28 14	07 ♏ 14 51	23 47	00 ♑ 00	29 ♐ 47	24 07	21 45	08 08	03 58	05 06
26	16 20 46	03 59 28	13 ♏ 58 27	20 ♏ 38 45	25 22	01 14	00 ♑ 33	24 15	21 42	08 06	03 57	05 07
27	16 24 43	05 00 12	27 ♏ 15 29	03 ♐ 48 23	26 57	02 29	01 18	24 22	21 39	08 03	03 57	05 08
28	16 28 39	06 00 57	10 ♐ 17 18	16 ♐ 42 06	28 ♏ 32	03 43	02 04	24 30	21 37	08 01	03 56	05 09
29	16 32 36	07 01 44	23 ♐ 02 45	29 ♐ 19 17	00 ♐ 06	04 57	02 49	24 38	21 34	07 58	03 55	05 10
30	16 36 33	08 ♐ 02 32	05 ♑ 31 48	11 ♑ 40 32	01 ♐ 41	06 ♑ 12	03 ♑ 35	24 ♍ 45	21 ♈ 31	07 ♊ 56	03 ♈ 55	05 ♒ 11

Date	Moon True ☊ ° '	Moon Mean ☊ ° '	Moon ☽ Latitude ° '
01	06 ♒ 30	06 ♒ 43	03 S 56
02	06 R 22	06 40	03 09
03	06 18	06 37	02 15
04	06 15	06 33	01 15
05	06 D 15	06 30	00 S 13
06	06 R 15	06 27	00 N 50
07	06 14	06 24	01 50
08	06 12	06 21	02 46
09	06 08	06 18	03 35
10	06 00	06 14	04 16
11	05 51	06 11	04 44
12	05 39	06 08	04 59
13	05 26	06 05	04 59
14	05 14	06 02	04 41
15	05 04	05 59	04 07
16	04 55	05 55	03 17
17	04 50	05 52	02 15
18	04 47	05 49	01 N 04
19	04 D 47	05 46	00 S 11
20	04 47	05 43	01 25
21	04 R 47	05 39	02 34
22	04 46	05 36	03 33
23	04 42	05 33	04 19
24	04 35	05 30	04 50
25	04 26	05 27	05 03
26	04 16	05 24	05 00
27	04 05	05 20	04 40
28	03 54	05 17	04 06
29	03 46	05 14	03 20
30	03 ♒ 39	05 ♒ 11	02 S 25

DECLINATIONS

Date	Sun ☉ ° '	Moon ☽ ° '	Mercury ☿ ° '	Venus ♀ ° '	Mars ♂ ° '	Jupiter ♃ ° '	Saturn ♄ ° '	Uranus ♅ ° '	Neptune ♆ ° '	Pluto ♇ ° '
01	14 S 27	26 S 30	06 S 13	20 S 33	23 S 04	04 N 46	06 N 28	21 N 43	00 N 24	23 S 41
02	14 46	26 34	06 24	20 51	23 10	04 42	06 27	21 43	00 23	23 41
03	15 05	25 19	06 40	21 08	23 16	04 38	06 25	21 42	00 23	23 41
04	15 23	22 53	06 59	21 25	23 22	04 34	06 23	21 42	00 22	23 41
05	15 41	19 30	07 21	21 42	23 28	04 30	06 22	21 42	00 22	23 40
06	16 00	15 21	07 46	21 57	23 33	04 26	06 21	21 41	00 21	23 40
07	16 17	10 38	08 13	22 12	23 38	04 22	06 19	21 41	00 21	23 40
08	16 35	05 30	08 42	22 27	23 43	04 19	06 18	21 41	00 21	23 40
09	16 52	00 S 05	09 13	22 40	23 47	04 15	06 16	21 40	00 20	23 40
10	17 09	05 N 26	09 45	22 53	23 52	04 11	06 15	21 40	00 20	23 39
11	17 26	10 51	10 18	23 06	23 56	04 08	06 14	21 40	00 19	23 39
12	17 42	15 57	10 52	23 17	24 00	04 04	06 12	21 39	00 19	23 39
13	17 58	20 25	11 26	23 29	24 03	04 01	06 11	21 39	00 19	23 39
14	18 14	23 54	12 01	23 39	24 06	03 57	06 10	21 39	00 18	23 38
15	18 30	26 04	12 36	23 49	24 10	03 54	06 08	21 38	00 18	23 38
16	18 45	26 36	13 10	23 58	24 12	03 50	06 07	21 38	00 17	23 38
17	18 59	25 25	13 45	24 06	24 15	03 47	06 06	21 38	00 17	23 38
18	19 14	22 35	14 20	24 13	24 17	03 43	06 05	21 37	00 17	23 37
19	19 28	18 24	14 54	24 20	24 19	03 40	06 04	21 37	00 17	23 37
20	19 42	13 10	15 27	24 26	24 21	03 37	06 03	21 36	00 16	23 37
21	19 55	07 17	16 01	24 32	24 23	03 34	06 02	21 36	00 16	23 37
22	20 08	01 N 04	16 33	24 36	24 24	03 31	06 01	21 36	00 16	23 36
23	20 21	05 S 08	17 05	24 40	24 25	03 27	06 00	21 35	00 15	23 36
24	20 33	11 03	17 37	24 43	24 26	03 24	05 59	21 35	00 15	23 36
25	20 45	16 22	18 07	24 46	24 26	03 21	05 58	21 34	00 15	23 35
26	20 57	20 47	18 37	24 47	24 27	03 19	05 57	21 34	00 15	23 35
27	21 08	24 05	19 06	24 48	24 27	03 16	05 56	21 34	00 14	23 35
28	21 19	26 02	19 34	24 48	24 26	03 13	05 55	21 33	00 14	23 34
29	21 29	26 35	20 01	24 48	24 26	03 10	05 54	21 33	00 14	23 34
30	21 S 39	25 S 44	20 S 27	24 S 46	24 S 25	03 N 07	05 N 54	21 N 32	00 N 14	23 S 34

ZODIAC SIGN ENTRIES

Date	h	m	Planets
01	07	34	♀ ♐
02	16	41	☽ ♑
05	04	13	☽ ♒
07	17	09	☽ ♓
10	00	26	☿ ♏
10	04	38	☽ ♈
12	12	58	☽ ♉
14	18	14	☽ ♊
16	21	39	☽ ♋
19	00	28	☽ ♌
21	03	25	☽ ♍
22	13	16	☉ ♐
23	06	52	☽ ♎
25	11	10	☽ ♏
25	11	59	♀ ♑
25	18	38	♂ ♑
27	17	01	☽ ♐
29	10	24	☿ ♐
30	01	18	☽ ♑

LATITUDES

Date	Mercury ☿ ° '	Venus ♀ ° '	Mars ♂ ° '	Jupiter ♃ ° '	Saturn ♄ ° '	Uranus ♅ ° '	Neptune ♆ ° '	Pluto ♇ ° '
01	02 N 00	00 S 22	00 S 51	01 N 02	02 S 46	00 S 06	01 S 28	04 S 46
04	02 11	00 30	00 53	01 02	02 46	00 06	01 27	04 46
07	02 12	00 37	00 54	01 03	02 46	00 05	01 27	04 45
10	02 05	00 45	00 55	01 04	02 45	00 05	01 27	04 45
13	01 53	00 52	00 56	01 04	02 45	00 05	01 27	04 45
16	01 37	01 00	00 57	01 05	02 44	00 05	01 27	04 45
19	01 18	01 07	00 58	01 05	02 44	00 05	01 27	04 45
22	00 58	01 13	00 59	01 06	02 43	00 05	01 27	04 45
25	00 37	01 20	01 00	01 07	02 42	00 05	01 27	04 45
28	00 N 16	01 25	01 01	01 07	02 42	00 05	01 27	04 44
31	00 S 04	01 S 31	01 S 02	01 N 08	02 S 41	00 S 05	01 S 27	04 S 44

DATA

Julian Date	2461711
Delta T	+73 seconds
Ayanamsa	24° 14' 57"
Synetic vernal point	04° ♓ 52' 02"
True obliquity of ecliptic	23° 26' 14"

MOON'S PHASES, APSIDES AND POSITIONS ☽

Date	h	m	Phase	Longitude ° '	Eclipse Indicator
06	08	00	☽	13 ♒ 41	
14	03	26	○	21 ♉ 31	
21	00	48	☾	28 ♌ 28	
28	03	24	●	05 ♐ 39	

Day	h	m		
06	11	37	Apogee	
19	00	22	Perigee	
02	01	13	Max dec	26° S 43'
09	12	22	0N	
16	07	34	Max dec	26° N 38'
22	16	07	0S	
29	09	06	Max dec	26° S 35'

LONGITUDES

Date	Chiron ⚷ ° '	Ceres ⚳ ° '	Pallas ⚴ ° '	Juno ⚵ ° '	Vesta ⚶ ° '	Black Moon Lilith ⚸ ° '
01	02 ♉ 01	12 ♎ 09	10 ♍ 54	28 ♋ 23	12 ♌ 20	15 ♒ 51
11	01 ♉ 33	16 ♎ 33	15 ♍ 21	00 ♌ 47	14 ♌ 39	16 ♒ 57
21	01 ♉ 08	20 ♎ 53	19 ♍ 29	02 ♌ 26	16 ♌ 31	18 ♒ 04
31	00 ♉ 45	25 ♎ 06	23 ♍ 16	03 ♌ 13	17 ♌ 49	19 ♒ 10

All ephemeris data is given at 12.00 UT and the Moon's longitude is additionally given for 24.00 UT

ASPECTARIAN

01 Monday
01 05 ☽ ☍ ♅
06 01 ☽ ☌ ♂
11 46 ☽ ⊥ ☉
17 33 ☉ ⊼ ♅
21 08 ☽ ∠ ♇
22 22 ☽ □ ♃
23 23 ☽ ⚹ ☿

02 Tuesday
03 40 ☽ △ ♄
06 17 ☽ ∠ ☉
14 22 ☽ ⊥ ♇
15 31 ☿ # ♄
20 22 ☽ ⊻ ♀
23 55 ☉ ± ♆

03 Wednesday
00 02 ☽ Q ☿
01 06 ☽ □ ♆
02 02 ☽ ⊻ ♇
09 25 ☽ ⊥ ♀
10 16 ☽ ⊼ ♅
12 10 ♀ Q ♃
14 02 ☽ ⚹ ☉
19 23 ☽ ⊻ ♂
22 06 ☽ ± ♅

04 Thursday
05 12 ☽ ∥ ♇
05 18 ☽ ∠ ♀
08 04 ☽ ∥ ♂
08 07 ☽ ⊥ ♂
08 55 ☿ ☍ ♄
09 51 ☽ △ ♃
12 42 ☽ Q ♆
14 12 ☽ □ ♄
14 44 ☽ □ ☿
16 02 ☽ bQ ♅
16 08 ☽ Q ☉
18 37 ♀ △ ♆
21 06 ☽ # ♅
22 16 ☽ ∥ ♀

05 Friday
03 25 ☽ ∠ ♂
04 50 ♀ ⚹ ♇
05 00 ☿ bQ ♅
12 21 ♀ # ♅
12 54 ☽ ⚹ ♆
14 00 ☽ ☌ ♇
15 05 ☽ ⚹ ♀
16 35 ☽ bQ ♃
22 19 ☽ △ ♅

06 Saturday
02 26 ☽ Q ♄
08 00 ☽ □ ☉
08 45 ☽ ∥ ☉
11 09 ☽ ± ♃
11 55 ☽ ⚹ ♂
18 21 ☽ Q ♀
19 21 ☽ ∠ ♆
23 32 ☽ ⊼ ♃

07 Sunday
02 42 ☽ ⚹ ♄
10 10 ☽ △ ☿
13 35 ☽ ⊥ ♆
13 55 ☽ Q ♂
15 00 ♀ bQ ♄
21 41 ♂ ∥ ♇
22 29 ☽ ∥ ☿

08 Monday
01 41 ☽ ⊻ ♆
02 56 ☽ ⊻ ♇
04 13 ☿ ⊥ ♃
08 20 ☽ # ♄
08 49 ☽ ∠ ♄
10 09 ♀ ☍ ♅
10 55 ☽ □ ♅
11 00 ☽ □ ♀
14 59 ☽ ⊥ ♇
17 23 ☽ # ♃
20 01 ☽ bQ ☿

09 Tuesday
02 05 ☽ △ ☉
02 39 ☽ ⊥ ♄
04 21 ☽ □ ♂
08 51 ☽ ∠ ♇
10 54 ☽ # ♆
12 31 ☽ ☍ ♃
13 50 ☽ ∥ ♆
14 25 ☽ ⊻ ♄
16 01 ☽ ± ☿
22 22 ☽ Q ♅

10 Wednesday
05 10 ☽ ⊼ ☿
06 40 ☽ ∥ ♃
10 00 ☽ bQ ☉
12 42 ☽ ☌ ♆
14 02 ☽ ⚹ ♇
15 34 ☽ ∥ ♄
21 23 ☽ ⚹ ♅

11 Thursday
04 01 ☽ △ ♀
04 38 ☽ ± ☉
09 14 ☽ # ☿
12 46 ☽ Q ♇
16 50 ☽ ⊼ ☉
17 36 ☽ △ ♂
19 09 ☉ bQ ♆
21 39 ☿ ± ♅
22 30 ☽ ⊼ ♃
23 16 ☽ ☌ ♄

12 Friday
01 23 ☽ ∠ ♅
07 13 ♂ ∠ ♇
09 40 ☽ ± ♃
10 49 ☽ bQ ♀
17 48 ☽ ⊥ ♅
20 23 ☽ ☍ ☿
20 29 ☽ ⊻ ♆
21 18 ☿ ⊼ ♆
21 30 ☽ # ☉
21 51 ☽ □ ♇
22 43 ☽ bQ ♂

13 Saturday
02 11 ☽ bQ ♃
04 11 ☉ ⊻ ♂
04 35 ☽ ⊻ ♅
04 47 ☽ ± ♀
07 15 ☽ ⊥ ♆
09 39 ☿ □ ♇
15 45 ☽ ± ♂
16 33 ☽ ⊼ ♀
19 21 ♃ ⊼ ♄
19 38 ☽ ∥ ♅
23 13 ☽ ∠ ♆

14 Sunday
02 58 ☽ ⊼ ♂
03 26 ☽ ☍ ☉
04 57 ☽ ⊻ ♄
05 07 ☽ △ ♃
09 47 ☽ # ♀
09 50 ☽ # ♇
10 47 ♀ ∥ ♇
10 55 ☿ ∠ ♂
13 46 ☽ # ♂
15 24 ☽ ⊥ ♄
23 02 ☉ ⊼ ♄

15 Monday
01 22 ☽ ⚹ ♆
02 47 ☽ △ ♇
04 43 ☿ ∠ ♃
06 08 ☉ ⚹ ♃
06 55 ☽ ∠ ♄
07 47 ☽ ⊼ ☿
09 02 ☽ ☌ ♅
12 53 ♂ △ ♄
13 14 ☉ Q ♇
18 52 ☿ ⊼ ♅
19 21 ☽ ± ☿
21 58 ☽ Q ♆

16 Tuesday
01 46 ☽ ☍ ♀
04 06 ♂ □ ♃
04 33 ☽ bQ ♇
08 29 ☽ ⚹ ♄
09 26 ☽ □ ♃
09 41 ☽ ☍ ♂
11 13 ☽ ⊼ ☉
12 37 ☽ bQ ☿
18 51 ☿ ± ♆
19 53 ☽ ± ♇
22 12 ☽ ± ☉

17 Wednesday
04 34 ☽ □ ♆
04 48 ☽ Q ♄
06 04 ☽ ⊼ ♇
09 40 ♀ ∠ ♇
12 01 ☽ ⊻ ♅
14 38 ☽ bQ ☉
16 19 ☽ Q ♃
17 12 ☽ △ ☿
22 09 ☽ ⊥ ♅
23 21 ☽ # ♂

18 Thursday
00 15 ☽ # ♀
04 35 ☽ # ♇
09 34 ☽ ⊼ ♀
11 08 ☽ □ ♄
12 50 ☽ ⚹ ♃
13 20 ☽ ∠ ♅
15 22 ☽ ⊼ ♂
17 58 ☽ △ ☉
18 15 ☽ ∥ ♅
20 41 ☽ ± ♀

19 Friday
02 05 ☽ ± ♂
02 51 ♀ △ ♄
06 44 ☽ # ☉
07 18 ☽ △ ♆
07 45 ♀ ∥ ♂
08 55 ☽ ☍ ♇
13 24 ☽ bQ ♀
14 30 ☽ ∠ ♃
14 40 ☽ ⚹ ♅
18 12 ☽ bQ ♂

20 Saturday
02 20 ☽ □ ☿
02 55 ☽ # ☿
03 35 ♀ □ ♃
05 56 ☽ ⊥ ♃
08 44 ☽ bQ ♆
10 59 ☽ Q ♅
13 48 ☽ △ ♄
13 56 ☉ ± ♄
16 15 ☽ ⊻ ♃
17 20 ☽ △ ♀
21 08 ☽ △ ♂

21 Sunday
00 03 ☽ ± ♆
00 48 ☽ □ ☉
10 16 ☽ ⊼ ♆
12 00 ☽ ⊼ ♇
12 51 ☽ Q ☿
15 17 ☽ bQ ♄
16 55 ☽ ∥ ♄
17 35 ☽ □ ♅
22 16 ☽ ± ♇

22 Monday
02 34 ☽ ∥ ♃
06 38 ☽ ± ♄
09 59 ☽ Q ☉
11 16 ☿ bQ ♆
12 01 ☽ ⚹ ☿
13 44 ☽ bQ ♇
15 07 ☽ ∥ ♆
16 54 ☽ ⊼ ♄
17 06 ☽ # ♆
20 06 ☽ ☌ ♃

23 Tuesday
01 41 ☽ □ ♀
03 27 ☽ □ ♂
05 29 ☽ # ♃
08 15 ☽ ⚹ ☉
13 46 ☽ ☍ ♆
15 22 ☽ # ♄
15 37 ☽ △ ♇
17 12 ☽ ∠ ☿
21 04 ☽ △ ♅

24 Wednesday
05 58 ☿ ⊼ ♄
10 52 ☽ ⊥ ☿
11 57 ☽ Q ♀
12 20 ☽ ∠ ☉
12 27 ☽ Q ♂
13 17 ♂ ⊥ ♇
18 09 ♀ ⊥ ♇
20 44 ☽ ☍ ♄
22 43 ☽ ⊻ ☿
23 08 ☽ bQ ♅

25 Thursday
00 42 ☽ ⊻ ♃
01 17 ☿ Q ♇
01 35 ♀ ☌ ♂
05 21 ☽ ⊥ ☉
10 44 ☽ ⚹ ♂
11 05 ☽ ⚹ ♀
11 22 ☽ ⊥ ♃
12 01 ☿ ⊥ ♂
14 56 ☽ ± ♅
16 47 ☽ ⊻ ☉
17 19 ☿ ⚹ ♃
18 10 ☽ ⊼ ♆
20 11 ☽ □ ♇
21 59 ☽ ∥ ☿

26 Friday
01 32 ☽ ⊼ ♅
02 47 ☿ ⊥ ♀
03 28 ☽ ∠ ♃
04 49 ☽ ± ♆
11 08 ☉ △ ♆
13 00 ☽ ∥ ☉
14 59 ☽ ∠ ♂
16 29 ☽ ∠ ♀
16 55 ☽ # ♅
20 56 ☽ bQ ♆

27 Saturday
01 51 ☽ ⊼ ♄
04 29 ☽ Q ♇
06 42 ☽ ⚹ ♃
07 40 ☽ ∥ ♇
08 14 ☽ ⊥ ♂
10 26 ☽ ⊥ ♀
11 22 ☽ ☌ ☿
12 43 ☽ ± ♄
15 01 ☉ ⚹ ♇
15 24 ☽ ∥ ♂
19 08 ☽ ∥ ♀
19 51 ☽ ⊻ ♂
22 24 ☿ ± ♄
22 33 ☽ ⊻ ♀

28 Sunday
00 15 ☽ △ ♆
02 27 ☽ ⚹ ♇
03 24 ☽ ☌ ☉
04 54 ☽ Q ♃
05 12 ☽ bQ ♄
07 47 ☽ ☍ ♅
16 10 ♀ □ ♆

29 Monday
01 08 ☉ Q ♃
01 27 ☉ bQ ♄
06 32 ☽ ∠ ♇
09 12 ☽ △ ♄
15 02 ☽ □ ♃
16 08 ♀ ⊻ ♇
21 09 ☉ # ♅
23 43 ☽ ⊥ ♇

30 Tuesday
03 28 ☽ ⊻ ☿
07 59 ☽ ☌ ♂
08 52 ☽ □ ♆
09 22 ☉ ☍ ♅
11 20 ☽ ⊻ ♇
13 26 ☽ ☌ ♀
16 39 ☽ ⊼ ♅
16 48 ☽ ⊥ ☿
17 20 ☽ ⊻ ☉
22 24 ♂ □ ♆

DECEMBER 2027

Sidereal time and LONGITUDES

Date	Sidereal time h m s	Sun ☉ ° ' "	Moon ☽ ° ' "	Moon ☽ 24.00 ° ' "	Mercury ☿ ° '	Venus ♀ ° '	Mars ♂ ° '	Jupiter ♃ ° '	Saturn ♄ ° '	Uranus ♅ ° '	Neptune ♆ ° '	Pluto ♇ ° '
01	16 40 29	09 ♐ 03 21	17 ♑ 45 45	23 ♑ 47 47	03 ♐ 15	07 ♑ 26	04 ♑ 21	24 ♍ 52	21 ♈ 29	07 ♊ 53	03 ♈ 54	05 ♒ 12
02	16 44 26	10 04 11	29 ♑ 47 04	05 ♒ 44 04	04 50	08 40	05 06	24 59	21 R 27	07 R 51	03 R 54	05 14
03	16 48 22	11 05 02	11 ♒ 39 21	17 ♒ 33 27	06 24	09 54	05 52	25 06	21 24	07 48	03 54	05 15
04	16 52 19	12 05 54	23 ♒ 27 01	29 ♒ 20 41	07 58	11 09	06 38	25 13	21 22	07 45	03 53	05 16
05	16 56 15	13 06 46	05 ♓ 15 07	11 ♓ 10 59	09 33	12 23	07 24	25 20	21 20	07 43	03 53	05 17
06	17 00 12	14 07 40	17 ♓ 08 59	23 ♓ 09 46	11 07	13 37	08 10	25 26	21 18	07 40	03 53	05 19
07	17 04 08	15 08 34	29 ♓ 13 59	05 ♈ 22 15	12 41	14 51	08 56	25 33	21 16	07 38	03 52	05 20
08	17 08 05	16 09 29	11 ♈ 35 07	17 ♈ 53 05	14 15	16 05	09 42	25 39	21 15	07 35	03 52	05 21
09	17 12 02	17 10 24	24 ♈ 16 36	00 ♉ 45 57	15 50	17 20	10 28	25 45	21 13	07 33	03 52	05 23
10	17 15 58	18 11 21	07 ♉ 21 21	14 ♉ 02 54	17 24	18 34	11 14	25 51	21 11	07 30	03 52	05 24
11	17 19 55	19 12 18	20 ♉ 50 31	27 ♉ 44 02	18 58	19 48	12 00	25 57	21 10	07 28	03 51	05 25
12	17 23 51	20 13 16	04 ♊ 43 06	11 ♊ 47 13	20 32	21 02	12 46	26 02	21 09	07 26	03 51	05 27
13	17 27 48	21 14 14	18 ♊ 55 48	26 ♊ 08 08	22 07	22 16	13 32	26 08	21 07	07 23	03 51	05 28
14	17 31 44	22 15 14	03 ♋ 23 27	10 ♋ 40 54	23 41	23 30	14 19	26 13	21 06	07 21	03 51	05 30
15	17 35 41	23 16 14	17 ♋ 59 39	25 ♋ 18 53	25 16	24 44	15 05	26 18	21 05	07 18	03 D 51	05 31
16	17 39 37	24 17 15	02 ♌ 37 48	09 ♌ 55 43	26 50	25 58	15 51	26 23	21 04	07 16	03 51	05 33
17	17 43 34	25 18 17	17 ♌ 11 59	24 ♌ 26 06	28 ♐ 25	27 12	16 38	26 28	21 04	07 13	03 51	05 34
18	17 47 31	26 19 20	01 ♍ 37 37	08 ♍ 46 12	00 ♑ 00	28 26	17 24	26 33	21 03	07 11	03 51	05 36
19	17 51 27	27 20 24	15 ♍ 51 37	22 ♍ 53 42	01 35	29 ♑ 39	18 11	26 37	21 02	07 09	03 51	05 37
20	17 55 24	28 21 29	29 ♍ 52 21	06 ♎ 47 31	03 10	00 ♒ 53	18 57	26 42	21 02	07 06	03 52	05 39
21	17 59 20	29 ♐ 22 34	13 ♎ 39 11	20 ♎ 27 22	04 46	02 07	19 44	26 46	21 02	07 04	03 52	05 40
22	18 03 17	00 ♑ 23 41	27 ♎ 12 07	03 ♏ 53 26	06 21	03 21	20 30	26 50	21 01	07 02	03 52	05 42
23	18 07 13	01 24 48	10 ♏ 31 24	17 ♏ 06 00	07 57	04 35	21 17	26 53	21 01	07 00	03 52	05 44
24	18 11 10	02 25 56	23 ♏ 37 19	00 ♐ 05 20	09 33	05 48	22 03	26 57	21 D 01	06 57	03 53	05 45
25	18 15 06	03 27 04	06 ♐ 30 07	12 ♐ 51 40	11 09	07 02	22 50	27 01	21 01	06 55	03 53	05 47
26	18 19 03	04 28 13	19 ♐ 10 03	25 ♐ 25 18	12 46	08 16	23 37	27 04	21 02	06 53	03 53	05 49
27	18 23 00	05 29 23	01 ♑ 37 31	07 ♑ 46 46	14 22	09 29	24 24	27 07	21 02	06 51	03 54	05 50
28	18 26 56	06 30 33	13 ♑ 53 13	19 ♑ 57 00	15 58	10 43	25 11	27 10	21 02	06 49	03 54	05 52
29	18 30 53	07 31 43	25 ♑ 58 21	01 ♒ 57 30	17 35	11 57	25 57	27 12	21 03	06 47	03 55	05 54
30	18 34 49	08 32 53	07 ♒ 54 44	13 ♒ 50 24	19 12	13 10	26 44	27 15	21 04	06 45	03 55	05 55
31	18 38 46	09 ♑ 34 03	19 ♒ 44 52	25 ♒ 38 33	20 ♑ 48	14 ♒ 24	27 ♑ 31	27 ♍ 17	21 ♈ 04	06 ♊ 43	03 ♈ 56	05 ♒ 57

Moon Nodes and Latitude

Date	Moon True ☊ ° '	Moon Mean ☊ ° '	Moon ☽ Latitude ° '
01	03 ♒ 35	05 ♒ 08	01 S 24
02	03 R 34	05 05	00 S 20
03	03 D 34	05 01	00 N 44
04	03 35	04 58	01 45
05	03 36	04 55	02 42
06	03 R 37	04 52	03 33
07	03 35	04 49	04 15
08	03 32	04 45	04 47
09	03 27	04 42	05 05
10	03 20	04 39	05 08
11	03 13	04 36	04 54
12	03 05	04 33	04 23
13	02 58	04 30	03 36
14	02 53	04 26	02 33
15	02 51	04 23	01 20
16	02 50	04 20	00 N 01
17	02 D 50	04 17	01 S 18
18	02 52	04 14	02 31
19	02 53	04 10	03 33
20	02 R 54	04 07	04 22
21	02 53	04 04	04 55
22	02 50	04 01	05 11
23	02 46	03 58	05 10
24	02 41	03 55	04 52
25	02 36	03 51	04 20
26	02 31	03 48	03 35
27	02 27	03 45	02 41
28	02 24	03 42	01 40
29	02 23	03 39	00 S 35
30	02 D 23	03 36	00 N 30
31	02 ♒ 24	03 ♒ 32	01 N 34

DECLINATIONS

Date	Sun ☉ ° '	Moon ☽ ° '	Mercury ☿ ° '	Venus ♀ ° '	Mars ♂ ° '	Jupiter ♃ ° '	Saturn ♄ ° '	Uranus ♅ ° '	Neptune ♆ ° '	Pluto ♇ ° '
01	21 S 48	23 S 39	20 S 53	24 S 44	24 S 24	03 N 05	05 N 53	21 N 32	00 N 14	23 S 34
02	21 57	20 32	21 17	24 41	24 22	03 02	05 52	21 32	00 14	23 33
03	22 06	16 35	21 40	24 38	24 21	02 59	05 52	21 31	00 14	23 33
04	22 14	12 03	22 02	24 33	24 19	02 57	05 51	21 31	00 13	23 33
05	22 22	07 04	22 23	24 28	24 16	02 55	05 51	21 30	00 13	23 32
06	22 30	01 S 48	22 43	24 23	24 14	02 52	05 50	21 30	00 13	23 32
07	22 37	03 N 36	23 02	24 16	24 11	02 50	05 50	21 30	00 13	23 31
08	22 43	08 59	23 20	24 09	24 08	02 48	05 49	21 29	00 13	23 31
09	22 49	14 08	23 37	24 01	24 05	02 45	05 49	21 29	00 13	23 31
10	22 55	18 49	23 52	23 52	24 01	02 43	05 49	21 29	00 13	23 30
11	23 00	22 41	24 06	23 42	23 57	02 41	05 48	21 28	00 13	23 30
12	23 05	25 24	24 19	23 32	23 53	02 39	05 48	21 28	00 13	23 30
13	23 09	26 33	24 30	23 21	23 49	02 37	05 48	21 27	00 13	23 29
14	23 13	25 57	24 41	23 10	23 44	02 35	05 48	21 27	00 13	23 29
15	23 16	23 33	24 49	22 57	23 39	02 34	05 48	21 27	00 13	23 28
16	23 19	19 35	24 57	22 45	23 34	02 32	05 48	21 26	00 13	23 28
17	23 21	14 27	25 03	22 31	23 29	02 30	05 47	21 26	00 13	23 28
18	23 23	08 33	25 08	22 17	23 23	02 29	05 47	21 25	00 13	23 27
19	23 25	02 N 18	25 12	22 02	23 17	02 27	05 48	21 25	00 13	23 27
20	23 26	03 S 58	25 14	21 46	23 11	02 26	05 48	21 25	00 13	23 27
21	23 26	09 55	25 14	21 30	23 04	02 24	05 48	21 25	00 13	23 26
22	23 26	15 18	25 14	21 13	22 57	02 23	05 48	21 24	00 14	23 26
23	23 26	19 52	25 11	20 56	22 50	02 22	05 48	21 24	00 14	23 25
24	23 25	23 23	25 07	20 38	22 43	02 20	05 48	21 23	00 14	23 25
25	23 24	25 39	25 02	20 20	22 35	02 19	05 49	21 23	00 14	23 24
26	23 22	26 34	24 55	20 00	22 27	02 18	05 49	21 22	00 14	23 24
27	23 19	26 07	24 47	19 41	22 19	02 17	05 49	21 22	00 14	23 24
28	23 17	24 22	24 37	19 21	22 11	02 16	05 50	21 22	00 15	23 23
29	23 13	21 32	24 26	19 00	22 02	02 16	05 50	21 21	00 15	23 23
30	23 10	17 48	24 13	18 39	21 53	02 15	05 51	21 21	00 15	23 22
31	23 S 06	13 S 24	23 S 58	18 S 17	21 S 44	02 N 14	05 N 51	21 N 21	00 N 15	23 S 22

ZODIAC SIGN ENTRIES

Date	h m	Planets
02	12 26	☽ ♒
05	01 20	☽ ♓
07	13 30	☽ ♈
09	22 36	☽ ♉
12	03 55	☽ ♊
14	06 24	☽ ♋
16	07 41	☽ ♌
18	09 17	☽ ♍
18	11 58	☿ ♑
19	18 40	♀ ♒
20	12 13	☽ ♎
22	02 42	☉ ♑
22	17 00	☽ ♏
24	23 50	☽ ♐
27	08 51	☽ ♑
29	20 04	☽ ♒

LATITUDES

Date	Mercury ☿ ° '	Venus ♀ ° '	Mars ♂ ° '	Jupiter ♃ ° '	Saturn ♄ ° '	Uranus ♅ ° '	Neptune ♆ ° '	Pluto ♇ ° '
01	00 S 04	01 S 31	01 S 02	01 N 08	02 S 41	00 S 05	01 S 27	04 S 44
04	00 25	01 36	01 03	01 09	02 40	00 05	01 26	04 44
07	00 44	01 40	01 03	01 09	02 40	00 05	01 26	04 44
10	01 02	01 44	01 04	01 10	02 39	00 05	01 26	04 44
13	01 18	01 47	01 04	01 11	02 38	00 05	01 26	04 44
16	01 33	01 49	01 05	01 12	02 37	00 05	01 26	04 44
19	01 46	01 51	01 05	01 12	02 36	00 05	01 26	04 44
22	01 57	01 52	01 06	01 13	02 36	00 05	01 26	04 44
25	02 05	01 53	01 06	01 14	02 35	00 05	01 26	04 44
28	02 09	01 52	01 06	01 15	02 34	00 05	01 25	04 44
31	02 S 10	01 S 51	01 S 06	01 N 16	02 S 33	00 S 05	01 S 25	04 S 44

DATA

Julian Date	2461741
Delta T	+73 seconds
Ayanamsa	24° 15' 03"
Synetic vernal point	04° ♓ 51' 56"
True obliquity of ecliptic	23° 26' 13"

MOON'S PHASES, APSIDES AND POSITIONS ☽

Date	h m	Phase	Longitude ° '	Eclipse Indicator
06	05 22	☽	13 ♓ 51	
13	16 09	○	21 ♊ 25	
20	09 11	☾	28 ♍ 14	
27	20 12	●	05 ♑ 50	

Day	h m		
04	08 43	Apogee	
16	02 36	Perigee	
06	20 02	ON	
13	15 57	Max dec	26° N 35'
19	20 45	OS	
26	15 50	Max dec	26° S 35'

LONGITUDES

Date	Chiron ⚷ ° '	Ceres ⚳ ° '	Pallas ⚴ ° '	Juno ⚵ ° '	Vesta ⚶ ° '	Black Moon Lilith ⚸ ° '
01	00 ♉ 45	25 ♎ 06	23 ♍ 16	03 ♌ 13	17 ♌ 49	19 ♒ 10
11	00 ♉ 26	29 ♎ 11	26 ♍ 37	03 ♌ 05	18 ♌ 29	20 ♒ 17
21	00 ♉ 12	03 ♏ 06	29 ♍ 29	02 ♌ 02	18 ♌ 27	21 ♒ 23
31	00 ♉ 04	06 ♏ 50	01 ♎ 46	00 ♌ 11	17 ♌ 38	22 ♒ 30

All ephemeris data is given at 12.00 UT and the Moon's longitude is additionally given for 24.00 UT

ASPECTARIAN

h m Aspects

01 Wednesday
00 50 ☽ ∥ ♀
04 22 ☽ ± ♅
04 41 ☽ ∥ ♂
06 10 ☽ ⊥ ☉
12 47 ☽ ∥ ♇
13 08 ☽ ∠ ☿
19 22 ☽ □ ♄
20 14 ☽ Q ♆
20 29 ♀ ⊼ ♅
21 53 ☿ △ ♆
22 09 ☽ ⚼ ♅

02 Thursday
01 40 ☽ ∠ ☉
02 17 ☽ △ ♃
02 18 ☽ ∥ ☉
05 00 ☽ # ♅
07 19 ☽ ∥ ☿
15 51 ♂ ⊻ ♇
18 06 ☿ ⚹ ♇
20 09 ☿ ⊻ ♂
20 17 ☽ ⚹ ♆
22 59 ☽ ☌ ♇
23 28 ☽ ⊻ ♂
23 44 ☽ ⚹ ☿

03 Friday
02 48 ☿ # ♅
04 12 ☽ △ ♅
07 27 ☽ Q ♄
08 02 ☽ ⊻ ♀
08 49 ☽ ⚼ ♃
10 44 ☽ ⚹ ☉
12 02 ☿ ⚼ ♄
12 28 ☽ ⊥ ♂
21 39 ☽ ⊥ ♀
23 34 ☿ Q ♃

04 Saturday
02 43 ☽ ∠ ♆
03 18 ☽ ± ♃
03 50 ☽ Q ☿
07 46 ☽ ⚹ ♄
08 02 ☽ ∠ ♂
08 46 ☿ ☍ ♅
13 27 ☽ Q ☉
15 38 ☽ ⊼ ♃
18 08 ☽ ∠ ♀
21 02 ☽ ⊥ ♆

05 Sunday
09 13 ☽ ⊻ ♆
09 37 ☉ ∥ ☿
12 04 ☽ ⊻ ♇
14 11 ☽ ∠ ♄
16 39 ☽ ⚹ ♂
16 58 ☽ □ ♅
17 39 ☽ # ♄
21 30 ♂ ⊼ ♅
22 01 ☽ □ ☿

06 Monday
00 14 ☽ ⊥ ♇
04 05 ☽ ⚹ ♀
05 22 ☽ □ ☉
07 09 ☽ # ♃
08 18 ☽ ⊥ ♄
13 03 ♀ ± ♅
18 20 ☽ ∠ ♇
18 26 ☽ Q ♂
19 04 ☽ # ♆
20 17 ☽ ⊻ ♄
21 01 ☽ ∥ ♆

07 Tuesday
04 40 ☽ ☍ ♃
04 55 ☽ Q ♅
06 47 ☽ Q ♀
08 37 ☽ ∥ ♃
21 05 ☽ ☌ ♆
21 53 ☽ ∥ ♄
23 57 ☽ ⚹ ♇

08 Wednesday
04 20 ☽ ⚹ ♅
08 08 ☽ □ ♂
14 08 ♀ ∥ ♂
17 50 ☽ △ ☿
19 26 ☉ ⊻ ♀
21 29 ☽ △ ☉
21 32 ☽ □ ♀
23 01 ☽ Q ♇

09 Thursday
03 34 ☿ ∥ ♇
06 17 ☽ ☌ ♄
08 47 ☽ ∠ ♅
14 46 ☽ ⊼ ♃

10 Friday
01 24 ☽ ⊥ ♅
01 46 ☽ ⚼ ☿
01 55 ☽ ± ♃
03 48 ☽ ⚼ ☉
05 39 ☽ ⊻ ♆
08 27 ☽ □ ♇
11 59 ☿ ∥ ♀
12 16 ☽ ⊻ ♅
16 31 ☽ ⊥ ♆
18 20 ☽ ⚼ ♃
19 24 ☽ △ ♂
20 14 ☽ ± ☿
21 24 ☽ ± ☉

11 Saturday
00 32 ☿ ∥ ♂
03 45 ☽ ∥ ♅
08 17 ☽ ⊼ ☿
08 31 ☽ ∠ ♆
08 54 ☽ ⊼ ☉
09 59 ☽ △ ♀
12 34 ☽ ⊻ ♄
14 17 ☽ # ☉
18 03 ☽ # ♇
19 18 ☽ # ♀
20 58 ☽ △ ♃
21 34 ☽ # ♂
22 17 ☉ ☌ ☿
23 00 ☽ ⊥ ♄
23 21 ☽ ⚼ ♂
23 57 ☽ # ☿

12 Sunday
10 32 ☽ ⚹ ♆
10 34 ☿ ∠ ♇
13 15 ☽ △ ♇
14 12 ♀ □ ♄
14 26 ☽ ∠ ♄
14 27 ☽ ⚼ ♀
15 42 ☽ ± ♂
16 36 ☽ ☌ ♅
17 27 ☉ ∠ ♇
17 58 ♀ ∥ ♇
21 07 ☿ △ ♄

13 Monday
02 26 ☽ ⊼ ♂
04 02 ♀ Q ♆
06 51 ☽ Q ♆
07 07 ☽ ± ♀
07 27 ♂ ± ♅
09 23 ☉ △ ♄
14 17 ♀ ⚼ ♅
14 35 ☽ ⚼ ♇
15 40 ☽ ⚹ ♄
16 09 ☽ ☍ ☉
17 58 ☽ ☍ ☿
18 05 ☽ ⊼ ♀
22 41 ☿ ⊻ ♀

14 Tuesday
00 04 ☽ □ ♃
05 33 ☽ ± ♇
07 30 ☉ ∥ ♀
11 32 ☽ Q ♄
12 46 ☽ □ ♆
15 28 ☽ ⊼ ♇
18 30 ☽ ⊻ ♅

15 Wednesday
01 53 ☽ # ☿
04 20 ☽ ⊥ ♅
05 55 ☽ Q ♃
06 57 ☽ ☍ ♂
09 09 ♆ St D
11 10 ☽ # ♂
12 32 ☽ # ♇
14 01 ☽ # ☉
16 28 ☽ # ♀
17 04 ☽ □ ♄
19 03 ☽ ∠ ♅
21 18 ☽ ⊼ ☉

16 Thursday
00 03 ☽ ☍ ♀
01 21 ☽ ⊼ ☿
01 42 ☽ ⚹ ♃
01 48 ☽ ∥ ♅
04 45 ☿ □ ♃
07 52 ☽ ± ☉
12 23 ☽ ± ☿
14 00 ☽ △ ♆
16 48 ☽ ☍ ♇
19 36 ☽ ⚹ ♅
20 52 ♀ △ ♃
23 46 ☽ ⚼ ☉

17 Friday
02 29 ☽ ∠ ♃
05 00 ☽ ⚼ ☿
11 00 ☽ ⊼ ♂
14 44 ☽ ⚼ ♆
15 21 ☽ Q ♅
16 24 ♂ ∥ ♇
17 27 ☽ ⊥ ♃
18 23 ☽ △ ♄
21 30 ☽ ± ♂

18 Saturday
02 28 ☽ △ ☉
03 28 ☽ ⊻ ♃
05 42 ☽ ± ♆
05 44 ☿ ⊥ ♇
06 09 ☽ ⊼ ♀
08 57 ☽ △ ☿
11 15 ☉ ∥ ♂
13 22 ☽ ⚼ ♂
15 44 ☽ ⊼ ♆
17 08 ☽ ± ♀
17 42 ☉ □ ♃
18 40 ☽ ⊼ ♇
19 25 ☽ ⚼ ♄
21 18 ☽ □ ♅
22 40 ☽ ∥ ♄

19 Sunday
04 48 ☽ ± ♇
09 46 ☽ ⚼ ♀
10 37 ☽ ± ♄
11 25 ☽ ∥ ♃
16 10 ☽ △ ♂
19 54 ☽ ∥ ♆
20 08 ☽ ⚼ ♇
20 49 ☽ ⊼ ♄
21 36 ☽ # ♆

20 Monday
06 05 ☽ # ♃
06 29 ☽ ☌ ♃
09 11 ☽ □ ☉
13 56 ☽ △ ♀
18 27 ☽ □ ☿
18 54 ☽ ☍ ♆
19 13 ☽ # ♄
22 02 ☽ △ ♇
22 22 ☿ □ ♆

21 Tuesday
00 31 ☽ △ ♅
11 01 ☉ ∥ ♇
19 05 ☽ Q ☉
19 11 ☉ ⊥ ♇
20 43 ♀ # ♅
23 21 ☽ □ ♂

22 Wednesday
01 00 ☽ ☍ ♄
01 55 ☿ ⊻ ♇
02 49 ☽ ⚼ ♅
06 15 ☽ Q ☿
11 20 ☽ ⊻ ♃
18 11 ☽ ⚹ ☉
18 50 ☽ ± ♅
21 54 ☿ ⊼ ♅
22 08 ☽ ⊥ ♃
22 08 ♀ ⚹ ♆
23 58 ☽ ⊼ ♆

23 Thursday
00 09 ☽ □ ♀
03 18 ☽ □ ♇
04 01 ♂ □ ♄
05 37 ☽ ⊼ ♅
06 42 ☽ ⚹ ☿
09 36 ☽ Q ♂
10 49 ☽ ± ♆
14 30 ☽ ∠ ♃
18 01 ☽ ∥ ♀
21 26 ☽ # ♅
23 39 ☽ ∠ ☉

24 Friday
02 46 ♄ St D
03 15 ☽ ⚼ ♆
06 21 ♂ Q ♆
06 55 ☽ ∥ ♂
07 12 ☽ ⊼ ♄
08 56 ☽ ⚹ ♂
09 01 ♂ ⚼ ♅
10 56 ♀ ☌ ♇
11 47 ☉ ∥ ♇
12 14 ☽ ∥ ☉
12 14 ☽ ∥ ♇
12 15 ☽ Q ♇
12 23 ☽ Q ♀
13 58 ☽ ∠ ☿
17 39 ☽ ⊥ ☉
18 12 ☽ ⚹ ♃
18 18 ☽ ± ♄

25 Saturday
04 14 ☽ ∥ ☿
05 47 ☽ ⊻ ☉
07 05 ☽ △ ♆
09 06 ☽ ⊥ ☿
09 48 ♀ △ ♅
10 39 ☽ ⚹ ♇
11 06 ☽ ⚼ ♄
12 47 ☽ ☍ ♅
13 07 ☽ ⚹ ♀
14 40 ☽ ∠ ♂
16 44 ☽ Q ♃
22 02 ☽ ⊻ ☿
22 12 ☉ □ ♆

26 Sunday
08 51 ☽ ⊥ ♂
13 50 ☿ ± ♅
15 09 ☽ ∠ ♇
15 33 ☽ △ ♄
20 42 ☽ ∠ ♀
21 06 ☽ ⊻ ♂

27 Monday
02 57 ♀ Q ♄
03 14 ☽ □ ♃
08 31 ☽ ⊥ ♇
16 02 ☽ ⊥ ♀
16 25 ☽ □ ♆
20 12 ☽ ☌ ☉
20 13 ☽ ⊻ ♇
20 25 ☉ ⊻ ♇
22 09 ☽ ⊼ ♅

28 Tuesday
05 04 ☽ ⊻ ♀
09 09 ☽ ∥ ☿
09 53 ☽ ± ♅
16 45 ☽ ☌ ☿
18 56 ☉ ⊼ ♅
21 20 ☽ ∥ ♇
22 28 ☽ ∥ ☉

29 Wednesday
02 10 ☽ □ ♄
03 40 ☽ ⚼ ♅
03 53 ☽ Q ♆
08 03 ☽ ∥ ♂
11 58 ☽ ☌ ♂
13 13 ☽ # ♅
14 29 ☽ △ ♃
17 21 ♀ ⚼ ♃

30 Thursday
03 56 ☽ ⚹ ♆
06 28 ☽ ∥ ♀
07 58 ☽ ☌ ♇
09 39 ☽ △ ♅
13 24 ☽ ⊻ ☉
14 19 ☽ Q ♄
20 48 ☽ ⚼ ♃
23 52 ☽ ☌ ♀

31 Friday
02 42 ☽ ⊥ ☉
04 32 ♂ △ ♃
10 20 ☽ ∠ ♆
14 30 ☽ ⊻ ☿
14 42 ☽ ⚹ ♄
15 09 ☽ ± ♃
15 59 ☿ □ ♄
22 44 ☽ ∠ ☉

JANUARY 2028

LONGITUDES

Date	Sidereal time h m s	Sun ☉ ° ' "	Moon ☽ ° ' "	Moon ☽ 24.00 ° ' "	Mercury ☿ ° '	Venus ♀ ° '	Mars ♂ ° '	Jupiter ♃ ° '	Saturn ♄ ° '	Uranus ♅ ° '	Neptune ♆ ° '	Pluto ♇ ° '
01	18 42 42	10 ♑ 35 13	01 ♓ 31 55	07 ♓ 25 28	22 ♑ 25	15 ♒ 37	28 ♑ 18	27 ♍ 19	21 ♈ 05	06 ♊ 41	03 ♈ 56	05 ♒ 59
02	18 46 39	11 36 23	13 ♓ 19 44	19 ♓ 15 16	24 01	16 50	29 05	27 21	21 06	06 R 39	03 57	06 01
03	18 50 35	12 37 32	25 ♓ 12 40	01 ♈ 12 32	25 38	18 04	29 ♑ 52	27 23	21 07	06 37	03 57	06 02
04	18 54 32	13 38 42	07 ♈ 15 28	13 ♈ 22 05	27 14	19 17	00 ♒ 39	27 25	21 08	06 35	03 58	06 04
05	18 58 29	14 39 51	19 ♈ 32 59	25 ♈ 48 44	28 ♑ 49	20 30	01 26	27 26	21 10	06 33	03 59	06 06
06	19 02 25	15 41 00	02 ♉ 09 53	08 ♉ 36 53	00 ♒ 24	21 44	02 13	27 27	21 11	06 31	04 00	06 08
07	19 06 22	16 42 08	15 ♉ 10 09	21 ♉ 49 58	01 58	22 57	03 00	27 28	21 13	06 29	04 00	06 09
08	19 10 18	17 43 17	28 ♉ 36 33	05 ♊ 29 56	03 31	24 10	03 47	27 29	21 14	06 28	04 01	06 11
09	19 14 15	18 44 25	12 ♊ 30 01	19 ♊ 36 32	05 02	25 23	04 35	27 30	21 16	06 26	04 02	06 13
10	19 18 11	19 45 32	26 ♊ 49 05	04 ♋ 07 04	06 33	26 36	05 22	27 30	21 18	06 24	04 03	06 15
11	19 22 08	20 46 39	11 ♋ 29 43	18 ♋ 56 09	08 01	27 49	06 09	27 31	21 20	06 23	04 04	06 17
12	19 26 04	21 47 46	26 ♋ 25 23	03 ♌ 56 21	09 26	29 ♒ 01	06 56	27 R 31	21 22	06 21	04 05	06 19
13	19 30 01	22 48 53	11 ♌ 27 55	18 ♌ 59 00	10 49	00 ♓ 14	07 43	27 31	21 24	06 20	04 06	06 20
14	19 33 58	23 49 59	26 ♌ 28 34	03 ♍ 55 37	12 09	01 27	08 31	27 30	21 27	06 18	04 07	06 22
15	19 37 54	24 51 05	11 ♍ 19 18	18 ♍ 38 54	13 25	02 39	09 18	27 30	21 29	06 17	04 08	06 24
16	19 41 51	25 52 11	25 ♍ 53 49	03 ♎ 03 38	14 36	03 52	10 05	27 29	21 31	06 15	04 09	06 26
17	19 45 47	26 53 16	10 ♎ 08 02	17 ♎ 06 53	15 41	05 04	10 53	27 28	21 34	06 14	04 10	06 28
18	19 49 44	27 54 22	24 ♎ 00 07	00 ♏ 47 50	16 41	06 17	11 40	27 27	21 37	06 13	04 11	06 30
19	19 53 40	28 55 27	07 ♏ 30 09	14 ♏ 07 17	17 33	07 29	12 27	27 26	21 40	06 11	04 12	06 32
20	19 57 37	29 ♑ 56 32	20 ♏ 39 31	27 ♏ 07 07	18 18	08 41	13 15	27 24	21 42	06 10	04 13	06 34
21	20 01 33	00 ♒ 57 36	03 ♐ 30 26	09 ♐ 49 46	18 54	09 54	14 02	27 23	21 45	06 09	04 15	06 35
22	20 05 30	01 58 41	16 ♐ 05 28	22 ♐ 17 50	19 21	11 06	14 49	27 21	21 49	06 08	04 16	06 37
23	20 09 27	02 59 45	28 ♐ 27 11	04 ♑ 33 50	19 37	12 18	15 37	27 19	21 52	06 07	04 17	06 39
24	20 13 23	04 00 48	10 ♑ 38 02	16 ♑ 40 05	19 R 42	13 30	16 24	27 17	21 55	06 06	04 19	06 41
25	20 17 20	05 01 50	22 ♑ 40 13	28 ♑ 38 43	19 36	14 41	17 12	27 14	21 58	06 05	04 20	06 43
26	20 21 16	06 02 52	04 ♒ 35 47	10 ♒ 31 42	19 18	15 53	17 59	27 12	22 02	06 04	04 21	06 45
27	20 25 13	07 03 53	16 ♒ 26 42	22 ♒ 21 02	18 49	17 05	18 46	27 09	22 06	06 03	04 23	06 47
28	20 29 09	08 04 53	28 ♒ 14 58	04 ♓ 08 48	18 09	18 16	19 34	27 06	22 09	06 02	04 24	06 49
29	20 33 06	09 05 52	10 ♓ 02 50	15 ♓ 57 24	17 19	19 28	20 21	27 03	22 13	06 01	04 26	06 51
30	20 37 02	10 06 50	21 ♓ 52 52	27 ♓ 49 35	16 21	20 39	21 09	26 59	22 17	06 01	04 27	06 52
31	20 40 59	11 ♒ 07 47	03 ♈ 48 00	09 ♈ 48 33	15 ♒ 15	21 ♓ 50	21 ♒ 56	26 ♍ 56	22 ♈ 21	06 ♊ 00	04 ♈ 29	06 ♒ 54

Moon True ☊, Moon Mean ☊, Moon ☽ Latitude and DECLINATIONS

Date	Moon True ☊ ° '	Moon Mean ☊ ° '	Moon ☽ Latitude ° '	Sun ☉ ° '	Moon ☽ ° '	Mercury ☿ ° '	Venus ♀ ° '	Mars ♂ ° '	Jupiter ♃ ° '	Saturn ♄ ° '	Uranus ♅ ° '	Neptune ♆ ° '	Pluto ♇ ° '
01	02 ♒ 25	03 ♒ 29	02 N 34	23 S 01	08 S 32	23 S 42	17 S 54	21 S 35	02 N 14	05 N 52	21 N 20	00 N 16	23 S 22
02	02 D 27	03 26	03 27	22 56	03 S 22	23 25	17 32	21 25	02 13	05 53	21 20	00 16	23 21
03	02 29	03 23	04 12	22 50	01 N 57	23 06	17 09	21 16	02 13	05 53	21 20	00 16	23 21
04	02 30	03 20	04 46	22 44	07 16	22 45	16 45	21 06	02 12	05 54	21 20	00 17	23 20
05	02 R 30	03 16	05 09	22 38	12 24	22 23	16 21	20 55	02 12	05 55	21 19	00 17	23 20
06	02 29	03 13	05 17	22 31	17 10	22 00	15 56	20 45	02 12	05 56	21 19	00 17	23 19
07	02 27	03 10	05 09	22 24	21 18	21 35	15 31	20 34	02 12	05 56	21 19	00 18	23 19
08	02 25	03 07	04 44	22 16	24 28	21 09	15 06	20 23	02 11	05 57	21 18	00 18	23 19
09	02 23	03 04	04 03	22 08	26 18	20 41	14 40	20 12	02 11	05 58	21 18	00 18	23 18
10	02 22	03 01	03 05	21 59	26 28	20 13	14 14	20 00	02 12	05 59	21 18	00 19	23 18
11	02 21	02 57	01 53	21 50	24 49	19 43	13 48	19 48	02 12	06 00	21 18	00 19	23 17
12	02 20	02 54	00 N 33	21 40	21 24	19 13	13 21	19 37	02 12	06 01	21 17	00 19	23 17
13	02 D 20	02 51	00 S 51	21 30	16 32	18 42	12 54	19 24	02 12	06 02	21 17	00 20	23 16
14	02 21	02 48	02 10	21 20	10 39	18 10	12 26	19 12	02 13	06 03	21 17	00 20	23 16
15	02 21	02 45	03 20	21 09	04 N 14	17 39	11 58	19 00	02 13	06 04	21 17	00 21	23 16
16	02 22	02 42	04 16	20 58	02 S 17	17 07	11 30	18 47	02 14	06 06	21 16	00 21	23 15
17	02 22	02 38	04 54	20 47	08 31	16 35	11 02	18 34	02 14	06 07	21 16	00 22	23 15
18	02 23	02 35	05 14	20 35	14 11	16 04	10 33	18 21	02 15	06 08	21 16	00 22	23 14
19	02 R 23	02 32	05 17	20 22	19 00	15 35	10 04	18 07	02 16	06 09	21 16	00 23	23 14
20	02 22	02 29	05 02	20 10	22 46	15 06	09 35	17 54	02 16	06 11	21 16	00 23	23 13
21	02 22	02 26	04 32	19 57	25 18	14 40	09 06	17 40	02 17	06 12	21 15	00 24	23 13
22	02 22	02 22	03 50	19 43	26 31	14 15	08 36	17 26	02 18	06 14	21 15	00 24	23 13
23	02 D 22	02 19	02 58	19 29	26 23	13 54	08 06	17 12	02 19	06 15	21 15	00 25	23 12
24	02 22	02 16	01 58	19 15	24 58	13 35	07 36	16 58	02 21	06 16	21 15	00 25	23 12
25	02 22	02 13	00 S 54	19 00	22 25	13 20	07 06	16 43	02 22	06 18	21 15	00 26	23 11
26	02 R 22	02 10	00 N 12	18 46	18 55	13 08	06 36	16 29	02 23	06 19	21 14	00 26	23 11
27	02 22	02 07	01 17	18 30	14 40	13 01	06 05	16 14	02 24	06 21	21 14	00 27	23 10
28	02 22	02 03	02 19	18 15	09 54	12 57	05 34	15 59	02 26	06 23	21 14	00 28	23 10
29	02 21	02 00	03 14	17 59	04 S 48	12 57	05 04	15 44	02 27	06 24	21 14	00 28	23 10
30	02 20	01 57	04 02	17 43	00 N 29	13 02	04 33	15 28	02 29	06 26	21 14	00 29	23 09
31	02 ♒ 19	01 ♒ 54	04 N 39	17 S 26	05 N 47	13 S 09	04 S 02	15 S 13	02 N 31	06 N 28	21 N 14	00 N 30	23 S 09

ZODIAC SIGN ENTRIES

Date	h	m	Planets
01	08	53	☽ ♓
03	16	02	♂ ♒
03	21	35	☽ ♈
06	05	58	☿ ♒
06	07	56	☽ ♉
08	14	26	☽ ♊
10	17	15	☽ ♋
12	17	43	☽ ♌
13	07	20	♀ ♓
14	17	40	☽ ♍
16	18	51	☽ ♎
18	22	35	☽ ♏
20	13	22	☉ ♒
21	05	24	☽ ♐
23	15	02	☽ ♑
26	02	44	☽ ♒
28	15	34	☽ ♓
31	04	22	☽ ♈

LATITUDES

Date	Mercury ☿ ° '	Venus ♀ ° '	Mars ♂ ° '	Jupiter ♃ ° '	Saturn ♄ ° '	Uranus ♅ ° '	Neptune ♆ ° '	Pluto ♇ ° '
01	02 S 10	01 S 50	01 S 06	01 N 16	02 S 33	00 S 05	01 S 25	04 S 44
04	02 05	01 48	01 07	01 17	02 32	00 05	01 25	04 44
07	01 54	01 45	01 07	01 18	02 31	00 05	01 25	04 44
10	01 38	01 41	01 07	01 18	02 30	00 05	01 25	04 44
13	01 14	01 36	01 06	01 19	02 29	00 05	01 25	04 44
16	00 S 41	01 31	01 06	01 20	02 29	00 05	01 25	04 45
19	00 00	01 25	01 06	01 21	02 28	00 05	01 24	04 45
22	00 N 48	01 18	01 06	01 22	02 27	00 05	01 24	04 45
25	01 42	01 10	01 06	01 23	02 26	00 05	01 24	04 45
28	02 34	01 01	01 05	01 23	02 25	00 05	01 24	04 45
31	03 N 15	00 S 52	01 S 05	01 N 24	02 S 25	00 S 05	01 S 24	04 S 46

DATA

Julian Date	2461772
Delta T	+73 seconds
Ayanamsa	24° 15' 08"
Synetic vernal point	04° ♓ 51' 50"
True obliquity of ecliptic	23° 26' 13"

LONGITUDES

Date	Chiron ⚷ ° '	Ceres ⚳ ° '	Pallas ⚴ ° '	Juno ⚵ ° '	Vesta ⚶ ° '	Black Moon Lilith ⚸ ° '
01	00 ♉ 03	07 ♏ 11	01 ♎ 58	29 ♋ 58	17 ♌ 31	22 ♒ 37
11	00 ♉ 01	10 ♏ 39	03 ♎ 30	27 ♋ 33	15 ♌ 54	23 ♒ 43
21	00 ♉ 05	13 ♏ 50	04 ♎ 15	25 ♋ 01	13 ♌ 40	24 ♒ 50
31	00 ♉ 14	16 ♏ 41	04 ♎ 07	22 ♋ 47	11 ♌ 05	25 ♒ 56

MOON'S PHASES, APSIDES AND POSITIONS ☽

Date	h	m	Phase	Longitude ° '	Eclipse Indicator
05	01	40	☽	14 ♈ 14	
12	04	03	○	21 ♋ 28	partial
18	19	26	☾	28 ♎ 13	
26	15	12	●	06 ♒ 11	Annular

Day	h	m		
01	03	57	Apogee	
13	07	53	Perigee	
28	15	43	Apogee	
03	03	12	ON	
10	02	24	Max dec	26° N 37'
16	03	32	OS	
22	21	34	Max dec	26° S 38'
30	09	49	ON	

All ephemeris data is given at 12.00 UT and the Moon's longitude is additionally given for 24.00 UT

ASPECTARIAN

01 Saturday: 01 13 ☿ Ꝑ ♅; 03 24 ☽ ⊼ ♃; 04 39 ☽ ⊥ ☿; 04 40 ☽ ⊥ ♆; 04 49 ☿ Q ♆; 04 57 ☽ ⊻ ♂; 16 54 ☽ ⊻ ♆; 18 03 ☽ ⊥ ♂; 21 05 ☽ ⊻ ♇; 21 17 ☽ ∠ ♄; 22 27 ☽ □ ♅

02 Sunday: 00 27 ☽ # ♄; 01 52 ☽ ∠ ☿; 08 10 ☽ ✶ ☉; 09 19 ☽ ⊥ ♇; 13 39 ☽ ∠ ♂; 15 36 ☽ ⊥ ♄; 16 45 ☿ ∥ ♇; 17 12 ☽ # ♃; 19 56 ☽ ⊻ ♀

03 Monday: 01 22 ♂ # ♅; 02 00 ☽ # ♆; 03 35 ☽ ∠ ♇; 03 45 ☽ ⊻ ♄; 04 25 ☽ ∥ ♆; 09 26 ☽ ⊥ ♀; 10 43 ☽ Q ☉; 10 48 ☽ Q ♅; 11 43 ☉ ± ♅; 12 58 ☽ ✶ ☿; 13 08 ☽ ∥ ♃; 16 22 ☽ ☍ ♃; 21 59 ☽ ✶ ♂

04 Tuesday: 05 27 ☽ ∠ ♀; 05 29 ☽ ☌ ♆; 05 44 ♀ ∠ ♆; 05 47 ☽ ∥ ♄; 09 39 ☽ ✶ ♇; 10 40 ☽ ✶ ♅; 13 14 ☉ ∥ ☿; 14 52 ☿ △ ♃; 16 28 ☽ Q ☿; 23 19 ☽ Q ♂

05 Wednesday: 01 40 ☽ □ ☉; 09 11 ☽ Q ♇; 14 03 ☽ ✶ ♀; 15 07 ☽ ☌ ♄; 15 51 ☽ ∠ ♅

06 Thursday: 01 11 ♀ ✶ ♄; 03 07 ☽ ⊼ ♃; 06 04 ☽ # ♀; 06 38 ♀ ± ♃; 06 54 ☉ ⊥ ♀; 08 13 ☽ □ ☿; 08 55 ☽ ⊥ ♅; 12 07 ☽ □ ♂; 14 25 ☽ ± ♃; 15 14 ☽ Q ♀; 15 25 ☽ ⊻ ♆; 19 25 ☽ □ ♇; 20 06 ☽ ⊻ ♅

07 Friday: 02 33 ☽ ⊥ ♆; 07 05 ☽ Ꝑ ♃; 07 34 ☽ # ♂; 12 04 ☽ ∥ ♅; 13 38 ☽ # ☿; 15 01 ☽ △ ☉; 18 57 ☽ ∠ ♆; 19 07 ☽ # ☉; 22 55 ☽ ⊻ ♄

08 Saturday: 02 12 ☽ # ♇; 03 03 ☿ # ♅; 03 22 ☽ □ ♀; 09 35 ☽ ⊥ ♄; 10 02 ☽ △ ♃; 19 07 ♂ ✶ ♆; 19 46 ☽ Ꝑ ☉; 20 00 ☿ ✶ ♆; 20 55 ☿ ☌ ♂; 21 27 ☽ ✶ ♆; 21 35 ☽ △ ♂; 21 38 ☽ △ ☿

09 Sunday: 01 13 ☽ △ ♇; 01 18 ☽ ∠ ♄; 01 38 ☽ ☌ ♅; 12 26 ☽ ± ☉; 18 00 ☽ Q ♆; 23 21 ☽ ⊼ ☉

10 Monday: 00 38 ☽ Ꝑ ♂; 02 13 ☽ Ꝑ ☿; 02 43 ☽ Ꝑ ♇; 02 49 ☽ ✶ ♄; 07 10 ☿ ☌ ♇; 09 51 ☿ △ ♅; 11 36 ☽ △ ♀; 13 08 ☽ □ ♃; 16 26 ☽ ± ♂; 17 40 ☽ ± ♇; 18 50 ☽ ± ☿; 22 41 ☽ Q ♄; 23 54 ☽ □ ♆

11 Tuesday: 02 50 ☽ ⊼ ♂; 03 31 ☽ ⊼ ♇; 03 42 ☽ ⊻ ♅; 05 04 ☿ ∥ ♂; 05 44 ☽ ⊼ ☿; 06 05 ♀ ⊼ ♃; 13 26 ☽ ⊥ ♅; 14 19 ☽ Ꝑ ♀; 16 06 ♂ ☌ ♇; 17 06 ♀ ⊥ ♆; 18 29 ☽ Q ♃; 18 47 ♂ △ ♅

12 Wednesday: 00 13 ☽ # ♇; 01 33 ☉ □ ♄; 01 50 ☉ Ꝑ ♅; 03 53 ☽ □ ♄; 03 54 ☽ ∠ ♅; 04 03 ☽ ☍ ☉; 06 05 ☽ ± ♀; 06 18 ♄ ∠ ♅; 08 53 ♃ St R; 10 23 ☽ # ☉; 10 44 ☿ Q ♄; 12 39 ☽ ∥ ♅; 13 44 ☽ ✶ ♃; 16 31 ☽ ⊼ ♀; 18 47 ☉ Q ♆; 22 00 ☽ # ♂

13 Thursday: 00 14 ☽ △ ♆; 00 54 ☽ # ☿; 03 49 ☽ ☍ ♇; 03 49 ☽ ✶ ♅; 05 42 ☽ ☍ ♂; 06 39 ♅ △ ♇; 10 52 ☽ ☍ ☿; 13 40 ☽ ∠ ♃; 22 56 ☽ Q ♅

14 Friday: 00 12 ☽ Ꝑ ♆; 03 55 ☽ △ ♄; 04 02 ☽ ⊥ ♃; 04 27 ☽ # ♀; 07 27 ☽ ⊼ ☉; 13 39 ☽ ⊻ ♃; 14 38 ☽ ± ♆; 17 48 ☽ ± ☉; 18 35 ☿ Ꝑ ♃; 19 38 ☉ # ♅; 20 42 ☽ ☍ ♀

15 Saturday: 00 19 ☽ ⊼ ♆; 03 49 ☽ □ ♅; 03 59 ☽ ⊼ ♇; 04 07 ☽ Ꝑ ♄; 05 14 ☽ ∥ ♄; 08 31 ☽ ⊼ ♂; 09 26 ☽ Ꝑ ☉; 13 46 ☽ ± ♇; 15 43 ☽ ⊼ ☿; 17 54 ♂ Q ♄; 18 49 ☽ ± ♄; 18 52 ☽ ± ♂; 19 23 ☽ ∥ ♃

16 Sunday: 02 15 ☽ ∥ ♆; 02 27 ☽ ± ☿; 04 35 ☽ Ꝑ ♇; 04 43 ☽ ⊼ ♄; 04 50 ☽ # ♆; 10 34 ☽ Ꝑ ♂; 11 47 ☽ # ♃; 11 57 ☽ △ ☉; 14 39 ☽ ☌ ♃; 17 41 ♀ ⊻ ♆; 18 42 ☽ Ꝑ ☿

17 Monday: 01 51 ☽ ☍ ♆; 02 30 ☽ # ♄; 02 36 ☽ ⊼ ♀; 05 22 ☽ △ ♅; 05 45 ☽ △ ♇; 13 21 ☽ △ ♂; 13 45 ☽ ± ♀; 21 29 ☽ ∥ ♀; 22 18 ☽ △ ☿

18 Tuesday: 01 29 ☉ △ ♃; 06 47 ☽ Ꝑ ♀; 07 08 ☽ Ꝑ ♅; 07 48 ☽ ☍ ♄; 10 39 ♀ □ ♅; 16 24 ♀ ⊻ ♇; 18 04 ☽ ⊻ ♃; 18 52 ♀ ∠ ♄; 19 26 ☽ □ ☉; 20 05 ☽ ∥ ☿; 22 56 ☽ ± ♅

19 Wednesday: 04 42 ☽ ⊥ ♃; 06 04 ☽ ⊼ ♆; 07 30 ☽ ∥ ♂; 09 39 ☽ ⊼ ♅; 10 14 ☽ □ ♇; 11 20 ♂ Ꝑ ♃; 11 58 ☽ △ ♀; 16 53 ☽ ± ♆; 19 34 ☽ ∥ ☉; 20 54 ☽ ∠ ♃; 21 32 ☽ □ ♂

20 Thursday: 01 32 ☽ # ♅; 06 34 ☽ Q ☉; 07 25 ☽ □ ☿; 09 21 ☽ Ꝑ ♆; 13 57 ☽ ⊼ ♄; 15 35 ☽ ∥ ♇; 19 15 ☽ Q ♇

21 Friday: 00 31 ☽ ✶ ♃; 01 09 ☽ ± ♄; 06 47 ☽ ✶ ☉; 09 02 ☽ Q ♂; 13 24 ☽ △ ♆; 17 00 ☽ ☍ ♅; 17 51 ☽ ✶ ♇; 18 11 ☽ Ꝑ ♄; 18 42 ☽ Q ☿; 23 07 ☽ Q ♃

22 Saturday: 01 24 ☽ □ ♀; 06 43 ☿ ∠ ♆; 09 24 ☽ ✶ ♂; 13 52 ☽ ∠ ☉; 18 27 ☽ ✶ ☿; 22 43 ☽ ∠ ♇; 23 06 ☽ △ ♄

23 Sunday: 08 54 ☽ ⊥ ☉; 09 47 ☽ □ ♃; 16 00 ☽ Q ♀; 16 19 ☽ ⊥ ♇; 16 31 ☽ ∠ ♂; 19 23 ♀ ⊥ ♇; 21 43 ☽ ⊻ ☉; 23 29 ☽ □ ♆

24 Monday: 00 14 ☽ ∠ ☿; 03 02 ☽ ⊼ ♅; 04 10 ☽ ⊻ ♇; 11 02 ☿ St R; 11 30 ☽ ⊥ ♂; 14 54 ☽ ± ♅; 18 04 ☽ ⊥ ☿; 18 18 ☽ ✶ ♀; 19 08 ☉ ✶ ♆

25 Tuesday: 00 17 ☽ ⊻ ♂; 05 40 ☽ ∥ ♇; 05 56 ☽ ⊻ ☿; 08 49 ☽ Ꝑ ♅; 10 36 ☽ □ ♄; 11 19 ☽ Q ♆; 20 42 ☽ # ♅; 21 08 ☽ △ ♃

26 Wednesday: 03 41 ☽ ∠ ♀; 08 46 ☿ ∠ ♆; 11 31 ☽ ✶ ♆; 12 23 ☉ △ ♅; 13 00 ☽ ∥ ☉; 14 58 ☽ △ ♅; 15 06 ♀ ⊥ ♄; 15 12 ☽ ☌ ☉; 16 22 ☽ ☌ ♇; 23 03 ☽ Q ♄; 23 54 ☽ ⊥ ♀

27 Thursday: 00 09 ♀ # ♄; 03 05 ☽ ∥ ♂; 03 19 ☽ Ꝑ ♃; 05 03 ☉ ☌ ♇; 12 44 ☿ ☌ ♂; 13 26 ☽ ⊻ ♀; 16 35 ☽ ☌ ☿; 17 04 ☽ ☌ ♂; 17 58 ☽ ∠ ♆; 20 47 ☽ ∥ ☿; 21 31 ☽ ± ♃; 23 32 ☽ ✶ ♄

28 Friday: 06 53 ♂ ∠ ♆; 09 40 ☽ ⊼ ♃; 10 29 ☿ ⊻ ♀; 12 19 ☽ ⊥ ♆

29 Saturday: 00 33 ☽ ⊻ ♆; 03 50 ☽ □ ♅; 04 37 ☽ # ♄; 05 28 ☽ ⊻ ♇; 06 13 ☽ ∠ ♄; 09 53 ☽ ⊻ ☉; 10 39 ☽ ∥ ♀; 17 42 ☽ ⊥ ♇; 22 39 ☽ # ♃; 23 13 ☽ ⊥ ☉

30 Sunday: 00 36 ☽ ⊥ ♄; 01 40 ☽ ⊻ ☿; 07 29 ♂ ± ♃; 07 39 ☽ # ♆; 09 14 ☽ ☌ ♀; 10 25 ☽ ⊻ ♂; 11 59 ☽ ∠ ♇; 12 00 ☽ ∥ ♆; 12 49 ☽ ⊻ ♄; 12 52 ☽ ⊥ ☿; 16 13 ☉ Q ♄; 16 18 ☽ Q ♅; 19 08 ☽ ∠ ☉; 21 04 ☽ ∥ ♃; 22 16 ☽ ☍ ♃; 23 23 ☽ ⊥ ♂

31 Monday: 04 45 ☽ # ♀; 05 30 ☽ ∠ ☿; 05 56 ☿ ⊥ ♀; 13 21 ☽ ☌ ♆; 13 22 ♀ ∠ ♇; 15 09 ☽ ∥ ♄; 16 24 ☽ ✶ ♅; 17 58 ♀ ⊻ ♂; 18 14 ☽ ✶ ♇; 18 43 ☽ ∠ ♂; 22 53 ♀ ⊻ ♄

FEBRUARY 2028

LONGITUDES

Date	Sidereal time h m s	Sun ☉ ° ′ ″	Moon ☽ ° ′ ″	Moon ☽ 24.00 ° ′ ″	Mercury ☿ ° ′	Venus ♀ ° ′	Mars ♂ ° ′	Jupiter ♃ ° ′	Saturn ♄ ° ′	Uranus ♅ ° ′	Neptune ♆ ° ′	Pluto ♇ ° ′
01	20 44 56	12 ♒ 08 43	15 ♈ 51 41	21 ♈ 57 53	14 ♒ 05	23 ♓ 02	22 ♒ 44	26 ♍ 52	22 ♈ 25	05 ♊ 59	04 ♈ 30	06 ♒ 56
02	20 48 52	13 09 37	28 ♈ 07 40	04 ♉ 21 33	12 R 52	24 13	23 31	26 R 48	22 29	05 R 59	04 32	06 58
03	20 52 49	14 10 30	10 ♉ 40 01	17 ♉ 03 35	11 37	25 23	24 19	26 44	22 33	05 58	04 33	07 00
04	20 56 45	15 11 22	23 ♉ 32 43	00 ♊ 07 51	10 25	26 34	25 06	26 40	22 38	05 58	04 35	07 02
05	21 00 42	16 12 12	06 ♊ 49 19	13 ♊ 37 25	09 15	27 45	25 54	26 35	22 42	05 57	04 36	07 04
06	21 04 38	17 13 01	20 ♊ 32 20	27 ♊ 34 05	08 11	28 ♓ 55	26 41	26 31	22 46	05 57	04 38	07 06
07	21 08 35	18 13 49	04 ♋ 42 35	11 ♋ 57 33	07 12	00 ♈ 06	27 29	26 26	22 51	05 57	04 40	07 07
08	21 12 31	19 14 35	19 ♋ 18 31	26 ♋ 44 52	06 20	01 16	28 16	26 21	22 56	05 56	04 41	07 09
09	21 16 28	20 15 20	04 ♌ 15 44	11 ♌ 50 10	05 37	02 26	29 04	26 16	23 00	05 56	04 43	07 11
10	21 20 25	21 16 03	19 ♌ 26 59	27 ♌ 05 00	05 01	03 36	29 ♒ 51	26 11	23 05	05 56	04 45	07 13
11	21 24 21	22 16 45	04 ♍ 42 52	12 ♍ 19 20	04 34	04 46	00 ♓ 39	26 06	23 10	05 56	04 47	07 15
12	21 28 18	23 17 25	19 ♍ 53 08	27 ♍ 23 08	04 14	05 56	01 26	26 00	23 15	05 56	04 49	07 17
13	21 32 14	24 18 04	04 ♎ 48 18	12 ♎ 07 51	04 03	07 05	02 13	25 55	23 20	05 D 56	04 50	07 19
14	21 36 11	25 18 43	19 ♎ 21 06	26 ♎ 27 38	03 59	08 15	03 01	25 49	23 25	05 56	04 52	07 20
15	21 40 07	26 19 20	03 ♏ 27 11	10 ♏ 19 43	04 D 02	09 24	03 48	25 43	23 31	05 56	04 54	07 22
16	21 44 04	27 19 56	17 ♏ 05 17	23 ♏ 44 07	04 13	10 33	04 36	25 37	23 36	05 56	04 56	07 24
17	21 48 00	28 20 30	00 ♐ 16 34	06 ♐ 43 01	04 29	11 42	05 23	25 31	23 41	05 56	04 58	07 26
18	21 51 57	29 ♒ 21 04	13 ♐ 03 59	19 ♐ 19 57	04 51	12 50	06 11	25 25	23 47	05 57	05 00	07 28
19	21 55 54	00 ♓ 21 37	25 ♐ 31 29	01 ♑ 39 07	05 19	13 59	06 58	25 18	23 52	05 57	05 02	07 29
20	21 59 50	01 22 08	07 ♑ 43 23	13 ♑ 44 49	05 51	15 07	07 45	25 12	23 58	05 57	05 04	07 31
21	22 03 47	02 22 38	19 ♑ 43 54	25 ♑ 41 07	06 29	16 16	08 33	25 05	24 03	05 58	05 06	07 33
22	22 07 43	03 23 06	01 ♒ 36 53	07 ♒ 31 38	07 10	17 24	09 20	24 58	24 09	05 58	05 08	07 35
23	22 11 40	04 23 33	13 ♒ 25 43	19 ♒ 19 28	07 56	18 32	10 08	24 51	24 15	05 59	05 10	07 36
24	22 15 36	05 23 58	25 ♒ 13 11	01 ♓ 07 09	08 45	19 39	10 55	24 45	24 21	05 59	05 12	07 38
25	22 19 33	06 24 21	07 ♓ 01 37	12 ♓ 56 48	09 37	20 47	11 42	24 37	24 27	06 00	05 14	07 40
26	22 23 29	07 24 43	18 ♓ 52 57	24 ♓ 50 16	10 33	21 54	12 30	24 30	24 33	06 01	05 16	07 42
27	22 27 26	08 25 03	00 ♈ 48 57	06 ♈ 49 14	11 31	23 01	13 17	24 23	24 39	06 01	05 18	07 43
28	22 31 23	09 25 22	12 ♈ 51 20	18 ♈ 55 30	12 33	24 08	14 04	24 16	24 45	06 02	05 20	07 45
29	22 35 19	10 ♓ 25 38	25 ♈ 02 01	01 ♉ 11 09	13 ♒ 36	25 ♈ 14	14 ♓ 51	24 ♍ 08	24 ♈ 51	06 ♊ 03	05 ♈ 22	07 ♒ 47

MOON

Date	Moon True ☊ ° ′	Moon Mean ☊ ° ′	Moon ☽ Latitude ° ′
01	02 ♒ 17	01 ♒ 51	05 N 04
02	02 R 17	01 48	05 16
03	02 16	01 44	05 14
04	02 D 16	01 41	04 56
05	02 17	01 38	04 21
06	02 18	01 35	03 31
07	02 19	01 32	02 27
08	02 20	01 28	01 N 12
09	02 R 21	01 25	00 S 11
10	02 20	01 22	01 33
11	02 18	01 19	02 50
12	02 16	01 16	03 53
13	02 13	01 13	04 40
14	02 10	01 09	05 07
15	02 07	01 06	05 15
16	02 06	01 03	05 04
17	02 D 05	01 00	04 38
18	02 06	00 57	03 58
19	02 07	00 54	03 08
20	02 09	00 50	02 10
21	02 11	00 47	01 08
22	02 12	00 44	00 S 03
23	02 R 11	00 41	01 N 01
24	02 09	00 38	02 03
25	02 04	00 34	02 59
26	01 59	00 31	03 47
27	01 52	00 28	04 26
28	01 45	00 25	04 54
29	01 ♒ 38	00 ♒ 22	05 N 08

DECLINATIONS

Date	Sun ☉ ° ′	Moon ☽ ° ′	Mercury ☿ ° ′	Venus ♀ ° ′	Mars ♂ ° ′	Jupiter ♃ ° ′	Saturn ♄ ° ′	Uranus ♅ ° ′	Neptune ♆ ° ′	Pluto ♇ ° ′
01	17 S 09	10 N 55	13 S 20	03 S 31	14 S 57	02 N 32	06 N 29	21 N 14	00 N 30	23 S 08
02	16 52	15 44	13 34	02 59	14 41	02 34	06 31	21 14	00 31	23 08
03	16 34	19 59	13 50	02 28	14 25	02 36	06 33	21 14	00 31	23 08
04	16 17	23 25	14 07	01 57	14 09	02 38	06 35	21 14	00 32	23 07
05	15 59	25 44	14 25	01 25	13 53	02 40	06 37	21 13	00 33	23 07
06	15 40	26 37	14 44	00 54	13 37	02 42	06 38	21 13	00 33	23 06
07	15 22	25 48	15 03	00 S 23	13 20	02 44	06 40	21 13	00 34	23 06
08	15 03	23 14	15 21	00 N 09	13 04	02 46	06 42	21 13	00 35	23 06
09	14 44	19 01	15 40	00 40	12 47	02 48	06 44	21 13	00 36	23 05
10	14 25	13 31	15 57	01 12	12 30	02 50	06 46	21 13	00 36	23 05
11	14 05	07 09	16 13	01 43	12 13	02 53	06 48	21 13	00 37	23 05
12	13 45	00 N 26	16 28	02 14	11 56	02 55	06 50	21 13	00 38	23 04
13	13 25	06 S 11	16 42	02 46	11 39	02 57	06 52	21 13	00 39	23 04
14	13 05	12 19	16 54	03 17	11 21	03 00	06 54	21 13	00 39	23 04
15	12 45	17 36	17 06	03 48	11 04	03 02	06 57	21 13	00 40	23 03
16	12 24	21 48	17 15	04 19	10 46	03 05	06 59	21 13	00 41	23 03
17	12 03	24 44	17 23	04 50	10 29	03 08	07 01	21 13	00 42	23 03
18	11 42	26 18	17 30	05 21	10 11	03 10	07 03	21 14	00 42	23 02
19	11 21	26 30	17 36	05 52	09 53	03 13	07 05	21 14	00 43	23 02
20	10 59	25 23	17 39	06 22	09 35	03 16	07 07	21 14	00 44	23 01
21	10 38	23 06	17 42	06 53	09 17	03 19	07 10	21 14	00 45	23 01
22	10 16	19 51	17 43	07 23	08 59	03 21	07 12	21 14	00 46	23 01
23	09 54	15 49	17 42	07 53	08 41	03 24	07 14	21 14	00 46	23 00
24	09 32	11 11	17 40	08 23	08 23	03 27	07 17	21 14	00 47	23 00
25	09 10	06 10	17 37	08 53	08 04	03 30	07 19	21 14	00 48	23 00
26	08 47	00 S 55	17 32	09 23	07 46	03 33	07 21	21 14	00 49	22 59
27	08 25	04 N 24	17 26	09 52	07 28	03 36	07 24	21 15	00 50	22 59
28	08 02	09 35	17 19	10 22	07 09	03 39	07 26	21 15	00 50	22 59
29	07 S 39	14 N 28	17 S 10	10 N 51	06 S 50	03 N 42	07 N 28	21 N 15	00 N 51	22 S 58

ZODIAC SIGN ENTRIES

Date	h	m	Planets
02	15	37	☽ ♉
04	23	46	☽ ♊
07	04	06	☽ ♋
07	10	01	♀ ♈
09	05	12	☽ ♌
10	16	32	♂ ♓
11	04	35	☽ ♍
13	04	13	☽ ♎
15	06	03	☽ ♏
17	11	29	☽ ♐
19	03	26	☉ ♓
19	20	45	☽ ♑
22	08	44	☽ ♒
24	21	43	☽ ♓
27	10	22	☽ ♈
29	21	42	☽ ♉

LATITUDES

Date	Mercury ☿ ° ′	Venus ♀ ° ′	Mars ♂ ° ′	Jupiter ♃ ° ′	Saturn ♄ ° ′	Uranus ♅ ° ′	Neptune ♆ ° ′	Pluto ♇ ° ′
01	03 N 24	00 S 48	01 S 04	01 N 24	02 S 24	00 S 05	01 S 24	04 S 46
04	03 39	00 38	01 04	01 25	02 24	00 04	01 24	04 46
07	03 32	00 27	01 03	01 26	02 23	00 04	01 24	04 46
10	03 09	00 16	01 03	01 27	02 22	00 04	01 24	04 47
13	02 37	00 S 03	01 02	01 27	02 22	00 04	01 24	04 47
16	02 00	00 N 09	01 01	01 28	02 21	00 04	01 24	04 47
19	01 23	00 23	01 00	01 28	02 20	00 04	01 24	04 48
22	00 47	00 36	00 59	01 29	02 20	00 04	01 24	04 48
25	00 N 14	00 50	00 58	01 29	02 19	00 04	01 23	04 48
28	00 S 17	01 05	00 57	01 30	02 19	00 04	01 23	04 49
31	00 S 44	01 N 19	00 S 56	01 N 30	02 S 18	00 S 04	01 S 23	04 S 49

DATA

Julian Date	2461803
Delta T	+73 seconds
Ayanamsa	24° 15′ 14″
Synetic vernal point	04° ♓ 51′ 45″
True obliquity of ecliptic	23° 26′ 13″

MOON'S PHASES, APSIDES AND POSITIONS ☽

Date	h	m	Phase	Longitude ° ′	Eclipse Indicator
03	19	10	☽	14 ♉ 29	
10	15	04	○	21 ♌ 24	
17	08	08	☾	28 ♏ 11	
25	10	37	●	06 ♓ 21	

Day	h	m		
10	20	02	Perigee	
24	16	46	Apogee	
06	12	51	Max dec	26° N 37′
12	13	32	0S	
19	03	23	Max dec	26° S 35′
26	16	07	0N	

LONGITUDES

Date	Chiron ⚷ ° ′	Ceres ⚳ ° ′	Pallas ⚴ ° ′	Juno ⚵ ° ′	Vesta ⚶ ° ′	Black Moon Lilith ⚸ ° ′
01	00 ♉ 15	16 ♏ 57	04 ♎ 03	22 ♋ 35	10 ♌ 50	26 ♒ 03
11	00 ♉ 31	19 ♏ 22	02 ♎ 53	21 ♋ 01	08 ♌ 14	27 ♒ 09
21	00 ♉ 51	21 ♏ 18	00 ♎ 50	20 ♋ 13	05 ♌ 57	28 ♒ 16
31	01 ♉ 16	22 ♏ 42	28 ♍ 03	20 ♋ 13	04 ♌ 16	29 ♒ 23

ASPECTARIAN

h m	Aspects
01 Tuesday	
01 33	♂ ✶ ♄
03 58	☽ ✶ ☉
05 48	☉ ⚼ ♃
08 48	☽ ✶ ☿
18 05	☽ Q ♇
22 05	☽ ∠ ♅
02 Wednesday	
00 23	☽ # ☿
00 57	☽ ☌ ♄
02 25	☽ ✶ ♂
03 34	☽ ⊻ ♀
05 43	☽ Q ☉
06 14	☽ Q ☿
06 55	☽ # ♂
07 22	♀ Q ♅
08 47	☉ ☌ ☿
09 27	☽ ⊼ ♃
15 35	☽ ⊥ ♅
16 27	☽ ⊥ ♀
17 42	☽ # ☉
20 58	☽ ± ♃
03 Thursday	
00 21	☽ ⊻ ♆
03 05	☽ ⊻ ♅
03 11	☽ Q ♂
05 01	☽ □ ♇
06 28	♀ # ♃
09 43	☿ ⚼ ♃
11 25	☽ ∠ ♀
11 47	☽ ⊥ ♆
13 39	☽ □ ☿
14 00	☽ ⚼ ♃
19 10	☽ □ ☉
19 57	☽ ∥ ♅
04 Friday	
00 20	☿ ∠ ♀
04 40	☽ ∠ ♆
07 57	☿ Q ♄
09 34	☽ # ♇
10 18	☽ ⊻ ♄
13 47	☿ ∥ ♂
13 48	♀ ☍ ♃
15 02	☽ □ ♂
17 41	☽ △ ♃
18 05	☽ ✶ ♀
21 20	☽ ⊥ ♄
05 Saturday	
08 02	☽ ✶ ♆
10 27	☽ ☌ ♅
12 26	☽ △ ♇
13 34	☽ ∠ ♄
15 59	☽ △ ☿
17 41	☽ Q ♀
06 Sunday	
05 14	☽ Q ♆
05 48	☽ △ ☉
07 18	♂ ⊼ ♃
14 41	☽ ⚼ ♇
15 52	☽ ✶ ♄
16 13	☽ ⚼ ☿
22 10	☽ □ ♃
23 08	☽ △ ♂
07 Monday	
03 24	♀ # ♆
03 35	☽ □ ♀
05 59	☽ ± ♇
06 29	☽ ± ☿
09 20	☽ ⚼ ☉
11 55	☽ □ ♆
12 14	☽ Q ♄
13 53	☿ ☌ ♇
14 03	☽ ⊻ ♅
15 54	☽ ⊼ ☿
16 02	☽ ⊼ ♇
23 58	☽ ⊥ ♅
08 Tuesday	
00 09	☉ ∥ ☿
01 22	☽ ± ☉
01 35	☽ ⚼ ♂
03 59	☽ Q ♃
11 53	☽ ⊼ ☉
12 56	☽ # ♇
14 39	☽ ∠ ♅
17 04	☽ ± ♂
17 53	☽ □ ♄
22 56	☉ ∠ ♆
23 19	☽ ✶ ♃
09 Wednesday	
00 30	☽ ∥ ♅
00 45	☿ △ ♅
01 18	♂ ⊥ ♆
03 15	☽ ⊼ ♂
08 25	♀ ∥ ♆
08 51	☽ △ ♀
12 22	☉ ± ♃
12 44	☽ △ ♆
14 03	☽ ☍ ☿
14 40	☽ ✶ ♅
16 39	☽ ☍ ♇
23 03	☽ ∠ ♃
10 Thursday	
02 26	☽ # ☿
08 09	☽ # ☉
09 37	☽ Q ♅
10 33	☽ ⚼ ♀
12 28	☽ ⚼ ♆
13 09	☽ ⊥ ♃
15 04	☽ ☍ ☉
16 10	☽ # ♂
17 45	☽ △ ♄
22 32	☽ ⊻ ♃
11 Friday	
00 27	☿ ✶ ♆
01 52	☽ ± ♀
02 39	☽ ± ♆
05 15	☽ ☍ ♂
08 51	☿ ✶ ♀
11 46	☽ ⊼ ☿
12 05	☽ ⊼ ♀
12 06	☽ ⊼ ♆
12 15	♀ ☌ ♆
13 15	☽ ∥ ♄
13 55	☽ □ ♅
16 00	☽ ⊼ ♇
17 28	☽ ⚼ ♄
21 00	☽ ± ☿
12 Saturday	
01 30	☽ ± ♇
03 12	☽ ∥ ♃
06 02	☽ ∥ ♀
07 47	☽ ± ♄
10 59	☽ ⚼ ☿
11 01	☉ ✶ ♄
11 17	☽ ∥ ♆
12 04	♀ ✶ ♅
15 47	☽ # ♆
15 50	☽ ⚼ ♇
17 24	☽ ⊼ ♄
17 50	☽ ⊼ ☉
21 43	☽ ☌ ♃
22 21	☽ # ♀
23 49	♅ St D
13 Sunday	
00 04	☽ # ♃
04 10	☽ ± ☉
07 34	☽ ⊼ ♂
10 47	☽ △ ☿
12 03	☽ ☍ ♆
13 50	☽ △ ♅
14 34	☽ # ♄
16 02	☽ ☍ ♀
16 06	☽ △ ♇
16 45	♀ ✶ ♇
17 54	☽ ± ♂
19 53	☽ ⚼ ☉
21 55	♀ ∥ ♃
14 Monday	
08 15	☽ ∥ ♂
09 38	☽ ⚼ ♂
12 39	☿ St D
14 39	☽ ⚼ ♅
15 05	☽ ∥ ☉
18 53	☽ ☍ ♄
22 49	☽ △ ☉
22 50	☽ ⊻ ♃
22 56	☉ ⊼ ♃
15 Tuesday	
05 56	☽ ± ♅
09 01	☽ ⊥ ♃
09 25	☽ ∥ ☿
12 39	☽ △ ♂
13 02	☽ □ ☿
14 31	☽ ⊼ ♆
16 18	☽ ⊼ ♅
18 50	☽ □ ♇
20 37	☿ ⊻ ♂
23 19	☽ ⊼ ♀
16 Wednesday	
00 36	☽ ∠ ♃
01 02	☽ ± ♆
08 17	☽ # ♅
10 57	☽ ± ♀
15 12	♀ ⊥ ♂
17 07	☽ ⚼ ♆
20 55	☽ ∥ ♇
21 24	☽ Q ☿
22 38	♂ ⊻ ♆
23 50	☽ ⊼ ♄
17 Thursday	
03 04	☽ Q ♇
03 19	☽ ✶ ♃
04 47	☽ ⚼ ♀
08 08	☽ □ ☉
10 54	☽ ± ♄
20 02	☽ ✶ ☿
20 44	☽ △ ♆
22 08	☽ □ ♂
22 33	☽ ☍ ♅
18 Friday	
01 22	☽ ✶ ♇
01 23	☽ Q ♃
03 15	☉ ⊥ ♆
03 49	☽ ⚼ ♄
04 55	♂ □ ♅
11 32	☽ △ ♀
20 33	☿ ✶ ♆
20 55	☽ Q ☉
19 Saturday	
01 29	☽ ∠ ☿
06 05	☽ ∠ ♇
08 45	☽ △ ♄
10 50	☽ Q ♂
11 34	☽ □ ♃
19 44	☽ ⊥ ☿
22 19	☽ ✶ ☉
23 43	☽ ⊥ ♇
20 Sunday	
04 30	♂ ⊻ ♇
06 43	☽ □ ♆
08 07	☽ ⊻ ☿
08 30	☽ ⊼ ♅
11 36	☽ ⊻ ♇
12 04	☽ ✶ ♂
16 04	☿ △ ♅
20 26	☽ ± ♅
21 Monday	
04 18	☽ □ ♀
06 50	☽ ∠ ☉
12 46	☽ ∥ ♇
14 29	☽ ⚼ ♅
18 47	☽ Q ♆
20 14	☽ ∠ ♂
20 47	☽ □ ♄
22 41	☽ △ ♃
22 Tuesday	
02 23	♀ ∥ ♄
02 35	☽ # ♅
02 38	☽ ⊥ ☉
05 35	♂ ∠ ♄
15 44	☽ ⊥ ♂
15 55	☽ ⊻ ☉
19 09	☽ ✶ ♆
20 29	☽ Q ♀
20 51	☽ △ ♅
23 Wednesday	
00 02	☽ ☌ ☿
00 08	☽ ☌ ♇
01 12	☽ ∥ ☿
01 37	☿ ☌ ♇
04 48	☽ ⊻ ♂
04 48	☽ ⚼ ♃
09 35	☽ Q ♄
22 57	☽ ± ♃
23 28	☽ ✶ ♀
24 Thursday	
01 44	☽ ∠ ♆
06 55	☉ ⊻ ♆
10 12	☽ ✶ ♄
11 02	☽ ⊼ ♃
11 37	♀ Q ♇
11 39	♀ # ♂
20 06	☽ ⊥ ♆
20 43	☽ ∥ ☉
25 Friday	
00 19	☽ # ♀
02 14	☉ □ ♅
02 28	☽ ∥ ♂
06 39	☽ # ♄
08 20	☽ ⊻ ♆
09 12	☽ ∠ ♀
09 55	☽ □ ♅
10 37	☽ ☌ ☉
12 02	☿ ⚼ ♃
13 18	☽ ⊻ ♇
16 51	♀ ∠ ♅
16 57	☽ ∠ ♄
17 42	☽ ⊻ ☿
19 32	☉ # ♀
22 10	☽ ☌ ♂
26 Saturday	
00 07	☽ # ♃
01 29	☽ ⊥ ♇
05 20	☽ ⊥ ♀
06 53	☽ ⊥ ☿
07 59	♃ ⊼ ♄
11 18	☽ ⊥ ♄
12 26	☽ # ♆
18 43	☽ ⊻ ♀
18 52	☉ ⊻ ♇
19 42	☽ ∠ ♇
19 48	☽ ∥ ♆
22 21	☽ Q ♅
23 13	☽ ☍ ♃
23 30	☽ ⊻ ♄
27 Sunday	
02 37	☽ ∠ ☿
08 22	☽ ∥ ♃
16 31	♂ # ♄
20 59	☽ ☌ ♆
22 25	☽ ✶ ♅
28 Monday	
01 16	☽ # ♂
01 49	☽ ✶ ♇
01 50	♂ ⊥ ♇
01 52	☽ ∥ ♄
04 34	☽ ⊻ ☉
05 14	☽ # ☉
11 19	☽ ✶ ☿
14 34	☽ ⊻ ♂
14 39	♀ ⊼ ♃
16 06	☽ ∥ ♀
17 06	☿ Q ♄
17 33	☽ ⊥ ☉
20 33	☉ ∠ ♄
29 Tuesday	
01 39	☽ Q ♇
02 41	♀ ☌ ♄
03 14	☽ ⊥ ♂
04 11	☽ ∠ ♅
10 16	☽ ⊼ ♃
11 38	☽ ☌ ♄
12 26	☽ ☌ ♀
12 50	☽ ∠ ☉
13 14	☽ Q ☿
21 48	☽ ⊥ ♅
21 52	☽ ± ♃
22 04	☽ ∠ ♂
22 29	☉ # ♄

All ephemeris data is given at 12.00 UT and the Moon's longitude is additionally given for 24.00 UT

MARCH 2028

LONGITUDES

Date	Sidereal time h m s	Sun ☉ ° ' "	Moon ☽ ° ' "	Moon ☽ 24.00 ° ' "	Mercury ☿ ° '	Venus ♀ ° '	Mars ♂ ° '	Jupiter ♃ ° '	Saturn ♄ ° '	Uranus ♅ ° '	Neptune ♆ ° '	Pluto ♇ ° '
01	22 39 16	11 ♓ 25 53	07 ♉ 23 13	13 ♉ 38 34	14 ♒ 42	26 ♈ 21	15 ♓ 39	24 ♍ 01	24 ♈ 57	06 ♊ 04	05 ♈ 24	07 ♒ 48
02	22 43 12	12 26 05	19 ♉ 57 33	26 ♉ 20 33	15 51	27 27	16 26	23 R 54	25 03	06 05	05 26	07 50
03	22 47 09	13 26 16	02 ♊ 47 58	09 ♊ 20 10	17 01	28 33	17 13	23 46	25 10	06 06	05 28	07 51
04	22 51 05	14 26 24	15 ♊ 57 32	22 ♊ 40 25	18 13	29 ♈ 38	18 00	23 38	25 16	06 07	05 30	07 53
05	22 55 02	15 26 31	29 ♊ 29 06	06 ♋ 23 50	19 27	00 ♉ 43	18 48	23 31	25 22	06 08	05 33	07 55
06	22 58 58	16 26 35	13 ♋ 24 43	20 ♋ 31 48	20 43	01 49	19 35	23 23	25 29	06 09	05 35	07 56
07	23 02 55	17 26 37	27 ♋ 44 54	05 ♌ 03 46	22 00	02 53	20 22	23 15	25 35	06 11	05 37	07 58
08	23 06 52	18 26 37	12 ♌ 27 52	19 ♌ 56 33	23 20	03 58	21 09	23 08	25 42	06 12	05 39	07 59
09	23 10 48	19 26 35	27 ♌ 28 57	05 ♍ 04 02	24 40	05 02	21 56	23 00	25 49	06 13	05 41	08 01
10	23 14 45	20 26 30	12 ♍ 40 35	20 ♍ 17 21	26 03	06 06	22 43	22 52	25 55	06 14	05 43	08 02
11	23 18 41	21 26 24	27 ♍ 53 00	05 ♎ 26 12	27 26	07 10	23 30	22 44	26 02	06 16	05 46	08 04
12	23 22 38	22 26 16	12 ♎ 55 42	20 ♎ 20 24	28 ♒ 51	08 13	24 17	22 36	26 09	06 17	05 48	08 05
13	23 26 34	23 26 06	27 ♎ 39 20	04 ♏ 51 44	00 ♓ 18	09 16	25 04	22 29	26 16	06 19	05 50	08 07
14	23 30 31	24 25 54	11 ♏ 57 04	18 ♏ 55 01	01 45	10 18	25 51	22 21	26 23	06 20	05 52	08 08
15	23 34 27	25 25 41	25 ♏ 45 28	02 ♐ 28 30	03 15	11 21	26 38	22 13	26 29	06 22	05 55	08 09
16	23 38 24	26 25 26	09 ♐ 04 19	15 ♐ 33 18	04 45	12 22	27 25	22 05	26 36	06 24	05 57	08 11
17	23 42 21	27 25 09	21 ♐ 55 57	28 ♐ 12 48	06 17	13 24	28 12	21 58	26 43	06 25	05 59	08 12
18	23 46 17	28 24 50	04 ♑ 24 27	10 ♑ 31 35	07 50	14 25	28 58	21 50	26 50	06 27	06 01	08 13
19	23 50 14	29 ♓ 24 30	16 ♑ 34 50	22 ♑ 34 52	09 24	15 26	29 ♓ 45	21 42	26 57	06 29	06 04	08 15
20	23 54 10	00 ♈ 24 08	28 ♑ 32 20	04 ♒ 27 52	11 00	16 26	00 ♈ 32	21 34	27 05	06 31	06 06	08 16
21	23 58 07	01 23 44	10 ♒ 22 04	16 ♒ 15 29	12 37	17 26	01 19	21 27	27 12	06 32	06 08	08 17
22	00 02 03	02 23 19	22 ♒ 08 37	28 ♒ 01 57	14 15	18 26	02 05	21 19	27 19	06 34	06 10	08 19
23	00 06 00	03 22 51	03 ♓ 55 54	09 ♓ 50 50	15 55	19 25	02 52	21 12	27 26	06 36	06 13	08 20
24	00 09 56	04 22 22	15 ♓ 47 04	21 ♓ 44 51	17 35	20 24	03 39	21 04	27 33	06 38	06 15	08 21
25	00 13 53	05 21 50	27 ♓ 44 27	03 ♈ 46 00	19 17	21 22	04 25	20 57	27 41	06 40	06 17	08 22
26	00 17 50	06 21 17	09 ♈ 49 41	15 ♈ 55 36	21 01	22 19	05 12	20 49	27 48	06 42	06 20	08 23
27	00 21 46	07 20 41	22 ♈ 03 51	28 ♈ 14 32	22 46	23 17	05 58	20 42	27 55	06 45	06 22	08 25
28	00 25 43	08 20 04	04 ♉ 27 44	10 ♉ 43 31	24 32	24 13	06 45	20 35	28 02	06 47	06 24	08 26
29	00 29 39	09 19 24	17 ♉ 02 00	23 ♉ 23 18	26 19	25 09	07 31	20 28	28 10	06 49	06 26	08 27
30	00 33 36	10 18 42	29 ♉ 47 33	06 ♊ 14 56	28 08	26 05	08 18	20 21	28 17	06 51	06 29	08 28
31	00 37 32	11 ♈ 17 58	12 ♊ 45 39	19 ♊ 19 54	29 ♓ 58	27 ♉ 00	09 ♈ 04	20 ♍ 14	28 ♈ 25	06 ♊ 54	06 ♈ 31	08 ♒ 29

Date	Moon True ☊ ° '	Moon Mean ☊ ° '	Moon ☽ Latitude ° '
01	01 ♒ 32	00 ♒ 19	05 N 08
02	01 R 28	00 15	04 54
03	01 25	00 12	04 25
04	01 D 25	00 09	03 41
05	01 25	00 06	02 44
06	01 27	00 ♒ 03	01 36
07	01 28	29 ♑ 59	00 N 20
08	01 R 27	29 56	00 S 59
09	01 25	29 53	02 16
10	01 20	29 50	03 24
11	01 13	29 47	04 17
12	01 05	29 44	04 52
13	00 57	29 40	05 06
14	00 50	29 37	05 01
15	00 44	29 34	04 38
16	00 40	29 31	04 01
17	00 39	29 28	03 12
18	00 D 39	29 25	02 16
19	00 39	29 21	01 15
20	00 40	29 18	00 S 11
21	00 R 39	29 15	00 N 52
22	00 37	29 12	01 52
23	00 32	29 09	02 48
24	00 24	29 05	03 36
25	00 14	29 02	04 16
26	00 02	28 59	04 44
27	29 ♑ 49	28 56	04 59
28	29 37	28 53	05 01
29	29 26	28 50	04 48
30	29 18	28 46	04 21
31	29 ♑ 12	28 ♑ 43	03 N 40

DECLINATIONS

Sun ☉ ° '	Moon ☽ ° '	Mercury ☿ ° '	Venus ♀ ° '	Mars ♂ ° '	Jupiter ♃ ° '	Saturn ♄ ° '	Uranus ♅ ° '	Neptune ♆ ° '	Pluto ♇ ° '
07 S 17	18 N 50	16 S 59	11 N 19	06 S 32	03 N 45	07 N 31	21 N 15	00 N 52	22 S 58
06 54	22 27	16 48	11 48	06 13	03 48	07 33	21 15	00 53	22 58
06 31	25 03	16 34	12 16	05 54	03 51	07 36	21 15	00 54	22 58
06 07	26 22	16 20	12 44	05 36	03 54	07 38	21 16	00 55	22 57
05 44	26 11	16 04	13 12	05 17	03 58	07 41	21 16	00 56	22 57
05 21	24 21	15 47	13 39	04 58	04 01	07 43	21 16	00 56	22 57
04 58	20 56	15 29	14 06	04 39	04 04	07 46	21 16	00 57	22 57
04 34	16 07	15 09	14 33	04 20	04 07	07 48	21 16	00 58	22 56
04 11	10 13	14 48	15 00	04 01	04 10	07 51	21 17	00 59	22 56
03 47	03 N 40	14 25	15 26	03 42	04 13	07 53	21 17	01 00	22 56
03 24	03 S 05	14 02	15 52	03 23	04 16	07 56	21 17	01 01	22 56
03 00	09 35	13 37	16 17	03 04	04 19	07 58	21 17	01 02	22 55
02 36	15 24	13 11	16 42	02 45	04 23	08 01	21 18	01 03	22 55
02 13	20 11	12 43	17 07	02 26	04 26	08 04	21 18	01 03	22 55
01 49	23 42	12 14	17 31	02 07	04 29	08 06	21 18	01 04	22 55
01 25	25 47	11 45	17 55	01 48	04 32	08 09	21 19	01 05	22 55
01 02	26 24	11 13	18 19	01 29	04 35	08 11	21 19	01 06	22 54
00 38	25 38	10 41	18 42	01 10	04 38	08 14	21 19	01 07	22 54
00 S 14	23 39	10 07	19 05	00 51	04 41	08 17	21 19	01 08	22 54
00 N 10	20 38	09 32	19 27	00 32	04 44	08 19	21 20	01 09	22 54
00 33	16 49	08 56	19 49	00 S 13	04 47	08 22	21 20	01 10	22 54
00 57	12 21	08 19	20 11	00 N 06	04 50	08 25	21 20	01 11	22 54
01 21	07 27	07 41	20 32	00 25	04 53	08 27	21 21	01 12	22 53
01 44	02 S 17	07 01	20 53	00 44	04 56	08 30	21 21	01 12	22 53
02 08	03 N 01	06 21	21 13	01 03	04 59	08 33	21 22	01 13	22 53
02 31	08 14	05 39	21 33	01 22	05 02	08 35	21 22	01 14	22 53
02 55	13 13	04 56	21 52	01 40	05 05	08 38	21 22	01 15	22 53
03 18	17 43	04 12	22 11	01 59	05 07	08 41	21 23	01 16	22 53
03 42	21 31	03 27	22 29	02 18	05 10	08 43	21 23	01 17	22 53
04 05	24 21	02 41	22 47	02 37	05 13	08 46	21 23	01 18	22 53
04 N 28	25 N 57	01 S 54	23 N 04	02 N 56	05 N 15	08 N 49	21 N 24	01 N 19	22 S 52

ZODIAC SIGN ENTRIES

Date	h	m	Planets
03	06	49	☽ ♊
04	20	01	♀ ♉
05	12	54	☽ ♋
07	15	43	☽ ♌
09	15	59	☽ ♍
11	15	21	☽ ♎
13	07	07	☿ ♓
13	15	53	☽ ♏
15	19	33	☽ ♐
18	03	27	☽ ♑
19	19	36	♂ ♈
20	02	17	☉ ♈
20	14	57	☽ ♒
23	04	00	☽ ♓
25	16	30	☽ ♈
28	03	24	☽ ♉
30	12	23	☽ ♊
31	12	28	☿ ♈

LATITUDES

Date	Mercury ☿ ° '	Venus ♀ ° '	Mars ♂ ° '	Jupiter ♃ ° '	Saturn ♄ ° '	Uranus ♅ ° '	Neptune ♆ ° '	Pluto ♇ ° '
01	00 S 36	01 N 15	00 S 57	01 N 30	02 S 18	00 S 04	01 S 23	04 S 49
04	01 01	01 29	00 56	01 30	02 18	00 04	01 23	04 50
07	01 23	01 44	00 54	01 31	02 17	00 04	01 23	04 50
10	01 41	01 59	00 53	01 31	02 17	00 04	01 23	04 51
13	01 56	02 15	00 52	01 31	02 16	00 04	01 23	04 51
16	02 07	02 30	00 51	01 31	02 16	00 04	01 23	04 52
19	02 15	02 45	00 49	01 31	02 16	00 04	01 23	04 52
22	02 18	02 59	00 48	01 31	02 15	00 04	01 23	04 53
25	02 17	03 14	00 47	01 31	02 15	00 04	01 23	04 53
28	02 13	03 28	00 45	01 31	02 15	00 04	01 23	04 54
31	02 S 03	03 N 41	00 S 44	01 N 31	02 S 14	00 S 04	01 S 23	04 S 54

DATA

Julian Date	2461832
Delta T	+73 seconds
Ayanamsa	24° 15' 17"
Synetic vernal point	04° ♓ 51' 42"
True obliquity of ecliptic	23° 26' 13"

LONGITUDES

Date	Chiron ⚷ ° '	Ceres ⚳ ° '	Pallas ⚴ ° '	Juno ⚵ ° '	Vesta ⚶ ° '	Black Moon Lilith ⚸ ° '
01	01 ♉ 13	22 ♏ 36	28 ♍ 21	20 ♋ 11	04 ♌ 24	29 ♒ 16
11	01 ♉ 41	23 ♏ 27	25 ♍ 12	20 ♋ 52	03 ♌ 23	00 ♓ 23
21	02 ♉ 13	23 ♏ 40	22 ♍ 00	22 ♋ 14	03 ♌ 09	01 ♓ 29
31	02 ♉ 47	23 ♏ 11	19 ♍ 09	24 ♋ 08	03 ♌ 40	02 ♓ 36

MOON'S PHASES, APSIDES AND POSITIONS ☽

Date	h	m	Phase	Longitude ° '	Eclipse Indicator
04	09	02	☽	14 ♊ 19	
11	01	06	○	20 ♍ 59	
17	23	23	☾	27 ♐ 53	
26	04	31	●	06 ♈ 03	

Day	h	m		
10	08	28	Perigee	
22	23	35	Apogee	
04	21	11	Max dec	26° N 29'
11	01	01	0S	
17	10	25	Max dec	26° S 24'
24	22	22	0N	

ASPECTARIAN

h m Aspects

01 Wednesday
01 52 ☽ ⋕ ☿
08 09 ☽ ⚺ ♆
09 27 ☽ ⚺ ♅
12 48 ☽ □ ♇
15 06 ☽ ⚼ ♃
19 44 ☽ ⊥ ♆
20 27 ☽ ✶ ☉

02 Thursday
03 23 ☽ ∥ ♅
03 25 ☽ □ ☿
04 52 ☽ ✶ ♂
09 33 ☉ ∠ ♀
12 54 ☽ ∠ ♆
16 03 ☽ ⋕ ♇
19 20 ☽ △ ♃
21 09 ☽ Q ☉
21 40 ☽ ⚺ ♄

03 Friday
03 22 ☽ ⚺ ♀
04 56 ☽ Q ♂
08 57 ☽ ⊥ ♄
15 31 ☽ ⊥ ♀
16 56 ☽ ✶ ♆
18 05 ☽ ☌ ♅
21 19 ☽ △ ♇
22 18 ☉ ⊥ ♇

04 Saturday
00 05 ☿ ⚺ ♂
01 37 ☽ ∠ ♄
01 39 ☿ ± ♃
09 02 ☽ □ ☉
09 24 ☽ ∠ ♀
12 05 ♀ ± ♃
14 47 ☽ Q ♆
15 54 ☽ □ ♂
16 28 ☽ △ ☿
22 47 ♀ ⊥ ♅

05 Sunday
00 24 ☽ ⚼ ♇
01 35 ☽ □ ♃
04 43 ☽ ✶ ♄
14 21 ☽ ✶ ♀
16 14 ☽ ± ♇
21 30 ☽ ⚼ ☿
22 33 ☽ □ ♆
23 34 ☽ ⚺ ♅

06 Monday
01 47 ☽ Q ♄
02 38 ☽ ⚻ ♇
08 35 ☽ Q ♃
08 35 ♂ ⊥ ♄
09 21 ☿ ∠ ♆
09 52 ☽ ⊥ ♅
12 44 ☽ Q ♀
14 26 ☽ ± ☿
17 31 ☽ △ ☉
23 01 ☽ △ ♂
23 15 ☽ ⋕ ♇

07 Tuesday
01 04 ☽ ∠ ♅
01 31 ☽ ⚻ ☿
04 37 ☽ ✶ ♃
08 24 ☽ □ ♄
10 04 ☽ ∥ ♅
20 17 ☽ ⚼ ☉
21 07 ☽ □ ♀

08 Wednesday
00 56 ☽ △ ♆
01 12 ☽ ⚼ ♂
01 50 ☽ ✶ ♅
04 45 ☽ ☍ ♇
05 03 ☽ ∠ ♃
08 42 ☿ ⚻ ♃
11 58 ☽ ± ☉
16 29 ☽ ⋕ ☿
16 34 ☽ ± ♂
18 15 ☽ ∥ ♀
19 26 ☽ ⊥ ♃
21 13 ☽ Q ♅
22 17 ☽ ⚻ ☉

09 Thursday
01 10 ☽ ⚼ ♆
02 29 ♂ ⋕ ♃
02 41 ☽ ⚻ ♂
04 56 ☽ ⚺ ♃
06 02 ☿ ⋕ ♀
07 06 ☽ ☍ ☿
09 20 ☽ △ ♄
09 43 ♃ ⚼ ♇
12 36 ☉ ⋕ ♃
15 30 ☽ ± ♆
20 50 ☽ ∥ ♄
21 57 ☉ ⊥ ♄

10 Friday
00 51 ☽ △ ♀
01 01 ☽ ⚻ ♆
01 50 ☽ □ ♅
03 15 ♀ ⚺ ♆
04 40 ☽ ⚻ ♇
09 13 ☽ ⚼ ♄
09 45 ☿ ✶ ♄
10 02 ☽ ∥ ♃
11 32 ☽ ⋕ ☉
11 51 ☽ ⋕ ♂
14 09 ☽ ± ♇
15 16 ♀ ⚺ ♅
15 57 ♂ ☍ ♃
21 27 ☽ ∥ ♆
22 07 ♂ ∠ ♇
23 30 ☽ ± ♄

11 Saturday
01 06 ☽ ☍ ☉
02 16 ☽ ⚼ ♀
03 56 ☽ ☌ ♃
04 22 ☽ ⚼ ♇
04 36 ☽ ⋕ ♆
04 42 ☽ ☍ ♂
09 03 ☽ ⚻ ♄
11 13 ☽ ⚻ ☿
13 02 ☽ ∥ ♂
13 03 ☽ ∥ ☉
13 23 ☉ ∥ ♂
16 18 ☽ ⋕ ♃
17 35 ☽ ± ♀
21 43 ☽ ± ☿
23 43 ♀ ⚼ ♃

12 Sunday
00 33 ☽ ☍ ♆
01 20 ☽ △ ♅
03 52 ☽ ⚻ ♀
04 13 ☽ △ ♇
05 51 ☽ ⋕ ♄
09 01 ♀ □ ♇
12 09 ♂ Q ♅
13 39 ☽ ⚼ ☿
15 37 ☉ ☍ ♃

13 Monday
01 34 ☽ ⚼ ♅
03 06 ☽ ∥ ☿
03 34 ☽ ⚺ ♃
03 59 ☉ ∠ ♇
04 12 ☿ ⊥ ♆
04 33 ☽ ⚻ ☉
07 30 ☽ ⚻ ♂
09 41 ☽ ☍ ♄
13 21 ☽ ⊥ ♃
15 10 ☽ ± ☉
16 25 ☽ ± ♅
16 52 ☽ △ ☿
17 59 ☽ ± ♂
18 33 ☽ ⋕ ♀

14 Tuesday
01 40 ☽ ⚻ ♆
02 28 ☽ ⚻ ♅
04 15 ☽ ∠ ♃
05 30 ☽ □ ♇
07 23 ☽ ⚼ ☉
08 58 ☽ ☍ ♀
09 43 ☉ Q ♅
10 01 ☽ ⚼ ♂
11 52 ☽ ± ♆
18 40 ☽ ⋕ ♅

15 Wednesday
03 27 ☽ ⚼ ♆
05 43 ☽ ∥ ♇
05 49 ☽ ✶ ♃
06 57 ♂ ⚺ ♄
11 22 ☽ △ ☉
12 42 ☽ Q ♇
13 19 ☽ ⚻ ♄
13 38 ☽ △ ♂

16 Thursday
00 08 ☽ ± ♄
01 41 ♀ ⊥ ♆
03 00 ☽ Q ♃
03 06 ☽ □ ☿
06 16 ☽ △ ♆
07 05 ☽ ☍ ♅
10 22 ☽ ✶ ♇
15 37 ♀ ∠ ♂
16 43 ☽ ⚼ ♄
16 59 ☉ ⚺ ♄
18 37 ☽ ⚻ ♀

17 Friday
06 48 ☽ ± ♀
07 16 ☿ ⚺ ♆
07 32 ☉ ⋕ ♆
12 03 ☽ □ ♃
14 13 ☿ □ ♅
14 25 ☽ ∠ ♇
17 05 ☽ Q ☿
21 13 ☽ △ ♄
23 23 ☽ □ ☉

18 Saturday
00 46 ☽ □ ♂
01 28 ☽ ⚼ ♀
07 45 ☽ ⊥ ♇
15 10 ☽ □ ♆
15 50 ♂ ⋕ ♆
16 00 ☽ ⚻ ♅
18 07 ☿ ⚺ ♇
19 29 ☽ ⚺ ♇
19 41 ☽ ✶ ☿

19 Sunday
03 50 ☽ ± ♅
09 30 ☽ △ ♀
13 48 ☽ Q ☉
14 30 ☽ Q ♂
18 43 ☽ ∥ ♇
21 49 ☽ ⚼ ♅
22 08 ☽ △ ♃

20 Monday
03 01 ☽ Q ♆
06 05 ☽ ∠ ☿
07 04 ☽ ⋕ ♅
09 01 ☽ □ ♄
16 07 ☽ ✶ ☉
16 19 ☽ ✶ ♂
19 12 ☽ ⋕ ♀

21 Tuesday
00 41 ☉ ⋕ ♂
02 36 ☉ ☌ ♂
03 09 ☽ ⊥ ☿
03 22 ☽ ✶ ♆
04 07 ☽ ⚼ ♃
04 12 ☽ △ ♅
05 18 ☿ ∠ ♄
07 46 ☽ ☌ ♇
17 18 ☽ ⚺ ☿
21 56 ☽ Q ♄
22 14 ☽ ± ♃

22 Wednesday
00 58 ☽ ∠ ♂
01 25 ☽ ∠ ☉
03 44 ☽ □ ♀
08 51 ☿ ⋕ ♄
10 01 ☽ ∠ ♆
10 20 ☽ ⚻ ♃
12 36 ♃ ± ♄
12 53 ☿ ⊥ ♇
20 37 ☽ ⊥ ♂
21 27 ☽ ⊥ ☉
22 39 ☽ ✶ ♄

23 Thursday
02 23 ☉ ∥ ♆
04 25 ☽ ⊥ ♆
07 18 ☽ ⋕ ♄
09 41 ☽ ⚺ ♂
10 47 ☽ ∥ ☿
10 47 ☽ ⚺ ☉
16 39 ☽ ⚺ ♆
17 27 ☽ □ ♅
19 43 ☽ Q ♀
20 57 ☽ ⚺ ♇
23 56 ☽ ⋕ ♃

24 Friday
05 25 ☽ ∠ ♄
09 06 ☽ ⊥ ♇
14 18 ☽ ⋕ ☉
16 14 ☽ ☌ ☿
16 52 ☽ ⋕ ♆
18 39 ☽ ⋕ ♂
22 06 ☽ ✶ ♀
22 32 ☽ ☍ ♃
23 44 ☽ ⊥ ♄

25 Saturday
02 32 ☽ ∥ ♂
02 51 ♀ △ ♃
03 15 ☽ ∠ ♇
03 52 ☽ ∥ ♆
05 51 ☽ Q ♅
07 42 ☽ ∥ ☉
10 04 ♀ ∠ ♆
11 52 ☽ ⚺ ♄
21 03 ☽ ∥ ♃
22 38 ♀ ∥ ♅

26 Sunday
01 27 ☽ ⋕ ☿
02 08 ♂ ∥ ♆
02 13 ☽ ☌ ♂
04 31 ☽ ☌ ☉
05 03 ☽ ☌ ♆
05 49 ☽ ✶ ♅
06 38 ☽ ∠ ♀
09 09 ☽ ✶ ♇
09 33 ☿ ☍ ♃
11 16 ☉ ☌ ♆
13 39 ☽ ∥ ♄
20 53 ☉ ✶ ♅
23 37 ☿ ⊥ ♄

27 Monday
01 52 ☽ ⊥ ♀
07 38 ☿ ⋕ ♃
08 46 ☽ Q ♇
09 22 ☽ ⚻ ♃
11 22 ☽ ∠ ♅
13 35 ☽ ⚺ ☿
14 34 ☽ ⚺ ♀
20 56 ☽ ± ♃
20 58 ☿ ∠ ♇
23 29 ☽ ☌ ♄

28 Tuesday
00 40 ♂ ☌ ♆
03 09 ☽ ⊥ ☿
03 14 ☿ ✶ ♀
04 53 ☽ ⊥ ♅
13 00 ♂ ✶ ♅
14 08 ☽ ⚼ ♃
14 20 ☉ ✶ ♇
15 29 ☿ Q ♅
15 44 ☽ ⚺ ♆
16 28 ☽ ⚺ ♅
16 41 ☽ ⚺ ♂
19 37 ☽ □ ♇
20 04 ☽ ⚺ ☉
23 19 ☽ ∠ ☿

29 Wednesday
03 15 ☽ ⊥ ♆
04 54 ☽ ⊥ ♂
06 56 ☉ ⋕ ☿
08 29 ☽ ⊥ ☉
11 02 ☽ ∥ ♅
18 26 ☽ △ ♃
20 00 ☽ ∥ ♀
20 21 ☽ ∠ ♆
22 22 ☽ ⋕ ♇
23 03 ☽ ∠ ♂

30 Thursday
02 55 ☽ ∠ ☉
04 31 ☽ ☌ ♀
08 23 ☽ ✶ ☿
09 10 ☽ ⚺ ♄
13 31 ☿ ⋕ ♂
14 14 ☿ ⚺ ♄
17 20 ♂ ✶ ♇
19 42 ♀ ⋕ ♇
20 27 ☽ ⊥ ♄

31 Friday
00 27 ☽ ✶ ♆
01 09 ☽ ☌ ♅
04 07 ☽ △ ♇
04 47 ☽ ✶ ♂
09 06 ☽ ✶ ☉
10 18 ☽ Q ☿
13 12 ☽ ∠ ♄
22 33 ☽ Q ♆

All ephemeris data is given at 12.00 UT and the Moon's longitude is additionally given for 24.00 UT

APRIL 2028

LONGITUDES

Date	Sidereal time h m s	Sun ☉ ° ' "	Moon ☽ ° ' "	Moon ☽ 24.00 ° ' "	Mercury ☿ ° '	Venus ♀ ° '	Mars ♂ ° '	Jupiter ♃ ° '	Saturn ♄ ° '	Uranus ♅ ° '	Neptune ♆ ° '	Pluto ♇ ° '
01	00 41 29	12 ♈ 17 12	25 ♊ 57 57	02 ♋ 40 01	01 ♈ 49	27 ♉ 54	09 ♈ 51	20 ♍ 07	28 ♈ 32	06 ♊ 56	06 ♈ 33	08 ♒ 30
02	00 45 25	13 16 23	09 ♋ 26 22	16 ♋ 17 13	03 42	28 48	10 37	20 R 00	28 40	06 58	06 35	08 31
03	00 49 22	14 15 32	23 ♋ 12 44	00 ♌ 13 03	05 37	29 ♉ 41	11 23	19 54	28 47	07 01	06 38	08 32
04	00 53 19	15 14 39	07 ♌ 18 11	14 ♌ 28 03	07 32	00 ♊ 34	12 09	19 47	28 55	07 03	06 40	08 33
05	00 57 15	16 13 43	21 ♌ 42 27	29 ♌ 00 59	09 29	01 26	12 56	19 41	29 02	07 06	06 42	08 34
06	01 01 12	17 12 45	06 ♍ 23 08	13 ♍ 48 12	11 27	02 17	13 42	19 35	29 10	07 08	06 44	08 35
07	01 05 08	18 11 45	21 ♍ 15 20	28 ♍ 43 31	13 27	03 07	14 28	19 29	29 17	07 11	06 47	08 36
08	01 09 05	19 10 42	06 ♎ 11 41	13 ♎ 38 40	15 28	03 57	15 14	19 23	29 25	07 13	06 49	08 36
09	01 13 01	20 09 37	21 ♎ 03 20	28 ♎ 24 33	17 30	04 45	16 00	19 17	29 32	07 16	06 51	08 37
10	01 16 58	21 08 31	05 ♏ 41 18	12 ♏ 52 44	19 33	05 33	16 46	19 11	29 40	07 18	06 53	08 38
11	01 20 54	22 07 22	19 ♏ 58 08	26 ♏ 56 58	21 37	06 20	17 32	19 05	29 47	07 21	06 56	08 39
12	01 24 51	23 06 12	03 ♐ 48 56	10 ♐ 33 53	23 42	07 06	18 18	19 00	29 ♈ 55	07 24	06 58	08 40
13	01 28 48	24 05 00	17 ♐ 11 52	23 ♐ 43 06	25 48	07 52	19 03	18 55	00 ♉ 03	07 27	07 00	08 40
14	01 32 44	25 03 46	00 ♑ 07 55	06 ♑ 26 47	27 ♈ 54	08 36	19 49	18 49	00 10	07 29	07 02	08 41
15	01 36 41	26 02 30	12 ♑ 40 13	18 ♑ 48 50	00 ♉ 01	09 19	20 35	18 44	00 18	07 32	07 04	08 42
16	01 40 37	27 01 13	24 ♑ 53 18	00 ♒ 54 18	02 08	10 02	21 21	18 40	00 26	07 35	07 07	08 42
17	01 44 34	27 59 54	06 ♒ 52 31	12 ♒ 48 38	04 14	10 43	22 06	18 35	00 33	07 38	07 09	08 43
18	01 48 30	28 58 33	18 ♒ 43 20	24 ♒ 37 17	06 20	11 23	22 52	18 30	00 41	07 41	07 11	08 44
19	01 52 27	29 ♈ 57 11	00 ♓ 31 06	06 ♓ 25 21	08 26	12 02	23 38	18 26	00 48	07 44	07 13	08 44
20	01 56 23	00 ♉ 55 46	12 ♓ 20 36	18 ♓ 17 19	10 30	12 39	24 23	18 22	00 56	07 47	07 15	08 45
21	02 00 20	01 54 20	24 ♓ 15 56	00 ♈ 16 49	12 33	13 16	25 09	18 18	01 04	07 50	07 17	08 45
22	02 04 17	02 52 53	06 ♈ 20 16	12 ♈ 26 31	14 34	13 51	25 54	18 14	01 11	07 53	07 19	08 46
23	02 08 13	03 51 23	18 ♈ 35 46	24 ♈ 48 07	16 33	14 25	26 39	18 10	01 19	07 56	07 21	08 46
24	02 12 10	04 49 52	01 ♉ 03 37	07 ♉ 22 18	18 30	14 58	27 25	18 06	01 27	07 59	07 23	08 46
25	02 16 06	05 48 19	13 ♉ 44 08	20 ♉ 09 03	20 24	15 29	28 10	18 03	01 34	08 02	07 26	08 47
26	02 20 03	06 46 44	26 ♉ 36 59	03 ♊ 07 50	22 15	15 58	28 55	18 00	01 42	08 05	07 28	08 47
27	02 23 59	07 45 07	09 ♊ 41 31	16 ♊ 17 59	24 03	16 26	29 ♈ 41	17 57	01 50	08 08	07 30	08 48
28	02 27 56	08 43 28	22 ♊ 57 09	29 ♊ 39 01	25 48	16 53	00 ♉ 26	17 54	01 57	08 11	07 32	08 48
29	02 31 52	09 41 47	06 ♋ 23 35	13 ♋ 10 52	27 29	17 17	01 11	17 51	02 05	08 15	07 34	08 48
30	02 35 49	10 ♉ 40 05	20 ♋ 00 55	26 ♋ 53 48	29 ♉ 06	17 ♊ 40	01 ♉ 56	17 ♍ 49	02 ♉ 12	08 ♊ 18	07 ♈ 36	08 ♒ 48

Date	Moon True ☊ ° '	Moon Mean ☊ ° '	Moon ☽ Latitude ° '
01	29 ♑ 09	28 ♑ 40	02 N 46
02	29 R 08	28 37	01 42
03	29 D 08	28 34	00 N 31
04	29 R 08	28 31	00 S 43
05	29 06	28 27	01 57
06	29 02	28 24	03 04
07	28 55	28 21	03 59
08	28 46	28 18	04 38
09	28 35	28 15	04 58
10	28 24	28 11	04 58
11	28 13	28 08	04 40
12	28 04	28 05	04 05
13	27 57	28 02	03 17
14	27 54	27 59	02 21
15	27 52	27 56	01 20
16	27 D 51	27 52	00 S 16
17	27 R 51	27 49	00 N 48
18	27 50	27 46	01 48
19	27 48	27 43	02 44
20	27 42	27 40	03 32
21	27 34	27 37	04 12
22	27 23	27 33	04 40
23	27 11	27 30	04 57
24	26 57	27 27	04 59
25	26 44	27 24	04 47
26	26 32	27 21	04 20
27	26 23	27 17	03 40
28	26 17	27 14	02 46
29	26 13	27 11	01 43
30	26 ♑ 12	27 ♑ 08	00 N 33

DECLINATIONS

Date	Sun ☉ ° '	Moon ☽ ° '	Mercury ☿ ° '	Venus ♀ ° '	Mars ♂ ° '	Jupiter ♃ ° '	Saturn ♄ ° '	Uranus ♅ ° '	Neptune ♆ ° '	Pluto ♇ ° '
01	04 N 51	26 N 09	01 S 06	23 N 21	03 N 14	05 N 18	08 N 51	21 N 24	01 N 20	22 S 52
02	05 14	24 48	00 S 17	23 37	03 33	05 21	08 54	21 25	01 20	22 52
03	05 37	21 57	00 N 33	23 53	03 52	05 23	08 57	21 25	01 21	22 52
04	06 00	17 45	01 24	24 09	04 10	05 26	09 00	21 25	01 22	22 52
05	06 23	12 26	02 16	24 24	04 28	05 28	09 02	21 26	01 23	22 52
06	06 46	06 N 19	03 08	24 38	04 47	05 30	09 05	21 26	01 24	22 52
07	07 08	00 S 12	04 01	24 52	05 05	05 33	09 08	21 27	01 25	22 52
08	07 30	06 43	04 55	25 05	05 24	05 35	09 10	21 27	01 26	22 52
09	07 53	12 49	05 49	25 18	05 42	05 37	09 13	21 28	01 27	22 52
10	08 15	18 06	06 44	25 30	06 00	05 39	09 16	21 28	01 27	22 52
11	08 37	22 13	07 39	25 42	06 18	05 41	09 18	21 28	01 28	22 52
12	08 59	24 55	08 34	25 53	06 36	05 43	09 21	21 29	01 29	22 52
13	09 20	26 06	09 29	26 04	06 54	05 45	09 24	21 29	01 30	22 52
14	09 42	25 47	10 24	26 14	07 12	05 47	09 27	21 30	01 31	22 52
15	10 03	24 09	11 19	26 24	07 29	05 49	09 29	21 30	01 32	22 52
16	10 25	21 24	12 13	26 33	07 47	05 51	09 32	21 31	01 33	22 52
17	10 46	17 47	13 06	26 41	08 05	05 52	09 35	21 31	01 33	22 52
18	11 07	13 30	13 58	26 50	08 22	05 54	09 37	21 32	01 34	22 52
19	11 27	08 44	14 50	26 57	08 39	05 56	09 40	21 32	01 35	22 52
20	11 48	03 S 40	15 40	27 04	08 57	05 57	09 43	21 33	01 36	22 52
21	12 08	01 N 34	16 28	27 11	09 14	05 59	09 45	21 33	01 37	22 52
22	12 28	06 48	17 14	27 17	09 31	06 00	09 48	21 34	01 37	22 52
23	12 48	11 51	17 59	27 22	09 48	06 01	09 51	21 34	01 38	22 53
24	13 08	16 31	18 42	27 27	10 05	06 03	09 53	21 35	01 39	22 53
25	13 27	20 31	19 22	27 31	10 22	06 04	09 56	21 35	01 40	22 53
26	13 47	23 37	20 00	27 35	10 38	06 05	09 58	21 36	01 41	22 53
27	14 06	25 31	20 35	27 39	10 55	06 06	10 01	21 36	01 41	22 53
28	14 24	26 01	21 09	27 41	11 11	06 07	10 04	21 37	01 42	22 53
29	14 43	25 00	21 39	27 44	11 27	06 08	10 06	21 37	01 43	22 53
30	15 N 01	22 N 29	22 N 07	27 N 45	11 N 44	06 N 08	10 N 09	21 N 38	01 N 44	22 S 53

ZODIAC SIGN ENTRIES

Date	h	m	Planets
01	19	14	☽ ♋
03	20	28	♀ ♊
03	23	38	☽ ♌
06	01	36	☽ ♍
08	02	03	☽ ♎
10	02	37	☽ ♏
12	05	18	☽ ♐
13	03	40	♄ ♉
14	11	45	☽ ♑
15	11	48	☿ ♉
16	22	11	☽ ♒
19	10	57	☽ ♓
19	13	09	☉ ♉
21	23	27	☽ ♈
24	09	58	☽ ♉
26	18	15	☽ ♊
27	22	21	♂ ♉
29	00	37	☽ ♋

LATITUDES

Date	Mercury ☿ ° '	Venus ♀ ° '	Mars ♂ ° '	Jupiter ♃ ° '	Saturn ♄ ° '	Uranus ♅ ° '	Neptune ♆ ° '	Pluto ♇ ° '
01	01 S 59	03 N 45	00 S 43	01 N 31	02 S 14	00 S 04	01 S 23	04 S 55
04	01 44	03 58	00 42	01 30	02 14	00 04	01 23	04 55
07	01 24	04 10	00 40	01 30	02 14	00 04	01 23	04 56
10	00 59	04 21	00 38	01 30	02 14	00 04	01 23	04 57
13	00 S 31	04 30	00 37	01 29	02 14	00 04	01 23	04 57
16	00 00	04 39	00 35	01 29	02 14	00 04	01 24	04 58
19	00 N 33	04 46	00 33	01 28	02 14	00 04	01 24	04 58
22	01 05	04 51	00 31	01 28	02 14	00 04	01 24	04 59
25	01 35	04 54	00 30	01 27	02 14	00 04	01 24	05 00
28	02 00	04 55	00 28	01 27	02 14	00 03	01 24	05 00
31	02 N 19	04 N 54	00 S 26	01 N 26	02 S 14	00 S 03	01 S 24	05 S 01

DATA

Julian Date	2461863
Delta T	+73 seconds
Ayanamsa	24° 15' 21"
Synetic vernal point	04° ♓ 51' 38"
True obliquity of ecliptic	23° 26' 13"

MOON'S PHASES, APSIDES AND POSITIONS ☽

Date	h	m	Phase	Longitude ° '	Eclipse Indicator
02	19	15	☽	13 ♋ 34	
09	10	27	○	20 ♎ 06	
16	16	37	☾	27 ♑ 12	
24	19	47	●	05 ♉ 09	

Day	h	m		
07	16	13	Perigee	
19	15	20	Apogee	
01	03	03	Max dec	26° N 15'
07	11	18	0S	
13	18	49	Max dec	26° S 09'
21	04	49	0N	
28	07	59	Max dec	26° N 02'

LONGITUDES

Date	Chiron ⚷ ° '	Ceres ⚳ ° '	Pallas ⚴ ° '	Juno ⚵ ° '	Vesta ⚶ ° '	Black Moon Lilith ⚸ ° '
01	02 ♉ 51	23 ♏ 06	18 ♍ 54	24 ♋ 22	03 ♌ 45	02 ♓ 43
11	03 ♉ 26	21 ♏ 54	16 ♍ 47	26 ♋ 47	05 ♌ 02	03 ♓ 49
21	04 ♉ 03	20 ♏ 09	15 ♍ 30	29 ♋ 34	06 ♌ 54	04 ♓ 56
31	04 ♉ 41	18 ♏ 01	15 ♍ 03	02 ♌ 38	09 ♌ 17	06 ♓ 03

ASPECTARIAN

h m	Aspects
01 Saturday	
01 31	☽ □ ♃
04 06	☽ Q ♂
05 20	☿ # ♆
07 33	☽ bQ ♇
08 44	☽ Q ☉
15 45	☽ ⊻ ♀
16 39	☽ ⚹ ♄
23 43	☽ ± ♇
02 Sunday	
00 12	☽ □ ☿
03 13	☽ ⊥ ♀
06 57	☽ □ ♆
07 27	♀ ⊻ ♄
07 38	☽ ⊻ ♅
09 30	☽ Q ♃
10 22	☽ ⊼ ♇
14 10	☽ Q ♄
14 12	☽ □ ♂
18 13	☽ ⊥ ♅
19 15	☽ □ ☉
19 17	☉ ∥ ♃
20 12	☽ ∠ ♀
22 28	☽ ∥ ♀
03 Monday	
05 30	☽ # ♇
06 19	☽ ⚹ ♃
09 55	☽ ∠ ♅
15 32	☽ ∥ ♅
21 38	☽ □ ♄
23 50	☽ ⚹ ♀
04 Tuesday	
00 59	☿ ☌ ♆
05 52	☿ ⚹ ♅
07 47	☽ ∠ ♃
10 55	☽ △ ♆
11 07	☿ ∥ ♆
11 34	☽ ⚹ ♅
12 27	☽ △ ☿
14 06	☽ ☍ ♇
20 36	☽ △ ♂
21 24	☽ Q ♀
22 47	☽ ⊥ ♃
05 Wednesday	
00 36	☿ ⚹ ♇
02 16	☽ △ ☉
07 40	☽ Q ♅
08 41	☽ ⊻ ♃
12 00	☽ bQ ♆
17 17	☽ bQ ☿
22 47	☽ bQ ♂
06 Thursday	
00 08	☽ △ ♄
01 31	☽ ∥ ♄
02 48	☽ ± ♆
04 44	☽ bQ ☉
04 55	☽ □ ♀
10 16	☽ ± ☿
10 27	☽ ∥ ☉
12 35	☽ ⊼ ♆
13 13	☽ □ ♅
14 14	☽ ± ♂
15 03	☽ ∥ ♃
15 34	☽ ⊼ ♇
17 30	☽ ∥ ♂
20 22	☽ ± ☉
21 29	☽ ⊼ ☿
22 26	☽ ∥ ☿
23 19	☉ ∠ ♀
07 Friday	
00 28	☽ ⊼ ♂
00 41	☽ bQ ♄
01 16	☽ ± ♇
06 09	☽ ∥ ♆
06 44	☽ ⊼ ☉
09 10	☽ ☌ ♃
15 17	☽ ± ♄
15 46	☽ bQ ♇
16 28	☽ # ♆
08 Saturday	
01 01	☽ ⊼ ♄
04 14	☽ # ☿
06 49	☽ # ♂
07 30	☿ ☌ ♂
07 45	☽ # ♃
08 10	☽ △ ♀
13 00	☽ ☍ ♆
13 39	☽ △ ♅
15 11	☽ # ☉
15 53	☽ △ ♇
16 25	☉ ⊼ ♃
21 28	☽ # ♄
09 Sunday	
03 21	☽ ☍ ♂
05 03	♂ ∥ ♃
05 19	☽ ☍ ☿
06 24	☿ ∥ ♃
07 00	☿ ∥ ♂
09 08	☽ ⊻ ♃
09 46	☽ bQ ♀
10 27	☽ ☍ ☉
13 58	☽ bQ ♅
18 50	☽ ⊥ ♃
23 26	☉ Q ♇
10 Monday	
01 18	☽ ± ♀
01 58	☽ ☍ ♄
04 44	☽ ± ♅
07 54	☿ ⊼ ♃
09 31	☽ ∠ ♃
11 46	☽ ⊼ ♀
14 00	☽ ⊼ ♆
14 42	☽ ⊼ ♅
15 58	♀ ⊥ ♄
16 54	☽ □ ♇
11 Tuesday	
00 03	☽ ± ♆
00 39	☿ Q ♇
06 45	☿ ∠ ♀
07 02	☽ # ♅
07 37	☽ ⊼ ♂
10 31	☽ ⚹ ♃
15 19	☽ ⊼ ☿
15 21	☽ bQ ♆
15 58	☽ ⊼ ☉
16 45	☽ ∥ ♇
17 54	☉ ∠ ♅
18 27	☽ ± ♂
20 37	☿ ∠ ♅
22 56	☉ ☌ ☿
23 29	☽ Q ♇
12 Wednesday	
03 07	☽ ± ☉
03 30	☽ ± ☿
05 06	☽ ⊼ ♄
07 05	☽ Q ♃
07 13	♀ ⚹ ♆
11 01	☽ bQ ♂
15 45	☽ ± ♄
17 35	☽ △ ♆
18 10	☽ ☍ ♀
18 22	☽ ☍ ♅
20 12	☽ bQ ☉
20 36	☽ ⚹ ♇
21 47	♀ ☌ ♅
22 17	☽ bQ ☿
13 Thursday	
02 13	☿ ± ♃
05 47	☉ ∥ ☿
07 53	♂ ⊼ ♃
08 03	☽ bQ ♄
09 37	☿ ∥ ♄
09 39	☽ # ♀
15 07	☽ □ ♃
15 37	☽ △ ♂
16 23	☉ ∥ ♄
22 21	☽ # ♀
23 55	☽ ∠ ♇
14 Friday	
01 43	☽ △ ☉
06 38	☉ ± ♃
06 59	☽ △ ☿
12 04	☽ △ ♄
14 49	♀ △ ♇
16 50	☽ ⊥ ♇
15 Saturday	
01 10	☽ □ ♆
02 03	☽ ⊼ ♅
04 19	☽ ⊻ ♇
05 08	☽ ⊼ ♀
13 41	☽ ± ♅
15 24	☿ ☌ ♄
15 34	♂ Q ♇
17 29	☽ ± ♀
23 47	☽ △ ♃
16 Sunday	
00 19	☽ ∥ ♇
04 31	☽ □ ♂
05 40	☿ ⊥ ♅
07 25	☽ bQ ♅
11 12	☽ # ♅
12 17	☽ bQ ♀
12 26	☽ Q ♆
16 37	☽ □ ☉
23 10	☽ □ ♄
17 Monday	
04 47	☿ bQ ♃
05 25	☽ bQ ♃
05 33	☽ □ ☿
12 33	☽ ⚹ ♆
13 32	☽ △ ♅
15 43	☽ ☌ ♇
18 58	☽ Q ♂
19 56	☿ ⊥ ♀
20 13	☽ △ ♀
23 28	☽ ± ♃
18 Tuesday	
05 45	♂ ∠ ♅
08 08	☽ Q ☉
09 53	☽ # ☿
11 34	☽ ⊼ ♃
11 55	☽ Q ♄
19 03	☽ ∠ ♆
21 01	☽ ⚹ ♂
21 48	☿ ⊻ ♆
23 27	☽ # ☉
19 Wednesday	
01 54	☽ Q ☿
03 45	☿ ⊻ ♅
07 30	☽ # ♄
10 45	☽ ⚹ ☉
12 22	☽ # ♂
12 36	☽ ⚹ ♄
13 25	☽ ⊥ ♆
15 32	☿ □ ♇
20 Thursday	
01 21	☽ # ♃
01 39	☽ ⊻ ♆
02 43	☽ □ ♅
04 42	☽ ⊻ ♇
05 36	☽ ∠ ♂
07 29	☽ ⚹ ☿
11 19	♂ ± ♃
12 10	☉ ☌ ♄
12 40	☽ □ ♀
16 51	☽ ⊥ ♇
19 20	☽ ∠ ♄
19 54	☽ ∠ ☉
21 29	☽ # ♆
21 Friday	
00 05	☽ ☍ ♃
01 01	☽ ⊥ ♂
10 01	☉ ⊥ ♅
10 58	☽ ∠ ♇
12 10	☽ ∥ ♆
13 37	☽ ⊥ ♄
13 52	☽ ⊻ ♂
15 08	☽ Q ♅
15 34	☽ ⊥ ☉
19 53	☽ ∠ ☿
20 54	☿ ⊥ ♆
22 Saturday	
00 02	☿ ⊻ ♀
01 42	☽ ⊻ ♄
02 41	☽ Q ♀
04 34	☽ ⊻ ☉
08 15	☽ ∥ ♃
13 57	☽ ☌ ♆
15 03	☽ ⚹ ♅
16 47	☽ ⚹ ♇
17 15	☽ ⊥ ☿
20 00	☉ bQ ♃
23 Sunday	
01 28	☽ ∥ ♂
02 11	☽ ∥ ♄
03 29	☽ ⚹ ♀
07 16	☽ ⊻ ☿
11 10	☽ ⊼ ♃
16 13	☽ Q ♇
16 24	♂ ∥ ♄
17 01	☽ ∥ ☉
20 26	☽ ∠ ♅
22 43	☽ ± ♃
24 Monday	
04 34	☽ ☌ ♂
07 17	☿ △ ♃
09 48	☽ ∠ ♀
12 45	☽ ☌ ♄
13 46	☽ ⊥ ♅
15 53	☽ bQ ♃
19 47	☽ ☌ ☉
25 Tuesday	
00 04	☽ ⊻ ♆
01 12	☽ ⊻ ♅
02 40	☽ □ ♇
03 03	☽ ∥ ☿
03 39	☽ ⊥ ♀
11 25	☽ ⊥ ♆
15 24	☽ ⊻ ♀
19 25	☽ ∥ ♅
20 03	☽ △ ♃
26 Wednesday	
02 34	☽ ☌ ☿
04 16	☽ ∠ ♆
05 28	☽ # ♇
14 49	☿ ∠ ♆
16 31	☽ ⊻ ♂
21 28	☽ ⊻ ♄
27 Thursday	
04 13	☽ ⊥ ♂
05 24	☉ ⊻ ♆
07 59	☽ ⚹ ♆
08 11	☽ ⊻ ☉
08 34	☽ ⊥ ♄
09 09	☽ ☌ ♅
10 22	☽ △ ♇
19 58	☽ ⊥ ☉
21 36	☽ ∠ ♂
22 02	☉ ⊻ ♅
28 Friday	
00 41	☽ ☌ ♀
01 05	☽ ∠ ♄
02 55	☽ □ ♃
05 49	☽ Q ♆
13 30	☽ ∠ ☉
13 31	☽ bQ ♇
13 50	☉ □ ♇
17 51	☽ ⊻ ☿
18 13	♀ ∠ ♄
29 Saturday	
02 11	☽ ⚹ ♂
04 16	☽ ⚹ ♄
05 37	☽ ± ♇
06 06	☽ ⊥ ☿
10 23	☿ ∥ ♅
11 03	☽ Q ♃
14 04	☽ □ ♆
15 17	☽ ⊻ ♅
16 16	☽ ⊼ ♇
18 18	☽ ⚹ ☉
30 Sunday	
00 14	☽ ∠ ☿
00 42	☽ Q ♂
01 43	☽ Q ♄
01 55	☽ ⊥ ♅
07 47	☽ ⊻ ♀
08 09	☽ ⚹ ♃
08 54	☽ # ♇
14 20	☽ ∥ ☿
16 59	☽ Q ☉
17 45	☽ ∠ ♅
18 03	☽ ∥ ♅
18 34	☽ ⊥ ♀
20 11	♀ □ ♃
22 41	♂ ☌ ♄

All ephemeris data is given at 12.00 UT and the Moon's longitude is additionally given for 24.00 UT

MAY 2028

LONGITUDES

Date	Sidereal time h m s	Sun ☉ ° ' "	Moon ☽ ° ' "	Moon ☽ 24.00 ° ' "	Mercury ☿ ° '	Venus ♀ ° '	Mars ♂ ° '	Jupiter ♃ ° '	Saturn ♄ ° '	Uranus ♅ ° '	Neptune ♆ ° '	Pluto ♇ ° '
01	02 39 46	11 ♉ 38 20	03 ♌ 49 34	10 ♌ 48 15	00 ♊ 40	18 ♊ 02	02 ♉ 41	17 ♍ 46	02 ♉ 20	08 ♊ 21	07 ♈ 38	08 ♒ 49
02	02 43 42	12 36 33	17 ♌ 49 52	24 ♌ 54 20	02 09	18 21	03 26	17 R 44	02 28	08 24	07 40	08 49
03	02 47 39	13 34 44	02 ♍ 01 34	09 ♍ 11 20	03 35	18 38	04 11	17 42	02 35	08 27	07 41	08 49
04	02 51 35	14 32 52	16 ♍ 23 18	23 ♍ 37 05	04 56	18 54	04 55	17 40	02 43	08 31	07 43	08 49
05	02 55 32	15 30 59	00 ♎ 52 06	08 ♎ 07 44	06 13	19 07	05 40	17 39	02 50	08 34	07 45	08 49
06	02 59 28	16 29 04	15 ♎ 23 16	22 ♎ 37 54	07 26	19 18	06 25	17 37	02 58	08 37	07 47	08 49
07	03 03 25	17 27 07	29 ♎ 50 47	07 ♏ 01 07	08 34	19 27	07 10	17 36	03 05	08 41	07 49	08 49
08	03 07 21	18 25 09	14 ♏ 08 07	21 ♏ 11 02	09 38	19 34	07 54	17 35	03 13	08 44	07 51	08 50
09	03 11 18	19 23 08	28 ♏ 09 15	05 ♐ 02 15	10 38	19 39	08 39	17 34	03 20	08 47	07 53	08 R 50
10	03 15 15	20 21 07	11 ♐ 49 41	18 ♐ 31 19	11 33	19 41	09 23	17 33	03 28	08 51	07 54	08 50
11	03 19 11	21 19 04	25 ♐ 07 04	01 ♑ 36 58	12 23	19 R 41	10 08	17 33	03 35	08 54	07 56	08 49
12	03 23 08	22 16 59	08 ♑ 01 13	14 ♑ 20 06	13 08	19 38	10 52	17 33	03 43	08 58	07 58	08 49
13	03 27 04	23 14 53	20 ♑ 34 01	26 ♑ 43 26	13 49	19 34	11 37	17 32	03 50	09 01	08 00	08 49
14	03 31 01	24 12 46	02 ♒ 48 53	08 ♒ 50 58	14 25	19 26	12 21	17 D 32	03 57	09 04	08 01	08 49
15	03 34 57	25 10 37	14 ♒ 50 18	20 ♒ 47 32	14 56	19 16	13 05	17 33	04 05	09 08	08 03	08 49
16	03 38 54	26 08 28	26 ♒ 43 21	02 ♓ 38 24	15 22	19 04	13 50	17 33	04 12	09 11	08 05	08 49
17	03 42 50	27 06 17	08 ♓ 33 20	14 ♓ 28 48	15 43	18 50	14 34	17 34	04 19	09 15	08 06	08 49
18	03 46 47	28 04 04	20 ♓ 25 24	26 ♓ 23 44	15 59	18 33	15 18	17 34	04 26	09 18	08 08	08 48
19	03 50 43	29 01 51	02 ♈ 24 18	08 ♈ 27 36	16 11	18 14	16 02	17 35	04 34	09 22	08 10	08 48
20	03 54 40	29 ♉ 59 37	14 ♈ 34 02	20 ♈ 43 59	16 17	17 52	16 46	17 36	04 41	09 25	08 11	08 48
21	03 58 37	00 ♊ 57 21	26 ♈ 57 43	03 ♉ 15 26	16 R 19	17 28	17 30	17 38	04 48	09 29	08 13	08 48
22	04 02 33	01 55 04	09 ♉ 37 15	16 ♉ 03 15	16 16	17 03	18 14	17 39	04 55	09 32	08 14	08 47
23	04 06 30	02 52 46	22 ♉ 33 23	29 ♉ 07 33	16 08	16 35	18 58	17 41	05 02	09 36	08 16	08 47
24	04 10 26	03 50 27	05 ♊ 45 35	12 ♊ 27 17	15 57	16 05	19 41	17 43	05 09	09 39	08 17	08 46
25	04 14 23	04 48 07	19 ♊ 12 22	26 ♊ 00 35	15 41	15 34	20 25	17 45	05 16	09 43	08 19	08 46
26	04 18 19	05 45 45	02 ♋ 51 36	09 ♋ 45 09	15 21	15 01	21 09	17 47	05 23	09 46	08 20	08 46
27	04 22 16	06 43 23	16 ♋ 40 55	23 ♋ 38 38	14 58	14 27	21 52	17 49	05 30	09 50	08 22	08 45
28	04 26 12	07 40 58	00 ♌ 38 03	07 ♌ 38 57	14 32	13 52	22 36	17 52	05 37	09 53	08 23	08 45
29	04 30 09	08 38 32	14 ♌ 41 07	21 ♌ 44 22	14 03	13 15	23 20	17 54	05 44	09 57	08 24	08 44
30	04 34 06	09 36 05	28 ♌ 48 31	05 ♍ 53 23	13 33	12 38	24 03	17 57	05 51	10 00	08 26	08 44
31	04 38 02	10 ♊ 33 37	12 ♍ 58 47	20 ♍ 04 31	13 ♊ 00	12 ♊ 01	24 ♉ 46	18 ♍ 00	05 ♉ 58	10 ♊ 04	08 ♈ 27	08 ♒ 43

Date	Moon True ☊ ° '	Moon Mean ☊ ° '	Moon ☽ Latitude ° '
01	26 ♑ 12	27 ♑ 05	00 S 40
02	26 R 12	27 02	01 52
03	26 11	26 58	02 58
04	26 07	26 55	03 53
05	26 01	26 52	04 34
06	25 53	26 49	04 57
07	25 43	26 46	05 01
08	25 33	26 43	04 47
09	25 23	26 39	04 15
10	25 15	26 36	03 29
11	25 09	26 33	02 33
12	25 05	26 30	01 30
13	25 04	26 27	00 S 24
14	25 D 04	26 23	00 N 41
15	25 04	26 20	01 44
16	25 R 05	26 17	02 41
17	25 04	26 14	03 31
18	25 01	26 11	04 12
19	24 57	26 08	04 43
20	24 50	26 04	05 01
21	24 41	26 01	05 05
22	24 32	25 58	04 55
23	24 23	25 55	04 30
24	24 15	25 52	03 50
25	24 08	25 48	02 56
26	24 04	25 45	01 52
27	24 02	25 42	00 N 40
28	24 D 02	25 39	00 S 36
29	24 03	25 36	01 49
30	24 04	25 33	02 57
31	24 ♑ 05	25 ♑ 29	03 S 54

DECLINATIONS

Date	Sun ☉ ° '	Moon ☽ ° '	Mercury ☿ ° '	Venus ♀ ° '	Mars ♂ ° '	Jupiter ♃ ° '	Saturn ♄ ° '	Uranus ♅ ° '	Neptune ♆ ° '	Pluto ♇ ° '
01	15 N 19	18 N 38	22 N 33	27 N 46	12 N 00	06 N 09	10 N 11	21 N 38	01 N 44	22 S 53
02	15 37	13 42	22 56	27 47	12 15	06 10	10 14	21 39	01 45	22 54
03	15 55	07 59	23 17	27 47	12 31	06 10	10 17	21 39	01 46	22 54
04	16 12	01 N 47	23 35	27 47	12 47	06 11	10 19	21 40	01 47	22 54
05	16 29	04 S 32	23 51	27 45	13 02	06 11	10 22	21 40	01 47	22 54
06	16 46	10 38	24 05	27 44	13 18	06 12	10 24	21 41	01 48	22 54
07	17 02	16 07	24 16	27 41	13 33	06 12	10 27	21 41	01 49	22 55
08	17 18	20 38	24 25	27 39	13 48	06 12	10 29	21 42	01 49	22 55
09	17 34	23 53	24 32	27 35	14 03	06 12	10 32	21 43	01 50	22 55
10	17 50	25 39	24 37	27 31	14 18	06 13	10 34	21 43	01 51	22 55
11	18 05	25 53	24 40	27 26	14 32	06 13	10 37	21 44	01 51	22 55
12	18 20	24 41	24 41	27 20	14 47	06 12	10 39	21 44	01 52	22 56
13	18 35	22 15	24 41	27 14	15 01	06 12	10 41	21 45	01 53	22 56
14	18 49	18 51	24 38	27 07	15 15	06 12	10 44	21 45	01 53	22 56
15	19 03	14 44	24 34	27 00	15 29	06 12	10 46	21 46	01 54	22 56
16	19 17	10 05	24 28	26 51	15 43	06 11	10 49	21 46	01 55	22 57
17	19 31	05 S 06	24 20	26 42	15 57	06 11	10 51	21 47	01 55	22 57
18	19 44	00 N 04	24 11	26 32	16 11	06 11	10 53	21 47	01 56	22 57
19	19 56	05 17	24 00	26 21	16 24	06 10	10 56	21 48	01 56	22 57
20	20 09	10 22	23 48	26 10	16 37	06 09	10 58	21 48	01 57	22 58
21	20 21	15 08	23 34	25 57	16 50	06 09	11 00	21 49	01 58	22 58
22	20 33	19 21	23 19	25 44	17 03	06 08	11 03	21 49	01 58	22 58
23	20 44	22 45	23 03	25 30	17 16	06 07	11 05	21 50	01 59	22 59
24	20 55	25 02	22 46	25 16	17 28	06 06	11 07	21 51	01 59	22 59
25	21 06	25 56	22 28	25 00	17 41	06 05	11 09	21 51	02 00	22 59
26	21 16	25 16	22 09	24 44	17 53	06 04	11 12	21 52	02 00	23 00
27	21 26	23 03	21 49	24 28	18 05	06 03	11 14	21 52	02 01	23 00
28	21 35	19 26	21 29	24 10	18 17	06 02	11 16	21 53	02 01	23 00
29	21 44	14 41	21 08	23 52	18 28	06 01	11 18	21 53	02 02	23 01
30	21 53	09 07	20 47	23 34	18 40	05 59	11 20	21 54	02 02	23 01
31	22 N 02	03 N 05	20 N 26	23 N 15	18 N 51	05 N 58	11 N 22	21 N 54	02 N 03	23 S 01

ZODIAC SIGN ENTRIES

Date	h	m	Planets
01	01	42	☿ ♊
01	05	23	☽ ♌
03	08	36	☽ ♍
05	10	34	☽ ♎
07	12	15	☽ ♏
09	15	12	☽ ♐
11	21	00	☽ ♑
14	06	26	☽ ♒
16	18	39	☽ ♓
19	07	12	☽ ♈
20	12	10	☉ ♊
21	17	48	☽ ♉
24	01	35	☽ ♊
26	07	00	☽ ♋
28	10	55	☽ ♌
30	14	01	☽ ♍

LATITUDES

Date	Mercury ☿ ° '	Venus ♀ ° '	Mars ♂ ° '	Jupiter ♃ ° '	Saturn ♄ ° '	Uranus ♅ ° '	Neptune ♆ ° '	Pluto ♇ ° '
01	02 N 19	04 N 54	00 S 26	01 N 26	02 S 14	00 S 03	01 S 24	05 S 01
04	02 31	04 49	00 24	01 25	02 14	00 03	01 24	05 02
07	02 34	04 41	00 22	01 25	02 14	00 03	01 24	05 02
10	02 29	04 29	00 20	01 24	02 14	00 03	01 24	05 03
13	02 14	04 13	00 19	01 24	02 14	00 03	01 24	05 04
16	01 50	03 53	00 17	01 23	02 14	00 03	01 24	05 04
19	01 17	03 27	00 15	01 22	02 14	00 03	01 25	05 05
22	00 N 36	02 57	00 13	01 22	02 15	00 03	01 25	05 06
25	00 S 12	02 22	00 11	01 21	02 15	00 03	01 25	05 06
28	01 04	01 43	00 09	01 20	02 15	00 03	01 25	05 07
31	01 S 56	01 N 02	00 S 07	01 N 20	02 S 16	00 S 03	01 S 25	05 S 08

DATA

Julian Date	2461893
Delta T	+73 seconds
Ayanamsa	24° 15' 25"
Synetic vernal point	04° ♓ 51' 34"
True obliquity of ecliptic	23° 26' 12"

MOON'S PHASES, APSIDES AND POSITIONS ☽

Date	h	m	Phase	Longitude ° '	Eclipse Indicator
02	02	26	☽	12 ♌ 13	
08	19	49	○	18 ♏ 44	
16	10	43	☾	26 ♒ 05	
24	08	16	●	03 ♊ 41	
31	07	37	☽	10 ♍ 23	

Day	h	m		
05	10	37	Perigee	
17	09	57	Apogee	
31	06	36	Perigee	
04	18	47	OS	
11	03	41	Max dec	25° S 59'
18	11	40	ON	
25	14	00	Max dec	25° N 56'
31	23	58	OS	

LONGITUDES

Date	Chiron ⚷ ° '	Ceres ⚳ ° '	Pallas ⚴ ° '	Juno ⚵ ° '	Vesta ⚶ ° '	Black Moon Lilith ⚸ ° '
01	04 ♉ 41	18 ♏ 01	15 ♍ 03	02 ♌ 38	09 ♌ 17	06 ♓ 03
11	05 ♉ 17	15 ♏ 47	15 ♍ 23	05 ♌ 57	12 ♌ 06	07 ♓ 10
21	05 ♉ 53	13 ♏ 40	16 ♍ 24	09 ♌ 27	15 ♌ 17	08 ♓ 16
31	06 ♉ 26	11 ♏ 56	18 ♍ 02	13 ♌ 05	18 ♌ 46	09 ♓ 23

ASPECTARIAN

h m	Aspects
01 Monday	
00 36	♂ ⊥ ♅
05 51	☽ ⚹ ☿
09 24	☽ □ ♄
09 55	☽ □ ♂
10 11	☽ ∠ ♃
10 35	☽ ∠ ♀
14 46	♂ bQ ♃
17 02	♄ ⊥ ♅
18 34	☽ △ ♆
19 49	☽ ⚹ ♅
20 35	☽ ☍ ♇
02 Tuesday	
01 37	☽ ⊥ ♃
02 26	☽ □ ☉
02 28	☉ ⊥ ♀
03 42	☽ ∥ ☉
05 00	☽ Q ☿
07 33	♀ ∠ ♂
09 18	☿ # ♇
11 50	☽ ⊻ ♃
12 54	☽ ⚹ ♀
16 23	☽ Q ♅
17 33	☿ ⊻ ♄
18 03	☽ ∥ ♂
20 13	☽ bQ ♆
03 Wednesday	
02 46	☽ ∥ ♄
09 37	☽ Q ♀
11 26	☽ ± ♆
12 57	☽ △ ♄
14 53	☉ ⊥ ♆
14 53	☽ □ ☿
15 49	☽ △ ♂
19 08	☽ ∥ ♃
21 31	☽ ⊼ ♆
22 49	☽ □ ♅
23 23	☽ ⊼ ♇
04 Thursday	
05 47	♃ bQ ♄
08 43	☽ △ ☉
09 23	☽ ± ♇
11 34	☿ ⊻ ♂
12 03	☽ ∥ ♆
14 08	☽ ☌ ♃
14 13	☽ bQ ♄
16 14	☽ □ ♀
18 12	☽ bQ ♂
05 Friday	
00 20	☽ bQ ♇
01 32	☽ # ♆
05 16	☽ ± ♄
09 55	☽ ± ♂
11 23	☽ bQ ☉
15 17	☽ ⊼ ♄
18 23	☽ # ♃
20 22	☽ ⊼ ♂
21 40	☽ △ ☿
23 24	☽ ☍ ♆
06 Saturday	
00 46	☽ △ ♅
01 09	☽ △ ♇
03 19	☽ ± ☉
11 05	☽ # ♄
13 57	☽ ⊼ ☉
15 42	☽ ⊻ ♃
18 33	☽ △ ♀
19 28	☿ ⚹ ♆
23 48	☽ # ♂
07 Sunday	
00 41	☽ bQ ☿
01 37	☽ ⊥ ♃
01 42	☽ bQ ♅
14 25	☿ ☌ ♅
15 38	☉ △ ♃
16 36	☽ ∠ ♃
16 44	☽ # ☉
16 45	☽ ± ♅
16 56	☽ ± ☿
17 28	☽ ☍ ♄
17 32	☿ △ ♇
19 46	☽ bQ ♀
08 Monday	
00 55	☽ ☍ ♂
00 55	☿ ⊥ ♄
01 22	☽ ⊼ ♆
02 51	☽ ⊼ ♅
03 02	☽ □ ♇
03 48	☽ ⊼ ☿
10 03	♂ ⊻ ♆
11 02	☽ ± ♀
11 31	☽ ± ♆
17 51	☽ ⚹ ♃
18 50	☽ # ♅
19 49	☽ ☍ ☉
21 18	☽ ⊼ ♀
09 Tuesday	
02 53	☽ bQ ♆
03 39	☽ ∥ ♇
09 32	♇ St R
09 42	☽ Q ♇
14 27	☽ Q ♃
16 58	♂ ⊻ ♅
17 44	♂ □ ♇
18 48	☉ ⊻ ♀
19 04	☽ # ☿
21 06	☽ ⊼ ♄
10 Wednesday	
03 06	♅ △ ♇
05 02	☽ △ ♆
06 40	☽ ⚹ ♇
06 41	☽ ☍ ♅
07 25	☽ ⊼ ♂
07 46	☽ ± ♄
11 27	☽ ☍ ☿
18 44	☽ ± ♂
22 15	☽ □ ♃
23 03	♀ St R
11 Thursday	
00 00	☽ bQ ♄
02 06	☽ ☍ ♀
04 31	☽ ⊼ ☉
09 38	☽ ∠ ♇
12 02	☽ bQ ♂
16 22	☽ ± ☉
12 Friday	
02 15	☽ ⊥ ♇
03 49	☽ △ ♄
10 30	☽ bQ ☉
11 54	☽ □ ♆
11 58	☽ # ☿
13 31	☽ ⊻ ♇
13 47	☽ ⊼ ♅
17 44	☽ △ ♂
22 17	☽ ⊼ ☿
13 Saturday	
01 15	☽ ± ♅
05 32	☉ ∠ ♆
06 09	☽ △ ♃
06 19	☽ ∥ ♇
10 04	☽ ⊼ ♀
10 28	☽ ± ☿
16 03	☽ # ♅
17 39	☽ △ ☉
18 44	☽ bQ ♅
20 00	♃ St D
21 38	☽ ± ♀
22 36	☽ Q ♆
14 Sunday	
04 57	☽ bQ ☿
11 27	☽ bQ ♃
12 12	☽ # ☉
14 17	☽ □ ♄
15 11	☽ bQ ♀
22 23	☽ ⚹ ♆
23 56	☽ ☌ ♇
15 Monday	
00 30	☽ △ ♅
05 23	☽ ± ♃
08 01	☽ # ♂
08 15	☽ □ ♂
12 11	☽ △ ☿
16 50	♀ ⊥ ♂
17 27	☽ ⊼ ♃
20 48	☽ △ ♀
16 Tuesday	
02 45	☽ Q ♄
02 58	♀ ∠ ♄
04 37	☽ ∠ ♆
08 24	☽ # ♄
10 43	☽ □ ☉
20 39	♂ ⊥ ♆
22 53	☽ ⊥ ♆
23 02	☽ Q ♂
17 Wednesday	
03 19	☽ ⚹ ♄
06 51	☽ # ♃
11 05	☽ ⊻ ♆
12 31	☽ ⊻ ♇
13 24	☽ □ ♅
18 Thursday	
00 40	☽ ⊥ ♇
00 58	☽ ⚹ ♂
02 26	☽ Q ☉
02 46	☽ # ♆
02 52	☽ □ ☿
06 15	☽ ☍ ♃
08 19	☽ □ ♀
10 00	☽ ∠ ♄
18 48	☽ ∠ ♇
20 34	☽ ∥ ♆
19 Friday	
01 53	☽ Q ♅
04 15	☽ ⊥ ♄
04 41	☽ ⚹ ☉
09 05	☽ ∠ ♂
15 34	☽ Q ☿
16 08	☽ ∥ ♃
16 20	☽ ⊻ ♄
17 54	☿ ⊻ ♂
19 22	☽ Q ♀
23 26	☽ ☌ ♆
20 Saturday	
00 40	☽ ⚹ ♇
01 51	☽ ⚹ ♅
04 04	☽ ⊥ ♂
12 54	☽ ∠ ☉
14 58	☽ ∥ ♄
15 22	☽ ⚹ ☿
16 34	☽ ⊻ ♂
17 56	☽ ⊼ ♃
18 15	☽ ⚹ ♀
21 Sunday	
00 07	☽ Q ♇
03 15	♀ □ ♃
05 35	☽ ± ♃
07 12	☽ ∠ ♅
07 50	☽ ⊥ ☉
08 43	☿ St R
11 29	♀ ⊻ ♂
16 28	♂ △ ♃
20 15	☽ ⊻ ☉
20 18	☽ ∠ ☿
21 43	☽ ∥ ♂
22 10	☽ ∠ ♀
22 50	☽ bQ ♃
22 Monday	
00 29	☽ ⊥ ♅
03 04	☽ ☌ ♄
09 24	☽ ⊻ ♆
10 26	☽ □ ♇
11 50	☽ ⊻ ♅
13 12	☽ ⊥ ☿
14 34	☽ ⊥ ♀
20 05	☽ ∥ ☉
20 39	☽ ⊥ ♆
23 Tuesday	
00 17	☽ ⊻ ☿
01 22	☽ ⊻ ♀
03 00	☽ △ ♃
04 44	☽ ∥ ♅
04 59	☽ ☌ ♂
13 18	☽ ∠ ♆
13 52	☽ # ♇
14 19	☽ ∥ ☿
18 44	☿ # ♇
24 Wednesday	
08 16	☽ ☌ ☉
10 54	☽ ⊻ ♄
15 05	☽ ∥ ♀
16 33	☽ ⚹ ♆
17 25	☽ △ ♇
19 01	☽ ☌ ♅
21 46	☽ ⊥ ♄
25 Thursday	
00 54	☿ ☌ ♀
05 47	☽ ☌ ♀
05 52	☽ ☌ ☿
09 24	☽ □ ♃
13 54	☽ ∠ ♄
13 58	☽ Q ♆
14 16	☽ ⊻ ♂
20 03	☽ bQ ♇
26 Friday	
01 21	☉ ⊻ ♄
01 26	☽ ⊥ ♂
11 49	☽ ± ♇
16 27	☽ ⚹ ♄
17 07	☽ Q ♃
17 26	☽ ⊻ ☉
18 03	☽ ∠ ♂
20 41	☽ ∥ ♀
21 33	☽ □ ♆
22 16	☽ ⊼ ♇
27 Saturday	
00 05	☽ ⊻ ♅
04 38	☽ ⊥ ☉
08 18	☽ ⊻ ♀
08 35	☿ ∥ ♅
09 07	☽ ⊻ ☿
10 31	☽ ⊥ ♅
12 25	☽ # ♇
13 26	☽ Q ♄
13 58	☽ ⚹ ♃
18 14	☽ ⊥ ♀
19 11	☽ ⊥ ☿
20 49	☽ ∥ ♅
21 20	☽ ∠ ☉
21 27	☽ ⚹ ♂
22 09	☽ ∥ ☿
23 19	☽ ∥ ☉
28 Sunday	
02 06	☽ ∠ ♅
06 51	☉ ∥ ☿
09 05	☽ ∠ ♀
10 10	☽ ∠ ☿
15 50	☽ ∠ ♃
18 05	☽ ∥ ♂
19 10	☽ Q ♂
20 36	☽ □ ♄
29 Monday	
00 56	☽ ⚹ ☉
01 17	☽ △ ♆
01 52	☽ ☍ ♇
03 53	☽ ⚹ ♅
05 57	☉ ⚹ ♆
07 15	☽ ⊥ ♃
09 40	☽ ⚹ ♀
10 58	☽ ⚹ ☿
14 18	☉ △ ♇
14 45	♂ ∠ ♆
17 30	☽ ⊻ ♃
22 52	☽ Q ☉
30 Tuesday	
00 24	☽ Q ♅
02 49	☽ ∥ ♄
02 51	☽ bQ ♆
03 29	☽ □ ♂
05 13	☽ Q ♀
06 40	☽ Q ☿
13 24	☉ ∥ ♅
18 09	☽ ± ♆
22 47	☉ ☌ ♅
31 Wednesday	
00 01	☽ △ ♄
00 40	☽ ∥ ♃
04 19	☽ ⊼ ♆
04 47	☽ ⊼ ♇
07 03	☽ □ ♅
07 37	☽ □ ☉
10 26	☽ □ ♀
12 03	☽ □ ☿
13 47	♀ ⊥ ♄
14 56	☽ ± ♇
16 02	☽ ∥ ♆
20 32	☽ ☌ ♃

All ephemeris data is given at 12.00 UT and the Moon's longitude is additionally given for 24.00 UT

JUNE 2028

Date	Sidereal time h m s	Sun ☉ ° ' "	Moon ☽ ° ' "	Moon ☽ 24.00 ° ' "	Mercury ☿ ° '	Venus ♀ ° '	Mars ♂ ° '	Jupiter ♃ ° '	Saturn ♄ ° '	Uranus ♅ ° '	Neptune ♆ ° '	Pluto ♇ ° '
01	04 41 59	11 ♊ 31 06	27 ♍ 10 19	04 ♎ 15 55	12 ♊ 27	11 ♊ 23	25 ♉ 30	18 ♍ 04	06 ♉ 04	10 ♊ 07	08 ♈ 28	08 ♒ 42
02	04 45 55	12 28 35	11 ♎ 21 01	18 ♎ 25 14	11 R 54	10 R 45	26 13	18 07	06 11	10 11	08 29	08 R 42
03	04 49 52	13 26 02	25 ♎ 28 11	02 ♏ 29 26	11 20	10 08	26 56	18 11	06 18	10 14	08 31	08 41
04	04 53 48	14 23 29	09 ♏ 28 34	16 ♏ 25 06	10 47	09 31	27 39	18 14	06 24	10 18	08 32	08 40
05	04 57 45	15 20 54	23 ♏ 18 38	00 ♐ 08 45	10 16	08 55	28 23	18 18	06 31	10 22	08 33	08 40
06	05 01 41	16 18 18	06 ♐ 55 05	13 ♐ 37 20	09 46	08 20	29 06	18 22	06 37	10 25	08 34	08 39
07	05 05 38	17 15 41	20 ♐ 15 16	26 ♐ 48 43	09 19	07 46	29 ♉ 49	18 26	06 44	10 29	08 35	08 38
08	05 09 35	18 13 04	03 ♑ 17 36	09 ♑ 41 57	08 55	07 13	00 ♊ 32	18 31	06 50	10 32	08 36	08 38
09	05 13 31	19 10 25	16 ♑ 01 51	22 ♑ 17 29	08 34	06 42	01 15	18 35	06 57	10 36	08 37	08 37
10	05 17 28	20 07 46	28 ♑ 29 05	04 ♒ 37 01	08 16	06 12	01 57	18 40	07 03	10 39	08 38	08 36
11	05 21 24	21 05 07	10 ♒ 41 38	16 ♒ 43 23	08 02	05 45	02 40	18 45	07 09	10 43	08 39	08 35
12	05 25 21	22 02 26	22 ♒ 42 47	28 ♒ 40 21	07 52	05 20	03 23	18 50	07 15	10 46	08 40	08 34
13	05 29 17	22 59 45	04 ♓ 36 38	10 ♓ 32 15	07 47	04 56	04 06	18 55	07 21	10 49	08 41	08 33
14	05 33 14	23 57 04	16 ♓ 27 48	22 ♓ 23 53	07 D 46	04 35	04 48	19 00	07 28	10 53	08 42	08 33
15	05 37 10	24 54 23	28 ♓ 21 07	04 ♈ 20 06	07 49	04 16	05 31	19 05	07 34	10 56	08 43	08 32
16	05 41 07	25 51 41	10 ♈ 21 25	16 ♈ 25 38	07 57	04 00	06 13	19 11	07 40	11 00	08 44	08 31
17	05 45 04	26 48 58	22 ♈ 33 15	28 ♈ 44 46	08 09	03 46	06 56	19 17	07 45	11 03	08 45	08 30
18	05 49 00	27 46 16	05 ♉ 00 35	11 ♉ 21 03	08 27	03 34	07 38	19 23	07 51	11 07	08 45	08 29
19	05 52 57	28 43 33	17 ♉ 46 27	24 ♉ 16 56	08 48	03 25	08 21	19 29	07 57	11 10	08 46	08 28
20	05 56 53	29 ♊ 40 50	00 ♊ 52 37	07 ♊ 33 29	09 15	03 18	09 03	19 35	08 03	11 13	08 47	08 27
21	06 00 50	00 ♋ 38 07	14 ♊ 19 25	21 ♊ 10 12	09 45	03 13	09 45	19 41	08 09	11 17	08 47	08 26
22	06 04 46	01 35 23	28 ♊ 05 31	05 ♋ 05 00	10 21	03 11	10 27	19 47	08 14	11 20	08 48	08 25
23	06 08 43	02 32 39	12 ♋ 08 09	19 ♋ 14 28	11 00	03 D 11	11 09	19 54	08 20	11 24	08 49	08 24
24	06 12 39	03 29 54	26 ♋ 23 21	03 ♌ 34 13	11 44	03 14	11 52	20 01	08 25	11 27	08 49	08 23
25	06 16 36	04 27 09	10 ♌ 46 28	17 ♌ 59 30	12 33	03 18	12 34	20 08	08 31	11 30	08 50	08 22
26	06 20 33	05 24 24	25 ♌ 12 45	02 ♍ 25 42	13 25	03 25	13 16	20 14	08 36	11 34	08 50	08 20
27	06 24 29	06 21 38	09 ♍ 37 49	16 ♍ 48 42	14 22	03 34	13 57	20 22	08 41	11 37	08 51	08 19
28	06 28 26	07 18 51	23 ♍ 57 57	01 ♎ 05 14	15 23	03 45	14 39	20 29	08 46	11 40	08 51	08 18
29	06 32 22	08 16 04	08 ♎ 10 16	15 ♎ 12 47	16 27	03 58	15 21	20 36	08 51	11 43	08 52	08 17
30	06 36 19	09 ♋ 13 16	22 ♎ 12 36	29 ♎ 09 32	17 ♊ 36	04 ♊ 14	16 ♊ 03	20 ♍ 44	08 ♉ 56	11 ♊ 47	08 ♈ 52	08 ♒ 16

Date	Moon True ☊ ° '	Moon Mean ☊ ° '	Moon ☽ Latitude ° '
01	24 ♑ 04	25 ♑ 26	04 S 36
02	24 R 01	25 23	05 02
03	23 56	25 20	05 09
04	23 51	25 17	04 58
05	23 45	25 14	04 30
06	23 39	25 10	03 46
07	23 34	25 07	02 51
08	23 31	25 04	01 48
09	23 29	25 01	00 S 41
10	23 D 29	24 58	00 N 27
11	23 30	24 54	01 33
12	23 32	24 51	02 33
13	23 33	24 48	03 26
14	23 35	24 45	04 10
15	23 R 35	24 42	04 44
16	23 34	24 39	05 05
17	23 32	24 35	05 14
18	23 29	24 32	05 07
19	23 25	24 29	04 46
20	23 21	24 26	04 09
21	23 18	24 23	03 18
22	23 16	24 20	02 14
23	23 14	24 16	01 N 01
24	23 D 14	24 13	00 S 17
25	23 15	24 10	01 35
26	23 16	24 07	02 47
27	23 17	24 04	03 49
28	23 18	24 00	04 36
29	23 18	23 57	05 05
30	23 ♑ 18	23 ♑ 54	05 S 16

DECLINATIONS

Date	Sun ☉ ° '	Moon ☽ ° '	Mercury ☿ ° '	Venus ♀ ° '	Mars ♂ ° '	Jupiter ♃ ° '	Saturn ♄ ° '	Uranus ♅ ° '	Neptune ♆ ° '	Pluto ♇ ° '
01	22 N 10	03 S 06	20 N 06	22 N 56	19 N 02	05 N 56	11 N 24	21 N 55	02 N 03	23 S 02
02	22 17	09 07	19 45	22 37	19 13	05 55	11 27	21 55	02 04	23 02
03	22 25	14 39	19 26	22 17	19 23	05 53	11 29	21 56	02 04	23 02
04	22 31	19 21	19 07	21 58	19 34	05 52	11 31	21 56	02 04	23 03
05	22 38	22 57	18 49	21 38	19 44	05 50	11 33	21 57	02 05	23 03
06	22 44	25 11	18 33	21 19	19 54	05 48	11 35	21 57	02 05	23 03
07	22 50	25 55	18 18	21 00	20 04	05 46	11 37	21 58	02 06	23 04
08	22 55	25 12	18 05	20 41	20 14	05 44	11 39	21 58	02 06	23 04
09	23 00	23 09	17 53	20 23	20 23	05 42	11 41	21 59	02 06	23 05
10	23 04	20 01	17 43	20 05	20 33	05 40	11 42	21 59	02 07	23 05
11	23 08	16 04	17 35	19 48	20 42	05 38	11 44	22 00	02 07	23 05
12	23 12	11 32	17 29	19 32	20 51	05 36	11 46	22 00	02 07	23 06
13	23 15	06 37	17 25	19 16	20 59	05 34	11 48	22 01	02 08	23 06
14	23 18	01 S 30	17 22	19 01	21 08	05 32	11 50	22 01	02 08	23 07
15	23 20	03 N 41	17 22	18 47	21 16	05 29	11 52	22 02	02 08	23 07
16	23 22	08 47	17 24	18 34	21 24	05 27	11 54	22 02	02 09	23 08
17	23 24	13 37	17 27	18 21	21 32	05 25	11 55	22 03	02 09	23 08
18	23 25	18 01	17 32	18 10	21 40	05 22	11 57	22 03	02 09	23 08
19	23 26	21 42	17 39	17 59	21 47	05 20	11 59	22 04	02 09	23 09
20	23 26	24 24	17 47	17 50	21 55	05 17	12 00	22 04	02 10	23 09
21	23 26	25 48	17 57	17 41	22 02	05 14	12 02	22 05	02 10	23 10
22	23 26	25 39	18 08	17 33	22 08	05 12	12 04	22 05	02 10	23 10
23	23 25	23 54	18 21	17 26	22 15	05 09	12 05	22 06	02 10	23 11
24	23 23	20 35	18 34	17 20	22 21	05 06	12 07	22 06	02 10	23 11
25	23 22	16 00	18 49	17 15	22 28	05 03	12 08	22 07	02 11	23 11
26	23 20	10 29	19 04	17 11	22 34	05 00	12 10	22 07	02 11	23 12
27	23 17	04 N 25	19 20	17 07	22 39	04 57	12 11	22 08	02 11	23 12
28	23 14	01 S 49	19 37	17 04	22 45	04 54	12 13	22 08	02 11	23 13
29	23 11	07 55	19 54	17 02	22 50	04 51	12 14	22 08	02 11	23 13
30	23 N 07	13 S 32	20 N 12	17 N 01	22 N 55	04 N 48	12 N 16	22 N 09	02 N 11	23 S 14

ZODIAC SIGN ENTRIES

Date	h m	Planets
01	16 47	☽ ♎
03	19 44	☽ ♏
05	23 45	☽ ♐
07	18 20	♂ ♊
08	05 53	☽ ♑
10	14 57	☽ ♒
13	02 41	☽ ♓
15	15 19	☽ ♈
18	02 25	☽ ♉
20	10 25	☽ ♊
20	20 02	☉ ♋
22	15 17	☽ ♋
24	18 02	☽ ♌
26	19 58	☽ ♍
28	22 10	☽ ♎

LATITUDES

Date	Mercury ☿ ° '	Venus ♀ ° '	Mars ♂ ° '	Jupiter ♃ ° '	Saturn ♄ ° '	Uranus ♅ ° '	Neptune ♆ ° '	Pluto ♇ ° '
01	02 S 13	00 N 48	00 S 06	01 N 20	02 S 16	00 S 03	01 S 25	05 S 08
04	02 58	00 N 05	00 04	01 19	02 16	00 03	01 25	05 08
07	03 35	00 S 37	00 S 02	01 18	02 16	00 03	01 26	05 09
10	04 01	01 16	00 00	01 18	02 17	00 03	01 26	05 10
13	04 15	01 53	00 N 02	01 17	02 17	00 03	01 26	05 10
16	04 18	02 25	00 04	01 16	02 18	00 03	01 26	05 11
19	04 10	02 54	00 06	01 16	02 18	00 03	01 26	05 11
22	03 54	03 18	00 08	01 15	02 19	00 03	01 26	05 12
25	03 31	03 38	00 10	01 15	02 19	00 03	01 26	05 12
28	03 02	03 54	00 12	01 14	02 20	00 03	01 27	05 13
31	02 S 29	04 S 06	00 N 14	01 N 14	02 S 20	00 S 03	01 S 27	05 S 13

DATA

Julian Date	2461924
Delta T	+73 seconds
Ayanamsa	24° 15' 29"
Synetic vernal point	04° ♓ 51' 30"
True obliquity of ecliptic	23° 26' 12"

MOON'S PHASES, APSIDES AND POSITIONS ☽

Date	h m	Phase	Longitude ° '	Eclipse Indicator
07	06 09	○	17 ♐ 02	
15	04 27	☾	24 ♓ 36	
22	18 27	●	01 ♋ 51	
29	12 11	☽	08 ♎ 16	

Day	h m		
14	04 44	Apogee	
26	04 17	Perigee	
07	11 53	Max dec	25° S 55'
14	18 55	0N	
21	22 02	Max dec	25° N 56'
28	04 59	0S	

LONGITUDES

Date	Chiron ⚷ ° '	Ceres ⚳ ° '	Pallas ⚴ ° '	Juno ⚵ ° '	Vesta ⚶ ° '	Black Moon Lilith ⚸ ° '
01	06 ♉ 29	11 ♏ 48	18 ♍ 13	13 ♌ 27	19 ♌ 08	09 ♓ 30
11	07 ♉ 00	10 ♏ 40	20 ♍ 23	17 ♌ 13	22 ♌ 55	10 ♓ 37
21	07 ♉ 27	10 ♏ 09	22 ♍ 57	21 ♌ 04	26 ♌ 54	11 ♓ 44
31	07 ♉ 50	10 ♏ 16	25 ♍ 52	24 ♌ 59	01 ♍ 06	12 ♓ 50

ASPECTARIAN

h m	Aspects
01 Thursday	
01 36	☽ Ꝑ ♄
05 13	♀ ⋕ ♇
06 09	☽ Ꝑ ♇
07 55	☽ ⋕ ♆
09 01	☽ △ ♂
10 00	☉ ☌ ♀
16 57	☽ ± ♄
23 07	☽ ⋕ ♃
02 Friday	
01 35	☿ ⊥ ♄
02 46	☉ ☌ ☿
03 11	☽ ⊼ ♄
03 42	☉ ⊥ ♄
07 09	☽ ☍ ♆
07 30	☽ △ ♇
10 01	☽ △ ♅
11 02	☽ △ ♀
11 46	☽ Ꝑ ♂
12 53	☽ △ ☿
14 03	☽ △ ☉
21 49	☽ ⋕ ♄
23 32	☽ ⊻ ♃
03 Saturday	
03 52	☽ ± ♂
05 26	☉ ∥ ♀
08 14	♀ ☌ ♅
09 47	☽ ⊥ ♃
11 27	☽ Ꝑ ♀
11 37	☽ Ꝑ ♅
13 25	☽ Ꝑ ☿
13 54	☿ ∥ ♂
14 38	☽ ⊼ ♂
17 26	☽ Ꝑ ☉
04 Sunday	
01 14	☽ ∠ ♃
02 11	☽ ± ♀
03 04	☽ ± ♅
04 14	☽ ± ☿
06 40	☽ ☍ ♄
10 00	☽ ± ☉
10 22	☽ ⊼ ♆
10 37	☽ □ ♇
10 45	☽ ⋕ ☿
12 04	☽ ⊼ ♀
13 16	☽ ⋕ ♂
13 26	☽ ⊼ ♅
13 42	♀ ∥ ♅
14 11	☽ ⊼ ☿
20 45	☽ ± ♆
21 07	☽ ⊼ ☉
05 Monday	
03 05	☽ ⋕ ♀
03 14	☽ ⚹ ♃
04 24	☽ ⋕ ♅
08 07	☿ ☌ ♅
09 24	☽ ⋕ ☉
12 25	☽ Ꝑ ♆
12 52	☽ ∥ ♇
17 52	☽ Q ♇
21 23	☽ ☍ ♂
22 25	♀ △ ♇
06 Tuesday	
00 20	☽ Q ♃
02 21	♀ ⚹ ♆
11 28	☽ ⊼ ♄
14 25	☽ ☍ ♀
14 57	☽ △ ♆
15 05	☽ ⚹ ♇
16 55	☽ ☍ ☿
18 16	☽ ☍ ♅
22 17	☽ ± ♄
07 Wednesday	
06 09	☽ ☍ ☉
08 41	☽ □ ♃
14 43	☽ Ꝑ ♄
18 10	☽ ∠ ♇
08 Thursday	
06 34	☽ ⊼ ♂
10 46	☽ ⊥ ♇
18 24	☽ ± ♂
18 41	☽ △ ♄
19 02	☽ ⊼ ♀
20 00	☉ □ ♃
21 57	☽ □ ♆
21 58	☽ ⊻ ♇
22 13	☽ ⊼ ☿
09 Friday	
01 38	☽ ⊼ ♅
02 25	♀ ⊻ ♄
04 37	♆ ⚹ ♇
05 55	☽ ± ♀
07 39	☿ ⚹ ♆
07 56	☿ △ ♇
09 16	☽ ± ☿
11 41	♀ ∥ ♂
12 26	☽ Ꝑ ♂
12 37	☽ ∥ ♇
13 05	☽ ± ♅
13 19	☽ ⋕ ☉
16 55	☽ △ ♃
18 30	☽ ⊼ ☉
21 47	☽ ⋕ ♅
22 26	☽ Ꝑ ♀
10 Saturday	
02 06	☽ Ꝑ ☿
06 28	☽ Ꝑ ♅
07 02	☽ ± ☉
08 24	☽ Q ♆
08 33	☽ ⋕ ♂
11 28	☽ ⋕ ♀
17 30	☉ ⋕ ♇
19 12	☽ △ ♂
22 12	☽ Ꝑ ♃
11 Sunday	
01 02	☉ Q ♆
02 07	☽ Ꝑ ☉
02 35	☽ △ ♀
02 59	☽ ⋕ ☿
04 56	☽ □ ♄
06 49	☽ △ ☿
07 50	☽ ☌ ♇
07 58	☽ ⚹ ♆
12 02	☽ △ ♅
16 06	☽ ± ♃
12 Monday	
04 09	☽ ⊼ ♃
10 32	☽ △ ☉
10 47	☽ ⋕ ♄
13 56	☽ ∠ ♆
17 09	☽ Q ♄
18 03	☉ ∠ ♄
13 Tuesday	
08 06	☽ ⊥ ♆
10 53	☽ □ ♂
12 38	☽ □ ♀
17 00	☽ ⋕ ♃
17 37	☽ ⚹ ♄
18 23	☽ □ ☿
19 59	☽ ⊻ ♇
20 16	☽ ⊻ ♆
14 Wednesday	
00 39	☽ □ ♅
01 54	☉ Ꝑ ♇
06 06	☿ St D
06 58	♀ ☌ ♂
08 07	☽ ⊥ ♇
09 02	☽ ⋕ ♆
17 10	☽ ☍ ♃
15 Thursday	
00 03	☽ Q ♀
00 14	☽ ∠ ♄
01 38	☽ Q ♂
02 18	☽ ∠ ♇
04 27	☽ □ ☉
04 49	☽ ∥ ♆
06 51	☽ Q ☿
13 11	☽ Q ♅
18 30	☽ ⊥ ♄
20 22	☽ ∥ ♃
23 36	☽ ⚹ ♀
16 Friday	
03 15	☽ ⚹ ♂
06 35	☽ ⊻ ♄
07 09	☽ ⚹ ☿
08 20	☽ ⚹ ♇
08 46	☽ ☌ ♆
13 17	☽ ⚹ ♅
19 32	☽ Q ☉
17 Saturday	
03 19	☽ ∥ ♄
04 43	☽ ∠ ♀
05 33	☽ ⊼ ♃
07 59	☽ Q ♇
10 43	☽ ∠ ♂
13 12	☽ ∠ ☿
17 20	☽ ± ♃
18 50	☽ ∠ ♅
20 58	☽ ⚹ ☉
21 56	☽ ⊥ ♀
18 Sunday	
05 10	☽ ⊥ ♂
06 58	☽ ⊥ ☿
09 10	☽ ∥ ☿
09 17	☽ ⊻ ♀
10 47	☽ Ꝑ ♃
12 12	☽ ⊥ ♅
12 52	☽ ∥ ♀
14 39	☿ △ ♇
17 17	☽ ⊻ ♂
17 27	☽ ☌ ♄
18 35	☽ □ ♇
18 42	☽ ⊻ ☿
19 07	☽ ⊻ ♆
20 36	♂ ⊻ ♄
23 36	☽ ⊻ ♅
19 Monday	
03 51	☽ ∠ ☉
06 24	☽ ⊥ ♆
09 39	☿ ⚹ ♆
12 41	☽ ∥ ♂
14 45	☽ ∥ ♅
15 11	☽ △ ♃
16 02	♂ △ ♇
21 52	☽ ⊥ ☉
23 04	☽ ∠ ♆
23 39	☽ ⋕ ♇
20 Tuesday	
02 13	☽ ∥ ☉
02 42	♂ ⚹ ♆
09 40	☽ ⊻ ☉
15 23	☿ ∥ ♀
16 20	☽ ☌ ♀
21 Wednesday	
00 58	☽ ⊻ ♄
01 34	☽ △ ♇
02 11	☽ ⚹ ♆
03 28	☽ ☌ ♂
03 35	☽ ☌ ☿
06 36	☽ ☌ ♅
11 41	☽ ⊥ ♄
12 51	☿ ☌ ♂
21 29	☽ □ ♃
23 21	☽ Q ♆
22 Thursday	
00 06	♂ ∥ ♅
03 32	☽ ∠ ♄
03 55	☽ Ꝑ ♇
18 27	☽ ☌ ☉
19 25	☽ ± ♇
20 45	☽ ⊻ ♀
22 13	♀ St D
23 Friday	
04 45	☽ Q ♃
05 29	☽ ⚹ ♄
05 39	☽ ⊼ ♇
06 21	☽ □ ♆
06 59	☽ ⊥ ♀
08 21	☉ ± ♇
09 59	☽ ⊻ ☿
10 15	☽ ⊻ ♂
10 44	☽ ⊻ ♅
16 18	☽ ∥ ☉
18 12	☽ ⋕ ♇
20 40	☽ ⊥ ☿
20 45	♂ ☌ ♅
20 55	☽ ⊥ ♅
20 56	☽ ⊥ ♂
22 15	☽ ∠ ♀
24 Saturday	
00 48	☽ ∥ ♂
01 13	☽ ⚹ ♃
01 55	☽ Q ♄
01 59	☿ ☌ ♅
02 16	☽ ∥ ♅
03 02	♄ □ ♇
04 46	☉ ⊻ ♀
12 06	☽ ∠ ♅
12 37	☽ ∠ ☿
12 50	☽ ∠ ♂
22 44	☽ ∥ ☿
23 29	☽ ⚹ ♀
25 Sunday	
00 43	☽ ⊻ ☉
02 31	☽ ∠ ♃
05 53	☽ ∥ ♀
07 59	☽ ☍ ♇
08 12	☽ □ ♄
08 46	☽ △ ♆
11 26	☽ ⊥ ☉
13 13	☽ ⚹ ♅
14 26	☿ ☌ ♂
15 07	☽ ⚹ ♂
15 08	☽ ⚹ ☿
17 37	☽ ⊥ ♃
19 35	☽ Q ♀
26 Monday	
03 27	☽ ∠ ☉
03 40	☽ ⊻ ♃
05 02	☽ ∥ ♄
09 15	☽ Q ♅
09 43	☽ Ꝑ ♆
12 05	☽ Q ♂
12 22	☽ Q ☿
27 Tuesday	
00 41	☽ ± ♆
01 47	☽ □ ♀
06 10	☽ ⚹ ☉
09 49	☽ ⊼ ♇
09 54	☽ ∥ ♃
10 25	☽ △ ♄
10 41	☽ ⊼ ♆
15 19	☽ □ ♅
19 36	☽ □ ♂
19 50	☽ ± ♇
20 29	☽ □ ☿
20 33	☿ ⊥ ♄
20 37	☽ ∥ ♆
28 Wednesday	
03 38	☽ Q ☉
06 06	☽ ☌ ♃
10 53	☽ Ꝑ ♇
11 40	☽ Ꝑ ♄
13 23	☽ ⋕ ♆
16 34	♂ ⊥ ♄
20 03	☉ ⋕ ♇
23 54	☽ ⋕ ♃
29 Thursday	
02 56	☽ ± ♄
04 46	☽ △ ♀
12 11	☽ □ ☉
12 12	☽ △ ♇
12 25	☉ ⊼ ♇
12 58	♄ ⊻ ♆
13 10	☽ ☍ ♆
13 10	☽ ⊼ ♄
18 04	☽ △ ♅
21 41	☉ Q ♃
30 Friday	
00 53	☽ △ ♂
03 01	☉ □ ♆
03 22	☽ △ ☿
04 17	☉ ⚹ ♄
06 19	☽ ⋕ ♄
06 47	☽ Ꝑ ♀
09 26	☽ ⊻ ♃
19 52	☽ ⊥ ♃
19 54	☽ Ꝑ ♅
22 36	☽ ± ♀

All ephemeris data is given at 12.00 UT and the Moon's longitude is additionally given for 24.00 UT

JULY 2028

LONGITUDES

Date	Sidereal time h m s	Sun ☉ ° ' "	Moon ☽ ° ' "	Moon ☽ 24.00 ° ' "	Mercury ☿ ° '	Venus ♀ ° '	Mars ♂ ° '	Jupiter ♃ ° '	Saturn ♄ ° '	Uranus ♅ ° '	Neptune ♆ ° '	Pluto ♇ ° '
01	06 40 15	10 ♋ 10 29	06 ♏ 03 25	12 ♏ 54 09	18 ♊ 49	04 ♊ 31	16 ♊ 45	20 ♍ 51	09 ♉ 01	11 ♊ 50	08 ♈ 52	08 ♒ 15
02	06 44 12	11 07 40	19 ♏ 41 38	26 ♏ 25 45	20 05	04 50	17 26	20 59	09 06	11 53	08 53	08 R 14
03	06 48 08	12 04 52	03 ♐ 06 27	09 ♐ 43 40	21 25	05 10	18 08	21 07	09 11	11 56	08 53	08 12
04	06 52 05	13 02 03	16 ♐ 17 22	22 ♐ 47 32	22 49	05 33	18 49	21 15	09 16	11 59	08 53	08 11
05	06 56 02	13 59 14	29 ♐ 14 10	05 ♑ 37 17	24 17	05 57	19 31	21 23	09 21	12 02	08 53	08 10
06	06 59 58	14 56 25	11 ♑ 56 57	18 ♑ 13 15	25 49	06 23	20 12	21 31	09 25	12 06	08 54	08 09
07	07 03 55	15 53 36	24 ♑ 26 18	00 ♒ 36 15	27 24	06 50	20 54	21 39	09 30	12 09	08 54	08 07
08	07 07 51	16 50 47	06 ♒ 43 18	12 ♒ 47 41	29 ♊ 02	07 19	21 35	21 48	09 34	12 12	08 54	08 06
09	07 11 48	17 47 59	18 ♒ 49 41	24 ♒ 49 36	00 ♋ 45	07 49	22 16	21 56	09 39	12 15	08 54	08 05
10	07 15 44	18 45 10	00 ♓ 47 46	06 ♓ 44 37	02 30	08 21	22 57	22 05	09 43	12 18	08 54	08 04
11	07 19 41	19 42 22	12 ♓ 40 33	18 ♓ 36 02	04 18	08 54	23 38	22 14	09 47	12 21	08 54	08 02
12	07 23 37	20 39 34	24 ♓ 31 33	00 ♈ 27 37	06 10	09 28	24 19	22 22	09 51	12 24	08 R 54	08 01
13	07 27 34	21 36 47	06 ♈ 24 46	12 ♈ 23 33	08 04	10 04	25 00	22 31	09 55	12 27	08 54	08 00
14	07 31 31	22 34 01	18 ♈ 24 33	24 ♈ 28 19	10 01	10 41	25 41	22 40	09 59	12 30	08 54	07 58
15	07 35 27	23 31 15	00 ♉ 35 25	06 ♉ 46 23	12 01	11 19	26 22	22 50	10 03	12 32	08 54	07 57
16	07 39 24	24 28 29	13 ♉ 01 45	19 ♉ 21 58	14 02	11 58	27 03	22 59	10 07	12 35	08 54	07 56
17	07 43 20	25 25 44	25 ♉ 47 29	02 ♊ 18 38	16 06	12 38	27 44	23 08	10 10	12 38	08 53	07 54
18	07 47 17	26 23 00	08 ♊ 55 42	15 ♊ 38 52	18 10	13 20	28 25	23 18	10 14	12 41	08 53	07 53
19	07 51 13	27 20 17	22 ♊ 28 12	29 ♊ 23 36	20 16	14 02	29 06	23 27	10 18	12 44	08 53	07 52
20	07 55 10	28 17 35	06 ♋ 24 54	13 ♋ 31 45	22 23	14 45	29 ♊ 46	23 37	10 21	12 46	08 53	07 50
21	07 59 06	29 ♋ 14 52	20 ♋ 43 40	28 ♋ 00 03	24 31	15 29	00 ♋ 27	23 47	10 24	12 49	08 52	07 49
22	08 03 03	00 ♌ 12 11	05 ♌ 20 08	12 ♌ 43 05	26 39	16 14	01 07	23 56	10 28	12 52	08 52	07 47
23	08 07 00	01 09 30	20 ♌ 08 00	27 ♌ 33 55	28 ♋ 46	17 00	01 48	24 06	10 31	12 54	08 52	07 46
24	08 10 56	02 06 49	04 ♍ 59 53	12 ♍ 24 56	00 ♌ 54	17 47	02 28	24 16	10 34	12 57	08 51	07 45
25	08 14 53	03 04 09	19 ♍ 48 12	27 ♍ 08 52	03 01	18 34	03 09	24 27	10 37	12 59	08 51	07 43
26	08 18 49	04 01 29	04 ♎ 26 15	11 ♎ 39 46	05 07	19 23	03 49	24 37	10 39	13 02	08 50	07 42
27	08 22 46	04 58 50	18 ♎ 48 58	25 ♎ 53 30	07 13	20 12	04 29	24 47	10 42	13 04	08 50	07 41
28	08 26 42	05 56 11	02 ♏ 53 12	09 ♏ 47 57	09 17	21 01	05 09	24 57	10 45	13 07	08 49	07 39
29	08 30 39	06 53 33	16 ♏ 37 46	23 ♏ 22 42	11 20	21 52	05 50	25 08	10 47	13 09	08 49	07 38
30	08 34 35	07 50 55	00 ♐ 02 54	06 ♐ 38 35	13 22	22 43	06 30	25 18	10 50	13 12	08 48	07 36
31	08 38 32	08 ♌ 48 18	13 ♐ 09 57	19 ♐ 37 15	15 ♌ 22	23 ♊ 34	07 ♋ 10	25 ♍ 29	10 ♉ 52	13 ♊ 14	08 ♈ 48	07 ♒ 35

Date	Moon True ☊ ° '	Moon Mean ☊ ° '	Moon ☽ Latitude ° '
01	23 ♑ 17	23 ♑ 51	05 S 08
02	23 R 16	23 48	04 43
03	23 14	23 45	04 03
04	23 13	23 41	03 10
05	23 12	23 38	02 09
06	23 11	23 35	01 S 02
07	23 D 11	23 32	00 N 07
08	23 11	23 29	01 14
09	23 12	23 26	02 17
10	23 12	23 22	03 14
11	23 13	23 19	04 01
12	23 13	23 16	04 38
13	23 13	23 13	05 04
14	23 13	23 10	05 16
15	23 13	23 06	05 15
16	23 13	23 03	04 59
17	23 13	23 00	04 28
18	23 14	22 57	03 42
19	23 14	22 54	02 43
20	23 14	22 51	01 32
21	23 15	22 47	00 N 14
22	23 R 14	22 44	01 S 07
23	23 14	22 41	02 24
24	23 13	22 38	03 31
25	23 12	22 35	04 24
26	23 11	22 32	05 00
27	23 10	22 28	05 16
28	23 10	22 25	05 12
29	23 D 10	22 22	04 50
30	23 11	22 19	04 13
31	23 ♑ 12	22 ♑ 16	03 S 24

DECLINATIONS

Sun ☉ ° '	Moon ☽ ° '	Mercury ☿ ° '	Venus ♀ ° '	Mars ♂ ° '	Jupiter ♃ ° '	Saturn ♄ ° '	Uranus ♅ ° '	Neptune ♆ ° '	Pluto ♇ ° '
23 N 03	18 S 23	20 N 30	17 N 01	23 N 00	04 N 45	12 N 17	22 N 09	02 N 11	23 S 14
22 58	22 12	20 48	17 01	23 05	04 42	12 19	22 10	02 11	23 15
22 53	24 45	21 05	17 01	23 09	04 39	12 20	22 10	02 11	23 15
22 48	25 53	21 22	17 03	23 14	04 35	12 21	22 11	02 11	23 16
22 42	25 35	21 39	17 04	23 18	04 32	12 22	22 11	02 11	23 16
22 36	23 56	21 55	17 07	23 21	04 29	12 24	22 11	02 11	23 17
22 29	21 07	22 10	17 10	23 25	04 25	12 25	22 12	02 11	23 17
22 22	17 23	22 24	17 13	23 28	04 22	12 26	22 12	02 11	23 17
22 15	13 00	22 37	17 17	23 32	04 18	12 27	22 13	02 11	23 18
22 07	08 10	22 49	17 21	23 35	04 15	12 28	22 13	02 11	23 18
21 59	03 S 05	22 58	17 25	23 37	04 11	12 30	22 14	02 11	23 19
21 51	02 N 05	23 06	17 30	23 40	04 08	12 31	22 14	02 11	23 19
21 42	07 12	23 12	17 35	23 42	04 04	12 32	22 14	02 11	23 20
21 33	12 05	23 15	17 40	23 44	04 00	12 33	22 15	02 11	23 20
21 23	16 35	23 16	17 45	23 46	03 56	12 34	22 15	02 11	23 21
21 13	20 29	23 15	17 51	23 48	03 53	12 35	22 15	02 11	23 21
21 03	23 32	23 11	17 57	23 49	03 49	12 36	22 16	02 11	23 22
20 52	25 26	23 04	18 03	23 51	03 45	12 37	22 16	02 11	23 22
20 41	25 56	22 55	18 09	23 52	03 41	12 37	22 17	02 11	23 23
20 30	24 49	22 43	18 15	23 53	03 37	12 38	22 17	02 10	23 23
20 18	22 04	22 28	18 21	23 53	03 33	12 39	22 17	02 10	23 24
20 06	17 51	22 11	18 28	23 54	03 29	12 40	22 18	02 10	23 24
19 54	12 30	21 51	18 34	23 54	03 25	12 41	22 18	02 10	23 24
19 41	06 N 24	21 29	18 40	23 54	03 21	12 41	22 18	02 10	23 25
19 28	00 S 01	21 04	18 46	23 54	03 17	12 42	22 19	02 09	23 25
19 15	06 21	20 37	18 52	23 53	03 13	12 43	22 19	02 09	23 26
19 01	12 14	20 08	18 58	23 53	03 08	12 43	22 19	02 09	23 26
18 47	17 21	19 37	19 04	23 52	03 04	12 44	22 20	02 09	23 27
18 33	21 26	19 05	19 10	23 51	03 00	12 44	22 20	02 08	23 27
18 18	24 17	18 31	19 16	23 50	02 56	12 45	22 20	02 08	23 28
18 N 03	25 S 45	17 N 55	19 N 21	23 N 48	02 N 51	12 N 45	22 N 20	02 N 08	23 S 28

ZODIAC SIGN ENTRIES

Date	h	m	Planets
01	01	28	☽ ♏
03	06	24	☽ ♐
05	13	26	☽ ♑
07	22	49	☽ ♒
09	01	37	☿ ♋
10	10	24	☽ ♓
12	23	04	☽ ♈
15	10	51	☽ ♉
17	19	46	☽ ♊
20	01	03	☽ ♋
20	20	10	♂ ♋
22	03	17	☽ ♌
22	06	54	☉ ♌
24	01	50	☿ ♌
24	03	56	☽ ♍
26	04	41	☽ ♎
28	07	02	☽ ♏
30	11	55	☽ ♐

LATITUDES

Date	Mercury ☿ ° '	Venus ♀ ° '	Mars ♂ ° '	Jupiter ♃ ° '	Saturn ♄ ° '	Uranus ♅ ° '	Neptune ♆ ° '	Pluto ♇ ° '
01	02 S 29	04 S 06	00 N 14	01 N 14	02 S 20	00 S 03	01 S 27	05 S 13
04	01 52	04 15	00 16	01 13	02 21	00 03	01 27	05 14
07	01 14	04 21	00 18	01 13	02 21	00 03	01 27	05 14
10	00 S 36	04 24	00 20	01 12	02 22	00 03	01 27	05 14
13	00 00	04 25	00 22	01 12	02 23	00 03	01 27	05 15
16	00 N 33	04 24	00 24	01 11	02 23	00 03	01 28	05 15
19	01 01	04 22	00 26	01 11	02 24	00 03	01 28	05 15
22	01 23	04 17	00 28	01 10	02 25	00 03	01 28	05 16
25	01 38	04 11	00 30	01 10	02 25	00 03	01 28	05 16
28	01 45	04 04	00 32	01 10	02 26	00 03	01 28	05 16
31	01 N 46	03 S 56	00 N 34	01 N 09	02 S 27	00 S 03	01 S 28	05 S 17

DATA

Julian Date	2461954
Delta T	+73 seconds
Ayanamsa	24° 15' 35"
Synetic vernal point	04° ♓ 51' 24"
True obliquity of ecliptic	23° 26' 11"

MOON'S PHASES, APSIDES AND POSITIONS ☽

Date	h	m	Phase	Longitude ° '	Eclipse Indicator
06	18	11	○	15 ♑ 11	partial
14	20	57	☾	22 ♈ 55	
22	03	02	●	29 ♋ 51	Total
28	17	40	☽	06 ♏ 10	

Day	h	m		
11	22	18	Apogee	
23	22	12	Perigee	
04	18	50	Max dec	25° S 57'
12	02	19	0N	
19	07	36	Max dec	25° N 57'
25	11	56	0S	

LONGITUDES

Date	Chiron ⚷ ° '	Ceres ⚳ ° '	Pallas ⚴ ° '	Juno ⚵ ° '	Vesta ⚶ ° '	Black Moon Lilith ⚸ ° '
01	07 ♉ 50	10 ♏ 16	25 ♍ 52	24 ♌ 59	01 ♍ 06	12 ♓ 50
11	08 ♉ 09	10 ♏ 59	29 ♍ 03	28 ♌ 56	05 ♍ 27	13 ♓ 57
21	08 ♉ 23	12 ♏ 14	02 ♎ 28	02 ♍ 56	09 ♍ 57	15 ♓ 04
31	08 ♉ 32	13 ♏ 58	06 ♎ 05	06 ♍ 57	14 ♍ 34	16 ♓ 11

All ephemeris data is given at 12.00 UT and the Moon's longitude is additionally given for 24.00 UT

ASPECTARIAN

01 Saturday
04 05 ☽ ⚼ ♂
04 45 ☽ # ♀
07 42 ☽ ⚼ ☿
09 15 ☽ ⊼ ♀
11 36 ☽ ± ♅
11 38 ☽ ∠ ♃
15 49 ☽ □ ♇
16 56 ☽ ⊼ ♆
17 13 ☽ ☍ ♄
18 42 ☉ ∥ ♂
19 45 ☽ △ ☉
20 39 ☽ ± ♂
22 09 ☽ ⊼ ♅

02 Sunday
00 30 ☉ ⊥ ♀
01 03 ☽ ± ☿
01 27 ☽ # ☿
03 29 ☽ ± ♆
07 47 ☽ ⊼ ♂
11 45 ☽ # ♅
12 46 ☽ ⊼ ☿
14 19 ☽ ✱ ♃
17 55 ☽ # ☉
19 13 ☽ # ♂
19 27 ☽ ⚼ ♆
20 26 ☽ ∥ ♇
23 37 ☽ Q ♇

03 Monday
00 20 ☽ ⚼ ☉
02 25 ☿ Q ♆
05 57 ☿ □ ♃
08 09 ☉ ⊻ ♅
12 01 ☽ Q ♃
15 50 ☽ ☍ ♀
17 48 ☽ ± ☉
21 13 ☽ ✱ ♇
22 28 ☽ △ ♆
23 05 ☽ ⊼ ♄

04 Tuesday
04 06 ☽ ☍ ♅
05 34 ☽ ⊼ ☉
10 07 ☽ ± ♄
16 55 ☽ ☍ ♂
17 55 ☿ ⚼ ♇
21 14 ☽ □ ♃

05 Wednesday
00 43 ☽ ∠ ♇
01 05 ♂ # ♇
01 35 ☽ ☍ ☿
02 49 ☽ ⚼ ♄
12 56 ☿ ∠ ♄
17 29 ☽ ⊥ ♇

06 Thursday
01 03 ☽ ⊼ ♀
04 47 ☽ ⊻ ♇
06 11 ☽ □ ♆
07 10 ☽ △ ♄
12 16 ☽ ⊼ ♅
12 51 ☽ ± ♀
17 32 ☽ # ♂
18 11 ☽ ☍ ☉
18 26 ☽ ∥ ♇
23 48 ☽ ± ♅

07 Friday
00 55 ☽ # ☉
03 46 ☽ # ♅
04 33 ☽ # ☿
04 44 ☽ ⊼ ♂
06 33 ☽ △ ♃
06 46 ☽ ⚼ ♀
12 08 ♂ Q ♆
14 32 ☿ ∥ ♅
16 46 ☽ Q ♆
17 03 ☽ ± ♂
17 17 ☽ ⚼ ♅
18 37 ☽ ⊼ ☿

08 Saturday
08 10 ☽ ± ☿
09 41 ☉ ∥ ☿
11 42 ☽ ⚼ ♂
12 09 ☽ ⚼ ♃
13 01 ☽ # ♀
13 13 ☽ △ ♀
14 43 ☽ ☌ ♇
16 17 ☽ ✱ ♆
17 39 ☽ □ ♄
21 27 ♂ □ ♃
22 51 ☽ △ ♅

09 Sunday
04 50 ☽ ⚼ ☿
06 10 ☽ ± ♃
09 46 ☽ ⊼ ☉
14 47 ☽ # ♄
18 17 ☽ ⊼ ♃
19 17 ☽ △ ♂
19 23 ☉ ∥ ♅
22 08 ☽ ∠ ♆
22 48 ☽ ± ☉
23 33 ♀ △ ♇
23 51 ☉ ⊥ ♅

10 Monday
05 46 ☽ Q ♄
06 09 ☿ ± ♇
15 35 ♂ ⚼ ♇
16 02 ☽ △ ☿
16 14 ☽ ⊥ ♆
18 29 ☽ ⚼ ☉

11 Tuesday
02 38 ☽ ⊻ ♇
03 59 ☽ □ ♀
04 22 ☽ ⊻ ♆
06 07 ☽ ✱ ♄
06 47 ☽ # ♃
11 20 ☽ □ ♅
12 02 ♀ ✱ ♆
13 06 ♆ St R
14 45 ☽ ⊥ ♇
16 09 ☽ # ♆

12 Wednesday
03 29 ☽ △ ☉
07 35 ☽ ☍ ♃
08 57 ☽ ∠ ♇
11 34 ☽ □ ♂
12 28 ☽ ∥ ♆
12 40 ☽ ∠ ♄
18 16 ☽ Q ♀
21 24 ☽ ∥ ♃
23 55 ☽ Q ♅

13 Thursday
05 22 ♀ ⊻ ♄
06 57 ☽ ⊥ ♄
08 33 ♂ ∠ ♄
11 02 ☿ ⊼ ♇
15 10 ☽ ✱ ♇
15 59 ☽ □ ☿
17 00 ☽ ☌ ♆
19 05 ☽ ⊻ ♄
19 44 ☽ ✱ ♀
20 17 ☉ Q ♄
22 14 ☿ □ ♆

14 Friday
00 09 ☽ ✱ ♅
02 02 ☽ Q ♂
11 32 ☿ ✱ ♄
14 22 ☽ ∥ ♄
15 06 ☽ Q ♇
15 13 ☉ ✱ ♃
20 34 ☽ ⊼ ♃
20 34 ☿ Q ♃
20 57 ☽ □ ☉
23 43 ☿ ⊻ ♀

15 Saturday
03 10 ☽ ∠ ♀
03 15 ☽ ✱ ♂
06 00 ☽ ∠ ♅
08 31 ☽ ± ♃
10 39 ☽ Q ☿
18 27 ☿ ⊻ ♅
18 57 ☽ ∥ ♀
21 41 ☽ ⊥ ♀
23 36 ☽ ⊥ ♅

16 Sunday
02 12 ☽ ⚼ ♃
02 15 ☽ □ ♇
04 05 ☽ ⊻ ♆
04 43 ♃ ⚼ ♇
06 24 ☽ ☌ ♄
09 52 ☽ ⊻ ♀
10 02 ☽ ∠ ♂
10 51 ☽ Q ☉
11 09 ☽ ⊻ ♅
14 17 ☽ ✱ ☿
15 33 ☽ ⊥ ♆
16 55 ☽ ∥ ☉

17 Monday
01 01 ☽ ∥ ♅
04 01 ☽ ⊥ ♂
07 00 ☽ △ ♃
08 28 ☽ ∠ ♆
08 54 ☽ ∥ ☿
10 24 ☽ # ♇
11 16 ☽ ✱ ☉
11 47 ♀ ☌ ♅
14 53 ☽ ∥ ♂
15 48 ☽ ⊻ ♂
23 36 ☽ ∠ ☿

18 Tuesday
10 07 ☽ △ ♇
11 56 ☽ ✱ ♆
14 21 ☽ ⊻ ♄
16 45 ☽ ∠ ☉
17 57 ☿ ⊥ ♅
18 45 ☽ ☌ ♅
18 53 ☽ ⊥ ☿
20 18 ☽ ☌ ♀

19 Wednesday
01 06 ☽ ⊥ ♄
07 27 ☽ ⊻ ☿
07 51 ☿ ⊥ ♀
09 14 ☽ Q ♆
09 52 ☽ ⊥ ☉
12 41 ☽ ⚼ ♇
13 44 ☽ □ ♃
16 56 ☽ ∠ ♄
21 05 ☽ ⊻ ☉
22 15 ☉ ∠ ♅

20 Thursday
00 04 ☽ ☌ ♂
04 12 ☽ ± ♇
11 32 ☿ Q ♄
14 24 ☽ ⊼ ♇
16 10 ☽ □ ♆
18 41 ☽ ✱ ♄
20 53 ☽ Q ♃
21 50 ☽ ∥ ♂
22 46 ☽ ⊻ ♅

21 Friday
02 07 ☽ # ♇
02 48 ☽ ⊻ ♀
03 00 ☿ ✱ ♃
08 49 ☽ ⊥ ♅
08 56 ☽ ∥ ☿
10 29 ☽ ∥ ♅
13 20 ☽ ⊥ ♀
14 47 ☽ Q ♄
17 06 ☽ ✱ ♃
19 20 ☽ ☌ ☿
23 32 ☽ ∥ ☉
23 44 ☽ ∠ ♅

22 Saturday
03 02 ☽ ☌ ☉
03 26 ☿ ∥ ♅
04 47 ☽ ⊻ ♂
04 57 ☽ ∠ ♀
09 02 ☽ ∥ ♀
15 03 ☽ ⊥ ♂
16 00 ☽ ☍ ♇
17 45 ☽ △ ♆
17 56 ☽ ∠ ♃
19 27 ♀ ⊥ ♄
20 22 ☽ □ ♄

23 Sunday
00 16 ☽ ✱ ♅
02 00 ☿ ∠ ♅
06 21 ☽ ∠ ♂
06 40 ☽ ✱ ♀
08 41 ☽ ⊥ ♃
11 01 ♂ ± ♇
11 16 ☽ ∥ ♄
18 01 ☽ ⚼ ♆
18 29 ☽ ⊻ ♃
19 44 ☽ Q ♅

24 Monday
03 07 ☽ Q ♀
04 17 ☽ ⊻ ☿
07 01 ☽ ⊻ ☉
07 43 ☽ ✱ ♂
08 32 ☽ ± ♆
15 35 ☽ ⊥ ☿
16 26 ☽ ⊼ ♇
17 23 ☽ ⊥ ☉
18 14 ☽ ⊼ ♆
21 02 ☽ △ ♄
23 35 ☽ ∥ ♃

25 Tuesday
00 54 ☽ □ ♅
02 08 ☽ ± ♇
03 54 ☽ ∥ ♆
04 04 ☽ Q ♂
08 36 ☽ ∠ ☿
08 59 ☽ ∠ ☉
09 53 ☽ □ ♀
13 07 ☉ ☌ ☿
14 09 ☿ ⊻ ♂
16 45 ☽ ⚼ ♇
18 17 ☉ ⊻ ♂
19 40 ☽ ☌ ♃
20 00 ☽ # ♆
21 31 ☽ ⚼ ♄
22 14 ☿ ∠ ♀

26 Wednesday
00 06 ☽ # ♃
10 55 ☽ □ ♂
11 16 ☽ ✱ ☉
12 22 ☽ ± ♄
13 19 ☽ ✱ ☿
17 24 ☽ △ ♇
19 18 ☽ ☍ ♆
22 21 ☽ ⊼ ♄

27 Thursday
02 20 ☽ △ ♅
08 41 ☽ Q ☉
12 47 ☽ Q ☿
14 08 ☽ # ♄
14 28 ☽ △ ♀
15 29 ☉ ∥ ♀
17 20 ☿ ☍ ♇
22 14 ☽ ⊻ ♃

28 Friday
03 46 ☽ ⚼ ♅
06 23 ♀ Q ♆
06 43 ☿ △ ♆
08 38 ☽ ⊥ ♃
16 07 ☽ △ ♂
17 40 ☽ □ ☉
17 46 ☽ ⚼ ♀
19 18 ☽ # ☉
19 21 ☽ ± ♅
20 15 ☽ □ ♇
20 37 ☿ ∠ ♃
21 31 ☽ # ♀
22 17 ☽ ⊼ ♆
23 04 ☽ # ☿

29 Saturday
00 26 ☽ ∠ ♃
01 03 ☽ □ ☿
01 42 ☽ ☍ ♄
05 01 ☉ ∠ ♀
05 31 ☿ □ ♄
05 52 ☽ ⊼ ♅
08 48 ☽ ± ♆
08 52 ☿ ∥ ♀
10 33 ☽ ± ♀
18 27 ☽ # ♅
19 50 ☽ ⚼ ♂
20 39 ☿ ⊥ ♂
21 55 ☽ ⊼ ♀

30 Sunday
00 46 ☽ ⚼ ♆
03 20 ☽ ✱ ♃
03 46 ☽ ∥ ♇
04 00 ☽ Q ♇
06 04 ☉ ☍ ♇
07 18 ☽ # ♂
09 10 ♀ ⚼ ♇
09 58 ☿ ✱ ♅
12 51 ☽ ± ♂

31 Monday
00 21 ☽ ⊼ ♂
01 24 ☽ Q ♃
01 44 ☽ ✱ ♇
02 43 ☉ ∥ ☿
03 20 ☽ △ ☉
03 57 ☽ △ ♆
07 45 ☽ ⊼ ♄
11 44 ☉ △ ♆
12 08 ☽ ☍ ♅
16 49 ☽ △ ☿
18 54 ☽ ± ♄

LONGITUDES

Date	Sidereal time h m s	Sun ☉ ° ' "	Moon ☽ ° ' "	Moon ☽ 24.00 ° ' "	Mercury ☿ ° '	Venus ♀ ° '	Mars ♂ ° '	Jupiter ♃ ° '	Saturn ♄ ° '	Uranus ♅ ° '	Neptune ♆ ° '	Pluto ♇ ° '
01	08 42 29	09 ♌ 45 41	26 ♐ 00 45	02 ♑ 20 41	17 ♌ 21	24 ♊ 26	07 ♋ 50	25 ♍ 40	10 ♉ 55	13 ♊ 16	08 ♈ 47	07 ♒ 34
02	08 46 25	10 43 05	08 ♑ 37 21	14 ♑ 50 59	19 19	25 19	08 30	25 50	10 57	13 19	08 R 46	07 R 32
03	08 50 22	11 40 30	21 ♑ 01 48	27 ♑ 10 04	21 14	26 13	09 09	26 01	10 59	13 21	08 46	07 31
04	08 54 18	12 37 55	03 ♒ 16 00	09 ♒ 19 48	23 09	27 07	09 49	26 12	11 01	13 23	08 45	07 29
05	08 58 15	13 35 22	15 ♒ 21 43	21 ♒ 21 56	25 02	28 01	10 29	26 23	11 03	13 25	08 44	07 28
06	09 02 11	14 32 49	27 ♒ 20 42	03 ♓ 18 14	26 53	28 56	11 09	26 34	11 05	13 27	08 43	07 27
07	09 06 08	15 30 18	09 ♓ 14 47	15 ♓ 10 39	28 ♌ 43	29 ♊ 52	11 48	26 45	11 06	13 29	08 43	07 25
08	09 10 04	16 27 47	21 ♓ 06 06	27 ♓ 01 27	00 ♍ 31	00 ♋ 48	12 28	26 56	11 08	13 31	08 42	07 24
09	09 14 01	17 25 18	02 ♈ 57 03	08 ♈ 53 18	02 17	01 45	13 07	27 08	11 09	13 33	08 41	07 23
10	09 17 58	18 22 50	14 ♈ 50 35	20 ♈ 49 21	04 02	02 42	13 47	27 19	11 11	13 35	08 40	07 21
11	09 21 54	19 20 23	26 ♈ 50 05	02 ♉ 53 15	05 46	03 39	14 26	27 30	11 12	13 37	08 39	07 20
12	09 25 51	20 17 58	08 ♉ 59 23	15 ♉ 09 01	07 28	04 37	15 06	27 42	11 13	13 39	08 38	07 18
13	09 29 47	21 15 34	21 ♉ 22 40	27 ♉ 40 53	09 08	05 35	15 45	27 53	11 14	13 41	08 37	07 17
14	09 33 44	22 13 12	04 ♊ 04 11	10 ♊ 33 03	10 47	06 34	16 24	28 05	11 15	13 43	08 36	07 16
15	09 37 40	23 10 52	17 ♊ 07 54	23 ♊ 49 07	12 24	07 33	17 04	28 16	11 16	13 44	08 35	07 14
16	09 41 37	24 08 33	00 ♋ 36 58	07 ♋ 31 37	14 00	08 33	17 43	28 28	11 17	13 46	08 34	07 13
17	09 45 33	25 06 15	14 ♋ 33 05	21 ♋ 41 14	15 35	09 33	18 22	28 40	11 17	13 48	08 33	07 12
18	09 49 30	26 03 59	28 ♋ 55 44	06 ♌ 16 07	17 08	10 33	19 01	28 52	11 18	13 49	08 32	07 11
19	09 53 27	27 01 45	13 ♌ 41 39	21 ♌ 11 28	18 39	11 34	19 40	29 03	11 18	13 51	08 31	07 09
20	09 57 23	27 59 31	28 ♌ 44 32	06 ♍ 19 40	20 09	12 35	20 19	29 15	11 19	13 52	08 30	07 08
21	10 01 20	28 57 19	13 ♍ 55 37	21 ♍ 31 06	21 38	13 36	20 58	29 27	11 19	13 54	08 28	07 07
22	10 05 16	29 ♌ 55 09	29 ♍ 04 50	06 ♎ 35 38	23 05	14 37	21 37	29 39	11 19	13 55	08 27	07 05
23	10 09 13	00 ♍ 53 00	14 ♎ 02 27	21 ♎ 24 21	24 30	15 39	22 15	29 ♍ 51	11 R 19	13 56	08 26	07 04
24	10 13 09	01 50 51	28 ♎ 40 39	05 ♏ 50 49	25 54	16 42	22 54	00 ♎ 03	11 19	13 58	08 25	07 03
25	10 17 06	02 48 45	12 ♏ 54 30	19 ♏ 51 35	27 16	17 44	23 33	00 16	11 18	13 59	08 23	07 02
26	10 21 02	03 46 39	26 ♏ 42 05	03 ♐ 26 10	28 37	18 47	24 11	00 28	11 18	14 00	08 22	07 00
27	10 24 59	04 44 35	10 ♐ 04 05	16 ♐ 36 12	29 ♍ 56	19 50	24 50	00 40	11 18	14 01	08 21	06 59
28	10 28 56	05 42 32	23 ♐ 02 56	29 ♐ 24 46	01 ♎ 14	20 53	25 28	00 52	11 17	14 03	08 19	06 58
29	10 32 52	06 40 30	05 ♑ 42 10	11 ♑ 55 37	02 29	21 57	26 07	01 05	11 17	14 04	08 18	06 57
30	10 36 49	07 38 30	18 ♑ 05 36	24 ♑ 12 35	03 43	23 01	26 45	01 17	11 16	14 05	08 17	06 56
31	10 40 45	08 ♍ 36 31	00 ♒ 16 59	06 ♒ 19 13	04 ♎ 55	24 ♋ 05	27 ♋ 23	01 ♎ 29	11 ♉ 15	14 ♊ 06	08 ♈ 15	06 ♒ 55

Moon nodes and latitude / DECLINATIONS

Date	Moon True ☊ ° '	Moon Mean ☊ ° '	Moon ☽ Latitude ° '	Sun ☉ ° '	Moon ☽ ° '	Mercury ☿ ° '	Venus ♀ ° '	Mars ♂ ° '	Jupiter ♃ ° '	Saturn ♄ ° '	Uranus ♅ ° '	Neptune ♆ ° '	Pluto ♇ ° '
01	23 ♑ 13	22 ♑ 12	02 S 25	17 N 48	25 S 47	17 N 18	19 N 26	23 N 47	02 N 47	12 N 46	22 N 21	02 N 08	23 S 29
02	23 D 14	22 09	01 20	17 33	24 29	16 40	19 31	23 45	02 43	12 46	22 21	02 07	23 29
03	23 15	22 06	00 S 12	17 17	22 00	16 02	19 36	23 43	02 38	12 47	22 21	02 07	23 29
04	23 R 14	22 03	00 N 55	17 01	18 32	15 22	19 41	23 41	02 34	12 47	22 22	02 07	23 30
05	23 13	22 00	01 59	16 45	14 20	14 41	19 45	23 38	02 29	12 48	22 22	02 06	23 30
06	23 11	21 57	02 57	16 28	09 37	14 00	19 49	23 36	02 25	12 48	22 22	02 06	23 31
07	23 08	21 53	03 47	16 11	04 S 36	13 18	19 53	23 33	02 20	12 48	22 22	02 06	23 31
08	23 04	21 50	04 26	15 54	00 N 33	12 35	19 57	23 30	02 16	12 48	22 23	02 05	23 31
09	23 00	21 47	04 55	15 37	05 41	11 53	20 00	23 27	02 11	12 49	22 23	02 05	23 32
10	22 56	21 44	05 10	15 19	10 37	11 10	20 03	23 23	02 07	12 49	22 23	02 04	23 32
11	22 53	21 41	05 13	15 01	15 12	10 26	20 05	23 20	02 02	12 49	22 23	02 04	23 33
12	22 52	21 37	05 01	14 43	19 15	09 43	20 07	23 16	01 57	12 49	22 24	02 04	23 33
13	22 D 51	21 34	04 36	14 25	22 32	08 59	20 09	23 12	01 53	12 49	22 24	02 03	23 33
14	22 52	21 31	03 56	14 06	24 49	08 16	20 11	23 08	01 48	12 49	22 24	02 03	23 34
15	22 53	21 28	03 03	13 47	25 51	07 32	20 12	23 04	01 43	12 49	22 24	02 02	23 34
16	22 54	21 25	01 59	13 28	25 25	06 48	20 12	23 00	01 39	12 49	22 24	02 02	23 35
17	22 55	21 22	00 N 46	13 09	23 24	06 05	20 12	22 55	01 34	12 49	22 25	02 01	23 35
18	22 R 56	21 18	00 S 33	12 50	19 50	05 22	20 12	22 50	01 29	12 49	22 25	02 01	23 35
19	22 54	21 15	01 51	12 30	14 56	04 39	20 12	22 45	01 24	12 49	22 25	02 00	23 36
20	22 51	21 12	03 03	12 10	09 03	03 56	20 11	22 40	01 20	12 49	22 25	02 00	23 36
21	22 47	21 09	04 02	11 50	02 N 36	03 13	20 09	22 35	01 15	12 49	22 25	01 59	23 36
22	22 42	21 06	04 45	11 30	03 S 59	02 31	20 07	22 30	01 10	12 48	22 26	01 59	23 37
23	22 36	21 03	05 07	11 10	10 15	01 50	20 05	22 24	01 05	12 48	22 26	01 58	23 37
24	22 32	20 59	05 09	10 49	15 48	01 08	20 02	22 18	01 00	12 48	22 26	01 58	23 38
25	22 28	20 56	04 51	10 28	20 20	00 N 27	19 58	22 12	00 55	12 48	22 26	01 57	23 38
26	22 27	20 53	04 17	10 07	23 35	00 S 13	19 54	22 06	00 50	12 47	22 26	01 57	23 38
27	22 D 26	20 50	03 30	09 46	25 25	00 53	19 50	22 00	00 45	12 47	22 26	01 56	23 38
28	22 27	20 47	02 33	09 25	25 48	01 32	19 45	21 54	00 41	12 47	22 27	01 56	23 39
29	22 29	20 43	01 30	09 04	24 48	02 10	19 40	21 47	00 36	12 46	22 27	01 55	23 39
30	22 30	20 40	00 24	08 42	22 36	02 48	19 34	21 41	00 31	12 46	22 27	01 55	23 39
31	22 ♑ 30	20 ♑ 37	00 N 42	08 N 20	19 S 24	03 S 25	19 N 28	21 N 34	00 N 26	12 N 45	22 N 27	01 N 54	23 S 40

ZODIAC SIGN ENTRIES

Date	h	m	Planets
01	19	33	☽ ♑
04	05	34	☽ ♒
06	17	21	☽ ♓
07	15	26	♀ ♋
08	05	10	☿ ♍
09	06	02	☽ ♈
11	18	17	☽ ♉
14	04	22	☽ ♊
16	10	55	☽ ♋
18	13	46	☽ ♌
20	14	00	☽ ♍
22	13	28	☽ ♎
22	14	01	☉ ♍
24	05	08	♃ ♎
24	14	12	☽ ♏
26	17	51	☽ ♐
27	13	08	☿ ♎
29	01	07	☽ ♑
31	11	26	☽ ♒

LATITUDES

Date	Mercury ☿ ° '	Venus ♀ ° '	Mars ♂ ° '	Jupiter ♃ ° '	Saturn ♄ ° '	Uranus ♅ ° '	Neptune ♆ ° '	Pluto ♇ ° '
01	01 N 46	03 S 53	00 N 34	01 N 09	02 S 27	00 S 03	01 S 28	05 S 17
04	01 39	03 43	00 36	01 09	02 28	00 03	01 29	05 17
07	01 28	03 33	00 38	01 09	02 29	00 03	01 29	05 17
10	01 13	03 22	00 40	01 08	02 29	00 03	01 29	05 17
13	00 55	03 10	00 42	01 08	02 30	00 03	01 29	05 17
16	00 33	02 58	00 44	01 08	02 31	00 03	01 29	05 17
19	00 N 10	02 45	00 46	01 07	02 32	00 03	01 29	05 18
22	00 S 15	02 32	00 48	01 07	02 32	00 03	01 29	05 18
25	00 41	02 19	00 50	01 07	02 33	00 02	01 30	05 18
28	01 08	02 05	00 52	01 07	02 34	00 02	01 30	05 18
31	01 S 35	01 S 51	00 N 54	01 N 07	02 S 35	00 S 02	01 S 30	05 S 18

DATA

Julian Date	2461985
Delta T	+73 seconds
Ayanamsa	24° 15' 40"
Synetic vernal point	04° ♓ 51' 19"
True obliquity of ecliptic	23° 26' 11"

MOON'S PHASES, APSIDES AND POSITIONS ☽

Date	h	m	Phase	Longitude ° '	Eclipse Indicator
05	08	10	○	13 ♒ 26	
13	11	45	☾	21 ♉ 15	
20	10	44	●	27 ♌ 56	
27	01	36	☽	04 ♐ 19	

Day	h	m		
08	12	41	Apogee	
21	04	07	Perigee	
01	00	43	Max dec	25° S 57'
08	09	25	ON	
15	17	13	Max dec	25° N 54'
21	21	25	OS	
28	06	28	Max dec	25° S 50'

LONGITUDES

Date	Chiron ⚷ ° '	Ceres ⚳ ° '	Pallas ⚴ ° '	Juno ⚵ ° '	Vesta ⚶ ° '	Black Moon Lilith ⚸ ° '
01	08 ♉ 33	14 ♏ 10	06 ♎ 27	07 ♍ 21	15 ♍ 02	16 ♓ 18
11	08 ♉ 35	16 ♏ 21	10 ♎ 15	11 ♍ 22	19 ♍ 46	17 ♓ 25
21	08 ♉ 33	18 ♏ 52	14 ♎ 11	15 ♍ 23	24 ♍ 37	18 ♓ 32
31	08 ♉ 24	21 ♏ 42	18 ♎ 14	19 ♍ 23	29 ♍ 32	19 ♓ 39

All ephemeris data is given at 12.00 UT and the Moon's longitude is additionally given for 24.00 UT

ASPECTARIAN

h m Aspects

01 Tuesday
02 42 ♂ ⊼ ♇
05 31 ☽ ∠ ♇
08 49 ☽ ☍ ♀
09 27 ☽ Ᵽ ☉
11 20 ☽ □ ♃
11 48 ☽ Ᵽ ♄
22 29 ☽ ⊥ ♇

02 Wednesday
02 14 ☽ Ᵽ ☿
03 55 ☽ ± ☉
09 55 ☽ ⊻ ♇
11 44 ☽ ☍ ♂
12 17 ☽ □ ♆
15 45 ☉ ∠ ♃
16 22 ☽ ⊼ ☉
16 29 ☽ △ ♄
17 57 ☉ □ ♄
19 14 ☿ ⊥ ♃
20 25 ☽ ⋕ ♂
21 03 ☽ ⊼ ♅
21 56 ♂ □ ♆
22 41 ☽ ± ☿
22 56 ☽ ∥ ♇

03 Thursday
05 30 ♀ □ ♃
05 32 ♀ ∠ ♄
05 40 ♃ Ᵽ ♄
08 43 ☽ ± ♅
09 04 ☽ ⋕ ♅
12 29 ☽ ⊼ ☿
21 54 ☽ △ ♃
22 56 ☽ ⊼ ♀
23 11 ☽ Q ♆

04 Friday
02 21 ☽ Ᵽ ♅
04 48 ☽ ⋕ ♀
11 40 ☽ ± ♀
19 35 ☿ Ᵽ ♆
20 20 ☽ ☌ ♇
21 44 ☽ ⋕ ☉
22 50 ☽ ⚹ ♆

05 Saturday
01 44 ☽ ⊼ ♂
03 23 ☽ □ ♄
03 58 ☽ Ᵽ ♃
06 57 ☽ Ᵽ ♀
07 33 ☉ ⚹ ♅
08 07 ☽ △ ♅
08 10 ☽ ☍ ☉
09 47 ☽ ⋕ ☿
14 22 ☽ ± ♂
17 08 ☿ Q ♅
20 05 ☽ ⋕ ♄
21 06 ☿ ∠ ♂
22 12 ☽ ± ♃

06 Sunday
04 44 ☽ ∠ ♆
07 28 ☿ ⊻ ♃
09 22 ♂ ⚹ ♄
09 27 ☽ Ᵽ ♂
10 25 ☽ ⊼ ♃
10 54 ☽ ☍ ☿
15 29 ☽ △ ♀
15 29 ☽ Q ♄
22 49 ☽ ⊥ ♆

07 Monday
08 19 ☽ ⊻ ♇
10 55 ☽ ⊻ ♆
15 46 ☽ ⚹ ♄
17 29 ☽ △ ♂
20 26 ☽ ⊥ ♇
20 36 ☽ □ ♅
22 43 ☽ ⋕ ♃
23 43 ☽ ⋕ ♆

08 Tuesday
01 19 ♂ ⋕ ♇
01 47 ☽ ⊼ ☉
04 46 ☿ ∥ ♄
14 37 ☽ ∠ ♇
15 00 ☽ ± ☉
19 07 ☽ ∥ ♆
19 50 ☽ ∥ ♃
20 16 ☿ ⚹ ♀
22 13 ☽ ∠ ♄

09 Wednesday
00 01 ☽ ☍ ♃
02 52 ♀ ± ♇
09 10 ☽ Q ♅
09 21 ☽ □ ♀
10 25 ☽ ⊼ ☿
10 50 ☽ Ᵽ ☉
16 28 ☽ ⊥ ♄
17 20 ☿ ± ♆
20 56 ☽ ⚹ ♇
23 34 ☽ ☌ ♆

10 Thursday
00 39 ☽ ± ☿
04 32 ♂ ⊻ ♅
04 36 ☽ ⊻ ♄
09 28 ☽ ⚹ ♅
09 44 ☽ □ ♂
14 23 ☽ ∥ ☿
19 44 ☽ △ ☉
21 02 ☽ Q ♇
21 51 ☽ Ᵽ ☿
23 14 ☽ ∥ ♄

11 Friday
00 46 ☽ Q ♀
00 47 ♃ ∥ ♆
11 05 ☽ ∥ ☉
13 21 ☽ ⊼ ♃
15 33 ☽ ∠ ♅
23 45 ☽ Q ♂

12 Saturday
01 26 ☽ ± ♃
02 40 ☽ ⚹ ♀
08 31 ☽ △ ☿
08 43 ☽ □ ♇
09 22 ☽ ⊥ ♅
09 53 ☿ ⊼ ♇
11 18 ☽ ⊻ ♆
16 22 ☽ ☌ ♄
17 54 ☽ ∥ ♀
19 21 ☽ Ᵽ ♃
21 07 ☽ ⊻ ♅
22 59 ☽ ⊥ ♆

13 Sunday
00 34 ☽ ⚹ ♂
04 39 ☿ ⊼ ♆
10 22 ☽ ∠ ♀
10 51 ☽ ∥ ♅
11 45 ☽ □ ☉
16 17 ☽ ∠ ♆
17 49 ☽ ∥ ♂
19 05 ♂ Q ♃
21 22 ☽ ⋕ ♇

14 Monday
00 35 ☽ △ ♃
04 54 ☽ ⊥ ♀
06 45 ☽ ∠ ♂
07 37 ☉ ⊥ ♃
17 02 ☽ ⊻ ♀
17 56 ☽ △ ♇
18 57 ☿ △ ♄
20 24 ☽ ⚹ ♆

15 Tuesday
00 18 ☽ Q ☉
00 21 ☽ ⊥ ♂
01 18 ☽ ⊻ ♄
02 11 ☽ □ ☿
02 30 ☉ ⊥ ♂
04 33 ♀ ⊼ ♇
05 49 ☽ ☌ ♅
11 52 ☽ ⊻ ♂
12 14 ☽ ⊥ ♄
18 13 ☽ Q ♆
21 10 ☽ Ᵽ ♇
21 53 ☉ Ᵽ ♆
23 42 ☽ ⚹ ☉

16 Wednesday
00 20 ☿ ± ♇
04 21 ☽ ∠ ♄
08 11 ☽ □ ♃
08 20 ☿ □ ♅
12 27 ♀ □ ♆
13 03 ☽ ± ♇
14 45 ☽ Q ☿
23 27 ☽ ⊼ ♇

17 Thursday
01 46 ☽ □ ♆
02 49 ☽ ☌ ♀
03 52 ☽ ∠ ☉
06 27 ☽ ⚹ ♄
10 24 ☽ ⋕ ♇
10 43 ☽ ⊻ ♅
13 57 ☽ ⚹ ☿
15 37 ☽ Q ♃
16 01 ☽ ∥ ♂
18 44 ☽ ☌ ♂
19 44 ☽ ∥ ♅
20 14 ☽ ⊥ ☉
20 51 ☽ ⊥ ♅

18 Friday
02 41 ☽ Q ♄
05 42 ☉ Q ♅
06 56 ☽ ⊻ ☉
09 55 ☽ ∥ ♀
11 49 ☽ ∠ ♅
11 53 ☽ ⚹ ♃
12 47 ☉ ∥ ♄
17 52 ☽ ∠ ☿

19 Saturday
01 27 ☽ ☍ ♇
03 39 ☽ △ ♆
05 52 ♀ ⚹ ♄
08 09 ☽ □ ♄
08 19 ☽ ⊻ ♀
10 08 ☽ ⊥ ☿
12 15 ☽ ⚹ ♅
12 36 ☽ ∠ ♃
18 39 ☽ ⊥ ♀
18 59 ♂ ⊥ ♅
20 50 ☽ ⊻ ☿
21 03 ☽ ∥ ♄
22 00 ☽ ⊻ ♂
22 58 ☽ ∥ ☉

20 Sunday
03 10 ☽ ⊥ ♃
03 40 ☽ Ᵽ ♆
07 26 ☽ Q ♅
07 59 ☽ ⊥ ♂
10 01 ☽ ∠ ♀
10 44 ☽ ☌ ☉
12 50 ☽ ⊻ ♃
16 36 ☿ ⚹ ♂
17 56 ☽ ± ♆
22 52 ☽ ∠ ♂

21 Monday
01 15 ☽ ⊼ ♇
03 24 ☽ ⊼ ♆
07 52 ☽ △ ♄
09 25 ☽ ∥ ☿
10 43 ☽ ± ♇
11 27 ☽ ⚹ ♀
11 57 ☽ □ ♅
14 12 ☽ ∥ ♆
16 57 ☽ ∥ ♃
19 09 ♀ ⊻ ♅
19 50 ☿ Ᵽ ♇
23 37 ☽ ⚹ ♂

22 Tuesday
00 55 ☽ Ᵽ ♇
01 28 ☽ ☌ ☿
01 47 ☽ ⋕ ♃
03 43 ☉ ⊻ ♃
04 39 ☽ ⋕ ♆
07 08 ☽ ⋕ ☿
07 36 ☽ Ᵽ ♄
07 49 ☽ Q ♀
12 56 ☽ ☌ ♃
13 26 ☽ ⊻ ☉
19 33 ☽ Q ♂
21 57 ☽ ± ♄
22 17 ♄ St R
23 40 ☽ ⊥ ☉

23 Wednesday
00 47 ☽ △ ♇
02 58 ☽ ☍ ♆
05 17 ♂ ∥ ♅
06 51 ☿ ∥ ♆
07 35 ☽ ⊼ ♄
11 50 ☽ △ ♅
14 49 ☽ □ ♀
15 12 ☽ ∠ ☉
15 29 ☽ ⋕ ☉
22 34 ☽ ⋕ ♄

24 Thursday
02 01 ☽ □ ♂
06 55 ☽ ⊻ ☿
12 29 ☽ Ᵽ ♅
14 20 ☽ ⊻ ♃
17 24 ☿ ∥ ♃
17 40 ☽ ⚹ ☉
17 57 ☽ ⊥ ☿
19 06 ☿ Ᵽ ♄

25 Friday
00 32 ☽ ⊥ ♃
01 43 ☉ ± ♆
02 01 ☽ □ ♇
03 15 ♂ Q ♄
03 36 ☽ ± ♅
04 19 ☽ ⊼ ♆
09 16 ☽ ☍ ♄
09 53 ☽ ⋕ ♀
10 48 ☽ ∠ ☿
13 51 ☽ ⊼ ♅
14 32 ☽ ± ♆
15 31 ☽ Q ☉
16 06 ☽ ∠ ♃
20 59 ☽ △ ♀

26 Saturday
00 15 ☽ ⋕ ♂
02 21 ☽ ⋕ ♅
02 59 ♀ Q ♃
06 08 ☽ Ᵽ ♆
07 21 ☽ △ ♂
09 01 ☽ Q ♇
11 12 ☉ ∠ ♀
12 33 ☽ ∥ ♇
15 46 ☽ ⚹ ☿
18 47 ☽ ⚹ ♃

27 Sunday
01 36 ☽ □ ☉
01 42 ☽ Ᵽ ♀
06 25 ☽ ⚹ ♇
08 10 ☿ ⋕ ♃
08 52 ☽ △ ♆
11 33 ☽ Ᵽ ♂
14 14 ☽ ⊼ ♄
15 48 ☽ Q ☿
16 28 ♀ ⊥ ♅
16 50 ☽ Q ♃
19 15 ☽ ☍ ♅
19 30 ☽ ± ♀

28 Monday
01 16 ☽ ± ♄
04 02 ☿ ☌ ♃
04 58 ☽ ± ♂
07 36 ☽ ⊼ ♀
09 59 ☽ ∠ ♇
16 47 ☽ ⊼ ♂
18 05 ☽ Ᵽ ♄

29 Tuesday
02 46 ☿ ⋕ ♆
02 56 ☽ ⊥ ♇
03 01 ☽ □ ♃
05 10 ☽ □ ☿
14 01 ☽ △ ☉
14 23 ☽ ⊻ ♇
16 59 ☽ □ ♆
18 40 ☉ ⊼ ♇
22 44 ☽ △ ♄

30 Wednesday
02 03 ☽ ∥ ♇
04 10 ☽ ⊼ ♅
13 24 ☽ ⋕ ♅
15 53 ☽ ± ♅
17 35 ♀ Q ♄
20 00 ☽ ⋕ ♂
21 41 ☽ Ᵽ ☉
22 34 ☽ ☍ ♀

31 Thursday
03 28 ☉ ⊼ ♆
04 03 ☽ Q ♆
05 57 ☽ ☍ ♂
09 38 ☽ Ᵽ ♅
11 35 ☽ ⋕ ♀
14 26 ☽ △ ♃
17 01 ☽ ± ☉
18 38 ☿ ± ♄
22 12 ☽ △ ☿

SEPTEMBER 2028

LONGITUDES

Date	Sidereal time h m s	Sun ☉ ° ' "	Moon ☽ ° ' "	Moon ☽ 24.00 ° ' "	Mercury ☿ ° '	Venus ♀ ° '	Mars ♂ ° '	Jupiter ♃ ° '	Saturn ♄ ° '	Uranus ♅ ° '	Neptune ♆ ° '	Pluto ♇ ° '
01	10 44 42	09 ♍ 34 33	12 ♒ 19 38	18 ♒ 18 36	06 ♎ 05	25 ♋ 09	28 ♋ 02	01 ♎ 42	11 ♉ 14	14 ♊ 07	08 ♈ 14	06 ♒ 53
02	10 48 38	10 32 37	24 ♒ 16 24	00 ♓ 13 18	07 13	26 14	28 40	01 54	11 R 13	14 07	08 R 13	06 R 52
03	10 52 35	11 30 43	06 ♓ 09 34	12 ♓ 05 25	08 19	27 19	29 18	02 07	11 12	14 08	08 11	06 51
04	10 56 31	12 28 50	18 ♓ 01 05	23 ♓ 56 44	09 22	28 24	29 ♋ 56	02 19	11 11	14 09	08 10	06 50
05	11 00 28	13 26 59	29 ♓ 52 35	05 ♈ 48 51	10 23	29 ♋ 29	00 ♌ 34	02 32	11 09	14 10	08 08	06 49
06	11 04 25	14 25 09	11 ♈ 45 45	17 ♈ 43 31	11 22	00 ♌ 35	01 12	02 44	11 08	14 10	08 07	06 48
07	11 08 21	15 23 22	23 ♈ 42 25	29 ♈ 42 45	12 18	01 41	01 50	02 57	11 06	14 11	08 05	06 47
08	11 12 18	16 21 36	05 ♉ 44 50	11 ♉ 49 01	13 11	02 47	02 28	03 10	11 04	14 12	08 04	06 46
09	11 16 14	17 19 53	17 ♉ 55 44	24 ♉ 05 23	14 01	03 53	03 05	03 22	11 03	14 12	08 02	06 45
10	11 20 11	18 18 12	00 ♊ 18 26	06 ♊ 35 21	14 48	04 59	03 43	03 35	11 01	14 13	08 01	06 44
11	11 24 07	19 16 32	12 ♊ 56 40	19 ♊ 22 53	15 31	06 06	04 21	03 48	10 59	14 13	07 59	06 43
12	11 28 04	20 14 55	25 ♊ 54 28	02 ♋ 31 53	16 10	07 13	04 58	04 01	10 57	14 13	07 58	06 42
13	11 32 00	21 13 20	09 ♋ 15 34	16 ♋ 05 50	16 46	08 20	05 36	04 13	10 55	14 14	07 56	06 41
14	11 35 57	22 11 47	23 ♋ 02 55	00 ♌ 06 55	17 17	09 27	06 13	04 26	10 52	14 14	07 54	06 40
15	11 39 54	23 10 16	07 ♌ 17 45	14 ♌ 35 09	17 43	10 34	06 51	04 39	10 50	14 14	07 53	06 40
16	11 43 50	24 08 48	21 ♌ 58 40	29 ♌ 27 35	18 05	11 42	07 28	04 52	10 48	14 14	07 51	06 39
17	11 47 47	25 07 21	07 ♍ 00 59	14 ♍ 37 46	18 21	12 50	08 05	05 05	10 45	14 14	07 50	06 38
18	11 51 43	26 05 56	22 ♍ 16 38	29 ♍ 56 12	18 31	13 58	08 43	05 18	10 42	14 14	07 48	06 37
19	11 55 40	27 04 33	07 ♎ 35 02	15 ♎ 11 40	18 36	15 06	09 20	05 30	10 40	14 R 14	07 46	06 36
20	11 59 36	28 03 12	22 ♎ 44 46	00 ♏ 13 09	18 R 33	16 14	09 57	05 43	10 37	14 14	07 45	06 36
21	12 03 33	29 ♍ 01 53	07 ♏ 35 48	14 ♏ 51 56	18 25	17 23	10 34	05 56	10 34	14 14	07 43	06 35
22	12 07 29	00 ♎ 00 36	22 ♏ 00 59	29 ♏ 02 38	18 09	18 31	11 11	06 09	10 31	14 14	07 41	06 34
23	12 11 26	00 59 21	05 ♐ 56 48	12 ♐ 43 34	17 46	19 40	11 48	06 22	10 28	14 14	07 40	06 33
24	12 15 23	01 58 07	19 ♐ 23 10	25 ♐ 55 59	17 16	20 49	12 25	06 35	10 25	14 14	07 38	06 33
25	12 19 19	02 56 55	02 ♑ 22 31	08 ♑ 43 19	16 38	21 58	13 01	06 48	10 21	14 13	07 36	06 32
26	12 23 16	03 55 45	14 ♑ 58 58	21 ♑ 10 05	15 54	23 07	13 38	07 01	10 18	14 13	07 35	06 31
27	12 27 12	04 54 36	27 ♑ 17 18	03 ♒ 21 14	15 03	24 17	14 15	07 14	10 15	14 13	07 33	06 31
28	12 31 09	05 53 29	09 ♒ 22 27	15 ♒ 21 32	14 06	25 26	14 51	07 27	10 11	14 12	07 31	06 30
29	12 35 05	06 52 24	21 ♒ 19 00	27 ♒ 15 19	13 04	26 36	15 28	07 40	10 08	14 12	07 30	06 30
30	12 39 02	07 ♎ 51 20	03 ♓ 10 54	09 ♓ 06 08	11 ♎ 59	27 ♌ 45	16 ♌ 04	07 ♎ 53	10 ♉ 04	14 ♊ 11	07 ♈ 28	06 ♒ 29

MOON NODES AND LATITUDE

Date	Moon True ☊ ° '	Moon Mean ☊ ° '	Moon ☽ Latitude ° '
01	22 ♑ 28	20 ♑ 34	01 N 45
02	22 R 23	20 31	02 43
03	22 17	20 28	03 33
04	22 09	20 24	04 14
05	21 59	20 21	04 43
06	21 49	20 18	05 01
07	21 39	20 15	05 05
08	21 31	20 12	04 56
09	21 25	20 09	04 34
10	21 21	20 05	03 58
11	21 19	20 02	03 10
12	21 D 19	19 59	02 12
13	21 20	19 56	01 N 04
14	21 R 20	19 53	00 S 09
15	21 19	19 49	01 24
16	21 16	19 46	02 36
17	21 10	19 43	03 38
18	21 02	19 40	04 26
19	20 53	19 37	04 55
20	20 43	19 34	05 02
21	20 34	19 30	04 50
22	20 27	19 27	04 18
23	20 22	19 24	03 33
24	20 19	19 21	02 36
25	20 19	19 18	01 34
26	20 D 19	19 15	00 S 28
27	20 R 19	19 11	00 N 37
28	20 17	19 08	01 39
29	20 14	19 05	02 36
30	20 ♑ 07	19 ♑ 02	03 N 26

DECLINATIONS

Date	Sun ☉ ° '	Moon ☽ ° '	Mercury ☿ ° '	Venus ♀ ° '	Mars ♂ ° '	Jupiter ♃ ° '	Saturn ♄ ° '	Uranus ♅ ° '	Neptune ♆ ° '	Pluto ♇ ° '
01	07 N 59	15 S 25	04 S 01	19 N 21	21 N 27	00 N 21	12 N 45	22 N 27	01 N 53	23 S 40
02	07 37	10 52	04 36	19 14	21 20	00 16	12 44	22 27	01 53	23 40
03	07 15	05 57	05 10	19 06	21 13	00 11	12 44	22 27	01 52	23 41
04	06 53	00 S 51	05 44	18 58	21 05	00 06	12 43	22 27	01 52	23 41
05	06 30	04 N 17	06 16	18 49	20 58	00 N 01	12 42	22 27	01 51	23 41
06	06 08	09 16	06 47	18 39	20 50	00 S 04	12 42	22 28	01 50	23 41
07	05 45	13 55	07 17	18 30	20 42	00 10	12 41	22 28	01 50	23 42
08	05 23	18 05	07 45	18 19	20 34	00 15	12 40	22 28	01 49	23 42
09	05 00	21 33	08 12	18 08	20 26	00 20	12 39	22 28	01 49	23 42
10	04 38	24 05	08 38	17 57	20 18	00 25	12 39	22 28	01 48	23 42
11	04 15	25 30	09 02	17 45	20 10	00 30	12 38	22 28	01 47	23 43
12	03 52	25 34	09 24	17 33	20 01	00 35	12 37	22 28	01 47	23 43
13	03 29	24 11	09 44	17 20	19 53	00 40	12 36	22 28	01 46	23 43
14	03 06	21 19	10 02	17 07	19 44	00 45	12 35	22 28	01 45	23 43
15	02 43	17 05	10 17	16 53	19 36	00 50	12 34	22 28	01 45	23 43
16	02 19	11 43	10 30	16 39	19 27	00 55	12 33	22 28	01 44	23 44
17	01 56	05 N 33	10 41	16 24	19 18	01 00	12 32	22 28	01 43	23 44
18	01 33	01 S 00	10 48	16 09	19 09	01 06	12 31	22 28	01 43	23 44
19	01 10	07 31	10 53	15 53	18 59	01 11	12 30	22 28	01 42	23 44
20	00 46	13 31	10 54	15 37	18 50	01 16	12 29	22 28	01 41	23 44
21	00 N 23	18 36	10 51	15 20	18 41	01 21	12 28	22 28	01 41	23 44
22	00 00	22 26	10 44	15 03	18 31	01 26	12 27	22 28	01 40	23 44
23	00 S 24	24 47	10 34	14 45	18 22	01 31	12 26	22 28	01 39	23 45
24	00 47	25 37	10 19	14 28	18 12	01 36	12 25	22 28	01 39	23 45
25	01 10	24 59	10 00	14 09	18 02	01 41	12 24	22 28	01 38	23 45
26	01 34	23 04	09 36	13 50	17 52	01 47	12 23	22 28	01 37	23 45
27	01 57	20 06	09 08	13 31	17 42	01 52	12 21	22 28	01 37	23 45
28	02 20	16 19	08 36	13 11	17 32	01 57	12 20	22 28	01 36	23 45
29	02 44	11 56	08 01	12 51	17 22	02 02	12 19	22 28	01 35	23 45
30	03 S 07	07 S 08	07 S 22	12 N 31	17 N 12	02 S 07	12 N 18	22 N 28	01 N 35	23 S 45

ZODIAC SIGN ENTRIES

Date	h	m	Planets
02	23	33	☽ ♓
04	14	36	♂ ♌
05	12	15	☽ ♈
05	23	18	♀ ♌
08	00	34	☽ ♉
10	11	25	☽ ♊
12	19	26	☽ ♋
14	23	48	☽ ♌
17	00	52	☽ ♍
19	00	06	☽ ♎
20	23	39	☽ ♏
22	11	45	☉ ♎
23	01	39	☽ ♐
25	07	33	☽ ♑
27	17	21	☽ ♒
30	05	33	☽ ♓

LATITUDES

Date	Mercury ☿ ° '	Venus ♀ ° '	Mars ♂ ° '	Jupiter ♃ ° '	Saturn ♄ ° '	Uranus ♅ ° '	Neptune ♆ ° '	Pluto ♇ ° '
01	01 S 44	01 S 47	00 N 55	01 N 07	02 S 35	00 S 02	01 S 30	05 S 18
04	02 12	01 33	00 57	01 06	02 35	00 02	01 30	05 18
07	02 38	01 19	00 59	01 06	02 36	00 02	01 30	05 18
10	03 02	01 06	01 01	01 06	02 37	00 02	01 30	05 18
13	03 24	00 53	01 03	01 06	02 37	00 02	01 30	05 18
16	03 42	00 39	01 05	01 06	02 38	00 02	01 30	05 17
19	03 53	00 27	01 07	01 06	02 39	00 02	01 30	05 17
22	03 55	00 14	01 09	01 06	02 39	00 02	01 30	05 17
25	03 45	00 S 02	01 11	01 06	02 40	00 02	01 30	05 17
28	03 18	00 N 10	01 13	01 06	02 40	00 02	01 30	05 17
31	02 S 35	00 N 21	01 N 15	01 N 06	02 S 41	00 S 02	01 S 30	05 S 17

DATA

Julian Date	2462016
Delta T	+73 seconds
Ayanamsa	24° 15' 44"
Synetic vernal point	04° ♓ 51' 15"
True obliquity of ecliptic	23° 26' 12"

MOON'S PHASES, APSIDES AND POSITIONS ☽

Date	h	m	Phase	Longitude ° '	Eclipse Indicator
03	23	48	○	11 ♓ 59	
12	00	46	☾	19 ♊ 48	
18	18	24	●	26 ♍ 22	
25	13	10	☽	03 ♑ 00	

Day	h	m		
04	19	46	Apogee	
18	14	14	Perigee	
04	15	56	0N	
12	01	17	Max dec	25° N 43'
18	08	21	0S	
24	13	14	Max dec	25° S 37'

LONGITUDES

Date	Chiron ⚷ ° '	Ceres ⚳ ° '	Pallas ⚴ ° '	Juno ⚵ ° '	Vesta ⚶ ° '	Black Moon Lilith ⚸ ° '
01	08 ♉ 23	21 ♏ 59	18 ♎ 39	19 ♍ 47	00 ♎ 02	19 ♓ 46
11	08 ♉ 10	25 ♏ 05	22 ♎ 48	23 ♍ 46	05 ♎ 02	20 ♓ 53
21	07 ♉ 51	28 ♏ 22	27 ♎ 02	27 ♍ 42	10 ♎ 06	22 ♓ 00
31	07 ♉ 28	01 ♐ 50	01 ♏ 20	01 ♎ 36	15 ♎ 14	23 ♓ 07

All ephemeris data is given at 12.00 UT and the Moon's longitude is additionally given for 24.00 UT

ASPECTARIAN

h m	Aspects
01 Friday	
01 09	☽ ☌ ♇
03 50	☽ ✶ ♆
06 01	☽ ⊼ ☉
09 49	☽ □ ♄
15 34	☽ △ ♅
20 55	☽ ⚼ ♃
02 Saturday	
02 26	☽ # ♄
04 44	☿ △ ♇
07 26	☽ ⚼ ☿
09 52	☽ ∠ ♆
15 21	☽ ± ♃
16 20	☽ ⊼ ♀
21 21	☽ ⊼ ♂
21 57	☽ Q ♄
03 Sunday	
03 27	☽ ± ☿
03 40	☽ ⊼ ♃
03 59	☽ ⊥ ♆
04 20	☉ △ ♄
05 19	☽ # ☉
05 40	☽ ± ♀
05 50	♂ ∠ ♅
09 13	☿ ☍ ♆
10 09	☽ ± ♂
13 24	☽ ⊻ ♇
15 19	☽ ∥ ☿
16 05	☽ ⊻ ♆
16 47	☽ ⊼ ☿
22 10	☽ ✶ ♄
23 48	☽ ☍ ☉
04 Monday	
01 32	☽ ⊥ ♇
01 42	☽ ⚼ ♀
04 10	☽ □ ♅
05 24	☽ ⚼ ♂
07 15	☽ # ♆
15 33	☽ # ♃
16 18	☽ ∥ ♃
19 43	☽ ∠ ♇
20 40	☉ ± ♇
05 Tuesday	
00 36	☽ ∥ ♆
04 29	☽ ∠ ♄
04 47	♀ ∠ ♅
11 08	☽ △ ♀
13 28	☽ △ ♂
16 38	☽ Q ♅
17 28	☽ ☍ ♃
18 26	☉ # ☿
21 51	☽ ∥ ☉
22 32	☽ # ☿
22 39	☽ ⊥ ♄
06 Wednesday	
02 00	☽ ✶ ♇
04 39	☽ ☌ ♆
05 50	☉ □ ♅
06 10	☿ ⊼ ♄
10 43	☽ ⊻ ♄
11 08	☽ ☍ ☿
16 52	☽ ✶ ♅
17 49	☽ ⊼ ☉
07 Thursday	
02 09	☽ Q ♇
05 25	☽ ∥ ♄
06 57	☽ ± ☉
19 51	♀ ☌ ♂
22 57	☽ ∠ ♅
08 Friday	
02 31	☽ ⚼ ☉
05 07	☽ □ ♂
05 30	☽ □ ♀
06 46	☽ ⊼ ♃
13 25	☽ ∥ ♀
14 01	☽ □ ♇
16 35	☽ ⊻ ♆
16 51	☽ ⊥ ♅
18 53	☽ ± ♃
22 22	♀ ✶ ♃
22 31	☽ ☌ ♄
09 Saturday	
03 47	☽ ⊼ ☿
04 00	☽ ∥ ♂
04 23	☽ ⊥ ♆
04 41	☽ ⊻ ♅
10 44	☽ △ ☉
12 53	☽ ⚼ ♃
16 22	☽ ± ☿
17 39	☿ △ ♅
18 30	☽ Q ♂
19 36	☽ ∥ ♅
20 28	☽ Q ♀
21 56	☽ ∠ ♆
10 Sunday	
04 19	♂ ✶ ♃
07 37	☽ # ♇
10 57	☽ ⚼ ☿
18 23	☽ △ ♃
18 52	☽ ✶ ♂
21 49	☽ ✶ ♀
11 Monday	
00 16	☽ △ ♇
02 40	☽ ✶ ♆
08 19	☽ ⊻ ♄
14 23	☽ ☌ ♅
16 48	☉ ∠ ♂
17 05	☽ △ ☿
19 31	☽ ⊥ ♄
12 Tuesday	
00 32	☽ ∠ ♂
00 46	☽ □ ☉
01 05	☽ Q ♆
01 12	♀ ☍ ♇
04 18	☽ ⚼ ♆
04 36	☽ ∠ ♀
12 04	☽ ∠ ♄
17 51	☽ ⊥ ♂
20 42	☽ ± ♇
22 30	☽ ⊥ ♀
13 Wednesday	
02 53	☽ □ ♃
03 41	♀ △ ♆
05 10	☽ ⊻ ♂
07 27	☽ ⊼ ♇
09 39	☽ □ ♆
10 12	☽ ⊻ ♀
11 56	☽ Q ☉
14 54	☽ ✶ ♄
16 52	☽ # ♇
20 44	☽ ⊻ ♅
23 19	☉ ⚼ ♇
14 Thursday	
01 41	☽ □ ☿
03 44	☽ ∥ ♅
07 10	☽ ⊥ ♅
10 26	☽ ✶ ☉
10 56	☽ Q ♃
11 42	☽ Q ♄
22 12	☽ ∥ ♂
22 31	☽ ∠ ♅
15 Friday	
04 59	♂ ☍ ♇
07 32	☽ ✶ ♃
09 18	☽ Q ☿
10 57	☽ ☍ ♇
11 13	☽ ☌ ♂
12 58	☽ △ ♆
13 04	☽ ∥ ♀
13 33	☽ ∠ ☉
17 20	♀ □ ♄
17 50	☽ □ ♄
17 52	☽ ☌ ♀
23 26	☽ ✶ ♅
16 Saturday	
05 17	♃ ± ♄
05 22	☽ ⊥ ☉
05 33	☽ ✶ ☿
08 32	☽ ∥ ♄
08 32	☽ ∠ ♃
13 25	☽ ⚼ ♆
15 44	☽ ⊻ ☉
16 48	☽ # ☿
18 51	☽ Q ♅
23 12	☽ ⊥ ♃
17 Sunday	
02 15	♂ △ ♆
03 47	☽ ± ♆
06 06	☽ ∠ ☿
08 53	☽ ⊻ ♃
11 23	☽ ⊼ ♇
13 17	☽ ⊼ ♆
13 46	☽ ⊻ ♂
17 53	☽ △ ♄
20 31	☽ ⊥ ☿
20 51	☽ ± ♇
21 54	☽ ⊻ ♀
23 23	☽ □ ♅
23 37	☽ ⊥ ♂
18 Monday	
01 39	☉ ∥ ♆
02 07	☽ ∥ ♆
02 08	☽ ∥ ☉
02 45	☉ ⚼ ♄
04 29	☽ # ♃
06 03	☽ ⊻ ☿
08 05	☽ ⊥ ♀
10 58	☽ ⚼ ♇
12 19	☽ ∥ ♃
13 52	☽ # ☉
14 20	☽ ∠ ♂
14 33	☽ # ♆
17 21	☽ ⚼ ♄
17 53	♀ ✶ ♅
18 24	☽ ☌ ☉
23 19	☽ ∠ ♀
19 Tuesday	
00 01	♅ St R
07 25	☽ ± ♄
08 41	☽ ☌ ♃
10 28	☽ △ ♇
11 14	☉ # ♃
12 18	☽ ☍ ♆
14 52	☽ ✶ ♂
16 33	☿ St R
16 50	☽ ⊼ ♄
22 30	☽ △ ♅
20 Wednesday	
00 49	☽ ✶ ♀
01 08	☽ ∥ ☿
05 22	☽ ☌ ☿
07 39	☽ # ♄
10 40	☽ Q ♂
20 48	☽ # ♀
21 06	☽ ⊻ ☉
21 32	☽ Q ♀
22 25	☽ ⚼ ♅
21 Thursday	
07 31	☽ ⊥ ☉
09 15	☽ ⊻ ♃
10 20	☽ □ ♇
11 59	♂ □ ♄
12 12	☽ ⊼ ♆
12 23	☽ # ♂
13 03	☽ ± ♅
16 52	☽ ☍ ♄
17 06	☽ □ ♂
19 15	☽ ⊥ ♃
22 04	☽ ± ♆
22 57	☽ ⊼ ♅
23 23	☽ ∠ ☉
22 Friday	
05 36	☽ □ ♀
05 37	☽ ⊻ ☿
05 52	☿ ✶ ♀
10 31	☽ ∠ ♃
12 19	☽ # ♅
13 08	☽ ⚼ ♆
15 32	☽ ⊥ ☿
16 20	☽ Q ♇
23 27	☽ ∥ ♇
23 Saturday	
02 42	☽ ✶ ☉
06 37	☽ ∠ ☿
12 45	☽ ✶ ♃
13 04	☽ ✶ ♇
15 01	☽ △ ♆
19 56	☽ ⊼ ♄
22 50	☽ △ ♂
24 Sunday	
01 27	☽ Q ☉
02 41	☽ ☍ ♅
06 38	☽ ± ♄
07 46	♃ △ ♇
08 19	☽ ✶ ☿
10 31	☽ Q ♃
14 51	☽ △ ♀
15 56	☽ ∠ ♇
22 27	♃ # ♆
22 59	☽ ⚼ ♄
25 Monday	
03 28	☽ ⚼ ♂
05 22	☽ Q ☿
07 46	♀ ∠ ♃
08 33	☽ ⊥ ♇
13 10	☽ □ ☉
19 50	☽ ⊻ ♇
20 30	☽ □ ♃
21 13	☽ ± ♂
21 32	☽ ⚼ ♀
21 51	☽ □ ♆
26 Tuesday	
01 01	♀ ⚼ ♆
03 03	☽ △ ♄
04 55	☽ ∥ ♇
09 16	☽ ⊼ ♂
10 32	☽ ⊼ ♅
13 40	☽ □ ☿
15 47	☉ # ♆
16 33	☽ ± ♀
17 31	☽ # ♅
20 36	☉ ± ♄
22 08	☽ ± ♅
27 Wednesday	
04 54	☉ ∥ ♃
05 28	☽ ⊼ ♀
08 35	☽ Q ♆
10 42	♂ ✶ ♅
15 47	☽ ⚼ ♅
28 Thursday	
00 35	☿ ✶ ♂
04 22	☽ # ♂
04 26	☽ △ ☉
06 17	☽ ☌ ♇
08 05	☽ △ ♃
08 19	☽ ✶ ♆
09 31	☿ △ ♅
13 37	☽ □ ♄
19 15	♃ ☍ ♆
20 44	☽ △ ☿
21 40	☽ △ ♅
23 34	☽ ☍ ♂
29 Friday	
02 52	☉ △ ♇
03 47	♀ Q ♅
06 43	☽ # ♀
09 58	☽ # ♄
13 13	☽ ⚼ ☉
14 22	☽ ∠ ♆
14 47	☽ ⚼ ♃
23 49	☽ ☍ ♀
30 Saturday	
00 30	☽ ⚼ ☿
01 42	☽ Q ♄
02 48	☉ ☍ ♆
03 45	☿ ∠ ♀
08 32	☽ ⊥ ♆
09 04	☽ ± ☉
09 19	☽ ± ♃
10 43	☽ ∥ ☿
12 51	☉ ☌ ♃
17 10	☽ ± ☿
18 42	☽ ⊻ ♇
20 40	☽ ⊻ ♆
21 42	☽ ⊼ ♃
22 20	☽ ⊼ ☉

OCTOBER 2028

LONGITUDES

Date	Sidereal time h m s	Sun ☉ ° ' "	Moon ☽ ° ' "	Moon ☽ 24.00 ° ' "	Mercury ☿ ° '	Venus ♀ ° '	Mars ♂ ° '	Jupiter ♃ ° '	Saturn ♄ ° '	Uranus ♅ ° '	Neptune ♆ ° '	Pluto ♇ ° '
01	12 42 58	08 ♎ 50 19	15 ♓ 01 22	20 ♓ 56 52	10 ♎ 51	28 ♌ 55	16 ♌ 40	08 ♎ 06	10 ♉ 00	14 ♊ 10	07 ♈ 26	06 ♒ 29
02	12 46 55	09 49 19	26 ♓ 52 54	02 ♈ 49 40	09 R 42	00 ♍ 05	17 17	08 19	09 R 56	14 R 10	07 R 25	06 R 28
03	12 50 52	10 48 21	08 ♈ 47 21	14 ♈ 46 06	08 35	01 16	17 53	08 32	09 53	14 09	07 23	06 28
04	12 54 48	11 47 25	20 ♈ 46 04	26 ♈ 47 23	07 30	02 26	18 29	08 45	09 49	14 08	07 21	06 27
05	12 58 45	12 46 32	02 ♉ 50 13	08 ♉ 54 41	06 30	03 36	19 05	08 58	09 45	14 08	07 20	06 27
06	13 02 41	13 45 40	15 ♉ 00 58	21 ♉ 09 16	05 37	04 47	19 41	09 11	09 41	14 07	07 18	06 27
07	13 06 38	14 44 51	27 ♉ 19 48	03 ♊ 32 51	04 52	05 58	20 17	09 24	09 37	14 06	07 16	06 26
08	13 10 34	15 44 04	09 ♊ 48 43	16 ♊ 07 44	04 15	07 08	20 53	09 37	09 32	14 05	07 15	06 26
09	13 14 31	16 43 20	22 ♊ 30 15	28 ♊ 56 42	03 49	08 19	21 29	09 50	09 28	14 04	07 13	06 26
10	13 18 27	17 42 37	05 ♋ 27 27	12 ♋ 02 58	03 34	09 31	22 04	10 03	09 24	14 03	07 12	06 25
11	13 22 24	18 41 57	18 ♋ 43 37	25 ♋ 29 46	03 D 29	10 42	22 40	10 16	09 20	14 02	07 10	06 25
12	13 26 21	19 41 20	02 ♌ 21 43	09 ♌ 19 41	03 35	11 53	23 15	10 28	09 15	14 01	07 08	06 25
13	13 30 17	20 40 44	16 ♌ 23 47	23 ♌ 33 55	03 52	13 04	23 51	10 41	09 11	13 59	07 07	06 25
14	13 34 14	21 40 11	00 ♍ 49 54	08 ♍ 11 17	04 18	14 16	24 26	10 54	09 06	13 58	07 05	06 25
15	13 38 10	22 39 40	15 ♍ 37 27	23 ♍ 07 33	04 55	15 28	25 02	11 07	09 02	13 57	07 03	06 25
16	13 42 07	23 39 12	00 ♎ 40 34	08 ♎ 15 18	05 40	16 39	25 37	11 20	08 57	13 55	07 02	06 25
17	13 46 03	24 38 45	15 ♎ 50 27	23 ♎ 24 39	06 33	17 51	26 12	11 33	08 53	13 54	07 00	06 24
18	13 50 00	25 38 21	00 ♏ 56 34	08 ♏ 24 56	07 34	19 03	26 47	11 46	08 48	13 53	06 59	06 24
19	13 53 56	26 37 58	15 ♏ 48 34	23 ♏ 06 33	08 40	20 15	27 22	11 58	08 43	13 51	06 57	06 D 24
20	13 57 53	27 37 38	00 ♐ 18 05	07 ♐ 22 38	09 53	21 27	27 57	12 11	08 39	13 50	06 56	06 24
21	14 01 50	28 37 19	14 ♐ 19 55	21 ♐ 09 48	11 10	22 39	28 32	12 24	08 34	13 48	06 54	06 25
22	14 05 46	29 ♎ 37 03	27 ♐ 52 22	04 ♑ 27 53	12 32	23 52	29 06	12 37	08 29	13 47	06 53	06 25
23	14 09 43	00 ♏ 36 48	10 ♑ 56 44	17 ♑ 19 23	13 57	25 04	29 ♌ 41	12 49	08 25	13 45	06 51	06 25
24	14 13 39	01 36 34	23 ♑ 36 26	29 ♑ 48 28	15 25	26 16	00 ♍ 16	13 02	08 20	13 43	06 50	06 25
25	14 17 36	02 36 23	05 ♒ 56 09	12 ♒ 00 10	16 56	27 29	00 50	13 15	08 15	13 42	06 48	06 25
26	14 21 32	03 36 13	18 ♒ 01 11	23 ♒ 59 49	18 28	28 42	01 24	13 27	08 10	13 40	06 47	06 25
27	14 25 29	04 36 04	29 ♒ 56 43	05 ♓ 52 29	20 03	29 ♍ 54	01 59	13 40	08 05	13 38	06 45	06 25
28	14 29 25	05 35 57	11 ♓ 47 39	17 ♓ 42 43	21 38	01 ♎ 07	02 33	13 52	08 00	13 36	06 44	06 26
29	14 33 22	06 35 52	23 ♓ 38 10	29 ♓ 34 23	23 15	02 20	03 07	14 05	07 56	13 34	06 42	06 26
30	14 37 19	07 35 49	05 ♈ 31 43	11 ♈ 30 29	24 52	03 33	03 41	14 17	07 51	13 33	06 41	06 26
31	14 41 15	08 ♏ 35 48	17 ♈ 30 54	23 ♈ 33 11	26 ♎ 30	04 ♎ 46	04 ♍ 15	14 ♎ 30	07 ♉ 46	13 ♊ 31	06 ♈ 40	06 ♒ 27

Date	Moon True ☊ ° '	Moon Mean ☊ ° '	Moon ☽ Latitude ° '
01	19 ♑ 58	18 ♑ 59	04 N 07
02	19 R 46	18 55	04 37
03	19 32	18 52	04 54
04	19 18	18 49	04 59
05	19 04	18 46	04 51
06	18 52	18 43	04 30
07	18 43	18 40	03 55
08	18 36	18 36	03 09
09	18 32	18 33	02 13
10	18 31	18 30	01 N 08
11	18 D 31	18 27	00 S 01
12	18 R 30	18 24	01 13
13	18 29	18 21	02 22
14	18 25	18 17	03 23
15	18 19	18 14	04 13
16	18 09	18 11	04 47
17	17 58	18 08	05 00
18	17 47	18 05	04 52
19	17 36	18 01	04 25
20	17 27	17 58	03 41
21	17 21	17 55	02 45
22	17 18	17 52	01 41
23	17 16	17 49	00 S 33
24	17 D 16	17 46	00 N 34
25	17 R 16	17 42	01 37
26	17 15	17 39	02 35
27	17 12	17 36	03 26
28	17 07	17 33	04 07
29	16 59	17 30	04 37
30	16 48	17 26	04 56
31	16 ♑ 36	17 ♑ 23	05 N 02

DECLINATIONS

Date	Sun ☉ ° '	Moon ☽ ° '	Mercury ☿ ° '	Venus ♀ ° '	Mars ♂ ° '	Jupiter ♃ ° '	Saturn ♄ ° '	Uranus ♅ ° '	Neptune ♆ ° '	Pluto ♇ ° '
01	03 S 30	02 S 06	06 S 40	12 N 10	17 N 01	02 S 12	12 N 16	22 N 28	01 N 34	23 S 45
02	03 53	02 N 59	05 57	11 49	16 51	02 17	12 15	22 28	01 34	23 45
03	04 17	08 00	05 13	11 27	16 41	02 22	12 14	22 28	01 33	23 46
04	04 40	12 44	04 29	11 06	16 30	02 27	12 12	22 27	01 32	23 46
05	05 03	17 00	03 47	10 43	16 19	02 32	12 11	22 27	01 32	23 46
06	05 26	20 38	03 07	10 21	16 09	02 38	12 10	22 27	01 31	23 46
07	05 49	23 22	02 31	09 58	15 58	02 43	12 08	22 27	01 30	23 46
08	06 12	25 02	01 58	09 34	15 47	02 48	12 07	22 27	01 30	23 46
09	06 34	25 26	01 30	09 11	15 36	02 53	12 05	22 27	01 29	23 46
10	06 57	24 28	01 07	08 47	15 25	02 58	12 04	22 27	01 28	23 46
11	07 20	22 07	00 50	08 23	15 14	03 03	12 03	22 27	01 28	23 46
12	07 42	18 27	00 39	07 58	15 03	03 08	12 01	22 27	01 27	23 46
13	08 04	13 40	00 32	07 34	14 52	03 13	12 00	22 26	01 26	23 46
14	08 27	08 00	00 32	07 09	14 41	03 18	11 58	22 26	01 26	23 46
15	08 49	01 N 46	00 36	06 43	14 30	03 23	11 57	22 26	01 25	23 46
16	09 11	04 S 39	00 46	06 18	14 19	03 28	11 55	22 26	01 25	23 46
17	09 33	10 51	01 00	05 52	14 07	03 33	11 54	22 26	01 24	23 46
18	09 55	16 22	01 18	05 26	13 56	03 38	11 52	22 26	01 23	23 45
19	10 16	20 48	01 40	05 00	13 45	03 43	11 51	22 26	01 23	23 45
20	10 38	23 49	02 05	04 34	13 33	03 48	11 49	22 25	01 22	23 45
21	10 59	25 15	02 33	04 07	13 22	03 53	11 48	22 25	01 21	23 45
22	11 20	25 06	03 03	03 41	13 10	03 58	11 46	22 25	01 21	23 45
23	11 41	23 32	03 36	03 14	12 59	04 02	11 45	22 25	01 20	23 45
24	12 02	20 49	04 10	02 47	12 47	04 07	11 43	22 25	01 20	23 45
25	12 23	17 13	04 46	02 20	12 36	04 12	11 41	22 24	01 19	23 45
26	12 43	12 58	05 23	01 52	12 24	04 17	11 40	22 24	01 19	23 45
27	13 03	08 17	06 01	01 25	12 13	04 22	11 38	22 24	01 18	23 45
28	13 23	03 S 20	06 40	00 58	12 01	04 27	11 37	22 24	01 18	23 44
29	13 43	01 N 43	07 19	00 30	11 49	04 31	11 35	22 24	01 17	23 44
30	14 03	06 43	07 58	00 02	11 38	04 36	11 34	22 23	01 16	23 44
31	14 S 22	11 N 31	08 S 38	00 S 25	11 N 26	04 S 41	11 N 32	22 N 23	01 N 16	23 S 44

ZODIAC SIGN ENTRIES

Date	h	m	Planets
02	10	08	♀ ♍
02	18	18	☽ ♈
05	06	23	☽ ♉
07	17	10	☽ ♊
10	01	57	☽ ♋
12	07	53	☽ ♌
14	10	38	☽ ♍
16	10	56	☽ ♎
18	10	30	☽ ♏
20	11	30	☽ ♐
22	15	51	☽ ♑
22	21	13	☉ ♏
24	01	10	♂ ♍
25	00	22	☽ ♒
27	12	07	☽ ♓
27	13	52	♀ ♎
30	00	52	☽ ♈

LATITUDES

Date	Mercury ☿ ° '	Venus ♀ ° '	Mars ♂ ° '	Jupiter ♃ ° '	Saturn ♄ ° '	Uranus ♅ ° '	Neptune ♆ ° '	Pluto ♇ ° '
01	02 S 35	00 N 21	01 N 15	01 N 06	02 S 41	00 S 02	01 S 30	05 S 17
04	01 39	00 31	01 17	01 06	02 41	00 02	01 30	05 17
07	00 S 38	00 41	01 19	01 06	02 41	00 02	01 30	05 16
10	00 N 19	00 50	01 21	01 06	02 42	00 02	01 30	05 16
13	01 05	00 59	01 23	01 06	02 42	00 02	01 30	05 16
16	01 37	01 07	01 25	01 06	02 42	00 02	01 30	05 16
19	01 56	01 14	01 27	01 06	02 42	00 02	01 30	05 16
22	02 04	01 21	01 29	01 07	02 42	00 02	01 30	05 15
25	02 03	01 27	01 31	01 07	02 42	00 02	01 30	05 15
28	01 55	01 32	01 33	01 07	02 42	00 02	01 30	05 15
31	01 N 42	01 N 36	01 N 36	01 N 07	02 S 42	00 S 02	01 S 30	05 S 15

DATA

Julian Date	2462046
Delta T	+73 seconds
Ayanamsa	24° 15' 47"
Synetic vernal point	04° ♓ 51' 12"
True obliquity of ecliptic	23° 26' 12"

MOON'S PHASES, APSIDES AND POSITIONS ☽

Date	h	m	Phase	Longitude ° '	Eclipse Indicator
03	16	25	○	10 ♈ 59	
11	11	57	☾	18 ♋ 42	
18	02	57	●	25 ♎ 16	
25	04	53	☽	02 ♒ 19	

Day	h	m		
01	21	26	Apogee	
17	00	33	Perigee	
29	06	49	Apogee	
01	21	55	0N	
09	07	11	Max dec	25° N 28'
15	18	38	0S	
21	21	38	Max dec	25° S 22'
29	03	50	0N	

LONGITUDES

Date	Chiron ⚷ ° '	Ceres ⚳ ° '	Pallas ⚴ ° '	Juno ⚵ ° '	Vesta ⚶ ° '	Black Moon Lilith ⚸ ° '
01	07 ♉ 28	01 ♐ 50	01 ♏ 20	01 ♎ 36	15 ♎ 14	23 ♓ 07
11	07 ♉ 03	05 ♐ 27	05 ♏ 41	05 ♎ 27	20 ♎ 24	24 ♓ 14
21	06 ♉ 35	09 ♐ 11	10 ♏ 04	09 ♎ 13	25 ♎ 37	25 ♓ 21
31	06 ♉ 06	13 ♐ 01	14 ♏ 29	12 ♎ 54	00 ♏ 51	26 ♓ 28

All ephemeris data is given at 12.00 UT and the Moon's longitude is additionally given for 24.00 UT

ASPECTARIAN

h m	Aspects
01 Sunday	
01 53	☽ ⚹ ♄
04 16	☽ ⊼ ☿
04 30	♀ ∥ ♄
05 52	☽ ∥ ☉
06 51	☽ ⊥ ♇
10 17	☽ □ ♅
11 33	☽ ∥ ♃
14 32	☽ # ♆
15 31	☽ ⊼ ♂
02 Monday	
01 04	☽ ∠ ♇
04 19	☽ ± ♂
05 16	☽ ∥ ♆
06 38	☿ ⊼ ♄
08 06	☽ ∠ ♄
08 38	☽ # ♃
10 37	☉ ☌ ☿
14 43	☉ ⊼ ♄
16 37	☽ # ☉
19 11	☽ ⊼ ♀
22 39	☽ Q ♅
23 28	☽ ⚼ ♂
03 Tuesday	
00 18	☽ # ☿
02 10	☽ ⊥ ♄
07 20	☽ ⚹ ♇
08 36	☽ ± ♀
09 11	☽ ☌ ♆
11 28	☽ ☍ ♃
11 36	☽ ☍ ☿
12 48	☿ ☌ ♃
14 10	☽ ⊻ ♄
14 29	♀ ± ♆
16 25	☽ ☍ ☉
22 45	☽ ⚹ ♅
04 Wednesday	
01 58	☿ ⊥ ♀
04 07	☽ ∥ ♀
04 37	☽ ⚼ ♀
07 12	☽ △ ♂
07 23	☽ Q ♇
08 18	☉ ∥ ☿
09 16	☽ ∥ ♄
15 25	☿ ☍ ♆
19 55	♀ ⊥ ♃
05 Thursday	
04 39	☽ ∠ ♅
08 07	☽ ∥ ♂
13 24	☿ △ ♇
13 41	☽ △ ♀
18 44	☽ ⊼ ☿
19 09	☽ □ ♇
20 52	☽ ⊻ ♆
22 26	☽ ⊥ ♅
06 Friday	
00 19	☽ ⊼ ♃
01 34	☽ ☌ ♄
05 45	☽ ± ☿
08 39	☽ ⊥ ♆
09 19	☽ ⊼ ☉
10 14	☽ ⊻ ♅
12 20	☽ ± ♃
20 23	☉ △ ♅
21 36	☽ □ ♂
22 06	☽ ± ☉
22 08	☿ ⊻ ♀
22 17	☽ ⚼ ☿
07 Saturday	
02 12	☽ ∠ ♆
02 56	☽ ∥ ♅
04 15	☿ ∠ ♂
04 47	☿ ∥ ♃
06 12	☽ ⚼ ♃
16 22	☽ # ♇
17 05	☽ ⚼ ☉
21 40	♀ ⊼ ♇
08 Sunday	
01 49	☽ △ ☿
05 33	☽ △ ♇
05 55	♃ ⊼ ♄
06 22	☽ □ ♀
07 06	☽ ⚹ ♆
10 08	☽ Q ♂
11 29	☽ ⊻ ♄
11 37	☽ △ ♃
14 05	♀ ⊼ ♆
20 07	☽ ☌ ♅
22 50	☽ ⊥ ♄
09 Monday	
00 12	☽ △ ☉
05 51	☽ Q ♆
09 59	☽ ⚹ ♂
09 59	☽ ⚼ ♇
13 06	☿ # ♆
15 39	☽ ∠ ♄
19 51	☽ Q ♀
10 Tuesday	
02 45	☽ ± ♇
08 34	☽ □ ☿
09 54	♀ △ ♄
13 46	☽ ⊼ ♇
15 05	☽ ∠ ♂
15 10	☽ □ ♆
16 40	♂ ⚼ ♆
19 09	☽ ⚹ ♄
20 07	☽ ⚹ ♀
20 30	☽ □ ♃
20 46	☽ # ♇
11 Wednesday	
01 15	♀ ⊻ ♃
03 35	☽ ⊻ ♅
08 08	☽ ⊥ ♂
09 16	☽ ∥ ♅
10 28	☿ St D
11 57	☽ □ ☉
14 19	☽ ⊥ ♅
16 36	☽ Q ♄
16 55	☽ Q ☿
19 19	☽ ⊻ ♂
12 Thursday	
01 32	☽ ∠ ♀
05 07	☽ Q ♃
06 10	☽ ∠ ♅
14 09	☽ ⚹ ☿
18 39	☽ ⊥ ♀
19 00	☽ ☍ ♇
20 14	☽ △ ♆
20 18	☉ # ♀
21 53	☽ Q ☉
22 45	♀ ± ♇
23 49	☽ □ ♄
13 Friday	
02 11	☽ ⚹ ♃
05 52	☽ ⊻ ♀
06 12	☽ ∥ ♂
07 56	☽ ⚹ ♅
16 15	☽ ∠ ☿
19 28	☽ ∥ ♄
19 43	☽ ⚹ ☉
21 34	☽ ⚼ ♆
14 Saturday	
01 00	☽ ☌ ♂
03 46	☽ ∠ ♃
04 00	☽ Q ♅
06 07	♀ □ ♅
07 42	☽ ⊥ ☿
10 20	☽ # ☉
12 25	☽ ± ♆
15 40	☽ ∥ ♀
17 54	☽ ⊻ ☿
18 46	☽ ⊥ ♃
21 07	☽ ⊼ ♇
22 11	☽ ⊼ ♆
22 13	☽ ∠ ☉
15 Sunday	
01 25	☽ △ ♄
04 38	☽ ⊻ ♃
06 00	☽ # ♃
06 50	☽ ± ♇
09 18	☽ □ ♅
11 43	☽ ☌ ♀
13 20	☽ ∥ ♆
13 47	☽ ⊥ ☉
16 17	☽ # ☿
21 05	☽ ∥ ☿
21 16	☽ ⚼ ♇
23 54	☽ # ♆
16 Monday	
00 03	☽ ⊻ ☉
01 22	☽ ⚼ ♄
03 38	☽ ⊻ ♂
07 30	☽ ∥ ♃
13 33	☽ ⊥ ♂
15 36	☽ ± ♄
17 50	☽ # ♀
20 22	☽ ☌ ☿
21 05	☽ △ ♇
22 03	☽ ☍ ♆
17 Tuesday	
00 15	♂ Q ♅
01 02	☽ ⊼ ♄
04 22	☽ ∠ ♂
05 07	☽ ☌ ♃
06 29	☽ ∥ ☉
08 17	☿ △ ♇
08 56	☽ △ ♅
15 27	☽ ⊻ ♀
16 17	☽ # ♄
22 49	☿ ☍ ♆
18 Wednesday	
01 20	☽ # ♂
01 48	☽ ⊥ ♀
02 57	☽ ☌ ☉
05 06	☽ ⚹ ♂
08 42	☽ ⚼ ♅
10 30	♂ ∠ ♃
17 25	☽ ∠ ♀
18 10	☿ # ♆
20 46	☽ □ ♇
21 40	☽ ⊼ ♆
23 07	☽ ± ♅
23 28	☽ ⊻ ☿
19 Thursday	
00 33	☽ ☍ ♄
01 07	☽ Q ♂
03 40	♇ St D
05 40	☽ ⊻ ♃
07 21	☽ ± ♆
08 49	☽ ⊼ ♅
09 59	☽ ⊥ ☿
12 56	☿ ⊼ ♄
15 36	☽ ⊥ ♃
19 56	☽ ⚹ ♀
22 04	☽ ⚼ ♆
23 26	☽ # ♅
20 Friday	
02 06	☽ ∠ ☿
02 09	☽ Q ♇
06 42	☽ ∠ ♃
07 11	☽ ⊻ ☉
07 53	☽ □ ♂
11 06	♀ ⚼ ♇
11 23	☽ ∥ ♇
17 49	☽ Q ♀
18 02	☽ ⊥ ☉
22 21	☽ ⚹ ♇
23 13	☽ △ ♆
21 Saturday	
02 06	☽ ⊼ ♄
05 57	☽ ⚹ ☿
06 36	☉ ⚹ ♂
08 05	☉ ⊥ ♀
08 36	☽ ⚹ ♃
10 40	☽ ∠ ☉
11 05	☽ ☍ ♅
12 24	☽ ± ♄
16 16	☉ ⚼ ♅
23 09	♀ # ♃
22 Sunday	
00 26	☽ ∠ ♇
04 05	☽ □ ♀
04 11	☽ ⚼ ♄
05 02	♀ ⚼ ♄
05 19	☽ Q ☿
06 03	☽ Q ♃
13 37	☿ ☌ ♃
14 20	☽ △ ♂
15 25	☽ ⚹ ☉
16 36	☽ ⊥ ♇
23 Monday	
03 17	☿ # ♀
03 35	☽ ⊻ ♇
04 25	☽ □ ♆
07 18	☽ △ ♄
08 43	☿ △ ♅
09 38	☽ ∥ ♇
15 23	☽ Q ☉
15 34	☽ □ ♃
15 39	☉ # ♄
17 15	☽ ⊼ ♅
18 21	☽ □ ☿
19 21	☽ ⚼ ♂
23 05	☽ # ♅
24 Tuesday	
04 35	☽ ± ♅
07 46	☿ ∠ ♂
09 52	☿ ∥ ♃
13 19	☽ ± ♂
14 21	☽ Q ♆
17 42	☽ △ ♀
21 52	☽ ⚼ ♅
25 Wednesday	
01 31	☽ ⊼ ♂
04 53	☽ □ ☉
10 45	♂ ± ♆
12 57	☽ ☌ ♇
13 42	☽ ⚹ ♆
16 32	☽ □ ♄
22 01	☉ # ♂
26 Thursday	
02 24	☽ ⚼ ♀
02 44	☽ △ ♃
03 20	☽ △ ♅
13 03	☽ △ ☿
13 13	☽ ∥ ☉
15 05	☽ # ♂
18 52	☽ # ♄
19 31	☽ ∠ ♆
22 26	☽ ± ♀
27 Friday	
04 16	☽ Q ♄
09 06	♃ △ ♅
09 22	☽ ⚼ ♃
11 55	☽ ⊼ ♀
13 38	☽ ⊥ ♆
16 19	☽ ☍ ♂
18 15	♀ ∥ ♆
21 49	☽ ∥ ☿
22 17	☽ △ ☉
23 55	☽ ⚼ ☿
28 Saturday	
01 07	☽ ⊻ ♇
01 45	☽ ⊻ ♆
03 54	☽ ± ♃
04 23	☽ ⚹ ♄
06 45	☽ ∥ ♃
13 17	☽ ⊥ ♇
15 40	☽ □ ♅
16 18	☽ ⊼ ♃
21 01	☽ ± ☿
21 43	☽ # ♆
29 Sunday	
00 25	☽ # ♀
04 29	♀ ± ♄
06 42	☽ ∥ ♀
07 30	☽ ⚼ ☉
07 32	☽ ∠ ♇
08 00	☉ □ ♇
09 56	☽ ∥ ♆
10 35	☽ ∠ ♄
11 06	☽ ⊼ ☿
14 34	☉ ⊼ ♆
30 Monday	
01 35	☽ # ♃
03 21	☽ ± ☉
04 00	☽ Q ♅
04 38	☽ ⊥ ♄
07 34	☽ ☍ ♀
08 05	☽ ⊼ ♂
10 44	☉ ± ♅
13 50	☽ ⚹ ♇
14 19	☽ ☌ ♆
16 32	☽ ⊼ ☉
16 38	☽ ⊻ ♄
16 47	♀ ⊻ ♂
17 34	☉ ☍ ♄
19 06	☽ # ☿
20 45	☽ ± ♂
21 29	♂ ∥ ♄
31 Tuesday	
04 02	☽ ⚹ ♅
05 52	☽ ☍ ♃
11 37	☽ ∥ ♂
12 07	☽ ∥ ♄
13 51	☽ Q ♇
15 37	☽ ⚼ ♂

NOVEMBER 2028

LONGITUDES

Date	Sidereal time h m s	Sun ☉ ° ' "	Moon ☽ ° ' "	Moon ☽ 24.00 ° ' "	Mercury ☿ ° '	Venus ♀ ° '	Mars ♂ ° '	Jupiter ♃ ° '	Saturn ♄ ° '	Uranus ♅ ° '	Neptune ♆ ° '	Pluto ⯓ ° '
01	14 45 12	09 ♏ 35 48	29 ♈ 37 28	05 ♉ 43 54	28 ♎ 09	05 ♎ 59	04 ♍ 48	14 ♎ 42	07 ♉ 41	13 ♊ 29	06 ♈ 38	06 ♒ 27
02	14 49 08	10 35 50	11 ♉ 52 33	18 ♉ 03 29	29 ♎ 47	07 12	05 22	14 54	07 R 36	13 R 27	06 R 37	06 27
03	14 53 05	11 35 54	24 ♉ 16 46	00 ♊ 32 27	01 ♏ 26	08 25	05 56	15 07	07 32	13 25	06 36	06 28
04	14 57 01	12 36 00	06 ♊ 50 35	13 ♊ 11 14	03 05	09 39	06 29	15 19	07 27	13 23	06 34	06 28
05	15 00 58	13 36 08	19 ♊ 34 32	26 ♊ 00 34	04 44	10 52	07 02	15 31	07 22	13 21	06 33	06 29
06	15 04 54	14 36 18	02 ♋ 29 29	09 ♋ 01 29	06 22	12 06	07 36	15 43	07 17	13 18	06 32	06 29
07	15 08 51	15 36 30	15 ♋ 36 46	22 ♋ 15 32	08 01	13 19	08 09	15 56	07 12	13 16	06 31	06 30
08	15 12 48	16 36 43	28 ♋ 58 02	05 ♌ 44 28	09 39	14 33	08 42	16 08	07 08	13 14	06 30	06 30
09	15 16 44	17 36 59	12 ♌ 35 02	19 ♌ 29 53	11 17	15 46	09 15	16 20	07 03	13 12	06 28	06 31
10	15 20 41	18 37 17	26 ♌ 29 07	03 ♍ 32 43	12 55	17 00	09 47	16 32	06 58	13 10	06 27	06 31
11	15 24 37	19 37 37	10 ♍ 40 36	17 ♍ 52 31	14 32	18 14	10 20	16 44	06 54	13 07	06 26	06 32
12	15 28 34	20 37 59	25 ♍ 08 06	02 ♎ 26 49	16 09	19 28	10 53	16 55	06 49	13 05	06 25	06 33
13	15 32 30	21 38 23	09 ♎ 48 00	17 ♎ 10 50	17 46	20 41	11 25	17 07	06 44	13 03	06 24	06 33
14	15 36 27	22 38 48	24 ♎ 34 26	01 ♏ 57 46	19 23	21 55	11 58	17 19	06 40	13 01	06 23	06 34
15	15 40 23	23 39 16	09 ♏ 19 49	16 ♏ 39 33	20 59	23 09	12 30	17 31	06 35	12 58	06 22	06 35
16	15 44 20	24 39 45	23 ♏ 56 00	01 ♐ 08 15	22 35	24 23	13 02	17 42	06 31	12 56	06 21	06 36
17	15 48 17	25 40 16	08 ♐ 15 34	15 ♐ 17 18	24 11	25 37	13 34	17 54	06 26	12 53	06 20	06 37
18	15 52 13	26 40 48	22 ♐ 12 13	29 ♐ 02 28	25 46	26 52	14 06	18 05	06 22	12 51	06 19	06 37
19	15 56 10	27 41 22	05 ♑ 45 28	12 ♑ 22 06	27 22	28 06	14 37	18 17	06 18	12 49	06 18	06 38
20	16 00 06	28 41 57	18 ♑ 52 33	25 ♑ 17 05	28 ♏ 57	29 ♎ 20	15 09	18 28	06 13	12 46	06 17	06 39
21	16 04 03	29 ♏ 42 34	01 ♒ 36 07	07 ♒ 50 07	00 ♐ 32	00 ♏ 34	15 40	18 39	06 09	12 44	06 16	06 40
22	16 07 59	00 ♐ 43 11	13 ♒ 59 40	20 ♒ 05 19	02 06	01 48	16 11	18 51	06 05	12 41	06 16	06 41
23	16 11 56	01 43 50	26 ♒ 07 43	02 ♓ 07 30	03 41	03 03	16 43	19 02	06 01	12 39	06 15	06 42
24	16 15 52	02 44 29	08 ♓ 05 18	14 ♓ 01 46	05 15	04 17	17 13	19 13	05 57	12 36	06 14	06 43
25	16 19 49	03 45 10	19 ♓ 57 30	25 ♓ 53 07	06 49	05 32	17 44	19 24	05 53	12 34	06 13	06 44
26	16 23 46	04 45 52	01 ♈ 49 11	07 ♈ 46 14	08 23	06 46	18 15	19 35	05 49	12 31	06 13	06 45
27	16 27 42	05 46 35	13 ♈ 44 43	19 ♈ 45 07	09 57	08 00	18 45	19 45	05 45	12 29	06 12	06 46
28	16 31 39	06 47 19	25 ♈ 47 48	01 ♉ 53 04	11 31	09 15	19 16	19 56	05 41	12 26	06 11	06 47
29	16 35 35	07 48 04	08 ♉ 01 14	14 ♉ 12 28	13 04	10 29	19 46	20 07	05 37	12 24	06 11	06 48
30	16 39 32	08 ♐ 48 51	20 ♉ 26 56	26 ♉ 44 45	14 ♐ 38	11 ♏ 44	20 ♍ 16	20 ♎ 17	05 ♉ 34	12 ♊ 21	06 ♈ 10	06 ♒ 49

Date	Moon True ☊ ° '	Moon Mean ☊ ° '	Moon ☽ Latitude ° '
01	16 ♑ 23	17 ♑ 20	04 N 54
02	16 R 10	17 17	04 32
03	16 00	17 14	03 58
04	15 51	17 11	03 11
05	15 45	17 07	02 15
06	15 42	17 04	01 N 10
07	15 41	17 01	00 00
08	15 D 42	16 58	01 S 10
09	15 42	16 55	02 19
10	15 R 42	16 52	03 20
11	15 40	16 48	04 11
12	15 36	16 45	04 47
13	15 30	16 42	05 04
14	15 22	16 39	05 02
15	15 13	16 36	04 39
16	15 05	16 32	03 59
17	14 58	16 29	03 04
18	14 54	16 26	01 59
19	14 51	16 23	00 S 49
20	14 D 51	16 20	00 N 22
21	14 52	16 17	01 29
22	14 54	16 13	02 31
23	14 55	16 10	03 25
24	14 R 55	16 07	04 08
25	14 53	16 04	04 41
26	14 50	16 01	05 02
27	14 44	15 58	05 10
28	14 38	15 54	05 04
29	14 31	15 51	04 44
30	14 ♑ 24	15 ♑ 48	04 N 11

DECLINATIONS

Date	Sun ☉ ° '	Moon ☽ ° '	Mercury ☿ ° '	Venus ♀ ° '	Mars ♂ ° '	Jupiter ♃ ° '	Saturn ♄ ° '	Uranus ♅ ° '	Neptune ♆ ° '	Pluto ⯓ ° '
01	14 S 41	15 N 55	09 S 18	00 S 53	11 N 14	04 S 46	11 N 31	22 N 23	01 N 15	23 S 44
02	15 00	19 43	09 58	01 21	11 03	04 50	11 29	22 23	01 15	23 44
03	15 19	22 41	10 37	01 49	10 51	04 55	11 28	22 22	01 14	23 43
04	15 37	24 36	11 17	02 17	10 39	05 00	11 26	22 22	01 14	23 43
05	15 55	25 16	11 56	02 45	10 28	05 04	11 25	22 22	01 14	23 43
06	16 13	24 35	12 34	03 12	10 16	05 09	11 23	22 22	01 13	23 43
07	16 31	22 32	13 12	03 40	10 04	05 13	11 22	22 21	01 13	23 43
08	16 48	19 13	13 50	04 08	09 53	05 18	11 20	22 21	01 12	23 42
09	17 05	14 49	14 26	04 36	09 41	05 22	11 19	22 21	01 12	23 42
10	17 22	09 33	15 03	05 04	09 29	05 27	11 18	22 21	01 11	23 42
11	17 38	03 N 41	15 38	05 31	09 17	05 31	11 16	22 20	01 11	23 42
12	17 54	02 S 27	16 13	05 59	09 06	05 36	11 15	22 20	01 10	23 42
13	18 10	08 32	16 47	06 27	08 54	05 40	11 13	22 20	01 10	23 41
14	18 26	14 12	17 20	06 54	08 43	05 45	11 12	22 20	01 10	23 41
15	18 41	19 01	17 52	07 21	08 31	05 49	11 11	22 19	01 09	23 41
16	18 56	22 36	18 23	07 48	08 19	05 53	11 09	22 19	01 09	23 41
17	19 10	24 42	18 54	08 15	08 08	05 58	11 08	22 19	01 09	23 40
18	19 25	25 11	19 24	08 42	07 56	06 02	11 07	22 18	01 08	23 40
19	19 39	24 08	19 52	09 09	07 45	06 06	11 05	22 18	01 08	23 40
20	19 52	21 45	20 20	09 36	07 33	06 10	11 04	22 18	01 08	23 39
21	20 05	18 21	20 46	10 02	07 22	06 15	11 03	22 18	01 07	23 39
22	20 18	14 13	21 12	10 28	07 10	06 19	11 02	22 17	01 07	23 39
23	20 30	09 36	21 37	10 54	06 59	06 23	11 00	22 17	01 07	23 38
24	20 42	04 S 41	22 00	11 20	06 48	06 27	10 59	22 17	01 07	23 38
25	20 54	00 N 20	22 23	11 45	06 36	06 31	10 58	22 16	01 06	23 38
26	21 05	05 20	22 44	12 10	06 25	06 35	10 57	22 16	01 06	23 38
27	21 16	10 11	23 04	12 35	06 14	06 39	10 56	22 16	01 06	23 37
28	21 26	14 41	23 23	13 00	06 03	06 43	10 55	22 15	01 06	23 37
29	21 36	18 40	23 41	13 24	05 52	06 47	10 54	22 15	01 05	23 37
30	21 S 46	21 N 53	23 S 57	13 S 48	05 N 41	06 S 50	10 N 53	22 N 15	01 N 05	23 S 36

ZODIAC SIGN ENTRIES

Date	h m	Planets
01	12 44	☽ ♉
02	15 04	☿ ♏
03	22 58	☽ ♊
06	07 24	☽ ♋
08	13 50	☽ ♌
10	17 59	☽ ♍
12	20 00	☽ ♎
14	20 49	☽ ♏
16	22 06	☽ ♐
19	01 42	☽ ♑
21	00 58	♀ ♏
21	04 00	☿ ♐
21	08 56	☽ ♒
21	18 54	☉ ♐
23	19 44	☽ ♓
26	08 19	☽ ♈
28	20 18	☽ ♉

LATITUDES

Date	Mercury ☿ ° '	Venus ♀ ° '	Mars ♂ ° '	Jupiter ♃ ° '	Saturn ♄ ° '	Uranus ♅ ° '	Neptune ♆ ° '	Pluto ⯓ ° '
01	01 N 37	01 N 37	01 N 36	01 N 07	02 S 42	00 S 02	01 S 30	05 S 15
04	01 20	01 41	01 39	01 07	02 42	00 02	01 30	05 14
07	01 02	01 43	01 41	01 08	02 42	00 02	01 30	05 14
10	00 42	01 45	01 43	01 08	02 42	00 02	01 30	05 14
13	00 22	01 46	01 45	01 08	02 41	00 02	01 30	05 14
16	00 N 01	01 46	01 48	01 09	02 41	00 02	01 30	05 13
19	00 S 19	01 46	01 50	01 09	02 41	00 02	01 29	05 13
22	00 38	01 44	01 52	01 09	02 40	00 02	01 29	05 13
25	00 57	01 42	01 55	01 10	02 40	00 02	01 29	05 13
28	01 14	01 40	01 57	01 10	02 39	00 02	01 29	05 13
31	01 S 30	01 N 37	02 N 00	01 N 10	02 S 38	00 S 02	01 S 29	05 S 13

DATA

Julian Date	2462077
Delta T	+73 seconds
Ayanamsa	24° 15' 50"
Synetic vernal point	04° ♓ 51' 09"
True obliquity of ecliptic	23° 26' 11"

MOON'S PHASES, APSIDES AND POSITIONS ☽

Date	h m	Phase	Longitude ° '	Eclipse Indicator
02	09 17	○	10 ♉ 29	
09	21 26	☾	18 ♌ 01	
16	13 18	●	24 ♏ 43	
24	00 15	☽	02 ♓ 15	

Day	h m		
14	05 44	Perigee	
26	00 07	Apogee	
05	12 03	Max dec	25° N 16'
12	02 28	0S	
18	07 10	Max dec	25° S 13'
25	10 24	0N	

LONGITUDES

Date	Chiron ⚷ ° '	Ceres ⚳ ° '	Pallas ⚴ ° '	Juno ⚵ ° '	Vesta ⚶ ° '	Black Moon Lilith ⚸ ° '
01	06 ♉ 03	13 ♐ 24	14 ♏ 55	13 ♎ 15	01 ♏ 23	26 ♓ 35
11	05 ♉ 35	17 ♐ 19	19 ♏ 21	16 ♎ 49	06 ♏ 38	27 ♓ 42
21	05 ♉ 08	21 ♐ 17	23 ♏ 45	20 ♎ 15	11 ♏ 54	28 ♓ 49
31	04 ♉ 44	25 ♐ 19	28 ♏ 09	23 ♎ 32	17 ♏ 10	29 ♓ 56

All ephemeris data is given at 12.00 UT and the Moon's longitude is additionally given for 24.00 UT

ASPECTARIAN

h m	Aspects
01 Wednesday	
04 27	☽ # ☉
08 38	☽ ☍ ☿
09 45	☽ ∠ ♅
16 45	☿ ♇ ♅
21 13	♀ △ ⯓
22 40	☽ △ ♂
02 Thursday	
00 40	♀ ☍ ♆
01 25	☽ □ ⯓
01 45	☽ ⊻ ♆
01 52	☽ ⊼ ♀
03 23	☽ ⊥ ♅
03 43	☽ ☌ ♄
06 58	♀ # ♆
09 17	☽ ☍ ☉
13 26	☽ ⊥ ♆
14 52	☽ ± ♀
15 03	☽ ⊻ ♅
18 00	☽ ⊼ ♃
19 26	♀ ⊼ ♄
03 Friday	
05 48	☽ ± ♃
06 50	☽ ∠ ♆
09 08	☽ ∥ ♅
10 10	☽ ♇ ♀
18 28	☿ # ♂
23 18	☽ # ⯓
23 22	☽ ♇ ♃
04 Saturday	
03 46	☽ ⊼ ☿
11 17	☽ □ ♂
11 18	☽ △ ⯓
11 23	☉ ± ♆
11 26	♂ ⊼ ⯓
11 29	☽ ✱ ♆
13 08	☽ ⊻ ♄
15 46	♂ ⊼ ♆
16 53	☽ ± ☿
17 43	☿ # ♄
17 53	☽ △ ♀
23 49	☽ ⊼ ☉
05 Sunday	
00 19	☽ ☌ ♅
00 24	☽ ⊥ ♄
04 16	☽ △ ♃
06 00	☉ ⊼ ♅
10 05	☽ Q ♆
12 03	☽ ± ☉
12 19	☽ ♇ ☿
15 34	☽ ♇ ⯓
17 11	☽ ∠ ♄
22 39	☽ Q ♂
23 59	♄ ⊥ ♅
06 Monday	
00 19	♂ △ ♄
06 13	☽ ♇ ☉
08 18	☽ ± ⯓
13 43	☿ □ ⯓
14 21	☿ ⊼ ♆
19 21	☽ ⊼ ⯓
19 25	☽ □ ♆
20 10	☽ △ ☿
20 46	☽ ✱ ♄
21 48	☽ ✱ ♂
07 Tuesday	
00 05	☽ # ⯓
00 46	☿ ☍ ♄
01 25	☿ ± ♅
07 24	☽ □ ♀
07 46	☽ ⊻ ♅
11 08	♀ △ ♅
11 59	☽ △ ☉
12 35	☽ □ ♃
13 30	☽ ∥ ♅
14 59	☿ ✱ ♂
18 28	☽ Q ♄
18 36	☽ ⊥ ♅
21 30	☉ ⊻ ♃
08 Wednesday	
01 34	♆ ✱ ⯓
02 11	☽ ∠ ♂
10 42	☽ ∠ ♅
17 25	♀ ⊥ ♂
18 54	☽ ⊥ ♂
18 59	☽ Q ♀
21 17	☽ Q ♃
09 Thursday	
00 58	☽ # ☉
01 18	☽ △ ♆
01 21	☽ ☍ ⯓
02 21	☽ □ ♄
05 55	☽ ⊻ ♂
09 25	☽ □ ☿
13 04	☽ ✱ ♅
13 37	☽ # ☿
18 05	☽ ✱ ♀
18 36	☽ ✱ ♃
21 26	☽ □ ☉
10 Friday	
00 59	♀ ☌ ♃
03 23	☽ ♇ ♆
04 21	☽ ∥ ♄
05 21	☿ ± ♆
09 45	☽ Q ♅
12 15	☽ ∥ ♂
15 38	☿ ⊼ ♅
18 45	☽ ± ♆
20 31	☽ Q ☿
20 42	☽ ∠ ♃
22 16	☽ ∠ ♀
11 Saturday	
04 45	☽ # ♃
04 53	☽ ⊼ ♆
05 02	☽ ⊼ ⯓
05 12	☽ # ♀
05 41	☽ △ ♄
06 30	☽ Q ☉
11 24	☽ ☌ ♂
12 02	♀ ∥ ♃
12 05	☽ ⊥ ♃
14 50	☽ ⊥ ♀
15 07	☽ ± ⯓
16 05	☽ □ ♅
19 16	☽ ✱ ☿
21 52	☽ ∥ ♆
22 14	☽ ⊻ ♃
12 Sunday	
01 45	☽ ⊻ ♀
04 01	☽ ✱ ☉
06 05	☽ ♇ ⯓
06 33	☽ ♇ ♄
07 02	☽ # ♆
15 01	♂ ⊥ ♃
21 17	☽ ± ♄
23 06	☽ ∠ ☿
13 Monday	
00 26	☽ ∥ ♃
01 01	☿ ⊻ ♃
02 56	☽ ∥ ♀
06 22	☉ ♇ ♆
06 28	☽ ☍ ♆
06 28	☽ ∠ ☉
06 43	☽ △ ⯓
07 02	☽ ⊼ ♄
13 25	☽ # ♂
14 44	☽ ⊻ ♂
15 36	☽ ⊥ ☿
17 16	☽ △ ♅
22 12	☽ ⊥ ☉
23 00	☽ # ♄
14 Tuesday	
00 04	☽ ☌ ♃
00 51	☽ ⊥ ♂
02 32	☽ ⊻ ☿
07 18	☽ ☌ ♀
08 39	☽ ⊻ ☉
16 01	☽ ∠ ♂
17 34	☽ ♇ ♅
15 Wednesday	
04 59	☽ ∥ ☿
07 10	☽ ⊼ ♆
07 31	☽ □ ⯓
07 33	☽ ☍ ♄
07 44	☉ ⊥ ♃
08 10	☽ ± ♅
10 05	☽ ∥ ☉
12 54	♄ □ ⯓
16 01	♂ ± ⯓
16 57	☽ ± ♆
17 22	☽ ✱ ♂
17 38	☿ ♇ ♆
17 56	☽ ⊼ ♅
16 Thursday	
01 35	☽ ⊻ ♃
07 44	☽ ♇ ♆
07 51	♂ □ ♅
09 29	☽ ☌ ☿
09 38	☽ # ♅
10 24	☉ Q ⯓
11 37	☽ ⊥ ♃
12 50	☽ ⊻ ♀
13 06	☽ Q ⯓
13 18	☽ ☌ ☉
13 53	☽ Q ♂
21 59	☽ ∥ ⯓
23 46	☽ ⊥ ♀
17 Friday	
02 50	☽ ∠ ♃
06 32	☉ Q ♂
07 07	☿ ⊥ ♃
07 11	♀ # ♂
08 44	☽ △ ♆
08 56	☽ ⊼ ♄
09 12	☽ ✱ ⯓
16 24	☽ ∠ ♀
17 04	☉ ⊻ ♀
18 30	☿ Q ⯓
19 05	☽ ± ♄
19 52	☽ ☍ ♅
21 24	☽ □ ♂
18 Saturday	
04 44	☽ ✱ ♃
10 31	☽ ♇ ♄
10 58	☽ ∠ ⯓
13 48	☉ ∥ ☿
19 03	☽ ⊻ ☿
19 13	☿ Q ♂
20 27	☽ ⊻ ☉
20 57	☽ ✱ ♀
19 Sunday	
02 03	☽ Q ♃
02 50	☽ ⊥ ⯓
06 40	♀ ♇ ♅
07 07	☽ ⊥ ☿
07 59	☽ ⊥ ☉
08 00	♄ ⊻ ♆
12 57	☽ △ ♄
12 59	☽ □ ♆
13 35	☽ ⊻ ⯓
17 48	☽ ∥ ⯓
20 40	☽ Q ♀
20 Monday	
00 46	☽ ⊼ ♅
01 38	☽ ∠ ☿
01 38	☽ ∠ ☉
01 40	☉ ☌ ☿
04 49	☽ △ ♂
07 21	☽ # ♅
11 14	☽ □ ♃
11 48	☽ ± ♅
21 30	☽ ∥ ☿
22 07	☽ Q ♆
21 Tuesday	
01 11	☽ ∥ ☉
04 39	☽ ♇ ♅
08 04	☽ ✱ ☉
09 39	☽ ✱ ☿
09 49	☽ □ ♀
10 08	☽ ♇ ♂
14 59	☿ ⊻ ♀
15 21	♀ ∠ ♂
20 41	☽ □ ♄
20 58	☽ ✱ ♆
21 45	☽ ☌ ⯓
22 Wednesday	
04 15	☽ ± ♂
09 17	☽ Q ☉
09 27	☽ △ ♅
12 15	☽ Q ☿
16 30	☽ ⊼ ♂
21 41	☽ △ ♃
23 Thursday	
02 18	☽ ∠ ♆
04 50	☽ # ♄
05 58	☽ ∥ ♀
07 48	☽ Q ♄
17 46	♀ # ♄
18 03	☿ ∠ ♃
20 13	☽ ⊥ ♆
24 Friday	
00 15	☽ □ ☉
01 25	☽ # ♂
03 27	☽ △ ♀
03 38	☽ ∥ ♃
04 04	☽ ♇ ♃
05 25	☽ □ ☿
07 42	☽ ✱ ♄
08 16	☽ ⊻ ♆
09 14	☽ ⊻ ⯓
21 05	☽ □ ♅
21 21	☽ ⊥ ⯓
22 11	☿ ⊼ ♄
22 31	☽ ± ♃
25 Saturday	
02 56	☿ △ ♆
05 08	☽ # ♆
05 14	☿ # ♅
07 18	☽ ☍ ♂
10 40	☿ ✱ ⯓
10 51	☽ ⊼ ♃
13 17	☽ ♇ ♀
13 51	☽ ∠ ♄
15 36	☽ ∠ ⯓
15 39	☽ ∥ ♆
18 28	♀ ☍ ♄
20 38	♂ # ♃
26 Sunday	
01 21	♀ ⊼ ♆
06 35	☉ ∠ ♃
07 27	♀ ± ♅
07 58	☽ ⊥ ♄
09 23	☽ Q ♅
09 37	☽ ± ♀
11 41	♀ □ ⯓
17 04	☽ ∥ ♂
18 08	☽ # ♃
18 30	☽ △ ☉
20 01	☽ ⊻ ♄
20 51	☽ ☌ ♆
21 58	☽ ✱ ⯓
23 08	☽ ⊼ ♀
27 Monday	
03 14	☽ △ ☿
09 28	☽ ✱ ♅
11 22	☉ ⊼ ♄
15 52	☽ ∥ ♄
21 55	☉ △ ♆
22 03	☽ Q ⯓
22 27	☽ ⊼ ♂
28 Tuesday	
00 12	☽ ☍ ♃
01 49	☽ # ♀
03 19	☽ ♇ ☉
10 54	☽ ± ♂
11 55	☉ ✱ ⯓
13 37	☽ ♇ ☿
14 34	☿ ± ♄
15 14	☽ ∠ ♅
22 44	☽ ± ☉
29 Wednesday	
01 53	☿ ☍ ♅
05 23	☽ ♇ ♂
06 22	☿ ∥ ⯓
07 21	☽ ☌ ♄
08 25	☽ ⊻ ♆
08 51	☽ ⊥ ♅
09 37	☽ □ ⯓
09 53	☽ ± ☿
11 32	☽ ⊼ ☉
17 20	☽ ☍ ♀
20 04	☽ ⊥ ♆
20 28	☽ ⊻ ♅
23 13	☽ ⊼ ☿
30 Thursday	
10 53	☽ # ☉
11 38	☽ △ ♂
11 42	☽ ⊼ ♃
13 23	☽ ∠ ♆
13 45	♂ ⊻ ♃
15 08	☽ ∥ ♅
20 20	♀ ± ♆
23 18	☽ ± ♃
23 35	♀ ⊼ ♅

DECEMBER 2028

LONGITUDES

Date	Sidereal time h m s	Sun ☉ ° ' "	Moon ☽ ° ' "	Moon ☽ 24.00 ° ' "	Mercury ☿ ° '	Venus ♀ ° '	Mars ♂ ° '	Jupiter ♃ ° '	Saturn ♄ ° '	Uranus ♅ ° '	Neptune ♆ ° '	Pluto ♇ ° '
01	16 43 28	09 ♐ 49 38	03 ♊ 05 56	09 ♊ 30 30	16 ♐ 11	12 ♏ 59	20 ♍ 46	20 ♎ 28	05 ♉ 30	12 ♊ 19	06 ♈ 10	06 ♒ 51
02	16 47 25	10 50 27	15 ♊ 58 23	22 ♊ 29 33	17 45	14 13	21 16	20 38	05 R 27	12 R 16	06 R 09	06 52
03	16 51 21	11 51 16	29 ♊ 03 54	05 ♋ 41 18	19 18	15 28	21 45	20 49	05 23	12 14	06 09	06 53
04	16 55 18	12 52 07	12 ♋ 21 40	19 ♋ 04 52	20 52	16 43	22 14	20 59	05 20	12 11	06 08	06 54
05	16 59 15	13 53 00	25 ♋ 50 47	02 ♌ 39 22	22 25	17 57	22 44	21 09	05 17	12 09	06 08	06 55
06	17 03 11	14 53 53	09 ♌ 30 28	16 ♌ 24 02	23 58	19 12	23 13	21 19	05 14	12 06	06 07	06 57
07	17 07 08	15 54 48	23 ♌ 19 58	00 ♍ 18 12	25 31	20 27	23 42	21 29	05 11	12 04	06 07	06 58
08	17 11 04	16 55 44	07 ♍ 18 36	14 ♍ 21 03	27 05	21 42	24 10	21 39	05 08	12 01	06 07	06 59
09	17 15 01	17 56 41	21 ♍ 25 23	28 ♍ 31 24	28 ♐ 38	22 57	24 39	21 49	05 05	11 58	06 06	07 01
10	17 18 57	18 57 39	05 ♎ 38 50	12 ♎ 47 22	00 ♑ 11	24 11	25 07	21 58	05 02	11 56	06 06	07 02
11	17 22 54	19 58 39	19 ♎ 56 36	27 ♎ 06 07	01 44	25 26	25 35	22 08	04 59	11 53	06 06	07 03
12	17 26 50	20 59 39	04 ♏ 15 24	11 ♏ 23 55	03 17	26 41	26 03	22 17	04 57	11 51	06 06	07 05
13	17 30 47	22 00 41	18 ♏ 31 05	25 ♏ 36 19	04 49	27 56	26 30	22 27	04 54	11 48	06 06	07 06
14	17 34 44	23 01 44	02 ♐ 39 03	09 ♐ 38 43	06 22	29 ♏ 11	26 58	22 36	04 52	11 46	06 06	07 08
15	17 38 40	24 02 47	16 ♐ 34 48	23 ♐ 26 52	07 54	00 ♐ 26	27 25	22 45	04 49	11 43	06 06	07 09
16	17 42 37	25 03 52	00 ♑ 14 32	06 ♑ 57 31	09 26	01 41	27 52	22 54	04 47	11 41	06 06	07 10
17	17 46 33	26 04 57	13 ♑ 35 38	20 ♑ 08 48	10 58	02 56	28 19	23 03	04 45	11 39	06 D 06	07 12
18	17 50 30	27 06 02	26 ♑ 37 01	03 ♒ 00 23	12 29	04 11	28 45	23 12	04 43	11 36	06 06	07 13
19	17 54 26	28 07 08	09 ♒ 19 06	15 ♒ 33 26	13 59	05 26	29 12	23 20	04 41	11 34	06 06	07 15
20	17 58 23	29 ♐ 08 15	21 ♒ 43 45	27 ♒ 50 26	15 29	06 41	29 ♍ 38	23 29	04 39	11 31	06 06	07 17
21	18 02 19	00 ♑ 09 21	03 ♓ 53 57	09 ♓ 54 50	16 58	07 56	00 ♎ 03	23 37	04 37	11 29	06 06	07 18
22	18 06 16	01 10 28	15 ♓ 53 37	21 ♓ 50 53	18 26	09 11	00 29	23 45	04 36	11 27	06 06	07 20
23	18 10 13	02 11 35	27 ♓ 47 13	03 ♈ 43 12	19 52	10 26	00 54	23 54	04 34	11 24	06 06	07 21
24	18 14 09	03 12 42	09 ♈ 39 28	15 ♈ 36 36	21 17	11 41	01 19	24 02	04 33	11 22	06 07	07 23
25	18 18 06	04 13 49	21 ♈ 35 11	27 ♈ 35 47	22 40	12 57	01 44	24 09	04 32	11 20	06 07	07 24
26	18 22 02	05 14 57	03 ♉ 38 55	09 ♉ 45 06	24 00	14 12	02 09	24 17	04 31	11 17	06 07	07 26
27	18 25 59	06 16 04	15 ♉ 54 45	22 ♉ 08 16	25 18	15 27	02 33	24 25	04 30	11 15	06 07	07 28
28	18 29 55	07 17 12	28 ♉ 25 59	04 ♊ 48 09	26 33	16 42	02 57	24 32	04 29	11 13	06 08	07 29
29	18 33 52	08 18 19	11 ♊ 14 57	17 ♊ 46 29	27 44	17 57	03 20	24 40	04 28	11 11	06 08	07 31
30	18 37 48	09 19 27	24 ♊ 22 45	01 ♋ 03 42	28 51	19 12	03 44	24 47	04 27	11 09	06 09	07 33
31	18 41 45	10 ♑ 20 35	07 ♋ 49 10	14 ♋ 38 54	29 ♑ 53	20 ♐ 27	04 ♎ 07	24 ♎ 54	04 ♉ 26	11 ♊ 06	06 ♈ 09	07 ♒ 34

Date	Moon True ☊ ° '	Moon Mean ☊ ° '	Moon ☽ Latitude ° '
01	14 ♑ 18	15 ♑ 45	03 N 25
02	14 R 14	15 42	02 28
03	14 11	15 38	01 22
04	14 10	15 35	00 N 10
05	14 D 11	15 32	01 S 04
06	14 12	15 29	02 14
07	14 14	15 26	03 18
08	14 15	15 23	04 11
09	14 R 15	15 19	04 49
10	14 14	15 16	05 10
11	14 12	15 13	05 12
12	14 09	15 10	04 55
13	14 06	15 07	04 19
14	14 03	15 04	03 28
15	14 00	15 00	02 25
16	13 58	14 57	01 15
17	13 58	14 54	00 S 02
18	13 D 58	14 51	01 N 09
19	13 59	14 48	02 16
20	14 01	14 44	03 14
21	14 02	14 41	04 02
22	14 03	14 38	04 39
23	14 04	14 35	05 04
24	14 R 04	14 32	05 16
25	14 03	14 29	05 14
26	14 02	14 25	04 58
27	14 01	14 22	04 29
28	14 00	14 19	03 46
29	13 59	14 16	02 52
30	13 59	14 13	01 47
31	13 ♑ 59	14 ♑ 10	00 N 34

DECLINATIONS

Date	Sun ☉ ° '	Moon ☽ ° '	Mercury ☿ ° '	Venus ♀ ° '	Mars ♂ ° '	Jupiter ♃ ° '	Saturn ♄ ° '	Uranus ♅ ° '	Neptune ♆ ° '	Pluto ♇ ° '
01	21 S 55	24 N 08	24 S 13	14 S 12	05 N 30	06 S 54	10 N 52	22 N 14	01 N 05	23 S 36
02	22 04	25 09	24 27	14 35	05 19	06 58	10 51	22 14	01 05	23 36
03	22 12	24 48	24 40	14 58	05 08	07 02	10 50	22 14	01 05	23 35
04	22 20	23 02	24 51	15 21	04 57	07 05	10 49	22 13	01 05	23 35
05	22 28	19 56	25 01	15 43	04 46	07 09	10 48	22 13	01 05	23 35
06	22 35	15 43	25 10	16 05	04 36	07 13	10 47	22 13	01 04	23 34
07	22 41	10 37	25 18	16 26	04 25	07 16	10 47	22 12	01 04	23 34
08	22 48	04 N 56	25 24	16 47	04 14	07 20	10 46	22 12	01 04	23 33
09	22 53	01 S 02	25 29	17 08	04 04	07 23	10 45	22 12	01 04	23 33
10	22 59	06 59	25 32	17 28	03 53	07 27	10 44	22 11	01 04	23 33
11	23 03	12 37	25 34	17 48	03 43	07 30	10 44	22 11	01 04	23 32
12	23 08	17 34	25 34	18 07	03 33	07 34	10 43	22 11	01 04	23 32
13	23 12	21 29	25 33	18 26	03 23	07 37	10 43	22 10	01 04	23 31
14	23 15	24 05	25 31	18 44	03 13	07 40	10 42	22 10	01 04	23 31
15	23 18	25 10	25 27	19 02	03 03	07 43	10 41	22 10	01 04	23 31
16	23 21	24 41	25 21	19 19	02 53	07 46	10 41	22 09	01 04	23 30
17	23 23	22 47	25 14	19 35	02 43	07 50	10 40	22 09	01 04	23 30
18	23 24	19 42	25 06	19 52	02 33	07 53	10 40	22 09	01 04	23 29
19	23 25	15 44	24 56	20 07	02 24	07 56	10 40	22 09	01 04	23 29
20	23 26	11 12	24 44	20 22	02 14	07 59	10 39	22 08	01 04	23 29
21	23 26	06 19	24 31	20 37	02 05	08 02	10 39	22 08	01 04	23 28
22	23 26	01 S 16	24 17	20 51	01 56	08 04	10 39	22 08	01 04	23 28
23	23 25	03 N 46	24 01	21 04	01 46	08 07	10 39	22 07	01 05	23 27
24	23 24	08 40	23 44	21 16	01 37	08 10	10 38	22 07	01 05	23 27
25	23 22	13 16	23 26	21 28	01 28	08 13	10 38	22 07	01 05	23 26
26	23 20	17 24	23 06	21 40	01 20	08 15	10 38	22 06	01 05	23 26
27	23 17	20 53	22 46	21 51	01 11	08 18	10 38	22 06	01 05	23 26
28	23 14	23 29	22 25	22 01	01 02	08 21	10 38	22 06	01 05	23 25
29	23 11	24 57	22 02	22 10	00 54	08 23	10 38	22 05	01 06	23 25
30	23 06	25 05	21 40	22 19	00 45	08 26	10 38	22 05	01 06	23 24
31	23 S 02	23 N 46	21 S 16	22 S 27	00 N 37	08 S 28	10 N 38	22 N 05	01 N 06	23 S 24

ZODIAC SIGN ENTRIES

Date	h	m	Planets
01	06	10	☽ ♊
03	13	42	☽ ♋
05	19	20	☽ ♌
07	23	29	☽ ♍
10	02	29	☽ ♎
10	09	12	☿ ♑
12	04	52	☽ ♏
14	07	29	☽ ♐
15	03	39	♀ ♐
16	11	34	☽ ♑
18	18	20	☽ ♒
21	04	16	☽ ♓
21	08	20	☉ ♑
21	08	46	♂ ♎
23	16	29	☽ ♈
26	04	47	☽ ♉
28	14	58	☽ ♊
30	22	06	☽ ♋
31	14	49	☿ ♒

LATITUDES

Date	Mercury ☿ ° '	Venus ♀ ° '	Mars ♂ ° '	Jupiter ♃ ° '	Saturn ♄ ° '	Uranus ♅ ° '	Neptune ♆ ° '	Pluto ♇ ° '
01	01 S 30	01 N 37	02 N 00	01 N 10	02 S 38	00 S 02	01 S 29	05 S 13
04	01 44	01 33	02 02	01 11	02 38	00 02	01 29	05 13
07	01 56	01 28	02 05	01 11	02 37	00 02	01 29	05 12
10	02 06	01 24	02 07	01 12	02 36	00 02	01 29	05 12
13	02 12	01 18	02 10	01 12	02 36	00 02	01 29	05 12
16	02 15	01 12	02 13	01 13	02 35	00 02	01 28	05 12
19	02 14	01 06	02 16	01 13	02 34	00 01	01 28	05 12
22	02 08	01 00	02 19	01 14	02 33	00 01	01 28	05 12
25	01 56	00 53	02 21	01 14	02 32	00 01	01 28	05 12
28	01 36	00 46	02 24	01 15	02 32	00 01	01 28	05 12
31	01 S 08	00 N 38	02 N 27	01 N 16	02 S 31	00 S 01	01 S 28	05 S 12

DATA

Julian Date	2462107
Delta T	+73 seconds
Ayanamsa	24° 15' 55"
Synetic vernal point	04° ♓ 51' 04"
True obliquity of ecliptic	23° 26' 10"

MOON'S PHASES, APSIDES AND POSITIONS ☽

Date	h	m	Phase	Longitude ° '	Eclipse Indicator
02	01	40	○	10 ♊ 24	
09	05	39	☾	17 ♍ 41	
16	02	06	●	24 ♐ 39	
23	21	45	☽	02 ♈ 36	
31	16	48	○	10 ♋ 33	total

Date	h	m		
11	12	19	Perigee	
23	21	01	Apogee	
02	18	01	Max dec	25° N 11'
09	07	52	0S	
15	16	26	Max dec	25° S 12'
22	18	01	0N	
30	02	19	Max dec	25° N 13'

LONGITUDES

Date	Chiron ⚷ ° '	Ceres ⚳ ° '	Pallas ⚴ ° '	Juno ⚵ ° '	Vesta ⚶ ° '	Black Moon Lilith ⚸ ° '
01	04 ♉ 44	25 ♐ 19	28 ♏ 09	23 ♎ 32	17 ♏ 10	29 ♓ 56
11	04 ♉ 24	29 ♐ 21	02 ♐ 30	26 ♎ 37	22 ♏ 24	01 ♈ 04
21	04 ♉ 08	03 ♑ 25	06 ♐ 48	29 ♎ 30	27 ♏ 37	02 ♈ 11
31	03 ♉ 58	07 ♑ 28	11 ♐ 02	02 ♏ 08	02 ♐ 47	03 ♈ 18

All ephemeris data is given at 12.00 UT and the Moon's longitude is additionally given for 24.00 UT

ASPECTARIAN

h m	Aspects
01 Friday	
00 42	♂ ⚼ ♄
05 06	☽ # ♇
13 28	☽ # ☿
16 30	☽ ⊻ ♄
16 30	☽ ⚼ ♃
17 44	☽ ✶ ♆
19 02	☽ △ ♇
02 Saturday	
01 40	☽ ☍ ☉
03 39	☽ ⊥ ♄
05 10	☽ ☌ ♅
08 25	☽ ⊼ ♀
15 43	☽ ☍ ☿
16 01	☽ Q ♆
20 12	☽ ∠ ♄
20 39	☽ ± ♀
20 43	☽ △ ♃
22 07	☽ □ ♂
22 52	☽ ⚼ ♇
03 Sunday	
01 32	☉ ± ♄
14 23	☽ # ☿
14 49	☽ ⚼ ♀
15 18	☽ ± ♇
15 46	☉ # ♅
18 36	♂ ⚼ ♇
20 29	☉ ☍ ♅
23 25	☽ ✶ ♄
04 Monday	
00 49	☽ □ ♆
02 10	☽ ⊼ ♇
04 08	☿ ⚼ ♄
06 10	☽ # ♇
08 03	☽ Q ♂
11 41	☽ ⊻ ♅
12 59	☽ ⊼ ☉
14 06	☿ ✶ ♃
18 00	☽ # ☉
19 16	☽ ∥ ♅
20 34	☽ △ ♀
20 51	☽ Q ♄
22 22	☽ ⊥ ♅
05 Tuesday	
00 34	☽ ± ☉
03 34	☽ □ ♃
04 17	☿ ∠ ♇
05 08	☽ ⊼ ☿
06 16	☽ ✶ ♂
14 17	☽ ∠ ♅
17 07	☽ ± ☿
17 48	☽ ⚼ ☉
19 01	☿ □ ♂
06 Wednesday	
04 32	☽ □ ♄
06 05	☽ △ ♆
07 31	☽ ☍ ♇
09 39	☽ ∠ ♂
10 15	☽ # ♀
10 56	☽ ⚼ ☿
11 40	☽ Q ♃
16 31	☽ ✶ ♅
22 08	☽ △ ☉
07 Thursday	
01 53	☽ ⊥ ♂
06 31	☽ □ ♀
08 10	☽ ⚼ ♆
08 46	☽ ✶ ♃
11 17	☽ ∥ ♄
12 39	☽ ⊻ ♂
13 15	☽ Q ♅
16 15	☽ △ ☿
23 41	☽ ± ♆
08 Friday	
00 50	♀ ⚼ ♆
02 12	☽ # ♃
08 17	☽ △ ♄
09 57	☽ ⊼ ♆
10 51	☽ ∠ ♃
10 56	♀ ⊻ ♃
11 27	☽ ⊼ ♇
14 55	☽ ∥ ♂
16 28	☽ Q ♀
20 00	☽ □ ♅
21 42	☽ ± ♇
09 Saturday	
02 22	☽ ⊥ ♃
03 36	☽ ∥ ♆
05 39	☽ □ ☉
09 44	☽ ⚼ ♄
12 08	☽ # ♆
12 40	☽ ⊻ ♃
13 00	☽ ⚼ ♇
14 49	☽ ✶ ♀
17 38	☽ ☌ ♂
23 47	☽ # ♂
10 Sunday	
00 53	☽ ± ♄
01 40	☽ □ ☿
10 58	☽ ⊼ ♄
12 44	☿ ⊥ ♀
12 46	☽ ☍ ♆
13 53	☽ ∥ ♃
14 20	☽ △ ♇
14 23	☽ Q ☉
18 32	☽ ∠ ♀
22 32	☽ △ ♅
11 Monday	
01 22	☿ ⊥ ♇
03 43	☽ # ♄
04 29	♀ Q ♇
11 04	☽ ⊥ ♀
11 36	☽ Q ☿
12 04	☽ ✶ ☉
12 11	☉ ⚼ ♄
15 42	☽ ☌ ♃
16 25	♀ ✶ ♂
21 46	☽ ⊻ ♂
22 05	☽ ⊻ ♀
23 37	☽ ⚼ ♅
12 Tuesday	
08 10	☽ ⊥ ♂
10 10	☽ ✶ ☿
13 09	☽ ☍ ♄
14 40	☽ ± ♅
15 05	☽ ⊼ ♆
15 08	☽ ∠ ☉
15 15	☽ ∥ ♀
16 45	☽ □ ♇
23 47	☽ ∠ ♂
13 Wednesday	
00 43	☽ ⊼ ♅
01 10	☽ ± ♆
05 24	☿ Q ♃
07 27	☽ ⊥ ☉
13 08	☿ △ ♄
14 10	☉ ∠ ♇
14 29	☽ ∠ ☿
16 21	☽ ⚼ ♆
17 13	☽ # ♅
18 22	☽ ⊻ ☉
18 43	☽ ⊻ ♃
23 06	♀ ⊥ ♃
23 10	☽ Q ♇
23 59	☉ ✶ ♃
14 Thursday	
01 59	☽ ✶ ♂
02 29	☽ ∥ ☉
05 00	☽ ⊥ ♃
05 24	☽ ∥ ♇
05 31	☽ ☌ ♀
07 37	☽ ⊥ ☿
07 45	☿ □ ♆
15 46	☽ ⊼ ♄
17 54	☽ △ ♆
19 09	☽ ⊻ ☿
19 41	☽ ✶ ♇
20 34	☽ ∠ ♃
23 12	☽ Q ♂
15 Friday	
00 01	☿ ⊻ ♇
02 03	☽ ± ♄
03 37	☽ ☍ ♅
17 38	☽ ⚼ ♄
21 44	☽ ∠ ♇
22 53	☽ ✶ ♃
16 Saturday	
02 06	☽ ☌ ☉
07 39	☽ □ ♂
13 40	☽ ⊥ ♇
14 50	☽ ⊻ ♀
20 05	☽ △ ♄
20 24	☽ Q ♃
20 45	♆ St D
22 27	☽ □ ♆
17 Sunday	
00 25	☽ ⊻ ♇
02 41	☽ ⊥ ♀
04 39	☽ ∥ ♇
06 03	☽ ∥ ☉
06 36	☽ ☌ ☿
08 28	☽ ⊼ ♅
17 36	☽ # ♅
19 22	☽ ± ♅
20 46	☽ ∠ ♀
22 26	☿ ⊼ ♅
18 Monday	
05 33	☽ □ ♃
07 18	☽ Q ♆
09 56	♂ ± ♄
10 57	☽ ∥ ♀
11 58	☽ ⚼ ♅
12 59	☽ ⊻ ☉
16 09	☽ △ ♂
21 54	♀ ⊼ ♄
19 Tuesday	
01 14	☽ ⊥ ☉
03 12	☽ □ ♄
03 47	☽ ✶ ♀
05 51	☽ ✶ ♆
08 03	☽ ☌ ♇
16 17	☽ △ ♅
19 57	☽ ∠ ☉
21 42	☽ ⚼ ♂
22 12	☽ ⊻ ☿
20 Wednesday	
00 38	♀ △ ♆
05 24	☽ Q ♀
10 46	☽ ∠ ♆
11 28	☽ ⊥ ☿
13 48	☽ Q ♄
14 45	☽ # ♄
15 28	☽ △ ♃
15 51	☽ ± ♂
23 32	♀ ✶ ♇
21 Thursday	
03 48	☽ ∥ ♃
03 54	☽ ✶ ☉
04 06	☽ ⊼ ♂
04 27	☽ ⊥ ♆
07 38	☽ ∠ ☿
08 00	☉ □ ♂
13 26	☽ ✶ ♄
16 23	☽ ⊻ ♆
18 48	☽ ⊻ ♇
20 13	☿ ± ♅
20 59	☽ □ ♀
21 31	☽ ⚼ ♃
22 Friday	
02 41	♀ ∠ ♃
03 06	☽ □ ♅
06 02	☽ Q ☉
06 50	☽ ⊥ ♇
08 48	☽ # ♂
12 56	☽ # ♆
15 42	☉ ⊥ ♇
15 48	☽ ± ♃
17 49	☽ ✶ ☿
19 27	☽ ∠ ♄
23 08	☽ ∥ ♆
23 Saturday	
01 00	☽ ∠ ♇
02 44	☽ ∥ ♂
04 02	☽ ⊼ ♃
13 35	☽ ⊥ ♄
14 30	♀ ± ♄
15 16	☽ Q ♅
18 32	☽ ☍ ♂
21 23	☽ Q ☿
21 45	☽ □ ☉
24 Sunday	
01 42	☽ ⊻ ♄
04 50	☽ ☌ ♆
05 56	♀ ☍ ♅
07 23	☽ ✶ ♇
09 30	☽ # ♃
15 26	☽ ✶ ♅
16 35	☽ △ ♀
22 05	☽ ∥ ♄
25 Monday	
07 37	☽ Q ♇
11 19	☿ ∥ ♇
14 26	☽ □ ☿
17 12	☽ ☍ ♃
17 26	☉ ∥ ☿
18 54	☉ △ ♄
21 27	☽ ∠ ♅
26 Tuesday	
02 10	☽ ⚼ ♀
08 55	☽ ⊼ ♂
10 30	♀ Q ♂
13 42	☽ ☌ ♄
14 02	☿ Q ♆
15 14	☽ ⊥ ♅
15 27	☽ △ ☉
16 52	☽ ⊻ ♆
17 39	☿ □ ♃
19 28	☽ □ ♇
21 09	☽ ± ♂
21 58	☽ ± ♀
27 Wednesday	
02 58	☽ ⊻ ♅
04 38	☽ ⊥ ♆
08 36	☉ □ ♆
11 00	☽ ⊼ ♀
15 16	☽ ⚼ ♂
15 57	☉ Q ♃
20 21	☽ # ♀
22 04	☽ ∥ ♅
22 04	☽ ∠ ♆
23 15	☽ ⚼ ☉
28 Thursday	
02 16	☽ # ☿
03 21	♂ ∥ ♆
04 31	☽ ⊼ ♃
05 34	☿ ⚼ ♅
08 03	☽ △ ☿
09 15	☽ # ☉
11 13	☽ # ♇
16 01	☽ ± ♃
16 55	☉ ⊻ ♇
17 52	☽ ± ☉
20 47	☽ △ ♂
23 23	☽ ⊻ ♄
23 55	♀ # ♅
29 Friday	
02 30	☽ ✶ ♆
05 03	☽ △ ♇
05 58	☿ ∥ ♀
06 04	☽ ⊼ ☉
08 37	☿ # ♅
09 02	☽ ⚼ ♃
10 33	☽ ⊥ ♄
11 52	☽ ☌ ♅
15 01	☽ ⚼ ☿
30 Saturday	
00 40	☽ Q ♆
01 37	☽ ☍ ♀
03 04	☽ ∠ ♄
08 41	☽ ⚼ ♇
09 00	☽ ± ☿
12 44	☽ △ ♃
16 41	♀ ⚼ ♄
20 44	☽ ⊼ ☿
31 Sunday	
00 53	☽ ± ♇
05 14	☽ □ ♂
06 01	☽ ✶ ♄
09 03	☽ □ ♆
11 34	☽ ⊼ ♇
16 09	☽ # ♇
16 48	☽ ☍ ☉
17 47	☽ ⊻ ♅
20 02	☽ # ☉

JANUARY 2029

LONGITUDES

Date	Sidereal time h m s	Sun ☉ ° ' "	Moon ☽ ° ' "	Moon ☽ 24.00 ° ' "	Mercury ☿ ° '	Venus ♀ ° '	Mars ♂ ° '	Jupiter ♃ ° '	Saturn ♄ ° '	Uranus ♅ ° '	Neptune ♆ ° '	Pluto ♇ ° '
01	18 45 42	11 ♑ 21 42	21 ♋ 32 36	28 ♋ 29 52	00 ♒ 49	21 ♐ 43	04 ♎ 29	25 ♎ 01	04 ♉ 26	11 ♊ 04	06 ♈ 10	07 ♒ 36
02	18 49 38	12 22 50	05 ♌ 30 17	12 ♌ 33 23	01 39	22 58	04 52	25 08	04 R 26	11 R 02	06 10	07 38
03	18 53 35	13 23 58	19 ♌ 38 39	26 ♌ 45 36	02 21	24 13	05 14	25 15	04 25	11 00	06 11	07 40
04	18 57 31	14 25 07	03 ♍ 53 42	11 ♍ 02 28	02 55	25 28	05 36	25 21	04 25	10 58	06 12	07 41
05	19 01 28	15 26 15	18 ♍ 11 26	25 ♍ 20 10	03 20	26 43	05 57	25 27	04 25	10 56	06 12	07 43
06	19 05 24	16 27 23	02 ♎ 28 16	09 ♎ 35 24	03 34	27 58	06 18	25 34	04 D 25	10 54	06 13	07 45
07	19 09 21	17 28 32	16 ♎ 41 15	23 ♎ 45 32	03 R 38	29 ♐ 14	06 39	25 40	04 25	10 52	06 14	07 47
08	19 13 17	18 29 41	00 ♏ 48 01	07 ♏ 48 30	03 30	00 ♑ 29	06 59	25 46	04 25	10 51	06 14	07 48
09	19 17 14	19 30 50	14 ♏ 46 47	21 ♏ 42 42	03 11	01 44	07 20	25 51	04 26	10 49	06 15	07 50
10	19 21 11	20 31 59	28 ♏ 36 05	05 ♐ 26 49	02 39	02 59	07 39	25 57	04 26	10 47	06 16	07 52
11	19 25 07	21 33 08	12 ♐ 14 45	18 ♐ 59 44	01 57	04 15	07 58	26 02	04 27	10 45	06 17	07 54
12	19 29 04	22 34 18	25 ♐ 41 39	02 ♑ 20 22	01 03	05 30	08 17	26 08	04 28	10 43	06 18	07 56
13	19 33 00	23 35 26	08 ♑ 55 47	15 ♑ 27 49	00 ♒ 01	06 45	08 36	26 13	04 29	10 42	06 19	07 58
14	19 36 57	24 36 35	21 ♑ 56 24	28 ♑ 21 29	28 ♑ 51	08 00	08 54	26 18	04 30	10 40	06 20	07 59
15	19 40 53	25 37 43	04 ♒ 43 04	11 ♒ 01 11	27 35	09 16	09 11	26 23	04 31	10 38	06 21	08 01
16	19 44 50	26 38 51	17 ♒ 15 54	23 ♒ 27 22	26 17	10 31	09 29	26 27	04 32	10 37	06 22	08 03
17	19 48 46	27 39 57	29 ♒ 35 44	05 ♓ 41 14	24 59	11 46	09 45	26 32	04 33	10 35	06 23	08 05
18	19 52 43	28 41 04	11 ♓ 44 10	17 ♓ 44 50	23 42	13 01	10 02	26 36	04 34	10 34	06 24	08 07
19	19 56 40	29 ♑ 42 09	23 ♓ 43 36	29 ♓ 40 56	22 29	14 17	10 18	26 40	04 36	10 33	06 25	08 09
20	20 00 36	00 ♒ 43 14	05 ♈ 37 16	11 ♈ 33 06	21 23	15 32	10 33	26 44	04 38	10 31	06 26	08 11
21	20 04 33	01 44 17	17 ♈ 28 58	23 ♈ 25 27	20 23	16 47	10 48	26 48	04 39	10 30	06 27	08 12
22	20 08 29	02 45 20	29 ♈ 23 07	05 ♉ 22 35	19 33	18 02	11 02	26 51	04 41	10 29	06 28	08 14
23	20 12 26	03 46 22	11 ♉ 24 26	17 ♉ 29 15	18 51	19 17	11 16	26 55	04 43	10 27	06 30	08 16
24	20 16 22	04 47 23	23 ♉ 37 40	29 ♉ 50 12	18 18	20 33	11 30	26 58	04 45	10 26	06 31	08 18
25	20 20 19	05 48 23	06 ♊ 07 24	12 ♊ 29 43	17 54	21 48	11 42	27 01	04 47	10 25	06 32	08 20
26	20 24 15	06 49 22	18 ♊ 57 34	25 ♊ 31 16	17 39	23 03	11 55	27 04	04 50	10 24	06 34	08 22
27	20 28 12	07 50 20	02 ♋ 11 03	08 ♋ 56 59	17 33	24 18	12 07	27 07	04 52	10 23	06 35	08 24
28	20 32 09	08 51 16	15 ♋ 49 04	22 ♋ 47 07	17 D 35	25 34	12 18	27 09	04 54	10 22	06 36	08 26
29	20 36 05	09 52 12	29 ♋ 50 49	06 ♌ 59 42	17 44	26 49	12 29	27 12	04 57	10 21	06 38	08 27
30	20 40 02	10 53 07	14 ♌ 13 09	21 ♌ 30 25	18 00	28 04	12 39	27 14	05 00	10 20	06 39	08 29
31	20 43 58	11 ♒ 54 00	28 ♌ 50 38	06 ♍ 12 54	18 ♑ 22	29 ♑ 19	12 ♎ 48	27 ♎ 16	05 ♉ 03	10 ♊ 19	06 ♈ 41	08 ♒ 31

Date	Moon True ☊ ° '	Moon Mean ☊ ° '	Moon ☽ Latitude ° '
01	13 ♑ 59	14 ♑ 06	00 S 42
02	13 D 59	14 03	01 57
03	13 59	14 00	03 05
04	13 R 59	13 57	04 03
05	13 59	13 54	04 46
06	13 59	13 50	05 11
07	13 58	13 47	05 17
08	13 D 59	13 44	05 04
09	13 59	13 41	04 33
10	13 59	13 38	03 46
11	14 00	13 35	02 47
12	14 01	13 31	01 40
13	14 01	13 28	00 S 28
14	14 R 01	13 25	00 N 44
15	14 01	13 22	01 52
16	13 59	13 19	02 54
17	13 57	13 15	03 46
18	13 55	13 12	04 27
19	13 53	13 09	04 56
20	13 50	13 06	05 12
21	13 49	13 03	05 15
22	13 48	13 00	05 04
23	13 D 48	12 56	04 40
24	13 49	12 53	04 03
25	13 51	12 50	03 13
26	13 52	12 47	02 13
27	13 54	12 44	01 N 04
28	13 R 54	12 41	00 S 11
29	13 53	12 37	01 26
30	13 51	12 34	02 39
31	13 ♑ 48	12 ♑ 31	03 S 42

DECLINATIONS

Date	Sun ☉ ° '	Moon ☽ ° '	Mercury ☿ ° '	Venus ♀ ° '	Mars ♂ ° '	Jupiter ♃ ° '	Saturn ♄ ° '	Uranus ♅ ° '	Neptune ♆ ° '	Pluto ♇ ° '
01	22 S 57	21 N 01	20 S 53	22 S 35	00 N 29	08 S 30	10 N 38	22 N 05	01 N 06	23 S 23
02	22 52	17 00	20 30	22 42	00 21	08 33	10 38	22 04	01 07	23 23
03	22 46	11 59	20 07	22 48	00 13	08 35	10 38	22 04	01 07	23 22
04	22 39	06 18	19 44	22 53	00 N 06	08 37	10 39	22 04	01 07	23 22
05	22 33	00 N 17	19 23	22 58	00 S 02	08 39	10 39	22 04	01 07	23 22
06	22 25	05 S 44	19 03	23 02	00 09	08 41	10 39	22 03	01 08	23 21
07	22 18	11 26	18 45	23 05	00 16	08 43	10 40	22 03	01 08	23 21
08	22 10	16 29	18 29	23 08	00 24	08 45	10 40	22 03	01 08	23 20
09	22 01	20 36	18 15	23 10	00 30	08 47	10 40	22 03	01 09	23 20
10	21 52	23 31	18 04	23 11	00 37	08 49	10 41	22 02	01 09	23 19
11	21 43	25 01	17 55	23 12	00 44	08 51	10 41	22 02	01 10	23 19
12	21 33	25 02	17 49	23 11	00 50	08 52	10 42	22 02	01 10	23 18
13	21 23	23 36	17 45	23 11	00 56	08 54	10 42	22 02	01 10	23 18
14	21 12	20 56	17 44	23 09	01 03	08 56	10 43	22 01	01 11	23 18
15	21 01	17 16	17 46	23 07	01 08	08 57	10 44	22 01	01 11	23 17
16	20 49	12 54	17 49	23 04	01 14	08 59	10 44	22 01	01 12	23 17
17	20 38	08 05	17 54	23 00	01 20	09 00	10 45	22 01	01 12	23 16
18	20 25	03 S 02	18 01	22 55	01 25	09 01	10 46	22 01	01 13	23 16
19	20 13	02 N 03	18 09	22 50	01 30	09 03	10 47	22 00	01 13	23 15
20	20 00	07 00	18 17	22 44	01 35	09 04	10 47	22 00	01 14	23 15
21	19 46	11 42	18 27	22 38	01 40	09 05	10 48	22 00	01 14	23 14
22	19 32	15 59	18 37	22 30	01 45	09 06	10 49	22 00	01 15	23 14
23	19 18	19 41	18 47	22 22	01 49	09 07	10 50	22 00	01 15	23 14
24	19 04	22 35	18 57	22 14	01 53	09 08	10 51	21 59	01 16	23 13
25	18 49	24 30	19 08	22 04	01 57	09 09	10 52	21 59	01 16	23 13
26	18 34	25 11	19 18	21 54	02 01	09 10	10 53	21 59	01 17	23 12
27	18 18	24 29	19 28	21 44	02 04	09 10	10 54	21 59	01 17	23 12
28	18 03	22 19	19 38	21 33	02 08	09 11	10 55	21 59	01 18	23 11
29	17 46	18 46	19 47	21 21	02 11	09 12	10 56	21 59	01 18	23 11
30	17 30	14 02	19 55	21 08	02 14	09 12	10 57	21 59	01 19	23 11
31	17 S 13	08 N 24	20 S 03	20 S 55	02 S 17	09 S 13	10 N 59	21 N 59	01 N 20	23 S 10

ZODIAC SIGN ENTRIES

Date	h m	Planets
02	02 35	☽ ♌
04	05 27	☽ ♍
06	07 50	☽ ♎
08	02 47	♀ ♑
08	10 38	☽ ♏
10	14 27	☽ ♐
12	19 46	☽ ♑
13	12 13	☿ ♑
15	03 05	☽ ♒
17	12 48	☽ ♓
19	19 01	☉ ♒
20	00 38	☽ ♈
22	13 14	☽ ♉
25	00 19	☽ ♊
27	08 05	☽ ♋
29	12 15	☽ ♌
31	13 53	☽ ♍

LATITUDES

Date	Mercury ☿ ° '	Venus ♀ ° '	Mars ♂ ° '	Jupiter ♃ ° '	Saturn ♄ ° '	Uranus ♅ ° '	Neptune ♆ ° '	Pluto ♇ ° '
01	00 S 56	00 N 36	02 N 28	01 N 16	02 S 30	00 S 01	01 S 28	05 S 12
04	00 S 14	00 28	02 31	01 17	02 29	00 01	01 28	05 12
07	00 N 36	00 21	02 35	01 17	02 29	00 01	01 27	05 12
10	01 33	00 13	02 38	01 18	02 28	00 01	01 27	05 12
13	02 27	00 N 05	02 41	01 18	02 27	00 01	01 27	05 12
16	03 08	00 S 02	02 44	01 19	02 26	00 01	01 27	05 13
19	03 28	00 10	02 48	01 20	02 25	00 01	01 27	05 13
22	03 26	00 17	02 51	01 21	02 24	00 01	01 27	05 13
25	03 08	00 24	02 54	01 21	02 23	00 01	01 27	05 13
28	02 41	00 31	02 58	01 22	02 22	00 01	01 27	05 13
31	02 N 09	00 S 38	03 N 01	01 N 23	02 S 22	00 S 01	01 S 26	05 S 13

DATA

Julian Date	2462138
Delta T	+74 seconds
Ayanamsa	24° 16' 01"
Synetic vernal point	04° ♓ 50' 58"
True obliquity of ecliptic	23° 26' 10"

MOON'S PHASES, APSIDES AND POSITIONS ☽

Date	h m	Phase	Longitude ° '	Eclipse Indicator
07	13 26	☾	17 ♎ 32	
14	17 24	●	24 ♑ 50	Partial
22	19 23	☽	03 ♉ 04	
30	06 04	○	10 ♌ 38	

Day	h m		
05	04 28	Perigee	
20	18 10	Apogee	
05	13 07	OS	
12	00 07	Max dec	25° S 13'
19	02 19	ON	
26	12 15	Max dec	25° N 11'

LONGITUDES

Date	Chiron ⚷ ° '	Ceres ⚳ ° '	Pallas ⚴ ° '	Juno ⚵ ° '	Vesta ⚶ ° '	Black Moon Lilith ⚸ ° '
01	03 ♉ 57	07 ♑ 52	11 ♐ 27	02 ♏ 23	03 ♐ 18	03 ♈ 25
11	03 ♉ 53	11 ♑ 54	15 ♐ 35	04 ♏ 40	08 ♐ 23	04 ♈ 32
21	03 ♉ 55	15 ♑ 54	19 ♐ 35	06 ♏ 37	13 ♐ 24	05 ♈ 39
31	04 ♉ 03	19 ♑ 50	23 ♐ 26	08 ♏ 09	18 ♐ 19	06 ♈ 46

All ephemeris data is given at 12.00 UT and the Moon's longitude is additionally given for 24.00 UT

ASPECTARIAN

h m Aspects

01 Monday
00 25 ☽ ⋕ ♀
03 07 ☽ Q ♄
04 06 ☽ ∥ ♅
04 15 ☽ ⊥ ♅
05 24 ☉ ⊼ ♅
08 18 ♂ ⊼ ♄
12 19 ☽ ⊼ ♀
13 06 ☽ ⋕ ☿
13 41 ☽ Q ♂
18 03 ☽ □ ♃
19 48 ☽ ∠ ♅
23 42 ☽ ± ♀

02 Tuesday
05 01 ☽ ☍ ☿
05 32 ♀ ∠ ♇
10 09 ☽ □ ♄
10 53 ☽ ⚹ ♂
13 08 ☽ △ ♆
15 38 ☽ ☍ ♇
16 36 ☽ ⚼ ♀
21 24 ☽ ⚹ ♅

03 Wednesday
00 37 ☽ ⊼ ☉
01 05 ☽ Q ♃
07 49 ☉ ∥ ♀
11 33 ☽ ± ☉
13 01 ☽ ∠ ♂
14 36 ☽ ⚼ ♆
17 39 ☽ Q ♅
17 54 ☽ ∥ ♄
20 27 ☽ △ ♀
21 31 ☽ ⚹ ♃

04 Thursday
02 33 ☽ ⋕ ♃
03 54 ☽ ⚼ ☉
04 35 ☽ ⊥ ♂
05 46 ☽ ± ♆
09 33 ♀ ⚹ ♃
10 18 ☽ ⊼ ☿
12 53 ☽ △ ♄
14 56 ☽ ⊻ ♂
15 52 ☽ ⊼ ♆
18 23 ☽ ⊼ ♇
20 43 ☽ ± ☿
22 55 ☽ ∠ ♃
23 51 ☽ □ ♅

05 Friday
04 29 ☽ ± ♇
07 01 ☽ △ ☉
08 41 ☽ ∥ ♆
12 14 ☽ ⚼ ☿
12 39 ♄ St D
12 59 ☽ ⋕ ♂
13 16 ☽ ∥ ♂
14 03 ☽ ⚼ ♄
14 08 ☽ ⊥ ♃
17 34 ☽ ⋕ ♆
19 37 ☽ ⚼ ♇

06 Saturday
00 18 ☽ ⊻ ♃
02 53 ☿ ⊥ ♀
03 42 ☽ □ ♀
05 11 ☽ ± ♄
05 33 ♂ ☍ ♆
13 52 ☽ △ ☿
15 17 ☽ ⊼ ♄
18 19 ☽ ☍ ♆
18 37 ☽ ☌ ♂
20 55 ☽ △ ♇
22 14 ☉ ± ♅

07 Sunday
00 16 ☽ ∥ ♃
02 11 ☽ △ ♅
07 56 ☿ St R
08 37 ☽ ⋕ ♄
13 00 ☽ Q ♀
13 26 ☽ □ ☉

08 Monday
03 21 ☽ ☌ ♃
03 34 ☽ ⚼ ♅
11 24 ☽ ⚹ ♀
16 32 ☽ □ ☿
18 12 ☽ ☍ ♄
18 54 ☽ ± ♅
21 19 ☽ ⊼ ♆
22 15 ☽ ∥ ☿
22 31 ☽ Q ☉
22 52 ☽ ⊻ ♂

09 Tuesday
00 02 ☽ □ ♇
03 23 ♃ ⚼ ♅
05 11 ☽ ⊼ ♅
07 38 ☽ ± ♆
07 45 ☉ ⋕ ♅
09 26 ☽ ⊥ ♂
14 01 ♀ ⊥ ♇
15 43 ☽ ∠ ♀
20 50 ☽ ⚹ ☉
21 49 ☽ ∥ ☉
22 29 ☽ ⋕ ♅
22 43 ☽ Q ☿
23 13 ☽ ⚼ ♆

10 Wednesday
01 23 ☽ ∠ ♂
07 13 ☽ Q ♇
07 20 ☽ ⊻ ♃
07 40 ☿ ⊻ ♀
08 34 ☽ ∥ ♀
08 54 ☽ ⊥ ♀
09 58 ☽ ∥ ♇
17 54 ☽ ⊥ ♃
18 47 ☽ ⚹ ☿
20 27 ☽ ⊻ ♀
22 14 ☽ ⊼ ♄

11 Thursday
01 08 ☽ ∠ ☉
01 27 ☽ △ ♆
04 16 ☽ ⚹ ♂
04 18 ☽ ⚹ ♇
05 43 ♂ △ ♇
08 49 ☽ ± ♄
09 22 ☽ ☍ ♅
09 51 ☽ ∠ ♃
16 00 ♀ △ ♄
18 21 ☽ ⊥ ☉
19 51 ☽ ∠ ☿

12 Friday
00 49 ☽ ⚼ ♄
02 05 ☽ Q ♂
05 56 ☽ ⊻ ☉
07 01 ☽ ∠ ♇
10 56 ☽ ⊥ ☿
12 47 ☽ ⚹ ♃
21 00 ☽ ⊻ ☿
23 17 ☽ ⊥ ♇

13 Saturday
03 29 ♀ □ ♆
03 52 ☽ △ ♄
07 13 ☽ □ ♆
07 36 ☽ ☌ ♀
10 13 ☽ ⊻ ♇
10 41 ☽ Q ♃
11 23 ☽ □ ♂
15 14 ☽ ⊼ ♅
15 22 ☽ ∥ ♇
16 42 ☽ ∥ ♀

14 Sunday
02 15 ☽ ± ♅
03 20 ☽ ⋕ ♅
05 15 ☉ Q ♆
09 51 ☽ ∥ ☉
11 42 ♀ ⊻ ♇
16 27 ☽ Q ♆
17 24 ☽ ☌ ☉
17 59 ♀ Q ♃
18 57 ☽ ⚼ ♅
20 11 ☽ □ ♃
23 46 ☽ ☌ ☿

15 Monday
09 05 ☽ ∥ ☿
10 18 ♀ □ ♂
11 36 ☽ □ ♄
12 18 ☉ ⚼ ♅
15 05 ☽ ⚹ ♆
18 17 ☽ ☌ ♇
20 42 ☽ △ ♂
21 36 ☽ ⊻ ♀
23 15 ☽ △ ♅

16 Tuesday
00 21 ♂ ⋕ ♆
07 05 ☉ □ ♃
08 16 ☉ ☌ ☿
09 05 ☿ □ ♃
10 23 ☽ ⊥ ♀
13 55 ♀ ⊼ ♅
19 56 ☽ ∠ ♆
22 13 ☽ Q ♄
22 56 ☽ ⋕ ♄

17 Wednesday
00 29 ☿ ⚼ ♅
02 19 ☽ ⚼ ♂
03 50 ☽ ⊻ ☿
05 50 ☽ ∠ ♀
05 57 ☽ △ ♃
07 34 ☽ ∥ ♃
07 53 ☽ ⊻ ☉
13 32 ☽ ⊥ ♆
14 27 ☽ ⊥ ☿
20 23 ☽ ± ♂
20 44 ☽ ⊥ ☉
21 46 ☽ ⚹ ♄
22 56 ☿ Q ♆

18 Thursday
01 23 ☽ ⊻ ♆
04 47 ☽ ⊻ ♇
06 32 ☽ ∠ ☿
07 35 ♄ ⊥ ♅
08 32 ☽ ⊼ ♂
09 41 ☽ □ ♅
11 44 ☽ ⚼ ♃
14 51 ☽ ⚹ ♀
16 14 ☽ ∠ ☉
16 45 ☽ ⊥ ♇
19 30 ☽ ∥ ♂
20 36 ☽ ⋕ ♆

19 Friday
03 42 ☽ ∠ ♄
05 49 ☽ ± ♃
08 04 ☽ ∥ ♆
09 23 ☽ ⋕ ♂
09 44 ☽ ⚹ ☿
10 50 ☽ ∠ ♇
17 44 ☽ Q ♀
17 57 ☽ ⊼ ♃
21 41 ☽ Q ♅
21 50 ☽ ⊥ ♄

20 Saturday
01 10 ☽ ⚹ ☉
07 50 ☽ Q ☿
09 26 ♂ △ ♅
09 59 ☽ ⊻ ♄
13 39 ☽ ☌ ♆
17 11 ☽ ⚹ ♇
21 53 ☽ ⚹ ♅
22 11 ☽ ☍ ♂
22 20 ☽ ⋕ ♃

21 Sunday
03 43 ☽ Q ☉
06 37 ♀ ± ♅
07 14 ☽ ∥ ♄
10 25 ☽ □ ♀
17 28 ☽ □ ☿
17 31 ☽ Q ♇

22 Monday
04 09 ☽ ∠ ♅
06 54 ☽ ☍ ♃
19 23 ☽ □ ☉
22 11 ☽ ⊥ ♅
22 39 ☽ ☌ ♄

23 Tuesday
02 13 ☽ ⊻ ♆
05 26 ☽ ⋕ ☿
05 45 ☽ □ ♇
06 20 ☿ ☌ ♀
09 31 ☽ ⋕ ☉
10 07 ☽ ⊻ ♅
11 43 ☽ ⊼ ♂
14 10 ☽ ⊥ ♆
23 47 ☽ ± ♂

24 Wednesday
02 00 ☽ △ ☿
05 18 ☽ △ ♀
06 24 ☽ ∥ ♅
07 53 ☽ ∠ ♆
08 45 ☽ ⋕ ♀
11 06 ☉ □ ♄
17 39 ☽ ⚼ ♂
18 05 ☉ ∥ ☿
18 30 ☽ ⊼ ♃
18 32 ☽ ⋕ ♇

25 Thursday
06 02 ☽ ⚼ ☿
06 04 ☽ ± ♃
09 28 ☽ ⊻ ♄
11 21 ☽ △ ☉
12 47 ☽ ⚹ ♆
13 25 ☽ ⚼ ♀
16 11 ☽ △ ♇
20 05 ☽ ☌ ♅
20 50 ☽ ⊥ ♄
22 39 ☽ ± ☿
22 42 ☽ △ ♂
23 09 ☽ ⚼ ♃

26 Friday
00 45 ♀ ⋕ ♅
05 39 ☉ ⚹ ♆
08 06 ☽ ± ♀
09 37 ☽ ⊼ ☿
11 16 ☽ Q ♆
13 36 ☽ ∠ ♄
17 42 ☽ ⚼ ☉
20 06 ☽ ⚼ ♇
20 18 ☽ ⊼ ♀

27 Saturday
02 51 ☽ △ ♃
11 20 ☽ ± ☉
12 23 ☽ ± ♇
16 48 ☽ ⚹ ♄
17 25 ♀ Q ♆
18 41 ☿ St D
19 50 ☽ □ ♆
22 51 ☽ ⊼ ☉
23 03 ☽ ⊼ ♇

28 Sunday
01 33 ☉ ☌ ♇
02 30 ☽ ⊻ ♅
04 08 ☽ ⋕ ♇
05 47 ☽ □ ♂
08 18 ♀ ⚼ ♅
12 57 ☽ ⊥ ♅
13 54 ☽ Q ♄
14 47 ☽ ∥ ♅
15 04 ☽ ☍ ☿
18 32 ☽ ⋕ ♀

29 Monday
04 23 ☽ ∠ ♅
06 14 ☽ ⋕ ☿
06 22 ☽ ☍ ♀
07 30 ☽ □ ♃
13 05 ☽ Q ♂
17 53 ☽ ⋕ ☉
19 34 ♀ □ ♃
20 37 ☽ □ ♄
23 06 ☉ △ ♅
23 24 ☽ △ ♆

30 Tuesday
02 28 ☽ ☍ ♇
05 34 ☽ ⚹ ♅
06 04 ☽ ☍ ☉
09 22 ☽ ⚹ ♂
13 41 ☽ Q ♃
18 23 ☽ ⊼ ☿

31 Wednesday
00 16 ☽ ⚼ ♆
01 20 ☽ Q ♅
01 26 ☽ ∥ ♄
04 29 ☽ ± ☿
08 44 ☽ ⋕ ♃
09 25 ☽ ⚹ ♃
10 17 ☽ ∠ ♂
12 51 ☽ ⊼ ♀
15 00 ☽ ± ♆
19 37 ☽ ⚼ ☿
22 08 ☽ △ ♄
23 31 ☽ ± ♀

FEBRUARY 2029

LONGITUDES

Date	Sidereal time h m s	Sun ☉ ° ' "	Moon ☽ ° ' "	Moon ☽ 24.00 ° ' "	Mercury ☿ ° '	Venus ♀ ° '	Mars ♂ ° '	Jupiter ♃ ° '	Saturn ♄ ° '	Uranus ♅ ° '	Neptune ♆ ° '	Pluto ♇ ° '
01	20 47 55	12 ♒ 54 53	13 ♍ 36 13	20 ♍ 59 38	18 ♑ 51	00 ♒ 34	12 ♎ 57	27 ♎ 18	05 ♉ 05	10 ♊ 18	06 ♈ 42	08 ♒ 33
02	20 51 51	13 55 45	28 ♍ 22 12	05 ♎ 43 02	19 25	01 49	13 06	27 19	05 08	10 R 17	06 44	08 35
03	20 55 48	14 56 36	13 ♎ 01 22	20 ♎ 16 33	20 03	03 05	13 14	27 21	05 11	10 17	06 45	08 37
04	20 59 44	15 57 26	27 ♎ 28 03	04 ♏ 35 30	20 47	04 20	13 21	27 22	05 15	10 16	06 47	08 39
05	21 03 41	16 58 15	11 ♏ 38 37	18 ♏ 37 16	21 34	05 35	13 27	27 23	05 18	10 16	06 48	08 41
06	21 07 38	17 59 03	25 ♏ 31 26	02 ♐ 21 10	22 25	06 50	13 33	27 24	05 21	10 15	06 50	08 42
07	21 11 34	18 59 51	09 ♐ 06 35	15 ♐ 47 51	23 19	08 05	13 38	27 25	05 25	10 14	06 52	08 44
08	21 15 31	20 00 38	22 ♐ 25 10	28 ♐ 58 46	24 17	09 21	13 43	27 25	05 28	10 14	06 53	08 46
09	21 19 27	21 01 23	05 ♑ 28 52	11 ♑ 55 41	25 17	10 36	13 47	27 25	05 32	10 14	06 55	08 48
10	21 23 24	22 02 08	18 ♑ 19 24	24 ♑ 40 13	26 20	11 51	13 50	27 25	05 36	10 13	06 57	08 50
11	21 27 20	23 02 51	00 ♒ 58 16	07 ♒ 13 44	27 25	13 06	13 52	27 R 25	05 40	10 13	06 58	08 52
12	21 31 17	24 03 33	13 ♒ 26 43	19 ♒ 37 20	28 33	14 21	13 54	27 25	05 44	10 13	07 00	08 53
13	21 35 13	25 04 14	25 ♒ 45 41	01 ♓ 51 54	29 ♑ 43	15 36	13 55	27 25	05 48	10 13	07 02	08 55
14	21 39 10	26 04 53	07 ♓ 56 05	13 ♓ 58 23	00 ♒ 54	16 52	13 R 55	27 24	05 52	10 12	07 04	08 57
15	21 43 07	27 05 31	19 ♓ 58 56	25 ♓ 57 57	02 08	18 07	13 55	27 23	05 56	10 12	07 06	08 59
16	21 47 03	28 06 07	01 ♈ 55 38	07 ♈ 52 16	03 23	19 22	13 54	27 22	06 00	10 D 12	07 07	09 01
17	21 51 00	29 ♒ 06 41	13 ♈ 48 08	19 ♈ 43 37	04 39	20 37	13 52	27 21	06 05	10 12	07 09	09 03
18	21 54 56	00 ♓ 07 14	25 ♈ 39 05	01 ♉ 35 00	05 57	21 52	13 49	27 19	06 09	10 12	07 11	09 04
19	21 58 53	01 07 45	07 ♉ 31 50	13 ♉ 30 08	07 17	23 07	13 45	27 18	06 14	10 13	07 13	09 06
20	22 02 49	02 08 15	19 ♉ 30 26	25 ♉ 33 22	08 38	24 22	13 41	27 16	06 19	10 13	07 15	09 08
21	22 06 46	03 08 42	01 ♊ 39 31	07 ♊ 49 31	10 00	25 37	13 36	27 14	06 23	10 13	07 17	09 10
22	22 10 42	04 09 08	14 ♊ 04 00	20 ♊ 23 34	11 24	26 52	13 30	27 12	06 28	10 13	07 19	09 11
23	22 14 39	05 09 32	26 ♊ 48 47	03 ♋ 20 11	12 48	28 07	13 24	27 09	06 33	10 14	07 21	09 13
24	22 18 36	06 09 54	09 ♋ 58 14	16 ♋ 43 14	14 14	29 ♒ 22	13 16	27 07	06 38	10 14	07 23	09 15
25	22 22 32	07 10 14	23 ♋ 35 26	00 ♌ 34 53	15 41	00 ♓ 37	13 08	27 04	06 43	10 15	07 25	09 17
26	22 26 29	08 10 32	07 ♌ 41 27	14 ♌ 54 49	17 09	01 52	12 59	27 01	06 48	10 15	07 27	09 18
27	22 30 25	09 10 48	22 ♌ 14 25	29 ♌ 39 32	18 39	03 07	12 49	26 58	06 53	10 16	07 29	09 20
28	22 34 22	10 ♓ 11 02	07 ♍ 09 10	14 ♍ 42 13	20 ♒ 09	04 ♓ 22	12 ♎ 39	26 ♎ 55	06 ♉ 59	10 ♊ 16	07 ♈ 31	09 ♒ 22

Date	Moon True ☊ ° '	Moon Mean ☊ ° '	Moon ☽ Latitude ° '
01	13 ♑ 43	12 ♑ 28	04 S 31
02	13 R 39	12 25	05 01
03	13 35	12 21	05 12
04	13 32	12 18	05 03
05	13 31	12 15	04 36
06	13 D 31	12 12	03 52
07	13 32	12 09	02 57
08	13 34	12 06	01 53
09	13 35	12 02	00 S 44
10	13 R 35	11 59	00 N 26
11	13 34	11 56	01 33
12	13 30	11 53	02 35
13	13 24	11 50	03 28
14	13 17	11 47	04 12
15	13 08	11 43	04 43
16	12 59	11 40	05 02
17	12 51	11 37	05 07
18	12 44	11 34	05 00
19	12 39	11 31	04 39
20	12 36	11 27	04 06
21	12 35	11 24	03 22
22	12 D 35	11 21	02 27
23	12 36	11 18	01 24
24	12 37	11 15	00 N 14
25	12 R 36	11 12	00 S 59
26	12 34	11 08	02 10
27	12 29	11 05	03 15
28	12 ♑ 21	11 ♑ 02	04 S 09

DECLINATIONS

Date	Sun ☉ ° '	Moon ☽ ° '	Mercury ☿ ° '	Venus ♀ ° '	Mars ♂ ° '	Jupiter ♃ ° '	Saturn ♄ ° '	Uranus ♅ ° '	Neptune ♆ ° '	Pluto ♇ ° '
01	16 S 56	02 N 17	20 S 10	20 S 41	02 S 19	09 S 13	11 N 00	21 N 58	01 N 20	23 S 10
02	16 39	03 S 58	20 17	20 26	02 21	09 13	11 01	21 58	01 21	23 09
03	16 21	09 56	20 22	20 11	02 23	09 13	11 02	21 58	01 22	23 09
04	16 03	15 17	20 27	19 56	02 25	09 14	11 04	21 58	01 22	23 08
05	15 45	19 41	20 30	19 39	02 26	09 14	11 05	21 58	01 23	23 08
06	15 26	22 54	20 33	19 23	02 28	09 14	11 06	21 58	01 24	23 08
07	15 08	24 44	20 34	19 05	02 29	09 14	11 08	21 58	01 24	23 07
08	14 49	25 06	20 35	18 47	02 29	09 14	11 09	21 58	01 25	23 07
09	14 29	24 03	20 34	18 29	02 30	09 14	11 11	21 58	01 26	23 06
10	14 10	21 45	20 32	18 10	02 30	09 14	11 12	21 58	01 26	23 06
11	13 50	18 26	20 29	17 50	02 30	09 14	11 14	21 58	01 27	23 06
12	13 30	14 19	20 25	17 30	02 30	09 13	11 15	21 58	01 28	23 05
13	13 10	09 40	20 20	17 10	02 29	09 13	11 17	21 58	01 28	23 05
14	12 49	04 S 42	20 13	16 49	02 29	09 12	11 18	21 58	01 29	23 04
15	12 29	00 N 22	20 05	16 27	02 27	09 12	11 20	21 58	01 30	23 04
16	12 08	05 23	19 56	16 05	02 26	09 11	11 22	21 58	01 31	23 04
17	11 47	10 10	19 46	15 43	02 24	09 11	11 23	21 58	01 31	23 03
18	11 26	14 34	19 35	15 20	02 23	09 10	11 25	21 58	01 32	23 03
19	11 04	18 25	19 22	14 57	02 20	09 09	11 27	21 58	01 33	23 03
20	10 43	21 33	19 08	14 33	02 18	09 08	11 29	21 58	01 34	23 02
21	10 21	23 47	18 52	14 09	02 15	09 07	11 30	21 58	01 35	23 02
22	09 59	24 55	18 36	13 44	02 12	09 06	11 32	21 58	01 35	23 02
23	09 37	24 48	18 18	13 20	02 09	09 05	11 34	21 58	01 36	23 01
24	09 15	23 18	17 59	12 54	02 06	09 04	11 36	21 58	01 37	23 01
25	08 53	20 25	17 38	12 29	02 02	09 03	11 38	21 58	01 38	23 01
26	08 30	16 15	17 16	12 03	01 58	09 02	11 40	21 58	01 39	23 00
27	08 08	11 01	16 53	11 37	01 53	09 01	11 42	21 58	01 39	23 00
28	07 S 45	05 N 02	16 S 29	11 S 10	01 S 49	08 S 59	11 N 44	21 N 58	01 N 40	23 S 00

ZODIAC SIGN ENTRIES

Date	h	m	Planets	
01	01	03	♀	♒
02	14	39	☽	♎
04	16	15	☽	♏
06	19	51	☽	♐
09	01	53	☽	♑
11	10	09	☽	♒
13	17	52	☿	♒
13	20	20	☽	♓
16	08	07	☽	♈
18	09	08	☉	♓
18	20	48	☽	♉
21	08	45	☽	♊
23	17	53	☽	♋
25	00	03	♀	♓
25	23	01	☽	♌
28	00	33	☽	♍

LATITUDES

Date	Mercury ☿ ° '	Venus ♀ ° '	Mars ♂ ° '	Jupiter ♃ ° '	Saturn ♄ ° '	Uranus ♅ ° '	Neptune ♆ ° '	Pluto ♇ ° '
01	01 N 57	00 S 40	03 N 02	01 N 23	02 S 21	00 S 01	01 S 26	05 S 14
04	01 24	00 47	03 06	01 24	02 20	00 01	01 26	05 14
07	00 52	00 53	03 09	01 24	02 20	00 01	01 26	05 14
10	00 N 21	00 58	03 12	01 25	02 19	00 01	01 26	05 14
13	00 S 07	01 03	03 15	01 26	02 18	00 01	01 26	05 15
16	00 33	01 08	03 18	01 27	02 17	00 01	01 26	05 15
19	00 57	01 12	03 21	01 27	02 16	00 01	01 26	05 15
22	01 17	01 16	03 23	01 28	02 16	00 01	01 26	05 16
25	01 35	01 19	03 25	01 29	02 15	00 01	01 26	05 16
28	01 49	01 21	03 27	01 29	02 14	00 01	01 26	05 17
31	02 S 00	01 S 23	03 N 28	01 N 30	02 S 14	00 S 01	01 S 26	05 S 17

DATA

Julian Date	2462169
Delta T	+74 seconds
Ayanamsa	24° 16' 07"
Synetic vernal point	04° ♓ 50' 52"
True obliquity of ecliptic	23° 26' 10"

LONGITUDES

Date	Chiron ⚷ ° '	Ceres ⚳ ° '	Pallas ⚴ ° '	Juno ⚵ ° '	Vesta ⚶ ° '	Black Moon Lilith ⚸ ° '
01	04 ♉ 04	20 ♑ 14	23 ♐ 49	08 ♏ 17	18 ♐ 48	06 ♈ 53
11	04 ♉ 18	24 ♑ 06	27 ♐ 29	09 ♏ 18	23 ♐ 33	08 ♈ 00
21	04 ♉ 37	27 ♑ 52	00 ♑ 55	09 ♏ 48	28 ♐ 10	09 ♈ 08
31	05 ♉ 01	01 ♒ 32	04 ♑ 06	09 ♏ 43	02 ♑ 34	10 ♈ 15

MOON'S PHASES, APSIDES AND POSITIONS ☽

Date	h	m	Phase	Longitude ° '	Eclipse Indicator
05	21	52	☾	17 ♏ 23	
13	10	31	●	25 ♒ 00	
21	15	10	☽	03 ♊ 17	
28	17	10	○	10 ♍ 24	

Day	h	m		
01	12	29	Perigee	
17	12	04	Apogee	
01	20	45	0S	
08	06	07	Max dec	25° S 08'
15	10	14	0N	
22	21	46	Max dec	25° N 02'

ASPECTARIAN

h m Aspects

01 Thursday
00 46 ☽ ⊼ ♆
01 06 ☽ ⊥ ♂
03 47 ☽ ⊼ ♇
06 39 ☽ □ ♅
09 52 ☽ ∠ ♃
10 48 ☽ ⊼ ☉
10 56 ☽ ⊻ ♂
11 52 ☽ # ♂
13 11 ☉ △ ♂
13 32 ☽ ± ♇
15 29 ☽ bQ ♀
15 37 ☽ ∥ ♆
20 50 ☽ △ ☿
21 15 ☽ ± ☉
22 34 ☽ bQ ♄

02 Friday
00 31 ☽ ⊥ ♃
01 54 ☽ # ♆
04 11 ☽ bQ ♇
05 45 ☽ ∥ ♂
10 17 ☽ ⊻ ♃
12 59 ☽ bQ ☉
13 16 ☽ ± ♄
18 10 ☽ △ ♀
23 06 ☽ ⊼ ♄
23 20 ☿ ∥ ♀

03 Saturday
01 41 ☽ ☍ ♆
04 44 ☽ △ ♇
07 29 ☽ △ ♅
09 04 ☽ ∥ ♃
12 20 ☽ ☌ ♂
15 24 ☽ △ ☉
16 43 ☽ # ♄

04 Sunday
00 14 ☽ □ ☿
08 19 ☽ bQ ♅
11 50 ☽ ☌ ♃
15 35 ☽ ∥ ☉
23 27 ☽ ± ♅

05 Monday
00 41 ☽ □ ♀
01 09 ☽ ☍ ♄
03 44 ☽ ⊼ ♆
06 18 ♀ □ ♄
06 55 ☽ □ ♇
08 14 ☽ Q ☿
09 38 ☽ ⊼ ♅
11 49 ☽ ∥ ♀
14 00 ☽ ± ♆
15 08 ☽ ⊻ ♂
17 21 ☽ ∥ ☿
21 52 ☽ □ ☉

06 Tuesday
01 33 ☽ ⊥ ♂
03 56 ☽ # ♅
05 33 ☽ bQ ♆
06 13 ☽ ✱ ☿
10 41 ☽ Q ♀
11 56 ♀ ✱ ♆
14 04 ☽ Q ♇
14 12 ☽ ∥ ♇
15 17 ☽ ⊻ ♃
17 21 ☽ ∠ ♂

07 Wednesday
01 52 ☽ ⊥ ♃
05 24 ☽ ⊼ ♄
07 56 ☽ Q ☉
07 59 ☽ △ ♆
10 00 ☽ ✱ ♀
10 29 ☽ ∠ ☿
11 20 ☽ ✱ ♇
14 01 ☽ ☍ ♅
16 08 ☽ ± ♄
17 55 ☽ ∠ ♃
20 10 ☽ ✱ ♂

08 Thursday
00 43 ♀ ☌ ♇
03 53 ☽ ⊥ ☿
07 15 ☽ ✱ ☉
08 27 ☽ bQ ♄
14 28 ☽ ∠ ♇
15 40 ☽ ⊻ ☿
15 53 ☽ ∠ ♀
18 03 ☽ Q ♂
21 08 ☽ ✱ ♃

09 Friday
03 04 ☿ Q ♆
04 58 ♀ △ ♅
07 02 ☽ ⊥ ♇
10 11 ☽ ⊥ ♀
10 38 ☿ bQ ♅
12 06 ☽ △ ♄
13 05 ☽ ∠ ☉
14 40 ☽ □ ♆
18 11 ☽ ⊻ ♇
19 20 ☽ Q ♃
20 49 ☽ ⊼ ♅
22 32 ☽ ⊻ ♀
23 27 ☽ ∥ ♇

10 Saturday
03 32 ☽ □ ♂
07 20 ☽ ⊥ ☉
08 03 ☽ ± ♅
09 48 ☉ ∠ ♆
10 15 ☽ # ♅
13 07 ♃ St R
19 37 ☽ ⊻ ☉
21 42 ☽ ∥ ☿

11 Sunday
00 33 ☽ Q ♆
01 02 ☽ bQ ♅
04 35 ☽ ☌ ☿
05 14 ☽ □ ♃
11 58 ☿ □ ♃
16 02 ☽ ∥ ♀
21 02 ☽ □ ♄
23 32 ☽ ✱ ♆

12 Monday
03 10 ♀ △ ♂
03 11 ☽ ☌ ♇
03 36 ☉ Q ♄
05 45 ☽ △ ♅
12 53 ☽ △ ♂
13 58 ☽ ☌ ♀
16 42 ☽ ∥ ☉

13 Tuesday
03 55 ☽ # ♄
04 41 ☽ ∠ ♆
08 08 ☽ Q ♄
10 31 ☽ ☌ ☉
14 12 ☽ ∥ ♃
15 14 ☽ △ ♃
18 12 ☽ bQ ♂
20 35 ☽ ⊻ ☿
22 23 ☽ ⊥ ♆

14 Wednesday
07 53 ☽ ✱ ♄
08 16 ♂ St R
09 44 ☽ ⊥ ☿
10 16 ☽ ⊻ ♆
11 59 ☽ ± ♂
14 01 ☽ ⊻ ♇
16 31 ☽ □ ♅
20 51 ☽ bQ ♃
22 34 ☽ ∥ ♂
23 54 ☽ ⊼ ♂

15 Thursday
01 59 ☽ ⊥ ♇
03 10 ☽ # ♆
05 38 ☽ ∠ ☿
07 49 ☽ ⊻ ♀
13 55 ☽ ∠ ♄
14 48 ☽ ± ♃
17 21 ☽ ∥ ♆
18 51 ☉ △ ♃
20 02 ☽ ∠ ♇
21 15 ☽ ⊥ ♀
21 52 ☽ # ♂

16 Friday
02 50 ☽ ⊼ ♃
03 35 ☽ ⊻ ☉
04 30 ☽ Q ♅
08 06 ☽ ⊥ ♄
10 51 ♅ St D
15 16 ☽ ✱ ☿
16 48 ☽ ⊥ ☉
17 30 ☽ ∠ ♀
20 17 ☽ ⊻ ♄
22 31 ☽ ☌ ♆

17 Saturday
02 21 ☽ ✱ ♇
04 43 ☽ ✱ ♅
06 15 ☉ bQ ♂
06 56 ☽ # ♃
12 07 ☽ ☍ ♂
12 41 ☽ ∠ ☉
18 29 ☽ Q ☿
18 30 ☽ ∥ ♄
19 56 ☽ # ☉

18 Sunday
02 42 ☽ Q ♇
03 26 ☽ ✱ ♀
11 06 ☽ ∠ ♅
12 31 ☉ # ♄
15 23 ☽ ☍ ♃
15 51 ☿ □ ♄
16 07 ☽ # ♀
18 19 ♀ ∠ ♆
21 53 ☽ ✱ ☉

19 Monday
05 18 ☽ ⊥ ♅
06 34 ☽ Q ♀
09 22 ☽ ☌ ♄
10 50 ☿ ✱ ♆
11 22 ☽ ⊻ ♆
11 26 ☽ □ ☿
14 13 ☉ ⊥ ♆
15 10 ☽ □ ♇
17 24 ☽ ⊻ ♅
18 12 ☽ # ☿
23 28 ☽ ⊥ ♆

20 Tuesday
00 17 ☽ Q ☉
00 26 ☽ ⊼ ♂
10 47 ♀ Q ♄
12 21 ☽ ± ♂
15 43 ☽ ∥ ♅
17 28 ☽ ∠ ♆
20 59 ☿ ☌ ♇
22 46 ☽ □ ♀

21 Wednesday
02 37 ☽ # ♇
03 20 ☽ ⊼ ♃
06 03 ☽ bQ ♂
15 04 ☽ ± ♃
15 10 ☽ □ ☉
15 45 ☿ △ ♅
21 17 ☽ ⊻ ♄
22 59 ☽ ✱ ♆

22 Thursday
02 37 ☽ △ ♇
04 37 ☽ ☌ ♅
06 14 ☽ △ ☿
08 26 ☽ bQ ♃
08 55 ☽ ⊥ ♄
10 56 ☽ △ ♂
18 05 ♀ △ ♃
22 00 ☽ Q ♆

23 Friday
02 07 ☽ ∠ ♄
07 10 ☽ bQ ♇
12 38 ☽ △ ♃
14 04 ☽ bQ ☿
14 41 ☽ △ ♀
16 46 ♀ bQ ♂
21 08 ☿ △ ♂
23 49 ☽ ± ♇

24 Saturday
04 35 ☽ △ ☉
05 57 ☽ ✱ ♄
07 20 ☽ □ ♆
08 30 ☽ ± ☿
10 42 ☽ ⊼ ♇
12 28 ☽ ⊻ ♅
14 59 ☽ # ♇
17 50 ☽ □ ♂
20 31 ☽ ⊼ ☿
20 39 ☽ bQ ♀
23 09 ☽ ⊥ ♅

25 Sunday
00 10 ☉ ∥ ♃
00 13 ☉ ✱ ♄
00 29 ☽ ∥ ♅
03 28 ☽ Q ♄
09 21 ☽ bQ ☉
11 14 ☉ ± ♂
13 58 ☽ ± ♀
14 51 ☽ ∠ ♅
17 59 ☽ □ ♃
18 04 ☉ ⊻ ♆

26 Monday
00 48 ☽ Q ♂
01 15 ☽ ⊼ ♀
02 00 ☽ ± ☉
03 40 ♀ ⊥ ♆
06 11 ☽ # ☿
10 30 ☽ □ ♄
11 36 ☽ △ ♆
12 52 ☽ ⊼ ☉
14 42 ☽ ☍ ♇
16 17 ☽ ✱ ♅
20 43 ☽ ✱ ♂

27 Tuesday
00 08 ☽ Q ♃
05 28 ☽ ☍ ☿
07 50 ♀ # ♄
09 07 ☽ ∥ ♄
09 14 ☽ # ♀
12 02 ☽ Q ♅
12 24 ☽ bQ ♆
15 43 ☉ ⊻ ♇
19 38 ☽ ✱ ♃
20 20 ☽ # ♃
20 56 ☽ ∠ ♂

28 Wednesday
00 39 ☽ # ☉
02 58 ☽ ± ♆
07 09 ☽ ☍ ♀
11 12 ☽ ⊥ ♂
11 43 ☽ △ ♄
12 35 ☽ ⊼ ♆
14 06 ☉ □ ♅
15 31 ☽ ⊼ ♇
16 58 ☽ □ ♅
17 10 ☽ ☍ ☉
19 33 ☽ ∠ ♃
20 38 ☽ ⊻ ♂

All ephemeris data is given at 12.00 UT and the Moon's longitude is additionally given for 24.00 UT

MARCH 2029

LONGITUDES

Date	Sidereal time h m s	Sun ☉ ° ' "	Moon ☽ ° ' "	Moon ☽ 24.00 ° ' "	Mercury ☿ ° '	Venus ♀ ° '	Mars ♂ ° '	Jupiter ♃ ° '	Saturn ♄ ° '	Uranus ♅ ° '	Neptune ♆ ° '	Pluto ♇ ° '
01	22 38 18	11 ♓ 11 14	22 ♍ 17 24	29 ♍ 53 22	21 ♒ 41	05 ♓ 37	12 ♎ 28	26 ♎ 52	07 ♉ 04	10 ♊ 17	07 ♈ 33	09 ♒ 23
02	22 42 15	12 11 25	07 ♎ 28 46	15 ♎ 02 17	23 13	06 52	12 R 16	26 R 48	07 10	10 18	07 35	09 25
03	22 46 11	13 11 34	22 ♎ 32 43	29 ♎ 59 01	24 47	08 07	12 03	26 44	07 15	10 18	07 37	09 26
04	22 50 08	14 11 41	07 ♏ 20 19	14 ♏ 35 57	26 22	09 22	11 50	26 40	07 21	10 19	07 39	09 28
05	22 54 05	15 11 47	21 ♏ 45 30	28 ♏ 48 42	27 58	10 37	11 35	26 36	07 26	10 20	07 42	09 30
06	22 58 01	16 11 52	05 ♐ 45 31	12 ♐ 36 01	29 ♒ 35	11 52	11 20	26 32	07 32	10 21	07 44	09 31
07	23 01 58	17 11 55	19 ♐ 20 27	25 ♐ 59 06	01 ♓ 13	13 07	11 05	26 28	07 38	10 22	07 46	09 33
08	23 05 54	18 11 56	02 ♑ 32 24	09 ♑ 00 45	02 52	14 22	10 48	26 23	07 44	10 23	07 48	09 34
09	23 09 51	19 11 56	15 ♑ 24 37	21 ♑ 44 27	04 32	15 37	10 31	26 18	07 50	10 24	07 50	09 36
10	23 13 47	20 11 54	28 ♑ 00 42	04 ♒ 13 47	06 13	16 51	10 14	26 13	07 56	10 25	07 52	09 37
11	23 17 44	21 11 50	10 ♒ 24 06	16 ♒ 31 58	07 56	18 06	09 56	26 08	08 02	10 26	07 55	09 39
12	23 21 40	22 11 45	22 ♒ 37 45	28 ♒ 41 40	09 40	19 21	09 37	26 03	08 08	10 28	07 57	09 40
13	23 25 37	23 11 38	04 ♓ 44 00	10 ♓ 44 55	11 24	20 36	09 17	25 58	08 14	10 29	07 59	09 42
14	23 29 34	24 11 29	16 ♓ 44 38	22 ♓ 43 17	13 10	21 51	08 57	25 52	08 20	10 30	08 01	09 43
15	23 33 30	25 11 18	28 ♓ 41 01	04 ♈ 38 01	14 58	23 06	08 37	25 47	08 26	10 32	08 03	09 45
16	23 37 27	26 11 05	10 ♈ 34 23	16 ♈ 30 20	16 46	24 20	08 16	25 41	08 33	10 33	08 06	09 46
17	23 41 23	27 10 49	22 ♈ 26 03	28 ♈ 21 45	18 36	25 35	07 55	25 35	08 39	10 35	08 08	09 48
18	23 45 20	28 10 32	04 ♉ 17 42	10 ♉ 14 13	20 26	26 50	07 33	25 29	08 46	10 36	08 10	09 49
19	23 49 16	29 ♓ 10 13	16 ♉ 11 37	22 ♉ 10 20	22 18	28 05	07 11	25 23	08 52	10 38	08 12	09 50
20	23 53 13	00 ♈ 09 52	28 ♉ 10 47	04 ♊ 13 28	24 12	29 ♓ 19	06 48	25 16	08 59	10 39	08 15	09 52
21	23 57 09	01 09 28	10 ♊ 18 55	16 ♊ 27 41	26 06	00 ♈ 34	06 26	25 10	09 05	10 41	08 17	09 53
22	00 01 06	02 09 02	22 ♊ 40 22	28 ♊ 57 35	28 02	01 49	06 03	25 04	09 12	10 43	08 19	09 54
23	00 05 03	03 08 34	05 ♋ 19 55	11 ♋ 47 59	29 ♓ 58	03 03	05 40	24 57	09 19	10 44	08 21	09 55
24	00 08 59	04 08 04	18 ♋ 22 19	25 ♋ 03 24	01 ♈ 56	04 18	05 16	24 50	09 25	10 46	08 24	09 57
25	00 12 56	05 07 31	01 ♌ 51 38	08 ♌ 47 17	03 55	05 33	04 53	24 43	09 32	10 48	08 26	09 58
26	00 16 52	06 06 56	15 ♌ 50 27	23 ♌ 01 02	05 55	06 47	04 30	24 37	09 39	10 50	08 28	09 59
27	00 20 49	07 06 18	00 ♍ 18 45	07 ♍ 43 02	07 56	08 02	04 06	24 30	09 46	10 52	08 30	10 00
28	00 24 45	08 05 38	15 ♍ 13 07	22 ♍ 47 58	09 58	09 16	03 43	24 22	09 53	10 54	08 33	10 01
29	00 28 42	09 04 56	00 ♎ 26 21	08 ♎ 06 52	12 00	10 31	03 20	24 15	10 00	10 56	08 35	10 03
30	00 32 38	10 04 12	15 ♎ 48 03	23 ♎ 28 21	14 03	11 45	02 57	24 08	10 07	10 58	08 37	10 04
31	00 36 35	11 ♈ 03 26	01 ♏ 06 20	08 ♏ 40 37	16 ♈ 06	13 ♈ 00	02 ♎ 34	24 ♎ 01	10 ♉ 14	11 ♊ 00	08 ♈ 40	10 ♒ 05

Moon True ☊, Moon Mean ☊, Moon ☽ Latitude — DECLINATIONS

Date	Moon True ☊ ° '	Moon Mean ☊ ° '	Moon ☽ Latitude ° '	Sun ☉ ° '	Moon ☽ ° '	Mercury ☿ ° '	Venus ♀ ° '	Mars ♂ ° '	Jupiter ♃ ° '	Saturn ♄ ° '	Uranus ♅ ° '	Neptune ♆ ° '	Pluto ♇ ° '
01	12 ♑ 13	10 ♑ 59	04 S 46	07 S 22	01 S 19	16 S 04	10 S 43	01 S 44	08 S 58	11 N 45	21 N 59	01 N 41	22 S 59
02	12 R 03	10 56	05 03	06 59	07 36	15 37	10 16	01 39	08 56	11 47	21 59	01 42	22 59
03	11 55	10 53	04 59	06 36	13 23	15 09	09 49	01 34	08 55	11 49	21 59	01 43	22 59
04	11 48	10 49	04 35	06 13	18 17	14 39	09 21	01 28	08 53	11 51	21 59	01 44	22 58
05	11 43	10 46	03 53	05 50	21 57	14 09	08 53	01 23	08 51	11 53	21 59	01 44	22 58
06	11 41	10 43	02 59	05 27	24 12	13 37	08 25	01 17	08 50	11 55	21 59	01 45	22 58
07	11 D 40	10 40	01 56	05 03	24 56	13 04	07 57	01 10	08 48	11 58	21 59	01 46	22 58
08	11 41	10 37	00 S 48	04 40	24 13	12 29	07 28	01 04	08 46	12 00	21 59	01 47	22 57
09	11 R 41	10 33	00 N 20	04 16	22 13	11 53	07 00	00 57	08 44	12 02	22 00	01 48	22 57
10	11 39	10 30	01 26	03 53	19 09	11 16	06 31	00 50	08 42	12 04	22 00	01 49	22 57
11	11 36	10 27	02 26	03 29	15 17	10 38	06 01	00 43	08 40	12 06	22 00	01 50	22 57
12	11 29	10 24	03 19	03 06	10 50	09 59	05 32	00 36	08 38	12 08	22 00	01 50	22 56
13	11 20	10 21	04 02	02 42	06 01	09 18	05 03	00 29	08 36	12 10	22 00	01 51	22 56
14	11 08	10 18	04 34	02 18	01 S 01	08 36	04 33	00 22	08 34	12 12	22 01	01 52	22 56
15	10 55	10 14	04 53	01 55	03 N 58	07 53	04 03	00 14	08 32	12 15	22 01	01 53	22 56
16	10 40	10 11	05 00	01 31	08 47	07 09	03 34	00 S 06	08 30	12 17	22 01	01 54	22 55
17	10 27	10 08	04 54	01 07	13 16	06 24	03 04	00 N 02	08 27	12 19	22 01	01 55	22 55
18	10 15	10 05	04 35	00 44	17 15	05 37	02 34	00 09	08 25	12 21	22 01	01 56	22 55
19	10 05	10 02	04 03	00 S 20	20 34	04 50	02 04	00 17	08 23	12 23	22 02	01 57	22 55
20	09 59	09 59	03 21	00 N 04	23 01	04 01	01 34	00 25	08 20	12 26	22 02	01 58	22 55
21	09 55	09 55	02 29	00 28	24 27	03 11	01 03	00 33	08 18	12 28	22 02	01 58	22 54
22	09 54	09 52	01 30	00 51	24 44	02 20	00 33	00 41	08 15	12 30	22 02	01 59	22 54
23	09 D 53	09 49	00 N 24	01 15	23 44	01 28	00 S 03	00 50	08 13	12 32	22 03	02 00	22 54
24	09 R 53	09 46	00 S 45	01 39	21 26	00 S 36	00 N 27	00 58	08 10	12 35	22 03	02 01	22 54
25	09 52	09 43	01 53	02 02	17 54	00 N 18	00 58	01 06	08 08	12 37	22 03	02 02	22 54
26	09 49	09 39	02 58	02 26	13 16	01 12	01 28	01 14	08 05	12 39	22 03	02 03	22 54
27	09 43	09 36	03 53	02 49	07 44	02 07	01 58	01 21	08 03	12 42	22 04	02 04	22 54
28	09 34	09 33	04 34	03 13	01 N 37	03 03	02 28	01 29	08 00	12 44	22 04	02 05	22 53
29	09 23	09 30	04 56	03 36	04 S 42	03 59	02 59	01 37	07 57	12 46	22 04	02 06	22 53
30	09 12	09 27	04 58	03 59	10 48	04 55	03 29	01 45	07 55	12 49	22 05	02 06	22 53
31	09 ♑ 01	09 ♑ 24	04 S 38	04 N 22	16 S 12	05 N 52	03 N 59	01 N 52	07 S 52	12 N 51	22 N 05	02 N 07	22 S 53

ZODIAC SIGN ENTRIES

Date	h	m	Planets
02	00	10	☽ ♎
04	00	02	☽ ♏
06	02	02	☽ ♐
06	18	15	☿ ♓
08	07	20	☽ ♑
10	15	50	☽ ♒
13	02	35	☽ ♓
15	14	39	☽ ♈
18	03	19	☽ ♉
20	08	02	☉ ♈
20	15	37	☽ ♊
21	01	04	♀ ♈
23	01	58	☽ ♋
23	12	18	☿ ♈
25	08	44	☽ ♌
27	11	29	☽ ♍
29	11	19	☽ ♎
31	10	15	☽ ♏

LATITUDES

Date	Mercury ☿ ° '	Venus ♀ ° '	Mars ♂ ° '	Jupiter ♃ ° '	Saturn ♄ ° '	Uranus ♅ ° '	Neptune ♆ ° '	Pluto ♇ ° '
01	01 S 53	01 S 22	03 N 28	01 N 29	02 S 14	00 S 01	01 S 26	05 S 17
04	02 03	01 24	03 29	01 30	02 13	00 01	01 26	05 17
07	02 09	01 25	03 29	01 31	02 13	00 01	01 26	05 18
10	02 12	01 26	03 29	01 31	02 12	00 01	01 26	05 18
13	02 11	01 26	03 29	01 32	02 12	00 01	01 26	05 19
16	02 06	01 26	03 27	01 32	02 11	00 01	01 26	05 19
19	01 56	01 25	03 25	01 32	02 11	00 01	01 26	05 20
22	01 42	01 23	03 22	01 33	02 10	00 01	01 26	05 20
25	01 22	01 21	03 18	01 33	02 10	00 S 01	01 26	05 21
28	00 59	01 18	03 14	01 33	02 09	00 00	01 26	05 22
31	00 S 30	01 S 15	03 N 09	01 N 34	02 S 09	00 00	01 S 26	05 S 22

DATA

Julian Date	2462197
Delta T	+74 seconds
Ayanamsa	24° 16' 10"
Synetic vernal point	04° ♓ 50' 49"
True obliquity of ecliptic	23° 26' 10"

MOON'S PHASES, APSIDES AND POSITIONS ☽

Date	h	m	Phase	Longitude ° '	Eclipse Indicator
07	07	52	☾	17 ♐ 02	
15	04	19	●	24 ♓ 52	
23	07	33	☽	02 ♋ 58	
30	02	26	○	09 ♎ 41	

Day	h	m		
01	18	36	Perigee	
16	21	45	Apogee	
30	05	46	Perigee	
01	07	02	0S	
07	11	51	Max dec	24° S 56'
14	16	54	0N	
22	05	16	Max dec	24° N 47'
28	18	09	0S	

LONGITUDES

Date	Chiron ⚷ ° '	Ceres ⚳ ° '	Pallas ⚴ ° '	Juno ⚵ ° '	Vesta ⚶ ° '	Black Moon Lilith ⚸ ° '
01	04 ♉ 56	00 ♒ 49	03 ♑ 29	09 ♏ 47	01 ♑ 43	10 ♈ 01
11	05 ♉ 23	04 ♒ 22	06 ♑ 26	09 ♏ 14	05 ♑ 56	11 ♈ 09
21	05 ♉ 54	07 ♒ 47	09 ♑ 01	08 ♏ 07	09 ♑ 54	12 ♈ 16
31	06 ♉ 28	11 ♒ 00	11 ♑ 12	06 ♏ 29	13 ♑ 33	13 ♈ 23

All ephemeris data is given at 12.00 UT and the Moon's longitude is additionally given for 24.00 UT

ASPECTARIAN

h m Aspects

01 Thursday
00 22 ☽ ⋕ ♂
00 44 ☽ ∥ ♆
01 04 ☽ ± ♇
09 45 ☽ ⊥ ♃
10 56 ☽ ⊼ ☿
11 39 ☽ bQ ♄
13 22 ☽ ⋕ ♆
13 32 ☽ ∥ ♂
15 19 ☽ bQ ♇
19 11 ☽ ⊻ ♃
21 28 ☽ ± ☿

02 Friday
00 42 ♂ ⋕ ♆
01 57 ☿ ∠ ♆
01 57 ☽ ± ♄
01 57 ♀ ± ♂
03 12 ☉ bQ ♃
09 44 ☽ ∥ ☉
10 57 ☽ ⊼ ♀
11 29 ☽ ⊼ ♄
12 10 ☽ ☍ ♆
13 19 ☽ bQ ☿
13 26 ☉ ⊼ ♂
15 04 ☽ △ ♇
16 28 ☽ △ ♅
17 16 ☽ ∥ ♃
18 00 ♀ ✶ ♄
19 29 ☽ ☌ ♂
20 00 ☽ ⊼ ☉
21 19 ☽ ± ♀
21 56 ☽ ∥ ♀

03 Saturday
02 11 ♀ ⊻ ♆
05 09 ☽ ⋕ ♄
06 15 ☽ ± ☉
13 00 ☽ bQ ♀
16 01 ☽ △ ☿
16 27 ☽ bQ ♅
18 43 ☽ ☌ ♃
19 20 ☽ ∥ ☿
19 35 ☿ Q ♄
21 46 ☽ bQ ☉

04 Sunday
07 03 ☽ ± ♅
12 01 ☽ ☍ ♄
12 32 ☽ ⊼ ♆
13 58 ♀ ⊻ ♇
15 31 ☽ □ ♇
15 39 ☽ △ ♀
16 29 ☿ △ ♃
16 55 ☽ ⊼ ♅
18 05 ☿ bQ ♂
19 17 ☽ ⊻ ♂
22 28 ☽ ± ♆

05 Monday
00 10 ☽ △ ☉
05 06 ☽ ⊥ ♂
06 31 ♀ □ ♅
12 14 ☽ ⋕ ♅
13 35 ☽ bQ ♆
13 40 ♀ ∥ ♃
19 20 ☉ ⊥ ♇
20 03 ☽ ∠ ♂
20 11 ☽ ⊻ ♃
20 57 ☽ ∥ ♇
21 46 ☽ Q ♇
23 55 ☽ □ ☿

06 Tuesday
03 36 ♀ ⊼ ♂
05 57 ♀ bQ ♃
06 26 ☽ ⊥ ♃
15 07 ☽ ⊼ ♄
15 27 ☽ △ ♆
18 35 ☽ ✶ ♇
20 03 ☽ ☍ ♅
21 36 ☽ ✶ ♂
22 04 ☽ ∠ ♃
23 47 ☽ □ ♀

07 Wednesday
01 45 ☽ ± ♄
07 52 ☽ □ ☉
11 44 ☽ Q ☿
17 58 ☽ bQ ♄
18 36 ☽ Q ♂
20 16 ☿ ⊥ ♆
21 24 ☽ ∠ ♇

08 Thursday
00 47 ☽ ✶ ♃
11 38 ☽ Q ♀
12 41 ☽ ✶ ☿
13 55 ☽ ⊥ ♇
19 20 ☽ Q ☉
21 41 ☽ △ ♄
21 46 ☽ □ ♆
22 46 ☽ Q ♃

09 Friday
01 04 ☽ ⊻ ♇
02 35 ☽ ⊼ ♅
03 01 ☽ □ ♂
04 41 ☽ ∥ ♇
06 48 ☿ ⋕ ♄
11 47 ♀ ⊥ ♇
11 53 ☿ ± ♂
12 25 ☽ ✶ ♀
13 52 ☽ ± ♅
14 03 ☽ ⋕ ♅
15 57 ♄ ⊻ ♆
19 47 ☽ ✶ ☉
21 00 ☽ ∠ ☿
21 28 ♂ △ ♅

10 Saturday
07 01 ☽ bQ ♅
07 53 ☽ Q ♆
08 35 ☽ □ ♃
12 31 ☉ ± ♃
16 56 ☽ ⊥ ☿
20 14 ☽ ∠ ♀

11 Sunday
03 06 ☽ ∠ ☉
06 25 ☽ ⊻ ☿
07 08 ☽ ✶ ♆
07 20 ☽ □ ♄
10 32 ☽ ☌ ♇
11 06 ☽ △ ♂
11 41 ☿ ⊻ ♆
12 05 ☽ △ ♅
13 25 ☿ ✶ ♄
15 42 ☽ ⊥ ♀
22 13 ☽ ⊥ ☉

12 Monday
04 48 ☽ ⊻ ♀
05 17 ☽ ⋕ ♄
07 40 ♂ △ ♇
11 04 ☽ ⊻ ☉
11 26 ☿ ⊼ ♂
12 12 ☿ ⊻ ♇
12 38 ☽ ∠ ♆
15 49 ☽ bQ ♂
17 03 ☽ ∥ ☿
18 43 ☽ △ ♃
18 59 ☽ Q ♄
23 10 ☽ ∥ ♃
23 11 ☿ □ ♅

13 Tuesday
00 32 ♀ ± ♃
06 12 ☿ bQ ♃
06 31 ☽ ⊥ ♆
09 12 ☽ ± ♂
13 01 ☉ ∠ ♄
17 13 ☽ ∥ ♀
18 30 ☽ ⊻ ♆
19 02 ☽ ✶ ♄
20 50 ☽ ⊼ ♂
21 55 ☽ ⊻ ♇
23 29 ☽ □ ♅

14 Wednesday
00 20 ☽ bQ ♃
03 37 ☽ ☌ ☿
05 20 ☽ ∥ ☉
07 58 ☽ ⋕ ♆
09 57 ☽ ⊥ ♇
13 22 ☿ ∥ ♃
15 16 ☽ ∥ ♂
18 13 ☽ ± ♃
18 27 ☽ ⋕ ♂
23 26 ☽ ☌ ♀

15 Thursday
01 05 ☉ ∠ ♇
01 21 ☽ ∠ ♄
01 55 ☽ ∥ ♆
02 48 ☽ ⋕ ☉
04 03 ☽ ∠ ♇
04 19 ☽ ☌ ☉
06 11 ☽ ⊼ ♃
11 41 ☽ Q ♅
12 25 ☽ ⋕ ♀
13 33 ☉ ⋕ ♆
19 18 ♀ ∠ ♄
19 38 ☽ ⊥ ♄
21 15 ♂ ⊼ ♄
22 36 ☿ ⊥ ♇

16 Friday
00 54 ☉ ⊼ ♃
04 51 ☽ ⋕ ☿
06 58 ☽ ☌ ♆
07 28 ☽ ☍ ♂
07 52 ☽ ⊻ ♄
10 22 ☽ ✶ ♇
10 32 ☽ ⋕ ♃
11 57 ☽ ✶ ♅
20 26 ♀ ∠ ♇
22 30 ♂ ☍ ♆

17 Saturday
02 48 ☽ ⊻ ☿
06 39 ☽ ∥ ♄
10 42 ☽ Q ♇
11 55 ♀ ⊼ ♃
12 51 ♃ bQ ♅
17 10 ☽ ⊥ ☿
18 19 ☽ ☍ ♃
18 22 ☽ ∠ ♅
19 08 ☽ ⊻ ♀
22 29 ☽ ⊻ ☉

18 Sunday
00 12 ☿ ± ♃
08 42 ☽ ⊥ ♀
11 44 ☽ ⊥ ☉
12 37 ☽ ⊥ ♅
14 45 ☽ ∠ ☿
18 22 ☽ ⊼ ♂
19 51 ☽ ⊻ ♆
21 06 ☽ ☌ ♄
22 33 ☉ Q ♅
23 10 ☽ □ ♇

19 Monday
00 46 ☽ ⊻ ♅
05 00 ☽ ∠ ♀
06 07 ☽ ± ♂
07 33 ☽ ∠ ☉
07 59 ☽ ⊥ ♆
13 48 ☉ ⋕ ♂
17 24 ♀ ⋕ ♆
22 51 ♀ Q ♅
23 39 ☽ bQ ♂

20 Tuesday
01 11 ☽ ∥ ♅
02 07 ☽ ∠ ♆
02 33 ☽ ✶ ☿
06 15 ☽ ⊼ ♃
09 05 ☿ ∠ ♄
10 40 ☽ ⋕ ♇
14 32 ☽ ✶ ♀
16 18 ☽ ✶ ☉
18 06 ☽ ± ♃
20 30 ☿ ∠ ♇

21 Wednesday
00 53 ☿ ⊼ ♃
04 35 ☽ △ ♂
06 50 ☽ Q ☿
08 00 ☽ ✶ ♆
09 34 ☽ ⊻ ♄
11 09 ☽ △ ♇
11 43 ☽ bQ ♃
12 43 ☽ ☌ ♅
16 55 ☽ Q ♀
18 03 ☽ Q ☉
20 55 ☉ ∥ ♂
21 25 ☽ ⊥ ♄

22 Thursday
03 53 ☉ ⋕ ♀
06 44 ♀ ⋕ ♂
07 27 ☽ Q ♆
14 57 ☽ ∠ ♄
16 17 ☽ bQ ♇
16 32 ☽ △ ♃
20 35 ☿ Q ♅
21 33 ☿ ⋕ ♆

23 Friday
00 05 ☽ □ ☿
07 16 ☽ □ ♀
07 33 ☽ □ ☉
09 22 ☽ ± ♇
12 36 ☽ □ ♂
16 14 ☉ ⋕ ☿
16 35 ☉ ⊥ ♄
17 23 ♀ ⊥ ♄
17 39 ☽ □ ♆
19 28 ☽ ✶ ♄
20 12 ☉ ☌ ♀
20 33 ☽ ⊼ ♇
22 04 ☽ ⊻ ♅
22 20 ☽ ⋕ ♇

24 Saturday
03 23 ☿ ⋕ ♂
06 45 ☽ ∥ ♅
09 05 ☽ ⊥ ♅
14 22 ☿ ⋕ ♀
17 33 ☽ Q ♄
20 34 ☽ Q ♂
23 31 ☽ □ ♃

25 Sunday
01 18 ☽ ∠ ♅
02 18 ♀ ☍ ♂
07 04 ☿ ⊥ ♄
07 49 ☉ ☍ ♂
11 50 ☉ ∥ ♆
16 12 ☽ △ ☿
17 07 ☽ ✶ ♂
18 07 ☽ △ ☉
19 02 ☽ △ ♀
20 30 ♀ ∥ ♂
21 41 ☿ ☍ ♂
23 25 ☽ △ ♆

26 Monday
01 24 ☽ □ ♄
02 02 ☽ ☍ ♇
03 29 ☽ ✶ ♅
06 34 ☽ Q ♃
12 39 ☿ ∥ ♂
14 47 ☽ ∥ ♄
16 34 ☉ ☌ ☿
17 58 ☽ ∠ ♂
21 29 ☽ bQ ☉
21 54 ☽ bQ ☿
22 54 ☽ bQ ♀
23 43 ☽ Q ♅

27 Tuesday
00 47 ☽ bQ ♆
02 31 ☽ ✶ ♃
03 15 ☿ ∥ ♀
08 29 ☽ ⊥ ♂
10 27 ☿ ∥ ♆
10 42 ☽ ⋕ ♃
13 23 ☽ ± ☉
14 55 ☿ ☌ ♀
15 03 ☽ ± ♀
15 04 ☽ ± ☿
15 35 ☽ ± ♆
16 35 ♀ ∥ ♆
18 01 ☽ ⊻ ♂
18 55 ☿ ☌ ♆
21 31 ♀ ☌ ♆
23 48 ☽ ⊼ ☉

28 Wednesday
01 18 ☽ ⊼ ♆
01 38 ☽ ⊼ ♀
02 17 ☽ ⊼ ☿
02 44 ☽ ∠ ♃
03 25 ☽ △ ♄
03 42 ☽ ⊼ ♇
04 11 ♂ ± ♄
05 06 ☽ □ ♅
06 14 ☽ ∥ ☉
07 13 ☽ ∥ ☿
08 57 ☽ ∥ ♀
10 14 ☽ ∥ ♆
10 58 ☿ ⊻ ♄
12 29 ☽ ∥ ♂
12 43 ☿ ✶ ♇
13 17 ☽ ± ♇
16 58 ☽ ⊥ ♃
19 09 ☉ ∥ ☿
23 12 ☿ ✶ ♅
23 25 ☉ ☌ ♆

29 Thursday
00 01 ☽ ⋕ ♂
00 57 ♀ ⊻ ♄
02 03 ☽ ⋕ ♆
02 22 ☽ ⊻ ♃
02 43 ♀ ✶ ♇
03 24 ☽ bQ ♄
03 31 ☽ bQ ♇
04 51 ☽ ⋕ ♀
07 30 ☽ ⋕ ☉
08 46 ☽ ⋕ ☿
16 25 ☽ ☌ ♂
17 37 ☽ ± ♄
20 17 ♀ ✶ ♅
23 11 ♄ □ ♇

30 Friday
00 29 ☽ ∥ ♃
00 46 ☽ ☍ ♆
02 26 ☽ ☍ ☉
03 02 ☽ △ ♇
03 03 ☽ ⊼ ♄
04 26 ☽ △ ♅
05 08 ☽ ☍ ♀
08 51 ☽ ☍ ☿
11 47 ☉ ✶ ♇
13 12 ☉ ⊻ ♄
20 32 ☽ ⋕ ♄

31 Saturday
00 56 ☽ ☌ ♃
03 57 ☽ bQ ♅
10 37 ☉ ✶ ♅
14 15 ☽ ⊻ ♂
18 11 ☽ ± ♅
23 31 ☽ ⊥ ♂

APRIL 2029

LONGITUDES

Date	Sidereal time h m s	Sun ☉ ° ' "	Moon ☽ ° ' "	Moon ☽ 24.00 ° ' "	Mercury ☿ ° '	Venus ♀ ° '	Mars ♂ ° '	Jupiter ♃ ° '	Saturn ♄ ° '	Uranus ♅ ° '	Neptune ♆ ° '	Pluto ♇ ° '
01	00 40 32	12 ♈ 02 39	16 ♏ 10 02	23 ♏ 33 38	18 ♈ 09	14 ♈ 14	02 ♎ 11	23 ♎ 53	10 ♉ 21	11 ♊ 02	08 ♈ 42	10 ♒ 06
02	00 44 28	13 01 49	00 ♐ 50 42	08 ♐ 00 43	20 13	15 29	01 R 48	23 R 46	10 28	11 04	08 44	10 07
03	00 48 25	14 00 57	15 ♐ 03 28	21 ♐ 58 53	22 15	16 43	01 26	23 39	10 35	11 07	08 46	10 08
04	00 52 21	15 00 04	28 ♐ 47 06	05 ♑ 28 26	24 17	17 58	01 04	23 31	10 42	11 09	08 49	10 09
05	00 56 18	15 59 09	12 ♑ 03 14	18 ♑ 32 01	26 18	19 12	00 43	23 24	10 50	11 11	08 51	10 10
06	01 00 14	16 58 13	24 ♑ 55 18	01 ♒ 13 38	28 ♈ 18	20 27	00 22	23 16	10 57	11 14	08 53	10 11
07	01 04 11	17 57 14	07 ♒ 27 36	13 ♒ 37 46	00 ♉ 15	21 41	00 ♎ 01	23 08	11 04	11 16	08 55	10 12
08	01 08 07	18 56 14	19 ♒ 44 41	25 ♒ 48 50	02 11	22 55	29 ♍ 41	23 01	11 12	11 18	08 58	10 13
09	01 12 04	19 55 12	01 ♓ 50 44	07 ♓ 50 48	04 04	24 10	29 21	22 53	11 19	11 21	09 00	10 13
10	01 16 01	20 54 08	13 ♓ 49 24	19 ♓ 46 56	05 54	25 24	29 02	22 45	11 26	11 23	09 02	10 14
11	01 19 57	21 53 02	25 ♓ 43 39	01 ♈ 39 51	07 41	26 38	28 43	22 38	11 34	11 26	09 04	10 15
12	01 23 54	22 51 54	07 ♈ 35 46	13 ♈ 31 35	09 24	27 53	28 25	22 30	11 41	11 29	09 07	10 16
13	01 27 50	23 50 45	19 ♈ 27 30	25 ♈ 23 40	11 04	29 ♈ 07	28 08	22 22	11 49	11 31	09 09	10 17
14	01 31 47	24 49 33	01 ♉ 20 17	07 ♉ 17 30	12 39	00 ♉ 21	27 51	22 15	11 56	11 34	09 11	10 17
15	01 35 43	25 48 20	13 ♉ 15 30	19 ♉ 14 29	14 10	01 36	27 35	22 07	12 04	11 36	09 13	10 18
16	01 39 40	26 47 04	25 ♉ 14 41	01 ♊ 16 22	15 37	02 50	27 20	21 59	12 11	11 39	09 15	10 19
17	01 43 36	27 45 47	07 ♊ 19 50	13 ♊ 25 25	16 59	04 04	27 06	21 52	12 19	11 42	09 18	10 19
18	01 47 33	28 44 27	19 ♊ 33 31	25 ♊ 44 31	18 15	05 18	26 52	21 44	12 26	11 45	09 20	10 20
19	01 51 30	29 ♈ 43 05	01 ♋ 58 53	08 ♋ 17 06	19 27	06 32	26 39	21 36	12 34	11 48	09 22	10 21
20	01 55 26	00 ♉ 41 41	14 ♋ 39 40	21 ♋ 07 05	20 34	07 47	26 26	21 29	12 41	11 50	09 24	10 21
21	01 59 23	01 40 15	27 ♋ 39 50	04 ♌ 18 23	21 35	09 01	26 15	21 21	12 49	11 53	09 26	10 22
22	02 03 19	02 38 46	11 ♌ 03 07	17 ♌ 54 21	22 31	10 15	26 04	21 14	12 57	11 56	09 28	10 22
23	02 07 16	03 37 16	24 ♌ 52 19	01 ♍ 57 02	23 21	11 29	25 54	21 06	13 04	11 59	09 31	10 23
24	02 11 12	04 35 43	09 ♍ 08 24	16 ♍ 26 07	24 06	12 43	25 45	20 59	13 12	12 02	09 33	10 23
25	02 15 09	05 34 08	23 ♍ 49 39	01 ♎ 18 15	24 45	13 57	25 37	20 52	13 20	12 05	09 35	10 24
26	02 19 05	06 32 31	08 ♎ 50 58	16 ♎ 26 41	25 18	15 11	25 29	20 45	13 27	12 08	09 37	10 24
27	02 23 02	07 30 51	24 ♎ 04 05	01 ♏ 41 50	25 46	16 25	25 22	20 37	13 35	12 11	09 39	10 25
28	02 26 59	08 29 11	09 ♏ 18 31	16 ♏ 52 48	26 08	17 39	25 16	20 30	13 43	12 14	09 41	10 25
29	02 30 55	09 27 28	24 ♏ 23 24	01 ♐ 49 14	26 24	18 53	25 11	20 23	13 50	12 17	09 43	10 25
30	02 34 52	10 ♉ 25 44	09 ♐ 09 22	16 ♐ 23 07	26 ♉ 35	20 ♉ 07	25 ♍ 06	20 ♎ 17	13 ♉ 58	12 ♊ 20	09 ♈ 45	10 ♒ 26

Date	Moon True ☊ ° '	Moon Mean ☊ ° '	Moon ☽ Latitude ° '
01	08 ♑ 53	09 ♑ 20	03 S 59
02	08 R 46	09 17	03 05
03	08 43	09 14	02 01
04	08 41	09 11	00 S 52
05	08 D 41	09 08	00 N 18
06	08 R 41	09 04	01 25
07	08 39	09 01	02 26
08	08 36	08 58	03 19
09	08 30	08 55	04 02
10	08 21	08 52	04 34
11	08 09	08 49	04 53
12	07 56	08 45	05 00
13	07 42	08 42	04 54
14	07 29	08 39	04 35
15	07 17	08 36	04 04
16	07 08	08 33	03 22
17	07 01	08 30	02 30
18	06 58	08 26	01 31
19	06 56	08 23	00 N 26
20	06 D 56	08 20	00 S 41
21	06 57	08 17	01 48
22	06 R 57	08 14	02 51
23	06 55	08 10	03 46
24	06 51	08 07	04 30
25	06 44	08 04	04 56
26	06 36	08 01	05 04
27	06 27	07 58	04 50
28	06 19	07 55	04 16
29	06 12	07 51	03 24
30	06 ♑ 07	07 ♑ 48	02 S 19

DECLINATIONS

Date	Sun ☉ ° '	Moon ☽ ° '	Mercury ☿ ° '	Venus ♀ ° '	Mars ♂ ° '	Jupiter ♃ ° '	Saturn ♄ ° '	Uranus ♅ ° '	Neptune ♆ ° '	Pluto ♇ ° '
01	04 N 46	20 S 29	06 N 49	04 N 29	01 N 59	07 S 49	12 N 53	22 N 05	02 N 08	22 S 53
02	05 09	23 21	07 45	04 58	02 06	07 46	12 56	22 06	02 09	22 53
03	05 32	24 36	08 41	05 28	02 13	07 44	12 58	22 06	02 10	22 53
04	05 55	24 18	09 37	05 58	02 20	07 41	13 00	22 06	02 11	22 53
05	06 17	22 36	10 31	06 27	02 27	07 38	13 03	22 07	02 12	22 53
06	06 40	19 45	11 25	06 57	02 33	07 35	13 05	22 07	02 13	22 53
07	07 03	16 03	12 17	07 26	02 39	07 32	13 07	22 07	02 14	22 53
08	07 25	11 45	13 08	07 55	02 45	07 30	13 10	22 08	02 14	22 52
09	07 47	07 03	13 58	08 24	02 51	07 27	13 12	22 08	02 15	22 52
10	08 09	02 S 09	14 45	08 53	02 56	07 24	13 14	22 08	02 16	22 52
11	08 32	02 N 47	15 30	09 21	03 02	07 21	13 17	22 09	02 17	22 52
12	08 53	07 37	16 14	09 49	03 06	07 18	13 19	22 09	02 18	22 52
13	09 15	12 09	16 55	10 17	03 11	07 15	13 21	22 09	02 19	22 52
14	09 37	16 14	17 33	10 45	03 15	07 13	13 24	22 10	02 20	22 52
15	09 58	19 42	18 09	11 13	03 19	07 10	13 26	22 10	02 20	22 52
16	10 19	22 20	18 42	11 40	03 23	07 07	13 29	22 10	02 21	22 52
17	10 41	24 00	19 13	12 07	03 26	07 04	13 31	22 11	02 22	22 52
18	11 02	24 32	19 41	12 34	03 30	07 01	13 33	22 11	02 23	22 53
19	11 22	23 52	20 06	13 00	03 32	06 59	13 36	22 12	02 24	22 53
20	11 43	21 57	20 29	13 26	03 35	06 56	13 38	22 12	02 25	22 53
21	12 03	18 52	20 49	13 52	03 37	06 53	13 40	22 12	02 25	22 53
22	12 23	14 42	21 06	14 17	03 39	06 50	13 43	22 13	02 26	22 53
23	12 43	09 40	21 20	14 42	03 40	06 48	13 45	22 13	02 27	22 53
24	13 03	03 N 58	21 32	15 07	03 42	06 45	13 47	22 13	02 28	22 53
25	13 23	02 S 05	21 41	15 31	03 42	06 42	13 50	22 14	02 29	22 53
26	13 42	08 10	21 48	15 55	03 43	06 40	13 52	22 14	02 29	22 53
27	14 01	13 49	21 52	16 19	03 43	06 37	13 54	22 15	02 30	22 53
28	14 20	18 38	21 53	16 42	03 43	06 35	13 57	22 15	02 31	22 53
29	14 38	22 09	21 51	17 05	03 43	06 32	13 59	22 16	02 32	22 53
30	14 N 57	24 S 06	21 N 47	17 N 27	03 N 42	06 S 30	14 N 01	22 N 16	02 N 33	22 S 54

ZODIAC SIGN ENTRIES

Date	h	m	Planets
02	10	36	☽ ♐
04	14	10	☽ ♑
06	21	39	☽ ♒
07	08	51	☿ ♉
07	13	09	♂ ♍
09	08	19	☽ ♓
11	20	38	☽ ♈
14	05	06	♀ ♉
14	09	18	☽ ♉
16	21	28	☽ ♊
19	08	12	☽ ♋
19	18	56	☉ ♉
21	16	14	☽ ♌
23	20	43	☽ ♍
25	21	55	☽ ♎
27	21	20	☽ ♏
29	21	03	☽ ♐

LATITUDES

Date	Mercury ☿ ° '	Venus ♀ ° '	Mars ♂ ° '	Jupiter ♃ ° '	Saturn ♄ ° '	Uranus ♅ ° '	Neptune ♆ ° '	Pluto ♇ ° '
01	00 S 20	01 S 14	03 N 07	01 N 34	02 S 09	00 00	01 S 26	05 S 22
04	00 N 12	01 10	03 01	01 34	02 08	00 00	01 26	05 23
07	00 47	01 06	02 54	01 34	02 08	00 00	01 26	05 24
10	01 20	01 01	02 47	01 34	02 08	00 00	01 26	05 24
13	01 51	00 56	02 40	01 34	02 07	00 00	01 26	05 25
16	02 18	00 50	02 32	01 34	02 07	00 00	01 26	05 26
19	02 37	00 44	02 24	01 33	02 07	00 00	01 26	05 26
22	02 48	00 38	02 16	01 33	02 06	00 00	01 26	05 27
25	02 49	00 31	02 08	01 33	02 06	00 00	01 26	05 28
28	02 40	00 25	02 00	01 33	02 06	00 00	01 26	05 28
31	02 N 20	00 S 18	01 N 52	01 N 32	02 S 06	00 00	01 S 26	05 S 29

DATA

Julian Date	2462228
Delta T	+74 seconds
Ayanamsa	24° 16' 13"
Synetic vernal point	04° ♓ 50' 46"
True obliquity of ecliptic	23° 26' 10"

MOON'S PHASES, APSIDES AND POSITIONS ☽

Date	h	m	Phase	Longitude ° '	Eclipse Indicator
05	19	52	☾	16 ♑ 18	
13	21	40	●	24 ♈ 14	
21	19	50	☽	01 ♌ 59	
28	10	37	○	08 ♏ 26	

Day	h	m		
12	23	22	Apogee	
27	16	29	Perigee	
03	19	02	Max dec	24° S 40'
10	22	25	0N	
18	10	47	Max dec	24° N 32'
25	03	49	0S	

LONGITUDES

Date	Chiron ⚷ ° '	Ceres ⚳ ° '	Pallas ⚴ ° '	Juno ⚵ ° '	Vesta ⚶ ° '	Black Moon Lilith ⚸ ° '
01	06 ♉ 31	11 ♒ 19	11 ♑ 23	06 ♏ 17	13 ♑ 54	13 ♈ 30
11	07 ♉ 07	14 ♒ 19	13 ♑ 01	04 ♏ 13	17 ♑ 08	14 ♈ 37
21	07 ♉ 45	17 ♒ 04	14 ♑ 05	01 ♏ 57	19 ♑ 54	15 ♈ 44
31	08 ♉ 23	19 ♒ 31	14 ♑ 31	29 ♎ 41	22 ♑ 08	16 ♈ 52

ASPECTARIAN

h m	Aspects
01 Sunday	
00 00	☽ ⊼ ♆
02 15	☽ □ ♇
02 36	☽ ☍ ♄
03 45	☽ ⊼ ♅
04 55	☽ ⊼ ☉
08 37	☽ ⊼ ♀
09 38	☽ ± ♆
13 36	☽ ∠ ♂
15 15	☽ ± ☉
15 44	☽ ⊼ ☿
19 12	☽ ± ♀
23 49	☽ # ♅
02 Monday	
00 15	☽ bQ ♆
00 26	☽ ⊻ ♃
03 06	☽ ± ☿
07 00	☽ bQ ☉
07 00	☽ ∥ ♇
07 28	☽ Q ♇
10 14	☽ ⊥ ♃
11 20	☽ bQ ♀
12 35	☿ # ♃
13 34	☽ ⚹ ♂
20 30	☽ bQ ☿
22 25	♂ ∥ ♆
03 Tuesday	
01 10	☽ ∠ ♃
01 15	☽ △ ♆
03 35	☽ ⚹ ♇
04 18	☽ ⊼ ♄
05 14	☽ ☍ ♅
09 17	☽ Q ♂
10 05	☽ △ ☉
10 32	☿ Q ♇
14 39	☽ ± ♄
15 09	☽ △ ♀
04 Wednesday	
00 09	☉ ∥ ♀
02 39	☽ △ ☿
02 47	☽ ⚹ ♃
03 26	☿ ☍ ♃
05 33	☽ ∠ ♇
06 30	☽ bQ ♄
15 58	☽ □ ♂
21 37	☽ ⊥ ♇
23 58	☽ Q ♃
05 Thursday	
06 07	☽ □ ♆
08 32	☽ ⊻ ♇
08 58	☽ ∥ ♇
09 44	☽ △ ♄
10 24	☽ ⊼ ♅
10 37	☿ ∠ ♅
16 46	☽ # ♅
19 52	☽ □ ☉
21 31	☽ ± ♅
06 Friday	
02 40	☽ □ ♀
08 54	☽ □ ♃
14 29	☽ bQ ♅
15 44	☽ Q ♆
19 35	☽ □ ☿
22 04	☽ △ ♂
07 Saturday	
08 50	☽ Q ☉
09 30	☿ ⊼ ♂
14 51	☽ ⚹ ♆
16 48	☽ Q ♀
16 49	♀ # ♃
17 19	☽ ☌ ♇
19 05	☽ □ ♄
19 25	☽ △ ♅
22 00	♀ Q ♇
08 Sunday	
02 19	☽ bQ ♂
04 28	☽ # ♄
05 43	☽ # ☿
10 16	☽ ⚹ ☉
12 40	☿ ∥ ♄
13 01	☽ Q ☿
13 32	♀ ☍ ♃
16 21	☉ # ♃
18 23	☽ △ ♃
18 59	☽ ⚹ ♀
19 34	☽ ± ♂
20 21	☽ ∠ ♆
09 Monday	
05 52	☽ # ♀
06 54	☽ Q ♄
07 10	☽ ⊼ ♂
08 35	☽ # ☉
10 00	☽ ∥ ♃
14 19	☽ ⊥ ♆
17 14	☽ ⚹ ☿
18 41	☽ ∠ ☉
21 54	♄ ⊻ ♅
23 57	☽ bQ ♃
10 Tuesday	
02 15	☿ ± ♂
02 21	☽ ⊻ ♆
04 20	☽ ∠ ♀
04 47	☽ ⊻ ♇
05 07	☿ ⊥ ♅
07 06	☽ □ ♅
07 09	☽ ⚹ ♄
08 13	☽ # ♂
11 25	☽ # ♆
14 22	☽ ⊥ ☉
16 52	☽ ⊥ ♇
17 50	☽ ± ♃
11 Wednesday	
00 32	☽ ⊥ ♀
03 32	☽ ⊻ ☉
04 47	☽ ∠ ☿
05 48	☽ ⊼ ♃
07 49	♀ ∠ ♅
09 31	☽ ∥ ♆
11 02	☽ ∠ ♇
13 11	☽ ∥ ♂
13 42	☽ ∠ ♄
14 04	☽ ⊻ ♀
17 54	☽ ☍ ♂
19 31	☽ Q ♅
21 08	☉ Q ♇
12 Thursday	
02 06	☽ ⊥ ☿
04 05	☉ ☍ ♃
07 44	☿ ⊻ ♆
08 06	☽ ⊥ ♄
10 28	☽ # ♃
15 04	☽ ☌ ♆
16 16	☽ ⊻ ☿
17 24	☽ ⚹ ♇
19 10	☽ ∥ ☉
19 53	☽ ⚹ ♅
20 22	☽ ⊻ ♄
20 32	♀ ⊼ ♂
13 Friday	
00 24	☿ □ ♇
00 47	☽ ∥ ♀
17 42	☽ Q ♇
17 50	☽ ☍ ♃
18 52	☽ ∥ ♄
18 56	☿ ⊻ ♅
21 40	☽ ☌ ☉
14 Saturday	
00 03	☿ ☌ ♄
02 20	☽ ∠ ♅
05 08	☽ ⊼ ♂
09 47	☽ ☌ ♀
14 39	☿ bQ ♂
16 58	☽ ± ♂
20 33	☽ ⊥ ♅
22 22	☽ ∥ ☿
15 Sunday	
03 51	☽ ⊻ ♆
06 03	☽ □ ♇
08 40	☽ ⊻ ♅
09 34	☽ ☌ ♄
10 41	☽ bQ ♂
14 06	☽ ☌ ☿
15 57	☽ ⊥ ♆
16 Monday	
05 34	☽ ⊼ ♃
05 46	☿ ⊥ ♆
08 37	☉ ∠ ♅
10 01	☽ ∠ ♆
10 13	☽ ∥ ♅
15 21	☽ ⊻ ☉
16 05	☽ △ ♂
17 24	☽ ± ♃
18 16	☽ # ♇
20 11	♀ ± ♂
21 42	♂ bQ ♄
22 48	☉ ⊼ ♂
17 Tuesday	
04 19	☽ ⊥ ☉
04 49	☽ ⊻ ♀
11 05	☽ bQ ♃
15 53	☽ ⚹ ♆
17 55	☽ △ ♇
18 01	☽ ⊥ ♀
20 39	☽ ☌ ♅
21 55	☽ ⊻ ♄
23 38	☽ ∠ ☉
18 Wednesday	
09 11	☽ ⊻ ☿
09 47	☽ ⊥ ♄
13 37	☽ ∠ ♀
15 27	☽ Q ♆
16 11	☽ △ ♃
20 54	♀ ⊥ ♅
22 07	☽ ⊥ ☿
23 13	☽ bQ ♇
19 Thursday	
01 55	☽ □ ♂
03 26	☽ ∠ ♄
07 18	☽ ⚹ ☉
16 31	☽ ± ♇
17 11	☽ ∠ ☿
21 38	☽ ⚹ ♀
20 Friday	
02 05	☽ □ ♆
02 12	☽ # ♇
03 54	☽ ⊼ ♇
06 41	☽ ⊻ ♅
08 01	☽ Q ☉
08 16	☽ ⚹ ♄
09 34	☽ ∥ ♅
11 35	☽ Q ♂
17 57	☽ ⊥ ♅
22 31	☽ Q ♀
23 15	☽ ∥ ☿
23 56	☽ ⚹ ☿
21 Saturday	
00 06	♀ ∥ ♄
00 33	☽ □ ♃
06 45	☽ Q ♄
07 07	☿ ⊼ ♃
09 27	☽ ⚹ ♂
10 35	☽ ∠ ♅
19 50	☽ □ ☉
20 32	♀ ⊻ ♆
23 31	☽ Q ☿
23 56	☉ ± ♂
22 Sunday	
08 48	☽ Q ♃
09 12	☽ △ ♆
10 26	☽ □ ♀
10 48	☽ ☍ ♇
12 02	☽ ∠ ♂
13 34	☽ ⚹ ♅
13 58	☽ ∥ ♀
14 29	♀ □ ♇
15 22	☽ □ ♄
17 01	☽ ∥ ♄
22 47	☽ ∥ ☉
23 Monday	
02 03	♀ bQ ♂
03 33	☽ ⊥ ♂
05 36	☽ ⚹ ♃
09 14	☽ □ ☿
10 29	☽ Q ♅
11 23	☽ bQ ♆
13 44	☽ ⊻ ♂
22 11	♀ ⊻ ♅
24 Tuesday	
00 29	☽ # ♃
02 39	☽ ± ♆
03 53	☽ △ ☉
06 48	☽ ∠ ♃
12 40	☽ ⊼ ♆
13 08	☽ ∥ ♂
14 04	☽ ⊼ ♇
16 48	☽ □ ♅
18 03	☽ ∥ ♆
18 27	☽ △ ♀
18 45	☽ △ ♄
21 33	☽ ⊥ ♃
22 30	♀ ☌ ♄
23 56	☽ ± ♇
25 Wednesday	
05 14	☿ ∠ ♆
06 21	☽ bQ ☉
07 15	☽ ⊻ ♃
13 32	☽ # ♆
13 32	☽ △ ☿
14 32	☽ bQ ♇
14 51	☽ ☌ ♂
18 21	☽ # ♂
19 18	☽ bQ ♄
20 58	☽ bQ ♀
21 52	☽ ± ☉
26 Thursday	
01 22	☉ ⊥ ♅
06 03	☽ ∥ ♃
08 05	☽ ⊼ ☉
09 46	☽ ± ♄
12 34	☽ ± ♀
13 13	☽ ☍ ♆
14 22	☽ bQ ☿
14 28	☽ △ ♇
17 13	☽ △ ♅
19 10	☿ △ ♂
19 21	☽ ⊼ ♄
20 37	♀ ⊥ ♆
22 54	☽ ⊼ ♀
27 Friday	
02 34	☉ ∥ ♄
03 22	♂ bQ ♇
05 02	☽ ± ☿
06 37	☽ ☌ ♃
12 23	☽ # ♄
12 56	☽ # ☉
14 02	☽ ⊻ ♂
14 44	☽ ⊼ ☿
16 55	☽ bQ ♅
23 24	☽ ⊥ ♂
28 Saturday	
00 48	☽ # ♀
07 08	☽ ± ♅
10 37	☽ ☍ ☉
12 35	☽ ⊼ ♆
13 30	☽ ∠ ♂
13 45	☽ □ ♇
16 39	☽ ⊼ ♅
19 02	☽ ☍ ♄
22 07	☽ ± ♆
29 Sunday	
02 24	☽ ☍ ♀
05 39	☽ ⊻ ♃
09 32	☽ # ☿
11 31	☿ ± ♃
12 31	☽ bQ ♆
12 56	☽ # ♅
13 16	☽ ⚹ ♂
15 12	☽ ⊥ ♃
15 17	☽ ☍ ☿
18 30	☽ Q ♇
18 36	☉ ⊻ ♆
18 58	☽ ∥ ♇
30 Monday	
04 34	☉ bQ ♂
05 41	☽ ∠ ♃
08 39	☽ Q ♂
11 57	☉ □ ♇
12 59	☽ △ ♆
14 06	☽ ⚹ ♇
14 15	☽ ⊼ ☉
14 54	♀ ⊼ ♃
17 17	☽ ☍ ♅
20 02	☽ ⊼ ♄

All ephemeris data is given at 12.00 UT and the Moon's longitude is additionally given for 24.00 UT

MAY 2029

LONGITUDES

Date	Sidereal time h m s	Sun ☉ ° ' "	Moon ☽ ° ' "	Moon ☽ 24.00 ° ' "	Mercury ☿ ° '	Venus ♀ ° '	Mars ♂ ° '	Jupiter ♃ ° '	Saturn ♄ ° '	Uranus ♅ ° '	Neptune ♆ ° '	Pluto ♇ ° '
01	02 38 48	11 ♉ 23 58	23 ♐ 30 00	00 ♑ 29 44	26 ♉ 40	21 ♉ 21	25 ♍ 03	20 ♎ 10	14 ♉ 06	12 ♊ 23	09 ♈ 47	10 ♒ 26
02	02 42 45	12 22 10	07 ♑ 22 15	14 ♑ 07 38	26 R 40	22 35	25 R 00	20 R 03	14 13	12 27	09 49	10 26
03	02 46 41	13 20 21	20 ♑ 46 08	27 ♑ 18 07	26 34	23 49	24 58	19 57	14 21	12 30	09 51	10 27
04	02 50 38	14 18 31	03 ♒ 44 02	10 ♒ 04 24	26 24	25 02	24 56	19 50	14 29	12 33	09 53	10 27
05	02 54 34	15 16 39	16 ♒ 19 46	22 ♒ 30 44	26 09	26 16	24 56	19 44	14 37	12 36	09 55	10 27
06	02 58 31	16 14 45	28 ♒ 37 53	04 ♓ 41 49	25 49	27 30	24 D 56	19 37	14 44	12 39	09 57	10 27
07	03 02 28	17 12 50	10 ♓ 43 05	16 ♓ 42 14	25 26	28 44	24 57	19 31	14 52	12 43	09 59	10 27
08	03 06 24	18 10 54	22 ♓ 39 48	28 ♓ 36 14	24 59	29 ♉ 58	24 58	19 25	15 00	12 46	10 01	10 27
09	03 10 21	19 08 56	04 ♈ 32 00	10 ♈ 27 28	24 29	01 ♊ 11	25 01	19 19	15 07	12 49	10 02	10 27
10	03 14 17	20 06 57	16 ♈ 23 02	22 ♈ 18 59	23 57	02 25	25 04	19 13	15 15	12 53	10 04	10 R 27
11	03 18 14	21 04 57	28 ♈ 15 37	04 ♉ 13 11	23 23	03 39	25 08	19 08	15 23	12 56	10 06	10 27
12	03 22 10	22 02 55	10 ♉ 11 54	16 ♉ 11 58	22 48	04 53	25 12	19 02	15 31	12 59	10 08	10 27
13	03 26 07	23 00 52	22 ♉ 13 34	28 ♉ 16 52	22 12	06 06	25 18	18 57	15 38	13 03	10 10	10 27
14	03 30 03	23 58 47	04 ♊ 22 03	10 ♊ 29 17	21 36	07 20	25 23	18 52	15 46	13 06	10 12	10 27
15	03 34 00	24 56 40	16 ♊ 38 44	22 ♊ 50 37	21 01	08 34	25 30	18 47	15 54	13 09	10 13	10 27
16	03 37 57	25 54 32	29 ♊ 05 08	05 ♋ 22 32	20 27	09 47	25 37	18 42	16 01	13 13	10 15	10 27
17	03 41 53	26 52 23	11 ♋ 43 03	18 ♋ 06 57	19 55	11 01	25 45	18 37	16 09	13 16	10 17	10 26
18	03 45 50	27 50 12	24 ♋ 34 33	01 ♌ 06 08	19 25	12 15	25 54	18 32	16 16	13 20	10 18	10 26
19	03 49 46	28 47 59	07 ♌ 41 59	14 ♌ 22 23	18 58	13 28	26 03	18 28	16 24	13 23	10 20	10 26
20	03 53 43	29 ♉ 45 44	21 ♌ 07 34	27 ♌ 57 45	18 34	14 42	26 13	18 23	16 32	13 27	10 22	10 26
21	03 57 39	00 ♊ 43 28	04 ♍ 53 03	11 ♍ 53 30	18 14	15 55	26 23	18 19	16 39	13 30	10 23	10 26
22	04 01 36	01 41 10	18 ♍ 59 02	26 ♍ 09 26	17 58	17 09	26 34	18 15	16 47	13 33	10 25	10 25
23	04 05 32	02 38 50	03 ♎ 24 24	10 ♎ 43 24	17 45	18 22	26 46	18 11	16 54	13 37	10 26	10 25
24	04 09 29	03 36 29	18 ♎ 05 50	25 ♎ 30 53	17 37	19 36	26 58	18 08	17 02	13 40	10 28	10 25
25	04 13 26	04 34 07	02 ♏ 57 41	10 ♏ 25 13	17 34	20 49	27 11	18 04	17 09	13 44	10 29	10 24
26	04 17 22	05 31 43	17 ♏ 52 27	25 ♏ 18 19	17 D 35	22 03	27 24	18 01	17 17	13 47	10 31	10 24
27	04 21 19	06 29 18	02 ♐ 41 47	10 ♐ 01 52	17 40	23 16	27 38	17 58	17 24	13 51	10 32	10 24
28	04 25 15	07 26 51	17 ♐ 17 44	24 ♐ 28 37	17 50	24 29	27 53	17 55	17 32	13 54	10 34	10 23
29	04 29 12	08 24 24	01 ♑ 33 58	08 ♑ 33 20	18 04	25 43	28 07	17 52	17 39	13 58	10 35	10 23
30	04 33 08	09 21 56	15 ♑ 26 26	22 ♑ 13 12	18 23	26 56	28 23	17 49	17 47	14 02	10 37	10 22
31	04 37 05	10 ♊ 19 26	28 ♑ 53 37	05 ♒ 27 52	18 ♉ 46	28 ♊ 09	28 ♍ 39	17 ♎ 47	17 ♉ 54	14 ♊ 05	10 ♈ 38	10 ♒ 22

Date	Moon True ☊ ° '	Moon Mean ☊ ° '	Moon ☽ Latitude ° '
01	06 ♑ 04	07 ♑ 45	01 S 07
02	06 D 03	07 42	00 N 07
03	06 04	07 39	01 18
04	06 05	07 36	02 23
05	06 R 05	07 32	03 19
06	06 04	07 29	04 04
07	06 02	07 26	04 38
08	05 57	07 23	04 59
09	05 50	07 20	05 07
10	05 42	07 16	05 02
11	05 34	07 13	04 44
12	05 25	07 10	04 13
13	05 18	07 07	03 31
14	05 13	07 04	02 39
15	05 09	07 01	01 39
16	05 07	06 57	00 N 33
17	05 D 07	06 54	00 S 36
18	05 08	06 51	01 44
19	05 10	06 48	02 48
20	05 11	06 45	03 45
21	05 R 11	06 42	04 30
22	05 10	06 38	05 00
23	05 08	06 35	05 12
24	05 04	06 32	05 04
25	05 00	06 29	04 36
26	04 56	06 26	03 49
27	04 52	06 22	02 47
28	04 50	06 19	01 35
29	04 49	06 16	00 S 18
30	04 D 49	06 13	00 N 58
31	04 ♑ 51	06 ♑ 10	02 N 08

DECLINATIONS

Date	Sun ☉ ° '	Moon ☽ ° '	Mercury ☿ ° '	Venus ♀ ° '	Mars ♂ ° '	Jupiter ♃ ° '	Saturn ♄ ° '	Uranus ♅ ° '	Neptune ♆ ° '	Pluto ♇ ° '
01	15 N 15	24 S 23	21 N 41	17 N 49	03 N 41	06 S 27	14 N 04	22 N 16	02 N 33	22 S 54
02	15 33	23 07	21 32	18 10	03 40	06 25	14 06	22 17	02 34	22 54
03	15 50	20 33	21 21	18 31	03 39	06 23	14 08	22 17	02 35	22 54
04	16 08	17 00	21 07	18 51	03 37	06 20	14 11	22 18	02 36	22 54
05	16 25	12 46	20 51	19 11	03 35	06 18	14 13	22 18	02 36	22 54
06	16 42	08 07	20 33	19 31	03 32	06 16	14 15	22 19	02 37	22 55
07	16 58	03 S 15	20 13	19 49	03 30	06 14	14 17	22 19	02 38	22 55
08	17 15	01 N 40	19 52	20 08	03 27	06 12	14 20	22 19	02 38	22 55
09	17 30	06 30	19 29	20 26	03 24	06 10	14 22	22 20	02 39	22 55
10	17 46	11 05	19 05	20 43	03 20	06 08	14 24	22 20	02 40	22 55
11	18 02	15 16	18 40	20 59	03 16	06 06	14 26	22 21	02 40	22 56
12	18 17	18 53	18 14	21 16	03 12	06 04	14 29	22 21	02 41	22 56
13	18 31	21 43	17 48	21 31	03 08	06 02	14 31	22 22	02 42	22 56
14	18 46	23 37	17 22	21 46	03 04	06 00	14 33	22 22	02 42	22 56
15	19 00	24 24	16 56	22 00	02 59	05 58	14 35	22 22	02 43	22 57
16	19 14	23 59	16 31	22 14	02 54	05 57	14 37	22 23	02 44	22 57
17	19 27	22 19	16 06	22 27	02 49	05 55	14 40	22 23	02 44	22 57
18	19 40	19 30	15 43	22 39	02 43	05 53	14 42	22 24	02 45	22 58
19	19 53	15 38	15 22	22 51	02 38	05 52	14 44	22 24	02 46	22 58
20	20 06	10 54	15 02	23 02	02 32	05 50	14 46	22 25	02 46	22 58
21	20 18	05 N 32	14 44	23 13	02 25	05 49	14 48	22 25	02 47	22 58
22	20 30	00 S 14	14 28	23 23	02 19	05 48	14 50	22 25	02 47	22 59
23	20 41	06 07	14 14	23 32	02 13	05 47	14 52	22 26	02 48	22 59
24	20 52	11 47	14 02	23 40	02 06	05 45	14 54	22 26	02 49	22 59
25	21 03	16 49	13 53	23 48	01 59	05 44	14 57	22 27	02 49	22 59
26	21 13	20 49	13 46	23 55	01 52	05 43	14 59	22 27	02 50	23 00
27	21 23	23 25	13 41	24 01	01 44	05 42	15 01	22 28	02 50	23 00
28	21 33	24 24	13 38	24 07	01 37	05 41	15 03	22 28	02 51	23 00
29	21 42	23 43	13 38	24 12	01 29	05 40	15 05	22 28	02 51	23 01
30	21 51	21 35	13 40	24 16	01 21	05 40	15 07	22 29	02 52	23 01
31	22 N 00	18 S 17	13 N 44	24 N 20	01 N 13	05 S 39	15 N 09	22 N 29	02 N 52	23 S 01

ZODIAC SIGN ENTRIES

Date	h	m	Planets
01	23	09	☽ ♑
04	05	01	☽ ♒
06	14	42	☽ ♓
08	12	46	♀ ♊
09	02	49	☽ ♈
11	15	30	☽ ♉
14	03	24	☽ ♊
16	13	45	☽ ♋
18	21	59	☽ ♌
20	17	56	☉ ♊
21	03	33	☽ ♍
23	06	23	☽ ♎
25	07	14	☽ ♏
27	07	37	☽ ♐
29	09	20	☽ ♑
31	14	00	☽ ♒

LATITUDES

Date	Mercury ☿ ° '	Venus ♀ ° '	Mars ♂ ° '	Jupiter ♃ ° '	Saturn ♄ ° '	Uranus ♅ ° '	Neptune ♆ ° '	Pluto ♇ ° '
01	02 N 20	00 S 18	01 N 52	01 N 32	02 S 06	00 00	01 S 26	05 S 29
04	01 49	00 10	01 45	01 32	02 06	00 00	01 26	05 30
07	01 08	00 S 03	01 37	01 31	02 06	00 00	01 26	05 31
10	00 N 20	00 N 04	01 30	01 31	02 06	00 00	01 26	05 31
13	00 S 32	00 12	01 23	01 30	02 06	00 00	01 27	05 32
16	01 24	00 19	01 16	01 30	02 06	00 00	01 27	05 33
19	02 11	00 27	01 09	01 29	02 06	00 00	01 27	05 33
22	02 50	00 34	01 03	01 28	02 06	00 00	01 27	05 34
25	03 19	00 41	00 56	01 27	02 06	00 00	01 27	05 35
28	03 39	00 48	00 50	01 27	02 06	00 00	01 27	05 35
31	03 S 49	00 N 54	00 N 45	01 N 26	02 S 06	00 00	01 S 27	05 S 36

DATA

Julian Date	2462258
Delta T	+74 seconds
Ayanamsa	24° 16' 17"
Synetic vernal point	04° ♓ 50' 42"
True obliquity of ecliptic	23° 26' 09"

LONGITUDES

Date	Chiron ⚷ ° '	Ceres ⚳ ° '	Pallas ⚴ ° '	Juno ⚵ ° '	Vesta ⚶ ° '	Black Moon Lilith ⚸ ° '
01	08 ♉ 23	19 ♒ 31	14 ♑ 31	29 ♎ 41	22 ♑ 08	16 ♈ 52
11	09 ♉ 00	21 ♒ 38	14 ♑ 13	27 ♎ 38	23 ♑ 45	17 ♈ 59
21	09 ♉ 37	23 ♒ 21	13 ♑ 11	25 ♎ 59	24 ♑ 38	19 ♈ 06
31	10 ♉ 12	24 ♒ 38	11 ♑ 26	24 ♎ 48	24 ♑ 44	20 ♈ 13

MOON'S PHASES, APSIDES AND POSITIONS ☽

Date	h	m	Phase	Longitude ° '	Eclipse Indicator
05	09	48	☾	15 ♒ 11	
13	13	42	●	23 ♉ 05	
21	04	16	☽	00 ♍ 25	
27	18	37	○	06 ♐ 45	

Day	h	m	
10	07	10	Apogee
25	22	32	Perigee

Day	h	m		
01	04	04	Max dec	24° S 29'
08	03	51	0N	
15	15	47	Max dec	24° N 25'
22	11	01	0S	
28	13	58	Max dec	24° S 24'

ASPECTARIAN

h m	Aspects
01 Tuesday	
00 57	☽ ± ☉
06 11	☽ ± ♄
06 24	☽ ✱ ♃
08 00	☽ ⊼ ♀
14 37	☽ □ ♂
15 18	☽ ∠ ♇
17 19	☽ ⚼ ☉
17 25	☽ ⊼ ☿
19 12	☽ ± ♀
21 40	☽ ⚼ ♄
23 06	☿ St R
02 Wednesday	
02 46	☽ Q ♃
03 46	☽ ± ☿
06 51	☽ ⊥ ♇
12 24	☽ ⚼ ♀
13 55	☉ ⊻ ♅
14 31	☽ ∥ ♇
16 20	☽ □ ♆
17 25	☽ ⊻ ♇
19 34	☽ ⚼ ☿
21 00	☽ # ♅
21 01	☽ ⊼ ♅
21 33	☽ △ ☉
03 Thursday	
00 18	☽ △ ♄
05 03	☽ # ☿
07 51	☽ ± ♅
10 30	☽ □ ♃
18 08	☽ △ ♀
19 40	☽ △ ♂
22 31	☽ △ ☿
04 Friday	
00 25	☽ ⚼ ♅
01 03	☽ Q ♆
01 08	☽ # ♀
08 51	♀ ∠ ♆
10 05	♀ △ ♂
16 51	☽ # ☉
16 57	☉ ☌ ♄
23 40	☽ ✱ ♆
23 44	☽ ⚼ ♂
05 Saturday	
00 42	☽ ☌ ♇
02 15	♀ ± ♃
04 12	☽ # ♄
04 48	☽ △ ♅
08 39	☽ □ ♄
09 48	☽ □ ☉
10 02	☿ ☌ ♀
17 02	☽ ± ♂
18 31	☽ △ ♃
19 00	♂ St D
06 Sunday	
04 19	☉ ⊥ ♆
04 44	☽ ⊼ ♂
04 44	☽ ∠ ♆
06 38	☽ □ ☿
09 31	☽ □ ♀
20 12	☽ Q ♄
21 19	☽ ∥ ♃
22 32	☽ ⊥ ♆
23 45	☽ ⚼ ♃
07 Monday	
00 04	☽ Q ☉
05 18	☿ ± ♃
10 31	☽ ⊻ ♆
10 49	☽ # ♂
11 28	☽ ⊻ ♇
15 02	☽ # ♆
16 00	☽ □ ♅
17 15	☽ Q ☿
17 34	☽ ± ♃
20 24	☽ ✱ ♄
23 29	☽ ⊥ ♇
08 Tuesday	
01 26	☽ Q ♀
02 10	☽ ✱ ☉
02 29	☿ ∥ ♀
05 31	☽ ⊼ ♃
10 53	☿ ∠ ♆
12 33	☿ △ ♂
16 30	☽ ✱ ☿
16 41	☽ ☍ ♂
16 46	☽ ∥ ♆
17 38	☽ ∠ ♇
20 38	☽ ∥ ♂
09 Wednesday	
02 59	☽ ∠ ♄
04 27	☽ Q ♅
04 27	☽ ✱ ♀
10 17	☽ # ♃
11 09	☽ ∠ ☉
15 52	☉ ⊼ ♃
21 24	☽ ⊥ ♄
21 37	☽ ∠ ☿
23 11	☽ ☌ ♆
23 59	☽ ✱ ♇
10 Thursday	
04 52	☽ ✱ ♅
07 00	☽ ⊥ ☉
09 41	☽ ⊻ ♄
14 20	☽ ∠ ♀
15 02	☽ ⊥ ☿
17 42	☽ ☍ ♃
20 13	☽ ⊻ ☉
11 Friday	
00 17	☽ Q ♇
02 37	☽ ⊻ ☿
04 17	♇ St R
05 39	☽ ⊼ ♂
06 58	☽ ∥ ♄
10 37	☽ ⊥ ♀
11 20	☽ ∠ ♅
17 49	☽ ± ♂
20 45	♀ ⚼ ♃
12 Saturday	
00 06	☽ ⊻ ♀
05 32	☽ ⊥ ♅
07 20	☽ ∥ ☉
07 52	☽ ∥ ☿
10 20	☉ ∥ ☿
11 52	☽ ⊻ ♆
12 01	☽ ⚼ ♂
12 31	☽ □ ♇
17 36	☽ ⊻ ♅
22 44	☽ ☌ ♄
23 27	☉ ☌ ☿
23 54	☽ ⊥ ♆
13 Sunday	
05 32	☽ ⊼ ♃
09 47	☽ ∥ ♀
11 57	☽ ☌ ☿
13 42	☽ ☌ ☉
17 22	☽ ± ♃
17 50	☽ ∠ ♆
18 08	☽ △ ♂
18 49	☽ ∥ ♅
14 Monday	
01 49	☽ # ♇
11 01	☽ ⚼ ♃
18 28	☽ ☌ ♀
23 27	☽ ✱ ♆
23 56	☽ △ ♇
15 Tuesday	
01 05	♂ ⚼ ♇
05 11	☽ ☌ ♅
08 09	☉ ± ♃
10 31	☽ ⊻ ♄
16 06	☽ △ ♃
19 05	☉ ∠ ♆
20 05	☽ ⊻ ☿
22 16	☽ ⊥ ♄
22 49	☽ Q ♆
16 Wednesday	
03 48	☉ △ ♂
05 01	☽ ⚼ ♇
05 17	☽ □ ♂
05 24	☽ ⊻ ☉
07 09	☽ ⊥ ☿
15 44	☽ ∠ ♄
17 51	☽ ⊥ ☉
21 13	♀ ✱ ♆
22 14	☽ ± ♇
23 38	☽ ∠ ☿
17 Thursday	
00 50	♀ △ ♇
04 46	☽ # ♇
04 54	♀ ∥ ♅
09 17	☽ □ ♆
09 36	☽ ⊼ ♇
10 32	☽ ⊻ ♀
10 46	☽ ∥ ♀
11 19	☽ ∥ ♅
12 19	☽ ∠ ☉
14 56	☽ ⊻ ♅
15 52	☽ Q ♂
20 24	☽ ✱ ♄
23 00	☽ ⊥ ♀
18 Friday	
00 51	☽ □ ♃
02 13	☽ ⊥ ♅
02 46	☽ ✱ ☿
04 56	♂ ∥ ♆
10 49	☽ ∥ ☉
14 28	☽ ✱ ♂
17 26	☽ ∠ ♀
18 29	☽ ✱ ☉
18 53	☽ Q ♄
18 57	☽ ∠ ♅
19 36	♄ ⊥ ♆
19 Saturday	
00 09	☽ Q ☿
09 46	☽ Q ♃
10 15	♀ ☌ ♅
13 38	☽ ∥ ☿
16 46	☽ △ ♆
16 51	☽ ∥ ♄
16 56	☽ ☍ ♇
18 02	☽ Q ☉
18 07	☽ ∠ ♂
22 17	☽ ✱ ♅
23 26	☽ ✱ ♀
20 Sunday	
02 12	♀ # ♇
03 46	☽ □ ♄
07 11	☽ ✱ ♃
07 36	☽ □ ☿
10 22	☽ ⊥ ♂
19 28	☽ ⚼ ♆
19 37	☽ Q ♅
21 04	☽ ⊻ ♂
22 45	☽ Q ♀
21 Monday	
03 37	☿ ⊼ ♃
04 16	☽ □ ☉
06 22	☿ ∥ ♄
09 19	☽ ∠ ♃
10 46	☽ # ♃
11 09	☽ ± ♆
21 27	☽ ⊼ ♆
21 30	☽ ⊼ ♇
23 34	☽ ∥ ♆
22 Tuesday	
00 40	☽ ⊥ ♃
01 18	☽ ∥ ♂
02 48	☽ □ ♅
04 00	♀ ⊻ ♄
07 41	☽ ± ♇
08 15	☽ △ ♄
08 37	☽ □ ♀
10 18	☽ △ ☿
10 47	☽ ⊻ ♃
19 27	♆ ✱ ♇
20 19	☽ # ♂
22 24	☽ # ♆
22 47	☽ ⚼ ♇
23 Wednesday	
00 52	☽ ☌ ♂
01 32	☿ ⊻ ♀
08 36	♀ △ ♃
09 30	☽ ⚼ ♄
10 35	☽ ∥ ♃
10 40	☽ △ ☉
10 57	☽ ⚼ ☿
23 30	☽ △ ♇
23 33	☽ ☍ ♆
24 Thursday	
00 24	☽ ± ♄
00 42	☉ ⚼ ♃
01 33	☽ ± ☿
04 47	☽ △ ♅
10 15	☽ ⊼ ♄
11 14	☽ ⊼ ☿
12 03	☽ ☌ ♃
12 53	☽ ⚼ ☉
14 39	☽ △ ♀
21 57	☽ # ☿
25 Friday	
02 33	☽ # ♄
02 33	☽ ⊻ ♂
04 26	☽ ± ☉
05 10	☽ ⚼ ♅
12 22	☽ ⊥ ♂
14 46	☽ ⊼ ☉
17 01	☽ ⚼ ♀
19 21	☿ St D
19 42	☽ ± ♅
23 58	☽ □ ♇
26 Saturday	
00 08	☽ ⊼ ♆
03 04	☽ ∠ ♂
05 24	☽ ⊼ ♅
08 47	☽ ± ♀
09 49	☽ ± ♆
11 02	☽ ☍ ♄
11 31	☽ ☍ ☿
12 13	☽ ⊻ ♃
15 08	☽ # ☉
19 20	☽ ⊼ ♀
21 27	♀ Q ♆
21 52	☽ ⊥ ♃
27 Sunday	
00 22	☽ ⚼ ♆
01 20	☽ # ♅
03 39	☽ ✱ ♂
05 00	☽ Q ♇
06 54	☽ ∥ ♇
12 26	☽ ∠ ♃
15 00	♀ ⊥ ♄
18 37	☽ ☍ ☉
20 54	☿ ⊥ ♀
22 42	☽ # ♀
23 32	☽ Q ♂
28 Monday	
00 35	☽ ✱ ♇
00 52	☽ △ ♆
06 22	☽ ☍ ♅
12 23	☽ ⊼ ♄
12 54	☽ ⊼ ☿
13 01	☽ ✱ ♃
19 01	☿ ⊼ ♃
22 30	☽ ± ♄
23 05	☽ ± ☿
29 Tuesday	
01 09	☽ ☍ ♀
01 31	☽ ∠ ♇
02 49	☽ # ♀
05 28	♀ ⚼ ♇
06 03	☽ □ ♂
09 07	☽ Q ♃
13 52	☽ ⚼ ♄
14 37	☽ ⚼ ☿
16 48	☽ ⊥ ♇
21 45	☽ ∥ ♇
30 Wednesday	
00 37	☽ ⊼ ☉
03 09	☽ ⊻ ♇
03 33	☽ □ ♆
03 37	☽ # ♅
09 30	☽ ⊼ ♅
09 44	☽ # ☉
11 51	☽ ± ☉
16 09	☽ △ ♄
16 10	☽ □ ♃
17 20	☽ △ ☿
18 15	♃ ⊼ ♄
20 08	☽ ± ♅
31 Thursday	
05 04	☽ ⚼ ☉
10 32	☽ ⊼ ♀
11 32	☽ Q ♆
11 33	☽ △ ♂
12 21	☽ ⚼ ♅
12 56	☉ △ ♇
19 58	☉ ✱ ♆
22 35	☽ ± ♀

All ephemeris data is given at 12.00 UT and the Moon's longitude is additionally given for 24.00 UT

JUNE 2029

LONGITUDES

Date	Sidereal time h m s	Sun ☉	Moon ☽	Moon ☽ 24.00	Mercury ☿	Venus ♀	Mars ♂	Jupiter ♃	Saturn ♄	Uranus ♅	Neptune ♆	Pluto ♇
01	04 41 01	11 ♊ 16 56	11 ♒ 56 13	18 ♒ 19 00	19 ♉ 14	29 ♊ 23	28 ♍ 55	17 ♎ 44	18 ♉ 01	14 ♊ 09	10 ♈ 39	10 ♒ 21
02	04 44 58	12 14 25	24 ♒ 36 40	00 ♓ 49 41	19 46	00 ♋ 36	29 12	17 R 42	18 08	14 12	10 41	10 R 21
03	04 48 55	13 11 53	06 ♓ 58 36	13 ♓ 03 58	20 21	01 49	29 29	17 40	18 16	14 16	10 42	10 20
04	04 52 51	14 09 21	19 ♓ 06 21	25 ♓ 06 19	21 01	03 03	29 ♍ 47	17 38	18 23	14 19	10 43	10 19
05	04 56 48	15 06 47	01 ♈ 04 27	07 ♈ 01 18	21 45	04 16	00 ♎ 06	17 37	18 30	14 23	10 44	10 19
06	05 00 44	16 04 13	12 ♈ 57 24	18 ♈ 53 17	22 32	05 29	00 24	17 35	18 37	14 26	10 46	10 18
07	05 04 41	17 01 39	24 ♈ 49 26	00 ♉ 46 17	23 24	06 42	00 43	17 34	18 44	14 30	10 47	10 17
08	05 08 37	17 59 03	06 ♉ 44 16	12 ♉ 43 46	24 19	07 55	01 03	17 33	18 52	14 33	10 48	10 17
09	05 12 34	18 56 28	18 ♉ 45 06	24 ♉ 48 35	25 17	09 09	01 23	17 32	18 59	14 37	10 49	10 16
10	05 16 30	19 53 51	00 ♊ 54 29	07 ♊ 03 00	26 19	10 22	01 43	17 31	19 06	14 40	10 50	10 15
11	05 20 27	20 51 14	13 ♊ 14 20	19 ♊ 28 38	27 24	11 35	02 04	17 31	19 13	14 44	10 51	10 14
12	05 24 24	21 48 36	25 ♊ 46 02	02 ♋ 06 37	28 33	12 48	02 26	17 31	19 20	14 47	10 52	10 14
13	05 28 20	22 45 58	08 ♋ 30 29	14 ♋ 57 40	29 ♉ 45	14 01	02 47	17 30	19 27	14 51	10 53	10 13
14	05 32 17	23 43 18	21 ♋ 28 13	28 ♋ 02 09	01 ♊ 01	15 14	03 09	17 D 30	19 33	14 55	10 54	10 12
15	05 36 13	24 40 38	04 ♌ 39 31	11 ♌ 20 18	02 19	16 27	03 32	17 31	19 40	14 58	10 55	10 11
16	05 40 10	25 37 57	18 ♌ 04 31	24 ♌ 52 07	03 41	17 40	03 54	17 31	19 47	15 02	10 56	10 11
17	05 44 06	26 35 15	01 ♍ 43 06	08 ♍ 37 24	05 06	18 53	04 18	17 32	19 54	15 05	10 57	10 09
18	05 48 03	27 32 32	15 ♍ 34 55	22 ♍ 35 33	06 34	20 06	04 41	17 32	20 00	15 08	10 58	10 08
19	05 51 59	28 29 49	29 ♍ 39 06	06 ♎ 45 21	08 05	21 19	05 05	17 33	20 07	15 12	10 59	10 07
20	05 55 56	29 ♊ 27 04	13 ♎ 54 02	21 ♎ 04 48	09 39	22 32	05 29	17 34	20 14	15 15	10 59	10 06
21	05 59 53	00 ♋ 24 19	28 ♎ 17 14	05 ♏ 30 51	11 16	23 45	05 54	17 36	20 20	15 19	11 00	10 06
22	06 03 49	01 21 33	12 ♏ 45 08	19 ♏ 59 31	12 56	24 58	06 18	17 37	20 27	15 22	11 01	10 05
23	06 07 46	02 18 47	27 ♏ 13 20	04 ♐ 25 59	14 39	26 11	06 44	17 39	20 33	15 26	11 01	10 03
24	06 11 42	03 16 00	11 ♐ 36 48	18 ♐ 45 09	16 25	27 23	07 09	17 41	20 39	15 29	11 02	10 02
25	06 15 39	04 13 13	25 ♐ 50 26	02 ♑ 52 07	18 14	28 36	07 35	17 43	20 46	15 33	11 03	10 01
26	06 19 35	05 10 25	09 ♑ 49 41	16 ♑ 42 46	20 05	29 ♋ 49	08 01	17 45	20 52	15 36	11 03	10 00
27	06 23 32	06 07 37	23 ♑ 31 02	00 ♒ 14 16	21 59	01 ♌ 02	08 27	17 47	20 58	15 39	11 04	09 59
28	06 27 28	07 04 49	06 ♒ 52 22	13 ♒ 25 18	23 56	02 14	08 54	17 49	21 05	15 43	11 04	09 58
29	06 31 25	08 02 01	19 ♒ 53 08	26 ♒ 16 03	25 55	03 27	09 21	17 52	21 11	15 46	11 05	09 57
30	06 35 22	08 ♋ 59 12	02 ♓ 34 17	08 ♓ 48 10	27 ♊ 57	04 ♌ 40	09 ♎ 48	17 ♎ 55	21 ♉ 17	15 ♊ 49	11 ♈ 05	09 ♒ 56

Moon nodes and latitude

Date	Moon True ☊	Moon Mean ☊	Moon ☽ Latitude
01	04 ♑ 52	06 ♑ 07	03 N 10
02	04 D 53	06 03	04 00
03	04 54	06 00	04 38
04	04 R 54	05 57	05 03
05	04 54	05 54	05 14
06	04 52	05 51	05 12
07	04 50	05 48	04 56
08	04 47	05 44	04 27
09	04 45	05 41	03 47
10	04 42	05 38	02 56
11	04 41	05 35	01 56
12	04 40	05 32	00 N 49
13	04 D 40	05 28	00 S 21
14	04 40	05 25	01 32
15	04 41	05 22	02 39
16	04 42	05 19	03 38
17	04 43	05 16	04 26
18	04 43	05 13	04 59
19	04 R 43	05 09	05 16
20	04 43	05 06	05 13
21	04 43	05 03	04 50
22	04 42	05 00	04 10
23	04 42	04 57	03 13
24	04 42	04 53	02 05
25	04 D 42	04 50	00 S 49
26	04 R 42	04 47	00 N 28
27	04 42	04 44	01 42
28	04 42	04 41	02 49
29	04 41	04 38	03 45
30	04 ♑ 41	04 ♑ 34	04 N 29

DECLINATIONS

Date	Sun ☉	Moon ☽	Mercury ☿	Venus ♀	Mars ♂	Jupiter ♃	Saturn ♄	Uranus ♅	Neptune ♆	Pluto ♇
01	22 N 08	14 S 10	13 N 50	24 N 23	01 N 05	05 S 38	15 N 11	22 N 30	02 N 53	23 S 02
02	22 15	09 32	13 58	24 25	00 56	05 38	15 13	22 30	02 53	23 02
03	22 23	04 S 38	14 08	24 26	00 48	05 37	15 15	22 31	02 54	23 03
04	22 30	00 N 20	14 20	24 27	00 39	05 37	15 17	22 31	02 54	23 03
05	22 36	05 14	14 33	24 27	00 30	05 36	15 18	22 31	02 54	23 03
06	22 42	09 54	14 48	24 26	00 21	05 36	15 20	22 32	02 55	23 03
07	22 48	14 12	15 05	24 25	00 12	05 36	15 22	22 32	02 55	23 04
08	22 54	17 58	15 23	24 22	00 N 03	05 36	15 24	22 33	02 56	23 04
09	22 59	21 02	15 42	24 20	00 S 07	05 36	15 26	22 33	02 56	23 04
10	23 03	23 12	16 02	24 16	00 17	05 36	15 28	22 34	02 56	23 05
11	23 07	24 18	16 23	24 12	00 26	05 36	15 30	22 34	02 57	23 06
12	23 11	24 11	16 45	24 07	00 36	05 36	15 31	22 34	02 57	23 06
13	23 14	22 49	17 08	24 01	00 46	05 36	15 33	22 35	02 58	23 07
14	23 17	20 13	17 32	23 54	00 57	05 36	15 35	22 35	02 58	23 07
15	23 20	16 32	17 56	23 47	01 07	05 37	15 37	22 36	02 58	23 07
16	23 22	11 57	18 21	23 39	01 17	05 37	15 38	22 36	02 58	23 08
17	23 23	06 43	18 46	23 31	01 28	05 37	15 40	22 36	02 59	23 08
18	23 25	01 N 05	19 11	23 22	01 39	05 38	15 42	22 37	02 59	23 09
19	23 26	04 S 41	19 36	23 12	01 50	05 39	15 43	22 37	02 59	23 09
20	23 26	10 17	20 01	23 01	02 00	05 39	15 45	22 38	03 00	23 10
21	23 26	15 23	20 26	22 50	02 12	05 40	15 47	22 38	03 00	23 10
22	23 26	19 38	20 50	22 38	02 23	05 41	15 48	22 38	03 00	23 10
23	23 25	22 40	21 14	22 26	02 34	05 42	15 50	22 39	03 00	23 11
24	23 24	24 14	21 37	22 13	02 45	05 43	15 52	22 39	03 00	23 11
25	23 22	24 11	21 59	21 59	02 57	05 44	15 53	22 39	03 01	23 12
26	23 20	22 36	22 19	21 45	03 08	05 45	15 55	22 40	03 01	23 12
27	23 18	19 42	22 39	21 30	03 20	05 46	15 56	22 40	03 01	23 13
28	23 15	15 49	22 57	21 14	03 32	05 47	15 58	22 41	03 01	23 13
29	23 12	11 17	23 13	20 58	03 43	05 48	15 59	22 41	03 01	23 14
30	23 N 08	06 S 23	23 N 27	20 N 41	03 S 55	05 S 50	16 N 01	22 N 41	03 N 01	23 S 14

ZODIAC SIGN ENTRIES

Date	h	m	Planets
02	00	11	♀ ♋
02	22	24	☽ ♓
05	04	49	♂ ♎
05	09	50	☽ ♈
07	22	27	☽ ♉
10	10	13	☽ ♊
12	20	01	☽ ♋
13	16	46	☿ ♊
15	03	34	☽ ♌
17	09	00	☽ ♍
19	12	35	☽ ♎
21	01	48	☉ ♋
21	14	51	☽ ♏
23	16	37	☽ ♐
25	19	05	☽ ♑
26	15	37	♀ ♌
27	23	34	☽ ♒
30	07	05	☽ ♓

LATITUDES

Date	Mercury ☿	Venus ♀	Mars ♂	Jupiter ♃	Saturn ♄	Uranus ♅	Neptune ♆	Pluto ♇
01	03 S 50	00 N 57	00 N 43	01 N 26	02 S 06	00 00	01 S 27	05 S 36
04	03 49	01 03	00 37	01 25	02 06	00 00	01 28	05 37
07	03 39	01 09	00 32	01 24	02 06	00 00	01 28	05 37
10	03 23	01 14	00 27	01 23	02 07	00 00	01 28	05 38
13	03 02	01 19	00 22	01 23	02 07	00 00	01 28	05 39
16	02 35	01 24	00 17	01 22	02 07	00 00	01 28	05 39
19	02 05	01 28	00 13	01 21	02 07	00 00	01 28	05 40
22	01 32	01 32	00 08	01 20	02 08	00 00	01 28	05 40
25	00 57	01 35	00 N 04	01 19	02 08	00 00	01 29	05 41
28	00 S 21	01 37	00 00	01 19	02 08	00 00	01 29	05 41
31	00 N 13	01 N 39	00 S 04	01 N 18	02 S 09	00 00	01 S 29	05 S 42

DATA

Julian Date	2462289
Delta T	+74 seconds
Ayanamsa	24° 16' 22"
Synetic vernal point	04° ♓ 50' 37"
True obliquity of ecliptic	23° 26' 08"

MOON'S PHASES, APSIDES AND POSITIONS ☽

Date	h	m	Phase	Longitude	Eclipse Indicator
04	01	19	☾	13 ♓ 44	
12	03	50	●	21 ♊ 29	Partial
19	09	54	☽	28 ♍ 25	
26	03	22	○	04 ♑ 50	total

Day	h	m		
06	22	15	Apogee	
22	15	39	Perigee	
04	10	22	0N	
11	21	52	Max dec	24° N 24'
18	16	30	0S	
24	23	18	Max dec	24° S 25'

LONGITUDES

Date	Chiron ⚷	Ceres ⚳	Pallas ⚴	Juno ⚵	Vesta ⚶	Black Moon Lilith ⚸
01	10 ♉ 15	24 ♒ 44	11 ♑ 13	24 ♎ 43	24 ♑ 42	20 ♈ 20
11	10 ♉ 48	25 ♒ 28	08 ♑ 49	24 ♎ 08	23 ♑ 54	21 ♈ 27
21	11 ♉ 17	25 ♒ 38	06 ♑ 01	24 ♎ 07	22 ♑ 21	22 ♈ 34
31	11 ♉ 43	25 ♒ 12	03 ♑ 09	24 ♎ 36	20 ♑ 15	23 ♈ 41

ASPECTARIAN

h m	Aspects
01 Friday	
00 20	♀ □ ♂
06 28	☽ # ♄
09 03	☽ ☌ ♇
09 37	☽ ✱ ♆
10 41	☽ △ ☉
13 44	☽ # ☿
15 48	☽ ꝑ ♂
16 09	☽ △ ♅
17 04	☽ ꝑ ♀
22 52	☽ △ ♃
23 33	☽ □ ♄
02 Saturday	
02 19	☽ □ ☿
09 14	☽ ± ♂
14 03	☽ ∠ ♆
21 03	☽ ⊼ ♂
03 Sunday	
00 50	☽ △ ♀
03 36	☽ ꝑ ♃
07 14	☽ ∥ ♃
07 32	☽ ⊥ ♆
10 35	☽ Q ♄
14 51	☽ Q ☿
18 36	☽ ⊻ ♇
19 20	☽ ⊻ ♆
20 24	☽ # ♆
21 13	☽ ± ♃
04 Monday	
01 19	☽ □ ☉
02 26	☽ □ ♅
06 28	☽ ⊥ ♇
07 04	☽ # ♂
09 05	☽ ⊼ ♃
10 33	☽ ✱ ♄
13 29	☽ ∥ ♂
16 03	☽ ✱ ☿
16 23	☉ ☌ ♅
16 54	☉ ∥ ♅
19 24	♀ ∠ ♄
05 Tuesday	
00 26	☽ ∠ ♇
00 30	☽ ∥ ♆
09 58	☽ ☍ ♂
12 56	♀ ± ♇
13 53	☽ # ♃
14 39	☽ Q ♅
16 28	☽ Q ☉
16 57	☽ ∠ ♄
19 10	☽ □ ♀
06 Wednesday	
00 15	☽ ∠ ☿
06 38	☽ ✱ ♇
07 33	☽ ☌ ♆
11 19	☽ ⊥ ♄
15 01	☽ ✱ ♅
18 51	☽ ✱ ☉
19 48	☽ ⊥ ☿
21 21	☽ ☍ ♃
23 35	☽ ⊻ ♄
07 Thursday	
06 53	☽ Q ♇
08 53	☽ ⊻ ☿
11 44	☽ Q ♀
16 35	☿ ± ♃
17 42	☽ ∥ ☿
19 08	☽ ∥ ♄
21 29	☽ ∠ ♅
08 Friday	
00 14	☽ ⊼ ♂
01 19	☉ △ ♃
03 48	☽ ∠ ☉
12 39	☽ ± ♂
14 13	☿ ∥ ♄
14 39	☽ ✱ ♀
15 40	☽ ⊥ ♅
19 05	☽ □ ♇
20 09	☽ ⊻ ♆
23 25	☽ ⊥ ☉
09 Saturday	
03 43	☽ ⊻ ♅
07 09	☽ ꝑ ♂
08 09	☽ ⊥ ♆
09 35	☽ ⊼ ♃
12 25	☽ ⊻ ☉
12 27	☽ ☌ ♄
13 03	☉ ⊻ ♄
21 28	☽ ± ♃
23 53	☽ ∠ ♀
10 Sunday	
00 47	☿ ∠ ♆
02 01	☽ ∠ ♆
02 07	☽ ☌ ☿
03 35	☽ ∥ ♅
09 47	☽ ∥ ☉
09 52	♀ ⊼ ♇
10 22	☽ # ♇
13 39	☽ △ ♂
15 10	☽ ꝑ ♃
19 30	☽ ⊥ ♀
21 28	♀ □ ♆
11 Monday	
01 48	☉ # ♇
06 12	☽ △ ♇
07 23	☽ ✱ ♆
08 18	☽ ∥ ♀
08 27	☽ ⊻ ♀
14 54	☽ ☌ ♅
20 14	☽ △ ♃
23 36	☽ ⊻ ♄
12 Tuesday	
03 50	☽ ☌ ☉
06 29	☽ Q ♆
10 58	☽ ꝑ ♇
11 09	☽ ⊥ ♄
14 31	☽ ∥ ♀
17 50	☽ ⊻ ☿
13 Wednesday	
00 58	☽ □ ♂
03 58	☽ ± ♇
04 19	☽ ∠ ♄
06 18	☽ ⊥ ☿
06 38	☽ ∥ ☉
08 14	☽ # ♇
14 40	☽ ∥ ♅
15 07	☉ Q ♆
15 11	☽ ⊼ ♇
16 27	☽ □ ♆
21 06	♃ St D
23 19	☽ ☌ ♀
23 51	☽ ⊻ ♅
14 Thursday	
00 51	☽ ∠ ☿
04 42	☽ □ ♃
05 13	♀ ⊻ ♅
08 27	☽ ✱ ♄
10 58	☽ ⊥ ♅
11 24	☽ Q ♂
16 27	☽ ⊻ ☉
15 Friday	
03 28	☽ ∠ ♅
04 14	☽ ⊥ ☉
04 24	☽ ∥ ☿
06 33	☽ Q ♄
07 18	☽ ✱ ☿
09 54	☽ ✱ ♂
13 32	☽ Q ♃
15 28	☿ ꝑ ♃
17 08	☽ ∥ ♄
21 43	☽ ∠ ☉
21 55	☽ ☍ ♇
23 16	☽ △ ♆
16 Saturday	
00 34	☉ ꝑ ♇
06 33	☽ ✱ ♅
07 16	☽ Q ☿
08 57	♀ □ ♃
11 01	☽ ✱ ♃
11 13	☽ ⊻ ♀
13 31	☽ ∠ ♂
15 03	☽ □ ♄
16 17	☉ ⊥ ♄
17 24	☿ △ ♂
22 52	☽ ⊥ ♀
17 Sunday	
01 53	☽ ꝑ ♆
02 21	☽ ✱ ☉
03 51	☽ Q ♅
05 50	☽ ⊥ ♂
13 25	☽ ∠ ♃
16 09	☽ ∠ ♀
16 37	☽ ⊻ ♂
16 45	☽ # ♃
17 38	☽ ± ♆
18 34	☽ □ ☿
18 Monday	
00 49	☽ Q ☉
02 38	☽ ⊼ ♇
04 01	☽ ∥ ♆
04 02	☽ ⊼ ♆
05 00	☉ ∥ ♀
05 02	☽ ⊥ ♃
09 43	☽ # ♂
09 54	♀ ✱ ♄
11 14	☽ □ ♅
12 57	☽ ± ♇
15 22	☽ ⊻ ♃
19 39	☽ △ ♄
20 29	☽ ✱ ♀
23 42	☽ ∥ ♂
19 Tuesday	
04 19	☽ ꝑ ♇
04 55	☽ # ♆
09 31	♀ ⊥ ♅
09 54	☽ □ ☉
14 59	♂ ꝑ ♄
16 01	☽ ∥ ♃
18 05	♀ # ♇
18 47	☽ Q ♀
21 19	☽ ꝑ ♄
21 27	☽ ☌ ♂
20 Wednesday	
03 58	☽ △ ☿
05 39	☽ △ ♇
07 07	☽ ☍ ♆
12 33	☽ ± ♄
14 17	☽ △ ♅
18 09	☽ ☌ ♃
18 54	☿ △ ♇
22 40	☽ ⊼ ♄
21 Thursday	
03 45	☽ □ ♀
08 11	☿ ✱ ♆
08 12	☽ ꝑ ☿
14 03	☽ # ♄
15 23	☽ ꝑ ♅
15 46	☽ △ ☉
16 18	♀ Q ♂
22 Friday	
01 00	☽ ⊻ ♂
01 05	☽ ± ☿
06 22	☽ ± ♅
07 34	☽ □ ♇
09 07	☽ ⊼ ♆
11 14	☽ ⊥ ♂
11 58	♀ ∥ ♅
12 20	☽ ⊼ ☿
16 22	☽ ⊼ ♅
18 24	☽ ꝑ ☉
19 04	☽ ± ♆
20 05	☽ ⊻ ♃
21 29	☽ # ☿
23 Saturday	
00 51	☽ ☍ ♄
02 36	☽ ∠ ♂
06 03	☽ ⊥ ♃
09 49	☽ # ♀
10 01	☽ ꝑ ♆
10 06	☽ △ ♀
10 23	☽ ± ☉
11 46	☽ # ♅
13 23	☽ Q ♇
17 44	☽ ∥ ♇
20 37	☽ # ☉
21 02	☽ ∠ ♃
21 04	☽ ⊼ ☉
23 04	☿ ☌ ♅
24 Sunday	
04 18	☽ ✱ ♂
09 22	☽ ✱ ♇
11 02	☽ △ ♆
13 25	☽ ꝑ ♀
18 32	☽ ☍ ♅
21 13	☽ ☍ ☿
22 12	☽ ✱ ♃
25 Monday	
01 04	☽ Q ♂
03 20	☽ ⊼ ♄
05 06	☿ △ ♃
06 00	☽ ± ♀
07 09	☉ ± ♇
10 37	☽ ∠ ♇
12 17	☿ ∥ ♀
13 35	☽ ± ♄
17 09	☽ ⊼ ♀
18 36	☽ Q ♃
20 21	♂ # ♆
26 Tuesday	
01 58	☽ ⊥ ♇
03 12	☽ # ☉
03 22	☽ ☍ ☉
05 06	☽ ∥ ♇
05 06	☽ ꝑ ♄
08 46	☽ □ ♂
11 20	☽ # ♅
12 18	☽ ⊻ ♇
14 08	☽ □ ♆
14 28	☽ # ☿
21 00	☽ # ♀
22 05	☽ ⊼ ♅
22 32	☿ ⊻ ♄
27 Wednesday	
01 51	☽ □ ♃
04 14	♀ ∠ ♅
07 28	☽ △ ♄
07 38	☉ ∠ ♄
08 41	☽ ± ♅
08 51	☽ ⊼ ☿
13 51	☿ ∥ ♅
21 19	☽ ± ☿
21 54	☽ Q ♆
28 Thursday	
00 48	☽ ꝑ ♅
01 25	☿ Q ♆
02 46	☽ ☍ ♀
11 14	☽ # ♄
12 24	☽ ⊼ ☉
15 50	☽ △ ♂
16 25	☽ ꝑ ☿
17 39	☽ ☌ ♇
19 41	☽ ✱ ♆
22 37	☉ # ♇
29 Friday	
00 16	☽ ± ☉
00 28	☿ ꝑ ♇
04 18	☽ △ ♅
06 03	♀ Q ♄
08 14	☽ △ ♃
10 28	☉ ∥ ☿
13 32	☿ # ♇
14 26	☽ □ ♄
18 23	☽ ꝑ ☉
20 41	☽ ꝑ ♂
23 39	☽ ∠ ♆
30 Saturday	
01 29	☽ △ ☿
03 46	☿ ⊥ ♄
12 40	☽ ꝑ ♃
14 27	☽ ± ♂
14 37	☽ ∥ ♃
16 27	☽ ⊼ ♀
16 50	☽ ⊥ ♆
18 31	♂ △ ♇
23 15	☽ ∥ ♂

All ephemeris data is given at 12.00 UT and the Moon's longitude is additionally given for 24.00 UT

JULY 2029

LONGITUDES

Date	Sidereal time h m s	Sun ☉ ° ' "	Moon ☽ ° ' "	Moon ☽ 24.00 ° ' "	Mercury ☿ ° '	Venus ♀ ° '	Mars ♂ ° '	Jupiter ♃ ° '	Saturn ♄ ° '	Uranus ♅ ° '	Neptune ♆ ° '	Pluto ♇ ° '
01	06 39 18	09 ♋ 56 24	14 ♓ 58 04	21 ♓ 04 27	00 ♋ 00	05 ♌ 52	10 ♎ 16	17 ♎ 58	21 ♉ 23	15 ♊ 53	11 ♈ 06	09 ♒ 55
02	06 43 15	10 53 36	27 ♓ 07 47	03 ♈ 08 35	02 05	07 05	10 43	18 01	21 29	15 56	11 06	09 R 54
03	06 47 11	11 50 48	09 ♈ 07 25	15 ♈ 04 50	04 12	08 18	11 12	18 04	21 35	15 59	11 07	09 52
04	06 51 08	12 48 00	21 ♈ 01 25	26 ♈ 57 44	06 20	09 30	11 40	18 08	21 40	16 03	11 07	09 51
05	06 55 04	13 45 13	02 ♉ 54 22	08 ♉ 51 53	08 29	10 43	12 08	18 12	21 46	16 06	11 07	09 50
06	06 59 01	14 42 26	14 ♉ 50 49	20 ♉ 51 41	10 38	11 55	12 37	18 15	21 52	16 09	11 08	09 49
07	07 02 57	15 39 39	26 ♉ 54 57	03 ♊ 01 06	12 48	13 08	13 06	18 19	21 57	16 12	11 08	09 48
08	07 06 54	16 36 52	09 ♊ 10 30	15 ♊ 23 31	14 58	14 20	13 36	18 24	22 03	16 15	11 08	09 46
09	07 10 51	17 34 06	21 ♊ 40 45	28 ♊ 01 26	17 08	15 32	14 05	18 28	22 08	16 19	11 08	09 45
10	07 14 47	18 31 20	04 ♋ 26 44	10 ♋ 56 23	19 18	16 45	14 35	18 32	22 14	16 22	11 08	09 44
11	07 18 44	19 28 34	17 ♋ 30 25	24 ♋ 08 46	21 27	17 57	15 05	18 37	22 19	16 25	11 08	09 43
12	07 22 40	20 25 49	00 ♌ 51 17	07 ♌ 37 48	23 35	19 10	15 35	18 42	22 25	16 28	11 08	09 41
13	07 26 37	21 23 03	14 ♌ 28 02	21 ♌ 21 42	25 42	20 22	16 06	18 47	22 30	16 31	11 08	09 40
14	07 30 33	22 20 18	28 ♌ 18 25	05 ♍ 17 49	27 47	21 34	16 37	18 52	22 35	16 34	11 R 08	09 39
15	07 34 30	23 17 33	12 ♍ 19 30	19 ♍ 23 03	29 ♋ 52	22 46	17 08	18 57	22 40	16 37	11 08	09 37
16	07 38 26	24 14 47	26 ♍ 28 03	03 ♎ 34 07	01 ♌ 55	23 59	17 39	19 02	22 45	16 40	11 08	09 36
17	07 42 23	25 12 02	10 ♎ 40 53	17 ♎ 47 57	03 56	25 11	18 10	19 08	22 50	16 43	11 08	09 35
18	07 46 20	26 09 17	24 ♎ 55 01	02 ♏ 01 46	05 56	26 23	18 42	19 14	22 55	16 46	11 08	09 33
19	07 50 16	27 06 33	09 ♏ 07 55	16 ♏ 13 10	07 54	27 35	19 14	19 20	22 59	16 49	11 08	09 32
20	07 54 13	28 03 48	23 ♏ 17 18	00 ♐ 20 02	09 50	28 47	19 46	19 26	23 04	16 52	11 08	09 31
21	07 58 09	29 01 04	07 ♐ 21 09	14 ♐ 20 22	11 44	29 ♌ 59	20 18	19 32	23 09	16 54	11 08	09 29
22	08 02 06	29 ♋ 58 20	21 ♐ 17 27	28 ♐ 12 10	13 37	01 ♍ 11	20 50	19 38	23 13	16 57	11 07	09 28
23	08 06 02	00 ♌ 55 36	05 ♑ 04 13	11 ♑ 53 23	15 28	02 23	21 23	19 44	23 18	17 00	11 07	09 27
24	08 09 59	01 52 53	18 ♑ 39 26	25 ♑ 22 06	17 18	03 35	21 56	19 51	23 22	17 03	11 07	09 25
25	08 13 55	02 50 10	02 ♒ 01 11	08 ♒ 36 36	19 05	04 47	22 29	19 57	23 26	17 06	11 06	09 24
26	08 17 52	03 47 28	15 ♒ 08 07	21 ♒ 35 42	20 51	05 58	23 02	20 04	23 31	17 08	11 06	09 23
27	08 21 49	04 44 47	27 ♒ 59 20	04 ♓ 19 01	22 35	07 10	23 35	20 11	23 35	17 11	11 06	09 21
28	08 25 45	05 42 06	10 ♓ 34 54	16 ♓ 47 07	24 17	08 22	24 09	20 18	23 39	17 13	11 05	09 20
29	08 29 42	06 39 27	22 ♓ 55 54	29 ♓ 01 34	25 58	09 33	24 43	20 25	23 43	17 16	11 05	09 18
30	08 33 38	07 36 48	05 ♈ 04 26	11 ♈ 04 56	27 37	10 45	25 17	20 33	23 47	17 19	11 04	09 17
31	08 37 35	08 ♌ 34 10	17 ♈ 03 32	23 ♈ 00 42	29 ♌ 14	11 ♍ 57	25 ♎ 51	20 ♎ 40	23 ♉ 50	17 ♊ 21	11 ♈ 04	09 ♒ 16

Date	Moon True ☊ ° '	Moon Mean ☊ ° '	Moon ☽ Latitude ° '
01	04 ♑ 40	04 ♑ 31	04 N 58
02	04 R 40	04 28	05 14
03	04 40	04 25	05 16
04	04 D 40	04 22	05 04
05	04 40	04 19	04 39
06	04 41	04 15	04 02
07	04 42	04 12	03 14
08	04 43	04 09	02 17
09	04 44	04 06	01 12
10	04 44	04 03	00 N 02
11	04 R 44	03 59	01 S 10
12	04 43	03 56	02 19
13	04 41	03 53	03 22
14	04 39	03 50	04 14
15	04 36	03 47	04 51
16	04 34	03 44	05 12
17	04 33	03 40	05 13
18	04 32	03 37	04 55
19	04 D 33	03 34	04 19
20	04 34	03 31	03 28
21	04 35	03 28	02 24
22	04 36	03 25	01 S 13
23	04 R 37	03 21	00 N 02
24	04 36	03 18	01 16
25	04 34	03 15	02 25
26	04 31	03 12	03 24
27	04 26	03 09	04 11
28	04 21	03 05	04 45
29	04 16	03 02	05 06
30	04 12	02 59	05 11
31	04 ♑ 08	02 ♑ 56	05 N 04

DECLINATIONS

Date	Sun ☉	Moon ☽	Mercury ☿	Venus ♀	Mars ♂	Jupiter ♃	Saturn ♄	Uranus ♅	Neptune ♆	Pluto ♇
01	23 N 04	01 S 20	23 N 39	20 N 24	04 S 07	05 S 51	16 N 02	22 N 42	03 N 02	23 S 15
02	22 59	03 N 40	23 49	20 06	04 19	05 52	16 03	22 42	03 02	23 15
03	22 54	08 27	23 56	19 48	04 32	05 54	16 05	22 42	03 02	23 16
04	22 49	12 54	24 01	19 29	04 44	05 55	16 06	22 43	03 02	23 16
05	22 44	16 51	24 03	19 09	04 56	05 57	16 07	22 43	03 02	23 17
06	22 37	20 09	24 02	18 49	05 08	05 59	16 09	22 43	03 02	23 17
07	22 31	22 37	23 58	18 29	05 21	06 00	16 10	22 44	03 02	23 17
08	22 24	24 04	23 52	18 08	05 33	06 02	16 11	22 44	03 02	23 18
09	22 17	24 22	23 43	17 46	05 46	06 04	16 13	22 44	03 02	23 18
10	22 09	23 23	23 31	17 24	05 58	06 06	16 14	22 45	03 02	23 19
11	22 01	21 08	23 17	17 02	06 11	06 08	16 15	22 45	03 02	23 19
12	21 53	17 42	23 00	16 39	06 24	06 10	16 16	22 45	03 02	23 20
13	21 44	13 16	22 40	16 16	06 37	06 12	16 17	22 46	03 02	23 20
14	21 35	08 05	22 19	15 52	06 49	06 14	16 19	22 46	03 02	23 21
15	21 26	02 N 27	21 55	15 28	07 02	06 17	16 20	22 46	03 02	23 21
16	21 16	03 S 22	21 29	15 04	07 15	06 19	16 21	22 47	03 02	23 22
17	21 05	09 01	21 02	14 39	07 28	06 21	16 22	22 47	03 02	23 22
18	20 55	14 13	20 33	14 14	07 41	06 24	16 23	22 47	03 02	23 23
19	20 44	18 38	20 02	13 48	07 54	06 26	16 24	22 47	03 01	23 23
20	20 33	21 57	19 30	13 22	08 07	06 28	16 25	22 48	03 01	23 24
21	20 21	23 54	18 57	12 56	08 20	06 31	16 26	22 48	03 01	23 24
22	20 09	24 21	18 22	12 29	08 34	06 34	16 27	22 48	03 01	23 25
23	19 57	23 18	17 46	12 03	08 47	06 36	16 28	22 49	03 01	23 25
24	19 44	20 53	17 10	11 35	09 00	06 39	16 29	22 49	03 01	23 26
25	19 31	17 22	16 33	11 08	09 13	06 42	16 30	22 49	03 01	23 26
26	19 18	13 03	15 55	10 40	09 26	06 44	16 30	22 49	03 00	23 27
27	19 04	08 14	15 16	10 12	09 40	06 47	16 31	22 50	03 00	23 27
28	18 51	03 S 12	14 37	09 44	09 53	06 50	16 32	22 50	03 00	23 27
29	18 36	01 N 52	13 58	09 15	10 06	06 53	16 33	22 50	03 00	23 28
30	18 22	06 47	13 18	08 47	10 19	06 56	16 34	22 50	02 59	23 28
31	18 N 07	11 N 22	12 N 38	08 N 18	10 S 33	06 S 59	16 N 34	22 N 51	02 N 59	23 S 29

ZODIAC SIGN ENTRIES

Date	h m	Planets
01	12 01	☿ ♋
02	17 43	☽ ♈
05	06 08	☽ ♉
07	18 05	☽ ♊
10	03 42	☽ ♋
12	10 29	☽ ♌
14	14 55	☽ ♍
15	13 37	☿ ♌
16	17 58	☽ ♎
18	20 34	☽ ♏
20	23 26	☽ ♐
21	12 21	♀ ♍
22	12 42	☉ ♌
23	03 08	☽ ♑
25	08 21	☽ ♒
27	15 48	☽ ♓
30	01 56	☽ ♈
31	23 35	☿ ♍

LATITUDES

Date	Mercury ☿	Venus ♀	Mars ♂	Jupiter ♃	Saturn ♄	Uranus ♅	Neptune ♆	Pluto ♇
01	00 N 13	01 N 39	00 S 04	01 N 18	02 S 09	00 00	01 S 29	05 S 42
04	00 44	01 40	00 08	01 17	02 09	00 00	01 29	05 42
07	01 09	01 41	00 11	01 16	02 10	00 00	01 29	05 43
10	01 29	01 40	00 15	01 15	02 10	00 00	01 29	05 43
13	01 42	01 39	00 18	01 15	02 10	00 00	01 30	05 44
16	01 48	01 38	00 21	01 14	02 11	00 N 01	01 30	05 44
19	01 48	01 35	00 25	01 13	02 11	00 01	01 30	05 44
22	01 42	01 32	00 28	01 13	02 12	00 01	01 30	05 45
25	01 31	01 29	00 30	01 12	02 12	00 01	01 30	05 45
28	01 16	01 24	00 33	01 11	02 12	00 01	01 30	05 45
31	00 N 57	01 N 19	00 S 36	01 N 10	02 S 13	00 N 01	01 S 31	05 S 45

DATA

Julian Date	2462319
Delta T	+74 seconds
Ayanamsa	24° 16' 27"
Synetic vernal point	04° ♓ 50' 32"
True obliquity of ecliptic	23° 26' 08"

MOON'S PHASES, APSIDES AND POSITIONS ☽

Date	h m	Phase	Longitude ° '	Eclipse Indicator
03	17 57	☾	12 ♈ 05	
11	15 51	●	19 ♋ 38	Partial
18	14 14	☽	26 ♎ 15	
25	13 36	○	02 ♒ 54	

Day	h m		
04	16 06	Apogee	
18	11 20	Perigee	
01	18 21	ON	
09	05 40	Max dec	24° N 25'
15	22 07	OS	
22	07 03	Max dec	24° S 23'
29	03 05	ON	

LONGITUDES

Date	Chiron ⚷ ° '	Ceres ⚳ ° '	Pallas ⚴ ° '	Juno ⚵ ° '	Vesta ⚶ ° '	Black Moon Lilith ⚸ ° '
01	11 ♉ 43	25 ♒ 12	03 ♑ 09	24 ♎ 36	20 ♑ 15	23 ♈ 41
11	12 ♉ 04	24 ♒ 12	00 ♑ 30	25 ♎ 32	17 ♑ 51	24 ♈ 49
21	12 ♉ 21	22 ♒ 40	28 ♐ 19	26 ♎ 54	15 ♑ 31	25 ♈ 56
31	12 ♉ 33	20 ♒ 43	26 ♐ 45	28 ♎ 38	13 ♑ 36	27 ♈ 03

ASPECTARIAN

h m	Aspects
01 Sunday	
01 02	☽ Q ♄
01 23	☽ △ ☉
02 10	☽ ⊻ ♇
02 28	☽ ⚻ ♂
03 58	☽ # ♆
04 27	☽ ⊻ ♆
05 18	☽ ± ♀
06 07	☽ ± ♃
08 31	☿ ⊥ ♀
11 20	☉ ⚻ ♇
13 47	☽ □ ♅
13 51	☽ ⊥ ♇
13 55	♀ Q ♃
17 54	☽ ⚻ ♃
02 Monday	
00 42	☽ ✶ ♄
00 53	☽ ⚼ ♀
03 42	☉ □ ♂
07 34	☽ ∠ ♇
08 55	☽ ∥ ♆
15 23	☽ # ♂
17 21	☉ □ ♆
22 58	☽ # ♃
23 59	☽ □ ☿
03 Tuesday	
01 39	☽ Q ♅
06 50	☽ ∠ ♄
07 47	♂ ☍ ♆
08 24	☿ ± ♇
10 09	☽ △ ♀
13 30	☽ ✶ ♇
16 00	☽ ☌ ♆
16 20	☽ ☍ ♂
17 57	☽ □ ☉
04 Wednesday	
01 06	☽ ⊥ ♄
01 54	☽ ✶ ♅
06 08	☽ ☍ ♃
13 19	☽ ⊻ ♄
13 41	☽ Q ♇
16 02	☿ ∠ ♄
18 52	♀ ☍ ♇
20 09	☽ Q ☿
05 Thursday	
07 19	☽ ∥ ♄
08 20	☽ ∠ ♅
09 28	☽ Q ☉
14 00	☉ ∥ ♅
20 10	♀ △ ♆
06 Friday	
01 42	☽ ✶ ☿
01 55	☽ □ ♇
02 32	☽ ⊥ ♅
02 41	☽ ∥ ♀
02 57	☿ ⚻ ♇
04 32	☽ ⊻ ♆
05 29	☽ □ ♀
07 21	☽ ⚻ ♂
11 42	☽ ✶ ☉
14 37	☽ ⊻ ♅
16 33	☽ ⊥ ♆
17 25	☿ □ ♆
18 51	☽ ⚻ ♃
19 51	☽ ± ♂
07 Saturday	
02 06	☽ ☌ ♄
06 51	☽ ± ♃
10 27	☽ ∠ ♆
10 54	☽ ∥ ☉
11 15	♀ ✶ ♂
13 23	☽ ∥ ♅
14 08	☽ ∠ ☿
14 27	☽ ⚼ ♂
16 17	☿ □ ♂
20 00	☽ ∠ ☉
20 03	☿ ⊻ ♀
21 03	☽ # ♇
21 12	☽ Q ♀
08 Sunday	
00 40	☽ ⚼ ♃
02 27	☉ ⊻ ♅
07 33	☽ ∥ ☿
11 32	☽ ⊥ ☿
13 09	☽ △ ♇
15 01	☽ ⊥ ☉
15 48	☽ ✶ ♆
20 54	☽ △ ♂
23 02	☽ ✶ ♀
09 Monday	
01 33	☽ ⊻ ☿
01 43	☽ ☌ ♅
02 33	☿ ⊻ ♅
03 32	☽ ⊻ ☉
05 51	☽ △ ♃
12 54	☽ ⊻ ♄
14 46	☽ Q ♆
17 49	☽ ⚼ ♇
20 32	☉ ☌ ☿
10 Tuesday	
00 18	☽ ⊥ ♄
03 15	☿ □ ♃
03 57	♀ ✶ ♅
06 27	☽ ∠ ♀
09 41	☽ ∥ ☿
10 40	☽ ± ♇
12 29	☉ □ ♃
13 03	☽ # ♇
17 12	☽ ∠ ♄
20 20	☽ ∥ ♅
21 46	☽ ⚻ ♇
11 Wednesday	
00 22	☽ □ ♆
00 50	☽ ⊥ ♀
03 33	☽ ∥ ☉
05 13	♂ ∥ ♃
07 25	☽ □ ♂
07 53	☿ # ♇
10 00	☽ ⊻ ♅
12 54	☽ ⊻ ♀
14 02	☽ □ ♃
15 51	☽ ☌ ☉
20 30	☽ ☌ ☿
20 46	☽ ✶ ♄
20 55	☽ ⊥ ♅
22 15	☿ ✶ ♄
23 07	☿ ⊥ ♅
12 Thursday	
02 08	♀ ✶ ♃
13 05	☽ ∠ ♅
17 03	☽ Q ♂
18 21	☽ Q ♄
18 43	☽ ∥ ♀
20 14	☽ ∥ ♄
22 25	☽ Q ♃
13 Friday	
03 36	☽ ☍ ♇
05 52	☿ ∥ ♅
06 11	☽ △ ♆
10 39	♀ ∥ ♄
14 57	☽ ✶ ♂
15 35	☽ ✶ ♅
19 34	☽ ✶ ♃
23 15	☽ ☌ ♀
14 Saturday	
00 56	☽ ⊻ ☉
02 03	☽ □ ♄
02 08	♆ St R
08 16	☽ ⚼ ♆
09 39	♂ △ ♅
10 20	♂ ± ♄
10 57	☽ ⊻ ☿
12 03	☽ ⊥ ☉
12 27	☽ Q ♅
17 18	☽ # ♂
17 54	☽ ∠ ♂
18 02	☉ ⊥ ♅
18 43	☉ ✶ ♄
19 58	☽ # ♃
21 36	☽ ∠ ♃
23 03	☽ ⊥ ☿
23 44	☽ ± ♆
15 Sunday	
04 37	☽ ∠ ☉
07 24	☽ ⚻ ♇
09 33	☽ ∥ ♆
09 42	♀ □ ♄
09 53	☽ ⊥ ♂
09 59	☽ ⚻ ♆
13 04	☽ ⊥ ♃
17 03	☽ ∠ ☿
17 36	☽ ± ♇
19 20	☽ □ ♅
20 29	☽ ⊻ ♂
23 20	☽ ⊻ ♃
16 Monday	
05 40	☽ △ ♄
07 24	☽ ⊻ ♀
07 58	☽ ✶ ☉
08 51	☽ ⚼ ♇
09 03	☿ ∠ ♅
10 37	☽ # ♆
18 29	☽ ⊥ ♀
22 44	☽ ✶ ☿
17 Tuesday	
00 26	☽ ∥ ♃
05 01	☽ ∥ ♂
05 42	☽ Q ☉
07 10	☽ ⚼ ♄
07 15	☉ ∥ ☿
10 09	☽ △ ♇
11 04	☽ ∠ ♀
12 46	☽ ☍ ♆
14 08	☉ ⊻ ♀
22 12	☽ △ ♅
22 18	☽ Q ☿
22 26	☽ ± ♄
23 14	☿ Q ♄
18 Wednesday	
01 07	☽ ☌ ♂
02 21	☽ ☌ ♃
07 07	♀ ⚼ ♆
08 36	☽ ⚻ ♄
12 03	☽ # ♀
14 14	☽ □ ☉
14 42	☽ ✶ ♀
23 11	☽ # ♄
23 35	☽ ⚼ ♅
19 Thursday	
00 50	☿ Q ♂
04 40	☿ Q ♃
09 34	☽ □ ☿
12 41	☽ □ ♇
12 50	☽ Q ♀
14 51	☽ ± ♅
15 23	☽ ⚻ ♆
17 21	♂ ☌ ♃
19 56	☽ # ☿
20 Friday	
01 03	☽ ⚻ ♅
01 22	☽ # ☉
01 33	☽ ± ♆
05 23	☽ ⊻ ♃
05 46	☽ ⊻ ♂
08 04	☿ ☍ ♇
11 37	☽ ☍ ♄
13 36	♀ Q ♅
15 40	☽ ⊥ ♃
16 23	☽ ⊥ ♂
16 50	☽ ⚼ ♆
19 11	☽ Q ♇
20 26	☽ # ♅
20 43	☽ △ ☉
22 14	☽ □ ♀
21 Saturday	
03 49	☽ ∥ ♇
04 15	☿ △ ♆
07 08	☽ ∠ ♃
08 21	☽ ∠ ♂
15 40	☽ ✶ ♇
18 28	☽ △ ♆
20 43	☽ △ ☿
22 Sunday	
00 17	☽ ⚼ ☉
04 29	☽ ☍ ♅
09 06	☽ ✶ ♃
11 11	☽ ✶ ♂
15 22	☽ ⚻ ♄
16 59	☽ ± ☉
17 30	☽ ∠ ♇
23 Monday	
01 51	☽ ± ♄
02 43	☽ ⚼ ☿
04 13	☽ ⚻ ☉
06 07	☽ Q ♃
06 50	☽ △ ♀
08 55	☽ Q ♂
09 09	☽ ⊥ ♇
10 19	☽ ∥ ♇
17 42	☽ ⚼ ♄
17 59	☽ # ♅
19 40	☽ ⊻ ♇
20 56	☽ ± ☿
22 38	☽ □ ♆
24 Tuesday	
08 38	☿ ✶ ♅
09 08	☽ ⚻ ♅
09 12	☽ ⚻ ☿
11 51	☽ ⚼ ♀
14 08	☽ □ ♃
16 21	☉ ∠ ♅
18 05	☽ □ ♂
19 52	☽ ± ♅
20 27	☽ △ ♄
21 06	☽ # ☉
25 Wednesday	
05 33	☽ ± ♀
06 44	☽ Q ♆
12 08	☽ ⚼ ♅
13 36	☽ ☍ ☉
13 56	☿ ∥ ♄
16 01	♀ ∠ ♃
17 06	☽ # ♄
17 30	☽ ⚻ ♀
17 41	☽ # ☿
18 35	♀ ± ♆
26 Thursday	
00 38	☿ ✶ ♃
01 25	☽ ☌ ♇
04 34	☽ ✶ ♆
15 43	☽ △ ♅
21 14	☽ △ ♃
27 Friday	
00 16	☽ ☍ ☿
01 23	☽ # ♀
03 21	☽ △ ♂
03 40	☽ □ ♄
05 23	☽ ∥ ♂
08 26	☽ ∠ ♆
11 23	♂ ⚻ ♄
18 54	☽ ∥ ♃
28 Saturday	
01 28	☽ ⊥ ♆
01 47	☽ ⚼ ♃
01 52	☽ ⚻ ☉
02 32	☿ □ ♄
06 51	♀ # ♂
07 17	☽ ☍ ♀
09 05	☿ ✶ ♂
09 07	☽ ⚼ ♂
09 36	☽ ⊻ ♇
10 28	☉ Q ♄
12 55	☽ # ♆
12 58	☽ ⊻ ♆
14 04	☽ Q ♄
14 20	☽ ± ☉
19 15	☽ ± ♃
21 09	☽ ⊥ ♇
29 Sunday	
00 54	☽ □ ♅
03 21	☽ ± ♂
07 03	☽ ⚻ ♃
07 03	♀ ⚻ ♇
09 18	☽ ⚼ ☉
13 32	☽ ✶ ♄
13 39	☿ ⚼ ♆
14 42	☽ ∠ ♇
15 40	☽ ⚻ ♂
17 23	☽ ∥ ♆
17 51	♀ ∠ ♂
18 54	☽ ⚻ ☿
30 Monday	
08 38	☽ ± ☿
12 28	☽ Q ♅
12 46	☽ # ♃
17 30	☽ △ ☉
18 20	♀ ⚻ ♆
19 25	☽ ∠ ♄
20 23	☽ ✶ ♇
21 14	☽ ∥ ♀
23 58	☽ ☌ ♆
31 Tuesday	
00 36	☽ ⚻ ♀
05 26	☽ ⚼ ☿
07 17	☽ # ♂
12 36	☽ ✶ ♅
13 35	☽ ⊥ ♄
13 54	☿ Q ♅
13 59	☽ ± ♀
14 56	☉ Q ♃
18 03	☽ ∥ ☿
19 21	☽ ☍ ♃
20 27	☽ Q ♇

All ephemeris data is given at 12.00 UT and the Moon's longitude is additionally given for 24.00 UT

AUGUST 2029

LONGITUDES

Date	Sidereal time h m s	Sun ☉ ° ' "	Moon ☽ ° ' "	Moon ☽ 24.00 ° ' "	Mercury ☿ ° '	Venus ♀ ° '	Mars ♂ ° '	Jupiter ♃ ° '	Saturn ♄ ° '	Uranus ♅ ° '	Neptune ♆ ° '	Pluto ♇ ° '
01	08 41 31	09 ♌ 31 33	28 ♈ 57 01	04 ♉ 53 01	00 ♍ 49	13 ♍ 08	26 ♎ 25	20 ♎ 48	23 ♉ 54	17 ♊ 24	11 ♈ 03	09 ♒ 14
02	08 45 28	10 28 58	10 ♉ 49 19	16 ♉ 46 31	02 23	14 20	26 59	20 56	23 58	17 26	11 R 02	09 R 13
03	08 49 24	11 26 24	22 ♉ 45 15	28 ♉ 46 07	03 55	15 31	27 34	21 03	24 01	17 28	11 02	09 11
04	08 53 21	12 23 51	04 ♊ 49 44	10 ♊ 56 41	05 25	16 43	28 09	21 11	24 04	17 31	11 01	09 10
05	08 57 18	13 21 19	17 ♊ 07 31	23 ♊ 22 44	06 53	17 54	28 44	21 19	24 08	17 33	11 00	09 09
06	09 01 14	14 18 48	29 ♊ 42 48	06 ♋ 08 04	08 20	19 05	29 19	21 28	24 11	17 36	11 00	09 07
07	09 05 11	15 16 19	12 ♋ 38 50	19 ♋ 15 16	09 45	20 17	29 ♎ 54	21 36	24 14	17 38	10 59	09 06
08	09 09 07	16 13 50	25 ♋ 57 26	02 ♌ 45 17	11 08	21 28	00 ♏ 29	21 44	24 17	17 40	10 58	09 05
09	09 13 04	17 11 23	09 ♌ 38 37	16 ♌ 37 05	12 30	22 39	01 05	21 53	24 20	17 42	10 57	09 03
10	09 17 00	18 08 57	23 ♌ 40 15	00 ♍ 47 32	13 49	23 50	01 41	22 01	24 23	17 44	10 57	09 02
11	09 20 57	19 06 32	07 ♍ 58 14	15 ♍ 11 37	15 06	25 01	02 17	22 10	24 26	17 46	10 56	09 01
12	09 24 53	20 04 08	22 ♍ 26 51	29 ♍ 43 08	16 22	26 12	02 53	22 19	24 28	17 48	10 55	08 59
13	09 28 50	21 01 45	06 ♎ 59 39	14 ♎ 15 40	17 35	27 23	03 29	22 28	24 31	17 51	10 54	08 58
14	09 32 47	21 59 23	21 ♎ 30 29	28 ♎ 43 31	18 47	28 34	04 05	22 37	24 33	17 52	10 53	08 56
15	09 36 43	22 57 02	05 ♏ 54 16	13 ♏ 02 22	19 56	29 ♍ 45	04 42	22 46	24 36	17 54	10 52	08 55
16	09 40 40	23 54 42	20 ♏ 07 32	27 ♏ 09 35	21 03	00 ♎ 56	05 19	22 56	24 38	17 56	10 51	08 54
17	09 44 36	24 52 23	04 ♐ 08 26	11 ♐ 04 02	22 07	02 06	05 55	23 05	24 40	17 58	10 50	08 52
18	09 48 33	25 50 05	17 ♐ 56 23	24 ♐ 45 34	23 10	03 17	06 32	23 14	24 42	18 00	10 49	08 51
19	09 52 29	26 47 49	01 ♑ 31 38	08 ♑ 14 40	24 09	04 28	07 09	23 24	24 44	18 02	10 48	08 50
20	09 56 26	27 45 33	14 ♑ 54 42	21 ♑ 31 49	25 06	05 38	07 47	23 34	24 46	18 03	10 47	08 49
21	10 00 22	28 43 18	28 ♑ 06 03	04 ♒ 37 24	26 00	06 49	08 24	23 44	24 48	18 05	10 46	08 47
22	10 04 19	29 ♌ 41 05	11 ♒ 05 52	17 ♒ 31 27	26 50	07 59	09 01	23 53	24 49	18 07	10 45	08 46
23	10 08 16	00 ♍ 38 53	23 ♒ 54 07	00 ♓ 13 53	27 38	09 10	09 39	24 03	24 51	18 08	10 43	08 45
24	10 12 12	01 36 42	06 ♓ 30 43	12 ♓ 44 38	28 22	10 20	10 17	24 13	24 52	18 10	10 42	08 43
25	10 16 09	02 34 32	18 ♓ 55 41	25 ♓ 03 58	29 03	11 30	10 55	24 24	24 54	18 11	10 41	08 42
26	10 20 05	03 32 25	01 ♈ 09 36	07 ♈ 12 45	29 ♍ 40	12 40	11 33	24 34	24 55	18 13	10 40	08 41
27	10 24 02	04 30 19	13 ♈ 13 39	19 ♈ 12 35	00 ♎ 13	13 50	12 11	24 44	24 56	18 14	10 39	08 40
28	10 27 58	05 28 14	25 ♈ 09 54	01 ♉ 06 00	00 41	15 00	12 49	24 55	24 57	18 16	10 37	08 39
29	10 31 55	06 26 12	07 ♉ 01 19	12 ♉ 56 21	01 05	16 10	13 28	25 05	24 58	18 17	10 36	08 37
30	10 35 51	07 24 11	18 ♉ 51 39	24 ♉ 47 48	01 24	17 20	14 06	25 16	24 59	18 18	10 35	08 36
31	10 39 48	08 ♍ 22 12	00 ♊ 45 25	06 ♊ 45 09	01 ♎ 38	18 ♎ 30	14 ♏ 45	25 ♎ 26	24 ♉ 59	18 ♊ 19	10 ♈ 33	08 ♒ 35

Date	Moon True ☊ ° '	Moon Mean ☊ ° '	Moon ☽ Latitude ° '
01	04 ♑ 06	02 ♑ 53	04 N 43
02	04 D 05	02 50	04 10
03	04 06	02 46	03 26
04	04 07	02 43	02 33
05	04 09	02 40	01 31
06	04 10	02 37	00 N 24
07	04 R 10	02 34	00 S 46
08	04 08	02 31	01 55
09	04 04	02 27	03 00
10	03 58	02 24	03 55
11	03 51	02 21	04 37
12	03 44	02 18	05 02
13	03 38	02 15	05 07
14	03 33	02 11	04 53
15	03 30	02 08	04 20
16	03 28	02 05	03 32
17	03 D 28	02 02	02 31
18	03 29	01 59	01 23
19	03 30	01 56	00 S 11
20	03 R 30	01 52	01 N 01
21	03 27	01 49	02 08
22	03 22	01 46	03 07
23	03 14	01 43	03 56
24	03 05	01 40	04 32
25	02 54	01 37	04 55
26	02 43	01 33	05 03
27	02 33	01 30	04 58
28	02 24	01 27	04 40
29	02 18	01 24	04 10
30	02 14	01 21	03 29
31	02 ♑ 12	01 ♑ 17	02 N 39

DECLINATIONS

Date	Sun ☉ ° '	Moon ☽ ° '	Mercury ☿ ° '	Venus ♀ ° '	Mars ♂ ° '	Jupiter ♃ ° '	Saturn ♄ ° '	Uranus ♅ ° '	Neptune ♆ ° '	Pluto ♇ ° '
01	17 N 52	15 N 30	11 N 58	07 N 49	10 S 46	07 S 02	16 N 35	22 N 51	02 N 59	23 S 29
02	17 36	19 01	11 17	07 19	10 59	07 05	16 36	22 51	02 59	23 30
03	17 21	21 47	10 37	06 50	11 12	07 08	16 37	22 51	02 58	23 30
04	17 05	23 36	09 56	06 20	11 26	07 11	16 37	22 52	02 58	23 31
05	16 49	24 20	09 16	05 50	11 39	07 15	16 38	22 52	02 58	23 31
06	16 32	23 50	08 36	05 20	11 52	07 18	16 38	22 52	02 57	23 32
07	16 15	22 04	07 56	04 50	12 05	07 21	16 39	22 52	02 57	23 32
08	15 58	19 04	07 16	04 20	12 19	07 24	16 40	22 53	02 57	23 32
09	15 41	14 56	06 36	03 49	12 32	07 28	16 40	22 53	02 56	23 33
10	15 23	09 55	05 57	03 19	12 45	07 31	16 41	22 53	02 56	23 33
11	15 05	04 N 17	05 19	02 48	12 58	07 35	16 41	22 53	02 56	23 34
12	14 47	01 S 37	04 40	02 17	13 11	07 38	16 41	22 53	02 55	23 34
13	14 29	07 28	04 03	01 47	13 24	07 42	16 42	22 54	02 55	23 34
14	14 11	12 54	03 26	01 16	13 37	07 45	16 42	22 54	02 54	23 35
15	13 52	17 34	02 49	00 45	13 51	07 49	16 43	22 54	02 54	23 35
16	13 33	21 10	02 14	00 N 14	14 04	07 52	16 43	22 54	02 54	23 36
17	13 14	23 27	01 39	00 S 17	14 16	07 56	16 43	22 54	02 53	23 36
18	12 54	24 16	01 05	00 48	14 29	08 00	16 44	22 54	02 53	23 36
19	12 35	23 36	00 N 32	01 19	14 42	08 03	16 44	22 55	02 52	23 37
20	12 15	21 36	00 00	01 50	14 55	08 07	16 44	22 55	02 52	23 37
21	11 55	18 27	00 S 31	02 20	15 08	08 11	16 44	22 55	02 51	23 38
22	11 35	14 26	01 00	02 51	15 21	08 15	16 44	22 55	02 51	23 38
23	11 15	09 50	01 29	03 22	15 33	08 18	16 45	22 55	02 50	23 38
24	10 54	04 S 54	01 55	03 53	15 46	08 22	16 45	22 55	02 50	23 39
25	10 33	00 N 08	02 21	04 24	15 58	08 26	16 45	22 56	02 49	23 39
26	10 12	05 06	02 44	04 54	16 11	08 30	16 45	22 56	02 49	23 39
27	09 51	09 48	03 06	05 25	16 23	08 34	16 45	22 56	02 48	23 40
28	09 30	14 05	03 25	05 55	16 35	08 38	16 45	22 56	02 48	23 40
29	09 09	17 47	03 43	06 26	16 48	08 42	16 45	22 56	02 47	23 40
30	08 47	20 47	03 58	06 56	17 00	08 46	16 45	22 56	02 47	23 41
31	08 N 26	22 N 54	04 S 10	07 S 26	17 S 12	08 S 50	16 N 45	22 N 56	02 N 46	23 S 41

ZODIAC SIGN ENTRIES

Date	h m	Planets
01	14 07	☽ ♉
04	02 27	☽ ♊
06	12 32	☽ ♋
07	16 03	♂ ♏
08	19 09	☽ ♌
10	22 40	☽ ♍
13	00 28	☽ ♎
15	02 08	☽ ♏
15	17 06	♀ ♎
17	04 52	☽ ♐
19	09 17	☽ ♑
21	15 29	☽ ♒
22	19 51	☉ ♍
23	23 34	☽ ♓
26	09 43	☽ ♈
27	02 21	☿ ♎
28	21 46	☽ ♉
31	10 29	☽ ♊

LATITUDES

Date	Mercury ☿ ° '	Venus ♀ ° '	Mars ♂ ° '	Jupiter ♃ ° '	Saturn ♄ ° '	Uranus ♅ ° '	Neptune ♆ ° '	Pluto ♇ ° '
01	00 N 50	01 N 17	00 S 37	01 N 10	02 S 14	00 N 01	01 S 31	05 S 46
04	00 27	01 11	00 39	01 10	02 14	00 01	01 31	05 46
07	00 N 01	01 04	00 42	01 09	02 15	00 01	01 31	05 46
10	00 S 27	00 56	00 44	01 08	02 15	00 01	01 31	05 46
13	00 56	00 48	00 47	01 08	02 16	00 01	01 31	05 46
16	01 26	00 40	00 49	01 07	02 17	00 01	01 31	05 46
19	01 57	00 30	00 51	01 07	02 17	00 01	01 32	05 46
22	02 28	00 20	00 53	01 06	02 18	00 01	01 32	05 47
25	02 58	00 N 10	00 55	01 06	02 18	00 01	01 32	05 47
28	03 26	00 S 01	00 57	01 05	02 19	00 01	01 32	05 47
31	03 S 50	00 S 12	00 S 59	01 N 05	02 S 20	00 N 01	01 S 32	05 S 47

DATA

Julian Date	2462350
Delta T	+74 seconds
Ayanamsa	24° 16' 31"
Synetic vernal point	04° ♓ 50' 27"
True obliquity of ecliptic	23° 26' 08"

MOON'S PHASES, APSIDES AND POSITIONS ☽

Date	h m	Phase	Longitude ° '	Eclipse Indicator
02	11 15	☾	10 ♉ 27	
10	01 56	●	17 ♌ 45	
16	18 55	☽	24 ♏ 11	
24	01 51	○	01 ♓ 12	

Day	h m		
01	10 39	Apogee	
13	09 57	Perigee	
29	04 37	Apogee	
05	14 38	Max dec	24° N 20'
12	05 27	OS	
18	13 08	Max dec	24° S 16'
25	11 20	ON	

LONGITUDES

Date	Chiron ⚷ ° '	Ceres ⚳ ° '	Pallas ⚴ ° '	Juno ⚵ ° '	Vesta ⚶ ° '	Black Moon Lilith ⚸ ° '
01	12 ♉ 33	20 ♒ 30	26 ♐ 38	28 ♎ 49	13 ♑ 26	27 ♈ 10
11	12 ♉ 39	18 ♒ 19	25 ♐ 49	00 ♏ 53	12 ♑ 14	28 ♈ 17
21	12 ♉ 39	16 ♒ 10	25 ♐ 42	03 ♏ 12	11 ♑ 48	29 ♈ 24
31	12 ♉ 33	14 ♒ 15	26 ♐ 13	05 ♏ 46	12 ♑ 10	00 ♉ 31

ASPECTARIAN

01 Wednesday
01 44 ☽ ⊻ ♄
04 56 ☉ ☍ ♇
06 37 ☽ ☍ ♂
10 10 ☽ Ꝑ ♀
16 21 ☽ △ ☿
18 56 ☽ ∥ ♄
18 59 ☽ ∠ ♅

02 Thursday
02 29 ☽ ∥ ☉
08 46 ☽ □ ♇
11 15 ☽ □ ☉
12 26 ☽ ⊻ ♆
13 14 ☽ ⊥ ♅
19 52 ☽ △ ♀
20 06 ☿ ⋕ ♂
22 25 ♀ ⋕ ♃

03 Friday
00 31 ☽ ⊥ ♆
01 22 ☽ ⊻ ♅
01 30 ♀ ⊥ ♃
01 51 ☉ △ ♆
05 31 ♀ ± ♇
08 34 ☽ ⊼ ♃
14 32 ☽ ☌ ♄
18 32 ☽ ∠ ♆
20 41 ☽ ± ♃
22 06 ☽ ⊼ ♂

04 Saturday
00 34 ☽ ∥ ♅
02 29 ☽ Q ☉
05 40 ☿ ± ♆
10 24 ☽ ⋕ ♇
10 35 ☽ ± ♂
13 19 ☽ □ ☿
14 42 ☽ Ꝑ ♃
20 31 ☽ △ ♇

05 Sunday
00 08 ☽ ⚹ ♆
01 46 ☿ ∠ ♃
04 05 ☽ ⚹ ☉
04 47 ♀ □ ♅
05 06 ☽ Ꝑ ♂
12 50 ☽ ☌ ♅
13 39 ☽ □ ♀
20 10 ☽ △ ♃
23 17 ☽ Q ♆

06 Monday
01 26 ☽ Ꝑ ♇
01 29 ☽ ⊻ ♄
03 07 ☉ ∥ ♄
04 48 ☽ Q ☿
11 11 ☽ ∠ ☉
11 13 ☽ △ ♂
12 53 ☽ ⊥ ♄
17 50 ☽ ⋕ ♇
18 23 ☽ ± ♇

07 Tuesday
01 04 ☿ ⊼ ♇
03 09 ☽ Q ♀
03 10 ☽ ∥ ♅
05 19 ☽ ⊥ ☉
05 30 ☽ ⊼ ♇
05 43 ☽ ∠ ♄
06 03 ☽ ⚹ ☿
08 57 ☽ □ ♆
17 10 ☽ ⊻ ☉
21 05 ☽ ⊻ ♅

08 Wednesday
03 11 ☽ ⚹ ♀
04 23 ☽ □ ♃
07 17 ☿ ⋕ ♃
07 54 ☽ ⊥ ♅
09 01 ☽ ⚹ ♄
09 07 ☿ ⊼ ♆
12 21 ☽ ∠ ☿
18 19 ♀ ⊻ ♃
20 23 ☽ □ ♂
23 53 ☽ ∠ ♅

09 Thursday
02 40 ☽ ∥ ♄
05 56 ☽ ⊥ ☿
06 14 ☽ Q ♄
07 49 ☽ ∥ ☉
08 13 ☽ ∠ ♀
10 59 ☽ ☍ ♇
12 25 ☽ Q ♃
14 16 ☽ △ ♆
17 26 ☽ ⊻ ☿
23 29 ☽ ⋕ ♂

10 Friday
01 10 ☽ ⊥ ♀
01 20 ☉ ⚹ ♅
01 53 ☽ ⚹ ♅
01 56 ☽ ☌ ☉
04 56 ☽ Q ♂
09 11 ☽ ⚹ ♃
12 18 ☽ ⊻ ♀
13 12 ☽ □ ♄
15 50 ☽ Ꝑ ♆
15 54 ♀ Ꝑ ♇
22 15 ☽ Q ♅
22 23 ☽ ⋕ ♃
23 33 ♀ △ ♄

11 Saturday
02 05 ☽ ⚹ ♂
06 03 ♀ ∥ ♆
06 56 ☽ ± ♆
07 15 ☽ ∥ ☿
10 11 ☿ ± ♇
10 39 ☽ ∠ ♃
13 44 ☽ ⊼ ♇
16 55 ☽ ⊼ ♆
17 36 ☽ ∥ ♆
18 41 ☽ ∥ ♀
21 36 ☉ ⊥ ♀
23 41 ☽ ± ♇

12 Sunday
01 00 ☽ ☌ ☿
01 45 ☽ ⊥ ♃
04 07 ☽ ∠ ♂
04 19 ☽ □ ♅
07 47 ☽ ⊻ ☉
08 59 ♂ Ꝑ ♅
10 57 ☿ ⊥ ♃
11 47 ☽ ⊻ ♃
14 29 ☽ ⋕ ♀
14 32 ☽ Ꝑ ♇
15 21 ☽ △ ♄
17 14 ☽ ⋕ ♆
18 24 ☽ ⊥ ☉
18 45 ☽ ☌ ♀
19 38 ☽ ⊥ ♂
23 11 ☽ ⋕ ☿

13 Monday
05 57 ☽ ⊻ ♂
10 18 ☽ ∠ ☉
12 56 ☽ ∥ ♃
15 15 ☽ △ ♇
16 03 ♀ ⊥ ♂
16 10 ☽ Ꝑ ♄
17 10 ☿ □ ♅
18 26 ☽ ☍ ♆

14 Tuesday
05 58 ☽ △ ♅
07 05 ☽ ⊻ ☿
07 06 ☽ ± ♄
12 51 ☽ ⚹ ☉
13 52 ☽ ☌ ♃
15 35 ☽ ∥ ♂
17 04 ☽ ⊼ ♄
17 44 ☽ ⋕ ☉
17 55 ☽ ⊥ ☿
18 47 ☉ Q ♂

15 Wednesday
00 47 ☽ ⊻ ♀
01 27 ☿ ∠ ♂
06 41 ☉ ⚹ ♃
06 58 ☽ Ꝑ ♅
07 10 ☽ ⋕ ♄
08 52 ☿ ∥ ♆
09 53 ☽ ☌ ♂
10 14 ☽ ∠ ☿
10 17 ☽ Q ☉
11 43 ☽ ⊥ ♀
13 00 ☉ ⋕ ♂
17 03 ☽ □ ♇
20 20 ☽ ⊼ ♆
22 07 ☽ ± ♅

16 Thursday
04 14 ☽ ∠ ♀
06 27 ☽ ± ♆
08 17 ☽ ⊼ ♅
13 42 ☽ ⚹ ☿
16 49 ☽ ⊻ ♃
18 55 ☽ □ ☉
19 42 ☽ ☍ ♄
21 45 ☽ Ꝑ ♆
23 32 ☽ Q ♇

17 Friday
03 12 ☽ ⊥ ♃
04 43 ☽ ⋕ ♅
06 41 ☉ □ ♄
08 10 ☽ ⚹ ♀
11 58 ☽ Q ☿
14 32 ☽ ∥ ♇
15 13 ☽ ⊻ ♂
18 54 ☽ ∠ ♃
20 11 ☽ ⚹ ♇
23 35 ☽ △ ♆

18 Saturday
02 07 ☽ ⊥ ♂
06 55 ☽ Q ♀
11 33 ☉ Ꝑ ♆
12 06 ☽ ☍ ♅
14 20 ☿ ⊻ ♃
18 23 ☿ ⋕ ♀
18 37 ☽ ∠ ♂
21 26 ☽ ⚹ ♃
21 54 ☽ □ ☿
22 23 ☽ ∠ ♇
23 56 ☽ ⊼ ♄

19 Sunday
02 58 ☽ △ ☉
04 20 ☿ Ꝑ ♇
10 35 ☽ ± ♄
11 50 ☽ ∥ ♇
14 19 ☽ ⊥ ♇
17 44 ☽ □ ♀
19 00 ☽ Q ♃
22 06 ☽ ⋕ ♅
22 32 ☽ ⚹ ♂

20 Monday
01 02 ☽ ⊻ ♇
02 42 ☽ Ꝑ ♄
03 12 ☿ △ ♄
04 34 ☽ □ ♆
07 49 ☽ Ꝑ ☉
17 42 ☽ ⊼ ♅
21 15 ☽ Q ♂

21 Tuesday
01 24 ☽ ± ☉
03 54 ☽ □ ♃
04 38 ☽ ± ♅
05 56 ☽ △ ♄
07 52 ☽ △ ☿
13 13 ☽ Q ♆
13 14 ☽ ⊼ ☉
21 11 ☽ Ꝑ ♅
22 47 ☽ ⋕ ♄

22 Wednesday
02 25 ♂ □ ♇
05 39 ☽ △ ♀
07 11 ☽ ∥ ♂
07 41 ☽ ☌ ♇
07 57 ☽ □ ♂
11 20 ☽ ⚹ ♆
11 33 ♀ ⋕ ♆
13 29 ☽ Ꝑ ☿
22 58 ☉ Q ♅

23 Thursday
01 08 ☽ △ ♅
03 41 ♀ △ ♇
04 22 ☽ ⋕ ☉
07 27 ☽ ± ☿
12 18 ☽ △ ♃
12 32 ☽ Ꝑ ♀
13 47 ☽ □ ♄
15 26 ☽ ∠ ♆
19 29 ☽ ∥ ♃
19 31 ☽ ⊼ ☿

24 Friday
01 51 ☽ ☍ ☉
02 23 ♀ Ꝑ ♄
07 23 ☽ ± ♀
08 32 ☽ ⊥ ♆
09 49 ♀ ⊻ ♂
16 15 ☽ ⊻ ♇
16 25 ☽ ∥ ♀
17 17 ☽ Ꝑ ♃
19 32 ♀ ☍ ♆
19 38 ☽ △ ♂
20 03 ☽ ⊻ ♆
20 06 ☽ ⊼ ♀
21 54 ☽ ⋕ ♆

25 Saturday
00 16 ☽ Q ♄
01 05 ☽ ∥ ☿
03 35 ♂ ⊼ ♆
03 48 ☽ ⊥ ♇
10 34 ☽ □ ♅
10 57 ☽ ± ♃
21 19 ☽ ∠ ♇
22 50 ☽ ⊼ ♃
23 29 ☽ ⋕ ☿
23 41 ☽ ⚹ ♄

26 Sunday
00 51 ☽ ∥ ♆
02 24 ☽ Ꝑ ♂
08 54 ☽ ☍ ☿
10 56 ☽ ⋕ ♀
16 54 ☿ ⋕ ♆
17 07 ☽ ⊼ ☉
21 10 ☽ ± ♂
22 02 ☽ Q ♅

27 Monday
02 54 ☽ ⚹ ♇
05 24 ☽ ∠ ♄
05 26 ☽ ⋕ ♃
06 05 ☽ ± ☉
06 51 ☽ ☌ ♆
09 48 ☽ ⊼ ♂
12 17 ☽ ∥ ☉
13 21 ☽ ☍ ♀
14 10 ♂ ± ♅
15 20 ☉ ± ♆
22 04 ☽ ⚹ ♅
23 27 ☽ ⊥ ♄

28 Tuesday
01 42 ☽ Ꝑ ☉
02 54 ☽ Q ♇
11 28 ☽ ☍ ♃
11 34 ☽ ⊻ ♄
18 02 ♃ ⊼ ♄
23 34 ☽ ⊼ ☿

29 Wednesday
04 24 ☽ ∠ ♅
04 42 ☽ ⋕ ♂
04 50 ☽ ∥ ♄
06 58 ♂ ⋕ ♄
10 42 ☽ △ ☉
12 08 ☽ ± ☿
15 14 ☽ □ ♇
19 14 ☽ ⊻ ♆
22 41 ☽ ⊥ ♅

30 Thursday
01 48 ☽ ☍ ♂
06 54 ☽ Ꝑ ☿
07 23 ☽ ⊥ ♆
08 34 ☽ ⊼ ♀
10 52 ☽ ⊻ ♅
13 26 ☉ ⋕ ♃
22 02 ☽ ± ♀

31 Friday
00 23 ☽ ☌ ♄
01 08 ☽ ⊼ ♃
01 33 ☽ ∠ ♆
08 22 ♀ △ ♅
12 34 ☽ ∥ ♅
13 23 ☽ ± ♃
13 48 ☽ △ ☿
17 12 ☉ ⊼ ♇
18 05 ☽ Ꝑ ♀
22 16 ♀ ± ♄

All ephemeris data is given at 12.00 UT and the Moon's longitude is additionally given for 24.00 UT

SEPTEMBER 2029

LONGITUDES

Date	Sidereal time h m s	Sun ☉	Moon ☽	Moon ☽ 24.00	Mercury ☿	Venus ♀	Mars ♂	Jupiter ♃	Saturn ♄	Uranus ♅	Neptune ♆	Pluto ♇
01	10 43 45	09 ♍ 20 15	12 ♊ 47 38	18 ♊ 53 33	01 ♎ 47	19 ♎ 39	15 ♏ 24	25 ♎ 37	25 ♉ 00	18 ♊ 20	10 ♈ 32	08 ♒ 34
02	10 47 41	10 18 20	25 ♊ 03 34	01 ♋ 18 18	01 50	20 49	16 02	25 48	25 00	18 22	10 R 31	08 R 33
03	10 51 38	11 16 27	07 ♋ 38 21	14 ♋ 04 15	01 R 47	21 59	16 41	25 59	25 01	18 23	10 29	08 32
04	10 55 34	12 14 36	20 ♋ 36 28	27 ♋ 15 20	01 38	23 08	17 21	26 10	25 01	18 24	10 28	08 31
05	10 59 31	13 12 47	04 ♌ 01 05	10 ♌ 53 47	01 23	24 17	18 00	26 21	25 01	18 25	10 26	08 29
06	11 03 27	14 10 59	17 ♌ 53 19	24 ♌ 59 24	01 01	25 27	18 39	26 32	25 R 01	18 26	10 25	08 28
07	11 07 24	15 09 14	02 ♍ 11 33	09 ♍ 29 04	00 ♎ 33	26 36	19 19	26 43	25 01	18 26	10 23	08 27
08	11 11 20	16 07 30	16 ♍ 51 05	24 ♍ 16 36	29 ♍ 58	27 45	19 58	26 54	25 01	18 27	10 22	08 26
09	11 15 17	17 05 48	01 ♎ 44 29	09 ♎ 13 33	29 18	28 ♎ 54	20 38	27 06	25 01	18 28	10 20	08 25
10	11 19 14	18 04 08	16 ♎ 42 37	24 ♎ 10 33	28 32	00 ♏ 03	21 18	27 17	25 00	18 29	10 19	08 24
11	11 23 10	19 02 29	01 ♏ 36 16	08 ♏ 58 54	27 41	01 12	21 58	27 28	25 00	18 29	10 17	08 23
12	11 27 07	20 00 53	16 ♏ 17 40	23 ♏ 32 01	26 45	02 21	22 38	27 40	24 59	18 30	10 16	08 22
13	11 31 03	20 59 17	00 ♐ 41 33	07 ♐ 46 01	25 46	03 29	23 18	27 52	24 58	18 30	10 14	08 21
14	11 35 00	21 57 44	14 ♐ 45 21	21 ♐ 39 36	24 45	04 38	23 59	28 03	24 58	18 31	10 13	08 20
15	11 38 56	22 56 12	28 ♐ 28 53	05 ♑ 13 26	23 42	05 46	24 39	28 15	24 57	18 31	10 11	08 19
16	11 42 53	23 54 41	11 ♑ 53 31	18 ♑ 29 24	22 41	06 55	25 20	28 27	24 56	18 32	10 10	08 19
17	11 46 49	24 53 12	25 ♑ 01 25	01 ♒ 29 51	21 41	08 03	26 00	28 39	24 54	18 32	10 08	08 18
18	11 50 46	25 51 45	07 ♒ 54 59	14 ♒ 17 04	20 44	09 11	26 41	28 50	24 53	18 33	10 06	08 17
19	11 54 43	26 50 19	20 ♒ 36 18	26 ♒ 52 55	19 53	10 19	27 22	29 02	24 52	18 33	10 05	08 16
20	11 58 39	27 48 55	03 ♓ 07 02	09 ♓ 18 48	19 07	11 27	28 03	29 14	24 50	18 33	10 03	08 15
21	12 02 36	28 47 33	15 ♓ 28 20	21 ♓ 35 44	18 29	12 35	28 44	29 26	24 49	18 33	10 02	08 15
22	12 06 32	29 ♍ 46 13	27 ♓ 41 04	03 ♈ 44 28	18 00	13 42	29 ♏ 25	29 39	24 47	18 33	10 00	08 14
23	12 10 29	00 ♎ 44 54	09 ♈ 46 01	15 ♈ 45 51	17 39	14 50	00 ♐ 06	29 ♎ 51	24 45	18 33	09 58	08 13
24	12 14 25	01 43 38	21 ♈ 44 09	27 ♈ 41 05	17 29	15 57	00 48	00 ♏ 03	24 44	18 R 33	09 57	08 12
25	12 18 22	02 42 24	03 ♉ 36 54	09 ♉ 31 54	17 D 28	17 05	01 29	00 15	24 42	18 33	09 55	08 12
26	12 22 18	03 41 12	15 ♉ 26 24	21 ♉ 20 48	17 37	18 12	02 11	00 27	24 40	18 33	09 53	08 11
27	12 26 15	04 40 02	27 ♉ 15 32	03 ♊ 11 05	17 56	19 19	02 53	00 40	24 37	18 33	09 52	08 10
28	12 30 12	05 38 54	09 ♊ 07 59	15 ♊ 06 49	18 25	20 25	03 34	00 52	24 35	18 33	09 50	08 10
29	12 34 08	06 37 49	21 ♊ 08 12	27 ♊ 12 46	19 03	21 32	04 16	01 05	24 33	18 32	09 48	08 09
30	12 38 05	07 ♎ 36 46	03 ♋ 21 11	09 ♋ 34 06	19 ♍ 50	22 ♏ 39	04 ♐ 58	01 ♏ 17	24 ♉ 30	18 ♊ 32	09 ♈ 47	08 ♒ 09

Date	Moon True ☊	Moon Mean ☊	Moon ☽ Latitude
01	02 ♑ 12	01 ♑ 14	01 N 41
02	02 D 12	01 11	00 N 38
03	02 R 12	01 08	00 S 29
04	02 11	01 05	01 36
05	02 07	01 02	02 40
06	02 01	00 58	03 37
07	01 53	00 55	04 22
08	01 42	00 52	04 51
09	01 32	00 49	05 01
10	01 21	00 46	04 51
11	01 13	00 42	04 20
12	01 07	00 39	03 33
13	01 04	00 36	02 33
14	01 03	00 33	01 25
15	01 D 03	00 30	00 S 14
16	01 R 02	00 27	00 N 57
17	01 01	00 23	02 03
18	00 57	00 20	03 01
19	00 50	00 17	03 50
20	00 40	00 14	04 26
21	00 28	00 11	04 50
22	00 14	00 08	04 59
23	00 00	00 04	04 55
24	29 ♐ 47	00 ♑ 01	04 38
25	29 35	29 ♐ 58	04 09
26	29 26	29 55	03 29
27	29 20	29 52	02 41
28	29 16	29 48	01 44
29	29 15	29 45	00 43
30	29 ♐ 15	29 ♐ 42	00 S 22

DECLINATIONS

Date	Sun ☉	Moon ☽	Mercury ☿	Venus ♀	Mars ♂	Jupiter ♃	Saturn ♄	Uranus ♅	Neptune ♆	Pluto ♇
01	08 N 04	24 N 00	04 S 20	07 S 56	17 S 24	08 S 54	16 N 45	22 N 56	02 N 45	23 S 41
02	07 42	23 59	04 27	08 26	17 36	08 58	16 45	22 57	02 45	23 42
03	07 20	22 44	04 31	08 56	17 47	09 02	16 45	22 57	02 44	23 42
04	06 58	20 16	04 31	09 25	17 59	09 06	16 45	22 57	02 44	23 42
05	06 36	16 39	04 28	09 54	18 11	09 10	16 44	22 57	02 43	23 42
06	06 13	12 01	04 21	10 23	18 22	09 15	16 44	22 57	02 43	23 43
07	05 51	06 37	04 11	10 52	18 34	09 19	16 44	22 57	02 42	23 43
08	05 28	00 N 43	03 56	11 21	18 45	09 23	16 44	22 57	02 41	23 43
09	05 06	05 S 18	03 38	11 50	18 56	09 27	16 44	22 57	02 41	23 43
10	04 43	11 02	03 15	12 18	19 07	09 31	16 43	22 57	02 40	23 44
11	04 20	16 06	02 49	12 46	19 18	09 35	16 43	22 57	02 40	23 44
12	03 57	20 07	02 20	13 14	19 29	09 40	16 43	22 57	02 39	23 44
13	03 34	22 47	01 47	13 41	19 40	09 44	16 42	22 57	02 38	23 44
14	03 11	23 58	01 12	14 08	19 50	09 48	16 42	22 57	02 38	23 45
15	02 48	23 39	00 S 35	14 35	20 01	09 53	16 41	22 58	02 37	23 45
16	02 25	21 57	00 N 04	15 02	20 11	09 57	16 41	22 58	02 36	23 45
17	02 02	19 06	00 43	15 28	20 21	10 01	16 40	22 58	02 36	23 45
18	01 39	15 22	01 22	15 54	20 31	10 06	16 40	22 58	02 35	23 45
19	01 15	10 59	01 59	16 20	20 41	10 10	16 40	22 58	02 34	23 45
20	00 52	06 13	02 35	16 45	20 51	10 14	16 39	22 58	02 34	23 46
21	00 29	01 S 16	03 09	17 10	21 00	10 19	16 38	22 58	02 33	23 46
22	00 N 05	03 N 39	03 39	17 34	21 10	10 23	16 38	22 58	02 32	23 46
23	00 S 18	08 23	04 05	17 59	21 19	10 27	16 37	22 58	02 32	23 46
24	00 41	12 46	04 27	18 22	21 28	10 32	16 37	22 58	02 31	23 46
25	01 05	16 37	04 44	18 46	21 37	10 36	16 36	22 58	02 30	23 46
26	01 28	19 48	04 57	19 09	21 46	10 40	16 35	22 58	02 30	23 47
27	01 51	22 09	05 04	19 31	21 54	10 45	16 35	22 58	02 29	23 47
28	02 15	23 32	05 07	19 53	22 03	10 49	16 34	22 58	02 29	23 47
29	02 38	23 51	05 05	20 15	22 11	10 54	16 33	22 58	02 28	23 47
30	03 S 01	23 N 02	04 N 58	20 S 36	22 S 19	10 S 58	16 N 32	22 N 58	02 N 27	23 S 47

ZODIAC SIGN ENTRIES

Date	h	m	Planets
02	21	30	☽ ♋
05	04	54	☽ ♌
07	08	22	☽ ♍
08	10	59	☿ ♍
09	09	12	☽ ♎
10	10	54	♀ ♏
11	09	24	☽ ♏
13	10	50	☽ ♐
15	14	41	☽ ♑
17	21	13	☽ ♒
20	05	59	☽ ♓
22	16	35	☽ ♈
22	17	38	☉ ♎
23	08	14	♂ ♐
24	06	24	♃ ♏
25	04	41	☽ ♉
27	17	33	☽ ♊
30	05	28	☽ ♋

LATITUDES

Date	Mercury ☿	Venus ♀	Mars ♂	Jupiter ♃	Saturn ♄	Uranus ♅	Neptune ♆	Pluto ♇
01	03 S 57	00 S 16	00 S 59	01 N 05	02 S 20	00 N 01	01 S 32	05 S 47
04	04 13	00 28	01 01	01 04	02 21	00 01	01 32	05 47
07	04 19	00 40	01 03	01 04	02 21	00 01	01 32	05 46
10	04 11	00 52	01 04	01 03	02 22	00 01	01 32	05 46
13	03 47	01 04	01 06	01 03	02 22	00 01	01 32	05 46
16	03 06	01 17	01 07	01 02	02 23	00 01	01 32	05 46
19	02 12	01 29	01 09	01 02	02 24	00 01	01 32	05 46
22	01 12	01 42	01 10	01 02	02 24	00 01	01 33	05 46
25	00 S 14	01 54	01 11	01 01	02 25	00 01	01 33	05 46
28	00 N 35	02 07	01 12	01 01	02 25	00 01	01 33	05 46
31	01 N 13	02 S 19	01 S 13	01 N 01	02 S 26	00 N 01	01 S 33	05 S 45

DATA

Julian Date	2462381
Delta T	+74 seconds
Ayanamsa	24° 16' 35"
Synetic vernal point	04° ♓ 50' 24"
True obliquity of ecliptic	23° 26' 08"

MOON'S PHASES, APSIDES AND POSITIONS ☽

Date	h	m	Phase	Longitude	Eclipse Indicator
01	04	33	☾	09 ♊ 02	
08	10	44	●	16 ♍ 04	
15	01	29	☽	22 ♐ 31	
22	16	29	○	29 ♓ 57	
30	20	57	☾	07 ♋ 59	

Day	h	m		
10	04	22	Perigee	
25	19	28	Apogee	
01	23	27	Max dec	24° N 08'
08	14	51	0S	
14	18	44	Max dec	24° S 02'
21	18	10	0N	
29	06	50	Max dec	23° N 53'

LONGITUDES

Date	Chiron ⚷	Ceres ⚳	Pallas ⚴	Juno ⚵	Vesta ⚶	Black Moon Lilith ⚸
01	12 ♉ 32	14 ♒ 05	26 ♐ 18	06 ♏ 02	12 ♑ 14	00 ♉ 37
11	12 ♉ 20	12 ♒ 38	27 ♐ 25	08 ♏ 47	13 ♑ 24	01 ♉ 45
21	12 ♉ 04	11 ♒ 44	29 ♐ 00	11 ♏ 43	15 ♑ 13	02 ♉ 52
31	11 ♉ 42	11 ♒ 27	00 ♑ 59	14 ♏ 46	17 ♑ 35	03 ♉ 59

ASPECTARIAN

h m	Aspects
01 Saturday	
02 11	☽ ⋕ ♇
03 37	☽ △ ♇
04 33	☽ □ ☉
07 32	☽ ✶ ♆
07 37	☽ ꝗ ♃
15 44	☉ ⋕ ♀
17 25	☽ ⊼ ♂
22 56	☽ ☌ ♅
02 Sunday	
02 54	☽ △ ♀
05 49	☽ ± ♂
07 04	☽ Q ♆
09 04	☽ ꝗ ♇
11 54	☽ ⊻ ♄
12 17	☿ St R
13 27	☽ △ ♃
16 56	☉ ⊼ ♆
18 47	☽ Q ☉
20 07	☽ ⋕ ♇
23 26	☽ ⊥ ♄
03 Monday	
00 07	☽ ꝗ ♂
00 59	☽ □ ☿
02 21	☽ ± ♇
02 57	☉ ∠ ♃
04 42	♂ ± ♆
09 12	☽ ‖ ♅
13 40	☽ ⊼ ♇
14 53	☿ ∠ ♂
16 27	☽ ∠ ♄
17 20	☽ □ ♆
18 13	♀ ‖ ♃
19 22	☽ ✶ ☉
04 Tuesday	
05 43	☽ △ ♂
07 57	☽ ⊻ ♅
10 15	☽ Q ☿
17 01	☽ □ ♀
18 52	☽ ⊥ ♅
19 59	☽ ✶ ♄
22 11	☽ □ ♃
05 Wednesday	
00 55	☽ ∠ ☉
03 11	☽ ⋕ ♂
07 26	☽ ✶ ☿
10 56	☽ ∠ ♅
11 29	☽ ‖ ♄
17 16	☽ Q ♄
18 01	☽ ⊥ ☉
19 49	☽ ☍ ♇
23 11	☽ △ ♆
06 Thursday	
03 07	♀ ⊼ ♄
03 26	♂ ⊼ ♅
03 43	☽ Q ♀
05 12	☽ ⊻ ☉
06 11	☽ Q ♃
08 34	♄ St R
08 54	☽ ∠ ☿
12 55	☽ ✶ ♅
13 22	☽ □ ♂
18 54	☽ ⋕ ♀
19 01	☉ ± ♇
23 41	☽ ⊥ ☿
07 Friday	
00 03	☽ □ ♄
00 31	☽ ⋕ ♃
00 41	☽ ꝗ ♆
01 53	☽ ✶ ♀
02 46	☽ ✶ ♃
09 06	☽ Q ♅
09 22	☽ ⊻ ☿
14 53	♀ ☌ ♃
15 24	☽ ‖ ☉
15 38	☽ ± ♆
20 50	☽ Q ♂
22 18	☽ ⊼ ♇
22 27	☽ ⋕ ☿
08 Saturday	
01 28	☽ ⊼ ♆
03 51	☽ ∠ ♃
04 05	☽ ‖ ♆
04 46	☽ ∠ ♀
08 05	☽ ± ♇
10 44	☽ ☌ ☉
14 36	☽ □ ♅
17 18	☽ ✶ ♂
18 39	☽ ⊥ ♃
20 36	☽ ⊥ ♀
22 38	☽ ꝗ ♇
09 Sunday	
01 11	☽ △ ♄
01 32	☽ ⋕ ♆
04 26	☽ ⊻ ♃
05 39	☽ ‖ ☿
07 04	☽ ⊻ ♀
08 16	☽ ☌ ☿
11 14	☽ ⋕ ☉
17 03	☿ ⊻ ♀
18 32	☽ ∠ ♂
22 42	☽ △ ♇
10 Monday	
01 15	☽ ꝗ ♄
01 46	☽ ☍ ♆
05 23	☽ ‖ ♃
09 38	☽ ⊥ ♂
14 20	☽ ⊻ ☉
14 50	☽ △ ♅
15 41	☽ ± ♄
18 11	☽ ‖ ♀
19 43	☽ ⊻ ♂
22 13	☉ □ ♅
11 Tuesday	
00 39	☽ ⊥ ☉
01 20	☽ ⊼ ♄
05 14	☽ ☌ ♃
06 00	☽ ⊻ ☿
11 17	☽ ☌ ♀
15 04	☽ ꝗ ♅
15 10	☽ ⊥ ☿
15 17	☽ ⋕ ♄
16 14	☽ ∠ ☉
16 26	☿ ⊻ ♃
20 34	☿ ⋕ ♆
23 01	☽ □ ♇
12 Wednesday	
02 07	☽ ⊼ ♆
04 59	☽ ∠ ☿
05 45	☽ ± ♅
07 29	☽ ‖ ♂
11 57	☽ ± ♆
15 39	☽ ⊼ ♅
16 35	☿ ⊥ ♀
18 36	☽ ✶ ☉
23 01	☽ ☌ ♂
13 Thursday	
02 25	☽ ☍ ♄
02 52	☽ ꝗ ♆
04 16	☽ ✶ ☿
04 43	☽ Q ♇
07 10	☽ ⊻ ♃
12 22	♀ ꝗ ♅
14 10	☽ ⋕ ♅
16 10	☽ Q ☉
17 09	☽ ⊻ ♀
17 26	☽ ⊥ ♃
23 11	☽ Q ☿
14 Friday	
00 59	☽ ✶ ♇
03 50	☽ ‖ ♇
04 12	☽ △ ♆
04 16	☽ ⊥ ♀
06 54	☿ △ ♄
09 01	☽ ∠ ♃
14 49	☉ ⊥ ♃
18 31	☽ ☍ ♅
21 14	☽ ∠ ♀
22 43	☿ ✶ ♂
15 Saturday	
01 29	☽ □ ☉
02 56	☽ ∠ ♇
04 11	☽ □ ☿
04 54	☽ ⊻ ♂
05 46	☽ ⊼ ♄
09 43	☽ ‖ ♇
11 35	☽ ✶ ♃
16 03	☽ ⊥ ♂
16 22	☽ ± ♄
18 49	☽ ⊥ ♇
20 57	☿ ꝗ ♇
21 11	☉ ☌ ☿
21 25	☉ ꝗ ♇
22 04	♂ ☍ ♄
23 57	☉ ‖ ♆
16 Sunday	
00 04	☽ ⋕ ♅
02 12	☽ ✶ ♀
05 33	☽ ⊻ ♇
06 15	♂ ꝗ ♆
08 27	☽ ꝗ ♄
08 53	☽ □ ♆
09 02	☽ ∠ ♂
09 21	☽ Q ♃
16 34	☿ ⊥ ♃
20 30	☿ ∠ ♀
17 Monday	
00 05	☽ ⊼ ♅
01 59	☽ Q ♀
03 02	☽ ‖ ♂
06 16	☽ △ ☿
11 06	☽ ± ♅
11 44	☽ △ ☉
11 47	☽ △ ♄
12 30	☉ △ ♄
13 55	☽ ✶ ♂
17 06	♀ □ ♇
17 44	☽ Q ♆
18 48	☽ □ ♃
22 01	♂ Q ♇
18 Tuesday	
03 48	☽ ꝗ ♅
04 08	☽ ⋕ ♄
08 11	☽ ꝗ ☿
09 07	☽ ‖ ♀
12 41	☽ ☌ ♇
13 32	☽ Q ♂
14 37	☽ □ ♀
16 07	☽ ✶ ♆
18 00	☽ ꝗ ☉
18 38	☉ ‖ ☿
19 Wednesday	
00 01	☽ ± ☿
07 03	♀ ⊼ ♆
08 05	☽ △ ♅
10 42	☽ ⊼ ☿
12 29	☽ ± ☉
16 11	☽ ‖ ♃
20 07	☽ □ ♄
20 32	☽ ∠ ♆
20 Thursday	
00 56	☽ ⊼ ☉
01 41	☽ □ ♂
04 24	☽ △ ♃
06 28	♀ ⋕ ♄
10 58	☿ ‖ ♆
13 48	☽ ⊥ ♆
21 56	☽ ⊻ ♇
21 Friday	
01 25	☽ ⊻ ♆
03 51	☽ ⋕ ☿
05 47	☽ △ ♀
05 49	☽ ⋕ ♆
06 49	☽ Q ♄
07 13	☉ ✶ ♂
09 26	☿ □ ♅
09 36	☽ ⊥ ♇
09 57	☽ ꝗ ♃
11 22	♀ ± ♅
16 10	☽ ⋕ ☉
17 39	☽ ☍ ☿
18 02	☽ □ ♅
19 52	☽ ‖ ☉
22 Saturday	
03 13	☽ ∠ ♇
03 53	☽ ± ♃
06 18	☽ ✶ ♄
06 32	☽ ‖ ♆
08 02	☉ ⊻ ♃
11 56	☽ ‖ ☿
14 14	☽ ꝗ ♀
15 38	☽ △ ♂
15 56	☽ ⊼ ♃
16 29	☽ ☍ ☉
22 57	♂ ⊻ ♃
23 Sunday	
01 00	☿ Q ♂
05 36	☽ Q ♅
08 55	☽ ✶ ♇
09 57	☽ ± ♀
11 59	☽ ∠ ♄
12 25	☽ ☌ ♆
16 19	♅ St R
23 11	☽ ⊼ ♀
23 12	☽ ⋕ ♃
23 20	☽ ꝗ ♂
24 Monday	
03 31	☽ ⊼ ☿
05 36	☽ ✶ ♅
05 58	☽ ⊥ ♄
08 55	☽ Q ♇
11 46	♀ ± ♆
15 29	☽ ± ☿
18 01	☽ ⊻ ♄
18 33	☽ ± ♂
25 Tuesday	
02 02	☿ St D
05 04	☽ ☍ ♃
07 26	☽ ⊼ ♂
09 39	☽ ꝗ ☿
10 00	☽ ⊼ ☉
11 51	☽ ‖ ♄
11 52	☽ ∠ ♅
21 06	☿ ✶ ♀
21 17	☽ □ ♇
23 16	☽ ± ☉
26 Wednesday	
00 45	☽ ⊻ ♆
05 45	☽ ⋕ ♀
06 08	☽ ⊥ ♅
12 55	☽ ⊥ ♆
16 31	☽ △ ☿
18 11	☽ ☍ ♀
18 19	☽ ⊻ ♅
19 12	☽ ꝗ ☉
19 37	♀ ⊼ ♅
27 Thursday	
06 40	☽ ☌ ♄
07 09	☽ ∠ ♆
08 48	☽ ⋕ ♂
19 01	☽ ⊼ ♃
23 54	☽ ‖ ♅
28 Friday	
00 05	☽ ☍ ♂
04 21	☽ △ ☉
07 21	☽ ± ♃
10 03	☽ △ ♇
13 24	☽ ✶ ♆
17 21	☿ □ ♅
20 46	☽ ⋕ ♇
29 Saturday	
01 45	☽ ꝗ ♃
01 55	☉ ⋕ ♆
06 50	☽ ☌ ♅
07 36	☽ □ ☿
12 52	☽ ⊼ ♀
13 20	☽ Q ♆
15 59	☽ ꝗ ♇
16 47	☽ ⋕ ♇
18 43	☽ ⊻ ♄
30 Sunday	
01 54	☽ ± ♀
06 08	☉ ∠ ♀
06 28	☽ ⊥ ♄
07 54	☽ △ ♃
09 39	☽ ± ♇
13 09	☽ ‖ ♅
15 19	☽ ⊼ ♂
20 57	☽ □ ☉
21 06	☽ ꝗ ♀
21 15	☽ ⊼ ♇
21 19	☽ Q ☿
21 41	☽ ⋕ ♂
23 50	☽ ∠ ♄

All ephemeris data is given at 12.00 UT and the Moon's longitude is additionally given for 24.00 UT

OCTOBER 2029

LONGITUDES

Date	Sidereal time h m s	Sun ☉ ° ' "	Moon ☽ ° ' "	Moon ☽ 24.00 ° ' "	Mercury ☿ ° '	Venus ♀ ° '	Mars ♂ ° '	Jupiter ♃ ° '	Saturn ♄ ° '	Uranus ♅ ° '	Neptune ♆ ° '	Pluto ♇ ° '
01	12 42 01	08 ♎ 35 45	15 ♋ 52 12	22 ♋ 16 04	20 ♍ 45	23 ♏ 45	05 ♐ 40	01 ♏ 30	24 ♉ 28	18 ♊ 32	09 ♈ 45	08 ♒ 08
02	12 45 58	09 34 47	28 ♋ 46 19	05 ♌ 23 24	21 47	24 51	06 22	01 42	24 R 25	18 R 31	09 R 43	08 R 08
03	12 49 54	10 33 50	12 ♌ 07 45	18 ♌ 59 35	22 56	25 57	07 05	01 55	24 22	18 31	09 42	08 07
04	12 53 51	11 32 57	25 ♌ 58 59	03 ♍ 05 52	24 11	27 03	07 47	02 07	24 19	18 30	09 40	08 07
05	12 57 47	12 32 05	10 ♍ 19 53	17 ♍ 40 29	25 31	28 09	08 29	02 20	24 17	18 30	09 38	08 06
06	13 01 44	13 31 15	25 ♍ 06 51	02 ♎ 38 00	26 56	29 ♏ 14	09 12	02 33	24 14	18 29	09 37	08 06
07	13 05 41	14 30 28	10 ♎ 12 45	17 ♎ 49 44	28 25	00 ♐ 20	09 55	02 45	24 10	18 28	09 35	08 05
08	13 09 37	15 29 43	25 ♎ 27 35	03 ♏ 04 53	29 ♍ 57	01 25	10 37	02 58	24 07	18 28	09 33	08 05
09	13 13 34	16 28 59	10 ♏ 40 16	18 ♏ 12 31	01 ♎ 32	02 30	11 20	03 11	24 04	18 27	09 32	08 05
10	13 17 30	17 28 18	25 ♏ 40 33	03 ♐ 03 31	03 10	03 34	12 03	03 24	24 01	18 26	09 30	08 04
11	13 21 27	18 27 39	10 ♐ 20 46	17 ♐ 31 50	04 49	04 39	12 46	03 37	23 57	18 25	09 28	08 04
12	13 25 23	19 27 01	24 ♐ 36 29	01 ♑ 34 41	06 29	05 43	13 29	03 50	23 54	18 24	09 27	08 04
13	13 29 20	20 26 25	08 ♑ 26 30	15 ♑ 12 10	08 11	06 47	14 12	04 02	23 50	18 23	09 25	08 04
14	13 33 16	21 25 51	21 ♑ 51 59	28 ♑ 26 20	09 53	07 51	14 55	04 15	23 46	18 22	09 24	08 03
15	13 37 13	22 25 19	04 ♒ 55 38	11 ♒ 20 19	11 37	08 55	15 39	04 28	23 43	18 21	09 22	08 03
16	13 41 10	23 24 48	17 ♒ 40 51	23 ♒ 57 39	13 20	09 58	16 22	04 41	23 39	18 20	09 20	08 03
17	13 45 06	24 24 19	00 ♓ 11 07	06 ♓ 21 39	15 03	11 01	17 06	04 54	23 35	18 19	09 19	08 03
18	13 49 03	25 23 52	12 ♓ 29 36	18 ♓ 35 16	16 47	12 04	17 49	05 07	23 31	18 18	09 17	08 03
19	13 52 59	26 23 26	24 ♓ 38 56	00 ♈ 40 52	18 30	13 07	18 33	05 20	23 27	18 16	09 16	08 03
20	13 56 56	27 23 03	06 ♈ 41 16	12 ♈ 40 21	20 14	14 09	19 17	05 33	23 23	18 15	09 14	08 03
21	14 00 52	28 22 41	18 ♈ 38 17	24 ♈ 35 15	21 56	15 11	20 00	05 46	23 19	18 14	09 12	08 D 03
22	14 04 49	29 ♎ 22 21	00 ♉ 31 25	06 ♉ 26 58	23 39	16 13	20 44	05 59	23 15	18 12	09 11	08 03
23	14 08 45	00 ♏ 22 04	12 ♉ 22 07	18 ♉ 17 03	25 21	17 14	21 28	06 13	23 10	18 11	09 09	08 03
24	14 12 42	01 21 48	24 ♉ 12 03	00 ♊ 07 21	27 03	18 15	22 12	06 26	23 06	18 09	09 08	08 03
25	14 16 39	02 21 35	06 ♊ 03 18	12 ♊ 00 14	28 ♎ 44	19 16	22 56	06 39	23 02	18 08	09 06	08 03
26	14 20 35	03 21 23	17 ♊ 58 34	23 ♊ 58 44	00 ♏ 25	20 16	23 41	06 52	22 57	18 06	09 05	08 03
27	14 24 32	04 21 14	00 ♋ 01 12	06 ♋ 06 30	02 05	21 16	24 25	07 05	22 53	18 05	09 03	08 03
28	14 28 28	05 21 07	12 ♋ 15 10	18 ♋ 27 46	03 45	22 15	25 09	07 18	22 48	18 03	09 02	08 04
29	14 32 25	06 21 02	24 ♋ 44 52	01 ♌ 07 04	05 24	23 14	25 54	07 31	22 44	18 01	09 00	08 04
30	14 36 21	07 21 00	07 ♌ 34 55	14 ♌ 08 54	07 03	24 13	26 38	07 44	22 39	18 00	08 59	08 04
31	14 40 18	08 ♏ 20 59	20 ♌ 49 29	27 ♌ 37 00	08 ♏ 41	25 ♐ 11	27 ♐ 23	07 ♏ 57	22 ♉ 35	17 ♊ 58	08 ♈ 58	08 ♒ 04

Date	Moon True ☊ ° '	Moon Mean ☊ ° '	Moon ☽ Latitude ° '
01	29 ♐ 15	29 ♐ 39	01 S 27
02	29 R 14	29 36	02 30
03	29 10	29 33	03 26
04	29 05	29 29	04 13
05	28 56	29 26	04 46
06	28 46	29 23	05 01
07	28 35	29 20	04 55
08	28 25	29 17	04 29
09	28 16	29 14	03 44
10	28 10	29 10	02 43
11	28 07	29 07	01 33
12	28 06	29 04	00 S 19
13	28 D 06	29 01	00 N 55
14	28 R 06	28 58	02 03
15	28 05	28 54	03 03
16	28 03	28 51	03 52
17	27 57	28 48	04 29
18	27 49	28 45	04 53
19	27 39	28 42	05 03
20	27 28	28 39	05 00
21	27 16	28 35	04 43
22	27 05	28 32	04 14
23	26 56	28 29	03 34
24	26 48	28 26	02 45
25	26 43	28 23	01 48
26	26 41	28 20	00 N 47
27	26 D 41	28 16	00 S 18
28	26 41	28 13	01 23
29	26 43	28 10	02 25
30	26 R 43	28 07	03 22
31	26 ♐ 42	28 ♐ 04	04 S 11

DECLINATIONS

Sun ☉ ° '	Moon ☽ ° '	Mercury ☿ ° '	Venus ♀ ° '	Mars ♂ ° '	Jupiter ♃ ° '	Saturn ♄ ° '	Uranus ♅ ° '	Neptune ♆ ° '	Pluto ♇ ° '
03 S 24	21 N 03	04 N 47	20 S 57	22 S 27	11 S 02	16 N 32	22 N 58	02 N 27	23 S 47
03 48	17 58	04 31	21 17	22 35	11 07	16 31	22 58	02 26	23 47
04 11	13 51	04 12	21 37	22 42	11 11	16 30	22 58	02 25	23 47
04 34	08 53	03 49	21 56	22 50	11 16	16 29	22 58	02 25	23 47
04 57	03 N 17	03 22	22 15	22 57	11 20	16 28	22 58	02 24	23 47
05 20	02 S 40	02 53	22 34	23 04	11 25	16 28	22 58	02 23	23 47
05 43	08 34	02 21	22 51	23 10	11 29	16 27	22 57	02 23	23 47
06 06	14 01	01 47	23 09	23 17	11 33	16 26	22 57	02 22	23 47
06 29	18 33	01 10	23 25	23 23	11 38	16 25	22 57	02 21	23 47
06 51	21 49	00 N 32	23 42	23 29	11 42	16 24	22 57	02 21	23 47
07 14	23 32	00 S 07	23 57	23 35	11 47	16 23	22 57	02 20	23 47
07 37	23 38	00 48	24 12	23 41	11 51	16 22	22 57	02 19	23 47
07 59	22 15	01 29	24 27	23 46	11 56	16 21	22 57	02 19	23 47
08 21	19 38	02 12	24 41	23 51	12 00	16 20	22 57	02 18	23 47
08 44	16 04	02 55	24 54	23 56	12 04	16 19	22 57	02 17	23 47
09 06	11 51	03 38	25 07	24 01	12 09	16 18	22 57	02 17	23 47
09 27	07 12	04 21	25 20	24 06	12 13	16 17	22 57	02 16	23 47
09 49	02 S 21	05 05	25 31	24 10	12 18	16 16	22 57	02 16	23 47
10 11	02 N 31	05 49	25 42	24 14	12 22	16 15	22 57	02 15	23 47
10 32	07 14	06 32	25 53	24 18	12 26	16 14	22 56	02 14	23 47
10 54	11 40	07 15	26 03	24 22	12 31	16 13	22 56	02 14	23 47
11 15	15 37	07 59	26 12	24 25	12 35	16 11	22 56	02 13	23 47
11 36	18 57	08 41	26 21	24 28	12 39	16 10	22 56	02 13	23 47
11 57	21 29	09 24	26 29	24 31	12 44	16 09	22 56	02 12	23 47
12 17	23 06	10 05	26 36	24 34	12 48	16 08	22 56	02 11	23 46
12 38	23 40	10 47	26 43	24 36	12 52	16 07	22 56	02 11	23 46
12 58	23 08	11 27	26 49	24 38	12 57	16 06	22 56	02 10	23 46
13 18	21 30	12 08	26 55	24 40	13 01	16 05	22 55	02 10	23 46
13 38	18 47	12 47	27 00	24 42	13 05	16 03	22 55	02 09	23 46
13 58	15 06	13 26	27 05	24 43	13 10	16 02	22 55	02 09	23 46
14 S 17	10 N 35	14 S 04	27 S 09	24 S 44	13 S 14	16 N 01	22 N 55	02 N 08	23 S 46

ZODIAC SIGN ENTRIES

Date	h	m	Planets
02	14	15	☽ ♌
04	18	48	☽ ♍
06	19	49	☽ ♎
07	04	47	♀ ♐
08	12	40	☿ ♎
08	19	09	☽ ♏
10	19	01	☽ ♐
12	21	16	☽ ♑
15	02	52	☽ ♒
17	11	38	☽ ♓
19	22	39	☽ ♈
22	10	56	☽ ♉
23	03	08	☉ ♏
24	23	45	☽ ♊
26	06	04	☿ ♏
27	11	58	☽ ♋
29	21	54	☽ ♌

LATITUDES

Date	Mercury ☿ ° '	Venus ♀ ° '	Mars ♂ ° '	Jupiter ♃ ° '	Saturn ♄ ° '	Uranus ♅ ° '	Neptune ♆ ° '	Pluto ♇ ° '
01	01 N 13	02 S 19	01 S 13	01 N 01	02 S 26	00 N 01	01 S 33	05 S 45
04	01 38	02 31	01 14	01 01	02 26	00 01	01 33	05 45
07	01 53	02 42	01 15	01 00	02 27	00 01	01 33	05 45
10	01 57	02 53	01 16	01 00	02 27	00 01	01 33	05 45
13	01 55	03 03	01 17	01 00	02 27	00 01	01 33	05 44
16	01 46	03 13	01 17	01 00	02 28	00 01	01 33	05 44
19	01 34	03 22	01 18	00 59	02 28	00 01	01 33	05 44
22	01 18	03 30	01 18	00 59	02 28	00 02	01 33	05 44
25	01 00	03 37	01 19	00 59	02 28	00 02	01 33	05 43
28	00 41	03 43	01 19	00 59	02 28	00 02	01 32	05 43
31	00 N 21	03 S 48	01 S 20	00 N 59	02 S 29	00 N 02	01 S 32	05 S 43

DATA

Julian Date	2462411
Delta T	+74 seconds
Ayanamsa	24° 16' 38"
Synetic vernal point	04° ♓ 50' 20"
True obliquity of ecliptic	23° 26' 08"

MOON'S PHASES, APSIDES AND POSITIONS ☽

Date	h	m	Phase	Longitude ° '	Eclipse Indicator
07	19	14	●	14 ♎ 48	
14	11	09	☽	21 ♑ 24	
22	09	28	○	29 ♈ 16	
30	11	32	☾	07 ♌ 20	

Day	h	m		
08	11	25	Perigee	
23	01	45	Apogee	
06	01	19	0S	
12	01	34	Max dec	23° S 47'
18	23	35	0N	
26	12	34	Max dec	23° N 40'

LONGITUDES

Date	Chiron ⚷ ° '	Ceres ⚳ ° '	Pallas ⚴ ° '	Juno ⚵ ° '	Vesta ⚶ ° '	Black Moon Lilith ⚸ ° '
01	11 ♉ 42	11 ♒ 27	00 ♑ 59	14 ♏ 46	17 ♑ 35	03 ♉ 59
11	11 ♉ 18	11 ♒ 45	03 ♑ 19	17 ♏ 56	20 ♑ 25	05 ♉ 05
21	10 ♉ 50	12 ♒ 37	05 ♑ 55	21 ♏ 11	23 ♑ 39	06 ♉ 12
31	10 ♉ 21	13 ♒ 59	08 ♑ 45	24 ♏ 31	27 ♑ 12	07 ♉ 19

All ephemeris data is given at 12.00 UT and the Moon's longitude is additionally given for 24.00 UT

ASPECTARIAN

h m Aspects

01 Monday
00 23 ☽ □ ♆
00 49 ☉ △ ♇
03 33 ☽ ± ♂
12 52 ☽ # ♀
17 00 ☽ ⚺ ♅
21 32 ☽ ⚼ ♂
21 56 ☽ ⚹ ☿

02 Tuesday
02 54 ♀ ☍ ♄
04 01 ☽ ⚹ ♄
04 07 ☽ △ ♀
04 11 ☽ ⊥ ♅
08 13 ☉ ⚼ ♄
09 15 ♀ ⚼ ♆
09 38 ☽ Q ☉
15 24 ☉ ☍ ♆
17 25 ☽ □ ♃
20 38 ☽ ∠ ♅
21 08 ☽ ∥ ♄

03 Wednesday
01 48 ☽ Q ♄
02 32 ☽ △ ♂
03 50 ☽ ∠ ☿
04 53 ☽ ☍ ♇
07 42 ☽ △ ♆
09 01 ☽ ⚹ ☉
12 29 ☉ # ☿
15 33 ♀ Q ♇
15 38 ☿ ⚼ ♇
21 14 ☽ ⊥ ☿
23 10 ☽ ⚹ ♅

04 Thursday
01 07 ☽ # ♃
01 48 ☽ Q ♃
08 37 ☽ ⚺ ☿
09 11 ☽ □ ♄
09 46 ☽ ⚼ ♆
13 02 ☽ ∠ ☉
13 58 ☽ □ ♀
14 31 ☿ △ ♄
19 39 ☽ Q ♅
22 31 ☽ ⚹ ♃
23 00 ♂ ⚹ ♇

05 Friday
00 56 ☽ ± ♆
04 27 ♂ ⊥ ♃
05 16 ☽ ⊥ ☉
05 29 ☽ # ☉
08 19 ☽ ⚻ ♇
08 48 ☽ □ ♂
10 52 ☽ ⚻ ♆
11 35 ☽ ∥ ☿
15 13 ♂ # ♅
15 37 ☽ ∥ ♆
15 53 ☽ ⚺ ☉
18 11 ☽ ± ♇
22 17 ☽ Q ♀
23 37 ☽ ∠ ♃

06 Saturday
01 19 ☽ □ ♅
04 19 ☿ ⊥ ♃
08 45 ☽ ⚼ ♇
10 35 ☽ △ ♄
10 54 ☽ # ♆
12 49 ☽ # ☿
14 19 ☽ ⊥ ♃
15 14 ☽ ☌ ☿
15 30 ☽ Q ♂
19 07 ☽ ⚹ ♀
20 19 ☿ Q ♂
23 29 ☽ ∥ ☉

07 Sunday
00 02 ☽ ⚺ ♃
01 27 ♂ △ ♆
08 39 ☽ △ ♇
10 22 ☽ ⚼ ♄
10 48 ☿ ∥ ♆
11 01 ☽ ☍ ♆
11 30 ☽ ⚹ ♂
19 14 ☽ ☌ ☉
20 21 ♀ # ♅
20 41 ☽ ∠ ♀

08 Monday
00 30 ☽ ± ♄
00 39 ☽ ∥ ♃
01 00 ☽ △ ♅
09 54 ☽ ⚻ ♄
11 55 ☽ ⊥ ♀
12 16 ☽ ∠ ♂
19 53 ☽ ⚺ ☿
22 05 ☽ ⚺ ♀
23 59 ☽ # ♄
23 59 ☽ ☌ ♃

09 Tuesday
00 35 ☽ ⚼ ♅
03 08 ☽ ⊥ ♂
06 27 ☽ ⊥ ☿
06 35 ♀ ∥ ♂
07 54 ☽ □ ♇
10 12 ☽ ⚻ ♆
13 06 ☽ ⚺ ♂
14 49 ☽ ± ♅
19 42 ☽ ± ♆
21 54 ☽ ⚺ ☉
22 27 ☽ ∠ ☿

10 Wednesday
00 22 ☽ ⚻ ♅
07 03 ♀ ⚺ ♃
08 12 ☽ ⊥ ☉
09 19 ☽ ☍ ♄
10 06 ☽ ⚼ ♆
12 38 ☽ Q ♇
15 47 ♃ ⚼ ♅
15 58 ☿ ⚺ ♃
20 36 ♀ ∥ ♇
23 50 ☽ ∠ ☉

11 Thursday
00 22 ☉ ± ♄
00 44 ☽ ⚺ ♃
01 11 ☽ # ♅
01 42 ☽ ⚹ ☿
01 52 ☽ ☌ ♀
05 21 ☿ ⚹ ♀
08 14 ☽ ⚹ ♇
10 33 ☽ △ ♆
10 46 ☽ ⊥ ♃
10 59 ☉ △ ♅
13 51 ☽ ∥ ♂
16 14 ☽ ☌ ♂

12 Friday
00 14 ☽ Q ☿
01 29 ☽ ☍ ♅
02 02 ☽ ∠ ♃
02 35 ☽ ⚹ ☉
09 22 ☽ ∠ ♇
10 29 ☽ ∥ ♂
10 47 ☽ ⚻ ♄
21 03 ☽ ± ♄

13 Saturday
00 41 ☽ Q ☉
00 50 ☽ ⊥ ♇
02 42 ☽ # ♅
04 10 ☽ ⚹ ♃
08 51 ☽ ⚺ ♀
10 15 ☿ △ ♇
11 20 ☽ ⚺ ♇
11 29 ☽ □ ☿
12 41 ☽ ⚼ ♄
13 43 ☽ □ ♆
16 56 ♂ ∥ ♇
20 22 ☽ ⊥ ♀
20 51 ☿ ⚼ ♄
22 48 ☽ ⚺ ♂

14 Sunday
01 43 ☽ Q ♃
05 06 ☿ ☍ ♆
05 41 ☽ ⚻ ♅
10 12 ☽ ⊥ ♂
11 09 ☽ □ ☉
13 57 ☽ ∠ ♀
15 27 ☽ △ ♄
15 34 ☿ # ♆
16 31 ♀ ⚹ ♇
16 33 ☽ ± ♅
22 03 ☽ Q ♆

15 Monday
03 36 ☽ ∠ ♂
09 05 ☽ ⚼ ♅
10 30 ☽ # ♄
11 08 ☽ □ ♃
17 50 ☽ ☌ ♇
20 07 ☽ ⚹ ♀
20 16 ☽ ⚹ ♆
21 57 ♀ △ ♆

16 Tuesday
02 28 ☽ △ ☿
09 21 ☽ ⚹ ♂
10 25 ☽ ∥ ♃
13 14 ☽ △ ♅
17 20 ☉ ⚻ ♄
20 56 ☽ Q ♀
23 21 ☽ □ ♄
23 54 ☽ △ ☉

17 Wednesday
00 42 ☽ ∠ ♆
01 22 ☽ ∥ ☉
08 34 ♀ ⊥ ♃
09 46 ☽ Q ♂
11 43 ☽ ⚼ ☿
18 03 ☽ ⊥ ♆
21 20 ☽ △ ♃

18 Thursday
00 20 ☽ ∥ ☿
03 18 ☽ ⚺ ♇
05 44 ☽ ⚺ ♆
07 32 ☽ ⚼ ☉
08 06 ☽ ± ☿
10 06 ☽ Q ♄
11 06 ☽ □ ♀
12 29 ☽ # ♆
15 03 ☽ ⊥ ♇
21 50 ☽ ⚻ ☿
21 52 ☿ ± ♄
23 09 ☽ □ ♂
23 24 ☽ □ ♅

19 Friday
02 49 ☽ ± ☉
03 09 ♂ ☍ ♅
03 18 ☽ ⚼ ♃
08 47 ☿ △ ♅
08 49 ☽ ∠ ♇
09 38 ☽ ⚹ ♄
10 42 ☽ ∥ ♆
13 02 ☿ ⚹ ♂
15 46 ☽ ⚻ ☉
21 30 ☽ ± ♃

20 Saturday
07 41 ☽ # ☿
09 42 ☽ ⚻ ♃
11 08 ☽ Q ♅
14 43 ☽ ⚹ ♇
15 23 ☽ ∠ ♄
17 05 ☽ ☌ ♆

21 Sunday
03 58 ♇ St D
04 23 ☽ △ ♀
07 18 ☽ # ☉
09 21 ☽ ⊥ ♄
11 10 ☽ ⚹ ♅
14 50 ☽ Q ♇
14 56 ☽ △ ♂
16 59 ☽ # ♃
19 47 ☽ ☍ ☿
21 22 ☽ ⚺ ♄

22 Monday
06 31 ☿ ⚻ ♄
09 28 ☽ ☍ ☉
13 31 ☽ ⚼ ♀
15 47 ☽ ∥ ♄
17 25 ☽ ∠ ♅
23 15 ☽ ⚼ ♂
23 17 ☽ ☍ ♃
23 47 ♂ ∠ ♃

23 Tuesday
03 14 ☽ □ ♇
05 30 ☽ ⚺ ♆
09 29 ☽ ± ♀
11 37 ☽ ⊥ ♅
17 38 ☽ ⊥ ♆
18 42 ☽ ± ♂
22 48 ☽ ⚻ ♀
23 46 ☽ ⚺ ♅

24 Wednesday
07 41 ☽ ⚻ ♂
09 47 ☽ ☌ ♄
09 52 ♀ ☍ ♅
11 51 ☽ ∠ ♆
18 44 ☽ ⚻ ☿

25 Thursday
03 51 ☽ ⚻ ☉
08 40 ☽ ∥ ♅
08 54 ☽ ± ☿
13 13 ☽ ⚻ ♃
14 38 ♂ ⚻ ♄
15 38 ♂ ∠ ♇
16 02 ☽ △ ♇
17 05 ☽ ± ☉
18 09 ☽ ⚹ ♆

26 Friday
01 32 ☽ ± ♃
06 01 ☽ ⚼ ☿
06 05 ☉ ⚼ ♅
12 15 ☽ ☌ ♅
12 50 ☽ ⚼ ☉
17 00 ☽ ☍ ♀
18 12 ☽ Q ♆
19 55 ☽ ⚼ ♃
21 54 ☽ ⚺ ♄
22 09 ☽ ⚼ ♇

27 Saturday
00 08 ☽ ☍ ♂
09 45 ☽ ⊥ ♄
09 51 ☉ ∥ ♃
16 01 ☽ ± ♇
16 13 ☽ ∥ ♅
16 44 ☽ △ ☿
21 19 ☽ △ ☉

28 Sunday
02 04 ☿ ⚼ ♅
02 10 ☽ △ ♃
03 23 ☽ ∠ ♄
03 49 ☽ ⚻ ♇
05 44 ☽ □ ♆
13 27 ♀ ∠ ♃
23 11 ☽ ⚺ ♅

29 Monday
00 26 ♀ ⚻ ♄
07 38 ♀ ∠ ♇
08 11 ☽ ⚹ ♄
08 54 ☽ ⚻ ♀
10 38 ☽ ⊥ ♅
14 18 ☽ ⚻ ♂
21 11 ☽ ± ♀

30 Tuesday
00 41 ☿ ∥ ♃
02 16 ☽ ± ♂
03 31 ☽ ∠ ♅
06 26 ☽ ∥ ♄
06 37 ☽ Q ♄
10 53 ☽ □ ☿
11 32 ☽ □ ☉
12 18 ☽ □ ♃
12 54 ☽ ☍ ♇
14 34 ☽ △ ♆
15 15 ☽ ⚼ ♀
18 02 ☽ # ☉
19 52 ☽ ⚼ ♂
20 13 ☽ # ☿
22 41 ☽ # ♃
23 14 ☉ ☌ ☿
23 38 ☿ ☌ ♃
23 57 ☉ ☌ ♃

31 Wednesday
02 55 ☿ □ ♇
05 17 ☉ □ ♇
06 54 ☽ ⚹ ♅
15 06 ☽ □ ♄
15 55 ☿ ⚻ ♆
17 33 ☽ ⚼ ♆
20 20 ☽ △ ♀
21 14 ☽ Q ♃
22 33 ☽ Q ☉
23 47 ☽ Q ☿

NOVEMBER 2029

LONGITUDES

Date	Sidereal time h m s	Sun ☉ ° ' "	Moon ☽ ° ' "	Moon ☽ 24.00 ° ' "	Mercury ☿ ° '	Venus ♀ ° '	Mars ♂ ° '	Jupiter ♃ ° '	Saturn ♄ ° '	Uranus ♅ ° '	Neptune ♆ ° '	Pluto ♇ ° '
01	14 44 14	09 ♏ 21 01	04 ♍ 31 42	11 ♍ 33 39	10 ♏ 19	26 ♐ 09	28 ♐ 07	08 ♏ 11	22 ♉ 30	17 ♊ 56	08 ♈ 56	08 ♒ 05
02	14 48 11	10 21 04	18 ♍ 42 45	25 ♍ 58 41	11 56	27 07	28 52	08 24	22 R 25	17 R 54	08 R 55	08 05
03	14 52 08	11 21 10	03 ♎ 20 56	10 ♎ 48 47	13 33	28 04	29 ♐ 37	08 37	22 20	17 52	08 53	08 05
04	14 56 04	12 21 18	18 ♎ 21 14	25 ♎ 57 11	15 10	29 00	00 ♑ 21	08 50	22 16	17 50	08 52	08 06
05	15 00 01	13 21 28	03 ♏ 35 20	11 ♏ 14 20	16 46	29 ♐ 56	01 06	09 03	22 11	17 48	08 51	08 06
06	15 03 57	14 21 40	18 ♏ 52 47	26 ♏ 29 19	18 21	00 ♑ 52	01 51	09 16	22 06	17 47	08 49	08 07
07	15 07 54	15 21 54	04 ♐ 02 40	11 ♐ 31 45	19 57	01 47	02 36	09 29	22 01	17 44	08 48	08 07
08	15 11 50	16 22 09	18 ♐ 55 35	26 ♐ 13 27	21 32	02 41	03 21	09 43	21 56	17 42	08 47	08 08
09	15 15 47	17 22 26	03 ♑ 24 50	10 ♑ 29 24	23 06	03 35	04 07	09 56	21 52	17 40	08 46	08 08
10	15 19 43	18 22 44	17 ♑ 27 01	24 ♑ 17 43	24 40	04 28	04 52	10 09	21 47	17 38	08 45	08 09
11	15 23 40	19 23 04	01 ♒ 01 40	07 ♒ 39 09	26 14	05 20	05 37	10 22	21 42	17 36	08 43	08 09
12	15 27 37	20 23 25	14 ♒ 10 34	20 ♒ 36 21	27 47	06 12	06 22	10 35	21 37	17 34	08 42	08 10
13	15 31 33	21 23 48	26 ♒ 56 59	03 ♓ 12 58	29 ♏ 21	07 03	07 08	10 48	21 32	17 32	08 41	08 10
14	15 35 30	22 24 12	09 ♓ 24 51	15 ♓ 33 07	00 ♐ 54	07 53	07 53	11 01	21 27	17 30	08 40	08 11
15	15 39 26	23 24 37	21 ♓ 38 17	27 ♓ 40 50	02 26	08 43	08 39	11 14	21 22	17 27	08 39	08 12
16	15 43 23	24 25 04	03 ♈ 41 13	09 ♈ 39 52	03 59	09 32	09 24	11 27	21 17	17 25	08 38	08 13
17	15 47 19	25 25 32	15 ♈ 37 10	21 ♈ 33 28	05 31	10 20	10 10	11 40	21 12	17 23	08 37	08 13
18	15 51 16	26 26 01	27 ♈ 29 07	03 ♉ 24 24	07 02	11 07	10 56	11 53	21 08	17 20	08 36	08 14
19	15 55 12	27 26 32	09 ♉ 19 36	15 ♉ 14 58	08 34	11 53	11 41	12 06	21 03	17 18	08 35	08 15
20	15 59 09	28 27 04	21 ♉ 10 45	27 ♉ 07 09	10 05	12 38	12 27	12 19	20 58	17 16	08 34	08 16
21	16 03 06	29 ♏ 27 38	03 ♊ 04 24	09 ♊ 02 44	11 36	13 22	13 13	12 32	20 53	17 13	08 33	08 17
22	16 07 02	00 ♐ 28 13	15 ♊ 02 22	21 ♊ 03 32	13 07	14 05	13 59	12 44	20 48	17 11	08 32	08 17
23	16 10 59	01 28 50	27 ♊ 06 30	03 ♋ 11 31	14 38	14 47	14 45	12 57	20 44	17 09	08 31	08 18
24	16 14 55	02 29 29	09 ♋ 18 55	15 ♋ 29 00	16 08	15 28	15 31	13 10	20 39	17 06	08 30	08 19
25	16 18 52	03 30 09	21 ♋ 42 05	27 ♋ 58 34	17 38	16 08	16 17	13 23	20 34	17 04	08 30	08 20
26	16 22 48	04 30 50	04 ♌ 18 47	10 ♌ 43 09	19 08	16 47	17 03	13 36	20 29	17 01	08 29	08 21
27	16 26 45	05 31 33	17 ♌ 12 01	23 ♌ 45 44	20 37	17 24	17 49	13 48	20 25	16 59	08 28	08 22
28	16 30 41	06 32 18	00 ♍ 24 40	07 ♍ 09 05	22 06	18 00	18 35	14 01	20 20	16 56	08 27	08 23
29	16 34 38	07 33 04	13 ♍ 59 12	20 ♍ 55 08	23 34	18 35	19 21	14 14	20 16	16 54	08 27	08 24
30	16 38 35	08 ♐ 33 51	27 ♍ 56 56	05 ♎ 04 27	25 ♐ 02	19 ♑ 09	20 ♑ 08	14 ♏ 26	20 ♉ 11	16 ♊ 51	08 ♈ 26	08 ♒ 25

Date	Moon True ☊ ° '	Moon Mean ☊ ° '	Moon ☽ Latitude ° '
01	26 ♐ 40	28 ♐ 00	04 S 46
02	26 R 35	27 57	05 06
03	26 29	27 54	05 06
04	26 22	27 51	04 46
05	26 15	27 48	04 06
06	26 10	27 45	03 08
07	26 06	27 41	01 57
08	26 04	27 38	00 S 39
09	26 D 04	27 35	00 N 40
10	26 05	27 32	01 54
11	26 07	27 29	02 59
12	26 08	27 26	03 52
13	26 R 08	27 22	04 33
14	26 06	27 19	04 59
15	26 03	27 16	05 11
16	25 59	27 13	05 09
17	25 54	27 10	04 54
18	25 48	27 06	04 26
19	25 43	27 03	03 47
20	25 38	27 00	02 58
21	25 35	26 57	02 00
22	25 33	26 54	00 N 58
23	25 D 33	26 51	00 S 09
24	25 33	26 47	01 15
25	25 35	26 44	02 19
26	25 37	26 41	03 18
27	25 38	26 38	04 08
28	25 39	26 35	04 46
29	25 R 39	26 31	05 10
30	25 ♐ 38	26 ♐ 28	05 S 16

DECLINATIONS

Date	Sun ☉ ° '	Moon ☽ ° '	Mercury ☿ ° '	Venus ♀ ° '	Mars ♂ ° '	Jupiter ♃ ° '	Saturn ♄ ° '	Uranus ♅ ° '	Neptune ♆ ° '	Pluto ♇ ° '
01	14 S 36	05 N 24	14 S 41	27 S 12	24 S 45	13 S 18	16 N 00	22 N 55	02 N 08	23 S 45
02	14 55	00 S 14	15 18	27 14	24 46	13 22	15 59	22 55	02 07	23 45
03	15 14	06 01	15 54	27 17	24 46	13 27	15 57	22 55	02 07	23 45
04	15 33	11 36	16 29	27 18	24 46	13 31	15 56	22 54	02 06	23 45
05	15 51	16 34	17 03	27 19	24 46	13 35	15 55	22 54	02 06	23 45
06	16 09	20 27	17 36	27 19	24 45	13 39	15 54	22 54	02 05	23 45
07	16 26	22 52	18 09	27 19	24 45	13 43	15 53	22 54	02 05	23 44
08	16 44	23 37	18 40	27 18	24 44	13 48	15 51	22 54	02 04	23 44
09	17 01	22 44	19 11	27 17	24 42	13 52	15 50	22 54	02 04	23 44
10	17 18	20 25	19 40	27 15	24 41	13 56	15 49	22 53	02 03	23 44
11	17 34	17 01	20 09	27 13	24 39	14 00	15 48	22 53	02 03	23 43
12	17 51	12 52	20 37	27 10	24 37	14 04	15 47	22 53	02 02	23 43
13	18 06	08 15	21 03	27 06	24 34	14 08	15 45	22 53	02 02	23 43
14	18 22	03 S 25	21 29	27 02	24 32	14 12	15 44	22 53	02 02	23 43
15	18 37	01 N 27	21 53	26 58	24 29	14 16	15 43	22 52	02 01	23 42
16	18 52	06 12	22 17	26 53	24 26	14 20	15 42	22 52	02 01	23 42
17	19 07	10 40	22 39	26 48	24 22	14 24	15 41	22 52	02 00	23 42
18	19 21	14 43	23 00	26 42	24 19	14 28	15 39	22 52	02 00	23 42
19	19 35	18 11	23 20	26 35	24 15	14 32	15 38	22 52	02 00	23 41
20	19 49	20 54	23 39	26 29	24 10	14 36	15 37	22 51	01 59	23 41
21	20 02	22 44	23 57	26 22	24 06	14 40	15 36	22 51	01 59	23 41
22	20 15	23 33	24 13	26 14	24 01	14 44	15 35	22 51	01 59	23 40
23	20 27	23 16	24 29	26 06	23 56	14 48	15 34	22 51	01 58	23 40
24	20 39	21 51	24 43	25 58	23 51	14 52	15 32	22 50	01 58	23 40
25	20 51	19 24	24 55	25 49	23 45	14 55	15 31	22 50	01 58	23 39
26	21 02	15 58	25 07	25 40	23 39	14 59	15 30	22 50	01 58	23 39
27	21 13	11 44	25 17	25 31	23 33	15 03	15 29	22 50	01 57	23 39
28	21 24	06 51	25 26	25 21	23 27	15 07	15 28	22 50	01 57	23 38
29	21 34	01 32	25 33	25 11	23 20	15 10	15 27	22 49	01 57	23 38
30	21 S 44	04 S 01	25 S 39	25 S 01	23 S 13	15 S 14	15 N 26	22 N 49	01 N 57	23 S 38

ZODIAC SIGN ENTRIES

Date	h	m	Planets
01	04	10	☽ ♍
03	06	34	☽ ♎
04	00	32	♂ ♑
05	06	22	☽ ♏
05	13	39	♀ ♑
07	05	34	☽ ♐
09	06	17	☽ ♑
11	10	09	☽ ♒
13	17	49	☽ ♓
13	22	09	☿ ♐
16	04	38	☽ ♈
18	17	06	☽ ♉
21	05	49	☽ ♊
22	00	49	☉ ♐
23	17	43	☽ ♋
26	03	51	☽ ♌
28	11	16	☽ ♍
30	15	28	☽ ♎

LATITUDES

Date	Mercury ☿ ° '	Venus ♀ ° '	Mars ♂ ° '	Jupiter ♃ ° '	Saturn ♄ ° '	Uranus ♅ ° '	Neptune ♆ ° '	Pluto ♇ ° '
01	00 N 14	03 S 49	01 S 20	00 N 59	02 S 29	00 N 02	01 S 32	05 S 43
04	00 S 06	03 52	01 20	00 59	02 29	00 02	01 32	05 43
07	00 26	03 54	01 20	00 59	02 29	00 02	01 32	05 42
10	00 46	03 54	01 20	00 59	02 28	00 02	01 32	05 42
13	01 04	03 52	01 20	00 58	02 28	00 02	01 32	05 42
16	01 22	03 48	01 20	00 58	02 28	00 02	01 32	05 42
19	01 38	03 42	01 20	00 58	02 28	00 02	01 32	05 41
22	01 52	03 34	01 19	00 58	02 28	00 02	01 32	05 41
25	02 04	03 23	01 19	00 58	02 27	00 02	01 32	05 41
28	02 14	03 10	01 19	00 59	02 27	00 02	01 32	05 41
31	02 S 20	02 S 53	01 S 18	00 N 59	02 S 26	00 N 02	01 S 31	05 S 41

DATA

Julian Date	2462442
Delta T	+74 seconds
Ayanamsa	24° 16' 42"
Synetic vernal point	04° ♓ 50' 17"
True obliquity of ecliptic	23° 26' 07"

MOON'S PHASES, APSIDES AND POSITIONS ☽

Date	h	m	Phase	Longitude ° '	Eclipse Indicator
06	04	24	●	14 ♏ 03	
13	00	35	☽	20 ♒ 55	
21	04	03	○	29 ♉ 08	
28	23	48	☾	07 ♍ 02	

Day	h	m		
05	23	01	Perigee	
19	02	41	Apogee	
02	11	03	0S	
08	10	41	Max dec	23° S 37'
15	04	49	0N	
22	17	48	Max dec	23° N 35'
29	18	40	0S	

LONGITUDES

Date	Chiron ⚷ ° '	Ceres ⚳ ° '	Pallas ⚴ ° '	Juno ⚵ ° '	Vesta ⚶ ° '	Black Moon Lilith ⚸ ° '
01	10 ♉ 18	14 ♒ 09	09 ♑ 03	24 ♏ 51	27 ♑ 34	07 ♉ 26
11	09 ♉ 49	16 ♒ 01	12 ♑ 05	28 ♏ 14	01 ♒ 26	08 ♉ 33
21	09 ♉ 21	18 ♒ 16	15 ♑ 17	01 ♐ 39	05 ♒ 30	09 ♉ 40
31	08 ♉ 55	20 ♒ 51	18 ♑ 36	05 ♐ 04	09 ♒ 45	10 ♉ 47

ASPECTARIAN

01 Thursday
00 15 ☽ △ ♂
00 44 ♃ □ ♇
02 17 ☉ ⊼ ♆
04 04 ☽ Q ♅
05 26 ☉ ∥ ☿
09 15 ☽ ± ♆
18 05 ☽ ⊼ ♇
18 21 ☽ ✶ ♃
19 32 ☽ ⊼ ♆
20 53 ☽ ✶ ☉
23 02 ♂ ± ♄
23 11 ☽ ✶ ☿

02 Friday
02 09 ☽ ∥ ♆
04 15 ☽ ± ♇
10 39 ☽ □ ♅
11 27 ☿ ± ♅
18 07 ☽ △ ♄
18 12 ☿ ∠ ♀
19 14 ☽ ⚼ ♇
19 50 ☽ # ♆
19 52 ☽ ∠ ♃
23 47 ☽ ∠ ☉

03 Saturday
02 49 ☽ □ ♀
03 15 ☽ ∠ ☿
05 37 ☽ □ ♂
10 48 ☽ ⊥ ♃
14 27 ☿ # ♄
15 28 ☽ ⊥ ☉
18 24 ☽ ⚼ ♄
18 31 ♀ ± ♄
19 36 ☽ ⊥ ☿
19 38 ☽ △ ♇
20 36 ☽ ⊻ ♃
20 54 ☽ ☍ ♆

04 Sunday
00 04 ☉ ± ♅
01 47 ☽ ⊻ ☉
06 20 ☽ ⊻ ☿
07 38 ☿ ± ♆
08 42 ☽ ± ♄
09 43 ☽ Q ♀
11 11 ☽ △ ♅
12 00 ☽ Q ♂
15 26 ♃ ⊼ ♆
17 26 ☿ ∠ ♂
18 09 ☽ ⊼ ♄
20 54 ☽ ∥ ♃

05 Monday
05 54 ☽ ✶ ♀
07 54 ☽ ✶ ♂
07 58 ☽ ∥ ☉
08 37 ☽ # ♄
10 47 ☽ ⚼ ♅
15 00 ☽ ∥ ☿
17 18 ☉ # ♄
19 05 ☽ □ ♇
20 14 ☽ ⊼ ♆
20 42 ☽ ☌ ♃

06 Tuesday
00 52 ☽ ± ♅
03 23 ☿ ⊼ ♅
04 24 ☽ ☌ ☉
05 38 ☽ ± ♆
06 57 ☽ ∠ ♀
08 39 ☽ ∠ ♂
10 16 ☽ ⊼ ♅
11 05 ☽ ☌ ☿
17 03 ☽ ☍ ♄
19 47 ☽ ⚼ ♆
20 12 ♂ ⊥ ♇
22 02 ☽ ⊥ ♀
22 51 ☉ ± ♆
23 24 ☽ Q ♇
23 34 ☽ ⊥ ♂

07 Wednesday
08 09 ☽ ⊻ ♀
09 35 ☽ ⊻ ♂
12 31 ☽ # ♅
18 31 ☽ ✶ ♇
19 36 ☽ △ ♆
20 51 ☽ ⊻ ♃
21 06 ♀ ⊥ ♇

08 Thursday
06 41 ☽ ⊥ ♃
07 32 ☽ ⊻ ☉
10 01 ☽ ☍ ♅
16 46 ☽ ⊻ ☿
16 54 ☽ ⊼ ♄
17 59 ☿ ☍ ♄
18 03 ☽ ⊥ ☉
18 53 ☽ ∠ ♇
21 38 ☽ ∠ ♃

09 Friday
02 46 ☽ ± ♄
03 54 ☽ ⊥ ☿
09 28 ☽ # ♅
09 51 ☽ ⊥ ♇
10 07 ☽ ∠ ☉
12 18 ☽ ☌ ♀
13 14 ☽ ☌ ♂
17 47 ☽ ⚼ ♄
18 54 ☉ ⊼ ♅
19 59 ☽ ⊻ ♇
20 55 ☽ ∠ ☿
21 02 ☽ □ ♆
21 59 ☿ ⚼ ♆
23 13 ☽ ✶ ♃

10 Saturday
12 20 ☽ ⊼ ♅
13 45 ☽ ✶ ☉
17 05 ☽ ∥ ☿
19 31 ☽ △ ♄
20 21 ☽ Q ♃
22 49 ☽ ± ♅

11 Sunday
02 19 ☽ ✶ ☿
04 19 ☽ Q ♆
08 42 ☽ ∥ ☉
10 46 ☿ Q ♇
12 42 ☽ Q ☉
14 50 ☽ ⚼ ♅
19 27 ☽ # ♄
20 20 ☽ ⊻ ♀
20 48 ☽ ⊻ ♂

12 Monday
00 55 ☽ ☌ ♇
01 56 ☽ ✶ ♆
02 49 ☽ Q ☿
05 15 ☽ □ ♃
05 28 ☽ ∥ ♃
08 05 ☽ ⊥ ♀
08 28 ☽ ⊥ ♂
18 17 ☽ △ ♅
18 52 ♂ ⚼ ♄
22 39 ♀ ⚼ ♄

13 Tuesday
00 35 ☽ □ ☉
01 49 ☽ □ ♄
02 03 ☽ ∠ ♀
02 18 ☽ ∠ ♂
05 49 ☽ ∠ ♆
15 00 ☉ ☍ ♄
17 13 ☽ □ ☿
22 58 ☽ ⊥ ♆

14 Wednesday
08 49 ☽ ✶ ♀
08 51 ☽ ✶ ♂
09 37 ☽ ⊻ ♇
10 33 ☽ ⊻ ♆
12 02 ♀ ☌ ♂
12 04 ☽ Q ♄
15 11 ☽ △ ♃
18 51 ☽ # ♆
20 42 ♀ ⊻ ♇
21 20 ☽ ⊥ ♇
21 31 ♂ ⊻ ♇

15 Thursday
03 46 ☽ □ ♅
09 55 ☽ Q ♂
10 02 ☽ Q ♀
10 05 ♀ □ ♆
11 28 ☽ ✶ ♄
12 01 ♂ □ ♆
14 49 ☽ ∥ ♆
15 05 ☽ ∠ ♇
15 50 ☽ △ ☉
17 34 ☉ ⚼ ♆
18 29 ☿ ⊥ ♂
21 14 ☿ ⊥ ♀
21 17 ☽ ⚼ ♃

16 Friday
09 01 ♃ ± ♅
10 57 ☉ ∠ ♂
12 40 ☽ △ ☿
15 27 ☽ Q ♅
15 36 ☽ ± ♃
17 11 ☽ ∠ ♄
21 05 ☽ ✶ ♇
21 54 ☽ ☌ ♆

17 Saturday
00 16 ☽ □ ♂
00 34 ☽ ⚼ ☉
00 34 ☽ □ ♀
00 46 ☉ ∠ ♀
03 53 ☽ ⊼ ♃
11 10 ☽ ⊥ ♄
15 32 ☽ ✶ ♅
20 24 ☽ ± ☉
21 18 ☽ Q ♇
23 13 ☽ ⊻ ♄
23 21 ☽ ⚼ ☿

18 Sunday
02 19 ☿ # ♅
07 11 ☉ Q ♇
09 40 ☽ ⊼ ☉
10 27 ☽ # ♃
18 05 ☽ ∥ ♄
20 16 ☽ ± ☿
21 48 ☽ ∠ ♅

19 Monday
06 56 ☿ ✶ ♇
09 49 ☽ □ ♇
10 14 ☽ ⊼ ☿
10 29 ☽ ⊻ ♆
12 13 ☿ △ ♆
15 59 ☽ ⊥ ♅
17 07 ☽ △ ♂
17 32 ☽ △ ♀
17 43 ☽ ☍ ♃
21 38 ♀ ✶ ♃
22 38 ☽ ⊥ ♆

20 Tuesday
00 31 ☽ # ☉
04 06 ☽ ⊻ ♅
05 47 ♂ ✶ ♃
11 34 ☽ ☌ ♄
14 15 ☿ ∥ ♇
16 49 ☽ ∠ ♆

21 Wednesday
01 33 ☽ ⚼ ♂
01 54 ☽ ⚼ ♀
04 03 ☽ ☍ ☉
14 06 ☽ ∥ ♅
20 54 ☽ ± ♂
21 12 ☽ ± ♀
21 57 ☿ ∥ ♂
22 28 ☽ △ ♇
22 59 ☽ ✶ ♆

22 Thursday
04 59 ☿ ⊻ ♃
07 19 ☽ ⊼ ♃
07 36 ☽ ☍ ☿
09 45 ☽ ⊼ ♂
09 59 ☽ ⊼ ♀
16 16 ☽ ☌ ♅
19 31 ☽ ± ♃
22 57 ☽ Q ♆
23 25 ☽ ⊻ ♄

23 Friday
04 27 ☽ ⚼ ♇
11 15 ☽ ⊥ ♄
13 42 ☽ ⚼ ♃
15 52 ☿ ⊻ ♂
16 46 ☿ ⊻ ♀
21 22 ☽ ∥ ♅
21 25 ☽ ⊼ ☉
22 16 ☽ ± ♇

24 Saturday
01 03 ♀ ☌ ♂
04 52 ☽ ∠ ♄
10 03 ☽ ⊼ ♇
10 15 ☽ ± ☉
10 25 ☽ □ ♆
19 38 ☽ △ ♃
23 59 ☽ # ☉

25 Sunday
00 40 ☽ ☍ ♀
00 51 ☽ ☍ ♂
03 05 ☽ ⊼ ☿
03 05 ☽ ⊻ ♅
03 06 ☿ ☍ ♅
05 18 ☽ ⚼ ☉
09 50 ☽ ✶ ♄
14 36 ☽ ⊥ ♅
16 12 ☽ ± ☿

26 Monday
07 41 ☽ ∠ ♅
08 35 ☽ Q ♄
11 09 ♂ ⊼ ♅
11 36 ☽ ⚼ ☿
12 25 ☽ △ ☉
12 50 ♂ ∥ ♇
14 53 ☽ ∥ ♄
17 53 ☽ # ♃
19 36 ☽ ☍ ♇
19 49 ☽ △ ♆
20 30 ♀ ⊼ ♅
20 46 ☿ ⊥ ♃

27 Tuesday
05 38 ☽ □ ♃
08 55 ☿ ⊼ ♄
11 36 ☽ ✶ ♅
12 24 ☽ ⊼ ♀
13 12 ☽ ⊼ ♂
17 52 ☽ □ ♄
19 04 ☽ △ ☿
23 27 ☽ ⚼ ♆
23 54 ☽ ± ♀

28 Wednesday
00 51 ☽ ± ♂
06 14 ☿ ∥ ♀
09 22 ☽ Q ♅
14 55 ☽ Q ♃
15 39 ☽ ± ♆
16 51 ☽ ⚼ ♀
18 01 ☽ ⚼ ♂
23 48 ☽ □ ☉

29 Thursday
02 12 ☽ ⊼ ♇
02 17 ☽ ⊼ ♆
09 18 ☿ ∠ ♇
10 11 ☽ ∥ ♆
12 25 ☽ ✶ ♃
12 44 ☽ ± ♇
17 02 ☽ □ ♅
20 19 ☽ △ ♀
21 51 ☽ △ ♂
22 48 ☽ △ ♄

30 Friday
03 05 ☽ # ♆
04 17 ☽ ⚼ ♇
06 28 ☽ □ ☿
08 35 ☉ ✶ ♇
08 56 ☉ △ ♆
09 28 ☽ Q ☉
11 45 ♃ ± ♆
13 36 ♂ △ ♄
14 33 ☽ ∠ ♃
20 57 ♆ ✶ ♇

All ephemeris data is given at 12.00 UT and the Moon's longitude is additionally given for 24.00 UT

DECEMBER 2029

LONGITUDES

Date	Sidereal time h m s	Sun ☉ ° ' "	Moon ☽ ° ' "	Moon ☽ 24.00 ° ' "	Mercury ☿ ° '	Venus ♀ ° '	Mars ♂ ° '	Jupiter ♃ ° '	Saturn ♄ ° '	Uranus ♅ ° '	Neptune ♆ ° '	Pluto ♇ ° '
01	16 42 31	09 ♐ 34 40	12 ♎ 17 28	19 ♎ 35 33	26 ♐ 29	19 ♑ 40	20 ♑ 54	14 ♏ 39	20 ♉ 07	16 ♊ 49	08 ♈ 25	08 ♒ 26
02	16 46 28	10 35 31	26 ♎ 58 07	04 ♏ 24 27	27 56	20 11	21 40	14 51	20 R 02	16 R 46	08 R 25	08 28
03	16 50 24	11 36 23	11 ♏ 53 40	19 ♏ 24 46	29 ♐ 22	20 40	22 27	15 04	19 58	16 44	08 24	08 29
04	16 54 21	12 37 16	26 ♏ 56 39	04 ♐ 28 12	00 ♑ 47	21 07	23 13	15 16	19 53	16 41	08 24	08 30
05	16 58 17	13 38 11	11 ♐ 58 15	19 ♐ 25 43	02 11	21 32	24 00	15 28	19 49	16 39	08 23	08 31
06	17 02 14	14 39 06	26 ♐ 49 35	04 ♑ 08 57	03 33	21 56	24 46	15 41	19 45	16 36	08 23	08 32
07	17 06 10	15 40 03	11 ♑ 23 02	18 ♑ 31 16	04 54	22 18	25 33	15 53	19 41	16 33	08 22	08 34
08	17 10 07	16 41 01	25 ♑ 33 12	02 ♒ 28 35	06 14	22 38	26 19	16 05	19 37	16 31	08 22	08 35
09	17 14 04	17 41 59	09 ♒ 17 18	15 ♒ 59 23	07 32	22 56	27 06	16 18	19 33	16 28	08 22	08 36
10	17 18 00	18 42 57	22 ♒ 35 02	29 ♒ 04 31	08 47	23 12	27 53	16 30	19 29	16 26	08 21	08 37
11	17 21 57	19 43 57	05 ♓ 28 13	11 ♓ 46 34	10 00	23 26	28 39	16 42	19 25	16 23	08 21	08 39
12	17 25 53	20 44 57	18 ♓ 00 04	24 ♓ 09 16	11 09	23 38	29 ♑ 26	16 54	19 21	16 21	08 21	08 40
13	17 29 50	21 45 57	00 ♈ 14 43	06 ♈ 16 59	12 16	23 47	00 ♒ 13	17 06	19 17	16 18	08 21	08 41
14	17 33 46	22 46 58	12 ♈ 16 40	18 ♈ 14 18	13 18	23 54	01 00	17 18	19 14	16 16	08 20	08 43
15	17 37 43	23 47 59	24 ♈ 10 27	00 ♉ 05 39	14 15	23 59	01 46	17 29	19 10	16 13	08 20	08 44
16	17 41 39	24 49 01	06 ♉ 00 23	11 ♉ 55 07	15 07	24 01	02 33	17 41	19 07	16 11	08 20	08 46
17	17 45 36	25 50 04	17 ♉ 50 18	23 ♉ 46 21	15 53	24 R 01	03 20	17 53	19 03	16 08	08 20	08 47
18	17 49 33	26 51 07	29 ♉ 43 36	05 ♊ 42 23	16 33	23 59	04 07	18 05	19 00	16 06	08 20	08 49
19	17 53 29	27 52 10	11 ♊ 43 01	17 ♊ 45 44	17 04	23 54	04 54	18 16	18 57	16 03	08 D 20	08 50
20	17 57 26	28 53 14	23 ♊ 50 46	29 ♊ 58 20	17 26	23 46	05 41	18 28	18 53	16 01	08 20	08 52
21	18 01 22	29 ♐ 54 19	06 ♋ 08 34	12 ♋ 21 38	17 39	23 36	06 28	18 39	18 50	15 58	08 20	08 53
22	18 05 19	00 ♑ 55 24	18 ♋ 37 39	24 ♋ 56 45	17 R 42	23 24	07 15	18 50	18 47	15 56	08 20	08 55
23	18 09 15	01 56 30	01 ♌ 19 01	07 ♌ 44 32	17 34	23 09	08 02	19 02	18 45	15 53	08 20	08 56
24	18 13 12	02 57 36	14 ♌ 13 23	20 ♌ 45 41	17 14	22 51	08 49	19 13	18 42	15 51	08 20	08 58
25	18 17 08	03 58 42	27 ♌ 21 28	04 ♍ 00 49	16 42	22 32	09 36	19 24	18 39	15 49	08 21	08 59
26	18 21 05	04 59 49	10 ♍ 43 49	17 ♍ 30 30	15 58	22 10	10 23	19 35	18 36	15 46	08 21	09 01
27	18 25 02	06 00 57	24 ♍ 20 53	01 ♎ 15 00	15 04	21 45	11 10	19 46	18 34	15 44	08 21	09 02
28	18 28 58	07 02 05	08 ♎ 12 48	15 ♎ 14 12	13 59	21 19	11 57	19 57	18 32	15 41	08 21	09 04
29	18 32 55	08 03 14	22 ♎ 19 04	29 ♎ 27 11	12 47	20 51	12 44	20 08	18 29	15 39	08 22	09 06
30	18 36 51	09 04 24	06 ♏ 38 17	13 ♏ 52 00	11 29	20 21	13 31	20 19	18 27	15 37	08 22	09 07
31	18 40 48	10 ♑ 05 34	21 ♏ 07 52	28 ♏ 25 21	10 ♑ 08	19 ♑ 49	14 ♒ 19	20 ♏ 29	18 ♉ 25	15 ♊ 35	08 ♈ 22	09 ♒ 09

Date	Moon True ☊ ° '	Moon Mean ☊ ° '	Moon ☽ Latitude ° '
01	25 ♐ 36	26 ♐ 25	05 S 03
02	25 R 34	26 22	04 30
03	25 33	26 19	03 39
04	25 31	26 16	02 32
05	25 30	26 12	01 S 14
06	25 D 30	26 09	00 N 07
07	25 30	26 06	01 27
08	25 31	26 03	02 39
09	25 32	26 00	03 40
10	25 32	25 57	04 27
11	25 33	25 53	04 58
12	25 33	25 50	05 15
13	25 R 33	25 47	05 17
14	25 33	25 44	05 04
15	25 33	25 41	04 39
16	25 32	25 37	04 02
17	25 D 32	25 34	03 15
18	25 32	25 31	02 19
19	25 33	25 28	01 16
20	25 33	25 25	00 N 09
21	25 R 33	25 22	00 S 59
22	25 32	25 18	02 05
23	25 32	25 15	03 06
24	25 31	25 12	03 59
25	25 30	25 09	04 40
26	25 29	25 06	05 07
27	25 28	25 03	05 17
28	25 D 28	24 59	05 09
29	25 28	24 56	04 43
30	25 29	24 53	03 59
31	25 ♐ 30	24 ♐ 50	02 S 59

DECLINATIONS

Date	Sun ☉ ° '	Moon ☽ ° '	Mercury ☿ ° '	Venus ♀ ° '	Mars ♂ ° '	Jupiter ♃ ° '	Saturn ♄ ° '	Uranus ♅ ° '	Neptune ♆ ° '	Pluto ♇ ° '
01	21 S 53	09 S 30	25 S 44	24 S 51	23 S 06	15 S 18	15 N 25	22 N 49	01 N 56	23 S 37
02	22 02	14 35	25 47	24 40	22 59	15 21	15 24	22 49	01 56	23 37
03	22 10	18 52	25 49	24 29	22 51	15 25	15 23	22 48	01 56	23 37
04	22 18	21 56	25 49	24 18	22 43	15 29	15 22	22 48	01 56	23 36
05	22 26	23 27	25 48	24 07	22 35	15 32	15 21	22 48	01 56	23 36
06	22 33	23 16	25 45	23 55	22 26	15 36	15 20	22 48	01 56	23 36
07	22 40	21 30	25 41	23 44	22 18	15 39	15 19	22 47	01 55	23 35
08	22 46	18 25	25 36	23 32	22 09	15 43	15 18	22 47	01 55	23 35
09	22 52	14 23	25 29	23 20	21 59	15 46	15 17	22 47	01 55	23 34
10	22 57	09 47	25 21	23 08	21 50	15 50	15 16	22 47	01 55	23 34
11	23 02	04 S 52	25 12	22 56	21 40	15 53	15 15	22 46	01 55	23 34
12	23 07	00 N 05	25 01	22 44	21 30	15 56	15 15	22 46	01 55	23 33
13	23 11	04 56	24 49	22 32	21 20	16 00	15 14	22 46	01 55	23 33
14	23 14	09 31	24 36	22 19	21 10	16 03	15 13	22 46	01 55	23 32
15	23 17	13 42	24 22	22 07	20 59	16 06	15 12	22 45	01 55	23 32
16	23 20	17 19	24 08	21 55	20 48	16 09	15 12	22 45	01 55	23 31
17	23 22	20 15	23 52	21 43	20 37	16 13	15 11	22 45	01 55	23 31
18	23 24	22 21	23 36	21 30	20 26	16 16	15 10	22 45	01 55	23 31
19	23 25	23 27	23 20	21 18	20 14	16 19	15 10	22 44	01 55	23 30
20	23 26	23 27	23 03	21 06	20 02	16 22	15 09	22 44	01 55	23 30
21	23 26	22 19	22 46	20 54	19 50	16 25	15 08	22 44	01 55	23 29
22	23 26	20 05	22 29	20 42	19 38	16 28	15 08	22 44	01 55	23 29
23	23 25	16 50	22 13	20 30	19 26	16 31	15 07	22 43	01 55	23 28
24	23 24	12 45	21 57	20 18	19 13	16 34	15 07	22 43	01 55	23 28
25	23 23	08 00	21 42	20 06	19 00	16 37	15 06	22 43	01 55	23 28
26	23 20	02 N 49	21 28	19 55	18 47	16 40	15 06	22 43	01 56	23 27
27	23 18	02 S 36	21 15	19 43	18 34	16 43	15 05	22 42	01 56	23 27
28	23 15	07 59	21 02	19 32	18 21	16 46	15 05	22 42	01 56	23 26
29	23 11	13 04	20 51	19 21	18 07	16 49	15 04	22 42	01 56	23 26
30	23 07	17 29	20 41	19 09	17 53	16 51	15 04	22 42	01 56	23 25
31	23 S 03	20 S 55	20 S 32	18 S 59	17 S 39	16 S 54	15 N 04	22 N 41	01 N 56	23 S 25

ZODIAC SIGN ENTRIES

Date	h	m	Planets
02	16	54	☽ ♏
03	22	47	☿ ♑
04	16	52	☽ ♐
06	17	11	☽ ♑
08	19	41	☽ ♒
11	01	43	☽ ♓
13	05	25	♂ ♒
13	11	31	☽ ♈
15	23	49	☽ ♉
18	12	33	☽ ♊
21	00	03	☽ ♋
21	14	14	☉ ♑
23	09	32	☽ ♌
25	16	47	☽ ♍
27	21	50	☽ ♎
30	00	55	☽ ♏

LATITUDES

Date	Mercury ☿ ° '	Venus ♀ ° '	Mars ♂ ° '	Jupiter ♃ ° '	Saturn ♄ ° '	Uranus ♅ ° '	Neptune ♆ ° '	Pluto ♇ ° '
01	02 S 20	02 S 53	01 S 18	00 N 59	02 S 26	00 N 02	01 S 31	05 S 41
04	02 23	02 33	01 18	00 59	02 26	00 02	01 31	05 40
07	02 21	02 10	01 17	00 59	02 25	00 02	01 31	05 40
10	02 13	01 43	01 17	00 59	02 25	00 02	01 31	05 40
13	01 58	01 12	01 16	00 59	02 24	00 02	01 31	05 40
16	01 34	00 S 37	01 15	00 59	02 24	00 02	01 31	05 40
19	00 59	00 N 01	01 14	00 59	02 23	00 02	01 31	05 40
22	00 S 14	00 43	01 13	00 59	02 22	00 02	01 30	05 40
25	00 N 42	01 28	01 12	01 00	02 22	00 02	01 30	05 40
28	01 40	02 15	01 11	01 00	02 21	00 02	01 30	05 40
31	02 N 32	03 N 02	01 S 10	01 N 00	02 S 20	00 N 02	01 S 30	05 S 39

DATA

Julian Date	2462472
Delta T	+74 seconds
Ayanamsa	24° 16' 46"
Synetic vernal point	04° ♓ 50' 13"
True obliquity of ecliptic	23° 26' 07"

MOON'S PHASES, APSIDES AND POSITIONS ☽

Date	h	m	Phase	Longitude ° '	Eclipse Indicator
05	14	52	●	13 ♐ 45	Partial
12	17	49	☽	21 ♓ 00	
20	22	46	○	29 ♊ 21	total
28	09	49	☾	06 ♎ 57	

Day	h	m		
04	10	31	Perigee	
16	13	53	Apogee	
05	21	29	Max dec	23° S 35'
12	11	33	0N	
20	00	09	Max dec	23° N 35'
27	00	30	0S	

LONGITUDES

Date	Chiron ⚷ ° '	Ceres ⚳ ° '	Pallas ⚴ ° '	Juno ⚵ ° '	Vesta ⚶ ° '	Black Moon Lilith ⚸ ° '
01	08 ♉ 55	20 ♒ 51	18 ♑ 36	05 ♐ 04	09 ♒ 45	10 ♉ 47
11	08 ♉ 33	23 ♒ 43	22 ♑ 00	08 ♐ 30	14 ♒ 10	11 ♉ 54
21	08 ♉ 15	26 ♒ 48	25 ♑ 28	11 ♐ 54	18 ♒ 41	13 ♉ 00
31	08 ♉ 02	00 ♓ 06	28 ♑ 59	15 ♐ 16	23 ♒ 18	14 ♉ 07

All ephemeris data is given at 12.00 UT and the Moon's longitude is additionally given for 24.00 UT

ASPECTARIAN

01 Saturday
00 07 ☽ ⚼ ♄
05 35 ☽ ☍ ♆
05 36 ☽ △ ♇
05 52 ☽ ⊥ ♃
06 04 ☿ ± ♄
07 10 ☽ ⚹ ☉
14 59 ☽ ± ♄
15 56 ☽ ⊻ ♃
16 01 ☽ Q ☿
19 25 ☽ △ ♅

02 Sunday
00 34 ☽ □ ♀
00 47 ☽ ⊼ ♄
02 55 ☽ □ ♂
05 54 ♀ △ ♄
09 36 ☽ ∠ ☉
13 44 ☽ ⚹ ☿
16 00 ☽ ∥ ♃
16 08 ☽ # ♄
19 44 ☽ ⚼ ♅

03 Monday
00 38 ♃ # ♄
01 12 ☽ ⊥ ☉
06 25 ☽ ⊼ ♆
06 32 ☽ □ ♇
06 40 ☽ Q ♀
09 33 ☽ Q ♂
10 08 ☽ ± ♅
11 30 ☽ ⊻ ☉
16 00 ☽ ± ♆
16 21 ☽ ∠ ☿
17 08 ☽ ☌ ♃
19 42 ☽ ⊼ ♅
20 02 ♂ # ♅
20 17 ♂ ± ♅

04 Tuesday
00 49 ☽ ☍ ♄
01 51 ☿ ∠ ♃
02 26 ☽ ⚹ ♀
05 45 ☽ ⚹ ♂
06 21 ☽ ⚼ ♆
08 12 ☽ ⊥ ☿
11 17 ☽ Q ♇
16 22 ☽ ∥ ☉
18 44 ☽ ⊻ ☿
20 37 ☽ ∥ ♂
22 30 ☽ # ♅

05 Wednesday
03 04 ☽ ∠ ♀
06 16 ☽ △ ♆
06 28 ☽ ⚹ ♇
06 58 ☽ ∠ ♂
14 52 ☽ ☌ ☉
17 42 ☽ ⊻ ♃
17 54 ☽ ⊥ ♀
18 02 ☿ ⊥ ♇
19 29 ☽ ☍ ♅
22 13 ☽ ⊥ ♂

06 Thursday
00 34 ☽ ⊼ ♄
01 27 ☉ ∥ ♂
03 31 ☽ ⊥ ♃
03 50 ☽ ⊻ ♀
06 39 ☽ ∠ ♇
08 28 ☽ ⊻ ♂
10 15 ☽ ± ♄
18 23 ☽ ∠ ♃
20 53 ☽ # ♅
21 22 ☽ ⊥ ♇
23 32 ☽ ∥ ☉

07 Friday
00 09 ☽ ☌ ☿
00 56 ☽ ⚼ ♄
02 53 ☽ ∥ ♂
06 59 ☽ □ ♆
07 17 ☽ ⊻ ♇
08 10 ☿ ⚼ ♄
18 26 ☉ ⊻ ♃
19 39 ☽ ⚹ ♃
19 44 ☽ ⊻ ☉
20 39 ☽ ⊼ ♅

08 Saturday
01 54 ☽ △ ♄
04 15 ♀ ± ♅
06 04 ♀ ∥ ♇
06 42 ☽ ⊥ ☉
06 49 ☽ ± ♅
06 53 ☽ ☌ ♀
08 11 ☉ ☍ ♅
10 16 ☉ ⊥ ♀
13 20 ♂ Q ♆
13 24 ☽ Q ♆
13 24 ☽ ☌ ♂
15 50 ☉ # ♅
16 26 ☽ Q ♃
22 17 ☽ ⚼ ♅
23 28 ☽ ∠ ☉

09 Sunday
04 22 ☽ ∥ ♃
07 00 ☽ # ♄
08 33 ☽ ⊻ ☿
10 21 ☽ ⚹ ♆
10 47 ☽ ☌ ♇
20 22 ☽ ⊥ ☿

10 Monday
00 44 ☽ □ ♃
00 50 ☽ △ ♅
03 46 ☿ □ ♆
04 21 ☽ ⚹ ☉
05 45 ♃ ⊼ ♅
06 21 ☽ □ ♄
08 52 ☿ ⊻ ♇
13 09 ☽ ⊻ ♀
13 25 ☽ ∠ ♆
14 26 ☽ ∠ ☿
22 24 ☽ ⊻ ♂

11 Tuesday
00 28 ☽ ⊥ ♀
03 03 ☉ ∥ ♀
04 22 ☽ Q ☉
04 58 ☉ ⊼ ♄
06 08 ☽ ⊥ ♆
10 22 ☽ ⊥ ♂
13 38 ♂ Q ♃
15 40 ☽ Q ♄
17 27 ☽ ⊻ ♆
17 43 ☽ ∠ ♀
18 02 ☽ ⊻ ♇
21 29 ☽ ⚹ ☿

12 Wednesday
02 16 ☽ # ♆
04 39 ☽ ∠ ♂
05 33 ☽ ⊥ ♇
07 21 ♀ # ♅
08 48 ☽ □ ♅
09 49 ☽ △ ♃
14 37 ☽ ⚹ ♄
17 49 ☽ □ ☉
20 56 ☽ ∥ ♆
23 03 ☽ Q ☿
23 04 ☽ ∠ ♇
23 07 ☽ ⚹ ♀

13 Thursday
11 56 ☽ ⚹ ♂
15 44 ☽ ⚼ ♃
19 59 ☽ ∠ ♄
20 02 ☽ Q ♅
23 07 ☽ Q ♀

14 Friday
04 06 ☽ ☌ ♆
04 51 ☽ ⚹ ♇
09 59 ☽ ± ♃
13 32 ☽ Q ♂
13 54 ☽ ⊥ ♄
14 14 ☽ □ ☿
19 48 ♂ ⚼ ♅
19 59 ☽ ⚹ ♅
22 16 ☽ ⊼ ♃

15 Saturday
01 56 ☽ ⊻ ♄
02 56 ☉ ⊥ ♃
05 02 ☽ Q ♇
10 29 ☉ ∠ ♇
11 10 ☽ △ ☉
11 37 ☽ □ ♀
16 35 ☉ ⊻ ♀
21 29 ☽ ∥ ♄

16 Sunday
02 14 ☽ ∠ ♅
03 43 ☽ # ♃
04 30 ☽ □ ♂
16 44 ☽ ⊻ ♆
17 36 ☽ □ ♇
18 34 ☉ ± ♄
20 26 ☽ ⊥ ♅
20 28 ☽ ⚼ ☉
23 47 ♀ St R

17 Monday
04 54 ☽ ⊥ ♆
07 48 ☽ △ ☿
08 34 ☽ ⊻ ♅
12 05 ☽ ☍ ♃
14 27 ☽ ☌ ♄
15 15 ☽ # ♂
16 25 ☽ ± ☉
19 59 ☿ ⊼ ♅
23 07 ☽ ∠ ♆

18 Tuesday
00 28 ☽ △ ♀
02 03 ☽ # ♀
05 40 ☽ ⊼ ☉
15 50 ☽ ⚼ ☿
18 27 ☽ ∥ ♅
20 05 ☿ ∥ ♇
21 26 ☽ △ ♂

19 Wednesday
04 27 ☉ ∥ ☿
05 15 ☽ ⚹ ♆
06 14 ☽ △ ♇
06 26 ☽ ⚼ ♀
08 24 ♆ St D
08 53 ☽ # ☿
10 39 ☽ ± ☿
10 55 ☽ # ☉
14 49 ☽ # ♇
20 35 ☽ ☌ ♅
23 00 ☽ ⊼ ☿

20 Thursday
00 09 ☽ ± ♀
01 13 ☽ ⊼ ♃
02 17 ☽ ⊻ ♄
05 05 ☽ Q ♆
05 20 ☽ ⚼ ♂
09 45 ☽ # ♇
11 51 ☽ ⊼ ♀
12 02 ☽ ⚼ ♇
12 46 ☽ # ☉
13 13 ☽ ± ♃
14 03 ☽ ⊥ ♄
22 46 ☽ ☍ ☉

21 Friday
00 13 ☽ ± ♂
03 05 ☽ # ☿
05 33 ☽ ∥ ♅
05 40 ☽ ± ♇
07 05 ☽ ⚼ ♃
07 33 ☽ ∠ ♄
12 40 ☽ ⊼ ♂
15 06 ☿ # ♅
16 14 ☽ □ ♆
17 19 ☽ ⊼ ♇

22 Saturday
05 50 ☿ St R
05 54 ☽ # ♀
06 52 ☽ ⊻ ♅
07 03 ♃ ☍ ♄
10 14 ☽ ☍ ☿
12 19 ☽ ⚹ ♄
12 25 ☽ △ ♃
16 02 ☽ # ♂
18 16 ☽ ⊥ ♅
20 54 ☽ ☍ ♀

23 Sunday
10 56 ☽ Q ♄
11 12 ☽ ∠ ♅
13 16 ☽ ⊼ ☉
14 01 ☽ # ♃
21 04 ☉ ⊥ ♂
21 25 ♂ ⚹ ♆
22 40 ☽ ∥ ♄

24 Monday
01 06 ☽ △ ♆
01 21 ☽ ☍ ♂
01 26 ☽ ± ☉
02 15 ☽ ☍ ♇
12 03 ☉ ⊥ ♇
14 59 ☽ ⚹ ♅
16 41 ♂ ☌ ♇
17 20 ☽ ⊼ ☿
19 27 ☽ ⚼ ☉
20 12 ☽ □ ♄
21 18 ☽ □ ♃

25 Tuesday
03 27 ☽ ⊼ ♀
03 54 ☽ ± ☿
04 36 ☉ ⚼ ♄
04 42 ☽ ⚼ ♆
12 49 ☽ Q ♅
14 04 ☽ ± ♀
19 28 ☽ ⚼ ☿
21 00 ☽ ± ♆

26 Wednesday
00 09 ☉ ∠ ♃
00 55 ☽ △ ☉
05 49 ☽ ⚼ ♀
06 19 ☽ Q ♃
07 45 ☽ ⊼ ♆
08 56 ☽ ⊼ ♇
11 21 ☽ ⊼ ♂
15 57 ☽ ∥ ♆
17 57 ☿ ⊼ ♅
19 37 ☽ ± ♇
20 44 ☽ △ ☿
20 54 ☽ □ ♅
22 38 ☽ ± ♂

27 Thursday
01 54 ☽ △ ♄
03 52 ☽ ⚹ ♃
07 36 ☽ △ ♀
09 01 ☽ # ♆
11 28 ☽ ⚼ ♇
13 36 ♀ ± ♅
15 22 ☽ ⚼ ♂

28 Friday
03 57 ☽ ⚼ ♄
06 19 ☽ ∠ ♃
09 49 ☽ □ ☉
12 15 ☽ ☍ ♆
13 28 ☽ △ ♇
18 47 ☽ △ ♂
19 22 ☽ ± ♄
21 07 ☽ □ ☿
21 56 ☽ ⊥ ♃

29 Saturday
00 44 ☽ △ ♅
05 32 ☽ ⊼ ♄
08 15 ☽ ⊻ ♃
09 36 ☽ □ ♀
12 37 ☿ ⊻ ♂
18 47 ☽ Q ☉
19 17 ☉ □ ♆
22 22 ☽ # ♄

30 Sunday
01 04 ☽ Q ☿
01 58 ☽ ⚼ ♅
08 15 ☽ ∥ ♃
13 13 ☉ ⊻ ♇
13 17 ♀ ⚹ ♃
14 18 ☽ ∥ ♂
14 45 ☽ Q ♀
14 53 ☽ ⊼ ♆
16 08 ☽ □ ♇
16 21 ☽ ⚹ ☉
16 56 ☽ ± ♅
19 23 ☽ ⚹ ☿
22 09 ☽ ∥ ♀

31 Monday
00 05 ☽ □ ♂
00 50 ☽ ± ♆
02 51 ☽ ⊼ ♅
07 32 ☽ ☍ ♄
08 55 ☽ ∥ ☿
09 55 ☽ ⚹ ♀
10 55 ☽ ☌ ♃
12 28 ☉ ☌ ☿
15 42 ☽ ⚼ ♆
18 02 ☽ ∠ ☿
19 01 ☽ ∠ ☉
21 56 ☽ Q ♇

JANUARY 2030

LONGITUDES

Date	Sidereal time h m s	Sun ☉ ° ' "	Moon ☽ ° ' "	Moon ☽ 24.00 ° ' "	Mercury ☿ ° '	Venus ♀ ° '	Mars ♂ ° '	Jupiter ♃ ° '	Saturn ♄ ° '	Uranus ♅ ° '	Neptune ♆ ° '	Pluto ♇ ° '
01	18 44 44	11 ♑ 06 44	05 ♐ 43 49	13 ♐ 02 37	08 ♑ 47	19 ♑ 16	15 ♒ 06	20 ♏ 40	18 ♉ 23	15 ♊ 32	08 ♈ 23	09 ♒ 11
02	18 48 41	12 07 55	20 ♐ 20 57	27 ♐ 38 06	07 R 28	18 R 42	15 53	20 50	18 R 21	15 R 30	08 23	09 12
03	18 52 37	13 09 05	04 ♑ 53 14	12 ♑ 05 35	06 12	18 07	16 40	21 01	18 19	15 28	08 24	09 14
04	18 56 34	14 10 16	19 ♑ 14 27	26 ♑ 19 09	05 04	17 31	17 27	21 11	18 18	15 26	08 24	09 16
05	19 00 30	15 11 27	03 ♒ 19 08	10 ♒ 13 55	04 04	16 55	18 14	21 21	18 16	15 24	08 25	09 18
06	19 04 27	16 12 38	17 ♒ 03 11	23 ♒ 46 41	03 13	16 18	19 02	21 31	18 14	15 22	08 26	09 19
07	19 08 24	17 13 48	00 ♓ 24 22	06 ♓ 56 14	02 32	15 41	19 49	21 41	18 13	15 20	08 26	09 21
08	19 12 20	18 14 58	13 ♓ 22 26	19 ♓ 43 15	02 01	15 05	20 36	21 51	18 12	15 17	08 27	09 23
09	19 16 17	19 16 07	25 ♓ 59 01	02 ♈ 10 09	01 40	14 29	21 23	22 01	18 11	15 15	08 28	09 25
10	19 20 13	20 17 16	08 ♈ 17 09	14 ♈ 20 33	01 29	13 54	22 11	22 10	18 10	15 14	08 28	09 26
11	19 24 10	21 18 24	20 ♈ 20 57	26 ♈ 18 57	01 D 27	13 20	22 58	22 20	18 09	15 12	08 29	09 28
12	19 28 06	22 19 32	02 ♉ 15 11	08 ♉ 10 16	01 33	12 47	23 45	22 30	18 08	15 10	08 30	09 30
13	19 32 03	23 20 39	14 ♉ 04 50	19 ♉ 59 31	01 47	12 15	24 32	22 39	18 07	15 08	08 31	09 32
14	19 35 59	24 21 46	25 ♉ 54 53	01 ♊ 51 32	02 09	11 45	25 20	22 48	18 07	15 06	08 32	09 34
15	19 39 56	25 22 52	07 ♊ 50 00	13 ♊ 50 45	02 37	11 17	26 07	22 57	18 06	15 04	08 33	09 35
16	19 43 53	26 23 58	19 ♊ 54 16	26 ♊ 00 55	03 11	10 51	26 54	23 06	18 06	15 02	08 34	09 37
17	19 47 49	27 25 03	02 ♋ 11 03	08 ♋ 24 55	03 50	10 27	27 41	23 15	18 06	15 01	08 35	09 39
18	19 51 46	28 26 07	14 ♋ 42 44	21 ♋ 04 37	04 35	10 05	28 29	23 24	18 05	14 59	08 36	09 41
19	19 55 42	29 ♑ 27 10	27 ♋ 30 37	04 ♌ 00 43	05 23	09 46	29 ♒ 16	23 33	18 D 05	14 57	08 37	09 43
20	19 59 39	00 ♒ 28 13	10 ♌ 34 49	17 ♌ 12 48	06 16	09 29	00 ♓ 03	23 41	18 06	14 56	08 38	09 45
21	20 03 35	01 29 16	23 ♌ 54 27	00 ♍ 39 29	07 12	09 14	00 50	23 50	18 06	14 54	08 39	09 46
22	20 07 32	02 30 17	07 ♍ 27 40	14 ♍ 18 40	08 11	09 02	01 37	23 58	18 06	14 53	08 40	09 48
23	20 11 28	03 31 18	21 ♍ 12 10	28 ♍ 07 53	09 13	08 52	02 25	24 06	18 06	14 51	08 41	09 50
24	20 15 25	04 32 19	05 ♎ 05 30	12 ♎ 04 46	10 18	08 45	03 12	24 14	18 07	14 50	08 42	09 52
25	20 19 22	05 33 19	19 ♎ 05 25	26 ♎ 07 14	11 25	08 40	03 59	24 22	18 08	14 49	08 44	09 54
26	20 23 18	06 34 19	03 ♏ 10 01	10 ♏ 13 36	12 34	08 38	04 46	24 30	18 08	14 47	08 45	09 56
27	20 27 15	07 35 18	17 ♏ 17 48	24 ♏ 22 28	13 46	08 D 38	05 33	24 37	18 09	14 46	08 46	09 58
28	20 31 11	08 36 17	01 ♐ 27 26	08 ♐ 32 28	14 59	08 41	06 21	24 45	18 10	14 45	08 47	09 59
29	20 35 08	09 37 15	15 ♐ 37 23	22 ♐ 41 53	16 14	08 46	07 08	24 52	18 11	14 44	08 49	10 01
30	20 39 04	10 38 12	29 ♐ 45 40	06 ♑ 48 22	17 30	08 53	07 55	25 00	18 13	14 43	08 50	10 03
31	20 43 01	11 ♒ 39 09	13 ♑ 49 37	20 ♑ 48 57	18 ♑ 48	09 ♑ 03	08 ♓ 42	25 ♏ 07	18 ♉ 14	14 ♊ 41	08 ♈ 51	10 ♒ 05

Date	Moon True ☊ ° '	Moon Mean ☊ ° '	Moon ☽ Latitude ° '
01	25 ♐ 31	24 ♐ 47	01 S 47
02	25 D 32	24 43	00 S 29
03	25 R 32	24 40	00 N 52
04	25 31	24 37	02 07
05	25 28	24 34	03 14
06	25 25	24 31	04 07
07	25 22	24 28	04 46
08	25 19	24 24	05 08
09	25 16	24 21	05 15
10	25 14	24 18	05 07
11	25 13	24 15	04 45
12	25 D 14	24 12	04 12
13	25 15	24 09	03 27
14	25 17	24 05	02 35
15	25 19	24 02	01 35
16	25 20	23 59	00 N 30
17	25 R 20	23 56	00 S 37
18	25 18	23 53	01 44
19	25 15	23 49	02 47
20	25 10	23 46	03 42
21	25 04	23 43	04 26
22	24 57	23 40	04 56
23	24 51	23 37	05 10
24	24 46	23 34	05 05
25	24 43	23 30	04 43
26	24 42	23 27	04 03
27	24 D 42	23 24	03 09
28	24 44	23 21	02 03
29	24 45	23 18	00 S 49
30	24 R 45	23 15	00 N 27
31	24 ♐ 43	23 ♐ 11	01 N 41

DECLINATIONS

Date	Sun ☉ ° '	Moon ☽ ° '	Mercury ☿ ° '	Venus ♀ ° '	Mars ♂ ° '	Jupiter ♃ ° '	Saturn ♄ ° '	Uranus ♅ ° '	Neptune ♆ ° '	Pluto ♇ ° '
01	22 S 58	23 S 01	20 S 24	18 S 48	17 S 25	16 S 57	15 N 04	22 N 41	01 N 57	23 S 24
02	22 53	23 34	20 18	18 37	17 11	17 00	15 03	22 41	01 57	23 24
03	22 47	22 29	20 13	18 27	16 56	17 02	15 03	22 41	01 57	23 23
04	22 41	19 57	20 10	18 17	16 41	17 05	15 03	22 41	01 57	23 23
05	22 34	16 16	20 08	18 08	16 26	17 07	15 03	22 40	01 58	23 23
06	22 27	11 47	20 08	17 59	16 11	17 10	15 03	22 40	01 58	23 22
07	22 19	06 52	20 09	17 50	15 56	17 13	15 02	22 40	01 58	23 22
08	22 11	01 S 47	20 12	17 41	15 41	17 15	15 02	22 40	01 59	23 21
09	22 03	03 N 13	20 17	17 33	15 25	17 17	15 02	22 40	01 59	23 21
10	21 54	07 59	20 22	17 25	15 09	17 20	15 02	22 39	01 59	23 20
11	21 45	12 21	20 29	17 18	14 54	17 22	15 02	22 39	02 00	23 20
12	21 35	16 11	20 36	17 11	14 37	17 24	15 02	22 39	02 00	23 19
13	21 25	19 22	20 44	17 05	14 21	17 27	15 02	22 39	02 00	23 19
14	21 14	21 44	20 53	16 59	14 05	17 29	15 03	22 38	02 01	23 18
15	21 04	23 10	21 01	16 54	13 49	17 31	15 03	22 38	02 01	23 18
16	20 52	23 33	21 10	16 50	13 32	17 33	15 03	22 38	02 02	23 17
17	20 40	22 48	21 19	16 45	13 15	17 35	15 03	22 38	02 02	23 17
18	20 28	20 54	21 27	16 42	12 59	17 38	15 03	22 38	02 02	23 16
19	20 16	17 56	21 35	16 39	12 42	17 40	15 04	22 38	02 03	23 16
20	20 03	14 01	21 43	16 36	12 25	17 42	15 04	22 37	02 03	23 16
21	19 50	09 22	21 50	16 34	12 07	17 44	15 04	22 37	02 04	23 15
22	19 36	04 N 11	21 56	16 32	11 50	17 46	15 05	22 37	02 04	23 15
23	19 22	01 S 15	22 02	16 31	11 33	17 47	15 05	22 37	02 05	23 14
24	19 07	06 42	22 07	16 30	11 15	17 49	15 06	22 37	02 05	23 14
25	18 53	11 50	22 11	16 30	10 58	17 51	15 06	22 37	02 06	23 13
26	18 38	16 22	22 14	16 30	10 40	17 53	15 06	22 37	02 06	23 13
27	18 22	20 01	22 16	16 30	10 22	17 55	15 07	22 36	02 07	23 12
28	18 06	22 27	22 18	16 31	10 04	17 56	15 08	22 36	02 07	23 12
29	17 50	23 29	22 18	16 32	09 46	17 58	15 08	22 36	02 08	23 11
30	17 34	22 59	22 17	16 33	09 28	18 00	15 09	22 36	02 09	23 11
31	17 S 17	21 S 02	22 S 15	16 S 35	09 S 10	18 S 01	15 N 09	22 N 36	02 N 09	23 S 11

ZODIAC SIGN ENTRIES

Date	h	m	Planets
01	02	36	☽ ♐
03	03	54	☽ ♑
05	06	18	☽ ♒
07	11	16	☽ ♓
09	19	47	☽ ♈
12	07	26	☽ ♉
14	20	15	☽ ♊
17	07	46	☽ ♋
19	16	37	☽ ♌
20	00	54	☉ ♒
20	10	27	♂ ♓
21	22	50	☽ ♍
24	03	14	☽ ♎
26	06	37	☽ ♏
28	09	32	☽ ♐
30	12	24	☽ ♑

LATITUDES

Date	Mercury ☿ ° '	Venus ♀ ° '	Mars ♂ ° '	Jupiter ♃ ° '	Saturn ♄ ° '	Uranus ♅ ° '	Neptune ♆ ° '	Pluto ♇ ° '
01	02 N 45	03 N 17	01 S 10	01 N 00	02 S 20	00 N 02	01 S 30	05 S 39
04	03 11	04 02	01 09	01 00	02 19	00 02	01 30	05 39
07	03 15	04 43	01 07	01 01	02 18	00 02	01 30	05 40
10	03 03	05 19	01 06	01 01	02 17	00 02	01 30	05 40
13	02 41	05 48	01 05	01 01	02 16	00 02	01 29	05 40
16	02 14	06 11	01 03	01 01	02 16	00 02	01 29	05 40
19	01 45	06 27	01 02	01 02	02 15	00 02	01 29	05 40
22	01 15	06 37	01 00	01 02	02 14	00 02	01 29	05 40
25	00 46	06 40	00 59	01 02	02 13	00 02	01 29	05 40
28	00 N 18	06 39	00 57	01 03	02 12	00 02	01 29	05 40
31	00 S 08	06 N 34	00 S 56	01 N 03	02 S 11	00 N 02	01 S 29	05 S 41

DATA

Julian Date	2462503
Delta T	+74 seconds
Ayanamsa	24° 16' 52"
Synetic vernal point	04° ♓ 50' 07"
True obliquity of ecliptic	23° 26' 06"

MOON'S PHASES, APSIDES AND POSITIONS ☽

Date	h	m	Phase	Longitude ° '	Eclipse Indicator
04	02	49	●	13 ♑ 47	
11	14	06	☽	21 ♈ 24	
19	15	54	○	29 ♋ 37	
26	18	14	☾	06 ♏ 50	

Day	h	m		
01	15	24	Perigee	
13	08	49	Apogee	
28	15	48	Perigee	
02	07	57	Max dec	23° S 35'
08	20	30	ON	
16	08	11	Max dec	23° N 34'
23	06	30	OS	
29	16	13	Max dec	23° S 30'

LONGITUDES

Date	Chiron ⚷ ° '	Ceres ⚳ ° '	Pallas ⚴ ° '	Juno ⚵ ° '	Vesta ⚶ ° '	Black Moon Lilith ⚸ ° '
01	08 ♉ 01	00 ♓ 26	29 ♑ 20	15 ♐ 36	23 ♒ 46	14 ♉ 14
11	07 ♉ 55	03 ♓ 55	02 ♒ 53	18 ♐ 53	28 ♒ 29	15 ♉ 21
21	07 ♉ 54	07 ♓ 31	06 ♒ 26	22 ♐ 05	03 ♓ 14	16 ♉ 27
31	08 ♉ 00	11 ♓ 14	09 ♒ 57	25 ♐ 10	08 ♓ 02	17 ♉ 34

ASPECTARIAN

h m Aspects

01 Tuesday
05 03 ☿ ⊻ ♇
06 45 ☽ # ♅
07 26 ☽ Q ♂
07 34 ☽ ⊥ ☿
08 27 ☿ ⊥ ♂
09 41 ☽ ∠ ♀
10 55 ☽ ⊥ ☉
11 11 ☽ ∥ ☉
16 21 ☽ △ ♆
16 35 ☽ ⊻ ☿
17 40 ☽ ✶ ♇
19 02 ☿ □ ♆
20 42 ☽ ∥ ♇
21 30 ☽ ⊻ ☉
23 55 ☽ ⊥ ♀

02 Wednesday
00 58 ♂ △ ♅
04 03 ☽ ☍ ♅
04 14 ☽ ✶ ♂
08 43 ☽ ⊼ ♄
09 23 ☽ ⊻ ♀
12 49 ☽ ⊻ ♃
18 22 ☽ ∠ ♇
18 34 ☽ ± ♄
19 27 ☽ ∥ ♇
22 49 ☽ ⊥ ♃

03 Thursday
03 11 ♀ △ ♄
03 16 ♂ ∥ ♃
06 22 ☽ ∠ ♂
07 32 ☽ ∥ ☉
09 15 ☽ ⊥ ♇
09 22 ☽ # ♅
09 24 ☽ bQ ♄
13 53 ☽ ∠ ♃
14 01 ☽ ☌ ☿
15 25 ☿ ∠ ♃
17 50 ☽ □ ♆
19 15 ☽ ⊻ ♇
22 11 ☽ ⊥ ♂

04 Friday
02 49 ☽ ☌ ☉
05 36 ☽ ⊼ ♅
08 49 ☽ ⊻ ♂
09 13 ☽ ☌ ♀
10 24 ☽ △ ♄
10 24 ☽ ∥ ☿
13 03 ♀ ⊻ ♂
13 08 ☉ # ♅
15 19 ☽ ✶ ♃
15 41 ☽ ± ♅

05 Saturday
00 09 ☽ ∥ ♀
00 10 ☽ Q ♆
06 59 ☽ ∥ ♃
06 59 ☽ bQ ♅
10 56 ☽ ∥ ♂
12 03 ☽ Q ♃
12 43 ♂ □ ♄
13 12 ☽ ⊻ ☿
16 38 ☉ ⊼ ♅
18 54 ☽ # ♄
20 50 ☽ ✶ ♆
22 23 ☽ ☌ ♇
22 59 ☽ ⊥ ☿
23 45 ☿ ∠ ♂

06 Sunday
08 46 ☿ ⊥ ♇
09 01 ☽ △ ♅
10 23 ☽ ⊻ ☉
10 43 ☽ ⊻ ♀
11 07 ☿ bQ ♄
13 18 ☉ ☌ ♀
13 57 ☽ ∠ ☿
14 06 ☽ □ ♄
15 43 ☽ ☌ ♂
20 03 ☽ □ ♃
20 56 ☽ ⊥ ♀
21 57 ☽ ⊥ ☉
23 23 ☽ ∠ ♆

07 Monday
12 29 ☽ ∠ ♀
15 37 ☽ ∠ ☉
15 43 ☽ ✶ ☿
15 43 ☽ ⊥ ♆
22 39 ☽ Q ♄

08 Tuesday
02 48 ☽ ⊻ ♆
03 10 ♀ ⊼ ♅
04 31 ☽ ⊻ ♇
10 48 ☉ △ ♄
11 07 ☽ # ♆
13 10 ☽ Q ☿
15 04 ☽ ✶ ♀
15 36 ☽ □ ♅
15 47 ☽ ⊥ ♇
20 16 ♀ ⊥ ♂
21 05 ☽ ✶ ♄
22 01 ☽ ✶ ☉

09 Wednesday
02 36 ☽ ⊻ ♂
04 17 ☽ △ ♃
05 59 ☽ ∥ ♆
08 58 ☽ ∠ ♇
12 55 ☽ Q ♀
14 54 ☽ ⊥ ♂
22 49 ☽ □ ☿
23 10 ☽ Q ☉

10 Thursday
01 57 ☽ ∠ ♄
02 05 ☽ Q ♅
09 40 ☽ ∠ ♂
09 47 ☽ bQ ♃
11 54 ♂ □ ♃
12 22 ☽ ☌ ♆
14 17 ☽ ✶ ♇
19 39 ☽ ⊥ ♄
22 37 ☽ □ ♀
22 42 ♂ # ♄

11 Friday
01 43 ☽ ✶ ♅
01 54 ♀ ∥ ♃
03 52 ☽ ± ♃
05 46 ☿ St D
07 36 ☽ ⊻ ♄
09 24 ☉ ± ♅
14 06 ☽ □ ☉
14 15 ☽ Q ♇
16 02 ☽ ⊼ ♃
17 37 ☽ ✶ ♂

12 Saturday
02 27 ☽ # ♂
04 11 ♂ ∠ ♆
04 25 ☽ ∥ ♄
07 47 ☽ ∠ ♅
10 33 ☽ △ ☿
16 38 ☉ ✶ ♃
18 49 ☽ # ♀
19 36 ☽ Q ♂
20 43 ☽ # ♃

13 Sunday
00 41 ☽ ⊻ ♆
01 58 ☽ ⊥ ♅
02 44 ☽ □ ♇
08 27 ☽ △ ♀
12 53 ☽ ⊥ ♆
14 08 ☽ ⊻ ♅
17 39 ☽ bQ ☿
20 12 ☽ ☌ ♄

14 Monday
01 37 ☽ # ☿
05 37 ☽ ☍ ♃
06 39 ☽ # ☉
07 10 ☽ ∠ ♆
08 34 ☽ △ ☉
10 44 ☽ □ ♂
12 29 ☽ ± ☿
13 38 ☽ bQ ♀

15 Tuesday
01 04 ☽ ⊼ ☿
01 06 ☽ ∥ ♅
07 05 ☽ ± ♀
13 26 ☽ ✶ ♆
14 50 ☉ ∥ ☿
15 24 ♀ ∠ ♂
15 31 ☽ △ ♇
15 42 ☽ # ♇
17 34 ☽ bQ ☉
18 40 ☽ ⊼ ♀

16 Wednesday
02 24 ☽ ☌ ♅
08 26 ☽ ⊻ ♄
08 43 ☿ bQ ♄
13 04 ☽ ± ☉
13 18 ☽ Q ♆
15 51 ☉ Q ♆
18 23 ☽ ⊼ ♃
20 14 ☽ ⊥ ♄
21 18 ☽ bQ ♇

17 Thursday
00 51 ☽ # ♇
01 54 ☽ ⊼ ☉
02 40 ☽ △ ♂
05 11 ☿ ⊥ ♇
06 15 ☽ ± ♃
13 45 ☽ ∠ ♄
14 47 ☽ ∥ ♅
14 51 ☽ ± ♇
15 23 ☽ ☍ ☿
23 50 ☽ bQ ♃

18 Friday
00 19 ☽ □ ♆
02 24 ☽ ⊼ ♇
03 27 ☽ ☍ ♀
06 39 ☽ # ☿
09 30 ☽ bQ ♂
12 31 ☽ ⊻ ♅
16 18 ☉ ⊻ ♂
16 24 ☽ # ☉
18 23 ☽ ✶ ♄
23 48 ☽ ⊥ ♅

19 Saturday
03 35 ☽ ± ♂
03 55 ♄ St D
04 32 ☽ △ ♃
13 50 ☽ # ♃
15 28 ☽ ⊼ ♂
15 45 ♀ ⊻ ♇
15 54 ☽ ☍ ☉
16 31 ☽ ∠ ♅
16 47 ☽ Q ♄
20 37 ☽ # ♀
23 37 ☉ bQ ♅

20 Sunday
00 44 ♃ bQ ♆
03 33 ☽ ⊼ ☿
06 04 ☽ ∥ ♄
08 26 ☽ △ ♆
10 02 ☽ ⊼ ♀
10 28 ☽ ☍ ♇
15 17 ☽ ± ☿
19 52 ☽ ✶ ♅
20 42 ☽ ± ♀
21 16 ☽ # ♂

21 Monday
01 35 ☽ □ ♄
08 42 ☽ bQ ☿
11 32 ☽ bQ ♆
11 51 ☽ □ ♃
12 34 ☽ bQ ♀
17 20 ☽ Q ♅

22 Tuesday
01 05 ☽ ☍ ♂
02 34 ☽ ⊼ ☉
03 32 ☽ ± ♆
13 22 ☽ △ ☿
13 59 ☽ ± ☉
14 07 ☽ ⊼ ♆
14 43 ☽ △ ♀
16 07 ☽ ⊼ ♇
16 51 ♀ ∠ ♃
19 59 ☽ Q ♃
21 23 ☽ ∥ ♆
23 31 ☿ □ ♆

23 Wednesday
00 58 ☽ □ ♅
02 38 ☽ ± ♇
04 57 ☿ ☌ ♀
06 37 ☽ △ ♄
06 58 ☽ bQ ☉
08 55 ☿ ∠ ♃
15 36 ☽ # ♆
17 05 ☽ ✶ ♃
18 19 ☽ bQ ♇
20 35 ♂ ⊥ ♆

24 Thursday
02 13 ☿ ⊻ ♇
08 33 ☽ ⊼ ♂
08 36 ☽ bQ ♄
10 58 ☽ △ ☉
18 13 ☽ ☍ ♆
18 14 ☽ □ ♀
19 11 ☽ ∠ ♃
19 29 ☽ ± ♂
20 13 ☽ △ ♇
21 00 ♀ □ ♆
21 43 ☽ □ ☿

25 Friday
00 04 ☽ ± ♄
04 41 ☽ △ ♅
08 00 ☽ ∥ ♂
10 21 ☽ ⊼ ♄
10 45 ☽ ⊥ ♃
11 48 ☽ bQ ♂
21 06 ☽ ⊻ ♃

26 Saturday
00 53 ☽ Q ♀
04 52 ☽ # ♄
06 16 ☽ bQ ♅
07 11 ☽ Q ☿
09 59 ☉ Q ♃
12 42 ☽ ∥ ♀
14 53 ☽ △ ♂
18 14 ☽ □ ☉
21 12 ☽ ∥ ♃
21 17 ☽ ✶ ♀
21 30 ☽ ⊼ ♆
21 33 ☽ ± ♅
21 33 ♀ St D
23 31 ☽ □ ♇

27 Sunday
01 04 ☽ ∥ ☉
05 27 ☽ ✶ ☿
07 42 ☽ ± ♆
07 43 ☽ ⊼ ♅
13 27 ☽ ☍ ♄
22 46 ☽ ∠ ♀
22 59 ☽ bQ ♆

28 Monday
00 32 ☽ ☌ ♃
03 09 ☽ Q ☉
06 07 ☽ Q ♇
06 41 ♂ Q ♄
07 31 ☿ ⊼ ♅
09 16 ☽ ∠ ☿
09 54 ☽ ∥ ☿
13 54 ☉ ⊻ ♀
14 05 ☽ ⊥ ♀
14 08 ☽ # ♅
16 28 ☉ ✶ ♆
20 46 ☽ □ ♂

29 Tuesday
00 18 ☽ ⊻ ♀
00 26 ☽ △ ♆
01 02 ☽ ∥ ♇
01 03 ☽ ✶ ☉
01 40 ☉ ∥ ♃
01 58 ☽ ⊥ ☿
02 29 ☽ ✶ ♇
10 29 ☽ ☍ ♅
13 07 ☽ ⊻ ☿
16 22 ☽ ⊼ ♄
21 47 ☉ ☌ ♇

30 Wednesday
00 32 ♀ □ ♆
02 33 ☽ ± ♄
03 50 ☽ ⊻ ♃
03 59 ☽ ∠ ♇
04 27 ☽ ∠ ☉
05 05 ☽ Q ♂
07 39 ☽ ∥ ♇
14 07 ☽ ⊥ ♃
17 53 ☽ bQ ♄
18 25 ☽ # ♅
19 19 ☽ ⊥ ♇
20 57 ☽ ⊥ ☉
22 56 ☽ ∥ ☿

31 Thursday
01 28 ☿ △ ♄
02 43 ☽ ✶ ♂
03 29 ☽ □ ♆
03 43 ☽ ☌ ♀
05 35 ☽ ⊻ ♇
05 36 ☽ ∠ ♃
07 59 ☽ ⊻ ☉
13 29 ☽ ⊼ ♅
17 00 ♂ ⊻ ♆
19 34 ☽ △ ♄
21 24 ☽ ☌ ☿
23 46 ☽ ± ♅

All ephemeris data is given at 12.00 UT and the Moon's longitude is additionally given for 24.00 UT

FEBRUARY 2030

LONGITUDES

Date	Sidereal time h m s	Sun ☉ ° ' "	Moon ☽ ° ' "	Moon ☽ 24.00 ° ' "	Mercury ☿ ° '	Venus ♀ ° '	Mars ♂ ° '	Jupiter ♃ ° '	Saturn ♄ ° '	Uranus ♅ ° '	Neptune ♆ ° '	Pluto ♇ ° '
01	20 46 57	12 ♒ 40 05	27 ♑ 45 57	04 ♒ 40 08	20 ♑ 07	09 ♑ 15	09 ♓ 29	25 ♏ 14	18 ♉ 15	14 ♊ 40	08 ♈ 53	10 ♒ 07
02	20 50 54	13 41 00	11 ♒ 31 04	18 ♒ 18 19	21 27	09 29	10 16	25 21	18 17	14 R 39	08 54	10 09
03	20 54 51	14 41 53	25 ♒ 01 32	01 ♓ 40 24	22 49	09 45	11 03	25 27	18 19	14 38	08 56	10 11
04	20 58 47	15 42 46	08 ♓ 14 42	14 ♓ 44 18	24 11	10 03	11 50	25 34	18 20	14 38	08 57	10 13
05	21 02 44	16 43 37	21 ♓ 09 09	27 ♓ 29 18	25 35	10 23	12 37	25 40	18 22	14 37	08 59	10 14
06	21 06 40	17 44 27	03 ♈ 44 56	09 ♈ 56 16	27 00	10 45	13 24	25 46	18 24	14 36	09 00	10 16
07	21 10 37	18 45 15	16 ♈ 03 39	22 ♈ 07 30	28 26	11 08	14 11	25 52	18 26	14 35	09 02	10 18
08	21 14 33	19 46 02	28 ♈ 08 17	04 ♉ 06 33	29 ♑ 53	11 34	14 58	25 58	18 29	14 35	09 04	10 20
09	21 18 30	20 46 48	10 ♉ 02 54	15 ♉ 57 56	01 ♒ 20	12 01	15 45	26 04	18 31	14 34	09 05	10 22
10	21 22 26	21 47 32	21 ♉ 52 19	27 ♉ 46 44	02 49	12 30	16 32	26 10	18 33	14 33	09 07	10 24
11	21 26 23	22 48 14	03 ♊ 41 51	09 ♊ 38 22	04 18	13 00	17 19	26 15	18 36	14 33	09 09	10 25
12	21 30 20	23 48 55	15 ♊ 36 56	21 ♊ 38 12	05 49	13 32	18 06	26 20	18 39	14 32	09 10	10 27
13	21 34 16	24 49 35	27 ♊ 42 47	03 ♋ 51 15	07 20	14 05	18 53	26 25	18 41	14 32	09 12	10 29
14	21 38 13	25 50 12	10 ♋ 04 06	16 ♋ 21 46	08 52	14 40	19 40	26 30	18 44	14 32	09 14	10 31
15	21 42 09	26 50 48	22 ♋ 44 34	29 ♋ 12 46	10 25	15 16	20 27	26 35	18 47	14 31	09 16	10 33
16	21 46 06	27 51 23	05 ♌ 46 29	12 ♌ 25 43	11 59	15 54	21 13	26 40	18 50	14 31	09 17	10 35
17	21 50 02	28 51 56	19 ♌ 10 20	26 ♌ 00 05	13 34	16 32	22 00	26 44	18 53	14 31	09 19	10 36
18	21 53 59	29 ♒ 52 27	02 ♍ 54 36	09 ♍ 53 22	15 10	17 12	22 47	26 49	18 56	14 31	09 21	10 38
19	21 57 55	00 ♓ 52 56	16 ♍ 55 49	24 ♍ 01 17	16 46	17 53	23 34	26 53	19 00	14 31	09 23	10 40
20	22 01 52	01 53 24	01 ♎ 09 03	08 ♎ 18 26	18 24	18 35	24 20	26 57	19 03	14 30	09 25	10 42
21	22 05 49	02 53 51	15 ♎ 28 43	22 ♎ 39 16	20 02	19 18	25 07	27 00	19 07	14 D 30	09 27	10 43
22	22 09 45	03 54 16	29 ♎ 49 28	06 ♏ 58 51	21 42	20 02	25 54	27 04	19 10	14 31	09 29	10 45
23	22 13 42	04 54 40	14 ♏ 07 00	21 ♏ 13 35	23 22	20 47	26 40	27 07	19 14	14 31	09 30	10 47
24	22 17 38	05 55 03	28 ♏ 18 23	05 ♐ 21 16	25 03	21 33	27 27	27 11	19 18	14 31	09 32	10 49
25	22 21 35	06 55 25	12 ♐ 22 07	19 ♐ 20 54	26 46	22 20	28 13	27 14	19 22	14 31	09 34	10 50
26	22 25 31	07 55 45	26 ♐ 17 38	03 ♑ 12 17	28 ♒ 29	23 08	29 00	27 17	19 26	14 31	09 36	10 52
27	22 29 28	08 56 04	10 ♑ 04 52	16 ♑ 55 22	00 ♓ 13	23 57	29 ♓ 46	27 19	19 30	14 32	09 38	10 54
28	22 33 24	09 ♓ 56 21	23 ♑ 43 42	00 ♒ 29 50	01 ♓ 58	24 ♑ 46	00 ♈ 33	27 ♏ 22	19 ♉ 34	14 ♊ 32	09 ♈ 40	10 ♒ 55

Date	Moon True ☊ ° '	Moon Mean ☊ ° '	Moon ☽ Latitude ° '
01	24 ♐ 39	23 ♐ 08	02 N 48
02	24 R 33	23 05	03 45
03	24 25	23 02	04 27
04	24 15	22 59	04 55
05	24 06	22 55	05 06
06	23 57	22 52	05 02
07	23 50	22 49	04 44
08	23 45	22 46	04 13
09	23 42	22 43	03 32
10	23 D 41	22 40	02 42
11	23 42	22 36	01 45
12	23 42	22 33	00 N 43
13	23 R 43	22 30	00 S 21
14	23 41	22 27	01 26
15	23 37	22 24	02 28
16	23 31	22 20	03 25
17	23 22	22 17	04 11
18	23 12	22 14	04 44
19	23 00	22 11	05 01
20	22 50	22 08	04 59
21	22 40	22 05	04 39
22	22 34	22 01	04 01
23	22 30	21 58	03 09
24	22 28	21 55	02 04
25	22 D 28	21 52	00 S 53
26	22 R 28	21 49	00 N 20
27	22 27	21 46	01 32
28	22 ♐ 24	21 ♐ 42	02 N 37

DECLINATIONS

Date	Sun ☉ ° '	Moon ☽ ° '	Mercury ☿ ° '	Venus ♀ ° '	Mars ♂ ° '	Jupiter ♃ ° '	Saturn ♄ ° '	Uranus ♅ ° '	Neptune ♆ ° '	Pluto ♇ ° '
01	17 S 00	17 S 51	22 S 12	16 S 37	08 S 52	18 S 03	15 N 10	22 N 36	02 N 10	23 S 10
02	16 43	13 43	22 07	16 39	08 33	18 04	15 11	22 36	02 10	23 10
03	16 25	08 58	22 02	16 41	08 15	18 06	15 12	22 36	02 11	23 09
04	16 07	03 S 55	21 55	16 43	07 57	18 07	15 12	22 35	02 12	23 09
05	15 49	01 N 11	21 47	16 45	07 38	18 09	15 13	22 35	02 12	23 08
06	15 31	06 06	21 38	16 48	07 20	18 10	15 14	22 35	02 13	23 08
07	15 12	10 41	21 27	16 50	07 01	18 11	15 15	22 35	02 13	23 07
08	14 53	14 45	21 16	16 53	06 42	18 12	15 16	22 35	02 14	23 07
09	14 34	18 11	21 03	16 55	06 24	18 14	15 17	22 35	02 15	23 07
10	14 14	20 50	20 48	16 58	06 05	18 15	15 18	22 35	02 15	23 06
11	13 55	22 37	20 33	17 00	05 46	18 16	15 18	22 35	02 16	23 06
12	13 35	23 23	20 16	17 02	05 27	18 17	15 19	22 35	02 17	23 05
13	13 15	23 04	19 58	17 05	05 08	18 18	15 20	22 35	02 18	23 05
14	12 54	21 37	19 38	17 07	04 49	18 19	15 22	22 35	02 18	23 05
15	12 34	19 05	19 17	17 09	04 30	18 20	15 23	22 35	02 19	23 04
16	12 13	15 31	18 55	17 10	04 11	18 21	15 24	22 35	02 20	23 04
17	11 52	11 06	18 32	17 12	03 52	18 22	15 25	22 35	02 20	23 03
18	11 31	06 01	18 07	17 13	03 33	18 23	15 26	22 35	02 21	23 03
19	11 10	00 N 33	17 41	17 14	03 14	18 24	15 27	22 35	02 22	23 03
20	10 48	05 S 02	17 13	17 15	02 55	18 24	15 28	22 35	02 23	23 02
21	10 26	10 23	16 44	17 15	02 36	18 25	15 29	22 35	02 24	23 02
22	10 05	15 10	16 14	17 15	02 17	18 26	15 31	22 35	02 24	23 02
23	09 43	19 04	15 43	17 15	01 58	18 26	15 32	22 35	02 25	23 01
24	09 20	21 48	15 10	17 15	01 39	18 27	15 33	22 35	02 26	23 01
25	08 58	23 09	14 36	17 14	01 20	18 28	15 35	22 35	02 27	23 00
26	08 36	23 03	14 00	17 13	01 01	18 28	15 36	22 35	02 27	23 00
27	08 13	21 31	13 23	17 11	00 42	18 29	15 37	22 35	02 28	23 00
28	07 S 50	18 S 46	12 S 45	17 S 09	00 S 23	18 S 29	15 N 39	22 N 35	02 N 29	22 S 59

ZODIAC SIGN ENTRIES

Date	h	m	Planets
01	15	52	☽ ♒
03	20	58	☽ ♓
06	04	48	☽ ♈
08	14	03	☿ ♒
08	15	44	☽ ♉
11	04	30	☽ ♊
13	16	29	☽ ♋
16	01	27	☽ ♌
18	06	58	☽ ♍
18	15	00	☉ ♓
20	10	04	☽ ♎
22	12	18	☽ ♏
24	14	53	☽ ♐
26	18	26	☽ ♑
27	09	00	☿ ♓
27	19	07	♂ ♈
28	23	07	☽ ♒

LATITUDES

Date	Mercury ☿ ° '	Venus ♀ ° '	Mars ♂ ° '	Jupiter ♃ ° '	Saturn ♄ ° '	Uranus ♅ ° '	Neptune ♆ ° '	Pluto ♇ ° '
01	00 S 16	06 N 31	00 S 55	01 N 03	02 S 11	00 N 03	01 S 29	05 S 41
04	00 39	06 21	00 53	01 04	02 10	00 03	01 29	05 41
07	01 00	06 09	00 52	01 04	02 09	00 03	01 28	05 41
10	01 19	05 55	00 50	01 04	02 08	00 03	01 28	05 41
13	01 35	05 39	00 48	01 05	02 07	00 03	01 28	05 42
16	01 48	05 21	00 46	01 05	02 06	00 03	01 28	05 42
19	01 58	05 03	00 45	01 05	02 06	00 03	01 28	05 43
22	02 05	04 44	00 43	01 06	02 05	00 03	01 28	05 43
25	02 08	04 25	00 41	01 06	02 04	00 03	01 28	05 43
28	02 07	04 05	00 39	01 07	02 03	00 03	01 28	05 44
31	02 S 03	03 N 45	00 S 37	01 N 07	02 S 02	00 N 03	01 S 28	05 S 44

DATA

Julian Date	2462534
Delta T	+74 seconds
Ayanamsa	24° 16' 57"
Synetic vernal point	04° ♓ 50' 02"
True obliquity of ecliptic	23° 26' 06"

MOON'S PHASES, APSIDES AND POSITIONS ☽

Date	h	m	Phase	Longitude ° '	Eclipse Indicator
02	16	07	●	13 ♒ 51	
10	11	49	☽	21 ♉ 47	
18	06	20	○	29 ♌ 38	
25	01	58	☾	06 ♐ 30	

Day	h	m		
10	06	06	Apogee	
22	10	23	Perigee	
05	06	25	0N	
12	17	09	Max dec	23° N 24'
19	14	20	0S	
25	22	11	Max dec	23° S 17'

LONGITUDES

Date	Chiron ⚷ ° '	Ceres ⚳ ° '	Pallas ⚴ ° '	Juno ⚵ ° '	Vesta ⚶ ° '	Black Moon Lilith ⚸ ° '
01	08 ♉ 01	11 ♓ 37	10 ♒ 19	25 ♐ 29	08 ♓ 31	17 ♉ 41
11	08 ♉ 13	15 ♓ 25	13 ♒ 48	28 ♐ 24	13 ♓ 20	18 ♉ 48
21	08 ♉ 30	19 ♓ 17	17 ♒ 15	01 ♑ 10	18 ♓ 10	19 ♉ 54
31	08 ♉ 53	23 ♓ 12	20 ♒ 37	03 ♑ 43	22 ♓ 59	21 ♉ 01

All ephemeris data is given at 12.00 UT and the Moon's longitude is additionally given for 24.00 UT

ASPECTARIAN

h m	Aspects
01 Friday	
02 03	♀ ⚹ ♂
05 59	☽ ∠ ♂
07 34	☽ ⚹ ♃
10 28	☽ Q ♆
10 44	☽ ∥ ♃
15 18	☽ ⚼ ♅
17 45	☽ ∥ ☉
19 39	☽ ∥ ♀
21 57	☿ ± ♅
22 32	☽ ⊥ ♂
02 Saturday	
04 02	☽ # ♄
04 37	☽ Q ♃
07 24	☽ ⚹ ♆
08 05	♂ ⊻ ♇
08 21	☽ ⊻ ♀
09 35	☽ ☌ ♇
09 40	☽ ⊻ ♂
16 07	☽ ☌ ☉
17 19	☉ ∥ ♀
17 32	☽ △ ♅
19 07	☽ ⊥ ♀
23 59	☽ □ ♄
03 Sunday	
07 35	☽ ⊻ ☿
10 02	☽ ∠ ♆
10 41	☉ △ ♅
11 29	☽ ∠ ♀
12 46	☽ □ ♃
15 43	☽ ∥ ♂
19 36	☽ ⊥ ☿
04 Monday	
02 19	☽ ⊥ ♆
08 30	☽ Q ♄
13 19	☽ ⊻ ♆
13 57	☽ ∠ ☿
15 24	☽ ⚹ ♀
15 37	☽ ⊻ ♇
19 03	☽ ☌ ♂
20 05	☽ # ♆
23 40	☉ ⊥ ♀
23 47	☽ □ ♅
05 Tuesday	
01 05	♀ ⊻ ♇
02 46	☽ ⊥ ♇
03 00	☽ ⊻ ☉
06 46	☽ ⚹ ♄
13 30	☿ ⚹ ♃
14 23	☽ Q ♀
15 14	☽ ⊥ ☉
16 54	☽ ∥ ♆
19 45	☽ ∠ ♇
20 36	☽ △ ♃
21 26	☽ ⚹ ☿
06 Wednesday	
09 47	☽ Q ♅
09 53	☽ ∠ ☉
11 20	☽ ∠ ♄
12 08	☿ Q ♆
14 10	♀ ∠ ♃
17 49	☽ # ♂
22 13	☽ ☌ ♆
23 30	☽ Q ☿
07 Thursday	
00 41	☽ ⚹ ♇
01 44	☽ ⚼ ♃
02 01	☽ □ ♀
04 17	☉ □ ♄
04 52	☽ ⊥ ♄
08 04	☽ ⊻ ♂
08 41	☉ # ♄
09 06	☽ ⚹ ♅
16 43	☽ ⊻ ♄
17 48	☽ ⚹ ☉
19 36	☽ ± ♃
20 43	☽ ⊥ ♂
23 59	♂ □ ♅
08 Friday	
00 23	☽ Q ♇
07 05	☿ ⚼ ♅
07 38	☽ ⊼ ♃
12 48	☽ # ☉
14 53	☽ ∠ ♅
15 19	☽ ∥ ♄
15 31	☿ ∠ ♂
15 56	☽ ∠ ♂
15 58	☽ □ ☿
19 58	☽ Q ☉
09 Saturday	
02 30	☽ # ♀
09 00	☽ ⊥ ♅
10 03	☽ ⊻ ♆
12 23	☽ # ♃
12 38	☽ □ ♇
16 09	☽ △ ♀
21 09	☽ ⊻ ♅
22 15	☽ ⊥ ♆
10 Sunday	
00 24	☽ ⚹ ♂
05 14	☽ ☌ ♄
07 22	♂ ⊥ ♇
11 39	☽ # ☿
11 49	☽ □ ☉
16 34	☽ ∠ ♆
20 47	☽ ☍ ♃
23 56	☽ ⚼ ♀
11 Monday	
02 30	☽ Q ♂
11 32	☽ ∥ ♅
13 25	☽ △ ☿
19 00	☽ ± ♀
23 02	☽ ⚹ ♆
23 06	☽ # ♇
12 Tuesday	
01 37	☽ △ ♇
07 38	☽ ⊼ ♀
09 51	☽ ☌ ♅
17 19	☽ □ ♂
18 04	☽ ⊻ ♄
20 43	☉ ∠ ♆
23 06	☽ Q ♆
23 51	☽ ⚼ ☿
13 Wednesday	
05 36	♂ ⚹ ♄
05 48	☽ △ ☉
06 01	☽ ⊥ ♄
07 36	☽ ⚼ ♇
09 27	☽ ⊼ ♃
11 18	☽ # ♇
20 06	☽ ± ☿
21 17	☽ ± ♃
22 15	☽ ∥ ♅
23 43	☽ ∠ ♄
14 Thursday	
01 15	☽ ± ♇
06 02	☿ Q ♃
06 14	♀ ⊼ ♅
09 22	☽ ⊼ ☿
10 23	☽ □ ♆
12 52	☽ ⊼ ♇
13 36	☽ ⚼ ☉
14 46	☽ ⚼ ♃
17 42	☿ ⚹ ♆
20 31	☽ ⊻ ♅
21 13	☽ ☍ ♀
15 Friday	
04 33	☽ ⚹ ♄
05 18	☉ □ ♃
07 25	☽ △ ♂
07 50	☽ ⊥ ♅
08 09	☽ ± ☉
10 07	☽ # ☿
13 57	☿ ☌ ♇
17 33	☽ # ♃
19 12	☽ △ ♃
20 17	☽ ⊼ ☉
16 Saturday	
00 34	☽ ∠ ♅
01 43	☽ # ♀
02 57	☽ Q ♄
12 44	☽ ∥ ♄
12 52	☽ ⚼ ♂
18 22	☽ △ ♆
20 42	☽ ☍ ♇
17 Sunday	
00 42	☽ ☍ ☿
03 44	☽ ⚹ ♅
06 02	☽ ± ♂
07 05	☽ ⊼ ♀
07 46	☽ # ☉
11 30	☽ □ ♄
17 18	☽ ⊼ ♂
18 14	☽ ± ♀
21 05	☽ ⚼ ♆
21 06	☿ ∥ ♃
18 Monday	
00 53	☽ Q ♅
01 21	☽ □ ♃
02 14	☿ △ ♅
06 20	☽ ☍ ☉
10 42	☽ ⚼ ♀
12 46	☽ ± ♆
23 06	☽ ⊼ ♆
23 36	☽ # ♂
19 Tuesday	
01 18	☽ ⊼ ♇
04 08	☽ ∥ ♆
07 53	☽ □ ♅
08 30	☽ Q ♃
11 33	☽ ± ♇
11 42	☽ ⊼ ☿
13 42	☽ △ ♀
15 31	☽ △ ♄
23 10	☽ ± ☿
23 52	☽ ☍ ♂
20 Wednesday	
00 31	☽ # ♆
02 48	☽ ⚼ ♇
03 22	☽ ∥ ♂
04 54	☽ ⚹ ♃
10 21	☿ ⊥ ♂
10 40	☿ ∥ ♀
13 20	☽ ⊼ ☉
16 15	☽ ⚼ ☿
16 47	☿ ⊻ ♀
16 54	☽ ⚼ ♄
21 59	☿ □ ♄
23 24	♅ St D
21 Thursday	
00 09	☽ ± ☉
01 53	☽ ☍ ♆
04 02	☽ △ ♇
05 14	♀ △ ♄
06 10	☽ ∠ ♃
08 02	☽ ± ♄
10 23	☽ △ ♅
12 16	☽ ∥ ☉
16 21	☽ ⚼ ☉
18 06	☽ ⊼ ♄
18 44	☽ □ ♀
20 37	☽ △ ☿
21 17	☽ ⊥ ♃
22 Friday	
01 27	☉ ⊥ ♆
03 25	♂ # ♆
05 02	☽ ⊼ ♂
07 22	☽ ⊻ ♃
07 30	♂ ∠ ♇
11 28	☽ ⚼ ♅
13 54	☽ # ♄
15 40	☽ ± ♂
17 19	☽ ∥ ☿
19 21	☽ △ ☉
23 Saturday	
00 02	☽ ∥ ♀
02 32	☽ Q ♀
02 34	☽ ± ♅
03 10	♀ ± ♅
04 14	☽ ⊼ ♆
06 23	☽ □ ♇
07 38	☽ ∥ ♃
07 39	☽ ⚼ ♂
12 40	☽ ⊼ ♅
14 21	☽ ± ♆
19 37	☿ # ♄
20 40	☽ ☍ ♄
23 54	☽ ⚹ ♀
24 Sunday	
03 07	♂ △ ♃
04 34	☿ ∠ ♆
05 36	☽ ⚼ ♆
05 44	☽ □ ☿
10 05	☽ ☌ ♃
10 27	☽ △ ♂
12 51	☽ Q ♇
22 40	☽ # ♅
25 Monday	
01 58	☽ □ ☉
02 53	☽ ∠ ♀
07 12	☽ △ ♆
07 26	☽ ∥ ♇
09 22	☽ ⚹ ♇
15 41	☽ ☍ ♅
16 41	☽ Q ☿
18 46	☿ □ ♃
19 14	☽ ⊥ ♀
23 15	☉ Q ♄
26 Tuesday	
00 05	☽ ⊼ ♄
06 12	☽ ⊻ ♀
10 30	☽ ± ♄
11 16	☽ ∠ ♇
11 19	☽ Q ☉
13 11	☽ ∥ ♇
13 43	☽ ⊻ ♃
16 20	☽ ⚹ ☿
16 58	☽ □ ♂
21 53	☽ # ♅
27 Wednesday	
00 10	☽ ⊥ ♃
00 53	☿ ⊻ ♂
02 12	☽ ⚼ ♄
02 56	☽ ⊥ ♇
05 00	☿ ⊥ ♀
09 50	☽ ⚹ ☉
11 14	☽ □ ♆
13 26	☽ ⊻ ♇
13 46	☉ ∠ ♀
15 56	☽ ∠ ♃
19 48	☽ ⊼ ♅
22 20	☽ ∠ ☿
28 Thursday	
02 18	☽ Q ♂
04 37	☽ △ ♄
05 26	☉ ⊻ ♆
06 21	☽ ± ♅
13 58	☽ ☌ ♀
14 01	☽ ∥ ♃
14 19	☽ ∠ ☉
16 34	☽ ⊥ ☿
18 28	☽ ⚹ ♃
19 00	☽ Q ♆
22 18	☽ ⚼ ♅
23 08	☽ ∥ ♀

MARCH 2030

LONGITUDES

Date	Sidereal time h m s	Sun ☉ ° ' "	Moon ☽ ° ' "	Moon ☽ 24.00 ° ' "	Mercury ☿ ° '	Venus ♀ ° '	Mars ♂ ° '	Jupiter ♃ ° '	Saturn ♄ ° '	Uranus ♅ ° '	Neptune ♆ ° '	Pluto ♇ ° '
01	22 37 21	10 ♓ 56 36	07 ♒ 13 37	13 ♒ 54 54	03 ♓ 45	25 ♑ 37	01 ♈ 19	27 ♏ 24	19 ♉ 38	14 ♊ 32	09 ♈ 42	10 ♒ 57
02	22 41 18	11 56 50	20 ♒ 33 32	27 ♒ 09 19	05 32	26 28	02 05	27 26	19 43	14 33	09 45	10 59
03	22 45 14	12 57 03	03 ♓ 42 04	10 ♓ 11 34	07 20	27 19	02 52	27 28	19 47	14 33	09 47	11 00
04	22 49 11	13 57 13	16 ♓ 37 42	23 ♓ 00 19	09 10	28 12	03 38	27 30	19 52	14 34	09 49	11 02
05	22 53 07	14 57 22	29 ♓ 19 21	05 ♈ 34 47	11 00	29 05	04 24	27 32	19 56	14 35	09 51	11 04
06	22 57 04	15 57 28	11 ♈ 46 39	17 ♈ 55 06	12 52	29 ♑ 58	05 11	27 33	20 01	14 35	09 53	11 05
07	23 01 00	16 57 33	24 ♈ 00 18	00 ♉ 02 33	14 44	00 ♒ 52	05 57	27 35	20 06	14 36	09 55	11 07
08	23 04 57	17 57 36	06 ♉ 02 10	11 ♉ 59 34	16 38	01 47	06 43	27 36	20 11	14 37	09 57	11 08
09	23 08 53	18 57 37	17 ♉ 55 14	23 ♉ 49 40	18 33	02 42	07 29	27 36	20 16	14 38	09 59	11 10
10	23 12 50	19 57 35	29 ♉ 43 29	05 ♊ 37 17	20 28	03 38	08 15	27 37	20 21	14 39	10 01	11 11
11	23 16 47	20 57 32	11 ♊ 31 44	17 ♊ 27 31	22 24	04 35	09 01	27 38	20 26	14 41	10 04	11 13
12	23 20 43	21 57 26	23 ♊ 25 19	29 ♊ 25 51	24 21	05 31	09 47	27 38	20 31	14 42	10 06	11 15
13	23 24 40	22 57 18	05 ♋ 29 49	11 ♋ 37 53	26 19	06 29	10 33	27 38	20 36	14 43	10 08	11 16
14	23 28 36	23 57 08	17 ♋ 50 41	24 ♋ 08 49	28 ♓ 18	07 26	11 19	27 R 38	20 42	14 44	10 10	11 18
15	23 32 33	24 56 56	00 ♌ 32 46	07 ♌ 02 57	00 ♈ 16	08 25	12 05	27 38	20 47	14 45	10 12	11 19
16	23 36 29	25 56 42	13 ♌ 39 41	20 ♌ 23 06	02 15	09 23	12 51	27 37	20 53	14 45	10 15	11 20
17	23 40 26	26 56 25	27 ♌ 13 12	04 ♍ 09 49	04 15	10 22	13 36	27 37	20 58	14 46	10 17	11 22
18	23 44 22	27 56 06	11 ♍ 12 37	18 ♍ 21 02	06 14	11 22	14 22	27 36	21 04	14 48	10 19	11 23
19	23 48 19	28 55 45	25 ♍ 34 24	02 ♎ 51 52	08 12	12 22	15 08	27 35	21 09	14 49	10 21	11 25
20	23 52 16	29 ♓ 55 22	10 ♎ 12 29	17 ♎ 35 14	10 13	13 22	15 53	27 34	21 15	14 51	10 23	11 26
21	23 56 12	00 ♈ 54 57	24 ♎ 59 03	02 ♏ 22 55	12 06	14 23	16 39	27 32	21 21	14 52	10 26	11 27
22	00 00 09	01 54 30	09 ♏ 45 53	17 ♏ 07 03	14 02	15 23	17 25	27 31	21 27	14 54	10 28	11 29
23	00 04 05	02 54 02	24 ♏ 25 43	01 ♐ 41 16	15 55	16 25	18 10	27 29	21 33	14 55	10 30	11 30
24	00 08 02	03 53 31	08 ♐ 53 14	16 ♐ 01 21	17 46	17 26	18 56	27 27	21 39	14 57	10 32	11 31
25	00 11 58	04 53 00	23 ♐ 05 23	00 ♑ 05 19	19 35	18 28	19 41	27 25	21 45	14 58	10 35	11 32
26	00 15 55	05 52 26	07 ♑ 01 08	13 ♑ 52 56	21 20	19 31	20 26	27 22	21 51	15 00	10 37	11 34
27	00 19 51	06 51 50	20 ♑ 40 52	27 ♑ 25 04	23 00	20 33	21 12	27 20	21 57	15 02	10 39	11 35
28	00 23 48	07 51 13	04 ♒ 05 43	10 ♒ 42 59	24 39	21 36	21 57	27 17	22 04	15 04	10 42	11 36
29	00 27 45	08 50 34	17 ♒ 17 01	23 ♒ 47 57	26 13	22 39	22 42	27 14	22 10	15 05	10 44	11 37
30	00 31 41	09 49 53	00 ♓ 15 53	06 ♓ 40 54	27 41	23 42	23 27	27 11	22 16	15 07	10 46	11 38
31	00 35 38	10 ♈ 49 11	13 ♓ 03 04	19 ♓ 22 25	29 ♈ 05	24 ♒ 46	24 ♈ 13	27 ♏ 08	22 ♉ 23	15 ♊ 09	10 ♈ 48	11 ♒ 39

Date	Moon True ☊ ° '	Moon Mean ☊ ° '	Moon ☽ Latitude ° '
01	22 ♐ 18	21 ♐ 39	03 N 33
02	22 R 08	21 36	04 16
03	21 57	21 33	04 45
04	21 43	21 30	04 59
05	21 29	21 26	04 57
06	21 16	21 23	04 42
07	21 05	21 20	04 13
08	20 57	21 17	03 33
09	20 51	21 14	02 45
10	20 48	21 11	01 49
11	20 D 47	21 07	00 N 49
12	20 R 47	21 04	00 S 14
13	20 46	21 01	01 17
14	20 45	20 58	02 18
15	20 41	20 55	03 14
16	20 35	20 52	04 01
17	20 25	20 48	04 37
18	20 14	20 45	04 57
19	20 03	20 42	04 59
20	19 51	20 39	04 42
21	19 41	20 36	04 06
22	19 34	20 32	03 14
23	19 29	20 29	02 08
24	19 27	20 26	00 S 56
25	19 D 27	20 23	00 N 19
26	19 R 27	20 20	01 32
27	19 27	20 17	02 38
28	19 24	20 13	03 33
29	19 19	20 10	04 17
30	19 11	20 07	04 46
31	19 ♐ 00	20 ♐ 04	05 N 01

DECLINATIONS

Date	Sun ☉ ° '	Moon ☽ ° '	Mercury ☿ ° '	Venus ♀ ° '	Mars ♂ ° '	Jupiter ♃ ° '	Saturn ♄ ° '	Uranus ♅ ° '	Neptune ♆ ° '	Pluto ♇ ° '
01	07 S 28	15 S 01	12 S 06	17 S 07	00 S 04	18 S 29	15 N 40	22 N 35	02 N 30	22 S 59
02	07 05	10 35	11 25	17 04	00 N 15	18 30	15 41	22 35	02 31	22 59
03	06 42	05 05	10 43	17 01	00 34	18 30	15 43	22 35	02 32	22 59
04	06 19	00 S 41	10 00	16 57	00 53	18 30	15 44	22 35	02 32	22 58
05	05 56	04 N 17	09 15	16 53	01 12	18 31	15 46	22 35	02 33	22 58
06	05 32	08 58	08 29	16 49	01 31	18 31	15 47	22 35	02 34	22 58
07	05 09	13 13	07 42	16 44	01 50	18 31	15 49	22 36	02 35	22 57
08	04 46	16 53	06 54	16 39	02 09	18 31	15 50	22 36	02 36	22 57
09	04 22	19 48	06 04	16 33	02 28	18 31	15 52	22 36	02 37	22 57
10	03 59	21 52	05 14	16 27	02 46	18 31	15 53	22 36	02 37	22 56
11	03 35	22 58	04 22	16 20	03 05	18 31	15 55	22 36	02 38	22 56
12	03 11	23 02	03 30	16 13	03 24	18 31	15 57	22 36	02 39	22 56
13	02 48	22 02	02 36	16 05	03 42	18 31	15 58	22 36	02 40	22 56
14	02 24	19 58	01 42	15 57	04 01	18 31	16 00	22 36	02 41	22 55
15	02 00	16 52	00 S 47	15 49	04 19	18 31	16 01	22 36	02 42	22 55
16	01 37	12 52	00 N 09	15 40	04 38	18 30	16 03	22 37	02 43	22 55
17	01 13	08 06	01 05	15 30	04 56	18 30	16 05	22 37	02 44	22 55
18	00 49	02 N 47	02 01	15 20	05 15	18 30	16 06	22 37	02 44	22 55
19	00 26	02 S 49	02 57	15 10	05 33	18 29	16 08	22 37	02 45	22 54
20	00 S 02	08 22	03 54	14 59	05 51	18 29	16 10	22 37	02 46	22 54
21	00 N 22	13 29	04 50	14 48	06 09	18 29	16 11	22 37	02 47	22 54
22	00 46	17 48	05 45	14 36	06 27	18 28	16 13	22 38	02 48	22 54
23	01 09	20 57	06 40	14 24	06 45	18 28	16 15	22 38	02 49	22 54
24	01 33	22 42	07 34	14 11	07 03	18 27	16 17	22 38	02 50	22 53
25	01 56	22 56	08 26	13 58	07 21	18 27	16 18	22 38	02 51	22 53
26	02 20	21 43	09 17	13 44	07 39	18 26	16 20	22 38	02 51	22 53
27	02 43	19 15	10 07	13 30	07 56	18 25	16 22	22 39	02 52	22 53
28	03 07	15 46	10 54	13 16	08 14	18 25	16 24	22 39	02 53	22 53
29	03 30	11 34	11 39	13 01	08 31	18 24	16 25	22 39	02 54	22 53
30	03 54	06 54	12 22	12 45	08 49	18 23	16 27	22 39	02 55	22 53
31	04 N 17	02 S 02	13 N 02	12 S 30	09 N 06	18 S 22	16 N 29	22 N 39	02 N 56	22 S 52

ZODIAC SIGN ENTRIES

Date	h	m	Planets
03	05	12	☽ ♓
05	13	18	☽ ♈
06	12	51	♀ ♒
07	23	55	☽ ♉
10	12	34	☽ ♊
13	01	08	☽ ♋
15	08	42	☿ ♈
15	10	59	☽ ♌
17	16	50	☽ ♍
19	19	18	☽ ♎
20	13	52	☉ ♈
21	20	08	☽ ♏
23	21	12	☽ ♐
25	23	51	☽ ♑
28	04	38	☽ ♒
30	11	30	☽ ♓

LATITUDES

Date	Mercury ☿ ° '	Venus ♀ ° '	Mars ♂ ° '	Jupiter ♃ ° '	Saturn ♄ ° '	Uranus ♅ ° '	Neptune ♆ ° '	Pluto ♇ ° '
01	02 S 06	03 N 58	00 S 38	01 N 07	02 S 03	00 N 03	01 S 28	05 S 44
04	02 00	03 38	00 36	01 07	02 02	00 03	01 28	05 44
07	01 50	03 18	00 35	01 08	02 02	00 03	01 28	05 45
10	01 35	02 58	00 33	01 08	02 01	00 03	01 28	05 45
13	01 15	02 39	00 31	01 09	02 00	00 03	01 28	05 46
16	00 49	02 19	00 29	01 09	01 59	00 03	01 28	05 46
19	00 S 19	02 00	00 27	01 09	01 59	00 03	01 28	05 47
22	00 N 14	01 42	00 25	01 10	01 58	00 03	01 28	05 48
25	00 51	01 24	00 23	01 10	01 58	00 03	01 28	05 48
28	01 27	01 06	00 21	01 10	01 57	00 03	01 28	05 49
31	02 N 01	00 N 49	00 S 19	01 N 11	01 S 56	00 N 03	01 S 28	05 S 49

DATA

Julian Date	2462562
Delta T	+74 seconds
Ayanamsa	24° 17' 01"
Synetic vernal point	04° ♓ 49' 58"
True obliquity of ecliptic	23° 26' 07"

MOON'S PHASES, APSIDES AND POSITIONS ☽

Date	h	m	Phase	Longitude ° '	Eclipse Indicator
04	06	35	●	13 ♓ 44	
12	08	48	☽	21 ♊ 49	
19	17	56	○	29 ♍ 11	
26	09	51	☾	05 ♑ 47	

Day	h	m		
10	02	25	Apogee	
21	22	10	Perigee	
04	15	16	ON	
12	01	37	Max dec	23° N 08'
18	23	58	OS	
25	03	45	Max dec	23° S 01'
31	21	54	ON	

LONGITUDES

Date	Chiron ⚷ ° '	Ceres ⚳ ° '	Pallas ⚴ ° '	Juno ⚵ ° '	Vesta ⚶ ° '	Black Moon Lilith ⚸ ° '
01	08 ♉ 48	22 ♓ 25	19 ♒ 57	03 ♑ 13	22 ♓ 01	20 ♉ 48
11	09 ♉ 14	26 ♓ 20	23 ♒ 15	05 ♑ 34	26 ♓ 50	21 ♉ 54
21	09 ♉ 44	00 ♈ 17	26 ♒ 27	07 ♑ 38	01 ♈ 38	23 ♉ 01
31	10 ♉ 17	04 ♈ 14	29 ♒ 31	09 ♑ 21	06 ♈ 23	24 ♉ 07

All ephemeris data is given at 12.00 UT and the Moon's longitude is additionally given for 24.00 UT

ASPECTARIAN

h m	Aspects
01 Friday	
00 49	☽ ✶ ♂
04 50	☽ ⊻ ☿
07 35	☽ ⊥ ☉
08 14	☽ # ♄
11 29	☿ ⊥ ♆
12 12	☉ ⊻ ♇
15 55	☽ Q ♃
16 27	☽ ✶ ♆
18 41	☽ ☌ ♇
19 12	☽ ⊻ ☉
02 Saturday	
01 08	☽ △ ♅
05 20	☽ ∠ ♂
06 56	☽ ∥ ☿
10 28	☽ □ ♄
19 37	☽ ∠ ♆
22 41	☽ ⊥ ♂
23 29	☽ ⊻ ♀
03 Sunday	
00 33	☽ □ ♃
02 24	♂ Q ♅
06 51	☽ ∥ ☉
10 22	☽ ⊻ ♂
11 15	☽ ⊥ ♀
12 08	☽ ⊥ ♆
16 25	♀ ✶ ♃
18 09	☿ Q ♄
19 35	☽ Q ♄
19 49	☽ ☌ ☿
23 15	☽ ⊻ ♆
04 Monday	
01 07	♀ Q ♆
01 32	☽ ⊻ ♇
03 11	☽ # ♆
05 07	☽ ∠ ♀
06 35	☽ ☌ ☉
08 09	☽ □ ♅
11 05	☽ # ♂
12 46	☽ ⊥ ♇
18 07	☽ ✶ ♄
20 01	☽ ∥ ♂
20 36	☿ ⊻ ♆
05 Tuesday	
02 51	☉ □ ♅
03 32	☽ ∥ ♆
05 47	☽ ∠ ♇
08 35	☽ △ ♃
11 30	☽ ✶ ♀
12 43	☿ ⊻ ♇
18 14	☽ Q ♅
19 36	☽ # ☉
22 23	☽ ☌ ♂
22 50	☽ ∠ ♄
06 Wednesday	
01 43	♀ bQ ♅
06 32	♂ ∠ ♄
08 18	☽ ☌ ♆
09 47	☽ # ☿
10 39	☽ ✶ ♇
12 24	☽ Q ♀
13 31	☽ bQ ♃
14 30	☽ ⊻ ☿
16 24	☽ ⊥ ♄
17 29	☽ ✶ ♅
20 53	☽ ⊻ ☉
07 Thursday	
04 14	☽ ⊻ ♄
04 22	☽ ⊥ ☿
07 11	☽ ± ♃
09 45	☽ ⊥ ☉
10 13	☿ □ ♅
10 14	☽ Q ♇
15 48	☉ ⊥ ♇
19 06	☽ ⚻ ♃
23 08	☽ ∠ ♅
08 Friday	
01 31	☽ ∠ ☿
02 46	☽ □ ♀
04 38	☽ ∥ ♄
05 16	☽ ∠ ☉
10 21	☽ # ♀
13 27	☽ ⊻ ♂
15 40	☿ ∠ ♀
17 12	☽ ⊥ ♅
18 29	☿ ⊥ ♇
19 54	☽ ⊻ ♆
22 18	☽ □ ♇
09 Saturday	
00 39	☽ # ♃
02 23	☽ ⊥ ♂
05 19	☽ ⊻ ♅
08 04	☽ ⊥ ♆
13 30	☽ ✶ ☿
14 18	☽ ✶ ☉
16 47	☽ ☌ ♄
21 55	☽ ∠ ♂
22 54	☉ ☌ ☿
10 Sunday	
00 02	♂ ∥ ♆
02 24	☽ ∠ ♆
07 43	☽ ☍ ♃
10 26	☿ ✶ ♄
16 58	☽ Q ☉
18 40	☽ Q ☿
20 39	☽ △ ♀
22 08	☉ ✶ ♄
11 Monday	
01 16	☽ ∥ ♅
06 33	☽ ✶ ♂
09 01	☽ ✶ ♆
10 49	☽ # ♇
11 22	☽ △ ♇
18 21	☽ ☌ ♅
12 Tuesday	
04 59	☉ # ♂
05 40	☽ bQ ♀
06 07	☽ ⊻ ♄
08 29	☽ Q ♂
08 48	☽ □ ☉
09 20	☽ Q ♆
13 59	☿ # ♂
14 14	☽ □ ☿
16 32	☽ # ♇
17 39	☽ bQ ♇
18 14	☽ ⊥ ♄
20 25	☽ ⚻ ♃
22 14	♂ ☌ ♆
13 Wednesday	
01 14	☽ ± ♀
01 37	☽ ∥ ♅
02 48	☉ ∥ ☿
08 20	☽ ± ♃
10 20	☿ # ♆
11 20	☿ ∠ ♇
11 33	☽ ± ♇
12 13	☽ ∠ ♄
14 06	☽ ⚻ ♀
14 33	♃ St R
19 34	☉ # ♆
21 06	☽ □ ♆
22 33	☽ □ ♂
23 19	☽ ⚻ ♇
14 Thursday	
01 57	☽ bQ ♃
03 59	☿ △ ♃
05 46	♀ # ♄
05 58	☽ ⊻ ♅
11 12	♂ ✶ ♇
17 29	☽ ✶ ♄
17 30	☽ ⊥ ♅
15 Friday	
00 13	☽ # ♃
00 37	☽ △ ☉
06 33	☽ △ ♃
10 29	☽ ∠ ♅
11 24	☽ △ ☿
16 11	☽ Q ♄
17 31	☽ ∥ ♄
19 10	☽ # ♀
16 Saturday	
03 39	☽ ☍ ♀
05 03	♂ bQ ♃
05 48	☽ △ ♆
06 42	☽ bQ ☉
07 48	☽ ☍ ♇
10 26	☽ △ ♂
13 58	☽ ✶ ♅
17 36	♀ Q ♃
18 03	☿ Q ♅
19 33	☽ bQ ☿
21 44	☉ ∠ ♇
17 Sunday	
00 07	☽ ± ☉
00 57	☽ □ ♄
08 36	☽ bQ ♆
09 37	♀ ✶ ♆
11 13	☽ Q ♅
11 28	☽ ⚻ ☉
12 41	☽ □ ♃
14 05	☽ ± ☿
14 31	☉ # ☿
14 33	☽ bQ ♂
18 Monday	
00 14	☽ ± ♆
01 42	☽ ∥ ♂
02 08	☽ ⚻ ☿
03 56	☉ △ ♃
06 54	☽ ± ♂
09 55	☿ ∠ ♄
10 29	☽ ⚻ ♆
12 10	☽ ∥ ♆
12 17	☽ ⚻ ♀
12 18	☽ ⚻ ♇
12 32	♀ ☌ ♇
14 50	☽ ∥ ☿
17 38	☽ ⚻ ♂
18 03	☽ □ ♅
19 23	☽ Q ♃
21 06	☽ # ☉
22 24	☽ ± ♇
23 07	☽ ± ♀
19 Tuesday	
01 53	♂ ✶ ♅
02 28	☽ ∥ ☉
04 38	☽ △ ♄
06 48	☿ ∥ ♆
11 43	☽ # ♆
12 41	☽ # ☿
12 59	♂ ⊥ ♄
13 23	☽ bQ ♇
15 10	☽ bQ ♀
15 18	☽ ✶ ♃
17 56	☽ ☍ ☉
20 Wednesday	
00 21	☽ # ♂
05 30	☽ bQ ♄
11 55	☽ ☍ ☿
12 18	☽ ☍ ♆
14 00	☽ △ ♇
14 51	☿ ☌ ♆
15 49	☽ ∠ ♃
17 31	☽ △ ♀
19 33	☽ △ ♅
20 16	☽ ± ♄
21 45	☽ ☍ ♂
21 Thursday	
03 48	☿ ✶ ♇
06 04	☽ ⚻ ♄
06 25	☽ ⊥ ♃
16 08	☽ ⊻ ♃
17 16	☿ bQ ♃
18 28	☽ ∥ ♀
19 56	☽ bQ ♅
22 19	☽ ⚻ ☉
23 55	♀ △ ♅
22 Friday	
02 29	☽ # ♄
08 46	☽ ± ☉
10 35	☽ ± ♅
13 09	☽ ⚻ ♆
14 47	☽ □ ♇
16 23	☽ ∥ ♃
19 59	☽ ⚻ ☿
20 23	☽ ⚻ ♅
21 52	☽ □ ♀
22 58	☽ ± ♆
23 05	☿ ✶ ♅
23 Saturday	
00 30	☽ bQ ☉
01 10	☽ ⚻ ♂
07 01	☿ ⊥ ♄
07 14	☽ ☍ ♄
07 16	☽ ± ☿
11 33	☽ ± ♂
12 25	☉ Q ♅
13 47	☽ bQ ♆
15 27	☿ ∥ ♂
17 01	☽ ☌ ♃
20 23	☽ Q ♇
24 Sunday	
00 18	☽ bQ ☿
02 15	☿ ✶ ♀
03 03	☽ △ ☉
03 16	☽ bQ ♂
05 48	☽ Q ♀
10 32	☽ # ♅
14 47	☽ △ ♆
16 25	☽ ✶ ♇
17 42	☽ ∥ ♇
22 12	☽ ☍ ♅
25 Monday	
03 32	☽ ✶ ♀
05 09	☽ △ ☿
05 52	☽ △ ♂
09 42	☽ ⚻ ♄
13 57	☽ ∥ ♇
14 33	☿ ☌ ♂
17 54	☽ ∠ ♇
19 23	☽ ⊻ ♃
20 02	☽ ± ♄
21 05	☽ # ♅
26 Tuesday	
05 42	☽ ⊥ ♃
07 17	☽ ∠ ♀
09 28	☽ ⊥ ♇
09 51	☽ □ ☉
11 42	☽ bQ ♄
12 37	☿ ± ♃
18 18	☽ □ ♆
19 46	☿ ⊻ ♄
19 56	☽ ⊻ ♇
21 20	☽ ∠ ♃
27 Wednesday	
00 17	☽ ⊥ ♀
02 00	☽ ⚻ ♅
11 45	☽ ⊻ ♀
12 37	☽ ± ♅
12 58	☽ □ ♂
14 17	☽ △ ♄
14 29	☉ ∠ ♄
16 09	♂ ± ♃
16 45	☽ □ ☿
18 19	☽ ∥ ♃
20 02	☽ Q ☉
20 09	☿ Q ♇
21 26	☉ ∥ ♆
23 49	☽ ✶ ♃
28 Thursday	
02 15	☽ Q ♆
04 43	☽ bQ ♅
08 06	☽ # ♄
16 04	♂ ⊻ ♄
19 21	☽ ✶ ☉
21 22	☽ Q ♃
23 15	☽ Q ♂
23 46	♀ □ ♄
23 59	☽ ✶ ♆
29 Friday	
01 38	☽ ☌ ♇
03 38	☽ ∥ ♀
05 38	☽ Q ☿
07 58	☽ △ ♅
11 38	☽ # ☿
16 28	♀ ✶ ♂
21 03	☽ □ ♄
22 36	☽ ✶ ♂
22 45	☽ ☌ ♀
30 Saturday	
01 05	☽ ∠ ☉
02 58	☽ # ♂
03 37	☽ ∠ ♆
04 03	☿ ⚻ ♃
06 18	☽ □ ♃
06 36	☽ ✶ ☿
17 53	♂ Q ♇
19 13	☽ ⊥ ☉
20 26	☽ ⊥ ♆
22 03	☿ # ♀
31 Sunday	
01 47	☽ # ☉
04 18	☽ ∠ ♂
06 55	☽ Q ♄
07 26	☽ ⊻ ☉
07 36	☽ # ♆
07 45	☽ ⊻ ♆
09 22	☽ ⊻ ♇
11 38	☉ ☌ ♆
14 11	☽ ∠ ☿
15 59	☽ □ ♅
20 45	☽ ⊥ ♇
22 24	☽ ⊥ ♂

APRIL 2030

LONGITUDES

Date	Sidereal time h m s	Sun ☉ ° ' "	Moon ☽ ° ' "	Moon ☽ 24.00 ° ' "	Mercury ☿ ° '	Venus ♀ ° '	Mars ♂ ° '	Jupiter ♃ ° '	Saturn ♄ ° '	Uranus ♅ ° '	Neptune ♆ ° '	Pluto ♇ ° '
01	00 39 34	11 ♈ 48 26	25 ♓ 38 59	01 ♈ 52 49	00 ♉ 23	25 ♒ 50	24 ♈ 58	27 ♏ 05	22 ♉ 29	15 ♊ 11	10 ♈ 51	11 ♒ 41
02	00 43 31	12 47 39	08 ♈ 03 56	14 ♈ 12 23	01 35	26 54	25 43	27 R 01	22 36	15 13	10 53	11 42
03	00 47 27	13 46 51	20 ♈ 18 16	26 ♈ 21 42	02 42	27 58	26 28	26 58	22 42	15 15	10 55	11 43
04	00 51 24	14 46 00	02 ♉ 22 49	08 ♉ 21 50	03 42	29 ♒ 03	27 13	26 54	22 49	15 17	10 57	11 44
05	00 55 20	15 45 07	14 ♉ 18 59	20 ♉ 14 34	04 36	00 ♓ 07	27 58	26 50	22 56	15 19	11 00	11 45
06	00 59 17	16 44 12	26 ♉ 08 57	02 ♊ 02 32	05 23	01 12	28 43	26 45	23 03	15 22	11 02	11 46
07	01 03 14	17 43 15	07 ♊ 55 46	13 ♊ 49 10	06 04	02 17	29 ♈ 27	26 41	23 09	15 24	11 04	11 47
08	01 07 10	18 42 16	19 ♊ 43 17	25 ♊ 38 41	06 38	03 23	00 ♉ 12	26 37	23 16	15 26	11 06	11 48
09	01 11 07	19 41 15	01 ♋ 36 01	07 ♋ 35 54	07 05	04 28	00 57	26 32	23 23	15 28	11 09	11 48
10	01 15 03	20 40 11	13 ♋ 39 02	19 ♋ 46 03	07 26	05 34	01 42	26 27	23 30	15 31	11 11	11 49
11	01 19 00	21 39 05	25 ♋ 57 38	02 ♌ 14 24	07 40	06 40	02 26	26 22	23 37	15 33	11 13	11 50
12	01 22 56	22 37 56	08 ♌ 36 57	15 ♌ 05 48	07 47	07 46	03 11	26 17	23 44	15 35	11 15	11 51
13	01 26 53	23 36 46	21 ♌ 41 23	28 ♌ 24 02	07 R 47	08 52	03 55	26 12	23 51	15 38	11 18	11 52
14	01 30 49	24 35 33	05 ♍ 13 56	12 ♍ 11 06	07 42	09 58	04 40	26 06	23 58	15 40	11 20	11 53
15	01 34 46	25 34 18	19 ♍ 15 21	26 ♍ 26 21	07 30	11 05	05 24	26 01	24 05	15 43	11 22	11 53
16	01 38 43	26 33 00	03 ♎ 43 31	11 ♎ 06 06	07 13	12 11	06 09	25 55	24 12	15 46	11 24	11 54
17	01 42 39	27 31 41	18 ♎ 33 09	26 ♎ 03 36	06 51	13 18	06 53	25 49	24 20	15 48	11 27	11 55
18	01 46 36	28 30 19	03 ♏ 36 16	11 ♏ 09 55	06 24	14 25	07 37	25 43	24 27	15 51	11 29	11 55
19	01 50 32	29 ♈ 28 56	18 ♏ 43 20	26 ♏ 15 19	05 52	15 32	08 21	25 37	24 34	15 53	11 31	11 56
20	01 54 29	00 ♉ 27 31	03 ♐ 44 49	11 ♐ 10 54	05 18	16 39	09 06	25 31	24 41	15 56	11 33	11 57
21	01 58 25	01 26 04	18 ♐ 32 46	25 ♐ 49 51	04 40	17 47	09 50	25 25	24 49	15 59	11 35	11 57
22	02 02 22	02 24 36	03 ♑ 01 40	10 ♑ 07 59	04 01	18 54	10 34	25 18	24 56	16 02	11 37	11 58
23	02 06 18	03 23 06	17 ♑ 08 40	24 ♑ 03 42	03 20	20 02	11 18	25 12	25 04	16 04	11 40	11 58
24	02 10 15	04 21 35	00 ♒ 53 12	07 ♒ 37 21	02 39	21 09	12 02	25 05	25 11	16 07	11 42	11 59
25	02 14 12	05 20 01	14 ♒ 16 22	20 ♒ 50 34	01 57	22 17	12 46	24 58	25 18	16 10	11 44	11 59
26	02 18 08	06 18 26	27 ♒ 20 14	03 ♓ 45 42	01 17	23 25	13 30	24 51	25 26	16 13	11 46	12 00
27	02 22 05	07 16 50	10 ♓ 07 16	16 ♓ 25 15	00 39	24 33	14 13	24 45	25 33	16 16	11 48	12 00
28	02 26 01	08 15 12	22 ♓ 39 56	28 ♓ 51 35	00 ♉ 03	25 42	14 57	24 38	25 41	16 19	11 50	12 01
29	02 29 58	09 13 32	05 ♈ 00 29	11 ♈ 06 51	29 ♈ 29	26 50	15 41	24 31	25 48	16 22	11 52	12 01
30	02 33 54	10 ♉ 11 51	17 ♈ 10 55	23 ♈ 12 54	28 ♈ 59	27 ♓ 58	16 ♉ 25	24 ♏ 23	25 ♉ 56	16 ♊ 25	11 ♈ 54	12 ♒ 02

Date	Moon True ☊ ° '	Moon Mean ☊ ° '	Moon ☽ Latitude ° '
01	18 ♐ 48	20 ♐ 01	05 N 00
02	18 R 36	19 58	04 45
03	18 24	19 54	04 17
04	18 14	19 51	03 38
05	18 06	19 48	02 50
06	18 01	19 45	01 54
07	17 59	19 42	00 N 54
08	17 D 58	19 38	00 S 09
09	17 59	19 35	01 13
10	18 00	19 32	02 13
11	18 R 00	19 29	03 09
12	17 58	19 26	03 58
13	17 55	19 23	04 36
14	17 50	19 19	05 00
15	17 42	19 16	05 07
16	17 35	19 13	04 55
17	17 27	19 10	04 23
18	17 20	19 07	03 33
19	17 15	19 04	02 28
20	17 12	19 00	01 S 12
21	17 D 12	18 57	00 N 07
22	17 12	18 54	01 25
23	17 14	18 51	02 35
24	17 14	18 48	03 35
25	17 R 14	18 44	04 21
26	17 12	18 41	04 52
27	17 08	18 38	05 08
28	17 03	18 35	05 09
29	16 56	18 32	04 55
30	16 ♐ 50	18 ♐ 29	04 N 28

DECLINATIONS

Date	Sun ☉ ° '	Moon ☽ ° '	Mercury ☿ ° '	Venus ♀ ° '	Mars ♂ ° '	Jupiter ♃ ° '	Saturn ♄ ° '	Uranus ♅ ° '	Neptune ♆ ° '	Pluto ♇ ° '
01	04 N 40	02 N 52	13 N 39	12 S 13	09 N 23	18 S 21	16 N 31	22 N 40	02 N 57	22 S 52
02	05 03	07 34	14 14	11 57	09 40	18 20	16 33	22 40	02 58	22 52
03	05 26	11 54	14 46	11 40	09 57	18 19	16 35	22 40	02 59	22 52
04	05 49	15 43	15 14	11 23	10 14	18 18	16 36	22 40	02 59	22 52
05	06 12	18 50	15 39	11 05	10 30	18 17	16 38	22 41	03 00	22 52
06	06 35	21 08	16 02	10 47	10 47	18 16	16 40	22 41	03 01	22 52
07	06 57	22 30	16 20	10 28	11 04	18 15	16 42	22 41	03 02	22 52
08	07 20	22 53	16 36	10 09	11 20	18 14	16 44	22 41	03 03	22 52
09	07 42	22 13	16 48	09 50	11 36	18 13	16 46	22 42	03 04	22 52
10	08 04	20 31	16 56	09 31	11 52	18 12	16 48	22 42	03 05	22 52
11	08 26	17 51	17 01	09 11	12 08	18 11	16 49	22 42	03 06	22 52
12	08 48	14 16	17 03	08 50	12 24	18 09	16 51	22 42	03 06	22 52
13	09 10	09 55	17 01	08 30	12 40	18 08	16 53	22 43	03 07	22 52
14	09 32	04 N 56	16 55	08 09	12 55	18 07	16 55	22 43	03 08	22 52
15	09 53	00 S 27	16 46	07 48	13 11	18 05	16 57	22 43	03 09	22 52
16	10 14	05 59	16 34	07 26	13 26	18 04	16 59	22 43	03 10	22 52
17	10 36	11 20	16 19	07 05	13 41	18 03	17 01	22 44	03 11	22 52
18	10 56	16 03	16 01	06 43	13 57	18 01	17 03	22 44	03 11	22 52
19	11 17	19 45	15 40	06 20	14 11	18 00	17 04	22 44	03 12	22 52
20	11 38	22 05	15 17	05 58	14 26	17 58	17 06	22 45	03 13	22 52
21	11 58	22 49	14 52	05 35	14 41	17 57	17 08	22 45	03 14	22 52
22	12 18	21 59	14 25	05 12	14 55	17 55	17 10	22 45	03 15	22 52
23	12 38	19 46	13 57	04 49	15 10	17 54	17 12	22 45	03 16	22 52
24	12 58	16 28	13 27	04 25	15 24	17 52	17 14	22 46	03 16	22 52
25	13 18	12 23	12 58	04 02	15 38	17 50	17 16	22 46	03 17	22 52
26	13 37	07 49	12 28	03 38	15 52	17 49	17 18	22 46	03 18	22 52
27	13 56	03 S 01	11 58	03 14	16 05	17 47	17 19	22 47	03 19	22 52
28	14 15	01 N 49	11 30	02 50	16 19	17 45	17 21	22 47	03 20	22 52
29	14 34	06 30	11 02	02 25	16 32	17 44	17 23	22 47	03 20	22 53
30	14 N 52	10 N 52	10 N 36	02 S 01	16 N 45	17 S 42	17 N 25	22 N 48	03 N 21	22 S 53

ZODIAC SIGN ENTRIES

Date	h m	Planets
01	04 47	☿ ♉
01	20 22	☽ ♈
04	07 15	☽ ♉
05	09 19	♀ ♓
06	19 50	☽ ♊
08	05 27	♂ ♉
09	08 47	☽ ♋
11	19 44	☽ ♌
14	02 50	☽ ♍
16	05 53	☽ ♎
18	06 16	☽ ♏
20	00 43	☉ ♉
20	05 59	☽ ♐
22	06 56	☽ ♑
24	10 26	☽ ♒
26	16 57	☽ ♓
28	13 44	☿ ♈
29	02 13	☽ ♈

LATITUDES

Date	Mercury ☿ ° '	Venus ♀ ° '	Mars ♂ ° '	Jupiter ♃ ° '	Saturn ♄ ° '	Uranus ♅ ° '	Neptune ♆ ° '	Pluto ♇ ° '
01	02 N 12	00 N 44	00 S 18	01 N 11	01 S 56	00 N 03	01 S 28	05 S 50
04	02 39	00 28	00 16	01 11	01 56	00 03	01 28	05 50
07	02 58	00 N 12	00 14	01 11	01 55	00 03	01 28	05 51
10	03 07	00 S 02	00 12	01 12	01 55	00 03	01 28	05 52
13	03 04	00 16	00 10	01 12	01 54	00 03	01 28	05 52
16	02 48	00 29	00 08	01 12	01 54	00 03	01 28	05 53
19	02 20	00 42	00 06	01 12	01 53	00 03	01 28	05 54
22	01 40	00 53	00 04	01 12	01 53	00 03	01 28	05 54
25	00 52	01 03	00 S 02	01 12	01 53	00 03	01 28	05 55
28	00 N 01	01 13	00 00	01 12	01 52	00 03	01 28	05 56
31	00 S 49	01 S 21	00 N 02	01 N 12	01 S 52	00 N 03	01 S 28	05 S 56

DATA

Julian Date	2462593
Delta T	+74 seconds
Ayanamsa	24° 17' 03"
Synetic vernal point	04° ♓ 49' 55"
True obliquity of ecliptic	23° 26' 06"

MOON'S PHASES, APSIDES AND POSITIONS ☽

Date	h m	Phase	Longitude ° '	Eclipse Indicator
02	22 02	●	13 ♈ 12	
11	02 57	☽	21 ♋ 17	
18	03 20	○	28 ♎ 09	
24	18 39	☾	04 ♒ 38	

Day	h m		
06	18 52	Apogee	
19	03 53	Perigee	
08	08 41	Max dec	22° N 53'
15	10 01	0S	
21	11 04	Max dec	22° S 49'
28	02 56	0N	

LONGITUDES

Date	Chiron ⚷ ° '	Ceres ⚳ ° '	Pallas ⚴ ° '	Juno ⚵ ° '	Vesta ⚶ ° '	Black Moon Lilith ⚸ ° '
01	10 ♉ 21	04 ♈ 38	29 ♒ 48	09 ♑ 30	06 ♈ 52	24 ♉ 14
11	10 ♉ 57	08 ♈ 33	02 ♓ 42	10 ♑ 48	11 ♈ 35	25 ♉ 21
21	11 ♉ 34	12 ♈ 27	05 ♓ 25	11 ♑ 38	16 ♈ 14	26 ♉ 27
31	12 ♉ 13	16 ♈ 18	07 ♓ 55	11 ♑ 58	20 ♈ 50	27 ♉ 34

ASPECTARIAN

h m	Aspects
01 Monday	
05 53	☽ ⚹ ♄
08 10	☿ ∠ ♅
08 44	☉ ⚹ ♇
09 18	☽ ⊥ ☿
10 36	☽ ⊻ ♂
12 21	♀ ∠ ♆
12 22	☽ ⊻ ♀
12 25	☽ ∥ ♆
13 58	☽ ∠ ♇
14 44	☽ △ ♃
18 18	☉ Ᵽ ♃
21 51	☽ ∥ ☉
22 07	☽ ⊻ ☿
02 Tuesday	
01 01	☽ ⊥ ♀
02 34	☽ Q ♅
11 05	☽ ∠ ♄
14 41	♀ □ ♃
17 30	☽ ☌ ♆
19 05	☽ ⚹ ♇
19 41	☽ Ᵽ ♃
20 11	☽ ∠ ♀
22 02	☽ ☌ ☉
03 Wednesday	
00 06	☽ ∥ ♂
02 02	☽ ⚹ ♅
04 51	☽ ⊥ ♄
10 42	☽ # ♀
13 17	☽ ± ♃
16 48	☽ ⊻ ♄
18 45	☽ Q ♇
04 Thursday	
01 01	☽ ☌ ♂
01 07	☽ ⊼ ♃
02 38	♂ ⊼ ♃
04 41	☽ ⚹ ♀
07 48	☽ ∠ ♅
08 19	☽ ∥ ☿
14 52	☽ ☌ ☿
18 24	☽ ∥ ♄
05 Friday	
01 10	☉ ⚹ ♅
01 54	☽ ⊥ ♅
05 17	☽ ⊻ ♆
06 48	☽ □ ♇
07 08	☽ Q ♀
07 25	☽ # ♃
14 03	☽ ⊻ ♅
15 10	☽ ⊻ ☉
17 26	☽ ⊥ ♆
06 Saturday	
04 26	☽ ⊥ ☉
05 37	☽ ☌ ♄
11 43	♀ # ♂
11 46	☽ ∠ ♆
13 14	☽ ☍ ♃
17 34	☽ ⊻ ♂
20 24	☉ ⊥ ♄
23 20	☽ □ ♀
07 Sunday	
00 25	☽ ∠ ☉
06 37	☽ ⊥ ♂
07 59	☽ ⊻ ☿
17 32	☽ ∥ ♅
18 25	☽ ⚹ ♆
19 51	☽ △ ♇
20 53	☽ ⊥ ☿
08 Monday	
02 12	☽ ∠ ♂
03 06	☽ # ♇
03 15	☽ ☌ ♅
09 45	☽ ⚹ ☉
14 18	☽ # ♇
16 03	☽ ∠ ☿
18 53	☽ Q ♆
19 16	☽ ⊻ ♄
19 49	♂ ∠ ♅
23 35	☽ ∥ ♅
09 Tuesday	
01 51	☽ ⊼ ♃
02 20	☽ Ᵽ ♇
06 43	☿ ∥ ♄
07 30	☽ ⊥ ♄
10 36	☽ ⚹ ♂
12 11	☽ Q ☉
13 51	☽ ± ♃
18 20	☽ △ ♀
20 26	☽ ± ♇
23 20	☽ ⚹ ☿
10 Wednesday	
01 42	☽ ∠ ♄
03 23	♀ ⊥ ♆
07 03	☉ ± ♃
07 06	☽ □ ♆
07 41	☽ Ᵽ ♃
08 23	☽ ⊼ ♇
12 05	☽ Q ♂
15 41	☽ ⊻ ♅
23 35	☽ Q ☿
11 Thursday	
02 52	☽ Ᵽ ♀
02 57	☽ □ ☉
03 27	☽ ⊥ ♅
07 26	☽ ⚹ ♄
09 26	☽ # ♃
10 12	☉ ∠ ♀
12 47	☽ △ ♃
18 03	☽ ∥ ☿
19 26	☽ ∥ ♄
20 49	☽ ∠ ♅
21 51	☽ ± ♀
12 Friday	
01 09	☽ □ ♂
06 33	☽ Q ♄
10 15	☽ ⊼ ♀
10 26	☽ □ ☿
12 27	☿ ⚹ ♀
13 15	☉ # ♀
16 33	♃ Ᵽ ♆
16 56	☽ △ ♆
18 01	☽ ☍ ♇
22 11	☽ ∥ ♂
13 Saturday	
00 57	☽ ⚹ ♅
02 34	☿ St R
15 32	☽ ∥ ☉
15 44	☽ △ ☉
15 55	☽ □ ♄
18 14	☉ Q ♇
18 37	☉ ⊻ ♄
19 40	☽ # ♀
20 01	☽ □ ♃
20 17	☽ Ᵽ ♆
22 40	☽ Q ♅
14 Sunday	
10 57	☽ △ ♂
12 10	☽ ± ♆
12 17	☿ ∥ ♄
16 14	☽ △ ☿
20 07	☽ Ᵽ ☉
20 10	☽ ∥ ♆
20 54	☽ ☍ ♀
22 34	☽ ⊼ ♆
23 29	☽ ⊼ ♇
15 Monday	
03 10	☽ Q ♃
06 00	☽ □ ♅
09 42	☽ ± ♇
12 34	☽ ± ☉
14 02	☽ Ᵽ ♂
17 21	☽ Ᵽ ☿
18 32	♀ ⊻ ♆
20 09	☽ △ ♄
21 49	☉ ⊼ ♃
23 13	☽ ⚹ ♃
23 20	☽ ⊼ ☉
23 42	☽ # ♆
16 Tuesday	
00 45	☽ Ᵽ ♇
05 45	♀ ⊻ ♇
05 48	☽ ± ♂
07 59	☽ ± ☿
12 28	♀ Q ♄
16 09	☽ ⊼ ♂
17 34	☽ ⊼ ☿
17 58	☽ ∥ ♀
21 00	☽ Ᵽ ♄
23 37	☽ ∠ ♃
17 Wednesday	
00 31	☽ ☍ ♆
01 18	☽ △ ♇
02 52	☽ ⊼ ♀
07 34	☽ △ ♅
08 20	☽ # ☉
11 15	☿ ☌ ♂
11 38	☽ ± ♄
13 18	☽ ± ♀
14 01	☽ ⊥ ♃
21 19	☽ ⊼ ♄
23 32	☽ ⊻ ♃
18 Thursday	
00 08	☽ # ♂
03 20	☽ ☍ ☉
04 49	☽ Ᵽ ♀
07 36	☽ Ᵽ ♅
11 49	☽ # ☿
16 17	☽ ☍ ☿
17 45	☽ # ♄
18 42	☽ ☍ ♂
21 56	☽ ± ♅
23 41	☽ ∥ ♃
19 Friday	
00 32	☽ ⊼ ♆
01 13	☽ □ ♇
06 32	☽ △ ♀
07 29	☽ ⊼ ♅
10 05	☽ ± ♆
19 56	♀ □ ♅
21 23	☽ ☍ ♄
22 54	☽ ☌ ♃
20 Saturday	
00 27	☽ Ᵽ ♆
05 54	☽ Q ♇
06 21	☽ ⊼ ☉
14 24	☽ ⊼ ☿
16 40	☽ ± ☉
21 04	☽ ⊼ ♂
23 42	☽ ± ☿
21 Sunday	
00 17	☉ ∠ ♅
00 38	☽ △ ♆
01 15	☽ ⚹ ♇
03 44	☽ # ♅
07 19	☽ ± ♂
07 48	☽ ☍ ♅
08 18	☽ Ᵽ ☉
10 38	☽ □ ♀
13 46	☽ Ᵽ ☿
15 50	♀ ⊥ ♇
17 18	♂ ⊥ ♅
18 17	☽ # ♅
18 32	☿ ∥ ♂
22 24	☽ ⊼ ♄
22 54	☽ Ᵽ ♂
23 13	☽ ⊻ ♃
22 Monday	
01 52	☽ ∠ ♇
08 28	☽ ± ♄
09 08	☽ ⊥ ♃
10 53	☽ △ ☉
13 29	☿ ∠ ♀
13 35	☽ △ ☿
16 57	☽ ⊥ ♇
19 05	☽ Q ♀
23 46	☽ Ᵽ ♄
23 Tuesday	
00 12	☽ ∠ ♃
01 26	☽ △ ♂
02 35	☽ □ ♆
03 08	☽ ⊻ ♇
10 09	☽ ⊼ ♅
11 14	☉ ☌ ☿
17 26	☽ ⚹ ♀
20 34	☽ ± ♅
24 Wednesday	
00 31	♂ ⊻ ♆
01 52	☽ △ ♄
01 52	☽ ⚹ ♃
01 57	♃ ☍ ♄
02 33	☽ ∥ ♃
07 01	☽ # ♄
09 53	☽ Q ♆
10 28	♂ □ ♇
12 25	☽ Ᵽ ♅
14 58	☽ □ ☿
18 17	☽ # ♂
18 39	☽ □ ☉
22 14	☽ ∠ ♀
22 57	☽ Q ♃
25 Thursday	
02 11	☉ ∥ ☿
07 16	☽ # ☉
07 23	☽ ⚹ ♆
07 52	☽ ☌ ♇
08 23	☽ # ☿
09 06	☽ □ ♂
15 28	☽ △ ♅
16 01	☽ ⊥ ♀
21 52	☽ Q ☿
26 Friday	
04 03	☽ ⊻ ♀
05 56	☽ Q ☉
07 27	☽ □ ♃
08 26	☽ □ ♄
10 56	☽ ∠ ♆
14 28	☿ ∠ ♅
19 00	☽ ⚹ ☿
20 13	☽ Q ♂
27 Saturday	
03 49	☽ ⊥ ♆
06 11	☽ ⚹ ☉
07 07	♀ # ♆
10 32	☽ # ♆
10 51	☽ ∥ ♀
13 13	☿ ⊥ ♀
15 12	☽ ⊻ ♆
15 34	♀ △ ♃
15 35	☽ ⊻ ♇
18 36	☽ Q ♄
20 17	☽ ⚹ ♂
22 02	☽ ∠ ☿
23 45	☽ □ ♅
28 Sunday	
03 03	☽ ⊥ ♇
11 43	♀ ⚹ ♄
13 14	☽ ∠ ☉
14 32	☽ ⊥ ☿
15 45	☽ △ ♃
16 40	☽ # ♀
17 53	☽ ⚹ ♄
18 27	☽ ☌ ♀
19 38	☽ ∥ ♆
20 25	☽ ∠ ♇
29 Monday	
01 40	☽ ⊻ ☿
03 01	☽ ∠ ♂
08 13	☽ ⊥ ☉
10 44	☽ Q ♅
15 59	♀ ∠ ♇
20 45	☽ Ᵽ ♃
21 00	☽ ⊻ ☉
21 46	☽ ⊥ ♂
23 31	☽ ∠ ♄
30 Tuesday	
01 32	☽ ☌ ♆
01 48	☽ ⚹ ♇
10 22	☽ ⊻ ♂
10 28	☽ ⚹ ♅
10 32	☽ ∥ ☿
12 08	♂ ⊻ ♅
14 22	☽ ± ♃
17 31	☽ ⊥ ♄
17 37	☉ ⊥ ♅

MAY 2030

LONGITUDES

Date	Sidereal time h m s	Sun ☉ ° ′ ″	Moon ☽ ° ′ ″	Moon ☽ 24.00 ° ′ ″	Mercury ☿ ° ′	Venus ♀ ° ′	Mars ♂ ° ′	Jupiter ♃ ° ′	Saturn ♄ ° ′	Uranus ♅ ° ′	Neptune ♆ ° ′	Pluto ♇ ° ′
01	02 37 51	11 ♉ 10 08	29 ♈ 13 00	05 ♉ 11 25	28 ♈ 33	29 ♓ 07	17 ♉ 08	24 ♏ 16	26 ♉ 04	16 ♊ 28	11 ♈ 56	12 ♒ 02
02	02 41 47	12 08 23	11 ♉ 08 23	17 ♉ 04 06	28 R 10	00 ♈ 15	17 52	24 R 09	26 11	16 31	11 58	12 02
03	02 45 44	13 06 37	22 ♉ 58 50	28 ♉ 52 49	27 53	01 24	18 35	24 02	26 19	16 34	12 00	12 02
04	02 49 41	14 04 49	04 ♊ 46 20	10 ♊ 39 43	27 39	02 33	19 19	23 54	26 26	16 37	12 02	12 03
05	02 53 37	15 02 59	16 ♊ 33 19	22 ♊ 27 29	27 30	03 42	20 02	23 47	26 34	16 40	12 04	12 03
06	02 57 34	16 01 07	28 ♊ 22 39	04 ♋ 19 16	27 26	04 51	20 46	23 39	26 42	16 43	12 06	12 03
07	03 01 30	16 59 14	10 ♋ 17 48	16 ♋ 18 46	27 D 27	06 00	21 29	23 32	26 50	16 46	12 08	12 03
08	03 05 27	17 57 18	22 ♋ 22 42	28 ♋ 30 10	27 33	07 09	22 12	23 24	26 57	16 50	12 10	12 03
09	03 09 23	18 55 21	04 ♌ 41 42	10 ♌ 57 52	27 43	08 18	22 55	23 17	27 05	16 53	12 12	12 03
10	03 13 20	19 53 22	17 ♌ 19 13	23 ♌ 46 16	27 58	09 27	23 38	23 09	27 13	16 56	12 14	12 04
11	03 17 16	20 51 20	00 ♍ 19 27	06 ♍ 59 10	28 17	10 36	24 22	23 01	27 20	16 59	12 16	12 04
12	03 21 13	21 49 17	13 ♍ 45 42	20 ♍ 39 14	28 40	11 46	25 05	22 54	27 28	17 03	12 18	12 04
13	03 25 10	22 47 13	27 ♍ 39 48	04 ♎ 47 15	29 08	12 55	25 48	22 46	27 36	17 06	12 20	12 R 04
14	03 29 06	23 45 06	12 ♎ 01 16	19 ♎ 21 21	29 ♈ 40	14 05	26 31	22 38	27 44	17 09	12 21	12 04
15	03 33 03	24 42 57	26 ♎ 46 49	04 ♏ 16 47	00 ♉ 16	15 14	27 13	22 31	27 51	17 13	12 23	12 04
16	03 36 59	25 40 48	11 ♏ 50 14	19 ♏ 26 02	00 56	16 24	27 56	22 23	27 59	17 16	12 25	12 03
17	03 40 56	26 38 36	27 ♏ 02 56	04 ♐ 39 42	01 40	17 34	28 39	22 15	28 07	17 19	12 27	12 03
18	03 44 52	27 36 23	12 ♐ 15 07	19 ♐ 48 00	02 27	18 43	29 ♉ 22	22 08	28 15	17 23	12 29	12 03
19	03 48 49	28 34 09	27 ♐ 17 18	04 ♑ 42 06	03 18	19 53	00 ♊ 04	22 00	28 22	17 26	12 30	12 03
20	03 52 45	29 ♉ 31 54	12 ♑ 01 40	19 ♑ 15 25	04 12	21 03	00 47	21 53	28 30	17 30	12 32	12 03
21	03 56 42	00 ♊ 29 37	26 ♑ 22 56	03 ♒ 24 01	05 09	22 13	01 30	21 45	28 38	17 33	12 34	12 03
22	04 00 39	01 27 20	10 ♒ 18 33	17 ♒ 06 36	06 10	23 23	02 12	21 38	28 46	17 36	12 35	12 02
23	04 04 35	02 25 01	23 ♒ 48 20	00 ♓ 24 01	07 13	24 33	02 55	21 30	28 53	17 40	12 37	12 02
24	04 08 32	03 22 41	06 ♓ 53 58	13 ♓ 18 35	08 20	25 44	03 37	21 23	29 01	17 43	12 39	12 02
25	04 12 28	04 20 20	19 ♓ 38 16	25 ♓ 53 29	09 29	26 54	04 20	21 16	29 09	17 47	12 40	12 02
26	04 16 25	05 17 59	02 ♈ 04 42	08 ♈ 12 20	10 42	28 04	05 02	21 08	29 17	17 50	12 42	12 01
27	04 20 21	06 15 36	14 ♈ 16 52	20 ♈ 18 43	11 57	29 ♈ 14	05 44	21 01	29 24	17 54	12 43	12 01
28	04 24 18	07 13 12	26 ♈ 18 18	02 ♉ 16 00	13 15	00 ♉ 25	06 26	20 54	29 32	17 57	12 45	12 00
29	04 28 14	08 10 48	08 ♉ 12 13	14 ♉ 07 17	14 35	01 35	07 09	20 47	29 40	18 01	12 46	12 00
30	04 32 11	09 08 22	20 ♉ 01 32	25 ♉ 55 17	15 59	02 46	07 51	20 40	29 48	18 04	12 48	12 00
31	04 36 08	10 ♊ 05 55	01 ♊ 48 51	07 ♊ 42 31	17 ♉ 25	03 ♉ 56	08 ♊ 33	20 ♏ 33	29 ♉ 55	18 ♊ 08	12 ♈ 49	11 ♒ 59

Date	Moon True ☊ ° ′	Moon Mean ☊ ° ′	Moon ☽ Latitude ° ′
01	16 ♐ 43	18 ♐ 25	03 N 50
02	16 R 38	18 22	03 01
03	16 34	18 19	02 05
04	16 32	18 16	01 N 04
05	16 D 31	18 13	00 00
06	16 31	18 09	01 S 05
07	16 33	18 06	02 07
08	16 35	18 03	03 04
09	16 36	18 00	03 55
10	16 37	17 57	04 35
11	16 R 36	17 54	05 02
12	16 35	17 50	05 15
13	16 33	17 47	05 09
14	16 30	17 44	04 44
15	16 27	17 41	04 01
16	16 24	17 38	02 59
17	16 23	17 35	01 45
18	16 22	17 31	00 S 23
19	16 D 22	17 28	01 N 00
20	16 23	17 25	02 17
21	16 24	17 22	03 24
22	16 25	17 19	04 16
23	16 26	17 15	04 53
24	16 R 26	17 12	05 13
25	16 26	17 09	05 16
26	16 25	17 06	05 05
27	16 23	17 03	04 40
28	16 22	17 00	04 03
29	16 21	16 56	03 16
30	16 20	16 53	02 21
31	16 ♐ 19	16 ♐ 50	01 N 20

DECLINATIONS

Date	Sun ☉ ° ′	Moon ☽ ° ′	Mercury ☿ ° ′	Venus ♀ ° ′	Mars ♂ ° ′	Jupiter ♃ ° ′	Saturn ♄ ° ′	Uranus ♅ ° ′	Neptune ♆ ° ′	Pluto ♇ ° ′
01	15 N 11	14 N 46	10 N 11	01 S 36	16 N 58	17 S 40	17 N 27	22 N 48	03 N 22	22 S 53
02	15 29	18 02	09 48	01 11	17 11	17 38	17 29	22 48	03 23	22 53
03	15 46	20 32	09 28	00 46	17 24	17 37	17 31	22 49	03 23	22 53
04	16 04	22 08	09 10	00 S 21	17 37	17 35	17 32	22 49	03 24	22 53
05	16 21	22 45	08 54	00 N 04	17 49	17 33	17 34	22 49	03 25	22 53
06	16 38	22 21	08 41	00 29	18 01	17 31	17 36	22 50	03 26	22 54
07	16 54	20 56	08 30	00 55	18 13	17 29	17 38	22 50	03 26	22 54
08	17 11	18 33	08 22	01 20	18 25	17 28	17 40	22 50	03 27	22 54
09	17 27	15 17	08 16	01 45	18 36	17 26	17 42	22 50	03 28	22 54
10	17 42	11 16	08 13	02 11	18 48	17 24	17 44	22 51	03 29	22 54
11	17 58	06 38	08 12	02 36	18 59	17 22	17 45	22 51	03 29	22 55
12	18 13	01 N 33	08 13	03 02	19 10	17 20	17 47	22 51	03 30	22 55
13	18 28	03 S 48	08 17	03 27	19 21	17 18	17 49	22 52	03 31	22 55
14	18 42	09 07	08 23	03 53	19 32	17 17	17 51	22 52	03 31	22 55
15	18 57	14 03	08 32	04 19	19 42	17 15	17 53	22 52	03 32	22 56
16	19 11	18 14	08 42	04 44	19 53	17 13	17 54	22 53	03 33	22 56
17	19 24	21 12	08 54	05 10	20 03	17 11	17 56	22 53	03 33	22 56
18	19 37	22 38	09 08	05 35	20 13	17 09	17 58	22 53	03 34	22 56
19	19 50	22 24	09 24	06 01	20 22	17 07	18 00	22 54	03 35	22 57
20	20 03	20 37	09 42	06 26	20 32	17 06	18 01	22 54	03 35	22 57
21	20 15	17 32	10 01	06 51	20 41	17 04	18 03	22 54	03 36	22 57
22	20 27	13 32	10 22	07 16	20 50	17 02	18 05	22 55	03 36	22 57
23	20 38	08 58	10 44	07 42	20 59	17 00	18 07	22 55	03 37	22 58
24	20 50	04 S 08	11 08	08 07	21 08	16 58	18 08	22 55	03 38	22 58
25	21 00	00 N 45	11 33	08 31	21 17	16 57	18 10	22 56	03 38	22 58
26	21 11	05 30	11 59	08 56	21 25	16 55	18 12	22 56	03 39	22 59
27	21 21	09 56	12 26	09 21	21 33	16 53	18 13	22 56	03 39	22 59
28	21 31	13 56	12 54	09 45	21 41	16 51	18 15	22 57	03 40	22 59
29	21 40	17 20	13 23	10 10	21 49	16 50	18 17	22 57	03 40	23 00
30	21 49	20 00	13 53	10 34	21 56	16 48	18 18	22 57	03 41	23 00
31	21 N 58	21 N 49	14 N 23	10 N 58	22 N 03	16 S 46	18 N 20	22 N 58	03 N 41	23 S 00

ZODIAC SIGN ENTRIES

Date	h	m	Planets
01	13	34	☽ ♉
02	06	37	♀ ♈
04	02	17	☽ ♊
06	15	17	☽ ♋
09	02	55	☽ ♌
11	11	25	☽ ♍
13	15	57	☽ ♎
15	01	30	☿ ♉
15	17	10	☽ ♏
17	16	39	☽ ♐
19	09	28	♂ ♊
19	16	22	☽ ♑
20	23	41	☉ ♊
21	18	10	☽ ♒
23	23	16	☽ ♓
26	07	57	☽ ♈
28	03	33	♀ ♉
28	19	26	☽ ♉
31	08	18	☽ ♊

LATITUDES

Date	Mercury ☿ ° ′	Venus ♀ ° ′	Mars ♂ ° ′	Jupiter ♃ ° ′	Saturn ♄ ° ′	Uranus ♅ ° ′	Neptune ♆ ° ′	Pluto ♇ ° ′
01	00 S 49	01 S 21	00 N 02	01 N 12	01 S 52	00 N 03	01 S 28	05 S 56
04	01 35	01 29	00 04	01 12	01 52	00 03	01 28	05 57
07	02 13	01 36	00 05	01 12	01 52	00 03	01 28	05 58
10	02 43	01 42	00 07	01 12	01 51	00 03	01 28	05 59
13	03 05	01 47	00 09	01 11	01 51	00 03	01 29	05 59
16	03 18	01 51	00 11	01 11	01 51	00 03	01 29	06 00
19	03 24	01 54	00 13	01 11	01 51	00 03	01 29	06 01
22	03 24	01 57	00 15	01 11	01 51	00 03	01 29	06 02
25	03 16	01 58	00 17	01 10	01 51	00 03	01 29	06 02
28	03 03	01 59	00 18	01 10	01 50	00 03	01 29	06 03
31	02 S 45	01 S 59	00 N 20	01 N 09	01 S 50	00 N 03	01 S 29	06 S 04

DATA

Julian Date	2462623
Delta T	+74 seconds
Ayanamsa	24° 17′ 06″
Synetic vernal point	04° ♓ 49′ 52″
True obliquity of ecliptic	23° 26′ 06″

MOON'S PHASES, APSIDES AND POSITIONS ☽

Date	h	m	Phase	Longitude ° ′	Eclipse Indicator
02	14	12	●	12 ♉ 14	
10	17	11	☽	20 ♌ 06	
17	11	19	○	26 ♏ 37	
24	04	57	☾	03 ♓ 06	

Day	h	m		
04	03	51	Apogee	
17	13	54	Perigee	
31	06	31	Apogee	
05	14	33	Max dec	22° N 45′
12	19	00	0S	
18	20	40	Max dec	22° S 45′
25	08	17	0N	

LONGITUDES

Date	Chiron ⚷ ° ′	Ceres ⚳ ° ′	Pallas ⚴ ° ′	Juno ⚵ ° ′	Vesta ⚶ ° ′	Black Moon Lilith ⚸ ° ′
01	12 ♉ 13	16 ♈ 18	07 ♓ 55	11 ♑ 58	20 ♈ 50	27 ♉ 34
11	12 ♉ 52	20 ♈ 06	10 ♓ 10	11 ♑ 46	25 ♈ 22	28 ♉ 40
21	13 ♉ 30	23 ♈ 49	12 ♓ 08	10 ♑ 59	29 ♈ 49	29 ♉ 47
31	14 ♉ 07	27 ♈ 27	13 ♓ 45	09 ♑ 39	04 ♉ 10	00 ♊ 53

All ephemeris data is given at 12.00 UT and the Moon's longitude is additionally given for 24.00 UT

ASPECTARIAN

01 Wednesday
01 37 ☽ Q ♇
02 12 ☽ ⊼ ♃
03 16 ☿ ⊻ ♀
05 37 ☽ ⊻ ♄
10 42 ☽ ☌ ☿
11 46 ☽ ⊻ ♀
15 00 ☽ ∥ ☉
16 32 ☽ ∠ ♅

02 Thursday
01 06 ☽ ⊥ ♀
04 44 ☽ ∥ ♂
07 27 ☽ ∥ ♄
07 44 ☉ ⊻ ♆
08 49 ☽ # ♃
09 25 ☉ □ ♇
10 44 ☽ ⊥ ♅
13 41 ☽ ⊻ ♆
13 49 ☽ □ ♇
14 12 ☽ ☌ ☉
15 51 ♂ ⊥ ♆
21 13 ☽ ∠ ♀
22 55 ☽ ⊻ ♅

03 Friday
01 52 ☽ ⊥ ♆
02 30 ☽ ☌ ♂
14 06 ☽ ☍ ♃
18 51 ☽ ☌ ♄
20 13 ☽ ∠ ♆
21 45 ☽ ⊻ ☿

04 Saturday
02 52 ♂ ∥ ♄
06 59 ☽ ⚹ ♀
09 12 ♂ # ♃
09 45 ☽ ⊥ ☿
15 40 ♆ ⚹ ♇

05 Sunday
02 49 ☽ △ ♇
02 51 ☽ ⚹ ♆
03 42 ♃ # ♄
03 51 ☽ ∠ ☿
08 40 ☽ ⊻ ☉
10 04 ☽ Q ♀
12 14 ☽ ☌ ♅
19 32 ☽ ⊻ ♂
21 57 ☽ ⊥ ☉

06 Monday
02 32 ☽ ⊼ ♃
03 19 ☽ Q ♆
08 31 ☽ ⊥ ♂
08 34 ☽ ⊻ ♄
09 19 ☽ bQ ♇
09 21 ♀ Q ♅
10 07 ☽ ⚹ ☿
14 33 ☽ ± ♃
17 49 ☽ ∠ ☉
20 15 ☿ St D
20 49 ☽ ⊥ ♄

07 Tuesday
02 27 ☽ □ ♀
03 29 ☽ ± ♇
03 51 ☽ ∠ ♂
06 26 ☉ ⊻ ♅
08 30 ☽ bQ ♃
10 18 ☽ Q ☿
15 05 ☽ ∠ ♄
15 31 ☽ ⊼ ♇
15 42 ☽ □ ♆

08 Wednesday
00 58 ☽ ⊻ ♅
02 30 ☽ ⚹ ☉
11 38 ☽ ⚹ ♂
12 53 ☽ ⊥ ♅
13 01 ☽ ∥ ♂
14 00 ☽ △ ♃
15 08 ♀ ∠ ♂
17 32 ☉ ⊥ ♆
19 03 ☽ ∥ ♄
20 46 ☽ # ♃
21 04 ☽ ⚹ ♄
21 58 ☽ ∥ ☉
22 15 ☽ □ ☿

09 Thursday
04 05 ☽ Q ☉
06 33 ☽ ∠ ♅
10 46 ☉ # ♃
11 35 ♀ bQ ♃
12 28 ☽ Q ♂
19 37 ☽ △ ♀
20 30 ☽ Q ♄
22 01 ♂ ☍ ♃

10 Friday
02 05 ☽ ☍ ♇
02 23 ☽ △ ♆
11 17 ☽ ⚹ ♅
13 49 ☉ ∥ ♄
17 11 ☽ □ ☉
17 47 ♄ ∠ ♆
22 45 ☽ □ ♃

11 Saturday
00 27 ☽ □ ♂
02 33 ☽ bQ ♀
04 13 ☽ ∥ ☿
06 25 ☽ bQ ♆
06 30 ☽ □ ♄
08 10 ☽ △ ☿
09 34 ☽ Q ♅
20 28 ☽ ± ♀
22 44 ☽ ± ♆

12 Sunday
03 00 ☽ ∥ ♆
05 40 ☽ ∥ ♀
07 00 ☽ Q ♃
08 09 ☽ ⊼ ♀
09 00 ☽ ⊼ ♇
09 25 ☽ ⊼ ♆
11 50 ☽ bQ ☿
17 46 ☽ □ ♅
18 11 ♀ ⚹ ♇
19 31 ☽ ± ♇
23 16 ♇ St R
23 22 ♀ ☌ ♆

13 Monday
03 03 ☽ △ ☉
03 43 ☽ ⚹ ♃
04 00 ☽ ± ☿
04 28 ♀ ∠ ♄
08 39 ☽ △ ♂
10 21 ☽ # ♀
10 43 ☽ # ♆
10 59 ☽ bQ ♇
11 33 ☉ ☍ ♃
11 53 ☽ △ ♄
14 35 ☽ ⊼ ☿
15 05 ♀ ∥ ♆

14 Tuesday
04 49 ☽ ∠ ♃
06 13 ☽ bQ ☉
08 36 ☽ # ☿
11 07 ☽ bQ ♂
12 04 ☽ △ ♇
12 33 ☽ ☍ ♆
13 10 ☽ bQ ♄
15 40 ☽ ☍ ♀
19 31 ☽ ⊥ ♃
20 27 ☽ △ ♅
22 03 ☽ ± ☉

15 Wednesday
02 35 ☽ ± ♂
03 59 ☽ ± ♄
05 11 ☽ ⊻ ♃
08 27 ☽ ⊼ ☉
12 45 ☽ ⊼ ♂
13 45 ☽ ⊼ ♄
17 44 ♂ ∠ ♆
17 51 ☽ ☍ ☿
20 44 ☽ bQ ♅

16 Thursday
05 39 ☽ ∥ ♃
09 54 ☽ # ♄
11 06 ☽ ± ♅
11 43 ♀ ± ♃
12 21 ☽ □ ♇
12 55 ☽ ⊼ ♆
13 57 ♂ ☌ ♄
19 05 ☽ # ☉
19 49 ☽ ⊼ ♀
20 37 ☽ ⊼ ♅
22 25 ☽ ± ♆

17 Friday
00 40 ☽ # ♂
04 31 ☽ ☌ ♃
06 03 ☽ ± ♀
06 48 ♀ ⚹ ♅
11 19 ☽ ☍ ☉
12 38 ☽ bQ ♆
13 42 ☽ ☍ ♄
14 39 ☽ ☍ ♂
16 44 ☽ Q ♇
19 39 ☽ ⊼ ☿
21 24 ☽ bQ ♀

18 Saturday
05 38 ☽ ± ☿
08 38 ☉ ∠ ♆
09 47 ☿ ∠ ♅
11 41 ☽ ⚹ ♇
12 21 ☽ △ ♆
20 10 ☽ ☍ ♅
20 44 ☽ bQ ☿
23 09 ☽ △ ♀

19 Sunday
03 36 ☽ ⊻ ♃
06 22 ☉ ☌ ♄
11 37 ☽ ∠ ♇
13 09 ☽ ⊥ ♃
13 46 ☽ ⊼ ♄
14 12 ☽ ⊼ ☉
16 43 ☽ ⊼ ♂
22 20 ☽ △ ☿
23 34 ☽ ± ♄

20 Monday
00 36 ☽ ± ☉
02 12 ☽ ⊥ ♇
02 58 ☽ ± ♂
03 37 ☽ ∠ ♃
12 02 ☽ ⊻ ♇
12 42 ☽ # ♂
12 50 ☽ □ ♆
14 27 ☽ bQ ♄
16 26 ☽ bQ ☉
16 47 ☽ # ☉
18 32 ☽ bQ ♂
21 05 ☽ ⊼ ♅

21 Tuesday
03 20 ♀ ⊼ ♃
04 15 ☽ ⚹ ♃
04 21 ☽ □ ♀
07 11 ☽ ± ♅
08 31 ☽ # ♄
15 07 ☽ ∥ ♃
15 52 ☽ △ ♄
19 08 ☽ Q ♆
19 31 ☽ △ ☉
21 11 ☽ △ ♂
21 31 ♀ ⊥ ♄
22 35 ☽ bQ ♅

22 Wednesday
00 30 ☽ Q ♃
04 12 ☽ □ ☿
14 04 ☽ Q ♀
15 02 ☽ ☌ ♇
16 00 ☽ ⚹ ♆

23 Thursday
00 56 ☽ △ ♅
01 21 ♀ Q ♇
03 39 ☽ # ☿
07 54 ☽ □ ♃
13 29 ☽ ⚹ ♀
14 47 ☽ Q ☿
17 56 ☽ # ♀
18 55 ☽ ∠ ♆
21 20 ☽ □ ♄

24 Friday
04 57 ☽ □ ☉
05 34 ☽ □ ♂
11 31 ☽ ⊥ ♆
14 29 ☽ # ♆
14 55 ☽ ⚹ ☿
19 52 ☽ ∠ ♀
21 35 ☽ ⊻ ♇
22 46 ☽ ⊻ ♆

25 Saturday
07 13 ☽ Q ♄
08 27 ☽ □ ♅
08 56 ☽ ⊥ ♇
10 50 ☉ ☌ ♂
14 39 ☽ ⊥ ♀
15 04 ☽ △ ♃
17 27 ☽ Q ♂
17 36 ☽ Q ☉
22 17 ☽ ∠ ☿

26 Sunday
02 11 ☽ ∠ ♇
02 29 ☽ ∥ ♆
03 24 ☽ ⊻ ♀
06 30 ☽ ⚹ ♄
17 41 ☽ ⊥ ☿
18 07 ☽ ⚹ ♂
18 50 ☽ ⚹ ☉
19 23 ☽ Q ♅
19 52 ☽ bQ ♃

27 Monday
06 51 ☽ ⊻ ☿
07 31 ☽ ⚹ ♇
08 21 ☽ ∥ ♀
08 54 ☽ ☌ ♆
10 56 ☿ ⊥ ♅
12 15 ☽ ∠ ♄
13 14 ☿ □ ♇
13 27 ☽ ± ♃
15 51 ♀ ⊻ ♄
19 13 ☽ ⚹ ♅

28 Tuesday
01 17 ☽ ⊼ ♃
01 39 ☽ ∠ ♂
02 37 ☿ ⊻ ♆
03 06 ☽ ∠ ☉
04 35 ☽ ∥ ☿
06 23 ☽ ⊥ ♄
07 24 ☽ Q ♇
13 26 ♀ ⊥ ♂
18 34 ☽ ⊻ ♄
20 51 ☽ ⊥ ♂
21 10 ☽ ☌ ♀
22 45 ☽ ⊥ ☉

29 Wednesday
01 27 ☽ ∠ ♅
08 10 ☽ # ♃
09 43 ☽ ⊻ ♂
11 57 ☽ ⊻ ☉
19 42 ☽ □ ♇
19 45 ☽ ⊥ ♅
19 47 ☽ ∥ ♄
21 16 ☽ ⊻ ♆

30 Thursday
02 40 ☽ ☌ ☿
08 00 ☽ ⊻ ♅
09 29 ☽ ⊥ ♆
13 18 ☽ ☍ ♃
18 37 ♀ ∠ ♅

31 Friday
03 51 ☽ ∠ ♆
08 06 ☽ ☌ ♄
14 52 ☽ ∥ ☉
16 48 ☽ ⊻ ♀
17 01 ☽ ∥ ♂

LONGITUDES

Date	Sidereal time h m s	Sun ☉ ° ' "	Moon ☽ ° ' "	Moon ☽ 24.00 ° ' "	Mercury ☿ ° '	Venus ♀ ° '	Mars ♂ ° '	Jupiter ♃ ° '	Saturn ♄ ° '	Uranus ♅ ° '	Neptune ♆ ° '	Pluto ♇ ° '
01	04 40 04	11 ♊ 03 28	13 ♊ 36 33	19 ♊ 31 15	18 ♉ 53	05 ♉ 07	09 ♊ 15	20 ♏ 26	00 ♊ 03	18 ♊ 11	12 ♈ 50	11 ♒ 59
02	04 44 01	12 00 59	25 ♊ 26 54	01 ♋ 23 47	20 24	06 17	09 57	20 R 20	00 11	18 15	12 52	11 R 58
03	04 47 57	12 58 29	07 ♋ 22 12	13 ♋ 22 27	21 58	07 28	10 39	20 13	00 18	18 18	12 53	11 58
04	04 51 54	13 55 58	19 ♋ 24 51	25 ♋ 29 45	23 34	08 39	11 21	20 07	00 26	18 22	12 55	11 57
05	04 55 50	14 53 26	01 ♌ 37 30	07 ♌ 48 28	25 13	09 50	12 03	20 00	00 34	18 25	12 56	11 56
06	04 59 47	15 50 53	14 ♌ 03 01	20 ♌ 21 32	26 54	11 00	12 45	19 54	00 41	18 29	12 57	11 56
07	05 03 43	16 48 19	26 ♌ 44 24	03 ♍ 11 59	28 ♉ 38	12 11	13 26	19 48	00 49	18 33	12 58	11 55
08	05 07 40	17 45 43	09 ♍ 44 38	16 ♍ 22 41	00 ♊ 24	13 22	14 08	19 42	00 56	18 36	13 00	11 55
09	05 11 37	18 43 07	23 ♍ 06 23	29 ♍ 55 57	02 13	14 33	14 50	19 36	01 04	18 40	13 01	11 54
10	05 15 33	19 40 29	06 ♎ 51 28	13 ♎ 52 59	04 05	15 44	15 31	19 30	01 11	18 43	13 02	11 53
11	05 19 30	20 37 50	21 ♎ 00 23	28 ♎ 13 24	05 58	16 55	16 13	19 24	01 19	18 47	13 03	11 52
12	05 23 26	21 35 10	05 ♏ 31 39	12 ♏ 54 37	07 54	18 06	16 54	19 19	01 26	18 50	13 04	11 52
13	05 27 23	22 32 30	20 ♏ 21 34	27 ♏ 51 40	09 52	19 17	17 36	19 13	01 34	18 54	13 05	11 51
14	05 31 19	23 29 48	05 ♐ 23 59	12 ♐ 57 26	11 53	20 28	18 17	19 08	01 41	18 57	13 06	11 50
15	05 35 16	24 27 06	20 ♐ 30 55	28 ♐ 03 18	13 55	21 39	18 59	19 03	01 49	19 01	13 07	11 49
16	05 39 12	25 24 23	05 ♑ 33 27	13 ♑ 00 20	16 00	22 51	19 40	18 58	01 56	19 04	13 08	11 48
17	05 43 09	26 21 39	20 ♑ 22 59	27 ♑ 40 35	18 06	24 02	20 21	18 53	02 03	19 08	13 09	11 48
18	05 47 06	27 18 55	04 ♒ 52 28	11 ♒ 58 06	20 14	25 13	21 02	18 49	02 11	19 12	13 10	11 47
19	05 51 02	28 16 11	18 ♒ 57 09	25 ♒ 49 28	22 22	26 25	21 44	18 44	02 18	19 15	13 11	11 46
20	05 54 59	29 ♊ 13 26	02 ♓ 34 59	09 ♓ 13 51	24 32	27 36	22 25	18 40	02 25	19 19	13 12	11 45
21	05 58 55	00 ♋ 10 41	15 ♓ 46 18	22 ♓ 12 39	26 43	28 47	23 06	18 35	02 32	19 22	13 13	11 44
22	06 02 52	01 07 56	28 ♓ 33 20	04 ♈ 48 49	28 ♊ 55	29 ♉ 59	23 47	18 31	02 40	19 26	13 13	11 43
23	06 06 48	02 05 11	10 ♈ 59 38	17 ♈ 06 20	01 ♋ 06	01 ♊ 10	24 28	18 28	02 47	19 29	13 14	11 42
24	06 10 45	03 02 25	23 ♈ 09 28	29 ♈ 09 38	03 18	02 22	25 09	18 24	02 54	19 33	13 15	11 41
25	06 14 41	03 59 40	05 ♉ 07 23	11 ♉ 03 16	05 29	03 33	25 50	18 20	03 01	19 36	13 16	11 40
26	06 18 38	04 56 54	16 ♉ 57 48	22 ♉ 51 32	07 39	04 45	26 31	18 17	03 08	19 39	13 16	11 39
27	06 22 35	05 54 09	28 ♉ 44 54	04 ♊ 38 22	09 49	05 57	27 11	18 14	03 15	19 43	13 17	11 38
28	06 26 31	06 51 23	10 ♊ 32 20	16 ♊ 27 12	11 58	07 08	27 52	18 11	03 22	19 46	13 18	11 37
29	06 30 28	07 48 37	22 ♊ 23 18	28 ♊ 20 57	14 06	08 20	28 33	18 08	03 29	19 50	13 18	11 36
30	06 34 24	08 ♋ 45 51	04 ♋ 20 25	10 ♋ 21 58	16 ♋ 12	09 ♊ 32	29 ♊ 14	18 ♏ 05	03 ♊ 36	19 ♊ 53	13 ♈ 19	11 ♒ 35

Date	Moon True ☊ ° '	Moon Mean ☊ ° '	Moon ☽ Latitude ° '
01	16 ♐ 19	16 ♐ 47	00 N 15
02	16 D 19	16 44	00 S 51
03	16 19	16 41	01 54
04	16 20	16 37	02 54
05	16 20	16 34	03 46
06	16 20	16 31	04 29
07	16 R 20	16 28	05 00
08	16 20	16 25	05 16
09	16 D 20	16 21	05 16
10	16 20	16 18	04 58
11	16 20	16 15	04 22
12	16 21	16 12	03 28
13	16 21	16 09	02 20
14	16 22	16 06	01 S 01
15	16 R 22	16 02	00 N 23
16	16 21	15 59	01 45
17	16 21	15 56	02 58
18	16 19	15 53	03 58
19	16 18	15 50	04 42
20	16 17	15 47	05 08
21	16 16	15 43	05 17
22	16 15	15 40	05 10
23	16 D 15	15 37	04 48
24	16 16	15 34	04 14
25	16 18	15 31	03 29
26	16 19	15 27	02 36
27	16 20	15 24	01 36
28	16 21	15 21	00 N 32
29	16 R 21	15 18	00 S 33
30	16 ♐ 20	15 ♐ 15	01 S 38

DECLINATIONS

Date	Sun ☉ ° '	Moon ☽ ° '	Mercury ☿ ° '	Venus ♀ ° '	Mars ♂ ° '	Jupiter ♃ ° '	Saturn ♄ ° '	Uranus ♅ ° '	Neptune ♆ ° '	Pluto ♇ ° '
01	22 N 06	22 N 41	14 N 54	11 N 21	22 N 11	16 S 45	18 N 22	22 N 58	03 N 42	23 S 01
02	22 14	22 31	15 26	11 45	22 17	16 43	18 23	22 58	03 42	23 01
03	22 21	21 20	15 58	12 08	22 24	16 42	18 25	22 59	03 43	23 01
04	22 28	19 10	16 30	12 31	22 31	16 40	18 26	22 59	03 43	23 02
05	22 35	16 07	17 03	12 54	22 37	16 38	18 28	22 59	03 44	23 02
06	22 41	12 19	17 36	13 17	22 43	16 37	18 30	23 00	03 44	23 03
07	22 47	07 54	18 08	13 39	22 49	16 36	18 31	23 00	03 45	23 03
08	22 52	03 N 02	18 41	14 01	22 54	16 34	18 33	23 00	03 45	23 03
09	22 57	02 S 06	19 13	14 23	23 00	16 33	18 34	23 00	03 46	23 04
10	23 02	07 17	19 44	14 44	23 05	16 31	18 36	23 01	03 46	23 04
11	23 06	12 14	20 15	15 05	23 10	16 30	18 37	23 01	03 46	23 05
12	23 10	16 38	20 46	15 26	23 14	16 29	18 39	23 01	03 47	23 05
13	23 13	20 04	21 15	15 47	23 19	16 27	18 40	23 02	03 47	23 05
14	23 16	22 12	21 43	16 07	23 23	16 26	18 42	23 02	03 47	23 06
15	23 19	22 43	22 09	16 27	23 27	16 25	18 43	23 02	03 48	23 06
16	23 21	21 35	22 34	16 46	23 31	16 24	18 45	23 03	03 48	23 07
17	23 23	18 57	22 57	17 05	23 35	16 23	18 46	23 03	03 48	23 07
18	23 24	15 11	23 18	17 24	23 38	16 22	18 47	23 03	03 49	23 08
19	23 25	10 40	23 37	17 42	23 42	16 21	18 49	23 04	03 49	23 08
20	23 26	05 45	23 54	18 00	23 45	16 20	18 50	23 04	03 49	23 09
21	23 26	00 S 44	24 08	18 17	23 47	16 19	18 51	23 04	03 50	23 09
22	23 26	04 N 10	24 19	18 34	23 50	16 18	18 53	23 04	03 50	23 09
23	23 25	08 46	24 28	18 51	23 52	16 17	18 54	23 05	03 50	23 10
24	23 24	12 55	24 34	19 07	23 55	16 16	18 56	23 05	03 50	23 10
25	23 22	16 30	24 37	19 22	23 56	16 16	18 57	23 05	03 51	23 11
26	23 21	19 23	24 37	19 37	23 58	16 15	18 58	23 06	03 51	23 11
27	23 18	21 26	24 34	19 52	24 00	16 14	18 59	23 06	03 51	23 12
28	23 15	22 33	24 29	20 06	24 01	16 14	19 01	23 06	03 51	23 12
29	23 12	22 40	24 21	20 20	24 02	16 13	19 02	23 06	03 51	23 13
30	23 N 09	21 N 44	24 N 11	20 N 33	24 N 03	16 S 12	19 N 03	23 N 07	03 N 51	23 S 13

ZODIAC SIGN ENTRIES

Date	h	m	Planets
01	02	34	♄ ♊
02	21	11	☽ ♋
05	08	50	☽ ♌
07	18	05	☽ ♍
08	06	31	☿ ♊
10	00	07	☽ ♎
12	02	56	☽ ♏
14	03	25	☽ ♐
16	03	06	☽ ♑
18	03	51	☽ ♒
20	07	23	☽ ♓
21	07	31	☉ ♋
22	12	23	♀ ♊
22	14	45	☽ ♈
22	23	56	☿ ♋
25	01	41	☽ ♉
27	14	33	☽ ♊
30	03	19	☽ ♋

LATITUDES

Date	Mercury ☿ ° '	Venus ♀ ° '	Mars ♂ ° '	Jupiter ♃ ° '	Saturn ♄ ° '	Uranus ♅ ° '	Neptune ♆ ° '	Pluto ♇ ° '
01	02 S 38	01 S 59	00 N 21	01 N 09	01 S 50	00 N 03	01 S 29	06 S 04
04	02 14	01 58	00 23	01 09	01 50	00 03	01 30	06 04
07	01 45	01 56	00 24	01 08	01 50	00 03	01 30	06 05
10	01 14	01 53	00 26	01 07	01 50	00 03	01 30	06 06
13	00 41	01 50	00 28	01 07	01 50	00 03	01 30	06 06
16	00 S 08	01 46	00 29	01 06	01 50	00 04	01 30	06 07
19	00 N 24	01 42	00 31	01 05	01 50	00 04	01 30	06 08
22	00 53	01 37	00 33	01 05	01 51	00 04	01 30	06 08
25	01 18	01 31	00 34	01 04	01 51	00 04	01 31	06 09
28	01 36	01 25	00 36	01 03	01 51	00 04	01 31	06 09
31	01 N 47	01 S 19	00 N 38	01 N 03	01 S 51	00 N 04	01 S 31	06 S 10

DATA

Julian Date	2462654
Delta T	+74 seconds
Ayanamsa	24° 17' 11"
Synetic vernal point	04° ♓ 49' 48"
True obliquity of ecliptic	23° 26' 05"

MOON'S PHASES, APSIDES AND POSITIONS ☽

Date	h	m	Phase	Longitude ° '	Eclipse Indicator
01	06	21	●	10 ♊ 50	Annular
09	03	36	☽	18 ♍ 23	
15	18	41	○	24 ♐ 43	partial
22	17	20	☾	01 ♈ 21	
30	21	34	●	09 ♋ 09	

Day	h	m	
14	23	39	Perigee
27	14	27	Apogee

Day	h	m		
01	20	12	Max dec	22° N 44'
09	02	17	0S	
15	07	27	Max dec	22° S 45'
21	15	33	0N	
29	02	34	Max dec	22° N 45'

LONGITUDES

Date	Chiron ⚷ ° '	Ceres ⚳ ° '	Pallas ⚴ ° '	Juno ⚵ ° '	Vesta ⚶ ° '	Black Moon Lilith ⚸ ° '
01	14 ♉ 10	27 ♈ 48	13 ♓ 54	09 ♑ 29	04 ♉ 35	01 ♊ 00
11	14 ♉ 45	01 ♉ 19	15 ♓ 06	07 ♑ 38	08 ♉ 49	02 ♊ 06
21	15 ♉ 16	04 ♉ 41	15 ♓ 51	05 ♑ 28	12 ♉ 55	03 ♊ 13
31	15 ♉ 44	07 ♉ 53	16 ♓ 06	03 ♑ 09	16 ♉ 52	04 ♊ 19

ASPECTARIAN

h m Aspects

01 Saturday
00 15 ☿ ⊻ ♅
02 35 ☽ ☌ ♂
05 45 ☉ ⊥ ♀
06 21 ☽ ☌ ☉
06 22 ☽ ⊥ ♀
08 41 ☽ △ ♇
10 26 ☽ ✶ ♆
11 15 ☿ ⊥ ♆
21 21 ☽ ☌ ♅

02 Sunday
00 17 ☽ ⊻ ☿
01 44 ☽ ⊼ ♃
02 40 ☽ ∠ ♀
10 49 ☽ Q ♆
10 49 ☉ △ ♇
10 53 ☿ ☍ ♃
13 46 ☽ ± ♃
14 13 ☽ ⊥ ☿
15 04 ☽ bQ ♇
17 51 ☽ ∥ ♂
19 12 ☽ ∥ ☉
21 39 ☽ ⊻ ♄

03 Monday
07 43 ☽ bQ ♃
09 10 ☽ ± ♇
09 44 ☉ ✶ ♆
09 51 ☽ ⊥ ♄
11 04 ☽ ∠ ☿
12 13 ☽ ✶ ♀
18 58 ☽ ⊻ ♂
21 10 ☽ ⊼ ♇
23 03 ☽ □ ♆

04 Tuesday
00 10 ☽ ⊻ ☉
04 01 ☽ ∠ ♄
07 39 ☽ ⊥ ♂
09 55 ☽ ⊻ ♅
13 07 ☽ ⊥ ☉
13 22 ☽ △ ♃
14 42 ☽ Q ♀
18 16 ☽ ∥ ♄
18 51 ☿ # ♃
21 29 ☽ ✶ ☿
21 49 ☽ ⊥ ♅

05 Wednesday
02 30 ☽ ∠ ♂
06 16 ☽ ∥ ☿
08 15 ☽ # ♃
08 20 ☽ ∠ ☉
08 25 ♂ △ ♇
09 54 ☽ ✶ ♄
15 31 ☽ ∠ ♅

06 Thursday
00 33 ☽ Q ☿
05 33 ☽ □ ♀
06 47 ☽ ∥ ♀
07 56 ☽ ☍ ♇
09 21 ☽ ✶ ♂
09 22 ☽ Q ♄
09 54 ☽ △ ♆
15 43 ☽ ✶ ☉
19 25 ♂ ✶ ♆
20 29 ☽ ✶ ♅
23 02 ☽ □ ♃

07 Friday
02 45 ☿ ∠ ♆
06 35 ♀ □ ♇
09 25 ☽ Q ♂
14 18 ☽ bQ ♆
16 06 ☽ □ ☿
16 10 ☽ Q ☉
19 07 ☽ Q ♅
19 35 ♀ ⊥ ♅
19 40 ☽ □ ♄

08 Saturday
04 12 ♀ ⊻ ♆
05 45 ☿ ∥ ♄
06 58 ☽ ± ♆
08 17 ☽ Q ♃
08 35 ☽ ∥ ♆
15 56 ☽ ⊼ ♇
17 54 ☽ ⊼ ♆
19 13 ☽ △ ♀
19 37 ☿ ☌ ♄
20 24 ☽ □ ♂

09 Sunday
02 44 ☽ ± ♇
03 36 ☽ □ ☉
04 03 ☽ □ ♅
05 48 ☽ ✶ ♃
10 27 ☉ ☌ ♅
16 15 ♂ ∥ ♅
18 41 ☽ bQ ♇
19 41 ☽ # ♆

10 Monday
00 24 ☽ bQ ♀
01 35 ♀ ⊻ ♂
02 06 ☽ △ ♄
05 18 ☉ ∥ ♅
06 27 ☽ △ ☿
07 58 ☽ ∠ ♃
08 01 ☉ ⊼ ♃
08 57 ♂ # ♇
17 23 ☽ ± ♀
20 36 ☽ △ ♇
22 34 ☽ ☍ ♆
23 16 ☽ ⊥ ♃

11 Tuesday
01 36 ☉ # ♇
03 32 ☽ △ ♂
04 03 ☽ bQ ♄
04 31 ☽ ⊼ ♀
08 15 ☽ △ ♅
09 20 ☽ ⊻ ♃
11 20 ☽ △ ☉
11 56 ☽ bQ ☿
19 15 ☽ ± ♄

12 Wednesday
04 27 ☽ # ♀
05 09 ☽ ± ☿
05 15 ☽ ⊼ ♄
05 46 ☽ bQ ♂
09 14 ☽ bQ ♅
11 06 ☽ ∥ ♃
13 51 ☽ bQ ☉
16 28 ☽ ⊼ ☿
21 11 ☽ ± ♂
22 18 ☽ □ ♇
23 56 ☽ ± ♅

13 Thursday
00 16 ☽ ⊼ ♆
01 08 ☽ # ♄
03 43 ♀ ⊻ ♅
05 26 ☽ ± ☉
07 21 ☽ ⊼ ♂
07 53 ♀ ⊥ ♆
09 39 ☽ ⊼ ♅
09 57 ☽ ± ♆
10 08 ☽ ☍ ♀
10 11 ☽ ☌ ♃
10 50 ♀ ☍ ♃
15 44 ☽ ⊼ ☉

14 Friday
00 22 ☽ bQ ♆
02 25 ☽ # ☿
03 09 ☽ Q ♇
06 03 ☽ ☍ ♄
11 26 ☿ △ ♇
19 42 ♃ ± ♆
22 13 ☽ ✶ ♇
23 54 ☽ ☍ ☿

15 Saturday
00 15 ☽ △ ♆
02 32 ☿ ✶ ♆
09 26 ☽ ☍ ♂
09 36 ☽ ☍ ♅
09 41 ☽ ⊻ ♃
10 06 ♀ # ♃
13 29 ♂ ☌ ♅
13 58 ☽ ⊼ ♀
14 19 ♂ ⊼ ♃
17 59 ♃ ⊼ ♅
18 41 ☽ ☍ ☉
19 10 ☽ ⊥ ♃
22 01 ☽ ∠ ♇
23 44 ☽ # ☿

16 Sunday
00 20 ☽ ± ♀
01 52 ☉ ± ♃
05 06 ☉ Q ♆
06 09 ☽ ⊼ ♄
09 28 ☽ ∠ ♃
12 24 ☽ ⊥ ♇
15 51 ☽ ± ♄
16 00 ☽ bQ ♀
22 03 ☽ ⊻ ♇

17 Monday
00 14 ☽ □ ♆
06 32 ☽ bQ ♄
07 39 ☽ ⊼ ☿
09 34 ☽ ✶ ♃
09 57 ☽ ⊼ ♅
11 57 ☽ ⊼ ♂
13 23 ☽ # ♄
18 30 ☿ ∥ ♅
18 31 ☽ △ ♀
19 08 ☽ ± ☿
19 50 ☽ ± ♅
20 37 ☿ ⊼ ♃
22 18 ☽ ± ♂
22 31 ☽ ⊼ ☉
22 41 ☉ bQ ♇
23 25 ☿ # ♇
23 43 ☽ # ♀

18 Tuesday
00 03 ☿ ☌ ♅
05 06 ☽ ∥ ♃
05 14 ☽ Q ♃
05 48 ☽ Q ♆
07 27 ☽ △ ♄
09 12 ☽ ± ☉
10 51 ☽ bQ ♅
12 42 ☽ bQ ☿
14 04 ☽ bQ ♂
20 07 ☉ ∥ ☿
23 40 ☽ ☌ ♇

19 Wednesday
01 25 ☿ ☌ ♂
01 31 ☽ bQ ☉
02 04 ☽ ✶ ♆
11 37 ☽ □ ♃
12 31 ☽ △ ♅
17 04 ☽ △ ♂
19 04 ☽ △ ☿
19 40 ☿ ∥ ♂

20 Thursday
02 17 ☽ □ ♀
04 11 ☽ ∠ ♆
05 33 ☽ △ ☉
11 42 ☽ □ ♄
13 17 ☿ ± ♃
19 17 ☿ Q ♆
20 19 ☽ ⊥ ♆
21 14 ☽ # ♆

21 Friday
00 12 ♀ ∠ ♆
04 34 ☽ ⊻ ♇
07 16 ☽ ⊻ ♆
12 07 ☿ bQ ♇
14 05 ☽ Q ♀
15 38 ☽ ⊥ ♇
17 12 ☽ △ ♃
18 43 ☽ □ ♅
20 57 ☽ Q ♄

22 Saturday
02 26 ☽ □ ♂
08 30 ☽ ∠ ♇
10 20 ☽ ∥ ♆
12 49 ☽ □ ☿
15 00 ☽ ✶ ♀
17 20 ☽ □ ☉
19 56 ☽ ✶ ♄
21 28 ☽ bQ ♃

23 Sunday
05 08 ☽ Q ♅
11 48 ♂ ± ♃
13 23 ☽ ✶ ♇
13 41 ☿ ⊻ ♀
14 51 ☽ ± ♃
15 03 ☽ Q ♂
16 24 ☽ ☌ ♆
17 46 ♀ ∥ ♄
23 16 ☽ ∠ ♀

24 Monday
01 28 ☽ ∠ ♄
02 36 ☽ ⊼ ♃
04 47 ☽ ✶ ♅
07 04 ☉ ☌ ☿
07 24 ☿ ⊻ ♄
07 26 ☽ Q ☉
07 28 ☽ Q ☿
07 53 ☉ ⊻ ♄
13 03 ☽ Q ♇
13 06 ☿ bQ ♃
15 38 ♂ Q ♆
16 12 ☽ ✶ ♂
19 06 ☽ ⊥ ♀
19 32 ☽ ⊥ ♄
20 26 ☉ bQ ♃
23 52 ♀ ☌ ♄

25 Tuesday
07 42 ☽ ⊻ ♄
08 30 ☽ ⊻ ♀
09 31 ☽ ✶ ☉
10 12 ☽ # ♃
10 56 ☽ ∠ ♅
12 53 ☽ ✶ ☿
14 01 ☿ ± ♇

26 Wednesday
00 15 ☽ ∠ ♂
01 13 ☽ □ ♇
04 30 ☽ ⊻ ♆
05 15 ☽ ⊥ ♅
08 04 ☽ ∥ ♄
14 37 ☽ ∥ ♀
14 40 ☽ ☍ ♃
16 42 ☽ ⊥ ♆
16 45 ♂ bQ ♇
17 31 ☽ ⊻ ♅
18 36 ☽ ∠ ☉
19 40 ☽ ⊥ ♂

27 Thursday
02 12 ☽ ∠ ☿
05 15 ☿ ⊥ ♄
05 19 ☉ ± ♇
07 38 ☉ ⊻ ♀
08 38 ☽ ⊻ ♂
11 03 ☽ ∠ ♆
14 34 ☽ ⊥ ☉
21 15 ☽ ☌ ♄

28 Friday
00 39 ☽ ⊥ ☿
03 51 ☽ ⊻ ☉
04 19 ☽ ☌ ♀
08 03 ☿ ⊼ ♇
14 11 ☽ △ ♇
15 33 ☽ ⊻ ☿
17 36 ☽ ✶ ♆

29 Saturday
02 59 ☿ □ ♆
03 25 ☽ ⊼ ♃
06 49 ☽ ☌ ♅
09 18 ☉ # ♇
15 30 ☽ ± ♃
17 53 ☽ Q ♆
18 20 ☿ ⊥ ♀
20 28 ☽ bQ ♇

30 Sunday
01 09 ☽ ☌ ♂
09 30 ☽ bQ ♃
10 30 ☽ ⊻ ♄
14 28 ☽ ± ♇
21 34 ☽ ☌ ☉
22 34 ☽ ⊥ ♄
23 29 ☽ ⊻ ♀

All ephemeris data is given at 12.00 UT and the Moon's longitude is additionally given for 24.00 UT

JULY 2030

LONGITUDES

Date	Sidereal time h m s	Sun ☉ ° ' "	Moon ☽ ° ' "	Moon ☽ 24.00 ° '	Mercury ☿ ° '	Venus ♀ ° '	Mars ♂ ° '	Jupiter ♃ ° '	Saturn ♄ ° '	Uranus ♅ ° '	Neptune ♆ ° '	Pluto ♇ ° '
01	06 38 21	09 ♋ 43 05	16 ♋ 25 49	22 ♋ 32 11	18 ♋ 17	10 ♊ 44	29 ♊ 54	18 ♏ 02	03 ♊ 42	19 ♊ 57	13 ♈ 19	11 ♒ 34
02	06 42 17	10 40 19	28 ♋ 41 14	04 ♌ 53 08	20 21	11 56	00 ♋ 35	18 R 00	03 49	20 00	13 20	11 R 32
03	06 46 14	11 37 33	11 ♌ 08 04	17 ♌ 26 09	22 21	13 08	01 16	17 58	03 56	20 03	13 21	11 31
04	06 50 10	12 34 46	23 ♌ 47 33	00 ♍ 12 25	24 20	14 19	01 56	17 56	04 03	20 07	13 21	11 30
05	06 54 07	13 31 59	06 ♍ 40 52	13 ♍ 13 04	26 18	15 31	02 37	17 54	04 09	20 10	13 21	11 29
06	06 58 04	14 29 12	19 ♍ 49 09	26 ♍ 29 16	28 ♋ 14	16 43	03 17	17 52	04 16	20 13	13 21	11 28
07	07 02 00	15 26 24	03 ♎ 13 31	10 ♎ 02 00	00 ♌ 08	17 55	03 57	17 51	04 22	20 17	13 22	11 27
08	07 05 57	16 23 37	16 ♎ 54 49	23 ♎ 51 59	01 59	19 08	04 38	17 50	04 29	20 20	13 22	11 25
09	07 09 53	17 20 49	00 ♏ 53 29	07 ♏ 59 13	03 49	20 20	05 18	17 49	04 35	20 23	13 22	11 24
10	07 13 50	18 18 01	15 ♏ 09 01	22 ♏ 22 36	05 37	21 32	05 58	17 48	04 42	20 27	13 23	11 23
11	07 17 46	19 15 13	29 ♏ 39 36	06 ♐ 59 31	07 23	22 44	06 38	17 47	04 48	20 30	13 23	11 22
12	07 21 43	20 12 25	14 ♐ 21 46	21 ♐ 45 37	09 07	23 56	07 19	17 46	04 54	20 33	13 23	11 20
13	07 25 39	21 09 37	29 ♐ 10 16	06 ♑ 34 51	10 49	25 08	07 59	17 46	05 00	20 36	13 23	11 19
14	07 29 36	22 06 49	13 ♑ 58 25	21 ♑ 20 02	12 29	26 21	08 39	17 46	05 07	20 39	13 23	11 18
15	07 33 33	23 04 01	28 ♑ 38 45	05 ♒ 53 44	14 07	27 33	09 19	17 D 46	05 13	20 42	13 23	11 17
16	07 37 29	24 01 14	13 ♒ 04 09	20 ♒ 09 21	15 43	28 45	09 59	17 46	05 19	20 46	13 23	11 15
17	07 41 26	24 58 27	27 ♒ 08 49	04 ♓ 02 08	17 17	29 ♊ 58	10 39	17 46	05 25	20 49	13 R 23	11 14
18	07 45 22	25 55 41	10 ♓ 49 05	17 ♓ 29 35	18 49	01 ♋ 10	11 19	17 47	05 31	20 52	13 23	11 13
19	07 49 19	26 52 55	24 ♓ 03 42	00 ♈ 31 37	20 19	02 23	11 59	17 48	05 36	20 55	13 23	11 11
20	07 53 15	27 50 10	06 ♈ 53 41	13 ♈ 10 16	21 47	03 35	12 38	17 48	05 42	20 58	13 23	11 10
21	07 57 12	28 47 25	19 ♈ 21 52	25 ♈ 29 03	23 13	04 48	13 18	17 49	05 48	21 01	13 23	11 09
22	08 01 08	29 ♋ 44 42	01 ♉ 32 23	07 ♉ 32 31	24 37	06 00	13 58	17 51	05 54	21 04	13 22	11 07
23	08 05 05	00 ♌ 41 59	13 ♉ 30 05	19 ♉ 25 43	25 59	07 13	14 38	17 52	05 59	21 07	13 22	11 06
24	08 09 02	01 39 17	25 ♉ 20 06	01 ♊ 13 49	27 18	08 26	15 17	17 54	06 05	21 10	13 22	11 05
25	08 12 58	02 36 36	07 ♊ 07 30	13 ♊ 01 44	28 36	09 38	15 57	17 56	06 10	21 12	13 22	11 03
26	08 16 55	03 33 56	18 ♊ 57 02	24 ♊ 53 55	29 ♌ 51	10 51	16 37	17 57	06 15	21 15	13 21	11 02
27	08 20 51	04 31 16	00 ♋ 52 50	06 ♋ 54 10	01 ♍ 04	12 04	17 16	18 00	06 21	21 18	13 21	11 01
28	08 24 48	05 28 38	12 ♋ 58 18	19 ♋ 05 28	02 15	13 17	17 56	18 02	06 26	21 21	13 21	10 59
29	08 28 44	06 26 00	25 ♋ 15 56	01 ♌ 29 50	03 23	14 30	18 35	18 04	06 31	21 24	13 20	10 58
30	08 32 41	07 23 24	07 ♌ 47 16	14 ♌ 08 19	04 29	15 42	19 15	18 07	06 36	21 26	13 20	10 56
31	08 36 37	08 ♌ 20 47	20 ♌ 32 56	27 ♌ 01 06	05 ♍ 32	16 ♋ 55	19 ♋ 54	18 ♏ 10	06 ♊ 41	21 ♊ 29	13 ♈ 19	10 ♒ 55

Date	Moon True ☊ ° '	Moon Mean ☊ ° '	Moon ☽ Latitude ° '
01	16 ♐ 18	15 ♐ 12	02 S 38
02	16 R 15	15 08	03 32
03	16 11	15 05	04 17
04	16 07	15 02	04 50
05	16 03	14 59	05 09
06	16 00	14 56	05 13
07	15 58	14 53	05 00
08	15 D 58	14 49	04 29
09	15 58	14 46	03 42
10	15 59	14 43	02 41
11	16 01	14 40	01 28
12	16 01	14 37	00 S 09
13	16 R 01	14 33	01 N 11
14	15 59	14 30	02 27
15	15 55	14 27	03 31
16	15 50	14 24	04 22
17	15 44	14 21	04 54
18	15 38	14 18	05 09
19	15 33	14 14	05 07
20	15 29	14 11	04 49
21	15 27	14 08	04 18
22	15 D 26	14 05	03 36
23	15 27	14 02	02 45
24	15 28	13 58	01 47
25	15 29	13 55	00 N 45
26	15 R 30	13 52	00 S 19
27	15 28	13 49	01 22
28	15 25	13 46	02 23
29	15 19	13 43	03 17
30	15 12	13 39	04 04
31	15 ♐ 03	13 ♐ 36	04 S 39

DECLINATIONS

Date	Sun ☉ ° '	Moon ☽ ° '	Mercury ☿ ° '	Venus ♀ ° '	Mars ♂ ° '	Jupiter ♃ ° '	Saturn ♄ ° '	Uranus ♅ ° '	Neptune ♆ ° '	Pluto ♇ ° '
01	23 N 05	19 N 48	23 N 58	20 N 45	24 N 04	16 S 12	19 N 04	23 N 07	03 N 52	23 S 14
02	23 00	16 57	23 42	20 57	24 04	16 12	19 06	23 07	03 52	23 14
03	22 56	13 18	23 25	21 09	24 04	16 11	19 07	23 07	03 52	23 15
04	22 50	09 01	23 05	21 20	24 04	16 11	19 08	23 08	03 52	23 15
05	22 45	04 N 16	22 44	21 30	24 04	16 11	19 09	23 08	03 52	23 16
06	22 39	00 S 46	22 21	21 40	24 04	16 10	19 10	23 08	03 52	23 16
07	22 32	05 52	21 56	21 49	24 03	16 10	19 11	23 08	03 52	23 17
08	22 26	10 47	21 29	21 57	24 02	16 10	19 12	23 09	03 52	23 17
09	22 19	15 15	21 01	22 05	24 01	16 10	19 14	23 09	03 52	23 18
10	22 11	18 56	20 32	22 13	24 00	16 10	19 15	23 09	03 52	23 18
11	22 03	21 31	20 02	22 19	23 59	16 10	19 16	23 09	03 52	23 19
12	21 55	22 40	19 31	22 26	23 57	16 10	19 17	23 10	03 52	23 19
13	21 46	22 19	18 59	22 31	23 55	16 10	19 18	23 10	03 52	23 20
14	21 37	20 16	18 26	22 36	23 53	16 11	19 19	23 10	03 52	23 20
15	21 28	16 58	17 52	22 40	23 51	16 11	19 20	23 10	03 52	23 20
16	21 18	12 42	17 18	22 44	23 49	16 11	19 21	23 11	03 52	23 21
17	21 08	07 51	16 44	22 47	23 46	16 11	19 22	23 11	03 52	23 21
18	20 57	02 S 44	16 08	22 49	23 43	16 12	19 23	23 11	03 52	23 22
19	20 47	02 N 21	15 33	22 51	23 40	16 12	19 24	23 11	03 52	23 22
20	20 35	07 10	14 57	22 52	23 37	16 13	19 24	23 11	03 52	23 23
21	20 24	11 33	14 21	22 52	23 34	16 13	19 25	23 12	03 52	23 23
22	20 12	15 23	13 45	22 52	23 30	16 14	19 26	23 12	03 52	23 24
23	20 00	18 30	13 09	22 51	23 26	16 15	19 27	23 12	03 52	23 24
24	19 47	20 50	12 33	22 50	23 22	16 15	19 28	23 12	03 51	23 25
25	19 34	22 14	11 58	22 47	23 18	16 16	19 29	23 12	03 51	23 25
26	19 21	22 40	11 22	22 45	23 14	16 17	19 30	23 13	03 51	23 26
27	19 08	22 04	10 47	22 41	23 09	16 18	19 30	23 13	03 51	23 26
28	18 54	20 26	10 12	22 37	23 05	16 18	19 31	23 13	03 51	23 27
29	18 40	17 51	09 37	22 32	23 00	16 19	19 32	23 13	03 51	23 27
30	18 25	14 23	09 03	22 26	22 55	16 20	19 33	23 13	03 50	23 28
31	18 N 10	10 N 14	08 N 30	22 N 20	22 N 49	16 S 21	19 N 33	23 N 14	03 N 50	23 S 28

ZODIAC SIGN ENTRIES

Date	h m	Planets
01	15 20	♂ ♋
02	14 33	☽ ♌
04	23 37	☽ ♍
07	06 16	☽ ♎
07	10 23	☿ ♌
09	10 29	☽ ♏
11	12 33	☽ ♐
13	13 21	☽ ♑
15	14 14	☽ ♒
17	12 46	♀ ♋
17	16 57	☽ ♓
19	23 01	☽ ♈
22	08 56	☽ ♉
22	18 25	☉ ♌
24	21 30	☽ ♊
26	14 54	☿ ♍
27	10 14	☽ ♋
29	21 08	☽ ♌

LATITUDES

Date	Mercury ☿ ° '	Venus ♀ ° '	Mars ♂ ° '	Jupiter ♃ ° '	Saturn ♄ ° '	Uranus ♅ ° '	Neptune ♆ ° '	Pluto ♇ ° '
01	01 N 47	01 S 19	00 N 38	01 N 03	01 S 51	00 N 04	01 S 31	06 S 10
04	01 52	01 12	00 39	01 02	01 51	00 04	01 31	06 10
07	01 51	01 05	00 41	01 01	01 51	00 04	01 31	06 11
10	01 44	00 57	00 42	01 00	01 52	00 04	01 32	06 11
13	01 31	00 50	00 44	00 59	01 52	00 04	01 32	06 12
16	01 14	00 42	00 45	00 59	01 52	00 04	01 32	06 12
19	00 53	00 34	00 47	00 58	01 52	00 04	01 32	06 12
22	00 N 28	00 26	00 48	00 57	01 53	00 04	01 32	06 13
25	00 00	00 18	00 50	00 56	01 53	00 04	01 32	06 13
28	00 S 31	00 10	00 51	00 56	01 53	00 04	01 33	06 13
31	01 S 03	00 S 02	00 N 52	00 N 55	01 S 54	00 N 04	01 S 33	06 S 14

DATA

Julian Date	2462684
Delta T	+74 seconds
Ayanamsa	24° 17' 17"
Synetic vernal point	04° ♓ 49' 42"
True obliquity of ecliptic	23° 26' 04"

MOON'S PHASES, APSIDES AND POSITIONS ☽

Date	h m	Phase	Longitude ° '	Eclipse Indicator
08	11 02	☽	16 ♎ 21	
15	02 12	○	22 ♑ 41	
22	08 08	☾	29 ♈ 35	
30	11 11	●	07 ♌ 21	

Day	h m		
13	05 20	Perigee	
25	05 01	Apogee	
06	08 23	0S	
12	17 39	Max dec	22° S 43'
19	00 50	0N	
26	10 01	Max dec	22° N 40'

LONGITUDES

Date	Chiron ⚷ ° '	Ceres ⚳ ° '	Pallas ⚴ ° '	Juno ⚵ ° '	Vesta ⚶ ° '	Black Moon Lilith ⚸ ° '
01	15 ♉ 44	07 ♉ 53	16 ♓ 06	03 ♑ 09	16 ♉ 58	04 ♊ 19
11	16 ♉ 08	10 ♉ 54	15 ♓ 49	00 ♑ 56	20 ♉ 39	05 ♊ 26
21	16 ♉ 28	13 ♉ 40	14 ♓ 57	29 ♐ 01	24 ♉ 15	06 ♊ 32
31	16 ♉ 42	16 ♉ 11	13 ♓ 31	27 ♐ 33	27 ♉ 36	07 ♊ 39

All ephemeris data is given at 12.00 UT and the Moon's longitude is additionally given for 24.00 UT

ASPECTARIAN

h m	Aspects
01 Monday	
00 04	☉ ∥ ♅
02 02	☿ ∥ ♂
02 23	☽ ⊼ ♇
02 40	☽ ∥ ♀
05 51	☽ □ ♆
09 18	☿ △ ♃
11 41	☉ ⊥ ♄
12 39	☽ ⊥ ♀
15 10	☽ △ ♃
16 23	☽ ☌ ☿
16 32	☽ ∠ ♄
17 18	☿ ∠ ♄
18 52	☽ ∥ ♄
18 57	☽ ⊻ ♅
02 Tuesday	
04 21	♀ △ ♇
06 45	☽ ⊥ ♅
08 02	☿ ⊻ ♅
08 12	☽ ∠ ♀
15 54	☽ ⊻ ♂
17 25	☽ # ♃
22 02	☽ ⚹ ♄
03 Wednesday	
00 17	☽ ∠ ♅
04 09	☽ ⊥ ♂
09 26	☉ ⊼ ♇
12 44	☽ ☍ ♇
13 01	☽ ⊻ ☉
16 12	☽ ⚹ ♀
16 12	☽ △ ♆
16 13	♀ ⚹ ♆
21 14	☽ Q ♄
22 19	☽ ∠ ♂
04 Thursday	
00 37	☿ # ♇
00 58	☽ □ ♃
01 22	☽ ⊥ ☉
05 02	☽ ⚹ ♅
09 18	☿ ∥ ♅
13 13	☽ ⊻ ☿
17 14	☽ Q ♀
19 40	☽ ∠ ☉
20 32	☽ ℙ ♆
05 Friday	
02 27	☽ ⊥ ☿
03 37	☽ Q ♅
04 03	☽ ⚹ ♂
07 17	☽ □ ♄
07 20	☉ □ ♆
10 18	☿ ⊥ ♅
10 34	☽ Q ♃
10 37	☉ ∥ ☿
13 14	☽ ± ♆
13 55	☽ ∥ ♆
20 49	☽ ⊼ ♇
21 58	♂ ℙ ♃
21 58	☽ ∠ ☿
06 Saturday	
00 15	☽ ⊼ ♆
01 34	☽ ⚹ ☉
03 19	☽ Q ♂
05 50	☽ □ ♀
07 44	☽ ± ♇
08 29	☽ ⚹ ♃
12 44	☽ □ ♅
23 56	☽ ℙ ♇
07 Sunday	
00 55	☽ Q ☉
02 35	☽ # ♆
05 36	☽ ⚹ ☿
10 31	♀ ⊼ ♃
11 20	☽ ∠ ♃
13 22	☽ □ ♂
14 03	☽ △ ♄
16 47	☿ ∥ ♀
08 Monday	
02 27	☽ △ ♇
03 09	☽ ⊥ ♃
05 42	♂ ⊻ ♄
05 49	☽ ☍ ♆
06 08	☽ Q ☿
11 02	☽ □ ☉
13 35	☽ ⊻ ♃
16 12	☽ △ ♀
16 29	☽ ℙ ♄
17 56	☽ △ ♅
09 Tuesday	
08 03	☽ ± ♄
13 16	♀ ☌ ♅
15 33	♂ ± ♇
17 27	☽ ∥ ♃
17 42	☽ □ ☿
18 19	☽ ⊼ ♄
19 39	☽ ℙ ♅
19 50	☽ △ ♂
20 13	☽ ℙ ♀
22 48	☿ ⚹ ♄
23 27	☉ △ ♃
10 Wednesday	
05 43	☽ □ ♇
09 02	☽ ⊼ ♆
09 21	☉ ∥ ♀
09 32	☿ ∠ ♅
10 49	☽ ± ♅
12 41	☽ ± ♀
14 20	☽ # ♄
16 24	☽ ☌ ♃
17 37	☽ △ ☉
19 02	☽ ± ♆
19 37	☿ ⊻ ♂
20 50	☽ ⊼ ♅
22 09	☽ ℙ ♂
23 15	☽ # ☿
23 33	☽ ⊼ ♀
11 Thursday	
09 53	☽ ℙ ♆
11 31	☽ Q ♇
13 41	☽ ± ♂
19 06	☽ # ☉
20 03	☽ ℙ ☉
20 29	☽ ☍ ♄
23 58	☽ ⊼ ♂
12 Friday	
02 18	☽ # ♀
02 20	☽ △ ☿
03 25	☉ ∠ ♄
03 34	☿ ∠ ♀
07 06	☽ ⚹ ♇
08 47	♀ ± ♃
10 24	☽ △ ♆
11 44	☽ ± ☉
17 32	☽ ⊻ ♃
21 07	☉ ⊻ ♅
22 04	☽ ☍ ♅
22 08	☽ ⊼ ☉
22 25	☿ ∥ ♄
13 Saturday	
03 15	☽ ⊥ ♃
04 54	☽ ☍ ♀
05 52	☽ ℙ ☿
06 06	☽ # ♀
07 23	☽ ∠ ♇
16 49	♀ Q ♆
17 49	☽ ∠ ♃
19 07	☿ ☍ ♇
20 27	☽ # ☉
21 31	☽ ⊼ ♄
21 56	☽ ⊥ ♇
22 18	☽ ± ☿
14 Sunday	
02 56	☽ ☍ ♂
07 19	☽ ± ♄
07 39	☽ ⊻ ♇
09 16	☽ ⊼ ☿
11 02	☽ □ ♆
11 06	♀ ℙ ♇
18 10	☽ ⚹ ♃
19 53	☽ # ♄
22 04	☽ ℙ ♄
22 56	☽ ⊼ ♅
15 Monday	
01 10	☿ △ ♆
01 26	♃ St D
02 12	☽ ☍ ☉
05 07	☽ # ☿
08 48	☽ ± ♅
10 02	☽ ⊼ ♀
13 50	☽ Q ♃
16 31	☽ Q ♆
16 50	☽ ∥ ♃
20 50	☽ ± ♀
22 56	☽ △ ♄
23 44	☽ ℙ ♅
16 Tuesday	
06 34	☽ ⊼ ♂
08 57	☽ ☌ ♇
12 32	☽ ⚹ ♆
13 16	☽ ℙ ♀
16 27	♆ St R
17 01	☽ ☍ ☿
17 09	☽ ± ♂
19 00	☿ ⊥ ♂
19 56	☽ □ ♃
17 Wednesday	
01 05	☽ △ ♅
07 59	☽ ⊼ ☉
09 17	☽ ℙ ♂
14 08	☿ Q ♄
14 08	☽ ∠ ♆
17 21	☽ △ ♀
19 08	☽ ± ☉
19 39	☿ □ ♃
18 Thursday	
02 31	☽ □ ♄
05 54	☽ ⊥ ♆
06 42	☽ # ♆
08 28	♂ ⊼ ♇
09 41	☿ # ♃
12 13	☽ ℙ ☉
12 42	☽ ⊻ ♇
12 56	☽ △ ♂
16 35	☽ ⊻ ♆
20 19	♂ ⊥ ♄
23 28	☽ ⊥ ♇
19 Friday	
00 32	☽ △ ♃
04 15	☽ ⊼ ☿
06 12	☽ □ ♅
11 09	☽ Q ♄
12 48	☉ ⊥ ♅
15 55	☽ ∠ ♇
16 42	☽ ± ☿
17 38	☽ △ ☉
19 25	☽ ∥ ♆
20 21	♀ ℙ ♃
22 02	☿ ⚹ ♅
20 Saturday	
04 16	☽ ℙ ♃
05 05	☽ □ ♀
09 43	☽ ⚹ ♄
11 46	☽ ℙ ☿
15 57	☽ Q ♅
20 08	☽ ⚹ ♇
21 23	☽ ± ♃
23 36	☽ □ ♂
21 Sunday	
00 24	☽ ☌ ♆
09 00	☽ ⊼ ♃
14 36	♂ □ ♆
14 49	☽ ∠ ♄
15 14	☽ ⚹ ♅
18 50	♀ ± ♇
19 23	☽ Q ♇
19 27	☽ Q ♀
20 31	☽ △ ☿
22 Monday	
02 48	☽ ∥ ☿
08 08	☽ □ ☉
08 42	☽ ⊥ ♄
09 35	♀ ⊻ ♄
12 54	☽ Q ♂
18 03	☽ # ♃
20 46	☽ ⊻ ♄
21 04	☽ ∠ ♅
21 55	☽ ⚹ ♀
23 Tuesday	
07 10	☽ □ ♇
11 44	☽ ⊻ ♆
14 25	☽ ⚹ ♂
15 16	☽ ⊥ ♅
20 49	☽ ∥ ♄
20 51	☽ ☍ ♃
21 57	♂ # ♇
23 27	☽ Q ☉
23 53	☽ ⊥ ♆
24 Wednesday	
01 09	☽ ∥ ☉
03 29	☽ ⊻ ♅
07 41	☽ ∠ ♀
16 30	☽ □ ☿
18 10	☽ ∠ ♆
22 41	☽ ∠ ♂
25 Thursday	
02 00	☽ ⚹ ☉
04 05	☽ ⊥ ♀
07 36	☿ ℙ ♆
10 02	☽ ☌ ♄
17 42	☽ ⊻ ♀
18 05	☽ ⊥ ♂
19 59	☽ △ ♇
21 38	☉ ∥ ♄
26 Friday	
00 40	☽ ⚹ ♆
06 59	☽ ⊻ ♂
09 31	☽ Q ☿
09 59	☽ ⊼ ♃
11 09	☽ ∠ ☉
15 31	♀ ⊼ ♇
16 40	☽ ☌ ♅
17 35	♂ ∥ ♅
22 08	☽ ± ♃
27 Saturday	
00 55	☽ Q ♆
02 15	☽ ℙ ♇
06 52	☽ ⊥ ☉
12 25	☽ ⚹ ☿
16 14	☽ ℙ ♃
18 00	♀ ⊥ ♄
19 53	☽ ⊻ ☉
20 13	☽ ± ♇
22 58	☽ ⊻ ♄
28 Sunday	
08 06	☽ ⊼ ♇
10 56	☽ ⊥ ♄
12 40	☽ ☌ ♀
12 44	☽ □ ♆
13 19	♀ □ ♆
15 58	♂ △ ♃
21 16	☽ ∠ ☿
21 27	☽ ∥ ♄
21 58	☽ △ ♃
22 17	☽ ☌ ♂
29 Monday	
04 28	☽ ⊻ ♅
04 39	☽ ∥ ☉
04 41	☽ ∠ ♄
10 58	☉ ∠ ♅
12 16	☿ Q ♅
14 22	☉ ⚹ ♄
16 08	☽ ⊥ ♅
16 29	☽ ⊥ ☿
23 02	☿ ∠ ♂
23 12	☽ # ♃
30 Tuesday	
05 06	☽ ⊻ ☿
09 26	☽ ∠ ♅
09 44	☽ ⚹ ♄
11 11	☽ ☌ ☉
17 58	☽ ☍ ♇
22 29	☽ △ ♆
31 Wednesday	
04 31	☽ ⊻ ♀
07 32	☽ □ ♃
08 30	☽ Q ♄
10 44	☽ ⊻ ♂
13 45	☽ ⚹ ♅
16 53	☽ ⊥ ♀
22 22	☽ ∥ ☿
22 28	☽ ⊥ ♂

AUGUST 2030

LONGITUDES

Date	Sidereal time h m s	Sun ☉ ° ' "	Moon ☽ ° ' "	Moon ☽ 24.00 ° ' "	Mercury ☿ ° '	Venus ♀ ° '	Mars ♂ ° '	Jupiter ♃ ° '	Saturn ♄ ° '	Uranus ♅ ° '	Neptune ♆ ° '	Pluto ⯓ ° '
01	08 40 34	09 ♌ 18 12	03 ♍ 32 43	10 ♍ 07 40	06 ♍ 32	18 ♋ 08	20 ♋ 34	18 ♏ 13	06 ♊ 46	21 ♊ 32	13 ♈ 19	10 ♒ 54
02	08 44 31	10 15 37	16 ♍ 45 49	23 ♍ 27 03	07 30	19 21	21 13	18 16	06 51	21 34	13 R 18	10 R 52
03	08 48 27	11 13 03	00 ♎ 11 12	06 ♎ 58 09	08 24	20 34	21 52	18 19	06 56	21 37	13 18	10 51
04	08 52 24	12 10 30	13 ♎ 47 47	20 ♎ 40 00	09 16	21 48	22 31	18 23	07 01	21 39	13 17	10 50
05	08 56 20	13 07 57	27 ♎ 34 43	04 ♏ 31 52	10 04	23 01	23 11	18 27	07 05	21 42	13 17	10 48
06	09 00 17	14 05 25	11 ♏ 31 23	18 ♏ 33 12	10 49	24 14	23 50	18 31	07 10	21 44	13 16	10 47
07	09 04 13	15 02 54	25 ♏ 37 12	02 ♐ 43 17	11 30	25 27	24 29	18 35	07 14	21 47	13 15	10 45
08	09 08 10	16 00 24	09 ♐ 51 17	17 ♐ 00 58	12 08	26 40	25 08	18 39	07 19	21 49	13 15	10 44
09	09 12 06	16 57 54	24 ♐ 12 02	01 ♑ 24 05	12 41	27 54	25 47	18 43	07 23	21 52	13 14	10 43
10	09 16 03	17 55 26	08 ♑ 36 41	15 ♑ 49 16	13 11	29 ♋ 07	26 26	18 48	07 27	21 54	13 13	10 41
11	09 20 00	18 52 58	23 ♑ 01 14	00 ♒ 11 55	13 36	00 ♌ 20	27 05	18 52	07 31	21 56	13 12	10 40
12	09 23 56	19 50 31	07 ♒ 20 37	14 ♒ 26 38	13 57	01 34	27 44	18 57	07 35	21 59	13 12	10 39
13	09 27 53	20 48 05	21 ♒ 29 17	28 ♒ 27 57	14 12	02 47	28 23	19 02	07 39	22 01	13 11	10 37
14	09 31 49	21 45 40	05 ♓ 22 03	12 ♓ 11 08	14 23	04 00	29 02	19 07	07 43	22 03	13 10	10 36
15	09 35 46	22 43 16	18 ♓ 54 53	25 ♓ 33 04	14 29	05 14	29 ♋ 41	19 13	07 47	22 05	13 09	10 35
16	09 39 42	23 40 54	02 ♈ 05 37	08 ♈ 32 33	14 R 29	06 27	00 ♌ 20	19 18	07 50	22 07	13 08	10 33
17	09 43 39	24 38 33	14 ♈ 54 04	21 ♈ 10 25	14 24	07 41	00 58	19 24	07 54	22 09	13 07	10 32
18	09 47 35	25 36 14	27 ♈ 21 59	03 ♉ 29 14	14 13	08 55	01 37	19 29	07 58	22 11	13 06	10 31
19	09 51 32	26 33 57	09 ♉ 32 42	15 ♉ 32 57	13 57	10 08	02 16	19 35	08 01	22 13	13 05	10 29
20	09 55 29	27 31 41	21 ♉ 30 38	27 ♉ 26 25	13 34	11 22	02 54	19 41	08 04	22 15	13 04	10 28
21	09 59 25	28 29 26	03 ♊ 20 58	09 ♊ 14 58	13 07	12 36	03 33	19 47	08 07	22 17	13 03	10 27
22	10 03 22	29 ♌ 27 14	15 ♊ 09 07	21 ♊ 04 04	12 34	13 49	04 12	19 54	08 11	22 19	13 02	10 25
23	10 07 18	00 ♍ 25 03	27 ♊ 00 30	02 ♋ 59 00	11 55	15 03	04 50	20 00	08 14	22 21	13 01	10 24
24	10 11 15	01 22 53	09 ♋ 00 09	15 ♋ 04 50	11 12	16 17	05 29	20 07	08 17	22 23	13 00	10 23
25	10 15 11	02 20 46	21 ♋ 12 26	27 ♋ 24 23	10 25	17 31	06 07	20 14	08 19	22 24	12 59	10 22
26	10 19 08	03 18 40	03 ♌ 40 40	10 ♌ 01 28	09 35	18 45	06 46	20 20	08 22	22 26	12 57	10 20
27	10 23 04	04 16 36	16 ♌ 26 54	22 ♌ 57 00	08 41	19 59	07 24	20 28	08 25	22 28	12 56	10 19
28	10 27 01	05 14 33	29 ♌ 31 41	06 ♍ 10 46	07 47	21 13	08 02	20 35	08 27	22 29	12 55	10 18
29	10 30 58	06 12 31	12 ♍ 54 00	19 ♍ 41 02	06 51	22 27	08 41	20 42	08 30	22 31	12 54	10 17
30	10 34 54	07 10 32	26 ♍ 31 30	03 ♎ 24 58	05 56	23 41	09 19	20 49	08 32	22 32	12 52	10 15
31	10 38 51	08 ♍ 08 34	10 ♎ 20 59	17 ♎ 19 06	05 ♍ 03	24 ♌ 55	09 ♌ 57	20 ♏ 57	08 ♊ 34	22 ♊ 34	12 ♈ 51	10 ♒ 14

Date	Moon True ☊ ° '	Moon Mean ☊ ° '	Moon ☽ Latitude ° '
01	14 ♐ 54	13 ♐ 33	05 S 00
02	14 R 45	13 30	05 05
03	14 37	13 27	04 55
04	14 32	13 24	04 27
05	14 29	13 20	03 44
06	14 28	13 17	02 47
07	14 D 28	13 14	01 39
08	14 28	13 11	00 S 25
09	14 R 28	13 08	00 N 52
10	14 26	13 04	02 05
11	14 21	13 01	03 10
12	14 14	12 58	04 03
13	14 05	12 55	04 40
14	13 54	12 52	05 00
15	13 44	12 49	05 02
16	13 34	12 45	04 47
17	13 26	12 42	04 19
18	13 20	12 39	03 39
19	13 17	12 36	02 49
20	13 16	12 33	01 53
21	13 D 15	12 30	00 N 52
22	13 R 16	12 26	00 S 10
23	13 15	12 23	01 12
24	13 12	12 20	02 12
25	13 07	12 17	03 07
26	13 00	12 14	03 54
27	12 50	12 10	04 30
28	12 38	12 07	04 53
29	12 25	12 04	05 01
30	12 13	12 01	04 51
31	12 ♐ 03	11 ♐ 58	04 S 25

DECLINATIONS

Date	Sun ☉ ° '	Moon ☽ ° '	Mercury ☿ ° '	Venus ♀ ° '	Mars ♂ ° '	Jupiter ♃ ° '	Saturn ♄ ° '	Uranus ♅ ° '	Neptune ♆ ° '	Pluto ⯓ ° '
01	17 N 55	05 N 33	07 N 57	22 N 13	22 N 44	16 S 22	19 N 34	23 N 14	03 N 50	23 S 29
02	17 40	00 N 32	07 25	22 06	22 38	16 24	19 35	23 14	03 50	23 29
03	17 24	04 S 35	06 54	21 58	22 33	16 25	19 35	23 14	03 49	23 30
04	17 08	09 32	06 24	21 49	22 27	16 26	19 36	23 14	03 49	23 30
05	16 52	14 05	05 55	21 40	22 21	16 27	19 37	23 15	03 49	23 31
06	16 36	17 56	05 27	21 30	22 14	16 29	19 37	23 15	03 48	23 31
07	16 19	20 46	05 01	21 19	22 08	16 30	19 38	23 15	03 48	23 31
08	16 02	22 20	04 36	21 08	22 01	16 31	19 39	23 15	03 48	23 32
09	15 45	22 27	04 12	20 56	21 54	16 33	19 39	23 15	03 48	23 32
10	15 27	21 04	03 50	20 43	21 48	16 34	19 40	23 15	03 47	23 33
11	15 10	18 21	03 30	20 30	21 40	16 36	19 40	23 15	03 47	23 33
12	14 52	14 31	03 12	20 17	21 33	16 37	19 41	23 16	03 46	23 34
13	14 33	09 55	02 56	20 02	21 26	16 39	19 41	23 16	03 46	23 34
14	14 15	04 S 54	02 42	19 47	21 18	16 40	19 42	23 16	03 46	23 34
15	13 56	00 N 14	02 31	19 32	21 11	16 42	19 42	23 16	03 45	23 35
16	13 37	05 14	02 22	19 16	21 03	16 44	19 43	23 16	03 45	23 35
17	13 18	09 51	02 17	19 00	20 55	16 45	19 43	23 16	03 45	23 36
18	12 59	13 56	02 14	18 43	20 47	16 47	19 43	23 16	03 44	23 36
19	12 39	17 20	02 14	18 25	20 38	16 49	19 44	23 17	03 44	23 36
20	12 20	19 57	02 17	18 07	20 30	16 51	19 44	23 17	03 43	23 37
21	12 00	21 41	02 24	17 48	20 21	16 53	19 45	23 17	03 43	23 37
22	11 40	22 26	02 34	17 29	20 13	16 55	19 45	23 17	03 42	23 38
23	11 19	22 12	02 47	17 09	20 04	16 57	19 45	23 17	03 42	23 38
24	10 59	20 56	03 03	16 49	19 55	16 59	19 46	23 17	03 42	23 38
25	10 38	18 41	03 23	16 29	19 46	17 01	19 46	23 17	03 41	23 39
26	10 17	15 32	03 45	16 08	19 36	17 03	19 46	23 17	03 40	23 39
27	09 56	11 37	04 11	15 46	19 27	17 05	19 46	23 18	03 40	23 39
28	09 35	07 03	04 38	15 24	19 17	17 07	19 47	23 18	03 39	23 40
29	09 14	02 N 05	05 07	15 02	19 08	17 09	19 47	23 18	03 39	23 40
30	08 52	03 S 05	05 38	14 39	18 58	17 11	19 47	23 18	03 38	23 40
31	08 N 31	08 S 10	06 N 10	14 N 16	18 N 48	17 S 13	19 N 47	23 N 18	03 N 38	23 S 41

ZODIAC SIGN ENTRIES

Date	h m	Planets	
01	05 30	☽	♍
03	11 40	☽	♎
05	16 11	☽	♏
07	19 24	☽	♐
09	21 40	☽	♑
11	05 24	♀	♌
11	23 40	☽	♒
14	02 39	☽	♓
15	23 56	♂	♌
16	08 08	☽	♈
18	17 09	☽	♉
21	05 12	☽	♊
23	01 36	☉	♍
23	18 01	☽	♋
26	04 59	☽	♌
28	12 51	☽	♍
30	18 04	☽	♎

LATITUDES

Date	Mercury ☿ ° '	Venus ♀ ° '	Mars ♂ ° '	Jupiter ♃ ° '	Saturn ♄ ° '	Uranus ♅ ° '	Neptune ♆ ° '	Pluto ⯓ ° '
01	01 S 15	00 N 01	00 N 53	00 N 55	01 S 54	00 N 04	01 S 33	06 S 14
04	01 50	00 09	00 54	00 54	01 54	00 04	01 33	06 14
07	02 25	00 16	00 56	00 54	01 54	00 04	01 33	06 14
10	03 00	00 24	00 57	00 53	01 55	00 04	01 33	06 14
13	03 33	00 31	00 58	00 52	01 55	00 04	01 33	06 14
16	04 03	00 38	00 59	00 51	01 56	00 04	01 33	06 15
19	04 25	00 45	01 00	00 50	01 56	00 04	01 34	06 15
22	04 38	00 51	01 02	00 50	01 56	00 04	01 34	06 15
25	04 37	00 56	01 03	00 49	01 57	00 04	01 34	06 15
28	04 20	01 02	01 04	00 48	01 57	00 04	01 34	06 15
31	03 S 45	01 N 07	01 N 06	00 N 48	01 S 58	00 N 04	01 S 34	06 S 15

DATA

Julian Date	2462715
Delta T	+74 seconds
Ayanamsa	24° 17' 21"
Synetic vernal point	04° ♓ 49' 37"
True obliquity of ecliptic	23° 26' 05"

MOON'S PHASES, APSIDES AND POSITIONS ☽

Date	h	m	Phase	Longitude ° '	Eclipse Indicator
06	16	43	☽	14 ♏ 17	
13	10	44	○	20 ♒ 45	
21	01	15	☾	28 ♉ 04	
28	23	07	●	05 ♍ 41	

Day	h	m		
09	22	51	Perigee	
21	22	51	Apogee	
02	14	30	0S	
09	01	51	Max dec	22° S 35'
15	10	52	0N	
22	18	12	Max dec	22° N 29'
29	21	44	0S	

LONGITUDES

Date	Chiron ⚷ ° '	Ceres ⚳ ° '	Pallas ⚴ ° '	Juno ⚵ ° '	Vesta ⚶ ° '	Black Moon Lilith ⚸ ° '
01	16 ♉ 43	16 ♉ 25	13 ♓ 20	27 ♐ 26	27 ♉ 55	07 ♊ 45
11	16 ♉ 52	18 ♉ 34	11 ♓ 21	26 ♐ 35	00 ♊ 58	08 ♊ 52
21	16 ♉ 55	20 ♉ 16	08 ♓ 59	26 ♐ 19	03 ♊ 42	09 ♊ 58
31	16 ♉ 51	21 ♉ 38	06 ♓ 26	26 ♐ 39	06 ♊ 02	11 ♊ 05

All ephemeris data is given at 12.00 UT and the Moon's longitude is additionally given for 24.00 UT

ASPECTARIAN

h m	Aspects
01 Thursday	
02 24	☽ ♇ ♆
03 51	☿ Q ♃
11 11	☽ ∠ ♀
11 58	☽ Q ♅
13 33	♀ △ ♃
15 53	☽ ∠ ♂
16 54	☽ Q ♃
17 54	☽ ☌ ☿
17 56	☽ □ ♄
18 19	☿ □ ♄
18 53	☽ ± ♆
20 20	☽ ∥ ♆
23 19	☽ ⊻ ☉
02 Friday	
01 22	☽ ⊼ ⯓
05 46	☽ ⊼ ♆
07 16	☿ ± ♆
11 01	☽ ⊥ ☉
12 12	☽ ± ⯓
14 43	☽ ⚹ ♃
17 08	☽ ⚹ ♀
20 25	☽ ⚹ ♂
20 40	☽ □ ♅
03 Saturday	
02 02	♂ ⊻ ♅
02 59	☉ ☍ ⯓
04 18	☽ ♇ ⯓
04 24	☽ ∠ ☉
08 27	☽ # ♆
14 33	♂ ∠ ♄
16 39	☽ Q ♀
17 35	☽ ∠ ♃
18 51	☽ Q ♂
22 04	☽ # ☿
04 Sunday	
00 00	☽ △ ♄
03 31	☽ ⊻ ☿
06 48	☽ △ ⯓
08 57	☽ ⚹ ☉
09 15	♀ ⊻ ♅
09 31	☽ ⊥ ♃
11 07	☽ ☍ ♆
14 44	☽ ⊥ ☿
16 33	♀ ∠ ♄
20 03	☽ ⊻ ♃
05 Monday	
01 46	☽ △ ♅
02 25	☽ ♇ ♄
03 19	☽ □ ♀
03 59	☽ □ ♂
07 23	☽ ∠ ☿
07 27	☽ Q ☉
15 36	☉ △ ♆
18 06	☽ ± ♄
19 02	♀ ☌ ♂
06 Tuesday	
02 12	☽ ∥ ♃
03 40	☽ # ☉
03 47	☽ ♇ ♅
04 29	☽ ⊼ ♄
10 43	☽ ⚹ ☿
10 44	☽ □ ⯓
10 53	☿ ⊼ ⯓
14 59	☽ ⊼ ♆
16 43	☽ □ ☉
19 14	☽ ± ♅
21 48	☉ # ♃
23 59	☽ ☌ ♃
07 Wednesday	
01 10	☽ # ♄
01 12	☽ ± ♆
05 28	☽ ⊼ ♅
08 14	☽ Q ☿
09 59	☽ △ ♂
11 41	☽ △ ♀
16 27	☽ ♇ ♆
17 18	☽ Q ⯓
17 50	☽ # ♀
08 Thursday	
05 36	☽ # ♂
07 42	☽ ☍ ♄
12 30	☽ ♇ ♂
13 28	☽ ⚹ ⯓
15 20	☽ ♇ ♀
15 59	☽ □ ☿
17 41	☽ △ ♆
23 03	☽ △ ☉
09 Friday	
02 48	☽ ⊻ ♃
04 17	☽ ± ♂
04 55	☿ ∠ ♀
07 47	☽ ± ♀
08 05	☽ ☍ ♅
11 23	♀ ⊥ ♅
12 52	☽ ⊥ ♃
14 31	☽ ∠ ⯓
14 46	☽ ⊼ ♂
18 44	☽ ⊼ ♀
10 Saturday	
01 40	☽ # ♂
01 52	☽ ♇ ☉
03 56	☽ ∠ ♃
05 29	☽ ⊥ ⯓
10 04	☽ ⊼ ♄
13 57	☿ ⊼ ♆
15 27	☽ ⊻ ⯓
15 27	☿ ∥ ♆
16 14	☽ # ♀
17 54	☽ ± ☉
19 40	☽ □ ♆
19 51	☽ △ ☿
20 06	☽ ± ♄
11 Sunday	
01 42	☽ # ♄
04 36	☽ ⊼ ☉
05 03	☽ ⚹ ♃
10 11	☽ ⊼ ♅
11 10	☽ ♇ ♄
11 41	☉ □ ♃
19 07	☽ ☍ ♂
20 14	☽ ± ♅
21 34	☽ ♇ ☿
23 37	☽ ∥ ♃
12 Monday	
01 12	☽ Q ♃
01 22	☽ ☍ ♀
01 41	☽ Q ♆
05 08	☉ Q ♄
09 52	☽ # ☉
11 23	☽ ♇ ♅
12 25	☽ △ ♄
13 02	☽ ± ☿
17 33	☽ ☌ ⯓
21 31	♂ ⊥ ♅
21 52	☽ ⚹ ♆
23 23	☽ ⊼ ☿
13 Tuesday	
07 47	☽ □ ♃
10 44	☽ ☍ ☉
12 54	☽ △ ♅
23 30	☽ ∠ ♆
14 Wednesday	
00 26	☽ ⊼ ♂
09 23	☽ ⊼ ♀
11 23	☽ ± ♂
15 09	☽ ⊥ ♆
16 08	☽ □ ♄
17 17	☽ # ♆
19 31	☉ ⚹ ♅
20 48	♀ ∥ ♄
20 57	☽ ± ♀
21 11	☽ ⊻ ⯓
22 20	♃ ± ♆
22 37	☽ # ☿
15 Thursday	
01 43	☽ ⊻ ♆
03 54	☿ ∠ ♂
04 02	☽ ☍ ☿
04 03	☽ ♇ ♂
07 49	☽ ⊥ ⯓
12 32	☽ △ ♃
14 36	☽ ♇ ♀
17 44	☽ □ ♅
19 24	☽ ⊼ ☉
22 29	☽ ∥ ☿
16 Friday	
00 02	☽ ∠ ⯓
00 28	☽ Q ♄
01 20	☿ St R
04 45	☽ ∥ ♆
07 12	☽ ± ☉
08 34	☽ △ ♂
16 07	☽ ♇ ♃
20 57	☽ △ ♀
22 44	☽ ⚹ ♄
17 Saturday	
01 15	☽ ♇ ☉
01 23	♀ ∠ ♅
03 00	☽ Q ♅
03 45	☽ ⚹ ⯓
08 37	☽ ☌ ♆
09 07	☽ ± ♃
11 03	☽ ⊼ ☿
16 29	♀ ⚹ ♄
20 38	☽ ⊼ ♃
22 23	☽ ± ☿
18 Sunday	
00 26	☿ ⊥ ♀
01 56	☽ ⚹ ♅
02 36	☽ Q ⯓
03 24	☽ ∠ ♄
06 32	☽ ∥ ☉
08 17	☽ △ ☉
15 33	☽ ♇ ☿
20 47	☽ □ ♂
21 02	☽ ⊥ ♄
19 Monday	
07 22	☽ ∠ ♅
07 55	☽ # ♃
08 57	☽ ⊻ ♄
13 19	☽ □ ♀
13 52	☽ □ ⯓
18 45	♀ ☍ ⯓
19 03	☽ ⊻ ♆
20 05	☽ ∥ ♀
20 32	☽ △ ☿
20 Tuesday	
01 23	☽ ⊥ ♅
07 05	☽ ⊥ ♆
08 17	☽ ☍ ♃
09 40	☽ ∥ ♄
10 43	☽ Q ♂
13 30	☽ ⊻ ♅
17 53	☽ ∥ ♂
21 Wednesday	
01 13	☉ ♇ ♆
01 15	☽ ∠ ♆
01 15	☽ □ ☉
05 45	☽ Q ♀
12 26	☽ ⚹ ♂
14 59	☿ ⊼ ♆
19 06	☿ ⊻ ♀
20 48	♀ △ ♆
21 45	☽ ☌ ♄
22 Thursday	
02 24	☽ △ ⯓
06 59	☽ □ ☿
07 42	☽ ⚹ ♆
08 59	☽ ⚹ ♀
17 05	☽ Q ☉
20 40	☽ ∠ ♂
21 43	☽ ⊼ ♃
23 Friday	
02 34	☽ ☌ ♅
07 59	☽ Q ♆
08 46	☽ ♇ ⯓
09 57	☽ ± ♃
15 53	☽ ⊥ ♂
17 32	☽ Q ☿
18 50	☽ ∠ ♀
19 27	☽ ⚹ ☉
24 Saturday	
01 59	♀ # ♃
02 49	☽ ± ⯓
04 11	☽ ♇ ♃
04 35	☽ ⊻ ♂
07 15	☿ ⊥ ♂
10 33	☽ ⊻ ♄
14 44	☽ ⊼ ⯓
14 50	☽ ⊥ ♀
16 06	☽ ⚹ ☿
19 53	☽ □ ♆
22 28	☽ ⊥ ♄
25 Sunday	
01 02	☽ ∥ ♂
01 44	☽ ∥ ♄
03 49	☽ ∠ ☉
03 59	☽ ⊻ ♀
10 04	☽ △ ♃
11 16	♂ ∥ ♄
13 46	☿ ⊼ ⯓
14 20	☽ ⊻ ♅
16 08	☽ ∠ ♄
19 39	☽ ∠ ☿
22 48	☽ ⊥ ☉
26 Monday	
01 26	☽ # ♃
01 57	☽ ⊥ ♅
07 04	☿ ∥ ♆
07 30	☽ ∥ ♀
11 15	☽ ⊻ ☉
11 49	☽ ⊥ ☿
18 10	☽ ☌ ♂
19 08	☽ ∠ ♅
20 55	☽ ⚹ ♄
22 26	☽ ⊻ ☿
27 Tuesday	
00 34	☽ ☍ ⯓
05 28	☽ △ ♆
14 17	♂ ∠ ♅
16 41	☉ Q ♅
17 25	☿ Q ♃
18 59	☿ □ ♄
19 13	☽ ☌ ♀
19 22	☽ Q ♄
19 29	☽ □ ♃
20 47	♀ Q ♄
21 54	☽ ∥ ☉
22 22	♀ □ ♃
23 07	☽ ⚹ ♅
28 Wednesday	
07 55	☿ ⊻ ♂
09 05	☽ ♇ ♆
20 58	☽ Q ♅
22 51	☽ ∥ ☿
23 07	☽ ☌ ☉
29 Thursday	
01 18	☽ ± ♆
01 54	☽ ☌ ☿
04 06	☽ ⊻ ♂
04 08	☽ □ ♄
04 27	☽ Q ♃
04 36	☽ ∥ ♆
04 37	♂ ⚹ ♄
07 20	☽ ⊼ ⯓
10 51	☿ ± ♆
11 59	☽ ⊼ ♆
13 21	♀ ⚹ ♅
15 19	☽ ⊥ ♂
17 59	☽ ± ⯓
20 11	☉ ☌ ☿
30 Friday	
01 55	☽ ⚹ ♃
04 40	☉ ± ♆
05 00	☽ □ ♅
06 31	☽ ⊻ ♀
07 57	☽ ∠ ♂
09 47	☽ ♇ ⯓
14 36	☽ # ♆
18 02	☽ ⊥ ♀
31 Saturday	
01 19	☽ # ☿
03 23	☽ ⊻ ☿
04 19	☽ ∠ ♃
07 54	☽ ⊻ ☉
08 55	☽ △ ♄
11 10	☽ ∠ ♀
11 17	☽ ⚹ ♂
11 48	☽ △ ⯓
13 08	☽ ⊥ ☿
13 36	☽ # ☉
16 19	☽ ☍ ♆
19 02	☽ ⊥ ☉
20 00	☽ ⊥ ♃
22 15	♂ ☍ ⯓
23 04	☉ □ ♄

SEPTEMBER 2030

LONGITUDES

Date	Sidereal time h m s	Sun ☉ ° ' "	Moon ☽ ° ' "	Moon ☽ 24.00 ° ' "	Mercury ☿ ° '	Venus ♀ ° '	Mars ♂ ° '	Jupiter ♃ ° '	Saturn ♄ ° '	Uranus ♅ ° '	Neptune ♆ ° '	Pluto ♇ ° '
01	10 42 47	09 ♍ 06 37	24 ♎ 18 54	01 ♏ 19 59	04 ♍ 13	26 ♌ 09	10 ♌ 36	21 ♏ 05	08 ♊ 36	22 ♊ 35	12 ♈ 50	10 ♒ 13
02	10 46 44	10 04 42	08 ♏ 22 02	15 ♏ 24 44	03 R 27	27 23	11 14	21 13	08 38	22 36	12 R 49	10 R 12
03	10 50 40	11 02 48	22 ♏ 27 52	29 ♏ 31 15	02 45	28 37	11 52	21 21	08 40	22 38	12 47	10 11
04	10 54 37	12 00 55	06 ♐ 34 43	13 ♐ 38 10	02 10	29 ♌ 51	12 30	21 29	08 42	22 39	12 46	10 10
05	10 58 33	12 59 04	20 ♐ 41 29	27 ♐ 44 35	01 42	01 ♍ 05	13 08	21 37	08 44	22 40	12 44	10 09
06	11 02 30	13 57 15	04 ♑ 47 19	11 ♑ 49 34	01 22	02 20	13 46	21 45	08 46	22 41	12 43	10 07
07	11 06 27	14 55 27	18 ♑ 51 07	25 ♑ 51 43	01 10	03 34	14 24	21 54	08 47	22 42	12 42	10 06
08	11 10 23	15 53 40	02 ♒ 51 06	09 ♒ 48 56	01 D 07	04 48	15 02	22 02	08 48	22 43	12 40	10 05
09	11 14 20	16 51 55	16 ♒ 44 49	23 ♒ 38 21	01 12	06 03	15 40	22 11	08 50	22 44	12 39	10 04
10	11 18 16	17 50 11	00 ♓ 29 07	07 ♓ 16 42	01 26	07 17	16 18	22 20	08 51	22 45	12 37	10 03
11	11 22 13	18 48 29	14 ♓ 00 43	20 ♓ 40 49	01 50	08 31	16 56	22 29	08 52	22 46	12 36	10 02
12	11 26 09	19 46 49	27 ♓ 16 43	03 ♈ 48 13	02 22	09 46	17 34	22 38	08 53	22 47	12 34	10 01
13	11 30 06	20 45 11	10 ♈ 15 10	16 ♈ 37 34	03 03	11 00	18 12	22 47	08 54	22 48	12 33	10 00
14	11 34 02	21 43 34	22 ♈ 55 27	29 ♈ 09 00	03 52	12 14	18 49	22 56	08 55	22 49	12 31	09 59
15	11 37 59	22 42 00	05 ♉ 18 28	11 ♉ 24 09	04 48	13 29	19 27	23 05	08 55	22 49	12 30	09 58
16	11 41 56	23 40 28	17 ♉ 26 30	23 ♉ 25 59	05 53	14 43	20 05	23 15	08 56	22 50	12 28	09 57
17	11 45 52	24 38 58	29 ♉ 23 07	05 ♊ 18 31	07 03	15 58	20 43	23 24	08 56	22 51	12 27	09 57
18	11 49 49	25 37 30	11 ♊ 12 49	17 ♊ 06 39	08 20	17 13	21 20	23 34	08 56	22 51	12 25	09 56
19	11 53 45	26 36 04	23 ♊ 00 42	28 ♊ 55 40	09 43	18 27	21 58	23 44	08 57	22 52	12 23	09 55
20	11 57 42	27 34 41	04 ♋ 52 15	10 ♋ 51 06	11 10	19 42	22 35	23 54	08 57	22 52	12 22	09 54
21	12 01 38	28 33 19	16 ♋ 52 54	22 ♋ 58 17	12 42	20 56	23 13	24 04	08 R 57	22 52	12 20	09 53
22	12 05 35	29 ♍ 32 00	29 ♋ 07 47	05 ♌ 21 58	14 18	22 11	23 50	24 14	08 57	22 53	12 19	09 52
23	12 09 31	00 ♎ 30 43	11 ♌ 41 13	18 ♌ 05 55	15 56	23 26	24 28	24 24	08 56	22 53	12 17	09 52
24	12 13 28	01 29 28	24 ♌ 36 17	01 ♍ 12 25	17 38	24 40	25 05	24 34	08 56	22 53	12 15	09 51
25	12 17 25	02 28 15	07 ♍ 54 20	14 ♍ 41 52	19 21	25 55	25 43	24 44	08 56	22 53	12 14	09 50
26	12 21 21	03 27 05	21 ♍ 34 43	28 ♍ 32 28	21 06	27 10	26 20	24 55	08 55	22 54	12 12	09 49
27	12 25 18	04 25 56	05 ♎ 34 37	12 ♎ 40 30	22 53	28 25	26 57	25 05	08 54	22 54	12 10	09 49
28	12 29 14	05 24 50	19 ♎ 49 26	27 ♎ 00 41	24 40	29 ♍ 40	27 35	25 16	08 54	22 R 54	12 09	09 48
29	12 33 11	06 23 45	04 ♏ 13 30	11 ♏ 27 09	26 28	00 ♎ 54	28 12	25 27	08 53	22 54	12 07	09 47
30	12 37 07	07 ♎ 22 42	18 ♏ 40 57	25 ♏ 54 17	28 ♍ 17	02 ♎ 09	28 ♌ 49	25 ♏ 38	08 ♊ 52	22 ♊ 54	12 ♈ 05	09 ♒ 47

Date	Moon True ☊ ° '	Moon Mean ☊ ° '	Moon ☽ Latitude ° '
01	11 ♐ 55	11 ♐ 55	03 S 43
02	11 R 50	11 51	02 47
03	11 48	11 48	01 40
04	11 D 48	11 45	00 S 28
05	11 R 48	11 42	00 N 47
06	11 47	11 39	01 58
07	11 44	11 36	03 03
08	11 38	11 32	03 55
09	11 30	11 29	04 33
10	11 19	11 26	04 55
11	11 07	11 23	05 00
12	10 55	11 20	04 49
13	10 43	11 16	04 22
14	10 33	11 13	03 43
15	10 26	11 10	02 54
16	10 22	11 07	01 58
17	10 20	11 04	00 N 58
18	10 D 19	11 01	00 S 05
19	10 19	10 57	01 07
20	10 R 19	10 54	02 07
21	10 18	10 51	03 02
22	10 14	10 48	03 49
23	10 08	10 45	04 27
24	09 59	10 41	04 52
25	09 49	10 38	05 03
26	09 38	10 35	04 57
27	09 27	10 32	04 33
28	09 18	10 29	03 51
29	09 12	10 26	02 55
30	09 ♐ 07	10 ♐ 22	01 S 47

DECLINATIONS

Date	Sun ☉ ° '	Moon ☽ ° '	Mercury ☿ ° '	Venus ♀ ° '	Mars ♂ ° '	Jupiter ♃ ° '	Saturn ♄ ° '	Uranus ♅ ° '	Neptune ♆ ° '	Pluto ♇ ° '
01	08 N 09	12 S 53	06 N 41	13 N 52	18 N 38	17 S 15	19 N 48	23 N 18	03 N 37	23 S 41
02	07 47	16 55	07 13	13 28	18 28	17 18	19 48	23 18	03 37	23 41
03	07 25	20 00	07 44	13 04	18 18	17 20	19 48	23 18	03 36	23 42
04	07 03	21 51	08 13	12 39	18 08	17 22	19 48	23 18	03 36	23 42
05	06 41	22 20	08 41	12 14	17 57	17 25	19 48	23 18	03 35	23 42
06	06 19	21 22	09 06	11 48	17 47	17 27	19 48	23 18	03 34	23 43
07	05 56	19 06	09 28	11 23	17 36	17 29	19 48	23 18	03 34	23 43
08	05 34	15 42	09 47	10 57	17 25	17 32	19 48	23 19	03 33	23 43
09	05 11	11 28	10 03	10 30	17 14	17 34	19 48	23 19	03 33	23 43
10	04 48	06 41	10 15	10 04	17 03	17 37	19 48	23 19	03 32	23 44
11	04 26	01 S 40	10 23	09 37	16 52	17 39	19 48	23 19	03 32	23 44
12	04 03	03 N 20	10 27	09 09	16 41	17 41	19 48	23 19	03 31	23 44
13	03 40	08 05	10 28	08 42	16 30	17 44	19 48	23 19	03 30	23 44
14	03 17	12 22	10 24	08 14	16 18	17 46	19 48	23 19	03 30	23 45
15	02 54	16 02	10 16	07 46	16 07	17 49	19 48	23 19	03 29	23 45
16	02 31	18 56	10 05	07 18	15 55	17 52	19 48	23 19	03 28	23 45
17	02 08	20 57	09 49	06 49	15 44	17 54	19 48	23 19	03 28	23 45
18	01 44	22 02	09 30	06 21	15 32	17 57	19 48	23 19	03 27	23 45
19	01 21	22 08	09 08	05 52	15 20	17 59	19 48	23 19	03 27	23 46
20	00 58	21 14	08 42	05 23	15 08	18 02	19 48	23 19	03 26	23 46
21	00 34	19 22	08 13	04 54	14 56	18 05	19 48	23 19	03 25	23 46
22	00 N 11	16 35	07 42	04 25	14 44	18 07	19 47	23 19	03 25	23 46
23	00 S 12	13 00	07 08	03 55	14 32	18 10	19 47	23 19	03 24	23 46
24	00 36	08 43	06 32	03 26	14 20	18 12	19 47	23 19	03 23	23 46
25	00 59	03 N 55	05 54	02 56	14 08	18 15	19 47	23 19	03 23	23 47
26	01 22	01 S 12	05 14	02 26	13 55	18 18	19 47	23 19	03 22	23 47
27	01 46	06 23	04 33	01 56	13 43	18 21	19 46	23 19	03 21	23 47
28	02 09	11 19	03 50	01 27	13 30	18 23	19 46	23 19	03 21	23 47
29	02 32	15 40	03 07	00 57	13 18	18 26	19 46	23 19	03 20	23 47
30	02 S 56	19 S 05	02 N 22	00 N 26	13 N 05	18 S 29	19 N 45	23 N 19	03 N 19	23 S 47

ZODIAC SIGN ENTRIES

Date	h	m	Planets
01	21	43	☽ ♏
04	00	49	☽ ♐
04	14	50	♀ ♍
06	03	51	☽ ♑
08	07	06	☽ ♒
10	11	09	☽ ♓
12	16	59	☽ ♈
15	01	39	☽ ♉
17	13	15	☽ ♊
20	02	10	☽ ♋
22	13	41	☽ ♌
22	23	27	☉ ♎
24	21	49	☽ ♍
27	02	30	☽ ♎
28	18	34	♀ ♎
29	04	59	☽ ♏

LATITUDES

Date	Mercury ☿ ° '	Venus ♀ ° '	Mars ♂ ° '	Jupiter ♃ ° '	Saturn ♄ ° '	Uranus ♅ ° '	Neptune ♆ ° '	Pluto ♇ ° '
01	03 S 31	01 N 08	01 N 06	00 N 48	01 S 58	00 N 04	01 S 34	06 S 15
04	02 39	01 12	01 07	00 47	01 58	00 04	01 34	06 15
07	01 42	01 16	01 08	00 46	01 59	00 04	01 34	06 15
10	00 S 46	01 19	01 10	00 46	01 59	00 05	01 34	06 15
13	00 N 05	01 22	01 11	00 45	02 00	00 05	01 35	06 14
16	00 47	01 24	01 12	00 45	02 00	00 05	01 35	06 14
19	01 18	01 25	01 13	00 44	02 01	00 05	01 35	06 14
22	01 39	01 26	01 14	00 44	02 01	00 05	01 35	06 14
25	01 50	01 26	01 15	00 43	02 02	00 05	01 35	06 14
28	01 52	01 25	01 16	00 43	02 02	00 05	01 35	06 14
31	01 N 48	01 N 24	01 N 17	00 N 42	02 S 03	00 N 05	01 S 35	06 S 13

DATA

Julian Date	2462746
Delta T	+74 seconds
Ayanamsa	24° 17' 25"
Synetic vernal point	04° ♓ 49' 34"
True obliquity of ecliptic	23° 26' 05"

MOON'S PHASES, APSIDES AND POSITIONS ☽

Date	h	m	Phase	Longitude ° '	Eclipse Indicator
04	21	56	☽	12 ♐ 25	
11	21	18	○	19 ♓ 11	
19	19	56	☾	26 ♊ 55	
27	09	55	●	04 ♎ 21	

Day	h	m		
04	16	58	Perigee	
18	18	07	Apogee	
30	15	30	Perigee	
05	07	54	Max dec	22° S 21'
11	19	56	0N	
19	02	17	Max dec	22° N 13'
26	06	25	0S	

LONGITUDES

Date	Chiron ⚷ ° '	Ceres ⚳ ° '	Pallas ⚴ ° '	Juno ⚵ ° '	Vesta ⚶ ° '	Black Moon Lilith ⚸ ° '
01	16 ♉ 51	21 ♉ 44	06 ♓ 10	26 ♐ 43	06 ♊ 15	11 ♊ 11
11	16 ♉ 42	22 ♉ 28	03 ♓ 40	27 ♐ 39	08 ♊ 05	12 ♊ 18
21	16 ♉ 27	22 ♉ 36	01 ♓ 25	29 ♐ 05	09 ♊ 22	13 ♊ 24
31	16 ♉ 07	22 ♉ 07	29 ♒ 37	00 ♑ 57	10 ♊ 02	14 ♊ 31

ASPECTARIAN

h m	Aspects
01 Sunday	
01 27	☿ Q ♅
03 44	☽ ∠ ☿
06 24	☽ ⊻ ♃
08 55	☽ Q ♂
09 02	☽ △ ♅
10 47	☽ ⚼ ♄
11 05	☉ Q ♃
11 37	☽ ∠ ☉
15 26	☽ ✶ ♀
17 00	☽ # ♀
02 Monday	
02 13	☽ ± ♄
04 01	☽ ✶ ☿
10 42	☽ ⚼ ♅
12 28	☽ ⊼ ♄
13 54	☽ Q ♀
14 32	☽ ∥ ♃
14 56	☉ ⊼ ♇
15 07	☽ □ ♇
15 08	☽ ✶ ☉
17 07	☽ □ ♂
19 33	☽ ⊼ ♆
20 08	♀ ⚼ ♆
22 26	☽ # ♂
23 28	☽ Q ☿
03 Tuesday	
02 03	☽ ± ♅
03 32	☉ ∥ ☿
05 45	☽ ± ♆
10 04	☽ ☌ ♃
10 05	☽ # ♄
12 17	☽ ⊼ ♅
13 04	☽ Q ☉
21 02	☽ ⚼ ♆
21 42	☽ Q ♇
23 28	☽ □ ♀
04 Wednesday	
04 47	☽ □ ☿
15 37	☽ ☍ ♄
18 05	☽ ✶ ♇
21 30	♂ △ ♆
21 56	☽ □ ☉
22 30	☽ △ ♆
22 33	☽ △ ♂
05 Thursday	
06 05	☉ ⊼ ♆
13 35	☽ ⊻ ♃
15 22	☽ ☍ ♅
19 34	☽ ∠ ♇
21 09	☿ ☌ ♀
22 58	☉ ⊻ ♂
23 54	☽ ⊥ ♃
06 Friday	
01 16	☽ ⚼ ♂
06 17	☽ △ ☿
07 24	☽ △ ♀
10 52	☽ ⊥ ♇
15 23	☽ ∠ ♃
17 19	☽ ± ♂
18 47	☽ ⊼ ♄
21 05	☽ ⊻ ♇
07 Saturday	
01 30	☽ □ ♆
04 03	☽ ⊼ ♂
04 47	☽ △ ☉
05 02	☽ ± ♄
05 47	☽ # ♄
07 27	☽ ⚼ ☿
11 28	☽ ⚼ ♀
17 15	☽ ✶ ♃
18 36	☽ ⊼ ♅
20 27	☽ ⚼ ♄
22 44	☽ ± ☿
08 Sunday	
00 01	♂ # ♃
00 05	☽ ∥ ♃
00 05	☽ # ♂
04 22	☽ ± ♀
04 54	☽ ± ♅
08 15	☽ Q ♆
08 23	☽ ⚼ ☉
09 00	☽ ⊼ ☿
09 27	☿ St D
10 25	♀ Q ♅
14 03	☽ Q ♃
15 41	☽ ⊼ ♀
16 42	☉ ± ♇
20 24	☽ ⚼ ♅
22 17	☽ △ ♄
09 Monday	
00 27	☽ ☌ ♇
01 03	☽ ± ☉
04 54	☽ ✶ ♆
10 03	☽ ☍ ♂
12 13	☽ ⊼ ☉
17 31	☽ # ♀
19 01	☽ # ☿
21 33	☽ □ ♃
22 27	☽ △ ♅
23 28	♀ ± ♆
10 Tuesday	
04 46	☿ ∥ ♀
06 59	☽ ∠ ♆
13 43	☽ ☍ ☿
21 48	☽ # ☉
22 49	☽ ⊥ ♆
11 Wednesday	
01 13	☽ ☍ ♀
02 48	☽ □ ♄
03 09	☽ # ♆
04 55	☽ ⊻ ♇
09 28	☽ ⊻ ♆
15 38	☽ ⊥ ♇
17 30	☽ ⊼ ♂
18 47	♀ □ ♄
21 18	☽ ☍ ☉
12 Thursday	
03 26	☽ △ ♃
03 48	☽ □ ♅
04 53	☽ ± ♂
07 53	☽ ∠ ♇
11 16	☽ Q ♄
12 53	☽ ∥ ♆
15 15	☽ ∥ ☉
16 58	♀ ⊼ ♇
21 50	☽ ⊼ ☿
22 12	☽ ⚼ ♂
13 Friday	
07 06	♀ Q ♃
07 19	☽ ⚼ ♃
09 28	☽ ✶ ♄
09 36	☽ ± ☿
11 32	☽ ✶ ♇
13 01	☽ Q ♅
13 33	☽ ⊼ ♀
14 59	☽ ∥ ♀
15 21	♃ ⊼ ♅
16 18	☽ ☌ ♆
22 17	☉ ∥ ♆
14 Saturday	
00 26	☽ ± ♃
00 49	☽ ∥ ☿
02 05	☽ ± ♀
03 41	☽ ⚼ ☿
03 46	☽ △ ♂
09 31	☽ ⊼ ☉
10 13	☽ Q ♇
11 47	☽ ✶ ♅
12 01	☽ ⊼ ♃
13 53	☽ ∠ ♄
17 17	♀ ⊼ ♆
21 14	☽ ⚼ ♀
22 02	☽ ± ☉
15 Sunday	
07 20	☽ ⊥ ♄
10 56	☽ △ ☿
12 20	☿ Q ♅
12 36	☽ ∥ ♂
15 02	☉ □ ♅
16 56	☽ ∠ ♅
17 06	☽ ⚼ ☉
19 06	☽ ⊻ ♄
21 10	☽ □ ♇
23 24	☉ ✶ ♃
16 Monday	
02 08	☽ ⊻ ♆
02 12	☽ # ♃
05 58	☽ △ ♀
10 47	☽ ⊥ ♅
14 03	☽ ⊥ ♆
17 34	☽ □ ♂
21 05	☽ ∥ ♄
22 48	☽ ⊻ ♅
23 47	☽ ☍ ♃
17 Tuesday	
00 03	☿ ± ♆
01 36	☽ △ ☉
08 05	☽ ∠ ♆
11 31	♀ ± ♇
19 05	☉ ⚼ ♇
20 42	♂ Q ♄
18 Wednesday	
05 25	☽ □ ☿
07 23	☽ ☌ ♄
07 58	☽ Q ♂
09 23	☽ △ ♇
14 26	☽ ✶ ♆
22 43	☿ □ ♄
19 Thursday	
01 38	☽ □ ♀
09 45	☽ ✶ ♂
11 41	☽ ☌ ♅
13 29	☽ ⊼ ♃
14 48	☽ Q ♆
15 18	☿ ⊼ ♇
15 51	☽ ⚼ ♇
19 56	☽ □ ☉
22 52	☽ Q ☿
20 Friday	
01 49	☽ ± ♃
10 03	☽ ± ♇
17 46	☽ ∠ ♂
18 20	☽ Q ♀
20 11	☽ ⊻ ♄
20 12	☽ ⚼ ♃
21 30	♄ St R
22 05	☽ ⊼ ♇
22 43	♂ ✶ ♅
21 Saturday	
00 50	☿ Q ♃
02 28	☽ ✶ ☿
02 59	☽ □ ♆
06 27	☿ ⊼ ♆
07 26	☽ ∥ ♄
08 10	☽ ⊥ ♄
11 18	☽ Q ☉
12 42	☽ ⊥ ♂
20 55	☽ ✶ ♀
23 49	☽ ⊻ ♅
23 54	☽ # ♃
22 Sunday	
01 09	☽ ⊻ ♂
01 54	☽ ∠ ♄
02 20	☽ △ ♃
11 31	☽ ⊥ ♅
12 22	☽ ∠ ☿
12 51	☽ ✶ ☉
23 Monday	
01 27	♀ □ ♅
01 46	☽ ∥ ♂
04 48	☽ ∠ ♅
05 10	☽ ∠ ♀
06 48	☽ ✶ ♄
08 12	☽ ⊥ ☿
08 22	♂ □ ♃
08 33	☽ ☍ ♇
08 54	☉ ⊥ ♂
10 51	☿ ± ♇
13 07	☽ △ ♆
19 46	☽ ∠ ☉
21 10	☽ ⊻ ☿
23 54	☽ ⊥ ♀
24 Tuesday	
05 15	☽ Q ♄
08 51	☽ ✶ ♅
09 38	♀ ✶ ♃
11 56	☽ □ ♃
12 08	☽ ⊻ ♀
12 56	☽ ☌ ♂
13 45	☽ ⊥ ☉
14 01	♀ ∥ ♆
15 17	♀ ⚼ ♇
16 50	☽ ⚼ ♆
25 Wednesday	
00 54	☽ ∥ ☿
01 30	☽ ⊻ ☉
04 02	♀ ⊻ ♂
06 37	☽ Q ♅
09 01	☽ ± ♆
13 49	☽ □ ♄
14 34	☽ ∥ ♆
15 25	☽ ⊼ ♇
17 11	☽ ∥ ♀
19 38	☽ ⊼ ♆
20 40	☽ Q ♃
26 Thursday	
00 53	☽ # ☉
01 59	☽ ± ♇
11 04	☽ ☌ ☿
12 50	☽ ∥ ☉
14 17	☽ □ ♅
17 11	☽ # ♀
17 36	☽ ⚼ ♇
17 50	☽ ✶ ♃
20 36	☽ ⊻ ♂
21 57	☽ # ♆
22 35	☽ ☌ ♀
27 Friday	
04 26	☽ # ☿
07 20	☽ ⊥ ♂
09 55	☽ ☌ ☉
12 12	☿ □ ♅
16 51	☉ # ♀
17 38	☽ △ ♄
19 10	☽ △ ♇
19 44	☽ ∠ ♃
19 57	♂ ⚼ ♆
23 08	☽ ☍ ♆
23 17	☽ ∠ ♂
28 Saturday	
08 27	♅ St R
11 03	☽ ⊥ ♃
13 44	☿ ⚼ ♇
17 08	☽ △ ♅
18 48	☽ ⚼ ♄
20 52	☿ ✶ ♃
21 12	☽ ⊻ ♃
21 15	☽ ⊻ ☿
23 01	☽ # ♂
29 Sunday	
01 32	☽ ✶ ♂
04 33	☿ ∥ ♆
05 58	☽ ⊻ ♀
08 40	☽ ⊥ ☿
09 46	☽ ± ♄
15 52	☽ ⊻ ☉
16 52	☽ ⊥ ♀
18 06	☽ ⚼ ♅
19 43	☽ ⊼ ♄
21 14	☽ □ ♇
22 22	☽ Q ♂
30 Monday	
00 12	☉ # ☿
01 05	☽ ⊼ ♆
01 45	☽ ∠ ☿
02 33	☽ ⊥ ☉
06 59	☽ ∥ ♃
09 02	☽ ± ♅
09 13	☽ ∠ ♀
11 01	☽ ± ♆
17 54	☽ # ♄
18 35	☽ ∠ ☉
18 59	☽ ⊼ ♅
22 56	☿ ⊻ ♂
23 41	☽ ☌ ♃

All ephemeris data is given at 12.00 UT and the Moon's longitude is additionally given for 24.00 UT

OCTOBER 2030

LONGITUDES

Date	Sidereal time h m s	Sun ☉ ° ′ ″	Moon ☽ ° ′ ″	Moon ☽ 24.00 ° ′ ″	Mercury ☿ ° ′	Venus ♀ ° ′	Mars ♂ ° ′	Jupiter ♃ ° ′	Saturn ♄ ° ′	Uranus ♅ ° ′	Neptune ♆ ° ′	Pluto ♇ ° ′
01	12 41 04	08 ♎ 21 42	03 ♐ 06 38	10 ♐ 17 32	00 ♎ 05	03 ♎ 24	29 ♌ 26	25 ♏ 49	08 ♊ 51	22 ♊ 53	12 ♈ 04	09 ♒ 46
02	12 45 00	09 20 43	17 ♐ 26 39	24 ♐ 33 43	01 54	04 39	00 ♍ 04	25 59	08 R 49	22 R 53	12 R 02	09 R 46
03	12 48 57	10 19 45	01 ♑ 38 32	08 ♑ 40 59	03 42	05 54	00 41	26 11	08 48	22 53	12 00	09 45
04	12 52 54	11 18 50	15 ♑ 40 57	22 ♑ 38 24	05 30	07 09	01 18	26 22	08 47	22 53	11 59	09 45
05	12 56 50	12 17 56	29 ♑ 33 18	06 ♒ 25 36	07 18	08 24	01 55	26 33	08 45	22 52	11 57	09 44
06	13 00 47	13 17 04	13 ♒ 15 15	20 ♒ 02 14	09 05	09 39	02 32	26 44	08 43	22 52	11 55	09 44
07	13 04 43	14 16 13	26 ♒ 46 26	03 ♓ 27 47	10 52	10 53	03 09	26 56	08 42	22 51	11 54	09 43
08	13 08 40	15 15 25	10 ♓ 06 10	16 ♓ 41 30	12 38	12 08	03 46	27 07	08 40	22 51	11 52	09 43
09	13 12 36	16 14 38	23 ♓ 13 38	29 ♓ 42 28	14 23	13 23	04 22	27 19	08 38	22 50	11 50	09 42
10	13 16 33	17 13 53	06 ♈ 07 56	12 ♈ 29 57	16 08	14 38	04 59	27 30	08 36	22 50	11 49	09 42
11	13 20 29	18 13 10	18 ♈ 48 29	25 ♈ 03 34	17 52	15 53	05 36	27 42	08 34	22 49	11 47	09 42
12	13 24 26	19 12 29	01 ♉ 15 15	07 ♉ 23 39	19 35	17 08	06 13	27 53	08 31	22 48	11 45	09 41
13	13 28 23	20 11 51	13 ♉ 28 57	19 ♉ 31 23	21 17	18 23	06 49	28 05	08 29	22 48	11 44	09 41
14	13 32 19	21 11 14	25 ♉ 31 15	01 ♊ 28 54	22 59	19 38	07 26	28 17	08 27	22 47	11 42	09 41
15	13 36 16	22 10 40	07 ♊ 24 45	13 ♊ 19 14	24 40	20 53	08 03	28 29	08 24	22 46	11 40	09 41
16	13 40 12	23 10 08	19 ♊ 12 53	25 ♊ 06 14	26 20	22 09	08 39	28 41	08 22	22 45	11 39	09 40
17	13 44 09	24 09 38	00 ♋ 59 52	06 ♋ 54 25	28 00	23 24	09 16	28 53	08 19	22 44	11 37	09 40
18	13 48 05	25 09 11	12 ♋ 50 30	18 ♋ 48 46	29 ♎ 39	24 39	09 52	29 05	08 16	22 43	11 36	09 40
19	13 52 02	26 08 45	24 ♋ 49 54	00 ♌ 54 31	01 ♏ 18	25 54	10 29	29 17	08 13	22 42	11 34	09 40
20	13 55 58	27 08 22	07 ♌ 03 16	13 ♌ 16 45	02 55	27 09	11 05	29 30	08 10	22 41	11 32	09 40
21	13 59 55	28 08 02	19 ♌ 35 30	26 ♌ 00 01	04 32	28 24	11 42	29 42	08 07	22 40	11 31	09 40
22	14 03 52	29 ♎ 07 43	02 ♍ 30 40	09 ♍ 07 46	06 09	29 ♎ 39	12 18	29 ♏ 54	08 04	22 39	11 29	09 40
23	14 07 48	00 ♏ 07 27	15 ♍ 51 29	22 ♍ 41 50	07 45	00 ♏ 54	12 54	00 ♐ 07	08 00	22 38	11 28	09 40
24	14 11 45	01 07 13	29 ♍ 38 41	06 ♎ 41 45	09 20	02 09	13 31	00 19	07 57	22 36	11 26	09 D 40
25	14 15 41	02 07 01	13 ♎ 50 34	21 ♎ 04 32	10 55	03 25	14 07	00 32	07 54	22 35	11 25	09 40
26	14 19 38	03 06 51	28 ♎ 22 52	05 ♏ 44 43	12 29	04 40	14 43	00 44	07 50	22 34	11 23	09 40
27	14 23 34	04 06 43	13 ♏ 09 08	20 ♏ 35 07	14 03	05 55	15 19	00 57	07 47	22 32	11 21	09 40
28	14 27 31	05 06 37	28 ♏ 01 38	05 ♐ 27 44	15 36	07 10	15 55	01 09	07 43	22 31	11 20	09 40
29	14 31 27	06 06 33	12 ♐ 52 29	20 ♐ 15 05	17 09	08 25	16 31	01 22	07 39	22 29	11 19	09 40
30	14 35 24	07 06 31	27 ♐ 34 48	04 ♑ 51 05	18 41	09 41	17 07	01 35	07 35	22 28	11 17	09 41
31	14 39 21	08 ♏ 06 31	12 ♑ 03 26	19 ♑ 11 33	20 ♏ 13	10 ♏ 56	17 ♍ 43	01 ♐ 48	07 ♊ 31	22 ♊ 26	11 ♈ 16	09 ♒ 41

Date	Moon True ☊ ° ′	Moon Mean ☊ ° ′	Moon ☽ Latitude ° ′
01	09 ♐ 06	10 ♐ 19	00 S 32
02	09 D 06	10 16	00 N 45
03	09 07	10 13	01 58
04	09 R 07	10 10	03 03
05	09 05	10 07	03 57
06	09 02	10 03	04 36
07	08 56	10 00	04 59
08	08 48	09 57	05 06
09	08 39	09 54	04 56
10	08 29	09 51	04 32
11	08 21	09 47	03 54
12	08 13	09 44	03 05
13	08 08	09 41	02 09
14	08 05	09 38	01 07
15	08 04	09 35	00 N 04
16	08 D 05	09 32	01 S 00
17	08 06	09 28	02 01
18	08 08	09 25	02 58
19	08 08	09 22	03 47
20	08 R 08	09 19	04 27
21	08 06	09 16	04 55
22	08 02	09 13	05 10
23	07 56	09 09	05 08
24	07 50	09 06	04 49
25	07 45	09 03	04 12
26	07 40	09 00	03 18
27	07 36	08 57	02 10
28	07 34	08 53	00 S 52
29	07 D 34	08 50	00 N 29
30	07 35	08 47	01 48
31	07 ♐ 36	08 ♐ 44	02 N 58

DECLINATIONS

Date	Sun ☉ ° ′	Moon ☽ ° ′	Mercury ☿ ° ′	Venus ♀ ° ′	Mars ♂ ° ′	Jupiter ♃ ° ′	Saturn ♄ ° ′	Uranus ♅ ° ′	Neptune ♆ ° ′	Pluto ♇ ° ′
01	03 S 19	21 S 18	01 N 37	00 S 04	12 N 52	18 S 31	19 N 45	23 N 19	03 N 19	23 S 47
02	03 42	22 06	00 52	00 34	12 40	18 34	19 45	23 19	03 18	23 47
03	04 05	21 28	00 N 06	01 04	12 27	18 37	19 44	23 19	03 17	23 47
04	04 28	19 29	00 S 40	01 34	12 14	18 40	19 44	23 19	03 17	23 47
05	04 52	16 23	01 26	02 04	12 01	18 42	19 44	23 19	03 16	23 47
06	05 15	12 25	02 12	02 34	11 48	18 45	19 43	23 19	03 15	23 48
07	05 38	07 53	02 58	03 04	11 35	18 48	19 43	23 19	03 15	23 48
08	06 00	03 S 03	03 44	03 34	11 22	18 51	19 43	23 19	03 14	23 48
09	06 23	01 N 51	04 30	04 04	11 09	18 53	19 42	23 19	03 13	23 48
10	06 46	06 36	05 15	04 34	10 56	18 56	19 42	23 19	03 13	23 48
11	07 09	10 58	06 00	05 04	10 43	18 59	19 41	23 19	03 12	23 48
12	07 31	14 48	06 45	05 34	10 29	19 02	19 41	23 19	03 12	23 48
13	07 54	17 56	07 29	06 03	10 16	19 04	19 40	23 19	03 11	23 48
14	08 16	20 14	08 13	06 33	10 03	19 07	19 40	23 19	03 10	23 48
15	08 38	21 36	08 56	07 02	09 49	19 10	19 39	23 19	03 10	23 48
16	09 00	22 00	09 39	07 31	09 36	19 13	19 39	23 19	03 09	23 48
17	09 22	21 25	10 21	08 00	09 22	19 15	19 38	23 19	03 08	23 48
18	09 44	19 52	11 02	08 29	09 09	19 18	19 38	23 19	03 08	23 47
19	10 06	17 26	11 43	08 58	08 55	19 21	19 37	23 19	03 08	23 47
20	10 27	14 12	12 23	09 26	08 42	19 24	19 37	23 19	03 07	23 47
21	10 49	10 16	13 02	09 55	08 28	19 26	19 36	23 19	03 06	23 47
22	11 10	05 45	13 41	10 23	08 15	19 29	19 35	23 19	03 05	23 47
23	11 31	00 N 50	14 19	10 50	08 01	19 32	19 35	23 19	03 05	23 47
24	11 52	04 S 17	14 56	11 18	07 48	19 35	19 34	23 19	03 04	23 47
25	12 12	09 20	15 32	11 45	07 34	19 37	19 34	23 19	03 03	23 47
26	12 33	13 59	16 07	12 12	07 20	19 40	19 33	23 19	03 03	23 47
27	12 53	17 51	16 42	12 39	07 07	19 43	19 32	23 19	03 02	23 47
28	13 13	20 34	17 16	13 05	06 53	19 46	19 32	23 19	03 02	23 46
29	13 33	21 51	17 49	13 31	06 39	19 48	19 31	23 18	03 01	23 46
30	13 53	21 37	18 21	13 57	06 25	19 51	19 30	23 18	03 01	23 46
31	14 S 12	19 S 56	18 S 52	14 S 23	06 N 12	19 S 54	19 N 30	23 N 18	03 N 00	23 S 46

ZODIAC SIGN ENTRIES

Date	h m	Planets
01	06 49	☽ ♐
01	10 50	☿ ♎
02	09 42	♂ ♍
03	09 13	☽ ♑
05	12 46	☽ ♒
07	17 47	☽ ♓
10	00 33	☽ ♈
12	09 34	☽ ♉
14	21 01	☽ ♊
17	09 58	☽ ♋
18	17 03	☿ ♏
19	22 13	☽ ♌
22	07 24	☽ ♍
22	18 40	♀ ♏
22	23 14	♃ ♐
23	09 00	☉ ♏
24	12 37	☽ ♎
26	14 39	☽ ♏
28	15 11	☽ ♐
30	15 59	☽ ♑

LATITUDES

Date	Mercury ☿ ° ′	Venus ♀ ° ′	Mars ♂ ° ′	Jupiter ♃ ° ′	Saturn ♄ ° ′	Uranus ♅ ° ′	Neptune ♆ ° ′	Pluto ♇ ° ′
01	01 N 48	01 N 24	01 N 17	00 N 42	02 S 03	00 N 05	01 S 35	06 S 13
04	01 39	01 23	01 18	00 42	02 03	00 05	01 35	06 13
07	01 26	01 21	01 20	00 41	02 03	00 05	01 35	06 13
10	01 10	01 18	01 21	00 41	02 04	00 05	01 35	06 13
13	00 53	01 15	01 22	00 40	02 04	00 05	01 35	06 12
16	00 33	01 11	01 23	00 40	02 04	00 05	01 35	06 12
19	00 N 13	01 07	01 24	00 40	02 05	00 05	01 35	06 12
22	00 S 07	01 02	01 25	00 39	02 05	00 05	01 35	06 11
25	00 27	00 57	01 26	00 39	02 05	00 05	01 35	06 11
28	00 47	00 52	01 26	00 38	02 05	00 05	01 35	06 11
31	01 S 07	00 N 46	01 N 27	00 N 38	02 S 06	00 N 05	01 S 35	06 S 11

DATA

Julian Date	2462776
Delta T	+74 seconds
Ayanamsa	24° 17′ 28″
Synetic vernal point	04° ♓ 49′ 31″
True obliquity of ecliptic	23° 26′ 05″

MOON'S PHASES, APSIDES AND POSITIONS ☽

Date	h m	Phase	Longitude ° ′	Eclipse Indicator
04	03 56	☽	10 ♑ 59	
11	10 47	○	18 ♈ 10	
19	14 50	☾	26 ♋ 16	
26	20 17	●	03 ♏ 28	

Day	h m		
16	13 14	Apogee	
28	11 57	Perigee	
02	13 15	Max dec	22° S 06′
09	02 54	0N	
16	09 37	Max dec	22° N 00′
23	15 58	0S	
29	20 13	Max dec	21° S 57′

LONGITUDES

Date	Chiron ⚷ ° ′	Ceres ⚳ ° ′	Pallas ⚴ ° ′	Juno ⚵ ° ′	Vesta ⚶ ° ′	Black Moon Lilith ⚸ ° ′
01	16 ♉ 07	22 ♉ 07	29 ♒ 37	00 ♑ 57	10 ♊ 02	14 ♊ 31
11	15 ♉ 43	21 ♉ 00	28 ♒ 23	03 ♑ 13	10 ♊ 00	15 ♊ 37
21	15 ♉ 17	19 ♉ 19	27 ♒ 42	05 ♑ 49	09 ♊ 14	16 ♊ 44
31	14 ♉ 47	17 ♉ 13	27 ♒ 38	08 ♑ 43	07 ♊ 45	17 ♊ 50

All ephemeris data is given at 12.00 UT and the Moon's longitude is additionally given for 24.00 UT

ASPECTARIAN

h m	Aspects
01 Tuesday	
01 57	☽ bQ ♆
03 07	☽ Q ♇
05 37	☽ □ ♂
06 14	☽ ⚹ ☿
11 47	☉ # ♆
12 32	☽ ⚹ ♀
21 25	☽ ⚹ ☉
21 34	☽ ☍ ♄
23 07	☽ ⚹ ♇
23 31	☉ △ ♄
02 Wednesday	
02 56	☽ △ ♆
05 11	☽ Q ☿
10 32	☽ Q ♀
17 45	☿ # ♀
19 03	☽ Q ☉
21 10	☽ ☍ ♅
22 03	☉ △ ♇
03 Thursday	
00 20	☽ ∠ ♇
02 36	☽ ⊻ ♃
10 17	☽ △ ♂
12 55	☽ ⊥ ♃
15 35	☽ ⊥ ♇
16 01	☽ □ ☿
19 57	☽ □ ♀
04 Friday	
00 11	☽ ⊼ ♄
01 49	☽ ⊻ ♇
03 56	☽ □ ☉
04 29	☽ ∠ ♃
05 39	☽ □ ♆
09 36	☽ # ♄
10 27	☽ ± ♄
13 06	☽ bQ ♂
13 26	☉ ∠ ♃
17 42	♀ ⊥ ♂
19 05	☽ ∥ ♃
05 Saturday	
00 24	☽ ⊼ ♅
01 57	☽ bQ ♄
03 46	☉ ☍ ♆
05 22	☽ ± ♂
06 42	☽ ⚹ ♃
10 49	☽ ± ♅
12 41	☽ Q ♆
16 18	☽ ⊼ ♂
18 44	♀ △ ♄
06 Sunday	
00 32	☿ ⊥ ♂
02 32	☽ bQ ♅
03 34	☽ △ ☿
03 56	☽ Q ♃
04 03	☽ △ ♄
05 00	☽ △ ♀
05 48	☽ ☌ ♇
07 13	☿ △ ♄
09 40	☽ ⚹ ♆
12 03	☽ △ ☉
13 38	♀ △ ♇
15 37	☽ # ♂
20 37	☿ △ ♇
07 Monday	
05 01	☽ △ ♅
08 37	♃ bQ ♆
10 07	☽ bQ ☿
10 16	☽ bQ ♀
12 13	☽ ∠ ♆
12 17	☽ □ ♃
13 21	☿ ☌ ♀
16 49	☽ bQ ☉
20 14	♀ # ♆
20 25	☿ # ♆
20 44	☿ ∥ ♀
22 31	☽ ∥ ☉
23 59	☽ ☍ ♂
08 Tuesday	
01 49	☿ ☍ ♆
04 05	☽ ± ♀
04 12	☿ ∠ ♃
04 21	☽ ⊥ ♆
04 45	☽ ± ☿
06 52	♀ ☍ ♆
09 05	☽ ∥ ☿
09 24	☽ □ ♄
09 42	☽ ∥ ♀
10 21	☽ ± ☉
11 06	☽ # ♆
11 18	☽ ⊻ ♇
11 28	♀ ∠ ♃
15 12	☽ ⊻ ♆
16 05	☽ ⊼ ♀
17 18	☽ ⊼ ☿
22 08	☽ ⊼ ☉
22 12	☽ ⊥ ♇
09 Wednesday	
11 17	☽ □ ♅
14 44	☽ ∠ ♇
18 17	☽ Q ♄
18 49	☽ ∥ ♆
19 40	☽ △ ♃
10 Thursday	
00 22	☽ # ♀
03 47	☽ # ☿
05 59	♂ Q ♅
09 45	☽ ⊼ ♂
13 00	☽ # ☉
16 37	☽ ⚹ ♄
18 43	☽ ⚹ ♇
20 50	☽ Q ♅
21 36	☽ ± ♂
22 41	☽ ☌ ♆
11 Friday	
00 11	☽ bQ ♃
05 49	☽ ☍ ♀
09 54	☽ ☍ ☿
10 34	☽ ∥ ♂
10 47	☽ ☍ ☉
11 59	♃ Q ♇
15 36	☽ bQ ♂
17 32	☽ Q ♇
17 37	☽ ± ♃
18 55	♂ ± ♆
19 41	☽ ⚹ ♅
21 05	☽ ∠ ♄
23 47	☉ ☌ ☿
12 Saturday	
05 22	☽ ⊼ ♃
14 28	☽ ⊥ ♄
22 11	☽ △ ♂
13 Sunday	
00 48	☽ ∠ ♅
02 11	☽ ⊻ ♄
04 31	☽ □ ♇
08 33	☽ ⊻ ♆
18 34	☽ ⊥ ♅
20 24	☽ ⊥ ♆
22 52	☽ ⊼ ♀
23 00	☽ # ♃
23 51	☿ ∠ ♂
14 Monday	
00 48	☿ ⊥ ♃
02 33	☽ ⊼ ☉
05 10	☽ ∥ ♄
06 05	☽ ⊼ ☿
06 31	☽ ⊻ ♅
09 10	☿ △ ♅
12 16	☽ ± ♀
14 22	☽ ∠ ♆
15 17	☉ ∥ ☿
15 39	☽ ± ☉
17 39	☽ ☍ ♃
18 24	☿ bQ ♄
20 07	☽ ± ☿
15 Tuesday	
08 33	☽ bQ ♀
11 29	☽ bQ ☉
13 21	☽ □ ♂
14 00	☽ ☌ ♄
16 36	☽ △ ♇
17 20	☽ bQ ☿
20 38	☽ ⚹ ♆
21 17	☉ ⊥ ♃
16 Wednesday	
01 06	♂ □ ♄
02 05	☉ △ ♅
10 48	☿ # ♂
16 24	☉ bQ ♄
18 40	☽ △ ♀
19 12	☽ ☌ ♅
20 48	☽ △ ☉
21 01	☽ Q ♆
23 07	☽ bQ ♇
23 35	♀ △ ♅
17 Thursday	
00 23	♀ ⊥ ♃
03 59	☽ Q ♂
04 54	☽ △ ☿
07 38	☽ ⊼ ♃
10 31	♀ bQ ♄
12 15	☉ # ♂
17 26	☽ ± ♇
18 32	☉ ∠ ♂
20 02	☽ ± ♃
18 Friday	
02 35	☿ ⊻ ♃
02 47	☽ ⊻ ♄
03 59	♂ ⊼ ♇
05 36	☽ ⊼ ♇
05 41	☽ ⚹ ♂
09 29	☽ □ ♆
14 33	☽ bQ ♃
14 48	☽ ∥ ♄
14 52	☽ ⊥ ♄
18 16	☽ # ♃
20 33	♀ ∠ ♂
19 Saturday	
07 46	☽ ⊻ ♅
08 48	☽ ∠ ♄
10 41	♀ # ♂
13 21	☽ ∠ ♂
14 21	☽ □ ♀
14 50	☽ □ ☉
19 39	☽ ⊥ ♅
20 58	☽ △ ♃
20 Sunday	
01 11	☿ ± ♄
02 43	☽ □ ☿
07 59	☽ ⊥ ♂
11 12	☉ ☌ ♀
13 13	☽ ∠ ♅
14 09	☽ ⚹ ♄
17 03	☽ ☍ ♇
20 12	☽ ⊻ ♂
20 39	☽ △ ♆
21 54	☽ # ☿
21 Monday	
04 53	☽ Q ☉
05 06	♂ ⊼ ♆
05 17	☽ Q ♀
09 07	☽ # ☉
12 59	☽ Q ♄
13 47	☽ # ♀
17 46	☽ ⚹ ♅
18 21	☽ Q ☿
22 22	☽ ∥ ♂
22 Tuesday	
00 56	☽ bQ ♆
05 17	☽ ⚹ ☉
06 12	☽ ⚹ ♀
07 08	☽ □ ♃
15 53	☽ Q ♅
17 25	☽ ± ♆
17 45	♀ ⊻ ♃
19 32	☽ ⚹ ☿
22 02	☽ □ ♄
23 Wednesday	
00 58	☽ ⊼ ♇
01 14	☽ ∥ ♆
03 05	♇ St D
04 12	☽ ⊼ ♆
06 31	☽ ☌ ♂
10 10	☿ bQ ♅
10 36	☽ ∠ ☉
11 34	☉ ⊻ ♃
11 40	☽ ± ♇
12 05	☽ ∠ ♀
15 46	☿ ⊼ ♄
16 02	☽ Q ♃
23 51	☽ □ ♅
24 Thursday	
01 40	☽ ∠ ☿
03 25	☽ bQ ♇
03 36	☽ ⊥ ☉
05 24	☽ ⊥ ♀
06 21	☽ # ♆
08 14	♀ ± ♄
08 51	♃ # ♄
13 10	☽ ⚹ ♃
14 43	☽ ⊻ ☉
16 43	☽ ⊻ ♀
16 57	☿ □ ♇
19 06	☽ ⊥ ☿
25 Friday	
02 04	☽ △ ♄
03 51	☽ # ♂
05 00	☽ △ ♇
06 30	☽ ⊻ ☿
06 57	☉ ± ♄
07 56	☽ ☍ ♆
12 28	☽ ⊻ ♂
14 51	☽ ∠ ♃
19 21	☿ ⊼ ♆
22 52	☽ ⊥ ♂
26 Saturday	
01 27	☽ ∥ ♀
02 28	☽ △ ♅
02 56	☽ bQ ♄
03 40	☽ ∥ ☉
05 56	☽ ⊥ ♃
14 17	☽ ∠ ♂
15 54	☽ ⊻ ♃
17 37	☽ ± ♄
20 17	☽ ☌ ☉
23 11	☽ ☌ ♀
27 Sunday	
02 50	☽ ∥ ☿
02 55	☽ bQ ♅
03 20	☽ ⊼ ♄
06 22	☽ □ ♇
09 06	☽ ⊼ ♆
13 38	☽ ☌ ☿
15 39	☽ ⚹ ♂
17 27	☽ ± ♅
18 47	☽ ± ♆
28 Monday	
01 29	☽ # ♄
02 01	♂ ± ♇
03 07	☽ ⊼ ♅
03 26	☽ ∥ ♃
09 16	☽ bQ ♆
11 26	☽ Q ♇
11 49	☽ Q ♂
17 07	☽ ☌ ♃
18 25	♀ bQ ♅
19 58	☿ ⚹ ♂
21 57	♀ ⊼ ♄
29 Tuesday	
00 15	☽ ⊻ ☉
01 49	☿ ± ♅
03 35	☽ ☍ ♄
04 07	☽ ⊻ ♀
06 49	☽ ⚹ ♇
09 28	☽ △ ♆
10 40	☽ ⊥ ☉
14 24	☿ ± ♆
14 45	☽ ⊥ ♀
18 10	☽ □ ♂
19 19	☉ ∥ ♀
19 46	☽ ⊻ ☿
30 Wednesday	
02 23	☽ ∠ ☉
03 37	☽ ☍ ♅
06 42	☽ ⊥ ☿
06 47	☽ ∠ ♀
07 14	☽ ∠ ♇
12 02	♀ □ ♇
18 41	☽ ⊻ ♃
20 14	☉ bQ ♅
22 04	☽ ⊥ ♇
22 49	☉ ⊼ ♄
23 16	☽ ∠ ☿
31 Thursday	
04 28	☽ ⊼ ♄
04 47	☽ ⊥ ♃
04 55	☽ ⚹ ☉
08 02	☽ ⊻ ♇
09 56	☽ ⚹ ♀
10 40	☽ □ ♆
12 22	☽ ∥ ♃
14 27	☽ ± ♄
16 17	☽ # ♄
18 12	♀ ⊼ ♆
20 04	☽ ∠ ♃
20 20	☽ ∥ ☿
21 56	☽ △ ♂

NOVEMBER 2030

LONGITUDES

Date	Sidereal time h m s	Sun ☉ ° ' "	Moon ☽ ° ' "	Moon ☽ 24.00 ° ' "	Mercury ☿ ° '	Venus ♀ ° '	Mars ♂ ° '	Jupiter ♃ ° '	Saturn ♄ ° '	Uranus ♅ ° '	Neptune ♆ ° '	Pluto ♇ ° '
01	14 43 17	09 ♏ 06 32	26 ♑ 15 12	03 ♒ 14 16	21 ♏ 44	12 ♏ 11	18 ♍ 19	02 ♐ 00	07 ♊ 27	22 ♊ 24	11 ♈ 14	09 ♒ 41
02	14 47 14	10 06 34	10 ♒ 08 44	16 ♒ 58 36	23 15	13 26	18 55	02 13	07 R 23	22 R 23	11 R 13	09 41
03	14 51 10	11 06 38	23 ♒ 43 59	00 ♓ 25 01	24 46	14 41	19 31	02 26	07 19	22 21	11 11	09 42
04	14 55 07	12 06 44	07 ♓ 01 51	13 ♓ 34 40	26 15	15 57	20 06	02 39	07 15	22 19	11 10	09 42
05	14 59 03	13 06 51	20 ♓ 03 39	26 ♓ 28 59	27 45	17 12	20 42	02 52	07 11	22 18	11 09	09 42
06	15 03 00	14 06 59	02 ♈ 50 50	09 ♈ 09 24	29 ♏ 14	18 27	21 18	03 05	07 07	22 16	11 07	09 43
07	15 06 56	15 07 10	15 ♈ 24 51	21 ♈ 37 20	00 ♐ 42	19 42	21 53	03 18	07 02	22 14	11 06	09 43
08	15 10 53	16 07 22	27 ♈ 47 02	03 ♉ 54 06	02 10	20 58	22 29	03 31	06 58	22 12	11 05	09 44
09	15 14 50	17 07 35	09 ♉ 58 44	16 ♉ 01 06	03 38	22 13	23 04	03 44	06 53	22 10	11 03	09 44
10	15 18 46	18 07 51	22 ♉ 01 24	27 ♉ 59 52	05 05	23 28	23 40	03 57	06 49	22 08	11 02	09 45
11	15 22 43	19 08 08	03 ♊ 56 43	09 ♊ 52 13	06 31	24 43	24 15	04 11	06 44	22 06	11 01	09 45
12	15 26 39	20 08 27	15 ♊ 46 42	21 ♊ 40 28	07 56	25 59	24 50	04 24	06 40	22 04	11 00	09 46
13	15 30 36	21 08 48	27 ♊ 33 53	03 ♋ 27 22	09 21	27 14	25 26	04 37	06 35	22 02	10 58	09 46
14	15 34 32	22 09 11	09 ♋ 21 19	15 ♋ 16 13	10 46	28 29	26 01	04 50	06 31	22 00	10 57	09 47
15	15 38 29	23 09 35	21 ♋ 12 34	27 ♋ 10 53	12 09	29 ♏ 44	26 36	05 04	06 26	21 58	10 56	09 48
16	15 42 25	24 10 01	03 ♌ 11 42	09 ♌ 15 36	13 32	01 ♐ 00	27 11	05 17	06 21	21 56	10 55	09 48
17	15 46 22	25 10 29	15 ♌ 23 08	21 ♌ 34 53	14 53	02 15	27 46	05 30	06 16	21 53	10 54	09 49
18	15 50 19	26 10 59	27 ♌ 51 24	04 ♍ 13 14	16 14	03 30	28 21	05 43	06 12	21 51	10 53	09 50
19	15 54 15	27 11 31	10 ♍ 40 51	17 ♍ 14 40	17 33	04 45	28 56	05 57	06 07	21 49	10 52	09 50
20	15 58 12	28 12 05	23 ♍ 55 04	00 ♎ 42 15	18 51	06 01	29 ♍ 31	06 10	06 02	21 47	10 51	09 51
21	16 02 08	29 ♏ 12 40	07 ♎ 36 23	14 ♎ 37 24	20 07	07 16	00 ♎ 06	06 24	05 57	21 44	10 50	09 52
22	16 06 05	00 ♐ 13 17	21 ♎ 45 08	28 ♎ 59 13	21 21	08 31	00 41	06 37	05 52	21 42	10 49	09 53
23	16 10 01	01 13 56	06 ♏ 19 06	13 ♏ 44 06	22 34	09 46	01 15	06 50	05 47	21 40	10 48	09 54
24	16 13 58	02 14 36	21 ♏ 13 18	28 ♏ 45 43	23 44	11 02	01 50	07 04	05 42	21 37	10 47	09 55
25	16 17 54	03 15 18	06 ♐ 20 12	13 ♐ 55 35	24 51	12 17	02 25	07 17	05 37	21 35	10 46	09 55
26	16 21 51	04 16 01	21 ♐ 30 39	29 ♐ 04 14	25 56	13 32	02 59	07 31	05 32	21 33	10 45	09 56
27	16 25 48	05 16 46	06 ♑ 35 12	14 ♑ 02 34	26 57	14 48	03 33	07 44	05 28	21 30	10 44	09 57
28	16 29 44	06 17 32	21 ♑ 25 29	28 ♑ 43 14	27 54	16 03	04 08	07 58	05 23	21 28	10 44	09 58
29	16 33 41	07 18 18	05 ♒ 55 17	13 ♒ 01 16	28 47	17 18	04 42	08 11	05 18	21 25	10 43	09 59
30	16 37 37	08 ♐ 19 06	20 ♒ 01 00	26 ♒ 54 24	29 ♐ 34	18 ♐ 34	05 ♎ 16	08 ♐ 24	05 ♊ 13	21 ♊ 23	10 ♈ 42	10 ♒ 00

Date	Moon True ☊ ° '	Moon Mean ☊ ° '	Moon ☽ Latitude ° '
01	07 ♐ 38	08 ♐ 41	03 N 56
02	07 R 38	08 38	04 39
03	07 38	08 34	05 05
04	07 36	08 31	05 14
05	07 33	08 28	05 07
06	07 29	08 25	04 44
07	07 25	08 22	04 08
08	07 22	08 19	03 21
09	07 19	08 15	02 25
10	07 17	08 12	01 23
11	07 17	08 09	00 N 18
12	07 D 17	08 06	00 S 47
13	07 18	08 03	01 50
14	07 19	07 59	02 49
15	07 20	07 56	03 40
16	07 22	07 53	04 23
17	07 22	07 50	04 55
18	07 R 23	07 47	05 13
19	07 22	07 44	05 17
20	07 22	07 40	05 05
21	07 21	07 37	04 35
22	07 20	07 34	03 47
23	07 19	07 31	02 44
24	07 18	07 28	01 28
25	07 18	07 25	00 S 05
26	07 D 18	07 21	01 N 18
27	07 19	07 18	02 36
28	07 19	07 15	03 42
29	07 19	07 12	04 32
30	07 ♐ 19	07 ♐ 09	05 N 04

DECLINATIONS

Sun ☉ ° '	Moon ☽ ° '	Mercury ☿ ° '	Venus ♀ ° '	Mars ♂ ° '	Jupiter ♃ ° '	Saturn ♄ ° '	Uranus ♅ ° '	Neptune ♆ ° '	Pluto ♇ ° '
14 S 32	17 S 02	19 S 22	14 S 48	05 N 58	19 S 56	19 N 29	23 N 18	03 N 00	23 S 46
14 51	13 13	19 52	15 12	05 44	19 59	19 28	23 18	02 59	23 46
15 10	08 48	20 20	15 37	05 30	20 02	19 28	23 18	02 58	23 45
15 28	04 S 03	20 47	16 01	05 17	20 04	19 27	23 18	02 58	23 45
15 46	00 N 46	21 13	16 24	05 03	20 07	19 26	23 18	02 57	23 45
16 04	05 29	21 39	16 47	04 49	20 09	19 25	23 18	02 57	23 45
16 22	09 53	22 03	17 10	04 35	20 12	19 25	23 18	02 56	23 45
16 40	13 49	22 26	17 32	04 22	20 15	19 24	23 18	02 56	23 44
16 57	17 06	22 48	17 54	04 08	20 17	19 23	23 17	02 56	23 44
17 14	19 37	23 09	18 15	03 54	20 20	19 22	23 17	02 55	23 44
17 30	21 14	23 28	18 35	03 40	20 22	19 22	23 17	02 55	23 44
17 47	21 54	23 47	18 56	03 27	20 25	19 21	23 17	02 54	23 44
18 03	21 35	24 04	19 15	03 13	20 27	19 20	23 17	02 54	23 43
18 18	20 18	24 20	19 35	02 59	20 30	19 19	23 17	02 53	23 43
18 34	18 08	24 35	19 53	02 46	20 32	19 19	23 17	02 53	23 43
18 49	15 10	24 48	20 11	02 32	20 35	19 18	23 17	02 52	23 42
19 03	11 31	25 01	20 29	02 18	20 37	19 17	23 17	02 52	23 42
19 18	07 18	25 12	20 46	02 05	20 40	19 16	23 16	02 52	23 42
19 32	02 N 40	25 21	21 02	01 51	20 42	19 15	23 16	02 51	23 42
19 45	02 S 15	25 29	21 18	01 37	20 44	19 15	23 16	02 51	23 41
19 59	07 13	25 36	21 33	01 24	20 47	19 14	23 16	02 51	23 41
20 12	11 59	25 41	21 48	01 10	20 49	19 13	23 16	02 50	23 41
20 24	16 12	25 45	22 01	00 57	20 51	19 12	23 16	02 50	23 40
20 36	19 29	25 48	22 15	00 43	20 54	19 11	23 16	02 50	23 40
20 48	21 27	25 49	22 27	00 30	20 56	19 11	23 16	02 49	23 40
21 00	21 52	25 49	22 39	00 16	20 58	19 10	23 15	02 49	23 39
21 11	20 40	25 47	22 50	00 N 03	21 01	19 09	23 15	02 49	23 39
21 21	18 05	25 44	23 01	00 S 10	21 03	19 08	23 15	02 48	23 39
21 31	14 24	25 39	23 11	00 24	21 05	19 07	23 15	02 48	23 38
21 S 41	10 S 00	25 S 33	23 S 20	00 S 37	21 S 07	19 N 07	23 N 15	02 N 48	23 S 38

ZODIAC SIGN ENTRIES

Date	h	m	Planets
01	18	25	☽ ♒
03	23	15	☽ ♓
06	06	37	☽ ♈
07	00	30	☿ ♐
08	16	20	☽ ♉
11	04	02	☽ ♊
13	16	58	☽ ♋
15	17	01	♀ ♐
16	05	38	☽ ♌
18	16	04	☽ ♍
20	22	46	☽ ♎
21	07	55	♂ ♎
22	06	44	☉ ♐
23	01	40	☽ ♏
25	01	58	☽ ♐
27	01	29	☽ ♑
29	02	07	☽ ♒

LATITUDES

Date	Mercury ☿ ° '	Venus ♀ ° '	Mars ♂ ° '	Jupiter ♃ ° '	Saturn ♄ ° '	Uranus ♅ ° '	Neptune ♆ ° '	Pluto ♇ ° '
01	01 S 13	00 N 44	01 N 28	00 N 38	02 S 06	00 N 05	01 S 35	06 S 10
04	01 31	00 38	01 29	00 38	02 06	00 05	01 35	06 10
07	01 48	00 31	01 30	00 37	02 06	00 05	01 34	06 10
10	02 02	00 24	01 31	00 37	02 06	00 05	01 34	06 10
13	02 15	00 17	01 31	00 37	02 06	00 05	01 34	06 09
16	02 24	00 10	01 32	00 36	02 06	00 05	01 34	06 09
19	02 31	00 N 03	01 33	00 36	02 06	00 05	01 34	06 09
22	02 33	00 S 04	01 34	00 36	02 06	00 05	01 34	06 08
25	02 29	00 12	01 35	00 36	02 06	00 06	01 34	06 08
28	02 19	00 19	01 36	00 35	02 05	00 06	01 34	06 08
31	01 S 59	00 S 26	01 N 37	00 N 35	02 S 05	00 N 06	01 S 34	06 S 08

DATA

Julian Date	2462807
Delta T	+74 seconds
Ayanamsa	24° 17' 31"
Synetic vernal point	04° ♓ 49' 28"
True obliquity of ecliptic	23° 26' 04"

MOON'S PHASES, APSIDES AND POSITIONS ☽

Date	h	m	Phase	Longitude ° '	Eclipse Indicator
02	11	56	☽	10 ♒ 06	
10	03	30	○	17 ♉ 47	
18	08	32	☾	26 ♌ 02	
25	06	46	●	03 ♐ 02	Total

Day	h	m		
13	04	46	Apogee	
25	21	03	Perigee	
05	08	09	ON	
12	16	09	Max dec	21° N 55'
20	01	07	OS	
26	06	04	Max dec	21° S 55'

LONGITUDES

Date	Chiron ⚷ ° '	Ceres ⚳ ° '	Pallas ⚴ ° '	Juno ⚵ ° '	Vesta ⚶ ° '	Black Moon Lilith ⚸ ° '
01	14 ♉ 44	16 ♉ 59	27 ♒ 39	09 ♑ 01	07 ♊ 33	17 ♊ 57
11	14 ♉ 15	14 ♉ 41	28 ♒ 12	12 ♑ 12	05 ♊ 24	19 ♊ 03
21	13 ♉ 45	12 ♉ 27	29 ♒ 14	15 ♑ 36	02 ♊ 53	20 ♊ 10
31	13 ♉ 18	10 ♉ 34	00 ♓ 43	19 ♑ 11	00 ♊ 16	21 ♊ 16

ASPECTARIAN

h m Aspects

01 Friday
02 35 ☽ Q ☉
03 23 ☽ ⚹ ☿
05 28 ☽ ⊼ ♅
05 34 ☽ Ᵽ ♄
08 07 ☽ Q ♀
15 41 ☽ ± ♅
17 06 ☽ Q ♆
17 18 ☿ # ♄
22 02 ☽ ⚹ ♃
22 22 ☿ ⊼ ♅

02 Saturday
00 41 ☽ Ᵽ ♂
01 12 ☽ ∥ ♀
01 54 ☉ □ ♇
02 27 ☽ Q ☿
03 01 ☽ ∥ ☉
07 11 ☽ Ᵽ ♅
07 13 ☽ △ ♄
11 12 ☽ ☌ ♇
11 56 ☽ □ ☉
13 52 ☽ ⚹ ♆
17 04 ☽ ± ♂
18 21 ☽ □ ♀
18 48 ☿ ∥ ♃
19 15 ☽ Q ♃

03 Sunday
04 08 ☽ ⊼ ♂
09 32 ☽ △ ♅
13 50 ☉ ⊼ ♆
14 04 ☽ □ ☿
16 23 ☽ ∠ ♆

04 Monday
03 55 ☽ □ ♃
05 36 ☽ # ♂
08 37 ☽ ⊥ ♆
10 32 ☿ Ᵽ ♆
12 24 ☽ □ ♄
16 53 ☽ ⊻ ♇
17 26 ☽ # ♆
19 04 ♀ ± ♅
19 33 ☽ ⊻ ♆
22 05 ☽ △ ☉

05 Tuesday
03 56 ☽ ⊥ ♇
06 07 ☽ △ ♀
10 58 ♀ ± ♆
11 20 ☿ Q ♇
13 15 ☽ ☍ ♂
16 09 ☽ □ ♅
20 41 ☽ ∠ ♇
21 30 ☽ Q ♄
22 32 ♂ Q ♃
22 59 ☽ ∥ ♆

06 Wednesday
04 16 ☽ △ ☿
04 21 ☽ Ᵽ ☉
08 43 ☽ ∥ ♂
12 27 ☽ △ ♃
13 16 ☽ Ᵽ ♀
20 03 ☽ ⚹ ♄
22 53 ☽ ± ☉

07 Thursday
01 04 ☽ ⚹ ♇
02 05 ☽ Q ♅
03 44 ☽ ☌ ♆
08 21 ☽ ± ♀
11 23 ☽ ⊼ ☉
12 38 ☽ Ᵽ ☿
17 40 ☽ Ᵽ ♃
21 13 ☽ ⊼ ♀

08 Friday
00 12 ☽ Q ♇
00 44 ☽ ∠ ♄
01 09 ♂ □ ♅
01 09 ☽ ⚹ ♅
01 09 ☽ ⊼ ♂
08 25 ☽ ± ☿
11 28 ☽ ± ♃
13 26 ☽ ± ♂
13 46 ☉ ± ♅
18 11 ☽ ⊥ ♄
21 46 ☽ ⊼ ☿
23 27 ☽ ⊼ ♃

09 Saturday
05 56 ☽ ⊻ ♄
06 27 ☽ ∠ ♅
08 02 ☽ Ᵽ ♂
10 20 ☉ ± ♆
10 36 ☽ # ☉
11 08 ♀ ⊼ ♅
11 31 ☽ □ ♇
14 08 ☽ ⊻ ♆
14 09 ☿ ☌ ♃
19 48 ☽ # ♀

10 Sunday
00 16 ☽ ⊥ ♅
02 03 ☽ ⊥ ♆
03 30 ☽ ☍ ☉
09 20 ☽ ∥ ♄
12 13 ☽ ⊻ ♅
15 14 ☽ ☍ ♀
15 27 ☽ △ ♂
19 05 ♀ ⚹ ♂
20 02 ☽ ∠ ♆
21 12 ☽ # ♃
22 22 ☿ # ♅

11 Monday
04 36 ☿ Q ♂
12 29 ☽ ☍ ♃
15 36 ☿ ☍ ♄
17 37 ☽ ☌ ♄
17 55 ☽ ☍ ☿
23 46 ☽ △ ♇

12 Tuesday
02 18 ☽ ⚹ ♆
07 36 ☿ ∥ ♇
08 44 ♂ Ᵽ ♇
12 20 ♀ Ᵽ ♆
21 42 ☽ ⊼ ☉

13 Wednesday
00 46 ☽ ☌ ♅
02 40 ☽ Q ♆
06 18 ☽ Ᵽ ♇
07 25 ☽ □ ♂
11 04 ☽ ± ☉
11 14 ☽ ⊼ ♀
17 35 ♀ # ♄
19 07 ☿ ⚹ ♇
22 28 ♀ Q ♇

14 Thursday
00 39 ☽ ± ♇
00 55 ☽ ± ♀
02 38 ☽ ⊼ ♃
06 15 ☽ ⊻ ♄
07 06 ☽ Ᵽ ☉
08 25 ☉ ⊼ ♅
09 16 ☽ # ♃
12 52 ☽ ⊼ ♇
15 04 ☽ ± ♃
15 14 ☽ ⊼ ☿
15 15 ☽ □ ♆
15 16 ☿ △ ♆
18 22 ☽ ⊥ ♄
19 54 ☽ # ♀
21 22 ☽ Ᵽ ♀
21 57 ☽ Q ♂
22 48 ♂ ∥ ♆

15 Friday
00 04 ☽ ∥ ♄
05 01 ☽ ± ☿
08 19 ☽ # ☉
09 38 ☽ Ᵽ ♃
12 27 ☽ ∠ ♄
13 31 ☽ ⊻ ♅
16 17 ☽ △ ☉
23 24 ☽ ⚹ ♂

16 Saturday
01 30 ☽ Ᵽ ☿
01 31 ☽ ⊥ ♅
07 06 ☽ △ ♀
16 13 ☽ △ ♃
18 13 ☽ ⚹ ♄
19 22 ☽ ∠ ♅

17 Sunday
01 05 ☽ ☍ ♇
03 14 ☽ △ ♆
06 38 ☽ ∠ ♂
10 55 ☽ △ ☿
17 34 ☽ Q ♄

18 Monday
00 33 ☽ ⚹ ♅
00 58 ☽ ⊥ ♂
01 37 ♀ ∥ ♃
04 55 ☉ Ᵽ ♆
08 15 ☽ Ᵽ ♆
08 32 ☽ □ ☉
09 41 ☉ # ♄
12 59 ☽ ⊻ ♂
23 17 ☽ Q ♅
23 49 ☽ □ ♀

19 Tuesday
01 13 ☽ ± ♆
03 04 ☽ □ ♃
03 35 ☽ □ ♄
10 27 ☽ ⊼ ♇
11 03 ☽ ∥ ♆
12 20 ☽ ⊼ ♆
16 15 ☽ ∥ ♂
20 57 ☽ Q ☉
21 28 ☽ ± ♇

20 Wednesday
01 05 ♃ ☍ ♄
01 55 ☽ □ ☿
03 37 ☉ Q ♇
08 11 ☽ □ ♅
09 07 ☽ # ♂
12 11 ☽ Q ♀
12 22 ♀ ☍ ♄
12 27 ☽ Q ♃
13 40 ☽ Ᵽ ♇
14 54 ☽ # ♆
15 41 ♀ ☌ ♃
20 12 ☽ ⚹ ☉
22 22 ☽ ☌ ♂

21 Thursday
09 09 ☽ △ ♄
09 52 ☽ ⚹ ♃
11 21 ☽ ⚹ ♀
12 58 ☽ Q ☿
15 53 ☽ △ ♇
17 32 ☽ ☍ ♆

22 Friday
00 10 ☽ ∠ ☉
10 32 ☽ Ᵽ ♄
11 17 ☽ ⚹ ☿
11 46 ☽ ∠ ♃
11 55 ☽ △ ♅
15 14 ☽ ∠ ♀
16 25 ☽ ⊥ ☉
18 34 ☿ ☍ ♅

23 Saturday
01 23 ☽ ± ♄
02 54 ☽ ⊥ ♃
03 04 ☽ ⊻ ☉
03 23 ☽ ⊻ ♂
07 28 ☽ ⊥ ♀
11 08 ☽ ⊼ ♄
12 33 ☽ Ᵽ ♅
12 52 ☽ ⊻ ♃
13 20 ☉ ⚹ ♂
13 35 ☽ ⊥ ♂
14 12 ☽ ∠ ☿
14 19 ♀ ⚹ ♇
17 48 ☽ □ ♇
18 08 ☽ ⊻ ♀
19 15 ☽ ⊼ ♆

24 Sunday
03 04 ☽ ± ♅
04 42 ☽ ∠ ♂
04 54 ☽ ± ♆
05 57 ☽ ⊥ ☿
07 20 ♀ △ ♆
09 28 ☽ # ♄
12 38 ☽ ⊼ ♅
16 20 ☽ ⊻ ☿
19 16 ☽ Ᵽ ♆
22 39 ☽ Q ♇

25 Monday
00 45 ☽ ∥ ☉
03 29 ☽ ∥ ♃
05 32 ☽ ⚹ ♂
06 46 ☽ ☌ ☉
10 53 ☽ ☍ ♄
13 29 ☿ ∠ ♇
13 31 ☽ ☌ ♃
17 41 ☽ ⚹ ♇
19 00 ☽ △ ♆
22 15 ☽ ☌ ♀

26 Tuesday
01 16 ☽ Q ♂
08 43 ☉ ∥ ♃
12 03 ☽ ☍ ♅
17 26 ☽ ∠ ♇
19 32 ☽ ☌ ☿

27 Wednesday
05 46 ☽ ∥ ☉
06 58 ☽ □ ♂
07 38 ☽ ∥ ♃
07 47 ☽ ⊥ ♇
09 45 ☽ ⊻ ☉
10 12 ☽ ⊼ ♄
13 52 ☽ ⊻ ♃
15 55 ☉ ☍ ♄
17 25 ☽ ⊻ ♇
18 40 ☽ □ ♆
19 47 ☽ ± ♄
20 05 ☽ ⊥ ☉
23 41 ☽ ⊥ ♃

28 Thursday
02 26 ☽ ⊻ ♀
03 29 ☽ # ♄
10 18 ☽ Ᵽ ♄
11 46 ☽ ∠ ☉
12 04 ☽ ⊼ ♅
13 07 ☽ ⊥ ♀
14 33 ☽ ∠ ♃
14 49 ♀ Q ♂
21 54 ☽ ± ♅
23 21 ☽ ⊻ ☿

29 Friday
00 00 ☽ Q ♆
05 23 ☽ ∠ ♀
09 52 ☽ △ ♂
09 58 ☽ ⊥ ☿
10 57 ☽ △ ♄
12 50 ☽ Ᵽ ♅
14 30 ☽ ⚹ ☉
15 52 ☽ ⚹ ♃
18 52 ☽ ☌ ♇
20 05 ☽ ⚹ ♆
22 05 ♀ # ♅

30 Saturday
02 06 ☽ ∠ ☿
09 14 ☽ ⚹ ♀
09 49 ♂ △ ♄
12 28 ☽ Ᵽ ♂
12 34 ☽ Q ☉
12 41 ☽ Q ♃
14 21 ☽ △ ♅
14 43 ☉ ☌ ♃
21 53 ☽ ∠ ♆

All ephemeris data is given at 12.00 UT and the Moon's longitude is additionally given for 24.00 UT

DECEMBER 2030

Date	Sidereal time h m s	Sun ☉ ° ' "	Moon ☽ ° ' "	Moon ☽ 24.00 ° ' "	Mercury ☿ ° '	Venus ♀ ° '	Mars ♂ ° '	Jupiter ♃ ° '	Saturn ♄ ° '	Uranus ♅ ° '	Neptune ♆ ° '	Pluto ♇ ° '
01	16 41 34	09 ♐ 19 54	03 ♓ 41 33	10 ♓ 22 37	00 ♑ 16	19 ♐ 49	05 ♎ 50	08 ♐ 38	05 ♊ 08	21 ♊ 20	10 ♈ 41	10 ♒ 01
02	16 45 30	10 20 44	16 ♓ 57 52	23 ♓ 27 38	00 51	21 04	06 24	08 51	05 R 03	21 R 18	10 R 41	10 02
03	16 49 27	11 21 34	29 ♓ 52 18	06 ♈ 12 17	01 19	22 19	06 58	09 05	04 58	21 15	10 40	10 04
04	16 53 23	12 22 25	12 ♈ 28 00	18 ♈ 39 53	01 39	23 35	07 32	09 18	04 53	21 13	10 40	10 05
05	16 57 20	13 23 17	24 ♈ 48 21	00 ♉ 53 49	01 50	24 50	08 06	09 32	04 48	21 10	10 39	10 06
06	17 01 17	14 24 10	06 ♉ 56 42	12 ♉ 57 22	01 R 51	26 05	08 40	09 45	04 43	21 08	10 38	10 07
07	17 05 13	15 25 03	18 ♉ 56 09	24 ♉ 53 25	01 42	27 21	09 14	09 59	04 39	21 05	10 38	10 08
08	17 09 10	16 25 58	00 ♊ 49 27	06 ♊ 44 35	01 21	28 36	09 47	10 12	04 34	21 03	10 37	10 09
09	17 13 06	17 26 54	12 ♊ 39 03	18 ♊ 33 08	00 50	29 ♐ 51	10 21	10 26	04 29	21 00	10 37	10 11
10	17 17 03	18 27 50	24 ♊ 27 06	00 ♋ 21 13	00 ♑ 07	01 ♑ 06	10 54	10 39	04 24	20 58	10 37	10 12
11	17 20 59	19 28 48	06 ♋ 15 43	12 ♋ 10 53	29 ♐ 13	02 22	11 27	10 52	04 20	20 55	10 36	10 13
12	17 24 56	20 29 46	18 ♋ 07 00	24 ♋ 04 20	28 09	03 37	12 01	11 06	04 15	20 53	10 36	10 14
13	17 28 52	21 30 45	00 ♌ 03 14	06 ♌ 04 00	26 57	04 52	12 34	11 19	04 11	20 50	10 36	10 16
14	17 32 49	22 31 46	12 ♌ 07 01	18 ♌ 12 39	25 38	06 07	13 07	11 32	04 06	20 47	10 35	10 17
15	17 36 46	23 32 47	24 ♌ 21 18	00 ♍ 33 23	24 16	07 23	13 40	11 46	04 01	20 45	10 35	10 18
16	17 40 42	24 33 49	06 ♍ 49 20	13 ♍ 09 36	22 53	08 38	14 13	11 59	03 57	20 42	10 35	10 20
17	17 44 39	25 34 52	19 ♍ 34 38	26 ♍ 04 49	21 32	09 53	14 46	12 12	03 53	20 40	10 35	10 21
18	17 48 35	26 35 56	02 ♎ 40 35	09 ♎ 22 17	20 15	11 08	15 18	12 26	03 48	20 37	10 35	10 23
19	17 52 32	27 37 01	16 ♎ 10 10	23 ♎ 04 28	19 05	12 23	15 51	12 39	03 44	20 35	10 34	10 24
20	17 56 28	28 38 07	00 ♏ 05 16	07 ♏ 12 31	18 04	13 39	16 24	12 52	03 40	20 32	10 34	10 26
21	18 00 25	29 ♐ 39 13	14 ♏ 26 02	21 ♏ 45 27	17 13	14 54	16 56	13 05	03 36	20 30	10 34	10 27
22	18 04 21	00 ♑ 40 21	29 ♏ 10 12	06 ♐ 39 35	16 32	16 09	17 28	13 19	03 31	20 27	10 D 34	10 29
23	18 08 18	01 41 29	14 ♐ 12 41	21 ♐ 48 26	16 02	17 24	18 00	13 32	03 27	20 25	10 34	10 30
24	18 12 15	02 42 38	29 ♐ 25 40	07 ♑ 03 05	15 43	18 40	18 33	13 45	03 23	20 22	10 35	10 32
25	18 16 11	03 43 47	14 ♑ 39 25	22 ♑ 13 22	15 34	19 55	19 05	13 58	03 19	20 20	10 35	10 33
26	18 20 08	04 44 57	29 ♑ 43 44	07 ♒ 09 28	15 D 35	21 10	19 36	14 11	03 16	20 17	10 35	10 35
27	18 24 04	05 46 06	14 ♒ 29 37	21 ♒ 43 28	15 45	22 25	20 08	14 24	03 12	20 15	10 35	10 36
28	18 28 01	06 47 16	28 ♒ 50 31	05 ♓ 50 24	16 04	23 40	20 40	14 37	03 08	20 12	10 35	10 38
29	18 31 57	07 48 25	12 ♓ 43 02	19 ♓ 28 25	16 30	24 56	21 11	14 50	03 05	20 10	10 35	10 39
30	18 35 54	08 49 35	26 ♓ 06 47	02 ♈ 38 27	17 03	26 11	21 43	15 03	03 01	20 07	10 36	10 41
31	18 39 50	09 ♑ 50 44	09 ♈ 03 50	15 ♈ 23 26	17 ♐ 42	27 ♑ 26	22 ♎ 14	15 ♐ 16	02 ♊ 58	20 ♊ 05	10 ♈ 36	10 ♒ 43

Date	Moon True ☊ ° '	Moon Mean ☊ ° '	Moon ☽ Latitude ° '
01	07 ♐ 19	07 ♐ 05	05 N 18
02	07 D 19	07 02	05 14
03	07 19	06 59	04 54
04	07 19	06 56	04 20
05	07 20	06 53	03 35
06	07 21	06 50	02 41
07	07 21	06 46	01 41
08	07 22	06 43	00 N 36
09	07 R 22	06 40	00 S 29
10	07 21	06 37	01 33
11	07 20	06 34	02 34
12	07 18	06 30	03 27
13	07 16	06 27	04 12
14	07 14	06 24	04 47
15	07 12	06 21	05 08
16	07 10	06 18	05 16
17	07 09	06 15	05 09
18	07 D 09	06 11	04 46
19	07 10	06 08	04 06
20	07 11	06 05	03 11
21	07 13	06 02	02 03
22	07 14	05 59	00 S 44
23	07 R 14	05 56	00 N 38
24	07 12	05 52	01 59
25	07 10	05 49	03 12
26	07 06	05 46	04 10
27	07 02	05 43	04 50
28	06 58	05 40	05 10
29	06 54	05 36	05 12
30	06 52	05 33	04 56
31	06 ♐ 51	05 ♐ 30	04 N 26

DECLINATIONS

Date	Sun ☉ ° '	Moon ☽ ° '	Mercury ☿ ° '	Venus ♀ ° '	Mars ♂ ° '	Jupiter ♃ ° '	Saturn ♄ ° '	Uranus ♅ ° '	Neptune ♆ ° '	Pluto ♇ ° '
01	21 S 51	05 S 13	25 S 25	23 S 29	00 S 50	21 S 09	19 N 06	23 N 15	02 N 48	23 S 38
02	22 00	00 S 20	25 17	23 36	01 03	21 12	19 05	23 14	02 47	23 37
03	22 08	04 N 27	25 06	23 43	01 17	21 14	19 04	23 14	02 47	23 37
04	22 16	08 55	24 55	23 50	01 30	21 16	19 04	23 14	02 47	23 36
05	22 24	12 57	24 42	23 55	01 43	21 18	19 03	23 14	02 47	23 36
06	22 31	16 22	24 28	24 00	01 56	21 20	19 02	23 14	02 47	23 36
07	22 38	19 04	24 13	24 04	02 09	21 22	19 01	23 14	02 47	23 35
08	22 45	20 55	23 56	24 08	02 22	21 24	19 01	23 14	02 46	23 35
09	22 51	21 49	23 38	24 11	02 35	21 26	19 00	23 13	02 46	23 34
10	22 56	21 46	23 20	24 12	02 47	21 28	18 59	23 13	02 46	23 34
11	23 01	20 44	23 00	24 14	03 00	21 30	18 58	23 13	02 46	23 34
12	23 06	18 47	22 39	24 14	03 13	21 32	18 58	23 13	02 46	23 33
13	23 10	16 01	22 17	24 14	03 26	21 34	18 57	23 13	02 46	23 33
14	23 13	12 34	21 55	24 13	03 38	21 36	18 56	23 13	02 46	23 32
15	23 17	08 33	21 34	24 11	03 51	21 38	18 56	23 12	02 46	23 32
16	23 19	04 N 06	21 13	24 08	04 03	21 39	18 55	23 12	02 46	23 32
17	23 22	00 S 37	20 53	24 05	04 16	21 41	18 54	23 12	02 46	23 31
18	23 23	05 26	20 35	24 01	04 28	21 43	18 54	23 12	02 46	23 31
19	23 25	10 08	20 19	23 56	04 40	21 45	18 53	23 12	02 46	23 30
20	23 26	14 29	20 05	23 50	04 52	21 46	18 52	23 12	02 46	23 30
21	23 26	18 07	19 55	23 44	05 05	21 48	18 52	23 11	02 46	23 29
22	23 26	20 41	19 47	23 37	05 17	21 50	18 51	23 11	02 46	23 29
23	23 25	21 52	19 43	23 29	05 29	21 52	18 51	23 11	02 46	23 28
24	23 24	21 27	19 41	23 21	05 41	21 53	18 50	23 11	02 46	23 28
25	23 23	19 27	19 42	23 12	05 53	21 55	18 50	23 11	02 46	23 27
26	23 21	16 08	19 45	23 02	06 04	21 56	18 49	23 11	02 46	23 27
27	23 18	11 51	19 50	22 51	06 16	21 58	18 49	23 10	02 46	23 27
28	23 16	07 01	19 58	22 40	06 28	21 59	18 48	23 10	02 46	23 26
29	23 12	01 S 59	20 06	22 28	06 39	22 01	18 48	23 10	02 47	23 26
30	23 08	02 N 59	20 16	22 15	06 51	22 02	18 47	23 10	02 47	23 25
31	23 S 04	07 N 40	20 S 27	22 S 02	07 S 02	22 S 04	18 N 47	23 N 10	02 N 47	23 S 25

ZODIAC SIGN ENTRIES

Date	h	m	Planets
01	02	27	☿ ♑
01	05	27	☽ ♓
03	12	14	☽ ♈
05	22	14	☽ ♉
08	10	20	☽ ♊
09	14	52	♀ ♑
10	15	11	☿ ♐
10	23	17	☽ ♋
13	11	54	☽ ♌
15	22	56	☽ ♍
18	07	09	☽ ♎
20	11	51	☽ ♏
21	20	10	☉ ♑
22	13	20	☽ ♐
24	12	54	☽ ♑
26	12	26	☽ ♒
28	13	58	☽ ♓
30	19	07	☽ ♈

LATITUDES

Date	Mercury ☿ ° '	Venus ♀ ° '	Mars ♂ ° '	Jupiter ♃ ° '	Saturn ♄ ° '	Uranus ♅ ° '	Neptune ♆ ° '	Pluto ♇ ° '
01	01 S 59	00 S 26	01 N 37	00 N 35	02 S 05	00 N 06	01 S 34	06 S 08
04	01 30	00 33	01 38	00 35	02 05	00 06	01 34	06 08
07	00 S 47	00 40	01 38	00 35	02 04	00 06	01 33	06 07
10	00 N 07	00 47	01 39	00 35	02 04	00 06	01 33	06 07
13	01 07	00 53	01 40	00 34	02 04	00 06	01 33	06 07
16	02 02	00 59	01 41	00 34	02 03	00 06	01 33	06 07
19	02 41	01 05	01 42	00 34	02 03	00 06	01 33	06 07
22	02 59	01 10	01 42	00 34	02 02	00 06	01 33	06 07
25	02 59	01 15	01 43	00 34	02 02	00 06	01 33	06 06
28	02 46	01 20	01 44	00 34	02 01	00 06	01 32	06 06
31	02 N 25	01 S 24	01 N 45	00 N 33	02 S 00	00 N 06	01 S 32	06 S 06

DATA

Julian Date	2462837
Delta T	+74 seconds
Ayanamsa	24° 17' 36"
Synetic vernal point	04° ♓ 49' 23"
True obliquity of ecliptic	23° 26' 03"

LONGITUDES

Date	Chiron ⚷ ° '	Ceres ⚳ ° '	Pallas ⚴ ° '	Juno ⚵ ° '	Vesta ⚶ ° '	Black Moon Lilith ⚸ ° '
01	13 ♉ 18	10 ♉ 34	00 ♓ 43	19 ♑ 11	00 ♊ 16	21 ♊ 16
11	12 ♉ 53	09 ♉ 13	02 ♓ 36	22 ♑ 57	27 ♉ 55	22 ♊ 23
21	12 ♉ 33	08 ♉ 29	04 ♓ 48	26 ♑ 51	26 ♉ 03	23 ♊ 29
31	12 ♉ 18	08 ♉ 26	07 ♓ 18	00 ♒ 52	24 ♉ 53	24 ♊ 36

MOON'S PHASES, APSIDES AND POSITIONS ☽

Date	h	m	Phase	Longitude ° '	Eclipse Indicator
01	22	57	☽	09 ♓ 48	
09	22	40	○	17 ♊ 54	
18	00	01	☾	26 ♍ 05	
24	17	32	●	02 ♑ 57	
31	13	36	☽	09 ♈ 55	

Day	h	m		
10	09	47	Apogee	
24	10	02	Perigee	
02	13	37	0N	
09	22	27	Max dec	21° N 55'
17	08	56	0S	
23	17	49	Max dec	21° S 55'
29	21	27	0N	

ASPECTARIAN

01 Sunday
04 52 ☽ ± ♂
05 36 ☽ ✶ ☿
08 19 ☽ Q ♀
13 47 ☽ ⊥ ♆
14 33 ☽ □ ♄
16 00 ☽ ⊼ ♂
21 00 ☽ □ ♃
22 57 ☽ □ ☉
23 23 ☽ ⊻ ♇
23 53 ☽ # ♆

02 Monday
00 33 ☽ ⊻ ♆
04 10 ☽ Q ☿
04 40 ☉ ✶ ♇
08 33 ☽ ∥ ♂
10 18 ☽ ⊥ ♇
14 29 ♀ ∥ ♇
16 15 ♀ ☍ ♅
19 12 ☽ # ♂
19 50 ☉ △ ♆
19 58 ☽ □ ♅
20 22 ☽ □ ♀
23 10 ☽ Q ♄

03 Tuesday
02 58 ☽ ∠ ♇
03 33 ☽ ∥ ♆
14 49 ☽ □ ☿
21 35 ☽ ✶ ♄

04 Wednesday
02 06 ☽ ☍ ♂
05 46 ☽ Q ♅
05 49 ☽ △ ♃
07 24 ☽ ✶ ♇
08 32 ☽ ☌ ♆
11 48 ☽ △ ☉

05 Thursday
02 17 ☽ ∠ ♄
04 55 ☽ ✶ ♅
06 41 ☽ Q ♇
11 27 ☽ ⅊ ♃
12 04 ☽ △ ♀
17 08 ♀ ∠ ♇
19 41 ☽ ⅊ ☉
19 49 ☽ ⊥ ♄

06 Friday
01 55 ☽ △ ☿
02 46 ☿ St R
05 32 ☽ ± ♃
07 37 ☽ ⊻ ♄
10 23 ☽ ∠ ♅
15 10 ☽ ± ☉
15 36 ☽ ⊼ ♂
17 42 ☽ ⊼ ♃
18 20 ☽ □ ♇
19 22 ☽ ⊻ ♆
21 14 ☽ ⅊ ♀

07 Saturday
04 11 ☽ ± ♂
04 17 ☽ ⊼ ☉
04 18 ☽ ⊥ ♅
07 22 ☽ ⊥ ♆
07 35 ☽ ⅊ ☿
11 35 ☽ ∥ ♄
16 19 ☽ ⊻ ♅
17 25 ☽ ± ♀
22 01 ☿ ∥ ♀
23 11 ☽ ⅊ ♂

08 Sunday
01 18 ☽ ± ☿
01 29 ☽ ∠ ♆
06 43 ♃ ✶ ♇
06 57 ☽ ⊼ ♀
13 02 ☽ ⊼ ☿
19 32 ☽ ☌ ♄
22 10 ☽ # ♃

09 Monday
04 32 ♂ △ ♇
06 58 ☽ △ ♇
07 05 ☽ △ ♂
07 23 ☽ ☍ ♃
07 52 ☽ ✶ ♆
17 14 ☿ ∥ ♇
17 49 ♂ ✶ ♃
22 40 ☽ ☍ ☉
23 37 ♂ ☍ ♆

10 Tuesday
00 10 ☿ ☌ ♀
04 55 ☽ ☌ ♅
08 01 ♃ △ ♆
08 15 ☽ Q ♆
09 39 ♂ # ♆
13 31 ☽ ⅊ ♇
19 45 ☿ # ♅
21 09 ☽ # ♃
22 44 ☽ ☍ ☿

11 Wednesday
03 08 ☽ ☍ ♀
07 51 ☽ ± ♇
08 06 ☽ ⊻ ♄
10 36 ☉ ∥ ☿
20 02 ☽ ⊼ ♇
20 12 ☽ ⊥ ♄
20 48 ☽ □ ♆
21 32 ☽ ⊼ ♃
23 03 ☽ □ ♂

12 Thursday
09 54 ☽ ± ♃
10 09 ☽ ∥ ♄
14 17 ☽ ∠ ♄
17 15 ☽ ⊼ ☉
17 33 ☽ ⊻ ♅
20 36 ☉ ☍ ♅
23 31 ♀ ⊼ ♄

13 Friday
00 14 ♀ ⊥ ♇
04 22 ☽ ⅊ ♃
05 34 ☽ ⊥ ♅
06 21 ☽ ⊼ ☿
06 26 ☽ ± ☉
13 04 ☽ Q ♂
17 13 ☽ ± ☿
20 11 ☽ ✶ ♄
22 44 ☽ ⊼ ♀
23 30 ☽ ∠ ♅

14 Saturday
02 04 ☽ ⅊ ☉
06 23 ☉ # ♅
08 22 ☽ ☍ ♇
08 59 ☽ △ ♆
09 22 ☽ ⅊ ☿
10 50 ☽ △ ♃
12 01 ☽ ± ♀
14 04 ☽ ✶ ♂
18 10 ☿ ∠ ♇
18 36 ☿ Q ♂
19 48 ☽ Q ♄

15 Sunday
04 59 ☽ ✶ ♅
07 42 ☽ ⅊ ♀
08 09 ☿ ∥ ♃
10 17 ☽ △ ☉
11 51 ☽ △ ☿
14 23 ☽ ⅊ ♆
19 14 ☉ ☌ ☿
20 44 ☽ ∠ ♂

16 Monday
04 09 ☽ Q ♅
06 33 ☽ □ ♄
07 44 ☽ ± ♆
12 15 ☽ # ♂
14 46 ☽ ⊥ ♂
15 49 ☽ △ ♀
18 40 ☽ ⊼ ♇
18 56 ☽ ∥ ♆
19 08 ☽ ⊼ ♆
21 58 ☽ □ ♃

17 Tuesday
02 36 ☽ ⊻ ♂
05 59 ☽ ± ♇
06 31 ☉ ∠ ♇
11 52 ♀ ± ♄
14 00 ☽ □ ♅
15 18 ☽ □ ☿
21 11 ♀ ⊻ ♇
22 41 ☽ ⅊ ♇
22 43 ☽ # ♆

18 Wednesday
00 01 ☽ □ ☉
01 17 ♀ □ ♆
04 43 ☿ ☍ ♅
06 59 ☽ ∥ ♂
07 52 ☽ Q ♃
14 01 ☽ △ ♄
21 11 ☽ Q ☿

19 Thursday
01 49 ☽ △ ♇
02 08 ☽ ☍ ♆
04 41 ☽ □ ♀
05 43 ☽ ✶ ♃
10 57 ☽ Q ☉
11 25 ☽ ☌ ♂
16 27 ☽ ⅊ ♄
16 44 ☽ ✶ ☿
18 02 ♀ ⊻ ♃
19 40 ☽ △ ♅
23 46 ☉ Q ♂

20 Friday
07 53 ☽ ± ♄
08 10 ☽ ∠ ♃
09 20 ☽ ✶ ☉
14 54 ☽ Q ♀
16 44 ☽ ∠ ☿
18 01 ☽ ⊼ ♄
21 10 ☽ ⅊ ♅
23 37 ☽ ⊥ ♃

21 Saturday
05 24 ☽ □ ♇
05 37 ☽ ⊼ ♆
06 57 ☽ ⊥ ☿
09 45 ☽ ⊻ ♃
12 06 ☽ ± ♅
12 23 ☽ ∠ ☉
12 50 ☽ ✶ ♀
15 31 ☽ ± ♆
16 16 ☽ ⊻ ♂
16 21 ☽ ⊻ ☿
17 12 ☿ ✶ ♂
17 54 ☽ # ♄
20 40 ♆ St D
21 55 ☽ ⊼ ♅

22 Sunday
02 26 ☽ ⊥ ♂
02 35 ☽ ∥ ☿
04 12 ☽ ⊥ ☉
06 12 ☽ ⅊ ♆
10 53 ☽ Q ♇
14 36 ☽ ⊻ ☉
15 29 ☽ ∠ ♀
17 00 ☿ ⊻ ♀
17 30 ☽ ∠ ♂
18 57 ☽ ☍ ♄

23 Monday
06 06 ☽ ✶ ♇
06 14 ☽ △ ♆
07 09 ☽ ⊥ ♀
10 54 ☽ ☌ ♃
11 39 ☽ ∥ ♃
14 49 ☽ ☌ ☿
15 11 ♀ ∥ ♇
17 31 ☽ ⊻ ♀
18 13 ☽ ✶ ♂
21 46 ☽ ☍ ♅
23 10 ☽ ∥ ♃

24 Tuesday
01 04 ☉ ∥ ♀
05 51 ☽ ∠ ♇
05 51 ♂ ⅊ ♄
07 05 ♀ ⅊ ♄
08 07 ♀ □ ♂
13 49 ☽ Q ♂
17 32 ☽ ☌ ☉
18 12 ☽ ⊼ ♄
20 02 ☽ ⊥ ♇

25 Wednesday
03 02 ☉ ⊼ ♄
03 37 ☽ ± ♄
05 30 ☽ ⊻ ♇
05 33 ☽ □ ♆
09 51 ☽ ∥ ☿
10 54 ☽ ⊻ ♃
13 17 ♀ ⊥ ♃
13 26 ☽ ⊻ ☿
14 47 ♀ # ♅
17 17 ☽ # ♄
17 47 ☽ ⅊ ♄
19 15 ☽ □ ♂
19 40 ♀ ⊼ ♅
20 32 ☽ ⊥ ♃
20 58 ☽ ⊼ ♅
21 05 ☽ ☌ ♀
21 15 ☿ St D
22 57 ☽ ⊥ ☿

26 Thursday
06 30 ☽ ± ♅
07 51 ☉ ⊥ ♇
10 09 ☽ Q ♆
11 07 ☽ ∠ ♃
13 23 ☽ ∠ ☿
14 20 ♆ ✶ ♇
17 40 ☽ △ ♄
20 41 ☽ ⊻ ☉
20 56 ☿ ⊥ ♀
20 56 ☽ ⅊ ♅

27 Friday
05 35 ☽ ✶ ♆
05 36 ☽ ☌ ♇
07 11 ☽ ⊥ ☉
11 51 ☽ ✶ ♃
14 07 ☽ ✶ ☿
16 34 ♂ △ ♅
21 30 ☽ △ ♅
21 42 ☽ △ ♂
23 12 ☽ ∠ ☉

28 Saturday
02 26 ☽ ⊻ ♀
06 29 ☽ ∠ ♆
08 11 ☽ Q ♃
10 39 ☽ Q ☿
13 33 ☽ ⊥ ♀
14 34 ☽ ∥ ♂
19 18 ☽ □ ♄
21 50 ☽ ⊥ ♆

29 Sunday
00 09 ☽ ⅊ ♂
02 44 ☽ ✶ ☉
06 37 ☽ ∠ ♀
08 13 ☽ # ♆
08 16 ☽ ⊻ ♆
08 23 ☽ ⊻ ♇
15 48 ☽ □ ♃
16 33 ☽ ± ♂
18 57 ☽ □ ☿
18 59 ☽ ⊥ ♇

30 Monday
01 12 ☽ □ ♅
01 38 ☽ Q ☉
02 49 ☽ Q ♄
02 58 ☉ # ♅
03 42 ☽ ⊼ ♂
10 58 ☽ ∥ ♆
11 00 ♀ ± ♅
11 13 ☽ ∠ ♇
12 08 ☽ ✶ ♀
16 16 ☉ ± ♄

31 Tuesday
00 39 ☽ ✶ ♄
08 31 ☽ # ♂
09 12 ♀ ∥ ♃
10 10 ☽ Q ♅
12 46 ☽ Q ♀
13 36 ☽ □ ☉
14 54 ☽ ☌ ♆
15 07 ☽ ✶ ♇
23 58 ☽ △ ♃

All ephemeris data is given at 12.00 UT and the Moon's longitude is additionally given for 24.00 UT

JANUARY 2031

LONGITUDES

Date	Sidereal time h m s	Sun ☉ ° ' "	Moon ☽ ° ' "	Moon ☽ 24.00 ° ' "	Mercury ☿ ° '	Venus ♀ ° '	Mars ♂ ° '	Jupiter ♃ ° '	Saturn ♄ ° '	Uranus ♅ ° '	Neptune ♆ ° '	Pluto ♇ ° '
01	18 43 47	10 ♑ 51 53	21 ♈ 37 49	27 ♈ 47 35	18 ♐ 27	28 ♑ 41	22 ♎ 45	15 ♐ 29	02 ♊ 54	20 ♊ 03	10 ♈ 36	10 ♒ 44
02	18 47 44	11 53 02	03 ♉ 53 19	09 ♉ 55 38	19 16	29 ♑ 56	23 16	15 42	02 R 51	20 R 00	10 37	10 46
03	18 51 40	12 54 11	15 ♉ 55 07	21 ♉ 52 21	20 10	01 ♒ 11	23 47	15 54	02 48	19 58	10 37	10 48
04	18 55 37	13 55 20	27 ♉ 47 54	03 ♊ 42 15	21 07	02 26	24 18	16 07	02 45	19 56	10 38	10 49
05	18 59 33	14 56 28	09 ♊ 35 53	15 ♊ 29 15	22 08	03 41	24 49	16 20	02 42	19 54	10 38	10 51
06	19 03 30	15 57 37	21 ♊ 22 44	27 ♊ 16 41	23 12	04 56	25 19	16 32	02 39	19 51	10 39	10 53
07	19 07 26	16 58 45	03 ♋ 11 25	09 ♋ 07 13	24 19	06 12	25 49	16 45	02 36	19 49	10 39	10 54
08	19 11 23	17 59 53	15 ♋ 04 19	21 ♋ 02 56	25 28	07 27	26 20	16 58	02 33	19 47	10 40	10 56
09	19 15 19	19 01 01	27 ♋ 03 15	03 ♌ 05 26	26 40	08 42	26 50	17 10	02 31	19 45	10 40	10 58
10	19 19 16	20 02 08	09 ♌ 09 40	15 ♌ 16 03	27 53	09 57	27 20	17 22	02 28	19 43	10 41	11 00
11	19 23 13	21 03 16	21 ♌ 24 47	27 ♌ 36 00	29 ♐ 08	11 12	27 49	17 35	02 26	19 41	10 42	11 01
12	19 27 09	22 04 23	03 ♍ 49 54	10 ♍ 06 38	00 ♑ 25	12 27	28 19	17 47	02 23	19 39	10 43	11 03
13	19 31 06	23 05 30	16 ♍ 26 27	22 ♍ 49 35	01 43	13 42	28 48	17 59	02 21	19 37	10 43	11 05
14	19 35 02	24 06 37	29 ♍ 16 16	05 ♎ 46 47	03 02	14 56	29 18	18 12	02 19	19 35	10 44	11 07
15	19 38 59	25 07 43	12 ♎ 21 25	19 ♎ 00 28	04 23	16 11	29 ♎ 47	18 24	02 17	19 33	10 45	11 09
16	19 42 55	26 08 50	25 ♎ 44 12	02 ♏ 32 52	05 45	17 26	00 ♏ 16	18 36	02 15	19 31	10 46	11 10
17	19 46 52	27 09 56	09 ♏ 26 40	16 ♏ 25 45	07 07	18 41	00 45	18 48	02 13	19 29	10 47	11 12
18	19 50 48	28 11 03	23 ♏ 30 09	00 ♐ 39 49	08 31	19 56	01 13	19 00	02 12	19 27	10 48	11 14
19	19 54 45	29 ♑ 12 09	07 ♐ 54 32	15 ♐ 13 56	09 56	21 11	01 42	19 12	02 10	19 25	10 49	11 16
20	19 58 42	00 ♒ 13 15	22 ♐ 37 31	00 ♑ 04 34	11 21	22 26	02 10	19 23	02 09	19 23	10 50	11 18
21	20 02 38	01 14 20	07 ♑ 34 14	15 ♑ 05 29	12 47	23 41	02 38	19 35	02 07	19 22	10 51	11 19
22	20 06 35	02 15 25	22 ♑ 37 12	00 ♒ 08 11	14 14	24 55	03 06	19 47	02 06	19 20	10 52	11 21
23	20 10 31	03 16 29	07 ♒ 37 12	15 ♒ 03 05	15 42	26 10	03 34	19 58	02 05	19 18	10 53	11 23
24	20 14 28	04 17 33	22 ♒ 24 43	29 ♒ 41 07	17 11	27 25	04 01	20 10	02 04	19 17	10 54	11 25
25	20 18 24	05 18 35	06 ♓ 51 31	13 ♓ 55 17	18 40	28 40	04 29	20 21	02 03	19 15	10 55	11 27
26	20 22 21	06 19 36	20 ♓ 52 01	27 ♓ 41 30	20 10	29 ♒ 54	04 56	20 33	02 02	19 14	10 56	11 29
27	20 26 17	07 20 37	04 ♈ 23 46	10 ♈ 58 55	21 40	01 ♓ 09	05 23	20 44	02 01	19 12	10 57	11 30
28	20 30 14	08 21 36	17 ♈ 27 19	23 ♈ 49 21	23 12	02 24	05 49	20 55	02 01	19 11	10 59	11 32
29	20 34 11	09 22 34	00 ♉ 05 33	06 ♉ 16 32	24 44	03 38	06 16	21 06	02 00	19 09	11 00	11 34
30	20 38 07	10 23 31	12 ♉ 22 56	18 ♉ 25 25	26 16	04 53	06 42	21 17	02 00	19 08	11 01	11 36
31	20 42 04	11 ♒ 24 27	24 ♉ 24 43	00 ♊ 21 29	27 ♑ 50	06 ♓ 07	07 ♏ 08	21 ♐ 28	02 ♊ 00	19 ♊ 07	11 ♈ 03	11 ♒ 38

MOON NODES AND LATITUDE

Date	Moon True ☊ ° '	Moon Mean ☊ ° '	Moon ☽ Latitude ° '
01	06 ♐ 52	05 ♐ 27	03 N 43
02	06 D 53	05 24	02 51
03	06 55	05 21	01 52
04	06 56	05 17	00 N 50
05	06 R 57	05 14	00 S 14
06	06 56	05 11	01 18
07	06 52	05 08	02 18
08	06 47	05 05	03 12
09	06 40	05 02	03 58
10	06 32	04 58	04 34
11	06 23	04 55	04 58
12	06 14	04 52	05 08
13	06 07	04 49	05 03
14	06 02	04 46	04 43
15	05 59	04 42	04 09
16	05 58	04 39	03 20
17	05 D 59	04 36	02 18
18	06 00	04 33	01 S 07
19	06 R 00	04 30	00 N 10
20	05 59	04 27	01 28
21	05 56	04 23	02 41
22	05 50	04 20	03 43
23	05 41	04 17	04 29
24	05 32	04 14	04 57
25	05 22	04 11	05 05
26	05 13	04 08	04 54
27	05 06	04 04	04 26
28	05 01	04 01	03 46
29	04 59	03 58	02 55
30	04 D 59	03 55	01 58
31	04 ♐ 59	03 ♐ 52	00 N 56

DECLINATIONS

Date	Sun ☉ ° '	Moon ☽ ° '	Mercury ☿ ° '	Venus ♀ ° '	Mars ♂ ° '	Jupiter ♃ ° '	Saturn ♄ ° '	Uranus ♅ ° '	Neptune ♆ ° '	Pluto ♇ ° '
01	22 S 59	11 N 52	20 S 39	21 S 48	07 S 13	22 S 05	18 N 46	23 N 10	02 N 47	23 S 24
02	22 54	15 30	20 51	21 34	07 25	22 07	18 46	23 09	02 47	23 24
03	22 49	18 24	21 04	21 18	07 36	22 08	18 45	23 09	02 47	23 23
04	22 42	20 28	21 17	21 03	07 47	22 10	18 45	23 09	02 48	23 23
05	22 36	21 39	21 29	20 46	07 58	22 11	18 45	23 09	02 48	23 22
06	22 29	21 52	21 42	20 29	08 09	22 12	18 44	23 09	02 48	23 22
07	22 21	21 06	21 54	20 11	08 19	22 13	18 44	23 09	02 48	23 21
08	22 13	19 24	22 05	19 53	08 30	22 15	18 44	23 08	02 49	23 21
09	22 05	16 51	22 16	19 34	08 41	22 16	18 44	23 08	02 49	23 20
10	21 56	13 33	22 26	19 15	08 51	22 17	18 43	23 08	02 49	23 20
11	21 47	09 39	22 36	18 55	09 02	22 18	18 43	23 08	02 50	23 19
12	21 38	05 19	22 45	18 34	09 12	22 20	18 43	23 08	02 50	23 19
13	21 28	00 N 41	22 53	18 13	09 22	22 21	18 43	23 08	02 50	23 18
14	21 17	04 S 03	23 00	17 52	09 32	22 22	18 43	23 07	02 51	23 18
15	21 06	08 42	23 06	17 30	09 42	22 23	18 42	23 07	02 51	23 17
16	20 55	13 02	23 11	17 07	09 52	22 24	18 42	23 07	02 51	23 17
17	20 43	16 49	23 15	16 44	10 02	22 25	18 42	23 07	02 52	23 16
18	20 31	19 43	23 18	16 21	10 12	22 26	18 42	23 07	02 52	23 16
19	20 19	21 27	23 20	15 57	10 22	22 27	18 42	23 07	02 53	23 15
20	20 06	21 46	23 20	15 33	10 31	22 28	18 42	23 07	02 53	23 15
21	19 53	20 32	23 20	15 08	10 41	22 29	18 42	23 06	02 53	23 15
22	19 39	17 52	23 18	14 43	10 50	22 30	18 42	23 06	02 54	23 14
23	19 25	14 01	23 15	14 17	10 59	22 31	18 42	23 06	02 54	23 14
24	19 11	09 21	23 11	13 51	11 08	22 32	18 42	23 06	02 55	23 13
25	18 56	04 S 16	23 05	13 25	11 17	22 32	18 42	23 06	02 55	23 13
26	18 41	00 N 53	22 58	12 58	11 26	22 33	18 42	23 06	02 56	23 12
27	18 26	05 49	22 50	12 31	11 35	22 34	18 42	23 06	02 56	23 12
28	18 10	10 20	22 41	12 04	11 43	22 35	18 43	23 06	02 57	23 11
29	17 54	14 22	22 30	11 36	11 52	22 36	18 43	23 06	02 57	23 11
30	17 38	17 25	22 18	11 08	12 00	22 36	18 43	23 05	02 58	23 10
31	17 S 21	19 N 47	22 S 05	10 S 40	12 S 09	22 S 37	18 N 43	23 N 05	02 N 58	23 S 10

ZODIAC SIGN ENTRIES

Date	h	m	Planets	
02	04	20	♀	♒
02	13	14	☽	♉
04	16	28	☽	♊
07	05	32	☽	♋
09	17	52	☽	♌
12	04	18	☿	♑
12	04	38	☽	♍
14	13	21	☽	♎
15	22	48	♂	♏
16	19	32	☽	♏
18	22	54	☽	♐
20	06	48	☉	♒
20	23	53	☽	♑
22	23	47	☽	♒
25	00	31	☽	♓
26	13	49	♀	♓
27	04	06	☽	♈
29	11	49	☽	♉
31	23	17	☽	♊

LATITUDES

Date	Mercury ☿ ° '	Venus ♀ ° '	Mars ♂ ° '	Jupiter ♃ ° '	Saturn ♄ ° '	Uranus ♅ ° '	Neptune ♆ ° '	Pluto ♇ ° '
01	02 N 17	01 S 25	01 N 45	00 N 33	02 S 00	00 N 06	01 S 32	06 S 06
04	01 52	01 28	01 46	00 33	01 59	00 06	01 32	06 06
07	01 25	01 31	01 46	00 33	01 59	00 06	01 32	06 06
10	00 59	01 33	01 47	00 33	01 58	00 06	01 32	06 06
13	00 32	01 35	01 48	00 33	01 57	00 06	01 32	06 06
16	00 N 08	01 36	01 48	00 33	01 56	00 06	01 32	06 06
19	00 S 16	01 36	01 49	00 33	01 56	00 06	01 31	06 06
22	00 38	01 36	01 49	00 33	01 55	00 06	01 31	06 07
25	00 58	01 35	01 50	00 33	01 54	00 06	01 31	06 07
28	01 15	01 33	01 50	00 33	01 53	00 06	01 31	06 07
31	01 S 31	01 S 31	01 N 51	00 N 33	01 S 52	00 N 06	01 S 31	06 S 07

DATA

Julian Date	2462868
Delta T	+74 seconds
Ayanamsa	24° 17' 41"
Synetic vernal point	04° ♓ 49' 18"
True obliquity of ecliptic	23° 26' 03"

MOON'S PHASES, APSIDES AND POSITIONS ☽

Date	h	m	Phase	Longitude ° '	Eclipse Indicator
08	18	26	○	18 ♋ 16	
16	12	47	☾	26 ♎ 11	
23	04	31	●	02 ♒ 57	
30	07	43	☽	10 ♉ 13	

Day	h	m		
06	10	48	Apogee	
21	21	36	Perigee	
06	05	13	Max dec	21° N 54'
13	15	31	0S	
20	04	54	Max dec	21° S 50'
26	07	51	0N	

LONGITUDES

Date	Chiron ⚷ ° '	Ceres ⚳ ° '	Pallas ⚴ ° '	Juno ⚵ ° '	Vesta ⚶ ° '	Black Moon Lilith ⚸ ° '
01	12 ♉ 16	08 ♉ 28	07 ♓ 33	01 ♒ 17	24 ♉ 48	24 ♊ 42
11	12 ♉ 07	09 ♉ 08	10 ♓ 19	05 ♒ 25	24 ♉ 26	25 ♊ 49
21	12 ♉ 04	10 ♉ 23	13 ♓ 17	09 ♒ 39	24 ♉ 48	26 ♊ 56
31	12 ♉ 07	12 ♉ 09	16 ♓ 25	13 ♒ 57	25 ♉ 49	28 ♊ 02

ASPECTARIAN

h m	Aspects
01 Wednesday	
04 51	☽ ∠ ♄
05 27	☽ △ ☿
05 53	☉ □ ♆
08 56	☉ ⊻ ♇
08 57	☽ ⚹ ♅
10 31	♀ Q ♆
14 09	☽ Q ♇
14 16	☽ ☍ ♂
22 13	☽ ⊥ ♄
02 Thursday	
03 19	☽ □ ♀
05 35	☽ ⚼ ♃
09 57	☽ ⊻ ♄
12 48	☽ ⚼ ☿
14 12	☽ ∠ ♅
23 45	☽ ± ♃
03 Friday	
01 23	☽ ⊻ ♆
01 42	☽ □ ♇
05 23	☽ △ ☉
05 33	♀ ∠ ♃
07 11	☿ ☍ ♅
08 06	☽ ⊥ ♅
08 11	☽ ± ☿
11 59	☽ ⊼ ♃
13 25	☽ ⊥ ♆
15 34	☽ ∥ ♄
20 08	☽ ⊻ ♅
21 17	☽ ⊼ ☿
04 Saturday	
00 13	☿ ∥ ♀
04 35	☽ ⊼ ♂
07 36	☽ ∠ ♆
14 30	☽ ⚼ ☉
17 18	☽ ± ♂
17 39	♀ △ ♄
19 45	☽ # ♀
22 00	☽ ☌ ♄
22 33	☽ △ ♀
05 Sunday	
05 45	☽ # ☿
10 32	☽ ± ☉
12 27	☽ ⚼ ♂
14 07	☽ ⚹ ♆
14 33	☽ △ ♇
23 55	☽ ⊼ ☉
06 Monday	
01 58	☽ ☍ ♃
08 44	☽ ⚼ ♀
08 55	☽ ☌ ♅
10 24	♀ ⚼ ♅
14 35	☽ Q ♆
16 06	☽ ☍ ☿
18 31	☽ # ☿
20 22	☽ △ ♂
21 11	☽ ⚼ ♇
07 Tuesday	
05 12	☽ ± ♀
05 14	☉ ⊻ ♃
10 49	☽ ⊻ ♄
15 29	☽ ± ♇
18 48	☽ ⊼ ♀
22 54	☽ ⊥ ♄
08 Wednesday	
02 00	☉ ⚼ ♄
03 07	☽ □ ♆
03 39	☽ ⊼ ♇
05 12	☽ # ♀
08 46	☉ ∥ ♃
15 52	☽ ⊼ ♃
16 58	☽ ∠ ♄
18 26	☽ ☍ ☉
19 05	☽ ∥ ♄
21 26	☽ ⊻ ♅
21 40	☿ ∠ ♇
22 04	♂ ± ♄
22 20	☉ ∥ ☿
09 Thursday	
04 06	☽ ± ♃
09 24	☽ ⊥ ♅
11 08	☽ ⊼ ☿
11 32	☽ □ ♂
11 34	☿ ∥ ♃
17 36	☿ ⚹ ♂
22 21	☽ ⚼ ♃
22 49	☽ ⚹ ♄
10 Friday	
00 24	☽ ± ☿
03 14	☽ ∠ ♅
04 37	☉ ⊼ ♅
13 43	☽ ☍ ♀
15 00	☽ △ ♆
15 37	☽ ☍ ♇
20 09	☽ ⚼ ☿
22 24	☽ Q ♄
11 Saturday	
00 38	☽ Q ♂
02 23	♀ ⚹ ♆
04 24	☽ △ ♃
08 38	☽ ⚹ ♅
08 40	♀ ☌ ♇
11 14	☽ ⊼ ☉
15 27	☽ # ♂
20 20	☽ ⚼ ♆
23 55	☽ ± ☉
12 Sunday	
00 57	☽ ⚹ ♂
01 59	♀ # ♄
04 40	☽ △ ☿
07 48	☽ Q ♅
09 14	☽ □ ♄
13 41	☽ ± ♆
18 45	☽ ⚼ ☉
13 Monday	
01 00	☽ ∥ ♆
01 09	☽ ⊼ ♆
01 49	☽ ⊼ ♇
06 14	☽ ⊼ ♀
06 49	☽ ∠ ♂
13 13	☽ ± ♇
14 58	☽ □ ♃
17 57	☽ □ ♅
18 47	☽ ± ♀
23 21	☿ ⊼ ♄
14 Tuesday	
00 26	☽ ⊥ ♂
01 34	☽ △ ☉
05 54	☽ # ♆
06 07	☽ ⚼ ♇
12 03	☽ ⊻ ♂
13 22	☽ ⚼ ♀
14 25	☉ ⊥ ♃
17 37	☽ △ ♄
19 45	☽ □ ☿
15 Wednesday	
00 57	☽ Q ♃
09 05	☽ ☍ ♆
09 47	☽ △ ♇
17 34	☿ # ♅
17 36	☽ ∥ ♂
19 39	☽ △ ♀
20 53	☽ ⚼ ♄
21 28	☉ ± ♅
23 04	☽ ⚹ ♃
16 Thursday	
00 56	☽ △ ♅
01 46	☿ ⊥ ♇
08 03	☽ Q ☿
12 47	☽ □ ☉
12 55	☽ ± ♄
20 17	☽ ☌ ♂
23 27	☽ ⊼ ♄
17 Friday	
02 02	☽ ∠ ♃
03 24	☽ ⚼ ♅
07 32	☽ ⚹ ☿
11 30	☽ ∥ ♀
14 18	☽ ⊼ ♆
14 30	♀ ⚹ ♃
15 02	☽ □ ♇
17 52	☽ ⊥ ♃
18 56	☽ ± ♅
21 07	☿ ∥ ♇
22 37	☽ Q ☉
18 Saturday	
00 37	☽ ± ♆
02 33	☽ # ♄
02 53	♀ △ ♅
04 16	☽ ⊻ ♃
05 09	☽ ⊼ ♅
05 23	☽ □ ♀
06 36	☿ ± ♄
12 02	☽ ∠ ☿
15 52	☽ ⚼ ♆
20 04	☽ ∥ ☉
20 28	☽ ⚹ ☉
21 38	☽ Q ♇
19 Sunday	
01 22	☽ ⊻ ♂
02 31	☽ ☍ ♄
02 36	☉ Q ♆
04 42	☽ ⊥ ☿
11 38	☽ ⊥ ♂
14 17	☽ Q ♀
15 40	☽ ⊻ ☿
16 46	☽ △ ♆
17 31	☽ ⚹ ♇
23 05	☽ ∠ ☉
20 Monday	
02 52	☽ ∠ ♂
03 05	☿ □ ♆
06 41	☽ ☌ ♃
06 46	☽ ☍ ♅
10 47	♂ ⊼ ♄
11 00	☿ ⊻ ♇
11 39	☽ ⚹ ♀
12 04	♃ ☍ ♅
14 46	☽ ⊥ ☉
17 56	☽ ∠ ♇
21 Tuesday	
01 08	☽ ⊻ ☉
03 18	☽ ⊼ ♄
03 51	☽ ⚹ ♂
08 24	☽ ⊥ ♇
12 53	☽ ± ♄
13 56	☽ ∠ ♀
17 14	☽ □ ♆
18 00	☽ ⊻ ♇
19 57	☽ ∥ ☉
21 13	☽ ☌ ☿
23 38	☽ Q ♂
22 Wednesday	
03 13	☽ ⚼ ♄
05 35	☽ ⊥ ♀
05 40	☽ # ♄
06 46	☽ ⊼ ♅
07 25	☽ ⊻ ♃
08 23	☉ △ ♄
16 00	☽ ⊻ ♀
16 19	☽ ± ♅
17 06	☽ ⊥ ♃
21 58	☽ Q ♆
23 Thursday	
03 07	☽ △ ♄
04 31	☽ ☌ ☉
05 17	☽ □ ♂
06 20	♀ ∠ ♆
06 41	☽ ⚼ ♅
07 41	☽ ∠ ♃
08 41	☿ Q ♂
10 19	☽ ∥ ♀
17 15	☽ ⚹ ♆
18 05	☽ ☌ ♇
21 55	☿ ∥ ♇
24 Friday	
00 25	☉ □ ♂
02 30	☽ ⊻ ☿
03 27	☽ ∥ ♂
06 53	☽ △ ♅
08 16	☽ ⚹ ♃
10 10	☿ ⚼ ♄
11 40	☉ ⚼ ♅
13 24	☽ ⊥ ☿
17 44	☽ ∠ ♆
21 01	☽ ☌ ♀
25 Saturday	
00 45	♂ ⚼ ♅
03 57	☽ □ ♄
04 20	☽ Q ♃
06 01	☽ ∠ ☿
07 52	☽ △ ♂
08 44	☽ ⊥ ♆
08 52	☿ # ♅
09 12	☽ ⊻ ☉
13 16	☉ ∠ ♃
18 15	☽ # ♆
18 53	☽ ⊻ ♆
19 47	☽ ⊻ ♇
20 08	☽ ⊥ ☉
21 16	☿ ⊼ ♅
26 Sunday	
06 06	☽ ⊥ ♇
09 09	☽ □ ♅
10 18	☉ # ♄
10 19	☽ ⚼ ♂
10 33	☽ Q ♄
10 38	☽ ⚹ ☿
11 26	☽ □ ♃
12 52	☽ ∠ ☉
18 54	☿ ⊻ ♃
21 47	☽ ∥ ♆
21 52	☽ ∠ ♇
27 Monday	
02 41	☽ ± ♂
05 34	☽ ⊻ ♀
07 44	☽ ⚹ ♄
10 32	☽ Q ☿
13 50	☽ ⊼ ♂
17 05	☽ Q ♅
17 31	☽ ⊥ ♀
17 47	☽ ⚹ ☉
23 58	☽ ☌ ♆
28 Tuesday	
01 00	☽ ⚹ ♇
04 42	♀ □ ♄
11 11	☽ ∠ ♄
11 53	☽ ∠ ♀
15 13	☽ ⚹ ♅
17 55	☽ Q ☉
18 36	☽ △ ♃
20 26	☽ # ♂
21 00	♂ ∠ ♃
21 08	☽ # ♀
23 29	☽ Q ♇
23 49	♀ Q ♃
29 Wednesday	
00 18	☽ □ ☿
00 50	☿ ∥ ♃
01 34	♀ ∥ ♂
04 10	☽ ⊥ ♄
15 42	☽ ⊻ ♄
18 33	☿ ± ♅
19 38	☽ ⚹ ♀
19 51	☽ ∠ ♅
23 50	☽ ⚼ ♃
30 Thursday	
00 25	☽ ☍ ♂
07 43	☽ □ ☉
09 18	☽ ⊻ ♆
10 27	☽ □ ♇
13 29	☽ ⊥ ♅
13 42	☽ # ☉
14 44	♀ ⊥ ♆
17 50	☽ ± ♃
21 13	☽ ⊥ ♆
21 57	☽ Q ♀
31 Friday	
00 07	☽ ∥ ♄
01 23	☽ ⊻ ♅
03 11	☉ ⚹ ♆
05 44	☿ ⊥ ♃
06 00	☽ ⊼ ♃
15 17	☽ ∠ ♆
17 27	☉ ☌ ♇
19 55	☽ △ ☿

FEBRUARY 2031

LONGITUDES

Date	Sidereal time h m s	Sun ☉ ° ' "	Moon ☽ ° ' "	Moon ☽ 24.00 ° ' "	Mercury ☿ ° '	Venus ♀ ° '	Mars ♂ ° '	Jupiter ♃ ° '	Saturn ♄ ° '	Uranus ♅ ° '	Neptune ♆ ° '	Pluto ♇ ° '
01	20 46 00	12 ♒ 25 22	06 ♊ 16 26	12 ♊ 10 12	29 ♑ 24	07 ♓ 22	07 ♏ 34	21 ♐ 39	02 ♊ 00	19 ♊ 05	11 ♈ 04	11 ♒ 40
02	20 49 57	13 26 15	18 ♊ 03 24	23 ♊ 56 39	00 ♒ 58	08 36	07 59	21 50	02 D 00	19 R 04	11 05	11 42
03	20 53 53	14 27 07	29 ♊ 50 28	05 ♋ 45 21	02 34	09 51	08 24	22 00	02 00	19 03	11 07	11 43
04	20 57 50	15 27 58	11 ♋ 41 43	17 ♋ 39 59	04 10	11 05	08 49	22 11	02 00	19 02	11 08	11 45
05	21 01 46	16 28 47	23 ♋ 40 25	29 ♋ 43 19	05 47	12 20	09 14	22 21	02 00	19 01	11 10	11 47
06	21 05 43	17 29 35	05 ♌ 48 52	11 ♌ 57 12	07 24	13 34	09 39	22 32	02 01	19 00	11 11	11 49
07	21 09 40	18 30 22	18 ♌ 08 25	24 ♌ 22 34	09 03	14 48	10 03	22 42	02 01	18 59	11 13	11 51
08	21 13 36	19 31 08	00 ♍ 39 39	06 ♍ 59 40	10 42	16 03	10 27	22 52	02 02	18 58	11 14	11 53
09	21 17 33	20 31 52	13 ♍ 22 33	19 ♍ 48 16	12 22	17 17	10 50	23 02	02 03	18 57	11 16	11 55
10	21 21 29	21 32 35	26 ♍ 16 48	02 ♎ 48 06	14 02	18 31	11 14	23 12	02 04	18 56	11 17	11 56
11	21 25 26	22 33 17	09 ♎ 22 10	15 ♎ 59 02	15 44	19 45	11 37	23 22	02 05	18 56	11 19	11 58
12	21 29 22	23 33 58	22 ♎ 38 44	29 ♎ 21 21	17 26	20 59	12 00	23 32	02 06	18 55	11 21	12 00
13	21 33 19	24 34 38	06 ♏ 07 00	12 ♏ 55 48	19 09	22 13	12 22	23 41	02 07	18 54	11 22	12 02
14	21 37 15	25 35 17	19 ♏ 47 52	26 ♏ 43 19	20 53	23 27	12 44	23 51	02 08	18 54	11 24	12 04
15	21 41 12	26 35 54	03 ♐ 42 14	10 ♐ 44 39	22 38	24 41	13 06	24 00	02 10	18 53	11 26	12 05
16	21 45 09	27 36 31	17 ♐ 50 31	24 ♐ 59 43	24 24	25 55	13 28	24 10	02 11	18 52	11 27	12 07
17	21 49 05	28 37 06	02 ♑ 12 01	09 ♑ 27 01	26 10	27 09	13 49	24 19	02 13	18 52	11 29	12 09
18	21 53 02	29 ♒ 37 41	16 ♑ 44 14	24 ♑ 03 03	27 58	28 23	14 10	24 28	02 15	18 52	11 31	12 11
19	21 56 58	00 ♓ 38 14	01 ♒ 22 41	08 ♒ 42 19	29 ♒ 46	29 ♓ 37	14 31	24 37	02 17	18 51	11 33	12 13
20	22 00 55	01 38 45	16 ♒ 01 00	23 ♒ 17 47	01 ♓ 35	00 ♈ 51	14 51	24 46	02 19	18 51	11 35	12 14
21	22 04 51	02 39 15	00 ♓ 31 43	07 ♓ 41 54	03 26	02 04	15 10	24 54	02 21	18 51	11 36	12 16
22	22 08 48	03 39 43	14 ♓ 47 32	21 ♓ 47 56	05 16	03 18	15 30	25 03	02 23	18 51	11 38	12 18
23	22 12 44	04 40 10	28 ♓ 42 34	05 ♈ 31 03	07 08	04 32	15 49	25 11	02 25	18 50	11 40	12 20
24	22 16 41	05 40 35	12 ♈ 13 11	18 ♈ 48 56	09 00	05 45	16 08	25 20	02 28	18 50	11 42	12 21
25	22 20 38	06 40 58	25 ♈ 18 23	01 ♉ 41 49	10 53	06 59	16 26	25 28	02 30	18 D 50	11 44	12 23
26	22 24 34	07 41 19	07 ♉ 59 35	14 ♉ 12 09	12 47	08 12	16 44	25 36	02 33	18 50	11 46	12 25
27	22 28 31	08 41 38	20 ♉ 20 05	26 ♉ 23 59	14 41	09 26	17 01	25 44	02 36	18 50	11 48	12 27
28	22 32 27	09 ♓ 41 55	02 ♊ 24 29	08 ♊ 22 18	16 ♓ 35	10 ♈ 39	17 ♏ 18	25 ♐ 52	02 ♊ 39	18 ♊ 51	11 ♈ 50	12 ♒ 28

ASPECTARIAN

01 Saturday
- 03 19 ☽ ☌ ♄
- 06 55 ☿ Q ♆
- 14 29 ☽ □ ♀
- 14 43 ☽ ⊼ ♂
- 17 40 ♀ △ ♂
- 21 46 ☽ ✶ ♆
- 23 00 ☽ △ ♇

02 Sunday
- 01 42 ☽ △ ☉
- 02 25 ♄ St D
- 02 47 ☽ # ☿
- 03 23 ☽ ± ♂
- 07 05 ☽ ⚼ ☿
- 14 04 ☽ ☌ ♅
- 19 48 ☽ ☍ ♃
- 22 17 ☽ Q ♆
- 22 25 ☽ ⚼ ♂

03 Monday
- 03 28 ☿ △ ♄
- 04 17 ☽ ± ☿
- 05 39 ☽ ⚼ ♇
- 11 08 ☽ ⚼ ☉
- 12 58 ☽ # ☿
- 16 23 ☽ ⊻ ♄
- 18 23 ☽ ⊼ ☿
- 23 58 ☽ ± ♇

04 Tuesday
- 04 32 ☽ ⊥ ♄
- 05 59 ☽ △ ♂
- 07 05 ☽ ± ☉
- 10 03 ☿ ⚼ ♅
- 10 38 ☽ △ ♀
- 10 52 ☽ □ ♆
- 12 07 ☽ ⊼ ♇
- 12 56 ♀ ⊻ ♆
- 20 17 ☽ ⊼ ☉
- 22 40 ☽ ∠ ♄

05 Wednesday
- 01 13 ♀ ⊻ ♇
- 01 14 ☽ ∥ ♄
- 02 43 ☽ ⊻ ♅
- 09 20 ☽ ⊼ ♃
- 14 40 ☽ ⊥ ♅
- 20 05 ☽ ⚼ ♀
- 21 26 ☽ ± ♃

06 Thursday
- 03 36 ☽ # ☉
- 04 31 ☽ ✶ ♄
- 08 26 ☽ ∠ ♅
- 14 01 ☿ ∠ ♃
- 15 24 ☽ ⚼ ♃
- 15 36 ☽ ☍ ☿
- 15 49 ☽ ± ♀
- 19 45 ☽ □ ♂
- 21 56 ☿ ⊥ ♀
- 22 23 ☽ # ♂
- 22 32 ☽ △ ♆
- 23 46 ☽ ☍ ♇

07 Friday
- 04 01 ☽ Q ♄
- 04 50 ☽ ⊼ ♀
- 12 46 ☽ ☍ ☉
- 13 37 ☽ ✶ ♅
- 20 54 ☽ △ ♃
- 23 05 ☉ △ ♅

08 Saturday
- 03 33 ☽ ⚼ ♆
- 07 16 ☿ □ ♂
- 07 39 ☽ Q ♂
- 10 45 ☽ # ♀
- 12 35 ☽ Q ♅
- 14 37 ☽ □ ♄
- 19 58 ☿ ✶ ♆
- 20 42 ☽ ± ♆

09 Sunday
- 05 25 ☿ ☌ ♇
- 06 21 ☽ ∥ ♆
- 07 06 ☽ ✶ ♂
- 08 02 ☽ ⊼ ♆
- 09 15 ☽ ⊼ ♇
- 09 48 ☽ ⊼ ☿
- 20 05 ☽ ☍ ♀
- 20 29 ☽ ± ♇
- 22 24 ☽ □ ♅
- 22 42 ☽ ± ☿

10 Monday
- 00 29 ♀ ⊥ ♇
- 01 40 ☿ # ♄
- 02 29 ☽ ⊼ ☉
- 06 14 ☽ □ ♃
- 11 54 ☽ ∠ ♂
- 13 13 ☽ ⚼ ♇
- 13 32 ☽ # ♆
- 14 32 ☽ ± ☉
- 15 54 ♂ ⊼ ♆
- 17 50 ☽ ⚼ ☿
- 20 05 ♀ □ ♅
- 22 39 ☽ △ ♄

11 Tuesday
- 01 37 ☽ ∥ ♀
- 04 57 ☽ ⊥ ♂
- 08 25 ☽ ⚼ ☉
- 15 33 ☽ ☍ ♆
- 15 40 ☽ Q ♃
- 16 12 ☽ ⊻ ♂
- 16 44 ☽ △ ♇
- 18 23 ♀ Q ♄

12 Wednesday
- 01 14 ☽ △ ☿
- 02 00 ☽ ⚼ ♄
- 05 18 ☽ △ ♅
- 08 43 ☽ ⊼ ♀
- 10 54 ☉ ✶ ♃
- 12 08 ☉ ∥ ♂
- 12 16 ♂ □ ♇
- 13 36 ☽ ✶ ♃
- 13 47 ☽ △ ☉
- 13 54 ♅ # ♇
- 18 11 ☽ ± ♄
- 20 34 ☽ ± ♀
- 21 49 ☽ ∥ ☉
- 22 57 ☽ ∥ ♂

13 Thursday
- 04 54 ☽ ⊼ ♄
- 08 05 ☽ ⚼ ♅
- 08 33 ☿ △ ♅
- 14 09 ☽ ⚼ ♀
- 16 36 ☽ ∠ ♃
- 20 07 ☽ ∥ ☿
- 21 17 ☽ ⊼ ♆
- 22 27 ☽ □ ♇
- 23 20 ☽ ☌ ♂
- 23 57 ☽ ± ♅

14 Friday
- 07 49 ☽ ± ♆
- 08 34 ☽ ⊥ ♃
- 10 25 ☽ ⊼ ♅
- 11 53 ☽ # ♄
- 14 10 ☽ □ ☿
- 18 58 ☽ △ ♀
- 19 07 ☽ ⊻ ♃
- 20 42 ♀ □ ♃
- 21 37 ♂ ± ♅
- 22 50 ☽ □ ☉
- 23 28 ☽ ⚼ ♆

15 Saturday
- 05 47 ☽ Q ♇
- 07 51 ☉ ∠ ♆
- 09 21 ☽ ☍ ♄
- 13 50 ♀ # ♆

16 Sunday
- 01 11 ☽ △ ♆
- 01 29 ☽ Q ☿
- 02 19 ☽ ✶ ♇
- 04 25 ☽ ⊻ ♂
- 07 57 ☽ Q ☉
- 08 28 ☿ ✶ ♃
- 13 44 ☽ ☍ ♅
- 14 48 ☽ ⊥ ♂
- 22 43 ☽ ☌ ♃

17 Monday
- 00 33 ☽ ✶ ☿
- 02 49 ☽ □ ♀
- 03 35 ☽ ∠ ♇
- 05 36 ☽ ✶ ☉
- 06 14 ☽ ∠ ♂
- 11 56 ♀ ∠ ♇
- 12 02 ☽ ⊼ ♄
- 16 17 ☿ ∠ ♆
- 18 34 ☽ ⊥ ♇
- 21 59 ☽ ± ♄

18 Tuesday
- 03 23 ☽ □ ♆
- 04 29 ☽ ⊻ ♇
- 04 47 ☿ ∥ ♂
- 04 55 ☽ ∠ ☿
- 07 40 ☽ ✶ ♂
- 08 16 ☽ ∠ ☉
- 11 22 ☽ Q ♀
- 12 50 ☽ ⚼ ♄
- 12 52 ☽ # ♄
- 15 29 ☽ ⊼ ♅
- 21 47 ☽ ⊥ ☿

19 Wednesday
- 00 09 ☽ ⊥ ☉
- 00 48 ☽ ⊻ ♃
- 01 19 ☽ ± ♅
- 03 50 ☽ Q ♂
- 05 36 ☿ ⊻ ♀
- 08 51 ☽ ✶ ♀
- 09 00 ☽ Q ♆
- 09 00 ☽ ⊻ ☿
- 09 07 ♀ ⚼ ♂
- 10 42 ☽ ⊻ ☉
- 10 44 ☽ ⊥ ♃
- 13 29 ☽ △ ♄
- 16 03 ☽ ⚼ ♅
- 19 33 ☽ ∥ ♂

20 Thursday
- 01 37 ☽ ∠ ♃
- 03 57 ☽ ∥ ☿
- 04 42 ☽ ✶ ♆
- 05 47 ☽ ☌ ♇
- 10 02 ☽ □ ♂
- 11 42 ☽ ∠ ♀
- 13 36 ☉ ☌ ☿
- 15 10 ☽ ∥ ☉
- 16 40 ☽ △ ♅
- 21 37 ☿ □ ♄

21 Friday
- 02 34 ☽ ✶ ♃
- 03 55 ☽ ⊥ ♀
- 04 24 ☉ □ ♄
- 05 28 ☽ ∠ ♆
- 14 49 ☽ ⊻ ♀
- 15 02 ☽ □ ♄
- 15 49 ☽ ☌ ☉
- 17 28 ♀ ✶ ♄
- 17 33 ☽ ☌ ☿
- 20 30 ☽ ⊥ ♆
- 22 46 ☽ Q ♃

22 Saturday
- 03 54 ☽ # ♆
- 06 38 ☽ ⊻ ♆
- 07 45 ☽ ⊻ ♇
- 13 14 ☽ △ ♂
- 16 13 ☽ # ♀
- 16 49 ☿ ⊥ ♆
- 18 00 ☽ ⊥ ♇
- 18 55 ☽ □ ♅
- 21 35 ☽ Q ♄
- 22 09 ☽ ∥ ♀

23 Sunday
- 05 48 ☽ □ ♃
- 09 35 ☽ ∠ ♇
- 10 11 ☽ ∥ ♆
- 12 46 ☿ Q ♃
- 15 47 ☽ ⚼ ♂
- 18 32 ☽ ✶ ♄
- 23 16 ☽ ☌ ♀
- 23 20 ☽ ⊻ ☉

24 Monday
- 02 21 ☽ Q ♅
- 03 17 ☉ ⊻ ♀
- 05 17 ☽ ⊻ ☿
- 08 08 ☽ ± ♂
- 10 56 ☽ ⊥ ☉
- 11 04 ☽ ☌ ♆
- 12 15 ☽ ✶ ♇
- 12 38 ☉ ⊥ ♆
- 17 16 ☽ # ☉
- 17 53 ☽ ⊥ ☿
- 18 18 ☽ # ☿
- 19 15 ☽ ⊼ ♂
- 21 33 ☽ ∠ ♄

25 Tuesday
- 00 03 ☽ ✶ ♅
- 04 43 ☽ ∠ ☉
- 05 53 ☉ ∥ ☿
- 09 12 ♀ Q ♅
- 10 17 ☽ Q ♇
- 11 24 ♅ St D
- 12 18 ☽ △ ♃
- 13 16 ☽ ∠ ☿
- 14 15 ☽ ⊥ ♄
- 22 56 ☿ ⊻ ♆

26 Wednesday
- 01 35 ☽ ⊻ ♄
- 04 00 ☽ # ♂
- 04 04 ☽ ∠ ♅
- 07 18 ☿ ⊻ ♇
- 09 31 ☉ Q ♃
- 11 22 ☽ ✶ ☉
- 12 27 ☽ ⊻ ♀
- 17 04 ☽ ⚼ ♃
- 19 18 ☽ ⊻ ♆
- 20 33 ☽ □ ♇
- 21 21 ☽ ⊥ ♅
- 22 55 ☽ ✶ ☿

27 Thursday
- 01 20 ☽ ⊥ ♀
- 05 20 ☽ ☍ ♂
- 07 00 ☽ ⊥ ♆
- 09 04 ☽ ⊻ ♅
- 10 48 ☽ ± ♃
- 12 46 ☽ Q ☉
- 13 37 ☽ ∥ ♄
- 21 00 ☽ ∠ ♀
- 21 10 ♀ ∥ ♆
- 22 47 ☽ ⊼ ♃

28 Friday
- 00 50 ☽ ∠ ♆
- 02 55 ☽ Q ☿
- 12 28 ☽ ☌ ♄
- 22 31 ☿ △ ♂

Moon True ☊, Mean ☊, Latitude

Date	Moon True ☊ ° '	Moon Mean ☊ ° '	Moon ☽ Latitude ° '
01	05 ♐ 00	03 ♐ 48	00 S 07
02	04 R 59	03 45	01 09
03	04 56	03 42	02 08
04	04 50	03 39	03 02
05	04 41	03 36	03 48
06	04 30	03 33	04 25
07	04 17	03 29	04 49
08	04 04	03 26	05 00
09	03 50	03 23	04 57
10	03 39	03 20	04 38
11	03 30	03 17	04 05
12	03 24	03 14	03 17
13	03 21	03 10	02 19
14	03 D 20	03 07	01 S 11
15	03 R 20	03 04	00 N 02
16	03 19	03 01	01 16
17	03 17	02 58	02 26
18	03 12	02 54	03 28
19	03 05	02 51	04 16
20	02 54	02 48	04 48
21	02 42	02 45	05 00
22	02 30	02 42	04 54
23	02 18	02 39	04 29
24	02 08	02 35	03 51
25	02 01	02 32	03 01
26	01 57	02 29	02 03
27	01 55	02 26	01 01
28	01 ♐ 55	02 ♐ 23	00 S 03

DECLINATIONS

Date	Sun ☉ ° '	Moon ☽ ° '	Mercury ☿ ° '	Venus ♀ ° '	Mars ♂ ° '	Jupiter ♃ ° '	Saturn ♄ ° '	Uranus ♅ ° '	Neptune ♆ ° '	Pluto ♇ ° '
01	17 S 04	21 N 14	21 S 50	10 S 12	12 S 17	22 S 38	18 N 43	23 N 05	02 N 59	23 S 09
02	16 47	21 45	21 34	09 43	12 25	22 39	18 44	23 05	03 00	23 09
03	16 30	21 18	21 16	09 14	12 33	22 39	18 44	23 05	03 00	23 08
04	16 12	19 54	20 57	08 44	12 41	22 40	18 44	23 05	03 01	23 08
05	15 54	17 37	20 37	08 15	12 49	22 40	18 45	23 05	03 01	23 07
06	15 35	14 32	20 15	07 45	12 56	22 41	18 45	23 05	03 02	23 07
07	15 17	10 48	19 52	07 15	13 04	22 42	18 45	23 05	03 03	23 07
08	14 58	06 33	19 28	06 45	13 11	22 42	18 46	23 05	03 03	23 06
09	14 39	01 N 58	19 02	06 15	13 18	22 43	18 46	23 05	03 04	23 06
10	14 19	02 S 47	18 34	05 45	13 26	22 43	18 47	23 05	03 05	23 05
11	14 00	07 28	18 06	05 14	13 33	22 44	18 47	23 04	03 05	23 05
12	13 40	11 52	17 36	04 43	13 40	22 44	18 48	23 04	03 06	23 04
13	13 20	15 44	17 04	04 12	13 46	22 45	18 48	23 04	03 07	23 04
14	12 59	18 49	16 31	03 41	13 53	22 45	18 49	23 04	03 07	23 04
15	12 39	20 51	15 57	03 10	14 00	22 46	18 49	23 04	03 08	23 03
16	12 18	21 37	15 21	02 39	14 06	22 46	18 50	23 04	03 09	23 03
17	11 57	20 59	14 43	02 08	14 12	22 46	18 50	23 04	03 09	23 02
18	11 36	18 57	14 05	01 37	14 18	22 47	18 51	23 04	03 10	23 02
19	11 15	15 41	13 25	01 06	14 24	22 47	18 52	23 04	03 11	23 02
20	10 53	11 27	12 43	00 34	14 30	22 48	18 52	23 04	03 12	23 01
21	10 32	06 36	12 01	00 S 03	14 36	22 48	18 53	23 04	03 12	23 01
22	10 10	01 S 28	11 17	00 N 29	14 42	22 48	18 54	23 04	03 13	23 00
23	09 48	03 N 36	10 31	01 00	14 47	22 49	18 54	23 04	03 14	23 00
24	09 26	08 22	09 45	01 31	14 53	22 49	18 55	23 04	03 15	23 00
25	09 03	12 35	08 57	02 03	14 58	22 49	18 56	23 04	03 15	22 59
26	08 41	16 07	08 08	02 34	15 03	22 49	18 57	23 04	03 16	22 59
27	08 19	18 48	07 18	03 05	15 08	22 50	18 57	23 04	03 17	22 59
28	07 S 56	20 N 36	06 S 27	03 N 37	15 S 13	22 S 50	18 N 58	23 N 04	03 N 18	22 S 58

ZODIAC SIGN ENTRIES

Date	h m	Planets
01	21 15	☿ ♒
03	12 19	☽ ♋
06	00 33	☽ ♌
08	10 45	☽ ♍
10	18 51	☽ ♎
13	01 09	☽ ♏
15	05 39	☽ ♐
17	08 21	☽ ♑
18	20 51	☉ ♓
19	09 45	☽ ♒
19	15 02	☿ ♓
19	19 30	♀ ♈
21	11 07	☽ ♓
23	14 16	☽ ♈
25	20 48	☽ ♉
28	07 11	☽ ♊

LATITUDES

Date	Mercury ☿ ° '	Venus ♀ ° '	Mars ♂ ° '	Jupiter ♃ ° '	Saturn ♄ ° '	Uranus ♅ ° '	Neptune ♆ ° '	Pluto ♇ ° '
01	01 S 36	01 S 30	01 N 51	00 N 32	01 S 52	00 N 06	01 S 31	06 S 07
04	01 48	01 27	01 51	00 32	01 51	00 06	01 31	06 07
07	01 57	01 23	01 51	00 32	01 51	00 06	01 31	06 08
10	02 03	01 18	01 52	00 32	01 50	00 06	01 31	06 08
13	02 06	01 13	01 52	00 32	01 49	00 06	01 30	06 08
16	02 05	01 08	01 52	00 32	01 48	00 06	01 30	06 09
19	01 59	01 01	01 52	00 32	01 47	00 06	01 30	06 09
22	01 50	00 55	01 52	00 32	01 46	00 06	01 30	06 09
25	01 35	00 47	01 51	00 32	01 46	00 06	01 30	06 10
28	01 15	00 40	01 51	00 32	01 45	00 06	01 30	06 10
31	00 S 50	00 S 31	01 N 50	00 N 32	01 S 44	00 N 06	01 S 30	06 S 11

DATA

Julian Date	2462899
Delta T	+74 seconds
Ayanamsa	24° 17' 46"
Synetic vernal point	04° ♓ 49' 13"
True obliquity of ecliptic	23° 26' 03"

MOON'S PHASES, APSIDES AND POSITIONS ☽

Date	h m	Phase	Longitude ° '	Eclipse Indicator
07	12 46	○	18 ♌ 32	
14	22 50	☾	26 ♏ 03	
21	15 49	●	02 ♓ 49	

Day	h m		
02	23 39	Apogee	
19	00 31	Perigee	
02	12 43	Max dec	21° N 45'
09	21 59	0S	
16	13 13	Max dec	21° S 37'
22	18 52	0N	

LONGITUDES

Date	Chiron ⚷ ° '	Ceres ⚳ ° '	Pallas ⚴ ° '	Juno ⚵ ° '	Vesta ⚶ ° '	Black Moon Lilith ⚸ ° '
01	12 ♉ 08	12 ♉ 21	16 ♓ 44	14 ♒ 23	25 ♉ 58	28 ♊ 09
11	12 ♉ 18	14 ♉ 36	20 ♓ 02	18 ♒ 46	27 ♉ 37	29 ♊ 15
21	12 ♉ 33	17 ♉ 14	23 ♓ 26	23 ♒ 11	29 ♉ 46	00 ♋ 22
31	12 ♉ 54	20 ♉ 11	26 ♓ 57	27 ♒ 38	02 ♊ 20	01 ♋ 29

All ephemeris data is given at 12.00 UT and the Moon's longitude is additionally given for 24.00 UT

MARCH 2031

LONGITUDES

Date	Sidereal time h m s	Sun ☉ ° ' "	Moon ☽ ° ' "	Moon ☽ 24.00 ° ' "	Mercury ☿ ° '	Venus ♀ ° '	Mars ♂ ° '	Jupiter ♃ ° '	Saturn ♄ ° '	Uranus ♅ ° '	Neptune ♆ ° '	Pluto ♇ ° '
01	22 36 24	10 ♓ 42 10	14 ♊ 18 08	20 ♊ 12 40	18 ♓ 30	11 ♈ 52	17 ♏ 34	25 ♐ 59	02 ♊ 41	18 ♊ 51	11 ♈ 52	12 ♒ 30
02	22 40 20	11 42 24	26 ♊ 06 37	02 ♋ 00 38	20 24	13 06	17 51	26 07	02 45	18 51	11 54	12 32
03	22 44 17	12 42 35	07 ♋ 55 22	13 ♋ 51 26	22 18	14 19	18 06	26 14	02 48	18 51	11 56	12 33
04	22 48 13	13 42 44	19 ♋ 49 24	25 ♋ 49 47	24 12	15 32	18 21	26 21	02 51	18 52	11 58	12 35
05	22 52 10	14 42 51	01 ♌ 53 00	07 ♌ 59 27	26 04	16 45	18 36	26 29	02 54	18 52	12 00	12 36
06	22 56 07	15 42 56	14 ♌ 09 26	20 ♌ 23 11	27 56	17 58	18 50	26 36	02 58	18 53	12 02	12 38
07	23 00 03	16 42 59	26 ♌ 40 50	03 ♍ 02 26	29 ♓ 46	19 11	19 04	26 42	03 01	18 53	12 04	12 40
08	23 04 00	17 43 00	09 ♍ 28 00	15 ♍ 57 27	01 ♈ 34	20 24	19 17	26 49	03 05	18 54	12 06	12 41
09	23 07 56	18 42 59	22 ♍ 30 36	29 ♍ 07 17	03 20	21 36	19 30	26 56	03 09	18 54	12 08	12 43
10	23 11 53	19 42 56	05 ♎ 47 15	12 ♎ 30 14	05 03	22 49	19 42	27 02	03 12	18 55	12 11	12 44
11	23 15 49	20 42 52	19 ♎ 15 57	26 ♎ 04 11	06 43	24 02	19 54	27 08	03 16	18 56	12 13	12 46
12	23 19 46	21 42 45	02 ♏ 54 38	09 ♏ 47 06	08 18	25 14	20 05	27 14	03 20	18 56	12 15	12 47
13	23 23 42	22 42 37	16 ♏ 41 25	23 ♏ 37 24	09 49	26 27	20 15	27 20	03 25	18 57	12 17	12 49
14	23 27 39	23 42 28	00 ♐ 34 58	07 ♐ 34 00	11 16	27 39	20 25	27 26	03 29	18 58	12 19	12 50
15	23 31 36	24 42 16	14 ♐ 34 27	21 ♐ 36 13	12 36	28 ♈ 52	20 35	27 31	03 33	18 59	12 21	12 52
16	23 35 32	25 42 03	28 ♐ 39 16	05 ♑ 43 27	13 51	00 ♉ 04	20 43	27 37	03 37	19 00	12 24	12 53
17	23 39 29	26 41 49	12 ♑ 48 39	19 ♑ 54 38	15 00	01 16	20 52	27 42	03 42	19 01	12 26	12 55
18	23 43 25	27 41 33	27 ♑ 01 10	04 ♒ 07 54	16 02	02 28	20 59	27 47	03 46	19 02	12 28	12 56
19	23 47 22	28 41 14	11 ♒ 14 25	18 ♒ 20 15	16 57	03 40	21 06	27 52	03 51	19 03	12 30	12 58
20	23 51 18	29 ♓ 40 55	25 ♒ 24 53	02 ♓ 27 43	17 44	04 52	21 12	27 57	03 56	19 04	12 32	12 59
21	23 55 15	00 ♈ 40 33	09 ♓ 28 12	16 ♓ 25 44	18 24	06 04	21 18	28 02	04 00	19 06	12 35	13 00
22	23 59 11	01 40 09	23 ♓ 19 48	00 ♈ 09 53	18 56	07 16	21 23	28 06	04 05	19 07	12 37	13 02
23	00 03 08	02 39 44	06 ♈ 55 34	13 ♈ 36 32	19 19	08 28	21 27	28 10	04 10	19 08	12 39	13 03
24	00 07 05	03 39 16	20 ♈ 12 33	26 ♈ 43 31	19 35	09 39	21 31	28 14	04 15	19 10	12 41	13 04
25	00 11 01	04 38 46	03 ♉ 09 25	09 ♉ 30 22	19 43	10 51	21 34	28 18	04 20	19 11	12 44	13 06
26	00 14 58	05 38 14	15 ♉ 46 34	21 ♉ 58 19	19 R 44	12 02	21 36	28 22	04 26	19 13	12 46	13 07
27	00 18 54	06 37 40	28 ♉ 06 01	04 ♊ 10 06	19 36	13 14	21 37	28 26	04 31	19 14	12 48	13 08
28	00 22 51	07 37 04	10 ♊ 11 06	16 ♊ 09 35	19 22	14 25	21 38	28 29	04 36	19 16	12 50	13 09
29	00 26 47	08 36 25	22 ♊ 06 09	28 ♊ 01 27	19 01	15 36	21 R 38	28 32	04 42	19 17	12 53	13 11
30	00 30 44	09 35 44	03 ♋ 56 07	09 ♋ 50 50	18 33	16 47	21 37	28 35	04 47	19 19	12 55	13 12
31	00 34 40	10 ♈ 35 01	15 ♋ 46 15	21 ♋ 43 01	18 ♈ 00	17 ♉ 58	21 ♏ 36	28 ♐ 38	04 ♊ 53	19 ♊ 21	12 ♈ 57	13 ♒ 13

Date	Moon True ☊ ° '	Moon Mean ☊ ° '	Moon ☽ Latitude ° '
01	01 ♐ 55	02 ♐ 19	01 S 05
02	01 R 54	02 16	02 05
03	01 52	02 13	02 59
04	01 47	02 10	03 45
05	01 39	02 07	04 22
06	01 28	02 04	04 48
07	01 16	02 00	05 00
08	01 03	01 57	04 58
09	00 51	01 54	04 40
10	00 40	01 51	04 07
11	00 31	01 48	03 20
12	00 26	01 45	02 20
13	00 23	01 41	01 S 12
14	00 D 22	01 38	00 N 01
15	00 23	01 35	01 15
16	00 R 23	01 32	02 25
17	00 22	01 29	03 26
18	00 18	01 25	04 15
19	00 13	01 22	04 48
20	00 04	01 19	05 03
21	29 ♏ 55	01 16	05 00
22	29 45	01 13	04 39
23	29 35	01 10	04 03
24	29 27	01 06	03 14
25	29 22	01 03	02 16
26	29 19	01 00	01 12
27	29 18	00 57	00 N 06
28	29 D 18	00 54	00 S 58
29	29 19	00 51	02 00
30	29 20	00 47	02 56
31	29 ♏ 20	00 ♐ 44	03 S 44

DECLINATIONS

Date	Sun ☉ ° '	Moon ☽ ° '	Mercury ☿ ° '	Venus ♀ ° '	Mars ♂ ° '	Jupiter ♃ ° '	Saturn ♄ ° '	Uranus ♅ ° '	Neptune ♆ ° '	Pluto ♇ ° '
01	07 S 33	21 N 26	05 S 35	04 N 08	15 S 18	22 S 50	18 N 59	23 N 04	03 N 19	22 S 58
02	07 10	21 18	04 42	04 39	15 23	22 50	19 00	23 04	03 19	22 58
03	06 47	20 14	03 49	05 10	15 27	22 51	19 01	23 04	03 20	22 57
04	06 24	18 15	02 55	05 41	15 31	22 51	19 02	23 04	03 21	22 57
05	06 01	15 28	02 01	06 11	15 36	22 51	19 02	23 04	03 22	22 57
06	05 38	11 59	01 06	06 42	15 40	22 51	19 03	23 04	03 23	22 56
07	05 15	07 55	00 S 12	07 12	15 44	22 51	19 04	23 04	03 23	22 56
08	04 51	03 N 25	00 N 42	07 43	15 48	22 52	19 05	23 04	03 24	22 56
09	04 28	01 S 19	01 35	08 13	15 51	22 52	19 06	23 04	03 25	22 55
10	04 04	06 05	02 27	08 43	15 55	22 52	19 07	23 05	03 26	22 55
11	03 41	10 37	03 19	09 12	15 58	22 52	19 08	23 05	03 27	22 55
12	03 17	14 41	04 08	09 42	16 02	22 52	19 09	23 05	03 28	22 54
13	02 54	17 58	04 56	10 11	16 05	22 52	19 10	23 05	03 29	22 54
14	02 30	20 15	05 42	10 40	16 08	22 52	19 11	23 05	03 29	22 54
15	02 06	21 18	06 26	11 09	16 11	22 53	19 12	23 05	03 30	22 54
16	01 42	21 01	07 07	11 38	16 13	22 53	19 13	23 05	03 31	22 53
17	01 19	19 24	07 46	12 06	16 16	22 53	19 14	23 05	03 32	22 53
18	00 55	16 35	08 21	12 34	16 19	22 53	19 15	23 05	03 33	22 53
19	00 31	12 47	08 53	13 02	16 21	22 53	19 16	23 05	03 34	22 53
20	00 S 08	08 17	09 21	13 29	16 23	22 53	19 18	23 05	03 35	22 53
21	00 N 16	03 S 23	09 46	13 56	16 25	22 53	19 19	23 05	03 35	22 52
22	00 40	01 N 38	10 06	14 23	16 27	22 53	19 20	23 06	03 36	22 52
23	01 04	06 28	10 23	14 50	16 29	22 53	19 21	23 06	03 37	22 52
24	01 27	10 53	10 36	15 16	16 30	22 53	19 22	23 06	03 38	22 52
25	01 51	14 41	10 44	15 42	16 32	22 53	19 23	23 06	03 39	22 52
26	02 14	17 42	10 48	16 07	16 33	22 53	19 24	23 06	03 40	22 51
27	02 38	19 50	10 48	16 32	16 34	22 53	19 26	23 06	03 41	22 51
28	03 01	21 01	10 43	16 57	16 36	22 53	19 27	23 06	03 42	22 51
29	03 25	21 12	10 35	17 21	16 36	22 54	19 28	23 06	03 42	22 51
30	03 48	20 27	10 22	17 45	16 37	22 54	19 29	23 07	03 43	22 51
31	04 N 11	18 N 47	10 N 06	18 N 08	16 S 38	22 S 54	19 N 30	23 N 07	03 N 44	22 S 51

ZODIAC SIGN ENTRIES

Date	h	m	Planets
02	19	55	☽ ♋
05	08	17	☽ ♌
07	15	01	☿ ♈
07	18	17	☽ ♍
10	01	35	☽ ♎
12	06	54	☽ ♏
14	11	00	☽ ♐
16	10	42	♀ ♉
16	14	17	☽ ♑
18	17	02	☽ ♒
20	19	41	☉ ♈
20	19	48	☽ ♓
22	23	43	☽ ♈
25	06	05	☽ ♉
27	15	45	☽ ♊
30	04	01	☽ ♋

LATITUDES

Date	Mercury ☿ ° '	Venus ♀ ° '	Mars ♂ ° '	Jupiter ♃ ° '	Saturn ♄ ° '	Uranus ♅ ° '	Neptune ♆ ° '	Pluto ♇ ° '
01	01 S 07	00 S 37	01 N 51	00 N 32	01 S 45	00 N 06	01 S 30	06 S 10
04	00 40	00 28	01 50	00 32	01 44	00 06	01 30	06 11
07	00 S 07	00 20	01 49	00 32	01 43	00 06	01 30	06 11
10	00 N 29	00 10	01 49	00 32	01 42	00 06	01 30	06 12
13	01 08	00 S 01	01 47	00 32	01 42	00 06	01 30	06 12
16	01 48	00 N 09	01 46	00 32	01 41	00 06	01 30	06 13
19	02 24	00 19	01 45	00 32	01 40	00 06	01 30	06 13
22	02 55	00 29	01 43	00 32	01 40	00 06	01 30	06 14
25	03 16	00 39	01 40	00 32	01 39	00 06	01 30	06 15
28	03 24	00 49	01 38	00 32	01 38	00 06	01 30	06 15
31	03 N 18	00 N 59	01 N 35	00 N 32	01 S 38	00 N 06	01 S 30	06 S 16

DATA

Julian Date	2462927
Delta T	+74 seconds
Ayanamsa	24° 17' 49"
Synetic vernal point	04° ♓ 49' 10"
True obliquity of ecliptic	23° 26' 03"

LONGITUDES

Date	Chiron ⚷ ° '	Ceres ⚳ ° '	Pallas ⚴ ° '	Juno ⚵ ° '	Vesta ⚶ ° '	Black Moon Lilith ⚸ ° '
01	12 ♉ 50	19 ♉ 34	26 ♓ 15	26 ♒ 45	01 ♊ 48	01 ♋ 15
11	13 ♉ 15	22 ♉ 44	29 ♓ 49	01 ♓ 14	04 ♊ 39	02 ♋ 22
21	13 ♉ 44	26 ♉ 08	03 ♈ 28	05 ♓ 43	07 ♊ 47	03 ♋ 29
31	14 ♉ 17	29 ♉ 43	07 ♈ 11	10 ♓ 14	11 ♊ 11	04 ♋ 35

MOON'S PHASES, APSIDES AND POSITIONS ☽

Date	h	m	Phase	Longitude ° '	Eclipse Indicator
01	04	02	☽	10 ♊ 22	
09	04	30	○	18 ♍ 24	
16	06	36	☾	25 ♐ 29	
23	03	49	●	02 ♈ 19	
31	00	32	☽	10 ♋ 07	

Day	h	m		
02	18	55	Apogee	
17	18	46	Perigee	
30	15	24	Apogee	
01	20	45	Max dec	21° N 30'
09	05	22	OS	
15	18	56	Max dec	21° S 21'
22	04	10	ON	
29	04	50	Max dec	21° N 15'

ASPECTARIAN

h m	Aspects
01 Saturday	
04 02	☽ □ ☉
04 25	♀ ± ♂
06 31	☽ ⚹ ♀
07 03	☽ ⚹ ♆
08 20	☽ △ ♇
11 50	♀ ☌ ♆
12 05	☿ ⊥ ♇
16 28	☿ □ ♅
18 48	☽ ⊼ ♂
21 14	☽ ☌ ♅
22 09	☽ □ ☿
02 Sunday	
00 34	♀ ⚹ ♇
07 17	☽ ± ♂
07 29	☽ Q ♆
09 42	☽ Q ♀
12 00	☽ ☍ ♃
12 58	☿ # ♀
14 53	☽ ⚼ ♇
16 27	☿ Q ♄
16 46	☉ ⊻ ♆
17 58	♂ ± ♆
03 Monday	
01 33	☽ ⊻ ♄
02 00	☽ ⚼ ♂
08 10	☉ ⊻ ♇
09 13	☽ ± ♇
13 46	☽ ⊥ ♄
20 08	☽ □ ♆
21 23	☽ ⊼ ♇
22 35	☽ △ ☉
04 Tuesday	
00 34	☿ # ♆
02 24	☽ □ ♀
03 52	☽ ∥ ♄
08 01	☽ ∠ ♄
08 59	☽ △ ♂
10 04	☽ ⊻ ♅
22 05	☽ ⊥ ♅
22 22	☽ △ ☿
05 Wednesday	
01 11	☽ ⊼ ♃
07 20	☽ ⚼ ☉
07 28	☉ # ♀
11 07	☽ # ♂
13 11	☽ ± ♃
14 01	☽ ⚹ ♄
15 55	☽ ∠ ♅
17 30	☿ □ ♃
06 Thursday	
02 36	☽ ± ☉
06 59	☽ ⚼ ♃
07 52	☽ △ ♆
08 03	☿ ∠ ♇
09 03	☽ ☍ ♇
09 13	☽ ⚼ ☿
11 55	♀ ∠ ♄
13 34	☽ Q ♄
15 17	☽ ⊼ ☉
16 06	♂ ⊼ ♅
20 09	☽ △ ♀
21 06	☽ ⚹ ♅
21 12	☽ □ ♂
07 Friday	
05 32	☽ ± ☿
06 05	♀ ⚹ ♅
09 12	♀ ⊼ ♂
12 03	☽ △ ♃
12 44	☽ ⚼ ♆
15 29	☽ ∥ ♀
18 49	☽ ⊼ ☿
19 57	☽ Q ♅
08 Saturday	
00 01	☽ □ ♄
03 37	☽ ⚼ ♀
03 48	☽ # ☉
05 43	☽ ± ♆
07 52	☽ Q ♂
12 02	☽ ∥ ♆
16 54	☽ ⊼ ♆
17 59	☽ ⊼ ♇
22 04	☽ ± ♀
23 41	☽ ∥ ☿
09 Sunday	
04 30	☽ ☍ ☉
05 03	☽ ± ♇
05 24	☽ □ ♅
06 25	☽ ⚹ ♂
09 15	☿ ⚹ ♄
10 11	☽ ⊼ ♀
11 57	☉ ⊥ ♇
13 35	☽ # ☿
16 32	☉ □ ♅
20 05	☽ □ ♃
21 28	☽ ⚼ ♇
22 33	☽ # ♆
10 Monday	
02 35	☽ ∥ ☉
06 23	☿ ⚼ ♂
07 21	☽ △ ♄
10 01	☽ ∠ ♂
10 29	☽ ☍ ☿
11 33	☉ △ ♂
23 27	☽ ☍ ♆
11 Tuesday	
00 27	☽ △ ♇
02 20	☽ ⊥ ♂
03 21	☽ # ♀
04 37	☽ Q ♃
10 14	☽ ⚼ ♄
11 24	☽ △ ♅
13 08	☽ ⊻ ♂
14 46	☽ ⊼ ☉
15 14	☿ Q ♅
16 02	☿ ∥ ♆
19 13	☉ # ☿
21 14	☽ ☍ ♀
12 Wednesday	
01 41	☉ # ♆
01 59	☽ ⚹ ♃
02 10	☽ ± ☉
02 11	☽ ± ♄
02 24	☉ Q ♄
02 54	♀ Q ♇
12 45	☽ ⊼ ♄
13 48	☽ ⚼ ♅
19 10	☽ ⚼ ☉
21 13	☽ ∥ ♂
22 36	☽ ⊼ ☿
13 Thursday	
04 20	☽ ⊼ ♆
04 23	☽ ∠ ♃
05 16	☽ □ ♇
05 30	☽ ± ♅
10 19	☽ ± ☿
14 46	☽ ± ♆
15 56	☽ ⊼ ♅
18 15	☽ ☌ ♂
20 06	☽ ⊥ ♃
23 12	☽ # ♄
23 14	☽ △ ☉
14 Friday	
03 43	☽ ⚼ ☿
06 22	☽ ⚼ ♆
06 29	☽ ⊼ ♀
06 32	☽ ⊻ ♃
07 09	♀ △ ♃
08 16	♀ ⊥ ♄
12 27	☽ Q ♇
17 00	☽ ☍ ♄
17 47	☽ ± ♀
15 Saturday	
07 14	☿ ☌ ♆
08 12	☽ △ ♆
08 18	☽ △ ☿
09 04	☽ ⚹ ♇
10 40	☽ ⚼ ♀
16 53	☿ ⚹ ♇
19 32	☽ ☍ ♅
22 22	☽ ⊻ ♂
16 Sunday	
06 36	☽ □ ☉
08 41	☽ ⊥ ♂
10 13	☽ ☌ ♃
10 42	☽ ∠ ♇
14 37	☽ △ ♀
20 29	☽ ⊼ ♄
17 Monday	
00 07	☽ ∠ ♂
02 00	☽ ⊥ ♇
06 42	☽ ± ♄
08 32	☿ ± ♂
11 21	☽ □ ♆
12 10	☽ ⊻ ♇
13 40	☽ # ♄
15 26	☽ Q ☉
16 00	☽ □ ☿
22 00	☽ ⚼ ♄
22 30	☽ ⊼ ♅
18 Tuesday	
01 44	☽ ⚹ ♂
08 39	☽ ± ♅
13 13	☽ ⚹ ☉
13 18	☽ ⊻ ♃
13 54	☽ ∥ ♂
14 30	☉ □ ♃
17 50	☽ Q ♆
18 03	☉ ∠ ♇
20 50	♃ ∥ ♇
22 03	☽ □ ♀
22 09	☽ Q ♂
23 27	☽ △ ♄
23 29	☽ ⊥ ♃
23 51	☽ ⚼ ♅
19 Wednesday	
00 40	☽ Q ☿
10 43	☽ # ♀
14 08	☽ ⚹ ♆
14 46	☽ ∠ ♃
14 55	☽ ☌ ♇
15 48	♀ ⊻ ♄
16 27	☽ ∠ ☉
19 46	♀ ∠ ♅
22 14	☽ ⚹ ☿
20 Thursday	
01 14	☽ △ ♅
04 49	☽ □ ♂
07 02	☽ # ☿
07 17	☽ Q ♀
08 50	☽ ⊥ ☉
15 37	☽ ∠ ♆
16 20	☽ ⚹ ♃
19 49	☽ ⊻ ☉
21 Friday	
01 06	☽ ∠ ☿
02 35	☽ □ ♄
03 03	♃ ∠ ♇
05 37	☽ ⚹ ♀
07 01	☽ ⊥ ♆
11 00	☽ # ♆
12 58	☽ Q ♃
17 15	☽ ⊥ ☿
17 22	☽ ⊻ ♆
18 06	☽ ⊻ ♇
22 Saturday	
01 48	☽ # ☉
04 03	☽ ⊻ ☿
04 30	☽ ⊥ ♇
04 39	☽ □ ♅
06 58	☽ ∥ ☉
08 35	☽ △ ♂
09 49	☽ Q ♄
09 58	☽ ∠ ♀
20 15	☽ ∠ ♇
20 24	☽ □ ♃
21 40	☽ ∥ ♆
23 03	☿ ⚹ ♅
23 11	☿ ∠ ♄
23 Sunday	
03 18	☽ ⊥ ♀
03 49	☽ ☌ ☉
07 04	☽ ⚹ ♄
11 09	☽ ⚼ ♂
12 23	☽ Q ♅
15 01	☽ ⊻ ♀
22 18	☽ ☌ ♆
23 01	☽ ⚹ ♇
24 Monday	
03 25	☽ ± ♂
10 05	☽ ⚹ ♅
10 14	☽ ∥ ☿
10 15	☽ ∠ ♄
10 51	☽ ☌ ☿
11 44	☉ ⊥ ♀
14 24	☽ ⊼ ♂
20 57	☽ Q ♇
25 Tuesday	
02 53	☽ △ ♃
02 56	☽ ⊥ ♄
03 56	☉ ⚹ ♄
13 56	☽ ∠ ♅
14 15	☽ ⊻ ♄
15 02	☽ ⊻ ☉
20 26	☽ ∥ ♀
26 Wednesday	
00 44	☿ St R
01 57	☽ # ♂
03 23	☽ ⊥ ☉
04 05	☽ ☌ ♀
06 12	☽ ⊻ ♆
06 53	☽ □ ♇
07 04	☽ ⊥ ♅
07 21	☽ ⚼ ♃
17 47	☽ ⊥ ♆
18 39	☽ ⊻ ♅
19 35	☽ ⊻ ☿
22 13	☽ ∠ ☉
23 18	☽ ☍ ♂
27 Thursday	
00 50	☽ ± ♃
03 04	♀ ⊻ ♆
06 21	☽ ∥ ♄
07 10	☽ ⊥ ☿
10 03	♀ □ ♇
11 25	☽ ∠ ♆
11 51	☉ ⚼ ♂
12 06	♀ ⊥ ♅
12 39	☽ ⊼ ♃
14 26	♀ # ♂
16 11	♀ ⚼ ♃
19 29	☿ ∠ ♄
28 Friday	
00 39	☽ ∠ ☿
00 47	☽ ☌ ♄
03 07	☉ Q ♅
06 24	☽ ⚹ ☉
17 20	☽ ⚹ ♆
17 58	☽ △ ♇
19 10	☿ ⚹ ♅
21 26	☽ ⊻ ♀
29 Saturday	
00 34	♂ St R
05 57	☽ ⚹ ☿
06 18	☽ ☌ ♅
08 42	☽ Q ☉
10 53	☽ ⊥ ♀
11 03	☽ ⊼ ♂
17 38	☽ Q ♆
23 12	☽ ± ♂
30 Sunday	
00 20	☽ ⚼ ♇
01 06	☽ ☍ ♃
05 24	☽ Q ☿
06 58	☉ ∥ ♆
07 10	☽ ∠ ♀
13 45	☽ ⊻ ♄
17 27	☽ ⚼ ♂
18 38	☽ ± ♇
31 Monday	
00 32	☽ □ ☉
02 01	☽ ⊥ ♄
03 13	☽ ∥ ♄
06 17	☽ □ ♆
06 49	☽ ⊼ ♇
12 19	☿ ⊻ ♀
16 17	☽ □ ☿
16 57	☽ ⚹ ♀
18 08	☽ ∥ ♀
19 14	☽ ⊻ ♅
20 22	☽ ∠ ♄
23 44	☽ △ ♂

All ephemeris data is given at 12.00 UT and the Moon's longitude is additionally given for 24.00 UT

APRIL 2031

LONGITUDES

Date	Sidereal time h m s	Sun ☉ ° ' "	Moon ☽ ° ' "	Moon ☽ 24.00 ° ' "	Mercury ☿ ° '	Venus ♀ ° '	Mars ♂ ° '	Jupiter ♃ ° '	Saturn ♄ ° '	Uranus ♅ ° '	Neptune ♆ ° '	Pluto ♇ ° '
01	00 38 37	11 ♈ 34 16	27 ♋ 41 46	03 ♌ 43 05	17 ♈ 22	19 ♉ 09	21 ♏ 34	28 ♐ 41	04 ♊ 58	19 ♊ 22	12 ♈ 59	13 ♒ 14
02	00 42 34	12 33 28	09 ♌ 47 33	15 ♌ 55 39	16 R 40	20 20	21 R 31	28 43	05 04	19 24	13 02	13 15
03	00 46 30	13 32 38	22 ♌ 07 50	28 ♌ 24 28	15 55	21 31	21 27	28 46	05 10	19 26	13 04	13 16
04	00 50 27	14 31 45	04 ♍ 45 50	11 ♍ 12 09	15 08	22 41	21 23	28 48	05 16	19 28	13 06	13 17
05	00 54 23	15 30 51	17 ♍ 43 30	24 ♍ 19 52	14 19	23 52	21 17	28 50	05 22	19 30	13 09	13 18
06	00 58 20	16 29 54	01 ♎ 01 10	07 ♎ 47 11	13 31	25 02	21 11	28 52	05 28	19 32	13 11	13 19
07	01 02 16	17 28 55	14 ♎ 37 37	21 ♎ 32 05	12 43	26 12	21 05	28 53	05 34	19 34	13 13	13 20
08	01 06 13	18 27 54	28 ♎ 30 09	05 ♏ 31 20	11 57	27 23	20 57	28 55	05 40	19 36	13 15	13 21
09	01 10 09	19 26 51	12 ♏ 35 05	19 ♏ 40 53	11 13	28 33	20 49	28 56	05 46	19 38	13 18	13 22
10	01 14 06	20 25 46	26 ♏ 48 12	03 ♐ 56 32	10 32	29 ♉ 42	20 40	28 57	05 52	19 40	13 20	13 23
11	01 18 03	21 24 40	11 ♐ 05 25	18 ♐ 14 26	09 56	00 ♊ 52	20 30	28 58	05 58	19 43	13 22	13 24
12	01 21 59	22 23 31	25 ♐ 23 10	02 ♑ 31 19	09 23	02 02	20 20	28 58	06 05	19 45	13 24	13 25
13	01 25 56	23 22 21	09 ♑ 38 35	16 ♑ 44 43	08 55	03 11	20 08	28 59	06 11	19 47	13 27	13 26
14	01 29 52	24 21 10	23 ♑ 49 29	00 ♒ 52 41	08 32	04 21	19 56	28 59	06 18	19 49	13 29	13 27
15	01 33 49	25 19 56	07 ♒ 54 07	14 ♒ 53 37	08 13	05 30	19 43	28 59	06 24	19 52	13 31	13 27
16	01 37 45	26 18 41	21 ♒ 50 59	28 ♒ 46 03	08 01	06 39	19 30	28 R 59	06 31	19 54	13 33	13 28
17	01 41 42	27 17 24	05 ♓ 38 36	12 ♓ 28 26	07 53	07 48	19 16	28 59	06 37	19 57	13 36	13 29
18	01 45 38	28 16 06	19 ♓ 15 21	25 ♓ 59 10	07 D 50	08 57	19 01	28 59	06 44	19 59	13 38	13 30
19	01 49 35	29 ♈ 14 45	02 ♈ 39 41	09 ♈ 16 45	07 53	10 06	18 45	28 58	06 51	20 02	13 40	13 30
20	01 53 32	00 ♉ 13 23	15 ♈ 50 13	22 ♈ 19 58	08 01	11 15	18 29	28 57	06 57	20 04	13 42	13 31
21	01 57 28	01 11 59	28 ♈ 45 57	05 ♉ 08 08	08 14	12 23	18 12	28 56	07 04	20 07	13 44	13 32
22	02 01 25	02 10 33	11 ♉ 26 34	17 ♉ 41 19	08 31	13 31	17 55	28 55	07 11	20 09	13 47	13 32
23	02 05 21	03 09 06	23 ♉ 52 34	00 ♊ 00 29	08 53	14 40	17 37	28 53	07 18	20 12	13 49	13 33
24	02 09 18	04 07 36	06 ♊ 05 20	12 ♊ 07 27	09 20	15 48	17 18	28 52	07 25	20 15	13 51	13 33
25	02 13 14	05 06 04	18 ♊ 07 12	24 ♊ 04 59	09 51	16 55	16 59	28 50	07 32	20 17	13 53	13 34
26	02 17 11	06 04 31	00 ♋ 01 16	05 ♋ 56 32	10 25	18 03	16 40	28 48	07 39	20 20	13 55	13 34
27	02 21 07	07 02 55	11 ♋ 51 21	17 ♋ 46 14	11 04	19 11	16 20	28 46	07 46	20 23	13 57	13 35
28	02 25 04	08 01 17	23 ♋ 41 48	29 ♋ 38 38	11 47	20 18	15 59	28 43	07 53	20 26	13 59	13 35
29	02 29 01	08 59 37	05 ♌ 37 20	11 ♌ 38 30	12 33	21 25	15 39	28 41	08 00	20 29	14 02	13 36
30	02 32 57	09 ♉ 57 55	17 ♌ 42 43	23 ♌ 50 34	13 ♈ 22	22 ♊ 32	15 ♏ 17	28 ♐ 38	08 ♊ 07	20 ♊ 31	14 ♈ 04	13 ♒ 36

Date	Moon True ☊ ° '	Moon Mean ☊ ° '	Moon ☽ Latitude ° '
01	29 ♏ 19	00 ♐ 41	04 S 23
02	29 R 15	00 38	04 51
03	29 10	00 35	05 06
04	29 03	00 31	05 07
05	28 56	00 28	04 52
06	28 48	00 25	04 22
07	28 42	00 22	03 36
08	28 37	00 19	02 36
09	28 34	00 16	01 26
10	28 33	00 12	00 S 09
11	28 D 33	00 09	01 N 08
12	28 35	00 06	02 21
13	28 36	00 03	03 25
14	28 37	00 ♐ 00	04 17
15	28 R 36	29 ♏ 57	04 52
16	28 34	29 53	05 10
17	28 30	29 50	05 10
18	28 26	29 47	04 52
19	28 21	29 44	04 19
20	28 17	29 41	03 32
21	28 13	29 37	02 34
22	28 11	29 34	01 31
23	28 10	29 31	00 N 24
24	28 D 10	29 28	00 S 44
25	28 11	29 25	01 48
26	28 13	29 22	02 47
27	28 14	29 18	03 38
28	28 16	29 15	04 20
29	28 16	29 12	04 52
30	28 ♏ 16	29 ♏ 09	05 S 11

DECLINATIONS

Date	Sun ☉ ° '	Moon ☽ ° '	Mercury ☿ ° '	Venus ♀ ° '	Mars ♂ ° '	Jupiter ♃ ° '	Saturn ♄ ° '	Uranus ♅ ° '	Neptune ♆ ° '	Pluto ♇ ° '
01	04 N 35	16 N 19	09 N 46	18 N 31	16 S 38	22 S 54	19 N 31	23 N 07	03 N 45	22 S 50
02	04 58	13 07	09 23	18 53	16 38	22 54	19 33	23 07	03 46	22 50
03	05 21	09 18	08 58	19 15	16 39	22 54	19 34	23 07	03 47	22 50
04	05 44	05 00	08 30	19 37	16 39	22 54	19 35	23 07	03 48	22 50
05	06 06	00 N 22	08 00	19 58	16 38	22 54	19 36	23 07	03 49	22 50
06	06 29	04 S 24	07 30	20 19	16 38	22 54	19 38	23 08	03 49	22 50
07	06 52	09 05	06 58	20 39	16 38	22 54	19 39	23 08	03 50	22 50
08	07 14	13 22	06 26	20 58	16 37	22 54	19 40	23 08	03 51	22 50
09	07 37	16 58	05 55	21 17	16 36	22 54	19 41	23 08	03 52	22 50
10	07 59	19 36	05 24	21 36	16 35	22 54	19 43	23 08	03 53	22 50
11	08 21	20 59	04 54	21 54	16 34	22 54	19 44	23 08	03 54	22 50
12	08 43	21 00	04 26	22 11	16 33	22 54	19 45	23 09	03 55	22 50
13	09 05	19 40	04 00	22 28	16 31	22 54	19 46	23 09	03 56	22 50
14	09 26	17 07	03 36	22 44	16 30	22 54	19 48	23 09	03 56	22 50
15	09 48	13 35	03 14	23 00	16 28	22 54	19 49	23 09	03 57	22 50
16	10 09	09 19	02 55	23 15	16 26	22 54	19 50	23 09	03 58	22 50
17	10 30	04 S 38	02 38	23 29	16 24	22 54	19 51	23 10	03 59	22 50
18	10 51	00 N 14	02 24	23 43	16 22	22 54	19 53	23 10	04 00	22 50
19	11 12	05 01	02 12	23 56	16 19	22 54	19 54	23 10	04 01	22 50
20	11 33	09 29	02 03	24 09	16 17	22 54	19 55	23 10	04 01	22 50
21	11 53	13 26	01 57	24 21	16 14	22 54	19 57	23 10	04 02	22 50
22	12 14	16 42	01 53	24 33	16 11	22 54	19 58	23 11	04 03	22 50
23	12 34	19 07	01 52	24 43	16 08	22 54	19 59	23 11	04 04	22 50
24	12 54	20 36	01 53	24 53	16 05	22 54	20 00	23 11	04 05	22 50
25	13 13	21 07	01 56	25 03	16 02	22 54	20 02	23 11	04 06	22 50
26	13 33	20 39	02 02	25 12	15 59	22 54	20 03	23 11	04 06	22 50
27	13 52	19 17	02 10	25 20	15 55	22 54	20 04	23 12	04 07	22 50
28	14 11	17 05	02 20	25 28	15 52	22 54	20 06	23 12	04 08	22 50
29	14 30	14 09	02 32	25 35	15 48	22 54	20 07	23 12	04 09	22 50
30	14 N 48	10 N 35	02 N 46	25 N 41	15 S 44	22 S 54	20 N 08	23 N 12	04 N 10	22 S 50

ZODIAC SIGN ENTRIES

Date	h	m	Planets	
01	16	36	☽	♌
04	03	01	☽	♍
06	10	11	☽	♎
08	14	34	☽	♏
10	17	23	☽	♐
10	18	01	♀	♊
12	19	45	☽	♑
14	22	30	☽	♒
17	02	09	☽	♓
19	07	12	☽	♈
20	06	31	☉	♉
21	14	19	☽	♉
23	23	59	☽	♊
26	11	57	☽	♋
29	00	43	☽	♌

LATITUDES

Date	Mercury ☿ ° '	Venus ♀ ° '	Mars ♂ ° '	Jupiter ♃ ° '	Saturn ♄ ° '	Uranus ♅ ° '	Neptune ♆ ° '	Pluto ♇ ° '
01	03 N 12	01 N 03	01 N 34	00 N 32	01 S 37	00 N 06	01 S 30	06 S 16
04	02 45	01 13	01 31	00 32	01 37	00 06	01 30	06 17
07	02 07	01 23	01 27	00 32	01 36	00 06	01 30	06 17
10	01 20	01 33	01 23	00 32	01 36	00 06	01 30	06 18
13	00 N 31	01 42	01 18	00 32	01 35	00 06	01 30	06 19
16	00 S 17	01 51	01 13	00 32	01 35	00 06	01 30	06 19
19	01 00	02 00	01 07	00 32	01 34	00 06	01 30	06 20
22	01 37	02 08	01 01	00 32	01 34	00 06	01 30	06 21
25	02 08	02 16	00 54	00 32	01 33	00 06	01 30	06 22
28	02 31	02 23	00 47	00 32	01 33	00 06	01 30	06 22
31	02 S 48	02 N 30	00 N 40	00 N 31	01 S 32	00 N 06	01 S 30	06 S 23

DATA

Julian Date	2462958
Delta T	+74 seconds
Ayanamsa	24° 17' 52"
Synetic vernal point	04° ♓ 49' 07"
True obliquity of ecliptic	23° 26' 03"

MOON'S PHASES, APSIDES AND POSITIONS ☽

Date	h	m	Phase	Longitude ° '	Eclipse Indicator
07	17	21	○	17 ♎ 42	
14	12	58	☾	24 ♑ 24	
21	16	57	●	01 ♉ 24	
29	19	19	☽	09 ♌ 17	

Day	h	m		
11	19	21	Perigee	
27	10	32	Apogee	
05	13	50	0S	
12	00	23	Max dec	21° S 10'
18	10	52	0N	
25	12	30	Max dec	21° N 07'

LONGITUDES

Date	Chiron ⚷ ° '	Ceres ⚳ ° '	Pallas ⚴ ° '	Juno ⚵ ° '	Vesta ⚶ ° '	Black Moon Lilith ⚸ ° '
01	14 ♉ 20	00 ♊ 05	07 ♈ 33	10 ♓ 41	11 ♊ 32	04 ♋ 42
11	14 ♉ 56	03 ♊ 51	11 ♈ 18	15 ♓ 15	13 ♊ 08	05 ♋ 49
21	15 ♉ 34	07 ♊ 44	15 ♈ 05	19 ♓ 41	15 ♊ 55	06 ♋ 55
31	16 ♉ 13	11 ♊ 44	18 ♈ 54	24 ♓ 09	22 ♊ 50	08 ♋ 02

All ephemeris data is given at 12.00 UT and the Moon's longitude is additionally given for 24.00 UT

ASPECTARIAN

h m Aspects

01 Tuesday
07 20 ☽ ⊥ ♅
08 32 ♀ ⊥ ♆
09 16 ☽ # ♂
13 59 ☽ ⊼ ♃
16 34 ♀ ⊻ ♅
19 39 ☽ Q ♀

02 Wednesday
01 20 ☽ ∠ ♅
01 58 ☽ ± ♃
02 36 ☽ ✶ ♄
17 54 ☽ △ ☉
18 22 ☽ △ ♆
18 48 ☽ ☍ ♇
19 44 ☽ P ♃
23 55 ☉ ☌ ♆

03 Thursday
00 40 ☽ △ ☿
02 20 ☽ Q ♄
05 13 ☉ ✶ ♇
06 47 ☽ ✶ ♅
10 41 ☽ □ ♀
10 42 ☽ □ ♂
10 48 ♀ ☍ ♂
14 10 ☽ ∥ ☿
16 55 ☿ ⊥ ♀
23 23 ☽ P ♆

04 Friday
00 43 ☽ △ ♃
01 18 ☽ P ☉
03 38 ☿ ± ♂
03 47 ☽ P ☿
05 47 ☽ Q ♅
08 22 ☽ ∥ ☉
09 48 ♀ ∥ ♄
12 57 ☽ □ ♄
14 15 ♀ ± ♃
16 24 ☽ ± ♆
18 19 ☽ ∥ ♆
19 37 ☽ ± ☉
19 40 ☽ ± ☿
20 02 ☉ ☌ ☿
20 33 ☽ Q ♂

05 Saturday
03 34 ☽ ⊼ ♆
03 52 ☽ ⊼ ♇
06 07 ☽ ⊼ ☿
07 00 ☉ ± ♂
07 37 ☽ ⊼ ☉
14 53 ☽ ± ♇
15 15 ☽ □ ♅
18 27 ☽ ✶ ♂

06 Sunday
00 14 ☽ △ ♀
07 10 ☽ P ♇
08 08 ☽ □ ♃
09 04 ☽ # ♆
17 34 ☿ ✶ ♇
19 57 ☽ △ ♄
21 07 ☽ ∠ ♂
21 31 ☿ ☌ ♆
23 28 ☽ # ☉

07 Monday
02 07 ☽ # ☿
05 27 ☽ P ♀
08 51 ☽ ☍ ☿
09 32 ☽ ☍ ♆
09 45 ☽ △ ♇
12 47 ☽ ⊥ ♂
14 45 ☉ ∥ ☿
15 57 ☽ Q ♃
17 21 ☽ ☍ ☉
20 37 ☽ △ ♅
22 23 ☽ P ♄
22 36 ☽ ± ♀
23 07 ☽ ⊻ ♂

08 Tuesday
06 39 ☿ ∠ ♀
09 53 ☽ ⊼ ♀
12 42 ☽ ✶ ♃
14 00 ☽ ± ♄
22 27 ☽ P ♅

09 Wednesday
00 20 ☽ ⊼ ♄
06 41 ♀ ∠ ♆
09 15 ☽ ∥ ♂
09 48 ☽ ⊼ ☿
13 12 ☽ ⊼ ♆
13 20 ☽ □ ♇
13 47 ☽ ± ♅
14 17 ☽ ∠ ♃
16 48 ☉ ✶ ♅
19 28 ☽ ± ☿
20 05 ♀ ⊼ ♃
23 22 ☽ ± ♆
23 57 ☽ ⊼ ♅

10 Thursday
00 28 ☽ ⊼ ☉
01 46 ☽ ☌ ♂
05 30 ☽ ⊥ ♃
09 58 ☽ P ☿
11 19 ☽ ± ☉
13 26 ☽ # ♄
14 35 ☽ P ♆
15 37 ☽ ⊻ ♃
16 56 ☉ ⊼ ♂
17 19 ☽ ☍ ♀
19 43 ☽ Q ♇

11 Friday
00 01 ☉ ∠ ♄
03 21 ☽ ☍ ♄
03 34 ☽ P ☉
10 07 ☽ △ ☿
15 50 ☽ △ ♆
15 53 ☽ ✶ ♇

12 Saturday
02 30 ☽ ☍ ♅
03 36 ☽ ⊻ ♂
06 36 ☽ △ ☉
13 33 ☽ ⊥ ♂
17 06 ☽ ∠ ♇
18 02 ☽ ☌ ♃
22 06 ♆ ✶ ♇

13 Sunday
00 10 ☽ ⊼ ♀
04 30 ☽ ∠ ♂
06 08 ☽ ⊼ ♄
08 16 ☽ ⊥ ♇
10 43 ☽ # ♄
10 48 ☽ □ ☿
11 10 ☽ ± ♀
16 18 ☿ ∥ ♆
16 19 ☽ ± ♄
18 24 ☽ ⊻ ♇
18 26 ☽ □ ♆

14 Monday
03 44 ☽ P ♀
05 12 ☽ ⊼ ♅
05 30 ☽ ✶ ♂
07 40 ☽ P ♄
12 58 ☽ □ ☉
15 24 ☽ ± ♅
16 29 ☽ Q ☿
16 43 ☽ ∥ ♂
20 28 ♀ # ♇
20 47 ☽ ⊻ ♃
22 53 ♂ ⊼ ♅

15 Tuesday
01 04 ☽ Q ♆
01 36 ☽ Q ♂
03 07 ♀ # ♃
06 47 ☽ P ♅
07 01 ☽ ⊥ ♃
07 32 ☽ △ ♀
09 25 ☽ △ ♄
12 04 ♃ St R
12 33 ☽ ✶ ☿
15 05 ☉ Q ♇
21 32 ☽ ☌ ♇
21 40 ☽ ✶ ♆
22 01 ☽ Q ☉
22 26 ☽ ∠ ♃

16 Wednesday
03 14 ♀ ∥ ♅
06 51 ♂ ± ♆
07 52 ☽ # ☉
08 00 ☽ □ ♂
08 37 ☽ △ ♅
08 40 ♀ ☌ ♄
13 59 ☽ ∠ ☿
20 19 ☽ ✶ ☉
23 40 ☽ ∠ ♆

17 Thursday
00 22 ☽ ✶ ♃
01 47 ☿ Q ♅
05 28 ☽ ⊥ ☿
13 30 ☿ ✶ ♀
13 44 ☽ □ ♄
15 13 ☽ # ♆
15 25 ☽ ⊥ ♆
15 54 ☽ ⊻ ☿
16 08 ☽ □ ♀
21 22 ☽ Q ♃
22 25 ☽ # ☿

18 Friday
00 35 ☽ ∠ ☉
01 47 ☽ ⊻ ♇
02 01 ☽ ⊻ ♆
11 16 ☿ St D
11 35 ☽ △ ♂
12 25 ☽ ⊥ ♇
13 18 ☽ □ ♅
17 47 ☽ ⊥ ☉
21 50 ☽ Q ♄
22 20 ☽ ∥ ☿

19 Saturday
03 01 ☽ Q ♀
04 31 ☽ ∠ ♇
05 05 ☉ △ ♃
05 20 ☽ □ ♃
05 22 ☽ ⊻ ☉
06 52 ☽ ∥ ♆
13 56 ☽ P ♂
19 38 ☽ ✶ ♄
21 32 ☽ ☌ ☿
21 45 ☽ Q ♅

20 Sunday
02 47 ☽ ✶ ♀
05 58 ☽ ± ♂
07 44 ☽ ✶ ♇
08 04 ☽ ☌ ♆
16 46 ☽ ⊼ ♂
19 50 ☽ ✶ ♅
20 42 ☿ Q ♅
23 24 ☽ ∠ ♄

21 Monday
01 13 ☽ ∥ ☉
05 56 ☽ Q ♇
08 21 ☉ ⊥ ♄
09 10 ☽ ∠ ♀
12 18 ☽ △ ♃
16 22 ☽ ⊥ ♄
16 57 ☽ ☌ ☉

22 Tuesday
00 00 ☽ ∠ ♅
03 47 ☽ ⊥ ♀
03 49 ☽ ⊻ ♄
06 17 ☽ ⊻ ☿
07 54 ☽ # ♂
12 18 ♀ △ ♇
16 01 ☽ □ ♇
16 23 ☽ ⊻ ♀
16 29 ☽ ⊻ ♆
16 43 ☽ P ♃
17 13 ☽ ⊥ ♅
17 30 ♀ ✶ ♆
18 04 ☽ ⊥ ☿

23 Wednesday
00 08 ☽ ☍ ♂
04 05 ☽ ⊥ ♆
04 50 ☽ ⊻ ♅
10 04 ☽ ± ♃
12 01 ☽ ∠ ☿
21 41 ☽ ∠ ♆
21 46 ☽ ⊼ ♃

24 Thursday
00 15 ☽ ∥ ♄
07 47 ☽ ⊻ ☉
14 39 ☽ ☌ ♄
18 42 ☽ ✶ ☿
20 43 ☽ ⊥ ☉

25 Friday
02 52 ☽ △ ♇
03 29 ☽ ✶ ♆
09 21 ☽ ☌ ♀
09 47 ☽ ⊼ ♂
12 59 ♀ ⊼ ♂
16 20 ☽ ∠ ☉
16 23 ☽ ☌ ♅
16 52 ☉ ∠ ♅
19 51 ☽ Q ☿
21 32 ☽ ± ♂

26 Saturday
03 41 ☽ Q ♆
09 04 ☽ P ♇
09 32 ☽ ☍ ♃
15 14 ☽ P ♂
17 56 ☿ ± ♂

27 Sunday
00 18 ☽ ∥ ♄
01 22 ☽ ✶ ☉
03 19 ☽ ± ♇
03 37 ☽ ⊻ ♄
10 18 ☽ □ ☿
15 30 ☽ ⊼ ♇
15 55 ☽ ⊥ ♄
16 16 ☽ □ ♆
20 49 ☽ △ ♂

28 Monday
03 54 ☽ Q ☉
04 25 ☽ ⊻ ♀
05 22 ☽ ⊻ ♅
08 08 ☉ ⊻ ♄
10 20 ☽ ∠ ♄
14 49 ♀ ☌ ♅
17 32 ☽ ⊥ ♅
17 49 ☽ ⊥ ♀
22 06 ☽ ⊼ ♃
22 50 ☽ # ♂

29 Tuesday
09 38 ☽ ∥ ☉
10 07 ☽ ± ♃
11 42 ☽ ∠ ♅
13 46 ☽ ∠ ♀
15 35 ♀ ± ♂
16 48 ☽ ✶ ♄
19 19 ☽ □ ☉

30 Wednesday
02 47 ☽ △ ☿
03 53 ☽ ☍ ♇
03 59 ☽ P ♃
04 46 ☽ △ ♆
07 22 ☽ □ ♂
16 47 ☽ Q ♄
17 32 ☽ ✶ ♅
18 32 ☿ ✶ ♇
22 24 ☽ ✶ ♀

MAY 2031

LONGITUDES

Date	Sidereal time h m s	Sun ☉ ° ' "	Moon ☽ ° ' "	Moon ☽ 24.00 ° ' "	Mercury ☿ ° '	Venus ♀ ° '	Mars ♂ ° '	Jupiter ♃ ° '	Saturn ♄ ° '	Uranus ♅ ° '	Neptune ♆ ° '	Pluto ♇ ° '
01	02 36 54	10 ♉ 56 11	00 ♍ 02 34	06 ♍ 19 14	14 ♈ 15	23 ♊ 39	14 ♏ 56	28 ♐ 35	08 ♊ 15	20 ♊ 34	14 ♈ 06	13 ♒ 37
02	02 40 50	11 54 24	12 ♍ 41 00	19 ♍ 08 13	15 11	24 46	14 R 35	28 R 32	08 22	20 37	14 08	13 37
03	02 44 47	12 52 36	25 ♍ 41 11	02 ♎ 20 03	16 10	25 52	14 13	28 29	08 29	20 40	14 10	13 37
04	02 48 43	13 50 46	09 ♎ 04 53	15 ♎ 55 40	17 12	26 59	13 51	28 26	08 36	20 43	14 12	13 38
05	02 52 40	14 48 54	22 ♎ 52 10	29 ♎ 54 07	18 17	28 05	13 29	28 22	08 44	20 46	14 14	13 38
06	02 56 36	15 47 00	07 ♏ 01 03	14 ♏ 12 26	19 24	29 ♊ 10	13 07	28 18	08 51	20 49	14 16	13 38
07	03 00 33	16 45 05	21 ♏ 27 34	28 ♏ 45 44	20 34	00 ♋ 16	12 45	28 14	08 59	20 52	14 18	13 38
08	03 04 30	17 43 08	06 ♐ 06 07	13 ♐ 27 53	21 47	01 22	12 23	28 10	09 06	20 55	14 20	13 38
09	03 08 26	18 41 09	20 ♐ 50 10	28 ♐ 12 09	23 02	02 27	12 01	28 06	09 13	20 59	14 22	13 39
10	03 12 23	19 39 09	05 ♑ 33 02	12 ♑ 52 08	24 19	03 32	11 39	28 02	09 21	21 02	14 24	13 39
11	03 16 19	20 37 08	20 ♑ 08 47	27 ♑ 22 28	25 39	04 36	11 17	27 57	09 28	21 05	14 26	13 39
12	03 20 16	21 35 05	04 ♒ 32 43	11 ♒ 39 12	27 01	05 41	10 56	27 52	09 36	21 08	14 28	13 39
13	03 24 12	22 33 01	18 ♒ 41 40	25 ♒ 39 57	28 26	06 45	10 34	27 48	09 44	21 11	14 29	13 39
14	03 28 09	23 30 56	02 ♓ 33 57	09 ♓ 23 39	29 ♈ 53	07 49	10 13	27 43	09 51	21 15	14 31	13 39
15	03 32 05	24 28 49	16 ♓ 09 06	22 ♓ 50 21	01 ♉ 22	08 53	09 52	27 37	09 59	21 18	14 33	13 R 39
16	03 36 02	25 26 41	29 ♓ 27 30	06 ♈ 00 42	02 53	09 56	09 32	27 32	10 06	21 21	14 35	13 39
17	03 39 59	26 24 32	12 ♈ 30 05	18 ♈ 55 49	04 26	11 00	09 12	27 27	10 14	21 24	14 37	13 39
18	03 43 55	27 22 22	25 ♈ 18 02	01 ♉ 36 56	06 02	12 03	08 53	27 21	10 22	21 28	14 39	13 39
19	03 47 52	28 20 11	07 ♉ 52 39	14 ♉ 05 23	07 39	13 05	08 34	27 15	10 29	21 31	14 40	13 39
20	03 51 48	29 ♉ 17 58	20 ♉ 15 18	26 ♉ 22 35	09 19	14 08	08 15	27 09	10 37	21 34	14 42	13 39
21	03 55 45	00 ♊ 15 44	02 ♊ 27 26	08 ♊ 30 04	11 01	15 10	07 57	27 03	10 45	21 38	14 44	13 38
22	03 59 41	01 13 29	14 ♊ 30 41	20 ♊ 29 33	12 45	16 12	07 40	26 57	10 53	21 41	14 46	13 38
23	04 03 38	02 11 12	26 ♊ 26 55	02 ♋ 23 07	14 32	17 13	07 23	26 51	11 00	21 44	14 47	13 38
24	04 07 34	03 08 54	08 ♋ 18 27	14 ♋ 13 16	16 20	18 14	07 07	26 45	11 08	21 48	14 49	13 38
25	04 11 31	04 06 34	20 ♋ 07 57	26 ♋ 02 56	18 11	19 15	06 52	26 38	11 16	21 51	14 51	13 38
26	04 15 28	05 04 13	01 ♌ 58 40	07 ♌ 55 35	20 04	20 15	06 37	26 32	11 24	21 55	14 52	13 37
27	04 19 24	06 01 51	13 ♌ 54 13	19 ♌ 55 05	21 58	21 16	06 23	26 25	11 31	21 58	14 54	13 37
28	04 23 21	06 59 27	25 ♌ 58 41	02 ♍ 05 36	23 55	22 15	06 10	26 18	11 39	22 02	14 55	13 37
29	04 27 17	07 57 01	08 ♍ 16 21	14 ♍ 31 29	25 54	23 15	05 57	26 12	11 47	22 05	14 57	13 36
30	04 31 14	08 54 34	20 ♍ 51 30	27 ♍ 16 53	27 55	24 13	05 46	26 05	11 55	22 09	14 58	13 36
31	04 35 10	09 ♊ 52 06	03 ♎ 48 03	10 ♎ 25 22	29 ♉ 58	25 ♋ 12	05 ♏ 35	25 ♐ 58	12 ♊ 02	22 ♊ 12	15 ♈ 00	13 ♒ 35

Date	Moon True ☊ ° '	Moon Mean ☊ ° '	Moon ☽ Latitude ° '
01	28 ♏ 15	29 ♏ 06	05 S 16
02	28 R 13	29 03	05 06
03	28 11	28 59	04 41
04	28 09	28 56	03 59
05	28 08	28 53	03 03
06	28 06	28 50	01 54
07	28 06	28 47	00 S 37
08	28 D 06	28 43	00 N 44
09	28 06	28 40	02 03
10	28 07	28 37	03 13
11	28 08	28 34	04 10
12	28 08	28 31	04 51
13	28 08	28 28	05 13
14	28 R 08	28 24	05 16
15	28 08	28 21	05 02
16	28 08	28 18	04 31
17	28 08	28 15	03 47
18	28 08	28 12	02 52
19	28 D 08	28 08	01 50
20	28 08	28 05	00 N 44
21	28 R 08	28 02	00 S 24
22	28 07	27 59	01 30
23	28 07	27 56	02 31
24	28 07	27 53	03 25
25	28 06	27 49	04 10
26	28 05	27 46	04 45
27	28 04	27 43	05 07
28	28 04	27 40	05 17
29	28 D 03	27 37	05 12
30	28 04	27 34	04 52
31	28 ♏ 04	27 ♏ 30	04 S 18

DECLINATIONS

Date	Sun ☉ ° '	Moon ☽ ° '	Mercury ☿ ° '	Venus ♀ ° '	Mars ♂ ° '	Jupiter ♃ ° '	Saturn ♄ ° '	Uranus ♅ ° '	Neptune ♆ ° '	Pluto ♇ ° '
01	15 N 06	06 N 31	03 N 02	25 N 46	15 S 41	22 S 54	20 N 09	23 N 12	04 N 10	22 S 51
02	15 24	02 N 05	03 20	25 51	15 37	22 54	20 11	23 13	04 11	22 51
03	15 42	02 S 35	03 39	25 56	15 33	22 54	20 12	23 13	04 12	22 51
04	15 59	07 16	04 00	25 59	15 29	22 54	20 13	23 13	04 13	22 51
05	16 17	11 43	04 23	26 02	15 25	22 54	20 14	23 13	04 13	22 51
06	16 34	15 39	04 47	26 05	15 21	22 54	20 16	23 13	04 14	22 51
07	16 50	18 43	05 12	26 06	15 16	22 54	20 17	23 14	04 15	22 52
08	17 07	20 36	05 39	26 07	15 12	22 54	20 18	23 14	04 16	22 52
09	17 23	21 04	06 07	26 08	15 08	22 54	20 19	23 14	04 16	22 52
10	17 39	20 06	06 37	26 07	15 04	22 54	20 21	23 14	04 17	22 52
11	17 54	17 48	07 07	26 06	15 00	22 54	20 22	23 15	04 18	22 52
12	18 09	14 24	07 39	26 05	14 56	22 54	20 23	23 15	04 19	22 53
13	18 24	10 15	08 12	26 03	14 52	22 54	20 24	23 15	04 19	22 53
14	18 39	05 38	08 45	26 00	14 48	22 54	20 26	23 15	04 20	22 53
15	18 53	00 S 49	09 20	25 57	14 44	22 54	20 27	23 15	04 21	22 53
16	19 07	03 N 56	09 55	25 52	14 40	22 54	20 28	23 16	04 21	22 53
17	19 21	08 25	10 31	25 48	14 36	22 54	20 29	23 16	04 22	22 54
18	19 34	12 28	11 08	25 43	14 32	22 54	20 30	23 16	04 23	22 54
19	19 47	15 52	11 46	25 37	14 29	22 54	20 32	23 16	04 23	22 54
20	20 00	18 30	12 24	25 30	14 26	22 54	20 33	23 17	04 24	22 55
21	20 12	20 15	13 03	25 23	14 22	22 54	20 34	23 17	04 24	22 55
22	20 24	21 03	13 41	25 16	14 19	22 54	20 35	23 17	04 25	22 55
23	20 36	20 52	14 21	25 08	14 16	22 54	20 36	23 17	04 26	22 55
24	20 47	19 46	15 00	24 59	14 13	22 54	20 37	23 17	04 26	22 56
25	20 58	17 48	15 40	24 50	14 11	22 53	20 39	23 18	04 27	22 56
26	21 08	15 05	16 19	24 40	14 08	22 53	20 40	23 18	04 27	22 56
27	21 19	11 44	16 58	24 30	14 06	22 53	20 41	23 18	04 28	22 57
28	21 28	07 53	17 37	24 19	14 04	22 53	20 42	23 18	04 29	22 57
29	21 38	03 N 38	18 16	24 08	14 02	22 53	20 43	23 18	04 29	22 57
30	21 47	00 S 52	18 54	23 56	14 01	22 53	20 44	23 19	04 30	22 58
31	21 N 55	05 S 27	19 N 31	23 N 44	13 S 59	22 S 53	20 N 45	23 N 19	04 N 30	22 S 58

ZODIAC SIGN ENTRIES

Date	h m	Planets
01	11 55	☽ ♍
03	19 48	☽ ♎
06	00 10	☽ ♏
07	06 06	♀ ♋
08	02 02	☽ ♐
10	02 56	☽ ♑
12	04 23	☽ ♒
14	07 31	☽ ♓
14	14 00	☿ ♉
16	12 59	☽ ♈
18	20 55	☽ ♉
21	05 28	☉ ♊
21	07 08	☽ ♊
23	19 10	☽ ♋
26	08 00	☽ ♌
28	19 54	☽ ♍
31	05 02	☽ ♎
31	12 23	☿ ♊

LATITUDES

Date	Mercury ☿ ° '	Venus ♀ ° '	Mars ♂ ° '	Jupiter ♃ ° '	Saturn ♄ ° '	Uranus ♅ ° '	Neptune ♆ ° '	Pluto ♇ ° '
01	02 S 48	02 N 30	00 N 40	00 N 31	01 S 32	00 N 06	01 S 30	06 S 23
04	02 59	02 35	00 32	00 31	01 32	00 06	01 30	06 24
07	03 03	02 40	00 24	00 31	01 32	00 07	01 30	06 25
10	03 02	02 44	00 16	00 31	01 31	00 07	01 30	06 25
13	02 55	02 47	00 N 08	00 31	01 31	00 07	01 31	06 26
16	02 43	02 49	00 00	00 31	01 31	00 07	01 31	06 27
19	02 26	02 50	00 S 08	00 30	01 30	00 07	01 31	06 28
22	02 04	02 50	00 16	00 30	01 30	00 07	01 31	06 28
25	01 39	02 48	00 24	00 30	01 30	00 07	01 31	06 29
28	01 10	02 46	00 31	00 30	01 29	00 07	01 31	06 30
31	00 S 39	02 N 41	00 S 39	00 N 29	01 S 29	00 N 07	01 S 31	06 S 31

DATA

Julian Date	2462988
Delta T	+74 seconds
Ayanamsa	24° 17' 55"
Synetic vernal point	04° ♓ 49' 04"
True obliquity of ecliptic	23° 26' 02"

MOON'S PHASES, APSIDES AND POSITIONS ☽

Date	h m	Phase	Longitude ° '	Eclipse Indicator
07	03 40	○	16 ♏ 25	
13	19 07	☾	22 ♒ 50	
21	07 17	●	00 ♊ 04	Annular
29	11 20	☽	07 ♍ 55	

Day	h m		
09	07 43	Perigee	
25	02 15	Apogee	
02	22 48	OS	
09	07 52	Max dec	21° S 06'
15	16 06	ON	
22	19 35	Max dec	21° N 06'
30	07 29	OS	

LONGITUDES

Date	Chiron ⚷ ° '	Ceres ⚳ ° '	Pallas ⚴ ° '	Juno ⚵ ° '	Vesta ⚶ ° '	Black Moon Lilith ⚸ ° '
01	16 ♉ 13	11 ♊ 44	18 ♈ 54	24 ♓ 09	22 ♊ 50	08 ♋ 02
11	16 ♉ 53	15 ♊ 49	22 ♈ 43	28 ♓ 34	26 ♊ 52	09 ♋ 09
21	17 ♉ 33	19 ♊ 59	26 ♈ 32	02 ♈ 57	01 ♋ 00	10 ♋ 16
31	18 ♉ 11	24 ♊ 12	00 ♉ 21	07 ♈ 15	05 ♋ 12	11 ♋ 22

All ephemeris data is given at 12.00 UT and the Moon's longitude is additionally given for 24.00 UT

ASPECTARIAN

h m	Aspects
01 Thursday	
01 38	♃ ∠ ♇
07 41	☿ ☌ ♆
09 12	☽ △ ♃
10 10	☽ ⚼ ♆
10 21	☽ ⚼ ☿
16 52	☽ Q ♅
17 24	☽ Q ♂
23 46	☽ Q ♀
02 Friday	
00 51	☿ ⚻ ♂
00 52	☽ ∥ ♆
03 24	☽ ± ♆
03 48	☽ □ ♄
04 53	☽ ± ☿
05 51	☽ ∥ ☿
09 24	♂ ± ♅
10 25	☽ △ ☉
13 45	☽ ⚻ ♇
14 43	☽ ⚻ ♆
15 27	☽ ⚹ ♂
17 03	☽ ⚻ ☿
03 Saturday	
00 53	☽ ± ♇
01 43	☉ # ♂
02 47	☽ □ ♅
12 22	☽ □ ♀
15 03	♂ ⚻ ♆
16 17	☽ ⚼ ☉
17 03	☽ □ ♃
17 19	☽ ⚼ ♇
17 53	☽ # ☿
18 13	☽ ∠ ♂
18 32	♀ Q ♆
20 17	☽ # ♆
04 Sunday	
02 10	☉ ⚼ ♃
06 31	☉ □ ♇
09 39	☽ ± ☉
09 53	☽ ⊥ ♂
11 09	☽ △ ♄
12 04	☉ ☍ ♂
19 59	☽ △ ♇
20 09	☽ ⊻ ♂
21 00	☽ ⚻ ☉
21 00	☽ ☍ ♆
21 01	☉ ⊻ ♆
05 Monday	
00 49	☽ Q ♃
02 07	☿ ∥ ♆
02 27	♂ □ ♇
03 24	☽ ☍ ☿
08 23	☽ △ ♅
10 50	☉ ⊥ ♅
13 29	☽ ⚼ ♄
17 58	♀ ☍ ♃
18 38	♀ ⚼ ♂
21 10	♂ ∠ ♃
21 21	☽ ⚹ ♃
21 39	☽ △ ♀
06 Tuesday	
00 08	♀ ⚼ ♇
04 56	☽ ± ♄
09 58	☽ ∥ ♂
09 59	☽ ⚼ ♅
15 06	☽ ⚻ ♄
18 57	☽ # ☉
21 56	☽ ☌ ♂
22 27	☽ ∠ ♃
23 03	☽ □ ♇
07 Wednesday	
00 07	☽ ⚻ ♆
00 55	☽ ⚼ ♀
01 04	☽ ± ♅
03 40	☽ ☍ ☉
10 05	☽ ± ♆
10 24	☽ ⚻ ☿
11 02	☽ ⚻ ♅
13 17	☽ ⊥ ♃
17 00	☽ ± ♀
18 23	☿ ⚹ ♅
21 09	☽ ± ☿
23 05	☽ ⊻ ♃
08 Thursday	
00 54	☽ ⚼ ♆
03 38	☽ ⚻ ♀
04 43	☽ Q ♇
06 46	☽ # ♄
13 12	☽ ⚼ ☿
16 56	☽ ☍ ♄
21 59	☽ ⊻ ♂
09 Friday	
00 17	☽ ⚹ ♇
01 26	☽ △ ♆
07 31	☽ ⊥ ♂
08 15	☽ ⚻ ☉
12 14	☽ ☍ ♅
15 55	☽ △ ☿
18 43	☽ ± ☉
21 49	☽ ∠ ♂
23 47	☽ ☌ ♃
10 Saturday	
00 43	☽ ∠ ♇
08 19	☽ # ♄
08 26	☽ ☍ ♀
10 26	☽ ⚼ ☉
12 32	☿ ∠ ♄
15 26	☽ ⊥ ♇
18 16	☽ ⚻ ♄
21 45	☽ ⚹ ♂
11 Sunday	
01 17	☽ ⊻ ♇
02 33	☽ □ ♆
04 13	☽ ± ♄
07 06	☉ ⊥ ♆
11 10	☽ # ☉
11 54	☿ Q ♇
12 50	☽ △ ☉
13 33	☽ ⚻ ♅
17 04	☽ Q ♂
19 14	☽ ⚼ ♄
22 05	☽ □ ☿
23 33	☽ ± ♅
12 Monday	
00 09	☉ ⊻ ♅
00 54	☽ ⊻ ♃
08 30	☽ Q ♆
08 38	☽ ∥ ♂
10 53	☽ ⊥ ♃
14 04	☽ ⚻ ♀
14 41	☽ ⚼ ♅
18 37	☉ ± ♃
20 36	☽ △ ♄
22 30	☽ □ ♂
13 Tuesday	
01 02	☽ ± ♀
01 46	☿ △ ♃
01 59	☽ ∠ ♃
03 23	☽ ☌ ♇
04 48	☽ ⚹ ♆
07 42	☽ Q ☿
16 18	☽ △ ♅
17 41	☽ ⚼ ♀
19 07	☽ □ ☉
21 43	☽ # ☿
14 Wednesday	
03 35	☽ ⚹ ♃
06 41	☽ ∠ ♆
06 45	☽ ⚹ ☿
08 11	♀ ± ♇
18 30	☽ # ♆
20 26	♇ St R
22 00	☽ △ ♀
22 29	☽ ⊥ ♆
15 Thursday	
00 29	☽ Q ♃
00 56	☽ □ ♄
01 07	☽ △ ♂
04 58	☽ Q ☉
06 38	♂ ⚻ ♄
07 33	☽ ⊻ ♇
09 09	☽ ⊻ ♆
12 25	☽ ∠ ☿
18 16	☽ ⊥ ♇
21 15	☽ □ ♅
16 Friday	
03 17	☽ ⚼ ♂
04 08	☽ ⚹ ☉
05 01	♀ △ ♂
06 42	☽ ⊥ ☿
08 31	☽ □ ♃
09 31	☽ Q ♄
10 32	☽ ∠ ♇
14 11	☽ ∥ ♆
16 18	♀ ⊻ ♄
19 05	☽ ⊻ ☿
19 16	☽ ± ♂
17 Saturday	
06 02	☽ ⚻ ♂
06 14	☽ Q ♅
07 45	☽ ⚹ ♄
08 38	☿ ⊥ ♄
08 57	☽ □ ♀
09 49	☽ ∠ ☉
14 08	☽ ⚹ ♇
15 56	☽ ☌ ♆
18 Sunday	
02 16	☽ ∥ ☿
03 59	☽ ⊥ ☉
04 43	☽ ⚹ ♅
05 23	♃ ∥ ♇
11 29	☉ ⚻ ♃
12 07	☽ ∠ ♄
12 39	☽ Q ♇
15 51	☽ △ ♃
16 15	☽ ⊻ ☉
18 39	☿ ∠ ♅
21 49	☽ Q ♀
19 Monday	
01 48	☽ # ♂
05 26	☽ ⊥ ♄
09 22	☽ ∠ ♅
11 31	☽ ☌ ☿
13 17	☽ ☍ ♂
17 05	☽ ⊻ ♄
20 23	☽ ⚼ ♃
22 59	☽ ⚹ ♀
23 03	☿ ☍ ♂
23 08	☽ □ ♇
20 Tuesday	
00 49	♀ ⚻ ♇
01 10	☽ ⊻ ♆
02 50	☽ ⊥ ♅
12 53	☽ ⊥ ♆
13 45	☽ ± ♃
14 35	☽ ⊻ ♅
22 22	☉ ∠ ♆
21 Wednesday	
01 26	☽ ⚻ ♃
01 42	♀ □ ♆
06 36	☽ ∠ ♆
07 03	☽ ∠ ♀
07 17	☽ ☌ ☉
07 51	☿ ⊻ ♄
10 50	☽ ∥ ☉
18 39	☽ ∥ ♄
22 39	☽ ⚻ ♂
22 Thursday	
01 36	☿ ⚼ ♃
02 34	☽ ⊥ ♀
04 39	☽ ☌ ♄
07 54	☽ ⊻ ☿
10 15	☽ △ ♇
10 21	☽ ± ♂
12 30	☽ ⚹ ♆
15 41	☽ ⊻ ♀
21 59	☽ ⊥ ☿
23 58	☿ □ ♇
23 Friday	
00 02	☉ ∠ ♀
02 28	☽ ☌ ♅
04 00	☽ ⚼ ♂
06 17	♀ ⊥ ♄
09 24	☿ # ♂
12 41	☽ Q ♆
12 48	☽ ☍ ♃
13 33	☉ ∥ ♄
15 31	☿ ⊻ ♆
16 25	☽ ⚼ ♇
18 51	☽ ∥ ☉
19 20	☽ ∠ ☿
19 47	☽ ∥ ♄
24 Saturday	
00 37	☽ ⊻ ☉
04 40	☿ ⊥ ♅
09 38	☽ △ ♂
10 38	☽ ± ♇
13 51	☽ ⊥ ☉
17 48	☽ ⊻ ♄
22 48	☽ ⚻ ♇
25 Sunday	
01 14	☽ □ ♆
06 07	☽ ⊥ ♄
07 18	☽ ⚹ ☿
09 44	☽ ∠ ☉
10 02	☽ ☌ ♀
12 32	♂ ⚼ ♅
15 31	☽ ⊻ ♅
26 Monday	
00 34	☽ ∠ ♄
01 05	☽ ⚻ ♃
03 44	☽ ⊥ ♅
03 46	☽ ∥ ☿
12 12	☽ Q ☿
13 06	☽ ± ♃
17 16	☿ ⚹ ♀
17 37	☿ ± ♃
18 48	☽ ⚹ ☉
19 19	☽ # ♂
21 11	☽ □ ♂
22 00	☽ ∠ ♅
22 21	☿ ⊥ ♆
27 Tuesday	
07 04	☽ ⚼ ♃
07 10	☽ ⚹ ♄
11 25	☽ ☍ ♇
11 57	☿ ⊻ ♅
13 59	☽ △ ♆
19 09	☉ ⚻ ♂
20 57	☽ Q ☉
28 Wednesday	
03 59	☽ ⊻ ♀
04 09	☽ ⚹ ♅
06 12	♀ ⊻ ♅
07 10	☽ □ ☿
07 21	☽ Q ♄
08 29	☽ Q ♂
12 39	☽ △ ♃
16 52	☽ ⊥ ♀
19 46	☽ ⚼ ♆
29 Thursday	
03 51	☽ Q ♅
07 21	☽ ∥ ♆
07 35	☽ ⚹ ♂
11 20	☽ □ ☉
11 56	☽ ∠ ♀
13 18	☽ ± ♆
15 16	☿ ⚻ ♃
18 49	☽ □ ♄
22 14	☽ ⚻ ♇
30 Friday	
00 43	♂ ± ♄
00 50	☽ ⚻ ♆
09 37	☽ ± ♇
11 49	☽ ∠ ♂
14 26	☽ □ ♅
18 50	☽ ⚹ ♀
21 41	☽ □ ♃
31 Saturday	
02 26	☽ ⚼ ♇
03 39	☽ △ ☿
04 21	☽ ⊥ ♂
07 03	☽ # ♆
12 22	☿ ∠ ♆
15 12	☽ ⊻ ♂
18 40	☽ Q ♀
23 51	☽ △ ☉

JUNE 2031

LONGITUDES

Date	Sidereal time h m s	Sun ☉ ° ' "	Moon ☽ ° ' "	Moon ☽ 24.00 ° ' "	Mercury ☿ ° '	Venus ♀ ° '	Mars ♂ ° '	Jupiter ♃ ° '	Saturn ♄ ° '	Uranus ♅ ° '	Neptune ♆ ° '	Pluto ♇ ° '
01	04 39 07	10 ♊ 49 37	17 ♎ 09 06	23 ♎ 59 26	02 ♊ 03	26 ♋ 10	05 ♏ 25	25 ♐ 51	12 ♊ 10	22 ♊ 16	15 ♈ 01	13 ♒ 35
02	04 43 03	11 47 06	00 ♏ 56 24	07 ♏ 59 53	04 09	27 08	05 R 15	25 R 43	12 18	22 19	15 03	13 R 34
03	04 47 00	12 44 34	15 ♏ 09 38	22 ♏ 25 13	06 17	28 05	05 07	25 36	12 26	22 23	15 04	13 34
04	04 50 57	13 42 01	29 ♏ 46 04	07 ♐ 11 24	08 26	29 02	04 59	25 29	12 34	22 26	15 06	13 33
05	04 54 53	14 39 26	14 ♐ 40 20	22 ♐ 11 50	10 36	29 ♋ 58	04 52	25 22	12 41	22 30	15 07	13 33
06	04 58 50	15 36 51	29 ♐ 44 47	07 ♑ 18 03	12 47	00 ♌ 53	04 46	25 14	12 49	22 33	15 08	13 32
07	05 02 46	16 34 16	14 ♑ 50 26	22 ♑ 20 51	14 58	01 49	04 41	25 07	12 57	22 37	15 10	13 32
08	05 06 43	17 31 39	29 ♑ 48 14	07 ♒ 11 43	17 10	02 43	04 36	24 59	13 05	22 40	15 11	13 31
09	05 10 39	18 29 02	14 ♒ 30 30	21 ♒ 44 00	19 23	03 37	04 33	24 52	13 13	22 44	15 12	13 30
10	05 14 36	19 26 24	28 ♒ 51 46	05 ♓ 53 32	21 34	04 31	04 30	24 44	13 20	22 47	15 13	13 30
11	05 18 32	20 23 45	12 ♓ 49 12	19 ♓ 38 45	23 46	05 24	04 28	24 36	13 28	22 51	15 15	13 29
12	05 22 29	21 21 06	26 ♓ 22 22	03 ♈ 00 14	25 57	06 16	04 27	24 29	13 36	22 55	15 16	13 28
13	05 26 26	22 18 27	09 ♈ 32 41	16 ♈ 00 04	28 ♊ 07	07 08	04 D 26	24 21	13 44	22 58	15 17	13 28
14	05 30 22	23 15 47	22 ♈ 22 45	28 ♈ 41 10	00 ♋ 15	07 59	04 27	24 14	13 51	23 02	15 18	13 27
15	05 34 19	24 13 07	04 ♉ 55 43	11 ♉ 06 49	02 22	08 49	04 28	24 06	13 59	23 05	15 19	13 26
16	05 38 15	25 10 26	17 ♉ 14 51	23 ♉ 20 12	04 28	09 39	04 30	23 58	14 07	23 09	15 20	13 25
17	05 42 12	26 07 45	29 ♉ 23 13	05 ♊ 24 15	06 32	10 28	04 33	23 51	14 15	23 12	15 21	13 25
18	05 46 08	27 05 04	11 ♊ 23 36	17 ♊ 21 33	08 34	11 16	04 36	23 43	14 22	23 16	15 22	13 24
19	05 50 05	28 02 22	23 ♊ 18 22	29 ♊ 14 19	10 35	12 04	04 41	23 35	14 30	23 20	15 23	13 23
20	05 54 01	28 59 39	05 ♋ 09 39	11 ♋ 04 35	12 33	12 50	04 46	23 28	14 38	23 23	15 24	13 22
21	05 57 58	29 ♊ 56 56	16 ♋ 59 24	22 ♋ 54 19	14 29	13 36	04 52	23 20	14 45	23 27	15 25	13 21
22	06 01 55	00 ♋ 54 13	28 ♋ 49 36	04 ♌ 45 32	16 23	14 21	04 58	23 12	14 53	23 30	15 26	13 20
23	06 05 51	01 51 29	10 ♌ 42 26	16 ♌ 40 36	18 15	15 05	05 05	23 05	15 00	23 34	15 27	13 19
24	06 09 48	02 48 44	22 ♌ 40 23	28 ♌ 42 11	20 04	15 48	05 14	22 57	15 08	23 37	15 28	13 18
25	06 13 44	03 45 59	04 ♍ 46 24	10 ♍ 53 27	21 51	16 30	05 22	22 50	15 16	23 41	15 28	13 17
26	06 17 41	04 43 13	17 ♍ 03 50	23 ♍ 18 00	23 36	17 12	05 32	22 43	15 23	23 44	15 29	13 16
27	06 21 37	05 40 26	29 ♍ 36 26	05 ♎ 59 38	25 19	17 52	05 42	22 35	15 31	23 48	15 30	13 15
28	06 25 34	06 37 39	12 ♎ 28 05	19 ♎ 02 14	27 00	18 30	05 53	22 28	15 38	23 51	15 31	13 14
29	06 29 30	07 34 52	25 ♎ 42 27	02 ♏ 29 07	28 ♋ 38	19 08	06 04	22 21	15 46	23 55	15 31	13 13
30	06 33 27	08 ♋ 32 04	09 ♏ 22 27	16 ♏ 22 34	00 ♌ 14	19 ♌ 45	06 ♏ 17	22 ♐ 14	15 ♊ 53	23 ♊ 58	15 ♈ 32	13 ♒ 12

Date	Moon True ☊ ° '	Moon Mean ☊ ° '	Moon ☽ Latitude ° '
01	28 ♏ 05	27 ♏ 27	03 S 28
02	28 D 06	27 24	02 25
03	28 07	27 21	01 S 11
04	28 R 07	27 18	00 N 09
05	28 07	27 14	01 30
06	28 06	27 11	02 46
07	28 04	27 08	03 50
08	28 01	27 05	04 38
09	27 59	27 02	05 07
10	27 57	26 59	05 15
11	27 56	26 55	05 05
12	27 D 56	26 52	04 37
13	27 57	26 49	03 56
14	27 58	26 46	03 04
15	28 00	26 43	02 04
16	28 01	26 40	00 N 59
17	28 R 01	26 36	00 S 07
18	28 01	26 33	01 13
19	27 58	26 30	02 14
20	27 55	26 27	03 09
21	27 50	26 24	03 56
22	27 44	26 20	04 33
23	27 38	26 17	04 58
24	27 32	26 14	05 10
25	27 28	26 11	05 09
26	27 25	26 08	04 53
27	27 23	26 05	04 24
28	27 D 23	26 01	03 40
29	27 24	25 58	02 44
30	27 ♏ 26	25 ♏ 55	01 S 37

DECLINATIONS

Date	Sun ☉ ° '	Moon ☽ ° '	Mercury ☿ ° '	Venus ♀ ° '	Mars ♂ ° '	Jupiter ♃ ° '	Saturn ♄ ° '	Uranus ♅ ° '	Neptune ♆ ° '	Pluto ♇ ° '
01	22 N 04	09 S 56	20 N 07	23 N 32	13 S 58	22 S 53	20 N 46	23 N 19	04 N 31	22 S 58
02	22 12	14 04	20 41	23 19	13 57	22 53	20 47	23 19	04 31	22 59
03	22 19	17 31	21 15	23 05	13 56	22 52	20 49	23 20	04 32	22 59
04	22 26	19 57	21 46	22 51	13 56	22 52	20 50	23 20	04 32	22 59
05	22 33	21 03	22 16	22 37	13 56	22 52	20 51	23 20	04 33	23 00
06	22 39	20 40	22 44	22 22	13 56	22 52	20 52	23 20	04 33	23 00
07	22 45	18 48	23 10	22 07	13 56	22 52	20 53	23 20	04 34	23 01
08	22 51	15 39	23 34	21 52	13 56	22 52	20 54	23 21	04 34	23 01
09	22 56	11 35	23 55	21 36	13 57	22 52	20 55	23 21	04 35	23 01
10	23 01	06 56	24 13	21 20	13 58	22 51	20 56	23 21	04 35	23 02
11	23 05	02 S 03	24 29	21 04	13 59	22 51	20 57	23 21	04 35	23 02
12	23 09	02 N 48	24 41	20 47	14 00	22 51	20 58	23 21	04 36	23 03
13	23 13	07 24	24 52	20 30	14 02	22 51	20 59	23 22	04 36	23 03
14	23 16	11 33	24 59	20 13	14 04	22 51	21 00	23 22	04 37	23 04
15	23 18	15 06	25 03	19 56	14 06	22 50	21 01	23 22	04 37	23 04
16	23 21	17 55	25 05	19 38	14 08	22 50	21 02	23 22	04 37	23 04
17	23 23	19 54	25 04	19 20	14 11	22 50	21 03	23 22	04 38	23 05
18	23 24	20 56	25 00	19 02	14 13	22 50	21 03	23 23	04 38	23 05
19	23 25	21 02	24 54	18 44	14 16	22 49	21 04	23 23	04 38	23 06
20	23 26	20 11	24 46	18 25	14 20	22 49	21 05	23 23	04 39	23 06
21	23 26	18 27	24 35	18 07	14 23	22 49	21 06	23 23	04 39	23 07
22	23 26	15 56	24 23	17 48	14 27	22 49	21 07	23 23	04 39	23 07
23	23 25	12 46	24 08	17 29	14 31	22 49	21 08	23 23	04 40	23 08
24	23 24	09 04	23 52	17 10	14 35	22 48	21 09	23 24	04 40	23 08
25	23 23	04 58	23 33	16 52	14 39	22 48	21 10	23 24	04 40	23 09
26	23 21	00 N 36	23 14	16 33	14 43	22 48	21 10	23 24	04 40	23 09
27	23 19	03 S 52	22 52	16 14	14 48	22 48	21 11	23 24	04 40	23 09
28	23 16	08 18	22 30	15 54	14 53	22 47	21 12	23 24	04 41	23 10
29	23 13	12 28	22 06	15 35	14 58	22 47	21 13	23 25	04 41	23 10
30	23 N 10	16 S 08	21 N 41	15 N 16	15 S 03	22 S 47	21 N 14	23 N 25	04 N 41	23 S 11

ZODIAC SIGN ENTRIES

Date	h m	Planets
02	10 23	☽ ♏
04	12 23	☽ ♐
05	12 57	♀ ♌
06	12 24	☽ ♑
08	12 19	☽ ♒
10	13 56	☽ ♓
12	18 33	☽ ♈
14	09 09	☿ ♋
15	02 31	☽ ♉
17	13 13	☽ ♊
20	01 33	☽ ♋
21	13 17	☉ ♋
22	14 23	☽ ♌
25	02 34	☽ ♍
27	12 45	☽ ♎
29	19 37	☽ ♏
30	08 25	☿ ♌

LATITUDES

Date	Mercury ☿ ° '	Venus ♀ ° '	Mars ♂ ° '	Jupiter ♃ ° '	Saturn ♄ ° '	Uranus ♅ ° '	Neptune ♆ ° '	Pluto ♇ ° '
01	00 S 28	02 N 40	00 S 41	00 N 29	01 S 29	00 N 07	01 S 31	06 S 31
04	00 N 04	02 34	00 48	00 29	01 29	00 07	01 31	06 31
07	00 35	02 26	00 54	00 29	01 29	00 07	01 32	06 32
10	01 03	02 17	01 00	00 28	01 29	00 07	01 32	06 33
13	01 26	02 05	01 06	00 28	01 28	00 07	01 32	06 34
16	01 43	01 52	01 11	00 28	01 28	00 07	01 32	06 34
19	01 54	01 37	01 16	00 27	01 28	00 07	01 32	06 35
22	01 58	01 20	01 21	00 27	01 28	00 07	01 32	06 35
25	01 55	01 01	01 25	00 26	01 28	00 07	01 33	06 36
28	01 46	00 40	01 29	00 26	01 28	00 07	01 33	06 37
31	01 N 32	00 N 16	01 S 33	00 N 25	01 S 28	00 N 07	01 S 33	06 S 37

DATA

Julian Date	2463019
Delta T	+74 seconds
Ayanamsa	24° 17' 59"
Synetic vernal point	04° ♓ 49' 00"
True obliquity of ecliptic	23° 26' 01"

MOON'S PHASES, APSIDES AND POSITIONS ☽

Date	h m	Phase	Longitude ° '	Eclipse Indicator
05	11 58	○	14 ♐ 39	
12	02 21	☾	20 ♓ 58	
19	22 25	●	28 ♊ 27	
28	00 19	☽	06 ♎ 10	

Day	h m		
06	12 18	Perigee	
21	11 37	Apogee	
05	17 51	Max dec	21° S 06'
11	22 03	0N	
19	02 15	Max dec	21° N 07'
26	15 16	0S	

LONGITUDES

Date	Chiron ⚷ ° '	Ceres ⚳ ° '	Pallas ⚴ ° '	Juno ⚵ ° '	Vesta ⚶ ° '	Black Moon Lilith ⚸ ° '
01	18 ♉ 15	24 ♊ 37	00 ♉ 43	07 ♈ 41	05 ♋ 37	11 ♋ 29
11	18 ♉ 51	28 ♊ 54	04 ♉ 31	11 ♈ 54	09 ♋ 54	12 ♋ 36
21	19 ♉ 25	03 ♋ 12	08 ♉ 16	15 ♈ 59	14 ♋ 14	13 ♋ 43
31	19 ♉ 56	07 ♋ 31	11 ♉ 57	19 ♈ 56	18 ♋ 36	14 ♋ 50

All ephemeris data is given at 12.00 UT and the Moon's longitude is additionally given for 24.00 UT

ASPECTARIAN

h m	Aspects
01 Sunday	
03 03	☽ △ ♄
04 49	♀ ⚻ ♃
05 40	☽ △ ♇
06 10	☽ Q ♃
08 13	☽ ☍ ♆
11 46	☽ ⚼ ☿
21 01	☽ △ ♅
02 Monday	
00 32	☉ ± ♂
03 05	☽ ✱ ♃
04 19	☽ ⚼ ☉
04 57	☽ □ ♀
05 41	☽ ⚼ ♄
06 20	☽ ± ☿
10 34	♀ ∥ ♅
11 19	☽ ∥ ♂
16 31	☿ ∥ ♄
17 03	♀ ∠ ♄
18 27	☽ ⚻ ☿
19 17	☽ ☌ ♂
20 51	☽ ± ☉
21 13	☽ ± ♄
22 54	☽ ⚼ ♅
23 43	☿ ⚻ ♂
03 Tuesday	
02 59	☉ ☌ ♄
04 27	☽ ∠ ♃
07 24	☽ ⚻ ♄
07 41	☽ ⚻ ☉
09 21	☽ □ ♇
11 51	☽ ⚻ ♆
14 02	☽ ± ♅
19 18	☽ ⊥ ♃
20 00	♀ ⊥ ♅
21 48	☽ ± ♆
22 08	♀ ⋕ ♇
23 59	☽ ⚻ ♅
04 Wednesday	
05 04	☽ ⚺ ♃
08 27	☉ △ ♇
10 03	♀ ⋕ ♃
10 43	☽ △ ♀
12 32	☽ ⚼ ♆
14 54	☽ Q ♇
20 23	☽ ⚺ ♂
05 Thursday	
03 41	☽ ⋕ ♄
04 22	☽ ☍ ☿
05 57	☽ ⊥ ♂
08 48	☽ ☍ ♄
10 12	☽ ✱ ♇
11 58	☽ ☍ ☉
12 30	☽ ⚼ ♀
12 43	☽ △ ♆
14 53	☿ ± ♂
16 24	☉ ∥ ♀
20 14	☽ ∠ ♂
23 24	☿ ∥ ♀
23 48	☉ ✱ ♆
06 Friday	
00 31	☽ ☍ ♅
03 47	☽ ± ♀
04 53	☽ ☌ ♃
06 16	☉ ∥ ☿
07 26	☽ ⋕ ♄
10 05	☽ ∠ ♇
12 30	☿ ☌ ♄
13 56	☽ ⚻ ♀
18 47	☿ ⋕ ♃
19 53	♀ ± ♃
19 56	☽ ✱ ♂
20 17	☿ △ ♇
07 Saturday	
00 22	☽ ⊥ ♇
02 39	☿ ⋕ ♇
08 58	☽ ⚻ ♄
09 55	☽ ⚺ ♇
12 15	☽ ⚻ ☿
12 31	☽ □ ♆
14 05	☿ ✱ ♆
14 55	☽ Q ♂
14 57	☽ ⚻ ☉
18 37	☽ ± ♄
22 08	☿ ∥ ♅
23 29	☽ ± ☿
08 Sunday	
00 29	☽ ⚻ ♅
01 12	☽ ± ☉
04 18	☽ ⚺ ♃
09 12	☽ ⚼ ♄
10 10	☽ ± ♅
13 54	☽ ⊥ ♃
15 21	☉ ⋕ ♃
16 30	☽ ⚼ ☿
16 43	☽ ⚼ ☉
17 02	☽ ☍ ♀
17 29	☽ Q ♆
18 50	☉ ☌ ☿
19 45	☽ □ ♂
22 10	☿ ∠ ♀
22 39	☽ ∥ ♂
09 Monday	
00 50	☽ ⚼ ♅
04 25	☽ ∠ ♃
09 51	☽ △ ♄
10 21	☽ ☌ ♇
13 09	☽ ✱ ♆
13 49	☿ ⚼ ♂
19 03	☽ △ ☉
21 31	☽ △ ☿
10 Tuesday	
01 44	☽ △ ♅
05 06	☽ ✱ ♃
11 32	♀ □ ♂
13 25	☉ ⚼ ♂
14 19	☽ ∠ ♆
17 59	☉ ⋕ ♇
21 35	☽ △ ♂
22 17	☽ ⚻ ♀
23 36	☽ ⋕ ♆
11 Wednesday	
01 20	☽ Q ♃
01 41	☿ ☌ ♅
05 47	☽ ⊥ ♆
09 21	☽ ± ♀
13 00	☉ ∠ ♀
13 09	☽ □ ♄
13 10	☽ ⚺ ♇
14 37	♄ △ ♇
16 15	☽ ⚺ ♆
20 45	☿ ☍ ♃
21 39	♀ ∥ ♄
23 40	☽ ⚼ ♂
23 42	☽ ⊥ ♇
12 Thursday	
02 15	☽ ⚼ ♀
02 21	☽ □ ☉
05 46	☽ □ ♅
08 38	☽ □ ♃
11 05	☽ □ ☿
15 44	☽ ± ♂
15 47	☽ ∠ ♇
21 11	☽ ∥ ♆
21 32	☽ Q ♄
13 Friday	
02 37	☽ ⚻ ♂
02 44	☿ Q ♆
07 15	☽ △ ♀
11 57	♂ St D
13 31	☽ Q ☉
14 39	☽ Q ♅
15 54	☿ ⚼ ♇
19 15	☽ ✱ ♇
19 50	☽ ✱ ♄
22 40	☽ ☌ ♆
14 Saturday	
02 40	☽ Q ☿
05 43	☉ ☌ ♅
13 14	☽ ✱ ♅
13 21	♀ ∠ ♅
13 49	☽ ✱ ☉
15 28	☽ △ ♃
17 49	☽ Q ♇
15 Sunday	
00 27	☽ ∠ ♄
04 39	☽ ⋕ ♂
06 04	☽ ✱ ☿
09 20	☉ ☍ ♃
11 06	☽ ☍ ♂
17 59	☽ ⊥ ♄
18 09	☽ ∠ ♅
18 50	♀ ⚼ ♃
20 00	☽ ⚼ ♃
20 06	☽ □ ♀
20 29	☿ ⊥ ♀
21 01	☽ ∠ ☉
16 Monday	
04 31	☽ □ ♇
05 48	☽ ⚺ ♄
08 15	☽ ⚺ ♆
11 48	☽ ⊥ ♅
12 20	☿ △ ♂
13 24	☽ ± ♃
16 07	☽ ⊥ ☉
17 16	☽ ∠ ☿
20 04	☽ ⊥ ♆
23 41	☽ ⚺ ♅
17 Tuesday	
01 07	☽ ⚻ ♃
04 59	☽ ⚺ ☉
05 09	☽ ∥ ♀
07 53	☉ ∥ ♅
10 03	☽ Q ♀
13 56	☽ ∠ ♆
14 45	☽ ⊥ ☿
22 10	☿ ± ♇
22 20	☽ ⚻ ♂
18 Wednesday	
05 11	☽ ⚺ ☿
10 24	☽ ± ♂
11 45	☽ ✱ ♀
16 01	☽ △ ♇
18 03	☽ ☌ ♄
18 38	☽ ∥ ♄
19 20	☉ Q ♆
20 00	☽ ✱ ♆
19 Thursday	
04 37	☽ ⚼ ♂
09 07	☽ ∥ ♄
12 02	☽ ☌ ♅
12 34	☽ ☍ ♃
20 08	☽ ∠ ♀
20 16	☽ Q ♆
20 28	☉ ⚼ ♇
22 15	☽ ⚼ ♇
22 25	☽ ☌ ☉
20 Friday	
11 11	☽ △ ♂
15 39	☽ ⊥ ♀
16 28	☽ ± ♇
18 01	☿ ⚺ ♀
21 47	♃ ☍ ♅
22 03	☿ ⚻ ♇
21 Saturday	
04 08	♀ ☍ ♇
04 38	☽ ⚻ ♇
04 40	☽ ⚺ ♀
05 55	☽ ☌ ☿
07 25	☽ ⚺ ♄
08 48	☽ □ ♆
15 41	☿ ⚺ ♄
16 14	☽ ∥ ♀
19 43	☽ ⊥ ♄
23 53	☿ □ ♆
22 Sunday	
00 44	☽ ⚻ ♃
01 10	☽ ⚺ ♅
12 46	☽ ± ♃
13 23	☽ ⊥ ♅
14 10	☽ ∠ ♄
16 34	☽ ⚺ ☉
23 36	☽ ⋕ ♂
23 Monday	
00 33	☽ □ ♂
05 45	☽ ⊥ ☉
06 46	☽ ⚼ ♃
07 39	☽ ∠ ♅
08 50	♀ ✱ ♄
17 15	☽ ☍ ♇
20 45	☽ ✱ ♄
21 23	☽ ☌ ♀
21 33	☽ △ ♆
24 Tuesday	
00 10	♀ △ ♆
01 26	☽ ∠ ☉
05 52	☽ ⚺ ☿
12 34	☽ △ ♃
13 07	☽ Q ♂
13 54	☽ ✱ ♅
19 57	☽ ⊥ ☿
20 59	☽ Q ♄
21 00	☉ ∥ ♅
25 Wednesday	
03 20	☿ ⊥ ♄
03 30	☽ ⚼ ♆
09 51	☽ ✱ ☉
13 12	☽ ✱ ♂
13 41	☽ ∥ ♆
13 48	☽ Q ♅
16 47	☽ ∠ ☿
21 14	☽ ± ♆
23 38	☿ ∥ ♅
26 Thursday	
00 27	☿ ⚻ ♃
02 15	☉ ∥ ☿
04 39	☽ ⚻ ♇
08 43	☽ □ ♄
08 56	☽ ⚻ ♆
11 17	☽ Q ☉
12 16	☽ ⚺ ♀
13 52	☿ ⚺ ♅
16 15	☽ ± ♇
17 07	☿ ⋕ ♇
18 47	☽ ∠ ♂
22 46	☽ □ ♃
27 Friday	
00 28	☽ ⊥ ♀
00 54	☽ □ ♅
02 35	☽ ✱ ☿
09 11	♄ ✱ ♆
09 27	☽ ⚼ ♇
12 11	☽ ⊥ ♂
12 47	☉ △ ♂
16 18	☽ ⋕ ♆
17 11	☿ ⋕ ♃
18 28	☽ ∠ ♀
23 37	☽ ⚺ ♂
28 Saturday	
00 19	☽ □ ☉
04 39	☽ Q ☿
08 21	☽ Q ♃
13 25	☽ △ ♇
17 35	☽ ☍ ♆
17 52	☽ △ ♄
23 36	☽ ✱ ♀
29 Sunday	
03 05	☉ ± ♇
06 02	☽ ✱ ♃
08 02	☿ ± ♃
08 47	☽ △ ♅
17 55	☽ □ ☿
21 03	☽ ⚼ ♄
22 05	☽ Q ♀
30 Monday	
03 48	☉ ⋕ ♇
04 14	☽ ∥ ♂
06 26	☽ ⋕ ♀
06 33	☽ ☌ ♂
07 50	☿ ⊥ ♅
08 19	☽ ∠ ♃
10 26	☽ △ ☉
11 18	☽ ⚼ ♅
12 53	☽ ± ♄
18 35	☽ □ ♇
22 34	☽ ⚻ ♆
22 43	☿ ∠ ♄
23 16	☽ ⚻ ♄
23 39	☽ ⊥ ♃

JULY 2031

LONGITUDES

Date	Sidereal time h m s	Sun ☉ ° ' "	Moon ☽ ° ' "	Moon ☽ 24.00 ° ' "	Mercury ☿ ° '	Venus ♀ ° '	Mars ♂ ° '	Jupiter ♃ ° '	Saturn ♄ ° '	Uranus ♅ ° '	Neptune ♆ ° '	Pluto ♇ ° '
01	06 37 24	09 ♋ 29 16	23 ♏ 29 28	00 ♐ 42 58	01 ♌ 48	20 ♌ 20	06 ♏ 29	22 ♐ 07	16 ♊ 00	24 ♊ 02	15 ♈ 32	13 ♒ 11
02	06 41 20	10 26 27	08 ♐ 02 41	15 ♐ 28 03	03 19	20 54	06 43	22 R 00	16 08	24 05	15 33	13 R 10
03	06 45 17	11 23 38	22 ♐ 58 18	00 ♑ 32 27	04 48	21 27	06 57	21 53	16 15	24 09	15 34	13 09
04	06 49 13	12 20 49	08 ♑ 09 23	15 ♑ 47 50	06 15	21 58	07 12	21 46	16 22	24 12	15 34	13 08
05	06 53 10	13 18 00	23 ♑ 26 27	01 ♒ 03 51	07 40	22 28	07 27	21 39	16 30	24 16	15 34	13 07
06	06 57 06	14 15 11	08 ♒ 38 42	16 ♒ 09 48	09 02	22 56	07 43	21 33	16 37	24 19	15 35	13 06
07	07 01 03	15 12 22	23 ♒ 36 01	00 ♓ 56 29	10 21	23 23	07 59	21 26	16 44	24 22	15 35	13 04
08	07 04 59	16 09 33	08 ♓ 10 30	15 ♓ 17 35	11 38	23 48	08 16	21 20	16 51	24 26	15 36	13 03
09	07 08 56	17 06 44	22 ♓ 17 29	29 ♓ 00 17	12 53	24 12	08 33	21 14	16 58	24 29	15 36	13 02
10	07 12 53	18 03 56	05 ♈ 55 36	12 ♈ 34 12	14 05	24 34	08 52	21 08	17 06	24 32	15 36	13 01
11	07 16 49	19 01 08	19 ♈ 06 15	25 ♈ 32 15	15 14	24 54	09 10	21 02	17 13	24 36	15 37	13 00
12	07 20 46	19 58 21	01 ♉ 52 41	08 ♉ 08 08	16 21	25 12	09 29	20 56	17 20	24 39	15 37	12 58
13	07 24 42	20 55 35	14 ♉ 19 10	20 ♉ 26 22	17 25	25 28	09 49	20 50	17 27	24 42	15 37	12 57
14	07 28 39	21 52 48	26 ♉ 30 19	02 ♊ 31 32	18 26	25 43	10 09	20 44	17 34	24 46	15 37	12 56
15	07 32 35	22 50 03	08 ♊ 30 34	14 ♊ 27 53	19 24	25 55	10 30	20 39	17 40	24 49	15 37	12 55
16	07 36 32	23 47 18	20 ♊ 23 56	26 ♊ 19 08	20 19	26 06	10 51	20 33	17 47	24 52	15 37	12 53
17	07 40 28	24 44 33	02 ♋ 13 49	08 ♋ 08 20	21 10	26 14	11 12	20 28	17 54	24 55	15 38	12 52
18	07 44 25	25 41 49	14 ♋ 02 58	19 ♋ 57 57	21 58	26 20	11 34	20 23	18 01	24 59	15 38	12 51
19	07 48 22	26 39 05	25 ♋ 53 32	01 ♌ 49 55	22 43	26 24	11 57	20 18	18 08	25 02	15 R 38	12 49
20	07 52 18	27 36 22	07 ♌ 47 17	13 ♌ 45 49	23 24	26 26	12 20	20 13	18 14	25 05	15 38	12 48
21	07 56 15	28 33 39	19 ♌ 45 42	25 ♌ 47 08	24 02	26 R 25	12 43	20 09	18 21	25 08	15 38	12 47
22	08 00 11	29 ♋ 30 56	01 ♍ 50 18	07 ♍ 55 27	24 35	26 22	13 07	20 04	18 27	25 11	15 37	12 45
23	08 04 08	00 ♌ 28 14	14 ♍ 02 49	20 ♍ 12 41	25 05	26 17	13 31	20 00	18 34	25 14	15 37	12 44
24	08 08 04	01 25 32	26 ♍ 25 22	02 ♎ 41 12	25 30	26 09	13 56	19 56	18 40	25 17	15 37	12 43
25	08 12 01	02 22 51	09 ♎ 00 35	15 ♎ 23 53	25 50	25 59	14 21	19 52	18 47	25 20	15 37	12 41
26	08 15 57	03 20 10	21 ♎ 51 31	28 ♎ 23 55	26 06	25 46	14 47	19 48	18 53	25 23	15 37	12 40
27	08 19 54	04 17 29	05 ♏ 01 28	11 ♏ 44 34	26 18	25 31	15 13	19 44	18 59	25 26	15 36	12 39
28	08 23 51	05 14 49	18 ♏ 33 32	25 ♏ 28 37	26 24	25 14	15 39	19 41	19 06	25 29	15 36	12 37
29	08 27 47	06 12 09	02 ♐ 29 58	09 ♐ 37 36	26 R 25	24 54	16 06	19 38	19 12	25 32	15 36	12 36
30	08 31 44	07 09 30	16 ♐ 51 23	24 ♐ 10 59	26 22	24 33	16 33	19 34	19 18	25 35	15 36	12 35
31	08 35 40	08 ♌ 06 51	01 ♑ 35 53	09 ♑ 05 20	26 ♌ 13	24 ♌ 09	17 ♏ 00	19 ♐ 31	19 ♊ 24	25 ♊ 38	15 ♈ 35	12 ♒ 33

Date	Moon True ☊ ° '	Moon Mean ☊ ° '	Moon ☽ Latitude ° '
01	27 ♏ 26	25 ♏ 52	00 S 21
02	27 R 26	25 49	00 N 57
03	27 24	25 46	02 14
04	27 20	25 42	03 22
05	27 14	25 39	04 16
06	27 07	25 36	04 52
07	27 01	25 33	05 07
08	26 55	25 30	05 02
09	26 50	25 26	04 38
10	26 48	25 23	03 59
11	26 D 47	25 20	03 09
12	26 47	25 17	02 10
13	26 48	25 14	01 07
14	26 49	25 11	00 N 02
15	26 R 48	25 07	01 S 02
16	26 45	25 04	02 03
17	26 40	25 01	02 58
18	26 33	24 58	03 45
19	26 23	24 55	04 22
20	26 12	24 52	04 48
21	26 00	24 48	05 01
22	25 49	24 45	05 01
23	25 40	24 42	04 48
24	25 32	24 39	04 21
25	25 28	24 36	03 41
26	25 25	24 32	02 48
27	25 D 25	24 29	01 46
28	25 25	24 26	00 S 36
29	25 R 25	24 23	00 N 38
30	25 24	24 20	01 51
31	25 ♏ 20	24 ♏ 17	02 N 59

DECLINATIONS

Date	Sun ☉ ° '	Moon ☽ ° '	Mercury ☿ ° '	Venus ♀ ° '	Mars ♂ ° '	Jupiter ♃ ° '	Saturn ♄ ° '	Uranus ♅ ° '	Neptune ♆ ° '	Pluto ♇ ° '
01	23 N 06	18 S 59	21 N 15	14 N 58	15 S 09	22 S 47	21 N 14	23 N 25	04 N 41	23 S 11
02	23 01	20 42	20 48	14 39	15 14	22 46	21 15	23 25	04 41	23 12
03	22 57	21 01	20 21	14 20	15 20	22 46	21 16	23 25	04 42	23 12
04	22 52	19 49	19 53	14 01	15 26	22 46	21 17	23 25	04 42	23 13
05	22 46	17 11	19 24	13 43	15 32	22 45	21 17	23 26	04 42	23 13
06	22 40	13 23	18 55	13 25	15 38	22 45	21 18	23 26	04 42	23 14
07	22 34	08 49	18 25	13 07	15 44	22 45	21 19	23 26	04 42	23 14
08	22 27	03 S 50	17 55	12 49	15 51	22 45	21 20	23 26	04 42	23 15
09	22 20	01 N 12	17 25	12 31	15 57	22 44	21 20	23 26	04 42	23 15
10	22 13	06 01	16 55	12 14	16 04	22 44	21 21	23 26	04 42	23 16
11	22 05	10 23	16 25	11 57	16 11	22 44	21 22	23 26	04 42	23 16
12	21 57	14 09	15 55	11 40	16 18	22 44	21 22	23 27	04 42	23 17
13	21 48	17 12	15 25	11 24	16 25	22 44	21 23	23 27	04 42	23 17
14	21 39	19 24	14 55	11 08	16 32	22 43	21 24	23 27	04 42	23 18
15	21 30	20 41	14 26	10 52	16 40	22 43	21 24	23 27	04 42	23 18
16	21 20	21 03	13 57	10 37	16 47	22 43	21 25	23 27	04 42	23 19
17	21 10	20 27	13 29	10 23	16 55	22 43	21 25	23 27	04 42	23 19
18	21 00	18 58	13 02	10 08	17 02	22 43	21 26	23 27	04 42	23 20
19	20 49	16 40	12 35	09 55	17 10	22 42	21 26	23 28	04 42	23 20
20	20 38	13 41	12 09	09 42	17 18	22 42	21 27	23 28	04 42	23 21
21	20 27	10 07	11 44	09 30	17 25	22 42	21 28	23 28	04 42	23 21
22	20 15	06 07	11 21	09 18	17 33	22 42	21 28	23 28	04 42	23 22
23	20 03	01 N 51	10 58	09 07	17 41	22 42	21 29	23 28	04 42	23 22
24	19 50	02 S 34	10 37	08 57	17 49	22 42	21 29	23 28	04 42	23 23
25	19 37	06 57	10 18	08 47	17 58	22 41	21 30	23 28	04 42	23 23
26	19 24	11 07	10 00	08 38	18 06	22 41	21 30	23 28	04 41	23 24
27	19 11	14 52	09 44	08 30	18 14	22 41	21 31	23 29	04 41	23 24
28	18 57	17 56	09 30	08 23	18 22	22 41	21 31	23 29	04 41	23 25
29	18 43	20 03	09 17	08 17	18 31	22 41	21 32	23 29	04 41	23 25
30	18 29	20 56	09 07	08 11	18 39	22 41	21 32	23 29	04 41	23 26
31	18 N 14	20 S 26	09 N 00	08 N 06	18 S 47	22 S 41	21 N 32	23 N 29	04 N 41	23 S 26

ZODIAC SIGN ENTRIES

Date	h	m	Planets
01	22	49	☽ ♐
03	23	09	☽ ♑
05	22	19	☽ ♒
07	22	27	☽ ♓
10	01	28	☽ ♈
12	08	26	☽ ♉
14	18	57	☽ ♊
17	07	28	☽ ♋
19	20	18	☽ ♌
22	08	22	☽ ♍
23	00	10	☉ ♌
24	18	52	☽ ♎
27	02	55	☽ ♏
29	07	45	☽ ♐
31	09	25	☽ ♑

LATITUDES

Date	Mercury ☿ ° '	Venus ♀ ° '	Mars ♂ ° '	Jupiter ♃ ° '	Saturn ♄ ° '	Uranus ♅ ° '	Neptune ♆ ° '	Pluto ♇ ° '
01	01 N 32	00 N 16	01 S 33	00 N 25	01 S 28	00 N 07	01 S 33	06 S 37
04	01 13	00 S 10	01 37	00 25	01 28	00 07	01 33	06 38
07	00 48	00 39	01 40	00 25	01 28	00 07	01 33	06 38
10	00 N 20	01 06	01 43	00 24	01 28	00 07	01 33	06 39
13	00 S 12	01 44	01 45	00 23	01 28	00 07	01 34	06 39
16	00 48	02 20	01 48	00 23	01 28	00 07	01 34	06 39
19	01 26	02 58	01 50	00 22	01 28	00 07	01 34	06 40
22	02 06	03 38	01 52	00 22	01 28	00 07	01 34	06 40
25	02 46	04 19	01 54	00 21	01 28	00 07	01 34	06 41
28	03 25	05 01	01 56	00 21	01 28	00 07	01 34	06 41
31	04 S 01	05 S 41	01 S 58	00 N 20	01 S 28	00 N 07	01 S 35	06 S 41

DATA

Julian Date	2463049
Delta T	+74 seconds
Ayanamsa	24° 18' 04"
Synetic vernal point	04° ♓ 48' 55"
True obliquity of ecliptic	23° 26' 01"

MOON'S PHASES, APSIDES AND POSITIONS ☽

Date	h	m	Phase	Longitude ° '	Eclipse Indicator
04	19	01	○	12 ♑ 38	
11	11	50	☾	19 ♈ 01	
19	13	40	●	26 ♋ 43	
27	10	35	☽	04 ♏ 14	

Day	h	m		
04	21	23	Perigee	
18	14	44	Apogee	
03	05	07	Max dec	21° S 05'
09	06	12	ON	
16	08	52	Max dec	21° N 03'
23	22	04	OS	
30	15	39	Max dec	20° S 57'

LONGITUDES

Date	Chiron ⚷ ° '	Ceres ⚳ ° '	Pallas ⚴ ° '	Juno ⚵ ° '	Vesta ⚶ ° '	Black Moon Lilith ⚸ ° '
01	19 ♉ 56	07 ♋ 31	11 ♉ 57	19 ♈ 56	18 ♋ 36	14 ♋ 50
11	20 ♉ 23	11 ♋ 52	15 ♉ 35	23 ♈ 42	23 ♋ 01	15 ♋ 57
21	20 ♉ 46	16 ♋ 13	19 ♉ 07	27 ♈ 14	27 ♋ 21	17 ♋ 04
31	21 ♉ 03	20 ♋ 33	22 ♉ 31	00 ♉ 28	01 ♌ 54	18 ♋ 11

All ephemeris data is given at 12.00 UT and the Moon's longitude is additionally given for 24.00 UT

ASPECTARIAN

01 Tuesday
01 05 ♀ # ♂
02 46 ☽ ± ♅
06 28 ☽ □ ♀
08 44 ☽ ± ♆
09 42 ☽ ⊻ ♃
12 32 ☿ ∥ ♄
12 54 ☽ ⊼ ♅
13 47 ☽ bQ ☉
23 43 ☽ bQ ♆

02 Wednesday
00 45 ☽ Q ♇
03 24 ☽ △ ☿
05 43 ☽ ± ☉
09 48 ☽ ⊻ ♂
13 41 ☽ # ☿
16 09 ☽ ⊼ ☉
19 41 ☽ ⊥ ♂
20 17 ☽ ⚹ ♇

03 Thursday
00 08 ☽ △ ♆
01 10 ☽ ☍ ♄
06 24 ☽ bQ ☿
07 33 ♂ ∠ ♃
09 29 ☽ △ ♀
10 17 ☽ ☌ ♃
10 21 ☽ ∠ ♂
13 52 ☽ ☍ ♅
20 12 ☽ ∠ ♇
22 15 ☽ ± ☿

04 Friday
04 23 ♀ △ ♃
08 42 ☽ ⊼ ☿
10 04 ☽ bQ ♀
10 23 ☽ ⊥ ♇
10 27 ☽ ⚹ ♂
11 04 ☽ # ☿
19 01 ☽ ☍ ☉
19 48 ☽ ⊻ ♇
20 04 ☿ bQ ♃
23 39 ☽ □ ♆

05 Saturday
00 41 ☽ ± ♀
01 00 ☽ ⊼ ♄
05 37 ☽ Q ♂
07 20 ☉ ⊼ ♇
07 27 ☿ □ ♂
09 13 ☽ ⊻ ♃
10 25 ☽ ⊼ ♀
10 30 ☽ ± ♄
13 17 ☽ ⊼ ♅
14 59 ☉ # ♃
18 35 ☽ ⊥ ♃
22 46 ☽ ± ♅
22 57 ☽ ∥ ♂

06 Sunday
00 47 ☽ bQ ♄
03 58 ☽ Q ♆
08 42 ☽ ∠ ♃
10 29 ☽ □ ♂
11 53 ☽ # ♀
12 40 ☽ ☍ ☿
13 04 ☽ bQ ♅
17 23 ☿ ∠ ♅
19 04 ☽ ☌ ♇
21 33 ☽ ⊼ ☉
23 04 ☽ ⚹ ♆

07 Monday
00 50 ☽ △ ♄
07 51 ☽ ± ☉
08 31 ☽ ⚹ ♃
11 38 ☽ ☍ ♀
13 16 ☽ △ ♅
21 42 ☉ □ ♆
23 25 ☽ ∠ ♆
23 33 ☽ bQ ☉

08 Tuesday
04 00 ☽ Q ♃
07 51 ☽ # ♆
12 09 ☽ △ ♂
14 23 ☽ ⊥ ♆
18 23 ☽ ⊼ ☿
20 11 ☽ ⊻ ♇

09 Wednesday
00 31 ☽ ⊻ ♆
02 27 ☽ △ ☉
02 47 ☽ □ ♄
05 34 ☽ ± ☿
06 24 ☽ ⊥ ♇
08 02 ☉ ⊻ ♄
10 11 ☽ □ ♃
14 15 ☽ bQ ♂
14 55 ☿ ☍ ♇
15 24 ☽ ⊼ ♀
15 49 ☽ □ ♅
21 59 ☽ ∠ ♇
22 41 ☽ bQ ☿

10 Thursday
02 12 ☽ ± ♀
05 17 ☽ ∥ ♆
06 24 ☽ ± ♂
10 19 ♀ ⚹ ♅
10 30 ☽ Q ♄
17 24 ☽ ⊼ ♂
18 43 ☽ bQ ♀

11 Friday
00 00 ☽ Q ♅
00 47 ☽ ⚹ ♇
04 12 ☽ △ ☿
05 33 ☽ ☌ ♆
07 19 ☉ ⊥ ♀
08 28 ☽ ⚹ ♄
11 50 ☽ □ ☉
15 32 ☽ △ ♃
19 57 ☿ △ ♆
20 47 ☽ ∥ ♀
20 52 ☿ # ♂
22 17 ☽ ⚹ ♅
22 58 ☽ Q ♇
23 04 ☽ △ ♀

12 Saturday
12 52 ☽ ∠ ♄
19 41 ☽ bQ ♃
23 14 ☽ ∥ ☿

13 Sunday
00 39 ☽ Q ☉
02 38 ♂ bQ ♅
03 00 ☽ ∠ ♅
03 00 ☽ ☍ ♂
04 56 ☽ # ♂
06 21 ☽ ⊥ ♄
09 20 ☽ □ ♇
09 51 ☉ ⊼ ♃
12 47 ☿ ⚹ ♄
13 00 ☽ ± ♃
14 32 ☽ ⊻ ♆
18 10 ☽ ⊻ ♄
18 37 ☽ □ ☿
20 38 ☽ ⊥ ♅

14 Monday
00 40 ☽ ⊼ ♃
02 04 ☽ ⚹ ☉
02 20 ☽ ⊥ ♆
08 31 ☽ ⊻ ♅
10 24 ☽ □ ♀
20 12 ☽ ∠ ♆

15 Tuesday
09 34 ☽ Q ☿
10 31 ☽ ∠ ☉
16 07 ☽ ⊼ ♂
20 51 ☽ △ ♇
23 04 ☽ Q ♀

16 Wednesday
02 00 ☉ ∥ ♄
02 20 ☽ ⚹ ♆
04 35 ☽ ± ♂
06 15 ☽ ⊥ ☉
06 40 ☽ ☌ ♄
11 48 ☽ ⚹ ☿
11 59 ☉ ⊥ ♄
12 19 ☽ ☍ ♃
18 08 ☿ △ ♃
19 28 ☽ ⊻ ☉
21 06 ☽ ☌ ♅
23 23 ☽ bQ ♂
23 41 ☽ ⚹ ♀

17 Thursday
02 39 ☽ Q ♆
03 09 ☽ bQ ♇
16 48 ☉ ⊻ ♅
20 36 ☽ ∠ ☿
21 24 ☽ ± ♇

18 Friday
06 27 ☽ ∠ ♀
06 48 ☽ △ ♂
09 34 ☽ ⊼ ♇
15 12 ☽ □ ♆
16 11 ☽ ⊥ ☿
20 07 ☽ ⊻ ♄

19 Saturday
00 46 ☽ ⊼ ♃
00 50 ☽ ⊥ ♀
03 55 ☉ ± ♃
05 10 ☽ ⊻ ☿
05 17 ☉ ⊻ ♀
06 10 ♆ St R
07 40 ☽ # ♂
08 23 ☽ ⊥ ♄
10 15 ☽ ⊻ ♅
12 49 ☽ ± ♃
13 02 ☽ ⊻ ♀
13 40 ☽ ☌ ☉
22 26 ☽ ⊥ ♅

20 Sunday
02 45 ☽ ∠ ♄
03 50 ♂ ± ♄
06 52 ☽ bQ ♃
16 38 ☽ ∠ ♅
17 08 ♀ St R
21 26 ☽ □ ♂
22 03 ☽ ☍ ♇

21 Monday
00 06 ☽ ∥ ☿
03 44 ☽ △ ♆
09 09 ☽ ⚹ ♄
12 46 ☽ △ ♃
15 26 ♂ □ ♇
16 05 ☽ ∥ ♀
20 56 ☽ ☌ ☿
22 45 ☽ ⚹ ♅

22 Tuesday
01 12 ☽ ☌ ♀
07 01 ☽ ⊻ ☉
09 14 ☽ Q ♄
09 36 ☽ bQ ♆
10 32 ☽ Q ♂
19 53 ☽ ⊥ ☉
20 06 ☽ ∥ ♆
22 35 ☽ Q ♅

23 Wednesday
03 20 ☽ ± ♆
09 27 ☽ ⊼ ♇
10 56 ☽ ⚹ ♂
15 01 ☽ ∠ ☉
15 04 ☽ ⊼ ♆
20 53 ☽ □ ♄
21 07 ☽ ± ♇
21 55 ☿ ⚹ ♅
23 31 ☽ □ ♃

24 Thursday
08 19 ☉ ⊥ ♅
09 48 ☽ □ ♅
10 09 ☽ ⊻ ☿
11 28 ☽ ⊻ ♀
11 45 ♂ ⊥ ♃
14 29 ☽ bQ ♇
16 59 ☽ ∠ ♂
22 01 ☽ ⊥ ☿
22 23 ☽ ⚹ ☉
22 50 ☽ ⊥ ♀
23 35 ☽ # ♆

25 Friday
09 51 ☽ Q ♃
10 43 ☽ ⊥ ♂
15 32 ☽ ∠ ☿
15 39 ☽ ∠ ♀
18 50 ☿ ☌ ♀
18 55 ☽ △ ♇
21 59 ☽ # ♀
22 23 ☽ ⊻ ♂
22 55 ☽ Q ☉

26 Saturday
00 24 ☽ ☍ ♆
05 48 ☽ # ☿
06 27 ☽ △ ♄
08 13 ☽ ⚹ ♃
18 31 ☽ △ ♅
19 04 ☽ ⚹ ♀
19 56 ☽ ⚹ ☿

27 Sunday
03 28 ☉ ∠ ♄
10 07 ☽ bQ ♄
10 35 ☽ □ ☉
11 29 ☽ ∠ ♃
16 23 ☽ Q ♀
17 55 ☽ Q ☿
18 11 ♀ ⚹ ♅
21 43 ☽ bQ ♅
22 35 ☉ bQ ♃

28 Monday
01 35 ☽ □ ♇
02 19 ☽ ± ♄
03 28 ☽ ⊥ ♃
06 44 ☽ ☌ ♂
06 49 ☽ ⊼ ♆
09 32 ♂ ⊼ ♆
12 56 ☽ ⊼ ♄
13 37 ☽ ± ♅
13 57 ☽ ⊻ ♃
16 24 ☽ ∥ ♂
17 18 ☽ ± ♆
21 10 ☽ # ☉
23 19 ☽ □ ♀

29 Tuesday
00 03 ☽ ⊼ ♅
01 38 ☽ □ ☿
06 47 ☿ St R
08 46 ☽ bQ ♆
08 47 ☽ Q ♇
18 42 ☽ △ ☉

30 Wednesday
01 20 ☉ # ♂
04 56 ☽ ⚹ ♇
09 55 ☽ △ ♆
11 28 ☽ ⊻ ♂
16 03 ☽ ☍ ♄
16 27 ☽ ☌ ♃
21 18 ☽ bQ ☉
21 38 ☽ ⊥ ♂

31 Thursday
00 16 ☽ △ ♀
02 19 ☽ ☍ ♅
03 24 ☽ △ ☿
05 29 ☽ ∠ ♇
12 40 ☽ ∠ ♂
12 53 ☽ ± ☉
19 57 ☽ ⊥ ♇
23 09 ☽ ⊼ ☉
23 45 ☽ bQ ♀

AUGUST 2031

Date	Sidereal time h m s	Sun ☉ ° ′ ″	Moon ☽ ° ′ ″	Moon ☽ 24.00 ° ′ ″	Mercury ☿ ° ′	Venus ♀ ° ′	Mars ♂ ° ′	Jupiter ♃ ° ′	Saturn ♄ ° ′	Uranus ♅ ° ′	Neptune ♆ ° ′	Pluto ♇ ° ′
01	08 39 37	09 ♌ 04 13	16 ♑ 38 25	24 ♑ 14 01	25 ♌ 59	23 ♌ 42	17 ♏ 28	19 ♐ 29	19 ♊ 30	25 ♊ 41	15 ♈ 35	12 ♒ 32
02	08 43 33	10 01 36	01 ♒ 50 52	09 ♒ 27 38	25 R 40	23 R 14	17 56	19 R 26	19 36	25 43	15 R 34	12 R 31
03	08 47 30	10 58 59	17 ♒ 02 55	24 ♒ 35 24	25 16	22 45	18 24	19 24	19 42	25 46	15 34	12 29
04	08 51 26	11 56 24	02 ♓ 03 49	09 ♓ 27 05	24 47	22 13	18 53	19 21	19 48	25 49	15 33	12 28
05	08 55 23	12 53 49	16 ♓ 44 17	23 ♓ 54 44	24 14	21 40	19 22	19 19	19 53	25 52	15 33	12 27
06	08 59 20	13 51 15	00 ♈ 57 58	07 ♈ 53 46	23 37	21 06	19 52	19 17	19 59	25 54	15 32	12 25
07	09 03 16	14 48 43	14 ♈ 42 04	21 ♈ 23 02	22 56	20 31	20 21	19 16	20 05	25 57	15 32	12 24
08	09 07 13	15 46 12	27 ♈ 56 57	04 ♉ 24 15	22 12	19 55	20 51	19 14	20 10	25 59	15 31	12 22
09	09 11 09	16 43 42	10 ♉ 45 28	17 ♉ 01 10	21 26	19 18	21 22	19 13	20 16	26 02	15 30	12 21
10	09 15 06	17 41 13	23 ♉ 11 59	29 ♉ 18 35	20 38	18 41	21 52	19 12	20 21	26 05	15 30	12 20
11	09 19 02	18 38 46	05 ♊ 21 37	11 ♊ 21 45	19 50	18 03	22 23	19 11	20 26	26 07	15 29	12 18
12	09 22 59	19 36 21	17 ♊ 19 36	23 ♊ 15 46	19 01	17 26	22 55	19 10	20 32	26 10	15 28	12 17
13	09 26 55	20 33 57	29 ♊ 10 50	05 ♋ 05 19	18 14	16 49	23 26	19 09	20 37	26 12	15 27	12 16
14	09 30 52	21 31 34	10 ♋ 59 41	16 ♋ 54 22	17 29	16 13	23 58	19 09	20 42	26 14	15 27	12 14
15	09 34 49	22 29 12	22 ♋ 49 44	28 ♋ 46 08	16 47	15 37	24 30	19 09	20 47	26 17	15 26	12 13
16	09 38 45	23 26 52	04 ♌ 43 48	10 ♌ 43 00	16 08	15 02	25 03	19 D 09	20 52	26 19	15 25	12 12
17	09 42 42	24 24 33	16 ♌ 43 53	22 ♌ 46 39	15 35	14 29	25 35	19 09	20 57	26 21	15 24	12 10
18	09 46 38	25 22 16	28 ♌ 51 23	04 ♍ 58 12	15 06	13 57	26 08	19 09	21 01	26 23	15 23	12 09
19	09 50 35	26 20 00	11 ♍ 07 11	17 ♍ 18 27	14 44	13 26	26 41	19 10	21 06	26 26	15 22	12 08
20	09 54 31	27 17 45	23 ♍ 32 04	29 ♍ 48 09	14 28	12 58	27 15	19 10	21 11	26 28	15 21	12 06
21	09 58 28	28 15 31	06 ♎ 06 50	12 ♎ 28 17	14 19	12 31	27 48	19 11	21 15	26 30	15 20	12 05
22	10 02 24	29 ♌ 13 18	18 ♎ 52 40	25 ♎ 20 14	14 D 18	12 06	28 22	19 12	21 20	26 32	15 19	12 04
23	10 06 21	00 ♍ 11 07	01 ♏ 51 12	08 ♏ 25 52	14 24	11 43	28 57	19 14	21 24	26 34	15 18	12 02
24	10 10 18	01 08 57	15 ♏ 04 29	21 ♏ 47 20	14 38	11 22	29 ♏ 31	19 15	21 29	26 36	15 17	12 01
25	10 14 14	02 06 48	28 ♏ 34 42	05 ♐ 26 48	15 00	11 04	00 ♐ 06	19 17	21 33	26 38	15 16	12 00
26	10 18 11	03 04 40	12 ♐ 23 48	19 ♐ 25 48	15 30	10 48	00 40	19 19	21 37	26 40	15 15	11 59
27	10 22 07	04 02 34	26 ♐ 32 45	03 ♑ 44 30	16 07	10 34	01 16	19 21	21 41	26 42	15 14	11 57
28	10 26 04	05 00 29	11 ♑ 00 46	18 ♑ 21 02	16 52	10 23	01 51	19 23	21 45	26 43	15 13	11 56
29	10 30 00	05 58 25	25 ♑ 44 41	03 ♒ 10 54	17 45	10 14	02 26	19 25	21 49	26 45	15 11	11 55
30	10 33 57	06 56 22	10 ♒ 38 43	18 ♒ 07 03	18 44	10 07	03 02	19 28	21 53	26 47	15 10	11 54
31	10 37 53	07 ♍ 54 21	25 ♒ 34 47	03 ♓ 00 45	19 ♌ 50	10 ♌ 03	03 ♐ 38	19 ♐ 30	21 ♊ 56	26 ♊ 49	15 ♈ 09	11 ♒ 52

Date	Moon True ☊ ° ′	Moon Mean ☊ ° ′	Moon ☽ Latitude ° ′
01	25 ♏ 13	24 ♏ 13	03 N 56
02	25 R 05	24 10	04 37
03	24 55	24 07	04 59
04	24 44	24 04	04 59
05	24 35	24 01	04 39
06	24 27	23 57	04 03
07	24 22	23 54	03 13
08	24 19	23 51	02 15
09	24 18	23 48	01 11
10	24 D 18	23 45	00 N 06
11	24 R 18	23 42	00 S 58
12	24 16	23 38	01 59
13	24 12	23 35	02 53
14	24 06	23 32	03 40
15	23 57	23 29	04 17
16	23 45	23 26	04 44
17	23 32	23 23	04 58
18	23 18	23 19	04 58
19	23 04	23 16	04 45
20	22 53	23 13	04 19
21	22 44	23 10	03 39
22	22 38	23 07	02 48
23	22 34	23 03	01 47
24	22 33	23 00	00 S 39
25	22 D 33	22 57	00 N 32
26	22 R 33	22 54	01 43
27	22 31	22 51	02 50
28	22 28	22 48	03 47
29	22 21	22 44	04 30
30	22 13	22 41	04 55
31	22 ♏ 02	22 ♏ 38	05 N 01

DECLINATIONS

Date	Sun ☉ ° ′	Moon ☽ ° ′	Mercury ☿ ° ′	Venus ♀ ° ′	Mars ♂ ° ′	Jupiter ♃ ° ′	Saturn ♄ ° ′	Uranus ♅ ° ′	Neptune ♆ ° ′	Pluto ♇ ° ′
01	17 N 59	18 S 29	08 N 54	08 N 03	18 S 56	22 S 41	21 N 33	23 N 29	04 N 41	23 S 27
02	17 44	15 14	08 51	08 00	19 04	22 41	21 33	23 29	04 40	23 27
03	17 28	10 58	08 51	07 58	19 12	22 41	21 34	23 29	04 40	23 28
04	17 12	06 05	08 53	07 57	19 21	22 41	21 34	23 29	04 40	23 28
05	16 56	00 S 57	08 58	07 57	19 29	22 41	21 34	23 30	04 40	23 29
06	16 40	04 N 06	09 06	07 57	19 38	22 41	21 35	23 30	04 39	23 29
07	16 23	08 46	09 16	07 59	19 46	22 41	21 35	23 30	04 39	23 30
08	16 06	12 50	09 28	08 01	19 54	22 41	21 36	23 30	04 39	23 30
09	15 49	16 10	09 43	08 04	20 03	22 41	21 36	23 30	04 38	23 31
10	15 32	18 40	10 00	08 08	20 11	22 41	21 36	23 30	04 38	23 31
11	15 14	20 14	10 18	08 12	20 20	22 41	21 37	23 30	04 38	23 31
12	14 56	20 52	10 39	08 17	20 28	22 41	21 37	23 30	04 37	23 32
13	14 38	20 33	11 00	08 23	20 36	22 41	21 37	23 30	04 37	23 32
14	14 20	19 19	11 22	08 29	20 45	22 42	21 37	23 30	04 37	23 33
15	14 01	17 16	11 45	08 36	20 53	22 42	21 38	23 30	04 36	23 33
16	13 42	14 29	12 08	08 43	21 01	22 42	21 38	23 30	04 36	23 34
17	13 23	11 05	12 31	08 51	21 09	22 42	21 38	23 31	04 36	23 34
18	13 04	07 12	12 54	08 58	21 17	22 42	21 39	23 31	04 35	23 34
19	12 44	03 N 00	13 15	09 07	21 25	22 43	21 39	23 31	04 35	23 35
20	12 24	01 S 24	13 36	09 15	21 33	22 43	21 39	23 31	04 34	23 35
21	12 05	05 47	13 55	09 24	21 41	22 43	21 39	23 31	04 34	23 36
22	11 45	09 59	14 13	09 32	21 49	22 43	21 39	23 31	04 34	23 36
23	11 24	13 47	14 28	09 41	21 57	22 44	21 40	23 31	04 33	23 36
24	11 04	16 59	14 41	09 50	22 04	22 44	21 40	23 31	04 33	23 37
25	10 43	19 19	14 52	09 58	22 12	22 44	21 40	23 31	04 32	23 37
26	10 22	20 34	15 00	10 07	22 19	22 44	21 40	23 31	04 32	23 38
27	10 01	20 34	15 06	10 15	22 27	22 45	21 40	23 31	04 31	23 38
28	09 40	19 13	15 08	10 24	22 34	22 45	21 41	23 31	04 31	23 38
29	09 19	16 34	15 08	10 32	22 41	22 45	21 41	23 31	04 30	23 39
30	08 58	12 49	15 04	10 40	22 48	22 46	21 41	23 31	04 30	23 39
31	08 N 36	08 S 16	14 N 58	10 N 48	22 S 55	22 S 46	21 N 41	23 N 31	04 N 29	23 S 39

ZODIAC SIGN ENTRIES

Date	h m	Planets
02	09 05	☽ ♒
04	08 40	☽ ♓
06	10 21	☽ ♈
08	15 47	☽ ♉
11	01 22	☽ ♊
13	13 40	☽ ♋
16	02 29	☽ ♌
18	14 15	☽ ♍
21	00 23	☽ ♎
23	07 23	☉ ♍
23	08 36	☽ ♏
25	08 08	♂ ♐
25	14 30	☽ ♐
27	17 47	☽ ♑
29	18 52	☽ ♒
31	19 08	☽ ♓

LATITUDES

Date	Mercury ☿ ° ′	Venus ♀ ° ′	Mars ♂ ° ′	Jupiter ♃ ° ′	Saturn ♄ ° ′	Uranus ♅ ° ′	Neptune ♆ ° ′	Pluto ♇ ° ′
01	04 S 12	05 S 54	01 S 58	00 N 20	01 S 29	00 N 07	01 S 35	06 S 41
04	04 38	06 30	01 59	00 20	01 29	00 07	01 35	06 41
07	04 52	07 02	02 01	00 19	01 29	00 07	01 35	06 42
10	04 52	07 27	02 02	00 19	01 29	00 07	01 35	06 42
13	04 35	07 46	02 03	00 18	01 29	00 07	01 35	06 42
16	04 02	07 57	02 03	00 18	01 29	00 08	01 35	06 42
19	03 18	08 00	02 04	00 17	01 30	00 08	01 36	06 42
22	02 26	07 56	02 04	00 17	01 30	00 08	01 36	06 42
25	01 32	07 46	02 05	00 16	01 30	00 08	01 36	06 42
28	00 S 40	07 30	02 05	00 16	01 30	00 08	01 36	06 42
31	00 N 06	07 S 11	02 S 05	00 N 15	01 S 30	00 N 08	01 S 36	06 S 42

DATA

Julian Date	2463080
Delta T	+74 seconds
Ayanamsa	24° 18' 09"
Synetic vernal point	04° ♓ 48' 50"
True obliquity of ecliptic	23° 26' 01"

MOON'S PHASES, APSIDES AND POSITIONS ☽

Date	h m	Phase	Longitude ° ′	Eclipse Indicator
03	01 46	○	10 ♒ 35	
10	00 24	☾	17 ♉ 13	
18	04 32	●	25 ♌ 04	
25	18 40	☽	02 ♐ 23	

Day	h m		
02	06 49	Perigee	
14	21 49	Apogee	
30	12 58	Perigee	
05	16 24	0N	
12	15 50	Max dec	20° N 52'
20	04 25	0S	
26	23 53	Max dec	20° S 44'

LONGITUDES

Date	Chiron ⚷ ° ′	Ceres ⚳ ° ′	Pallas ⚴ ° ′	Juno ⚵ ° ′	Vesta ⚶ ° ′	Black Moon Lilith ⚸ ° ′
01	21 ♉ 05	20 ♋ 59	22 ♉ 51	00 ♉ 46	02 ♌ 21	18 ♋ 17
11	21 ♉ 16	25 ♋ 18	26 ♉ 04	03 ♉ 36	06 ♌ 48	19 ♋ 24
21	21 ♉ 22	29 ♋ 36	29 ♉ 04	05 ♉ 59	11 ♌ 16	20 ♋ 31
31	21 ♉ 22	03 ♌ 50	01 ♊ 46	07 ♉ 47	15 ♌ 43	21 ♋ 38

All ephemeris data is given at 12.00 UT and the Moon's longitude is additionally given for 24.00 UT

ASPECTARIAN

h m	Aspects
01 Friday	
03 10	☽ ⚼ ☿
05 30	☽ ⊻ ♇
08 03	☽ ∥ ♂
08 27	♃ ☍ ♄
10 19	☽ □ ♆
13 21	☽ ⚹ ♂
13 38	☽ ± ♀
16 29	☽ ⊻ ♃
16 33	☽ ⊼ ♄
16 48	☽ ⋕ ☉
17 12	☽ ± ☿
22 51	☽ ⊼ ♀
02 Saturday	
01 55	☽ ⊥ ♃
02 05	☽ ± ♄
02 19	☽ ⊼ ♅
02 29	☽ ⊼ ☿
08 40	☿ ⚹ ♅
08 53	☽ Q ♂
11 48	☽ ± ♅
14 43	☽ Q ♆
16 04	☽ ∠ ♃
16 22	☽ ⚼ ♄
03 Sunday	
01 46	☽ ☍ ☉
02 02	☽ ⚼ ♅
04 47	☽ ☌ ♇
06 22	☉ ∠ ♅
09 39	☽ ⚹ ♆
14 13	☽ □ ♂
15 43	☽ ⚹ ♃
16 14	☽ △ ♄
20 45	☽ ☍ ♀
22 37	☽ ⋕ ☿
04 Monday	
00 42	☽ ☍ ☿
01 56	☽ △ ♅
03 04	☽ ⋕ ♀
09 34	☽ ∠ ♆
10 52	☽ Q ♃
18 40	☽ ⋕ ♆
05 Tuesday	
00 10	☽ ⊥ ♆
00 51	☉ ☍ ♇
04 55	☽ ⊻ ♇
05 12	☽ ⊼ ☉
09 41	♂ ⊻ ♃
10 02	☽ ⊻ ♆
14 50	☽ ⊥ ♇
15 51	☽ ± ☉
16 17	☽ □ ♃
16 32	☽ △ ♂
17 17	☽ □ ♄
19 55	☽ ⊼ ♀
06 Wednesday	
00 02	☽ ⊼ ☿
03 20	☽ □ ♅
05 39	☽ ± ♀
05 57	☽ ∠ ♇
08 07	☽ ⚼ ☉
09 47	☽ ± ☿
14 16	♂ ± ♅
14 44	☽ ∥ ♆
18 58	☽ ⚼ ♂
19 14	♂ ⊼ ♄
20 31	☽ ⚼ ♀
07 Thursday	
00 14	☽ Q ♄
00 39	☽ ⚼ ☿
07 44	☽ ∥ ♀
07 55	☽ ⚹ ♇
10 39	☽ Q ♅
11 22	☽ ± ♂
12 13	☽ △ ☉
13 28	☽ ☌ ♆
14 53	☽ ∥ ☿
15 23	♀ □ ♂
15 42	♅ ⋕ ♇
20 09	☽ △ ♃
21 42	☽ ⚹ ♄
21 58	☽ △ ♀
22 32	☽ ⊼ ♂
08 Friday	
02 03	☽ △ ☿
03 07	♀ ⚹ ♄
05 27	☽ Q ♇
05 44	☉ △ ♆
08 23	☽ ⚹ ♅
23 40	☽ ⚼ ♃
09 Saturday	
01 32	☽ ∠ ♄
09 23	☽ ∥ ☉
12 32	☽ ∠ ♅
13 17	☿ □ ♂
15 02	☽ □ ♇
15 20	♀ △ ♃
16 41	☽ ± ♃
18 33	♂ ± ♆
18 44	☽ ⊥ ♄
21 05	☽ ⊻ ♆
10 Sunday	
00 24	☽ □ ☉
03 37	☽ □ ♀
04 13	☽ ⊼ ♃
05 54	☽ ⊥ ♅
06 24	☽ ⊻ ♄
07 19	☽ □ ☿
08 41	☽ ⊥ ♆
09 18	☽ ☍ ♂
17 39	☽ ⊻ ♅
19 44	☿ ⚹ ♄
11 Monday	
02 20	☽ ∠ ♆
03 01	☉ ☌ ♀
13 19	☽ Q ♀
14 24	☽ ⋕ ♂
14 47	☽ Q ☉
16 37	☽ Q ☿
12 Tuesday	
01 07	☉ △ ♃
01 52	☽ △ ♇
04 05	☉ ☌ ☿
07 43	☿ △ ♃
08 16	☽ ⚹ ♆
12 12	☽ ⚹ ♀
15 13	☽ ⚹ ☿
15 42	☽ ☍ ♃
17 00	☽ ⚹ ☉
18 30	☽ ☌ ♄
23 49	☽ ⊼ ♂
13 Wednesday	
05 56	☽ ☌ ♅
08 07	☽ ⚼ ♇
08 30	☽ Q ♆
10 21	☽ ⋕ ♂
12 33	☽ ± ♂
13 16	☉ ⚹ ♄
17 05	☽ ∠ ♀
19 44	☽ ∠ ☿
14 Thursday	
02 07	☽ ∠ ☉
02 21	☽ ± ♇
07 41	☽ ⚼ ♂
10 29	☽ ⊥ ♀
12 56	☽ ⊥ ☿
14 31	☽ ⊼ ♇
21 01	☽ □ ♆
22 01	☽ ⊥ ☉
22 05	☽ ⊻ ♀
15 Friday	
00 25	☽ ⊻ ☿
04 32	☽ ⊼ ♃
07 49	☽ ⊻ ♄
11 15	☽ ⊻ ☉
15 33	☽ △ ♂
16 41	☽ ± ♃
19 00	☽ ⊻ ♅
19 51	♀ △ ♆
20 03	☽ ⊥ ♄
16 Saturday	
04 58	♃ St D
07 08	☽ ⊥ ♅
10 49	☽ ⚼ ♃
14 18	☽ ∠ ♄
18 31	☽ ∥ ☉
17 Sunday	
01 14	☽ ∠ ♅
02 55	☽ ☍ ♇
03 16	☽ ∥ ☿
07 43	☽ ☌ ♀
09 21	☽ △ ♆
09 48	☽ ☌ ☿
16 48	☽ △ ♃
20 26	☽ ⚹ ♄
20 42	☿ △ ♆
18 Monday	
01 42	☽ ∥ ♀
04 32	☽ ☌ ☉
06 23	☽ □ ♂
07 08	☽ ⚹ ♅
15 00	☽ ⚼ ♆
17 43	☉ ∥ ☿
20 15	☽ Q ♄
23 55	♂ ⊼ ♅
19 Tuesday	
03 06	☽ ∥ ♆
06 44	☽ Q ♅
08 36	☽ ± ♆
13 57	☽ ⊼ ♇
14 26	☉ ⚹ ♅
16 20	☽ ⊻ ♀
18 51	☽ ⊻ ☿
19 15	☽ Q ♂
20 14	☽ ⊼ ♆
20 Wednesday	
01 33	☽ ± ♇
03 31	☽ ⊥ ♀
03 36	☽ □ ♃
06 12	☽ ⊥ ☿
07 27	☽ □ ♄
08 59	☉ □ ♂
17 38	☽ □ ♅
18 50	☽ ⚼ ♇
19 27	☽ ⚹ ♂
19 49	☽ ⊻ ☉
20 11	☽ ∠ ♀
23 12	☽ ∠ ☿
21 Thursday	
05 19	☽ ⋕ ♆
06 15	♂ ⋕ ♄
08 11	☽ ⊥ ☉
14 02	☽ Q ♃
23 15	☽ △ ♇
23 41	☽ ⚹ ♀
22 Friday	
01 13	☽ ∠ ♂
02 35	☽ ∠ ☉
03 25	☽ ⚹ ☿
04 28	☿ St D
05 21	☽ ☍ ♆
09 17	☽ ⋕ ♀
12 37	☽ ⚹ ♃
14 15	♀ ☍ ♇
16 36	☽ △ ♄
18 48	☽ ⊥ ♂
21 25	☽ Q ♀
21 51	☽ ⋕ ☉
23 Saturday	
01 52	☽ Q ☿
02 15	☽ △ ♅
06 24	☽ ⊻ ♂
08 42	☽ ⚹ ☉
14 53	☉ ⚼ ♆
16 21	☽ ∠ ♃
17 00	☽ ⋕ ☿
20 22	☽ ⚼ ♄
21 14	♀ ∠ ♅
24 Sunday	
05 29	☽ □ ♀
05 43	☽ ⚼ ♅
06 30	☽ □ ♇
08 16	☽ Q ☉
08 43	☽ ⊥ ♃
11 12	☽ □ ☿
12 23	☽ ⊼ ♆
12 43	☽ ± ♄
19 30	☽ ⊻ ♃
21 54	☽ ± ♅
23 05	☽ ± ♆
23 30	☽ ⊼ ♄
25 Monday	
08 08	♂ Q ♇
08 34	☽ ⊼ ♅
14 29	☽ Q ♇
14 47	☽ ☌ ♂
14 58	☽ ⚼ ♆
18 40	☽ □ ☉
18 58	♂ ⚼ ♆
26 Tuesday	
01 11	☿ △ ♆
09 18	☽ △ ♀
11 17	☽ ⚹ ♇
16 53	☽ △ ♆
17 32	☽ △ ☿
23 49	☽ ☌ ♃
27 Wednesday	
00 36	☉ ∥ ♀
02 22	☉ Q ♄
03 46	☽ ☍ ♄
10 23	☽ ⚼ ♀
12 15	☽ ☍ ♅
12 41	☽ ∠ ♇
20 02	☽ ⚼ ☿
20 13	☽ ⊻ ♂
28 Thursday	
01 11	☽ ± ♀
01 23	☽ △ ☉
03 39	☽ ⊥ ♇
06 34	☽ ⊥ ♂
10 58	☽ ⊼ ♀
11 45	☽ ± ☿
13 31	☽ ⊻ ♇
18 52	☽ □ ♆
21 57	☽ ∠ ♂
22 10	☽ ⊼ ☿
29 Friday	
01 43	☽ ⊻ ♃
03 44	☽ ⚼ ☉
05 36	☽ ⊼ ♄
11 28	☽ ⊥ ♃
13 38	☽ ⊼ ♅
15 21	☽ ± ♄
19 18	☽ ± ☉
22 05	☽ ⋕ ☿
23 15	☽ ⚹ ♂
23 20	☽ ± ♅
30 Saturday	
00 00	☽ Q ♆
02 02	☽ ∠ ♃
02 20	♂ ∥ ♃
05 38	☽ ⊼ ☉
05 55	☽ ⚼ ♄
11 10	☽ ☍ ♀
13 50	☽ ⚼ ♅
14 00	☽ ☌ ♇
19 15	☽ ⚹ ♆
19 21	☽ Q ♂
23 27	☽ ⋕ ♀
31 Sunday	
02 00	☽ ☍ ☿
02 12	☽ ⚹ ♃
04 40	☿ △ ♃
06 07	☽ △ ♄
10 11	☽ ⋕ ☉
13 59	☽ △ ♅
19 22	☽ ∠ ♆
21 36	☽ Q ♃

SEPTEMBER 2031

LONGITUDES

Date	Sidereal time h m s	Sun ☉ ° ′ ″	Moon ☽ ° ′ ″	Moon ☽ 24.00 ° ′ ″	Mercury ☿ ° ′	Venus ♀ ° ′	Mars ♂ ° ′	Jupiter ♃ ° ′	Saturn ♄ ° ′	Uranus ♅ ° ′	Neptune ♆ ° ′	Pluto ♇ ° ′
01	10 41 50	08 ♍ 52 21	10 ♓ 23 47	17 ♓ 42 52	21 ♌ 03	10 ♌ 01	04 ♐ 14	19 ♐ 33	22 ♊ 00	26 ♊ 50	15 ♈ 08	11 ♒ 51
02	10 45 46	09 50 23	24 ♓ 57 03	02 ♈ 05 34	22 22	10 D 02	04 51	19 36	22 03	26 52	15 R 06	11 R 50
03	10 49 43	10 48 26	09 ♈ 07 50	16 ♈ 03 28	23 46	10 05	05 27	19 40	22 07	26 53	15 05	11 49
04	10 53 40	11 46 32	22 ♈ 52 16	29 ♈ 34 11	25 16	10 10	06 04	19 43	22 10	26 55	15 04	11 48
05	10 57 36	12 44 39	06 ♉ 09 22	12 ♉ 38 07	26 49	10 17	06 41	19 47	22 13	26 56	15 03	11 47
06	11 01 33	13 42 49	19 ♉ 00 47	25 ♉ 17 54	28 ♌ 27	10 27	07 18	19 50	22 16	26 58	15 01	11 45
07	11 05 29	14 41 00	01 ♊ 30 01	07 ♊ 37 44	00 ♍ 09	10 39	07 55	19 54	22 19	26 59	15 00	11 44
08	11 09 26	15 39 13	13 ♊ 41 42	19 ♊ 42 35	01 53	10 53	08 33	19 59	22 22	27 00	14 58	11 43
09	11 13 22	16 37 29	25 ♊ 41 03	01 ♋ 37 45	03 40	11 09	09 11	20 03	22 25	27 01	14 57	11 42
10	11 17 19	17 35 46	07 ♋ 33 19	13 ♋ 28 22	05 29	11 26	09 49	20 07	22 28	27 03	14 56	11 41
11	11 21 15	18 34 06	19 ♋ 23 28	25 ♋ 19 10	07 20	11 46	10 27	20 12	22 30	27 04	14 54	11 40
12	11 25 12	19 32 27	01 ♌ 15 56	07 ♌ 14 13	09 12	12 08	11 05	20 17	22 33	27 05	14 53	11 39
13	11 29 09	20 30 51	13 ♌ 14 22	19 ♌ 16 44	11 05	12 31	11 43	20 22	22 35	27 06	14 51	11 38
14	11 33 05	21 29 16	25 ♌ 21 33	01 ♍ 29 03	12 58	12 56	12 22	20 27	22 38	27 07	14 50	11 37
15	11 37 02	22 27 44	07 ♍ 39 22	13 ♍ 52 37	14 52	13 23	13 01	20 32	22 40	27 08	14 48	11 36
16	11 40 58	23 26 13	20 ♍ 08 51	26 ♍ 28 04	16 45	13 51	13 40	20 38	22 42	27 09	14 47	11 35
17	11 44 55	24 24 44	02 ♎ 50 16	09 ♎ 15 26	18 39	14 21	14 19	20 43	22 44	27 10	14 45	11 34
18	11 48 51	25 23 18	15 ♎ 43 30	22 ♎ 14 27	20 32	14 52	14 58	20 49	22 46	27 10	14 44	11 33
19	11 52 48	26 21 53	28 ♎ 48 15	05 ♏ 24 52	22 24	15 25	15 37	20 55	22 48	27 11	14 42	11 32
20	11 56 44	27 20 29	12 ♏ 04 18	18 ♏ 46 35	24 16	15 59	16 17	21 01	22 49	27 12	14 40	11 31
21	12 00 41	28 19 08	25 ♏ 31 45	02 ♐ 19 51	26 08	16 34	16 57	21 07	22 51	27 12	14 39	11 31
22	12 04 38	29 ♍ 17 48	09 ♐ 10 55	16 ♐ 05 02	27 58	17 11	17 36	21 14	22 52	27 13	14 37	11 30
23	12 08 34	00 ♎ 16 30	23 ♐ 02 11	00 ♑ 02 24	29 ♍ 48	17 49	18 16	21 20	22 54	27 13	14 36	11 29
24	12 12 31	01 15 14	07 ♑ 05 37	14 ♑ 11 40	01 ♎ 37	18 28	18 57	21 27	22 55	27 14	14 34	11 28
25	12 16 27	02 13 59	21 ♑ 20 23	28 ♑ 31 26	03 25	19 08	19 37	21 34	22 56	27 14	14 32	11 27
26	12 20 24	03 12 46	05 ♒ 44 25	12 ♒ 58 48	05 12	19 50	20 17	21 41	22 57	27 15	14 31	11 27
27	12 24 20	04 11 35	20 ♒ 14 01	27 ♒ 29 21	06 59	20 32	20 58	21 48	22 58	27 15	14 29	11 26
28	12 28 17	05 10 25	04 ♓ 44 03	11 ♓ 57 20	08 44	21 15	21 39	21 55	22 59	27 15	14 28	11 25
29	12 32 13	06 09 17	19 ♓ 08 24	26 ♓ 16 31	10 28	22 00	22 20	22 02	23 00	27 16	14 26	11 25
30	12 36 10	07 ♎ 08 12	03 ♈ 20 56	10 ♈ 21 03	12 ♎ 12	22 ♌ 45	23 ♐ 00	22 ♐ 10	23 ♊ 00	27 ♊ 16	14 ♈ 24	11 ♒ 24

Date	Moon True ☊ ° ′	Moon Mean ☊ ° ′	Moon ☽ Latitude ° ′
01	21 ♏ 52	22 ♏ 35	04 N 46
02	21 R 42	22 32	04 13
03	21 34	22 29	03 24
04	21 28	22 25	02 25
05	21 25	22 22	01 20
06	21 24	22 19	00 N 13
07	21 D 25	22 16	00 S 54
08	21 25	22 13	01 56
09	21 R 25	22 09	02 52
10	21 23	22 06	03 40
11	21 18	22 03	04 19
12	21 11	22 00	04 46
13	21 02	21 57	05 01
14	20 52	21 54	05 03
15	20 41	21 50	04 51
16	20 30	21 47	04 25
17	20 21	21 44	03 45
18	20 14	21 41	02 54
19	20 10	21 38	01 52
20	20 08	21 35	00 S 43
21	20 D 07	21 31	00 N 29
22	20 08	21 28	01 41
23	20 09	21 25	02 48
24	20 R 09	21 22	03 46
25	20 08	21 19	04 30
26	20 04	21 15	04 59
27	19 59	21 12	05 08
28	19 52	21 09	04 58
29	19 45	21 06	04 29
30	19 ♏ 39	21 ♏ 03	03 N 44

DECLINATIONS

Date	Sun ☉ ° ′	Moon ☽ ° ′	Mercury ☿ ° ′	Venus ♀ ° ′	Mars ♂ ° ′	Jupiter ♃ ° ′	Saturn ♄ ° ′	Uranus ♅ ° ′	Neptune ♆ ° ′	Pluto ♇ ° ′
01	08 N 14	03 S 15	14 N 48	10 N 55	23 S 02	22 S 47	21 N 41	23 N 32	04 N 29	23 S 40
02	07 53	01 N 52	14 34	11 02	23 09	22 47	21 41	23 32	04 28	23 40
03	07 31	06 45	14 18	11 09	23 16	22 47	21 41	23 32	04 28	23 40
04	07 09	11 09	13 58	11 15	23 22	22 48	21 42	23 32	04 27	23 41
05	06 46	14 50	13 36	11 21	23 29	22 48	21 42	23 32	04 27	23 41
06	06 24	17 40	13 10	11 27	23 35	22 49	21 42	23 32	04 26	23 41
07	06 02	19 35	12 42	11 32	23 41	22 49	21 42	23 32	04 25	23 42
08	05 39	20 31	12 11	11 37	23 47	22 50	21 42	23 32	04 25	23 42
09	05 17	20 30	11 37	11 41	23 53	22 50	21 42	23 32	04 24	23 42
10	04 54	19 33	11 02	11 45	23 59	22 51	21 42	23 32	04 24	23 42
11	04 31	17 46	10 24	11 48	24 04	22 51	21 42	23 32	04 23	23 43
12	04 08	15 13	09 45	11 51	24 10	22 52	21 42	23 32	04 23	23 43
13	03 45	12 01	09 04	11 53	24 15	22 52	21 42	23 32	04 22	23 43
14	03 22	08 18	08 22	11 55	24 20	22 53	21 42	23 32	04 21	23 43
15	02 59	04 N 11	07 38	11 56	24 25	22 53	21 42	23 32	04 21	23 44
16	02 36	00 S 10	06 54	11 57	24 29	22 54	21 42	23 32	04 20	23 44
17	02 13	04 34	06 08	11 58	24 34	22 54	21 42	23 32	04 20	23 44
18	01 50	08 51	05 22	11 57	24 38	22 55	21 42	23 32	04 19	23 44
19	01 27	12 47	04 36	11 57	24 43	22 55	21 42	23 32	04 18	23 44
20	01 03	16 08	03 49	11 56	24 47	22 56	21 42	23 32	04 18	23 45
21	00 40	18 40	03 02	11 54	24 50	22 56	21 42	23 32	04 17	23 45
22	00 N 17	20 10	02 14	11 51	24 54	22 57	21 42	23 32	04 16	23 45
23	00 S 07	20 28	01 27	11 49	24 57	22 58	21 42	23 32	04 16	23 45
24	00 30	19 29	00 N 39	11 45	25 01	22 58	21 42	23 32	04 15	23 45
25	00 53	17 17	00 S 08	11 41	25 04	22 59	21 42	23 32	04 15	23 45
26	01 17	14 00	00 55	11 37	25 07	22 59	21 42	23 32	04 14	23 46
27	01 40	09 52	01 42	11 32	25 09	23 00	21 42	23 33	04 13	23 46
28	02 03	05 09	02 29	11 26	25 12	23 00	21 42	23 33	04 13	23 46
29	02 27	00 10	03 15	11 20	25 14	23 01	21 42	23 33	04 12	23 46
30	02 S 50	04 N 45	04 S 01	11 N 14	25 S 16	23 S 02	21 N 42	23 N 33	04 N 11	23 S 46

ZODIAC SIGN ENTRIES

Date	h	m	Planets
02	20	28	☽ ♈
05	00	47	☽ ♉
07	09	05	☽ ♊
07	09	56	☿ ♍
09	20	42	☽ ♋
12	09	27	☽ ♌
14	21	06	☽ ♍
17	06	40	☽ ♎
19	14	11	☽ ♏
21	19	54	☽ ♐
23	05	15	☉ ♎
23	14	35	☿ ♎
23	23	56	☽ ♑
26	02	28	☽ ♒
28	04	09	☽ ♓
30	06	18	☽ ♈

LATITUDES

Date	Mercury ☿ ° ′	Venus ♀ ° ′	Mars ♂ ° ′	Jupiter ♃ ° ′	Saturn ♄ ° ′	Uranus ♅ ° ′	Neptune ♆ ° ′	Pluto ♇ ° ′
01	00 N 20	07 S 04	02 S 05	00 N 15	01 S 31	00 N 08	01 S 36	06 S 42
04	00 56	06 41	02 05	00 14	01 31	00 08	01 36	06 42
07	01 22	06 16	02 05	00 14	01 31	00 08	01 36	06 42
10	01 39	05 49	02 05	00 13	01 31	00 08	01 36	06 42
13	01 47	05 22	02 05	00 13	01 32	00 08	01 37	06 42
16	01 48	04 55	02 04	00 13	01 32	00 08	01 37	06 42
19	01 43	04 28	02 04	00 12	01 32	00 08	01 37	06 42
22	01 34	04 01	02 03	00 12	01 32	00 08	01 37	06 41
25	01 20	03 34	02 03	00 11	01 33	00 08	01 37	06 41
28	01 04	03 08	02 02	00 11	01 33	00 08	01 37	06 41
31	00 N 46	02 S 43	02 S 01	00 N 10	01 S 33	00 N 08	01 S 37	06 S 41

DATA

Julian Date	2463111
Delta T	+74 seconds
Ayanamsa	24° 18′ 13″
Synetic vernal point	04° ♓ 48′ 46″
True obliquity of ecliptic	23° 26′ 02″

MOON'S PHASES, APSIDES AND POSITIONS ☽

Date	h	m	Phase	Longitude ° ′	Eclipse Indicator
01	09	20	○	08 ♓ 46	
08	16	14	☾	15 ♊ 50	
16	18	47	●	23 ♍ 43	
24	01	20	☽	00 ♑ 49	
30	18	58	○	07 ♈ 25	

Day	h	m		
11	12	00	Apogee	
27	07	11	Perigee	
02	03	13	0N	
08	23	22	Max dec	20° N 38′
16	11	08	0S	
23	05	43	Max dec	20° S 30′
29	12	49	0N	

LONGITUDES

Date	Chiron ⚷ ° ′	Ceres ⚳ ° ′	Pallas ⚴ ° ′	Juno ⚵ ° ′	Vesta ⚶ ° ′	Black Moon Lilith ⚸ ° ′
01	21 ♉ 22	04 ♌ 16	02 ♊ 02	07 ♉ 56	16 ♌ 10	21 ♋ 45
11	21 ♉ 15	08 ♌ 27	04 ♊ 20	09 ♉ 00	20 ♌ 35	22 ♋ 52
21	21 ♉ 03	12 ♌ 33	06 ♊ 08	09 ♉ 17	24 ♌ 59	23 ♋ 59
31	20 ♉ 45	16 ♌ 33	07 ♊ 21	08 ♉ 44	29 ♌ 20	25 ♋ 06

ASPECTARIAN

h m	Aspects
01 Monday	
01 34	☽ □ ♂
06 13	☽ # ♆
09 20	☽ ☍ ☉
09 56	☽ ⊥ ♆
11 03	☉ Q ♅
11 23	☽ ⊼ ♀
14 23	☽ ⊻ ♇
17 57	♀ St D
18 14	☉ ± ♆
19 44	☽ ⊻ ♆
21 13	☽ ± ♀
21 15	♅ bQ ♇
02 Tuesday	
00 13	☽ ⊥ ♇
03 05	☽ □ ♃
06 13	☿ ⚹ ♄
07 10	☽ □ ♄
07 15	☽ ⊼ ☿
12 08	☽ bQ ♀
15 09	☽ ∠ ♇
15 12	☽ □ ♅
16 55	☉ ⊻ ♀
18 19	☽ ± ☿
03 Wednesday	
00 33	☽ ∥ ♆
05 25	☽ △ ♂
11 19	☽ bQ ☿
13 38	☽ △ ♀
13 42	☽ Q ♄
15 06	☽ ⊼ ☉
15 39	☽ ∥ ☉
16 37	☽ ⚹ ♇
21 59	☽ Q ♅
22 17	☽ ☌ ♆
04 Thursday	
02 19	☽ ± ☉
06 24	☽ △ ♃
08 39	☽ bQ ♂
10 45	☽ ⚹ ♄
12 28	☉ ⊼ ♇
12 40	☽ ∥ ♀
13 38	☽ Q ♇
16 48	☽ △ ☿
19 14	☽ ⚹ ♅
19 31	☽ bQ ☉
05 Friday	
01 32	☽ ± ♂
04 13	☽ ∥ ☿
09 28	☽ bQ ♃
13 01	☽ ⊼ ♂
13 42	☿ ⚹ ♅
13 58	☽ ∠ ♄
19 43	☽ □ ♀
22 23	☽ □ ♇
22 43	☽ ∠ ♅
23 30	♂ # ♅
06 Saturday	
01 12	☽ △ ☉
02 12	☽ ± ♃
04 29	☽ ⊻ ♆
06 48	☽ ⊥ ♄
13 35	☽ ⊼ ♃
15 42	☽ ⊥ ♅
15 48	☽ ⊥ ♆
18 14	☽ ⊻ ♄
07 Sunday	
03 13	☽ ⊻ ♅
06 22	☽ Q ♀
08 57	☽ □ ☿
09 05	☽ ∠ ♆
09 55	☿ bQ ♆
13 36	♂ ∥ ♇
19 34	☉ ⊼ ♆
20 08	♅ Q ♆
08 Monday	
01 16	☽ ☍ ♂
06 18	☽ ⚹ ♀
08 05	☽ △ ♇
14 32	☽ ⚹ ♆
16 14	☽ □ ☉
09 Tuesday	
00 36	☽ ☍ ♃
02 31	☽ Q ☿
05 24	☽ ☌ ♄
09 51	☿ ∥ ♀
12 57	☽ ∠ ♀
14 03	☽ bQ ♇
14 33	☽ Q ♆
14 42	☽ ☌ ♅
22 10	☿ Q ♄
10 Wednesday	
06 20	☉ ⊥ ♀
07 03	☽ ⚹ ☿
07 36	☽ ⊥ ♀
07 41	☽ Q ☉
08 13	☽ ± ♇
14 09	☉ ± ♇
16 50	☽ ⊼ ♂
20 06	☽ ⊻ ♀
20 22	☽ ⊼ ♇
11 Thursday	
02 55	☽ □ ♆
05 13	♀ ☍ ♇
05 41	☽ ± ♂
10 11	☽ ⚹ ☉
13 39	☽ ⊼ ♃
18 20	☽ ⊻ ♄
19 04	☽ ∠ ☿
20 41	☉ ∥ ♆
12 Friday	
00 57	☽ bQ ♂
01 52	☽ ± ♃
03 33	☽ ⊻ ♅
06 30	☽ ⊥ ♄
07 57	☿ ± ♆
08 47	♀ ∠ ♅
10 29	☿ Q ♅
15 40	☽ ⊥ ♅
16 37	☽ ⊥ ☿
19 10	☽ ∠ ☉
20 08	☽ bQ ♃
13 Saturday	
00 40	☽ ∠ ♄
06 53	☽ ⊻ ☿
07 52	☉ □ ♃
08 48	☽ △ ♂
08 48	☽ ☍ ♇
08 49	♂ ⚹ ♇
09 43	☽ ∠ ♅
10 31	☽ ☌ ♀
12 56	☽ ∥ ♀
14 46	☽ ⊥ ☉
15 12	☽ △ ♆
19 01	☿ ⊼ ♇
14 Sunday	
00 25	☿ □ ♂
02 14	☽ △ ♃
03 42	☽ ⊻ ☉
06 36	☽ ⚹ ♄
11 29	☿ ⊻ ♀
11 35	☽ ∥ ☿
15 27	☽ ⚹ ♅
20 45	☽ bQ ♆
15 Monday	
06 11	☽ Q ♄
11 07	☽ ∥ ♆
11 18	☿ ⊼ ♆
14 13	☽ ± ♆
14 51	☽ Q ♅
17 12	☉ □ ♄
19 23	☽ ∥ ☉
19 37	☽ ⊼ ♇
22 54	☽ □ ♂
23 28	☽ ⊻ ♀
16 Tuesday	
01 45	☽ ⊼ ♆
04 22	☽ ☌ ☿
07 07	☽ ± ♇
11 25	☽ ⊥ ♀
12 55	☽ □ ♃
16 52	☽ □ ♄
18 47	☽ ☌ ☉
22 28	☿ ± ♇
17 Wednesday	
00 12	☽ bQ ♇
00 13	☽ # ☉
01 18	☽ □ ♅
05 10	☽ ∠ ♀
10 39	☽ # ♆
10 57	☽ Q ♂
18 19	♀ △ ♂
19 20	☽ # ☿
23 05	☽ Q ♃
18 Thursday	
03 36	♂ △ ♆
04 17	☽ △ ♇
05 51	♀ △ ♆
10 09	☽ ☍ ♆
10 21	☽ ⚹ ♀
10 31	☽ ⚹ ♂
15 51	☿ □ ♃
18 03	☿ ⊥ ♀
21 27	☽ ⚹ ♃
22 21	☽ ⊻ ☿
19 Friday	
01 00	☽ △ ♄
06 37	☽ # ♀
07 12	☽ ⊻ ☉
09 03	☽ △ ♅
09 21	☽ Q ♀
11 09	☽ ⊥ ☿
15 29	☽ ∠ ♂
16 14	☉ bQ ♇
17 04	☿ □ ♄
18 59	☽ ⊥ ☉
21 04	☿ ∥ ♆
20 Saturday	
01 00	☽ ∠ ♃
04 20	☽ bQ ♄
06 09	☽ ∠ ☿
08 22	☉ □ ♅
08 37	☽ ⊥ ♂
11 01	☽ □ ♇
12 13	☽ bQ ♅
12 31	☽ ∠ ☉
16 39	☽ ⊼ ♆
17 19	☽ ⊥ ♃
19 20	☽ □ ♀
19 56	☽ ⊻ ♂
20 32	☽ ± ♄
21 Sunday	
03 21	☽ ± ♆
04 07	☽ ⊻ ♃
04 19	☽ ± ♅
07 14	☽ ⊼ ♄
13 14	☽ ⚹ ☿
14 58	☽ ⊼ ♅
16 54	☿ bQ ♇
17 19	☽ ⚹ ☉
19 02	☽ Q ♇
19 16	☽ bQ ♆
22 Monday	
02 03	☿ □ ♅
13 36	☽ Q ☿
15 58	☽ Q ☉
16 02	☽ ⚹ ♇
21 27	☽ △ ♆
23 Tuesday	
02 34	☽ △ ♀
03 23	☽ ☌ ♂
09 03	☽ ☌ ♃
11 46	☽ ☍ ♄
17 55	☽ ∠ ♇
19 11	☽ ☍ ♅
24 Wednesday	
01 19	☽ □ ☿
01 20	☽ □ ☉
01 27	☉ ☌ ☿
05 32	☽ bQ ♀
09 15	☽ ⊥ ♇
15 15	☉ # ☿
19 24	☽ ⊻ ♇
21 32	☽ ± ♀
25 Thursday	
00 36	☽ □ ♆
08 08	☽ ⊼ ♀
08 58	☽ ⊻ ♂
12 22	☽ ⊻ ♃
14 41	☽ ⊼ ♄
19 30	☽ ⊥ ♂
21 52	☽ ⊼ ♅
22 29	☽ ⊥ ♃
26 Friday	
00 42	☽ ± ♄
03 36	☿ ∠ ♀
06 39	☽ Q ♆
07 30	☽ △ ☉
07 51	☽ ± ♅
10 59	☽ △ ☿
11 13	☽ ∠ ♂
13 34	☽ ∠ ♃
15 41	☽ bQ ♄
21 27	☽ ☌ ♇
22 47	☽ bQ ♅
27 Saturday	
02 31	☽ ⚹ ♆
02 37	☽ # ♀
09 53	☉ ∥ ☿
10 09	☽ bQ ☉
12 31	☽ ☍ ♀
13 16	☽ ⚹ ♂
14 36	☽ ⚹ ♃
15 17	☽ bQ ☿
16 32	☽ △ ♄
23 37	☽ △ ♅
28 Sunday	
02 07	☽ ± ☉
03 17	☽ ∠ ♆
08 14	☽ ± ☿
10 06	☽ Q ♂
10 38	☽ Q ♃
12 47	☽ ⊼ ☉
16 34	☽ # ♆
18 10	☽ ⊥ ♆
19 33	☽ ⊼ ☿
23 06	☽ ⊻ ♇
23 11	☽ ∥ ☿
23 44	♂ ☌ ♃
29 Monday	
01 53	☽ ∥ ☉
04 09	☽ ⊻ ♆
05 33	☿ Q ♃
08 38	☿ Q ♂
09 06	☽ ⊥ ♇
13 36	♀ △ ♃
16 55	☽ □ ♃
17 04	☽ ⊼ ♀
17 37	☽ □ ♂
18 29	☽ □ ♄
30 Tuesday	
00 13	☽ ∠ ♇
00 56	☿ △ ♇
01 40	☽ □ ♅
01 42	☽ # ☉
03 45	☽ ± ♀
07 41	☽ # ☿
09 12	☽ ∥ ♆
11 54	♂ ☍ ♄
17 15	☿ # ♆
18 58	☽ ☍ ☉
19 55	♀ ⚹ ♄
19 58	☽ bQ ♀

All ephemeris data is given at 12.00 UT and the Moon's longitude is additionally given for 24.00 UT

OCTOBER 2031

LONGITUDES

Date	Sidereal time h m s	Sun ☉ ° ' "	Moon ☽ ° ' "	Moon ☽ 24.00 ° ' "	Mercury ☿ ° '	Venus ♀ ° '	Mars ♂ ° '	Jupiter ♃ ° '	Saturn ♄ ° '	Uranus ♅ ° '	Neptune ♆ ° '	Pluto ♇ ° '
01	12 40 07	08 ♎ 07 08	17 ♈ 16 21	24 ♈ 06 24	13 ♎ 54	23 ♌ 32	23 ♐ 42	22 ♐ 18	23 ♊ 01	27 ♊ 16	14 ♈ 23	11 ♒ 23
02	12 44 03	09 06 06	00 ♉ 50 58	07 ♉ 29 52	15 36	24 19	24 23	22 25	23 01	27 16	14 R 21	11 R 23
03	12 48 00	10 05 06	14 ♉ 03 08	20 ♉ 30 50	17 17	25 07	25 04	22 33	23 01	27 R 16	14 19	11 22
04	12 51 56	11 04 09	26 ♉ 53 13	03 ♊ 10 35	18 57	25 56	25 46	22 41	23 02	27 16	14 18	11 22
05	12 55 53	12 03 14	09 ♊ 23 21	15 ♊ 31 57	20 36	26 46	26 27	22 50	23 R 02	27 16	14 16	11 21
06	12 59 49	13 02 21	21 ♊ 36 56	27 ♊ 38 52	22 14	27 36	27 09	22 58	23 02	27 16	14 14	11 21
07	13 03 46	14 01 31	03 ♋ 38 20	09 ♋ 35 57	23 52	28 27	27 51	23 07	23 01	27 15	14 13	11 20
08	13 07 42	15 00 42	15 ♋ 32 22	21 ♋ 28 10	25 29	29 ♌ 19	28 33	23 15	23 01	27 15	14 11	11 20
09	13 11 39	15 59 56	27 ♋ 24 01	03 ♌ 20 28	27 05	00 ♍ 12	29 15	23 24	23 01	27 15	14 09	11 19
10	13 15 36	16 59 13	09 ♌ 18 08	15 ♌ 17 32	28 ♎ 40	01 05	29 ♐ 57	23 33	23 00	27 14	14 08	11 19
11	13 19 32	17 58 31	21 ♌ 19 11	27 ♌ 23 31	00 ♏ 14	01 59	00 ♑ 39	23 42	23 00	27 14	14 06	11 18
12	13 23 29	18 57 52	03 ♍ 30 57	09 ♍ 41 49	01 48	02 54	01 22	23 51	22 59	27 14	14 04	11 18
13	13 27 25	19 57 15	15 ♍ 56 24	22 ♍ 14 54	03 21	03 49	02 04	24 00	22 58	27 13	14 03	11 18
14	13 31 22	20 56 41	28 ♍ 37 27	05 ♎ 04 07	04 54	04 45	02 47	24 09	22 57	27 13	14 01	11 17
15	13 35 18	21 56 08	11 ♎ 34 53	18 ♎ 09 42	06 25	05 41	03 29	24 19	22 56	27 12	13 59	11 17
16	13 39 15	22 55 38	24 ♎ 48 26	01 ♏ 30 52	07 56	06 38	04 12	24 28	22 55	27 11	13 58	11 17
17	13 43 11	23 55 09	08 ♏ 16 49	15 ♏ 06 00	09 27	07 36	04 55	24 38	22 54	27 10	13 56	11 17
18	13 47 08	24 54 43	21 ♏ 58 08	28 ♏ 52 55	10 56	08 34	05 38	24 48	22 52	27 10	13 54	11 16
19	13 51 05	25 54 18	05 ♐ 50 04	12 ♐ 49 16	12 25	09 32	06 21	24 58	22 51	27 09	13 53	11 16
20	13 55 01	26 53 56	19 ♐ 50 13	26 ♐ 52 40	13 54	10 31	07 04	25 08	22 49	27 08	13 51	11 16
21	13 58 58	27 53 35	03 ♑ 56 20	11 ♑ 00 58	15 21	11 31	07 48	25 18	22 47	27 07	13 49	11 16
22	14 02 54	28 53 16	18 ♑ 06 17	25 ♑ 12 03	16 48	12 30	08 31	25 28	22 46	27 06	13 48	11 16
23	14 06 51	29 ♎ 52 59	02 ♒ 18 00	09 ♒ 23 52	18 14	13 31	09 15	25 38	22 44	27 05	13 46	11 16
24	14 10 47	00 ♏ 52 43	16 ♒ 29 23	23 ♒ 34 13	19 39	14 31	09 58	25 49	22 42	27 04	13 45	11 16
25	14 14 44	01 52 29	00 ♓ 38 05	07 ♓ 40 38	21 04	15 32	10 42	25 59	22 40	27 03	13 43	11 D 16
26	14 18 40	02 52 16	14 ♓ 41 30	21 ♓ 40 21	22 28	16 34	11 26	26 10	22 37	27 02	13 41	11 16
27	14 22 37	03 52 05	28 ♓ 36 47	05 ♈ 30 29	23 51	17 36	12 10	26 21	22 35	27 01	13 40	11 16
28	14 26 34	04 51 56	12 ♈ 21 05	19 ♈ 08 17	25 13	18 38	12 53	26 32	22 33	26 59	13 38	11 16
29	14 30 30	05 51 49	25 ♈ 51 49	02 ♉ 31 27	26 34	19 41	13 37	26 42	22 30	26 58	13 37	11 16
30	14 34 27	06 51 44	09 ♉ 07 00	15 ♉ 38 23	27 54	20 43	14 21	26 53	22 27	26 57	13 35	11 16
31	14 38 23	07 ♏ 51 41	22 ♉ 05 31	28 ♉ 28 28	29 ♏ 13	21 ♍ 47	15 ♑ 06	27 ♐ 05	22 ♊ 25	26 ♊ 55	13 ♈ 34	11 ♒ 17

Date	Moon True ☊ ° '	Moon Mean ☊ ° '	Moon ☽ Latitude ° '
01	19 ♏ 33	21 ♏ 00	02 N 46
02	19 R 30	20 56	01 39
03	19 28	20 53	00 N 29
04	19 D 28	20 50	00 S 40
05	19 30	20 47	01 46
06	19 31	20 44	02 46
07	19 33	20 41	03 38
08	19 R 33	20 37	04 19
09	19 33	20 34	04 49
10	19 30	20 31	05 07
11	19 27	20 28	05 12
12	19 22	20 25	05 03
13	19 17	20 21	04 39
14	19 12	20 18	04 02
15	19 08	20 15	03 11
16	19 05	20 12	02 09
17	19 03	20 09	00 S 59
18	19 D 03	20 06	00 N 16
19	19 03	20 02	01 31
20	19 05	19 59	02 42
21	19 06	19 56	03 43
22	19 07	19 53	04 31
23	19 R 08	19 50	05 02
24	19 07	19 46	05 15
25	19 06	19 43	05 09
26	19 04	19 40	04 45
27	19 01	19 37	04 04
28	18 59	19 34	03 09
29	18 58	19 31	02 04
30	18 57	19 27	00 54
31	18 ♏ 57	19 ♏ 24	00 S 17

DECLINATIONS

Date	Sun ☉ ° '	Moon ☽ ° '	Mercury ☿ ° '	Venus ♀ ° '	Mars ♂ ° '	Jupiter ♃ ° '	Saturn ♄ ° '	Uranus ♅ ° '	Neptune ♆ ° '	Pluto ♇ ° '
01	03 S 13	09 N 20	04 S 47	11 N 07	25 S 18	23 S 02	21 N 42	23 N 33	04 N 11	23 S 46
02	03 36	13 19	05 32	10 59	25 19	23 03	21 42	23 33	04 10	23 46
03	04 00	16 31	06 17	10 51	25 21	23 03	21 42	23 33	04 09	23 46
04	04 23	18 48	07 01	10 42	25 22	23 04	21 42	23 33	04 09	23 46
05	04 46	20 06	07 44	10 33	25 23	23 05	21 42	23 33	04 08	23 46
06	05 09	20 24	08 28	10 23	25 23	23 05	21 41	23 33	04 07	23 46
07	05 32	19 45	09 10	10 13	25 24	23 06	21 41	23 33	04 07	23 47
08	05 55	18 14	09 52	10 02	25 24	23 06	21 41	23 33	04 06	23 47
09	06 18	15 57	10 33	09 50	25 24	23 07	21 41	23 33	04 05	23 47
10	06 40	12 59	11 14	09 39	25 24	23 07	21 41	23 33	04 05	23 47
11	07 03	09 28	11 54	09 26	25 23	23 08	21 41	23 33	04 04	23 47
12	07 26	05 30	12 33	09 14	25 22	23 08	21 41	23 33	04 03	23 47
13	07 48	01 N 15	13 12	09 00	25 21	23 09	21 41	23 33	04 03	23 47
14	08 10	03 S 09	13 50	08 47	25 20	23 10	21 41	23 33	04 02	23 47
15	08 33	07 31	14 27	08 32	25 19	23 10	21 40	23 33	04 02	23 47
16	08 55	11 37	15 03	08 18	25 17	23 11	21 40	23 33	04 01	23 47
17	09 17	15 11	15 38	08 03	25 15	23 11	21 40	23 33	04 00	23 47
18	09 39	18 00	16 13	07 47	25 13	23 12	21 40	23 33	04 00	23 46
19	10 00	19 47	16 47	07 31	25 10	23 12	21 40	23 33	03 59	23 46
20	10 22	20 22	17 20	07 15	25 08	23 13	21 40	23 33	03 58	23 46
21	10 43	19 40	17 52	06 58	25 05	23 13	21 40	23 33	03 58	23 46
22	11 05	17 44	18 23	06 41	25 02	23 14	21 39	23 33	03 57	23 46
23	11 26	14 44	18 53	06 23	24 58	23 14	21 39	23 33	03 57	23 46
24	11 47	10 52	19 23	06 05	24 54	23 15	21 39	23 33	03 56	23 46
25	12 07	06 25	19 51	05 47	24 50	23 15	21 39	23 33	03 55	23 46
26	12 28	01 S 39	20 18	05 28	24 46	23 15	21 39	23 33	03 55	23 46
27	12 48	03 N 10	20 45	05 09	24 42	23 16	21 39	23 33	03 54	23 46
28	13 08	07 47	21 10	04 50	24 37	23 16	21 38	23 33	03 54	23 46
29	13 28	11 55	21 34	04 30	24 32	23 17	21 38	23 33	03 53	23 45
30	13 48	15 23	21 57	04 10	24 27	23 17	21 38	23 33	03 52	23 45
31	14 S 08	18 N 00	22 S 19	03 N 50	24 S 21	23 S 17	21 N 38	23 N 33	03 N 52	23 S 45

ZODIAC SIGN ENTRIES

Date	h	m	Planets
02	10	29	☽ ♉
04	17	55	☽ ♊
07	04	42	☽ ♋
09	06	33	♀ ♍
09	17	15	☽ ♌
10	13	47	♂ ♑
11	08	18	☿ ♏
12	05	07	☽ ♍
14	14	34	☽ ♎
16	21	18	☽ ♏
19	01	56	☽ ♐
21	05	19	☽ ♑
23	08	07	☽ ♒
23	14	49	☉ ♏
25	10	55	☽ ♓
27	14	24	☽ ♈
29	19	26	☽ ♉

LATITUDES

Date	Mercury ☿ ° '	Venus ♀ ° '	Mars ♂ ° '	Jupiter ♃ ° '	Saturn ♄ ° '	Uranus ♅ ° '	Neptune ♆ ° '	Pluto ♇ ° '
01	00 N 46	02 S 43	02 S 01	00 N 10	01 S 33	00 N 08	01 S 37	06 S 41
04	00 26	02 18	02 00	00 10	01 33	00 08	01 37	06 40
07	00 N 06	01 55	01 59	00 10	01 34	00 08	01 37	06 40
10	00 S 15	01 32	01 58	00 09	01 34	00 08	01 37	06 40
13	00 36	01 11	01 56	00 09	01 34	00 08	01 37	06 40
16	00 57	00 50	01 55	00 08	01 34	00 09	01 37	06 39
19	01 17	00 30	01 54	00 08	01 35	00 09	01 37	06 39
22	01 36	00 S 12	01 52	00 08	01 35	00 09	01 37	06 39
25	01 54	00 N 06	01 51	00 07	01 35	00 09	01 37	06 38
28	02 10	00 22	01 49	00 07	01 35	00 09	01 37	06 38
31	02 S 24	00 N 37	01 S 47	00 N 07	01 S 35	00 N 09	01 S 37	06 S 38

DATA

Julian Date	2463141
Delta T	+74 seconds
Ayanamsa	24° 18' 15"
Synetic vernal point	04° ♓ 48' 44"
True obliquity of ecliptic	23° 26' 01"

MOON'S PHASES, APSIDES AND POSITIONS ☽

Date	h	m	Phase	Longitude ° '	Eclipse Indicator
08	10	50	☾	14 ♋ 58	
16	08	21	●	22 ♎ 47	
23	07	36	☽	29 ♑ 42	
30	07	33	○	06 ♉ 41	

Day	h	m		
09	06	24	Apogee	
22	20	17	Perigee	
06	07	28	Max dec	20° N 25'
13	18	51	0S	
20	10	59	Max dec	20° S 22'
26	20	10	0N	

LONGITUDES

Date	Chiron ⚷ ° '	Ceres ⚳ ° '	Pallas ⚴ ° '	Juno ⚵ ° '	Vesta ⚶ ° '	Black Moon Lilith ⚸ ° '
01	20 ♉ 45	16 ♌ 33	07 ♊ 21	08 ♉ 44	29 ♌ 20	25 ♋ 06
11	20 ♉ 23	20 ♌ 25	07 ♊ 48	07 ♉ 23	03 ♍ 38	26 ♋ 13
21	19 ♉ 57	24 ♌ 08	07 ♊ 22	05 ♉ 24	07 ♍ 50	27 ♋ 20
31	19 ♉ 28	27 ♌ 39	05 ♊ 57	03 ♉ 08	11 ♍ 56	28 ♋ 27

All ephemeris data is given at 12.00 UT and the Moon's longitude is additionally given for 24.00 UT

ASPECTARIAN

h m Aspects

01 Wednesday
01 08 ☽ Q ♄
01 48 ☽ ⚹ ♇
05 20 ☽ ☍ ☿
06 58 ☽ ☌ ♆
08 30 ☽ Q ♅
18 30 ☿ ☍ ♆
20 53 ☽ △ ♃
21 53 ☽ ∥ ♀
22 04 ☽ ⚹ ♄
22 43 ☽ Q ♇
23 39 ☽ △ ♀
23 52 ☽ △ ♂

02 Thursday
05 36 ☽ ⚹ ♅
23 59 ☽ bQ ♃

03 Friday
00 57 ☽ ∠ ♄
02 19 ♀ △ ♂
02 42 ♅ St R
04 08 ☽ ⊼ ☉
04 17 ☽ bQ ♂
07 04 ☽ □ ♇
08 43 ☽ ∠ ♅
12 30 ☽ ⊻ ♆
16 04 ☽ ± ☉
16 12 ☉ ∠ ♀
16 41 ☽ ± ♃
17 30 ☽ ⊥ ♄
18 52 ☽ ⊼ ☿
21 50 ☉ ⋕ ♆
21 50 ☽ ± ♂
23 37 ☽ ⊥ ♆

04 Saturday
01 19 ☉ Q ♃
01 24 ☽ ⊥ ♅
04 00 ☽ ⊼ ♃
04 43 ☽ ⊻ ♄
07 47 ☽ ± ☿
09 45 ☽ ⊼ ♂
10 04 ☽ □ ♀
10 19 ☽ bQ ☉
12 43 ☽ ⊻ ♅
16 34 ☽ ∠ ♆
19 01 ☉ △ ♇

05 Sunday
03 33 ☽ bQ ☿
08 31 ♂ ∠ ♇
10 50 ♄ St R
15 49 ☽ △ ♇
17 38 ☽ △ ☉
21 30 ☽ ⚹ ♆
23 15 ☽ Q ♀

06 Monday
02 20 ♀ ⚹ ♅
13 26 ☽ △ ☿
14 43 ☽ ☍ ♃
14 48 ☽ ☌ ♄
15 50 ♂ ☍ ♅
21 10 ☽ Q ♆
21 23 ☽ bQ ♇
21 54 ♃ ☍ ♄
23 13 ☽ ☌ ♅
23 33 ☿ △ ♄
23 41 ☽ ☍ ♂
23 43 ☿ ⚹ ♃

07 Tuesday
00 49 ☽ ⚹ ♀
15 24 ☽ ± ♇
16 21 ☉ ☍ ♆

08 Wednesday
03 30 ☽ ⊼ ♇
08 14 ♀ bQ ♆
09 16 ☽ □ ♆
09 21 ☽ ∠ ♀
10 50 ☽ □ ☉
16 27 ☿ ⋕ ♀

09 Thursday
03 08 ☽ ⊻ ♄
03 48 ☽ ⊼ ♃
05 00 ☽ ⊥ ♀
11 15 ☽ □ ☿
11 41 ☽ ⊻ ♅
14 31 ☿ △ ♅
15 15 ☽ ⊥ ♄
15 58 ☽ ⊼ ♂
16 05 ☽ ± ♃
18 07 ☽ ⊻ ♀
23 48 ☽ ⊥ ♅

10 Friday
02 32 ☽ Q ☉
04 50 ☽ ± ♂
09 24 ☽ ∠ ♄
10 28 ☽ bQ ♃
16 02 ☽ ☍ ♇
17 54 ☽ ∠ ♅
21 39 ☽ △ ♆
22 21 ☽ ⋕ ☿

11 Saturday
00 01 ☽ bQ ♂
04 45 ☽ ⚹ ☉
04 58 ☽ Q ☿
12 08 ☽ ∥ ♀
15 19 ☽ ⚹ ♄
16 46 ☽ △ ♃
23 28 ☿ ⚹ ♂
23 41 ☽ ⚹ ♅

12 Sunday
01 36 ☽ ⋕ ☉
03 19 ☽ bQ ♆
07 32 ☽ △ ♂
08 10 ☽ ⚹ ☿
10 42 ☽ ☌ ♀
12 57 ☽ ∠ ☉
14 51 ☽ Q ♄
20 20 ☽ ∥ ♆
20 50 ☽ ± ♆
23 05 ☽ Q ♅

13 Monday
03 05 ☽ ⊼ ♇
07 52 ☽ ⊥ ☉
08 22 ☽ ⊼ ♆
14 35 ☽ ± ♇
17 15 ☽ ∠ ☿
20 18 ☽ ⊻ ☉
21 40 ☉ Q ♂

14 Tuesday
01 21 ☽ □ ♄
03 30 ☽ □ ♃
06 16 ☿ ⚹ ♀
07 37 ☽ bQ ♇
09 21 ☽ □ ♅
12 35 ☽ ⊥ ☿
16 48 ☽ ⋕ ♆
17 03 ♀ Q ♄
20 12 ☽ □ ♂

15 Wednesday
00 18 ☽ ⊻ ♀
01 15 ☽ ⊻ ☿
11 27 ☽ △ ♇
11 55 ☉ ⋕ ♀
12 13 ☽ ⊥ ♀
13 21 ☽ Q ♃
16 23 ☽ ☍ ♆
17 31 ☽ ⋕ ♀
18 25 ☽ ∥ ☉

16 Thursday
05 51 ☽ ∠ ♀
07 03 ☽ Q ♂
08 21 ☽ ☌ ☉
08 36 ☽ △ ♄
11 24 ☽ ⚹ ♃
11 37 ☿ bQ ♄
11 42 ☉ △ ♄
16 16 ☽ △ ♅

17 Friday
05 43 ☽ ⚹ ♂
10 42 ☽ ⚹ ♀
11 19 ☽ bQ ♄
14 19 ☽ ☌ ☿
14 25 ☽ ∠ ♃
15 25 ☿ ∠ ♃
16 09 ☽ ∥ ☿
17 17 ☽ □ ♇
18 51 ☽ bQ ♅
20 07 ♀ ± ♆
21 56 ☽ ⊼ ♆
23 59 ♂ ⊥ ♇

18 Saturday
03 07 ☽ ± ♄
06 24 ☽ ⊥ ♃
08 25 ☽ ± ♆
08 41 ☉ ⚹ ♃
09 22 ☽ Q ♀
09 33 ☽ ∠ ♂
10 36 ☽ ± ♅
13 34 ☽ ⊼ ♄
16 59 ☽ ⊻ ♃
17 25 ☿ □ ♇
17 31 ☽ ⊻ ☉
21 01 ☽ ⊼ ♅

19 Sunday
00 01 ☽ bQ ♆
00 41 ☽ Q ♇
02 02 ☽ ⊥ ♂
02 35 ♀ Q ♅
04 42 ☽ ⊥ ☉
07 37 ☿ bQ ♅
12 57 ☽ ⊻ ♂
18 51 ☽ □ ♀
21 21 ☽ ⚹ ♇
21 23 ☽ ∠ ☉

20 Monday
00 39 ☽ ⊻ ☿
01 47 ☽ △ ♆
11 20 ☿ ⊼ ♆
12 06 ☽ ⊥ ☿
17 05 ☽ ☍ ♄
17 35 ☉ △ ♅
21 08 ☽ ☌ ♃
22 58 ☽ ∠ ♇

21 Tuesday
00 25 ☽ ☍ ♅
00 57 ☽ ⚹ ☉
05 12 ☽ ∠ ☿
06 12 ♀ ⊼ ♇
14 15 ☽ ⊥ ♇
18 54 ☽ ☌ ♂
22 52 ☽ Q ☉

22 Wednesday
00 26 ☽ ⊻ ♇
01 48 ☽ △ ♀
04 43 ☽ □ ♆
06 38 ☽ ∥ ☿
09 32 ☽ ⚹ ☿
11 22 ☿ ± ♄
19 51 ☽ ⊼ ♄

23 Thursday
00 36 ☽ ⊻ ♃
03 12 ☽ ⊼ ♅
05 06 ☽ bQ ♀
05 59 ☽ ± ♄
07 36 ☽ □ ☉
08 07 ☽ Q ☿
10 52 ☽ ⊥ ♃
11 06 ☽ Q ♆
13 20 ☽ ± ♅
18 02 ♀ ⊼ ♆
21 09 ☽ bQ ♄
21 29 ☽ ± ♀

24 Friday
00 22 ☽ ⊻ ♂
02 17 ☽ ∠ ♃
03 10 ☽ ☌ ♇
04 32 ☽ bQ ♅
07 10 ☽ ∥ ☉
07 22 ☽ ⚹ ♆
08 25 ☽ ⊼ ♀
11 04 ☽ ⊥ ♂
13 27 ☿ ± ♆
15 03 ☿ ⊥ ♃
17 58 ☽ □ ☿
22 29 ☽ △ ♄
23 12 ♇ St D

25 Saturday
03 09 ☽ ∠ ♂
04 00 ☽ ⚹ ♃
05 55 ☽ △ ♅
08 45 ☽ ∠ ♆
11 42 ☿ ± ♅
14 16 ☽ △ ☉
15 29 ☽ ⋕ ♀

26 Sunday
00 03 ☽ ⊥ ♆
00 42 ☽ Q ♃
00 42 ☽ ⋕ ♆
06 06 ☽ ⚹ ♂
06 08 ☽ ⊻ ♇
06 40 ♂ ⊻ ♇
10 17 ☽ ⊻ ♆
14 41 ☿ ⊼ ♄
15 28 ☽ ☍ ♀
16 25 ☽ ⊥ ♇
17 53 ☽ bQ ☉

27 Monday
01 22 ♃ ∠ ♇
01 36 ☽ □ ♄
02 50 ☽ △ ☿
03 52 ☽ Q ♂
04 23 ♀ ± ♇
07 56 ☽ ∠ ♇
08 01 ☽ □ ♃
09 13 ☽ □ ♅
10 36 ☽ ± ☉
15 42 ☽ ∥ ♆
21 28 ☽ ∥ ♀
21 51 ☽ ⊼ ☉

28 Tuesday
07 49 ☽ bQ ☿
08 50 ☽ Q ♄
10 06 ☽ ⚹ ♇
13 00 ☽ □ ♂
14 16 ☽ ☌ ♆
16 39 ☽ Q ♅

29 Wednesday
00 02 ☽ ⊼ ♀
01 29 ☽ ± ☿
06 00 ☽ ⚹ ♄
07 22 ☽ Q ♇
11 38 ☽ ± ♀
11 43 ♂ □ ♆
13 24 ☽ ⊼ ☿
13 32 ☽ △ ♃
13 59 ☽ ⚹ ♅
15 01 ☿ ⊻ ♃
16 06 ☿ ⋕ ♄
19 08 ☿ ⊼ ♅
23 09 ☽ ⋕ ☉

30 Thursday
05 16 ☽ bQ ♀
07 33 ☽ ☍ ☉
08 59 ☽ ∠ ♄
15 57 ☽ □ ♇
17 10 ☽ bQ ♃
17 11 ☽ ∠ ♅
18 11 ♃ ☍ ♅
20 12 ☽ ⊻ ♆
22 12 ☽ △ ♂

31 Friday
00 23 ☿ bQ ♆
01 28 ☽ ⊥ ♄
01 41 ☉ bQ ♄
07 18 ☽ ⊥ ♆
09 27 ♀ ∥ ♆
09 49 ☽ ⊥ ♅
10 04 ☽ ± ♃
11 22 ☽ △ ♀
12 36 ☽ ⊻ ♄
13 11 ☿ Q ♇
21 03 ☽ ⊻ ♅
21 30 ☽ ⊼ ♃

NOVEMBER 2031

LONGITUDES

Date	Sidereal time h m s	Sun ☉ ° ' "	Moon ☽ ° ' "	Moon ☽ 24.00 ° ' "	Mercury ☿ ° '	Venus ♀ ° '	Mars ♂ ° '	Jupiter ♃ ° '	Saturn ♄ ° '	Uranus ♅ ° '	Neptune ♆ ° '	Pluto ♇ ° '
01	14 42 20	08 ♏ 51 39	04 ♊ 47 18	11 ♊ 02 11	00 ♐ 30	22 ♍ 50	15 ♑ 50	27 ♐ 16	22 ♊ 22	26 ♊ 54	13 ♈ 32	11 ♒ 17
02	14 46 16	09 51 40	17 ♊ 13 20	23 ♊ 21 04	01 47	23 54	16 34	27 27	22 R 19	26 R 52	13 R 31	11 17
03	14 50 13	10 51 43	29 ♊ 25 42	05 ♋ 27 39	03 02	24 59	17 18	27 38	22 16	26 51	13 29	11 17
04	14 54 09	11 51 48	11 ♋ 27 22	17 ♋ 25 19	04 15	26 03	18 03	27 50	22 13	26 49	13 28	11 17
05	14 58 06	12 51 55	23 ♋ 22 02	29 ♋ 18 03	05 26	27 08	18 47	28 01	22 10	26 48	13 27	11 18
06	15 02 03	13 52 04	05 ♌ 13 58	11 ♌ 10 21	06 36	28 13	19 32	28 13	22 07	26 46	13 25	11 18
07	15 05 59	14 52 15	17 ♌ 07 48	23 ♌ 06 54	07 43	29 ♍ 19	20 16	28 25	22 03	26 44	13 24	11 18
08	15 09 56	15 52 28	29 ♌ 08 15	05 ♍ 12 23	08 48	00 ♎ 25	21 01	28 36	22 00	26 43	13 23	11 19
09	15 13 52	16 52 43	11 ♍ 19 52	17 ♍ 31 11	09 50	01 31	21 46	28 48	21 56	26 41	13 21	11 19
10	15 17 49	17 53 00	23 ♍ 46 48	00 ♎ 07 06	10 49	02 37	22 31	29 00	21 53	26 39	13 20	11 20
11	15 21 45	18 53 19	06 ♎ 32 25	13 ♎ 02 59	11 45	03 44	23 16	29 12	21 49	26 37	13 19	11 20
12	15 25 42	19 53 40	19 ♎ 38 58	26 ♎ 20 23	12 37	04 50	24 00	29 24	21 45	26 35	13 17	11 21
13	15 29 38	20 54 02	03 ♏ 07 11	09 ♏ 59 13	13 24	05 57	24 45	29 36	21 41	26 33	13 16	11 21
14	15 33 35	21 54 27	16 ♏ 56 11	23 ♏ 57 41	14 07	07 05	25 31	29 ♐ 48	21 37	26 31	13 15	11 22
15	15 37 32	22 54 53	01 ♐ 03 14	08 ♐ 12 15	14 44	08 12	26 16	00 ♑ 01	21 34	26 29	13 14	11 22
16	15 41 28	23 55 22	15 ♐ 24 05	22 ♐ 38 02	15 14	09 20	27 01	00 13	21 29	26 27	13 12	11 23
17	15 45 25	24 55 51	29 ♐ 53 24	07 ♑ 09 26	15 38	10 28	27 46	00 26	21 25	26 25	13 11	11 24
18	15 49 21	25 56 22	14 ♑ 25 26	21 ♑ 40 45	15 55	11 36	28 31	00 38	21 21	26 23	13 10	11 24
19	15 53 18	26 56 54	28 ♑ 54 45	06 ♒ 06 55	16 03	12 44	29 ♑ 17	00 51	21 17	26 21	13 09	11 25
20	15 57 14	27 57 28	13 ♒ 16 48	20 ♒ 24 01	16 R 01	13 52	00 ♒ 02	01 03	21 13	26 19	13 08	11 26
21	16 01 11	28 58 02	27 ♒ 28 17	04 ♓ 29 22	15 51	15 01	00 47	01 16	21 08	26 17	13 07	11 26
22	16 05 07	29 ♏ 58 38	11 ♓ 27 07	18 ♓ 21 28	15 29	16 10	01 33	01 29	21 04	26 15	13 06	11 27
23	16 09 04	00 ♐ 59 15	25 ♓ 12 22	01 ♈ 59 47	14 58	17 19	02 18	01 41	20 59	26 12	13 05	11 28
24	16 13 01	01 59 53	08 ♈ 43 46	15 ♈ 24 21	14 15	18 28	03 04	01 54	20 55	26 10	13 04	11 29
25	16 16 57	03 00 32	22 ♈ 01 34	28 ♈ 35 29	13 22	19 37	03 49	02 07	20 50	26 08	13 03	11 30
26	16 20 54	04 01 12	05 ♉ 06 10	11 ♉ 33 38	12 20	20 47	04 35	02 20	20 46	26 06	13 02	11 31
27	16 24 50	05 01 54	17 ♉ 57 59	24 ♉ 19 15	11 09	21 57	05 21	02 33	20 41	26 03	13 01	11 31
28	16 28 47	06 02 37	00 ♊ 37 30	06 ♊ 52 50	09 52	23 06	06 06	02 46	20 36	26 01	13 00	11 32
29	16 32 43	07 03 21	13 ♊ 05 19	19 ♊ 15 05	08 31	24 16	06 52	02 59	20 32	25 59	12 59	11 33
30	16 36 40	08 ♐ 04 07	25 ♊ 22 15	01 ♋ 27 00	07 ♐ 09	25 ♎ 27	07 ♒ 38	03 ♑ 12	20 ♊ 27	25 ♊ 56	12 ♈ 59	11 ♒ 34

Date	Moon True ☊ ° '	Moon Mean ☊ ° '	Moon ☽ Latitude ° '
01	18 ♏ 57	19 ♏ 21	01 S 27
02	18 D 58	19 18	02 30
03	18 59	19 15	03 26
04	19 00	19 12	04 12
05	19 01	19 08	04 46
06	19 01	19 05	05 08
07	19 R 01	19 02	05 17
08	19 01	18 59	05 12
09	19 01	18 56	04 54
10	19 00	18 52	04 21
11	19 00	18 49	03 35
12	19 D 00	18 46	02 36
13	19 01	18 43	01 27
14	19 01	18 40	00 S 12
15	19 R 01	18 37	01 N 06
16	19 00	18 33	02 21
17	19 00	18 30	03 28
18	18 59	18 27	04 21
19	18 58	18 24	04 58
20	18 57	18 21	05 15
21	18 57	18 18	05 13
22	18 D 57	18 14	04 53
23	18 58	18 11	04 16
24	18 59	18 08	03 25
25	19 00	18 05	02 24
26	19 01	18 02	01 16
27	19 02	17 58	00 N 06
28	19 R 01	17 55	01 S 04
29	19 00	17 52	02 09
30	18 ♏ 58	17 ♏ 49	03 S 07

DECLINATIONS

Date	Sun ☉ ° '	Moon ☽ ° '	Mercury ☿ ° '	Venus ♀ ° '	Mars ♂ ° '	Jupiter ♃ ° '	Saturn ♄ ° '	Uranus ♅ ° '	Neptune ♆ ° '	Pluto ♇ ° '
01	14 S 27	19 N 40	22 S 40	03 N 29	24 S 16	23 S 18	21 N 38	23 N 33	03 N 51	23 S 45
02	14 46	20 19	22 59	03 08	24 10	23 18	21 37	23 33	03 51	23 45
03	15 05	20 00	23 18	02 47	24 03	23 18	21 37	23 33	03 50	23 45
04	15 23	18 45	23 34	02 25	23 57	23 19	21 37	23 33	03 50	23 44
05	15 42	16 43	23 50	02 04	23 50	23 19	21 37	23 33	03 49	23 44
06	16 00	13 58	24 04	01 41	23 43	23 19	21 36	23 32	03 49	23 44
07	16 18	10 39	24 17	01 19	23 36	23 20	21 36	23 32	03 48	23 44
08	16 35	06 53	24 29	00 57	23 29	23 20	21 36	23 32	03 48	23 44
09	16 52	02 N 47	24 39	00 34	23 21	23 20	21 36	23 32	03 47	23 43
10	17 09	01 S 31	24 47	00 N 11	23 13	23 20	21 36	23 32	03 47	23 43
11	17 26	05 53	24 54	00 S 12	23 05	23 20	21 35	23 32	03 46	23 43
12	17 43	10 05	24 59	00 35	22 56	23 21	21 35	23 32	03 46	23 43
13	17 59	13 55	25 02	00 59	22 48	23 21	21 35	23 32	03 45	23 42
14	18 14	17 04	25 04	01 23	22 39	23 21	21 35	23 32	03 45	23 42
15	18 30	19 17	25 03	01 46	22 30	23 21	21 34	23 32	03 44	23 42
16	18 45	20 18	25 01	02 10	22 20	23 21	21 34	23 32	03 44	23 42
17	19 00	19 58	24 57	02 34	22 11	23 21	21 34	23 32	03 44	23 41
18	19 14	18 19	24 50	02 58	22 01	23 21	21 33	23 32	03 43	23 41
19	19 28	15 31	24 41	03 22	21 51	23 21	21 33	23 32	03 43	23 41
20	19 42	11 47	24 30	03 47	21 40	23 21	21 33	23 32	03 42	23 40
21	19 55	07 26	24 16	04 11	21 30	23 21	21 33	23 32	03 42	23 40
22	20 08	02 S 45	24 00	04 35	21 19	23 21	21 32	23 32	03 42	23 40
23	20 21	02 N 01	23 41	05 00	21 08	23 21	21 32	23 32	03 41	23 39
24	20 33	06 36	23 19	05 24	20 57	23 21	21 32	23 32	03 41	23 39
25	20 45	10 48	22 54	05 49	20 46	23 21	21 32	23 32	03 41	23 39
26	20 57	14 25	22 27	06 13	20 34	23 21	21 31	23 32	03 40	23 38
27	21 08	17 16	21 57	06 38	20 22	23 21	21 31	23 32	03 40	23 38
28	21 19	19 14	21 26	07 02	20 10	23 21	21 31	23 32	03 40	23 38
29	21 29	20 14	20 54	07 27	19 58	23 20	21 30	23 31	03 39	23 37
30	21 S 39	20 N 14	20 S 21	07 S 51	19 S 45	23 S 20	21 N 30	23 N 31	03 N 39	23 S 37

ZODIAC SIGN ENTRIES

Date	h	m	Planets
01	02	35	☿ ♐
01	02	53	☽ ♊
03	13	08	☽ ♋
06	01	25	☽ ♌
06	02	59	♀ ♎
08	13	43	☽ ♍
10	23	47	☽ ♎
13	06	30	☽ ♏
15	10	13	☽ ♐
15	10	29	♃ ♑
17	12	11	☽ ♑
19	13	48	☽ ♒
20	10	57	♂ ♒
21	16	19	☽ ♓
22	12	33	☉ ♐
23	20	28	☽ ♈
26	02	35	☽ ♉
28	10	48	☽ ♊
30	21	08	☽ ♋

LATITUDES

Date	Mercury ☿ ° '	Venus ♀ ° '	Mars ♂ ° '	Jupiter ♃ ° '	Saturn ♄ ° '	Uranus ♅ ° '	Neptune ♆ ° '	Pluto ♇ ° '
01	02 S 28	00 N 42	01 S 47	00 N 07	01 S 35	00 N 09	01 S 37	06 S 38
04	02 38	00 56	01 45	00 06	01 36	00 09	01 37	06 37
07	02 44	01 09	01 43	00 06	01 36	00 09	01 37	06 37
10	02 45	01 20	01 41	00 06	01 36	00 09	01 37	06 36
13	02 39	01 30	01 39	00 05	01 36	00 09	01 36	06 36
16	02 25	01 40	01 37	00 05	01 36	00 09	01 36	06 36
19	02 00	01 48	01 35	00 05	01 36	00 09	01 36	06 35
22	01 21	01 55	01 33	00 04	01 36	00 09	01 36	06 35
25	00 S 30	02 01	01 31	00 04	01 36	00 09	01 36	06 35
28	00 N 30	02 05	01 28	00 04	01 36	00 09	01 36	06 35
31	01 N 28	02 N 09	01 S 26	00 N 03	01 S 35	00 N 09	01 S 36	06 S 34

DATA

Julian Date	2463172
Delta T	+74 seconds
Ayanamsa	24° 18' 18"
Synetic vernal point	04° ♓ 48' 41"
True obliquity of ecliptic	23° 26' 01"

MOON'S PHASES, APSIDES AND POSITIONS ☽

Date	h	m	Phase	Longitude ° '	Eclipse Indicator
07	07	02	☾	14 ♌ 50	
14	21	10	●	22 ♏ 18	Ann-Total
21	14	45	☽	29 ♒ 05	
28	23	18	○	06 ♊ 31	

Day	h	m		
06	02	42	Apogee	
17	21	56	Perigee	
02	15	51	Max dec	20° N 20'
10	03	35	OS	
16	18	16	Max dec	20° S 20'
23	01	50	ON	
30	00	02	Max dec	20° N 21'

LONGITUDES

Date	Chiron ⚷ ° '	Ceres ⚳ ° '	Pallas ⚴ ° '	Juno ⚵ ° '	Vesta ⚶ ° '	Black Moon Lilith ⚸ ° '
01	19 ♉ 25	27 ♌ 59	05 ♊ 45	02 ♉ 54	12 ♍ 20	28 ♋ 34
11	18 ♉ 54	01 ♍ 14	03 ♊ 18	00 ♉ 46	16 ♍ 18	29 ♋ 41
21	18 ♉ 24	04 ♍ 11	00 ♊ 08	29 ♈ 05	20 ♍ 05	00 ♌ 58
31	17 ♉ 55	06 ♍ 46	26 ♉ 44	28 ♈ 25	23 ♍ 39	01 ♌ 56

ASPECTARIAN

01 Saturday
00 09 ☽ ∠ ♆
01 43 ♀ □ ♄
02 55 ☽ ☍ ☿
04 00 ☽ ⯓ ♂
20 29 ☽ ⊼ ☉
22 17 ☽ ± ♂

02 Sunday
00 28 ☽ △ ♇
02 25 ☿ ∠ ♂
04 49 ☽ ✶ ♆
09 07 ☽ ± ☉
10 39 ☽ ⊼ ♂
21 56 ☽ ☌ ♄

03 Monday
02 21 ☽ □ ♀
04 14 ☽ Q ♆
04 19 ☽ ⯓ ☉
05 47 ☽ ⯓ ♇
06 54 ☽ ☌ ♅
08 24 ☽ ☍ ♃
13 15 ☿ ∥ ♃
19 58 ☽ ⊼ ☿
22 14 ☉ □ ♇
23 39 ☽ ± ♇

04 Tuesday
09 09 ☿ # ♅
09 18 ☽ ± ☿
11 01 ☉ ⯓ ♅
11 40 ☽ ⊼ ♇
12 54 ☽ △ ☉
16 02 ☽ □ ♆
17 15 ♀ ⯓ ♇
17 45 ☽ Q ♀

05 Wednesday
02 08 ☽ ☍ ♂
02 45 ☿ ∥ ♇
04 35 ♀ □ ♅
05 26 ☽ ⯓ ☿
09 35 ☽ ⊻ ♄
12 10 ☿ ∥ ♂
16 39 ☉ ∠ ♃
18 55 ☽ ⊻ ♅
20 23 ☽ ✶ ♀
20 34 ☽ # ☉
21 34 ☽ ⊼ ♃
21 39 ☽ ⊥ ♄

06 Thursday
01 33 ☉ ⊼ ♆
07 01 ☽ ⊥ ♅
09 41 ♂ ∥ ♇
09 54 ☽ ± ♃
11 45 ♀ □ ♃
15 03 ☽ △ ☿
15 47 ☽ ∠ ♄

07 Friday
00 16 ☽ ☍ ♇
01 10 ☽ ∠ ♅
04 23 ☽ ⯓ ♃
04 30 ☽ △ ♆
05 46 ☽ ∠ ♀
07 02 ☽ □ ☉
18 44 ☽ ⊼ ♂
21 50 ☽ ✶ ♄
23 53 ♂ # ♅

08 Saturday
01 39 ☽ ⊥ ♀
07 11 ☽ ✶ ♅
07 31 ☽ ± ♂
10 29 ☽ ⯓ ♆
10 56 ☽ △ ♃
14 45 ☉ ± ♄
14 47 ☽ ⊻ ♀
21 34 ☽ Q ♄
22 13 ☽ Q ☉

09 Sunday
02 29 ☽ ⯓ ♂
04 14 ☽ ± ♆
06 16 ☽ ∥ ♆
06 50 ☽ Q ♅
08 50 ☽ □ ☿
11 59 ☽ ⊼ ♇
14 48 ♂ ∥ ♃
15 56 ☽ ⊼ ♆
17 09 ♂ ⊼ ♄
23 37 ☽ ± ♇
23 43 ☽ ✶ ☉

10 Monday
01 39 ☽ ∥ ♀
05 12 ☽ # ♀
08 23 ☽ □ ♄
09 26 ☽ △ ♂
16 51 ☽ ⯓ ♇
17 26 ☽ □ ♅
22 03 ☽ □ ♃
22 20 ☽ Q ☿

11 Tuesday
00 22 ☽ # ♆
00 58 ☿ ✶ ♇
06 16 ☽ ☌ ♀
06 39 ☽ ∠ ☉
20 52 ☽ △ ♇
21 50 ☉ ± ♆
22 19 ☽ ✶ ☿

12 Wednesday
00 27 ☽ ☍ ♆
00 41 ☽ ⊥ ☉
07 52 ☽ Q ♃
12 29 ☽ ⊻ ☉
15 46 ☽ △ ♄
20 18 ☽ □ ♂
23 59 ☉ ∠ ♀

13 Thursday
00 25 ☽ △ ♅
03 10 ☽ ∠ ☿
04 03 ☉ ± ♅
05 42 ☽ ✶ ♃
07 49 ☿ △ ♆
17 25 ☽ ⊻ ♀
18 14 ☽ ⯓ ♄
19 56 ☽ ⊥ ☿

14 Friday
02 23 ☽ □ ♇
02 41 ☽ ⯓ ♅
04 46 ☽ ⊥ ♀
05 39 ☽ ⊼ ♆
05 40 ☉ ⊼ ♄
05 46 ☽ Q ♂
06 53 ☽ ⊻ ☿
08 17 ☽ ∠ ♃
09 45 ☽ ± ♄
15 57 ☽ ± ♆
18 08 ☽ ± ♅
19 59 ☽ ⊼ ♄
21 10 ☽ ☌ ☉
21 33 ☽ ∠ ♀
23 55 ☽ ⊥ ♃

15 Saturday
00 40 ☽ ∥ ☉
03 27 ☽ ✶ ♂
04 19 ☽ ⊼ ♅
07 14 ☽ ⯓ ♆
09 10 ☽ Q ♇
10 13 ☽ ⊻ ♃
19 02 ♂ ⊼ ♅

16 Sunday
01 01 ☽ ✶ ♀
05 18 ☽ ✶ ♇
06 03 ☽ ∠ ♂
08 21 ☽ △ ♆
11 43 ☽ ☌ ☿
20 52 ☉ ⊥ ♃
21 50 ☽ ⊥ ♂
22 04 ☽ ☍ ♄
22 40 ☽ Q ♀

17 Monday
01 57 ♂ ± ♄
03 11 ☽ ⊻ ☉
06 13 ☽ ∠ ♇
06 17 ☽ ☍ ♅
08 18 ☽ ⊻ ♂
12 54 ☽ ☌ ♃
13 51 ☽ ⊥ ☉
21 06 ☽ ⊥ ♇

18 Tuesday
02 19 ☽ ∥ ☉
03 25 ♅ ⯓ ♇
05 49 ☽ ∠ ☉
06 56 ☽ □ ♀
07 01 ☽ ⊻ ♇
07 56 ♀ △ ♇
09 56 ☽ □ ♆
14 29 ☽ ⊻ ☿
22 19 ☉ ⊼ ♅
23 24 ☽ ⊼ ♄

19 Wednesday
00 32 ☽ ⊥ ☿
07 46 ☽ ⊼ ♅
08 30 ☽ ✶ ☉
09 18 ☽ ± ♄
12 38 ☽ ☌ ♂
15 15 ☽ ⊻ ♃
15 33 ☽ ∠ ☿
15 43 ☽ Q ♆
17 43 ☽ ± ♅
20 38 ♀ ☍ ♆
21 14 ☿ St R

20 Thursday
00 13 ☽ ⯓ ♄
01 25 ☽ ⊥ ♃
06 00 ☽ Q ☉
07 48 ♀ # ♆
08 43 ☽ ⯓ ♅
08 53 ☽ ☌ ♇
11 45 ☽ ✶ ♆
13 05 ☽ △ ♀
16 05 ☉ ⯓ ♆
16 35 ☽ ✶ ☿
16 44 ☽ ∠ ♃

21 Friday
01 18 ☽ △ ♄
05 34 ♂ # ♄
09 59 ☽ △ ♅
12 37 ☽ Q ☿
13 06 ☽ ∠ ♆
13 14 ☿ ∠ ♂
14 45 ☽ □ ☉
16 44 ☽ ⯓ ♀
17 59 ☽ ⊻ ♂
18 34 ☽ ✶ ♃
22 04 ♂ Q ♆
23 24 ☉ Q ♇

22 Saturday
01 33 ☿ ✶ ♀
03 26 ☽ ∥ ♀
04 30 ☽ ⊥ ♆
04 52 ☽ ⊥ ♂
07 14 ☽ # ♆
08 49 ♂ ⊻ ♃
09 34 ☽ ± ♀
12 00 ☽ ⊻ ♇
14 51 ☽ ⊻ ♆
15 34 ☽ Q ♃
18 47 ☽ □ ☿
20 55 ☽ ⊼ ♀
21 22 ☽ ∠ ♂
22 26 ☽ ⊥ ♇

23 Sunday
04 38 ☽ □ ♄
09 02 ♂ ± ♅
13 36 ☿ ∥ ♇
13 45 ☽ □ ♅
14 13 ☽ ∠ ♇
20 36 ☽ ∥ ♆
22 07 ☿ # ♅
23 02 ☽ △ ☉
23 38 ☽ □ ♃

24 Monday
01 18 ☽ ✶ ♂
04 58 ☽ # ♀
09 05 ☉ ⊻ ♃
09 35 ☿ ∥ ♃
12 20 ☽ Q ♄
16 56 ☽ ✶ ♇
19 46 ☽ ☌ ♆
21 20 ☽ △ ☿
21 45 ☽ Q ♅

25 Tuesday
00 04 ☽ Q ♂
04 06 ☽ ⯓ ☉
07 13 ☽ ☍ ♀
09 51 ☽ ✶ ♄
12 21 ☉ ∥ ♂
14 41 ☽ Q ♇
19 28 ☽ ✶ ♅
19 59 ☿ △ ♆
21 52 ☽ ± ☉
22 47 ☽ ⯓ ☿

26 Wednesday
00 32 ♀ Q ♃
06 48 ☽ △ ♃
09 50 ☽ ⊼ ☉
10 59 ☽ □ ♂
11 37 ♀ △ ♄
13 13 ☽ ∠ ♄
14 06 ☽ ± ☿
23 06 ☽ ∠ ♅
23 55 ☽ □ ♇

27 Thursday
00 20 ☽ ⊼ ☿
02 44 ☽ ⊻ ♆
04 51 ☿ ✶ ♇
05 53 ☽ ⊥ ♄
11 12 ☽ ⯓ ♃
13 59 ☽ ⊥ ♆
15 55 ☽ ⊥ ♅
17 06 ☽ ⊻ ♄
20 15 ☽ ⊼ ♀
21 44 ♂ ⯓ ♄

28 Friday
03 15 ☽ ⊻ ♅
04 31 ☽ ± ♃
07 00 ☽ ∠ ♆
08 22 ☿ # ♄
08 48 ☽ ± ♀
16 07 ☉ ∥ ☿
16 10 ☽ ⊼ ♃
18 00 ☉ ✶ ♂
23 12 ☽ △ ♂
23 18 ☽ ☍ ☉

29 Saturday
03 51 ☽ ⯓ ♀
04 02 ☽ ☍ ☿
04 32 ☽ # ♂
04 52 ☿ ∠ ♀
09 01 ☽ △ ♇
11 49 ☽ ✶ ♆
15 21 ☉ # ♄

30 Sunday
02 24 ☽ ☌ ♄
02 43 ☉ ☌ ☿
06 15 ☽ ⯓ ♂
06 32 ☿ ✶ ♂
11 14 ☽ Q ♆
12 09 ☽ △ ♀
13 06 ☽ ☌ ♅
14 22 ☽ ⯓ ♇
21 48 ♀ △ ♅

All ephemeris data is given at 12.00 UT and the Moon's longitude is additionally given for 24.00 UT

LONGITUDES

Date	Sidereal time h m s	Sun ☉ ° ' "	Moon ☽ ° ' "	Moon ☽ 24.00 ° ' "	Mercury ☿ ° '	Venus ♀ ° '	Mars ♂ ° '	Jupiter ♃ ° '	Saturn ♄ ° '	Uranus ♅ ° '	Neptune ♆ ° '	Pluto ♇ ° '
01	16 40 36	09 ♐ 04 53	07 ♋ 29 31	13 ♋ 30 04	05 ♐ 47	26 ♎ 37	08 ♒ 24	03 ♑ 25	20 ♊ 22	25 ♊ 54	12 ♈ 58	11 ♒ 35
02	16 44 33	10 05 42	19 ♋ 28 53	25 ♋ 26 19	04 R 30	27 47	09 09	03 38	20 R 17	25 R 51	12 R 57	11 36
03	16 48 30	11 06 31	01 ♌ 22 43	07 ♌ 18 30	03 20	28 ♎ 58	09 55	03 52	20 13	25 49	12 56	11 37
04	16 52 26	12 07 22	13 ♌ 14 04	19 ♌ 09 57	02 17	00 ♏ 08	10 41	04 05	20 08	25 46	12 56	11 38
05	16 56 23	13 08 14	25 ♌ 06 37	01 ♍ 04 38	01 25	01 19	11 27	04 18	20 03	25 44	12 55	11 40
06	17 00 19	14 09 07	07 ♍ 04 34	13 ♍ 07 00	00 44	02 30	12 13	04 32	19 58	25 41	12 54	11 41
07	17 04 16	15 10 02	19 ♍ 12 32	25 ♍ 21 46	00 ♐ 15	03 41	12 59	04 45	19 53	25 39	12 54	11 42
08	17 08 12	16 10 58	01 ♎ 35 16	07 ♎ 53 37	29 ♏ 56	04 52	13 45	04 59	19 48	25 36	12 53	11 43
09	17 12 09	17 11 55	14 ♎ 17 19	20 ♎ 46 51	29 49	06 04	14 31	05 12	19 43	25 34	12 53	11 44
10	17 16 05	18 12 54	27 ♎ 22 35	04 ♏ 04 50	29 ♏ 52	07 15	15 17	05 26	19 38	25 31	12 52	11 45
11	17 20 02	19 13 54	10 ♏ 53 44	17 ♏ 49 21	00 ♐ 05	08 27	16 03	05 39	19 33	25 29	12 52	11 47
12	17 23 59	20 14 54	24 ♏ 51 32	01 ♐ 59 59	00 D 27	09 38	16 49	05 53	19 28	25 26	12 51	11 48
13	17 27 55	21 15 56	09 ♐ 14 14	16 ♐ 33 38	00 57	10 50	17 35	06 06	19 23	25 24	12 51	11 49
14	17 31 52	22 16 59	23 ♐ 57 19	01 ♑ 24 21	01 34	12 02	18 21	06 20	19 18	25 21	12 51	11 50
15	17 35 48	23 18 02	08 ♑ 53 39	16 ♑ 24 02	02 18	13 14	19 07	06 34	19 13	25 19	12 50	11 52
16	17 39 45	24 19 07	23 ♑ 54 21	01 ♒ 23 27	03 07	14 26	19 53	06 47	19 08	25 16	12 50	11 53
17	17 43 41	25 20 11	08 ♒ 50 16	16 ♒ 13 50	04 01	15 38	20 39	07 01	19 03	25 13	12 50	11 54
18	17 47 38	26 21 16	23 ♒ 33 22	00 ♓ 48 12	05 00	16 50	21 26	07 15	18 59	25 11	12 50	11 56
19	17 51 34	27 22 22	07 ♓ 57 52	15 ♓ 02 05	06 02	18 02	22 12	07 28	18 54	25 08	12 49	11 57
20	17 55 31	28 23 27	22 ♓ 00 43	28 ♓ 53 45	07 08	19 14	22 58	07 42	18 49	25 06	12 49	11 59
21	17 59 28	29 ♐ 24 33	05 ♈ 41 19	12 ♈ 23 36	08 17	20 27	23 44	07 56	18 44	25 03	12 49	12 00
22	18 03 24	00 ♑ 25 39	19 ♈ 00 54	25 ♈ 33 32	09 28	21 39	24 30	08 10	18 39	25 01	12 49	12 01
23	18 07 21	01 26 45	02 ♉ 01 52	08 ♉ 26 14	10 42	22 52	25 17	08 24	18 34	24 58	12 49	12 03
24	18 11 17	02 27 52	14 ♉ 47 00	21 ♉ 04 32	11 58	24 04	26 03	08 37	18 30	24 56	12 D 49	12 04
25	18 15 14	03 28 59	27 ♉ 19 07	03 ♊ 31 05	13 15	25 17	26 49	08 51	18 25	24 53	12 49	12 06
26	18 19 10	04 30 06	09 ♊ 40 39	15 ♊ 48 05	14 35	26 30	27 35	09 05	18 20	24 50	12 49	12 07
27	18 23 07	05 31 13	21 ♊ 53 35	27 ♊ 57 20	15 55	27 42	28 21	09 19	18 16	24 48	12 49	12 09
28	18 27 03	06 32 20	03 ♋ 59 29	10 ♋ 00 13	17 17	28 ♏ 55	29 08	09 33	18 11	24 45	12 49	12 10
29	18 31 00	07 33 28	15 ♋ 59 40	21 ♋ 58 00	18 39	00 ♐ 08	29 ♒ 54	09 47	18 07	24 43	12 49	12 12
30	18 34 57	08 34 36	27 ♋ 55 23	03 ♌ 52 00	20 03	01 21	00 ♓ 40	10 01	18 02	24 40	12 50	12 14
31	18 38 53	09 ♑ 35 44	09 ♌ 48 04	15 ♌ 43 49	21 ♐ 28	02 ♐ 34	01 ♓ 26	10 ♑ 14	17 ♊ 58	24 ♊ 38	12 ♈ 50	12 ♒ 15

Moon

Date	Moon True ☊ ° '	Moon Mean ☊ ° '	Moon ☽ Latitude ° '
01	18 ♏ 55	17 ♏ 46	03 S 56
02	18 R 51	17 43	04 34
03	18 47	17 39	05 00
04	18 44	17 36	05 13
05	18 42	17 33	05 12
06	18 40	17 30	04 58
07	18 D 40	17 27	04 31
08	18 41	17 24	03 50
09	18 42	17 20	02 58
10	18 44	17 17	01 55
11	18 45	17 14	00 S 43
12	18 R 45	17 11	00 N 32
13	18 44	17 08	01 50
14	18 41	17 04	03 00
15	18 36	17 01	04 00
16	18 31	16 58	04 43
17	18 25	16 55	05 06
18	18 20	16 52	05 09
19	18 17	16 49	04 53
20	18 15	16 45	04 19
21	18 D 15	16 42	03 31
22	18 16	16 39	02 32
23	18 18	16 36	01 27
24	18 19	16 33	00 N 19
25	18 R 19	16 30	00 S 49
26	18 16	16 26	01 53
27	18 11	16 23	02 51
28	18 04	16 20	03 41
29	17 55	16 17	04 20
30	17 45	16 14	04 48
31	17 ♏ 34	16 ♏ 10	05 S 03

DECLINATIONS

Date	Sun ☉ ° '	Moon ☽ ° '	Mercury ☿ ° '	Venus ♀ ° '	Mars ♂ ° '	Jupiter ♃ ° '	Saturn ♄ ° '	Uranus ♅ ° '	Neptune ♆ ° '	Pluto ♇ ° '
01	21 S 48	19 N 17	19 S 49	08 S 15	19 S 33	23 S 20	21 N 30	23 N 31	03 N 39	23 S 37
02	21 57	17 30	19 19	08 39	19 20	23 20	21 30	23 31	03 39	23 36
03	22 06	14 58	18 51	09 03	19 07	23 19	21 29	23 31	03 38	23 36
04	22 14	11 50	18 26	09 28	18 54	23 19	21 29	23 31	03 38	23 35
05	22 22	08 14	18 06	09 51	18 40	23 19	21 29	23 31	03 38	23 35
06	22 30	04 18	17 49	10 15	18 27	23 18	21 28	23 31	03 38	23 35
07	22 37	00 N 07	17 37	10 39	18 13	23 18	21 28	23 31	03 38	23 34
08	22 43	04 S 09	17 29	11 02	17 59	23 18	21 28	23 31	03 37	23 34
09	22 49	08 22	17 25	11 26	17 45	23 17	21 27	23 31	03 37	23 33
10	22 55	12 19	17 25	11 49	17 31	23 17	21 27	23 31	03 37	23 33
11	23 00	15 46	17 29	12 12	17 16	23 16	21 27	23 31	03 37	23 33
12	23 04	18 26	17 36	12 34	17 01	23 16	21 27	23 31	03 37	23 32
13	23 09	20 01	17 45	12 57	16 47	23 15	21 26	23 31	03 37	23 32
14	23 13	20 18	17 56	13 19	16 32	23 15	21 26	23 30	03 37	23 31
15	23 16	19 09	18 10	13 41	16 16	23 14	21 26	23 30	03 37	23 31
16	23 19	16 41	18 25	14 03	16 01	23 14	21 25	23 30	03 36	23 30
17	23 21	13 07	18 41	14 24	15 46	23 13	21 25	23 30	03 36	23 30
18	23 23	08 47	18 58	14 45	15 30	23 12	21 25	23 30	03 36	23 29
19	23 24	04 S 03	19 16	15 06	15 14	23 12	21 25	23 30	03 36	23 29
20	23 25	00 N 48	19 34	15 27	14 58	23 11	21 24	23 30	03 36	23 29
21	23 26	05 29	19 53	15 47	14 42	23 10	21 24	23 30	03 36	23 28
22	23 26	09 47	20 11	16 07	14 26	23 10	21 24	23 30	03 36	23 28
23	23 26	13 32	20 30	16 26	14 10	23 09	21 23	23 30	03 36	23 27
24	23 25	16 34	20 48	16 45	13 54	23 08	21 23	23 30	03 36	23 27
25	23 23	18 46	21 06	17 04	13 37	23 07	21 23	23 29	03 36	23 26
26	23 21	20 02	21 23	17 22	13 20	23 06	21 23	23 29	03 37	23 26
27	23 19	20 20	21 40	17 40	13 04	23 05	21 22	23 29	03 37	23 25
28	23 16	19 42	21 56	17 57	12 47	23 05	21 22	23 29	03 37	23 25
29	23 13	18 10	22 12	18 14	12 30	23 04	21 22	23 29	03 37	23 24
30	23 09	15 52	22 26	18 31	12 13	23 03	21 21	23 29	03 37	23 24
31	23 S 05	12 N 55	22 S 40	18 S 47	11 S 55	23 S 02	21 N 21	23 N 29	03 N 37	23 S 23

ZODIAC SIGN ENTRIES

Date	h	m	Planets
03	09	13	☽ ♌
04	09	09	♀ ♏
05	21	50	☽ ♍
08	05	36	☿ ♏
08	08	57	☽ ♎
10	16	43	☽ ♏
11	04	21	☿ ♐
12	20	39	☽ ♐
14	21	44	☽ ♑
16	21	46	☽ ♒
18	22	40	☽ ♓
21	01	56	☽ ♈
22	01	55	☉ ♑
23	08	13	☽ ♉
25	17	11	☽ ♊
28	04	04	☽ ♋
29	09	17	♀ ♐
29	15	15	♂ ♓
30	16	11	☽ ♌

LATITUDES

Date	Mercury ☿ ° '	Venus ♀ ° '	Mars ♂ ° '	Jupiter ♃ ° '	Saturn ♄ ° '	Uranus ♅ ° '	Neptune ♆ ° '	Pluto ♇ ° '
01	01 N 28	02 N 09	01 S 26	00 N 03	01 S 35	00 N 09	01 S 36	06 S 34
04	02 13	02 12	01 24	00 03	01 35	00 09	01 36	06 34
07	02 38	02 14	01 21	00 03	01 35	00 09	01 36	06 34
10	02 45	02 14	01 19	00 03	01 35	00 09	01 35	06 33
13	02 39	02 14	01 17	00 02	01 35	00 09	01 35	06 33
16	02 24	02 13	01 14	00 02	01 34	00 09	01 35	06 33
19	02 05	02 11	01 12	00 02	01 34	00 09	01 35	06 33
22	01 42	02 08	01 09	00 01	01 34	00 09	01 35	06 33
25	01 18	02 05	01 07	00 01	01 33	00 09	01 35	06 33
28	00 54	02 00	01 04	00 01	01 33	00 09	01 35	06 33
31	00 N 29	01 N 55	01 S 02	00 00	01 S 32	00 N 09	01 S 34	06 S 32

DATA

Julian Date	2463202
Delta T	+74 seconds
Ayanamsa	24° 18' 23"
Synetic vernal point	04° ♓ 48' 36"
True obliquity of ecliptic	23° 26' 00"

MOON'S PHASES, APSIDES AND POSITIONS ☽

Date	h	m	Phase	Longitude ° '	Eclipse Indicator
07	03	20	☾	14 ♍ 48	
14	09	06	●	22 ♐ 10	
21	00	00	☽	28 ♓ 54	
28	17	33	○	06 ♋ 46	

Day	h	m		
03	22	27	Apogee	
15	21	20	Perigee	
31	13	02	Apogee	
07	12	39	0S	
14	04	39	Max dec	20° S 22'
20	08	03	0N	
27	07	34	Max dec	20° N 21'

LONGITUDES

Date	Chiron ⚷ ° '	Ceres ⚳ ° '	Pallas ⚴ ° '	Juno ⚵ ° '	Vesta ⚶ ° '	Black Moon Lilith ⚸ ° '
01	17 ♉ 55	06 ♍ 46	26 ♉ 44	28 ♈ 25	23 ♍ 39	01 ♌ 56
11	17 ♉ 28	08 ♍ 55	23 ♉ 39	28 ♈ 36	26 ♍ 57	03 ♌ 03
21	17 ♉ 06	10 ♍ 33	21 ♉ 22	29 ♈ 42	29 ♍ 56	04 ♌ 10
31	16 ♉ 48	11 ♍ 35	20 ♉ 13	01 ♉ 39	02 ♎ 32	05 ♌ 17

ASPECTARIAN

h m Aspects

01 Monday
01 11 ☽ ± ♂
03 45 ☽ ☍ ♃
06 16 ☽ ⋕ ♂
08 12 ☽ ± ♇
08 57 ☽ ⊼ ☿
13 55 ☽ ⊼ ♂
15 28 ☽ ⊼ ☉
19 44 ☽ ± ☿
20 11 ☽ ⊼ ♇
22 55 ☽ □ ♆

02 Tuesday
04 34 ☽ ± ☉
10 09 ☿ ∥ ♂
12 02 ☽ bQ ☿
13 37 ☽ ⊻ ♄
19 11 ☿ ⊥ ♀

03 Wednesday
00 22 ☽ bQ ☉
00 48 ☽ ⊻ ♅
01 37 ☽ ⊥ ♄
02 33 ☿ ⊻ ♃
06 35 ☽ □ ♀
09 24 ♂ ⊥ ♃
12 53 ☽ ⊥ ♅
15 36 ☽ △ ☿
17 07 ☽ ⊼ ♃
19 42 ☽ ∠ ♄

04 Thursday
00 24 ☉ ✶ ♇
05 30 ☽ ± ♃
06 29 ☽ ☍ ♂
07 02 ☽ ∠ ♅
08 46 ☽ ☍ ♇
09 32 ☽ △ ☉
11 23 ☽ △ ♆
14 38 ♂ bQ ♅
23 01 ☽ Q ♀

05 Friday
00 04 ☽ bQ ♃
01 51 ☽ ✶ ♄
02 35 ☽ ⋕ ♀
06 52 ☉ △ ♆
13 15 ☽ ✶ ♅
13 15 ☿ ⊻ ♀
17 39 ☽ bQ ♆
18 44 ♂ ☌ ♇
23 58 ☽ □ ☿

06 Saturday
01 51 ☽ Q ♄
01 51 ☽ ✶ ♀
06 49 ☽ △ ♃
11 40 ☽ ± ♆
13 13 ☽ Q ♅
15 53 ☽ ∥ ♆
21 10 ☽ ⊼ ♇
21 31 ☿ Q ♂
22 54 ☽ ⊼ ♂
23 35 ☽ ⊼ ♆

07 Sunday
03 20 ☽ □ ☉
09 02 ☽ ± ♇
09 27 ♂ ✶ ♆
10 10 ☽ Q ☿
10 52 ☽ ∠ ♀
11 31 ☽ ± ♂
13 19 ☽ □ ♄

08 Monday
00 31 ☽ □ ♅
02 36 ☽ bQ ♇
06 11 ☽ bQ ♂
06 14 ☽ ⊥ ♀
08 53 ☽ ✶ ☿
09 01 ☽ ⋕ ♆
10 37 ♀ bQ ♄
14 34 ♀ ✶ ♃
17 24 ☽ Q ☉
18 35 ☽ □ ♃
18 55 ☽ ⊻ ♀

09 Tuesday
07 13 ☽ △ ♇
09 22 ☽ ☍ ♆
12 27 ☽ △ ♂
12 59 ☽ ∠ ☿
16 24 ☿ St D
17 52 ☽ ✶ ☉
21 59 ☽ △ ♄

10 Wednesday
04 43 ☽ Q ♃
05 35 ☽ ⊥ ☿
08 26 ☽ ∥ ♀
08 39 ☽ △ ♅
16 32 ☽ ⊻ ☿
19 16 ☿ ∥ ♂
23 19 ☽ ∠ ☉

11 Thursday
00 54 ☽ bQ ♄
02 38 ☽ ✶ ♃
07 18 ☽ ☌ ♀
11 17 ☽ bQ ♅
13 32 ☽ □ ♇
15 26 ☽ ⊼ ♆
16 23 ☽ ⊥ ☉
16 36 ☽ ± ♄
19 00 ☉ ☍ ♄
21 28 ☽ □ ♂
23 22 ☽ ∥ ♂

12 Friday
01 47 ☽ ± ♆
02 35 ♃ ⊥ ♇
02 48 ☽ ± ♅
02 53 ☽ ⊼ ♄
02 57 ☽ ∥ ☿
03 33 ☽ ⊻ ☉
05 07 ☽ ∠ ♃
12 59 ☽ ⊼ ♅
17 03 ☽ bQ ♆
20 20 ☽ Q ♇
20 35 ☽ ⊥ ♃
21 43 ☽ ☌ ☿

13 Saturday
03 32 ♀ bQ ♅
05 37 ☽ Q ♂
06 45 ☽ ⊻ ♃
14 52 ☽ ⊻ ♀
16 15 ☽ ✶ ♇
17 56 ☽ △ ♆

14 Sunday
01 33 ☽ ⊥ ♀
02 25 ☽ ✶ ♂
04 30 ☽ ☍ ♄
08 09 ♀ □ ♇
09 06 ☽ ☌ ☉
14 15 ☽ ☍ ♅
16 40 ☽ ∠ ♇
17 24 ☽ ∠ ♀

15 Monday
00 52 ☽ ⊻ ☿
01 49 ☉ ∥ ♃
03 56 ☽ ∠ ♂
04 16 ♀ ⊼ ♆
07 08 ☽ ⊥ ♇
08 12 ☽ ☌ ♃
10 59 ☽ ⊥ ☿
11 54 ♀ ± ♄
14 53 ♂ △ ♄
16 45 ☽ ⊻ ♇
18 18 ☽ □ ♆
19 07 ☽ ⊥ ♂
19 32 ☽ ✶ ♀
22 04 ☽ ∥ ☿

16 Tuesday
02 11 ☽ ∠ ☿
04 25 ☽ ⊼ ♄
05 14 ☽ ⊻ ♂
12 43 ☽ ⊻ ☉
13 58 ☽ ± ♄
14 10 ☽ ⊼ ♅
16 23 ☽ Q ♀
17 23 ☽ ∥ ♂
17 49 ♅ ⋕ ♇
23 02 ☽ ⊥ ☉
23 06 ☽ Q ♆
23 46 ☽ ± ♅

17 Wednesday
03 42 ☽ ✶ ☿
04 20 ☽ bQ ♄
04 42 ☽ ∥ ♀
09 01 ☽ ⊻ ♃
09 27 ☉ ☍ ♅
14 14 ☽ bQ ♅
14 36 ☽ ∠ ☉
16 59 ☽ ☌ ♇
18 28 ☽ ✶ ♆
18 53 ☽ ⊥ ♃

18 Thursday
00 00 ☽ □ ♀
00 28 ☽ Q ☿
04 31 ☽ △ ♄
08 18 ☽ ☌ ♂
09 49 ☽ ∠ ♃
14 40 ☽ △ ♅
16 58 ☽ ✶ ☉
19 03 ☽ ∠ ♆

19 Friday
01 52 ☉ ∠ ♇
08 30 ☽ □ ☿
10 05 ☽ ⊥ ♆
11 10 ☽ ✶ ♃
14 12 ☽ ⋕ ♆
14 34 ☽ Q ☉
17 29 ♀ ∥ ♂
18 46 ☽ ⊻ ♇
20 14 ☽ ⊻ ♆

20 Saturday
00 20 ♂ ∠ ♃
03 42 ♀ ± ♆
04 03 ♀ ⊼ ♄
05 02 ☽ ⊥ ♇
06 31 ☽ □ ♄
06 46 ☽ △ ♀
07 57 ☽ Q ♃
09 14 ♀ ± ♅
13 45 ☽ ⊻ ♂
17 20 ☽ □ ♅
20 39 ☽ ∠ ♇

21 Sunday
00 00 ☽ □ ☉
00 51 ☽ ⊥ ♂
02 14 ☽ ∥ ♆
02 59 ☿ ⊻ ♃
11 32 ☽ bQ ♀
13 51 ☽ Q ♄
16 04 ☽ □ ♃
17 04 ☽ △ ☿
17 46 ☽ ∠ ♂
23 19 ☽ ✶ ♇

22 Monday
00 46 ☽ ☌ ♆
01 09 ☽ Q ♅
05 17 ☽ ± ♀
11 21 ☽ ✶ ♄
17 19 ☽ ⊼ ♀
21 11 ☽ Q ♇
22 41 ☽ ✶ ♂
22 57 ☽ ✶ ♅
23 02 ☽ bQ ☿

23 Tuesday
02 54 ♂ △ ♅
10 49 ☽ △ ☉
14 52 ☽ ∠ ♄
16 09 ☽ ⋕ ♂
17 32 ☽ ± ☿
22 27 ☽ Q ♂

24 Wednesday
00 08 ☽ △ ♃
01 02 ♀ ∠ ♃
02 50 ☽ ∠ ♅
06 04 ☽ ⊼ ☿
06 51 ☽ □ ♇
07 39 ♆ St D
07 41 ☽ ⊥ ♄
08 16 ☽ ⊻ ♆
13 52 ☽ ⋕ ♀
14 03 ☿ ✶ ♇
17 33 ☽ bQ ☉
19 01 ☽ ⊻ ♄
19 41 ☽ ⊥ ♆
19 52 ☽ ⊥ ♅

25 Thursday
03 52 ☿ △ ♆
04 21 ♀ ⊼ ♅
05 12 ☽ bQ ♃
07 19 ☽ ⊻ ♅
07 39 ☽ ☍ ♀
10 58 ☽ □ ♂
12 21 ☽ ± ☉
12 58 ☽ ∠ ♆
22 55 ☽ ± ♃

26 Friday
01 00 ☽ ⊼ ☉
10 49 ☽ ⊼ ♃
11 17 ☿ ⋕ ♄
16 48 ☽ △ ♇
18 09 ☽ ✶ ♆
19 16 ♂ ∠ ♆
22 46 ☽ ☍ ☿

27 Saturday
00 33 ♅ Q ♆
04 53 ☽ ☌ ♄
14 13 ♀ bQ ♆
17 44 ☽ ☌ ♅
17 47 ☽ Q ♆
22 25 ☽ bQ ♇

28 Sunday
00 48 ☽ ⊼ ♀
01 40 ☽ △ ♂
03 12 ☉ ⊥ ♇
05 55 ☿ Q ♂
14 04 ☽ ± ♀
16 22 ☽ ± ♇
17 33 ☽ ☍ ☉
22 57 ♀ □ ♂
23 18 ☽ ☍ ♃

29 Monday
03 00 ☿ ☍ ♄
04 23 ☽ ⊼ ♇
05 39 ☽ □ ♆
09 39 ☽ bQ ♂
10 05 ☽ bQ ♀
11 15 ☽ ⋕ ♀
13 16 ♀ Q ♇
16 13 ☽ ⊻ ♄
18 03 ☽ ⊼ ☿

30 Tuesday
04 13 ☽ ⊥ ♄
04 59 ☽ ± ♂
05 28 ☽ ⊻ ♅
07 43 ☽ ± ☿
17 32 ☽ ⊥ ♅
17 55 ☽ ⊼ ♂
19 43 ☽ △ ♀
22 15 ☽ ∠ ♄

31 Wednesday
04 20 ☽ bQ ☿
11 33 ☽ ⊼ ☉
11 40 ☽ ∠ ♅
12 55 ☽ ⊼ ♃
16 58 ☽ ☍ ♇
18 08 ☽ △ ♆
19 59 ☽ ⋕ ♂

All ephemeris data is given at 12.00 UT and the Moon's longitude is additionally given for 24.00 UT

JANUARY 2032

LONGITUDES

Date	Sidereal time h m s	Sun ☉ ° ' "	Moon ☽ ° ' "	Moon ☽ 24.00 ° ' "	Mercury ☿ ° '	Venus ♀ ° '	Mars ♂ ° '	Jupiter ♃ ° '	Saturn ♄ ° '	Uranus ♅ ° '	Neptune ♆ ° '	Pluto ♇ ° '
01	18 42 50	10 ♑ 36 53	21 ♌ 39 33	27 ♌ 35 36	22 ♐ 53	03 ♐ 47	02 ♓ 12	10 ♑ 28	17 ♊ 53	24 ♊ 36	12 ♈ 50	12 ♒ 17
02	18 46 46	11 38 01	03 ♍ 32 18	09 ♍ 30 06	24 19	05 01	02 58	10 42	17 R 49	24 R 33	12 50	12 18
03	18 50 43	12 39 10	15 ♍ 29 26	21 ♍ 30 48	25 46	06 14	03 45	10 56	17 45	24 31	12 51	12 20
04	18 54 39	13 40 20	27 ♍ 34 45	03 ♎ 41 50	27 14	07 27	04 31	11 10	17 40	24 28	12 51	12 22
05	18 58 36	14 41 29	09 ♎ 52 40	16 ♎ 07 50	28 ♐ 42	08 40	05 17	11 24	17 36	24 26	12 52	12 23
06	19 02 32	15 42 39	22 ♎ 27 58	28 ♎ 53 37	00 ♑ 10	09 54	06 03	11 38	17 32	24 24	12 52	12 25
07	19 06 29	16 43 49	05 ♏ 25 22	12 ♏ 03 42	01 39	11 07	06 49	11 52	17 28	24 21	12 53	12 27
08	19 10 26	17 44 59	18 ♏ 49 02	25 ♏ 41 38	03 09	12 21	07 35	12 06	17 24	24 19	12 53	12 28
09	19 14 22	18 46 09	02 ♐ 41 40	09 ♐ 49 05	04 39	13 34	08 22	12 19	17 20	24 17	12 54	12 30
10	19 18 19	19 47 19	17 ♐ 03 39	24 ♐ 24 54	06 10	14 48	09 08	12 33	17 17	24 14	12 54	12 32
11	19 22 15	20 48 30	01 ♑ 52 06	09 ♑ 24 21	07 41	16 01	09 54	12 47	17 13	24 12	12 55	12 34
12	19 26 12	21 49 40	17 ♑ 00 30	24 ♑ 39 12	09 13	17 15	10 40	13 01	17 09	24 10	12 55	12 35
13	19 30 08	22 50 50	02 ♒ 19 04	09 ♒ 58 35	10 45	18 28	11 26	13 15	17 06	24 08	12 56	12 37
14	19 34 05	23 51 59	17 ♒ 36 18	25 ♒ 10 51	12 17	19 42	12 12	13 29	17 02	24 06	12 57	12 39
15	19 38 01	24 53 08	02 ♓ 41 02	10 ♓ 05 49	13 50	20 56	12 58	13 42	16 59	24 04	12 58	12 41
16	19 41 58	25 54 16	17 ♓ 24 25	24 ♓ 36 17	15 24	22 10	13 44	13 56	16 56	24 02	12 58	12 42
17	19 45 55	26 55 23	01 ♈ 41 07	08 ♈ 38 47	16 58	23 23	14 30	14 10	16 52	23 59	12 59	12 44
18	19 49 51	27 56 30	15 ♈ 29 23	22 ♈ 13 08	18 32	24 37	15 16	14 24	16 49	23 57	13 00	12 46
19	19 53 48	28 57 36	28 ♈ 50 24	05 ♉ 21 39	20 07	25 51	16 02	14 37	16 46	23 55	13 01	12 48
20	19 57 44	29 ♑ 58 41	11 ♉ 47 21	18 ♉ 08 05	21 43	27 05	16 48	14 51	16 43	23 54	13 02	12 50
21	20 01 41	00 ♒ 59 45	24 ♉ 24 23	00 ♊ 36 48	23 19	28 19	17 34	15 05	16 41	23 52	13 03	12 51
22	20 05 37	02 00 48	06 ♊ 45 52	12 ♊ 52 05	24 56	29 ♐ 32	18 20	15 18	16 38	23 50	13 04	12 53
23	20 09 34	03 01 50	18 ♊ 55 55	24 ♊ 57 47	26 33	00 ♑ 46	19 06	15 32	16 35	23 48	13 05	12 55
24	20 13 30	04 02 52	00 ♋ 58 04	06 ♋ 57 06	28 11	02 00	19 52	15 45	16 33	23 46	13 06	12 57
25	20 17 27	05 03 52	12 ♋ 55 09	18 ♋ 52 29	29 ♑ 49	03 14	20 38	15 59	16 30	23 44	13 07	12 59
26	20 21 24	06 04 52	24 ♋ 49 18	00 ♌ 45 47	01 ♒ 28	04 28	21 24	16 13	16 28	23 43	13 08	13 00
27	20 25 20	07 05 50	06 ♌ 42 07	12 ♌ 38 26	03 08	05 42	22 10	16 26	16 26	23 41	13 09	13 02
28	20 29 17	08 06 48	18 ♌ 34 53	24 ♌ 31 38	04 48	06 56	22 55	16 39	16 23	23 39	13 10	13 04
29	20 33 13	09 07 45	00 ♍ 28 50	06 ♍ 26 41	06 29	08 10	23 41	16 53	16 21	23 38	13 11	13 06
30	20 37 10	10 08 41	12 ♍ 25 25	18 ♍ 25 16	08 10	09 24	24 27	17 06	16 20	23 36	13 13	13 08
31	20 41 06	11 ♒ 09 36	24 ♍ 26 34	00 ♎ 29 38	09 ♒ 53	10 ♑ 38	25 ♓ 13	17 ♑ 20	16 ♊ 18	23 ♊ 35	13 ♈ 14	13 ♒ 10

Date	Moon True ☊ ° '	Moon Mean ☊ ° '	Moon ☽ Latitude ° '
01	17 ♏ 24	16 ♏ 07	05 S 05
02	17 R 16	16 04	04 53
03	17 10	16 01	04 29
04	17 06	15 58	03 53
05	17 04	15 55	03 05
06	17 D 04	15 51	02 08
07	17 05	15 48	01 S 02
08	17 R 06	15 45	00 N 09
09	17 04	15 42	01 22
10	17 01	15 39	02 33
11	16 55	15 35	03 35
12	16 46	15 32	04 23
13	16 36	15 29	04 53
14	16 26	15 26	05 02
15	16 16	15 23	04 50
16	16 09	15 20	04 19
17	16 04	15 16	03 32
18	16 01	15 13	02 34
19	16 01	15 10	01 30
20	16 D 01	15 07	00 N 22
21	16 R 01	15 04	00 S 44
22	15 59	15 01	01 48
23	15 55	14 57	02 45
24	15 47	14 54	03 34
25	15 37	14 51	04 13
26	15 24	14 48	04 41
27	15 09	14 45	04 56
28	14 54	14 41	04 59
29	14 40	14 38	04 48
30	14 27	14 35	04 25
31	14 ♏ 18	14 ♏ 32	03 S 50

DECLINATIONS

Date	Sun ☉ ° '	Moon ☽ ° '	Mercury ☿ ° '	Venus ♀ ° '	Mars ♂ ° '	Jupiter ♃ ° '	Saturn ♄ ° '	Uranus ♅ ° '	Neptune ♆ ° '	Pluto ♇ ° '
01	23 S 01	09 N 28	22 S 53	19 S 03	11 S 38	23 S 01	21 N 21	23 N 29	03 N 37	23 S 23
02	22 55	05 39	23 05	19 18	11 21	23 00	21 21	23 29	03 37	23 22
03	22 50	01 N 35	23 16	19 33	11 03	22 59	21 21	23 29	03 38	23 22
04	22 44	02 S 36	23 26	19 47	10 45	22 58	21 21	23 29	03 38	23 21
05	22 37	06 45	23 34	20 00	10 28	22 57	21 20	23 28	03 38	23 21
06	22 31	10 43	23 42	20 13	10 10	22 56	21 20	23 28	03 38	23 20
07	22 23	14 18	23 48	20 26	09 52	22 54	21 20	23 28	03 38	23 20
08	22 15	17 16	23 54	20 38	09 34	22 53	21 20	23 28	03 39	23 19
09	22 07	19 21	23 58	20 49	09 16	22 52	21 20	23 28	03 39	23 19
10	21 58	20 16	24 01	21 00	08 58	22 51	21 19	23 28	03 39	23 18
11	21 49	19 51	24 02	21 11	08 40	22 50	21 19	23 28	03 39	23 18
12	21 40	18 00	24 03	21 20	08 21	22 48	21 19	23 28	03 40	23 17
13	21 30	14 52	24 02	21 29	08 03	22 47	21 19	23 28	03 40	23 17
14	21 20	10 45	23 59	21 38	07 45	22 46	21 19	23 28	03 40	23 16
15	21 09	06 00	23 56	21 46	07 26	22 45	21 19	23 28	03 41	23 16
16	20 58	01 S 00	23 51	21 53	07 08	22 43	21 19	23 27	03 41	23 15
17	20 46	03 N 55	23 45	22 00	06 49	22 42	21 19	23 27	03 41	23 15
18	20 34	08 28	23 37	22 06	06 31	22 41	21 18	23 27	03 42	23 14
19	20 22	12 28	23 28	22 11	06 12	22 39	21 18	23 27	03 42	23 14
20	20 09	15 43	23 17	22 16	05 53	22 38	21 18	23 27	03 43	23 13
21	19 56	18 09	23 05	22 20	05 35	22 36	21 18	23 27	03 43	23 13
22	19 42	19 40	22 52	22 23	05 16	22 35	21 18	23 27	03 43	23 12
23	19 29	20 14	22 37	22 26	04 57	22 33	21 18	23 27	03 44	23 12
24	19 14	19 52	22 21	22 28	04 38	22 32	21 18	23 27	03 44	23 11
25	19 00	18 37	22 03	22 29	04 20	22 30	21 18	23 27	03 45	23 11
26	18 45	16 33	21 44	22 30	04 01	22 29	21 18	23 27	03 45	23 10
27	18 30	13 49	21 24	22 30	03 42	22 27	21 18	23 27	03 46	23 10
28	18 14	10 31	21 01	22 30	03 23	22 26	21 18	23 27	03 46	23 09
29	17 58	06 48	20 38	22 28	03 04	22 24	21 18	23 26	03 47	23 09
30	17 42	02 49	20 13	22 26	02 45	22 23	21 18	23 26	03 47	23 08
31	17 S 25	01 S 19	19 S 46	22 S 24	02 S 26	22 S 21	21 N 18	23 N 26	03 N 48	23 S 08

ZODIAC SIGN ENTRIES

Date	h	m	Planets
02	04	52	☽ ♍
04	16	46	☽ ♎
06	09	14	☿ ♑
07	02	03	☽ ♏
09	07	24	☽ ♐
11	09	00	☽ ♑
13	08	22	☽ ♒
15	07	42	☽ ♓
17	09	08	☽ ♈
19	14	07	☽ ♉
20	12	31	☉ ♒
21	22	49	☽ ♊
22	20	56	♀ ♑
24	10	04	☽ ♋
25	14	41	☿ ♒
26	22	27	☽ ♌
29	11	02	☽ ♍
31	23	01	☽ ♎

LATITUDES

Date	Mercury ☿ ° '	Venus ♀ ° '	Mars ♂ ° '	Jupiter ♃ ° '	Saturn ♄ ° '	Uranus ♅ ° '	Neptune ♆ ° '	Pluto ♇ ° '
01	00 N 22	01 N 53	01 S 01	00 00	01 S 32	00 N 10	01 S 34	06 S 32
04	00 S 01	01 48	00 58	00 00	01 32	00 10	01 34	06 32
07	00 23	01 41	00 56	00 00	01 31	00 10	01 34	06 32
10	00 43	01 34	00 53	00 00	01 30	00 10	01 34	06 32
13	01 02	01 27	00 50	00 S 01	01 30	00 10	01 34	06 32
16	01 19	01 19	00 48	00 01	01 29	00 10	01 34	06 32
19	01 33	01 11	00 45	00 01	01 29	00 10	01 33	06 32
22	01 45	01 03	00 43	00 02	01 28	00 10	01 33	06 33
25	01 55	00 54	00 40	00 02	01 27	00 10	01 33	06 33
28	02 01	00 46	00 38	00 02	01 27	00 10	01 33	06 33
31	02 S 05	00 N 37	00 S 35	00 S 02	01 S 26	00 N 10	01 S 33	06 S 33

DATA

Julian Date	2463233
Delta T	+75 seconds
Ayanamsa	24° 18' 28"
Synetic vernal point	04° ♓ 48' 31"
True obliquity of ecliptic	23° 26' 00"

MOON'S PHASES, APSIDES AND POSITIONS ☽

Date	h	m	Phase	Longitude ° '	Eclipse Indicator
05	22	04	☾	15 ♎ 07	
12	20	07	●	22 ♑ 10	
19	12	14	☽	28 ♈ 58	
27	12	52	○	07 ♌ 08	

Day	h	m		
13	07	48	Perigee	
27	15	56	Apogee	
03	21	05	0S	
10	16	41	Max dec	20° S 18'
16	16	47	0N	
23	14	27	Max dec	20° N 14'
31	04	24	0S	

LONGITUDES

Date	Chiron ⚷ ° '	Ceres ⚳ ° '	Pallas ⚴ ° '	Juno ⚵ ° '	Vesta ⚶ ° '	Black Moon Lilith ⚸ ° '
01	16 ♉ 46	11 ♍ 39	20 ♉ 10	01 ♉ 53	02 ♎ 46	05 ♌ 24
11	16 ♉ 34	11 ♍ 56	20 ♉ 18	04 ♉ 37	04 ♎ 50	06 ♌ 31
21	16 ♉ 28	11 ♍ 29	21 ♉ 33	07 ♉ 58	06 ♎ 18	07 ♌ 38
31	16 ♉ 29	10 ♍ 19	23 ♉ 46	11 ♉ 47	07 ♎ 07	08 ♌ 46

All ephemeris data is given at 12.00 UT and the Moon's longitude is additionally given for 24.00 UT

ASPECTARIAN

01 Thursday
00 50 ☽ ± ☉
01 18 ☽ ± ♃
04 25 ☽ ✶ ♄
07 41 ☉ ☌ ♃
10 21 ☉ ∥ ♃
14 49 ☽ △ ☿
17 55 ☽ ✶ ♅
19 52 ☽ bQ ♃
20 45 ☽ bQ ☉
22 35 ☉ ∥ ☿

02 Friday
00 30 ☽ bQ ♆
02 20 ☿ ∥ ♃
04 32 ☽ Q ♄
04 36 ♀ ⊥ ♃
10 47 ☽ ☍ ♂
15 18 ☽ □ ♀
15 44 ☿ ☍ ♅
18 03 ☽ Q ♅
18 39 ☽ ± ♆

03 Saturday
00 03 ☽ ∥ ♆
02 42 ☽ △ ♃
04 18 ☉ ⊻ ♇
05 40 ☽ ⊼ ♇
05 48 ☽ △ ☉
06 42 ☽ ⊼ ♆
16 28 ☽ □ ♄
16 36 ☉ □ ♆
17 41 ☽ ± ♇

04 Sunday
01 44 ☿ ∥ ♇
05 53 ☽ □ ♅
07 20 ☽ Q ♀
11 13 ☽ □ ☿
11 34 ☽ bQ ♇
14 16 ☿ ∠ ♇
17 54 ☽ # ♆
19 42 ☿ # ♅

05 Monday
02 30 ☽ ⊼ ♂
09 25 ☽ ✶ ♀
14 18 ☉ ⊥ ♀
14 53 ☽ ± ♂
14 59 ☽ □ ♃
16 51 ☽ △ ♇
17 45 ☽ ☍ ♆
22 04 ☽ □ ☉

06 Tuesday
02 44 ☽ △ ♄
02 48 ☽ Q ☿
08 47 ☽ ∥ ♂
09 10 ☽ bQ ♂
15 36 ☽ △ ♅
17 02 ☽ ∠ ♀

07 Wednesday
01 36 ☽ Q ♃
04 13 ☽ ✶ ☿
06 38 ☽ bQ ♄
10 38 ☽ Q ☉
11 23 ☽ ⊥ ♀
13 41 ♂ ⊥ ♆
14 42 ☽ △ ♂
19 07 ☽ bQ ♅
22 53 ☽ ± ♄
23 21 ☽ ⊻ ♀
23 51 ☽ ✶ ♃

08 Thursday
00 43 ☽ □ ♇
01 28 ☽ ⊼ ♆
04 22 ☉ ⊼ ♄
05 58 ♀ ⊻ ♃
09 31 ☽ ⊼ ♄
09 58 ☽ ✶ ☉
10 41 ☽ ∠ ☿
11 07 ☽ ± ♅
12 07 ☽ ± ♆
14 38 ♀ ✶ ♇
21 35 ☽ ⊼ ♅
22 41 ♀ △ ♆

09 Friday
02 39 ☽ ∠ ♃
03 47 ☽ bQ ♆
04 16 ☽ ⊥ ☿
08 15 ☽ Q ♇
13 58 ☽ ∠ ☉
15 43 ☽ ⊻ ☿
18 14 ☽ ⊥ ♃
22 06 ☽ □ ♂

10 Saturday
04 26 ☽ ⊻ ♃
04 30 ☽ ✶ ♇
05 08 ☽ △ ♆
06 12 ☽ ⊥ ☉
07 55 ☽ ☌ ♀
09 14 ♃ ⊻ ♇
12 21 ☽ ☍ ♄
16 48 ☽ ⊻ ☉
17 55 ☿ ⊥ ♇
23 41 ☽ ☍ ♅

11 Sunday
05 04 ☽ ∠ ♇
05 17 ☽ Q ♂
19 30 ☽ ⊥ ♇
22 18 ☽ ☌ ☿

12 Monday
01 28 ☽ ✶ ♂
01 54 ♃ □ ♆
05 01 ☽ ⊻ ♇
05 34 ☽ □ ♆
05 37 ☽ ☌ ♃
09 19 ♀ # ♄
10 18 ♀ ☍ ♄
12 14 ☽ ⊼ ♄
12 24 ☽ ⊻ ♀
20 07 ☽ ☌ ☉
21 37 ☽ ± ♄
22 39 ☽ ⊥ ♀
23 13 ☽ ⊼ ♅

13 Tuesday
02 18 ☽ ∠ ♂
08 35 ☽ ± ♅
09 50 ☽ Q ♆
11 39 ☽ bQ ♄
12 47 ☉ ∥ ♀
13 58 ☽ ∠ ♀
17 08 ☽ ⊥ ♂
17 33 ☉ ± ♄
22 39 ☽ bQ ♅

14 Wednesday
02 41 ☽ ⊻ ☿
03 03 ☽ ⊻ ♂
04 11 ☽ ☌ ♇
04 40 ☽ ✶ ♆
05 24 ☽ ⊻ ♃
09 24 ☿ ✶ ♂
11 07 ☽ △ ♄
13 12 ☽ ⊥ ☿
13 36 ☉ # ♄
15 00 ☽ ⊥ ♃
15 36 ☽ ✶ ♀
17 12 ☉ ⊼ ♅
17 42 ☿ ⊻ ♇
22 15 ☽ △ ♅
22 19 ☿ □ ♆
22 38 ☽ ⊻ ☉

15 Thursday
02 27 ♂ ⊻ ♇
04 25 ☽ ∠ ♆
04 27 ☽ ∥ ♂
05 07 ☽ ∠ ☿
05 31 ☽ ∠ ♃
08 54 ☽ ⊥ ☉
09 37 ☿ ☌ ♃
11 38 ♂ ⊻ ♆
12 26 ☽ Q ♀
18 54 ☽ ⊥ ♆
23 09 ☽ # ♆

16 Friday
00 31 ☽ ∠ ☉
04 15 ☽ ⊻ ♇
04 42 ☽ ⊻ ♆
05 37 ☽ ☌ ♂
06 11 ☽ ✶ ♃
08 17 ☽ ✶ ☿
11 13 ☽ □ ♄
14 09 ☽ ⊥ ♇
20 38 ☽ □ ♀
20 48 ♂ ✶ ♃
23 00 ☽ □ ♅

17 Saturday
02 28 ☽ Q ♃
03 17 ☽ ✶ ☉
05 16 ☽ ∠ ♇
06 47 ☽ Q ☿
10 41 ☿ ⊼ ♄
10 52 ☽ ∥ ♆
17 27 ☽ Q ♄
23 27 ♀ ☍ ♅

18 Sunday
01 29 ☽ Q ☉
02 03 ☽ # ♂
05 48 ☽ Q ♅
07 11 ☽ ✶ ♇
07 36 ☽ ☌ ♆
10 02 ☽ □ ♃
11 36 ☽ ⊻ ♂
14 21 ☽ ✶ ♄
18 08 ☽ □ ☿
22 56 ☽ ⊥ ♂

19 Monday
03 06 ☽ ✶ ♅
04 38 ☽ Q ♇
06 00 ☽ △ ♀
12 14 ☽ □ ☉
13 16 ☿ # ♅
16 17 ☽ ∠ ♂
17 21 ☽ ∠ ♄

20 Tuesday
06 35 ☽ ∠ ♅
09 33 ♂ □ ♄
10 00 ☽ ⊥ ♄
10 02 ☉ ± ♅
12 36 ☽ bQ ♀
13 57 ☽ □ ♇
14 20 ☽ ⊻ ♆
17 53 ☽ △ ♃
20 37 ☿ ∥ ♇
21 17 ☽ ⊻ ♄
22 05 ☽ ✶ ♂
23 31 ☽ ⊥ ♅

21 Wednesday
01 43 ☽ ⊥ ♆
02 42 ☿ ± ♄
02 56 ♀ ∠ ♇
07 32 ☽ ± ♀
09 36 ☽ △ ☿
10 57 ☽ ⊻ ♅
13 12 ☉ Q ♆
19 02 ☽ ∠ ♆
19 59 ☿ ⊼ ♅
20 22 ☽ ⊼ ♀
22 38 ☽ Q ♂
23 10 ☽ bQ ♃

22 Thursday
01 53 ☽ △ ☉
03 20 ☉ bQ ♄
12 47 ☽ # ☉
17 05 ☽ ± ♃
19 09 ☽ bQ ☿

23 Friday
00 04 ☽ △ ♇
00 24 ☽ ✶ ♆
05 08 ☽ ⊼ ♃
05 55 ♂ ⊥ ♇
07 22 ☽ ☌ ♄
10 03 ☽ bQ ☉
12 22 ☽ □ ♂
15 42 ☽ ± ☿
18 23 ☿ ∥ ♃
21 39 ☽ ☌ ♅

24 Saturday
00 15 ☽ Q ♆
03 07 ☿ ∥ ♀
05 32 ☽ ⊼ ☿
05 37 ☽ ± ☉
05 56 ☽ bQ ♇
14 19 ☽ ☍ ♀
18 45 ☽ ⊼ ☉

25 Sunday
00 01 ☽ ± ♇
04 56 ☽ # ☉
10 54 ☿ ± ♅
12 07 ☽ ⊼ ♇
12 24 ☽ □ ♆
18 18 ☽ ☍ ♃
19 12 ☽ ⊻ ♄
23 44 ♀ ∥ ♃

26 Monday
04 37 ☽ △ ♂
07 06 ☿ Q ♆
07 15 ☽ ⊥ ♄
09 46 ☽ ⊻ ♅
11 58 ☿ bQ ♄
21 51 ☽ ⊥ ♅

27 Tuesday
01 22 ☽ ∠ ♄
03 36 ☽ ☍ ☿
07 21 ♂ # ♆
09 45 ☽ ⊼ ♀
11 19 ♃ ⊼ ♄
12 52 ☽ ☍ ☉
12 59 ☽ bQ ♂
16 00 ☽ ∠ ♅
17 52 ☉ ∠ ♂
18 13 ☿ # ♄
23 17 ☽ ± ♀

28 Wednesday
00 50 ☽ ☍ ♇
01 03 ☽ △ ♆
07 35 ☽ ✶ ♄
08 02 ☽ ⊼ ♃
08 25 ☽ ± ♂
14 37 ♀ ⊥ ♇
19 34 ☽ bQ ♀
20 23 ☽ ± ♃
21 22 ☽ ⊼ ♂
22 13 ☽ ✶ ♅

29 Thursday
00 29 ☉ bQ ♅
07 23 ☽ bQ ♆
07 44 ☽ Q ♄
10 18 ♂ □ ♅
14 53 ☽ bQ ♃
22 20 ☽ Q ♅

30 Friday
01 31 ☽ ± ♆
02 04 ☽ ⊼ ☿
05 15 ☽ △ ♀
06 15 ☽ ∥ ♆
07 00 ☽ ⊼ ☉
12 21 ☽ # ♂
13 25 ☽ ⊼ ♇
13 35 ☽ ⊼ ♆
16 05 ☽ ± ☿
17 57 ☿ bQ ♅
19 47 ☽ □ ♄
20 08 ☽ ± ☉
21 33 ☽ △ ♃

31 Saturday
01 27 ☽ ± ♇
10 17 ☽ □ ♅
13 01 ☽ bQ ☿
13 37 ☽ ☍ ♂
15 44 ☽ bQ ☉
18 05 ☽ ∥ ♂
19 24 ☽ bQ ♇
20 18 ☿ ∠ ♂

FEBRUARY 2032

LONGITUDES

Date	Sidereal time h m s	Sun ☉ ° ' "	Moon ☽ ° ' "	Moon ☽ 24.00 ° ' "	Mercury ☿ ° '	Venus ♀ ° '	Mars ♂ ° '	Jupiter ♃ ° '	Saturn ♄ ° '	Uranus ♅ ° '	Neptune ♆ ° '	Pluto ♇ ° '
01	20 45 03	12 ♒ 10 31	06 ♎ 34 52	12 ♎ 42 42	11 ♒ 36	11 ♑ 53	25 ♓ 58	17 ♑ 33	16 ♊ 16	23 ♊ 33	13 ♈ 15	13 ♒ 11
02	20 48 59	13 11 24	18 53 36	25 08 06	13 19	13 07	26 44	17 46	16 R 14	23 R 32	13 17	13 13
03	20 52 56	14 12 17	01 ♏ 26 43	07 ♏ 50 00	15 04	14 21	27 29	17 59	16 13	23 30	13 18	13 15
04	20 56 53	15 13 09	14 ♏ 18 31	20 ♏ 52 47	16 49	15 35	28 15	18 13	16 11	23 29	13 19	13 17
05	21 00 49	16 14 00	27 ♏ 31 08	04 ♐ 20 28	18 35	16 49	29 01	18 26	16 10	23 28	13 21	13 19
06	21 04 46	17 14 51	11 ♐ 14 37	18 ♐ 15 54	20 21	18 03	29 ♓ 46	18 39	16 09	23 27	13 22	13 21
07	21 08 42	18 15 40	25 ♐ 24 21	02 ♑ 39 46	22 08	19 18	00 ♈ 32	18 52	16 08	23 25	13 24	13 23
08	21 12 39	19 16 29	10 ♑ 01 44	17 ♑ 29 36	23 55	20 32	01 17	19 05	16 07	23 24	13 25	13 24
09	21 16 35	20 17 17	25 ♑ 02 28	02 ♒ 39 12	25 44	21 46	02 03	19 18	16 06	23 23	13 27	13 26
10	21 20 32	21 18 03	10 ♒ 18 32	17 ♒ 58 59	27 32	23 00	02 48	19 31	16 06	23 22	13 28	13 28
11	21 24 28	22 18 48	25 ♒ 39 05	03 ♓ 17 19	29 ♒ 21	24 14	03 33	19 43	16 05	23 21	13 30	13 30
12	21 28 25	23 19 32	10 ♓ 52 16	18 ♓ 22 40	01 ♓ 10	25 29	04 19	19 56	16 04	23 20	13 31	13 32
13	21 32 22	24 20 14	25 ♓ 47 26	03 ♈ 05 43	03 00	26 43	05 04	20 09	16 04	23 19	13 33	13 33
14	21 36 18	25 20 54	10 ♈ 16 56	17 ♈ 20 43	04 49	27 57	05 49	20 22	16 04	23 18	13 34	13 35
15	21 40 15	26 21 33	24 ♈ 16 55	01 ♉ 05 37	06 38	29 ♑ 11	06 35	20 34	16 04	23 18	13 36	13 37
16	21 44 11	27 22 11	07 ♉ 47 04	14 ♉ 23 36	08 27	00 ♒ 26	07 20	20 47	16 D 04	23 17	13 38	13 39
17	21 48 08	28 22 46	20 ♉ 49 43	27 ♉ 11 57	10 15	01 40	08 05	20 59	16 04	23 16	13 40	13 41
18	21 52 04	29 ♒ 23 20	03 ♊ 28 55	09 ♊ 41 14	12 02	02 54	08 50	21 12	16 04	23 16	13 41	13 43
19	21 56 01	00 ♓ 23 52	15 ♊ 49 31	21 ♊ 54 24	13 48	04 09	09 35	21 24	16 04	23 15	13 43	13 44
20	21 59 57	01 24 22	27 ♊ 56 28	03 ♋ 56 18	15 33	05 23	10 20	21 36	16 05	23 14	13 45	13 46
21	22 03 54	02 24 51	09 ♋ 54 24	15 ♋ 51 17	17 15	06 37	11 05	21 48	16 05	23 14	13 47	13 48
22	22 07 51	03 25 17	21 ♋ 47 21	27 ♋ 43 00	18 55	07 51	11 50	22 01	16 06	23 14	13 48	13 50
23	22 11 47	04 25 42	03 ♌ 38 36	09 ♌ 34 25	20 31	09 06	12 35	22 13	16 07	23 13	13 50	13 51
24	22 15 44	05 26 05	15 ♌ 30 43	21 ♌ 27 42	22 04	10 20	13 20	22 25	16 07	23 13	13 52	13 53
25	22 19 40	06 26 26	27 ♌ 25 35	03 ♍ 24 30	23 33	11 34	14 05	22 37	16 08	23 13	13 54	13 55
26	22 23 37	07 26 46	09 ♍ 24 36	15 ♍ 26 02	24 57	12 49	14 50	22 48	16 10	23 12	13 56	13 57
27	22 27 33	08 27 03	21 ♍ 28 57	27 ♍ 33 28	26 15	14 03	15 34	23 00	16 11	23 12	13 58	13 58
28	22 31 30	09 27 19	03 ♎ 39 45	09 ♎ 48 01	27 27	15 17	16 19	23 12	16 12	23 12	14 00	14 00
29	22 35 26	10 ♓ 27 34	15 ♎ 58 29	22 ♎ 11 22	28 ♓ 33	16 ♒ 32	17 ♈ 04	23 ♑ 23	16 ♊ 13	23 ♊ 12	14 ♈ 02	14 ♒ 02

MOON NODES AND LATITUDE

Date	Moon True ☊ ° '	Moon Mean ☊ ° '	Moon ☽ Latitude ° '
01	14 ♏ 11	14 ♏ 29	03 S 04
02	14 R 07	14 26	02 09
03	14 05	14 22	01 S 06
04	14 D 05	14 19	00 N 01
05	14 R 05	14 16	01 11
06	14 03	14 13	02 18
07	14 00	14 10	03 20
08	13 53	14 07	04 10
09	13 44	14 03	04 45
10	13 33	14 00	05 00
11	13 21	13 57	04 54
12	13 11	13 54	04 26
13	13 02	13 51	03 41
14	12 55	13 47	02 43
15	12 52	13 44	01 37
16	12 51	13 41	00 N 27
17	12 D 51	13 38	00 S 42
18	12 R 51	13 35	01 47
19	12 50	13 32	02 45
20	12 47	13 28	03 35
21	12 41	13 25	04 14
22	12 32	13 22	04 43
23	12 21	13 19	04 58
24	12 08	13 16	05 01
25	11 55	13 13	04 51
26	11 43	13 09	04 28
27	11 32	13 06	03 52
28	11 23	13 03	03 06
29	11 ♏ 18	13 ♏ 00	02 S 11

DECLINATIONS

Date	Sun ☉ ° '	Moon ☽ ° '	Mercury ☿ ° '	Venus ♀ ° '	Mars ♂ ° '	Jupiter ♃ ° '	Saturn ♄ ° '	Uranus ♅ ° '	Neptune ♆ ° '	Pluto ♇ ° '
01	17 S 08	05 S 26	19 S 18	22 S 20	02 S 08	22 S 20	21 N 18	23 N 26	03 N 48	23 S 07
02	16 51	09 23	18 49	22 16	01 49	22 18	21 18	23 26	03 49	23 07
03	16 34	13 01	18 18	22 12	01 30	22 16	21 18	23 26	03 49	23 06
04	16 16	16 07	17 45	22 07	01 11	22 15	21 18	23 26	03 50	23 06
05	15 58	18 28	17 11	22 01	00 52	22 13	21 18	23 26	03 51	23 06
06	15 40	19 50	16 36	21 54	00 33	22 11	21 19	23 26	03 51	23 05
07	15 21	20 02	15 59	21 47	00 S 14	22 10	21 19	23 26	03 52	23 05
08	15 02	18 54	15 20	21 39	00 N 05	22 08	21 19	23 26	03 52	23 04
09	14 43	16 27	14 41	21 30	00 23	22 06	21 19	23 26	03 53	23 04
10	14 24	12 50	13 59	21 21	00 42	22 04	21 19	23 26	03 54	23 03
11	14 04	08 21	13 17	21 11	01 01	22 03	21 19	23 26	03 54	23 03
12	13 44	03 S 23	12 33	21 01	01 20	22 01	21 20	23 26	03 55	23 02
13	13 24	01 N 43	11 48	20 50	01 38	21 59	21 20	23 26	03 56	23 02
14	13 04	06 34	11 02	20 38	01 57	21 57	21 20	23 26	03 56	23 01
15	12 44	10 55	10 15	20 26	02 16	21 56	21 20	23 25	03 57	23 01
16	12 23	14 32	09 27	20 13	02 34	21 54	21 20	23 25	03 58	23 01
17	12 02	17 17	08 39	19 59	02 53	21 52	21 21	23 25	03 58	23 00
18	11 41	19 06	07 49	19 45	03 11	21 50	21 21	23 25	03 59	23 00
19	11 20	19 56	06 59	19 30	03 30	21 48	21 21	23 25	04 00	22 59
20	10 58	19 50	06 09	19 15	03 48	21 46	21 21	23 25	04 00	22 59
21	10 37	18 50	05 19	18 59	04 07	21 45	21 22	23 25	04 01	22 59
22	10 15	17 01	04 29	18 43	04 25	21 43	21 22	23 25	04 02	22 58
23	09 53	14 30	03 40	18 26	04 43	21 41	21 22	23 25	04 03	22 58
24	09 31	11 23	02 51	18 09	05 01	21 39	21 23	23 25	04 03	22 57
25	09 09	07 48	02 04	17 51	05 20	21 37	21 23	23 25	04 04	22 57
26	08 46	03 N 54	01 18	17 32	05 38	21 35	21 23	23 25	04 05	22 57
27	08 24	00 S 11	00 S 33	17 13	05 56	21 34	21 24	23 25	04 06	22 56
28	08 01	04 18	00 N 08	16 54	06 14	21 32	21 24	23 25	04 06	22 56
29	07 S 39	08 S 17	00 N 48	16 S 34	06 N 31	21 S 30	21 N 24	23 N 25	04 N 07	22 S 55

ZODIAC SIGN ENTRIES

Date	h m	Planets
03	09 16	☽ ♏
05	16 21	☽ ♐
06	19 19	♂ ♈
07	19 37	☽ ♑
09	19 50	☽ ♒
11	18 49	☽ ♓
11	20 34	☿ ♓
13	18 54	☽ ♈
15	22 04	☽ ♉
16	03 40	♀ ♒
18	05 20	☽ ♊
19	02 32	☉ ♓
20	16 07	☽ ♋
23	04 37	☽ ♌
25	17 10	☽ ♍
28	04 48	☽ ♎

LATITUDES

Date	Mercury ☿ ° '	Venus ♀ ° '	Mars ♂ ° '	Jupiter ♃ ° '	Saturn ♄ ° '	Uranus ♅ ° '	Neptune ♆ ° '	Pluto ♇ ° '
01	02 S 05	00 N 34	00 S 34	00 S 03	01 S 26	00 N 10	01 S 33	06 S 33
04	02 03	00 25	00 32	00 03	01 25	00 10	01 33	06 33
07	01 57	00 16	00 29	00 03	01 24	00 10	01 33	06 33
10	01 46	00 N 07	00 27	00 04	01 24	00 10	01 33	06 34
13	01 30	00 S 01	00 24	00 04	01 23	00 10	01 32	06 34
16	01 08	00 10	00 22	00 04	01 22	00 10	01 32	06 34
19	00 40	00 18	00 19	00 04	01 22	00 10	01 32	06 35
22	00 S 06	00 26	00 17	00 05	01 21	00 10	01 32	06 35
25	00 N 33	00 33	00 15	00 05	01 20	00 10	01 32	06 35
28	01 15	00 41	00 12	00 05	01 20	00 10	01 32	06 36
31	01 N 59	00 S 48	00 S 10	00 S 06	01 S 19	00 N 10	01 S 32	06 S 36

DATA

Julian Date	2463264
Delta T	+75 seconds
Ayanamsa	24° 18' 33"
Synetic vernal point	04° ♓ 48' 26"
True obliquity of ecliptic	23° 26' 00"

MOON'S PHASES, APSIDES AND POSITIONS ☽

Date	h m	Phase	Longitude ° '	Eclipse Indicator
04	13 49	☾	15 ♏ 18	
11	06 24	●	22 ♒ 05	
18	03 29	☽	29 ♉ 08	
26	07 43	○	07 ♍ 16	

Day	h m		
10	20 41	Perigee	
23	18 32	Apogee	
07	03 37	Max dec	20° S 06'
13	03 52	0N	
19	21 17	Max dec	20° N 01'
27	10 55	0S	

LONGITUDES

Date	Chiron ⚷ ° '	Ceres ⚳ ° '	Pallas ⚴ ° '	Juno ⚵ ° '	Vesta ⚶ ° '	Black Moon Lilith ⚸ ° '
01	16 ♉ 29	10 ♍ 10	24 ♉ 02	12 ♉ 12	07 ♎ 09	08 ♌ 52
11	16 ♉ 36	08 ♍ 20	27 ♉ 06	16 ♉ 27	07 ♎ 07	10 ♌ 00
21	16 ♉ 50	06 ♍ 07	00 ♊ 49	21 ♉ 00	06 ♎ 17	11 ♌ 07
31	17 ♉ 09	03 ♍ 48	05 ♊ 00	25 ♉ 47	04 ♎ 41	12 ♌ 14

All ephemeris data is given at 12.00 UT and the Moon's longitude is additionally given for 24.00 UT

ASPECTARIAN

h m Aspects

01 Sunday
02 27 ☽ ⧺ ♆
22 16 ♀ ∥ ♃
23 26 ☽ △ ☿
23 32 ☽ □ ♀
23 56 ☽ △ ☉

02 Monday
00 58 ☽ △ ♇
01 05 ☽ ☍ ♆
01 43 ☿ ⊻ ♀
06 52 ☽ △ ♄
07 32 ☉ ☌ ☿
09 47 ☽ □ ♃
10 34 ☿ ☌ ♇
11 20 ☿ ✶ ♆
12 47 ☉ ☌ ♇
14 04 ☉ ✶ ♆
14 12 ♀ ⊻ ♇
15 15 ♀ □ ♆
20 33 ☉ ⊻ ♀
20 55 ☽ △ ♅

03 Tuesday
04 01 ☽ ⊼ ♂
11 34 ☽ ⚼ ♄
13 53 ☽ Q ♀
16 06 ☽ ± ♂
20 42 ☽ Q ♃

04 Wednesday
01 14 ☽ ⚼ ♅
03 36 ☿ △ ♄
04 24 ☽ ± ♄
09 56 ☽ ⚼ ♂
10 06 ☽ □ ♇
10 11 ☽ ⊼ ♆
13 05 ♂ ∠ ♇
13 15 ☽ ∥ ☉
13 49 ☽ □ ☉
14 35 ☽ ✶ ♀
15 27 ☽ ⊼ ♄
17 18 ☽ □ ☿
17 49 ☽ ± ♅
19 16 ☽ ✶ ♃
21 11 ☽ ± ♆
23 37 ♀ ⊼ ♄

05 Thursday
00 38 ☽ ∥ ☿
04 41 ☽ ⊼ ♅
09 43 ☿ ⊻ ♃
10 32 ☉ △ ♄
13 25 ☽ ⚼ ♆
14 45 ☽ △ ♂
18 41 ☽ Q ♇
20 19 ☽ ∠ ♀
21 35 ♀ Q ♂
22 34 ☽ ∠ ♃

06 Friday
00 45 ☽ Q ☉
06 15 ☽ Q ☿
13 32 ☽ ⊥ ♀
14 27 ☽ ⊥ ♃
15 37 ☽ ✶ ♇
15 40 ☽ △ ♆
20 24 ☽ ☍ ♄
23 04 ☽ ✶ ☉

07 Saturday
00 46 ☽ ⊻ ♀
00 51 ☽ ⊻ ♃
01 55 ♀ ☌ ♃
05 44 ☽ ✶ ☿
08 42 ☽ ☍ ♅
16 57 ☽ ∠ ♇
20 57 ☽ □ ♂

08 Sunday
01 57 ☽ ∠ ☉
02 56 ♂ Q ♃
05 08 ☿ △ ♅
06 09 ☉ ⊻ ♃
07 44 ☽ ⊥ ♇
09 58 ☽ ∠ ☿
17 28 ☽ ⊻ ♇
17 28 ☽ □ ♆
17 37 ☽ ⊥ ☉
21 47 ☽ ⊼ ♄

09 Monday
02 25 ☽ ⊥ ☿
02 45 ☽ ☌ ♃
03 39 ☽ Q ♂
03 55 ☽ ⊻ ☉
05 31 ☿ ⊥ ♃
06 21 ☽ ☌ ♀
07 21 ☽ ± ♄
09 02 ☉ ∥ ☿
09 23 ☽ ⊼ ♅
13 14 ☽ ⊻ ☿
18 29 ♀ ± ♄
18 51 ☽ ± ♅
19 15 ☿ ⊥ ♂
21 33 ☽ ⚼ ♄
22 07 ☽ Q ♆
23 37 ☽ ✶ ♂

10 Tuesday
01 33 ☽ ∥ ☉
03 29 ☽ ∥ ☿
08 58 ☽ ⚼ ♅
15 02 ♆ ✶ ♇
16 37 ♀ ⧺ ♄
16 57 ☽ ✶ ♆
16 57 ☽ ☌ ♇
18 59 ♀ ⊼ ♅
21 02 ☽ △ ♄

11 Wednesday
00 19 ☽ ∠ ♂
00 31 ☿ ∠ ♆
02 36 ☽ ⊻ ♃
06 24 ☽ ☌ ☉
08 24 ☽ △ ♅
09 36 ☽ ⊻ ♀
12 07 ☽ ⊥ ♃
15 08 ☽ ⊥ ♂
16 28 ☽ ∠ ♆
18 35 ☽ ☌ ☿
19 50 ☽ ⊥ ♀

12 Thursday
01 04 ☽ ⊻ ♂
02 28 ☽ ∠ ♃
04 32 ♂ Q ♄
06 40 ☽ ⊥ ♆
09 28 ☽ ⧺ ♆
11 19 ☽ ∠ ♀
12 15 ☉ △ ♅
16 14 ☽ ⊻ ♆
16 14 ☽ ⊻ ♇
20 18 ☽ □ ♄
21 04 ☽ ⧺ ♂

13 Friday
00 37 ☿ ⊥ ♀
01 53 ☽ ⊥ ♇
02 43 ☽ ✶ ♃
07 59 ☽ □ ♅
09 28 ☽ ⊻ ☉
11 37 ☽ ∥ ♂
13 39 ☽ ✶ ♀
16 32 ☽ ∠ ♇
20 00 ☽ ⊥ ☉
22 35 ☽ Q ♃
22 43 ☽ ∥ ♆

14 Saturday
01 32 ☽ ⊻ ☿
01 36 ☽ Q ♄
04 07 ☽ ☌ ♂
11 24 ☽ Q ♀
12 07 ☽ ∠ ☉
13 02 ☽ ⊥ ☿
13 44 ☽ Q ♅
17 35 ☽ ☌ ♆
17 36 ☽ ✶ ♇
20 05 ☿ ∠ ♃
21 48 ☽ ✶ ♄

15 Sunday
05 27 ☽ □ ♃
06 42 ☽ ∠ ☿
08 45 ☽ ⧺ ☿
10 17 ☽ ✶ ♅
10 36 ☿ ⊻ ♂
13 58 ♀ ± ♅
14 21 ☽ Q ♇
15 56 ☽ ✶ ☉
18 19 ☉ ⊥ ♃
21 30 ☽ □ ♀
22 25 ☽ ⧺ ☉
23 56 ☽ ∠ ♄

16 Monday
00 57 ☿ ⊥ ♆
07 00 ♄ St D
11 08 ☽ ⊻ ♂
12 54 ☽ ∠ ♅
13 24 ☽ ✶ ☿
15 07 ☽ Q ☉
16 08 ☽ ⊥ ♄
22 41 ☽ ⊻ ♆
22 43 ☽ □ ♇
22 43 ☽ ⊥ ♂

17 Tuesday
00 14 ♀ ⚼ ♄
03 08 ☽ ⊻ ♄
05 23 ☽ ⊥ ♅
09 49 ☽ ⊥ ♆
11 49 ♀ Q ♆
12 18 ☽ △ ♃
15 06 ☽ Q ☿
16 29 ☽ ∠ ♂
16 34 ☽ ⊻ ♅
18 49 ☉ ∠ ♆

18 Wednesday
02 48 ☽ ∠ ♆
03 29 ☽ □ ☉
10 46 ☽ △ ♀
17 19 ☽ ⚼ ♃
23 01 ☽ ✶ ♂
23 51 ☽ ⧺ ♀

19 Thursday
07 23 ☽ □ ☿
07 51 ☽ ✶ ♆
07 54 ☽ △ ♇
10 44 ☿ ⊻ ♆
11 03 ☿ ⊻ ♇
11 09 ☽ ± ♃
12 29 ☽ ☌ ♄
19 17 ☽ ⚼ ♀
23 11 ☽ ⊼ ♃

20 Friday
00 07 ☽ Q ♂
02 39 ☽ ☌ ♅
07 37 ☽ Q ♆
13 39 ☽ ⚼ ♇
15 12 ☽ ± ♀
19 28 ☿ □ ♄
19 34 ☽ △ ☉

21 Saturday
04 21 ♂ ∥ ♆
04 37 ☽ ⊼ ♀
07 45 ☽ ± ♇
08 38 ☽ ⧺ ♀
14 32 ☽ □ ♂
16 39 ♂ Q ♅
19 49 ☽ □ ♆
19 52 ☽ ⊼ ♇

22 Sunday
00 29 ☽ ⊻ ♄
04 33 ☽ ⚼ ☉
05 14 ☽ △ ☿
12 27 ☽ ☍ ♃
12 37 ☽ ⊥ ♄
13 25 ☿ ⧺ ♂
14 54 ☽ ⊻ ♅
19 06 ♀ ⚼ ♅
23 04 ♃ ± ♄

23 Monday
00 28 ☽ ± ☉
00 59 ☿ ⧺ ♆
01 49 ☿ ⊥ ♇
03 03 ☽ ⊥ ♅
06 52 ☽ ∠ ♄
13 44 ☽ ⊼ ☉
16 23 ☽ ⚼ ☿
21 15 ☽ ∠ ♅

24 Tuesday
00 19 ☽ ☍ ♀
07 18 ☽ △ ♂
08 40 ☽ △ ♆
08 42 ☽ ☍ ♇
13 14 ☽ ✶ ♄
13 17 ☽ ± ☿
18 17 ☿ ✶ ♃

25 Wednesday
02 09 ☽ ⊼ ♃
02 18 ☽ ⧺ ☉
03 07 ☽ ⊼ ☿
03 31 ☽ ✶ ♅
05 56 ♂ ☌ ♆
06 26 ♂ ✶ ♇
06 27 ☿ □ ♅
13 26 ☽ Q ♄
14 25 ☽ ± ♃
14 58 ☽ ⚼ ♆
15 33 ☽ ⚼ ♂

26 Thursday
02 19 ☽ ∥ ♂
03 36 ☽ Q ♅
07 43 ☽ ☍ ☉
08 45 ☽ ⚼ ♃
09 02 ☽ ± ♆
10 45 ☽ ± ♂
10 56 ☽ ∥ ♆
19 33 ☽ ⊼ ♀
21 02 ☽ ⊼ ♆
21 03 ☽ ⊼ ♇
22 41 ☉ ∠ ♃
23 30 ☽ ⊼ ♂
23 57 ☉ ⊥ ♆

27 Friday
01 28 ☽ □ ♄
06 45 ☽ ⧺ ☿
08 50 ☽ ± ♀
09 00 ☽ ± ♇
10 17 ♀ ✶ ♆
10 27 ♀ ☌ ♇
13 51 ☽ ∥ ☿
15 03 ☽ △ ♃
15 24 ☽ □ ♅
22 29 ☽ ☍ ☿

28 Saturday
02 49 ☽ ⚼ ♇
04 37 ☽ ⚼ ♀
08 06 ♂ ✶ ♄
10 50 ☽ ⧺ ♆
12 37 ♃ ⊼ ♅

29 Sunday
00 20 ☽ ⧺ ♂
00 20 ☽ ⊼ ☉
06 03 ♀ △ ♄
08 13 ☽ ☍ ♆
08 13 ☽ △ ♇
08 22 ☽ ∥ ☉
11 15 ♆ ✶ ♇
12 29 ☽ △ ♄
13 01 ☽ ± ☉
13 11 ☽ △ ♀
14 14 ☽ ☍ ♂
23 51 ☿ ∠ ♇

MARCH 2032

LONGITUDES

Date	Sidereal time h m s	Sun ☉ ° ' "	Moon ☽ ° ' "	Moon ☽ 24.00 ° ' "	Mercury ☿ ° '	Venus ♀ ° '	Mars ♂ ° '	Jupiter ♃ ° '	Saturn ♄ ° '	Uranus ♅ ° '	Neptune ♆ ° '	Pluto ♇ ° '
01	22 39 23	11 ♓ 27 47	28 ♎ 26 59	04 ♏ 45 39	29 ♓ 31	17 ♒ 46	17 ♈ 48	23 ♑ 35	16 ♊ 15	23 ♊ 12	14 ♈ 04	14 ♒ 03
02	22 43 20	12 27 58	11 ♏ 07 43	17 ♏ 33 32	00 ♈ 21	19 00	18 33	23 46	16 17	23 D 12	14 06	14 05
03	22 47 16	13 28 08	24 ♏ 03 31	00 ♐ 38 03	01 04	20 14	19 18	23 58	16 18	23 12	14 08	14 07
04	22 51 13	14 28 16	07 ♐ 17 30	14 ♐ 02 13	01 37	21 29	20 02	24 09	16 20	23 12	14 10	14 08
05	22 55 09	15 28 23	20 ♐ 52 29	27 ♐ 48 30	02 02	22 43	20 47	24 20	16 22	23 13	14 12	14 10
06	22 59 06	16 28 28	04 ♑ 50 20	11 ♑ 57 58	02 17	23 57	21 31	24 31	16 24	23 13	14 14	14 11
07	23 03 02	17 28 32	19 ♑ 11 10	26 ♑ 29 35	02 23	25 12	22 15	24 42	16 27	23 13	14 16	14 13
08	23 06 59	18 28 34	03 ♒ 52 36	11 ♒ 19 29	02 R 21	26 26	23 00	24 53	16 29	23 14	14 18	14 15
09	23 10 55	19 28 34	18 ♒ 49 18	26 ♒ 20 57	02 09	27 40	23 44	25 04	16 31	23 14	14 20	14 16
10	23 14 52	20 28 33	03 ♓ 53 16	11 ♓ 24 59	01 49	28 ♒ 55	24 28	25 15	16 34	23 14	14 22	14 18
11	23 18 49	21 28 29	18 ♓ 54 52	26 ♓ 21 43	01 20	00 ♓ 09	25 13	25 25	16 36	23 15	14 24	14 19
12	23 22 45	22 28 24	03 ♈ 44 28	11 ♈ 02 10	00 45	01 23	25 57	25 36	16 39	23 16	14 26	14 21
13	23 26 42	23 28 17	18 ♈ 14 03	25 ♈ 19 35	00 ♈ 03	02 38	26 41	25 46	16 42	23 16	14 28	14 22
14	23 30 38	24 28 07	02 ♉ 18 21	09 ♉ 10 13	29 ♓ 16	03 52	27 25	25 56	16 45	23 17	14 31	14 24
15	23 34 35	25 27 56	15 ♉ 55 09	22 ♉ 33 19	28 25	05 06	28 09	26 07	16 48	23 18	14 33	14 25
16	23 38 31	26 27 42	29 ♉ 05 00	05 ♊ 30 36	27 31	06 21	28 53	26 17	16 51	23 18	14 35	14 27
17	23 42 28	27 27 26	11 ♊ 50 35	18 ♊ 05 29	26 35	07 35	29 ♈ 37	26 27	16 54	23 19	14 37	14 28
18	23 46 24	28 27 08	24 ♊ 15 53	00 ♋ 22 24	25 39	08 49	00 ♉ 21	26 37	16 58	23 20	14 39	14 30
19	23 50 21	29 ♓ 26 48	06 ♋ 25 38	12 ♋ 26 12	24 43	10 03	01 05	26 46	17 01	23 21	14 41	14 31
20	23 54 18	00 ♈ 26 25	18 ♋ 24 43	24 ♋ 21 44	23 49	11 18	01 49	26 56	17 05	23 22	14 44	14 32
21	23 58 14	01 26 00	00 ♌ 17 50	06 ♌ 13 30	22 58	12 32	02 32	27 06	17 08	23 23	14 46	14 34
22	00 02 11	02 25 33	12 ♌ 09 13	18 ♌ 05 26	22 11	13 46	03 16	27 15	17 12	23 24	14 48	14 35
23	00 06 07	03 25 04	24 ♌ 02 32	00 ♍ 00 53	21 28	15 00	04 00	27 24	17 16	23 25	14 50	14 37
24	00 10 04	04 24 32	06 ♍ 00 45	12 ♍ 02 26	20 50	16 15	04 43	27 33	17 20	23 27	14 53	14 38
25	00 14 00	05 23 58	18 ♍ 06 08	24 ♍ 12 02	20 18	17 29	05 27	27 42	17 24	23 28	14 55	14 39
26	00 17 57	06 23 22	00 ♎ 20 18	06 ♎ 31 04	19 51	18 43	06 10	27 51	17 28	23 29	14 57	14 40
27	00 21 53	07 22 44	12 ♎ 44 25	19 ♎ 00 28	19 30	19 57	06 54	28 00	17 32	23 31	14 59	14 42
28	00 25 50	08 22 04	25 ♎ 19 17	01 ♏ 40 58	19 15	21 12	07 37	28 09	17 36	23 32	15 02	14 43
29	00 29 47	09 21 22	08 ♏ 05 36	14 ♏ 33 17	19 06	22 26	08 21	28 17	17 40	23 33	15 04	14 44
30	00 33 43	10 20 38	21 ♏ 04 07	27 ♏ 38 13	19 02	23 40	09 04	28 26	17 45	23 35	15 06	14 45
31	00 37 40	11 ♈ 19 52	04 ♐ 15 41	10 ♐ 56 39	19 ♓ 04	24 ♓ 54	09 ♉ 47	28 ♑ 34	17 ♊ 49	23 ♊ 37	15 ♈ 08	14 ♒ 46

Date	Moon True ☊ ° '	Moon Mean ☊ ° '	Moon ☽ Latitude ° '
01	11 ♏ 14	12 ♏ 57	01 S 08
02	11 R 14	12 53	00 S 01
03	11 D 14	12 50	01 N 08
04	11 15	12 47	02 15
05	11 R 15	12 44	03 16
06	11 13	12 41	04 07
07	11 10	12 38	04 44
08	11 04	12 34	05 04
09	10 56	12 31	05 03
10	10 48	12 28	04 42
11	10 41	12 25	04 01
12	10 34	12 22	03 05
13	10 30	12 19	01 57
14	10 28	12 15	00 N 44
15	10 D 28	12 12	00 S 29
16	10 29	12 09	01 39
17	10 30	12 06	02 42
18	10 31	12 03	03 35
19	10 R 31	11 59	04 17
20	10 29	11 56	04 47
21	10 26	11 53	05 05
22	10 21	11 50	05 10
23	10 14	11 47	05 01
24	10 07	11 44	04 39
25	10 01	11 40	04 05
26	09 55	11 37	03 19
27	09 51	11 34	02 22
28	09 49	11 31	01 18
29	09 48	11 28	00 S 09
30	09 D 48	11 24	01 N 01
31	09 ♏ 49	11 ♏ 21	02 N 10

DECLINATIONS

Date	Sun ☉ ° '	Moon ☽ ° '	Mercury ☿ ° '	Venus ♀ ° '	Mars ♂ ° '	Jupiter ♃ ° '	Saturn ♄ ° '	Uranus ♅ ° '	Neptune ♆ ° '	Pluto ♇ ° '
01	07 S 16	11 S 59	01 N 24	16 S 13	06 N 49	21 S 28	21 N 25	23 N 25	04 N 08	22 S 55
02	06 53	15 10	01 58	15 53	07 07	21 26	21 25	23 25	04 09	22 55
03	06 30	17 41	02 27	15 31	07 24	21 24	21 26	23 25	04 10	22 54
04	06 07	19 18	02 53	15 10	07 42	21 23	21 26	23 25	04 10	22 54
05	05 43	19 52	03 14	14 47	07 59	21 21	21 26	23 25	04 11	22 54
06	05 20	19 14	03 31	14 25	08 17	21 19	21 27	23 25	04 12	22 53
07	04 57	17 22	03 44	14 02	08 34	21 17	21 27	23 25	04 13	22 53
08	04 33	14 21	03 52	13 38	08 51	21 15	21 28	23 25	04 14	22 53
09	04 10	10 22	03 55	13 15	09 08	21 13	21 28	23 25	04 14	22 52
10	03 46	05 42	03 53	12 51	09 25	21 12	21 29	23 25	04 15	22 52
11	03 23	00 S 41	03 46	12 26	09 42	21 10	21 29	23 25	04 16	22 52
12	02 59	04 N 19	03 35	12 01	09 59	21 08	21 30	23 25	04 17	22 52
13	02 35	08 57	03 20	11 36	10 16	21 06	21 30	23 25	04 18	22 51
14	02 12	12 58	03 01	11 11	10 32	21 04	21 31	23 25	04 19	22 51
15	01 48	16 08	02 38	10 45	10 49	21 02	21 31	23 26	04 19	22 51
16	01 24	18 20	02 12	10 19	11 05	21 01	21 32	23 26	04 20	22 50
17	01 01	19 32	01 44	09 53	11 21	20 59	21 32	23 26	04 21	22 50
18	00 37	19 44	01 15	09 26	11 37	20 57	21 33	23 26	04 22	22 50
19	00 S 13	19 00	00 43	09 00	11 53	20 55	21 33	23 26	04 23	22 50
20	00 N 10	17 25	00 N 12	08 32	12 09	20 54	21 34	23 26	04 24	22 50
21	00 34	15 07	00 S 20	08 05	12 25	20 52	21 34	23 26	04 25	22 49
22	00 58	12 11	00 51	07 38	12 40	20 50	21 35	23 26	04 25	22 49
23	01 21	08 46	01 21	07 10	12 56	20 49	21 36	23 26	04 26	22 49
24	01 45	04 58	01 49	06 42	13 11	20 47	21 36	23 26	04 27	22 49
25	02 09	00 N 57	02 16	06 14	13 26	20 45	21 37	23 26	04 28	22 48
26	02 32	03 S 10	02 41	05 46	13 42	20 44	21 37	23 26	04 29	22 48
27	02 56	07 13	03 04	05 17	13 57	20 42	21 38	23 26	04 30	22 48
28	03 19	11 00	03 24	04 49	14 11	20 40	21 39	23 26	04 31	22 48
29	03 42	14 21	03 41	04 20	14 26	20 39	21 39	23 26	04 31	22 48
30	04 06	17 02	03 57	03 51	14 41	20 37	21 40	23 26	04 32	22 48
31	04 N 29	18 S 52	04 S 09	03 S 22	14 N 55	20 S 36	21 N 40	23 N 26	04 N 33	22 S 48

ZODIAC SIGN ENTRIES

Date	h	m	Planets
01	14	57	☽ ♏
02	01	19	☿ ♈
03	22	51	☽ ♐
06	03	46	☽ ♑
08	05	43	☽ ♒
10	05	49	☽ ♓
11	09	04	♀ ♓
12	05	54	☽ ♈
13	13	47	☿ ♓
14	08	01	☽ ♉
16	13	42	☽ ♊
18	00	35	♂ ♉
18	23	16	☽ ♋
20	01	22	☉ ♈
21	11	24	☽ ♌
23	23	58	☽ ♍
26	11	20	☽ ♎
28	20	50	☽ ♏
31	04	18	☽ ♐

LATITUDES

Date	Mercury ☿ ° '	Venus ♀ ° '	Mars ♂ ° '	Jupiter ♃ ° '	Saturn ♄ ° '	Uranus ♅ ° '	Neptune ♆ ° '	Pluto ♇ ° '
01	01 N 44	00 S 45	00 S 11	00 S 06	01 S 19	00 N 10	01 S 32	06 S 36
04	02 26	00 52	00 08	00 06	01 18	00 10	01 32	06 37
07	03 02	00 58	00 06	00 06	01 18	00 10	01 32	06 37
10	03 27	01 04	00 04	00 07	01 17	00 10	01 32	06 38
13	03 37	01 09	00 S 02	00 07	01 16	00 10	01 32	06 38
16	03 29	01 13	00 00	00 07	01 16	00 10	01 32	06 39
19	03 04	01 18	00 N 03	00 08	01 15	00 10	01 32	06 39
22	02 27	01 21	00 05	00 08	01 14	00 10	01 32	06 40
25	01 43	01 24	00 07	00 09	01 14	00 10	01 32	06 41
28	00 56	01 26	00 09	00 09	01 13	00 10	01 32	06 41
31	00 N 11	01 S 28	00 N 11	00 S 09	01 S 12	00 N 10	01 S 32	06 S 42

DATA

Julian Date	2463293
Delta T	+75 seconds
Ayanamsa	24° 18' 36"
Synetic vernal point	04° ♓ 48' 23"
True obliquity of ecliptic	23° 26' 00"

MOON'S PHASES, APSIDES AND POSITIONS ☽

Date	h	m	Phase	Longitude ° '	Eclipse Indicator
05	01	47	☾	15 ♐ 03	
11	16	25	●	21 ♓ 39	
18	20	57	☽	28 ♊ 49	
27	00	46	○	06 ♎ 55	

Day	h	m		
10	06	45	Perigee	
22	08	23	Apogee	
05	11	33	Max dec	19° S 52'
11	15	15	ON	
18	04	54	Max dec	19° N 46'
25	17	32	OS	

LONGITUDES

Date	Chiron ⚷ ° '	Ceres ⚳ ° '	Pallas ⚴ ° '	Juno ⚵ ° '	Vesta ⚶ ° '	Black Moon Lilith ⚸ ° '
01	17 ♉ 07	04 ♍ 01	04 ♊ 34	25 ♉ 17	04 ♎ 52	12 ♌ 07
11	17 ♉ 31	01 ♍ 52	09 ♊ 07	00 ♊ 14	02 ♎ 43	13 ♌ 15
21	17 ♉ 59	00 ♍ 10	13 ♊ 58	05 ♊ 19	00 ♎ 11	14 ♌ 22
31	18 ♉ 32	29 ♌ 06	19 ♊ 03	10 ♊ 29	27 ♍ 39	15 ♌ 29

All ephemeris data is given at 12.00 UT and the Moon's longitude is additionally given for 24.00 UT

ASPECTARIAN

h m	Aspects
01 Monday	
01 35	♅ St D
01 57	☽ △ ♅
02 32	☽ □ ♃
07 52	☽ bQ ☉
14 03	♀ ✶ ♂
14 12	☽ ⚻ ☿
17 21	☽ bQ ♄
02 Tuesday	
02 25	☽ ± ☿
03 45	☉ # ♂
06 30	☽ bQ ♅
10 24	☽ ± ♄
13 14	☽ Q ♃
14 43	☽ △ ☉
17 21	☽ ∥ ♀
17 32	☽ □ ♇
17 34	☽ ⚻ ♆
19 43	☉ ⊥ ♂
20 24	☽ bQ ☿
21 38	☽ ⚻ ♄
22 45	♃ # ♄
23 20	☽ ± ♅
03 Wednesday	
02 41	☽ ⚻ ♂
04 14	☽ □ ♀
04 44	☽ ± ♆
10 26	☽ ⚻ ♅
11 49	☽ ✶ ♃
14 24	☽ ± ♂
21 17	☽ bQ ♆
04 Thursday	
01 22	☽ △ ☿
02 42	☽ Q ♇
03 47	☉ ⚺ ♇
04 17	☉ ⚺ ♆
07 43	☽ bQ ♂
15 22	☽ ∠ ♃
16 18	☽ Q ♀
05 Friday	
00 12	☽ ✶ ♇
00 15	☽ △ ♆
01 47	☽ □ ☉
04 06	☽ ☍ ♄
07 30	☽ ⊥ ♃
11 49	☽ △ ♂
15 31	☽ ✶ ♀
16 04	☽ ☍ ♅
18 06	☽ ⚺ ♃
21 32	♀ △ ♅
06 Saturday	
02 21	☽ ∠ ♇
07 36	☽ □ ☿
10 18	☉ □ ♄
11 20	☽ Q ☉
17 41	☽ ⊥ ♇
19 37	☽ ∠ ♀
07 Sunday	
00 48	♀ ⚺ ♃
03 45	☽ ⚺ ♇
03 49	☽ □ ♆
07 27	☽ ⚻ ♄
08 58	☽ ✶ ☉
12 01	☽ ⊥ ♀
13 59	☽ Q ☿
16 22	☿ St R
17 20	☽ □ ♂
17 23	☽ ± ♄
18 39	☽ ⚻ ♅
21 11	☽ ☌ ♃
22 47	☽ ⚺ ♀
08 Monday	
04 27	☽ ± ♅
08 07	☽ bQ ♄
09 26	☽ Q ♆
09 32	☽ ✶ ☿
10 21	☿ ⊥ ♀
11 18	☽ ∠ ☉
17 15	☽ ∥ ♀
19 01	☽ bQ ♅
19 33	♂ ✶ ♅
09 Tuesday	
00 04	☽ Q ♂
02 50	☽ ⊥ ☉
04 43	☽ ☌ ♇
04 48	☽ ✶ ♆
07 35	☉ # ♆
08 19	☽ △ ♄
09 22	☽ ∠ ☿
13 07	☽ ⚺ ☉
18 13	☽ # ♂
19 02	☽ △ ♅
20 14	☽ ✶ ♂
22 05	☽ ⚺ ♃
23 27	☽ ⊥ ☿
10 Wednesday	
03 22	☽ ☌ ♀
04 14	☉ # ☿
04 47	☽ ∠ ♆
07 36	☉ ⊥ ♇
07 44	☽ ⊥ ♃
08 46	☽ ⚺ ☿
18 59	☽ # ♆
19 09	☽ ⊥ ♆
20 56	☽ # ☿
21 04	♀ ∠ ♆
21 21	☽ ∠ ♂
22 06	☽ ∥ ☉
22 15	☽ ∠ ♃
11 Thursday	
04 38	☽ ⚺ ♇
04 45	☽ ⚺ ♆
08 17	☽ □ ♄
12 30	☽ ⊥ ♂
14 16	☽ ⊥ ♇
16 25	☽ ☌ ☉
18 59	☽ □ ♅
21 04	♂ □ ♃
22 36	☽ ✶ ♃
22 40	☽ ⚺ ♂
12 Friday	
03 46	☿ ⚺ ♀
04 50	☽ ∠ ♇
06 01	☽ # ☉
07 19	☽ ☌ ☿
07 49	☽ ⚺ ♀
08 37	☽ ∥ ☿
11 51	☽ ∥ ♆
13 30	☽ Q ♄
16 39	♀ ⊥ ♃
18 24	☽ Q ♃
18 32	☽ ⊥ ♀
13 Saturday	
00 23	☽ Q ♅
01 36	♂ Q ♇
05 32	☽ ✶ ♇
05 42	☽ ☌ ♆
07 06	☉ □ ♅
09 25	☽ ✶ ♄
10 53	☽ ∠ ♀
19 53	☽ ∥ ♂
20 30	☽ ✶ ♅
21 31	☽ ⚺ ☉
14 Sunday	
00 55	☽ □ ♃
01 49	☽ Q ♇
01 53	☽ # ♀
03 06	☽ ☌ ♂
07 03	☽ ⚺ ☿
08 25	☿ ∠ ♇
08 35	☽ ⊥ ☉
11 02	☽ ∠ ♄
14 59	☽ ✶ ♀
16 52	☽ ⊥ ☿
22 27	☽ ∠ ♅
15 Monday	
01 31	☽ ∠ ☉
02 50	☽ ⊥ ♄
07 48	☽ ∠ ☿
09 19	☽ □ ♇
09 32	☽ ⚺ ♆
10 06	♀ # ♂
13 35	☽ ⚺ ♄
14 21	☽ Q ♀
14 28	☽ ⊥ ♅
16 00	☿ ⚺ ♂
20 22	☽ ⊥ ♆
16 Tuesday	
01 22	☽ ⚺ ♅
06 42	☉ ✶ ♃
06 45	☽ △ ♃
06 45	☽ ✶ ☉
09 18	☽ ✶ ☿
11 36	☽ ⚺ ♂
12 56	☽ ∠ ♆
23 29	☽ ⊥ ♂
17 Wednesday	
01 11	☉ ☌ ☿
03 02	☽ □ ♀
06 14	☽ Q ☿
07 04	☽ Q ☉
11 14	☽ bQ ♃
15 04	☿ ✶ ♃
17 02	☽ △ ♇
17 20	☽ ✶ ♆
17 38	☽ ∠ ♂
21 45	☽ ☌ ♄
18 Thursday	
04 47	☽ ± ♃
08 43	♀ ⊥ ♆
10 11	☽ ☌ ♅
14 31	☽ □ ☿
16 39	☽ ⚻ ♃
16 42	☽ Q ♆
20 57	☽ □ ☉
22 17	☽ bQ ♇
19 Friday	
00 43	☽ ✶ ♂
06 47	☿ ⊥ ♂
13 46	☉ ∠ ♇
16 10	☽ ± ♇
20 04	☽ △ ♀
20 Saturday	
02 09	☽ Q ♂
04 12	☽ ⚻ ♇
04 34	☽ □ ♆
09 18	☽ ⚺ ♄
12 36	☉ ∥ ☿
21 27	☽ ⊥ ♄
21 31	♂ ∠ ♄
22 01	☽ ⚺ ♅
22 10	☽ △ ☿
21 Sunday	
00 25	☿ □ ♅
02 13	♀ ∠ ♃
05 26	☽ ☍ ♃
05 45	☽ bQ ♀
10 09	☽ ⊥ ♅
14 31	☽ △ ☉
15 44	☽ ∠ ♄
16 50	☽ □ ♂
22 Monday	
02 06	☽ ± ♀
02 34	☽ bQ ☿
04 24	☽ ∠ ♅
08 33	☽ ∥ ♂
15 39	☽ ⚻ ♀
16 56	☽ ☍ ♇
17 22	☽ △ ♆
19 41	☽ ± ☿
22 15	☽ ✶ ♄
23 38	☽ bQ ☉
23 Tuesday	
04 08	♀ ⚺ ♇
07 06	☽ ⚻ ☿
08 39	♀ ⚺ ♆
10 45	☽ ✶ ♅
15 19	☉ # ☿
18 51	☽ ⚻ ♃
19 24	☽ ± ☉
22 33	☽ Q ♄
23 41	☽ bQ ♆
23 48	☽ # ♀
24 Wednesday	
07 02	☽ ± ♃
08 31	☽ ⚻ ☉
09 15	☽ △ ♂
10 52	☽ Q ♅
15 09	☽ ∥ ♆
17 44	☽ ± ♆
20 29	☿ ⊥ ♇
25 Thursday	
01 11	☽ bQ ♃
04 58	☽ # ☿
05 10	☽ ⚻ ♇
05 34	☽ ∥ ☉
05 41	☽ ⚻ ♆
09 03	☿ ∠ ♂
10 10	♀ □ ♄
10 35	☽ □ ♄
10 38	☽ ☍ ♀
11 48	☉ Q ♄
16 10	☽ ☍ ☿
16 29	☉ ⚺ ♂
16 55	☽ bQ ♂
17 02	☽ ± ♇
22 35	☽ □ ♅
26 Friday	
07 06	☽ △ ♃
07 55	☽ # ☉
08 51	☽ ∥ ☿
10 42	☽ bQ ♇
11 40	☽ ± ♂
19 42	☽ # ♆
27 Saturday	
00 02	☽ ⚻ ♂
00 46	☽ ☍ ☉
01 39	☽ ∥ ♀
04 57	☿ ☌ ♀
15 45	☽ △ ♇
16 20	☽ ☍ ♆
21 14	☽ △ ♄
28 Sunday	
00 40	☽ ⚻ ☿
02 34	♀ ⊥ ♇
03 19	☽ ⚻ ♀
08 36	☽ △ ♅
11 52	☽ ± ☿
15 55	☽ ± ♀
17 24	☽ □ ♃
29 Monday	
01 48	☽ bQ ♄
02 42	♀ # ♆
04 36	☽ bQ ☿
08 23	☉ # ☿
10 38	☽ bQ ♀
12 30	☽ ☍ ♂
12 43	☽ # ♂
12 52	☽ bQ ♅
14 33	☽ ⚻ ☉
18 42	☽ ± ♄
19 17	♂ ∠ ♅
30 Tuesday	
00 21	☽ □ ♇
00 59	☽ ⚻ ♆
02 35	☽ ± ☉
03 22	☽ Q ♃
05 19	☉ # ♀
05 35	☽ ± ♅
05 51	☽ ⚻ ♄
08 16	☽ △ ☿
09 00	☿ ∥ ♀
10 22	♀ □ ♅
12 04	☽ ± ♆
14 29	☉ Q ♃
14 29	☿ St D
16 37	☽ ⚻ ♅
17 15	☽ △ ♀
20 27	☽ bQ ☉
31 Wednesday	
01 35	☽ ✶ ♃
04 31	☽ bQ ♆
06 48	♀ ∠ ♂
09 19	☽ Q ♇
16 41	☉ ∥ ♆
18 56	☉ Q ♅
22 30	☽ ⚻ ♂

APRIL 2032

Date	Sidereal time h m s	Sun ☉ LONGITUDES	Moon ☽	Moon ☽ 24.00	Mercury ☿	Venus ♀	Mars ♂	Jupiter ♃	Saturn ♄	Uranus ♅	Neptune ♆	Pluto ⯓
01	00 41 36	12 ♈ 19 04	17 ♐ 41 13	24 ♐ 29 31	19 ♓ 12	26 ♓ 08	10 ♉ 31	28 ♑ 42	17 ♊ 54	23 ♊ 38	15 ♈ 11	14 ♒ 48
02	00 45 33	13 18 15	01 ♑ 21 35	08 ♑ 17 28	19 D 25	27 22	11 14	28 50	17 59	23 40	15 13	14 49
03	00 49 29	14 17 36	15 ♑ 17 09	22 ♑ 20 32	19 43	28 37	11 57	28 58	18 03	23 41	15 15	14 50
04	00 53 26	15 16 31	29 ♑ 27 27	06 ♒ 37 37	20 06	29 ♓ 51	12 40	29 06	18 08	23 43	15 17	14 51
05	00 57 22	16 15 37	13 ♒ 50 40	21 ♒ 06 08	20 33	01 ♈ 05	13 23	29 14	18 13	23 45	15 20	14 52
06	01 01 19	17 14 41	28 ♒ 23 26	05 ♓ 41 55	21 05	02 19	14 06	29 21	18 18	23 47	15 22	14 53
07	01 05 16	18 13 42	13 ♓ 00 48	20 ♓ 19 19	21 41	03 33	14 49	29 29	18 23	23 49	15 24	14 54
08	01 09 12	19 12 42	27 ♓ 36 37	04 ♈ 51 54	22 21	04 47	15 32	29 36	18 28	23 51	15 27	14 55
09	01 13 09	20 11 40	12 ♈ 04 21	19 ♈ 13 15	23 05	06 02	16 15	29 43	18 34	23 53	15 29	14 56
10	01 17 05	21 10 36	26 ♈ 17 56	03 ♉ 17 53	23 52	07 16	16 58	29 50	18 39	23 55	15 31	14 57
11	01 21 02	22 09 31	10 ♉ 12 38	17 ♉ 01 55	24 43	08 30	17 41	29 ♑ 56	18 44	23 57	15 33	14 58
12	01 24 58	23 08 23	23 ♉ 45 33	00 ♊ 23 29	25 36	09 44	18 23	00 ♒ 03	18 50	23 59	15 36	14 59
13	01 28 55	24 07 13	06 ♊ 55 47	13 ♊ 22 37	26 33	10 58	19 06	00 09	18 55	24 01	15 38	14 59
14	01 32 51	25 06 00	19 ♊ 44 15	26 ♊ 01 03	27 33	12 12	19 49	00 16	19 01	24 03	15 40	15 00
15	01 36 48	26 04 46	02 ♋ 13 25	08 ♋ 21 51	28 35	13 26	20 31	00 22	19 07	24 05	15 42	15 01
16	01 40 45	27 03 29	14 ♋ 26 50	20 ♋ 28 57	29 ♓ 40	14 40	21 14	00 28	19 12	24 08	15 45	15 02
17	01 44 41	28 02 10	26 ♋ 28 44	02 ♌ 26 48	00 ♈ 48	15 54	21 57	00 34	19 18	24 10	15 47	15 03
18	01 48 38	29 00 49	08 ♌ 23 43	14 ♌ 20 03	01 58	17 08	22 39	00 39	19 24	24 12	15 49	15 03
19	01 52 34	29 ♈ 59 26	20 ♌ 16 23	26 ♌ 13 15	03 10	18 22	23 21	00 45	19 30	24 15	15 51	15 04
20	01 56 31	00 ♉ 58 00	02 ♍ 11 11	08 ♍ 10 40	04 25	19 36	24 04	00 50	19 36	24 17	15 53	15 05
21	02 00 27	01 56 32	14 ♍ 12 09	20 ♍ 16 03	05 42	20 50	24 46	00 55	19 42	24 19	15 56	15 05
22	02 04 24	02 55 03	26 ♍ 22 44	02 ♎ 32 30	07 00	22 04	25 28	01 00	19 48	24 22	15 58	15 06
23	02 08 20	03 53 31	08 ♎ 45 38	15 ♎ 02 20	08 21	23 18	26 11	01 05	19 54	24 24	16 00	15 07
24	02 12 17	04 51 57	21 ♎ 22 47	27 ♎ 47 02	09 44	24 32	26 53	01 10	20 01	24 27	16 02	15 07
25	02 16 14	05 50 21	04 ♏ 15 11	10 ♏ 47 11	11 10	25 46	27 35	01 14	20 07	24 30	16 04	15 08
26	02 20 10	06 48 43	17 ♏ 23 00	24 ♏ 02 32	12 37	27 00	28 17	01 19	20 13	24 32	16 07	15 08
27	02 24 07	07 47 04	00 ♐ 45 39	07 ♐ 32 10	14 05	28 14	28 59	01 23	20 20	24 35	16 09	15 09
28	02 28 03	08 45 23	14 ♐ 21 52	21 ♐ 14 33	15 36	29 ♈ 28	29 ♉ 41	01 27	20 26	24 38	16 11	15 09
29	02 32 00	09 43 40	28 ♐ 09 58	05 ♑ 07 52	17 09	00 ♉ 42	00 ♊ 23	01 31	20 33	24 40	16 13	15 10
30	02 35 56	10 ♉ 41 56	12 ♑ 07 58	19 ♑ 10 00	18 ♈ 44	01 ♉ 56	01 ♊ 05	01 ♒ 34	20 ♊ 39	24 ♊ 43	16 ♈ 15	15 ♒ 10

Date	Moon True ☊	Moon Mean ☊	Moon ☽ Latitude
01	09 ♏ 51	11 ♏ 18	03 N 13
02	09 D 52	11 15	04 06
03	09 R 53	11 12	04 46
04	09 52	11 09	05 09
05	09 50	11 05	05 13
06	09 48	11 02	04 58
07	09 45	10 59	04 23
08	09 42	10 56	03 31
09	09 40	10 53	02 26
10	09 38	10 50	01 N 13
11	09 D 38	10 46	00 S 03
12	09 38	10 43	01 17
13	09 39	10 40	02 25
14	09 41	10 37	03 24
15	09 42	10 34	04 11
16	09 43	10 30	04 46
17	09 43	10 27	05 08
18	09 R 43	10 24	05 16
19	09 42	10 21	05 11
20	09 41	10 18	04 53
21	09 40	10 15	04 21
22	09 39	10 11	03 37
23	09 38	10 08	02 43
24	09 38	10 05	01 39
25	09 38	10 02	00 S 30
26	09 D 38	09 59	00 N 43
27	09 38	09 56	01 55
28	09 38	09 52	03 01
29	09 R 38	09 49	03 58
30	09 ♏ 38	09 ♏ 46	04 N 41

DECLINATIONS

Date	Sun ☉	Moon ☽	Mercury ☿	Venus ♀	Mars ♂	Jupiter ♃	Saturn ♄	Uranus ♅	Neptune ♆	Pluto ⯓
01	04 N 52	19 S 40	04 S 19	02 S 53	15 N 09	20 S 34	21 N 41	23 N 27	04 N 34	22 S 47
02	05 15	19 20	04 27	02 24	15 23	20 33	21 42	23 27	04 35	22 47
03	05 38	17 49	04 32	01 55	15 37	20 31	21 42	23 27	04 36	22 47
04	06 01	15 13	04 34	01 26	15 51	20 30	21 43	23 27	04 37	22 47
05	06 24	11 40	04 35	00 56	16 05	20 28	21 43	23 27	04 38	22 47
06	06 46	07 22	04 32	00 S 27	16 18	20 27	21 44	23 27	04 38	22 47
07	07 09	02 S 37	04 28	00 N 03	16 31	20 25	21 45	23 27	04 39	22 47
08	07 31	02 N 17	04 22	00 32	16 45	20 24	21 45	23 27	04 40	22 47
09	07 53	07 01	04 13	01 01	16 58	20 23	21 46	23 27	04 41	22 47
10	08 16	11 17	04 03	01 31	17 10	20 21	21 47	23 27	04 42	22 47
11	08 38	14 50	03 50	02 00	17 23	20 20	21 47	23 27	04 43	22 47
12	08 59	17 28	03 36	02 30	17 36	20 19	21 48	23 28	04 44	22 46
13	09 21	19 05	03 20	02 59	17 48	20 18	21 49	23 28	04 45	22 46
14	09 43	19 39	03 02	03 28	18 00	20 16	21 49	23 28	04 45	22 46
15	10 04	19 14	02 43	03 58	18 12	20 15	21 50	23 28	04 46	22 46
16	10 25	17 54	02 22	04 27	18 24	20 14	21 50	23 28	04 47	22 46
17	10 46	15 48	02 01	04 56	18 36	20 13	21 51	23 28	04 48	22 46
18	11 07	13 04	01 35	05 25	18 47	20 12	21 52	23 28	04 49	22 46
19	11 28	09 48	01 09	05 54	18 58	20 11	21 52	23 28	04 50	22 46
20	11 48	06 08	00 42	06 22	19 10	20 10	21 53	23 28	04 50	22 46
21	12 09	02 N 12	00 S 14	06 51	19 21	20 09	21 54	23 28	04 51	22 46
22	12 29	01 S 53	00 N 16	07 19	19 31	20 08	21 54	23 29	04 52	22 47
23	12 49	05 58	00 47	07 48	19 42	20 07	21 55	23 29	04 53	22 47
24	13 08	09 52	01 19	08 16	19 52	20 06	21 55	23 29	04 54	22 47
25	13 28	13 24	01 52	08 44	20 02	20 05	21 56	23 29	04 55	22 47
26	13 47	16 20	02 26	09 11	20 12	20 05	21 57	23 29	04 55	22 47
27	14 06	18 26	03 02	09 39	20 22	20 04	21 57	23 29	04 56	22 47
28	14 25	19 31	03 38	10 06	20 32	20 03	21 58	23 29	04 57	22 47
29	14 43	19 27	04 16	10 33	20 41	20 02	21 59	23 29	04 58	22 47
30	15 N 02	18 S 13	04 N 54	11 N 00	20 N 51	20 S 02	21 N 59	23 N 29	04 N 58	22 S 47

ZODIAC SIGN ENTRIES

Date	h	m	Planets
02	09	38	☽ ♑
04	12	55	☽ ♒
04	14	58	♀ ♈
06	14	39	☽ ♓
08	15	57	☽ ♈
10	18	20	☽ ♉
12	00	58	♃ ♒
12	23	17	☽ ♊
15	07	41	☽ ♋
16	19	08	☿ ♈
17	19	04	☽ ♌
19	12	14	☉ ♉
20	07	36	☽ ♍
22	19	04	☽ ♎
25	04	07	☽ ♏
27	10	39	☽ ♐
28	12	18	♀ ♉
28	22	45	♂ ♊
29	15	10	☽ ♑

LATITUDES

Date	Mercury ☿	Venus ♀	Mars ♂	Jupiter ♃	Saturn ♄	Uranus ♅	Neptune ♆	Pluto ⯓
01	00 S 03	01 S 28	00 N 11	00 S 09	01 S 12	00 N 10	01 S 32	06 S 42
04	00 42	01 29	00 13	00 10	01 12	00 10	01 32	06 43
07	01 17	01 30	00 15	00 10	01 11	00 10	01 32	06 43
10	01 45	01 29	00 17	00 11	01 11	00 10	01 32	06 44
13	02 08	01 28	00 19	00 11	01 10	00 10	01 32	06 45
16	02 26	01 27	00 21	00 12	01 09	00 10	01 32	06 46
19	02 38	01 25	00 23	00 12	01 09	00 10	01 32	06 46
22	02 44	01 22	00 24	00 13	01 08	00 10	01 32	06 47
25	02 46	01 19	00 26	00 13	01 08	00 10	01 32	06 48
28	02 43	01 16	00 28	00 13	01 08	00 10	01 32	06 49
31	02 S 34	01 S 12	00 N 30	00 S 14	01 S 07	00 N 10	01 S 32	06 S 49

DATA

Julian Date	2463324
Delta T	+75 seconds
Ayanamsa	24° 18' 38"
Synetic vernal point	04° ♓ 48' 20"
True obliquity of ecliptic	23° 26' 00"

MOON'S PHASES, APSIDES AND POSITIONS ☽

Date	h	m	Phase	Longitude	Eclipse Indicator
03	10	10	☾	14 ♑ 13	
10	02	39	●	20 ♈ 48	
17	15	24	☽	28 ♋ 10	
25	15	10	○	05 ♏ 58	total

Day	h	m		
07	06	50	Perigee	
19	03	01	Apogee	
01	17	06	Max dec	19° S 41'
08	00	50	0N	
14	13	31	Max dec	19° N 39'
22	00	58	0S	
28	22	43	Max dec	19° S 38'

LONGITUDES

Date	Chiron ⚷	Ceres ⚳	Pallas ⚴	Juno ⚵	Vesta ⚶	Black Moon Lilith ⚸
01	18 ♉ 35	29 ♌ 01	19 ♊ 34	11 ♊ 01	27 ♍ 24	15 ♌ 36
11	19 ♉ 11	28 ♌ 43	24 ♊ 50	16 ♊ 15	25 ♍ 15	16 ♌ 43
21	19 ♉ 50	29 ♌ 08	00 ♋ 12	21 ♊ 32	23 ♍ 45	17 ♌ 50
31	20 ♉ 30	00 ♍ 11	05 ♋ 37	26 ♊ 50	23 ♍ 02	18 ♌ 57

ASPECTARIAN

h m	Aspects
01 Thursday	
01 42	☽ △ ☉
04 51	☽ ∠ ♃
06 51	☽ ⚹ ⯓
07 32	☽ △ ♆
09 48	☽ ± ♂
12 23	☽ ☍ ♄
14 43	☽ □ ☿
20 57	☽ ⊥ ♃
22 31	☽ ☍ ♅
02 Friday	
02 33	☽ ♇ ♂
04 22	☽ □ ♀
07 34	☽ ⊻ ♃
09 18	☽ ∠ ⯓
22 42	☽ Q ☿
03 Saturday	
00 55	☽ ⊥ ⯓
05 59	☽ △ ♂
10 10	☽ □ ☉
11 13	☽ ⊻ ⯓
11 57	☽ □ ♆
14 29	☽ Q ♀
15 57	♂ ⊥ ♄
16 45	☽ ⊼ ♄
19 45	☽ ⚹ ☿
19 50	♀ ⚹ ♃
04 Sunday	
01 23	☉ ⚹ ⯓
02 19	☽ ⊼ ♅
02 59	☽ ± ♄
07 20	☽ # ♂
11 24	☽ ☌ ♃
12 00	♀ ∠ ⯓
12 24	☉ ☌ ♆
12 27	☽ ± ♅
12 43	☽ ⚹ ♀
18 12	☽ ♇ ♄
18 27	☽ Q ♆
18 52	☽ Q ☉
21 45	☽ ∠ ☿
05 Monday	
03 31	☽ ♇ ♅
11 12	☽ □ ♂
13 13	☽ ⊥ ☿
13 42	☽ ☌ ⯓
14 28	☽ ⚹ ♆
16 03	☽ ∠ ♀
16 18	☽ ⚹ ☉
19 17	☽ △ ♄
23 30	☽ ⊻ ☿
06 Tuesday	
02 45	☿ ⊥ ⯓
04 24	☽ △ ♅
08 17	☽ ⊥ ♀
13 36	☽ ⊻ ♃
14 53	☽ # ☉
15 15	☽ ∠ ♆
18 25	☽ Q ♂
18 47	☽ ∠ ☉
19 03	☽ ⊻ ♀
23 32	☽ ⊥ ♃
07 Wednesday	
01 55	☽ # ♆
02 40	☽ ∥ ☿
06 04	☽ ⊥ ♆
10 37	☽ ⊥ ☉
14 25	☽ ∠ ♃
14 41	♂ □ ⯓
15 06	☽ ⊻ ⯓
15 07	☽ ⚹ ♂
15 56	☽ ⊻ ♆
16 15	☉ ⚹ ♄
20 52	☽ □ ♄
21 11	☽ ⊻ ☉
23 28	☽ # ♀
08 Thursday	
00 58	☽ ⊥ ⯓
02 29	☽ ∥ ♀
02 55	☽ ☌ ☿
05 47	☽ □ ♅
08 40	♂ ⊻ ♆
15 18	☽ ⚹ ♃
15 49	☽ ∠ ⯓
17 05	☽ ∠ ♂
22 05	☽ # ☿
23 57	☽ ∥ ♆
09 Friday	
00 59	☽ ☌ ♀
02 46	☽ Q ♄
08 48	☽ ⊥ ♂
11 24	☽ Q ♃
11 40	☽ Q ♅
16 47	☽ ⚹ ⯓
17 06	☽ ∥ ☉
17 43	☽ ☌ ♆
19 22	☽ ⊻ ♂
22 57	☽ ⚹ ♄
23 10	♀ Q ♄
10 Saturday	
02 39	☽ ☌ ☉
07 37	☽ ⊻ ☿
07 56	☽ ⚹ ♅
13 06	☽ Q ⯓
13 18	☿ □ ♅
18 05	☽ □ ♃
18 29	☽ ⊥ ☿
11 Sunday	
00 41	☽ ∠ ♄
08 43	☽ ⊻ ♀
09 47	☽ ∠ ♅
11 04	☽ ∠ ☿
13 15	♃ ± ♅
16 28	☽ ⊥ ♄
20 16	☽ ⊥ ♀
20 21	☽ □ ⯓
21 23	♂ ⊥ ♅
21 25	☽ ⊻ ♆
12 Monday	
01 39	☽ ⊥ ♅
01 53	☽ ☌ ♂
03 08	☽ ⊻ ♄
08 06	☽ ⊥ ♆
10 48	☽ ⊻ ☉
12 24	☽ ⊻ ♅
13 42	☽ ∥ ♂
13 56	☽ ∠ ♀
15 34	☽ ⚹ ☿
22 30	☽ ⊥ ☉
23 28	☽ △ ♃
13 Tuesday	
00 24	☽ ∠ ♆
04 59	♂ ⊻ ♄
09 19	☉ ⚹ ♅
15 15	☽ Q ☿
16 24	☽ ∠ ☉
20 18	☽ ⚹ ♀
22 45	☿ # ♀
14 Wednesday	
03 03	☽ △ ⯓
03 28	☽ ♇ ♃
04 17	☽ ⚹ ♆
08 58	♀ Q ♅
10 37	☽ ☌ ♄
12 09	☽ ⊻ ♂
13 16	♀ Q ♃
20 15	☽ ☌ ♅
20 42	☽ ± ♃
21 27	☽ Q ♀
23 06	☽ ⚹ ☉
15 Thursday	
00 19	☽ ⊥ ♂
03 13	☽ Q ♆
04 18	☽ □ ☿
07 43	☽ ♇ ⯓
08 22	☽ ⊼ ♃
18 50	☽ ∠ ♂
16 Friday	
00 26	☽ Q ☉
01 18	☽ ± ⯓
05 28	☽ ∥ ♂
11 20	☉ Q ⯓
12 30	☽ □ ♀
13 10	☽ ⊼ ⯓
14 35	☽ □ ♆
19 04	♀ ⚹ ⯓
19 54	☿ ∠ ⯓
21 32	☽ ⊻ ♄
17 Saturday	
02 21	☽ ⚹ ♂
05 20	♀ ∥ ♆
06 11	♂ ⊥ ♆
06 37	☿ ⚹ ♃
07 21	☽ ⊻ ♅
09 29	♀ ☌ ♆
09 37	☽ ⊥ ♄
13 41	♀ ⊥ ♂
15 24	☽ □ ☉
19 26	☽ ⊥ ♅
20 16	☽ ☍ ♃
21 36	☽ △ ☿
18 Sunday	
03 52	☽ ∠ ♄
03 58	☽ Q ♂
13 38	☽ ∠ ♅
19 Monday	
01 18	☽ ∥ ☉
01 28	☽ ☍ ⯓
03 03	☽ △ ♆
07 16	☽ ♇ ☿
07 43	☽ △ ♀
10 26	☽ ⚹ ♄
18 37	☽ □ ♂
20 02	☽ ⚹ ♅
20 Tuesday	
03 31	☽ ± ☿
08 29	☉ □ ♃
09 16	☽ ⊼ ♃
09 20	☽ △ ☉
09 23	☽ ♇ ♆
10 42	☽ ∥ ♀
10 49	☽ Q ♄
11 51	♀ ⚹ ♄
17 00	☽ ⊼ ☿
17 25	☽ ♇ ♀
19 53	♂ ⊻ ♅
20 01	☽ ∥ ♆
20 14	☽ Q ♅
21 23	☽ ± ♃
21 Wednesday	
03 28	☽ ± ♆
13 25	☽ ± ♀
13 46	☽ ⊼ ⯓
15 26	☽ ⊼ ♆
15 26	☽ ♇ ♃
17 54	☽ ♇ ☉
22 59	☽ □ ♄
22 Thursday	
00 48	☽ ∥ ☿
01 12	☽ # ☿
01 38	☽ ± ⯓
02 37	☽ ⊼ ♀
08 03	☽ □ ♅
10 07	☽ △ ♂
13 09	☽ ± ☉
19 16	☽ ♇ ⯓
21 05	☽ △ ♃
23 Friday	
01 49	☽ ⊼ ☉
03 24	☿ Q ♄
05 32	☽ # ♆
11 08	☽ ☍ ☿
16 55	☽ ♇ ♂
24 Saturday	
00 09	☽ △ ⯓
00 31	☽ # ♀
01 52	☽ ☍ ♆
09 24	☽ △ ♄
10 11	♀ ⚹ ♅
11 00	☽ ± ♂
16 01	☉ ∠ ♄
17 47	☽ △ ♅
18 34	☽ ☍ ♀
22 55	☽ ⊼ ♂
25 Sunday	
06 24	☽ □ ♃
12 31	☽ # ☉
13 36	☽ ♇ ♄
15 10	☽ ☍ ☉
18 23	♂ # ♃
21 40	☽ ♇ ♅
26 Monday	
02 15	☽ ⊼ ☿
06 13	☽ ± ♄
07 56	☽ □ ⯓
09 41	☽ ⊼ ♆
10 48	☿ Q ♅
14 06	☽ ± ♅
14 29	☽ ± ☿
14 36	♀ Q ⯓
15 30	☽ Q ♃
17 10	☽ ⊼ ♄
20 33	☽ ± ♆
27 Tuesday	
00 02	☿ Q ♃
00 56	☽ ⊼ ♅
07 03	☽ ⊼ ♀
08 39	☽ ♇ ☿
08 40	☽ ☍ ♂
08 52	☿ ∠ ♂
12 41	☽ ♇ ♆
13 07	☽ ⚹ ♃
16 14	☽ Q ⯓
18 47	☽ ± ♀
28 Wednesday	
01 24	☽ ⊼ ☉
04 54	☿ ⚹ ⯓
12 12	☽ ♇ ♀
12 44	☽ ± ☉
13 23	☽ ⚹ ⯓
14 26	☽ △ ☿
15 11	☽ △ ♆
15 40	☽ ∠ ♃
21 14	☿ ☌ ♆
21 42	♀ ⊻ ♂
22 41	☽ ☍ ♄
29 Thursday	
05 36	☽ ♇ ☉
05 56	☽ ☍ ♅
07 23	☽ ⊥ ♃
10 35	☉ ∠ ♅
15 27	☽ ∠ ⯓
16 02	☽ ⊼ ♂
16 48	☽ △ ♀
17 48	☽ ⊻ ♃
30 Friday	
02 54	☽ ± ♂
04 33	♀ □ ♃
06 56	☽ ⊥ ⯓
09 22	☽ △ ☉
14 40	☿ ∥ ♆
17 11	☽ ⊻ ⯓
18 02	♂ ∠ ♆
19 03	☽ □ ♆
19 06	☽ ♇ ♂

All ephemeris data is given at 12.00 UT and the Moon's longitude is additionally given for 24.00 UT

MAY 2032

LONGITUDES

Date	Sidereal time h m s	Sun ☉ ° ' "	Moon ☽ ° ' "	Moon ☽ 24.00 ° ' "	Mercury ☿ ° '	Venus ♀ ° '	Mars ♂ ° '	Jupiter ♃ ° '	Saturn ♄ ° '	Uranus ♅ ° '	Neptune ♆ ° '	Pluto ♇ ° '
01	02 39 53	11 ♉ 40 10	26 ♑ 13 42	03 ♒ 18 46	20 ♈ 20	03 ♉ 10	01 ♊ 47	01 ♒ 38	20 ♊ 46	24 ♊ 46	16 ♈ 17	15 ♒ 11
02	02 43 49	12 38 23	10 ♒ 24 55	17 ♒ 31 50	21 58	04 24	02 29	01 41	20 53	24 49	16 19	15 11
03	02 47 46	13 36 34	24 ♒ 39 14	01 ♓ 46 47	23 39	05 38	03 11	01 44	20 59	24 52	16 21	15 11
04	02 51 43	14 34 44	08 ♓ 54 08	16 ♓ 00 56	25 21	06 52	03 52	01 47	21 06	24 55	16 23	15 12
05	02 55 39	15 32 52	23 ♓ 06 50	00 ♈ 11 26	27 04	08 06	04 34	01 50	21 13	24 57	16 25	15 12
06	02 59 36	16 30 59	07 ♈ 14 21	14 ♈ 15 12	28 ♈ 50	09 20	05 16	01 52	21 20	25 00	16 27	15 12
07	03 03 32	17 29 05	21 ♈ 13 36	28 ♈ 09 10	00 ♉ 38	10 34	05 57	01 55	21 27	25 03	16 30	15 12
08	03 07 29	18 27 09	05 ♉ 01 33	11 ♉ 50 26	02 27	11 47	06 39	01 57	21 34	25 06	16 32	15 13
09	03 11 25	19 25 11	18 ♉ 35 31	25 ♉ 16 36	04 19	13 01	07 20	01 59	21 41	25 09	16 34	15 13
10	03 15 22	20 23 12	01 ♊ 53 29	08 ♊ 26 05	06 12	14 15	08 02	02 01	21 48	25 13	16 35	15 13
11	03 19 18	21 21 12	14 ♊ 54 20	21 ♊ 18 17	08 07	15 29	08 43	02 02	21 55	25 16	16 37	15 13
12	03 23 15	22 19 09	27 ♊ 38 01	03 ♋ 53 43	10 04	16 43	09 25	02 04	22 02	25 19	16 39	15 13
13	03 27 12	23 17 05	10 ♋ 05 36	16 ♋ 14 00	12 03	17 57	10 06	02 05	22 09	25 22	16 41	15 13
14	03 31 08	24 14 59	22 ♋ 19 14	28 ♋ 21 46	14 04	19 11	10 48	02 06	22 16	25 25	16 43	15 13
15	03 35 05	25 12 52	04 ♌ 22 00	10 ♌ 20 29	16 06	20 24	11 29	02 07	22 24	25 28	16 45	15 13
16	03 39 01	26 10 43	16 ♌ 17 43	22 ♌ 14 16	18 10	21 38	12 10	02 08	22 31	25 31	16 47	15 R 13
17	03 42 58	27 08 32	28 ♌ 10 43	04 ♍ 07 38	20 16	22 52	12 51	02 08	22 38	25 35	16 49	15 13
18	03 46 54	28 06 19	10 ♍ 05 38	16 ♍ 05 16	22 22	24 06	13 32	02 09	22 46	25 38	16 51	15 13
19	03 50 51	29 ♉ 04 05	22 ♍ 07 08	28 ♍ 11 47	24 31	25 20	14 13	02 09	22 53	25 41	16 52	15 13
20	03 54 47	00 ♊ 01 49	04 ♎ 19 43	10 ♎ 31 26	26 40	26 34	14 55	02 R 09	23 00	25 44	16 54	15 13
21	03 58 44	00 59 31	16 ♎ 47 21	23 ♎ 07 51	28 ♉ 50	27 47	15 36	02 08	23 08	25 48	16 56	15 13
22	04 02 41	01 57 12	29 ♎ 33 13	06 ♏ 03 41	01 ♊ 01	29 ♉ 01	16 17	02 08	23 15	25 51	16 58	15 13
23	04 06 37	02 54 51	12 ♏ 39 20	19 ♏ 20 14	03 13	00 ♊ 15	16 58	02 07	23 23	25 54	16 59	15 12
24	04 10 34	03 52 30	26 ♏ 06 16	02 ♐ 57 15	05 25	01 29	17 38	02 06	23 30	25 58	17 01	15 12
25	04 14 30	04 50 06	09 ♐ 52 53	16 ♐ 52 47	07 36	02 42	18 19	02 05	23 38	26 01	17 03	15 12
26	04 18 27	05 47 42	23 ♐ 56 25	01 ♑ 03 15	09 48	03 56	19 00	02 04	23 45	26 05	17 04	15 12
27	04 22 23	06 45 17	08 ♑ 12 38	15 ♑ 23 55	11 59	05 10	19 41	02 03	23 53	26 08	17 06	15 11
28	04 26 20	07 42 51	22 ♑ 36 25	29 ♑ 49 28	14 09	06 24	20 22	02 01	24 00	26 11	17 08	15 11
29	04 30 16	08 40 23	07 ♒ 02 26	14 ♒ 14 43	16 17	07 37	21 02	02 00	24 08	26 15	17 09	15 11
30	04 34 13	09 37 55	21 ♒ 25 49	28 ♒ 35 18	18 25	08 51	21 43	01 58	24 16	26 18	17 11	15 10
31	04 38 10	10 ♊ 35 26	05 ♓ 42 46	12 ♓ 47 56	20 ♊ 31	10 ♊ 05	22 ♊ 24	01 ♒ 55	24 ♊ 23	26 ♊ 22	17 ♈ 12	15 ♒ 10

Date	Moon True ☊ ° '	Moon Mean ☊ ° '	Moon ☽ Latitude ° '
01	09 ♏ 38	09 ♏ 43	05 N 08
02	09 R 38	09 40	05 17
03	09 D 38	09 36	05 06
04	09 38	09 33	04 37
05	09 39	09 30	03 50
06	09 39	09 27	02 50
07	09 40	09 24	01 40
08	09 40	09 21	00 N 26
09	09 R 40	09 17	00 S 49
10	09 39	09 14	02 00
11	09 38	09 11	03 03
12	09 37	09 08	03 55
13	09 35	09 05	04 35
14	09 33	09 02	05 02
15	09 31	08 58	05 14
16	09 30	08 55	05 13
17	09 D 30	08 52	04 59
18	09 30	08 49	04 32
19	09 31	08 46	03 53
20	09 33	08 42	03 02
21	09 34	08 39	02 02
22	09 35	08 36	00 S 55
23	09 R 36	08 33	00 N 17
24	09 35	08 30	01 30
25	09 33	08 27	02 39
26	09 30	08 23	03 40
27	09 26	08 20	04 28
28	09 22	08 17	05 00
29	09 19	08 14	05 12
30	09 16	08 11	05 06
31	09 ♏ 15	08 ♏ 08	04 N 40

DECLINATIONS

Date	Sun ☉ ° '	Moon ☽ ° '	Mercury ☿ ° '	Venus ♀ ° '	Mars ♂ ° '	Jupiter ♃ ° '	Saturn ♄ ° '	Uranus ♅ ° '	Neptune ♆ ° '	Pluto ♇ ° '
01	15 N 20	15 S 51	05 N 34	11 N 27	21 N 00	20 S 01	22 N 00	23 N 30	04 N 59	22 S 47
02	15 38	12 32	06 14	11 53	21 09	20 01	22 00	23 30	05 00	22 48
03	15 55	08 29	06 55	12 19	21 17	20 00	22 01	23 30	05 01	22 48
04	16 12	03 S 57	07 37	12 45	21 26	20 00	22 02	23 30	05 02	22 48
05	16 30	00 N 48	08 19	13 10	21 34	19 59	22 02	23 30	05 02	22 48
06	16 46	05 29	09 02	13 35	21 42	19 59	22 03	23 30	05 03	22 48
07	17 03	09 50	09 46	14 00	21 50	19 58	22 03	23 30	05 04	22 48
08	17 19	13 36	10 30	14 25	21 58	19 58	22 04	23 30	05 05	22 48
09	17 35	16 34	11 14	14 49	22 05	19 58	22 05	23 30	05 05	22 49
10	17 50	18 35	11 59	15 13	22 12	19 57	22 05	23 31	05 06	22 49
11	18 06	19 33	12 44	15 36	22 19	19 57	22 06	23 31	05 07	22 49
12	18 21	19 30	13 30	15 59	22 26	19 57	22 06	23 31	05 07	22 49
13	18 35	18 29	14 15	16 22	22 33	19 57	22 07	23 31	05 08	22 50
14	18 50	16 37	15 00	16 44	22 39	19 57	22 07	23 31	05 09	22 50
15	19 04	14 04	15 45	17 06	22 46	19 57	22 08	23 31	05 10	22 50
16	19 18	10 57	16 29	17 27	22 52	19 57	22 08	23 31	05 10	22 50
17	19 31	07 25	17 13	17 48	22 57	19 57	22 09	23 31	05 11	22 51
18	19 44	03 N 35	17 56	18 08	23 03	19 57	22 09	23 32	05 12	22 51
19	19 57	00 S 26	18 38	18 28	23 08	19 57	22 10	23 32	05 12	22 51
20	20 09	04 30	19 19	18 47	23 14	19 57	22 11	23 32	05 13	22 51
21	20 21	08 29	19 59	19 06	23 19	19 58	22 11	23 32	05 13	22 52
22	20 33	12 10	20 37	19 25	23 23	19 58	22 12	23 32	05 14	22 52
23	20 44	15 22	21 14	19 43	23 28	19 58	22 12	23 32	05 15	22 52
24	20 55	17 49	21 48	20 00	23 32	19 59	22 12	23 32	05 15	22 53
25	21 06	19 19	22 20	20 17	23 36	19 59	22 13	23 32	05 16	22 53
26	21 16	19 38	22 50	20 33	23 40	19 59	22 13	23 32	05 17	22 53
27	21 26	18 43	23 18	20 49	23 44	20 00	22 14	23 32	05 17	22 53
28	21 35	16 37	23 43	21 04	23 48	20 00	22 14	23 33	05 18	22 54
29	21 45	13 28	24 05	21 19	23 51	20 01	22 15	23 33	05 18	22 54
30	21 53	09 31	24 25	21 33	23 54	20 02	22 15	23 33	05 19	22 55
31	22 N 02	05 S 04	24 N 41	21 N 46	23 N 57	20 S 02	22 N 16	23 N 33	05 N 19	22 S 55

ZODIAC SIGN ENTRIES

Date	h	m	Planets
01	18	24	☽ ♒
03	21	00	☽ ♓
05	23	41	☽ ♈
07	03	35	☿ ♉
08	03	13	☽ ♉
10	08	33	☽ ♊
12	16	31	☽ ♋
15	03	16	☽ ♌
17	15	41	☽ ♍
20	03	32	☽ ♎
20	11	15	☉ ♊
22	00	45	☿ ♊
22	12	50	☽ ♏
23	07	10	♀ ♊
24	18	51	☽ ♐
26	22	14	☽ ♑
29	00	18	☽ ♒
31	02	22	☽ ♓

LATITUDES

Date	Mercury ☿ ° '	Venus ♀ ° '	Mars ♂ ° '	Jupiter ♃ ° '	Saturn ♄ ° '	Uranus ♅ ° '	Neptune ♆ ° '	Pluto ♇ ° '
01	02 S 34	01 S 12	00 N 30	00 S 14	01 S 07	00 N 10	01 S 32	06 S 49
04	02 21	01 07	00 31	00 14	01 07	00 10	01 32	06 50
07	02 03	01 02	00 33	00 15	01 06	00 10	01 32	06 51
10	01 41	00 57	00 34	00 15	01 06	00 10	01 32	06 52
13	01 16	00 51	00 36	00 16	01 05	00 10	01 32	06 52
16	00 47	00 45	00 37	00 17	01 05	00 10	01 32	06 53
19	00 S 16	00 39	00 39	00 17	01 05	00 10	01 33	06 54
22	00 N 16	00 32	00 40	00 18	01 04	00 10	01 33	06 55
25	00 46	00 25	00 41	00 18	01 04	00 10	01 33	06 56
28	01 14	00 18	00 43	00 19	01 04	00 10	01 33	06 56
31	01 N 36	00 S 11	00 N 44	00 S 19	01 S 03	00 N 10	01 S 33	06 S 57

DATA

Julian Date	2463354
Delta T	+75 seconds
Ayanamsa	24° 18' 42"
Synetic vernal point	04° ♓ 48' 17"
True obliquity of ecliptic	23° 25' 59"

MOON'S PHASES, APSIDES AND POSITIONS ☽

Date	h	m	Phase	Longitude ° '	Eclipse Indicator
02	16	02	☾	12 ♒ 48	
09	13	36	●	19 ♉ 29	Annular
17	09	43	☽	27 ♌ 03	
25	02	37	○	04 ♐ 28	
31	20	51	☾	10 ♓ 57	

Day	h	m		
03	20	35	Perigee	
16	22	19	Apogee	
29	02	57	Perigee	
05	08	00	0N	
11	22	35	Max dec	19° N 39'
19	09	25	0S	
26	06	26	Max dec	19° S 40'

LONGITUDES

Date	Chiron ⚷ ° '	Ceres ⚳ ° '	Pallas ⚴ ° '	Juno ⚵ ° '	Vesta ⚶ ° '	Black Moon Lilith ⚸ ° '
01	20 ♉ 30	00 ♍ 11	05 ♋ 37	26 ♊ 50	23 ♍ 02	18 ♌ 57
11	21 ♉ 11	01 ♍ 49	11 ♋ 05	02 ♋ 07	23 ♍ 08	20 ♌ 05
21	21 ♉ 52	03 ♍ 57	16 ♋ 34	07 ♋ 24	24 ♍ 01	21 ♌ 12
31	22 ♉ 32	06 ♍ 29	22 ♋ 02	12 ♋ 38	25 ♍ 37	22 ♌ 19

ASPECTARIAN

h m	Aspects
01 Saturday	
00 41	☽ □ ☿
02 39	☽ ⚻ ♄
06 17	♂ △ ♃
09 31	☽ ⚻ ♅
12 55	☽ ± ♄
15 50	☽ # ☉
18 51	☿ ⚹ ♄
19 43	☽ ± ♅
21 11	☽ ☌ ♃
21 54	☽ △ ♂
02 Sunday	
00 52	☽ □ ♀
01 41	☽ Q ♆
04 17	☽ ⚼ ♄
10 59	☽ ⚼ ♅
11 09	☽ Q ☿
15 42	☽ # ♀
16 02	☽ □ ☉
20 03	☽ ☌ ♇
21 59	☽ ⚹ ♆
03 Monday	
05 47	☽ △ ♄
10 04	☽ ⚹ ☿
10 07	☽ Q ♀
12 21	☽ △ ♅
19 24	☽ # ☿
19 40	♀ ∠ ♄
23 19	☽ ∠ ♆
23 58	☽ ⚺ ♃
04 Tuesday	
00 34	☽ Q ☉
03 05	☽ □ ♂
05 44	☿ ⚹ ♅
06 28	☽ # ♆
08 14	☽ ⚹ ♀
10 07	☽ ⊥ ♃
14 31	☽ ⊥ ♆
14 46	☽ ∠ ☿
22 16	☽ ⚹ ☉
22 37	☽ ⚺ ♇
05 Wednesday	
00 40	☽ ⚺ ♆
01 21	☽ ∠ ♃
02 42	☉ ⊥ ♄
03 18	☉ □ ♇
08 04	☽ ⊥ ☿
08 45	☽ ⊥ ♇
08 46	☽ □ ♄
11 02	☽ Q ♂
11 58	☽ ∠ ♀
13 42	☿ Q ♇
15 08	☽ □ ♅
19 40	☽ ⚺ ☿
06 Thursday	
00 01	☽ ∠ ♇
01 32	☽ ∠ ☉
02 50	☽ ⚹ ♃
04 42	☽ ⊥ ♀
08 27	☽ ⚹ ♂
09 46	☽ ∥ ♆
10 30	☉ ⚺ ♆
15 36	☽ Q ♄
15 55	☽ ⚺ ♀
18 01	☽ ⊥ ☉
21 20	☿ ⊥ ♂
21 54	☽ Q ♅
23 23	☽ Q ♃
07 Friday	
01 38	☽ ⚹ ♇
01 47	♀ ∠ ♅
03 49	☽ ☌ ♆
05 04	☽ ⚺ ☉
11 30	☽ ∠ ♂
11 31	☽ ∥ ☿
12 23	☽ ⚹ ♄
18 39	☽ ⚹ ♅
22 21	☽ Q ♇
08 Saturday	
03 56	☽ ⊥ ♂
05 14	☿ □ ♃
06 36	☽ □ ♃
06 49	☽ ☌ ☿
14 43	☽ ∠ ♄
15 00	☽ ⚺ ♂
18 48	☽ ∥ ♀
20 58	☽ ∠ ♅
09 Sunday	
01 06	☽ ☌ ♀
05 08	☉ ⊥ ♅
05 59	☽ □ ♇
06 46	☽ ⊥ ♄
08 22	☽ ⚺ ♆
10 13	♂ ∥ ♄
13 01	☽ ⊥ ♅
13 36	☽ ☌ ☉
17 35	☽ ⚺ ♄
19 07	☽ ⊥ ♆
23 50	☽ ⚺ ♅
10 Monday	
00 12	☽ ∥ ☉
11 27	☽ ∠ ♆
12 13	☽ △ ♃
19 58	☿ ∠ ♄
21 15	☽ ⚺ ☿
23 53	☽ ☌ ♂
11 Tuesday	
06 47	♀ □ ♇
10 17	☽ ⊥ ☿
12 35	☽ △ ♇
13 12	☽ ⚺ ♀
15 13	☽ ⚹ ♆
16 00	☽ ⚼ ♃
21 17	♀ ⊥ ♄
23 32	☿ ⚺ ♂
12 Wednesday	
01 05	☽ ⚺ ☉
01 17	☽ ☌ ♄
01 40	☽ ⊥ ♀
03 55	☉ ⚺ ♄
06 14	☽ ∠ ☿
07 34	☽ ☌ ♅
09 00	☽ ± ♃
10 48	♀ ⚺ ♆
13 25	☽ ⊥ ☉
13 57	☽ Q ♆
15 00	☿ ∠ ♅
16 56	☽ ⚼ ♇
20 30	☽ ⚻ ♃
20 39	☉ ⊥ ♆
20 40	☽ ∠ ♀
13 Thursday	
08 11	☽ ∠ ☉
10 18	☽ ± ♇
10 24	☽ ∥ ☉
12 01	☽ ⚺ ♂
16 33	☽ ⚹ ☿
22 01	☽ ⚻ ♇
14 Friday	
00 27	☽ ⊥ ♂
00 56	☽ □ ♆
05 06	☽ ⚹ ♀
11 00	☽ ∥ ♀
11 54	☽ ⚺ ♄
16 09	☽ ⚹ ☉
16 53	♀ ⊥ ♅
18 10	☽ ⚺ ♅
19 18	☽ ∠ ♂
20 55	☽ Q ☿
23 56	☽ ⊥ ♄
15 Saturday	
00 22	☽ ∥ ☿
01 41	☿ □ ♇
06 11	☽ ⊥ ♅
07 30	☽ ☍ ♃
07 38	☽ Q ♀
15 38	☿ ⊥ ♄
15 54	♇ St R
18 08	☽ ∠ ♄
18 13	☽ Q ☉
18 44	☉ ⚺ ♅
19 42	☿ ⚺ ♆
16 Sunday	
00 19	☽ ∠ ♅
03 10	☽ ⚹ ♂
06 15	♂ # ♇
09 50	☽ ☍ ♇
12 59	☽ △ ♆
16 35	☽ □ ☿
17 Monday	
00 02	☽ □ ♀
00 41	☽ ⚹ ♄
04 00	☿ ⊥ ♅
04 52	☽ Q ♂
06 43	☽ ⚹ ♅
06 59	♀ ⚺ ♄
09 43	☽ □ ☉
10 53	♀ ⊥ ♆
19 21	☽ ⚼ ♆
20 00	☽ ⚻ ♃
18 Tuesday	
01 10	☽ Q ♄
02 08	☽ ∥ ♆
07 02	☽ Q ♅
08 05	☽ ± ♃
13 30	☽ ± ♆
16 36	☿ ⚺ ♄
17 21	☿ ⊥ ♆
19 20	☽ □ ♂
22 16	☽ ⚻ ♇
19 Wednesday	
00 42	☿ ∥ ♀
01 32	☽ ⚻ ♆
02 07	☽ ⚼ ♃
10 13	☽ ± ♇
13 02	☉ # ♃
13 32	☽ □ ♄
14 47	♃ St R
17 45	☽ △ ☿
19 04	☽ △ ♀
19 05	☽ □ ♅
19 18	♀ ⚺ ♅
20 Thursday	
01 26	☿ ⚺ ♅
02 53	☽ △ ☉
03 58	☽ ⚼ ♇
07 44	☽ △ ♃
09 10	☿ ☌ ♀
16 13	☽ # ♆
22 44	♂ △ ♇
21 Friday	
03 31	☽ ⚼ ♀
05 11	☽ ⚼ ☿
09 00	☽ △ ♇
09 35	☽ △ ♂
10 21	☽ ⚼ ☉
11 06	☿ # ♃
12 16	☽ ☍ ♆
22 29	☽ ± ♀
22 Saturday	
00 07	☽ △ ♄
01 49	☽ ± ☿
04 45	☽ ± ☉
05 04	☽ △ ♅
07 56	☉ ∥ ☿
10 54	☽ ⚻ ♀
12 12	☉ ∠ ♆
15 17	☽ ⚻ ☿
15 22	☽ ⚼ ♂
16 25	☉ △ ♃
16 46	☽ □ ♃
16 48	☽ ⚻ ☉
22 24	☿ ∠ ♆
23 Sunday	
00 05	☿ △ ♃
04 10	☽ ⚼ ♄
06 07	☉ ☌ ☿
08 46	☽ ± ♂
08 49	☽ ⚼ ♅
13 09	♂ ⚹ ♆
16 36	☽ □ ♇
17 38	♂ ⚼ ♃
19 49	☽ ⚻ ♆
20 10	☽ ⚻ ♂
20 35	☽ ± ♄
24 Monday	
01 04	☽ ± ♅
01 23	☽ Q ♃
06 32	☽ ± ♆
07 22	☽ ⚻ ♄
09 45	♀ # ♃
11 03	♂ ∥ ♅
11 45	☽ ⚻ ♅
22 21	☽ ☍ ♀
22 23	☽ ⚼ ♆
22 31	☽ ⚹ ♃
22 49	♀ ∠ ♆
25 Tuesday	
00 09	♀ △ ♃
00 26	☽ Q ♇
02 37	☽ ☍ ☉
06 23	☿ ∥ ♄
07 20	☽ ☍ ☿
21 08	☽ ⚹ ♇
26 Wednesday	
00 18	☽ △ ♆
00 21	☽ ∠ ♃
03 11	☽ ☍ ♂
10 20	♃ ± ♅
11 41	☽ ☍ ♄
14 32	☿ # ♇
15 36	☽ ⊥ ♃
15 38	☽ ☍ ♅
22 33	☽ ∠ ♇
27 Thursday	
01 41	☽ ⚺ ♃
06 25	☽ ⚻ ♀
09 23	☽ ⚻ ☉
13 38	☽ ⊥ ♇
17 24	☽ ± ♀
19 25	☽ ⚻ ☿
20 08	☽ ± ☉
23 39	☽ ⚺ ♇
28 Friday	
02 00	☿ ∥ ♅
02 52	☽ □ ♆
07 11	☽ ± ☿
08 05	☽ ⚻ ♂
09 48	☽ ⚼ ♀
12 11	☽ ⚼ ☉
14 21	☽ ⚻ ♄
17 59	☽ ⚻ ♅
18 01	☿ ∥ ♂
18 33	☽ ± ♂
23 36	☿ △ ♇
29 Saturday	
00 25	☽ ± ♄
00 46	☽ ⚼ ☿
03 37	☽ ☌ ♃
04 00	☽ ± ♅
08 51	☽ Q ♆
10 15	☽ ⚼ ♂
13 04	☽ △ ♀
14 55	☽ △ ☉
15 31	☽ ⚼ ♄
19 02	☽ ⚼ ♅
19 48	☿ ⚼ ♃
21 52	☿ ⚹ ♆
30 Sunday	
01 33	☽ ☌ ♇
04 53	☽ ⚹ ♆
06 05	☽ △ ☿
12 30	☽ △ ♂
16 47	☽ △ ♄
20 12	☽ △ ♅
31 Monday	
05 38	☽ ⚺ ♃
06 04	☽ ∠ ♆
10 41	☽ # ♆
15 44	☽ ⊥ ♃
20 06	☽ □ ♀
20 51	☽ □ ☉
21 19	☽ ⊥ ♆

All ephemeris data is given at 12.00 UT and the Moon's longitude is additionally given for 24.00 UT

JUNE 2032

Date	Sidereal time h m s	Sun ☉ ° ' "	Moon ☽ ° ' "	Moon ☽ 24.00 ° ' "	Mercury ☿ ° '	Venus ♀ ° '	Mars ♂ ° '	Jupiter ♃ ° '	Saturn ♄ ° '	Uranus ♅ ° '	Neptune ♆ ° '	Pluto ♇ ° '
01	04 42 06	11 ♊ 32 56	19 ♓ 50 35	26 ♓ 50 34	22 ♊ 35	11 ♊ 19	23 ♊ 04	01 ♒ 53	24 ♊ 31	26 ♊ 25	17 ♈ 14	15 ♒ 10
02	04 46 03	12 30 26	03 ♈ 47 45	10 ♈ 42 04	24 37	12 32	23 45	01 R 51	24 39	26 29	17 15	15 R 09
03	04 49 59	13 27 54	17 ♈ 33 29	24 ♈ 21 58	26 37	13 46	24 25	01 48	24 46	26 32	17 17	15 09
04	04 53 56	14 25 22	01 ♉ 07 30	07 ♉ 50 03	28 ♊ 34	15 00	25 06	01 45	24 54	26 36	17 18	15 08
05	04 57 52	15 22 50	14 ♉ 29 36	21 ♉ 06 07	00 ♋ 30	16 14	25 46	01 42	25 02	26 39	17 20	15 08
06	05 01 49	16 20 16	27 ♉ 39 35	04 ♊ 09 56	02 23	17 27	26 27	01 39	25 10	26 43	17 21	15 07
07	05 05 45	17 17 42	10 ♊ 37 09	17 ♊ 01 13	04 14	18 41	27 07	01 35	25 17	26 46	17 22	15 07
08	05 09 42	18 15 07	23 ♊ 22 06	29 ♊ 39 50	06 02	19 55	27 47	01 32	25 25	26 50	17 24	15 06
09	05 13 39	19 12 31	05 ♋ 54 27	12 ♋ 06 02	07 47	21 09	28 28	01 28	25 33	26 53	17 25	15 05
10	05 17 35	20 09 55	18 ♋ 14 42	24 ♋ 20 37	09 31	22 22	29 08	01 24	25 41	26 57	17 26	15 05
11	05 21 32	21 07 17	00 ♌ 24 01	06 ♌ 25 11	11 11	23 36	29 ♊ 48	01 20	25 49	27 01	17 28	15 04
12	05 25 28	22 04 38	12 ♌ 24 24	18 ♌ 22 05	12 49	24 50	00 ♋ 28	01 16	25 56	27 04	17 29	15 03
13	05 29 25	23 01 59	24 ♌ 18 39	00 ♍ 14 33	14 24	26 04	01 08	01 11	26 04	27 08	17 30	15 03
14	05 33 21	23 59 18	06 ♍ 10 20	12 ♍ 06 31	15 57	27 17	01 49	01 07	26 12	27 11	17 31	15 02
15	05 37 18	24 56 37	18 ♍ 03 42	24 ♍ 02 29	17 26	28 31	02 29	01 02	26 20	27 15	17 32	15 01
16	05 41 14	25 53 55	00 ♎ 03 30	06 ♎ 07 22	18 54	29 ♊ 45	03 09	00 57	26 28	27 18	17 33	15 00
17	05 45 11	26 51 12	12 ♎ 14 43	18 ♎ 26 09	20 18	00 ♋ 58	03 49	00 52	26 35	27 22	17 34	15 00
18	05 49 08	27 48 28	24 ♎ 42 15	01 ♏ 03 33	21 40	02 12	04 29	00 47	26 43	27 26	17 35	14 59
19	05 53 04	28 45 43	07 ♏ 30 31	14 ♏ 03 32	22 59	03 26	05 09	00 41	26 51	27 29	17 36	14 58
20	05 57 01	29 ♊ 42 58	20 ♏ 42 53	27 ♏ 28 44	24 15	04 39	05 49	00 36	26 59	27 33	17 37	14 57
21	06 00 57	00 ♋ 40 12	04 ♐ 21 06	11 ♐ 19 50	25 28	05 53	06 28	00 30	27 07	27 36	17 38	14 56
22	06 04 54	01 37 25	18 ♐ 24 38	25 ♐ 34 59	26 38	07 07	07 08	00 25	27 15	27 40	17 39	14 55
23	06 08 50	02 34 38	02 ♑ 50 16	10 ♑ 09 40	27 45	08 21	07 48	00 19	27 22	27 43	17 40	14 54
24	06 12 47	03 31 51	17 ♑ 32 14	24 ♑ 56 57	28 49	09 34	08 28	00 13	27 30	27 47	17 41	14 53
25	06 16 43	04 29 04	02 ♒ 22 45	09 ♒ 48 33	29 ♋ 50	10 48	09 08	00 07	27 38	27 51	17 42	14 52
26	06 20 40	05 26 16	17 ♒ 13 20	24 ♒ 36 08	00 ♌ 48	12 02	09 47	00 ♒ 00	27 46	27 54	17 43	14 52
27	06 24 37	06 23 28	01 ♓ 56 09	09 ♓ 12 40	01 42	13 15	10 27	29 ♑ 54	27 53	27 58	17 43	14 51
28	06 28 33	07 20 40	16 ♓ 25 11	23 ♓ 33 18	02 33	14 29	11 07	29 47	28 01	28 01	17 44	14 50
29	06 32 30	08 17 52	00 ♈ 36 48	07 ♈ 35 35	03 21	15 43	11 46	29 41	28 09	28 05	17 45	14 49
30	06 36 26	09 ♋ 15 05	14 ♈ 29 40	21 ♈ 19 08	04 ♌ 04	16 ♋ 57	12 ♋ 26	29 ♑ 34	28 ♊ 17	28 ♊ 08	17 ♈ 46	14 ♒ 47

Date	Moon True ☊ ° '	Moon Mean ☊ ° '	Moon ☽ Latitude ° '
01	09 ♏ 15	08 ♏ 04	03 N 58
02	09 D 16	08 01	03 02
03	09 18	07 58	01 56
04	09 19	07 55	00 N 45
05	09 R 19	07 52	00 S 28
06	09 18	07 48	01 38
07	09 15	07 45	02 42
08	09 10	07 42	03 36
09	09 03	07 39	04 19
10	08 56	07 36	04 49
11	08 48	07 33	05 05
12	08 42	07 29	05 08
13	08 36	07 26	04 57
14	08 33	07 23	04 34
15	08 31	07 20	03 59
16	08 D 31	07 17	03 12
17	08 32	07 13	02 17
18	08 33	07 10	01 14
19	08 34	07 07	00 S 06
20	08 R 33	07 04	01 N 05
21	08 30	07 01	02 14
22	08 26	06 58	03 17
23	08 19	06 54	04 09
24	08 11	06 51	04 46
25	08 02	06 48	05 04
26	07 55	06 45	05 01
27	07 49	06 42	04 39
28	07 45	06 39	03 59
29	07 43	06 35	03 05
30	07 ♏ 43	06 ♏ 32	02 N 01

DECLINATIONS

Date	Sun ☉ ° '	Moon ☽ ° '	Mercury ☿ ° '	Venus ♀ ° '	Mars ♂ ° '	Jupiter ♃ ° '	Saturn ♄ ° '	Uranus ♅ ° '	Neptune ♆ ° '	Pluto ♇ ° '
01	22 N 10	00 S 22	24 N 56	21 N 59	24 N 00	20 S 03	22 N 16	23 N 33	05 N 20	22 S 55
02	22 17	04 N 17	25 07	22 11	24 02	20 04	22 16	23 33	05 20	22 56
03	22 25	08 41	25 16	22 23	24 04	20 05	22 17	23 33	05 21	22 56
04	22 31	12 34	25 22	22 34	24 06	20 05	22 17	23 33	05 21	22 56
05	22 38	15 44	25 26	22 44	24 08	20 06	22 18	23 33	05 22	22 57
06	22 44	18 02	25 27	22 54	24 10	20 07	22 18	23 34	05 22	22 57
07	22 50	19 22	25 26	23 02	24 11	20 08	22 18	23 34	05 23	22 58
08	22 55	19 40	25 22	23 11	24 12	20 09	22 19	23 34	05 23	22 58
09	23 00	18 59	25 17	23 18	24 13	20 10	22 19	23 34	05 23	22 58
10	23 04	17 25	25 10	23 25	24 14	20 11	22 19	23 34	05 24	22 58
11	23 08	15 05	25 00	23 32	24 15	20 12	22 19	23 34	05 24	22 59
12	23 12	12 09	24 49	23 37	24 15	20 13	22 20	23 34	05 25	22 59
13	23 15	08 44	24 37	23 42	24 15	20 14	22 20	23 34	05 25	23 00
14	23 18	05 00	24 23	23 46	24 15	20 16	22 21	23 34	05 26	23 00
15	23 20	01 N 03	24 07	23 50	24 15	20 17	22 21	23 34	05 26	23 01
16	23 22	02 S 58	23 50	23 52	24 14	20 18	22 21	23 34	05 27	23 01
17	23 24	06 56	23 33	23 55	24 14	20 19	22 22	23 34	05 27	23 02
18	23 25	10 43	23 14	23 56	24 13	20 21	22 22	23 35	05 27	23 02
19	23 26	14 06	22 54	23 57	24 12	20 22	22 22	23 35	05 28	23 03
20	23 26	16 53	22 33	23 57	24 10	20 23	22 22	23 35	05 28	23 03
21	23 26	18 49	22 12	23 56	24 09	20 25	22 23	23 35	05 28	23 04
22	23 25	19 39	21 50	23 54	24 07	20 26	22 23	23 35	05 29	23 04
23	23 24	19 15	21 28	23 52	24 05	20 28	22 23	23 35	05 29	23 05
24	23 23	17 34	21 05	23 49	24 03	20 29	22 23	23 35	05 29	23 05
25	23 21	14 41	20 42	23 46	24 01	20 31	22 24	23 35	05 29	23 06
26	23 19	10 53	20 19	23 41	23 58	20 32	22 24	23 35	05 30	23 06
27	23 17	06 26	19 55	23 36	23 56	20 34	22 24	23 35	05 30	23 07
28	23 14	01 S 41	19 32	23 31	23 53	20 35	22 24	23 35	05 30	23 07
29	23 10	03 N 04	19 09	23 24	23 50	20 37	22 24	23 35	05 30	23 07
30	23 N 07	07 N 34	18 N 46	23 N 17	23 N 46	20 S 38	22 N 24	23 N 35	05 N 31	23 S 08

ZODIAC SIGN ENTRIES

Date	h	m	Planets
02	05	26	☽ ♈
04	10	00	☽ ♉
05	05	43	☿ ♋
06	16	18	☽ ♊
09	00	39	☽ ♋
11	11	12	☽ ♌
11	19	06	♂ ♋
13	23	31	☽ ♍
16	11	53	☽ ♎
16	17	00	♀ ♋
18	22	01	☽ ♏
20	19	09	☉ ♋
21	04	25	☽ ♐
23	07	19	☽ ♑
25	08	10	☽ ♒
25	15	57	☿ ♌
26	12	57	♃ ♑
27	08	49	☽ ♓
29	10	57	☽ ♈

LATITUDES

Date	Mercury ☿ ° '	Venus ♀ ° '	Mars ♂ ° '	Jupiter ♃ ° '	Saturn ♄ ° '	Uranus ♅ ° '	Neptune ♆ ° '	Pluto ♇ ° '
01	01 N 42	00 S 09	00 N 45	00 S 19	01 S 03	00 N 10	01 S 33	06 S 57
04	01 56	00 S 02	00 46	00 20	01 03	00 10	01 33	06 58
07	02 04	00 N 05	00 47	00 21	01 03	00 10	01 33	06 59
10	02 04	00 13	00 48	00 21	01 02	00 10	01 34	06 59
13	01 58	00 20	00 49	00 22	01 02	00 10	01 34	07 00
16	01 45	00 26	00 51	00 22	01 02	00 10	01 34	07 01
19	01 26	00 33	00 52	00 23	01 02	00 10	01 34	07 01
22	01 02	00 40	00 53	00 23	01 01	00 10	01 34	07 02
25	00 N 32	00 46	00 54	00 24	01 01	00 10	01 34	07 03
28	00 S 03	00 52	00 55	00 24	01 01	00 10	01 35	07 03
31	00 S 42	00 N 58	00 N 56	00 S 25	01 S 01	00 N 10	01 S 35	07 S 04

DATA

Julian Date	2463385
Delta T	+75 seconds
Ayanamsa	24° 18' 46"
Synetic vernal point	04° ♓ 48' 13"
True obliquity of ecliptic	23° 25' 59"

MOON'S PHASES, APSIDES AND POSITIONS ☽

Date	h	m	Phase	Longitude ° '	Eclipse Indicator
08	01	32	●	17 ♊ 50	
16	03	00	☽	25 ♍ 32	
23	11	32	○	02 ♑ 34	
30	02	12	☾	08 ♈ 52	

Day	h	m		
13	16	31	Apogee	
25	14	59	Perigee	
01	13	54	0N	
08	07	13	Max dec	19° N 41'
15	18	19	0S	
22	16	30	Max dec	19° S 41'
28	20	27	0N	

LONGITUDES

Date	Chiron ⚷ ° '	Ceres ⚳ ° '	Pallas ⚴ ° '	Juno ⚵ ° '	Vesta ⚶ ° '	Black Moon Lilith ⚸ ° '
01	22 ♉ 36	06 ♍ 46	22 ♋ 35	13 ♋ 10	25 ♍ 48	22 ♌ 26
11	23 ♉ 14	09 ♍ 42	28 ♋ 01	18 ♋ 22	28 ♍ 03	23 ♌ 33
21	23 ♉ 51	12 ♍ 55	03 ♌ 24	23 ♋ 31	00 ♎ 49	24 ♌ 40
31	24 ♉ 24	16 ♍ 23	08 ♌ 45	28 ♋ 37	04 ♎ 01	25 ♌ 47

ASPECTARIAN

01 Tuesday
04 01 ☽ ⊻ ♇; 06 58 ☽ ∠ ♃; 07 32 ☽ ⊻ ♆; 14 15 ☽ ⊥ ♇; 17 29 ☽ □ ☿; 17 48 ☽ □ ♂; 20 05 ☽ □ ♄; 20 38 ☿ ☌ ♂; 23 19 ☽ □ ♅

02 Wednesday
05 42 ☽ ∠ ♇; 05 49 ☽ Q ♀; 05 54 ☽ Q ☉; 08 38 ☽ ✶ ♃; 09 00 ☉ ∥ ♄; 09 07 ☉ ☌ ♀; 12 26 ☿ ☌ ♄; 17 35 ☽ ∥ ♆; 23 15 ♀ ∥ ♄

03 Thursday
02 25 ☿ ± ♃; 02 32 ☽ Q ♂; 03 32 ☽ Q ♄; 04 17 ☽ ✶ ☉; 04 42 ☽ ✶ ♀; 05 26 ☽ Q ♃; 05 57 ☽ Q ☿; 06 41 ☽ Q ♅; 07 46 ☽ ✶ ♇; 11 04 ☿ ☌ ♅; 11 31 ☽ ☌ ♆; 22 49 ☉ ∥ ♀

04 Friday
00 44 ☽ ✶ ♂; 00 51 ☽ ✶ ♄; 03 33 ♂ ☌ ♄; 03 55 ☽ ✶ ♅; 04 55 ☽ Q ♇; 06 42 ☽ ✶ ☿; 08 44 ☽ ∠ ☉; 09 47 ☽ ∠ ♀; 13 07 ☽ □ ♃; 14 41 ♀ △ ♇; 21 08 ☿ Q ♆

05 Saturday
02 04 ☽ ⊥ ☉; 03 31 ☽ ⊥ ♀; 03 52 ☽ ∠ ♄; 04 55 ☽ ∠ ♂; 05 43 ☉ △ ♇; 06 51 ☽ ∠ ♅; 07 20 ☿ Ᵽ ♇; 09 41 ♂ ± ♃; 13 09 ☽ □ ♇; 13 44 ☽ ⊻ ☉; 14 08 ☽ ∠ ☿; 15 28 ☽ ⊻ ♀; 17 09 ☽ ⊻ ♆; 20 19 ☽ ⊥ ♄; 20 51 ♀ Ᵽ ♃; 22 05 ☽ ⊥ ♂; 23 14 ☽ ⊥ ♅

06 Sunday
02 48 ☿ ⊼ ♃; 04 05 ☽ ⊥ ♆; 07 22 ☽ ⊻ ♄; 09 16 ☽ ⊥ ☿; 09 39 ☽ ⊻ ♂; 09 54 ♀ ✶ ♆; 10 15 ☽ ⊻ ♅; 19 18 ☉ Ᵽ ♃; 19 19 ☽ △ ♃; 20 39 ☽ ∠ ♆; 21 55 ♀ ♯ ♇; 22 09 ☽ ⊻ ☿; 22 33 ♂ ☌ ♅

07 Monday
13 59 ☉ ✶ ♆; 20 24 ☽ △ ♇; 23 08 ☽ Ᵽ ♃

08 Tuesday
00 41 ☽ ✶ ♆; 01 32 ☽ ☌ ☉; 04 45 ☽ ☌ ♀; 15 56 ☽ ☌ ♄; 16 05 ☽ ± ♃; 18 37 ☽ ☌ ♅; 20 53 ☽ ☌ ♂; 23 30 ☽ Q ♆

09 Wednesday
00 49 ☽ Ᵽ ♇; 01 35 ♃ ± ♄; 03 30 ☽ ⊼ ♃; 04 37 ☉ ♯ ♇; 16 14 ☽ ☌ ☿; 18 09 ☽ ± ♇

10 Thursday
05 49 ☽ ⊼ ♇; 05 57 ☿ ± ♇; 10 25 ☽ □ ♆; 16 05 ☽ ⊻ ☉; 21 01 ☽ ⊻ ♀; 23 19 ♂ Q ♆

11 Friday
02 48 ☽ ⊻ ♄; 04 56 ☽ ⊥ ☉; 05 14 ☽ ⊻ ♅; 10 14 ☽ ⊥ ♀; 10 44 ☽ ⊻ ♂; 13 51 ☽ ☍ ♃; 14 50 ☽ ⊥ ♄; 17 13 ☽ ⊥ ♅; 21 20 ♂ Ᵽ ♇; 22 07 ♀ ∥ ♅; 23 24 ☽ ⊥ ♂

12 Saturday
00 23 ☽ ∠ ☉; 06 14 ☽ ∠ ♀; 09 01 ☽ ∠ ♄; 11 19 ☽ ∠ ♅; 12 57 ☽ ⊻ ☿; 17 19 ☽ ☍ ♇; 18 32 ☽ ∠ ♂; 19 58 ♀ ± ♃; 22 13 ☽ △ ♆

13 Sunday
02 54 ☽ ⊥ ☿; 09 11 ☽ ✶ ☉; 12 16 ♀ ☌ ♄; 13 31 ♂ ⊼ ♃; 15 36 ☽ ✶ ♄; 15 57 ☽ ✶ ♀; 17 44 ☽ ✶ ♅; 21 51 ☿ ⊼ ♇; 23 51 ☽ ∠ ☿

14 Monday
01 50 ☽ ⊼ ♃; 02 39 ☽ ✶ ♂; 04 36 ☽ Ᵽ ♆; 09 19 ☽ ∥ ♆; 09 57 ♀ ☌ ♅; 11 36 ☽ Q ☉; 13 53 ☽ ± ♃; 16 09 ☽ Q ♄; 18 08 ☽ Q ♅; 19 02 ☽ Q ♀; 22 49 ☽ ± ♆

15 Tuesday
00 12 ☿ ∥ ♂; 04 21 ☽ Q ♂; 05 53 ☽ ⊼ ♇; 07 57 ☽ Ᵽ ♃; 10 34 ☽ ✶ ☿; 10 57 ☽ ⊼ ♆; 13 35 ☿ □ ♆; 14 04 ☉ ± ♃; 17 56 ☽ ± ♇

16 Wednesday
03 00 ☽ □ ☉; 04 45 ☽ □ ♄; 06 30 ☽ □ ♅; 08 15 ♀ Q ♆; 09 29 ☿ ∥ ♀; 11 18 ☽ □ ♀; 11 54 ☽ Ᵽ ♇; 13 46 ☽ △ ♃; 13 53 ☽ Q ☿; 17 03 ♀ Ᵽ ♇; 18 29 ☽ □ ♂

17 Thursday
02 53 ☽ ♯ ♆; 04 23 ☉ ☌ ♄; 09 29 ☿ ∥ ♅; 10 03 ♀ ⊼ ♃; 17 20 ☽ △ ♇; 22 21 ☽ ☍ ♆; 22 38 ☉ ∥ ☿

18 Friday
01 45 ☉ ☌ ♅; 05 30 ☽ □ ☿; 15 52 ☽ △ ♄; 17 11 ☽ △ ♅; 18 22 ☽ △ ☉; 23 24 ☽ □ ♃

19 Saturday
01 38 ☿ ♯ ♇; 03 38 ☽ △ ♀; 07 23 ☽ △ ♂; 20 03 ☽ Ᵽ ♄; 21 11 ☽ Ᵽ ♅

20 Sunday
00 21 ☽ Ᵽ ☉; 01 38 ☽ □ ♇; 06 27 ☽ ⊼ ♆; 08 14 ☽ Q ♃; 09 38 ☉ Q ♆; 09 55 ☽ Ᵽ ♀; 12 11 ☽ Ᵽ ♂; 12 29 ☽ ± ♄; 13 29 ☽ ± ♅; 17 11 ☽ ± ♆; 17 45 ☽ ± ☉; 17 49 ☉ Ᵽ ♇; 18 55 ☽ △ ☿; 23 14 ☽ ⊼ ♄

21 Monday
00 09 ☿ ∥ ♄; 00 10 ☽ ⊼ ♅; 03 28 ☽ ± ♀; 04 55 ☽ ± ♂; 05 07 ☽ ⊼ ☉; 05 21 ☽ ✶ ♃; 08 14 ☉ ⊼ ♃; 09 01 ☽ Ᵽ ♆; 09 33 ☽ Q ♇; 14 55 ☽ ⊼ ♀; 15 51 ☽ ⊼ ♂; 23 29 ☽ Ᵽ ☿

22 Tuesday
06 07 ☽ ✶ ♇; 06 58 ☽ ∠ ♃; 10 44 ☽ △ ♆; 12 58 ♀ ☌ ♂; 16 04 ☽ ± ☿; 21 59 ☽ ⊥ ♃

23 Wednesday
02 36 ☿ ⊻ ♄; 02 53 ☽ ☍ ♄; 02 55 ☽ ⊼ ☿; 03 31 ☽ ☍ ♅; 07 10 ☽ ∠ ♇; 07 52 ☽ ⊻ ♃; 11 18 ☿ ⊻ ♅; 11 32 ☽ ☍ ☉; 20 32 ☽ ☍ ♂; 21 51 ☽ ☍ ♀; 21 56 ☽ ⊥ ♇; 22 52 ♀ ± ♇

24 Thursday
07 42 ☽ ⊻ ♇; 12 14 ☽ □ ♆

25 Friday
03 06 ♂ ± ♇; 04 16 ☽ ⊼ ♄; 04 39 ☽ ⊼ ♅; 07 36 ☽ ☍ ☿; 08 22 ☽ ☌ ♃; 14 02 ☽ ± ♄; 14 22 ☽ ± ♅; 15 38 ☽ ⊼ ☉; 17 22 ☽ Q ♆; 17 59 ☿ ☍ ♃; 23 03 ☿ ♯ ♃; 23 24 ☽ ⊼ ♂

26 Saturday
01 59 ☽ ± ☉; 02 50 ☽ ⊼ ♀; 04 43 ☽ Ᵽ ♄; 04 58 ☽ Ᵽ ♅; 08 10 ☽ ☌ ♇; 09 34 ☽ ± ♂; 12 48 ☽ ✶ ♆; 13 26 ☽ ± ♀; 17 35 ☽ Ᵽ ☉

27 Sunday
00 53 ☽ Ᵽ ♂; 05 19 ☽ △ ♄; 05 25 ☽ Ᵽ ♀; 05 28 ☽ △ ♅; 08 41 ☽ ⊻ ♃; 11 36 ☽ ⊼ ☿; 13 18 ☽ ∠ ♆; 16 47 ♀ ∥ ♅; 16 50 ☽ ♯ ♆; 18 28 ☽ ⊥ ♃; 19 51 ☽ △ ☉; 22 07 ☽ ± ☿

28 Monday
02 44 ☽ △ ♂; 04 11 ☽ ⊥ ♆; 08 28 ☽ △ ♀; 09 18 ☽ ∠ ♃; 09 20 ☽ ⊻ ♇; 12 03 ♄ ☌ ♅; 14 01 ☽ Ᵽ ☿; 14 12 ☽ ⊻ ♆; 18 33 ♀ ⊼ ♇; 19 23 ☽ ⊥ ♇

29 Tuesday
07 39 ☽ □ ♅; 07 45 ☽ □ ♄; 10 25 ☽ ✶ ♃; 10 38 ☽ ∠ ♇; 16 56 ☽ △ ☿

30 Wednesday
00 37 ☉ ± ♇; 00 46 ☽ ∥ ♆; 02 12 ☽ □ ☉; 04 35 ☉ ♯ ♇; 06 56 ☽ Q ♃; 08 13 ☽ □ ♂; 12 31 ☽ ✶ ♇; 14 36 ☿ ⊥ ♅; 14 53 ☽ Q ♅; 15 09 ☽ Q ♄; 16 43 ☽ □ ♀; 17 44 ☽ ☌ ♆; 21 01 ☿ ⊥ ♄

All ephemeris data is given at 12.00 UT and the Moon's longitude is additionally given for 24.00 UT

JULY 2032

Date	Sidereal time h m s	LONGITUDES Sun ☉ ° ' "	Moon ☽ ° ' "	Moon ☽ 24.00 ° ' "	Mercury ☿ ° '	Venus ♀ ° '	Mars ♂ ° '	Jupiter ♃ ° '	Saturn ♄ ° '	Uranus ♅ ° '	Neptune ♆ ° '	Pluto ♇ ° '
01	06 40 23	10 ♋ 12 17	28 ♈ 04 11	04 ♉ 45 00	04 ♌ 44	18 ♋ 10	13 ♋ 05	29 ♑ 27	28 ♊ 24	28 ♊ 12	17 ♈ 46	14 ♒ 46
02	06 44 19	11 09 30	11 ♉ 21 52	17 ♉ 55 00	05 20	19 24	13 45	29 R 22	28 32	28 15	17 47	14 R 45
03	06 48 16	12 06 43	24 ♉ 24 40	00 ♊ 51 05	05 52	20 38	14 25	29 13	28 40	28 19	17 47	14 44
04	06 52 12	13 03 56	07 ♊ 14 28	13 ♊ 35 01	06 20	21 51	15 04	29 06	28 47	28 22	17 48	14 43
05	06 56 09	14 01 09	19 ♊ 52 52	26 ♊ 08 09	06 43	23 05	15 43	28 59	28 55	28 26	17 48	14 42
06	07 00 06	14 58 23	02 ♋ 21 00	08 ♋ 31 30	07 02	24 19	16 23	28 52	29 03	28 29	17 49	14 41
07	07 04 02	15 55 36	14 ♋ 39 45	20 ♋ 45 50	07 17	25 33	17 02	28 44	29 10	28 33	17 49	14 40
08	07 07 59	16 52 50	26 ♋ 49 52	02 ♌ 51 58	07 27	26 46	17 42	28 37	29 18	28 36	17 50	14 39
09	07 11 55	17 50 04	08 ♌ 52 17	14 ♌ 51 00	07 32	28 00	18 21	28 30	29 26	28 40	17 50	14 37
10	07 15 52	18 47 17	20 ♌ 48 21	26 ♌ 44 34	07 R 32	29 ♋ 14	19 00	28 22	29 33	28 43	17 51	14 36
11	07 19 48	19 44 31	02 ♍ 40 00	08 ♍ 34 59	07 28	00 ♌ 27	19 39	28 15	29 41	28 46	17 51	14 35
12	07 23 45	20 41 45	14 ♍ 29 57	20 ♍ 25 21	07 19	01 41	20 19	28 07	29 48	28 50	17 51	14 34
13	07 27 41	21 38 58	26 ♍ 21 42	02 ♎ 19 33	07 05	02 55	20 58	27 59	29 ♊ 56	28 53	17 51	14 33
14	07 31 38	22 36 12	08 ♎ 19 30	14 ♎ 22 08	06 47	04 09	21 37	27 52	00 ♋ 03	28 57	17 52	14 31
15	07 35 35	23 33 26	20 ♎ 28 07	26 ♎ 38 06	06 25	05 22	22 16	27 44	00 10	29 00	17 52	14 30
16	07 39 31	24 30 40	02 ♏ 52 44	09 ♏ 12 38	05 58	06 36	22 55	27 36	00 18	29 03	17 52	14 29
17	07 43 28	25 27 54	15 ♏ 38 24	22 ♏ 10 34	05 28	07 50	23 35	27 29	00 25	29 07	17 52	14 28
18	07 47 24	26 25 08	28 ♏ 49 35	05 ♐ 35 45	04 54	09 03	24 14	27 21	00 32	29 10	17 52	14 26
19	07 51 21	27 22 23	12 ♐ 29 17	19 ♐ 30 11	04 18	10 17	24 53	27 13	00 40	29 13	17 52	14 25
20	07 55 17	28 19 37	26 ♐ 38 17	03 ♑ 53 10	03 39	11 31	25 32	27 05	00 47	29 16	17 52	14 24
21	07 59 14	29 ♋ 16 52	11 ♑ 14 15	18 ♑ 40 40	02 58	12 45	26 11	26 58	00 54	29 20	17 R 52	14 22
22	08 03 10	00 ♌ 14 08	26 ♑ 11 24	03 ♒ 45 15	02 16	13 58	26 50	26 50	01 01	29 23	17 52	14 21
23	08 07 07	01 11 24	11 ♒ 20 54	18 ♒ 56 59	01 34	15 12	27 29	26 42	01 09	29 26	17 52	14 20
24	08 11 04	02 08 40	26 ♒ 32 10	04 ♓ 05 08	00 53	16 26	28 08	26 34	01 16	29 29	17 52	14 18
25	08 15 00	03 05 58	11 ♓ 34 46	19 ♓ 00 04	00 ♌ 12	17 39	28 46	26 27	01 23	29 32	17 52	14 17
26	08 18 57	04 03 16	26 ♓ 20 15	03 ♈ 34 46	29 ♋ 33	18 53	29 ♋ 25	26 19	01 30	29 35	17 52	14 16
27	08 22 53	05 00 35	10 ♈ 43 14	17 ♈ 45 28	28 57	20 07	00 ♌ 04	26 12	01 37	29 38	17 52	14 14
28	08 26 50	05 57 55	24 ♈ 41 29	01 ♉ 31 23	28 25	21 20	00 43	26 04	01 44	29 41	17 51	14 13
29	08 30 46	06 55 16	08 ♉ 15 27	14 ♉ 53 58	27 56	22 34	01 22	25 57	01 51	29 44	17 51	14 12
30	08 34 43	07 52 38	21 ♉ 27 20	27 ♉ 55 58	27 32	23 48	02 01	25 49	01 58	29 48	17 51	14 10
31	08 38 39	08 ♌ 50 02	04 ♊ 20 16	10 ♊ 40 41	27 ♋ 13	25 ♌ 01	02 ♌ 39	25 ♑ 42	02 ♋ 04	29 ♊ 50	17 ♈ 50	14 ♒ 09

Date	Moon True ☊ ° '	Moon Mean ☊ ° '	Moon ☽ Latitude ° '
01	07 ♏ 43	06 ♏ 29	00 N 51
02	07 R 44	06 26	00 S 19
03	07 42	06 23	01 28
04	07 39	06 19	02 30
05	07 33	06 16	03 24
06	07 24	06 13	04 08
07	07 13	06 10	04 39
08	07 01	06 07	04 57
09	06 48	06 04	05 01
10	06 37	06 00	04 52
11	06 27	05 57	04 31
12	06 19	05 54	03 58
13	06 14	05 51	03 14
14	06 11	05 48	02 22
15	06 11	05 45	01 22
16	06 D 11	05 41	00 S 17
17	06 R 11	05 38	00 N 50
18	06 09	05 35	01 57
19	06 05	05 32	02 59
20	05 59	05 29	03 53
21	05 50	05 25	04 34
22	05 40	05 22	04 56
23	05 29	05 19	04 59
24	05 19	05 16	04 40
25	05 11	05 13	04 03
26	05 05	05 10	03 09
27	05 02	05 06	02 05
28	05 D 01	05 03	00 N 54
29	05 R 01	05 00	00 S 17
30	05 01	04 57	01 26
31	04 ♏ 59	04 ♏ 54	02 S 28

DECLINATIONS

Date	Sun ☉ ° '	Moon ☽ ° '	Mercury ☿ ° '	Venus ♀ ° '	Mars ♂ ° '	Jupiter ♃ ° '	Saturn ♄ ° '	Uranus ♅ ° '	Neptune ♆ ° '	Pluto ♇ ° '
01	23 N 02	11 N 35	18 N 24	23 N 09	23 N 43	20 S 40	22 N 25	23 N 35	05 N 31	23 S 08
02	22 58	14 56	18 01	23 01	23 39	20 41	22 25	23 35	05 31	23 09
03	22 53	17 27	17 40	22 52	23 35	20 43	22 25	23 36	05 31	23 09
04	22 47	19 03	17 19	22 42	23 31	20 45	22 25	23 36	05 31	23 10
05	22 42	19 39	16 59	22 31	23 27	20 46	22 25	23 36	05 31	23 10
06	22 36	19 17	16 40	22 20	23 23	20 48	22 25	23 36	05 32	23 11
07	22 29	18 00	16 21	22 09	23 18	20 50	22 25	23 36	05 32	23 12
08	22 22	15 56	16 04	21 56	23 13	20 51	22 25	23 36	05 32	23 12
09	22 15	11 55	15 48	21 43	23 08	20 53	22 26	23 36	05 32	23 13
10	22 07	09 56	15 34	21 29	23 03	20 55	22 26	23 36	05 32	23 13
11	21 59	06 18	15 21	21 15	22 58	20 56	22 26	23 36	05 32	23 14
12	21 50	02 N 26	15 09	21 00	22 52	20 58	22 26	23 36	05 32	23 14
13	21 42	01 S 32	14 59	20 45	22 47	21 00	22 26	23 36	05 32	23 15
14	21 32	05 28	14 51	20 29	22 41	21 01	22 26	23 36	05 32	23 15
15	21 23	09 16	14 44	20 12	22 35	21 03	22 26	23 36	05 32	23 16
16	21 13	12 44	14 40	19 55	22 29	21 05	22 26	23 36	05 32	23 16
17	21 02	15 43	14 37	19 37	22 22	21 06	22 26	23 36	05 32	23 17
18	20 52	17 59	14 36	19 18	22 16	21 08	22 26	23 36	05 32	23 17
19	20 41	19 19	14 37	19 00	22 09	21 10	22 26	23 36	05 32	23 18
20	20 29	19 30	14 39	18 40	22 02	21 11	22 26	23 36	05 32	23 18
21	20 18	18 25	14 44	18 20	21 55	21 13	22 26	23 36	05 32	23 19
22	20 06	16 03	14 50	18 00	21 48	21 15	22 26	23 36	05 32	23 19
23	19 53	12 35	14 57	17 39	21 40	21 16	22 26	23 36	05 32	23 20
24	19 41	08 16	15 06	17 17	21 33	21 18	22 26	23 36	05 32	23 20
25	19 28	03 S 29	15 17	16 56	21 25	21 19	22 26	23 36	05 32	23 21
26	19 14	01 N 26	15 28	16 33	21 17	21 21	22 26	23 36	05 32	23 21
27	19 01	06 09	15 40	16 10	21 09	21 22	22 26	23 36	05 31	23 22
28	18 47	10 24	15 54	15 47	21 01	21 24	22 26	23 36	05 31	23 22
29	18 32	13 59	16 08	15 24	20 53	21 26	22 26	23 37	05 31	23 23
30	18 18	16 44	16 22	14 59	20 45	21 27	22 26	23 37	05 31	23 23
31	18 N 03	18 N 35	16 N 37	14 N 35	20 N 36	21 S 29	22 N 25	23 N 37	05 N 31	23 S 24

ZODIAC SIGN ENTRIES

Date	h	m	Planets
01	15	27	☽ ♉
03	22	24	☽ ♊
06	07	27	☽ ♋
08	18	18	☽ ♌
11	03	04	♀ ♌
11	06	36	☽ ♍
13	19	20	☽ ♎
14	02	16	♄ ♋
16	06	29	☽ ♏
18	14	06	☽ ♐
20	17	35	☽ ♑
22	06	05	☉ ♌
22	18	03	☽ ♒
24	17	30	☽ ♓
25	19	20	☿ ♋
26	18	03	☽ ♈
27	09	23	♂ ♌
28	21	19	☽ ♉
31	03	52	☽ ♊

LATITUDES

Date	Mercury ☿ ° '	Venus ♀ ° '	Mars ♂ ° '	Jupiter ♃ ° '	Saturn ♄ ° '	Uranus ♅ ° '	Neptune ♆ ° '	Pluto ♇ ° '
01	00 S 42	00 N 58	00 N 56	00 S 25	01 S 01	00 N 10	01 S 35	07 S 04
04	01 25	01 03	00 57	00 25	01 00	00 10	01 35	07 04
07	02 10	01 08	00 58	00 26	01 00	00 10	01 35	07 05
10	02 55	01 13	00 59	00 26	01 00	00 10	01 35	07 05
13	03 37	01 17	00 59	00 27	01 00	00 10	01 35	07 06
16	04 14	01 20	01 00	00 27	01 00	00 10	01 36	07 06
19	04 42	01 23	01 01	00 28	01 00	00 10	01 36	07 07
22	04 57	01 26	01 02	00 28	01 00	00 10	01 36	07 07
25	04 56	01 28	01 03	00 28	01 00	00 11	01 36	07 07
28	04 40	01 29	01 03	00 29	01 00	00 11	01 36	07 08
31	04 S 11	01 N 30	01 N 04	00 S 29	01 S 00	00 N 11	01 S 36	07 S 08

DATA

Julian Date	2463415
Delta T	+75 seconds
Ayanamsa	24° 18' 51"
Synetic vernal point	04° ♓ 48' 08"
True obliquity of ecliptic	23° 25' 58"

MOON'S PHASES, APSIDES AND POSITIONS ☽

Date	h	m	Phase	Longitude ° '	Eclipse Indicator
07	14	41	●	16 ♋ 02	
15	18	32	☽	23 ♎ 49	
22	18	52	○	00 ♒ 30	
29	09	25	☾	06 ♉ 49	

Day	h	m		
11	07	57	Apogee	
23	18	51	Perigee	
05	14	50	Max dec	19° N 39'
13	02	47	0S	
20	03	37	Max dec	19° S 35'
26	04	57	0N	

LONGITUDES

Date	Chiron ⚷ ° '	Ceres ⚳ ° '	Pallas ⚴ ° '	Juno ⚵ ° '	Vesta ⚶ ° '	Black Moon Lilith ⚸ ° '
01	24 ♉ 24	16 ♍ 23	08 ♌ 45	28 ♋ 37	04 ♎ 01	25 ♌ 47
11	24 ♉ 53	20 ♍ 03	14 ♌ 03	03 ♌ 40	07 ♎ 35	26 ♌ 54
21	25 ♉ 19	23 ♍ 52	19 ♌ 18	08 ♌ 38	11 ♎ 27	28 ♌ 02
31	25 ♉ 39	27 ♍ 50	24 ♌ 29	13 ♌ 33	15 ♎ 34	29 ♌ 09

ASPECTARIAN

h m	Aspects
01 Thursday	
04 07	♀ □ ♆
09 41	☽ Q ♇
12 14	☽ ✶ ♅
12 16	☽ Q ☉
12 36	☽ ✶ ♄
14 19	♀ # ♇
14 27	☽ □ ♃
17 41	☽ Q ♂
02 Friday	
00 33	☽ □ ☿
04 03	☽ Q ♀
11 36	☽ ✶ ☉
15 28	☽ ∠ ♅
16 00	☽ ∠ ♄
16 35	☽ ✶ ♂
18 11	☽ □ ♇
23 46	☽ ⊻ ♆
03 Saturday	
04 16	☽ ✶ ♀
05 29	☉ ∥ ♀
08 06	☽ ⊥ ♅
08 44	☽ ⊥ ♄
10 33	♂ ∥ ♅
10 51	☽ ⊥ ♆
10 57	☽ Q ☿
14 09	☽ ∥ ☿
17 25	☽ ∠ ☉
19 18	☽ ⊻ ♅
19 59	☽ ⊻ ♄
20 52	☽ △ ♃
21 48	☽ ∠ ♂
23 43	♂ ⊼ ♇
04 Sunday	
03 39	☽ ∠ ♆
10 13	☽ ✶ ☿
11 12	☽ ∠ ♀
11 39	☽ ⊥ ☉
15 38	☽ ⊥ ♂
23 55	☽ ⊻ ☉
05 Monday	
00 52	☽ ⚼ ♃
02 08	☽ △ ♇
03 38	☽ ⊻ ♂
06 05	☽ ⊥ ♀
08 02	☽ ✶ ♆
15 37	☽ ∠ ☿
17 53	☽ ± ♃
18 20	♃ ⊼ ♄
18 48	☽ ⊻ ♀
06 Tuesday	
01 33	♀ ∥ ♄
04 30	☽ ☌ ♅
04 48	☉ ⊼ ♇
05 19	☽ ⊼ ♃
05 33	☽ ☌ ♄
06 51	☽ ⚼ ♇
07 06	☽ Q ♆
09 24	☽ ⊥ ☿
21 18	☽ ⊻ ☿
07 Wednesday	
00 17	☽ ± ♇
12 00	☽ ⊼ ♇
14 41	☽ ☌ ☉
16 56	☽ ☌ ♂
18 13	☽ □ ♆
08 Thursday	
00 23	☉ ∥ ♄
10 23	☽ ∥ ☿
11 52	☽ ☌ ♀
13 57	♃ ⊼ ♅
15 31	☽ ☍ ♃
15 32	☽ ⊻ ♅
16 57	☽ ⊻ ♄
17 07	♂ □ ♆
17 28	♂ # ♇
09 Friday	
03 32	☽ ⊥ ♅
05 02	☽ ⊥ ♄
09 18	☽ ☌ ☿
12 05	☉ □ ♆
20 45	♀ ☍ ♃
21 39	☽ ∠ ♅
23 16	☽ ∠ ♄
23 32	☽ ☍ ♇
10 Saturday	
01 32	♀ ⊻ ♅
02 33	☿ St R
06 01	☽ △ ♆
07 35	☽ ⊻ ☉
08 09	☽ ⊻ ♂
19 01	♀ ⊻ ♄
20 37	♄ ⚼ ♇
20 45	☽ ⊥ ☉
20 59	☽ ⊥ ♂
11 Sunday	
03 08	☽ ⊼ ♃
04 05	☽ ✶ ♅
05 16	☉ ☌ ♂
05 53	☽ ✶ ♄
07 00	☽ ⊻ ♀
12 22	☽ ⚼ ♆
15 10	☽ ± ♃
16 16	☽ ∠ ♂
16 35	☽ ∠ ☉
16 53	☽ ∥ ♆
20 35	☽ ⊥ ♀
21 38	☽ ⊻ ☿
12 Monday	
04 31	☽ Q ♅
06 28	☽ Q ♄
06 38	☽ ± ♆
09 14	☽ ⚼ ♃
09 39	☽ ⊥ ☿
12 08	☽ ⊼ ♇
15 05	♀ # ♃
16 57	☽ ∠ ♀
18 48	☽ ⊼ ♆
22 18	♄ Q ♆
13 Tuesday	
00 16	☽ ± ♇
00 28	☽ ✶ ♂
01 39	☽ ✶ ☉
03 33	☽ ∠ ☿
15 15	☽ △ ♃
17 07	☽ □ ♅
18 24	☽ ⚼ ♇
19 15	☽ □ ♄
14 Wednesday	
02 03	☽ Q ♂
02 41	☽ ✶ ♀
03 55	☽ Q ☉
09 01	☽ ✶ ☿
12 23	☽ # ♆
15 Thursday	
00 17	☽ △ ♇
04 22	♀ ⊥ ♅
05 15	☽ Q ♀
06 53	☽ ☍ ♆
08 06	☽ Q ☿
15 43	☽ □ ♂
18 32	☽ □ ☉
22 31	☿ ⊥ ♄
16 Friday	
01 59	☽ □ ♃
03 04	☿ ☌ ♀
04 38	☽ △ ♅
05 25	♀ ⊥ ♄
07 00	☽ △ ♄
17 40	☽ □ ☿
19 50	☽ □ ♀
21 56	♂ ∥ ♄
17 Saturday	
02 38	☽ # ☿
04 19	☉ # ♃
09 09	☽ ⚼ ♅
09 49	☽ □ ♇
11 35	☽ ⚼ ♄
11 42	☽ Q ♃
16 07	☽ ⊼ ♆
18 Sunday	
01 45	☽ ± ♅
02 05	☿ ⊥ ♅
03 05	☽ ± ♆
03 18	☽ △ ♂
04 13	☽ ± ♄
07 21	☽ △ ☉
09 22	☽ ✶ ♃
12 36	☽ ⊼ ♅
15 05	☽ ⊼ ♄
18 25	☽ Q ♇
19 12	☽ ⚼ ♆
22 20	☽ △ ☿
19 Monday	
05 43	☽ # ♀
07 15	☽ ⚼ ♂
07 49	☽ △ ♀
08 34	☉ ☍ ♃
11 32	☽ ∠ ♃
11 47	☽ ⚼ ☉
15 19	☽ ✶ ♇
21 14	☽ △ ♆
23 08	☽ ⚼ ☿
23 28	☽ ± ♂
20 Tuesday	
02 46	☽ ⊥ ♃
04 15	☽ ± ☉
10 04	☽ ⊼ ♂
11 46	☽ ⚼ ♀
12 45	☽ ⊻ ♃
13 36	☽ ± ☿
15 01	☽ ⊼ ☉
16 24	☽ ☍ ♅
16 35	☽ ∠ ♇
18 57	☽ ☍ ♄
20 45	♆ St R
23 05	☽ ⊼ ☿
21 Wednesday	
04 01	☽ ± ♀
07 21	☽ ⊥ ♇
13 12	☉ ⊻ ♅
13 16	☽ # ♀
14 39	☽ ⊼ ♀
17 04	☽ ⊻ ♇
22 42	☽ □ ♆
22 Thursday	
03 13	♅ ⚼ ♇
12 05	♂ ☍ ♃
13 01	☽ ☌ ♃
13 04	☽ ☍ ♂
17 05	☽ ⊼ ♅
18 52	☽ ☍ ☉
19 20	♀ ☍ ♇
19 44	☽ ⊼ ♄
20 22	♀ ∠ ♅
20 58	☽ # ☿
21 13	☽ ☍ ☿
23 Friday	
02 37	☽ ± ♅
03 21	☽ Q ♆
05 18	☽ ± ♄
10 39	☉ ⊻ ♄
16 42	☽ ☌ ♇
16 53	☽ ⚼ ♅
17 31	☉ ☌ ☿
18 37	☽ ☍ ♀
19 38	☽ ⚼ ♄
22 18	☽ ✶ ♆
24 Saturday	
00 33	☿ ⊻ ♄
08 27	♀ ∠ ♄
12 04	☽ ⊻ ♃
14 38	☽ ⊼ ♂
16 42	☽ △ ♅
18 35	☽ ⊼ ☿
19 34	☽ △ ♄
21 31	☽ ⊼ ☉
21 31	☽ ⊥ ♃
22 04	☽ ∠ ♆
25 Sunday	
00 37	☽ ± ♂
01 54	☽ # ♆
03 45	☽ ± ☿
07 45	☽ ± ☉
11 47	☽ ∠ ♃
12 28	☽ ⊥ ♆
15 42	☽ ⚼ ♂
16 08	♀ △ ♆
16 21	☽ ⊻ ♇
17 35	☽ ⚼ ☿
22 09	☽ ⊻ ♆
22 42	☽ ⊼ ♀
23 16	☽ ⚼ ☉
26 Monday	
02 04	☽ ⊥ ♇
03 14	♂ # ♃
09 23	☽ ± ♀
10 53	☿ ⊻ ♅
11 58	☽ ✶ ♃
14 33	☿ ☌ ♂
16 49	☽ ∠ ♇
17 06	☽ △ ☿
17 20	☽ △ ♂
17 23	☽ □ ♅
18 41	♂ ⊻ ♅
20 36	☽ □ ♄
27 Tuesday	
01 40	☽ ⚼ ♀
01 42	☽ △ ☉
07 46	☽ Q ♃
08 41	☽ ∥ ♆
17 58	☽ ✶ ♇
23 50	☽ Q ♅
28 Wednesday	
00 10	☽ ☌ ♆
03 19	☽ Q ♄
04 44	☉ ⊥ ♅
05 36	☽ △ ♀
07 46	☿ ∥ ♀
14 23	☽ □ ♃
14 40	☽ Q ♇
18 17	☽ □ ☿
20 48	☽ ✶ ♅
23 06	☽ □ ♂
29 Thursday	
00 28	☽ ✶ ♄
09 25	☽ □ ☉
21 54	☽ ∥ ♀
22 42	☽ □ ♇
23 45	☽ ∠ ♅
30 Friday	
01 27	☽ Q ☿
03 41	☽ ∠ ♄
05 23	☽ ⊻ ♆
07 49	☽ ∥ ☿
09 12	☽ Q ♂
09 36	♂ ⊻ ♄
14 19	☉ ⊥ ♄
16 20	☽ ⊥ ♅
16 25	☽ ⊥ ♆
16 46	☽ □ ♀
19 59	☽ △ ♃
20 24	☽ ⊥ ♄
20 50	☽ Q ☉
22 58	☽ ✶ ☿
31 Saturday	
03 32	☽ ⊻ ♅
04 44	☽ ∥ ☉
07 42	☽ ⊻ ♄
08 40	☽ ✶ ♂
09 11	☽ ∠ ♆
11 58	♅ Q ♆
21 12	☽ ✶ ☉
23 55	☽ ⚼ ♃
23 58	♀ ⊼ ♃

All ephemeris data is given at 12.00 UT and the Moon's longitude is additionally given for 24.00 UT

AUGUST 2032

Date	Sidereal time h m s	Sun ☉ ° ' "	Moon ☽ ° ' "	Moon ☽ 24.00 ° ' "	Mercury ☿ ° '	Venus ♀ ° '	Mars ♂ ° '	Jupiter ♃ ° '	Saturn ♄ ° '	Uranus ♅ ° '	Neptune ♆ ° '	Pluto ♇ ° '
01	08 42 36	09 ♌ 47 26	16 ♊ 57 36	23 ♊ 11 24	26 ♋ 59	26 ♌ 15	03 ♌ 18	25 ♑ 34	02 ♋ 11	29 ♊ 53	17 ♈ 50	14 ♒ 08
02	08 46 33	10 44 52	29 ♊ 22 26	05 ♋ 30 59	26 R 51	27 29	03 57	25 R 27	02 18	29 56	17 R 50	14 R 06
03	08 50 29	11 42 18	11 ♋ 37 20	17 ♋ 41 45	26 D 49	28 42	04 36	25 20	02 24	29 ♊ 59	17 49	14 05
04	08 54 26	12 39 46	23 ♋ 44 24	29 ♋ 45 30	26 54	29 ♌ 56	05 14	25 13	02 31	00 ♋ 02	17 49	14 04
05	08 58 22	13 37 15	05 ♌ 45 13	11 ♌ 43 41	27 05	01 ♍ 10	05 53	25 06	02 38	00 05	17 48	14 02
06	09 02 19	14 34 44	17 ♌ 41 05	23 ♌ 37 35	27 23	02 23	06 32	24 59	02 44	00 08	17 48	14 01
07	09 06 15	15 32 15	29 ♌ 33 20	05 ♍ 28 34	27 47	03 37	07 10	24 52	02 51	00 10	17 47	14 00
08	09 10 12	16 29 46	11 ♍ 23 30	17 ♍ 18 24	28 18	04 51	07 49	24 45	02 57	00 13	17 47	13 58
09	09 14 08	17 27 19	23 ♍ 13 35	29 ♍ 09 23	28 56	06 04	08 28	24 39	03 03	00 16	17 46	13 57
10	09 18 05	18 24 52	05 ♎ 06 13	11 ♎ 04 31	29 ♋ 40	07 18	09 06	24 32	03 10	00 19	17 45	13 55
11	09 22 02	19 22 27	17 ♎ 04 46	23 ♎ 07 30	00 ♌ 31	08 31	09 45	24 26	03 16	00 21	17 45	13 54
12	09 25 58	20 20 02	29 ♎ 13 16	05 ♏ 22 40	01 28	09 45	10 23	24 19	03 22	00 24	17 44	13 53
13	09 29 55	21 17 38	11 ♏ 36 19	17 ♏ 54 49	02 32	10 59	11 02	24 13	03 28	00 26	17 43	13 51
14	09 33 51	22 15 16	24 ♏ 18 46	00 ♐ 48 45	03 41	12 12	11 40	24 07	03 34	00 29	17 43	13 50
15	09 37 48	23 12 54	07 ♐ 25 15	14 ♐ 08 43	04 56	13 26	12 19	24 01	03 40	00 31	17 42	13 49
16	09 41 44	24 10 33	20 ♐ 59 28	27 ♐ 57 40	06 17	14 39	12 57	23 56	03 46	00 34	17 41	13 47
17	09 45 41	25 08 13	05 ♑ 03 20	12 ♑ 16 14	07 43	15 53	13 35	23 50	03 52	00 36	17 40	13 46
18	09 49 37	26 05 55	19 ♑ 35 57	27 ♑ 01 50	09 14	17 06	14 14	23 45	03 58	00 39	17 39	13 45
19	09 53 34	27 03 37	04 ♒ 33 00	12 ♒ 08 19	10 49	18 20	14 52	23 39	04 04	00 41	17 38	13 43
20	09 57 31	28 01 21	19 ♒ 46 32	27 ♒ 26 16	12 29	19 33	15 31	23 34	04 09	00 43	17 37	13 42
21	10 01 27	28 59 05	05 ♓ 06 04	12 ♓ 44 29	14 12	20 47	16 09	23 29	04 15	00 46	17 37	13 41
22	10 05 24	29 ♌ 56 52	20 ♓ 20 11	27 ♓ 51 56	15 59	22 00	16 47	23 24	04 21	00 48	17 36	13 39
23	10 09 20	00 ♍ 54 39	05 ♈ 18 44	12 ♈ 39 43	17 48	23 14	17 26	23 19	04 26	00 50	17 35	13 38
24	10 13 17	01 52 29	19 ♈ 54 18	27 ♈ 02 07	19 40	24 27	18 04	23 15	04 31	00 52	17 34	13 37
25	10 17 13	02 50 20	04 ♉ 02 59	10 ♉ 56 53	21 34	25 41	18 42	23 10	04 37	00 54	17 32	13 36
26	10 21 10	03 48 13	17 ♉ 43 59	24 ♉ 24 33	23 30	26 54	19 20	23 06	04 42	00 56	17 31	13 34
27	10 25 06	04 46 08	00 ♊ 58 59	07 ♊ 27 42	25 26	28 08	19 59	23 02	04 47	00 58	17 30	13 33
28	10 29 03	05 44 05	13 ♊ 51 12	20 ♊ 09 58	27 24	29 ♍ 21	20 37	22 58	04 52	01 00	17 29	13 32
29	10 33 00	06 42 03	26 ♊ 24 31	02 ♋ 35 23	29 ♌ 22	00 ♎ 35	21 15	22 54	04 57	01 02	17 28	13 31
30	10 36 56	07 40 04	08 ♋ 43 01	14 ♋ 47 54	01 ♍ 20	01 48	21 53	22 51	05 02	01 04	17 27	13 29
31	10 40 53	08 ♍ 38 06	20 ♋ 50 27	26 ♋ 51 04	03 ♍ 19	03 ♎ 02	22 ♌ 32	22 ♑ 47	05 ♋ 07	01 ♋ 06	17 ♈ 26	13 ♒ 28

Date	Moon True ☊ ° '	Moon Mean ☊ ° '	Moon ☽ Latitude ° '
01	04 ♏ 55	04 ♏ 51	03 S 22
02	04 R 48	04 47	04 05
03	04 38	04 44	04 37
04	04 26	04 41	04 55
05	04 13	04 38	05 00
06	03 59	04 35	04 52
07	03 46	04 31	04 31
08	03 35	04 28	03 58
09	03 27	04 25	03 15
10	03 21	04 22	02 24
11	03 18	04 19	01 25
12	03 17	04 16	00 S 22
13	03 D 17	04 12	00 N 44
14	03 R 17	04 09	01 49
15	03 16	04 06	02 51
16	03 13	04 03	03 45
17	03 08	04 00	04 28
18	03 01	03 57	04 55
19	02 52	03 53	05 03
20	02 43	03 50	04 50
21	02 34	03 47	04 16
22	02 27	03 44	03 24
23	02 22	03 41	02 19
24	02 19	03 37	01 N 06
25	02 D 18	03 34	00 S 09
26	02 19	03 31	01 22
27	02 20	03 28	02 27
28	02 R 19	03 25	03 24
29	02 17	03 22	04 08
30	02 13	03 18	04 41
31	02 ♏ 07	03 ♏ 15	05 S 00

DECLINATIONS

Date	Sun ☉ ° '	Moon ☽ ° '	Mercury ☿ ° '	Venus ♀ ° '	Mars ♂ ° '	Jupiter ♃ ° '	Saturn ♄ ° '	Uranus ♅ ° '	Neptune ♆ ° '	Pluto ♇ ° '
01	17 N 48	19 N 26	16 N 51	14 N 10	20 N 27	21 S 30	22 N 25	23 N 37	05 N 31	23 S 24
02	17 32	19 20	17 06	13 45	20 18	21 31	22 25	23 37	05 30	23 25
03	17 16	18 20	17 20	13 19	20 09	21 33	22 25	23 37	05 30	23 25
04	17 00	16 30	17 34	12 53	20 00	21 34	22 25	23 37	05 30	23 26
05	16 44	13 59	17 47	12 27	19 51	21 36	22 25	23 37	05 30	23 26
06	16 27	10 54	18 00	12 00	19 42	21 37	22 25	23 37	05 29	23 27
07	16 10	07 24	18 11	11 34	19 32	21 38	22 25	23 37	05 29	23 27
08	15 53	03 N 37	18 21	11 06	19 22	21 40	22 25	23 37	05 29	23 28
09	15 36	00 S 18	18 29	10 39	19 12	21 41	22 24	23 37	05 29	23 28
10	15 18	04 13	18 36	10 11	19 02	21 42	22 24	23 37	05 28	23 29
11	15 00	08 01	18 42	09 43	18 52	21 43	22 24	23 37	05 28	23 29
12	14 42	11 32	18 45	09 15	18 42	21 45	22 24	23 37	05 28	23 29
13	14 24	14 37	18 46	08 46	18 32	21 46	22 24	23 37	05 27	23 30
14	14 05	17 05	18 45	08 17	18 21	21 47	22 24	23 37	05 27	23 30
15	13 47	18 44	18 42	07 48	18 11	21 48	22 24	23 37	05 27	23 31
16	13 28	19 23	18 36	07 19	18 00	21 49	22 23	23 37	05 26	23 31
17	13 08	18 52	18 27	06 50	17 49	21 50	22 23	23 37	05 26	23 32
18	12 49	17 08	18 16	06 20	17 38	21 51	22 23	23 37	05 26	23 32
19	12 29	14 13	18 02	05 50	17 27	21 52	22 23	23 37	05 25	23 32
20	12 09	10 17	17 45	05 20	17 16	21 53	22 23	23 37	05 25	23 33
21	11 49	05 40	17 25	04 50	17 05	21 54	22 23	23 37	05 24	23 33
22	11 29	00 S 42	17 03	04 20	16 53	21 55	22 22	23 37	05 24	23 34
23	11 09	04 N 14	16 38	03 50	16 42	21 56	22 22	23 37	05 24	23 34
24	10 48	08 48	16 10	03 19	16 30	21 57	22 22	23 37	05 23	23 34
25	10 28	12 43	15 40	02 49	16 19	21 57	22 22	23 37	05 23	23 35
26	10 07	15 48	15 07	02 18	16 07	21 58	22 22	23 37	05 22	23 35
27	09 46	17 57	14 32	01 47	15 55	21 59	22 21	23 37	05 22	23 36
28	09 24	19 05	13 56	01 16	15 43	22 00	22 21	23 37	05 21	23 36
29	09 03	19 15	13 18	00 46	15 31	22 00	22 21	23 37	05 21	23 36
30	08 41	18 29	12 38	00 15	15 19	22 01	22 21	23 37	05 20	23 37
31	08 N 20	16 N 52	11 N 56	00 S 16	15 N 06	22 S 02	22 N 21	23 N 37	05 N 20	23 S 37

ZODIAC SIGN ENTRIES

Date	h	m	Planets
02	13	13	☽ ♋
03	18	20	♅ ♋
04	13	20	♀ ♍
05	00	29	☽ ♌
07	12	54	☽ ♍
10	01	42	☽ ♎
10	21	47	☿ ♌
12	13	32	☽ ♏
14	22	31	☽ ♐
17	03	28	☽ ♑
19	04	45	☽ ♒
21	04	01	☽ ♓
22	13	18	☉ ♍
23	03	25	☽ ♈
25	05	03	☽ ♉
27	10	12	☽ ♊
29	00	40	♀ ♎
29	18	58	☽ ♋
29	19	41	☿ ♍

LATITUDES

Date	Mercury ☿ ° '	Venus ♀ ° '	Mars ♂ ° '	Jupiter ♃ ° '	Saturn ♄ ° '	Uranus ♅ ° '	Neptune ♆ ° '	Pluto ♇ ° '
01	03 S 58	01 N 30	01 N 04	00 S 29	01 S 00	00 N 11	01 S 36	07 S 08
04	03 16	01 30	01 05	00 30	01 00	00 11	01 37	07 08
07	02 28	01 29	01 06	00 30	00 59	00 11	01 37	07 09
10	01 39	01 27	01 06	00 30	00 59	00 11	01 37	07 09
13	00 50	01 26	01 07	00 30	00 59	00 11	01 37	07 09
16	00 S 06	01 23	01 07	00 31	00 59	00 11	01 37	07 09
19	00 N 32	01 20	01 08	00 31	00 59	00 11	01 37	07 09
22	01 03	01 16	01 09	00 31	00 59	00 11	01 38	07 09
25	01 26	01 12	01 09	00 31	00 59	00 11	01 38	07 09
28	01 40	01 07	01 10	00 31	00 59	00 11	01 38	07 09
31	01 N 46	01 N 01	01 N 10	00 S 32	00 S 59	00 N 11	01 S 38	07 S 09

DATA

Julian Date	2463446
Delta T	+75 seconds
Ayanamsa	24° 18' 55"
Synetic vernal point	04° ♓ 48' 03"
True obliquity of ecliptic	23° 25' 58"

MOON'S PHASES, APSIDES AND POSITIONS ☽

Date	h	m	Phase	Longitude ° '	Eclipse Indicator
06	05	11	●	14 ♌ 18	
14	07	51	☽	22 ♏ 05	
21	01	47	○	28 ♒ 34	
27	19	33	☾	05 ♊ 04	

Day	h	m		
07	17	30	Apogee	
21	03	58	Perigee	
01	21	27	Max dec	19° N 31'
09	10	10	0S	
16	13	52	Max dec	19° S 23'
22	15	20	0N	
29	03	52	Max dec	19° N 18'

LONGITUDES

Date	Chiron ⚷ ° '	Ceres ⚳ ° '	Pallas ⚴ ° '	Juno ⚵ ° '	Vesta ⚶ ° '	Black Moon Lilith ⚸ ° '
01	25 ♉ 41	28 ♍ 14	24 ♌ 59	14 ♌ 02	16 ♎ 00	29 ♌ 15
11	25 ♉ 55	02 ♎ 20	00 ♍ 06	18 ♌ 51	20 ♎ 22	00 ♍ 22
21	26 ♉ 04	06 ♎ 31	05 ♍ 10	23 ♌ 36	24 ♎ 55	01 ♍ 30
31	26 ♉ 07	10 ♎ 47	10 ♍ 09	28 ♌ 15	29 ♎ 38	02 ♍ 37

ASPECTARIAN

h m	Aspects
01 Sunday	
02 38	☽ ∠ ☿
06 15	☽ Q ♀
06 35	☽ △ ♇
13 41	☽ ✶ ♆
14 43	☽ ∠ ♂
16 58	☽ ± ♃
19 38	☽ ⊥ ☿
02 Monday	
00 46	☿ ⊻ ♀
04 22	☽ ∠ ☉
04 27	☽ ⊼ ♃
07 08	☽ ⊻ ☿
07 54	☽ ✶ ♀
09 05	☽ ⊥ ♂
11 29	☽ bQ ♇
12 53	☽ Q ♆
13 06	☽ ☌ ♅
17 45	☽ ☌ ♄
21 26	☽ ⊻ ♂
23 23	☽ ⊥ ☉
03 Tuesday	
05 03	☽ ± ♇
06 52	☿ St D
08 49	☉ ∥ ☿
12 11	☽ ⊻ ☉
16 34	☽ ∠ ♀
16 51	☽ ⊼ ♇
04 Wednesday	
00 15	☽ □ ♆
00 40	☽ ∥ ☿
05 27	☽ ∥ ☉
12 26	☽ ⊥ ♀
14 06	♀ ✶ ♅
14 54	☽ ☍ ♃
18 22	☽ ☌ ☿
05 Thursday	
00 36	☽ ⊻ ♅
01 45	☽ ⊻ ♀
05 41	☽ ⊻ ♄
10 52	♀ ± ♃
12 17	☽ ☌ ♂
12 40	☽ ⊥ ♅
17 49	☽ ⊥ ♄
19 58	♂ ⊥ ♅
22 13	☉ ☍ ♇
06 Friday	
02 24	☽ ∥ ♀
04 37	☽ ☍ ♇
05 11	☽ ☌ ☉
06 50	☽ ∠ ♅
12 06	☽ ∠ ♄
12 14	☽ △ ♆
19 30	♀ ✶ ♄
19 58	♀ bQ ♆
07 Saturday	
02 28	☉ ∠ ♅
02 36	☽ ⊼ ♃
08 16	☽ ⊻ ☿
13 16	☽ ✶ ♅
14 38	☽ ± ♃
17 23	☿ ⊥ ♀
18 33	☽ bQ ♆
18 44	☽ ✶ ♄
20 56	☽ ⊥ ☿
21 11	☽ ☌ ♀
08 Sunday	
00 18	☽ ∥ ♆
04 20	☽ ⊻ ♂
08 43	☽ bQ ♃
12 47	☽ ± ♆
13 41	☽ Q ♅
16 04	☽ ∠ ☿
17 12	☽ ⊥ ♂
17 13	☽ ⊼ ♇
19 17	☽ Q ♄
23 16	☽ ⊻ ☉
09 Monday	
00 57	☽ ⊼ ♆
05 22	☽ ± ♇
12 30	☽ ∠ ♂
12 30	☽ ⊥ ☉
14 51	☽ △ ♃
19 45	☉ △ ♆
23 33	☽ bQ ♇
10 Tuesday	
00 17	☽ ✶ ☿
02 18	☽ □ ♅
04 56	☉ ∠ ♄
08 03	☽ □ ♄
08 18	☽ ∠ ☉
14 44	♂ ⊥ ♄
16 55	☽ ⊻ ♀
19 46	☽ # ♆
20 30	☽ ✶ ♂
11 Wednesday	
02 10	☽ Q ☿
05 40	☽ △ ♇
06 19	☽ ⊥ ♀
07 26	☿ ⊻ ♅
13 20	☽ ☍ ♆
16 58	☽ ✶ ☉
21 47	☽ Q ♂
21 59	☽ # ♀
12 Thursday	
02 13	☽ ∠ ♀
02 27	☽ □ ♃
04 20	♀ bQ ♃
06 30	☿ ∥ ♂
14 19	☽ △ ♅
16 47	☽ □ ☿
18 36	☽ Q ☉
20 10	☽ △ ♄
13 Friday	
10 23	☽ # ☉
10 40	☽ ✶ ♀
10 50	☽ □ ♂
13 10	☽ Q ♃
14 05	♀ ⊻ ♂
16 18	☽ □ ♇
19 20	☽ bQ ♅
23 38	☽ ⊼ ♆
14 Saturday	
01 10	☽ bQ ♄
02 27	♀ ± ♆
07 51	☽ □ ☉
09 36	☿ ⊻ ♄
10 53	☽ ± ♆
11 39	☽ ✶ ♃
11 46	☽ Q ♀
12 19	☽ ± ♅
17 39	♀ Q ♅
18 05	☽ ± ♄
23 26	☽ ⊼ ♅
15 Sunday	
01 51	☽ Q ♇
03 24	☽ # ♂
03 27	☽ bQ ♆
05 10	☽ ⊼ ♄
07 02	☽ △ ☿
11 23	☽ # ☿
14 52	☽ ∠ ♃
19 21	♀ ⊼ ♇
21 11	☽ △ ♂
23 23	☽ ✶ ♇
23 48	☽ □ ♀
16 Monday	
06 14	☽ △ ♆
06 21	☉ ⊼ ♃
06 42	☽ ⊥ ♃
12 34	☽ bQ ☿
16 57	☿ ⊥ ♅
17 03	☽ ⊻ ♃
17 55	☽ △ ☉
17 Tuesday	
00 33	☽ bQ ♂
01 23	☽ ∠ ♇
04 28	☽ ☍ ♅
05 44	☽ ± ☿
10 00	☽ ☍ ♄
11 45	♀ Q ♄
16 26	☽ ± ♂
16 32	☽ ⊥ ♇
16 58	☽ ⊼ ☿
18 23	♂ ☍ ♇
20 36	☽ # ☿
21 04	☽ bQ ☉
18 Wednesday	
02 27	☽ ⊻ ♇
02 50	☽ ⊼ ♂
06 00	☽ # ♂
07 34	☽ △ ♀
08 50	☽ □ ♆
12 52	☽ ± ☉
18 40	☽ ☌ ♃
22 36	♀ ⊼ ♆
23 14	☽ ⊼ ☉
23 57	☿ ⊥ ♄
19 Thursday	
05 50	☽ ⊼ ♅
09 54	☽ bQ ♀
11 13	☽ ⊼ ♄
13 44	☽ Q ♆
15 24	☽ ± ♅
20 47	☽ ± ♄
23 08	☽ ☍ ☿
20 Friday	
00 13	☽ # ☉
01 23	☽ ± ♀
02 28	☽ ☌ ♇
05 01	☽ ☍ ♂
05 37	☽ bQ ♅
08 15	♀ ∥ ♆
08 38	☽ ✶ ♆
11 01	☽ bQ ♄
11 38	☽ ⊼ ♀
14 45	♀ ± ♇
17 54	☽ ⊻ ♃
20 27	♂ ∠ ♅
21 Saturday	
01 47	☽ ☍ ☉
03 15	☽ ⊥ ♃
04 51	☿ ☍ ♇
05 11	☽ △ ♅
08 06	☽ ∠ ♆
10 40	☽ △ ♄
13 15	☽ # ♆
16 31	☽ # ♀
17 17	☽ ∠ ♃
22 12	☽ ⊥ ♆
23 27	☉ ± ♃
22 Sunday	
01 27	☽ ⊻ ♇
04 11	☽ ⊼ ☿
06 08	☽ ⊼ ♂
07 40	☽ ⊻ ♆
09 29	☿ ∠ ♅
10 56	☽ ⊥ ♇
13 06	☿ ⊥ ♀
14 53	☽ ☍ ♀
14 58	☽ ± ☿
16 04	☽ ± ♂
16 51	☽ ✶ ♃
23 Monday	
01 15	☽ ∠ ♇
04 24	☽ ⊼ ☉
04 25	☿ ☌ ♂
04 45	☽ □ ♅
04 50	☿ ∥ ♂
07 07	☽ bQ ♂
07 22	☽ bQ ☿
09 02	☿ △ ♆
09 59	☉ ✶ ♅
10 09	☽ ∥ ♀
10 34	☽ □ ♄
12 01	☽ Q ♃
13 39	♀ △ ♃
14 47	☽ ± ☉
17 28	♂ △ ♆
17 50	☽ ∥ ♆
19 55	♀ ⊥ ♂
24 Tuesday	
01 35	☽ ✶ ♇
06 36	☽ bQ ☉
08 06	☽ ☌ ♆
08 48	☽ △ ♂
10 02	☿ ∠ ♄
10 16	☽ Q ♅
11 33	☽ △ ☿
16 25	☽ Q ♄
17 34	☽ □ ♃
20 22	☽ ⊼ ♀
21 35	☽ Q ♇
22 46	☽ ∥ ☉
25 Wednesday	
04 43	☉ bQ ♆
06 35	☽ ✶ ♅
07 32	☽ ± ♀
09 46	☽ △ ☉
12 59	☽ ✶ ♄
26 Thursday	
00 40	☽ bQ ♀
04 38	☽ □ ♇
06 58	☽ ∥ ☿
07 17	☿ ⊼ ♃
08 48	☽ ∠ ♅
11 38	☽ ⊻ ♆
14 38	☽ ∥ ♂
15 01	☽ □ ♂
15 32	☽ ∠ ♄
21 35	☽ △ ♃
22 23	☽ ⊥ ♆
27 Friday	
00 07	☽ □ ☿
01 00	☽ ⊥ ♅
03 43	♂ ∠ ♄
06 14	☽ △ ♀
07 57	☽ ⊥ ♄
11 59	☽ ⊻ ♅
12 31	☉ ✶ ♄
14 48	☽ ∠ ♆
19 04	☽ ⊻ ♄
19 33	☽ □ ☉
20 07	♀ bQ ♇
28 Saturday	
01 00	☽ bQ ♃
01 38	☽ Q ♂
11 24	☽ △ ♇
15 28	☽ Q ☿
17 53	☽ ± ♃
18 53	☽ ✶ ♆
29 Sunday	
01 33	☽ ✶ ♂
05 17	☽ ⊼ ♃
06 33	☿ ± ♃
08 26	☽ Q ☉
16 04	☽ bQ ♇
17 55	☽ Q ♆
18 49	☽ ✶ ☿
20 58	☽ □ ♀
21 00	☽ ☌ ♅
21 14	♀ □ ♅
30 Monday	
04 44	☽ ☌ ♄
08 13	☽ ∠ ♂
08 38	☿ ✶ ♅
09 36	☽ ± ♇
09 46	☽ ✶ ☉
16 13	☉ bQ ♃
21 24	☽ ⊼ ♇
23 49	♅ # ♇
31 Tuesday	
01 20	☿ bQ ♆
02 46	☿ ⊻ ♀
02 57	☽ ⊥ ♂
05 13	☽ □ ♆
05 59	☽ ∠ ☿
12 25	☽ Q ♀
15 33	☽ ⊻ ♂
15 52	☽ ☍ ♃
18 04	☽ ∠ ☉
21 14	♂ ⊼ ♃

All ephemeris data is given at 12.00 UT and the Moon's longitude is additionally given for 24.00 UT

SEPTEMBER 2032

LONGITUDES

Date	Sidereal time h m s	Sun ☉ ° ' "	Moon ☽ ° ' "	Moon ☽ 24.00 ° ' "	Mercury ☿ ° '	Venus ♀ ° '	Mars ♂ ° '	Jupiter ♃ ° '	Saturn ♄ ° '	Uranus ♅ ° '	Neptune ♆ ° '	Pluto ♇ ° '
01	10 44 49	09 ♍ 36 10	02 ♌ 50 07	08 ♌ 47 54	05 ♍ 17	04 ♎ 15	23 ♌ 10	22 ♑ 44	05 ♋ 12	01 ♋ 08	17 ♈ 24	13 ♒ 27
02	10 48 46	10 34 16	14 ♌ 44 43	20 ♌ 40 50	07 14	05 28	23 48	22 R 41	05 17	01 10	17 R 23	13 R 26
03	10 52 42	11 32 23	26 ♌ 36 30	02 ♍ 31 54	09 11	06 42	24 26	22 38	05 21	01 11	17 22	13 25
04	10 56 39	12 30 32	08 ♍ 27 17	14 ♍ 22 50	11 08	07 55	25 04	22 36	05 26	01 13	17 21	13 23
05	11 00 35	13 28 43	20 ♍ 18 45	26 ♍ 15 16	13 03	09 08	25 42	22 33	05 30	01 15	17 19	13 22
06	11 04 32	14 26 56	02 ♎ 12 36	08 ♎ 11 01	14 58	10 22	26 20	22 31	05 35	01 16	17 18	13 21
07	11 08 29	15 25 10	14 ♎ 10 48	20 ♎ 12 14	16 51	11 35	26 58	22 29	05 39	01 18	17 17	13 20
08	11 12 25	16 23 26	26 ♎ 15 42	02 ♏ 21 32	18 44	12 48	27 37	22 27	05 43	01 19	17 15	13 19
09	11 16 22	17 21 44	08 ♏ 30 10	14 ♏ 42 02	20 35	14 02	28 15	22 25	05 47	01 21	17 14	13 18
10	11 20 18	18 20 03	20 ♏ 57 36	27 ♏ 17 20	22 26	15 15	28 53	22 24	05 51	01 22	17 12	13 17
11	11 24 15	19 18 24	03 ♐ 41 42	10 ♐ 11 12	24 15	16 28	29 ♌ 31	22 22	05 55	01 23	17 11	13 16
12	11 28 11	20 16 46	16 ♐ 46 15	23 ♐ 27 14	26 03	17 41	00 ♍ 09	22 21	05 59	01 25	17 10	13 15
13	11 32 08	21 15 10	00 ♑ 14 30	07 ♑ 08 15	27 50	18 55	00 47	22 20	06 03	01 26	17 08	13 14
14	11 36 04	22 13 35	14 ♑ 08 35	21 ♑ 15 28	29 ♍ 36	20 08	01 25	22 20	06 07	01 27	17 07	13 13
15	11 40 01	23 12 02	28 ♑ 28 39	05 ♒ 47 44	01 ♎ 21	21 21	02 03	22 19	06 10	01 28	17 05	13 12
16	11 43 58	24 10 31	13 ♒ 12 07	20 ♒ 41 00	03 05	22 34	02 41	22 19	06 14	01 29	17 04	13 11
17	11 47 54	25 09 01	28 ♒ 13 24	05 ♓ 48 11	04 48	23 47	03 18	22 19	06 17	01 30	17 02	13 10
18	11 51 51	26 07 33	13 ♓ 24 08	20 ♓ 59 58	06 29	25 00	03 56	22 D 19	06 20	01 31	17 01	13 09
19	11 55 47	27 06 07	28 ♓ 34 24	06 ♈ 06 11	08 10	26 14	04 34	22 19	06 24	01 32	16 59	13 08
20	11 59 44	28 04 43	13 ♈ 34 13	20 ♈ 57 31	09 50	27 27	05 12	22 19	06 27	01 33	16 58	13 07
21	12 03 40	29 ♍ 03 21	28 ♈ 15 17	05 ♉ 26 53	11 28	28 40	05 50	22 20	06 30	01 34	16 56	13 06
22	12 07 37	00 ♎ 02 00	12 ♉ 31 54	19 ♉ 30 06	13 06	29 ♎ 53	06 28	22 21	06 33	01 35	16 54	13 05
23	12 11 33	01 00 43	26 ♉ 21 25	03 ♊ 05 55	14 42	01 ♏ 06	07 06	22 22	06 35	01 35	16 53	13 05
24	12 15 30	01 59 27	09 ♊ 43 51	16 ♊ 15 30	16 18	02 19	07 44	22 23	06 38	01 36	16 51	13 04
25	12 19 27	02 58 14	22 ♊ 41 17	29 ♊ 01 41	17 53	03 32	08 21	22 24	06 41	01 37	16 50	13 03
26	12 23 23	03 57 03	05 ♋ 17 12	11 ♋ 28 22	19 27	04 45	08 59	22 26	06 43	01 37	16 48	13 02
27	12 27 20	04 55 54	17 ♋ 35 45	23 ♋ 39 52	21 00	05 58	09 37	22 28	06 46	01 38	16 46	13 01
28	12 31 16	05 54 47	29 ♋ 41 18	05 ♌ 40 32	22 32	07 11	10 15	22 30	06 48	01 38	16 45	13 01
29	12 35 13	06 53 43	11 ♌ 38 05	17 ♌ 34 25	24 03	08 24	10 53	22 32	06 50	01 39	16 43	13 00
30	12 39 09	07 ♎ 52 41	23 ♌ 29 59	29 ♌ 25 11	25 ♎ 33	09 ♏ 36	11 ♍ 30	22 ♑ 34	06 ♋ 52	01 ♋ 39	16 ♈ 41	12 ♒ 59

Date	Moon True ☊ ° '	Moon Mean ☊ ° '	Moon ☽ Latitude ° '
01	01 ♏ 58	03 ♏ 12	05 S 06
02	01 R 49	03 09	04 58
03	01 39	03 06	04 38
04	01 30	03 02	04 05
05	01 22	02 59	03 22
06	01 16	02 56	02 30
07	01 12	02 53	01 31
08	01 11	02 50	00 S 27
09	01 D 11	02 47	00 N 40
10	01 12	02 43	01 45
11	01 13	02 40	02 47
12	01 14	02 37	03 42
13	01 R 14	02 34	04 27
14	01 12	02 31	04 57
15	01 08	02 28	05 11
16	01 04	02 24	05 04
17	00 59	02 21	04 37
18	00 54	02 18	03 50
19	00 50	02 15	02 47
20	00 47	02 12	01 33
21	00 46	02 08	00 N 14
22	00 D 46	02 05	01 S 04
23	00 48	02 02	02 16
24	00 49	01 59	03 18
25	00 50	01 56	04 07
26	00 R 51	01 53	04 44
27	00 50	01 49	05 06
28	00 48	01 46	05 14
29	00 45	01 43	05 09
30	00 ♏ 42	01 ♏ 40	04 S 50

DECLINATIONS

Date	Sun ☉ ° '	Moon ☽ ° '	Mercury ☿ ° '	Venus ♀ ° '	Mars ♂ ° '	Jupiter ♃ ° '	Saturn ♄ ° '	Uranus ♅ ° '	Neptune ♆ ° '	Pluto ♇ ° '
01	07 N 58	14 N 33	11 N 14	00 S 47	14 N 54	22 S 02	22 N 20	23 N 37	05 N 19	23 S 37
02	07 36	11 39	10 30	01 18	14 41	22 03	22 20	23 37	05 19	23 38
03	07 14	08 17	09 45	01 49	14 29	22 03	22 20	23 37	05 18	23 38
04	06 52	04 36	09 00	02 20	14 16	22 04	22 20	23 37	05 18	23 38
05	06 30	00 N 44	08 14	02 51	14 03	22 04	22 20	23 37	05 17	23 39
06	06 07	03 S 10	07 27	03 22	13 51	22 05	22 19	23 37	05 17	23 39
07	05 45	06 59	06 40	03 52	13 38	22 05	22 19	23 37	05 16	23 39
08	05 22	10 33	05 53	04 23	13 25	22 05	22 19	23 37	05 16	23 40
09	04 59	13 42	05 06	04 54	13 12	22 06	22 19	23 37	05 15	23 40
10	04 37	16 18	04 18	05 24	12 59	22 06	22 19	23 37	05 15	23 40
11	04 14	18 09	03 31	05 55	12 45	22 06	22 19	23 37	05 14	23 40
12	03 51	19 05	02 43	06 25	12 32	22 06	22 18	23 37	05 13	23 41
13	03 28	18 59	01 56	06 55	12 19	22 06	22 18	23 37	05 13	23 41
14	03 05	17 45	01 08	07 25	12 05	22 07	22 18	23 37	05 12	23 41
15	02 42	15 23	00 N 21	07 55	11 52	22 07	22 18	23 37	05 12	23 41
16	02 19	11 59	00 S 26	08 25	11 38	22 07	22 18	23 37	05 11	23 42
17	01 56	07 45	01 12	08 54	11 25	22 07	22 17	23 37	05 10	23 42
18	01 32	02 S 59	01 58	09 24	11 11	22 07	22 17	23 37	05 10	23 42
19	01 09	01 N 59	02 44	09 53	10 57	22 07	22 17	23 37	05 09	23 42
20	00 46	06 47	03 29	10 22	10 44	22 07	22 17	23 37	05 09	23 42
21	00 N 23	11 04	04 14	10 51	10 30	22 07	22 17	23 37	05 08	23 43
22	00 S 01	14 35	04 59	11 19	10 16	22 07	22 17	23 37	05 07	23 43
23	00 24	17 08	05 43	11 48	10 02	22 06	22 16	23 37	05 07	23 43
24	00 47	18 39	06 26	12 16	09 48	22 06	22 16	23 37	05 06	23 43
25	01 11	19 07	07 09	12 43	09 34	22 06	22 16	23 37	05 05	23 43
26	01 34	18 36	07 52	13 11	09 20	22 06	22 16	23 37	05 05	23 43
27	01 58	17 13	08 34	13 38	09 05	22 06	22 16	23 37	05 04	23 43
28	02 21	15 05	09 15	14 05	08 51	22 05	22 16	23 37	05 04	23 44
29	02 44	12 21	09 55	14 31	08 37	22 05	22 16	23 37	05 03	23 44
30	03 S 07	09 N 07	10 S 35	14 S 58	08 N 23	22 S 05	22 N 15	23 N 37	05 N 02	23 S 44

ZODIAC SIGN ENTRIES

Date	h m	Planets
01	06 19	☽ ♌
03	18 52	☽ ♍
06	07 33	☽ ♎
08	19 22	☽ ♏
11	05 06	☽ ♐
12	06 32	♂ ♍
13	11 35	☽ ♑
14	17 24	☿ ♎
15	14 31	☽ ♒
17	14 49	☽ ♓
19	14 16	☽ ♈
21	14 54	☽ ♉
22	11 11	☉ ♎
22	14 23	♀ ♏
23	18 28	☽ ♊
26	01 51	☽ ♋
28	12 37	☽ ♌

LATITUDES

Date	Mercury ☿ ° '	Venus ♀ ° '	Mars ♂ ° '	Jupiter ♃ ° '	Saturn ♄ ° '	Uranus ♅ ° '	Neptune ♆ ° '	Pluto ♇ ° '
01	01 N 47	00 N 59	01 N 10	00 S 32	00 S 59	00 N 11	01 S 38	07 S 09
04	01 44	00 53	01 11	00 32	00 59	00 11	01 38	07 09
07	01 37	00 46	01 11	00 32	01 00	00 11	01 38	07 09
10	01 25	00 39	01 11	00 32	01 00	00 11	01 38	07 09
13	01 10	00 31	01 12	00 32	01 00	00 11	01 38	07 09
16	00 52	00 24	01 12	00 32	01 00	00 11	01 38	07 09
19	00 33	00 15	01 12	00 32	01 00	00 11	01 39	07 08
22	00 N 12	00 N 07	01 13	00 32	01 00	00 12	01 39	07 08
25	00 S 09	00 S 02	01 13	00 32	01 00	00 12	01 39	07 08
28	00 31	00 11	01 13	00 32	01 00	00 12	01 39	07 08
31	00 S 53	00 S 20	01 N 13	00 S 32	01 S 00	00 N 12	01 S 39	07 S 07

DATA

Julian Date	2463477
Delta T	+75 seconds
Ayanamsa	24° 18' 59"
Synetic vernal point	04° ♓ 48' 00"
True obliquity of ecliptic	23° 25' 59"

MOON'S PHASES, APSIDES AND POSITIONS ☽

Date	h m	Phase	Longitude ° '	Eclipse Indicator
04	20 57	●	12 ♍ 52	
12	18 49	☽	20 ♐ 33	
19	09 30	○	27 ♓ 00	
26	09 12	☾	03 ♋ 50	

Day	h m		
03	20 14	Apogee	
18	14 14	Perigee	
05	16 31	OS	
12	21 48	Max dec	19° S 11'
19	02 25	ON	
25	11 09	Max dec	19° N 07'

LONGITUDES

Date	Chiron ⚷ ° '	Ceres ⚳ ° '	Pallas ⚴ ° '	Juno ⚵ ° '	Vesta ⚶ ° '	Black Moon Lilith ⚸ ° '
01	26 ♉ 07	11 ♎ 12	10 ♍ 39	28 ♌ 43	00 ♏ 07	02 ♍ 43
11	26 ♉ 02	15 ♎ 32	15 ♍ 34	03 ♍ 16	04 ♏ 59	03 ♍ 50
21	25 ♉ 52	19 ♎ 54	20 ♍ 25	07 ♍ 42	09 ♏ 57	04 ♍ 57
31	25 ♉ 37	24 ♎ 18	25 ♍ 11	12 ♍ 01	15 ♏ 02	06 ♍ 04

All ephemeris data is given at 12.00 UT and the Moon's longitude is additionally given for 24.00 UT

ASPECTARIAN

h m Aspects

01 Wednesday
03 28 ☽ ⊥ ☿
08 32 ☽ ∥ ♂
08 34 ☽ ⊻ ♅
10 59 ☿ ✶ ♄
13 41 ☽ ⊥ ☉
15 10 ☽ ✶ ♀
16 47 ☽ ⊻ ♄
17 53 ☽ ⊻ ☿
20 40 ☽ ⊥ ♅

02 Thursday
02 50 ☽ ⊻ ☉
04 57 ☽ ⊥ ♄
07 58 ♀ □ ♄
09 21 ☽ ☍ ♇
14 52 ☽ ∠ ♅
17 20 ☽ △ ♆
17 21 ☿ bQ ♃
22 58 ☽ ∥ ☿
23 16 ☽ ∠ ♄

03 Friday
00 55 ☽ ∠ ♀
04 00 ☽ ⚻ ♃
07 21 ☽ ☌ ♂
07 46 ☉ ± ♆
16 06 ☽ ± ♃
19 49 ☽ ∥ ☉
21 14 ☽ ⊥ ♀
21 18 ☽ ✶ ♅
23 38 ☽ bQ ♆

04 Saturday
05 50 ☽ ✶ ♄
07 35 ☽ ∥ ♆
10 16 ☽ bQ ♃
10 47 ☽ ⊻ ♀
14 38 ☿ ± ♆
17 50 ☽ ± ♆
18 28 ☽ ☌ ☿
20 57 ☽ ☌ ☉
21 40 ☽ Q ♅
21 59 ☽ ⚻ ♇

05 Sunday
00 32 ☽ # ♀
05 58 ☽ ⚻ ♆
05 59 ☉ Q ♅
06 17 ☽ Q ♄
09 24 ☉ ⚻ ♇
10 06 ☽ ± ♇
14 23 ☿ Q ♅
15 56 ☿ ⚻ ♇
16 31 ☽ △ ♃
22 48 ☉ ☌ ☿
23 30 ☽ ⊻ ♂

06 Monday
04 15 ☽ bQ ♇
10 06 ☽ □ ♅
12 17 ☽ ⊥ ♂
13 20 ☽ ∥ ♀
18 49 ☽ □ ♄

07 Tuesday
01 06 ☽ # ♆
04 48 ☽ # ☉
06 13 ☽ ☌ ♀
07 21 ☽ ∠ ♂
10 19 ☽ △ ♇
10 20 ☽ # ☿
14 41 ☽ ⊻ ☉
17 18 ☿ ⚻ ♆
18 10 ☽ ☍ ♆
18 20 ☽ ⊻ ☿
22 32 ☿ Q ♄

08 Wednesday
03 40 ☽ ⊥ ☉
03 57 ♀ ∠ ♂
04 29 ☽ □ ♃
08 26 ☽ ⊥ ☿
14 48 ☽ ✶ ♂
19 05 ☉ ∥ ♆
19 26 ☿ ± ♇
21 52 ♀ △ ♇
21 59 ☽ △ ♅
22 58 ☽ ∠ ☉

09 Thursday
05 19 ☽ ∠ ☿
06 41 ☽ △ ♄
07 11 ☿ ∥ ♆
08 07 ☽ # ♂
08 51 ☉ ⚻ ♆
14 38 ☉ # ♀
15 34 ☽ Q ♂
15 42 ☿ # ♀
15 43 ☽ Q ♃
18 00 ☉ ∥ ☿
18 30 ♂ ± ♃
21 17 ☽ □ ♇
23 22 ☉ Q ♄
23 52 ☽ ⊻ ♀

10 Friday
03 11 ☽ bQ ♅
04 32 ♀ # ♆
04 50 ☽ ⚻ ♆
06 34 ☽ ✶ ☉
11 34 ☿ △ ♃
11 48 ☽ bQ ♄
12 37 ☽ ⊥ ♀
14 44 ☽ ✶ ♃
15 16 ☽ ✶ ☿
16 16 ☽ ± ♆
20 23 ☽ ± ♅

11 Saturday
03 46 ☽ □ ♂
04 55 ☽ ± ♄
07 11 ☽ Q ☉
07 25 ☽ ∠ ♀
07 28 ☽ Q ♇
07 41 ☽ ⚻ ♅
09 11 ☽ bQ ♆
10 56 ☉ ± ♇
16 10 ☽ ⚻ ♄
17 31 ☽ Q ☿
18 49 ☽ ∠ ♃

12 Sunday
01 46 ♀ ☍ ♆
05 36 ☽ ✶ ♇
11 15 ☽ ⊥ ♃
12 42 ☽ △ ♆
13 50 ☽ ✶ ♀
18 49 ☽ □ ☉
22 02 ☽ ⊻ ♃

13 Monday
07 08 ☽ □ ☿
08 28 ☽ ∠ ♇
12 59 ☽ △ ♂
13 17 ☽ Q ♀
14 05 ☽ ☍ ♅
17 13 ☿ bQ ♇
22 10 ☽ ☍ ♄

14 Tuesday
00 09 ☽ ⊥ ♇
10 25 ☽ ⊻ ♇
13 30 ♂ ✶ ♅
14 29 ☉ △ ♃
16 01 ☽ bQ ♂
17 01 ☽ □ ♆
23 03 ☽ □ ♀

15 Wednesday
01 47 ☽ ☌ ♃
02 37 ☽ △ ☉
07 47 ☽ ± ♂
13 36 ☿ □ ♅
13 36 ♂ bQ ♆
16 56 ☽ ⚻ ♅
17 22 ☽ △ ☿
18 08 ☽ ⚻ ♂
22 50 ☽ Q ♆

16 Thursday
00 40 ☽ ⚻ ♄
02 44 ☽ ± ♅
03 03 ☿ ⊻ ♂
05 02 ☽ bQ ☉
06 57 ♀ □ ♃
10 26 ☽ ± ♄
11 58 ☽ ☌ ♇
14 16 ☽ # ♂
17 17 ☽ bQ ♅
18 12 ☽ ✶ ♆
20 33 ☽ ± ☉
20 51 ☽ bQ ☿

17 Friday
00 55 ☽ bQ ♄
02 36 ☽ ⊻ ♃
04 20 ☽ △ ♀
06 27 ☽ ∥ ♀
06 47 ☽ ⚻ ☉
12 08 ☽ ⊥ ♃
13 01 ☽ ± ☿
17 12 ☽ △ ♅
18 02 ☽ ∠ ♆
19 52 ♃ St D
20 24 ☽ ☍ ♂
23 43 ☽ ⚻ ☿

18 Saturday
00 49 ☽ △ ♄
01 14 ☽ # ♆
02 23 ☽ ∠ ♃
03 04 ☉ # ☿
06 10 ☽ bQ ♀
08 14 ☽ ⊥ ♆
09 49 ☿ □ ♄
11 36 ☽ ⊻ ♇
16 15 ☽ ∥ ☿
17 41 ☽ ⊻ ♆
19 35 ☽ # ☉
21 04 ☽ ⊥ ♇
21 37 ☽ ± ♀

19 Sunday
02 05 ☽ ✶ ♃
07 57 ☽ ⚻ ♀
08 16 ☽ ∥ ☉
09 30 ☽ ☍ ☉
11 18 ☽ ∠ ♇
16 19 ☽ # ☿
16 43 ☽ □ ♅
21 09 ☽ Q ♃
21 58 ☽ ⚻ ♂

20 Monday
00 31 ☽ □ ♄
03 38 ☽ ∥ ♆
05 13 ☽ ☍ ☿
08 01 ☽ ± ♂
11 16 ☽ ✶ ♇
12 57 ☉ bQ ♇
17 29 ☽ ☌ ♆
21 43 ☽ Q ♅
23 15 ☽ bQ ♂

21 Tuesday
00 06 ♀ # ♂
02 14 ☽ □ ♃
05 47 ☽ Q ♄
06 48 ☽ Q ♇
08 45 ☽ ∥ ♂
10 30 ☽ # ♀
12 44 ☽ ☍ ♀
13 25 ☽ ⚻ ☉
17 30 ☽ ✶ ♅
20 45 ☿ ⊥ ♂

22 Wednesday
00 10 ☽ ± ☉
01 14 ☽ △ ♂
01 49 ☽ ✶ ♄
11 54 ☿ △ ♇
12 57 ☽ □ ♇
13 05 ☽ ⚻ ☿
15 14 ♂ ✶ ♄
16 32 ☿ # ♆
16 36 ☽ bQ ☉
18 57 ☽ ∠ ♅
19 30 ☽ ⊻ ♆

23 Thursday
00 47 ☽ ± ☿
03 31 ☉ ⊻ ♀
03 36 ☽ ∠ ♄
04 58 ☽ △ ♃
05 54 ☽ ⊥ ♆
10 39 ☽ ⊥ ♅
18 44 ☽ bQ ☿
19 32 ☽ ⊥ ♄
20 55 ☽ △ ☉
21 15 ☽ ⚻ ♀
21 18 ☽ ⊻ ♅
21 48 ☽ ∠ ♆
21 48 ♀ △ ♅
22 27 ♂ bQ ♃

24 Friday
02 18 ☉ □ ♅
06 21 ☽ ⊻ ♄
07 43 ☽ bQ ♃
08 10 ☽ □ ♂
09 09 ☽ ± ♀
18 06 ☽ △ ♇
20 13 ☿ ☍ ♆

25 Saturday
00 15 ☽ ± ♃
01 05 ☽ ✶ ♆
01 46 ☽ △ ☿
03 24 ☽ bQ ♀
11 28 ☽ ⚻ ♃
19 17 ☽ Q ♂
22 08 ☽ bQ ♇
23 36 ☽ Q ♆

26 Sunday
04 57 ☽ ☌ ♅
09 12 ☽ □ ☉
10 51 ☽ △ ♀
14 47 ☽ ☌ ♄
15 23 ☽ ± ♇
19 33 ☽ ✶ ♂

27 Monday
03 02 ☽ ⚻ ♇
10 23 ☽ □ ♆
19 41 ☽ □ ☿
21 38 ☽ ☍ ♃
23 29 ☽ Q ☉

28 Tuesday
01 47 ☿ # ♂
02 39 ☽ ∠ ♂
04 18 ♀ △ ♄
11 27 ☿ □ ♃
15 54 ☽ ⊻ ♅
20 03 ☽ # ♀
21 38 ☽ ⊥ ♂

29 Wednesday
01 36 ☽ ✶ ☉
02 18 ☽ ⊻ ♄
03 57 ☽ ⊥ ♅
04 43 ☽ □ ♀
06 12 ♂ ± ♆
10 23 ☽ ⊻ ♂
10 28 ☉ □ ♄
12 57 ☽ Q ☿
14 26 ☽ ⊥ ♄
14 45 ☽ ☍ ♇
22 07 ☽ ∠ ♅
22 15 ☽ △ ♆

30 Thursday
03 13 ☽ # ☿
08 41 ☽ ∠ ♄
10 06 ☽ ⚻ ♃
10 38 ☽ ∠ ☉
16 45 ☽ ✶ ☿
17 30 ☽ ∥ ♂
21 17 ☽ Q ♀
22 19 ☽ ± ♃

OCTOBER 2032

Date	Sidereal time h m s	Sun ☉ ° ' "	Moon ☽ ° ' "	Moon ☽ 24.00 ° ' "	Mercury ☿ ° '	Venus ♀ ° '	Mars ♂ ° '	Jupiter ♃ ° '	Saturn ♄ ° '	Uranus ♅ ° '	Neptune ♆ ° '	Pluto ♇ ° '
01	12 43 06	08 ♎ 51 41	05 ♍ 20 24	11 ♍ 15 58	27 ♎ 02	10 ♏ 49	12 ♍ 08	22 ♑ 37	06 ♋ 54	01 ♋ 39	16 ♈ 40	12 ♒ 59
02	12 47 02	09 50 43	17 ♍ 12 11	23 ♍ 09 21	28 31	12 02	12 46	22 39	06 56	01 40	16 R 38	12 R 58
03	12 50 59	10 49 47	29 ♍ 07 43	05 ♎ 07 32	29 ♎ 58	13 15	13 24	22 42	06 58	01 40	16 37	12 58
04	12 54 56	11 48 54	11 ♎ 08 59	17 ♎ 12 19	01 ♏ 25	14 28	14 01	22 45	06 59	01 40	16 35	12 57
05	12 58 52	12 48 02	23 ♎ 17 42	29 ♎ 25 21	02 50	15 41	14 39	22 49	07 01	01 40	16 33	12 56
06	13 02 49	13 47 13	05 ♏ 35 27	11 ♏ 48 13	04 15	16 53	15 17	22 52	07 02	01 40	16 32	12 56
07	13 06 45	14 46 25	18 ♏ 03 51	24 ♏ 22 36	05 39	18 06	15 55	22 56	07 04	01 R 40	16 30	12 55
08	13 10 42	15 45 39	00 ♐ 44 39	07 ♐ 10 17	07 02	19 19	16 32	23 00	07 05	01 40	16 28	12 55
09	13 14 38	16 44 56	13 ♐ 39 42	20 ♐ 13 10	08 23	20 32	17 10	23 04	07 06	01 40	16 26	12 54
10	13 18 35	17 44 14	26 ♐ 50 54	03 ♑ 33 06	09 44	21 44	17 47	23 08	07 07	01 40	16 25	12 54
11	13 22 31	18 43 34	10 ♑ 19 57	17 ♑ 11 33	11 03	22 57	18 25	23 12	07 08	01 39	16 23	12 54
12	13 26 28	19 42 55	24 ♑ 07 58	01 ♒ 09 09	12 22	24 09	19 03	23 17	07 09	01 39	16 21	12 53
13	13 30 25	20 42 18	08 ♒ 15 01	15 ♒ 25 18	13 38	25 22	19 40	23 21	07 09	01 39	16 20	12 53
14	13 34 21	21 41 43	22 ♒ 39 39	29 ♒ 57 38	14 54	26 35	20 18	23 26	07 10	01 38	16 18	12 53
15	13 38 18	22 41 10	07 ♓ 18 36	14 ♓ 41 52	16 08	27 47	20 55	23 31	07 10	01 38	16 16	12 52
16	13 42 14	23 40 39	22 ♓ 06 36	29 ♓ 31 56	17 21	29 ♏ 00	21 33	23 37	07 11	01 38	16 15	12 52
17	13 46 11	24 40 09	06 ♈ 56 56	14 ♈ 20 36	18 32	00 ♐ 12	22 11	23 42	07 11	01 37	16 13	12 52
18	13 50 07	25 39 41	21 ♈ 42 03	29 ♈ 00 21	19 41	01 25	22 48	23 48	07 11	01 36	16 12	12 52
19	13 54 04	26 39 15	06 ♉ 14 42	13 ♉ 24 25	20 48	02 37	23 26	23 53	07 R 11	01 36	16 10	12 51
20	13 58 00	27 38 52	20 ♉ 28 54	27 ♉ 27 42	21 53	03 49	24 03	23 59	07 11	01 35	16 08	12 51
21	14 01 57	28 38 30	04 ♊ 20 31	11 ♊ 07 12	22 55	05 02	24 41	24 05	07 11	01 34	16 07	12 51
22	14 05 54	29 ♎ 38 11	17 ♊ 47 42	24 ♊ 22 07	23 55	06 14	25 18	24 12	07 10	01 34	16 05	12 51
23	14 09 50	00 ♏ 37 54	00 ♋ 50 40	07 ♋ 13 39	24 52	07 26	25 56	24 18	07 10	01 33	16 03	12 51
24	14 13 47	01 37 39	13 ♋ 31 27	19 ♋ 44 30	25 45	08 38	26 33	24 25	07 09	01 32	16 02	12 51
25	14 17 43	02 37 27	25 ♋ 53 20	01 ♌ 58 29	26 35	09 51	27 11	24 31	07 09	01 31	16 00	12 51
26	14 21 40	03 37 16	08 ♌ 00 30	13 ♌ 59 59	27 22	11 03	27 48	24 38	07 08	01 30	15 59	12 D 51
27	14 25 36	04 37 08	19 ♌ 57 31	25 ♌ 53 40	28 03	12 15	28 25	24 45	07 07	01 29	15 57	12 51
28	14 29 33	05 37 02	01 ♍ 49 01	07 ♍ 44 07	28 40	13 27	29 03	24 52	07 06	01 28	15 55	12 51
29	14 33 29	06 36 58	13 ♍ 39 31	19 ♍ 35 41	29 12	14 39	29 ♍ 40	25 00	07 05	01 27	15 54	12 51
30	14 37 26	07 36 57	25 ♍ 33 05	01 ♎ 32 11	29 38	15 51	00 ♎ 18	25 07	07 04	01 26	15 52	12 51
31	14 41 23	08 ♏ 36 57	07 ♎ 33 19	13 ♎ 36 52	29 ♏ 57	17 ♐ 03	00 ♎ 55	25 ♑ 15	07 ♋ 02	01 ♋ 24	15 ♈ 51	12 ♒ 51

Date	Moon True ☊ ° '	Moon Mean ☊ ° '	Moon ☽ Latitude ° '
01	00 ♏ 38	01 ♏ 37	04 S 19
02	00 R 34	01 34	03 36
03	00 31	01 30	02 44
04	00 29	01 27	01 45
05	00 28	01 24	00 S 40
06	00 D 28	01 21	00 N 28
07	00 29	01 18	01 36
08	00 30	01 14	02 40
09	00 31	01 11	03 37
10	00 32	01 08	04 24
11	00 33	01 05	04 58
12	00 R 33	01 02	05 15
13	00 33	00 59	05 14
14	00 32	00 55	04 54
15	00 31	00 52	04 14
16	00 30	00 49	03 17
17	00 30	00 46	02 07
18	00 29	00 43	00 N 49
19	00 D 29	00 40	00 S 32
20	00 29	00 36	01 49
21	00 30	00 33	02 57
22	00 30	00 30	03 54
23	00 R 30	00 27	04 36
24	00 30	00 24	05 04
25	00 30	00 20	05 17
26	00 D 30	00 17	05 15
27	00 30	00 14	05 00
28	00 30	00 11	04 32
29	00 31	00 08	03 52
30	00 32	00 05	03 02
31	00 ♏ 32	00 ♏ 01	02 S 04

DECLINATIONS

Date	Sun ☉ ° '	Moon ☽ ° '	Mercury ☿ ° '	Venus ♀ ° '	Mars ♂ ° '	Jupiter ♃ ° '	Saturn ♄ ° '	Uranus ♅ ° '	Neptune ♆ ° '	Pluto ♇ ° '
01	03 S 31	05 N 32	11 S 14	15 S 23	08 N 08	22 S 04	22 N 15	23 N 37	05 N 02	23 S 44
02	03 54	01 N 44	11 53	15 49	07 54	22 04	22 15	23 37	05 01	23 44
03	04 17	02 S 10	12 31	16 14	07 39	22 03	22 15	23 37	05 00	23 44
04	04 40	06 01	13 08	16 39	07 25	22 03	22 15	23 37	05 00	23 44
05	05 03	09 40	13 44	17 03	07 11	22 02	22 15	23 37	04 59	23 44
06	05 26	12 56	14 20	17 27	06 56	22 02	22 15	23 37	04 59	23 44
07	05 49	15 40	14 54	17 50	06 41	22 01	22 15	23 37	04 58	23 44
08	06 12	17 42	15 28	18 14	06 27	22 00	22 15	23 37	04 57	23 44
09	06 35	18 51	16 01	18 36	06 12	22 00	22 15	23 37	04 56	23 44
10	06 58	19 00	16 33	18 58	05 58	21 59	22 14	23 37	04 56	23 44
11	07 20	18 05	17 05	19 20	05 43	21 58	22 14	23 37	04 55	23 44
12	07 43	16 06	17 35	19 41	05 28	21 58	22 14	23 37	04 55	23 44
13	08 05	13 08	18 04	20 02	05 14	21 57	22 14	23 37	04 54	23 44
14	08 27	09 19	18 32	20 22	04 59	21 56	22 14	23 37	04 53	23 44
15	08 49	04 53	18 59	20 42	04 44	21 55	22 14	23 37	04 53	23 44
16	09 11	00 S 06	19 25	21 01	04 29	21 54	22 14	23 37	04 52	23 44
17	09 33	04 N 52	19 50	21 19	04 14	21 53	22 14	23 37	04 51	23 44
18	09 55	09 12	20 13	21 37	04 00	21 52	22 14	23 37	04 51	23 44
19	10 17	13 06	20 36	21 55	03 45	21 51	22 14	23 37	04 50	23 44
20	10 38	16 07	20 57	22 12	03 30	21 50	22 14	23 37	04 49	23 44
21	10 59	18 06	21 16	22 28	03 15	21 49	22 14	23 37	04 49	23 44
22	11 20	19 00	21 34	22 44	03 00	21 48	22 14	23 38	04 48	23 44
23	11 41	18 50	21 51	22 59	02 45	21 47	22 14	23 38	04 48	23 44
24	12 02	17 42	22 06	23 13	02 31	21 46	22 14	23 38	04 47	23 44
25	12 23	15 47	22 19	23 27	02 16	21 45	22 14	23 38	04 46	23 44
26	12 43	13 11	22 30	23 40	02 01	21 43	22 14	23 38	04 46	23 44
27	13 03	10 05	22 40	23 53	01 46	21 42	22 14	23 38	04 45	23 43
28	13 23	06 36	22 47	24 04	01 31	21 41	22 14	23 38	04 45	23 43
29	13 43	02 N 51	22 52	24 16	01 16	21 40	22 14	23 38	04 44	23 43
30	14 03	01 S 01	22 55	24 26	01 01	21 38	22 14	23 38	04 44	23 43
31	14 S 22	04 S 54	22 S 55	24 S 36	00 N 46	21 S 37	22 N 14	23 N 38	04 N 43	23 S 43

ZODIAC SIGN ENTRIES

Date	h m	Planets
01	01 11	☽ ♍
03	12 29	☿ ♏
03	13 45	☽ ♎
06	01 08	☽ ♏
08	10 36	☽ ♐
10	17 40	☽ ♑
12	22 02	☽ ♒
15	00 04	☽ ♓
17	00 45	☽ ♈
17	07 59	♀ ♐
19	01 38	☽ ♉
21	04 24	☽ ♊
22	20 46	☉ ♏
23	10 26	☽ ♋
25	20 06	☽ ♌
28	08 19	☽ ♍
30	00 38	♂ ♎
30	20 56	☽ ♎
31	17 33	☿ ♐

LATITUDES

Date	Mercury ☿ ° '	Venus ♀ ° '	Mars ♂ ° '	Jupiter ♃ ° '	Saturn ♄ ° '	Uranus ♅ ° '	Neptune ♆ ° '	Pluto ♇ ° '
01	00 S 53	00 S 20	01 N 13	00 S 32	01 S 00	00 N 12	01 S 39	07 S 07
04	01 15	00 30	01 14	00 32	01 00	00 12	01 39	07 07
07	01 36	00 39	01 14	00 32	01 00	00 12	01 39	07 07
10	01 56	00 48	01 14	00 32	01 00	00 12	01 39	07 06
13	02 14	00 58	01 14	00 32	01 00	00 12	01 39	07 06
16	02 31	01 07	01 14	00 32	01 00	00 12	01 39	07 06
19	02 45	01 16	01 14	00 33	01 00	00 12	01 39	07 05
22	02 55	01 24	01 14	00 33	01 00	00 12	01 39	07 05
25	03 01	01 33	01 14	00 33	01 00	00 12	01 39	07 05
28	03 00	01 41	01 14	00 33	01 00	00 12	01 39	07 04
31	02 S 51	01 S 48	01 N 14	00 S 33	01 S 00	00 N 12	01 S 39	07 S 04

DATA

Julian Date	2463507
Delta T	+75 seconds
Ayanamsa	24° 19' 01"
Synetic vernal point	04° ♓ 47' 57"
True obliquity of ecliptic	23° 25' 59"

MOON'S PHASES, APSIDES AND POSITIONS ☽

Date	h m	Phase	Longitude ° '	Eclipse Indicator
04	13 26	●	11 ♎ 52	
12	03 48	☽	19 ♑ 23	
18	18 58	○	25 ♈ 57	total
26	02 29	☾	03 ♌ 14	

Day	h m		
01	03 14	Apogee	
16	21 32	Perigee	
28	18 28	Apogee	
02	22 40	0S	
10	03 35	Max dec	19° S 04'
16	12 32	0N	
22	20 01	Max dec	19° N 03'
30	05 42	0S	

LONGITUDES

Date	Chiron ⚷ ° '	Ceres ⚳ ° '	Pallas ⚴ ° '	Juno ⚵ ° '	Vesta ⚶ ° '	Black Moon Lilith ⚸ ° '
01	25 ♉ 37	24 ♎ 18	25 ♍ 11	12 ♍ 01	15 ♏ 02	06 ♍ 04
11	25 ♉ 16	28 ♎ 43	29 ♍ 53	16 ♍ 13	20 ♏ 12	07 ♍ 11
21	24 ♉ 50	03 ♏ 09	04 ♎ 30	20 ♍ 15	25 ♏ 26	08 ♍ 18
31	24 ♉ 22	07 ♏ 34	09 ♎ 01	24 ♍ 08	00 ♐ 43	09 ♍ 25

All ephemeris data is given at 12.00 UT and the Moon's longitude is additionally given for 24.00 UT

ASPECTARIAN

h m	Aspects
01 Friday	
04 32	☽ ✶ ♅
04 34	☽ ⚼ ♆
06 31	☽ ⊥ ☉
07 40	♀ Q ♃
14 46	☿ ∠ ♂
15 10	☽ ✶ ♄
15 17	☽ ∥ ♆
16 37	☽ ⚼ ♃
19 47	☽ ⊻ ☉
22 45	☽ ± ♆
23 42	☽ ⧺ ☉
02 Saturday	
00 22	☽ ✶ ♀
02 32	☽ ☌ ♂
03 27	☽ ⊼ ♇
03 29	☽ ∠ ☿
04 50	☽ Q ♅
10 51	☽ ⊼ ♆
15 30	☽ Q ♄
15 34	☽ ± ♇
19 37	♂ ⊼ ♇
23 02	☽ △ ♃
03 Sunday	
00 12	☽ ⊥ ☿
06 17	♀ □ ♇
09 15	♄ ± ♇
09 39	☽ ⚼ ♇
10 02	☽ ∠ ♀
13 55	☽ ⊻ ☿
17 05	☽ □ ♅
17 57	♀ ✶ ♂
22 15	♂ Q ♅
04 Monday	
02 34	☽ ∥ ☉
03 42	☽ □ ♄
05 34	☽ ⧺ ♆
06 03	☽ ⊥ ♀
13 26	☽ ☌ ☉
15 34	☽ △ ♇
16 12	☿ △ ♅
18 01	☽ ⊻ ♂
19 18	☽ ⊻ ♀
20 27	☽ ⧺ ♂
22 44	☽ ☍ ♆
05 Tuesday	
06 31	☽ ⊥ ♂
07 42	☉ ⧺ ♆
11 03	☽ □ ♃
15 21	☉ △ ♇
06 Wednesday	
01 07	☽ ∠ ♂
04 23	☽ △ ♅
04 57	♀ ⊼ ♆
07 36	♀ ⚼ ♅
09 04	☽ ☌ ☿
14 49	☽ △ ♄
19 52	♅ St R
22 15	☽ Q ♃
07 Thursday	
02 09	☽ □ ♇
02 40	☽ ∥ ☿
05 10	☽ ⊻ ☉
07 40	☽ ✶ ♂
09 01	☽ ⊼ ♆
09 20	☽ ⚼ ♅
12 05	☽ ☌ ♀
17 36	☽ ⊥ ☉
19 37	☽ ⚼ ♄
20 25	☽ ± ♆
21 18	☽ ✶ ♃
08 Friday	
02 26	☽ ± ♅
07 38	☽ Q ♂
09 33	♂ ⊼ ♆
12 02	☽ ∠ ☉
12 19	☽ Q ♇
12 38	☽ ± ♄
12 59	☿ △ ♄
13 21	☽ ⚼ ♆
13 44	☽ ⊼ ♅
21 30	☉ ⧺ ♂
23 51	☽ ⊼ ♄
09 Saturday	
00 56	☽ ∥ ♀
01 07	☽ ⊻ ☿
01 36	☽ ∠ ♃
04 44	☉ ☍ ♆
10 37	☽ ✶ ♇
13 29	☽ ⊥ ☿
17 05	☽ △ ♆
18 08	☽ ✶ ☉
18 16	☽ ⊥ ♃
18 45	☽ □ ♂
10 Sunday	
01 50	☽ ⊻ ♀
05 15	☽ ⊻ ♃
07 45	☽ ∠ ☿
12 48	☽ ∥ ♀
13 45	☽ ⊥ ♀
13 53	☽ ∠ ♇
15 36	☉ ⊻ ♂
17 36	☽ Q ☉
19 38	♀ ⚼ ♄
20 38	☽ ☍ ♅
11 Monday	
01 06	♀ ± ♆
05 56	☽ ⊥ ♇
06 21	☽ ☍ ♄
07 23	☽ ∠ ♀
13 24	☽ ✶ ☿
14 52	☿ Q ♃
16 30	☽ ⊻ ♇
17 23	♀ ✶ ♃
22 34	☽ □ ♆
23 00	☽ ∥ ☿
12 Tuesday	
02 48	☽ △ ♂
03 48	☽ □ ☉
06 00	♂ ± ♇
10 31	☽ ☌ ♃
12 03	☽ ✶ ♀
12 26	☽ Q ☿
15 53	♂ Q ♄
21 48	☿ □ ♇
13 Wednesday	
00 51	☽ ⊼ ♅
05 24	☽ Q ♆
05 41	☽ ⚼ ♂
10 09	☽ ⊼ ♄
10 23	☽ Q ♀
10 59	☽ ± ♅
17 31	♀ ± ♅
19 46	☽ ☌ ♇
20 14	☽ ± ♄
21 30	☽ ± ♂
21 54	☽ □ ☿
14 Thursday	
01 29	☽ ✶ ♆
02 02	☽ ⚼ ♅
07 55	☽ ⊼ ♂
10 17	☽ △ ☉
11 11	☽ ⚼ ♄
13 17	☽ ⊻ ♃
16 32	☽ ∥ ☉
19 02	☽ □ ♀
21 23	♂ ∥ ♆
23 13	☽ ⊥ ♃
15 Friday	
02 10	☽ ∠ ♆
02 45	☽ △ ♅
11 47	☽ △ ♄
12 03	☽ ⧺ ♆
12 39	☽ ⚼ ☉
12 50	☽ ⧺ ♂
13 59	☽ ∠ ♃
14 39	☿ ⊼ ♆
16 49	☽ ⊥ ♆
21 02	☽ ⊻ ♇
21 45	☿ ⚼ ♅
16 Saturday	
02 32	☽ ⊻ ♆
03 37	☽ △ ☿
04 19	☽ ± ☉
06 45	☽ ⊥ ♇
10 14	☉ □ ♃
11 03	☽ ☍ ♂
14 27	☽ ✶ ♃
14 43	☽ ⊼ ☉
21 18	☽ ∠ ♇
17 Sunday	
00 07	☽ △ ♀
03 23	☽ □ ♅
05 59	☽ ⚼ ☿
09 46	☽ ∥ ♂
09 58	☽ Q ♃
12 23	☽ □ ♄
12 47	☽ ∥ ♆
21 35	☽ ✶ ♇
21 49	☽ ± ☿
18 Monday	
01 06	♀ Q ♇
02 35	☽ ⚼ ♀
03 01	☽ ☌ ♆
07 31	♀ ± ♄
07 47	♀ ⚼ ♆
08 25	☽ ⊼ ☿
08 35	☽ Q ♅
13 53	☽ ⊼ ♂
15 27	☽ □ ♃
15 55	♀ ⊼ ♅
16 27	☽ ⧺ ☉
17 11	☽ Q ♇
17 43	☽ Q ♄
18 38	☽ ± ♀
18 58	☽ ☍ ☉
22 26	♄ St R
19 Tuesday	
00 11	☽ ± ♂
04 17	☽ ✶ ♅
05 25	☽ ⊼ ♀
07 26	♀ ∥ ♃
13 34	☽ ✶ ♄
15 48	☽ ⚼ ♂
16 24	☉ ⊥ ♀
23 04	☽ □ ♇
20 Wednesday	
04 38	☽ ⊻ ♆
05 23	☽ ∠ ♅
09 04	♂ △ ♃
14 35	☽ ☍ ☿
14 50	☽ ⊥ ♆
14 54	☽ ∠ ♄
15 34	♀ ⧺ ♄
17 48	☿ ± ♆
18 03	☽ △ ♃
18 24	☽ △ ♂
18 56	☿ ⚼ ♄
20 45	☽ ⊥ ♅
21 Thursday	
01 17	☽ ⊼ ☉
06 21	☽ ∠ ♆
06 28	☽ ⊥ ♄
07 09	☽ ⊻ ♅
12 34	☽ ± ☉
13 19	☽ ☍ ♀
17 00	☽ ⊻ ♄
20 27	☽ ⚼ ♃
22 Friday	
03 06	☽ △ ♇
05 50	☽ ⚼ ☉
08 55	☽ ✶ ♆
12 44	☽ ± ♃
19 51	☿ ✶ ♃
23 46	☽ ⊼ ♃
23 Saturday	
00 03	☽ ⊼ ☿
02 25	☽ □ ♂
06 26	☽ ⚼ ♇
06 37	♀ ⊼ ♄
06 49	☽ Q ♆
06 50	☿ ∥ ♃
11 34	☽ △ ☉
12 02	☽ ± ☿
13 19	☽ ☌ ♅
23 17	☽ ± ♇
23 52	☽ ☌ ♄
24 Sunday	
01 42	☽ ⊼ ♀
05 57	☿ ± ♅
06 19	☽ ⚼ ☿
08 22	♀ Q ♂
09 43	☉ △ ♅
10 42	☽ ⊼ ♇
14 05	☽ Q ♂
14 22	☽ ± ♀
16 48	☽ □ ♆
25 Monday	
03 08	☿ ⧺ ♄
04 56	♀ ∠ ♃
09 18	☽ ☍ ♃
09 44	☽ ⚼ ♀
13 28	☽ △ ☿
14 40	☽ ✶ ♂
21 12	♇ St D
23 05	☽ ⊻ ♅
26 Tuesday	
02 29	☽ □ ☉
07 32	♀ ⧺ ♅
10 15	☽ ⊻ ♄
10 59	☽ ⊥ ♅
13 47	♂ ⚼ ♇
15 26	☽ ⧺ ☉
18 35	♀ ∥ ♇
18 45	☽ △ ♀
21 41	☽ ☍ ♇
22 07	☽ ∠ ♂
22 15	☽ ⊥ ♄
23 31	☉ ⊥ ♂
27 Wednesday	
03 56	☽ △ ♆
05 00	☽ ∠ ♅
16 21	☽ ∠ ♄
17 15	☽ ⊥ ♂
17 52	☽ Q ☉
21 47	☽ ⊼ ♃
28 Thursday	
00 00	♀ ✶ ♇
05 18	☽ □ ☿
06 05	☽ ⊻ ♂
10 04	☽ ± ♃
10 12	☽ ⚼ ♆
11 17	☽ ✶ ♅
20 25	☽ ✶ ☉
22 42	☽ ✶ ♄
29 Friday	
00 03	☽ ∥ ♆
04 24	☽ ± ♆
04 30	☽ ⚼ ♃
10 22	☽ ⊼ ♇
11 34	☽ Q ♅
14 14	☽ □ ♀
16 31	☽ ⊼ ♆
19 27	☽ Q ☿
22 30	☽ ± ♇
22 32	☽ ∥ ♂
22 55	☉ △ ♄
22 57	☽ Q ♄
30 Saturday	
05 33	☽ ∠ ☉
11 07	☽ △ ♃
12 00	☽ ⧺ ♂
12 27	♀ △ ♆
16 37	☽ ⚼ ♇
20 25	☽ ✶ ☿
22 02	☽ ☌ ♂
23 46	☽ □ ♅
31 Sunday	
01 16	☽ ⊥ ☉
06 28	☽ Q ♀
10 52	☽ ⧺ ♆
10 58	☽ □ ♄
14 18	☽ ⊻ ☉
22 30	☽ △ ♇

NOVEMBER 2032

LONGITUDES

Date	Sidereal time h m s	Sun ☉	Moon ☽	Moon ☽ 24.00	Mercury ☿	Venus ♀	Mars ♂	Jupiter ♃	Saturn ♄	Uranus ♅	Neptune ♆	Pluto ♇
		° ′ ″	° ′ ″	° ′ ″	° ′	° ′	° ′	° ′	° ′	° ′	° ′	° ′
01	14 45 19	09 ♏ 36 59	19 ♎ 43 08	25 ♎ 52 20	00 ♐ 08	18 ♐ 15	01 ♎ 32	25 ♑ 23	07 ♋ 01	01 ♋ 23	15 ♈ 49	12 ♒ 51
02	14 49 16	10 37 04	02 ♏ 04 41	08 ♏ 20 21	00 R 13	19 27	02 10	25 30	06 R 59	01 R 22	15 R 48	12 52
03	14 53 12	11 37 10	14 ♏ 39 27	21 ♏ 02 02	00 ♐ 08	20 39	02 47	25 39	06 58	01 20	15 46	12 52
04	14 57 09	12 37 18	27 ♏ 28 09	03 ♐ 57 47	29 ♏ 55	21 50	03 24	25 47	06 56	01 19	15 45	12 52
05	15 01 05	13 37 28	10 ♐ 30 55	17 ♐ 07 27	29 32	23 02	04 02	25 55	06 54	01 17	15 44	12 52
06	15 05 02	14 37 40	23 ♐ 47 21	00 ♑ 30 29	29 00	24 14	04 39	26 04	06 52	01 16	15 42	12 53
07	15 08 58	15 37 53	07 ♑ 16 46	14 ♑ 06 05	28 17	25 26	05 16	26 12	06 50	01 14	15 41	12 53
08	15 12 55	16 38 08	20 ♑ 58 17	27 ♑ 53 16	27 26	26 37	05 54	26 21	06 48	01 13	15 39	12 53
09	15 16 52	17 38 24	04 ♒ 50 52	11 ♒ 50 57	26 25	27 49	06 31	26 30	06 45	01 11	15 38	12 54
10	15 20 48	18 38 42	18 ♒ 53 20	25 ♒ 57 50	25 16	29 ♐ 00	07 08	26 39	06 43	01 10	15 37	12 54
11	15 24 45	19 39 01	03 ♓ 04 14	10 ♓ 12 15	24 02	00 ♑ 12	07 45	26 48	06 40	01 08	15 35	12 55
12	15 28 41	20 39 22	17 ♓ 21 37	24 ♓ 31 59	22 43	01 23	08 23	26 57	06 38	01 06	15 34	12 55
13	15 32 38	21 39 43	01 ♈ 42 57	08 ♈ 54 04	21 23	02 34	09 00	27 07	06 35	01 04	15 33	12 56
14	15 36 34	22 40 06	16 ♈ 04 51	23 ♈ 14 46	20 03	03 46	09 37	27 16	06 32	01 02	15 31	12 56
15	15 40 31	23 40 31	00 ♉ 23 17	07 ♉ 29 49	18 47	04 57	10 14	27 26	06 30	01 01	15 30	12 57
16	15 44 27	24 40 57	14 ♉ 33 47	21 ♉ 34 38	17 37	06 08	10 51	27 36	06 27	00 59	15 29	12 57
17	15 48 24	25 41 25	28 ♉ 31 52	05 ♊ 25 01	16 36	07 19	11 28	27 46	06 23	00 57	15 28	12 58
18	15 52 21	26 41 55	12 ♊ 13 41	18 ♊ 57 34	15 43	08 30	12 05	27 56	06 20	00 55	15 27	12 59
19	15 56 17	27 42 26	25 ♊ 36 27	02 ♋ 10 13	15 02	09 41	12 42	28 06	06 17	00 53	15 25	12 59
20	16 00 14	28 42 58	08 ♋ 38 51	15 ♋ 02 27	14 32	10 52	13 19	28 16	06 14	00 51	15 24	13 00
21	16 04 10	29 ♏ 43 33	21 ♋ 21 12	27 ♋ 35 22	14 14	12 02	13 57	28 26	06 10	00 49	15 23	13 01
22	16 08 07	00 ♐ 44 09	03 ♌ 45 19	09 ♌ 51 27	14 08	13 13	14 34	28 37	06 07	00 47	15 22	13 01
23	16 12 03	01 44 46	15 ♌ 54 16	21 ♌ 54 19	14 D 12	14 23	15 11	28 47	06 03	00 44	15 21	13 02
24	16 16 00	02 45 26	27 ♌ 52 10	03 ♍ 48 26	14 27	15 34	15 48	28 58	06 00	00 42	15 20	13 03
25	16 19 56	03 46 07	09 ♍ 43 43	15 ♍ 38 41	14 51	16 44	16 25	29 09	05 56	00 40	15 19	13 04
26	16 23 53	04 46 49	21 ♍ 33 57	27 ♍ 30 11	15 24	17 55	17 02	29 19	05 52	00 38	15 18	13 05
27	16 27 50	05 47 33	03 ♎ 27 58	09 ♎ 27 54	16 05	19 05	17 39	29 30	05 48	00 36	15 17	13 05
28	16 31 46	06 48 19	15 ♎ 30 33	21 ♎ 36 26	16 52	20 15	18 15	29 42	05 44	00 33	15 16	13 06
29	16 35 43	07 49 06	27 ♎ 45 59	03 ♏ 59 37	17 46	21 25	18 52	29 ♑ 53	05 40	00 31	15 15	13 07
30	16 39 39	08 ♐ 49 55	10 ♏ 17 40	16 ♏ 40 21	18 ♏ 44	22 ♑ 35	19 ♎ 29	00 ♒ 04	05 ♋ 36	00 ♋ 29	15 ♈ 15	13 ♒ 08

DECLINATIONS

Date	Moon True ☊	Moon Mean ☊	Moon ☽ Latitude	Sun ☉	Moon ☽	Mercury ☿	Venus ♀	Mars ♂	Jupiter ♃	Saturn ♄	Uranus ♅	Neptune ♆	Pluto ♇
	° ′	° ′	° ′	° ′	° ′	° ′	° ′	° ′	° ′	° ′	° ′	° ′	° ′
01	00 ♏ 33	29 ♎ 58	01 S 00	14 S 41	08 S 38	22 S 52	24 S 45	00 N 32	21 S 36	22 N 15	23 N 38	04 N 42	23 S 43
02	00 R 33	29 55	00 N 08	15 00	12 04	22 46	24 54	00 17	21 34	22 15	23 38	04 42	23 42
03	00 33	29 52	01 17	15 19	15 00	22 37	25 01	00 N 02	21 33	22 15	23 38	04 41	23 42
04	00 32	29 49	02 24	15 37	17 15	22 25	25 08	00 S 13	21 31	22 15	23 38	04 41	23 42
05	00 30	29 46	03 24	15 55	18 39	22 09	25 15	00 28	21 30	22 15	23 38	04 40	23 42
06	00 28	29 42	04 14	16 13	19 04	21 49	25 20	00 43	21 28	22 15	23 38	04 40	23 42
07	00 26	29 39	04 51	16 31	18 24	21 26	25 25	00 57	21 26	22 15	23 38	04 39	23 41
08	00 24	29 36	05 12	16 48	16 40	20 58	25 29	01 12	21 25	22 15	23 38	04 39	23 41
09	00 23	29 33	05 15	17 05	13 57	20 27	25 33	01 27	21 23	22 15	23 38	04 38	23 41
10	00 D 23	29 30	05 00	17 22	10 24	19 52	25 36	01 42	21 21	22 15	23 38	04 38	23 41
11	00 23	29 26	04 26	17 39	06 14	19 15	25 38	01 56	21 20	22 16	23 38	04 37	23 40
12	00 24	29 23	03 36	17 55	01 S 41	18 35	25 39	02 11	21 18	22 16	23 38	04 37	23 40
13	00 26	29 20	02 32	18 11	03 N 00	17 54	25 39	02 26	21 16	22 16	23 38	04 36	23 40
14	00 27	29 17	01 N 18	18 26	07 32	17 14	25 39	02 40	21 14	22 16	23 38	04 36	23 40
15	00 28	29 14	00 00	18 41	11 37	16 35	25 38	02 55	21 12	22 16	23 38	04 35	23 39
16	00 R 27	29 11	01 S 17	18 56	14 59	15 58	25 37	03 10	21 10	22 16	23 38	04 35	23 39
17	00 25	29 07	02 38	19 11	17 25	15 24	25 34	03 24	21 08	22 16	23 38	04 34	23 39
18	00 22	29 04	03 30	19 25	18 47	14 55	25 31	03 39	21 06	22 17	23 38	04 34	23 39
19	00 18	29 01	04 18	19 39	19 04	14 31	25 27	03 53	21 04	22 17	23 38	04 34	23 38
20	00 13	28 58	04 51	19 52	18 18	14 12	25 23	04 08	21 02	22 17	23 38	04 33	23 38
21	00 08	28 55	05 09	20 05	16 39	13 58	25 17	04 22	21 00	22 17	23 38	04 33	23 38
22	00 04	28 51	05 12	20 18	14 15	13 50	25 11	04 36	20 58	22 17	23 38	04 33	23 37
23	00 02	28 48	05 01	20 30	11 16	13 46	25 05	04 51	20 56	22 18	23 39	04 32	23 37
24	00 00	28 45	04 37	20 42	07 52	13 48	24 57	05 05	20 54	22 18	23 39	04 32	23 37
25	00 D 00	28 42	04 01	20 54	04 12	13 53	24 49	05 19	20 52	22 18	23 39	04 31	23 36
26	00 01	28 39	03 15	21 05	00 N 21	14 02	24 41	05 34	20 49	22 18	23 39	04 31	23 36
27	00 03	28 36	02 20	21 16	03 S 31	14 15	24 31	05 48	20 47	22 18	23 39	04 31	23 36
28	00 05	28 32	01 19	21 26	07 19	14 30	24 21	06 02	20 45	22 19	23 39	04 30	23 35
29	00 06	28 29	00 13	21 36	10 52	14 48	24 10	06 16	20 42	22 19	23 39	04 30	23 35
30	00 ♏ 05	28 ♎ 26	00 N 55	21 S 46	14 S 02	15 S 07	23 S 59	06 S 30	20 S 40	22 N 19	23 N 39	04 N 30	23 S 34

ZODIAC SIGN ENTRIES

Date	h m	Planets
02	08 00	☽ ♏
04	04 38	☿ ♏
04	16 41	☽ ♐
06	23 06	☽ ♑
09	03 39	☽ ♒
11	06 49	☽ ♓
11	08 04	♀ ♑
13	09 08	☽ ♈
15	11 21	☽ ♉
17	14 33	☽ ♊
19	20 01	☽ ♋
21	18 31	☉ ♐
22	04 41	☽ ♌
24	16 18	☽ ♍
27	05 02	☽ ♎
29	16 19	☽ ♏
30	03 32	♃ ♒

LATITUDES

Date	Mercury ☿	Venus ♀	Mars ♂	Jupiter ♃	Saturn ♄	Uranus ♅	Neptune ♆	Pluto ♇
	° ′	° ′	° ′	° ′	° ′	° ′	° ′	° ′
01	02 S 45	01 S 51	01 N 14	00 S 33	01 S 00	00 N 12	01 S 39	07 S 04
04	02 21	01 58	01 14	00 33	01 00	00 12	01 39	07 03
07	01 42	02 04	01 14	00 33	01 00	00 12	01 39	07 03
10	00 S 49	02 10	01 14	00 33	01 00	00 12	01 38	07 03
13	00 N 12	02 15	01 14	00 33	01 00	00 13	01 38	07 02
16	01 10	02 19	01 14	00 33	01 00	00 13	01 38	07 02
19	01 55	02 23	01 14	00 33	01 00	00 13	01 38	07 02
22	02 21	02 26	01 14	00 33	01 00	00 13	01 38	07 01
25	02 31	02 27	01 13	00 33	01 00	00 13	01 38	07 01
28	02 28	02 28	01 13	00 33	01 00	00 13	01 38	07 01
31	02 N 17	02 S 28	01 N 13	00 S 33	01 S 00	00 N 13	01 S 38	07 S 00

DATA

Julian Date	2463538
Delta T	+75 seconds
Ayanamsa	24° 19′ 04″
Synetic vernal point	04° ♓ 47′ 55″
True obliquity of ecliptic	23° 25′ 58″

MOON'S PHASES, APSIDES AND POSITIONS ☽

Date	h m	Phase	Longitude ° ′	Eclipse Indicator
03	05 45	●	11 ♏ 22	Partial
10	11 33	☽	18 ♒ 38	
17	06 42	○	25 ♉ 28	
24	22 48	☾	03 ♍ 13	

Day	h m		
13	15 34	Perigee	
25	14 17	Apogee	
06	09 12	Max dec	19° S 04′
12	20 36	0N	
19	06 08	Max dec	19° N 06′
26	14 13	0S	

LONGITUDES

Date	Chiron ⚷	Ceres ⚳	Pallas ⚴	Juno ⚵	Vesta ⚶	Black Moon Lilith ⚸
	° ′	° ′	° ′	° ′	° ′	° ′
01	24 ♉ 19	08 ♏ 00	09 ♎ 28	24 ♍ 31	01 ♐ 15	09 ♍ 32
11	23 ♉ 48	12 ♏ 25	13 ♎ 52	28 ♍ 10	06 ♐ 36	10 ♍ 39
21	23 ♉ 17	16 ♏ 47	18 ♎ 08	01 ♎ 36	11 ♐ 59	11 ♍ 46
31	22 ♉ 46	21 ♏ 06	22 ♎ 16	04 ♎ 46	17 ♐ 23	12 ♍ 52

All ephemeris data is given at 12.00 UT and the Moon's longitude is additionally given for 24.00 UT

ASPECTARIAN

h m	Aspects
01 Monday	
02 53	☽ ∠ ☿
04 22	☽ ☍ ♆
06 09	♂ □ ♅
08 48	☽ ✶ ♀
20 43	☽ ⊥ ☿
23 09	☽ □ ♃
02 Tuesday	
08 24	☽ ⊻ ☿
10 37	☽ △ ♅
11 56	☿ St R
12 10	☽ ⊻ ♂
13 23	♀ ⊥ ♃
17 02	☽ ∠ ♀
21 24	☽ △ ♄
03 Wednesday	
00 16	☽ ⊥ ♂
05 45	☽ ☌ ☉
08 36	☽ □ ♇
10 04	☽ Q ♃
11 58	☽ ⊥ ♀
14 06	☽ ⊼ ♆
15 10	☽ ⅊ ♅
15 21	☽ ∥ ☉
18 12	☽ ∠ ♂
04 Thursday	
00 26	☽ ⊻ ♀
01 22	☽ ± ♆
01 42	☽ ⅊ ♄
08 00	☽ ± ♅
08 50	☽ ✶ ♃
16 26	☽ ☌ ☿
17 55	☉ □ ♇
18 04	☽ ⅊ ♆
18 18	☽ Q ♇
18 24	☽ ± ♄
19 06	☽ ⊼ ♅
23 32	☽ ✶ ♂
05 Friday	
04 07	☿ ⋕ ♄
05 24	☽ ⊼ ♄
12 45	☽ ∠ ♃
16 18	☽ ✶ ♇
18 07	☽ ⊻ ☉
20 11	☉ Q ♃
21 27	☽ △ ♆
22 31	☽ Q ♂
06 Saturday	
04 07	♄ ± ♇
05 14	☽ ⊥ ♃
05 51	☽ ⊥ ☉
12 52	☽ ☌ ♀
16 07	☽ ⊻ ♃
19 19	☽ ∠ ♇
20 52	☽ ⊻ ☿
23 16	☽ ∠ ☉
07 Sunday	
01 19	☽ ☍ ♅
07 00	☽ ⊥ ☿
08 17	☽ □ ♂
11 13	☽ ☍ ♄
11 18	☽ ⊥ ♇
11 26	☿ ∥ ♃
13 05	☉ ⊼ ♆
21 52	☽ ⊻ ♇
21 59	☽ ∠ ☿
08 Monday	
02 11	☉ ⅊ ♅
02 44	☽ □ ♆
03 50	☽ ✶ ☉
05 47	♀ ⊻ ♃
10 39	☽ ∥ ☉
20 57	☿ ⊻ ♀
21 26	☽ ☌ ♃
22 28	☽ ✶ ☿
22 44	☽ ⊻ ♀
09 Tuesday	
02 20	☽ Q ☉
05 43	☽ ⊼ ♅
09 55	☽ Q ♆
10 03	☽ ⊥ ♀
10 22	☿ ✶ ♃
13 41	♀ ∠ ♇
15 00	☽ △ ♂
15 16	☽ ⊼ ♄
16 01	☽ ± ♅
17 41	☽ Q ☿
20 46	♂ □ ♄
10 Wednesday	
01 31	☽ ± ♄
01 48	☽ ☌ ♇
02 55	☽ ∠ ♀
06 26	☽ ✶ ♆
07 22	☽ ⅊ ♅
11 33	☽ □ ☉
14 21	☿ ± ♅
16 47	☽ ⅊ ♄
17 46	☽ ⅊ ♂
21 59	☽ □ ☿
11 Thursday	
01 18	☽ ⊻ ♃
06 42	☽ ✶ ♀
07 49	☽ ∠ ♆
08 44	☽ △ ♅
09 41	☽ ± ♂
10 25	☿ ⊥ ♀
11 32	☽ ⊥ ♃
18 03	☽ △ ♄
20 15	☽ ⊼ ♂
20 41	☽ ⋕ ♆
22 57	☽ ⊥ ♆
12 Friday	
02 50	☽ ∠ ♃
03 56	☿ ∠ ♂
04 33	☽ ⊻ ♇
04 44	☽ Q ♀
06 26	♀ ☍ ♅
09 00	☽ ⊻ ♆
09 31	☽ ∥ ♂
14 37	☽ ⊥ ♇
17 56	☽ △ ☉
20 12	☽ △ ☿
13 Saturday	
04 13	☽ ✶ ♃
05 11	☉ ∥ ☿
05 40	☽ ∠ ♇
08 13	☿ ⅊ ♄
08 52	☽ ⋕ ♂
09 03	☿ ± ♆
09 08	☉ ☌ ☿
09 15	☉ ± ♆
10 16	☉ ⅊ ♄
10 56	☽ □ ♅
13 34	☽ □ ♀
19 08	☽ ⅊ ☿
20 06	☽ □ ♄
20 17	☽ ∥ ♆
20 53	☽ ⅊ ☉
14 Sunday	
00 29	☽ Q ♃
00 42	☽ ☍ ♂
06 44	☽ ✶ ♇
08 54	☽ ± ☿
11 04	☽ ☌ ♆
13 03	☽ ± ☉
16 57	☽ Q ♅
18 06	☽ ⊼ ☿
23 52	☽ ⊼ ☉
15 Monday	
00 36	☿ ∠ ♀
02 07	☽ Q ♄
02 51	☽ Q ♇
06 58	☽ □ ♃
13 03	☽ ✶ ♅
20 23	☽ △ ♀
22 16	☽ ✶ ♄
16 Tuesday	
05 24	☽ ⊼ ♂
09 16	☽ □ ♇
13 34	☽ ⊻ ♆
14 25	☽ ∠ ♅
16 05	☽ ± ♂
16 51	☽ ☍ ☿
18 02	♀ ☍ ♄
18 50	☉ ± ♅
19 00	☽ ⋕ ☿
22 58	☿ ⊥ ♂
23 44	☽ ∠ ♄
23 49	☽ ⊥ ♆
17 Wednesday	
00 16	☽ ⅊ ♀
04 50	♀ ⊥ ♇
05 49	☽ ⊥ ♅
06 42	☽ ☍ ☉
08 16	☽ ⅊ ♂
10 39	☽ △ ♃
15 13	☽ ⊥ ♄
15 21	☽ ∠ ♆
16 11	☽ ⊻ ♅
17 17	☽ ± ♀
18 Thursday	
01 39	☽ ⊻ ♄
04 47	☽ ⊼ ♀
06 03	☿ ⅊ ♅
06 59	☿ Q ♃
11 45	☽ △ ♂
13 15	☽ ⅊ ♃
13 20	☽ △ ♇
17 42	☽ ✶ ♆
17 53	☽ ⊼ ☿
21 14	☿ ⊼ ♆
19 Friday	
02 15	♅ ⋕ ♇
04 06	☽ ± ☿
05 33	☽ ± ♃
12 00	☉ ∠ ♂
15 18	☽ Q ♆
16 08	☽ ⊼ ☉
16 20	☽ ⅊ ♇
16 35	☽ ⊼ ♃
19 45	☽ ⅊ ☿
21 36	☽ ☌ ♅
23 03	☉ ✶ ♃
23 04	♂ △ ♇
20 Saturday	
04 05	☽ ± ☉
07 31	☽ ☌ ♄
08 56	☽ ± ♇
16 33	☽ ☍ ♀
20 09	☽ ⊼ ♇
21 13	☽ □ ♂
22 19	☽ ⅊ ☉
22 45	☽ △ ☿
21 Sunday	
00 40	☽ □ ♆
21 02	☿ ⊻ ♂
22 03	☉ ± ♄
22 Monday	
01 50	☽ ☍ ♃
03 27	☉ ⅊ ♆
05 35	☽ △ ☉
05 38	♂ ⋕ ♆
06 12	☽ ⊻ ♅
08 02	♀ ⊻ ♇
09 32	☽ Q ♂
12 57	☉ ⊼ ♅
14 02	☿ St D
15 42	☽ ⋕ ☿
16 36	☽ ⊻ ♄
17 55	☽ ⊥ ♅
18 54	☉ Q ♇
23 Tuesday	
04 23	☽ ⊥ ♄
06 17	☽ ☍ ♇
07 39	☿ ✶ ♀
08 34	☽ □ ☿
08 40	☽ ⊼ ♀
10 28	☽ ✶ ♂
10 54	☽ △ ♆
11 41	☽ ∠ ♅
18 38	♂ ☍ ♆
21 56	☽ ± ♀
22 15	☽ ∠ ♄
24 Wednesday	
07 22	♀ □ ♆
14 15	☽ ⊼ ♃
16 58	☽ ⅊ ♆
17 43	☽ ✶ ♅
18 02	☽ ⅊ ♀
18 14	☽ ∠ ♂
21 33	☽ Q ☿
21 51	♀ □ ♂
22 48	☽ □ ☉
25 Thursday	
02 34	☽ ± ♃
04 21	☽ ✶ ♄
05 14	☽ ⋕ ♂
07 57	☉ ∥ ♃
09 55	☽ ∥ ♆
11 10	☽ ± ♆
13 27	☽ ⊥ ♂
17 57	☽ Q ♅
18 46	☽ ⊼ ♇
21 06	☽ ⅊ ♃
22 52	☽ ✶ ☿
23 19	☽ ⊼ ♆
26 Friday	
02 18	☽ ⊻ ♂
03 47	☽ △ ♀
04 33	☽ Q ♄
06 57	☽ ± ♇
08 08	☿ ⊼ ♆
14 41	☽ Q ☉
20 11	☿ ⅊ ♅
27 Saturday	
01 10	☽ ⅊ ♇
03 55	☽ △ ♃
06 15	☽ □ ♅
06 54	☽ ∠ ☿
12 14	☉ ⊼ ♄
16 40	☽ □ ♄
17 06	☽ ✶ ☉
18 10	☽ ⋕ ♆
28 Sunday	
02 07	☽ ⊥ ☿
03 13	☽ ∥ ♂
07 14	☽ △ ♇
11 32	☽ ☍ ♆
14 53	☽ ⊻ ☿
17 43	☽ ☌ ♂
22 19	☽ □ ♀
29 Monday	
01 30	☽ ∠ ☉
15 48	☿ Q ♃
16 09	☽ □ ♃
17 18	☽ △ ♅
20 31	☽ ⊥ ☉
30 Tuesday	
03 07	☽ △ ♄
08 59	☽ ⊻ ☉
12 36	☽ Q ♀
17 22	☽ □ ♇
21 19	☽ ⊼ ♆
21 44	☽ ⅊ ♅
22 57	☽ ∥ ☿

DECEMBER 2032

LONGITUDES

Date	Sidereal time h m s	Sun ☉ ° ' "	Moon ☽ ° ' "	Moon ☽ 24.00 ° ' "	Mercury ☿ ° '	Venus ♀ ° '	Mars ♂ ° '	Jupiter ♃ ° '	Saturn ♄ ° '	Uranus ♅ ° '	Neptune ♆ ° '	Pluto ♇ ° '
01	16 43 36	09 ♐ 50 45	23 ♏ 07 49	29 ♏ 40 08	19 ♏ 48	23 ♑ 45	20 ♎ 06	00 ♒ 15	05 ♋ 32	00 ♋ 26	15 ♈ 14	13 ♒ 09
02	16 47 32	10 51 36	06 ♐ 17 15	12 ♐ 59 01	20 55	24 54	20 43	00 27	05 R 28	00 R 24	15 R 13	13 10
03	16 51 29	11 52 29	19 ♐ 45 12	26 ♐ 35 28	22 06	26 04	21 20	00 38	05 23	00 22	15 12	13 11
04	16 55 25	12 53 22	03 ♑ 29 25	10 ♑ 26 34	23 19	27 13	21 57	00 50	05 19	00 19	15 12	13 12
05	16 59 22	13 54 17	17 ♑ 26 26	24 ♑ 28 29	24 36	28 23	22 33	01 02	05 15	00 17	15 11	13 13
06	17 03 19	14 55 13	01 ♒ 32 11	08 ♒ 37 02	25 54	29 ♑ 32	23 10	01 14	05 10	00 14	15 10	13 14
07	17 07 15	15 56 09	15 ♒ 42 34	22 ♒ 48 21	27 14	00 ♒ 41	23 47	01 26	05 06	00 12	15 10	13 16
08	17 11 12	16 57 06	29 ♒ 54 02	06 ♓ 59 18	28 ♏ 36	01 50	24 24	01 38	05 01	00 09	15 09	13 17
09	17 15 08	17 58 03	14 ♓ 03 55	21 ♓ 07 40	00 ♐ 00	02 59	25 00	01 50	04 56	00 07	15 08	13 18
10	17 19 05	18 59 01	28 ♓ 10 24	05 ♈ 12 01	01 24	04 07	25 37	02 02	04 52	00 04	15 08	13 19
11	17 23 01	20 00 00	12 ♈ 12 24	19 ♈ 11 26	02 50	05 16	26 14	02 14	04 47	00 ♋ 02	15 07	13 20
12	17 26 58	21 00 59	26 ♈ 09 02	03 ♉ 05 03	04 16	06 24	26 50	02 26	04 42	29 ♊ 59	15 07	13 21
13	17 30 54	22 01 59	09 ♉ 59 20	16 ♉ 51 42	05 44	07 32	27 27	02 39	04 38	29 57	15 06	13 23
14	17 34 51	23 03 00	23 ♉ 41 56	00 ♊ 29 47	07 12	08 40	28 03	02 51	04 33	29 54	15 06	13 24
15	17 38 48	24 04 01	07 ♊ 15 00	13 ♊ 57 18	08 40	09 48	28 40	03 04	04 28	29 52	15 06	13 25
16	17 42 44	25 05 03	20 ♊ 36 25	27 ♊ 12 07	10 09	10 55	29 16	03 16	04 23	29 49	15 05	13 27
17	17 46 41	26 06 05	03 ♋ 44 11	10 ♋ 12 25	11 39	12 03	29 ♎ 53	03 29	04 18	29 47	15 05	13 28
18	17 50 37	27 07 08	16 ♋ 36 44	22 ♋ 57 03	13 09	13 10	00 ♏ 29	03 42	04 13	29 44	15 05	13 29
19	17 54 34	28 08 12	29 ♋ 13 26	05 ♌ 25 58	14 39	14 17	01 06	03 54	04 09	29 42	15 04	13 31
20	17 58 30	29 ♐ 09 16	11 ♌ 34 49	17 ♌ 40 17	16 09	15 24	01 42	04 07	04 04	29 39	15 04	13 32
21	18 02 27	00 ♑ 10 22	23 ♌ 42 41	29 ♌ 42 25	17 40	16 30	02 18	04 20	03 59	29 36	15 04	13 33
22	18 06 23	01 11 27	05 ♍ 39 58	11 ♍ 35 52	19 12	17 37	02 55	04 33	03 54	29 34	15 04	13 35
23	18 10 20	02 12 34	17 ♍ 30 40	23 ♍ 25 02	20 43	18 43	03 31	04 46	03 49	29 31	15 04	13 36
24	18 14 17	03 13 41	29 ♍ 19 35	05 ♎ 15 00	22 15	19 49	04 07	04 59	03 44	29 29	15 04	13 38
25	18 18 13	04 14 49	11 ♎ 11 58	17 ♎ 11 11	23 47	20 54	04 44	05 13	03 39	29 26	15 04	13 39
26	18 22 10	05 15 57	23 ♎ 13 21	29 ♎ 19 06	25 19	22 00	05 20	05 26	03 34	29 24	15 D 04	13 41
27	18 26 06	06 17 06	05 ♏ 29 05	11 ♏ 43 52	26 52	23 05	05 56	05 39	03 29	29 21	15 04	13 42
28	18 30 03	07 18 16	18 ♏ 03 57	24 ♏ 29 46	28 25	24 10	06 32	05 52	03 24	29 18	15 04	13 44
29	18 33 59	08 19 26	01 ♐ 01 38	07 ♐ 39 44	29 ♐ 58	25 14	07 08	06 06	03 19	29 16	15 04	13 45
30	18 37 56	09 20 37	14 ♐ 24 07	21 ♐ 14 41	01 ♑ 31	26 19	07 44	06 19	03 14	29 13	15 04	13 47
31	18 41 52	10 ♑ 21 47	28 ♐ 11 09	05 ♑ 13 07	03 ♑ 05	27 ♒ 23	08 ♏ 20	06 ♒ 33	03 ♋ 09	29 ♊ 11	15 ♈ 04	13 ♒ 48

Date	Moon True ☊ ° '	Moon Mean ☊ ° '	Moon ☽ Latitude ° '
01	00 ♏ 03	28 ♎ 23	02 N 02
02	29 ♎ 58	28 20	03 04
03	29 52	28 17	03 57
04	29 45	28 13	04 37
05	29 37	28 10	05 01
06	29 31	28 07	05 08
07	29 25	28 04	04 56
08	29 22	28 01	04 26
09	29 20	27 57	03 40
10	29 D 20	27 54	02 40
11	29 22	27 51	01 31
12	29 22	27 48	00 N 17
13	29 R 22	27 45	00 S 57
14	29 19	27 42	02 07
15	29 14	27 38	03 09
16	29 06	27 35	03 59
17	28 56	27 32	04 36
18	28 45	27 29	04 57
19	28 34	27 26	05 04
20	28 24	27 23	04 56
21	28 16	27 19	04 35
22	28 10	27 16	04 02
23	28 06	27 13	03 19
24	28 05	27 10	02 27
25	28 D 05	27 07	01 29
26	28 06	27 03	00 S 26
27	28 R 05	27 00	00 N 39
28	28 04	26 57	01 44
29	27 59	26 54	02 46
30	27 52	26 51	03 40
31	27 ♎ 42	26 ♎ 48	04 N 23

DECLINATIONS

Date	Sun ☉ ° '	Moon ☽ ° '	Mercury ☿ ° '	Venus ♀ ° '	Mars ♂ ° '	Jupiter ♃ ° '	Saturn ♄ ° '	Uranus ♅ ° '	Neptune ♆ ° '	Pluto ♇ ° '
01	21 S 55	16 S 35	15 S 29	23 S 47	06 S 44	20 S 38	22 N 19	23 N 39	04 N 30	23 S 34
02	22 04	18 20	15 51	23 34	06 58	20 35	22 19	23 39	04 29	23 34
03	22 12	19 06	16 15	23 21	07 12	20 33	22 20	23 39	04 29	23 33
04	22 20	18 46	16 39	23 07	07 26	20 30	22 20	23 39	04 29	23 33
05	22 28	17 19	17 04	22 52	07 39	20 28	22 20	23 39	04 29	23 32
06	22 35	14 48	17 29	22 37	07 53	20 25	22 20	23 39	04 28	23 32
07	22 41	11 24	17 54	22 21	08 07	20 23	22 21	23 39	04 28	23 32
08	22 48	07 21	18 19	22 05	08 20	20 20	22 21	23 39	04 28	23 31
09	22 53	02 S 53	18 44	21 48	08 34	20 17	22 21	23 39	04 28	23 31
10	22 59	01 N 43	19 08	21 31	08 47	20 15	22 21	23 39	04 28	23 30
11	23 03	06 13	19 32	21 13	09 01	20 12	22 21	23 39	04 28	23 30
12	23 08	10 22	19 56	20 54	09 14	20 09	22 22	23 39	04 27	23 29
13	23 12	13 54	20 19	20 35	09 27	20 06	22 22	23 39	04 27	23 29
14	23 15	16 39	20 41	20 16	09 40	20 03	22 22	23 39	04 27	23 29
15	23 18	18 25	21 03	19 56	09 54	20 01	22 23	23 39	04 27	23 28
16	23 20	19 08	21 24	19 35	10 07	19 58	22 23	23 39	04 27	23 28
17	23 22	18 47	21 44	19 15	10 20	19 55	22 23	23 39	04 27	23 27
18	23 23	17 29	22 03	18 53	10 33	19 52	22 23	23 39	04 27	23 27
19	23 25	15 21	22 21	18 31	10 45	19 49	22 24	23 39	04 27	23 26
20	23 26	12 33	22 38	18 09	10 58	19 46	22 24	23 39	04 27	23 26
21	23 26	09 17	22 54	17 47	11 11	19 43	22 24	23 39	04 27	23 25
22	23 26	05 41	23 09	17 24	11 23	19 40	22 24	23 39	04 27	23 25
23	23 25	01 N 53	23 23	17 00	11 36	19 37	22 24	23 39	04 27	23 24
24	23 24	01 S 59	23 35	16 37	11 48	19 34	22 25	23 39	04 27	23 24
25	23 22	05 48	23 47	16 13	12 01	19 30	22 25	23 39	04 27	23 23
26	23 20	09 25	23 57	15 48	12 13	19 27	22 25	23 39	04 27	23 23
27	23 17	12 44	24 06	15 23	12 25	19 24	22 25	23 39	04 27	23 22
28	23 14	15 32	24 14	14 58	12 37	19 21	22 26	23 39	04 27	23 22
29	23 10	17 39	24 21	14 33	12 49	19 17	22 26	23 39	04 27	23 21
30	23 06	18 53	24 26	14 07	13 01	19 14	22 26	23 39	04 27	23 21
31	23 S 02	19 S 03	24 S 30	13 S 42	13 S 13	19 S 11	22 N 26	23 N 39	04 N 27	23 S 20

ZODIAC SIGN ENTRIES

Date	h m	Planets
02	00 36	☽ ♐
04	05 57	☽ ♑
06	09 24	☽ ♒
06	21 48	♀ ♒
08	12 10	☽ ♓
09	12 05	☿ ♐
10	15 07	☽ ♈
12	06 22	♅ ♊
12	18 39	☽ ♉
14	23 07	☽ ♊
17	05 07	☽ ♋
17	16 47	♂ ♏
19	13 30	☽ ♌
21	07 56	☉ ♑
22	00 35	☽ ♍
24	13 22	☽ ♎
27	01 20	☽ ♏
29	10 08	☽ ♐
29	12 34	☿ ♑
31	15 07	☽ ♑

LATITUDES

Date	Mercury ☿ ° '	Venus ♀ ° '	Mars ♂ ° '	Jupiter ♃ ° '	Saturn ♄ ° '	Uranus ♅ ° '	Neptune ♆ ° '	Pluto ♇ ° '
01	02 N 17	02 S 28	01 N 13	00 S 33	01 S 00	00 N 13	01 S 38	07 S 00
04	02 01	02 27	01 12	00 33	01 00	00 13	01 38	07 00
07	01 41	02 25	01 12	00 33	00 59	00 13	01 37	07 00
10	01 20	02 22	01 12	00 33	00 59	00 13	01 37	06 59
13	00 57	02 17	01 11	00 33	00 59	00 13	01 37	06 59
16	00 35	02 11	01 11	00 33	00 59	00 13	01 37	06 59
19	00 N 12	02 04	01 10	00 34	00 59	00 13	01 37	06 59
22	00 S 09	01 56	01 09	00 34	00 58	00 13	01 37	06 58
25	00 30	01 47	01 09	00 34	00 58	00 13	01 37	06 58
28	00 49	01 36	01 08	00 34	00 58	00 13	01 36	06 58
31	01 S 07	01 S 24	01 N 07	00 S 34	00 S 57	00 N 13	01 S 36	06 S 58

DATA

Julian Date	2463568
Delta T	+75 seconds
Ayanamsa	24° 19' 08"
Synetic vernal point	04° ♓ 47' 51"
True obliquity of ecliptic	23° 25' 57"

MOON'S PHASES, APSIDES AND POSITIONS ☽

Date	h m	Phase	Longitude ° '	Eclipse Indicator
02	20 53	●	11 ♐ 14	
09	19 08	☽	18 ♓ 16	
16	20 49	○	25 ♊ 27	
24	20 39	☾	03 ♎ 36	

Day	h m		
08	19 09	Perigee	
23	11 31	Apogee	
03	16 58	Max dec	19° S 08'
10	03 03	0N	
16	16 05	Max dec	19° N 08'
23	23 41	0S	
31	03 24	Max dec	19° S 07'

LONGITUDES

Date	Chiron ⚷ ° '	Ceres ⚳ ° '	Pallas ⚴ ° '	Juno ⚵ ° '	Vesta ⚶ ° '	Black Moon Lilith ⚸ ° '
01	22 ♉ 46	21 ♏ 06	22 ♎ 16	04 ♎ 46	17 ♐ 23	12 ♍ 52
11	22 ♉ 18	25 ♏ 21	26 ♎ 14	07 ♎ 38	22 ♐ 48	13 ♍ 59
21	21 ♉ 53	29 ♏ 32	00 ♏ 00	10 ♎ 09	28 ♐ 12	15 ♍ 06
31	21 ♉ 33	03 ♐ 37	03 ♏ 32	12 ♎ 15	03 ♑ 37	16 ♍ 13

All ephemeris data is given at 12.00 UT and the Moon's longitude is additionally given for 24.00 UT

ASPECTARIAN

01 Wednesday
02 49 ☽ Q ♃
05 15 ☽ ☌ ☿
06 07 ☽ ⊻ ♂
07 13 ☽ bQ ♄
08 29 ☽ ± ♆
13 14 ☽ ⚹ ♀
14 25 ☽ ± ♅
17 45 ☽ ⊥ ♂
23 41 ☽ ± ♄

02 Thursday
01 01 ☽ bQ ♆
01 16 ☽ ⚹ ♃
01 22 ☽ ⊼ ♅
02 43 ☽ Q ♇
03 01 ☿ ⊻ ♂
03 02 ☿ bQ ♄
03 19 ♀ # ♅
07 12 ♃ ⊼ ♅
10 31 ☽ ⊼ ♄
10 55 ☽ ∠ ♂
12 45 ♀ ∥ ♇
18 12 ☿ ± ♆
19 07 ☽ ∠ ♀
20 53 ☽ ☌ ☉

03 Friday
00 21 ☽ ⚹ ♇
03 57 ☽ △ ♆
04 37 ☽ ∠ ♃
12 36 ☽ ⊥ ♀
14 55 ☽ ⚹ ♂
16 31 ☽ ⊻ ☿
20 43 ☽ ⊥ ♃

04 Saturday
00 06 ☽ ⊻ ♀
02 48 ☽ ∠ ♇
04 03 ☽ ⊥ ☿
06 31 ☽ ☍ ♅
07 20 ☽ ⊻ ♃
10 56 ☉ # ♄
12 49 ☽ Q ♂
15 09 ☽ ☍ ♄
18 26 ☽ ⊥ ♇
19 34 ☉ ⚹ ♇
21 11 ☽ ∠ ☿

05 Sunday
04 46 ☽ ⊻ ♇
05 28 ☽ ⊻ ☉
06 22 ☿ ± ♅
08 08 ☽ □ ♆
14 27 ☽ ∥ ☿
16 32 ☽ ⊥ ☉
21 08 ☽ □ ♂

06 Monday
01 26 ☽ ⚹ ☿
08 17 ☽ ☌ ♀
09 03 ☽ ∠ ☉
09 48 ☽ ⊼ ♅
11 28 ☽ ☌ ♃
14 24 ☉ ∥ ♀
14 46 ☽ Q ♆
17 49 ☉ △ ♆
18 08 ☽ ⊼ ♄
19 57 ☽ ± ♅
23 54 ☽ Q ☿

07 Tuesday
02 18 ♀ ⊼ ♅
04 14 ☽ ± ♄
07 51 ☽ ☌ ♇
11 04 ☽ ⚹ ♆
11 08 ☽ bQ ♅
12 25 ☽ ⚹ ☉
12 59 ♀ # ♄
19 23 ☽ bQ ♄

08 Wednesday
02 16 ☽ △ ♂
02 24 ☉ ∠ ♃
06 43 ☽ ∥ ♂
06 50 ♀ ☌ ♃
09 34 ☽ □ ☿
10 16 ☽ Q ☉
12 25 ☽ ∠ ♆
12 26 ☽ △ ♅
14 58 ☽ ⊻ ♃
15 33 ☽ ⊻ ♀
18 47 ☿ ± ♄
20 37 ☽ △ ♄

09 Thursday
01 16 ☽ ⊥ ♃
02 36 ☽ ⊥ ♀
03 39 ☽ ⊥ ♆
03 41 ☽ # ♆
04 48 ☽ bQ ♂
10 30 ☉ ∠ ♀
10 42 ☽ ⊻ ♇
13 49 ☽ ⊻ ♆
14 00 ☿ ⊼ ♅
14 27 ☿ bQ ♆
15 24 ♀ Q ♆
16 45 ☽ ∠ ♃
19 08 ☽ □ ☉
19 14 ☽ ∠ ♀
20 46 ☽ ± ♂
20 54 ☽ ⊥ ♇

10 Friday
07 27 ☽ ⊼ ♂
10 29 ☿ Q ♇
12 15 ☽ ∠ ♇
15 14 ☽ □ ♅
18 08 ☽ △ ☿
18 14 ☿ ⊥ ♂
18 41 ☽ ⚹ ♃
23 03 ☽ ⚹ ♀
23 22 ☽ □ ♄

11 Saturday
00 17 ☿ ⚹ ♃
02 27 ☽ ∥ ♆
02 39 ♀ ⊼ ♄
13 57 ☽ ⚹ ♇
15 32 ☽ Q ♃
17 00 ☽ ☌ ♆
21 27 ☽ Q ♀
21 59 ☽ Q ♅
22 46 ☽ bQ ☿

12 Sunday
02 27 ☽ △ ☉
03 41 ♀ ± ♅
04 46 ☽ # ♂
06 05 ☽ Q ♄
10 38 ☽ Q ♇
13 14 ☽ ☍ ♂
16 06 ☽ ± ☿
18 37 ☽ ⚹ ♅
18 48 ☿ ⊼ ♄
23 03 ☽ □ ♃

13 Monday
00 04 ☿ ∥ ♃
02 44 ☽ ⚹ ♄
03 43 ☽ ⊼ ☿
06 27 ☽ bQ ☉
07 21 ☽ □ ♀
17 55 ☽ □ ♇
20 38 ☽ ∠ ♅
20 56 ☽ ⊻ ♆
21 19 ☿ ∥ ♀
23 24 ☽ ± ☉

14 Tuesday
04 45 ☽ ∠ ♄
07 26 ☽ ⊥ ♆
10 46 ☽ ⊼ ☉
12 22 ☽ ⊥ ♅
20 02 ☽ ⊼ ♂
20 30 ☽ ⊥ ♄
22 55 ☽ ⊻ ♅
23 18 ☽ ∠ ♆

15 Wednesday
04 26 ☽ △ ♃
05 26 ♀ ∥ ♃
07 05 ☽ ⊻ ♄
07 11 ☽ ± ♂
14 51 ☽ ☍ ☿
15 39 ♃ Q ♆
16 58 ☽ △ ♀
23 03 ☽ △ ♇

16 Thursday
00 01 ☽ bQ ♂
01 20 ♀ ± ♄
02 02 ☽ ⚹ ♆
07 42 ☽ bQ ♃
20 49 ☽ ☍ ☉
22 34 ☽ bQ ♀
23 47 ☽ Q ♆

17 Friday
00 19 ☽ ± ♃
02 18 ☽ bQ ♇
04 33 ☽ △ ♂
04 44 ☽ ☌ ♅
08 15 ♂ △ ♅
11 31 ☽ ⊼ ♃
13 03 ☽ ☌ ♄
16 40 ☽ ± ♀
18 55 ☽ ± ♇

18 Saturday
04 37 ☽ ⊼ ☿
04 55 ☽ ⊼ ♀
06 07 ☽ ⊼ ♇
09 07 ☽ □ ♆
13 30 ☿ ⚹ ♀
17 25 ☽ ± ☿
17 35 ☿ ⚹ ♇
19 01 ♀ ☌ ♇

19 Sunday
05 10 ☉ ⊥ ♃
09 44 ☽ ⊼ ☉
12 54 ☽ ⊻ ♅
12 56 ☽ bQ ☿
15 47 ☽ □ ♂
15 48 ☿ # ♄
18 47 ☿ △ ♆
20 29 ♀ bQ ♅
21 00 ☉ ∠ ♇
21 12 ☽ ☍ ♃
21 26 ☽ ⊻ ♄
22 20 ☽ ± ☉

20 Monday
00 28 ☽ ⊥ ♅
05 01 ♀ ⚹ ♆
06 57 ♃ ⊼ ♄
09 02 ☽ ⊥ ♄
10 57 ☉ ∥ ♇
15 51 ☽ ☍ ♇
17 31 ☽ bQ ☉
18 01 ☽ ∠ ♅
18 52 ☽ △ ♆
20 16 ☽ ☍ ♀
22 17 ☽ △ ☿
23 11 ☉ ☍ ♅
23 16 ☽ # ♂

21 Tuesday
02 21 ☿ ∠ ♂
02 39 ☽ ∠ ♄
04 52 ☽ Q ♂
23 45 ☽ ⚹ ♅

22 Wednesday
00 43 ☽ bQ ♆
02 08 ☽ △ ☉
06 09 ☽ ⚹ ♂
08 27 ☽ ⚹ ♄
09 43 ☽ ⊼ ♃
18 39 ☿ ∠ ♃
18 52 ☽ ± ♆
19 52 ☽ ∥ ♆
22 04 ☽ ± ♃
23 53 ☽ Q ♅

23 Thursday
04 03 ☽ ⊼ ♇
07 02 ☽ ⊼ ♆
08 34 ☽ Q ♄
14 01 ♀ bQ ♄
14 09 ☽ ∠ ♂
14 41 ☽ ⊼ ♀
14 54 ☿ ∥ ♇
15 41 ☉ ∥ ☿
16 16 ☽ ± ♇
16 41 ☽ bQ ♃
19 29 ☽ □ ☿
22 18 ♂ △ ♄

24 Friday
04 08 ☽ ± ♀
05 51 ☉ ∥ ♇
09 25 ☽ ⊥ ♂
10 35 ☽ bQ ♇
12 18 ☽ □ ♅
19 08 ☿ # ♅
20 39 ☽ □ ☉
20 52 ☽ □ ♄
22 14 ☽ ⊻ ♂
22 55 ☉ ☍ ♄
23 42 ☽ △ ♃

25 Saturday
00 14 ☽ bQ ♀
03 26 ☽ # ♆
13 21 ☽ Q ☿
16 56 ☽ △ ♇
19 45 ☽ ☍ ♆
21 01 ♆ St D

26 Sunday
08 28 ♃ ± ♅
09 20 ☽ △ ♀
12 06 ☽ Q ☉
15 40 ☉ ⚹ ♂
16 45 ☽ ⚹ ☿
16 57 ☉ ⊻ ♃
18 21 ♂ □ ♃

27 Monday
00 06 ☽ △ ♅
08 09 ☽ △ ♄
09 27 ☽ ∥ ♂
12 20 ☽ □ ♃
12 55 ☽ ☌ ♂
13 41 ☽ ⚹ ☉

28 Tuesday
01 58 ☽ ∠ ☿
03 47 ☽ □ ♇
04 55 ☽ bQ ♅
06 20 ☽ ⊼ ♆
07 27 ☽ ∥ ♀
12 37 ☽ bQ ♄
17 00 ☿ ∠ ♇
17 37 ☽ ± ♆
20 37 ☽ ∠ ☉
21 14 ☽ ⊥ ☿
21 46 ☽ ± ♅
22 13 ☉ ⊥ ♇
23 02 ☽ Q ♃

29 Wednesday
00 25 ☽ □ ♀
01 30 ☿ ☍ ♅
05 15 ☽ ± ♄
08 47 ☽ ⊼ ♅
09 48 ☽ ⊻ ☿
10 15 ☽ bQ ♆
13 20 ☽ Q ♇
14 26 ☿ ⊥ ♃
14 33 ☽ ⊥ ☉
16 08 ☽ ⊼ ♄
21 21 ☽ ⚹ ♃
23 35 ☽ ⊻ ♂

30 Thursday
02 16 ☽ ⊻ ☉
10 46 ☽ ⊥ ♂
10 54 ☽ ⚹ ♇
11 49 ☽ Q ♀
13 11 ☽ △ ♆

31 Friday
00 20 ☽ ∠ ♃
03 16 ☽ ∠ ♂
10 29 ☽ ⚹ ♀
13 01 ☿ ☍ ♄
13 04 ☽ ∠ ♇
13 42 ☽ ☍ ♅
16 07 ☽ ⊥ ♃
20 26 ☽ ☍ ♄
21 25 ☽ ☌ ☿

JANUARY 2033

LONGITUDES

Date	Sidereal time h m s	Sun ☉ ° ' "	Moon ☽ ° ' "	Moon ☽ 24.00 ° ' "	Mercury ☿ ° '	Venus ♀ ° '	Mars ♂ ° '	Jupiter ♃ ° '	Saturn ♄ ° '	Uranus ♅ ° '	Neptune ♆ ° '	Pluto ♇ ° '
01	18 45 49	11 ♑ 22 59	12 ♑ 19 58	19 ♑ 30 59	04 ♑ 39	28 ♒ 26	08 ♏ 56	06 ♒ 46	03 ♋ 04	29 ♊ 08	15 ♈ 05	13 ♒ 50
02	18 49 46	12 24 10	26 ♑ 45 22	04 ♒ 02 10	06 13	29 ♒ 30	09 32	07 00	02 R 59	29 R 06	15 05	13 52
03	18 53 42	13 25 21	11 ♒ 20 28	18 ♒ 39 19	07 48	00 ♓ 33	10 08	07 14	02 54	29 03	15 05	13 53
04	18 57 39	14 26 31	25 ♒ 57 50	03 ♓ 15 13	09 23	01 35	10 44	07 27	02 50	29 01	15 05	13 55
05	19 01 35	15 27 42	10 ♓ 30 45	17 ♓ 43 52	10 58	02 38	11 20	07 41	02 45	28 58	15 06	13 56
06	19 05 32	16 28 52	24 ♓ 54 08	02 ♈ 01 14	12 34	03 40	11 56	07 55	02 40	28 56	15 06	13 58
07	19 09 28	17 30 02	09 ♈ 04 58	16 ♈ 05 16	14 10	04 41	12 32	08 08	02 35	28 54	15 07	14 00
08	19 13 25	18 31 11	23 ♈ 02 08	29 ♈ 55 37	15 47	05 42	13 07	08 22	02 31	28 51	15 07	14 01
09	19 17 21	19 32 20	06 ♉ 45 50	13 ♉ 32 55	17 24	06 43	13 43	08 36	02 26	28 49	15 08	14 03
10	19 21 18	20 33 28	20 ♉ 17 00	26 ♉ 58 12	19 01	07 43	14 19	08 50	02 21	28 46	15 08	14 05
11	19 25 15	21 34 36	03 ♊ 36 38	10 ♊ 12 22	20 39	08 43	14 54	09 04	02 17	28 44	15 09	14 06
12	19 29 11	22 35 43	16 ♊ 45 26	23 ♊ 15 50	22 17	09 43	15 30	09 18	02 12	28 42	15 09	14 08
13	19 33 08	23 36 50	29 ♊ 43 35	06 ♋ 08 38	23 56	10 41	16 05	09 32	02 08	28 40	15 10	14 10
14	19 37 04	24 37 56	12 ♋ 30 54	18 ♋ 50 22	25 35	11 40	16 41	09 46	02 04	28 37	15 11	14 12
15	19 41 01	25 39 02	25 ♋ 06 58	01 ♌ 20 40	27 15	12 38	17 16	10 00	01 59	28 35	15 11	14 13
16	19 44 57	26 40 08	07 ♌ 31 30	13 ♌ 39 29	28 ♑ 55	13 35	17 51	10 14	01 55	28 33	15 12	14 15
17	19 48 54	27 41 13	19 ♌ 44 44	25 ♌ 47 22	00 ♒ 36	14 32	18 27	10 28	01 51	28 31	15 13	14 17
18	19 52 50	28 42 17	01 ♍ 47 38	07 ♍ 45 47	02 17	15 28	19 02	10 42	01 47	28 29	15 14	14 19
19	19 56 47	29 ♑ 43 21	13 ♍ 42 09	19 ♍ 37 09	03 58	16 24	19 37	10 56	01 42	28 27	15 14	14 20
20	20 00 44	00 ♒ 44 25	25 ♍ 31 12	01 ♎ 24 50	05 40	17 18	20 12	11 11	01 38	28 24	15 15	14 22
21	20 04 40	01 45 28	07 ♎ 18 37	13 ♎ 13 09	07 23	18 13	20 47	11 25	01 34	28 22	15 16	14 24
22	20 08 37	02 46 30	19 ♎ 09 04	25 ♎ 07 03	09 06	19 06	21 23	11 39	01 31	28 20	15 17	14 26
23	20 12 33	03 47 33	01 ♏ 07 48	07 ♏ 12 01	10 49	19 59	21 58	11 53	01 27	28 18	15 18	14 28
24	20 16 30	04 48 35	13 ♏ 20 24	19 ♏ 33 37	12 33	20 52	22 33	12 07	01 23	28 16	15 19	14 29
25	20 20 26	05 49 36	25 ♏ 52 19	02 ♐ 17 05	14 17	21 43	23 07	12 22	01 19	28 15	15 20	14 31
26	20 24 23	06 50 37	08 ♐ 48 24	15 ♐ 26 40	16 01	22 34	23 42	12 36	01 16	28 13	15 21	14 33
27	20 28 19	07 51 37	22 ♐ 12 08	29 ♐ 04 53	17 45	23 24	24 17	12 50	01 12	28 11	15 22	14 35
28	20 32 16	08 52 37	06 ♑ 04 49	13 ♑ 11 39	19 30	24 13	24 52	13 04	01 09	28 09	15 23	14 37
29	20 36 13	09 53 36	20 ♑ 24 52	27 ♑ 43 47	21 14	25 01	25 26	13 19	01 06	28 07	15 24	14 39
30	20 40 09	10 54 34	05 ♒ 07 27	12 ♒ 34 51	22 57	25 48	26 01	13 33	01 02	28 06	15 26	14 40
31	20 44 06	11 ♒ 55 31	20 ♒ 04 47	27 ♒ 36 02	24 ♒ 40	26 ♓ 35	26 ♏ 36	13 ♒ 47	00 ♋ 59	28 ♊ 04	15 ♈ 27	14 ♒ 42

Date	Moon True ☊ ° '	Moon Mean ☊ ° '	Moon ☽ Latitude ° '
01	27 ♎ 31	26 ♎ 44	04 N 51
02	27 R 19	26 41	05 01
03	27 08	26 38	04 52
04	26 58	26 35	04 24
05	26 52	26 32	03 39
06	26 48	26 29	02 40
07	26 46	26 25	01 32
08	26 D 46	26 22	00 N 20
09	26 R 46	26 19	00 S 53
10	26 45	26 16	02 01
11	26 41	26 13	03 01
12	26 33	26 09	03 51
13	26 23	26 06	04 28
14	26 10	26 03	04 51
15	25 56	26 00	05 00
16	25 42	25 57	04 53
17	25 28	25 54	04 34
18	25 17	25 50	04 02
19	25 08	25 47	03 20
20	25 02	25 44	02 29
21	24 59	25 41	01 32
22	24 58	25 38	00 S 31
23	24 D 58	25 35	00 N 33
24	24 R 57	25 31	01 36
25	24 56	25 28	02 36
26	24 52	25 25	03 30
27	24 46	25 22	04 15
28	24 37	25 19	04 46
29	24 26	25 15	05 01
30	24 14	25 12	04 56
31	24 ♎ 03	25 ♎ 09	04 N 32

DECLINATIONS

Date	Sun ☉ ° '	Moon ☽ ° '	Mercury ☿ ° '	Venus ♀ ° '	Mars ♂ ° '	Jupiter ♃ ° '	Saturn ♄ ° '	Uranus ♅ ° '	Neptune ♆ ° '	Pluto ♇ ° '
01	22 S 57	18 S 02	24 S 33	13 S 15	13 S 25	19 S 07	22 N 27	23 N 39	04 N 27	23 S 20
02	22 51	15 52	24 35	12 49	13 36	19 04	22 27	23 39	04 28	23 19
03	22 45	12 41	24 35	12 23	13 48	19 01	22 27	23 39	04 28	23 19
04	22 39	08 43	24 33	11 56	13 59	18 57	22 27	23 39	04 28	23 18
05	22 32	04 S 15	24 30	11 29	14 11	18 54	22 28	23 39	04 28	23 18
06	22 25	00 N 25	24 26	11 02	14 22	18 50	22 28	23 39	04 28	23 17
07	22 17	05 01	24 21	10 34	14 33	18 47	22 28	23 39	04 29	23 17
08	22 09	09 15	24 14	10 07	14 44	18 43	22 28	23 39	04 29	23 16
09	22 01	12 56	24 05	09 39	14 55	18 40	22 28	23 39	04 29	23 16
10	21 52	15 52	23 55	09 12	15 06	18 36	22 29	23 39	04 29	23 15
11	21 42	17 54	23 44	08 44	15 17	18 32	22 29	23 39	04 30	23 15
12	21 32	18 56	23 31	08 16	15 27	18 29	22 29	23 39	04 30	23 14
13	21 22	18 58	23 16	07 48	15 38	18 25	22 29	23 39	04 30	23 14
14	21 11	18 01	23 00	07 20	15 48	18 21	22 30	23 39	04 30	23 13
15	21 00	16 11	22 43	06 52	15 59	18 18	22 30	23 39	04 31	23 13
16	20 49	13 39	22 24	06 24	16 09	18 14	22 30	23 39	04 31	23 12
17	20 37	10 33	22 03	05 55	16 19	18 10	22 30	23 39	04 31	23 12
18	20 25	07 02	21 41	05 27	16 29	18 06	22 30	23 39	04 32	23 11
19	20 12	03 N 20	21 18	04 59	16 39	18 02	22 31	23 39	04 32	23 11
20	19 59	00 S 30	20 52	04 31	16 49	17 59	22 31	23 39	04 33	23 10
21	19 46	04 19	20 26	04 03	16 58	17 55	22 31	23 39	04 33	23 10
22	19 32	07 58	19 57	03 35	17 08	17 51	22 31	23 39	04 33	23 09
23	19 18	11 21	19 28	03 07	17 17	17 47	22 31	23 39	04 34	23 09
24	19 03	14 19	18 56	02 39	17 26	17 43	22 32	23 39	04 34	23 08
25	18 49	16 41	18 24	02 12	17 36	17 39	22 32	23 38	04 35	23 08
26	18 33	18 18	17 49	01 44	17 45	17 35	22 32	23 38	04 35	23 07
27	18 18	18 58	17 14	01 17	17 54	17 32	22 32	23 38	04 35	23 06
28	18 02	18 32	16 37	00 50	18 03	17 28	22 32	23 38	04 36	23 06
29	17 46	16 55	15 59	00 S 22	18 11	17 24	22 33	23 38	04 36	23 05
30	17 29	14 11	15 19	00 N 04	18 20	17 20	22 33	23 38	04 37	23 05
31	17 S 13	10 S 29	14 S 39	00 N 31	18 S 28	17 S 15	22 N 33	23 N 38	04 N 37	23 S 04

ZODIAC SIGN ENTRIES

Date	h	m	Planets
02	17	21	☽ ♒
02	23	35	♀ ♓
04	18	38	☽ ♓
06	20	35	☽ ♈
09	00	08	☽ ♉
11	05	28	☽ ♊
13	12	31	☽ ♋
15	21	24	☽ ♌
17	03	32	☿ ♒
18	08	24	☽ ♍
19	18	33	☉ ♒
20	21	07	☽ ♎
23	09	45	☽ ♏
25	19	45	☽ ♐
28	01	35	☽ ♑
30	03	42	☽ ♒

LATITUDES

Date	Mercury ☿ ° '	Venus ♀ ° '	Mars ♂ ° '	Jupiter ♃ ° '	Saturn ♄ ° '	Uranus ♅ ° '	Neptune ♆ ° '	Pluto ♇ ° '
01	01 S 12	01 S 20	01 N 07	00 S 34	00 S 57	00 N 13	01 S 36	06 S 58
04	01 27	01 06	01 06	00 34	00 57	00 13	01 36	06 58
07	01 41	00 50	01 06	00 34	00 56	00 13	01 36	06 58
10	01 51	00 34	01 05	00 35	00 56	00 13	01 36	06 58
13	01 59	00 S 16	01 04	00 35	00 56	00 13	01 36	06 58
16	02 04	00 N 04	01 03	00 35	00 55	00 13	01 35	06 58
19	02 06	00 25	01 02	00 35	00 55	00 13	01 35	06 58
22	02 03	00 47	01 00	00 35	00 54	00 13	01 35	06 58
25	01 56	01 11	00 59	00 36	00 54	00 13	01 35	06 58
28	01 44	01 36	00 58	00 36	00 53	00 13	01 35	06 58
31	01 S 26	02 N 03	00 N 56	00 S 36	00 S 53	00 N 13	01 S 35	06 S 58

DATA

Julian Date	2463599
Delta T	+75 seconds
Ayanamsa	24° 19' 14"
Synetic vernal point	04° ♓ 47' 45"
True obliquity of ecliptic	23° 25' 57"

MOON'S PHASES, APSIDES AND POSITIONS ☽

Date	h	m	Phase	Longitude ° '	Eclipse Indicator
01	10	17	●	11 ♑ 19	
08	03	34	☽	18 ♈ 10	
15	13	07	○	25 ♋ 42	
23	17	46	☾	04 ♏ 02	
30	22	00	●	11 ♒ 20	

Day	h	m		
04	05	19	Perigee	
20	06	56	Apogee	
06	09	49	ON	
13	00	29	Max dec	19° N 04'
20	08	52	OS	
27	14	52	Max dec	18° S 58'

LONGITUDES

Date	Chiron ⚷ ° '	Ceres ⚳ ° '	Pallas ⚴ ° '	Juno ⚵ ° '	Vesta ⚶ ° '	Black Moon Lilith ⚸ ° '
01	21 ♉ 31	04 ♐ 01	03 ♏ 52	12 ♎ 26	04 ♑ 09	16 ♍ 20
11	21 ♉ 16	07 ♐ 57	07 ♏ 05	14 ♎ 00	09 ♑ 32	17 ♍ 26
21	21 ♉ 08	11 ♐ 45	09 ♏ 58	15 ♎ 02	14 ♑ 53	18 ♍ 33
31	21 ♉ 06	15 ♐ 22	12 ♏ 27	15 ♎ 28	20 ♑ 11	19 ♍ 40

ASPECTARIAN

h m	Aspects
01 Saturday	
02 29	☽ ⊻ ♃
04 25	☽ ⊥ ♇
06 02	☽ ⚹ ♂
06 10	♀ ∥ ♂
10 17	☽ ☌ ☉
14 00	☽ ∠ ♀
14 31	☽ ⊻ ♇
16 36	☽ □ ♆
02 Sunday	
02 59	☽ Q ♂
03 20	♀ △ ♅
06 11	☽ ⊥ ♀
15 51	☽ ⚻ ♅
16 52	☽ ⊻ ♀
22 13	☽ ⚻ ♄
22 26	☽ Q ♆
03 Monday	
01 28	♀ ∠ ♆
01 42	☽ ± ♅
01 49	☿ ⊻ ♃
04 47	☽ ∥ ♂
05 08	☽ ☌ ♃
05 29	☽ ⊻ ☿
08 02	☽ ± ♄
09 56	☽ □ ♂
13 20	☿ ⊥ ♇
14 19	☽ ∥ ♀
15 40	☽ ⊻ ☉
16 11	☽ ☌ ♇
16 26	☽ ⚼ ♅
16 31	☽ ⊥ ☿
18 09	☽ ⚹ ♆
22 43	☽ ⚼ ♄
23 12	☉ ⊻ ♇
04 Tuesday	
02 15	☽ ⊥ ☉
09 05	☽ ∠ ☿
17 00	☽ △ ♅
18 09	☽ ∠ ☉
18 47	☽ ∠ ♆
21 58	☽ ☌ ♀
23 14	☽ △ ♄
05 Wednesday	
03 21	☉ □ ♆
07 14	☽ ⊻ ♃
09 39	☽ ⊥ ♆
10 51	☽ # ♆
12 51	☽ ⚹ ☿
13 25	☽ △ ♂
14 36	♀ △ ♄
17 21	☽ ⊥ ♃
17 42	☽ ⊻ ♇
19 37	☽ ⊻ ♆
20 42	☿ ⚹ ♂
20 51	☽ ⚹ ☉
23 51	♅ ⚼ ♇
06 Thursday	
03 09	☉ # ♄
03 43	☽ ⊥ ♇
08 36	☽ ∠ ♃
11 22	☽ Q ☿
15 33	☽ ⚼ ♂
18 29	☽ Q ☉
18 46	☽ □ ♅
18 51	☽ ∠ ♇
07 Friday	
01 01	☽ □ ♄
03 56	☽ ⊻ ♀
07 27	☽ ± ♂
09 07	☽ ∥ ♆
09 21	☿ ⊻ ♇
10 22	☽ ⚹ ♃
14 57	☽ ⊥ ♀
18 09	☽ ⚻ ♂
20 25	☽ ⚹ ♇
21 50	☽ □ ☿
22 19	☽ ☌ ♆
08 Saturday	
01 21	☽ Q ♅
02 07	☿ □ ♆
03 34	☽ □ ☉
07 19	☽ Q ♃
07 39	☽ ∠ ♀
07 39	☽ Q ♄
16 40	☽ # ♀
17 12	☽ Q ♇
22 06	☽ ⚹ ♅
22 54	♃ ± ♄
09 Sunday	
04 26	☽ ⚹ ♄
11 55	☽ ⚹ ♀
15 18	☽ □ ♃
15 44	♂ ⚼ ♅
10 Monday	
00 26	☽ ∠ ♅
00 52	☽ ☍ ♂
00 55	☽ □ ♇
02 15	♂ □ ♇
02 49	☽ ⊻ ♆
04 29	☽ # ♂
06 48	☽ ∠ ♄
09 25	☽ △ ☿
10 55	☽ Q ♀
12 32	☽ △ ☉
13 31	☽ ⊥ ♆
16 27	☽ ⊥ ♅
22 50	☽ ⊥ ♄
11 Tuesday	
03 13	☽ ⊻ ♅
05 43	☽ ∠ ♆
09 36	☽ ⊻ ♄
16 13	☽ ⚼ ☿
17 50	☽ ⚼ ☉
21 40	☿ # ♅
21 57	♂ ⚻ ♆
22 03	☽ □ ♀
22 06	☽ △ ♃
22 22	♀ ⊥ ♆
22 57	♀ ⊻ ♃
23 16	☽ # ♃
12 Wednesday	
07 11	☽ △ ♇
09 03	☽ ⚹ ♆
09 34	☽ ⚻ ♂
11 00	☽ ± ☿
11 41	☽ ± ☉
21 09	☽ ± ♂
23 40	☽ ⚻ ☿
23 41	☽ ⚻ ☉
23 56	☉ ☌ ☿
13 Thursday	
02 10	☽ ⚼ ♃
07 14	☽ Q ♆
10 01	☽ ☌ ♅
10 57	☽ ⚼ ♇
14 39	☽ ⚼ ♂
16 24	☿ ∥ ♇
16 28	☽ ☌ ♄
19 14	☽ ± ♃
14 Friday	
03 50	☽ ± ♇
05 07	☽ # ♃
06 43	☽ ⚻ ♃
10 16	☽ △ ♀
15 11	☽ ⚻ ♇
17 03	☽ □ ♆
20 16	☽ △ ♂
15 Saturday	
01 50	♂ ⚼ ♄
13 07	☽ ☍ ☉
14 08	☽ # ♂
16 43	☽ ☍ ☿
17 14	☽ ⚼ ♀
18 39	☽ ⊻ ♅
16 Sunday	
00 56	☿ ∠ ♀
01 10	☽ ⊻ ♄
04 40	☿ # ♄
06 14	☽ ⊥ ♅
06 52	☿ ⚻ ♅
12 07	☽ ± ♀
12 45	☽ ⊥ ♄
17 24	☽ ☍ ♃
23 45	☽ ∠ ♅
17 Monday	
00 51	☽ ⚻ ♀
01 12	☽ ☍ ♇
03 03	☽ △ ♆
05 30	♀ ⊻ ♇
06 18	☽ ∠ ♄
08 46	☿ Q ♂
09 18	☽ □ ♂
18 Tuesday	
05 08	☿ ⚻ ♄
05 15	☽ ⚻ ☉
05 23	☽ ⚹ ♅
05 44	♀ ⊻ ♆
06 49	☉ ⚻ ♅
08 51	☽ ⚼ ♆
11 58	☽ ⚹ ♄
13 08	☽ ⚻ ☿
18 23	☽ ± ☉
23 04	☽ Q ♂
23 59	☽ # ♀
19 Wednesday	
01 32	☿ Q ♆
02 58	☽ ± ♆
03 12	☽ ± ☿
04 24	☽ ∥ ♆
05 26	☽ Q ♅
06 18	☽ ⚻ ♃
12 01	☽ Q ♄
13 18	☽ ⚻ ♇
14 16	☽ ⚼ ☉
15 07	☽ ⚻ ♆
17 55	☽ ☍ ♀
18 30	☿ ± ♅
18 42	☽ ± ♃
20 Thursday	
00 29	☽ ⚼ ☿
00 38	☽ ⚹ ♂
01 30	☽ ± ♇
07 17	♀ ⊥ ♃
10 55	♀ # ♆
13 22	☽ ⚼ ♃
17 52	☽ □ ♅
19 52	☽ ⚼ ♇
23 38	☽ △ ☉
21 Friday	
00 24	☽ □ ♄
07 57	☉ ⚻ ♄
08 45	☽ ∠ ♂
10 32	☽ ∥ ♀
12 11	☽ △ ☿
13 31	☽ # ♆
14 34	☿ ± ♄
20 30	☽ △ ♃
22 Saturday	
02 26	☽ △ ♇
03 58	☽ ⊥ ♂
04 10	☽ ☍ ♆
08 08	♂ ± ♆
11 54	☽ ⚻ ♀
16 43	☽ ⊻ ♂
23 Sunday	
00 12	☉ Q ♆
00 56	☽ ± ♀
06 24	☽ △ ♅
12 38	☽ △ ♄
17 46	☽ □ ☉
20 14	☽ ⚼ ♀
21 13	☉ Q ♂
23 46	☉ ± ♅
24 Monday	
01 25	♀ ⊥ ♇
01 34	♂ ± ♅
02 06	☉ ∥ ☿
05 06	☿ ☌ ♃
09 35	☽ □ ♃
10 13	☽ □ ☿
11 52	☽ ⚼ ♅
14 14	☽ □ ♇
15 50	☽ ⚻ ♆
17 52	☽ ⚼ ♄
21 51	☿ ⚼ ♅
25 Tuesday	
03 23	☽ ± ♆
03 33	☽ △ ♀
05 08	☽ ± ♅
06 33	☽ ☌ ♂
07 48	☽ Q ☉
10 58	☽ ± ♄
15 21	☿ ☌ ♇
16 27	☽ ⚻ ♅
18 53	♂ ∥ ♃
20 23	☽ ⚼ ♆
20 35	☽ Q ♃
22 10	☽ ⚻ ♄
26 Wednesday	
00 18	☽ ∥ ♃
00 28	☽ Q ♇
01 07	☽ ∥ ♂
01 51	☽ Q ☿
02 41	☿ ⚹ ♆
05 52	☽ ∥ ☿
08 06	☽ ⚹ ☉
14 33	☿ ∥ ♂
15 16	☿ ⚼ ♄
16 38	☽ ∥ ☉
19 00	☽ ⚹ ♃
21 23	☉ ± ♄
22 25	☽ ⚹ ♇
22 43	☿ ∥ ♃
23 51	☽ △ ♆
27 Thursday	
02 29	☿ ⊥ ♀
02 58	☽ ⚹ ☿
13 15	☽ ∠ ☉
14 14	☽ □ ♀
15 49	☽ ⊻ ♂
22 01	☽ ∠ ♃
22 25	☽ ☍ ♅
28 Friday	
00 54	☽ ∠ ♇
02 41	☽ ⊥ ♂
03 36	☽ ☍ ♄
06 06	☽ ⊥ ☉
08 55	☽ ∠ ☿
11 30	☉ ∥ ♂
13 43	☽ ⊥ ♃
16 18	☽ ⊥ ♇
17 06	☽ ⊻ ☉
18 40	☽ ∠ ♂
19 09	♃ ⚼ ♅
20 29	☽ ∥ ♂
22 52	♂ ± ♄
22 59	☽ Q ♀
23 29	☽ ∥ ☉
29 Saturday	
00 00	☽ ⊻ ♃
02 13	☽ ⊥ ☿
02 24	☽ ⊻ ♇
03 41	☽ □ ♆
06 19	☽ ∥ ♃
13 31	☽ ⊻ ☿
20 00	☽ ⚹ ♀
20 36	☽ ⚹ ♂
30 Sunday	
00 25	☽ ∥ ☿
00 37	☽ ⚻ ♅
01 16	☉ ∠ ♀
05 25	☽ ⚻ ♄
09 15	☽ Q ♆
10 20	☽ ± ♅
15 05	☽ ± ♄
16 51	☽ Q ♂
21 39	☽ ∠ ♀
22 00	☽ ☌ ☉
31 Monday	
00 48	☽ ⚼ ♅
01 46	☽ ☌ ♃
03 23	☽ ☌ ♇
04 35	☽ ⚹ ♆
05 29	☽ ⚼ ♄
06 39	☉ ∥ ♃
12 50	☽ ⊥ ♀
13 41	♀ △ ♂
20 16	☽ ☌ ☿
22 48	☽ □ ♂
22 55	☽ ⊻ ♀

All ephemeris data is given at 12.00 UT and the Moon's longitude is additionally given for 24.00 UT

FEBRUARY 2033

LONGITUDES

Date	Sidereal time h m s	Sun ☉	Moon ☽	Moon ☽ 24.00	Mercury ☿	Venus ♀	Mars ♂	Jupiter ♃	Saturn ♄	Uranus ♅	Neptune ♆	Pluto ♇
01	20 48 02	12 ♒ 56 27	05 ♓ 07 20	12 ♓ 37 30	26 ♒ 22	27 ♓ 20	27 ♏ 10	14 ♒ 02	00 ♋ 56	28 ♊ 02	15 ♈ 28	14 ♒ 44
02	20 51 59	13 57 21	20 ♓ 05 27	27 ♓ 30 14	28 04	28 05	27 44	14 16	00 R 53	28 R 01	15 29	14 46
03	20 55 55	14 58 15	04 ♈ 51 05	12 ♈ 07 22	29 ♒ 43	28 48	28 19	14 30	00 50	27 59	15 31	14 48
04	20 59 52	15 59 07	19 ♈ 18 41	26 ♈ 24 48	01 ♓ 21	29 ♓ 30	28 53	14 44	00 48	27 58	15 32	14 50
05	21 03 48	16 59 58	03 ♉ 25 36	10 ♉ 21 06	02 57	00 ♈ 11	29 ♏ 27	14 59	00 45	27 56	15 33	14 51
06	21 07 45	18 00 47	17 ♉ 11 28	23 ♉ 56 53	04 30	00 51	00 ♐ 01	15 13	00 42	27 55	15 35	14 53
07	21 11 42	19 01 34	00 ♊ 37 37	07 ♊ 13 57	05 59	01 30	00 35	15 27	00 40	27 54	15 36	14 55
08	21 15 38	20 02 21	13 ♊ 46 12	20 ♊ 14 39	07 25	02 07	01 09	15 42	00 38	27 52	15 38	14 57
09	21 19 35	21 03 05	26 ♊ 39 36	03 ♋ 01 19	08 46	02 43	01 43	15 56	00 35	27 51	15 39	14 59
10	21 23 31	22 03 49	09 ♋ 20 01	15 ♋ 35 56	10 01	03 17	02 17	16 10	00 33	27 50	15 41	15 00
11	21 27 28	23 04 30	21 ♋ 49 13	28 ♋ 00 04	11 11	03 50	02 50	16 24	00 31	27 49	15 42	15 02
12	21 31 24	24 05 10	04 ♌ 08 35	10 ♌ 14 54	12 14	04 22	03 24	16 39	00 29	27 47	15 44	15 04
13	21 35 21	25 05 49	16 ♌ 19 09	22 ♌ 21 26	13 10	04 52	03 58	16 53	00 27	27 46	15 45	15 06
14	21 39 17	26 06 26	28 ♌ 21 53	04 ♍ 20 40	13 57	05 20	04 31	17 07	00 26	27 45	15 47	15 08
15	21 43 14	27 07 02	10 ♍ 17 57	16 ♍ 13 57	14 36	05 47	05 04	17 21	00 24	27 44	15 48	15 10
16	21 47 11	28 07 36	22 ♍ 08 53	28 ♍ 03 03	15 05	06 12	05 38	17 35	00 22	27 43	15 50	15 11
17	21 51 07	29 ♒ 08 09	03 ♎ 56 47	09 ♎ 50 28	15 24	06 35	06 11	17 50	00 21	27 43	15 52	15 13
18	21 55 04	00 ♓ 08 40	15 ♎ 44 30	21 ♎ 39 23	15 33	06 56	06 44	18 04	00 20	27 42	15 53	15 15
19	21 59 00	01 09 10	27 ♎ 35 36	03 ♏ 33 43	15 R 32	07 15	07 17	18 18	00 18	27 41	15 55	15 17
20	22 02 57	02 09 39	09 ♏ 34 18	15 ♏ 37 59	15 21	07 32	07 50	18 32	00 17	27 40	15 57	15 18
21	22 06 53	03 10 07	21 ♏ 45 24	27 ♏ 57 10	14 59	07 48	08 23	18 46	00 16	27 40	15 59	15 20
22	22 10 50	04 10 33	04 ♐ 13 56	10 ♐ 36 17	14 29	08 01	08 55	19 00	00 16	27 39	16 00	15 22
23	22 14 46	05 10 58	17 ♐ 04 47	23 ♐ 39 55	13 49	08 11	09 28	19 14	00 15	27 38	16 02	15 24
24	22 18 43	06 11 21	00 ♑ 22 05	07 ♑ 11 34	13 02	08 20	10 01	19 28	00 14	27 38	16 04	15 25
25	22 22 40	07 11 44	14 ♑ 08 27	21 ♑ 12 43	12 09	08 26	10 33	19 42	00 14	27 37	16 06	15 27
26	22 26 36	08 12 04	28 ♑ 24 06	05 ♒ 42 08	11 11	08 30	11 05	19 56	00 13	27 37	16 08	15 29
27	22 30 33	09 12 24	13 ♒ 06 07	20 ♒ 35 11	10 09	08 32	11 37	20 10	00 13	27 37	16 10	15 31
28	22 34 29	10 ♓ 12 41	28 ♒ 08 16	05 ♓ 44 09	09 ♓ 05	08 ♈ 31	12 ♐ 10	20 ♒ 24	00 ♋ 13	27 ♊ 36	16 ♈ 12	15 ♒ 32

Date	Moon True ☊	Moon Mean ☊	Moon ☽ Latitude
01	23 ♎ 54	25 ♎ 06	03 N 48
02	23 R 48	25 03	02 49
03	23 44	25 00	01 39
04	23 42	24 56	00 N 23
05	23 D 43	24 53	00 S 52
06	23 43	24 50	02 02
07	23 R 42	24 47	03 03
08	23 40	24 44	03 54
09	23 34	24 40	04 31
10	23 27	24 37	04 55
11	23 17	24 34	05 04
12	23 05	24 31	04 58
13	22 54	24 28	04 39
14	22 43	24 25	04 08
15	22 34	24 21	03 26
16	22 27	24 18	02 35
17	22 22	24 15	01 37
18	22 20	24 12	00 S 35
19	22 D 20	24 09	00 N 28
20	22 21	24 06	01 32
21	22 22	24 02	02 32
22	22 23	23 59	03 27
23	22 R 22	23 56	04 13
24	22 19	23 53	04 47
25	22 15	23 50	05 06
26	22 09	23 46	05 07
27	22 02	23 43	04 49
28	21 ♎ 55	23 ♎ 40	04 N 10

DECLINATIONS

Date	Sun ☉	Moon ☽	Mercury ☿	Venus ♀	Mars ♂	Jupiter ♃	Saturn ♄	Uranus ♅	Neptune ♆	Pluto ♇
01	16 S 55	06 S 06	13 S 57	00 N 58	18 S 37	17 S 11	22 N 33	23 N 38	04 N 38	23 S 04
02	16 38	01 S 20	13 15	01 24	18 45	17 07	22 33	23 38	04 38	23 03
03	16 20	03 N 26	12 31	01 50	18 53	17 03	22 34	23 38	04 39	23 03
04	16 02	07 55	11 48	02 15	19 01	16 59	22 34	23 38	04 40	23 03
05	15 44	11 51	11 04	02 40	19 09	16 55	22 34	23 38	04 40	23 02
06	15 26	15 01	10 19	03 05	19 17	16 51	22 34	23 38	04 41	23 02
07	15 07	17 17	09 35	03 30	19 24	16 47	22 34	23 38	04 41	23 01
08	14 48	18 35	08 52	03 54	19 32	16 43	22 35	23 38	04 42	23 01
09	14 29	18 52	08 09	04 18	19 39	16 38	22 35	23 38	04 42	23 00
10	14 09	18 12	07 28	04 41	19 47	16 34	22 35	23 38	04 43	23 00
11	13 49	16 40	06 48	05 04	19 54	16 30	22 35	23 38	04 44	22 59
12	13 29	14 23	06 10	05 26	20 01	16 26	22 35	23 38	04 44	22 59
13	13 09	11 30	05 34	05 48	20 08	16 22	22 35	23 38	04 45	22 58
14	12 49	08 10	05 01	06 09	20 14	16 17	22 36	23 38	04 46	22 58
15	12 28	04 32	04 32	06 30	20 21	16 13	22 36	23 38	04 46	22 57
16	12 07	00 N 45	04 06	06 50	20 28	16 09	22 36	23 38	04 47	22 57
17	11 46	03 S 03	03 44	07 09	20 34	16 05	22 36	23 38	04 48	22 56
18	11 25	06 44	03 27	07 27	20 40	16 01	22 36	23 38	04 48	22 56
19	11 04	10 10	03 14	07 45	20 47	15 56	22 37	23 38	04 49	22 56
20	10 42	13 13	03 06	08 02	20 53	15 52	22 37	23 38	04 50	22 55
21	10 20	15 45	03 03	08 19	20 59	15 48	22 37	23 38	04 50	22 55
22	09 59	17 36	03 04	08 34	21 04	15 43	22 37	23 38	04 51	22 54
23	09 37	18 37	03 11	08 48	21 10	15 39	22 37	23 38	04 52	22 54
24	09 14	18 39	03 22	09 02	21 16	15 35	22 37	23 38	04 52	22 54
25	08 52	17 36	03 38	09 14	21 21	15 31	22 38	23 38	04 53	22 53
26	08 30	15 27	03 57	09 26	21 26	15 26	22 38	23 38	04 54	22 53
27	08 07	12 16	04 19	09 36	21 32	15 22	22 38	23 38	04 55	22 52
28	07 S 44	08 S 12	04 S 44	09 N 45	21 S 37	15 S 18	22 N 38	23 N 38	04 N 55	22 S 52

ZODIAC SIGN ENTRIES

Date	h m	Planets
01	03 50	☽ ♓
03	04 04	☽ ♈
03	16 04	☿ ♓
05	05 27	♀ ♈
05	06 07	☽ ♉
06	11 12	♂ ♐
07	10 52	☽ ♊
09	18 17	☽ ♋
12	03 54	☽ ♌
14	15 17	☽ ♍
17	03 58	☽ ♎
18	08 34	☉ ♓
19	16 51	☽ ♏
22	03 56	☽ ♐
24	11 21	☽ ♑
26	14 39	☽ ♒
28	14 57	☽ ♓

LATITUDES

Date	Mercury ☿	Venus ♀	Mars ♂	Jupiter ♃	Saturn ♄	Uranus ♅	Neptune ♆	Pluto ♇
01	01 S 18	02 N 12	00 N 56	00 S 36	00 S 53	00 N 13	01 S 35	06 S 58
04	00 52	02 40	00 54	00 36	00 52	00 13	01 35	06 59
07	00 S 18	03 10	00 53	00 37	00 52	00 13	01 34	06 59
10	00 N 22	03 41	00 51	00 37	00 51	00 13	01 34	06 59
13	01 08	04 13	00 49	00 37	00 50	00 13	01 34	06 59
16	01 56	04 45	00 47	00 37	00 50	00 13	01 34	07 00
19	02 41	05 19	00 45	00 38	00 49	00 13	01 34	07 00
22	03 17	05 52	00 43	00 38	00 49	00 13	01 34	07 00
25	03 39	06 25	00 41	00 38	00 48	00 13	01 34	07 01
28	03 41	06 56	00 38	00 39	00 48	00 13	01 34	07 01
31	03 N 25	07 N 26	00 N 36	00 S 39	00 S 47	00 N 13	01 S 34	07 S 02

DATA

Julian Date	2463630
Delta T	+75 seconds
Ayanamsa	24° 19' 19"
Synetic vernal point	04° ♓ 47' 40"
True obliquity of ecliptic	23° 25' 58"

MOON'S PHASES, APSIDES AND POSITIONS ☽

Date	h m	Phase	Longitude	Eclipse Indicator
06	13 34	☽	18 ♉ 05	
14	07 04	○	25 ♌ 54	
22	11 53	☾	04 ♐ 10	

Day	h m	
01	07 23	Perigee
16	19 43	Apogee

02	18 41	0N	
09	07 07	Max dec	18° N 53'
16	16 40	0S	
24	00 57	Max dec	18° S 46'

LONGITUDES

Date	Chiron ⚷	Ceres ⚳	Pallas ⚴	Juno ⚵	Vesta ⚶	Black Moon Lilith ⚸
01	21 ♉ 06	15 ♐ 43	12 ♏ 40	15 ♎ 29	20 ♑ 43	19 ♍ 47
11	21 ♉ 12	19 ♐ 06	14 ♏ 37	15 ♎ 12	25 ♑ 57	20 ♍ 53
21	21 ♉ 23	22 ♐ 15	15 ♏ 59	14 ♎ 15	01 ♒ 07	22 ♍ 00
31	21 ♉ 41	25 ♐ 06	16 ♏ 41	12 ♎ 42	06 ♒ 12	23 ♍ 07

All ephemeris data is given at 12.00 UT and the Moon's longitude is additionally given for 24.00 UT

ASPECTARIAN

h m	Aspects
01 Tuesday	
00 43	☽ △ ♅
04 34	☽ ∠ ♆
05 21	☽ △ ♄
14 17	☉ ⚼ ♅
18 57	☽ ⊥ ♆
19 29	☽ # ♆
02 Wednesday	
01 25	☽ ⚺ ☉
02 29	☽ ⚺ ♃
03 25	☽ ⚺ ♇
04 35	☽ ⚺ ♆
05 02	☿ □ ♂
10 03	♀ □ ♅
11 20	☿ △ ♅
11 45	☽ # ♀
11 46	☽ ⊥ ☉
12 17	☽ ⊥ ♃
12 23	☿ ⚺ ♀
13 05	☽ ⊥ ♇
21 30	☉ ☌ ♃
23 00	♂ ⚻ ♅
03 Thursday	
00 48	☽ □ ♅
00 53	☽ △ ♂
01 36	☽ ☌ ♀
02 33	☽ ⚺ ☿
03 01	☽ ∥ ♀
03 06	☽ ∠ ♃
03 26	☽ ∠ ☉
03 43	☽ ∠ ♇
05 27	☽ □ ♄
07 43	☉ ☌ ♇
13 37	☽ ⊥ ☿
18 18	☽ ∥ ♆
23 43	☿ ∠ ♆
04 Friday	
01 04	☉ ⚹ ♆
02 33	☽ ⚼ ♂
03 57	☿ △ ♄
04 13	☽ ⚹ ♃
04 29	☽ ⚹ ♇
05 40	☽ ☌ ♆
06 00	☽ ⚹ ☉
06 24	☽ Q ♅
06 25	☽ ∠ ☿
07 40	☉ ⚼ ♄
11 08	☽ Q ♄
18 16	☽ ± ♂
21 50	♃ ☌ ♇
23 50	♀ ∠ ♇
05 Saturday	
00 44	☽ Q ♇
00 46	♀ ∠ ♃
00 46	☽ Q ♃
02 37	☽ ⚹ ♅
03 49	☽ Q ☉
04 53	☽ ⚻ ♂
06 09	☽ ⚺ ♀
07 25	☽ ⚹ ♄
07 42	☽ # ☿
11 04	☽ ⚹ ☿
17 00	☽ ⊥ ♀
06 Sunday	
04 30	☽ ∠ ♅
07 07	♀ □ ♄
07 56	☽ □ ♇
08 28	☽ □ ♃
09 09	☽ ⚺ ♆
09 23	☽ ∠ ♄
09 31	☽ ∠ ♀
10 37	☽ Q ☿
13 34	☽ □ ☉
15 19	☽ # ☉
19 47	☽ ⊥ ♆
20 22	☽ ⊥ ♅
07 Monday	
01 19	☽ ⊥ ♄
05 53	☽ # ♃
07 05	☽ ⚺ ♅
11 55	☽ ☍ ♂
11 57	☽ ∠ ♆
12 04	☽ ⚺ ♄
12 45	♂ ⚼ ♆
13 39	☽ ⚹ ♀
15 10	♂ ⚻ ♄
22 55	☽ □ ☿
08 Tuesday	
04 34	♃ ⚹ ♆
06 13	♃ ⚼ ♄
12 40	☽ Q ♀
14 11	☽ △ ♇
15 26	☽ ⚹ ♆
15 37	☽ △ ♃
09 Wednesday	
00 37	☽ △ ☉
13 52	☽ Q ♆
14 14	☽ ☌ ♅
18 16	☽ ⚼ ♇
19 23	☽ ☌ ♄
20 12	☽ ⚼ ♃
21 58	☽ ⚻ ♂
23 58	☽ □ ♀
10 Thursday	
05 06	☿ ⊥ ♆
07 18	☽ ⚼ ☉
09 54	☽ ± ♂
11 23	☽ ± ♇
13 27	☽ △ ☿
13 38	☽ ± ♃
14 05	♀ ∥ ♆
22 53	☽ ⚻ ♇
11 Friday	
00 10	☽ □ ♆
01 21	☽ ⚻ ♃
02 02	☽ ± ☉
03 57	☽ ⚼ ♂
14 03	☽ # ♃
14 39	☽ ⚻ ☉
20 57	♂ Q ♇
21 17	☽ ⚼ ☿
23 37	☽ ⚺ ♅
12 Saturday	
04 52	☽ ⚺ ♄
10 29	☽ △ ♂
11 19	☽ ⊥ ♅
12 27	☽ △ ♀
16 28	☽ ± ☿
16 35	☽ ⊥ ♄
20 58	☽ # ☉
13 Sunday	
05 00	☽ ∠ ♅
05 16	☽ ⚻ ☿
06 08	☿ # ♀
09 35	☽ ☍ ♇
10 18	☽ ∠ ♄
10 53	☽ △ ♆
13 08	☽ ☍ ♃
19 20	☽ ⚼ ♀
22 55	♅ Q ♆
14 Monday	
07 04	☽ ☍ ☉
10 47	☽ ⚹ ♅
14 01	☽ ± ♀
16 07	☽ ⚹ ♄
16 51	☽ ⚼ ♆
15 Tuesday	
00 16	☿ # ♆
00 17	☽ ∥ ♀
00 57	☽ □ ♂
02 33	☽ ⚻ ♀
10 26	☽ ∥ ♆
10 52	☽ Q ♅
11 00	☽ ± ♆
11 58	☽ # ☿
16 14	☽ Q ♄
21 05	☽ ☍ ☿
21 51	☽ ⚻ ♇
23 10	☽ ⚻ ♆
16 Wednesday	
02 34	☽ ⚻ ♃
02 35	☉ △ ♅
09 15	♂ Q ♃
10 03	☽ ± ♇
15 00	☽ ± ♃
15 09	☽ Q ♂
19 36	☿ ⚺ ♇
23 19	☽ □ ♅
17 Thursday	
01 17	☽ ⚻ ☉
04 23	☽ ⚼ ♇
04 42	☽ □ ♄
09 41	☽ ⚼ ♃
14 39	☽ ± ☉
16 00	☽ ∥ ☿
16 46	☽ ⚹ ♂
17 32	☽ ☍ ♀
23 12	☽ # ♆
18 Friday	
10 40	☽ ⚼ ☉
11 00	☽ △ ♇
11 37	☽ ⚻ ☿
12 18	☽ ☍ ♆
16 17	☉ △ ♄
16 49	☽ △ ♃
17 18	☽ # ♀
21 20	☿ St R
23 49	☽ ± ☿
19 Saturday	
00 45	☽ ∠ ♂
06 16	☉ ∠ ♆
08 58	♀ △ ♂
12 11	☽ △ ♅
15 23	☉ ⊥ ♀
17 28	☽ △ ♄
17 52	☽ ⚼ ☿
17 58	☽ ∥ ☉
19 47	☽ ⊥ ♂
19 50	☽ △ ☉
20 Sunday	
03 27	♄ ⚼ ♇
07 52	☽ ⚻ ♀
08 22	☽ ⚺ ♂
14 50	☿ ⚺ ♇
18 09	☽ ⚼ ♅
20 03	☽ ± ♀
23 09	☽ △ ☿
23 18	☽ ⚼ ♄
23 23	☽ □ ♇
21 Monday	
00 39	☽ ⚻ ♆
06 03	☽ □ ♃
11 49	☽ ± ♅
12 26	☽ ± ♆
12 30	☽ ∥ ♃
14 03	☽ ⚼ ♀
16 53	☽ ± ♄
23 26	☽ ⚻ ♅
22 Tuesday	
04 26	☽ ⚻ ♄
05 51	☽ ⚼ ♆
10 21	☽ Q ♇
11 53	☽ □ ☉
17 20	☽ Q ♃
19 15	☽ △ ♀
21 15	☽ ☌ ♂
23 Wednesday	
06 18	☽ □ ☿
08 53	☽ ⚹ ♇
10 05	☽ △ ♆
16 02	☽ ⚹ ♃
24 Thursday	
00 02	☽ Q ☉
07 08	☽ ☍ ♅
11 46	☽ ☍ ♄
12 06	☽ ∠ ♇
13 07	☽ Q ☿
19 22	☽ ∠ ♃
20 37	☉ # ♀
23 04	☽ ⚹ ☉
25 Friday	
02 06	☽ □ ♀
03 55	☽ ⊥ ♇
05 35	☽ ⚺ ♂
08 48	☽ ⚹ ☿
11 15	☽ ⊥ ♃
14 15	☽ ⚺ ♇
15 21	☽ □ ♆
16 16	☽ ⊥ ♂
21 37	☽ ⚺ ♃
26 Saturday	
02 41	☽ ∠ ☉
08 01	☽ ∠ ♂
08 33	☽ ∠ ☿
08 50	☽ Q ♀
10 42	☽ ⚻ ♅
12 09	☽ ∥ ♃
13 24	☿ □ ♂
15 00	☽ ⚻ ♄
18 44	☽ ⊥ ☉
19 32	☉ ⚺ ♀
20 35	☽ ± ♅
21 27	☽ Q ♆
22 25	☽ ⊥ ☿
27 Sunday	
00 50	☽ ± ♄
04 36	☽ ⚹ ♀
05 15	☽ ⚺ ☉
07 33	☽ ⚺ ☿
09 32	☽ ⚹ ♂
11 13	☽ ⚼ ♅
11 45	☿ ⊥ ♆
15 24	☽ ⚼ ♄
15 41	♀ St R
15 53	☽ ☌ ♇
16 56	☽ ⚹ ♆
23 00	☉ ☌ ☿
23 31	☽ ☌ ♃
28 Monday	
03 38	☽ # ♀
04 41	☽ ∠ ♀
05 27	☽ Q ♂
11 10	☽ △ ♅
11 33	☉ ⊥ ♆
14 45	☽ ∥ ☉
15 17	☽ △ ♄
16 51	☽ ∠ ♆
18 54	☽ ⊥ ♀
22 34	☿ # ♆

MARCH 2033

LONGITUDES

Date	Sidereal time h m s	Sun ☉ ° ' "	Moon ☽ ° ' "	Moon ☽ 24.00 ° ' "	Mercury ☿ ° '	Venus ♀ ° '	Mars ♂ ° '	Jupiter ♃ ° '	Saturn ♄ ° '	Uranus ♅ ° '	Neptune ♆ ° '	Pluto ♇ ° '
01	22 38 26	11 ♓ 12 57	13 ♓ 21 31	20 ♓ 59 03	08 ♓ 02	08 ♈ 28	12 ♐ 41	20 ♒ 38	00 ♋ 13	27 ♊ 36	16 ♈ 13	15 ♒ 34
02	22 42 22	12 13 11	28 ♓ 35 24	06 ♈ 09 22	06 R 59	08 R 22	13 13	20 52	00 D 13	27 R 36	16 15	15 36
03	22 46 19	13 13 23	13 ♈ 39 50	21 ♈ 05 49	05 59	08 14	13 45	21 06	00 13	27 36	16 17	15 37
04	22 50 15	14 13 33	28 ♈ 26 36	05 ♉ 41 35	05 03	08 03	14 17	21 19	00 13	27 36	16 19	15 39
05	22 54 12	15 13 41	12 ♉ 50 24	19 ♉ 52 51	04 11	07 49	14 48	21 33	00 14	27 36	16 21	15 41
06	22 58 09	16 13 47	26 ♉ 48 52	03 ♊ 38 33	03 25	07 34	15 19	21 47	00 14	27 D 36	16 23	15 42
07	23 02 05	17 13 51	10 ♊ 22 07	16 ♊ 59 49	02 45	07 15	15 50	22 01	00 15	27 36	16 25	15 44
08	23 06 02	18 13 53	23 ♊ 32 01	29 ♊ 59 06	02 11	06 55	16 21	22 14	00 15	27 36	16 27	15 45
09	23 09 58	19 13 52	06 ♋ 21 29	12 ♋ 39 36	01 44	06 32	16 52	22 28	00 16	27 36	16 30	15 47
10	23 13 55	20 13 50	18 ♋ 53 52	25 ♋ 04 41	01 24	06 07	17 23	22 41	00 17	27 36	16 32	15 49
11	23 17 51	21 13 45	01 ♌ 12 27	07 ♌ 17 32	01 10	05 39	17 54	22 55	00 18	27 37	16 34	15 50
12	23 21 48	22 13 38	13 ♌ 20 18	19 ♌ 21 03	01 03	05 10	18 24	23 08	00 19	27 37	16 36	15 52
13	23 25 44	23 13 29	25 ♌ 20 06	01 ♍ 17 43	01 D 02	04 39	18 54	23 21	00 21	27 37	16 38	15 53
14	23 29 41	24 13 18	07 ♍ 14 11	13 ♍ 09 44	01 08	04 06	19 24	23 35	00 22	27 38	16 40	15 55
15	23 33 38	25 13 04	19 ♍ 04 38	24 ♍ 59 05	01 19	03 32	19 54	23 48	00 23	27 38	16 42	15 56
16	23 37 34	26 12 49	00 ♎ 53 21	06 ♎ 47 42	01 35	02 57	20 24	24 01	00 25	27 39	16 44	15 58
17	23 41 31	27 12 32	12 ♎ 42 22	18 ♎ 37 38	01 57	02 21	20 54	24 14	00 27	27 39	16 46	15 59
18	23 45 27	28 12 13	24 ♎ 33 50	00 ♏ 31 17	02 24	01 43	21 23	24 27	00 29	27 40	16 49	16 01
19	23 49 24	29 ♓ 11 52	06 ♏ 30 19	12 ♏ 31 21	02 55	01 06	21 53	24 40	00 31	27 41	16 51	16 02
20	23 53 20	00 ♈ 11 29	18 ♏ 34 46	24 ♏ 41 01	03 31	00 ♈ 28	22 22	24 54	00 33	27 42	16 53	16 03
21	23 57 17	01 11 04	00 ♐ 50 33	07 ♐ 03 51	04 11	29 ♓ 50	22 51	25 06	00 35	27 42	16 55	16 05
22	00 01 13	02 10 38	13 ♐ 21 23	19 ♐ 43 37	04 54	29 13	23 20	25 19	00 37	27 43	16 57	16 06
23	00 05 10	03 10 10	26 ♐ 11 02	02 ♑ 44 03	05 41	28 36	23 48	25 32	00 39	27 44	17 00	16 08
24	00 09 07	04 09 40	09 ♑ 23 02	16 ♑ 08 17	06 32	28 00	24 17	25 45	00 42	27 45	17 02	16 09
25	00 13 03	05 09 09	22 ♑ 58 57	29 ♑ 58 23	07 25	27 24	24 45	25 58	00 44	27 46	17 04	16 10
26	00 17 00	06 08 35	07 ♒ 03 17	14 ♒ 14 30	08 22	26 50	25 13	26 10	00 47	27 47	17 06	16 11
27	00 20 56	07 08 00	21 ♒ 31 43	28 ♒ 54 20	09 22	26 18	25 41	26 23	00 50	27 48	17 08	16 13
28	00 24 53	08 07 23	06 ♓ 21 40	13 ♓ 52 47	10 24	25 47	26 08	26 35	00 53	27 50	17 11	16 14
29	00 28 49	09 06 44	21 ♓ 26 41	29 ♓ 02 12	11 29	25 18	26 36	26 48	00 56	27 51	17 13	16 15
30	00 32 46	10 06 04	06 ♈ 38 08	14 ♈ 13 15	12 36	24 50	27 03	27 00	00 59	27 52	17 15	16 16
31	00 36 42	11 ♈ 05 21	21 ♈ 46 22	29 ♈ 16 21	13 ♓ 45	24 ♓ 25	27 ♐ 29	27 ♒ 12	01 ♋ 02	27 ♊ 53	17 ♈ 17	16 ♒ 18

MOON NODES AND LATITUDE / DECLINATIONS

Date	Moon True ☊ ° '	Moon Mean ☊ ° '	Moon Latitude ° '	Sun ☉ ° '	Moon ☽ ° '	Mercury ☿ ° '	Venus ♀ ° '	Mars ♂ ° '	Jupiter ♃ ° '	Saturn ♄ ° '	Uranus ♅ ° '	Neptune ♆ ° '	Pluto ♇ ° '
01	21 ♎ 50	23 ♎ 37	03 N 13	07 S 21	03 S 34	05 S 11	09 N 53	21 S 42	15 S 13	22 N 38	23 N 38	04 N 56	22 S 52
02	21 R 46	23 34	02 03	06 59	01 N 19	05 39	10 00	21 46	15 09	22 39	23 38	04 57	22 51
03	21 44	23 31	00 N 44	06 35	06 04	06 08	10 05	21 51	15 05	22 39	23 38	04 58	22 51
04	21 D 44	23 27	00 S 37	06 12	10 21	06 37	10 09	21 56	15 00	22 39	23 38	04 59	22 51
05	21 45	23 24	01 53	05 49	13 54	07 06	10 12	22 00	14 56	22 39	23 38	04 59	22 50
06	21 47	23 21	03 00	05 26	16 31	07 33	10 13	22 05	14 52	22 39	23 38	05 00	22 50
07	21 48	23 18	03 55	05 03	18 07	07 59	10 13	22 09	14 48	22 39	23 38	05 01	22 49
08	21 R 48	23 15	04 36	04 39	18 41	08 24	10 11	22 13	14 43	22 40	23 38	05 02	22 49
09	21 46	23 12	05 02	04 16	18 16	08 46	10 08	22 17	14 39	22 40	23 38	05 03	22 49
10	21 43	23 08	05 12	03 52	16 57	09 07	10 04	22 21	14 35	22 40	23 38	05 03	22 49
11	21 39	23 05	05 08	03 29	14 52	09 25	09 57	22 25	14 30	22 40	23 38	05 04	22 48
12	21 34	23 02	04 51	03 05	12 10	09 41	09 50	22 29	14 26	22 40	23 38	05 05	22 48
13	21 29	22 59	04 20	02 41	08 59	09 55	09 41	22 32	14 22	22 41	23 38	05 06	22 48
14	21 24	22 56	03 39	02 18	05 28	10 06	09 30	22 36	14 18	22 41	23 38	05 07	22 47
15	21 20	22 52	02 48	01 54	01 N 45	10 15	09 18	22 39	14 13	22 41	23 38	05 07	22 47
16	21 17	22 49	01 50	01 30	02 S 02	10 22	09 05	22 42	14 09	22 41	23 38	05 08	22 47
17	21 16	22 46	00 S 47	01 07	05 44	10 27	08 50	22 46	14 05	22 41	23 38	05 09	22 46
18	21 D 15	22 43	00 N 18	00 43	09 14	10 29	08 35	22 49	14 01	22 41	23 38	05 10	22 46
19	21 16	22 40	01 23	00 S 19	12 22	10 30	08 18	22 52	13 56	22 42	23 38	05 11	22 46
20	21 18	22 37	02 25	00 N 05	15 01	10 28	08 00	22 55	13 52	22 42	23 38	05 12	22 46
21	21 19	22 33	03 22	00 28	17 02	10 24	07 42	22 58	13 48	22 42	23 38	05 13	22 45
22	21 21	22 30	04 10	00 52	18 16	10 19	07 22	23 00	13 44	22 42	23 38	05 13	22 45
23	21 21	22 27	04 47	01 16	18 36	10 11	07 02	23 03	13 40	22 42	23 38	05 14	22 45
24	21 R 22	22 24	05 10	01 39	17 57	10 02	06 42	23 06	13 35	22 42	23 38	05 15	22 45
25	21 21	22 21	05 17	02 03	16 16	09 51	06 21	23 08	13 31	22 43	23 38	05 16	22 45
26	21 19	22 18	05 05	02 27	13 30	09 38	06 00	23 11	13 27	22 43	23 38	05 17	22 44
27	21 18	22 14	04 34	02 50	10 00	09 23	05 39	23 13	13 23	22 43	23 38	05 18	22 44
28	21 16	22 11	03 44	03 13	05 42	09 07	05 18	23 15	13 19	22 43	23 38	05 19	22 44
29	21 15	22 08	02 38	03 37	00 S 58	08 50	04 57	23 17	13 15	22 43	23 38	05 19	22 44
30	21 14	22 05	01 20	04 00	03 N 52	08 30	04 36	23 19	13 11	22 43	23 38	05 20	22 44
31	21 ♎ 14	22 ♎ 02	00 S 03	04 N 23	08 N 26	08 S 10	04 N 16	23 S 22	13 S 07	22 N 43	23 N 38	05 N 21	22 S 44

ZODIAC SIGN ENTRIES

Date	h m	Planets
02	14 14	☽ ♈
04	14 34	☽ ♉
06	17 35	☽ ♊
09	00 02	☽ ♋
11	09 38	☽ ♌
13	21 23	☽ ♍
16	10 12	☽ ♎
18	22 57	☽ ♏
20	07 23	☉ ♈
21	05 49	♀ ♓
21	10 22	☽ ♐
23	19 01	☽ ♑
26	00 03	☽ ♒
28	01 46	☽ ♓
30	01 31	☽ ♈

LATITUDES

Date	Mercury ☿ ° '	Venus ♀ ° '	Mars ♂ ° '	Jupiter ♃ ° '	Saturn ♄ ° '	Uranus ♅ ° '	Neptune ♆ ° '	Pluto ♇ ° '
01	03 N 38	07 N 07	00 N 38	00 S 39	00 S 48	00 N 13	01 S 34	07 S 01
04	03 16	07 35	00 35	00 39	00 47	00 13	01 34	07 02
07	02 41	08 00	00 32	00 40	00 46	00 13	01 34	07 03
10	02 00	08 19	00 29	00 40	00 46	00 13	01 33	07 03
13	01 16	08 32	00 26	00 40	00 45	00 13	01 33	07 04
16	00 N 34	08 38	00 23	00 41	00 45	00 13	01 33	07 04
19	00 S 04	08 35	00 19	00 41	00 44	00 13	01 33	07 05
22	00 39	08 23	00 15	00 42	00 44	00 13	01 33	07 05
25	01 09	08 03	00 11	00 42	00 43	00 13	01 33	07 06
28	01 34	07 36	00 07	00 42	00 43	00 13	01 33	07 06
31	01 S 55	07 N 04	00 N 03	00 S 43	00 S 42	00 N 13	01 S 33	07 S 07

DATA

Julian Date	2463658
Delta T	+75 seconds
Ayanamsa	24° 19' 22"
Synetic vernal point	04° ♓ 47' 37"
True obliquity of ecliptic	23° 25' 58"

MOON'S PHASES, APSIDES AND POSITIONS ☽

Date	h m	Phase	Longitude ° '	Eclipse Indicator
01	08 23	●	11 ♓ 04	
08	01 27	☽	17 ♊ 47	
16	01 37	○	25 ♍ 47	
24	01 50	☾	03 ♑ 44	
30	17 52	●	10 ♈ 21	Total

Day	h m		
01	18 10	Perigee	
15	21 41	Apogee	
30	06 04	Perigee	
02	05 33	0N	
08	13 22	Max dec	18° N 41'
15	23 05	0S	
23	08 27	Max dec	18° S 37'
29	16 49	0N	

LONGITUDES

Date	Chiron ⚷ ° '	Ceres ⚳ ° '	Pallas ⚴ ° '	Juno ⚵ ° '	Vesta ⚶ ° '	Black Moon Lilith ⚸ ° '
01	21 ♉ 37	24 ♐ 33	16 ♏ 36	13 ♎ 03	05 ♒ 11	22 ♍ 53
11	21 ♉ 59	27 ♐ 08	16 ♏ 42	11 ♎ 06	10 ♒ 11	24 ♍ 00
21	22 ♉ 27	29 ♐ 20	16 ♏ 00	08 ♎ 48	15 ♒ 04	25 ♍ 06
31	22 ♉ 58	01 ♑ 06	14 ♏ 28	06 ♎ 23	19 ♒ 50	26 ♍ 13

All ephemeris data is given at 12.00 UT and the Moon's longitude is additionally given for 24.00 UT

ASPECTARIAN

h m Aspects

01 Tuesday
01 31 ☿ ⚺ ♀
04 09 ☽ ☌ ☿
04 20 ☽ ⚺ ♀
04 34 ☽ ∥ ☿
05 07 ☽ ⋕ ♆
07 04 ☽ ⊥ ♆
08 23 ☽ ☌ ☉
10 55 ☽ □ ♂
15 29 ☽ ⚺ ♇
16 02 ♄ St D
16 31 ☽ ⚺ ♆
23 37 ☽ ⚺ ♃

02 Wednesday
00 56 ☽ ⊥ ♇
09 14 ☽ ⊥ ♃
10 26 ☽ □ ♅
14 34 ☽ □ ♄
15 11 ☽ ∠ ♇
23 43 ☽ ∠ ♃

03 Thursday
00 29 ☽ ⚺ ☿
03 23 ☽ ☌ ♀
06 16 ☽ ∥ ♆
09 28 ☽ ⊥ ☿
11 14 ☽ ⚺ ☉
12 09 ☽ △ ♂
12 26 ☽ ⋕ ☿
14 34 ☽ ⋕ ☉
15 06 ☽ Q ♅
15 09 ☽ ⚹ ♇
16 14 ☽ ☌ ♆
19 20 ☽ Q ♄
21 37 ☽ ⊥ ☉
23 06 ☽ ∠ ☿

04 Friday
00 11 ☽ ⚹ ♃
00 33 ☉ ∥ ☿
10 36 ☽ ⚹ ♅
10 42 ☽ Q ♇
10 48 ☽ ∥ ♀
13 23 ☽ ∠ ☉
13 25 ☽ ⅊ ♂
14 28 ☉ □ ♂
14 55 ☽ ⚹ ♄
20 11 ☽ Q ♃
22 17 ☽ ⚹ ☿

05 Saturday
03 42 ☽ ⚺ ♀
04 56 ☽ ± ♂
11 35 ☽ ∠ ♅
13 38 ☽ ⊥ ♀
14 41 ♅ St D
15 27 ☽ ⊼ ♂
16 03 ☽ ∠ ♄
16 22 ☽ ⚹ ☉
16 49 ☽ □ ♇
17 22 ☽ Q ☿
17 59 ☽ ⚺ ♆
20 15 ☽ ⋕ ♃
23 02 ☉ ⚺ ♇

06 Sunday
02 57 ☽ ⊥ ♅
03 07 ☽ □ ♃
04 18 ☽ ⊥ ♆
04 46 ☽ ∠ ♀
07 30 ☽ ⊥ ♄
13 22 ☽ ⚺ ♅
14 40 ☽ Q ☉
15 58 ☉ ⚺ ♆
17 59 ☽ ⚺ ♄
20 02 ☽ ∠ ♆
23 02 ☽ □ ☿

07 Monday
06 33 ☽ ⚹ ♀
06 40 ♂ ⚹ ♇
13 37 ☉ ⋕ ♆
21 43 ☽ △ ♇
22 18 ☽ ☍ ♂
22 38 ♀ ∠ ♃
22 59 ☽ ⚹ ♆

08 Tuesday
01 27 ☽ □ ☉
03 43 ☽ Q ♀
09 34 ☽ △ ♃
17 03 ♂ △ ♆
19 33 ☽ ☌ ♅
21 10 ☽ Q ♆

09 Wednesday
00 31 ☽ ☌ ♄
01 28 ☽ ⅊ ♇
03 33 ☽ △ ☿
12 19 ☽ □ ♀
14 08 ☽ ⅊ ♃
18 31 ☽ ± ♇

10 Thursday
02 57 ☿ ∠ ♆
06 02 ☽ ⊼ ♇
07 17 ☽ ⅊ ☿
07 25 ☽ □ ♆
07 39 ☽ ± ♃
08 57 ☽ ⊼ ♂
14 48 ☽ △ ☉
19 29 ☽ ⊼ ♃
21 04 ☽ ± ♂

11 Friday
00 22 ☽ ± ☿
04 57 ☽ ⚺ ♅
10 13 ☽ ⚺ ♄
11 56 ☽ ⊼ ☿
15 28 ☽ ⅊ ♂
15 37 ☽ ⋕ ♃
16 44 ☽ ⊥ ♅
20 26 ☽ △ ♀
22 04 ☽ ⊥ ♄
22 47 ☽ ⅊ ☉

12 Saturday
02 58 ☉ ⊥ ♇
10 34 ☽ ∠ ♅
15 58 ☽ ∠ ♄
17 02 ☽ ☍ ♇
18 17 ☽ ± ☉
18 31 ☽ △ ♆
21 05 ☿ ⋕ ♀
22 33 ☽ △ ♂

13 Sunday
01 05 ☽ ⅊ ♀
02 58 ☿ St D
05 43 ☽ ⋕ ☿
06 45 ☽ ∥ ♀
07 23 ☽ ⊼ ☉
07 57 ☽ ☍ ♃
16 06 ☉ ⚺ ♃
16 36 ☽ ⚹ ♅
18 23 ☽ ± ♀
19 15 ☿ Q ♂
22 06 ☽ ⚹ ♄
23 33 ☽ ☍ ☿

14 Monday
00 43 ☽ ⅊ ♆
05 57 ☽ ⊼ ♀
14 19 ☽ ∥ ♆
16 51 ☽ Q ♅
18 58 ☽ ± ♆
22 25 ☽ Q ♄

15 Tuesday
05 37 ☽ ⊼ ♇
07 10 ☽ ⊼ ♆
10 54 ☽ ⋕ ☉
13 46 ☽ □ ♂
17 49 ☽ ± ♇
21 47 ☽ ⊼ ♃

16 Wednesday
01 16 ♂ ⋕ ♄
01 37 ☽ ☍ ☉
05 24 ☽ □ ♅
08 57 ☽ ∥ ☉
10 12 ☽ ± ♃
11 02 ☽ □ ♄
12 09 ☽ ⅊ ♇
13 28 ☽ ⊼ ☿
15 59 ☽ ☍ ♀
23 33 ☿ ∠ ♆

17 Thursday
02 02 ☽ ± ☿
03 56 ☽ Q ♂
04 50 ☽ ⅊ ♃
08 07 ☽ ⋕ ♆
17 25 ♂ ∥ ♇
18 40 ☽ △ ♇
20 16 ☽ ☍ ♆
20 56 ☽ ⅊ ☿
21 01 ☿ ⚺ ♀
22 57 ☉ □ ♅

18 Friday
05 19 ☽ ⚹ ♂
07 41 ☽ ⋕ ♀
11 47 ☽ △ ♃
18 16 ☽ △ ♅
20 00 ☽ ⊼ ☉
21 16 ☽ ∥ ☿
23 57 ☽ △ ♄

19 Saturday
01 42 ☽ ⊼ ♀
04 28 ☽ △ ☿
09 09 ☽ ± ☉
12 47 ☽ ∠ ♂
13 08 ☽ ± ♀
14 23 ♀ ∠ ♇
23 59 ♀ ⊥ ♃

20 Sunday
00 20 ☽ ⅊ ♅
01 12 ☽ ∥ ♃
04 42 ☽ ⅊ ☉
05 59 ☽ ⅊ ♄
06 09 ☽ ⅊ ♀
07 00 ☽ □ ♇
07 26 ☽ ⊥ ♂
08 38 ☽ ⊼ ♆
09 17 ♀ □ ♄
16 05 ☉ ☌ ♀
18 08 ☽ ± ♅
19 46 ☽ ⚺ ♂
20 30 ☽ ± ♆
20 49 ☉ □ ♄
23 46 ☽ ± ♄

21 Monday
00 38 ☽ □ ♃
05 54 ☽ ⊼ ♅
09 25 ☉ ∠ ♇
09 37 ☉ ⊥ ♃
10 09 ☽ △ ♀
11 29 ☽ ⊼ ♄
12 43 ☽ △ ☉
14 06 ☽ ⅊ ♆
18 16 ☽ Q ♇
18 50 ☽ □ ☿

22 Tuesday
11 56 ☽ Q ♃
17 12 ☽ ⚹ ♇
18 49 ☽ △ ♆

23 Wednesday
07 04 ☽ Q ☿
07 26 ☽ ☌ ♂
10 47 ☽ ⚹ ♃
14 52 ☽ ☍ ♅
16 14 ☽ □ ♀
20 02 ☿ Q ♂
20 14 ☽ ☍ ♄
21 05 ☽ ∠ ♇

24 Thursday
01 50 ☽ □ ☉
06 31 ☽ ⚹ ☿
13 22 ☽ ⊥ ♇
14 29 ☽ ∠ ♃
21 24 ♀ □ ♅
23 15 ☽ Q ♀

25 Friday
00 02 ☽ ⚺ ♇
01 36 ☽ □ ♆
06 37 ☽ ⊥ ♃
10 56 ☽ ∠ ☿
12 17 ☽ Q ☉
15 08 ☽ ⚺ ♂
17 12 ☽ ⚺ ♃
19 18 ☽ ⚹ ♀
20 14 ☽ ⊼ ♅

26 Saturday
01 21 ☽ ⊼ ♄
01 47 ☽ ⊥ ♂
03 30 ☽ ⊥ ☿
06 29 ☽ ± ♅
08 42 ☽ Q ♆
10 21 ☽ ⚹ ☉
11 33 ☽ ± ♄
13 01 ☽ ∥ ♃
14 22 ☽ ⚺ ☿
17 28 ☽ ∠ ♂
19 43 ☽ ∠ ♀
21 36 ☽ ⅊ ♅

27 Sunday
02 36 ☽ ⅊ ♄
03 15 ☽ ☌ ♇
04 47 ☽ ⚹ ♆
09 19 ♀ ⚺ ♃
10 03 ☽ ⊥ ♀
13 04 ☽ ∠ ☉
15 51 ☽ ∥ ☿
18 59 ☽ ⚹ ♂
19 30 ☽ ⚺ ♀
20 01 ☽ ☌ ♃
22 14 ☽ △ ♅

28 Monday
03 08 ♀ □ ♂
03 09 ☽ △ ♄
04 43 ☽ ⊥ ☉
05 16 ☽ ∠ ♆
11 21 ♀ ∥ ♆
14 04 ☽ ⋕ ♆
14 17 ☽ ⋕ ♀
14 56 ☽ Q ♂
15 01 ☽ ⚺ ☉
18 57 ☽ ☌ ☿
19 43 ☽ ⊥ ♆
23 48 ☽ ⋕ ☉

29 Tuesday
03 46 ☽ ⚺ ♇
05 17 ☽ ⚺ ♆
06 07 ☿ ⊥ ♆
13 17 ☽ ⊥ ♇
17 54 ☽ ☌ ♀
20 23 ☽ □ ♂
20 34 ☽ ⚺ ♃
22 08 ☽ □ ♅

30 Wednesday
03 02 ☽ □ ♄
03 31 ☽ ∠ ♇
06 11 ☽ ⊥ ♃
07 38 ♂ ⚹ ♃
12 46 ☽ ∥ ☉
15 31 ☽ ∥ ♀
17 52 ☽ ☌ ☉
19 32 ☽ ∥ ♆
20 36 ☽ ∠ ♃
22 12 ☽ ⚺ ☿

31 Thursday
02 38 ☽ Q ♅
03 17 ☽ ⚹ ♇
04 51 ☽ ☌ ♆
07 37 ☽ Q ♄
07 59 ☉ ∥ ♀
08 31 ☽ ⊥ ☿
10 36 ☽ ⋕ ☿
15 57 ♄ ⋕ ♇
16 07 ☽ ⚺ ♀
20 48 ☽ ⚹ ♃
21 25 ☽ △ ♂
21 48 ☽ ⚹ ♅
22 27 ☽ Q ♇

APRIL 2033

LONGITUDES

Date	Sidereal time h m s	Sun ☉ ° ' "	Moon ☽ ° ' "	Moon ☽ 24.00 ° ' "	Mercury ☿ ° '	Venus ♀ ° '	Mars ♂ ° '	Jupiter ♃ ° '	Saturn ♄ ° '	Uranus ♅ ° '	Neptune ♆ ° '	Pluto ♇ ° '
01	00 40 39	12 ♈ 04 36	06 ♉ 42 13	14 ♉ 03 04	14 ♓ 57	24 ♓ 02	27 ♐ 56	27 ♒ 24	01 ♋ 05	27 ♊ 55	17 ♈ 20	16 ♒ 19
02	00 44 36	13 03 49	21 ♉ 18 14	28 ♉ 27 10	16 11	23 R 41	28 22	27 36	01 09	27 56	17 22	16 20
03	00 48 32	14 03 00	05 ♊ 29 33	12 ♊ 25 11	17 26	23 23	28 49	27 48	01 12	27 58	17 24	16 21
04	00 52 29	15 02 08	19 ♊ 14 03	25 ♊ 56 17	18 44	23 07	29 14	28 00	01 16	27 59	17 27	16 22
05	00 56 25	16 01 14	02 ♋ 32 04	09 ♋ 01 47	20 04	22 53	29 ♐ 40	28 12	01 19	28 01	17 29	16 23
06	01 00 22	17 00 18	15 ♋ 25 47	21 ♋ 44 32	21 25	22 42	00 ♑ 05	28 24	01 23	28 02	17 31	16 24
07	01 04 18	17 59 20	27 ♋ 58 33	04 ♌ 08 19	22 48	22 33	00 30	28 36	01 27	28 04	17 33	16 25
08	01 08 15	18 58 19	10 ♌ 14 23	16 ♌ 17 16	24 13	22 27	00 55	28 47	01 31	28 06	17 36	16 26
09	01 12 11	19 57 16	22 ♌ 17 29	28 ♌ 15 32	25 40	22 23	01 20	28 59	01 35	28 08	17 38	16 27
10	01 16 08	20 56 10	04 ♍ 11 55	10 ♍ 07 04	27 09	22 21	01 44	29 10	01 39	28 09	17 40	16 28
11	01 20 05	21 55 03	16 ♍ 01 25	21 ♍ 55 24	28 ♓ 39	22 D 22	02 08	29 22	01 43	28 11	17 42	16 29
12	01 24 01	22 53 53	27 ♍ 49 21	03 ♎ 43 39	00 ♈ 10	22 25	02 32	29 33	01 47	28 13	17 45	16 30
13	01 27 58	23 52 41	09 ♎ 38 36	15 ♎ 34 31	01 44	22 31	02 55	29 44	01 52	28 15	17 47	16 31
14	01 31 54	24 51 27	21 ♎ 31 39	27 ♎ 30 18	03 19	22 38	03 18	29 ♒ 55	01 56	28 17	17 49	16 32
15	01 35 51	25 50 11	03 ♏ 30 41	09 ♏ 33 03	04 55	22 48	03 41	00 ♓ 06	02 01	28 19	17 51	16 33
16	01 39 47	26 48 54	15 ♏ 37 38	21 ♏ 44 40	06 34	23 00	04 03	00 17	02 05	28 21	17 54	16 34
17	01 43 44	27 47 34	27 ♏ 54 22	04 ♐ 06 58	08 13	23 15	04 25	00 28	02 10	28 23	17 56	16 34
18	01 47 40	28 46 12	10 ♐ 22 42	16 ♐ 41 49	09 55	23 31	04 47	00 38	02 15	28 25	17 58	16 35
19	01 51 37	29 ♈ 44 49	23 ♐ 04 33	29 ♐ 31 09	11 38	23 49	05 08	00 49	02 20	28 28	18 00	16 36
20	01 55 33	00 ♉ 43 24	06 ♑ 01 51	12 ♑ 36 53	13 23	24 09	05 29	00 59	02 25	28 30	18 03	16 37
21	01 59 30	01 41 58	19 ♑ 16 28	26 ♑ 00 46	15 09	24 31	05 49	01 10	02 30	28 32	18 05	16 37
22	02 03 27	02 40 30	02 ♒ 49 56	09 ♒ 44 04	16 57	24 54	06 10	01 20	02 35	28 35	18 07	16 38
23	02 07 23	03 39 00	16 ♒ 43 09	23 ♒ 47 10	18 47	25 20	06 29	01 30	02 40	28 37	18 09	16 38
24	02 11 20	04 37 29	00 ♓ 55 55	08 ♓ 09 09	20 38	25 47	06 49	01 40	02 45	28 39	18 12	16 39
25	02 15 16	05 35 56	15 ♓ 26 27	22 ♓ 47 20	22 31	26 15	07 08	01 50	02 51	28 42	18 14	16 40
26	02 19 13	06 34 21	00 ♈ 11 08	07 ♈ 37 06	24 26	26 45	07 26	02 00	02 56	28 44	18 16	16 41
27	02 23 09	07 32 45	15 ♈ 04 23	22 ♈ 32 03	26 22	27 17	07 44	02 10	03 01	28 47	18 18	16 41
28	02 27 06	08 31 07	29 ♈ 59 06	07 ♉ 24 34	28 ♈ 20	27 50	08 02	02 19	03 07	28 49	18 20	16 42
29	02 31 02	09 29 27	14 ♉ 47 28	22 ♉ 06 51	00 ♉ 19	28 24	08 19	02 29	03 13	28 52	18 22	16 42
30	02 34 59	10 ♉ 27 46	29 ♉ 21 54	06 ♊ 31 54	02 ♉ 20	28 ♓ 59	08 ♑ 36	02 ♓ 38	03 ♋ 18	28 ♊ 54	18 ♈ 25	16 ♒ 43

DECLINATIONS

Date	Moon True ☊ ° '	Moon Mean ☊ ° '	Moon ☽ Latitude ° '	Sun ☉ ° '	Moon ☽ ° '	Mercury ☿ ° '	Venus ♀ ° '	Mars ♂ ° '	Jupiter ♃ ° '	Saturn ♄ ° '	Uranus ♅ ° '	Neptune ♆ ° '	Pluto ♇ ° '
01	21 ♎ 14	21 ♎ 58	01 S 25	04 N 46	12 N 25	07 S 47	03 N 56	23 S 24	13 S 03	22 N 44	23 N 38	05 N 22	22 S 43
02	21 D 14	21 55	02 40	05 09	15 31	07 24	03 37	23 25	12 59	22 44	23 38	05 23	22 43
03	21 15	21 52	03 42	05 32	17 34	06 59	03 18	23 27	12 55	22 44	23 38	05 24	22 43
04	21 16	21 49	04 30	05 55	18 31	06 32	03 00	23 29	12 51	22 44	23 38	05 25	22 43
05	21 16	21 46	05 01	06 18	18 23	06 05	02 43	23 31	12 47	22 44	23 38	05 25	22 43
06	21 16	21 43	05 16	06 41	17 18	05 36	02 26	23 33	12 43	22 44	23 38	05 26	22 43
07	21 R 16	21 39	05 16	07 03	15 24	05 05	02 10	23 34	12 39	22 44	23 38	05 27	22 43
08	21 16	21 36	05 01	07 26	12 50	04 34	01 56	23 36	12 35	22 45	23 38	05 28	22 43
09	21 16	21 33	04 33	07 48	09 47	04 01	01 42	23 37	12 31	22 45	23 38	05 29	22 43
10	21 16	21 30	03 53	08 10	06 21	03 27	01 29	23 39	12 27	22 45	23 38	05 30	22 42
11	21 D 16	21 27	03 04	08 32	02 N 41	02 52	01 17	23 40	12 24	22 45	23 38	05 31	22 42
12	21 16	21 24	02 07	08 54	01 S 04	02 16	01 06	23 42	12 20	22 45	23 38	05 31	22 42
13	21 16	21 20	01 S 04	09 16	04 48	01 39	00 56	23 43	12 16	22 45	23 38	05 32	22 42
14	21 R 16	21 17	00 N 01	09 37	08 22	01 01	00 47	23 45	12 12	22 45	23 38	05 33	22 42
15	21 16	21 14	01 08	09 59	11 37	00 S 21	00 39	23 46	12 09	22 45	23 38	05 34	22 42
16	21 16	21 11	02 11	10 20	14 25	00 N 19	00 32	23 48	12 05	22 45	23 38	05 35	22 42
17	21 15	21 08	03 10	10 41	16 36	01 01	00 26	23 49	12 01	22 45	23 38	05 36	22 42
18	21 14	21 04	04 00	11 02	18 02	01 43	00 21	23 50	11 58	22 45	23 39	05 37	22 42
19	21 13	21 01	04 40	11 23	18 36	02 26	00 17	23 52	11 54	22 46	23 39	05 37	22 42
20	21 12	20 58	05 06	11 43	18 12	03 10	00 14	23 53	11 51	22 46	23 39	05 38	22 42
21	21 11	20 55	05 17	12 04	16 49	03 55	00 12	23 54	11 47	22 46	23 39	05 39	22 42
22	21 D 11	20 52	05 11	12 24	14 29	04 41	00 11	23 56	11 44	22 46	23 39	05 40	22 42
23	21 11	20 49	04 46	12 44	11 16	05 28	00 11	23 57	11 40	22 46	23 39	05 41	22 42
24	21 12	20 45	04 04	13 04	07 20	06 15	00 11	23 58	11 37	22 46	23 39	05 41	22 42
25	21 13	20 42	03 06	13 23	02 S 53	07 03	00 12	24 00	11 34	22 46	23 39	05 42	22 42
26	21 14	20 39	01 54	13 42	01 N 49	07 51	00 15	24 01	11 30	22 46	23 39	05 43	22 42
27	21 15	20 36	00 N 34	14 01	06 28	08 40	00 18	24 03	11 27	22 46	23 39	05 44	22 43
28	21 R 15	20 33	00 S 48	14 20	10 43	09 29	00 21	24 04	11 24	22 46	23 39	05 45	22 43
29	21 14	20 29	02 07	14 39	14 15	10 19	00 26	24 06	11 21	22 46	23 39	05 46	22 43
30	21 ♎ 12	20 ♎ 26	03 S 15	14 N 57	16 N 50	11 N 09	00 N 31	24 S 07	11 S 17	22 N 46	23 N 39	05 N 46	22 S 43

ZODIAC SIGN ENTRIES

Date	h	m	Planets
01	01	10	☽ ♉
03	02	37	☽ ♊
05	07	22	☽ ♋
06	06	51	♂ ♑
07	15	56	☽ ♌
10	03	31	☽ ♍
12	09	18	☿ ♈
12	16	26	☽ ♎
14	22	45	♃ ♓
15	05	00	☽ ♏
17	16	03	☽ ♐
19	18	13	☉ ♉
20	00	53	☽ ♑
22	07	02	☽ ♒
24	10	27	☽ ♓
26	11	42	☽ ♈
28	12	01	☽ ♉
29	08	09	☿ ♉
30	13	03	☽ ♊

LATITUDES

Date	Mercury ☿ ° '	Venus ♀ ° '	Mars ♂ ° '	Jupiter ♃ ° '	Saturn ♄ ° '	Uranus ♅ ° '	Neptune ♆ ° '	Pluto ♇ ° '
01	02 S 01	06 N 52	00 N 01	00 S 43	00 S 42	00 N 13	01 S 33	07 S 07
04	02 16	06 14	00 S 03	00 44	00 42	00 13	01 33	07 08
07	02 26	05 35	00 08	00 44	00 41	00 13	01 33	07 09
10	02 32	04 55	00 14	00 45	00 41	00 13	01 33	07 09
13	02 33	04 15	00 19	00 45	00 40	00 13	01 33	07 10
16	02 29	03 36	00 25	00 46	00 40	00 13	01 33	07 11
19	02 21	02 59	00 32	00 46	00 39	00 13	01 33	07 12
22	02 08	02 24	00 38	00 47	00 39	00 13	01 33	07 12
25	01 51	01 51	00 45	00 48	00 38	00 13	01 33	07 13
28	01 29	01 20	00 53	00 48	00 38	00 13	01 34	07 14
31	01 S 04	00 N 51	01 S 01	00 S 49	00 S 37	00 N 13	01 S 34	07 S 15

DATA

Julian Date	2463689
Delta T	+75 seconds
Ayanamsa	24° 19' 24"
Synetic vernal point	04° ♓ 47' 35"
True obliquity of ecliptic	23° 25' 58"

MOON'S PHASES, APSIDES AND POSITIONS ☽

Date	h	m	Phase	Longitude ° '	Eclipse Indicator
06	15	14	☽	17 ♋ 08	
14	19	17	○	25 ♎ 09	total
22	11	42	☾	02 ♒ 40	
29	02	46	●	09 ♉ 07	

Day	h	m		
12	02	18	Apogee	
27	14	30	Perigee	
04	21	01	Max dec	18° N 35'
12	05	10	0S	
19	14	16	Max dec	18° S 36'
26	02	46	0N	

LONGITUDES

Date	Chiron ⚷ ° '	Ceres ⚳ ° '	Pallas ⚴ ° '	Juno ⚵ ° '	Vesta ⚶ ° '	Black Moon Lilith ⚸ ° '
01	23 ♉ 02	01 ♑ 15	14 ♏ 16	06 ♎ 09	20 ♒ 18	26 ♍ 20
11	23 ♉ 38	02 ♑ 27	11 ♏ 56	03 ♎ 54	24 ♒ 52	27 ♍ 26
21	24 ♉ 16	03 ♑ 06	09 ♏ 24	02 ♎ 01	29 ♒ 12	28 ♍ 33
31	24 ♉ 57	03 ♑ 08	06 ♏ 00	00 ♎ 38	03 ♓ 28	29 ♍ 39

All ephemeris data is given at 12.00 UT and the Moon's longitude is additionally given for 24.00 UT

ASPECTARIAN

h m Aspects

01 Friday
00 08 ☽ ∠ ☿
01 29 ☽ ⊥ ♀
02 53 ☽ ✶ ♄
10 48 ♂ ☍ ♅
15 42 ☽ ∠ ♀
16 16 ☽ # ♃
16 27 ☽ Q ♃
21 24 ☽ ⊻ ☉
22 04 ☉ ∠ ♃
22 09 ☽ ∠ ♅
22 29 ☽ bQ ♂

02 Saturday
02 42 ☽ ✶ ☿
03 25 ☽ ∠ ♄
03 45 ☽ □ ♇
05 27 ☽ ⊻ ♆
08 00 ☽ ⊥ ☉
13 04 ☽ ⊥ ♅
13 51 ☽ ± ♂
15 04 ☿ ⊻ ♇
15 27 ☽ ⊥ ♆
15 54 ☽ ✶ ♀
18 27 ☽ ⊥ ♄
22 43 ☽ □ ♃
23 09 ☽ ⊻ ♅

03 Sunday
00 11 ☽ ∠ ☉
00 15 ☽ ⊼ ♂
00 39 ☽ Q ☿
02 33 ☉ ∥ ♆
04 38 ☽ ⊻ ♄
06 42 ☽ ∠ ♆
11 21 ☿ ⊻ ♆
11 49 ☽ Q ♀

04 Monday
04 00 ☽ ✶ ☉
06 56 ☽ △ ♇
08 49 ☽ ✶ ♆
09 20 ♃ △ ♅
11 01 ☽ □ ☿
18 48 ☽ □ ♀

05 Tuesday
03 06 ☽ Q ☉
03 44 ☽ ☌ ♅
03 59 ☽ △ ♃
05 43 ☉ # ☿
06 24 ☽ Q ♆
06 35 ☽ ☍ ♂
09 46 ☽ ☌ ♄
09 54 ☽ bQ ♇
11 50 ☉ Q ♅
21 08 ☉ ✶ ♇

06 Wednesday
02 33 ☽ ± ♇
08 07 ☽ bQ ♃
13 51 ☽ ⊼ ♇
15 14 ☽ □ ☉
15 58 ☽ □ ♆
19 23 ☿ # ♆

07 Thursday
00 21 ♄ bQ ♇
00 47 ☽ △ ☿
01 01 ☉ ☌ ♆
01 28 ☽ ± ♃
01 39 ☽ △ ♀
05 20 ☿ ⊥ ♇
07 57 ☿ ☌ ♀
12 11 ☽ ⊻ ♅
13 13 ☽ ⊼ ♃
17 05 ☽ ⊼ ♂
18 47 ☽ ⊻ ♄
23 54 ☽ ⊥ ♅

08 Friday
05 14 ☽ ± ♂
06 32 ☽ bQ ♀
06 35 ☽ ⊥ ♄
09 44 ☽ bQ ☿
12 35 ♀ ⊥ ♇
14 12 ☽ # ♃
17 40 ☽ ∠ ♅
23 40 ☽ bQ ♂

09 Saturday
00 14 ☽ ± ♀
00 19 ☽ ☍ ♇
00 31 ☽ ∠ ♄
02 09 ☉ Q ♄
02 39 ☽ △ ♆
06 02 ☽ ± ☿
06 54 ☽ △ ☉
12 10 ☽ ⊼ ♀
19 44 ☽ ⊼ ☿
19 50 ♂ ∠ ♇
23 46 ☽ ✶ ♅

10 Sunday
00 43 ☽ ∥ ☉
01 19 ♂ # ♅
01 41 ☽ ☍ ♃
05 47 ♂ ☍ ♄
06 49 ☽ ✶ ♄
06 50 ☽ △ ♂
08 54 ☽ bQ ♆
15 27 ♀ St D
15 50 ☽ bQ ☉
17 40 ☽ ∥ ♆

11 Monday
00 07 ☽ Q ♅
03 12 ☽ ± ♆
04 36 ☿ □ ♅
07 17 ☽ Q ♄
10 35 ☽ # ☿
11 46 ☽ ± ☉
12 57 ☽ ⊼ ♇
15 26 ☽ ⊼ ♆
21 27 ☽ ∥ ♀
23 28 ☉ ⊻ ♀

12 Tuesday
00 52 ☿ ⊻ ♃
00 57 ☽ ☍ ♀
01 05 ☽ ⊼ ☉
01 10 ☽ ± ♇
12 11 ☽ # ♀
12 49 ☽ □ ♅
15 34 ☽ ⊼ ♃
17 30 ☽ ☍ ☿
18 34 ☽ ∥ ☿
19 30 ☽ bQ ♇
20 07 ☽ □ ♄
21 53 ☽ □ ♂

13 Wednesday
03 57 ☽ ± ♃
08 45 ☿ ∠ ♇
14 07 ☿ □ ♄
14 14 ♀ ⊥ ♇
16 51 ☽ # ♆
22 28 ☽ bQ ♃

14 Thursday
01 55 ☽ △ ♇
04 30 ☽ ☍ ♆
11 31 ☽ Q ♂
11 43 ☿ □ ♂
14 16 ☽ ⊼ ♀
19 17 ☽ ☍ ☉
22 01 ☽ # ☉
22 21 ☿ # ♀

15 Friday
01 36 ☽ △ ♅
02 28 ☽ ± ♀
05 05 ☽ △ ♃
08 59 ☽ △ ♄
12 20 ☽ ✶ ♂
15 15 ☽ ⊼ ☿
16 04 ☽ ∥ ♃
20 40 ☽ bQ ♀

16 Saturday
05 00 ☽ ± ☿
07 27 ☿ ⊥ ♃
07 30 ☽ bQ ♅
13 50 ☽ □ ♇
14 54 ☽ bQ ♄
16 28 ☽ ⊼ ♆
18 40 ☿ ∥ ♀
18 56 ☽ ∠ ♂

17 Sunday
01 14 ☽ ± ♅
01 27 ☽ bQ ☿
02 44 ☽ △ ♀
04 15 ☽ ± ♆
08 36 ☽ ± ♄
11 46 ☽ ⊼ ☉
12 56 ☽ ⊼ ♅
13 01 ☽ ⊥ ♂
17 01 ☽ □ ♃
20 18 ☽ ⊼ ♄
21 45 ☽ bQ ♆

18 Monday
00 21 ☽ ± ☉
00 54 ☽ Q ♇
00 57 ☽ ⊻ ♂
03 11 ☉ ✶ ♅
07 28 ☉ Q ♇
10 59 ☽ △ ☿
19 00 ☽ bQ ☉
23 48 ☽ ✶ ♇

19 Tuesday
02 27 ☽ △ ♆
03 53 ☽ Q ♃
13 25 ☽ □ ♀
14 27 ☉ ⊥ ♀
22 04 ☽ ☍ ♅

20 Wednesday
01 26 ☽ △ ☉
02 36 ☽ ✶ ♃
03 52 ☽ ∠ ♇
05 18 ☽ ☍ ♄
10 58 ☽ ☌ ♂
19 23 ☉ # ♃
19 57 ☉ ✶ ♃
20 22 ☽ ⊥ ♇
23 28 ☽ Q ♀

21 Thursday
03 26 ☽ □ ☿
06 20 ☽ ∠ ♃
07 14 ☽ ⊻ ♇
09 51 ☽ □ ♆
21 37 ☽ ✶ ♀
22 38 ☽ ⊥ ♃

22 Friday
02 57 ☿ ∠ ♃
04 30 ☽ ⊼ ♅
06 56 ☿ Q ♅
07 48 ☿ ✶ ♇
09 20 ☽ ⊻ ♃
09 26 ☉ ✶ ♄
11 33 ☽ ⊼ ♄
11 42 ☽ □ ☉
15 03 ☽ ± ♅
16 16 ☽ Q ☿
17 45 ☽ Q ♆
17 57 ☽ ⊻ ♂
22 04 ☽ ± ♄

23 Saturday
00 41 ☽ ∠ ♀
02 43 ☽ # ☉
03 42 ☿ ☌ ♆
04 35 ☽ ⊥ ♂
06 40 ☽ bQ ♅
09 12 ☽ ∥ ♃
11 53 ☽ ☌ ♇
13 38 ☽ bQ ♄
14 27 ☽ ✶ ♆
16 02 ☽ ✶ ☿
16 35 ☽ ⊥ ♀
18 49 ☿ ∥ ♆
20 18 ☽ ∠ ♂
21 00 ☽ Q ☉

24 Sunday
03 04 ☽ ⊻ ♀
08 11 ☽ △ ♅
13 15 ☽ ☌ ♃
13 38 ☿ Q ♄
15 03 ☽ △ ♄
15 47 ☽ ∠ ♆
17 08 ☽ # ☿
18 36 ☽ ✶ ☉
20 59 ☽ ∠ ☿
21 04 ☽ # ♆
22 00 ☽ ✶ ♂

25 Monday
06 43 ☽ ⊥ ♆
14 01 ☽ ⊻ ♇
14 01 ☽ ⊥ ☿
16 34 ☽ ⊻ ♆
18 10 ☽ Q ♂
21 02 ☽ ∠ ☉
23 49 ☽ ⊥ ♇

26 Tuesday
01 16 ☽ ⊻ ☿
01 37 ☽ # ♀
03 56 ☽ ∥ ♀
06 15 ☽ ☌ ♀
09 39 ☽ □ ♅
12 40 ☽ ⊥ ☉
14 25 ☽ ∠ ♇
14 58 ☽ ⊻ ♃
16 28 ☽ □ ♄
23 02 ☽ ⊻ ☉
23 57 ☽ □ ♂

27 Wednesday
00 45 ☽ ⊥ ♃
08 08 ☽ ∥ ♆
14 36 ☽ ✶ ♇
14 45 ☽ Q ♅
15 23 ☽ ∠ ♃
17 12 ☽ ☌ ♆
18 47 ☉ △ ♂
21 38 ☽ Q ♄

28 Thursday
02 58 ☽ ∥ ☿
03 30 ☿ ⊻ ♀
08 23 ☽ ⊻ ♀
08 55 ☽ ☌ ☿
09 55 ☽ Q ♇
10 07 ☽ ✶ ♅
15 48 ☽ ✶ ♃
16 10 ☽ # ♃
16 26 ☿ Q ♇
17 05 ☽ ✶ ♄
18 05 ☿ ✶ ♅
18 27 ☽ ⊥ ♀

29 Friday
01 16 ☽ △ ♂
02 46 ☽ ☌ ☉
09 38 ☽ ∠ ♀
10 29 ☽ ∠ ♅
11 29 ☽ Q ♃
15 07 ☽ □ ♇
15 31 ☽ ∥ ☉
17 38 ☽ ∠ ♄
17 52 ☽ ⊻ ♆

30 Saturday
01 16 ☽ ⊥ ♅
02 15 ☽ bQ ♂
03 46 ☽ ⊥ ♆
08 28 ♀ □ ♅
08 33 ☽ ⊥ ♄
11 14 ☽ ⊻ ♅
11 21 ☽ ✶ ♀
15 43 ☿ ✶ ♃
15 53 ☿ # ♃
17 30 ☽ ± ♂
17 31 ☽ □ ♃
17 47 ☽ ⊻ ☿
18 37 ☽ ⊻ ♄
18 46 ☽ ∠ ♆
23 55 ☿ ✶ ♄

MAY 2033

LONGITUDES

Date	Sidereal time h m s	Sun ☉ ° ' "	Moon ☽ ° ' "	Moon ☽ 24.00 ° ' "	Mercury ☿ ° '	Venus ♀ ° '	Mars ♂ ° '	Jupiter ♃ ° '	Saturn ♄ ° '	Uranus ♅ ° '	Neptune ♆ ° '	Pluto ♇ ° '
01	02 38 56	11 ♉ 26 02	13 ♊ 36 15	20 ♊ 34 32	04 ♉ 23	29 ♓ 36	08 ♑ 52	02 ♓ 47	03 ♋ 24	28 ♊ 57	18 ♈ 27	16 ♒ 43
02	02 42 52	12 24 17	27 ♊ 26 28	04 ♋ 11 54	06 27	00 ♈ 14	09 08	02 56	03 30	29 00	18 29	16 43
03	02 46 49	13 22 30	10 ♋ 50 52	17 ♋ 23 30	08 33	00 53	09 23	03 05	03 36	29 03	18 31	16 44
04	02 50 45	14 20 41	23 ♋ 50 05	00 ♌ 10 57	10 39	01 33	09 37	03 14	03 42	29 05	18 33	16 44
05	02 54 42	15 18 50	06 ♌ 26 34	12 ♌ 37 25	12 47	02 14	09 52	03 23	03 48	29 08	18 35	16 45
06	02 58 38	16 16 57	18 ♌ 44 03	24 ♌ 47 04	14 56	02 57	10 05	03 31	03 54	29 11	18 37	16 45
07	03 02 35	17 15 01	00 ♍ 47 04	06 ♍ 44 39	17 05	03 40	10 18	03 40	04 00	29 14	18 39	16 45
08	03 06 31	18 13 04	12 ♍ 40 26	18 ♍ 35 01	19 16	04 24	10 31	03 48	04 06	29 17	18 41	16 45
09	03 10 28	19 11 05	24 ♍ 28 58	00 ♎ 22 51	21 26	05 09	10 43	03 56	04 12	29 20	18 43	16 46
10	03 14 25	20 09 05	06 ♎ 17 10	12 ♎ 12 26	23 37	05 55	10 54	04 04	04 19	29 22	18 45	16 46
11	03 18 21	21 07 02	18 ♎ 09 05	24 ♎ 07 32	25 47	06 41	11 05	04 12	04 25	29 25	18 47	16 46
12	03 22 18	22 04 58	00 ♏ 08 09	06 ♏ 11 13	27 ♉ 57	07 29	11 15	04 20	04 31	29 28	18 49	16 46
13	03 26 14	23 02 52	12 ♏ 17 03	18 ♏ 25 49	00 ♊ 06	08 17	11 25	04 27	04 38	29 31	18 51	16 46
14	03 30 11	24 00 45	24 ♏ 37 45	00 ♐ 52 56	02 14	09 06	11 34	04 35	04 45	29 35	18 53	16 46
15	03 34 07	24 58 36	07 ♐ 11 29	13 ♐ 33 26	04 20	09 56	11 42	04 42	04 51	29 38	18 55	16 46
16	03 38 04	25 56 26	19 ♐ 58 49	26 ♐ 27 37	06 25	10 47	11 50	04 49	04 58	29 41	18 57	16 46
17	03 42 00	26 54 14	02 ♑ 59 50	09 ♑ 35 24	08 28	11 38	11 57	04 56	05 04	29 44	18 59	16 46
18	03 45 57	27 52 01	16 ♑ 14 16	22 ♑ 56 24	10 29	12 29	12 04	05 03	05 11	29 47	19 01	16 R 46
19	03 49 54	28 49 47	29 ♑ 41 44	06 ♒ 30 13	12 28	13 22	12 10	05 10	05 18	29 50	19 03	16 46
20	03 53 50	29 ♉ 47 32	13 ♒ 21 48	20 ♒ 16 26	14 25	14 15	12 15	05 17	05 25	29 53	19 04	16 46
21	03 57 47	00 ♊ 45 16	27 ♒ 14 03	04 ♓ 14 35	16 18	15 08	12 19	05 23	05 32	29 ♊ 57	19 06	16 46
22	04 01 43	01 42 59	11 ♓ 17 54	18 ♓ 23 53	18 10	16 02	12 23	05 29	05 39	00 ♋ 00	19 08	16 46
23	04 05 40	02 40 40	25 ♓ 32 20	02 ♈ 43 01	19 58	16 57	12 26	05 35	05 45	00 03	19 10	16 46
24	04 09 36	03 38 21	09 ♈ 55 35	17 ♈ 09 40	21 43	17 52	12 28	05 41	05 52	00 06	19 12	16 46
25	04 13 33	04 36 01	24 ♈ 24 46	01 ♉ 40 21	23 26	18 48	12 29	05 47	06 00	00 10	19 13	16 46
26	04 17 29	05 33 39	08 ♉ 55 47	16 ♉ 10 22	25 06	19 44	12 30	05 53	06 07	00 13	19 15	16 45
27	04 21 26	06 31 17	23 ♉ 23 24	00 ♊ 34 09	26 42	20 40	12 R 30	05 58	06 14	00 16	19 17	16 45
28	04 25 23	07 28 54	07 ♊ 41 52	14 ♊ 45 54	28 16	21 37	12 29	06 03	06 21	00 20	19 18	16 45
29	04 29 19	08 26 29	21 ♊ 45 36	28 ♊ 40 28	29 ♊ 46	22 35	12 28	06 09	06 28	00 23	19 20	16 45
30	04 33 16	09 24 03	05 ♋ 30 04	12 ♋ 14 06	01 ♋ 13	23 33	12 26	06 14	06 35	00 27	19 22	16 44
31	04 37 12	10 ♊ 21 36	18 ♋ 52 25	25 ♋ 24 59	02 ♋ 37	24 ♈ 31	12 ♑ 23	06 ♓ 18	06 ♋ 43	00 ♋ 30	19 ♈ 23	16 ♒ 44

Moon Nodes and Latitude

Date	Moon True ☊ ° '	Moon Mean ☊ ° '	Moon ☽ Latitude ° '
01	21 ♎ 09	20 ♎ 23	04 S 10
02	21 R 06	20 20	04 49
03	21 03	20 17	05 10
04	21 01	20 14	05 15
05	21 00	20 10	05 04
06	21 D 00	20 07	04 39
07	21 00	20 04	04 03
08	21 02	20 01	03 16
09	21 03	19 58	02 21
10	21 05	19 55	01 21
11	21 06	19 51	00 S 16
12	21 R 06	19 48	00 N 49
13	21 04	19 45	01 54
14	21 00	19 42	02 53
15	20 56	19 39	03 46
16	20 50	19 35	04 27
17	20 44	19 32	04 56
18	20 39	19 29	05 10
19	20 34	19 26	05 07
20	20 31	19 23	04 47
21	20 30	19 20	04 10
22	20 D 30	19 16	03 17
23	20 32	19 13	02 12
24	20 33	19 10	00 N 57
25	20 R 33	19 07	00 S 21
26	20 32	19 04	01 38
27	20 28	19 01	02 48
28	20 23	18 57	03 47
29	20 16	18 54	04 31
30	20 08	18 51	04 58
31	20 ♎ 00	18 ♎ 48	05 S 07

DECLINATIONS

Date	Sun ☉ ° '	Moon ☽ ° '	Mercury ☿ ° '	Venus ♀ ° '	Mars ♂ ° '	Jupiter ♃ ° '	Saturn ♄ ° '	Uranus ♅ ° '	Neptune ♆ ° '	Pluto ♇ ° '
01	15 N 15	18 N 17	11 N 59	00 N 37	24 S 09	11 S 14	22 N 46	23 N 39	05 N 47	22 S 43
02	15 33	18 36	12 49	00 44	24 11	11 11	22 46	23 39	05 48	22 43
03	15 51	17 50	13 38	00 51	24 12	11 08	22 46	23 39	05 49	22 43
04	16 08	16 10	14 28	00 59	24 14	11 05	22 46	23 39	05 49	22 43
05	16 25	13 45	15 17	01 08	24 16	11 02	22 46	23 39	05 50	22 44
06	16 42	10 47	16 05	01 17	24 18	10 59	22 46	23 39	05 51	22 44
07	16 59	07 24	16 52	01 27	24 20	10 57	22 46	23 39	05 52	22 44
08	17 15	03 47	17 39	01 38	24 22	10 54	22 46	23 39	05 52	22 44
09	17 31	00 N 02	18 24	01 48	24 24	10 51	22 45	23 39	05 53	22 44
10	17 47	03 S 44	19 07	02 00	24 26	10 48	22 45	23 39	05 54	22 44
11	18 02	07 22	19 49	02 12	24 28	10 46	22 45	23 39	05 55	22 45
12	18 17	10 45	20 29	02 24	24 31	10 43	22 45	23 39	05 55	22 45
13	18 32	13 43	21 07	02 37	24 33	10 41	22 45	23 39	05 56	22 45
14	18 46	16 07	21 42	02 51	24 36	10 38	22 45	23 39	05 57	22 45
15	19 00	17 48	22 16	03 05	24 38	10 35	22 45	23 39	05 57	22 45
16	19 14	18 37	22 46	03 19	24 41	10 33	22 45	23 39	05 58	22 46
17	19 28	18 28	23 15	03 34	24 44	10 31	22 45	23 39	05 59	22 46
18	19 41	17 19	23 40	03 49	24 47	10 28	22 45	23 39	05 59	22 46
19	19 54	15 12	24 04	04 04	24 50	10 26	22 44	23 39	06 00	22 47
20	20 06	12 13	24 24	04 20	24 53	10 24	22 44	23 39	06 01	22 47
21	20 18	08 31	24 42	04 36	24 56	10 22	22 44	23 39	06 01	22 47
22	20 30	04 S 17	24 57	04 52	25 00	10 20	22 44	23 39	06 02	22 47
23	20 41	00 N 15	25 10	05 09	25 03	10 18	22 44	23 39	06 03	22 48
24	20 52	04 49	25 20	05 26	25 07	10 16	22 44	23 39	06 03	22 48
25	21 03	09 08	25 28	05 43	25 11	10 14	22 43	23 39	06 04	22 48
26	21 14	12 55	25 34	06 00	25 15	10 12	22 43	23 39	06 05	22 49
27	21 24	15 54	25 38	06 18	25 19	10 10	22 43	23 39	06 05	22 49
28	21 33	17 51	25 39	06 36	25 23	10 09	22 43	23 39	06 06	22 49
29	21 42	18 40	25 39	06 54	25 27	10 07	22 43	23 39	06 06	22 50
30	21 51	18 22	25 37	07 12	25 31	10 05	22 42	23 39	06 07	22 50
31	22 N 00	17 N 02	25 N 33	07 N 30	25 S 35	10 S 04	22 N 42	23 N 39	06 N 07	22 S 50

ZODIAC SIGN ENTRIES

Date	h	m	Planets	
02	03	12	♀	♈
02	16	31	☽	♋
04	23	39	☽	♌
07	10	26	☽	♍
09	23	14	☽	♎
12	11	44	☽	♏
13	10	57	☿	♊
14	22	19	☽	♐
17	06	31	☽	♑
19	12	32	☽	♒
20	17	11	☉	♊
21	16	45	☽	♓
22	13	14	♅	♋
23	19	28	☽	♈
25	21	14	☽	♉
27	23	03	☽	♊
29	15	50	☿	♋
30	02	19	☽	♋

LATITUDES

Date	Mercury ☿ ° '	Venus ♀ ° '	Mars ♂ ° '	Jupiter ♃ ° '	Saturn ♄ ° '	Uranus ♅ ° '	Neptune ♆ ° '	Pluto ♇ ° '
01	01 S 04	00 N 51	01 S 01	00 S 49	00 S 37	00 N 13	01 S 34	07 S 15
04	00 35	00 N 24	01 09	00 49	00 37	00 13	01 34	07 16
07	00 S 04	00 00	01 18	00 50	00 37	00 13	01 34	07 16
10	00 N 28	00 S 23	01 27	00 51	00 36	00 13	01 34	07 17
13	00 58	00 43	01 37	00 52	00 36	00 13	01 34	07 18
16	01 25	01 02	01 47	00 52	00 35	00 13	01 34	07 19
19	01 47	01 18	01 58	00 53	00 35	00 13	01 34	07 19
22	02 03	01 33	02 09	00 54	00 35	00 13	01 34	07 20
25	02 12	01 47	02 21	00 55	00 34	00 13	01 34	07 21
28	02 14	01 58	02 33	00 55	00 34	00 13	01 35	07 22
31	02 N 08	02 S 08	02 S 45	00 S 56	00 S 34	00 N 13	01 S 35	07 S 23

DATA

Julian Date	2463719
Delta T	+75 seconds
Ayanamsa	24° 19' 27"
Synetic vernal point	04° ♓ 47' 32"
True obliquity of ecliptic	23° 25' 57"

MOON'S PHASES, APSIDES AND POSITIONS ☽

Date	h	m	Phase	Longitude ° '	Eclipse Indicator
06	06	45	☽	16 ♌ 04	
14	10	43	○	23 ♏ 58	
21	18	29	☾	01 ♓ 01	
28	11	36	●	07 ♊ 28	

Day	h	m		
09	16	21	Apogee	
25	12	47	Perigee	
02	06	37	Max dec	18° N 37'
09	12	12	0S	
16	20	25	Max dec	18° S 40'
23	10	44	0N	
29	17	10	Max dec	18° N 42'

LONGITUDES

Date	Chiron ⚷ ° '	Ceres ⚳ ° '	Pallas ⚴ ° '	Juno ⚵ ° '	Vesta ⚶ ° '	Black Moon Lilith ⚸ ° '
01	24 ♉ 57	03 ♑ 08	06 ♏ 00	00 ♎ 38	03 ♓ 28	29 ♍ 39
11	25 ♉ 39	02 ♑ 32	03 ♏ 25	29 ♍ 50	07 ♓ 25	00 ♎ 46
21	26 ♉ 21	01 ♑ 21	00 ♏ 39	29 ♍ 38	11 ♓ 05	01 ♎ 52
31	27 ♉ 03	29 ♐ 38	28 ♎ 55	29 ♍ 59	14 ♓ 25	02 ♎ 59

All ephemeris data is given at 12.00 UT and the Moon's longitude is additionally given for 24.00 UT

ASPECTARIAN

h m	Aspects
01 Sunday	
03 47	☽ ⚻ ♂
05 35	☽ ⊥ ☿
08 02	☽ ⚺ ☉
08 26	☽ Q ♀
17 20	☽ △ ♇
19 04	☽ ⊥ ☉
20 20	☽ ⚹ ♆
23 40	☽ ∠ ☿
02 Monday	
08 21	☿ ⊥ ♀
11 56	☽ ∠ ☉
14 45	☽ ☌ ♅
17 11	☽ □ ♀
17 23	☽ Q ♆
19 35	☽ ⚼ ♇
21 51	☽ △ ♃
22 50	☽ ☌ ♄
03 Tuesday	
07 02	☽ ⚹ ☿
09 17	☽ ☍ ♂
11 47	☽ ± ♇
16 59	☽ ⚹ ☉
22 47	☽ ⚻ ♇
22 47	☿ △ ♂
04 Wednesday	
01 26	☽ ⚼ ♃
02 07	☽ □ ♆
05 20	☉ ∠ ♅
09 21	☽ Q ☿
12 12	☽ ∥ ☉
17 07	☽ Q ☉
18 29	☽ ± ♃
18 30	♀ ∠ ♇
21 57	☽ ⚺ ♅
05 Thursday	
01 11	☽ ∥ ☿
03 27	☽ △ ♀
06 02	☽ ⚻ ♃
06 52	☽ ⚺ ♄
09 28	☽ ⊥ ♅
13 50	☉ Q ♃
18 33	☽ ⊥ ♄
18 45	☽ ⚻ ♂
06 Friday	
02 55	☽ □ ☿
03 01	☽ ∠ ♅
03 26	☿ ∠ ♅
06 41	☽ ± ♂
06 45	☽ □ ☉
08 05	☽ ☍ ♇
10 21	☽ ⚼ ♀
10 22	☽ # ♃
11 46	☽ △ ♆
12 19	☽ ∠ ♄
19 00	☿ Q ♃
23 35	☉ □ ♇
07 Saturday	
00 50	☽ ⚼ ♂
05 21	☽ ± ♀
08 14	☿ □ ♇
08 52	☽ ⚹ ♅
10 26	♃ ∠ ♆
11 53	♀ ⚺ ♃
15 11	☉ ☌ ☿
16 59	☉ ∥ ☿
17 47	☽ ⚼ ♆
17 51	☽ ☍ ♃
18 10	☽ ⚻ ♀
18 31	☽ ⚹ ♄
22 22	☽ ∥ ♆
08 Sunday	
00 50	♀ □ ♄
05 34	☿ ⚺ ♆
07 33	☽ △ ♂
09 09	☽ Q ♅
10 10	☿ ∠ ♄
12 02	☽ ± ♆
14 20	☿ ∠ ♀
19 01	☽ Q ♄
20 17	☽ ⚻ ♇
09 Monday	
00 04	☉ ⚺ ♆
00 15	☽ ⚻ ♆
00 16	☽ △ ☉
01 13	☽ ∥ ♀
04 24	☽ △ ☿
08 30	☽ ± ♇
12 37	☉ ∠ ♄
17 40	♂ ⊥ ♇
21 54	☽ □ ♅
10 Tuesday	
00 19	☽ # ♀
02 48	☽ ⚼ ♇
07 27	☽ ⚻ ♃
07 57	☽ □ ♄
09 21	☿ ⊥ ♅
09 29	☽ ⚼ ☉
11 11	☽ ☍ ♀
17 46	☽ ⚼ ☿
19 45	☽ ± ♃
21 31	☽ □ ♂
11 Wednesday	
00 51	☿ ⊥ ♆
02 13	☽ # ♆
05 20	☽ ± ☉
09 12	☽ △ ♇
13 17	☽ ☍ ♆
14 08	☽ ⚼ ♃
15 41	☿ ⚼ ♂
16 00	☽ ± ☿
18 29	☽ ⚻ ☉
12 Thursday	
06 41	☽ ⚻ ☿
10 14	☽ Q ♂
10 41	☽ △ ♅
11 47	☽ ∥ ♃
18 49	☿ ⊥ ♄
20 25	☽ △ ♃
20 47	☽ △ ♄
13 Friday	
03 35	☽ ⚻ ♀
05 29	☿ ⚺ ♅
10 17	☽ ⚹ ♂
16 12	☽ ± ♀
16 24	☽ ⚼ ♅
20 46	☽ □ ♇
14 Saturday	
00 31	☉ ⊥ ♅
00 51	☽ ⚻ ♆
02 28	☽ ⚼ ♄
03 38	♄ # ♇
09 58	☽ ± ♅
10 43	☽ ☍ ☉
10 55	☽ ⚼ ♀
12 30	☽ ± ♆
15 47	☽ ∠ ♂
19 58	☽ ± ♄
21 32	☽ ⚻ ♅
15 Sunday	
04 15	☉ ∠ ♀
05 30	☽ ☍ ☿
05 46	☽ ⚼ ♆
07 09	☿ ∠ ♆
07 14	☽ □ ♃
07 25	☽ Q ♇
07 31	☽ ⚻ ♄
09 09	☽ ⊥ ♂
10 29	☉ ⊥ ♆
16 27	☿ □ ♃
17 33	☽ △ ♀
18 14	☿ ⚺ ♄
20 37	☽ ⚺ ♂
16 Monday	
04 50	☿ ± ♂
06 01	☽ ⚹ ♇
10 05	☽ △ ♆
10 43	☿ ∥ ♄
11 22	☿ # ♇
13 33	♀ ⊥ ♃
17 19	☽ Q ♃
23 55	☽ ⚻ ☉
17 Tuesday	
06 00	☽ ☍ ♅
09 46	☽ ∠ ♇
11 49	☽ ± ☉
12 59	♇ St R
13 29	☉ ⚼ ♂
15 35	☽ ⚹ ♃
15 49	☽ ☍ ♄
22 33	♀ □ ♂
23 47	☽ ⚻ ☿
18 Wednesday	
02 09	☽ ⊥ ♇
04 25	☽ ☌ ♂
04 47	☽ □ ♀
05 27	☽ ⚼ ☉
10 38	☿ ∥ ♅
12 32	☽ ± ☿
12 58	☽ ⚺ ♇
16 59	☽ □ ♆
18 54	☽ ∠ ♃
19 Thursday	
07 23	☽ ⚼ ☿
08 02	☿ ⚻ ♂
10 21	☽ △ ☉
11 03	☽ ⊥ ♃
12 15	☽ ⚻ ♅
15 09	☽ Q ♀
21 44	☽ ⚺ ♃
21 58	☽ ⚻ ♄
22 52	☽ ± ♅
20 Friday	
00 59	☽ Q ♆
01 15	☉ ⊥ ♄
08 10	☿ ⚹ ♀
08 34	☽ ± ♄
10 02	☽ ⚺ ♂
13 38	☽ ⚹ ♀
14 07	☽ △ ☿
14 34	☉ ⚺ ♅
14 40	☽ ⚼ ♅
17 56	☽ ☌ ♇
20 32	☽ ⊥ ♂
21 57	☽ ⚹ ♆
21 Saturday	
00 20	☽ ⚼ ♄
00 25	☽ ∥ ♃
12 09	☽ ∠ ♂
16 40	☽ △ ♅
17 19	☽ ∠ ♀
17 56	☿ △ ♇
18 29	☽ □ ☉
23 47	☽ ∠ ♆
22 Sunday	
02 03	☽ ☌ ♃
02 18	☽ △ ♄
02 23	☽ # ♆
09 00	☽ # ♀
09 43	☽ ⊥ ♀
13 50	☽ ⚹ ♂
15 07	☽ ⊥ ♆
18 29	☿ # ♂
20 34	☽ ⚺ ♀
21 15	☽ ⚺ ♇
23 Monday	
01 05	☿ ⚹ ♆
01 16	☽ ⚺ ♆
01 18	☽ □ ☿
03 15	☽ Q ☉
07 12	♀ ⚹ ♇
07 21	☽ ⊥ ♇
10 08	☽ Q ♂
19 35	☽ □ ♅
22 25	☽ ∠ ♇
24 Tuesday	
00 47	☽ ⚹ ☉
04 54	☽ ⚺ ♃
05 12	☽ □ ♄
11 37	☽ Q ☿
14 57	☽ ⊥ ♃
15 33	☽ ∥ ♀
16 14	☽ □ ♂
18 35	♀ Q ♅
18 43	☽ ∥ ♆
23 20	☽ ⚹ ♇
25 Wednesday	
01 37	☽ Q ♅
02 04	☽ ☌ ♀
02 15	☉ ∠ ♆
03 24	☽ ☌ ♆
03 28	☽ ∠ ☉
05 58	☽ ∠ ♃
10 10	☽ ⚹ ☿
11 18	☽ Q ♄
18 29	☽ # ♃
19 11	☽ Q ♇
19 25	☽ ⊥ ☉
21 32	☽ ⚹ ♅
23 20	♀ ☌ ♆
26 Thursday	
06 02	☽ ⚺ ☉
06 55	☽ ⚹ ♃
07 18	☽ ⚹ ♄
14 10	☽ ∠ ☿
17 55	☽ △ ♂
17 57	♀ ∥ ♆
20 47	☉ □ ♃
22 27	☽ ∠ ♅
23 47	♂ St R
27 Friday	
00 58	☽ □ ♇
02 56	☽ Q ♃
03 39	☉ ⚺ ♄
05 09	☽ ⚺ ♆
06 58	☽ ⊥ ☿
07 09	☽ ⚺ ♀
08 22	☽ ∠ ♄
11 33	☉ ± ♂
13 29	☽ ⊥ ♅
15 09	☽ ⊥ ♆
17 52	☽ ⊥ ♀
18 12	☽ ⚺ ☿
18 52	☽ ⚼ ♂
20 19	♀ ∠ ♃
23 31	☽ ⊥ ♄
23 33	☽ ⚺ ♅
28 Saturday	
06 16	☽ ∠ ♆
09 13	☽ □ ♃
09 42	☽ ⚺ ♄
09 58	☽ ± ♂
10 03	☽ ∠ ♀
11 36	☽ ☌ ☉
20 07	☽ ⚻ ♂
29 Sunday	
03 23	☽ △ ♇
07 49	☽ ⚹ ♆
13 31	☽ ⚹ ♀
22 34	☿ ☌ ♅
30 Monday	
03 03	☽ ☌ ♅
03 34	☽ ☌ ☿
04 41	☽ Q ♆
05 22	☽ ⚼ ♇
12 05	☽ Q ♀
13 18	☽ △ ♃
13 57	☽ ☌ ♄
14 28	☿ Q ♆
19 28	☽ ⚺ ☉
20 46	☿ ⚼ ♇
21 19	☽ ± ♇
31 Tuesday	
00 18	☽ ☍ ♂
04 24	☿ # ♂
07 05	☽ ⊥ ☉
08 07	☽ ⚻ ♇
12 56	☽ □ ♆
16 28	☽ ⚼ ♃
17 35	♀ Q ♄
23 10	☽ □ ♀

JUNE 2033

LONGITUDES

Date	Sidereal time h m s	Sun ☉ ° ' "	Moon ☽ ° ' "	Moon ☽ 24.00 ° ' "	Mercury ☿ ° '	Venus ♀ ° '	Mars ♂ ° '	Jupiter ♃ ° '	Saturn ♄ ° '	Uranus ♅ ° '	Neptune ♆ ° '	Pluto ♇ ° '
01	04 41 09	11 ♊ 19 08	01 ♌ 51 52	08 ♌ 13 17	03 ♋ 58	25 ♈ 29	12 ♑ 19	06 ♓ 23	06 ♋ 50	00 ♋ 33	19 ♈ 25	16 ♒ 43
02	04 45 05	12 16 39	14 ♌ 29 33	20 ♌ 41 04	05 15	26 28	12 R 14	06 27	06 57	00 37	19 26	16 R 43
03	04 49 02	13 14 08	26 ♌ 48 18	02 ♍ 51 47	06 30	27 27	12 09	06 32	07 05	00 40	19 28	16 43
04	04 52 58	14 11 36	08 ♍ 52 08	14 ♍ 49 58	07 40	28 27	12 03	06 36	07 12	00 44	19 29	16 42
05	04 56 55	15 09 03	20 ♍ 45 55	26 ♍ 40 40	08 48	29 ♈ 27	11 56	06 40	07 19	00 47	19 31	16 42
06	05 00 52	16 06 28	02 ♎ 34 53	08 ♎ 29 12	09 52	00 ♉ 27	11 49	06 43	07 27	00 51	19 32	16 41
07	05 04 48	17 03 52	14 ♎ 24 16	20 ♎ 20 42	10 52	01 28	11 40	06 47	07 34	00 54	19 34	16 41
08	05 08 45	18 01 16	26 ♎ 19 05	02 ♏ 19 57	11 49	02 29	11 32	06 50	07 42	00 58	19 35	16 40
09	05 12 41	18 58 38	08 ♏ 23 47	14 ♏ 31 00	12 42	03 30	11 22	06 54	07 49	01 01	19 36	16 40
10	05 16 38	19 55 59	20 ♏ 41 58	26 ♏ 56 58	13 31	04 31	11 12	06 57	07 57	01 05	19 38	16 39
11	05 20 34	20 53 20	03 ♐ 16 12	09 ♐ 39 47	14 16	05 33	11 01	06 59	08 04	01 08	19 39	16 38
12	05 24 31	21 50 39	16 ♐ 07 45	22 ♐ 40 04	14 58	06 35	10 49	07 02	08 12	01 12	19 40	16 38
13	05 28 27	22 47 58	29 ♐ 16 34	05 ♑ 57 03	15 35	07 37	10 37	07 05	08 20	01 15	19 42	16 37
14	05 32 24	23 45 16	12 ♑ 41 56	19 ♑ 28 53	16 08	08 40	10 24	07 07	08 27	01 19	19 43	16 36
15	05 36 21	24 42 34	26 ♑ 19 33	03 ♒ 12 54	16 37	09 43	10 10	07 09	08 35	01 23	19 44	16 36
16	05 40 17	25 39 51	10 ♒ 08 34	17 ♒ 06 11	17 01	10 46	09 56	07 11	08 43	01 26	19 45	16 35
17	05 44 14	26 37 08	24 ♒ 05 25	01 ♓ 06 00	17 21	11 49	09 42	07 13	08 50	01 30	19 46	16 34
18	05 48 10	27 34 24	08 ♓ 07 39	15 ♓ 10 09	17 37	12 53	09 27	07 14	08 58	01 33	19 48	16 33
19	05 52 07	28 31 41	22 ♓ 13 21	29 ♓ 17 04	17 48	13 56	09 11	07 15	09 06	01 37	19 49	16 33
20	05 56 03	29 ♊ 28 57	06 ♈ 21 12	13 ♈ 25 36	17 54	15 00	08 55	07 17	09 13	01 40	19 50	16 32
21	06 00 00	00 ♋ 26 12	20 ♈ 30 07	27 ♈ 34 37	17 R 56	16 04	08 38	07 17	09 21	01 44	19 51	16 31
22	06 03 56	01 23 28	04 ♉ 38 53	11 ♉ 42 40	17 54	17 09	08 21	07 18	09 29	01 48	19 52	16 30
23	06 07 53	02 20 44	18 ♉ 45 41	25 ♉ 47 35	17 47	18 13	08 04	07 19	09 37	01 51	19 53	16 29
24	06 11 50	03 17 59	02 ♊ 47 58	09 ♊ 46 24	17 35	19 18	07 46	07 19	09 45	01 55	19 54	16 28
25	06 15 46	04 15 14	16 ♊ 42 25	23 ♊ 35 34	17 20	20 23	07 29	07 19	09 52	01 58	19 55	16 27
26	06 19 43	05 12 30	00 ♋ 25 23	07 ♋ 11 27	17 00	21 28	07 10	07 R 19	10 00	02 02	19 55	16 27
27	06 23 39	06 09 44	13 ♋ 53 25	20 ♋ 30 58	16 37	22 33	06 52	07 19	10 08	02 06	19 56	16 26
28	06 27 36	07 06 59	27 ♋ 03 54	03 ♌ 32 06	16 10	23 39	06 34	07 19	10 16	02 09	19 57	16 25
29	06 31 32	08 04 13	09 ♌ 55 32	16 ♌ 14 17	15 41	24 45	06 15	07 18	10 23	02 13	19 58	16 24
30	06 35 29	09 ♋ 01 27	22 ♌ 28 32	28 ♌ 38 34	15 ♋ 09	25 ♉ 50	05 ♑ 57	07 ♓ 17	10 ♋ 31	02 ♋ 16	19 ♈ 59	16 ♒ 23

Moon Nodes and Latitude

Date	Moon True ☊ ° '	Moon Mean ☊ ° '	Moon ☽ Latitude ° '
01	19 ♎ 53	18 ♎ 45	05 S 01
02	19 R 47	18 41	04 40
03	19 44	18 38	04 06
04	19 42	18 35	03 22
05	19 D 42	18 32	02 29
06	19 43	18 29	01 31
07	19 44	18 26	00 S 29
08	19 R 44	18 22	00 N 35
09	19 42	18 19	01 38
10	19 38	18 16	02 38
11	19 31	18 13	03 31
12	19 23	18 10	04 14
13	19 13	18 07	04 45
14	19 02	18 03	05 02
15	18 52	18 00	05 01
16	18 44	17 57	04 43
17	18 38	17 54	04 08
18	18 34	17 51	03 18
19	18 33	17 47	02 15
20	18 D 33	17 44	01 N 04
21	18 R 33	17 41	00 S 10
22	18 32	17 38	01 24
23	18 29	17 35	02 33
24	18 24	17 32	03 32
25	18 15	17 28	04 17
26	18 05	17 25	04 47
27	17 53	17 22	05 01
28	17 41	17 19	04 57
29	17 30	17 16	04 39
30	17 ♎ 21	17 ♎ 13	04 S 07

DECLINATIONS

Date	Sun ☉ ° '	Moon ☽ ° '	Mercury ☿ ° '	Venus ♀ ° '	Mars ♂ ° '	Jupiter ♃ ° '	Saturn ♄ ° '	Uranus ♅ ° '	Neptune ♆ ° '	Pluto ♇ ° '
01	22 N 08	14 N 51	25 N 27	07 N 49	25 S 40	10 S 02	22 N 42	23 N 39	06 N 08	22 S 51
02	22 16	12 01	25 20	08 08	25 44	10 01	22 42	23 39	06 09	22 51
03	22 23	08 43	25 11	08 26	25 49	10 00	22 41	23 39	06 09	22 51
04	22 30	05 07	25 02	08 45	25 54	09 58	22 41	23 39	06 10	22 52
05	22 36	01 N 22	24 51	09 04	25 59	09 57	22 41	23 39	06 10	22 52
06	22 42	02 S 25	24 39	09 23	26 04	09 56	22 40	23 39	06 11	22 53
07	22 48	06 07	24 26	09 42	26 08	09 55	22 40	23 39	06 11	22 53
08	22 54	09 36	24 12	10 01	26 13	09 54	22 40	23 39	06 12	22 53
09	22 59	12 45	23 57	10 20	26 19	09 53	22 39	23 39	06 12	22 54
10	23 03	15 23	23 42	10 39	26 24	09 52	22 39	23 39	06 13	22 54
11	23 07	17 21	23 26	10 58	26 29	09 51	22 39	23 39	06 13	22 55
12	23 11	18 29	23 10	11 17	26 34	09 51	22 38	23 39	06 13	22 55
13	23 14	18 40	22 53	11 36	26 39	09 50	22 38	23 39	06 14	22 55
14	23 17	17 49	22 36	11 55	26 44	09 50	22 38	23 39	06 14	22 56
15	23 20	15 57	22 19	12 14	26 49	09 49	22 37	23 39	06 15	22 56
16	23 22	13 09	22 02	12 33	26 55	09 49	22 37	23 39	06 15	22 57
17	23 23	09 35	21 44	12 52	27 00	09 48	22 36	23 39	06 15	22 57
18	23 25	05 28	21 27	13 11	27 05	09 48	22 36	23 39	06 16	22 58
19	23 25	01 S 01	21 10	13 29	27 10	09 48	22 36	23 39	06 16	22 58
20	23 26	03 N 30	20 53	13 48	27 15	09 47	22 35	23 39	06 17	22 59
21	23 26	07 51	20 37	14 06	27 20	09 47	22 35	23 39	06 17	22 59
22	23 25	11 44	20 21	14 24	27 25	09 47	22 34	23 39	06 17	23 00
23	23 25	14 57	20 06	14 42	27 29	09 47	22 34	23 39	06 18	23 00
24	23 23	17 15	19 51	15 00	27 34	09 48	22 33	23 38	06 18	23 00
25	23 22	18 30	19 37	15 17	27 38	09 48	22 33	23 38	06 18	23 01
26	23 20	18 39	19 24	15 35	27 43	09 48	22 32	23 38	06 18	23 01
27	23 17	17 44	19 11	15 52	27 47	09 48	22 32	23 38	06 19	23 02
28	23 14	15 53	19 00	16 09	27 51	09 49	22 31	23 38	06 19	23 02
29	23 11	13 16	18 50	16 25	27 55	09 49	22 31	23 38	06 19	23 03
30	23 N 08	10 N 07	18 N 40	16 N 42	27 S 59	09 S 50	22 N 30	23 N 38	06 N 19	23 S 03

ZODIAC SIGN ENTRIES

Date	h	m	Planets
01	08	31	☽ ♌
03	18	19	☽ ♍
06	01	08	♀ ♉
06	06	45	☽ ♎
08	19	21	☽ ♏
11	05	49	☽ ♐
13	13	18	☽ ♑
15	18	25	☽ ♒
17	22	07	☽ ♓
20	01	13	☽ ♈
21	01	01	☉ ♋
22	04	07	☽ ♉
24	07	12	☽ ♊
26	11	15	☽ ♋
28	17	26	☽ ♌

LATITUDES

Date	Mercury ☿ ° '	Venus ♀ ° '	Mars ♂ ° '	Jupiter ♃ ° '	Saturn ♄ ° '	Uranus ♅ ° '	Neptune ♆ ° '	Pluto ♇ ° '
01	02 N 05	02 S 11	02 S 49	00 S 56	00 S 34	00 N 13	01 S 35	07 S 23
04	01 49	02 19	03 01	00 57	00 33	00 13	01 35	07 24
07	01 27	02 26	03 14	00 58	00 33	00 13	01 35	07 24
10	00 58	02 31	03 27	00 59	00 33	00 13	01 35	07 25
13	00 N 22	02 35	03 39	01 00	00 32	00 13	01 35	07 26
16	00 S 20	02 38	03 52	01 01	00 32	00 13	01 35	07 27
19	01 06	02 39	04 04	01 01	00 32	00 13	01 36	07 27
22	01 54	02 39	04 15	01 02	00 31	00 13	01 36	07 28
25	02 43	02 39	04 26	01 03	00 31	00 13	01 36	07 29
28	03 29	02 37	04 35	01 04	00 31	00 13	01 36	07 29
31	04 S 07	02 S 34	04 S 44	01 S 05	00 S 31	00 N 13	01 S 36	07 S 30

DATA

Julian Date	2463750
Delta T	+75 seconds
Ayanamsa	24° 19' 31"
Synetic vernal point	04° ♓ 47' 28"
True obliquity of ecliptic	23° 25' 57"

MOON'S PHASES, APSIDES AND POSITIONS ☽

Date	h	m	Phase	Longitude ° '	Eclipse Indicator
04	23	39	☽	14 ♍ 39	
12	23	19	○	22 ♐ 18	
19	23	29	☾	28 ♓ 59	
26	21	07	●	05 ♋ 34	

Day	h	m		
06	10	12	Apogee	
21	01	28	Perigee	
05	20	40	0S	
13	04	20	Max dec	18° S 43'
19	17	22	0N	
26	03	07	Max dec	18° N 43'

LONGITUDES

Date	Chiron ⚷ ° '	Ceres ⚳ ° '	Pallas ⚴ ° '	Juno ⚵ ° '	Vesta ⚶ ° '	Black Moon Lilith ⚸ ° '
01	27 ♉ 08	29 ♐ 26	28 ♒ 47	00 ♎ 03	14 ♓ 43	03 ♎ 06
11	27 ♉ 48	27 ♐ 21	27 ♒ 55	00 ♎ 58	17 ♓ 37	04 ♎ 12
21	28 ♉ 27	25 ♐ 09	27 ♒ 50	02 ♎ 19	20 ♓ 04	05 ♎ 19
31	29 ♉ 04	23 ♐ 03	28 ♒ 28	04 ♎ 22	21 ♓ 57	06 ♎ 25

ASPECTARIAN

h m Aspects

01 Wednesday
00 51 ☽ ∠ ☉
09 13 ☽ ± ♃
09 32 ☽ ⊻ ♅
16 24 ☽ ⊻ ☿
20 34 ☽ ⊼ ♃
20 53 ☽ ⊥ ♅
21 27 ☽ ⊻ ♄

02 Thursday
05 05 ☽ ⊥ ☿
07 23 ☽ ✶ ☉
07 42 ☽ ⊼ ♂
09 01 ☽ ⊥ ♄
11 05 ☉ ⊼ ♂
14 10 ☽ ∠ ♅
16 18 ☽ ☍ ♇
19 12 ☽ ± ♂
21 36 ☽ △ ♆

03 Friday
00 26 ☽ ∠ ☿
02 37 ☽ ∠ ♄
02 58 ☽ # ♃
08 39 ☽ Q ☉
12 40 ☽ bQ ♂
12 46 ☿ △ ♃
13 24 ☽ △ ♀
13 47 ☽ ∥ ♀
19 41 ☽ ✶ ♅

04 Saturday
01 06 ☿ ☌ ♄
03 13 ☽ bQ ♆
05 14 ☽ ∥ ♆
07 25 ☽ ☍ ♃
08 37 ☽ ✶ ♄
09 21 ☽ ✶ ☿
18 01 ♀ Q ♇
18 20 ☽ △ ♂
19 48 ☽ Q ♅
21 19 ☽ ± ♆
22 03 ☽ bQ ♀
23 39 ☽ □ ☉

05 Sunday
03 46 ☽ ⊼ ♇
09 03 ☽ Q ♄
09 28 ☽ ⊼ ♆
12 04 ☽ Q ☿
15 55 ☽ ± ♇
17 57 ☽ ± ♀

06 Monday
03 58 ☉ ∥ ♄
07 16 ☽ ⊼ ♀
08 27 ☽ □ ♅
10 11 ☽ bQ ♇
20 28 ☽ ⊼ ♃
21 51 ♀ ✶ ♅
22 00 ☽ □ ♄

07 Tuesday
02 24 ☉ △ ♇
04 11 ☽ □ ☿
06 32 ☽ □ ♂
07 25 ☿ ± ♇
08 42 ☽ ± ♃
12 27 ☽ # ♆
16 36 ☽ △ ♇
17 51 ☽ △ ☉
22 26 ☽ ☍ ♆

08 Wednesday
02 58 ☽ bQ ♃
03 40 ♀ # ♃
05 36 ☿ ☍ ♂
10 46 ☉ # ♇
14 07 ☽ ∥ ♃
15 18 ☽ # ♀
18 20 ☽ Q ♂
21 19 ☽ △ ♅

09 Thursday
01 25 ☽ ☍ ♀
02 31 ☽ bQ ☉
09 01 ☽ △ ♃
10 51 ☽ △ ♄
17 45 ☽ ✶ ♂
21 04 ☽ △ ☿
21 45 ☽ ± ☉

10 Friday
03 00 ☽ bQ ♅
04 10 ☽ □ ♇
04 13 ☉ ✶ ♆
09 56 ☽ ⊼ ♆
10 24 ☽ ⊼ ☉
16 23 ☽ bQ ♄
16 44 ☿ ∥ ♅
20 28 ☽ ± ♅
21 30 ☽ ± ♆
22 24 ☽ ∠ ♂

11 Saturday
03 57 ☽ bQ ☿
07 57 ☽ ⊼ ♅
09 43 ☽ ± ♄
14 35 ☽ Q ♇
14 37 ☽ bQ ♆
15 14 ☽ ⊥ ♂
16 41 ☽ ⊼ ♀
19 01 ☽ □ ♃
21 07 ☽ ⊼ ♄
21 57 ☽ ± ☿

12 Sunday
02 18 ☽ ⊻ ♂
04 52 ☽ ± ♀
09 44 ☽ ⊼ ☿
10 40 ☉ ∥ ☿
12 55 ☽ ✶ ♇
18 32 ☽ △ ♆
22 48 ♀ ✶ ♃
22 53 ☽ bQ ♀
23 19 ☽ ☍ ☉

13 Monday
04 22 ☽ Q ♃
08 42 ☿ # ♇
11 20 ♂ ⊥ ♇
15 35 ☽ ☍ ♅
16 13 ☽ ∠ ♇

14 Tuesday
02 03 ☽ ✶ ♃
04 15 ☽ △ ♀
04 24 ☽ ☍ ♄
06 27 ♀ ✶ ♄
08 00 ☽ ☌ ♂
08 18 ☽ ⊥ ♇
09 47 ☿ ∥ ♄
18 20 ☽ ☍ ☿
18 56 ☽ ⊻ ♇

15 Wednesday
00 26 ☽ □ ♆
04 40 ☽ ∠ ♃
08 58 ☽ ⊼ ☉
10 58 ☉ ∠ ♀
10 59 ☿ ⊼ ♇
20 13 ☽ ± ☉
20 26 ☽ ⊥ ♃
20 36 ♀ △ ♂
20 50 ☽ ⊼ ♅

16 Thursday
06 52 ☽ ⊻ ♃
07 18 ☽ ± ♅
07 52 ☽ Q ♆
09 30 ☽ ⊼ ♄
11 39 ☽ ⊻ ♂
12 58 ☽ bQ ☉
13 10 ☽ □ ♀
16 02 ☽ # ♀
19 57 ☽ ± ♄
21 50 ☽ ⊥ ♂
22 54 ☽ bQ ♅
23 06 ☽ ☌ ♇

17 Friday
00 10 ☽ ⊼ ☿
04 35 ☽ ✶ ♆
10 40 ☽ ∥ ♃
10 43 ☽ ± ☿
11 34 ☽ bQ ♄
13 01 ☽ ∠ ♂
16 39 ☽ △ ☉
22 36 ☽ Q ♀

18 Saturday
00 44 ☽ △ ♅
02 26 ☽ bQ ☿
06 18 ☽ ∠ ♆
07 32 ☽ # ♆
10 28 ☽ ☌ ♃
12 42 ♅ bQ ♇
13 27 ☽ △ ♄
14 12 ☽ ✶ ♂
20 45 ☽ ✶ ♀
21 40 ☽ ⊥ ♆

19 Sunday
02 21 ☽ ⊻ ♇
04 24 ☽ △ ☿
07 54 ☽ ⊻ ♆
10 16 ☽ Q ♂
12 33 ☽ ⊥ ♇
17 14 ♂ ☍ ♄
23 29 ☽ □ ☉

20 Monday
00 21 ☽ ∠ ♀
03 49 ☽ ∠ ♇
04 01 ☽ □ ♅
13 34 ☽ ⊻ ♃
16 16 ☽ □ ♂
16 52 ☽ ⊥ ♀
16 55 ☽ □ ♄
23 45 ☽ ⊥ ♃

21 Tuesday
03 07 ☽ ∥ ♆
03 53 ☽ ⊻ ♀
05 15 ☽ ✶ ♇
07 39 ☽ □ ☿
08 15 ☽ Q ☉
10 05 ☿ St R
10 42 ☽ Q ♅
10 53 ☽ ☌ ♆
15 02 ☽ ∠ ♃
21 46 ♀ □ ♇
23 33 ☽ # ♃
23 44 ☽ Q ♄

22 Wednesday
01 35 ☽ Q ♇
03 39 ♀ ∠ ♅
06 04 ☽ ✶ ☉
07 08 ☽ ✶ ♅
14 06 ☽ Q ☿
14 45 ☉ bQ ♇
16 31 ☽ ✶ ♃
18 10 ☽ △ ♂
20 17 ☽ ✶ ♄
22 49 ☉ ☌ ♅

23 Thursday
00 04 ☉ Q ♆
03 14 ☿ ✶ ♀
08 08 ☽ □ ♇
08 44 ☽ ∠ ♅
09 25 ☽ ∠ ☉
09 39 ☽ ∥ ♀
10 21 ☽ ✶ ☿
11 00 ☽ ☌ ♀
12 56 ☽ Q ♃
13 54 ☽ ⊻ ♆
19 12 ☽ bQ ♂
22 04 ☽ ∠ ♄

24 Friday
00 09 ☽ ⊥ ♅
00 10 ☽ ⊥ ♆
01 33 ♅ Q ♆
01 53 ☽ ⊥ ☉
10 17 ☽ ± ♂
10 28 ☽ ⊻ ♅
11 39 ☽ ∠ ☿
12 24 ♀ Q ♃
12 55 ☽ ⊻ ☉
13 38 ☽ ⊥ ♄
15 36 ☽ ∠ ♆
19 46 ☽ □ ♃
20 23 ☽ ⊼ ♂

25 Saturday
00 03 ☽ ⊻ ♄
01 21 ♀ ⊻ ♆
02 52 ☽ ⊥ ☿
11 34 ☽ △ ♇
13 03 ☽ ⊻ ☿
17 35 ☽ ✶ ♆
18 57 ☽ ⊻ ♀
21 52 ♃ St R

26 Sunday
00 15 ♂ ✶ ♃
06 21 ☽ ⊥ ♀
13 48 ☽ bQ ♇
14 39 ☽ Q ♆
14 51 ☽ ☌ ♅
21 07 ☽ ☌ ☉
23 39 ☽ ∠ ♀
23 42 ☽ ☍ ♂

27 Monday
00 11 ♀ bQ ♂
00 14 ☽ △ ♃
05 11 ☽ ☌ ♄
05 47 ☽ ± ♇
16 34 ☽ ⊼ ♇
16 46 ☽ ☌ ☿
22 55 ☿ ⊼ ♇
22 58 ☽ □ ♆

28 Tuesday
01 30 ☉ ☍ ♂
03 17 ☽ bQ ♃
05 09 ☽ ✶ ♀
09 26 ☽ ∥ ♀
16 53 ☉ △ ♃
19 51 ☽ ± ♃
21 28 ☽ ⊻ ♅

29 Wednesday
05 15 ☽ ⊼ ♂
05 26 ☽ Q ♀
07 04 ☽ ⊼ ♃
08 13 ☽ ⊻ ☉
08 45 ☽ ⊥ ♅
12 18 ♄ ± ♇
12 53 ☽ ⊻ ♄
16 19 ☽ ± ♂
20 30 ☽ ⊥ ☉
22 29 ☽ ⊻ ☿

30 Thursday
00 17 ☽ ☍ ♇
00 25 ☽ ⊥ ♄
01 56 ☽ ∠ ♅
04 09 ♀ ∠ ♄
07 10 ☽ △ ♆
09 07 ☽ bQ ♂
09 32 ☽ ⊥ ☿
14 02 ☽ # ♃
15 06 ♀ ⊥ ♆
15 15 ☽ ∠ ☉
17 58 ☽ ∠ ♄
19 10 ☽ □ ♀
22 03 ♀ ⊥ ♅

All ephemeris data is given at 12.00 UT and the Moon's longitude is additionally given for 24.00 UT

JULY 2033

LONGITUDES

Date	Sidereal time h m s	Sun ☉ ° ' "	Moon ☽ ° ' "	Moon ☽ 24.00 ° ' "	Mercury ☿ ° '	Venus ♀ ° '	Mars ♂ ° '	Jupiter ♃ ° '	Saturn ♄ ° '	Uranus ♅ ° '	Neptune ♆ ° '	Pluto ♇ ° '
01	06 39 25	09 ♋ 58 40	04 ♍ 44 44	10 ♍ 47 29	14 ♋ 34	26 ♉ 56	05 ♑ 38	07 ♓ 16	10 ♋ 39	02 ♋ 20	19 ♈ 59	16 ♒ 22
02	06 43 22	10 55 54	16 ♍ 47 18	22 ♍ 44 45	13 R 59	28 02	05 R 20	07 R 15	10 47	02 23	20 00	16 R 21
03	06 47 19	11 53 06	28 ♍ 40 27	04 ♎ 35 03	13 22	29 ♉ 09	05 02	07 14	10 55	02 27	20 01	16 19
04	06 51 15	12 50 19	10 ♎ 29 11	16 ♎ 23 34	12 46	00 ♊ 15	04 44	07 12	11 03	02 31	20 01	16 18
05	06 55 12	13 47 31	22 ♎ 18 52	28 ♎ 15 47	12 09	01 21	04 26	07 10	11 10	02 34	20 02	16 17
06	06 59 08	14 44 43	04 ♏ 14 59	10 ♏ 17 05	11 34	02 28	04 08	07 08	11 18	02 38	20 03	16 16
07	07 03 05	15 41 55	16 ♏ 22 43	22 ♏ 32 24	11 00	03 35	03 51	07 06	11 26	02 41	20 03	16 15
08	07 07 01	16 39 06	28 ♏ 46 38	05 ♐ 05 49	10 29	04 42	03 34	07 04	11 34	02 45	20 04	16 14
09	07 10 58	17 36 18	11 ♐ 30 14	18 ♐ 00 07	10 00	05 49	03 17	07 02	11 42	02 48	20 04	16 13
10	07 14 54	18 33 29	24 ♐ 35 31	01 ♑ 16 25	09 34	06 56	03 01	06 59	11 49	02 52	20 04	16 12
11	07 18 51	19 30 41	08 ♑ 02 38	14 ♑ 53 52	09 13	08 03	02 45	06 56	11 57	02 55	20 05	16 11
12	07 22 48	20 27 53	21 ♑ 49 43	28 ♑ 49 39	08 55	09 11	02 30	06 53	12 05	02 59	20 05	16 09
13	07 26 44	21 25 05	05 ♒ 53 05	12 ♒ 59 20	08 42	10 19	02 15	06 50	12 13	03 02	20 06	16 08
14	07 30 41	22 22 17	20 ♒ 07 45	27 ♒ 17 38	08 34	11 26	02 01	06 46	12 20	03 06	20 06	16 07
15	07 34 37	23 19 30	04 ♓ 28 20	11 ♓ 39 13	08 31	12 34	01 47	06 43	12 28	03 09	20 06	16 06
16	07 38 34	24 16 43	18 ♓ 49 48	25 ♓ 59 34	08 D 33	13 42	01 34	06 39	12 36	03 12	20 06	16 04
17	07 42 30	25 13 57	03 ♈ 08 12	10 ♈ 15 23	08 40	14 50	01 22	06 35	12 43	03 16	20 07	16 03
18	07 46 27	26 11 11	17 ♈ 20 54	24 ♈ 24 37	08 53	15 58	01 10	06 31	12 51	03 19	20 07	16 02
19	07 50 23	27 08 27	01 ♉ 26 26	08 ♉ 26 15	09 12	17 07	00 59	06 26	12 59	03 23	20 07	16 01
20	07 54 20	28 05 43	15 ♉ 24 02	22 ♉ 19 44	09 36	18 15	00 49	06 22	13 06	03 26	20 07	15 59
21	07 58 17	29 ♋ 03 00	29 ♉ 13 18	06 ♊ 04 37	10 06	19 24	00 39	06 17	13 14	03 29	20 07	15 58
22	08 02 13	00 ♌ 00 17	12 ♊ 53 35	19 ♊ 40 05	10 42	20 32	00 31	06 12	13 22	03 33	20 07	15 57
23	08 06 10	00 57 36	26 ♊ 23 57	03 ♋ 05 00	11 23	21 41	00 23	06 08	13 29	03 36	20 R 07	15 56
24	08 10 06	01 54 55	09 ♋ 43 02	16 ♋ 17 52	12 10	22 50	00 15	06 02	13 37	03 39	20 07	15 54
25	08 14 03	02 52 15	22 ♋ 49 20	29 ♋ 17 16	13 02	23 59	00 09	05 57	13 44	03 42	20 07	15 53
26	08 17 59	03 49 36	05 ♌ 41 34	12 ♌ 02 08	14 00	25 08	00 ♑ 03	05 52	13 52	03 46	20 07	15 52
27	08 21 56	04 46 57	18 ♌ 19 00	24 ♌ 32 11	15 03	26 17	29 ♐ 59	05 46	13 59	03 49	20 07	15 50
28	08 25 52	05 44 19	00 ♍ 41 50	06 ♍ 48 06	16 12	27 27	29 55	05 40	14 07	03 52	20 07	15 49
29	08 29 49	06 41 41	12 ♍ 51 17	18 ♍ 51 41	17 25	28 36	29 52	05 34	14 14	03 55	20 07	15 48
30	08 33 46	07 39 04	24 ♍ 49 43	00 ♎ 45 49	18 44	29 ♊ 46	29 49	05 28	14 22	03 58	20 06	15 46
31	08 37 42	08 ♌ 36 28	06 ♎ 40 29	12 ♎ 34 18	20 ♋ 07	00 ♋ 55	29 ♐ 48	05 ♓ 22	14 ♋ 29	04 ♋ 01	20 ♈ 06	15 ♒ 45

Date	Moon True ☊ ° '	Moon Mean ☊ ° '	Moon ☽ Latitude ° '
01	17 ♎ 14	17 ♎ 09	03 S 25
02	17 R 10	17 06	02 34
03	17 08	17 03	01 36
04	17 D 08	17 00	00 S 35
05	17 R 08	16 57	00 N 27
06	17 07	16 53	01 29
07	17 05	16 50	02 28
08	17 00	16 47	03 22
09	16 53	16 44	04 06
10	16 44	16 41	04 39
11	16 32	16 38	04 58
12	16 20	16 34	04 59
13	16 09	16 31	04 43
14	15 59	16 28	04 10
15	15 52	16 25	03 20
16	15 48	16 22	02 17
17	15 46	16 18	01 N 06
18	15 D 46	16 15	00 S 08
19	15 R 46	16 12	01 22
20	15 45	16 09	02 30
21	15 42	16 06	03 29
22	15 37	16 03	04 15
23	15 29	15 59	04 46
24	15 18	15 56	05 00
25	15 07	15 53	04 59
26	14 55	15 50	04 43
27	14 44	15 47	04 12
28	14 34	15 44	03 31
29	14 28	15 40	02 40
30	14 24	15 37	01 42
31	14 ♎ 22	15 ♎ 34	00 S 41

DECLINATIONS

Date	Sun ☉ ° '	Moon ☽ ° '	Mercury ☿ ° '	Venus ♀ ° '	Mars ♂ ° '	Jupiter ♃ ° '	Saturn ♄ ° '	Uranus ♅ ° '	Neptune ♆ ° '	Pluto ♇ ° '
01	23 N 03	06 N 35	18 N 32	16 N 58	28 S 02	09 S 50	22 N 30	23 N 38	06 N 20	23 S 04
02	22 59	02 N 51	18 25	17 14	28 06	09 51	22 29	23 38	06 20	23 04
03	22 54	00 S 57	18 20	17 29	28 09	09 52	22 29	23 38	06 20	23 05
04	22 49	04 41	18 15	17 44	28 12	09 53	22 28	23 38	06 20	23 05
05	22 43	08 16	18 12	17 59	28 15	09 54	22 28	23 38	06 20	23 06
06	22 37	11 32	18 11	18 14	28 18	09 55	22 27	23 38	06 21	23 06
07	22 31	14 22	18 10	18 28	28 20	09 56	22 26	23 38	06 21	23 07
08	22 24	16 36	18 11	18 42	28 23	09 57	22 26	23 38	06 21	23 08
09	22 16	18 06	18 13	18 55	28 25	09 58	22 25	23 38	06 21	23 08
10	22 09	18 41	18 17	19 08	28 27	09 59	22 25	23 38	06 21	23 09
11	22 01	18 14	18 21	19 21	28 29	10 01	22 24	23 38	06 21	23 09
12	21 52	16 44	18 27	19 33	28 30	10 02	22 23	23 37	06 21	23 10
13	21 44	14 13	18 34	19 45	28 32	10 04	22 23	23 37	06 21	23 10
14	21 35	10 49	18 41	19 57	28 33	10 05	22 22	23 37	06 21	23 11
15	21 25	06 46	18 50	20 08	28 34	10 07	22 21	23 37	06 21	23 11
16	21 15	02 S 19	18 59	20 18	28 35	10 08	22 21	23 37	06 21	23 12
17	21 05	02 N 16	19 09	20 28	28 36	10 10	22 20	23 37	06 21	23 12
18	20 54	06 41	19 20	20 38	28 37	10 12	22 20	23 37	06 21	23 13
19	20 43	10 41	19 31	20 47	28 37	10 14	22 19	23 37	06 21	23 13
20	20 32	14 03	19 42	20 56	28 38	10 16	22 18	23 37	06 21	23 14
21	20 21	16 35	19 53	21 04	28 38	10 17	22 17	23 37	06 21	23 14
22	20 09	18 08	20 04	21 12	28 38	10 19	22 17	23 37	06 21	23 15
23	19 56	18 38	20 15	21 19	28 38	10 22	22 16	23 37	06 21	23 15
24	19 44	18 05	20 25	21 25	28 38	10 24	22 15	23 37	06 21	23 16
25	19 31	16 35	20 35	21 32	28 37	10 26	22 15	23 37	06 21	23 16
26	19 17	14 16	20 44	21 37	28 37	10 28	22 14	23 36	06 21	23 17
27	19 04	11 20	20 52	21 42	28 36	10 30	22 13	23 36	06 21	23 17
28	18 50	07 56	20 59	21 47	28 36	10 33	22 13	23 36	06 21	23 18
29	18 36	04 16	21 05	21 51	28 35	10 35	22 12	23 36	06 21	23 18
30	18 21	00 29	21 09	21 54	28 34	10 37	22 11	23 36	06 21	23 19
31	18 N 06	03 S 17	21 N 11	21 N 57	28 S 34	10 S 40	22 N 10	23 N 36	06 N 21	23 S 19

ZODIAC SIGN ENTRIES

Date	h	m	Planets
01	02	39	☽ ♍
03	14	41	☽ ♎
04	06	37	♀ ♊
06	03	29	☽ ♏
08	14	20	☽ ♐
10	21	44	☽ ♑
13	02	00	☽ ♒
15	04	32	☽ ♓
17	06	44	☽ ♈
19	09	32	☽ ♉
21	13	22	☽ ♊
22	11	53	☉ ♌
23	18	27	☽ ♋
26	01	20	☽ ♌
27	04	35	♂ ♐
28	10	38	☽ ♍
30	17	00	♀ ♋
30	22	27	☽ ♎

LATITUDES

Date	Mercury ☿ ° '	Venus ♀ ° '	Mars ♂ ° '	Jupiter ♃ ° '	Saturn ♄ ° '	Uranus ♅ ° '	Neptune ♆ ° '	Pluto ♇ ° '
01	04 S 07	02 S 34	04 S 44	01 S 05	00 S 31	00 N 13	01 S 36	07 S 30
04	04 35	02 31	04 52	01 06	00 30	00 13	01 36	07 30
07	04 49	02 27	04 58	01 07	00 30	00 13	01 37	07 31
10	04 49	02 21	05 03	01 07	00 30	00 14	01 37	07 31
13	04 36	02 16	05 07	01 08	00 30	00 14	01 37	07 32
16	04 11	02 09	05 10	01 09	00 29	00 14	01 37	07 32
19	03 37	02 02	05 11	01 10	00 29	00 14	01 37	07 33
22	02 57	01 54	05 12	01 11	00 29	00 14	01 38	07 33
25	02 13	01 46	05 11	01 11	00 29	00 14	01 38	07 33
28	01 29	01 38	05 10	01 12	00 29	00 14	01 38	07 34
31	00 S 45	01 S 29	05 S 08	01 S 13	00 S 28	00 N 14	01 S 38	07 S 34

DATA

Julian Date	2463780
Delta T	+75 seconds
Ayanamsa	24° 19' 36"
Synetic vernal point	04° ♓ 47' 23"
True obliquity of ecliptic	23° 25' 56"

LONGITUDES

Date	Chiron ⚷ ° '	Ceres ⚳ ° '	Pallas ⚴ ° '	Juno ⚵ ° '	Vesta ⚶ ° '	Black Moon Lilith ⚸ ° '
01	29 ♉ 04	23 ♐ 03	28 ♎ 28	04 ♎ 02	21 ♓ 57	06 ♎ 25
11	29 ♉ 36	21 ♐ 18	29 ♎ 44	06 ♎ 05	23 ♓ 13	07 ♎ 32
21	00 ♊ 05	20 ♐ 03	01 ♏ 33	08 ♎ 25	23 ♓ 47	08 ♎ 38
31	00 ♊ 29	19 ♐ 22	03 ♏ 49	10 ♎ 58	23 ♓ 34	09 ♎ 44

MOON'S PHASES, APSIDES AND POSITIONS ☽

Date	h	m	Phase	Longitude ° '	Eclipse Indicator
04	17	12	☽	13 ♎ 03	
12	09	29	○	20 ♑ 22	
19	04	07	☾	26 ♈ 50	
26	08	13	●	03 ♌ 41	

Day	h	m		
04	04	55	Apogee	
16	09	33	Perigee	
31	23	16	Apogee	
03	06	02	OS	
10	14	01	Max dec	18° S 41'
17	00	06	ON	
23	11	18	Max dec	18° N 38'
30	15	05	OS	

ASPECTARIAN

h m	Aspects
01 Friday	
02 17	☽ ∠ ☿
07 13	☽ ⚹ ♅
09 29	☉ # ♇
12 29	☽ bQ ♆
13 44	☽ △ ♂
13 44	☽ ∥ ♆
17 00	☽ ☍ ♃
21 26	☉ ± ♇
23 16	☽ ⚹ ☉
23 51	☽ ⚹ ♄
02 Saturday	
06 24	☽ ± ♆
06 38	☽ ⚹ ☿
07 10	☽ Q ♅
07 38	☉ ☌ ♄
11 06	☽ ⚻ ♇
18 28	☽ ⚻ ♆
23 10	☽ ± ♇
03 Sunday	
00 12	☽ Q ♄
01 14	☿ ∠ ♀
01 27	☽ Q ☉
05 38	☽ Q ☿
10 05	♀ ± ♂
13 03	☽ △ ♀
17 22	☽ bQ ♇
19 42	☽ □ ♅
04 Monday	
00 35	☽ □ ♂
05 20	☽ ⚻ ♃
10 48	☉ ☌ ☿
13 09	☽ □ ♄
16 24	☽ □ ☿
17 12	☽ □ ☉
17 31	☽ ± ♃
22 41	☽ bQ ♀
22 53	☽ # ♆
23 48	☽ △ ♇
05 Tuesday	
07 23	☽ ☍ ♆
11 43	☽ bQ ♃
12 14	☽ Q ♂
18 47	☽ ± ♀
23 44	☽ ∥ ♃
06 Wednesday	
07 15	☿ ∥ ♀
08 04	☽ ⚻ ♀
08 45	☽ △ ♅
11 47	☽ ⚹ ♂
15 38	♀ ⊻ ♅
17 45	☽ △ ♃
20 56	☿ ☌ ♄
07 Thursday	
01 53	☽ △ ☿
02 10	☽ △ ♄
10 33	☽ △ ☉
11 45	☽ □ ♇
14 34	☽ bQ ♅
16 32	♀ ⚻ ♂
16 43	☽ ∠ ♂
19 11	☽ ⚻ ♆
08 Friday	
01 39	☉ ⚻ ♇
04 16	☉ ∥ ♄
05 55	☽ bQ ☿
06 47	☽ ± ♆
07 43	☽ bQ ♄
08 05	☽ ± ♅
08 43	☿ ⊥ ♀
09 44	☽ ⊥ ♂
17 56	☽ bQ ☉
19 35	☽ ⚻ ♅
19 50	♀ ∠ ♆
20 54	☽ ⊻ ♂
22 21	☽ Q ♇
22 25	☽ ± ☿
23 56	☽ bQ ♆
09 Saturday	
00 18	☿ ± ♇
00 20	☽ ☍ ♀
01 01	☽ ± ♄
03 39	☽ □ ♃
09 00	♀ ⊥ ♄
09 17	☽ ⚻ ☿
12 12	☽ ± ☉
12 21	☽ ⚻ ♄
15 15	☽ # ☿
20 42	☽ ⚹ ♇
10 Sunday	
00 09	☽ ⚻ ☉
03 47	☽ △ ♆
12 42	☽ Q ♃
12 55	♀ □ ♃
23 25	♂ ☍ ♅
23 51	☽ ∠ ♇
11 Monday	
02 49	☽ ☌ ♂
02 54	☽ ☍ ♅
09 12	☽ # ☿
10 03	☽ ⚹ ♃
12 02	☽ ⚻ ♀
14 00	☽ ☍ ☿
15 44	☽ ⊥ ♇
18 55	☽ ☍ ♄
23 28	☽ ± ♀
12 Tuesday	
02 12	☽ ⊻ ♇
02 27	☉ □ ♆
07 23	☿ ⊻ ♀
09 00	☽ □ ♆
09 29	☽ ☍ ☉
12 05	☽ ∠ ♃
16 24	☽ bQ ♀
13 Wednesday	
03 27	☽ ⊥ ♃
05 56	☽ ⊻ ♂
07 09	☽ ⚻ ♅
13 35	☽ ⊻ ♃
15 44	☽ Q ♆
15 56	☽ ⊥ ♂
16 42	☽ ⚻ ☿
17 21	☽ ± ♅
20 08	☽ △ ♀
21 43	☉ bQ ♃
22 47	☽ ⚻ ♄
14 Thursday	
02 43	☽ ± ☿
05 16	☽ ☌ ♇
06 52	☽ ∠ ♂
08 34	☽ bQ ♅
08 58	☽ ± ♄
11 57	☽ ⚹ ♆
16 02	☽ ⚻ ☉
16 34	☽ ∥ ♃
17 43	☽ bQ ☿
15 Friday	
00 11	☽ bQ ♄
02 47	☽ ± ☉
07 35	☽ ⚹ ♂
09 34	♀ ⊻ ♄
09 47	☽ △ ♅
13 03	☽ ∠ ♆
14 15	☽ # ♆
14 19	☿ St D
15 43	☽ ☌ ♃
18 45	☽ △ ☿
18 54	☽ bQ ☉
16 Saturday	
01 29	☽ △ ♄
02 41	☽ □ ♀
03 20	☽ Q ♂
04 06	☽ ⊥ ♆
07 24	☽ ⊻ ♇
14 08	☽ ⊻ ♆
17 25	☽ ⊥ ♇
21 47	☽ △ ☉
17 Sunday	
08 30	☽ ∠ ♇
09 04	☽ □ ♂
11 27	☽ Q ♀
12 13	☽ □ ♅
17 46	☽ ⊻ ♃
21 27	☽ □ ☿
18 Monday	
03 51	☽ ⊥ ♃
04 19	☽ □ ♄
09 28	☽ ⚹ ♀
09 46	☽ ⚹ ♇
10 11	☽ ∥ ♆
13 13	♀ △ ♇
16 42	☽ ☌ ♆
18 46	☽ Q ♅
19 02	☽ ∠ ♃
19 Tuesday	
04 07	☽ □ ☉
04 35	☽ Q ☿
06 09	☽ Q ♇
07 41	☉ ∥ ♀
08 26	♂ ∠ ♇
09 02	☽ # ♃
11 12	☽ Q ♄
11 14	☽ △ ♂
13 15	☽ ∠ ♀
15 20	☽ ⚹ ♅
20 31	☽ ⚹ ♃
20 Wednesday	
01 42	☽ ⚹ ☿
06 05	☽ ⊥ ♀
08 00	☽ ⚹ ♄
12 43	☽ bQ ♂
13 01	☽ □ ♇
13 17	☽ Q ☉
17 06	☽ Q ♃
17 16	☽ ∠ ♅
17 23	☽ ⊻ ♀
20 10	☽ ⊻ ♆
21 Thursday	
04 08	☽ ± ♂
04 33	☽ ∠ ☿
06 01	☿ ± ♇
06 35	☽ ⊥ ♆
08 58	☽ ⊥ ♅
10 16	☽ ∠ ♄
11 41	☽ ⚹ ☉
14 29	☽ ⚻ ♂
19 29	☽ ⊻ ♅
20 55	☽ ⊥ ☿
22 19	☽ ∠ ♆
22 Friday	
00 18	☽ □ ♃
02 10	☽ ⊥ ♄
03 10	♀ ⚹ ♆
07 57	☽ ⊻ ☿
12 50	☽ ⊻ ♄
16 01	☽ ∠ ☉
16 42	☉ ± ♃
17 02	☉ ∥ ☿
17 24	☽ △ ♇
23 07	☉ ⚻ ♂
23 Saturday	
00 48	☽ ⚹ ♆
02 49	☽ ☌ ♀
09 14	☽ ⊥ ☉
10 26	♆ St R
19 04	☽ ☍ ♂
20 06	☽ bQ ♇
20 49	☽ ⊻ ☉
22 16	☽ Q ♆
24 Sunday	
00 59	☽ ☌ ♅
05 23	☽ △ ♃
12 20	☽ ± ♇
16 46	☽ ☌ ☿
19 10	☽ ☌ ♄
23 16	☽ ⚻ ♇
25 Monday	
07 01	☽ □ ♆
08 34	☽ bQ ♃
14 22	☽ ⊻ ♀
26 Tuesday	
01 09	☽ ± ♃
01 30	☽ ⚻ ♂
02 37	☽ ⊥ ♀
08 12	☿ ☌ ♄
08 13	☽ ☌ ☉
08 21	☽ ⊻ ♅
10 15	☉ ⊻ ♅
12 19	☽ ⚻ ♃
12 41	☽ ± ♂
19 43	☽ ⊥ ♅
21 14	☽ ∠ ♀
27 Wednesday	
03 38	☽ ⊻ ♄
05 10	☽ ⊻ ☿
05 39	☽ bQ ♂
07 16	☽ ☍ ♇
12 58	☽ ∠ ♅
15 15	☽ ⊥ ♄
15 28	☽ △ ♆
17 47	☽ ⊥ ☿
18 01	☽ # ♃
28 Thursday	
04 23	☿ ⚻ ♇
05 00	☽ ⚹ ♀
08 52	☽ ∠ ♄
10 28	☽ △ ♂
10 30	☉ ⚻ ♃
13 05	☽ ∠ ☿
16 06	☉ ± ♂
18 15	☽ ⚹ ♅
20 40	☽ bQ ♆
21 42	☽ ☍ ♃
22 34	☽ ∥ ♆
22 45	☽ ⊻ ☉
29 Friday	
07 03	☽ Q ♀
11 39	☽ ⊥ ☉
14 30	☽ ± ♆
14 47	☽ ⚹ ♄
17 51	☽ ⚻ ♇
18 08	☽ Q ♅
22 12	☽ ⚹ ☿
30 Saturday	
02 30	☽ ⚻ ♆
05 51	☽ ± ♇
07 14	☽ ∠ ☉
13 18	♀ ☍ ♂
15 08	☽ Q ♄
22 04	☽ □ ♂
23 03	☽ □ ♀
31 Sunday	
00 00	☽ bQ ♇
01 29	☽ Q ☿
06 36	☽ □ ♅
08 35	♀ bQ ♇
09 23	☽ ⚻ ♃
11 45	☿ □ ♆
15 59	☿ bQ ♃
16 17	☽ ⚹ ☉
21 28	☽ ± ♃

All ephemeris data is given at 12.00 UT and the Moon's longitude is additionally given for 24.00 UT

AUGUST 2033

LONGITUDES

Date	Sidereal time h m s	Sun ☉	Moon ☽	Moon ☽ 24.00	Mercury ☿	Venus ♀	Mars ♂	Jupiter ♃	Saturn ♄	Uranus ♅	Neptune ♆	Pluto ⯓
01	08 41 39	09 ♌ 33 52	18 ♎ 27 50	24 ♎ 21 43	21 ♋ 35	02 ♋ 05	29 ♐ 48	05 ♓ 16	14 ♋ 36	04 ♋ 05	20 ♈ 06	15 ♒ 44
02	08 45 35	10 31 17	00 ♏ 16 37	06 ♏ 13 12	23 08	03 14	29 D 48	05 R 10	14 44	04 08	20 R 06	15 R 42
03	08 49 32	11 28 43	12 ♏ 12 08	18 ♏ 14 07	24 44	04 24	29 49	05 03	14 51	04 11	20 05	15 41
04	08 53 28	12 26 09	24 ♏ 19 46	00 ♐ 29 45	26 25	05 34	29 51	04 57	14 58	04 14	20 05	15 40
05	08 57 25	13 23 36	06 ♐ 44 37	13 ♐ 04 55	28 10	06 44	29 54	04 50	15 06	04 17	20 04	15 38
06	09 01 21	14 21 04	19 ♐ 31 03	26 ♐ 03 24	29 ♋ 58	07 54	29 ♐ 58	04 43	15 13	04 20	20 04	15 37
07	09 05 18	15 18 32	02 ♑ 42 08	09 ♑ 27 22	01 ♌ 49	09 05	00 ♑ 02	04 36	15 20	04 23	20 04	15 36
08	09 09 15	16 16 01	16 ♑ 19 02	23 ♑ 16 52	03 42	10 15	00 08	04 29	15 27	04 25	20 03	15 34
09	09 13 11	17 13 32	00 ♒ 20 31	07 ♒ 29 26	05 38	11 25	00 14	04 22	15 34	04 28	20 02	15 33
10	09 17 08	18 11 03	14 ♒ 42 54	22 ♒ 00 07	07 36	12 36	00 21	04 15	15 41	04 31	20 02	15 31
11	09 21 04	19 08 35	29 ♒ 20 13	06 ♓ 42 13	09 36	13 46	00 28	04 07	15 48	04 34	20 01	15 30
12	09 25 01	20 06 08	14 ♓ 05 12	21 ♓ 28 12	11 37	14 57	00 37	04 00	15 55	04 37	20 01	15 29
13	09 28 57	21 03 43	28 ♓ 50 23	06 ♈ 10 56	13 39	16 07	00 46	03 52	16 02	04 39	20 00	15 27
14	09 32 54	22 01 19	13 ♈ 29 13	20 ♈ 44 39	15 41	17 18	00 56	03 45	16 09	04 42	19 59	15 26
15	09 36 50	22 58 56	27 ♈ 56 48	05 ♉ 05 22	17 43	18 29	01 07	03 37	16 16	04 45	19 59	15 25
16	09 40 47	23 56 35	12 ♉ 10 08	19 ♉ 10 59	19 46	19 40	01 19	03 30	16 22	04 47	19 58	15 23
17	09 44 44	24 54 16	26 ♉ 07 52	03 ♊ 00 47	21 48	20 51	01 31	03 22	16 29	04 50	19 57	15 22
18	09 48 40	25 51 58	09 ♊ 49 49	16 ♊ 35 01	23 50	22 02	01 44	03 14	16 36	04 53	19 56	15 21
19	09 52 37	26 49 42	23 ♊ 16 31	29 ♊ 54 23	25 51	23 14	01 57	03 07	16 42	04 55	19 56	15 19
20	09 56 33	27 47 28	06 ♋ 28 44	12 ♋ 59 40	27 51	24 25	02 12	02 59	16 49	04 58	19 55	15 18
21	10 00 30	28 45 15	19 ♋ 27 16	25 ♋ 51 35	29 ♌ 51	25 36	02 27	02 51	16 55	05 00	19 54	15 17
22	10 04 26	29 ♌ 43 04	02 ♌ 12 43	08 ♌ 30 44	01 ♍ 49	26 48	02 42	02 43	17 02	05 03	19 53	15 15
23	10 08 23	00 ♍ 40 54	14 ♌ 45 42	20 ♌ 57 42	03 46	27 59	02 59	02 35	17 08	05 05	19 52	15 14
24	10 12 19	01 38 46	27 ♌ 06 50	03 ♍ 13 14	05 42	29 ♋ 11	03 16	02 27	17 15	05 07	19 51	15 13
25	10 16 16	02 36 39	09 ♍ 17 02	15 ♍ 18 26	07 37	00 ♌ 22	03 33	02 19	17 21	05 10	19 50	15 12
26	10 20 13	03 34 31	21 ♍ 17 39	27 ♍ 14 56	09 30	01 34	03 52	02 11	17 27	05 12	19 49	15 10
27	10 24 09	04 32 30	03 ♎ 10 36	09 ♎ 05 00	11 23	02 46	04 11	02 03	17 33	05 14	19 48	15 09
28	10 28 06	05 30 27	14 ♎ 58 30	20 ♎ 51 32	13 14	03 58	04 30	01 56	17 39	05 16	19 47	15 08
29	10 32 02	06 28 26	26 ♎ 44 36	02 ♏ 38 12	15 03	05 10	04 50	01 48	17 45	05 18	19 46	15 06
30	10 35 59	07 26 26	08 ♏ 32 53	14 ♏ 29 12	16 52	06 22	05 11	01 40	17 51	05 21	19 45	15 05
31	10 39 55	08 ♍ 24 28	20 ♏ 27 47	26 ♏ 29 13	18 ♍ 39	07 ♌ 34	05 ♑ 32	01 ♓ 32	17 ♋ 57	05 ♋ 23	19 ♈ 44	15 ♒ 04

Date	Moon True ☊	Moon Mean ☊	Moon ☽ Latitude
01	14 ♎ 21	15 ♎ 31	00 N 22
02	14 D 22	15 28	01 24
03	14 R 22	15 24	02 23
04	14 22	15 21	03 17
05	14 19	15 18	04 03
06	14 14	15 15	04 38
07	14 07	15 12	05 00
08	13 59	15 09	05 05
09	13 50	15 05	04 53
10	13 41	15 02	04 22
11	13 34	14 59	03 34
12	13 29	14 56	02 31
13	13 26	14 53	01 N 18
14	13 25	14 50	00 00
15	13 D 26	14 46	01 S 18
16	13 27	14 43	02 29
17	13 R 27	14 40	03 30
18	13 26	14 37	04 18
19	13 23	14 34	04 50
20	13 19	14 30	05 07
21	13 12	14 27	05 07
22	13 05	14 24	04 52
23	12 57	14 21	04 23
24	12 50	14 18	03 42
25	12 44	14 15	02 51
26	12 40	14 11	01 54
27	12 38	14 08	00 S 51
28	12 D 37	14 05	00 N 13
29	12 38	14 02	01 16
30	12 40	13 59	02 17
31	12 ♎ 42	13 ♎ 56	03 N 12

DECLINATIONS

Date	Sun ☉	Moon ☽	Mercury ☿	Venus ♀	Mars ♂	Jupiter ♃	Saturn ♄	Uranus ♅	Neptune ♆	Pluto ⯓
01	17 N 51	06 S 54	21 N 12	21 N 59	28 S 33	10 S 42	22 N 10	23 N 36	06 N 20	23 S 20
02	17 36	10 15	21 10	22 01	28 32	10 45	22 09	23 36	06 20	23 20
03	17 20	13 13	21 06	22 02	28 31	10 48	22 08	23 36	06 20	23 21
04	17 04	15 40	21 00	22 02	28 29	10 50	22 08	23 36	06 20	23 21
05	16 48	17 26	20 51	22 02	28 28	10 53	22 07	23 36	06 20	23 22
06	16 31	18 24	20 40	22 01	28 27	10 56	22 06	23 36	06 19	23 22
07	16 14	18 25	20 26	22 00	28 26	10 58	22 05	23 36	06 19	23 23
08	15 57	17 23	20 10	21 58	28 24	11 01	22 05	23 35	06 19	23 23
09	15 40	15 18	19 51	21 56	28 23	11 04	22 04	23 35	06 19	23 24
10	15 23	12 14	19 29	21 53	28 21	11 07	22 03	23 35	06 18	23 24
11	15 05	08 21	19 05	21 49	28 20	11 09	22 02	23 35	06 18	23 25
12	14 47	03 S 56	18 38	21 45	28 18	11 12	22 02	23 35	06 18	23 25
13	14 28	00 N 44	18 09	21 40	28 16	11 15	22 01	23 35	06 18	23 26
14	14 10	05 19	17 38	21 34	28 15	11 18	22 00	23 35	06 17	23 26
15	13 51	09 32	17 04	21 28	28 13	11 21	21 59	23 35	06 17	23 27
16	13 32	13 07	16 29	21 21	28 11	11 24	21 59	23 35	06 17	23 27
17	13 13	15 53	15 52	21 14	28 09	11 27	21 58	23 35	06 16	23 28
18	12 54	17 40	15 13	21 06	28 07	11 30	21 57	23 35	06 16	23 28
19	12 34	18 26	14 33	20 58	28 05	11 33	21 56	23 35	06 16	23 28
20	12 14	18 10	13 52	20 49	28 03	11 36	21 56	23 35	06 15	23 29
21	11 54	16 57	13 10	20 39	28 01	11 38	21 55	23 34	06 15	23 29
22	11 34	14 55	12 27	20 29	27 59	11 41	21 54	23 34	06 14	23 30
23	11 14	12 12	11 43	20 18	27 56	11 44	21 53	23 34	06 14	23 30
24	10 53	08 59	10 58	20 06	27 54	11 47	21 53	23 34	06 14	23 30
25	10 32	05 26	10 13	19 54	27 52	11 50	21 52	23 34	06 13	23 31
26	10 12	01 N 43	09 27	19 42	27 49	11 53	21 51	23 34	06 13	23 31
27	09 51	02 S 03	08 41	19 29	27 47	11 56	21 50	23 34	06 12	23 32
28	09 29	05 42	07 54	19 15	27 44	11 59	21 50	23 34	06 12	23 32
29	09 08	09 07	07 08	19 01	27 41	12 02	21 49	23 34	06 11	23 32
30	08 47	12 11	06 21	18 46	27 38	12 05	21 48	23 34	06 11	23 33
31	08 N 25	14 S 46	05 N 34	18 N 31	27 S 35	12 S 08	21 N 47	23 N 34	06 N 11	23 S 33

ZODIAC SIGN ENTRIES

Date	h	m	Planets
02	11	26	☽ ♏
04	23	02	☽ ♐
06	12	31	☿ ♌
07	00	47	♂ ♑
07	07	09	☽ ♑
09	11	25	☽ ♒
11	13	05	☽ ♓
13	13	54	☽ ♈
15	15	26	☽ ♉
17	18	44	☽ ♊
20	00	10	☽ ♋
21	13	55	☿ ♍
22	07	49	☽ ♌
22	19	02	☉ ♍
24	17	40	☽ ♍
25	04	29	♀ ♌
27	05	34	☽ ♎
29	18	38	☽ ♏

LATITUDES

Date	Mercury ☿	Venus ♀	Mars ♂	Jupiter ♃	Saturn ♄	Uranus ♅	Neptune ♆	Pluto ⯓
01	00 S 31	01 S 26	05 S 07	01 S 13	00 S 28	00 N 14	01 S 38	07 S 34
04	00 N 08	01 17	05 03	01 14	00 28	00 14	01 38	07 34
07	00 42	01 07	05 00	01 14	00 28	00 14	01 38	07 35
10	01 10	00 58	04 55	01 15	00 28	00 14	01 39	07 35
13	01 29	00 48	04 51	01 15	00 28	00 14	01 39	07 35
16	01 41	00 38	04 46	01 16	00 27	00 14	01 39	07 35
19	01 46	00 29	04 40	01 16	00 27	00 14	01 39	07 35
22	01 44	00 19	04 35	01 16	00 27	00 14	01 39	07 35
25	01 37	00 S 10	04 29	01 17	00 27	00 14	01 39	07 35
28	01 26	00 00	04 23	01 17	00 27	00 14	01 39	07 35
31	01 N 11	00 N 09	04 S 17	01 S 17	00 S 27	00 N 14	01 S 40	07 S 35

DATA

Julian Date	2463811
Delta T	+75 seconds
Ayanamsa	24° 19' 40"
Synetic vernal point	04° ♓ 47' 19"
True obliquity of ecliptic	23° 25' 57"

MOON'S PHASES, APSIDES AND POSITIONS ☽

Date	h	m	Phase	Longitude	Eclipse Indicator
03	10	26	☽	11 ♏ 25	
10	18	08	○	18 ♒ 26	
17	09	43	☾	24 ♉ 49	
24	21	40	●	02 ♍ 02	

Day	h	m		
12	21	24	Perigee	
28	15	34	Apogee	
07	00	21	Max dec	18° S 32'
13	08	16	0N	
19	17	42	Max dec	18° N 28'
26	22	53	0S	

LONGITUDES

Date	Chiron ⚷	Ceres ⚳	Pallas ⚴	Juno ⚵	Vesta ⚶	Black Moon Lilith ⚸
01	00 ♊ 31	19 ♐ 20	04 ♏ 04	11 ♎ 14	23 ♓ 30	09 ♎ 51
11	00 ♊ 50	19 ♐ 19	06 ♏ 45	14 ♎ 00	22 ♓ 26	10 ♎ 58
21	01 ♊ 02	19 ♐ 27	09 ♏ 36	16 ♎ 55	20 ♓ 39	12 ♎ 04
31	01 ♊ 09	20 ♐ 58	11 ♏ 00	19 ♎ 59	18 ♓ 22	13 ♎ 10

ASPECTARIAN

h m Aspects

01 Monday
04 04 ☽ □ ♄
06 26 ☽ △ ⯓
08 13 ☽ # ♆
10 38 ☽ Q ♂
12 23 ♀ Q ♆
14 24 ♂ St D
15 19 ☽ ☍ ♆
15 38 ☽ ♇ ♃
18 52 ☽ Q ☉
19 17 ☽ □ ☿

02 Tuesday
01 33 ☉ ⊥ ♅
11 02 ☽ ⚹ ♂
15 51 ☽ ∥ ♃
18 39 ☽ △ ♀
19 49 ☽ △ ♅
21 47 ☽ △ ♃

03 Wednesday
04 07 ♃ ∠ ♆
07 06 ♀ ☌ ♅
10 26 ☽ □ ☉
17 14 ☽ ∠ ♂
17 20 ☽ △ ♄
18 55 ☽ □ ⯓

04 Thursday
00 11 ♀ △ ♃
01 55 ☽ ♇ ♅
03 39 ☽ ⊼ ♆
03 50 ☽ ♇ ♀
11 04 ☽ ⊥ ♂
15 25 ☽ ± ♆
16 45 ☽ △ ☿
19 38 ☽ ± ♅
22 47 ☽ ⊻ ♂
23 06 ☽ ♇ ♄
23 16 ☽ ± ♀

05 Friday
03 28 ☽ # ☉
06 04 ☽ Q ⯓
07 16 ☽ ⊼ ♅
08 23 ☽ □ ♃
08 49 ☽ ♇ ♆
11 59 ☽ ⊼ ♀
16 31 ☽ ± ♄
20 28 ☿ ± ♃

06 Saturday
01 36 ☽ △ ☉
02 07 ☽ ♇ ☿
03 55 ☽ ⊼ ♄
04 45 ☽ ⚹ ⯓
12 01 ☿ ⊼ ♂
13 01 ☽ △ ♆
17 51 ☽ Q ♃
21 30 ☽ ± ☿

07 Sunday
04 34 ☉ ♇ ♂
07 11 ☽ ☌ ♂
07 22 ☽ ♇ ☉
08 13 ☽ ∠ ⯓
10 08 ☽ ⊼ ☿
12 37 ☉ ⊻ ♄
15 00 ☽ ☍ ♅
15 22 ☽ ⚹ ♃
18 55 ☉ ☍ ⯓
22 23 ♀ ± ⯓

08 Monday
00 13 ☽ ⊥ ⯓
00 24 ☽ ☍ ♀
00 38 ☽ ± ☉
10 29 ☽ ☍ ♄
10 42 ☽ ⊻ ⯓
11 54 ☽ ⊼ ☉
14 16 ☉ ⊥ ♀
17 26 ☽ ∠ ♃
18 27 ☽ □ ♆
20 29 ♃ △ ♅
21 09 ☿ ⊼ ♃
21 12 ☿ ⊻ ♅

09 Tuesday
07 59 ☽ # ☉
08 33 ♄ ⊼ ⯓
08 41 ☽ ⊥ ♃
11 48 ☽ ⊻ ♂
18 43 ☽ ⊻ ♃
18 58 ☽ ⊼ ♅
19 40 ☿ ± ♂
21 58 ☽ ⊥ ♂
22 19 ☽ ☍ ☿

10 Wednesday
00 55 ☽ Q ♆
05 01 ☽ ± ♅
08 11 ☽ ⊼ ♀
13 03 ☽ ∠ ♂
13 20 ☽ ☌ ⯓
13 37 ☽ ⊼ ♄
18 08 ☽ ☍ ☉
18 58 ☽ ± ♀
19 18 ☽ ∥ ♃
19 57 ☽ ♇ ♅
20 46 ☽ ⚹ ♆
23 34 ☽ ± ♄

11 Thursday
11 00 ☽ ♇ ♀
13 52 ☽ ⚹ ♂
14 24 ☽ ♇ ♄
16 08 ♂ ∠ ⯓
19 44 ☽ ☌ ♃
20 33 ☽ △ ♅
21 15 ☽ ∠ ♆
23 07 ☉ ∠ ♅
23 26 ☽ # ♆
23 48 ☿ ⊥ ♅

12 Friday
07 21 ☽ ⊼ ☿
09 35 ☽ Q ♂
09 46 ☉ △ ♆
11 53 ☽ ⊥ ♆
13 31 ☽ △ ♀
14 16 ☽ ⊻ ⯓
15 00 ☽ △ ♄
18 39 ☽ ± ☿
21 37 ☽ ⊻ ♆
22 27 ☽ ⊼ ☉
22 39 ♀ ⊼ ⯓

13 Saturday
00 00 ☽ ⊥ ⯓
08 54 ☽ ± ☉
09 53 ♀ ☌ ♄
11 38 ☽ ♇ ☿
14 38 ☽ ∠ ⯓
15 11 ☽ □ ♂
20 09 ☽ ⊻ ♃
21 32 ☽ □ ♅

14 Sunday
00 38 ☽ ♇ ☉
05 54 ☽ ⊥ ♃
09 08 ☿ ☍ ⯓
15 12 ☽ ⚹ ⯓
15 18 ☿ ♇ ♂
15 30 ☉ ⊥ ♄
16 13 ☽ △ ☿
16 25 ☽ □ ♄
17 17 ☽ ∥ ♆
17 47 ☿ ⊻ ♄
18 52 ☽ □ ♀
20 37 ☽ ∠ ♃
22 44 ☽ ☌ ♆

15 Monday
03 08 ☽ △ ☉
03 18 ☽ Q ♅
11 06 ☽ Q ⯓
14 30 ♀ ♇ ♃
17 23 ☽ △ ♂
21 27 ☽ ⚹ ♃
22 41 ☽ Q ♄
23 28 ☽ ⚹ ♅
23 41 ☽ # ♃

16 Tuesday
03 40 ☽ Q ♀
09 20 ☿ ⊻ ♀
12 20 ☿ ∠ ♅
14 21 ☿ △ ♆
14 53 ☽ ∥ ☉
17 29 ☽ □ ⯓
17 38 ☽ Q ♃
17 57 ♀ □ ♆
19 10 ☽ ♇ ♂
19 14 ☽ ⚹ ♄

17 Wednesday
01 05 ☽ ∠ ♅
01 20 ☽ ⊻ ♆
02 02 ☽ ⚹ ♀
03 14 ☽ □ ☿
09 43 ☽ □ ☉
10 55 ☽ ± ♂
11 41 ☽ ⊥ ♆
11 53 ☽ ∥ ☿
16 43 ☽ ⊥ ♅
20 31 ☿ ⊥ ♄
21 24 ☽ ∠ ♄
21 31 ☽ ⊼ ♂

18 Thursday
00 30 ☽ □ ♃
03 15 ☽ ⊻ ♅
03 23 ☽ ∠ ♆
06 36 ☽ ∠ ♀
13 22 ☽ ⊥ ♄
16 10 ☽ Q ☿
19 42 ☽ Q ☉
21 46 ☽ △ ⯓

19 Friday
00 06 ☽ ⊥ ♀
00 07 ☽ ⊻ ♄
05 59 ☽ ⚹ ♆
11 54 ☽ ⊻ ♀
17 28 ☽ ⚹ ☿
18 55 ☽ ⚹ ☉

20 Saturday
00 44 ☽ ♇ ⯓
03 40 ☽ Q ♆
04 01 ☽ ☍ ♂
05 39 ☽ △ ♃
09 13 ☽ ☌ ♅
10 33 ☉ ☌ ☿
17 11 ☽ ± ⯓

21 Sunday
00 33 ☽ ∠ ☉
01 52 ☽ ∠ ☿
04 15 ☽ ⊼ ⯓
07 15 ☽ ☌ ♄
09 02 ☽ ♇ ♃
12 49 ☽ □ ♆
18 40 ☽ ⊥ ☉
21 43 ☽ ⊥ ☿

22 Monday
00 42 ☽ ☌ ♀
01 43 ☽ ± ♃
04 22 ☉ # ♃
06 53 ☽ ⊻ ☉
10 35 ♀ ± ♃
11 06 ☽ ⊻ ☿
12 36 ♂ ⚹ ♃
12 57 ☽ ⊼ ♃
12 57 ☽ ⊼ ♂
14 48 ☿ ∠ ♄
17 24 ☽ ⊻ ♅
22 21 ☿ ☍ ♃

23 Tuesday
00 39 ☽ ± ♂
00 41 ☿ △ ♂
04 54 ☽ ⊥ ♅
11 05 ☿ # ♃
12 55 ☽ ☍ ⯓
15 37 ☽ # ♃
16 38 ☽ ⊻ ♄
17 09 ☽ ∥ ☿
18 22 ☽ ♇ ♂
19 02 ☿ ⊥ ♀
20 33 ☽ ∥ ☉
21 52 ☽ △ ♆
22 20 ☽ ∠ ♅

24 Wednesday
01 30 ☿ ♇ ♆
04 22 ☽ ⊥ ♄
04 38 ☿ ⚹ ♅
16 29 ☽ ⊻ ♀
16 38 ☉ ∥ ☿
21 40 ☽ ☌ ☉
22 10 ☽ ∠ ♄
22 22 ☽ ☍ ♃

25 Thursday
00 23 ☽ △ ♂
03 12 ☽ ♇ ♆
03 48 ☽ ⚹ ♅
04 41 ☉ ∠ ♄
05 36 ☽ ⊥ ♀
05 40 ☉ ☍ ♃
06 50 ☽ ∥ ♆
08 04 ☽ ☌ ☿
09 16 ♃ ♇ ♄
21 03 ☽ ± ♆
23 45 ☽ ⊼ ⯓

26 Friday
01 29 ☽ ∠ ♀
03 45 ☽ Q ♅
04 13 ☽ ⚹ ♄
09 02 ☽ ⊼ ♆
11 45 ☽ ± ⯓
22 29 ☉ △ ♂
23 12 ♀ ⊼ ♃

27 Saturday
04 36 ☽ Q ♄
05 53 ☽ ♇ ⯓
09 45 ☽ ⊼ ♃
11 05 ☽ ⚹ ♀
14 05 ☽ □ ♂
15 01 ☽ ⊻ ☉
16 11 ☽ □ ♅
18 17 ☉ ♇ ♆
21 48 ☽ ± ♃

28 Sunday
04 18 ☽ ⊥ ☉
05 55 ☉ ⚹ ♅
07 47 ☽ ⊻ ☿
12 19 ☽ △ ⯓
14 15 ☽ Q ♀
15 22 ☽ # ♆
15 56 ☽ ♇ ♃
17 31 ☽ □ ♄
19 11 ☿ ± ♆
21 47 ☽ ☍ ♆
22 16 ☽ ⊥ ☿

29 Monday
00 18 ☽ ∠ ☉
00 25 ☽ # ☿
02 52 ♀ ⊼ ♂
03 48 ☽ Q ♂
12 06 ☽ # ☉
12 42 ☿ ⊼ ⯓
14 57 ♀ ⊻ ♅
19 58 ☽ ∠ ☿
22 11 ☽ △ ♃

30 Tuesday
04 58 ☽ ⚹ ♂
05 29 ☽ △ ♅
07 04 ☽ □ ♀
09 33 ☽ ⚹ ☉
11 07 ☽ ∥ ♃
17 18 ☿ ∥ ♆
18 35 ☿ Q ♅

31 Wednesday
00 13 ♂ ☍ ♅
01 11 ☽ □ ⯓
02 10 ☿ ⚹ ♄
06 56 ☽ △ ♄
07 44 ☽ ⚹ ☿
10 32 ☽ ⊼ ♆
11 50 ☽ ♇ ♅
11 53 ☽ Q ☉
12 09 ☽ ∠ ♂
22 28 ☽ ± ♆

All ephemeris data is given at 12.00 UT and the Moon's longitude is additionally given for 24.00 UT

SEPTEMBER 2033

LONGITUDES

Date	Sidereal time h m s	Sun ☉ ° ' "	Moon ☽ ° ' "	Moon ☽ 24.00 ° ' "	Mercury ☿ ° '	Venus ♀ ° '	Mars ♂ ° '	Jupiter ♃ ° '	Saturn ♄ ° '	Uranus ♅ ° '	Neptune ♆ ° '	Pluto ♇ ° '
01	10 43 52	09 ♍ 22 31	02 ♐ 34 09	08 ♐ 43 11	20 ♍ 25	08 ♌ 46	05 ♑ 54	01 ♓ 24	18 ♋ 03	05 ♋ 25	19 ♈ 42	15 ♒ 03
02	10 47 48	10 20 35	14 ♐ 56 55	21 ♐ 15 56	22 09	09 58	06 16	01 R 17	18 09	05 27	19 R 41	15 R 02
03	10 51 45	11 18 41	27 ♐ 40 44	04 ♑ 11 47	23 52	11 11	06 39	01 09	18 15	05 29	19 40	15 00
04	10 55 42	12 16 48	10 ♑ 49 26	17 ♑ 33 56	25 34	12 23	07 03	01 01	18 20	05 30	19 39	14 59
05	10 59 38	13 14 57	24 ♑ 25 24	01 ♒ 23 48	27 15	13 35	07 26	00 54	18 26	05 32	19 37	14 58
06	11 03 35	14 13 07	08 ♒ 28 56	15 ♒ 40 25	28 ♍ 55	14 48	07 51	00 46	18 32	05 34	19 36	14 57
07	11 07 31	15 11 18	22 ♒ 57 42	00 ♓ 20 03	00 ♎ 33	16 00	08 16	00 39	18 37	05 36	19 35	14 56
08	11 11 28	16 09 32	07 ♓ 46 35	15 ♓ 16 18	02 11	17 13	08 41	00 32	18 42	05 38	19 34	14 55
09	11 15 24	17 07 47	22 ♓ 48 06	00 ♈ 20 51	03 47	18 26	09 07	00 24	18 48	05 39	19 32	14 54
10	11 19 21	18 06 03	07 ♈ 53 24	15 ♈ 24 40	05 22	19 39	09 33	00 17	18 53	05 41	19 31	14 52
11	11 23 17	19 04 22	22 ♈ 53 38	00 ♉ 19 24	06 56	20 51	10 00	00 10	18 58	05 42	19 30	14 51
12	11 27 14	20 02 42	07 ♉ 41 10	14 ♉ 58 21	08 28	22 04	10 27	00 ♓ 03	19 03	05 44	19 28	14 50
13	11 31 11	21 01 05	22 ♉ 10 26	29 ♉ 17 08	10 00	23 17	10 54	29 ♒ 56	19 08	05 45	19 27	14 49
14	11 35 07	21 59 30	06 ♊ 18 15	13 ♊ 13 42	11 31	24 30	11 22	29 49	19 13	05 47	19 25	14 48
15	11 39 04	22 57 57	20 ♊ 03 32	26 ♊ 47 53	13 00	25 43	11 51	29 43	19 18	05 48	19 24	14 47
16	11 43 00	23 56 27	03 ♋ 26 56	10 ♋ 00 57	14 28	26 56	12 19	29 36	19 23	05 50	19 22	14 46
17	11 46 57	24 54 58	16 ♋ 30 12	22 ♋ 55 00	15 55	28 10	12 49	29 30	19 28	05 51	19 21	14 45
18	11 50 53	25 53 32	29 ♋ 15 40	05 ♌ 32 31	17 21	29 ♌ 23	13 18	29 23	19 32	05 52	19 19	14 44
19	11 54 50	26 52 07	11 ♌ 45 52	17 ♌ 56 02	18 46	00 ♍ 36	13 48	29 17	19 37	05 53	19 18	14 43
20	11 58 46	27 50 45	24 ♌ 03 18	00 ♍ 07 59	20 10	01 50	14 18	29 11	19 41	05 54	19 16	14 42
21	12 02 43	28 49 25	06 ♍ 10 20	12 ♍ 10 38	21 32	03 03	14 49	29 05	19 45	05 55	19 15	14 42
22	12 06 40	29 ♍ 48 07	18 ♍ 09 07	24 ♍ 06 05	22 53	04 17	15 20	28 59	19 50	05 56	19 13	14 41
23	12 10 36	00 ♎ 46 51	00 ♎ 01 45	05 ♎ 56 23	24 13	05 30	15 51	28 54	19 54	05 57	19 12	14 40
24	12 14 33	01 45 36	11 ♎ 50 16	17 ♎ 43 41	25 31	06 44	16 22	28 48	19 58	05 58	19 10	14 39
25	12 18 29	02 44 24	23 ♎ 36 55	29 ♎ 30 19	26 49	07 57	16 54	28 43	20 02	05 59	19 09	14 38
26	12 22 26	03 43 14	05 ♏ 24 11	11 ♏ 18 55	28 04	09 11	17 27	28 38	20 06	06 00	19 07	14 37
27	12 26 22	04 42 05	17 ♏ 14 54	23 ♏ 12 33	29 ♎ 18	10 25	17 59	28 33	20 10	06 01	19 05	14 37
28	12 30 19	05 40 59	29 ♏ 12 19	05 ♐ 14 40	00 ♏ 31	11 39	18 32	28 28	20 14	06 02	19 04	14 36
29	12 34 15	06 39 54	11 ♐ 20 06	17 ♐ 29 06	01 42	12 52	19 05	28 23	20 17	06 02	19 02	14 35
30	12 38 12	07 ♎ 38 51	23 ♐ 42 11	29 ♐ 59 51	02 ♏ 51	14 ♍ 06	19 ♑ 39	28 ♒ 19	20 ♋ 21	06 ♋ 03	19 ♈ 01	14 ♒ 34

MOON

Date	Moon True ☊ ° '	Moon Mean ☊ ° '	Moon ☽ Latitude ° '
01	12 ♎ 43	13 ♎ 52	04 N 00
02	12 R 43	13 49	04 38
03	12 42	13 46	05 03
04	12 39	13 43	05 14
05	12 36	13 40	05 07
06	12 32	13 36	04 42
07	12 28	13 33	03 59
08	12 24	13 30	02 59
09	12 22	13 27	01 46
10	12 21	13 24	00 N 25
11	12 D 21	13 21	00 S 58
12	12 22	13 17	02 15
13	12 24	13 14	03 23
14	12 25	13 11	04 16
15	12 25	13 08	04 53
16	12 R 25	13 05	05 12
17	12 24	13 02	05 15
18	12 22	12 58	05 03
19	12 20	12 55	04 36
20	12 18	12 52	03 56
21	12 16	12 49	03 07
22	12 14	12 46	02 10
23	12 13	12 42	01 07
24	12 13	12 39	00 S 02
25	12 D 13	12 36	01 N 03
26	12 14	12 33	02 05
27	12 15	12 30	03 03
28	12 16	12 27	03 53
29	12 16	12 23	04 33
30	12 ♎ 17	12 ♎ 20	05 N 02

DECLINATIONS

Date	Sun ☉ ° '	Moon ☽ ° '	Mercury ☿ ° '	Venus ♀ ° '	Mars ♂ ° '	Jupiter ♃ ° '	Saturn ♄ ° '	Uranus ♅ ° '	Neptune ♆ ° '	Pluto ♇ ° '
01	08 N 03	16 S 44	04 N 48	18 N 15	27 S 32	12 S 10	21 N 47	23 N 34	06 N 10	23 S 33
02	07 41	17 59	04 01	17 59	27 29	12 13	21 46	23 34	06 10	23 34
03	07 19	18 22	03 15	17 42	27 26	12 16	21 45	23 34	06 09	23 34
04	06 57	17 47	02 28	17 24	27 23	12 19	21 45	23 34	06 09	23 35
05	06 35	16 11	01 42	17 07	27 19	12 21	21 44	23 33	06 08	23 35
06	06 13	13 35	00 56	16 48	27 16	12 24	21 43	23 33	06 08	23 35
07	05 50	10 05	00 N 11	16 29	27 12	12 27	21 42	23 33	06 07	23 35
08	05 28	05 53	00 S 34	16 10	27 08	12 29	21 42	23 33	06 07	23 36
09	05 05	01 S 14	01 19	15 50	27 04	12 32	21 41	23 33	06 06	23 36
10	04 42	03 N 30	02 04	15 30	27 00	12 34	21 40	23 33	06 06	23 36
11	04 19	08 00	02 48	15 10	26 56	12 37	21 40	23 33	06 05	23 37
12	03 57	11 56	03 31	14 48	26 52	12 39	21 39	23 33	06 04	23 37
13	03 34	15 02	04 14	14 27	26 48	12 42	21 38	23 33	06 04	23 37
14	03 11	17 09	04 57	14 05	26 43	12 44	21 38	23 33	06 03	23 37
15	02 47	18 12	05 39	13 43	26 38	12 47	21 37	23 33	06 03	23 38
16	02 24	18 11	06 21	13 20	26 34	12 49	21 36	23 33	06 02	23 38
17	02 01	17 12	07 01	12 57	26 29	12 51	21 36	23 33	06 01	23 38
18	01 38	15 22	07 42	12 33	26 24	12 53	21 35	23 33	06 01	23 38
19	01 15	12 50	08 21	12 10	26 18	12 55	21 35	23 33	06 00	23 39
20	00 51	09 47	09 00	11 45	26 13	12 57	21 34	23 33	06 00	23 39
21	00 28	06 21	09 38	11 21	26 08	12 59	21 34	23 33	05 59	23 39
22	00 N 05	02 N 42	10 16	10 56	26 02	13 01	21 33	23 33	05 58	23 39
23	00 S 19	01 S 02	10 53	10 31	25 56	13 03	21 32	23 33	05 58	23 39
24	00 42	04 43	11 29	10 05	25 50	13 05	21 32	23 33	05 57	23 40
25	01 05	08 11	12 04	09 40	25 44	13 07	21 31	23 33	05 57	23 40
26	01 29	11 21	12 38	09 13	25 38	13 09	21 31	23 33	05 56	23 40
27	01 52	14 03	13 11	08 47	25 31	13 10	21 30	23 33	05 55	23 40
28	02 15	16 11	13 43	08 21	25 25	13 12	21 30	23 33	05 55	23 40
29	02 39	17 37	14 15	07 54	25 18	13 13	21 29	23 33	05 54	23 40
30	03 S 02	18 S 16	14 S 45	07 N 27	25 S 11	13 S 15	21 N 29	23 N 33	05 N 53	23 S 40

ZODIAC SIGN ENTRIES

Date	h	m	Planets
01	06	57	☽ ♐
03	16	18	☽ ♑
05	21	37	☽ ♒
07	03	49	☿ ♎
07	23	28	☽ ♓
09	23	27	☽ ♈
11	23	29	☽ ♉
12	22	28	♃ ♒
14	01	13	☽ ♊
16	05	45	☽ ♋
18	13	24	☽ ♌
19	00	09	♀ ♍
20	23	44	☽ ♍
22	16	52	☉ ♎
23	11	56	☽ ♎
26	01	00	☽ ♏
28	01	44	☿ ♏
28	13	35	☽ ♐

LATITUDES

Date	Mercury ☿ ° '	Venus ♀ ° '	Mars ♂ ° '	Jupiter ♃ ° '	Saturn ♄ ° '	Uranus ♅ ° '	Neptune ♆ ° '	Pluto ♇ ° '
01	01 N 05	00 N 12	04 S 14	01 S 17	00 S 27	00 N 14	01 S 40	07 S 35
04	00 47	00 20	04 08	01 17	00 26	00 14	01 40	07 35
07	00 26	00 29	04 02	01 17	00 26	00 15	01 40	07 35
10	00 N 04	00 37	03 56	01 17	00 26	00 15	01 40	07 35
13	00 S 18	00 44	03 49	01 17	00 26	00 15	01 40	07 35
16	00 42	00 51	03 43	01 17	00 26	00 15	01 40	07 35
19	01 05	00 58	03 37	01 17	00 26	00 15	01 40	07 34
22	01 29	01 04	03 30	01 17	00 25	00 15	01 40	07 34
25	01 51	01 10	03 24	01 17	00 25	00 15	01 40	07 34
28	02 13	01 14	03 18	01 16	00 25	00 15	01 40	07 34
31	02 S 33	01 N 19	03 S 11	01 S 16	00 S 25	00 N 15	01 S 41	07 S 33

DATA

Julian Date	2463842
Delta T	+75 seconds
Ayanamsa	24° 19' 44"
Synetic vernal point	04° ♓ 47' 15"
True obliquity of ecliptic	23° 25' 57"

MOON'S PHASES, APSIDES AND POSITIONS ☽

Date	h	m	Phase	Longitude ° '	Eclipse Indicator
02	02	24	☽	09 ♐ 57	
09	02	21	○	16 ♓ 44	
15	17	34	☾	23 ♊ 11	
23	13	40	●	00 ♎ 51	Partial

Day	h	m	
10	01	53	Perigee
25	01	45	Apogee

Day	h	m		
03	09	50	Max dec	18° S 22'
09	18	14	0N	
15	23	36	Max dec	18° N 19'
23	05	19	0S	
30	17	29	Max dec	18° S 17'

LONGITUDES

Date	Chiron ⚷ ° '	Ceres ⚳ ° '	Pallas ⚴ ° '	Juno ⚵ ° '	Vesta ⚶ ° '	Black Moon Lilith ° '
01	01 ♊ 09	21 ♐ 07	13 ♏ 20	20 ♎ 18	18 ♓ 07	13 ♎ 17
11	01 ♊ 08	22 ♐ 43	16 ♏ 49	23 ♎ 28	15 ♓ 36	14 ♎ 24
21	01 ♊ 02	24 ♐ 45	20 ♏ 29	26 ♎ 44	13 ♓ 12	15 ♎ 30
31	00 ♊ 49	27 ♐ 07	24 ♏ 17	00 ♏ 04	11 ♓ 16	16 ♎ 36

All ephemeris data is given at 12.00 UT and the Moon's longitude is additionally given for 24.00 UT

ASPECTARIAN

h m Aspects

01 Thursday
02 30 ☿ ⊼ ♆
05 46 ☽ ± ♅
06 35 ☽ ⊥ ♂
09 44 ☽ □ ♃
11 38 ☽ Q ☿
12 56 ☽ Q ♇
12 58 ☽ bQ ♄
16 11 ☽ bQ ♆
17 35 ☽ ⊼ ♅
18 43 ☽ ⊻ ♂
20 39 ☿ ± ♇
23 26 ♅ # ♇

02 Friday
01 24 ☽ △ ♀
02 24 ☽ □ ☉
06 35 ☽ ± ♄
11 56 ☽ # ♀
12 09 ☽ ⚹ ♇
18 09 ☽ ⊼ ♄
20 10 ☽ Q ♃
21 00 ☽ △ ♆

03 Saturday
03 48 ☽ □ ☿
08 55 ☽ bQ ♀
16 18 ☽ ∠ ♇
18 06 ♀ ⊥ ♅
18 21 ☽ ⚹ ♃

04 Sunday
01 33 ☉ ⊻ ♀
02 22 ☽ ☍ ♅
03 11 ☽ ± ♀
04 58 ☽ ☌ ♂
08 42 ☽ ⊥ ♇
14 49 ☽ △ ☉
15 04 ☽ ⊼ ♀
19 25 ☽ ⊻ ♇
21 09 ☽ # ♀
21 11 ☽ ∠ ♃

05 Monday
01 28 ☽ ☍ ♄
03 38 ☽ □ ♆
07 33 ♀ ± ♂
12 49 ☽ ⊥ ♃
17 34 ☽ △ ☿
19 06 ☽ bQ ☉
21 06 ☉ ± ♆
23 03 ☽ ⊻ ♃

06 Tuesday
07 05 ☽ ⊼ ♅
10 31 ☽ Q ♆
10 54 ☽ ⊻ ♂
11 31 ☽ ± ☉
14 56 ♀ ☍ ♇
17 11 ☽ ± ♅
17 28 ☉ ∥ ♆
20 42 ☽ ∥ ♃
21 14 ☽ ⊥ ♂
22 15 ☽ bQ ☿
22 16 ☽ ⊼ ☉
22 47 ☽ ☌ ♇
23 31 ☽ ☍ ♀

07 Wednesday
02 53 ☿ bQ ♇
04 49 ☽ ⊼ ♄
05 43 ☉ ⊼ ♇
06 28 ☽ ⚹ ♆
08 07 ☽ bQ ♅
12 30 ☽ ∠ ♂
12 56 ☿ Q ♄
13 14 ☿ ⊼ ♃
14 43 ☽ ± ♄
14 56 ☽ ± ☿

08 Thursday
00 24 ☽ ☌ ♃
01 53 ☽ ⊼ ☿
05 25 ☽ bQ ♄
06 50 ☽ ∠ ♆
08 32 ☽ △ ♅
10 46 ☽ # ♆
13 30 ☽ ⚹ ♂
14 20 ☿ ∠ ♀
14 27 ☽ # ☉
21 15 ☽ ⊥ ♆
23 25 ☽ ⊻ ♇

09 Friday
00 07 ♂ ⊥ ♇
02 21 ☽ ☍ ☉
04 26 ☽ ⊼ ♀
05 35 ☽ △ ♄
06 49 ☽ ⊻ ♆
08 58 ☽ ⊥ ♇
09 14 ☽ Q ♂
11 38 ☽ ∥ ☿
14 49 ☽ ± ♀
19 49 ♀ ⊻ ♄
23 16 ☽ ∠ ♇

10 Saturday
00 00 ☽ ⊻ ♃
01 19 ☉ Q ♅
03 17 ☽ # ☿
06 23 ☽ bQ ♀
07 31 ☽ ☍ ☿
08 29 ☽ □ ♅
09 28 ☽ ⊥ ♃
09 32 ♀ △ ♆
14 44 ☽ □ ♂
16 54 ☿ □ ♅
17 43 ☽ ∥ ☉
23 08 ☽ ⚹ ♇
23 42 ☽ ∠ ♃

11 Sunday
01 06 ☿ ± ♃
01 29 ☽ ∥ ♆
05 26 ☽ ⊼ ☉
05 40 ☽ □ ♄
06 33 ☽ ☌ ♆
08 26 ☽ △ ♀
08 59 ♀ ∠ ♅
09 12 ☉ ⚹ ♄
13 19 ☽ Q ♅
15 45 ☽ ± ☉
18 23 ☽ Q ♇
22 07 ☉ ⊼ ♆
23 39 ☽ ⚹ ♃

12 Monday
07 23 ☽ bQ ☉
08 48 ☽ ⚹ ♅
10 58 ☽ Q ♄
13 27 ☽ ⊼ ☿
16 41 ☽ △ ♂
17 04 ☽ # ♃
19 07 ☽ Q ♃
21 08 ☉ # ☿
23 46 ☽ □ ♇

13 Tuesday
00 30 ☽ ± ☿
06 54 ☽ ⚹ ♄
07 13 ☉ ± ♇
07 23 ☽ ∥ ♀
07 27 ☽ ⊻ ♆
09 37 ☽ ∠ ♅
09 55 ☽ △ ☉
14 02 ☽ □ ♀
17 19 ☽ bQ ☿
17 29 ☽ ⊥ ♆
18 29 ☽ bQ ♂

14 Wednesday
00 49 ☽ ⊥ ♅
01 00 ☽ □ ♃
08 24 ☽ ∠ ♄
08 46 ☽ ∠ ♆
08 50 ☿ □ ♂
10 21 ☽ ± ♂
11 06 ☽ ⊻ ♅
21 05 ☽ ⊼ ♂
22 06 ☽ △ ☿
23 47 ☽ Q ♀

15 Thursday
00 03 ☽ ⊥ ♄
02 44 ☽ △ ♇
03 08 ♀ ⊥ ♄
10 39 ☽ ⊻ ♄
10 50 ☽ ⚹ ♆
17 34 ☽ □ ☉
23 05 ☽ ⚹ ♀

16 Friday
01 25 ☿ # ♆
05 06 ☽ △ ♃
05 21 ☽ bQ ♇
08 15 ☽ Q ♆
10 14 ♄ □ ♆
14 03 ☿ bQ ♃
16 20 ☽ ☌ ♅
16 55 ☿ △ ♇
21 42 ☽ ± ♇

17 Saturday
00 29 ♀ bQ ♂
04 48 ☽ Q ☉
04 53 ☽ ☍ ♂
05 09 ☽ ∠ ♀
08 18 ☽ bQ ♃
08 45 ☽ ⊼ ♇
10 47 ☽ □ ☿
17 18 ☽ □ ♆
17 26 ♀ # ♃
17 33 ☽ ☌ ♄
23 42 ☽ ⊥ ♀

18 Sunday
00 59 ☽ ± ♃
05 05 ☽ ⚹ ☉
12 09 ♀ ☍ ♃
12 15 ☽ ⊼ ♃
12 15 ☽ ⊻ ♀

19 Monday
00 39 ☽ ⊻ ♅
01 07 ☽ Q ☿
11 16 ☽ # ♃
12 13 ☽ ∠ ☉
12 14 ☽ ⊥ ♅
16 07 ☽ ⊼ ♂
17 44 ☽ ☍ ♇
18 32 ☽ ∥ ♀
20 59 ☿ ☍ ♆

20 Tuesday
02 38 ☽ △ ♆
03 21 ☿ □ ♄
03 22 ☽ ⊻ ♄
03 23 ☽ ⚹ ☿
04 19 ☽ ± ♂
05 48 ☽ ∠ ♅
07 17 ☽ ⊥ ☉
07 26 ♂ ∠ ♃
15 14 ☽ ⊥ ♄
16 42 ☽ # ☿
20 08 ☽ ⊻ ☉
22 03 ☽ ☍ ♃
22 48 ☽ bQ ♂

21 Wednesday
05 05 ☽ ☌ ♀
06 35 ♂ ⊻ ♇
08 11 ☽ bQ ♆
09 10 ☽ ∠ ♄
11 30 ☽ ⚹ ♅
12 48 ☽ ∠ ☿
14 26 ☽ ∥ ♆
17 54 ☉ ⊼ ♃

22 Thursday
02 07 ☽ ± ♆
05 01 ☽ ⊼ ♇
06 04 ☽ △ ♂
08 59 ☉ bQ ♇
09 07 ☽ ⊥ ☿
10 58 ♀ bQ ♆
11 35 ☽ Q ♅
14 09 ☽ ⊼ ♆
15 24 ☽ ⚹ ♄
17 05 ☽ ± ♇
22 45 ☽ ⊻ ☿
23 31 ♀ ∠ ♄

23 Friday
03 30 ☿ # ♀
04 09 ☽ # ☉
06 46 ☽ ∥ ☉
09 43 ☽ ⊼ ♃
11 16 ☽ bQ ♇
13 40 ☽ ☌ ☉
15 49 ☽ Q ♄
21 03 ♀ ⚹ ♅
21 48 ☽ ± ♃

24 Saturday
00 03 ☽ □ ♅
00 24 ☽ ⊻ ♀
14 01 ☽ ⊥ ♀
15 59 ☽ bQ ♃
17 27 ☉ Q ♄
17 43 ☽ △ ♇
20 21 ☽ # ♆
21 41 ☽ □ ♂

25 Sunday
02 54 ☽ ☍ ♆
04 40 ☽ □ ♄
10 30 ☽ ∠ ♀
19 18 ☽ ☌ ☿
21 30 ☽ # ♀
22 19 ☽ △ ♃

26 Monday
08 16 ☽ ⊻ ☉
12 05 ☽ Q ♂
13 13 ☽ △ ♅
20 34 ☽ ⚹ ♀
21 34 ☽ ⊥ ☉
22 10 ☿ △ ♃

27 Tuesday
01 37 ☽ ∥ ☿
03 34 ☽ ∥ ♃
06 40 ☽ □ ♇
08 33 ☉ ± ♃
11 22 ☿ ∥ ♃
13 34 ☽ ⚹ ♂
15 42 ☽ ⊼ ♆
17 23 ☽ ∠ ☉
17 54 ☽ △ ♄
19 36 ☽ bQ ♅
23 36 ☽ Q ♀

28 Wednesday
03 44 ☽ ± ♆
10 32 ☽ □ ♃
13 38 ☽ ± ♅
14 54 ☽ ⊻ ☿
18 45 ☽ Q ♇
20 29 ☉ □ ♅
21 01 ☽ ∠ ♂
21 38 ☽ bQ ♆

29 Thursday
00 01 ☽ bQ ♄
01 34 ☽ ⊼ ♅
02 00 ☽ ⚹ ☉
04 05 ☽ ⊥ ☿
09 54 ♂ □ ♆
15 06 ♀ ± ♆
15 21 ☽ □ ♀
15 36 ☽ ⊥ ♂
17 48 ☽ ± ♄
18 21 ☽ ⚹ ♇
21 49 ☽ Q ♃
23 33 ☽ ∠ ☿

30 Friday
02 58 ☽ △ ♆
03 31 ☽ Q ☉
03 49 ☽ ⊻ ♂
05 30 ☽ ⊼ ♄
20 46 ☽ ⚹ ♃
21 02 ♀ ⊼ ♇
23 11 ☽ ∠ ♇

OCTOBER 2033

Date	Sidereal time h m s	Sun ☉ ° ' "	Moon ☽ ° ' "	Moon ☽ 24.00 ° ' "	Mercury ☿ ° '	Venus ♀ ° '	Mars ♂ ° '	Jupiter ♃ ° '	Saturn ♄ ° '	Uranus ♅ ° '	Neptune ♆ ° '	Pluto ♇ ° '
01	12 42 09	08 ♎ 37 50	06 ♑ 22 36	12 ♑ 50 54	03 ♏ 58	15 ♍ 20	20 ♑ 12	28 ♒ 15	20 ♋ 24	06 ♋ 03	18 ♈ 59	14 ♒ 34
02	12 46 05	09 36 50	19 ♑ 25 09	26 ♑ 05 43	05 03	16 34	20 46	28 R 11	20 27	06 04	18 R 57	14 R 33
03	12 50 02	10 35 52	02 ♒ 52 52	09 ♒ 46 45	06 05	17 48	21 21	28 07	20 31	06 04	18 56	14 32
04	12 53 58	11 34 56	16 ♒ 47 25	23 ♒ 54 45	07 06	19 02	21 55	28 03	20 34	06 05	18 54	14 32
05	12 57 55	12 34 02	01 ♓ 08 28	08 ♓ 28 09	08 03	20 16	22 30	27 59	20 37	06 05	18 52	14 31
06	13 01 51	13 33 09	15 ♓ 53 09	23 ♓ 22 40	08 58	21 31	23 05	27 56	20 40	06 05	18 51	14 31
07	13 05 48	14 32 19	00 ♈ 55 46	08 ♈ 31 22	09 49	22 45	23 40	27 53	20 43	06 06	18 49	14 30
08	13 09 44	15 31 30	16 ♈ 08 16	23 ♈ 45 15	10 38	23 59	24 16	27 50	20 46	06 06	18 47	14 30
09	13 13 41	16 30 43	01 ♉ 21 06	08 ♉ 54 36	11 22	25 13	24 51	27 47	20 48	06 06	18 46	14 29
10	13 17 38	17 29 59	16 ♉ 24 39	23 ♉ 50 16	12 03	26 28	25 27	27 44	20 51	06 06	18 44	14 29
11	13 21 34	18 29 16	01 ♊ 10 39	08 ♊ 25 07	12 39	27 42	26 04	27 42	20 53	06 06	18 42	14 28
12	13 25 31	19 28 36	15 ♊ 33 12	22 ♊ 34 37	13 10	28 ♍ 56	26 40	27 39	20 56	06 R 06	18 41	14 28
13	13 29 27	20 27 59	29 ♊ 29 14	06 ♋ 17 03	13 35	00 ♎ 11	27 16	27 37	20 58	06 06	18 39	14 27
14	13 33 24	21 27 23	12 ♋ 58 14	19 ♋ 33 04	13 55	01 25	27 53	27 35	21 00	06 06	18 37	14 27
15	13 37 20	22 26 50	26 ♋ 01 53	02 ♌ 25 06	14 09	02 40	28 30	27 34	21 02	06 06	18 36	14 27
16	13 41 17	23 26 19	08 ♌ 43 11	14 ♌ 56 38	14 15	03 54	29 07	27 32	21 04	06 06	18 34	14 26
17	13 45 13	24 25 51	21 ♌ 05 59	27 ♌ 11 43	14 R 15	05 09	29 ♑ 45	27 31	21 06	06 05	18 32	14 26
18	13 49 10	25 25 25	03 ♍ 14 21	09 ♍ 14 22	14 06	06 24	00 ♒ 22	27 30	21 07	06 05	18 31	14 26
19	13 53 07	26 25 00	15 ♍ 12 16	21 ♍ 08 29	13 50	07 38	01 00	27 29	21 09	06 05	18 29	14 25
20	13 57 03	27 24 38	27 ♍ 03 25	02 ♎ 57 28	13 24	08 53	01 38	27 28	21 10	06 04	18 27	14 25
21	14 01 00	28 24 19	08 ♎ 51 01	14 ♎ 44 23	12 50	10 08	02 16	27 28	21 12	06 04	18 26	14 25
22	14 04 56	29 ♎ 24 01	20 ♎ 37 52	26 ♎ 31 46	12 07	11 22	02 54	27 28	21 13	06 03	18 24	14 25
23	14 08 53	00 ♏ 23 45	02 ♏ 26 22	08 ♏ 21 54	11 16	12 37	03 32	27 D 28	21 14	06 02	18 22	14 25
24	14 12 49	01 23 32	14 ♏ 18 38	20 ♏ 16 47	10 16	13 52	04 11	27 28	21 15	06 02	18 21	14 25
25	14 16 46	02 23 20	26 ♏ 16 37	02 ♐ 18 23	09 10	15 07	04 50	27 28	21 16	06 01	18 19	14 25
26	14 20 42	03 23 10	08 ♐ 22 20	14 ♐ 28 44	07 59	16 22	05 29	27 29	21 17	06 00	18 17	14 25
27	14 24 39	04 23 02	20 ♐ 37 52	26 ♐ 50 02	06 44	17 37	06 08	27 29	21 18	06 00	18 16	14 25
28	14 28 36	05 22 56	03 ♑ 05 34	09 ♑ 24 47	05 27	18 52	06 47	27 30	21 18	05 59	18 14	14 D 25
29	14 32 32	06 22 51	15 ♑ 48 01	22 ♑ 15 37	04 11	20 07	07 26	27 31	21 19	05 58	18 13	14 25
30	14 36 29	07 22 48	28 ♑ 47 56	05 ♒ 25 16	02 58	21 22	08 05	27 33	21 19	05 57	18 11	14 25
31	14 40 25	08 ♏ 22 47	12 ♒ 07 55	18 ♒ 56 08	01 ♏ 51	22 ♎ 37	08 ♒ 45	27 ♒ 34	21 ♋ 19	05 ♋ 56	18 ♈ 10	14 ♒ 25

Date	Moon True ☊ ° '	Moon Mean ☊ ° '	Moon ☽ Latitude ° '
01	12 ♎ 17	12 ♎ 17	05 N 17
02	12 R 17	12 14	05 16
03	12 17	12 11	04 58
04	12 16	12 07	04 22
05	12 D 16	12 04	03 30
06	12 16	12 01	02 22
07	12 17	11 58	01 N 03
08	12 R 17	11 55	00 S 21
09	12 16	11 52	01 44
10	12 16	11 48	02 59
11	12 15	11 45	04 00
12	12 15	11 42	04 44
13	12 14	11 39	05 10
14	12 13	11 36	05 18
15	12 D 13	11 33	05 09
16	12 14	11 29	04 45
17	12 15	11 26	04 08
18	12 16	11 23	03 20
19	12 17	11 20	02 25
20	12 18	11 17	01 24
21	12 19	11 13	00 S 19
22	12 R 19	11 10	00 N 46
23	12 17	11 07	01 49
24	12 15	11 04	02 48
25	12 12	11 01	03 40
26	12 09	10 58	04 22
27	12 05	10 54	04 53
28	12 02	10 51	05 11
29	12 00	10 48	05 14
30	11 58	10 45	05 01
31	11 ♎ 58	10 ♎ 42	04 N 32

DECLINATIONS

Date	Sun ☉ ° '	Moon ☽ ° '	Mercury ☿ ° '	Venus ♀ ° '	Mars ♂ ° '	Jupiter ♃ ° '	Saturn ♄ ° '	Uranus ♅ ° '	Neptune ♆ ° '	Pluto ♇ ° '
01	03 S 25	18 S 01	15 S 14	06 N 59	25 S 04	13 S 16	21 N 28	23 N 33	05 N 53	23 S 41
02	03 48	16 49	15 42	06 32	24 57	13 18	21 28	23 33	05 52	23 41
03	04 12	14 40	16 09	06 04	24 49	13 19	21 27	23 33	05 52	23 41
04	04 35	11 38	16 34	05 36	24 42	13 20	21 27	23 33	05 51	23 41
05	04 58	07 48	16 58	05 08	24 34	13 21	21 27	23 33	05 50	23 41
06	05 21	03 S 23	17 20	04 40	24 26	13 22	21 26	23 33	05 50	23 41
07	05 44	01 N 20	17 41	04 11	24 18	13 22	21 26	23 33	05 49	23 41
08	06 07	06 01	18 00	03 43	24 09	13 24	21 25	23 33	05 48	23 41
09	06 29	10 19	18 18	03 14	24 01	13 25	21 25	23 33	05 48	23 41
10	06 52	13 53	18 33	02 45	23 52	13 26	21 25	23 33	05 47	23 41
11	07 15	16 28	18 46	02 16	23 44	13 27	21 24	23 33	05 46	23 41
12	07 37	17 56	18 57	01 47	23 35	13 27	21 24	23 33	05 46	23 41
13	08 00	18 16	19 06	01 18	23 25	13 28	21 24	23 33	05 45	23 41
14	08 22	17 32	19 11	00 49	23 16	13 28	21 24	23 33	05 45	23 41
15	08 44	15 53	19 14	00 N 20	23 07	13 29	21 23	23 33	05 44	23 41
16	09 06	13 30	19 14	00 S 09	22 57	13 29	21 23	23 33	05 43	23 41
17	09 28	10 33	19 10	00 39	22 47	13 30	21 23	23 33	05 43	23 41
18	09 50	07 12	19 03	01 08	22 37	13 30	21 23	23 33	05 42	23 41
19	10 11	03 N 36	18 51	01 37	22 27	13 30	21 22	23 33	05 41	23 41
20	10 33	00 S 07	18 36	02 07	22 16	13 30	21 22	23 33	05 41	23 41
21	10 54	03 48	18 16	02 36	22 06	13 30	21 22	23 33	05 40	23 41
22	11 15	07 21	17 52	03 05	21 55	13 30	21 22	23 33	05 40	23 41
23	11 36	10 36	17 23	03 35	21 44	13 30	21 22	23 33	05 39	23 41
24	11 57	13 27	16 50	04 04	21 33	13 30	21 22	23 33	05 38	23 41
25	12 18	15 45	16 13	04 33	21 22	13 29	21 22	23 33	05 38	23 40
26	12 38	17 23	15 32	05 02	21 11	13 29	21 22	23 33	05 37	23 40
27	12 59	18 14	14 49	05 31	20 59	13 29	21 22	23 33	05 37	23 40
28	13 19	18 13	14 05	06 00	20 47	13 28	21 22	23 33	05 36	23 40
29	13 39	17 18	13 20	06 29	20 35	13 28	21 21	23 34	05 35	23 40
30	13 58	15 28	12 36	06 58	20 23	13 27	21 21	23 34	05 35	23 40
31	14 S 18	12 S 47	11 S 53	07 S 26	20 S 11	13 S 26	21 N 21	23 N 34	05 N 34	23 S 40

ZODIAC SIGN ENTRIES

Date	h	m	Planets
01	00	00	☽ ♑
03	06	56	☽ ♒
05	10	07	☽ ♓
07	10	32	☽ ♈
09	09	52	☽ ♉
11	10	04	☽ ♊
13	08	32	♀ ♎
13	12	54	☽ ♋
15	19	26	☽ ♌
17	21	52	♂ ♒
18	05	33	☽ ♍
20	17	59	☽ ♎
23	02	27	☉ ♏
23	07	03	☽ ♏
25	19	25	☽ ♐
28	06	05	☽ ♑
30	14	11	☽ ♒

LATITUDES

Date	Mercury ☿ ° '	Venus ♀ ° '	Mars ♂ ° '	Jupiter ♃ ° '	Saturn ♄ ° '	Uranus ♅ ° '	Neptune ♆ ° '	Pluto ♇ ° '
01	02 S 33	01 N 19	03 S 11	01 S 16	00 S 25	00 N 15	01 S 41	07 S 33
04	02 51	01 23	03 05	01 16	00 25	00 15	01 41	07 33
07	03 06	01 26	02 59	01 15	00 25	00 15	01 41	07 33
10	03 16	01 28	02 53	01 15	00 25	00 15	01 41	07 32
13	03 20	01 30	02 47	01 15	00 24	00 15	01 41	07 32
16	03 16	01 31	02 41	01 14	00 24	00 15	01 41	07 32
19	03 01	01 32	02 35	01 14	00 24	00 15	01 41	07 31
22	02 31	01 32	02 29	01 13	00 24	00 16	01 41	07 31
25	01 45	01 31	02 23	01 13	00 24	00 16	01 41	07 31
28	00 S 47	01 30	02 17	01 12	00 24	00 16	01 41	07 30
31	00 N 14	01 N 28	02 S 12	01 S 12	00 S 23	00 N 16	01 S 40	07 S 30

DATA

Julian Date	2463872
Delta T	+75 seconds
Ayanamsa	24° 19' 46"
Synetic vernal point	04° ♓ 47' 13"
True obliquity of ecliptic	23° 25' 57"

MOON'S PHASES, APSIDES AND POSITIONS ☽

Date	h	m	Phase	Longitude ° '	Eclipse Indicator
01	16	33	☽	08 ♑ 49	
08	10	58	○	15 ♈ 29	total
15	04	47	☾	22 ♋ 09	
23	07	28	●	00 ♏ 12	
31	04	46	☽	08 ♒ 05	

Day	h	m		
08	12	19	Perigee	
22	03	32	Apogee	
07	05	18	0N	
13	06	59	Max dec	18° N 17'
20	11	18	0S	
27	23	41	Max dec	18° S 20'

LONGITUDES

Date	Chiron ⚷ ° '	Ceres ⚳ ° '	Pallas ⚴ ° '	Juno ⚵ ° '	Vesta ⚶ ° '	Black Moon Lilith ⚸ ° '
01	00 ♊ 49	27 ♐ 07	24 ♏ 17	00 ♏ 04	11 ♓ 16	16 ♎ 36
11	00 ♊ 30	29 ♐ 48	28 ♏ 12	03 ♏ 27	09 ♓ 59	17 ♎ 43
21	00 ♊ 06	02 ♑ 44	02 ♐ 13	06 ♏ 52	09 ♓ 27	18 ♎ 49
31	29 ♉ 39	05 ♑ 54	06 ♐ 18	10 ♏ 17	09 ♓ 41	19 ♎ 56

ASPECTARIAN

h m	Aspects
01 Saturday	
07 03	☽ ✱ ☿
11 24	☽ ☍ ♅
16 04	☽ ⊥ ♇
16 33	☽ □ ☉
21 13	♂ ☍ ♄
02 Sunday	
00 40	☽ ∠ ♃
03 08	☽ ⊻ ♇
06 17	☽ △ ♀
07 18	☽ Q ☿
11 09	☽ □ ♆
13 53	☽ ☍ ♄
14 34	☽ ☌ ♂
16 57	☽ ⊥ ♃
23 25	☽ ∥ ☿
03 Monday	
03 37	☽ ⊻ ♃
11 39	☿ △ ♅
11 51	☽ bQ ♀
17 17	♀ Q ♅
17 35	☽ ⊼ ♅
18 03	☽ □ ☿
19 03	☽ Q ♆
23 02	♀ ∥ ♆
23 27	☽ ∥ ♃
04 Tuesday	
02 26	☽ △ ☉
03 57	☽ ± ♅
04 59	☽ ± ♀
08 09	☽ ☌ ♇
09 22	♀ ⊼ ♆
15 34	☽ ✱ ♆
16 10	☽ ⊼ ♀
16 45	♂ ⊥ ♃
18 24	☽ ⊼ ♄
19 15	☽ bQ ♅
21 02	☽ ⊻ ♂
05 Wednesday	
04 29	☽ ± ♄
05 39	☽ bQ ☉
06 48	☽ ☌ ♃
07 27	☽ ⊥ ♂
16 29	☽ ∠ ♆
16 45	♀ ± ♇
16 49	☉ # ♀
18 56	♀ ✱ ♄
19 22	☽ bQ ♄
20 07	☽ △ ♅
21 32	☽ ± ☉
21 39	☉ bQ ♃
22 51	☽ ∠ ♂
23 00	☽ # ♆
06 Thursday	
00 05	☽ △ ☿
02 27	☽ ∥ ☉
04 30	☽ # ♀
07 06	☽ ⊥ ♆
07 58	☽ ⊼ ☉
09 47	☽ ⊻ ♇
16 44	☽ ⊻ ♆
19 25	☽ ⊥ ♇
19 41	☽ △ ♄
21 49	☽ ☍ ♀
07 Friday	
00 00	☽ ✱ ♂
01 44	☽ bQ ☿
07 10	☽ ⊻ ♃
09 44	☽ ∠ ♇
11 05	☉ △ ♇
16 39	☽ ⊥ ♃
16 51	☽ ± ☿
17 24	☉ # ♆
19 48	☽ Q ♂
20 10	☽ □ ♅
08 Saturday	
01 11	☽ ∥ ♀
02 51	☽ ⊼ ☿
06 48	☽ ∠ ♃
09 25	☽ ✱ ♇
10 54	☽ ∥ ♆
10 58	☽ ☍ ☉
12 33	☽ # ☉
16 10	☽ ☌ ♆
19 18	☽ □ ♄
22 27	♀ △ ♂
09 Sunday	
00 33	☽ Q ♅
01 19	☽ □ ♂
01 27	☽ ⊼ ♀
04 19	☽ Q ♇
06 22	☽ ✱ ♃
11 46	☽ ± ♀
19 32	☽ ✱ ♅
23 52	☽ Q ♄
10 Monday	
01 21	☽ Q ♃
03 21	☽ bQ ♀
04 41	☽ ☍ ☿
08 35	☽ # ♃
08 54	☽ □ ♇
13 53	☽ ⊼ ☉
15 44	☽ ⊻ ♆
19 10	☽ ✱ ♄
19 34	☽ ∠ ♅
11 Tuesday	
00 16	☽ ± ☉
01 26	☽ ⊥ ♆
03 15	☽ △ ♂
05 46	☽ △ ♀
06 18	☽ □ ♃
10 00	☿ ∠ ♀
10 14	☽ ⊥ ♅
11 54	♀ ⊼ ♃
16 02	♅ St R
16 05	☽ bQ ☉
16 10	☽ ∠ ♆
17 06	☉ ☍ ♆
18 24	♂ ∥ ♇
19 48	☽ ∠ ♄
20 09	☽ ⊻ ♅
12 Wednesday	
05 09	☽ bQ ♂
07 49	☽ ⊼ ☿
10 09	☽ △ ♇
10 56	☽ ⊥ ♄
16 35	♂ # ♅
17 18	☽ ✱ ♆
18 21	☽ ± ☿
19 11	☽ △ ☉
21 07	☽ ± ♂
21 11	☽ ⊻ ♄
22 04	♀ bQ ♇
13 Thursday	
07 57	☽ ⊼ ♂
08 45	☽ △ ♃
10 23	☽ bQ ☿
11 57	☽ bQ ♇
13 20	☽ □ ♀
14 02	☽ Q ♆
23 40	☽ ☌ ♅
14 Friday	
00 29	☉ □ ♄
00 56	♂ ⊻ ♃
03 52	☽ ± ♇
11 19	☽ bQ ♃
13 45	☽ △ ☿
14 41	☽ ⊼ ♇
22 16	☽ □ ♆
15 Saturday	
01 00	☽ Q ♀
02 42	☽ ☌ ♄
03 43	☽ ± ♃
04 47	☽ □ ☉
14 51	☽ ⊼ ♃
16 51	☽ ☍ ♂
19 19	♀ Q ♄
16 Sunday	
01 49	☽ ✱ ♀
05 01	♀ ± ♃
06 59	☽ ⊻ ♅
12 03	☽ # ♃
17 41	☽ Q ☉
18 29	☽ ⊥ ♅
22 02	☿ St R
22 42	☽ □ ☿
23 01	☽ ☍ ♇
17 Monday	
07 00	☽ △ ♆
09 56	☽ ∠ ♀
11 59	☽ ∠ ♅
11 59	☽ ⊻ ♄
19 07	☽ ✱ ☉
19 12	☽ # ☉
23 50	☽ ⊥ ♄
18 Tuesday	
00 37	☽ ☍ ♃
05 41	☽ ⊥ ♀
05 59	☽ ⊼ ♂
06 01	♀ □ ♅
09 47	☽ Q ☿
12 32	☽ bQ ♆
17 40	☽ ✱ ♅
17 46	☽ ∠ ♄
18 36	☽ ± ♂
19 02	☽ ⊻ ♀
22 10	☽ ∥ ♆
19 Wednesday	
03 41	☽ ∠ ☉
06 32	☽ ± ♆
09 18	☽ ✱ ☿
10 26	☽ ⊼ ♇
13 41	☽ bQ ♂
14 49	☿ ⊥ ♀
17 48	☽ Q ♅
18 36	☽ ⊼ ♆
22 33	☽ ± ♇
23 20	☽ # ♀
23 30	☽ ⊥ ☉
20 Thursday	
00 03	☽ ✱ ♄
12 47	☽ ⊻ ☉
12 51	☽ ⊼ ♃
13 28	☉ △ ♃
14 37	☽ ∠ ☿
16 48	☽ bQ ♇
21 49	☽ △ ♂
21 Friday	
00 28	☽ Q ♄
01 02	☽ ± ♃
02 56	☽ ∥ ♀
06 19	☽ □ ♅
08 06	☽ ⊥ ☿
14 55	☽ ☌ ♀
19 22	☽ bQ ♃
19 41	☽ ⊻ ☿
23 21	☽ △ ♇
22 Saturday	
00 26	☽ # ♆
07 28	☽ ☍ ♆
13 12	☽ □ ♄
20 40	☿ ⊻ ♀
23 Sunday	
01 53	☽ △ ♃
07 19	♃ St D
07 28	☽ ☌ ☉
08 52	♀ bQ ♃
14 22	☽ □ ♂
19 17	☽ △ ♅
21 04	☽ ∥ ☉
24 Monday	
04 30	☽ ☌ ☿
11 00	☽ ⊻ ♀
12 12	☽ □ ♇
12 22	☽ ∥ ♃
20 06	☽ ⊼ ♆
22 26	♀ △ ♇
25 Tuesday	
00 29	☽ ⊥ ♀
01 29	☽ bQ ♅
01 58	☽ △ ♄
04 43	☽ Q ♂
08 06	☽ ± ♆
12 36	♂ # ♄
14 22	☽ □ ♃
16 16	☽ ∥ ☿
19 27	☽ ± ♅
20 32	☽ ∠ ♀
26 Wednesday	
00 12	☽ Q ♇
01 15	☽ ⊻ ☉
01 59	☽ bQ ♆
05 57	☽ ✱ ♂
07 20	☽ ⊼ ♅
07 52	☽ bQ ♄
11 18	☽ ⊻ ☿
14 11	☽ ⊥ ☉
22 01	☽ ⊥ ☿
23 52	☽ ✱ ♇
27 Thursday	
01 35	☽ ± ♄
01 58	☽ Q ♃
05 28	☽ ✱ ♀
07 10	♂ ⊼ ♅
07 24	☽ △ ♆
09 22	☽ ∠ ☉
13 01	☽ ∠ ♂
13 18	☽ ⊼ ♄
13 56	☽ ∠ ☿
16 21	♀ # ♆
16 45	♇ St D
16 55	♂ Q ♆
19 33	☿ □ ♂
28 Friday	
00 16	♀ ☍ ♆
01 17	☽ ✱ ♃
02 04	☿ △ ♅
04 57	☽ ∠ ♇
07 16	☽ Q ♀
07 20	☽ ⊥ ♂
12 45	☉ ☌ ☿
16 05	☽ ✱ ☿
16 44	☽ ✱ ☉
17 29	☽ ☍ ♅
19 24	☽ ⊻ ♂
22 06	☽ ⊥ ♇
23 03	☉ ∥ ♃
29 Saturday	
02 06	☉ △ ♅
05 00	☉ ∥ ☿
05 51	☽ ∠ ♃
07 44	☿ ∥ ♃
09 24	☽ ⊻ ♇
12 40	☽ Q ☿
16 29	☽ □ ♆
17 13	☽ Q ☉
20 53	☽ □ ♀
22 15	☽ ☍ ♄
22 39	☽ ⊥ ♃
30 Sunday	
09 43	☽ ⊻ ♃
11 13	♀ □ ♄
18 58	☽ □ ☿
31 Monday	
00 48	☽ ∥ ☉
00 56	☽ ⊼ ♅
01 21	☽ Q ♆
04 46	☽ □ ☉
05 40	☽ ☌ ♂
06 48	☽ ∥ ♃
11 38	☽ ± ♅
16 02	☽ ☌ ♇
20 26	☽ ∥ ☿
22 37	☽ ✱ ♆

All ephemeris data is given at 12.00 UT and the Moon's longitude is additionally given for 24.00 UT

NOVEMBER 2033

LONGITUDES

Date	Sidereal time h m s	Sun ☉ ° ' "	Moon ☽ ° ' "	Moon ☽ 24.00 ° ' "	Mercury ☿ ° '	Venus ♀ ° '	Mars ♂ ° '	Jupiter ♃ ° '	Saturn ♄ ° '	Uranus ♅ ° '	Neptune ♆ ° '	Pluto ♇ ° '
01	14 44 22	09 ♏ 22 47	25 ♒ 50 05	02 ♓ 49 53	00 ♏ 51	23 ♎ 52	09 ♒ 25	27 ♒ 36	21 ♋ 20	05 ♋ 55	18 ♈ 08	14 ♒ 25
02	14 48 18	10 22 49	09 ♓ 55 31	17 ♓ 06 50	00 ♏ 00	25 07	10 05	27 38	21 R 20	05 R 54	18 R 07	14 25
03	14 52 15	11 22 52	24 ♓ 23 34	01 ♈ 45 16	29 ♎ 20	26 22	10 45	27 40	21 20	05 52	18 05	14 25
04	14 56 11	12 22 57	09 ♈ 11 20	16 ♈ 41 00	28 R 51	27 37	11 25	27 43	21 19	05 51	18 04	14 25
05	15 00 08	13 23 03	24 ♈ 13 21	01 ♉ 47 20	28 33	28 ♎ 52	12 05	27 45	21 19	05 50	18 02	14 26
06	15 04 05	14 23 12	09 ♉ 21 47	16 ♉ 55 30	28 27	00 ♏ 07	12 45	27 48	21 19	05 49	18 01	14 26
07	15 08 01	15 23 22	24 ♉ 27 15	01 ♊ 55 53	28 D 32	01 22	13 25	27 51	21 18	05 47	17 59	14 26
08	15 11 58	16 23 34	09 ♊ 20 18	16 ♊ 39 33	28 48	02 37	14 06	27 54	21 17	05 46	17 58	14 26
09	15 15 54	17 23 48	23 ♊ 52 51	00 ♋ 59 36	29 14	03 53	14 46	27 57	21 17	05 45	17 56	14 27
10	15 19 51	18 24 03	07 ♋ 59 22	14 ♋ 51 58	29 ♎ 49	05 08	15 27	28 01	21 16	05 43	17 55	14 27
11	15 23 47	19 24 21	21 ♋ 37 21	28 ♋ 15 39	00 ♏ 32	06 23	16 08	28 05	21 15	05 42	17 54	14 28
12	15 27 44	20 24 41	04 ♌ 47 08	11 ♌ 12 11	01 23	07 38	16 49	28 08	21 14	05 40	17 52	14 28
13	15 31 40	21 25 02	17 ♌ 31 18	23 ♌ 45 01	02 20	08 53	17 30	28 13	21 13	05 38	17 51	14 28
14	15 35 37	22 25 26	29 ♌ 53 56	05 ♍ 58 42	03 22	10 09	18 11	28 17	21 11	05 37	17 50	14 29
15	15 39 34	23 25 51	11 ♍ 59 57	17 ♍ 58 19	04 30	11 24	18 52	28 21	21 10	05 35	17 48	14 29
16	15 43 30	24 26 18	23 ♍ 54 27	29 ♍ 48 58	05 42	12 39	19 33	28 26	21 08	05 33	17 47	14 30
17	15 47 27	25 26 48	05 ♎ 42 26	11 ♎ 35 25	06 57	13 55	20 14	28 31	21 07	05 32	17 46	14 30
18	15 51 23	26 27 18	17 ♎ 28 25	23 ♎ 21 55	08 15	15 10	20 56	28 36	21 05	05 30	17 44	14 31
19	15 55 20	27 27 51	29 ♎ 16 19	05 ♏ 11 59	09 37	16 25	21 37	28 41	21 03	05 28	17 43	14 32
20	15 59 16	28 28 25	11 ♏ 09 16	17 ♏ 08 26	11 00	17 41	22 19	28 46	21 01	05 26	17 42	14 32
21	16 03 13	29 ♏ 29 01	23 ♏ 09 43	29 ♏ 13 17	12 25	18 56	23 00	28 52	20 59	05 24	17 41	14 33
22	16 07 09	00 ♐ 29 38	05 ♐ 19 19	11 ♐ 27 55	13 52	20 11	23 42	28 58	20 57	05 22	17 40	14 34
23	16 11 06	01 30 17	17 ♐ 39 10	23 ♐ 53 10	15 20	21 27	24 24	29 04	20 55	05 20	17 39	14 34
24	16 15 03	02 30 57	00 ♑ 09 57	06 ♑ 29 36	16 49	22 42	25 05	29 10	20 52	05 18	17 38	14 35
25	16 18 59	03 31 39	12 ♑ 52 11	19 ♑ 17 46	18 19	23 58	25 47	29 16	20 50	05 16	17 37	14 36
26	16 22 56	04 32 21	25 ♑ 46 27	02 ♒ 18 20	19 49	25 13	26 29	29 22	20 47	05 14	17 35	14 37
27	16 26 52	05 33 05	08 ♒ 53 33	15 ♒ 32 16	21 21	26 29	27 11	29 29	20 45	05 12	17 34	14 38
28	16 30 49	06 33 50	22 ♒ 14 37	29 ♒ 00 48	22 52	27 44	27 53	29 36	20 42	05 10	17 34	14 38
29	16 34 45	07 34 35	05 ♓ 50 57	12 ♓ 45 13	24 25	28 ♏ 59	28 35	29 42	20 39	05 08	17 33	14 39
30	16 38 42	08 ♐ 35 22	19 ♓ 43 41	26 ♓ 46 24	25 ♏ 57	00 ♐ 15	29 ♒ 18	29 ♒ 50	20 ♋ 36	05 ♋ 05	17 ♈ 32	14 ♒ 40

Date	Moon True ☊ ° '	Moon Mean ☊ ° '	Moon ☽ Latitude ° '
01	11 ♎ 59	10 ♎ 39	03 N 47
02	12 D 00	10 35	02 47
03	12 02	10 32	01 35
04	12 02	10 29	00 N 16
05	12 R 02	10 26	01 S 06
06	12 00	10 23	02 24
07	11 56	10 19	03 31
08	11 51	10 16	04 23
09	11 46	10 13	04 57
10	11 40	10 10	05 11
11	11 36	10 07	05 07
12	11 33	10 04	04 47
13	11 32	10 00	04 13
14	11 D 32	09 57	03 28
15	11 33	09 54	02 34
16	11 35	09 51	01 35
17	11 36	09 48	00 S 32
18	11 R 36	09 45	00 N 32
19	11 34	09 41	01 34
20	11 30	09 38	02 33
21	11 24	09 35	03 25
22	11 15	09 32	04 09
23	11 06	09 29	04 41
24	10 56	09 25	05 01
25	10 46	09 22	05 06
26	10 38	09 19	04 56
27	10 33	09 16	04 30
28	10 30	09 13	03 49
29	10 29	09 10	02 55
30	10 ♎ 29	09 ♎ 06	01 N 49

DECLINATIONS

Date	Sun ☉ ° '	Moon ☽ ° '	Mercury ☿ ° '	Venus ♀ ° '	Mars ♂ ° '	Jupiter ♃ ° '	Saturn ♄ ° '	Uranus ♅ ° '	Neptune ♆ ° '	Pluto ♇ ° '
01	14 S 37	09 S 20	11 S 15	07 S 54	19 S 59	13 S 26	21 N 21	23 N 34	05 N 34	23 S 39
02	14 56	05 16	10 40	08 23	19 46	13 25	21 22	23 34	05 33	23 39
03	15 14	00 S 46	10 10	08 51	19 33	13 24	21 22	23 34	05 33	23 39
04	15 33	03 N 53	09 46	09 18	19 20	13 23	21 22	23 34	05 32	23 39
05	15 51	08 22	09 28	09 46	19 07	13 22	21 22	23 34	05 31	23 39
06	16 09	12 20	09 15	10 14	18 54	13 21	21 22	23 34	05 31	23 39
07	16 27	15 28	09 08	10 41	18 41	13 19	21 22	23 34	05 30	23 38
08	16 44	17 30	09 07	11 08	18 27	13 18	21 22	23 34	05 30	23 38
09	17 01	18 21	09 10	11 34	18 14	13 17	21 22	23 34	05 29	23 38
10	17 18	18 01	09 19	12 01	18 00	13 16	21 23	23 34	05 29	23 38
11	17 35	16 38	09 31	12 27	17 46	13 14	21 23	23 35	05 28	23 37
12	17 51	14 25	09 47	12 53	17 32	13 13	21 23	23 35	05 28	23 37
13	18 07	11 34	10 06	13 18	17 18	13 11	21 23	23 35	05 27	23 37
14	18 22	08 16	10 28	13 43	17 04	13 09	21 24	23 35	05 27	23 37
15	18 38	04 41	10 52	14 08	16 49	13 08	21 24	23 35	05 26	23 36
16	18 52	00 N 58	11 18	14 33	16 35	13 06	21 24	23 35	05 26	23 36
17	19 07	02 S 45	11 46	14 57	16 20	13 04	21 24	23 35	05 26	23 36
18	19 21	06 22	12 15	15 21	16 05	13 02	21 25	23 35	05 25	23 35
19	19 35	09 44	12 45	15 44	15 50	13 00	21 25	23 35	05 25	23 35
20	19 49	12 45	13 15	16 07	15 35	12 58	21 25	23 35	05 24	23 35
21	20 02	15 15	13 46	16 29	15 20	12 56	21 26	23 35	05 24	23 34
22	20 15	17 06	14 18	16 52	15 04	12 54	21 26	23 36	05 23	23 34
23	20 27	18 11	14 49	17 13	14 49	12 52	21 27	23 36	05 23	23 34
24	20 39	18 25	15 21	17 34	14 33	12 50	21 27	23 36	05 23	23 33
25	20 51	17 44	15 52	17 55	14 18	12 47	21 28	23 36	05 22	23 33
26	21 02	16 08	16 23	18 15	14 02	12 45	21 28	23 36	05 22	23 33
27	21 13	13 41	16 53	18 35	13 46	12 42	21 29	23 36	05 22	23 32
28	21 24	10 29	17 24	18 54	13 30	12 40	21 29	23 36	05 21	23 32
29	21 34	06 39	17 53	19 13	13 14	12 37	21 30	23 36	05 21	23 32
30	21 S 44	02 S 24	18 S 22	19 S 31	12 S 57	12 S 35	21 N 30	23 N 36	05 N 21	23 S 31

ZODIAC SIGN ENTRIES

Date	h	m	Planets
01	19	10	☽ ♓
02	12	07	☿ ♎
03	21	09	☽ ♈
05	21	10	☽ ♉
06	09	45	♀ ♏
07	20	53	☽ ♊
09	22	19	☽ ♋
10	18	34	☿ ♏
12	03	11	☽ ♌
14	12	12	☽ ♍
17	00	22	☽ ♎
19	13	29	☽ ♏
22	00	16	☉ ♐
22	01	32	☽ ♐
24	11	41	☽ ♑
26	19	47	☽ ♒
29	01	44	☽ ♓
30	07	17	♀ ♐

LATITUDES

Date	Mercury ☿ ° '	Venus ♀ ° '	Mars ♂ ° '	Jupiter ♃ ° '	Saturn ♄ ° '	Uranus ♅ ° '	Neptune ♆ ° '	Pluto ♇ ° '
01	00 N 34	01 N 27	02 S 10	01 S 12	00 S 23	00 N 16	01 S 40	07 S 30
04	01 23	01 24	02 04	01 11	00 23	00 16	01 40	07 29
07	01 57	01 21	01 59	01 11	00 23	00 16	01 40	07 29
10	02 14	01 17	01 53	01 10	00 23	00 16	01 40	07 28
13	02 19	01 13	01 48	01 10	00 23	00 16	01 40	07 28
16	02 14	01 08	01 43	01 09	00 22	00 16	01 40	07 28
19	02 03	01 03	01 38	01 09	00 22	00 16	01 40	07 27
22	01 47	00 58	01 32	01 08	00 22	00 16	01 40	07 27
25	01 28	00 52	01 28	01 08	00 22	00 16	01 40	07 26
28	01 08	00 46	01 23	01 07	00 22	00 16	01 40	07 26
31	00 N 47	00 N 39	01 S 18	01 S 07	00 S 21	00 N 16	01 S 40	07 S 26

DATA

Julian Date	2463903
Delta T	+75 seconds
Ayanamsa	24° 19' 49"
Synetic vernal point	04° ♓ 47' 10"
True obliquity of ecliptic	23° 25' 57"

LONGITUDES

Date	Chiron ⚷ ° '	Ceres ⚳ ° '	Pallas ⚴ ° '	Juno ⚵ ° '	Vesta ⚶ ° '	Black Moon Lilith ⚸ ° '
01	29 ♉ 36	06 ♑ 13	06 ♐ 42	10 ♏ 38	09 ♓ 45	20 ♎ 02
11	29 ♉ 05	09 ♑ 35	10 ♐ 50	14 ♏ 04	10 ♓ 46	21 ♎ 09
21	28 ♉ 33	13 ♑ 05	15 ♐ 01	17 ♏ 28	12 ♓ 25	22 ♎ 15
31	28 ♉ 01	16 ♑ 43	19 ♐ 12	20 ♏ 50	14 ♓ 35	23 ♎ 22

MOON'S PHASES, APSIDES AND POSITIONS ☽

Date	h	m	Phase	Longitude ° '	Eclipse Indicator
06	20	32	○	14 ♉ 45	
13	20	09	☾	21 ♌ 46	
22	01	39	●	00 ♐ 03	
29	15	15	☽	07 ♓ 43	

Day	h	m		
06	00	05	Perigee	
18	10	54	Apogee	
03	15	58	0N	
09	16	56	Max dec	18° N 22'
16	18	13	0S	
24	06	06	Max dec	18° S 27'

ASPECTARIAN

h m	Aspects
01 Tuesday	
03 28	☽ bQ ♅
04 11	☽ ⚻ ♄
08 15	☽ △ ♀
14 20	☉ □ ♂
14 34	☽ ± ♄
15 03	☽ ☌ ♃
19 51	☽ ∥ ♀
20 06	☽ △ ☿
02 Wednesday	
00 30	☽ ∠ ♆
05 13	☽ △ ♅
05 56	☽ bQ ♄
07 04	♄ St R
10 22	☽ # ♆
12 16	☽ ⚺ ♂
12 21	☽ bQ ♀
12 49	☽ △ ☉
15 39	☽ ⊥ ♆
19 31	☽ ⚺ ♇
20 05	☽ bQ ☿
22 46	☽ ⊥ ♂
03 Thursday	
01 38	☽ ⚺ ♆
04 45	☽ ± ♀
05 28	☽ ⊥ ♇
06 58	☽ △ ♄
10 19	☽ ± ☿
14 19	☽ ∠ ♂
15 30	☽ bQ ☉
15 31	☽ ⚻ ♀
17 22	☽ ⚺ ♃
19 46	☽ ⚻ ☿
20 13	☽ ∠ ♇
04 Friday	
03 09	☽ ⊥ ♃
06 38	☽ □ ♅
07 09	☽ ± ☉
13 55	♀ △ ♃
15 44	☽ ⚹ ♂
17 29	☽ ⚻ ☉
17 40	☽ ∠ ♃
20 23	☽ ⚹ ♇
20 38	☽ ∥ ♆
05 Saturday	
01 56	☿ ∥ ♀
02 10	☽ ☌ ♆
03 29	♂ ± ♅
06 50	☿ ☌ ♀
07 23	☽ □ ♄
11 23	☽ Q ♅
11 46	☽ Q ♂
15 30	☽ Q ♇
17 37	☽ ⚹ ♃
17 54	☽ # ☿
18 47	☽ ☍ ☿
20 02	☽ ☍ ♀
21 05	☽ # ♀
06 Sunday	
06 23	☽ ⚹ ♅
11 55	☽ Q ♄
12 40	☿ St D
12 42	☽ Q ♃
13 04	☉ □ ♇
17 37	☽ □ ♂
18 59	☽ # ♃
20 02	☽ □ ♇
20 32	☽ ☍ ☉
07 Monday	
01 42	☽ ⚺ ♆
06 10	☽ ∠ ♅
06 58	☽ ⚹ ♄
11 15	☽ ⊥ ♆
17 27	☽ □ ♃
18 38	☽ ⚻ ☿
20 32	☽ ⊥ ♅
23 30	☽ # ☉
08 Tuesday	
00 07	☽ ⚻ ♀
01 41	☽ ∠ ♆
04 28	☽ ± ☿
06 12	☽ ⚺ ♅
07 03	☽ ∠ ♄
10 44	☽ ± ♀
19 29	☽ bQ ☿
20 10	☽ △ ♂
20 21	☽ △ ♇
21 44	☽ ⊥ ♄
09 Wednesday	
00 17	♂ ☌ ♇
00 25	☽ ⚻ ☉
02 08	☽ ⚹ ♆
02 53	☽ bQ ♀
07 12	☽ # ♂
07 40	☽ ⚺ ♄
11 08	☽ ± ☉
18 53	☽ △ ♃
21 22	☽ △ ☿
21 23	☽ bQ ♇
22 11	☽ Q ♆
22 26	☽ bQ ♂
10 Thursday	
00 39	☉ ⚻ ♆
03 30	☽ bQ ☉
06 35	☽ △ ♀
08 05	☽ ☌ ♅
12 30	☽ # ♂
12 48	☽ ± ♇
14 40	☽ ± ♂
20 47	☽ bQ ♃
23 05	♀ △ ♅
23 17	☽ ⚻ ♇
11 Friday	
00 07	☽ # ☉
01 43	☽ ⚻ ♂
05 22	☽ □ ♆
07 43	☽ △ ☉
11 20	☽ ☌ ♄
12 49	☽ ± ♃
21 14	☉ ∥ ♂
23 43	☽ ⚻ ♃
12 Saturday	
05 15	☽ □ ☿
13 38	☽ ⚺ ♅
17 53	☽ □ ♀
17 58	☉ bQ ♅
22 50	☽ # ♃
23 46	☽ # ♀
13 Sunday	
00 51	☽ ⊥ ♅
05 41	♀ ∥ ♃
06 11	☽ ☍ ♇
07 10	☉ △ ♄
11 57	☽ ☍ ♂
12 37	☽ △ ♆
17 52	☽ Q ☿
17 59	☽ ∠ ♅
19 05	☽ ⚺ ♄
20 09	☽ □ ☉
21 53	☽ # ☿
14 Monday	
00 00	♂ ⚹ ♆
06 42	☽ ⊥ ♄
08 10	☽ Q ♀
08 49	☽ ☍ ♃
17 45	☽ bQ ♆
19 31	☽ ⚹ ☿
23 15	☽ ⚹ ♅
15 Tuesday	
00 24	☽ ∠ ♄
07 01	☽ ∥ ♆
10 40	☽ ⚹ ♀
10 46	☽ Q ☉
11 37	☽ ± ♆
17 00	☽ ⚻ ♇
20 41	☉ ± ♆
23 12	☽ Q ♅
23 38	☽ ⚻ ♆
16 Wednesday	
02 39	☽ ⚻ ♂
04 45	☽ ∠ ☿
05 06	☽ ± ♇
06 25	☽ ⚹ ♄
09 24	☿ △ ♅
13 11	☽ ⚹ ☉
15 32	☽ ± ♂
20 31	☽ ∠ ♀
21 15	☽ ⚻ ♃
23 22	☽ bQ ♇
17 Thursday	
01 08	☽ ⊥ ☿
06 44	☽ Q ♄
09 33	☽ ± ♃
10 59	☽ bQ ♂
11 38	☽ □ ♅
14 51	☽ ⚺ ☿
17 02	☽ ⊥ ♀
21 39	♂ bQ ♅
22 34	☽ ∠ ☉
23 30	♀ □ ♇
18 Friday	
04 02	☽ bQ ♃
05 35	☽ # ♆
05 58	☽ △ ♇
06 44	☽ ⚺ ♀
12 33	☽ ☍ ♆
17 13	♂ ⚻ ♄
18 39	☽ ⊥ ☉
19 20	☽ □ ♄
19 29	☽ △ ♂
19 Saturday	
01 58	♅ # ♇
07 59	☽ ⚺ ☉
10 48	☽ △ ♃
15 49	♀ ∥ ♂
23 27	☿ ∥ ♃
20 Sunday	
00 30	☽ △ ♅
11 39	☽ ☌ ☿
12 26	♀ ⚻ ♆
13 56	☽ ∥ ♃
17 34	☽ ∥ ☿
18 48	☽ □ ♇
19 49	☉ □ ♃
21 Monday	
01 06	☽ ⚻ ♆
02 36	☽ ☌ ♀
06 32	☽ bQ ♅
07 41	☽ △ ♄
10 09	☉ ± ♅
11 40	☽ □ ♂
12 49	☽ ∥ ♂
13 02	☽ ± ♆
23 23	☽ □ ♃
22 Tuesday	
00 20	☽ ± ♅
01 39	☽ ☌ ☉
06 34	☽ Q ♇
06 47	☽ bQ ♆
07 12	☽ ∥ ♀
12 06	☽ ⚻ ♅
13 14	☽ bQ ♄
15 21	♀ bQ ♅
23 33	☿ □ ♇
23 Wednesday	
01 12	☽ Q ♂
02 05	♀ △ ♄
06 02	☽ ⚹ ♇
06 43	☽ ± ♄
06 54	☽ ⚺ ☿
10 51	☽ Q ♃
11 48	☿ ∥ ♂
11 59	☽ △ ♆
18 16	☽ ⚻ ♄
20 03	☽ ⊥ ☿
20 08	☽ ⚺ ♀
24 Thursday	
01 44	☽ ⚹ ♂
08 54	☽ ⊥ ♀
10 04	☽ ⚹ ♃
10 54	☽ ∠ ♇
13 40	☉ Q ♇
14 34	☉ bQ ♆
15 33	☽ ∠ ☿
16 51	☽ ⚺ ☉
21 44	☽ ☍ ♅
25 Friday	
00 52	☿ ⚻ ♆
03 51	☽ ∠ ♀
03 58	☽ ⊥ ♇
05 11	☽ ⊥ ☉
05 22	♀ ± ♆
07 52	☽ ∠ ♂
08 50	☽ ∥ ♀
14 38	☽ ∠ ♃
15 14	☽ ⚺ ♇
20 51	☽ □ ♆
23 28	☽ ∠ ☉
23 31	☽ ⚹ ☿
26 Saturday	
01 39	☽ ⊥ ♂
02 48	☽ ☍ ♄
07 31	☽ ⊥ ♃
09 40	☽ ∥ ☿
10 52	☽ ⚹ ♀
13 23	☽ ⚺ ♂
18 21	☿ bQ ♅
18 41	☽ ⚺ ♃
21 45	♂ ± ♄
27 Sunday	
00 35	☽ Q ☿
02 49	☿ △ ♄
03 57	☉ ⚻ ♅
05 18	☽ ⚻ ♅
05 25	☽ ⚹ ☉
05 59	☽ Q ♆
11 10	☽ Q ♀
11 16	☽ ∥ ♂
16 10	☽ ± ♅
16 24	☉ bQ ♄
20 01	☽ ∥ ♃
22 22	☽ ☌ ♇
28 Monday	
03 38	☽ ⚹ ♆
04 54	☽ Q ☉
08 18	☽ bQ ♅
09 15	☽ ⚻ ♄
13 16	☽ □ ☿
18 43	♀ □ ♂
19 53	☽ ± ♄
22 33	☽ ☌ ♂
22 36	☿ ± ♆
22 44	☽ □ ♀
29 Tuesday	
00 32	☉ # ♄
01 08	☽ ☌ ♃
06 13	☽ ∠ ♆
10 45	☽ △ ♅
11 39	☽ bQ ♄
14 34	♀ ± ♅
15 15	☽ □ ☉
19 35	☽ # ♆
21 54	☽ ⊥ ♆
30 Wednesday	
03 07	♀ □ ♃
03 18	☽ ⚺ ♇
08 14	☽ ⚺ ♆
13 29	☽ △ ♄
13 37	☽ ⊥ ♇
23 54	☽ △ ☿

All ephemeris data is given at 12.00 UT and the Moon's longitude is additionally given for 24.00 UT

DECEMBER 2033

LONGITUDES

Date	Sidereal time h m s	Sun ☉ ° ' "	Moon ☽ ° ' "	Moon ☽ 24.00 ° ' "	Mercury ☿ ° '	Venus ♀ ° '	Mars ♂ ° '	Jupiter ♃ ° '	Saturn ♄ ° '	Uranus ♅ ° '	Neptune ♆ ° '	Pluto ♇ ° '
01	16 42 38	09 ♐ 36 09	03 ♈ 53 19	11 ♈ 04 17	27 ♏ 30	01 ♐ 30	00 ♓ 00	29 ♒ 57	20 ♋ 33	05 ♋ 03	17 ♈ 31	14 ♒ 41
02	16 46 35	10 36 58	18 ♈ 19 02	25 ♈ 37 11	29 ♏ 02	02 46	00 42	00 ♓ 04	20 R 30	05 R 01	17 R 30	14 42
03	16 50 32	11 37 47	02 ♉ 58 12	10 ♉ 21 23	00 ♐ 35	04 01	01 24	00 12	20 27	04 59	17 29	14 43
04	16 54 28	12 38 37	17 ♉ 45 55	25 ♉ 10 53	02 09	05 17	02 07	00 19	20 23	04 56	17 28	14 44
05	16 58 25	13 39 29	02 ♊ 35 16	09 ♊ 58 02	03 42	06 32	02 49	00 27	20 20	04 54	17 27	14 45
06	17 02 21	14 40 21	17 ♊ 18 08	24 ♊ 34 33	05 15	07 48	03 31	00 35	20 17	04 52	17 27	14 46
07	17 06 18	15 41 14	01 ♋ 46 24	08 ♋ 52 53	06 48	09 03	04 14	00 43	20 13	04 49	17 26	14 47
08	17 10 14	16 42 09	15 ♋ 53 24	22 ♋ 47 28	08 22	10 18	04 56	00 52	20 09	04 47	17 25	14 48
09	17 14 11	17 43 04	29 ♋ 34 51	06 ♌ 15 26	09 55	11 34	05 39	01 00	20 06	04 44	17 25	14 49
10	17 18 07	18 44 01	12 ♌ 49 18	19 ♌ 16 41	11 29	12 49	06 21	01 09	20 02	04 42	17 24	14 50
11	17 22 04	19 44 58	25 ♌ 37 55	01 ♍ 53 28	13 02	14 05	07 04	01 17	19 58	04 40	17 24	14 52
12	17 26 01	20 45 57	08 ♍ 03 53	14 ♍ 09 47	14 36	15 20	07 46	01 26	19 54	04 37	17 23	14 53
13	17 29 57	21 46 57	20 ♍ 11 50	26 ♍ 10 42	16 10	16 36	08 29	01 35	19 50	04 35	17 22	14 54
14	17 33 54	22 47 58	02 ♎ 07 06	08 ♎ 01 45	17 44	17 51	09 12	01 44	19 46	04 32	17 22	14 55
15	17 37 50	23 49 00	13 ♎ 55 20	19 ♎ 48 32	19 17	19 07	09 54	01 54	19 42	04 30	17 22	14 56
16	17 41 47	24 50 03	25 ♎ 41 59	01 ♏ 36 18	20 51	20 22	10 37	02 03	19 38	04 27	17 21	14 58
17	17 45 43	25 51 07	07 ♏ 32 03	13 ♏ 29 43	22 25	21 38	11 20	02 12	19 33	04 25	17 21	14 59
18	17 49 40	26 52 12	19 ♏ 29 47	25 ♏ 32 37	24 00	22 54	12 02	02 22	19 29	04 22	17 20	15 00
19	17 53 36	27 53 17	01 ♐ 38 32	07 ♐ 47 46	25 34	24 09	12 45	02 32	19 25	04 19	17 20	15 02
20	17 57 33	28 54 23	14 ♐ 00 30	20 ♐ 16 49	27 09	25 25	13 28	02 42	19 20	04 17	17 20	15 03
21	18 01 30	29 ♐ 55 30	26 ♐ 36 44	03 ♑ 00 13	28 ♐ 43	26 40	14 11	02 52	19 16	04 14	17 19	15 04
22	18 05 26	00 ♑ 56 38	09 ♑ 27 10	15 ♑ 57 28	00 ♑ 18	27 56	14 54	03 02	19 11	04 12	17 19	15 06
23	18 09 23	01 57 46	22 ♑ 30 55	29 ♑ 07 20	01 53	29 ♐ 11	15 36	03 12	19 06	04 09	17 19	15 07
24	18 13 19	02 58 54	05 ♒ 46 32	12 ♒ 28 20	03 29	00 ♑ 27	16 19	03 23	19 02	04 07	17 19	15 08
25	18 17 16	04 00 02	19 ♒ 12 34	25 ♒ 59 05	05 04	01 42	17 02	03 33	18 57	04 04	17 19	15 10
26	18 21 12	05 01 11	02 ♓ 47 49	09 ♓ 38 40	06 40	02 58	17 45	03 44	18 52	04 01	17 19	15 11
27	18 25 09	06 02 19	16 ♓ 31 38	23 ♓ 26 42	08 16	04 13	18 28	03 55	18 48	03 59	17 19	15 13
28	18 29 05	07 03 28	00 ♈ 23 53	07 ♈ 23 11	09 52	05 29	19 11	04 05	18 43	03 56	17 D 19	15 14
29	18 33 02	08 04 36	14 ♈ 24 37	21 ♈ 28 09	11 29	06 44	19 54	04 16	18 38	03 54	17 19	15 16
30	18 36 59	09 05 45	28 ♈ 33 40	05 ♉ 41 02	13 06	08 00	20 37	04 27	18 33	03 51	17 19	15 17
31	18 40 55	10 ♑ 06 53	12 ♉ 49 59	20 ♉ 00 11	14 ♑ 43	09 ♑ 15	21 ♓ 20	04 ♓ 39	18 ♋ 28	03 ♋ 48	17 ♈ 19	15 ♒ 19

Date	Moon True ☊ ° '	Moon Mean ☊ ° '	Moon ☽ Latitude ° '
01	10 ♎ 30	09 ♎ 03	00 N 35
02	10 R 30	09 00	00 S 42
03	10 28	08 57	01 57
04	10 23	08 54	03 05
05	10 16	08 51	04 01
06	10 06	08 47	04 40
07	09 55	08 44	05 00
08	09 44	08 41	05 01
09	09 35	08 38	04 45
10	09 27	08 35	04 14
11	09 22	08 31	03 31
12	09 19	08 28	02 38
13	09 18	08 25	01 40
14	09 D 19	08 22	00 S 38
15	09 R 19	08 19	00 N 25
16	09 17	08 16	01 26
17	09 14	08 12	02 24
18	09 08	08 09	03 16
19	08 59	08 06	04 00
20	08 47	08 03	04 33
21	08 34	08 00	04 54
22	08 19	07 56	05 00
23	08 06	07 53	04 51
24	07 55	07 50	04 26
25	07 46	07 47	03 46
26	07 40	07 44	02 53
27	07 37	07 41	01 49
28	07 D 37	07 37	00 N 38
29	07 R 37	07 34	00 S 36
30	07 36	07 31	01 48
31	07 ♎ 33	07 ♎ 28	02 S 55

DECLINATIONS

Date	Sun ☉ ° '	Moon ☽ ° '	Mercury ☿ ° '	Venus ♀ ° '	Mars ♂ ° '	Jupiter ♃ ° '	Saturn ♄ ° '	Uranus ♅ ° '	Neptune ♆ ° '	Pluto ♇ ° '
01	21 S 53	02 N 05	18 S 50	19 S 49	12 S 41	12 S 32	21 N 31	23 N 36	05 N 20	23 S 31
02	22 02	06 32	19 18	20 06	12 25	12 29	21 31	23 37	05 20	23 30
03	22 10	10 40	19 45	20 23	12 08	12 26	21 32	23 37	05 20	23 30
04	22 18	14 09	20 10	20 39	11 52	12 24	21 32	23 37	05 20	23 30
05	22 26	16 44	20 35	20 54	11 35	12 21	21 33	23 37	05 19	23 29
06	22 33	18 11	20 59	21 09	11 18	12 18	21 33	23 37	05 19	23 29
07	22 40	18 25	21 22	21 23	11 01	12 15	21 34	23 37	05 19	23 28
08	22 46	17 30	21 45	21 36	10 44	12 12	21 35	23 37	05 19	23 28
09	22 52	15 35	22 06	21 49	10 27	12 08	21 35	23 37	05 19	23 27
10	22 57	12 54	22 26	22 02	10 10	12 05	21 36	23 37	05 18	23 27
11	23 02	09 40	22 44	22 13	09 53	12 02	21 37	23 37	05 18	23 27
12	23 07	06 06	23 02	22 24	09 36	11 59	21 37	23 38	05 18	23 26
13	23 11	02 N 21	23 19	22 34	09 18	11 55	21 38	23 38	05 18	23 26
14	23 14	01 S 26	23 34	22 44	09 01	11 52	21 39	23 38	05 18	23 25
15	23 17	05 07	23 49	22 53	08 44	11 49	21 39	23 38	05 17	23 25
16	23 20	08 36	24 02	23 01	08 26	11 45	21 40	23 38	05 17	23 24
17	23 22	11 45	24 13	23 08	08 09	11 42	21 41	23 38	05 17	23 24
18	23 24	14 27	24 24	23 15	07 51	11 38	21 41	23 38	05 17	23 23
19	23 25	16 34	24 33	23 21	07 33	11 34	21 42	23 38	05 17	23 23
20	23 26	17 57	24 41	23 27	07 16	11 31	21 43	23 38	05 17	23 22
21	23 26	18 30	24 48	23 31	06 58	11 27	21 44	23 38	05 17	23 22
22	23 26	18 06	24 53	23 35	06 40	11 23	21 44	23 39	05 17	23 21
23	23 25	16 46	24 57	23 39	06 22	11 19	21 45	23 39	05 17	23 21
24	23 24	14 31	25 00	23 41	06 04	11 16	21 46	23 39	05 17	23 20
25	23 22	11 28	25 01	23 43	05 47	11 12	21 46	23 39	05 17	23 20
26	23 20	07 47	25 00	23 44	05 29	11 08	21 47	23 39	05 17	23 19
27	23 18	03 S 38	24 59	23 44	05 11	11 04	21 48	23 39	05 17	23 19
28	23 15	00 N 44	24 56	23 44	04 53	11 00	21 49	23 39	05 17	23 18
29	23 11	05 08	24 51	23 42	04 35	10 56	21 50	23 39	05 17	23 18
30	23 07	09 16	24 45	23 40	04 17	10 51	21 50	23 39	05 17	23 17
31	23 S 03	12 N 55	24 S 37	23 S 38	03 S 58	10 S 47	21 N 51	23 N 39	05 N 17	23 S 17

ZODIAC SIGN ENTRIES

Date	h	m	Planets	
01	05	27	☽	♈
01	12	10	♂	♓
01	22	34	♃	♓
03	02	51	☿	♐
03	07	10	☽	♉
05	07	48	☽	♊
07	09	02	☽	♋
09	12	45	☽	♌
11	20	21	☽	♍
14	07	43	☽	♎
16	20	45	☽	♏
19	08	47	☽	♐
21	13	46	☉	♑
21	18	22	☽	♑
22	07	24	☿	♑
24	01	35	☽	♒
24	03	31	♀	♑
26	07	05	☽	♓
28	11	19	☽	♈
30	14	26	☽	♉

LATITUDES

Date	Mercury ☿ ° '	Venus ♀ ° '	Mars ♂ ° '	Jupiter ♃ ° '	Saturn ♄ ° '	Uranus ♅ ° '	Neptune ♆ ° '	Pluto ♇ ° '
01	00 N 47	00 N 39	01 S 18	01 S 07	00 S 21	00 N 16	01 S 40	07 S 26
04	00 25	00 32	01 13	01 07	00 21	00 16	01 39	07 25
07	00 N 04	00 25	01 09	01 06	00 21	00 16	01 39	07 25
10	00 S 17	00 18	01 04	01 06	00 21	00 16	01 39	07 25
13	00 36	00 11	01 00	01 05	00 20	00 16	01 39	07 24
16	00 55	00 N 04	00 55	01 05	00 20	00 17	01 39	07 24
19	01 12	00 S 03	00 51	01 05	00 20	00 17	01 39	07 24
22	01 27	00 10	00 47	01 04	00 20	00 17	01 39	07 24
25	01 41	00 17	00 43	01 04	00 19	00 17	01 38	07 23
28	01 52	00 24	00 39	01 04	00 19	00 17	01 38	07 23
31	02 S 01	00 S 31	00 S 35	01 S 03	00 S 19	00 N 17	01 S 38	07 S 23

DATA

Julian Date	2463933
Delta T	+75 seconds
Ayanamsa	24° 19' 53"
Synetic vernal point	04° ♓ 47' 06"
True obliquity of ecliptic	23° 25' 56"

MOON'S PHASES, APSIDES AND POSITIONS ☽

Date	h	m	Phase	Longitude ° '	Eclipse Indicator
06	07	22	○	14 ♊ 29	
13	15	28	☾	21 ♍ 56	
21	18	46	●	00 ♑ 13	
29	00	20	☽	07 ♈ 35	

Day	h	m		
04	08	14	Perigee	
16	03	34	Apogee	
01	00	54	0N	
07	04	40	Max dec	18° N 28'
14	02	54	0S	
21	14	10	Max dec	18° S 30'
28	07	59	0N	

LONGITUDES

Date	Chiron ⚷ ° '	Ceres ⚳ ° '	Pallas ⚴ ° '	Juno ⚵ ° '	Vesta ⚶ ° '	Black Moon Lilith ⚸ ° '
01	28 ♉ 01	16 ♑ 43	19 ♐ 12	20 ♏ 50	14 ♓ 35	23 ♎ 22
11	27 ♉ 31	20 ♑ 27	23 ♐ 22	24 ♏ 09	17 ♓ 13	24 ♎ 28
21	27 ♉ 03	24 ♑ 16	27 ♐ 32	27 ♏ 23	20 ♓ 13	25 ♎ 35
31	26 ♉ 40	28 ♑ 08	01 ♑ 39	00 ♐ 31	23 ♓ 32	26 ♎ 41

All ephemeris data is given at 12.00 UT and the Moon's longitude is additionally given for 24.00 UT

ASPECTARIAN

h m	Aspects
01 Thursday	
04 55	☽ ∠ ♇
05 06	☽ ⊻ ♂
05 19	☽ ⊻ ♃
07 36	☽ △ ♀
09 58	♂ ☌ ♃
13 57	☽ □ ♅
15 29	☽ ⊥ ♃
15 43	☽ ⊥ ♂
22 17	☽ △ ☉
02 Friday	
03 58	♂ ∥ ♃
04 05	☽ ⚼ ☿
05 25	☽ ∥ ♆
06 01	☽ ⚹ ♇
06 35	☽ ∠ ♃
07 02	♀ ⚼ ♆
07 27	☽ ∠ ♂
10 39	☽ ☌ ♆
10 48	♀ Q ♇
11 00	☽ ⚼ ♀
11 37	☿ ± ♅
15 35	☽ □ ♄
19 43	☽ Q ♅
20 42	☽ ± ☿
03 Saturday	
00 53	☽ ⚼ ☉
01 47	☽ Q ♇
03 10	☽ ± ♀
05 19	☿ □ ♃
07 26	☽ ⚹ ♃
07 40	☽ ⚻ ☿
09 19	☽ ⚹ ♂
13 52	☽ ⚻ ♀
15 15	☽ ⚹ ♅
16 39	☽ ± ☉
20 48	☽ ⋕ ♂
20 52	☽ Q ♄
23 23	☽ ⋕ ♃
04 Sunday	
03 06	☽ ⚻ ☉
03 07	☽ Q ♃
05 45	♀ ⚻ ♅
05 47	☽ Q ♂
07 05	☽ □ ♇
11 03	☿ □ ♂
11 31	☽ ⊻ ♆
14 05	♀ ⚼ ♄
15 31	☽ ∠ ♅
16 14	☽ ⚹ ♄
17 02	☿ ⚼ ♆
21 13	☿ Q ♇
21 13	☽ ⊥ ♆
05 Monday	
00 05	♂ ∠ ♆
06 02	☽ ⊥ ♅
08 30	☽ □ ♃
11 47	☽ ∠ ♆
12 23	☽ □ ♂
14 01	☽ ☍ ☿
15 45	☽ ⊻ ♅
16 27	☽ ∠ ♄
19 00	☽ ☍ ♀
06 Tuesday	
03 08	☉ ± ♄
06 08	☿ ⚻ ♅
07 03	☽ ⊥ ♄
07 22	☽ ☍ ☉
07 50	☽ △ ♇
12 14	☽ ⚹ ♆
12 22	☿ ⚼ ♄
14 16	☉ ⚹ ♇
16 52	☽ ⊻ ♄
07 Wednesday	
08 05	☽ Q ♆
08 40	☽ ⚼ ♇
10 13	☽ △ ♃
13 02	☿ ∥ ♀
16 21	☽ △ ♂
17 07	☽ ☌ ♅
21 32	☽ ⚻ ☿
23 51	☽ ± ♇
08 Thursday	
00 44	☿ ⋕ ♄
01 29	☽ ⚻ ♀
07 01	♂ △ ♅
08 39	♀ ⋕ ♄
09 02	☽ ± ☿
10 07	☽ ⚻ ♇
11 57	☽ ⚼ ♃
12 48	☽ ± ♀
13 31	☽ ⚻ ☉
14 39	☽ □ ♆
18 51	♂ ⚼ ♄
19 22	☽ ☌ ♄
19 24	☽ ⚼ ♂
09 Friday	
00 48	☽ ± ☉
02 41	☽ ⚼ ☿
03 48	☽ ± ♃
04 51	☉ △ ♆
06 06	☽ ⚼ ♀
12 07	☽ ± ♂
14 34	☽ ⚻ ♃
18 05	☽ ⚼ ☉
21 14	☽ ⊻ ♅
23 30	☽ ⚻ ♂
10 Saturday	
08 07	☽ ⊥ ♅
09 12	☽ △ ☿
12 00	☽ △ ♀
15 44	☽ ☍ ♇
18 26	☽ ⋕ ♃
20 29	☽ △ ♆
23 19	☉ Q ♃
23 56	☽ △ ☉
11 Sunday	
00 45	☽ ∠ ♅
01 21	☽ ⊻ ♄
09 54	♀ ± ♄
10 20	☽ ⋕ ♂
12 38	☽ ⊥ ♄
16 49	☉ ⚻ ♄
22 58	☽ ☍ ♃
12 Monday	
00 57	☽ ⚼ ♆
01 41	☿ ± ♄
03 03	♀ ⚹ ♇
05 18	☽ ⚹ ♅
05 52	☽ ∠ ♄
11 24	☽ ☍ ♂
16 20	☿ ⚹ ♇
17 10	☽ ∥ ♆
18 30	☽ ± ♆
20 11	☉ ∥ ☿
13 Tuesday	
01 26	☽ ⚻ ♇
02 46	☽ □ ☿
04 00	☽ □ ♀
04 49	☽ Q ♅
06 23	☽ ⚻ ♆
11 17	☽ ⚹ ♄
13 24	☽ ± ♇
15 28	☽ □ ☉
22 03	☿ ∥ ♇
14 Wednesday	
02 41	♀ △ ♆
06 31	☿ △ ♆
07 32	☽ ⚼ ♇
11 13	☽ ⚻ ♃
11 17	☽ Q ♄
16 53	☽ □ ♅
17 35	☿ ⋕ ♅
20 26	☽ Q ☿
20 29	☽ Q ♀
22 25	☿ ☌ ♀
23 34	☽ ± ♃
15 Thursday	
03 17	☽ ⚻ ♂
07 18	☽ Q ☉
13 11	☽ ⋕ ♆
14 05	☽ △ ♇
16 18	☽ ± ♂
17 58	☿ ⚻ ♄
18 08	☽ ⚼ ♃
19 00	☽ ☍ ♆
22 16	☿ Q ♃
22 29	♀ ⚻ ♄
23 42	☽ □ ♄
23 51	☽ ⚹ ♀
16 Friday	
00 37	☽ ⚹ ☿
04 55	♀ Q ♃
10 04	☽ ⚹ ☉
10 57	☽ ∥ ♂
11 49	☽ ⚼ ♂
17 Saturday	
01 05	☽ △ ♃
05 42	☽ △ ♅
09 58	☽ ∠ ♀
11 32	☽ ∥ ♃
11 45	☽ ∠ ☿
12 35	♂ ⊥ ♆
19 19	☽ ∠ ☉
20 08	☽ △ ♂
18 Sunday	
03 00	☽ □ ♇
06 12	☽ ⊥ ♀
07 40	♃ ∠ ♆
07 42	☽ ⚻ ♆
08 00	☉ ∥ ♇
08 33	☽ ⊥ ☿
11 44	☽ ⚼ ♅
11 58	☽ △ ♄
14 59	☽ ⊥ ☉
19 32	☽ ⊻ ♀
19 38	☽ ± ♆
22 16	☽ ⊻ ☿
19 Monday	
03 57	☽ ⊻ ☉
05 30	☽ ± ♅
13 21	☽ ⚼ ♆
13 46	☽ □ ♃
14 43	☽ Q ♇
17 13	☽ ⚻ ♅
17 23	☽ ⚼ ♄
18 01	♀ ∥ ♇
20 Tuesday	
06 16	☉ ∥ ♀
10 43	☽ ± ♄
10 54	☽ □ ♂
14 00	☽ ⚹ ♇
18 22	☽ △ ♆
22 08	☽ ⚻ ♄
21 Wednesday	
00 58	☽ Q ♃
12 07	☽ ☌ ♀
15 31	☉ ∠ ♇
16 32	☽ ☌ ☿
18 31	☽ ∠ ♇
18 46	☽ ☌ ☉
23 04	☽ Q ♂
23 54	☽ ⚹ ♃
22 Thursday	
02 15	☽ ☍ ♅
08 47	☿ ∠ ♇
11 20	☽ ⊥ ♇
18 59	♂ ⊻ ♇
22 26	☽ ⊻ ♇
22 37	☽ ⚹ ♂
23 Friday	
02 30	☽ □ ♆
04 01	☽ ∠ ♃
05 49	☽ ☍ ♄
12 51	♀ ⋕ ♅
15 09	☉ ☌ ☿
20 38	☽ ⊥ ♃
24 Saturday	
01 23	☽ ⊻ ♀
03 31	☽ ∠ ♂
06 05	♀ ∠ ♇
06 33	☽ ⊻ ☉
07 18	☽ ⊻ ☿
07 38	☽ ⊻ ♃
09 01	☽ ⚻ ♅
10 20	☿ ⚹ ♃
11 10	☽ Q ♆
13 20	☽ ⊥ ♀
18 14	☽ ⊥ ☉
19 32	☽ ⊥ ☿
19 45	☽ ± ♅
20 37	☽ ⊥ ♂
21 17	☿ ☍ ♅
23 16	☉ ⚹ ♃
25 Sunday	
04 47	☽ ☌ ♇
07 05	☽ ∠ ♀
07 55	☽ ⊻ ♂
08 38	☽ ⚹ ♆
11 05	☿ Q ♂
11 33	☽ ⚻ ♄
11 36	☽ ∠ ☉
11 45	☽ ⚼ ♅
13 28	☉ ☍ ♅
13 44	☽ ∠ ☿
14 00	☽ ∥ ♃
21 20	♂ ⊻ ♆
22 07	☽ ± ♄
26 Monday	
11 09	☽ ∠ ♆
12 19	☽ ⚹ ♀
13 40	☽ ☌ ♃
13 53	☽ ⚼ ♄
14 09	☽ △ ♅
16 13	☽ ⚹ ☉
19 41	☽ ⚹ ☿
21 47	☉ ∥ ♇
27 Tuesday	
01 27	♃ ⚼ ♄
02 40	☽ ∥ ♂
02 44	☽ ⋕ ♆
02 55	☽ ⊥ ♆
03 36	♂ ⋕ ♆
05 03	♀ ⚹ ♃
07 33	♀ ☍ ♅
09 43	☽ ⊻ ♇
11 25	☽ Q ♀
13 22	☽ ⊻ ♆
14 50	☽ Q ☉
15 33	☽ ☌ ♂
15 55	☽ △ ♄
19 21	☽ Q ☿
19 38	♃ △ ♅
20 09	☽ ⊥ ♇
21 59	♂ △ ♄
28 Wednesday	
02 24	☿ ⊥ ♇
07 38	♆ St D
11 43	☽ ∠ ♇
18 04	☽ □ ♅
18 26	☽ ⊻ ♃
21 36	☽ □ ♀
21 38	☉ Q ♂
29 Thursday	
00 20	☽ □ ☉
04 51	☽ ⊥ ♃
06 21	☽ □ ☿
09 08	☽ ⋕ ♂
12 52	☽ ∥ ♆
13 27	☽ ⚹ ♇
16 56	☽ ☌ ♆
19 09	☽ □ ♄
20 23	☽ ∠ ♃
21 50	☽ ⊻ ♂
30 Friday	
00 41	☽ Q ♅
08 32	☽ ⊥ ♂
09 51	☽ Q ♇
16 39	☉ ⊥ ♇
20 53	☽ ⚹ ♅
21 46	☽ ⋕ ♃
22 04	☽ ⚹ ♃
22 55	♀ ⋕ ♅
31 Saturday	
00 30	☽ ∠ ♂
01 23	☽ Q ♄
05 25	☽ △ ♀
07 06	☽ △ ☉
11 37	♂ ⊥ ♇
13 08	♀ ⊥ ♇
15 05	♀ Q ♂
15 33	☽ △ ☿
16 10	☽ □ ♇
18 28	☽ Q ♃
19 30	☽ ⊻ ♆
21 01	☿ ⊻ ♇
21 23	☽ ⚹ ♄
21 58	☽ ∠ ♅

JANUARY 2034

LONGITUDES

Date	Sidereal time h m s	Sun ☉ ° ' "	Moon ☽ ° ' "	Moon ☽ 24.00 ° ' "	Mercury ☿ ° '	Venus ♀ ° '	Mars ♂ ° '	Jupiter ♃ ° '	Saturn ♄ ° '	Uranus ♅ ° '	Neptune ♆ ° '	Pluto ⯓ ° '
01	18 44 52	11 ♑ 08 01	27 ♉ 11 11	04 ♊ 22 28	16 ♑ 20	10 ♑ 31	22 ♓ 02	04 ♓ 50	18 ♋ 24	03 ♋ 46	17 ♈ 19	15 ♒ 20
02	18 48 48	12 09 09	11 ♊ 33 24	18 ♊ 43 17	17 58	11 46	22 45	05 01	18 R 19	03 R 43	17 19	15 22
03	18 52 45	13 10 17	25 ♊ 51 24	02 ♋ 57 00	19 36	13 02	23 28	05 13	18 14	03 41	17 19	15 24
04	18 56 41	14 11 25	09 ♋ 59 20	16 ♋ 57 46	21 14	14 17	24 11	05 24	18 09	03 38	17 20	15 25
05	19 00 38	15 12 33	23 ♋ 51 41	00 ♌ 40 36	22 53	15 33	24 54	05 36	18 04	03 36	17 20	15 27
06	19 04 34	16 13 41	07 ♌ 24 09	14 ♌ 02 06	24 31	16 48	25 37	05 48	17 59	03 33	17 20	15 28
07	19 08 31	17 14 49	20 ♌ 34 22	27 ♌ 00 59	26 10	18 04	26 20	05 59	17 54	03 31	17 21	15 30
08	19 12 28	18 15 56	03 ♍ 22 08	09 ♍ 38 05	27 49	19 19	27 03	06 11	17 49	03 28	17 21	15 32
09	19 16 24	19 17 04	15 ♍ 49 13	21 ♍ 56 01	29 ♑ 29	20 35	27 46	06 23	17 44	03 26	17 21	15 33
10	19 20 21	20 18 12	27 ♍ 59 02	03 ♎ 58 52	01 ♒ 08	21 50	28 28	06 35	17 39	03 23	17 22	15 35
11	19 24 17	21 19 20	09 ♎ 56 10	15 ♎ 51 36	02 47	23 06	29 11	06 48	17 34	03 21	17 22	15 37
12	19 28 14	22 20 27	21 ♎ 45 53	27 ♎ 39 42	04 26	24 21	29 ♓ 54	07 00	17 29	03 19	17 23	15 38
13	19 32 10	23 21 35	03 ♏ 33 45	09 ♏ 28 45	06 05	25 37	00 ♈ 37	07 12	17 24	03 16	17 23	15 40
14	19 36 07	24 22 43	15 ♏ 25 19	21 ♏ 24 07	07 43	26 52	01 20	07 25	17 19	03 14	17 24	15 42
15	19 40 03	25 23 50	27 ♏ 25 43	03 ♐ 30 38	09 21	28 07	02 03	07 37	17 14	03 11	17 24	15 43
16	19 44 00	26 24 58	09 ♐ 39 20	15 ♐ 52 12	10 58	29 ♑ 23	02 46	07 50	17 10	03 09	17 25	15 45
17	19 47 57	27 26 05	22 ♐ 09 32	28 ♐ 31 30	12 34	00 ♒ 38	03 28	08 02	17 05	03 07	17 26	15 47
18	19 51 53	28 27 12	04 ♑ 58 15	11 ♑ 29 45	14 08	01 54	04 11	08 15	17 00	03 05	17 27	15 49
19	19 55 50	29 ♑ 28 18	18 ♑ 05 54	24 ♑ 46 30	15 41	03 09	04 54	08 28	16 55	03 02	17 27	15 50
20	19 59 46	00 ♒ 29 24	01 ♒ 31 16	08 ♒ 19 51	17 12	04 25	05 37	08 41	16 50	03 00	17 28	15 52
21	20 03 43	01 30 29	15 ♒ 11 48	22 ♒ 06 43	18 40	05 40	06 20	08 54	16 46	02 58	17 29	15 54
22	20 07 39	02 31 33	29 ♒ 04 05	06 ♓ 03 27	20 06	06 55	07 02	09 07	16 41	02 56	17 30	15 56
23	20 11 36	03 32 37	13 ♓ 04 23	20 ♓ 06 29	21 27	08 11	07 45	09 20	16 36	02 54	17 31	15 57
24	20 15 32	04 33 39	27 ♓ 09 23	04 ♈ 12 47	22 45	09 26	08 28	09 33	16 32	02 52	17 32	15 59
25	20 19 29	05 34 41	11 ♈ 16 26	18 ♈ 20 08	23 58	10 42	09 11	09 46	16 27	02 49	17 32	16 01
26	20 23 26	06 35 41	25 ♈ 23 44	02 ♉ 27 05	25 05	11 57	09 53	10 00	16 23	02 47	17 33	16 03
27	20 27 22	07 36 41	09 ♉ 30 05	16 ♉ 32 36	26 05	13 12	10 36	10 13	16 18	02 45	17 34	16 05
28	20 31 19	08 37 39	23 ♉ 33 24	00 ♊ 35 40	26 58	14 28	11 19	10 26	16 14	02 43	17 36	16 06
29	20 35 15	09 38 36	07 ♊ 35 53	14 ♊ 34 55	27 43	15 43	12 01	10 40	16 09	02 42	17 37	16 08
30	20 39 12	10 39 32	21 ♊ 32 30	28 ♊ 28 21	28 19	16 58	12 44	10 53	16 05	02 40	17 38	16 10
31	20 43 08	11 ♒ 40 26	05 ♋ 22 06	12 ♋ 13 24	28 ♒ 46	18 ♒ 14	13 ♈ 27	11 ♓ 07	16 ♋ 01	02 ♋ 38	17 ♈ 39	16 ♒ 12

Moon nodes and latitude; DECLINATIONS

Date	Moon True ☊ ° '	Moon Mean ☊ ° '	Moon ☽ Latitude ° '	Sun ☉ ° '	Moon ☽ ° '	Mercury ☿ ° '	Venus ♀ ° '	Mars ♂ ° '	Jupiter ♃ ° '	Saturn ♄ ° '	Uranus ♅ ° '	Neptune ♆ ° '	Pluto ⯓ ° '
01	07 ♎ 28	07 ♎ 25	03 S 50	22 S 58	15 N 47	24 S 28	23 S 34	03 S 40	10 S 43	21 N 52	23 N 39	05 N 17	23 S 16
02	07 R 19	07 22	04 31	22 53	17 41	24 17	23 30	03 22	10 39	21 53	23 39	05 17	23 16
03	07 08	07 18	04 55	22 47	18 27	24 05	23 25	03 04	10 34	21 53	23 40	05 17	23 15
04	06 56	07 15	05 00	22 41	18 04	23 51	23 20	02 46	10 30	21 54	23 40	05 18	23 15
05	06 43	07 12	04 47	22 34	16 36	23 36	23 13	02 28	10 26	21 55	23 40	05 18	23 14
06	06 31	07 09	04 19	22 27	14 15	23 19	23 06	02 10	10 21	21 56	23 40	05 18	23 14
07	06 22	07 06	03 37	22 19	11 12	23 00	22 59	01 52	10 17	21 56	23 40	05 18	23 13
08	06 15	07 02	02 44	22 11	07 43	22 40	22 50	01 34	10 12	21 57	23 40	05 18	23 13
09	06 11	06 59	01 46	22 03	03 58	22 19	22 41	01 15	10 08	21 58	23 40	05 19	23 12
10	06 09	06 56	00 S 43	21 54	00 N 08	21 56	22 31	00 57	10 03	21 59	23 40	05 19	23 12
11	06 D 09	06 53	00 N 20	21 45	03 S 38	21 31	22 21	00 39	09 59	21 59	23 40	05 19	23 11
12	06 09	06 50	01 22	21 35	07 13	21 05	22 09	00 21	09 54	22 00	23 40	05 19	23 11
13	06 R 08	06 47	02 20	21 25	10 30	20 37	21 58	00 S 03	09 49	22 01	23 40	05 19	23 10
14	06 06	06 43	03 13	21 14	13 23	20 09	21 45	00 N 15	09 45	22 02	23 40	05 20	23 10
15	06 02	06 40	03 57	21 03	15 43	19 38	21 32	00 33	09 40	22 02	23 40	05 20	23 09
16	05 54	06 37	04 32	20 52	17 24	19 07	21 18	00 51	09 35	22 03	23 40	05 20	23 08
17	05 45	06 34	04 54	20 40	18 18	18 34	21 03	01 09	09 31	22 04	23 40	05 21	23 08
18	05 33	06 31	05 03	20 28	18 18	18 00	20 48	01 27	09 26	22 05	23 41	05 21	23 07
19	05 21	06 28	04 55	20 15	17 20	17 26	20 33	01 45	09 21	22 05	23 41	05 21	23 07
20	05 09	06 24	04 32	20 02	15 24	16 50	20 16	02 03	09 16	22 06	23 41	05 22	23 06
21	04 59	06 21	03 52	19 49	12 34	16 14	19 59	02 21	09 11	22 07	23 41	05 22	23 06
22	04 52	06 18	02 58	19 35	09 00	15 38	19 42	02 39	09 06	22 07	23 41	05 22	23 05
23	04 47	06 15	01 53	19 21	04 54	15 01	19 24	02 57	09 01	22 08	23 41	05 23	23 05
24	04 45	06 12	00 N 41	19 07	00 S 31	14 24	19 05	03 15	08 56	22 09	23 41	05 23	23 04
25	04 D 45	06 08	00 S 35	18 52	03 N 55	13 48	18 46	03 32	08 51	22 09	23 41	05 24	23 04
26	04 46	06 05	01 48	18 37	08 08	13 13	18 26	03 50	08 46	22 10	23 41	05 24	23 03
27	04 R 46	06 02	02 55	18 22	11 53	12 39	18 06	04 08	08 41	22 11	23 41	05 24	23 03
28	04 45	05 59	03 51	18 06	14 56	12 07	17 45	04 25	08 36	22 11	23 41	05 25	23 02
29	04 42	05 56	04 33	17 50	17 05	11 36	17 24	04 43	08 31	22 12	23 41	05 25	23 02
30	04 36	05 53	04 58	17 33	18 12	11 08	17 02	05 01	08 25	22 13	23 41	05 26	23 01
31	04 ♎ 28	05 ♎ 49	05 S 06	17 S 17	18 N 14	10 S 43	16 S 40	05 N 18	08 S 20	22 N 13	23 N 41	05 N 26	23 S 01

ZODIAC SIGN ENTRIES

Date	h	m	Planets
01	16	42	☽ ♊
03	19	00	☽ ♋
05	22	48	☽ ♌
08	05	37	☽ ♍
09	19	35	☿ ♒
10	16	01	☽ ♎
12	15	15	♂ ♈
13	04	45	☽ ♏
15	17	05	☽ ♐
16	23	49	♀ ♒
18	02	46	☽ ♑
20	00	27	☉ ♒
20	09	18	☽ ♒
22	13	36	☽ ♓
24	16	50	☽ ♈
26	19	50	☽ ♉
28	22	59	☽ ♊
31	02	39	☽ ♋

LATITUDES

Date	Mercury ☿ ° '	Venus ♀ ° '	Mars ♂ ° '	Jupiter ♃ ° '	Saturn ♄ ° '	Uranus ♅ ° '	Neptune ♆ ° '	Pluto ⯓ ° '
01	02 S 03	00 S 33	00 S 34	01 S 03	00 S 19	00 N 17	01 S 38	07 S 23
04	02 02	00 40	00 30	01 03	00 18	00 17	01 38	07 23
07	02 08	00 46	00 26	01 03	00 18	00 17	01 38	07 23
10	02 05	00 52	00 23	01 02	00 18	00 17	01 38	07 23
13	01 56	00 58	00 19	01 02	00 17	00 17	01 37	07 23
16	01 42	01 03	00 16	01 02	00 17	00 17	01 37	07 23
19	01 22	01 08	00 13	01 02	00 17	00 17	01 37	07 23
22	00 54	01 12	00 09	01 02	00 16	00 17	01 37	07 23
25	00 S 18	01 16	00 06	01 01	00 16	00 17	01 37	07 23
28	00 N 26	01 19	00 03	01 01	00 16	00 17	01 37	07 23
31	01 N 16	01 S 22	00 00	01 S 01	00 S 15	00 N 17	01 S 37	07 S 23

DATA

Julian Date	2463964
Delta T	+75 seconds
Ayanamsa	24° 19' 58"
Synetic vernal point	04° ♓ 47' 01"
True obliquity of ecliptic	23° 25' 56"

MOON'S PHASES, APSIDES AND POSITIONS ☽

Date	h	m	Phase	Longitude ° '	Eclipse Indicator
04	19	47	○	14 ♋ 31	
12	13	17	☾	22 ♎ 24	
20	10	01	●	00 ♒ 24	
27	08	32	☽	07 ♉ 28	

Day	h	m		
01	00	20	Perigee	
13	00	23	Apogee	
25	21	14	Perigee	
03	15	52	Max dec	18° N 28'
10	12	53	0S	
17	23	57	Max dec	18° S 25'
24	14	45	0N	
31	00	32	Max dec	18° N 21'

LONGITUDES

Date	Chiron ⚷ ° '	Ceres ⚳ ° '	Pallas ⚴ ° '	Juno ⚵ ° '	Vesta ⚶ ° '	Black Moon Lilith ° '
01	26 ♉ 38	28 ♑ 32	02 ♑ 04	00 ♐ 50	23 ♓ 53	26 ♎ 48
11	26 ♉ 20	02 ♒ 27	06 ♑ 08	03 ♐ 49	27 ♓ 29	27 ♎ 55
21	26 ♉ 08	06 ♒ 23	10 ♑ 07	06 ♐ 39	01 ♈ 17	29 ♎ 01
31	26 ♉ 03	10 ♒ 21	14 ♑ 01	09 ♐ 16	05 ♈ 15	00 ♏ 08

ASPECTARIAN

h m Aspects

01 Sunday
02 57 ☽ ✶ ♂
05 32 ☽ ⊥ ♆
08 56 ☽ ♇ ♀
10 06 ☽ ♇ ☉
12 58 ☽ ⊥ ♅
19 49 ☽ ♇ ☿
20 34 ☽ ∠ ♆
22 18 ☽ ∠ ♄
22 57 ☽ ⊻ ♅

02 Monday
00 02 ☽ Q ♂
00 56 ☽ □ ♃
01 24 ☽ ± ♀
02 17 ☽ ± ☉
02 28 ☿ □ ♆
12 24 ☽ ⊼ ♀
12 46 ☽ ± ☿
13 04 ☽ ⊼ ☉
13 15 ☽ ⊥ ♄
16 49 ☿ ☍ ♄
18 23 ☽ △ ⯓
21 39 ☽ ✶ ♆
23 15 ☽ ⊻ ♄

03 Tuesday
00 07 ☽ ⊼ ☿
07 46 ☽ □ ♂
17 51 ☽ Q ♆
19 41 ☽ ♇ ⯓
22 09 ☿ ∠ ♃

04 Wednesday
01 12 ☽ ☌ ♅
02 10 ☉ ☌ ♀
04 04 ☽ △ ♃
11 01 ☽ ± ⯓
19 47 ☽ ☍ ☉
20 07 ☽ ☍ ♀
21 21 ☽ ⊼ ⯓

05 Thursday
00 38 ☽ □ ♆
01 58 ☽ ☌ ♄
05 58 ☿ # ♅
06 13 ☽ ♇ ♃
08 34 ♀ ∥ ⯓
10 02 ♀ ⊻ ⯓
10 03 ☽ ☍ ☿
13 55 ☽ △ ♂
17 43 ☉ ⊻ ⯓
22 14 ☽ ± ♃

06 Friday
05 08 ☽ ⊻ ♅
09 04 ☽ ⊼ ♃
15 52 ☽ ⊥ ♅
18 07 ☽ ♇ ♂
18 57 ☿ ∥ ⯓
22 11 ♀ □ ♆

07 Saturday
02 39 ☽ ☍ ⯓
05 21 ☽ ⊼ ☉
06 03 ☽ △ ♆
06 53 ☽ ⊼ ♀
07 06 ☽ ⊻ ♄
08 13 ☽ ∠ ♅
09 04 ♀ ☍ ♄
11 32 ☽ ± ♂
14 15 ☉ □ ♆
15 37 ☿ ∥ ♀
16 02 ☿ ✶ ♂
17 23 ☽ ± ☉
18 08 ☽ ⊥ ♄
18 43 ☽ # ♃
19 11 ☽ ± ♀
23 21 ☽ ⊼ ♂
23 58 ☽ ⊼ ☿

08 Sunday
02 12 ☉ ☍ ♄
10 04 ☽ ♇ ♆
10 57 ☽ ∠ ♄
11 47 ☽ ♇ ☉
12 12 ☽ ✶ ♅
13 00 ☽ ± ☿
14 01 ☽ ♇ ♀
17 28 ☽ ☍ ♃

09 Monday
03 19 ☽ ± ♆
03 31 ☽ ∥ ♆
08 59 ☽ ♇ ☿
11 15 ☽ Q ♅
11 29 ☽ ⊼ ⯓
15 00 ☽ ⊼ ♆
15 43 ☽ ✶ ♄
19 24 ☽ △ ☉
22 24 ☽ △ ♀
23 17 ☽ ± ⯓

10 Tuesday
00 15 ☉ # ♄
03 03 ☿ ⊥ ♃
06 25 ♀ ∠ ♃
06 27 ☽ # ♂
09 04 ☿ # ♄
13 02 ☽ ☍ ♂
14 53 ☉ ∥ ☿
15 18 ☽ Q ♄
17 12 ☽ ♇ ⯓
18 24 ☽ ∥ ♂
19 17 ☽ △ ☿
22 47 ☽ □ ♅

11 Wednesday
05 33 ☽ ⊼ ♃
17 53 ☽ ± ♃
20 01 ☿ ⊼ ♅
23 08 ☽ # ♆
23 31 ☽ △ ⯓

12 Thursday
01 52 ☉ ∠ ♃
03 05 ☽ ☍ ♆
03 22 ☽ □ ♄
12 29 ☽ ♇ ♃
13 17 ☽ □ ☉
17 53 ☽ □ ♀

13 Friday
01 50 ☿ Q ♆
05 38 ☽ ⊼ ♂
05 47 ♀ # ♄
06 59 ☽ ∥ ♃
11 24 ☽ △ ♅
13 43 ♂ ∠ ⯓
16 06 ♄ □ ♆
17 56 ☽ □ ☿
18 36 ☽ ± ♂
19 31 ☽ △ ♃

14 Saturday
05 17 ☽ Q ☉
06 48 ☿ ⊻ ♃
10 45 ☽ Q ♀
12 33 ☽ □ ⯓
13 57 ☽ ♇ ♂
15 48 ☽ △ ♄
15 59 ☽ ⊼ ♆
17 38 ☽ ♇ ♅

15 Sunday
04 00 ☽ ± ♆
07 36 ☽ ✶ ☉
09 44 ☿ ± ♅
11 32 ☽ ± ♅
11 49 ☽ Q ☿
13 32 ☽ ✶ ♀
21 26 ☽ ♇ ♄
21 41 ☽ △ ♂
21 50 ☽ ♇ ♆
23 20 ☽ ⊼ ♅

16 Monday
00 27 ☽ Q ⯓
08 23 ☽ □ ♃
14 54 ☽ ± ♄
14 55 ☽ ✶ ☿
15 43 ☽ ∠ ☉
22 10 ☽ ∠ ♀
23 48 ☽ ✶ ⯓

17 Tuesday
00 32 ♂ □ ♅
02 23 ☽ ⊼ ♄
02 59 ☽ △ ♆
10 30 ☽ ⊥ ☉
17 12 ☽ ⊥ ♀
19 16 ☽ ∥ ☿
19 27 ☽ Q ♃
22 49 ☽ ⊻ ☉
23 38 ☽ ∠ ☿

18 Wednesday
04 15 ☽ ∠ ⯓
05 41 ☽ ⊻ ♀
08 30 ☽ ☍ ♅
10 28 ☽ □ ♂
18 09 ☽ ✶ ♃
18 38 ☽ ⊥ ☿
20 11 ♀ ⊥ ♃
20 56 ☽ ⊥ ⯓

19 Thursday
07 04 ☽ ⊻ ☿
07 54 ☽ ⊻ ⯓
09 06 ☽ ∥ ☿
09 53 ☽ ☍ ♄
09 53 ♀ ⊼ ♅
10 50 ☽ □ ♆
14 27 ☿ ☌ ⯓
21 08 ☽ Q ♂
21 49 ☽ ∠ ♃

20 Friday
06 29 ☿ ⊼ ♄
10 01 ☽ ☌ ☉
14 05 ☽ ⊥ ♃
14 37 ☽ ⊼ ♅
16 20 ☿ ✶ ♆
17 37 ☽ ☌ ♀
18 58 ☽ Q ♆
19 37 ☽ ✶ ♂

21 Saturday
00 39 ☿ ♇ ♅
00 49 ☽ ⊻ ♃
01 09 ☽ ± ♅
08 24 ♀ Q ♆
13 13 ☽ ☌ ⯓
14 42 ☽ ⊼ ♄
15 59 ☽ ✶ ♆
16 48 ☽ ♇ ♅
18 45 ☽ ☌ ☿
23 13 ☽ ∠ ♂

22 Sunday
01 03 ☽ ± ♄
11 24 ☽ ∥ ♃
15 34 ☽ ⊥ ♂
16 28 ☽ ♇ ♄
17 07 ♀ ✶ ♂
17 54 ☽ ∠ ♆
18 25 ☽ ⊻ ☉
18 37 ☽ △ ♅
21 12 ☉ ⊼ ♅

23 Monday
02 25 ☽ ⊻ ♂
02 49 ☽ ⊻ ♀
05 30 ☽ ⊥ ☉
05 30 ☽ ☌ ♃
05 38 ☉ ⊥ ♃
09 20 ☽ ⊥ ♆
09 21 ☽ # ♆
14 05 ☽ ⊥ ♀
16 56 ☽ ⊻ ⯓
18 00 ☽ △ ♄
19 35 ☽ ⊻ ♆
22 04 ☽ ∠ ☉
22 08 ☽ # ♂

24 Tuesday
01 15 ♀ ± ♅
01 48 ☉ ∥ ♀
03 11 ☽ ⊥ ⯓
03 46 ☽ ⊻ ☿
06 55 ☽ ∠ ♀
08 01 ☿ ± ♄
14 38 ♀ ⊻ ♃
14 58 ☽ ⊥ ☿
18 32 ☽ ∠ ⯓
21 40 ☽ □ ♅

25 Wednesday
01 34 ☽ ✶ ☉
07 43 ☽ ∠ ☿
08 15 ☽ ☌ ♂
09 24 ☽ ⊻ ♃
09 45 ☽ ∥ ♂
10 55 ☽ ✶ ♀
11 07 ☉ Q ♆
19 46 ☽ ⊥ ♃
20 05 ☽ ✶ ⯓
20 10 ☽ ∥ ♆
20 45 ☽ □ ♄
22 40 ☽ ☌ ♆
23 33 ☽ Q ☉

26 Thursday
00 12 ☿ ∠ ♂
04 11 ☽ Q ♅
09 18 ☽ Q ♀
11 18 ☽ ∠ ♃
11 25 ☽ ✶ ☿
15 42 ☽ # ♃
16 31 ☽ Q ⯓
17 02 ♂ ⊻ ♃

27 Friday
00 33 ☽ ✶ ♅
03 12 ☽ Q ♄
08 32 ☽ □ ☉
09 25 ☽ Q ☿
13 14 ☽ ✶ ♃
13 58 ☽ ⊻ ♂
16 42 ☽ # ☿
18 56 ☽ □ ♀
23 14 ☽ □ ⯓
23 32 ☽ ✶ ♄

28 Saturday
00 45 ☽ ⊥ ♂
01 47 ☽ ⊻ ♆
02 02 ☽ ∠ ♅
10 02 ☽ Q ♃
12 02 ☽ ⊥ ♆
14 14 ☉ ± ♅
16 56 ☽ ∠ ♂
17 22 ☽ ⊥ ♅
18 09 ☽ □ ☿

29 Sunday
01 01 ☽ ∠ ♄
03 26 ☽ ∠ ♆
03 37 ☽ ⊻ ♅
15 47 ☽ △ ☉
15 57 ☽ # ♀
16 22 ☽ ⊥ ♄
16 26 ♄ ⊼ ⯓
17 21 ☽ □ ♃
19 58 ♀ ⊼ ♄
20 00 ☽ ✶ ♂
20 15 ♀ ☌ ⯓
22 39 ☽ # ☉

30 Monday
02 38 ☽ ⊻ ♄
02 43 ☽ △ ⯓
03 20 ☽ △ ♀
05 14 ☽ ✶ ♆
17 49 ☽ Q ♂
18 55 ☉ ⊻ ♃
19 41 ☽ ♇ ☉

31 Tuesday
00 10 ☽ △ ☿
00 45 ♀ ✶ ♆
00 53 ♀ ♇ ♅
02 02 ☽ Q ♆
04 43 ☽ ♇ ⯓
07 14 ☽ ☌ ♅
07 53 ☽ ♇ ♀
12 35 ☽ ± ☉
20 28 ☽ ± ⯓
22 13 ☽ △ ♃
23 41 ♂ ∥ ♆
23 55 ☽ ⊼ ☉

All ephemeris data is given at 12.00 UT and the Moon's longitude is additionally given for 24.00 UT

FEBRUARY 2034

LONGITUDES

Date	Sidereal time h m s	Sun ☉ ° ' "	Moon ☽ ° ' "	Moon ☽ 24.00 ° ' "	Mercury ☿ ° '	Venus ♀ ° '	Mars ♂ ° '	Jupiter ♃ ° '	Saturn ♄ ° '	Uranus ♅ ° '	Neptune ♆ ° '	Pluto ♇ ° '
01	20 47 05	12 ♒ 41 20	19 ♋ 01 54	25 ♋ 47 14	29 ♒ 02	19 ♒ 29	14 ♈ 09	11 ♓ 20	15 ♋ 57	02 ♋ 36	17 ♈ 40	16 ♒ 14
02	20 51 01	13 42 12	02 ♌ 29 04	09 ♌ 07 06	29 R 07	20 44	14 52	11 34	15 R 53	02 R 34	17 41	16 15
03	20 54 58	14 43 03	15 ♌ 41 07	22 ♌ 10 56	29 01	21 59	15 34	11 48	15 49	02 33	17 42	16 17
04	20 58 55	15 43 53	28 ♌ 36 28	04 ♍ 57 41	28 44	23 15	16 17	12 01	15 45	02 31	17 44	16 19
05	21 02 51	16 44 42	11 ♍ 14 41	17 ♍ 27 35	28 16	24 30	16 59	12 15	15 41	02 29	17 45	16 21
06	21 06 48	17 45 30	23 ♍ 36 38	29 ♍ 42 10	27 38	25 45	17 42	12 29	15 37	02 28	17 46	16 23
07	21 10 44	18 46 17	05 ♎ 44 31	11 ♎ 44 10	26 50	27 00	18 24	12 43	15 33	02 26	17 48	16 25
08	21 14 41	19 47 03	17 ♎ 41 35	23 ♎ 37 21	25 55	28 15	19 07	12 57	15 30	02 24	17 49	16 26
09	21 18 37	20 47 47	29 ♎ 32 01	05 ♏ 26 13	24 52	29 ♒ 31	19 49	13 11	15 26	02 23	17 50	16 28
10	21 22 34	21 48 31	11 ♏ 20 35	17 ♏ 15 46	23 45	00 ♓ 46	20 31	13 25	15 22	02 22	17 52	16 30
11	21 26 30	22 49 14	23 ♏ 12 27	29 ♏ 11 15	22 35	02 01	21 14	13 39	15 19	02 20	17 53	16 32
12	21 30 27	23 49 56	05 ♐ 12 50	11 ♐ 17 48	21 25	03 16	21 56	13 53	15 16	02 19	17 55	16 34
13	21 34 24	24 50 36	17 ♐ 26 44	23 ♐ 40 08	20 15	04 31	22 38	14 07	15 12	02 18	17 56	16 35
14	21 38 20	25 51 16	29 ♐ 58 29	06 ♑ 22 08	19 08	05 46	23 21	14 21	15 09	02 16	17 58	16 37
15	21 42 17	26 51 54	12 ♑ 51 22	19 ♑ 26 22	18 05	07 01	24 03	14 35	15 06	02 15	17 59	16 39
16	21 46 13	27 52 31	26 ♑ 07 10	02 ♒ 53 44	17 08	08 16	24 45	14 50	15 03	02 14	18 01	16 41
17	21 50 10	28 53 07	09 ♒ 45 49	16 ♒ 43 08	16 17	09 31	25 27	15 04	15 00	02 13	18 03	16 43
18	21 54 06	29 ♒ 53 41	23 ♒ 45 12	00 ♓ 51 29	15 33	10 47	26 10	15 18	14 58	02 12	18 04	16 44
19	21 58 03	00 ♓ 54 14	08 ♓ 01 20	15 ♓ 14 02	14 56	12 02	26 52	15 32	14 55	02 11	18 06	16 46
20	22 01 59	01 54 45	22 ♓ 28 52	29 ♓ 45 02	14 27	13 17	27 34	15 47	14 52	02 10	18 08	16 48
21	22 05 56	02 55 15	07 ♈ 01 50	14 ♈ 18 33	14 06	14 32	28 16	16 01	14 50	02 09	18 09	16 50
22	22 09 53	03 55 42	21 ♈ 34 31	28 ♈ 49 12	13 52	15 46	28 58	16 15	14 47	02 08	18 11	16 51
23	22 13 49	04 56 08	06 ♉ 02 04	13 ♉ 12 44	13 45	17 01	29 ♈ 40	16 30	14 45	02 07	18 13	16 53
24	22 17 46	05 56 32	20 ♉ 20 50	27 ♉ 26 08	13 D 46	18 16	00 ♉ 22	16 44	14 43	02 06	18 14	16 55
25	22 21 42	06 56 54	04 ♊ 28 26	11 ♊ 27 36	13 53	19 31	01 04	16 59	14 41	02 06	18 16	16 57
26	22 25 39	07 57 14	18 ♊ 23 31	25 ♊ 16 08	14 06	20 46	01 46	17 13	14 39	02 05	18 18	16 58
27	22 29 35	08 57 32	02 ♋ 05 25	08 ♋ 51 20	14 24	22 01	02 28	17 27	14 37	02 04	18 20	17 00
28	22 33 32	09 ♓ 57 49	15 ♋ 33 54	22 ♋ 13 06	14 ♒ 49	23 ♓ 16	03 ♉ 10	17 ♓ 42	14 ♋ 35	02 ♋ 04	18 ♈ 22	17 ♒ 02

MOON

Date	Moon True ☊ ° '	Moon Mean ☊ ° '	Moon ☽ Latitude ° '
01	04 ♎ 19	05 ♎ 46	04 S 56
02	04 R 10	05 43	04 30
03	04 02	05 40	03 50
04	03 55	05 37	02 58
05	03 50	05 34	01 59
06	03 47	05 30	00 S 55
07	03 D 47	05 27	00 N 11
08	03 47	05 24	01 15
09	03 49	05 21	02 15
10	03 50	05 18	03 10
11	03 R 51	05 14	03 57
12	03 50	05 11	04 33
13	03 48	05 08	04 59
14	03 44	05 05	05 11
15	03 38	05 02	05 07
16	03 32	04 59	04 48
17	03 26	04 55	04 11
18	03 21	04 52	03 19
19	03 17	04 49	02 14
20	03 15	04 46	00 N 59
21	03 D 15	04 43	00 S 21
22	03 16	04 40	01 39
23	03 17	04 36	02 50
24	03 19	04 33	03 50
25	03 19	04 30	04 36
26	03 R 19	04 27	05 04
27	03 17	04 24	05 15
28	03 ♎ 15	04 ♎ 20	05 S 07

DECLINATIONS

Date	Sun ☉ ° '	Moon ☽ ° '	Mercury ☿ ° '	Venus ♀ ° '	Mars ♂ ° '	Jupiter ♃ ° '	Saturn ♄ ° '	Uranus ♅ ° '	Neptune ♆ ° '	Pluto ♇ ° '
01	17 S 00	17 N 11	10 S 21	16 S 17	05 N 35	08 S 15	22 N 14	23 N 41	05 N 27	23 S 00
02	16 42	15 13	10 03	15 54	05 53	08 10	22 14	23 41	05 27	23 00
03	16 25	12 28	09 49	15 30	06 10	08 05	22 15	23 41	05 28	22 59
04	16 07	09 10	09 39	15 06	06 27	07 59	22 16	23 41	05 28	22 59
05	15 49	05 31	09 34	14 42	06 44	07 54	22 16	23 41	05 29	22 58
06	15 30	01 N 42	09 34	14 17	07 01	07 49	22 17	23 41	05 29	22 58
07	15 12	02 S 07	09 37	13 52	07 18	07 43	22 17	23 41	05 30	22 57
08	14 53	05 47	09 45	13 26	07 35	07 38	22 18	23 41	05 30	22 57
09	14 33	09 12	09 57	13 00	07 52	07 33	22 18	23 41	05 31	22 56
10	14 14	12 13	10 13	12 34	08 09	07 27	22 19	23 41	05 32	22 56
11	13 54	14 45	10 31	12 07	08 26	07 22	22 19	23 41	05 32	22 55
12	13 34	16 41	10 51	11 40	08 42	07 16	22 20	23 41	05 33	22 55
13	13 14	17 53	11 13	11 13	08 59	07 11	22 20	23 42	05 33	22 54
14	12 54	18 15	11 36	10 45	09 15	07 05	22 21	23 42	05 34	22 54
15	12 33	17 43	12 00	10 18	09 32	07 00	22 21	23 42	05 35	22 53
16	12 12	16 13	12 23	09 49	09 48	06 54	22 22	23 42	05 35	22 53
17	11 52	13 46	12 46	09 21	10 04	06 49	22 22	23 42	05 36	22 52
18	11 30	10 28	13 07	08 52	10 20	06 43	22 23	23 42	05 37	22 52
19	11 09	06 29	13 28	08 23	10 36	06 38	22 23	23 42	05 37	22 51
20	10 47	02 S 05	13 47	07 54	10 52	06 32	22 24	23 42	05 38	22 51
21	10 26	02 N 28	14 05	07 25	11 08	06 27	22 24	23 42	05 39	22 50
22	10 04	06 53	14 21	06 56	11 24	06 21	22 24	23 42	05 39	22 50
23	09 42	10 51	14 35	06 26	11 39	06 16	22 25	23 42	05 40	22 50
24	09 20	14 07	14 48	05 56	11 55	06 10	22 25	23 42	05 41	22 49
25	08 57	16 31	14 59	05 26	12 10	06 04	22 25	23 42	05 41	22 49
26	08 35	17 53	15 08	04 56	12 26	05 59	22 26	23 42	05 42	22 48
27	08 13	18 10	15 15	04 26	12 41	05 53	22 26	23 42	05 43	22 48
28	07 S 50	17 N 26	15 S 20	03 S 55	12 N 56	05 S 48	22 N 26	23 N 42	05 N 44	22 S 48

ZODIAC SIGN ENTRIES

Date	h	m	Planets
02	07	32	☽ ♌
04	14	37	☽ ♍
07	00	35	☽ ♎
09	12	57	☽ ♏
09	21	23	♀ ♓
12	01	37	☽ ♐
14	12	03	☽ ♑
16	18	54	☽ ♒
18	14	30	☉ ♓
18	22	33	☽ ♓
21	00	25	☽ ♈
23	01	58	☽ ♉
23	23	24	♂ ♉
25	04	22	☽ ♊
27	08	19	☽ ♋

LATITUDES

Date	Mercury ☿ ° '	Venus ♀ ° '	Mars ♂ ° '	Jupiter ♃ ° '	Saturn ♄ ° '	Uranus ♅ ° '	Neptune ♆ ° '	Pluto ♇ ° '
01	01 N 33	01 S 23	00 N 01	01 S 01	00 S 15	00 N 17	01 S 36	07 S 23
04	02 24	01 25	00 03	01 01	00 15	00 17	01 36	07 23
07	03 08	01 27	00 06	01 01	00 14	00 17	01 36	07 23
10	03 35	01 28	00 09	01 01	00 14	00 17	01 36	07 24
13	03 42	01 28	00 11	01 01	00 14	00 17	01 36	07 24
16	03 29	01 28	00 14	01 01	00 13	00 17	01 36	07 24
19	03 01	01 27	00 16	01 01	00 13	00 17	01 36	07 25
22	02 24	01 26	00 19	01 01	00 12	00 17	01 36	07 25
25	01 45	01 24	00 21	01 01	00 12	00 17	01 36	07 25
28	01 06	01 21	00 23	01 01	00 12	00 17	01 35	07 26
31	00 N 29	01 S 18	00 N 25	01 S 01	00 S 11	00 N 17	01 S 35	07 S 26

DATA

Julian Date	2463995
Delta T	+75 seconds
Ayanamsa	24° 20' 03"
Synetic vernal point	04° ♓ 46' 56"
True obliquity of ecliptic	23° 25' 56"

MOON'S PHASES, APSIDES AND POSITIONS ☽

Date	h	m	Phase	Longitude ° '	Eclipse Indicator
03	10	04	○	14 ♌ 38	
11	11	09	☾	22 ♏ 47	
18	23	10	●	00 ♓ 22	
25	16	34	☽	07 ♊ 08	

Day	h	m		
09	21	33	Apogee	
21	15	20	Perigee	
06	22	37	0S	
14	10	05	Max dec	18° S 16'
20	22	59	0N	
27	06	42	Max dec	18° N 12'

LONGITUDES

Date	Chiron ⚷ ° '	Ceres ⚳ ° '	Pallas ⚴ ° '	Juno ⚵ ° '	Vesta ⚶ ° '	Black Moon Lilith ⚸ ° '
01	26 ♉ 03	10 ♒ 44	14 ♑ 24	09 ♐ 32	05 ♈ 39	00 ♏ 14
11	26 ♉ 05	14 ♒ 41	18 ♑ 10	11 ♐ 54	09 ♈ 46	01 ♏ 21
21	26 ♉ 14	18 ♒ 35	21 ♑ 47	14 ♐ 00	13 ♈ 46	02 ♏ 28
31	26 ♉ 30	22 ♒ 28	25 ♑ 15	15 ♐ 46	18 ♈ 16	03 ♏ 34

ASPECTARIAN

h m	Aspects
01 Wednesday	
01 13	☽ ± ♀
02 55	☽ □ ♂
03 02	☽ ⚼ ☿
06 09	☿ ∠ ♂
06 35	☽ ☌ ♄
07 02	☽ ⊼ ♇
09 35	☽ □ ♆
12 53	☽ ⊼ ♀
15 33	☽ # ☉
19 10	☽ ± ☿
02 Thursday	
01 12	☽ ⚼ ♃
03 07	☽ # ♀
05 57	☽ ⊼ ☿
11 22	☿ St R
12 09	☽ ⊻ ♅
17 39	☽ ± ♃
22 59	☽ ⊥ ♅
03 Friday	
01 53	♃ ⊥ ♆
04 45	☽ ⊼ ♃
08 44	♀ ± ♄
10 04	☽ ☍ ☉
11 47	☽ △ ♂
12 14	☽ ⊻ ♄
13 07	☽ ☍ ♇
15 25	☽ ∠ ♅
15 44	☽ △ ♆
19 21	♂ □ ♄
23 15	☽ ⊥ ♄
04 Saturday	
00 54	☽ ☍ ♀
08 30	☽ # ☿
12 14	☽ ☍ ☿
12 16	☉ ⊼ ♄
13 21	♂ ⚹ ♇
16 00	☽ ∠ ♄
17 20	☽ ⚼ ♂
19 21	☽ ⚹ ♅
19 47	☽ ⚼ ♆
20 08	☽ # ♃
05 Sunday	
02 20	☉ ☌ ♇
04 41	☽ ∥ ♂
11 29	☽ ± ♂
12 13	☽ ∥ ♆
12 58	☽ ± ♆
13 59	☽ ☍ ♃
18 14	☽ Q ♅
20 30	☽ ⚹ ♄
21 52	☽ ⊼ ♇
23 34	☽ ⊼ ☉
23 46	☽ ⊼ ♂
06 Monday	
00 35	☽ ⊼ ♆
05 06	☉ ⚼ ♅
07 09	☉ ⚹ ♂
09 35	☽ ± ♇
12 19	☉ ⚹ ♆
12 19	☽ ± ☉
14 37	♂ ☌ ♆
16 41	☽ ⊼ ♀
19 27	☽ ⊼ ☿
19 50	☽ Q ♄
07 Tuesday	
03 22	☽ ⚼ ♇
05 26	☽ □ ♅
05 55	☽ ± ♀
06 36	☽ ± ☿
07 43	☽ ⚼ ☉
10 07	☿ ☌ ♀
23 21	☽ ⚼ ☿
08 Wednesday	
02 01	☽ ⚼ ♀
02 15	☽ ⊼ ♃
03 44	♂ ⊥ ♃
07 35	☽ □ ♄
09 28	☽ △ ♇
10 06	☽ # ♆
12 15	☽ ☍ ♆
14 35	☽ ± ♃
14 45	♂ # ♃
15 03	☽ ☍ ♂
16 37	☽ △ ☉
09 Thursday	
00 23	☽ ∥ ♃
01 32	☽ # ♂
03 19	☽ △ ☿
09 12	☽ ⚼ ♃
11 57	☽ △ ♀
17 47	☽ △ ♅
18 12	☽ ∥ ☿
10 Friday	
02 15	☉ ± ♄
04 52	♀ ⚼ ♄
06 35	♂ Q ♅
14 36	☽ ∥ ♀
16 17	☽ △ ♃
20 08	☽ △ ♄
22 29	☽ □ ♇
11 Saturday	
00 10	☽ ⚼ ♅
01 15	☽ ⊼ ♆
04 24	☽ ∥ ☉
07 46	☽ ⊼ ♂
09 29	☉ ☌ ☿
10 52	☽ □ ☿
11 09	☽ □ ☉
13 22	☽ ± ♆
18 02	♀ △ ♅
18 17	☽ ± ♅
20 35	☽ ± ♂
12 Sunday	
02 11	☽ ⚼ ♄
05 03	♀ ∠ ♆
05 20	☿ ⚹ ♂
06 15	☽ ⊼ ♅
07 25	☽ ⚼ ♆
07 41	☽ □ ♀
10 42	☽ Q ♇
15 14	☿ ± ♄
15 37	☽ ⚼ ♂
19 34	☽ Q ☿
19 58	☽ ± ♄
13 Monday	
02 13	☽ Q ☉
05 24	☽ □ ♃
07 40	☽ ⊼ ♄
10 20	☽ ⚹ ♇
11 57	☿ ∥ ♀
12 57	☽ △ ♆
16 58	☽ ⚹ ☿
22 38	☽ △ ♂
22 53	☽ Q ♀
14 Tuesday	
03 30	☽ ⚹ ☉
15 07	☽ ∠ ♇
16 19	☽ ☍ ♅
16 34	☽ Q ♃
19 13	☽ ∠ ☿
15 Wednesday	
00 03	☽ ⚹ ♀
07 56	☽ ⊥ ♇
10 02	☽ ∠ ☉
10 41	☽ ⊥ ☿
14 14	☿ ⚹ ♆
15 14	☽ ⚹ ♃
16 06	☽ ☍ ♄
18 57	☽ ⊻ ♇
20 53	☽ ⊻ ☿
21 23	☽ □ ♆
16 Thursday	
03 46	☽ ⊥ ☉
06 22	☉ ∥ ☿
06 23	☽ ∠ ♀
09 13	☿ ⚼ ♅
09 25	☽ □ ♂
12 43	♀ # ♂
15 23	☽ ⊻ ☉
18 42	☽ ∠ ♃
22 49	☽ ⊼ ♅
23 50	☿ ☌ ♇
17 Friday	
00 00	☽ ⊥ ♀
05 30	☽ Q ♆
07 11	♃ △ ♄
09 19	☽ ± ♅
10 46	☽ ⊥ ♃
11 33	☽ ⊻ ♀
18 44	☽ Q ♂
19 05	☽ ∥ ☿
21 02	☽ ⊼ ♄
21 19	☽ ⊻ ♃
22 40	☽ ☌ ☿
18 Saturday	
00 01	☽ ☌ ♇
00 50	☽ ⚼ ♅
02 18	☽ ⚹ ♆
04 11	☽ ∥ ☉
07 16	☽ ± ♄
12 45	☽ # ♂
13 28	☉ ⚼ ♄
16 17	☽ ⚹ ♂
18 34	☿ ⊻ ♃
22 27	☽ ⚼ ♄
23 10	☽ ☌ ☉
23 25	☽ ∥ ♀
19 Sunday	
02 14	☽ △ ♅
03 45	☽ ∠ ♆
04 48	♀ ∠ ♂
11 11	☽ ∥ ♃
13 01	☿ ⊼ ♄
13 24	♀ ⊥ ♆
13 26	☿ Q ♂
16 52	☽ # ♆
18 43	☽ ∠ ♂
18 48	☽ ⊥ ♆
19 18	☽ ☌ ♀
23 06	☽ ⊻ ☿
23 26	☽ △ ♄
20 Monday	
00 43	☽ ☌ ♃
02 35	☽ ⊻ ♇
04 47	☽ ⊻ ♆
08 45	☽ ⊥ ☿
08 55	☉ # ♂
10 24	☽ ⊥ ♂
12 32	☽ ⊥ ♇
17 52	☉ △ ♅
20 49	☽ ⊻ ♂
23 13	☽ ∠ ☿
21 Tuesday	
03 24	☽ ∠ ♇
03 58	☽ □ ♅
04 43	☽ ⊻ ☉
05 25	☿ ⊻ ♀
15 21	☽ ⊥ ☉
17 39	♀ △ ♄
17 42	☉ ∠ ♆
23 27	☽ ⚹ ☿
22 Wednesday	
00 49	☽ □ ♄
01 31	☽ ⊻ ♀
03 04	☽ ⊻ ♃
04 11	☽ ⚹ ♇
05 08	☽ ∥ ♆
06 23	☽ ☌ ♆
07 18	☽ ∠ ☉
08 05	♂ Q ♇
09 06	☽ # ♃
09 37	☽ Q ♅
12 15	☽ # ♀
12 22	☽ ⊥ ♀
13 09	☽ ⊥ ♃
19 02	☽ Q ☿
23 26	♀ ☌ ♃
23 Thursday	
00 05	☽ Q ♇
00 52	☽ ☌ ♂
04 19	☽ ∠ ♃
04 41	☽ ∠ ♀
05 19	☽ # ☉
05 29	☽ ⚹ ♅
06 33	☽ Q ♄
09 18	♀ ⊻ ♇
10 02	☽ ⚹ ☉
17 52	☽ ∥ ♂
22 10	♀ ∥ ♃
22 55	☿ St D
24 Friday	
00 54	☽ □ ☿
02 33	☽ ⚹ ♄
05 49	☽ ⚹ ♃
06 12	☽ □ ♇
06 33	☽ ∠ ♅
07 38	☽ Q ☉
08 10	☽ ⚹ ♀
08 27	☽ ⊻ ♆
11 23	♀ ⊻ ♆
18 22	☽ # ☿
18 36	☽ ⊥ ♆
21 44	☽ ⊥ ♅
25 Saturday	
00 01	♀ # ♆
02 27	☽ Q ♃
03 50	☽ ∠ ♄
05 53	☽ ⊻ ♂
06 28	☽ Q ♀
07 56	☽ ⊻ ♅
08 28	♃ ⊻ ♇
09 56	☽ ∠ ♆
16 34	☽ □ ☉
16 41	☽ ⊥ ♂
19 12	☽ ⊥ ♄
20 01	☿ ⊥ ♀
26 Sunday	
04 25	☽ △ ☿
05 31	☽ ⊻ ♄
09 02	☽ ∠ ♂
09 32	☽ △ ♇
09 55	☽ □ ♃
11 51	☽ ⚹ ♆
16 33	☽ □ ♀
22 46	♂ ⚹ ♅
27 Monday	
07 09	☽ ⚼ ☿
08 53	☽ Q ♆
11 31	♂ ∠ ♃
11 51	☽ ⚼ ♇
11 58	☽ ☌ ♅
12 42	☽ ⚹ ♂
16 57	♂ Q ♄
23 32	☽ ± ☿
28 Tuesday	
00 09	☿ ⊼ ♄
01 10	☽ △ ☉
03 52	☽ ± ♇
07 25	♀ ⊥ ♇
10 15	☽ ☌ ♄
10 36	☽ ⊼ ☿
11 14	☽ Q ♂
14 38	☽ ⊼ ♇
15 54	☽ △ ♃
17 03	☽ □ ♆

All ephemeris data is given at 12.00 UT and the Moon's longitude is additionally given for 24.00 UT

MARCH 2034

LONGITUDES

Date	Sidereal time h m s	Sun ☉ ° ' "	Moon ☽ ° ' "	Moon ☽ 24.00 ° ' "	Mercury ☿ ° '	Venus ♀ ° '	Mars ♂ ° '	Jupiter ♃ ° '	Saturn ♄ ° '	Uranus ♅ ° '	Neptune ♆ ° '	Pluto ♇ ° '
01	22 37 28	10 ♓ 58 03	28 ♋ 48 56	05 ♌ 21 25	15 ♒ 18	24 ♓ 31	03 ♉ 52	17 ♓ 56	14 ♋ 34	02 ♋ 03	18 ♈ 24	17 ♒ 03
02	22 41 25	11 58 15	11 ♌ 50 32	18 ♌ 16 20	15 52	25 45	04 34	18 11	14 R 32	02 R 03	18 26	17 05
03	22 45 22	12 58 25	24 ♌ 38 49	00 ♍ 58 02	16 30	27 00	05 15	18 25	14 31	02 03	18 27	17 07
04	22 49 18	13 58 33	07 ♍ 14 04	13 ♍ 26 59	17 13	28 15	05 57	18 40	14 29	02 02	18 29	17 08
05	22 53 15	14 58 39	19 ♍ 36 54	25 ♍ 43 59	17 59	29 ♓ 29	06 39	18 54	14 28	02 02	18 31	17 10
06	22 57 11	15 58 44	01 ♎ 48 24	07 ♎ 50 23	18 49	00 ♈ 44	07 20	19 09	14 27	02 02	18 33	17 12
07	23 01 08	16 58 46	13 ♎ 50 12	19 ♎ 48 09	19 42	01 59	08 02	19 23	14 26	02 02	18 35	17 13
08	23 05 04	17 58 47	25 ♎ 44 35	01 ♏ 39 52	20 38	03 13	08 44	19 38	14 25	02 01	18 37	17 15
09	23 09 01	18 58 47	07 ♏ 34 28	13 ♏ 28 48	21 36	04 28	09 25	19 52	14 24	02 01	18 39	17 17
10	23 12 57	19 58 44	19 ♏ 23 24	25 ♏ 18 46	22 38	05 42	10 07	20 07	14 23	02 D 01	18 41	17 18
11	23 16 54	20 58 40	01 ♐ 15 28	07 ♐ 14 03	23 42	06 57	10 48	20 21	14 23	02 01	18 43	17 20
12	23 20 51	21 58 35	13 ♐ 15 07	19 ♐ 19 15	24 48	08 11	11 30	20 36	14 22	02 02	18 45	17 21
13	23 24 47	22 58 27	25 ♐ 27 02	01 ♑ 39 02	25 57	09 26	12 11	20 50	14 22	02 02	18 48	17 23
14	23 28 44	23 58 18	07 ♑ 55 47	14 ♑ 17 48	27 08	10 40	12 53	21 05	14 22	02 02	18 50	17 24
15	23 32 40	24 58 08	20 ♑ 45 30	27 ♑ 19 16	28 20	11 55	13 34	21 19	14 22	02 02	18 52	17 26
16	23 36 37	25 57 55	03 ♒ 59 21	10 ♒ 45 56	29 ♒ 35	13 09	14 16	21 34	14 D 22	02 02	18 54	17 27
17	23 40 33	26 57 41	17 ♒ 39 01	24 ♒ 38 32	00 ♓ 52	14 23	14 57	21 48	14 22	02 03	18 56	17 29
18	23 44 30	27 57 25	01 ♓ 44 11	08 ♓ 55 35	02 10	15 38	15 38	22 03	14 22	02 03	18 58	17 30
19	23 48 26	28 57 07	16 ♓ 12 09	23 ♓ 33 09	03 30	16 52	16 20	22 17	14 22	02 04	19 00	17 32
20	23 52 23	29 ♓ 56 48	00 ♈ 57 45	08 ♈ 25 01	04 51	18 06	17 01	22 32	14 23	02 04	19 02	17 33
21	23 56 20	00 ♈ 56 26	15 ♈ 53 56	23 ♈ 23 26	06 14	19 21	17 42	22 46	14 23	02 05	19 05	17 35
22	00 00 16	01 56 02	00 ♉ 52 29	08 ♉ 20 05	07 39	20 35	18 23	23 01	14 24	02 05	19 07	17 36
23	00 04 13	02 55 36	15 ♉ 45 19	23 ♉ 07 20	09 05	21 49	19 05	23 15	14 25	02 06	19 09	17 37
24	00 08 09	03 55 08	00 ♊ 25 27	07 ♊ 39 05	10 33	23 03	19 46	23 30	14 26	02 07	19 11	17 39
25	00 12 06	04 54 37	14 ♊ 47 50	21 ♊ 51 23	12 02	24 17	20 27	23 44	14 27	02 08	19 13	17 40
26	00 16 02	05 54 05	28 ♊ 49 35	05 ♋ 42 23	13 33	25 31	21 08	23 58	14 28	02 08	19 16	17 41
27	00 19 59	06 53 30	12 ♋ 29 51	19 ♋ 12 05	15 05	26 45	21 49	24 13	14 29	02 09	19 18	17 43
28	00 23 55	07 52 52	25 ♋ 49 18	02 ♌ 21 45	16 38	27 59	22 30	24 27	14 30	02 10	19 20	17 44
29	00 27 52	08 52 12	08 ♌ 49 42	15 ♌ 13 28	18 13	29 ♈ 13	23 11	24 41	14 32	02 11	19 22	17 45
30	00 31 49	09 51 30	21 ♌ 33 20	27 ♌ 49 38	19 49	00 ♉ 27	23 52	24 56	14 33	02 12	19 25	17 47
31	00 35 45	10 ♈ 50 46	04 ♍ 02 40	10 ♍ 12 43	21 ♓ 27	01 ♉ 41	24 ♉ 33	25 ♓ 10	14 ♋ 35	02 ♋ 13	19 ♈ 27	17 ♒ 48

Date	Moon True ☊ ° '	Moon Mean ☊ ° '	Moon ☽ Latitude ° '
01	03 ♎ 11	04 ♎ 17	04 S 44
02	03 R 08	04 14	04 06
03	03 05	04 11	03 16
04	03 02	04 08	02 18
05	03 01	04 05	01 13
06	03 00	04 01	00 S 07
07	03 D 01	03 58	01 N 00
08	03 02	03 55	02 03
09	03 03	03 52	03 00
10	03 04	03 49	03 50
11	03 05	03 45	04 30
12	03 06	03 42	04 59
13	03 R 06	03 39	05 15
14	03 06	03 36	05 16
15	03 05	03 33	05 03
16	03 04	03 30	04 33
17	03 03	03 26	03 46
18	03 03	03 23	02 45
19	03 02	03 20	01 32
20	03 02	03 17	00 N 12
21	03 D 02	03 14	01 S 11
22	03 02	03 11	02 29
23	03 03	03 07	03 36
24	03 R 02	03 04	04 28
25	03 02	03 01	05 02
26	03 02	02 58	05 17
27	03 D 02	02 55	05 14
28	03 02	02 51	04 53
29	03 03	02 48	04 18
30	03 04	02 45	03 31
31	03 ♎ 04	02 ♎ 42	02 S 34

DECLINATIONS

Date	Sun ☉ ° '	Moon ☽ ° '	Mercury ☿ ° '	Venus ♀ ° '	Mars ♂ ° '	Jupiter ♃ ° '	Saturn ♄ ° '	Uranus ♅ ° '	Neptune ♆ ° '	Pluto ♇ ° '
01	07 S 27	15 N 45	15 S 24	03 S 25	13 N 11	05 S 42	22 N 27	23 N 42	05 N 44	22 S 47
02	07 04	13 17	15 25	02 54	13 25	05 36	22 27	23 42	05 45	22 47
03	06 41	10 13	15 25	02 23	13 40	05 31	22 27	23 42	05 46	22 46
04	06 18	06 43	15 24	01 53	13 55	05 25	22 28	23 42	05 47	22 46
05	05 55	02 N 59	15 21	01 22	14 09	05 19	22 28	23 42	05 47	22 46
06	05 32	00 S 49	15 16	00 51	14 23	05 14	22 28	23 42	05 48	22 45
07	05 08	04 32	15 09	00 S 20	14 38	05 08	22 28	23 42	05 49	22 45
08	04 45	08 02	15 01	00 N 11	14 52	05 02	22 28	23 42	05 50	22 45
09	04 22	11 12	14 52	00 42	15 06	04 57	22 29	23 42	05 50	22 44
10	03 58	13 53	14 41	01 13	15 19	04 51	22 29	23 42	05 51	22 44
11	03 35	16 00	14 28	01 44	15 33	04 45	22 29	23 42	05 52	22 44
12	03 11	17 27	14 14	02 14	15 47	04 40	22 29	23 42	05 53	22 43
13	02 47	18 07	13 58	02 45	16 00	04 34	22 29	23 42	05 54	22 43
14	02 24	17 56	13 42	03 16	16 13	04 28	22 30	23 42	05 54	22 43
15	02 00	16 51	13 23	03 47	16 26	04 23	22 30	23 42	05 55	22 42
16	01 36	14 50	13 04	04 17	16 39	04 17	22 30	23 42	05 56	22 42
17	01 12	11 56	12 43	04 48	16 52	04 11	22 30	23 42	05 57	22 42
18	00 49	08 17	12 20	05 18	17 05	04 05	22 30	23 42	05 58	22 41
19	00 25	04 S 01	11 56	05 49	17 17	04 00	22 30	23 42	05 59	22 41
20	00 S 01	00 N 34	11 31	06 19	17 30	03 54	22 30	23 42	05 59	22 41
21	00 N 22	05 10	11 05	06 49	17 42	03 48	22 30	23 42	06 00	22 41
22	00 46	09 27	10 38	07 19	17 54	03 43	22 30	23 42	06 01	22 40
23	01 10	13 06	10 09	07 49	18 06	03 37	22 30	23 42	06 02	22 40
24	01 33	15 52	09 39	08 18	18 17	03 32	22 30	23 42	06 03	22 40
25	01 57	17 34	09 07	08 48	18 29	03 26	22 30	23 41	06 04	22 40
26	02 21	18 09	08 35	09 17	18 40	03 20	22 30	23 41	06 04	22 40
27	02 44	17 38	08 01	09 46	18 52	03 15	22 30	23 41	06 05	22 39
28	03 08	16 10	07 26	10 15	19 03	03 09	22 30	23 41	06 06	22 39
29	03 31	13 54	06 50	10 43	19 14	03 03	22 30	23 41	06 07	22 39
30	03 54	10 59	06 13	11 11	19 24	02 58	22 30	23 41	06 08	22 39
31	04 N 17	07 N 37	05 S 35	11 N 39	19 N 35	02 S 52	22 N 30	23 N 41	06 N 09	22 S 39

ZODIAC SIGN ENTRIES

Date	h	m	Planets
01	14	10	☽ ♌
03	22	09	☽ ♍
05	21	51	♀ ♈
06	08	25	☽ ♎
08	20	37	☽ ♏
11	09	28	☽ ♐
13	20	49	☽ ♑
16	04	51	☽ ♒
16	19	52	☿ ♓
18	09	05	☽ ♓
20	10	27	☽ ♈
20	13	17	☉ ♈
22	10	36	☽ ♉
24	11	18	☽ ♊
26	14	02	☽ ♋
28	19	39	☽ ♌
30	03	12	♀ ♉
31	04	11	☽ ♍

LATITUDES

Date	Mercury ☿ ° '	Venus ♀ ° '	Mars ♂ ° '	Jupiter ♃ ° '	Saturn ♄ ° '	Uranus ♅ ° '	Neptune ♆ ° '	Pluto ♇ ° '
01	00 N 53	01 S 20	00 N 24	01 S 01	00 S 12	00 N 17	01 S 35	07 S 26
04	00 N 17	01 17	00 26	01 01	00 11	00 17	01 35	07 26
07	00 S 16	01 13	00 28	01 01	00 11	00 17	01 35	07 27
10	00 45	01 09	00 30	01 01	00 11	00 17	01 35	07 27
13	01 11	01 04	00 32	01 01	00 10	00 17	01 35	07 28
16	01 33	00 59	00 34	01 01	00 10	00 17	01 35	07 28
19	01 51	00 53	00 36	01 01	00 09	00 17	01 35	07 29
22	02 05	00 47	00 37	01 01	00 09	00 17	01 35	07 29
25	02 15	00 40	00 39	01 02	00 09	00 17	01 35	07 30
28	02 21	00 33	00 41	01 02	00 08	00 17	01 35	07 31
31	02 S 23	00 S 26	00 N 42	01 S 02	00 S 08	00 N 17	01 S 35	07 S 31

DATA

Julian Date	2464023
Delta T	+75 seconds
Ayanamsa	24° 20' 06"
Synetic vernal point	04° ♓ 46' 53"
True obliquity of ecliptic	23° 25' 57"

MOON'S PHASES, APSIDES AND POSITIONS ☽

Date	h	m	Phase	Longitude ° '	Eclipse Indicator
05	02	10	○	14 ♍ 34	
13	06	44	☾	22 ♐ 45	
20	10	15	●	29 ♓ 52	Total
27	01	18	☽	06 ♋ 27	

Day	h	m		
09	15	56	Apogee	
21	18	09	Perigee	
06	06	49	OS	
13	19	07	Max dec	18° S 09'
20	09	07	ON	
26	12	34	Max dec	18° N 09'

LONGITUDES

Date	Chiron ⚷ ° '	Ceres ⚳ ° '	Pallas ⚴ ° '	Juno ⚵ ° '	Vesta ⚶ ° '	Black Moon Lilith ⚸ ° '
01	26 ♉ 26	21 ♒ 42	24 ♑ 34	15 ♐ 26	17 ♈ 25	03 ♏ 21
11	26 ♉ 46	25 ♒ 31	27 ♑ 53	16 ♐ 55	21 ♈ 45	04 ♏ 27
21	27 ♉ 12	29 ♒ 16	00 ♒ 58	17 ♐ 58	26 ♈ 09	05 ♏ 34
31	27 ♉ 43	02 ♓ 56	03 ♒ 47	18 ♐ 32	00 ♉ 35	06 ♏ 41

All ephemeris data is given at 12.00 UT and the Moon's longitude is additionally given for 24.00 UT

ASPECTARIAN

01 Wednesday
03 01 ♃ # ♆
03 20 ☽ △ ♀
06 23 ☽ bQ ☉
11 42 ♅ bQ ♇
15 58 ☽ # ☿
17 56 ☽ ⊻ ♅
19 42 ☽ bQ ♃
21 46 ☽ □ ♂

02 Thursday
00 13 ☽ ± ☉
04 58 ☽ ⊥ ♅
09 46 ☽ bQ ♀
10 57 ☽ ∥ ♂
12 16 ☽ ⊼ ☉
12 38 ☽ ± ♃
17 00 ☽ ⊻ ♄
19 53 ☽ ☍ ☿
21 42 ☽ ∠ ♅
21 48 ☽ ☍ ♇
23 15 ☉ ⊥ ♆

03 Friday
00 03 ☽ ⊼ ♃
00 19 ☽ △ ♆
04 13 ☽ ⊥ ♄
04 23 ☽ ± ♀
16 10 ♃ ⊻ ♆
16 57 ☽ ⊼ ♀
21 12 ☽ ∠ ♄

04 Saturday
02 03 ☽ ⚹ ♅
04 48 ☽ bQ ♆
06 18 ☿ bQ ♅
09 23 ☽ △ ♂
09 35 ☿ ☌ ♇
15 04 ☽ # ☉
18 09 ☽ ∥ ♆
20 43 ☽ # ♃
22 10 ☽ ± ♆
23 59 ☉ △ ♄

05 Sunday
01 08 ☽ Q ♅
01 59 ☽ ⚹ ♄
02 10 ☽ ☍ ☉
07 13 ☽ ⊼ ♇
08 36 ☽ ⊼ ☿
09 52 ☽ ⊼ ♆
10 35 ☽ ☍ ♃
16 13 ☽ bQ ♂
18 59 ☽ ± ♇
19 41 ☉ # ♆
21 10 ☽ ± ☿
23 49 ☽ # ♀

06 Monday
01 25 ☽ Q ♄
04 28 ☿ ⚹ ♆
09 38 ☽ ☍ ♀
11 01 ☽ ± ♂
12 10 ☽ ∥ ♀
12 27 ☽ □ ♅
12 46 ☽ bQ ♇
16 17 ☽ bQ ☿
23 41 ☽ ⊼ ♂

07 Tuesday
00 49 ☿ ⊻ ♃
12 38 ☉ ∥ ♃
12 57 ♀ □ ♅
13 11 ☽ □ ♄
14 33 ♀ ⊥ ♂
15 36 ☽ ∥ ☉
15 50 ☽ ∥ ♃
16 51 ♀ ∠ ♇
18 01 ☉ ⊻ ♇
18 49 ☽ △ ♇
18 54 ☽ ⊼ ☉
20 33 ☽ # ♆
21 35 ☽ ☍ ♆
23 24 ☽ ⊼ ♃

08 Wednesday
00 46 ☽ △ ☿
06 44 ☿ ± ♄
08 06 ☽ ± ☉
11 46 ☽ ± ♃
21 54 ☿ # ♂

09 Thursday
00 44 ☽ △ ♅
03 56 ☉ ⊻ ♆
04 01 ☽ bQ ☉
04 56 ☽ ⊼ ♀
06 24 ☽ bQ ♃
15 59 ☽ ☍ ♂
18 34 ☽ ± ♀

10 Friday
01 52 ☽ △ ♄
06 48 ♅ St D
07 12 ☽ bQ ♅
07 45 ☽ □ ♇
10 34 ☽ ⊼ ♆
13 18 ☽ △ ☉
13 30 ☽ △ ♃
14 59 ☽ bQ ♀
16 17 ☉ ☌ ♃
19 13 ☽ □ ☿
19 32 ☽ ∥ ☿
22 46 ☽ ± ♆

11 Saturday
01 26 ☽ ± ♅
05 31 ☽ # ♂
08 13 ☽ bQ ♄
13 33 ☽ ⊼ ♅
16 58 ☽ bQ ♆
20 12 ☽ Q ♇

12 Sunday
00 45 ☽ △ ♀
02 17 ☽ ± ♄
08 18 ☽ ⊼ ♂
11 01 ☽ Q ☿
14 13 ☽ ⊼ ♄
20 09 ☽ ⚹ ♇
20 55 ☽ ± ♂
22 55 ☽ △ ♆

13 Monday
02 48 ☽ □ ♃
06 44 ☽ □ ☉
12 52 ☉ # ♀
13 04 ☽ ⚹ ☿
15 35 ☽ bQ ♂
22 03 ☉ ⊥ ♇

14 Tuesday
00 44 ☽ ☍ ♅
01 26 ☽ ∠ ♇
14 14 ☽ Q ♃
17 45 ☽ □ ♀
18 35 ☽ ⊥ ♇
20 17 ☽ Q ☉
20 45 ☽ ∠ ☿
21 53 ☽ △ ♂

15 Wednesday
00 07 ☽ ☍ ♄
05 50 ☽ ⊻ ♇
08 29 ☽ □ ♆
13 04 ☽ ⚹ ♃
15 13 ☽ ⊥ ☿
17 13 ☽ # ♂
20 21 ☽ ⚹ ☉

16 Thursday
02 30 ♄ St D
03 16 ☽ ⊻ ☿
06 24 ☽ Q ♀
07 44 ☿ bQ ♄
08 31 ☽ ⊼ ♅
11 37 ♀ # ♃
15 29 ♂ ⚹ ♄
16 40 ☽ ∠ ♃
17 11 ☽ Q ♆
19 12 ☽ ± ♅

17 Friday
01 19 ☽ ∠ ☉
05 30 ☽ ∥ ☿
05 47 ☽ ⚹ ♀
06 18 ☽ ⊼ ♄
07 04 ☽ □ ♂
08 45 ☽ ⊥ ♃
10 57 ☽ bQ ♅
11 28 ♀ □ ♄
11 42 ☽ ☌ ♇
14 13 ☽ ⚹ ♆
16 41 ☽ ± ♄
18 09 ☽ ⊥ ☉
19 17 ☽ ⊻ ♃

18 Saturday
05 09 ☽ ⊻ ☉
08 01 ☽ bQ ♄
09 58 ☽ ∠ ♀
09 59 ☿ △ ♅
12 27 ♀ ⊻ ♂
12 32 ☽ △ ♅
12 47 ☽ ☌ ☿
15 21 ☽ Q ♂
15 45 ☽ ∠ ♆

19 Sunday
01 21 ☽ # ♆
02 24 ☽ ⊥ ♀
03 15 ☽ # ♀
06 44 ☽ ⊥ ♆
09 00 ☽ △ ♄
12 09 ☽ ∥ ♃
12 13 ☽ ⚹ ♂
13 11 ☽ ⊻ ♀
14 11 ☽ ⊻ ♇
16 36 ☽ ⊻ ♆
19 59 ♀ ∥ ♆
21 17 ☿ ∠ ♆
22 07 ☽ ☌ ♃
23 59 ☽ ⊥ ♇

20 Monday
01 06 ♀ ⚹ ♇
08 43 ☽ ∥ ☉
09 26 ☽ # ☉
10 15 ☽ ☌ ☉
13 47 ☽ ∠ ♂
13 47 ☽ □ ♅
13 56 ♂ ∠ ♅
14 34 ☽ ∠ ♇
17 35 ☿ Q ♂
18 55 ☽ ⊻ ☿

21 Tuesday
04 57 ☽ ⊥ ♂
04 59 ☽ # ♃
05 32 ☽ ⊥ ☿
06 41 ♀ ☌ ♆
07 28 ♂ □ ♇
09 35 ☽ □ ♄
14 42 ☽ ⚹ ♇
15 02 ☽ ⊻ ♂
16 32 ☽ ∥ ♆
17 06 ☽ ☌ ♆
18 01 ☽ ☌ ♀
18 42 ☽ Q ♅
21 26 ☽ ∠ ☿
22 07 ☽ ∥ ♀
23 11 ☽ ⊻ ♃

22 Wednesday
02 26 ♀ Q ♅
08 58 ☽ ⊥ ♃
09 57 ☽ Q ♇
13 49 ☽ ⊻ ☉
13 57 ☽ ⚹ ♅
14 27 ☽ Q ♄
15 49 ☉ □ ♅
18 25 ☽ # ☿
23 40 ☽ ∠ ♃

23 Thursday
00 03 ☽ ⚹ ☿
00 10 ☽ ⊥ ☉
04 30 ☉ ∠ ♇
09 49 ☽ ⚹ ♄
14 11 ☽ ∠ ♅
14 44 ♂ ⊻ ♆
15 02 ☽ □ ♇
15 47 ☽ ∠ ☉
17 32 ☽ ⊻ ♆
17 40 ☽ ☌ ♂
21 38 ☽ Q ☿
22 46 ☽ ⊻ ♀

24 Friday
00 25 ☽ ⚹ ♃
03 22 ☽ ⊥ ♆
04 54 ☽ ⊥ ♅
09 31 ☽ ⊥ ♀
10 21 ☽ ∠ ♄
14 48 ☽ ⊻ ♅
18 13 ☽ ⚹ ☉
18 15 ☽ ∠ ♆
20 33 ☽ Q ♃
22 41 ♀ ⊻ ♃

25 Saturday
01 19 ☽ ⊥ ♄
01 52 ☽ ∠ ♀
04 48 ♃ ⊥ ♇
06 48 ☽ □ ☿
11 24 ☽ ⊻ ♄
15 51 ☽ Q ☉
16 52 ☽ △ ♇
19 32 ☽ ⚹ ♆
19 46 ☿ # ♀
22 05 ☽ ⊻ ♂

26 Sunday
03 29 ☽ □ ♃
05 44 ☽ ⚹ ♀
07 23 ☿ ⊥ ♆
08 55 ☽ ⊥ ♂
16 14 ☽ Q ♆
17 46 ☽ ☌ ♅
18 44 ☽ bQ ♇

27 Monday
01 18 ☽ □ ☉
01 25 ☽ ∠ ♂
02 36 ☿ △ ♄
04 42 ☽ Q ♀
05 58 ☉ ∠ ♂
10 36 ☽ ± ♇
15 33 ☽ ☌ ♄
17 12 ☽ △ ☿
21 20 ☽ ⊼ ♇

28 Tuesday
00 12 ☽ □ ♆
05 38 ☽ ⚹ ♂
09 27 ☽ △ ♃
13 13 ☉ # ♃
16 22 ☽ □ ♀
23 40 ☽ ⊻ ♅

29 Wednesday
00 07 ☽ bQ ☿
04 50 ☽ Q ♂
04 58 ☿ ⊻ ♇
10 48 ☽ ⊥ ♅
12 05 ☽ △ ☉
13 38 ☽ bQ ♃
19 14 ☽ ± ☿
22 36 ♀ Q ♇
22 43 ☽ ⊻ ♄

30 Thursday
03 44 ☽ ∠ ♅
04 49 ☽ ☍ ♇
05 47 ☿ ⊻ ♆
06 55 ☽ ± ♃
07 54 ☽ △ ♆
08 13 ☽ ⊼ ☿
10 05 ☽ ⊥ ♄
10 39 ☽ ∥ ♀
15 22 ☿ # ♆
16 39 ☽ □ ♂
18 34 ☽ ⊼ ♃
18 51 ☽ bQ ☉
23 30 ♀ ⊥ ♃

31 Friday
03 21 ☽ ∠ ♄
06 56 ☽ △ ♀
08 28 ☽ ⚹ ♅
12 47 ☽ bQ ♆
13 41 ☽ ± ☉
21 52 ☽ ∥ ♆
22 43 ♀ ⚹ ♅

APRIL 2034

LONGITUDES

Date	Sidereal time h m s	Sun ☉ ° ' "	Moon ☽ ° ' "	Moon ☽ 24.00 ° ' "	Mercury ☿ ° '	Venus ♀ ° '	Mars ♂ ° '	Jupiter ♃ ° '	Saturn ♄ ° '	Uranus ♅ ° '	Neptune ♆ ° '	Pluto ♇ ° '
01	00 39 42	11 ♈ 49 59	16 ♍ 20 06	22 ♍ 25 04	23 ♓ 06	02 ♉ 55	25 ♉ 14	25 ♓ 24	14 ♋ 37	02 ♋ 15	19 ♈ 29	17 ♒ 49
02	00 43 38	12 49 10	28 ♍ 27 53	04 ♎ 28 48	24 46	04 09	25 54	25 38	14 39	02 16	19 31	17 50
03	00 47 35	13 48 19	10 ♎ 28 05	16 ♎ 25 57	26 28	05 22	26 35	25 53	14 41	02 17	19 34	17 51
04	00 51 31	14 47 26	22 ♎ 22 40	28 ♎ 18 28	28 11	06 36	27 16	26 07	14 43	02 18	19 36	17 52
05	00 55 28	15 46 31	04 ♏ 13 38	10 ♏ 08 24	29 ♓ 56	07 50	27 57	26 21	14 45	02 20	19 38	17 54
06	00 59 24	16 45 34	16 ♏ 03 06	21 ♏ 58 02	01 ♈ 42	09 03	28 37	26 35	14 47	02 21	19 40	17 55
07	01 03 21	17 44 36	27 ♏ 53 31	03 ♐ 49 56	03 29	10 17	29 18	26 49	14 49	02 23	19 43	17 56
08	01 07 18	18 43 35	09 ♐ 47 40	15 ♐ 47 08	05 18	11 31	29 ♉ 59	27 03	14 52	02 24	19 45	17 57
09	01 11 14	19 42 33	21 ♐ 48 46	27 ♐ 53 04	07 09	12 44	00 ♊ 39	27 17	14 55	02 26	19 47	17 58
10	01 15 11	20 41 29	04 ♑ 00 28	10 ♑ 11 31	09 00	13 58	01 20	27 31	14 57	02 27	19 49	17 59
11	01 19 07	21 40 23	16 ♑ 26 41	22 ♑ 46 30	10 54	15 11	02 00	27 45	15 00	02 29	19 52	18 00
12	01 23 04	22 39 15	29 ♑ 11 26	05 ♒ 41 56	12 49	16 24	02 41	27 59	15 03	02 31	19 54	18 01
13	01 27 00	23 38 06	12 ♒ 18 26	19 ♒ 01 15	14 45	17 38	03 21	28 12	15 06	02 32	19 56	18 02
14	01 30 57	24 36 54	25 ♒ 50 39	02 ♓ 46 47	16 43	18 51	04 02	28 26	15 09	02 34	19 58	18 03
15	01 34 53	25 35 42	09 ♓ 49 38	16 ♓ 59 05	18 42	20 04	04 42	28 40	15 12	02 36	20 01	18 03
16	01 38 50	26 34 27	24 ♓ 14 48	01 ♈ 36 18	20 43	21 18	05 23	28 54	15 15	02 38	20 03	18 04
17	01 42 47	27 33 10	09 ♈ 02 54	16 ♈ 33 44	22 45	22 31	06 03	29 07	15 19	02 40	20 05	18 05
18	01 46 43	28 31 52	24 ♈ 07 47	01 ♉ 43 53	24 48	23 44	06 43	29 21	15 22	02 42	20 07	18 06
19	01 50 40	29 ♈ 30 32	09 ♉ 20 49	16 ♉ 57 16	26 53	24 57	07 24	29 34	15 26	02 44	20 10	18 07
20	01 54 36	00 ♉ 29 10	24 ♉ 31 59	02 ♊ 03 46	28 ♈ 58	26 10	08 04	29 ♓ 48	15 30	02 46	20 12	18 08
21	01 58 33	01 27 46	09 ♊ 31 31	16 ♊ 54 18	01 ♉ 05	27 23	08 44	00 ♈ 01	15 33	02 48	20 14	18 08
22	02 02 29	02 26 20	24 ♊ 11 21	01 ♋ 22 06	03 12	28 36	09 24	00 15	15 37	02 50	20 16	18 09
23	02 06 26	03 24 51	08 ♋ 26 12	15 ♋ 23 26	05 20	29 ♉ 49	10 04	00 28	15 41	02 52	20 19	18 10
24	02 10 22	04 23 21	22 ♋ 13 49	28 ♋ 57 28	07 28	01 ♊ 02	10 45	00 41	15 45	02 54	20 21	18 10
25	02 14 19	05 21 48	05 ♌ 34 39	12 ♌ 05 44	09 36	02 15	11 25	00 55	15 49	02 57	20 23	18 11
26	02 18 16	06 20 13	18 ♌ 31 07	24 ♌ 51 18	11 44	03 28	12 05	01 08	15 54	02 59	20 25	18 11
27	02 22 12	07 18 36	01 ♍ 06 47	07 ♍ 18 06	13 52	04 41	12 45	01 21	15 58	03 01	20 28	18 12
28	02 26 09	08 16 57	13 ♍ 25 46	19 ♍ 30 18	15 59	05 53	13 25	01 34	16 02	03 04	20 30	18 13
29	02 30 05	09 15 15	25 ♍ 32 11	01 ♎ 31 53	18 06	07 06	14 05	01 47	16 07	03 06	20 32	18 13
30	02 34 02	10 ♉ 13 32	07 ♎ 29 50	13 ♎ 26 27	20 ♉ 10	08 ♊ 19	14 ♊ 45	02 ♈ 00	16 ♋ 11	03 ♋ 08	20 ♈ 34	18 ♒ 14

Moon nodes and latitude

Date	Moon True ☊ ° '	Moon Mean ☊ ° '	Moon ☽ Latitude ° '
01	03 ♎ 05	02 ♎ 39	01 S 32
02	03 D 05	02 36	00 S 26
03	03 R 05	02 32	00 N 41
04	03 05	02 29	01 45
05	03 03	02 26	02 44
06	03 02	02 23	03 36
07	02 59	02 20	04 19
08	02 57	02 17	04 51
09	02 55	02 13	05 10
10	02 53	02 10	05 16
11	02 52	02 07	05 07
12	02 D 52	02 04	04 43
13	02 53	02 01	04 04
14	02 55	01 57	03 11
15	02 56	01 54	02 04
16	02 57	01 51	00 N 48
17	02 R 57	01 48	00 S 34
18	02 56	01 45	01 54
19	02 54	01 42	03 07
20	02 50	01 38	04 06
21	02 46	01 35	04 48
22	02 42	01 32	05 10
23	02 39	01 29	05 12
24	02 37	01 26	04 55
25	02 D 37	01 23	04 23
26	02 37	01 19	03 38
27	02 39	01 16	02 44
28	02 40	01 13	01 43
29	02 41	01 10	00 39
30	02 ♎ 42	01 ♎ 07	00 N 26

DECLINATIONS

Date	Sun ☉ ° '	Moon ☽ ° '	Mercury ☿ ° '	Venus ♀ ° '	Mars ♂ ° '	Jupiter ♃ ° '	Saturn ♄ ° '	Uranus ♅ ° '	Neptune ♆ ° '	Pluto ♇ ° '
01	04 N 41	03 N 59	04 S 55	12 N 07	19 N 45	02 S 47	22 N 30	23 N 41	06 N 09	22 S 38
02	05 04	00 N 13	04 15	12 34	19 56	02 41	22 30	23 41	06 10	22 38
03	05 27	03 S 31	03 33	13 02	20 06	02 36	22 30	23 41	06 11	22 38
04	05 50	07 05	02 51	13 28	20 16	02 30	22 30	23 41	06 12	22 38
05	06 12	10 21	02 07	13 55	20 25	02 24	22 30	23 41	06 13	22 38
06	06 35	13 11	01 22	14 21	20 35	02 19	22 29	23 41	06 14	22 38
07	06 58	15 29	00 S 37	14 47	20 44	02 13	22 29	23 41	06 15	22 38
08	07 20	17 07	00 N 10	15 12	20 53	02 08	22 29	23 41	06 15	22 37
09	07 42	18 01	00 58	15 37	21 02	02 02	22 29	23 41	06 16	22 37
10	08 05	18 06	01 46	16 02	21 11	01 57	22 29	23 41	06 17	22 37
11	08 27	17 20	02 35	16 26	21 20	01 52	22 29	23 41	06 18	22 37
12	08 49	15 41	03 25	16 50	21 28	01 46	22 28	23 41	06 19	22 37
13	09 10	13 11	04 16	17 13	21 37	01 41	22 28	23 41	06 20	22 37
14	09 32	09 54	05 08	17 36	21 45	01 35	22 28	23 41	06 21	22 37
15	09 54	05 58	06 00	17 59	21 53	01 30	22 28	23 41	06 21	22 37
16	10 15	01 S 33	06 52	18 21	22 00	01 25	22 27	23 41	06 22	22 37
17	10 36	03 N 04	07 45	18 43	22 08	01 19	22 27	23 41	06 23	22 37
18	10 57	07 35	08 39	19 04	22 15	01 14	22 27	23 41	06 24	22 37
19	11 18	11 39	09 32	19 25	22 22	01 09	22 26	23 41	06 25	22 37
20	11 38	14 55	10 26	19 45	22 29	01 04	22 26	23 41	06 25	22 37
21	11 59	17 07	11 19	20 04	22 36	00 58	22 26	23 41	06 26	22 37
22	12 19	18 09	12 13	20 23	22 43	00 53	22 25	23 41	06 27	22 37
23	12 39	17 59	13 05	20 42	22 49	00 48	22 25	23 40	06 28	22 37
24	12 59	16 44	13 58	21 00	22 55	00 43	22 25	23 40	06 29	22 37
25	13 18	14 37	14 49	21 17	23 01	00 38	22 24	23 40	06 30	22 37
26	13 38	11 48	15 39	21 34	23 07	00 32	22 24	23 40	06 30	22 37
27	13 57	08 31	16 28	21 51	23 13	00 27	22 24	23 40	06 31	22 37
28	14 16	04 55	17 16	22 06	23 18	00 22	22 23	23 40	06 32	22 37
29	14 34	01 11	18 02	22 21	23 23	00 17	22 23	23 40	06 33	22 37
30	14 N 53	02 S 34	18 N 46	22 N 36	23 N 28	00 S 12	22 N 22	23 N 40	06 N 34	22 S 37

ZODIAC SIGN ENTRIES

Date	h m	Planets
02	15 03	☽ ♎
05	03 26	☽ ♏
05	13 01	☿ ♈
07	16 16	☽ ♐
08	12 49	♂ ♊
10	04 10	☽ ♑
12	13 30	☽ ♒
14	19 13	☽ ♓
16	21 24	☽ ♈
18	21 16	☽ ♉
20	00 04	☉ ♉
20	20 42	☽ ♊
20	23 46	☿ ♉
21	09 39	♃ ♈
22	21 42	☽ ♋
23	15 30	♀ ♊
25	01 53	☽ ♌
27	09 51	☽ ♍
29	20 56	☽ ♎

LATITUDES

Date	Mercury ☿ ° '	Venus ♀ ° '	Mars ♂ ° '	Jupiter ♃ ° '	Saturn ♄ ° '	Uranus ♅ ° '	Neptune ♆ ° '	Pluto ♇ ° '
01	02 S 23	00 S 23	00 N 43	01 S 02	00 S 08	00 N 17	01 S 35	07 S 32
04	02 19	00 15	00 44	01 02	00 08	00 17	01 35	07 32
07	02 10	00 S 08	00 46	01 03	00 07	00 16	01 35	07 33
10	01 58	00 00	00 47	01 03	00 07	00 16	01 35	07 34
13	01 40	00 N 09	00 48	01 03	00 07	00 16	01 35	07 35
16	01 18	00 17	00 50	01 04	00 06	00 16	01 35	07 35
19	00 53	00 25	00 51	01 04	00 06	00 16	01 35	07 36
22	00 S 23	00 34	00 52	01 04	00 06	00 16	01 35	07 37
25	00 N 08	00 42	00 53	01 05	00 05	00 16	01 35	07 38
28	00 40	00 50	00 54	01 05	00 05	00 16	01 35	07 38
31	01 N 11	00 N 58	00 N 55	01 S 05	00 S 05	00 N 16	01 S 35	07 S 39

DATA

Julian Date	2464054
Delta T	+75 seconds
Ayanamsa	24° 20' 08"
Synetic vernal point	04° ♓ 46' 50"
True obliquity of ecliptic	23° 25' 57"

LONGITUDES

Date	Chiron ⚷ ° '	Ceres ⚳ ° '	Pallas ⚴ ° '	Juno ⚵ ° '	Vesta ⚶ ° '	Black Moon Lilith ⚸ ° '
01	27 ♉ 46	03 ♓ 18	04 ♒ 03	18 ♐ 34	01 ♉ 01	06 ♏ 47
11	28 ♉ 21	06 ♓ 50	06 ♒ 32	18 ♐ 34	05 ♉ 28	07 ♏ 54
21	29 ♉ 00	10 ♓ 15	08 ♒ 40	18 ♐ 01	09 ♉ 55	09 ♏ 01
31	29 ♉ 41	13 ♓ 30	10 ♒ 23	16 ♐ 56	14 ♉ 22	10 ♏ 08

MOON'S PHASES, APSIDES AND POSITIONS ☽

Date	h m	Phase	Longitude ° '	Eclipse Indicator
03	19 19	○	14 ♎ 06	
11	22 45	☾	22 ♑ 07	
18	19 26	●	28 ♈ 50	
25	11 35	☽	05 ♌ 21	

Day	h m		
06	03 30	Apogee	
19	03 58	Perigee	
02	13 23	0S	
10	02 33	Max dec	18° S 10'
16	20 06	0N	
22	20 28	Max dec	18° N 13'
29	19 30	0S	

ASPECTARIAN

h m	Aspects
01 Saturday	
02 24	☽ ⊼ ☉
04 34	☽ # ☿
05 57	♀ Q ♄
06 23	☽ ± ♆
07 53	☽ Q ♅
07 56	☽ ∥ ☉
08 36	☽ ⚹ ♄
14 55	☽ ⊼ ♇
15 27	☽ ₽ ♀
17 38	☉ # ☿
18 13	☽ ⊼ ♆
19 55	☽ # ♃
21 34	♂ ⚹ ♃
21 38	♂ ⊥ ♆
22 33	☿ ⊥ ♇
02 Sunday	
02 48	☽ ± ♇
03 28	☽ ☍ ☿
06 16	☽ ☍ ♃
06 37	☽ △ ♂
08 22	☽ Q ♄
11 17	☽ ± ♀
19 35	☽ □ ♅
20 44	☽ ₽ ♇
03 Monday	
00 37	☽ ⊼ ♀
01 01	♂ ⊥ ♅
02 25	☿ ☌ ♃
06 08	☽ ∥ ♃
12 13	☽ ∥ ☿
14 23	☽ ₽ ♂
14 54	☿ ⚹ ♂
19 19	☽ ☍ ☉
20 29	☽ □ ♄
04 Tuesday	
02 19	☽ # ☉
02 53	☽ △ ♇
05 53	☽ # ♆
06 22	☽ ☍ ♆
09 37	☽ ± ♂
09 57	☉ □ ♄
19 42	☽ ⊼ ♃
22 29	☽ ⊼ ♂
05 Wednesday	
01 07	☿ ∥ ♃
01 46	☽ ⊼ ☿
08 06	☽ ± ♃
08 08	☽ △ ♅
12 33	☉ ∥ ♆
16 03	☽ ± ☿
20 09	☽ ☍ ♀
06 Thursday	
02 44	☽ ₽ ♃
09 25	☽ △ ♄
13 32	☽ ₽ ☿
13 34	☽ ⊼ ☉
14 39	☽ ₽ ♅
15 47	☽ □ ♇
19 22	☽ ⊼ ♆
20 59	☿ □ ♅
07 Friday	
02 05	☽ # ♀
02 50	☽ ± ☉
04 30	☿ ∠ ♇
07 34	☽ ± ♆
08 56	☽ ± ♅
09 47	☽ △ ♃
15 01	☽ ☍ ♂
15 55	☽ ₽ ♄
16 36	☉ ⚹ ♇
21 05	☽ ⊼ ♅
22 41	☽ ₽ ☉
08 Saturday	
01 20	☽ △ ☿
01 49	☽ ₽ ♆
04 15	☽ Q ♇
07 47	♂ ∠ ♄
10 08	☽ ± ♄
15 50	☽ ⊼ ♀
20 13	☿ ⊥ ♀
22 12	☽ ⊼ ♄
09 Sunday	
01 02	♀ ∠ ♃
04 20	☽ ⚹ ♇
05 11	☽ ± ♀
07 27	☽ △ ☉
07 58	☽ △ ♆
13 56	☉ ☌ ♆
23 01	☽ □ ♃
10 Monday	
01 01	☽ ₽ ♀
06 01	☉ Q ♅
06 28	☽ ⊼ ♂
08 57	☽ ☍ ♅
09 59	☽ ∠ ♇
16 52	☿ # ♃
18 50	☽ ± ♂
23 27	☽ □ ☿
11 Tuesday	
03 28	☽ ⊥ ♇
08 15	♀ ⚹ ♄
09 14	☽ ☍ ♄
09 20	☽ △ ♀
10 38	☽ Q ♃
13 08	☽ ₽ ♂
14 58	☽ ⊻ ♇
18 31	☽ □ ♆
22 45	☽ □ ☉
23 50	☽ # ♀
12 Wednesday	
05 38	♂ ⊻ ♅
09 42	☽ ⚹ ♃
15 32	☽ Q ☿
18 09	☽ ⊼ ♅
18 49	☽ △ ♂
13 Thursday	
04 04	☽ Q ♆
05 10	☽ ± ♅
10 08	♀ ∠ ♅
10 41	☽ Q ☉
13 39	☽ ∠ ♃
16 22	☿ □ ♄
17 02	☽ ⊼ ♄
17 09	☽ ⚹ ☿
19 54	♀ □ ♇
21 23	☽ ₽ ♅
22 15	☽ ☌ ♇
22 29	☽ □ ♀
14 Friday	
01 40	☽ ⚹ ♆
03 44	☽ ± ♄
05 56	☽ ⊥ ♃
09 41	☽ ⚹ ☉
14 13	☽ # ☉
16 35	☽ ⊻ ♃
19 30	☽ ₽ ♄
23 40	☽ △ ♅
23 51	☽ ∠ ☿
15 Saturday	
02 51	☽ □ ♂
03 48	☽ ∠ ♆
04 11	☿ ⚹ ♇
08 45	☽ Q ♀
09 46	☽ # ♆
10 45	♀ ⊻ ♆
11 51	☽ # ☿
13 23	☽ ∠ ☉
17 38	☽ ⊥ ☿
19 03	☽ ⊥ ♆
21 04	☽ △ ♄
22 02	☿ ∥ ♆
23 42	♂ ∠ ♆
16 Sunday	
01 48	☽ ⊻ ♇
03 57	☿ ☌ ♆
05 04	☽ ⊻ ♆
05 15	☽ ⊻ ☿
05 31	☽ ⊥ ☉
05 58	☿ ∠ ♂
06 42	☽ ⚹ ♀
10 30	☽ Q ♂
10 58	☿ Q ♅
11 43	☽ ⊥ ♇
12 45	☽ ∥ ♃
16 05	☽ ⊻ ☉
19 43	☽ ☌ ♃
17 Monday	
01 41	☽ □ ♅
02 23	☽ ∠ ♇
03 07	☽ # ♃
05 13	☿ ⊻ ♀
06 57	☽ ⚹ ♂
09 19	☽ ∠ ♀
22 03	☽ □ ♄
18 Tuesday	
00 58	☽ ⊥ ♀
02 26	☽ ⚹ ♇
05 30	☽ ∥ ♆
05 39	☽ ☌ ♆
06 33	☽ Q ♅
08 01	☽ ∠ ♂
11 19	☽ ⊻ ♀
13 14	☽ ☌ ☿
19 26	☽ ☌ ☉
19 31	☽ ∥ ☿
20 22	☽ ⊻ ♃
21 26	☽ Q ♇
22 53	☽ ⊥ ♂
19 Wednesday	
01 33	☽ ⚹ ♅
02 39	☽ Q ♄
05 58	☽ ⊥ ♃
08 47	☽ ⊻ ♂
09 32	☽ ∥ ☉
14 02	☉ ⊻ ♃
20 22	☽ ∠ ♃
21 38	☽ ⚹ ♄
20 Thursday	
01 15	☽ ∠ ♅
01 26	♂ ∥ ♄
01 51	☽ □ ♇
03 02	☉ Q ♇
05 07	☽ ⊻ ♆
12 34	♀ ⊥ ♆
14 39	☽ ⊥ ♆
14 50	☽ ☌ ♀
15 33	☽ ⊥ ♅
20 12	☽ ⊻ ☿
20 30	☽ ⚹ ♃
21 32	☽ ∠ ♄
22 08	☽ ⊻ ☉
22 36	☿ ⊻ ♃
23 57	♀ ⊥ ♅
21 Friday	
01 09	☽ ⊻ ♅
01 17	☿ Q ♇
05 04	☽ ∠ ♆
07 24	☽ ⊥ ☿
08 26	☽ ⊥ ☉
10 40	☽ ☌ ♂
12 03	☽ ⊥ ♄
15 08	♂ # ♇
16 06	☽ Q ♃
20 08	☉ ☌ ☿
21 50	☽ ⊻ ♄
22 Saturday	
00 05	☽ ∠ ☉
00 26	☽ ∠ ☿
02 02	☽ △ ♇
05 31	☽ ⚹ ♆
07 48	☿ ⚹ ♅
15 25	☿ ⊥ ♂
16 39	☉ ∥ ☿
16 56	☿ Q ♄
20 03	☽ ⊻ ♀
20 35	♂ ⊥ ♄
22 02	☉ ⚹ ♅
22 16	☽ □ ♃
23 Sunday	
01 34	☽ Q ♆
02 30	☽ ☌ ♅
02 49	☽ ⚹ ☉
03 01	☽ ₽ ♇
05 45	☽ ⚹ ☿
07 07	☽ ⊥ ♀
14 57	☽ ⊻ ♂
18 24	☽ ± ♇
19 11	☉ Q ♄
24 Monday	
00 05	☽ ∠ ♀
00 35	☽ ☌ ♄
00 57	☽ Q ☉
01 52	☽ ⊥ ♂
02 17	☿ ⊥ ♃
03 34	♀ ⚹ ♃
04 51	☽ ⊼ ♇
06 03	♀ ∠ ♄
06 13	☽ Q ☿
08 40	☽ □ ♆
18 34	☽ ∠ ♂
25 Tuesday	
03 22	☽ △ ♃
05 20	☽ ⚹ ♀
07 11	☽ ⊻ ♅
10 33	☽ ∥ ☿
11 35	☽ □ ☉
15 45	☉ ⊥ ♂
18 11	☽ ⊥ ♅
20 51	☽ □ ☿
22 35	☽ ∥ ☉
23 19	☽ ⚹ ♂
26 Wednesday	
02 06	♀ ⊻ ♅
05 41	☽ Q ♀
07 03	☽ ⊻ ♄
07 26	☽ ₽ ♃
10 59	☽ ∠ ♅
11 23	☽ ☍ ♇
15 36	☽ △ ♆
17 32	☿ ⊻ ♂
18 24	☽ ⊥ ♄
23 07	☽ Q ♂
27 Thursday	
00 45	☽ ± ♃
11 43	☽ ∠ ♄
12 28	☽ ⊼ ♃
13 10	☉ ⊥ ♃
15 42	☽ ⚹ ♅
19 39	☽ □ ♀
20 26	☽ ₽ ♆
28 Friday	
01 03	☽ △ ☉
01 27	☽ ∥ ♆
03 56	♀ ∠ ♆
11 58	☽ □ ♂
12 32	☿ ⚹ ♄
14 06	☽ ± ♆
15 13	☽ Q ♅
17 10	☽ ⚹ ♄
18 06	☽ △ ☿
19 16	☿ ∠ ♃
20 01	♂ Q ♃
21 27	☽ ⊼ ♇
29 Saturday	
02 00	☽ ⊼ ♆
09 13	☽ ₽ ☉
09 23	☽ ± ♇
12 05	☿ ∠ ♅
13 29	☿ □ ♇
14 17	♀ ∥ ♄
17 11	☽ Q ♄
17 47	☽ # ♃
21 08	☽ ∥ ♃
30 Sunday	
00 44	☽ ☍ ♃
03 12	☽ □ ♅
03 24	☽ ₽ ♇
04 50	☽ ± ☉
06 20	☽ ₽ ☿
13 50	☽ △ ♀
14 58	♀ # ♇
16 40	☿ ⊻ ♆
18 00	☽ ⊼ ☉

All ephemeris data is given at 12.00 UT and the Moon's longitude is additionally given for 24.00 UT

MAY 2034

LONGITUDES

Date	Sidereal time h m s	Sun ☉ ° ' "	Moon ☽ ° ' "	Moon ☽ 24.00 ° ' "	Mercury ☿ ° '	Venus ♀ ° '	Mars ♂ ° '	Jupiter ♃ ° '	Saturn ♄ ° '	Uranus ♅ ° '	Neptune ♆ ° '	Pluto ⯓ ° '
01	02 37 58	11 ♉ 11 47	19 ♎ 22 05	25 ♎ 17 04	22 ♉ 14	09 ♊ 31	15 ♊ 25	02 ♈ 13	16 ♋ 16	03 ♋ 11	20 ♈ 36	18 ♒ 14
02	02 41 55	12 10 00	01 ♏ 11 44	07 ♏ 06 19	24 15	10 44	16 05	02 25	16 21	03 13	20 38	18 15
03	02 45 51	13 08 11	13 ♏ 01 07	18 ♏ 56 20	26 14	11 56	16 44	02 38	16 25	03 16	20 40	18 15
04	02 49 48	14 06 21	24 ♏ 52 12	00 ♐ 48 57	28 11	13 09	17 24	02 51	16 30	03 19	20 43	18 16
05	02 53 45	15 04 29	06 ♐ 46 47	12 ♐ 45 55	00 ♊ 04	14 21	18 04	03 03	16 35	03 21	20 45	18 16
06	02 57 41	16 02 35	18 ♐ 46 36	24 ♐ 49 05	01 56	15 33	18 44	03 16	16 40	03 24	20 47	18 16
07	03 01 38	17 00 40	00 ♑ 53 38	07 ♑ 00 33	03 44	16 46	19 24	03 28	16 45	03 27	20 49	18 17
08	03 05 34	17 58 43	13 ♑ 10 09	19 ♑ 22 47	05 28	17 58	20 03	03 41	16 50	03 29	20 51	18 17
09	03 09 31	18 56 45	25 ♑ 38 51	01 ♒ 58 43	07 10	19 10	20 43	03 53	16 56	03 32	20 53	18 17
10	03 13 27	19 54 46	08 ♒ 22 48	14 ♒ 51 32	08 48	20 22	21 23	04 05	17 01	03 35	20 55	18 17
11	03 17 24	20 52 45	21 ♒ 25 18	28 ♒ 04 31	10 22	21 34	22 02	04 17	17 06	03 38	20 57	18 18
12	03 21 20	21 50 43	04 ♓ 49 31	11 ♓ 40 34	11 53	22 46	22 42	04 29	17 12	03 41	20 59	18 18
13	03 25 17	22 48 39	18 ♓ 37 53	25 ♓ 41 32	13 20	23 58	23 21	04 41	17 17	03 44	21 01	18 18
14	03 29 14	23 46 34	02 ♈ 51 29	10 ♈ 07 29	14 44	25 10	24 01	04 53	17 23	03 46	21 03	18 18
15	03 33 10	24 44 29	17 ♈ 29 08	24 ♈ 55 51	16 03	26 22	24 41	05 05	17 29	03 49	21 05	18 18
16	03 37 07	25 42 21	02 ♉ 26 50	10 ♉ 00 41	17 19	27 33	25 20	05 17	17 34	03 52	21 07	18 18
17	03 41 03	26 40 13	17 ♉ 37 26	25 ♉ 14 39	18 31	28 45	26 00	05 28	17 40	03 55	21 09	18 18
18	03 45 00	27 38 03	02 ♊ 51 22	10 ♊ 26 14	19 39	29 ♊ 57	26 39	05 40	17 46	03 58	21 11	18 18
19	03 48 56	28 35 52	17 ♊ 57 58	25 ♊ 25 22	20 42	01 ♋ 08	27 18	05 51	17 52	04 02	21 13	18 R 18
20	03 52 53	29 ♉ 33 39	02 ♋ 47 24	10 ♋ 03 13	21 42	02 20	27 58	06 03	17 58	04 05	21 15	18 18
21	03 56 49	00 ♊ 31 25	17 ♋ 12 14	24 ♋ 14 00	22 38	03 31	28 37	06 14	18 04	04 08	21 17	18 18
22	04 00 46	01 29 09	01 ♌ 08 22	07 ♌ 55 19	23 29	04 43	29 17	06 25	18 10	04 11	21 18	18 18
23	04 04 43	02 26 52	14 ♌ 35 03	21 ♌ 07 51	24 16	05 54	29 ♊ 56	06 36	18 16	04 14	21 20	18 18
24	04 08 39	03 24 32	27 ♌ 34 12	03 ♍ 54 35	24 58	07 05	00 ♋ 35	06 47	18 23	04 17	21 22	18 18
25	04 12 36	04 22 12	10 ♍ 09 35	16 ♍ 19 51	25 36	08 17	01 15	06 58	18 29	04 21	21 24	18 18
26	04 16 32	05 19 49	22 ♍ 26 00	28 ♍ 28 41	26 10	09 28	01 54	07 09	18 35	04 24	21 26	18 18
27	04 20 29	06 17 26	04 ♎ 28 31	10 ♎ 26 08	26 39	10 39	02 33	07 19	18 42	04 27	21 27	18 18
28	04 24 25	07 15 01	16 ♎ 22 06	22 ♎ 16 57	27 03	11 50	03 12	07 30	18 48	04 30	21 29	18 17
29	04 28 22	08 12 34	28 ♎ 11 11	04 ♏ 05 15	27 23	13 01	03 51	07 40	18 55	04 34	21 31	18 17
30	04 32 18	09 10 06	09 ♏ 59 34	15 ♏ 54 29	27 38	14 12	04 31	07 50	19 01	04 37	21 32	18 17
31	04 36 15	10 ♊ 07 37	21 ♏ 50 20	27 ♏ 47 22	27 ♊ 48	15 ♋ 22	05 ♋ 08	08 ♈ 01	19 ♋ 08	04 ♋ 40	21 ♈ 34	18 ♒ 16

Date	Moon True ☊ ° '	Moon Mean ☊ ° '	Moon ☽ Latitude ° '
01	02 ♎ 40	01 ♎ 03	01 N 30
02	02 R 37	01 00	02 29
03	02 32	00 57	03 21
04	02 25	00 54	04 05
05	02 17	00 51	04 39
06	02 09	00 48	05 00
07	02 01	00 44	05 08
08	01 55	00 41	05 03
09	01 50	00 38	04 42
10	01 48	00 35	04 08
11	01 D 47	00 32	03 20
12	01 48	00 29	02 20
13	01 49	00 25	01 N 10
14	01 R 49	00 22	00 S 06
15	01 48	00 19	01 23
16	01 45	00 16	02 37
17	01 40	00 13	03 40
18	01 32	00 09	04 28
19	01 24	00 06	04 57
20	01 15	00 03	05 05
21	01 08	00 ♎ 00	04 54
22	01 02	29 ♍ 57	04 25
23	00 58	29 54	03 42
24	00 57	29 50	02 49
25	00 D 57	29 47	01 49
26	00 57	29 44	00 S 45
27	00 R 58	29 41	00 N 19
28	00 56	29 38	01 21
29	00 53	29 34	02 20
30	00 47	29 31	03 12
31	00 ♎ 39	29 ♍ 28	03 N 56

DECLINATIONS

Date	Sun ☉	Moon ☽	Mercury ☿	Venus ♀	Mars ♂	Jupiter ♃	Saturn ♄	Uranus ♅	Neptune ♆	Pluto ⯓
01	15 N 11	06 S 12	19 N 28	22 N 50	23 N 33	00 S 07	22 N 22	23 N 40	06 N 34	22 S 38
02	15 29	09 34	20 08	23 03	23 38	00 S 02	22 21	23 40	06 35	22 38
03	15 47	12 32	20 46	23 15	23 42	00 N 03	22 21	23 40	06 35	22 38
04	16 04	15 00	21 21	23 27	23 46	00 07	22 20	23 40	06 37	22 38
05	16 21	16 51	21 54	23 39	23 50	00 12	22 20	23 40	06 38	22 38
06	16 38	17 58	22 24	23 49	23 54	00 17	22 19	23 40	06 38	22 38
07	16 55	18 17	22 52	23 59	23 57	00 22	22 19	23 40	06 39	22 38
08	17 11	17 46	23 17	24 08	24 01	00 27	22 18	23 40	06 40	22 39
09	17 27	16 23	23 40	24 17	24 04	00 31	22 18	23 39	06 41	22 39
10	17 43	14 10	24 00	24 25	24 07	00 36	22 17	23 39	06 41	22 39
11	17 58	11 12	24 18	24 32	24 10	00 41	22 17	23 39	06 42	22 39
12	18 13	07 34	24 34	24 39	24 12	00 45	22 16	23 39	06 43	22 39
13	18 28	03 S 25	24 47	24 45	24 15	00 50	22 15	23 39	06 44	22 40
14	18 43	01 N 03	24 58	24 50	24 17	00 55	22 15	23 39	06 44	22 40
15	18 57	05 35	25 07	24 54	24 19	00 59	22 14	23 39	06 45	22 40
16	19 11	09 52	25 14	24 58	24 21	01 04	22 13	23 39	06 46	22 40
17	19 24	13 33	25 19	25 01	24 22	01 08	22 13	23 39	06 46	22 40
18	19 38	16 20	25 22	25 03	24 24	01 12	22 12	23 39	06 47	22 41
19	19 50	17 57	25 23	25 04	24 25	01 17	22 11	23 39	06 48	22 41
20	20 03	18 19	25 23	25 06	24 26	01 21	22 11	23 38	06 48	22 41
21	20 15	17 28	25 21	25 06	24 26	01 25	22 10	23 38	06 49	22 41
22	20 27	15 36	25 17	25 05	24 27	01 30	22 09	23 38	06 50	22 42
23	20 39	12 55	25 12	25 04	24 28	01 34	22 09	23 38	06 50	22 42
24	20 50	09 40	25 06	25 02	24 28	01 38	22 08	23 38	06 51	22 42
25	21 01	06 05	24 58	24 59	24 27	01 42	22 07	23 38	06 52	22 43
26	21 11	02 N 18	24 49	24 56	24 27	01 46	22 06	23 38	06 52	22 43
27	21 21	01 S 29	24 38	24 52	24 27	01 50	22 05	23 38	06 53	22 43
28	21 31	05 11	24 27	24 47	24 26	01 54	22 05	23 38	06 53	22 44
29	21 40	08 39	24 15	24 42	24 25	01 58	22 04	23 38	06 54	22 44
30	21 49	11 46	24 02	24 36	24 24	02 02	22 03	23 37	06 55	22 44
31	21 N 58	14 S 25	23 N 48	24 N 29	24 N 23	02 N 06	22 N 02	23 N 37	06 N 55	22 S 45

ZODIAC SIGN ENTRIES

Date	h m	Planets
02	09 34	☽ ♏
04	22 21	☽ ♐
05	11 03	☿ ♊
07	10 14	☽ ♑
09	20 16	☽ ♒
12	03 26	☽ ♓
14	07 14	☽ ♈
16	08 06	☽ ♉
18	07 30	☽ ♊
18	13 05	♀ ♋
20	07 26	☽ ♋
20	22 57	☉ ♊
22	10 00	☽ ♌
23	14 26	♂ ♋
24	16 35	☽ ♍
27	03 02	☽ ♎
29	15 41	☽ ♏

LATITUDES

Date	Mercury ☿	Venus ♀	Mars ♂	Jupiter ♃	Saturn ♄	Uranus ♅	Neptune ♆	Pluto ⯓
01	01 N 11	00 N 58	00 N 55	01 S 05	00 S 05	00 N 16	01 S 35	07 S 39
04	01 39	01 06	00 56	01 06	00 04	00 16	01 35	07 40
07	02 01	01 13	00 57	01 06	00 04	00 16	01 35	07 41
10	02 16	01 20	00 58	01 07	00 04	00 16	01 35	07 42
13	02 25	01 27	00 59	01 07	00 04	00 16	01 35	07 43
16	02 25	01 33	01 00	01 08	00 03	00 16	01 36	07 43
19	02 17	01 39	01 00	01 08	00 03	00 16	01 36	07 44
22	02 01	01 44	01 01	01 09	00 03	00 16	01 36	07 45
25	01 36	01 49	01 02	01 09	00 02	00 16	01 36	07 46
28	01 03	01 53	01 03	01 10	00 02	00 16	01 36	07 47
31	00 N 23	01 N 57	01 N 03	01 S 10	00 S 02	00 N 16	01 S 36	07 S 47

DATA

Julian Date	2464084
Delta T	+75 seconds
Ayanamsa	24° 20' 11"
Synetic vernal point	04° ♓ 46' 48"
True obliquity of ecliptic	23° 25' 56"

MOON'S PHASES, APSIDES AND POSITIONS ☽

Date	h m	Phase	Longitude ° '	Eclipse Indicator
03	12 16	○	13 ♏ 09	
11	10 56	☾	20 ♒ 50	
18	03 13	●	27 ♉ 17	
24	23 58	☽	03 ♍ 53	

Day	h m		
03	06 07	Apogee	
17	14 23	Perigee	
30	11 13	Apogee	
07	09 08	Max dec	18° S 18'
14	06 26	0N	
20	06 54	Max dec	18° N 21'
27	02 32	0S	

LONGITUDES

Date	Chiron ⚷ ° '	Ceres ⚳ ° '	Pallas ⚴ ° '	Juno ⚵ ° '	Vesta ⚶ ° '	Black Moon Lilith ⚸ ° '
01	29 ♉ 41	13 ♓ 30	10 ♒ 23	16 ♐ 56	14 ♉ 22	10 ♏ 08
11	00 ♊ 24	16 ♓ 34	11 ♒ 38	15 ♐ 21	18 ♉ 48	11 ♏ 14
21	01 ♊ 08	19 ♓ 24	12 ♒ 22	13 ♐ 22	23 ♉ 12	12 ♏ 21
31	01 ♊ 52	21 ♓ 59	12 ♒ 27	11 ♐ 10	27 ♉ 35	13 ♏ 28

All ephemeris data is given at 12.00 UT and the Moon's longitude is additionally given for 24.00 UT

ASPECTARIAN

h m	Aspects
01 Monday	
03 31	☽ △ ♂
04 19	☽ ± ☿
05 40	☽ □ ♄
09 42	☽ △ ⯓
14 31	☽ ☍ ♆
14 37	☽ # ♆
19 00	☽ ⊼ ☿
23 38	☽ ♇ ♀
02 Tuesday	
03 47	♀ ⊥ ♄
11 45	☽ ♇ ♂
14 32	☽ ⊼ ♃
16 08	☽ △ ♅
20 00	☽ ± ♀
22 56	♂ ⊻ ♄
03 Wednesday	
01 07	♂ ∥ ♅
02 24	♅ ♇ ⯓
02 57	☽ ± ♃
07 06	☽ ± ♂
09 33	☽ ⊼ ♀
12 16	☽ ☍ ☉
17 32	☿ ⊥ ♆
18 57	☽ △ ♄
20 00	☽ ⊼ ♂
21 32	☽ ♇ ♃
22 37	☽ □ ⯓
22 41	☽ ♇ ♅
04 Thursday	
01 00	☿ ⊥ ♅
03 34	☽ ⊼ ♆
15 44	☽ ± ♆
16 57	☽ ± ♅
19 57	☽ ☍ ☿
05 Friday	
01 29	☽ ♇ ♄
03 05	☽ # ☉
04 23	☽ △ ♃
05 05	☽ ⊼ ♅
09 55	☽ ♇ ♆
10 58	☽ Q ⯓
14 33	♀ ∥ ♅
19 14	♂ △ ⯓
19 41	☽ ± ♄
06 Saturday	
04 52	☽ ☍ ♀
04 59	♀ Q ♃
06 04	☽ ⊼ ☉
07 46	☽ ⊼ ♄
08 08	☿ ∥ ♄
08 29	☿ ∠ ♄
11 00	☽ ⚹ ⯓
11 54	☽ ☍ ♂
12 58	♃ ∠ ⯓
16 00	☽ △ ♆
19 04	☽ ± ☉
23 55	☿ # ⯓
07 Sunday	
05 02	☉ ⚹ ♄
05 21	♀ ∥ ♂
08 01	♃ □ ♅
08 05	☿ ⊻ ♅
08 06	☿ ⚹ ♃
11 55	♀ ⊻ ♄
14 23	☽ ♇ ☉
16 41	☽ ∠ ⯓
17 02	☽ ☍ ♅
17 09	☽ □ ♃
18 30	☽ ⊼ ☿
08 Monday	
08 10	☽ ± ☿
10 17	☽ ⊥ ⯓
13 42	☉ ⊻ ♀
17 23	☿ ∠ ♆
18 25	♀ △ ⯓
19 09	☽ ☍ ♄
19 34	☉ □ ⯓
21 53	☽ ⊻ ⯓
21 54	☽ # ☉
22 05	☽ △ ☉
22 16	☽ ⊼ ♀
09 Tuesday	
01 18	☉ ∠ ♅
02 03	☽ ⊼ ♂
02 52	☽ □ ♆
04 19	☽ ♇ ☿
04 41	☽ Q ♃
09 58	☉ ∠ ♃
10 59	☽ ± ♀
11 16	☿ ∥ ♅
14 09	☽ ± ♂
18 26	♂ ⚹ ♆
10 Wednesday	
02 59	☽ ⊼ ♅
03 50	☽ ⚹ ♃
05 47	☽ ♇ ♀
08 03	☽ ♇ ♂
12 53	☽ △ ☿
13 00	☽ Q ♆
14 15	☽ ± ♅
21 54	☿ ∥ ♂
23 22	♀ ⚹ ♆
11 Thursday	
04 04	☽ ⊼ ♄
06 18	☽ ☌ ⯓
06 54	☽ ♇ ♅
08 03	☽ ∠ ♃
10 56	☽ □ ☉
11 09	☽ ⚹ ♆
12 17	☽ △ ♀
13 11	☽ △ ♂
13 53	☉ ⊻ ♆
15 05	☽ ± ♄
12 Friday	
00 16	☿ ⊥ ♄
00 34	☽ ⊥ ♃
07 19	☽ ♇ ♄
08 58	♀ ☌ ♂
09 58	☽ △ ♅
11 24	☽ ⊻ ♃
14 03	☽ ∠ ♆
17 06	☽ # ♆
21 29	☽ Q ☉
13 Saturday	
01 50	☽ □ ☿
02 41	☿ ∥ ♀
05 47	☽ ⊥ ♆
09 41	☽ △ ♄
11 26	☽ ⊻ ⯓
16 05	☽ ⊻ ♆
19 39	☽ ⚹ ☉
20 27	☽ □ ♂
21 39	☽ ⊥ ⯓
21 55	☽ □ ♀
14 Sunday	
01 46	☽ # ♃
11 15	☽ ∥ ♃
11 46	☽ Q ☿
12 44	☽ ∠ ⯓
13 32	☽ □ ♅
15 25	☽ ☌ ♃
22 29	☽ ∠ ☉
15 Monday	
03 49	☽ Q ♂
06 28	☽ Q ♀
06 55	☉ ⊻ ♂
09 27	☽ ⚹ ☿
11 59	☽ □ ♄
13 20	☽ ⚹ ⯓
14 10	☽ ⊥ ☉
17 50	☽ ☌ ♆
18 22	☽ ∥ ♆
19 02	☽ Q ♅
16 Tuesday	
00 08	☽ ⚹ ♂
00 30	☽ ⊻ ☉
03 32	☽ ⚹ ♀
08 35	☽ Q ⯓
11 03	☿ Q ♃
11 47	☽ ∠ ☿
14 16	☽ ⚹ ♅
16 33	☽ ⊻ ♃
17 00	☽ Q ♄
17 27	☿ ⊻ ♄
17 Wednesday	
01 04	☽ ∠ ♂
02 10	☽ ⊥ ♃
03 16	☽ ⊥ ☿
05 23	☽ ∠ ♀
07 43	☿ △ ⯓
12 04	☽ ⚹ ♄
13 05	☽ □ ⯓
13 31	☽ ⊻ ☿
14 03	☽ ∠ ♅
15 54	☽ ⊥ ♂
16 33	☽ ∠ ♃
17 34	☽ ⊻ ♆
20 46	☽ ⊥ ♀
18 Thursday	
00 22	☉ ⊥ ♆
01 46	☽ ⊻ ♂
03 02	☽ ⊥ ♆
03 13	☽ ☌ ☉
04 17	☽ ⊥ ♅
07 01	☽ ⊻ ♀
11 52	☽ ∠ ♄
13 46	☽ ⊻ ♅
16 30	☽ ⚹ ♃
17 16	☽ ∠ ♆
20 57	☉ ⊥ ♅
19 Friday	
02 13	☽ ⊥ ♄
07 00	⯓ St R
11 49	☽ Q ♃
11 51	☽ ⊻ ♄
12 33	☽ △ ⯓
16 43	☽ ☌ ☿
17 13	☽ ⚹ ♆
20 Saturday	
00 24	☿ ⚹ ♆
03 45	☽ ☌ ♂
06 21	☽ ⊻ ☉
11 11	☽ ☌ ♀
12 45	☽ Q ♆
12 51	☽ ♇ ⯓
14 07	☽ ☌ ♅
16 53	☽ ⊥ ☉
17 25	☽ □ ♃
21 Sunday	
03 46	☽ ± ⯓
06 53	♀ Q ♆
07 37	♀ ♇ ⯓
08 57	☽ ∠ ☉
13 29	☽ ☌ ♄
13 52	☽ ⊼ ⯓
18 57	☽ □ ♆
21 51	☽ ⊻ ☿
22 Monday	
00 47	♀ ☌ ♅
08 35	☽ ⊻ ♂
08 55	☽ ⊥ ☿
12 39	☽ ⚹ ☉
17 23	☽ ⊻ ♅
18 54	☽ ⊻ ♀
19 40	☽ ⊥ ♂
21 27	☽ △ ♃
23 Tuesday	
01 49	☽ ∠ ☿
04 07	☽ ⊥ ♅
06 40	☽ ⊥ ♀
11 44	☽ Q ☉
12 40	☽ ∠ ♂
18 41	♄ ⊼ ⯓
18 47	☽ ☍ ⯓
18 48	☽ ⊻ ♄
20 32	☽ ∠ ♅
24 Wednesday	
00 25	☽ △ ♆
00 45	☽ ∠ ♀
01 03	☽ ♇ ♃
04 42	♀ □ ♃
05 59	☽ ⊥ ♄
06 52	☽ ⚹ ☿
11 08	☉ ∠ ♄
18 00	☽ ⚹ ♂
18 09	☽ ± ♃
23 05	☽ ∠ ♄
23 58	☽ □ ☉
25 Thursday	
00 47	☽ ⚹ ♅
04 24	☿ ∥ ♀
04 44	☽ ♇ ♆
05 46	☽ ⊼ ♃
06 50	☽ Q ☿
06 55	☽ ∥ ♆
07 59	☽ ⚹ ♀
11 15	☉ ⊻ ♅
18 19	☽ Q ♂
22 12	☽ ± ♆
26 Friday	
00 04	☽ Q ♅
03 51	☽ ⊼ ⯓
04 22	☽ ⚹ ♄
09 53	☽ Q ♀
10 00	☽ ⊼ ♆
15 19	☽ ∥ ♃
15 41	☽ ± ⯓
19 43	☽ □ ☿
27 Saturday	
04 22	☽ Q ♄
07 55	☽ □ ♂
09 38	☽ ♇ ⯓
11 57	☽ □ ♅
14 16	☽ # ♃
15 58	☽ △ ☉
16 16	☉ ∠ ♆
17 49	☽ ☍ ♃
28 Sunday	
01 48	☽ □ ♀
14 18	☿ ∥ ♂
15 03	♂ ♇ ⯓
15 54	☽ △ ⯓
16 59	☽ □ ♄
19 39	☉ ⚹ ♃
21 14	♀ ± ⯓
22 24	☽ ☍ ♆
22 44	♂ Q ♆
23 36	☽ # ♆
29 Monday	
00 59	☽ ♇ ☉
10 20	☽ △ ☿
20 54	☽ ± ☉
30 Tuesday	
00 13	☽ △ ♂
01 01	☽ △ ♅
07 35	☽ ⊼ ♃
10 11	☽ ⊼ ☉
16 12	♂ ☌ ♅
17 27	☽ ♇ ☿
19 56	☽ ± ♃
21 28	☽ △ ♀
31 Wednesday	
04 48	☽ □ ⯓
06 28	☽ △ ♄
07 36	☽ ♇ ♅
08 25	☽ ♇ ♂
11 27	☽ ⊼ ♆
11 56	☽ ± ☿
14 25	☽ ♇ ♃
23 35	☽ ± ♆
23 50	☉ ∥ ♄

JUNE 2034

Date	Sidereal time h m s	Sun ☉ ° ' "	Moon ☽ ° ' "	Moon ☽ 24.00 ° ' "	Mercury ☿ ° '	Venus ♀ ° '	Mars ♂ ° '	Jupiter ♃ ° '	Saturn ♄ ° '	Uranus ♅ ° '	Neptune ♆ ° '	Pluto ♇ ° '
01	04 40 12	11 ♊ 05 07	03 ♐ 45 48	09 ♐ 45 51	27 ♊ 54	16 ♋ 33	05 ♋ 49	08 ♈ 11	19 ♋ 15	04 ♋ 44	21 ♈ 36	18 ♒ 16
02	04 44 08	12 02 36	15 ♐ 47 40	21 ♐ 51 23	27 R 55	17 44	06 28	08 21	19 21	04 47	21 37	18 R 16
03	04 48 05	13 00 04	27 ♐ 57 08	04 ♑ 05 03	27 51	18 54	07 07	08 31	19 28	04 50	21 39	18 15
04	04 52 01	13 57 31	10 ♑ 15 13	16 ♑ 27 47	27 44	20 05	07 46	08 41	19 35	04 54	21 40	18 15
05	04 55 58	14 54 57	22 ♑ 42 54	29 ♑ 00 43	27 32	21 15	08 25	08 51	19 42	04 57	21 42	18 15
06	04 59 54	15 52 22	05 ♒ 21 26	11 ♒ 45 16	27 15	22 26	09 04	09 00	19 49	05 01	21 44	18 14
07	05 03 51	16 49 47	18 ♒ 12 28	24 ♒ 43 18	26 56	23 36	09 43	09 10	19 55	05 04	21 45	18 14
08	05 07 47	17 47 11	01 ♓ 18 04	07 ♓ 57 04	26 33	24 46	10 22	09 19	20 02	05 08	21 46	18 13
09	05 11 44	18 44 34	14 ♓ 40 30	21 ♓ 28 51	26 06	25 56	11 01	09 28	20 09	05 11	21 48	18 13
10	05 15 41	19 41 57	28 ♓ 22 09	05 ♈ 20 37	25 38	27 06	11 40	09 37	20 16	05 15	21 49	18 12
11	05 19 37	20 39 19	12 ♈ 24 19	19 ♈ 33 12	25 07	28 16	12 19	09 46	20 24	05 18	21 51	18 11
12	05 23 34	21 36 41	26 ♈ 47 04	04 ♉ 05 35	24 35	29 ♋ 26	12 58	09 55	20 31	05 22	21 52	18 11
13	05 27 30	22 34 03	11 ♉ 28 13	18 ♉ 54 16	24 01	00 ♌ 35	13 37	10 04	20 38	05 25	21 53	18 10
14	05 31 27	23 31 24	26 ♉ 22 50	03 ♊ 52 56	23 28	01 45	14 15	10 12	20 45	05 29	21 55	18 10
15	05 35 23	24 28 44	11 ♊ 23 24	18 ♊ 53 04	22 54	02 55	14 54	10 21	20 52	05 32	21 56	18 09
16	05 39 20	25 26 04	26 ♊ 20 41	03 ♋ 45 06	22 21	04 04	15 33	10 29	21 00	05 36	21 57	18 08
17	05 43 16	26 23 24	11 ♋ 05 13	18 ♋ 20 07	21 49	05 13	16 12	10 37	21 07	05 39	21 58	18 08
18	05 47 13	27 20 42	25 ♋ 29 03	02 ♌ 31 25	21 19	06 23	16 51	10 45	21 14	05 43	22 00	18 07
19	05 51 10	28 18 00	09 ♌ 26 52	16 ♌ 15 14	20 52	07 32	17 29	10 53	21 22	05 47	22 01	18 06
20	05 55 06	29 ♊ 15 17	22 ♌ 56 33	29 ♌ 30 58	20 27	08 41	18 08	11 01	21 29	05 50	22 02	18 05
21	05 59 03	00 ♋ 12 34	05 ♍ 58 51	12 ♍ 20 36	20 06	09 50	18 47	11 08	21 36	05 54	22 03	18 05
22	06 02 59	01 09 50	18 ♍ 36 46	24 ♍ 47 56	19 48	10 59	19 26	11 16	21 44	05 57	22 04	18 04
23	06 06 56	02 07 05	00 ♎ 54 44	06 ♎ 57 51	19 33	12 07	20 04	11 23	21 51	06 01	22 05	18 03
24	06 10 52	03 04 19	12 ♎ 57 57	18 ♎ 55 42	19 23	13 16	20 43	11 30	21 59	06 04	22 06	18 02
25	06 14 49	04 01 33	24 ♎ 51 45	00 ♏ 46 46	19 18	14 25	21 22	11 37	22 06	06 08	22 07	18 01
26	06 18 45	04 58 46	06 ♏ 41 19	12 ♏ 35 58	19 D 17	15 33	22 00	11 44	22 14	06 12	22 08	18 00
27	06 22 42	05 55 59	18 ♏ 31 14	24 ♏ 27 34	19 21	16 41	22 39	11 51	22 21	06 15	22 09	17 59
28	06 26 39	06 53 11	00 ♐ 25 24	06 ♐ 25 04	19 29	17 49	23 17	11 57	22 29	06 19	22 10	17 58
29	06 30 35	07 50 23	12 ♐ 26 52	18 ♐ 31 02	19 42	18 57	23 56	12 04	22 37	06 22	22 11	17 57
30	06 34 32	08 ♋ 47 35	24 ♐ 37 46	00 ♑ 47 11	20 ♊ 00	20 ♌ 05	24 ♋ 35	12 ♈ 10	22 ♋ 44	06 ♋ 26	22 ♈ 12	17 ♒ 56

Date	Moon True ☊ ° '	Moon Mean ☊ ° '	Moon ☽ Latitude ° '
01	00 ♎ 28	29 ♍ 25	04 N 30
02	00 R 16	29 22	04 52
03	00 03	29 19	05 01
04	29 ♍ 51	29 15	04 56
05	29 40	29 12	04 38
06	29 32	29 09	04 05
07	29 27	29 06	03 20
08	29 24	29 03	02 23
09	29 23	29 00	01 17
10	29 D 23	28 56	00 N 05
11	29 R 23	28 53	01 S 08
12	29 21	28 50	02 20
13	29 17	28 47	03 23
14	29 10	28 44	04 13
15	29 00	28 40	04 47
16	28 49	28 37	05 00
17	28 38	28 34	04 54
18	28 28	28 31	04 28
19	28 20	28 28	03 47
20	28 14	28 25	02 55
21	28 11	28 21	01 55
22	28 10	28 18	00 S 50
23	28 D 10	28 15	00 N 15
24	28 R 10	28 12	01 18
25	28 08	28 09	02 16
26	28 05	28 06	03 09
27	27 59	28 02	03 53
28	27 51	27 59	04 27
29	27 40	27 56	04 50
30	27 ♍ 27	27 ♍ 53	05 N 00

DECLINATIONS

Date	Sun ☉ ° '	Moon ☽ ° '	Mercury ☿ ° '	Venus ♀ ° '	Mars ♂ ° '	Jupiter ♃ ° '	Saturn ♄ ° '	Uranus ♅ ° '	Neptune ♆ ° '	Pluto ♇ ° '
01	22 N 06	16 S 29	23 N 33	24 N 21	24 N 22	02 N 10	22 N 01	23 N 37	06 N 56	22 S 45
02	22 14	17 50	23 17	24 13	24 20	02 14	22 00	23 37	06 56	22 45
03	22 21	18 24	23 01	24 04	24 18	02 17	22 00	23 37	06 57	22 46
04	22 28	18 07	22 44	23 55	24 16	02 21	21 59	23 37	06 57	22 46
05	22 35	16 57	22 27	23 45	24 14	02 25	21 58	23 37	06 58	22 46
06	22 41	14 58	22 10	23 34	24 11	02 28	21 57	23 37	06 58	22 47
07	22 47	12 12	21 52	23 23	24 09	02 32	21 56	23 37	06 59	22 47
08	22 52	08 47	21 34	23 10	24 06	02 35	21 55	23 36	06 59	22 48
09	22 57	04 51	21 16	22 58	24 03	02 38	21 54	23 36	07 00	22 48
10	23 02	00 S 34	20 59	22 45	24 00	02 42	21 53	23 36	07 00	22 48
11	23 06	03 N 51	20 41	22 31	23 57	02 45	21 52	23 36	07 01	22 49
12	23 10	08 09	20 24	22 16	23 53	02 48	21 51	23 36	07 01	22 49
13	23 13	12 03	20 08	22 01	23 49	02 52	21 50	23 36	07 02	22 50
14	23 16	15 14	19 53	21 46	23 45	02 55	21 49	23 36	07 02	22 50
15	23 19	17 24	19 38	21 30	23 41	02 58	21 48	23 35	07 03	22 51
16	23 21	18 23	19 25	21 13	23 37	03 01	21 47	23 35	07 03	22 51
17	23 23	18 05	19 13	20 56	23 33	03 04	21 46	23 35	07 03	22 52
18	23 24	16 38	19 02	20 38	23 28	03 07	21 45	23 35	07 04	22 52
19	23 25	14 14	18 52	20 20	23 23	03 10	21 44	23 35	07 04	22 52
20	23 26	11 07	18 44	20 01	23 18	03 12	21 43	23 35	07 05	22 53
21	23 26	07 32	18 38	19 42	23 13	03 15	21 42	23 35	07 05	22 53
22	23 26	03 N 44	18 33	19 22	23 08	03 18	21 41	23 34	07 05	22 54
23	23 25	00 S 08	18 30	19 02	23 02	03 20	21 40	23 34	07 06	22 54
24	23 24	03 56	18 29	18 41	22 56	03 23	21 39	23 34	07 06	22 55
25	23 22	07 31	18 29	18 20	22 50	03 25	21 38	23 34	07 06	22 55
26	23 20	10 46	18 31	17 59	22 44	03 28	21 36	23 34	07 07	22 56
27	23 18	13 36	18 35	17 37	22 38	03 30	21 35	23 34	07 07	22 56
28	23 15	15 52	18 39	17 14	22 32	03 33	21 34	23 33	07 07	22 57
29	23 12	17 29	18 46	16 52	22 25	03 35	21 33	23 33	07 07	22 57
30	23 N 08	18 S 19	18 N 54	16 N 28	22 N 19	03 N 37	21 N 32	23 N 33	07 N 08	22 S 58

ZODIAC SIGN ENTRIES

Date	h	m	Planets	
01	04	27	☽	♐
03	16	01	☽	♑
06	01	52	☽	♒
08	09	38	☽	♓
10	14	49	☽	♈
12	17	18	☽	♉
12	23	49	♀	♌
14	17	48	☽	♊
16	17	55	☽	♋
18	19	41	☽	♌
21	00	53	☽	♍
21	06	44	☉	♋
23	10	12	☽	♎
25	22	25	☽	♏
28	11	09	☽	♐
30	22	28	☽	♑

LATITUDES

Date	Mercury ☿ ° '	Venus ♀ ° '	Mars ♂ ° '	Jupiter ♃ ° '	Saturn ♄ ° '	Uranus ♅ ° '	Neptune ♆ ° '	Pluto ♇ ° '
01	00 N 08	01 N 58	01 N 03	01 S 11	00 S 02	00 N 16	01 S 36	07 S 48
04	00 S 41	02 00	01 04	01 11	00 02	00 16	01 36	07 49
07	01 32	02 02	01 05	01 12	00 01	00 16	01 36	07 49
10	02 23	02 03	01 05	01 13	00 01	00 16	01 37	07 50
13	03 10	02 03	01 05	01 13	00 S 01	00 16	01 37	07 51
16	03 48	02 02	01 06	01 14	00 00	00 16	01 37	07 52
19	04 15	02 01	01 06	01 15	00 00	00 16	01 37	07 52
22	04 30	01 58	01 07	01 15	00 00	00 17	01 37	07 53
25	04 32	01 55	01 07	01 16	00 00	00 17	01 37	07 54
28	04 22	01 50	01 07	01 17	00 N 01	00 17	01 38	07 54
31	04 S 03	01 N 45	01 N 08	01 S 18	00 N 01	00 N 17	01 S 38	07 S 55

DATA

Julian Date	2464115
Delta T	+75 seconds
Ayanamsa	24° 20' 15"
Synetic vernal point	04° ♓ 46' 44"
True obliquity of ecliptic	23° 25' 56"

MOON'S PHASES, APSIDES AND POSITIONS ☽

Date	h	m	Phase	Longitude ° '	Eclipse Indicator
02	03	54	○	11 ♐ 43	
09	19	44	☾	19 ♓ 03	
16	10	26	●	25 ♊ 22	
23	14	35	☽	02 ♎ 13	

Day	h	m		
14	21	34	Perigee	
27	00	16	Apogee	
03	16	00	Max dec	18° S 25'
10	15	05	0N	
16	18	27	Max dec	18° N 25'
23	11	08	0S	
30	23	49	Max dec	18° S 26'

LONGITUDES

Date	Chiron ⚷ ° '	Ceres ⚳ ° '	Pallas ⚴ ° '	Juno ⚵ ° '	Vesta ⚶ ° '	Black Moon Lilith ⚸ ° '
01	01 ♊ 57	22 ♓ 14	12 ♒ 26	10 ♐ 56	28 ♉ 01	13 ♏ 35
11	02 ♊ 40	24 ♓ 29	11 ♒ 51	08 ♐ 42	02 ♊ 21	14 ♏ 42
21	03 ♊ 22	26 ♓ 22	10 ♒ 37	06 ♐ 39	06 ♊ 38	15 ♏ 48
31	04 ♊ 01	27 ♓ 51	08 ♒ 46	04 ♐ 58	10 ♊ 50	16 ♏ 55

ASPECTARIAN

01 Thursday
00 09 ☽ ⊼ ☿
01 50 ☽ ± ♅
03 37 ☽ ± ♂
04 45 ☿ ∥ ♅
07 05 ☽ ⅊ ♀
10 52 ♀ ∥ ♂
12 58 ☽ ⅊ ♄
13 56 ☽ ⊼ ♅
16 21 ☽ ⊼ ♂
17 01 ☽ Q ♇
17 41 ☽ ⅊ ♆
20 59 ☽ △ ♃

02 Friday
03 03 ☽ ± ♀
03 54 ☽ ☍ ☉
05 24 ☿ St R
07 06 ☽ ± ♄
16 15 ☽ ⊼ ♀
16 54 ☽ ⚹ ♇
19 07 ☽ ⊼ ♄
22 49 ♀ ⊼ ♇
23 34 ☽ △ ♆

03 Saturday
11 49 ☽ ☍ ☿
22 23 ☽ ∠ ♇
22 32 ☉ ⊥ ♀

04 Sunday
00 40 ♀ ☌ ♄
01 13 ☉ ⊥ ♄
01 32 ☽ ☍ ♅
06 54 ☽ ☍ ♂
08 55 ☽ □ ♃
09 22 ☿ ⋕ ♇
15 52 ☽ ⊥ ♇
19 46 ☽ ⊼ ☉

05 Monday
03 26 ☽ ⊻ ♇
04 12 ☉ ∥ ☿
06 09 ☽ ☍ ♄
08 16 ☽ ± ☉
08 55 ☽ ☍ ♀
10 03 ☽ □ ♆
19 59 ☽ Q ♃
21 00 ☽ ⊼ ☿
21 21 ♀ □ ♆

06 Tuesday
02 50 ☽ ⅊ ☉
06 00 ♀ ∥ ♅
08 08 ☽ ± ☿
08 49 ♂ □ ♃
11 21 ☽ ⊼ ♅
18 56 ☽ ⚹ ♃
19 21 ☽ ⊼ ♂
20 13 ☽ Q ♆
22 40 ☽ ± ♅

07 Wednesday
00 38 ☽ ⅊ ☿
06 01 ☿ ∥ ♄
07 08 ☽ ± ♂
09 15 ☽ △ ☉
12 02 ☽ ☌ ♇
13 58 ☉ ⋕ ♇
15 12 ☽ ⊼ ♄
15 27 ☽ ⅊ ♅
18 33 ☽ ⚹ ♆
22 55 ☽ ⊼ ♀
23 06 ☽ ∠ ♃

08 Thursday
00 37 ☽ ⅊ ♂
02 20 ☽ ± ♄
03 35 ☽ △ ☿
10 56 ☽ ± ♀
15 42 ☽ ⊥ ♃
18 49 ☽ ⅊ ♄
18 57 ☽ △ ♅
21 54 ☽ ∠ ♆
22 46 ☉ △ ♇
23 14 ☽ ⋕ ♆

09 Friday
02 37 ☽ ⊻ ♃
04 42 ☽ ⅊ ♀
05 09 ☽ △ ♂
12 43 ☉ ∥ ♀
14 00 ☽ ⊥ ♆
14 34 ☿ ⊻ ♀
18 15 ☽ ⊻ ♇
19 44 ☽ □ ☉
21 46 ☽ △ ♄

10 Saturday
00 22 ☽ ⋕ ♃
00 35 ☽ ⊻ ♆
04 46 ☽ ⊥ ♇
05 17 ♀ ⋕ ♇
07 24 ☽ □ ☿
09 36 ☽ △ ♀
20 19 ☽ ∠ ♇
23 53 ☽ □ ♅

11 Sunday
04 29 ☉ ⊻ ♄
05 11 ☽ Q ☉
05 57 ☽ ∥ ♃
07 30 ☽ ☌ ♃
07 33 ♂ ± ♇
11 50 ☽ □ ♂
13 09 ☽ Q ☿
21 43 ☽ ⚹ ♇

12 Monday
01 31 ☽ □ ♄
02 49 ☽ ⚹ ☉
03 51 ☽ ☌ ♆
05 32 ☽ ∥ ♆
06 19 ☽ Q ♅
08 29 ☽ ⚹ ☿
16 44 ☽ □ ♀
17 35 ☽ Q ♇
18 36 ☉ ⚹ ♆
19 11 ☽ Q ♂
21 03 ☉ Q ♃

13 Tuesday
02 07 ☽ ⚹ ♅
04 05 ☿ ⊥ ♀
05 14 ☽ ∠ ☉
07 21 ☽ Q ♄
08 10 ☽ ∠ ☿
09 42 ☽ ⊻ ♃
15 37 ☽ ⚹ ♂
19 30 ☽ ⊥ ♃
20 48 ☽ ⊥ ☉
22 12 ☽ ⊥ ☿
22 49 ☽ □ ♇

14 Wednesday
00 28 ☽ Q ♀
02 30 ☽ ∠ ♅
02 54 ☽ ⚹ ♄
04 49 ☽ ⊻ ♆
06 21 ♀ ∥ ♄
07 06 ☽ ⊻ ☉
07 29 ☽ ⊻ ☿
10 06 ☽ ∠ ♃
10 59 ☉ ☌ ☿
14 27 ☽ ⊥ ♆
16 49 ☽ ∠ ♂
16 59 ☽ ⊥ ♅
21 19 ☽ ⚹ ♀

15 Thursday
02 36 ☽ ⊻ ♅
03 06 ☽ ∠ ♄
04 52 ☽ ∠ ♆
07 51 ☽ ⊥ ♂
10 19 ☽ ⚹ ♃
17 37 ☽ ⊥ ♄
17 53 ☽ ⊻ ♂
22 49 ☽ △ ♇
23 19 ☽ ∠ ♀

16 Friday
03 19 ☽ ⊻ ♄
04 55 ☽ ⚹ ♆
05 43 ☽ Q ♃
05 47 ☽ ☌ ☿
07 16 ☿ Q ♃
10 26 ☽ ☌ ☉
15 01 ☽ ⊥ ♀
21 55 ♂ ∥ ♅
23 00 ☽ ⅊ ♇

17 Saturday
00 21 ☽ Q ♆
01 35 ☽ ⊻ ♀
03 04 ☽ ☌ ♅
05 11 ☿ ⚹ ♆
11 13 ☽ □ ♃
13 43 ☽ ± ♇
20 50 ☽ ☌ ♂
21 29 ♀ ⊻ ♅
23 39 ☽ ⊼ ♇

18 Sunday
04 47 ☽ ☌ ♄
05 13 ☽ ⊻ ☿
06 07 ☽ □ ♆
11 13 ☿ ∠ ♀
15 01 ☽ ⊥ ☿
15 23 ☽ ⊻ ☉
15 28 ☿ ⊻ ♄

19 Monday
02 24 ☽ ⊥ ☉
02 56 ☉ ∥ ♂
05 35 ☽ ⊻ ♅
05 57 ☽ ∠ ☿
08 21 ☽ ☌ ♀
14 32 ☽ △ ♃
16 06 ☽ ⊥ ♅
19 17 ☽ ∠ ☉

20 Tuesday
02 55 ☽ ⊻ ♂
03 17 ☽ ☍ ♇
07 38 ☽ ⚹ ☿
08 11 ☽ ∠ ♅
09 20 ☽ ⊻ ♄
10 17 ♂ ⊼ ♇
10 21 ☽ △ ♆
14 16 ☽ ⊥ ♂
17 38 ☽ ⅊ ♃
20 21 ☽ ⊥ ♄

21 Wednesday
00 26 ☽ ⚹ ☉
04 57 ☽ Q ☿
07 41 ☽ ∠ ♂
10 25 ☽ ± ♃
11 50 ☽ ⚹ ♅
13 11 ☽ ∠ ♄
14 01 ☽ ⅊ ♆
14 55 ☽ ∥ ♆
19 57 ☽ ⊻ ♀
21 49 ☽ ⊼ ♃

22 Thursday
00 42 ☽ Q ☉
07 06 ☽ ± ♆
08 32 ☽ ⊥ ♀
10 44 ☽ Q ♅
10 56 ☽ ⊼ ♇
13 39 ☽ ⚹ ♂
14 14 ☽ □ ☿
14 39 ☽ ∥ ♃
18 05 ☽ ⚹ ♄
18 42 ☽ ⊼ ♆
18 42 ♀ △ ♃
21 46 ☿ ⊻ ♂
22 33 ☽ ± ♇

23 Friday
03 47 ☽ ∠ ♀
09 36 ♀ ⊥ ♅
14 25 ☽ Q ♂
14 35 ☽ □ ☉
16 13 ☽ ⅊ ♇
17 53 ☽ Q ♄
22 10 ☽ □ ♅

24 Saturday
08 27 ☽ ⋕ ♃
09 02 ☽ ☍ ♃
11 03 ☉ ⅊ ♇
12 40 ☽ ⚹ ♀
17 43 ♂ ⋕ ♇
22 11 ☽ △ ♇

25 Sunday
00 49 ☽ △ ☿
02 03 ☿ ∥ ♀
04 30 ☽ □ ♂
06 22 ☽ □ ♄
06 26 ☽ ☍ ♆
09 12 ☽ ⋕ ♆
14 25 ☉ Q ♆
15 05 ♄ □ ♆
15 28 ☽ Q ♀

26 Monday
05 23 ☿ St D
07 06 ☽ ⅊ ☿
08 13 ☽ △ ☉
11 00 ☽ △ ♅
17 03 ♂ □ ♆
22 21 ☽ ⊼ ♃
22 34 ♂ ☌ ♄

27 Tuesday
01 26 ☽ ± ☿
07 53 ☽ □ ♀
10 37 ☽ ± ♃
10 56 ☽ □ ♇
13 41 ☽ ⊼ ☿
17 18 ☽ ⅊ ☉
17 33 ☽ ⅊ ♅
19 21 ☽ ⊼ ♆
19 51 ☽ △ ♄
20 38 ☉ ☌ ♅
20 49 ☽ △ ♂

28 Wednesday
04 58 ☽ ⅊ ♃
07 28 ☽ ± ♆
11 47 ☽ ± ♅
13 01 ☽ ± ☉
15 11 ♀ ☍ ♇
23 06 ☽ Q ♇
23 51 ☽ ⊼ ♅

29 Thursday
01 31 ☽ ⅊ ♆
02 03 ☽ ⊼ ☉
02 17 ☽ ⅊ ♄
03 30 ☽ ⋕ ♀
04 37 ☽ ⅊ ♂
11 14 ☽ △ ♃
20 20 ☽ ± ♄
22 53 ☽ ⚹ ♇
23 27 ☽ ± ♂

30 Friday
02 11 ☽ △ ♀
02 41 ☽ ☍ ☿
07 14 ☽ △ ♆
08 15 ☽ ⊼ ♄
09 33 ☿ ⚹ ♀
11 53 ☽ ⊼ ♂

All ephemeris data is given at 12.00 UT and the Moon's longitude is additionally given for 24.00 UT

JULY 2034

LONGITUDES

Date	Sidereal time h	m	s	Sun ☉ °	'	"	Moon ☽ °	'	"	Moon ☽ 24.00 °	'	"	Mercury ☿ °	'	Venus ♀ °	'	Mars ♂ °	'	Jupiter ♃ °	'	Saturn ♄ °	'	Uranus ♅ °	'	Neptune ♆ °	'	Pluto ⯓ °	'
01	06	38	28	09 ♋	44	46	06 ♑	59	22	13 ♑	14	24	20 ♊	23	21 ♌	13	25 ♋	13	12 ♈	16	22 ♋	52	06 ♋	30	22 ♈	12	17 ♒	56
02	06	42	25	10	41	57	19 ♑	32	16	25 ♑	52	58	20	51	22	20	25	52	12	22	23	00	06	33	22	13	17 R	55
03	06	46	21	11	39	08	02 ♒	16	30	08 ♒	42	52	21	24	23	28	26	30	12	28	23	07	06	37	22	14	17	53
04	06	50	18	12	36	20	15 ♒	12	01	21 ♒	44	00	22	01	24	35	27	09	12	33	23	15	06	40	22	15	17	52
05	06	54	14	13	33	31	28 ♒	18	50	04 ♓	56	34	22	43	25	42	27	47	12	39	23	23	06	44	22	15	17	51
06	06	58	11	14	30	42	11 ♓	37	16	18 ♓	21	03	23	30	26	49	28	26	12	44	23	30	06	48	22	16	17	50
07	07	02	08	15	27	54	25 ♓	08	02	01 ♈	58	18	24	22	27	56	29	04	12	49	23	38	06	51	22	17	17	49
08	07	06	04	16	25	06	08 ♈	51	59	15 ♈	49	09	25	18	29 ♌	03	29 ♋	43	12	54	23	46	06	55	22	17	17	48
09	07	10	01	17	22	19	22 ♈	49	50	29 ♈	53	59	26	18	00 ♍	10	00 ♌	21	12	59	23	54	06	58	22	18	17	47
10	07	13	57	18	19	32	07 ♉	01	31	14 ♉	12	11	27	23	01	16	01	00	13	03	24	01	07	02	22	18	17	46
11	07	17	54	19	16	45	21 ♉	25	39	28 ♉	41	29	28	33	02	22	01	38	13	08	24	09	07	05	22	19	17	45
12	07	21	50	20	13	59	05 ♊	58	35	13 ♊	17	44	29 ♊	47	03	28	02	16	13	12	24	17	07	09	22	19	17	44
13	07	25	47	21	11	14	20 ♊	36	41	27 ♊	55	02	01 ♋	05	04	34	02	55	13	16	24	25	07	13	22	20	17	42
14	07	29	43	22	08	28	05 ♋	11	53	12 ♋	26	20	02	27	05	40	03	33	13	20	24	32	07	16	22	20	17	41
15	07	33	40	23	05	44	19 ♋	37	31	26 ♋	44	38	03	53	06	45	04	12	13	23	24	40	07	20	22	20	17	40
16	07	37	37	24	02	59	03 ♌	47	00	10 ♌	44	03	05	24	07	51	04	50	13	27	24	48	07	23	22	21	17	39
17	07	41	33	25	00	15	17 ♌	35	22	24 ♌	20	42	06	58	08	56	05	28	13	30	24	56	07	27	22	21	17	38
18	07	45	30	25	57	31	00 ♍	59	57	07 ♍	33	08	08	36	10	01	06	07	13	33	25	04	07	30	22	21	17	36
19	07	49	26	26	54	47	14 ♍	00	26	20 ♍	22	08	10	18	11	06	06	45	13	36	25	11	07	34	22	22	17	35
20	07	53	23	27	52	04	26 ♍	38	38	02 ♎	50	23	12	04	12	10	07	23	13	39	25	19	07	37	22	22	17	34
21	07	57	19	28	49	20	08 ♎	57	56	15 ♎	01	51	13	52	13	15	08	02	13	41	25	27	07	40	22	22	17	33
22	08	01	16	29 ♋	46	37	21 ♎	02	47	27 ♎	01	22	15	44	14	19	08	40	13	44	25	35	07	44	22	22	17	31
23	08	05	12	00 ♌	43	55	02 ♏	58	15	08 ♏	54	04	17	39	15	23	09	18	13	46	25	42	07	47	22	22	17	30
24	08	09	09	01	41	12	14 ♏	49	29	20 ♏	45	06	19	36	16	27	09	57	13	48	25	50	07	51	22	22	17	29
25	08	13	06	02	38	30	26 ♏	41	30	02 ♐	39	16	21	35	17	30	10	35	13	50	25	58	07	54	22	22	17	28
26	08	17	02	03	35	49	08 ♐	38	53	14 ♐	40	49	23	36	18	33	11	13	13	51	26	06	07	57	22 R	22	17	26
27	08	20	59	04	33	08	20 ♐	45	28	26 ♐	53	12	25	39	19	36	11	52	13	53	26	13	08	01	22	22	17	25
28	08	24	55	05	30	27	03 ♑	04	16	09 ♑	18	53	27	43	20	39	12	30	13	54	26	21	08	04	22	22	17	24
29	08	28	52	06	27	47	15 ♑	37	13	21 ♑	59	18	29 ♋	48	21	41	13	08	13	55	26	29	08	07	22	22	17	22
30	08	32	48	07	25	08	28 ♑	25	11	04 ♒	54	47	01 ♌	54	22	44	13	46	13	56	26	37	08	10	22	22	17	21
31	08	36	45	08 ♌	22	29	11 ♒	28	02	18 ♒	04	46	04 ♌	00	23 ♍	45	14 ♌	25	13 ♈	56	26 ♋	44	08 ♋	14	22 ♈	22	17 ♒	20

Date	Moon True ☊ °	'	Moon Mean ☊ °	'	Moon ☽ Latitude °	'
01	27 ♍	15	27 ♍	50	04 N	56
02	27 R	02	27	46	04	38
03	26	52	27	43	04	06
04	26	43	27	40	03	21
05	26	38	27	37	02	24
06	26	35	27	34	01	19
07	26	34	27	31	00 N	08
08	26 D	35	27	27	01 S	05
09	26 R	35	27	24	02	15
10	26	33	27	21	03	18
11	26	30	27	18	04	09
12	26	24	27	15	04	44
13	26	16	27	12	05	02
14	26	07	27	08	05	00
15	25	57	27	05	04	38
16	25	48	27	02	04	00
17	25	41	26	59	03	08
18	25	36	26	56	02	07
19	25	33	26	52	01 S	01
20	25 D	33	26	49	00 N	06
21	25	33	26	46	01	11
22	25	34	26	43	02	12
23	25 R	35	26	40	03	07
24	25	33	26	37	03	53
25	25	31	26	33	04	29
26	25	26	26	30	04	54
27	25	19	26	27	05	06
28	25	11	26	24	05	04
29	25	02	26	21	04	48
30	24	54	26	18	04	16
31	24 ♍	47	26 ♍	14	03 N	32

DECLINATIONS

Date	Sun ☉ °	'	Moon ☽ °	'	Mercury ☿ °	'	Venus ♀ °	'	Mars ♂ °	'	Jupiter ♃ °	'	Saturn ♄ °	'	Uranus ♅ °	'	Neptune ♆ °	'	Pluto ⯓ °	'
01	23 N	04	18 S	19	19 N	03	16 N	05	22 N	12	03 N	39	21 N	31	23 N	33	07 N	08	22 S	58
02	23	00	17	25	19	13	15	41	22	05	03	41	21	29	23	33	07	08	22	59
03	22	55	15	39	19	24	15	17	21	57	03	43	21	28	23	33	07	08	22	59
04	22	50	13	04	19	36	14	52	21	50	03	45	21	27	23	32	07	09	23	00
05	22	45	09	48	19	48	14	28	21	43	03	47	21	26	23	32	07	09	23	00
06	22	39	06	00	20	02	14	02	21	35	03	49	21	24	23	32	07	09	23	01
07	22	32	01 S	49	20	15	13	37	21	27	03	50	21	23	23	32	07	09	23	01
08	22	25	02 N	31	20	29	13	11	21	19	03	52	21	22	23	32	07	09	23	02
09	22	18	06	47	20	44	12	45	21	11	03	54	21	21	23	32	07	10	23	02
10	22	11	10	44	20	58	12	19	21	03	03	55	21	20	23	31	07	10	23	03
11	22	03	14	06	21	12	11	52	20	54	03	57	21	18	23	31	07	10	23	04
12	21	54	16	38	21	25	11	26	20	46	03	58	21	17	23	31	07	10	23	04
13	21	46	18	05	21	38	10	59	20	37	03	59	21	16	23	31	07	10	23	05
14	21	37	18	18	21	51	10	31	20	28	04	00	21	14	23	31	07	10	23	05
15	21	27	17	24	22	02	10	04	20	19	04	01	21	13	23	30	07	10	23	06
16	21	18	15	25	22	12	09	36	20	10	04	03	21	12	23	30	07	10	23	07
17	21	07	12	34	22	21	09	08	20	00	04	04	21	11	23	30	07	10	23	07
18	20	57	09	08	22	29	08	40	19	51	04	05	21	09	23	30	07	10	23	08
19	20	46	05	21	22	35	08	12	19	41	04	06	21	08	23	30	07	10	23	08
20	20	35	01 N	25	22	38	07	44	19	32	04	06	21	07	23	30	07	10	23	09
21	20	23	02 S	28	22	40	07	15	19	22	04	07	21	05	23	29	07	10	23	09
22	20	11	06	10	22	40	06	47	19	12	04	08	21	04	23	29	07	10	23	09
23	19	59	09	34	22	37	06	18	19	02	04	08	21	02	23	29	07	10	23	10
24	19	47	12	34	22	32	05	49	18	51	04	09	21	01	23	29	07	10	23	10
25	19	34	15	03	22	24	05	20	18	41	04	09	21	00	23	29	07	10	23	11
26	19	21	16	54	22	13	04	51	18	31	04	10	20	58	23	28	07	10	23	11
27	19	07	18	01	22	00	04	22	18	20	04	10	20	57	23	28	07	10	23	12
28	18	53	18	20	21	44	03	53	18	09	04	10	20	56	23	28	07	10	23	13
29	18	39	17	45	21	25	03	24	17	58	04	10	20	54	23	28	07	10	23	13
30	18	25	16	17	21	04	02	55	17	47	04	10	20	53	23	28	07	10	23	14
31	18 N	10	13 S	57	20 N	40	02 N	26	17 N	36	04 N	10	20 N	52	23 N	28	07 N	10	23 S	14

ZODIAC SIGN ENTRIES

Date	h	m	Planets
03	07	44	☽ ♒
05	15	04	☽ ♓
07	20	33	☽ ♈
08	22	51	♂ ♌
09	08	34	♀ ♍
10	00	10	☽ ♉
12	02	09	☽ ♊
12	16	13	☿ ♋
14	03	26	☽ ♋
16	05	32	☽ ♌
18	10	11	☽ ♍
20	18	29	☽ ♎
22	17	36	☉ ♌
23	04	20	☽ ♏
25	18	40	☽ ♐
28	06	03	☽ ♑
29	14	13	☿ ♌
30	14	56	☽ ♒

LATITUDES

Date	Mercury ☿ °	'	Venus ♀ °	'	Mars ♂ °	'	Jupiter ♃ °	'	Saturn ♄ °	'	Uranus ♅ °	'	Neptune ♆ °	'	Pluto ⯓ °	'
01	04 S	03	01 N	45	01 N	08	01 S	18	00 N	01	00 N	17	01 S	38	07 S	55
04	03	36	01	39	01	08	01	19	00	01	00	17	01	38	07	55
07	03	04	01	31	01	08	01	19	00	01	00	17	01	38	07	56
10	02	27	01	23	01	08	01	20	00	02	00	17	01	38	07	57
13	01	47	01	14	01	09	01	21	00	02	00	17	01	38	07	57
16	01	07	01	05	01	09	01	22	00	02	00	17	01	39	07	58
19	00 S	27	00	52	01	09	01	23	00	03	00	17	01	39	07	58
22	00 N	09	00	40	01	09	01	24	00	03	00	17	01	39	07	58
25	00	42	00	26	01	09	01	25	00	03	00	17	01	39	07	59
28	01	09	00 N	12	01	09	01	25	00	03	00	17	01	39	07	59
31	01 N	28	00 S	03	01 N	09	01 S	26	00 N	03	00 N	17	01 S	39	08 S	00

DATA

Julian Date	2464145
Delta T	+75 seconds
Ayanamsa	24° 20' 20"
Synetic vernal point	04° ♓ 46' 39"
True obliquity of ecliptic	23° 25' 55"

MOON'S PHASES, APSIDES AND POSITIONS ☽

Date	h	m	Phase	Longitude °	'	Eclipse Indicator
01	17	44	○	09 ♑	58	
09	01	59	☾	16 ♈	58	
15	18	15	●	23 ♋	21	
23	07	05	☽	00 ♏	32	
31	05	54	○	08 ♒	08	

Day	h	m		
12	19	24	Perigee	
24	17	19	Apogee	
07	22	05	0N	
14	05	07	Max dec	18° N 23'
20	20	44	0S	
28	08	33	Max dec	18° S 20'

LONGITUDES

Date	Chiron ⚷ °	'	Ceres ⚳ °	'	Pallas ⚴ °	'	Juno ⚵ °	'	Vesta ⚶ °	'	Black Moon Lilith ⚸ °	'
01	04 ♊	01	27 ♓	51	08 ♒	46	04 ♐	58	10 ♊	50	16 ♏	55
11	04 ♊	38	28 ♓	51	06 ♒	27	03 ♐	45	14 ♊	59	18 ♏	02
21	05 ♊	10	29 ♓	20	03 ♒	49	03 ♐	04	19 ♊	02	19 ♏	09
31	05 ♊	38	29 ♓	14	01 ♒	08	02 ♐	57	22 ♊	59	20 ♏	16

ASPECTARIAN

01 Saturday
04 09 ☽ ∠ ⯓
10 21 ☽ ♇ ♀
11 03 ☽ ☍ ♅
17 44 ☽ ☍ ☉
18 19 ♀ ∠ ♅
21 28 ☽ ⊥ ⯓
22 13 ☽ □ ♃

02 Sunday
05 19 ☽ ± ♀
08 55 ☽ ⊻ ⯓
09 26 ♀ △ ♆
14 36 ☽ ⊼ ☿
17 05 ☽ □ ♆
17 49 ☉ # ⯓
17 50 ☽ ⊼ ♀
18 37 ☽ ☍ ♄

03 Monday
00 36 ☽ ☍ ♂
02 25 ☽ ± ☿
03 43 ♀ ⊻ ♄
08 35 ☽ Q ♃
16 45 ☽ # ♀
17 55 ☉ ± ⯓
20 04 ☽ ♇ ☿
20 08 ☽ ⊼ ♅

04 Tuesday
02 50 ☽ Q ♆
06 50 ☽ ⊼ ☉
07 05 ☽ ⚹ ♃
07 19 ☽ ± ♅
10 34 ☉ □ ♃
16 55 ☽ ☌ ⯓
18 45 ☽ ± ☉
20 14 ☿ ⚹ ♆
23 57 ☽ ♇ ♅

05 Wednesday
00 57 ☽ ⚹ ♆
01 13 ☽ △ ☿
02 55 ☽ ⊼ ♄
06 49 ☽ ☍ ♀
10 46 ☽ ∠ ♃
11 00 ☽ ⊼ ♂
12 29 ☽ ♇ ☉
13 57 ☽ ± ♄
22 25 ☽ ± ♂

06 Thursday
03 10 ☽ ⊥ ♃
03 18 ☽ △ ♅
04 11 ☽ ∠ ♆
05 00 ☽ # ♆
06 22 ☽ ♇ ♄
12 10 ☿ ⊻ ♄
14 00 ☽ ⊻ ♃
15 24 ☽ ♇ ♂
17 34 ☽ △ ☉
20 18 ☽ ⊥ ♆
23 05 ☽ ⊻ ⯓

07 Friday
00 38 ☽ # ♃
06 57 ☽ ⊻ ♆
09 11 ♀ ♇ ♃
09 20 ☽ △ ♄
09 41 ☽ ⊥ ⯓
10 32 ☽ □ ☿
17 22 ☽ ⊼ ♀
19 16 ☽ △ ♂

08 Saturday
01 08 ☿ Q ♃
01 26 ♂ ∥ ♄
01 28 ☽ ∠ ⯓
04 47 ☽ ± ♀
08 36 ☽ □ ♅
19 01 ☽ ☌ ♃
19 32 ☽ ∥ ♃
20 14 ☽ Q ☿
21 44 ☽ ♇ ♀

09 Sunday
01 59 ☽ □ ☉
03 23 ☽ ⚹ ⯓
05 29 ♀ ⊥ ♄
11 05 ☽ ☌ ♆
13 50 ☽ □ ♄
14 09 ☽ ∥ ♆
15 40 ☽ Q ♅
18 23 ☽ ⚹ ☿
21 54 ♀ ⊻ ♂
22 10 ☉ ⊼ ⯓
23 47 ☽ Q ⯓

10 Monday
01 22 ☽ □ ♂
01 29 ☽ △ ♀
10 44 ☽ Q ☉
12 01 ☽ ⚹ ♅
17 14 ☿ ∥ ♂
20 26 ☽ Q ♄
21 24 ☽ ∥ ♀
21 45 ☽ ∠ ☿
22 08 ☽ ⊻ ♃

11 Tuesday
05 54 ☽ □ ⯓
08 10 ☽ ⊥ ♃
08 11 ☽ ⚹ ☉
08 53 ☽ Q ♂
13 06 ☽ ∠ ♅
13 28 ☽ ⊻ ♆
14 01 ☽ ⊥ ☿
16 33 ☽ ⚹ ♄
22 30 ☿ ∥ ♄
23 07 ☽ ∠ ♃
23 23 ☽ ⊥ ♆

12 Wednesday
00 50 ☽ ⊻ ☿
04 01 ☽ ⊥ ♅
05 37 ☽ ⚹ ♂
07 32 ☽ □ ♀
10 41 ☽ ∠ ☉
13 55 ☽ ⊻ ♅
14 12 ☽ ∠ ♆
17 28 ☽ ∠ ♄
23 54 ☽ ⚹ ♃

13 Thursday
02 29 ☽ ⊥ ☉
07 15 ☽ △ ⯓
07 22 ☽ ∠ ♂
08 21 ☽ ⊥ ♄
13 01 ☽ ⊻ ☉
14 49 ☽ ⚹ ♆
15 28 ☽ Q ♀
18 18 ☽ ⊻ ♄
19 40 ☽ Q ♃
20 20 ☉ ∥ ☿
22 49 ☽ ⊥ ♂

14 Friday
06 59 ☽ ☌ ☿
07 52 ☽ ♇ ⯓
09 10 ☽ ⊻ ♂
10 34 ☽ Q ♆
12 50 ☽ ⚹ ♀
15 26 ☽ ☌ ♅
15 59 ☿ ♇ ⯓
16 54 ☉ □ ♆
22 44 ☽ ± ⯓

15 Saturday
01 32 ☽ □ ♃
08 44 ☽ ⊼ ⯓
15 53 ☽ ∠ ♀
16 34 ☽ □ ♆
18 15 ☽ ☌ ☉
19 19 ☿ Q ♆
20 34 ☽ ☌ ♄
20 35 ☿ ⊻ ♂

16 Sunday
00 54 ♀ ♇ ♆
01 14 ♀ ⚹ ♅
02 40 ♀ ± ♃
08 24 ☽ ⊥ ♀
13 53 ☽ ☌ ♂
15 07 ☽ ⊻ ☿
18 13 ☽ ⊻ ♅
19 36 ☽ ⊻ ♀

17 Monday
02 50 ☽ ⊥ ☿
03 54 ☉ ∥ ♄
04 41 ☽ ⊥ ♅
04 48 ☽ △ ♃
09 49 ☉ ☌ ♄
12 04 ☽ ☍ ⯓
19 17 ☿ ☌ ♅
20 27 ☽ △ ♆
20 38 ☽ ∠ ♅
20 49 ☽ ∠ ☿

18 Tuesday
01 10 ☽ ⊻ ♄
02 12 ☽ ⊻ ☉
07 33 ☽ ♇ ♃
12 07 ☽ ⊥ ♄
13 02 ♀ ∠ ♄
13 53 ☽ ⊥ ☉
15 25 ☽ ∥ ♀
21 50 ☽ ⊻ ♂
23 38 ☽ ♇ ♆
23 58 ☽ ⚹ ♅

19 Wednesday
00 03 ☽ ± ♃
00 36 ☽ ∥ ♆
04 02 ☽ ⚹ ☿
04 48 ☽ ∠ ♄
06 04 ☽ ☌ ♀
07 46 ☽ ∠ ☉
09 32 ☽ ⊥ ♂
11 14 ☽ ⊼ ♃
16 25 ☽ ± ♆
18 43 ☽ ⊼ ⯓
19 40 ☽ ∥ ♃
22 31 ☽ Q ♅

20 Thursday
03 25 ☽ ∠ ♂
03 48 ☽ ⊼ ♆
05 23 ☿ ± ⯓
06 06 ☽ ± ⯓
06 13 ☽ Q ☿
09 25 ☽ ⚹ ♄
14 33 ☽ ⚹ ☉
15 49 ☿ ⚹ ♀
21 20 ♂ ⊻ ♅
23 27 ☽ ♇ ⯓

21 Friday
08 59 ☽ Q ♄
09 27 ☽ □ ♅
09 35 ☿ □ ♃
10 03 ☽ ⚹ ♂
15 58 ☽ Q ☉
16 16 ♀ ∥ ♆
21 17 ☽ ⊻ ♀
21 22 ☽ ☍ ♃
22 19 ♀ ⊼ ♃
22 36 ☽ # ♃
23 27 ☽ □ ☿

22 Saturday
04 58 ☽ △ ⯓
10 24 ☽ ⊥ ♀
11 12 ☽ Q ♂
14 39 ☽ ☍ ♆
15 41 ☽ # ♀
18 52 ☽ # ♆
21 11 ☽ □ ♄

23 Sunday
06 16 ☽ ∠ ♀
07 05 ☽ □ ☉
07 42 ♀ ⊥ ♂
10 16 ☿ ⊼ ⯓
21 47 ☽ △ ♅

24 Monday
01 33 ☽ □ ♂
09 55 ☽ ⊼ ♃
10 20 ♀ ± ♆
15 36 ☽ ⚹ ♀
17 22 ☽ □ ⯓
22 06 ☽ ± ♃
23 36 ☽ △ ☿

25 Tuesday
03 16 ☽ ⊼ ♆
04 18 ☽ ♇ ♅
10 31 ☽ △ ♄
11 03 ♀ ⊼ ⯓
15 23 ☽ ± ♆
16 19 ☽ ♇ ♃
18 13 ☽ Q ♀
21 24 ☿ □ ♆
22 32 ☽ ± ♅
22 33 ♆ St R

26 Wednesday
01 01 ☽ △ ☉
05 36 ☽ Q ⯓
09 27 ☽ ♇ ♆
10 37 ☽ ⊼ ♅
11 54 ☽ ♇ ☿
16 56 ☽ ♇ ♄
17 25 ☽ △ ♂
22 23 ☽ △ ♃
22 26 ☉ ∠ ♀

27 Thursday
05 25 ☽ ⚹ ⯓
09 23 ☽ ± ☿
09 26 ☽ ♇ ☉
09 31 ☽ □ ♀
10 56 ☽ ± ♄
15 10 ☽ △ ♆
19 05 ☿ ☌ ♄
21 49 ♀ Q ♅
22 05 ☽ # ♂
22 12 ♀ ∥ ♃
22 49 ☽ ⊼ ♄
23 32 ☽ ⊼ ☿

28 Friday
00 36 ☽ ♇ ♂
04 31 ☽ ± ☉
10 42 ☽ ∠ ⯓
17 05 ☽ ⊼ ☉
18 57 ☽ ± ♂
21 39 ☽ ☍ ♅

29 Saturday
03 57 ☽ ⊥ ⯓
04 41 ☽ # ♂
07 02 ☽ ⊼ ♂
08 46 ☽ □ ♃
15 18 ☽ ⊻ ⯓

30 Sunday
00 27 ☽ △ ♀
00 42 ☽ □ ♆
03 38 ♀ ⊼ ♆
08 36 ☽ ☍ ♄
18 03 ♂ △ ♃
18 30 ☽ Q ♃
19 42 ☽ ☍ ☿

31 Monday
00 22 ☿ ∥ ♄
02 12 ♀ ± ⯓
04 36 ♂ ⊥ ♅
05 54 ☽ ☍ ☉
06 04 ☽ ⊼ ♅
06 38 ☽ ♇ ♀
08 07 ☉ ⊻ ♅
09 59 ☽ Q ♆
16 30 ☽ ⚹ ♃
17 03 ☽ ± ♅
17 37 ☽ ☍ ♂
22 37 ☽ ☌ ⯓

All ephemeris data is given at 12.00 UT and the Moon's longitude is additionally given for 24.00 UT

AUGUST 2034

LONGITUDES

Date	Sidereal time h m s	Sun ☉ ° ' "	Moon ☽ ° ' "	Moon ☽ 24.00 ° ' "	Mercury ☿ ° '	Venus ♀ ° '	Mars ♂ ° '	Jupiter ♃ ° '	Saturn ♄ ° '	Uranus ♅ ° '	Neptune ♆ ° '	Pluto ♇ ° '
01	08 40 41	09 ♌ 19 51	24 ♒ 44 50	01 ♓ 28 01	06 ♌ 06	24 ♍ 47	15 ♌ 03	13 ♈ 57	26 ♋ 52	08 ♋ 17	22 ♈ 22	17 ♒ 18
02	08 44 38	10 17 14	08 ♓ 14 07	15 ♓ 02 57	08 12	25 48	15 41	13 57	27 00	08 20	22 R 21	17 R 17
03	08 48 35	11 14 38	21 ♓ 54 17	28 ♓ 47 57	10 17	26 49	16 19	13 R 57	27 07	08 23	22 21	17 16
04	08 52 31	12 12 03	05 ♈ 43 46	12 ♈ 41 35	12 22	27 50	16 57	13 57	27 15	08 26	22 21	17 14
05	08 56 28	13 09 30	19 ♈ 41 14	26 ♈ 42 35	14 26	28 50	17 36	13 57	27 22	08 30	22 20	17 13
06	09 00 24	14 06 57	03 ♉ 45 29	10 ♉ 49 46	16 29	29 ♍ 50	18 14	13 56	27 30	08 33	22 20	17 12
07	09 04 21	15 04 26	17 ♉ 55 15	25 ♉ 01 43	18 30	00 ♎ 50	18 52	13 55	27 38	08 36	22 20	17 10
08	09 08 17	16 01 56	02 ♊ 08 35	09 ♊ 16 28	20 31	01 49	19 30	13 54	27 45	08 39	22 19	17 09
09	09 12 14	16 59 28	16 ♊ 24 05	23 ♊ 31 20	22 30	02 48	20 09	13 53	27 53	08 42	22 19	17 08
10	09 16 10	17 57 01	00 ♋ 37 45	07 ♋ 42 51	24 28	03 47	20 47	13 52	28 00	08 45	22 18	17 06
11	09 20 07	18 54 35	14 ♋ 46 06	21 ♋ 46 59	26 24	04 45	21 25	13 50	28 08	08 48	22 18	17 05
12	09 24 04	19 52 11	28 ♋ 45 00	05 ♌ 39 40	28 ♌ 19	05 43	22 03	13 48	28 15	08 51	22 17	17 04
13	09 28 00	20 49 48	12 ♌ 30 33	19 ♌ 17 16	00 ♍ 13	06 40	22 41	13 46	28 22	08 54	22 17	17 02
14	09 31 57	21 47 26	25 ♌ 59 32	02 ♍ 37 09	02 05	07 37	23 20	13 44	28 30	08 57	22 16	17 01
15	09 35 53	22 45 05	09 ♍ 10 01	15 ♍ 38 04	03 56	08 34	23 58	13 42	28 37	08 59	22 15	16 59
16	09 39 50	23 42 46	22 ♍ 01 25	28 ♍ 20 11	05 45	09 30	24 36	13 39	28 44	09 02	22 15	16 58
17	09 43 46	24 40 27	04 ♎ 34 39	10 ♎ 45 06	07 33	10 25	25 14	13 37	28 52	09 05	22 14	16 57
18	09 47 43	25 38 10	16 ♎ 51 56	22 ♎ 55 35	09 19	11 20	25 52	13 34	28 59	09 08	22 13	16 55
19	09 51 39	26 35 53	28 ♎ 56 33	04 ♏ 55 22	11 04	12 15	26 30	13 31	29 06	09 10	22 13	16 54
20	09 55 36	27 33 38	10 ♏ 52 35	16 ♏ 48 48	12 47	13 09	27 09	13 27	29 13	09 13	22 12	16 53
21	09 59 33	28 31 24	22 ♏ 44 36	28 ♏ 40 36	14 29	14 02	27 47	13 24	29 21	09 16	22 11	16 51
22	10 03 29	29 ♌ 29 11	04 ♐ 37 25	10 ♐ 35 37	16 09	14 55	28 25	13 20	29 28	09 18	22 10	16 50
23	10 07 26	00 ♍ 27 00	16 ♐ 35 48	22 ♐ 38 31	17 49	15 48	29 03	13 16	29 35	09 21	22 09	16 49
24	10 11 22	01 24 49	28 ♐ 44 17	04 ♑ 53 33	19 26	16 39	29 ♌ 41	13 12	29 42	09 23	22 08	16 48
25	10 15 19	02 22 40	11 ♑ 06 45	17 ♑ 24 14	21 03	17 31	00 ♍ 20	13 08	29 49	09 26	22 07	16 46
26	10 19 15	03 20 32	23 ♑ 46 17	00 ♒ 13 06	22 38	18 21	00 58	13 04	29 ♋ 56	09 28	22 07	16 45
27	10 23 12	04 18 25	06 ♒ 44 48	13 ♒ 21 26	24 11	19 11	01 36	12 59	00 ♌ 03	09 31	22 06	16 44
28	10 27 08	05 16 20	20 ♒ 02 55	26 ♒ 49 06	25 44	20 00	02 14	12 54	00 10	09 33	22 05	16 42
29	10 31 05	06 14 16	03 ♓ 39 45	10 ♓ 34 34	27 15	20 48	02 52	12 50	00 16	09 36	22 04	16 41
30	10 35 02	07 12 14	17 ♓ 33 08	24 ♓ 35 02	28 ♍ 44	21 35	03 30	12 45	00 23	09 38	22 02	16 40
31	10 38 58	08 ♍ 10 13	01 ♈ 39 45	08 ♈ 46 46	00 ♎ 13	22 ♎ 22	04 ♍ 09	12 ♈ 39	00 ♌ 30	09 ♋ 40	22 ♈ 01	16 ♒ 39

DECLINATIONS

Date	Moon True ☊ ° '	Moon Mean ☊ ° '	Moon ☽ Latitude ° '	Sun ☉ ° '	Moon ☽ ° '	Mercury ☿ ° '	Venus ♀ ° '	Mars ♂ ° '	Jupiter ♃ ° '	Saturn ♄ ° '	Uranus ♅ ° '	Neptune ♆ ° '	Pluto ♇ ° '
01	24 ♍ 41	26 ♍ 11	02 N 34	17 N 55	10 S 51	20 N 15	01 N 56	17 N 25	04 N 10	20 N 50	23 N 27	07 N 10	23 S 15
02	24 R 38	26 08	01 28	17 40	07 28	19 46	01 27	17 14	04 10	20 49	23 27	07 09	23 15
03	24 37	26 05	00 N 15	17 24	02 S 59	19 16	00 58	17 02	04 10	20 47	23 27	07 09	23 16
04	24 D 37	26 02	01 S 00	17 08	01 N 21	18 44	00 N 29	16 51	04 09	20 46	23 27	07 09	23 16
05	24 38	25 58	02 12	16 52	05 40	18 10	00 S 01	16 39	04 09	20 45	23 27	07 09	23 17
06	24 40	25 55	03 16	16 35	09 41	17 34	00 30	16 27	04 09	20 43	23 26	07 09	23 17
07	24 R 40	25 52	04 09	16 19	13 10	16 57	00 59	16 15	04 08	20 42	23 26	07 09	23 18
08	24 39	25 49	04 47	16 02	15 53	16 19	01 28	16 03	04 07	20 40	23 26	07 08	23 18
09	24 36	25 46	05 08	15 44	17 38	15 40	01 57	15 51	04 07	20 39	23 26	07 08	23 19
10	24 32	25 43	05 09	15 27	18 17	14 59	02 26	15 39	04 06	20 38	23 26	07 08	23 19
11	24 27	25 39	04 52	15 09	17 47	14 18	02 55	15 27	04 05	20 36	23 26	07 08	23 20
12	24 22	25 36	04 17	14 51	16 13	13 35	03 24	15 14	04 04	20 35	23 25	07 07	23 20
13	24 17	25 33	03 28	14 33	13 43	12 53	03 52	15 02	04 03	20 33	23 25	07 07	23 21
14	24 13	25 30	02 28	14 14	10 32	12 09	04 21	14 49	04 02	20 32	23 25	07 07	23 21
15	24 10	25 27	01 21	13 56	06 53	11 25	04 49	14 37	04 01	20 31	23 25	07 07	23 21
16	24 09	25 23	00 S 12	13 37	02 N 59	10 41	05 18	14 24	04 00	20 29	23 25	07 06	23 22
17	24 D 10	25 20	00 N 57	13 18	00 S 57	09 56	05 46	14 11	03 58	20 28	23 25	07 06	23 22
18	24 11	25 17	02 01	12 58	04 45	09 12	06 14	13 58	03 57	20 26	23 24	07 06	23 23
19	24 13	25 14	02 59	12 39	08 18	08 27	06 42	13 45	03 56	20 25	23 24	07 05	23 23
20	24 14	25 11	03 49	12 19	11 27	07 42	07 09	13 32	03 54	20 24	23 24	07 05	23 24
21	24 15	25 08	04 28	11 59	14 07	06 57	07 37	13 19	03 52	20 22	23 24	07 05	23 24
22	24 R 15	25 04	04 56	11 39	16 12	06 12	08 04	13 05	03 51	20 21	23 24	07 04	23 25
23	24 14	25 01	05 12	11 19	17 35	05 27	08 31	12 52	03 49	20 19	23 24	07 04	23 25
24	24 12	24 58	05 14	10 58	18 12	04 42	08 58	12 39	03 47	20 18	23 23	07 04	23 25
25	24 09	24 55	05 01	10 37	17 58	03 58	09 24	12 25	03 45	20 17	23 23	07 03	23 26
26	24 06	24 52	04 34	10 17	16 51	03 13	09 51	12 11	03 44	20 15	23 23	07 03	23 26
27	24 03	24 49	03 52	09 56	14 50	02 29	10 17	11 58	03 42	20 14	23 23	07 02	23 27
28	24 01	24 45	02 56	09 35	12 00	01 45	10 42	11 44	03 39	20 13	23 23	07 02	23 27
29	23 59	24 42	01 50	09 13	08 27	01 02	11 08	11 30	03 37	20 11	23 23	07 02	23 27
30	23 58	24 39	00 35	08 52	04 22	00 19	11 33	11 16	03 35	20 10	23 22	07 01	23 28
31	23 ♍ 58	24 ♍ 36	00 S 42	08 N 30	00 N 01	00 S 24	11 S 58	11 N 02	03 N 33	20 N 08	23 N 22	07 N 01	23 S 28

ZODIAC SIGN ENTRIES

Date	h	m	Planets
01	21	23	☽ ♓
04	02	05	☽ ♈
06	05	36	☽ ♉
06	15	54	♀ ♎
08	08	23	☽ ♊
10	10	56	☽ ♋
12	14	10	☽ ♌
13	09	14	☿ ♍
14	19	14	☽ ♍
17	03	11	☽ ♎
19	14	07	☽ ♏
22	02	40	☽ ♐
23	00	48	☉ ♍
24	14	28	☽ ♑
24	23	42	♂ ♍
26	23	36	☽ ♒
27	02	46	♄ ♌
29	05	36	☽ ♓
31	08	33	☿ ♎
31	09	11	☽ ♈

LATITUDES

Date	Mercury ☿ ° '	Venus ♀ ° '	Mars ♂ ° '	Jupiter ♃ ° '	Saturn ♄ ° '	Uranus ♅ ° '	Neptune ♆ ° '	Pluto ♇ ° '
01	01 N 33	00 S 09	01 N 09	01 S 27	00 N 04	00 N 17	01 S 40	08 S 00
04	01 43	00 25	01 09	01 27	00 04	00 17	01 40	08 00
07	01 46	00 43	01 09	01 28	00 04	00 17	01 40	08 00
10	01 43	01 01	01 09	01 29	00 04	00 17	01 40	08 00
13	01 35	01 20	01 09	01 30	00 05	00 17	01 40	08 01
16	01 23	01 40	01 09	01 31	00 05	00 17	01 40	08 01
19	01 07	02 01	01 09	01 32	00 05	00 17	01 41	08 01
22	00 48	02 22	01 08	01 32	00 05	00 17	01 41	08 01
25	00 27	02 45	01 08	01 33	00 06	00 17	01 41	08 01
28	00 N 04	03 07	01 08	01 34	00 06	00 17	01 41	08 01
31	00 S 20	03 S 31	01 N 08	01 S 34	00 N 06	00 N 18	01 S 41	08 S 01

DATA

Julian Date	2464176
Delta T	+75 seconds
Ayanamsa	24° 20' 25"
Synetic vernal point	04° ♓ 46' 34"
True obliquity of ecliptic	23° 25' 56"

MOON'S PHASES, APSIDES AND POSITIONS ☽

Date	h	m	Phase	Longitude ° '	Eclipse Indicator
07	06	50	☾	14 ♉ 52	
14	03	53	●	21 ♌ 28	
22	00	43	☽	29 ♏ 02	
29	16	49	○	06 ♓ 26	

Day	h	m		
08	08	29	Perigee	
21	12	03	Apogee	
04	04	32	ON	
10	13	29	Max dec	18° N 17'
17	06	11	OS	
24	17	37	Max dec	18° S 13'
31	11	56	ON	

LONGITUDES

Date	Chiron ⚷ ° '	Ceres ⚳ ° '	Pallas ⚴ ° '	Juno ⚵ ° '	Vesta ⚶ ° '	Black Moon Lilith ⚸ ° '
01	05 ♊ 41	29 ♓ 12	00 ♒ 53	02 ♐ 58	23 ♊ 22	20 ♏ 23
11	06 ♊ 03	28 ♓ 26	28 ♑ 25	03 ♐ 26	27 ♊ 10	21 ♏ 30
21	06 ♊ 20	27 ♓ 07	26 ♑ 23	04 ♐ 23	00 ♋ 50	22 ♏ 37
31	06 ♊ 31	25 ♓ 18	24 ♑ 54	05 ♐ 47	04 ♋ 18	23 ♏ 44

All ephemeris data is given at 12.00 UT and the Moon's longitude is additionally given for 24.00 UT

ASPECTARIAN

h m	Aspects
01 Tuesday	
00 23	☽ ± ♀
07 43	☽ ⚹ ♆
09 22	☽ bQ ♅
12 04	☽ ⊼ ♀
15 50	☽ ⊼ ♄
19 31	☽ ∠ ♃
02 Wednesday	
02 37	☽ ± ♄
10 27	☽ ∠ ♆
11 30	☽ ⊥ ♃
11 48	☽ # ♆
11 56	☽ ⊼ ☿
12 11	☽ △ ♅
13 35	☿ ⊻ ♅
15 54	☽ ⊼ ☉
18 41	☽ bQ ♄
22 04	☽ ⊻ ♃
03 Thursday	
00 25	☽ ± ☿
01 45	☽ ⊼ ♂
02 17	☽ ⊥ ♆
03 14	☽ ± ☉
03 41	♃ St R
03 54	☽ ⊻ ♇
05 20	☽ # ♃
12 46	☽ ± ♂
12 47	☽ ⊻ ♆
14 22	☽ ⊥ ♇
18 57	☽ bQ ☿
20 04	♀ ⚹ ♄
20 07	☽ bQ ☉
21 10	☽ △ ♄
21 15	☽ ☍ ♀
04 Friday	
00 38	☽ # ♀
05 10	☽ bQ ♂
05 58	☽ ∠ ♇
07 39	☽ ∥ ♀
08 24	☉ ☌ ☿
16 42	☽ □ ♅
22 15	♂ ☍ ♇
22 30	☿ ∠ ♀
23 59	☽ △ ☉
05 Saturday	
01 26	☽ △ ☿
02 09	☽ ☌ ♃
03 28	☽ ∥ ♃
06 18	☿ △ ♃
07 46	☽ ⚹ ♇
08 15	☽ △ ♂
12 44	☿ ⊥ ♅
16 32	☽ ☌ ♆
20 35	☽ ∥ ♆
23 40	☽ Q ♅
06 Sunday	
01 15	☽ □ ♄
04 15	☽ Q ♇
04 49	☽ ⊼ ♀
07 27	☉ △ ♃
15 48	☽ ± ♀
20 09	☽ ⚹ ♅
20 20	☿ ☍ ♇
23 21	☉ ⊥ ♅
07 Monday	
05 14	☽ ⊻ ♃
06 50	☽ □ ☉
08 05	☽ Q ♄
08 12	☽ bQ ♀
10 44	☽ □ ♇
13 09	☽ □ ☿
13 41	☽ □ ♂
15 22	☽ ⊥ ♃
18 16	☿ ☌ ♂
19 26	☽ ⊻ ♆
21 37	☽ ∠ ♅
08 Tuesday	
03 26	☉ ∥ ♂
04 31	☽ ⚹ ♄
05 33	☽ ⊥ ♆
06 32	☽ ∠ ♃
11 24	☽ △ ♀
12 51	☽ ⊥ ♅
13 18	☽ ∥ ☉
13 39	☽ ∥ ♂
15 24	☽ Q ☉
15 37	☽ ∥ ☿
19 49	♀ bQ ♇
20 42	☽ ∠ ♆
21 27	☽ Q ♂
22 59	☽ ⊻ ♅
09 Wednesday	
00 28	☽ Q ☿
01 48	☿ ∥ ♂
06 01	☽ ∠ ♄
06 53	☉ ∥ ☿
07 46	☽ ⚹ ♃
09 43	☿ △ ♆
13 04	☽ ⚹ ☉
13 13	☽ △ ♇
15 18	☉ ☍ ♇
18 36	☽ ⚹ ♂
21 18	☽ ⊥ ♄
21 57	☽ ⚹ ♆
23 56	☽ ⚹ ☿
10 Thursday	
02 57	☿ ∠ ♅
03 58	☽ Q ♃
07 31	☽ ⊻ ♄
14 29	☽ bQ ♇
16 13	☽ ∠ ☉
17 43	☽ □ ♀
18 13	☽ Q ♆
21 08	☽ ∠ ♂
11 Friday	
01 48	☽ ☌ ♅
05 22	☽ ∠ ☿
05 44	☽ ± ♇
08 36	☽ ⊥ ☉
10 25	☽ □ ♃
13 09	☽ ⊥ ♂
15 57	☽ ⊼ ♇
19 36	☽ ⊻ ☉
23 11	☽ ⊥ ☿
23 55	☽ ⊻ ♂
12 Saturday	
00 52	☽ □ ♆
02 40	☽ Q ♀
11 00	☿ ⊻ ♄
11 08	☽ ☌ ♄
11 09	☽ ⊻ ☿
18 00	☿ bQ ♃
20 46	♂ △ ♆
23 16	☽ ∥ ♂
13 Sunday	
01 00	☽ ⚹ ♀
03 54	☽ ∥ ☉
05 38	☽ ⊻ ♅
14 13	☽ △ ♃
16 13	☽ ⊥ ♅
19 59	☽ ☍ ♇
20 40	♀ # ♃
20 53	☽ ∥ ☿
23 46	☿ ⊥ ♀
14 Monday	
03 53	☽ ☌ ☉
05 19	☽ △ ♆
05 29	☽ ∠ ♀
06 58	☽ ☌ ♂
08 18	☽ ∠ ♅
16 34	☽ ⊻ ♄
16 56	☽ bQ ♃
22 58	☽ ⊥ ♀
23 48	☉ △ ♆
15 Tuesday	
00 50	☽ ☌ ☿
03 34	☽ ⊥ ♄
08 29	☽ bQ ♆
09 18	☽ ± ♃
10 31	☽ ∥ ♆
10 48	☽ ⊻ ♀
11 40	☽ ⚹ ♅
13 07	♂ ∠ ♅
20 19	☽ ∠ ♄
20 22	☽ ⊼ ♃
21 44	☿ ⊥ ♄
23 23	☽ # ♀
23 35	♀ □ ♅
16 Wednesday	
01 09	☽ ± ♆
02 31	☽ ⊼ ♇
05 49	☽ ∥ ♃
10 08	☽ Q ♅
12 25	☽ ⊼ ♆
13 47	☽ ± ♇
15 28	☽ ⊻ ☉
17 08	☽ ⊻ ♂
20 30	☉ ∠ ♅
20 33	♀ ∠ ♂
17 Thursday	
00 54	☽ ⚹ ♄
03 51	☽ ⊥ ☉
05 13	☽ ⊥ ♂
06 56	☽ bQ ♇
07 54	☿ bQ ♆
12 55	☿ ± ♃
18 43	☽ ⊻ ☿
20 47	☽ □ ♅
22 44	☽ ∠ ☉
23 36	☽ ∠ ♂
18 Friday	
00 16	☽ ☌ ♀
00 20	☽ Q ♄
01 16	♀ Q ♄
05 32	☽ ☍ ♃
06 50	☽ # ♃
08 26	☽ ⊥ ☿
09 26	☿ ⚹ ♅
12 07	☽ △ ♇
22 35	☽ ☍ ♆
23 11	☽ ∥ ♀
19 Saturday	
03 34	☽ # ♆
05 16	☽ ∠ ☿
05 22	☉ ☌ ♂
06 52	☽ ⚹ ♂
06 54	☽ ⚹ ☉
12 20	☽ □ ♄
12 52	☽ # ☿
20 Sunday	
08 18	☽ Q ♂
08 22	♀ # ♆
08 38	☽ △ ♅
09 07	☽ Q ☉
16 30	☽ ⚹ ☿
16 58	☽ ⊻ ♀
17 11	☽ ⊼ ♃
18 29	☽ # ☉
19 44	♀ ☍ ♃
21 11	☿ ⊼ ♃
22 39	♅ # ♇
22 45	☿ # ♀
22 54	☿ ⊻ ♀
21 Monday	
00 07	☽ □ ♇
04 47	☽ # ♂
05 16	☽ ± ♃
06 05	☽ ⊥ ♀
07 43	☿ ∥ ♆
09 04	☉ bQ ♃
09 54	☿ ∠ ♄
10 52	☽ ⊼ ♆
15 05	☽ bQ ♅
20 49	☽ Q ☿
22 46	☽ □ ♂
22 59	☽ ± ♆
23 23	☽ bQ ♃
22 Tuesday	
00 43	☽ □ ☉
01 29	☽ △ ♄
01 46	☽ ∠ ♀
09 15	♂ bQ ♃
09 20	☽ ± ♅
11 17	☉ ⊻ ♄
12 12	☿ ± ♆
12 26	☽ Q ♇
17 07	☽ bQ ♆
21 27	☽ ⊼ ♅
21 42	☿ ⊼ ♇
23 Wednesday	
05 24	☽ △ ♃
07 56	☽ bQ ♄
10 17	☽ ⚹ ♀
12 26	☽ ⚹ ♇
14 48	☽ □ ☿
23 01	☽ △ ♆
24 Thursday	
01 59	☽ ± ♄
11 50	☽ Q ♀
12 19	♂ ⊻ ♄
13 54	☽ ⊼ ♄
13 58	☽ △ ♂
15 41	♀ △ ♇
17 41	☽ △ ☉
17 58	☽ ∠ ♇
25 Friday	
08 46	☽ ☍ ♅
11 21	☽ ⊥ ♇
15 51	☽ □ ♃
18 00	☿ Q ♅
18 49	☿ ∥ ♃
20 29	☽ bQ ♂
22 47	☽ ⊻ ♇
26 Saturday	
00 56	☽ bQ ☉
01 04	☽ □ ♀
04 10	☿ ⊼ ♆
08 53	☽ □ ♆
09 33	☽ △ ☿
13 19	☉ ∠ ♀
13 49	☿ ± ♇
14 21	☽ ± ♂
19 12	☽ ± ☉
23 34	☽ ☍ ♄
27 Sunday	
01 21	☉ # ♀
01 29	☽ Q ♃
02 04	☽ ⊼ ♂
07 11	☽ ⊼ ☉
17 03	☽ bQ ☿
17 03	☽ ⊼ ♅
18 05	☽ Q ♆
23 16	☽ ⚹ ♃
28 Monday	
03 56	☽ ± ♅
06 02	☽ ☌ ♇
11 21	☽ ± ☿
11 54	☽ △ ♀
14 08	☽ # ♂
15 36	☽ ⚹ ♆
20 02	☽ bQ ♅
20 17	☽ ∥ ♀
23 21	☽ ⊼ ☿
29 Tuesday	
01 50	☽ ∠ ♃
06 01	☽ ⊼ ♄
06 38	☽ # ☉
10 33	☽ ☍ ♂
12 58	☉ ⊥ ♄
15 57	☽ bQ ♀
16 35	☽ ± ♄
16 49	☽ ☍ ☉
17 28	☽ ⊥ ♃
17 54	☽ ∠ ♆
20 43	☽ # ♆
22 20	☽ △ ♅
30 Wednesday	
01 28	☉ ± ♃
01 48	♀ # ♂
03 47	☽ ⊻ ♃
08 02	☉ bQ ♆
08 15	☽ bQ ♄
08 26	☽ ± ♀
09 25	☽ ⊥ ♆
10 29	☽ ⊻ ♇
16 26	☽ # ♃
19 19	☽ ⊼ ♀
19 40	☽ ⊻ ♆
20 43	☽ ⊥ ♇
31 Thursday	
01 32	♀ ☍ ♆
09 16	☽ ☍ ☿
10 01	☽ △ ♄
10 06	☽ ∥ ☿
11 58	☽ ∠ ♇
14 28	☽ # ☿
16 23	☽ ⊼ ♂
17 08	☿ ⚹ ♄
23 46	☽ ⊼ ☉

SEPTEMBER 2034

LONGITUDES

Date	Sidereal time h	m	s	Sun ☉ °	'	"	Moon ☽ °	'	"	Moon ☽ 24.00 °	'	"	Mercury ☿ °	'	Venus ♀ °	'	Mars ♂ °	'	Jupiter ♃ °	'	Saturn ♄ °	'	Uranus ♅ °	'	Neptune ♆ °	'	Pluto ♇ °	'
01	10	42	55	09 ♍	08	14	15 ♈	55	35	23 ♈	05	38	01 ♎	39	23 ♎	08	04 ♍	47	12 ♈	34	00 ♌	37	09 ♋	42	22 ♈	00	16 ♒	37
02	10	46	51	10	06	17	00 ♉	16	24	07 ♉	27	24	03	05	23	53	05	25	12 R	28	00	43	09	45	21 R	59	16 R	36
03	10	50	48	11	04	21	14 ♉	38	10	21 ♉	48	16	04	29	24	37	06	03	12	23	00	50	09	47	21	58	16	35
04	10	54	44	12	02	28	28 ♉	57	18	06 ♊	04	57	05	52	25	20	06	41	12	17	00	56	09	49	21	57	16	34
05	10	58	41	13	00	37	13 ♊	10	53	20 ♊	14	50	07	13	26	03	07	20	12	11	01	03	09	51	21	56	16	33
06	11	02	37	13	58	48	27 ♊	16	34	04 ♋	15	52	08	33	26	44	07	58	12	05	01	09	09	53	21	54	16	31
07	11	06	34	14	57	01	11 ♋	12	32	18 ♋	06	24	09	51	27	24	08	36	11	59	01	16	09	55	21	53	16	30
08	11	10	31	15	55	16	24 ♋	57	18	01 ♌	45	07	11	07	28	03	09	14	11	52	01	22	09	57	21	52	16	29
09	11	14	27	16	53	33	08 ♌	29	42	15 ♌	10	57	12	22	28	41	09	52	11	46	01	28	09	59	21	51	16	28
10	11	18	24	17	51	51	21 ♌	48	45	28 ♌	23	04	13	35	29	18	10	31	11	39	01	34	10	01	21	49	16	27
11	11	22	20	18	50	12	04 ♍	53	48	11 ♍	20	58	14	47	29 ♎	54	11	09	11	32	01	40	10	02	21	48	16	26
12	11	26	17	19	48	35	17 ♍	44	32	24 ♍	04	34	15	56	00 ♏	28	11	47	11	25	01	47	10	04	21	47	16	25
13	11	30	13	20	46	59	00 ♎	21	07	06 ♎	34	20	17	04	01	01	12	25	11	18	01	53	10	06	21	45	16	24
14	11	34	10	21	45	25	12 ♎	44	21	18 ♎	51	23	18	09	01	33	13	04	11	11	01	59	10	08	21	44	16	23
15	11	38	06	22	43	53	24 ♎	55	41	00 ♏	57	32	19	12	02	03	13	42	11	04	02	04	10	09	21	43	16	22
16	11	42	03	23	42	23	06 ♏	57	17	12 ♏	55	18	20	13	02	32	14	20	10	57	02	10	10	11	21	41	16	21
17	11	46	00	24	40	54	18 ♏	52	01	24 ♏	47	53	21	12	03	00	14	58	10	49	02	16	10	12	21	40	16	20
18	11	49	56	25	39	27	00 ♐	43	24	06 ♐	39	03	22	07	03	25	15	36	10	42	02	22	10	14	21	38	16	19
19	11	53	53	26	38	02	12 ♐	35	25	18 ♐	33	02	23	00	03	49	16	15	10	34	02	27	10	15	21	37	16	18
20	11	57	49	27	36	38	24 ♐	32	29	00 ♑	34	21	23	49	04	12	16	53	10	27	02	33	10	17	21	35	16	17
21	12	01	46	28	35	17	06 ♑	39	11	12 ♑	47	34	24	35	04	32	17	31	10	19	02	38	10	18	21	34	16	16
22	12	05	42	29 ♍	33	56	19 ♑	00	02	25 ♑	17	05	25	18	04	51	18	10	10	11	02	44	10	19	21	32	16	15
23	12	09	39	00 ♎	32	38	01 ♒	39	10	08 ♒	06	40	25	57	05	08	18	48	10	03	02	49	10	20	21	31	16	14
24	12	13	35	01	31	21	14 ♒	39	56	21 ♒	19	09	26	31	05	23	19	26	09	55	02	54	10	22	21	29	16	13
25	12	17	32	02	30	06	28 ♒	04	27	04 ♓	55	50	27	01	05	36	20	04	09	48	03	00	10	23	21	28	16	12
26	12	21	29	03	28	53	11 ♓	53	10	18 ♓	56	10	27	25	05	46	20	43	09	40	03	05	10	24	21	26	16	11
27	12	25	25	04	27	41	26 ♓	04	26	03 ♈	17	23	27	45	05	55	21	21	09	32	03	10	10	25	21	24	16	10
28	12	29	22	05	26	32	10 ♈	34	23	17 ♈	54	38	27	58	06	02	21	59	09	24	03	15	10	26	21	23	16	10
29	12	33	18	06	25	24	25 ♈	17	15	02 ♉	41	19	28	06	06	06	22	38	09	16	03	20	10	27	21	21	16	09
30	12	37	15	07 ♎	24	19	10 ♉	05	54	17 ♉	30	02	28 ♎	07	06 ♏	08	23 ♍	16	09 ♈	08	03 ♌	24	10 ♋	28	21 ♈	20	16 ♒	08

Date	Moon True ☊ °	'	Moon Mean ☊ °	'	Moon ☽ Latitude °	'
01	23 ♍	59	24 ♍	33	01 S	59
02	24 D	00	24	29	03	08
03	24	01	24	26	04	05
04	24	02	24	23	04	47
05	24	02	24	20	05	11
06	24 R	02	24	17	05	16
07	24	01	24	14	05	03
08	24	00	24	10	04	32
09	23	59	24	07	03	46
10	23	58	24	04	02	49
11	23	58	24	01	01	44
12	23	58	23	58	00 S	34
13	23 D	58	23	55	00 N	35
14	23	58	23	51	01	42
15	23	58	23	48	02	44
16	23	58	23	45	03	37
17	23 R	58	23	42	04	20
18	23	58	23	39	04	52
19	23	58	23	35	05	11
20	23	58	23	32	05	17
21	23 D	58	23	29	05	10
22	23	58	23	26	04	48
23	23	58	23	23	04	11
24	23	59	23	20	03	21
25	24	00	23	16	02	19
26	24	01	23	13	01 N	07
27	24 R	01	23	10	00 S	11
28	24	00	23	07	01	30
29	23	59	23	04	02	45
30	23 ♍	58	23 ♍	01	03 S	48

DECLINATIONS

Date	Sun ☉ °	'	Moon ☽ °	'	Mercury ☿ °	'	Venus ♀ °	'	Mars ♂ °	'	Jupiter ♃ °	'	Saturn ♄ °	'	Uranus ♅ °	'	Neptune ♆ °	'	Pluto ♇ °	'
01	08 N	09	04 N	26	01 S	06	12 S	22	10 N	48	03 N	31	20 N	07	23 N	22	07 N	00	23 S	29
02	07	47	08	38	01	48	12	46	10	34	03	28	20	06	23	22	07	00	23	29
03	07	25	12	19	02	29	13	10	10	20	03	26	20	04	23	22	06	59	23	29
04	07	03	15	15	03	09	13	33	10	06	03	24	20	03	23	22	06	59	23	30
05	06	40	17	14	03	49	13	56	09	52	03	21	20	02	23	22	06	58	23	30
06	06	18	18	08	04	29	14	19	09	37	03	18	20	00	23	21	06	58	23	30
07	05	56	17	56	05	07	14	41	09	23	03	16	19	59	23	21	06	57	23	31
08	05	33	16	40	05	45	15	03	09	08	03	13	19	58	23	21	06	57	23	31
09	05	10	14	29	06	23	15	24	08	54	03	11	19	57	23	21	06	56	23	31
10	04	48	11	34	06	59	15	45	08	39	03	08	19	55	23	21	06	56	23	32
11	04	25	08	06	07	34	16	05	08	25	03	05	19	54	23	21	06	55	23	32
12	04	02	04	19	08	09	16	25	08	10	03	02	19	53	23	21	06	55	23	32
13	03	39	00 N	24	08	43	16	44	07	55	02	59	19	52	23	21	06	54	23	32
14	03	16	03 S	28	09	15	17	03	07	40	02	56	19	50	23	21	06	54	23	33
15	02	53	07	07	09	47	17	21	07	26	02	54	19	49	23	20	06	53	23	33
16	02	30	10	25	10	17	17	39	07	11	02	51	19	48	23	20	06	53	23	33
17	02	07	13	16	10	46	17	56	06	56	02	48	19	47	23	20	06	52	23	33
18	01	44	15	32	11	14	18	12	06	41	02	45	19	46	23	20	06	51	23	34
19	01	20	17	09	11	40	18	28	06	26	02	41	19	44	23	20	06	51	23	34
20	00	57	18	02	12	05	18	42	06	11	02	38	19	43	23	20	06	50	23	34
21	00	34	18	07	12	28	18	57	05	56	02	35	19	42	23	20	06	50	23	34
22	00 N	10	17	20	12	50	19	10	05	41	02	32	19	41	23	20	06	49	23	35
23	00 S	13	15	42	13	09	19	22	05	26	02	29	19	40	23	20	06	49	23	35
24	00	36	13	13	13	26	19	34	05	11	02	26	19	39	23	20	06	48	23	35
25	01	00	09	58	13	41	19	45	04	55	02	23	19	38	23	20	06	47	23	35
26	01	23	06	04	13	54	19	55	04	40	02	20	19	36	23	20	06	47	23	35
27	01	46	01 S	44	14	04	20	03	04	25	02	16	19	35	23	20	06	46	23	35
28	02	10	02 N	48	14	11	20	11	04	10	02	13	19	34	23	20	06	45	23	36
29	02	33	07	13	14	15	20	18	03	54	02	10	19	33	23	19	06	45	23	36
30	02 S	56	11 N	14	14 S	15	20 S	24	03 N	39	02 N	07	19 N	32	23 N	19	06 N	44	23 S	36

ZODIAC SIGN ENTRIES

Date	h	m	Planets
02	11	33	☽ ♉
04	13	45	☽ ♊
06	16	40	☽ ♋
08	20	54	☽ ♌
11	02	58	☽ ♍
11	16	18	♀ ♏
13	11	19	☽ ♎
15	22	05	☽ ♏
18	10	32	☽ ♐
20	22	52	☽ ♑
22	22	39	☉ ♎
23	08	54	☽ ♒
25	15	23	☽ ♓
27	18	33	☽ ♈
29	19	39	☽ ♉

LATITUDES

Date	Mercury ☿ °	'	Venus ♀ °	'	Mars ♂ °	'	Jupiter ♃ °	'	Saturn ♄ °	'	Uranus ♅ °	'	Neptune ♆ °	'	Pluto ♇ °	'
01	00 S	29	03 S	39	01 N	08	01 S	35	00 N	06	00 N	18	01 S	41	08 S	01
04	00	54	04	02	01	07	01	35	00	07	00	18	01	41	08	01
07	01	20	04	27	01	07	01	36	00	07	00	18	01	41	08	01
10	01	46	04	51	01	07	01	36	00	07	00	18	01	42	08	01
13	02	11	05	15	01	06	01	37	00	08	00	18	01	42	08	00
16	02	35	05	39	01	06	01	37	00	08	00	18	01	42	08	00
19	02	57	06	02	01	06	01	37	00	08	00	18	01	42	08	00
22	03	16	06	25	01	05	01	38	00	08	00	18	01	42	08	00
25	03	31	06	46	01	05	01	38	00	09	00	18	01	42	07	59
28	03	41	07	04	01	04	01	38	00	09	00	18	01	42	07	59
31	03 S	41	07 S	20	01 N	04	01 S	38	00 N	09	00 N	18	01 S	42	07 S	59

DATA

Julian Date	2464207
Delta T	+75 seconds
Ayanamsa	24° 20' 28"
Synetic vernal point	04° ♓ 46' 31"
True obliquity of ecliptic	23° 25' 56"

MOON'S PHASES, APSIDES AND POSITIONS ☽

Date	h	m	Phase	Longitude °	'	Eclipse Indicator
05	11	41	☾	13 ♊	00	
12	16	14	●	19 ♍	59	Annular
20	18	39	☽	27 ♐	53	
28	02	57	○	05 ♈	04	partial

Day	h	m		
02	15	38	Perigee	
18	07	07	Apogee	
30	04	22	Perigee	
06	19	33	Max dec	18° N 11'
13	14	28	OS	
21	02	13	Max dec	18° S 11'
27	21	14	ON	

LONGITUDES

Date	Chiron ⚷ °	'	Ceres ⚳ °	'	Pallas ⚴ °	'	Juno ⚵ °	'	Vesta ⚶ °	'	Black Moon Lilith ⚸ °	'
01	06 ♊	32	25 ♓	06	24 ♑	47	05 ♐	56	04 ♋	38	23 ♏	51
11	06 ♊	35	22 ♓	57	23 ♑	59	07 ♐	46	07 ♋	52	24 ♏	58
21	06 ♊	32	20 ♓	44	23 ♑	47	09 ♐	56	10 ♋	51	26 ♏	05
31	06 ♊	23	18 ♓	39	24 ♑	10	12 ♐	23	13 ♋	30	27 ♏	12

ASPECTARIAN

h m	Aspects
01 Friday	
01 32	☽ □ ♅
02 58	☽ ± ♂
06 24	☽ ☌ ♃
06 57	☽ ∥ ♃
10 35	☽ ± ☉
11 27	☿ ⚼ ♇
13 10	☽ ✶ ♇
18 45	☽ ⚼ ♂
22 10	☽ ☌ ♆
02 Saturday	
00 44	☽ ☍ ♀
02 23	☽ ∥ ♆
02 42	☉ ✶ ♅
02 44	☽ ⚼ ☉
07 19	☽ ∥ ☉
07 46	☽ Q ♅
09 13	☽ Q ♇
12 45	☽ □ ♄
17 12	☽ ⊼ ☿
20 59	☽ △ ♂
23 22	☽ ∥ ♂
03 Sunday	
03 52	☽ ✶ ♅
04 19	☽ ± ☿
05 37	☽ △ ☉
08 15	☽ ⊻ ♃
15 15	☽ □ ♇
18 13	☽ ⊥ ♃
19 04	☽ Q ♄
19 11	☽ # ♀
20 59	☽ ⚼ ☿
22 41	♂ ± ♃
04 Monday	
00 15	☽ ⊻ ♆
05 02	☽ ∠ ♅
05 36	☽ ⊼ ♀
09 12	☽ ∠ ♃
10 19	☽ ⊥ ♆
15 22	☽ ✶ ♄
16 08	☉ ∥ ♆
16 13	☽ ± ♀
17 25	☉ ⊼ ♃
19 57	☿ # ♃
20 12	☽ ⊥ ♅
21 27	♂ ⚼ ♆
23 19	♂ ⊥ ♄
05 Tuesday	
00 51	☽ △ ☿
01 26	☽ ∠ ♆
01 38	☽ □ ♂
06 21	☽ ⊻ ♅
08 12	☽ ⚼ ♀
10 19	☽ ✶ ♃
11 41	☽ □ ☉
15 47	☿ ⊻ ♂
16 54	☽ ∠ ♄
17 42	☽ △ ♇
06 Wednesday	
02 50	☽ ✶ ♆
06 34	☽ Q ♃
08 20	☽ ⊥ ♄
09 39	☽ Q ♂
11 01	☽ △ ♀
18 42	☽ ⊻ ♄
19 17	☽ ⚼ ♇
20 40	☽ Q ☉
23 22	☽ Q ♆
07 Thursday	
07 16	☽ ✶ ♂
09 24	☽ □ ☿
09 45	☽ ☌ ♅
10 47	☽ ± ♇
13 19	☽ □ ♃
13 20	☿ □ ♅
18 59	☽ ✶ ☉
21 12	☽ ⊼ ♇
08 Friday	
06 35	☽ □ ♆
07 06	☉ # ☿
10 39	☉ ± ♆
10 40	☽ ∠ ♂
17 44	☽ □ ♀
20 06	☽ Q ☿
23 20	☽ ∠ ☉
23 24	☽ ☌ ♄
09 Saturday	
00 17	☉ ∠ ♄
01 09	☿ ☍ ♃
01 42	☉ ⊼ ♇
03 21	☽ ⊥ ♂
04 13	☽ # ♀
14 35	☽ ⊻ ♂
14 40	☽ ⊻ ♅
16 15	♂ ✶ ♅
16 38	☽ ⊥ ☉
17 48	☽ △ ♃
19 39	☽ ✶ ☿
10 Sunday	
01 28	☽ ⊥ ♅
02 18	☽ ☍ ♇
03 26	☽ Q ♀
04 17	☽ ⊻ ☉
09 57	☿ # ♆
11 36	☿ Q ♄
12 01	☽ △ ♆
17 50	☽ ∠ ♅
20 45	☽ ⚼ ♃
11 Monday	
01 38	☽ ∠ ☿
02 20	☽ ✶ ♀
06 00	☽ ⊻ ♄
09 49	☽ ∥ ♂
13 10	☽ ± ♃
14 58	☽ # ☿
15 31	☽ ⚼ ♆
17 12	☽ ⊥ ♄
19 39	☽ ∥ ♆
19 56	☽ ⊥ ☿
21 35	☽ ✶ ♅
12 Tuesday	
00 14	☽ ☌ ♂
00 14	☽ ⊼ ♃
00 26	♂ ⊼ ♃
07 31	☽ ∠ ♀
08 16	☽ ⊻ ☿
08 18	☽ ± ♆
09 30	☽ ⊼ ♇
10 10	☽ ∠ ♄
12 26	☿ # ♂
13 55	☽ ∥ ☉
16 14	☽ ☌ ☉
19 37	☽ ⊼ ♆
19 58	☽ ∥ ♃
20 12	☽ Q ♅
20 49	☽ ± ♇
21 51	☿ △ ♇
13 Wednesday	
01 20	☽ ⊥ ♀
13 21	☽ ⊻ ♀
14 00	☽ ⚼ ♇
14 57	☽ ✶ ♄
14 Thursday	
06 54	☽ □ ♅
08 46	☽ # ♃
09 00	☽ ☍ ♃
10 54	☽ # ☉
11 24	☉ ⊼ ♆
12 39	☽ ⊻ ♂
14 26	☽ Q ♄
19 07	☽ △ ♇
21 22	☉ Q ♅
23 38	☽ ☌ ☿
15 Friday	
01 05	☽ ⊥ ♂
03 00	☉ ± ♇
05 38	☽ ☍ ♆
07 16	☽ ⊻ ☉
10 28	☽ # ♆
11 26	☉ ∥ ♃
13 07	♀ □ ♄
14 02	☽ # ♂
19 54	☽ ∠ ♂
20 13	☽ ⊥ ☉
16 Saturday	
02 21	☽ □ ♄
02 47	☽ ☌ ♀
10 48	☽ ∥ ☿
15 50	☽ ∠ ☉
18 30	☽ △ ♅
19 43	☿ ⊥ ♂
19 56	☽ ⊼ ♃
17 Sunday	
03 41	☽ ✶ ♂
06 52	☽ □ ♇
07 55	☽ ± ♃
17 07	☽ ⊻ ☿
17 38	☽ ⊼ ♆
18 32	♂ ∥ ♆
23 41	☿ ☍ ♆
18 Monday	
00 49	☽ ✶ ☉
00 51	☽ ⚼ ♅
01 56	☽ ⚼ ♃
05 20	☽ Q ♂
05 46	☽ ± ♆
06 17	☽ ⊥ ☿
13 01	♂ ± ♆
15 21	☽ △ ♄
17 40	☽ ⊻ ♀
19 07	☽ ± ♅
19 15	☽ Q ♇
23 57	☽ ⚼ ♆
19 Tuesday	
02 00	☽ ∠ ☿
03 18	☽ Q ☉
06 13	☽ ⊥ ♀
07 16	☽ ⊼ ♅
07 58	☽ △ ♃
13 43	♂ ⊼ ♇
19 27	☽ ✶ ♇
19 47	☽ □ ♂
21 53	☽ ⚼ ♄
20 Wednesday	
00 58	☽ ∠ ♀
06 06	☽ △ ♆
10 28	☽ ✶ ☿
12 40	♅ ± ♇
16 02	☽ ± ♄
18 39	☽ □ ☉
21 Thursday	
01 23	☽ ∠ ♇
04 02	☽ ⊼ ♄
07 43	☽ ✶ ♀
10 12	☉ ⊥ ♀
11 52	☽ Q ☿
14 43	♃ □ ♅
17 12	♂ ∠ ♄
19 04	☽ ⊥ ♇
19 06	☽ □ ♃
19 09	☽ ☍ ♅
22 Friday	
06 42	☽ ⊻ ♇
07 46	☽ Q ♀
10 18	☽ △ ♂
16 51	☽ □ ♆
23 Saturday	
00 41	☽ □ ☿
05 19	☽ Q ♃
09 45	☽ △ ☉
14 12	☽ ☍ ♄
16 13	☽ ⚼ ♂
18 37	☽ □ ♀
24 Sunday	
02 33	☽ Q ♆
03 25	☽ ✶ ♃
04 07	☽ ⊼ ♅
04 37	☉ ⚼ ♇
09 39	☽ ± ♂
10 18	☽ ∥ ☿
14 49	☽ ☌ ♇
15 05	☽ ± ♅
15 38	☽ ⚼ ☉
21 03	☽ ⊼ ♂
21 17	♀ # ♄
25 Monday	
00 16	☽ ✶ ♆
06 15	☽ ∠ ♃
07 14	☽ ⚼ ♅
09 00	☽ ± ☉
10 03	☽ △ ☿
20 22	☽ ⊼ ☉
20 41	☽ ⊼ ♄
21 56	☽ ⊥ ♃
26 Tuesday	
01 12	☉ ✶ ♄
01 20	☽ △ ♀
02 38	☽ ∠ ♆
07 09	☽ ± ♄
07 53	☽ # ♆
08 13	☽ ⊻ ♃
09 26	☽ △ ♅
12 56	☽ ⚼ ☿
15 11	♀ ∠ ♂
18 03	☽ ⊥ ♆
19 20	☽ ⊻ ♇
20 29	☽ # ♂
22 37	☽ ⚼ ♄
27 Wednesday	
03 16	☽ ⚼ ♀
03 42	☽ ☍ ♂
04 11	☽ ⊻ ♆
04 35	☽ ± ☿
05 28	☽ ⊥ ♇
09 04	☽ # ♃
11 49	☽ ∥ ☉
14 06	♂ ⊼ ♆
14 51	☽ ⊼ ☿
18 28	☽ ± ♀
20 29	☽ ∠ ♇
23 51	☽ △ ♄
28 Thursday	
02 57	☽ ☍ ☉
04 28	☽ ⊼ ♀
08 19	☽ # ☉
09 00	☽ ∥ ♃
10 05	☽ ☌ ♃
11 07	☿ ⊥ ♂
11 46	☽ □ ♅
15 18	☉ # ♃
18 26	♂ ± ♇
18 52	☽ ∥ ♂
21 08	☽ ✶ ♇
29 Friday	
03 28	☉ ⊻ ♀
05 01	♂ Q ♅
05 37	☽ ☌ ♆
07 29	☽ ⊼ ♂
09 21	☽ ∥ ♆
16 34	☽ ☍ ☿
16 38	☽ Q ♇
17 08	☽ Q ♅
17 40	☽ ± ♂
30 Saturday	
01 06	☽ □ ♄
02 59	☿ St R
05 34	☽ ☍ ♀
07 20	☽ ⊼ ☉
08 54	☽ ⚼ ♂
10 26	☽ ⊻ ♃
12 35	☽ ✶ ♅
17 44	☽ ± ☉
19 37	♀ St R
20 04	☽ ⊥ ♃
21 47	☽ □ ♇

All ephemeris data is given at 12.00 UT and the Moon's longitude is additionally given for 24.00 UT

OCTOBER 2034

Date	Sidereal time h m s	Sun ☉ ° ' "	Moon ☽ ° ' "	Moon ☽ 24.00 ° ' "	Mercury ☿ ° '	Venus ♀ ° '	Mars ♂ ° '	Jupiter ♃ ° '	Saturn ♄ ° '	Uranus ♅ ° '	Neptune ♆ ° '	Pluto ♇ ° '
01	12 41 11	08 ♎ 23 16	24 ♉ 52 51	02 ♊ 13 31	28 ♎ 00	06 ♏ 07	23 ♍ 54	08 ♈ 59	03 ♌ 29	10 ♋ 28	21 ♈ 18	16 ♒ 07
02	12 45 08	09 22 15	09 ♊ 31 19	16 ♊ 45 39	27 R 47	06 R 04	24 33	08 R 51	03 34	10 29	21 R 16	16 R 07
03	12 49 04	10 21 17	23 ♊ 56 02	01 ♋ 02 06	27 26	05 59	25 11	08 43	03 38	10 30	21 15	16 06
04	12 53 01	11 20 21	08 ♋ 03 37	15 ♋ 00 27	26 57	05 52	25 49	08 35	03 43	10 31	21 13	16 05
05	12 56 58	12 19 27	21 ♋ 52 35	28 ♋ 40 04	26 20	05 42	26 28	08 27	03 47	10 31	21 11	16 05
06	13 00 54	13 18 35	05 ♌ 23 00	12 ♌ 01 34	25 36	05 29	27 06	08 19	03 51	10 32	21 10	16 04
07	13 04 51	14 17 46	18 ♌ 35 57	25 ♌ 06 23	24 45	05 15	27 44	08 11	03 56	10 32	21 08	16 04
08	13 08 47	15 16 59	01 ♍ 33 05	07 ♍ 56 17	23 46	04 58	28 23	08 03	04 00	10 33	21 07	16 03
09	13 12 44	16 16 15	14 ♍ 16 13	20 ♍ 33 05	22 43	04 38	29 01	07 55	04 04	10 33	21 05	16 03
10	13 16 40	17 15 32	26 ♍ 47 06	02 ♎ 58 26	21 35	04 17	29 ♍ 40	07 48	04 08	10 33	21 03	16 02
11	13 20 37	18 14 51	09 ♎ 07 18	15 ♎ 13 52	20 24	03 53	00 ♎ 18	07 40	04 12	10 34	21 01	16 02
12	13 24 33	19 14 13	21 ♎ 18 18	27 ♎ 20 47	19 12	03 28	00 56	07 32	04 15	10 34	21 00	16 01
13	13 28 30	20 13 37	03 ♏ 21 31	09 ♏ 20 40	18 01	03 00	01 35	07 24	04 19	10 34	20 58	16 01
14	13 32 27	21 13 03	15 ♏ 18 29	21 ♏ 15 12	16 53	02 31	02 13	07 17	04 23	10 34	20 56	16 00
15	13 36 23	22 12 30	27 ♏ 11 05	03 ♐ 06 28	15 51	02 00	02 52	07 09	04 26	10 34	20 55	16 00
16	13 40 20	23 12 00	09 ♐ 01 39	14 ♐ 57 01	14 55	01 27	03 30	07 02	04 30	10 R 34	20 53	15 59
17	13 44 16	24 11 31	20 ♐ 53 00	26 ♐ 50 02	14 07	00 54	04 09	06 54	04 33	10 34	20 51	15 59
18	13 48 13	25 11 04	02 ♑ 48 36	08 ♑ 49 13	13 30	00 ♏ 19	04 47	06 47	04 36	10 34	20 50	15 59
19	13 52 09	26 10 39	14 ♑ 52 26	20 ♑ 58 48	13 03	29 ♎ 43	05 26	06 40	04 39	10 34	20 48	15 58
20	13 56 06	27 10 16	27 ♑ 08 53	03 ♒ 23 16	12 47	29 07	06 04	06 33	04 42	10 34	20 46	15 58
21	14 00 02	28 09 55	09 ♒ 42 31	16 ♒ 07 11	12 D 42	28 30	06 43	06 26	04 45	10 34	20 45	15 58
22	14 03 59	29 ♎ 09 35	22 ♒ 37 45	29 ♒ 14 40	12 48	27 54	07 21	06 19	04 48	10 33	20 43	15 58
23	14 07 56	00 ♏ 09 16	05 ♓ 58 16	12 ♓ 48 47	13 05	27 17	08 00	06 12	04 50	10 33	20 41	15 58
24	14 11 52	01 09 00	19 ♓ 46 21	26 ♓ 50 54	13 32	26 41	08 38	06 06	04 53	10 33	20 40	15 57
25	14 15 49	02 08 45	04 ♈ 02 13	11 ♈ 19 50	14 09	26 05	09 17	05 59	04 55	10 32	20 38	15 57
26	14 19 45	03 08 32	18 ♈ 43 09	26 ♈ 11 20	14 54	25 30	09 55	05 53	04 58	10 32	20 37	15 57
27	14 23 42	04 08 21	03 ♉ 43 23	11 ♉ 18 06	15 47	24 57	10 34	05 47	05 00	10 31	20 35	15 57
28	14 27 38	05 08 12	18 ♉ 54 15	26 ♉ 30 30	16 46	24 24	11 13	05 41	05 02	10 30	20 33	15 57
29	14 31 35	06 08 05	04 ♊ 05 32	11 ♊ 38 06	17 52	23 54	11 51	05 35	05 04	10 30	20 32	15 57
30	14 35 31	07 08 00	19 ♊ 07 03	26 ♊ 31 24	19 04	23 24	12 30	05 29	05 06	10 29	20 30	15 D 57
31	14 39 28	08 ♏ 07 58	03 ♋ 50 22	11 ♋ 03 21	20 ♎ 19	22 ♎ 57	13 ♎ 08	05 ♈ 24	05 ♌ 08	10 ♋ 28	20 ♈ 29	15 ♒ 57

DECLINATIONS

Date	Moon True ☊ ° '	Moon Mean ☊ ° '	Moon ☽ Latitude ° '	Sun ☉ ° '	Moon ☽ ° '	Mercury ☿ ° '	Venus ♀ ° '	Mars ♂ ° '	Jupiter ♃ ° '	Saturn ♄ ° '	Uranus ♅ ° '	Neptune ♆ ° '	Pluto ♇ ° '
01	23 ♍ 56	22 ♍ 57	04 S 36	03 S 20	14 N 31	14 S 12	20 S 28	03 N 24	02 N 04	19 N 31	23 N 19	06 N 44	23 S 36
02	23 R 54	22 54	05 06	03 43	16 50	14 05	20 31	03 08	02 01	19 30	23 19	06 43	23 36
03	23 53	22 51	05 16	04 06	18 02	13 54	20 34	02 53	01 57	19 29	23 19	06 42	23 36
04	23 53	22 48	05 06	04 29	18 05	13 39	20 34	02 38	01 54	19 28	23 19	06 42	23 36
05	23 D 53	22 45	04 39	04 52	17 04	13 19	20 34	02 22	01 51	19 28	23 19	06 41	23 36
06	23 54	22 41	03 57	05 15	15 05	12 55	20 32	02 07	01 48	19 27	23 19	06 41	23 36
07	23 55	22 38	03 03	05 38	12 21	12 26	20 29	01 51	01 45	19 26	23 19	06 40	23 37
08	23 57	22 35	02 00	06 01	09 03	11 53	20 24	01 36	01 42	19 25	23 19	06 39	23 37
09	23 58	22 32	00 S 53	06 24	05 22	11 16	20 18	01 21	01 39	19 24	23 19	06 39	23 37
10	23 R 58	22 29	00 N 16	06 47	01 N 31	10 35	20 10	01 05	01 36	19 23	23 19	06 38	23 37
11	23 57	22 26	01 22	07 09	02 S 21	09 52	20 01	00 50	01 33	19 22	23 19	06 37	23 37
12	23 55	22 22	02 25	07 32	06 04	09 08	19 51	00 34	01 30	19 22	23 19	06 37	23 37
13	23 51	22 19	03 20	07 54	09 30	08 22	19 39	00 19	01 27	19 21	23 19	06 36	23 37
14	23 47	22 16	04 05	08 16	12 31	07 37	19 26	00 N 03	01 24	19 20	23 19	06 35	23 37
15	23 42	22 13	04 40	08 39	14 59	06 54	19 11	00 S 12	01 21	19 19	23 19	06 35	23 37
16	23 37	22 10	05 02	09 01	16 49	06 14	18 55	00 28	01 18	19 19	23 19	06 34	23 37
17	23 32	22 06	05 12	09 23	17 56	05 37	18 37	00 43	01 15	19 18	23 19	06 34	23 37
18	23 28	22 03	05 08	09 45	18 16	05 05	18 19	00 59	01 13	19 17	23 19	06 33	23 37
19	23 26	22 00	04 51	10 06	17 47	04 38	17 59	01 14	01 10	19 17	23 19	06 32	23 37
20	23 D 26	21 57	04 20	10 28	16 28	04 17	17 38	01 29	01 07	19 16	23 19	06 32	23 37
21	23 26	21 54	03 36	10 49	14 20	04 01	17 17	01 45	01 05	19 16	23 19	06 31	23 37
22	23 28	21 51	02 40	11 10	11 26	03 51	16 54	02 00	01 02	19 15	23 20	06 30	23 36
23	23 29	21 47	01 34	11 31	07 52	03 47	16 31	02 16	01 00	19 15	23 20	06 30	23 36
24	23 30	21 44	00 N 20	11 52	03 S 44	03 48	16 07	02 31	00 57	19 14	23 20	06 29	23 36
25	23 R 31	21 41	00 S 57	12 13	00 N 44	03 55	15 43	02 46	00 55	19 14	23 20	06 29	23 36
26	23 28	21 38	02 13	12 33	05 17	04 06	15 18	03 02	00 52	19 13	23 20	06 28	23 36
27	23 23	21 35	03 21	12 54	09 36	04 21	14 54	03 17	00 50	19 13	23 20	06 27	23 36
28	23 18	21 32	04 15	13 14	13 21	04 41	14 29	03 32	00 48	19 12	23 20	06 27	23 36
29	23 11	21 28	04 52	13 34	16 11	05 03	14 05	03 48	00 46	19 12	23 20	06 26	23 36
30	23 05	21 25	05 08	13 53	17 52	05 29	13 41	04 03	00 44	19 12	23 20	06 26	23 36
31	22 ♍ 59	21 ♍ 22	05 S 03	14 S 13	18 N 19	05 S 57	13 S 17	04 S 18	00 N 42	19 N 11	23 N 20	06 N 25	23 S 35

ZODIAC SIGN ENTRIES

Date	h	m	Planets
01	20	21	☽ ♊
03	22	15	☽ ♋
06	02	22	☽ ♌
08	09	06	☽ ♍
10	18	13	☽ ♎
11	00	44	♂ ♎
13	05	17	☽ ♏
15	17	42	☽ ♐
18	06	22	☽ ♑
19	00	40	♀ ♎
20	17	30	☽ ♒
23	01	21	☽ ♓
23	08	16	☉ ♏
25	05	17	☽ ♈
27	06	05	☽ ♉
29	05	31	☽ ♊
31	05	41	☽ ♋

LATITUDES

Date	Mercury ☿ ° '	Venus ♀ ° '	Mars ♂ ° '	Jupiter ♃ ° '	Saturn ♄ ° '	Uranus ♅ ° '	Neptune ♆ ° '	Pluto ♇ ° '
01	03 S 41	07 S 20	01 N 04	01 S 38	00 N 09	00 N 18	01 S 42	07 S 59
04	03 30	07 33	01 03	01 38	00 10	00 18	01 42	07 59
07	03 04	07 40	01 03	01 38	00 10	00 18	01 42	07 58
10	02 21	07 42	01 02	01 38	00 10	00 19	01 42	07 58
13	01 25	07 37	01 02	01 37	00 11	00 19	01 42	07 57
16	00 S 23	07 25	01 01	01 37	00 11	00 19	01 42	07 57
19	00 N 34	07 05	01 01	01 37	00 11	00 19	01 42	07 57
22	01 18	06 38	01 00	01 36	00 12	00 19	01 42	07 56
25	01 48	06 04	00 59	01 36	00 12	00 19	01 42	07 56
28	02 04	05 25	00 58	01 35	00 13	00 19	01 42	07 55
31	02 N 09	04 S 42	00 N 57	01 S 35	00 N 13	00 N 19	01 S 42	07 S 55

DATA

Julian Date	2464237
Delta T	+75 seconds
Ayanamsa	24° 20' 30"
Synetic vernal point	04° ♓ 46' 29"
True obliquity of ecliptic	23° 25' 56"

MOON'S PHASES, APSIDES AND POSITIONS ☽

Date	h	m	Phase	Longitude ° '	Eclipse Indicator
04	18	05	☾	11 ♋ 35	
12	07	33	●	19 ♎ 03	
20	12	03	☽	27 ♑ 10	
27	12	42	○	04 ♉ 10	

Day	h	m		
16	00	14	Apogee	
28	10	20	Perigee	
04	01	10	Max dec	18° N 12'
10	21	21	0S	
18	09	57	Max dec	18° S 16'
25	08	08	0N	
31	08	49	Max dec	18° N 20'

LONGITUDES

Date	Chiron ⚷ ° '	Ceres ⚳ ° '	Pallas ⚴ ° '	Juno ⚵ ° '	Vesta ⚶ ° '	Black Moon Lilith ⚸ ° '
01	06 ♊ 23	18 ♓ 39	24 ♑ 10	12 ♐ 23	13 ♋ 30	27 ♏ 12
11	06 ♊ 07	16 ♓ 57	25 ♑ 04	15 ♐ 05	15 ♋ 46	28 ♏ 19
21	05 ♊ 46	15 ♓ 46	26 ♑ 25	18 ♐ 01	17 ♋ 36	29 ♏ 26
31	05 ♊ 20	15 ♓ 11	28 ♑ 10	21 ♐ 08	18 ♋ 53	00 ♐ 33

All ephemeris data is given at 12.00 UT and the Moon's longitude is additionally given for 24.00 UT

ASPECTARIAN

h m	Aspects
01 Sunday	
06 11	☽ ⊻ ♆
06 27	☽ Q ♄
09 24	☽ ⚼ ☉
09 32	☽ # ☿
10 20	☽ △ ♂
10 34	☽ ∠ ♃
12 58	☽ ∠ ♅
14 40	☉ # ♂
15 56	☽ ⊥ ♆
17 02	☽ ⊼ ☿
02 Monday	
00 58	☉ ☍ ♃
02 08	☽ ✶ ♄
02 43	☽ ± ☿
03 42	☽ ⊥ ♅
06 20	☽ ⊼ ♀
06 39	☽ ∠ ♆
10 55	☽ ✶ ♃
11 44	☽ △ ☉
13 36	☽ ⊻ ♅
16 12	☽ ± ♀
17 17	☽ ⚼ ☿
22 55	☽ △ ♇
03 Tuesday	
03 05	☽ ∠ ♄
06 40	☽ Q ♃
07 05	☽ ⚼ ♀
07 30	☽ ✶ ♆
14 12	☽ □ ♂
15 32	☉ □ ♅
17 43	☽ △ ☿
18 17	☽ ⊥ ♄
04 Wednesday	
00 06	☽ ⚼ ♇
03 44	☽ Q ♆
04 31	☽ ⊻ ♄
08 16	☽ △ ♀
12 54	☽ □ ♃
15 29	☽ ± ♇
16 13	☽ ☌ ♅
18 05	☽ □ ☉
22 25	☽ Q ♂
05 Thursday	
01 52	☽ ⊼ ♇
09 47	☿ ⊻ ♂
10 48	☽ □ ♆
19 29	☽ □ ☿
20 29	☽ ✶ ♂
06 Friday	
04 08	☽ Q ☉
09 15	☽ ☌ ♄
12 11	☽ □ ♀
17 14	☽ △ ♃
21 17	☽ ⊻ ♅
07 Saturday	
00 45	☽ ∠ ♂
01 59	☽ Q ☿
03 30	☽ ✶ ☉
07 21	☽ ☍ ♇
08 13	☽ ⊥ ♅
11 13	☽ # ☿
16 39	☽ △ ♆
18 05	☽ ⊥ ♂
20 22	☽ ⚼ ♃
20 23	☽ Q ♀
22 34	☽ ✶ ☿
08 Sunday	
00 37	♂ ∥ ♃
00 48	☽ ∠ ♅
05 46	☽ ⊻ ♂
09 26	☽ ∠ ☉
12 56	☽ ± ♃
16 36	☽ ⊻ ♄
18 15	☽ ✶ ♀
20 32	☽ ⚼ ♆
09 Monday	
00 06	☽ ⊼ ♃
00 33	☽ ∠ ☿
02 40	♀ ⊥ ♂
03 46	☽ ⊥ ☉
03 51	☽ ∥ ♆
03 59	☽ ⊥ ♄
04 56	☽ ✶ ♅
06 04	☽ # ☉
06 29	☉ △ ♇
06 38	☉ Q ♄
13 32	☽ ± ♆
15 22	☽ ⊼ ♇
16 08	☽ ⊻ ☉
16 17	☽ ⊥ ☿
21 12	☽ ∠ ♄
21 59	☽ ∠ ♀
10 Tuesday	
00 59	☽ ⊼ ♆
02 49	☽ ⊻ ☿
02 51	☽ ± ♇
03 12	☉ # ♆
03 51	☽ Q ♅
11 29	☽ ∥ ♃
14 49	☽ ⊥ ♀
14 50	☽ ∥ ♂
17 52	☽ ☌ ♂
20 12	♀ □ ♄
20 13	☽ ⚼ ♇
23 00	☿ ☍ ♆
11 Wednesday	
02 06	☽ ⊻ ♀
02 20	☽ ✶ ♄
03 05	☽ # ♂
07 01	☽ # ♃
09 11	☽ ☍ ♃
14 49	☽ □ ♅
12 Thursday	
01 33	☽ △ ♇
01 58	☽ Q ♄
07 33	☽ ☌ ☉
08 12	☽ ☌ ☿
11 23	☽ ☍ ♆
11 35	☉ ☌ ☿
14 49	♂ ⚼ ♇
15 37	☽ # ♆
23 06	☽ ∥ ☉
13 Friday	
05 18	☽ ∥ ☿
08 15	☽ ⊻ ♂
11 19	☽ ☌ ♀
13 56	☽ □ ♄
20 01	☽ ⊼ ♃
20 56	☽ ⊥ ♂
22 00	☉ ∥ ☿
14 Saturday	
02 28	☽ △ ♅
05 29	☉ ☍ ♆
07 57	☽ ± ♃
13 24	☽ □ ♇
14 55	☽ ⊻ ☿
16 05	☽ ∠ ♂
18 03	♀ ⊻ ♂
21 49	♀ # ♄
22 53	☿ Q ♄
23 21	☽ ⊼ ♆
15 Sunday	
01 01	☽ ⊻ ☉
01 56	☽ ⚼ ♃
02 03	☽ ⊥ ☿
08 22	☿ △ ♇
08 44	☽ ⚼ ♅
11 27	☽ ± ♆
14 16	☽ ⊥ ☉
18 51	☽ ∠ ☿
21 19	☽ ⊻ ♀
23 28	☿ # ♆
16 Monday	
00 10	☽ ✶ ♂
01 48	☽ Q ♇
02 46	☽ △ ♄
02 58	☽ ± ♅
05 39	☽ ⚼ ♆
07 59	☽ △ ♃
08 57	☽ ⊥ ♀
10 10	☽ ∠ ☉
10 16	♅ St R
15 08	☽ ⊼ ♅
23 09	☽ ✶ ☿
17 Tuesday	
01 52	☽ Q ♂
02 06	☽ ✶ ♇
02 22	☽ ∠ ♀
09 17	☽ ⚼ ♄
11 57	☽ △ ♆
19 17	☽ ✶ ☉
22 00	☽ Q ☿
18 Wednesday	
03 31	☽ ± ♄
04 22	♂ ✶ ♄
07 14	☽ ✶ ♀
08 20	☽ ∠ ♇
15 36	☽ ⊼ ♄
16 11	☽ □ ♂
19 52	☽ □ ♃
21 32	☽ Q ☉
19 Thursday	
02 18	☽ ⊥ ♇
03 29	☽ ☍ ♅
06 03	☽ Q ♀
06 36	♂ # ♃
08 29	☽ □ ☿
14 10	☽ ⊻ ♇
23 38	☽ □ ♆
20 Friday	
07 00	☽ Q ♃
12 03	☽ □ ☉
15 37	☽ □ ♀
21 Saturday	
02 34	☽ ☍ ♄
03 08	♂ ☍ ♃
05 51	☽ ✶ ♃
06 02	☽ △ ♂
10 11	☽ Q ♆
10 23	☿ St D
13 36	☽ ⊼ ♅
17 04	☉ ☌ ♀
17 38	☽ △ ☿
23 43	☽ ☌ ♇
22 Sunday	
00 49	☽ ± ♅
08 30	☽ ✶ ♆
09 37	☽ ∠ ♃
11 28	☽ ⚼ ♂
13 46	☽ ∥ ☉
17 20	☽ ⚼ ♅
21 09	☽ △ ♀
21 34	☽ ⚼ ☿
23 Monday	
00 48	☽ △ ☉
01 49	☽ ⊥ ♃
04 35	☽ ± ♂
09 59	☽ ⊼ ♄
11 30	☽ ∠ ♆
12 25	☽ ⊻ ♃
14 01	☽ ± ☿
15 45	☽ ⊼ ♂
20 03	☽ △ ♅
20 16	☽ # ♆
20 35	☽ ± ♄
22 37	☽ ⚼ ♀
24 Tuesday	
00 51	☽ ⊼ ☿
03 14	☽ ⊥ ♆
05 18	☽ ⚼ ☉
05 27	☽ ⊻ ♇
11 37	☽ ∥ ☿
12 11	☽ ⚼ ♄
13 29	☽ ± ♀
13 31	☽ ⊻ ♆
15 44	☽ ⊥ ♇
18 19	☽ ∥ ♂
21 50	☽ ± ☉
23 15	☽ ⊼ ♀
25 Wednesday	
03 14	☽ # ♃
06 53	☽ ∠ ♇
08 38	☽ ⊼ ☉
12 56	☽ ∥ ♃
13 28	☽ △ ♄
15 12	☽ ☌ ♃
21 02	☽ ☍ ♂
22 42	☽ □ ♅
23 22	☽ # ♂
26 Thursday	
05 24	☽ # ☿
05 27	☽ ☍ ☿
07 32	☽ ✶ ♇
15 03	☽ ☌ ♆
18 21	☽ ∥ ♆
22 31	☽ ☍ ♀
27 Friday	
02 49	☽ Q ♇
03 43	☽ Q ♅
10 14	♂ □ ♅
12 42	☽ ☍ ☉
14 02	☽ □ ♄
15 14	☽ ⊻ ♃
16 26	☿ △ ♇
22 45	☽ ✶ ♅
23 19	☽ ⊼ ♂
28 Saturday	
00 40	☽ ⊥ ♃
07 21	☽ □ ♇
08 23	☽ ⊼ ☿
09 12	☽ ± ♂
09 32	☉ □ ♄
11 07	☽ # ☉
14 36	☽ ⊻ ♆
14 47	☽ ∠ ♃
18 10	☿ Q ♄
18 32	☽ Q ♄
18 34	☽ ± ☿
19 38	☽ # ♀
20 24	☽ ⊼ ♀
22 25	☽ ∠ ♅
23 53	☉ ⊼ ♃
29 Sunday	
00 02	☽ ⚼ ♂
00 03	☽ ⊥ ♆
05 34	☽ ± ♀
09 54	☽ ⚼ ☿
12 38	☽ ⊥ ♅
13 33	☽ ✶ ♄
13 46	♇ St D
14 17	☽ ∠ ♆
14 21	☽ ✶ ♃
15 28	☽ ⊼ ☉
19 23	☽ ⚼ ♀
22 11	☽ ⊻ ♅
30 Monday	
00 54	☽ △ ♂
01 43	☽ ± ☉
04 55	☉ ∥ ♀
06 55	☽ △ ♇
09 23	☽ Q ♃
11 54	☽ △ ☿
13 36	☽ ∠ ♄
14 14	☽ ✶ ♆
17 13	☽ ⚼ ☉
18 43	☽ △ ♀
31 Tuesday	
04 15	☽ ⊥ ♄
07 15	☽ ⚼ ♇
09 45	☽ Q ♆
14 09	☽ ⊻ ♄
14 33	☽ □ ♃
14 46	☿ ☍ ♆
19 39	☽ △ ☉
22 09	☽ ± ♇
23 01	☽ ☌ ♅

NOVEMBER 2034

LONGITUDES

Date	Sidereal time h m s	Sun ☉ ° ' "	Moon ☽ ° ' "	Moon ☽ 24.00 ° ' "	Mercury ☿ ° '	Venus ♀ ° '	Mars ♂ ° '	Jupiter ♃ ° '	Saturn ♄ ° '	Uranus ♅ ° '	Neptune ♆ ° '	Pluto ♇ ° '
01	14 43 25	09 ♏ 07 57	18 ♋ 09 59	25 ♋ 10 04	21 ♎ 39	22 ♎ 32	13 ♎ 47	05 ♈ 18	05 ♌ 10	10 ♋ 28	20 ♈ 27	15 ♒ 57
02	14 47 21	10 07 59	02 ♌ 03 34	08 ♌ 50 38	23 02	22 R 09	14 26	05 R 13	05 11	10 R 27	20 R 25	15 57
03	14 51 18	11 08 02	15 ♌ 31 30	22 ♌ 06 32	24 28	21 48	15 04	05 08	05 13	10 26	20 24	15 57
04	14 55 14	12 08 08	28 ♌ 36 07	05 ♍ 00 44	25 56	21 30	15 43	05 03	05 14	10 25	20 22	15 58
05	14 59 11	13 08 16	11 ♍ 20 50	17 ♍ 36 54	27 26	21 13	16 22	04 58	05 16	10 24	20 21	15 58
06	15 03 07	14 08 26	23 ♍ 49 25	29 ♍ 58 49	28 ♎ 58	21 00	17 00	04 53	05 17	10 23	20 19	15 58
07	15 07 04	15 08 38	06 ♎ 05 32	12 ♎ 09 56	00 ♏ 31	20 48	17 39	04 49	05 18	10 22	20 18	15 58
08	15 11 00	16 08 52	18 ♎ 12 23	24 ♎ 13 12	02 05	20 40	18 18	04 45	05 19	10 20	20 16	15 58
09	15 14 57	17 09 08	00 ♏ 12 38	06 ♏ 10 57	03 39	20 33	18 56	04 41	05 20	10 19	20 15	15 59
10	15 18 54	18 09 25	12 ♏ 08 22	18 ♏ 05 04	05 15	20 29	19 35	04 37	05 21	10 18	20 13	15 59
11	15 22 50	19 09 45	24 ♏ 01 15	29 ♏ 57 04	06 50	20 28	20 14	04 33	05 21	10 17	20 12	15 59
12	15 26 47	20 10 06	05 ♐ 52 43	11 ♐ 48 22	08 26	20 D 29	20 53	04 30	05 22	10 15	20 11	16 00
13	15 30 43	21 10 29	17 ♐ 44 15	23 ♐ 40 34	10 02	20 32	21 31	04 26	05 22	10 14	20 09	16 00
14	15 34 40	22 10 53	29 ♐ 37 35	05 ♑ 35 36	11 39	20 38	22 10	04 23	05 22	10 12	20 08	16 01
15	15 38 36	23 11 19	11 ♑ 35 57	17 ♑ 36 00	13 15	20 46	22 49	04 20	05 22	10 11	20 06	16 01
16	15 42 33	24 11 46	23 ♑ 39 11	29 ♑ 44 56	14 51	20 56	23 28	04 18	05 R 23	10 09	20 05	16 02
17	15 46 29	25 12 15	05 ♒ 53 47	12 ♒ 06 13	16 28	21 09	24 07	04 15	05 22	10 08	20 04	16 02
18	15 50 26	26 12 45	18 ♒ 22 49	24 ♒ 44 08	18 04	21 23	24 45	04 13	05 22	10 06	20 03	16 03
19	15 54 23	27 13 16	01 ♓ 10 44	07 ♓ 43 09	19 40	21 40	25 24	04 11	05 22	10 05	20 01	16 03
20	15 58 19	28 13 48	14 ♓ 21 53	21 ♓ 07 22	21 16	21 59	26 03	04 09	05 22	10 03	20 00	16 04
21	16 02 16	29 ♏ 14 22	27 ♓ 59 54	04 ♈ 59 43	22 52	22 19	26 42	04 07	05 21	10 01	19 59	16 05
22	16 06 12	00 ♐ 14 57	12 ♈ 06 51	19 ♈ 21 08	24 27	22 42	27 21	04 06	05 20	09 59	19 58	16 05
23	16 10 09	01 15 33	26 ♈ 42 11	04 ♉ 09 24	26 03	23 06	28 00	04 05	05 20	09 57	19 56	16 06
24	16 14 05	02 16 10	11 ♉ 41 57	19 ♉ 18 44	27 38	23 32	28 39	04 03	05 19	09 56	19 55	16 06
25	16 18 02	03 16 48	26 ♉ 58 29	04 ♊ 39 47	29 ♏ 13	24 00	29 17	04 03	05 18	09 54	19 54	16 07
26	16 21 58	04 17 28	12 ♊ 21 06	20 ♊ 00 57	00 ♐ 48	24 29	29 ♎ 56	04 02	05 17	09 52	19 53	16 08
27	16 25 55	05 18 10	27 ♊ 37 51	05 ♋ 10 29	02 23	25 00	00 ♏ 35	04 02	05 16	09 50	19 52	16 09
28	16 29 52	06 18 53	12 ♋ 37 42	19 ♋ 58 37	03 58	25 33	01 14	04 01	05 14	09 48	19 51	16 10
29	16 33 48	07 19 37	27 ♋ 12 32	04 ♌ 19 03	05 32	26 07	01 53	04 D 01	05 13	09 46	19 50	16 10
30	16 37 45	08 ♐ 20 23	11 ♌ 17 58	18 ♌ 09 18	07 ♐ 07	26 ♎ 43	02 ♏ 32	04 ♈ 01	05 ♌ 12	09 ♋ 44	19 ♈ 49	16 ♒ 11

Moon Nodes and Latitude

Date	Moon True ☊ ° '	Moon Mean ☊ ° '	Moon ☽ Latitude ° '
01	22 ♍ 55	21 ♍ 19	04 S 40
02	22 R 53	21 16	04 00
03	22 D 53	21 12	03 08
04	22 54	21 09	02 07
05	22 55	21 06	01 S 02
06	22 R 56	21 03	00 N 05
07	22 55	21 00	01 10
08	22 51	20 57	02 11
09	22 46	20 53	03 06
10	22 37	20 50	03 52
11	22 27	20 47	04 28
12	22 15	20 44	04 51
13	22 03	20 41	05 03
14	21 52	20 38	05 01
15	21 43	20 34	04 46
16	21 36	20 31	04 18
17	21 31	20 28	03 38
18	21 29	20 25	02 47
19	21 D 29	20 22	01 46
20	21 30	20 18	00 N 38
21	21 R 30	20 15	00 S 35
22	21 28	20 12	01 47
23	21 24	20 09	02 55
24	21 17	20 06	03 53
25	21 08	20 03	04 35
26	20 57	19 59	04 58
27	20 46	19 56	04 59
28	20 36	19 53	04 40
29	20 29	19 50	04 02
30	20 ♍ 24	19 ♍ 47	03 S 11

DECLINATIONS

Date	Sun ☉ ° '	Moon ☽ ° '	Mercury ☿ ° '	Venus ♀ ° '	Mars ♂ ° '	Jupiter ♃ ° '	Saturn ♄ ° '	Uranus ♅ ° '	Neptune ♆ ° '	Pluto ♇ ° '
01	14 S 32	17 N 35	06 S 27	12 S 54	04 S 34	00 N 40	19 N 11	23 N 20	06 N 25	23 S 35
02	14 51	15 48	06 59	12 31	04 49	00 38	19 11	23 20	06 24	23 35
03	15 10	13 11	07 32	12 09	05 04	00 36	19 10	23 20	06 23	23 35
04	15 28	09 58	08 07	11 48	05 19	00 34	19 10	23 21	06 23	23 35
05	15 47	06 21	08 42	11 28	05 34	00 32	19 10	23 21	06 22	23 34
06	16 05	02 N 32	09 19	11 09	05 49	00 31	19 10	23 21	06 22	23 34
07	16 22	01 S 21	09 55	10 51	06 04	00 29	19 10	23 21	06 21	23 34
08	16 40	05 07	10 32	10 34	06 19	00 28	19 10	23 21	06 21	23 34
09	16 57	08 38	11 09	10 18	06 34	00 26	19 09	23 21	06 20	23 34
10	17 14	11 47	11 46	10 03	06 49	00 25	19 09	23 21	06 20	23 33
11	17 31	14 27	12 23	09 50	07 04	00 24	19 09	23 21	06 19	23 33
12	17 47	16 30	13 00	09 37	07 19	00 23	19 09	23 21	06 19	23 33
13	18 03	17 51	13 36	09 26	07 34	00 22	19 09	23 22	06 18	23 33
14	18 19	18 25	14 12	09 16	07 48	00 21	19 09	23 22	06 18	23 32
15	18 34	18 11	14 47	09 07	08 03	00 20	19 09	23 22	06 17	23 32
16	18 49	17 08	15 22	08 59	08 18	00 19	19 10	23 22	06 17	23 32
17	19 04	15 16	15 56	08 52	08 32	00 18	19 10	23 22	06 16	23 31
18	19 18	12 40	16 30	08 47	08 47	00 18	19 10	23 22	06 16	23 31
19	19 32	09 24	17 03	08 42	09 01	00 17	19 10	23 22	06 15	23 31
20	19 46	05 34	17 35	08 39	09 16	00 17	19 10	23 23	06 15	23 31
21	19 59	01 S 19	18 06	08 36	09 30	00 16	19 11	23 23	06 14	23 30
22	20 12	03 N 08	18 36	08 35	09 44	00 16	19 11	23 23	06 14	23 30
23	20 24	07 34	19 06	08 35	09 58	00 16	19 11	23 23	06 14	23 30
24	20 37	11 38	19 35	08 35	10 13	00 15	19 12	23 23	06 13	23 29
25	20 48	15 00	20 02	08 36	10 27	00 15	19 12	23 23	06 13	23 29
26	21 00	17 21	20 29	08 38	10 41	00 15	19 12	23 23	06 13	23 28
27	21 11	18 26	20 55	08 41	10 55	00 15	19 13	23 24	06 12	23 28
28	21 21	18 12	21 19	08 45	11 08	00 16	19 13	23 24	06 12	23 28
29	21 32	16 45	21 43	08 50	11 22	00 16	19 14	23 24	06 11	23 27
30	21 S 41	14 N 19	22 S 06	08 S 56	11 S 36	00 N 16	19 N 14	23 N 24	06 N 11	23 S 27

ZODIAC SIGN ENTRIES

Date	h m	Planets
02	08 24	☽ ♌
04	14 36	☽ ♍
07	00 02	☽ ♎
07	04 05	☿ ♏
09	11 35	☽ ♏
12	00 06	☽ ♐
14	12 45	☽ ♑
17	00 30	☽ ♒
19	09 49	☽ ♓
21	15 27	☽ ♈
22	06 05	☉ ♐
23	17 20	☽ ♉
25	16 43	☽ ♊
25	23 50	☿ ♐
26	14 16	♂ ♏
27	15 45	☽ ♋
29	16 41	☽ ♌

LATITUDES

Date	Mercury ☿ ° '	Venus ♀ ° '	Mars ♂ ° '	Jupiter ♃ ° '	Saturn ♄ ° '	Uranus ♅ ° '	Neptune ♆ ° '	Pluto ♇ ° '
01	02 N 09	04 S 27	00 N 57	01 S 34	00 N 13	00 N 19	01 S 42	07 S 55
04	02 02	03 42	00 56	01 33	00 13	00 19	01 42	07 54
07	01 51	02 57	00 55	01 33	00 14	00 19	01 42	07 54
10	01 35	02 13	00 55	01 32	00 14	00 19	01 42	07 53
13	01 17	01 32	00 54	01 32	00 14	00 19	01 42	07 53
16	00 58	00 53	00 53	01 31	00 15	00 19	01 42	07 53
19	00 37	00 S 17	00 52	01 30	00 15	00 20	01 42	07 52
22	00 N 17	00 N 16	00 51	01 29	00 16	00 20	01 42	07 52
25	00 S 04	00 46	00 50	01 28	00 16	00 20	01 41	07 51
28	00 24	01 12	00 49	01 27	00 16	00 20	01 41	07 51
31	00 S 43	01 N 36	00 N 48	01 S 27	00 N 17	00 N 20	01 S 41	07 S 50

DATA

Julian Date	2464268
Delta T	+75 seconds
Ayanamsa	24° 20' 33"
Synetic vernal point	04° ♓ 46' 25"
True obliquity of ecliptic	23° 25' 56"

MOON'S PHASES, APSIDES AND POSITIONS ☽

Date	h m	Phase	Longitude ° '	Eclipse Indicator
03	03 27	☾	10 ♌ 47	
11	01 16	●	18 ♏ 43	
19	04 01	☽	26 ♒ 53	
25	22 32	○	03 ♊ 43	

Day	h m		
12	09 51	Apogee	
25	22 12	Perigee	
07	03 38	0S	
14	17 06	Max dec	18° S 26'
21	19 11	0N	
27	19 38	Max dec	18° N 29'

LONGITUDES

Date	Chiron ⚷ ° '	Ceres ⚳ ° '	Pallas ⚴ ° '	Juno ⚵ ° '	Vesta ⚶ ° '	Black Moon Lilith ⚸ ° '
01	05 ♊ 17	15 ♓ 10	28 ♑ 21	21 ♐ 28	18 ♋ 59	00 ♐ 40
11	04 ♊ 47	15 ♓ 16	00 ♒ 28	24 ♐ 45	19 ♋ 35	01 ♐ 47
21	04 ♊ 14	15 ♓ 58	02 ♒ 50	28 ♐ 15	19 ♋ 29	02 ♐ 54
31	03 ♊ 41	17 ♓ 13	05 ♒ 27	01 ♑ 44	18 ♋ 39	04 ♐ 01

ASPECTARIAN

h m Aspects

01 Wednesday
04 14 ☽ □ ♂
08 15 ☽ ⊼ ♇
10 10 ☿ # ♆
15 53 ☽ □ ♆
18 36 ☽ □ ☿
19 15 ☽ □ ♀
23 59 ☿ ☌ ♀

02 Thursday
12 41 ☽ Q ♂
16 49 ♃ △ ♄
17 31 ☽ △ ♃
17 32 ☽ ☌ ♄
19 21 ☉ △ ♅
20 26 ☽ # ☉

03 Friday
01 58 ☽ Q ♀
02 50 ☽ ⊻ ♅
03 27 ☽ □ ☉
05 49 ☽ Q ☿
11 08 ☽ ✱ ♂
11 53 ☉ ± ♃
12 47 ☽ ☍ ♇
13 38 ☽ ⊥ ♅
20 19 ☽ Ꝑ ♃
20 51 ☽ △ ♆
21 07 ☽ # ♀
23 10 ☽ ✱ ♀

04 Saturday
06 06 ☽ ∠ ♅
06 25 ☽ ✱ ☿
12 49 ☽ ± ♃
15 06 ☽ Q ☉
16 09 ☽ ∠ ♂
21 08 ♂ △ ♇
22 47 ☽ # ☿
23 59 ☽ ⊼ ♃

05 Sunday
00 27 ☽ ⊻ ♄
00 39 ☽ Ꝑ ♆
02 28 ☽ ∠ ♀
10 01 ☽ ⊥ ♂
10 12 ☽ ✱ ♅
11 50 ☽ ⊥ ♄
11 52 ☽ ∥ ♆
14 22 ☽ ∠ ☿
15 43 ☽ ✱ ☉
16 40 ☽ # ♂
17 43 ☽ ± ♆
19 16 ☽ ⊥ ♀
20 50 ☽ ⊼ ♇
22 07 ☽ ⊻ ♂

06 Monday
05 08 ☽ ∠ ♄
05 14 ☽ ⊼ ♆
06 37 ☽ ⊻ ♀
08 24 ☽ ± ♇
09 12 ☽ Q ♅
10 06 ☽ ⊥ ☿
22 35 ♂ Q ♄
23 17 ☽ ∠ ☉
23 27 ☽ ⊻ ☿

07 Tuesday
00 32 ☽ ∥ ♃
01 56 ☽ Ꝑ ♇
06 42 ☽ # ♃
09 30 ☽ ☍ ♃
10 26 ☽ ✱ ♄
18 34 ☽ ⊥ ☉
20 25 ☽ □ ♅

08 Wednesday
07 32 ☽ ⊻ ☉
07 33 ☽ △ ♇
07 50 ☉ □ ♇
10 13 ☽ Q ♄
12 11 ☽ ☌ ♂
12 57 ☿ ∥ ♀
14 14 ♂ # ♆
16 06 ☽ ☍ ♆
16 50 ☽ ☌ ♀
20 09 ☽ # ♆
20 37 ☽ ∥ ♂

09 Thursday
19 59 ☽ ☌ ☿
20 55 ☽ ⊼ ♃
22 18 ☽ □ ♄
23 23 ☽ ∥ ♀

10 Friday
02 52 ☿ ⊼ ♃
08 18 ☽ △ ♅
08 56 ☽ ± ♃
11 48 ☽ ∥ ☿
13 30 ☿ □ ♄
19 46 ☽ □ ♇

11 Saturday
01 16 ☽ ☌ ☉
03 01 ☽ Ꝑ ♃
03 54 ☽ ⊻ ♂
04 17 ☽ ⊼ ♆
04 49 ☽ ⊻ ♀
10 52 ♂ ☍ ♆
14 02 ♀ St D
14 32 ☽ Ꝑ ♅
16 24 ☽ ± ♆
16 44 ☽ ⊥ ♂
16 57 ☽ ⊥ ♀
20 42 ♀ ☌ ♂
20 47 ☉ Ꝑ ♃

12 Sunday
08 11 ☽ Q ♇
08 43 ☽ ± ♅
09 13 ☽ △ ♃
10 35 ☽ Ꝑ ♆
10 57 ☽ △ ♄
11 12 ☽ ∠ ♀
12 00 ☽ ∠ ♂
12 12 ☉ ⊼ ♆
17 59 ☽ ⊻ ☿
19 48 ☉ ⊻ ♀
20 50 ☽ ⊼ ♅

13 Monday
08 02 ☽ ⊥ ☿
08 29 ☽ ✱ ♇
14 48 ☿ △ ♅
16 52 ☽ △ ♆
17 19 ☽ Ꝑ ♄
17 42 ☽ ✱ ♀
17 46 ☿ ± ♃
19 35 ☽ ⊻ ☉
20 06 ☽ ✱ ♂
20 41 ☽ ∥ ☉

14 Tuesday
05 03 ☽ ∠ ☿
08 49 ☽ ⊥ ☉
11 15 ☉ ⊻ ♂
11 29 ☽ ± ♄
14 47 ☽ ∠ ♇
18 07 ☽ Q ♀
21 32 ☽ □ ♃
21 40 ☽ Q ♂
22 12 ☽ ∥ ☉
23 33 ☽ ⊼ ♄

15 Wednesday
04 35 ☽ ∠ ☉
08 52 ☽ ⊥ ♇
09 12 ☽ ☍ ♅
15 51 ☽ ✱ ☿
20 51 ☽ ⊻ ♇

16 Thursday
04 57 ☽ □ ♆
06 33 ☽ □ ♀
09 19 ☽ Q ♃
11 36 ☽ □ ♂
12 00 ♄ St R
13 10 ☽ ✱ ☉
19 17 ☽ Q ☿

17 Friday
05 32 ☿ □ ♇
06 05 ☽ ∥ ☿
08 49 ☽ ✱ ♃
10 17 ☉ Ꝑ ♅
10 59 ☽ ☍ ♄
14 46 ☽ Q ☉
16 12 ☽ Q ♆
20 11 ☽ ⊼ ♅
22 20 ☉ # ♄

18 Saturday
07 33 ☽ ☌ ♇
07 40 ☽ ± ♅
08 03 ☉ ± ♆
11 19 ☽ □ ☿
12 03 ♀ ∥ ♂
13 35 ☽ ∠ ♃
15 09 ☽ ✱ ♆
17 49 ☽ △ ♀

19 Sunday
00 40 ☽ Ꝑ ♅
00 41 ☽ △ ♂
04 01 ☽ □ ☉
04 51 ☿ Ꝑ ♃
06 27 ☽ ⊥ ♃
14 23 ☽ ∥ ♂
16 42 ☽ ∥ ♀
17 15 ☿ ⊼ ♆
17 31 ☽ ⊻ ♃
19 04 ☽ ∠ ♆
19 42 ☽ ⊼ ♄
22 19 ☽ Ꝑ ♀

20 Monday
03 08 ☉ ⊥ ♀
03 13 ♅ ± ♇
04 15 ☽ △ ♅
05 44 ☽ Ꝑ ♂
06 36 ☽ ± ♄
07 57 ☽ # ♆
11 21 ☽ ⊥ ♆
14 57 ☽ ± ♀
15 02 ☽ ⊻ ♇
22 00 ☽ ⊻ ♆
22 37 ☽ ± ♂
22 39 ☽ Ꝑ ♄

21 Tuesday
01 32 ☿ ⊻ ♀
01 40 ☽ ⊥ ♇
01 50 ☽ ⊼ ♀
01 52 ☽ △ ☿
09 38 ☽ ⊼ ♂
14 19 ☽ △ ☉
17 18 ☽ ∠ ♇
17 44 ☽ # ♃
20 37 ☽ ∥ ♃
22 30 ☽ ☌ ♃

22 Wednesday
00 36 ☽ △ ♄
06 59 ☽ Ꝑ ☿
08 27 ☽ □ ♅
17 37 ☽ Ꝑ ☉
18 36 ☽ ✱ ♇
19 54 ☿ Ꝑ ♅
23 49 ☽ ± ☿

23 Thursday
00 59 ☽ ☌ ♆
04 37 ☽ ∥ ♆
05 58 ☽ ☍ ♀
09 29 ☽ ± ☉
10 27 ☿ ± ♆
10 48 ☽ ⊼ ☿
14 02 ☽ Q ♅
14 11 ☽ ☍ ♂
14 15 ☽ Q ♇
16 18 ☿ # ♄
17 39 ☽ # ♀
19 53 ☽ ⊼ ☉
23 51 ☽ ⊻ ♃

24 Friday
01 52 ☽ □ ♄
02 38 ☽ # ♂
09 12 ☽ ✱ ♅
09 24 ☽ ⊥ ♃
18 58 ☽ □ ♇
23 35 ☽ ∠ ♃

25 Saturday
00 57 ☽ ⊻ ♆
06 15 ☽ Q ♄
07 12 ☽ ⊼ ♀
08 45 ☽ ∠ ♅
10 20 ☽ ⊥ ♆
13 50 ☿ ⊻ ♂
15 47 ☽ ⊼ ♂
15 54 ☽ ☍ ☿
16 52 ☽ ± ♀
22 32 ☽ ☍ ☉
22 47 ☽ ⊥ ♅
23 01 ☽ ✱ ♃

26 Sunday
00 22 ☽ ∠ ♆
00 59 ☽ ✱ ♄
01 33 ☽ ± ♂
02 10 ☉ ± ♅
05 04 ☿ ⊥ ♀
05 55 ☉ △ ♃
07 23 ☽ Ꝑ ♀
08 06 ☉ Q ♇
08 07 ☽ ⊻ ♅
16 13 ☽ Ꝑ ♂
17 45 ☽ Q ♃
17 55 ☽ △ ♇
23 47 ☽ ✱ ♆

27 Monday
00 24 ☽ ∠ ♄
01 51 ☉ Ꝑ ♆
07 42 ☽ △ ♀
11 03 ☉ △ ♄
14 35 ☽ ⊥ ♄
16 54 ☽ △ ♂
17 34 ☽ Ꝑ ♇
18 43 ☽ Q ♆
20 26 ☽ ⊼ ☿
22 10 ☽ □ ♃

28 Tuesday
00 07 ☽ ⊻ ♄
01 05 ☽ ⊼ ☉
07 10 ☽ ± ☿
07 26 ☽ ☌ ♅
08 00 ☽ ± ♇
09 35 ☿ ± ♅
11 27 ☽ ± ☉
12 56 ☿ △ ♃
15 00 ☿ Q ♇
15 28 ☉ ∥ ☿
17 44 ☽ ⊼ ♇
23 35 ☽ Ꝑ ☿
23 47 ☽ □ ♆

29 Wednesday
01 24 ☿ Ꝑ ♆
02 25 ♃ St D
03 16 ☽ Ꝑ ☉
07 13 ☿ △ ♄
10 06 ☽ □ ♀
20 15 ☽ □ ♂
23 30 ☽ △ ♃

30 Thursday
01 31 ☽ ☌ ♄
03 52 ☽ △ ☿
06 29 ☽ △ ☉
09 17 ☽ ⊻ ♅
18 13 ☽ Q ♀
19 43 ☽ ⊥ ♅
20 32 ☽ ☍ ♇

All ephemeris data is given at 12.00 UT and the Moon's longitude is additionally given for 24.00 UT

DECEMBER 2034

Date	Sidereal time h m s	LONGITUDES Sun ☉ ° ' "	Moon ☽ ° ' "	Moon ☽ 24.00 ° ' "	Mercury ☿ ° '	Venus ♀ ° '	Mars ♂ ° '	Jupiter ♃ ° '	Saturn ♄ ° '	Uranus ♅ ° '	Neptune ♆ ° '	Pluto ♇ ° '
01	16 41 41	09 ♐ 21 10	24 ♌ 53 15	01 ♍ 30 11	08 ♐ 41	27 ♎ 20	03 ♏ 11	04 ♈ 02	05 ♌ 10	09 ♋ 42	19 ♈ 48	16 ♒ 12
02	16 45 38	10 21 58	08 ♍ 00 32	14 ♍ 24 50	10 15	27 58	03 50	04 02	05 R 08	09 R 40	19 R 47	16 13
03	16 49 34	11 22 48	20 ♍ 43 42	26 ♍ 57 44	11 50	28 37	04 29	04 03	05 06	09 37	19 46	16 14
04	16 53 31	12 23 40	03 ♎ 07 34	09 ♎ 13 47	13 24	29 ♎ 18	05 08	04 04	05 04	09 35	19 45	16 15
05	16 57 27	13 24 33	15 ♎ 17 01	21 ♎ 17 48	14 58	00 ♏ 00	05 47	04 05	05 02	09 33	19 44	16 16
06	17 01 24	14 25 27	27 ♎ 16 40	03 ♏ 14 04	16 32	00 43	06 26	04 07	05 00	09 31	19 44	16 17
07	17 05 20	15 26 22	09 ♏ 10 27	15 ♏ 06 11	18 06	01 27	07 05	04 09	04 58	09 28	19 43	16 18
08	17 09 17	16 27 19	21 ♏ 01 36	26 ♏ 56 57	19 40	02 12	07 44	04 10	04 56	09 26	19 42	16 19
09	17 13 14	17 28 17	02 ♐ 52 31	08 ♐ 48 28	21 15	02 58	08 23	04 12	04 53	09 24	19 41	16 20
10	17 17 10	18 29 16	14 ♐ 45 00	20 ♐ 42 15	22 49	03 45	09 02	04 15	04 51	09 21	19 41	16 21
11	17 21 07	19 30 15	26 ♐ 40 21	02 ♑ 39 28	24 23	04 33	09 41	04 17	04 48	09 19	19 40	16 22
12	17 25 03	20 31 16	08 ♑ 39 43	14 ♑ 41 16	25 58	05 22	10 21	04 20	04 45	09 17	19 39	16 23
13	17 29 00	21 32 17	20 ♑ 44 17	26 ♑ 49 00	27 32	06 12	11 00	04 23	04 42	09 14	19 39	16 24
14	17 32 56	22 33 19	02 ♒ 55 38	09 ♒ 04 29	29 ♐ 06	07 02	11 39	04 26	04 40	09 12	19 38	16 25
15	17 36 53	23 34 22	15 ♒ 15 54	21 ♒ 30 13	00 ♑ 41	07 54	12 18	04 29	04 37	09 09	19 38	16 27
16	17 40 49	24 35 25	27 ♒ 47 51	04 ♓ 09 16	02 16	08 46	12 57	04 32	04 33	09 07	19 37	16 28
17	17 44 46	25 36 28	10 ♓ 34 54	17 ♓ 05 14	03 50	09 39	13 36	04 36	04 30	09 04	19 37	16 29
18	17 48 43	26 37 32	23 ♓ 40 44	00 ♈ 21 51	05 25	10 32	14 15	04 40	04 27	09 02	19 36	16 30
19	17 52 39	27 38 36	07 ♈ 08 57	14 ♈ 02 22	07 00	11 26	14 55	04 44	04 24	08 59	19 36	16 31
20	17 56 36	28 39 40	21 ♈ 02 16	28 ♈ 08 43	08 35	12 21	15 34	04 48	04 20	08 57	19 36	16 33
21	18 00 32	29 ♐ 40 45	05 ♉ 21 37	12 ♉ 40 37	10 10	13 16	16 13	04 53	04 17	08 54	19 35	16 34
22	18 04 29	00 ♑ 41 50	20 ♉ 05 13	27 ♉ 34 39	11 45	14 12	16 52	04 57	04 13	08 52	19 35	16 35
23	18 08 25	01 42 55	05 ♊ 07 56	12 ♊ 43 55	13 20	15 09	17 31	05 02	04 09	08 49	19 35	16 37
24	18 12 22	02 44 01	20 ♊ 21 16	27 ♊ 58 37	14 55	16 06	18 11	05 07	04 05	08 47	19 34	16 38
25	18 16 18	03 45 07	05 ♋ 34 32	13 ♋ 07 36	16 29	17 04	18 50	05 12	04 02	08 44	19 34	16 40
26	18 20 15	04 46 13	20 ♋ 36 35	28 ♋ 00 20	18 04	18 02	19 29	05 17	03 58	08 42	19 34	16 41
27	18 24 12	05 47 20	05 ♌ 17 59	12 ♌ 28 48	19 38	19 01	20 08	05 23	03 54	08 39	19 34	16 42
28	18 28 08	06 48 27	19 ♌ 32 23	26 ♌ 28 28	21 12	20 00	20 48	05 29	03 50	08 36	19 34	16 44
29	18 32 05	07 49 34	03 ♍ 17 03	09 ♍ 58 18	22 45	21 00	21 27	05 35	03 46	08 34	19 34	16 45
30	18 36 01	08 50 42	16 ♍ 32 31	23 ♍ 00 08	24 17	22 00	22 06	05 41	03 41	08 31	19 34	16 47
31	18 39 58	09 ♑ 51 51	29 ♍ 21 42	05 ♎ 37 47	25 ♑ 49	23 ♏ 01	22 ♏ 45	05 ♈ 47	03 ♌ 37	08 ♋ 29	19 ♈ 34	16 ♒ 48

Date	Moon True ☊ ° '	Moon Mean ☊ ° '	Moon ☽ Latitude ° '
01	20 ♍ 21	19 ♍ 44	02 S 10
02	20 D 21	19 40	01 05
03	20 R 21	19 37	00 N 02
04	20 20	19 34	01 07
05	20 18	19 31	02 08
06	20 13	19 28	03 02
07	20 05	19 24	03 47
08	19 55	19 21	04 23
09	19 41	19 18	04 47
10	19 26	19 15	04 58
11	19 11	19 12	04 57
12	18 56	19 09	04 42
13	18 44	19 05	04 15
14	18 34	19 02	03 35
15	18 28	18 59	02 45
16	18 24	18 56	01 46
17	18 23	18 53	00 N 41
18	18 D 23	18 50	00 S 28
19	18 R 22	18 46	01 38
20	18 21	18 43	02 44
21	18 17	18 40	03 41
22	18 10	18 37	04 26
23	18 01	18 34	04 54
24	17 50	18 30	05 01
25	17 39	18 27	04 47
26	17 29	18 24	04 13
27	17 20	18 21	03 23
28	17 15	18 18	02 21
29	17 12	18 15	01 13
30	17 11	18 11	00 03
31	17 ♍ 11	18 ♍ 08	01 N 04

DECLINATIONS

Date	Sun ☉ ° '	Moon ☽ ° '	Mercury ☿ ° '	Venus ♀ ° '	Mars ♂ ° '	Jupiter ♃ ° '	Saturn ♄ ° '	Uranus ♅ ° '	Neptune ♆ ° '	Pluto ♇ ° '
01	21 S 51	11 N 10	22 S 27	09 S 02	11 S 50	00 N 17	19 N 15	23 N 24	06 N 11	23 S 27
02	22 00	07 34	22 47	09 09	12 03	00 17	19 15	23 25	06 10	23 26
03	22 08	03 N 42	23 07	09 16	12 17	00 18	19 16	23 25	06 10	23 26
04	22 16	00 S 13	23 25	09 24	12 30	00 18	19 16	23 25	06 10	23 25
05	22 24	04 03	23 42	09 33	12 43	00 19	19 17	23 25	06 10	23 25
06	22 31	07 40	23 57	09 42	12 57	00 20	19 17	23 25	06 09	23 25
07	22 38	10 57	24 12	09 52	13 10	00 21	19 18	23 25	06 09	23 24
08	22 45	13 47	24 25	10 03	13 23	00 22	19 19	23 26	06 09	23 24
09	22 51	16 02	24 37	10 14	13 36	00 23	19 20	23 26	06 09	23 23
10	22 56	17 37	24 47	10 25	13 49	00 24	19 20	23 26	06 08	23 23
11	23 01	18 27	24 56	10 36	14 01	00 25	19 21	23 26	06 08	23 22
12	23 06	18 28	25 04	10 49	14 14	00 27	19 22	23 26	06 08	23 22
13	23 10	17 38	25 11	11 01	14 27	00 28	19 23	23 27	06 08	23 21
14	23 13	16 00	25 16	11 14	14 39	00 30	19 23	23 27	06 08	23 21
15	23 17	13 37	25 20	11 27	14 52	00 31	19 24	23 27	06 07	23 20
16	23 19	10 34	25 22	11 40	15 04	00 33	19 25	23 27	06 07	23 20
17	23 22	06 58	25 23	11 54	15 16	00 34	19 26	23 27	06 07	23 20
18	23 23	02 S 56	25 22	12 08	15 28	00 36	19 27	23 27	06 07	23 19
19	23 25	01 N 21	25 21	12 22	15 40	00 38	19 28	23 28	06 07	23 19
20	23 26	05 41	25 17	12 36	15 52	00 40	19 29	23 28	06 07	23 18
21	23 26	09 50	25 12	12 50	16 04	00 42	19 30	23 28	06 07	23 18
22	23 26	13 29	25 05	13 05	16 15	00 44	19 31	23 28	06 07	23 17
23	23 25	16 20	24 57	13 19	16 27	00 46	19 32	23 28	06 07	23 17
24	23 24	18 05	24 47	13 34	16 38	00 48	19 33	23 29	06 07	23 16
25	23 23	18 32	24 36	13 49	16 49	00 51	19 34	23 29	06 07	23 16
26	23 21	17 41	24 24	14 04	17 01	00 53	19 35	23 29	06 07	23 15
27	23 18	15 40	24 09	14 19	17 12	00 56	19 36	23 29	06 07	23 15
28	23 15	12 43	23 54	14 33	17 22	00 58	19 37	23 29	06 07	23 14
29	23 12	09 10	23 37	14 48	17 33	01 01	19 38	23 29	06 07	23 13
30	23 08	05 15	23 18	15 03	17 44	01 03	19 39	23 30	06 07	23 13
31	23 S 04	01 N 14	22 S 58	15 S 18	17 S 54	01 N 06	19 N 40	23 N 30	06 N 07	23 S 12

ZODIAC SIGN ENTRIES

Date	h m	Planets
01	21 15	☽ ♍
04	05 54	☽ ♎
05	12 04	♀ ♏
06	17 29	☽ ♏
09	06 11	☽ ♐
11	18 41	☽ ♑
14	06 16	☽ ♒
15	01 35	☿ ♑
16	16 10	☽ ♓
18	23 21	☽ ♈
20	03 06	☽ ♉
21	19 34	☉ ♑
23	03 51	☽ ♊
25	03 11	☽ ♋
27	03 16	☽ ♌
29	06 11	☽ ♍
31	13 13	☽ ♎

LATITUDES

Date	Mercury ☿ ° '	Venus ♀ ° '	Mars ♂ ° '	Jupiter ♃ ° '	Saturn ♄ ° '	Uranus ♅ ° '	Neptune ♆ ° '	Pluto ♇ ° '
01	00 S 43	01 N 36	00 N 48	01 S 27	00 N 17	00 N 20	01 S 41	07 S 50
04	01 01	01 56	00 46	01 26	00 17	00 20	01 41	07 50
07	01 18	02 14	00 45	01 25	00 18	00 20	01 41	07 50
10	01 33	02 30	00 44	01 24	00 18	00 20	01 41	07 49
13	01 46	02 43	00 43	01 23	00 18	00 20	01 41	07 49
16	01 57	02 54	00 41	01 22	00 19	00 20	01 41	07 49
19	02 06	03 03	00 40	01 21	00 19	00 20	01 40	07 48
22	02 11	03 10	00 39	01 21	00 19	00 20	01 40	07 48
25	02 12	03 15	00 37	01 20	00 20	00 20	01 40	07 48
28	02 09	03 18	00 36	01 19	00 20	00 20	01 40	07 48
31	02 S 01	03 N 20	00 N 34	01 S 18	00 N 21	00 N 20	01 S 40	07 S 47

DATA

Julian Date	2464298
Delta T	+75 seconds
Ayanamsa	24° 20' 38"
Synetic vernal point	04° ♓ 46' 21"
True obliquity of ecliptic	23° 25' 56"

MOON'S PHASES, APSIDES AND POSITIONS ☽

Date	h m	Phase	Longitude ° '	Eclipse Indicator
02	16 46	☾	10 ♍ 34	
10	20 14	●	18 ♐ 50	
18	17 45	☽	26 ♓ 52	
25	08 54	○	03 ♋ 37	

Day	h m		
09	10 18	Apogee	
24	10 43	Perigee	
04	10 40	0S	
12	00 20	Max dec	18° S 33'
19	04 34	0N	
25	08 15	Max dec	18° N 33'
31	19 25	0S	

LONGITUDES

Date	Chiron ⚷ ° '	Ceres ⚳ ° '	Pallas ⚴ ° '	Juno ⚵ ° '	Vesta ⚶ ° '	Black Moon Lilith ⚸ ° '
01	03 ♊ 41	17 ♓ 13	05 ♒ 27	01 ♑ 46	18 ♋ 39	04 ♐ 01
11	03 ♊ 08	18 ♓ 57	08 ♒ 15	05 ♑ 22	17 ♋ 05	05 ♐ 08
21	02 ♊ 38	21 ♓ 06	11 ♒ 12	09 ♑ 05	14 ♋ 55	06 ♐ 16
31	02 ♊ 12	23 ♓ 36	14 ♒ 17	12 ♑ 52	12 ♋ 23	07 ♐ 23

All ephemeris data is given at 12.00 UT and the Moon's longitude is additionally given for 24.00 UT

ASPECTARIAN

h m Aspects

01 Friday
00 54 ☉ ⊥ ♂
01 33 ☽ bQ ♃
02 55 ☽ △ ♆
05 02 ☽ Q ♂
07 39 ☽ # ♂
11 39 ☽ ∠ ♅
16 37 ☽ ✱ ♀
17 41 ☽ ± ♃
19 49 ☉ ⊼ ♅

02 Saturday
01 01 ☿ ⊥ ♂
02 04 ☽ # ♀
03 05 ☿ ⊼ ♅
03 52 ☽ ✱ ♂
04 39 ☽ ⊼ ♃
06 02 ☽ bQ ♆
06 41 ☽ ⊻ ♄
15 04 ☽ ✱ ♅
16 45 ☉ ☌ ☿
16 46 ☽ □ ☉
16 46 ☽ □ ☿
17 49 ☽ ⊥ ♄
19 46 ♂ ⊼ ♃
20 45 ☽ ∥ ♆
21 46 ☽ ∠ ♀
22 48 ☽ ± ♆

03 Sunday
03 25 ☽ ⊼ ♇
09 30 ☽ ∠ ♂
10 10 ☽ ⊼ ♆
10 49 ☽ ∠ ♄
13 42 ☽ Q ♅
14 53 ☽ ± ♇
15 50 ☽ ⊥ ♀

04 Monday
03 47 ☽ ⊥ ♂
04 06 ☽ ⊻ ♀
06 11 ☽ Q ☉
08 08 ☽ Q ☿
08 19 ☽ bQ ♇
08 48 ☽ ∥ ♃
09 54 ♂ □ ♄
12 19 ☿ # ♅
12 33 ☽ # ♃
12 57 ☿ ∥ ♇
13 51 ☽ ☍ ♃
15 49 ☽ ✱ ♄
16 09 ☽ ⊻ ♂

05 Tuesday
00 40 ☽ □ ♅
07 12 ♅ # ♇
07 56 ☽ ✱ ☉
11 16 ☽ ✱ ☿
12 53 ☿ ∠ ♀
13 57 ☽ △ ♇
15 29 ☽ Q ♄
20 53 ☽ ☍ ♆

06 Wednesday
01 44 ☽ # ♆
07 59 ☿ ✱ ♇
16 43 ☽ ∠ ☉
19 22 ☽ ☌ ♀
21 52 ☽ ∠ ☿

07 Thursday
01 48 ☽ ⊼ ♃
03 19 ☽ ∥ ♀
03 32 ☽ □ ♄
07 32 ☽ ☌ ♂
12 35 ☽ ⊥ ☉
12 36 ☽ △ ♅
13 58 ☽ ± ♃
18 50 ☽ ⊥ ☿

08 Friday
01 52 ☽ ⊻ ☉
02 26 ☽ □ ♇
07 59 ☽ ∥ ♂
08 14 ☽ bQ ♃
08 30 ☉ ✱ ♇
08 50 ☽ ⊻ ☿
09 19 ☽ ⊼ ♆
12 25 ☿ △ ♆
15 47 ☿ bQ ♄
18 53 ☽ bQ ♅
21 28 ☽ ± ♆

09 Saturday
12 12 ☽ ⊻ ♀
13 03 ☽ ± ♅
14 42 ☽ △ ♃
14 57 ☽ Q ♇
15 40 ☽ bQ ♆
16 03 ☽ △ ♄
23 48 ☽ ⊻ ♂

10 Sunday
01 09 ☽ ⊼ ♅
01 12 ☽ ⊥ ♀
12 37 ☽ ⊥ ♂
15 13 ☽ ✱ ♇
20 14 ☽ ☌ ☉
20 39 ☽ ∠ ♀
21 56 ☽ △ ♆
22 14 ☽ bQ ♄
23 01 ♂ △ ♅

11 Monday
03 31 ♀ ⊼ ♃
06 43 ☽ ☌ ☿
07 47 ☽ ∠ ♂
15 49 ☉ △ ♆
16 15 ☽ ± ♄
17 52 ☉ ∠ ♀
18 43 ☉ bQ ♄
18 55 ♀ □ ♄
19 57 ☿ ∠ ♂
21 25 ☽ ∠ ♇

12 Tuesday
03 19 ☽ □ ♃
04 14 ☽ ⊼ ♄
04 57 ☽ ✱ ♀
11 29 ♂ ± ♃
13 13 ☽ ☍ ♅
15 26 ☽ ⊥ ♇
15 33 ☽ ✱ ♂

13 Wednesday
03 24 ☽ ⊻ ♇
06 36 ☽ Q ♀
09 50 ☽ □ ♆
13 44 ☽ ⊻ ☉
15 15 ☽ Q ♃
16 43 ☽ Q ♂

14 Thursday
02 38 ☽ ⊥ ☉
03 24 ☽ ⊻ ☿
05 23 ☿ ± ♄
14 57 ☽ ✱ ♃
15 23 ☽ ☍ ♄
16 54 ☽ ⊥ ☿
20 38 ☽ □ ♀
21 12 ☽ Q ♆
21 51 ☽ ∠ ☉

15 Friday
00 12 ☽ ⊼ ♅
01 20 ☽ ∥ ♂
05 57 ☽ □ ♂
11 47 ☽ ± ♅
12 56 ☽ ∠ ☿
14 16 ☽ ☌ ♇
20 10 ☽ ∠ ♃
20 24 ☽ ✱ ♆
23 41 ☿ ∠ ♇

16 Saturday
04 25 ☽ ∥ ♀
05 01 ☽ bQ ♅
05 22 ☽ ✱ ☉
13 25 ☽ ⊥ ♃
15 35 ♃ △ ♄
17 58 ☉ ∥ ♇
21 12 ♀ △ ♅
21 38 ☽ ✱ ☿

17 Sunday
00 42 ☽ ⊼ ♄
00 47 ☽ ⊻ ♃
00 52 ☽ ∠ ♆
05 59 ☽ Q ☉
09 12 ☽ △ ♅
10 08 ☽ △ ♀
11 51 ☽ ± ♄
17 13 ☽ # ♆
17 36 ☽ ⊥ ♆
17 53 ☽ △ ♂
21 44 ☿ ⊼ ♄
22 55 ☽ ⊻ ♇
23 03 ☽ Q ☿

18 Monday
00 02 ☿ □ ♃
04 21 ☽ bQ ♄
04 36 ☽ ⊻ ♆
09 52 ☽ ⊥ ♇
15 35 ☽ bQ ♀
15 47 ♀ ± ♃
17 45 ☽ □ ☉
22 32 ☽ bQ ♂

19 Tuesday
01 07 ☽ # ♃
02 03 ☽ ∠ ♇
07 10 ☽ △ ♄
07 43 ☽ ☌ ♃
08 03 ☽ ∥ ♃
08 46 ☽ ± ♀
11 42 ☽ □ ☿
15 13 ☽ □ ♅
15 14 ☽ ± ♂
20 01 ☽ ⊼ ♀

20 Wednesday
02 10 ☽ ⊼ ♂
04 19 ☽ ✱ ♇
04 43 ☉ ± ♄
09 32 ☽ ☌ ♆
14 26 ☽ ∥ ♆
17 22 ☿ ☍ ♅
21 58 ☽ Q ♅

21 Thursday
00 41 ☽ Q ♇
01 51 ☽ △ ☉
10 13 ☽ □ ♄
11 12 ☽ ⊻ ♃
17 49 ☽ ✱ ♅
18 11 ☿ ⊥ ♇
20 51 ☽ △ ☿
21 07 ☽ ⊥ ♃

22 Friday
01 26 ♂ □ ♇
01 51 ☽ ☍ ♀
04 23 ☽ bQ ☉
06 21 ☽ □ ♇
06 34 ☽ ☍ ♂
08 53 ☽ # ♀
11 11 ☽ ⊻ ♆
11 47 ☽ ∠ ♃
15 24 ☽ Q ♄
18 03 ☽ ∠ ♅
19 56 ☽ ± ☉
20 49 ☽ ⊥ ♆
23 56 ☽ bQ ☿

23 Saturday
06 12 ☽ ⊼ ☉
08 21 ☽ ⊥ ♅
09 32 ☉ ∠ ♇
10 27 ☽ ✱ ♄
11 07 ☽ ∠ ♆
11 51 ☽ ✱ ♃
13 16 ☽ # ♂
15 53 ☽ ± ☿
17 49 ☽ ⊻ ♅

24 Sunday
02 26 ☽ ⊼ ☿
04 52 ☽ ⊼ ♀
06 08 ☽ △ ♇
06 53 ☽ Q ♃
08 25 ☽ ⊼ ♂
10 01 ☽ ∠ ♄
10 46 ☽ ✱ ♆
14 56 ☽ ± ♀
18 17 ☽ ± ♂

25 Monday
00 08 ☽ ⊥ ♄
01 39 ♀ □ ♇
05 40 ☽ Q ♆
05 48 ☽ bQ ♇
06 04 ☽ bQ ♀
08 54 ☽ ☍ ☉
09 07 ☽ bQ ♂
09 34 ☽ ⊻ ♄
11 24 ☽ □ ♃
14 39 ☿ ⊻ ♇
17 00 ☽ ☌ ♅
17 06 ☉ ∠ ♂
18 05 ☉ ⊼ ♄
20 05 ☽ ± ♇

26 Tuesday
05 41 ☽ ⊼ ♇
07 25 ☽ ☍ ☿
07 34 ☽ △ ♀
10 06 ☽ △ ♂
10 19 ☽ □ ♆
10 57 ☿ ✱ ♀
15 04 ♂ ⊼ ♆
20 42 ☽ # ♂

27 Wednesday
01 28 ☉ □ ♃
09 41 ☽ ☌ ♄
10 59 ☿ □ ♆
12 08 ☽ △ ♃
12 52 ☽ ⊼ ☉
17 33 ☽ ⊻ ♅
22 30 ♂ bQ ♃
22 53 ☽ # ♀
23 40 ☽ ± ☉

28 Thursday
01 22 ☿ ✱ ♂
01 22 ♀ ⊼ ♆
03 37 ☽ ⊥ ♅
07 12 ☽ ☍ ♇
12 02 ☽ △ ♆
12 51 ☽ □ ♀
13 37 ☽ bQ ♃
14 15 ☽ □ ♂
15 12 ☽ ⊼ ☿
16 13 ☽ bQ ☉
18 59 ☽ ∠ ♅

29 Friday
00 30 ☉ ∥ ♇
00 43 ♀ bQ ♃
02 57 ☽ ± ☿
05 23 ☽ ± ♃
12 50 ☽ ⊻ ♄
14 17 ☽ bQ ♆
16 07 ☽ ⊼ ♃
20 48 ☽ △ ☉
21 02 ☽ bQ ☿
21 15 ☿ # ♅
21 25 ☽ ✱ ♅
23 04 ☽ Q ♀
23 33 ☽ ⊥ ♄
23 37 ☽ Q ♂

30 Saturday
01 46 ☿ Q ♃
04 38 ☉ ☍ ♅
06 32 ☽ ± ♆
06 52 ☽ ∥ ♆
12 26 ☽ ⊼ ♇
15 57 ☽ ∠ ♄
17 35 ☽ ⊼ ♆
18 16 ☿ ∥ ♇
18 51 ♀ ☌ ♂
19 20 ☽ Q ♅
20 09 ♆ St D
22 52 ☽ ✱ ♂
22 59 ☽ ✱ ♀
23 36 ☽ ± ♇

31 Sunday
02 57 ☉ ∥ ☿
04 23 ☽ △ ☿
12 49 ☽ ∥ ♃
16 40 ☽ bQ ♇
20 05 ☽ ✱ ♄
22 33 ♀ bQ ♅

JANUARY 2035

Date	Sidereal time (h m s)	Sun ☉ (° ' ")	Moon ☽ (° ' ")	Moon ☽ 24.00 (° ' ")	Mercury ☿ (° ')	Venus ♀ (° ')	Mars ♂ (° ')	Jupiter ♃ (° ')	Saturn ♄ (° ')	Uranus ♅ (° ')	Neptune ♆ (° ')	Pluto ♇ (° ')
01	18 43 54	10 ♑ 53 00	11 ♎ 49 01	17 ♎ 56 04	27 ♑ 20	24 ♏ 02	23 ♏ 25	05 ♈ 53	03 ♌ 33	08 ♋ 26	19 ♈ 34	16 ♒ 50
02	18 47 51	11 54 09	23 ♎ 59 36	00 ♏ 00 15	28 ♑ 49	25 03	24 04	06 00	03 R 28	08 R 23	19 D 34	16 51
03	18 51 47	12 55 18	05 ♏ 58 39	11 ♏ 55 24	00 ♒ 16	26 05	24 43	06 06	03 24	08 21	19 34	16 53
04	18 55 44	13 56 28	17 ♏ 51 03	23 ♏ 46 08	01 42	27 07	25 23	06 13	03 19	08 18	19 34	16 54
05	18 59 41	14 57 38	29 ♏ 41 07	05 ♐ 36 24	03 05	28 10	26 02	06 20	03 15	08 16	19 34	16 56
06	19 03 37	15 58 49	11 ♐ 32 22	17 ♐ 29 20	04 26	29 ♏ 13	26 41	06 28	03 10	08 13	19 34	16 58
07	19 07 34	16 59 59	23 ♐ 27 34	29 ♐ 27 16	05 43	00 ♐ 16	27 21	06 35	03 06	08 11	19 35	16 59
08	19 11 30	18 01 09	05 ♑ 28 39	11 ♑ 31 51	06 57	01 19	28 00	06 42	03 01	08 08	19 35	17 01
09	19 15 27	19 02 20	17 ♑ 36 58	23 ♑ 44 07	08 05	02 23	28 40	06 50	02 56	08 05	19 35	17 02
10	19 19 23	20 03 30	29 ♑ 53 23	06 ♒ 04 52	09 09	03 27	29 19	06 58	02 52	08 03	19 36	17 04
11	19 23 20	21 04 39	12 ♒ 18 38	18 ♒ 34 48	10 06	04 32	29 ♏ 58	07 06	02 47	08 00	19 36	17 06
12	19 27 16	22 05 49	24 ♒ 53 30	01 ♓ 14 53	10 57	05 37	00 ♐ 38	07 14	02 42	07 58	19 37	17 07
13	19 31 13	23 06 57	07 ♓ 39 08	14 ♓ 06 27	11 40	06 42	01 17	07 22	02 37	07 55	19 37	17 09
14	19 35 10	24 08 06	20 ♓ 37 05	27 ♓ 11 16	12 14	07 47	01 57	07 31	02 32	07 53	19 37	17 11
15	19 39 06	25 09 13	03 ♈ 49 18	10 ♈ 31 24	12 39	08 52	02 36	07 39	02 28	07 51	19 38	17 12
16	19 43 03	26 10 20	17 ♈ 17 51	24 ♈ 08 50	12 53	09 58	03 15	07 48	02 23	07 48	19 39	17 14
17	19 46 59	27 11 26	01 ♉ 04 31	08 ♉ 04 57	12 R 56	11 04	03 55	07 57	02 18	07 46	19 39	17 16
18	19 50 56	28 12 32	15 ♉ 10 06	22 ♉ 19 49	12 48	12 10	04 34	08 06	02 13	07 43	19 40	17 17
19	19 54 52	29 ♑ 13 36	29 ♉ 33 49	06 ♊ 51 37	12 28	13 17	05 14	08 15	02 08	07 41	19 40	17 19
20	19 58 49	00 ♒ 14 40	14 ♊ 12 38	21 ♊ 36 06	11 56	14 24	05 53	08 24	02 03	07 39	19 41	17 21
21	20 02 45	01 15 43	29 ♊ 01 09	06 ♋ 26 48	11 13	15 30	06 33	08 34	01 58	07 36	19 42	17 23
22	20 06 42	02 16 45	13 ♋ 51 58	21 ♋ 15 37	10 20	16 37	07 12	08 43	01 53	07 34	19 43	17 24
23	20 10 39	03 17 46	28 ♋ 36 40	05 ♌ 54 10	09 19	17 45	07 52	08 53	01 48	07 32	19 43	17 26
24	20 14 35	04 18 47	13 ♌ 07 14	20 ♌ 15 07	08 10	18 52	08 31	09 03	01 43	07 30	19 44	17 28
25	20 18 32	05 19 47	27 ♌ 17 15	04 ♍ 13 13	06 57	20 00	09 11	09 13	01 38	07 27	19 45	17 30
26	20 22 28	06 20 46	11 ♍ 02 48	17 ♍ 45 53	05 41	21 08	09 50	09 23	01 33	07 25	19 46	17 32
27	20 26 25	07 21 44	24 ♍ 22 33	00 ♎ 53 02	04 25	22 16	10 30	09 33	01 29	07 23	19 47	17 33
28	20 30 21	08 22 42	07 ♎ 17 37	13 ♎ 36 43	03 11	23 24	11 09	09 43	01 24	07 21	19 48	17 35
29	20 34 18	09 23 39	19 ♎ 50 50	26 ♎ 00 31	02 00	24 32	11 49	09 54	01 19	07 19	19 49	17 37
30	20 38 14	10 24 35	02 ♏ 06 20	08 ♏ 08 55	00 ♒ 56	25 41	12 28	10 04	01 14	07 17	19 50	17 39
31	20 42 11	11 ♒ 25 31	14 ♏ 08 52	20 ♏ 06 49	29 ♑ 58	26 ♐ 49	13 ♐ 08	10 ♈ 15	01 ♌ 09	07 ♋ 15	19 ♈ 51	17 ♒ 40

Date	Moon True ☊ (° ')	Moon Mean ☊ (° ')	Moon ☽ Latitude (° ')
01	17 ♍ 12	18 ♍ 05	02 N 07
02	17 R 11	18 02	03 03
03	17 08	17 59	03 49
04	17 02	17 55	04 25
05	16 54	17 52	04 50
06	16 44	17 49	05 02
07	16 32	17 46	05 01
08	16 20	17 43	04 47
09	16 09	17 40	04 20
10	15 59	17 36	03 40
11	15 51	17 33	02 49
12	15 47	17 30	01 49
13	15 44	17 27	00 N 43
14	15 D 44	17 24	00 S 26
15	15 45	17 21	01 36
16	15 46	17 17	02 42
17	15 R 46	17 14	03 40
18	15 45	17 11	04 26
19	15 41	17 08	04 56
20	15 36	17 05	05 08
21	15 29	17 01	05 00
22	15 22	16 58	04 31
23	15 16	16 55	03 45
24	15 11	16 52	02 45
25	15 07	16 49	01 35
26	15 06	16 46	00 S 22
27	15 D 06	16 42	00 N 50
28	15 07	16 39	01 58
29	15 09	16 36	02 58
30	15 10	16 33	03 48
31	15 ♍ 11	16 ♍ 30	04 N 28

DECLINATIONS

Date	Sun ☉ (° ')	Moon ☽ (° ')	Mercury ☿ (° ')	Venus ♀ (° ')	Mars ♂ (° ')	Jupiter ♃ (° ')	Saturn ♄ (° ')	Uranus ♅ (° ')	Neptune ♆ (° ')	Pluto ♇ (° ')
01	22 S 59	02 S 43	22 S 37	15 S 33	18 S 05	01 N 09	19 N 42	23 N 30	06 N 07	23 S 12
02	22 54	06 28	22 14	15 47	18 15	01 12	19 43	23 30	06 07	23 11
03	22 48	09 54	21 50	16 02	18 25	01 14	19 44	23 30	06 07	23 11
04	22 42	12 54	21 25	16 16	18 35	01 17	19 45	23 30	06 07	23 10
05	22 36	15 21	20 59	16 31	18 45	01 20	19 46	23 31	06 07	23 10
06	22 29	17 10	20 32	16 45	18 55	01 24	19 47	23 31	06 07	23 09
07	22 21	18 15	20 05	16 59	19 04	01 27	19 48	23 31	06 07	23 09
08	22 13	18 32	19 37	17 12	19 14	01 30	19 50	23 31	06 08	23 08
09	22 05	17 59	19 08	17 26	19 23	01 33	19 51	23 31	06 08	23 08
10	21 56	16 35	18 40	17 39	19 32	01 36	19 52	23 31	06 08	23 07
11	21 47	14 24	18 12	17 52	19 41	01 40	19 53	23 32	06 08	23 07
12	21 37	11 30	17 45	18 05	19 50	01 43	19 54	23 32	06 08	23 06
13	21 27	08 02	17 19	18 17	19 59	01 47	19 56	23 32	06 09	23 05
14	21 17	04 S 07	16 54	18 29	20 07	01 50	19 57	23 32	06 09	23 05
15	21 06	00 N 03	16 31	18 41	20 16	01 54	19 58	23 32	06 09	23 04
16	20 55	04 18	16 10	18 53	20 24	01 58	19 59	23 32	06 09	23 04
17	20 43	08 25	15 52	19 04	20 32	02 01	20 01	23 33	06 10	23 03
18	20 31	12 08	15 37	19 15	20 40	02 05	20 02	23 33	06 10	23 03
19	20 18	15 14	15 25	19 25	20 47	02 09	20 03	23 33	06 10	23 02
20	20 06	17 24	15 16	19 35	20 55	02 13	20 04	23 33	06 11	23 02
21	19 52	18 26	15 11	19 45	21 02	02 17	20 05	23 33	06 11	23 01
22	19 39	18 13	15 09	19 54	21 10	02 21	20 07	23 33	06 11	23 00
23	19 25	16 46	15 11	20 03	21 17	02 25	20 08	23 33	06 12	23 00
24	19 11	14 15	15 15	20 11	21 24	02 29	20 09	23 33	06 12	22 59
25	18 56	10 55	15 23	20 19	21 31	02 33	20 10	23 34	06 12	22 59
26	18 41	07 05	15 33	20 27	21 37	02 37	20 11	23 34	06 13	22 58
27	18 26	03 N 00	15 45	20 34	21 44	02 41	20 13	23 34	06 13	22 58
28	18 10	01 S 05	15 58	20 40	21 50	02 45	20 14	23 34	06 13	22 57
29	17 54	05 01	16 12	20 46	21 56	02 49	20 15	23 34	06 14	22 57
30	17 38	08 37	16 27	20 52	22 02	02 54	20 16	23 34	06 14	22 56
31	17 S 21	11 S 49	16 S 42	20 S 57	22 S 08	02 N 58	20 N 17	23 N 34	06 N 15	22 S 56

ZODIAC SIGN ENTRIES

Date	h	m	Planets
03	00	00	☽ ♏
03	07	28	☿ ♒
05	12	38	☽ ♐
07	05	59	♀ ♐
08	01	05	☽ ♑
10	12	13	☽ ♒
11	13	01	♂ ♐
12	21	39	☽ ♓
15	05	06	☽ ♈
17	10	09	☽ ♉
19	12	43	☽ ♊
20	06	14	☉ ♒
21	13	35	☽ ♋
23	14	17	☽ ♌
25	16	40	☽ ♍
27	22	22	☽ ♎
30	07	51	☽ ♏
31	10	58	☿ ♑

LATITUDES

Date	Mercury ☿ (° ')	Venus ♀ (° ')	Mars ♂ (° ')	Jupiter ♃ (° ')	Saturn ♄ (° ')	Uranus ♅ (° ')	Neptune ♆ (° ')	Pluto ♇ (° ')
01	01 S 57	03 N 20	00 N 34	01 S 18	00 N 21	00 N 20	01 S 40	07 S 47
04	01 41	03 19	00 32	01 17	00 21	00 20	01 40	07 47
07	01 17	03 18	00 30	01 16	00 21	00 20	01 39	07 47
10	00 44	03 15	00 28	01 16	00 22	00 20	01 39	07 47
13	00 S 02	03 11	00 27	01 15	00 22	00 20	01 39	07 47
16	00 N 48	03 06	00 25	01 14	00 23	00 20	01 39	07 47
19	01 43	02 59	00 23	01 13	00 23	00 20	01 39	07 47
22	02 35	02 52	00 21	01 13	00 23	00 20	01 39	07 47
25	03 15	02 44	00 19	01 12	00 23	00 20	01 38	07 47
28	03 34	02 36	00 17	01 11	00 24	00 20	01 38	07 47
31	03 N 32	02 N 27	00 N 15	01 S 11	00 N 24	00 N 20	01 S 38	07 S 47

DATA

Julian Date	2464329
Delta T	+76 seconds
Ayanamsa	24° 20' 43"
Synetic vernal point	04° ♓ 46' 16"
True obliquity of ecliptic	23° 25' 56"

MOON'S PHASES, APSIDES AND POSITIONS ☽

Date	h	m	Phase	Longitude (° ')	Eclipse Indicator
01	10	01	☾	10 ♎ 48	
09	15	03	●	19 ♑ 10	
17	04	45	☽	26 ♈ 53	
23	20	17	○	03 ♌ 39	
31	06	02	☾	11 ♏ 10	

Day	h	m		
05	19	02	Apogee	
21	18	05	Perigee	
08	08	11	Max dec	18° S 33'
15	11	42	0N	
21	19	50	Max dec	18° N 30'
28	05	34	0S	

LONGITUDES

Date	Chiron ⚷ (° ')	Ceres ⚳ (° ')	Pallas ⚴ (° ')	Juno ⚵ (° ')	Vesta ⚶ (° ')	Black Moon Lilith ⚸ (° ')
01	02 ♊ 09	23 ♓ 52	14 ♒ 35	13 ♑ 15	12 ♋ 07	07 ♐ 30
11	01 ♊ 48	26 ♓ 43	17 ♒ 46	17 ♑ 04	09 ♋ 30	08 ♐ 37
21	01 ♊ 33	29 ♓ 48	21 ♒ 02	20 ♑ 55	07 ♋ 09	09 ♐ 44
31	01 ♊ 24	03 ♈ 06	24 ♒ 20	24 ♑ 46	05 ♋ 21	10 ♐ 51

All ephemeris data is given at 12.00 UT and the Moon's longitude is additionally given for 24.00 UT

ASPECTARIAN

h m	Aspects
01 Monday	
00 24	☽ ☍ ♃
02 14	☽ ⋕ ♃
05 00	☽ ∠ ♂
05 26	☽ □ ♅
06 05	☽ ∠ ♀
10 01	☽ □ ☉
10 43	☉ ⊥ ♇
12 45	♂ ⊡ ♅
19 15	☽ Q ♄
21 51	☽ △ ♇
23 36	☽ ⊥ ♂
02 Tuesday	
01 18	☽ ⊥ ♀
03 13	☽ ☍ ♆
09 38	☽ ⋕ ♆
12 09	☽ ⚺ ♂
14 18	☽ ⚺ ♀
22 57	☽ □ ☿
23 58	♀ ± ♆
03 Wednesday	
00 54	☽ Q ☉
06 51	☽ □ ♄
12 16	☽ ⚻ ♃
16 46	☽ △ ♅
04 Thursday	
00 29	☽ ± ♃
03 20	☽ ⚹ ☉
10 05	☽ □ ♇
15 29	☽ ⚻ ♆
16 15	☽ Q ☿
18 54	☽ ⊡ ♃
18 59	♂ ± ♆
23 01	☽ ⊡ ♅
05 Friday	
03 39	☽ ± ♆
04 10	☽ ☌ ♂
08 37	☽ ☌ ♀
12 37	☽ ∠ ☉
14 39	☿ ☍ ♄
17 12	☽ ± ♅
19 11	☽ △ ♄
19 48	☽ ⚹ ☿
21 54	☽ ⊡ ♆
22 40	☽ Q ♇
06 Saturday	
01 37	☽ △ ♃
04 25	☽ ∥ ♀
05 19	☽ ⚻ ♅
08 33	☽ ⊥ ☉
21 48	☽ ⚺ ☉
22 57	☽ ⚹ ♇
07 Sunday	
01 18	☽ ⊡ ♄
04 12	☽ △ ♆
05 51	☽ ∠ ☿
11 40	☉ ⚺ ♇
19 14	☽ ± ♄
20 14	☽ ⚺ ♂
08 Monday	
01 22	☿ ⋕ ♄
01 58	☽ ⊥ ☿
02 56	☽ ⚺ ♀
05 05	☽ ∠ ♇
06 42	☿ ⚹ ♃
07 08	☽ ⚻ ♄
08 54	☽ ⊥ ♂
14 28	☽ □ ♃
15 14	☽ ⚺ ☿
16 01	☽ ⊥ ♀
17 15	☽ ☍ ♅
23 00	☽ ⊥ ♇
09 Tuesday	
01 15	☿ Q ♆
02 44	☿ ∥ ♂
03 46	☽ ∠ ♂
05 34	♀ ± ♅
10 52	☽ ⚺ ♇
11 31	☽ ∠ ♀
12 01	☿ ⚻ ♅
15 03	☽ ☌ ☉
15 53	☽ □ ♆
21 39	☽ ∥ ♀
23 32	♀ △ ♄
10 Wednesday	
01 02	☉ □ ♆
02 18	☽ Q ♃
10 49	☽ ⚹ ♂
17 44	☽ ☍ ♄
19 35	☽ ⚹ ♀
11 Thursday	
01 51	☽ ⚹ ♃
02 56	☽ Q ♆
03 45	☽ ⚻ ♅
07 26	☽ ☌ ☿
11 19	☽ Q ♂
13 33	♀ ⊡ ♆
15 15	☽ ± ♅
20 51	☽ Q ♀
21 11	☽ ☌ ♇
12 Friday	
00 08	☿ ∥ ♀
00 50	♀ Q ♇
01 57	☽ ⚹ ♆
06 14	☽ ⚺ ☉
06 54	☽ ∠ ♃
08 22	☽ ⊡ ♅
18 35	☽ ⊥ ☉
23 25	☽ □ ♂
13 Saturday	
00 06	☽ ⊥ ♃
02 36	♂ ⋕ ♄
02 38	☽ ⚻ ♄
06 19	☽ ∠ ♆
10 03	☽ □ ♀
11 28	☽ ⚺ ♃
12 30	☽ △ ♅
12 56	☽ ∠ ☉
13 48	☽ ± ♄
19 51	☽ ⚺ ☿
23 06	☽ ⊥ ♆
23 50	☽ ⋕ ♆
14 Sunday	
05 12	♀ △ ♃
05 40	☽ ⚺ ♇
06 23	☽ ⊡ ♄
07 26	☽ ⊥ ☿
09 57	♂ ± ♅
10 10	☽ ⚺ ♆
14 08	♀ ⚻ ♅
16 42	☽ ⊥ ♇
18 59	☽ ⚹ ☉
15 Monday	
00 30	☽ ∠ ☿
01 03	☽ ⋕ ♃
07 27	♂ △ ♄
09 05	☽ ∠ ♇
09 34	☽ △ ♄
09 41	☽ △ ♂
18 28	☽ Q ☉
18 57	☽ ☌ ♃
19 12	☽ □ ♅
21 52	☽ △ ♀
22 34	☽ ∥ ♃
16 Tuesday	
01 49	☉ Q ♃
04 05	☽ ⚹ ☿
11 53	☽ ⚹ ♇
11 57	♃ □ ♅
13 46	☽ ⊡ ♂
16 08	☽ ☌ ♆
22 40	☽ ∥ ♆
17 Wednesday	
01 22	☽ Q ☿
02 36	☽ ⊡ ♀
02 50	☽ Q ♅
04 45	☽ □ ☉
06 16	☽ ± ♂
06 44	☿ St R
08 52	☽ Q ♇
14 05	☽ □ ♄
17 07	☽ ⚻ ♂
19 27	☽ ± ♀
23 25	☽ ⚹ ♅
23 54	☽ ⚺ ♃
18 Thursday	
01 22	☉ ∥ ♂
06 31	☽ ⚻ ♀
08 03	☽ □ ☿
10 11	☽ ⊥ ♃
15 25	♂ ⊡ ♆
15 35	☽ □ ♇
19 33	☽ ⚺ ♆
20 25	☽ Q ♄
22 44	☿ ⚹ ♀
19 Friday	
00 37	☽ ∠ ♅
01 25	☽ ∠ ♃
05 33	☽ ⊥ ♆
11 24	☽ △ ☉
13 34	☽ ⋕ ☿
15 29	☽ ⊥ ♅
15 30	♂ Q ♇
16 13	☽ ⚹ ♄
20 25	☽ ∠ ♆
21 46	☽ ☍ ♂
20 Saturday	
01 19	☽ ⚺ ♅
02 26	☽ ⚹ ♃
08 26	☽ △ ☿
12 19	☽ ☍ ♀
13 48	☽ ⊡ ☉
14 16	☉ ⋕ ♄
16 35	☽ ∠ ♄
17 07	☽ △ ♇
20 54	☽ ⚹ ♆
22 10	☽ Q ♃
21 Sunday	
05 29	☽ ± ☉
07 06	☽ ⊥ ♄
07 42	☽ ⊡ ☿
15 53	☽ ⚻ ☉
16 20	☽ Q ♆
16 44	☽ ⚺ ♄
17 26	☽ ⊡ ♇
19 48	☉ ∥ ♀
21 29	☽ ± ☿
22 Monday	
00 43	☽ ⚻ ♂
01 50	☽ ☌ ♅
03 25	☉ ☍ ♄
03 35	☽ □ ♃
06 38	☽ ⚻ ☿
08 01	☽ ± ♇
10 52	☽ ± ♂
16 50	☽ ⚻ ♀
17 14	♀ ⊡ ♄
17 45	☽ ⚻ ♇
21 29	☽ □ ♆
23 Tuesday	
00 36	♂ ⚻ ♅
02 10	☽ ⊡ ♂
03 23	☽ ± ♀
05 14	♀ ⚹ ♇
17 13	☽ ☌ ♄
19 21	☽ ⊡ ♀
20 09	☿ ⚹ ♃
20 17	☽ ☍ ☉
24 Wednesday	
02 39	☽ ⚺ ♅
03 40	☽ ⋕ ☿
03 58	☽ △ ♂
03 59	♀ ⋕ ♄
04 23	☽ ☍ ☿
05 08	☽ △ ♃
07 31	☿ ⚹ ♂
12 37	☽ ⊥ ♅
19 19	☽ ☍ ♇
20 36	☿ Q ♆
22 30	☽ △ ♀
23 09	☽ △ ♆
25 Thursday	
01 58	☿ ⚻ ♅
03 46	☽ ∠ ♅
06 40	☽ ⊡ ♃
06 46	♀ △ ♆
13 43	♂ △ ♃
19 28	☽ ⚺ ♄
22 22	☽ ± ♃
26 Friday	
00 57	☽ ⊡ ♆
03 03	☽ ⚻ ☉
03 22	☽ ⚻ ☿
05 06	☉ ☌ ☿
05 37	☽ ⚹ ♅
05 53	☽ ⊥ ♄
07 37	☿ ∠ ♀
09 01	☽ ⚻ ♃
09 45	☽ □ ♂
13 02	☽ ± ☿
14 30	☽ ± ☉
16 51	☽ ± ♆
17 11	☽ ∥ ♆
21 46	☽ ∠ ♄
23 36	☽ ⚻ ♇
27 Saturday	
02 57	☽ Q ♅
03 38	☽ ⚻ ♆
03 46	☽ ⊡ ☿
07 46	☽ □ ♀
08 01	☽ ⊡ ☉
10 30	☽ ± ♇
12 28	☉ ⚻ ♅
13 50	☽ ∥ ♃
19 58	☽ Q ♂
22 07	☉ Q ♆
28 Sunday	
01 01	☽ ⚹ ♄
03 09	☽ ⊡ ♇
04 57	☽ △ ☿
08 39	☉ ∠ ♀
12 06	☽ □ ♅
14 14	☽ △ ☉
16 39	☽ ☍ ♃
19 43	☽ ⚹ ♂
20 33	☽ Q ♀
22 12	☽ ⋕ ♃
23 31	☽ Q ♄
29 Monday	
07 40	☽ △ ♇
11 56	☽ ☍ ♆
19 51	☽ ⋕ ♆
22 03	☽ ⚹ ♀
30 Tuesday	
02 12	☉ ⚹ ♃
02 21	☽ ∠ ♂
03 18	♀ ± ♄
03 48	☿ ⊥ ♀
04 25	☿ ☍ ♄
09 51	☽ □ ☿
10 17	☽ □ ♄
21 09	☽ ⊥ ♂
22 14	☽ △ ♅
31 Wednesday	
04 04	☽ ⚻ ♃
06 02	☽ □ ☉
06 50	☽ ∠ ♀
09 50	☽ ⚺ ♂
16 16	☽ ± ♃
19 06	☽ □ ♇
19 08	☽ Q ☿
23 29	☽ ⚻ ♆

FEBRUARY 2035

Date	Sidereal time h m s	LONGITUDES Sun ☉ ° ' "	Moon ☽ ° ' "	Moon ☽ 24.00 ° ' "	Mercury ☿ ° '	Venus ♀ ° '	Mars ♂ ° '	Jupiter ♃ ° '	Saturn ♄ ° '	Uranus ♅ ° '	Neptune ♆ ° '	Pluto ♇ ° '
01	20 46 08	12 ♒ 26 26	26 ♏ 03 23	01 ♐ 59 10	29 ♑ 07	27 ♐ 58	13 ♐ 47	10 ♈ 25	01 ♌ 04	07 ♋ 13	19 ♈ 52	17 ♒ 42
02	20 50 04	13 27 20	07 ♐ 54 45	13 ♐ 50 41	28 R 26	29 ♐ 07	14 27	10 36	01 R 00	07 R 11	19 53	17 44
03	20 54 01	14 28 14	19 ♐ 47 27	25 ♐ 45 31	27 52	00 ♑ 16	15 06	10 47	00 55	07 09	19 54	17 46
04	20 57 57	15 29 07	01 ♑ 45 19	07 ♑ 47 13	27 27	01 25	15 46	10 58	00 50	07 07	19 56	17 48
05	21 01 54	16 29 58	13 ♑ 51 31	19 ♑ 58 31	27 11	02 34	16 25	11 09	00 45	07 05	19 57	17 49
06	21 05 50	17 30 49	26 ♑ 08 24	02 ♒ 21 20	27 03	03 44	17 05	11 21	00 41	07 03	19 58	17 51
07	21 09 47	18 31 39	08 ♒ 37 27	14 ♒ 56 47	27 D 03	04 53	17 45	11 32	00 36	07 02	19 59	17 53
08	21 13 43	19 32 27	21 ♒ 19 25	27 ♒ 45 18	27 09	06 03	18 24	11 43	00 32	07 00	20 01	17 55
09	21 17 40	20 33 14	04 ♓ 14 27	10 ♓ 46 47	27 23	07 13	19 04	11 55	00 27	06 58	20 02	17 57
10	21 21 37	21 34 00	17 ♓ 22 17	24 ♓ 00 51	27 43	08 23	19 43	12 07	00 23	06 56	20 03	17 58
11	21 25 33	22 34 44	00 ♈ 42 25	07 ♈ 26 57	28 08	09 33	20 23	12 18	00 18	06 55	20 05	18 00
12	21 29 30	23 35 27	14 ♈ 14 22	21 ♈ 04 36	28 39	10 43	21 02	12 30	00 14	06 53	20 06	18 02
13	21 33 26	24 36 08	27 ♈ 57 36	04 ♉ 53 16	29 15	11 53	21 42	12 42	00 10	06 52	20 08	18 04
14	21 37 23	25 36 48	11 ♉ 51 32	18 ♉ 52 17	29 ♑ 56	13 03	22 22	12 54	00 05	06 50	20 09	18 06
15	21 41 19	26 37 26	25 ♉ 55 21	03 ♊ 00 33	00 ♒ 40	14 13	23 01	13 06	00 ♌ 01	06 49	20 11	18 07
16	21 45 16	27 38 02	10 ♊ 07 39	17 ♊ 16 20	01 28	15 24	23 41	13 18	29 ♋ 57	06 48	20 12	18 09
17	21 49 12	28 38 36	24 ♊ 26 15	01 ♋ 36 58	02 20	16 34	24 20	13 31	29 53	06 46	20 14	18 11
18	21 53 09	29 ♒ 39 09	08 ♋ 48 00	15 ♋ 58 49	03 15	17 45	25 00	13 43	29 49	06 45	20 15	18 13
19	21 57 06	00 ♓ 39 40	23 ♋ 08 51	00 ♌ 17 30	04 13	18 56	25 39	13 55	29 45	06 44	20 17	18 15
20	22 01 02	01 40 09	07 ♌ 24 09	14 ♌ 28 14	05 14	20 06	26 19	14 08	29 42	06 43	20 18	18 16
21	22 04 59	02 40 37	21 ♌ 29 10	28 ♌ 26 27	06 17	21 17	26 59	14 20	29 38	06 41	20 20	18 18
22	22 08 55	03 41 02	05 ♍ 19 39	12 ♍ 08 23	07 23	22 28	27 38	14 33	29 34	06 40	20 22	18 20
23	22 12 52	04 41 26	18 ♍ 52 23	25 ♍ 31 28	08 30	23 39	28 18	14 46	29 31	06 39	20 23	18 22
24	22 16 48	05 41 49	02 ♎ 05 33	08 ♎ 34 40	09 40	24 50	28 57	14 58	29 27	06 38	20 25	18 23
25	22 20 45	06 42 10	14 ♎ 58 53	21 ♎ 18 26	10 52	26 01	29 ♐ 37	15 11	29 24	06 37	20 27	18 25
26	22 24 41	07 42 29	27 ♎ 33 33	03 ♏ 44 37	12 06	27 13	00 ♑ 17	15 24	29 20	06 36	20 29	18 27
27	22 28 38	08 42 47	09 ♏ 52 01	15 ♏ 56 12	13 21	28 24	00 56	15 37	29 17	06 36	20 30	18 29
28	22 32 35	09 ♓ 43 04	21 ♏ 57 41	27 ♏ 57 01	14 ♒ 38	29 ♑ 35	01 ♑ 36	15 ♈ 50	29 ♋ 14	06 ♋ 35	20 ♈ 32	18 ♒ 30

Date	Moon True ☊ ° '	Moon Mean ☊ ° '	Moon ☽ Latitude ° '
01	15 ♍ 10	16 ♍ 27	04 N 55
02	15 R 07	16 23	05 10
03	15 03	16 20	05 11
04	14 58	16 17	04 59
05	14 53	16 14	04 34
06	14 48	16 11	03 55
07	14 44	16 07	03 05
08	14 41	16 04	02 05
09	14 39	16 01	00 N 57
10	14 D 39	15 58	00 S 15
11	14 39	15 55	01 27
12	14 41	15 52	02 36
13	14 42	15 48	03 37
14	14 43	15 45	04 25
15	14 44	15 42	04 59
16	14 R 43	15 39	05 15
17	14 42	15 36	05 11
18	14 40	15 33	04 48
19	14 39	15 29	04 07
20	14 37	15 26	03 11
21	14 36	15 23	02 04
22	14 35	15 20	00 S 51
23	14 D 35	15 17	00 N 24
24	14 35	15 13	01 36
25	14 36	15 10	02 41
26	14 37	15 07	03 36
27	14 37	15 04	04 21
28	14 ♍ 38	15 ♍ 01	04 N 53

DECLINATIONS

Date	Sun ☉ ° '	Moon ☽ ° '	Mercury ☿ ° '	Venus ♀ ° '	Mars ♂ ° '	Jupiter ♃ ° '	Saturn ♄ ° '	Uranus ♅ ° '	Neptune ♆ ° '	Pluto ♇ ° '
01	17 S 04	14 S 29	16 S 57	21 S 02	22 S 13	03 N 02	20 N 18	23 N 34	06 N 15	22 S 55
02	16 47	16 31	17 12	21 06	22 19	03 07	20 20	23 35	06 16	22 55
03	16 29	17 52	17 27	21 09	22 24	03 11	20 21	23 35	06 16	22 54
04	16 11	18 26	17 40	21 12	22 29	03 16	20 22	23 35	06 17	22 54
05	15 53	18 11	17 53	21 14	22 34	03 20	20 23	23 35	06 17	22 53
06	15 35	17 04	18 05	21 16	22 38	03 25	20 24	23 35	06 18	22 53
07	15 16	15 07	18 16	21 18	22 43	03 30	20 25	23 35	06 18	22 52
08	14 57	12 25	18 26	21 18	22 47	03 34	20 26	23 35	06 19	22 52
09	14 38	09 04	18 35	21 18	22 51	03 39	20 27	23 35	06 19	22 51
10	14 19	05 13	18 43	21 18	22 55	03 44	20 28	23 35	06 20	22 50
11	13 59	01 S 03	18 50	21 17	22 59	03 48	20 29	23 35	06 20	22 50
12	13 39	03 N 13	18 55	21 15	23 03	03 53	20 30	23 35	06 21	22 50
13	13 19	07 22	18 59	21 13	23 06	03 58	20 31	23 35	06 22	22 49
14	12 59	11 11	19 02	21 10	23 09	04 03	20 32	23 36	06 22	22 49
15	12 38	14 23	19 04	21 07	23 12	04 07	20 33	23 36	06 23	22 48
16	12 17	16 46	19 04	21 03	23 15	04 12	20 34	23 36	06 23	22 48
17	11 57	18 08	19 04	20 59	23 18	04 17	20 35	23 36	06 24	22 47
18	11 35	18 21	19 01	20 54	23 20	04 22	20 36	23 36	06 25	22 47
19	11 14	17 23	18 58	20 48	23 23	04 27	20 37	23 36	06 25	22 46
20	10 53	15 20	18 53	20 42	23 25	04 32	20 38	23 36	06 26	22 46
21	10 31	12 22	18 47	20 35	23 27	04 37	20 39	23 36	06 26	22 45
22	10 09	08 46	18 40	20 27	23 28	04 42	20 40	23 36	06 27	22 45
23	09 47	04 46	18 31	20 19	23 30	04 47	20 40	23 36	06 28	22 44
24	09 25	00 N 38	18 21	20 11	23 31	04 52	20 41	23 36	06 29	22 44
25	09 03	03 S 26	18 10	20 02	23 32	04 57	20 42	23 36	06 29	22 43
26	08 41	07 14	17 58	19 52	23 33	05 02	20 43	23 36	06 30	22 43
27	08 18	10 38	17 44	19 42	23 34	05 07	20 43	23 36	06 31	22 43
28	07 S 55	13 S 32	17 S 29	19 S 31	23 S 35	05 N 13	20 N 44	23 N 36	06 N 31	22 S 42

ZODIAC SIGN ENTRIES

Date	h	m	Planets
01	19	59	☽ ♐
03	06	29	♀ ♑
04	08	30	☽ ♑
06	19	28	☽ ♒
09	04	10	☽ ♓
11	10	44	☽ ♈
13	15	32	☽ ♉
14	14	31	☿ ♒
15	18	55	☽ ♊
15	19	35	♄ ♋
17	21	18	☽ ♋
18	20	16	☉ ♓
19	23	31	☽ ♌
22	02	42	☽ ♍
24	08	10	☽ ♎
26	01	58	♂ ♑
26	16	43	☽ ♏
28	20	23	♀ ♒

LATITUDES

Date	Mercury ☿ ° '	Venus ♀ ° '	Mars ♂ ° '	Jupiter ♃ ° '	Saturn ♄ ° '	Uranus ♅ ° '	Neptune ♆ ° '	Pluto ♇ ° '
01	03 N 27	02 N 23	00 N 14	01 S 11	00 N 24	00 N 20	01 S 38	07 S 47
04	03 03	02 13	00 12	01 10	00 24	00 20	01 38	07 47
07	02 31	02 03	00 09	01 10	00 25	00 20	01 38	07 47
10	01 56	01 52	00 07	01 09	00 25	00 20	01 38	07 48
13	01 20	01 41	00 04	01 08	00 25	00 20	01 38	07 48
16	00 46	01 30	00 N 02	01 08	00 26	00 20	01 37	07 48
19	00 N 14	01 19	00 S 01	01 08	00 26	00 20	01 37	07 48
22	00 S 15	01 07	00 04	01 07	00 26	00 20	01 37	07 49
25	00 42	00 56	00 07	01 07	00 26	00 20	01 37	07 49
28	01 05	00 44	00 09	01 06	00 26	00 20	01 37	07 50
31	01 S 26	00 N 33	00 S 13	01 S 06	00 N 27	00 N 20	01 S 37	07 S 50

DATA

Julian Date	2464360
Delta T	+76 seconds
Ayanamsa	24° 20' 47"
Synetic vernal point	04° ♓ 46' 12"
True obliquity of ecliptic	23° 25' 56"

MOON'S PHASES, APSIDES AND POSITIONS ☽

Date	h	m	Phase	Longitude ° '	Eclipse Indicator
08	08	22	●	19 ♒ 23	
15	13	17	☽	26 ♉ 41	
22	08	54	○	03 ♍ 33	

Day	h	m		
02	12	53	Apogee	
18	05	46	Perigee	
04	16	35	Max dec	18° S 27'
11	17	56	0N	
18	04	18	Max dec	18° N 24'
24	15	40	0S	

LONGITUDES

Date	Chiron ⚷ ° '	Ceres ⚳ ° '	Pallas ⚴ ° '	Juno ⚵ ° '	Vesta ⚶ ° '	Black Moon Lilith ⚸ ° '
01	01 ♊ 23	03 ♈ 27	24 ♒ 40	25 ♑ 09	05 ♋ 13	10 ♐ 58
11	01 ♊ 22	06 ♈ 56	27 ♒ 59	29 ♑ 00	04 ♋ 11	12 ♐ 05
21	01 ♊ 28	10 ♈ 34	01 ♓ 20	02 ♒ 49	03 ♋ 55	13 ♐ 12
31	01 ♊ 41	14 ♈ 19	04 ♓ 39	06 ♒ 36	04 ♋ 22	14 ♐ 20

All ephemeris data is given at 12.00 UT and the Moon's longitude is additionally given for 24.00 UT

ASPECTARIAN

h m	Aspects
01 Thursday	
02 51	☽ ⊥ ♀
04 15	☽ ⚼ ♅
10 42	☽ ⚼ ♃
11 37	☽ ± ♆
16 16	☽ ⊻ ♀
16 49	☉ ∥ ☿
17 43	☿ ∠ ♂
17 50	☽ ✶ ☿
21 42	☽ Q ☉
22 05	☽ △ ♄
22 24	☽ ± ♅
02 Friday	
02 50	☿ ⊻ ♀
05 40	☉ ± ♅
05 52	☽ ⚼ ♆
06 50	☿ Q ♃
07 35	☽ Q ♇
10 31	☽ ⊼ ♅
15 07	☽ ∥ ☉
17 32	☽ △ ♃
22 37	☽ ∠ ☿
03 Saturday	
00 16	☽ ✶ ☉
00 48	☽ ∥ ☿
02 00	☽ ☌ ♂
04 14	☽ ⚼ ♄
07 54	☽ ✶ ♇
12 14	☽ △ ♆
16 02	☽ ⊥ ☿
22 14	☽ ± ♄
04 Sunday	
00 38	♀ ⊼ ♄
03 40	☽ ⊻ ☿
09 14	☽ ∠ ☉
10 10	☽ ⊼ ♄
11 15	☽ ☌ ♀
14 05	☽ ∠ ♇
14 15	♂ ⚼ ♄
22 38	☽ ☍ ♅
05 Monday	
04 46	☽ ⊥ ☉
06 35	☽ □ ♃
06 57	☉ ✶ ♂
07 59	☽ ⊥ ♇
17 20	☽ ⊻ ♂
17 20	♀ ∠ ♇
17 40	☽ ⊻ ☉
18 50	☽ ∥ ☿
19 49	☽ ⊻ ♇
23 58	☽ □ ♆
06 Tuesday	
05 44	☽ ⊥ ♂
13 45	☽ ☌ ☿
18 17	☽ Q ♃
20 19	☉ ☌ ♇
20 43	☽ ☍ ♄
07 Wednesday	
00 07	☽ ∠ ♂
01 26	☿ St D
04 08	☽ ⊻ ♀
08 57	☽ ⊼ ♅
10 15	☽ ∥ ☉
10 47	☽ Q ♆
16 45	☽ ⊥ ♀
17 21	♂ ✶ ♇
17 37	☽ ✶ ♃
20 21	☽ ± ♅
08 Thursday	
05 35	☽ ☌ ♇
06 13	☽ ✶ ♂
08 22	☽ ☌ ☉
09 32	☽ ✶ ♆
11 26	☽ ∠ ♀
13 15	☽ ⚼ ♅
22 14	☽ ∠ ♃
23 03	☽ ⊻ ☿
23 23	☉ ✶ ♆
09 Friday	
05 03	☽ ⊼ ♄
05 49	☽ Q ♂
07 04	♀ ☍ ♅
10 06	♂ ∥ ♇
10 23	☽ ⊥ ☿
13 28	☽ ∠ ♆
15 08	☽ ⊥ ♃
16 03	☽ ± ♄
17 00	☽ △ ♅
18 00	☽ ✶ ♀
10 Saturday	
02 17	☽ ⊻ ♃
03 17	☽ ∠ ☿
05 20	☽ ⋕ ♆
05 58	☽ ⊥ ♆
08 24	☽ ⚼ ♄
13 06	☽ ⊻ ♇
16 29	☽ □ ♂
16 52	☽ ⊻ ♆
17 58	☽ Q ♀
20 13	☽ ⊻ ☉
20 35	☽ ⋕ ♃
20 39	☉ ⚼ ♅
23 57	☽ ⊥ ♇
11 Sunday	
00 33	♂ △ ♆
07 14	☽ ✶ ☿
07 53	☽ ⊥ ☉
11 17	☽ △ ♄
16 06	☽ ∠ ♇
23 02	☽ □ ♅
12 Monday	
01 13	☽ ∠ ☉
05 11	☽ □ ♀
05 25	☽ Q ☿
08 53	☽ ☌ ♃
15 51	☽ ∥ ♃
18 41	☽ ✶ ♇
22 19	☽ ☌ ♆
13 Tuesday	
00 32	☽ △ ♂
05 42	☽ ✶ ☉
06 00	☽ ∥ ♆
06 37	☽ Q ♅
14 21	☽ □ ☿
15 40	☽ Q ♇
15 48	☽ □ ♄
15 51	♀ ⊥ ♇
14 Wednesday	
03 23	☽ ✶ ♅
03 53	☽ ⚼ ♂
04 08	☽ Q ☉
08 14	♀ □ ♃
13 49	☽ ⊻ ♃
14 14	☽ △ ♀
17 04	☿ ☍ ♄
20 05	☽ ± ♂
22 37	☽ Q ♄
22 42	☽ □ ♇
23 35	☽ ⋕ ☉
15 Thursday	
00 13	☽ ⊥ ♃
02 12	☽ ⊻ ♆
05 02	☽ ∠ ♅
06 50	☽ ⊼ ♂
12 26	☽ ⊥ ♆
13 17	☽ □ ☉
15 45	☽ ∠ ♃
18 06	☽ ⚼ ♀
18 55	☽ ✶ ♄
20 17	☽ ⊥ ♅
20 31	☽ △ ☿
16 Friday	
03 41	☽ ∠ ♆
05 34	☿ Q ♃
06 24	☽ ⊻ ♅
10 40	☽ ± ♀
17 25	☽ ✶ ♃
20 04	☽ ∠ ♄
21 02	♂ ± ♄
21 39	☽ ⊼ ♀
23 20	☽ ⚼ ☿
17 Saturday	
01 30	☽ △ ♇
04 56	☽ ✶ ♆
07 58	☉ ∠ ♃
11 05	☽ ⊥ ♄
11 50	☽ ☍ ♂
13 49	☽ Q ♃
15 23	☽ ± ☿
19 34	☽ △ ☉
21 04	☽ ⊻ ♄
18 Sunday	
01 03	☽ Q ♆
02 05	☽ ⊼ ☿
02 39	☽ ⚼ ♇
08 35	☽ ☌ ♅
15 45	☉ ⊼ ♄
17 43	☽ ± ♇
20 20	☽ □ ♃
21 41	♀ ⊻ ♇
22 31	☽ ⚼ ☉
19 Monday	
03 46	☽ ⊼ ♇
04 18	☽ ☍ ♀
07 11	☽ □ ♆
14 44	☽ ± ☉
16 25	☽ ⊼ ♂
23 03	☽ ☌ ♄
20 Tuesday	
01 35	☽ ⊼ ☉
03 00	☽ ± ♂
08 02	☽ ☍ ☿
10 50	☽ ⊻ ♅
16 09	♀ □ ♆
18 58	☽ ⚼ ♂
20 59	☽ ⊥ ♅
23 35	☽ △ ♃
21 Wednesday	
00 13	♀ ⋕ ♄
06 32	☽ ☍ ♇
10 01	☽ △ ♆
11 38	☽ ⊼ ♀
12 21	☽ ∠ ♅
20 54	☿ ⊼ ♅
21 56	☽ △ ♂
22 56	☽ ± ♀
22 Thursday	
01 46	☽ ⚼ ♃
02 00	☽ ⊻ ♄
02 11	☽ ⋕ ☉
08 54	☽ ☍ ☉
12 04	☽ ⚼ ♆
12 25	☽ ⊥ ♄
14 21	☽ ✶ ♅
15 55	☽ ⊼ ☿
16 07	☽ ⚼ ♀
17 45	☽ ± ♃
23 Friday	
02 01	☽ ∥ ♆
03 29	☽ ± ☿
03 59	☽ ± ♆
04 14	☽ ∠ ♄
04 32	☽ ⊼ ♃
09 31	☿ Q ♆
11 05	☽ ⊼ ♇
11 36	☽ Q ♅
11 53	☽ ∥ ♃
14 44	☽ ⊼ ♆
21 08	☽ ⚼ ☿
21 27	☽ △ ♀
21 55	☽ ± ♇
24 Saturday	
05 10	☉ ∠ ♆
05 57	☽ □ ♂
06 28	☉ ± ♄
07 11	☽ ✶ ♄
14 24	☽ ⚼ ♇
19 13	☽ ⊼ ☉
20 23	☽ □ ♅
25 Sunday	
03 28	☽ △ ☿
04 34	♂ ⊼ ♄
05 17	☽ Q ♄
07 21	☽ ± ☉
10 06	☉ △ ♅
12 24	☽ ☍ ♃
17 15	☽ Q ♂
18 31	☽ △ ♇
21 35	☽ ⋕ ♃
22 23	☽ ☍ ♆
26 Monday	
01 52	☽ ⚼ ☉
07 10	☽ ⋕ ♆
11 15	☽ □ ♀
15 25	☽ □ ♄
17 33	☽ ✶ ♂
20 49	☽ ∥ ☉
21 44	☿ ± ♅
27 Tuesday	
05 35	☽ △ ♅
09 32	☽ △ ☉
19 41	☽ □ ☿
23 35	☽ ⊼ ♃
28 Wednesday	
00 41	☽ ∠ ♂
02 19	☽ Q ♀
05 05	☽ □ ♇
05 11	♀ ☍ ♄
09 09	☽ ⊼ ♆
11 14	☽ ⚼ ♅
11 45	☽ ± ♃
15 36	☉ ⊥ ♃
19 42	☽ ⊥ ♂
21 11	☽ ± ♆

MARCH 2035

LONGITUDES

Date	Sidereal time h m s	Sun ☉ ° ' "	Moon ☽ ° ' "	Moon ☽ 24.00 ° ' "	Mercury ☿ ° '	Venus ♀ ° '	Mars ♂ ° '	Jupiter ♃ ° '	Saturn ♄ ° '	Uranus ♅ ° '	Neptune ♆ ° '	Pluto ♇ ° '
01	22 36 31	10 ♓ 43 19	03 ♐ 54 44	09 ♐ 51 26	15 ♒ 57	00 ♒ 46	02 ♑ 15	16 ♈ 03	29 ♋ 11	06 ♋ 34	20 ♈ 34	18 ♒ 32
02	22 40 28	11 43 32	15 ♐ 47 42	21 ♐ 44 08	17 17	01 58	02 55	16 16	29 R 08	06 R 33	20 36	18 34
03	22 44 24	12 43 45	27 ♐ 41 18	03 ♑ 39 48	18 38	03 09	03 34	16 30	29 05	06 33	20 38	18 35
04	22 48 21	13 43 55	09 ♑ 40 09	15 ♑ 42 54	20 01	04 21	04 14	16 43	29 02	06 32	20 40	18 37
05	22 52 17	14 44 04	21 ♑ 48 31	27 ♑ 57 27	21 26	05 33	04 54	16 56	28 59	06 32	20 42	18 39
06	22 56 14	15 44 12	04 ♒ 10 05	10 ♒ 26 45	22 51	06 44	05 33	17 10	28 57	06 31	20 43	18 40
07	23 00 10	16 44 17	16 ♒ 47 43	23 ♒ 13 09	24 18	07 56	06 13	17 23	28 54	06 31	20 45	18 42
08	23 04 07	17 44 21	29 ♒ 43 12	06 ♓ 17 52	25 46	09 08	06 52	17 37	28 52	06 30	20 47	18 44
09	23 08 04	18 44 23	12 ♓ 57 07	19 ♓ 40 48	27 16	10 20	07 32	17 50	28 50	06 30	20 49	18 45
10	23 12 00	19 44 24	26 ♓ 28 44	03 ♈ 20 37	28 ♒ 46	11 31	08 11	18 04	28 47	06 30	20 51	18 47
11	23 15 57	20 44 22	10 ♈ 16 06	17 ♈ 14 48	00 ♓ 18	12 43	08 51	18 17	28 45	06 30	20 53	18 48
12	23 19 53	21 44 18	24 ♈ 16 16	01 ♉ 20 01	01 51	13 55	09 30	18 31	28 43	06 29	20 55	18 50
13	23 23 50	22 44 12	08 ♉ 25 36	15 ♉ 32 30	03 25	15 07	10 10	18 45	28 41	06 29	20 57	18 51
14	23 27 46	23 44 05	22 ♉ 40 16	29 ♉ 48 26	05 01	16 19	10 49	18 59	28 40	06 29	20 59	18 53
15	23 31 43	24 43 55	06 ♊ 56 35	14 ♊ 04 20	06 37	17 31	11 29	19 12	28 38	06 D 29	21 02	18 55
16	23 35 39	25 43 42	21 ♊ 11 20	28 ♊ 17 16	08 15	18 43	12 08	19 26	28 36	06 29	21 04	18 56
17	23 39 36	26 43 28	05 ♋ 21 52	12 ♋ 24 53	09 54	19 55	12 48	19 40	28 35	06 29	21 06	18 58
18	23 43 33	27 43 11	19 ♋ 26 06	26 ♋ 25 18	11 35	21 07	13 27	19 54	28 33	06 30	21 08	18 59
19	23 47 29	28 42 52	03 ♌ 22 19	10 ♌ 16 58	13 16	22 20	14 07	20 08	28 32	06 30	21 10	19 00
20	23 51 26	29 ♓ 42 30	17 ♌ 09 05	23 ♌ 58 29	14 59	23 32	14 46	20 22	28 31	06 30	21 12	19 02
21	23 55 22	00 ♈ 42 07	00 ♍ 45 02	07 ♍ 28 33	16 43	24 44	15 25	20 36	28 30	06 30	21 14	19 03
22	23 59 19	01 41 41	14 ♍ 08 55	20 ♍ 46 00	18 28	25 56	16 05	20 50	28 29	06 31	21 16	19 05
23	00 03 15	02 41 13	27 ♍ 19 39	03 ♎ 49 49	20 14	27 08	16 44	21 04	28 28	06 31	21 19	19 06
24	00 07 12	03 40 43	10 ♎ 16 25	16 ♎ 39 27	22 02	28 21	17 24	21 18	28 28	06 32	21 21	19 08
25	00 11 08	04 40 10	22 ♎ 58 55	29 ♎ 14 54	23 51	29 ♒ 33	18 03	21 33	28 27	06 32	21 23	19 09
26	00 15 05	05 39 37	05 ♏ 27 31	11 ♏ 36 57	25 41	00 ♓ 45	18 42	21 47	28 26	06 33	21 25	19 10
27	00 19 02	06 39 01	17 ♏ 43 24	23 ♏ 47 11	27 33	01 58	19 22	22 01	28 26	06 33	21 27	19 12
28	00 22 58	07 38 23	29 ♏ 48 36	05 ♐ 48 04	29 ♓ 26	03 10	20 01	22 15	28 26	06 34	21 29	19 13
29	00 26 55	08 37 44	11 ♐ 46 00	17 ♐ 42 51	01 ♈ 20	04 23	20 40	22 29	28 26	06 35	21 32	19 14
30	00 30 51	09 37 02	23 ♐ 39 10	29 ♐ 35 29	03 16	05 35	21 19	22 44	28 25	06 36	21 34	19 15
31	00 34 48	10 ♈ 36 19	05 ♑ 32 21	11 ♑ 30 22	05 ♈ 12	06 ♓ 48	21 ♑ 59	22 ♈ 58	28 ♋ 26	06 ♋ 37	21 ♈ 36	19 ♒ 17

Moon nodes and latitude

Date	Moon True ☊ ° '	Moon Mean ☊ ° '	Moon ☽ Latitude ° '
01	14 ♍ 38	14 ♍ 58	05 N 12
02	14 R 38	14 54	05 17
03	14 38	14 51	05 09
04	14 38	14 48	04 48
05	14 D 38	14 45	04 13
06	14 38	14 42	03 26
07	14 38	14 39	02 28
08	14 38	14 35	01 22
09	14 38	14 32	00 N 09
10	14 R 38	14 29	01 S 05
11	14 38	14 26	02 17
12	14 37	14 23	03 22
13	14 36	14 19	04 16
14	14 35	14 16	04 54
15	14 34	14 13	05 14
16	14 34	14 10	05 14
17	14 D 34	14 07	04 56
18	14 35	14 04	04 20
19	14 36	14 00	03 29
20	14 37	13 57	02 26
21	14 38	13 54	01 16
22	14 38	13 51	00 S 03
23	14 R 38	13 48	01 N 09
24	14 36	13 44	02 17
25	14 34	13 41	03 16
26	14 31	13 38	04 04
27	14 27	13 35	04 41
28	14 24	13 32	05 04
29	14 21	13 29	05 14
30	14 18	13 25	05 10
31	14 ♍ 17	13 ♍ 22	04 N 53

DECLINATIONS

Date	Sun ☉ ° '	Moon ☽ ° '	Mercury ☿ ° '	Venus ♀ ° '	Mars ♂ ° '	Jupiter ♃ ° '	Saturn ♄ ° '	Uranus ♅ ° '	Neptune ♆ ° '	Pluto ♇ ° '
01	07 S 33	15 S 49	17 S 12	19 S 19	23 S 35	05 N 18	20 N 45	23 N 36	06 N 32	22 S 42
02	07 10	17 25	16 54	19 07	23 36	05 23	20 46	23 36	06 33	22 41
03	06 47	18 16	16 35	18 55	23 36	05 28	20 46	23 36	06 33	22 41
04	06 24	18 18	16 15	18 42	23 35	05 33	20 47	23 36	06 34	22 41
05	06 01	17 30	15 54	18 28	23 35	05 39	20 47	23 36	06 35	22 40
06	05 37	15 52	15 31	18 14	23 35	05 44	20 48	23 36	06 36	22 40
07	05 14	13 26	15 06	17 59	23 34	05 49	20 49	23 36	06 36	22 39
08	04 51	10 17	14 41	17 44	23 33	05 54	20 49	23 36	06 37	22 39
09	04 27	06 33	14 14	17 28	23 32	06 00	20 50	23 36	06 38	22 39
10	04 04	02 S 24	13 46	17 12	23 31	06 05	20 50	23 36	06 39	22 38
11	03 40	01 N 57	13 17	16 56	23 29	06 10	20 51	23 36	06 39	22 38
12	03 17	06 16	12 47	16 38	23 28	06 15	20 51	23 36	06 40	22 38
13	02 53	10 16	12 15	16 21	23 26	06 21	20 52	23 36	06 41	22 37
14	02 29	13 42	11 42	16 03	23 24	06 26	20 52	23 36	06 42	22 37
15	02 06	16 18	11 08	15 44	23 22	06 31	20 53	23 36	06 43	22 37
16	01 42	17 55	10 33	15 25	23 19	06 37	20 53	23 36	06 43	22 36
17	01 18	18 24	09 56	15 06	23 17	06 42	20 53	23 36	06 44	22 36
18	00 54	17 44	09 19	14 46	23 14	06 47	20 54	23 36	06 45	22 36
19	00 31	16 00	08 40	14 26	23 11	06 53	20 54	23 36	06 46	22 35
20	00 S 07	13 22	07 59	14 05	23 08	06 58	20 54	23 36	06 47	22 35
21	00 N 17	10 01	07 18	13 44	23 05	07 03	20 54	23 36	06 47	22 35
22	00 40	06 12	06 36	13 23	23 02	07 09	20 55	23 36	06 48	22 35
23	01 04	02 N 08	05 52	13 01	22 58	07 14	20 55	23 36	06 49	22 34
24	01 28	01 S 58	05 08	12 39	22 54	07 19	20 55	23 36	06 50	22 34
25	01 51	05 54	04 22	12 16	22 50	07 25	20 55	23 36	06 51	22 34
26	02 15	09 30	03 35	11 53	22 46	07 30	20 55	23 36	06 52	22 34
27	02 38	12 37	02 47	11 30	22 42	07 35	20 56	23 36	06 52	22 33
28	03 02	15 09	01 58	11 06	22 38	07 41	20 56	23 36	06 53	22 33
29	03 25	17 01	01 08	10 43	22 33	07 46	20 56	23 36	06 54	22 33
30	03 49	18 07	00 17	10 18	22 28	07 52	20 56	23 36	06 55	22 33
31	04 N 12	18 S 26	00 N 34	09 S 54	22 S 23	07 N 57	20 N 56	23 N 36	06 N 56	22 S 33

ZODIAC SIGN ENTRIES

Date	h m	Planets
01	04 07	☽ ♐
03	16 39	☽ ♑
06	03 58	☽ ♒
08	12 31	☽ ♓
10	18 10	☽ ♈
11	07 16	☿ ♓
12	21 44	☽ ♉
15	00 19	☽ ♊
17	02 54	☽ ♋
19	06 10	☽ ♌
20	19 03	☉ ♈
21	10 40	☽ ♍
23	16 55	☽ ♎
25	20 56	♀ ♓
26	01 27	☽ ♏
28	12 23	☽ ♐
28	19 13	☿ ♈
31	00 50	☽ ♑

LATITUDES

Date	Mercury ☿ ° '	Venus ♀ ° '	Mars ♂ ° '	Jupiter ♃ ° '	Saturn ♄ ° '	Uranus ♅ ° '	Neptune ♆ ° '	Pluto ♇ ° '
01	01 S 12	00 N 40	00 S 10	01 S 06	00 N 26	00 N 20	01 S 37	07 S 50
04	01 32	00 29	00 14	01 06	00 27	00 20	01 37	07 50
07	01 48	00 18	00 17	01 05	00 27	00 20	01 37	07 51
10	02 00	00 N 07	00 20	01 05	00 27	00 20	01 37	07 51
13	02 09	00 S 03	00 23	01 05	00 27	00 20	01 37	07 52
16	02 14	00 13	00 27	01 04	00 27	00 20	01 37	07 52
19	02 16	00 23	00 30	01 04	00 28	00 20	01 36	07 53
22	02 13	00 32	00 34	01 04	00 28	00 20	01 36	07 53
25	02 06	00 41	00 38	01 04	00 28	00 20	01 36	07 54
28	01 54	00 49	00 42	01 03	00 28	00 20	01 36	07 55
31	01 S 38	00 S 57	00 S 46	01 S 03	00 N 28	00 N 20	01 S 36	07 S 55

DATA

Julian Date	2464388
Delta T	+76 seconds
Ayanamsa	24° 20' 50"
Synetic vernal point	04° ♓ 46' 09"
True obliquity of ecliptic	23° 25' 56"

MOON'S PHASES, APSIDES AND POSITIONS ☽

Date	h m	Phase	Longitude ° '	Eclipse Indicator
02	03 01	☾	11 ♐ 21	
09	23 09	●	19 ♓ 12	Annular
16	20 15	☽	26 ♊ 04	
23	22 42	○	03 ♎ 08	
31	23 07	☾	11 ♑ 04	

Day	h m		
02	09 34	Apogee	
15	01 38	Perigee	
30	05 35	Apogee	
04	01 07	Max dec	18° S 23'
11	01 16	0N	
17	10 05	Max dec	18° N 24'
24	00 24	0S	
31	09 25	Max dec	18° S 27'

LONGITUDES

Date	Chiron ⚷ ° '	Ceres ⚳ ° '	Pallas ⚴ ° '	Juno ⚵ ° '	Vesta ⚶ ° '	Black Moon Lilith ⚸ ° '
01	01 ♊ 38	13 ♈ 34	04 ♓ 00	05 ♒ 51	04 ♋ 13	14 ♐ 06
11	01 ♊ 55	17 ♈ 23	07 ♓ 18	09 ♒ 35	05 ♋ 13	15 ♐ 13
21	02 ♊ 19	21 ♈ 17	10 ♓ 34	13 ♒ 15	06 ♋ 50	16 ♐ 21
31	02 ♊ 49	25 ♈ 14	13 ♓ 46	16 ♒ 49	08 ♋ 57	17 ♐ 28

All ephemeris data is given at 12.00 UT and the Moon's longitude is additionally given for 24.00 UT

ASPECTARIAN

h m	Aspects
01 Thursday	
02 31	☽ △ ♄
04 59	☽ ⚹ ♀
05 16	☽ ± ♅
06 08	☽ Ꝑ ♃
08 28	☽ ⊻ ♂
12 05	☽ Q ☿
14 22	☿ ⚹ ♃
15 21	☽ Ꝑ ♆
17 18	☽ Q ♇
17 21	☽ ⊼ ♅
02 Friday	
03 01	☽ □ ☉
04 34	☽ ∥ ☿
08 39	☽ Ꝑ ♄
12 59	☽ △ ♃
14 38	☽ ∠ ♀
15 23	☽ ⚹ ☿
17 36	☽ ⚹ ♇
21 44	☽ △ ♆
03 Saturday	
02 45	☽ ± ♄
09 47	☿ ∠ ♂
10 49	☽ ⊥ ♀
11 04	☿ ☌ ♇
12 31	♂ ∠ ♇
14 48	☽ ⊼ ♄
18 40	☽ Q ☉
23 53	☽ ∠ ♇
04 Sunday	
00 12	☽ ⊻ ♀
00 31	☽ ☌ ♂
01 30	☉ # ♆
01 30	☽ ∠ ☿
05 45	☽ ☍ ♅
06 48	♀ ⊻ ♂
17 52	☽ ⊥ ♇
18 58	☉ Ꝑ ♄
20 48	☽ ⚹ ☉
21 02	♀ Q ♃
21 47	☽ ⊥ ☿
23 12	☿ ⚹ ♆
05 Monday	
02 14	☽ □ ♃
05 46	☽ ⊻ ♇
09 48	☽ □ ♆
10 57	☉ ⊥ ♆
11 09	☽ ⊻ ☿
13 40	☿ Ꝑ ♅
06 Tuesday	
01 57	☽ ☍ ♄
04 48	☽ ∠ ☉
06 36	☉ # ♃
07 38	♀ ⊼ ♅
13 56	☽ Q ♃
14 48	☽ ⊻ ♂
16 30	☽ ⊼ ♅
16 59	☽ ∥ ☿
17 27	☽ ☌ ♀
20 45	☽ Q ♆
23 34	☽ ⊥ ☉
07 Wednesday	
02 53	☽ ⊥ ♂
03 55	☽ ± ♅
11 53	☽ ⊻ ☉
13 08	☽ ⚹ ♃
15 35	☽ ☌ ♇
19 26	☽ ⚹ ♆
20 43	☽ ∠ ♂
20 49	☽ Ꝑ ♅
22 48	♂ ☍ ♅
08 Thursday	
03 48	☽ ☌ ☿
05 00	♀ Q ♆
08 00	☉ ⊻ ♃
10 26	☽ ⊼ ♄
17 23	☽ ∠ ♃
21 22	☽ ± ♄
23 06	☽ ∠ ♆
09 Friday	
00 22	☽ △ ♅
01 43	☽ ⚹ ♂
06 49	☽ ⊻ ♀
09 58	☽ ⊥ ♃
11 31	☽ # ♆
12 18	☉ ⊻ ♇
13 34	☽ Ꝑ ♄
15 17	☽ # ♃
15 21	☽ ⊥ ♆
18 37	☽ ⊥ ♀
20 52	☽ ⊻ ♃
22 22	☽ ⊻ ♇
23 09	☽ ☌ ☉
10 Saturday	
00 20	☽ Q ♂
01 37	☽ ∥ ☉
02 04	☽ ⊻ ♆
09 00	☽ ⊥ ♇
12 05	☽ ∠ ♀
12 16	☿ ⊼ ♄
16 03	☽ △ ♄
16 32	☽ ⊻ ☿
11 Sunday	
00 47	☽ ∠ ♇
04 17	☽ ⊥ ☿
05 28	☽ □ ♅
07 26	♀ ± ♅
09 25	☽ □ ♂
15 43	☉ ⊻ ♆
16 37	☽ ⚹ ♀
19 37	☉ Q ♂
20 38	☽ # ☉
21 45	☽ ∠ ☿
12 Monday	
02 01	☽ ☌ ♃
02 42	☽ ⚹ ♇
06 16	☽ ☌ ♆
07 21	☽ ⊻ ☉
11 56	☽ ∥ ♃
12 22	☽ Q ♅
14 19	☽ ∥ ♆
15 04	☽ Q ♀
18 21	☽ ⊥ ☉
19 33	☽ □ ♄
23 10	☽ Q ♇
13 Tuesday	
02 29	☽ ⚹ ☿
08 43	☽ ⚹ ♅
10 45	☽ ∠ ☉
15 05	☽ △ ♂
17 43	☿ ∠ ♃
23 28	☽ # ☿
14 Wednesday	
00 19	☽ □ ♀
00 57	♃ ⚹ ♇
01 17	☽ Q ☿
01 54	☽ Q ♄
05 37	☽ □ ♇
05 41	☽ ⊻ ♃
06 46	☿ ± ♄
09 10	☽ ⊻ ♆
10 00	☽ ∠ ♅
13 55	☽ ⚹ ☉
15 57	☽ ⊥ ♃
17 33	☽ Ꝑ ♂
19 17	☽ ⊥ ♆
21 31	♅ St D
22 03	☽ ⚹ ♄
15 Thursday	
01 09	☽ ⊥ ♅
02 54	☿ ∠ ♆
06 41	☽ # ♀
07 19	☽ ∠ ♃
09 25	☽ ± ♂
09 57	☿ △ ♅
10 18	♀ ⊥ ♂
10 27	☽ ∠ ♆
11 14	☽ ⊻ ♅
11 24	☽ □ ☿
11 37	☽ Q ☉
16 22	☉ ⊥ ♇
20 00	☽ ⊼ ♂
23 14	☽ ∠ ♄
16 Friday	
07 27	☽ △ ♀
08 11	☽ △ ♇
09 00	☽ ⚹ ♃
11 47	☽ ⚹ ♆
14 23	☽ ⊥ ♄
16 22	♀ ☌ ♇
20 15	☽ □ ☉
17 Saturday	
00 31	☽ ⊻ ♄
05 38	☽ Q ♃
05 47	♀ ⚹ ♃
08 08	☽ Q ♆
09 37	☽ Ꝑ ♇
11 11	☽ Ꝑ ♀
13 55	☽ ☌ ♅
18 12	♂ ⊥ ♇
20 45	☽ △ ☿
23 57	♃ ∥ ♆
18 Sunday	
00 57	☽ ± ♇
01 16	☽ ☍ ♂
03 56	☽ ± ♀
11 13	☽ ⊼ ♇
12 09	♀ ⚹ ♆
12 49	☽ □ ♃
14 55	☽ □ ♆
15 10	☽ ⊼ ♀
19 24	♀ Ꝑ ♅
19 Monday	
01 57	☽ Ꝑ ☿
03 20	☽ △ ☉
03 39	☽ ☌ ♄
07 48	☉ △ ♄
15 46	☿ Ꝑ ♄
17 25	☽ ⊻ ♅
19 42	☽ ± ☿
20 Tuesday	
02 09	☿ ⊥ ♃
03 52	☽ ⊥ ♅
05 20	☽ # ♀
07 14	☿ ⚹ ♂
07 23	☽ Ꝑ ☉
07 37	☽ ⊼ ♂
07 39	☽ ⊼ ☿
15 11	☿ ⊥ ♆
15 18	☽ ☍ ♇
17 45	☽ △ ♃
18 40	☽ ± ♂
19 08	☽ △ ♆
19 39	☽ ∠ ♅
21 Wednesday	
00 18	☽ ☍ ♀
00 26	☽ ± ☉
08 01	☽ ⊻ ♄
08 34	☉ ⊥ ♀
11 23	☽ Ꝑ ♂
11 54	☽ ⊼ ☉
18 40	☽ ⊥ ♄
19 33	☿ # ♃
20 48	☽ Ꝑ ♃
21 48	☽ Ꝑ ♆
22 16	☽ ⚹ ♅
22 Thursday	
05 09	☿ # ♆
06 21	☽ ∥ ♃
08 19	☽ ∥ ♆
09 02	☽ # ☿
10 48	☽ ∠ ♄
13 16	☽ ± ♃
14 02	☽ ± ♆
15 40	☽ △ ♂
19 54	☽ Q ♅
20 30	☿ ⊻ ♇
20 57	☽ ⊼ ♇
21 01	☽ ☍ ☿
23 Friday	
00 21	☽ ⊼ ♃
00 57	☽ ⊼ ♆
07 55	☽ ± ♇
11 37	☽ ⊼ ♀
14 06	☽ ⚹ ♄
17 37	☽ ∥ ☉
22 42	☽ ☍ ☉
23 49	☽ ± ♀
24 Saturday	
00 32	☽ Ꝑ ♇
00 54	☿ ⊻ ♃
02 40	☿ ⊻ ♆
05 00	☽ □ ♅
08 40	☽ # ☉
12 21	☽ Q ♄
14 14	♀ ⊼ ♄
16 32	♃ ☌ ♆
18 22	☽ Ꝑ ♀
23 05	☉ ∠ ♇
25 Sunday	
02 07	☽ □ ♂
04 00	☽ ∥ ☿
04 42	☽ △ ♇
08 57	☽ ☍ ♆
09 12	☽ ☍ ♃
13 56	☽ ⊼ ☿
18 06	☽ # ♆
22 02	☽ # ♃
22 27	☽ □ ♄
26 Monday	
01 56	☽ △ ♀
03 26	☽ ± ☿
05 13	☿ ⊥ ♇
12 26	☽ ⊼ ☉
14 07	☽ △ ♅
14 33	☽ Q ♂
27 Tuesday	
00 00	☽ Ꝑ ☿
01 09	☽ ± ☉
03 57	☽ ∥ ♀
05 43	♂ ⊻ ♇
09 44	☉ □ ♅
14 53	☉ # ☿
14 54	☽ □ ♇
15 25	☽ ⚹ ♂
19 24	☽ ⊼ ♆
19 35	☽ Ꝑ ♅
20 27	☽ Ꝑ ☉
20 39	☽ ⊼ ♃
23 18	☿ △ ♄
28 Wednesday	
07 22	☽ ± ♆
08 50	☽ ± ♃
09 15	☽ △ ♄
11 06	☽ △ ☿
13 31	☽ ± ♅
19 29	☽ □ ♀
23 01	☽ ∠ ♂
29 Thursday	
01 26	☽ Ꝑ ♆
01 33	☽ ⊼ ♅
02 52	☽ Q ♇
03 13	☽ Ꝑ ♃
05 07	☽ △ ☉
11 26	♂ ∥ ♇
12 57	♀ ± ♄
15 21	☽ Ꝑ ♄
18 12	☽ ⊥ ♂
30 Friday	
03 06	☽ ⚹ ♇
07 01	☽ ⊻ ♂
07 46	☽ △ ♆
09 31	☽ ± ♄
10 06	☽ △ ♃
11 51	☽ Q ♀
13 12	☿ Q ♂
13 13	♄ St D
21 25	♂ □ ♆
21 39	☽ ⊼ ♄
31 Saturday	
00 28	☿ ∠ ♇
08 05	♀ ∠ ♆
08 18	♀ △ ♅
09 27	☽ ∠ ♇
11 12	☽ □ ☿
14 10	☽ ☍ ♅
14 49	☽ ⚹ ♀
19 58	♀ ∠ ♂
23 07	☽ □ ☉

APRIL 2035

Date	Sidereal time h m s	Sun ☉ ° ' "	Moon ☽ ° ' "	Moon ☽ 24.00 ° ' "	Mercury ☿ ° '	Venus ♀ ° '	Mars ♂ ° '	Jupiter ♃ ° '	Saturn ♄ ° '	Uranus ♅ ° '	Neptune ♆ ° '	Pluto ♇ ° '
01	00 38 44	11 ♈ 35 34	17 ♑ 30 09	23 ♑ 32 17	07 ♈ 11	08 ♓ 00	22 ♑ 38	23 ♈ 12	28 ♋ 26	06 ♋ 38	21 ♈ 38	19 ♒ 18
02	00 42 41	12 34 48	29 ♑ 37 23	05 ♒ 46 00	09 10	09 13	23 17	23 27	28 D 26	06 38	21 41	19 19
03	00 46 37	13 33 59	11 ♒ 58 44	18 ♒ 16 03	11 10	10 25	23 56	23 41	28 26	06 39	21 43	19 20
04	00 50 34	14 33 09	24 ♒ 38 26	01 ♓ 06 17	13 12	11 38	24 35	23 55	28 27	06 41	21 45	19 21
05	00 54 31	15 32 17	07 ♓ 39 52	14 ♓ 19 23	15 14	12 50	25 15	24 10	28 27	06 42	21 47	19 23
06	00 58 27	16 31 23	21 ♓ 04 56	27 ♓ 56 27	17 18	14 03	25 54	24 24	28 28	06 43	21 50	19 24
07	01 02 24	17 30 27	04 ♈ 53 44	11 ♈ 56 27	19 22	15 16	26 33	24 39	28 29	06 44	21 52	19 25
08	01 06 20	18 29 29	19 ♈ 04 06	26 ♈ 16 05	21 27	16 28	27 12	24 53	28 30	06 45	21 54	19 26
09	01 10 17	19 28 30	03 ♉ 31 38	10 ♉ 49 56	23 32	17 41	27 51	25 07	28 31	06 47	21 56	19 27
10	01 14 13	20 27 28	18 ♉ 10 05	25 ♉ 31 08	25 37	18 54	28 30	25 22	28 32	06 48	21 59	19 28
11	01 18 10	21 26 24	02 ♊ 52 11	10 ♊ 12 21	27 43	20 06	29 09	25 36	28 33	06 49	22 01	19 29
12	01 22 06	22 25 17	17 ♊ 30 49	24 ♊ 46 55	29 ♈ 48	21 19	29 ♑ 48	25 51	28 35	06 51	22 03	19 30
13	01 26 03	23 24 09	02 ♋ 00 03	09 ♋ 09 45	01 ♉ 52	22 32	00 ♒ 27	26 05	28 36	06 52	22 06	19 31
14	01 30 00	24 22 58	16 ♋ 15 42	23 ♋ 17 41	03 55	23 45	01 05	26 20	28 38	06 54	22 08	19 32
15	01 33 56	25 21 45	00 ♌ 15 35	07 ♌ 09 23	05 57	24 57	01 44	26 34	28 39	06 56	22 10	19 33
16	01 37 53	26 20 30	13 ♌ 59 10	20 ♌ 45 00	07 58	26 10	02 23	26 49	28 41	06 57	22 12	19 34
17	01 41 49	27 19 12	27 ♌ 27 04	04 ♍ 05 30	09 56	27 23	03 02	27 03	28 43	06 59	22 15	19 35
18	01 45 46	28 17 52	10 ♍ 40 30	17 ♍ 12 13	11 52	28 36	03 40	27 17	28 45	07 01	22 17	19 36
19	01 49 42	29 ♈ 16 30	23 ♍ 40 50	00 ♎ 06 29	13 46	29 ♓ 48	04 19	27 32	28 47	07 03	22 19	19 36
20	01 53 39	00 ♉ 15 06	06 ♎ 29 16	12 ♎ 49 19	15 36	01 ♈ 01	04 58	27 46	28 49	07 04	22 21	19 37
21	01 57 35	01 13 40	19 ♎ 06 43	25 ♎ 21 33	17 23	02 14	05 36	28 01	28 52	07 06	22 24	19 38
22	02 01 32	02 12 11	01 ♏ 33 54	07 ♏ 43 50	19 07	03 27	06 15	28 15	28 54	07 08	22 26	19 39
23	02 05 29	03 10 41	13 ♏ 51 27	19 ♏ 56 51	20 47	04 39	06 53	28 30	28 57	07 10	22 28	19 39
24	02 09 25	04 09 09	26 ♏ 00 11	02 ♐ 01 36	22 23	05 52	07 32	28 44	28 59	07 12	22 30	19 40
25	02 13 22	05 07 36	08 ♐ 01 19	13 ♐ 59 34	23 55	07 05	08 10	28 58	29 02	07 14	22 33	19 41
26	02 17 18	06 06 00	19 ♐ 56 39	25 ♐ 52 55	25 23	08 18	08 48	29 13	29 05	07 16	22 35	19 41
27	02 21 15	07 04 23	01 ♑ 48 45	07 ♑ 44 35	26 46	09 31	09 27	29 27	29 08	07 18	22 37	19 42
28	02 25 11	08 02 45	13 ♑ 40 53	19 ♑ 38 11	28 05	10 44	10 05	29 42	29 11	07 21	22 39	19 43
29	02 29 08	09 01 05	25 ♑ 37 03	01 ♒ 38 04	29 ♉ 20	11 57	10 43	29 ♈ 56	29 14	07 23	22 41	19 43
30	02 33 04	09 ♉ 59 23	07 ♒ 41 51	13 ♒ 49 02	00 ♊ 29	13 ♈ 10	11 ♒ 21	00 ♉ 10	29 ♋ 17	07 ♋ 25	22 ♈ 44	19 ♒ 44

Date	Moon True ☊ ° '	Moon Mean ☊ ° '	Moon ☽ Latitude ° '
01	14 ♍ 17	13 ♍ 19	04 N 23
02	14 D 18	13 16	03 41
03	14 20	13 13	02 48
04	14 22	13 10	01 46
05	14 23	13 06	00 N 37
06	14 R 23	13 03	00 S 37
07	14 21	13 00	01 50
08	14 18	12 57	02 58
09	14 13	12 54	03 56
10	14 08	12 50	04 39
11	14 03	12 47	05 05
12	13 58	12 44	05 10
13	13 54	12 41	04 55
14	13 52	12 38	04 22
15	13 D 52	12 35	03 34
16	13 53	12 31	02 35
17	13 55	12 28	01 28
18	13 55	12 25	00 S 18
19	13 R 55	12 22	00 N 53
20	13 53	12 19	01 59
21	13 48	12 16	02 58
22	13 41	12 12	03 48
23	13 33	12 09	04 26
24	13 23	12 06	04 52
25	13 14	12 03	05 05
26	13 05	12 00	05 04
27	12 58	11 56	04 50
28	12 52	11 53	04 23
29	12 49	11 50	03 45
30	12 ♍ 48	11 ♍ 47	02 N 57

DECLINATIONS

Date	Sun ☉ ° '	Moon ☽ ° '	Mercury ☿ ° '	Venus ♀ ° '	Mars ♂ ° '	Jupiter ♃ ° '	Saturn ♄ ° '	Uranus ♅ ° '	Neptune ♆ ° '	Pluto ♇ ° '
01	04 N 35	17 S 56	01 N 27	09 S 29	22 S 18	08 N 02	20 N 56	23 N 36	06 N 57	22 S 32
02	04 58	16 37	02 20	09 04	22 13	08 08	20 56	23 36	06 57	22 32
03	05 21	14 30	03 14	08 39	22 08	08 13	20 56	23 36	06 58	22 32
04	05 44	11 38	04 09	08 13	22 02	08 18	20 56	23 36	06 59	22 32
05	06 07	08 07	05 04	07 48	21 56	08 24	20 56	23 36	07 00	22 32
06	06 30	04 S 06	06 00	07 22	21 51	08 29	20 56	23 36	07 01	22 32
07	06 52	00 N 16	06 56	06 55	21 45	08 34	20 55	23 35	07 02	22 32
08	07 15	04 43	07 52	06 29	21 38	08 40	20 55	23 35	07 02	22 31
09	07 37	08 59	08 47	06 02	21 32	08 45	20 55	23 35	07 03	22 31
10	07 59	12 45	09 43	05 36	21 26	08 50	20 55	23 35	07 04	22 31
11	08 21	15 45	10 38	05 09	21 19	08 55	20 55	23 35	07 05	22 31
12	08 43	17 42	11 33	04 42	21 13	09 01	20 54	23 35	07 06	22 31
13	09 05	18 30	12 27	04 14	21 06	09 06	20 54	23 35	07 07	22 31
14	09 27	18 06	13 19	03 47	20 59	09 11	20 54	23 35	07 08	22 31
15	09 48	16 36	14 11	03 19	20 52	09 16	20 54	23 35	07 08	22 31
16	10 10	14 09	15 01	02 52	20 44	09 22	20 53	23 35	07 09	22 31
17	10 31	10 58	15 49	02 24	20 37	09 27	20 53	23 35	07 10	22 31
18	10 52	07 17	16 35	01 56	20 30	09 32	20 53	23 35	07 11	22 31
19	11 13	03 N 19	17 20	01 28	20 22	09 37	20 52	23 34	07 12	22 31
20	11 33	00 S 45	18 02	01 00	20 14	09 42	20 52	23 34	07 13	22 31
21	11 54	04 44	18 42	00 32	20 06	09 48	20 51	23 34	07 13	22 31
22	12 14	08 27	19 20	00 S 04	19 58	09 53	20 51	23 34	07 14	22 31
23	12 34	11 45	19 55	00 N 24	19 50	09 58	20 50	23 34	07 15	22 31
24	12 54	14 31	20 27	00 52	19 42	10 03	20 50	23 34	07 16	22 31
25	13 14	16 38	20 57	01 20	19 34	10 08	20 49	23 34	07 17	22 31
26	13 33	18 00	21 25	01 48	19 25	10 13	20 49	23 34	07 17	22 31
27	13 52	18 36	21 50	02 16	19 17	10 18	20 48	23 33	07 18	22 31
28	14 11	18 22	22 12	02 45	19 08	10 23	20 48	23 33	07 19	22 31
29	14 30	17 19	22 32	03 13	18 59	10 28	20 47	23 33	07 20	22 31
30	14 N 48	15 S 30	22 N 50	03 N 41	18 S 51	10 N 33	20 N 46	23 N 33	07 N 21	22 S 31

ZODIAC SIGN ENTRIES

Date	h m	Planets
02	12 44	☽ ♒
04	21 58	☽ ♓
07	03 34	☽ ♈
09	06 11	☽ ♉
11	07 19	☽ ♊
12	14 22	☿ ♉
12	19 35	♂ ♒
13	08 40	☽ ♋
15	11 33	☽ ♌
17	16 36	☽ ♍
19	15 52	♀ ♈
19	23 48	☽ ♎
20	05 49	☉ ♉
22	08 58	☽ ♏
24	19 57	☽ ♐
27	08 20	☽ ♑
29	18 57	♃ ♉
29	20 45	☽ ♒
30	01 45	☿ ♊

LATITUDES

Date	Mercury ☿ ° '	Venus ♀ ° '	Mars ♂ ° '	Jupiter ♃ ° '	Saturn ♄ ° '	Uranus ♅ ° '	Neptune ♆ ° '	Pluto ♇ ° '
01	01 S 31	00 S 59	00 S 47	01 S 03	00 N 28	00 N 20	01 S 36	07 S 55
04	01 09	01 06	00 51	01 03	00 28	00 20	01 36	07 56
07	00 42	01 13	00 55	01 03	00 29	00 20	01 36	07 57
10	00 S 12	01 18	01 00	01 03	00 29	00 20	01 36	07 58
13	00 N 21	01 23	01 04	01 03	00 29	00 20	01 36	07 58
16	00 54	01 28	01 09	01 02	00 29	00 20	01 36	07 59
19	01 26	01 31	01 14	01 02	00 29	00 20	01 36	08 00
22	01 54	01 34	01 18	01 02	00 29	00 20	01 36	08 01
25	02 17	01 36	01 23	01 02	00 29	00 20	01 36	08 02
28	02 33	01 38	01 29	01 02	00 29	00 20	01 36	08 02
31	02 N 40	01 S 39	01 S 34	01 S 02	00 N 30	00 N 20	01 S 37	08 S 03

DATA

Julian Date	2464419
Delta T	+76 seconds
Ayanamsa	24° 20' 53"
Synetic vernal point	04° ♓ 46' 06"
True obliquity of ecliptic	23° 25' 57"

MOON'S PHASES, APSIDES AND POSITIONS ☽

Date	h m	Phase	Longitude ° '	Eclipse Indicator
08	10 58	●	18 ♈ 27	
15	02 55	☽	25 ♋ 00	
22	13 21	○	02 ♏ 15	
30	16 54	☾	10 ♒ 11	

Day	h m		
11	00 55	Perigee	
26	22 25	Apogee	
07	10 34	ON	
13	15 51	Max dec	18° N 31'
20	07 32	OS	
27	17 21	Max dec	18° S 37'

LONGITUDES

Date	Chiron ⚷ ° '	Ceres ⚳ ° '	Pallas ⚴ ° '	Juno ⚵ ° '	Vesta ⚶ ° '	Black Moon Lilith ⚸ ° '
01	02 ♊ 52	25 ♈ 37	14 ♓ 05	17 ♒ 10	09 ♋ 11	17 ♐ 35
11	03 ♊ 26	29 ♈ 37	17 ♓ 12	20 ♒ 36	11 ♋ 47	18 ♐ 42
21	04 ♊ 05	03 ♉ 38	20 ♓ 13	23 ♒ 52	14 ♋ 45	19 ♐ 49
31	04 ♊ 46	07 ♉ 40	23 ♓ 07	26 ♒ 58	18 ♋ 01	20 ♐ 56

ASPECTARIAN

h m	Aspects
01 Sunday	
03 35	☽ ⊥ ♇
05 16	☿ □ ♅
15 35	☽ ⊻ ♇
17 02	♀ ∠ ♃
20 16	☽ □ ♆
22 47	☽ ☌ ♂
23 34	☽ □ ♃
02 Monday	
00 09	☽ ∠ ♀
06 13	☽ Q ☿
09 40	☽ ☍ ♄
13 27	☿ ⊻ ♀
14 03	☽ Q ☉
19 47	☽ ⊥ ♀
21 15	♂ □ ♃
03 Tuesday	
01 43	☽ ⊼ ♅
07 38	☽ Q ♆
08 41	☽ ⊻ ♀
10 09	☽ ⚹ ☿
11 25	☽ Q ♃
13 18	☽ ± ♅
15 18	☽ ⚹ ☉
04 Wednesday	
02 03	☽ ☌ ♇
06 26	☽ ⚼ ♅
06 34	☽ ⚹ ♆
08 12	♀ # ♃
10 38	☽ ⚹ ♃
11 54	☽ ⊻ ♂
19 06	☽ ⊼ ♄
19 52	☽ ∠ ☿
21 53	☽ ∠ ☉
23 38	☽ ⊥ ♂
05 Thursday	
06 09	☽ ± ♄
10 14	☽ △ ♅
10 20	☽ # ♃
10 24	☽ ∠ ♆
14 21	☽ ∥ ♀
14 46	☽ ∠ ♃
15 23	☽ ⊥ ☿
15 40	☽ ⊥ ☉
16 55	☽ ∠ ♂
18 41	☉ ☌ ☿
18 57	☽ # ♆
22 16	☽ ☌ ♀
22 28	☽ ⚼ ♄
23 14	☽ # ☉
06 Friday	
00 21	♀ ⚼ ♄
02 40	☽ ⊥ ♆
03 02	☽ # ☿
03 17	☽ ⊻ ☉
04 06	☽ ⊻ ☿
07 11	☽ ⊥ ♃
09 01	☽ ⊻ ♇
13 19	☽ ⊻ ♆
17 56	☽ ⊻ ♃
19 35	☽ ⊥ ♇
20 52	☽ ⚹ ♂
07 Saturday	
00 56	☽ △ ♄
06 28	♀ # ♆
09 32	☉ ∥ ☿
11 10	☽ ∠ ♇
11 55	☿ # ♀
12 32	☿ ⚹ ♇
13 33	☉ # ♀
14 36	☿ ∥ ♆
15 09	☽ □ ♅
18 32	☽ Q ♂
21 56	♃ Q ♅
22 25	☉ ∥ ♆
08 Sunday	
00 21	♀ ⊥ ♆
07 14	☽ ⊻ ♀
10 58	☽ ☌ ☉
12 37	☽ ⚹ ♇
16 39	☽ ☌ ☿
16 45	☽ ☌ ♆
17 20	☿ ☌ ♆
18 13	☽ ⊥ ♀
20 48	☽ # ♀
21 30	☽ Q ♅
21 52	☽ ☌ ♃
09 Monday	
00 52	☽ ∥ ♆
02 11	☽ □ ♂
03 19	☽ ∥ ☉
03 43	☽ □ ♄
08 34	☽ Q ♇
10 29	☽ ∠ ♀
10 31	☽ ∥ ☿
10 34	☽ ∥ ♃
10 45	☿ ∥ ♃
11 23	☉ ⚹ ♇
16 04	☿ ⊥ ♀
17 21	☽ ⚹ ♅
10 Tuesday	
02 25	☿ Q ♅
08 37	☿ ☌ ♃
09 20	☽ Q ♄
13 18	☽ ⚹ ♀
13 25	♂ ☍ ♄
14 07	☽ □ ♇
16 00	☽ ⊻ ☉
17 56	☽ ∠ ♅
18 14	☽ ⊻ ♆
23 30	♀ ⊻ ♇
23 37	♀ ⊥ ♃
23 57	☽ ⊻ ♃
11 Wednesday	
02 11	☽ ⊻ ☿
02 30	☽ ⊥ ☉
04 03	☽ ⊥ ♆
04 57	☽ ⚹ ♄
05 38	☽ △ ♂
08 39	☽ ⊥ ♅
09 54	☽ ⊥ ♃
10 38	☽ Q ♀
13 36	☽ ⊥ ☿
18 15	☽ ∠ ☉
18 28	☽ ⊻ ♅
18 48	☽ ∠ ♆
21 48	☿ □ ♄
12 Thursday	
00 52	☽ ∠ ♃
02 40	☉ ☌ ♆
05 31	☽ ∠ ♄
06 47	☽ ∠ ☿
07 19	☽ ⚼ ♂
11 59	☿ □ ♂
15 17	☽ △ ♇
18 51	☽ □ ♀
19 30	☽ ⚹ ♆
20 22	☽ ⊥ ♄
20 41	☽ ⚹ ☉
22 51	☽ ± ♂
13 Friday	
02 00	☽ ⚹ ♃
03 03	♀ ⊻ ♆
06 20	☽ ⊻ ♄
07 54	☿ Q ♇
09 17	☽ ⊼ ♂
11 44	☽ ⚹ ☿
13 00	☉ ∥ ♃
15 30	☽ Q ♆
16 13	☽ ⚼ ♇
18 06	☽ Q ☉
20 10	☽ ☌ ♅
22 22	☽ Q ♃
14 Saturday	
07 22	☽ ± ♇
11 20	☽ Q ☿
17 34	☽ ⊼ ♇
22 02	☽ □ ♆
15 Sunday	
01 01	☉ Q ♅
01 59	☽ △ ♀
02 55	☽ □ ☉
04 49	♂ # ♄
05 31	☽ □ ♃
09 13	☽ ☌ ♄
14 41	☽ ☍ ♂
23 36	☽ □ ☿
23 37	☽ ⊻ ♅
23 43	☿ ⚹ ♅
23 54	♀ ⊥ ♇
16 Monday	
06 13	☽ ∥ ☿
06 33	☽ ⚼ ♀
10 11	☽ ⊥ ♅
21 54	☽ ☍ ♇
17 Tuesday	
00 03	☽ ± ♀
02 11	☽ ∠ ♅
02 38	☽ △ ♆
03 13	☉ ☌ ♃
03 52	♀ ⊻ ♃
05 56	☉ ⊻ ♀
11 16	☽ △ ♃
11 45	☽ △ ☉
11 51	☽ ⊼ ♀
14 17	☽ ⊻ ♄
14 52	☽ ∥ ☉
22 02	☽ ∥ ♃
22 35	☽ ⊼ ♂
18 Wednesday	
01 10	☽ ⊥ ♄
05 18	☽ ⚹ ♅
05 47	☽ ⚼ ♆
10 04	☽ ± ♂
12 41	☽ ∥ ♆
14 34	☽ △ ☿
15 01	☽ ⚼ ♃
15 16	♀ △ ♄
17 12	☽ ⚼ ☉
17 40	☽ ∠ ♄
22 20	☽ ± ♆
23 35	☉ □ ♄
19 Thursday	
03 23	☽ Q ♅
03 29	☽ ⚼ ♂
04 26	☽ ⊼ ♇
07 56	☽ ± ♃
09 28	☽ ⊼ ♆
11 11	☽ ± ☉
15 35	☽ ± ♇
19 19	☽ ⊼ ♃
21 33	☽ ⚹ ♄
23 05	☽ ⚼ ☿
23 18	☽ ⊼ ☉
20 Friday	
00 15	☽ # ♀
00 38	☽ ☍ ♀
08 28	☽ ⚼ ♇
08 58	☽ △ ♂
13 06	☽ □ ♅
13 20	☽ ∥ ♀
18 52	☽ ± ☿
20 14	☽ Q ♄
21 Saturday	
04 43	☿ Q ♄
05 12	☿ ∠ ♀
08 10	☽ ⊼ ☿
13 00	☽ △ ♇
18 19	☽ ☍ ♆
22 04	☉ Q ♇
22 Sunday	
03 53	☽ # ♆
05 28	☽ ☍ ♃
06 50	☽ □ ♄
13 21	☽ ☍ ☉
16 03	☽ ⊼ ♀
19 33	☿ □ ♇
21 36	☽ □ ♂
22 14	☽ # ♃
22 52	☽ △ ♅
23 Monday	
05 02	☽ ± ♀
09 28	☿ # ♂
11 58	♀ ∠ ♇
19 26	☽ # ☉
23 11	♂ ⊼ ♅
23 26	☽ □ ♇
24 Tuesday	
00 42	☽ ⚼ ♀
03 45	☽ ☍ ☿
04 26	☽ ⚼ ♅
05 02	☽ ⊼ ♆
09 09	☿ ∠ ♅
11 00	☽ Q ♂
13 53	☿ ⊻ ♆
17 00	☽ ± ♆
17 33	☽ ⊼ ♃
17 58	☽ △ ♄
22 23	☽ ± ♅
25 Wednesday	
05 34	☿ ∥ ♄
05 41	☽ ⊼ ☉
05 46	☽ ± ♃
09 55	☽ △ ♀
10 25	☽ ⊼ ♅
11 02	☽ ⚼ ♆
11 19	☽ Q ♇
12 18	☽ ⚹ ♂
15 05	♀ □ ♅
18 47	☽ ± ☉
19 35	♃ □ ♄
26 Thursday	
00 08	☽ ⚼ ♄
00 12	☽ ⚼ ♃
11 29	☽ ⚹ ♇
14 33	☽ ⚼ ☉
17 20	☽ △ ♆
18 22	☽ ± ♄
20 15	☽ ∠ ♂
27 Friday	
00 28	☽ ⊼ ☿
06 33	☽ ⊼ ♄
07 08	☽ △ ♃
09 04	♀ ⚹ ♂
14 11	☽ ± ☿
15 29	☽ ⊥ ♂
17 51	☽ ∠ ♇
18 01	☉ ⚹ ♅
23 09	☽ ☍ ♅
23 36	☽ △ ☉
28 Saturday	
04 19	☽ ⊻ ♂
05 21	☽ □ ♀
10 39	☽ ⚼ ☿
12 04	☽ ⊥ ♇
23 07	☿ ⊥ ♆
29 Sunday	
00 10	☽ ⊻ ♇
06 07	☽ □ ♆
10 02	☿ ⚹ ♄
10 29	☿ # ♇
10 55	♂ Q ♆
19 15	☽ ☍ ♄
20 13	☽ △ ☿
20 47	☽ □ ♃
21 36	☽ Q ♀
30 Monday	
03 31	☿ ⊻ ♃
11 27	☽ ⊼ ♅
16 54	☽ □ ☉
17 58	☽ Q ♆
18 19	☽ # ☉
19 34	☽ ☌ ♂
23 16	☽ ± ♅
23 53	☽ ⚹ ♀

All ephemeris data is given at 12.00 UT and the Moon's longitude is additionally given for 24.00 UT

MAY 2035

LONGITUDES

Date	Sidereal time h m s	Sun ☉ ° ' "	Moon ☽ ° ' "	Moon ☽ 24.00 ° ' "	Mercury ☿ ° '	Venus ♀ ° '	Mars ♂ ° '	Jupiter ♃ ° '	Saturn ♄ ° '	Uranus ♅ ° '	Neptune ♆ ° '	Pluto ♇ ° '
01	02 37 01	10 ♉ 57 39	20 ♒ 00 14	26 ♒ 16 05	01 ♊ 34	14 ♈ 22	11 ♒ 59	00 ♉ 24	29 ♋ 20	07 ♋ 28	22 ♈ 46	19 ♒ 44
02	02 40 58	11 55 55	02 ♓ 37 12	09 ♓ 04 05	02 34	15 35	12 37	00 39	29 24	07 30	22 48	19 45
03	02 44 54	12 54 08	15 ♓ 37 16	22 ♓ 17 06	03 29	16 48	13 15	00 53	29 27	07 32	22 50	19 45
04	02 48 51	13 52 20	29 ♓ 03 54	05 ♈ 57 46	04 19	18 01	13 53	01 07	29 31	07 35	22 52	19 46
05	02 52 47	14 50 31	12 ♈ 58 40	20 ♈ 06 24	05 04	19 14	14 31	01 22	29 35	07 37	22 54	19 46
06	02 56 44	15 48 40	27 ♈ 20 31	04 ♉ 40 23	05 44	20 27	15 09	01 36	29 38	07 40	22 57	19 47
07	03 00 40	16 46 48	12 ♉ 05 10	19 ♉ 33 52	06 19	21 40	15 46	01 50	29 42	07 42	22 59	19 47
08	03 04 37	17 44 54	27 ♉ 05 18	04 ♊ 38 14	06 49	22 53	16 24	02 04	29 46	07 45	23 01	19 47
09	03 08 33	18 42 58	12 ♊ 11 23	19 ♊ 43 27	07 13	24 06	17 01	02 18	29 50	07 47	23 03	19 48
10	03 12 30	19 41 01	27 ♊ 13 17	04 ♋ 39 47	07 32	25 19	17 39	02 32	29 54	07 50	23 05	19 48
11	03 16 27	20 39 01	12 ♋ 02 05	19 ♋ 19 27	07 46	26 32	18 16	02 46	29 ♋ 58	07 53	23 07	19 48
12	03 20 23	21 37 00	26 ♋ 31 23	03 ♌ 37 34	07 55	27 45	18 53	03 00	00 ♌ 03	07 55	23 09	19 48
13	03 24 20	22 34 57	10 ♌ 37 50	17 ♌ 32 14	07 59	28 ♈ 58	19 30	03 14	00 07	07 58	23 11	19 49
14	03 28 16	23 32 53	24 ♌ 20 53	01 ♍ 04 01	07 R 58	00 ♉ 11	20 07	03 28	00 12	08 01	23 13	19 49
15	03 32 13	24 30 46	07 ♍ 41 57	14 ♍ 15 04	07 52	01 24	20 44	03 42	00 16	08 04	23 15	19 49
16	03 36 09	25 28 37	20 ♍ 43 44	27 ♍ 08 21	07 42	02 37	21 21	03 56	00 21	08 07	23 17	19 49
17	03 40 06	26 26 27	03 ♎ 29 19	09 ♎ 47 00	07 27	03 50	21 58	04 10	00 25	08 10	23 19	19 49
18	03 44 02	27 24 15	16 ♎ 01 43	22 ♎ 13 49	07 08	05 03	22 34	04 24	00 30	08 13	23 21	19 49
19	03 47 59	28 22 02	28 ♎ 23 32	04 ♏ 31 09	06 46	06 16	23 11	04 38	00 35	08 15	23 23	19 49
20	03 51 56	29 ♉ 19 47	10 ♏ 36 50	16 ♏ 40 47	06 20	07 29	23 47	04 52	00 40	08 18	23 25	19 49
21	03 55 52	00 ♊ 17 30	22 ♏ 43 10	28 ♏ 44 08	05 52	08 42	24 24	05 05	00 45	08 21	23 27	19 R 49
22	03 59 49	01 15 12	04 ♐ 43 48	10 ♐ 42 20	05 21	09 55	25 00	05 19	00 50	08 24	23 29	19 49
23	04 03 45	02 12 53	16 ♐ 39 53	22 ♐ 36 37	04 48	11 08	25 36	05 33	00 55	08 28	23 31	19 49
24	04 07 42	03 10 33	28 ♐ 32 45	04 ♑ 28 32	04 15	12 21	26 12	05 46	01 00	08 31	23 32	19 49
25	04 11 38	04 08 12	10 ♑ 24 12	16 ♑ 20 07	03 40	13 34	26 48	06 00	01 06	08 34	23 34	19 49
26	04 15 35	05 05 49	22 ♑ 16 36	28 ♑ 14 06	03 06	14 47	27 24	06 13	01 11	08 37	23 36	19 49
27	04 19 31	06 03 25	04 ♒ 13 03	10 ♒ 13 58	02 33	16 00	27 59	06 27	01 17	08 40	23 38	19 49
28	04 23 28	07 01 01	16 ♒ 17 23	22 ♒ 23 53	02 00	17 13	28 35	06 40	01 22	08 43	23 40	19 49
29	04 27 25	07 58 35	28 ♒ 34 04	04 ♓ 48 32	01 30	18 26	29 10	06 53	01 28	08 46	23 41	19 48
30	04 31 21	08 56 09	11 ♓ 07 56	17 ♓ 32 50	01 02	19 39	29 ♒ 45	07 07	01 33	08 50	23 43	19 48
31	04 35 18	09 ♊ 53 41	24 ♓ 03 49	00 ♈ 41 24	00 ♊ 36	20 ♉ 53	00 ♓ 20	07 ♉ 20	01 ♌ 39	08 ♋ 53	23 ♈ 45	19 ♒ 48

Date	Moon True ☊ ° '	Moon Mean ☊ ° '	Moon ☽ Latitude ° '
01	12 ♍ 48	11 ♍ 44	01 N 59
02	12 D 49	11 41	00 N 54
03	12 R 49	11 37	00 S 15
04	12 48	11 34	01 26
05	12 45	11 31	02 34
06	12 39	11 28	03 34
07	12 31	11 25	04 22
08	12 22	11 22	04 52
09	12 12	11 18	05 03
10	12 03	11 15	04 52
11	11 56	11 12	04 22
12	11 51	11 09	03 36
13	11 49	11 06	02 38
14	11 D 49	11 02	01 31
15	11 49	10 59	00 S 22
16	11 R 49	10 56	00 N 47
17	11 47	10 53	01 52
18	11 43	10 50	02 51
19	11 36	10 47	03 40
20	11 26	10 43	04 18
21	11 14	10 40	04 45
22	11 01	10 37	04 58
23	10 47	10 34	04 58
24	10 34	10 31	04 46
25	10 23	10 28	04 20
26	10 15	10 24	03 44
27	10 09	10 21	02 57
28	10 06	10 18	02 02
29	10 D 05	10 15	01 N 01
30	10 R 05	10 12	00 S 06
31	10 ♍ 04	10 ♍ 08	01 S 13

DECLINATIONS

Date	Sun ☉ ° '	Moon ☽ ° '	Mercury ☿ ° '	Venus ♀ ° '	Mars ♂ ° '	Jupiter ♃ ° '	Saturn ♄ ° '	Uranus ♅ ° '	Neptune ♆ ° '	Pluto ♇ ° '
01	15 N 07	12 S 56	23 N 05	04 N 09	18 S 42	10 N 38	20 N 46	23 N 33	07 N 21	22 S 31
02	15 25	09 42	23 17	04 37	18 33	10 43	20 45	23 33	07 22	22 31
03	15 42	05 54	23 28	05 05	18 24	10 48	20 44	23 33	07 23	22 31
04	16 00	01 S 41	23 36	05 32	18 14	10 53	20 44	23 33	07 24	22 32
05	16 17	02 N 46	23 41	06 00	18 05	10 58	20 43	23 32	07 25	22 32
06	16 34	07 12	23 45	06 28	17 56	11 03	20 42	23 32	07 25	22 32
07	16 51	11 18	23 46	06 55	17 46	11 08	20 42	23 32	07 26	22 32
08	17 07	14 45	23 46	07 22	17 37	11 13	20 41	23 32	07 27	22 32
09	17 23	17 15	23 43	07 50	17 27	11 18	20 40	23 32	07 28	22 32
10	17 39	18 32	23 38	08 17	17 18	11 23	20 39	23 32	07 28	22 32
11	17 55	18 32	23 31	08 43	17 08	11 27	20 38	23 31	07 29	22 33
12	18 10	17 18	23 22	09 10	16 58	11 32	20 37	23 31	07 30	22 33
13	18 25	15 02	23 11	09 37	16 49	11 37	20 36	23 31	07 30	22 33
14	18 39	11 58	22 59	10 03	16 39	11 42	20 36	23 31	07 31	22 33
15	18 54	08 20	22 45	10 29	16 29	11 46	20 35	23 31	07 32	22 33
16	19 08	04 24	22 29	10 55	16 19	11 51	20 34	23 31	07 33	22 34
17	19 21	00 N 20	22 11	11 20	16 09	11 56	20 33	23 30	07 33	22 34
18	19 34	03 S 41	21 52	11 46	15 59	12 00	20 32	23 30	07 34	22 34
19	19 47	07 28	21 32	12 11	15 49	12 05	20 31	23 30	07 35	22 34
20	20 00	10 54	21 11	12 35	15 38	12 09	20 30	23 30	07 35	22 35
21	20 12	13 51	20 49	13 00	15 28	12 14	20 29	23 30	07 36	22 35
22	20 24	16 11	20 26	13 24	15 18	12 18	20 28	23 29	07 37	22 35
23	20 36	17 49	20 03	13 48	15 08	12 23	20 27	23 29	07 37	22 36
24	20 47	18 40	19 40	14 12	14 58	12 27	20 25	23 29	07 38	22 36
25	20 58	18 42	19 17	14 35	14 47	12 32	20 24	23 29	07 39	22 36
26	21 09	17 54	18 54	14 58	14 37	12 36	20 23	23 29	07 39	22 36
27	21 19	16 19	18 32	15 20	14 27	12 41	20 22	23 28	07 40	22 37
28	21 29	14 00	18 10	15 43	14 16	12 45	20 21	23 28	07 41	22 37
29	21 38	11 01	17 50	16 04	14 06	12 49	20 20	23 28	07 41	22 37
30	21 47	07 28	17 31	16 26	13 56	12 54	20 18	23 28	07 42	22 38
31	21 N 56	03 S 29	17 N 14	16 N 47	13 S 46	12 N 58	20 N 17	23 N 28	07 N 42	22 S 38

ZODIAC SIGN ENTRIES

Date	h	m	Planets
02	07	04	☽ ♓
04	13	38	☽ ♈
06	16	22	☽ ♉
08	16	38	☽ ♊
10	16	28	☽ ♋
11	20	45	♄ ♌
12	17	51	☽ ♌
14	08	26	♀ ♉
14	22	05	☽ ♍
17	05	24	☽ ♎
19	15	09	☽ ♏
21	04	43	☉ ♊
22	02	32	☽ ♐
24	14	57	☽ ♑
27	03	33	☽ ♒
29	14	46	☽ ♓
30	22	08	♂ ♓
31	22	46	☽ ♈

LATITUDES

Date	Mercury ☿ ° '	Venus ♀ ° '	Mars ♂ ° '	Jupiter ♃ ° '	Saturn ♄ ° '	Uranus ♅ ° '	Neptune ♆ ° '	Pluto ♇ ° '
01	02 N 40	01 S 39	01 S 34	01 S 02	00 N 30	00 N 20	01 S 37	08 S 03
04	02 38	01 39	01 39	01 02	00 30	00 20	01 37	08 04
07	02 27	01 39	01 45	01 02	00 30	00 20	01 37	08 05
10	02 06	01 38	01 51	01 02	00 30	00 20	01 37	08 06
13	01 35	01 36	01 56	01 02	00 30	00 20	01 37	08 07
16	00 54	01 34	02 02	01 02	00 30	00 20	01 37	08 07
19	00 N 06	01 31	02 09	01 02	00 30	00 20	01 37	08 08
22	00 S 46	01 27	02 15	01 02	00 30	00 20	01 37	08 09
25	01 38	01 23	02 21	01 03	00 31	00 20	01 37	08 10
28	02 26	01 19	02 28	01 03	00 31	00 20	01 37	08 11
31	03 S 07	01 S 14	02 S 35	01 S 03	00 N 31	00 N 20	01 S 37	08 S 12

DATA

Julian Date	2464449
Delta T	+76 seconds
Ayanamsa	24° 20' 56"
Synetic vernal point	04° ♓ 46' 03"
True obliquity of ecliptic	23° 25' 56"

MOON'S PHASES, APSIDES AND POSITIONS ☽

Date	h	m	Phase	Longitude ° '	Eclipse Indicator
07	20	04	●	17 ♉ 06	
14	10	28	☽	23 ♌ 29	
22	04	26	○	00 ♐ 57	
30	07	31	☾	08 ♓ 45	

Day	h	m		
09	03	00	Perigee	
24	09	05	Apogee	
04	21	09	0N	
10	23	58	Max dec	18° N 41'
17	13	57	0S	
25	00	56	Max dec	18° S 47'

LONGITUDES

Date	Chiron ⚷ ° '	Ceres ⚳ ° '	Pallas ⚴ ° '	Juno ⚵ ° '	Vesta ⚶ ° '	Black Moon Lilith ⚸ ° '
01	04 ♊ 46	07 ♉ 40	23 ♓ 07	26 ♒ 58	18 ♋ 01	20 ♐ 56
11	05 ♊ 30	11 ♉ 41	25 ♓ 51	29 ♒ 51	21 ♋ 32	22 ♐ 03
21	06 ♊ 16	15 ♉ 42	28 ♓ 25	02 ♓ 28	25 ♋ 16	23 ♐ 11
31	07 ♊ 02	19 ♉ 41	00 ♈ 47	04 ♓ 45	29 ♋ 12	24 ♐ 18

All ephemeris data is given at 12.00 UT and the Moon's longitude is additionally given for 24.00 UT

ASPECTARIAN

h m	Aspects
01 Tuesday	
08 51	☽ Q ♃
09 25	☿ ⊥ ♅
11 29	☽ ☌ ♇
16 44	☽ ⚻ ♅
17 19	☽ ⚹ ♆
02 Wednesday	
04 59	☽ # ♃
05 54	☽ ⊼ ♄
06 31	☽ Q ☉
07 47	☽ ∠ ♀
08 13	☽ ⚹ ♃
11 54	☽ □ ☿
17 13	☽ ± ♄
21 07	☽ △ ♅
21 41	☽ ∠ ♆
03 Thursday	
02 18	☽ ⊥ ♀
03 00	☽ # ♆
06 39	☽ ⚹ ☉
07 28	☽ ⊻ ♂
09 52	☽ ⚻ ♄
12 29	☽ ∠ ♃
14 12	☽ ⊥ ♆
14 22	☽ ⊻ ♀
16 21	☽ # ♀
18 53	☽ ⊥ ♂
19 28	☽ ⊻ ♇
23 18	☽ Q ☿
23 34	♂ ± ♅
04 Friday	
01 01	☽ ⊻ ♆
01 44	☿ ∥ ♅
04 55	☽ ⊥ ♃
06 11	☽ ⊥ ♇
11 38	☽ ∠ ☉
11 40	☽ ∠ ♂
12 47	☽ △ ♄
12 48	☉ □ ♂
15 40	☽ ⊻ ♃
21 42	☽ ⚹ ☿
21 56	☽ ∠ ♇
05 Saturday	
02 49	☽ □ ♅
04 25	☽ ⊥ ☉
14 43	☽ ⚹ ♂
15 23	☽ ⊻ ☉
22 36	♀ ⚹ ♇
23 26	☽ ⚹ ♇
23 31	☽ ☌ ♀
06 Sunday	
00 32	☽ ∠ ☿
04 42	☽ ☌ ♆
07 28	☽ ∥ ♀
09 13	☽ Q ♅
11 39	☽ Q ♂
13 16	☽ ∥ ♆
15 47	☽ □ ♄
16 07	☽ ⊥ ☿
19 06	☽ ☌ ♃
19 17	☽ Q ♇
23 13	☿ ∠ ♀
07 Monday	
02 19	☽ ⊻ ☿
04 54	☽ ⚹ ♅
06 43	♃ Q ♇
10 56	☽ ∥ ♃
18 11	☽ □ ♂
20 04	☽ ☌ ☉
21 04	☽ Q ♄
08 Tuesday	
00 21	☽ □ ♇
04 43	☽ ⊻ ♀
05 04	☽ ∠ ♅
05 30	☽ ⊻ ♆
12 32	☉ Q ♄
14 38	♀ ☌ ♆
15 04	☽ ⊥ ♆
15 06	☽ ⊥ ♀
16 03	♀ ∥ ♆
16 17	☽ ⚹ ♄
19 26	☽ ⊥ ♅
20 03	☽ ⊻ ♃
09 Wednesday	
03 54	☽ ☌ ☿
04 59	☽ ⊻ ♅
05 24	☽ ∠ ♆
05 44	☽ ⊥ ♃
06 40	☽ ∠ ♀
14 10	☽ ∥ ☉
14 33	☽ # ♂
15 55	☉ # ♂
16 14	☽ ∠ ♄
20 01	☽ △ ♂
20 16	☽ ∠ ♃
23 06	☽ ⊻ ☉
10 Thursday	
00 07	☽ △ ♇
05 21	☽ ⚹ ♆
06 39	☽ ⊥ ♄
08 40	☽ ⚹ ♀
09 22	☽ ⊥ ☉
14 51	☉ □ ♇
16 20	☽ ⊻ ♄
20 42	☽ ⚹ ♃
21 07	☽ ⚻ ♂
22 38	♀ Q ♅
11 Friday	
00 13	☽ ⚻ ♇
00 43	☽ Q ♆
00 52	☽ ∠ ☉
04 57	☽ ⊻ ☿
05 12	☽ ☌ ♅
05 46	☽ Q ♀
10 25	☿ ∥ ♅
12 24	☽ ± ♂
14 53	☽ ⊥ ☿
14 54	☽ ± ♇
16 34	☽ Q ♃
22 42	☽ ⊼ ♂
12 Saturday	
00 25	☽ ∥ ☉
00 48	☽ ⊼ ♇
03 13	☽ ⚹ ☉
05 56	☽ ∠ ☿
06 21	☽ □ ♆
12 23	☿ ⊻ ♅
14 15	☽ □ ♀
16 39	☽ # ♂
17 58	☽ ☌ ♄
23 08	☽ □ ♃
13 Sunday	
00 52	☽ Q ☉
07 24	☽ ⊻ ♅
07 27	☽ ⚹ ☿
17 48	☽ ⊥ ♅
18 41	☿ St R
22 08	☉ ∠ ♅
22 15	☿ ⊻ ♅
23 58	♂ ☌ ♇
14 Monday	
03 32	☉ ⊻ ♆
03 59	☽ ☍ ♇
04 11	☽ ☍ ♂
04 18	☽ Q ☿
09 38	☽ ∠ ♅
10 00	☽ △ ♆
10 28	☽ □ ☉
12 15	♀ □ ♄
13 52	☽ ∥ ♃
22 29	☽ ⊻ ♄
23 27	☽ △ ♀
23 40	☽ ∥ ♀
15 Tuesday	
04 38	☽ △ ♃
09 23	☽ ⊥ ♄
12 19	☽ □ ☿
12 40	☽ ⚹ ♅
13 01	☽ ⚻ ♆
17 01	☽ ∥ ♆
20 16	♀ Q ♇
20 21	☿ ⊥ ♀
16 Wednesday	
01 58	☽ ∠ ♄
04 45	☿ # ♇
05 35	☽ ± ♆
05 37	☽ ⚻ ♀
08 37	☽ ⚻ ♃
10 18	☽ ⊼ ♇
10 51	☽ Q ♅
13 13	☽ ⊼ ♂
16 47	☽ ⊼ ♆
21 31	☽ ± ♇
21 36	☽ △ ☉
17 Thursday	
00 10	☽ ± ♀
01 01	☽ ± ♂
01 45	☽ ± ♃
06 09	☽ ⚹ ♄
12 43	☽ ⊼ ♀
13 19	☽ ⊼ ♃
14 32	☽ ⚻ ♇
17 13	♀ Q ♂
18 57	☽ ⚻ ♂
19 22	☽ △ ☿
20 16	♀ ☌ ♃
20 56	☽ □ ♅
18 Friday	
01 07	♂ Q ♃
04 27	☽ ⚻ ☉
05 10	☽ Q ♄
19 20	☽ △ ♇
23 16	☽ ± ☉
23 29	☽ ⚻ ☿
19 Saturday	
01 19	☽ △ ♂
02 13	☽ ☍ ♆
05 10	♀ ∥ ♃
11 57	☽ ⊼ ☉
12 42	☽ # ♆
15 15	♂ ⚻ ♅
16 19	☽ □ ♄
16 29	☽ ± ☿
19 22	☿ ⊻ ♀
20 26	♂ ⚹ ♆
20 Sunday	
00 27	☽ ☍ ♃
03 52	☽ ⊼ ☿
05 08	☽ ☍ ♀
07 26	☽ △ ♅
14 14	☉ ⊥ ♆
21 54	☽ # ♃
22 58	♇ St R
21 Monday	
03 20	☽ # ♀
05 00	♀ ⚹ ♅
06 14	☽ □ ♇
13 16	☽ ⚻ ♅
13 27	☽ ⊼ ♆
15 31	☽ □ ♂
22 Tuesday	
00 32	☉ ⚹ ♄
01 28	☽ ± ♆
02 46	☽ ∥ ♂
04 09	☽ △ ♄
04 26	☽ ☍ ☉
07 20	☽ ± ♅
10 41	☿ ∥ ♄
12 57	☿ ⊻ ♃
13 11	☽ ☍ ☿
13 12	☽ ⊼ ♃
13 27	☉ ∥ ☿
18 11	☉ ∥ ♄
18 12	☽ Q ♇
19 25	☽ ⊼ ♅
19 33	☽ ⚻ ♆
23 36	☽ ⊼ ♀
23 Wednesday	
01 29	☽ ± ♃
05 30	☽ Q ♂
10 29	☽ ⚻ ♄
13 03	☽ ± ♀
18 22	☽ ⚹ ♇
18 26	☉ ⊥ ♅
19 59	☽ ⚻ ♃
24 Thursday	
01 51	☽ △ ♆
04 48	☽ ± ♄
07 00	☽ ⚹ ♂
09 18	☽ ⚻ ♀
17 01	☽ ⊼ ♄
22 12	☽ ⊼ ☉
23 00	☽ ⊼ ☿
25 Friday	
00 42	☽ ∠ ♇
02 54	☽ △ ♃
04 45	☉ ☌ ☿
08 15	☽ ☍ ♅
10 35	☽ ± ☿
11 25	☽ ± ☉
14 58	☽ ∠ ♂
18 54	☽ ⊥ ♇
19 08	☽ △ ♀
20 57	♀ # ♂
26 Saturday	
03 58	☽ ⚻ ☿
07 02	☽ ⊻ ♇
07 13	☽ ⚻ ☉
10 07	☽ ⊥ ♂
14 41	☽ □ ♆
22 51	☽ ⊻ ♂
27 Sunday	
06 04	☽ ☍ ♄
07 12	☿ ⊥ ♅
08 48	☽ △ ☿
16 00	☽ △ ☉
16 32	☽ □ ♃
20 55	☽ ⊼ ♅
21 31	☽ # ♀
28 Monday	
00 38	☉ ⊻ ♃
02 49	☽ Q ♆
08 53	☽ ± ♅
09 21	☽ ∥ ♂
14 02	☽ □ ♀
18 56	☽ ☌ ♇
22 28	☽ # ♃
29 Tuesday	
02 30	☽ ⚹ ♆
02 39	☽ ⚻ ♅
04 44	☽ Q ♃
13 13	☽ ☌ ♂
13 30	☿ ⚹ ♄
17 26	☽ □ ☿
17 37	☽ ⊼ ♄
30 Wednesday	
04 15	☽ ⚹ ♃
04 43	☽ Q ♀
05 11	☽ ± ♄
06 26	☉ ∠ ♆
07 26	☽ ∠ ♆
07 31	☽ □ ☉
07 38	☽ △ ♅
09 07	☉ ⊻ ♅
09 50	♀ Q ♄
10 36	☽ # ♆
14 51	♀ □ ♇
22 14	☽ ⚻ ♄
31 Thursday	
00 21	☽ ⊥ ♆
02 16	☽ Q ☿
04 10	☽ ⊻ ♇
05 33	☽ ⚹ ♀
08 47	☽ ∠ ♃
11 25	☽ ⊻ ♆
15 10	☽ ⊥ ♇
18 26	☿ □ ♂
19 30	☽ Q ☉
23 30	☽ ⚹ ☿
23 53	☽ ⊻ ♂

JUNE 2035

LONGITUDES

Date	Sidereal time h m s	Sun ☉ ° ' "	Moon ☽ ° ' "	Moon ☽ 24.00 ° ' "	Mercury ☿ ° '	Venus ♀ ° '	Mars ♂ ° '	Jupiter ♃ ° '	Saturn ♄ ° '	Uranus ♅ ° '	Neptune ♆ ° '	Pluto ♇ ° '
01	04 39 14	10 ♊ 51 13	07 ♈ 25 59	14 ♈ 17 51	00 ♊ 13	22 ♉ 06	00 ♓ 55	07 ♉ 33	01 ♌ 45	08 ♋ 56	23 ♈ 47	19 ♒ 48
02	04 43 11	11 48 44	21 ♈ 17 11	28 ♈ 23 54	29 ♉ 54	23 19	01 30	07 46	01 51	08 59	23 48	19 R 47
03	04 47 07	12 46 15	05 ♉ 37 47	12 ♉ 58 22	29 R 39	24 32	02 04	07 59	01 57	09 03	23 50	19 47
04	04 51 04	13 43 44	20 ♉ 24 54	27 ♉ 56 30	29 28	25 45	02 39	08 12	02 03	09 06	23 52	19 47
05	04 55 00	14 41 13	05 ♊ 31 58	13 ♊ 10 02	29 21	26 58	03 13	08 25	02 09	09 09	23 53	19 46
06	04 58 57	15 38 41	20 ♊ 49 16	28 ♊ 28 12	29 18	28 11	03 47	08 38	02 15	09 13	23 55	19 46
07	05 02 54	16 36 08	06 ♋ 05 25	13 ♋ 39 36	29 D 20	29 ♉ 25	04 21	08 51	02 21	09 16	23 56	19 45
08	05 06 50	17 33 34	21 ♋ 09 32	28 ♋ 34 17	29 26	00 ♊ 38	04 55	09 04	02 27	09 20	23 58	19 45
09	05 10 47	18 30 59	05 ♌ 53 03	13 ♌ 05 19	29 37	01 51	05 28	09 17	02 33	09 23	23 59	19 44
10	05 14 43	19 28 22	20 ♌ 10 47	27 ♌ 09 19	29 ♉ 52	03 04	06 01	09 29	02 40	09 27	24 01	19 44
11	05 18 40	20 25 45	04 ♍ 01 01	10 ♍ 46 03	00 ♊ 12	04 17	06 34	09 42	02 46	09 30	24 02	19 43
12	05 22 36	21 23 07	17 ♍ 24 47	23 ♍ 57 36	00 36	05 31	07 07	09 54	02 52	09 33	24 04	19 43
13	05 26 33	22 20 27	00 ♎ 24 59	06 ♎ 47 26	01 04	06 44	07 40	10 07	02 59	09 37	24 05	19 42
14	05 30 29	23 17 47	13 ♎ 05 27	19 ♎ 19 32	01 37	07 57	08 12	10 19	03 05	09 40	24 06	19 42
15	05 34 26	24 15 05	25 ♎ 30 13	01 ♏ 37 56	02 14	09 10	08 44	10 31	03 12	09 44	24 08	19 41
16	05 38 23	25 12 23	07 ♏ 43 07	13 ♏ 46 10	02 55	10 24	09 16	10 44	03 19	09 47	24 09	19 40
17	05 42 19	26 09 40	19 ♏ 47 27	25 ♏ 47 16	03 40	11 37	09 48	10 56	03 25	09 51	24 10	19 40
18	05 46 16	27 06 56	01 ♐ 45 55	07 ♐ 43 39	04 30	12 50	10 19	11 08	03 32	09 55	24 12	19 39
19	05 50 12	28 04 12	13 ♐ 40 40	19 ♐ 37 12	05 23	14 03	10 51	11 20	03 39	09 58	24 13	19 38
20	05 54 09	29 01 27	25 ♐ 33 25	01 ♑ 29 30	06 20	15 17	11 22	11 32	03 46	10 02	24 14	19 38
21	05 58 05	29 ♊ 58 41	07 ♑ 25 38	13 ♑ 22 00	07 21	16 30	11 52	11 43	03 52	10 05	24 15	19 37
22	06 02 02	00 ♋ 55 55	19 ♑ 18 50	25 ♑ 16 21	08 26	17 43	12 23	11 55	03 59	10 09	24 16	19 36
23	06 05 58	01 53 09	01 ♒ 14 48	07 ♒ 14 30	09 34	18 57	12 53	12 07	04 06	10 12	24 18	19 35
24	06 09 55	02 50 23	13 ♒ 15 47	19 ♒ 19 00	10 46	20 10	13 23	12 18	04 13	10 16	24 19	19 34
25	06 13 52	03 47 36	25 ♒ 24 36	01 ♓ 32 59	12 02	21 23	13 53	12 30	04 20	10 20	24 20	19 34
26	06 17 48	04 44 49	07 ♓ 44 41	14 ♓ 00 10	13 21	22 37	14 22	12 41	04 27	10 23	24 21	19 33
27	06 21 45	05 42 02	20 ♓ 19 58	26 ♓ 44 37	14 44	23 50	14 51	12 52	04 34	10 27	24 22	19 32
28	06 25 41	06 39 15	03 ♈ 14 37	09 ♈ 50 27	16 10	25 03	15 19	13 04	04 41	10 30	24 23	19 31
29	06 29 38	07 36 28	16 ♈ 32 32	23 ♈ 21 13	17 39	26 17	15 48	13 15	04 49	10 34	24 24	19 30
30	06 33 34	08 ♋ 33 41	00 ♉ 16 43	07 ♉ 19 07	19 ♊ 12	27 ♊ 30	16 ♓ 16	13 ♉ 26	04 ♌ 56	10 ♋ 38	24 ♈ 25	19 ♒ 29

Date	Moon True ☊ ° '	Moon Mean ☊ ° '	Moon ☽ Latitude ° '
01	10 ♍ 03	10 ♍ 05	02 S 19
02	09 R 59	10 02	03 19
03	09 53	09 59	04 09
04	09 44	09 56	04 44
05	09 34	09 53	05 00
06	09 23	09 49	04 55
07	09 13	09 46	04 29
08	09 05	09 43	03 44
09	08 59	09 40	02 45
10	08 56	09 37	01 37
11	08 55	09 33	00 S 26
12	08 D 55	09 30	00 N 45
13	08 R 55	09 27	01 51
14	08 54	09 24	02 50
15	08 50	09 21	03 40
16	08 44	09 18	04 19
17	08 35	09 14	04 46
18	08 24	09 11	04 59
19	08 12	09 08	05 00
20	07 59	09 05	04 48
21	07 47	09 02	04 23
22	07 37	08 59	03 46
23	07 29	08 55	03 00
24	07 24	08 52	02 05
25	07 21	08 49	01 N 03
26	07 D 21	08 46	00 S 02
27	07 21	08 43	01 09
28	07 22	08 39	02 14
29	07 R 21	08 36	03 14
30	07 ♍ 19	08 ♍ 33	04 S 04

DECLINATIONS

Date	Sun ☉ ° '	Moon ☽ ° '	Mercury ☿ ° '	Venus ♀ ° '	Mars ♂ ° '	Jupiter ♃ ° '	Saturn ♄ ° '	Uranus ♅ ° '	Neptune ♆ ° '	Pluto ♇ ° '
01	22 N 04	00 N 49	16 N 58	17 N 08	13 S 35	13 N 02	20 N 16	23 N 27	07 N 43	22 S 38
02	22 12	05 13	16 44	17 28	13 25	13 06	20 15	23 27	07 43	22 39
03	22 19	09 29	16 32	17 47	13 15	13 10	20 13	23 27	07 44	22 39
04	22 26	13 17	16 22	18 07	13 04	13 15	20 12	23 27	07 45	22 40
05	22 33	16 18	16 14	18 26	12 54	13 19	20 11	23 26	07 45	22 40
06	22 40	18 13	16 08	18 44	12 44	13 23	20 10	23 26	07 46	22 40
07	22 45	18 49	16 05	19 02	12 34	13 27	20 08	23 26	07 46	22 41
08	22 51	18 05	16 03	19 19	12 24	13 31	20 07	23 26	07 47	22 41
09	22 56	16 08	16 03	19 36	12 14	13 35	20 05	23 26	07 47	22 41
10	23 01	13 13	16 06	19 52	12 03	13 39	20 04	23 25	07 48	22 42
11	23 05	09 38	16 10	20 08	11 53	13 43	20 03	23 25	07 48	22 42
12	23 09	05 40	16 17	20 24	11 43	13 46	20 01	23 25	07 49	22 43
13	23 13	01 N 32	16 25	20 38	11 33	13 50	20 00	23 25	07 49	22 43
14	23 16	02 S 33	16 34	20 52	11 24	13 54	19 58	23 24	07 50	22 44
15	23 18	06 26	16 46	21 06	11 14	13 58	19 57	23 24	07 50	22 44
16	23 21	10 00	16 58	21 19	11 04	14 01	19 55	23 24	07 50	22 44
17	23 23	13 05	17 13	21 32	10 54	14 05	19 54	23 24	07 51	22 45
18	23 24	15 37	17 28	21 43	10 45	14 09	19 52	23 23	07 51	22 45
19	23 25	17 28	17 44	21 55	10 35	14 12	19 51	23 23	07 52	22 46
20	23 26	18 34	18 02	22 05	10 26	14 16	19 49	23 23	07 52	22 46
21	23 26	18 51	18 20	22 15	10 16	14 20	19 48	23 22	07 53	22 47
22	23 26	18 18	18 39	22 25	10 07	14 23	19 46	23 22	07 53	22 47
23	23 25	16 57	18 59	22 34	09 58	14 27	19 45	23 22	07 53	22 48
24	23 24	14 50	19 19	22 42	09 49	14 30	19 43	23 22	07 54	22 48
25	23 23	12 03	19 40	22 49	09 40	14 33	19 41	23 21	07 54	22 49
26	23 21	08 42	20 00	22 56	09 31	14 37	19 40	23 21	07 54	22 49
27	23 19	04 53	20 21	23 02	09 22	14 40	19 38	23 21	07 55	22 50
28	23 16	00 S 46	20 42	23 08	09 13	14 43	19 36	23 21	07 55	22 50
29	23 13	03 N 31	21 02	23 13	09 05	14 46	19 35	23 20	07 55	22 51
30	23 N 09	07 N 45	21 N 22	23 N 17	08 S 56	14 N 50	19 N 33	23 N 20	07 N 55	22 S 51

ZODIAC SIGN ENTRIES

Date	h	m	Planets
02	04	33	☿ ♉
03	02	40	☽ ♉
05	03	16	☽ ♊
07	02	24	☽ ♋
07	23	37	♀ ♊
09	02	20	☽ ♌
10	22	32	☿ ♊
11	04	57	☽ ♍
13	11	13	☽ ♎
15	20	48	☽ ♏
18	08	27	☽ ♐
20	20	59	☽ ♑
21	12	33	☉ ♋
23	09	30	☽ ♒
25	20	59	☽ ♓
28	06	02	☽ ♈
30	11	31	☽ ♉

LATITUDES

Date	Mercury ☿ ° '	Venus ♀ ° '	Mars ♂ ° '	Jupiter ♃ ° '	Saturn ♄ ° '	Uranus ♅ ° '	Neptune ♆ ° '	Pluto ♇ ° '
01	03 S 18	01 S 12	02 S 37	01 S 03	00 N 31	00 N 20	01 S 37	08 S 12
04	03 45	01 07	02 44	01 03	00 31	00 20	01 38	08 13
07	04 01	01 01	02 51	01 03	00 31	00 20	01 38	08 14
10	04 06	00 55	02 58	01 03	00 31	00 20	01 38	08 14
13	04 02	00 48	03 05	01 04	00 31	00 20	01 38	08 15
16	03 50	00 41	03 13	01 04	00 32	00 20	01 38	08 16
19	03 31	00 34	03 20	01 04	00 32	00 20	01 38	08 17
22	03 05	00 27	03 28	01 04	00 32	00 20	01 39	08 17
25	02 35	00 20	03 36	01 05	00 32	00 20	01 39	08 18
28	02 02	00 13	03 44	01 05	00 32	00 20	01 39	08 19
31	01 S 26	00 S 05	03 S 52	01 S 05	00 N 32	00 N 20	01 S 39	08 S 19

DATA

Julian Date	2464480
Delta T	+76 seconds
Ayanamsa	24° 21' 00"
Synetic vernal point	04° ♓ 45' 59"
True obliquity of ecliptic	23° 25' 56"

MOON'S PHASES, APSIDES AND POSITIONS ☽

Date	h	m	Phase	Longitude ° '	Eclipse Indicator
06	03	21	●	15 ♊ 18	
12	19	50	☽	21 ♍ 42	
20	19	37	○	29 ♐ 20	
28	18	43	☾	06 ♈ 55	

Day	h	m		
06	11	27	Perigee	
20	12	14	Apogee	
01	07	31	ON	
07	10	39	Max dec	18° N 49'
13	20	57	OS	
21	08	16	Max dec	18° S 52'
28	16	17	ON	

LONGITUDES

Date	Chiron ⚷ ° '	Ceres ⚳ ° '	Pallas ⚴ ° '	Juno ⚵ ° '	Vesta ⚶ ° '	Black Moon Lilith ⚸ ° '
01	07 ♊ 06	20 ♉ 05	01 ♈ 00	04 ♓ 58	29 ♋ 36	24 ♐ 25
11	07 ♊ 52	24 ♉ 02	03 ♈ 06	06 ♓ 50	03 ♌ 42	25 ♐ 32
21	08 ♊ 37	27 ♉ 55	04 ♈ 53	08 ♓ 14	07 ♌ 55	26 ♐ 39
31	08 ♊ 20	01 ♊ 44	06 ♈ 19	09 ♓ 06	12 ♌ 16	27 ♐ 46

ASPECTARIAN

h m Aspects

01 Friday
01 22 ☽ ⊥ ♃
01 49 ☽ △ ♄
05 27 ☿ ∥ ♀
07 20 ☽ ∠ ♇
11 03 ☽ ⊥ ♂
11 21 ☽ ∠ ♀
12 13 ☽ ⊻ ♃
14 39 ☽ □ ♅
18 27 ☽ ✶ ☉

02 Saturday
01 17 ☽ ∠ ☿
03 27 ☽ ∠ ♂
04 33 ☽ ⊥ ♀
09 27 ☽ ✶ ♇
15 46 ☽ ⊻ ♀
16 17 ☽ ☌ ♆
16 22 ☽ ⊥ ☿
20 12 ☿ ⊥ ♆
21 41 ☽ Q ♅
21 54 ♀ ⊻ ♆
22 01 ☽ ∠ ☉

03 Sunday
01 57 ☽ ∥ ♆
01 58 ♀ ∠ ♅
02 15 ☽ ⊻ ☿
05 24 ♂ ⊼ ♄
05 39 ☽ Q ♇
05 52 ☽ □ ♄
05 53 ☽ ✶ ♂
14 00 ☽ ⊥ ☉
15 56 ☽ ☌ ♃
17 38 ☽ ✶ ♅
19 07 ♂ # ♃

04 Monday
00 29 ☽ ⊻ ☉
02 20 ☽ Q ♂
10 38 ☽ # ♂
10 59 ☽ □ ♇
11 24 ☽ Q ♄
11 43 ☽ ∥ ♃
17 31 ☽ ⊻ ♆
17 55 ☽ ∠ ♅
21 16 ☽ ☌ ♀

05 Tuesday
02 18 ☽ ☌ ☿
03 04 ☽ ⊥ ♆
03 26 ☉ ⊥ ♃
06 37 ☽ ✶ ♄
08 12 ☽ □ ♂
08 14 ☽ ⊥ ♅
11 24 ☽ ∥ ☿
16 37 ☽ ⊻ ♃
17 17 ☽ ∠ ♆
17 43 ☽ ⊻ ♅

06 Wednesday
02 10 ☽ ⊥ ♃
03 21 ☽ ☌ ☉
06 21 ☽ ∠ ♄
10 21 ☽ △ ♇
14 52 ☿ St D
15 06 ☉ # ♇
16 29 ☽ ∠ ♃
16 51 ☽ ✶ ♆
20 34 ☽ ⊥ ♄

07 Thursday
00 34 ☽ ⊻ ♀
01 19 ☽ ⊻ ☿
06 03 ☽ ⊻ ♄
09 09 ☽ △ ♂
09 54 ☽ Ᵽ ♇
10 23 ☿ ☌ ♀
10 48 ☽ ⊥ ☿
10 50 ☽ ⊥ ♀
11 46 ☽ Q ♆
16 26 ☽ ✶ ♃
17 03 ☽ ☌ ♅
22 37 ♀ ⊥ ♆

08 Friday
00 09 ☽ ± ♇
01 09 ☽ ∠ ☿
02 21 ☽ ∠ ♀
05 50 ☽ ⊻ ☉
08 56 ☉ ∠ ♄
09 44 ☽ ⊼ ♇
09 55 ☽ Ᵽ ♂
11 51 ☽ Q ♃
16 08 ☽ ⊥ ☉
16 32 ☽ □ ♆

09 Saturday
01 03 ☽ ± ♂
01 33 ☽ ✶ ☿
04 45 ☽ ✶ ♀
06 29 ☽ ☌ ♄
07 49 ☽ ∠ ☉
11 17 ☽ ⊼ ♂
12 38 ☽ ∥ ☿
17 43 ☽ □ ♃
17 50 ☽ ⊻ ♅
21 41 ☽ Q ☿

10 Sunday
02 31 ☽ Q ♀
03 10 ♀ ✶ ♄
03 56 ☽ ⊥ ♅
05 02 ♃ ✶ ♅
08 52 ☽ ∥ ♃
10 42 ☽ ✶ ☉
11 14 ☽ ☍ ♇
18 26 ☉ △ ♇
18 35 ☽ △ ♆
19 20 ☽ ∠ ♅
19 41 ♀ ⊥ ♅
20 33 ☽ # ♂

11 Monday
00 25 ☿ ⊥ ♆
04 09 ♀ ∥ ♄
05 07 ☽ □ ☿
09 00 ☽ Q ☉
09 47 ☽ ⊻ ♄
12 32 ☽ □ ♀
16 43 ☽ ☍ ♂
20 29 ☽ ⊥ ♄
20 55 ☽ Ᵽ ♆
21 46 ☽ ✶ ♅
22 14 ☽ △ ♃
23 13 ☽ ∥ ♆

12 Tuesday
12 51 ☽ ∠ ♄
13 11 ☽ ± ♆
16 12 ☽ ⊼ ♇
19 37 ☽ Q ♅
19 50 ☽ □ ☉

13 Wednesday
00 12 ☽ ⊼ ♆
01 58 ☽ Ᵽ ♃
03 14 ☽ ± ♇
13 16 ☽ △ ☿
16 51 ☽ ✶ ♄
19 03 ☽ ± ♃
20 03 ☽ Ᵽ ♇

14 Thursday
01 09 ☽ △ ♀
02 16 ☽ ⊼ ♂
05 27 ☽ □ ♅
06 37 ☽ ⊼ ♃
14 14 ☽ ± ♂
15 52 ☽ Q ♄
19 06 ☽ Ᵽ ☿
20 50 ♀ □ ♂

15 Friday
00 42 ☽ △ ♇
08 24 ☽ Ᵽ ♂
08 52 ☉ ✶ ♆
09 07 ☽ Ᵽ ♀
09 19 ☽ ☍ ♆
09 21 ☽ △ ☉
11 10 ♀ ∠ ♆
13 30 ☽ ± ☿
21 08 ☽ # ♆
23 35 ♀ ⊻ ♅

16 Saturday
01 57 ☽ ⊼ ☿
03 13 ☽ □ ♄
04 42 ☽ ± ♀
06 21 ♂ ∠ ♆
14 11 ♂ ± ♄
15 13 ☽ △ ♂
16 07 ☽ △ ♅
17 21 ☽ Ᵽ ☉
17 53 ☽ ⊼ ♀
18 03 ☽ ☍ ♃
19 30 ☽ ∥ ♂
19 51 ♀ ⊻ ♃

17 Sunday
02 54 ☿ ✶ ♄
04 33 ☉ ∠ ♃
11 45 ☽ □ ♇
12 48 ☽ ± ☉
14 35 ♂ △ ♅
17 47 ☿ ⊥ ♅
20 47 ☽ ⊼ ♆
20 59 ☽ # ♃
22 10 ☽ Ᵽ ♅

18 Monday
00 30 ☉ ∥ ♅
01 51 ☽ ⊼ ☉
08 50 ☽ ± ♆
15 35 ☽ △ ♄
16 20 ☽ ± ♅
17 55 ☽ ☍ ☿
23 50 ☽ Q ♇
23 53 ☉ ⊥ ♄

19 Tuesday
02 59 ☽ Ᵽ ♆
04 29 ☽ ⊼ ♅
06 01 ☽ □ ♂
07 11 ☽ ⊼ ♃
12 51 ☽ ☍ ♀
17 49 ☽ # ☿
19 30 ☽ ± ♃
22 08 ☽ Ᵽ ♄

20 Wednesday
00 02 ☽ ✶ ♇
09 19 ☽ △ ♆
14 00 ☽ Ᵽ ♃
16 30 ☽ ± ♄
19 37 ☽ ☍ ☉
20 02 ☽ Q ♂

21 Thursday
00 31 ♂ ✶ ♃
04 45 ☽ ⊼ ♄
06 19 ☽ ∠ ♇
11 50 ☽ ⊼ ☿
17 24 ☽ ☍ ♅
20 50 ☽ △ ♃
21 23 ☽ ✶ ♂

22 Friday
00 29 ☽ ⊥ ♇
01 08 ☽ ± ☿
04 49 ☽ # ☿
08 25 ☽ ⊼ ♀
12 35 ☽ ⊻ ♇
16 37 ♀ ⊥ ♃
21 09 ☽ Ᵽ ☿
21 54 ☽ ± ♀
22 00 ☽ □ ♆

23 Saturday
04 57 ☽ ∠ ♂
06 15 ☿ ∠ ♆
13 24 ☽ ⊼ ☉
15 28 ♀ ∠ ♄
17 47 ☽ ☍ ♄
18 01 ☽ Ᵽ ♀
23 46 ☽ ⊥ ♂

24 Sunday
00 31 ♀ △ ♇
01 37 ☿ ⊻ ♅
02 26 ☽ ± ☉
06 00 ☽ ⊼ ♅
06 28 ☽ △ ☿
10 04 ☽ □ ♃
10 06 ☽ Q ♆
12 15 ☽ ⊻ ♂
15 12 ☽ # ♃
18 00 ☽ ± ♅
21 51 ☽ Ᵽ ☉

25 Monday
00 30 ☽ ☌ ♇
03 12 ☽ △ ♀
09 52 ☽ ✶ ♆
10 22 ♀ # ♇
11 50 ☽ Ᵽ ♅
13 52 ☿ ∥ ♄
22 06 ☽ Q ♃
22 08 ☿ ⊻ ♃

26 Tuesday
03 35 ☉ ⊻ ♄
05 35 ☽ ⊼ ♄
05 43 ☽ △ ☉
06 13 ☽ ∥ ♂
07 01 ☉ Ᵽ ♇
07 36 ☉ ∥ ♅
15 05 ☽ ∠ ♆
17 06 ☽ △ ♅
17 11 ☽ # ♆
17 16 ☽ ± ♄
21 38 ☽ ✶ ♃

27 Wednesday
00 03 ☽ □ ☿
01 12 ☽ ☌ ♂
08 17 ☽ ⊥ ♆
10 29 ☽ ⊻ ♇
10 33 ☽ Ᵽ ♄
15 08 ☿ □ ♂
19 16 ☽ □ ♀
19 34 ☽ ⊻ ♆
21 44 ☽ ⊥ ♇
22 30 ♀ ✶ ♆

28 Thursday
02 18 ☽ ∠ ♃
04 57 ☉ Q ♆
13 54 ☽ Q ☿
14 20 ☽ ∠ ♇
14 40 ☽ △ ♄
18 43 ☽ □ ☉
19 04 ☽ ⊥ ♃

29 Friday
01 16 ☽ □ ♅
06 02 ☽ ⊻ ♃
07 34 ☽ Q ♀
10 38 ☽ ⊻ ♂
12 56 ☉ ∥ ♀
14 14 ☽ ✶ ☿
17 14 ☽ ✶ ♇
21 36 ☽ ⊥ ♂

30 Saturday
01 50 ☽ ☌ ♆
05 07 ☽ Q ☉
06 45 ☽ ✶ ♀
09 09 ☽ Q ♅
13 01 ☽ ∥ ♆
13 45 ☽ ∠ ♂
14 04 ☽ Q ♇
15 51 ☿ ⊥ ♃
16 12 ☿ △ ♇
18 46 ☽ # ♂
19 34 ☽ ∠ ☿
20 01 ☽ □ ♄
23 48 ☿ ∠ ♄

All ephemeris data is given at 12.00 UT and the Moon's longitude is additionally given for 24.00 UT

JULY 2035

LONGITUDES

Date	Sidereal time h m s	Sun ☉ ° ' "	Moon ☽ ° ' "	Moon ☽ 24.00 ° ' "	Mercury ☿ ° '	Venus ♀ ° '	Mars ♂ ° '	Jupiter ♃ ° '	Saturn ♄ ° '	Uranus ♅ ° '	Neptune ♆ ° '	Pluto ♇ ° '
01	06 37 31	09 ♋ 30 54	14 ♉ 28 20	21 ♉ 44 07	20 ♊ 49	28 ♊ 44	16 ♓ 43	13 ♉ 37	05 ♌ 03	10 ♋ 41	24 ♈ 25	19 ♒ 28
02	06 41 27	10 28 08	29 ♉ 05 56	06 ♊ 33 06	22 28	29 ♊ 57	17 11	13 48	05 10	10 45	24 26	19 R 27
03	06 45 24	11 25 21	14 ♊ 04 43	21 ♊ 39 39	24 11	01 ♋ 11	17 38	13 58	05 18	10 48	24 27	19 26
04	06 49 21	12 22 35	29 ♊ 16 41	06 ♋ 54 27	25 57	02 24	18 04	14 09	05 25	10 52	24 28	19 25
05	06 53 17	13 19 49	14 ♋ 31 34	22 ♋ 06 42	27 46	03 38	18 30	14 19	05 32	10 56	24 29	19 24
06	06 57 14	14 17 02	29 ♋ 38 35	07 ♌ 06 06	29 ♊ 38	04 51	18 56	14 30	05 40	10 59	24 29	19 23
07	07 01 10	15 14 16	14 ♌ 28 17	21 ♌ 44 25	01 ♋ 33	06 05	19 21	14 40	05 47	11 03	24 30	19 22
08	07 05 07	16 11 29	28 ♌ 53 57	05 ♍ 56 36	03 30	07 18	19 46	14 50	05 54	11 07	24 31	19 21
09	07 09 03	17 08 43	12 ♍ 52 13	19 ♍ 40 51	05 30	08 32	20 11	15 00	06 02	11 10	24 31	19 20
10	07 13 00	18 05 56	26 ♍ 22 42	02 ♎ 58 05	07 32	09 46	20 35	15 10	06 09	11 14	24 32	19 19
11	07 16 56	19 03 09	09 ♎ 27 23	15 ♎ 51 04	09 35	10 59	20 58	15 20	06 17	11 17	24 33	19 18
12	07 20 53	20 00 22	22 ♎ 09 41	28 ♎ 23 45	11 40	12 13	21 21	15 30	06 24	11 21	24 33	19 17
13	07 24 50	20 57 34	04 ♏ 33 50	10 ♏ 40 28	13 47	13 27	21 44	15 40	06 32	11 25	24 34	19 16
14	07 28 46	21 54 47	16 ♏ 44 12	22 ♏ 45 32	15 55	14 40	22 06	15 49	06 40	11 28	24 34	19 14
15	07 32 43	22 52 00	28 ♏ 44 57	04 ♐ 42 55	18 03	15 54	22 28	15 59	06 47	11 32	24 35	19 13
16	07 36 39	23 49 13	10 ♐ 39 50	16 ♐ 36 05	20 12	17 08	22 49	16 08	06 55	11 35	24 35	19 12
17	07 40 36	24 46 26	22 ♐ 32 01	28 ♐ 27 56	22 21	18 21	23 09	16 17	07 02	11 39	24 35	19 11
18	07 44 32	25 43 40	04 ♑ 24 08	10 ♑ 20 52	24 29	19 35	23 30	16 26	07 10	11 42	24 36	19 10
19	07 48 29	26 40 53	16 ♑ 18 21	22 ♑ 16 49	26 38	20 49	23 49	16 35	07 18	11 46	24 36	19 08
20	07 52 25	27 38 08	28 ♑ 16 27	04 ♒ 17 28	28 ♋ 45	22 02	24 08	16 44	07 25	11 49	24 36	19 07
21	07 56 22	28 35 22	10 ♒ 20 04	16 ♒ 24 28	00 ♌ 52	23 16	24 27	16 52	07 33	11 53	24 37	19 06
22	08 00 19	29 ♋ 32 37	22 ♒ 30 52	28 ♒ 39 32	02 58	24 30	24 45	17 01	07 41	11 56	24 37	19 05
23	08 04 15	00 ♌ 29 53	04 ♓ 50 42	11 ♓ 04 39	05 03	25 44	25 02	17 09	07 48	12 00	24 37	19 03
24	08 08 12	01 27 09	17 ♓ 21 41	23 ♓ 42 08	07 06	26 58	25 18	17 18	07 56	12 03	24 37	19 02
25	08 12 08	02 24 27	00 ♈ 06 18	06 ♈ 34 33	09 08	28 11	25 35	17 26	08 04	12 07	24 37	19 01
26	08 16 05	03 21 45	13 ♈ 07 12	19 ♈ 44 34	11 08	29 ♋ 25	25 50	17 34	08 11	12 10	24 37	19 00
27	08 20 01	04 19 04	26 ♈ 26 56	03 ♉ 14 32	13 07	00 ♌ 39	26 05	17 42	08 19	12 14	24 37	18 58
28	08 23 58	05 16 24	10 ♉ 07 32	17 ♉ 06 01	15 05	01 53	26 19	17 49	08 27	12 17	24 37	18 57
29	08 27 54	06 13 45	24 ♉ 09 57	01 ♊ 19 09	17 01	03 07	26 32	17 57	08 35	12 21	24 R 37	18 56
30	08 31 51	07 11 07	08 ♊ 33 18	15 ♊ 51 59	18 55	04 21	26 45	18 04	08 42	12 24	24 37	18 54
31	08 35 48	08 ♌ 08 30	23 ♊ 14 31	00 ♋ 40 11	20 ♌ 47	05 ♌ 35	26 ♓ 57	18 ♉ 12	08 ♌ 50	12 ♋ 27	24 ♈ 37	18 ♒ 53

Date	Moon True ☊ ° '	Moon Mean ☊ ° '	Moon ☽ Latitude ° '
01	07 ♍ 15	08 ♍ 30	04 S 42
02	07 R 09	08 27	05 02
03	07 02	08 24	05 03
04	06 53	08 20	04 43
05	06 46	08 17	04 02
06	06 40	08 14	03 05
07	06 36	08 11	01 56
08	06 34	08 08	00 S 41
09	06 D 34	08 05	00 N 34
10	06 35	08 01	01 45
11	06 36	07 58	02 48
12	06 R 36	07 55	03 41
13	06 35	07 52	04 22
14	06 32	07 49	04 51
15	06 28	07 45	05 06
16	06 21	07 42	05 08
17	06 14	07 39	04 56
18	06 07	07 36	04 32
19	05 59	07 33	03 56
20	05 54	07 30	03 10
21	05 49	07 26	02 14
22	05 47	07 23	01 12
23	05 46	07 20	00 N 05
24	05 D 46	07 17	01 S 03
25	05 48	07 14	02 09
26	05 49	07 11	03 10
27	05 50	07 07	04 02
28	05 R 50	07 04	04 42
29	05 49	07 01	05 07
30	05 47	06 58	05 13
31	05 ♍ 44	06 ♍ 55	04 S 59

DECLINATIONS

Date	Sun ☉ ° '	Moon ☽ ° '	Mercury ☿ ° '	Venus ♀ ° '	Mars ♂ ° '	Jupiter ♃ ° '	Saturn ♄ ° '	Uranus ♅ ° '	Neptune ♆ ° '	Pluto ♇ ° '
01	23 N 05	11 N 41	21 N 41	23 N 20	08 S 48	14 N 53	19 N 31	23 N 20	07 N 56	22 S 52
02	23 01	15 02	22 00	23 23	08 40	14 56	19 30	23 19	07 56	22 52
03	22 56	17 28	22 17	23 25	08 32	14 59	19 28	23 19	07 56	22 53
04	22 51	18 43	22 33	23 27	08 24	15 02	19 26	23 19	07 56	22 53
05	22 46	18 38	22 48	23 27	08 16	15 05	19 25	23 19	07 57	22 54
06	22 40	17 12	23 01	23 27	08 09	15 08	19 23	23 18	07 57	22 54
07	22 34	14 38	23 13	23 27	08 01	15 11	19 21	23 18	07 57	22 55
08	22 27	11 13	23 22	23 25	07 54	15 14	19 19	23 18	07 57	22 55
09	22 20	07 15	23 30	23 23	07 47	15 17	19 17	23 17	07 57	22 56
10	22 13	03 N 03	23 34	23 21	07 40	15 19	19 16	23 17	07 58	22 56
11	22 05	01 S 10	23 37	23 17	07 33	15 22	19 14	23 17	07 58	22 57
12	21 57	05 12	23 37	23 13	07 27	15 25	19 12	23 17	07 58	22 58
13	21 48	08 55	23 34	23 08	07 21	15 28	19 10	23 16	07 58	22 58
14	21 39	12 11	23 29	23 03	07 14	15 30	19 08	23 16	07 58	22 59
15	21 30	14 54	23 20	22 57	07 08	15 33	19 07	23 16	07 58	22 59
16	21 20	16 58	23 09	22 50	07 03	15 35	19 05	23 15	07 58	23 00
17	21 10	18 17	22 56	22 43	06 57	15 38	19 03	23 15	07 58	23 00
18	21 00	18 49	22 39	22 34	06 52	15 40	19 01	23 15	07 59	23 01
19	20 49	18 32	22 21	22 26	06 46	15 43	18 59	23 14	07 59	23 01
20	20 38	17 24	21 59	22 16	06 41	15 45	18 57	23 14	07 59	23 02
21	20 26	15 30	21 36	22 06	06 37	15 47	18 55	23 14	07 59	23 02
22	20 14	12 53	21 10	21 55	06 32	15 50	18 53	23 14	07 59	23 03
23	20 02	09 39	20 42	21 44	06 28	15 52	18 52	23 13	07 59	23 03
24	19 50	05 57	20 13	21 32	06 24	15 54	18 50	23 13	07 59	23 04
25	19 37	01 S 56	19 42	21 19	06 20	15 56	18 48	23 13	07 59	23 05
26	19 24	02 N 16	19 09	21 06	06 16	15 58	18 46	23 13	07 59	23 05
27	19 10	06 26	18 35	20 52	06 13	16 00	18 44	23 12	07 59	23 06
28	18 57	10 23	17 59	20 37	06 10	16 02	18 42	23 12	07 59	23 06
29	18 43	13 51	17 22	20 22	06 07	16 04	18 40	23 11	07 59	23 07
30	18 28	16 34	16 45	20 06	06 04	16 06	18 38	23 11	07 58	23 07
31	18 N 14	18 N 17	16 N 06	19 N 50	06 S 02	16 N 08	18 N 36	23 N 11	07 N 58	23 S 08

ZODIAC SIGN ENTRIES

Date	h	m	Planets
02	12	53	♀ ♋
02	13	27	☽ ♊
04	13	08	☽ ♋
06	12	34	☽ ♌
06	16	36	☿ ♋
08	13	52	☽ ♍
10	18	34	☽ ♎
13	03	07	☽ ♏
15	14	31	☽ ♐
18	03	06	☽ ♑
20	15	27	☽ ♒
21	02	06	☿ ♌
22	23	29	☉ ♌
23	02	37	☽ ♓
25	11	48	☽ ♈
26	23	18	♀ ♌
27	18	18	☽ ♉
29	21	48	☽ ♊
31	22	55	☽ ♋

LATITUDES

Date	Mercury ☿ ° '	Venus ♀ ° '	Mars ♂ ° '	Jupiter ♃ ° '	Saturn ♄ ° '	Uranus ♅ ° '	Neptune ♆ ° '	Pluto ♇ ° '
01	01 S 26	00 S 05	03 S 52	01 S 05	00 N 32	00 N 20	01 S 39	08 S 19
04	00 49	00 N 02	04 00	01 06	00 33	00 20	01 39	08 20
07	00 S 13	00 09	04 09	01 06	00 33	00 20	01 39	08 21
10	00 N 21	00 16	04 17	01 06	00 33	00 20	01 40	08 21
13	00 51	00 23	04 25	01 07	00 33	00 20	01 40	08 22
16	01 15	00 30	04 34	01 07	00 33	00 20	01 40	08 22
19	01 33	00 37	04 42	01 08	00 34	00 20	01 40	08 23
22	01 44	00 43	04 51	01 08	00 34	00 20	01 40	08 23
25	01 48	00 49	04 59	01 08	00 34	00 20	01 40	08 24
28	01 45	00 55	05 07	01 09	00 34	00 20	01 41	08 24
31	01 N 38	01 N 00	05 S 15	01 S 09	00 N 35	00 N 20	01 S 41	08 S 24

DATA

Julian Date	2464510
Delta T	+76 seconds
Ayanamsa	24° 21' 04"
Synetic vernal point	04° ♓ 45' 54"
True obliquity of ecliptic	23° 25' 56"

MOON'S PHASES, APSIDES AND POSITIONS ☽

Date	h	m	Phase	Longitude ° '	Eclipse Indicator
05	09	59	●	13 ♋ 15	
12	07	33	☽	19 ♎ 50	
20	10	37	○	27 ♑ 35	
28	02	56	☾	04 ♉ 55	

Day	h	m		
04	20	58	Perigee	
17	17	24	Apogee	
04	22	24	Max dec	18° N 51'
11	05	16	0S	
18	15	31	Max dec	18° S 50'
25	23	07	0N	

LONGITUDES

Date	Chiron ⚷ ° '	Ceres ⚳ ° '	Pallas ⚴ ° '	Juno ⚵ ° '	Vesta ⚶ ° '	Black Moon Lilith ⚸ ° '
01	09 ♊ 20	01 ♊ 44	06 ♈ 19	09 ♓ 06	12 ♌ 16	27 ♐ 46
11	10 ♊ 00	05 ♊ 29	07 ♈ 21	09 ♓ 22	16 ♌ 43	28 ♐ 53
21	10 ♊ 37	09 ♊ 06	07 ♈ 55	08 ♓ 57	21 ♌ 15	00 ♑ 00
31	11 ♊ 09	12 ♊ 37	07 ♈ 56	07 ♓ 52	25 ♌ 52	01 ♑ 08

ASPECTARIAN

h m Aspects

01 Sunday
03 07 ☽ ⚹ ☉
05 39 ☽ ⚹ ♅
09 06 ♀ ∥ ♅
09 18 ♀ ∠ ♃
10 33 ☽ ☌ ♃
10 39 ☽ ∠ ♀
12 38 ☽ ⊥ ☿
15 52 ☽ ⚹ ♂
18 57 ♀ ⊥ ♄
20 16 ☽ □ ♇
23 49 ☽ ⊻ ☿

02 Monday
02 16 ☽ Q ♄
02 52 ☽ ⊥ ♀
04 25 ☽ ⊻ ♆
05 42 ☽ ∠ ☉
06 32 ☽ ∠ ♅
11 13 ☽ ∥ ♃
12 08 ☽ Q ♂
13 31 ☽ ⊻ ♀
14 10 ☽ ⊥ ♆
19 30 ☉ ☌ ♅
21 09 ☽ ⊥ ♅
21 15 ☽ ⊥ ☉
21 52 ☽ ⚹ ♄

03 Tuesday
04 38 ☽ ∠ ♆
06 47 ☽ ⊻ ♅
07 30 ☽ ⊻ ☉
11 50 ☽ ⊻ ♃
15 41 ☿ ⚹ ♆
17 48 ☽ □ ♂
20 29 ☽ △ ♇
21 27 ☽ ⊥ ♃
21 55 ☽ ∠ ♄

04 Wednesday
04 02 ☉ # ♇
04 25 ☽ ⚹ ♆
06 04 ☽ ☌ ☿
11 48 ☽ ∠ ♃
12 13 ☽ ⊥ ♄
17 21 ☽ ☌ ♀
20 05 ☽ ⊡ ♇
21 44 ☽ ⊻ ♄
23 19 ☽ Q ♆

05 Thursday
06 18 ☽ ☌ ♅
09 16 ☉ ∥ ☿
09 59 ☽ ☌ ☉
10 14 ☽ ± ♇
10 33 ♂ ⊥ ♆
11 41 ☽ ⚹ ♃
13 49 ☉ ± ♇
18 28 ☽ △ ♂
19 42 ☽ ⊼ ♇
22 18 ☿ # ♇

06 Friday
02 56 ♀ ⊡ ♇
03 46 ☽ □ ♆
06 55 ☽ Q ♃
10 04 ☿ ∠ ♃
11 59 ☽ ⊻ ☿
12 19 ☿ ⊥ ♄
18 35 ☉ ⚹ ♃
19 05 ☽ ⊡ ♂
21 07 ☽ ⊻ ♀
21 45 ☽ ☌ ♄
23 03 ☽ ⊥ ☿

07 Saturday
05 30 ♀ ⊻ ♄
06 23 ☽ ⊻ ♅
07 39 ☽ ∥ ♃
07 44 ☽ ⊥ ♀
10 07 ☽ ± ♂
12 20 ☽ □ ♃
12 49 ♂ ⊻ ♇
13 21 ☽ ⊻ ☉
15 56 ☽ ∠ ☿
16 15 ☽ ⊥ ♅
20 03 ☽ ☍ ♇
20 17 ☽ ⊼ ♂
20 19 ♀ Q ♆
23 55 ☽ ∠ ♀
23 57 ☽ ⊥ ☉

08 Sunday
00 15 ☿ ∥ ♅
01 51 ♂ # ♆
04 37 ☽ △ ♆
07 17 ☽ ∠ ♅
16 10 ☽ ∠ ☉
19 41 ☿ ∥ ♀
21 06 ☽ ⚹ ☿
22 11 ☿ ⊡ ♇

09 Monday
00 03 ☽ ⊻ ♄
03 44 ☽ ⚹ ♀
06 10 ☽ ⊡ ♆
07 52 ☽ ∥ ♆
08 47 ☽ # ♂
09 01 ☽ ⚹ ♅
10 31 ☽ ⊥ ♄
15 47 ☽ △ ♃
18 47 ☿ ⊻ ♄
20 05 ☽ ⚹ ☉
21 33 ☽ Q ☿
21 57 ☽ ± ♆
23 22 ☽ ⊼ ♇

10 Tuesday
00 16 ☿ Q ♆
01 17 ☽ ☍ ♂
02 32 ☽ ∠ ♄
02 52 ☽ Q ♀
06 19 ☽ Q ♅
08 40 ☽ ⊼ ♆
10 05 ☽ ± ♇
18 59 ☽ ⊡ ♃
19 17 ☽ Q ☉

11 Wednesday
02 27 ☽ ⊡ ♇
06 03 ☽ ⚹ ♄
11 47 ☽ ± ♃
12 17 ☽ □ ☿
14 47 ♀ ∥ ♅
15 10 ☽ □ ♀
15 26 ☽ □ ♅
18 01 ☉ ⊼ ♇
18 13 ♀ ☌ ♅
23 11 ☽ ⊼ ♃

12 Thursday
04 46 ☽ Q ♄
06 30 ☽ △ ♇
07 33 ☽ □ ☉
08 11 ☿ ☌ ♅
10 24 ☽ ⊼ ♂
16 35 ☽ ☍ ♆
17 01 ♂ ⊡ ♄
22 18 ☽ ± ♂

13 Friday
01 47 ☽ ∥ ♂
02 45 ☿ ☌ ♀
05 35 ☽ # ♆
06 05 ☿ ± ♇
08 28 ♀ ± ♇
15 54 ☽ □ ♄
16 23 ☽ ⊡ ♂

14 Saturday
01 31 ☽ △ ♅
07 26 ☽ △ ♀
10 01 ☽ △ ☿
10 09 ☽ ☍ ♃
10 54 ☿ ⚹ ♃
16 58 ☽ □ ♇
19 35 ☉ △ ♂
23 01 ☽ △ ♂
23 12 ☽ △ ☉

15 Sunday
03 38 ☽ ⊼ ♆
03 57 ♀ # ♇
07 31 ☽ ⊡ ♅
13 48 ♀ ⚹ ♃
15 40 ☽ ± ♆
16 49 ☽ ⊡ ♀
18 53 ☽ # ♃
22 33 ☽ ⊡ ☿
23 24 ☿ ∥ ♅

16 Monday
00 58 ☿ ⊼ ♇
01 43 ☽ ± ♅
04 21 ☽ △ ♄
05 01 ☽ Q ♇
07 57 ☽ ⊡ ☉
09 49 ☽ ⊡ ♆
13 02 ☽ ± ♀
13 53 ☽ ⊼ ♅
20 43 ☽ ± ☿
23 12 ☽ ⊼ ♃

17 Tuesday
02 34 ☽ ⊼ ♀
03 44 ☽ ± ☉
04 59 ☿ # ♇
05 14 ☽ ⚹ ♇
07 20 ☉ □ ♆
10 59 ☽ ⊡ ♄
11 29 ☽ ± ♃
11 32 ☽ ⊼ ☿
13 18 ☽ □ ♂
16 10 ☽ △ ♆
16 56 ☽ ⊼ ☉
22 47 ☿ △ ♂

18 Wednesday
03 54 ♀ ⊼ ♇
05 23 ☽ ± ♄
05 56 ☽ ⊡ ♃
11 31 ☽ ∠ ♇
13 11 ☿ □ ♆
17 38 ☽ ⊼ ♄

19 Thursday
00 52 ☿ ∥ ♀
02 43 ☽ Q ♂
02 49 ☽ ☍ ♅
05 38 ☽ ⊥ ♇
12 34 ☽ △ ♃
13 05 ☉ ☌ ☿
17 41 ☽ ⊻ ♇
22 05 ☽ ☍ ♀

20 Friday
03 30 ☽ ⚹ ♂
04 39 ☽ □ ♆
10 37 ☽ ☍ ☉
11 41 ☿ Q ♃
13 10 ☽ ☍ ☿

21 Saturday
06 25 ☽ ☍ ♄
08 54 ☽ # ♃
10 11 ☽ ∠ ♂
15 05 ☽ ⊼ ♅
16 30 ☽ Q ♆
20 25 ☉ Q ♃

22 Sunday
01 04 ☽ □ ♃
01 21 ♂ ⊻ ♆
02 59 ☽ ± ♅
04 25 ☽ ⊥ ♂
05 16 ☽ ☌ ♇
14 14 ♀ □ ♆
16 06 ☽ ⚹ ♆
16 19 ☽ ⊼ ♀
16 28 ☽ ⊻ ♂
18 14 ♀ △ ♂
20 42 ☽ ⊡ ♅

23 Monday
02 52 ☽ ⊼ ☉
05 18 ☽ ± ♀
12 28 ☽ ⊼ ☿
12 36 ☽ Q ♃
14 07 ♂ ⊥ ♇
15 27 ☽ ± ☉
17 46 ☽ ⊼ ♄
21 12 ☽ ∠ ♆
23 13 ☽ # ♆

24 Tuesday
00 34 ☽ ⊡ ♀
01 50 ☽ △ ♅
02 17 ☽ ± ☿
05 24 ☽ ± ♄
09 15 ☽ ∥ ♂
10 08 ☽ ⊡ ☉
11 52 ☽ ⚹ ♃
14 23 ☽ ⊥ ♆
15 11 ☽ ⊻ ♇
22 27 ☿ ☌ ♄
22 39 ☽ ⊡ ♄
22 42 ☽ ⊡ ☿

25 Wednesday
01 44 ☽ ⊻ ♆
02 29 ☽ ⊥ ♇
03 21 ☽ ☌ ♂
08 03 ☽ △ ♀
16 22 ☽ ∠ ♃
16 38 ☽ △ ☉
17 48 ☉ ∥ ☿
19 15 ☽ ∠ ♇

26 Thursday
02 53 ☽ △ ♄
07 44 ☽ △ ☿
07 47 ☿ ⊡ ♂
09 08 ☽ ⊥ ♃
10 16 ☽ □ ♅
15 06 ♀ Q ♃
20 09 ☽ ⊻ ♃
22 38 ☽ ⚹ ♇

27 Friday
00 49 ☿ ⊻ ♅
05 12 ☿ ∥ ♄
08 45 ☽ ☌ ♆
10 43 ☽ # ♂
11 20 ☽ ⊻ ♂
18 44 ☽ Q ♅
20 00 ☽ Q ♇
20 11 ☽ □ ♀
21 09 ☽ ∥ ♆
22 08 ☽ ⊥ ♂

28 Saturday
02 56 ☽ □ ☉
09 04 ☽ □ ♄
12 30 ♆ St R
14 06 ☽ ∠ ♂
15 45 ☽ ⚹ ♅
21 55 ☽ □ ☿

29 Sunday
01 21 ☽ ☌ ♃
03 08 ☽ □ ♇
06 21 ☽ Q ♀
12 07 ☽ Q ☉
12 46 ☽ ⊻ ♆
16 04 ☽ ⚹ ♂
16 06 ☽ Q ♄
17 00 ☉ ∥ ♄
17 22 ☽ ∠ ♅
22 50 ☽ ⊥ ♆

30 Monday
00 37 ☿ □ ♃
04 23 ☽ ⚹ ♀
05 18 ☿ ⊥ ♅
07 09 ☽ ∥ ♃
08 25 ☽ ⊥ ♅
08 53 ☽ Q ☿
09 35 ☽ ⚹ ☉
11 57 ☿ ☍ ♇
12 15 ☽ ⚹ ♄
12 20 ☽ Q ♂
13 28 ☽ ∥ ☿
13 45 ☽ ∠ ♆
18 21 ☽ ⊻ ♅

31 Tuesday
03 44 ☽ ⊻ ♃
04 56 ☽ △ ♇
07 17 ☽ ∠ ♀
07 26 ☽ ⚹ ☿
10 58 ☽ ∥ ☉
10 59 ☿ ∥ ♃
11 50 ☽ ∠ ☉
12 58 ☽ ∠ ♄
13 33 ☽ ⊥ ♃
14 14 ☽ ⚹ ♆
14 24 ☿ ± ♂
18 05 ☽ □ ♂
19 56 ☽ ∥ ♄
23 10 ☽ ⊥ ♀

All ephemeris data is given at 12.00 UT and the Moon's longitude is additionally given for 24.00 UT

AUGUST 2035

Date	Sidereal time h m s	Sun ☉ ° ' "	Moon ☽ ° ' "	Moon ☽ 24.00 ° ' "	Mercury ☿ ° '	Venus ♀ ° '	Mars ♂ ° '	Jupiter ♃ ° '	Saturn ♄ ° '	Uranus ♅ ° '	Neptune ♆ ° '	Pluto ♇ ° '
01	08 39 44	09 ♌ 05 54	08 ♋ 08 04	15 ♋ 37 09	22 ♌ 38	06 ♌ 49	27 ♓ 09	18 ♉ 19	08 ♌ 58	12 ♋ 31	24 ♈ 37	18 ♒ 52
02	08 43 41	10 03 20	23 ♋ 06 22	00 ♌ 34 39	24 27	08 03	27 19	18 26	09 05	12 34	24 R 37	18 R 51
03	08 47 37	11 00 46	08 ♌ 00 53	15 ♌ 24 04	26 15	09 17	27 29	18 32	09 13	12 37	24 37	18 49
04	08 51 34	11 58 13	22 ♌ 43 16	29 ♌ 57 43	28 01	10 31	27 38	18 39	09 21	12 41	24 37	18 48
05	08 55 30	12 55 41	07 ♍ 06 44	14 ♍ 09 50	29 ♌ 45	11 45	27 46	18 46	09 29	12 44	24 36	18 47
06	08 59 27	13 53 09	21 ♍ 06 43	27 ♍ 57 12	01 ♍ 28	12 59	27 54	18 52	09 36	12 47	24 36	18 45
07	09 03 23	14 50 39	04 ♎ 41 15	11 ♎ 19 01	03 09	14 13	28 01	18 58	09 44	12 50	24 36	18 44
08	09 07 20	15 48 09	17 ♎ 50 42	24 ♎ 16 38	04 48	15 27	28 07	19 04	09 52	12 53	24 35	18 42
09	09 11 17	16 45 40	00 ♏ 37 13	06 ♏ 52 55	06 26	16 41	28 12	19 10	10 00	12 57	24 35	18 41
10	09 15 13	17 43 12	13 ♏ 04 15	19 ♏ 11 44	08 02	17 55	28 16	19 16	10 07	13 00	24 35	18 40
11	09 19 10	18 40 45	25 ♏ 15 57	01 ♐ 17 26	09 37	19 09	28 20	19 21	10 15	13 03	24 34	18 38
12	09 23 06	19 38 18	07 ♐ 16 45	13 ♐ 14 28	11 10	20 23	28 23	19 27	10 23	13 06	24 34	18 37
13	09 27 03	20 35 53	19 ♐ 11 05	25 ♐ 07 07	12 42	21 38	28 25	19 32	10 30	13 09	24 33	18 36
14	09 30 59	21 33 29	01 ♑ 03 03	06 ♑ 59 19	14 12	22 52	28 26	19 37	10 38	13 12	24 33	18 34
15	09 34 56	22 31 05	12 ♑ 56 21	18 ♑ 54 30	15 40	24 06	28 R 26	19 42	10 46	13 15	24 32	18 33
16	09 38 52	23 28 43	24 ♑ 54 09	00 ♒ 55 34	17 07	25 20	28 26	19 46	10 53	13 18	24 32	18 32
17	09 42 49	24 26 21	06 ♒ 59 04	13 ♒ 04 51	18 32	26 34	28 24	19 51	11 01	13 21	24 31	18 30
18	09 46 46	25 24 01	19 ♒ 13 08	25 ♒ 24 07	19 55	27 48	28 22	19 55	11 08	13 24	24 30	18 29
19	09 50 42	26 21 42	01 ♓ 37 56	07 ♓ 54 44	21 17	29 ♌ 03	28 19	19 59	11 16	13 27	24 30	18 28
20	09 54 39	27 19 25	14 ♓ 14 38	20 ♓ 37 43	22 37	00 ♍ 17	28 15	20 03	11 23	13 30	24 29	18 26
21	09 58 35	28 17 09	27 ♓ 04 06	03 ♈ 33 50	23 55	01 31	28 11	20 07	11 31	13 32	24 28	18 25
22	10 02 32	29 ♌ 14 54	10 ♈ 07 00	16 ♈ 43 40	25 11	02 45	28 06	20 11	11 39	13 35	24 27	18 24
23	10 06 28	00 ♍ 12 41	23 ♈ 23 53	00 ♉ 07 42	26 26	04 00	27 59	20 14	11 46	13 38	24 27	18 22
24	10 10 25	01 10 30	06 ♉ 55 06	13 ♉ 46 08	27 39	05 14	27 53	20 17	11 53	13 41	24 26	18 21
25	10 14 21	02 08 20	20 ♉ 40 44	27 ♉ 38 50	28 49	06 28	27 45	20 20	12 01	13 43	24 25	18 20
26	10 18 18	03 06 12	04 ♊ 40 20	11 ♊ 45 02	29 ♍ 58	07 43	27 36	20 23	12 08	13 46	24 24	18 19
27	10 22 15	04 04 06	18 ♊ 52 43	26 ♊ 03 03	01 ♎ 04	08 57	27 27	20 26	12 16	13 49	24 23	18 17
28	10 26 11	05 02 02	03 ♋ 15 41	10 ♋ 30 07	02 08	10 11	27 17	20 29	12 23	13 51	24 22	18 16
29	10 30 08	06 00 00	17 ♋ 45 51	25 ♋ 02 17	03 10	11 26	27 06	20 31	12 30	13 54	24 21	18 15
30	10 34 04	06 58 00	02 ♌ 18 46	09 ♌ 34 36	04 09	12 40	26 55	20 33	12 38	13 56	24 20	18 13
31	10 38 01	07 ♍ 56 01	16 ♌ 49 05	24 ♌ 01 32	05 ♎ 05	13 ♍ 54	26 ♓ 43	20 ♉ 35	12 ♌ 45	13 ♋ 59	24 ♈ 19	18 ♒ 12

Date	Moon True ☊ ° '	Moon Mean ☊ ° '	Moon ☽ Latitude ° '
01	05 ♍ 40	06 ♍ 51	04 S 25
02	05 R 37	06 48	03 32
03	05 34	06 45	02 26
04	05 32	06 42	01 S 10
05	05 D 32	06 39	00 N 09
06	05 33	06 36	01 25
07	05 34	06 32	02 34
08	05 35	06 29	03 33
09	05 36	06 26	04 19
10	05 37	06 23	04 52
11	05 R 37	06 20	05 11
12	05 36	06 17	05 16
13	05 35	06 13	05 07
14	05 33	06 10	04 45
15	05 31	06 07	04 11
16	05 29	06 04	03 26
17	05 28	06 01	02 31
18	05 27	05 57	01 29
19	05 26	05 54	00 N 21
20	05 D 26	05 51	00 S 49
21	05 27	05 48	01 57
22	05 27	05 45	03 01
23	05 28	05 42	03 56
24	05 28	05 38	04 39
25	05 29	05 35	05 07
26	05 R 29	05 32	05 17
27	05 29	05 29	05 09
28	05 29	05 26	04 41
29	05 D 29	05 22	03 55
30	05 29	05 19	02 54
31	05 ♍ 29	05 ♍ 16	01 S 42

DECLINATIONS

Date	Sun ☉ ° '	Moon ☽ ° '	Mercury ☿ ° '	Venus ♀ ° '	Mars ♂ ° '	Jupiter ♃ ° '	Saturn ♄ ° '	Uranus ♅ ° '	Neptune ♆ ° '	Pluto ♇ ° '
01	17 N 59	18 N 47	15 N 27	19 N 33	06 S 00	16 N 10	18 N 34	23 N 11	07 N 58	23 S 08
02	17 43	17 58	14 47	19 16	05 58	16 12	18 32	23 10	07 58	23 09
03	17 28	15 55	14 06	18 58	05 56	16 13	18 30	23 10	07 58	23 09
04	17 12	12 50	13 25	18 39	05 55	16 15	18 28	23 10	07 58	23 10
05	16 56	09 02	12 44	18 20	05 54	16 17	18 26	23 09	07 58	23 10
06	16 39	04 49	12 02	18 01	05 53	16 18	18 24	23 09	07 58	23 11
07	16 23	00 N 30	11 20	17 41	05 52	16 20	18 22	23 09	07 57	23 11
08	16 06	03 S 43	10 37	17 20	05 52	16 21	18 20	23 09	07 57	23 12
09	15 48	07 38	09 55	16 59	05 52	16 23	18 18	23 08	07 57	23 12
10	15 31	11 07	09 13	16 38	05 52	16 24	18 16	23 08	07 57	23 13
11	15 13	14 03	08 30	16 16	05 53	16 25	18 14	23 08	07 57	23 13
12	14 55	16 20	07 48	15 54	05 54	16 27	18 12	23 07	07 56	23 14
13	14 37	17 54	07 06	15 31	05 55	16 28	18 10	23 07	07 56	23 14
14	14 19	18 40	06 24	15 07	05 56	16 29	18 08	23 07	07 56	23 15
15	14 00	18 38	05 42	14 44	05 57	16 30	18 06	23 07	07 56	23 15
16	13 41	17 46	05 00	14 20	05 59	16 31	18 04	23 06	07 55	23 16
17	13 22	16 05	04 19	13 55	06 01	16 32	18 02	23 06	07 55	23 16
18	13 03	13 39	03 38	13 30	06 04	16 33	18 00	23 06	07 55	23 17
19	12 44	10 34	02 58	13 05	06 06	16 34	17 58	23 05	07 55	23 17
20	12 24	06 57	02 18	12 40	06 09	16 35	17 56	23 05	07 54	23 18
21	12 04	02 S 57	01 38	12 14	06 12	16 36	17 54	23 05	07 54	23 18
22	11 44	01 N 14	00 59	11 47	06 15	16 37	17 53	23 05	07 54	23 19
23	11 24	05 26	00 N 21	11 21	06 18	16 37	17 51	23 04	07 53	23 19
24	11 03	09 25	00 S 16	10 54	06 22	16 38	17 49	23 04	07 53	23 19
25	10 43	12 59	00 53	10 27	06 25	16 39	17 47	23 04	07 53	23 20
26	10 22	15 52	01 29	09 59	06 29	16 39	17 45	23 04	07 52	23 20
27	10 01	17 50	02 04	09 32	06 33	16 40	17 43	23 03	07 52	23 21
28	09 40	18 43	02 39	09 04	06 38	16 40	17 41	23 03	07 51	23 21
29	09 18	18 22	03 12	08 35	06 42	16 41	17 39	23 03	07 51	23 21
30	08 57	16 49	03 44	08 07	06 46	16 41	17 37	23 03	07 51	23 22
31	08 N 36	14 N 10	04 S 15	07 N 38	06 S 51	16 N 42	17 N 35	23 N 02	07 N 50	23 S 22

ZODIAC SIGN ENTRIES

Date	h	m	Planets
02	23	04	☽ ♌
05	00	04	☽ ♍
05	15	28	☿ ♍
07	03	38	☽ ♎
09	10	49	☽ ♏
11	21	25	☽ ♐
14	09	52	☽ ♑
16	22	10	☽ ♒
19	08	52	☽ ♓
20	06	33	♀ ♍
21	17	26	☽ ♈
23	06	44	☉ ♍
23	23	46	☽ ♉
26	04	02	☽ ♊
26	12	51	☿ ♎
28	06	35	☽ ♋
30	08	11	☽ ♌

LATITUDES

Date	Mercury ☿ ° '	Venus ♀ ° '	Mars ♂ ° '	Jupiter ♃ ° '	Saturn ♄ ° '	Uranus ♅ ° '	Neptune ♆ ° '	Pluto ♇ ° '
01	01 N 34	01 N 01	05 S 18	01 S 09	00 N 35	00 N 20	01 S 41	08 S 24
04	01 20	01 06	05 25	01 10	00 35	00 20	01 41	08 25
07	01 03	01 10	05 32	01 10	00 35	00 20	01 41	08 25
10	00 43	01 14	05 39	01 11	00 35	00 20	01 41	08 25
13	00 N 20	01 17	05 45	01 11	00 36	00 20	01 42	08 25
16	00 S 06	01 20	05 51	01 12	00 36	00 20	01 42	08 26
19	00 32	01 22	05 55	01 12	00 36	00 20	01 42	08 26
22	01 00	01 24	05 59	01 13	00 37	00 20	01 42	08 26
25	01 29	01 25	06 02	01 13	00 37	00 21	01 42	08 26
28	01 58	01 25	06 03	01 14	00 37	00 21	01 42	08 26
31	02 S 26	01 N 25	06 S 03	01 S 15	00 N 38	00 N 21	01 S 42	08 S 26

DATA

Julian Date	2464541
Delta T	+76 seconds
Ayanamsa	24° 21' 09"
Synetic vernal point	04° ♓ 45' 50"
True obliquity of ecliptic	23° 25' 56"

MOON'S PHASES, APSIDES AND POSITIONS ☽

Date	h	m	Phase	Longitude ° '	Eclipse Indicator
03	17	12	●	11 ♌ 13	
10	21	52	☽	18 ♏ 07	
19	01	00	○	25 ♒ 55	partial
26	09	08	☾	02 ♊ 59	

Day	h	m		
02	04	02	Perigee	
14	06	05	Apogee	
30	02	25	Perigee	
01	09	08	Max dec	18° N 47'
07	14	45	0S	
14	22	54	Max dec	18° S 46'
22	04	59	0N	
28	17	21	Max dec	18° N 45'

LONGITUDES

Date	Chiron ⚷ ° '	Ceres ⚳ ° '	Pallas ⚴ ° '	Juno ⚵ ° '	Vesta ⚶ ° '	Black Moon Lilith ⚸ ° '
01	11 ♊ 12	12 ♊ 57	07 ♈ 55	07 ♓ 43	26 ♌ 20	01 ♑ 14
11	11 ♊ 40	16 ♊ 17	07 ♈ 18	05 ♓ 56	01 ♍ 01	02 ♑ 21
21	12 ♊ 01	19 ♊ 26	06 ♈ 05	03 ♓ 41	05 ♍ 45	03 ♑ 28
31	12 ♊ 17	22 ♊ 21	04 ♈ 16	01 ♓ 13	10 ♍ 33	04 ♑ 35

All ephemeris data is given at 12.00 UT and the Moon's longitude is additionally given for 24.00 UT

ASPECTARIAN

h m	Aspects
01 Wednesday	
03 21	☽ ⊥ ☉
03 37	☽ ⊥ ♄
04 11	☽ ∠ ♃
05 09	☽ ⚼ ♇
08 03	☉ ☌ ♄
09 34	☽ Q ♆
09 41	☽ ⊻ ♀
11 05	☽ ∠ ☿
13 20	☽ ⊻ ♄
13 39	☽ ⊻ ☉
19 03	☽ ☌ ♅
19 34	☽ ± ♇
23 28	☽ ∥ ♄
02 Thursday	
03 31	☽ ⊥ ☿
04 26	☽ ⚹ ♃
05 10	☽ ⊼ ♇
14 08	☿ △ ♆
14 25	☽ □ ♆
14 28	☽ ⊻ ☿
16 28	☽ ∥ ☉
18 50	☽ △ ♂
23 51	☽ Q ♃
03 Friday	
09 08	☽ ∥ ♃
10 43	♀ ☌ ♄
13 58	☽ ☌ ♄
14 14	☽ ☌ ♀
17 12	☽ ☌ ☉
19 19	☽ ⚼ ♂
19 30	☽ ⊻ ♅
04 Saturday	
05 16	☽ □ ♃
05 20	☽ ⊥ ♅
05 34	☽ ☍ ♇
06 21	☿ ⊼ ♂
06 52	☽ ∥ ☿
07 14	☿ ∠ ♅
09 44	♅ # ♇
10 12	☽ ± ♂
15 07	☽ △ ♆
20 12	☽ ⊼ ♂
20 14	☽ ∠ ♅
21 58	☽ ☌ ☿
05 Sunday	
03 52	♀ ∥ ♄
06 44	☉ ⊻ ♅
07 29	☉ ⚼ ♂
14 48	♃ □ ♇
16 03	☽ ⊻ ♄
16 13	☽ ⚼ ♆
18 15	☽ ∥ ♆
20 37	☽ ⊻ ♀
21 35	☽ ⚹ ♅
22 37	☽ ⊻ ☉
06 Monday	
02 15	♅ ± ♇
02 23	☽ ⊥ ♄
06 04	☽ # ♂
07 39	☽ ± ♆
07 55	☽ ⊼ ♇
07 56	☽ ⊥ ♀
08 00	♀ ⊻ ♅
08 04	☽ △ ♃
09 43	☽ ⊥ ☉
10 14	♀ ⚼ ♂
18 06	☽ ⊼ ♆
18 10	☽ ∠ ♄
18 21	☽ ± ♇
18 27	☽ Q ♅
07 Tuesday	
00 00	☽ ☍ ♂
01 15	☽ ∠ ♀
02 41	☽ ∠ ☉
08 51	☽ ⊻ ☿
10 17	☽ ⚼ ♇
10 42	☽ ⚼ ♃
15 55	☉ ∥ ♃
21 13	☽ ⚹ ♄
21 13	☽ ⊥ ☿
08 Wednesday	
02 50	☽ □ ♅
03 08	☽ ± ♃
07 07	☽ ⚹ ♀
07 56	☽ ⚹ ☉
13 36	☽ △ ♇
14 17	☽ ⊼ ♃
16 10	☽ ∠ ☿
19 33	☽ Q ♄
09 Thursday	
00 35	☽ ☍ ♆
00 52	☽ ∥ ♂
07 22	☽ ⊼ ♂
07 55	☽ Q ♀
08 11	☽ Q ☉
14 01	☽ # ♆
18 40	☉ ☌ ♀
18 53	☽ ± ♂
10 Friday	
00 42	☽ # ☿
00 48	☽ ⚹ ☿
06 12	☽ □ ♄
11 51	☽ △ ♅
12 23	☽ ⚼ ♂
21 52	☽ □ ☉
22 33	☽ □ ♀
22 56	☽ □ ♇
11 Saturday	
00 13	☽ ☍ ♃
02 12	♀ ☍ ♇
02 26	♀ ∥ ♃
03 43	☽ Q ☿
09 49	♀ ⊥ ♅
10 37	☽ ⊼ ♆
11 04	☉ ☍ ♇
11 14	☿ ⚼ ♆
16 11	♀ □ ♃
17 33	☽ ⚼ ♅
18 07	☽ △ ♂
21 44	☉ ⊥ ♅
22 11	☽ # ☉
22 33	☿ ⊻ ♄
22 33	☽ ± ♆
12 Sunday	
06 38	☉ □ ♃
07 06	☿ ∥ ♆
07 36	☽ # ♀
10 41	☽ Q ♇
11 38	☽ ± ♅
13 27	☽ # ♃
16 35	☽ ⚼ ♆
18 18	☽ △ ♄
20 59	☽ □ ☿
23 46	☽ ⊼ ♅
13 Monday	
10 49	☽ ⚹ ♇
12 42	☽ ⊼ ♃
15 07	☽ △ ☉
17 31	☽ △ ♀
18 03	☽ # ♄
19 28	☿ ⚹ ♅
22 51	☽ △ ♆
14 Tuesday	
00 55	☽ ⚼ ♄
00 55	☽ ± ♃
03 29	♀ ± ♂
06 41	☽ □ ♂
17 05	☽ ∠ ♇
19 15	☽ ⚼ ♃
19 19	☽ ± ♄
15 Wednesday	
00 07	☽ ⚼ ☉
03 17	☿ # ♂
03 22	☽ ⚼ ♀
07 34	☽ ⊼ ♄
09 54	☉ ± ♂
10 01	♂ St R
11 13	☽ ⊥ ♇
12 38	☽ ☍ ♅
18 15	☽ △ ☿
19 02	☽ Q ♂
19 50	☽ ± ☉
20 26	♀ △ ♆
23 16	☽ ⊻ ♇
23 34	☽ ± ♀
16 Thursday	
01 40	☽ △ ♃
05 29	☽ # ♄
07 50	☿ ⊥ ♄
08 54	☽ ⊼ ☉
11 15	☽ □ ♆
12 58	☽ ⊼ ♀
19 01	☽ ⚹ ♂
17 Friday	
04 16	☽ ⚼ ☿
06 33	☽ # ♃
11 36	☿ ⊼ ♇
11 45	☿ ± ♆
13 53	☉ △ ♆
20 01	☽ ☍ ♄
22 53	☽ Q ♆
18 Saturday	
00 20	☽ ± ☿
00 35	☽ ⊼ ♅
00 36	☽ ∠ ♂
10 34	☽ ☌ ♇
12 00	☿ △ ♃
12 21	☽ ± ♅
13 22	☽ □ ♃
13 27	☽ # ♀
13 32	☽ ⊼ ☿
17 41	☽ # ☉
18 06	☽ ⊥ ♂
22 15	☽ ⚹ ♆
22 31	♀ ⊼ ♂
23 55	♀ ∠ ♅
19 Sunday	
01 00	☽ ☍ ☉
05 40	☽ ⊻ ♂
05 51	☽ ⚼ ♅
06 29	☽ ☍ ♀
20 Monday	
00 13	☽ Q ♃
03 00	☽ ∠ ♆
05 55	☽ # ♆
06 33	☽ ⊼ ♄
10 35	☽ △ ♅
16 55	☽ ∥ ♂
17 59	☽ ± ♄
19 53	☽ ⊻ ♇
19 58	☽ ⊥ ♆
22 59	☽ ⚹ ♃
21 Tuesday	
05 29	☽ ☍ ☿
07 05	☽ ⊥ ♇
07 10	☽ ⊻ ♆
09 38	☉ ⊼ ♂
10 58	☽ ⚼ ♄
14 03	☽ ☌ ♂
14 26	☽ ⊼ ☉
18 40	☉ ∠ ♅
21 02	☽ # ☿
21 06	☽ ⊼ ♀
21 12	☿ ± ♇
22 13	☿ ⊼ ♆
23 43	☽ ∠ ♇
22 Wednesday	
02 23	☽ ± ☉
02 56	☽ ∠ ♃
09 16	☽ ± ♀
10 48	☽ ∥ ☿
14 48	☽ △ ♄
18 20	☽ □ ♅
19 25	☽ ⊥ ♃
19 52	☿ Q ♅
20 06	☽ ⚼ ☉
23 Thursday	
01 28	☉ ∥ ♀
02 59	☽ ⚹ ♇
03 17	☽ ⚼ ♀
06 18	☽ ⊻ ♃
13 52	☽ ☌ ♆
17 09	☽ # ♂
17 58	☽ ⊼ ☿
19 18	☿ ∠ ♄
20 08	☽ ⊻ ♂
24 Friday	
00 25	☽ Q ♇
01 05	☽ △ ☉
02 30	☽ ∥ ♆
02 43	☽ Q ♅
05 40	☽ ± ☿
06 41	☽ ⊥ ♂
08 44	☽ △ ♀
16 15	☿ ☍ ♂
20 28	☽ ∥ ♀
20 48	☽ □ ♄
21 37	☽ ∥ ☉
22 21	☽ ∠ ♂
22 59	☽ ⚼ ☿
23 53	☽ ⚹ ♅
25 Saturday	
07 56	☽ □ ♇
11 25	☽ ☌ ♃
18 26	☽ ⊻ ♆
26 Sunday	
00 03	☽ ⚹ ♂
01 53	☽ ∠ ♅
03 15	☽ △ ☿
04 12	☽ Q ♄
04 43	☽ ⊥ ♆
09 08	☽ □ ☉
17 16	☽ ⊥ ♅
17 39	☽ □ ♀
20 01	☽ ∠ ♆
20 16	☽ ∥ ♃
20 17	☽ Q ♂
27 Monday	
00 46	☽ ⚹ ♄
03 27	☽ ⊻ ♅
09 58	☽ ∥ ♄
11 01	☽ △ ♇
14 37	☽ ⊻ ♃
17 44	☽ Q ☉
20 23	♀ ⚼ ♆
21 13	☽ ⚹ ♆
28 Tuesday	
00 41	☽ ⊥ ♃
02 08	☽ ∠ ♄
02 11	☽ □ ♂
02 46	☽ Q ♀
04 03	♂ ⚼ ♄
09 59	☽ □ ☿
12 00	☽ ⚼ ♇
15 09	☽ ⚹ ☉
15 41	☽ ∠ ♃
17 09	☽ Q ♆
17 14	☽ ⊥ ♄
29 Wednesday	
00 33	☽ ⚹ ♀
02 54	☽ ± ♇
03 14	☽ ⊻ ♄
05 36	☽ ☌ ♅
12 48	☽ ⊼ ♇
14 01	☿ ⚼ ♇
16 33	☽ ⚹ ♃
17 43	☽ ∠ ☉
18 01	☽ Q ☿
22 52	☽ □ ♆
30 Thursday	
01 41	☽ ∥ ♄
03 13	☽ △ ♂
03 37	☽ ∠ ♀
09 37	☽ ⊥ ☉
11 13	♀ ⊻ ♄
12 23	☽ Q ♃
13 24	☽ ∥ ♃
15 14	☽ ⚹ ☿
19 51	☽ ⊥ ♀
20 14	☽ ⊻ ☉
31 Friday	
01 53	♀ ∥ ♆
03 40	☽ ⚼ ♂
05 12	☽ ☌ ♄
06 43	☽ ⊻ ♀
07 17	☽ ⊻ ♅
13 30	♀ ⚹ ♅
14 18	☽ ☍ ♇
17 16	☽ ⊥ ♅
17 48	☽ ∠ ☿
18 16	☽ □ ♃
18 24	☽ ± ♂

SEPTEMBER 2035

LONGITUDES

Date	Sidereal time h m s	Sun ☉ ° ' "	Moon ☽ ° ' "	Moon ☽ 24.00 ° ' "	Mercury ☿ ° '	Venus ♀ ° '	Mars ♂ ° '	Jupiter ♃ ° '	Saturn ♄ ° '	Uranus ♅ ° '	Neptune ♆ ° '	Pluto ♇ ° '
01	10 41 57	08 ♍ 54 04	01 ♍ 11 13	08 ♍ 17 33	05 ♎ 58	15 ♍ 09	26 ♓ 31	20 ♉ 37	12 ♌ 52	14 ♋ 01	24 ♈ 18	18 ♒ 11
02	10 45 54	09 52 09	15 ♍ 19 56	22 ♍ 17 52	06 49	16 23	26 R 18	20 38	12 59	14 04	24 R 17	18 R 10
03	10 49 50	10 50 15	29 ♍ 10 57	05 ♎ 58 54	07 36	17 37	26 04	20 39	13 07	14 06	24 16	18 08
04	10 53 47	11 48 23	12 ♎ 41 30	19 ♎ 18 42	08 20	18 52	25 50	20 41	13 14	14 08	24 15	18 07
05	10 57 44	12 46 32	25 ♎ 50 31	02 ♏ 17 04	09 00	20 06	25 35	20 41	13 21	14 11	24 14	18 06
06	11 01 40	13 44 43	08 ♏ 38 34	14 ♏ 55 20	09 37	21 21	25 20	20 42	13 28	14 13	24 13	18 05
07	11 05 37	14 42 55	21 ♏ 07 43	27 ♏ 16 10	10 09	22 35	25 05	20 43	13 35	14 15	24 11	18 04
08	11 09 33	15 41 09	03 ♐ 21 10	09 ♐ 23 15	10 36	23 50	24 49	20 43	13 42	14 17	24 10	18 03
09	11 13 30	16 39 25	15 ♐ 22 57	21 ♐ 20 51	10 59	25 04	24 33	20 43	13 49	14 19	24 09	18 01
10	11 17 26	17 37 42	27 ♐ 17 32	03 ♑ 13 36	11 17	26 18	24 17	20 R 43	13 56	14 21	24 08	18 00
11	11 21 23	18 36 00	09 ♑ 09 37	15 ♑ 06 10	11 30	27 33	24 00	20 43	14 03	14 23	24 06	17 59
12	11 25 19	19 34 20	21 ♑ 03 47	27 ♑ 03 01	11 36	28 ♍ 47	23 44	20 42	14 09	14 25	24 05	17 58
13	11 29 16	20 32 42	03 ♒ 04 20	09 ♒ 08 13	11 R 37	00 ♎ 02	23 27	20 42	14 16	14 27	24 04	17 57
14	11 33 13	21 31 05	15 ♒ 15 02	21 ♒ 25 11	11 31	01 16	23 10	20 41	14 23	14 29	24 02	17 56
15	11 37 09	22 29 30	27 ♒ 38 57	03 ♓ 56 34	11 19	02 31	22 53	20 40	14 30	14 31	24 01	17 55
16	11 41 06	23 27 57	10 ♓ 18 14	16 ♓ 44 03	11 00	03 45	22 37	20 38	14 36	14 33	24 00	17 54
17	11 45 02	24 26 26	23 ♓ 14 03	29 ♓ 48 13	10 34	05 00	22 20	20 37	14 43	14 35	23 58	17 53
18	11 48 59	25 24 56	06 ♈ 26 27	13 ♈ 08 36	10 01	06 14	22 04	20 35	14 49	14 36	23 57	17 52
19	11 52 55	26 23 28	19 ♈ 54 27	26 ♈ 43 45	09 21	07 29	21 47	20 33	14 56	14 38	23 56	17 51
20	11 56 52	27 22 03	03 ♉ 36 11	10 ♉ 31 28	08 35	08 43	21 31	20 31	15 02	14 40	23 54	17 50
21	12 00 48	28 20 39	17 ♉ 29 13	24 ♉ 29 07	07 44	09 58	21 16	20 29	15 08	14 41	23 53	17 49
22	12 04 45	29 ♍ 19 18	01 ♊ 30 47	08 ♊ 33 55	06 47	11 12	21 00	20 26	15 15	14 43	23 51	17 48
23	12 08 42	00 ♎ 17 59	15 ♊ 38 10	22 ♊ 43 14	05 45	12 27	20 45	20 24	15 21	14 44	23 50	17 47
24	12 12 38	01 16 42	29 ♊ 48 51	06 ♋ 54 44	04 41	13 41	20 30	20 21	15 27	14 46	23 48	17 46
25	12 16 35	02 15 28	14 ♋ 00 38	21 ♋ 06 17	03 35	14 56	20 16	20 18	15 33	14 47	23 47	17 45
26	12 20 31	03 14 16	28 ♋ 11 28	05 ♌ 15 54	02 30	16 10	20 02	20 15	15 39	14 48	23 45	17 44
27	12 24 28	04 13 06	12 ♌ 19 20	19 ♌ 21 29	01 25	17 25	19 49	20 11	15 45	14 50	23 44	17 43
28	12 28 24	05 11 59	26 ♌ 22 04	03 ♍ 20 45	00 ♎ 25	18 39	19 36	20 08	15 51	14 51	23 42	17 43
29	12 32 21	06 10 53	10 ♍ 17 12	17 ♍ 11 07	29 ♍ 29	19 54	19 24	20 04	15 57	14 52	23 40	17 42
30	12 36 17	07 ♎ 09 50	24 ♍ 02 08	00 ♎ 49 57	28 ♍ 40	21 ♎ 08	19 ♓ 12	20 ♉ 00	16 ♌ 03	14 ♋ 53	23 ♈ 39	17 ♒ 41

MOON NODES AND LATITUDE

Date	Moon True ☊ ° '	Moon Mean ☊ ° '	Moon ☽ Latitude ° '
01	05 ♍ 29	05 ♍ 13	00 S 24
02	05 R 29	05 10	00 N 54
03	05 29	05 07	02 08
04	05 28	05 03	03 12
05	05 27	05 00	04 04
06	05 26	04 57	04 43
07	05 25	04 54	05 07
08	05 25	04 51	05 16
09	05 24	04 48	05 12
10	05 D 24	04 44	04 54
11	05 25	04 41	04 24
12	05 26	04 38	03 42
13	05 28	04 35	02 50
14	05 29	04 32	01 50
15	05 30	04 28	00 N 43
16	05 R 30	04 25	00 S 27
17	05 29	04 22	01 36
18	05 27	04 19	02 42
19	05 25	04 16	03 41
20	05 22	04 13	04 27
21	05 18	04 09	04 59
22	05 16	04 06	05 13
23	05 14	04 03	05 09
24	05 13	04 00	04 46
25	05 D 14	03 57	04 05
26	05 15	03 54	03 10
27	05 16	03 50	02 03
28	05 17	03 47	00 S 49
29	05 R 18	03 44	00 N 27
30	05 ♍ 16	03 ♍ 41	01 N 41

DECLINATIONS

Date	Sun ☉ ° '	Moon ☽ ° '	Mercury ☿ ° '	Venus ♀ ° '	Mars ♂ ° '	Jupiter ♃ ° '	Saturn ♄ ° '	Uranus ♅ ° '	Neptune ♆ ° '	Pluto ♇ ° '
01	08 N 14	10 N 41	04 S 45	07 N 09	06 S 56	16 N 42	17 N 33	23 N 02	07 N 50	23 S 23
02	07 52	06 37	05 13	06 40	07 00	16 42	17 31	23 02	07 49	23 23
03	07 30	02 N 17	05 40	06 11	07 05	16 42	17 29	23 02	07 49	23 23
04	07 08	02 S 04	06 06	05 42	07 10	16 42	17 27	23 02	07 48	23 24
05	06 46	06 11	06 29	05 12	07 15	16 42	17 25	23 01	07 48	23 24
06	06 23	09 55	06 51	04 42	07 20	16 42	17 23	23 01	07 48	23 24
07	06 01	13 06	07 11	04 12	07 25	16 42	17 21	23 01	07 47	23 25
08	05 39	15 38	07 28	03 42	07 29	16 42	17 19	23 01	07 47	23 25
09	05 16	17 28	07 44	03 12	07 34	16 42	17 18	23 01	07 46	23 25
10	04 53	18 30	07 56	02 42	07 39	16 42	17 16	23 00	07 46	23 26
11	04 30	18 44	08 06	02 11	07 44	16 42	17 14	23 00	07 45	23 26
12	04 08	18 08	08 13	01 41	07 48	16 42	17 12	23 00	07 45	23 26
13	03 45	16 43	08 17	01 11	07 53	16 41	17 10	23 00	07 44	23 27
14	03 22	14 31	08 17	00 40	07 57	16 41	17 08	23 00	07 44	23 27
15	02 59	11 37	08 14	00 N 09	08 01	16 40	17 06	22 59	07 43	23 27
16	02 36	08 07	08 07	00 S 21	08 05	16 40	17 05	22 59	07 42	23 27
17	02 12	04 S 10	07 56	00 52	08 09	16 39	17 03	22 59	07 42	23 28
18	01 49	00 N 04	07 41	01 22	08 12	16 39	17 01	22 59	07 41	23 28
19	01 26	04 22	07 21	01 53	08 16	16 38	16 59	22 59	07 41	23 28
20	01 03	08 31	06 58	02 24	08 19	16 37	16 58	22 59	07 40	23 28
21	00 39	12 15	06 30	02 54	08 22	16 37	16 56	22 58	07 40	23 29
22	00 N 16	15 20	05 58	03 25	08 24	16 36	16 54	22 58	07 39	23 29
23	00 S 07	17 32	05 23	03 55	08 27	16 35	16 52	22 58	07 39	23 29
24	00 31	18 40	04 45	04 26	08 29	16 34	16 51	22 58	07 38	23 29
25	00 54	18 38	04 05	04 56	08 31	16 33	16 49	22 58	07 37	23 30
26	01 17	17 25	03 23	05 26	08 32	16 32	16 47	22 58	07 37	23 30
27	01 41	15 08	02 41	05 56	08 34	16 31	16 46	22 58	07 36	23 30
28	02 04	11 58	01 59	06 26	08 35	16 30	16 44	22 58	07 36	23 30
29	02 27	08 08	01 19	06 56	08 35	16 29	16 42	22 57	07 35	23 30
30	02 S 51	03 N 55	00 S 41	07 S 26	08 S 36	16 N 28	16 N 41	22 N 57	07 N 34	23 S 30

ZODIAC SIGN ENTRIES

Date	h	m	Planets
01	10	00	☽ ♍
03	13	26	☽ ♎
05	19	44	☽ ♏
08	05	22	☽ ♐
10	17	28	☽ ♑
13	05	53	☽ ♒
13	11	24	♀ ♎
15	16	30	☽ ♓
18	00	21	☽ ♈
20	05	43	☽ ♉
22	09	25	☽ ♊
23	04	39	☉ ♎
24	12	19	☽ ♋
26	15	04	☽ ♌
28	18	14	☽ ♍
28	22	18	☿ ♍
30	22	31	☽ ♎

LATITUDES

Date	Mercury ☿ ° '	Venus ♀ ° '	Mars ♂ ° '	Jupiter ♃ ° '	Saturn ♄ ° '	Uranus ♅ ° '	Neptune ♆ ° '	Pluto ♇ ° '
01	02 S 35	01 N 25	06 S 02	01 S 15	00 N 38	00 N 21	01 S 43	08 S 26
04	03 02	01 24	06 00	01 15	00 38	00 21	01 43	08 26
07	03 26	01 22	05 57	01 16	00 38	00 21	01 43	08 26
10	03 47	01 20	05 52	01 16	00 39	00 21	01 43	08 25
13	04 01	01 18	05 45	01 17	00 39	00 21	01 43	08 25
16	04 06	01 14	05 37	01 17	00 39	00 21	01 43	08 25
19	03 58	01 11	05 27	01 18	00 40	00 21	01 43	08 25
22	03 34	01 06	05 16	01 18	00 40	00 21	01 43	08 25
25	02 54	01 02	05 04	01 18	00 41	00 21	01 43	08 24
28	01 59	00 57	04 51	01 19	00 41	00 21	01 43	08 24
31	00 S 59	00 N 51	04 S 38	01 S 19	00 N 42	00 N 21	01 S 44	08 S 24

DATA

Julian Date	2464572
Delta T	+76 seconds
Ayanamsa	24° 21' 13"
Synetic vernal point	04° ♓ 45' 46"
True obliquity of ecliptic	23° 25' 57"

MOON'S PHASES, APSIDES AND POSITIONS ☽

Date	h	m	Phase	Longitude ° '	Eclipse Indicator
02	01	59	●	09 ♍ 28	Total
09	14	47	☽	16 ♐ 46	
17	14	23	○	24 ♓ 32	
24	14	40	☾	01 ♋ 23	

Day	h	m		
10	23	24	Apogee	
25	13	37	Perigee	
04	00	31	0S	
11	06	36	Max dec	18° S 45'
18	11	36	0N	
24	23	15	Max dec	18° N 48'

LONGITUDES

Date	Chiron ⚷ ° '	Ceres ⚳ ° '	Pallas ⚴ ° '	Juno ⚵ ° '	Vesta ⚶ ° '	Black Moon Lilith ⚸ ° '
01	12 ♊ 18	22 ♊ 38	04 ♈ 04	00 ♓ 58	11 ♍ 02	04 ♑ 42
11	12 ♊ 27	25 ♊ 14	01 ♈ 43	28 ♒ 37	15 ♍ 53	05 ♑ 49
21	12 ♊ 28	27 ♊ 31	29 ♓ 46	26 ♒ 41	20 ♍ 45	06 ♑ 56
31	12 ♊ 23	29 ♊ 23	26 ♓ 21	25 ♒ 24	25 ♍ 40	08 ♑ 03

All ephemeris data is given at 12.00 UT and the Moon's longitude is additionally given for 24.00 UT

ASPECTARIAN

01 Saturday
00 29 ☽ △ ♆
01 41 ☿ ⚼ ♃
04 16 ☽ ⊼ ♂
08 21 ☽ ∠ ♅
09 50 ☽ ⊥ ☿
20 36 ☽ ⊻ ☿
21 46 ♀ # ♂
21 47 ☉ ⚼ ♆

02 Sunday
01 42 ☽ ⚼ ♆
01 59 ☽ ☌ ☉
04 10 ☽ ∥ ☉
05 06 ☽ ∥ ♆
07 58 ☽ ⊻ ♄
09 49 ☽ ⚹ ♅
09 49 ☽ # ♂
11 39 ☽ ∥ ♀
13 59 ☽ ☌ ♀
14 56 ☉ ∥ ♆
16 51 ☽ ⊼ ♇
17 04 ☽ ± ♆
18 20 ☽ ⊥ ♄
19 03 ☽ # ☿
21 08 ☽ △ ♃

03 Monday
03 13 ☽ ± ♇
03 26 ☽ ⊼ ♆
06 36 ☽ Q ♅
06 38 ☽ ☍ ♂
10 06 ☽ ∠ ♄
18 57 ☽ ⚼ ♇
21 50 ♀ ⊼ ♇
23 26 ☽ ⚼ ♃

04 Tuesday
00 15 ♀ ± ♆
01 24 ☿ # ♀
03 45 ☽ ☌ ☿
10 09 ☉ # ♂
10 17 ☽ ⊻ ☉
12 59 ☽ ⚹ ♄
14 37 ☽ □ ♅
15 35 ☽ ± ♃
19 47 ♀ ⊥ ♄
21 49 ☽ △ ♇
21 59 ☽ ⊥ ☉

05 Wednesday
00 21 ☽ ⊻ ♀
02 31 ☽ ⊼ ♃
06 43 ☽ # ♀
09 02 ☽ ☍ ♆
11 05 ☽ Q ♄
11 32 ☽ ⊼ ♂
12 32 ☽ ⊥ ♀
14 02 ☽ ∥ ☿
15 14 ☽ # ☉
15 53 ☽ ∠ ☉
18 42 ☽ ∥ ♂
20 55 ☉ # ☿
22 01 ☽ # ♆
22 29 ☽ ± ♂
23 26 ♀ △ ♃

06 Thursday
04 06 ☉ ⊻ ♄
07 11 ☽ ∠ ♀
13 56 ☽ ⊻ ☿
15 09 ☽ ⚼ ♂
21 17 ☽ □ ♄
22 34 ☽ ⚹ ☉
22 40 ☽ △ ♅

07 Friday
00 03 ☉ ⚹ ♅
01 56 ☽ ⊥ ☿
06 04 ☽ □ ♇
11 11 ☽ ☍ ♃
15 09 ☽ ⚹ ♀
17 57 ☽ ⊼ ♆
19 33 ☽ △ ♂
20 10 ☽ ∠ ☿
23 51 ☽ Q ☉

08 Saturday
03 57 ☽ ⚼ ♅
05 43 ☽ ± ♆
14 25 ☿ ∥ ♂
16 06 ♀ ± ♇
17 20 ☽ Q ♇
17 28 ☽ Q ♀
18 32 ♀ ⊼ ♆
21 50 ☽ ± ♅
23 33 ☽ ⚼ ♆

09 Sunday
00 48 ☽ # ♃
02 56 ☽ ⚹ ☿
03 45 ♀ ☍ ♂
08 49 ☽ △ ♄
09 19 ☽ # ♄
09 52 ☽ ⊼ ♅
14 15 ♃ St R
14 47 ☽ □ ☉
16 14 ☿ # ♆
17 18 ☽ ⚹ ♇
22 44 ☽ ⊼ ♃

10 Monday
03 43 ☽ Q ☿
05 37 ☽ △ ♆
06 03 ☽ □ ♂
09 47 ☽ □ ♀
10 50 ☽ ± ♃
12 57 ♀ Q ♅
15 20 ☽ ⚼ ♄
21 06 ☉ ⊼ ♇
23 32 ☽ ∠ ♇

11 Tuesday
00 06 ☉ ± ♆
02 08 ♂ ⊻ ♆
05 02 ☽ ⚼ ♃
09 43 ☽ ± ♄
13 40 ♂ ⊥ ♇
16 46 ☽ □ ☿
17 37 ☽ Q ♂
17 42 ☽ ⊥ ♇
19 00 ♂ # ♆
21 57 ☽ ⊼ ♄
22 35 ☽ ☍ ♅

12 Wednesday
05 47 ☽ ⊻ ♇
08 44 ☽ △ ☉
11 17 ☽ △ ♃
17 13 ☽ ⚹ ♂
18 03 ☽ □ ♆
19 48 ♀ ∠ ♄

13 Thursday
02 27 ☿ St R
04 18 ☉ ⊥ ♄
05 15 ☽ △ ♀
05 28 ☽ # ♄
12 18 ☽ # ♃
15 36 ☉ △ ♃
17 20 ☽ ⚼ ☉
22 24 ☽ ∠ ♂

14 Friday
04 46 ☽ △ ☿
05 44 ☽ Q ♆
10 17 ☽ ☍ ♄
10 30 ☽ ⊼ ♅
12 34 ☽ ± ☉
14 13 ☽ ⚼ ♀
15 40 ☽ ⊥ ♂
17 13 ☽ ☌ ♇
22 13 ☽ ± ♅
22 33 ☽ □ ♃

15 Saturday
01 14 ☽ ⊼ ☉
03 03 ☽ ⊻ ♂
05 02 ☽ ⚹ ♆
09 29 ☽ ⚼ ☿
09 35 ☽ ± ♀
15 35 ☽ ⚼ ♅
19 29 ♄ ⊻ ♅
19 37 ♀ ⚼ ♇
19 39 ☉ ☍ ♂
22 18 ☽ ⊼ ♀

16 Sunday
02 16 ☽ ± ☿
08 53 ☽ Q ♃
09 33 ☽ ∠ ♆
11 58 ☽ ∥ ☿
12 11 ☽ ∥ ♂
13 16 ☽ ⊼ ☿
14 34 ☽ # ♆
15 52 ☿ ∥ ♂
19 57 ☽ △ ♅
20 06 ☽ ⊼ ♄
22 24 ☉ ± ♇

17 Monday
00 45 ☉ ⊼ ♆
02 08 ☽ ⊻ ♇
02 19 ☽ ⊥ ♆
07 11 ☽ ⚹ ♃
07 19 ☽ ± ♄
10 23 ☽ ☌ ♂
13 11 ☽ ⊥ ♇
13 21 ☽ ⊻ ♆
14 23 ☽ ☍ ☉
23 42 ♀ ⚼ ♃
23 56 ☽ ⚼ ♄

18 Tuesday
00 20 ☽ # ☉
04 47 ☽ ∥ ♀
05 33 ☽ ∠ ♇
10 28 ☽ ∠ ♃
10 58 ☿ # ♆
11 36 ☽ ☍ ♀
18 08 ☽ ☍ ☿
20 13 ☽ # ♀
20 57 ☽ ∥ ☉
23 59 ☉ # ♀

19 Wednesday
02 32 ☽ ⊥ ♃
02 38 ☽ □ ♅
03 06 ☽ △ ♄
08 21 ☽ ⚹ ♇
13 08 ☽ ⊻ ♃
15 16 ☽ ⊻ ♂
18 07 ☉ Q ♅
19 04 ☽ ☌ ♆

20 Thursday
00 17 ☽ ⊼ ☉
01 36 ☽ ⊥ ♂
03 37 ☽ # ☿
05 26 ☽ Q ♇
06 58 ☽ ∥ ♆
10 21 ☽ Q ♅
10 30 ☿ ☌ ♀
10 45 ☽ # ♂
11 34 ☽ ± ☉
16 59 ☽ ∠ ♂
20 10 ☽ ⊼ ☿
21 45 ☽ ⊼ ♀

21 Friday
04 20 ☽ ⚼ ☉
05 55 ☽ ± ☿
07 10 ☽ ⚹ ♅
07 56 ☽ □ ♄
09 07 ☽ ± ♀
12 33 ☽ □ ♇
17 08 ☽ ☌ ♃
18 22 ☽ ⚹ ♂
19 48 ♂ ± ♄
20 26 ☽ ⚼ ☿
22 56 ☽ ⊻ ♆

22 Saturday
02 03 ☽ ⚼ ♀
07 59 ☽ △ ☉
08 55 ☽ ∠ ♅
09 10 ☽ ⊥ ♆
14 30 ☽ Q ♂
14 59 ☽ Q ♄
20 22 ☽ △ ☿

23 Sunday
00 16 ☽ ⊥ ♅
00 19 ☽ ∥ ♃
00 28 ☽ ∠ ♆
03 36 ☽ ∥ ♄
06 04 ☽ △ ♀
10 28 ☽ ⊻ ♅
11 31 ☽ ⚹ ♄
13 24 ☉ ∠ ♄
15 38 ☽ △ ♇
20 02 ☽ ⊻ ♃
20 31 ☽ □ ♂
20 32 ☿ ⚼ ♃

24 Monday
01 51 ☽ ⚹ ♆
06 10 ☽ ⊥ ♃
13 05 ☽ ∠ ♄
14 40 ☽ □ ☉
16 59 ☽ ⚼ ♇
18 47 ☿ ∥ ♀
19 39 ☽ □ ☿
21 19 ☽ ∠ ♃
22 06 ☽ Q ♆

25 Tuesday
00 18 ♀ ± ♃
04 25 ☽ ⊥ ♄
07 52 ♂ ⚹ ♃
08 11 ☽ ± ♇
09 09 ♀ □ ♅
13 18 ☽ ☌ ♅
13 42 ☽ □ ♀
14 38 ☽ ⊻ ♄
18 19 ☽ ⊼ ♇
22 25 ☽ △ ♂
22 36 ☽ ⚹ ♃
23 21 ☽ Q ☉
23 54 ☽ Q ☿
23 55 ☉ ⚼ ♇

26 Wednesday
01 14 ♀ ⚹ ♄
03 22 ☉ ☌ ☿
04 30 ☽ □ ♆
06 34 ☿ ⚼ ♇
18 47 ☽ ⚹ ☿
18 51 ☽ Q ♃
19 56 ☽ ∥ ♄
21 12 ☽ ⚹ ☉
22 40 ☽ ∥ ♃
23 07 ☽ Q ♀
23 26 ☽ ⚼ ♂

27 Thursday
14 30 ☽ ± ♂
16 16 ☽ ⊻ ♅
17 54 ☽ ☌ ♄
17 57 ♀ △ ♇
18 31 ☽ ∠ ☿
21 12 ☽ ☍ ♇
21 31 ☽ ⚹ ♀

28 Friday
00 35 ☽ ⊼ ♂
00 39 ☽ ∠ ☉
01 21 ☽ □ ♃
02 12 ☿ ∠ ♄
02 32 ☽ ⊥ ♅
07 26 ☽ △ ♆
08 52 ☽ ⊥ ☿
10 17 ☉ ⚼ ♃
10 19 ☉ ∥ ☿
17 14 ☽ ⊥ ☉
17 59 ☽ ∠ ♅
18 30 ☽ ⊻ ☿

29 Saturday
01 46 ☽ ∠ ♀
03 40 ♀ ⊼ ♂
04 21 ☽ ⊻ ☉
09 13 ☽ ⚼ ♆
09 16 ☽ # ♂
15 02 ♀ ⊼ ♃
15 13 ☽ ∥ ♆
18 13 ☽ # ♀
18 53 ☽ ⊥ ♀
19 58 ☽ ⚹ ♅
21 55 ☽ ⊻ ♄

30 Sunday
00 50 ☽ ± ♆
00 53 ☽ ⊼ ♇
03 38 ☽ ☍ ♂
04 57 ☽ △ ♃
06 24 ☽ ⊻ ♀
08 29 ☽ ⊥ ♄
11 19 ☽ ⊼ ♆
11 23 ☽ ± ♇
17 02 ☽ Q ♅
17 25 ☽ # ☉
18 54 ♀ # ♆
19 44 ☽ ☌ ☿

OCTOBER 2035

LONGITUDES

Date	Sidereal time h m s	Sun ☉ ° ' "	Moon ☽ ° ' "	Moon ☽ 24.00 ° ' "	Mercury ☿ ° '	Venus ♀ ° '	Mars ♂ ° '	Jupiter ♃ ° '	Saturn ♄ ° '	Uranus ♅ ° '	Neptune ♆ ° '	Pluto ♇ ° '
01	12 40 14	08 ♎ 08 48	07 ♎ 34 15	14 ♎ 14 47	27 ♍ 58	22 ♎ 23	19 ♓ 01	19 ♉ 55	16 ♌ 09	14 ♋ 54	23 ♈ 37	17 ♒ 40
02	12 44 11	09 07 49	20 ♎ 51 18	27 ♎ 23 40	27 R 25	23 37	18 R 51	19 R 51	16 15	14 55	23 R 36	17 R 39
03	12 48 07	10 06 52	03 ♏ 51 47	10 ♏ 15 36	27 03	24 52	18 41	19 47	16 20	14 56	23 34	17 39
04	12 52 04	11 05 57	16 ♏ 35 12	22 ♏ 50 41	26 50	26 06	18 32	19 42	16 26	14 57	23 32	17 38
05	12 56 00	12 05 03	29 ♏ 02 16	05 ♐ 10 14	26 D 48	27 21	18 24	19 37	16 31	14 58	23 31	17 37
06	12 59 57	13 04 12	11 ♐ 14 56	17 ♐ 16 46	26 56	28 35	18 17	19 32	16 37	14 59	23 29	17 37
07	13 03 53	14 03 22	23 ♐ 16 13	29 ♐ 13 47	27 15	29 ♎ 50	18 10	19 27	16 42	15 00	23 27	17 36
08	13 07 50	15 02 34	05 ♑ 10 02	11 ♑ 05 34	27 43	01 ♏ 05	18 04	19 21	16 47	15 00	23 26	17 35
09	13 11 46	16 01 48	17 ♑ 01 00	22 ♑ 56 57	28 21	02 19	17 59	19 16	16 52	15 01	23 24	17 35
10	13 15 43	17 01 04	28 ♑ 54 04	04 ♒ 53 01	29 ♍ 08	03 34	17 55	19 10	16 58	15 02	23 22	17 34
11	13 19 40	18 00 21	10 ♒ 54 23	16 ♒ 58 48	00 ♎ 03	04 48	17 51	19 04	17 03	15 02	23 21	17 34
12	13 23 36	18 59 40	23 ♒ 06 51	29 ♒ 19 02	01 05	06 03	17 49	18 58	17 08	15 03	23 19	17 33
13	13 27 33	19 59 01	05 ♓ 35 50	11 ♓ 57 39	02 13	07 17	17 47	18 52	17 12	15 03	23 17	17 33
14	13 31 29	20 58 24	18 ♓ 24 46	24 ♓ 57 25	03 28	08 32	17 46	18 46	17 17	15 03	23 16	17 32
15	13 35 26	21 57 49	01 ♈ 35 42	08 ♈ 19 33	04 47	09 46	17 D 45	18 39	17 22	15 04	23 14	17 32
16	13 39 22	22 57 15	15 ♈ 08 51	22 ♈ 03 17	06 10	11 01	17 46	18 33	17 27	15 04	23 12	17 31
17	13 43 19	23 56 44	29 ♈ 02 26	06 ♉ 05 46	07 38	12 15	17 47	18 26	17 31	15 04	23 11	17 31
18	13 47 15	24 56 15	13 ♉ 12 40	20 ♉ 22 25	09 08	13 30	17 49	18 19	17 36	15 04	23 09	17 31
19	13 51 12	25 55 48	27 ♉ 34 16	04 ♊ 47 26	10 41	14 44	17 52	18 12	17 40	15 05	23 07	17 30
20	13 55 09	26 55 23	12 ♊ 01 11	19 ♊ 14 47	12 15	15 59	17 56	18 05	17 44	15 05	23 06	17 30
21	13 59 05	27 55 00	26 ♊ 27 36	03 ♋ 39 05	13 52	17 13	18 00	17 58	17 48	15 R 05	23 04	17 30
22	14 03 02	28 54 40	10 ♋ 48 46	17 ♋ 56 17	15 30	18 28	18 05	17 51	17 53	15 05	23 02	17 30
23	14 06 58	29 ♎ 54 22	25 ♋ 01 24	02 ♌ 03 54	17 09	19 42	18 11	17 43	17 57	15 05	23 01	17 29
24	14 10 55	00 ♏ 54 06	09 ♌ 03 44	16 ♌ 00 50	18 48	20 57	18 18	17 36	18 01	15 04	22 59	17 29
25	14 14 51	01 53 52	22 ♌ 55 12	29 ♌ 46 52	20 29	22 11	18 25	17 28	18 04	15 04	22 57	17 29
26	14 18 48	02 53 41	06 ♍ 35 52	13 ♍ 22 14	22 09	23 26	18 33	17 20	18 08	15 04	22 56	17 29
27	14 22 44	03 53 32	20 ♍ 05 59	26 ♍ 47 08	23 50	24 40	18 42	17 13	18 12	15 04	22 54	17 29
28	14 26 41	04 53 25	03 ♎ 25 37	10 ♎ 01 25	25 31	25 55	18 52	17 05	18 15	15 03	22 52	17 29
29	14 30 38	05 53 20	16 ♎ 34 26	23 ♎ 04 34	27 11	27 09	19 02	16 57	18 19	15 03	22 51	17 29
30	14 34 34	06 53 17	29 ♎ 31 46	05 ♏ 55 53	28 ♎ 52	28 24	19 13	16 49	18 22	15 02	22 49	17 29
31	14 38 31	07 ♏ 53 16	12 ♏ 16 53	18 ♏ 34 42	00 ♏ 32	29 ♏ 38	19 ♓ 24	16 ♉ 41	18 ♌ 25	15 ♋ 02	22 ♈ 48	17 ♒ 28

Date	Moon True ☊ ° '	Moon Mean ☊ ° '	Moon ☽ Latitude ° '
01	05 ♍ 13	03 ♍ 38	02 N 47
02	05 R 09	03 34	03 43
03	05 03	03 31	04 26
04	04 57	03 28	04 55
05	04 51	03 25	05 09
06	04 45	03 22	05 08
07	04 41	03 19	04 54
08	04 39	03 15	04 28
09	04 D 38	03 12	03 50
10	04 39	03 09	03 02
11	04 41	03 06	02 05
12	04 42	03 03	01 N 02
13	04 R 43	03 00	00 S 05
14	04 42	02 56	01 13
15	04 39	02 53	02 20
16	04 33	02 50	03 20
17	04 26	02 47	04 10
18	04 18	02 44	04 46
19	04 10	02 40	05 04
20	04 02	02 37	05 03
21	03 57	02 34	04 43
22	03 53	02 31	04 05
23	03 51	02 28	03 13
24	03 D 52	02 25	02 09
25	03 52	02 21	00 S 58
26	03 R 53	02 18	00 N 15
27	03 51	02 15	01 26
28	03 48	02 12	02 31
29	03 41	02 09	03 27
30	03 32	02 06	04 11
31	03 ♍ 21	02 ♍ 02	04 N 42

DECLINATIONS

Date	Sun ☉ ° '	Moon ☽ ° '	Mercury ☿ ° '	Venus ♀ ° '	Mars ♂ ° '	Jupiter ♃ ° '	Saturn ♄ ° '	Uranus ♅ ° '	Neptune ♆ ° '	Pluto ♇ ° '
01	03 S 14	00 S 27	00 S 05	07 S 55	08 S 36	16 N 27	16 N 39	22 N 57	07 N 34	23 S 30
02	03 37	04 41	00 N 26	08 25	08 36	16 25	16 38	22 57	07 33	23 31
03	04 00	08 37	00 53	08 54	08 35	16 24	16 36	22 57	07 33	23 31
04	04 23	12 05	01 15	09 23	08 34	16 23	16 34	22 57	07 32	23 31
05	04 47	14 55	01 32	09 52	08 33	16 21	16 33	22 57	07 31	23 31
06	05 10	17 02	01 43	10 20	08 32	16 20	16 31	22 57	07 31	23 31
07	05 33	18 22	01 50	10 49	08 30	16 18	16 30	22 57	07 30	23 31
08	05 55	18 52	01 51	11 17	08 28	16 17	16 29	22 57	07 30	23 31
09	06 18	18 33	01 47	11 45	08 25	16 15	16 27	22 57	07 29	23 31
10	06 41	17 24	01 38	12 12	08 22	16 13	16 26	22 57	07 28	23 31
11	07 04	15 29	01 24	12 40	08 19	16 12	16 24	22 57	07 28	23 31
12	07 26	12 50	01 07	13 07	08 16	16 10	16 23	22 57	07 27	23 31
13	07 49	09 32	00 45	13 33	08 12	16 08	16 22	22 57	07 26	23 31
14	08 11	05 42	00 N 20	14 00	08 08	16 07	16 20	22 57	07 26	23 31
15	08 33	01 S 30	00 S 07	14 26	08 04	16 05	16 19	22 57	07 25	23 31
16	08 55	02 N 53	00 38	14 51	07 59	16 03	16 18	22 57	07 25	23 31
17	09 17	07 14	01 11	15 17	07 54	16 01	16 17	22 57	07 24	23 31
18	09 39	11 15	01 46	15 41	07 49	15 59	16 15	22 57	07 23	23 31
19	10 01	14 40	02 22	16 06	07 44	15 57	16 14	22 57	07 23	23 31
20	10 22	17 13	03 00	16 30	07 38	15 55	16 13	22 57	07 22	23 31
21	10 44	18 40	03 39	16 54	07 32	15 53	16 12	22 57	07 21	23 31
22	11 05	18 55	04 19	17 17	07 26	15 51	16 11	22 57	07 21	23 31
23	11 26	17 58	05 00	17 40	07 19	15 49	16 10	22 57	07 20	23 31
24	11 47	15 55	05 41	18 02	07 12	15 47	16 09	22 57	07 20	23 31
25	12 08	12 57	06 22	18 24	07 05	15 45	16 08	22 57	07 19	23 31
26	12 28	09 19	07 04	18 46	06 58	15 43	16 07	22 57	07 18	23 31
27	12 49	05 14	07 45	19 07	06 51	15 41	16 06	22 57	07 18	23 31
28	13 09	00 N 57	08 27	19 27	06 43	15 38	16 05	22 57	07 17	23 31
29	13 29	03 S 20	09 08	19 47	06 35	15 36	16 04	22 57	07 17	23 30
30	13 49	07 23	09 49	20 06	06 27	15 34	16 03	22 57	07 16	23 30
31	14 S 08	11 S 02	10 S 30	20 S 25	06 S 19	15 N 32	16 N 02	22 N 57	07 N 15	23 S 30

ZODIAC SIGN ENTRIES

Date	h	m	Planets
03	04	49	☽ ♏
05	13	53	☽ ♐
07	15	13	♀ ♏
08	01	33	☽ ♑
10	14	13	☽ ♒
11	10	49	☿ ♎
13	01	19	☽ ♓
15	09	08	☽ ♈
17	13	38	☽ ♉
19	16	02	☽ ♊
21	17	54	☽ ♋
23	14	16	☉ ♏
23	20	28	☽ ♌
26	00	23	☽ ♍
28	05	48	☽ ♎
30	12	53	☽ ♏
31	04	18	☿ ♏
31	19	06	♀ ♐

LATITUDES

Date	Mercury ☿ ° '	Venus ♀ ° '	Mars ♂ ° '	Jupiter ♃ ° '	Saturn ♄ ° '	Uranus ♅ ° '	Neptune ♆ ° '	Pluto ♇ ° '
01	00 S 59	00 N 51	04 S 38	01 S 19	00 N 42	00 N 21	01 S 44	08 S 24
04	00 S 01	00 45	04 23	01 20	00 42	00 22	01 44	08 23
07	00 N 48	00 38	04 09	01 20	00 42	00 22	01 44	08 23
10	01 24	00 32	03 54	01 20	00 43	00 22	01 44	08 23
13	01 47	00 25	03 40	01 20	00 43	00 22	01 44	08 22
16	01 59	00 17	03 25	01 21	00 44	00 22	01 44	08 22
19	02 01	00 10	03 11	01 21	00 44	00 22	01 44	08 21
22	01 56	00 N 02	02 57	01 21	00 45	00 22	01 44	08 21
25	01 45	00 S 06	02 44	01 21	00 45	00 22	01 44	08 20
28	01 31	00 14	02 31	01 21	00 46	00 22	01 44	08 20
31	01 N 14	00 S 22	02 S 18	01 S 21	00 N 46	00 N 22	01 S 44	08 S 19

DATA

Julian Date	2464602
Delta T	+76 seconds
Ayanamsa	24° 21' 15"
Synetic vernal point	04° ♓ 45' 44"
True obliquity of ecliptic	23° 25' 57"

MOON'S PHASES, APSIDES AND POSITIONS ☽

Date	h	m	Phase	Longitude ° '	Eclipse Indicator
01	13	07	●	08 ♎ 12	
09	09	49	☽	15 ♑ 56	
17	02	35	○	23 ♈ 33	
23	20	57	☾	00 ♌ 17	
31	02	59	●	07 ♏ 31	

Day	h	m		
08	19	02	Apogee	
20	19	50	Perigee	
01	09	33	0S	
08	14	43	Max dec	18° S 53'
15	20	17	0N	
22	04	52	Max dec	18° N 58'
28	17	16	0S	

LONGITUDES

Date	Chiron ⚷ ° '	Ceres ⚳ ° '	Pallas ⚴ ° '	Juno ⚵ ° '	Vesta ⚶ ° '	Black Moon Lilith ⚸ ° '
01	12 ♊ 23	29 ♊ 23	26 ♓ 21	25 ♒ 24	25 ♍ 40	08 ♑ 03
11	12 ♊ 11	00 ♋ 48	23 ♓ 49	24 ♒ 53	00 ♎ 35	09 ♑ 10
21	11 ♊ 53	01 ♋ 39	21 ♓ 41	25 ♒ 10	05 ♎ 31	10 ♑ 17
31	11 ♊ 29	01 ♋ 53	20 ♓ 08	26 ♒ 13	10 ♎ 26	11 ♑ 24

All ephemeris data is given at 12.00 UT and the Moon's longitude is additionally given for 24.00 UT

ASPECTARIAN

h m	Aspects
01 Monday	
00 29	☽ ∠ ♄
03 16	☽ Ꝑ ♇
07 18	☽ Ꝑ ♃
08 36	☽ # ☿
10 17	☽ ‖ ☿
13 07	☽ ☌ ☉
23 21	☽ ± ♃
02 Tuesday	
01 12	☽ □ ♅
03 33	☽ ✶ ♄
05 11	☽ ‖ ☉
06 11	☽ △ ♇
08 23	☽ ⊼ ♂
10 11	☽ ⊼ ♃
11 27	♀ ☍ ♆
17 00	☽ ☍ ♆
17 36	☽ ☌ ♀
19 13	☽ ± ♂
20 53	♀ ‖ ♂
23 40	☽ ⊻ ☿
03 Wednesday	
01 40	☽ Q ♄
05 10	☽ # ♆
08 56	♀ ± ♂
10 30	☽ ⊥ ☿
11 41	☽ Ꝑ ♂
11 45	☽ ‖ ♂
14 03	☽ ‖ ♀
21 00	☿ Q ♅
04 Thursday	
00 43	☽ ⊻ ☉
03 05	☽ ∠ ☿
08 53	☽ △ ♅
11 42	☽ □ ♄
13 04	☽ ⊥ ☉
14 00	☽ □ ♇
15 41	☽ △ ♂
17 55	☽ ☍ ♃
05 Friday	
01 11	☿ ⊻ ♀
01 19	☽ ⊼ ♆
04 53	☿ St D
07 38	☽ ✶ ☿
07 52	☽ ∠ ☉
08 21	☽ ⊻ ♀
12 55	☽ ± ♆
13 49	☽ Ꝑ ♅
21 23	☽ ⊥ ♀
06 Saturday	
00 53	☽ Q ♇
03 11	☽ # ♃
05 32	☽ # ♄
06 33	☽ Ꝑ ♆
07 20	☽ Q ☿
07 30	☽ ± ♅
12 24	♀ Q ♄
15 56	☽ ✶ ☉
16 43	☿ Q ♅
17 11	☽ ∠ ♀
19 25	☽ ⊼ ♅
22 18	☉ ± ♃
22 45	☽ △ ♄
07 Sunday	
00 39	☽ ✶ ♇
01 52	☽ □ ♂
04 23	☽ ⊼ ♃
12 22	☽ △ ♆
16 20	☽ ± ♃
18 07	☽ Q ☉
20 18	☽ □ ☿
08 Monday	
02 46	☽ ✶ ♀
05 07	☽ Ꝑ ♄
06 48	☽ ∠ ♇
10 22	☽ Ꝑ ♃
11 03	☉ □ ♅
13 49	☽ Q ♂
23 28	☽ ± ♄
09 Tuesday	
01 00	☽ ⊥ ♇
05 54	☽ Q ♀
07 57	☽ ☍ ♅
09 49	☽ □ ☉
11 42	☽ ⊼ ♄
13 09	☽ ⊻ ♇
13 57	☽ ✶ ♂
16 30	☽ △ ♃
10 Wednesday	
00 10	♀ Ꝑ ♂
00 53	☽ □ ♆
10 26	☉ ✶ ♄
12 30	☽ △ ☿
20 01	☽ ∠ ♂
22 26	☽ □ ♀
11 Thursday	
01 21	☉ △ ♇
01 30	☽ # ♄
04 00	☽ # ♃
08 33	☉ ⊼ ♂
12 52	☽ Q ♆
13 53	☽ ⊥ ♂
20 10	☽ ⊼ ♅
20 56	☽ Ꝑ ☿
12 Friday	
00 12	☽ ☍ ♄
01 08	☽ ☌ ♇
01 40	☽ ⊻ ♂
03 15	☽ △ ☉
03 58	☽ □ ♃
07 58	☽ ± ♅
10 01	☽ ‖ ♀
11 24	☉ ⊼ ♃
12 24	☽ ✶ ♆
12 53	☉ # ♆
16 12	☽ ± ☿
13 Saturday	
01 24	☽ Ꝑ ♅
04 54	☽ ⊼ ☿
10 44	☽ Ꝑ ☉
11 38	☿ ∠ ♄
14 23	☽ Q ♃
15 33	☽ △ ♀
17 05	☽ ∠ ♆
18 24	☿ Ꝑ ♇
20 51	☽ ‖ ♂
22 10	☽ ‖ ☉
14 Sunday	
01 31	☽ # ♆
05 06	☽ ± ☉
05 47	☽ △ ♅
09 17	☉ ‖ ♂
09 53	☽ ⊥ ♆
09 54	☽ ⊼ ♄
10 23	☽ ⊻ ♇
10 48	☽ ☌ ♂
12 38	☽ ✶ ♃
17 06	☽ ⊼ ☉
17 08	☿ Ꝑ ♃
20 53	☽ ⊻ ♆
21 00	☽ ± ♄
21 25	☽ ⊥ ♇
22 22	☽ Ꝑ ♀
15 Monday	
08 32	♂ St D
13 23	☽ Ꝑ ♄
13 41	☽ ∠ ♇
15 39	☽ ∠ ♃
16 17	☽ ± ♀
18 21	☽ ☍ ☿
18 50	☽ ‖ ☿
22 07	☽ # ☿
16 Tuesday	
04 01	☽ ⊼ ♀
07 28	☽ ⊥ ♃
11 52	☽ □ ♅
16 01	☽ △ ♄
16 09	☽ ✶ ♇
16 34	☽ ⊻ ♂
17 52	☽ ⊻ ♃
17 57	☉ ☍ ♆
17 Wednesday	
01 58	☽ ☌ ♆
02 35	☽ ☍ ☉
02 58	☽ ⊥ ♂
08 04	☉ ± ♂
12 12	♄ ☍ ♇
12 49	☽ Q ♇
12 59	☽ ‖ ♆
15 47	☽ # ♂
18 24	☽ ∠ ♂
18 52	☽ Q ♅
18 Thursday	
01 08	☽ # ☉
04 18	☽ ⊼ ☿
12 31	☽ ☍ ♀
15 08	☽ ✶ ♅
15 36	☽ ± ☿
19 13	☽ □ ♇
19 23	☽ □ ♄
19 45	☽ ✶ ♂
20 30	☽ ☌ ♃
19 Friday	
03 58	♀ # ♃
04 36	☽ ⊻ ♆
08 28	☽ Ꝑ ☿
09 04	☽ ⊼ ☉
14 35	☽ ⊥ ♆
15 50	☽ Q ♂
16 10	☽ ∠ ♅
18 35	♀ △ ♅
19 43	♀ # ♄
19 47	☽ ± ☉
22 48	☽ ‖ ♃
20 Saturday	
01 30	☽ ‖ ♄
01 31	☽ Q ♄
02 41	☽ # ♀
05 30	☽ ∠ ♆
07 07	☽ ⊥ ♅
09 33	☿ ± ♃
11 50	☽ Ꝑ ☉
12 27	☽ △ ☿
17 05	☽ ⊻ ♅
19 11	☽ ⊼ ♀
21 06	☽ △ ♇
21 32	☽ ✶ ♄
21 52	☽ □ ♂
21 59	☽ ⊻ ♃
21 Sunday	
06 06	☽ ± ♀
06 22	☽ ✶ ♆
07 11	♂ ✶ ♃
07 53	☽ ⊥ ♃
08 03	♅ St R
14 36	☽ △ ☉
17 22	♀ □ ♇
22 04	☽ Ꝑ ♇
22 31	☽ Ꝑ ♀
22 38	☽ ∠ ♄
22 46	☽ ∠ ♃
22 Monday	
00 03	♀ □ ♄
01 07	♀ ☍ ♃
02 20	☽ Q ♆
04 16	♀ △ ♂
05 50	☿ □ ♅
07 47	♃ □ ♄
13 09	☽ ± ♇
13 48	☽ ⊥ ♄
19 10	☽ ☌ ♅
20 55	☽ □ ☿
23 15	☽ ⊼ ♇
23 44	☽ ✶ ♃
23 57	☽ ⊻ ♄
23 Tuesday	
00 20	☽ △ ♂
02 07	☽ △ ♀
07 31	♂ # ♆
08 36	☽ □ ♆
12 57	☉ Q ♄
15 36	☽ # ♀
16 58	☿ △ ♇
19 44	☿ ⊼ ♃
19 55	☽ Q ♃
20 57	☽ □ ☉
24 Wednesday	
00 01	☿ ✶ ♄
02 02	☽ Ꝑ ♂
04 07	☿ ⊼ ♂
07 36	☽ Q ☿
09 46	☽ ‖ ♄
13 13	☽ ‖ ♃
17 37	☽ ± ♂
22 22	☽ ⊻ ♅
25 Thursday	
02 33	☽ ☍ ♇
02 36	☽ □ ♃
03 32	☽ ☌ ♄
04 06	☽ ⊼ ♂
06 20	☽ Q ☉
07 09	☽ ✶ ☿
08 47	☽ ⊥ ♅
09 16	♃ □ ♇
10 36	☽ □ ♀
12 04	☽ △ ♆
17 16	☽ # ☉
26 Friday	
00 30	☽ ∠ ♅
02 36	♀ ⊼ ♆
04 57	☽ ✶ ☉
09 20	☿ ‖ ♂
13 07	☽ ∠ ☿
14 21	☽ Ꝑ ♆
20 25	☿ # ♆
21 25	☽ Q ♀
22 57	☿ ☍ ♆
23 32	☽ # ☿
27 Saturday	
00 04	☽ ‖ ♆
02 26	☽ # ♂
03 01	☽ ✶ ♅
06 18	☽ ± ♆
06 37	☉ Ꝑ ♂
06 53	☽ △ ♃
07 19	☽ ⊼ ♇
07 22	☽ ⊥ ☿
08 35	☽ ⊻ ♄
09 28	☽ ☍ ♂
09 40	☽ ∠ ☉
17 01	☽ ⊼ ♆
18 03	☽ ± ♇
19 23	☽ ⊥ ♄
19 39	☽ ⊻ ☿
21 02	☽ ✶ ♀
28 Sunday	
00 29	☽ Q ♅
01 45	☿ ± ♂
03 08	☽ ⊥ ☉
09 35	☽ Ꝑ ♃
10 17	☽ Ꝑ ♇
11 41	☽ ∠ ♄
14 52	☽ ⊻ ☉
29 Monday	
01 48	☽ ± ♃
03 02	☽ ∠ ♀
09 12	☽ □ ♅
09 59	☿ ⊻ ♀
12 41	☽ ⊼ ♃
13 40	☽ △ ♇
15 13	☽ ✶ ♄
16 35	☽ ⊼ ♂
21 20	☽ ⊥ ♀
23 33	☽ ☍ ♆
30 Tuesday	
03 51	☽ ± ♂
06 31	☽ ‖ ♂
09 39	☽ ⊻ ♀
10 34	☽ ☌ ☿
11 19	☽ # ♆
13 35	☽ Q ♄
20 07	♀ ± ♆
20 54	☽ Ꝑ ♂
31 Wednesday	
02 59	☽ ☌ ☉
07 23	☽ ‖ ☿
10 21	☿ Q ♄
11 36	♇ St D
17 13	☽ △ ♅
19 38	♀ Ꝑ ♅
20 18	☽ ☍ ♃
21 53	☽ □ ♇
23 45	☽ □ ♄

NOVEMBER 2035

LONGITUDES

Date	Sidereal time h m s	Sun ☉ ° ' "	Moon ☽ ° ' "	Moon ☽ 24.00 ° ' "	Mercury ☿ ° '	Venus ♀ ° '	Mars ♂ ° '	Jupiter ♃ ° '	Saturn ♄ ° '	Uranus ♅ ° '	Neptune ♆ ° '	Pluto ♇ ° '
01	14 42 27	08 ♏ 53 17	24 ♏ 49 19	01 ♐ 00 47	02 ♏ 12	00 ♐ 52	19 ♓ 37	16 ♉ 33	18 ♌ 29	15 ♋ 01	22 ♈ 46	17 ♒ 29
02	14 46 24	09 53 20	07 ♐ 09 09	13 ♐ 14 36	03 52	02 07	19 50	16 R 25	18 32	15 R 01	22 R 44	17 D 29
03	14 50 20	10 53 25	19 ♐ 17 19	25 ♐ 17 35	05 32	03 21	20 03	16 17	18 35	15 00	22 43	17 29
04	14 54 17	11 53 31	01 ♑ 15 45	07 ♑ 12 11	07 11	04 36	20 17	16 09	18 38	14 59	22 41	17 29
05	14 58 13	12 53 39	13 ♑ 07 22	19 ♑ 01 48	08 49	05 50	20 32	16 01	18 40	14 58	22 40	17 29
06	15 02 10	13 53 48	24 ♑ 56 04	00 ♒ 50 44	10 28	07 05	20 47	15 53	18 43	14 58	22 38	17 29
07	15 06 07	14 54 00	06 ♒ 46 29	12 ♒ 43 57	12 06	08 19	21 03	15 44	18 46	14 57	22 37	17 29
08	15 10 03	15 54 12	18 ♒ 43 50	24 ♒ 46 50	13 43	09 34	21 20	15 36	18 48	14 56	22 35	17 29
09	15 14 00	16 54 26	00 ♓ 53 37	07 ♓ 04 51	15 21	10 48	21 37	15 28	18 50	14 55	22 34	17 30
10	15 17 56	17 54 42	13 ♓ 21 11	19 ♓ 43 11	16 58	12 02	21 55	15 20	18 53	14 54	22 32	17 30
11	15 21 53	18 54 59	26 ♓ 11 20	02 ♈ 46 01	18 34	13 17	22 13	15 12	18 55	14 53	22 31	17 30
12	15 25 49	19 55 17	09 ♈ 27 32	16 ♈ 16 00	20 10	14 31	22 32	15 04	18 57	14 52	22 29	17 30
13	15 29 46	20 55 37	23 ♈ 11 22	00 ♉ 13 23	21 46	15 46	22 51	14 56	18 59	14 50	22 28	17 31
14	15 33 42	21 55 59	07 ♉ 21 39	14 ♉ 35 33	23 22	17 00	23 11	14 48	19 01	14 49	22 26	17 31
15	15 37 39	22 56 22	21 ♉ 54 16	29 ♉ 16 51	24 57	18 14	23 31	14 40	19 02	14 48	22 25	17 32
16	15 41 36	23 56 47	06 ♊ 42 14	14 ♊ 09 17	26 32	19 29	23 52	14 32	19 04	14 47	22 24	17 32
17	15 45 32	24 57 13	21 ♊ 36 49	29 ♊ 03 44	28 06	20 43	24 13	14 24	19 06	14 45	22 22	17 32
18	15 49 29	25 57 41	06 ♋ 28 57	13 ♋ 51 35	29 ♏ 41	21 57	24 35	14 16	19 07	14 44	22 21	17 33
19	15 53 25	26 58 12	21 ♋ 10 51	28 ♋ 26 09	01 ♐ 15	23 12	24 57	14 08	19 08	14 42	22 20	17 33
20	15 57 22	27 58 43	05 ♌ 37 03	12 ♌ 43 16	02 49	24 26	25 19	14 00	19 09	14 41	22 18	17 34
21	16 01 18	28 59 17	19 ♌ 44 42	26 ♌ 41 21	04 23	25 40	25 42	13 53	19 10	14 39	22 17	17 35
22	16 05 15	29 ♏ 59 52	03 ♍ 33 18	10 ♍ 20 46	05 56	26 55	26 05	13 45	19 11	14 38	22 16	17 35
23	16 09 11	01 ♐ 00 29	17 ♍ 03 57	23 ♍ 43 08	07 30	28 09	26 29	13 38	19 12	14 36	22 15	17 36
24	16 13 08	02 01 08	00 ♎ 18 33	06 ♎ 50 30	09 03	29 ♐ 23	26 53	13 30	19 13	14 34	22 13	17 36
25	16 17 05	03 01 49	13 ♎ 19 12	19 ♎ 44 53	10 36	00 ♑ 38	27 17	13 23	19 14	14 33	22 12	17 37
26	16 21 01	04 02 31	26 ♎ 07 43	02 ♏ 27 51	12 09	01 52	27 42	13 16	19 14	14 31	22 11	17 38
27	16 24 58	05 03 14	08 ♏ 45 24	15 ♏ 00 28	13 42	03 06	28 07	13 09	19 14	14 29	22 10	17 38
28	16 28 54	06 03 59	21 ♏ 13 08	27 ♏ 23 25	15 14	04 20	28 33	13 02	19 15	14 27	22 09	17 39
29	16 32 51	07 04 46	03 ♐ 31 25	09 ♐ 37 09	16 47	05 35	28 59	12 55	19 15	14 25	22 08	17 40
30	16 36 47	08 ♐ 05 33	15 ♐ 40 44	21 ♐ 42 14	18 ♐ 19	06 ♑ 49	29 ♓ 25	12 ♉ 48	19 ♌ 15	14 ♋ 24	22 ♈ 07	17 ♒ 41

MOON NODES AND LATITUDE; DECLINATIONS

Date	Moon True ☊ ° '	Moon Mean ☊ ° '	Moon ☽ Latitude ° '	Sun ☉ ° '	Moon ☽ ° '	Mercury ☿ ° '	Venus ♀ ° '	Mars ♂ ° '	Jupiter ♃ ° '	Saturn ♄ ° '	Uranus ♅ ° '	Neptune ♆ ° '	Pluto ♇ ° '
01	03 ♍ 09	01 ♍ 59	04 N 59	14 S 27	14 S 08	11 S 10	20 S 43	06 S 10	15 N 30	16 N 01	22 N 57	07 N 15	23 S 30
02	02 R 56	01 56	05 01	14 46	16 32	11 50	21 01	06 01	15 27	16 00	22 57	07 14	23 30
03	02 45	01 53	04 50	15 05	18 11	12 29	21 18	05 52	15 25	16 00	22 58	07 14	23 30
04	02 36	01 50	04 26	15 24	19 00	13 08	21 35	05 43	15 23	15 59	22 58	07 13	23 29
05	02 29	01 46	03 50	15 42	18 58	13 46	21 51	05 34	15 21	15 58	22 58	07 13	23 29
06	02 25	01 43	03 05	16 00	18 06	14 23	22 06	05 24	15 18	15 58	22 58	07 12	23 29
07	02 23	01 40	02 12	16 18	16 27	15 00	22 21	05 15	15 16	15 57	22 58	07 12	23 29
08	02 D 23	01 37	01 12	16 36	14 04	15 36	22 35	05 05	15 14	15 56	22 58	07 11	23 29
09	02 23	01 34	00 N 08	16 53	11 02	16 11	22 48	04 55	15 12	15 56	22 58	07 10	23 28
10	02 R 23	01 31	00 S 58	17 10	07 26	16 45	23 01	04 45	15 09	15 55	22 58	07 10	23 28
11	02 21	01 27	02 02	17 27	03 S 23	17 19	23 13	04 34	15 07	15 55	22 59	07 09	23 28
12	02 16	01 24	03 03	17 43	00 N 57	17 51	23 24	04 23	15 05	15 54	22 59	07 09	23 28
13	02 09	01 21	03 55	17 59	05 22	18 23	23 35	04 13	15 03	15 54	22 59	07 08	23 27
14	01 59	01 18	04 34	18 15	09 39	18 54	23 45	04 02	15 00	15 53	22 59	07 08	23 27
15	01 47	01 15	04 56	18 30	13 28	19 24	23 54	03 51	14 58	15 53	22 59	07 07	23 27
16	01 36	01 11	04 59	18 45	16 30	19 53	24 03	03 40	14 56	15 53	22 59	07 07	23 26
17	01 25	01 08	04 42	19 00	18 28	20 21	24 11	03 28	14 54	15 52	23 00	07 06	23 26
18	01 16	01 05	04 06	19 14	19 11	20 47	24 18	03 17	14 52	15 52	23 00	07 06	23 26
19	01 10	01 02	03 14	19 29	18 34	21 13	24 24	03 05	14 49	15 52	23 00	07 06	23 26
20	01 06	00 59	02 10	19 42	16 45	21 38	24 30	02 54	14 47	15 52	23 00	07 05	23 25
21	01 D 05	00 56	01 S 00	19 56	13 57	22 02	24 35	02 42	14 45	15 52	23 00	07 05	23 25
22	01 R 05	00 52	00 N 13	20 09	10 24	22 25	24 39	02 30	14 43	15 51	23 00	07 04	23 25
23	01 03	00 49	01 23	20 21	06 23	22 46	24 43	02 18	14 41	15 51	23 01	07 04	23 24
24	01 03	00 46	02 28	20 34	02 N 08	23 07	24 46	02 06	14 39	15 51	23 01	07 03	23 24
25	00 58	00 43	03 25	20 45	02 S 08	23 26	24 48	01 54	14 37	15 51	23 01	07 03	23 24
26	00 50	00 40	04 07	20 57	06 15	23 44	24 49	01 41	14 35	15 51	23 01	07 03	23 23
27	00 40	00 37	04 39	21 08	10 01	24 01	24 49	01 29	14 33	15 51	23 02	07 02	23 23
28	00 26	00 33	04 56	21 19	13 18	24 16	24 49	01 16	14 31	15 51	23 02	07 02	23 23
29	00 12	00 30	04 59	21 29	15 57	24 31	24 48	01 04	14 30	15 52	23 02	07 01	23 22
30	29 ♌ 57	00 ♍ 27	04 N 49	21 S 39	17 S 53	24 S 44	24 S 46	00 S 51	14 N 28	15 N 52	23 N 02	07 N 01	23 S 22

ZODIAC SIGN ENTRIES

Date	h	m	Planets
01	22	02	☽ ♐
04	09	27	☽ ♑
06	22	17	☽ ♒
09	10	15	☽ ♓
11	18	59	☽ ♈
13	23	37	☽ ♉
16	01	10	☽ ♊
18	01	31	☽ ♋
18	16	53	☿ ♐
20	02	36	☽ ♌
22	05	46	☽ ♍
22	12	03	☉ ♐
24	11	26	☽ ♎
24	23	51	♀ ♑
26	19	19	☽ ♏
29	05	06	☽ ♐

LATITUDES

Date	Mercury ☿ ° '	Venus ♀ ° '	Mars ♂ ° '	Jupiter ♃ ° '	Saturn ♄ ° '	Uranus ♅ ° '	Neptune ♆ ° '	Pluto ♇ ° '
01	01 N 08	00 S 24	02 S 14	01 S 21	00 N 46	00 N 22	01 S 44	08 S 19
04	00 49	00 32	02 05	01 20	00 47	00 22	01 44	08 19
07	00 29	00 40	01 51	01 20	00 47	00 22	01 44	08 18
10	00 N 09	00 47	01 40	01 20	00 48	00 23	01 43	08 18
13	00 S 11	00 55	01 30	01 19	00 49	00 23	01 43	08 17
16	00 31	01 02	01 20	01 19	00 49	00 23	01 43	08 17
19	00 50	01 09	01 11	01 19	00 50	00 23	01 43	08 16
22	01 08	01 16	01 02	01 18	00 50	00 23	01 43	08 16
25	01 21	01 22	00 53	01 18	00 51	00 23	01 43	08 15
28	01 40	01 28	00 45	01 17	00 51	00 23	01 43	08 15
31	01 S 53	01 S 33	00 S 38	01 S 16	00 N 52	00 N 23	01 S 43	08 S 15

DATA

Julian Date	2464633
Delta T	+76 seconds
Ayanamsa	24° 21' 18"
Synetic vernal point	04° ♓ 45' 41"
True obliquity of ecliptic	23° 25' 57"

MOON'S PHASES, APSIDES AND POSITIONS ☽

Date	h	m	Phase	Longitude ° '	Eclipse Indicator
08	05	50	☽	15 ♒ 39	
15	13	49	○	23 ♉ 01	
22	05	16	☾	29 ♌ 43	
29	19	38	●	07 ♐ 24	

Day	h	m		
05	15	05	Apogee	
17	11	41	Perigee	
04	23	06	Max dec	19° S 05'
12	06	51	ON	
18	12	47	Max dec	19° N 11'
24	23	57	OS	

LONGITUDES

Date	Chiron ⚷ ° '	Ceres ⚳ ° '	Pallas ⚴ ° '	Juno ⚵ ° '	Vesta ⚶ ° '	Black Moon Lilith ⚸ ° '
01	11 ♊ 26	01 ♋ 53	20 ♓ 01	26 ♒ 22	10 ♎ 56	11 ♑ 31
11	10 ♊ 57	01 ♋ 23	19 ♓ 12	28 ♒ 11	15 ♎ 50	12 ♑ 38
21	10 ♊ 25	00 ♋ 50	19 ♓ 04	00 ♓ 38	20 ♎ 43	13 ♑ 45
31	09 ♊ 51	28 ♊ 29	19 ♓ 35	03 ♓ 36	25 ♎ 32	14 ♑ 51

All ephemeris data is given at 12.00 UT and the Moon's longitude is additionally given for 24.00 UT

ASPECTARIAN

h m	Aspects
01 Thursday	
01 48	☽ △ ♂
08 03	☽ ⚻ ♆
15 17	☽ ∥ ☉
19 37	☽ ± ♆
22 04	☽ ⚼ ♅
02 Friday	
00 29	☽ # ♃
01 03	☽ ☌ ♀
04 34	☽ ⚺ ☿
06 02	☽ # ♄
08 43	☽ Q ♇
13 09	☽ ⚼ ♆
15 39	☽ ± ♅
17 52	☽ ⚺ ☉
18 11	☽ ⊥ ☿
03 Saturday	
03 29	☽ ⚻ ♅
04 01	☿ ⚼ ♂
06 05	☽ ⚻ ♃
06 48	☽ ⊥ ☉
08 24	☽ ⚹ ♇
10 35	☽ △ ♄
13 33	☽ □ ♂
14 52	☽ ∠ ☿
17 55	☽ ± ♃
18 49	☽ △ ♆
04 Sunday	
02 24	☽ ∠ ☉
10 49	☉ # ♃
11 46	☽ ⚼ ♃
14 27	☽ ∠ ♇
16 47	☽ ⚼ ♄
19 31	☽ ⚺ ♀
05 Monday	
01 53	☽ ⚹ ☿
02 30	☽ Q ♂
05 04	♀ Q ♇
08 40	☽ ⊥ ♇
09 05	☽ ⊥ ♀
11 05	☽ ± ♄
11 30	☽ ⚹ ☉
15 45	☽ ☍ ♅
17 48	☽ △ ♃
20 51	☽ ⚺ ♇
23 19	☽ ⚻ ♄
06 Tuesday	
03 23	☽ ⚹ ♂
05 31	☽ ∠ ♀
06 10	☽ Q ☿
07 20	☽ □ ♆
08 28	☉ # ♄
14 08	☽ Q ☉
22 33	♀ ⚼ ♆
07 Wednesday	
10 31	☽ ∠ ♂
13 05	☉ △ ♅
13 31	☽ ∥ ☉
15 29	☽ ⚹ ♀
17 43	☽ # ♄
19 43	☽ Q ♆
22 08	☿ # ♃
23 58	♀ ± ♅
08 Thursday	
00 22	☽ ∥ ☿
00 25	☽ □ ☿
00 58	☽ # ♃
04 25	☽ ⚻ ♅
05 03	☽ ⊥ ♂
05 42	☉ ☍ ♃
05 50	☽ □ ♃
05 50	☽ □ ☉
09 31	☽ ☌ ♇
12 09	☽ ☍ ♄
16 22	☽ ± ♅
17 18	☽ ⚺ ♂
18 16	☽ Q ♀
19 39	☽ ⚹ ♆
09 Friday	
01 45	☿ # ♄
05 40	☿ △ ♅
10 05	☽ ⚼ ♅
13 41	☿ ☍ ♃
16 57	☽ Q ♃
10 Saturday	
00 54	☽ ∠ ♆
02 04	☉ □ ♇
07 32	♀ # ♅
09 14	☽ □ ♀
13 40	☽ # ♆
14 55	☽ △ ♅
15 43	☽ ⚹ ♃
18 01	☽ ⊥ ♆
19 48	☽ △ ☿
19 50	☽ ⚺ ♇
20 03	☿ □ ♇
21 21	☽ △ ☉
22 27	☽ ⚻ ♄
11 Sunday	
04 29	☽ ☌ ♂
04 57	☽ ∥ ♂
05 13	☽ ⚺ ♆
07 02	☽ ⊥ ♇
09 39	☽ ± ♄
11 57	☉ □ ♄
17 17	☿ □ ♄
19 16	☽ ∠ ♃
23 28	☉ ∥ ☿
23 32	☽ ∠ ♇
12 Monday	
01 56	☉ ☌ ☿
02 07	☽ ⚼ ♄
03 13	☽ ⚼ ☉
03 17	☽ ⚼ ☿
08 56	♂ ⚺ ♆
11 18	☽ ⊥ ♃
18 27	♀ ⚻ ♅
19 29	♀ ∥ ♇
20 31	☽ ± ☉
21 26	☽ ± ☿
21 26	♀ ⚻ ♃
21 31	☽ □ ♅
21 48	☽ ⚺ ♃
21 50	☽ △ ♀
13 Tuesday	
02 10	☽ ⚹ ♇
04 43	☽ △ ♄
05 55	☽ # ♂
07 48	☽ ⚻ ☉
09 14	☽ ⚻ ☿
10 45	☽ ☌ ♆
11 24	☽ ⚺ ♂
21 42	☽ ∥ ♆
21 54	☽ ⊥ ♂
22 19	☿ ⚻ ♆
22 48	☽ Q ♇
14 Wednesday	
02 09	☽ ⚼ ♀
04 24	☽ Q ♅
06 16	♃ ⚹ ♅
08 34	☿ △ ♂
13 24	☽ ∠ ♂
18 38	☽ ± ♀
22 06	♀ ⚹ ♇
23 49	☉ ⚻ ♆
15 Thursday	
00 13	☽ ☌ ♃
00 21	☽ ⚹ ♅
04 50	☽ □ ♇
05 27	☽ ⚻ ♀
07 18	☽ □ ♄
12 38	♂ ⊥ ♇
12 50	☽ ⚺ ♆
13 49	☽ ☍ ☉
14 42	☽ ⚹ ♂
17 34	☽ ☍ ☿
22 35	☽ ⊥ ♆
22 50	☽ ∥ ♃
16 Friday	
00 49	☽ ∠ ♅
03 50	♀ △ ♄
06 23	☽ ∥ ♄
08 57	☉ △ ♂
10 37	☽ Q ♂
12 35	☽ Q ♄
13 07	☽ ∠ ♆
15 20	☽ ⊥ ♅
17 Saturday	
00 29	☽ ⚺ ♃
00 59	☽ ⚺ ♅
05 27	☽ △ ♇
06 20	♀ ± ♃
07 56	☽ ⚹ ♄
10 03	☽ ⊥ ♃
10 26	☽ ☍ ♀
13 13	☽ ⚹ ♆
15 58	☿ ± ♆
16 17	☽ □ ♂
17 46	☽ ⚻ ☉
23 42	☽ ⚻ ☿
18 Sunday	
00 26	☽ ∠ ♃
04 09	☽ ± ☉
05 38	☽ ⚼ ♇
08 10	☽ ∠ ♄
08 33	☽ Q ♆
10 33	☽ ± ☿
12 45	☿ ⚼ ♅
19 27	♀ △ ♆
19 49	☽ ⚼ ☉
20 14	☽ ± ♇
22 48	☽ ⊥ ♄
19 Monday	
00 33	☽ ⚹ ♃
01 24	☽ ☌ ♅
02 56	☽ ⚼ ☿
06 03	☽ ⚻ ♇
08 38	☽ ⚺ ♄
13 53	☽ □ ♆
15 38	☽ ⚻ ♀
18 22	☽ △ ♂
20 07	☽ Q ♃
22 17	☽ △ ☉
20 Tuesday	
01 14	♂ ± ♄
02 31	☽ ± ♀
06 44	☽ △ ☿
19 03	☽ ⚼ ♀
19 36	☉ ± ♆
20 08	☽ ⚼ ♂
20 30	☽ ∥ ♄
21 Wednesday	
02 03	☽ □ ♃
03 18	☽ ⚺ ♅
05 43	☽ ∥ ♃
08 17	☽ ☍ ♇
11 01	☽ ☌ ♄
11 55	☽ ± ♂
12 40	♀ □ ♂
13 34	☽ ⊥ ♅
16 22	☽ △ ♆
22 34	☽ ⚻ ♂
23 15	☽ △ ♀
22 Thursday	
03 28	☉ ⚼ ♅
05 08	☽ ∠ ♅
05 16	☽ □ ☉
06 33	☿ Q ♇
16 44	☽ □ ☿
18 32	☽ ⚼ ♆
23 Friday	
05 54	☽ △ ♃
07 36	☽ ⚹ ♅
08 05	☽ ∥ ♆
08 11	☿ ⚼ ♆
10 31	☽ ± ♆
12 57	☽ ⚻ ♇
15 47	☽ Q ☉
15 51	☽ ⚺ ♄
20 26	♀ ⚼ ♃
21 19	☽ ⚻ ♆
23 47	☽ ± ♇
24 Saturday	
02 43	☽ ⊥ ♄
04 49	☿ ± ♅
05 09	☿ # ♅
05 12	☽ Q ♅
05 15	☽ Q ☿
05 33	☽ ☍ ♂
08 44	☽ ⚼ ♃
10 09	☽ □ ♀
12 14	☽ # ♂
15 24	☽ ⚹ ☉
16 13	☽ ⚼ ♇
19 10	☽ ∠ ♄
25 Sunday	
01 06	☽ ± ♃
06 16	☽ ⚹ ☿
09 18	☿ ∥ ♇
10 42	☽ ∥ ♂
12 07	☽ ⚻ ♃
14 17	☽ □ ♅
20 01	☽ △ ♇
21 32	☽ ∠ ☉
22 57	☽ Q ♀
23 02	☽ ⚹ ♄
26 Monday	
04 35	☽ ☍ ♆
14 11	☽ ∠ ☿
15 04	☽ ⚻ ♂
15 56	☽ ⊥ ☉
16 53	☽ # ♆
21 40	☽ Q ♄
27 Tuesday	
00 02	☽ ⚹ ♀
02 51	☽ ± ♂
02 58	♀ ∠ ♇
04 08	☿ ⚻ ♃
04 19	☽ ⚺ ☉
09 41	☽ ⊥ ☿
13 10	♂ ∠ ♃
20 21	☽ ☍ ♃
20 40	☽ ⚼ ♂
22 48	☽ ⚺ ☿
22 58	☽ △ ♅
28 Wednesday	
00 06	☿ ⚻ ♅
02 05	☉ Q ♇
05 06	☽ □ ♇
07 58	☽ ∠ ♀
08 11	☽ □ ♄
10 11	♀ ⚼ ♄
13 48	☽ ⚻ ♆
22 13	☽ # ♃
29 Thursday	
01 27	☽ ± ♆
02 47	☽ △ ♂
03 24	☽ ⊥ ♀
04 00	☽ ⚼ ♅
11 02	☽ # ♄
13 07	☉ ⚼ ♆
16 13	☽ Q ♇
16 29	☽ ⚺ ♀
19 05	☽ ⚼ ♆
19 38	☽ ☌ ☉
21 37	☽ ± ♅
30 Friday	
01 57	☿ ⚹ ♇
06 21	☽ ⚻ ♃
09 27	☽ ⚻ ♅
11 07	♄ St R
15 59	☽ ⚹ ♇
16 08	☿ ∥ ♀
18 01	☽ ☌ ☿
18 10	☽ ± ♃
18 53	☉ ± ♅
19 06	☽ △ ♄
19 07	☿ ± ♃

DECEMBER 2035

LONGITUDES

Date	Sidereal time h m s	Sun ☉ ° ' "	Moon ☽ ° ' "	Moon ☽ 24.00 ° ' "	Mercury ☿ ° '	Venus ♀ ° '	Mars ♂ ° '	Jupiter ♃ ° '	Saturn ♄ ° '	Uranus ♅ ° '	Neptune ♆ ° '	Pluto ⯓ ° '
01	16 40 44	09 ♐ 06 22	27 ♐ 41 49	03 ♑ 39 39	19 ♐ 51	08 ♑ 03	29 ♓ 51	12 ♉ 42	19 ♌ 15	14 ♋ 22	22 ♈ 06	17 ♒ 42
02	16 44 40	10 07 13	09 ♑ 35 58	15 ♑ 31 01	21 23	09 17	00 ♈ 18	12 R 35	19 R 15	14 R 20	22 R 05	17 42
03	16 48 37	11 08 04	21 ♑ 25 10	27 ♑ 18 46	22 55	10 31	00 46	12 29	19 14	14 18	22 04	17 43
04	16 52 34	12 08 56	03 ♒ 12 17	09 ♒ 06 12	24 27	11 46	01 13	12 23	19 14	14 16	22 03	17 44
05	16 56 30	13 09 49	15 ♒ 01 04	20 ♒ 57 27	25 59	13 00	01 41	12 17	19 14	14 13	22 02	17 45
06	17 00 27	14 10 42	26 ♒ 56 00	02 ♓ 57 21	27 30	14 14	02 09	12 11	19 13	14 11	22 01	17 46
07	17 04 23	15 11 37	09 ♓ 02 11	15 ♓ 11 13	29 ♐ 01	15 28	02 37	12 06	19 12	14 09	22 00	17 47
08	17 08 20	16 12 32	21 ♓ 25 06	27 ♓ 44 30	00 ♑ 32	16 42	03 06	12 00	19 11	14 07	21 59	17 48
09	17 12 16	17 13 28	04 ♈ 10 02	10 ♈ 42 15	02 03	17 56	03 35	11 55	19 10	14 05	21 58	17 49
10	17 16 13	18 14 25	17 ♈ 21 36	24 ♈ 08 24	03 33	19 10	04 04	11 50	19 09	14 03	21 58	17 50
11	17 20 09	19 15 22	01 ♉ 02 51	08 ♉ 04 54	05 03	20 24	04 33	11 45	19 08	14 00	21 57	17 51
12	17 24 06	20 16 20	15 ♉ 14 21	22 ♉ 30 46	06 32	21 38	05 03	11 40	19 07	13 58	21 56	17 52
13	17 28 03	21 17 19	29 ♉ 53 27	07 ♊ 21 30	08 00	22 52	05 33	11 35	19 05	13 56	21 55	17 53
14	17 31 59	22 18 18	14 ♊ 53 51	22 ♊ 29 13	09 28	24 06	06 03	11 31	19 04	13 53	21 55	17 54
15	17 35 56	23 19 18	00 ♋ 06 15	07 ♋ 43 35	10 54	25 20	06 33	11 26	19 02	13 51	21 54	17 56
16	17 39 52	24 20 19	15 ♋ 19 49	22 ♋ 53 41	12 19	26 34	07 04	11 22	19 01	13 49	21 54	17 57
17	17 43 49	25 21 21	00 ♌ 24 04	07 ♌ 49 59	13 43	27 48	07 35	11 18	18 59	13 46	21 53	17 58
18	17 47 45	26 22 24	15 ♌ 10 41	22 ♌ 25 37	15 06	29 ♑ 02	08 06	11 15	18 57	13 44	21 53	17 59
19	17 51 42	27 23 27	29 ♌ 34 27	06 ♍ 37 00	16 26	00 ♒ 15	08 37	11 11	18 55	13 41	21 52	18 00
20	17 55 38	28 24 31	13 ♍ 33 17	20 ♍ 23 27	17 44	01 29	09 08	11 08	18 53	13 39	21 52	18 02
21	17 59 35	29 ♐ 25 36	27 ♍ 07 42	03 ♎ 46 23	19 00	02 43	09 40	11 05	18 50	13 36	21 51	18 03
22	18 03 32	00 ♑ 26 42	10 ♎ 19 52	16 ♎ 48 33	20 12	03 57	10 11	11 02	18 48	13 34	21 51	18 04
23	18 07 28	01 27 49	23 ♎ 12 50	29 ♎ 33 08	21 20	05 10	10 43	10 59	18 46	13 31	21 50	18 05
24	18 11 25	02 28 56	05 ♏ 49 51	12 ♏ 03 21	22 25	06 24	11 15	10 57	18 43	13 29	21 50	18 07
25	18 15 21	03 30 04	18 ♏ 13 58	24 ♏ 22 02	23 24	07 38	11 47	10 54	18 41	13 26	21 50	18 08
26	18 19 18	04 31 13	00 ♐ 27 47	06 ♐ 31 31	24 18	08 51	12 20	10 52	18 38	13 24	21 50	18 10
27	18 23 14	05 32 22	12 ♐ 33 44	18 ♐ 33 41	25 06	10 05	12 52	10 50	18 35	13 21	21 49	18 11
28	18 27 11	06 33 32	24 ♐ 32 31	00 ♑ 30 04	25 46	11 18	13 25	10 48	18 32	13 19	21 49	18 12
29	18 31 07	07 34 42	06 ♑ 26 32	12 ♑ 22 05	26 18	12 32	13 58	10 47	18 29	13 16	21 49	18 14
30	18 35 04	08 35 52	18 ♑ 16 55	24 ♑ 11 14	26 40	13 45	14 31	10 46	18 26	13 14	21 49	18 15
31	18 39 01	09 ♑ 37 02	00 ♒ 05 17	05 ♒ 59 21	26 ♑ 53	14 ♒ 58	15 ♈ 04	10 ♉ 45	18 ♌ 23	13 ♋ 11	21 ♈ 49	18 ♒ 17

Date	Moon True ☊ ° '	Moon Mean ☊ ° '	Moon ☽ Latitude ° '
01	29 ♌ 44	00 ♍ 24	04 N 25
02	29 R 32	00 21	03 51
03	29 24	00 17	03 06
04	29 18	00 14	02 13
05	29 15	00 11	01 15
06	29 14	00 08	00 N 12
07	29 D 14	00 05	00 S 52
08	29 R 14	00 ♍ 02	01 55
09	29 13	29 ♌ 58	02 54
10	29 10	29 55	03 47
11	29 04	29 52	04 28
12	28 55	29 49	04 54
13	28 45	29 46	05 03
14	28 35	29 43	04 51
15	28 25	29 39	04 18
16	28 17	29 36	03 27
17	28 11	29 33	02 22
18	28 08	29 30	01 S 09
19	28 D 08	29 27	00 N 08
20	28 08	29 23	01 22
21	28 09	29 20	02 29
22	28 R 08	29 17	03 26
23	28 06	29 14	04 11
24	28 01	29 11	04 43
25	27 54	29 08	05 01
26	27 44	29 04	05 05
27	27 34	29 01	04 55
28	27 23	28 58	04 33
29	27 13	28 55	03 58
30	27 05	28 52	03 13
31	26 ♌ 59	28 ♌ 49	02 N 20

DECLINATIONS

Date	Sun ☉ ° '	Moon ☽ ° '	Mercury ☿ ° '	Venus ♀ ° '	Mars ♂ ° '	Jupiter ♃ ° '	Saturn ♄ ° '	Uranus ♅ ° '	Neptune ♆ ° '	Pluto ⯓ ° '
01	21 S 49	18 S 59	24 S 56	24 S 44	00 S 38	14 N 26	15 N 52	23 N 02	07 N 01	23 S 21
02	21 58	19 15	25 06	24 41	00 25	14 24	15 52	23 03	07 00	23 21
03	22 06	18 40	25 15	24 37	00 S 12	14 23	15 52	23 03	07 00	23 21
04	22 15	17 16	25 23	24 32	00 N 01	14 21	15 53	23 03	07 00	23 20
05	22 22	15 08	25 30	24 26	00 14	14 19	15 53	23 03	06 59	23 20
06	22 30	12 20	25 35	24 20	00 27	14 18	15 53	23 04	06 59	23 19
07	22 37	08 59	25 38	24 13	00 40	14 16	15 54	23 04	06 59	23 19
08	22 43	05 10	25 40	24 06	00 54	14 15	15 54	23 04	06 59	23 18
09	22 49	01 S 01	25 41	23 57	01 07	14 14	15 55	23 04	06 58	23 18
10	22 55	03 N 19	25 40	23 48	01 21	14 12	15 55	23 05	06 58	23 18
11	23 00	07 39	25 38	23 38	01 34	14 11	15 56	23 05	06 58	23 17
12	23 05	11 42	25 35	23 28	01 48	14 10	15 56	23 05	06 58	23 17
13	23 09	15 11	25 29	23 16	02 02	14 09	15 57	23 05	06 57	23 16
14	23 13	17 46	25 23	23 05	02 15	14 07	15 58	23 06	06 57	23 16
15	23 16	19 08	25 15	22 52	02 29	14 06	15 58	23 06	06 57	23 15
16	23 19	19 07	25 05	22 39	02 43	14 05	15 59	23 06	06 57	23 15
17	23 21	17 45	24 54	22 25	02 57	14 04	16 00	23 06	06 57	23 14
18	23 23	15 11	24 42	22 10	03 10	14 04	16 01	23 07	06 57	23 14
19	23 24	11 44	24 29	21 55	03 24	14 03	16 01	23 07	06 56	23 13
20	23 25	07 43	24 14	21 39	03 38	14 02	16 02	23 07	06 56	23 13
21	23 26	03 N 25	23 58	21 23	03 52	14 01	16 03	23 08	06 56	23 12
22	23 26	00 S 56	23 41	21 06	04 06	14 01	16 04	23 08	06 56	23 12
23	23 25	05 07	23 23	20 48	04 20	14 00	16 05	23 08	06 56	23 11
24	23 25	09 00	23 04	20 30	04 34	14 00	16 06	23 08	06 56	23 11
25	23 23	12 25	22 45	20 11	04 48	13 59	16 07	23 09	06 56	23 10
26	23 21	15 16	22 25	19 52	05 02	13 59	16 08	23 09	06 56	23 10
27	23 19	17 25	22 05	19 32	05 16	13 58	16 09	23 09	06 56	23 09
28	23 16	18 47	21 45	19 11	05 31	13 58	16 10	23 09	06 56	23 09
29	23 13	19 19	21 25	18 50	05 45	13 58	16 11	23 10	06 56	23 08
30	23 09	19 00	21 05	18 29	05 59	13 58	16 12	23 10	06 56	23 08
31	23 S 05	17 S 51	20 S 47	18 S 07	06 N 13	13 N 58	16 N 13	23 N 10	06 N 56	23 S 07

ZODIAC SIGN ENTRIES

Date	h	m	Planets
01	16	38	☽ ♑
01	19	37	♂ ♈
04	05	28	☽ ♒
06	18	07	☽ ♓
08	03	27	☿ ♑
09	04	14	☽ ♈
11	10	12	☽ ♉
13	12	11	☽ ♊
15	11	50	☽ ♋
17	11	21	☽ ♌
19	07	00	♀ ♒
19	12	43	☽ ♍
21	17	10	☽ ♎
22	01	31	☉ ♑
24	00	51	☽ ♏
26	11	05	☽ ♐
28	22	59	☽ ♑
31	11	49	☽ ♒

LATITUDES

Date	Mercury ☿ ° '	Venus ♀ ° '	Mars ♂ ° '	Jupiter ♃ ° '	Saturn ♄ ° '	Uranus ♅ ° '	Neptune ♆ ° '	Pluto ⯓ ° '
01	01 S 53	01 S 33	00 S 38	01 S 16	00 N 52	00 N 23	01 S 43	08 S 15
04	02 04	01 38	00 31	01 16	00 52	00 23	01 43	08 14
07	02 12	01 42	00 24	01 15	00 53	00 23	01 42	08 14
10	02 17	01 45	00 18	01 14	00 54	00 23	01 42	08 13
13	02 18	01 48	00 12	01 13	00 54	00 23	01 42	08 13
16	02 14	01 51	00 06	01 12	00 55	00 23	01 42	08 13
19	02 04	01 52	00 S 01	01 12	00 55	00 23	01 42	08 12
22	01 47	01 53	00 N 05	01 11	00 56	00 23	01 42	08 12
25	01 22	01 53	00 09	01 10	00 56	00 23	01 42	08 12
28	00 46	01 53	00 14	01 09	00 57	00 24	01 41	08 11
31	00 00	01 S 51	00 N 18	01 S 08	00 N 57	00 N 24	01 S 41	08 S 11

DATA

Julian Date	2464663
Delta T	+76 seconds
Ayanamsa	24° 21' 22"
Synetic vernal point	04° ♓ 45' 37"
True obliquity of ecliptic	23° 25' 56"

MOON'S PHASES, APSIDES AND POSITIONS ☽

Date	h	m	Phase	Longitude ° '	Eclipse Indicator
08	01	05	☽	15 ♓ 45	
15	00	33	○	22 ♊ 50	
21	16	28	☾	29 ♍ 37	
29	14	31	●	07 ♑ 41	

Day	h	m		
03	08	17	Apogee	
15	19	48	Perigee	
30	15	59	Apogee	
02	07	21	Max dec	19° S 16'
09	17	39	ON	
15	23	50	Max dec	19° N 18'
22	06	49	OS	
29	15	01	Max dec	19° S 19'

LONGITUDES

Date	Chiron ⚷ ° '	Ceres ⚳ ° '	Pallas ⚴ ° '	Juno ⚵ ° '	Vesta ⚶ ° '	Black Moon Lilith ⚸ ° '
01	09 ♊ 51	28 ♊ 29	19 ♓ 35	03 ♓ 36	25 ♎ 32	14 ♑ 51
11	09 ♊ 16	26 ♊ 18	20 ♓ 41	07 ♓ 03	00 ♏ 18	15 ♑ 58
21	08 ♊ 44	23 ♊ 57	22 ♓ 20	10 ♓ 54	04 ♏ 59	17 ♑ 05
31	08 ♊ 14	21 ♊ 42	24 ♓ 26	15 ♓ 05	09 ♏ 32	18 ♑ 12

ASPECTARIAN

h m	Aspects
01 Saturday	
00 48	☽ △ ♆
02 32	☿ △ ♄
12 00	☽ ♇ ♃
16 31	☽ □ ♂
22 04	☽ ∠ ⯓
02 Sunday	
01 11	☽ ♇ ♄
11 18	☽ ☌ ♀
13 09	☽ ⊻ ☉
16 16	☽ ⊥ ⯓
18 01	☽ △ ♃
19 23	☽ ± ♄
21 33	☽ ☍ ♅
22 36	☿ △ ♆
03 Monday	
02 28	☽ ⊥ ☉
04 28	☽ ⊻ ⯓
06 22	☽ Q ♂
07 34	☽ ⊼ ♄
13 18	☽ □ ♆
15 31	☽ ⊻ ☿
22 30	☽ ∠ ☉
04 Tuesday	
05 34	☽ ⊥ ☿
07 47	☽ ⚹ ♂
11 31	♀ ⊥ ⯓
17 04	☉ ⊼ ♃
23 12	♀ △ ♃
05 Wednesday	
01 54	☽ Q ♆
02 36	☽ ∠ ☿
04 26	☽ ⋕ ♄
06 30	☽ □ ♃
07 26	☽ ⊻ ♀
07 53	☽ ⚹ ☉
10 24	☽ ⊼ ♅
15 30	☽ ∠ ♂
16 25	♀ ± ♄
17 32	☽ ☌ ⯓
19 37	☽ ⋕ ♃
20 30	☽ ☍ ♄
20 59	☽ ⊥ ♀
22 29	☽ ± ♅
06 Thursday	
02 08	☽ ⚹ ♆
06 09	☉ ⊻ ♀
07 17	☿ ♇ ♃
10 21	☽ Q ☉
10 22	☽ ⊥ ♂
11 11	♀ ☍ ♅
12 15	☉ ⊼ ♅
13 19	☽ ⚹ ☿
16 30	☽ ♇ ♅
17 07	☽ ∠ ♀
18 27	☽ Q ♃
22 49	☽ ⊻ ♂
07 Friday	
08 00	☽ ∠ ♆
16 27	☽ Q ☿
17 56	☽ ⚹ ♃
20 26	♂ ∠ ⯓
21 58	☽ △ ♅
08 Saturday	
00 54	☽ ⋕ ♆
01 05	☽ □ ☉
01 34	☽ ⊥ ♆
01 56	☽ ⚹ ♀
05 03	☽ ⊻ ⯓
07 44	☽ ⊼ ♄
13 05	☽ ⊻ ♆
16 33	☽ ⊥ ⯓
19 10	☽ ± ♄
22 32	☽ ∠ ♃
09 Sunday	
03 17	☽ Q ♀
07 33	☽ □ ☿
09 29	☽ ∠ ⯓
09 38	♀ ⊻ ⯓
10 52	☽ ☌ ♂
11 25	☽ ⋕ ♂
12 01	☽ ♇ ♄
15 12	☽ ⊥ ♃
10 Monday	
00 25	☿ ∠ ⯓
00 31	☽ ∥ ♂
02 06	☽ ⊻ ♃
02 14	☉ ⚹ ⯓
02 59	☉ ± ♃
06 04	☽ □ ♅
11 42	♀ ⊼ ♄
12 51	☽ ⚹ ⯓
13 42	☽ △ ☉
15 12	☽ △ ♄
15 33	☽ □ ♀
16 10	♂ ♇ ♄
20 09	☽ ☌ ♆
21 35	☿ ♇ ♄
11 Tuesday	
00 20	☿ □ ♂
08 09	☽ ∥ ♆
09 12	☉ △ ♄
09 56	☽ Q ⯓
13 39	☽ Q ♅
17 56	☽ ♇ ☉
18 14	☽ ⊻ ♂
19 39	☽ △ ☿
12 Wednesday	
04 45	☽ ⊥ ♂
06 04	☽ ☌ ♃
09 53	☽ ⚹ ♅
10 16	☽ ± ☉
15 38	☉ ⋕ ♅
16 22	☽ □ ⯓
17 45	♀ □ ♆
18 24	☽ □ ♄
20 14	☽ ∠ ♂
20 57	☽ ⊼ ☉
23 03	☽ ⊻ ♆
23 32	☽ △ ♀
23 33	☽ ♇ ☿
13 Thursday	
04 16	☽ ∥ ♃
08 49	☽ ⊥ ♆
10 27	☽ ∠ ♅
12 37	♀ ∥ ⯓
13 31	♂ ⊥ ♃
15 46	☽ ± ☿
18 12	☽ ∥ ♄
21 25	☽ ⚹ ♂
23 18	☽ ∠ ♆
23 33	☽ Q ♄
23 50	☉ ∥ ♀
14 Friday	
00 53	☽ ⊥ ♅
01 58	☽ ♇ ♀
02 26	☽ ⊼ ☿
02 51	☉ △ ♆
06 39	☽ ⊻ ♃
09 50	♀ ⋕ ♅
10 19	♀ Q ♂
10 24	☽ ⊻ ♅
16 07	☽ ⊥ ♃
16 46	☽ △ ⯓
17 10	☽ Q ♂
17 31	☽ ± ♀
18 35	☽ ⚹ ♄
23 05	☽ ⚹ ♆
15 Saturday	
00 33	☽ ☍ ☉
03 50	☽ ⊼ ♀
06 15	☽ ∠ ♃
08 00	☉ ∥ ⯓
16 27	☽ ♇ ⯓
17 59	☽ Q ♆
18 11	☽ ∠ ♄
20 39	☿ △ ♃
22 30	☽ □ ♂
16 Sunday	
05 30	☿ ⊥ ⯓
05 47	☽ ⚹ ♃
06 39	☽ ± ⯓
06 46	☽ ☍ ☿
08 20	☽ ⊥ ♄
09 36	☽ ☌ ♅
16 09	☽ ⊼ ⯓
17 49	☽ ⊻ ♄
22 24	☽ □ ♆
23 29	☿ ± ♄
17 Monday	
00 42	☽ Q ♃
03 20	☽ ⊼ ☉
07 27	☽ ☍ ♀
12 48	☿ ☍ ♅
13 39	☽ ± ☉
18 Tuesday	
00 00	☽ △ ♂
05 16	☽ ∥ ♄
05 18	☽ ♇ ☉
05 35	☽ □ ♃
09 09	☉ ♇ ♃
09 38	☽ ⊻ ♅
11 51	☽ ⊼ ☿
16 38	☽ ☍ ⯓
18 12	☽ ☌ ♄
19 30	☽ ⊥ ♅
20 28	☽ ∥ ♃
22 47	☽ ± ☿
23 04	☽ △ ♆
19 Wednesday	
01 36	☽ ♇ ♂
08 02	☽ △ ☉
10 31	☽ ∠ ♅
13 16	☽ ⊼ ♀
15 29	☽ ♇ ☿
17 21	☽ ± ♂
20 Thursday	
00 25	☽ ♇ ♆
00 29	☽ ± ♀
04 02	☽ ⊼ ♂
07 48	☽ △ ♃
12 10	☽ ⚹ ♅
16 01	☽ ± ♆
16 26	☽ ∥ ♆
17 35	☿ ⊻ ⯓
17 38	☽ ♇ ♀
19 51	☽ ⊼ ⯓
20 04	☽ △ ☿
21 18	☽ ⊻ ♄
21 Friday	
02 36	☽ ⊼ ♆
06 29	☽ ± ⯓
07 55	☽ ⊥ ♄
09 09	☿ ⊼ ♄
09 17	☽ Q ♅
09 38	☽ ∥ ♂
10 08	☽ ♇ ♃
16 28	☽ □ ☉
22 42	☽ ♇ ⯓
23 07	☽ △ ♀
22 Saturday	
00 05	☽ ∠ ♄
02 19	☽ ± ♃
11 44	☽ ☍ ♂
13 17	☽ ⊼ ♃
17 57	☽ □ ♅
23 Sunday	
02 23	☽ △ ⯓
03 40	☽ ⚹ ♄
04 21	☽ Q ☉
07 06	☽ ⋕ ♂
08 08	☽ □ ☿
08 44	☉ ∥ ☿
09 25	☽ ☍ ♆
22 53	☿ □ ♆
22 54	☽ ⋕ ♆
23 02	♂ ⊻ ♃
24 Monday	
02 15	☽ Q ♄
03 34	☿ ∥ ⯓
05 01	☽ ⚹ ☉
06 54	☿ ⋕ ♅
13 13	☽ □ ♀
21 37	☽ Q ☿
21 49	☽ ☍ ♃
22 55	☽ ⊼ ♂
25 Tuesday	
02 43	☽ △ ♅
03 13	☉ ∠ ⯓
11 06	☽ ± ♂
11 49	☽ □ ⯓
12 34	☽ ∠ ☉
12 52	☽ □ ♄
15 56	☉ ♇ ♄
19 01	☽ ⊼ ♆
22 57	☽ ⚹ ☿
26 Wednesday	
00 30	☽ ⋕ ♃
04 05	☽ Q ♀
05 32	☽ ♇ ♂
06 48	☽ ± ♆
07 49	☽ ⊥ ☉
07 56	☽ ♇ ♅
20 46	☽ ⊻ ☉
20 49	☽ ⋕ ♄
23 18	☽ Q ⯓
27 Thursday	
00 36	☽ ♇ ♆
01 41	☽ ± ♅
06 30	☽ ⚹ ♀
06 46	☽ ∠ ☿
07 02	♀ Q ♆
08 35	☽ ⊼ ♃
12 40	☽ △ ♂
12 46	♅ ⋕ ⯓
13 35	☽ ⊼ ♅
20 32	☽ ± ♃
23 16	☽ ⚹ ⯓
28 Friday	
00 00	☽ △ ♄
01 52	☽ ⊥ ☿
02 32	♀ □ ♃
06 32	☽ △ ♆
07 34	♂ □ ♅
14 32	☽ ♇ ♃
14 35	☽ ⊻ ☿
15 57	☽ ∠ ♀
20 23	☽ ∥ ♀
29 Saturday	
05 30	☽ ∠ ⯓
06 03	☽ ♇ ♄
12 11	☽ ⊥ ♀
14 31	☽ ☌ ☉
20 46	☽ △ ♃
23 45	☽ ⊥ ⯓
30 Sunday	
00 11	☽ ± ♄
01 44	☽ ⊻ ♀
01 47	☽ ☍ ♅
02 04	♀ ⊼ ♅
03 59	☽ □ ♂
07 37	☉ ⋕ ♅
11 57	☽ ⊻ ⯓
12 18	☽ ⊼ ♄
19 11	☽ □ ♆
22 58	☉ ∥ ⯓
31 Monday	
05 24	☽ ☌ ☿
06 11	☽ ∥ ♀
15 35	♀ ⚹ ♂
18 22	☽ Q ♂

All ephemeris data is given at 12.00 UT and the Moon's longitude is additionally given for 24.00 UT

JANUARY 2036

LONGITUDES

Date	Sidereal time h m s	Sun ☉ ° ' "	Moon ☽ ° ' "	Moon ☽ 24.00 ° ' "	Mercury ☿ ° '	Venus ♀ ° '	Mars ♂ ° '	Jupiter ♃ ° '	Saturn ♄ ° '	Uranus ♅ ° '	Neptune ♆ ° '	Pluto ♇ ° '
01	18 42 57	10 ♑ 38 12	11 ♒ 53 45	17 ♒ 48 48	26 ♑ 55	16 ♒ 12	15 ♈ 38	10 ♉ 44	18 ♌ 19	13 ♋ 08	21 ♈ 49	18 ♒ 18
02	18 46 54	11 39 23	23 ♒ 44 55	29 ♒ 42 32	26 R 46	17 25	16 11	10 R 43	18 R 16	13 R 06	21 D 49	18 20
03	18 50 50	12 40 33	05 ♓ 42 08	11 ♓ 44 13	26 25	18 38	16 45	10 43	18 13	13 03	21 49	18 21
04	18 54 47	13 41 43	17 ♓ 49 19	23 ♓ 58 02	25 52	19 51	17 19	10 42	18 09	13 01	21 49	18 23
05	18 58 43	14 42 52	00 ♈ 10 55	06 ♈ 28 35	25 07	21 04	17 52	10 D 42	18 05	12 58	21 49	18 24
06	19 02 40	15 44 01	12 ♈ 51 36	19 ♈ 20 31	24 12	22 18	18 26	10 43	18 02	12 55	21 49	18 26
07	19 06 36	16 45 10	25 ♈ 55 49	02 ♉ 37 56	23 07	23 31	19 01	10 43	17 58	12 53	21 49	18 27
08	19 10 33	17 46 19	09 ♉ 27 09	16 ♉ 23 41	21 55	24 43	19 35	10 44	17 54	12 50	21 49	18 29
09	19 14 30	18 47 27	23 ♉ 27 31	00 ♊ 38 30	20 38	25 56	20 09	10 44	17 50	12 48	21 50	18 30
10	19 18 26	19 48 35	07 ♊ 56 14	15 ♊ 20 10	19 18	27 09	20 44	10 45	17 46	12 45	21 50	18 32
11	19 22 23	20 49 42	22 ♊ 49 28	00 ♋ 23 08	17 58	28 22	21 18	10 47	17 42	12 42	21 50	18 34
12	19 26 19	21 50 49	08 ♋ 00 00	15 ♋ 38 47	16 41	29 ♒ 35	21 53	10 48	17 38	12 40	21 51	18 35
13	19 30 16	22 51 56	23 ♋ 18 06	00 ♌ 56 35	15 27	00 ♓ 47	22 28	10 50	17 34	12 37	21 51	18 37
14	19 34 12	23 53 02	08 ♌ 32 55	16 ♌ 05 53	14 21	02 00	23 02	10 51	17 30	12 35	21 51	18 39
15	19 38 09	24 54 08	23 ♌ 34 25	00 ♍ 57 38	13 22	03 12	23 37	10 53	17 25	12 32	21 52	18 40
16	19 42 05	25 55 13	08 ♍ 14 49	15 ♍ 25 31	12 32	04 25	24 12	10 56	17 21	12 30	21 52	18 42
17	19 46 02	26 56 18	22 ♍ 29 26	29 ♍ 26 25	11 52	05 37	24 47	10 58	17 17	12 27	21 53	18 44
18	19 49 59	27 57 23	06 ♎ 16 34	13 ♎ 00 00	11 21	06 49	25 23	11 01	17 12	12 25	21 53	18 45
19	19 53 55	28 58 28	19 ♎ 37 04	26 ♎ 08 05	11 00	08 01	25 58	11 04	17 08	12 22	21 54	18 47
20	19 57 52	29 ♑ 59 32	02 ♏ 33 30	08 ♏ 53 48	10 48	09 13	26 33	11 07	17 03	12 20	21 55	18 49
21	20 01 48	01 ♒ 00 36	15 ♏ 09 28	21 ♏ 21 01	10 D 44	10 25	27 09	11 10	16 59	12 17	21 55	18 50
22	20 05 45	02 01 40	27 ♏ 28 56	03 ♐ 33 43	10 49	11 37	27 44	11 13	16 54	12 15	21 56	18 52
23	20 09 41	03 02 44	09 ♐ 35 50	15 ♐ 35 44	11 01	12 49	28 20	11 17	16 49	12 13	21 57	18 54
24	20 13 38	04 03 47	21 ♐ 33 49	27 ♐ 30 30	11 21	14 01	28 56	11 21	16 45	12 10	21 57	18 56
25	20 17 34	05 04 49	03 ♑ 26 06	09 ♑ 20 59	11 47	15 12	29 ♈ 31	11 25	16 40	12 08	21 58	18 57
26	20 21 31	06 05 51	15 ♑ 15 26	21 ♑ 09 44	12 18	16 24	00 ♉ 07	11 29	16 35	12 06	21 58	18 59
27	20 25 28	07 06 52	27 ♑ 04 09	02 ♒ 58 56	12 55	17 35	00 43	11 33	16 30	12 03	22 00	19 01
28	20 29 24	08 07 52	08 ♒ 54 20	14 ♒ 50 35	13 37	18 47	01 19	11 38	16 26	12 01	22 01	19 03
29	20 33 21	09 08 51	20 ♒ 47 54	26 ♒ 46 34	14 23	19 58	01 55	11 42	16 21	11 59	22 02	19 04
30	20 37 17	10 09 49	02 ♓ 46 50	08 ♓ 48 58	15 13	21 09	02 31	11 47	16 16	11 57	22 03	19 06
31	20 41 14	11 ♒ 10 46	14 ♓ 53 16	21 ♓ 00 02	16 ♑ 06	22 ♓ 20	03 ♉ 07	11 ♉ 52	16 ♌ 11	11 ♋ 55	22 ♈ 04	19 ♒ 08

Date	Moon True ☊ ° '	Moon Mean ☊ ° '	Moon ☽ Latitude ° '
01	26 ♌ 55	28 ♌ 45	01 N 20
02	26 R 54	28 42	00 N 17
03	26 D 54	28 39	00 S 48
04	26 55	28 36	01 51
05	26 57	28 33	02 51
06	26 58	28 29	03 44
07	26 R 57	28 26	04 27
08	26 55	28 23	04 57
09	26 52	28 20	05 11
10	26 47	28 17	05 05
11	26 41	28 14	04 40
12	26 36	28 10	03 54
13	26 31	28 07	02 51
14	26 28	28 04	01 37
15	26 27	28 01	00 S 16
16	26 D 28	27 58	01 N 04
17	26 29	27 55	02 18
18	26 31	27 51	03 21
19	26 32	27 48	04 12
20	26 R 32	27 45	04 47
21	26 31	27 42	05 08
22	26 29	27 39	05 14
23	26 26	27 35	05 06
24	26 22	27 32	04 45
25	26 19	27 29	04 11
26	26 15	27 26	03 27
27	26 12	27 23	02 34
28	26 10	27 20	01 34
29	26 09	27 16	00 N 30
30	26 D 09	27 13	00 S 37
31	26 ♌ 10	27 ♌ 10	01 S 42

DECLINATIONS

Date	Sun ☉ ° '	Moon ☽ ° '	Mercury ☿ ° '	Venus ♀ ° '	Mars ♂ ° '	Jupiter ♃ ° '	Saturn ♄ ° '	Uranus ♅ ° '	Neptune ♆ ° '	Pluto ♇ ° '
01	23 S 00	15 S 56	20 S 29	17 S 44	06 N 27	13 N 58	16 N 15	23 N 10	06 N 56	23 S 07
02	22 55	13 20	20 13	17 21	06 41	13 58	16 16	23 11	06 56	23 06
03	22 50	10 09	19 58	16 58	06 55	13 58	16 17	23 11	06 56	23 05
04	22 44	06 31	19 45	16 34	07 09	13 58	16 18	23 11	06 56	23 05
05	22 37	02 S 32	19 35	16 10	07 24	13 59	16 20	23 12	06 56	23 04
06	22 30	01 N 38	19 26	15 45	07 38	13 59	16 21	23 12	06 56	23 04
07	22 23	05 52	19 19	15 20	07 52	13 59	16 22	23 12	06 56	23 03
08	22 15	09 56	19 14	14 54	08 06	14 00	16 23	23 12	06 57	23 03
09	22 07	13 37	19 10	14 28	08 20	14 00	16 25	23 13	06 57	23 02
10	21 58	16 36	19 09	14 02	08 34	14 01	16 26	23 13	06 57	23 02
11	21 49	18 35	19 09	13 35	08 48	14 02	16 28	23 13	06 57	23 01
12	21 40	19 18	19 11	13 08	09 02	14 02	16 29	23 13	06 57	23 00
13	21 30	18 36	19 14	12 41	09 16	14 03	16 30	23 14	06 57	23 00
14	21 19	16 34	19 18	12 13	09 30	14 04	16 32	23 14	06 58	22 59
15	21 09	13 25	19 23	11 45	09 44	14 05	16 33	23 14	06 58	22 59
16	20 57	09 28	19 29	11 17	09 58	14 06	16 35	23 14	06 58	22 58
17	20 46	05 06	19 35	10 49	10 12	14 07	16 36	23 15	06 58	22 58
18	20 34	00 N 36	19 43	10 20	10 25	14 08	16 38	23 15	06 58	22 57
19	20 21	03 S 47	19 51	09 51	10 39	14 09	16 39	23 15	06 59	22 57
20	20 09	07 51	19 59	09 22	10 53	14 10	16 41	23 15	06 59	22 56
21	19 56	11 27	20 07	08 52	11 07	14 12	16 42	23 15	06 59	22 55
22	19 42	14 29	20 16	08 22	11 20	14 13	16 44	23 16	07 00	22 55
23	19 28	16 50	20 24	07 53	11 34	14 14	16 45	23 16	07 00	22 54
24	19 14	18 26	20 33	07 22	11 47	14 16	16 47	23 16	07 00	22 54
25	18 59	19 12	20 41	06 52	12 01	14 17	16 48	23 16	07 01	22 53
26	18 45	19 08	20 48	06 22	12 14	14 18	16 50	23 17	07 01	22 53
27	18 29	18 13	20 56	05 51	12 28	14 20	16 51	23 17	07 01	22 52
28	18 14	16 31	21 02	05 21	12 41	14 22	16 53	23 17	07 02	22 51
29	17 58	14 05	21 08	04 50	12 54	14 24	16 55	23 17	07 02	22 51
30	17 42	11 03	21 13	04 19	13 07	14 26	16 56	23 17	07 02	22 50
31	17 S 25	07 S 31	21 S 17	03 S 48	13 N 20	14 N 27	16 N 58	23 N 18	07 N 03	22 S 50

ZODIAC SIGN ENTRIES

Date	h	m	Planets
03	00	35	☽ ♓
05	11	39	☽ ♈
07	19	19	☽ ♉
09	22	56	☽ ♊
11	23	23	☽ ♋
12	20	23	♀ ♓
13	22	31	☽ ♌
15	22	26	☽ ♍
18	00	58	☽ ♎
20	07	12	☽ ♏
20	12	11	☉ ♒
22	16	57	☽ ♐
25	05	02	☽ ♑
26	07	15	♂ ♉
27	17	57	☽ ♒
30	06	27	☽ ♓

LATITUDES

Date	Mercury ☿ ° '	Venus ♀ ° '	Mars ♂ ° '	Jupiter ♃ ° '	Saturn ♄ ° '	Uranus ♅ ° '	Neptune ♆ ° '	Pluto ♇ ° '
01	00 N 17	01 S 51	00 N 20	01 S 08	00 N 57	00 N 24	01 S 41	08 S 11
04	01 14	01 48	00 24	01 07	00 58	00 24	01 41	08 11
07	02 10	01 45	00 27	01 06	00 58	00 24	01 41	08 11
10	02 55	01 41	00 31	01 05	00 59	00 24	01 41	08 11
13	03 20	01 36	00 34	01 04	00 59	00 24	01 41	08 11
16	03 23	01 30	00 37	01 03	01 00	00 24	01 40	08 11
19	03 09	01 23	00 40	01 02	01 00	00 24	01 40	08 10
22	02 44	01 16	00 43	01 02	01 01	00 24	01 40	08 10
25	02 14	01 08	00 46	01 01	01 01	00 24	01 40	08 10
28	01 43	00 59	00 48	01 00	01 02	00 24	01 40	08 10
31	01 N 11	00 S 49	00 N 51	00 S 59	01 N 02	00 N 24	01 S 40	08 S 10

DATA

Julian Date	2464694
Delta T	+76 seconds
Ayanamsa	24° 21' 28"
Synetic vernal point	04° ♓ 45' 31"
True obliquity of ecliptic	23° 25' 56"

MOON'S PHASES, APSIDES AND POSITIONS ☽

Date	h	m	Phase	Longitude ° '	Eclipse Indicator
06	17	48	☽	15 ♈ 59	
13	11	16	○	22 ♋ 50	
20	06	46	☾	29 ♎ 46	
28	10	17	●	08 ♒ 04	

Day	h	m		
13	08	48	Perigee	
26	16	22	Apogee	
06	02	41	ON	
12	12	19	Max dec	19° N 18'
18	15	11	OS	
25	22	01	Max dec	19° S 17'

LONGITUDES

Date	Chiron ⚷ ° '	Ceres ⚳ ° '	Pallas ⚴ ° '	Juno ⚵ ° '	Vesta ⚶ ° '	Black Moon Lilith ⚸ ° '
01	08 ♊ 11	21 ♊ 30	24 ♓ 40	15 ♓ 31	09 ♏ 59	18 ♑ 19
11	07 ♊ 46	19 ♊ 41	27 ♓ 13	20 ♓ 03	14 ♏ 23	19 ♑ 25
21	07 ♊ 27	18 ♊ 27	00 ♈ 07	24 ♓ 49	18 ♏ 17	20 ♑ 32
31	07 ♊ 14	17 ♊ 54	03 ♈ 19	29 ♓ 49	22 ♏ 36	21 ♑ 39

All ephemeris data is given at 12.00 UT and the Moon's longitude is additionally given for 24.00 UT

ASPECTARIAN

h m	Aspects
01 Tuesday	
04 24	☿ St R
07 46	☽ Q ♆
08 40	☽ # ♄
09 12	☽ ⊻ ☉
09 38	☽ □ ♃
14 09	☉ △ ♃
14 31	☽ ⊼ ♅
18 31	♄ ☍ ♇
19 57	☽ ✶ ♂
21 43	☽ ☌ ♀
22 31	☽ ⊥ ☉
02 Wednesday	
00 58	☽ ☍ ♄
01 01	☽ ☌ ♇
02 38	☽ ± ♅
06 35	♆ St D
06 39	☽ # ♃
08 05	☽ ✶ ♆
17 57	☽ ⊻ ☿
18 25	☽ ∠ ☉
20 44	☽ bQ ♅
22 00	☽ Q ♃
03 Thursday	
01 38	☉ ± ♄
03 42	☽ ∠ ♂
04 00	♀ ☍ ♄
04 11	☉ ⊥ ♇
05 39	☽ ⊥ ☿
06 17	♀ ☌ ♇
13 15	♂ ∥ ♆
14 13	☽ ∠ ♆
16 42	☉ ⊥ ♀
19 54	♀ ± ♅
20 30	☉ ☍ ♅
21 58	☽ ✶ ♃
22 32	☽ ⊥ ♂
22 55	☽ ∠ ☿
04 Friday	
02 33	☽ △ ♅
03 08	☽ ✶ ☉
08 03	☽ ⊥ ♆
08 12	☽ # ♂
09 23	☽ # ♆
10 57	☽ ⊻ ♂
12 39	☽ ⊼ ♄
13 06	☽ ⊻ ♇
16 26	☽ ⊻ ♀
19 49	☽ ⊻ ♆
05 Saturday	
00 18	☽ ± ♄
00 49	☽ ⊥ ♇
02 43	♀ # ♄
02 50	☽ ✶ ☿
03 22	☽ ∠ ♃
03 59	♃ St D
04 44	☽ Q ☉
05 22	☽ ⊥ ♀
17 32	☽ bQ ♄
18 10	☽ ∠ ♇
20 20	♂ △ ♄
20 39	☽ ⊥ ♃
06 Sunday	
00 22	☽ Q ☿
00 26	☽ ∠ ♀
02 37	♀ ✶ ♆
07 58	☽ ⊻ ♃
11 30	♂ ✶ ♇
12 07	☽ □ ♅
17 48	☽ □ ☉
21 32	☽ △ ♄
22 20	☽ ✶ ♇
22 49	☽ ☌ ♂
07 Monday	
04 32	☽ ☌ ♆
07 10	☽ ✶ ♀
07 19	☽ □ ☿
08 04	☿ ⊻ ♀
18 12	☽ ∥ ♆
20 08	☽ Q ♇
20 51	☽ Q ♅
08 Tuesday	
00 17	☽ ∥ ♂
06 46	☽ Q ♀
13 54	☿ □ ♆
14 13	☽ ☌ ♃
14 55	☉ ⊼ ♄
17 51	☽ ✶ ♅
09 Wednesday	
02 31	☽ □ ♄
03 29	☽ △ ☉
03 36	☽ □ ♇
05 09	☉ ⊻ ♇
06 10	☽ ⊻ ♂
07 38	☽ △ ☿
09 15	☽ ⊻ ♆
14 49	☽ ∥ ♃
16 33	☽ □ ♀
16 43	☽ ⊥ ♂
17 30	☽ # ♀
18 08	☿ □ ♂
18 36	☿ ⊥ ♀
19 15	☽ ∠ ♅
19 19	☽ ⊥ ♆
10 Thursday	
06 30	☽ bQ ☉
06 33	☽ bQ ☿
06 52	☉ ☌ ☿
08 14	☽ ∠ ♂
08 28	☽ Q ♄
10 04	☽ ⊥ ♅
10 12	☽ ∠ ♆
10 27	☽ ∥ ♄
12 51	♀ # ♃
16 36	☽ ⊻ ♃
19 48	☽ ⊻ ♅
20 00	☽ ± ☿
22 14	☽ ± ☉
23 25	♀ bQ ♅
11 Friday	
01 33	☿ ⊻ ♇
02 18	☽ ⊥ ♃
03 51	☽ ✶ ♄
04 52	☽ ⊼ ☿
05 10	☽ △ ♇
08 35	☽ ⊼ ☉
09 28	☽ ✶ ♂
10 26	☽ ✶ ♆
16 42	☽ ∠ ♃
17 07	☿ ⊼ ♄
20 16	♀ Q ♃
21 34	☽ △ ♀
12 Saturday	
01 59	☽ # ☿
03 36	☽ ∠ ♄
05 03	☽ bQ ♇
05 16	☽ Q ♂
05 27	☽ Q ♆
10 28	♂ ☌ ♆
11 55	☉ □ ♆
13 48	☉ □ ♂
16 24	☽ ✶ ♃
17 41	☽ ⊥ ♄
19 13	☽ ± ♇
19 18	☽ ☌ ♅
21 23	☽ # ☿
23 13	☽ bQ ♀
13 Sunday	
00 36	☽ ☍ ☿
03 03	☽ ⊻ ♄
04 38	☽ ⊼ ♇
08 42	☿ ∠ ♀
09 43	☽ □ ♆
10 38	☽ □ ♂
11 15	☽ Q ♃
11 16	☽ ☍ ☉
14 27	♅ ± ♇
14 32	☽ ± ♀
14 Monday	
00 46	☽ ⊼ ♀
12 20	☽ ∥ ♄
15 40	☽ □ ♃
18 23	☽ ⊻ ♅
20 37	☽ ⊼ ☿
15 Tuesday	
02 10	☽ ☌ ♄
03 56	☽ ⊥ ♅
04 06	☽ ☍ ♇
05 38	☽ ± ☿
07 27	☽ ∥ ♃
09 15	☽ △ ♆
12 05	☽ △ ♂
14 18	☽ ⊼ ☉
18 24	☽ ∠ ♅
19 20	☽ bQ ☿
23 57	☽ # ♀
16 Wednesday	
00 48	☽ ± ☉
05 05	☽ ☍ ♀
07 12	☿ ⊥ ♇
09 19	☽ ∥ ♂
09 43	☽ bQ ♆
13 27	☿ ☍ ♅
13 40	☽ bQ ♂
16 28	☽ △ ♃
16 47	☽ bQ ☉
18 48	☽ △ ☿
19 04	☽ ✶ ♅
17 Thursday	
00 46	☽ ± ♆
01 53	☽ ∥ ♆
03 11	☽ ⊻ ♄
05 25	☽ ± ♂
05 34	☽ ⊼ ♇
10 57	☽ ⊼ ♆
13 21	☽ ⊥ ♄
15 22	☽ Q ♅
15 51	☽ ± ♇
16 07	☽ ⊼ ♂
18 00	☽ bQ ♃
20 16	☽ △ ☉
18 Friday	
04 52	☽ ∠ ♄
07 32	☽ bQ ♇
08 56	♀ # ♂
09 45	☽ ± ♃
13 03	☽ ⊼ ♀
13 26	♀ ∠ ♆
20 28	☽ ⊼ ♃
20 47	☽ □ ☿
22 54	☿ ± ♄
22 55	☽ □ ♅
19 Saturday	
00 50	☽ ± ♀
07 30	☽ ✶ ♄
07 37	☿ △ ♃
10 28	☽ △ ♇
16 11	☽ ☍ ♆
18 53	☽ bQ ♀
20 Sunday	
00 15	☽ ☍ ♂
05 02	☽ Q ☿
05 28	☽ Q ♄
06 40	☽ # ♆
06 46	☽ □ ☉
20 29	☽ ∥ ♀
22 59	☉ ∥ ☿
21 Monday	
01 57	☽ △ ♀
03 31	☽ ✶ ☿
04 18	☽ ☍ ♃
06 30	☽ △ ♅
09 20	☽ # ♂
09 43	☿ St D
15 30	☽ □ ♄
18 30	☿ ✶ ♀
19 08	☽ □ ♇
20 07	☽ Q ☉
22 Tuesday	
01 07	☽ ⊼ ♆
03 34	♀ ✶ ♃
08 42	☽ ∠ ☿
09 37	☽ # ♃
11 33	☽ bQ ♅
12 32	☽ ⊼ ♂
12 53	☽ ± ♆
20 13	☿ ± ♄
21 47	☽ ✶ ☉
23 Wednesday	
00 15	♀ △ ♅
00 59	☽ ± ♂
02 43	☽ ⊥ ☿
05 17	☽ ± ♅
06 37	☽ Q ♇
06 42	☽ bQ ♆
11 02	☽ # ♄
14 55	☽ ⊻ ☿
15 23	☽ ⊼ ♃
17 12	☽ ⊼ ♅
19 09	☽ □ ♀
19 51	☽ bQ ♂
24 Thursday	
02 22	☽ △ ♄
03 28	☽ ± ♃
06 30	☽ ∠ ☉
06 41	☽ ✶ ♇
08 31	♀ ∠ ♂
11 43	☿ △ ♃
12 47	☽ △ ♆
21 42	☽ bQ ♃
25 Friday	
02 21	☽ ⊥ ☉
03 39	☽ △ ♂
05 02	☽ ∥ ☉
05 26	♀ # ♆
08 26	☽ bQ ♄
11 29	☽ Q ♀
13 04	☽ ∠ ♇
15 39	☽ ⊻ ☉
26 Saturday	
02 35	☽ ± ♄
03 32	♀ ⊥ ♆
03 36	☿ ☍ ♅
04 17	☽ △ ♃
05 36	☽ ☍ ♅
05 42	☽ ☌ ☿
07 22	☽ ⊥ ♇
11 06	♂ Q ♅
14 35	☽ ✶ ♀
14 41	☽ ⊼ ♄
15 33	♀ ⊼ ♄
19 36	☽ ⊻ ♇
27 Sunday	
01 41	☽ □ ♆
05 14	☽ ∥ ☉
15 40	☿ ⊥ ♇
19 48	☽ □ ♂
28 Monday	
00 27	☽ ∠ ♀
00 33	♂ Q ♇
07 39	☽ # ♄
10 17	☽ ☌ ☉
14 14	☽ Q ♆
17 30	♀ ⊻ ♇
17 33	☽ □ ♃
18 17	☽ ⊼ ♅
20 42	☽ ⊥ ♀
22 09	☽ ⊻ ☿
29 Tuesday	
03 06	☽ ☍ ♄
06 21	☽ ± ♅
08 31	☽ ☌ ♇
09 19	☽ # ♃
10 08	☽ Q ♂
10 08	☽ ⊻ ♀
11 06	☽ ⊥ ☿
14 28	☽ ✶ ♆
21 14	☽ # ♂
30 Wednesday	
00 22	☽ bQ ♅
05 59	☽ Q ♃
06 29	☽ ∠ ☿
09 06	☉ Q ♆
11 27	☽ ✶ ♂
20 30	☽ ∠ ♆
31 Thursday	
04 01	☽ ⊻ ☉
06 01	☽ ✶ ♃
06 08	☽ △ ♅
06 20	♀ ⊻ ♆
09 08	♀ ± ♄
13 50	☿ ⊼ ♄
14 19	☽ ⊥ ♆
14 32	☽ ⊼ ♄
14 36	☽ ✶ ☿
15 00	☽ # ♆
16 55	☽ ⊥ ☉
18 41	☽ ∠ ♂
18 48	♃ ✶ ♅
20 22	☽ ⊻ ♇

FEBRUARY 2036

Date	Sidereal time h m s	Sun ☉ ° ' "	Moon ☽ ° ' "	Moon ☽ 24.00 ° ' "	Mercury ☿ ° '	Venus ♀ ° '	Mars ♂ ° '	Jupiter ♃ ° '	Saturn ♄ ° '	Uranus ♅ ° '	Neptune ♆ ° '	Pluto ♇ ° '
01	20 45 10	12 ♒ 11 42	27 ♓ 09 38	03 ♈ 22 24	17 ♑ 03	23 ♓ 31	03 ♉ 43	11 ♉ 58	16 ♌ 06	11 ♋ 52	22 ♈ 05	19 ♒ 10
02	20 49 07	13 12 37	09 ♈ 38 41	15 ♈ 58 55	18 03	24 42	04 20	12 03	16 R 01	11 R 50	22 06	19 12
03	20 53 03	14 13 31	22 ♈ 23 26	28 ♈ 52 38	19 06	25 52	04 56	12 09	15 56	11 48	22 07	19 13
04	20 57 00	15 14 23	05 ♉ 26 52	12 ♉ 06 28	20 11	27 03	05 32	12 15	15 51	11 46	22 08	19 15
05	21 00 57	16 15 14	18 ♉ 51 42	25 ♉ 42 45	21 18	28 13	06 09	12 21	15 47	11 44	22 09	19 17
06	21 04 53	17 16 03	02 ♊ 39 44	09 ♊ 42 39	22 27	29 ♓ 24	06 45	12 27	15 42	11 42	22 10	19 19
07	21 08 50	18 16 51	16 ♊ 51 22	24 ♊ 05 36	23 39	00 ♈ 34	07 22	12 33	15 37	11 40	22 11	19 20
08	21 12 46	19 17 37	01 ♋ 24 55	08 ♋ 48 42	24 52	01 44	07 58	12 39	15 32	11 38	22 13	19 22
09	21 16 43	20 18 22	16 ♋ 16 13	23 ♋ 46 34	26 06	02 54	08 35	12 46	15 27	11 36	22 14	19 24
10	21 20 39	21 19 06	01 ♌ 18 44	08 ♌ 51 38	27 23	04 03	09 12	12 53	15 22	11 35	22 15	19 26
11	21 24 36	22 19 48	16 ♌ 24 07	23 ♌ 55 03	28 ♑ 41	05 13	09 48	13 00	15 17	11 33	22 16	19 28
12	21 28 32	23 20 28	01 ♍ 23 20	08 ♍ 47 55	00 ♒ 00	06 22	10 25	13 07	15 12	11 31	22 18	19 29
13	21 32 29	24 21 07	16 ♍ 07 57	23 ♍ 22 37	01 20	07 31	11 02	13 14	15 08	11 29	22 19	19 31
14	21 36 26	25 21 46	00 ♎ 31 21	07 ♎ 33 42	02 42	08 40	11 39	13 22	15 03	11 28	22 20	19 33
15	21 40 22	26 22 22	14 ♎ 29 24	21 ♎ 18 21	04 05	09 49	12 15	13 29	14 58	11 26	22 22	19 35
16	21 44 19	27 22 58	28 ♎ 00 35	04 ♏ 36 17	05 29	10 58	12 52	13 37	14 53	11 24	22 23	19 37
17	21 48 15	28 23 32	11 ♏ 05 43	17 ♏ 29 16	06 55	12 07	13 29	13 45	14 49	11 23	22 25	19 38
18	21 52 12	29 ♒ 24 06	23 ♏ 47 23	00 ♐ 00 34	08 21	13 15	14 06	13 53	14 44	11 21	22 26	19 40
19	21 56 08	00 ♓ 24 38	06 ♐ 09 21	12 ♐ 14 19	09 48	14 23	14 43	14 01	14 39	11 20	22 28	19 42
20	22 00 05	01 25 08	18 ♐ 16 02	24 ♐ 15 05	11 17	15 31	15 20	14 09	14 35	11 18	22 29	19 44
21	22 04 01	02 25 38	00 ♑ 12 02	06 ♑ 07 26	12 47	16 39	15 57	14 18	14 30	11 17	22 31	19 45
22	22 07 58	03 26 06	12 ♑ 01 50	17 ♑ 55 43	14 17	17 47	16 34	14 26	14 26	11 16	22 33	19 47
23	22 11 55	04 26 33	23 ♑ 49 35	29 ♑ 43 51	15 49	18 54	17 11	14 35	14 21	11 15	22 34	19 49
24	22 15 51	05 26 58	05 ♒ 38 56	11 ♒ 35 13	17 21	20 01	17 48	14 44	14 17	11 13	22 36	19 51
25	22 19 48	06 27 21	17 ♒ 33 02	23 ♒ 32 40	18 55	21 08	18 25	14 53	14 12	11 12	22 38	19 52
26	22 23 44	07 27 43	29 ♒ 34 23	05 ♓ 38 25	20 29	22 15	19 03	15 02	14 08	11 11	22 39	19 54
27	22 27 41	08 28 04	11 ♓ 44 59	17 ♓ 54 15	22 05	23 22	19 40	15 11	14 04	11 10	22 41	19 56
28	22 31 37	09 28 22	24 ♓ 06 23	00 ♈ 21 30	23 41	24 28	20 17	15 21	14 00	11 09	22 43	19 58
29	22 35 34	10 ♓ 28 39	06 ♈ 39 43	13 ♈ 01 10	25 ♒ 19	25 ♈ 34	20 ♉ 54	15 ♉ 30	13 ♌ 56	11 ♋ 08	22 ♈ 45	19 ♒ 59

Date	Moon True ☊ ° '	Moon Mean ☊ ° '	Moon ☽ Latitude ° '
01	26 ♌ 11	27 ♌ 07	02 S 44
02	26 D 13	27 04	03 39
03	26 14	27 00	04 24
04	26 14	26 57	04 57
05	26 R 14	26 54	05 15
06	26 14	26 51	05 16
07	26 13	26 48	04 57
08	26 13	26 45	04 20
09	26 12	26 41	03 24
10	26 11	26 38	02 14
11	26 11	26 35	00 S 54
12	26 D 11	26 32	00 N 29
13	26 11	26 29	01 49
14	26 11	26 26	03 00
15	26 R 11	26 22	03 58
16	26 11	26 19	04 40
17	26 11	26 16	05 07
18	26 11	26 13	05 18
19	26 D 11	26 10	05 13
20	26 11	26 06	04 55
21	26 12	26 03	04 24
22	26 12	26 00	03 42
23	26 13	25 57	02 51
24	26 14	25 54	01 52
25	26 15	25 51	00 N 48
26	26 R 15	25 47	00 S 18
27	26 14	25 44	01 25
28	26 13	25 41	02 28
29	26 ♌ 11	25 ♌ 38	03 S 26

DECLINATIONS

Date	Sun ☉ ° '	Moon ☽ ° '	Mercury ☿ ° '	Venus ♀ ° '	Mars ♂ ° '	Jupiter ♃ ° '	Saturn ♄ ° '	Uranus ♅ ° '	Neptune ♆ ° '	Pluto ♇ ° '
01	17 S 08	03 S 38	21 S 20	03 S 17	13 N 34	14 N 29	16 N 59	23 N 18	07 N 03	22 S 49
02	16 51	00 N 28	21 23	02 45	13 47	14 31	17 01	23 18	07 04	22 49
03	16 33	04 38	21 24	02 14	13 59	14 33	17 02	23 18	07 04	22 48
04	16 16	08 40	21 25	01 43	14 12	14 35	17 04	23 18	07 05	22 48
05	15 58	12 22	21 24	01 11	14 25	14 37	17 06	23 18	07 05	22 47
06	15 39	15 31	21 22	00 40	14 38	14 39	17 07	23 19	07 06	22 47
07	15 21	17 51	21 19	00 S 09	14 50	14 42	17 09	23 19	07 06	22 46
08	15 02	19 06	21 15	00 N 23	15 03	14 44	17 10	23 19	07 07	22 46
09	14 43	19 04	21 10	00 54	15 15	14 46	17 12	23 19	07 07	22 45
10	14 23	17 41	21 04	01 26	15 28	14 48	17 13	23 19	07 08	22 45
11	14 04	15 03	20 56	01 57	15 40	14 51	17 15	23 19	07 08	22 44
12	13 44	11 26	20 47	02 28	15 52	14 53	17 16	23 20	07 09	22 44
13	13 24	07 08	20 37	02 59	16 04	14 55	17 18	23 20	07 09	22 43
14	13 04	02 N 32	20 26	03 31	16 16	14 58	17 19	23 20	07 10	22 43
15	12 43	02 S 03	20 13	04 02	16 28	15 00	17 21	23 20	07 10	22 42
16	12 23	06 24	19 59	04 33	16 40	15 03	17 22	23 20	07 11	22 42
17	12 02	10 17	19 44	05 04	16 52	15 06	17 24	23 20	07 12	22 41
18	11 41	13 35	19 28	05 34	17 03	15 08	17 25	23 20	07 12	22 41
19	11 19	16 11	19 10	06 05	17 15	15 11	17 27	23 20	07 13	22 40
20	10 58	18 01	18 51	06 36	17 26	15 14	17 28	23 21	07 13	22 40
21	10 36	19 02	18 31	07 06	17 37	15 16	17 29	23 21	07 14	22 39
22	10 15	19 12	18 09	07 36	17 48	15 19	17 31	23 21	07 15	22 39
23	09 53	18 32	17 47	08 06	17 59	15 22	17 32	23 21	07 15	22 38
24	09 31	17 03	17 22	08 36	18 10	15 25	17 33	23 21	07 16	22 38
25	09 08	14 48	16 57	09 06	18 21	15 27	17 35	23 21	07 17	22 37
26	08 46	11 54	16 30	09 36	18 32	15 30	17 36	23 21	07 17	22 37
27	08 24	08 28	16 02	10 05	18 42	15 33	17 37	23 21	07 18	22 36
28	08 01	04 37	15 33	10 34	18 53	15 36	17 39	23 21	07 19	22 36
29	07 S 38	00 S 30	15 S 02	11 N 03	19 N 03	15 N 39	17 N 40	23 N 21	07 N 19	22 S 36

ZODIAC SIGN ENTRIES

Date	h	m	Planets
01	17	30	☽ ♈
04	02	04	☽ ♉
06	07	25	☽ ♊
07	00	26	♀ ♈
08	09	41	☽ ♋
10	09	55	☽ ♌
12	09	46	☽ ♍
12	12	03	☿ ♒
14	11	07	☽ ♎
16	15	36	☽ ♏
18	23	59	☽ ♐
19	02	14	☉ ♓
21	11	36	☽ ♑
24	00	33	☽ ♒
26	12	51	☽ ♓
28	23	19	☽ ♈

LATITUDES

Date	Mercury ☿ ° '	Venus ♀ ° '	Mars ♂ ° '	Jupiter ♃ ° '	Saturn ♄ ° '	Uranus ♅ ° '	Neptune ♆ ° '	Pluto ♇ ° '
01	01 N 01	00 S 46	00 N 51	00 S 59	01 N 02	00 N 24	01 S 40	08 S 11
04	00 31	00 35	00 53	00 58	01 02	00 24	01 39	08 11
07	00 N 03	00 24	00 55	00 57	01 02	00 24	01 39	08 11
10	00 S 23	00 S 12	00 57	00 56	01 03	00 24	01 39	08 11
13	00 47	00 00	00 59	00 56	01 03	00 24	01 39	08 11
16	01 08	00 N 13	01 01	00 55	01 03	00 24	01 39	08 11
19	01 26	00 27	01 02	00 54	01 03	00 24	01 39	08 12
22	01 41	00 41	01 04	00 53	01 04	00 24	01 39	08 12
25	01 54	00 55	01 05	00 53	01 04	00 24	01 38	08 12
28	02 03	01 10	01 07	00 52	01 04	00 24	01 38	08 13
31	02 S 08	01 N 25	01 N 08	00 S 51	01 N 04	00 N 23	01 S 38	08 S 13

DATA

Julian Date	2464725
Delta T	+76 seconds
Ayanamsa	24° 21' 32"
Synetic vernal point	04° ♓ 45' 27"
True obliquity of ecliptic	23° 25' 57"

MOON'S PHASES, APSIDES AND POSITIONS ☽

Date	h	m	Phase	Longitude ° '	Eclipse Indicator
05	07	01	☽	16 ♉ 03	
11	22	09	○	22 ♌ 45	total
18	23	47	☾	29 ♏ 54	
27	04	59	●	08 ♓ 10	Partial

Day	h	m		
10	21	06	Perigee	
23	03	25	Apogee	
02	09	17	ON	
08	23	26	Max dec	19° N 15'
15	01	10	OS	
22	04	47	Max dec	19° S 15'
29	14	54	ON	

LONGITUDES

Date	Chiron ⚷ ° '	Ceres ⚳ ° '	Pallas ⚴ ° '	Juno ⚵ ° '	Vesta ⚶ ° '	Black Moon Lilith ⚸ ° '
01	07 ♊ 13	17 ♊ 53	03 ♈ 39	00 ♈ 20	22 ♏ 59	21 ♑ 45
11	07 ♊ 08	18 ♊ 07	07 ♈ 09	05 ♈ 32	26 ♏ 41	22 ♑ 52
21	07 ♊ 10	18 ♊ 59	10 ♈ 53	10 ♈ 54	00 ♐ 01	23 ♑ 59
31	07 ♊ 19	20 ♊ 28	14 ♈ 49	16 ♈ 25	02 ♐ 57	25 ♑ 05

All ephemeris data is given at 12.00 UT and the Moon's longitude is additionally given for 24.00 UT

ASPECTARIAN

h m Aspects

01 Friday
02 05 ☽ ⊻ ♆
02 13 ☽ ± ♄
04 10 ☽ ☌ ♀
04 38 ☉ ⊼ ♅
05 57 ☉ □ ♃
08 06 ☽ ⊥ ♇
11 37 ☽ ∠ ♃
12 04 ☽ ∠ ☉
13 09 ☽ ⊥ ♂
14 26 ☽ ‖ ♀
15 59 ☽ Q ☿
19 35 ☽ bQ ♄
23 15 ☉ # ♄

02 Saturday
01 19 ☽ ⊻ ♂
01 33 ☽ ∠ ♇
05 05 ☽ ⊥ ♃
16 09 ☽ □ ♅
16 36 ☽ ⊻ ♃
19 21 ☽ ⚹ ☉
22 21 ♀ ⊥ ♇
23 43 ☽ # ♀

03 Sunday
00 00 ☽ △ ♄
05 17 ☽ □ ☿
06 04 ☽ ⚹ ♇
11 29 ☽ ☌ ♆
14 57 ☿ ⊻ ♇
19 06 ☽ ⊻ ♀
19 43 ☽ Q ☉

04 Monday
01 40 ☽ Q ♅
02 24 ☽ ‖ ♆
04 20 ☽ Q ♇
07 13 ☽ ⊥ ♀
12 10 ☽ ☌ ♂
16 19 ♀ ∠ ♃
23 22 ☽ ⚹ ♅

05 Tuesday
00 20 ☽ ☌ ♃
01 02 ☽ ∠ ♀
01 32 ☉ ☍ ♄
06 34 ☽ □ ♄
07 01 ☽ □ ☉
12 44 ☽ □ ♇
16 41 ☽ △ ☿
17 47 ☽ ⊻ ♆

06 Wednesday
01 45 ☽ ∠ ♅
04 03 ☽ ‖ ♂
04 15 ☽ ⊥ ♆
04 44 ☽ ‖ ♃
05 52 ☽ ⚹ ♀
06 01 ☿ □ ♆
13 01 ☽ # ☉
13 45 ☽ Q ♄
15 48 ♂ ‖ ♃
17 11 ☽ ⊥ ♅
19 18 ☽ ⊻ ♂
19 42 ☽ ∠ ♆
20 55 ☽ bQ ☿
21 58 ☉ ± ♅

07 Thursday
03 19 ☽ ⊻ ♅
03 28 ☽ ‖ ♄
04 10 ☽ Q ♀
04 44 ☽ ⊻ ♃
05 54 ☽ ⊥ ♂
09 56 ☽ ⚹ ♄
12 57 ♀ bQ ♄
13 26 ☽ ± ☿
14 33 ☽ △ ☉
14 51 ☽ ⊥ ♃
16 09 ☽ △ ♇
20 52 ☽ ⚹ ♆
21 33 ☽ ∠ ♂

08 Friday
00 17 ☽ ⊼ ☿
05 49 ☽ ∠ ♃
10 34 ☽ ∠ ♄
11 15 ☉ # ♂
12 33 ☽ □ ♀
13 53 ☉ ☌ ♇
16 33 ☽ Q ♆
16 49 ☽ bQ ♇
17 02 ☽ bQ ☉
22 32 ♀ ⊥ ♂
23 06 ☽ ⚹ ♂

09 Saturday
01 05 ☽ ⊥ ♄
04 31 ☽ ☌ ♅
06 20 ☽ ⚹ ♃
07 23 ☽ ± ♇
08 23 ☉ # ♃
08 37 ☽ ± ☉
10 41 ☽ ⊻ ♄
17 01 ☽ ⊼ ♇
18 56 ☽ ⊼ ☉
19 12 ☽ Q ♂
21 33 ☽ □ ♆

10 Sunday
01 41 ☽ Q ♃
05 10 ☽ ☍ ☿
16 43 ☽ △ ♀
17 03 ☽ ‖ ♄
19 58 ♀ ∠ ♇

11 Monday
01 04 ☽ □ ♂
04 17 ☽ ⊻ ♅
06 32 ☽ □ ♃
07 32 ☽ ‖ ♂
10 14 ☽ ☌ ♄
10 37 ☉ ⚹ ♆
13 33 ☽ ‖ ♃
13 49 ☽ ⊥ ♅
16 53 ☽ ☍ ♇
18 35 ☽ bQ ♀
19 54 ☽ # ☉
21 23 ☽ △ ♆
22 09 ☽ ☍ ☉

12 Tuesday
04 11 ☽ ∠ ♅
09 32 ☽ ⊼ ☿
10 13 ☽ ± ♀
20 11 ☽ ± ☿
20 44 ☽ ⊼ ♀
21 34 ☽ bQ ♆

13 Wednesday
03 16 ☽ △ ♂
04 24 ☽ ⚹ ♅
05 20 ♀ ⊥ ♃
07 12 ☽ △ ♃
10 21 ☽ ⊻ ♄
11 55 ☽ ‖ ♆
12 18 ☽ ± ♆
12 23 ☽ bQ ☿
17 36 ☽ ⊼ ♇
20 13 ☽ ⊥ ♄
22 15 ☽ ⊼ ♆

14 Thursday
00 10 ☽ Q ♅
02 40 ☽ ⊼ ☉
03 37 ☽ ± ♇
05 10 ♂ ⚹ ♅
05 10 ☽ bQ ♂
07 29 ☽ ‖ ♀
08 19 ☽ bQ ♃
11 12 ☽ ∠ ♄
13 32 ☽ ± ☉
16 06 ☽ △ ☿
18 51 ☽ bQ ♇
21 07 ☽ ± ♂
23 46 ☽ ± ♃

15 Friday
03 10 ☽ ☍ ♀
06 09 ☽ bQ ☉
06 42 ☽ □ ♅
07 56 ☽ ⊼ ♂
10 14 ☽ ⊼ ♃
12 50 ☽ ⚹ ♄
13 24 ☉ bQ ♅
20 58 ☽ △ ♇

16 Saturday
00 06 ☽ # ♀
01 54 ☽ ☍ ♆
09 59 ☽ Q ♄
10 47 ☽ △ ☉
16 38 ☽ # ♆
21 00 ♀ □ ♅

17 Sunday
03 17 ☽ □ ☿
12 32 ☽ △ ♅
14 05 ☽ ⊼ ♀
16 41 ☽ ☍ ♂
17 01 ☽ ☍ ♃
18 55 ☽ □ ♄
22 58 ☽ ‖ ☉

18 Monday
01 02 ♂ ☌ ♃
02 29 ☽ ± ♀
04 07 ☽ □ ♇
09 25 ☽ ⊼ ♆
16 55 ☽ bQ ♅
17 34 ☽ Q ☿
20 58 ☽ ± ♆
21 27 ☽ bQ ♀
23 47 ☽ □ ☉

19 Tuesday
01 44 ☽ # ♃
03 07 ♀ ⊻ ♃
09 52 ♂ □ ♄
10 23 ☽ ± ♅
14 35 ☽ bQ ♆
15 02 ☽ Q ♇
17 19 ♀ △ ♄
20 11 ☽ ⚹ ☿
22 11 ☽ ⊼ ♅
22 55 ☿ Q ♆

20 Wednesday
02 00 ☽ # ♂
03 19 ♀ ⊻ ♂
03 26 ☽ # ♄
03 43 ☽ ⊼ ♃
04 42 ☽ △ ♄
05 50 ☽ ⊼ ♂
05 57 ☽ △ ♀
12 24 ☿ ⊼ ♅
14 31 ☽ Q ☉
14 56 ☽ ⚹ ♇
15 49 ☽ ± ♃
16 49 ♂ ‖ ♄
18 28 ☽ ± ♂
20 29 ☽ △ ♆

21 Thursday
00 58 ☽ ‖ ☿
06 24 ☽ ∠ ☿
08 22 ☉ Q ♃
10 09 ☽ bQ ♃
10 36 ☽ bQ ♄
13 36 ☽ bQ ♂
16 55 ☽ ⚹ ☉
18 24 ♀ ‖ ♆
21 15 ☽ ∠ ♇

22 Friday
03 16 ☽ ⊥ ☿
04 43 ☽ ± ♄
10 27 ☽ ☍ ♅
10 44 ♃ □ ♄
14 09 ☿ ☍ ♄
14 42 ☿ □ ♃
15 35 ☽ ⊥ ♇
16 51 ☽ ⊼ ♄
16 58 ☽ △ ♃
17 16 ☽ ⊻ ☿
21 44 ☽ △ ♂

23 Saturday
00 55 ☽ □ ♀
02 15 ☽ ∠ ☉
03 00 ☿ # ♂
03 49 ☽ ⊻ ♇
09 26 ☽ □ ♆
21 07 ☽ # ♂
22 15 ☽ ⊥ ☉

24 Sunday
01 49 ☿ # ♄
05 07 ☽ # ♄
06 14 ☽ ‖ ☿
08 07 ♀ ⚹ ♇
10 00 ☿ ± ♅
11 33 ☽ ⊻ ☉
17 18 ☽ Q ♀
22 02 ☽ Q ♆
23 15 ☽ ⊼ ♅
23 34 ☿ □ ♂

25 Monday
05 19 ☽ ☍ ♄
05 51 ☽ # ♃
06 34 ☽ □ ♃
09 56 ☉ Q ♂
11 18 ☽ ± ♅
13 06 ☉ # ♀
13 51 ☽ □ ♂
15 09 ☽ ☌ ☿
16 40 ☽ ☌ ♇
19 55 ☽ ⚹ ♀
22 11 ☽ ⚹ ♆

26 Tuesday
02 57 ☿ ☌ ♇
05 16 ☽ bQ ♅
16 45 ☉ ∠ ♆
18 56 ☽ Q ♃
20 59 ♀ ☌ ♆

27 Wednesday
02 28 ☽ # ♀
03 33 ☽ Q ♂
04 00 ☽ ∠ ♆
04 41 ☽ ∠ ♀
04 59 ☽ ☌ ☉
10 51 ☽ △ ♅
12 31 ☽ ‖ ☉
16 30 ☽ ⊼ ♄
18 48 ☽ ⚹ ♃
19 28 ☽ # ♆
21 13 ☿ ⚹ ♆
21 39 ☽ ⊥ ♆
22 54 ♂ □ ♇

28 Thursday
00 01 ☽ ⊥ ♀
03 58 ☽ ⊻ ♇
04 06 ☽ ± ♄
04 13 ☽ ⚹ ♂
09 18 ☽ ⊻ ♆
09 44 ☿ # ♃
11 04 ☽ ⊻ ☿
12 45 ☽ ⊻ ♀
14 01 ☉ ∠ ♀
15 34 ☽ ⊥ ♇
21 20 ☽ bQ ♄

29 Friday
00 07 ☽ ∠ ♃
00 19 ☽ ⊥ ☿
08 49 ☽ ∠ ♇
10 29 ☽ ∠ ♂
17 26 ☽ ⊥ ♃
19 50 ☽ ⊻ ☉
19 56 ☽ ∠ ☿
20 26 ☽ □ ♅
23 11 ☿ ⚹ ♀
23 51 ☿ bQ ♅

MARCH 2036

LONGITUDES

Date	Sidereal time h m s	Sun ☉ ° ' "	Moon ☽ ° ' "	Moon ☽ 24.00 ° ' "	Mercury ☿ ° '	Venus ♀ ° '	Mars ♂ ° '	Jupiter ♃ ° '	Saturn ♄ ° '	Uranus ♅ ° '	Neptune ♆ ° '	Pluto ♇ ° '
01	22 39 30	11 ♓ 28 54	19 ♈ 25 56	25 ♈ 54 08	26 ♒ 57	26 ♈ 40	21 ♉ 31	15 ♉ 40	13 ♌ 51	11 ♋ 07	22 ♈ 46	20 ♒ 01
02	22 43 27	12 29 07	02 ♉ 25 50	09 ♉ 01 08	28 ♒ 37	27 45	22 09	15 49	13 R 47	11 R 06	22 48	20 03
03	22 47 24	13 29 18	15 ♉ 40 07	22 ♉ 22 53	00 ♓ 18	28 51	22 46	15 59	13 44	11 05	22 50	20 04
04	22 51 20	14 29 27	29 ♉ 09 28	05 ♊ 59 55	01 59	29 ♈ 56	23 23	16 09	13 40	11 04	22 52	20 06
05	22 55 17	15 29 34	12 ♊ 54 16	19 ♊ 52 30	03 42	01 ♉ 01	24 01	16 19	13 36	11 04	22 54	20 08
06	22 59 13	16 29 39	26 ♊ 54 31	04 ♋ 00 12	05 26	02 05	24 38	16 30	13 32	11 03	22 56	20 09
07	23 03 10	17 29 41	11 ♋ 09 20	18 ♋ 21 39	07 11	03 09	25 15	16 40	13 29	11 02	22 58	20 11
08	23 07 06	18 29 42	25 ♋ 36 46	02 ♌ 54 12	08 57	04 13	25 53	16 50	13 25	11 02	22 59	20 12
09	23 11 03	19 29 40	10 ♌ 13 24	17 ♌ 33 42	10 44	05 17	26 30	17 01	13 21	11 01	23 01	20 14
10	23 14 59	20 29 36	24 ♌ 54 24	02 ♍ 14 41	12 32	06 20	27 08	17 11	13 18	11 01	23 03	20 16
11	23 18 56	21 29 30	09 ♍ 33 44	16 ♍ 50 43	14 22	07 23	27 45	17 22	13 15	11 01	23 05	20 17
12	23 22 53	22 29 22	24 ♍ 04 50	01 ♎ 15 18	16 12	08 25	28 22	17 33	13 12	11 00	23 07	20 19
13	23 26 49	23 29 12	08 ♎ 21 25	15 ♎ 22 36	18 04	09 28	29 00	17 44	13 08	11 00	23 09	20 20
14	23 30 46	24 29 00	22 ♎ 18 22	29 ♎ 08 23	19 57	10 29	29 ♉ 37	17 55	13 05	11 00	23 11	20 22
15	23 34 42	25 28 47	05 ♏ 52 24	12 ♏ 30 23	21 51	11 31	00 ♊ 15	18 06	13 02	10 59	23 13	20 23
16	23 38 39	26 28 31	19 ♏ 02 21	25 ♏ 28 30	23 46	12 32	00 52	18 17	12 59	10 59	23 16	20 25
17	23 42 35	27 28 14	01 ♐ 49 05	08 ♐ 04 30	25 42	13 33	01 30	18 29	12 57	10 59	23 18	20 26
18	23 46 32	28 27 56	14 ♐ 15 12	20 ♐ 21 42	27 39	14 33	02 07	18 40	12 54	10 59	23 20	20 28
19	23 50 28	29 ♓ 27 35	26 ♐ 24 33	02 ♑ 24 22	29 ♓ 38	15 33	02 45	18 51	12 51	10 D 59	23 22	20 29
20	23 54 25	00 ♈ 27 13	08 ♑ 21 47	14 ♑ 17 26	01 ♈ 37	16 32	03 22	19 03	12 49	10 59	23 24	20 31
21	23 58 22	01 26 49	20 ♑ 11 58	26 ♑ 06 00	03 37	17 31	04 00	19 15	12 46	10 59	23 26	20 32
22	00 02 18	02 26 23	02 ♒ 00 09	07 ♒ 55 02	05 37	18 30	04 37	19 26	12 44	11 00	23 28	20 34
23	00 06 15	03 25 56	13 ♒ 51 13	19 ♒ 49 12	07 38	19 28	05 15	19 38	12 42	11 00	23 30	20 35
24	00 10 11	04 25 26	25 ♒ 49 29	01 ♓ 52 29	09 40	20 25	05 52	19 50	12 40	11 00	23 32	20 37
25	00 14 08	05 24 55	07 ♓ 58 36	14 ♓ 08 08	11 41	21 23	06 30	20 02	12 38	11 00	23 35	20 38
26	00 18 04	06 24 22	20 ♓ 21 19	26 ♓ 38 21	13 43	22 19	07 07	20 14	12 36	11 01	23 37	20 39
27	00 22 01	07 23 47	02 ♈ 59 20	09 ♈ 24 19	15 44	23 15	07 45	20 26	12 34	11 01	23 39	20 41
28	00 25 57	08 23 09	15 ♈ 53 16	22 ♈ 26 05	17 45	24 11	08 22	20 39	12 32	11 02	23 41	20 42
29	00 29 54	09 22 30	29 ♈ 02 38	05 ♉ 42 45	19 44	25 05	09 00	20 51	12 31	11 02	23 43	20 44
30	00 33 51	10 21 49	12 ♉ 26 10	19 ♉ 12 41	21 42	26 00	09 37	21 03	12 29	11 03	23 46	20 44
31	00 37 47	11 ♈ 21 05	26 ♉ 02 00	02 ♊ 53 53	23 ♈ 39	26 ♉ 53	10 ♊ 15	21 ♉ 16	12 ♌ 28	11 ♋ 04	23 ♈ 48	20 ♒ 46

Date	Moon True ☊ ° '	Moon Mean ☊ ° '	Moon ☽ Latitude ° '
01	26 ♌ 09	25 ♌ 35	04 S 14
02	26 R 07	25 32	04 50
03	26 05	25 28	05 11
04	26 03	25 25	05 16
05	26 03	25 22	05 02
06	26 D 03	25 19	04 31
07	26 04	25 16	03 43
08	26 05	25 12	02 40
09	26 07	25 09	01 27
10	26 07	25 06	00 S 07
11	26 R 07	25 03	01 N 13
12	26 05	25 00	02 28
13	26 02	24 57	03 32
14	25 57	24 53	04 21
15	25 53	24 50	04 54
16	25 48	24 47	05 11
17	25 45	24 44	05 11
18	25 42	24 41	04 57
19	25 41	24 38	04 29
20	25 D 41	24 34	03 51
21	25 42	24 31	03 02
22	25 44	24 28	02 06
23	25 46	24 25	01 N 05
24	25 R 46	24 22	00 00
25	25 46	24 18	01 S 06
26	25 43	24 15	02 06
27	25 38	24 12	03 08
28	25 32	24 09	03 58
29	25 24	24 06	04 37
30	25 17	24 03	05 01
31	25 ♌ 09	23 ♌ 59	05 S 08

DECLINATIONS

Date	Sun ☉ ° '	Moon ☽ ° '	Mercury ☿ ° '	Venus ♀ ° '	Mars ♂ ° '	Jupiter ♃ ° '	Saturn ♄ ° '	Uranus ♅ ° '	Neptune ♆ ° '	Pluto ♇ ° '
01	07 S 15	03 N 41	14 S 31	11 N 31	19 N 13	15 N 42	17 N 41	23 N 21	07 N 20	22 S 35
02	06 52	07 47	13 57	12 00	19 23	15 45	17 42	23 22	07 21	22 35
03	06 29	11 34	13 23	12 28	19 33	15 48	17 43	23 22	07 21	22 34
04	06 06	14 50	12 47	12 56	19 43	15 51	17 45	23 22	07 22	22 34
05	05 43	17 20	12 10	13 23	19 53	15 54	17 46	23 22	07 23	22 33
06	05 20	18 52	11 32	13 51	20 03	15 57	17 47	23 22	07 24	22 33
07	04 56	19 15	10 52	14 18	20 12	16 01	17 48	23 22	07 24	22 33
08	04 33	18 23	10 12	14 44	20 21	16 04	17 49	23 22	07 25	22 32
09	04 09	16 17	09 30	15 11	20 30	16 07	17 50	23 22	07 26	22 32
10	03 46	13 07	08 46	15 36	20 39	16 10	17 51	23 22	07 27	22 32
11	03 22	09 07	08 02	16 02	20 48	16 13	17 52	23 22	07 27	22 31
12	02 59	04 N 37	07 16	16 27	20 57	16 16	17 53	23 22	07 28	22 31
13	02 35	00 S 04	06 29	16 52	21 06	16 20	17 54	23 22	07 29	22 31
14	02 11	04 38	05 41	17 17	21 14	16 23	17 54	23 22	07 30	22 30
15	01 48	08 50	04 52	17 41	21 22	16 26	17 55	23 22	07 30	22 30
16	01 24	12 29	04 02	18 05	21 30	16 30	17 56	23 22	07 31	22 30
17	01 00	15 26	03 11	18 28	21 38	16 33	17 57	23 22	07 32	22 29
18	00 37	17 35	02 18	18 51	21 46	16 36	17 58	23 22	07 33	22 29
19	00 S 13	18 54	01 25	19 14	21 54	16 39	17 59	23 22	07 33	22 29
20	00 N 11	19 20	00 S 31	19 36	22 01	16 43	17 59	23 22	07 34	22 28
21	00 35	18 55	00 N 24	19 58	22 09	16 46	18 00	23 22	07 35	22 28
22	00 58	17 39	01 19	20 19	22 16	16 49	18 01	23 22	07 36	22 28
23	01 22	15 38	02 15	20 40	22 23	16 53	18 01	23 22	07 37	22 28
24	01 45	12 55	03 11	21 00	22 30	16 56	18 02	23 22	07 38	22 27
25	02 09	09 36	04 08	21 20	22 37	17 00	18 02	23 22	07 38	22 27
26	02 33	05 48	05 05	21 40	22 43	17 03	18 03	23 22	07 39	22 27
27	02 56	01 S 41	06 01	21 59	22 50	17 06	18 03	23 22	07 40	22 27
28	03 19	02 N 35	06 58	22 18	22 56	17 10	18 04	23 22	07 41	22 26
29	03 43	06 49	07 54	22 36	23 02	17 13	18 04	23 22	07 42	22 26
30	04 06	10 47	08 49	22 53	23 08	17 17	18 05	23 21	07 42	22 26
31	04 N 29	14 N 16	09 N 43	23 N 10	23 N 14	17 N 20	18 N 05	23 N 21	07 N 43	22 S 26

ZODIAC SIGN ENTRIES

Date	h	m	Planets
02	07	53	☽ ♉
03	07	48	☿ ♓
04	13	29	☽ ♊
04	13	34	♀ ♉
06	17	14	☽ ♋
08	19	14	☽ ♌
10	20	20	☽ ♍
12	21	54	☽ ♎
15	01	31	☽ ♏
15	02	37	♂ ♊
17	08	33	☽ ♐
19	16	31	☿ ♈
19	19	10	☽ ♑
20	01	03	☉ ♈
22	07	56	☽ ♒
24	20	18	☽ ♓
27	06	22	☽ ♈
29	13	44	☽ ♉
31	18	57	☽ ♊

LATITUDES

Date	Mercury ☿ ° '	Venus ♀ ° '	Mars ♂ ° '	Jupiter ♃ ° '	Saturn ♄ ° '	Uranus ♅ ° '	Neptune ♆ ° '	Pluto ♇ ° '
01	02 S 07	01 N 20	01 N 07	00 S 52	01 N 04	00 N 24	01 S 38	08 S 13
04	02 10	01 35	01 09	00 51	01 04	00 23	01 38	08 13
07	02 09	01 51	01 10	00 50	01 04	00 23	01 38	08 14
10	02 05	02 06	01 11	00 50	01 04	00 23	01 38	08 15
13	01 56	02 21	01 12	00 49	01 04	00 23	01 38	08 15
16	01 42	02 37	01 12	00 48	01 04	00 23	01 38	08 16
19	01 23	02 52	01 13	00 48	01 04	00 23	01 38	08 16
22	01 00	03 07	01 14	00 47	01 04	00 23	01 38	08 17
25	00 S 32	03 22	01 15	00 47	01 04	00 23	01 38	08 17
28	00 00	03 36	01 15	00 46	01 04	00 23	01 38	08 18
31	00 N 35	03 N 49	01 N 16	00 S 46	01 N 04	00 N 23	01 S 38	08 S 19

DATA

Julian Date	2464754
Delta T	+76 seconds
Ayanamsa	24° 21' 35"
Synetic vernal point	04° ♓ 45' 24"
True obliquity of ecliptic	23° 25' 57"

MOON'S PHASES, APSIDES AND POSITIONS ☽

Date	h	m	Phase	Longitude ° '	Eclipse Indicator
05	16	49	☽	15 ♊ 42	
12	09	09	○	22 ♍ 22	
19	18	39	☾	29 ♐ 44	
27	20	57	●	07 ♈ 46	

Day	h	m		
10	02	36	Perigee	
21	21	46	Apogee	
07	07	27	Max dec	19° N 17'
13	11	38	0S	
20	12	08	Max dec	19° S 20'
27	21	32	0N	

LONGITUDES

Date	Chiron ⚷ ° '	Ceres ⚳ ° '	Pallas ⚴ ° '	Juno ⚵ ° '	Vesta ⚶ ° '	Black Moon Lilith ⚸ ° '
01	07 ♊ 18	20 ♊ 18	14 ♈ 25	15 ♈ 51	02 ♐ 41	24 ♑ 59
11	07 ♊ 33	22 ♊ 18	18 ♈ 31	21 ♈ 28	05 ♐ 10	26 ♑ 05
21	07 ♊ 55	24 ♊ 38	22 ♈ 47	27 ♈ 11	07 ♐ 05	27 ♑ 12
31	08 ♊ 24	27 ♊ 24	27 ♈ 12	02 ♉ 59	08 ♐ 18	28 ♑ 19

All ephemeris data is given at 12.00 UT and the Moon's longitude is additionally given for 24.00 UT

ASPECTARIAN

h m	Aspects
01 Saturday	
01 38	☽ △ ♄
03 24	☉ △ ♅
04 20	☽ ⊥ ♂
04 52	☽ ⊻ ♃
07 19	☉ # ♆
08 03	☽ ⊥ ☉
13 05	☽ ⚹ ♇
16 05	☽ ⊻ ♂
18 13	☽ ☌ ♆
02 Sunday	
02 10	☽ ∠ ☉
02 38	☽ ☌ ♀
03 59	☽ ⚹ ☿
05 54	☽ Q ♅
07 04	☽ # ☉
09 23	☽ ∥ ♆
11 17	☽ Q ♇
03 Monday	
03 45	☽ ⚹ ♅
05 03	☽ Q ☿
07 46	☽ ⚹ ☉
08 31	☽ □ ♄
12 35	☽ ☌ ♃
14 43	♂ ⊻ ♆
17 18	♀ Q ♅
17 19	☉ ⊼ ♄
19 11	☽ ∥ ♀
19 54	☽ □ ♇
22 51	☽ # ☿
04 Tuesday	
00 50	☽ ⊻ ♆
01 18	☽ ☌ ♂
06 34	☽ ∠ ♅
06 55	☽ Q ☉
08 49	☿ # ♀
11 29	☽ ⊥ ♆
13 29	☽ ⊻ ♀
16 23	☽ Q ♄
17 42	☽ □ ☿
21 02	☽ ∥ ♃
22 23	☽ ⊥ ♅
05 Wednesday	
00 54	☽ ⊥ ♀
03 17	☽ ∠ ♆
08 49	☽ ⊻ ♅
13 12	☽ ⚹ ♄
16 49	☽ □ ☉
17 15	☽ ∥ ♄
17 48	☽ ∠ ♀
17 58	☽ ⊻ ♃
21 34	☿ Q ♃
06 Thursday	
00 27	☽ △ ♇
04 23	☽ ⊥ ♃
05 12	☽ ⚹ ♆
07 57	☽ ⊻ ♂
11 57	☉ ⚹ ♃
13 37	♀ Q ♇
14 45	☽ ∠ ♄
18 36	☽ ⊥ ♂
19 52	☽ ∠ ♃
21 29	☽ ⚹ ♀
22 43	☉ ⊥ ♆
07 Friday	
01 35	☽ Q ♆
01 58	☽ ⚼ ♇
04 25	☽ △ ☿
05 52	☽ ⊥ ♄
10 26	☽ ∠ ♂
11 49	☽ ☌ ♅
15 51	☽ ⊻ ♄
17 04	☽ ± ♇
19 12	☽ Q ♀
21 18	☽ ⚹ ♃
22 46	☿ ∠ ♆
23 21	☽ △ ☉
08 Saturday	
03 03	☽ ⊼ ♇
07 40	☽ □ ♆
08 52	☽ ⚼ ☿
12 28	☽ ⚹ ♂
17 23	☽ Q ♃
17 49	♂ ∠ ♅
19 58	☽ ∥ ♄
09 Sunday	
01 45	☽ ± ☿
01 55	☽ ⚼ ☉
03 16	☽ □ ♀
08 54	☉ ± ♄
09 03	☽ Q ♂
12 57	☽ ⊼ ☿
13 19	☽ ⊻ ♅
13 29	☽ ∥ ♃
15 49	☿ △ ♅
17 06	☽ ☌ ♄
17 45	☽ ± ☉
20 07	☽ ∥ ♀
23 07	☽ ⊥ ♅
23 14	☽ □ ♃
10 Monday	
04 16	☽ ⊼ ☉
04 24	☽ ☍ ♇
06 17	☉ ⊻ ♇
08 58	☽ △ ♆
13 49	☽ ∠ ♅
15 47	☽ □ ♂
21 44	☿ ⊼ ♄
11 Tuesday	
08 08	☽ △ ♀
09 35	☽ ⚼ ♆
14 23	☽ ⚹ ♅
18 02	☽ ⊻ ♄
19 11	☽ # ☿
21 03	☽ ☍ ☿
21 03	☽ ∥ ♆
12 Wednesday	
00 04	♀ ∥ ♃
00 26	☽ ± ♆
01 02	☽ △ ♃
03 55	☽ ⊥ ♄
05 44	☽ ⊼ ♇
05 56	☿ # ♆
09 09	☽ ☍ ☉
10 12	☽ Q ♅
10 24	☽ ⊼ ♆
10 49	☽ ⚼ ♀
15 13	☿ Q ♂
15 44	☽ ± ♇
18 50	☽ ∠ ♄
19 29	☽ △ ♂
21 10	☽ # ☉
13 Thursday	
00 03	☿ ⊥ ♆
02 22	☽ ⚼ ♃
03 04	☽ ± ♀
03 46	☉ ⊻ ♆
06 53	☽ ⚼ ♇
07 12	☿ ⚹ ♃
14 01	☽ ⊼ ♀
16 30	☽ □ ♅
17 50	☽ ± ♃
18 31	♂ ⊥ ♆
20 08	☽ ⚹ ♄
22 05	☽ ⚼ ♂
14 Friday	
00 01	☽ ∥ ☉
01 19	☿ ± ♄
04 16	☽ ⊼ ♃
07 15	☽ ⊼ ☿
08 37	☽ △ ♇
13 33	☽ ☍ ♆
14 24	☽ ± ♂
16 06	☽ ⊼ ☉
16 51	☽ Q ♄
16 51	☽ ∥ ☿
17 20	☿ ⊻ ♇
19 24	☽ ± ☿
23 44	♀ ⚹ ♅
15 Saturday	
01 28	☽ ⊼ ♂
03 31	☽ ± ☉
04 03	☽ # ♆
14 03	☽ ⚼ ☿
20 59	☽ ⚼ ☉
21 14	☽ △ ♅
23 03	☽ ☍ ♀
16 Sunday	
00 55	☽ □ ♄
02 55	♀ ∥ ♄
05 32	☿ ⊻ ♆
10 33	☉ ⊥ ♇
10 35	☽ ☍ ♃
14 33	☽ □ ♇
16 22	♂ Q ♄
19 52	☽ ⊼ ♆
22 21	♀ □ ♄
22 22	☽ △ ☿
17 Monday	
00 58	☽ ⚼ ♅
03 03	☽ △ ☉
07 11	☽ ± ♆
11 21	☽ ☍ ♂
18 03	☽ ± ♅
21 12	☿ ⊥ ♇
22 55	☉ ⚼ ♄
23 34	☽ # ♃
18 Tuesday	
00 27	☽ ⚼ ♆
00 44	☽ Q ♇
05 38	☽ ⊼ ♅
09 22	☽ △ ♄
12 38	☽ ⊼ ♀
14 53	☿ ⚼ ♄
15 29	♅ St D
17 34	☽ # ♄
20 48	☽ ⊼ ♃
19 Wednesday	
00 14	☽ ⚹ ♇
01 29	☽ ± ♀
05 55	☽ △ ♆
07 54	☉ ☌ ☿
08 52	☽ ± ♃
14 53	☽ ⚼ ♄
18 39	☽ □ ☉
19 42	☽ □ ☿
21 01	☽ ⚼ ♀
20 Thursday	
01 22	☽ ⊼ ♂
03 10	☽ ⚼ ♃
06 15	☽ ∠ ♇
08 53	☽ ± ♄
10 14	☿ ∠ ♀
14 08	☽ ± ♂
17 18	☽ ☍ ♅
18 15	☉ # ☿
20 59	☽ ⊼ ♄
21 Friday	
00 29	☽ ⊥ ♇
06 04	☽ △ ♀
09 25	☽ ⚼ ♂
10 02	☽ △ ♃
10 20	☽ Q ☉
12 41	☽ ⊻ ♇
15 27	☽ Q ☿
18 36	☽ □ ♆
18 39	☿ ⚹ ♂
20 20	☉ ∥ ☿
20 25	☿ ∠ ♃
22 Saturday	
06 37	☽ # ♄
11 19	☿ ∠ ♇
12 58	☽ ⚹ ☉
17 36	☽ △ ♂
20 51	☽ ⚹ ☿
22 40	☽ # ♃
23 Sunday	
02 30	♂ ⊥ ♅
06 14	☽ ⊼ ♅
07 15	☽ Q ♆
09 41	☽ ☍ ♄
17 24	♀ ☌ ♃
18 20	☽ ± ♅
22 03	☽ ∠ ☉
23 50	☽ □ ♃
24 Monday	
00 16	☽ □ ♀
01 34	☽ ☌ ♇
02 53	♂ # ♇
07 26	☽ ⚹ ♆
09 13	☽ ∠ ☿
12 21	☽ ⚼ ♅
16 45	♀ □ ♇
17 38	☽ ⊥ ☉
25 Tuesday	
00 28	☉ ∠ ♃
03 55	☿ □ ♅
06 32	☽ ⊻ ☉
06 37	☽ ⊥ ☿
08 56	☽ □ ♂
12 07	☽ Q ♃
13 11	☽ ∠ ♆
14 58	☽ Q ♀
17 22	☉ ∠ ♇
17 55	☽ △ ♅
20 40	☽ ⊻ ☿
21 03	☽ ⊼ ♄
22 58	☿ △ ♄
26 Wednesday	
00 40	☽ # ♆
06 43	☽ ⊥ ♆
08 38	☽ ± ♄
11 46	☽ ⚹ ♃
12 34	☽ ⊻ ♇
15 32	☽ # ☿
16 04	☽ ⚹ ♀
18 15	☽ ⊻ ♆
18 53	☿ ⊥ ♃
21 35	☽ Q ♂
27 Thursday	
00 03	☽ ⊥ ♇
01 47	☽ ⚼ ♄
05 30	☽ # ☉
16 40	☽ ∠ ♃
17 03	☽ ∠ ♇
20 57	☽ ☌ ☉
21 22	☽ ⚹ ♂
22 37	☽ ∠ ♀
22 43	♀ ⊻ ♆
28 Friday	
03 01	☽ □ ♅
05 50	☽ △ ♄
09 40	☽ ⊥ ♃
11 01	☉ ⚹ ♂
16 02	☽ ☌ ☿
16 32	☽ ⊥ ♀
16 34	☽ ∥ ☉
19 24	♃ □ ♇
20 51	☽ ⚹ ♇
20 52	☽ ⊻ ♃
21 38	☿ ⊥ ♀
23 26	♀ # ♇
29 Saturday	
00 51	♂ ∠ ♆
02 19	☽ ☌ ♆
02 23	☽ ∠ ♂
04 18	☽ ⊻ ♀
06 44	☿ ∥ ♆
11 59	☽ Q ♅
17 08	☽ ∥ ♆
18 38	☽ Q ♇
19 29	☽ ⊥ ♂
20 10	☽ ∥ ☿
30 Sunday	
00 04	☿ ⚹ ♇
03 03	☿ ⊻ ♃
06 45	☽ ⊻ ♂
08 01	☽ ⊻ ☉
09 32	☽ ⚹ ♅
12 05	☽ □ ♄
13 26	♀ ∠ ♅
19 31	☽ ⊥ ☉
31 Monday	
02 43	☽ □ ♇
03 29	☽ ☌ ♃
04 50	☉ □ ♅
07 08	☽ ⊻ ☿
08 04	☽ ⊻ ♆
12 03	☽ ∠ ♅
12 36	☽ ∠ ☉
13 36	☽ ☌ ♀
13 52	☿ ☌ ♆
18 36	☽ ⊥ ♆
19 01	♀ ∥ ♂
19 22	☽ ⊥ ☿
19 44	☽ Q ♄

APRIL 2036

Date	Sidereal time h m s	Sun ☉ ° ' "	Moon ☽ ° ' "	Moon ☽ 24.00 ° ' "	Mercury ☿ ° '	Venus ♀ ° '	Mars ♂ ° '	Jupiter ♃ ° '	Saturn ♄ ° '	Uranus ♅ ° '	Neptune ♆ ° '	Pluto ♇ ° '
01	00 41 44	12 ♈ 20 20	09 ♊ 48 06	16 ♊ 44 23	25 ♈ 34	27 ♉ 46	10 ♊ 53	21 ♉ 28	12 ♌ 27	11 ♋ 04	23 ♈ 50	20 ♒ 47
02	00 45 40	13 19 32	23 ♊ 42 32	00 ♋ 42 22	27 26	28 39	11 30	21 41	12 R 25	11 05	23 52	20 48
03	00 49 37	14 18 41	07 ♋ 43 44	14 ♋ 46 26	29 ♈ 15	29 ♉ 30	12 08	21 53	12 24	11 06	23 55	20 49
04	00 53 33	15 17 49	21 ♋ 50 22	28 ♋ 55 21	01 ♉ 01	00 ♊ 21	12 45	22 06	12 23	11 07	23 57	20 51
05	00 57 30	16 16 54	06 ♌ 01 14	13 ♌ 07 49	02 43	01 12	13 23	22 19	12 22	11 08	23 59	20 52
06	01 01 26	17 15 56	20 ♌ 14 51	27 ♌ 22 05	04 21	02 01	14 00	22 32	12 22	11 09	24 01	20 53
07	01 05 23	18 14 56	04 ♍ 29 11	11 ♍ 35 43	05 55	02 49	14 38	22 44	12 21	11 10	24 04	20 54
08	01 09 20	19 13 54	18 ♍ 41 17	25 ♍ 45 23	07 25	03 37	15 16	22 57	12 21	11 11	24 06	20 55
09	01 13 16	20 12 50	02 ♎ 47 30	09 ♎ 47 06	08 50	04 24	15 53	23 10	12 20	11 12	24 08	20 56
10	01 17 13	21 11 44	16 ♎ 43 39	23 ♎ 36 38	10 09	05 10	16 31	23 23	12 20	11 13	24 10	20 57
11	01 21 09	22 10 35	00 ♏ 25 37	07 ♏ 10 12	11 23	05 55	17 08	23 36	12 20	11 14	24 13	20 58
12	01 25 06	23 09 25	13 ♏ 50 04	20 ♏ 25 01	12 32	06 39	17 46	23 49	12 19	11 16	24 15	20 59
13	01 29 02	24 08 13	26 ♏ 54 57	03 ♐ 19 52	13 35	07 22	18 24	24 03	12 D 19	11 17	24 17	21 00
14	01 32 59	25 06 59	09 ♐ 39 51	15 ♐ 55 08	14 32	08 04	19 01	24 16	12 20	11 18	24 19	21 01
15	01 36 55	26 05 43	22 ♐ 06 00	28 ♐ 12 50	15 24	08 45	19 39	24 29	12 20	11 20	24 22	21 02
16	01 40 52	27 04 26	04 ♑ 16 08	10 ♑ 16 24	16 09	09 25	20 16	24 42	12 20	11 21	24 24	21 03
17	01 44 49	28 03 07	16 ♑ 14 12	22 ♑ 10 11	16 49	10 03	20 54	24 56	12 21	11 23	24 26	21 04
18	01 48 45	29 ♈ 01 46	28 ♑ 04 59	03 ♒ 59 16	17 22	10 41	21 31	25 09	12 21	11 24	24 28	21 05
19	01 52 42	00 ♉ 00 24	09 ♒ 53 44	15 ♒ 49 02	17 49	11 17	22 09	25 23	12 22	11 26	24 31	21 06
20	01 56 38	00 58 59	21 ♒ 45 53	27 ♒ 44 53	18 11	11 52	22 47	25 36	12 23	11 28	24 33	21 06
21	02 00 35	01 57 33	03 ♓ 46 41	09 ♓ 51 51	18 26	12 25	23 24	25 50	12 23	11 29	24 35	21 07
22	02 04 31	02 56 06	16 ♓ 00 53	22 ♓ 14 16	18 35	12 57	24 02	26 03	12 24	11 31	24 37	21 08
23	02 08 28	03 54 36	28 ♓ 32 21	04 ♈ 55 25	18 38	13 28	24 39	26 17	12 25	11 33	24 40	21 09
24	02 12 24	04 53 05	11 ♈ 23 38	17 ♈ 57 05	18 R 35	13 57	25 17	26 31	12 27	11 35	24 42	21 09
25	02 16 21	05 51 33	24 ♈ 35 42	01 ♉ 19 19	18 27	14 24	25 54	26 44	12 28	11 37	24 44	21 10
26	02 20 18	06 49 58	08 ♉ 07 40	15 ♉ 00 21	18 14	14 50	26 32	26 58	12 29	11 39	24 46	21 11
27	02 24 14	07 48 22	21 ♉ 56 53	28 ♉ 56 45	17 56	15 15	27 10	27 12	12 31	11 41	24 49	21 11
28	02 28 11	08 46 44	05 ♊ 59 20	13 ♊ 04 02	17 34	15 37	27 47	27 25	12 32	11 43	24 51	21 12
29	02 32 07	09 45 03	20 ♊ 10 14	27 ♊ 17 22	17 07	15 58	28 25	27 39	12 34	11 45	24 53	21 13
30	02 36 04	10 ♉ 43 21	04 ♋ 24 54	11 ♋ 32 23	16 ♉ 37	16 ♊ 16	29 ♊ 02	27 ♉ 53	12 ♌ 36	11 ♋ 47	24 ♈ 55	21 ♒ 13

Date	Moon True ☊ ° '	Moon Mean ☊ ° '	Moon ☽ Latitude ° '
01	25 ♌ 04	23 ♌ 56	04 S 58
02	25 R 00	23 53	04 31
03	24 58	23 50	03 47
04	24 D 58	23 47	02 49
05	24 59	23 44	01 40
06	25 00	23 40	00 S 26
07	25 R 00	23 37	00 N 51
08	24 57	23 34	02 04
09	24 53	23 31	03 08
10	24 46	23 28	04 01
11	24 36	23 24	04 38
12	24 26	23 21	05 00
13	24 16	23 18	05 04
14	24 07	23 15	04 54
15	24 00	23 12	04 29
16	23 55	23 09	03 53
17	23 52	23 05	03 07
18	23 52	23 02	02 13
19	23 D 52	22 59	01 14
20	23 52	22 56	00 N 11
21	23 R 52	22 53	00 S 53
22	23 50	22 49	01 55
23	23 45	22 46	02 53
24	23 38	22 43	03 44
25	23 28	22 40	04 25
26	23 17	22 37	04 51
27	23 05	22 34	05 01
28	22 54	22 30	04 54
29	22 45	22 27	04 28
30	22 ♌ 39	22 ♌ 24	03 S 46

DECLINATIONS

Date	Sun ☉ ° '	Moon ☽ ° '	Mercury ☿ ° '	Venus ♀ ° '	Mars ♂ ° '	Jupiter ♃ ° '	Saturn ♄ ° '	Uranus ♅ ° '	Neptune ♆ ° '	Pluto ♇ ° '
01	04 N 52	17 N 00	10 N 36	23 N 27	23 N 19	17 N 23	18 N 06	23 N 21	07 N 44	22 S 26
02	05 16	18 47	11 27	23 43	23 25	17 27	18 06	23 21	07 45	22 25
03	05 38	19 26	12 17	23 59	23 30	17 30	18 06	23 21	07 46	22 25
04	06 01	18 53	13 05	24 14	23 35	17 34	18 06	23 21	07 47	22 25
05	06 24	17 08	13 51	24 29	23 40	17 37	18 07	23 21	07 47	22 25
06	06 47	14 20	14 35	24 43	23 45	17 40	18 07	23 21	07 48	22 25
07	07 09	10 39	15 17	24 56	23 49	17 44	18 07	23 21	07 49	22 25
08	07 32	06 22	15 56	25 09	23 54	17 47	18 07	23 21	07 50	22 24
09	07 54	01 N 46	16 32	25 22	23 58	17 51	18 07	23 21	07 51	22 24
10	08 16	02 S 52	17 06	25 34	24 02	17 54	18 07	23 20	07 52	22 24
11	08 38	07 16	17 36	25 46	24 06	17 57	18 07	23 20	07 52	22 24
12	09 00	11 13	18 04	25 57	24 09	18 01	18 07	23 20	07 53	22 24
13	09 22	14 31	18 30	26 07	24 13	18 04	18 07	23 20	07 54	22 24
14	09 43	17 03	18 52	26 17	24 16	18 07	18 07	23 20	07 55	22 24
15	10 04	18 43	19 11	26 26	24 19	18 11	18 07	23 20	07 56	22 24
16	10 26	19 29	19 27	26 35	24 22	18 14	18 07	23 20	07 56	22 24
17	10 47	19 21	19 41	26 44	24 25	18 18	18 07	23 20	07 57	22 24
18	11 08	18 22	19 51	26 51	24 27	18 21	18 07	23 19	07 58	22 24
19	11 28	16 35	19 58	26 59	24 30	18 24	18 07	23 19	07 59	22 24
20	11 49	14 04	20 02	27 05	24 32	18 28	18 06	23 19	08 00	22 24
21	12 09	10 56	20 04	27 12	24 34	18 31	18 06	23 19	08 01	22 24
22	12 29	07 17	20 02	27 17	24 36	18 34	18 06	23 19	08 01	22 24
23	12 49	03 S 14	19 58	27 22	24 38	18 37	18 05	23 19	08 02	22 24
24	13 09	01 N 04	19 50	27 27	24 39	18 41	18 05	23 18	08 03	22 24
25	13 28	05 25	19 40	27 31	24 40	18 44	18 05	23 18	08 04	22 24
26	13 48	09 37	19 27	27 34	24 42	18 47	18 04	23 18	08 05	22 24
27	14 07	13 23	19 11	27 37	24 43	18 51	18 04	23 18	08 05	22 24
28	14 25	16 29	18 53	27 40	24 43	18 54	18 03	23 18	08 06	22 24
29	14 44	18 37	18 33	27 42	24 44	18 57	18 03	23 18	08 07	22 24
30	15 N 02	19 N 36	18 N 11	27 N 43	24 N 44	19 N 00	18 N 02	23 N 17	08 N 08	22 S 24

ZODIAC SIGN ENTRIES

Date	h	m	Planets
02	22	47	☽ ♋
03	22	09	☿ ♉
04	01	54	♀ ♊
05	01	49	☽ ♌
07	04	26	☽ ♍
09	07	14	☽ ♎
11	11	15	☽ ♏
13	17	45	☽ ♐
16	03	32	☽ ♑
18	15	54	☽ ♒
19	11	50	☉ ♉
21	04	30	☽ ♓
23	14	46	☽ ♈
25	21	39	☽ ♉
28	01	48	☽ ♊
30	04	34	☽ ♋

LATITUDES

Date	Mercury ☿ ° '	Venus ♀ ° '	Mars ♂ ° '	Jupiter ♃ ° '	Saturn ♄ ° '	Uranus ♅ ° '	Neptune ♆ ° '	Pluto ♇ ° '
01	00 N 46	03 N 54	01 N 16	00 S 46	01 N 04	00 N 23	01 S 38	08 S 19
04	01 21	04 06	01 16	00 45	01 04	00 23	01 38	08 20
07	01 54	04 18	01 17	00 45	01 04	00 23	01 38	08 20
10	02 21	04 29	01 17	00 44	01 04	00 23	01 38	08 21
13	02 42	04 39	01 17	00 44	01 04	00 23	01 38	08 22
16	02 55	04 47	01 18	00 44	01 04	00 23	01 38	08 23
19	02 57	04 54	01 18	00 43	01 04	00 23	01 38	08 23
22	02 48	04 59	01 18	00 43	01 04	00 23	01 38	08 24
25	02 27	05 02	01 18	00 42	01 04	00 23	01 38	08 25
28	01 54	05 02	01 18	00 42	01 04	00 23	01 38	08 26
31	01 N 11	05 N 00	01 N 19	00 S 42	01 N 03	00 N 23	01 S 38	08 S 27

DATA

Julian Date	2464785
Delta T	+76 seconds
Ayanamsa	24° 21' 38"
Synetic vernal point	04° ♓ 45' 21"
True obliquity of ecliptic	23° 25' 57"

MOON'S PHASES, APSIDES AND POSITIONS ☽

Date	h	m	Phase	Longitude ° '	Eclipse Indicator
04	00	03	☽	14 ♋ 48	
10	20	22	○	21 ♎ 32	
18	14	06	☾	29 ♑ 07	
26	09	33	●	06 ♉ 44	

Day	h	m		
06	09	47	Perigee	
18	17	40	Apogee	
03	13	09	Max dec	19° N 26'
09	21	07	0S	
16	20	26	Max dec	19° S 32'
24	06	08	0N	
30	19	03	Max dec	19° N 39'

LONGITUDES

Date	Chiron ⚷ ° '	Ceres ⚳ ° '	Pallas ⚴ ° '	Juno ⚵ ° '	Vesta ⚶ ° '	Black Moon Lilith ⚸ ° '
01	08 ♊ 27	27 ♊ 41	27 ♈ 39	03 ♉ 34	08 ♐ 23	28 ♑ 25
11	09 ♊ 01	00 ♋ 48	02 ♉ 13	09 ♉ 26	08 ♐ 46	29 ♑ 32
21	09 ♊ 39	04 ♋ 11	06 ♉ 56	15 ♉ 21	08 ♐ 20	00 ♒ 38
31	10 ♊ 21	07 ♋ 47	11 ♉ 46	21 ♉ 19	07 ♐ 06	01 ♒ 45

ASPECTARIAN

h m	Aspects
01 Tuesday	
03 42	♀ ∥ ♅
03 47	☽ ⊥ ♅
10 19	☽ ∠ ♆
13 31	☽ ∠ ☿
13 57	☽ ☌ ♂
14 12	☽ ⊻ ♅
14 27	☉ △ ♄
16 24	☽ ∥ ♃
16 34	☽ ✶ ♄
16 44	☽ ✶ ☉
18 02	☿ ∠ ♂
19 40	♂ ⊻ ♅
20 58	♂ ∥ ♅
02 Wednesday	
01 04	☽ ∥ ♄
07 00	☽ △ ♇
08 27	☽ ⊻ ♃
12 17	☽ ✶ ♆
14 59	☽ Q ☉
18 22	☽ ∠ ♄
18 55	☽ ⊥ ♃
19 21	☽ ✶ ☿
21 02	☽ ⊻ ♀
03 Thursday	
07 58	☽ ⊥ ♀
08 45	☽ ⚻ ♇
08 53	☽ Q ♆
09 45	☽ ⊥ ♄
10 00	☿ Q ♅
10 33	☽ ∠ ♃
17 45	☽ ☌ ♅
18 42	☿ ⊻ ♀
18 52	☽ Q ☿
19 51	☽ ⊻ ♂
19 57	☽ ⊻ ♄
22 17	♂ ✶ ♄
23 51	♀ ⊥ ♆
04 Friday	
00 03	☽ □ ☉
00 06	☽ ± ♇
00 17	☽ ∠ ♀
06 31	☽ ⊥ ♂
10 18	☽ ⊼ ♇
12 27	☽ ✶ ♃
12 54	♀ Q ♄
15 35	☽ □ ♆
21 47	☉ ∠ ♀
22 29	☽ ∠ ♂
05 Saturday	
00 37	☽ ∥ ♄
03 20	☽ ✶ ♀
05 40	☽ □ ☿
06 54	☽ ∥ ♃
09 04	☽ Q ♃
12 56	☉ ⊥ ♃
14 06	☿ Q ♇
20 38	☽ ⊻ ♅
22 43	☽ ☌ ♄
06 Sunday	
00 51	☽ Q ♀
01 00	☽ ✶ ♂
06 36	☽ △ ☉
06 46	☽ ⊥ ♅
10 25	☽ ∥ ☿
13 04	☽ ☍ ♇
15 54	☽ □ ♃
18 23	☽ △ ♆
21 57	☽ ∠ ♅
22 09	☽ Q ♂
07 Monday	
09 02	☽ □ ♀
09 46	☽ ⚻ ☉
14 43	☽ △ ☿
19 44	☽ ⚻ ♆
23 17	☽ ✶ ♅
08 Tuesday	
01 16	☽ ⊻ ♄
02 05	☽ ± ☉
04 07	☽ ∥ ♆
05 56	☽ □ ♂
06 15	☽ ∥ ☉
11 00	☽ ± ♆
11 25	☽ ⊥ ♄
12 59	☽ ⊼ ☉
15 47	☽ ⊼ ♇
19 02	☽ ⚻ ☿
19 21	☽ △ ♃
19 38	☽ Q ♅
21 12	☽ ⊼ ♆
09 Wednesday	
02 00	☽ ± ♇
02 41	☽ ∠ ♄
08 26	☉ ∥ ♆
12 04	☽ ± ☿
14 55	☽ △ ♀
17 23	☽ ⚻ ♇
21 22	☽ ⚻ ♃
23 27	☽ ⊼ ☿
10 Thursday	
02 27	☽ □ ♅
04 23	☽ ✶ ♄
05 57	☉ ✶ ♇
11 37	☽ △ ♂
13 10	☽ ± ♃
13 42	♀ ⊥ ♅
18 20	☽ ⚻ ♀
19 22	☽ △ ♇
20 22	☽ ☍ ☉
23 48	☽ ⊼ ♃
11 Friday	
01 01	☽ ☍ ♆
01 15	☽ Q ♄
01 49	☿ ⊥ ♂
08 59	☿ ✶ ♅
11 03	☽ ± ♀
15 11	☽ ⚻ ♂
15 29	☽ # ♆
20 43	☽ # ☉
22 20	☽ ⊼ ♀
12 Saturday	
07 20	☽ △ ♅
07 28	☿ □ ♄
08 04	☽ ± ♂
08 10	☿ ∥ ♃
09 16	☽ □ ♄
09 26	☽ ☍ ☿
14 37	☿ ∥ ♄
19 30	☽ ⊼ ♂
22 37	♄ St D
13 Sunday	
01 04	☽ □ ♇
06 26	☽ ⊼ ☉
06 35	☽ ☍ ♃
07 07	☽ ⊼ ♆
09 04	☉ ⊻ ♃
10 49	☽ ⚻ ♅
15 47	☉ ☌ ♆
18 18	☽ ± ♆
18 30	☽ ± ☉
14 Monday	
03 43	☽ ± ♅
08 47	☽ ☍ ♀
10 30	♃ ∥ ♄
10 46	☽ Q ♇
11 21	☽ ⚻ ♆
12 56	☽ ⚻ ☉
15 09	☽ ⊼ ♅
17 05	☽ △ ♄
19 41	♃ ⊻ ♆
22 04	☽ ⊼ ☿
15 Tuesday	
01 50	☽ # ♄
02 24	☽ # ♃
06 58	☽ ☍ ♂
09 55	☽ ✶ ♇
10 32	☽ ± ☿
16 26	☽ △ ♆
16 45	☽ ⊼ ♃
20 31	☽ △ ☉
22 16	☽ ⚻ ♄
16 Wednesday	
04 48	☽ ± ♃
05 26	☽ ⚻ ☿
07 24	☽ # ☿
11 31	♀ ∠ ♆
15 33	☽ ∠ ♇
16 07	☽ ± ♄
20 27	☽ # ☿
22 52	☽ ⊼ ♀
23 04	☽ ⚻ ♃
17 Thursday	
02 12	☽ ☍ ♅
04 09	☽ ⊼ ♄
09 38	☽ ⊥ ♇
11 37	☽ ± ♀
13 14	☽ △ ☿
18 33	♂ △ ♇
21 46	☽ ⊻ ♇
21 57	☽ ⊼ ♂
18 Friday	
04 39	☽ □ ♆
05 56	☽ △ ♃
06 51	☽ ⚻ ♀
10 48	☽ ± ♂
12 15	☽ # ♃
14 06	☽ □ ☉
16 07	☽ # ♄
21 31	☉ Q ♅
19 Saturday	
06 07	☽ ⚻ ♂
14 57	☽ △ ♀
15 08	☽ ⊼ ♅
17 01	☽ ☍ ♄
17 19	☽ Q ♆
18 35	♀ ⊻ ♅
20 Sunday	
03 18	☽ ± ♅
04 34	☽ □ ☿
05 54	☽ Q ☉
10 40	☽ ☌ ♇
14 09	☽ △ ♂
17 37	☽ ✶ ♆
19 51	☽ □ ♃
21 27	☽ ⚻ ♅
21 Monday	
04 02	☽ # ☉
08 04	☽ ✶ ☉
10 44	♀ ✶ ♄
17 19	☽ Q ☿
23 30	☽ ∠ ♆
22 Tuesday	
03 13	☽ △ ♅
04 58	☽ ⊼ ♄
05 47	☽ □ ♀
07 23	☽ # ♆
08 07	☽ Q ♃
16 02	☽ ∠ ☉
16 38	☽ ± ♄
16 55	☉ Q ♇
17 00	☽ ✶ ☿
17 04	☽ ⊥ ♆
21 53	☽ ⊻ ♇
23 Wednesday	
04 14	☽ □ ♂
04 37	☽ ⊻ ♆
07 38	☽ ✶ ♃
09 21	☽ ⊥ ♇
09 53	☽ ⚻ ♄
10 43	☽ ⊥ ☉
11 04	☿ ⊥ ♂
12 19	♂ ✶ ♆
13 18	☿ St R
17 44	☽ Q ♀
21 34	☽ ∠ ☿
22 56	☽ ⊻ ☉
24 Thursday	
02 17	☽ ∠ ♇
12 13	☽ ∠ ♃
12 21	☽ □ ♅
13 56	☽ △ ♄
14 11	☽ ⊥ ☿
15 39	☽ Q ♂
16 52	☽ ✶ ♀
20 52	♃ ∠ ♅
25 Friday	
01 03	☽ ⊻ ☿
04 56	☽ ⊥ ♃
05 49	☽ ✶ ♇
12 15	☽ ☌ ♆
14 28	☽ ✶ ♂
15 54	☽ ⊻ ♃
20 53	☽ ∠ ♀
20 59	☽ Q ♅
26 Saturday	
02 59	☽ ∥ ♆
03 17	☽ Q ♇
09 33	☽ ☌ ☉
13 17	☽ ⊥ ♀
18 10	☽ ✶ ♅
18 14	☽ ∠ ♂
19 38	☽ □ ♄
27 Sunday	
00 04	☽ ⊻ ♀
05 14	☽ ☌ ☿
10 35	☽ ⊥ ♂
10 42	☽ □ ♇
14 02	♂ ⊻ ♃
16 56	☽ ⊻ ♆
17 36	☽ ∥ ☉
20 08	☽ ∠ ♅
21 09	☽ ☌ ♃
21 22	☽ ⊻ ♂
28 Monday	
02 12	♂ ∠ ♄
02 42	☽ Q ♄
03 14	☽ ⊥ ♆
11 30	☿ ∥ ♃
11 32	☽ ⊥ ♅
17 05	☽ ⊻ ☉
18 34	☽ ∠ ♆
21 44	☽ ⊻ ♅
23 08	☽ ✶ ♄
29 Tuesday	
03 59	☽ ⊥ ☉
04 22	☽ ∥ ♄
04 43	☽ ☌ ♀
07 01	☽ ⊻ ☿
11 16	☽ ∥ ☿
13 45	☽ △ ♇
16 48	☽ ⊥ ☿
17 50	☽ ∥ ♃
19 47	☉ ⊥ ♀
19 58	☽ ✶ ♆
20 17	☽ ∠ ☉
30 Wednesday	
00 30	☽ ∠ ♄
00 49	☽ ⊻ ♃
02 32	☽ ☌ ♂
07 27	☽ ∠ ☿
11 05	☽ ⊥ ♃
15 02	☽ ⚻ ♇
15 41	☽ ⊥ ♄
16 14	☽ Q ♆
21 01	☿ ∥ ♄
21 58	☿ ⊻ ♀
23 24	☽ ✶ ☉

All ephemeris data is given at 12.00 UT and the Moon's longitude is additionally given for 24.00 UT

MAY 2036

LONGITUDES

Date	Sidereal time h m s	Sun ☉ ° ' "	Moon ☽ ° ' "	Moon ☽ 24.00 ° ' "	Mercury ☿ ° '	Venus ♀ ° '	Mars ♂ ° '	Jupiter ♃ ° '	Saturn ♄ ° '	Uranus ♅ ° '	Neptune ♆ ° '	Pluto ♇ ° '
01	02 40 00	11 ♉ 41 37	18 ♋ 39 25	25 ♋ 45 41	16 ♉ 04	16 ♊ 33	29 ♊ 40	28 ♉ 07	12 ♌ 38	11 ♋ 49	24 ♈ 58	21 ♒ 14
02	02 43 57	12 39 51	02 ♌ 50 57	09 ♌ 55 04	15 R 29	16 48	00 ♋ 17	28 21	12 40	11 51	25 00	21 14
03	02 47 53	13 38 03	16 ♌ 57 52	23 ♌ 59 17	14 52	17 01	00 55	28 35	12 42	11 54	25 02	21 15
04	02 51 50	14 36 13	00 ♍ 59 15	07 ♍ 57 41	14 14	17 12	01 33	28 49	12 44	11 56	25 04	21 15
05	02 55 47	15 34 20	14 ♍ 54 33	21 ♍ 49 43	13 36	17 20	02 10	29 02	12 47	11 58	25 06	21 16
06	02 59 43	16 32 26	28 ♍ 43 05	05 ♎ 34 29	12 59	17 26	02 48	29 16	12 49	12 01	25 08	21 16
07	03 03 40	17 30 30	12 ♎ 23 43	19 ♎ 10 33	12 22	17 30	03 25	29 30	12 52	12 03	25 10	21 16
08	03 07 36	18 28 32	25 ♎ 54 45	02 ♏ 36 03	11 47	17 32	04 03	29 44	12 54	12 05	25 13	21 17
09	03 11 33	19 26 32	09 ♏ 14 11	15 ♏ 48 56	11 15	17 R 31	04 40	29 ♉ 58	12 57	12 08	25 15	21 17
10	03 15 29	20 24 30	22 ♏ 20 03	28 ♏ 47 23	10 45	17 28	05 18	00 ♊ 12	13 00	12 10	25 17	21 17
11	03 19 26	21 22 27	05 ♐ 10 49	11 ♐ 30 20	10 18	17 22	05 55	00 26	13 03	12 13	25 19	21 18
12	03 23 22	22 20 23	17 ♐ 45 55	23 ♐ 57 43	09 55	17 14	06 33	00 40	13 06	12 16	25 21	21 18
13	03 27 19	23 18 17	00 ♑ 05 54	06 ♑ 10 43	09 36	17 03	07 11	00 54	13 09	12 18	25 23	21 18
14	03 31 16	24 16 10	12 ♑ 12 32	18 ♑ 11 45	09 21	16 51	07 48	01 09	13 12	12 21	25 25	21 18
15	03 35 12	25 14 02	24 ♑ 08 49	00 ♒ 04 18	09 11	16 35	08 26	01 23	13 16	12 24	25 27	21 19
16	03 39 09	26 11 52	05 ♒ 58 44	11 ♒ 52 46	09 05	16 17	09 03	01 37	13 19	12 26	25 29	21 19
17	03 43 05	27 09 41	17 ♒ 47 03	23 ♒ 42 14	09 D 03	15 57	09 41	01 51	13 22	12 29	25 31	21 19
18	03 47 02	28 07 29	29 ♒ 39 01	05 ♓ 38 06	09 06	15 35	10 18	02 05	13 26	12 32	25 33	21 19
19	03 50 58	29 ♉ 05 16	11 ♓ 40 09	17 ♓ 45 51	09 13	15 11	10 56	02 19	13 30	12 35	25 35	21 19
20	03 54 55	00 ♊ 03 01	23 ♓ 55 47	00 ♈ 10 34	09 26	14 44	11 33	02 33	13 33	12 38	25 37	21 19
21	03 58 51	01 00 46	06 ♈ 30 41	12 ♈ 56 33	09 42	14 16	12 11	02 47	13 37	12 40	25 39	21 19
22	04 02 48	01 58 29	19 ♈ 28 29	26 ♈ 06 41	10 03	13 46	12 48	03 01	13 41	12 43	25 41	21 R 19
23	04 06 45	02 56 11	02 ♉ 51 11	09 ♉ 41 52	10 28	13 13	13 26	03 15	13 45	12 46	25 43	21 19
24	04 10 41	03 53 52	16 ♉ 38 30	23 ♉ 40 39	10 58	12 40	14 03	03 29	13 49	12 49	25 45	21 19
25	04 14 38	04 51 32	00 ♊ 47 45	07 ♊ 59 05	11 31	12 06	14 41	03 44	13 54	12 52	25 47	21 19
26	04 18 34	05 49 11	15 ♊ 13 51	22 ♊ 31 11	12 09	11 30	15 18	03 58	13 58	12 55	25 48	21 19
27	04 22 31	06 46 49	29 ♊ 50 08	07 ♋ 09 50	12 50	10 54	15 56	04 12	14 02	12 58	25 50	21 18
28	04 26 27	07 44 25	14 ♋ 28 35	21 ♋ 48 00	13 36	10 17	16 33	04 26	14 07	13 02	25 52	21 18
29	04 30 24	08 42 00	29 ♋ 04 59	06 ♌ 19 46	14 25	09 39	17 11	04 40	14 11	13 05	25 54	21 18
30	04 34 20	09 39 34	13 ♌ 31 53	20 ♌ 41 00	15 17	09 02	17 48	04 54	14 16	13 08	25 56	21 18
31	04 38 17	10 ♊ 37 06	27 ♌ 46 53	04 ♍ 49 23	16 ♉ 13	08 ♊ 24	18 ♋ 26	05 ♊ 08	14 ♌ 20	13 ♋ 11	25 ♈ 57	21 ♒ 18

DECLINATIONS

Date	Moon True ☊ ° '	Moon Mean ☊ ° '	Moon ☽ Latitude ° '	Sun ☉ ° '	Moon ☽ ° '	Mercury ☿ ° '	Venus ♀ ° '	Mars ♂ ° '	Jupiter ♃ ° '	Saturn ♄ ° '	Uranus ♅ ° '	Neptune ♆ ° '	Pluto ♇ ° '
01	22 ♌ 35	22 ♌ 21	02 S 49	15 N 20	19 N 20	17 N 47	27 N 43	24 N 44	19 N 03	18 N 02	23 N 17	08 N 09	22 S 24
02	22 R 33	22 18	01 42	15 38	17 51	17 21	27 43	24 45	19 07	18 01	23 17	08 09	22 24
03	22 D 33	22 15	00 S 30	15 56	15 17	16 55	27 43	24 44	19 10	18 00	23 17	08 10	22 24
04	22 R 33	22 11	00 N 45	16 13	11 49	16 27	27 42	24 44	19 13	18 00	23 17	08 11	22 24
05	22 32	22 08	01 55	16 30	07 43	15 59	27 40	24 43	19 16	17 59	23 16	08 12	22 25
06	22 28	22 05	02 59	16 47	03 N 15	15 31	27 37	24 43	19 19	17 58	23 16	08 13	22 25
07	22 22	22 02	03 51	17 03	01 S 21	15 04	27 34	24 42	19 22	17 58	23 16	08 13	22 25
08	22 13	21 59	04 30	17 19	05 49	14 37	27 30	24 41	19 25	17 57	23 16	08 14	22 25
09	22 01	21 55	04 53	17 35	09 56	14 11	27 25	24 40	19 28	17 56	23 15	08 15	22 25
10	21 48	21 52	05 00	17 51	13 30	13 46	27 20	24 38	19 31	17 55	23 15	08 15	22 25
11	21 36	21 49	04 52	18 06	16 22	13 23	27 14	24 37	19 34	17 54	23 15	08 16	22 26
12	21 24	21 46	04 30	18 21	18 24	13 02	27 07	24 35	19 37	17 53	23 15	08 17	22 26
13	21 14	21 43	03 55	18 36	19 31	12 42	27 00	24 33	19 40	17 52	23 14	08 18	22 26
14	21 07	21 40	03 10	18 50	19 43	12 25	26 51	24 31	19 43	17 52	23 14	08 18	22 26
15	21 02	21 36	02 17	19 04	19 02	12 10	26 42	24 28	19 46	17 51	23 14	08 19	22 26
16	21 00	21 33	01 19	19 18	17 30	11 57	26 32	24 26	19 49	17 50	23 14	08 20	22 27
17	20 D 59	21 30	00 N 17	19 31	15 14	11 47	26 21	24 23	19 52	17 49	23 13	08 21	22 27
18	20 R 59	21 27	00 S 46	19 44	12 18	11 39	26 09	24 20	19 55	17 47	23 13	08 21	22 27
19	20 59	21 24	01 47	19 57	08 50	11 34	25 57	24 17	19 58	17 46	23 13	08 22	22 27
20	20 57	21 21	02 45	20 09	04 56	11 31	25 43	24 14	20 01	17 45	23 13	08 22	22 28
21	20 53	21 17	03 37	20 21	00 S 44	11 30	25 29	24 10	20 04	17 44	23 12	08 23	22 28
22	20 46	21 14	04 18	20 33	03 N 38	11 32	25 14	24 07	20 06	17 43	23 12	08 24	22 28
23	20 37	21 11	04 47	20 44	07 57	11 36	24 58	24 03	20 09	17 42	23 12	08 24	22 28
24	20 26	21 08	05 01	20 55	12 00	11 42	24 42	23 59	20 12	17 41	23 11	08 25	22 29
25	20 15	21 05	04 57	21 06	15 28	11 50	24 25	23 55	20 15	17 39	23 11	08 26	22 29
26	20 04	21 01	04 33	21 16	18 05	12 00	24 07	23 51	20 17	17 38	23 11	08 26	22 29
27	19 55	20 58	03 52	21 26	19 34	12 12	23 48	23 46	20 20	17 37	23 11	08 27	22 30
28	19 49	20 55	02 56	21 36	19 44	12 26	23 29	23 42	20 23	17 36	23 10	08 28	22 30
29	19 45	20 52	01 47	21 45	18 35	12 41	23 10	23 37	20 25	17 34	23 10	08 28	22 30
30	19 44	20 49	00 30	21 54	16 14	12 58	22 50	23 32	20 28	17 33	23 10	08 29	22 31
31	19 ♌ 44	20 ♌ 46	00 N 43	22 N 02	12 N 55	13 N 17	22 N 30	23 N 27	20 N 31	17 N 31	23 N 09	08 N 29	22 S 31

ZODIAC SIGN ENTRIES

Date	h	m	Planets	
02	00	50	♂	♋
02	07	10	☽	♌
04	10	18	☽	♍
06	14	14	☽	♎
08	19	19	☽	♏
09	14	52	♃	♊
11	02	16	☽	♐
13	11	48	☽	♑
15	23	51	☽	♒
18	12	42	☽	♓
20	10	45	☉	♊
20	23	40	☽	♈
23	06	57	☽	♉
25	10	40	☽	♊
27	12	16	☽	♋
29	13	31	☽	♌
31	15	46	☽	♍

LATITUDES

Date	Mercury ☿ ° '	Venus ♀ ° '	Mars ♂ ° '	Jupiter ♃ ° '	Saturn ♄ ° '	Uranus ♅ ° '	Neptune ♆ ° '	Pluto ♇ ° '
01	01 N 11	05 N 00	01 N 19	00 S 42	01 N 03	00 N 23	01 S 38	08 S 27
04	00 N 22	04 54	01 19	00 41	01 03	00 23	01 38	08 28
07	00 S 30	04 45	01 19	00 41	01 03	00 23	01 38	08 29
10	01 21	04 31	01 19	00 41	01 03	00 23	01 38	08 29
13	02 05	04 13	01 19	00 40	01 03	00 23	01 38	08 30
16	02 42	03 50	01 18	00 40	01 03	00 23	01 38	08 31
19	03 10	03 22	01 18	00 40	01 03	00 23	01 38	08 32
22	03 28	02 49	01 18	00 40	01 03	00 23	01 38	08 33
25	03 37	02 12	01 18	00 39	01 03	00 23	01 38	08 34
28	03 39	01 31	01 18	00 39	01 03	00 23	01 39	08 35
31	03 S 33	00 N 49	01 N 18	00 S 39	01 N 03	00 N 23	01 S 39	08 S 35

DATA

Julian Date	2464815
Delta T	+76 seconds
Ayanamsa	24° 21' 41"
Synetic vernal point	04° ♓ 45' 18"
True obliquity of ecliptic	23° 25' 57"

MOON'S PHASES, APSIDES AND POSITIONS ☽

Date	h	m	Phase	Longitude ° '	Eclipse Indicator
03	05	54	☽	13 ♌ 23	
10	08	09	○	20 ♏ 15	
18	08	39	☾	27 ♒ 59	
25	19	17	●	05 ♊ 09	

Day	h	m		
01	08	17	Perigee	
16	12	37	Apogee	
28	09	10	Perigee	
07	04	57	0S	
14	05	16	Max dec	19° S 45'
21	16	04	0N	
28	03	07	Max dec	19° N 50'

LONGITUDES

Date	Chiron ⚷ ° '	Ceres ⚳ ° '	Pallas ⚴ ° '	Juno ⚵ ° '	Vesta ⚶ ° '	Black Moon Lilith ⚸ ° '
01	10 ♊ 21	07 ♋ 47	11 ♉ 46	21 ♉ 19	07 ♐ 06	01 ♒ 45
11	11 ♊ 06	11 ♋ 34	16 ♉ 43	27 ♉ 17	05 ♐ 11	02 ♒ 52
21	11 ♊ 53	15 ♋ 31	21 ♉ 46	03 ♊ 15	02 ♐ 52	03 ♒ 58
31	12 ♊ 41	19 ♋ 35	26 ♉ 56	09 ♊ 13	00 ♐ 28	05 ♒ 05

ASPECTARIAN

h m	Aspects
01 Thursday	
00 26	☽ ☌ ♅
01 49	☽ ⊻ ♄
02 30	☽ ∠ ♃
06 13	☽ ± ♇
07 48	☽ ✱ ☿
08 23	☽ ⊻ ♀
15 12	☉ ✱ ♅
16 21	☽ ⊼ ♇
18 12	☽ ∥ ♃
18 42	☽ ⊥ ♀
21 08	☽ Q ☉
22 40	☽ □ ♆
02 Friday	
03 17	☽ Q ☿
04 15	☽ ✱ ♃
07 28	☽ ⊻ ♂
10 04	☽ ∥ ♄
10 12	☽ ∠ ♀
12 04	☉ □ ♄
15 44	☿ ∠ ♂
18 07	☽ ⊥ ♂
18 46	☽ ∥ ☿
03 Saturday	
00 56	☽ Q ♃
03 20	☽ ⊻ ♅
04 43	☽ ☌ ♄
05 54	☽ □ ☉
07 13	☽ ∥ ☉
08 35	☽ □ ☿
10 08	☽ ∠ ♂
12 05	☽ ✱ ♀
13 35	☽ ⊥ ♅
19 19	☽ ☍ ♇
04 Sunday	
01 49	☽ △ ♆
05 01	☽ ∠ ♅
06 34	☉ ☌ ☿
08 12	☽ □ ♃
08 53	☽ Q ♀
13 00	☽ ✱ ♂
19 40	☉ ∥ ☿
05 Monday	
03 40	☽ bQ ♆
06 54	☽ ✱ ♅
08 18	☽ ⊻ ♄
09 21	☽ ∥ ♆
09 51	☽ △ ☿
10 40	☽ Q ♂
13 14	☽ △ ☉
16 14	☽ □ ♀
18 43	☽ ⊥ ♄
19 17	☽ ± ♆
23 01	☽ ⊼ ♇
06 Tuesday	
03 46	☽ Q ♅
05 45	☽ ⊼ ♆
09 28	☽ ± ♇
10 26	☽ ∠ ♄
10 46	☽ bQ ☿
12 59	☽ △ ♃
17 18	☽ bQ ☉
17 52	☿ □ ♄
19 28	☽ □ ♂
07 Wednesday	
01 13	☽ bQ ♇
01 51	☽ ± ☿
10 19	☽ ± ☉
11 23	☽ □ ♅
11 50	☉ ⊻ ♀
11 58	☽ ⊼ ☿
12 50	☽ ✱ ♄
15 48	☽ bQ ♃
21 03	☽ △ ♀
21 44	☽ ⊼ ☉
08 Thursday	
00 18	☿ ✱ ♅
03 44	☽ △ ♇
08 03	☽ ± ♃
10 12	☽ Q ♄
10 44	☽ ☍ ♆
15 58	♀ St R
18 59	☽ ⊼ ♃
23 28	☿ ⊥ ♀
23 52	☽ bQ ♀
09 Friday	
01 48	☽ ⋕ ♆
03 20	☽ △ ♂
15 31	☽ ☍ ☿
16 08	☽ ± ♀
17 17	☽ △ ♅
18 47	☽ □ ♄
10 Saturday	
03 04	☽ ⊼ ♀
04 15	☉ ∠ ♂
08 03	☽ bQ ♂
08 09	☽ ☍ ☉
10 04	☽ □ ♇
13 45	☽ ⋕ ☿
17 29	☽ ⊼ ♆
18 31	☉ ∥ ♄
21 01	☽ bQ ♅
11 Sunday	
01 37	☽ ± ♂
02 55	☽ ☍ ♃
04 42	☽ ± ♆
10 00	☉ □ ♇
13 29	☽ ⊼ ♂
13 58	☽ ± ♅
19 48	☽ Q ♇
21 25	☽ ⊼ ☿
21 46	☽ bQ ♆
12 Monday	
01 24	☽ ⊼ ♅
02 18	♂ bQ ♇
03 00	☽ △ ♄
05 00	☽ ⋕ ♄
08 33	☽ ± ☿
10 59	☽ ☍ ♀
11 13	☽ ⋕ ☉
18 50	☽ ✱ ♇
19 35	♂ ⊥ ♃
21 36	☽ ⊼ ☉
13 Tuesday	
01 30	☽ bQ ☿
02 45	☽ △ ♆
08 10	☽ bQ ♄
10 19	☽ ± ☉
10 56	♂ ⊥ ♄
13 37	☽ ⊼ ♃
20 18	☽ ⋕ ♃
20 29	♂ Q ♆
14 Wednesday	
00 15	☽ ∠ ♇
01 43	☽ ± ♃
01 59	☽ ± ♄
02 44	☽ ☍ ♂
05 38	☽ bQ ☉
06 25	☽ △ ☿
11 39	☽ ⋕ ♃
12 17	☽ ☍ ♅
14 00	☽ ⊼ ♄
18 12	☽ ⊥ ♇
20 02	☽ bQ ♃
20 10	♃ Q ♄
21 06	☽ ⊼ ♀
15 Thursday	
06 16	☽ ⊻ ♇
08 55	☽ ± ♀
11 15	☽ ⋕ ☉
14 24	☽ △ ☉
14 39	☽ □ ♆
17 38	☉ ⊻ ♆
21 02	♃ ⊥ ♆
16 Friday	
02 43	☽ bQ ♀
02 57	☽ △ ♃
07 44	☽ ⋕ ♄
12 50	☿ ✱ ♂
18 16	☽ □ ☿
18 36	☽ ⊼ ♂
17 Saturday	
01 11	☽ ⊼ ♅
03 00	☽ ☍ ♄
03 19	☽ Q ♆
07 29	☽ ± ♂
08 24	☽ △ ♀
08 29	☿ St D
13 26	☽ ± ♅
19 10	☽ ☌ ♇
20 28	☉ ∠ ♅
18 Sunday	
02 45	☽ bQ ♂
03 43	☽ ✱ ♆
06 49	☽ Q ☿
07 43	☽ bQ ♅
08 39	☽ □ ☉
16 56	☽ ⋕ ☿
16 59	☽ □ ♃
19 Monday	
07 05	☽ ✱ ☿
09 51	☽ ∠ ♆
10 09	☿ ⊥ ♀
10 27	☽ △ ♂
13 48	☽ △ ♅
14 33	☉ ∥ ♃
15 03	☽ ⋕ ♆
15 37	☽ ⊼ ♄
18 41	☽ □ ♀
23 35	☽ Q ☉
20 Tuesday	
03 28	☽ ± ♄
03 36	☽ ⊥ ♆
05 19	☽ Q ♃
06 56	☽ ⊻ ♇
12 59	☽ ∠ ☿
15 16	☽ ⊻ ♆
18 32	☽ ⊥ ♇
20 57	☽ bQ ♄
21 Wednesday	
00 44	☽ ✱ ☉
04 16	☽ Q ♀
04 50	☽ ✱ ♃
06 34	☽ ⊥ ☿
11 38	☽ ∠ ♇
15 35	♇ St R
18 08	☽ ⊻ ☿
23 07	☽ □ ♂
23 33	☽ □ ♅
22 Thursday	
01 19	☽ △ ♄
01 55	☽ ✱ ♀
04 16	☉ Q ♄
04 27	☉ ⊥ ♆
07 04	☽ ∠ ☉
08 39	♂ ☌ ♅
09 18	☽ ∠ ♃
15 01	♀ ✱ ♄
15 21	☽ ✱ ♇
23 15	☽ ☌ ♆
23 Friday	
00 40	☽ ⊥ ☉
01 52	☽ ⊥ ♃
04 06	☽ ∠ ♀
07 53	♀ ⊻ ♂
08 18	☽ Q ♅
09 22	☽ Q ♂
12 10	☽ ⊻ ☉
12 43	☽ ⊻ ♃
12 49	☽ Q ♇
14 36	☽ ∥ ♆
19 24	☽ ⊥ ♀
22 33	☉ ☌ ♃
24 Saturday	
01 49	☽ ☌ ☿
01 57	♂ ⊻ ♄
05 24	☽ ✱ ♅
05 26	☽ ⊻ ♀
06 13	♀ ⊻ ♅
07 07	☽ □ ♄
07 20	☽ ✱ ♂
10 06	☽ ∥ ☿
19 59	☽ □ ♇
25 Sunday	
03 32	☽ ⊻ ♆
07 04	☽ ∠ ♅
10 03	☽ ∠ ♂
13 39	☽ ⊥ ♆
13 51	☽ Q ♄
16 59	☽ ☌ ♃
19 17	☽ ☌ ☉
22 11	☽ ⊥ ♅
23 30	☿ ⊻ ♀
26 Monday	
01 45	☽ ⊥ ♂
04 40	☽ ∠ ♆
06 05	☽ ☌ ♀
06 40	☽ ⊻ ☿
07 09	☽ ∥ ♄
08 10	☽ ⊻ ♅
09 54	☽ ✱ ♄
12 08	☽ ⊻ ♂
12 14	♂ ± ♇
17 03	☽ ⊥ ☿
22 01	☽ △ ♇
27 Tuesday	
05 26	☽ ✱ ♆
08 34	☽ ∠ ☿
10 41	☽ ∠ ♄
14 13	♀ ∠ ♆
15 27	♀ ∥ ♂
16 44	☿ ✱ ♅
17 08	☉ ⊥ ♅
19 15	☽ ⊻ ♃
22 36	☽ bQ ♇
28 Wednesday	
00 10	☽ ⊻ ☉
01 08	☽ Q ♆
01 30	☽ ⊥ ♄
05 14	☽ ⊥ ♃
05 23	☽ ⊻ ♀
06 38	♀ ⊥ ♂
09 36	☽ ☌ ♅
10 27	☽ ✱ ☿
10 41	☽ ⊥ ☉
11 22	☽ ⊻ ♄
13 20	☽ ± ♇
14 49	☽ ⊥ ♀
15 33	☽ ☌ ♂
20 14	☽ ∠ ♃
23 11	☽ ⊼ ♇
29 Thursday	
02 30	☽ ∠ ☉
04 55	☿ □ ♄
05 00	☽ ∠ ♀
06 44	☽ □ ♆
07 19	☽ Q ☿
12 09	♀ ∥ ♅
21 23	☽ ✱ ♃
23 51	☽ ∥ ♄
30 Friday	
02 25	☉ ☌ ♀
04 48	☽ ✱ ♀
05 05	☽ ✱ ☉
11 20	☽ ⊻ ♅
13 14	☽ ☌ ♄
15 08	☽ □ ☿
17 44	☽ Q ♃
19 30	☽ ⊻ ♂
21 25	☽ ⊥ ♅
31 Saturday	
00 03	☽ Q ♀
01 02	☽ ☍ ♇
02 38	☽ Q ☉
06 04	☽ ⊥ ♂
08 54	☽ △ ♆
09 47	☽ ∥ ☿
11 30	♀ ⋕ ♇
12 41	☽ ∠ ♅
20 43	☉ ∠ ♆
22 04	☽ ∠ ♂

All ephemeris data is given at 12.00 UT and the Moon's longitude is additionally given for 24.00 UT

JUNE 2036

LONGITUDES

Date	Sidereal time h m s	Sun ☉ ° ' "	Moon ☽ ° ' "	Moon ☽ 24.00 ° ' "	Mercury ☿ ° '	Venus ♀ ° '	Mars ♂ ° '	Jupiter ♃ ° '	Saturn ♄ ° '	Uranus ♅ ° '	Neptune ♆ ° '	Pluto ♇ ° '
01	04 42 14	11 ♊ 34 37	11 ♍ 48 27	18 ♍ 44 04	17 ♉ 12	07 ♊ 47	19 ♋ 04	05 ♊ 22	14 ♌ 25	13 ♋ 14	25 ♈ 59	21 ♒ 18
02	04 46 10	12 32 06	25 ♍ 36 19	02 ♎ 25 14	18 15	07 R 10	19 41	05 36	14 30	13 17	26 01	21 R 17
03	04 50 07	13 29 34	09 ♎ 10 53	15 ♎ 53 23	19 20	06 34	20 19	05 50	14 35	13 21	26 02	21 17
04	04 54 03	14 27 01	22 ♎ 32 45	29 ♎ 09 03	20 29	05 59	20 56	06 04	14 40	13 24	26 04	21 17
05	04 58 00	15 24 27	05 ♏ 42 19	12 ♏ 12 33	21 41	05 26	21 34	06 18	14 45	13 27	26 06	21 16
06	05 01 56	16 21 52	18 ♏ 39 44	25 ♏ 03 53	22 56	04 53	22 11	06 32	14 50	13 31	26 07	21 16
07	05 05 53	17 19 15	01 ♐ 24 58	07 ♐ 43 00	24 14	04 23	22 49	06 46	14 55	13 34	26 09	21 15
08	05 09 49	18 16 38	13 ♐ 57 59	20 ♐ 09 57	25 35	03 54	23 26	07 00	15 01	13 37	26 11	21 15
09	05 13 46	19 14 00	26 ♐ 18 59	02 ♑ 25 11	26 59	03 27	24 04	07 14	15 06	13 41	26 12	21 15
10	05 17 43	20 11 21	08 ♑ 28 44	14 ♑ 29 48	28 26	03 02	24 41	07 28	15 11	13 44	26 14	21 14
11	05 21 39	21 08 42	20 ♑ 28 40	26 ♑ 25 39	29 ♉ 56	02 40	25 19	07 42	15 17	13 47	26 15	21 14
12	05 25 36	22 06 01	02 ♒ 21 06	08 ♒ 15 27	01 ♊ 28	02 19	25 56	07 56	15 22	13 51	26 17	21 13
13	05 29 32	23 03 21	14 ♒ 09 08	20 ♒ 02 42	03 04	02 01	26 34	08 09	15 28	13 54	26 18	21 12
14	05 33 29	24 00 39	25 ♒ 56 40	01 ♓ 51 38	04 42	01 45	27 11	08 23	15 34	13 58	26 19	21 12
15	05 37 25	24 57 58	07 ♓ 48 14	13 ♓ 47 05	06 23	01 32	27 49	08 37	15 39	14 01	26 21	21 11
16	05 41 22	25 55 16	19 ♓ 48 51	25 ♓ 54 11	08 07	01 21	28 26	08 51	15 45	14 05	26 22	21 11
17	05 45 18	26 52 33	02 ♈ 03 43	08 ♈ 18 05	09 53	01 12	29 04	09 04	15 51	14 08	26 23	21 10
18	05 49 15	27 49 50	14 ♈ 37 51	21 ♈ 03 32	11 42	01 06	29 ♋ 41	09 18	15 57	14 12	26 25	21 09
19	05 53 12	28 47 07	27 ♈ 35 35	04 ♉ 14 18	13 34	01 03	00 ♌ 19	09 32	16 03	14 15	26 26	21 09
20	05 57 08	29 ♊ 44 24	10 ♉ 59 54	17 ♉ 52 26	15 28	01 01	00 56	09 45	16 09	14 19	26 27	21 08
21	06 01 05	00 ♋ 41 41	24 ♉ 51 48	01 ♊ 57 41	17 25	01 D 02	01 34	09 59	16 15	14 22	26 29	21 07
22	06 05 01	01 38 57	09 ♊ 09 39	16 ♊ 27 01	19 24	01 05	02 11	10 13	16 21	14 26	26 30	21 06
23	06 08 58	02 36 14	23 ♊ 48 59	01 ♋ 14 36	21 26	01 11	02 49	10 26	16 28	14 29	26 31	21 06
24	06 12 54	03 33 29	08 ♋ 42 48	16 ♋ 12 30	23 29	01 18	03 27	10 40	16 34	14 33	26 32	21 05
25	06 16 51	04 30 45	23 ♋ 42 35	01 ♌ 11 56	25 34	01 28	04 04	10 53	16 40	14 37	26 33	21 04
26	06 20 47	05 28 00	08 ♌ 39 34	16 ♌ 04 35	27 41	01 40	04 42	11 06	16 47	14 40	26 34	21 03
27	06 24 44	06 25 15	23 ♌ 26 12	00 ♍ 43 48	29 ♊ 49	01 54	05 19	11 20	16 53	14 44	26 35	21 02
28	06 28 41	07 22 29	07 ♍ 56 55	15 ♍ 05 12	01 ♋ 58	02 10	05 57	11 33	16 59	14 47	26 36	21 01
29	06 32 37	08 19 42	22 ♍ 08 28	29 ♍ 06 38	04 08	02 27	06 34	11 46	17 06	14 51	26 37	21 01
30	06 36 34	09 ♋ 16 55	05 ♎ 59 43	12 ♎ 47 47	06 ♋ 19	02 ♊ 47	07 ♌ 12	12 ♊ 00	17 ♌ 13	14 ♋ 55	26 ♈ 38	21 ♒ 00

MOON / DECLINATIONS

Date	Moon True ☊ ° '	Moon Mean ☊ ° '	Moon ☽ Latitude ° '	Sun ☉ ° '	Moon ☽ ° '	Mercury ☿ ° '	Venus ♀ ° '	Mars ♂ ° '	Jupiter ♃ ° '	Saturn ♄ ° '	Uranus ♅ ° '	Neptune ♆ ° '	Pluto ♇ ° '
01	19 ♌ 44	20 ♌ 42	01 N 55	22 N 10	08 N 54	13 N 37	22 N 10	23 N 21	20 N 33	17 N 30	23 N 09	08 N 30	22 S 31
02	19 R 43	20 39	02 59	22 18	04 N 29	13 58	21 50	23 16	20 36	17 29	23 09	08 31	22 32
03	19 41	20 36	03 52	22 25	00 S 05	14 20	21 30	23 10	20 38	17 27	23 08	08 31	22 32
04	19 36	20 33	04 31	22 32	04 35	14 44	21 11	23 04	20 41	17 26	23 08	08 32	22 32
05	19 28	20 30	04 55	22 38	08 47	15 08	20 51	22 59	20 43	17 24	23 08	08 32	22 33
06	19 19	20 27	05 04	22 44	12 30	15 33	20 32	22 52	20 46	17 23	23 07	08 33	22 33
07	19 08	20 23	04 57	22 50	15 35	15 59	20 13	22 46	20 48	17 21	23 07	08 33	22 34
08	18 57	20 20	04 36	22 55	17 54	16 26	19 55	22 40	20 50	17 20	23 07	08 34	22 34
09	18 47	20 17	04 02	23 00	19 21	16 53	19 37	22 33	20 53	17 18	23 06	08 34	22 34
10	18 39	20 14	03 17	23 04	19 53	17 21	19 21	22 26	20 55	17 16	23 06	08 35	22 35
11	18 33	20 11	02 24	23 08	19 30	17 49	19 04	22 19	20 57	17 15	23 06	08 35	22 35
12	18 30	20 07	01 26	23 12	18 14	18 18	18 49	22 12	21 00	17 13	23 05	08 36	22 36
13	18 28	20 04	00 N 23	23 15	16 12	18 46	18 35	22 05	21 02	17 11	23 05	08 36	22 36
14	18 D 28	20 01	00 S 40	23 18	13 30	19 14	18 21	21 58	21 04	17 10	23 05	08 37	22 36
15	18 29	19 58	01 42	23 20	10 13	19 42	18 09	21 50	21 06	17 08	23 04	08 37	22 37
16	18 30	19 55	02 41	23 22	06 30	20 10	17 57	21 42	21 09	17 06	23 04	08 38	22 37
17	18 R 30	19 52	03 33	23 24	02 S 26	20 37	17 46	21 35	21 11	17 05	23 03	08 38	22 38
18	18 29	19 48	04 17	23 25	01 N 49	21 04	17 36	21 27	21 13	17 03	23 03	08 39	22 38
19	18 25	19 45	04 48	23 26	06 07	21 29	17 27	21 18	21 15	17 01	23 03	08 39	22 39
20	18 20	19 42	05 06	23 26	10 16	21 54	17 20	21 10	21 17	16 59	23 02	08 39	22 39
21	18 13	19 39	05 06	23 26	14 01	22 17	17 13	21 02	21 19	16 57	23 02	08 40	22 40
22	18 06	19 36	04 48	23 25	17 04	22 39	17 06	20 53	21 21	16 56	23 02	08 40	22 40
23	17 59	19 33	04 11	23 24	19 07	22 59	17 01	20 44	21 23	16 54	23 01	08 40	22 41
24	17 54	19 29	03 16	23 23	19 53	23 17	16 57	20 36	21 25	16 52	23 01	08 41	22 41
25	17 50	19 26	02 07	23 21	19 16	23 33	16 53	20 27	21 27	16 50	23 00	08 41	22 42
26	17 48	19 23	00 S 49	23 19	17 18	23 47	16 51	20 17	21 29	16 48	23 00	08 42	22 42
27	17 D 48	19 20	00 N 31	23 17	14 11	23 58	16 49	20 08	21 31	16 46	23 00	08 42	22 43
28	17 49	19 17	01 47	23 14	10 15	24 07	16 47	19 59	21 33	16 44	22 59	08 42	22 43
29	17 50	19 13	02 56	23 10	05 49	24 14	16 47	19 49	21 35	16 42	22 59	08 42	22 44
30	17 ♌ 51	19 ♌ 10	03 N 52	23 N 06	01 N 11	24 N 17	16 N 47	19 N 39	21 N 36	16 N 40	22 N 58	08 N 43	22 S 44

ZODIAC SIGN ENTRIES

Date	h m	Planets
02	19 44	☽ ♎
05	01 33	☽ ♏
07	09 19	☽ ♐
09	19 14	☽ ♑
11	13 06	☿ ♊
12	07 14	☽ ♒
14	20 14	☽ ♓
17	08 00	☽ ♈
18	23 57	♂ ♌
19	16 22	☽ ♉
20	18 32	☉ ♋
21	20 42	☽ ♊
23	22 00	☽ ♋
25	22 05	☽ ♌
27	14 02	☿ ♋
27	22 48	☽ ♍
30	01 33	☽ ♎

LATITUDES

Date	Mercury ☿ ° '	Venus ♀ ° '	Mars ♂ ° '	Jupiter ♃ ° '	Saturn ♄ ° '	Uranus ♅ ° '	Neptune ♆ ° '	Pluto ♇ ° '
01	03 S 30	00 N 35	01 N 17	00 S 39	01 N 03	00 N 23	01 S 39	08 S 36
04	03 15	00 S 08	01 17	00 39	01 03	00 23	01 39	08 37
07	02 55	00 48	01 17	00 39	01 03	00 23	01 39	08 37
10	02 31	01 27	01 17	00 38	01 03	00 23	01 39	08 38
13	02 02	02 01	01 16	00 38	01 03	00 23	01 39	08 39
16	01 30	02 32	01 16	00 38	01 03	00 23	01 39	08 40
19	00 57	02 58	01 15	00 38	01 03	00 23	01 40	08 41
22	00 S 22	03 20	01 15	00 38	01 03	00 23	01 40	08 41
25	00 N 11	03 38	01 15	00 38	01 03	00 23	01 40	08 42
28	00 42	03 52	01 14	00 37	01 03	00 23	01 40	08 43
31	01 N 08	04 S 03	01 N 14	00 S 37	01 N 03	00 N 23	01 S 40	08 S 43

DATA

Julian Date	2464846
Delta T	+76 seconds
Ayanamsa	24° 21' 45"
Synetic vernal point	04° ♓ 45' 13"
True obliquity of ecliptic	23° 25' 57"

MOON'S PHASES, APSIDES AND POSITIONS ☽

Date	h m	Phase	Longitude ° '	Eclipse Indicator
01	11 34	☽	11 ♍ 34	
08	21 02	○	18 ♐ 38	
17	01 03	☾	26 ♓ 26	
24	03 10	●	03 ♋ 12	
30	18 13	☽	09 ♎ 32	

Day	h m		
13	04 57	Apogee	
25	10 23	Perigee	
03	11 32	OS	
10	13 43	Max dec	19° S 53'
18	01 50	ON	
24	13 25	Max dec	19° N 54'
30	18 06	OS	

LONGITUDES

Date	Chiron ⚷ ° '	Ceres ⚳ ° '	Pallas ⚴ ° '	Juno ⚵ ° '	Vesta ⚶ ° '	Black Moon Lilith ⚸ ° '
01	12 ♊ 46	20 ♋ 00	27 ♉ 27	09 ♊ 49	00 ♐ 14	05 ♒ 11
11	13 ♊ 34	24 ♋ 11	02 ♊ 43	15 ♊ 45	28 ♏ 10	06 ♒ 18
21	14 ♊ 22	28 ♋ 28	08 ♊ 05	21 ♊ 40	26 ♏ 43	07 ♒ 24
31	15 ♊ 08	02 ♌ 49	13 ♊ 32	27 ♊ 31	26 ♏ 02	08 ♒ 31

ASPECTARIAN

h m	Aspects
01 Sunday	
00 45	☽ □ ♃
05 22	☽ □ ♀
10 35	☽ bQ ♆
11 34	☽ □ ☉
12 27	☉ ∥ ♀
14 14	☽ ∥ ♆
14 29	☽ ✶ ♅
16 32	☽ ⊻ ♄
22 05	☽ △ ☿
02 Monday	
01 10	☽ ✶ ♂
02 12	☽ ± ♆
03 01	☽ ⊥ ♄
04 27	☽ ⊼ ♇
07 30	♀ ⊥ ♅
11 27	☽ Q ♅
12 43	☽ ⊼ ♆
14 57	☽ ± ♇
18 53	☽ ∠ ♄
23 13	☽ Q ♂
03 Tuesday	
02 38	☽ bQ ☿
05 57	☽ △ ♃
06 51	☽ bQ ♇
07 33	☽ △ ♀
08 05	☉ ⊻ ♅
19 28	☽ □ ♅
20 07	☽ ± ☿
20 18	☽ △ ☉
20 43	♂ ∥ ♅
21 43	☽ ✶ ♄
04 Wednesday	
07 55	☽ ⊼ ☿
08 57	☽ □ ♂
09 17	☽ bQ ♃
09 18	☽ bQ ♀
09 36	♀ ☌ ♃
09 42	☽ △ ♇
13 04	♀ ∠ ♂
14 58	☉ # ♇
17 54	☉ ✶ ♄
18 24	☽ ☍ ♆
19 31	☽ Q ♄
20 13	♂ ∠ ♃
05 Thursday	
00 58	☽ ± ♀
00 59	♂ ⊼ ♇
01 32	☽ bQ ☉
01 55	☽ ± ♃
03 46	☿ □ ♇
06 47	☿ ✶ ♂
10 34	☽ # ♆
11 31	☽ ⊼ ♀
13 07	☽ ⊼ ♃
19 22	☽ ± ☉
20 41	♀ ∥ ♃
23 02	☉ ⊥ ♂
06 Friday	
02 22	☽ △ ♅
04 49	☽ □ ♄
07 22	☽ ⊼ ☉
16 52	☽ □ ♇
18 56	☽ △ ♂
20 54	☽ ☍ ☿
07 Saturday	
02 01	☽ ⊼ ♆
04 43	☉ ∥ ♂
06 35	☽ bQ ♅
13 24	☽ ± ♆
16 20	☽ # ☿
17 25	☽ ☍ ♀
22 23	☽ ☍ ♃
23 46	☽ ± ♅
08 Sunday	
00 49	☽ bQ ♂
02 56	☽ Q ♇
05 11	☽ # ♄
06 37	☽ bQ ♆
11 20	☽ ⊼ ♅
14 02	☽ △ ♄
19 04	☽ ± ♂
21 02	☽ ☍ ☉
22 19	☿ ⊻ ♆
09 Monday	
02 06	☽ ✶ ♇
07 21	☽ ⊼ ♂
07 35	♂ # ♇
11 46	☽ △ ♆
13 30	☽ ⊼ ☿
17 34	☽ # ♀
19 29	☽ bQ ♄
10 Tuesday	
01 34	☽ ⊼ ♀
02 53	☽ ± ☿
04 42	♀ Q ♄
07 33	☽ ∠ ♇
08 06	☿ ∥ ♄
09 57	☽ ⊼ ♃
13 05	☽ ± ♀
13 25	☽ ± ♄
17 02	☿ ∠ ♅
21 28	☉ ∥ ♅
22 08	☽ ± ♃
22 31	☽ ☍ ♅
23 16	☽ bQ ☿
11 Wednesday	
01 28	☽ ⊥ ♇
01 29	☽ ⊼ ♄
06 30	☽ bQ ♀
13 28	☽ ⊼ ☉
13 30	☽ ⊻ ♇
14 00	☉ △ ♇
16 33	☽ bQ ♃
22 17	☽ ☍ ♂
23 40	☽ □ ♆
12 Thursday	
00 36	☽ # ♀
01 14	♃ ⊥ ♅
02 38	☽ ± ☉
09 57	☽ △ ☿
11 24	☽ # ☿
11 56	☽ △ ♀
15 10	♀ ⊥ ♆
22 30	☽ bQ ☉
22 46	☿ ☌ ♀
23 33	☽ △ ♃
13 Friday	
00 23	☿ ⊥ ♆
01 18	☽ # ♄
01 32	♂ □ ♆
05 40	☿ ∥ ♀
11 30	☽ ⊼ ♅
12 18	☽ Q ♆
14 42	☽ ☍ ♄
18 21	☿ Q ♄
23 46	☽ ± ♅
14 Saturday	
02 21	☽ ☌ ♇
07 43	☽ △ ☉
12 46	☽ ✶ ♆
14 40	☽ ⊼ ♂
18 09	☽ bQ ♅
23 34	☽ □ ♀
15 Sunday	
03 30	☽ ± ♂
08 39	☽ □ ☿
13 40	☽ □ ♃
19 08	☽ ∠ ♆
22 36	☽ # ♆
22 36	☽ bQ ♂
16 Monday	
00 32	☽ △ ♅
03 52	☽ ⊼ ♄
11 05	☽ Q ♀
11 32	☿ ⊥ ♅
13 06	☽ ⊥ ♆
14 42	☽ ⊻ ♇
15 52	☽ ± ♄
23 29	☿ ☌ ♃
23 31	☉ ✶ ♆
17 Tuesday	
00 56	☽ ⊻ ♆
01 03	☽ □ ☉
02 06	☽ Q ♃
02 29	☽ ⊥ ♇
02 30	☽ Q ☿
05 52	☽ △ ♂
09 38	☽ bQ ♄
10 22	☽ ✶ ♀
19 54	☽ ∠ ♇
18 Wednesday	
01 43	☽ ✶ ♃
05 32	☽ ✶ ☿
08 08	☿ ∠ ♆
11 11	☽ □ ♅
14 26	☽ Q ☉
14 30	☽ △ ♄
14 45	☽ ∠ ♀
21 23	☿ ∥ ♃
19 Thursday	
00 10	☽ ✶ ♇
04 16	☿ ∥ ♂
06 18	☽ ∠ ♃
07 21	☽ ⊥ ♀
09 53	☽ ☌ ♆
14 04	☽ ∠ ☿
14 20	☽ ✶ ☉
17 11	☽ □ ♂
18 14	☽ ⊻ ♀
19 55	♂ ∥ ♃
20 29	☽ Q ♅
20 59	☿ ⊻ ♅
22 02	☽ Q ♇
22 55	☽ ⊥ ♃
20 Friday	
02 27	☽ ∥ ♆
08 52	☽ ⊥ ☿
09 46	☽ ⊻ ♃
14 12	♀ St D
15 02	♀ ✶ ♂
17 50	☽ ✶ ♅
19 03	☽ ∠ ☉
20 33	☿ ∠ ♂
20 52	☿ ✶ ♄
21 05	☽ □ ♄
21 07	☽ ⊻ ☿
21 Saturday	
02 30	☽ Q ♂
05 36	☽ □ ♇
11 42	☽ ⊥ ☉
14 45	☽ ⊻ ♆
19 41	☽ ∠ ♅
20 56	☉ ⊻ ♀
22 28	☽ ☌ ♀
22 35	☽ ⊻ ☉
23 51	☽ ✶ ♂
22 Sunday	
00 53	☽ ⊥ ♆
03 43	☉ ∠ ♄
03 57	☽ Q ♄
10 41	☽ ∥ ♄
10 47	☽ ⊥ ♅
12 19	☽ ∥ ♀
13 46	☽ ☌ ♃
13 55	☿ # ♇
15 52	☽ ∠ ♆
20 44	☽ ⊻ ♅
23 56	☽ ✶ ♄
23 Monday	
01 48	☽ ∠ ♂
07 30	☽ ☌ ☿
07 35	☽ △ ♇
08 06	☿ △ ♇
15 12	☿ ∥ ♅
16 23	☽ ✶ ♆
17 04	☽ ⊥ ♂
24 Tuesday	
00 00	☽ ⊻ ♀
00 26	☽ ∠ ♄
03 10	☽ ☌ ☉
03 10	☽ ⊻ ♂
03 35	☉ ⊻ ♂
07 47	☽ bQ ♇
09 43	☽ ⊥ ♀
11 43	☽ Q ♆
14 59	☽ ⊥ ♄
15 10	☽ ⊻ ♃
20 09	☉ ∥ ☿
21 23	☽ ☌ ♅
22 11	☽ ± ♇
25 Wednesday	
00 17	☽ ∠ ♀
00 39	☽ ⊻ ♄
00 55	☽ ⊥ ♃
07 47	☽ ⊼ ♇
15 28	☽ ⊻ ☿
15 32	☽ ∠ ♃
16 33	☽ □ ♆
23 18	☿ ✶ ♆
26 Thursday	
00 36	☽ ✶ ♀
02 40	☽ ⊥ ☿
05 20	☽ ☌ ♂
06 30	☽ ⊻ ☉
16 01	☽ ✶ ♃
16 07	☽ ∥ ♀
16 29	☽ ∥ ♄
16 51	☽ ⊥ ☉
19 35	☽ ∠ ☿
20 13	☽ Q ♀
21 46	☽ ⊻ ♅
27 Friday	
01 14	☽ ☌ ♄
02 32	☉ bQ ♇
04 08	☿ ⊥ ♂
07 33	☽ ⊥ ♅
08 05	☽ ☍ ♇
08 28	☽ ∠ ☉
11 49	☽ Q ♃
17 11	☽ △ ♆
22 23	☽ ∠ ♅
28 Saturday	
00 19	☽ ✶ ☿
02 11	☽ □ ♀
08 31	☽ ⊻ ♂
10 58	☽ ✶ ☉
12 13	☿ ∠ ♄
14 26	☿ ⊻ ♀
18 08	♃ ∠ ♆
18 08	☽ bQ ♆
18 08	☽ □ ♃
19 01	☽ ⊥ ♂
20 34	☽ ∥ ♆
23 33	☽ ✶ ♅
23 56	☽ Q ☿
29 Sunday	
03 21	☽ ⊻ ♄
08 41	☽ Q ☉
09 24	☽ ± ♆
10 04	☽ ⊼ ♇
10 59	☽ ∠ ♂
13 39	☽ ⊥ ♄
16 50	☉ ⊥ ♀
19 30	☉ Q ♆
19 42	☽ ⊼ ♆
20 08	☽ Q ♅
20 21	☽ ± ♇
30 Monday	
01 02	♀ ⊥ ♆
05 20	☽ ∠ ♄
06 14	☽ △ ♀
08 29	☿ bQ ♇
12 00	☽ bQ ♇
12 40	☽ □ ☿
14 13	☽ ✶ ♂
18 13	☽ □ ☉
22 45	☽ △ ♃

All ephemeris data is given at 12.00 UT and the Moon's longitude is additionally given for 24.00 UT

JULY 2036

LONGITUDES

Date	Sidereal time h m s	Sun ☉ ° ′ ″	Moon ☽ ° ′ ″	Moon ☽ 24.00 ° ′ ″	Mercury ☿ ° ′	Venus ♀ ° ′	Mars ♂ ° ′	Jupiter ♃ ° ′	Saturn ♄ ° ′	Uranus ♅ ° ′	Neptune ♆ ° ′	Pluto ♇ ° ′
01	06 40 30	10 ♋ 14 08	19 ♎ 31 01	26 ♎ 09 36	08 ♋ 30	03 ♊ 08	07 ♌ 50	12 ♊ 13	17 ♌ 19	14 ♋ 58	26 ♈ 39	20 ♒ 59
02	06 44 27	11 11 20	02 ♏ 43 46	09 ♏ 13 45	10 41	03 31	08 27	12 26	17 26	15 02	26 40	20 R 58
03	06 48 23	12 08 32	15 ♏ 39 48	22 ♏ 02 09	12 51	03 56	09 05	12 39	17 33	15 06	26 41	20 57
04	06 52 20	13 05 43	28 ♏ 21 02	04 ♐ 36 41	15 01	04 22	09 42	12 52	17 39	15 09	26 42	20 56
05	06 56 16	14 02 55	10 ♐ 49 20	16 ♐ 59 10	17 10	04 50	10 20	13 05	17 46	15 13	26 43	20 55
06	07 00 13	15 00 06	23 ♐ 06 23	29 ♐ 11 11	19 18	05 19	10 58	13 18	17 53	15 16	26 43	20 54
07	07 04 10	15 57 17	05 ♑ 13 47	11 ♑ 14 21	21 25	05 50	11 35	13 31	18 00	15 20	26 44	20 53
08	07 08 06	16 54 29	17 ♑ 13 07	23 ♑ 10 19	23 31	06 22	12 13	13 44	18 07	15 24	26 45	20 52
09	07 12 03	17 51 40	29 ♑ 06 11	05 ♒ 00 59	25 35	06 56	12 50	13 57	18 14	15 27	26 46	20 51
10	07 15 59	18 48 52	10 ♒ 55 01	16 ♒ 48 37	27 38	07 31	13 28	14 09	18 21	15 31	26 46	20 50
11	07 19 56	19 46 04	22 ♒ 42 09	28 ♒ 35 59	29 ♋ 39	08 07	14 06	14 22	18 28	15 35	26 47	20 48
12	07 23 52	20 43 16	04 ♓ 30 33	10 ♓ 26 20	01 ♌ 38	08 44	14 43	14 35	18 35	15 38	26 47	20 47
13	07 27 49	21 40 28	16 ♓ 23 47	22 ♓ 23 25	03 35	09 22	15 21	14 47	18 42	15 42	26 48	20 46
14	07 31 45	22 37 41	28 ♓ 25 48	04 ♈ 31 29	05 31	10 02	15 59	15 00	18 49	15 45	26 48	20 45
15	07 35 42	23 34 55	10 ♈ 41 00	16 ♈ 54 55	07 25	10 42	16 36	15 12	18 56	15 49	26 49	20 44
16	07 39 39	24 32 09	23 ♈ 13 47	29 ♈ 38 08	09 17	11 24	17 14	15 24	19 03	15 53	26 49	20 43
17	07 43 35	25 29 24	06 ♉ 08 24	12 ♉ 45 00	11 07	12 07	17 52	15 37	19 11	15 56	26 50	20 42
18	07 47 32	26 26 39	19 ♉ 28 15	26 ♉ 18 20	12 56	12 50	18 29	15 49	19 18	16 00	26 50	20 40
19	07 51 28	27 23 56	03 ♊ 15 20	10 ♊ 19 10	14 42	13 35	19 07	16 01	19 25	16 03	26 51	20 39
20	07 55 25	28 21 13	17 ♊ 29 35	24 ♊ 46 08	16 27	14 20	19 45	16 13	19 33	16 07	26 51	20 38
21	07 59 21	29 ♋ 18 31	02 ♋ 08 14	09 ♋ 35 05	18 10	15 06	20 22	16 25	19 40	16 11	26 51	20 37
22	08 03 18	00 ♌ 15 49	17 ♋ 05 46	24 ♋ 39 11	19 51	15 53	21 00	16 37	19 47	16 14	26 52	20 36
23	08 07 14	01 13 08	02 ♌ 14 13	09 ♌ 49 39	21 30	16 41	21 38	16 49	19 55	16 18	26 52	20 34
24	08 11 11	02 10 28	17 ♌ 24 18	24 ♌ 57 01	23 07	17 29	22 16	17 01	20 02	16 21	26 52	20 33
25	08 15 07	03 07 48	02 ♍ 26 45	09 ♍ 52 33	24 43	18 19	22 53	17 13	20 10	16 25	26 52	20 32
26	08 19 04	04 05 09	17 ♍ 13 38	24 ♍ 29 22	26 16	19 08	23 31	17 24	20 17	16 28	26 52	20 31
27	08 23 01	05 02 30	01 ♎ 39 17	08 ♎ 43 06	27 48	19 59	24 09	17 36	20 25	16 32	26 52	20 29
28	08 26 57	05 59 51	15 ♎ 40 39	22 ♎ 31 55	29 ♌ 18	20 50	24 47	17 47	20 32	16 35	26 52	20 28
29	08 30 54	06 57 13	29 ♎ 17 00	05 ♏ 56 07	00 ♍ 47	21 42	25 25	17 59	20 40	16 39	26 52	20 27
30	08 34 50	07 54 35	12 ♏ 29 31	18 ♏ 57 33	02 13	22 35	26 02	18 10	20 47	16 42	26 R 52	20 25
31	08 38 47	08 ♌ 51 58	25 ♏ 20 36	01 ♐ 39 03	03 ♍ 37	23 ♊ 28	26 ♌ 40	18 ♊ 21	20 ♌ 55	16 ♋ 46	26 ♈ 52	20 ♒ 24

DECLINATIONS

Date	Moon True ☊ ° ′	Moon Mean ☊ ° ′	Moon ☽ Latitude ° ′	Sun ☉ ° ′	Moon ☽ ° ′	Mercury ☿ ° ′	Venus ♀ ° ′	Mars ♂ ° ′	Jupiter ♃ ° ′	Saturn ♄ ° ′	Uranus ♅ ° ′	Neptune ♆ ° ′	Pluto ♇ ° ′
01	17 ♌ 50	19 ♌ 07	04 N 35	23 N 02	03 S 24	24 N 18	16 N 48	19 N 30	21 N 38	16 N 38	22 N 58	08 N 43	22 S 45
02	17 R 48	19 04	05 01	22 58	07 42	24 16	16 49	19 20	21 40	16 36	22 58	08 43	22 45
03	17 45	19 01	05 11	22 53	11 33	24 11	16 51	19 10	21 42	16 34	22 57	08 44	22 46
04	17 40	18 58	05 06	22 47	14 48	24 03	16 53	18 59	21 43	16 32	22 57	08 44	22 46
05	17 35	18 54	04 47	22 42	17 20	23 53	16 56	18 49	21 45	16 30	22 56	08 44	22 47
06	17 29	18 51	04 14	22 35	19 02	23 40	17 00	18 39	21 47	16 28	22 56	08 44	22 47
07	17 24	18 48	03 30	22 29	19 50	23 25	17 04	18 28	21 48	16 26	22 56	08 45	22 48
08	17 20	18 45	02 37	22 22	19 43	23 07	17 08	18 17	21 50	16 24	22 55	08 45	22 48
09	17 17	18 42	01 38	22 14	18 44	22 47	17 13	18 07	21 51	16 22	22 55	08 45	22 49
10	17 16	18 38	00 N 35	22 07	16 56	22 25	17 17	17 56	21 53	16 19	22 54	08 45	22 49
11	17 D 16	18 35	00 S 30	21 59	14 25	22 01	17 23	17 45	21 54	16 17	22 54	08 45	22 50
12	17 16	18 32	01 34	21 50	11 18	21 35	17 28	17 33	21 56	16 15	22 54	08 45	22 51
13	17 18	18 29	02 34	21 41	07 44	21 08	17 34	17 22	21 57	16 13	22 53	08 46	22 51
14	17 20	18 26	03 28	21 32	03 S 48	20 38	17 40	17 11	21 59	16 11	22 53	08 46	22 52
15	17 21	18 23	04 14	21 22	00 N 20	20 08	17 46	16 59	22 00	16 08	22 52	08 46	22 52
16	17 22	18 19	04 48	21 12	04 33	19 36	17 52	16 48	22 01	16 06	22 52	08 46	22 53
17	17 R 21	18 16	05 10	21 02	08 41	19 04	17 59	16 36	22 03	16 04	22 52	08 46	22 53
18	17 20	18 13	05 16	20 51	12 32	18 30	18 05	16 24	22 04	16 02	22 51	08 46	22 54
19	17 18	18 10	05 04	20 40	15 50	17 55	18 12	16 12	22 05	16 00	22 51	08 46	22 54
20	17 15	18 07	04 33	20 29	18 19	17 19	18 18	16 00	22 07	15 57	22 50	08 46	22 55
21	17 13	18 04	03 44	20 17	19 41	16 43	18 25	15 48	22 08	15 55	22 50	08 46	22 55
22	17 11	18 00	02 39	20 05	19 43	16 06	18 32	15 36	22 09	15 53	22 49	08 46	22 56
23	17 10	17 57	01 S 22	19 53	18 20	15 29	18 38	15 23	22 10	15 50	22 49	08 46	22 57
24	17 D 10	17 54	00 N 01	19 40	15 38	14 51	18 45	15 11	22 12	15 48	22 48	08 46	22 57
25	17 10	17 51	01 24	19 27	11 54	14 13	18 51	14 58	22 13	15 46	22 48	08 46	22 58
26	17 11	17 48	02 39	19 14	07 29	13 34	18 58	14 46	22 14	15 43	22 48	08 46	22 58
27	17 12	17 44	03 43	19 00	02 N 45	12 56	19 04	14 33	22 15	15 41	22 47	08 46	22 59
28	17 13	17 41	04 31	18 46	02 S 00	12 17	19 10	14 20	22 16	15 39	22 47	08 46	22 59
29	17 13	17 38	05 02	18 32	06 30	11 38	19 16	14 07	22 17	15 36	22 47	08 46	23 00
30	17 R 13	17 35	05 16	18 17	10 33	11 00	19 22	13 54	22 18	15 34	22 46	08 46	23 00
31	17 ♌ 13	17 ♌ 32	05 N 14	18 N 02	14 S 00	10 N 21	19 N 27	13 N 41	22 N 19	15 N 32	22 N 46	08 N 46	23 S 01

ZODIAC SIGN ENTRIES

Date	h	m	Planets
02	07	00	☽ ♏
04	15	09	☽ ♐
07	01	37	☽ ♑
09	13	49	☽ ♒
11	16	16	☿ ♌
12	02	51	☽ ♓
14	15	06	☽ ♈
17	00	41	☽ ♉
19	06	24	☽ ♊
21	08	32	☽ ♋
22	05	23	☉ ♌
23	08	28	☽ ♌
25	08	04	☽ ♍
27	09	13	☽ ♎
28	23	16	☿ ♍
29	13	17	☽ ♏
31	20	51	☽ ♐

LATITUDES

Date	Mercury ☿ ° ′	Venus ♀ ° ′	Mars ♂ ° ′	Jupiter ♃ ° ′	Saturn ♄ ° ′	Uranus ♅ ° ′	Neptune ♆ ° ′	Pluto ♇ ° ′
01	01 N 08	04 S 03	01 N 14	00 S 37	01 N 03	00 N 23	01 S 40	08 S 43
04	01 29	04 11	01 13	00 37	01 03	00 23	01 40	08 44
07	01 42	04 17	01 13	00 37	01 03	00 23	01 41	08 45
10	01 49	04 19	01 12	00 37	01 03	00 23	01 41	08 45
13	01 50	04 20	01 12	00 37	01 03	00 23	01 41	08 46
16	01 45	04 19	01 11	00 37	01 03	00 23	01 41	08 46
19	01 34	04 15	01 10	00 37	01 03	00 23	01 41	08 47
22	01 19	04 11	01 10	00 37	01 03	00 23	01 41	08 47
25	00 59	04 05	01 09	00 37	01 03	00 23	01 42	08 48
28	00 37	03 58	01 08	00 37	01 04	00 23	01 42	08 48
31	00 N 11	03 S 49	01 N 08	00 S 36	01 N 04	00 N 23	01 S 42	08 S 48

DATA

Julian Date	2464876
Delta T	+76 seconds
Ayanamsa	24° 21′ 50″
Synetic vernal point	04° ♓ 45′ 09″
True obliquity of ecliptic	23° 25′ 57″

MOON'S PHASES, APSIDES AND POSITIONS ☽

Date	h	m	Phase	Longitude ° ′	Eclipse Indicator
08	11	19	○	16 ♑ 53	
16	14	40	☾	24 ♈ 38	
23	10	17	●	01 ♌ 09	Partial
30	02	56	☽	07 ♏ 33	

Day	h	m	
10	16	06	Apogee
23	18	28	Perigee

Day	h	m		
07	21	08	Max dec	19° S 54′
15	10	04	0N	
22	00	35	Max dec	19° N 52′
28	01	48	0S	

LONGITUDES

Date	Chiron ⚷ ° ′	Ceres ⚳ ° ′	Pallas ⚴ ° ′	Juno ⚵ ° ′	Vesta ⚶ ° ′	Black Moon Lilith ⚸ ° ′
01	15 ♊ 08	02 ♌ 49	13 ♊ 32	27 ♊ 31	26 ♏ 02	08 ♒ 31
11	15 ♊ 52	07 ♌ 14	19 ♊ 03	03 ♋ 18	26 ♏ 01	09 ♒ 37
21	16 ♊ 33	11 ♌ 41	24 ♊ 37	09 ♋ 06	27 ♏ 06	10 ♒ 44
31	17 ♊ 10	16 ♌ 11	00 ♋ 14	14 ♋ 39	27 ♏ 43	11 ♒ 50

ASPECTARIAN

h m Aspects

01 Tuesday
01 40 ☿ ⊻ ♂
03 50 ☽ □ ♅
08 02 ☽ ⚹ ♄
09 27 ☽ bQ ♀
12 35 ☽ Q ♂
13 44 ☿ Q ♆
14 11 ♅ ± ♇
14 38 ☽ △ ♇
20 32 ☿ ⊥ ♀

02 Wednesday
00 55 ☽ ☍ ♆
02 09 ☽ bQ ♃
02 11 ☽ ± ♀
05 54 ☽ Q ♄
12 01 ☉ ∥ ♅
13 30 ☽ ⊼ ♀
18 05 ☽ # ♆
18 54 ☉ ⊥ ♄
18 56 ☽ ± ♃
20 46 ☿ ⊥ ♄
22 04 ☉ ☌ ☿
23 06 ☽ □ ♂

03 Thursday
04 54 ☽ △ ☉
05 40 ☽ △ ☿
06 16 ☽ ⊼ ♃
09 33 ☿ ⊻ ♃
10 56 ☽ △ ♅
15 34 ☽ □ ♄
21 56 ☽ □ ♇

04 Friday
04 38 ☉ ⊻ ♃
08 51 ☽ ⊼ ♆
11 03 ☿ ± ♇
11 28 ☽ bQ ☉
13 34 ☿ ☌ ♅
15 28 ☽ bQ ♅
15 51 ☽ bQ ☿
15 59 ☉ # ♇
20 20 ☽ ± ♆
23 59 ☽ ☍ ♀

05 Saturday
03 20 ☽ # ♄
06 11 ☽ ± ☉
07 37 ☽ # ♀
08 18 ☽ Q ♇
08 52 ☽ ± ♅
11 00 ☽ △ ♂
12 49 ☽ ± ☿
13 44 ☽ bQ ♆
16 29 ☽ ☍ ♃
18 48 ☽ ⊼ ☉
19 06 ☿ ⊻ ♄
20 35 ☽ ⊼ ♅

06 Sunday
01 39 ☽ △ ♄
02 58 ☽ ⊼ ☿
06 02 ☽ # ♂
07 40 ☽ ⚹ ♇
09 24 ☉ ± ♇
11 58 ☿ ⊥ ♃
17 56 ☽ bQ ♂
19 08 ☽ △ ♆
19 19 ☉ ☌ ♅

07 Monday
03 11 ☿ ∠ ♀
05 53 ☿ ⊼ ♇
07 31 ☽ bQ ♄
12 45 ☽ ± ♂
13 16 ☽ ⊼ ♀
13 18 ☽ ∠ ♇
21 18 ♀ Q ♄

08 Tuesday
01 24 ☽ ⊼ ♂
01 39 ☽ ± ♄
01 49 ☽ ± ♀
04 52 ☽ ⊼ ♃
07 16 ☽ ⊥ ♇
08 19 ☽ ☍ ♅
11 19 ☽ ☍ ☉
13 49 ☽ ⊼ ♄
17 09 ☽ ± ♃
19 20 ☽ ⊻ ♇
20 47 ☽ bQ ♀

09 Wednesday
02 44 ☿ ∥ ♅
03 23 ☽ ☍ ☿
07 15 ☽ □ ♆
09 52 ☿ # ♇
11 40 ☽ bQ ♃
22 33 ☉ ⊻ ♄
22 38 ☽ # ♂

10 Thursday
01 50 ☿ □ ♆
04 43 ☽ △ ♀
08 04 ☽ # ♀
17 29 ☽ ☍ ♂
18 31 ☽ # ♄
18 43 ☽ △ ♃
19 51 ☽ Q ♆
21 25 ☽ ⊼ ♅

11 Friday
03 17 ☽ ☍ ♄
05 30 ☽ ⊼ ☉
08 09 ☽ ☌ ♇
08 17 ☿ ∠ ♃
09 42 ☽ ± ♅
15 17 ☉ ∥ ☿
17 54 ☿ ∥ ♃
18 47 ☽ ± ☉
20 18 ☽ ⚹ ♆
22 24 ☉ ∥ ♃

12 Saturday
03 44 ♂ ⚹ ♃
04 06 ☽ bQ ♅
04 59 ☽ ⊼ ☿
07 21 ☉ ⊥ ♃
13 41 ☉ ⊼ ♇
14 40 ☽ bQ ☉
19 23 ♀ ∥ ♂
19 35 ☽ ± ☿
21 02 ☽ □ ♀

13 Sunday
02 44 ☽ ∠ ♆
05 21 ☽ # ♆
08 42 ☽ □ ♃
09 47 ☽ ⊼ ♂
10 35 ☽ △ ♅
16 40 ☽ ⊼ ♄
17 15 ☽ bQ ☿
20 45 ☽ ⊻ ♇
20 50 ☽ ⊥ ♆
22 28 ☽ ± ♂
23 29 ☽ △ ☉

14 Monday
01 08 ♀ ⊥ ♅
02 47 ♂ ⊻ ♅
04 46 ☽ ± ♄
08 41 ☽ ⊥ ♇
08 47 ☽ ⊻ ♆
11 10 ☽ Q ♀
17 18 ☽ bQ ♂
21 09 ☽ Q ♃
22 43 ☽ bQ ♄

15 Tuesday
02 23 ☽ ∠ ♇
04 30 ☽ △ ☿
12 03 ☽ ⚹ ♀
16 39 ♅ # ♇
20 52 ☽ ⚹ ♃
21 57 ☽ □ ♅

16 Wednesday
00 00 ☽ △ ♂
04 01 ☽ △ ♄
07 15 ☽ ⚹ ♇
14 40 ☽ □ ☉
18 19 ☽ ∠ ♀
18 46 ☽ ☌ ♆

17 Thursday
01 39 ☽ ∠ ♃
02 32 ♀ ∠ ♆
05 40 ☽ Q ♇
07 56 ☽ Q ♅
11 57 ☽ ⊥ ♀
12 29 ☽ ∥ ♆
18 26 ☽ ⊥ ♃
22 30 ☽ □ ☿
23 28 ☽ ⊻ ♀

18 Friday
02 21 ☽ Q ☉
05 24 ☽ ⊻ ♃
05 48 ☽ ⚹ ♅
09 56 ☿ ⚹ ♀
10 10 ☽ □ ♂
11 42 ☽ □ ♄
14 07 ☽ □ ♇
21 58 ☉ □ ♆

19 Saturday
00 56 ☽ ⊻ ♆
01 09 ☽ ⚹ ☉
02 17 ☿ ∥ ♀
08 13 ☽ ∠ ♅
10 55 ☽ Q ☿
11 18 ☽ ⊥ ♆
13 19 ☽ ∥ ♄
14 52 ☽ ∥ ♂
18 33 ♃ ⊻ ♅
18 53 ☽ Q ♂
19 10 ☽ Q ♄
23 36 ☽ ⊥ ♅

20 Sunday
02 28 ♂ ☌ ♄
02 34 ☽ ∠ ♆
03 24 ☽ ∥ ☿
04 36 ☽ ∠ ☉
06 26 ☽ ☌ ♀
07 16 ☿ ⊻ ♅
08 27 ☿ ⚹ ♃
09 42 ☽ ⊻ ♅
09 51 ☽ ☌ ♃
10 01 ☽ ⚹ ☿
11 55 ☽ ∥ ♀
15 26 ☽ ⚹ ♄
15 54 ☽ ⚹ ♂
17 12 ☽ △ ♇
18 42 ♂ ∥ ♄
20 36 ☽ ⊥ ☉

21 Monday
03 24 ☽ ⚹ ♆
07 06 ☽ ⊻ ☉
13 52 ☽ ∠ ☿
16 07 ☽ ∠ ♄
17 27 ☽ ∠ ♂
17 36 ☽ bQ ♇
20 53 ♂ ☍ ♇
22 50 ☽ Q ♆

22 Tuesday
06 10 ☽ ⊥ ☿
06 41 ☽ ⊥ ♄
08 01 ☽ ± ♇
08 31 ☽ ⊥ ♂
09 58 ☽ ⊻ ♀
10 38 ☽ ☌ ♅
11 07 ☿ ☌ ♄
11 14 ☽ ⊻ ♃
16 19 ☽ ⊻ ♄
16 55 ☽ ⊻ ☿
17 33 ☽ ⊼ ♇
18 29 ☽ ⊻ ♂
20 02 ☽ ⊥ ♀
20 54 ☽ ⊥ ♃
21 17 ☿ ∥ ♄
22 40 ☿ ☍ ♇
23 32 ♀ ⊻ ♅

23 Wednesday
03 30 ☽ □ ♆
08 27 ☽ ∥ ♀
10 17 ☽ ☌ ☉
11 04 ☽ ∠ ♀
11 20 ☽ ∠ ♃
15 10 ☿ ☌ ♂
17 19 ☿ ∥ ♂
17 27 ♀ ☌ ♃

24 Thursday
00 11 ☿ ⊥ ♅
06 57 ☉ ∠ ♃
10 20 ☽ ⊻ ♅
10 45 ☽ ∥ ♄
11 22 ☽ ⚹ ♃
12 08 ☽ ⚹ ♀
15 29 ☽ ∥ ♂
15 59 ♂ ⊥ ♅
16 13 ☽ ☌ ♄
16 59 ☽ ☍ ♇
18 47 ☽ ∥ ☿
19 54 ☽ ⊥ ♅
20 03 ☽ ☌ ♂
22 10 ☽ ☌ ☿

25 Friday
03 04 ☽ △ ♆
06 44 ☽ Q ♃
08 22 ☽ Q ♀
10 20 ☽ ∠ ♅
13 11 ☽ ⊻ ☉
23 32 ☽ ⊥ ☉

26 Saturday
03 15 ☽ bQ ♆
05 16 ☽ ∥ ♆
10 45 ☽ ⚹ ♅
12 18 ☽ □ ♃
15 16 ☽ ∠ ☉
15 20 ☽ □ ♀
17 05 ☽ ⊻ ♄
17 24 ☽ ⊼ ♇
18 00 ☽ ± ♆
21 18 ☿ △ ♆
22 52 ☽ ⊻ ♂
23 22 ☉ ∠ ♀

27 Sunday
03 07 ☽ ⊥ ♄
03 21 ☽ ± ♇
03 58 ☽ ⊼ ♆
04 46 ☽ ⊻ ☿
06 44 ☽ Q ♅
07 31 ☉ ∥ ♀
09 21 ☽ ⊥ ♂
16 04 ☽ ⊥ ☿
18 09 ☽ ⚹ ☉
18 25 ☽ ∠ ♄
18 29 ☽ bQ ♇

28 Monday
00 53 ♄ ☍ ♇
01 20 ☽ ∠ ♂
01 52 ♀ △ ♇
02 04 ♀ ⚹ ♄
09 20 ☽ ∠ ☿
13 35 ☽ □ ♅
15 44 ☽ △ ♃
16 21 ☽ Q ☉
20 21 ☽ △ ♇
20 34 ☽ ⚹ ♄
20 58 ☿ Q ♃
21 37 ☽ △ ♀

29 Tuesday
04 45 ☽ ⚹ ♂
07 42 ☽ ☍ ♆
15 00 ☽ ⚹ ☿
18 08 ☽ Q ♄
18 44 ☽ bQ ♃

30 Wednesday
00 19 ♆ St R
01 03 ☽ # ♆
02 21 ☽ bQ ♀
02 56 ☽ □ ☉
03 05 ☿ ∠ ♅
03 25 ☽ Q ♂
11 23 ☽ ± ♃
14 25 ☽ # ☿
15 34 ☽ Q ☿
19 50 ☽ △ ♅
20 07 ☽ ± ♀
22 40 ☽ ⊼ ♃

31 Thursday
02 43 ☽ □ ♇
03 34 ☽ □ ♄
08 11 ☽ ⊼ ♀
09 42 ☽ # ♂
14 39 ☽ □ ♂
14 54 ☽ ⊼ ♆
19 47 ♂ △ ♆

All ephemeris data is given at 12.00 UT and the Moon's longitude is additionally given for 24.00 UT

AUGUST 2036

LONGITUDES

Date	Sidereal time h m s	Sun ☉ ° ' "	Moon ☽ ° ' "	Moon ☽ 24.00 ° ' "	Mercury ☿ ° '	Venus ♀ ° '	Mars ♂ ° '	Jupiter ♃ ° '	Saturn ♄ ° '	Uranus ♅ ° '	Neptune ♆ ° '	Pluto ♇ ° '
01	08 42 43	09 ♌ 49 22	07 ♐ 53 21	14 ♐ 03 54	05 ♍ 00	24 ♊ 21	27 ♌ 18	18 ♊ 32	21 ♌ 02	16 ♋ 49	26 ♈ 52	20 ♒ 23
02	08 46 40	10 46 46	20 ♐ 11 07	26 ♐ 15 26	06 20	25 15	27 56	18 43	21 10	16 53	26 R 52	20 R 21
03	08 50 36	11 44 11	02 ♑ 17 14	08 ♑ 16 54	07 39	26 10	28 34	18 54	21 17	16 56	26 52	20 20
04	08 54 33	12 41 36	14 ♑ 14 47	20 ♑ 11 14	08 55	27 05	29 12	19 05	21 25	16 59	26 52	20 19
05	08 58 30	13 39 03	26 ♑ 06 34	02 ♒ 01 07	10 09	28 01	29 ♌ 49	19 16	21 33	17 03	26 52	20 18
06	09 02 26	14 36 30	07 ♒ 55 09	13 ♒ 48 59	11 21	28 57	00 ♍ 27	19 27	21 40	17 06	26 52	20 16
07	09 06 23	15 33 58	19 ♒ 42 54	25 ♒ 37 10	12 31	29 ♊ 54	01 05	19 37	21 48	17 09	26 51	20 15
08	09 10 19	16 31 28	01 ♓ 32 05	07 ♓ 27 57	13 39	00 ♋ 51	01 43	19 48	21 56	17 13	26 51	20 14
09	09 14 16	17 28 58	13 ♓ 25 03	19 ♓ 23 43	14 44	01 48	02 21	19 58	22 03	17 16	26 51	20 12
10	09 18 12	18 26 30	25 ♓ 24 17	01 ♈ 27 06	15 46	02 46	02 59	20 08	22 11	17 19	26 50	20 11
11	09 22 09	19 24 03	07 ♈ 32 31	13 ♈ 40 56	16 46	03 45	03 37	20 18	22 19	17 22	26 50	20 09
12	09 26 05	20 21 37	19 ♈ 52 44	26 ♈ 08 19	17 43	04 44	04 15	20 28	22 26	17 25	26 50	20 08
13	09 30 02	21 19 12	02 ♉ 28 06	08 ♉ 52 29	18 38	05 43	04 53	20 38	22 34	17 29	26 49	20 07
14	09 33 59	22 16 49	15 ♉ 21 50	21 ♉ 56 32	19 29	06 43	05 31	20 48	22 42	17 32	26 49	20 05
15	09 37 55	23 14 28	28 ♉ 36 52	05 ♊ 23 07	20 17	07 43	06 09	20 58	22 49	17 35	26 48	20 04
16	09 41 52	24 12 08	12 ♊ 15 27	19 ♊ 13 56	21 02	08 43	06 47	21 08	22 57	17 38	26 48	20 03
17	09 45 48	25 09 50	26 ♊ 18 32	03 ♋ 29 04	21 43	09 44	07 25	21 17	23 05	17 41	26 47	20 01
18	09 49 45	26 07 34	10 ♋ 45 14	18 ♋ 06 32	22 20	10 45	08 03	21 27	23 12	17 44	26 46	20 00
19	09 53 41	27 05 19	25 ♋ 32 21	03 ♌ 01 51	22 54	11 46	08 41	21 36	23 20	17 47	26 46	19 59
20	09 57 38	28 03 05	10 ♌ 34 09	18 ♌ 08 09	23 23	12 48	09 19	21 45	23 28	17 50	26 45	19 57
21	10 01 34	29 00 53	25 ♌ 42 46	03 ♍ 16 48	23 48	13 50	09 57	21 54	23 35	17 53	26 44	19 56
22	10 05 31	29 ♌ 58 42	10 ♍ 49 07	18 ♍ 18 33	24 08	14 52	10 35	22 03	23 43	17 56	26 44	19 55
23	10 09 28	00 ♍ 56 33	25 ♍ 44 07	03 ♎ 04 52	24 23	15 54	11 14	22 12	23 51	17 59	26 43	19 54
24	10 13 24	01 54 25	10 ♎ 20 05	17 ♎ 29 09	24 33	16 57	11 52	22 20	23 58	18 02	26 42	19 52
25	10 17 21	02 52 18	24 ♎ 31 40	01 ♏ 27 24	24 38	18 00	12 30	22 29	24 06	18 05	26 41	19 51
26	10 21 17	03 50 12	08 ♏ 16 16	14 ♏ 58 21	24 R 37	19 04	13 08	22 37	24 13	18 07	26 40	19 50
27	10 25 14	04 48 08	21 ♏ 33 51	28 ♏ 03 05	24 30	20 07	13 46	22 46	24 21	18 10	26 40	19 48
28	10 29 10	05 46 05	04 ♐ 26 27	10 ♐ 44 24	24 17	21 11	14 25	22 54	24 29	18 13	26 39	19 47
29	10 33 07	06 44 03	16 ♐ 57 28	23 ♐ 06 12	23 58	22 15	15 03	23 02	24 36	18 16	26 38	19 46
30	10 37 03	07 42 03	29 ♐ 11 09	05 ♑ 12 54	23 33	23 20	15 41	23 10	24 44	18 18	26 37	19 45
31	10 41 00	08 ♍ 40 04	11 ♑ 12 00	17 ♑ 09 00	23 ♍ 02	24 ♋ 25	16 ♍ 19	23 ♊ 17	24 ♌ 51	18 ♋ 21	26 ♈ 36	19 ♒ 43

Date	Moon True ☊ ° '	Moon Mean ☊ ° '	Moon ☽ Latitude ° '
01	17 ♌ 12	17 ♌ 29	04 N 57
02	17 R 11	17 25	04 26
03	17 11	17 22	03 44
04	17 10	17 19	02 53
05	17 10	17 16	01 54
06	17 09	17 13	00 N 51
07	17 D 09	17 10	00 S 14
08	17 09	17 06	01 19
09	17 R 09	17 03	02 21
10	17 09	17 00	03 17
11	17 09	16 57	04 05
12	17 09	16 54	04 43
13	17 08	16 50	05 07
14	17 08	16 47	05 18
15	17 D 08	16 44	05 11
16	17 08	16 41	04 48
17	17 09	16 38	04 07
18	17 10	16 35	03 09
19	17 10	16 31	01 57
20	17 11	16 28	00 S 37
21	17 R 11	16 25	00 N 47
22	17 10	16 22	02 07
23	17 09	16 19	03 18
24	17 07	16 16	04 14
25	17 05	16 12	04 52
26	17 04	16 09	05 13
27	17 02	16 06	05 15
28	17 02	16 03	05 02
29	17 D 02	16 00	04 34
30	17 03	15 56	03 55
31	17 ♌ 05	15 ♌ 53	03 N 06

DECLINATIONS

Date	Sun ☉	Moon ☽	Mercury ☿	Venus ♀	Mars ♂	Jupiter ♃	Saturn ♄	Uranus ♅	Neptune ♆	Pluto ♇
01	17 N 47	16 S 44	09 N 43	19 N 33	13 N 28	22 N 20	15 N 29	22 N 45	08 N 46	23 S 01
02	17 32	18 38	09 04	19 38	13 14	22 21	15 27	22 45	08 46	23 02
03	17 16	19 40	08 26	19 43	13 01	22 22	15 25	22 45	08 46	23 02
04	17 00	19 48	07 49	19 47	12 48	22 23	15 22	22 44	08 46	23 03
05	16 43	19 03	07 12	19 52	12 34	22 24	15 20	22 44	08 46	23 04
06	16 27	17 28	06 35	19 56	12 21	22 24	15 17	22 43	08 46	23 04
07	16 10	15 08	05 59	20 00	12 07	22 25	15 15	22 43	08 45	23 05
08	15 53	12 09	05 23	20 03	11 53	22 26	15 13	22 42	08 45	23 05
09	15 35	08 41	04 48	20 06	11 39	22 27	15 10	22 42	08 45	23 06
10	15 18	04 50	04 14	20 09	11 25	22 28	15 08	22 42	08 45	23 06
11	15 00	00 S 46	03 41	20 11	11 11	22 28	15 05	22 41	08 45	23 07
12	14 42	03 N 24	03 09	20 13	10 57	22 29	15 03	22 41	08 44	23 07
13	14 23	07 31	02 38	20 15	10 43	22 30	15 00	22 41	08 44	23 08
14	14 05	11 22	02 08	20 16	10 29	22 31	14 58	22 40	08 44	23 08
15	13 46	14 47	01 39	20 17	10 15	22 31	14 55	22 40	08 44	23 09
16	13 27	17 30	01 11	20 18	10 00	22 32	14 53	22 39	08 44	23 09
17	13 08	19 16	00 45	20 18	09 46	22 33	14 51	22 39	08 43	23 10
18	12 48	19 52	00 N 21	20 17	09 31	22 33	14 48	22 39	08 43	23 10
19	12 29	19 06	00 S 02	20 17	09 17	22 34	14 46	22 38	08 43	23 11
20	12 09	17 00	00 23	20 15	09 02	22 34	14 43	22 38	08 42	23 11
21	11 49	13 41	00 42	20 13	08 48	22 35	14 41	22 37	08 42	23 11
22	11 29	09 28	00 58	20 11	08 33	22 35	14 38	22 37	08 42	23 12
23	11 08	04 N 43	01 13	20 09	08 18	22 36	14 36	22 37	08 42	23 12
24	10 48	00 S 12	01 24	20 05	08 03	22 36	14 33	22 36	08 41	23 13
25	10 27	04 58	01 33	20 02	07 49	22 37	14 31	22 36	08 41	23 13
26	10 06	09 19	01 40	19 58	07 34	22 37	14 29	22 36	08 41	23 14
27	09 45	13 04	01 43	19 53	07 19	22 38	14 26	22 35	08 40	23 14
28	09 24	16 04	01 42	19 48	07 04	22 38	14 24	22 35	08 40	23 14
29	09 02	18 14	01 39	19 42	06 49	22 39	14 21	22 35	08 39	23 15
30	08 41	19 31	01 31	19 36	06 34	22 39	14 19	22 34	08 39	23 15
31	08 N 19	19 S 53	01 S 21	19 N 30	06 N 18	22 N 39	14 N 16	22 N 34	08 N 39	23 S 16

ZODIAC SIGN ENTRIES

Date	h	m	Planets
03	07	26	☽ ♑
05	18	43	♂ ♍
05	19	54	☽ ♒
07	14	37	♀ ♋
08	08	53	☽ ♓
10	21	08	☽ ♈
13	07	20	☽ ♉
15	14	28	☽ ♊
17	18	12	☽ ♋
19	19	09	☽ ♌
21	18	48	☽ ♍
22	12	32	☉ ♍
23	18	57	☽ ♎
25	21	28	☽ ♏
28	03	38	☽ ♐
30	13	37	☽ ♑

LATITUDES

Date	Mercury ☿	Venus ♀	Mars ♂	Jupiter ♃	Saturn ♄	Uranus ♅	Neptune ♆	Pluto ♇
01	00 N 02	03 S 46	01 N 07	00 S 36	01 N 04	00 N 23	01 S 42	08 S 49
04	00 S 27	03 37	01 07	00 36	01 04	00 23	01 42	08 49
07	00 57	03 26	01 06	00 36	01 04	00 23	01 42	08 49
10	01 29	03 15	01 05	00 36	01 04	00 23	01 43	08 49
13	02 02	03 04	01 04	00 36	01 05	00 23	01 43	08 50
16	02 35	02 51	01 04	00 36	01 05	00 23	01 43	08 50
19	03 06	02 39	01 03	00 36	01 05	00 23	01 43	08 50
22	03 36	02 26	01 02	00 36	01 05	00 23	01 43	08 50
25	04 01	02 13	01 01	00 36	01 06	00 23	01 43	08 50
28	04 20	01 59	01 00	00 36	01 06	00 24	01 44	08 50
31	04 S 28	01 S 46	00 N 59	00 S 36	01 N 06	00 N 24	01 S 44	08 S 50

DATA

Julian Date	2464907
Delta T	+76 seconds
Ayanamsa	24° 21' 55"
Synetic vernal point	04° ♓ 45' 04"
True obliquity of ecliptic	23° 25' 57"

MOON'S PHASES, APSIDES AND POSITIONS ☽

Date	h	m	Phase	Longitude ° '	Eclipse Indicator
07	02	49	○	15 ♒ 12	total
15	01	36	☾	22 ♉ 49	
21	17	35	●	29 ♌ 14	Partial
28	14	43	☽	05 ♐ 53	

Day	h	m		
06	19	43	Apogee	
21	04	22	Perigee	
04	03	29	Max dec	19° S 52'
11	16	24	0N	
18	10	46	Max dec	19° N 52'
24	11	02	0S	
31	09	33	Max dec	19° S 53'

LONGITUDES

Date	Chiron ⚷	Ceres ⚳	Pallas ⚴	Juno ⚵	Vesta ⚶	Black Moon Lilith ⚸
01	17 ♊ 14	16 ♌ 39	00 ♋ 48	15 ♋ 12	28 ♏ 55	11 ♒ 57
11	17 ♊ 46	21 ♌ 10	06 ♋ 27	20 ♋ 43	01 ♐ 12	13 ♒ 03
21	18 ♊ 12	25 ♌ 43	12 ♋ 06	26 ♋ 06	04 ♐ 30	14 ♒ 10
31	18 ♊ 33	00 ♍ 16	17 ♋ 43	01 ♌ 22	07 ♐ 14	15 ♒ 16

All ephemeris data is given at 12.00 UT and the Moon's longitude is additionally given for 24.00 UT

ASPECTARIAN

h m	Aspects
01 Friday	
00 16	☽ bQ ♅
00 23	☽ # ♄
02 20	☽ ± ♆
05 44	☽ □ ☿
12 57	☽ Q ♇
16 04	☽ △ ☉
17 42	☽ ± ♅
19 44	☽ bQ ♆
22 31	☽ # ☉
02 Saturday	
05 28	☽ ⊼ ♅
09 05	☽ ☍ ♃
12 20	☽ ⚹ ♇
13 57	☽ △ ♄
22 50	☽ ☍ ♀
23 37	☿ ∥ ♆
03 Sunday	
00 00	☽ bQ ☉
01 13	☽ △ ♆
04 10	☽ △ ♂
14 00	☽ # ♀
18 05	☽ ∠ ♇
19 30	☽ ± ☉
20 06	☽ bQ ♄
04 Monday	
00 00	☽ △ ☿
06 17	♀ ⚹ ♆
08 36	☽ ⊼ ☉
11 53	☽ bQ ♂
12 08	☽ ⊥ ♇
12 53	☽ # ♀
14 23	☽ ± ♄
17 33	☽ ☍ ♅
21 55	☽ ⊼ ♃
05 Tuesday	
00 14	☽ ⊻ ♇
00 56	☿ Q ♀
02 39	☽ ⊼ ♄
07 06	☽ ± ♂
09 50	☽ bQ ☿
10 16	☽ ± ♃
13 32	☽ □ ♆
16 12	☽ ⊼ ♀
19 58	☽ ⊼ ♂
06 Wednesday	
04 49	☽ bQ ♃
05 27	☽ ± ♀
06 12	☽ ± ☿
19 46	☽ ⊼ ☿
22 17	☿ bQ ♆
07 Thursday	
00 51	☽ # ☉
01 21	☽ bQ ♀
02 07	☽ Q ♆
02 49	☽ ☍ ☉
06 46	☽ ⊼ ♅
10 52	☽ # ♄
11 48	☽ △ ♃
13 05	☽ ☌ ♇
16 17	☽ ☍ ♄
19 01	☽ ± ♅
08 Friday	
02 30	☽ ⚹ ♆
10 29	☽ △ ♀
12 24	☽ ☍ ♂
13 22	☽ bQ ♅
14 10	☽ # ♂
15 55	♂ Q ♃
09 Saturday	
06 11	☉ ⊻ ♅
08 24	♂ ∠ ♅
08 50	☽ ∠ ♆
11 34	☽ # ♆
14 54	☽ ☍ ☿
19 46	☽ △ ♅
20 53	☽ ⊼ ☉
10 Sunday	
01 20	☽ □ ♃
01 35	☽ ⊻ ♇
02 54	☽ ⊥ ♆
05 30	☽ ⊼ ♄
09 55	☽ ± ☉
13 32	☽ ⊥ ♇
14 51	☽ ⊻ ♆
16 11	☽ # ☿
17 25	♃ △ ♇
17 35	☽ ± ♄
11 Monday	
02 41	♀ ⚹ ♂
03 43	☉ ∥ ♄
03 51	☽ ⊼ ♂
03 53	☽ □ ♀
05 18	☽ bQ ☉
07 19	☽ ∠ ♇
11 32	☽ bQ ♄
13 31	☽ Q ♃
16 17	☽ ± ♂
12 Tuesday	
03 54	☿ ⚹ ♅
06 31	☉ ☍ ♇
07 14	☽ □ ♅
07 30	☽ ⊼ ☿
10 41	☽ ∥ ☿
10 43	☽ bQ ♂
12 30	☽ ⚹ ♇
13 00	☽ △ ☉
13 10	☽ ⚹ ♃
15 27	☉ ⚹ ♃
16 58	☽ △ ♄
17 57	☽ Q ♀
19 58	☽ ± ☿
21 42	♀ bQ ♇
13 Wednesday	
01 18	☽ ☌ ♆
11 20	☽ Q ♇
14 21	☽ bQ ☿
16 46	☽ △ ♂
17 41	☽ Q ♅
18 02	☽ ∠ ♃
18 37	☽ ⚹ ♀
19 25	☽ ∥ ♆
14 Thursday	
06 37	☽ ∥ ♂
10 57	☽ ⊥ ♃
15 59	☽ ⚹ ♅
20 02	☽ △ ☿
20 38	☽ □ ♇
22 04	☽ ⊻ ♃
23 53	☉ ☌ ♄
15 Friday	
00 31	☽ ∠ ♀
01 29	☽ □ ♄
01 36	☽ □ ☉
05 06	☽ ∥ ☉
05 35	☿ ⊼ ♇
08 46	☽ ⊻ ♆
13 07	☽ ∥ ♄
15 04	♀ ∠ ♄
17 57	☽ ⊥ ♀
19 05	☽ ∠ ♅
19 26	☽ ⊥ ♆
21 00	☉ ⊥ ♅
16 Saturday	
01 59	☽ □ ♂
04 16	☿ ± ♆
05 21	☽ ⊻ ♀
09 42	☽ Q ♄
10 55	☽ ⊥ ♅
11 12	☽ ∠ ♆
11 54	☽ Q ☉
13 51	♀ Q ♆
16 15	☿ □ ♃
21 18	☽ ⊻ ♅
17 Sunday	
01 22	☽ △ ♇
03 24	☽ ☌ ♃
03 51	☽ □ ☿
06 29	☽ ⚹ ♄
09 56	☽ ⚹ ☉
10 26	☽ Q ♂
11 17	☿ Q ♀
12 48	☽ ⚹ ♆
18 Monday	
02 32	☽ bQ ♇
07 21	☽ ⚹ ♂
07 46	☽ ∠ ♄
08 45	☽ Q ♆
11 17	☽ Q ☿
11 59	☽ ☌ ♀
12 39	☽ ∠ ☉
17 18	☽ ± ♇
22 37	☽ ⊥ ♄
23 26	☽ ☌ ♅
19 Tuesday	
03 03	☽ ⊼ ♇
03 58	☉ △ ♆
04 20	☽ ⊥ ☉
05 35	☽ ⊻ ♃
07 35	☽ ⚹ ☿
08 25	☽ ⊻ ♄
08 53	☽ ∠ ♂
13 58	☽ □ ♆
14 40	☽ ⊻ ☉
15 20	☽ ⊥ ♃
23 57	☽ ⊥ ♂
20 Wednesday	
05 52	☽ ∠ ♃
08 25	☽ ∠ ☿
09 56	☽ ⊻ ♂
15 47	☽ ⊻ ♀
17 49	☿ ⊻ ♄
23 08	☽ ⊥ ☿
23 34	☽ ⊻ ♅
21 Thursday	
02 00	☽ ⊥ ♀
02 52	☽ ☍ ♇
05 28	☽ ∥ ♄
05 54	☽ ⚹ ♃
08 36	☽ ☌ ♄
08 53	☽ ⊻ ☿
09 06	☽ ⊥ ♅
13 38	☽ △ ♆
14 29	♀ ± ♇
17 18	☽ ∠ ♀
17 35	☽ ☌ ☉
21 19	♂ ∥ ♆
23 25	☽ ∠ ♅
22 Friday	
00 09	☽ ∥ ☉
01 07	☽ Q ♃
11 37	☽ ☌ ♂
13 27	☽ bQ ♆
16 04	☽ ∥ ♆
17 05	☽ ∥ ♂
18 57	☽ ⚹ ♀
23 26	☽ ⚹ ♅
23 Saturday	
02 34	☽ ⊼ ♇
03 53	☽ ± ♆
06 12	☽ □ ♃
08 54	☽ ⊻ ♄
09 47	☽ ☌ ☿
12 15	☽ ± ♇
13 35	☽ ⊼ ♆
15 48	☽ Q ♀
18 45	☽ ⊥ ♄
18 57	☽ Q ♅
21 05	☽ ⊻ ☉
22 28	☉ ∠ ♀
24 Sunday	
02 58	☽ bQ ♇
04 26	☽ # ☿
06 02	♂ bQ ♆
07 41	☽ ⊥ ☉
09 42	☽ ∠ ♄
11 40	♃ ∥ ♅
14 40	☽ ⊻ ♂
18 10	☽ ∥ ☿
23 49	☽ ∠ ☉
23 59	☽ □ ♀
25 Monday	
00 58	☽ □ ♅
01 14	☽ ⊥ ♂
04 01	☽ △ ♇
06 19	♄ ⊥ ♅
08 27	☽ △ ♃
11 15	☽ ⚹ ♄
12 11	☽ ⊻ ☿
13 42	♀ ☌ ♅
14 20	♀ ⊥ ♄
15 43	☽ ☍ ♆
17 22	☽ ∠ ♂
17 22	☉ ∠ ♅
19 47	☿ St R
22 34	☽ ⊥ ☿
26 Tuesday	
02 30	☽ # ♂
03 34	☽ ⚹ ☉
08 16	☽ # ♆
08 20	☽ Q ♄
10 50	☽ bQ ♃
14 23	☽ ∠ ☿
16 17	☽ # ☉
21 08	☽ ⚹ ♂
27 Wednesday	
02 38	☽ Q ☉
03 08	☽ ± ♃
04 57	♀ ⊼ ♇
05 46	☽ △ ♅
08 47	☽ □ ♇
09 08	☽ △ ♀
10 45	☉ Q ♃
14 13	☽ ⊼ ♃
17 11	☽ □ ♄
17 20	☽ ⚹ ☿
20 10	☽ Q ♂
21 24	☽ ⊼ ♆
22 01	☽ # ♄
23 28	☿ ⊻ ♄
28 Thursday	
08 37	☽ ± ♆
09 41	☽ bQ ♅
14 43	☽ □ ☉
15 25	☽ Q ☿
15 37	☽ bQ ♀
18 20	☽ Q ♇
29 Friday	
01 43	☽ bQ ♆
02 53	☽ ± ♅
08 06	☽ □ ♂
10 31	☽ ± ♀
14 33	☽ ⊼ ♅
17 27	☽ ⚹ ♇
23 20	☽ ⊼ ♀
23 59	☽ ☍ ♃
30 Saturday	
01 17	☽ □ ☿
03 06	☽ △ ♄
06 55	☽ △ ♆
07 38	♀ ⊻ ♃
13 56	☉ ∥ ♆
14 27	☽ # ♀
15 27	☿ ⚹ ♀
23 02	☽ ∠ ♇
31 Sunday	
03 07	☿ □ ♃
06 28	☽ △ ☉
09 16	☽ bQ ♄
17 04	☽ ⊥ ♇
22 55	☽ △ ♂
23 11	♀ ⊻ ♄

SEPTEMBER 2036

LONGITUDES

Date	Sidereal time h m s	Sun ☉ ° ' "	Moon ☽ ° ' "	Moon ☽ 24.00 ° ' "	Mercury ☿ ° '	Venus ♀ ° '	Mars ♂ ° '	Jupiter ♃ ° '	Saturn ♄ ° '	Uranus ♅ ° '	Neptune ♆ ° '	Pluto ♇ ° '
01	10 44 57	09 ♍ 38 07	23 ♑ 04 25	28 ♑ 58 46	22 ♍ 25	25 ♋ 29	16 ♍ 58	23 ♊ 25	24 ♌ 59	18 ♋ 24	26 ♈ 35	19 ♒ 42
02	10 48 53	10 36 10	04 ♒ 52 31	10 ♒ 46 05	21 R 43	26 35	17 36	23 32	25 06	18 26	26 R 34	19 R 41
03	10 52 50	11 34 16	16 ♒ 39 54	22 ♒ 34 20	20 56	27 40	18 14	23 40	25 14	18 29	26 33	19 40
04	10 56 46	12 32 23	28 ♒ 29 41	04 ♓ 26 18	20 04	28 45	18 53	23 47	25 21	18 31	26 32	19 38
05	11 00 43	13 30 31	10 ♓ 24 26	16 ♓ 24 20	19 09	29 ♋ 51	19 31	23 54	25 29	18 34	26 31	19 37
06	11 04 39	14 28 41	22 ♓ 26 13	28 ♓ 30 18	18 11	00 ♌ 57	20 10	24 01	25 36	18 36	26 29	19 36
07	11 08 36	15 26 53	04 ♈ 36 46	10 ♈ 45 47	17 12	02 03	20 48	24 07	25 43	18 38	26 28	19 35
08	11 12 32	16 25 07	16 ♈ 57 32	23 ♈ 12 12	16 13	03 10	21 26	24 14	25 51	18 41	26 27	19 34
09	11 16 29	17 23 23	29 ♈ 29 56	05 ♉ 50 55	15 13	04 17	22 05	24 20	25 58	18 43	26 26	19 32
10	11 20 26	18 21 40	12 ♉ 15 21	18 ♉ 43 24	14 19	05 23	22 43	24 27	26 05	18 45	26 25	19 31
11	11 24 22	19 20 00	25 ♉ 15 18	01 ♊ 51 13	13 27	06 30	23 22	24 33	26 13	18 47	26 23	19 30
12	11 28 19	20 18 22	08 ♊ 31 21	15 ♊ 15 53	12 40	07 38	24 00	24 39	26 20	18 50	26 22	19 29
13	11 32 15	21 16 46	22 ♊ 04 58	28 ♊ 58 44	12 00	08 45	24 39	24 44	26 27	18 52	26 21	19 28
14	11 36 12	22 15 12	05 ♋ 57 13	13 ♋ 00 26	11 26	09 53	25 18	24 50	26 34	18 54	26 20	19 27
15	11 40 08	23 13 41	20 ♋ 08 16	27 ♋ 20 32	11 02	11 00	25 56	24 55	26 42	18 56	26 18	19 26
16	11 44 05	24 12 11	04 ♌ 36 55	11 ♌ 56 57	10 45	12 08	26 35	25 01	26 49	18 58	26 17	19 25
17	11 48 01	25 10 44	19 ♌ 20 04	26 ♌ 45 33	10 39	13 16	27 13	25 06	26 56	19 00	26 16	19 24
18	11 51 58	26 09 19	04 ♍ 12 33	11 ♍ 40 08	10 D 41	14 25	27 52	25 11	27 03	19 02	26 14	19 23
19	11 55 55	27 07 55	19 ♍ 07 17	26 ♍ 32 57	10 53	15 33	28 31	25 15	27 10	19 03	26 13	19 22
20	11 59 51	28 06 34	03 ♎ 56 06	11 ♎ 15 45	11 15	16 42	29 10	25 20	27 17	19 05	26 11	19 21
21	12 03 48	29 ♍ 05 14	18 ♎ 30 58	25 ♎ 40 58	11 46	17 50	29 ♍ 48	25 24	27 24	19 07	26 10	19 20
22	12 07 44	00 ♎ 03 57	02 ♏ 45 08	09 ♏ 42 58	12 26	18 59	00 ♎ 27	25 29	27 31	19 09	26 08	19 19
23	12 11 41	01 02 41	16 ♏ 34 11	23 ♏ 18 39	13 15	20 08	01 06	25 33	27 38	19 10	26 07	19 18
24	12 15 37	02 01 27	29 ♏ 56 23	06 ♐ 27 35	14 12	21 18	01 45	25 37	27 44	19 12	26 06	19 17
25	12 19 34	03 00 15	12 ♐ 52 32	19 ♐ 11 40	15 15	22 27	02 23	25 40	27 51	19 14	26 04	19 16
26	12 23 30	03 59 05	25 ♐ 25 28	01 ♑ 34 29	16 26	23 36	03 02	25 44	27 58	19 15	26 02	19 15
27	12 27 27	04 57 56	07 ♑ 39 21	13 ♑ 40 42	17 42	24 46	03 41	25 47	28 05	19 17	26 01	19 14
28	12 31 24	05 56 49	19 ♑ 39 11	25 ♑ 35 29	19 04	25 56	04 20	25 50	28 11	19 18	25 59	19 13
29	12 35 20	06 55 44	01 ♒ 30 13	07 ♒ 24 03	20 31	27 06	04 59	25 53	28 18	19 19	25 58	19 13
30	12 39 17	07 ♎ 54 40	13 ♒ 17 35	19 ♒ 11 24	22 ♍ 02	28 ♌ 16	05 ♎ 38	25 ♊ 56	28 ♌ 24	19 ♋ 21	25 ♈ 56	19 ♒ 12

DECLINATIONS

Date	Moon True ☊ ° '	Moon Mean ☊ ° '	Moon ☽ Latitude ° '	Sun ☉ ° '	Moon ☽ ° '	Mercury ☿ ° '	Venus ♀ ° '	Mars ♂ ° '	Jupiter ♃ ° '	Saturn ♄ ° '	Uranus ♅ ° '	Neptune ♆ ° '	Pluto ♇ ° '
01	17 ♌ 06	15 ♌ 50	02 N 09	07 N 57	19 S 20	01 S 06	19 N 23	06 N 03	22 N 40	14 N 14	22 N 34	08 N 38	23 S 16
02	17 D 08	15 47	01 07	07 35	17 57	00 48	19 15	05 48	22 40	14 12	22 33	08 38	23 16
03	17 08	15 44	00 N 03	07 13	15 48	00 26	19 07	05 33	22 41	14 09	22 33	08 37	23 17
04	17 R 08	15 41	01 S 02	06 51	12 58	00 S 01	18 58	05 17	22 41	14 07	22 33	08 37	23 17
05	17 06	15 37	02 05	06 29	09 35	00 N 27	18 49	05 02	22 41	14 04	22 32	08 37	23 18
06	17 03	15 34	03 02	06 07	05 47	00 58	18 39	04 47	22 42	14 02	22 32	08 36	23 18
07	16 58	15 31	03 52	05 44	01 S 43	01 32	18 29	04 31	22 42	13 59	22 32	08 36	23 18
08	16 54	15 28	04 31	05 21	02 N 29	02 07	18 19	04 16	22 42	13 57	22 31	08 35	23 19
09	16 49	15 25	04 59	04 59	06 38	02 43	18 07	04 00	22 42	13 55	22 31	08 35	23 19
10	16 44	15 22	05 12	04 36	10 34	03 19	17 56	03 45	22 43	13 52	22 31	08 34	23 19
11	16 41	15 18	05 09	04 13	14 04	03 55	17 44	03 29	22 43	13 50	22 31	08 34	23 20
12	16 38	15 15	04 50	03 50	16 56	04 30	17 31	03 14	22 43	13 48	22 30	08 33	23 20
13	16 D 38	15 12	04 15	03 27	18 57	05 03	17 18	02 58	22 43	13 45	22 30	08 33	23 20
14	16 39	15 09	03 24	03 04	19 54	05 33	17 04	02 43	22 44	13 43	22 30	08 32	23 20
15	16 40	15 06	02 20	02 41	19 37	06 01	16 50	02 27	22 44	13 41	22 30	08 32	23 21
16	16 41	15 02	01 S 05	02 18	18 03	06 25	16 35	02 11	22 44	13 38	22 29	08 31	23 21
17	16 R 42	14 59	00 N 14	01 55	15 15	06 46	16 20	01 56	22 44	13 36	22 29	08 31	23 21
18	16 40	14 56	01 34	01 32	11 25	07 02	16 05	01 40	22 44	13 34	22 29	08 30	23 22
19	16 37	14 53	02 47	01 08	06 52	07 14	15 49	01 24	22 44	13 31	22 28	08 30	23 22
20	16 32	14 50	03 48	00 45	01 N 56	07 22	15 32	01 09	22 45	13 29	22 28	08 29	23 22
21	16 26	14 47	04 34	00 N 22	03 S 02	07 25	15 15	00 53	22 45	13 27	22 28	08 29	23 22
22	16 19	14 43	05 01	00 S 02	07 43	07 23	14 58	00 37	22 45	13 25	22 28	08 28	23 23
23	16 12	14 40	05 09	00 25	11 51	07 17	14 40	00 21	22 45	13 22	22 28	08 28	23 23
24	16 07	14 37	05 00	00 48	15 14	07 07	14 22	00 N 06	22 45	13 20	22 28	08 27	23 23
25	16 03	14 34	04 36	01 12	17 46	06 52	14 03	00 S 10	22 45	13 18	22 27	08 26	23 23
26	16 01	14 31	03 59	01 35	19 22	06 34	13 44	00 26	22 45	13 16	22 27	08 26	23 23
27	16 D 00	14 27	03 12	01 58	20 01	06 12	13 24	00 42	22 45	13 14	22 27	08 25	23 23
28	16 01	14 24	02 18	02 22	19 43	05 47	13 04	00 57	22 46	13 11	22 27	08 25	23 24
29	16 02	14 21	01 18	02 45	18 33	05 18	12 44	01 13	22 46	13 09	22 27	08 24	23 24
30	16 ♌ 03	14 ♌ 18	00 N 15	03 S 08	16 S 35	04 N 47	12 N 23	01 S 29	22 N 46	13 N 07	22 N 26	08 N 23	23 S 24

ZODIAC SIGN ENTRIES

Date	h	m	Planets
02	02	05	☽ ♒
04	15	03	☽ ♓
05	15	12	♀ ♌
07	02	57	☽ ♈
09	12	57	☽ ♉
11	20	39	☽ ♊
14	01	46	☽ ♋
16	04	24	☽ ♌
18	05	13	☽ ♍
20	05	36	☽ ♎
21	19	16	♂ ♎
22	07	18	☽ ♏
22	10	23	☉ ♎
24	12	07	☽ ♐
26	20	55	☽ ♑
29	08	57	☽ ♒

LATITUDES

Date	Mercury ☿ ° '	Venus ♀ ° '	Mars ♂ ° '	Jupiter ♃ ° '	Saturn ♄ ° '	Uranus ♅ ° '	Neptune ♆ ° '	Pluto ♇ ° '
01	04 S 28	01 S 41	00 N 59	00 S 36	01 N 06	00 N 24	01 S 44	08 S 50
04	04 17	01 28	00 58	00 36	01 07	00 24	01 44	08 50
07	03 50	01 14	00 57	00 36	01 07	00 24	01 44	08 50
10	03 06	01 01	00 56	00 36	01 07	00 24	01 44	08 50
13	02 11	00 48	00 55	00 36	01 08	00 24	01 44	08 49
16	01 12	00 35	00 54	00 36	01 08	00 24	01 44	08 49
19	00 S 16	00 23	00 53	00 36	01 08	00 24	01 45	08 49
22	00 N 32	00 S 10	00 52	00 36	01 09	00 24	01 45	08 49
25	01 09	00 N 02	00 51	00 36	01 09	00 24	01 45	08 48
28	01 35	00 13	00 50	00 36	01 10	00 24	01 45	08 48
31	01 N 49	00 N 24	00 N 49	00 S 36	01 N 10	00 N 24	01 S 45	08 S 48

DATA

Julian Date	2464938
Delta T	+76 seconds
Ayanamsa	24° 21' 58"
Synetic vernal point	04° ♓ 45' 00"
True obliquity of ecliptic	23° 25' 58"

MOON'S PHASES, APSIDES AND POSITIONS ☽

Date	h	m	Phase	Longitude ° '	Eclipse Indicator
05	18	46	○	13 ♓ 47	
13	10	29	☾	21 ♊ 13	
20	01	52	●	27 ♍ 42	
27	06	12	☽	04 ♑ 44	

Day	h	m		
03	00	21	Apogee	
18	12	29	Perigee	
30	13	12	Apogee	
07	21	50	0N	
14	18	43	Max dec	19° N 57'
20	21	15	0S	
27	16	26	Max dec	20° S 02'

LONGITUDES

Date	Chiron ⚷ ° '	Ceres ⚳ ° '	Pallas ⚴ ° '	Juno ⚵ ° '	Vesta ⚶ ° '	Black Moon Lilith ⚸ ° '
01	18 ♊ 35	00 ♍ 43	18 ♋ 17	01 ♌ 53	07 ♐ 35	15 ♒ 23
11	18 ♊ 48	05 ♍ 16	23 ♋ 29	06 ♌ 58	11 ♐ 12	16 ♒ 29
21	18 ♊ 54	09 ♍ 48	29 ♋ 18	11 ♌ 53	15 ♐ 07	17 ♒ 36
31	18 ♊ 53	14 ♍ 18	04 ♌ 37	16 ♌ 36	19 ♐ 17	18 ♒ 42

All ephemeris data is given at 12.00 UT and the Moon's longitude is additionally given for 24.00 UT

ASPECTARIAN

h m	Aspects
01 Monday	
02 29	☽ ☍ ♅
03 37	☽ ± ♄
05 10	☽ ⊻ ♇
10 45	☽ △ ☿
10 59	☽ # ♀
12 42	☽ ⊼ ♃
15 27	☽ bQ ☉
15 55	☽ ⊼ ♄
17 24	☽ ☍ ♀
19 07	☽ □ ♆
02 Tuesday	
01 01	☽ ± ♃
07 06	☽ bQ ♂
11 24	☽ ± ☉
11 45	♀ □ ♆
15 32	☽ bQ ☿
19 32	☽ bQ ♃
03 Wednesday	
00 42	☽ ⊼ ☉
02 29	☽ ± ♂
07 42	☽ Q ♆
08 42	☽ ± ☿
11 24	☉ bQ ♆
15 23	☽ ⊼ ♂
15 42	☽ ⊼ ♅
18 04	☽ ☌ ♇
20 06	☽ ⊼ ☿
21 29	♂ ✱ ♅
23 09	☿ ± ♆
04 Thursday	
02 21	☽ △ ♃
02 46	☽ # ♄
03 55	☽ ± ♅
05 34	☽ ☍ ♄
08 02	☽ ✱ ♆
12 35	☽ ⊼ ♀
22 11	☽ bQ ♅
23 38	☿ ⊼ ♇
05 Friday	
01 55	☽ ± ♀
06 22	☿ ☌ ♂
13 02	♀ ⊥ ♃
14 12	☽ ∠ ♆
15 36	♂ ⊼ ♇
18 24	☽ # ♆
18 46	☽ ☍ ☉
21 48	☽ bQ ♀
06 Saturday	
02 12	☿ ✱ ♅
04 11	☽ ☍ ☿
04 21	☽ △ ♅
06 22	☽ ⊻ ♇
07 13	☽ ☍ ♂
08 09	☽ ⊥ ♆
09 50	☽ # ☉
15 09	☽ □ ♃
18 15	☽ ⊥ ♇
18 20	☽ ⊼ ♄
18 28	☽ # ♂
20 01	☽ ⊻ ♆
07 Sunday	
00 04	♂ ± ♆
06 17	☽ ± ♄
06 30	☽ △ ♀
11 56	☽ ∠ ♇
12 56	☽ # ☿
13 37	☿ ∠ ♀
08 Monday	
00 03	☽ bQ ♄
02 46	☽ Q ♃
09 26	☉ ☌ ☿
09 32	☽ ∥ ☿
10 40	☽ ⊼ ☿
10 52	☽ ⊼ ☉
15 19	☽ □ ♅
17 00	☽ ✱ ♇
21 05	☽ ⊼ ♂
21 22	☽ ± ☿
21 38	☽ ∥ ♂
23 23	☽ ± ☉
09 Tuesday	
02 05	☽ ✱ ♃
03 10	☽ ∥ ☉
05 13	☽ △ ♄
06 11	☽ ☌ ♆
09 10	☽ ± ♂
13 18	☽ bQ ☿
15 52	☽ Q ♇
17 56	☽ bQ ☉
21 54	☽ □ ♀
23 37	☽ ∥ ♆
10 Wednesday	
01 40	☽ Q ♅
03 04	☽ bQ ♂
06 42	☽ ∠ ♃
15 35	☽ △ ☿
22 03	☉ ✱ ♅
23 34	☽ ⊥ ♃
11 Thursday	
00 05	☽ ✱ ♅
00 06	☿ ∥ ♂
00 15	☽ △ ☉
01 27	☽ □ ♇
08 22	☽ △ ♂
10 17	☽ ∥ ♄
10 30	☽ Q ♀
10 41	☽ ⊻ ♃
13 46	☽ □ ♄
14 04	☽ ⊻ ♆
16 05	☉ ⊼ ♇
19 36	☉ ∥ ☿
23 44	☿ ⊥ ♀
12 Friday	
00 57	☽ ⊥ ♆
03 32	☽ ∠ ♅
10 15	☽ ✱ ♀
13 33	☉ ± ♆
17 05	☽ ∠ ♆
17 18	☽ ∥ ♀
18 22	♄ △ ♆
19 01	☽ □ ☿
19 42	☽ ⊥ ♅
22 27	☽ Q ♄
13 Saturday	
06 20	☽ ⊻ ♅
07 25	☽ △ ♇
10 29	☽ □ ☉
15 10	☽ ∠ ♀
15 55	♂ □ ♃
16 40	☽ ☌ ♃
16 42	☽ □ ♂
19 26	☽ ✱ ♆
19 41	☽ ✱ ♄
14 Sunday	
01 12	☽ Q ☿
08 08	☽ ⊥ ♀
09 25	☽ bQ ♇
10 59	♀ ∠ ♃
16 03	☽ Q ♆
17 37	♂ ± ♇
18 07	☿ bQ ♆
19 16	☽ ⊻ ♀
19 52	☽ Q ☉
21 03	☽ ✱ ☿
21 39	☽ ∠ ♄
15 Monday	
00 44	☽ ± ♇
01 05	☽ Q ♂
08 33	♀ ∠ ♂
09 58	☽ ☌ ♅
10 49	☽ ⊼ ♇
12 19	☿ ⊻ ♀
12 56	☽ ⊥ ♄
17 32	☽ ✱ ☉
20 02	☽ ⊻ ♃
21 36	☽ ∠ ☿
22 07	☽ ✱ ♂
22 16	☽ □ ♆
23 01	☽ ⊻ ♄
16 Tuesday	
01 18	♂ ⊼ ♆
06 02	☽ ⊥ ♃
12 14	☽ ⊥ ☿
20 03	☽ ∠ ☉
20 53	☽ ∠ ♃
21 57	☽ ⊻ ☿
22 35	♂ ⊻ ♄
23 55	☽ ∠ ♂
17 Wednesday	
01 20	☽ ☌ ♀
02 58	☽ ∥ ♀
09 25	☉ ∥ ♂
09 46	☉ □ ♃
11 27	☽ ⊻ ♅
11 44	☽ ⊥ ☉
12 06	☽ ☍ ♇
15 12	☽ ⊥ ♂
17 12	☉ ± ♇
17 20	☿ St D
21 10	☽ ⊥ ♅
21 22	☽ ✱ ♃
22 07	☽ ⊻ ☉
23 09	☽ ∥ ♄
23 11	☽ △ ♆
18 Thursday	
00 22	☽ ☌ ♄
01 20	☽ ⊻ ♂
11 42	☽ ∠ ♅
13 58	☉ ⊼ ♆
16 48	☽ Q ♃
22 32	☽ ☌ ☿
23 17	☽ bQ ♆
19 Friday	
03 44	☽ ∥ ♆
05 46	☽ ⊻ ♀
10 12	☽ ∥ ☿
11 54	☽ ✱ ♅
12 23	☽ ⊼ ♇
12 54	☉ ⊻ ♄
13 46	☽ ± ♆
16 15	☽ ⊥ ♀
21 58	☽ □ ♃
22 04	☽ ± ♇
23 26	☽ ⊼ ♆
20 Saturday	
01 06	☽ ⊻ ♄
01 52	☽ ☌ ☉
03 53	☽ ☌ ♂
07 21	☽ Q ♅
08 03	☽ ∠ ♀
08 34	☿ bQ ♆
10 55	☽ ⊥ ♄
12 40	☽ bQ ♇
15 59	☽ ∥ ♂
18 07	☽ ∥ ☉
23 57	☽ # ☉
21 Sunday	
00 24	☽ ⊻ ☿
01 47	☽ ∠ ♄
02 02	☽ # ♂
10 43	☽ ⊥ ☿
10 47	☽ ✱ ♀
13 00	☽ □ ♅
13 21	☽ △ ♇
23 36	☽ △ ♃
22 Monday	
00 48	☽ ☍ ♆
02 29	☽ ∠ ☿
03 01	☽ ✱ ♄
07 05	☽ ⊻ ☉
07 53	☽ ⊻ ♂
08 43	☽ Q ♀
10 16	☽ # ☿
15 20	♀ ⊻ ♅
16 07	☽ # ♆
18 07	☽ ⊥ ☉
18 38	♀ ☍ ♇
18 40	☽ ⊥ ♂
23 44	☽ Q ♄
23 Tuesday	
01 23	☽ bQ ♃
05 46	☽ ✱ ☿
09 51	☉ # ♂
11 00	☽ ∠ ☉
11 07	☽ ∠ ♂
14 59	♂ Q ♅
15 15	☉ Q ♅
15 45	☉ ☌ ♂
16 37	☽ △ ♅
16 49	☽ □ ♇
17 18	☽ ± ♃
18 55	☽ □ ♀
22 01	☽ # ♄
24 Wednesday	
04 06	☽ ⊼ ♃
04 39	☽ Q ☿
05 01	☽ ⊼ ♆
05 49	☽ # ♀
07 58	☽ □ ♄
15 28	☽ ✱ ♂
15 56	☽ ± ♆
16 07	☽ ✱ ☉
19 50	☽ bQ ♅
25 Thursday	
01 31	☽ Q ♇
08 36	☽ bQ ♆
12 40	☽ ± ♅
15 01	☽ Q ♂
16 22	☽ Q ☉
16 57	☽ □ ☿
26 Friday	
00 05	☽ ⊼ ♅
00 07	☽ ✱ ♇
08 07	☽ △ ♀
11 24	♅ ⊼ ♇
11 26	☉ ⊥ ♄
12 36	☽ ☍ ♃
13 12	☽ △ ♆
16 59	☽ △ ♄
18 26	☉ bQ ♇
27 Saturday	
03 43	☽ □ ♂
05 15	☽ ∠ ♇
06 12	☽ □ ☉
16 39	☽ bQ ♀
22 44	♀ ⊥ ♅
22 54	☽ bQ ♄
23 06	☽ ⊥ ♇
28 Sunday	
02 40	♀ ∥ ♄
05 21	♂ ⊥ ♄
07 57	♂ bQ ♇
09 59	♀ ✱ ♃
10 40	☽ △ ☿
11 08	☽ ⊻ ♇
11 17	☽ ☍ ♅
12 37	☽ ± ♀
13 13	♀ △ ♆
14 32	☿ ⊼ ♇
15 57	☿ ✱ ♅
17 09	☽ ± ♄
29 Monday	
00 33	☽ ⊼ ♃
00 47	☽ □ ♆
02 04	☽ ⊼ ♀
03 08	☿ ± ♆
05 25	☽ ⊼ ♄
12 47	☽ ± ♃
19 29	☽ △ ♂
21 21	☽ bQ ☿
30 Tuesday	
00 03	☽ △ ☉
07 10	☽ bQ ♃
13 19	☽ Q ♆
14 17	♃ ✱ ♆
15 13	♀ ☌ ♄
18 24	☽ ± ☿

OCTOBER 2036

LONGITUDES

Date	Sidereal time h m s	Sun ☉ ° ' "	Moon ☽ ° ' "	Moon ☽ 24.00 ° ' "	Mercury ☿ ° '	Venus ♀ ° '	Mars ♂ ° '	Jupiter ♃ ° '	Saturn ♄ ° '	Uranus ♅ ° '	Neptune ♆ ° '	Pluto ♇ ° '
01	12 43 13	08 ♎ 53 38	25 ♒ 06 01	01 ♓ 01 57	23 ♍ 36	29 ♌ 26	06 ♎ 17	25 ♊ 58	28 ♌ 31	19 ♋ 22	25 ♈ 55	19 ♒ 11
02	12 47 10	09 52 38	06 ♓ 59 38	12 ♓ 59 27	25 13	00 ♍ 36	06 56	26 01	28 37	19 23	25 R 53	19 R 10
03	12 51 06	10 51 40	19 ♓ 01 45	25 ♓ 06 47	26 52	01 47	07 35	26 03	28 44	19 24	25 52	19 09
04	12 55 03	11 50 44	01 ♈ 14 48	07 ♈ 25 55	28 ♍ 33	02 57	08 14	26 05	28 50	19 26	25 50	19 09
05	12 58 59	12 49 50	13 ♈ 40 16	19 ♈ 57 54	00 ♎ 16	04 08	08 53	26 07	28 56	19 27	25 48	19 08
06	13 02 56	13 48 58	26 ♈ 18 49	02 ♉ 42 59	02 00	05 19	09 32	26 08	29 02	19 28	25 47	19 07
07	13 06 53	14 48 08	09 ♉ 10 22	15 ♉ 40 51	03 44	06 29	10 11	26 10	29 08	19 29	25 45	19 07
08	13 10 49	15 47 20	22 ♉ 14 24	28 ♉ 50 53	05 29	07 40	10 50	26 11	29 15	19 30	25 43	19 06
09	13 14 46	16 46 34	05 ♊ 30 16	12 ♊ 12 28	07 15	08 52	11 30	26 12	29 21	19 30	25 42	19 06
10	13 18 42	17 45 51	18 ♊ 57 27	25 ♊ 45 13	09 01	10 03	12 09	26 13	29 27	19 31	25 40	19 05
11	13 22 39	18 45 10	02 ♋ 35 45	09 ♋ 29 04	10 46	11 14	12 48	26 13	29 32	19 32	25 38	19 04
12	13 26 35	19 44 32	16 ♋ 25 12	23 ♋ 24 09	12 32	12 26	13 27	26 14	29 38	19 33	25 37	19 04
13	13 30 32	20 43 55	00 ♌ 25 54	07 ♌ 30 25	14 17	13 37	14 07	26 14	29 44	19 33	25 35	19 03
14	13 34 28	21 43 21	14 ♌ 37 33	21 ♌ 47 09	16 02	14 49	14 46	26 R 14	29 50	19 34	25 33	19 03
15	13 38 25	22 42 50	28 ♌ 58 56	06 ♍ 12 31	17 46	16 01	15 25	26 14	29 ♌ 55	19 34	25 32	19 02
16	13 42 22	23 42 20	13 ♍ 27 26	20 ♍ 43 05	19 30	17 13	16 05	26 14	00 ♍ 01	19 35	25 30	19 02
17	13 46 18	24 41 53	27 ♍ 58 48	05 ♎ 13 49	21 14	18 25	16 44	26 13	00 07	19 35	25 28	19 02
18	13 50 15	25 41 28	12 ♎ 27 21	19 ♎ 38 33	22 57	19 37	17 24	26 12	00 12	19 36	25 27	19 01
19	13 54 11	26 41 05	26 ♎ 46 37	03 ♏ 50 49	24 39	20 49	18 03	26 11	00 17	19 36	25 25	19 01
20	13 58 08	27 40 44	10 ♏ 50 26	17 ♏ 44 56	26 21	22 01	18 43	26 10	00 23	19 36	25 23	19 01
21	14 02 04	28 40 26	24 ♏ 33 51	01 ♐ 16 55	28 02	23 13	19 22	26 09	00 28	19 37	25 22	19 00
22	14 06 01	29 ♎ 40 09	07 ♐ 53 59	14 ♐ 25 01	29 ♎ 42	24 26	20 02	26 07	00 33	19 37	25 20	19 00
23	14 09 57	00 ♏ 39 53	20 ♐ 50 12	27 ♐ 09 46	01 ♏ 22	25 38	20 41	26 05	00 38	19 37	25 18	19 00
24	14 13 54	01 39 40	03 ♑ 24 06	09 ♑ 33 40	03 02	26 51	21 21	26 03	00 43	19 37	25 17	19 00
25	14 17 51	02 39 28	15 ♑ 39 00	21 ♑ 40 43	04 40	28 04	22 00	26 01	00 48	19 R 37	25 15	18 59
26	14 21 47	03 39 18	27 ♑ 39 28	03 ♒ 35 55	06 19	29 ♍ 16	22 40	25 58	00 53	19 37	25 13	18 59
27	14 25 44	04 39 10	09 ♒ 30 46	15 ♒ 24 44	07 56	00 ♎ 29	23 20	25 56	00 58	19 37	25 12	18 59
28	14 29 40	05 39 03	21 ♒ 18 29	27 ♒ 12 44	09 33	01 42	23 59	25 53	01 02	19 37	25 10	18 59
29	14 33 37	06 38 58	03 ♓ 08 07	09 ♓ 05 15	11 10	02 55	24 39	25 50	01 07	19 36	25 08	18 59
30	14 37 33	07 38 55	15 ♓ 04 43	21 ♓ 07 03	12 46	04 08	25 19	25 47	01 11	19 36	25 07	18 59
31	14 41 30	08 ♏ 38 53	27 ♓ 12 41	03 ♈ 22 01	14 ♏ 22	05 ♎ 21	25 ♎ 59	25 ♊ 44	01 ♍ 16	19 ♋ 36	25 ♈ 05	18 ♒ 59

Date	Moon True ☊ ° '	Moon Mean ☊ ° '	Moon ☽ Latitude ° '
01	16 ♌ 03	14 ♌ 15	00 S 49
02	16 R 00	14 12	01 50
03	15 56	14 08	02 48
04	15 49	14 05	03 38
05	15 39	14 02	04 19
06	15 29	13 59	04 48
07	15 18	13 56	05 03
08	15 08	13 53	05 02
09	15 00	13 49	04 45
10	14 54	13 46	04 13
11	14 51	13 43	03 25
12	14 50	13 40	02 25
13	14 D 50	13 37	01 16
14	14 50	13 33	00 S 01
15	14 R 50	13 30	01 N 15
16	14 47	13 27	02 26
17	14 41	13 24	03 28
18	14 33	13 21	04 16
19	14 22	13 18	04 48
20	14 10	13 14	05 01
21	13 59	13 11	04 57
22	13 48	13 08	04 36
23	13 40	13 05	04 01
24	13 35	13 02	03 16
25	13 31	12 59	02 22
26	13 30	12 55	01 23
27	13 D 30	12 52	00 N 21
28	13 R 30	12 49	00 S 41
29	13 29	12 46	01 42
30	13 25	12 43	02 39
31	13 ♌ 19	12 ♌ 39	03 S 29

DECLINATIONS

Date	Sun ☉ ° '	Moon ☽ ° '	Mercury ☿ ° '	Venus ♀ ° '	Mars ♂ ° '	Jupiter ♃ ° '	Saturn ♄ ° '	Uranus ♅ ° '	Neptune ♆ ° '	Pluto ♇ ° '
01	03 S 32	13 S 55	04 N 13	12 N 02	01 S 45	22 N 46	13 N 05	22 N 26	08 N 23	23 S 24
02	03 55	10 39	03 37	11 41	02 00	22 46	13 03	22 26	08 22	23 24
03	04 18	06 55	02 59	11 19	02 16	22 46	13 01	22 26	08 22	23 24
04	04 41	02 S 50	02 20	10 57	02 32	22 46	12 59	22 26	08 21	23 24
05	05 04	01 N 25	01 39	10 34	02 48	22 46	12 57	22 26	08 20	23 25
06	05 27	05 41	00 57	10 11	03 03	22 46	12 55	22 26	08 20	23 25
07	05 50	09 46	00 N 15	09 48	03 19	22 46	12 53	22 26	08 19	23 25
08	06 13	13 27	00 S 29	09 25	03 35	22 46	12 51	22 25	08 19	23 25
09	06 35	16 32	01 13	09 01	03 50	22 46	12 49	22 25	08 18	23 25
10	06 58	18 47	01 58	08 37	04 06	22 46	12 47	22 25	08 17	23 25
11	07 21	19 59	02 42	08 12	04 22	22 46	12 45	22 25	08 17	23 25
12	07 43	20 01	03 27	07 48	04 37	22 46	12 43	22 25	08 16	23 25
13	08 06	18 49	04 12	07 23	04 53	22 46	12 41	22 25	08 16	23 25
14	08 28	16 25	04 57	06 58	05 09	22 46	12 39	22 25	08 15	23 25
15	08 50	12 59	05 41	06 32	05 24	22 46	12 37	22 25	08 14	23 25
16	09 12	08 45	06 26	06 06	05 40	22 46	12 36	22 25	08 14	23 25
17	09 34	03 N 59	07 10	05 41	05 55	22 46	12 34	22 25	08 13	23 25
18	09 56	00 S 59	07 53	05 14	06 11	22 46	12 32	22 25	08 12	23 25
19	10 17	05 51	08 37	04 48	06 26	22 46	12 30	22 25	08 12	23 25
20	10 39	10 18	09 19	04 22	06 41	22 46	12 29	22 25	08 11	23 25
21	11 00	14 06	10 02	03 55	06 57	22 46	12 27	22 25	08 11	23 25
22	11 21	17 05	10 43	03 28	07 12	22 46	12 25	22 25	08 10	23 25
23	11 42	19 06	11 24	03 01	07 27	22 46	12 24	22 25	08 09	23 25
24	12 03	20 08	12 04	02 34	07 43	22 46	12 22	22 25	08 09	23 25
25	12 23	20 10	12 44	02 07	07 58	22 46	12 21	22 25	08 08	23 25
26	12 44	19 16	13 22	01 39	08 13	22 46	12 19	22 25	08 08	23 24
27	13 04	17 32	14 01	01 12	08 28	22 46	12 18	22 25	08 07	23 24
28	13 24	15 03	14 39	00 44	08 43	22 46	12 16	22 25	08 06	23 24
29	13 44	11 56	15 16	00 N 17	08 58	22 46	12 15	22 25	08 06	23 24
30	14 03	08 19	15 52	00 S 11	09 13	22 46	12 13	22 25	08 05	23 24
31	14 S 23	04 S 19	16 S 27	00 S 39	09 S 28	22 N 46	12 N 12	22 N 25	08 N 05	23 S 24

ZODIAC SIGN ENTRIES

Date	h m	Planets
01	21 55	☽ ♓
01	23 39	♀ ♍
04	09 34	☽ ♈
05	08 21	☿ ♎
06	18 55	☽ ♉
09	02 05	☽ ♊
11	07 27	☽ ♋
13	11 16	☽ ♌
15	13 42	☽ ♍
16	07 34	♄ ♍
17	15 20	☽ ♎
19	17 27	☽ ♏
21	21 42	☽ ♐
22	16 16	☿ ♏
22	19 59	☉ ♏
24	05 26	☽ ♑
26	16 43	☽ ♒
27	02 21	♀ ♎
29	05 39	☽ ♓
31	17 27	☽ ♈

LATITUDES

Date	Mercury ☿ ° '	Venus ♀ ° '	Mars ♂ ° '	Jupiter ♃ ° '	Saturn ♄ ° '	Uranus ♅ ° '	Neptune ♆ ° '	Pluto ♇ ° '
01	01 N 49	00 N 24	00 N 49	00 S 36	01 N 10	00 N 24	01 S 45	08 S 48
04	01 55	00 34	00 48	00 37	01 11	00 25	01 45	08 47
07	01 53	00 44	00 47	00 37	01 11	00 25	01 45	08 47
10	01 45	00 53	00 45	00 36	01 11	00 25	01 45	08 47
13	01 33	01 01	00 44	00 36	01 12	00 25	01 45	08 46
16	01 18	01 09	00 43	00 36	01 13	00 25	01 45	08 46
19	01 00	01 16	00 42	00 36	01 13	00 25	01 45	08 45
22	00 41	01 22	00 40	00 36	01 14	00 25	01 45	08 45
25	00 22	01 28	00 39	00 36	01 14	00 25	01 45	08 44
28	00 N 01	01 33	00 38	00 36	01 15	00 25	01 45	08 44
31	00 S 19	01 N 37	00 N 37	00 S 36	01 N 15	00 N 25	01 S 45	08 S 43

DATA

Julian Date	2464968
Delta T	+76 seconds
Ayanamsa	24° 22' 01"
Synetic vernal point	04° ♓ 44' 58"
True obliquity of ecliptic	23° 25' 58"

MOON'S PHASES, APSIDES AND POSITIONS ☽

Date	h m	Phase	Longitude ° '	Eclipse Indicator
05	10 15	○	12 ♈ 46	
12	18 09	☾	20 ♋ 00	
19	11 50	●	26 ♎ 41	
27	01 14	☽	04 ♒ 12	

Day	h m	
16	11 54	Perigee
28	07 32	Apogee

Day	h m		
05	04 04	0N	
12	00 40	Max dec	20° N 09'
18	07 15	0S	
25	00 49	Max dec	20° S 16'

LONGITUDES

Date	Chiron ⚷ ° '	Ceres ⚳ ° '	Pallas ⚴ ° '	Juno ⚵ ° '	Vesta ⚶ ° '	Black Moon Lilith ⚸ ° '
01	18 ♊ 53	14 ♍ 18	04 ♌ 37	16 ♌ 36	19 ♐ 17	18 ♒ 42
11	18 ♊ 45	18 ♍ 45	09 ♌ 44	21 ♌ 05	23 ♐ 39	19 ♒ 48
21	18 ♊ 30	23 ♍ 08	14 ♌ 36	25 ♌ 19	28 ♐ 12	20 ♒ 55
31	18 ♊ 09	27 ♍ 26	19 ♌ 09	29 ♌ 15	02 ♑ 53	22 ♒ 01

All ephemeris data is given at 12.00 UT and the Moon's longitude is additionally given for 24.00 UT

ASPECTARIAN

h m	Aspects
01 Wednesday	
00 00	☽ ☌ ♇
00 20	☽ ⊼ ♅
03 48	☽ ⚼ ♂
08 28	☽ ⊼ ☿
09 20	☽ ⚼ ☉
12 32	☽ ± ♅
13 38	☽ ⚹ ♆
13 47	☽ △ ♃
18 34	☽ ⋕ ♄
18 58	☽ ☍ ♄
21 43	☽ ☍ ♀
23 05	☽ ± ♂
02 Thursday	
03 58	☽ ⋕ ♀
04 58	☉ ⋕ ☿
05 10	☽ ± ☉
06 45	☽ ⚼ ♅
11 27	☿ ± ♇
11 52	☽ ⊼ ♂
18 18	☽ ⊼ ☉
19 47	☽ ∠ ♆
21 43	☿ ⊼ ♆
03 Friday	
00 01	☿ □ ♃
02 58	☽ ⋕ ♆
03 07	♀ ⊥ ♂
12 15	☽ ⊻ ♇
12 45	☽ △ ♅
13 38	☽ ⊥ ♆
04 Saturday	
00 05	☽ ⊥ ♇
01 26	☽ ⊻ ♆
01 53	☽ □ ♃
02 14	☽ ∥ ☉
05 54	☽ ☍ ☿
06 52	☿ ⋕ ♂
07 15	☽ ⊼ ♄
13 39	☽ ∥ ♂
15 25	☽ ⋕ ☿
15 40	☽ ⊼ ♀
16 12	☿ ⊻ ♄
17 38	☽ ∠ ♇
19 01	☽ ± ♄
05 Sunday	
02 18	☽ ☍ ♂
04 29	☽ ± ♀
10 15	☽ ☍ ☉
12 31	☽ ⚼ ♄
12 51	☽ Q ♃
13 09	☽ ∥ ☿
18 30	♀ ∠ ♅
20 13	☽ ⋕ ♂
22 25	☽ ⚹ ♇
23 01	☽ □ ♅
23 29	☽ ⚼ ♀
06 Monday	
04 34	☿ Q ♅
10 33	☽ ⋕ ☉
11 00	☽ ☌ ♆
11 40	☽ ⚹ ♃
17 10	☽ △ ♄
18 02	☉ ∠ ♄
21 01	☽ Q ♇
07 Tuesday	
00 19	☽ ⊼ ☿
03 20	☽ ∥ ♆
06 32	☽ △ ♀
08 51	☽ Q ♅
12 13	☽ ∥ ♀
13 12	☽ ± ☿
13 59	☽ ⊼ ♂
15 41	☽ ∠ ♃
17 07	☿ ⚼ ♇
23 14	☽ ⊼ ☉
08 Wednesday	
01 37	☽ ± ♂
06 17	☽ □ ♇
06 59	☽ ⚹ ♅
07 49	☽ ∥ ♄
08 15	☽ ⊥ ♃
08 19	☽ ⚼ ☿
08 24	☿ ⊥ ♄
11 07	☽ ± ☉
18 19	☽ ⊻ ♆
18 53	☽ ⚼ ♂
19 11	☽ ⊻ ♃
22 29	♀ Q ♃
09 Thursday	
00 49	☽ □ ♄
04 45	☽ ⚼ ☉
05 09	☽ ⊥ ♆
10 12	☽ ∠ ♅
15 36	☽ △ ☿
18 36	☽ □ ♀
21 17	☽ ∠ ♆
23 16	☽ △ ♂
10 Friday	
02 20	☽ ⊥ ♅
09 18	☽ Q ♄
09 43	☽ △ ☉
12 13	☽ △ ♇
13 00	☽ ⊻ ♅
23 50	☽ ⚹ ♆
11 Saturday	
00 15	♀ ⚼ ♆
00 49	☽ ☌ ♃
05 34	☽ Q ♀
06 37	☽ ⚹ ♄
07 38	♀ ∥ ♆
14 35	☽ ⚼ ♇
19 43	☉ △ ♇
20 47	☽ Q ♆
12 Sunday	
04 18	☽ □ ☿
04 27	☽ ⚹ ♀
06 13	☽ ± ♇
06 38	☽ □ ♂
07 09	☉ □ ♅
07 40	☿ ⊻ ♀
08 54	☽ ∠ ♄
14 18	☉ ⋕ ♀
16 33	☽ ⊼ ♇
17 23	☽ ☌ ♅
18 09	☽ □ ☉
13 Monday	
00 29	☽ ⊥ ♄
03 45	☽ □ ♆
04 51	☽ ⊻ ♃
08 13	☿ ☌ ♂
08 38	☽ ∠ ♀
10 48	☽ ⊻ ♄
15 00	☽ Q ♂
15 04	☽ ⊥ ♃
15 36	☽ Q ☿
18 33	☿ ∠ ♄
22 25	☉ ⋕ ♆
14 Tuesday	
01 19	☽ ⊥ ♀
01 38	♃ St R
03 07	☽ Q ☉
06 18	☽ ∠ ♃
09 47	♀ ⊻ ♂
12 15	☽ ⚹ ♂
12 21	☽ ⊻ ♀
14 41	☽ ⚹ ☿
14 45	♂ ∠ ♄
19 25	☽ ☍ ♇
20 17	☽ ⊻ ♅
21 40	☿ ∥ ♂
15 Wednesday	
00 47	☽ ⚹ ☉
06 16	☽ △ ♆
06 19	☽ ⊥ ♅
07 25	☽ ⚹ ♃
13 35	☽ ☌ ♄
14 15	☽ ∥ ♄
14 30	☽ ∠ ♂
19 09	☽ ∠ ☿
21 18	☽ ∠ ♅
16 Thursday	
03 21	☽ Q ♃
03 34	☽ ∠ ☉
05 23	☿ ⋕ ♀
05 31	☿ △ ♇
06 09	☽ ⊥ ♂
07 07	☽ ⚼ ♆
09 47	☽ ⋕ ☉
12 05	☽ ⊥ ☿
13 06	☿ □ ♅
14 44	☽ ∥ ♆
16 32	☽ ⊻ ♂
18 46	☽ ☌ ♀
19 32	☽ ⊥ ☉
21 13	☽ ⊼ ♇
21 58	☽ ± ♆
22 08	☽ ⚹ ♅
22 19	☽ ⋕ ☿
23 20	☽ ⊻ ☿
17 Friday	
02 51	☽ ∥ ♀
02 57	☽ ⋕ ♂
03 35	♀ ⋕ ♂
06 11	☽ ⊻ ☉
07 07	☽ ± ♇
07 52	☽ ⊼ ♆
09 05	☽ □ ♃
15 33	☽ ⊻ ♄
17 58	☽ Q ♅
22 00	☽ ⚼ ♇
18 Saturday	
00 16	♀ ⊼ ♇
01 32	☽ ⊥ ♄
06 13	☉ ☍ ♆
08 45	♀ ± ♆
11 42	♀ ⚹ ♅
16 36	☽ ∠ ♄
20 38	☽ ☌ ♂
21 09	☉ ⊥ ♀
22 23	☿ ⋕ ♆
22 57	☽ △ ♇
23 56	☽ □ ♅
19 Sunday	
00 09	☉ △ ♃
01 03	☽ ⊻ ♀
07 11	☽ ⋕ ♀
07 55	☽ ☌ ☿
09 43	☽ ☍ ♆
11 00	☽ △ ♃
11 50	☽ ☌ ☉
12 04	☽ ⊥ ♀
15 13	☽ ∥ ♂
17 59	☽ ⚹ ♄
22 42	☿ ☍ ♆
20 Monday	
00 18	☽ ⋕ ♆
04 48	☽ ∠ ♀
05 24	☽ ∥ ☿
09 31	☿ △ ♃
12 34	☽ ⚼ ♃
14 12	☽ ∥ ☉
14 40	☽ Q ♄
22 54	♂ △ ♇
21 Tuesday	
01 03	☽ ⋕ ♄
02 12	☽ □ ♇
02 22	☽ ⊻ ♂
03 16	☽ △ ♅
04 13	☽ ± ♃
09 24	☽ ⚹ ♀
13 25	☽ ⊼ ♆
13 30	☽ ⊥ ♂
14 48	☽ ⊼ ♃
19 03	☽ ⊻ ☿
19 54	☽ ⊻ ☉
20 55	♂ □ ♅
22 36	☽ □ ♄
22 Wednesday	
00 07	☽ ± ♆
06 01	☽ ⚼ ♅
06 30	☽ ∠ ♂
07 25	☽ ⊥ ☿
07 36	☽ ⊥ ☉
09 03	☽ Q ♀
10 22	☽ Q ♇
10 46	☉ ☌ ☿
16 27	☽ ⚼ ♆
22 31	☽ ± ♅
23 15	♀ ± ♇
23 Thursday	
00 50	☿ ⚹ ♄
01 31	☽ ∠ ☉
02 23	☽ ∠ ☿
05 29	♀ ⊼ ♆
08 33	☽ ⚹ ♇
09 42	☽ ⊼ ♅
11 12	☉ ⚹ ♄
11 42	☽ ⚹ ♂
20 26	☽ △ ♆
20 36	♀ □ ♃
21 55	☽ ☍ ♃
22 04	☽ □ ♀
24 Friday	
02 31	☿ ⊥ ♀
06 47	☽ △ ♄
08 21	☽ ⚹ ☉
10 01	☉ ∥ ☿
11 10	☽ ⚹ ☿
11 53	☽ Q ♂
13 09	☽ ∠ ♇
22 12	☿ ⋕ ♄
25 Saturday	
03 24	♅ St R
06 44	☽ ⊥ ♇
08 49	☉ ⋕ ♄
09 52	☽ Q ☉
12 18	☽ ⚼ ♄
14 21	☽ Q ☿
18 38	☽ ⊻ ♇
19 53	☽ ☍ ♅
26 Sunday	
01 24	☽ □ ♂
03 55	♂ ⋕ ♆
06 23	☽ ± ♄
07 07	☽ □ ♆
08 38	☽ ⊼ ♃
15 38	☽ △ ♀
18 33	☽ ⊼ ♄
20 41	☽ ± ♃
27 Monday	
01 14	☽ □ ☉
08 17	☽ □ ☿
14 52	☽ ⚼ ♃
19 28	☽ Q ♆
21 57	♀ ⊻ ♄
28 Tuesday	
01 33	☽ ⚼ ♀
07 16	☽ ☌ ♇
08 33	☽ ⊼ ♅
10 12	♀ Q ♅
14 45	☽ ∥ ☿
17 47	☽ △ ♂
19 50	☽ ⚹ ♆
20 45	☽ ± ♅
21 16	☽ △ ♃
21 58	☽ ± ♀
23 59	☽ ∥ ☉
29 Wednesday	
07 13	☿ ⚼ ♃
07 53	☽ ☍ ♄
09 49	☽ ⋕ ♄
11 31	☽ ⊼ ♀
14 58	☽ ⚼ ♅
19 45	☽ △ ☉
30 Thursday	
01 55	☽ ⚼ ♂
02 05	☽ ∠ ♆
04 53	♂ ☍ ♆
06 40	☽ △ ☿
06 40	☽ ∥ ♂
08 54	♀ ⚼ ♇
13 27	☽ ⋕ ♆
18 38	☿ Q ♄
19 46	☽ ⊻ ♇
20 01	☽ ⊥ ♆
20 55	☽ ± ♂
21 00	☽ △ ♅
31 Friday	
03 32	♂ △ ♃
04 22	☽ ⚼ ☉
07 37	☽ ⊥ ♇
07 50	☽ ⊻ ♆
09 06	☽ □ ♃
09 27	☽ ⊼ ♂
16 50	☽ ⚼ ☿
19 58	☽ ⊼ ♄
23 32	☽ ± ☉

NOVEMBER 2036

LONGITUDES

Date	Sidereal time h m s	Sun ☉ ° ' "	Moon ☽ ° ' "	Moon ☽ 24.00 ° ' "	Mercury ☿ ° '	Venus ♀ ° '	Mars ♂ ° '	Jupiter ♃ ° '	Saturn ♄ ° '	Uranus ♅ ° '	Neptune ♆ ° '	Pluto ♇ ° '
01	14 45 26	09 ♏ 38 53	09 ♈ 35 21	15 ♈ 52 54	15 ♏ 57	06 ♎ 35	26 ♎ 39	25 ♊ 40	01 ♍ 20	19 ♋ 36	25 ♈ 03	18 ♒ 59
02	14 49 23	10 38 55	22 ♈ 14 48	28 ♈ 41 05	17 32	07 48	27 19	25 R 36	01 24	19 R 35	25 R 02	18 D 59
03	14 53 20	11 38 58	05 ♉ 11 42	11 ♉ 46 29	19 06	09 01	27 58	25 32	01 28	19 35	25 00	18 59
04	14 57 16	12 39 04	18 ♉ 25 14	25 ♉ 07 40	20 40	10 15	28 38	25 28	01 33	19 34	24 59	18 59
05	15 01 13	13 39 11	01 ♊ 53 27	08 ♊ 42 13	22 13	11 28	29 18	25 24	01 37	19 34	24 57	18 59
06	15 05 09	14 39 20	15 ♊ 33 34	22 ♊ 27 09	23 46	12 42	29 ♎ 58	25 19	01 40	19 33	24 56	18 59
07	15 09 06	15 39 31	29 ♊ 22 36	06 ♋ 19 37	25 19	13 55	00 ♏ 38	25 15	01 44	19 32	24 54	18 59
08	15 13 02	16 39 44	13 ♋ 17 54	20 ♋ 17 14	26 51	15 09	01 18	25 10	01 48	19 31	24 52	18 59
09	15 16 59	17 40 00	27 ♋ 17 27	04 ♌ 18 24	28 23	16 23	01 58	25 05	01 52	19 31	24 51	19 00
10	15 20 55	18 40 17	11 ♌ 20 00	18 ♌ 22 10	29 ♏ 55	17 36	02 39	25 00	01 55	19 30	24 49	19 00
11	15 24 52	19 40 36	25 ♌ 24 50	02 ♍ 27 55	01 ♐ 26	18 50	03 19	24 54	01 59	19 29	24 48	19 00
12	15 28 49	20 40 57	09 ♍ 31 17	16 ♍ 34 49	02 57	20 04	03 59	24 49	02 02	19 28	24 46	19 00
13	15 32 45	21 41 20	23 ♍ 38 17	00 ♎ 41 27	04 28	21 18	04 39	24 43	02 05	19 27	24 45	19 01
14	15 36 42	22 41 45	07 ♎ 43 58	14 ♎ 45 27	05 58	22 32	05 19	24 37	02 08	19 26	24 44	19 01
15	15 40 38	23 42 12	21 ♎ 45 27	28 ♎ 43 29	07 28	23 46	06 00	24 31	02 11	19 25	24 42	19 01
16	15 44 35	24 42 41	05 ♏ 39 03	12 ♏ 31 38	08 58	25 00	06 40	24 25	02 14	19 24	24 41	19 02
17	15 48 31	25 43 11	19 ♏ 20 45	26 ♏ 05 59	10 27	26 14	07 20	24 19	02 17	19 23	24 39	19 02
18	15 52 28	26 43 43	02 ♐ 46 56	09 ♐ 23 19	11 56	27 29	08 01	24 13	02 20	19 21	24 38	19 03
19	15 56 24	27 44 17	15 ♐ 54 56	22 ♐ 21 43	13 24	28 43	08 41	24 06	02 22	19 20	24 37	19 03
20	16 00 21	28 44 52	28 ♐ 43 38	05 ♑ 00 51	14 52	29 ♎ 57	09 21	23 59	02 25	19 19	24 35	19 04
21	16 04 18	29 ♏ 45 28	11 ♑ 13 33	17 ♑ 22 03	16 19	01 ♏ 11	10 02	23 53	02 27	19 17	24 34	19 04
22	16 08 14	00 ♐ 46 06	23 ♑ 26 46	29 ♑ 28 09	17 46	02 26	10 42	23 46	02 30	19 16	24 33	19 05
23	16 12 11	01 46 45	05 ♒ 26 45	11 ♒ 23 10	19 12	03 40	11 23	23 39	02 32	19 14	24 31	19 05
24	16 16 07	02 47 25	17 ♒ 18 00	23 ♒ 11 56	20 38	04 55	12 04	23 32	02 34	19 13	24 30	19 06
25	16 20 04	03 48 06	29 ♒ 05 39	04 ♓ 59 50	22 03	06 09	12 44	23 24	02 36	19 11	24 29	19 07
26	16 24 00	04 48 48	10 ♓ 55 12	16 ♓ 52 25	23 26	07 24	13 25	23 17	02 38	19 10	24 28	19 07
27	16 27 57	05 49 31	22 ♓ 52 10	28 ♓ 55 04	24 49	08 38	14 05	23 09	02 40	19 08	24 27	19 08
28	16 31 53	06 50 15	05 ♈ 01 42	11 ♈ 12 37	26 11	09 53	14 46	23 02	02 41	19 06	24 25	19 09
29	16 35 50	07 51 00	17 ♈ 28 15	23 ♈ 49 00	27 31	11 07	15 27	22 54	02 43	19 05	24 24	19 10
30	16 39 47	08 ♐ 51 47	00 ♉ 15 06	06 ♉ 46 44	28 ♐ 50	12 ♏ 22	16 ♏ 08	22 ♊ 47	02 ♍ 44	19 ♋ 03	24 ♈ 23	19 ♒ 10

Date	Moon True ☊ ° '	Moon Mean ☊ ° '	Moon ☽ Latitude ° '
01	13 ♌ 10	12 ♌ 36	04 S 11
02	12 R 59	12 33	04 41
03	12 46	12 30	04 58
04	12 33	12 27	04 59
05	12 20	12 24	04 43
06	12 09	12 20	04 11
07	12 02	12 17	03 25
08	11 57	12 14	02 25
09	11 55	12 11	01 17
10	11 D 54	12 08	00 S 03
11	11 R 54	12 05	01 N 11
12	11 53	12 01	02 21
13	11 50	11 58	03 22
14	11 44	11 55	04 10
15	11 35	11 52	04 44
16	11 24	11 49	05 00
17	11 12	11 45	04 58
18	10 59	11 42	04 40
19	10 48	11 39	04 07
20	10 38	11 36	03 22
21	10 32	11 33	02 29
22	10 28	11 30	01 29
23	10 27	11 26	00 N 27
24	10 D 27	11 23	00 S 37
25	10 27	11 20	01 38
26	10 R 27	11 17	02 35
27	10 25	11 14	03 26
28	10 22	11 11	04 09
29	10 15	11 07	04 41
30	10 ♌ 07	11 ♌ 04	05 S 00

DECLINATIONS

Date	Sun ☉	Moon ☽	Mercury ☿	Venus ♀	Mars ♂	Jupiter ♃	Saturn ♄	Uranus ♅	Neptune ♆	Pluto ♇
01	14 S 42	00 S 03	17 S 01	01 S 07	09 S 43	22 N 46	12 N 10	22 N 25	08 N 04	23 S 24
02	15 01	04 N 18	17 34	01 35	09 57	22 46	12 09	22 25	08 04	23 23
03	15 19	08 34	18 07	02 03	10 12	22 45	12 08	22 26	08 03	23 23
04	15 38	12 31	18 38	02 31	10 27	22 45	12 07	22 26	08 02	23 23
05	15 56	15 54	19 09	02 58	10 41	22 45	12 05	22 26	08 02	23 23
06	16 14	18 29	19 39	03 26	10 56	22 45	12 04	22 26	08 01	23 23
07	16 31	20 01	20 08	03 54	11 10	22 45	12 03	22 26	08 01	23 22
08	16 49	20 22	20 35	04 22	11 25	22 45	12 02	22 26	08 00	23 22
09	17 06	19 27	21 02	04 50	11 39	22 45	12 01	22 26	08 00	23 22
10	17 23	17 20	21 28	05 18	11 53	22 45	12 00	22 26	07 59	23 22
11	17 39	14 10	21 53	05 45	12 07	22 44	11 59	22 27	07 59	23 22
12	17 55	10 10	22 16	06 13	12 21	22 44	11 58	22 27	07 58	23 21
13	18 11	05 37	22 39	06 41	12 35	22 44	11 57	22 27	07 58	23 21
14	18 26	00 N 46	23 00	07 08	12 49	22 44	11 56	22 27	07 57	23 21
15	18 42	04 S 05	23 20	07 35	13 03	22 44	11 55	22 27	07 57	23 20
16	18 56	08 41	23 39	08 02	13 17	22 44	11 54	22 27	07 56	23 20
17	19 11	12 46	23 57	08 29	13 30	22 43	11 53	22 28	07 56	23 20
18	19 25	16 08	24 14	08 56	13 44	22 43	11 52	22 28	07 55	23 20
19	19 39	18 35	24 29	09 23	13 57	22 43	11 52	22 28	07 55	23 19
20	19 52	20 03	24 44	09 49	14 11	22 43	11 51	22 28	07 54	23 19
21	20 06	20 29	24 56	10 16	14 24	22 43	11 50	22 29	07 54	23 19
22	20 18	19 56	25 08	10 42	14 37	22 42	11 50	22 29	07 53	23 18
23	20 31	18 28	25 18	11 08	14 50	22 42	11 49	22 29	07 53	23 18
24	20 43	16 14	25 27	11 33	15 03	22 42	11 49	22 29	07 52	23 18
25	20 54	13 19	25 35	11 59	15 16	22 42	11 48	22 29	07 52	23 17
26	21 06	09 52	25 41	12 24	15 29	22 41	11 48	22 30	07 52	23 17
27	21 16	05 59	25 46	12 48	15 42	22 41	11 47	22 30	07 51	23 16
28	21 27	01 S 49	25 49	13 13	15 54	22 41	11 47	22 30	07 51	23 16
29	21 37	02 N 32	25 51	13 37	16 06	22 40	11 46	22 30	07 51	23 16
30	21 S 46	06 N 52	25 S 52	14 S 01	16 S 19	22 N 40	11 N 46	22 N 31	07 N 50	23 S 15

ZODIAC SIGN ENTRIES

Date	h	m	Planets
03	02	26	☽ ♉
05	08	39	☽ ♊
06	13	03	♂ ♏
07	13	05	☽ ♋
09	16	38	☽ ♌
10	13	20	☿ ♐
11	19	48	☽ ♍
13	22	49	☽ ♎
16	02	12	☽ ♏
18	06	59	☽ ♐
20	12	57	♀ ♏
20	14	25	☽ ♑
21	17	45	☉ ♐
23	01	04	☽ ♒
25	13	51	☽ ♓
28	02	08	☽ ♈
30	11	32	☽ ♉

LATITUDES

Date	Mercury ☿	Venus ♀	Mars ♂	Jupiter ♃	Saturn ♄	Uranus ♅	Neptune ♆	Pluto ♇
01	00 S 26	01 N 38	00 N 36	00 S 36	01 N 16	00 N 25	01 S 45	08 S 43
04	00 45	01 41	00 35	00 36	01 16	00 26	01 45	08 43
07	01 04	01 43	00 33	00 36	01 17	00 26	01 45	08 42
10	01 22	01 45	00 32	00 36	01 17	00 26	01 45	08 41
13	01 39	01 45	00 30	00 36	01 18	00 26	01 45	08 41
16	01 54	01 45	00 29	00 35	01 19	00 26	01 45	08 40
19	02 06	01 45	00 27	00 35	01 19	00 26	01 45	08 40
22	02 16	01 43	00 26	00 35	01 20	00 26	01 44	08 39
25	02 23	01 41	00 24	00 35	01 21	00 26	01 44	08 39
28	02 27	01 38	00 23	00 34	01 21	00 26	01 44	08 38
31	02 S 25	01 N 35	00 N 21	00 S 34	01 N 22	00 N 26	01 S 44	08 S 38

DATA

Julian Date	2464999
Delta T	+76 seconds
Ayanamsa	24° 22' 04"
Synetic vernal point	04° ♓ 44' 55"
True obliquity of ecliptic	23° 25' 58"

MOON'S PHASES, APSIDES AND POSITIONS ☽

Date	h	m	Phase	Longitude ° '	Eclipse Indicator
04	00	44	○	12 ♉ 11	
11	01	29	☾	19 ♌ 14	
18	00	14	●	26 ♏ 14	
25	22	28	☽	04 ♓ 15	

Day	h	m		
11	18	38	Perigee	
25	04	24	Apogee	
01	12	16	0N	
08	06	31	Max dec	20° N 24'
14	15	46	0S	
21	10	18	Max dec	20° S 29'
28	22	06	0N	

LONGITUDES

Date	Chiron ⚷ ° '	Ceres ⚳ ° '	Pallas ⚴ ° '	Juno ⚵ ° '	Vesta ⚶ ° '	Black Moon Lilith ⚸ ° '
01	18 ♊ 06	27 ♍ 52	19 ♌ 35	29 ♌ 37	03 ♑ 21	22 ♒ 08
11	17 ♊ 39	02 ♎ 03	23 ♌ 41	03 ♍ 10	08 ♑ 10	23 ♒ 14
21	17 ♊ 07	06 ♎ 07	27 ♌ 17	06 ♍ 18	13 ♑ 04	24 ♒ 21
31	16 ♊ 33	10 ♎ 00	00 ♍ 15	08 ♍ 58	18 ♑ 03	25 ♒ 27

All ephemeris data is given at 12.00 UT and the Moon's longitude is additionally given for 24.00 UT

ASPECTARIAN

h m Aspects

01 Saturday
01 11 ☽ ∠ ♇
05 34 ☽ ☍ ♀
06 40 ☽ ∥ ♀
07 39 ☽ ± ♄
07 55 ♇ St D
12 07 ☽ ⊼ ☉
12 47 ☽ ± ☿
19 10 ☽ ⋕ ♀
19 45 ☽ Q ♃

02 Sunday
00 56 ☽ Ᵽ ♄
01 51 ☽ ⊼ ☿
03 50 ♀ ⊥ ♄
05 51 ☽ ⚹ ♇
07 00 ☽ □ ♅
10 59 ☉ Ᵽ ♃
17 12 ☽ ☌ ♆
18 15 ☽ ⚹ ♃
21 57 ☽ ☍ ♂

03 Monday
04 15 ☽ Q ♇
05 07 ☽ △ ♄
09 00 ☽ ∥ ♆
10 12 ☿ □ ♇
16 21 ☽ Q ♅
18 28 ☿ ± ♃
19 18 ☿ △ ♅
19 42 ☽ ⊼ ♀
21 42 ☽ ∠ ♃
22 14 ☽ ⋕ ♂

04 Tuesday
00 44 ☽ ☍ ☉
07 41 ☽ ± ♀
09 23 ☽ ∥ ♄
13 00 ☽ □ ♇
13 52 ☽ ⊥ ♃
14 04 ☽ ⚹ ♅
16 33 ☽ ☍ ☿
23 43 ☽ ⊻ ♆

05 Wednesday
00 33 ☽ ⊻ ♃
01 25 ☽ Ᵽ ♀
07 11 ☽ ⊼ ♂
10 21 ☽ ⊥ ♆
10 52 ☉ Q ♄
11 30 ☽ □ ♄
12 13 ☽ ⋕ ☉
16 42 ☽ ∠ ♅
18 20 ☽ ± ♂

06 Thursday
02 10 ☽ ∠ ♆
06 30 ☽ △ ♀
08 29 ☽ ⊥ ♅
10 18 ☽ ⊼ ☉
10 55 ☽ Ᵽ ♂
17 58 ☽ △ ♇
18 57 ☽ ⊻ ♅
19 12 ☽ Q ♄
21 34 ☽ ± ☉

07 Friday
04 05 ☽ ⊼ ☿
04 16 ☽ ⚹ ♆
04 53 ☽ ☌ ♃
05 38 ☿ ⊼ ♆
10 57 ☿ ⊼ ♃
14 17 ☽ △ ♂
14 23 ☽ Ᵽ ☉
15 46 ☽ ± ☿
16 06 ☽ ⚹ ♄
19 58 ☽ Ᵽ ♇

08 Saturday
00 58 ☽ Q ♆
09 13 ☽ Ᵽ ☿
11 28 ☽ ± ♇
15 29 ☽ □ ♀
18 02 ☽ ∠ ♄
18 14 ☽ △ ☉
21 47 ☽ ⊼ ♇
22 41 ☽ ☌ ♅

09 Sunday
07 30 ♂ ⚹ ♄
07 50 ☽ □ ♆
08 14 ☽ ⊻ ♃
09 32 ☽ ⊥ ♄
14 06 ☽ △ ☿
18 27 ☽ ⊥ ♃
19 51 ☽ ⊻ ♄
20 25 ☽ □ ♂
21 56 ♀ ∠ ♄

10 Monday
01 17 ☽ Q ♀
09 44 ☽ ∠ ♃
11 36 ☽ ⋕ ☉
19 07 ☉ ± ♃
19 50 ☉ □ ♇
22 10 ♂ ⋕ ♄
23 43 ☽ ⚹ ♀

11 Tuesday
01 05 ☽ ☍ ♇
01 29 ☽ □ ☉
01 54 ☽ ⊻ ♅
02 06 ☿ ± ♆
04 40 ☽ Q ♂
07 26 ☉ △ ♅
10 57 ☽ △ ♆
11 09 ☽ ⚹ ♃
12 07 ☽ ⊥ ♅
15 15 ♀ △ ♇
20 51 ☿ □ ♄
23 13 ☽ ☌ ♄
23 29 ☽ □ ☿

12 Wednesday
00 04 ☽ ⋕ ♂
00 27 ♀ □ ♅
01 42 ☽ ∥ ♄
02 07 ☽ ⚹ ♂
03 25 ☽ ∠ ♅
03 42 ☽ ∠ ♀
07 26 ☽ Q ♃
10 28 ☽ Q ☉
12 26 ☽ Ᵽ ♆
20 28 ☽ ⊥ ♀
23 11 ☿ ⋕ ♅
23 54 ☽ ∥ ♆

13 Thursday
01 56 ♃ ⚹ ♆
03 42 ☽ ± ♆
04 08 ☽ ⊼ ♇
04 53 ☽ ∠ ♂
04 53 ☽ ⚹ ♅
07 04 ☽ ⋕ ♀
07 39 ☽ ⊻ ♀
08 26 ☽ ⚹ ☉
09 46 ☽ Q ☿
11 48 ☿ Ᵽ ♅
13 50 ☽ □ ♃
13 53 ☽ ⊼ ♆
14 20 ☽ ± ♇
17 24 ☿ ⊻ ♂
17 56 ☿ ⋕ ♃
20 57 ☽ ⊥ ♂

14 Friday
01 17 ☽ Q ♅
02 26 ☽ ⊻ ♄
05 40 ☽ Ᵽ ♇
07 41 ☽ ⊻ ♂
08 38 ☽ ⚹ ☿
11 56 ☽ ∠ ☉
12 42 ☽ ⊥ ♄

15 Saturday
04 08 ☽ ∠ ♄
04 30 ☽ ⊥ ☉
04 53 ☿ Q ♇
05 09 ☉ ⊻ ♀
07 18 ☽ △ ♇
07 59 ☽ □ ♅
12 11 ☿ ∥ ♇
13 22 ☽ ∠ ☿
15 36 ☽ ⊻ ☉
15 48 ☽ ☌ ♀
16 43 ☽ △ ♃
17 03 ☽ ☍ ♆

16 Sunday
01 34 ♀ △ ♃
05 45 ☉ ⊼ ♃
05 49 ♀ ☍ ♆
06 03 ☽ ⚹ ♄
06 27 ♀ ⋕ ♆
06 46 ☽ ⊥ ☿
07 56 ☽ ⋕ ♆
08 06 ☽ ∥ ♀
11 15 ☉ ⊼ ♆
13 51 ☽ ☌ ♂
18 28 ☽ ⊻ ☿
18 31 ☽ Ᵽ ♃
23 25 ☿ Ᵽ ♆

17 Monday
03 03 ☽ Q ♄
06 28 ☽ ⋕ ♄
10 12 ☽ ± ♃
11 27 ☽ □ ♇
12 03 ☽ △ ♅
17 07 ☽ ∥ ♂
20 45 ☽ ⊼ ♃
21 24 ☽ ⊼ ♆

18 Tuesday
00 14 ☽ ☌ ☉
01 29 ☽ ⊻ ♀
08 08 ☽ ± ♆
11 11 ☽ □ ♄
13 23 ☽ ⊥ ♀
14 50 ☽ Ᵽ ♅
19 44 ☽ Q ♇
22 00 ☽ ⊻ ♂

19 Wednesday
00 25 ☽ Ᵽ ♆
06 46 ☽ ☌ ☿
07 15 ☽ ± ♅
07 30 ☽ ∠ ♀
09 36 ☽ ⊥ ♂
10 57 ☿ ± ♅
17 50 ☽ ⚹ ♇
18 20 ☽ ⊼ ♅

20 Thursday
00 46 ♂ Ᵽ ♃
03 08 ☽ ☍ ♃
03 17 ☽ ∠ ♂
04 12 ☽ △ ♆
06 54 ☽ ∥ ☉
12 03 ☽ ⊻ ☉
14 35 ☽ ⚹ ♀
19 03 ☽ △ ♄
21 29 ☿ ∠ ♀
22 11 ☽ ∠ ♇

21 Friday
00 30 ☽ ⊥ ☉
03 08 ☿ ⊥ ♂
09 33 ☽ ⚹ ♂
15 36 ☽ ⊥ ♇
16 15 ☽ Q ♀
19 30 ☽ ∠ ☉
23 17 ☽ ⊻ ☿

22 Saturday
00 13 ☽ Ᵽ ♄
03 22 ☽ ⊻ ♇
03 38 ☽ ∥ ☉
03 45 ☽ ☍ ♅
06 48 ☉ ± ♆
10 27 ☽ Q ♂
12 37 ☽ ⊼ ♃
12 44 ☽ ⊥ ☿
13 18 ♀ ⚹ ♄
14 11 ☽ □ ♆
18 05 ☽ ± ♄

23 Sunday
00 28 ☽ ± ♃
03 57 ☽ ⚹ ☉
06 07 ☽ ⊼ ♄
08 01 ☽ □ ♀
09 10 ☽ ∠ ☿
10 04 ☿ ⚹ ♇
12 34 ☿ ⊼ ♅
18 23 ☽ Ᵽ ♃

24 Monday
00 43 ☽ □ ♂
02 17 ☽ Q ♆
06 26 ☽ Q ☉
06 30 ☉ □ ♄
15 40 ☽ ☌ ♇
15 53 ☽ ⊼ ♅
19 42 ☽ ⚹ ☿
21 33 ☽ ∥ ♂

25 Tuesday
00 32 ☽ △ ♃
02 12 ♀ ⋕ ♄
02 38 ☽ ⚹ ♆
04 04 ☽ ± ♅
19 09 ☽ ☍ ♄
20 39 ☽ ∥ ♀
20 57 ☉ Ᵽ ♅
22 20 ☽ Ᵽ ♅
22 28 ☽ □ ☉
22 58 ☽ ⋕ ♄
23 25 ☽ Q ☿

26 Wednesday
04 01 ☽ △ ♀
09 03 ☽ ∠ ♆
09 29 ☿ ☍ ♃
17 20 ☽ △ ♂

27 Thursday
00 42 ☽ ⋕ ♆
03 10 ☽ ⊥ ♆
03 38 ♀ Ᵽ ♃
04 32 ☽ ⊻ ♇
04 33 ☽ △ ♅
05 29 ☿ △ ♆
12 16 ♅ ⊼ ♇
12 34 ☽ □ ♃
13 42 ☽ Ᵽ ♀
15 08 ☽ ⊻ ♆
16 23 ☽ □ ☿
16 31 ☽ ⊥ ♇

28 Friday
01 04 ☽ Ᵽ ♂
07 25 ☽ ⊼ ♄
09 04 ♂ Q ♄
09 30 ☽ ± ♀
10 16 ☽ ∠ ♇
15 51 ☽ △ ☉
19 08 ☽ ± ♄
19 24 ☉ Q ♇
19 42 ☽ ± ♂
22 28 ☽ ⊼ ♀
23 32 ☽ Q ♃

29 Saturday
07 55 ☽ ⊼ ♂
12 28 ☽ Ᵽ ♄
15 03 ☽ □ ♅
15 13 ☽ ⚹ ♇
22 11 ☽ ⚹ ♃
23 04 ☽ Ᵽ ☉

30 Sunday
01 05 ☽ ☌ ♆
09 05 ☽ △ ☿
13 42 ☽ Q ♇
16 36 ☽ △ ♄
17 14 ☽ ± ☉
17 26 ☽ ∥ ♆

DECEMBER 2036

LONGITUDES

Date	Sidereal time h m s	Sun ☉ ° ' "	Moon ☽ ° ' "	Moon ☽ 24.00 ° ' "	Mercury ☿ ° '	Venus ♀ ° '	Mars ♂ ° '	Jupiter ♃ ° '	Saturn ♄ ° '	Uranus ♅ ° '	Neptune ♆ ° '	Pluto ♇ ° '
01	16 43 43	09 ♐ 52 34	13 ♉ 23 55	20 ♉ 06 36	00 ♑ 07	13 ♏ 37	16 ♏ 48	22 ♊ 39	02 ♍ 46	19 ♋ 01	24 ♈ 22	19 ♒ 11
02	16 47 40	10 53 22	26 ♉ 54 32	03 ♊ 47 24	01 23	14 51	17 29	22 R 31	02 47	18 R 59	24 R 21	19 12
03	16 51 36	11 54 12	10 ♊ 44 44	17 ♊ 46 01	02 35	16 06	18 10	22 23	02 48	18 57	24 20	19 13
04	16 55 33	12 55 03	24 ♊ 50 38	01 ♋ 57 55	03 45	17 21	18 51	22 15	02 49	18 55	24 19	19 14
05	16 59 29	13 55 54	09 ♋ 07 11	16 ♋ 17 45	04 53	18 35	19 32	22 07	02 50	18 53	24 18	19 14
06	17 03 26	14 56 47	23 ♋ 29 00	00 ♌ 40 21	05 56	19 50	20 13	21 59	02 51	18 51	24 17	19 15
07	17 07 22	15 57 42	07 ♌ 51 15	15 ♌ 01 17	06 56	21 05	20 54	21 51	02 51	18 49	24 16	19 16
08	17 11 19	16 58 37	22 ♌ 10 05	29 ♌ 17 22	07 50	22 20	21 35	21 43	02 52	18 47	24 15	19 17
09	17 15 16	17 59 34	06 ♍ 22 53	13 ♍ 26 31	08 40	23 35	22 16	21 35	02 53	18 45	24 15	19 18
10	17 19 12	19 00 31	20 ♍ 28 06	27 ♍ 27 35	09 24	24 50	22 57	21 27	02 53	18 43	24 14	19 19
11	17 23 09	20 01 30	04 ♎ 24 52	11 ♎ 19 53	10 00	26 05	23 38	21 18	02 53	18 41	24 13	19 20
12	17 27 05	21 02 31	18 ♎ 12 33	25 ♎ 02 47	10 29	27 20	24 19	21 10	02 53	18 39	24 12	19 21
13	17 31 02	22 03 32	01 ♏ 50 29	08 ♏ 35 31	10 50	28 35	25 00	21 02	02 R 53	18 37	24 11	19 22
14	17 34 58	23 04 34	15 ♏ 17 44	21 ♏ 56 59	11 01	29 ♏ 50	25 42	20 54	02 53	18 34	24 11	19 24
15	17 38 55	24 05 37	28 ♏ 33 06	05 ♐ 05 58	11 R 02	01 ♐ 05	26 23	20 46	02 53	18 32	24 10	19 25
16	17 42 51	25 06 42	11 ♐ 35 25	18 ♐ 01 21	10 52	02 20	27 04	20 38	02 53	18 30	24 09	19 26
17	17 46 48	26 07 46	24 ♐ 23 43	00 ♑ 42 27	10 30	03 35	27 45	20 30	02 52	18 27	24 09	19 27
18	17 50 45	27 08 52	06 ♑ 57 37	13 ♑ 09 15	09 57	04 50	28 27	20 21	02 52	18 25	24 08	19 28
19	17 54 41	28 09 58	19 ♑ 17 32	25 ♑ 22 38	09 13	06 05	29 08	20 13	02 51	18 23	24 08	19 29
20	17 58 38	29 ♐ 11 05	01 ♒ 24 50	07 ♒ 24 28	08 17	07 20	29 ♏ 50	20 05	02 50	18 20	24 07	19 31
21	18 02 34	00 ♑ 12 12	13 ♒ 21 53	19 ♒ 17 33	07 12	08 35	00 ♐ 31	19 57	02 49	18 18	24 07	19 32
22	18 06 31	01 13 19	25 ♒ 11 56	01 ♓ 05 34	05 59	09 50	01 13	19 50	02 48	18 15	24 06	19 33
23	18 10 27	02 14 26	06 ♓ 59 01	12 ♓ 52 53	04 40	11 05	01 54	19 42	02 47	18 13	24 06	19 34
24	18 14 24	03 15 34	18 ♓ 47 47	24 ♓ 44 23	03 18	12 20	02 36	19 34	02 46	18 11	24 06	19 36
25	18 18 20	04 16 41	00 ♈ 43 18	06 ♈ 45 13	01 56	13 35	03 17	19 26	02 45	18 08	24 05	19 37
26	18 22 17	05 17 49	12 ♈ 50 44	19 ♈ 00 29	00 ♑ 36	14 51	03 59	19 19	02 43	18 06	24 05	19 38
27	18 26 14	06 18 57	25 ♈ 15 02	01 ♉ 34 54	29 ♐ 20	16 06	04 40	19 11	02 42	18 03	24 05	19 40
28	18 30 10	07 20 04	08 ♉ 00 32	14 ♉ 32 16	28 12	17 21	05 22	19 04	02 40	18 00	24 04	19 41
29	18 34 07	08 21 12	21 ♉ 10 21	27 ♉ 54 56	27 13	18 36	06 04	18 57	02 38	17 58	24 04	19 43
30	18 38 03	09 22 20	04 ♊ 45 58	11 ♊ 43 18	26 23	19 51	06 46	18 49	02 37	17 55	24 04	19 44
31	18 42 00	10 ♑ 23 28	18 ♊ 46 36	25 ♊ 55 24	25 ♐ 43	21 ♐ 06	07 ♐ 27	18 ♊ 42	02 ♍ 35	17 ♋ 53	24 ♈ 04	19 ♒ 45

Date	Moon True ☊ ° '	Moon Mean ☊ ° '	Moon ☽ Latitude ° '
01	09 ♌ 57	11 ♌ 01	05 S 04
02	09 R 46	10 58	04 51
03	09 36	10 55	04 21
04	09 28	10 51	03 35
05	09 21	10 48	02 35
06	09 18	10 45	01 24
07	09 17	10 42	00 S 08
08	09 D 17	10 39	01 N 09
09	09 18	10 36	02 21
10	09 R 19	10 32	03 23
11	09 18	10 29	04 13
12	09 15	10 26	04 48
13	09 09	10 23	05 06
14	09 02	10 20	05 06
15	08 54	10 16	04 50
16	08 46	10 13	04 20
17	08 38	10 10	03 36
18	08 32	10 07	02 43
19	08 28	10 04	01 42
20	08 26	10 01	00 N 38
21	08 D 26	09 57	00 S 27
22	08 27	09 54	01 30
23	08 29	09 51	02 30
24	08 31	09 48	03 23
25	08 32	09 45	04 08
26	08 R 32	09 42	04 43
27	08 30	09 38	05 05
28	08 27	09 35	05 13
29	08 23	09 32	05 06
30	08 18	09 29	04 41
31	08 ♌ 13	09 ♌ 26	03 S 59

DECLINATIONS

Date	Sun ☉ ° '	Moon ☽ ° '	Mercury ☿ ° '	Venus ♀ ° '	Mars ♂ ° '	Jupiter ♃ ° '	Saturn ♄ ° '	Uranus ♅ ° '	Neptune ♆ ° '	Pluto ♇ ° '
01	21 S 56	11 N 01	25 S 51	14 S 24	16 S 31	22 N 40	11 N 46	22 N 31	07 N 50	23 S 15
02	22 04	14 44	25 49	14 48	16 43	22 39	11 46	22 31	07 49	23 14
03	22 13	17 44	25 45	15 10	16 55	22 39	11 45	22 32	07 49	23 14
04	22 20	19 45	25 40	15 33	17 07	22 39	11 45	22 32	07 49	23 14
05	22 28	20 33	25 34	15 55	17 18	22 38	11 45	22 32	07 48	23 13
06	22 35	20 01	25 26	16 17	17 30	22 38	11 45	22 33	07 48	23 13
07	22 42	18 11	25 17	16 38	17 41	22 38	11 45	22 33	07 48	23 12
08	22 48	15 12	25 07	16 59	17 53	22 37	11 45	22 33	07 48	23 12
09	22 53	11 21	24 56	17 19	18 04	22 37	11 45	22 33	07 47	23 11
10	22 59	06 53	24 44	17 39	18 15	22 37	11 45	22 34	07 47	23 11
11	23 03	02 N 07	24 31	17 58	18 26	22 36	11 45	22 34	07 47	23 11
12	23 08	02 S 42	24 17	18 17	18 36	22 36	11 45	22 34	07 47	23 10
13	23 12	07 20	24 02	18 36	18 47	22 35	11 46	22 35	07 46	23 10
14	23 15	11 32	23 47	18 54	18 57	22 35	11 46	22 35	07 46	23 09
15	23 18	15 06	23 31	19 11	19 08	22 35	11 46	22 35	07 46	23 09
16	23 21	17 53	23 15	19 28	19 18	22 34	11 47	22 36	07 46	23 08
17	23 23	19 43	22 59	19 45	19 28	22 34	11 47	22 36	07 46	23 08
18	23 24	20 33	22 43	20 01	19 37	22 33	11 47	22 36	07 45	23 07
19	23 25	20 21	22 26	20 16	19 47	22 33	11 48	22 37	07 45	23 07
20	23 26	19 13	22 10	20 31	19 57	22 32	11 48	22 37	07 45	23 06
21	23 26	17 14	21 54	20 45	20 06	22 32	11 49	22 37	07 45	23 06
22	23 26	14 32	21 38	20 58	20 15	22 32	11 49	22 38	07 45	23 05
23	23 25	11 16	21 23	21 11	20 24	22 31	11 50	22 38	07 45	23 05
24	23 24	07 33	21 09	21 24	20 33	22 31	11 51	22 38	07 45	23 04
25	23 22	03 S 30	20 55	21 35	20 42	22 30	11 51	22 39	07 45	23 03
26	23 20	00 N 44	20 44	21 47	20 50	22 30	11 52	22 39	07 45	23 03
27	23 17	05 02	20 33	21 57	20 58	22 29	11 53	22 40	07 45	23 02
28	23 14	09 14	20 25	22 07	21 06	22 29	11 54	22 40	07 45	23 02
29	23 10	13 08	20 19	22 16	21 14	22 29	11 54	22 40	07 45	23 01
30	23 06	16 29	20 14	22 24	21 22	22 28	11 55	22 41	07 45	23 01
31	23 S 02	19 N 00	20 S 12	22 S 32	21 S 30	22 N 28	11 N 56	22 N 41	07 N 45	23 S 00

ZODIAC SIGN ENTRIES

Date	h	m	Planets	
01	09	40	☿	♑
02	17	24	☽	♊
04	20	42	☽	♋
06	22	53	☽	♌
09	01	12	☽	♍
11	04	23	☽	♎
13	08	44	☽	♏
14	15	21	♀	♐
15	14	39	☽	♐
17	22	39	☽	♑
20	09	11	☽	♒
20	18	01	♂	♐
21	07	13	☉	♑
22	21	46	☽	♓
25	10	33	☽	♈
26	23	09	☿	♐
27	21	01	☽	♉
30	03	40	☽	♊

LATITUDES

Date	Mercury ☿ ° '	Venus ♀ ° '	Mars ♂ ° '	Jupiter ♃ ° '	Saturn ♄ ° '	Uranus ♅ ° '	Neptune ♆ ° '	Pluto ♇ ° '
01	02 S 25	01 N 35	00 N 21	00 S 34	01 N 22	00 N 26	01 S 44	08 S 38
04	02 17	01 31	00 20	00 34	01 23	00 26	01 44	08 38
07	02 02	01 27	00 18	00 33	01 23	00 26	01 44	08 37
10	01 38	01 22	00 16	00 33	01 24	00 26	01 44	08 37
13	01 03	01 16	00 14	00 32	01 25	00 27	01 44	08 36
16	00 S 16	01 10	00 13	00 32	01 26	00 27	01 43	08 36
19	00 N 41	01 04	00 11	00 32	01 26	00 27	01 43	08 36
22	01 40	00 57	00 09	00 31	01 27	00 27	01 43	08 35
25	02 30	00 50	00 07	00 31	01 28	00 27	01 43	08 35
28	03 00	00 43	00 05	00 30	01 28	00 27	01 43	08 35
31	03 N 10	00 N 36	00 N 03	00 S 30	01 N 29	00 N 27	01 S 43	08 S 34

DATA

Julian Date	2465029
Delta T	+76 seconds
Ayanamsa	24° 22' 08"
Synetic vernal point	04° ♓ 44' 51"
True obliquity of ecliptic	23° 25' 58"

MOON'S PHASES, APSIDES AND POSITIONS ☽

Date	h	m	Phase	Longitude ° '	Eclipse Indicator
03	14	08	○	12 ♊ 00	
10	09	18	☾	18 ♍ 54	
17	15	34	●	26 ♐ 17	
25	19	45	☽	04 ♈ 36	

Day	h	m		
07	02	02	Perigee	
23	01	11	Apogee	
05	14	27	Max dec	20° N 33'
11	22	31	0S	
18	19	29	Max dec	20° S 36'
26	07	54	0N	

LONGITUDES

Date	Chiron ⚷ ° '	Ceres ⚳ ° '	Pallas ⚴ ° '	Juno ⚵ ° '	Vesta ⚶ ° '	Black Moon Lilith ⚸ ° '
01	16 ♊ 33	10 ♎ 00	00 ♍ 15	08 ♍ 58	18 ♑ 03	25 ♒ 27
11	15 ♊ 57	13 ♎ 42	02 ♍ 28	11 ♍ 05	23 ♑ 06	26 ♒ 34
21	15 ♊ 22	17 ♎ 09	03 ♍ 48	12 ♍ 34	28 ♑ 11	27 ♒ 40
31	14 ♊ 49	20 ♎ 19	04 ♍ 06	13 ♍ 20	03 ♒ 18	28 ♒ 47

All ephemeris data is given at 12.00 UT and the Moon's longitude is additionally given for 24.00 UT

ASPECTARIAN

h m Aspects

01 Monday
00 12 ☉ ♇ ♆
00 28 ☽ Q ♅
01 41 ☽ ∠ ♃
05 06 ☽ ⊼ ☉
07 19 ♂ ± ♃
12 25 ☽ ☍ ♀
15 25 ☽ ♇ ☿
16 32 ☽ ∥ ♄
17 46 ☽ ⊥ ♃
18 26 ☽ ☍ ♂
22 02 ☽ ✶ ♅
22 22 ☽ □ ⯓

02 Tuesday
04 20 ☽ ⊻ ♃
07 31 ☽ ⊻ ♆
09 02 ☽ ± ☿
10 37 ♀ Q ♄
12 28 ☽ # ♀
18 01 ☽ ⊥ ♆
20 34 ☽ ⊼ ☿
22 16 ☽ □ ♄

03 Wednesday
00 19 ☽ ∠ ♅
04 07 ☽ # ♂
09 35 ☽ ∠ ♆
14 08 ☽ ☍ ☉
15 47 ☽ ⊥ ♅
16 21 ☿ △ ♄
17 00 ♀ ± ♃
22 03 ☽ ⊼ ♀

04 Thursday
01 19 ☽ ⊼ ♂
01 59 ☽ ⊻ ♅
02 28 ☽ △ ⯓
05 11 ☽ Q ♄
07 39 ☽ ☌ ♃
09 13 ☽ ± ♀
11 07 ☽ ✶ ♆
12 00 ☽ ± ♂
12 09 ☉ ± ♅
14 33 ♂ △ ♅
16 44 ☿ ∠ ♂
22 04 ☿ ∠ ⯓

05 Friday
01 27 ☽ ✶ ♄
01 38 ♂ □ ⯓
01 51 ☽ ♇ ♀
03 49 ☽ ♇ ⯓
03 55 ☽ ♇ ♂
04 18 ☽ ☍ ☿
07 17 ☽ Q ♆
17 38 ♀ △ ♅
18 54 ☽ ± ⯓
20 40 ☽ ⊼ ☉

06 Saturday
00 41 ♀ □ ⯓
02 35 ☽ ∠ ♄
03 01 ☉ # ♅
04 18 ☽ ☌ ♅
04 56 ☽ ⊼ ⯓
05 20 ☽ △ ♀
06 16 ☽ △ ♂
07 27 ☽ ± ☉
09 31 ☽ ⊻ ♃
13 20 ☽ □ ♆
17 37 ☽ ⊥ ♄
19 27 ☽ ⊥ ♃
22 23 ☉ # ♃
23 36 ☽ ♇ ☉

07 Sunday
03 39 ☽ ⊻ ♄
03 58 ♀ ☌ ♂
10 20 ☽ ⊼ ☿
10 20 ☽ ∠ ♃
16 17 ☽ # ♂
21 05 ☽ ± ☿

08 Monday
00 05 ☽ # ♀
01 19 ♀ ⊼ ♃
02 37 ☽ △ ☉
06 20 ☽ ⊻ ♅
07 09 ☽ ☍ ⯓
10 58 ☽ □ ♂
11 15 ☽ ✶ ♃
12 18 ☽ □ ♀
13 12 ☽ ♇ ☿
15 31 ☽ △ ♆
16 01 ♂ ⊼ ♃
16 24 ☽ ⊥ ♅

09 Tuesday
06 04 ☽ ☌ ♄
07 18 ☽ Q ♃
07 34 ☽ ∠ ♅
09 40 ☽ ∥ ♄
16 06 ☽ △ ☿
16 22 ☿ ∠ ♀
16 51 ☽ ♇ ♆
18 56 ☽ Q ♂
21 41 ☽ Q ♀

10 Wednesday
00 37 ♀ ⊼ ♆
05 24 ☉ ⊼ ♅
07 21 ☽ ∥ ♆
08 10 ☽ ± ♆
09 01 ☽ ✶ ♅
09 18 ☽ □ ☉
10 02 ☽ ⊼ ⯓
13 39 ☽ □ ♃
16 28 ☽ ✶ ♂
18 27 ☽ ⊼ ♆
19 31 ☉ ✶ ⯓
20 13 ☽ ✶ ♀
20 20 ☽ ± ⯓

11 Thursday
05 34 ☽ Q ♅
09 21 ☽ ⊻ ♄
11 52 ☽ ♇ ⯓
18 45 ☽ Q ☉
19 42 ☽ ∠ ♂
19 45 ☽ ⊥ ♄
22 05 ☽ □ ☿

12 Friday
00 43 ☽ ∠ ♀
08 02 ♂ ⊼ ♆
11 26 ☽ ∠ ♄
12 12 ☽ ⊥ ♂
12 46 ☽ □ ♅
14 01 ☽ △ ⯓
14 42 ☉ ☍ ♃
17 08 ☽ △ ♃
17 22 ☽ ✶ ☉
18 01 ☽ ⊥ ♀
22 30 ☽ ☍ ♆
23 17 ☽ ⊻ ♂

13 Saturday
00 04 ☉ ∥ ⯓
02 58 ♄ St R
05 38 ☽ ⊻ ♀
06 34 ☽ Q ☿
13 51 ☽ ✶ ♄
14 26 ☽ # ♆
19 22 ☽ ♇ ♃
22 01 ☽ ∠ ☉

14 Sunday
04 15 ☽ ✶ ☿
08 15 ♃ ∥ ♅
11 16 ☽ Q ♄
11 18 ☽ ± ♃
13 27 ☽ # ♄
14 22 ♃ Q ♄
15 28 ☽ ⊥ ☉
17 53 ☽ △ ♅
18 44 ♀ ± ♆
19 23 ☽ □ ⯓
22 00 ☽ ⊼ ♃
23 01 ♀ ∥ ♂

15 Monday
00 22 ☿ ∠ ♂
02 05 ☿ St R
03 13 ☽ ⊻ ☉
04 02 ☽ ⊼ ♆
07 26 ☽ ∠ ☿
07 50 ☽ ☌ ♂
13 45 ☉ △ ♆
14 57 ☽ ± ♆
17 06 ☽ ☌ ♀
19 55 ☽ □ ♄
21 06 ☽ ♇ ♅
23 46 ☽ ⊥ ☿

16 Tuesday
04 17 ☽ Q ⯓
05 07 ☉ ∥ ☿
07 30 ☽ ♇ ♆
10 41 ☽ ⊻ ☿
13 41 ☽ ± ♅
22 30 ♀ □ ♄
22 55 ☿ ∥ ⯓

17 Wednesday
00 51 ☽ ⊼ ♅
02 40 ☽ ✶ ⯓
04 43 ☽ ☍ ♃
07 10 ☽ ∥ ♂
09 46 ♀ ♇ ♅
11 32 ☽ △ ♆
12 40 ☽ ∥ ♀
15 34 ☽ ☌ ☉
18 45 ☽ ⊻ ♂

18 Thursday
00 42 ☿ ⊥ ♀
04 08 ☽ △ ♄
06 53 ☽ ⊥ ♂
07 12 ☽ ∠ ⯓
07 26 ☽ ⊻ ♀
17 29 ☽ ☌ ☿
20 19 ☽ ⊥ ♀
20 56 ☿ # ♅

19 Friday
00 38 ☽ ⊥ ⯓
01 19 ☽ ∠ ♂
02 05 ☿ # ♃
09 10 ☽ ♇ ♄
10 13 ☽ ☍ ♅
12 23 ☽ ⊻ ⯓
13 49 ☽ ⊼ ♃
14 16 ☽ ∥ ♀
15 55 ☽ ∠ ♀
21 32 ☽ □ ♆

20 Saturday
00 41 ☽ ∥ ♂
01 32 ☽ ± ♃
02 54 ☽ ± ♄
07 09 ☽ ⊻ ☉
08 39 ☽ ✶ ♂
14 50 ☽ ⊼ ♄
15 30 ♀ Q ⯓
19 16 ☽ ♇ ♃
20 14 ☽ ⊥ ☉
21 59 ☿ ⊻ ♀
22 07 ♂ ± ♆

21 Sunday
00 39 ☽ ⊻ ☿
01 14 ☽ ✶ ♀
09 29 ☽ Q ♆
10 11 ☽ Q ♂
11 42 ☽ ⊥ ☿
16 04 ☽ ∠ ☉
20 44 ☿ ⊥ ♂
21 57 ☽ ⊼ ♅
22 07 ♀ ♇ ♆

22 Monday
00 30 ☽ ☌ ⯓
01 12 ☽ △ ♃
04 15 ☽ ∠ ☿
04 21 ☽ Q ♀
09 47 ☽ ✶ ♆
10 06 ☽ ± ♅
11 05 ☉ ⊻ ♂

23 Tuesday
01 00 ☽ □ ♂
01 25 ☽ ✶ ☉
03 28 ☽ ☍ ♄
04 21 ☽ ♇ ♅
07 46 ☽ ✶ ☿
08 04 ☽ # ♄
13 37 ☿ ∠ ⯓
16 18 ☽ ∠ ♆
21 21 ☽ □ ♀
22 20 ☿ ∥ ♀

24 Wednesday
00 38 ☉ △ ♄
04 09 ☽ Q ☉
05 39 ☽ Q ☿
07 22 ♃ △ ⯓
08 58 ♀ ± ♅
10 35 ☽ ⊥ ♆
10 44 ☽ # ♆
10 45 ☽ △ ♅
12 25 ☉ ☌ ☿
13 33 ☽ □ ♃
13 37 ☽ ⊻ ⯓
17 50 ♂ □ ♄
20 12 ☿ ⊻ ♂
21 27 ☿ △ ♄
22 42 ☽ ⊻ ♆

25 Thursday
01 45 ☽ ⊥ ⯓
07 00 ♂ ♇ ♅
14 10 ☽ □ ☿
16 02 ☽ ⊼ ♄
17 26 ☽ △ ♂
19 45 ☽ □ ☉
19 47 ☽ ∠ ⯓
20 12 ☉ ∠ ⯓

26 Friday
01 13 ☽ Q ♃
03 54 ☽ ± ♄
04 09 ☿ ∥ ♂
16 21 ☽ △ ♀
21 29 ☽ ♇ ♄
22 11 ☽ □ ♅

27 Saturday
00 28 ☽ ✶ ♃
00 39 ☽ ♇ ♂
01 15 ☽ ✶ ⯓
09 46 ☽ ☌ ♆
18 53 ☽ ± ♂
19 07 ☽ △ ☿

28 Sunday
00 11 ☽ Q ⯓
00 18 ☽ ♇ ♀
02 04 ☽ △ ♄
03 23 ☽ ∥ ♆
04 44 ☽ ∠ ♃
06 49 ☽ ⊼ ♂
08 18 ☽ Q ♅
10 39 ☽ △ ☉
18 49 ☽ ± ♀
20 52 ☽ ♇ ☿
21 13 ☽ ⊥ ♃

29 Monday
00 13 ♀ ⊼ ♅
04 10 ☽ ∥ ♄
06 15 ☽ ✶ ♅
06 53 ☽ ⊼ ♀
08 01 ☽ ⊻ ♃
09 22 ☽ □ ⯓
12 04 ☽ ± ☿
16 13 ☽ ♇ ☉
17 11 ☽ ⊻ ♆
17 58 ♀ ☍ ♃
22 06 ☽ ⊼ ☿

30 Tuesday
03 48 ☽ ⊥ ♆
08 15 ☽ □ ♄
08 48 ☽ ∠ ♅
09 23 ☽ ± ☉
09 39 ♀ ✶ ⯓
15 38 ☽ ☍ ♂
19 27 ☽ ∠ ♆
20 36 ☽ ⊼ ☉
22 30 ♀ # ♃

31 Wednesday
00 18 ☽ ⊥ ♅
10 29 ☽ ⊻ ♅
11 53 ☽ ☌ ♃
13 39 ☽ △ ⯓
15 02 ☽ Q ♄
16 18 ☽ ☍ ♀
19 24 ☉ ∥ ⯓
20 54 ☽ ✶ ♆
22 46 ♂ Q ⯓
23 14 ☽ ☍ ☿

JANUARY 2037

LONGITUDES

Date	Sidereal time h m s	Sun ☉ ° ' "	Moon ☽ ° ' "	Moon ☽ 24.00 ° ' "	Mercury ☿ ° '	Venus ♀ ° '	Mars ♂ ° '	Jupiter ♃ ° '	Saturn ♄ ° '	Uranus ♅ ° '	Neptune ♆ ° '	Pluto ♇ ° '
01	18 45 56	11 ♑ 24 36	03 ♋ 09 05	10 ♋ 26 54	25 ♐ 14	22 ♐ 22	08 ♐ 09	18 ♊ 35	02 ♍ 33	17 ♋ 50	24 ♈ 04	19 ♒ 47
02	18 49 53	12 25 44	17 ♋ 48 00	25 ♋ 11 28	24 R 55	23 37	08 51	18 R 28	02 R 30	17 R 48	24 R 04	19 48
03	18 53 49	13 26 52	02 ♌ 36 20	10 ♌ 01 38	24 46	24 52	09 33	18 22	02 28	17 45	24 04	19 50
04	18 57 46	14 28 00	17 ♌ 26 27	24 ♌ 49 55	24 D 46	26 07	10 15	18 15	02 26	17 42	24 D 04	19 51
05	19 01 43	15 29 08	02 ♍ 11 16	09 ♍ 29 47	24 55	27 22	10 57	18 09	02 23	17 40	24 04	19 53
06	19 05 39	16 30 16	16 ♍ 44 58	23 ♍ 56 19	25 12	28 38	11 39	18 02	02 21	17 37	24 04	19 54
07	19 09 36	17 31 24	01 ♎ 03 33	08 ♎ 06 26	25 37	29 ♐ 53	12 21	17 56	02 18	17 35	24 04	19 56
08	19 13 32	18 32 33	15 ♎ 04 51	21 ♎ 58 45	26 08	01 ♑ 08	13 03	17 50	02 15	17 32	24 04	19 57
09	19 17 29	19 33 41	28 ♎ 48 09	05 ♏ 33 10	26 44	02 23	13 45	17 44	02 13	17 29	24 04	19 59
10	19 21 25	20 34 50	12 ♏ 13 53	18 ♏ 50 28	27 27	03 39	14 27	17 38	02 10	17 27	24 05	20 01
11	19 25 22	21 35 59	25 ♏ 23 05	01 ♐ 51 54	28 14	04 54	15 09	17 33	02 07	17 24	24 05	20 02
12	19 29 18	22 37 08	08 ♐ 17 05	14 ♐ 38 49	29 ♐ 05	06 09	15 51	17 27	02 03	17 22	24 05	20 04
13	19 33 15	23 38 16	20 ♐ 57 17	27 ♐ 12 38	00 ♑ 00	07 24	16 34	17 22	02 00	17 19	24 05	20 05
14	19 37 12	24 39 25	03 ♑ 25 03	09 ♑ 34 41	00 59	08 40	17 16	17 17	01 57	17 17	24 06	20 07
15	19 41 08	25 40 33	15 ♑ 41 42	21 ♑ 46 17	02 01	09 55	17 58	17 12	01 54	17 14	24 06	20 09
16	19 45 05	26 41 40	27 ♑ 48 37	03 ♒ 48 55	03 05	11 10	18 40	17 07	01 50	17 11	24 07	20 10
17	19 49 01	27 42 48	09 ♒ 47 23	15 ♒ 44 16	04 12	12 25	19 23	17 03	01 47	17 09	24 07	20 12
18	19 52 58	28 43 54	21 ♒ 39 51	27 ♒ 34 25	05 21	13 41	20 05	16 58	01 43	17 06	24 08	20 14
19	19 56 54	29 ♑ 45 00	03 ♓ 28 19	09 ♓ 21 54	06 32	14 56	20 47	16 54	01 39	17 04	24 08	20 15
20	20 00 51	00 ♒ 46 05	15 ♓ 15 35	21 ♓ 09 48	07 46	16 11	21 30	16 50	01 36	17 01	24 09	20 17
21	20 04 47	01 47 09	27 ♓ 05 00	03 ♈ 01 42	09 00	17 27	22 12	16 46	01 32	16 59	24 09	20 19
22	20 08 44	02 48 13	09 ♈ 00 25	15 ♈ 01 41	10 17	18 42	22 55	16 42	01 28	16 56	24 10	20 20
23	20 12 41	03 49 15	21 ♈ 06 04	27 ♈ 14 09	11 34	19 57	23 37	16 39	01 24	16 54	24 11	20 22
24	20 16 37	04 50 16	03 ♉ 26 28	09 ♉ 43 35	12 53	21 12	24 20	16 36	01 20	16 51	24 11	20 24
25	20 20 34	05 51 17	16 ♉ 06 01	22 ♉ 34 16	14 14	22 27	25 03	16 33	01 16	16 49	24 12	20 26
26	20 24 30	06 52 16	29 ♉ 08 43	05 ♊ 49 43	15 35	23 43	25 45	16 30	01 12	16 47	24 13	20 27
27	20 28 27	07 53 14	12 ♊ 37 30	19 ♊ 32 11	16 58	24 58	26 28	16 27	01 07	16 44	24 14	20 29
28	20 32 23	08 54 11	26 ♊ 33 43	03 ♋ 41 55	18 21	26 13	27 11	16 24	01 03	16 42	24 14	20 31
29	20 36 20	09 55 07	10 ♋ 56 24	18 ♋ 16 39	19 46	27 28	27 53	16 22	00 59	16 39	24 15	20 33
30	20 40 16	10 56 02	25 ♋ 41 56	03 ♌ 11 22	21 11	28 44	28 36	16 20	00 55	16 37	24 16	20 34
31	20 44 13	11 ♒ 56 56	10 ♌ 43 56	18 ♌ 18 32	22 ♑ 38	29 ♑ 59	29 ♐ 19	16 ♊ 18	00 ♍ 50	16 ♋ 35	24 ♈ 17	20 ♒ 36

Date	Moon True ☊ ° '	Moon Mean ☊ ° '	Moon ☽ Latitude ° '
01	08 ♌ 10	09 ♌ 22	03 S 01
02	08 R 07	09 19	01 49
03	08 06	09 16	00 S 30
04	08 D 06	09 13	00 N 51
05	08 07	09 10	02 09
06	08 09	09 07	03 18
07	08 10	09 03	04 12
08	08 10	09 00	04 51
09	08 R 10	08 57	05 12
10	08 09	08 54	05 15
11	08 07	08 51	05 02
12	08 05	08 48	04 34
13	08 03	08 44	03 52
14	08 01	08 41	03 00
15	07 59	08 38	02 00
16	07 58	08 35	00 N 56
17	07 D 58	08 32	00 S 10
18	07 58	08 28	01 15
19	07 59	08 25	02 17
20	08 00	08 22	03 13
21	08 01	08 19	04 01
22	08 01	08 16	04 38
23	08 02	08 13	05 04
24	08 02	08 09	05 17
25	08 R 02	08 06	05 15
26	08 02	08 03	04 57
27	08 02	08 00	04 22
28	08 D 02	07 57	03 31
29	08 02	07 54	02 25
30	08 02	07 50	01 08
31	08 ♌ 02	07 ♌ 47	00 N 15

DECLINATIONS

Date	Sun ☉ ° '	Moon ☽ ° '	Mercury ☿ ° '	Venus ♀ ° '	Mars ♂ ° '	Jupiter ♃ ° '	Saturn ♄ ° '	Uranus ♅ ° '	Neptune ♆ ° '	Pluto ♇ ° '
01	22 S 57	20 N 23	20 S 12	22 S 39	21 S 37	22 N 27	11 N 57	22 N 41	07 N 45	23 S 00
02	22 51	20 26	20 14	22 46	21 44	22 27	11 58	22 42	07 45	22 59
03	22 45	19 05	20 18	22 52	21 51	22 26	11 59	22 42	07 45	22 59
04	22 39	16 25	20 23	22 57	21 58	22 26	12 00	22 42	07 45	22 58
05	22 32	12 42	20 30	23 01	22 05	22 26	12 01	22 43	07 45	22 57
06	22 25	08 16	20 38	23 05	22 11	22 25	12 02	22 43	07 45	22 57
07	22 17	03 N 26	20 46	23 08	22 17	22 25	12 04	22 43	07 45	22 56
08	22 09	01 S 28	20 56	23 10	22 24	22 25	12 05	22 44	07 45	22 56
09	22 00	06 11	21 05	23 11	22 29	22 24	12 06	22 44	07 45	22 55
10	21 51	10 30	21 15	23 12	22 35	22 24	12 07	22 44	07 45	22 55
11	21 42	14 13	21 25	23 12	22 40	22 24	12 09	22 45	07 45	22 54
12	21 32	17 11	21 35	23 12	22 46	22 23	12 10	22 45	07 46	22 54
13	21 22	19 16	21 45	23 11	22 51	22 23	12 11	22 45	07 46	22 53
14	21 11	20 23	21 54	23 09	22 56	22 23	12 13	22 46	07 46	22 52
15	21 00	20 31	22 03	23 06	23 00	22 22	12 14	22 46	07 46	22 52
16	20 49	19 41	22 11	23 02	23 05	22 22	12 15	22 46	07 46	22 51
17	20 37	17 57	22 18	22 58	23 09	22 22	12 17	22 47	07 47	22 51
18	20 25	15 28	22 25	22 53	23 13	22 22	12 18	22 47	07 47	22 50
19	20 12	12 22	22 31	22 48	23 17	22 21	12 20	22 47	07 47	22 50
20	19 59	08 46	22 36	22 42	23 20	22 21	12 21	22 48	07 47	22 49
21	19 45	04 50	22 41	22 35	23 23	22 21	12 23	22 48	07 48	22 48
22	19 32	00 S 42	22 44	22 27	23 27	22 21	12 24	22 48	07 48	22 48
23	19 18	03 N 31	22 46	22 19	23 29	22 21	12 26	22 49	07 48	22 47
24	19 03	07 41	22 47	22 10	23 32	22 21	12 28	22 49	07 48	22 47
25	18 48	11 37	22 47	22 00	23 35	22 21	12 29	22 49	07 49	22 46
26	18 33	15 08	22 46	21 50	23 37	22 20	12 31	22 50	07 49	22 46
27	18 17	17 58	22 44	21 39	23 39	22 20	12 32	22 50	07 49	22 45
28	18 02	19 52	22 40	21 27	23 41	22 20	12 34	22 50	07 50	22 44
29	17 45	20 34	22 36	21 15	23 42	22 20	12 36	22 50	07 50	22 44
30	17 29	19 53	22 30	21 02	23 43	22 20	12 37	22 51	07 51	22 43
31	17 S 12	17 N 47	22 S 23	20 S 48	23 S 45	22 N 20	12 N 39	22 N 51	07 N 51	22 S 43

ZODIAC SIGN ENTRIES

Date	h	m	Planets
01	06	47	☽ ♋
03	07	47	☽ ♌
05	08	25	☽ ♍
07	10	12	☽ ♎
07	14	16	♀ ♑
09	14	07	☽ ♏
11	20	32	☽ ♐
13	11	52	☿ ♑
14	05	23	☽ ♑
16	16	22	☽ ♒
19	04	56	☽ ♓
19	17	54	☉ ♒
21	17	54	☽ ♈
24	05	22	☽ ♉
26	13	33	☽ ♊
28	17	48	☽ ♋
30	18	54	☽ ♌
31	12	22	♀ ♒

LATITUDES

Date	Mercury ☿ ° '	Venus ♀ ° '	Mars ♂ ° '	Jupiter ♃ ° '	Saturn ♄ ° '	Uranus ♅ ° '	Neptune ♆ ° '	Pluto ♇ ° '
01	03 N 09	00 N 33	00 N 03	00 S 29	01 N 29	00 N 27	01 S 43	08 S 34
04	02 56	00 26	00 N 01	00 29	01 30	00 27	01 42	08 34
07	02 35	00 18	00 S 01	00 28	01 31	00 27	01 42	08 34
10	02 09	00 11	00 03	00 28	01 31	00 27	01 42	08 34
13	01 41	00 N 03	00 06	00 27	01 32	00 27	01 42	08 34
16	01 13	00 S 05	00 08	00 27	01 32	00 27	01 42	08 33
19	00 45	00 12	00 10	00 26	01 33	00 27	01 42	08 33
22	00 N 18	00 19	00 12	00 25	01 33	00 27	01 41	08 33
25	00 S 07	00 27	00 14	00 25	01 34	00 27	01 41	08 33
28	00 30	00 33	00 16	00 24	01 34	00 27	01 41	08 33
31	00 S 51	00 S 40	00 S 19	00 S 24	01 N 35	00 N 27	01 S 41	08 S 33

DATA

Julian Date	2465060
Delta T	+76 seconds
Ayanamsa	24° 22' 13"
Synetic vernal point	04° ♓ 44' 45"
True obliquity of ecliptic	23° 25' 58"

MOON'S PHASES, APSIDES AND POSITIONS ☽

Date	h	m	Phase	Longitude ° '	Eclipse Indicator
02	02	35	○	12 ♋ 02	
08	18	29	☾	18 ♎ 49	
16	09	34	●	26 ♑ 35	Partial
24	14	55	☽	04 ♉ 58	
31	14	04	○	12 ♌ 02	total

Day	h	m	
03	21	35	Perigee
19	17	50	Apogee

Day	h	m		
02	00	57	Max dec	20° N 35'
08	04	48	0S	
15	03	04	Max dec	20° S 35'
22	15	58	0N	
29	12	24	Max dec	20° N 34'

LONGITUDES

Date	Chiron ⚷ ° '	Ceres ⚳ ° '	Pallas ⚴ ° '	Juno ⚵ ° '	Vesta ⚶ ° '	Black Moon Lilith ⚸ ° '
01	14 ♊ 46	20 ♎ 37	04 ♍ 04	13 ♍ 22	03 ♒ 49	28 ♒ 54
11	14 ♊ 18	23 ♎ 23	03 ♍ 08	13 ♍ 16	08 ♒ 57	00 ♓ 00
21	13 ♊ 55	25 ♎ 44	01 ♍ 08	12 ♍ 23	14 ♒ 06	01 ♓ 07
31	13 ♊ 38	27 ♎ 36	28 ♌ 15	10 ♍ 46	19 ♒ 14	02 ♓ 13

ASPECTARIAN

h m	Aspects
01 Thursday	
07 02	☽ # ☿
11 00	☽ ⚹ ♄
14 42	☽ bQ ♇
16 48	☽ Q ♆
19 18	♀ # ♅
20 39	☽ ⚻ ♂
02 Friday	
02 35	☽ ☍ ☉
05 29	☽ ± ♇
06 57	☽ ± ♂
11 31	☽ ∠ ♄
11 59	☽ ☌ ♅
13 05	☽ ⚺ ♃
15 16	☽ ⚻ ♇
17 32	☽ # ☿
19 21	♂ bQ ♆
20 37	♀ △ ♆
22 10	☽ □ ♆
22 19	☽ bQ ♂
22 19	☽ ⚻ ♀
22 45	☽ ⊥ ♃
22 51	☉ ∥ ♀
23 24	☽ ⚻ ☿
03 Saturday	
02 06	☽ ⊥ ♄
08 56	☽ ± ♀
09 02	☽ ± ☿
10 07	☿ ☌ ♀
11 47	☽ ⚺ ♄
13 13	☽ ∠ ♃
17 45	♆ St D
21 15	☉ ⊥ ♇
22 58	☿ St D
23 33	☽ bQ ☿
23 44	☉ # ♅
23 47	☽ △ ♂
04 Sunday	
00 49	☽ bQ ♀
06 50	☽ ⚻ ☉
12 26	☽ ⚺ ♅
13 18	☽ ⚹ ♃
15 55	☽ ☍ ♇
17 16	☽ ± ☉
18 20	♀ ∥ ♇
22 09	☽ ⊥ ♅
22 45	☽ △ ♆
23 59	☽ △ ☿
05 Monday	
03 24	☽ △ ♀
08 41	☽ Q ♃
09 01	☽ bQ ☉
12 20	☽ ☌ ♄
12 47	☽ ∠ ♅
15 52	☽ ∥ ♄
23 17	☽ bQ ♆
06 Tuesday	
03 07	☽ □ ♂
10 17	☉ # ♃
11 15	♂ ± ♅
11 34	☽ △ ☉
13 27	☽ ⚹ ♅
14 08	☽ □ ♃
14 11	☽ ± ♆
14 37	☽ ∥ ♆
17 16	☽ ⚻ ♇
07 Wednesday	
00 13	☽ ⚻ ♆
02 31	☽ □ ☿
03 20	☽ ± ♇
07 00	☉ bQ ♄
09 30	☽ Q ♅
09 49	☽ □ ♀
10 44	☽ Q ♂
11 20	☉ ∥ ♂
13 14	☉ ☍ ♅
14 06	☽ ⚺ ♄
18 35	☽ bQ ♇
20 46	☉ ⚻ ♃
08 Thursday	
00 18	☽ ⊥ ♄
08 18	☽ ⚹ ♂
10 17	☽ Q ☿
15 45	☽ ∠ ♄
15 56	♂ # ♃
16 14	☽ □ ♅
16 44	☽ △ ♃
18 29	☽ □ ☉
19 45	☽ Q ♀
20 29	☽ △ ♇
09 Friday	
03 40	☽ ☍ ♆
08 11	☽ ⚹ ☿
08 41	♀ △ ♄
11 54	☽ ∠ ♂
18 01	☽ ⚹ ♄
18 55	☽ bQ ♃
19 01	☽ ⚹ ♀
20 26	☽ # ♆
22 13	☉ ⚺ ♇
10 Saturday	
02 03	☉ ⊥ ♂
04 49	☽ ⊥ ♂
04 53	☽ Q ☉
10 56	☽ ± ♃
12 25	☽ ∠ ☿
15 29	☽ Q ♄
16 14	☽ ⚺ ♂
21 26	☽ △ ♅
21 44	☽ ⚻ ♃
22 02	☽ # ♄
11 Sunday	
00 52	☽ ∠ ♀
02 10	☽ □ ♇
04 27	☽ ⚹ ☉
05 49	☽ ⊥ ☿
09 36	☽ ⚻ ♆
14 43	♀ ∠ ♇
17 37	☽ ⚺ ☿
19 11	☽ ⊥ ♀
20 41	☽ ± ♆
12 Monday	
00 24	☽ □ ♄
00 58	☽ bQ ♅
07 34	☽ ⚺ ♀
08 32	☉ ∥ ☿
08 53	♂ # ♅
10 39	☽ ∠ ☉
11 35	☽ Q ♇
13 30	☽ bQ ♆
17 46	☽ ± ♅
13 Tuesday	
03 08	☽ ☌ ♂
05 06	☽ ⚻ ♅
05 07	☽ ⊥ ☉
05 12	☽ ☍ ♃
06 05	☉ ± ♃
10 21	☽ ⚹ ♇
17 35	☽ ⚺ ☉
18 00	☽ △ ♆
21 31	♂ ∥ ♇
22 44	☉ □ ♆
14 Wednesday	
06 52	☽ ☌ ☿
09 10	☽ △ ♄
12 25	♂ ⚻ ♅
12 32	♂ ☍ ♃
14 44	♃ ⚺ ♅
15 19	☽ ∠ ♇
23 22	☽ ☌ ♀
15 Thursday	
08 57	☽ ⊥ ♇
09 29	☿ △ ♄
14 21	☽ bQ ♄
14 56	☽ ⚻ ♃
15 01	☽ ☍ ♅
16 45	☽ ⚺ ♂
16 53	☉ ± ♄
20 48	☽ ⚺ ♇
16 Friday	
02 44	☽ ± ♃
04 38	☽ □ ♆
05 20	♀ ∥ ♂
05 22	☽ ⊥ ♂
08 05	☽ ± ♄
09 34	☽ ☌ ☉
20 00	☽ ⚻ ♄
20 33	☽ bQ ♃
23 36	☽ ⚺ ☿
17 Saturday	
00 27	☽ ∠ ♂
12 55	☽ ⊥ ☿
16 42	☽ Q ♆
17 56	☽ ⚺ ♀
23 13	☿ # ♃
18 Sunday	
02 33	☽ △ ♃
02 48	☽ ⚻ ♅
07 30	☽ ⊥ ♀
08 36	☽ ⚹ ♂
09 03	☽ ∠ ☿
09 05	☽ ☌ ♇
09 22	☿ ∠ ♇
14 55	☽ ± ♅
17 00	☽ ⚹ ♆
17 06	♂ ⚹ ♇
22 46	♀ ⊥ ♇
19 Monday	
03 43	☽ ⚺ ☉
03 56	☽ ∠ ♀
04 20	♀ ∥ ♇
08 19	☽ ☍ ♄
09 09	☽ bQ ♅
10 32	☽ Q ♂
12 13	☽ # ♄
13 33	♀ # ♅
17 05	☽ ⊥ ☉
18 58	☽ ⚹ ☿
23 32	☽ ∠ ♆
20 Tuesday	
13 08	☽ ∠ ☉
14 07	☽ ⚹ ♀
15 11	☽ □ ♃
15 34	☽ △ ♅
17 52	☽ ⊥ ♆
18 09	☽ # ♆
19 24	♀ bQ ♄
22 13	☽ Q ☿
22 14	☽ ⚺ ♇
23 01	☿ ∥ ♀
23 42	♀ ⚻ ♃
21 Wednesday	
01 29	☽ □ ♂
03 26	♀ ☍ ♅
06 04	☽ ⚺ ♆
06 20	☉ ⚻ ♄
10 26	☽ ⊥ ♇
11 34	☉ bQ ♃
17 20	☽ Q ♀
19 55	♅ # ♇
20 56	☽ ⚻ ♄
22 23	☽ ⚹ ☉
22 Thursday	
03 25	☽ Q ♃
04 38	☽ ∠ ♇
08 56	☽ ± ♄
14 51	☽ □ ☿
23 Friday	
00 37	☽ Q ☉
02 46	☽ bQ ♄
03 15	☽ ⚹ ♃
03 44	☽ □ ♅
06 01	♀ # ♃
09 28	☽ □ ♀
10 33	☽ ⚹ ♇
17 15	☽ △ ♂
18 02	☽ ☌ ♆
20 14	♀ ⚺ ♇
24 Saturday	
03 18	☿ ∥ ♇
07 01	♂ △ ♆
07 58	☽ △ ♄
08 27	☽ ∠ ♃
09 59	☽ Q ♇
12 44	☽ ∥ ♆
14 42	☽ Q ♅
14 55	☽ □ ☉
23 55	☽ bQ ♂
25 Sunday	
01 35	☽ ⊥ ♃
08 05	☽ △ ☿
12 49	☽ ⚺ ♃
13 20	☽ ⚹ ♅
13 32	♀ ± ♃
15 36	☿ ⊥ ♇
17 39	☽ ∥ ♄
17 48	☽ ± ♂
20 04	☽ □ ♇
26 Monday	
01 03	☽ △ ♀
03 00	☽ ⚺ ♆
05 29	☽ ⚻ ♂
13 56	☽ ⊥ ♆
14 54	☽ bQ ☿
15 41	☽ □ ♄
16 44	☽ ∠ ♅
21 41	♀ □ ♆
21 56	☿ ∥ ♇
22 07	☿ bQ ♄
27 Tuesday	
02 59	☽ △ ☉
03 21	☿ ⚻ ♃
06 01	☽ ∠ ♆
06 51	☽ bQ ♀
08 11	☿ ☍ ♅
08 42	☽ ⊥ ♅
08 45	☽ ± ☿
14 52	♀ ± ♄
14 56	☽ # ☉
18 39	☽ ☌ ♃
19 08	☽ ⚺ ♅
20 24	☽ ⚻ ☿
23 14	☽ Q ♄
28 Wednesday	
00 06	☽ ± ♀
01 40	☽ △ ♇
07 08	☽ bQ ☉
08 03	☽ ⚹ ♆
11 22	☽ ⚻ ♀
13 06	☽ ☍ ♂
19 32	☽ ⚹ ♄
23 29	☽ ± ☉
29 Thursday	
03 03	☽ bQ ♇
04 15	☽ Q ♆
10 11	☽ ⚻ ☉
17 55	☽ ± ♇
20 13	☽ ∠ ♄
20 52	☽ ⚺ ♃
21 20	☽ ☌ ♅
30 Friday	
01 27	☿ ⚺ ♇
03 43	☽ ⚻ ♇
03 57	☽ ☍ ☿
06 21	♀ ⚺ ♂
06 35	☽ ⊥ ♃
09 42	☽ □ ♆
10 44	☽ ⊥ ♄
16 53	☽ ⚻ ♂
17 18	☽ ☍ ♀
20 19	☽ ⚺ ♄
21 01	☽ ∠ ♃
31 Saturday	
02 57	☽ ± ♂
06 43	☿ ± ♃
14 04	☽ ☍ ☉
17 15	☽ # ☉
17 57	☽ bQ ♂
19 44	☿ # ♃
20 05	☉ Q ♆
20 49	☽ ⚹ ♃
21 15	☽ ⚺ ♅

All ephemeris data is given at 12.00 UT and the Moon's longitude is additionally given for 24.00 UT

FEBRUARY 2037

LONGITUDES

Date	Sidereal time h m s	Sun ☉ ° ' "	Moon ☽ ° ' "	Moon ☽ 24.00 ° ' "	Mercury ☿ ° '	Venus ♀ ° '	Mars ♂ ° '	Jupiter ♃ ° '	Saturn ♄ ° '	Uranus ♅ ° '	Neptune ♆ ° '	Pluto ♇ ° '
01	20 48 10	12 ♒ 57 49	25 ♌ 53 57	03 ♍ 28 59	24 ♑ 05	01 ♒ 14	00 ♑ 02	16 ♊ 17	00 ♍ 46	16 ♋ 33	24 ♈ 18	20 ♒ 38
02	20 52 06	13 58 40	11 ♍ 02 26	18 ♍ 33 13	25 33	02 29	00 44	16 R 15	00 R 41	16 R 30	24 19	20 40
03	20 56 03	14 59 31	26 ♍ 00 19	03 ♎ 22 51	27 02	03 44	01 27	16 14	00 37	16 28	24 20	20 41
04	20 59 59	16 00 21	10 ♎ 40 09	17 ♎ 51 40	28 ♑ 32	05 00	02 10	16 13	00 32	16 26	24 21	20 43
05	21 03 56	17 01 10	24 ♎ 57 03	01 ♏ 56 07	00 ♒ 02	06 15	02 53	16 12	00 27	16 24	24 22	20 45
06	21 07 52	18 01 58	08 ♏ 48 47	15 ♏ 35 11	01 34	07 30	03 36	16 11	00 23	16 22	24 23	20 47
07	21 11 49	19 02 45	22 ♏ 15 28	28 ♏ 49 57	03 06	08 45	04 19	16 11	00 18	16 20	24 25	20 49
08	21 15 45	20 03 31	05 ♐ 18 58	11 ♐ 42 55	04 39	10 00	05 02	16 10	00 13	16 18	24 26	20 50
09	21 19 42	21 04 16	18 ♐ 02 12	24 ♐ 17 18	06 12	11 16	05 45	16 D 10	00 09	16 16	24 27	20 52
10	21 23 39	22 05 01	00 ♑ 28 37	06 ♑ 36 35	07 47	12 31	06 28	16 10	00 ♍ 04	16 14	24 28	20 54
11	21 27 35	23 05 44	12 ♑ 41 39	18 ♑ 44 11	09 22	13 46	07 11	16 11	29 ♌ 59	16 12	24 29	20 56
12	21 31 32	24 06 26	24 ♑ 44 35	00 ♒ 43 10	10 58	15 01	07 55	16 11	29 54	16 10	24 31	20 57
13	21 35 28	25 07 07	06 ♒ 40 18	12 ♒ 36 16	12 35	16 16	08 38	16 12	29 49	16 08	24 32	20 59
14	21 39 25	26 07 46	18 ♒ 31 21	24 ♒ 25 50	14 13	17 31	09 21	16 13	29 44	16 06	24 33	21 01
15	21 43 21	27 08 24	00 ♓ 19 57	06 ♓ 13 58	15 52	18 47	10 04	16 14	29 40	16 04	24 35	21 03
16	21 47 18	28 09 00	12 ♓ 08 08	18 ♓ 02 42	17 32	20 02	10 47	16 16	29 35	16 03	24 36	21 05
17	21 51 14	29 ♒ 09 35	23 ♓ 57 55	29 ♓ 54 04	19 12	21 17	11 31	16 17	29 30	16 01	24 38	21 06
18	21 55 11	00 ♓ 10 09	05 ♈ 51 26	11 ♈ 50 21	20 53	22 32	12 14	16 19	29 25	15 59	24 39	21 08
19	21 59 08	01 10 40	17 ♈ 51 07	23 ♈ 54 08	22 36	23 47	12 57	16 21	29 20	15 58	24 41	21 10
20	22 03 04	02 11 10	29 ♈ 59 46	06 ♉ 08 26	24 19	25 02	13 41	16 23	29 15	15 56	24 42	21 12
21	22 07 01	03 11 38	12 ♉ 20 35	18 ♉ 36 38	26 03	26 17	14 24	16 25	29 10	15 54	24 44	21 13
22	22 10 57	04 12 04	24 ♉ 57 03	01 ♊ 22 18	27 48	27 32	15 08	16 28	29 06	15 53	24 45	21 15
23	22 14 54	05 12 28	07 ♊ 52 50	14 ♊ 29 02	29 ♒ 34	28 ♒ 47	15 51	16 30	29 01	15 52	24 47	21 17
24	22 18 50	06 12 51	21 ♊ 11 17	27 ♊ 59 52	01 ♓ 21	00 ♓ 02	16 35	16 33	28 56	15 50	24 48	21 19
25	22 22 47	07 13 11	04 ♋ 55 01	11 ♋ 56 48	03 09	01 17	17 18	16 37	28 51	15 49	24 50	21 20
26	22 26 43	08 13 30	19 ♋ 05 10	26 ♋ 19 55	04 58	02 32	18 02	16 40	28 46	15 47	24 52	21 22
27	22 30 40	09 13 46	03 ♌ 40 40	11 ♌ 06 50	06 48	03 47	18 45	16 43	28 42	15 46	24 53	21 24
28	22 34 37	10 ♓ 14 01	18 ♌ 37 38	26 ♌ 12 08	08 ♓ 38	05 ♓ 02	19 ♑ 29	16 ♊ 47	28 ♌ 37	15 ♋ 45	24 ♈ 55	21 ♒ 25

Date	Moon True ☊ ° '	Moon Mean ☊ ° '	Moon ☽ Latitude ° '
01	08 ♌ 02	07 ♌ 44	01 N 38
02	08 R 01	07 41	02 54
03	08 01	07 38	03 56
04	08 00	07 34	04 42
05	07 59	07 31	05 09
06	07 58	07 28	05 18
07	07 D 58	07 25	05 08
08	07 59	07 22	04 42
09	08 00	07 19	04 03
10	08 01	07 15	03 14
11	08 02	07 12	02 16
12	08 03	07 09	01 13
13	08 04	07 06	00 N 08
14	08 R 03	07 03	00 S 58
15	08 02	07 00	02 00
16	08 00	06 56	02 57
17	07 56	06 53	03 47
18	07 52	06 50	04 27
19	07 49	06 47	04 56
20	07 45	06 44	05 12
21	07 42	06 40	05 13
22	07 41	06 37	05 00
23	07 D 40	06 34	04 32
24	07 41	06 31	03 48
25	07 42	06 28	02 51
26	07 44	06 25	01 41
27	07 45	06 21	00 22
28	07 ♌ 45	06 ♌ 18	00 N 59

DECLINATIONS

Date	Sun ☉ ° '	Moon ☽ ° '	Mercury ☿ ° '	Venus ♀ ° '	Mars ♂ ° '	Jupiter ♃ ° '	Saturn ♄ ° '	Uranus ♅ ° '	Neptune ♆ ° '	Pluto ♇ ° '
01	16 S 55	14 N 25	22 S 14	20 S 34	23 S 45	22 N 20	12 N 41	22 N 51	07 N 51	22 S 42
02	16 38	10 06	22 05	20 19	23 46	22 20	12 43	22 52	07 52	22 42
03	16 20	05 12	21 54	20 04	23 47	22 20	12 44	22 52	07 52	22 41
04	16 02	00 N 07	21 42	19 48	23 47	22 20	12 46	22 52	07 53	22 40
05	15 44	04 S 51	21 28	19 31	23 47	22 21	12 48	22 52	07 53	22 40
06	15 25	09 25	21 13	19 14	23 46	22 21	12 50	22 53	07 54	22 39
07	15 07	13 22	20 57	18 56	23 46	22 21	12 51	22 53	07 54	22 39
08	14 47	16 33	20 40	18 38	23 45	22 21	12 53	22 53	07 54	22 38
09	14 28	18 51	20 21	18 19	23 44	22 21	12 55	22 53	07 55	22 38
10	14 09	20 12	20 01	18 00	23 43	22 21	12 57	22 54	07 55	22 37
11	13 49	20 34	19 39	17 40	23 42	22 22	12 59	22 54	07 56	22 37
12	13 29	19 58	19 16	17 20	23 40	22 22	13 00	22 54	07 56	22 36
13	13 09	18 29	18 52	16 59	23 38	22 22	13 02	22 54	07 57	22 36
14	12 48	16 11	18 27	16 38	23 36	22 22	13 04	22 54	07 58	22 35
15	12 28	13 13	18 00	16 16	23 34	22 23	13 06	22 55	07 58	22 35
16	12 07	09 44	17 31	15 54	23 31	22 23	13 08	22 55	07 59	22 34
17	11 46	05 52	17 02	15 31	23 28	22 23	13 09	22 55	07 59	22 34
18	11 25	01 S 46	16 31	15 08	23 25	22 24	13 11	22 55	08 00	22 33
19	11 03	02 N 27	15 58	14 44	23 22	22 24	13 13	22 55	08 00	22 33
20	10 42	06 36	15 24	14 20	23 18	22 25	13 15	22 56	08 01	22 32
21	10 20	10 33	14 49	13 56	23 15	22 25	13 17	22 56	08 02	22 32
22	09 58	14 08	14 13	13 31	23 11	22 25	13 18	22 56	08 02	22 31
23	09 36	17 09	13 35	13 06	23 06	22 26	13 20	22 56	08 03	22 31
24	09 14	19 21	12 56	12 41	23 02	22 26	13 22	22 56	08 03	22 30
25	08 51	20 30	12 15	12 15	22 57	22 27	13 24	22 56	08 04	22 30
26	08 29	20 25	11 33	11 49	22 53	22 27	13 25	22 57	08 05	22 29
27	08 06	18 58	10 50	11 23	22 47	22 28	13 27	22 57	08 05	22 29
28	07 S 44	16 N 11	10 S 06	10 S 56	22 S 42	22 N 28	13 N 29	22 N 57	08 N 06	22 S 28

ZODIAC SIGN ENTRIES

Date	h m	Planets
01	11 08	♂ ♑
01	18 29	☽ ♍
03	18 29	☽ ♎
05	11 26	☿ ♒
05	20 39	☽ ♏
08	02 09	☽ ♐
10	11 04	☽ ♑
11	06 46	♄ ♌
12	22 33	☽ ♒
15	11 19	☽ ♓
18	00 12	☽ ♈
18	07 59	☉ ♓
20	12 00	☽ ♉
22	21 27	☽ ♊
23	17 52	☿ ♓
24	11 15	♀ ♓
25	03 30	☽ ♋
27	06 01	☽ ♌

LATITUDES

Date	Mercury ☿ ° '	Venus ♀ ° '	Mars ♂ ° '	Jupiter ♃ ° '	Saturn ♄ ° '	Uranus ♅ ° '	Neptune ♆ ° '	Pluto ♇ ° '
01	00 S 58	00 S 42	00 S 19	00 S 23	01 N 35	00 N 27	01 S 41	08 S 33
04	01 16	00 48	00 22	00 23	01 35	00 27	01 41	08 33
07	01 32	00 54	00 24	00 22	01 36	00 27	01 41	08 34
10	01 45	01 00	00 27	00 22	01 36	00 27	01 40	08 34
13	01 56	01 05	00 29	00 21	01 37	00 27	01 40	08 34
16	02 03	01 09	00 31	00 20	01 37	00 27	01 40	08 34
19	02 06	01 13	00 34	00 20	01 37	00 27	01 40	08 35
22	02 07	01 17	00 36	00 19	01 37	00 27	01 40	08 35
25	02 03	01 20	00 39	00 19	01 38	00 27	01 40	08 35
28	01 54	01 22	00 41	00 18	01 38	00 27	01 40	08 36
31	01 S 41	01 S 24	00 S 44	00 S 18	01 N 38	00 N 27	01 S 39	08 S 36

DATA

Julian Date	2465091
Delta T	+76 seconds
Ayanamsa	24° 22' 19"
Synetic vernal point	04° ♓ 44' 40"
True obliquity of ecliptic	23° 25' 58"

LONGITUDES

Date	Chiron ⚷ ° '	Ceres ⚳ ° '	Pallas ⚴ ° '	Juno ⚵ ° '	Vesta ⚶ ° '	Black Moon Lilith ⚸ ° '
01	13 ♊ 36	27 ♎ 45	27 ♌ 56	10 ♍ 34	19 ♒ 45	02 ♓ 20
11	13 ♊ 27	28 ♎ 59	24 ♌ 34	08 ♍ 19	24 ♒ 51	03 ♓ 27
21	13 ♊ 26	29 ♎ 34	21 ♌ 18	05 ♍ 46	29 ♒ 56	04 ♓ 33
31	13 ♊ 32	29 ♎ 26	18 ♌ 36	03 ♍ 15	04 ♓ 59	05 ♓ 40

MOON'S PHASES, APSIDES AND POSITIONS ☽

Date	h m	Phase	Longitude ° '	Eclipse Indicator
07	05 43	☾	18 ♏ 47	
15	04 54	●	26 ♒ 50	
23	06 41	☽	04 ♊ 59	

Day	h m		
01	07 06	Perigee	
16	00 03	Apogee	
04	12 31	OS	
11	08 58	Max dec	20° S 35'
18	22 04	ON	
25	22 29	Max dec	20° N 38'

ASPECTARIAN

01 Sunday
03 27 ♀ ⊼ ♄
03 39 ☽ ☍ ♇
06 43 ☽ ⊥ ♅
08 49 ☽ ⊼ ☿
09 28 ☽ △ ♆
12 46 ♀ bQ ♃
15 39 ☿ □ ♆
15 45 ☽ Q ♃
18 51 ☽ △ ♂
19 19 ☽ ± ☿
19 39 ☽ ☌ ♄
20 54 ☽ ∠ ♅
21 12 ☽ ⊼ ♀
22 06 ☽ ∥ ♄
22 35 ☿ ± ♄

02 Monday
07 34 ☽ ± ♀
09 15 ☽ bQ ♆
10 21 ♂ △ ♄
11 08 ☽ bQ ☿
17 01 ☽ ⊼ ☉
20 18 ☽ □ ♃
20 42 ☽ ⚹ ♅
23 11 ☽ ∥ ♆
23 14 ☽ bQ ♀
23 38 ☽ ± ♆

03 Tuesday
03 20 ☽ ± ☉
03 25 ☽ ⊼ ♇
09 18 ☽ ⊼ ♆
13 07 ☽ ± ♇
13 51 ☽ △ ☿
15 59 ☽ Q ♅
18 57 ☽ bQ ☉
19 26 ☽ ⊻ ♄
21 18 ☽ □ ♂

04 Wednesday
01 46 ☽ △ ♀
03 49 ☽ bQ ♇
05 12 ☽ ⊥ ♄
16 47 ☉ △ ♃
20 03 ☽ ∠ ♄
21 13 ☽ △ ♃
21 34 ☽ △ ☉
21 35 ☽ □ ♅
21 47 ☉ ⊼ ♅

05 Thursday
04 44 ☽ Q ♂
04 51 ☽ △ ♇
11 01 ☽ ☍ ♆
18 19 ☿ ⊼ ♄
21 23 ☽ ⚹ ♄
21 47 ☽ □ ☿
22 43 ☽ bQ ♃

06 Friday
02 24 ☽ ⚹ ♂
03 40 ☽ # ♆
06 10 ☿ bQ ♃
09 28 ☽ □ ♀
14 25 ☽ ± ♃
18 15 ☽ Q ♄

07 Saturday
01 04 ☽ ⊼ ♃
01 21 ☽ △ ♅
05 43 ☽ □ ☉
06 23 ☽ ∠ ♂
08 38 ☽ # ♄
09 22 ☽ □ ♇
09 37 ☽ Q ☿
15 55 ☽ ⊼ ♆
21 03 ☽ Q ♀
23 15 ☽ ∥ ☉
23 42 ☽ ⊥ ♂

08 Sunday
02 37 ☽ □ ♄
02 55 ☽ ± ♆
04 33 ☽ bQ ♅
10 06 ☉ ∠ ♂
10 35 ☽ ⚹ ☿
11 27 ☽ ⊻ ♂
17 34 ☽ Q ☉
18 36 ☽ Q ♇
19 42 ☽ bQ ♆
21 18 ☽ ± ♅
21 44 ☽ ⚹ ♀
23 11 ☿ ⊻ ♂

09 Monday
06 20 ☽ ∥ ♀
07 04 ☉ ☌ ♇
07 41 ♃ St D
08 27 ☽ ☍ ♃
08 37 ☽ ⊼ ♅
16 02 ♂ ∠ ♇
17 26 ☽ ⚹ ♇
18 19 ☽ ⚹ ☉
18 56 ☽ ∠ ☿

10 Tuesday
00 20 ☽ △ ♆
05 35 ☽ ∠ ♀
08 27 ☽ ∥ ☿
10 09 ♀ ⊥ ♂
11 09 ♀ Q ♆
11 12 ☽ △ ♄
14 55 ☽ ⊥ ☿
15 18 ☉ ± ♅
22 38 ☽ ∠ ♇

11 Wednesday
00 28 ☽ ☌ ♂
01 10 ☽ ⊥ ♀
02 06 ☽ ∠ ☉
04 27 ☽ ⊻ ☿
14 22 ☽ ⊻ ♀
16 26 ☽ ⊥ ♇
16 30 ☽ bQ ♄
18 55 ☽ ⊼ ♃
18 56 ☽ ☍ ♅
21 18 ♃ ⊻ ♅
21 32 ☽ ⊥ ☉

12 Thursday
04 25 ☽ ⊻ ♇
06 53 ☽ ± ♃
10 20 ☽ ± ♄
10 37 ☽ ⊻ ☉
11 32 ☽ □ ♆
21 49 ☉ ⚹ ♆
22 17 ☽ ⊼ ♄

13 Friday
00 57 ☽ bQ ♃
05 10 ☽ ∥ ☿
09 24 ♀ ⊼ ♅
10 39 ♀ △ ♃
11 09 ☿ Q ♆
16 13 ☽ ⊻ ♂
19 00 ☉ # ♄
23 53 ☽ Q ♆

14 Saturday
01 53 ☽ ☌ ☿
05 09 ☽ ⊥ ♂
07 06 ☽ ⊼ ♅
07 13 ☽ ∥ ♀
07 19 ☽ △ ♃
09 44 ☽ ☌ ♀
17 05 ☽ ☌ ♇
19 15 ☽ ± ♅

15 Sunday
00 17 ☽ ⚹ ♆
00 36 ☽ ∠ ♂
04 54 ☽ ☌ ☉
10 39 ☽ ☍ ♄
12 55 ☽ # ♄
13 30 ☽ bQ ♅
14 55 ☿ ⊼ ♅
17 14 ☿ ⊥ ♂
17 25 ☿ △ ♃
18 10 ☽ ∥ ☉

16 Monday
06 51 ☽ ∠ ♆
09 05 ☽ ⚹ ♂
19 55 ☽ △ ♅
20 24 ☽ □ ♃
23 08 ☽ # ♆

17 Tuesday
00 45 ☽ ⊻ ☿
01 10 ☽ ⊥ ♆
05 55 ☽ ⊻ ♀
06 12 ☽ ⊻ ♇
08 35 ♀ ☌ ♇
11 01 ☽ Q ♂
13 21 ☽ ⊻ ♆
14 55 ☽ ⊥ ☿
18 22 ☽ ⊥ ♇
19 27 ☉ ☍ ♄
19 30 ☽ ⊥ ♀
23 07 ☽ ⊼ ♄
23 29 ☽ ⊻ ☉

18 Wednesday
01 46 ♀ ± ♅
08 53 ☽ Q ♃
11 07 ☽ ± ♄
12 05 ☽ ∠ ☿
12 34 ☽ ∠ ♇
12 41 ☽ ⊥ ☉
15 32 ☿ ☌ ♇
15 46 ☽ ∠ ♀

19 Thursday
01 37 ☽ □ ♂
03 14 ☿ ± ♅
05 02 ☽ bQ ♄
06 57 ☉ bQ ♅
08 14 ☽ □ ♅
08 22 ☽ ∠ ☉
09 00 ☽ ⚹ ♃
18 36 ☽ ⚹ ♇
22 58 ☽ ⚹ ☿

20 Friday
01 07 ☽ ⚹ ♀
01 34 ☽ ☌ ♆
05 29 ♀ ⚹ ♆
10 33 ☽ △ ♄
14 43 ☽ ∠ ♃
16 40 ☽ ⚹ ☉
17 29 ☿ ⚹ ♆
18 16 ☽ Q ♇
19 41 ☽ Q ♅
20 25 ☽ ∥ ♆

21 Saturday
02 21 ☽ Q ☿
03 17 ☽ Q ♀
05 10 ♂ bQ ♄
08 17 ☽ ⊥ ♃
10 42 ☽ # ☉
16 12 ☽ △ ♂
17 57 ☽ Q ☉
18 50 ☽ ⚹ ♅
19 51 ☽ ⊻ ♃
23 32 ☿ ☌ ♀

22 Sunday
05 00 ☽ □ ♇
06 04 ☽ ∥ ♄
08 06 ☽ # ♀
11 38 ☽ ⊻ ♆
12 28 ☽ # ☿
16 18 ♂ ⊥ ♇
17 23 ☽ □ ♀
18 11 ☽ □ ☿
19 43 ☽ □ ♄
22 16 ☽ bQ ♂
22 53 ☽ ⊥ ♆
23 04 ☽ ∠ ♅
23 43 ♀ # ♄

23 Monday
04 51 ☿ ☍ ♄
06 41 ☽ □ ☉
12 14 ♂ ☍ ♅
15 29 ☽ ∠ ♆
15 36 ☽ ⊥ ♅
15 49 ☽ ± ♂
16 04 ♀ ☍ ♄
20 48 ☿ # ♄

24 Tuesday
02 27 ☽ ⊻ ♅
03 18 ☽ ⊼ ♂
03 42 ☽ ☌ ♃
04 27 ☽ Q ♄
05 12 ☿ bQ ♅
11 19 ♂ ⊼ ♃
12 13 ☽ △ ♇
17 07 ☿ ∠ ♂
18 25 ☽ ⚹ ♆

25 Wednesday
01 33 ☽ ⚹ ♄
03 01 ♀ bQ ♅
05 07 ☽ △ ♀
08 30 ☽ △ ☿
11 59 ☿ ∥ ♀
14 27 ☽ bQ ♇
15 18 ☽ Q ♆
16 15 ☽ △ ☉
16 33 ♂ # ♅

26 Thursday
03 08 ☽ ∠ ♄
05 46 ☽ ± ♇
06 30 ☽ ☌ ♅
07 56 ☽ ⊻ ♃
09 10 ☽ bQ ♀
10 08 ☽ ☍ ♂
13 40 ☽ bQ ☿
15 48 ☽ ⊼ ♇
17 58 ☽ ⊥ ♃
18 06 ☽ ⊥ ♄
19 23 ☽ bQ ☉
21 36 ☽ □ ♆

27 Friday
01 30 ☽ ± ♀
03 55 ☽ ⊻ ♄
06 38 ☽ ± ☿
07 32 ♃ Q ♄
08 48 ☽ ∠ ♃
10 18 ♀ ∠ ♂
11 13 ☽ ± ☉
12 12 ☽ ⊼ ♀
13 05 ☉ # ♆
17 45 ☽ ⊼ ☿
21 37 ☽ ⊼ ☉

28 Saturday
04 17 ☉ ∠ ♆
07 25 ☽ ⊻ ♅
09 03 ☽ ⚹ ♃
12 48 ♃ # ♇
13 25 ☽ ⊼ ♂
16 27 ☽ ☍ ♇
16 57 ☽ ⊥ ♅
22 00 ☽ △ ♆
23 24 ☽ ± ♂

MARCH 2037

LONGITUDES

Date	Sidereal time h m s	Sun ☉ ° ' "	Moon ☽ ° ' "	Moon ☽ 24.00 ° ' "	Mercury ☿ ° '	Venus ♀ ° '	Mars ♂ ° '	Jupiter ♃ ° '	Saturn ♄ ° '	Uranus ♅ ° '	Neptune ♆ ° '	Pluto ♇ ° '
01	22 38 33	11 ♓ 14 13	03 ♍ 49 12	11 ♍ 27 36	10 ♓ 30	06 ♓ 17	20 ♑ 12	16 ♊ 51	28 ♌ 32	15 ♋ 44	24 ♈ 57	21 ♒ 27
02	22 42 30	12 14 24	19 ♍ 06 00	26 ♍ 43 04	12 23	07 32	20 56	16 55	28 R 28	15 R 43	24 59	21 29
03	22 46 26	13 14 33	04 ♎ 17 29	11 ♎ 48 02	14 16	08 47	21 40	16 59	28 23	15 42	25 00	21 31
04	22 50 23	14 14 40	19 ♎ 13 39	26 ♎ 33 26	16 11	10 02	22 23	17 03	28 18	15 41	25 02	21 32
05	22 54 19	15 14 46	03 ♏ 46 39	10 ♏ 52 52	18 06	11 17	23 07	17 08	28 14	15 40	25 04	21 34
06	22 58 16	16 14 50	17 ♏ 51 45	24 ♏ 43 16	20 02	12 32	23 51	17 13	28 09	15 39	25 06	21 36
07	23 02 12	17 14 53	01 ♐ 27 29	08 ♐ 04 40	21 58	13 47	24 35	17 18	28 05	15 38	25 08	21 37
08	23 06 09	18 14 54	14 ♐ 35 09	20 ♐ 59 26	23 55	15 02	25 18	17 23	28 01	15 37	25 10	21 39
09	23 10 06	19 14 53	27 ♐ 18 01	03 ♑ 31 30	25 52	16 17	26 02	17 28	27 56	15 36	25 12	21 40
10	23 14 02	20 14 51	09 ♑ 40 29	15 ♑ 45 33	27 49	17 32	26 46	17 33	27 52	15 36	25 14	21 42
11	23 17 59	21 14 47	21 ♑ 47 21	27 ♑ 46 27	29 ♓ 46	18 46	27 30	17 39	27 48	15 35	25 16	21 44
12	23 21 55	22 14 42	03 ♒ 43 24	09 ♒ 38 46	01 ♈ 43	20 01	28 14	17 45	27 43	15 34	25 17	21 45
13	23 25 52	23 14 35	15 ♒ 33 01	21 ♒ 26 37	03 39	21 16	28 58	17 51	27 39	15 34	25 19	21 47
14	23 29 48	24 14 26	27 ♒ 19 59	03 ♓ 13 29	05 34	22 31	29 ♑ 42	17 57	27 35	15 33	25 21	21 48
15	23 33 45	25 14 15	09 ♓ 07 26	15 ♓ 02 08	07 28	23 46	00 ♒ 26	18 03	27 31	15 33	25 23	21 50
16	23 37 41	26 14 02	20 ♓ 57 50	26 ♓ 54 46	09 20	25 00	01 10	18 09	27 27	15 32	25 25	21 51
17	23 41 38	27 13 47	02 ♈ 53 06	08 ♈ 53 03	11 09	26 15	01 54	18 16	27 23	15 32	25 27	21 53
18	23 45 35	28 13 30	14 ♈ 54 46	20 ♈ 58 24	12 56	27 30	02 38	18 23	27 20	15 32	25 29	21 54
19	23 49 31	29 ♓ 13 11	27 ♈ 04 07	03 ♉ 12 06	14 40	28 45	03 22	18 29	27 16	15 32	25 32	21 56
20	23 53 28	00 ♈ 12 50	09 ♉ 22 32	15 ♉ 35 36	16 21	29 ♓ 59	04 06	18 36	27 12	15 31	25 34	21 57
21	23 57 24	01 12 27	21 ♉ 51 34	28 ♉ 10 40	17 57	01 ♈ 14	04 50	18 44	27 08	15 31	25 36	21 59
22	00 01 21	02 12 01	04 ♊ 33 11	10 ♊ 59 26	19 29	02 29	05 34	18 51	27 05	15 31	25 38	22 00
23	00 05 17	03 11 34	17 ♊ 29 43	24 ♊ 04 24	20 56	03 44	06 18	18 58	27 01	15 D 31	25 40	22 02
24	00 09 14	04 11 04	00 ♋ 43 47	07 ♋ 28 12	22 18	04 58	07 02	19 06	26 58	15 31	25 42	22 03
25	00 13 10	05 10 32	14 ♋ 17 54	21 ♋ 13 07	23 34	06 13	07 46	19 14	26 55	15 31	25 44	22 04
26	00 17 07	06 09 57	28 ♋ 13 58	05 ♌ 20 29	24 44	07 27	08 30	19 21	26 52	15 31	25 46	22 06
27	00 21 04	07 09 21	12 ♌ 32 33	19 ♌ 49 55	25 48	08 42	09 14	19 29	26 49	15 32	25 49	22 07
28	00 25 00	08 08 41	27 ♌ 12 09	04 ♍ 38 36	26 45	09 57	09 58	19 38	26 45	15 32	25 51	22 09
29	00 28 57	09 08 00	12 ♍ 08 28	19 ♍ 40 47	27 35	11 11	10 43	19 46	26 43	15 32	25 53	22 10
30	00 32 53	10 07 16	27 ♍ 14 25	04 ♎ 48 08	28 18	12 26	11 27	19 54	26 40	15 32	25 55	22 11
31	00 36 50	11 ♈ 06 30	12 ♎ 20 41	19 ♎ 50 46	28 ♈ 54	13 ♈ 40	12 ♒ 11	20 ♊ 03	26 ♌ 37	15 ♋ 33	25 ♈ 57	22 ♒ 12

Date	Moon True ☊ ° '	Moon Mean ☊ ° '	Moon ☽ Latitude ° '
01	07 ♌ 43	06 ♌ 15	02 N 18
02	07 R 39	06 12	03 27
03	07 34	06 09	04 21
04	07 28	06 05	04 56
05	07 23	06 02	05 11
06	07 19	05 59	05 06
07	07 16	05 56	04 44
08	07 14	05 53	04 08
09	07 D 15	05 50	03 20
10	07 16	05 46	02 24
11	07 17	05 43	01 23
12	07 18	05 40	00 N 19
13	07 R 18	05 37	00 S 45
14	07 16	05 34	01 46
15	07 11	05 31	02 43
16	07 04	05 27	03 33
17	06 56	05 24	04 14
18	06 46	05 21	04 44
19	06 35	05 18	05 02
20	06 26	05 15	05 05
21	06 18	05 11	04 55
22	06 12	05 08	04 30
23	06 08	05 05	03 50
24	06 07	05 02	02 58
25	06 D 07	04 59	01 54
26	06 08	04 56	00 S 42
27	06 R 08	04 52	00 N 34
28	06 06	04 49	01 50
29	06 02	04 46	03 00
30	05 56	04 43	03 57
31	05 ♌ 47	04 ♌ 40	04 N 38

DECLINATIONS

Date	Sun ☉ ° '	Moon ☽ ° '	Mercury ☿ ° '	Venus ♀ ° '	Mars ♂ ° '	Jupiter ♃ ° '	Saturn ♄ ° '	Uranus ♅ ° '	Neptune ♆ ° '	Pluto ♇ ° '
01	07 S 21	12 N 15	09 S 20	10 S 29	22 S 37	22 N 29	13 N 30	22 N 57	08 N 07	22 S 28
02	06 58	07 29	08 33	10 02	22 31	22 30	13 32	22 57	08 07	22 28
03	06 35	02 N 17	07 45	09 34	22 25	22 30	13 34	22 57	08 08	22 27
04	06 12	02 S 58	06 55	09 06	22 19	22 31	13 35	22 57	08 09	22 27
05	05 49	07 54	06 05	08 38	22 12	22 31	13 37	22 57	08 09	22 26
06	05 25	12 15	05 13	08 10	22 06	22 32	13 38	22 58	08 10	22 26
07	05 02	15 48	04 21	07 42	21 59	22 33	13 40	22 58	08 11	22 26
08	04 39	18 26	03 27	07 13	21 52	22 33	13 42	22 58	08 12	22 25
09	04 15	20 04	02 33	06 44	21 44	22 34	13 43	22 58	08 12	22 25
10	03 52	20 41	01 39	06 15	21 37	22 35	13 45	22 58	08 13	22 24
11	03 28	20 18	00 S 43	05 46	21 29	22 35	13 46	22 58	08 14	22 24
12	03 05	19 00	00 N 12	05 16	21 21	22 36	13 48	22 58	08 14	22 24
13	02 41	16 53	01 08	04 47	21 13	22 37	13 49	22 58	08 15	22 23
14	02 17	14 03	02 04	04 17	21 05	22 37	13 50	22 58	08 16	22 23
15	01 54	10 40	02 59	03 47	20 56	22 38	13 52	22 58	08 17	22 23
16	01 30	06 51	03 54	03 17	20 48	22 39	13 53	22 58	08 17	22 22
17	01 06	02 S 45	04 48	02 48	20 39	22 40	13 55	22 58	08 18	22 22
18	00 42	01 N 30	05 41	02 17	20 30	22 40	13 56	22 58	08 19	22 22
19	00 S 19	05 44	06 33	01 47	20 20	22 41	13 57	22 58	08 20	22 21
20	00 N 05	09 47	07 24	01 17	20 11	22 42	13 58	22 58	08 21	22 21
21	00 29	13 29	08 12	00 47	20 01	22 43	14 00	22 58	08 21	22 21
22	00 52	16 38	08 59	00 S 17	19 51	22 43	14 01	22 58	08 22	22 20
23	01 16	19 01	09 43	00 N 14	19 41	22 44	14 02	22 58	08 23	22 20
24	01 40	20 28	10 25	00 44	19 31	22 45	14 03	22 58	08 24	22 20
25	02 03	20 46	11 04	01 14	19 21	22 46	14 04	22 58	08 24	22 20
26	02 27	19 49	11 40	01 44	19 10	22 46	14 05	22 58	08 25	22 19
27	02 50	17 35	12 13	02 15	18 59	22 47	14 06	22 58	08 26	22 19
28	03 14	14 10	12 43	02 45	18 48	22 48	14 07	22 58	08 27	22 19
29	03 37	09 46	13 10	03 15	18 37	22 49	14 08	22 58	08 28	22 19
30	04 00	04 44	13 33	03 45	18 26	22 50	14 09	22 58	08 28	22 18
31	04 N 24	00 S 36	13 N 52	04 N 15	18 S 14	22 N 50	14 N 10	22 N 58	08 N 29	22 S 18

ZODIAC SIGN ENTRIES

Date	h	m	Planets
01	05	59	☽ ♍
03	05	11	☽ ♎
05	05	42	☽ ♏
07	09	23	☽ ♐
09	17	11	☽ ♑
11	14	48	☿ ♈
12	04	29	☽ ♒
14	17	26	☽ ♓
14	22	02	♂ ♒
17	06	13	☽ ♈
19	17	45	☽ ♉
20	06	50	☉ ♈
20	12	10	♀ ♈
22	03	26	☽ ♊
24	10	41	☽ ♋
26	15	00	☽ ♌
28	16	31	☽ ♍
30	16	23	☽ ♎

LATITUDES

Date	Mercury ☿ ° '	Venus ♀ ° '	Mars ♂ ° '	Jupiter ♃ ° '	Saturn ♄ ° '	Uranus ♅ ° '	Neptune ♆ ° '	Pluto ♇ ° '
01	01 S 50	01 S 23	00 S 42	00 S 18	01 N 38	00 N 27	01 S 40	08 S 36
04	01 36	01 24	00 45	00 18	01 38	00 27	01 39	08 36
07	01 16	01 26	00 47	00 17	01 38	00 27	01 39	08 37
10	00 51	01 26	00 50	00 17	01 38	00 27	01 39	08 37
13	00 S 21	01 26	00 53	00 16	01 38	00 27	01 39	08 38
16	00 N 13	01 26	00 55	00 16	01 38	00 27	01 39	08 38
19	00 50	01 24	00 58	00 15	01 38	00 27	01 39	08 39
22	01 28	01 23	01 01	00 15	01 38	00 26	01 39	08 39
25	02 04	01 20	01 03	00 14	01 38	00 26	01 39	08 40
28	02 35	01 18	01 06	00 14	01 38	00 26	01 39	08 41
31	02 N 59	01 S 14	01 S 09	00 S 13	01 N 37	00 N 26	01 S 39	08 S 41

DATA

Julian Date	2465119
Delta T	+76 seconds
Ayanamsa	24° 22' 22"
Synetic vernal point	04° ♓ 44' 37"
True obliquity of ecliptic	23° 25' 59"

MOON'S PHASES, APSIDES AND POSITIONS ☽

Date	h	m	Phase	Longitude ° '	Eclipse Indicator
02	00	28	○	11 ♍ 45	
08	19	25	☾	18 ♐ 33	
16	23	36	●	26 ♓ 43	
24	18	40	☽	04 ♋ 28	
31	09	53	○	11 ♎ 01	

Day	h	m		
01	19	51	Perigee	
15	01	36	Apogee	
30	06	38	Perigee	
03	22	23	0S	
10	14	36	Max dec	20° S 41'
18	03	32	0N	
25	06	03	Max dec	20° N 49'
31	09	18	0S	

LONGITUDES

Date	Chiron ⚷ ° '	Ceres ⚳ ° '	Pallas ⚴ ° '	Juno ⚵ ° '	Vesta ⚶ ° '	Black Moon Lilith ⚸ ° '
01	13 ♊ 30	29 ♎ 31	19 ♌ 05	03 ♍ 44	03 ♓ 59	05 ♓ 26
11	13 ♊ 42	28 ♎ 50	17 ♌ 03	01 ♍ 27	08 ♓ 59	06 ♓ 33
21	14 ♊ 01	27 ♎ 29	15 ♌ 58	29 ♌ 40	13 ♓ 55	07 ♓ 40
31	14 ♊ 27	25 ♎ 35	15 ♌ 49	28 ♌ 32	18 ♓ 48	08 ♓ 46

All ephemeris data is given at 12.00 UT and the Moon's longitude is additionally given for 24.00 UT

ASPECTARIAN

h m	Aspects
01 Sunday	
03 44	☽ ☌ ♄
04 08	☽ Q ♃
04 46	☿ ∠ ♆
04 59	☽ ∥ ♄
07 09	☽ ∠ ♅
14 17	☽ ⚼ ♂
16 14	☽ ☍ ♀
21 39	☽ ⚼ ♆
22 15	☽ # ♀
23 58	☽ ☍ ☿
02 Monday	
00 28	☽ ☍ ☉
05 55	☽ # ☿
06 41	☽ ⚹ ♅
08 09	☉ ☌ ☿
08 33	☽ □ ♃
08 57	☽ ∥ ♆
11 48	☽ ± ♆
14 39	☽ # ☉
15 02	☽ △ ♂
15 45	☽ ⚻ ♇
16 41	♂ # ♃
21 16	☽ ⚻ ♆
03 Tuesday	
00 33	☿ # ♆
01 14	☽ ± ♇
01 33	☽ Q ♅
02 05	♂ ∥ ♇
02 41	☽ ⊻ ♄
06 48	♂ ⊻ ♇
12 09	☽ ⊥ ♄
15 32	☽ ⚼ ♇
19 49	☽ ⚻ ♀
04 Wednesday	
02 28	☽ ∠ ♄
03 21	☽ ⚻ ☉
05 44	☿ △ ♅
06 15	☽ □ ♅
06 20	☽ ⚻ ☿
06 21	☽ ± ♀
08 28	☽ △ ♃
09 36	♂ ± ♄
12 02	♀ ∠ ♆
13 47	☽ ± ☉
15 46	☽ △ ♇
17 25	☽ □ ♂
17 32	☽ ± ☿
21 31	☽ ☍ ♆
22 23	☽ ⚼ ♀
23 26	☿ □ ♃
05 Thursday	
02 21	☽ ∥ ☉
02 49	☽ ⚹ ♄
04 15	☽ ∥ ☿
05 40	☽ ⚼ ☉
09 14	☽ ⚼ ♃
10 41	☽ ⚼ ☿
12 34	♂ ± ♃
13 21	☽ # ♆
15 29	☽ ∥ ♀
21 48	☉ △ ♅
22 50	☽ Q ♄
06 Friday	
00 15	☿ ⊥ ♆
00 30	☽ ± ♃
01 05	☽ Q ♂
01 42	☉ ∥ ☿
01 56	☽ △ ♀
08 10	☽ △ ♅
08 59	☽ △ ☉
10 52	☽ ⚻ ♃
11 56	♀ # ♆
16 23	☽ △ ☿
18 31	☽ □ ♇
20 47	☽ # ♄
23 03	☽ ⚹ ♂
07 Saturday	
00 42	☽ ⚻ ♆
06 00	☽ □ ♄
07 37	☿ ⊻ ♇
10 31	☽ ⚼ ♅
11 25	☽ ± ♆
13 11	☉ □ ♃
19 10	♅ ± ♇
08 Sunday	
02 50	☽ ± ♅
02 51	☽ Q ♇
03 37	☽ ∠ ♂
03 48	☽ ⚼ ♆
07 02	♂ □ ♆
12 55	☽ □ ♀
13 55	☽ ⚻ ♅
17 15	☽ ☍ ♃
19 25	☽ □ ☉
21 22	☽ ⊥ ♂
23 11	♀ △ ♅
09 Monday	
01 16	☽ ⚹ ♇
03 34	☿ ⊻ ♆
07 58	☽ △ ♆
08 45	☽ □ ☿
09 26	☽ ⊻ ♂
10 38	☉ ⊥ ♆
13 13	☽ △ ♄
15 18	☿ ⚹ ♂
10 Tuesday	
02 59	☽ Q ♀
06 10	☽ ∠ ♇
08 57	☽ Q ☉
10 29	☿ ⊥ ♇
12 31	☿ ⚻ ♄
12 37	♀ □ ♃
18 14	☽ ⚼ ♄
23 40	☽ ☍ ♅
23 55	☽ ⊥ ♇
11 Wednesday	
02 27	☽ Q ☿
03 41	☽ ⚻ ♃
05 18	☽ ⚹ ♀
10 49	☽ ⚹ ☉
11 52	☽ ⊻ ♇
12 01	☽ ± ♄
15 45	☽ ± ♃
18 58	☽ □ ♆
20 52	♂ ⚻ ♄
21 34	♀ ⊥ ♆
23 51	☉ ⊻ ♇
23 58	☽ ⚻ ♄
12 Thursday	
00 12	☽ ☌ ♂
07 09	☽ ⚹ ☿
10 00	☽ ⚼ ♃
14 56	☽ ∠ ♀
19 47	☽ ∠ ☉
13 Friday	
07 27	☽ Q ♆
11 21	☽ ⊥ ♀
12 02	☽ ⚻ ♅
12 03	☿ ± ♄
15 46	☽ ⊥ ☉
16 42	☽ △ ♃
19 33	☽ ∠ ☿
22 03	♀ ⊻ ♇
14 Saturday	
00 14	☽ ± ♅
00 43	☽ ☌ ♇
01 01	☽ ⊻ ♀
05 07	☽ ⊻ ☉
07 58	☽ ⚹ ♆
12 31	☽ ☍ ♄
13 39	☽ # ♄
16 07	☉ # ☿
17 01	☿ Q ♃
17 08	☽ ⊻ ♂
17 26	☽ ⊥ ☿
18 33	☽ ⚼ ♅
15 Sunday	
03 51	☿ ∠ ♇
06 09	☽ ⊥ ♂
07 59	☽ ⊻ ☿
14 35	☽ ∠ ♆
15 47	☉ ⊻ ♆
16 Monday	
01 02	☽ △ ♅
01 38	☽ ∠ ♂
01 38	☿ # ♀
03 13	☽ # ♆
06 16	☽ □ ♃
08 53	☽ ⊥ ♆
13 48	☽ ⊻ ♇
20 13	♀ ⊻ ♆
21 02	☽ ⊻ ♆
21 07	☽ ☌ ♀
23 36	☽ ☌ ☉
17 Tuesday	
01 01	☽ ⚻ ♄
01 56	☽ ⊥ ♇
02 14	☽ # ☿
09 53	☽ ⚹ ♂
11 41	☽ ∥ ♀
13 00	☽ ± ♄
15 36	☉ ⚻ ♄
18 50	☽ Q ♃
20 01	☽ ∠ ♇
22 17	☽ ∥ ☉
18 Wednesday	
03 56	☿ ⚼ ♄
04 07	☉ ⊥ ♇
06 53	☽ ⚼ ♄
07 25	☽ ☌ ☿
07 54	☽ # ☉
08 47	♀ ⚻ ♄
11 24	☽ Q ♂
13 13	☽ □ ♅
15 58	☽ # ♀
18 56	☽ ⚹ ♃
19 59	♀ ⊥ ♇
19 Thursday	
01 52	☽ ⚹ ♇
08 58	☽ ☌ ♆
12 23	☽ △ ♄
15 40	☽ ⊻ ♀
16 35	☽ ⊻ ☉
17 05	♂ ⚼ ♃
18 01	☽ ∥ ☿
20 Friday	
00 04	☿ □ ♅
00 38	☽ Q ♅
00 41	☽ ∠ ♃
01 05	☽ □ ♂
01 27	☽ Q ♇
03 16	☽ ∥ ♆
04 42	☽ ⊥ ♀
05 19	☿ Q ♂
05 20	☽ ⊥ ☉
18 18	☽ ⊥ ♃
23 52	☽ ⚹ ♅
21 Saturday	
00 03	☽ ∠ ♀
00 15	☽ ∠ ☉
03 27	☽ ⊻ ☿
05 57	☽ ⊻ ♃
09 16	☉ ☌ ♀
12 14	☽ □ ♇
15 39	☽ ∥ ♄
16 34	☽ ⊥ ☿
16 35	☿ ∥ ♆
19 08	☽ ⊻ ♆
20 03	☉ # ♀
22 00	☽ □ ♄
22 Sunday	
00 56	☿ ⚹ ♃
04 25	☽ ∠ ♅
06 30	☽ ⊥ ♆
07 13	☽ ⚹ ☉
07 42	☽ ⚹ ♀
11 52	☽ ∠ ☿
14 00	☽ △ ♂
21 16	☽ ⊥ ♅
23 05	♀ ± ♄
23 22	☽ ∠ ♆
23 Monday	
07 25	☽ Q ☉
07 29	☽ Q ♄
08 10	☉ ± ♄
08 22	☽ ⊻ ♅
08 24	☽ Q ♀
08 25	♅ St D
14 44	☽ ☌ ♃
19 03	☽ ⚹ ☿
19 22	☽ ⚼ ♂
20 07	☽ # ♂
20 18	☽ △ ♇
24 Tuesday	
02 55	☽ ⚹ ♆
05 16	☽ ⚹ ♄
07 23	☿ ⚹ ♇
12 34	☽ ± ♂
18 40	☽ □ ☉
19 04	☽ Q ☿
20 20	☽ □ ♀
23 17	☽ ⚼ ♇
23 52	☽ ⚻ ♂
25 Wednesday	
00 27	☽ Q ♆
07 51	☽ ∠ ♄
14 08	☽ ☌ ♅
15 06	☽ ± ♇
20 39	☽ ⊻ ♃
23 26	☽ ⊥ ♄
26 Thursday	
01 30	☽ ⚻ ♇
04 56	♀ ∠ ♇
05 31	☽ □ ☿
07 03	☽ ⊥ ♃
07 48	☽ □ ♆
09 40	☽ ⊻ ♄
09 52	♀ Q ♃
21 24	☽ # ♂
22 27	☽ ∠ ♃
27 Friday	
02 22	☽ △ ☉
05 01	☽ △ ♀
06 13	☽ ☍ ♂
11 06	☉ ∠ ♇
12 21	☿ ☌ ♆
16 56	☽ ⊻ ♅
21 26	☉ Q ♃
23 33	☽ ⚹ ♃
28 Saturday	
02 46	☽ ⊥ ♅
03 46	☽ ☍ ♇
04 56	☽ ⚼ ☉
08 00	☽ ⚼ ♀
09 48	☽ △ ♆
11 13	☽ △ ☿
11 17	☽ ☌ ♄
12 15	☽ ∥ ♄
12 21	☿ △ ♄
13 33	♀ ⚹ ♂
17 23	☽ ∠ ♅
19 13	☽ Q ♃
19 36	☽ ∥ ☿
20 33	☽ ± ☉
23 52	☽ ± ♀
29 Sunday	
06 51	☽ ⚻ ☉
09 36	☽ ⚻ ♂
09 59	☽ ⚼ ♆
10 20	☽ ⚻ ♀
12 44	☽ ⚼ ☿
17 25	☽ ⚹ ♅
18 27	☽ ∥ ♆
19 39	☽ ± ♂
21 45	♀ ⚼ ♄
30 Monday	
00 15	☽ □ ♃
00 21	☽ ± ♆
03 47	☽ ± ☿
03 58	☽ ⚻ ♇
09 54	☽ ⚻ ♆
10 41	☽ ⚼ ♂
11 05	☽ ⊻ ♄
12 29	☽ Q ♅
13 30	☽ ± ♇
13 45	☽ ⚻ ☿
15 03	☽ ∥ ☉
16 03	☽ ∥ ♀
20 34	☽ ⊥ ♄
31 Tuesday	
03 49	☽ ⚼ ♇
09 53	☽ ☍ ☉
10 50	☽ ∠ ♄
11 44	☽ △ ♂
14 18	☽ ☍ ♀
17 07	☽ □ ♅
23 47	☉ ⚼ ♄

APRIL 2037

LONGITUDES

Date	Sidereal time h m s	Sun ☉ ° ' "	Moon ☽ ° ' "	Moon ☽ 24.00 ° ' "	Mercury ☿ ° '	Venus ♀ ° '	Mars ♂ ° '	Jupiter ♃ ° '	Saturn ♄ ° '	Uranus ♅ ° '	Neptune ♆ ° '	Pluto ♇ ° '
01	00 40 46	12 ♈ 05 42	27 ♎ 17 11	04 ♏ 38 51	29 ♈ 23	14 ♈ 55	12 ♒ 55	20 ♊ 11	26 ♌ 34	15 ♋ 33	25 ♈ 59	22 ♒ 14
02	00 44 43	13 04 52	11 ♏ 54 50	19 ♏ 04 23	29 44	16 09	13 40	20 20	26 R 32	15 34	26 02	22 15
03	00 48 39	14 04 01	26 ♏ 06 58	03 ♐ 02 17	29 ♈ 58	17 24	14 24	20 29	26 29	15 34	26 04	22 16
04	00 52 36	15 03 07	09 ♐ 50 13	16 ♐ 30 50	00 ♉ 05	18 38	15 08	20 38	26 27	15 35	26 06	22 17
05	00 56 33	16 02 12	23 ♐ 04 23	29 ♐ 31 13	00 ♉ 05	19 52	15 52	20 47	26 24	15 36	26 08	22 19
06	01 00 29	17 01 15	05 ♑ 51 50	12 ♑ 06 48	29 ♈ 59	21 07	16 37	20 57	26 22	15 36	26 11	22 20
07	01 04 26	18 00 17	18 ♑ 16 44	24 ♑ 22 17	29 R 45	22 21	17 21	21 06	26 20	15 37	26 13	22 21
08	01 08 22	18 59 16	00 ♒ 24 10	06 ♒ 23 03	29 26	23 36	18 05	21 15	26 18	15 38	26 15	22 22
09	01 12 19	19 58 14	12 ♒ 19 36	18 ♒ 14 30	29 01	24 50	18 50	21 25	26 16	15 39	26 17	22 23
10	01 16 15	20 57 10	24 ♒ 08 20	00 ♓ 01 44	28 32	26 04	19 34	21 35	26 14	15 40	26 20	22 24
11	01 20 12	21 56 04	05 ♓ 55 13	11 ♓ 49 17	27 57	27 19	20 18	21 44	26 13	15 41	26 22	22 25
12	01 24 08	22 54 56	17 ♓ 44 22	23 ♓ 40 53	27 20	28 33	21 03	21 54	26 11	15 42	26 24	22 26
13	01 28 05	23 53 47	29 ♓ 39 09	05 ♈ 39 27	26 39	29 ♈ 47	21 47	22 04	26 10	15 43	26 26	22 27
14	01 32 02	24 52 35	11 ♈ 41 59	17 ♈ 46 57	25 56	01 ♉ 01	22 31	22 15	26 08	15 44	26 29	22 28
15	01 35 58	25 51 22	23 ♈ 54 28	00 ♉ 04 37	25 12	02 16	23 16	22 25	26 07	15 45	26 31	22 29
16	01 39 55	26 50 07	06 ♉ 17 26	12 ♉ 32 59	24 28	03 30	24 00	22 35	26 06	15 47	26 33	22 30
17	01 43 51	27 48 50	18 ♉ 51 17	25 ♉ 12 19	23 44	04 44	24 44	22 46	26 05	15 48	26 35	22 31
18	01 47 48	28 47 30	01 ♊ 36 07	08 ♊ 02 43	23 02	05 58	25 29	22 56	26 04	15 49	26 38	22 32
19	01 51 44	29 ♈ 46 09	14 ♊ 32 12	21 ♊ 04 36	22 21	07 12	26 13	23 07	26 03	15 51	26 40	22 33
20	01 55 41	00 ♉ 44 46	27 ♊ 40 04	04 ♋ 18 42	21 44	08 27	26 58	23 17	26 02	15 52	26 42	22 34
21	01 59 37	01 43 20	11 ♋ 00 42	17 ♋ 46 12	21 09	09 41	27 42	23 28	26 01	15 54	26 45	22 34
22	02 03 34	02 41 52	24 ♋ 35 23	01 ♌ 28 25	20 38	10 55	28 26	23 39	26 01	15 55	26 47	22 35
23	02 07 31	03 40 23	08 ♌ 25 24	15 ♌ 26 26	20 11	12 09	29 11	23 50	26 00	15 57	26 49	22 36
24	02 11 27	04 38 50	22 ♌ 31 30	29 ♌ 40 29	19 48	13 23	29 ♒ 55	24 01	26 00	15 58	26 51	22 37
25	02 15 24	05 37 16	06 ♍ 53 09	14 ♍ 09 11	19 30	14 37	00 ♓ 39	24 12	26 00	16 00	26 54	22 38
26	02 19 20	06 35 39	21 ♍ 28 03	28 ♍ 49 07	19 17	15 51	01 24	24 24	26 00	16 02	26 56	22 38
27	02 23 17	07 34 01	06 ♎ 11 36	13 ♎ 34 36	19 09	17 05	02 08	24 35	26 D 00	16 04	26 58	22 39
28	02 27 13	08 32 20	20 ♎ 57 09	28 ♎ 18 14	19 06	18 19	02 52	24 46	26 00	16 05	27 00	22 40
29	02 31 10	09 30 38	05 ♏ 36 49	12 ♏ 51 55	19 D 07	19 33	03 37	24 58	26 00	16 07	27 03	22 40
30	02 35 06	10 ♉ 28 53	20 ♏ 02 40	27 ♏ 08 16	19 ♈ 14	20 ♉ 47	04 ♓ 21	25 ♊ 09	26 ♌ 00	16 ♋ 09	27 ♈ 05	22 ♒ 41

DECLINATIONS

Date	Moon True ☊ ° '	Moon Mean ☊ ° '	Moon ☽ Latitude ° '	Sun ☉ ° '	Moon ☽ ° '	Mercury ☿ ° '	Venus ♀ ° '	Mars ♂ ° '	Jupiter ♃ ° '	Saturn ♄ ° '	Uranus ♅ ° '	Neptune ♆ ° '	Pluto ♇ ° '
01	05 ♌ 37	04 ♌ 37	05 N 00	04 N 47	05 S 50	14 N 08	04 N 45	18 S 03	22 N 51	14 N 11	22 N 58	08 N 30	22 S 18
02	05 R 26	04 33	05 01	05 10	10 37	14 19	05 15	17 51	22 52	14 12	22 58	08 31	22 18
03	05 17	04 30	04 44	05 33	14 41	14 28	05 45	17 39	22 53	14 13	22 58	08 32	22 18
04	05 11	04 27	04 10	05 56	17 48	14 32	06 14	17 27	22 53	14 13	22 58	08 33	22 17
05	05 06	04 24	03 24	06 18	19 52	14 32	06 44	17 14	22 54	14 14	22 57	08 33	22 17
06	05 04	04 21	02 29	06 41	20 50	14 29	07 13	17 02	22 55	14 15	22 57	08 34	22 17
07	05 D 03	04 17	01 28	07 04	20 44	14 22	07 42	16 49	22 56	14 15	22 57	08 35	22 17
08	05 03	04 14	00 N 25	07 26	19 39	14 11	08 11	16 36	22 57	14 16	22 57	08 36	22 17
09	05 R 03	04 11	00 S 39	07 48	17 43	13 57	08 40	16 24	22 57	14 17	22 57	08 37	22 17
10	05 02	04 08	01 40	08 11	15 02	13 39	09 08	16 11	22 58	14 17	22 57	08 37	22 16
11	04 58	04 05	02 36	08 33	11 45	13 19	09 37	15 57	22 59	14 18	22 57	08 38	22 16
12	04 51	04 02	03 26	08 54	08 00	12 56	10 05	15 44	23 00	14 18	22 57	08 39	22 16
13	04 42	03 58	04 07	09 16	03 S 55	12 30	10 33	15 31	23 00	14 19	22 56	08 40	22 16
14	04 30	03 55	04 37	09 38	00 N 22	12 02	11 01	15 17	23 01	14 19	22 56	08 41	22 16
15	04 17	03 52	04 55	09 59	04 42	11 33	11 28	15 03	23 02	14 19	22 56	08 42	22 16
16	04 03	03 49	05 00	10 21	08 53	11 03	11 55	14 49	23 02	14 20	22 56	08 42	22 16
17	03 50	03 46	04 50	10 42	12 46	10 32	12 22	14 35	23 03	14 20	22 56	08 43	22 16
18	03 39	03 43	04 26	11 03	16 08	10 02	12 48	14 21	23 04	14 20	22 56	08 44	22 16
19	03 30	03 39	03 47	11 23	18 46	09 31	13 15	14 07	23 05	14 20	22 55	08 45	22 16
20	03 25	03 36	02 57	11 44	20 28	09 01	13 41	13 53	23 05	14 21	22 55	08 46	22 16
21	03 22	03 33	01 55	12 04	21 04	08 32	14 06	13 38	23 06	14 21	22 55	08 46	22 16
22	03 D 21	03 30	00 S 46	12 24	20 27	08 05	14 32	13 24	23 07	14 21	22 55	08 47	22 16
23	03 R 21	03 27	00 N 27	12 44	18 35	07 40	14 56	13 09	23 07	14 21	22 55	08 48	22 16
24	03 20	03 23	01 40	13 04	15 34	07 16	15 21	12 54	23 08	14 21	22 54	08 49	22 16
25	03 18	03 20	02 47	13 24	11 34	06 55	15 45	12 39	23 08	14 21	22 54	08 50	22 16
26	03 13	03 17	03 45	13 43	06 50	06 36	16 09	12 24	23 09	14 21	22 54	08 50	22 16
27	03 06	03 14	04 28	14 02	01 N 39	06 20	16 32	12 09	23 10	14 21	22 54	08 51	22 16
28	02 56	03 11	04 54	14 21	03 S 38	06 06	16 55	11 54	23 10	14 21	22 54	08 52	22 16
29	02 45	03 08	05 00	14 39	08 40	05 55	17 17	11 39	23 11	14 21	22 53	08 53	22 16
30	02 ♌ 33	03 ♌ 04	04 N 47	14 N 58	13 S 08	05 N 46	17 N 39	11 S 23	23 N 11	14 N 20	22 N 53	08 N 54	22 S 16

ZODIAC SIGN ENTRIES

Date	h	m	Planets
01	16	24	☽ ♏
03	16	23	☿ ♉
03	18	42	☽ ♐
06	00	54	☽ ♑
06	08	17	☿ ♈
08	11	12	☽ ♒
10	23	56	☽ ♓
13	12	42	☽ ♈
13	16	10	♀ ♉
15	23	51	☽ ♉
18	09	00	☽ ♊
19	17	40	☉ ♉
20	16	13	☽ ♋
22	21	26	☽ ♌
24	14	44	♂ ♓
25	00	33	☽ ♍
27	01	55	☽ ♎
29	02	47	☽ ♏

LATITUDES

Date	Mercury ☿ ° '	Venus ♀ ° '	Mars ♂ ° '	Jupiter ♃ ° '	Saturn ♄ ° '	Uranus ♅ ° '	Neptune ♆ ° '	Pluto ♇ ° '
01	03 N 05	01 S 13	01 S 10	00 S 13	01 N 37	00 N 26	01 S 39	08 S 41
04	03 14	01 09	01 12	00 13	01 37	00 26	01 39	08 42
07	03 11	01 04	01 15	00 12	01 37	00 26	01 39	08 43
10	02 54	01 00	01 18	00 12	01 37	00 26	01 39	08 44
13	02 23	00 54	01 20	00 11	01 37	00 26	01 39	08 44
16	01 41	00 49	01 23	00 11	01 36	00 26	01 39	08 45
19	00 53	00 42	01 26	00 11	01 36	00 26	01 39	08 46
22	00 N 02	00 36	01 28	00 10	01 36	00 26	01 39	08 47
25	00 S 46	00 29	01 31	00 10	01 36	00 26	01 39	08 48
28	01 29	00 22	01 33	00 10	01 35	00 26	01 39	08 49
31	02 S 05	00 S 15	01 S 36	00 S 09	01 N 35	00 N 26	01 S 39	08 S 49

DATA

Julian Date	2465150
Delta T	+76 seconds
Ayanamsa	24° 22' 24"
Synetic vernal point	04° ♓ 44' 35"
True obliquity of ecliptic	23° 25' 59"

MOON'S PHASES, APSIDES AND POSITIONS ☽

Date	h	m	Phase	Longitude ° '	Eclipse Indicator
07	11	25	☾	17 ♑ 59	
15	16	08	●	26 ♈ 01	
23	03	11	☽	03 ♌ 19	
29	18	54	○	09 ♏ 47	

Date	h	m	
11	13	14	Apogee
27	09	54	Perigee

Date	h	m		
06	21	43	Max dec	20° S 55'
14	09	56	0N	
21	11	56	Max dec	21° N 04'
27	19	29	0S	

LONGITUDES

Date	Chiron ⚷ ° '	Ceres ⚳ ° '	Pallas ⚴ ° '	Juno ⚵ ° '	Vesta ⚶ ° '	Black Moon Lilith ⚸ ° '
01	14 ♊ 30	25 ♎ 23	15 ♌ 51	28 ♌ 28	19 ♓ 17	08 ♓ 53
11	15 ♊ 02	23 ♎ 09	16 ♌ 37	28 ♌ 05	24 ♓ 04	10 ♓ 00
21	15 ♊ 40	20 ♎ 54	18 ♌ 04	28 ♌ 21	28 ♓ 46	11 ♓ 07
31	16 ♊ 22	18 ♎ 56	20 ♌ 06	29 ♌ 13	03 ♈ 21	12 ♓ 13

All ephemeris data is given at 12.00 UT and the Moon's longitude is additionally given for 24.00 UT

ASPECTARIAN

h m Aspects

01 Wednesday
00 26 ☽ △ ♃
03 49 ☽ △ ♇
06 22 ☽ # ♀
06 39 ☽ # ☉
09 54 ☽ ☍ ♆
10 51 ☽ ⚹ ♄
15 29 ☽ ☍ ☿
18 08 ☉ ∥ ♀
18 30 ☿ ∥ ♄

02 Thursday
00 35 ♀ □ ♅
01 01 ☽ ꝑ ♃
01 01 ☽ # ♆
06 24 ☽ Q ♄
14 05 ☽ ⊼ ☉
15 04 ☽ □ ♂
16 05 ☽ ± ♃
18 06 ☽ △ ♅
19 45 ☽ ⊼ ♀

03 Friday
00 39 ♂ Q ♆
00 55 ☽ ± ☉
02 17 ☽ ⊼ ♃
05 25 ☽ □ ♇
06 53 ☽ ± ♀
08 55 ☽ # ♄
10 30 ☽ # ☿
11 55 ☽ ⊼ ♆
12 38 ☽ □ ♄
17 29 ☽ ꝑ ☉
18 44 ☽ ⊼ ☿
19 43 ☽ ꝑ ♅
22 20 ☽ ± ♆
23 30 ☽ Q ♂
23 57 ☽ ꝑ ♀

04 Saturday
05 19 ☽ ± ☿
09 02 ☽ ∥ ♂
11 33 ☽ ± ♅
12 48 ☽ Q ♇
14 16 ☽ ꝑ ♆
20 03 ☉ ⚹ ♂
21 27 ☽ ꝑ ☿
22 04 ☽ ⚹ ♂
22 06 ☽ △ ☉
22 20 ☽ ⊼ ♅

05 Sunday
00 06 ☿ St R
01 08 ☉ □ ♅
02 51 ♂ ⊼ ♅
05 30 ☽ △ ♀
07 45 ☽ ☍ ♃
10 35 ☽ ⚹ ♇
17 42 ☽ △ ♆
18 10 ☽ △ ♄

06 Monday
00 59 ☽ △ ☿
03 26 ☽ ∠ ♂
08 13 ♀ ⚹ ♃
14 48 ☽ ∠ ♇
21 40 ☽ ⊥ ♂
22 32 ☽ ꝑ ♄

07 Tuesday
06 48 ☽ ☍ ♅
08 13 ☽ ⊥ ♇
10 04 ☽ ⊻ ♂
11 25 ☽ □ ☉
11 53 ♀ ⚹ ♇
16 02 ☽ ± ♄
17 37 ☽ ⊼ ♃
20 01 ☽ ⊻ ♇
20 55 ☽ □ ♀
21 28 ☿ Q ♂

08 Wednesday
02 22 ☿ ∥ ♄
03 42 ☽ □ ♆
03 51 ☽ ⊼ ♄
05 38 ☽ ± ♃
10 07 ☽ □ ☿
23 54 ☽ ꝑ ♃

09 Thursday
02 24 ☽ Q ☉
02 27 ♃ ∥ ♅
05 15 ♄ △ ♆
09 15 ♀ ∥ ♆
13 08 ☽ Q ♀
15 59 ☽ Q ♆
18 45 ☽ ⊼ ♅
21 09 ☽ Q ☿

10 Friday
01 41 ☽ ∥ ♂
02 04 ☽ ☌ ♂
04 55 ☽ ⚹ ☉
06 43 ☽ △ ♃
06 57 ☽ ± ♅
08 27 ☽ ☌ ♇
15 13 ♀ △ ♄
16 16 ☽ ☍ ♄
16 24 ☽ ⚹ ♀
16 28 ☽ ⚹ ♆
17 09 ♀ ☌ ♆
17 50 ☽ # ♄
20 32 ☽ ⚹ ☿
23 46 ☽ # ☿

11 Saturday
01 19 ☽ ꝑ ♅
06 19 ☉ ⚹ ♃
14 15 ☽ ∠ ☉
18 26 ☉ ∥ ♆
20 25 ☿ ☌ ♀
23 07 ☽ ∠ ♆

12 Sunday
00 04 ☉ ⚹ ♇
00 27 ☽ # ♀
01 36 ☽ ∠ ☿
02 31 ☽ ∠ ♀
06 54 ☽ # ☉
07 52 ☽ △ ♅
08 01 ☽ # ♆
10 11 ☽ ⊥ ☉
17 24 ☽ ⊥ ♆
18 52 ☽ ⊥ ☿
19 07 ☽ ⊻ ♂
20 32 ☽ □ ♃
21 30 ☽ ⊻ ♇
22 50 ☽ ⊥ ♀
23 24 ☽ ⊻ ☉

13 Monday
05 00 ☽ ⊼ ♄
05 32 ☽ ⊻ ♆
06 18 ☽ ⊻ ☿
08 00 ☽ ⊥ ♂
09 36 ☽ ⊥ ♇
09 48 ♂ ± ♅
12 18 ☽ ⊻ ♀
17 01 ☽ ± ♄
18 49 ☿ ☌ ♆

14 Tuesday
00 11 ♂ △ ♃
03 10 ☽ ∠ ♂
03 36 ☽ ∠ ♇
05 13 ☿ △ ♄
09 04 ☽ Q ♃
10 14 ♂ ☌ ♇
10 53 ☽ ꝑ ♄
19 59 ☽ □ ♅

15 Wednesday
02 56 ☉ ☌ ☿
09 02 ☽ ⚹ ♃
09 13 ☽ ⚹ ♇
10 40 ☽ ⚹ ♂
14 15 ☿ ∥ ♀
14 24 ☽ ☌ ☿
16 08 ☽ ☌ ☉
16 18 ☽ △ ♄
17 06 ☽ ☌ ♆
18 13 ☉ △ ♄
23 16 ♃ △ ♇

16 Thursday
04 49 ☉ ☌ ♆
06 01 ☽ ☌ ♀
07 09 ☽ Q ♅
08 33 ☽ Q ♇
10 54 ☽ ∥ ♆
11 25 ☽ Q ♂
14 31 ☽ ∠ ♃
17 32 ♀ Q ♅
19 39 ☿ ⚹ ♂
21 32 ☽ ∥ ☉
23 32 ☽ ∥ ☿

17 Friday
06 11 ☽ ⚹ ♅
07 44 ☉ ∥ ☿
07 45 ♀ Q ♇
07 58 ☽ ⊥ ♃
08 59 ☽ ∥ ♀
18 56 ☽ □ ♇
19 29 ☽ ⊻ ♃
20 45 ☽ ⊻ ☿
22 36 ☽ ∥ ♄
23 36 ☽ # ♂
23 49 ☽ □ ♂

18 Saturday
01 37 ☽ □ ♄
02 39 ☽ ⊻ ♆
06 18 ☽ ⊻ ☉
07 26 ☽ ⊥ ☿
10 32 ☽ ∠ ♅
13 37 ♂ # ♄
13 55 ☽ ⊥ ♆
14 44 ☿ ⚹ ♃
18 26 ☽ ⊥ ☉
21 00 ☽ ⊻ ♀
23 22 ☽ ∠ ☿

19 Sunday
03 19 ☽ ⊥ ♅
05 17 ☿ ⚹ ♇
06 27 ♂ ☍ ♄
06 41 ☽ ∠ ♆
09 17 ☽ ⊥ ♀
11 06 ☽ Q ♄
12 28 ☽ ∠ ☉
14 25 ☽ ⊻ ♅

20 Monday
01 40 ☽ ⚹ ☿
02 42 ☽ △ ♇
03 18 ♂ ⚹ ♆
03 31 ☽ ∠ ♀
03 56 ☽ ☌ ♃
08 33 ♀ ∠ ♃
09 02 ☽ ⚹ ♄
10 15 ☽ ⚹ ♆
10 38 ☽ △ ♂
18 01 ☽ ⚹ ☉
19 10 ♀ # ♂
22 29 ☽ Q ☿

21 Tuesday
00 34 ☿ ∥ ♆
05 51 ☽ ꝑ ♇
07 56 ☽ Q ♆
09 23 ☽ ⚹ ♀
12 01 ☽ ∠ ♄
13 02 ♀ Q ♂
15 11 ☽ ꝑ ♂
17 12 ☽ Q ☉
20 42 ☽ ☌ ♅
21 54 ☽ ± ♇

22 Wednesday
01 50 ♀ ∥ ♄
03 58 ☽ ⊥ ♄
05 17 ☽ □ ☿
08 01 ☽ ± ♂
08 29 ☽ ⊼ ♇
08 46 ☽ Q ♀
10 20 ☽ ⊻ ♃
14 29 ☽ ⊻ ♄
15 50 ☽ □ ♆
19 06 ☽ ⊼ ♂
20 57 ☽ ⊥ ♃

23 Thursday
03 11 ☽ □ ☉
12 43 ☽ ∠ ♃
18 56 ☉ Q ♅
19 00 ☽ □ ♀

24 Friday
00 53 ☽ ⊻ ♅
05 08 ☉ # ♂
07 31 ☽ △ ☿
11 04 ☽ ⊥ ♅
11 09 ☉ Q ♇
12 09 ☽ ☍ ♇
13 20 ☽ ∥ ♀
14 33 ☽ ⚹ ♃
17 51 ☽ ☌ ♄
19 18 ☽ △ ♆
19 54 ☽ ∥ ♄

25 Saturday
01 05 ☽ ☍ ♂
02 12 ☽ ∠ ♅
02 28 ☽ ∥ ☉
05 36 ☽ # ♂
08 07 ☽ ꝑ ☿
09 45 ☽ △ ☉
10 51 ☽ Q ♃
20 18 ☽ ꝑ ♆
22 45 ☽ ± ☿
23 44 ♂ ꝑ ♅

26 Sunday
01 56 ☽ △ ♀
02 12 ☽ ∥ ♆
03 04 ☽ ⚹ ♅
08 29 ☽ ⊼ ☿
11 07 ☽ ± ♆
12 13 ☽ ꝑ ☉
13 08 ☽ ∥ ☿
13 55 ☽ ⊼ ♇
15 39 ♀ ⚹ ♅
16 51 ☽ □ ♃
19 24 ☽ ⊻ ♄
20 57 ☽ ⊼ ♆
22 44 ☽ Q ♅
23 43 ☽ ± ♇

27 Monday
03 57 ☽ ± ☉
04 42 ☽ ꝑ ♀
04 57 ♄ St D
05 03 ☽ ⊼ ♂
05 10 ☽ ⊥ ♄
14 22 ☽ ꝑ ♇
14 23 ☽ ⊼ ☉
15 19 ☽ ± ♂
19 48 ☽ ∠ ♄
20 40 ☽ ± ♀

28 Tuesday
04 04 ☽ □ ♅
06 43 ☽ ꝑ ♂
07 19 ☽ ⊼ ♀
08 59 ☽ ☍ ☿
12 01 ☉ ∥ ♄
14 47 ☽ △ ♇
15 56 ☿ St D
18 18 ☽ △ ♃
20 14 ☽ ⚹ ♄
21 54 ☽ ☍ ♆
22 31 ♀ ⊥ ♃
23 06 ☽ # ☿

29 Wednesday
03 24 ☿ ⊻ ♀
08 32 ☽ △ ♂
13 04 ☽ # ♆
15 56 ☽ Q ♄
18 54 ☽ ☍ ☉
19 17 ☽ ꝑ ♃

30 Thursday
01 53 ☉ ∠ ♃
02 41 ☽ ∥ ♂
05 28 ☽ △ ♅
07 22 ☿ ∠ ♂
10 29 ☽ ± ♃
10 37 ☽ ⊼ ☿
13 21 ☽ ☍ ♀
16 27 ☽ □ ♇
19 17 ☽ # ♄
20 45 ☽ ⊼ ♃
20 52 ☽ ± ☿
22 04 ☽ □ ♄
23 56 ☽ ⊼ ♆

MAY 2037

LONGITUDES

Date	Sidereal time h m s	Sun ☉ ° ' "	Moon ☽ ° ' "	Moon ☽ 24.00 ° '	Mercury ☿ ° '	Venus ♀ ° '	Mars ♂ ° '	Jupiter ♃ ° '	Saturn ♄ ° '	Uranus ♅ ° '	Neptune ♆ ° '	Pluto ♇ ° '
01	02 39 03	11 ♉ 27 08	04 ♐ 08 08	11 ♐ 01 49	19 ♈ 25	22 ♉ 01	05 ♓ 05	25 ♊ 21	26 ♌ 01	16 ♋ 11	27 ♈ 07	22 ♒ 41
02	02 43 00	12 25 20	17 ♐ 49 02	24 ♐ 29 42	19 41	23 14	05 49	25 33	26 01	16 13	27 09	22 42
03	02 46 56	13 23 31	01 ♑ 03 51	07 ♑ 31 43	20 02	24 28	06 34	25 44	26 02	16 15	27 11	22 42
04	02 50 53	14 21 40	13 ♑ 53 38	20 ♑ 10 00	20 27	25 42	07 18	25 56	26 02	16 17	27 14	22 43
05	02 54 49	15 19 48	26 ♑ 21 23	02 ♒ 28 21	20 56	26 56	08 02	26 08	26 03	16 19	27 16	22 43
06	02 58 46	16 17 55	08 ♒ 31 32	14 ♒ 31 36	21 29	28 10	08 47	26 20	26 04	16 22	27 18	22 44
07	03 02 42	17 16 00	20 ♒ 29 14	26 ♒ 25 07	22 07	29 ♉ 24	09 31	26 32	26 05	16 24	27 20	22 44
08	03 06 39	18 14 04	02 ♓ 19 55	08 ♓ 14 18	22 48	00 ♊ 37	10 15	26 44	26 06	16 26	27 22	22 45
09	03 10 35	19 12 06	14 ♓ 08 54	20 ♓ 04 18	23 32	01 51	10 59	26 56	26 08	16 28	27 24	22 45
10	03 14 32	20 10 07	26 ♓ 01 03	01 ♈ 59 39	24 21	03 05	11 43	27 08	26 09	16 31	27 26	22 45
11	03 18 29	21 08 06	08 ♈ 00 34	14 ♈ 04 09	25 12	04 19	12 28	27 21	26 10	16 33	27 29	22 46
12	03 22 25	22 06 04	20 ♈ 10 44	26 ♈ 20 34	26 07	05 32	13 12	27 33	26 12	16 36	27 31	22 46
13	03 26 22	23 04 01	02 ♉ 33 50	08 ♉ 50 37	27 05	06 46	13 56	27 46	26 14	16 38	27 33	22 46
14	03 30 18	24 01 56	15 ♉ 10 59	21 ♉ 34 56	28 07	08 00	14 40	27 58	26 15	16 41	27 35	22 47
15	03 34 15	24 59 50	28 ♉ 02 22	04 ♊ 33 11	29 ♈ 11	09 13	15 24	28 10	26 17	16 43	27 37	22 47
16	03 38 11	25 57 43	11 ♊ 07 15	17 ♊ 44 24	00 ♉ 18	10 27	16 08	28 23	26 19	16 46	27 39	22 47
17	03 42 08	26 55 34	24 ♊ 24 28	01 ♋ 07 17	01 27	11 40	16 52	28 36	26 21	16 48	27 41	22 47
18	03 46 04	27 53 23	07 ♋ 52 43	14 ♋ 40 37	02 40	12 54	17 36	28 48	26 24	16 51	27 43	22 47
19	03 50 01	28 51 11	21 ♋ 30 53	28 ♋ 23 26	03 55	14 08	18 20	29 01	26 26	16 54	27 45	22 47
20	03 53 58	29 ♉ 48 57	05 ♌ 18 12	12 ♌ 15 09	05 13	15 21	19 04	29 14	26 28	16 56	27 47	22 48
21	03 57 54	00 ♊ 46 42	19 ♌ 14 13	26 ♌ 15 22	06 33	16 35	19 48	29 27	26 31	16 59	27 49	22 48
22	04 01 51	01 44 25	03 ♍ 18 31	10 ♍ 23 32	07 56	17 48	20 32	29 39	26 33	17 02	27 51	22 48
23	04 05 47	02 42 06	17 ♍ 30 15	24 ♍ 38 27	09 21	19 02	21 16	29 ♊ 52	26 36	17 05	27 53	22 R 48
24	04 09 44	03 39 46	01 ♎ 47 47	08 ♎ 57 53	10 48	20 15	21 59	00 ♋ 05	26 39	17 08	27 55	22 48
25	04 13 40	04 37 24	16 ♎ 08 15	23 ♎ 18 22	12 18	21 29	22 43	00 18	26 41	17 10	27 57	22 48
26	04 17 37	05 35 00	00 ♏ 27 37	07 ♏ 35 22	13 51	22 42	23 27	00 31	26 44	17 13	27 59	22 48
27	04 21 33	06 32 36	14 ♏ 40 56	21 ♏ 43 40	15 26	23 55	24 10	00 44	26 47	17 16	28 01	22 47
28	04 25 30	07 30 10	28 ♏ 42 58	05 ♐ 38 16	17 03	25 09	24 54	00 57	26 51	17 19	28 02	22 47
29	04 29 27	08 27 42	12 ♐ 29 06	19 ♐ 15 05	18 42	26 22	25 38	01 11	26 54	17 22	28 04	22 47
30	04 33 23	09 25 14	25 ♐ 55 57	02 ♑ 31 34	20 24	27 35	26 21	01 24	26 57	17 25	28 06	22 47
31	04 37 20	10 ♊ 22 45	09 ♑ 01 53	15 ♑ 26 59	22 ♉ 09	28 ♊ 49	27 ♓ 05	01 ♋ 37	27 ♌ 01	17 ♋ 28	28 ♈ 08	22 ♒ 47

DECLINATIONS

Date	Moon True ☊ ° '	Moon Mean ☊ ° '	Moon ☽ Latitude ° '	Sun ☉ ° '	Moon ☽ ° '	Mercury ☿ ° '	Venus ♀ ° '	Mars ♂ ° '	Jupiter ♃ ° '	Saturn ♄ ° '	Uranus ♅ ° '	Neptune ♆ ° '	Pluto ♇ ° '
01	02 ♌ 23	03 ♌ 01	04 N 16	15 N 16	16 S 46	05 N 40	18 N 01	11 S 08	23 N 12	14 N 20	22 N 53	08 N 54	22 S 16
02	02 R 15	02 58	03 31	15 34	19 22	05 37	18 22	10 52	23 12	14 20	22 53	08 55	22 16
03	02 09	02 55	02 36	15 51	20 49	05 35	18 43	10 37	23 13	14 20	22 52	08 56	22 16
04	02 06	02 52	01 35	16 09	21 08	05 37	19 03	10 21	23 13	14 19	22 52	08 57	22 16
05	02 06	02 49	00 N 30	16 26	20 23	05 40	19 22	10 05	23 14	14 19	22 52	08 57	22 16
06	02 D 04	02 45	00 S 34	16 43	18 41	05 46	19 42	09 49	23 14	14 18	22 52	08 58	22 17
07	02 R 04	02 42	01 36	16 59	16 11	05 55	20 00	09 34	23 15	14 18	22 51	08 59	22 17
08	02 04	02 39	02 33	17 15	13 02	06 05	20 18	09 18	23 15	14 18	22 51	09 00	22 17
09	02 01	02 36	03 24	17 31	09 22	06 17	20 36	09 02	23 16	14 17	22 51	09 00	22 17
10	01 56	02 33	04 06	17 47	05 20	06 31	20 53	08 45	23 16	14 17	22 50	09 01	22 17
11	01 48	02 29	04 37	18 02	01 S 04	06 47	21 09	08 29	23 16	14 16	22 50	09 02	22 17
12	01 38	02 26	04 56	18 17	03 N 18	07 05	21 25	08 13	23 17	14 16	22 50	09 03	22 18
13	01 27	02 23	05 02	18 32	07 38	07 25	21 40	07 57	23 17	14 15	22 49	09 03	22 18
14	01 15	02 20	04 53	18 47	11 54	07 46	21 54	07 41	23 17	14 14	22 49	09 04	22 18
15	01 03	02 17	04 30	19 01	15 20	08 08	22 08	07 24	23 18	14 13	22 49	09 05	22 18
16	00 54	02 14	03 52	19 14	18 17	08 32	22 22	07 08	23 18	14 12	22 48	09 05	22 18
17	00 46	02 10	03 00	19 28	20 19	08 58	22 35	06 51	23 18	14 12	22 48	09 06	22 19
18	00 42	02 07	01 58	19 41	21 14	09 24	22 47	06 35	23 18	14 11	22 48	09 07	22 19
19	00 40	02 04	00 S 49	19 54	20 55	09 52	22 58	06 18	23 19	14 10	22 47	09 08	22 19
20	00 D 39	02 01	00 N 25	20 06	19 20	10 21	23 09	06 02	23 19	14 09	22 47	09 08	22 19
21	00 40	01 58	01 38	20 18	16 36	10 51	23 19	05 45	23 19	14 08	22 47	09 09	22 20
22	00 R 40	01 54	02 45	20 30	12 52	11 21	23 28	05 29	23 19	14 07	22 46	09 10	22 20
23	00 39	01 51	03 44	20 42	08 22	11 53	23 37	05 12	23 19	14 06	22 46	09 10	22 20
24	00 36	01 48	04 28	20 53	03 N 23	12 26	23 45	04 56	23 19	14 05	22 46	09 11	22 20
25	00 31	01 45	04 56	21 03	01 S 47	12 59	23 53	04 39	23 19	14 04	22 45	09 12	22 21
26	00 24	01 42	05 06	21 14	06 51	13 33	23 59	04 22	23 19	14 03	22 45	09 12	22 21
27	00 16	01 39	04 56	21 24	11 31	14 07	24 05	04 06	23 20	14 02	22 45	09 13	22 21
28	00 07	01 35	04 29	21 33	15 30	14 42	24 11	03 49	23 20	14 01	22 44	09 13	22 22
29	00 00	01 32	03 46	21 43	18 33	15 17	24 15	03 32	23 20	14 00	22 44	09 14	22 22
30	29 ♋ 53	01 29	02 52	21 51	20 30	15 53	24 19	03 16	23 20	13 58	22 43	09 15	22 22
31	29 ♋ 49	01 ♌ 26	01 N 50	22 N 00	21 S 18	16 N 29	24 N 22	02 S 59	23 N 19	13 N 57	22 N 43	09 N 15	22 S 23

ZODIAC SIGN ENTRIES

Date	h m	Planets
01	04 53	☽ ♐
03	10 03	☽ ♑
05	19 08	☽ ♒
07	23 51	♀ ♊
08	07 16	☽ ♓
10	20 00	☽ ♈
13	07 04	☽ ♉
15	15 37	☽ ♊
16	05 46	☿ ♉
17	22 00	☽ ♋
20	02 48	☽ ♌
20	16 35	☉ ♊
22	06 23	☽ ♍
24	02 12	♃ ♋
24	08 59	☽ ♎
26	11 14	☽ ♏
28	14 13	☽ ♐
30	19 23	☽ ♑

LATITUDES

Date	Mercury ☿ ° '	Venus ♀ ° '	Mars ♂ ° '	Jupiter ♃ ° '	Saturn ♄ ° '	Uranus ♅ ° '	Neptune ♆ ° '	Pluto ♇ ° '
01	02 S 05	00 S 15	01 S 36	00 S 09	01 N 35	00 N 26	01 S 39	08 S 49
04	02 34	00 08	01 38	00 09	01 35	00 26	01 39	08 50
07	02 54	00 S 01	01 41	00 08	01 35	00 26	01 39	08 51
10	03 08	00 N 07	01 43	00 08	01 34	00 26	01 39	08 52
13	03 14	00 14	01 46	00 08	01 34	00 26	01 39	08 53
16	03 14	00 22	01 48	00 08	01 34	00 26	01 39	08 54
19	03 08	00 29	01 50	00 07	01 33	00 26	01 39	08 55
22	02 57	00 36	01 53	00 07	01 33	00 26	01 39	08 56
25	02 40	00 43	01 55	00 07	01 33	00 26	01 39	08 57
28	02 19	00 50	01 57	00 06	01 33	00 26	01 40	08 57
31	01 S 53	00 N 57	01 S 59	00 S 06	01 N 32	00 N 26	01 S 40	08 S 58

DATA

Julian Date	2465180
Delta T	+76 seconds
Ayanamsa	24° 22' 27"
Synetic vernal point	04° ♓ 44' 32"
True obliquity of ecliptic	23° 25' 59"

MOON'S PHASES, APSIDES AND POSITIONS ☽

Date	h m	Phase	Longitude ° '	Eclipse Indicator
07	04 56	☾	16 ♒ 59	
15	05 54	●	24 ♉ 45	
22	09 08	☽	01 ♍ 38	
29	04 24	○	08 ♐ 09	

Day	h m		
09	06 37	Apogee	
24	14 52	Perigee	
04	06 45	Max dec	21° S 10'
11	17 52	0N	
18	18 01	Max dec	21° N 16'
25	03 43	0S	
31	16 34	Max dec	21° S 19'

LONGITUDES

Date	Chiron ⚷ ° '	Ceres ⚳ ° '	Pallas ⚴ ° '	Juno ⚵ ° '	Vesta ⚶ ° '	Black Moon Lilith ⚸ ° '
01	16 ♊ 22	18 ♎ 56	20 ♌ 06	29 ♌ 13	03 ♈ 21	12 ♓ 13
11	17 ♊ 07	17 ♎ 25	22 ♌ 35	00 ♍ 36	07 ♈ 50	13 ♓ 20
21	17 ♊ 55	16 ♎ 32	25 ♌ 27	02 ♍ 25	12 ♈ 10	14 ♓ 27
31	18 ♊ 45	16 ♎ 18	28 ♌ 38	04 ♍ 36	16 ♈ 20	15 ♓ 34

All ephemeris data is given at 12.00 UT and the Moon's longitude is additionally given for 24.00 UT

ASPECTARIAN

h m Aspects

01 Friday
00 17 ☽ # ☉
06 55 ☽ ⚼ ♅
10 14 ☽ ± ♆
12 30 ☽ ⚼ ☿
13 44 ☽ □ ♂
22 33 ☽ ± ♅
23 25 ☽ Q ♇
23 53 ☽ # ♀

02 Saturday
01 22 ♀ □ ♇
01 43 ☽ ⊼ ☉
01 57 ☽ ⚼ ♆
09 09 ☽ ⊼ ♅
13 10 ☽ ± ☉
15 25 ☽ △ ☿
20 46 ☽ ⚹ ♇
22 44 ☽ ⊼ ♀
23 25 ☽ Q ♂

03 Sunday
02 07 ☽ ☍ ♃
02 47 ☽ △ ♄
04 53 ☽ △ ♆
06 42 ☽ ⚼ ☉
10 48 ☽ ± ♀
22 49 ☽ ⚹ ♂

04 Monday
00 21 ☽ ∠ ♇
05 19 ☽ ⚼ ♀
06 36 ☽ ⚼ ♄
12 58 ☽ △ ☉
16 35 ☽ ☍ ♅
17 23 ☽ ⊥ ♇
17 24 ♀ ⊻ ♃
18 40 ♀ □ ♄
23 46 ☽ ± ♄

05 Tuesday
01 02 ☽ □ ☿
01 43 ♃ ⚹ ♄
04 56 ☽ ⊻ ♇
05 08 ☽ ∠ ♂
11 25 ☽ ⊼ ♄
11 33 ☽ ⊼ ♃
11 58 ☿ ⊥ ♀
13 15 ☽ △ ♀
13 46 ☽ □ ♆
18 37 ♀ ⊻ ♆
23 31 ☽ ± ♃
23 52 ☽ ⊥ ♂

06 Wednesday
01 12 ☽ # ♀
12 32 ☽ ⊻ ♂
13 37 ☉ ⚹ ♅
14 01 ☽ Q ☿
17 42 ☽ ⚼ ♃

07 Thursday
01 35 ☽ Q ♆
03 44 ☽ ⊼ ♅
04 56 ☽ □ ☉
05 37 ☽ # ☉
15 28 ☽ ⚹ ☿
15 52 ☽ ± ♅
16 33 ☽ ☌ ♇
23 21 ☽ ☍ ♄

08 Friday
00 27 ☽ △ ♃
01 54 ☽ ⚹ ♆
02 50 ☽ # ♄
08 08 ☽ □ ♀
10 10 ☽ ⚼ ♅
10 22 ☿ ⚹ ♇
20 38 ☽ Q ☉
23 50 ☽ ∠ ☿

09 Saturday
04 23 ♀ ∠ ♅
05 10 ☽ ☌ ♂
08 27 ☽ ∠ ♆
13 38 ♂ # ♆
14 14 ☽ # ♆
14 17 ☽ ∥ ♂
16 44 ☽ △ ♅
19 22 ☽ ⊥ ☿
23 09 ☽ ⚹ ☉

10 Sunday
00 53 ☽ Q ♀
02 45 ☽ ⊥ ♆
05 25 ☽ ⊻ ♇
05 31 ☽ # ☿
08 23 ☽ ⊻ ☿
12 16 ☽ ⊼ ♄
14 18 ☽ □ ♃
14 52 ☽ ⊻ ♆
17 31 ☽ ⊥ ♇
19 16 ♀ ⊥ ♆

11 Monday
00 20 ☽ ± ♄
03 47 ☽ ⚹ ♀
07 57 ☽ ∠ ☉
11 31 ☽ ∠ ♇
12 33 ♂ ∠ ♆
18 17 ☽ ⚼ ♄
18 39 ☉ ⊥ ♃
21 24 ☽ ⊻ ♂

12 Tuesday
02 46 ☽ Q ♃
03 19 ☽ ⊥ ☉
04 57 ☽ □ ♅
06 25 ♃ ⚹ ♆
09 57 ☽ ⊥ ♂
12 47 ☽ ∠ ♀
14 02 ☿ △ ♄
16 04 ☽ ⊻ ☉
17 03 ☽ ⚹ ♇
23 45 ☽ △ ♄

13 Wednesday
00 32 ☽ ☌ ☿
02 18 ☽ ☌ ♆
02 35 ☽ ⚹ ♃
04 34 ☽ ∠ ♂
04 39 ☉ □ ♇
08 10 ☽ ⊥ ♀
10 41 ☽ ∥ ☿
13 43 ☽ # ♂
15 59 ☽ Q ♅
16 14 ☽ Q ♇
20 13 ☽ ∥ ♆
20 55 ☽ ⊻ ♀
23 16 ☿ ☌ ♆

14 Thursday
07 45 ☽ ∠ ♃
07 51 ☿ ⚹ ♃
08 44 ☿ # ♂
10 58 ☽ ⚹ ♂
14 49 ☽ ⚹ ♅

15 Friday
00 56 ☽ ⊥ ♃
02 14 ☽ □ ♇
04 16 ☽ ∥ ♄
05 54 ☽ ☌ ☉
08 45 ☽ □ ♄
10 45 ☽ Q ♂
11 13 ☽ ⊻ ♆
12 15 ☽ ⊻ ♃
14 18 ☽ ⊻ ☿
18 49 ☽ ∠ ♅
22 18 ☽ ⊥ ♆

16 Saturday
02 22 ☽ ⊥ ☿
05 31 ♀ # ♇
10 39 ☽ ☌ ♀
11 21 ☽ ⊥ ♅
14 47 ☽ ∠ ♆
17 50 ☽ Q ♄
18 21 ♀ ⊥ ♅
20 18 ☽ ∠ ☿
21 16 ☉ □ ♄
21 38 ☽ □ ♂
22 16 ☽ ⊻ ♅
23 07 ☽ ∥ ☉

17 Sunday
01 22 ♅ ± ♇
09 05 ☽ △ ♇
09 45 ♂ △ ♅
15 30 ☽ ⚹ ♄
16 52 ☽ ⊻ ☉
17 53 ☽ ⚹ ♆
19 37 ☽ ☌ ♃
20 08 ☿ ∥ ♆

18 Monday
01 50 ☽ ⚹ ☿
04 23 ☽ ⊥ ☉
07 34 ☉ ⊻ ♆
08 19 ♀ ∠ ♆
08 58 ☿ ∠ ♂
11 50 ☽ ⚼ ♇
14 22 ♀ ∥ ♅
15 16 ☽ Q ♆
18 14 ☽ ∠ ♄
21 32 ☽ ∠ ☉
21 45 ☽ ⊻ ♀

19 Tuesday
01 11 ☽ Q ☿
03 43 ☽ ± ♇
03 52 ☽ ☌ ♅
06 07 ☽ △ ♂
09 20 ☽ ⊥ ♀
10 06 ☽ ⊥ ♄
14 14 ☽ ⊼ ♇
17 15 ☉ ⊻ ♃
18 07 ♀ Q ♄
20 37 ☽ ⊻ ♄
22 55 ☽ □ ♆

20 Wednesday
01 18 ☽ ⊻ ♃
01 46 ☽ ⚹ ☉
02 35 ☽ ∠ ♀
03 33 ☽ ∥ ☉
04 19 ☿ Q ♇
06 50 ☿ Q ♅
09 44 ☽ ⚼ ♂
11 49 ☽ □ ☿
11 52 ☽ ⊥ ♃

21 Thursday
00 05 ☽ Q ☉
02 09 ☽ ± ♂
03 39 ☽ ∠ ♃
07 00 ☽ ⚹ ♀
08 07 ☽ ⊻ ♅
12 05 ♀ ∥ ♃
13 01 ☽ ⊼ ♂
18 05 ☽ ☍ ♇
18 26 ☽ ⊥ ♅
20 16 ♀ ⊻ ♅

22 Friday
00 28 ☽ ☌ ♄
02 42 ☽ △ ♆
04 28 ☽ ∥ ♄
05 29 ☽ Q ♀
05 42 ☽ ⚹ ♃
09 08 ☽ □ ☉
09 49 ☽ ∠ ♅
19 30 ☽ ∥ ☿
19 38 ☉ ∠ ♅
20 41 ☽ △ ☿

23 Saturday
02 21 ☽ Q ♃
04 11 ☽ ⚼ ♆
07 51 ♇ St R
07 56 ☽ ∥ ♆
11 17 ☽ ⚹ ♅
14 48 ☽ □ ♀
18 39 ☽ ☍ ♂
19 23 ☽ ± ♆
20 54 ☽ ⊼ ♇

24 Sunday
00 48 ☽ ⚼ ☿
03 20 ☽ ⊻ ♄
04 19 ☽ # ♂
05 29 ☽ ⊼ ♆
06 58 ☽ ± ♇
07 31 ☽ Q ♅
09 06 ☽ □ ♃
09 30 ♂ ⊥ ♆
13 25 ☽ ⊥ ♄
15 21 ☽ △ ☉
17 37 ☽ ± ☿
18 32 ☉ ⊥ ♆
22 02 ☽ ⚼ ♇

25 Monday
04 32 ☽ ∠ ♄
04 50 ☽ ⊼ ☿
13 44 ☽ □ ♅
14 30 ♂ ⊻ ♇
18 15 ☽ ⚼ ☉
21 41 ☉ Q ♂
21 46 ☽ △ ♀
23 08 ☽ △ ♇
23 36 ☽ ⊼ ♂

26 Tuesday
00 42 ☽ ∥ ♂
05 44 ☽ ⚹ ♄
07 49 ☽ ☍ ♆
10 12 ☽ ± ♂
10 25 ☽ ± ☉
12 06 ☽ △ ♃
13 49 ♀ △ ♇
21 14 ☽ ⊼ ☉
23 45 ☽ # ♆

27 Wednesday
01 20 ☽ ⚼ ♀
02 00 ☽ Q ♄
02 10 ☽ ⚼ ♂
08 22 ☿ ∥ ♄
13 26 ☽ ☍ ☿
13 49 ☽ ⚼ ♃
16 25 ☽ △ ♅
17 24 ☿ ∠ ♃
18 02 ☽ ± ♀

28 Thursday
00 08 ♀ □ ♂
01 49 ☽ □ ♇
02 30 ☽ # ♄
05 04 ☽ △ ♂
05 16 ☽ ⊼ ♀
05 26 ☽ ± ♃
05 50 ☽ # ☿
08 46 ☽ □ ♄
10 50 ☽ ⊼ ♆
15 56 ☽ ⊼ ♃
16 07 ☿ ⚹ ♅
18 15 ☽ ⚼ ♅
21 14 ☽ ± ♆

29 Friday
04 24 ☽ ☍ ☉
09 01 ☽ Q ♇
10 02 ☽ ± ♅
13 02 ☽ ⚼ ♆
20 41 ☽ ⊼ ♅
22 55 ♀ ⚹ ♄

30 Saturday
00 37 ☽ ⊼ ☿
06 19 ☽ ⚹ ♇
12 48 ☽ □ ♂
12 59 ☽ ± ☿
13 51 ☽ △ ♄
15 18 ☽ ☍ ♀
15 56 ☽ △ ♆
22 06 ☽ ☍ ♃
22 20 ♀ ⚹ ♆

31 Sunday
07 57 ☽ ⚼ ☿
09 36 ♂ ⊼ ♄
09 41 ☽ ∠ ♇
14 42 ☽ ⊼ ☉
17 34 ☽ ⚼ ♄
20 38 ☿ □ ♇
23 59 ☽ Q ♂

JUNE 2037

LONGITUDES

Date	Sidereal time h m s	Sun ☉ ° ' "	Moon ☽ ° ' "	Moon ☽ 24.00 ° ' "	Mercury ☿ ° '	Venus ♀ ° '	Mars ♂ ° '	Jupiter ♃ ° '	Saturn ♄ ° '	Uranus ♅ ° '	Neptune ♆ ° '	Pluto ♇ ° '
01	04 41 16	11 ♊ 20 15	21 ♑ 47 04	28 ♑ 02 24	23 ♉ 55	00 ♋ 02	27 ♓ 48	01 ♋ 50	27 ♌ 04	17 ♋ 31	28 ♈ 10	22 ♒ 47
02	04 45 13	12 17 44	04 ♒ 13 22	10 ♒ 20 24	25 44	01 15	28 31	02 03	27 08	17 35	28 11	22 R 46
03	04 49 09	13 15 12	16 ♒ 24 01	22 ♒ 24 44	27 36	02 28	29 15	02 17	27 11	17 38	28 13	22 46
04	04 53 06	14 12 40	28 ♒ 23 11	04 ♓ 19 57	29 ♉ 29	03 42	29 ♓ 58	02 30	27 15	17 41	28 15	22 46
05	04 57 02	15 10 06	10 ♓ 15 41	16 ♓ 11 01	01 ♊ 25	04 55	00 ♈ 41	02 43	27 19	17 44	28 17	22 45
06	05 00 59	16 07 32	22 ♓ 06 36	28 ♓ 03 02	03 23	06 08	01 25	02 57	27 23	17 47	28 18	22 45
07	05 04 56	17 04 57	04 ♈ 00 57	10 ♈ 00 54	05 23	07 21	02 08	03 10	27 27	17 50	28 20	22 45
08	05 08 52	18 02 22	16 ♈ 03 27	22 ♈ 09 04	07 25	08 34	02 51	03 23	27 31	17 54	28 22	22 44
09	05 12 49	18 59 46	28 ♈ 18 13	04 ♉ 31 14	09 29	09 47	03 34	03 37	27 35	17 57	28 23	22 44
10	05 16 45	19 57 10	10 ♉ 48 28	17 ♉ 10 05	11 35	11 01	04 17	03 50	27 39	18 00	28 25	22 43
11	05 20 42	20 54 33	23 ♉ 36 16	00 ♊ 07 04	13 42	12 14	05 00	04 04	27 44	18 04	28 26	22 43
12	05 24 38	21 51 55	06 ♊ 42 25	13 ♊ 22 14	15 51	13 27	05 42	04 17	27 48	18 07	28 28	22 43
13	05 28 35	22 49 17	20 ♊ 06 18	26 ♊ 54 22	18 01	14 40	06 25	04 31	27 53	18 10	28 29	22 42
14	05 32 31	23 46 38	03 ♋ 46 07	10 ♋ 41 10	20 12	15 53	07 08	04 44	27 57	18 14	28 31	22 41
15	05 36 28	24 43 59	17 ♋ 39 07	24 ♋ 39 35	22 23	17 06	07 50	04 58	28 02	18 17	28 32	22 41
16	05 40 25	25 41 18	01 ♌ 42 06	08 ♌ 46 18	24 35	18 19	08 33	05 12	28 07	18 21	28 34	22 40
17	05 44 21	26 38 37	15 ♌ 51 46	22 ♌ 58 09	26 47	19 32	09 15	05 25	28 12	18 24	28 35	22 40
18	05 48 18	27 35 55	00 ♍ 05 05	07 ♍ 12 16	28 ♊ 59	20 45	09 58	05 39	28 17	18 27	28 36	22 39
19	05 52 14	28 33 13	14 ♍ 19 24	21 ♍ 26 12	01 ♋ 10	21 57	10 40	05 52	28 22	18 31	28 38	22 38
20	05 56 11	29 ♊ 30 29	28 ♍ 32 25	05 ♎ 37 48	03 21	23 10	11 22	06 06	28 27	18 34	28 39	22 38
21	06 00 07	00 ♋ 27 44	12 ♎ 42 05	19 ♎ 45 01	05 30	24 23	12 04	06 20	28 32	18 38	28 40	22 37
22	06 04 04	01 24 59	26 ♎ 46 20	03 ♏ 45 45	07 39	25 36	12 46	06 33	28 37	18 41	28 42	22 36
23	06 08 00	02 22 13	10 ♏ 43 01	17 ♏ 37 49	09 46	26 49	13 28	06 47	28 42	18 45	28 43	22 36
24	06 11 57	03 19 27	24 ♏ 29 54	01 ♐ 18 58	11 52	28 02	14 10	07 00	28 47	18 48	28 44	22 35
25	06 15 54	04 16 40	08 ♐ 04 46	14 ♐ 47 04	13 56	29 ♋ 14	14 52	07 14	28 53	18 52	28 45	22 34
26	06 19 50	05 13 53	21 ♐ 25 41	28 ♐ 00 27	15 58	00 ♌ 27	15 34	07 28	28 58	18 56	28 46	22 33
27	06 23 47	06 11 05	04 ♑ 31 15	10 ♑ 58 03	17 59	01 40	16 15	07 41	29 04	18 59	28 48	22 32
28	06 27 43	07 08 17	17 ♑ 20 50	23 ♑ 39 41	19 57	02 52	16 57	07 55	29 09	19 03	28 49	22 32
29	06 31 40	08 05 29	29 ♑ 54 43	06 ♒ 06 08	21 54	04 05	17 38	08 09	29 15	19 06	28 50	22 31
30	06 35 36	09 ♋ 02 41	12 ♒ 14 10	18 ♒ 19 08	23 ♋ 48	05 ♌ 17	18 ♈ 19	08 ♋ 22	29 ♌ 21	19 ♋ 10	28 ♈ 51	22 ♒ 30

Date	Moon True ☊ ° '	Moon Mean ☊ ° '	Moon ☽ Latitude ° '
01	29 ♋ 47	01 ♌ 23	00 N 43
02	29 D 47	01 20	00 S 24
03	29 48	01 16	01 29
04	29 49	01 13	02 28
05	29 50	01 10	03 21
06	29 R 50	01 07	04 05
07	29 49	01 04	04 39
08	29 45	01 00	05 01
09	29 40	00 57	05 10
10	29 34	00 54	05 04
11	29 28	00 51	04 43
12	29 22	00 48	04 07
13	29 16	00 45	03 17
14	29 12	00 41	02 14
15	29 10	00 38	01 S 02
16	29 D 10	00 35	00 N 14
17	29 11	00 32	01 30
18	29 12	00 29	02 41
19	29 13	00 26	03 42
20	29 14	00 22	04 30
21	29 R 13	00 19	05 00
22	29 12	00 16	05 13
23	29 09	00 13	05 07
24	29 05	00 10	04 43
25	29 02	00 06	04 04
26	28 58	00 03	03 12
27	28 56	00 ♌ 00	02 10
28	28 54	29 ♋ 57	01 N 03
29	28 D 54	29 54	00 S 06
30	28 ♋ 54	29 ♋ 51	01 S 13

DECLINATIONS

Date	Sun ☉ ° '	Moon ☽ ° '	Mercury ☿ ° '	Venus ♀ ° '	Mars ♂ ° '	Jupiter ♃ ° '	Saturn ♄ ° '	Uranus ♅ ° '	Neptune ♆ ° '	Pluto ♇ ° '
01	22 N 08	20 S 58	17 N 04	24 N 25	02 S 42	23 N 19	13 N 56	22 N 43	09 N 16	22 S 23
02	22 16	19 35	17 40	24 26	02 26	23 19	13 54	22 42	09 16	22 23
03	22 23	17 20	18 15	24 27	02 09	23 19	13 53	22 42	09 17	22 24
04	22 30	14 21	18 50	24 28	01 52	23 19	13 52	22 41	09 18	22 24
05	22 36	10 49	19 24	24 27	01 36	23 19	13 50	22 41	09 18	22 24
06	22 43	06 53	19 58	24 26	01 19	23 19	13 49	22 40	09 19	22 25
07	22 48	02 S 41	20 30	24 24	01 03	23 19	13 47	22 40	09 19	22 25
08	22 54	01 N 41	21 02	24 22	00 46	23 18	13 46	22 40	09 20	22 26
09	22 59	06 02	21 32	24 18	00 29	23 18	13 44	22 39	09 20	22 26
10	23 03	10 15	22 01	24 14	00 S 13	23 18	13 43	22 39	09 21	22 26
11	23 07	14 06	22 28	24 10	00 N 04	23 17	13 41	22 38	09 21	22 27
12	23 11	17 22	22 53	24 04	00 20	23 17	13 40	22 38	09 22	22 27
13	23 14	19 48	23 16	23 58	00 36	23 17	13 38	22 37	09 22	22 28
14	23 17	21 09	23 37	23 51	00 53	23 16	13 36	22 37	09 23	22 28
15	23 20	21 14	23 56	23 44	01 09	23 16	13 35	22 36	09 23	22 29
16	23 22	20 00	24 11	23 36	01 25	23 16	13 33	22 36	09 24	22 29
17	23 23	17 31	24 25	23 27	01 41	23 15	13 31	22 35	09 24	22 30
18	23 25	13 57	24 35	23 17	01 58	23 15	13 30	22 35	09 25	22 30
19	23 26	09 35	24 43	23 07	02 14	23 14	13 28	22 35	09 25	22 30
20	23 26	04 N 42	24 47	22 56	02 30	23 14	13 26	22 34	09 25	22 31
21	23 26	00 S 24	24 49	22 45	02 46	23 13	13 24	22 34	09 26	22 31
22	23 26	05 27	24 49	22 32	03 02	23 13	13 22	22 33	09 26	22 32
23	23 25	10 10	24 45	22 20	03 17	23 12	13 20	22 33	09 27	22 32
24	23 23	14 18	24 39	22 06	03 33	23 12	13 18	22 32	09 27	22 33
25	23 22	17 38	24 30	21 52	03 49	23 11	13 16	22 32	09 27	22 33
26	23 20	19 58	24 19	21 37	04 04	23 10	13 15	22 31	09 28	22 34
27	23 17	21 11	24 06	21 22	04 20	23 09	13 13	22 31	09 28	22 34
28	23 14	21 16	23 50	21 06	04 36	23 09	13 11	22 30	09 28	22 35
29	23 11	20 15	23 32	20 50	04 51	23 08	13 08	22 30	09 29	22 35
30	23 N 07	18 S 17	23 N 13	20 N 33	05 N 06	23 N 07	13 N 06	22 N 29	09 N 29	22 S 36

ZODIAC SIGN ENTRIES

Date	h	m	Planets
01	11	22	♀ ♋
02	03	47	☽ ♒
04	13	03	♂ ♈
04	15	15	☽ ♓
04	18	27	☿ ♊
07	03	56	☽ ♈
09	15	17	☽ ♉
11	23	47	☽ ♊
14	05	25	☽ ♋
16	09	06	☽ ♌
18	11	51	☽ ♍
18	23	11	☿ ♋
20	14	28	☽ ♎
21	00	22	☉ ♋
22	17	32	☽ ♏
24	21	41	☽ ♐
26	03	06	♀ ♌
27	03	39	☽ ♑
29	12	10	☽ ♒

LATITUDES

Date	Mercury ☿ ° '	Venus ♀ ° '	Mars ♂ ° '	Jupiter ♃ ° '	Saturn ♄ ° '	Uranus ♅ ° '	Neptune ♆ ° '	Pluto ♇ ° '
01	01 S 44	00 N 59	02 S 00	00 S 06	01 N 32	00 N 26	01 S 40	08 S 59
04	01 14	01 05	02 02	00 06	01 32	00 26	01 40	08 59
07	00 42	01 11	02 03	00 05	01 32	00 26	01 40	09 00
10	00 S 09	01 16	02 05	00 05	01 32	00 26	01 40	09 01
13	00 N 23	01 21	02 07	00 05	01 31	00 26	01 40	09 02
16	00 52	01 26	02 09	00 04	01 31	00 26	01 40	09 03
19	01 17	01 30	02 10	00 04	01 31	00 26	01 41	09 04
22	01 36	01 33	02 11	00 04	01 31	00 26	01 41	09 04
25	01 48	01 36	02 13	00 03	01 31	00 26	01 41	09 05
28	01 54	01 38	02 14	00 03	01 30	00 26	01 41	09 06
31	01 N 54	01 N 40	02 S 15	00 S 03	01 N 30	00 N 26	01 S 41	09 S 06

DATA

Julian Date	2465211
Delta T	+76 seconds
Ayanamsa	24° 22' 32"
Synetic vernal point	04° ♓ 44' 27"
True obliquity of ecliptic	23° 25' 59"

MOON'S PHASES, APSIDES AND POSITIONS ☽

Date	h	m	Phase	Longitude ° '	Eclipse Indicator
05	22	49	☾	15 ♓ 36	
13	17	10	●	23 ♊ 02	
20	13	45	☽	29 ♍ 35	
27	15	20	○	06 ♑ 19	

Day	h	m		
06	01	23	Apogee	
18	16	28	Perigee	
08	02	47	0N	
15	01	41	Max dec	21° N 22'
21	10	07	0S	
28	01	34	Max dec	21° S 22'

LONGITUDES

Date	Chiron ⚷ ° '	Ceres ⚳ ° '	Pallas ⚴ ° '	Juno ⚵ ° '	Vesta ⚶ ° '	Black Moon Lilith ⚸ ° '
01	18 ♊ 50	16 ♎ 19	28 ♌ 57	04 ♍ 50	16 ♈ 45	15 ♓ 40
11	19 ♊ 41	16 ♎ 48	02 ♍ 23	07 ♍ 21	20 ♈ 14	16 ♓ 47
21	20 ♊ 32	17 ♎ 52	06 ♍ 01	10 ♍ 06	24 ♈ 30	17 ♓ 54
31	21 ♊ 22	19 ♎ 28	09 ♍ 49	13 ♍ 03	28 ♈ 02	19 ♓ 01

ASPECTARIAN

h m	Aspects
01 Monday	
02 30	☽ ⊥ ♇
02 52	☽ ± ☉
03 53	☽ ☍ ♅
10 38	☽ ± ♄
13 54	☽ ⊻ ♇
16 36	☿ ⊥ ♀
16 46	☽ △ ☿
16 55	☉ ⊥ ♅
21 27	☽ ⚼ ☉
22 10	☽ ⊼ ♄
02 Tuesday	
00 15	☽ ⚹ ♂
00 16	☽ □ ♆
00 28	♂ ⊻ ♆
05 35	☽ ⊼ ♀
07 42	☽ ⊼ ♃
16 44	☿ ⊥ ♃
18 35	☽ ± ♀
19 39	☽ ± ♃
20 11	♂ ⊥ ♇
03 Wednesday	
04 54	☽ # ☿
05 13	☽ △ ☉
06 39	☿ □ ♄
07 17	♀ ☌ ♃
07 27	☽ ∠ ♂
11 07	☉ ∠ ♆
11 38	☽ Q ♆
13 47	☽ ⚼ ♃
14 06	☉ # ♇
14 23	☽ ⚼ ♀
14 27	☽ ⊼ ♅
20 08	☿ ⊻ ♆
04 Thursday	
00 42	☽ ☌ ♇
02 30	☽ ± ♅
02 33	☽ ⊥ ♂
09 42	☽ ☍ ♄
11 43	☽ ⚹ ♆
14 38	☽ □ ☿
15 24	☽ ⊻ ♂
15 34	☽ # ♄
20 27	☽ △ ♃
20 42	☽ ⚼ ♅
21 41	☿ ⚹ ♂
23 56	☽ △ ♀
05 Friday	
15 59	☉ Q ♄
18 07	☽ ∠ ♆
21 30	☽ # ♆
22 49	☽ □ ☉
06 Saturday	
03 12	☽ △ ♅
03 51	☉ ∥ ♅
04 35	☿ ∠ ♅
06 01	☿ ⊻ ♃
10 14	☽ Q ☿
12 24	☽ ⊥ ♆
13 18	☽ ⊻ ♇
22 43	☽ ⊼ ♄
23 15	☿ ⊥ ♆
07 Sunday	
00 32	☽ ⊻ ♆
01 24	☽ ⊥ ♇
07 58	☽ ☌ ♂
10 16	☽ □ ♃
10 51	☽ ± ♄
14 20	☽ Q ☉
15 18	☽ ⚹ ☿
19 26	☽ □ ♀
19 28	☽ ∠ ♇
19 42	♀ ⚼ ♇
21 40	☽ ∥ ♂
08 Monday	
04 56	☽ ⚼ ♄
07 17	☽ # ♂
08 09	☉ ⊻ ♅
15 39	☽ □ ♅
16 15	☽ ⚹ ☉
22 42	☽ Q ♃
09 Tuesday	
01 09	☽ ⚹ ♇
03 03	☽ ∠ ☿
10 36	☽ △ ♄
10 54	☽ Q ♀
12 10	☽ ☌ ♆
12 53	♂ ± ♄
14 31	♂ □ ♃
20 23	☿ ⊻ ♀
22 27	☽ ⚹ ♃
22 47	☽ ⊻ ♂
23 54	☽ ∠ ☉
23 59	♀ Q ♆
10 Wednesday	
00 02	☽ ⊥ ☿
00 24	☽ Q ♇
02 48	☽ Q ♅
06 44	☽ ∥ ♆
10 56	☽ ⊥ ♂
12 25	☽ ⚹ ♀
13 46	☽ ⊻ ☿
16 55	☿ ⊥ ♅
18 26	☽ ⊥ ☉
11 Thursday	
01 38	☽ ⚹ ♅
03 24	☽ ∠ ♃
04 53	☽ ∠ ♂
06 35	☽ ⊻ ☉
08 57	☿ ∠ ♆
09 19	☽ ∥ ♄
10 21	☽ □ ♇
10 44	☿ # ♇
19 23	☽ ∠ ♀
19 40	☽ □ ♄
20 23	☽ ⊥ ♃
20 56	☽ ⊻ ♆
21 15	☿ ∥ ♅
22 35	♀ ∠ ♄
12 Friday	
05 27	☽ ∠ ♅
07 32	☽ ⊻ ♃
07 55	☽ ⊥ ♆
10 05	☽ ⚹ ♂
11 28	☿ Q ♄
13 28	☽ ⊥ ♀
21 47	☽ ⊥ ♅
13 Saturday	
00 11	☽ ∠ ♆
01 21	☽ ⊻ ♀
04 27	☽ Q ♄
07 35	☽ ☌ ☿
08 33	☽ ⊻ ♅
08 51	☽ Q ♂
08 59	☉ △ ♇
09 20	☉ ∥ ☿
12 22	☿ ∥ ♃
13 46	☿ ⊻ ♅
16 35	☽ △ ♇
17 10	☽ ☌ ☉
18 38	☿ Q ♂
14 Sunday	
01 47	☽ ⚹ ♄
02 48	☽ ⚹ ♆
05 42	☉ ∥ ♃
13 43	☽ ☌ ♃
18 10	☽ □ ♂
18 49	☽ ⚼ ♇
23 43	☽ Q ♆
15 Monday	
00 56	☿ ∥ ♀
03 54	♀ ± ♇
04 01	☽ ∠ ♄
06 40	♂ ∠ ♇
10 20	☽ ± ♇
10 57	☽ ☌ ♀
13 05	☽ ☌ ♅
15 12	☿ △ ♇
19 33	☽ ⊥ ♄
20 37	☽ ⊼ ♇
21 37	☽ ⊻ ☿
16 Tuesday	
01 01	☽ ⊻ ☉
05 51	☽ ⊻ ♄
06 39	☽ □ ♆
09 45	☽ ⊥ ☿
11 59	☽ ⊥ ☉
12 38	♀ ☌ ♅
18 02	☽ ⊻ ♃
17 Wednesday	
00 14	☽ △ ♂
03 50	☽ ∠ ☿
04 21	☽ ∠ ☉
04 22	☽ ⊥ ♃
09 18	☉ ☌ ☿
16 18	☽ ⊻ ♅
18 46	☽ ⊻ ♀
19 40	☉ ∥ ♀
19 49	☽ ∠ ♃
23 28	☽ ☍ ♇
18 Thursday	
02 28	☽ ⊥ ♅
02 55	☽ ⚼ ♂
04 00	☿ ⚹ ♄
05 50	☽ ⊥ ♀
07 30	☽ ⚹ ☉
07 53	☿ ⚹ ♆
08 56	☽ ☌ ♄
09 30	☽ △ ♆
09 48	☽ ⚹ ☿
14 44	☽ ∥ ♄
17 42	☽ ∠ ♅
18 41	♀ ∥ ♃
18 53	☽ ± ♂
21 31	☽ ⚹ ♃
22 26	☽ ∠ ♀
19 Friday	
05 11	☽ Q ☉
05 31	☽ ⊼ ♂
06 38	☉ ⚹ ♄
09 42	☽ Q ☿
10 50	☽ ⚼ ♆
12 51	☽ ∥ ♆
13 57	☉ ⚹ ♆
18 05	☽ Q ♃
19 06	☽ ⚹ ♅
20 31	♀ ⊥ ♄
20 Saturday	
01 22	♀ ⊼ ♇
02 01	☽ ⊼ ♇
02 02	☽ ± ♆
02 05	☽ ⚹ ♀
11 50	☽ ⊻ ♄
12 09	☽ ± ♇
12 11	☽ ⊼ ♆
13 45	☽ □ ☉
15 27	☽ Q ♅
21 36	☽ □ ☿
21 55	☽ ∥ ♂
22 03	☽ ⊥ ♄
21 Sunday	
00 17	☽ Q ♀
01 00	☽ □ ♃
03 23	☽ ⚼ ♇
10 53	☽ ☍ ♂
13 25	☽ ∠ ♄
22 08	☽ □ ♅
22 14	☿ ☌ ♃
23 43	☽ # ♂
22 Monday	
04 53	☽ △ ♇
09 48	☽ □ ♀
10 43	♀ ∥ ♅
11 31	☿ ⚼ ♇
13 19	♀ # ♇
15 11	☽ ⚹ ♄
15 18	☽ ☍ ♆
20 33	☽ △ ☉
23 Tuesday	
05 05	☽ △ ♃
08 09	☽ # ♆
10 04	☽ △ ☿
11 58	☽ Q ♄
16 33	♄ △ ♆
17 02	☽ ⊼ ♂
21 04	♂ ⚼ ♄
21 19	♅ # ♇
22 54	☿ Q ♆
24 Wednesday	
00 24	☽ ⚼ ☉
02 00	☽ △ ♅
04 01	☽ ± ♂
05 51	☽ # ♄
07 34	☽ ⚼ ♃
08 39	☽ □ ♇
16 54	☽ ⚼ ☿
17 20	☽ ± ☉
18 48	☽ △ ♀
19 27	☽ ⊼ ♆
19 36	☽ □ ♄
20 39	☽ ⚼ ♂
23 39	☽ ± ♃
25 Thursday	
02 17	♀ □ ♆
04 22	♀ ⊻ ♄
04 29	☽ ⚼ ♅
04 44	☽ ⊼ ☉
06 05	☽ ± ♆
10 28	☽ ⊼ ♃
11 22	☿ ∠ ♄
11 41	☽ ± ☿
16 26	☽ Q ♇
20 36	☽ ± ♅
22 10	☽ ⚼ ♆
26 Friday	
00 07	☽ ⚼ ♀
00 22	☽ ⊼ ☿
00 49	☽ △ ♂
04 37	☿ □ ♂
07 27	☽ ⊼ ♅
14 03	☽ ⚹ ♇
18 03	☽ ± ♀
18 54	☿ ± ♇
21 17	♃ ⚼ ♇
27 Saturday	
01 26	☽ △ ♆
01 52	☽ △ ♄
06 10	☽ ⊼ ♀
15 20	☽ ☍ ☉
17 36	☽ ∠ ♇
17 41	☽ # ♀
17 59	☽ ☍ ♃
28 Sunday	
00 34	☿ ☌ ♅
05 56	☽ ⚼ ♄
10 27	☽ ⊥ ♇
11 12	☽ □ ♂
15 14	☽ ☍ ♅
17 51	☽ ☍ ☿
21 08	☽ # ♀
21 39	☉ ⚼ ♇
21 49	☽ ⊻ ♇
23 07	☽ ± ♄
29 Monday	
09 55	☽ □ ♆
10 43	☽ ⊼ ♄
13 43	☉ ☌ ♃
19 38	☿ ⊼ ♇
20 57	☽ ☍ ♀
23 45	☽ Q ♂
30 Tuesday	
04 17	☽ ⊼ ♃
05 13	☽ ⊼ ☉
05 52	☿ ⊥ ♄
14 16	☉ ∥ ♃
16 17	☽ ± ♃
18 00	☽ ± ☉
18 53	☿ ∥ ♃
19 45	☉ ∥ ☿
21 06	☽ Q ♆

All ephemeris data is given at 12.00 UT and the Moon's longitude is additionally given for 24.00 UT

JULY 2037

LONGITUDES

Date	Sidereal time h m s	Sun ☉ ° ' "	Moon ☽ ° ' "	Moon ☽ 24.00 ° ' "	Mercury ☿ ° '	Venus ♀ ° '	Mars ♂ ° '	Jupiter ♃ ° '	Saturn ♄ ° '	Uranus ♅ ° '	Neptune ♆ ° '	Pluto ♇ ° '
01	06 39 33	09 ♋ 59 53	24 ♒ 21 23	00 ♓ 21 20	25 ♋ 41	06 ♌ 30	19 ♈ 00	08 ♋ 36	29 ♌ 27	19 ♋ 14	28 ♈ 52	22 ♒ 29
02	06 43 29	10 57 05	06 ♓ 19 26	12 ♓ 16 09	27 31	07 43	19 42	08 50	29 32	19 17	28 53	22 R 28
03	06 47 26	11 54 17	18 ♓ 12 02	24 ♓ 07 36	29 ♋ 20	08 55	20 22	09 03	29 38	19 21	28 54	22 27
04	06 51 23	12 51 29	00 ♈ 03 26	06 ♈ 00 07	01 ♌ 06	10 08	21 03	09 17	29 44	19 24	28 55	22 26
05	06 55 19	13 48 41	11 ♈ 58 13	17 ♈ 58 21	02 50	11 20	21 44	09 30	29 50	19 28	28 56	22 25
06	06 59 16	14 45 54	24 ♈ 01 05	00 ♉ 06 57	04 33	12 32	22 25	09 44	29 ♌ 56	19 32	28 56	22 24
07	07 03 12	15 43 07	06 ♉ 16 31	12 ♉ 30 15	06 13	13 45	23 05	09 58	00 ♍ 02	19 35	28 57	22 23
08	07 07 09	16 40 20	18 ♉ 48 36	25 ♉ 11 56	07 51	14 57	23 46	10 11	00 09	19 39	28 58	22 22
09	07 11 05	17 37 34	01 ♊ 40 33	08 ♊ 14 41	09 26	16 09	24 26	10 25	00 15	19 43	28 59	22 21
10	07 15 02	18 34 48	14 ♊ 54 26	21 ♊ 39 48	11 00	17 22	25 06	10 38	00 21	19 46	29 00	22 20
11	07 18 58	19 32 02	28 ♊ 30 41	05 ♋ 26 53	12 32	18 34	25 46	10 52	00 27	19 50	29 00	22 19
12	07 22 55	20 29 17	12 ♋ 28 02	19 ♋ 33 42	14 01	19 46	26 26	11 05	00 34	19 54	29 01	22 18
13	07 26 52	21 26 32	26 ♋ 43 21	03 ♌ 56 19	15 29	20 58	27 05	11 19	00 40	19 57	29 02	22 17
14	07 30 48	22 23 47	11 ♌ 11 56	18 ♌ 29 27	16 54	22 11	27 45	11 32	00 47	20 01	29 02	22 16
15	07 34 45	23 21 02	25 ♌ 48 07	03 ♍ 07 08	18 17	23 23	28 24	11 46	00 53	20 04	29 03	22 15
16	07 38 41	24 18 18	10 ♍ 25 49	17 ♍ 43 26	19 38	24 35	29 04	11 59	01 00	20 08	29 03	22 14
17	07 42 38	25 15 33	24 ♍ 59 23	02 ♎ 13 06	20 56	25 47	29 ♈ 43	12 13	01 07	20 12	29 04	22 12
18	07 46 34	26 12 49	09 ♎ 24 07	16 ♎ 32 03	22 13	26 59	00 ♉ 22	12 26	01 13	20 15	29 04	22 11
19	07 50 31	27 10 05	23 ♎ 36 35	00 ♏ 37 30	23 26	28 11	01 00	12 40	01 20	20 19	29 05	22 10
20	07 54 27	28 07 21	07 ♏ 34 38	14 ♏ 27 54	24 38	29 ♌ 23	01 39	12 53	01 27	20 23	29 05	22 09
21	07 58 24	29 ♋ 04 37	21 ♏ 17 15	28 ♏ 02 42	25 47	00 ♍ 35	02 17	13 06	01 33	20 26	29 05	22 08
22	08 02 21	00 ♌ 01 53	04 ♐ 44 16	11 ♐ 22 01	26 53	01 47	02 56	13 20	01 40	20 30	29 06	22 06
23	08 06 17	00 59 10	17 ♐ 56 03	24 ♐ 26 27	27 56	02 58	03 34	13 33	01 47	20 34	29 06	22 05
24	08 10 14	01 56 27	00 ♑ 53 19	07 ♑ 16 46	28 57	04 10	04 12	13 46	01 54	20 37	29 06	22 04
25	08 14 10	02 53 45	13 ♑ 36 56	19 ♑ 53 56	29 ♌ 55	05 22	04 49	14 00	02 01	20 41	29 07	22 03
26	08 18 07	03 51 03	26 ♑ 07 55	02 ♒ 19 02	00 ♍ 50	06 34	05 27	14 13	02 08	20 44	29 07	22 02
27	08 22 03	04 48 22	08 ♒ 27 27	14 ♒ 33 21	01 42	07 45	06 04	14 26	02 15	20 48	29 07	22 00
28	08 26 00	05 45 42	20 ♒ 36 56	26 ♒ 38 25	02 30	08 57	06 42	14 39	02 22	20 51	29 07	21 59
29	08 29 56	06 43 02	02 ♓ 38 05	08 ♓ 36 12	03 15	10 08	07 19	14 52	02 29	20 55	29 07	21 58
30	08 33 53	07 40 23	14 ♓ 33 04	20 ♓ 29 04	03 56	11 20	07 55	15 05	02 36	20 59	29 07	21 56
31	08 37 50	08 ♌ 37 45	26 ♓ 24 33	02 ♈ 19 56	04 ♍ 34	12 ♍ 31	08 ♉ 32	15 ♋ 18	02 ♍ 43	21 ♋ 02	29 ♈ 08	21 ♒ 55

Date	Moon True ☊ ° '	Moon Mean ☊ ° '	Moon ☽ Latitude ° '
01	28 ♋ 55	29 ♋ 47	02 S 16
02	28 D 57	29 44	03 12
03	28 58	29 41	04 00
04	28 59	29 38	04 37
05	29 00	29 35	05 02
06	29 R 00	29 32	05 15
07	28 59	29 28	05 14
08	28 57	29 25	04 58
09	28 56	29 22	04 26
10	28 55	29 19	03 40
11	28 54	29 16	02 40
12	28 53	29 12	01 30
13	28 53	29 09	00 S 12
14	28 D 53	29 06	01 N 08
15	28 53	29 03	02 24
16	28 53	29 00	03 31
17	28 54	28 57	04 24
18	28 54	28 53	04 59
19	28 54	28 50	05 16
20	28 R 54	28 47	05 14
21	28 D 54	28 44	04 53
22	28 54	28 41	04 17
23	28 54	28 38	03 28
24	28 55	28 34	02 29
25	28 55	28 31	01 24
26	28 55	28 28	00 N 15
27	28 R 55	28 25	00 S 53
28	28 55	28 22	01 57
29	28 54	28 18	02 56
30	28 53	28 15	03 47
31	28 ♋ 51	28 ♋ 12	04 S 27

DECLINATIONS

Date	Sun ☉ ° '	Moon ☽ ° '	Mercury ☿ ° '	Venus ♀ ° '	Mars ♂ ° '	Jupiter ♃ ° '	Saturn ♄ ° '	Uranus ♅ ° '	Neptune ♆ ° '	Pluto ♇ ° '
01	23 N 03	15 S 32	22 N 52	20 N 15	05 N 21	23 N 06	13 N 04	22 N 29	09 N 29	22 S 36
02	22 59	12 10	22 29	19 57	05 37	23 06	13 02	22 28	09 30	22 37
03	22 54	08 20	22 04	19 38	05 52	23 05	13 00	22 28	09 30	22 37
04	22 49	04 S 13	21 39	19 19	06 07	23 04	12 58	22 27	09 30	22 38
05	22 43	00 N 05	21 12	18 59	06 22	23 03	12 56	22 27	09 30	22 38
06	22 37	04 26	20 44	18 39	06 36	23 02	12 54	22 26	09 31	22 39
07	22 30	08 40	20 14	18 18	06 51	23 01	12 51	22 26	09 31	22 39
08	22 24	12 38	19 44	17 57	07 06	23 00	12 49	22 25	09 31	22 40
09	22 16	16 08	19 14	17 36	07 20	22 59	12 47	22 25	09 31	22 41
10	22 09	18 56	18 42	17 13	07 34	22 58	12 45	22 24	09 32	22 41
11	22 01	20 45	18 10	16 51	07 49	22 57	12 43	22 23	09 32	22 42
12	21 52	21 22	17 38	16 28	08 03	22 56	12 40	22 23	09 32	22 42
13	21 43	20 37	17 05	16 04	08 17	22 55	12 38	22 22	09 32	22 43
14	21 34	18 30	16 31	15 40	08 31	22 54	12 36	22 22	09 32	22 43
15	21 25	15 10	15 58	15 16	08 45	22 53	12 33	22 21	09 32	22 44
16	21 15	10 54	15 24	14 52	08 58	22 52	12 31	22 21	09 33	22 44
17	21 05	06 01	14 50	14 27	09 12	22 51	12 28	22 20	09 33	22 45
18	20 54	00 N 52	14 17	14 01	09 25	22 50	12 26	22 20	09 33	22 45
19	20 43	04 S 16	13 43	13 35	09 39	22 49	12 24	22 19	09 33	22 46
20	20 32	09 05	13 10	13 09	09 52	22 48	12 21	22 19	09 33	22 47
21	20 20	13 21	12 37	12 43	10 05	22 46	12 19	22 18	09 33	22 47
22	20 08	16 51	12 04	12 16	10 18	22 45	12 16	22 18	09 33	22 48
23	19 56	19 26	11 32	11 49	10 31	22 44	12 14	22 17	09 33	22 48
24	19 43	20 56	11 00	11 22	10 43	22 43	12 11	22 17	09 33	22 49
25	19 30	21 21	10 29	10 54	10 56	22 41	12 09	22 16	09 33	22 49
26	19 17	20 40	09 58	10 26	11 08	22 40	12 06	22 16	09 33	22 50
27	19 03	19 00	09 28	09 58	11 21	22 39	12 04	22 15	09 33	22 50
28	18 50	16 28	09 00	09 30	11 33	22 37	12 01	22 14	09 33	22 51
29	18 35	13 16	08 32	09 01	11 45	22 36	11 59	22 14	09 33	22 52
30	18 21	09 34	08 06	08 32	11 57	22 35	11 56	22 13	09 33	22 52
31	18 N 06	05 S 31	07 N 40	08 N 03	12 N 09	22 N 33	11 N 54	22 N 13	09 N 33	22 S 53

ZODIAC SIGN ENTRIES

Date	h	m	Planets
01	23	17	☽ ♓
03	21	01	☿ ♌
04	11	53	☽ ♈
06	23	46	☽ ♉
07	02	31	♄ ♍
09	08	55	☽ ♊
11	14	35	☽ ♋
13	17	28	☽ ♌
15	18	53	☽ ♍
17	20	19	☽ ♎
17	22	42	♂ ♉
19	22	56	☽ ♏
21	00	23	♀ ♍
22	03	30	☽ ♐
22	11	12	☉ ♌
24	10	20	☽ ♑
25	14	09	☿ ♍
26	19	30	☽ ♒
29	06	43	☽ ♓
31	19	17	☽ ♈

LATITUDES

Date	Mercury ☿ ° '	Venus ♀ ° '	Mars ♂ ° '	Jupiter ♃ ° '	Saturn ♄ ° '	Uranus ♅ ° '	Neptune ♆ ° '	Pluto ♇ ° '
01	01 N 54	01 N 40	02 S 15	00 S 03	01 N 30	00 N 26	01 S 41	09 S 06
04	01 47	01 41	02 16	00 03	01 30	00 26	01 41	09 07
07	01 35	01 41	02 17	00 02	01 30	00 26	01 42	09 08
10	01 17	01 41	02 18	00 02	01 30	00 26	01 42	09 08
13	00 56	01 40	02 18	00 02	01 30	00 26	01 42	09 09
16	00 30	01 38	02 19	00 01	01 30	00 26	01 42	09 10
19	00 N 01	01 35	02 19	00 01	01 30	00 26	01 42	09 10
22	00 S 31	01 32	02 19	00 01	01 30	00 26	01 42	09 11
25	01 05	01 28	02 20	00 S 01	01 30	00 26	01 43	09 11
28	01 42	01 23	02 19	00 00	01 30	00 26	01 43	09 11
31	02 S 19	01 N 18	02 S 19	00 00	01 N 30	00 N 26	01 S 43	09 S 12

DATA

Julian Date	2465241
Delta T	+76 seconds
Ayanamsa	24° 22' 37"
Synetic vernal point	04° ♓ 44' 22"
True obliquity of ecliptic	23° 25' 59"

MOON'S PHASES, APSIDES AND POSITIONS ☽

Date	h	m	Phase	Longitude ° '	Eclipse Indicator
05	16	00	☾	13 ♈ 58	
13	02	32	●	21 ♋ 04	Total
19	18	31	☽	27 ♎ 26	
27	04	15	○	04 ♒ 30	partial

Day	h	m		
03	19	46	Apogee	
15	16	47	Perigee	
31	12	22	Apogee	
05	11	31	ON	
12	11	00	Max dec	21° N 22'
18	15	59	OS	
25	08	46	Max dec	21° S 21'

LONGITUDES

Date	Chiron ⚷ ° '	Ceres ⚳ ° '	Pallas ⚴ ° '	Juno ⚵ ° '	Vesta ⚶ ° '	Black Moon Lilith ⚸ ° '
01	21 ♊ 22	19 ♎ 28	09 ♍ 49	13 ♍ 03	28 ♈ 02	19 ♓ 01
11	22 ♊ 11	21 ♎ 31	13 ♍ 44	16 ♍ 10	01 ♉ 17	20 ♓ 08
21	22 ♊ 57	23 ♎ 57	17 ♍ 47	19 ♍ 25	04 ♉ 12	21 ♓ 15
31	23 ♊ 39	26 ♎ 42	21 ♍ 54	22 ♍ 47	06 ♉ 43	22 ♓ 22

ASPECTARIAN

h m Aspects

01 Wednesday
00 44 ☽ ✱ ♂
01 45 ☽ ⊼ ♅
08 16 ☽ ☌ ♇
10 28 ☽ ⚼ ♃
13 23 ☽ ⚼ ☉
13 45 ☽ ± ♅
15 08 ☽ ⊼ ☿
20 22 ♂ □ ♅
21 01 ☽ ✱ ♆
22 15 ☽ ☍ ♄

02 Thursday
03 55 ☿ ⋕ ♇
05 21 ☽ ± ☿
06 04 ☽ ⋕ ♄
07 52 ☽ ⚼ ♅
08 31 ☽ ∠ ♂
10 10 ☉ Q ♆
12 38 ☿ ∥ ♅
15 06 ☽ ⊼ ♀
17 09 ☽ △ ♃
22 09 ☽ △ ☉

03 Friday
02 46 ☽ ⚼ ☿
03 17 ☽ ∠ ♆
03 47 ☽ ⊥ ♂
04 36 ☽ ± ♀
04 58 ☽ ⋕ ♆
06 08 ☿ □ ♆
14 20 ☽ △ ♅
15 18 ♀ ⊻ ♃
16 22 ☿ ⊻ ♄
16 40 ☽ ⊻ ♂
20 36 ☽ ⊻ ♇
21 31 ☽ ⊥ ♆

04 Saturday
00 53 ☽ ⚼ ♀
01 45 ☽ ⋕ ♂
08 44 ☽ ⊥ ♇
09 41 ☽ ⊻ ♆
11 21 ☽ ⊼ ♄
14 29 ☽ △ ☿
23 34 ☽ ± ♄

05 Sunday
02 52 ☽ ∠ ♇
06 57 ☽ □ ♃
10 35 ☽ △ ♀
16 00 ☽ □ ☉
17 47 ☽ ⚼ ♄

06 Monday
03 03 ☽ □ ♅
04 55 ☉ ⋕ ♇
08 38 ☽ ☌ ♂
08 49 ☽ ✱ ♇
11 45 ♂ ✱ ♇
16 53 ☉ ∠ ♄
19 28 ☽ Q ♃
21 42 ☽ ☌ ♆
23 45 ☽ △ ♄

07 Tuesday
00 57 ☽ ∥ ♂
06 37 ☽ Q ☉
08 20 ☽ Q ♇
11 51 ☽ □ ☿
14 33 ☽ Q ♅
16 58 ☽ ∥ ♆
19 15 ☽ ✱ ♃

08 Wednesday
03 54 ☽ □ ♀
04 33 ☉ ± ♇
06 36 ☉ ∥ ♅
07 37 ☽ ✱ ☉
13 09 ☽ ∥ ♄
13 36 ☽ ✱ ♅
18 42 ☽ □ ♇
21 49 ☽ ⊻ ♂

09 Thursday
00 11 ☽ ∠ ♃
03 05 ☽ Q ☿
07 01 ☽ ⊻ ♆
09 21 ☽ □ ♄
09 35 ☽ ⊥ ♂
13 53 ☽ ∠ ☉
17 01 ☽ Q ♀
17 07 ☽ ⊥ ♃
17 35 ☽ ∠ ♅
18 04 ☽ ⊥ ♆
18 16 ♀ ⊥ ♃
22 16 ☽ ∥ ♀

10 Friday
02 54 ☽ ∠ ♂
04 04 ☽ ✱ ☿
04 12 ☽ ⊻ ♃
05 23 ☿ ⊻ ♃
07 30 ☽ ⊥ ☉
09 57 ☽ ⊥ ♅
10 07 ☽ ∥ ☿
10 22 ☽ ∠ ♆
16 49 ☽ ✱ ♀
18 11 ☽ Q ♄
19 02 ☽ ⊻ ☉
20 42 ☽ ⊻ ♅

11 Saturday
01 10 ☽ △ ♇
06 58 ☽ ✱ ♂
10 05 ☽ ∠ ☿
12 51 ☽ ✱ ♆
15 25 ☽ ✱ ♄
19 59 ☉ ☌ ♅
21 35 ☽ ∠ ♀

12 Sunday
03 11 ☽ ⚼ ♇
03 31 ☽ ⊥ ☿
03 32 ♃ Q ♆
04 46 ☽ Q ♂
09 32 ☽ Q ♆
09 37 ☽ ☌ ♃
14 25 ☽ ⊥ ♀
14 32 ♀ ⊻ ♅
14 57 ☽ ⊻ ☿
17 18 ☽ ∠ ♄
18 29 ☽ ± ♇

13 Monday
00 37 ☽ ☌ ♅
01 29 ☽ ⊻ ♀
02 32 ☽ ☌ ☉
04 35 ☽ ⊼ ♇
08 33 ☽ ⊥ ♄
12 39 ☽ □ ♂
15 51 ☽ □ ♆
18 38 ☽ ⊻ ♄

14 Tuesday
08 43 ☉ ⊼ ♇
12 34 ☽ ⊻ ♃
13 41 ♀ ☍ ♇
22 23 ☽ ☌ ☿
22 36 ☽ ⊥ ♃

15 Wednesday
01 09 ☿ ⊥ ♃
02 34 ☽ ⊻ ♅
06 00 ☽ ∥ ☿
06 10 ☽ ☍ ♇
07 40 ☽ ☌ ♀
07 42 ☽ ⊻ ☉
09 10 ☉ ⊻ ♀
11 19 ☽ ∥ ♀
12 27 ☽ ⊥ ♅
13 36 ☽ ∠ ♃
16 28 ☽ △ ♂
17 19 ☽ △ ♆
18 14 ☽ ⊥ ☉
20 24 ☽ ☌ ♄

16 Thursday
03 16 ☽ ∠ ♅
03 22 ☽ ∥ ♄
10 01 ☽ ∠ ☉
11 50 ♂ ☌ ♆
14 36 ☽ ✱ ♃
17 58 ☽ ⚼ ♆
18 15 ☽ ⚼ ♂
18 55 ☽ ∥ ♆
21 19 ☽ ∥ ♂
21 36 ☿ ⊻ ♅

17 Friday
04 03 ☽ ✱ ♅
04 39 ☽ ⊻ ☿
07 24 ☽ ⊼ ♇
07 43 ☉ ⊥ ♄
08 49 ☽ ± ♆
09 47 ☽ ± ♂
10 42 ☽ Q ♃
12 29 ☽ ✱ ☉
13 26 ☽ ⊻ ♀
15 33 ☽ ⊥ ☿
17 19 ☽ ± ♇
18 45 ☽ ⊼ ♆
20 12 ☽ ⊼ ♂
20 42 ♀ ⊥ ♅
22 14 ☽ ⊻ ♄

18 Saturday
00 01 ☽ Q ♅
00 18 ☽ ⊥ ♀
07 59 ☽ ∠ ☿
08 18 ☽ ⚼ ♇
08 19 ☽ ⊥ ♄
09 52 ☽ Q ☉
11 34 ☿ ☍ ♇
16 30 ♂ Q ♃
16 44 ☽ ∠ ♀
17 11 ☽ □ ♃
23 11 ♀ ∠ ♃
23 34 ☽ ∠ ♄

19 Sunday
01 31 ♂ ∥ ♆
06 23 ☽ □ ♅
09 33 ☽ △ ♇
11 41 ☽ ✱ ☿
18 31 ☽ □ ☉
20 33 ☽ ✱ ♀
21 21 ☽ ☍ ♆

20 Monday
01 16 ☽ ☍ ♂
01 19 ☽ ✱ ♄
02 46 ♂ △ ♄
06 02 ♀ △ ♆
10 12 ☽ Q ☿
13 39 ☿ ∥ ♀
14 27 ☽ ⋕ ♆
16 20 ☽ ⋕ ♂
19 15 ☽ Q ♀
21 23 ☽ △ ♃
22 18 ☽ Q ♄

21 Tuesday
00 50 ♃ ⋕ ♇
05 51 ☽ ⋕ ♄
08 04 ☽ ⋕ ☿
08 32 ☽ ⋕ ♀
10 30 ☽ △ ♅
12 22 ☉ □ ♆
13 29 ☽ □ ♇
20 41 ☽ □ ☿

22 Wednesday
00 19 ☽ ⚼ ♃
01 53 ☽ ⊼ ♆
02 05 ☿ ∥ ♄
02 54 ☽ △ ☉
03 06 ☿ ⊥ ♅
06 09 ☽ □ ♀
06 26 ☽ □ ♄
08 35 ☽ ⊼ ♂
09 39 ♀ ☌ ♄
11 47 ♀ ∥ ♄
12 39 ☽ ± ♆
13 23 ☽ ⚼ ♅
16 45 ☽ ± ♃
19 57 ☽ ± ♂
21 42 ☽ Q ♇

23 Thursday
03 50 ☽ ⊼ ♃
04 59 ☽ ⚼ ♆
05 47 ☽ ± ♅
08 09 ☽ ⚼ ☉
13 13 ☽ ⚼ ♂
16 51 ☽ ⊼ ♅
17 41 ☽ ⋕ ☉
19 38 ☽ ✱ ♇

24 Friday
02 03 ☽ ± ☉
06 25 ☿ ∠ ♃
07 20 ♂ Q ♇
08 05 ☽ △ ☿
08 40 ☽ △ ♆
10 52 ☉ ⊻ ♄
12 57 ♀ △ ♂
13 55 ☽ △ ♄
14 08 ☽ ⊼ ☉
15 48 ☿ △ ♆
18 31 ☽ △ ♂
18 47 ☽ △ ♀
20 52 ☿ ∥ ♂
23 35 ☽ ∠ ♇

25 Saturday
10 51 ♀ ∥ ♂
12 44 ☽ ☍ ♃
14 40 ☽ ⚼ ☿
16 37 ☽ ⊥ ♇
18 33 ☽ ⚼ ♄
18 35 ♀ ∠ ♅

26 Sunday
01 34 ☽ ☍ ♅
02 16 ☽ ⚼ ♀
04 06 ☽ ⊻ ♇
09 18 ☽ ± ☿
12 00 ☽ ± ♄
17 47 ☽ □ ♆
21 30 ☽ ± ♀
21 48 ☽ ⊼ ☿
23 45 ☽ ⊼ ♄

27 Monday
04 15 ☽ ☍ ☉
07 05 ☽ □ ♂
08 02 ☿ ∥ ♆
10 28 ☽ ⊼ ♀
11 09 ☽ ⋕ ☉
23 58 ☽ ⊼ ♃

28 Tuesday
05 04 ☽ Q ♆
07 16 ☿ ☌ ♄
08 58 ♀ ∥ ♆
12 04 ☽ ± ♃
12 29 ☽ ⊼ ♅
14 43 ☽ ☌ ♇
20 33 ☽ Q ♂

29 Wednesday
00 30 ☽ ± ♅
04 58 ☽ ✱ ♆
06 21 ☽ ⚼ ♃
11 42 ☽ ☍ ♄
13 19 ☽ ☍ ☿
18 37 ☽ ⚼ ♅
20 48 ☽ ⋕ ♄
20 55 ☽ ⊼ ☉
21 40 ☽ ⋕ ♂
21 54 ☽ ✱ ♂

30 Thursday
04 47 ☽ ☍ ♀
10 04 ☽ ± ☉
10 47 ♂ ∥ ♄
11 08 ☽ ∠ ♆
12 04 ☽ ⋕ ♆
13 06 ☽ △ ♃
19 06 ☽ ⋕ ♀
21 58 ☽ ⋕ ☿

31 Friday
01 03 ☽ △ ♅
02 55 ☽ ⊻ ♇
05 21 ☽ ⊥ ♆
05 22 ☉ □ ♂
05 51 ☽ ∠ ♂
05 52 ☽ ⚼ ☉
15 03 ☽ ⊥ ♇
17 30 ☽ ⊻ ♆

All ephemeris data is given at 12.00 UT and the Moon's longitude is additionally given for 24.00 UT

AUGUST 2037

LONGITUDES

Date	Sidereal time h m s	Sun ☉ ° ' "	Moon ☽ ° ' "	Moon ☽ 24.00 ° ' "	Mercury ☿ ° '	Venus ♀ ° '	Mars ♂ ° '	Jupiter ♃ ° '	Saturn ♄ ° '	Uranus ♅ ° '	Neptune ♆ ° '	Pluto ⯓ ° '
01	08 41 46	09 ♌ 35 08	08 ♈ 15 40	14 ♈ 12 13	05 ♍ 08	13 ♍ 43	09 ♉ 08	15 ♋ 31	02 ♍ 51	21 ♋ 06	29 ♈ 08	21 ♒ 54
02	08 45 43	10 32 32	20 ♈ 10 05	26 ♈ 09 47	05 38	14 54	09 45	15 44	02 58	21 09	29 R 08	21 R 53
03	08 49 39	11 29 58	02 ♉ 11 51	08 ♉ 16 50	06 03	16 05	10 21	15 57	03 05	21 13	29 07	21 51
04	08 53 36	12 27 24	14 ♉ 25 18	20 ♉ 37 47	06 24	17 17	10 56	16 10	03 12	21 16	29 07	21 50
05	08 57 32	13 24 52	26 ♉ 54 48	03 ♊ 16 52	06 40	18 28	11 32	16 23	03 20	21 19	29 07	21 49
06	09 01 29	14 22 21	09 ♊ 44 27	16 ♊ 17 54	06 52	19 39	12 07	16 36	03 27	21 23	29 07	21 47
07	09 05 25	15 19 51	22 ♊ 57 34	29 ♊ 43 39	06 58	20 50	12 42	16 48	03 34	21 26	29 07	21 46
08	09 09 22	16 17 23	06 ♋ 36 15	13 ♋ 35 19	06 R 59	22 01	13 17	17 01	03 42	21 30	29 07	21 45
09	09 13 19	17 14 56	20 ♋ 40 41	27 ♋ 52 00	06 55	23 12	13 52	17 14	03 49	21 33	29 07	21 43
10	09 17 15	18 12 30	05 ♌ 08 44	12 ♌ 30 14	06 46	24 23	14 26	17 26	03 56	21 36	29 06	21 42
11	09 21 12	19 10 05	19 ♌ 55 39	27 ♌ 24 03	06 31	25 34	15 00	17 39	04 04	21 40	29 06	21 41
12	09 25 08	20 07 41	04 ♍ 54 22	12 ♍ 25 29	06 11	26 45	15 34	17 51	04 11	21 43	29 06	21 39
13	09 29 05	21 05 19	19 ♍ 56 17	27 ♍ 25 40	05 45	27 56	16 07	18 04	04 19	21 46	29 05	21 38
14	09 33 01	22 02 57	04 ♎ 52 35	12 ♎ 16 08	05 15	29 ♍ 07	16 41	18 16	04 26	21 50	29 05	21 37
15	09 36 58	23 00 36	19 ♎ 35 31	26 ♎ 50 06	04 39	00 ♎ 17	17 14	18 28	04 34	21 53	29 04	21 35
16	09 40 54	23 58 16	03 ♏ 59 25	11 ♏ 03 08	03 59	01 28	17 46	18 41	04 41	21 56	29 04	21 34
17	09 44 51	24 55 58	18 ♏ 01 07	24 ♏ 53 19	03 15	02 39	18 19	18 53	04 49	21 59	29 03	21 33
18	09 48 48	25 53 40	01 ♐ 39 51	08 ♐ 20 52	02 27	03 49	18 51	19 05	04 56	22 03	29 03	21 31
19	09 52 44	26 51 23	14 ♐ 56 38	21 ♐ 27 29	01 37	04 59	19 22	19 17	05 04	22 06	29 02	21 30
20	09 56 41	27 49 08	27 ♐ 53 45	04 ♑ 15 47	00 ♍ 46	06 10	19 54	19 29	05 11	22 09	29 02	21 29
21	10 00 37	28 46 54	10 ♑ 33 59	16 ♑ 48 43	29 ♌ 53	07 20	20 25	19 41	05 19	22 12	29 01	21 27
22	10 04 34	29 ♌ 44 40	23 ♑ 00 19	29 ♑ 09 08	29 01	08 30	20 56	19 53	05 27	22 15	29 00	21 26
23	10 08 30	00 ♍ 42 28	05 ♒ 15 29	11 ♒ 19 39	28 10	09 41	21 26	20 05	05 34	22 18	29 00	21 25
24	10 12 27	01 40 18	17 ♒ 21 56	23 ♒ 22 34	27 21	10 51	21 57	20 16	05 42	22 21	28 59	21 23
25	10 16 23	02 38 08	29 ♒ 21 47	05 ♓ 19 50	26 36	12 01	22 26	20 28	05 49	22 24	28 58	21 22
26	10 20 20	03 36 01	11 ♓ 16 55	17 ♓ 13 16	25 56	13 11	22 56	20 40	05 57	22 27	28 58	21 21
27	10 24 17	04 33 54	23 ♓ 09 05	29 ♓ 04 38	25 20	14 20	23 25	20 51	06 05	22 30	28 57	21 20
28	10 28 13	05 31 49	05 ♈ 00 08	10 ♈ 55 54	24 51	15 30	23 54	21 03	06 12	22 33	28 56	21 18
29	10 32 10	06 29 46	16 ♈ 52 12	22 ♈ 49 22	24 29	16 40	24 22	21 14	06 20	22 36	28 55	21 17
30	10 36 06	07 27 45	28 ♈ 47 47	04 ♉ 47 49	24 14	17 49	24 50	21 25	06 27	22 39	28 54	21 16
31	10 40 03	08 ♍ 25 46	10 ♉ 49 54	16 ♉ 54 31	24 ♌ 07	18 ♎ 59	25 ♉ 18	21 ♋ 37	06 ♍ 35	22 ♋ 42	28 ♈ 53	21 ♒ 14

MOON NODES AND LATITUDE / DECLINATIONS

Date	Moon True ☊ ° '	Moon Mean ☊ ° '	Moon ☽ Latitude ° '	Sun ☉ ° '	Moon ☽ ° '	Mercury ☿ ° '	Venus ♀ ° '	Mars ♂ ° '	Jupiter ♃ ° '	Saturn ♄ ° '	Uranus ♅ ° '	Neptune ♆ ° '	Pluto ⯓ ° '
01	28 ♋ 50	28 ♋ 09	04 S 56	17 N 51	01 S 16	07 N 16	07 N 34	12 N 20	22 N 32	11 N 51	22 N 12	09 N 33	22 S 53
02	28 R 49	28 06	05 13	17 35	03 N 03	06 54	07 05	12 32	22 31	11 48	22 12	09 33	22 54
03	28 48	28 03	05 16	17 20	07 17	06 33	06 35	12 43	22 29	11 46	22 11	09 33	22 54
04	28 D 48	27 59	05 05	17 04	11 18	06 14	06 05	12 54	22 28	11 43	22 11	09 33	22 55
05	28 48	27 56	04 39	16 47	14 56	05 57	05 35	13 05	22 26	11 41	22 10	09 33	22 55
06	28 49	27 53	03 59	16 31	17 58	05 42	05 05	13 16	22 25	11 38	22 10	09 33	22 56
07	28 51	27 50	03 06	16 14	20 09	05 29	04 35	13 27	22 23	11 35	22 09	09 33	22 56
08	28 52	27 47	02 00	15 57	21 16	05 19	04 05	13 38	22 22	11 33	22 09	09 33	22 57
09	28 53	27 43	00 S 45	15 40	21 06	05 11	03 34	13 48	22 20	11 30	22 08	09 32	22 57
10	28 R 53	27 40	00 N 35	15 22	19 32	05 06	03 04	13 59	22 19	11 27	22 08	09 32	22 58
11	28 52	27 37	01 54	15 04	16 38	05 04	02 33	14 09	22 17	11 25	22 07	09 32	22 58
12	28 50	27 34	03 06	14 46	12 35	05 04	02 02	14 19	22 16	11 22	22 07	09 32	22 59
13	28 47	27 31	04 05	14 28	07 44	05 07	01 32	14 29	22 14	11 19	22 06	09 32	23 00
14	28 44	27 28	04 48	14 09	02 N 28	05 14	01 01	14 38	22 13	11 16	22 06	09 32	23 00
15	28 41	27 24	05 10	13 51	02 S 52	05 23	00 N 30	14 48	22 11	11 14	22 05	09 31	23 01
16	28 39	27 21	05 13	13 32	07 56	05 36	00 S 01	14 57	22 09	11 11	22 05	09 31	23 01
17	28 38	27 18	04 56	13 12	12 27	05 51	00 32	15 07	22 08	11 08	22 04	09 31	23 02
18	28 D 38	27 15	04 24	12 53	16 11	06 09	01 03	15 16	22 06	11 06	22 04	09 31	23 02
19	28 39	27 12	03 37	12 33	18 59	06 30	01 34	15 25	22 05	11 03	22 03	09 30	23 02
20	28 40	27 09	02 41	12 14	20 44	06 53	02 05	15 34	22 03	11 00	22 03	09 30	23 03
21	28 42	27 05	01 38	11 54	21 23	07 18	02 35	15 42	22 01	10 57	22 02	09 30	23 03
22	28 43	27 02	00 N 31	11 34	20 58	07 44	03 06	15 51	22 00	10 55	22 02	09 30	23 04
23	28 R 43	26 59	00 S 36	11 13	19 32	08 12	03 37	15 59	21 58	10 52	22 01	09 29	23 04
24	28 41	26 56	01 40	10 53	17 13	08 40	04 08	16 07	21 57	10 49	22 01	09 29	23 05
25	28 38	26 53	02 39	10 32	14 11	09 08	04 39	16 16	21 55	10 46	22 00	09 29	23 05
26	28 33	26 49	03 31	10 11	10 35	09 36	05 09	16 24	21 53	10 44	22 00	09 28	23 06
27	28 27	26 46	04 13	09 50	06 36	10 03	05 40	16 31	21 52	10 41	21 59	09 28	23 06
28	28 21	26 43	04 45	09 29	02 S 22	10 29	06 10	16 39	21 50	10 38	21 59	09 28	23 07
29	28 14	26 40	05 04	09 08	01 N 57	10 54	06 40	16 47	21 48	10 35	21 58	09 27	23 07
30	28 08	26 37	05 10	08 46	06 13	11 16	07 10	16 54	21 47	10 33	21 58	09 27	23 07
31	28 ♋ 03	26 ♋ 34	05 S 02	08 N 24	10 N 17	11 N 36	07 S 40	17 N 01	21 N 45	10 N 30	21 N 58	09 N 27	23 S 08

ZODIAC SIGN ENTRIES

Date	h	m	Planets
03	07	38	☽ ♉
05	17	50	☽ ♊
08	00	29	☽ ♋
10	03	32	☽ ♌
12	04	10	☽ ♍
14	04	08	☽ ♎
15	06	06	♀ ♎
16	05	17	☽ ♏
18	09	02	☽ ♐
20	15	57	☽ ♑
21	08	56	☿ ♌
22	18	22	☉ ♍
23	01	40	☽ ♒
25	13	17	☽ ♓
28	01	52	☽ ♈
30	14	25	☽ ♉

LATITUDES

Date	Mercury ☿ ° '	Venus ♀ ° '	Mars ♂ ° '	Jupiter ♃ ° '	Saturn ♄ ° '	Uranus ♅ ° '	Neptune ♆ ° '	Pluto ⯓ ° '
01	02 S 32	01 N 16	02 S 19	00 00	01 N 30	00 N 26	01 S 43	09 S 12
04	03 09	01 09	02 19	00 00	01 30	00 26	01 43	09 12
07	03 43	01 02	02 18	00 N 01	01 30	00 26	01 43	09 13
10	04 13	00 54	02 18	00 01	01 30	00 26	01 44	09 13
13	04 36	00 46	02 17	00 01	01 30	00 26	01 44	09 13
16	04 46	00 37	02 16	00 02	01 30	00 26	01 44	09 13
19	04 42	00 28	02 14	00 02	01 30	00 26	01 44	09 13
22	04 21	00 17	02 13	00 02	01 30	00 26	01 44	09 13
25	03 44	00 N 07	02 11	00 03	01 30	00 26	01 44	09 13
28	02 55	00 S 04	02 09	00 03	01 31	00 26	01 45	09 13
31	02 S 00	00 S 16	02 S 07	00 N 03	01 N 31	00 N 26	01 S 45	09 S 13

DATA

Julian Date	2465272
Delta T	+76 seconds
Ayanamsa	24° 22' 41"
Synetic vernal point	04° ♓ 44' 17"
True obliquity of ecliptic	23° 26' 00"

MOON'S PHASES, APSIDES AND POSITIONS ☽

Date	h	m	Phase	Longitude ° '	Eclipse Indicator
04	07	51	☾	12 ♉ 17	
11	10	42	●	19 ♌ 07	
18	01	00	☽	25 ♏ 27	
25	19	09	○	02 ♓ 55	

Day	h	m	
12	17	35	Perigee
28	00	12	Apogee

01	19	02	0N	
08	21	01	Max dec	21° N 22'
14	23	01	0S	
21	14	19	Max dec	21° S 24'
29	01	11	0N	

LONGITUDES

Date	Chiron ⚷ ° '	Ceres ⚳ ° '	Pallas ⚴ ° '	Juno ⚵ ° '	Vesta ⚶ ° '	Black Moon Lilith ⚸ ° '
01	23 ♊ 43	27 ♎ 00	22 ♍ 19	23 ♍ 07	06 ♉ 57	22 ♓ 28
11	24 ♊ 21	00 ♏ 21	26 ♍ 32	26 ♍ 34	08 ♉ 58	23 ♓ 35
21	24 ♊ 54	03 ♏ 20	00 ♎ 49	00 ♎ 05	10 ♉ 07	24 ♓ 42
31	25 ♊ 21	06 ♏ 49	05 ♎ 08	03 ♎ 39	11 ♉ 18	25 ♓ 49

ASPECTARIAN

01 Saturday
00 55 ☽ ⊼ ♄
01 05 ☽ ⊥ ♂
05 22 ☽ ⊼ ☿
09 15 ☽ ∠ ⯓
09 54 ♂ Q ♅
13 11 ☽ ± ♄
13 52 ☽ ⊻ ♂
13 56 ♆ St R
14 55 ☽ △ ☉
18 04 ☽ ± ☿
20 19 ♀ ♇ ♆

02 Sunday
00 14 ☽ ⊼ ♀
02 55 ☽ □ ♃
07 32 ☽ ♇ ♄
12 58 ☽ ♇ ☿
13 38 ☽ ± ♀
13 59 ☽ □ ♅
15 25 ☽ ✶ ⯓

03 Monday
02 13 ♃ ± ⯓
05 54 ☽ ☌ ♆
08 06 ☽ ∥ ☿
08 22 ☽ ∥ ♀
08 33 ♀ ✶ ♃
09 34 ☽ ♇ ♀
13 46 ☽ △ ♄
15 16 ☽ Q ⯓
15 32 ☽ Q ♃
16 07 ☿ ∥ ♀
19 51 ☽ △ ☿

04 Tuesday
00 11 ☿ ∠ ♅
01 18 ☽ ∥ ♆
01 53 ☽ Q ♅
04 52 ☽ ☌ ♂
07 51 ☽ □ ☉
14 34 ☽ ∥ ♄
15 27 ☽ ✶ ♃
18 08 ☽ △ ♀
22 45 ☽ ∥ ♂

05 Wednesday
01 17 ☽ ✶ ♅
02 17 ☽ □ ⯓
16 11 ☽ ⊻ ♆
20 35 ☽ ∠ ♃
21 11 ☽ Q ☉

06 Thursday
00 12 ☽ □ ♄
00 52 ☽ ∥ ☉
03 26 ☽ ⊥ ♆
05 45 ☽ ∠ ♅
06 36 ☽ □ ☿
13 36 ☽ ⊥ ♃
16 35 ☽ ⊻ ♂
20 02 ☽ ∠ ♆
21 10 ☽ ✶ ☉
22 23 ☽ ⊥ ♅

07 Friday
00 44 ☽ ⊻ ♃
04 00 ☽ ⊥ ♂
07 50 ☽ □ ♀
09 16 ☽ ⊻ ♅
09 30 ☽ Q ♄
09 52 ☽ △ ⯓
15 36 ☽ Q ☿
20 49 ☽ ∠ ♂
22 55 ☽ ✶ ♆

08 Saturday
00 47 ♀ ✶ ♅
02 03 ☽ ∠ ☉
05 45 ☿ St R
06 29 ♀ ⊼ ⯓
06 54 ☽ ✶ ♄
12 15 ☽ ♇ ⯓
12 40 ☽ ✶ ☿
18 26 ☽ Q ♀
18 49 ☽ ⊥ ☉
19 46 ☽ Q ♆
23 58 ☽ ✶ ♂

09 Sunday
03 39 ☽ ± ⯓
05 48 ☽ ⊻ ☉
06 06 ☽ ☌ ♃
08 50 ☽ ∠ ♄
10 02 ♀ ± ♆
11 19 ☉ ⊻ ♃
13 28 ☽ ☌ ♅
13 45 ☽ ⊼ ⯓
14 04 ☽ ∠ ☿
16 36 ☉ ⊥ ♀
16 37 ☽ ✶ ♀
21 02 ☽ Q ♂

10 Monday
00 01 ☽ ⊥ ♄
02 03 ☽ □ ♆
04 53 ☽ ⊥ ☿
10 00 ☽ ⊻ ♄
14 37 ☽ ⊻ ☿
19 32 ☽ ∠ ♀

11 Tuesday
01 23 ☿ ∠ ♅
03 44 ☽ □ ♂
08 16 ☽ ⊻ ♃
10 42 ☽ ☌ ☉
11 23 ☽ ⊥ ♀
14 48 ☽ ⊻ ♅
14 49 ☽ ☍ ⯓
16 37 ♅ ⊼ ⯓
18 04 ☽ ⊥ ♃
21 51 ☽ ⊻ ♀
22 47 ☽ ∥ ☉

12 Wednesday
00 28 ☽ ⊥ ♅
02 43 ☽ △ ♆
02 49 ☽ ∥ ♂
08 41 ☽ ∠ ♃
10 51 ☽ ☌ ♄
13 59 ☽ ☌ ☿
14 54 ☽ ∠ ♅
18 26 ☽ ∥ ♄

13 Thursday
02 39 ☽ ♇ ♆
03 29 ☽ ∥ ♆
05 40 ☽ △ ♂
06 03 ♀ ± ⯓
08 58 ☽ ✶ ♃
11 25 ☉ ∥ ♂
13 58 ☽ ⊻ ☉
14 43 ☽ ⊼ ⯓
14 57 ☽ ✶ ♅
17 02 ☽ ± ♆
23 49 ☽ ∥ ☿

14 Friday
00 15 ☽ ⊥ ☉
00 19 ☽ ± ⯓
01 19 ☉ ☍ ⯓
01 55 ☽ ☌ ♀
02 40 ☽ ⊼ ♆
04 28 ☽ Q ♃
06 10 ☉ ⊻ ♅
06 38 ☽ ♇ ♂
10 18 ☽ Q ♅
11 17 ☽ ⊻ ♄
11 23 ♀ ⊼ ♆
12 34 ☽ ⊻ ☿
14 48 ☽ ♇ ⯓
15 46 ☽ ∠ ☉
19 10 ☽ ∥ ♀
21 06 ☽ ⊥ ♄
21 46 ☽ ± ♂
21 57 ☽ ⊥ ☿

15 Saturday
02 12 ☽ ⋕ ♀
07 58 ☽ ⊼ ♂
10 08 ☽ □ ♃
11 57 ☽ ∠ ♄
12 05 ☽ ∠ ☿
14 43 ☿ ☌ ♄
15 17 ☽ △ ⯓
15 48 ☽ □ ♅
16 32 ♀ Q ♃
18 03 ☽ ✶ ☉

16 Sunday
00 09 ☽ ⋕ ☿
03 44 ☽ ☍ ♆
07 22 ☽ ⊻ ♀
11 59 ☽ ✶ ☿
13 11 ☽ ✶ ♄
15 36 ☽ Q ☉
18 25 ☽ ⊥ ♀
19 57 ☿ ∠ ♃
20 02 ☽ ⋕ ♆

17 Monday
04 45 ☽ ⋕ ♄
07 27 ☽ Q ☿
09 54 ☽ Q ♄
10 21 ☉ ⊥ ♃
11 17 ☽ ∠ ♀
12 32 ☽ ☍ ♂
13 31 ☽ △ ♃
16 09 ☽ ⋕ ☉
18 08 ☽ □ ⯓
18 57 ☽ △ ♅
19 27 ☿ ⊻ ♀

18 Tuesday
01 00 ☽ □ ☉
05 15 ☽ ⋕ ♂
07 21 ☽ ⊼ ♆
13 01 ♀ ♇ ♂
13 20 ☽ □ ☿
16 13 ☽ ✶ ♀
16 23 ☽ ♇ ♃
16 50 ♀ Q ♅
17 55 ☽ □ ♄
18 03 ☽ ± ♆
21 41 ☽ ♇ ♅

19 Wednesday
02 06 ☽ Q ⯓
05 23 ♂ ✶ ♃
08 55 ☽ ± ♃
10 21 ☽ ♇ ♆
13 41 ♀ ⊻ ♄
14 07 ☽ ± ♅
16 08 ☽ Q ♀
20 06 ☽ ⊼ ♃
20 29 ☽ ⊼ ♂

20 Thursday
00 04 ☽ ✶ ⯓
01 14 ☽ ⊼ ♅
08 06 ☽ ± ♂
11 51 ☽ △ ☉
14 08 ☽ △ ♆
17 03 ☽ △ ☿
18 19 ♀ ♇ ⯓
19 02 ☿ ⊥ ♀
20 42 ☉ ⊥ ♅
21 30 ♃ ∥ ♅

21 Friday
01 46 ☽ ♇ ♂
01 54 ☽ △ ♄
04 11 ☽ ∠ ⯓
05 12 ☽ □ ♀
17 51 ☉ △ ♆
18 41 ☽ ♇ ☉
19 45 ☽ ♇ ☿
21 22 ☽ ⊥ ⯓

22 Saturday
02 28 ☉ ☌ ☿
05 50 ☽ ☍ ♃
06 58 ☽ ♇ ♄
07 48 ☽ △ ♂
08 57 ☽ ⊻ ⯓
10 32 ☽ ☍ ♅
12 01 ☽ ± ☿
12 15 ☿ △ ♆
13 34 ☽ ± ☉
22 58 ☽ ⊼ ☿
23 42 ☽ □ ♆

23 Sunday
00 42 ☽ ± ♄
02 17 ☽ ⊼ ☉
08 21 ☿ ⊥ ♅
10 48 ♂ □ ⯓
12 37 ☽ ⊼ ♄
21 39 ☽ △ ♀

24 Monday
11 15 ☽ Q ♆
16 39 ☉ ∥ ♄
17 54 ☽ ⊼ ♃
20 01 ☽ ☌ ⯓
20 50 ☽ ⋕ ♂
21 32 ☽ □ ♂
22 00 ☽ ⊼ ♅

25 Tuesday
06 06 ☽ ± ♃
06 46 ☽ ♇ ♀
06 47 ☽ ☍ ☿
06 53 ☿ ∠ ♀
07 40 ♀ ⊥ ♄
10 04 ☽ ± ♅
10 05 ♂ ✶ ♅
11 13 ☽ ✶ ♆
15 38 ☿ ⊥ ♃
19 09 ☽ ☍ ☉

26 Wednesday
00 29 ☽ ♇ ♃
01 08 ☽ ☍ ♄
02 49 ☽ ± ♀
04 15 ☽ ♇ ♅
05 29 ☿ ∥ ♆
11 06 ☽ ⋕ ♄
11 16 ☽ Q ♂
14 46 ☽ ⋕ ☉
16 14 ☽ ⊼ ♀
17 24 ☽ ∠ ♆
17 28 ☽ ⋕ ☿
18 54 ☽ ⋕ ♆

27 Thursday
05 24 ☉ ∥ ☿
07 16 ☽ △ ♃
08 19 ☽ ⊻ ⯓
10 41 ☽ △ ♅
11 35 ☽ ⊥ ♆
12 33 ☽ ✶ ♂
16 14 ☽ ⊼ ☿
16 49 ☽ ∥ ♀
20 26 ☽ ⊥ ⯓
23 43 ☽ ⊻ ♆

28 Friday
03 54 ☽ ± ☿
13 10 ☽ ⊼ ☉
13 19 ☉ ∥ ♆
14 27 ☽ ⊼ ♄
14 38 ☽ ∠ ⯓
19 45 ☿ ∥ ♄
20 12 ☽ ∠ ♂
21 29 ☽ ♇ ☿

29 Saturday
02 23 ☽ ± ☉
02 44 ☽ ± ♄
03 54 ☉ ∠ ♃
07 12 ☉ ☌ ♄
11 32 ☽ ☍ ♀
15 09 ☽ ⊥ ♂
15 32 ☿ □ ♂
17 35 ♃ ⊼ ⯓
20 53 ☽ ✶ ⯓
20 57 ☽ □ ♃
21 05 ☽ ♇ ♄
22 09 ☽ ♇ ☉
23 36 ☽ □ ♅

30 Sunday
03 00 ☽ △ ☿
03 43 ☽ ⊻ ♂
12 13 ☽ ☌ ♆
16 47 ☉ ∠ ♅
18 18 ☽ ⋕ ♀
20 55 ☽ Q ⯓

31 Monday
00 48 ♃ ∠ ♄
01 40 ☽ ∥ ☉
03 28 ☽ △ ♄
06 49 ☽ △ ☉
06 56 ☽ ∥ ♆
09 32 ☽ Q ♃
11 43 ☽ Q ♅
13 19 ☽ ∥ ♄
20 07 ☿ St D
20 55 ☽ ∥ ☿
22 36 ♀ ± ♂

All ephemeris data is given at 12.00 UT and the Moon's longitude is additionally given for 24.00 UT

SEPTEMBER 2037

LONGITUDES

Date	Sidereal time h m s	Sun ☉ ° ' "	Moon ☽ ° ' "	Moon ☽ 24.00 ° ' "	Mercury ☿ ° '	Venus ♀ ° '	Mars ♂ ° '	Jupiter ♃ ° '	Saturn ♄ ° '	Uranus ♅ ° '	Neptune ♆ ° '	Pluto ♇ ° '
01	10 43 59	09 ♍ 23 48	23 ♉ 02 07	29 ♉ 13 15	24 ♌ 08	20 ♎ 08	25 ♉ 45	21 ♋ 48	06 ♍ 42	22 ♋ 44	28 ♈ 52	21 ♒ 13
02	10 47 56	10 21 52	05 ♊ 28 24	11 ♊ 48 07	24 D 18	21 18	26 12	21 59	06 50	22 47	28 R 51	21 R 12
03	10 51 52	11 19 58	18 ♊ 12 56	24 ♊ 43 20	24 37	22 27	26 38	22 10	06 58	22 50	28 50	21 11
04	10 55 49	12 18 07	01 ♋ 19 46	08 ♋ 02 39	25 03	23 36	27 04	22 21	07 05	22 52	28 49	21 09
05	10 59 46	13 16 17	14 ♋ 52 17	21 ♋ 48 51	25 39	24 45	27 29	22 31	07 13	22 55	28 48	21 08
06	11 03 42	14 14 29	28 ♋ 52 24	06 ♌ 02 49	26 22	25 54	27 54	22 42	07 20	22 58	28 47	21 07
07	11 07 39	15 12 43	13 ♌ 19 48	20 ♌ 42 49	27 13	27 03	28 19	22 53	07 28	23 00	28 46	21 06
08	11 11 35	16 10 59	28 ♌ 11 10	05 ♍ 43 53	28 12	28 12	28 43	23 03	07 36	23 03	28 45	21 05
09	11 15 32	17 09 17	13 ♍ 19 53	20 ♍ 57 54	29 ♌ 18	29 ♎ 21	29 07	23 14	07 43	23 05	28 44	21 03
10	11 19 28	18 07 37	28 ♍ 36 33	06 ♎ 14 28	00 ♍ 30	00 ♏ 30	29 30	23 24	07 51	23 08	28 43	21 02
11	11 23 25	19 05 58	13 ♎ 50 16	21 ♎ 22 39	01 48	01 38	29 ♉ 52	23 34	07 58	23 10	28 42	21 01
12	11 27 21	20 04 21	28 ♎ 50 31	06 ♏ 12 54	03 12	02 47	00 ♊ 14	23 44	08 06	23 13	28 40	21 00
13	11 31 18	21 02 46	13 ♏ 29 04	20 ♏ 38 29	04 41	03 55	00 36	23 54	08 13	23 15	28 39	20 59
14	11 35 15	22 01 12	27 ♏ 40 52	04 ♐ 36 06	06 15	05 03	00 57	24 04	08 21	23 17	28 38	20 58
15	11 39 11	22 59 40	11 ♐ 24 16	18 ♐ 05 36	07 52	06 11	01 17	24 14	08 28	23 19	28 37	20 57
16	11 43 08	23 58 10	24 ♐ 40 27	01 ♑ 09 15	09 32	07 19	01 37	24 24	08 35	23 22	28 35	20 56
17	11 47 04	24 56 41	07 ♑ 32 30	13 ♑ 50 45	11 15	08 27	01 56	24 34	08 43	23 24	28 34	20 55
18	11 51 01	25 55 14	20 ♑ 04 33	26 ♑ 14 27	13 00	09 35	02 15	24 43	08 50	23 26	28 33	20 54
19	11 54 57	26 53 48	02 ♒ 21 01	08 ♒ 24 46	14 47	10 43	02 33	24 52	08 58	23 28	28 31	20 52
20	11 58 54	27 52 24	14 ♒ 26 12	20 ♒ 25 45	16 36	11 50	02 51	25 02	09 05	23 30	28 30	20 51
21	12 02 50	28 51 02	26 ♒ 23 50	02 ♓ 20 51	18 25	12 58	03 07	25 11	09 12	23 32	28 29	20 50
22	12 06 47	29 ♍ 49 42	08 ♓ 17 06	14 ♓ 12 54	20 15	14 05	03 24	25 20	09 20	23 34	28 27	20 49
23	12 10 44	00 ♎ 48 23	20 ♓ 08 30	26 ♓ 04 07	22 06	15 12	03 39	25 29	09 27	23 36	28 26	20 49
24	12 14 40	01 47 06	01 ♈ 59 58	07 ♈ 56 14	23 57	16 19	03 54	25 38	09 34	23 38	28 24	20 48
25	12 18 37	02 45 51	13 ♈ 53 04	19 ♈ 50 41	25 47	17 26	04 09	25 46	09 41	23 40	28 23	20 47
26	12 22 33	03 44 39	25 ♈ 49 14	01 ♉ 48 54	27 38	18 33	04 22	25 55	09 48	23 41	28 21	20 46
27	12 26 30	04 43 28	07 ♉ 49 56	13 ♉ 52 33	29 ♍ 28	19 39	04 35	26 03	09 56	23 43	28 20	20 45
28	12 30 26	05 42 20	19 ♉ 57 01	26 ♉ 03 40	01 ♎ 18	20 45	04 47	26 12	10 03	23 45	28 18	20 44
29	12 34 23	06 41 13	02 ♊ 12 50	08 ♊ 24 55	03 07	21 52	04 59	26 20	10 10	23 46	28 17	20 43
30	12 38 19	07 ♎ 40 10	14 ♊ 40 18	20 ♊ 59 28	04 ♎ 55	22 ♏ 58	05 ♊ 10	26 ♋ 28	10 ♍ 17	23 ♋ 48	28 ♈ 15	20 ♒ 42

Date	Moon True ☊ ° '	Moon Mean ☊ ° '	Moon ☽ Latitude ° '
01	28 ♋ 00	26 ♋ 30	04 S 41
02	27 R 58	26 27	04 06
03	27 D 58	26 24	03 19
04	27 59	26 21	02 19
05	28 00	26 18	01 S 11
06	28 R 01	26 15	00 N 05
07	28 00	26 11	01 22
08	27 57	26 08	02 35
09	27 52	26 05	03 39
10	27 45	26 02	04 28
11	27 37	25 59	04 58
12	27 30	25 55	05 06
13	27 23	25 52	04 55
14	27 19	25 49	04 25
15	27 16	25 46	03 40
16	27 D 16	25 43	02 46
17	27 16	25 40	01 44
18	27 17	25 36	00 N 39
19	27 R 17	25 33	00 S 27
20	27 15	25 30	01 30
21	27 11	25 27	02 29
22	27 05	25 24	03 20
23	26 56	25 21	04 03
24	26 45	25 17	04 35
25	26 32	25 14	04 55
26	26 20	25 11	05 02
27	26 08	25 08	04 56
28	25 58	25 05	04 36
29	25 50	25 01	04 04
30	25 ♋ 46	24 ♋ 58	03 S 19

DECLINATIONS

Date	Sun ☉ ° '	Moon ☽ ° '	Mercury ☿ ° '	Venus ♀ ° '	Mars ♂ ° '	Jupiter ♃ ° '	Saturn ♄ ° '	Uranus ♅ ° '	Neptune ♆ ° '	Pluto ♇ ° '
01	08 N 03	14 N 00	11 N 53	08 S 10	17 N 08	21 N 43	10 N 27	21 N 57	09 N 26	23 S 08
02	07 41	17 10	12 07	08 40	17 15	21 42	10 24	21 57	09 26	23 09
03	07 19	19 37	12 19	09 10	17 22	21 40	10 22	21 56	09 25	23 09
04	06 57	21 06	12 27	09 39	17 29	21 38	10 19	21 56	09 25	23 09
05	06 34	21 26	12 31	10 08	17 35	21 37	10 16	21 55	09 25	23 10
06	06 12	20 27	12 32	10 37	17 42	21 35	10 13	21 55	09 24	23 10
07	05 50	18 07	12 29	11 06	17 48	21 33	10 11	21 55	09 24	23 11
08	05 27	14 32	12 22	11 35	17 54	21 32	10 08	21 54	09 23	23 11
09	05 04	09 55	12 12	12 03	18 00	21 30	10 05	21 54	09 23	23 11
10	04 42	04 N 39	11 59	12 31	18 06	21 29	10 02	21 53	09 23	23 12
11	04 19	00 S 53	11 41	12 59	18 12	21 27	10 00	21 53	09 22	23 12
12	03 56	06 17	11 21	13 27	18 17	21 25	09 57	21 53	09 22	23 12
13	03 33	11 12	10 57	13 54	18 23	21 24	09 54	21 52	09 21	23 13
14	03 10	15 20	10 30	14 21	18 28	21 22	09 52	21 52	09 21	23 13
15	02 47	18 30	10 01	14 48	18 34	21 20	09 49	21 52	09 20	23 13
16	02 24	20 34	09 28	15 14	18 39	21 19	09 46	21 51	09 20	23 13
17	02 00	21 30	08 54	15 40	18 44	21 17	09 43	21 51	09 19	23 14
18	01 37	21 18	08 17	16 06	18 49	21 16	09 41	21 51	09 19	23 14
19	01 14	20 04	07 38	16 32	18 53	21 14	09 38	21 50	09 18	23 14
20	00 51	17 56	06 58	16 57	18 58	21 13	09 35	21 50	09 18	23 14
21	00 27	15 03	06 16	17 22	19 03	21 11	09 33	21 50	09 17	23 15
22	00 N 04	11 33	05 33	17 46	19 07	21 10	09 30	21 49	09 17	23 15
23	00 S 19	07 37	04 49	18 10	19 11	21 08	09 28	21 49	09 16	23 15
24	00 43	03 S 24	04 04	18 33	19 15	21 07	09 25	21 49	09 15	23 15
25	01 06	00 N 57	03 19	18 57	19 20	21 05	09 22	21 48	09 15	23 16
26	01 29	05 17	02 33	19 19	19 24	21 04	09 20	21 48	09 14	23 16
27	01 53	09 27	01 47	19 42	19 27	21 02	09 17	21 48	09 14	23 16
28	02 16	13 17	01 00	20 04	19 31	21 01	09 15	21 48	09 13	23 16
29	02 39	16 37	00 13	20 25	19 35	20 59	09 12	21 47	09 13	23 16
30	03 S 03	19 N 15	00 S 34	20 S 46	19 N 39	20 N 58	09 N 09	21 N 47	09 N 12	23 S 17

ZODIAC SIGN ENTRIES

Date	h	m	Planets
02	01	30	☽ ♊
04	09	36	☽ ♋
06	13	54	☽ ♌
08	14	54	☽ ♍
10	01	38	♀ ♏
10	02	17	☿ ♍
10	14	11	☽ ♎
11	20	29	♂ ♊
12	13	52	☽ ♏
14	16	00	☽ ♐
16	21	51	☽ ♑
19	07	22	☽ ♒
21	19	16	☽ ♓
22	16	13	☉ ♎
24	07	57	☽ ♈
26	20	22	☽ ♉
27	18	59	☿ ♎
29	07	42	☽ ♊

LATITUDES

Date	Mercury ☿ ° '	Venus ♀ ° '	Mars ♂ ° '	Jupiter ♃ ° '	Saturn ♄ ° '	Uranus ♅ ° '	Neptune ♆ ° '	Pluto ♇ ° '
01	01 S 41	00 S 20	02 S 07	00 N 03	01 N 31	00 N 26	01 S 45	09 S 13
04	00 S 46	00 32	02 04	00 04	01 31	00 27	01 45	09 13
07	00 N 03	00 44	02 02	00 04	01 31	00 27	01 45	09 13
10	00 44	00 56	01 59	00 04	01 31	00 27	01 45	09 13
13	01 15	01 09	01 56	00 05	01 32	00 27	01 45	09 13
16	01 36	01 22	01 52	00 05	01 32	00 27	01 45	09 13
19	01 47	01 34	01 49	00 05	01 32	00 27	01 46	09 12
22	01 51	01 47	01 45	00 06	01 33	00 27	01 46	09 12
25	01 47	02 00	01 40	00 06	01 33	00 27	01 46	09 12
28	01 39	02 12	01 36	00 06	01 33	00 27	01 46	09 12
31	01 N 26	02 S 24	01 S 31	00 N 07	01 N 34	00 N 27	01 S 46	09 S 11

DATA

Julian Date	2465303
Delta T	+76 seconds
Ayanamsa	24° 22' 45"
Synetic vernal point	04° ♓ 44' 14"
True obliquity of ecliptic	23° 26' 00"

MOON'S PHASES, APSIDES AND POSITIONS ☽

Date	h	m	Phase	Longitude ° '	Eclipse Indicator
02	22	03	☾	10 ♊ 46	
09	18	25	●	17 ♍ 25	
16	10	36	☽	23 ♐ 55	
24	11	32	○	01 ♈ 46	

Day	h	m		
10	02	03	Perigee	
24	03	28	Apogee	
05	06	20	Max dec	21° N 28'
11	08	11	0S	
17	19	40	Max dec	21° S 33'
25	06	48	0N	

LONGITUDES

Date	Chiron ⚷ ° '	Ceres ⚳ ° '	Pallas ⚴ ° '	Juno ⚵ ° '	Vesta ⚶ ° '	Black Moon Lilith ⚸ ° '
01	25 ♊ 23	07 ♏ 10	05 ♎ 35	04 ♎ 01	11 ♉ 20	25 ♓ 56
11	25 ♊ 43	10 ♏ 49	09 ♎ 57	07 ♎ 37	11 ♉ 25	27 ♓ 03
21	25 ♊ 56	14 ♏ 36	14 ♎ 22	11 ♎ 15	10 ♉ 45	28 ♓ 10
31	26 ♊ 01	18 ♏ 29	18 ♎ 48	14 ♎ 52	09 ♉ 21	29 ♓ 17

All ephemeris data is given at 12.00 UT and the Moon's longitude is additionally given for 24.00 UT

ASPECTARIAN

h m	Aspects
01 Tuesday	
05 45	☽ ⊼ ♀
08 25	☉ ⋕ ♀
08 28	☽ □ ♇
09 33	☽ ⚹ ♃
11 25	☽ ⚹ ♅
14 10	☽ □ ☿
17 28	☽ ☌ ♂
18 40	☽ ± ♀
23 19	☽ ⊻ ♆
02 Wednesday	
10 01	♀ △ ♇
10 50	☽ ⊥ ♆
12 44	☽ ∥ ♂
13 44	☽ ꝑ ♀
14 37	☽ □ ♄
14 55	☽ ∠ ♃
16 25	☽ ∠ ♅
22 03	☽ □ ☉
03 Thursday	
00 33	♀ ∠ ♄
01 14	☽ Q ☿
03 51	☽ ∠ ♆
04 52	♀ □ ♃
08 07	☽ ⊥ ♃
09 25	☽ ⊥ ♅
17 29	☽ △ ♇
19 24	☽ ⊻ ♃
20 11	♀ □ ♅
20 33	☽ ⊻ ♅
20 36	☽ △ ♀
04 Friday	
00 11	☽ ⚹ ☿
00 33	☽ Q ♄
00 43	♀ ⋕ ♆
04 01	☽ ⊻ ♂
07 29	☽ ⚹ ♆
10 00	☽ Q ☉
15 13	☽ ⊥ ♂
20 38	☽ ꝑ ♇
22 24	☽ ⚹ ♄
05 Saturday	
04 14	☽ ∠ ☿
04 53	☽ Q ♆
07 42	☽ ∠ ♂
09 00	☽ ⚹ ☉
12 28	☽ ± ♇
17 55	♀ ⋕ ♄
20 42	☽ ⊥ ☿
22 49	☽ ⊼ ♇
06 Sunday	
00 48	☽ ∠ ♄
01 01	☉ ꝑ ♆
01 23	☽ ☌ ♃
01 56	☽ ☌ ♅
06 32	☽ □ ♀
07 31	☽ ⊻ ☿
10 19	☽ ⚹ ♂
11 51	☽ □ ♆
12 40	☽ ∠ ☉
16 11	☽ ⊥ ♄
07 Monday	
02 16	☽ ⊻ ♄
04 45	☽ ⊥ ☉
06 55	☽ Q ♂
14 25	☽ ∥ ♂
15 04	☽ Q ♀
15 17	☽ ⊻ ☉
08 Tuesday	
00 36	☽ ☍ ♇
03 41	☽ ⊻ ♃
03 44	☽ ⊻ ♅
10 33	♃ ☌ ♅
10 50	☿ ⚹ ♀
12 01	☽ ☌ ☿
12 02	☽ ⚹ ♀
12 52	☽ □ ♂
12 54	☽ △ ♆
13 23	☽ ⊥ ♅
13 24	☽ ⊥ ♃
14 06	♂ ⊻ ♆
23 18	♀ ☍ ♆
09 Wednesday	
00 12	☽ ∥ ☿
00 13	☿ △ ♆
02 25	☽ ⋕ ♀
03 04	☽ ☌ ♄
03 42	☽ ∠ ♅
03 51	☽ ∠ ♃
04 20	♀ ⊼ ♂
06 02	☿ □ ♂
07 30	☿ ⊥ ♅
10 23	☿ ⊥ ♃
11 12	☽ ∥ ♄
12 38	☽ ꝑ ♆
13 44	☽ ∠ ♀
14 35	☽ ∥ ♆
17 28	☿ ⋕ ♀
18 25	☽ ☌ ☉
10 Thursday	
00 08	☽ ⊼ ♇
02 45	☽ ± ♆
03 22	☽ ⚹ ♅
03 44	☽ ⚹ ♃
05 01	☽ ⊥ ♀
09 32	☽ ± ♇
10 41	☿ ⚹ ♀
11 48	☽ ∥ ☉
12 10	☽ ⊼ ♆
13 25	☽ △ ♂
15 12	☽ ⊻ ♀
15 14	☽ ⊻ ☿
22 17	☽ Q ♅
22 48	☽ Q ♃
23 40	☽ ꝑ ♇
11 Friday	
01 33	☽ ⊥ ☿
02 39	☽ ⊻ ♄
12 13	☽ ⊥ ♄
13 41	☽ ꝑ ♂
17 11	☽ ∠ ☿
20 56	☽ ⊻ ☉
23 25	☽ △ ♇
12 Saturday	
02 04	☽ ⋕ ☉
02 40	☽ ∠ ♄
02 54	☽ □ ♅
03 41	☽ □ ♃
04 24	☽ ± ♂
07 13	☽ ⊥ ☉
11 44	☽ ☍ ♆
14 19	☽ ⊼ ♂
18 55	☽ ☌ ♀
19 52	☽ ⚹ ☿
22 51	☽ ∠ ☉
13 Sunday	
02 38	☽ ⋕ ♆
03 13	☽ ⚹ ♄
05 26	☽ ⋕ ♄
10 26	☉ ⊼ ♇
10 50	☽ ⋕ ☿
17 59	☽ Q ☿
20 19	♄ ∠ ♅
23 23	☽ Q ♄
14 Monday	
00 34	☽ □ ♇
01 37	☽ ⚹ ☉
04 28	☽ △ ♅
04 59	☽ ∥ ♀
05 45	☽ △ ♃
13 38	☽ ⊼ ♆
17 47	☽ ☍ ♂
23 49	☽ Q ☉
15 Tuesday	
00 02	☽ ± ♆
01 57	☽ ⊻ ♀
02 45	☉ ± ♆
04 52	☽ □ ☿
06 32	☽ ꝑ ♅
06 45	☽ □ ♄
07 39	☽ Q ♇
08 06	☽ ꝑ ♃
12 33	☽ ⋕ ♂
13 32	☽ ⊥ ♀
15 56	☽ ꝑ ♆
18 54	☿ ∠ ♅
20 26	☉ ⚹ ♅
21 31	☿ ☌ ♄
21 45	☿ ∥ ♄
22 38	☽ ± ♅
16 Wednesday	
00 25	☽ ± ♃
05 09	☽ ⚹ ♇
07 17	☽ ∠ ♀
09 35	☽ ⊼ ♅
09 56	☿ ∠ ♃
10 36	☽ □ ☉
11 29	☽ ⊼ ♃
18 15	☿ ∥ ♆
19 13	☽ △ ♆
17 Thursday	
00 40	☉ ⚹ ♃
01 12	☽ ⊼ ♂
03 34	☽ ⋕ ♃
08 55	☽ ∠ ♇
12 46	☽ ± ♂
13 54	☽ ⚹ ♀
14 15	☽ △ ♄
18 09	♀ ⚹ ♄
20 10	☽ △ ☿
18 Friday	
02 01	☽ ⊥ ♇
06 24	☽ ꝑ ♂
13 10	☽ ⋕ ♃
13 35	☽ ⊻ ♇
15 13	☽ Q ♀
18 32	☽ ☍ ♅
19 14	☿ ꝑ ♆
19 23	☽ ꝑ ♄
21 09	☽ ☍ ♃
19 Saturday	
00 21	☽ △ ☉
04 29	☽ □ ♆
06 05	☽ ꝑ ☿
11 28	☉ ± ♇
12 24	☽ △ ♂
13 13	☽ ± ♄
20 Sunday	
01 13	☽ ⊼ ♄
01 55	☽ ⋕ ♂
02 59	☽ ± ☿
06 17	☽ □ ♀
08 36	☽ ꝑ ☉
16 07	☽ Q ♆
17 05	☽ ⊼ ☿
19 45	☽ ∥ ♀
21 Monday	
00 51	☽ ☌ ♇
03 01	☉ ⊼ ♆
04 14	☽ ± ☉
06 13	☽ ⊼ ♅
09 31	☽ ⊼ ♃
16 11	☽ ⚹ ♆
17 23	☽ ⊼ ☉
18 20	☽ ± ♅
21 46	☽ ± ♃
22 Tuesday	
01 53	☽ □ ♂
12 34	☽ ꝑ ♅
14 08	☽ ☍ ♄
16 12	☽ ꝑ ♃
19 21	☿ ⊼ ♇
22 26	☽ ∠ ♆
23 Wednesday	
00 57	☽ ⋕ ♄
00 57	☽ △ ♀
02 13	☽ ⋕ ♆
13 21	☽ ⊻ ♇
15 08	☽ Q ♂
16 14	☿ ± ♆
16 37	☽ ⊥ ♆
16 42	☽ ☍ ☿
19 01	☽ △ ♅
22 57	☽ △ ♃
24 Thursday	
01 29	☽ ⊥ ♇
04 45	☽ ⊻ ♆
07 28	☽ ⋕ ☿
07 50	☿ ⚹ ♅
10 29	☽ ꝑ ♀
11 32	☽ ☍ ☉
15 56	☽ ⚹ ♂
19 39	☽ ∠ ♇
25 Friday	
01 40	☽ ∥ ☉
03 27	☽ ⊼ ♄
06 33	☽ ± ♀
11 46	☿ ⚹ ♃
12 55	☽ ⋕ ☉
15 40	☽ ± ♄
19 53	☽ ⊼ ♀
22 48	☽ ∠ ♂
23 05	☽ ∥ ☿
26 Saturday	
00 45	☿ ± ♇
01 51	☽ ⚹ ♇
07 43	☽ □ ♅
09 57	☽ ꝑ ♄
10 37	♂ ⊥ ♆
12 12	☽ □ ♃
16 17	☽ ⊼ ☿
17 04	☽ ☌ ♆
17 12	☽ ⊥ ♂
17 18	♀ ⋕ ♂
21 20	☿ ⊼ ♆
27 Sunday	
01 52	☽ Q ♇
05 16	☽ ⊼ ☉
05 25	☽ ⊻ ♂
06 27	☽ ± ☿
07 39	☉ △ ♂
09 54	☉ ⋕ ☿
10 41	☽ ∥ ♆
11 01	☽ ∥ ♄
16 12	☽ △ ♄
18 15	☽ ± ☉
19 44	☽ Q ♅
28 Monday	
00 30	☽ Q ♃
02 03	☉ ∠ ♀
03 31	☽ ꝑ ☿
11 29	♀ □ ♇
12 39	☉ ꝑ ♇
13 01	☉ Q ♅
13 32	☽ □ ♇
13 37	☽ ꝑ ☉
13 45	☽ ☍ ♀
19 29	☽ ⚹ ♅
29 Tuesday	
00 24	☽ ⚹ ♃
03 15	♄ ∥ ♆
04 21	☽ ⊻ ♆
14 03	☽ △ ☿
16 00	☽ ⊥ ♆
17 27	☽ ☌ ♂
19 25	♀ Q ♄
21 25	☽ △ ☉
30 Wednesday	
00 43	☽ ∠ ♅
03 31	☽ □ ♄
05 48	☽ ∠ ♃
09 18	☽ ∠ ♆
15 28	☿ △ ♂
16 30	☽ ∥ ♂
17 58	☽ ⊥ ♅
22 21	☿ ꝑ ♇
23 08	☽ ⊥ ♃
23 27	☽ △ ♇
23 51	☿ Q ♅

OCTOBER 2037

LONGITUDES

Date	Sidereal time h m s	Sun ☉ ° ' "	Moon ☽ ° ' "	Moon ☽ 24.00 ° ' "	Mercury ☿ ° '	Venus ♀ ° '	Mars ♂ ° '	Jupiter ♃ ° '	Saturn ♄ ° '	Uranus ♅ ° '	Neptune ♆ ° '	Pluto ♇ ° '
01	12 42 16	08 ♎ 39 08	27 ♊ 22 53	03 ♋ 51 01	06 ♎ 43	24 ♏ 04	05 ♊ 20	26 ♋ 36	10 ♍ 24	23 ♋ 49	28 ♈ 14	20 ♒ 41
02	12 46 13	09 38 09	10 ♋ 24 20	17 ♋ 03 18	08 30	25 09	05 29	26 44	10 31	23 51	28 R 12	20 R 41
03	12 50 09	10 37 12	23 ♋ 48 20	00 ♌ 39 44	10 17	26 15	05 37	26 51	10 38	23 52	28 11	20 40
04	12 54 06	11 36 17	07 ♌ 37 45	14 ♌ 42 28	12 02	27 20	05 45	26 59	10 45	23 54	28 09	20 39
05	12 58 02	12 35 24	21 ♌ 53 50	29 ♌ 11 34	13 47	28 25	05 52	27 06	10 52	23 55	28 07	20 38
06	13 01 59	13 34 34	06 ♍ 35 14	14 ♍ 04 06	15 31	29 ♏ 30	05 58	27 14	10 58	23 56	28 06	20 38
07	13 05 55	14 33 46	21 ♍ 37 17	29 ♍ 13 38	17 14	00 ♐ 35	06 03	27 21	11 05	23 58	28 04	20 37
08	13 09 52	15 33 00	06 ♎ 51 54	14 ♎ 30 39	18 56	01 39	06 07	27 28	11 12	23 59	28 03	20 36
09	13 13 48	16 32 17	22 ♎ 08 27	29 ♎ 43 53	20 38	02 44	06 11	27 34	11 19	24 00	28 01	20 36
10	13 17 45	17 31 35	07 ♏ 15 36	14 ♏ 42 24	22 19	03 48	06 13	27 41	11 25	24 01	27 59	20 35
11	13 21 42	18 30 55	22 ♏ 03 18	29 ♏ 17 31	23 59	04 52	06 15	27 48	11 32	24 02	27 58	20 35
12	13 25 38	19 30 17	06 ♐ 24 32	13 ♐ 24 03	25 38	05 56	06 16	27 54	11 39	24 03	27 56	20 34
13	13 29 35	20 29 41	20 ♐ 15 59	27 ♐ 00 27	27 17	06 59	06 R 16	28 00	11 45	24 04	27 54	20 33
14	13 33 31	21 29 07	03 ♑ 37 43	10 ♑ 08 11	28 ♎ 55	08 02	06 15	28 06	11 52	24 05	27 53	20 33
15	13 37 28	22 28 35	16 ♑ 32 24	22 ♑ 50 54	00 ♏ 32	09 05	06 13	28 12	11 58	24 06	27 51	20 32
16	13 41 24	23 28 04	29 ♑ 04 20	05 ♒ 13 20	02 08	10 08	06 10	28 18	12 04	24 06	27 49	20 32
17	13 45 21	24 27 35	11 ♒ 18 34	17 ♒ 20 41	03 44	11 10	06 07	28 23	12 11	24 07	27 48	20 32
18	13 49 17	25 27 08	23 ♒ 20 18	29 ♒ 18 00	05 19	12 12	06 02	28 29	12 17	24 08	27 46	20 31
19	13 53 14	26 26 42	05 ♓ 14 20	11 ♓ 09 49	06 54	13 14	05 57	28 34	12 23	24 08	27 44	20 31
20	13 57 11	27 26 18	17 ♓ 04 55	23 ♓ 00 01	08 28	14 15	05 51	28 39	12 29	24 09	27 43	20 30
21	14 01 07	28 25 56	28 ♓ 55 30	04 ♈ 51 39	10 01	15 16	05 43	28 44	12 35	24 09	27 41	20 30
22	14 05 04	29 ♎ 25 36	10 ♈ 48 44	16 ♈ 46 58	11 34	16 17	05 35	28 49	12 41	24 10	27 39	20 30
23	14 09 00	00 ♏ 25 18	22 ♈ 46 32	28 ♈ 47 35	13 06	17 17	05 27	28 53	12 47	24 10	27 38	20 29
24	14 12 57	01 25 02	04 ♉ 50 13	10 ♉ 54 34	14 37	18 17	05 17	28 57	12 53	24 10	27 36	20 29
25	14 16 53	02 24 48	17 ♉ 00 44	23 ♉ 08 48	16 08	19 17	05 06	29 02	12 59	24 11	27 34	20 29
26	14 20 50	03 24 36	29 ♉ 18 55	05 ♊ 31 12	17 39	20 16	04 55	29 06	13 05	24 11	27 32	20 29
27	14 24 46	04 24 25	11 ♊ 45 50	18 ♊ 03 00	19 09	21 15	04 43	29 10	13 10	24 11	27 31	20 28
28	14 28 43	05 24 18	24 ♊ 22 56	00 ♋ 45 55	20 38	22 14	04 30	29 13	13 16	24 11	27 29	20 28
29	14 32 40	06 24 12	07 ♋ 12 14	13 ♋ 42 14	22 07	23 12	04 16	29 17	13 22	24 11	27 27	20 28
30	14 36 36	07 24 08	20 ♋ 16 14	26 ♋ 54 35	23 35	24 09	04 01	29 20	13 27	24 R 11	27 26	20 28
31	14 40 33	08 ♏ 24 07	03 ♌ 37 38	10 ♌ 25 41	25 ♏ 03	25 ♐ 06	03 ♊ 46	29 ♋ 23	13 ♍ 33	24 ♋ 11	27 ♈ 24	20 ♒ 28

Date	Moon True ☊ ° '	Moon Mean ☊ ° '	Moon ☽ Latitude ° '
01	25 ♋ 44	24 ♋ 55	02 S 24
02	25 D 43	24 52	01 20
03	25 43	24 49	00 S 10
04	25 R 43	24 46	01 N 03
05	25 41	24 42	02 14
06	25 36	24 39	03 18
07	25 29	24 36	04 10
08	25 19	24 33	04 45
09	25 08	24 30	05 00
10	24 57	24 27	04 54
11	24 47	24 23	04 28
12	24 39	24 20	03 45
13	24 34	24 17	02 50
14	24 32	24 14	01 48
15	24 D 31	24 11	00 N 42
16	24 R 31	24 07	00 S 24
17	24 30	24 04	01 28
18	24 28	24 01	02 26
19	24 23	23 58	03 17
20	24 16	23 55	04 00
21	24 05	23 52	04 32
22	23 53	23 48	04 52
23	23 38	23 45	04 59
24	23 24	23 42	04 54
25	23 11	23 39	04 35
26	22 59	23 36	04 02
27	22 50	23 32	03 18
28	22 45	23 29	02 24
29	22 42	23 26	01 21
30	22 41	23 23	00 13
31	22 ♋ 41	23 ♋ 20	00 N 58

DECLINATIONS

Date	Sun ☉ ° '	Moon ☽ ° '	Mercury ☿ ° '	Venus ♀ ° '	Mars ♂ ° '	Jupiter ♃ ° '	Saturn ♄ ° '	Uranus ♅ ° '	Neptune ♆ ° '	Pluto ♇ ° '
01	03 S 26	21 N 00	01 S 21	21 S 07	19 N 42	20 N 56	09 N 07	21 N 47	09 N 12	23 S 17
02	03 49	21 41	02 07	21 27	19 45	20 55	09 04	21 47	09 11	23 17
03	04 12	21 10	02 54	21 47	19 49	20 54	09 02	21 47	09 10	23 17
04	04 35	19 22	03 40	22 06	19 52	20 52	08 59	21 46	09 10	23 17
05	04 58	16 19	04 26	22 24	19 55	20 51	08 57	21 46	09 09	23 17
06	05 21	12 09	05 12	22 42	19 58	20 50	08 54	21 46	09 09	23 17
07	05 44	07 09	05 57	23 00	20 01	20 49	08 52	21 46	09 08	23 17
08	06 07	01 N 39	06 41	23 17	20 03	20 47	08 50	21 46	09 07	23 17
09	06 30	03 S 59	07 26	23 34	20 06	20 46	08 47	21 45	09 07	23 18
10	06 53	09 18	08 09	23 50	20 09	20 45	08 45	21 45	09 06	23 18
11	07 15	13 58	08 52	24 05	20 11	20 44	08 42	21 45	09 06	23 18
12	07 38	17 41	09 35	24 20	20 14	20 43	08 40	21 45	09 05	23 18
13	08 00	20 15	10 17	24 35	20 16	20 42	08 38	21 45	09 04	23 18
14	08 23	21 35	10 58	24 48	20 18	20 41	08 35	21 45	09 04	23 18
15	08 45	21 43	11 39	25 02	20 20	20 39	08 33	21 45	09 03	23 18
16	09 07	20 44	12 19	25 14	20 22	20 38	08 31	21 44	09 03	23 18
17	09 29	18 47	12 58	25 26	20 24	20 37	08 29	21 44	09 02	23 18
18	09 50	16 02	13 37	25 38	20 26	20 37	08 26	21 44	09 01	23 18
19	10 12	12 39	14 14	25 49	20 27	20 36	08 24	21 44	09 01	23 18
20	10 34	08 47	14 51	25 59	20 29	20 35	08 22	21 44	09 00	23 18
21	10 55	04 35	15 28	26 09	20 30	20 34	08 20	21 44	09 00	23 18
22	11 16	00 S 12	16 03	26 18	20 31	20 33	08 18	21 44	08 59	23 18
23	11 37	04 N 13	16 37	26 27	20 33	20 32	08 15	21 44	08 58	23 18
24	11 58	08 31	17 11	26 35	20 34	20 31	08 13	21 44	08 58	23 17
25	12 18	12 31	17 44	26 42	20 35	20 31	08 11	21 44	08 57	23 17
26	12 39	16 03	18 16	26 49	20 35	20 30	08 09	21 44	08 57	23 17
27	12 59	18 55	18 47	26 55	20 36	20 29	08 07	21 44	08 56	23 17
28	13 19	20 55	19 17	27 01	20 37	20 29	08 05	21 44	08 55	23 17
29	13 39	21 53	19 46	27 06	20 37	20 28	08 03	21 44	08 55	23 17
30	13 59	21 42	20 15	27 10	20 37	20 28	08 01	21 44	08 54	23 17
31	14 S 18	20 N 16	20 S 42	27 S 14	20 N 37	20 N 27	07 N 59	21 N 44	08 N 54	23 S 17

ZODIAC SIGN ENTRIES

Date	h m	Planets
01	16 53	☽ ♋
03	22 51	☽ ♌
06	01 19	☽ ♍
06	23 03	♀ ♐
08	01 13	☽ ♎
10	00 26	☽ ♏
12	01 11	☽ ♐
14	05 24	☽ ♑
15	04 07	☿ ♏
16	13 48	☽ ♒
19	01 25	☽ ♓
21	14 11	☽ ♈
23	01 50	☉ ♏
24	02 24	☽ ♉
26	13 20	☽ ♊
28	22 34	☽ ♋
31	05 32	☽ ♌

LATITUDES

Date	Mercury ☿ ° '	Venus ♀ ° '	Mars ♂ ° '	Jupiter ♃ ° '	Saturn ♄ ° '	Uranus ♅ ° '	Neptune ♆ ° '	Pluto ♇ ° '
01	01 N 26	02 S 24	01 S 31	00 N 07	01 N 34	00 N 27	01 S 46	09 S 11
04	01 11	02 36	01 25	00 07	01 34	00 27	01 46	09 11
07	00 53	02 48	01 19	00 08	01 35	00 28	01 46	09 10
10	00 34	02 59	01 13	00 08	01 35	00 28	01 46	09 10
13	00 N 14	03 09	01 06	00 08	01 36	00 28	01 46	09 10
16	00 S 06	03 19	00 59	00 09	01 36	00 28	01 46	09 09
19	00 27	03 28	00 51	00 09	01 37	00 28	01 46	09 09
22	00 47	03 36	00 43	00 10	01 37	00 28	01 46	09 08
25	01 07	03 43	00 35	00 10	01 38	00 28	01 46	09 08
28	01 26	03 49	00 26	00 11	01 38	00 28	01 46	09 07
31	01 S 44	03 S 53	00 S 17	00 N 11	01 N 39	00 N 28	01 S 46	09 S 06

DATA

Julian Date	2465333
Delta T	+76 seconds
Ayanamsa	24° 22' 48"
Synetic vernal point	04° ♓ 44' 11"
True obliquity of ecliptic	23° 26' 00"

MOON'S PHASES, APSIDES AND POSITIONS ☽

Date	h m	Phase	Longitude ° '	Eclipse Indicator
02	10 29	☾	09 ♋ 34	
09	02 34	●	16 ♎ 09	
16	00 15	☽	22 ♑ 59	
24	04 36	○	01 ♉ 07	
31	21 06	☾	08 ♌ 47	

Day	h m	
08	12 55	Perigee
21	07 24	Apogee

Day	h m		
02	13 59	Max dec	21° N 42'
08	19 00	OS	
15	02 37	Max dec	21° S 48'
22	13 05	ON	
29	20 08	Max dec	21° N 57'

LONGITUDES

Date	Chiron ⚷ ° '	Ceres ⚳ ° '	Pallas ⚴ ° '	Juno ⚵ ° '	Vesta ⚶ ° '	Black Moon Lilith ⚸ ° '
01	26 ♊ 01	18 ♏ 29	18 ♎ 48	14 ♎ 52	09 ♉ 21	29 ♓ 17
11	25 ♊ 59	22 ♏ 28	23 ♎ 15	18 ♎ 29	07 ♉ 19	00 ♈ 24
21	25 ♊ 49	26 ♏ 31	27 ♎ 42	22 ♎ 05	04 ♉ 51	01 ♈ 32
31	25 ♊ 32	00 ♐ 37	02 ♏ 08	25 ♎ 39	02 ♉ 15	02 ♈ 39

ASPECTARIAN

h m Aspects

01 Thursday
00 35 ♀ # ♃
05 12 ☽ ⊼ ♀
05 19 ☽ ⊻ ♅
06 46 ♀ △ ♅
10 30 ☉ Q ♃
10 31 ☽ ⊻ ♃
10 50 ☽ ∥ ♃
13 35 ☽ ✶ ♆
13 55 ☽ Q ♄
15 08 ☽ # ♀
17 27 ☽ ± ♀

02 Friday
02 54 ☽ ⊻ ♂
03 22 ☽ ⚼ ♇
08 00 ☽ □ ☿
10 29 ☽ □ ☉
11 30 ☽ ⚼ ♀
11 38 ☽ Q ♆
12 12 ☽ ✶ ♄
13 58 ☽ ⊥ ♂
15 16 ☿ Q ♃
19 43 ☽ ± ♇
23 17 ☽ # ♀

03 Saturday
06 18 ☽ ∠ ♂
06 27 ☽ ⊼ ♇
11 57 ♀ # ♅
12 07 ☽ ☌ ♅
12 19 ☉ ⊻ ♄
15 14 ☽ ∠ ♄
16 40 ☽ △ ♀
17 09 ☿ ⊻ ♄
17 10 ☽ ∥ ♃
17 25 ☽ ☌ ♃
19 39 ☽ □ ♆
21 00 ☽ Q ☿
21 06 ☽ Q ☉
22 34 ☉ ☌ ☿

04 Sunday
03 13 ♀ △ ♃
06 58 ☽ ∥ ♂
07 01 ☽ ⊥ ♄
08 45 ☽ ✶ ♂
17 21 ☽ ⊻ ♄
19 16 ☽ ✶ ☉
20 33 ☽ ✶ ☿
22 47 ☿ ∠ ♀

05 Monday
05 14 ☽ Q ♂
05 35 ♀ ⊼ ♆
09 55 ☽ ☍ ♇
15 21 ☽ ⊻ ♅
20 39 ☽ ⊻ ♃
22 03 ☽ ∠ ☉
22 14 ☽ △ ♆
23 36 ☽ □ ♀

06 Tuesday
00 51 ☽ ∠ ☿
01 12 ☽ ⊥ ♅
06 31 ☽ ⊥ ♃
10 59 ☽ □ ♂
13 42 ☽ ⊥ ☉
15 48 ☽ ∠ ♅
17 20 ☽ ⊥ ☿
19 06 ☽ ☌ ♄
21 08 ☽ ∠ ♃
22 26 ☽ ⚼ ♆
22 33 ☉ ∥ ☿

07 Wednesday
00 00 ☽ ⊻ ☉
02 52 ☽ ∥ ♆
04 05 ☽ ∥ ♄
04 09 ☽ ⊻ ☿
06 49 ☽ Q ♀
09 48 ☿ ⊥ ♄
10 25 ☽ ⊼ ♇
12 42 ☽ ± ♆
15 42 ☽ ✶ ♅
16 44 ☽ # ☿
17 54 ☽ # ☉
19 53 ☽ ± ♇
21 06 ☽ ✶ ♃
22 10 ☽ ⊼ ♆

08 Thursday
03 12 ☽ ✶ ♀
10 02 ☽ ⚼ ♇
10 37 ☽ Q ♅
10 49 ☽ △ ♂
12 24 ♀ ∥ ♇
16 06 ☽ Q ♃
18 51 ☽ ⊻ ♄

09 Friday
02 34 ☽ ☌ ☉
04 21 ☽ ⊥ ♄
04 32 ☽ ∠ ♀
09 20 ☽ ☌ ☿
09 34 ☽ △ ♇
10 28 ☽ ⚼ ♂
11 27 ☿ △ ♇
14 56 ☽ □ ♅
18 38 ☽ ∠ ♄
19 48 ☽ ⊥ ♀
19 57 ☿ ⚼ ♂
20 39 ☽ □ ♃
21 16 ☽ ☍ ♆
23 57 ☽ ∥ ☉

10 Saturday
00 45 ☽ ± ♂
05 46 ☽ ∥ ☿
06 03 ☽ ⊻ ♀
09 10 ☉ ⊥ ♄
09 24 ☽ # ♄
10 20 ☽ ⊼ ♂
11 04 ☽ # ♆
16 07 ♀ ± ♆
18 44 ☽ ✶ ♄

11 Sunday
05 47 ☽ ⊻ ☉
06 38 ☿ # ♄
09 34 ☽ □ ♇
12 45 ☿ □ ♅
14 27 ☽ Q ♄
15 16 ☽ △ ♅
15 35 ☽ ⊻ ☿
16 21 ☽ ⊥ ☉
19 19 ☿ # ♆
21 34 ☽ △ ♃
21 45 ☽ ⊼ ♆

12 Monday
02 53 ☽ ⊥ ☿
07 49 ☽ ± ♆
08 32 ☽ ∠ ☉
11 07 ☽ ☌ ♀
11 45 ☽ ☍ ♂
15 41 ☽ Q ♇
16 31 ☽ ⚼ ♅
18 15 ♃ □ ♆
19 39 ♀ ☍ ♂
20 12 ☽ ∠ ☿
21 02 ☽ □ ♄
23 09 ♂ St R
23 10 ☽ ⚼ ♆
23 13 ☽ ⚼ ♃

13 Tuesday
03 42 ☿ ∠ ♄
08 07 ☽ ± ♅
12 15 ☽ # ♂
12 26 ☽ ✶ ☉
12 31 ☽ ✶ ♇
13 30 ☉ △ ♇
15 05 ☽ ± ♃
17 55 ☽ # ♃
18 45 ☽ ⊼ ♅
21 01 ☿ ☍ ♆
23 18 ☿ □ ♃

14 Wednesday
01 35 ☽ △ ♆
01 54 ☽ ⊼ ♃
02 14 ☽ ✶ ☿
06 19 ☉ ⚼ ♂
11 43 ☽ Q ☉
15 31 ☽ ∠ ♇
16 48 ☽ ⊼ ♂
19 05 ☽ # ♅
20 49 ☽ ⊻ ♀
23 38 ♀ Q ♇

15 Thursday
00 32 ☉ # ♄
03 20 ☽ △ ♄
03 23 ☽ Q ☿
03 54 ☽ ± ♂
07 26 ☿ ± ♂
08 14 ☽ ⊥ ♇
09 01 ☽ ⊥ ♀
10 15 ☽ # ♅
12 12 ♀ ⚼ ♅
19 35 ☽ ⊻ ♇
20 51 ☽ ⚼ ♂

16 Friday
00 15 ☽ □ ☉
02 24 ☽ ☍ ♅
03 41 ☽ ∠ ♀
07 39 ☉ # ♆
08 06 ☽ ⚼ ♄
09 35 ☽ □ ♆
10 29 ☽ ☍ ♃
13 27 ☽ # ♃
17 24 ☽ # ♂
18 52 ☽ □ ☿

17 Saturday
01 47 ☽ ± ♄
01 48 ☽ △ ♂
03 37 ☉ □ ♅
11 41 ☽ ✶ ♀
13 44 ☽ ⊼ ♄
20 53 ☽ Q ♆

18 Sunday
06 21 ☽ ☌ ♇
13 35 ☽ ⊼ ♅
13 54 ☽ Q ♀
14 04 ♀ □ ♄
16 38 ☽ △ ☉
20 53 ☽ ✶ ♆
22 20 ☿ ⊼ ♂
22 25 ☽ ⊼ ♃

19 Monday
00 48 ♀ ⚼ ♆
01 41 ☽ ± ♅
02 52 ☽ ∥ ☿
10 38 ☽ ± ♃
13 25 ☽ □ ♂
15 52 ☽ △ ☿
19 54 ☽ ⚼ ♅
20 32 ♀ ⚼ ♃

20 Tuesday
01 43 ☽ ⚼ ☉
02 09 ☽ ∥ ☉
02 36 ☽ ☍ ♄
02 44 ☿ ⊥ ♀
03 09 ☽ ∠ ♆
04 59 ☽ ⚼ ♃
05 43 ☽ □ ♀
10 40 ☽ # ♆
13 16 ☉ ∠ ♄
14 28 ☽ # ♄
18 22 ☉ ☍ ♆
18 56 ☽ ⊻ ♇
21 22 ☽ ⊥ ♆
21 39 ☽ ± ☉

21 Wednesday
01 35 ☽ Q ♂
02 20 ☽ △ ♅
02 54 ☽ ⚼ ☿
07 06 ☽ ⊥ ♇
09 29 ☽ ⊻ ♆
10 55 ☽ ⊼ ☉
11 36 ☽ △ ♃
19 49 ☉ □ ♃
23 50 ☽ ± ☿

22 Thursday
01 17 ☽ ∠ ♇
01 36 ☽ ✶ ♂
05 36 ♄ ⚼ ♆
13 44 ☽ ⊼ ☿
15 29 ☉ ± ♂
15 48 ☽ ⊼ ♄

23 Friday
00 01 ☽ △ ♀
03 57 ☽ ± ♄
06 47 ☿ ✶ ♄
07 24 ☽ ∠ ♂
07 26 ☽ ✶ ♇
08 21 ♂ ∥ ♃
14 47 ☽ □ ♅
21 39 ☽ ☌ ♆
22 05 ☽ ⚼ ♄

24 Saturday
00 15 ☽ □ ♃
01 07 ☽ ⊥ ♂
04 36 ☽ ☍ ☉
07 20 ☽ Q ♇
08 40 ☽ ⚼ ♀
09 11 ♀ ± ♅
10 21 ☽ ∥ ♄
12 52 ☽ ⊻ ♂
14 35 ☽ ∥ ♆

25 Sunday
02 30 ☽ Q ♅
04 01 ☽ △ ♄
04 02 ☽ ± ♀
10 03 ☽ ☍ ☿
10 32 ☽ # ☉
12 02 ☽ Q ♃
16 51 ☽ ⊼ ♀
18 47 ☽ □ ♇

26 Monday
02 01 ☽ ✶ ♅
08 34 ☽ ⊻ ♆
11 34 ☽ ✶ ♃
16 58 ♀ ✶ ♇
20 10 ☽ ⊥ ♆
20 37 ☽ ⊼ ☉
22 40 ☽ ☌ ♂

27 Tuesday
07 03 ☽ ∠ ♅
09 10 ☽ ± ☉
10 24 ☽ # ☿
13 26 ☽ ∠ ♆
14 43 ☽ □ ♄
16 36 ☽ ∠ ♃
18 00 ☉ ⊼ ♂

28 Wednesday
00 16 ☽ ⊥ ♅
03 50 ☽ ⚼ ☉
03 58 ☽ ⊼ ☿
04 36 ☽ △ ♇
05 38 ☽ ∥ ♃
07 21 ☽ ∥ ♂
07 35 ☽ ☍ ♀
09 20 ☿ □ ♇
09 48 ☽ ⊥ ♃
11 38 ☽ ⊻ ♅
16 49 ☽ ± ☿
17 50 ☽ ✶ ♆
21 09 ☽ ⊻ ♃

29 Thursday
01 02 ☽ Q ♄
05 23 ☽ ∥ ♅
06 38 ☽ ⊻ ♂
08 47 ☽ ⚼ ♇
10 23 ☽ △ ☉
11 49 ☽ ⚼ ☿
14 13 ♀ ± ♃
16 10 ☽ Q ♆
17 33 ☽ ⊥ ♂
23 27 ☽ ✶ ♄

30 Friday
01 24 ☽ ± ♇
01 59 ♅ St R
09 46 ☽ ∠ ♂
10 43 ☽ ∥ ♅
12 21 ☽ ⊼ ♇
12 51 ♀ ⊼ ♅
18 45 ☽ △ ☿
19 06 ☽ ☌ ♅
19 35 ☽ ⊼ ♀
21 52 ☿ △ ♅
23 08 ☿ # ♃

31 Saturday
00 55 ☽ □ ♆
02 52 ☽ ∠ ♄
04 25 ☽ ☌ ♃
07 10 ☽ ± ♀
07 37 ☽ ∥ ♂
07 41 ☽ # ☿
07 59 ☿ # ♂
09 48 ☽ ∥ ♃
12 14 ☽ ✶ ♂
14 46 ☿ ⊻ ♀
18 58 ☽ ⊥ ♄
20 42 ☿ Q ♄
21 06 ☽ □ ☉

All ephemeris data is given at 12.00 UT and the Moon's longitude is additionally given for 24.00 UT

NOVEMBER 2037

LONGITUDES

Date	Sidereal time h m s	Sun ☉ ° ' "	Moon ☽ ° ' "	Moon ☽ 24.00 ° ' "	Mercury ☿ ° '	Venus ♀ ° '	Mars ♂ ° '	Jupiter ♃ ° '	Saturn ♄ ° '	Uranus ♅ ° '	Neptune ♆ ° '	Pluto ♇ ° '
01	14 44 29	09 ♏ 24 08	17 ♌ 18 59	24 ♌ 17 41	26 ♏ 30	26 ♐ 03	03 ♊ 29	29 ♋ 26	13 ♍ 38	24 ♋ 11	27 ♈ 23	20 ♒ 28
02	14 48 26	10 24 10	01 ♍ 21 51	08 ♍ 31 24	27 56	26 59	03 R 13	29 29	13 43	24 R 11	27 R 21	20 R 28
03	14 52 22	11 24 15	15 ♍ 46 06	23 ♍ 05 31	29 ♏ 22	27 55	02 55	29 31	13 48	24 11	27 19	20 D 28
04	14 56 19	12 24 23	00 ♎ 29 05	07 ♎ 55 59	00 ♐ 47	28 50	02 37	29 34	13 54	24 11	27 18	20 28
05	15 00 15	13 24 32	15 ♎ 25 16	22 ♎ 55 50	02 12	29 ♐ 44	02 18	29 36	13 59	24 10	27 16	20 28
06	15 04 12	14 24 43	00 ♏ 26 30	07 ♏ 56 01	03 35	00 ♑ 38	01 59	29 38	14 04	24 10	27 14	20 28
07	15 08 09	15 24 56	15 ♏ 23 10	22 ♏ 46 46	04 58	01 32	01 39	29 40	14 09	24 09	27 13	20 28
08	15 12 05	16 25 11	00 ♐ 05 47	07 ♐ 19 21	06 20	02 24	01 19	29 41	14 13	24 09	27 11	20 28
09	15 16 02	17 25 27	14 ♐ 26 46	21 ♐ 27 31	07 41	03 16	00 58	29 43	14 18	24 08	27 10	20 28
10	15 19 58	18 25 45	28 ♐ 21 19	05 ♑ 08 05	09 00	04 08	00 37	29 44	14 23	24 08	27 08	20 29
11	15 23 55	19 26 05	11 ♑ 47 53	18 ♑ 20 57	10 19	04 58	00 ♊ 16	29 45	14 27	24 07	27 07	20 29
12	15 27 51	20 26 26	24 ♑ 47 39	01 ♒ 08 26	11 36	05 48	29 ♉ 54	29 46	14 32	24 06	27 05	20 29
13	15 31 48	21 26 49	07 ♒ 23 51	13 ♒ 34 29	12 52	06 37	29 32	29 46	14 36	24 06	27 04	20 29
14	15 35 44	22 27 13	19 ♒ 41 00	25 ♒ 44 02	14 06	07 26	29 10	29 47	14 41	24 05	27 02	20 30
15	15 39 41	23 27 38	01 ♓ 44 16	07 ♓ 42 20	15 18	08 13	28 48	29 47	14 45	24 04	27 01	20 30
16	15 43 38	24 28 05	13 ♓ 38 52	19 ♓ 34 29	16 28	09 00	28 26	29 R 47	14 49	24 03	26 59	20 30
17	15 47 34	25 28 33	25 ♓ 29 46	01 ♈ 25 14	17 36	09 45	28 04	29 47	14 53	24 02	26 58	20 31
18	15 51 31	26 29 02	07 ♈ 21 22	13 ♈ 18 37	18 40	10 30	27 42	29 46	14 57	24 01	26 56	20 31
19	15 55 27	27 29 32	19 ♈ 17 21	25 ♈ 17 54	19 42	11 14	27 20	29 46	15 01	24 00	26 55	20 32
20	15 59 24	28 30 04	01 ♉ 20 33	07 ♉ 25 30	20 40	11 56	26 58	29 45	15 05	23 59	26 54	20 32
21	16 03 20	29 ♏ 30 38	13 ♉ 32 56	19 ♉ 42 58	21 35	12 38	26 36	29 44	15 09	23 58	26 52	20 32
22	16 07 17	00 ♐ 31 12	25 ♉ 55 42	02 ♊ 11 09	22 25	13 18	26 15	29 43	15 12	23 57	26 51	20 33
23	16 11 13	01 31 49	08 ♊ 29 22	14 ♊ 50 21	23 09	13 58	25 54	29 42	15 16	23 55	26 50	20 34
24	16 15 10	02 32 26	21 ♊ 14 06	27 ♊ 40 38	23 48	14 36	25 33	29 40	15 19	23 54	26 48	20 34
25	16 19 07	03 33 06	04 ♋ 09 56	10 ♋ 42 02	24 21	15 12	25 12	29 38	15 22	23 53	26 47	20 35
26	16 23 03	04 33 46	17 ♋ 16 59	23 ♋ 54 51	24 46	15 48	24 52	29 36	15 26	23 51	26 46	20 35
27	16 27 00	05 34 29	00 ♌ 35 43	07 ♌ 19 40	25 04	16 22	24 33	29 34	15 29	23 50	26 45	20 36
28	16 30 56	06 35 12	14 ♌ 06 48	20 ♌ 57 13	25 13	16 54	24 13	29 32	15 32	23 48	26 43	20 37
29	16 34 53	07 35 58	27 ♌ 50 59	04 ♍ 48 10	25 R 12	17 25	23 55	29 29	15 35	23 47	26 42	20 37
30	16 38 49	08 ♐ 36 44	11 ♍ 48 45	18 ♍ 52 39	25 ♐ 01	17 ♑ 55	23 ♉ 37	29 ♋ 27	15 ♍ 38	23 ♋ 45	26 ♈ 41	20 ♒ 38

MOON NODES AND LATITUDE

Date	Moon True ☊ ° '	Moon Mean ☊ ° '	Moon ☽ Latitude ° '
01	22 ♋ 41	23 ♋ 17	02 N 07
02	22 R 39	23 13	03 10
03	22 35	23 10	04 03
04	22 29	23 07	04 40
05	22 20	23 04	05 00
06	22 09	23 01	04 59
07	21 59	22 58	04 38
08	21 49	22 54	03 58
09	21 42	22 51	03 04
10	21 37	22 48	02 00
11	21 35	22 45	00 N 52
12	21 D 34	22 42	00 S 17
13	21 35	22 38	01 24
14	21 36	22 35	02 24
15	21 R 35	22 32	03 18
16	21 33	22 29	04 02
17	21 28	22 26	04 35
18	21 22	22 23	04 56
19	21 13	22 19	05 05
20	21 03	22 16	05 01
21	20 53	22 13	04 42
22	20 43	22 10	04 10
23	20 35	22 07	03 26
24	20 29	22 04	02 31
25	20 25	22 00	01 26
26	20 24	21 57	00 S 17
27	20 D 24	21 54	00 N 55
28	20 25	21 51	02 05
29	20 27	21 48	03 09
30	20 ♋ 27	21 ♋ 44	04 N 03

DECLINATIONS

Date	Sun ☉ ° '	Moon ☽ ° '	Mercury ☿ ° '	Venus ♀ ° '	Mars ♂ ° '	Jupiter ♃ ° '	Saturn ♄ ° '	Uranus ♅ ° '	Neptune ♆ ° '	Pluto ♇ ° '
01	14 S 37	17 N 39	21 S 08	27 S 17	20 N 37	20 N 27	07 N 58	21 N 44	08 N 53	23 S 17
02	14 56	13 56	21 33	27 19	20 37	20 26	07 56	21 44	08 53	23 16
03	15 15	09 20	21 57	27 22	20 37	20 26	07 54	21 44	08 52	23 16
04	15 33	04 N 06	22 20	27 23	20 37	20 26	07 52	21 44	08 51	23 16
05	15 52	01 S 27	22 42	27 24	20 36	20 25	07 50	21 44	08 51	23 16
06	16 10	06 57	23 03	27 24	20 35	20 25	07 49	21 45	08 50	23 16
07	16 27	12 01	23 23	27 24	20 35	20 25	07 47	21 45	08 50	23 15
08	16 45	16 17	23 41	27 23	20 34	20 25	07 45	21 45	08 49	23 15
09	17 02	19 29	23 58	27 22	20 33	20 24	07 44	21 45	08 49	23 15
10	17 19	21 25	24 14	27 20	20 31	20 24	07 42	21 45	08 48	23 15
11	17 35	22 03	24 29	27 18	20 30	20 24	07 40	21 45	08 48	23 14
12	17 51	21 27	24 42	27 15	20 29	20 24	07 39	21 45	08 47	23 14
13	18 07	19 46	24 54	27 12	20 28	20 24	07 37	21 46	08 47	23 14
14	18 23	17 12	25 05	27 08	20 26	20 24	07 36	21 46	08 46	23 14
15	18 38	13 56	25 14	27 03	20 24	20 25	07 34	21 46	08 46	23 13
16	18 53	10 09	25 22	26 59	20 22	20 25	07 33	21 46	08 45	23 13
17	19 08	06 00	25 28	26 53	20 20	20 25	07 32	21 46	08 45	23 13
18	19 22	01 S 37	25 33	26 48	20 18	20 25	07 30	21 46	08 44	23 13
19	19 36	02 N 50	25 36	26 42	20 16	20 25	07 29	21 47	08 44	23 12
20	19 49	07 14	25 38	26 35	20 14	20 26	07 28	21 47	08 43	23 12
21	20 02	11 25	25 38	26 28	20 12	20 26	07 27	21 47	08 43	23 12
22	20 15	15 11	25 37	26 21	20 10	20 26	07 25	21 47	08 42	23 11
23	20 28	18 19	25 34	26 13	20 07	20 27	07 24	21 48	08 42	23 11
24	20 40	20 38	25 30	26 05	20 05	20 27	07 23	21 48	08 41	23 11
25	20 52	21 56	25 24	25 57	20 03	20 28	07 22	21 48	08 41	23 10
26	21 03	22 02	25 16	25 48	20 01	20 29	07 21	21 48	08 41	23 10
27	21 14	20 55	25 06	25 39	19 58	20 29	07 20	21 49	08 40	23 10
28	21 24	18 35	24 55	25 30	19 56	20 30	07 19	21 49	08 40	23 09
29	21 34	15 10	24 42	25 20	19 54	20 31	07 18	21 49	08 39	23 09
30	21 S 44	10 N 52	24 S 27	25 S 10	19 N 52	20 N 31	07 N 17	21 N 50	08 N 39	23 S 08

ZODIAC SIGN ENTRIES

Date	h	m	Planets
02	09	42	☽ ♍
03	22	40	☿ ♐
04	11	13	☽ ♎
05	18	58	♀ ♑
06	11	18	☽ ♏
08	11	50	☽ ♐
10	14	54	☽ ♑
12	05	40	♂ ♉
12	21	50	☽ ♒
15	08	31	☽ ♓
17	21	07	☽ ♈
20	09	20	☽ ♉
21	23	38	☉ ♐
22	19	49	☽ ♊
25	04	18	☽ ♋
27	10	56	☽ ♌
29	15	43	☽ ♍

LATITUDES

Date	Mercury ☿ ° '	Venus ♀ ° '	Mars ♂ ° '	Jupiter ♃ ° '	Saturn ♄ ° '	Uranus ♅ ° '	Neptune ♆ ° '	Pluto ♇ ° '
01	01 S 49	03 S 55	00 S 14	00 N 11	01 N 39	00 N 28	01 S 46	09 S 06
04	02 05	03 57	00 S 04	00 12	01 40	00 28	01 46	09 06
07	02 18	03 59	00 N 06	00 12	01 40	00 29	01 46	09 05
10	02 28	03 58	00 15	00 13	01 41	00 29	01 46	09 05
13	02 35	03 56	00 25	00 13	01 42	00 29	01 46	09 04
16	02 38	03 52	00 34	00 14	01 42	00 29	01 46	09 04
19	02 35	03 45	00 44	00 14	01 43	00 29	01 46	09 03
22	02 24	03 36	00 53	00 15	01 44	00 29	01 46	09 03
25	02 05	03 24	01 01	00 15	01 45	00 29	01 46	09 02
28	01 34	03 10	01 09	00 16	01 45	00 29	01 45	09 02
31	00 S 51	02 S 52	01 N 16	00 N 16	01 N 46	00 N 29	01 S 45	09 S 01

DATA

Julian Date	2465364
Delta T	+76 seconds
Ayanamsa	24° 22' 51"
Synetic vernal point	04° ♓ 44' 08"
True obliquity of ecliptic	23° 26' 00"

MOON'S PHASES, APSIDES AND POSITIONS ☽

Date	h	m	Phase	Longitude ° '	Eclipse Indicator
07	12	03	●	15 ♏ 25	
14	17	59	☽	22 ♒ 42	
22	21	35	○	00 ♊ 55	
30	06	06	☾	08 ♍ 22	

Day	h	m		
05	22	04	Perigee	
17	21	02	Apogee	
05	05	45	0S	
11	11	55	Max dec	22° S 03'
18	20	44	0N	
26	02	13	Max dec	22° N 08'

LONGITUDES

Date	Chiron ⚷ ° '	Ceres ⚳ ° '	Pallas ⚴ ° '	Juno ⚵ ° '	Vesta ⚶ ° '	Black Moon Lilith ⚸ ° '
01	25 ♊ 30	01 ♐ 02	02 ♏ 35	26 ♎ 00	02 ♉ 00	02 ♈ 45
11	25 ♊ 05	05 ♐ 10	07 ♏ 01	29 ♎ 30	29 ♈ 38	03 ♈ 53
21	24 ♊ 36	09 ♐ 20	11 ♏ 24	02 ♏ 55	27 ♈ 46	05 ♈ 00
31	24 ♊ 02	13 ♐ 31	15 ♏ 45	06 ♏ 14	26 ♈ 34	06 ♈ 07

All ephemeris data is given at 12.00 UT and the Moon's longitude is additionally given for 24.00 UT

ASPECTARIAN

h m Aspects

01 Sunday
00 17 ☽ ♇ ♀
05 34 ☽ ⊻ ♄
08 54 ☽ Q ♂
17 26 ☽ ☍ ♇
23 03 ♂ ⊥ ♆
23 49 ☽ ⊻ ♅

02 Monday
02 20 ☿ ⊼ ♆
04 04 ☽ △ ♀
05 13 ☽ △ ♄
05 33 ☽ □ ☿
06 35 ☽ # ☉
06 37 ☽ Q ☉
08 48 ☽ ⊥ ♄
10 00 ☽ ⊥ ♀
15 03 ☽ □ ♂
18 56 ☽ ⊥ ♃
21 08 ☽ △ ♆
22 48 ☿ # ♅

03 Tuesday
01 06 ☽ ∠ ♅
03 35 ♆ St D
04 15 ☽ ⚹ ☉
06 19 ☽ ♇ ♆
08 45 ☽ ☌ ♄
09 56 ☽ ∠ ♃
14 15 ☽ ∥ ♆
14 42 ☿ △ ♃
14 55 ☽ Q ☿
18 51 ☽ ∥ ♄
19 43 ☽ ⊼ ♆
21 06 ☽ ± ♆

04 Wednesday
01 46 ☽ ⚹ ♅
05 29 ☽ ± ♇
06 39 ☽ ∠ ☉
06 51 ☽ ⊼ ♆
09 09 ☽ □ ♀
10 30 ☽ ⚹ ♃
12 32 ☽ ⚹ ☿
15 22 ☽ △ ♂
20 02 ☽ Q ♄
21 10 ☽ Q ♅
22 14 ☽ ⊥ ☉

05 Thursday
05 52 ☽ Q ♃
08 10 ♀ ⊼ ♃
08 33 ☽ ⊻ ☉
09 41 ☽ ⊻ ♄
13 34 ☿ ☍ ♂
14 57 ☽ ♇ ♂
15 07 ☽ ∠ ☿
15 56 ☽ Q ♀
19 19 ☽ ⊥ ♄
20 04 ☽ △ ♆

06 Friday
01 58 ☽ □ ♅
02 50 ☉ ⚹ ♄
05 02 ☽ ± ♂
06 10 ☿ ± ♆
06 54 ☽ ☍ ♆
06 58 ☽ ⊥ ☿
09 47 ☽ ∠ ♄
10 42 ☽ □ ♃
12 20 ☽ ⚹ ♀
14 25 ☽ ⊼ ♂
15 52 ☽ # ♄
17 33 ☽ ⊻ ☿
20 36 ☽ # ♆

07 Saturday
03 07 ☿ ∥ ♆
09 59 ☽ ⚹ ♄
12 03 ☽ ☌ ☉
13 58 ☽ ♇ ♅
14 31 ♀ ⊼ ♂
20 14 ☽ □ ♆

08 Sunday
02 14 ☽ △ ♅
05 32 ☽ ± ♀
05 36 ☽ Q ♄
07 13 ☽ ⊼ ♆
11 20 ☽ △ ♃
13 58 ☽ ☍ ♂
15 13 ☽ ∥ ☉
16 04 ☽ ⊻ ♀
17 06 ☽ ± ♆
23 25 ☽ ☌ ☿

09 Monday
01 55 ☽ Q ♆
03 03 ☽ ♇ ♄
08 08 ☽ ♇ ♆
11 45 ☽ □ ♄
12 27 ☽ ♇ ♃
17 28 ☽ ⊻ ☉
18 17 ☽ ± ☉
21 34 ☽ # ♃
22 18 ☽ ⚹ ♆
23 05 ☽ # ♂

10 Tuesday
02 22 ☿ Q ♆
03 55 ☽ ± ♃
04 36 ☽ ⊥ ☉
04 38 ☽ ⊼ ☿
09 52 ☽ △ ♆
14 13 ☽ ♇ ☿
14 25 ☽ ⊼ ♃
15 53 ☽ ⊼ ☉
19 30 ☽ # ♅
21 41 ☽ ∠ ☉
22 53 ☽ ☌ ♆

11 Wednesday
00 37 ☽ ∠ ♆
02 17 ☽ ± ☉
09 01 ☽ ⊻ ☿
16 52 ☽ △ ♄
16 53 ☽ ⊥ ♆
18 09 ☽ ♇ ♂
21 10 ☽ ⊥ ☿

12 Thursday
02 42 ♀ ∠ ♆
03 11 ☽ ⚹ ☉
03 57 ☽ ⊻ ♆
04 35 ☽ # ♅
10 43 ☽ ☍ ♅
13 03 ☉ □ ♆
14 02 ♀ ± ♂
15 47 ☽ ∠ ☿
16 18 ☽ □ ♆
20 58 ☿ ♇ ♀
21 00 ☽ ♇ ♄
21 06 ♂ ⚹ ♃
21 23 ☽ △ ♂
21 23 ☽ ☍ ♃

13 Friday
03 38 ☽ # ♂
03 45 ☽ Q ☉
04 22 ☽ ⊻ ♀
10 24 ☽ ⊻ ♀
14 21 ☽ ± ♄
22 51 ☽ ⊥ ♀
23 48 ☽ ⚹ ☿

14 Saturday
02 06 ☽ ⊼ ♄
02 44 ☽ ∥ ☉
02 53 ☽ Q ♆
08 58 ☉ ∠ ♀
13 36 ☽ ☌ ♀
17 49 ☽ ∠ ♀
17 59 ☽ □ ☉
20 42 ☽ ⊼ ☿

15 Sunday
00 13 ☿ □ ♄
01 35 ☿ ⊻ ♄
02 09 ☽ Q ☿
02 25 ♂ ∥ ♃
02 34 ☽ ⚹ ♆
06 18 ☽ □ ♂
08 05 ☽ ⊼ ♃
08 39 ☽ ± ♅
20 08 ☽ △ ♀

16 Monday
00 14 ♃ ∠ ♄
01 57 ☽ ⚹ ♀
02 14 ☉ △ ♅
02 17 ♃ St R
02 44 ☽ ♇ ♅
08 39 ☽ ∠ ♆
14 18 ☽ ♇ ♃
14 23 ☽ ∠ ♆
17 28 ☽ Q ♂
18 19 ☽ □ ☿
20 16 ☽ # ♆

17 Tuesday
01 54 ☽ ⊻ ♇
02 50 ☽ ⊥ ♆
03 16 ☽ # ♄
03 54 ☽ Q ☿
09 03 ☽ △ ☉
11 57 ☽ △ ☉
14 04 ☽ ⊥ ♆
14 58 ☽ ⊻ ♆
17 03 ☽ ⚹ ♂
20 41 ☽ △ ♄
21 34 ☿ ± ♄

18 Wednesday
08 17 ☽ ∠ ♆
18 46 ☽ ∠ ☿
21 06 ☽ ♇ ☉
22 27 ☽ ∠ ♆
22 37 ☉ ⊼ ♆
23 55 ☉ Q ♄

19 Thursday
03 23 ☽ ⊼ ♄
09 10 ☉ ☍ ♂
12 54 ☽ △ ☿
14 29 ☽ ⚹ ♆
15 29 ☽ ± ♄
15 58 ☽ ⊥ ♂
16 49 ☽ ± ☉
21 24 ☽ □ ♆

20 Friday
03 11 ☽ ☌ ♆
03 34 ☽ ± ♆
05 51 ☽ ⊼ ☉
08 23 ☿ ⚹ ♆
08 51 ☽ ♇ ♆
09 29 ☽ ♇ ♀
12 35 ♀ ♇ ♂
13 16 ☽ ∥ ♄
14 21 ☽ Q ♆
17 00 ♂ ⊻ ♆
20 19 ☽ ∥ ♆
21 16 ☽ ♇ ♅

21 Saturday
08 54 ☽ Q ♅
10 06 ☽ △ ♀
15 07 ☽ △ ♄
16 16 ☽ ± ♅
17 15 ☉ △ ♃
20 09 ☽ Q ♃

22 Sunday
01 36 ☽ □ ♆
02 44 ☉ # ♂
04 45 ☽ ⊻ ♆
08 11 ☽ ⚹ ♆
12 36 ☽ ☌ ♂
13 46 ☽ ⊻ ♆
16 49 ☽ ♇ ♀
19 16 ☽ ⚹ ♃
21 35 ☽ ☍ ☉

23 Monday
01 15 ☽ ⊥ ♆
10 23 ☉ # ♃
10 56 ☽ ± ♀
12 49 ☽ ∠ ♅
18 18 ☽ ∠ ♆
22 53 ☽ ⊼ ♀
23 42 ☽ ∠ ♃

24 Tuesday
00 51 ☽ □ ♄
05 22 ☽ ∥ ♂
05 46 ☽ ⊥ ♅
06 45 ☿ ± ♃
09 40 ☽ ∥ ♃
10 45 ☽ △ ♆
11 03 ♀ ⊥ ♇
12 21 ☽ # ☉
15 42 ☿ ⊼ ♅
16 32 ☽ ⊥ ♃
16 58 ☽ ⊻ ♅
17 02 ☽ ☍ ☿
18 09 ☉ ± ♆
19 50 ☽ ⊻ ♂
22 22 ☽ ⚹ ♆

25 Wednesday
03 39 ☽ ⊻ ♃
06 40 ☽ ⊥ ♂
08 17 ☽ ∥ ♅
10 32 ☽ Q ♄
10 46 ☽ ⊼ ♆
14 36 ☽ ♇ ♆
19 31 ♀ △ ♄
20 29 ☽ Q ♆
22 43 ☽ ± ☉
22 49 ☽ ∠ ♂

26 Thursday
07 06 ☽ ± ♆
08 37 ☽ ⚹ ♄
09 10 ☽ ☍ ♀
15 21 ☿ ⊼ ♂
16 29 ☽ ♇ ☉
18 00 ☽ ⊼ ♆
19 49 ☽ ∥ ♅
23 52 ☽ ☌ ♅

27 Friday
01 23 ☽ ⚹ ♂
01 53 ☽ ⊼ ♆
05 06 ☽ □ ♆
07 49 ☽ # ☉
10 10 ☽ ☌ ♃
11 48 ☽ ∠ ♄
12 51 ☽ ± ☿
17 25 ☽ ∥ ♃
21 36 ☽ △ ☉
22 22 ☽ Q ♂
23 19 ☽ ∥ ♂

28 Saturday
03 53 ☽ ⊥ ♄
05 04 ☽ △ ☿
14 30 ☽ ⊻ ♄
17 06 ☽ ⊻ ♆
22 00 ☿ St R
23 25 ☽ ☍ ♇

29 Sunday
04 01 ☽ ± ♆
04 57 ☽ ⊻ ♅
05 19 ☽ □ ♂
07 25 ☽ △ ♀
10 01 ☽ △ ♆
14 50 ☽ ⊻ ♆
15 20 ☽ ⊥ ♆
20 12 ☽ ♇ ♀
23 29 ♂ ⚹ ♄

30 Monday
01 08 ☽ □ ☿
06 06 ☽ □ ☉
06 47 ☽ ∠ ♅
11 47 ☽ ⊻ ♆
12 30 ☉ Q ♆
15 18 ☉ ♇ ♆
16 28 ☽ ∠ ♃
18 31 ☽ ☌ ♄
22 44 ☽ △ ♀
23 04 ☽ ∥ ♄

DECEMBER 2037

LONGITUDES

Date	Sidereal time h m s	Sun ☉ ° ' "	Moon ☽ ° ' "	Moon ☽ 24.00 ° ' "	Mercury ☿ ° '	Venus ♀ ° '	Mars ♂ ° '	Jupiter ♃ ° '	Saturn ♄ ° '	Uranus ♅ ° '	Neptune ♆ ° '	Pluto ♇ ° '
01	16 42 46	09 ♐ 37 33	25 ♍ 59 43	03 ♎ 09 42	24 ♐ 39	18 ♑ 23	23 ♉ 19	29 ♋ 24	15 ♍ 40	23 ♋ 44	26 ♈ 40	20 ♒ 39
02	16 46 42	10 38 23	10 ♎ 22 14	17 ♎ 36 51	24 R 06	18 49	23 R 03	29 R 20	15 43	23 R 42	26 R 39	20 40
03	16 50 39	11 39 14	24 ♎ 52 57	02 ♏ 09 53	23 22	19 14	22 46	29 17	15 46	23 40	26 38	20 40
04	16 54 36	12 40 06	09 ♏ 26 53	16 ♏ 43 08	22 27	19 36	22 31	29 14	15 48	23 39	26 37	20 41
05	16 58 32	13 41 00	23 ♏ 57 48	01 ♐ 10 03	21 23	19 57	22 16	29 10	15 50	23 37	26 36	20 42
06	17 02 29	14 41 56	08 ♐ 19 06	15 ♐ 24 14	20 11	20 16	22 02	29 06	15 53	23 35	26 35	20 43
07	17 06 25	15 42 52	22 ♐ 24 48	29 ♐ 20 20	18 52	20 33	21 49	29 02	15 55	23 33	26 34	20 44
08	17 10 22	16 43 49	06 ♑ 10 25	12 ♑ 54 51	17 30	20 48	21 37	28 58	15 57	23 31	26 33	20 45
09	17 14 18	17 44 47	19 ♑ 33 30	26 ♑ 06 23	16 07	21 01	21 25	28 53	15 59	23 29	26 32	20 46
10	17 18 15	18 45 46	02 ♒ 33 41	08 ♒ 55 37	14 46	21 12	21 14	28 49	16 01	23 27	26 31	20 47
11	17 22 11	19 46 46	15 ♒ 12 33	21 ♒ 24 53	13 30	21 20	21 04	28 44	16 02	23 25	26 30	20 48
12	17 26 08	20 47 46	27 ♒ 33 07	03 ♓ 37 49	12 20	21 26	20 55	28 39	16 04	23 23	26 29	20 49
13	17 30 05	21 48 46	09 ♓ 39 31	15 ♓ 38 50	11 20	21 30	20 46	28 34	16 05	23 21	26 28	20 50
14	17 34 01	22 49 47	21 ♓ 36 24	27 ♓ 32 49	10 29	21 31	20 39	28 29	16 07	23 19	26 28	20 51
15	17 37 58	23 50 49	03 ♈ 27 42	09 ♈ 24 41	09 49	21 R 30	20 32	28 23	16 08	23 17	26 27	20 52
16	17 41 54	24 51 51	15 ♈ 21 19	21 ♈ 19 10	09 21	21 26	20 26	28 18	16 09	23 15	26 26	20 53
17	17 45 51	25 52 53	27 ♈ 18 46	03 ♉ 20 34	09 03	21 20	20 21	28 12	16 10	23 13	26 25	20 54
18	17 49 47	26 53 56	09 ♉ 25 02	15 ♉ 32 30	08 57	21 12	20 17	28 06	16 11	23 10	26 25	20 55
19	17 53 44	27 55 00	21 ♉ 43 19	27 ♉ 57 44	09 D 00	21 01	20 13	28 00	16 12	23 08	26 24	20 56
20	17 57 40	28 56 04	04 ♊ 15 56	10 ♊ 38 03	09 12	20 47	20 11	27 54	16 13	23 06	26 24	20 57
21	18 01 37	29 ♐ 57 08	17 ♊ 04 08	23 ♊ 34 12	09 33	20 31	20 09	27 47	16 13	23 03	26 23	20 59
22	18 05 34	00 ♑ 58 13	00 ♋ 08 09	06 ♋ 45 55	10 02	20 12	20 08	27 41	16 14	23 01	26 22	21 00
23	18 09 30	01 59 18	13 ♋ 27 18	20 ♋ 12 06	10 37	19 51	20 D 08	27 34	16 14	22 59	26 22	21 01
24	18 13 27	03 00 24	27 ♋ 00 06	03 ♌ 51 01	11 19	19 28	20 08	27 28	16 14	22 56	26 22	21 02
25	18 17 23	04 01 30	10 ♌ 44 38	17 ♌ 40 37	12 06	19 03	20 10	27 21	16 15	22 54	26 21	21 04
26	18 21 20	05 02 37	24 ♌ 38 45	01 ♍ 38 44	12 58	18 36	20 12	27 14	16 R 15	22 52	26 21	21 05
27	18 25 16	06 03 44	08 ♍ 40 18	15 ♍ 43 12	13 54	18 06	20 14	27 07	16 15	22 49	26 20	21 06
28	18 29 13	07 04 52	22 ♍ 47 12	29 ♍ 52 01	14 54	17 36	20 18	27 00	16 14	22 47	26 20	21 08
29	18 33 09	08 06 00	06 ♎ 57 26	14 ♎ 03 09	15 58	17 03	20 22	26 53	16 14	22 44	26 20	21 09
30	18 37 06	09 07 09	21 ♎ 08 56	28 ♎ 14 28	17 04	16 30	20 27	26 45	16 14	22 42	26 20	21 10
31	18 41 03	10 ♑ 08 18	05 ♏ 19 28	12 ♏ 23 35	18 ♐ 13	15 ♑ 55	20 ♉ 33	26 ♋ 38	16 ♍ 13	22 ♋ 39	26 ♈ 19	21 ♒ 12

Date	Moon True ☊ ° '	Moon Mean ☊ ° '	Moon ☽ Latitude ° '
01	20 ♋ 26	21 ♋ 41	04 N 42
02	20 R 23	21 38	05 05
03	20 18	21 35	05 09
04	20 12	21 32	04 53
05	20 06	21 29	04 18
06	20 01	21 25	03 27
07	19 56	21 22	02 24
08	19 54	21 19	01 14
09	19 53	21 16	00 N 02
10	19 D 54	21 13	01 S 09
11	19 55	21 10	02 14
12	19 57	21 06	03 12
13	19 59	21 03	04 00
14	19 59	21 00	04 36
15	19 R 59	20 57	05 01
16	19 57	20 54	05 13
17	19 54	20 50	05 11
18	19 51	20 47	04 56
19	19 47	20 44	04 27
20	19 43	20 41	03 44
21	19 41	20 38	02 50
22	19 39	20 35	01 45
23	19 38	20 31	00 S 34
24	19 D 38	20 28	00 N 41
25	19 38	20 25	01 54
26	19 40	20 22	03 02
27	19 41	20 19	03 59
28	19 42	20 16	04 42
29	19 42	20 12	05 09
30	19 R 42	20 09	05 16
31	19 ♋ 41	20 ♋ 06	05 N 05

DECLINATIONS

Date	Sun ☉ ° '	Moon ☽ ° '	Mercury ☿ ° '	Venus ♀ ° '	Mars ♂ ° '	Jupiter ♃ ° '	Saturn ♄ ° '	Uranus ♅ ° '	Neptune ♆ ° '	Pluto ♇ ° '
01	21 S 53	05 N 55	24 S 10	25 S 00	19 N 50	20 N 32	07 N 17	21 N 50	08 N 39	23 S 08
02	22 02	00 N 35	23 51	24 50	19 48	20 33	07 16	21 50	08 38	23 07
03	22 11	04 S 50	23 31	24 40	19 46	20 34	07 15	21 51	08 38	23 07
04	22 19	10 00	23 09	24 29	19 44	20 35	07 14	21 51	08 38	23 07
05	22 26	14 35	22 45	24 18	19 42	20 36	07 14	21 51	08 37	23 06
06	22 33	18 17	22 19	24 07	19 41	20 37	07 13	21 52	08 37	23 06
07	22 40	20 49	21 53	23 56	19 39	20 38	07 12	21 52	08 37	23 05
08	22 46	22 03	21 27	23 44	19 38	20 39	07 12	21 52	08 36	23 05
09	22 52	21 59	21 00	23 33	19 37	20 40	07 11	21 53	08 36	23 04
10	22 57	20 42	20 35	23 21	19 36	20 41	07 11	21 53	08 36	23 04
11	23 02	18 24	20 12	23 09	19 35	20 42	07 10	21 53	08 36	23 03
12	23 07	15 19	19 51	22 58	19 34	20 43	07 10	21 54	08 35	23 03
13	23 11	11 39	19 33	22 46	19 33	20 44	07 10	21 54	08 35	23 03
14	23 14	07 34	19 19	22 34	19 32	20 46	07 09	21 54	08 35	23 02
15	23 17	03 S 13	19 08	22 22	19 32	20 47	07 09	21 55	08 34	23 02
16	23 20	01 N 14	19 00	22 10	19 32	20 48	07 09	21 55	08 34	23 01
17	23 22	05 40	18 56	21 58	19 32	20 49	07 09	21 56	08 34	23 01
18	23 24	09 57	18 55	21 46	19 32	20 51	07 09	21 56	08 34	23 00
19	23 25	13 54	18 58	21 34	19 32	20 52	07 09	21 56	08 34	23 00
20	23 26	17 19	19 03	21 22	19 33	20 54	07 09	21 57	08 34	22 59
21	23 26	19 59	19 10	21 10	19 33	20 55	07 09	21 57	08 33	22 59
22	23 26	21 41	19 19	20 58	19 34	20 56	07 09	21 58	08 33	22 58
23	23 25	22 11	19 30	20 46	19 35	20 58	07 09	21 58	08 33	22 57
24	23 24	21 25	19 42	20 34	19 36	20 59	07 09	21 58	08 33	22 57
25	23 22	19 22	19 55	20 22	19 37	21 01	07 09	21 59	08 33	22 56
26	23 20	16 09	20 09	20 10	19 39	21 02	07 09	21 59	08 33	22 56
27	23 18	12 01	20 24	19 59	19 40	21 04	07 10	22 00	08 33	22 55
28	23 15	07 11	20 39	19 47	19 42	21 05	07 10	22 00	08 33	22 55
29	23 11	01 N 58	20 54	19 36	19 44	21 07	07 10	22 01	08 33	22 54
30	23 07	03 S 21	21 09	19 24	19 46	21 09	07 11	22 01	08 33	22 54
31	23 S 03	08 S 30	21 S 23	19 S 13	19 N 48	21 N 10	07 N 11	22 N 01	08 N 33	22 S 53

ZODIAC SIGN ENTRIES

Date	h	m	Planets
01	18	43	☽ ♎
03	20	26	☽ ♏
05	22	03	☽ ♐
08	01	09	☽ ♑
10	07	13	☽ ♒
12	16	49	☽ ♓
15	04	58	☽ ♈
17	17	21	☽ ♉
20	03	54	☽ ♊
21	13	08	☉ ♑
22	11	45	☽ ♋
24	17	16	☽ ♌
26	21	11	☽ ♍
29	00	14	☽ ♎
31	02	59	☽ ♏

LATITUDES

Date	Mercury ☿ ° '	Venus ♀ ° '	Mars ♂ ° '	Jupiter ♃ ° '	Saturn ♄ ° '	Uranus ♅ ° '	Neptune ♆ ° '	Pluto ♇ ° '
01	00 S 51	02 S 52	01 N 16	00 N 16	01 N 46	00 N 29	01 S 45	09 S 01
04	00 N 05	02 30	01 23	00 17	01 47	00 29	01 45	09 00
07	01 05	02 05	01 29	00 17	01 48	00 29	01 45	09 00
10	01 59	01 37	01 35	00 18	01 48	00 30	01 45	09 00
13	02 36	01 04	01 40	00 18	01 49	00 30	01 45	08 59
16	02 53	00 S 27	01 45	00 19	01 50	00 30	01 45	08 59
19	02 52	00 N 14	01 48	00 19	01 51	00 30	01 44	08 58
22	02 40	00 58	01 52	00 20	01 52	00 30	01 44	08 58
25	02 20	01 44	01 55	00 20	01 52	00 30	01 44	08 58
28	01 57	02 31	01 57	00 21	01 53	00 30	01 44	08 57
31	01 N 32	03 N 17	01 N 59	00 N 21	01 N 54	00 N 30	01 S 44	08 S 57

DATA

Julian Date	2465394
Delta T	+76 seconds
Ayanamsa	24° 22' 55"
Synetic vernal point	04° ♓ 44' 03"
True obliquity of ecliptic	23° 26' 00"

MOON'S PHASES, APSIDES AND POSITIONS ☽

Date	h	m	Phase	Longitude ° '	Eclipse Indicator
06	23	38	●	15 ♐ 11	
14	14	42	☽	22 ♓ 57	
22	13	39	○	01 ♋ 02	
29	14	05	☾	08 ♎ 11	

Day	h	m		
03	21	16	Perigee	
15	16	32	Apogee	
29	18	47	Perigee	
02	14	33	0S	
08	22	31	Max dec	22° S 11'
16	05	24	0N	
23	09	46	Max dec	22° N 12'
29	20	51	0S	

LONGITUDES

Date	Chiron ⚷ ° '	Ceres ⚳ ° '	Pallas ⚴ ° '	Juno ⚵ ° '	Vesta ⚶ ° '	Black Moon Lilith ⚸ ° '
01	24 ♊ 02	13 ♐ 31	15 ♏ 45	06 ♏ 14	26 ♈ 34	06 ♈ 07
11	23 ♊ 26	17 ♐ 41	20 ♏ 03	09 ♏ 27	26 ♈ 06	07 ♈ 14
21	22 ♊ 49	21 ♐ 51	24 ♏ 15	12 ♏ 30	26 ♈ 21	08 ♈ 21
31	22 ♊ 13	25 ♐ 58	28 ♏ 21	15 ♏ 23	27 ♈ 21	09 ♈ 29

All ephemeris data is given at 12.00 UT and the Moon's longitude is additionally given for 24.00 UT

ASPECTARIAN

h m Aspects

01 Tuesday
02 41 ☉ # ♅
02 59 ☽ ⊼ ♇
03 02 ☽ ± ♆
05 37 ☽ ∥ ♄
07 36 ☽ △ ♂
08 12 ☽ ✶ ♅
09 48 ☽ □ ☿
13 06 ☽ ± ♇
13 07 ☽ ⊼ ♆
14 57 ☽ Q ☉
17 41 ☽ ✶ ♃

02 Wednesday
04 09 ☽ ⚼ ♇
04 15 ☽ Q ♅
08 12 ☽ ⚼ ♂
12 29 ☽ ✶ ☉
13 36 ☽ Q ♃
14 45 ☽ Q ☿
20 53 ☽ ⊻ ♄
22 51 ☽ ± ♂

03 Thursday
02 15 ☿ ⊼ ♅
02 24 ☽ □ ♀
05 03 ☽ △ ♇
06 50 ☽ ⊥ ♄
08 35 ☽ ⊼ ♂
09 38 ☽ ✶ ☿
10 01 ☽ □ ♅
11 23 ☉ ⚼ ♆
14 27 ☿ ± ♃
14 52 ☽ ☍ ♆
15 08 ☽ ∠ ☉
19 14 ☽ □ ♃
21 43 ☽ ∠ ♄
22 57 ☽ # ♄

04 Friday
05 25 ☽ # ♆
07 05 ☽ ⊥ ☉
08 53 ☽ Q ♀
08 55 ☽ ∠ ☿
09 57 ☿ ⊼ ♂
13 57 ☿ ∥ ♇
17 43 ☽ ⊻ ☉
22 31 ☽ ✶ ♄
22 48 ☽ ⊥ ☿

05 Saturday
05 12 ☽ ✶ ♀
06 35 ☽ □ ♇
08 02 ☽ ⊻ ☿
09 14 ☽ ☍ ♂
11 25 ☽ △ ♅
16 22 ☽ ⊼ ♆
18 28 ☽ Q ♄
20 37 ☽ △ ♃
21 31 ☉ ⊥ ♀
22 43 ☉ ⚼ ♃

06 Sunday
01 37 ☉ ∥ ☿
01 44 ☿ ✶ ♇
02 22 ☽ ± ♆
06 46 ☽ ∠ ♀
10 31 ☿ ⊻ ♀
12 27 ☽ ⚼ ♅
12 40 ☽ Q ♇
17 30 ☽ ⚼ ♆
21 44 ☽ ⚼ ♃
22 17 ☽ ⊥ ♀
23 38 ☽ ☌ ☉
23 40 ☽ # ♂

07 Monday
00 50 ☽ □ ♄
03 41 ☽ ± ♅
06 27 ☽ ☌ ☿
08 45 ☽ ⊻ ♀
09 06 ☽ ✶ ♇
09 39 ☽ # ♃
10 59 ☽ ⊼ ♂
13 04 ☿ # ♅
13 04 ☽ ± ♃
13 58 ☽ ⊼ ♅
16 52 ☉ □ ♄
19 09 ☽ △ ♆
21 12 ☽ ± ♂
23 25 ☽ ⊼ ♃

08 Tuesday
00 53 ☽ ∥ ☿
05 19 ♀ ⊻ ♇
05 59 ☽ # ♅
11 14 ☽ ∠ ♇
11 39 ☿ ± ♅
12 45 ☽ ⚼ ♂
19 42 ☉ ☌ ☿

09 Wednesday
03 18 ☽ ⊥ ♇
05 30 ☽ △ ♄
06 06 ☉ ± ♅
06 21 ☽ ⊻ ☿
08 26 ☽ ⊻ ☉
14 12 ☽ ⊻ ♇
14 23 ☿ □ ♄
14 42 ☽ ☌ ♀
15 08 ☽ # ♅
15 20 ☽ △ ♂
16 14 ☽ ⊥ ☿
19 10 ☽ ☍ ♅
20 18 ☽ ⊥ ☉

10 Thursday
00 46 ☽ □ ♆
05 03 ☽ ☍ ♃
05 07 ☿ ⊥ ♀
06 50 ☿ # ♃
07 16 ☽ ∠ ☿
09 06 ☽ ⚼ ♄
12 16 ☽ # ♃
13 51 ☽ ∥ ☿
14 27 ☽ ∠ ☉
14 37 ♀ △ ♂

11 Friday
00 52 ☽ # ♂
02 05 ☽ ± ♄
07 04 ☿ ⚼ ♃
09 00 ☽ ✶ ☿
10 38 ☽ Q ♆
13 36 ☽ ⊼ ♄
17 09 ☉ ∥ ♇
21 37 ☽ ✶ ☉
22 24 ☉ ∥ ♀
22 48 ☽ ☌ ♇
23 11 ☽ □ ♂
23 57 ☽ ⊻ ♀

12 Saturday
00 43 ♀ ∥ ♇
03 52 ☽ ⊼ ♅
06 13 ☽ Q ☿
09 54 ☽ ✶ ♆
11 46 ☽ ⊥ ♀
12 19 ☉ ✶ ♇
14 09 ☽ ⊼ ♃
14 22 ☉ ⊼ ♂
15 36 ☽ ± ♅
23 18 ☽ Q ☉

13 Sunday
01 56 ☽ ± ♃
03 16 ♂ □ ♇
04 14 ☉ ⊻ ♀
05 41 ☽ ∠ ♀
08 16 ☿ ⚼ ♆
09 24 ☽ ⚼ ♅
10 15 ☽ Q ♂
12 18 ☿ # ♂
15 06 ☽ □ ☿
15 37 ☽ ∠ ♆
19 46 ☽ ⚼ ♃

14 Monday
00 55 ☽ ☍ ♄
04 20 ☉ ± ♃
06 11 ☽ # ♆
09 41 ☽ ⊥ ♆
10 05 ☽ ✶ ♂
10 28 ☽ ⊻ ♇
11 50 ☽ ✶ ♀
12 46 ♀ St R
14 17 ☽ # ♄
14 42 ☽ □ ☉
15 27 ☽ △ ♅
21 47 ☽ ⊻ ♆
22 36 ☽ ⊥ ♇
23 08 ☉ ⊼ ♅

15 Tuesday
01 47 ☽ △ ♃
12 03 ☽ Q ♀
16 07 ☽ ∠ ♂
16 50 ☽ ∠ ♇

16 Wednesday
00 18 ☽ △ ☿
10 09 ☽ ⊥ ♂
13 37 ☽ ⊼ ♄
22 09 ☽ ⊻ ♂
23 08 ☽ ✶ ♇

17 Thursday
00 09 ☽ □ ♀
01 41 ☽ ± ♄
03 49 ☽ □ ♅
05 36 ☽ ⚼ ☿
08 53 ☽ △ ☉
10 14 ☽ ☌ ♆
13 45 ☽ □ ♃
16 36 ♀ # ♅
19 42 ☽ ⚼ ♄
20 09 ☽ ∥ ♄
22 16 ☉ ± ♂
23 08 ☽ Q ♇
23 17 ☽ ± ☿

18 Friday
00 39 ☉ △ ♆
04 07 ☽ ∥ ♆
11 04 ☽ ⊼ ☿
15 26 ☽ Q ♅
16 13 ☿ St D
17 19 ☽ ⚼ ☉

19 Saturday
00 59 ☽ Q ♃
01 16 ☽ △ ♄
09 06 ☽ ☌ ♂
10 29 ☽ □ ♇
10 39 ☽ △ ♀
12 25 ☽ ± ☉
13 46 ☉ ⊼ ♃
14 43 ☽ ✶ ♅
19 18 ♀ ⊻ ♇
21 00 ☽ ⊻ ♆
23 58 ☽ ✶ ♃

20 Sunday
00 58 ☽ ⊼ ☉
08 27 ☽ ⊥ ♆
14 49 ☽ ⚼ ♀
19 13 ☽ ∠ ♅
21 33 ☽ ☍ ☿

21 Monday
01 25 ☽ ∠ ♆
03 16 ☽ # ☿
04 06 ☽ ∠ ♃
07 21 ☽ ± ♀
07 30 ☽ ∥ ♂
10 25 ☽ □ ♄
11 59 ☽ ⊥ ♅
17 41 ☽ ⊻ ♂
18 14 ☽ ⊼ ♀
19 14 ☽ △ ♇
20 39 ☽ ⊥ ♃
23 02 ☽ ⊻ ♅
23 27 ☽ ∥ ♃

22 Tuesday
01 15 ☽ # ♀
04 42 ☽ ⊥ ♂
05 09 ☽ ✶ ♆
07 34 ☽ ⊻ ♃
13 39 ☽ ☍ ☉
14 34 ♀ # ♃
17 20 ♀ △ ♂
19 16 ☽ ∥ ♅
19 26 ☽ Q ♄
21 03 ☽ ∠ ♂
22 38 ☽ ⚼ ♇

23 Wednesday
02 53 ☽ Q ♆
06 31 ♂ St D
06 40 ☽ ⊼ ☿
14 48 ☽ ± ♇
16 58 ☽ ✶ ♄
17 56 ☽ ± ☿
23 05 ☽ ☍ ♀
23 18 ☿ # ♂
23 52 ☽ ✶ ♂
23 53 ☽ ∥ ♅

24 Thursday
01 28 ☽ ⊼ ♇
04 52 ☽ ☌ ♅
10 43 ☽ ⚼ ☿
10 52 ☽ □ ♆
12 48 ☽ ☌ ♃
13 27 ☿ ⚼ ♆
18 22 ☽ ∥ ♃
19 26 ☽ ∠ ♄
21 01 ☽ Q ♂
23 22 ☽ ⊼ ☉

25 Friday
01 01 ☽ # ♀
07 11 ☽ # ☿
09 39 ☽ ∥ ♂
10 39 ☽ ± ☉
11 08 ☽ ⊥ ♄
14 30 ☽ △ ☿
18 16 ☿ ⚼ ♃
21 31 ☽ ⊻ ♄

26 Saturday
01 55 ☽ ⊼ ♀
03 27 ☽ ⚼ ☉
04 19 ☽ □ ♂
05 24 ☿ ⊥ ♀
05 52 ☽ ☍ ♇
08 56 ☽ ⊻ ♅
11 36 ♄ St R
11 55 ☽ ± ♀
12 57 ☿ ∥ ♀
14 55 ☽ △ ♆
15 41 ☉ ⚼ ♂
16 24 ☽ ⊻ ♃
19 13 ☽ ⊥ ♅

27 Sunday
02 36 ☽ ⊥ ♃
02 49 ☽ ⚼ ♀
07 12 ☽ △ ☉
10 33 ☽ ∠ ♅
13 02 ☉ ∠ ♇
16 33 ☽ ⚼ ♆
17 49 ☽ ∠ ♃
21 35 ☽ □ ☿

28 Monday
00 53 ☽ ☌ ♄
03 30 ☽ △ ♀
05 29 ☽ ∥ ♆
07 46 ☽ △ ♂
07 50 ☽ ± ♆
09 11 ☽ ⊼ ♇
11 59 ☽ ✶ ♅
12 06 ☽ ∥ ♄
18 01 ☽ ⊼ ♆
19 05 ☽ ✶ ♃
19 22 ☽ ± ♇
20 57 ♀ # ♂

29 Tuesday
06 31 ☽ Q ☿
08 15 ☽ Q ♅
09 18 ☽ ⚼ ♂
10 38 ☽ ⚼ ♇
14 05 ☽ □ ☉
15 13 ☽ Q ♃
17 54 ☿ □ ♄

30 Wednesday
00 37 ☽ ± ♂
03 41 ☽ ⊻ ♄
03 46 ☿ ⊻ ♀
04 12 ☿ ± ♅
04 26 ☽ □ ♀
04 31 ☽ ✶ ☿
10 49 ☽ ⊼ ♂
11 57 ☿ # ♃
12 02 ☽ △ ♇
13 50 ☽ ⊥ ♄
14 36 ☽ □ ♅
20 45 ☽ ☍ ♆
21 24 ☽ □ ♃
22 53 ☽ Q ☉
23 11 ♀ △ ♄

31 Thursday
05 03 ☽ ∠ ♄
05 42 ☽ # ♄
08 07 ☽ ∠ ☿
09 42 ☽ Q ♀
12 13 ☽ # ♆
20 48 ☽ ✶ ☉

JANUARY 2038

Date	Sidereal time h m s	LONGITUDES Sun ☉ ° ' "	Moon ☽ ° ' "	Moon ☽ 24.00 ° ' "	Mercury ☿ ° '	Venus ♀ ° '	Mars ♂ ° '	Jupiter ♃ ° '	Saturn ♄ ° '	Uranus ♅ ° '	Neptune ♆ ° '	Pluto ♇ ° '
01	18 44 59	11 ♑ 09 28	19 ♏ 26 28	26 ♏ 27 47	19 ♐ 25	15 ♑ 19	20 ♉ 40	26 ♋ 30	16 ♍ 13	22 ♋ 37	26 ♈ 19	21 ♒ 13
02	18 48 56	12 10 38	03 ♐ 27 09	10 ♐ 24 11	20 38	14 R 43	20 47	26 R 23	16 R 12	22 R 34	26 R 19	21 15
03	18 52 52	13 11 48	17 ♐ 18 32	24 ♐ 09 51	21 54	14 07	20 55	26 15	16 11	22 32	26 19	21 16
04	18 56 49	14 12 59	00 ♑ 57 47	07 ♑ 42 04	23 11	13 30	21 03	26 07	16 10	22 29	26 19	21 18
05	19 00 45	15 14 10	14 ♑ 22 28	20 ♑ 58 47	24 29	12 54	21 12	25 59	16 09	22 26	26 19	21 19
06	19 04 42	16 15 20	27 ♑ 30 54	03 ♒ 58 46	25 49	12 17	21 22	25 52	16 08	22 24	26 D 19	21 21
07	19 08 38	17 16 31	10 ♒ 22 23	16 ♒ 41 51	27 10	11 42	21 32	25 44	16 07	22 21	26 19	21 22
08	19 12 35	18 17 41	22 ♒ 57 18	29 ♒ 08 58	28 32	11 07	21 43	25 36	16 05	22 19	26 19	21 24
09	19 16 32	19 18 51	05 ♓ 17 09	11 ♓ 22 10	29 ♐ 55	10 34	21 55	25 28	16 04	22 16	26 19	21 25
10	19 20 28	20 20 00	17 ♓ 24 27	23 ♓ 24 26	01 ♑ 19	10 02	22 07	25 20	16 02	22 13	26 19	21 27
11	19 24 25	21 21 09	29 ♓ 22 36	05 ♈ 19 29	02 43	09 31	22 20	25 12	16 01	22 11	26 19	21 28
12	19 28 21	22 22 17	11 ♈ 15 38	17 ♈ 11 38	04 09	09 02	22 33	25 04	15 59	22 08	26 19	21 30
13	19 32 18	23 23 25	23 ♈ 08 05	29 ♈ 05 33	05 35	08 35	22 47	24 56	15 57	22 06	26 20	21 31
14	19 36 14	24 24 32	05 ♉ 04 39	11 ♉ 05 58	07 02	08 10	23 02	24 47	15 55	22 03	26 20	21 33
15	19 40 11	25 25 38	17 ♉ 10 04	23 ♉ 17 28	08 30	07 47	23 17	24 39	15 53	22 00	26 20	21 35
16	19 44 07	26 26 44	29 ♉ 28 43	05 ♊ 44 14	09 58	07 26	23 32	24 31	15 51	21 58	26 21	21 36
17	19 48 04	27 27 49	12 ♊ 04 24	18 ♊ 29 35	11 27	07 08	23 48	24 23	15 48	21 55	26 21	21 38
18	19 52 01	28 28 53	24 ♊ 59 58	01 ♋ 35 45	12 57	06 52	24 05	24 15	15 46	21 53	26 21	21 39
19	19 55 57	29 ♑ 29 57	08 ♋ 16 57	15 ♋ 03 30	14 27	06 39	24 22	24 07	15 43	21 50	26 22	21 41
20	19 59 54	00 ♒ 31 00	21 ♋ 55 14	28 ♋ 51 52	15 58	06 28	24 39	23 59	15 41	21 47	26 22	21 43
21	20 03 50	01 32 03	05 ♌ 52 59	12 ♌ 58 06	17 29	06 19	24 57	23 51	15 38	21 45	26 23	21 44
22	20 07 47	02 33 04	20 ♌ 06 38	27 ♌ 17 54	19 02	06 13	25 15	23 43	15 35	21 42	26 23	21 46
23	20 11 43	03 34 05	04 ♍ 31 14	11 ♍ 45 54	20 34	06 09	25 34	23 35	15 32	21 40	26 24	21 48
24	20 15 40	04 35 06	19 ♍ 01 09	26 ♍ 16 19	22 07	06 D 08	25 53	23 28	15 30	21 37	26 25	21 50
25	20 19 36	05 36 05	03 ♎ 30 43	10 ♎ 43 47	23 41	06 10	26 12	23 20	15 26	21 35	26 25	21 51
26	20 23 33	06 37 05	17 ♎ 55 00	25 ♎ 03 55	25 16	06 14	26 32	23 12	15 23	21 32	26 26	21 53
27	20 27 30	07 38 04	02 ♏ 10 12	09 ♏ 13 35	26 51	06 20	26 53	23 05	15 20	21 30	26 27	21 55
28	20 31 26	08 39 02	16 ♏ 13 53	23 ♏ 10 58	28 ♑ 26	06 28	27 13	22 57	15 17	21 27	26 27	21 56
29	20 35 23	09 40 00	00 ♐ 04 47	06 ♐ 55 19	00 ♒ 03	06 39	27 34	22 50	15 13	21 25	26 28	21 58
30	20 39 19	10 40 57	13 ♐ 42 33	20 ♐ 26 31	01 40	06 52	27 56	22 42	15 10	21 22	26 29	22 00
31	20 43 16	11 ♒ 41 53	27 ♐ 07 16	03 ♑ 44 50	03 ♒ 17	07 ♑ 07	28 ♉ 17	22 ♋ 35	15 ♍ 06	21 ♋ 20	26 ♈ 30	22 ♒ 02

Date	Moon True ☊ ° '	Moon Mean ☊ ° '	Moon ☽ Latitude ° '
01	19 ♋ 40	20 ♋ 03	04 N 35
02	19 R 39	20 00	03 49
03	19 38	19 56	02 50
04	19 38	19 53	01 42
05	19 37	19 50	00 N 29
06	19 D 37	19 47	00 S 44
07	19 38	19 44	01 53
08	19 38	19 41	02 55
09	19 38	19 37	03 47
10	19 R 38	19 34	04 29
11	19 38	19 31	04 58
12	19 38	19 28	05 14
13	19 37	19 25	05 17
14	19 D 38	19 21	05 06
15	19 38	19 18	04 41
16	19 39	19 15	04 04
17	19 39	19 12	03 14
18	19 40	19 09	02 13
19	19 41	19 06	01 S 03
20	19 R 41	19 02	00 N 12
21	19 41	18 59	01 29
22	19 39	18 56	02 41
23	19 38	18 53	03 43
24	19 36	18 50	04 32
25	19 34	18 47	05 03
26	19 32	18 43	05 15
27	19 32	18 40	05 08
28	19 D 32	18 37	04 42
29	19 33	18 34	04 00
30	19 34	18 31	03 05
31	19 ♋ 36	18 ♋ 27	02 N 01

DECLINATIONS

Date	Sun ☉ ° '	Moon ☽ ° '	Mercury ☿ ° '	Venus ♀ ° '	Mars ♂ ° '	Jupiter ♃ ° '	Saturn ♄ ° '	Uranus ♅ ° '	Neptune ♆ ° '	Pluto ♇ ° '
01	22 S 58	13 S 10	21 S 38	19 S 02	19 N 50	21 N 12	07 N 12	22 N 02	08 N 33	22 S 53
02	22 53	17 06	21 52	18 52	19 53	21 13	07 12	22 02	08 33	22 52
03	22 47	20 01	22 05	18 41	19 56	21 15	07 13	22 03	08 33	22 51
04	22 40	21 44	22 18	18 31	19 58	21 16	07 13	22 03	08 33	22 51
05	22 34	22 11	22 30	18 21	20 01	21 18	07 14	22 04	08 33	22 50
06	22 27	21 22	22 41	18 12	20 04	21 20	07 15	22 04	08 33	22 50
07	22 19	19 27	22 52	18 03	20 07	21 21	07 15	22 04	08 33	22 49
08	22 11	16 37	23 01	17 55	20 11	21 23	07 16	22 05	08 33	22 49
09	22 03	13 05	23 10	17 46	20 14	21 25	07 17	22 05	08 33	22 48
10	21 54	09 06	23 17	17 39	20 18	21 26	07 18	22 06	08 33	22 47
11	21 44	04 48	23 24	17 32	20 21	21 28	07 19	22 06	08 33	22 47
12	21 35	00 S 22	23 29	17 25	20 25	21 29	07 20	22 07	08 33	22 46
13	21 24	04 N 05	23 34	17 19	20 29	21 31	07 21	22 07	08 33	22 46
14	21 14	08 24	23 37	17 13	20 33	21 33	07 22	22 07	08 34	22 45
15	21 03	12 27	23 39	17 08	20 37	21 34	07 23	22 08	08 34	22 45
16	20 52	16 04	23 40	17 04	20 41	21 36	07 24	22 08	08 34	22 44
17	20 40	19 02	23 39	17 00	20 45	21 37	07 25	22 09	08 34	22 43
18	20 28	21 08	23 38	16 56	20 49	21 39	07 26	22 09	08 34	22 43
19	20 15	22 08	23 35	16 54	20 53	21 40	07 27	22 10	08 34	22 42
20	20 02	21 51	23 30	16 51	20 58	21 42	07 28	22 10	08 35	22 42
21	19 49	20 14	23 25	16 49	21 02	21 43	07 30	22 10	08 35	22 41
22	19 35	17 19	23 18	16 48	21 07	21 45	07 31	22 11	08 35	22 40
23	19 21	13 19	23 10	16 47	21 11	21 46	07 32	22 11	08 35	22 40
24	19 07	08 31	23 00	16 47	21 16	21 48	07 34	22 12	08 36	22 39
25	18 52	03 N 15	22 49	16 47	21 21	21 49	07 35	22 12	08 36	22 39
26	18 37	02 S 10	22 37	16 47	21 25	21 51	07 36	22 12	08 36	22 38
27	18 21	07 24	22 23	16 48	21 30	21 52	07 38	22 13	08 37	22 38
28	18 06	12 11	22 08	16 49	21 35	21 53	07 39	22 13	08 37	22 37
29	17 50	16 15	21 52	16 50	21 40	21 55	07 41	22 14	08 37	22 36
30	17 33	19 23	21 34	16 52	21 45	21 56	07 42	22 14	08 38	22 36
31	17 S 16	21 S 23	21 S 15	16 S 53	21 N 49	21 N 57	07 N 44	22 N 14	08 N 38	22 S 35

ZODIAC SIGN ENTRIES

Date	h	m	Planets
02	06	04	☽ ♐
04	10	18	☽ ♑
06	16	36	☽ ♒
09	01	39	☽ ♓
09	13	32	☿ ♑
11	13	15	☽ ♈
14	01	49	☽ ♉
16	13	00	☽ ♊
18	21	07	☽ ♋
19	23	49	☉ ♒
21	01	57	☽ ♌
23	04	30	☽ ♍
25	06	11	☽ ♎
27	08	20	☽ ♏
29	11	22	☿ ♒
29	11	52	☽ ♐
31	17	12	☽ ♑

LATITUDES

Date	Mercury ☿ ° '	Venus ♀ ° '	Mars ♂ ° '	Jupiter ♃ ° '	Saturn ♄ ° '	Uranus ♅ ° '	Neptune ♆ ° '	Pluto ♇ ° '
01	01 N 23	03 N 32	02 N 00	00 N 21	01 N 54	00 N 30	01 S 44	08 S 57
04	00 58	04 15	02 02	00 22	01 55	00 30	01 44	08 57
07	00 32	04 53	02 03	00 22	01 56	00 30	01 43	08 56
10	00 N 08	05 25	02 04	00 23	01 56	00 30	01 43	08 56
13	00 S 15	05 51	02 05	00 23	01 57	00 30	01 43	08 56
16	00 36	06 10	02 06	00 23	01 58	00 30	01 43	08 56
19	00 56	06 23	02 06	00 24	01 59	00 30	01 43	08 56
22	01 14	06 30	02 06	00 24	01 59	00 30	01 43	08 56
25	01 29	06 31	02 06	00 24	02 00	00 30	01 42	08 56
28	01 42	06 28	02 06	00 25	02 01	00 30	01 42	08 56
31	01 S 53	06 N 22	02 N 06	00 N 25	02 N 01	00 N 30	01 S 42	08 S 56

DATA

Julian Date	2465425
Delta T	+77 seconds
Ayanamsa	24° 23' 01"
Synetic vernal point	04° ♓ 43' 58"
True obliquity of ecliptic	23° 26' 00"

MOON'S PHASES, APSIDES AND POSITIONS ☽

Date	h	m	Phase	Longitude ° '	Eclipse Indicator
05	13	41	●	15 ♑ 18	Annular
13	12	34	☽	23 ♈ 25	
21	04	00	○	01 ♌ 12	
27	22	00	☾	08 ♏ 03	

Day	h	m		
12	13	58	Apogee	
24	10	04	Perigee	
05	08	14	Max dec	22° S 12'
12	13	56	0N	
19	19	03	Max dec	22° N 11'
26	02	23	0S	

LONGITUDES

Date	Chiron ⚷ ° '	Ceres ⚳ ° '	Pallas ⚴ ° '	Juno ⚵ ° '	Vesta ⚶ ° '	Black Moon Lilith ⚸ ° '
01	22 ♊ 09	26 ♐ 22	28 ♏ 45	15 ♏ 40	27 ♈ 29	09 ♈ 35
11	21 ♊ 36	00 ♑ 26	02 ♐ 42	18 ♏ 19	29 ♈ 06	10 ♈ 42
21	21 ♊ 08	04 ♑ 25	06 ♐ 30	20 ♏ 43	01 ♉ 14	11 ♈ 50
31	20 ♊ 46	08 ♑ 20	10 ♐ 06	22 ♏ 49	03 ♉ 46	12 ♈ 57

All ephemeris data is given at 12.00 UT and the Moon's longitude is additionally given for 24.00 UT

ASPECTARIAN

h m Aspects

01 Friday
00 47 ☽ ⊥ ☿
05 16 ☽ ⚹ ♀
06 30 ☽ ⚹ ♄
11 57 ☽ ⊻ ☿
14 06 ☽ ☍ ♂
15 02 ☽ □ ♇
15 56 ♀ ⊥ ♇
17 24 ☽ △ ♅
23 45 ☽ ⊼ ♆
23 58 ☽ △ ♃

02 Saturday
00 23 ☽ ∠ ☉
02 59 ☽ Q ♄
05 51 ☽ ∠ ♀
07 25 ☿ ± ♃
10 03 ☽ ± ♆
14 35 ☉ ∥ ♇
15 01 ☿ ⊼ ♂
17 04 ☽ ⊥ ☉
19 05 ☽ bQ ♅
20 42 ☽ ⊥ ♀
22 01 ☽ Q ♇
23 48 ♃ □ ♆
23 52 ☿ ⚹ ♇

03 Sunday
00 38 ☽ ∥ ♀
01 34 ☽ bQ ♃
01 35 ☽ bQ ♆
04 16 ☽ ⊻ ☉
06 40 ☽ ⊻ ♀
07 26 ☿ # ♅
10 03 ☽ □ ♄
10 38 ☽ ± ♅
11 07 ☽ # ♂
17 05 ☽ ± ♃
18 21 ☽ ⊼ ♂
18 56 ☽ ⚹ ♇
20 50 ☽ ☌ ☿
21 06 ☽ ⊼ ♅
23 30 ☿ ⊼ ♅

04 Monday
01 27 ☉ ☌ ♀
03 25 ☽ # ♃
03 31 ☽ ⊼ ♃
03 47 ☽ △ ♆
05 00 ☽ ± ♂
21 07 ☽ # ♅
21 09 ☽ bQ ♂
21 30 ☽ ∠ ♇

05 Tuesday
09 26 ☽ ☌ ♀
13 41 ☽ ☌ ☉
13 42 ☽ ⊥ ♇
13 57 ☉ ⊥ ♇
15 13 ☽ △ ♄
16 51 ☉ ∥ ☿
19 08 ☽ # ♅

06 Wednesday
00 34 ☽ △ ♂
00 38 ☽ ⊻ ♇
02 37 ☽ ☍ ♅
05 35 ♆ St D
08 00 ♂ □ ♇
08 30 ☽ ⊻ ☿
08 59 ☽ ☍ ♃
09 09 ☉ △ ♄
09 47 ☽ □ ♆
12 38 ☽ # ♃
12 47 ☿ ⊼ ♃
18 41 ☽ bQ ♄
20 53 ☽ ⊥ ☿
20 57 ☿ △ ♆

07 Thursday
05 00 ☽ # ♂
06 04 ☿ ∥ ♇
11 30 ☽ ± ♄
14 24 ☽ ⊻ ♀
15 47 ☽ ∠ ☿
19 28 ☽ Q ♆
19 41 ☿ ± ♂
22 52 ☽ ⊼ ♄

08 Friday
01 18 ☽ ∥ ♀
01 18 ☽ ⊥ ♀
02 16 ☽ ⊻ ☉
08 59 ☽ ☌ ♇
09 35 ☽ □ ♂
10 46 ☽ ⊼ ♅
14 49 ☽ ⊥ ☉
17 03 ☽ ⊼ ♃
17 51 ☽ ∠ ♀
18 30 ☽ ⚹ ♆
22 20 ☽ ± ♅

09 Saturday
00 09 ☽ ⚹ ☿
04 35 ☽ ± ♃
04 45 ☉ # ♅
09 55 ☽ ∠ ☉
15 53 ☽ bQ ♅
21 17 ☽ Q ♂
21 58 ☽ ⚹ ♀
22 06 ☽ bQ ♃
23 54 ☽ ∠ ♆

10 Sunday
02 47 ☽ Q ☿
09 17 ☽ ☍ ♄
15 06 ☽ # ♆
17 49 ☽ ⊥ ♆
18 23 ☽ ⚹ ☉
20 05 ☽ ⊻ ♇
20 51 ☽ Q ♀
21 35 ☽ ⚹ ♂
21 36 ☽ △ ♅
21 50 ♂ ⚹ ♅
22 10 ☽ # ♄

11 Monday
03 41 ☽ △ ♃
05 51 ☽ ⊻ ♆
08 09 ☽ ⊥ ♇
14 50 ☉ ⊻ ♇
19 40 ☽ □ ☿
20 46 ☽ Q ☉

12 Tuesday
02 21 ☽ ∠ ♇
04 22 ☽ ∠ ♂
06 42 ☉ ☍ ♅
07 40 ☽ □ ♀
17 36 ☉ △ ♂
21 31 ☽ ⊼ ♄
22 55 ☉ # ♃
22 55 ☽ ⊥ ♂

13 Wednesday
08 44 ☽ ⚹ ♇
09 37 ☽ ± ♄
09 54 ☽ □ ♅
11 17 ☽ ⊻ ♂
12 34 ☽ □ ☉
15 34 ☽ □ ♃
18 26 ☽ ☌ ♆

14 Thursday
03 41 ☽ bQ ♄
03 45 ☿ ∠ ♇
06 04 ☽ ∥ ♄
08 56 ☽ Q ♇
12 53 ☽ ∥ ♆
16 28 ☽ △ ☿
17 13 ♀ bQ ♂
17 58 ☽ △ ♀
19 58 ☉ ☍ ♃
21 53 ☽ Q ♅

15 Friday
02 39 ☿ ☌ ♀
03 11 ☽ Q ♃
07 35 ☿ bQ ♂
09 28 ☽ △ ♄
20 40 ☽ □ ♇
21 28 ☽ ⚹ ♅
22 42 ☽ bQ ♀

16 Saturday
00 14 ☽ ☌ ♂
02 05 ☽ bQ ☿
02 30 ☽ ⚹ ♃
05 36 ☽ △ ☉
05 56 ☽ ⊻ ♆
09 34 ☉ □ ♆
15 41 ☽ ± ♀
17 31 ☽ ⊥ ♆
19 17 ☽ # ♀
21 47 ☽ ± ☿

17 Sunday
02 17 ☽ ∠ ♅
02 52 ☽ ⊼ ♀
04 24 ☉ # ♂
06 59 ☽ ∠ ♃
10 38 ☽ ∠ ♆
10 41 ☽ ⊼ ☿
12 48 ☽ bQ ☉
18 58 ☽ □ ♄
19 11 ☽ ⊥ ♅
23 41 ☽ ⊥ ♃

18 Monday
03 50 ☽ # ☉
05 51 ☽ △ ♇
06 17 ☽ ⊻ ♅
06 59 ☽ ± ☉
07 22 ☽ ∥ ♂
10 16 ☽ ⊻ ♂
10 39 ☽ ⊻ ♃
14 29 ☽ ⚹ ♆
18 53 ☽ ⊼ ☉
21 13 ☽ ∥ ♃
21 27 ☽ ⊥ ♂
22 17 ♂ ⚹ ♃

19 Tuesday
03 52 ☽ Q ♄
07 46 ♀ ∠ ♇
09 07 ☽ ☍ ♀
09 09 ☽ bQ ♇
12 09 ☽ Q ♆
13 57 ☽ ∠ ♂
14 11 ☽ ∥ ♅
23 41 ☽ ∥ ♅

20 Wednesday
00 18 ☽ ☍ ☿
01 08 ☽ ⚹ ♄
01 08 ☽ ± ♇
07 34 ☿ △ ♄
07 53 ☿ ⊥ ♇
11 38 ☽ ⊼ ♇
11 47 ☽ ☌ ♅
15 30 ☽ ∥ ♃
15 33 ☽ ☌ ♃
15 41 ☉ bQ ♄
16 50 ☽ ⚹ ♂
19 43 ☽ □ ♆

21 Thursday
02 47 ☽ ∥ ♂
03 04 ☽ ∠ ♄
04 00 ☽ ☍ ☉
12 44 ☽ ⊼ ♀
13 51 ☽ Q ♂
14 35 ♅ ⊼ ♇
16 31 ☽ # ☉
18 21 ☽ ⊥ ♄
22 49 ☽ ± ♀

22 Friday
04 26 ☽ ⊻ ♄
09 58 ☽ ⊼ ☿
13 50 ☽ bQ ♀
14 40 ☽ ⊻ ♅
14 47 ☽ ☍ ♇
15 27 ☽ # ♀
17 59 ☽ ⊻ ♃
20 47 ☽ □ ♂
21 12 ☽ ± ☿
22 30 ☽ △ ♆

23 Saturday
00 38 ☽ ⊥ ♅
03 53 ☽ ⊥ ♃
10 18 ☽ ⊼ ☉
13 57 ☽ bQ ☿
14 42 ☽ △ ♀
15 33 ☽ ∠ ♅
18 41 ☽ ∠ ♃
21 00 ☽ ± ☉
23 24 ☽ bQ ♆

24 Sunday
04 30 ☿ ☍ ♅
06 11 ☽ ☌ ♄
07 21 ☿ ⊻ ♇
10 20 ♀ St D
11 37 ☽ ∥ ♆
13 00 ☽ bQ ☉
14 18 ☽ ± ♆
16 18 ☽ ⚹ ♅
16 27 ☽ ∥ ♄
16 39 ☽ ⊼ ♇
17 45 ☽ △ ☿
19 17 ☽ ⚹ ♃
23 37 ☽ △ ♂

25 Monday
00 14 ☽ ⊼ ♆
02 36 ☽ ± ♇
07 00 ☿ ☍ ♃
12 07 ☽ Q ♅
15 00 ☽ Q ♃
15 44 ☽ △ ☉
16 25 ☽ □ ♀
17 34 ☽ bQ ♇

26 Tuesday
01 06 ☽ bQ ♂
02 01 ☉ ⊻ ♀
04 09 ♂ ⊻ ♆
07 47 ☽ ⊻ ♄
09 40 ☿ ∥ ♇
16 30 ☽ ± ♂
17 48 ☽ ⊥ ♄
18 03 ☽ □ ♅
18 40 ☽ △ ♇
20 47 ☽ □ ♃
22 39 ☽ Q ♀

27 Wednesday
01 52 ☽ □ ☿
02 19 ☽ ☍ ♆
02 50 ☽ ⊼ ♂
05 55 ☿ □ ♆
08 54 ☽ ∠ ♄
12 37 ☿ △ ♂
13 06 ☽ # ♄
17 49 ☽ # ♆
19 08 ☽ ⚹ ♀
22 00 ☽ □ ☉

28 Thursday
04 26 ☿ # ♅
10 22 ☽ ⚹ ♄
12 24 ☽ Q ☿
20 59 ☽ △ ♅
21 09 ☽ ∠ ♀
21 52 ☽ □ ♇
23 30 ☽ △ ♃

29 Friday
02 06 ☉ ± ♄
05 42 ☽ ⊼ ♆
07 02 ☽ Q ♄
07 27 ☽ Q ☉
07 30 ☽ ☍ ♂
08 08 ☿ # ♃
11 56 ☽ ⚹ ☿
13 01 ☽ ⊥ ♀
14 37 ☿ bQ ♄
15 59 ☽ ∥ ♀
16 11 ☽ ± ♆
22 14 ☽ ∥ ☉
23 04 ☽ bQ ♅
23 42 ☽ ⊻ ♀

30 Saturday
00 53 ☿ # ♂
01 28 ☽ bQ ♃
05 25 ☽ Q ♇
06 12 ☽ ⚹ ☉
08 03 ☽ bQ ♆
14 35 ☽ □ ♄
14 57 ☽ ± ♅
17 17 ☽ ± ♃
17 58 ☽ ∠ ☿

31 Sunday
01 38 ☽ ⊼ ♅
02 49 ☽ ⚹ ♇
02 50 ♀ ∠ ♇
03 54 ☽ ⊼ ♃
10 01 ☽ ∥ ☿
10 52 ☽ △ ♆
11 10 ☽ ∠ ☉
12 20 ☽ ⊥ ☿
14 10 ☽ ⊼ ♂
21 33 ☽ # ♂

FEBRUARY 2038

Date	Sidereal time h m s	Sun ☉ ° ' "	Moon ☽ ° ' "	Moon ☽ 24.00 ° ' "	Mercury ☿ ° '	Venus ♀ ° '	Mars ♂ ° '	Jupiter ♃ ° '	Saturn ♄ ° '	Uranus ♅ ° '	Neptune ♆ ° '	Pluto ♇ ° '
01	20 47 12	12 ♒ 42 49	10 ♑ 19 16	16 ♑ 50 36	04 ♒ 56	07 ♑ 24	28 ♉ 39	22 ♋ 28	15 ♍ 03	21 ♋ 18	26 ♈ 31	22 ♒ 03
02	20 51 09	13 43 44	23 ♑ 18 52	29 ♑ 44 05	06 35	07 43	29 02	22 R 21	14 R 59	21 R 15	26 32	22 05
03	20 55 05	14 44 37	06 ♒ 06 19	12 ♒ 25 33	08 14	08 04	29 25	22 14	14 55	21 13	26 33	22 07
04	20 59 02	15 45 30	18 ♒ 41 53	24 ♒ 55 20	09 55	08 27	29 ♉ 48	22 07	14 52	21 11	26 34	22 09
05	21 02 59	16 46 22	01 ♓ 06 01	07 ♓ 14 01	11 36	08 52	00 ♊ 11	22 00	14 48	21 08	26 35	22 10
06	21 06 55	17 47 12	13 ♓ 19 31	19 ♓ 22 41	13 18	09 18	00 35	21 53	14 44	21 06	26 36	22 12
07	21 10 52	18 48 01	25 ♓ 23 44	01 ♈ 22 58	15 01	09 46	00 59	21 47	14 40	21 04	26 37	22 14
08	21 14 48	19 48 48	07 ♈ 20 42	13 ♈ 17 18	16 44	10 16	01 23	21 41	14 36	21 02	26 38	22 16
09	21 18 45	20 49 34	19 ♈ 13 10	25 ♈ 08 46	18 29	10 47	01 48	21 34	14 32	21 00	26 39	22 17
10	21 22 41	21 50 19	01 ♉ 04 35	07 ♉ 01 11	20 14	11 20	02 13	21 28	14 27	20 57	26 40	22 19
11	21 26 38	22 51 02	12 ♉ 59 06	18 ♉ 58 56	22 00	11 54	02 38	21 22	14 23	20 55	26 41	22 21
12	21 30 34	23 51 43	25 ♉ 01 18	01 ♊ 06 48	23 47	12 29	03 03	21 16	14 19	20 53	26 42	22 23
13	21 34 31	24 52 23	07 ♊ 16 03	13 ♊ 29 40	25 34	13 06	03 29	21 11	14 14	20 51	26 44	22 25
14	21 38 28	25 53 01	19 ♊ 48 12	26 ♊ 12 11	27 22	13 44	03 55	21 05	14 10	20 49	26 45	22 26
15	21 42 24	26 53 38	02 ♋ 42 05	09 ♋ 18 17	29 ♒ 11	14 23	04 21	21 00	14 06	20 47	26 46	22 28
16	21 46 21	27 54 13	16 ♋ 01 04	22 ♋ 50 33	01 ♓ 01	15 03	04 47	20 55	14 01	20 45	26 48	22 30
17	21 50 17	28 54 46	29 ♋ 46 47	06 ♌ 49 34	02 51	15 45	05 14	20 50	13 57	20 43	26 49	22 32
18	21 54 14	29 ♒ 55 18	13 ♌ 58 35	21 ♌ 13 17	04 42	16 28	05 41	20 45	13 52	20 41	26 50	22 33
19	21 58 10	00 ♓ 55 48	28 ♌ 32 59	05 ♍ 56 49	06 34	17 11	06 08	20 40	13 47	20 40	26 52	22 35
20	22 02 07	01 56 16	13 ♍ 23 46	20 ♍ 52 43	08 25	17 56	06 35	20 36	13 43	20 38	26 53	22 37
21	22 06 03	02 56 42	28 ♍ 22 31	05 ♎ 51 59	10 18	18 41	07 03	20 31	13 38	20 36	26 55	22 39
22	22 10 00	03 57 08	13 ♎ 20 00	20 ♎ 45 32	12 10	19 28	07 30	20 27	13 33	20 34	26 56	22 40
23	22 13 57	04 57 31	28 ♎ 07 40	05 ♏ 25 40	14 02	20 15	07 58	20 23	13 29	20 33	26 58	22 42
24	22 17 53	05 57 54	12 ♏ 38 57	19 ♏ 47 06	15 54	21 04	08 26	20 19	13 24	20 31	26 59	22 44
25	22 21 50	06 58 15	26 ♏ 49 54	03 ♐ 47 15	17 46	21 53	08 54	20 16	13 19	20 30	27 01	22 46
26	22 25 46	07 58 35	10 ♐ 39 11	17 ♐ 25 53	19 37	22 43	09 23	20 12	13 14	20 28	27 03	22 47
27	22 29 43	08 58 53	24 ♐ 07 33	00 ♑ 44 30	21 26	23 33	09 51	20 09	13 10	20 27	27 04	22 49
28	22 33 39	09 ♓ 59 10	07 ♑ 17 01	13 ♑ 45 29	23 ♓ 15	24 ♑ 25	10 ♊ 20	20 ♋ 06	13 ♍ 05	20 ♋ 25	27 ♈ 06	22 ♒ 51

Date	Moon True ☊ ° '	Moon Mean ☊ ° '	Moon ☽ Latitude ° '
01	19 ♋ 37	18 ♋ 24	00 N 51
02	19 R 37	18 21	00 S 20
03	19 36	18 18	01 29
04	19 33	18 15	02 33
05	19 30	18 12	03 28
06	19 25	18 08	04 13
07	19 19	18 05	04 45
08	19 14	18 02	05 05
09	19 09	17 59	05 12
10	19 06	17 56	05 05
11	19 04	17 53	04 45
12	19 D 03	17 49	04 13
13	19 04	17 46	03 28
14	19 05	17 43	02 33
15	19 07	17 40	01 28
16	19 08	17 37	00 S 17
17	19 R 07	17 33	00 N 58
18	19 05	17 30	02 11
19	19 01	17 27	03 17
20	18 55	17 24	04 11
21	18 48	17 21	04 49
22	18 41	17 18	05 07
23	18 35	17 14	05 04
24	18 31	17 11	04 41
25	18 28	17 08	04 02
26	18 D 28	17 05	03 09
27	18 29	17 02	02 07
28	18 ♋ 30	16 ♋ 59	01 N 00

DECLINATIONS

Date	Sun ☉ ° '	Moon ☽ ° '	Mercury ☿ ° '	Venus ♀ ° '	Mars ♂ ° '	Jupiter ♃ ° '	Saturn ♄ ° '	Uranus ♅ ° '	Neptune ♆ ° '	Pluto ♇ ° '
01	16 S 59	22 S 11	20 S 54	16 S 55	21 N 54	21 N 59	07 N 45	22 N 15	08 N 38	22 S 35
02	16 42	21 45	20 32	16 57	21 59	22 00	07 47	22 15	08 39	22 34
03	16 24	20 11	20 08	17 00	22 04	22 01	07 49	22 15	08 39	22 34
04	16 07	17 38	19 43	17 02	22 09	22 02	07 50	22 16	08 40	22 33
05	15 48	14 19	19 16	17 05	22 14	22 03	07 52	22 16	08 40	22 32
06	15 30	10 26	18 49	17 07	22 19	22 05	07 54	22 16	08 40	22 32
07	15 11	06 12	18 19	17 09	22 24	22 06	07 55	22 17	08 41	22 31
08	14 52	01 S 46	17 48	17 12	22 28	22 07	07 57	22 17	08 41	22 31
09	14 33	02 N 42	17 16	17 14	22 33	22 08	07 59	22 18	08 42	22 30
10	14 13	07 04	16 42	17 17	22 38	22 09	08 01	22 18	08 42	22 30
11	13 54	11 12	16 07	17 19	22 43	22 10	08 02	22 18	08 43	22 29
12	13 34	14 56	15 30	17 21	22 48	22 11	08 04	22 18	08 43	22 29
13	13 14	18 06	14 52	17 23	22 53	22 11	08 06	22 19	08 44	22 28
14	12 53	20 30	14 13	17 25	22 57	22 13	08 08	22 19	08 44	22 28
15	12 33	21 56	13 32	17 27	23 02	22 14	08 10	22 19	08 45	22 27
16	12 12	22 12	12 50	17 28	23 07	22 15	08 12	22 20	08 45	22 26
17	11 51	21 08	12 06	17 30	23 11	22 15	08 14	22 20	08 46	22 26
18	11 30	18 43	11 22	17 31	23 16	22 16	08 15	22 20	08 46	22 25
19	11 08	15 03	10 36	17 31	23 21	22 17	08 17	22 20	08 47	22 25
20	10 47	10 23	09 48	17 32	23 25	22 17	08 19	22 21	08 47	22 24
21	10 25	05 N 04	09 00	17 32	23 30	22 18	08 21	22 21	08 48	22 24
22	10 03	00 S 33	08 11	17 32	23 34	22 19	08 23	22 21	08 49	22 23
23	09 41	06 04	07 20	17 31	23 38	22 20	08 25	22 21	08 49	22 23
24	09 19	11 09	06 29	17 30	23 42	22 20	08 27	22 22	08 50	22 22
25	08 57	15 31	05 37	17 29	23 47	22 21	08 29	22 22	08 50	22 22
26	08 35	18 55	04 45	17 28	23 51	22 22	08 31	22 22	08 51	22 21
27	08 12	21 11	03 52	17 26	23 55	22 22	08 33	22 22	08 52	22 21
28	07 S 49	22 S 14	03 S 00	17 S 23	23 N 59	22 N 23	08 N 35	22 N 23	08 N 52	22 S 21

ZODIAC SIGN ENTRIES

Date	h	m	Planets
03	00	30	☽ ♒
05	00	33	♂ ♊
05	09	51	☽ ♓
07	21	13	☽ ♈
10	09	49	☽ ♉
12	21	49	☽ ♊
15	07	02	☽ ♋
15	22	41	☿ ♓
17	12	23	☽ ♌
18	13	52	☉ ♓
19	14	22	☽ ♍
21	14	36	☽ ♎
23	15	04	☽ ♏
25	17	27	☽ ♐
27	22	39	☽ ♑

LATITUDES

Date	Mercury ☿ ° '	Venus ♀ ° '	Mars ♂ ° '	Jupiter ♃ ° '	Saturn ♄ ° '	Uranus ♅ ° '	Neptune ♆ ° '	Pluto ♇ ° '
01	01 S 55	06 N 19	02 N 06	00 N 25	02 N 02	00 N 30	01 S 42	08 S 56
04	02 02	06 08	02 06	00 25	02 02	00 30	01 42	08 56
07	02 05	05 56	02 05	00 26	02 03	00 30	01 42	08 56
10	02 04	05 41	02 05	00 26	02 03	00 30	01 42	08 56
13	02 00	05 26	02 04	00 26	02 04	00 30	01 41	08 56
16	01 51	05 09	02 04	00 26	02 04	00 30	01 41	08 56
19	01 36	04 51	02 03	00 27	02 04	00 30	01 41	08 57
22	01 17	04 32	02 02	00 27	02 05	00 30	01 41	08 57
25	00 51	04 13	02 01	00 27	02 05	00 30	01 41	08 57
28	00 S 20	03 54	02 01	00 27	02 05	00 30	01 41	08 58
31	00 N 16	03 N 35	02 N 00	00 N 27	02 N 06	00 N 30	01 S 41	08 S 58

DATA

Julian Date	2465456
Delta T	+77 seconds
Ayanamsa	24° 23' 06"
Synetic vernal point	04° ♓ 43' 53"
True obliquity of ecliptic	23° 26' 01"

MOON'S PHASES, APSIDES AND POSITIONS ☽

Date	h	m	Phase	Longitude ° '	Eclipse Indicator
04	05	52	●	15 ♒ 30	
12	09	30	☽	23 ♉ 45	
19	16	09	○	01 ♍ 06	
26	06	56	☾	07 ♐ 46	

Day	h	m		
09	10	02	Apogee	
21	08	09	Perigee	
01	15	26	Max dec	22° S 12'
08	21	26	0N	
16	04	54	Max dec	22° N 15'
22	09	40	0S	
28	20	36	Max dec	22° S 19'

LONGITUDES

Date	Chiron ⚷ ° '	Ceres ⚳ ° '	Pallas ⚴ ° '	Juno ⚵ ° '	Vesta ⚶ ° '	Black Moon Lilith ⚸ ° '
01	20 ♊ 44	08 ♑ 43	10 ♐ 26	23 ♏ 01	04 ♉ 03	13 ♈ 04
11	20 ♊ 30	12 ♑ 30	13 ♐ 46	24 ♏ 44	06 ♉ 59	14 ♈ 11
21	20 ♊ 23	16 ♑ 09	16 ♐ 49	26 ♏ 02	10 ♉ 12	15 ♈ 18
31	20 ♊ 24	19 ♑ 39	19 ♐ 30	26 ♏ 54	13 ♉ 39	16 ♈ 25

All ephemeris data is given at 12.00 UT and the Moon's longitude is additionally given for 24.00 UT

ASPECTARIAN

01 Monday
00 44 ☽ # ♃
00 45 ☽ ⊻ ☿
01 22 ☽ ± ♂
04 51 ☽ ⊥ ☉
06 01 ☽ ∠ ♇
06 32 ☽ ☌ ♀
16 46 ☽ ⊻ ☉
17 04 ☉ ∥ ♀
18 18 ☽ ⅊ ♂
20 39 ☽ △ ♄
22 34 ☽ ⊥ ♇

02 Tuesday
05 25 ☽ # ♃
06 16 ☽ # ♂
08 11 ☽ ☍ ♅
09 42 ☽ ⊻ ♇
10 13 ☽ ☍ ♃
11 39 ☉ ⊥ ♀
16 45 ♂ ∥ ♃
18 00 ☽ □ ♆
23 00 ☽ △ ♂

03 Wednesday
00 25 ☽ ⅊ ♄
07 10 ☉ Q ♆
08 51 ☿ ⊻ ♀
12 43 ☽ ∥ ☿
15 50 ☽ ⊻ ♀
16 01 ☉ ⊼ ♄
16 39 ☽ ☌ ☿
17 19 ☽ ± ♄
21 28 ☿ ± ♄

04 Thursday
03 36 ☽ ⊥ ♀
04 04 ☽ Q ♆
04 41 ☽ ⊼ ♄
05 52 ☽ ☌ ☉
06 51 ♃ ⊼ ♇
16 42 ☽ ∥ ♀
16 45 ☽ ⊼ ♅
18 31 ☽ ⊼ ♃
18 39 ☽ ☌ ♇
21 28 ☽ ∠ ♀

05 Friday
00 50 ☽ ∥ ☉
03 11 ☽ ⚹ ♆
04 19 ☽ ± ♅
06 01 ☽ ± ♃
10 10 ☽ □ ♂
21 49 ☽ ⅊ ♅
23 26 ☽ ⅊ ♃

06 Saturday
00 20 ♂ ∥ ♅
03 46 ☽ ⚹ ♀
08 34 ☽ ∠ ♆
11 57 ☽ ⊻ ☿
14 46 ☽ ☍ ♄
21 39 ☽ ⊻ ☉
22 10 ☽ # ♆
22 46 ☽ Q ♂

07 Sunday
01 48 ☽ ⊥ ☿
02 26 ☽ ⊥ ♆
02 28 ☽ # ♄
03 23 ☽ △ ♅
04 28 ☽ Q ♀
04 51 ☽ △ ♃
05 40 ☽ ⊻ ♇
06 17 ☿ Q ♆
07 15 ☿ ⊼ ♄
10 42 ☽ ⊥ ☉
14 26 ☽ ⊻ ♆
17 41 ☽ ⊥ ♇
22 49 ☽ ∠ ☿
23 35 ☽ ⚹ ♂

08 Monday
02 38 ☿ ⊥ ♀
06 26 ☽ ∠ ☉
11 50 ☽ ∠ ♇
18 09 ☽ □ ♀
22 00 ♂ # ♇

09 Tuesday
02 33 ☽ ⊼ ♄
06 55 ☽ ∠ ♂
10 15 ☽ ⚹ ☿
13 06 ☿ ∥ ♀
14 38 ☽ ± ♄
15 33 ☽ ⚹ ☉
15 35 ☽ □ ♅
15 48 ☉ ⊼ ♅
16 43 ☽ □ ♃
18 14 ☽ ⚹ ♇

10 Wednesday
01 48 ☽ ⊥ ♂
03 04 ☽ ☌ ♆
04 03 ☉ ⊼ ♃
08 44 ☽ ⅊ ♄
14 23 ☽ ⊻ ♂
14 45 ☽ Q ☿
17 21 ☽ ∥ ♄
18 06 ☽ Q ☉
18 34 ☽ Q ♇
21 18 ☽ ∥ ♆
21 40 ☿ ⊼ ♅
23 46 ☉ ☌ ♇

11 Thursday
03 51 ☽ Q ♅
03 57 ☿ ⊼ ♃
04 48 ☽ Q ♃
09 42 ☽ △ ♀
14 47 ☽ △ ♄
15 23 ♂ ⊥ ♆
16 51 ☿ ☌ ♇

12 Friday
03 38 ☽ # ☉
03 49 ☽ ⚹ ♅
04 37 ☽ ⚹ ♃
06 45 ☽ □ ♇
09 06 ☽ □ ☿
09 30 ☽ □ ☉
14 40 ☉ ☌ ☿
15 20 ☽ ⊻ ♆
15 26 ☽ # ☿
17 07 ☽ ⅊ ♀

13 Saturday
03 09 ☽ ⊥ ♆
04 22 ☽ ☌ ♂
06 07 ☽ # ♀
09 16 ☽ ∠ ♅
09 54 ☽ ∠ ♃
11 39 ☽ ± ♀
20 38 ☽ ∠ ♆
23 50 ☽ ⊼ ♀

14 Sunday
01 21 ☽ □ ♄
02 33 ☽ ⊥ ♅
03 07 ☽ ⊥ ♃
03 40 ☿ ⚹ ♆
04 48 ☿ ± ♅
08 25 ☿ ± ♃
13 55 ☽ ⊻ ♅
14 24 ☽ ⊻ ♃
16 58 ☽ △ ♇

15 Monday
00 23 ☽ △ ☉
01 02 ☽ ⚹ ♆
02 30 ♀ △ ♄
04 29 ☽ △ ☿
09 03 ☉ ⚹ ♆
09 31 ☉ ± ♅
10 53 ☽ Q ♄
14 18 ☉ ± ♃
15 07 ☽ ⊻ ♂
16 03 ☿ ∠ ♀
20 42 ☽ ⅊ ♇
23 04 ☽ Q ♆

16 Tuesday
01 33 ☽ ∥ ♃
02 21 ☽ ⊥ ♂
06 00 ☽ ⅊ ☉
07 46 ☽ ∥ ♃
08 28 ☽ ⚹ ♄
10 12 ☽ ☍ ♀
12 00 ☽ ⅊ ☿
12 51 ☽ ± ♇
18 53 ☽ ∠ ♂
20 20 ☽ ☌ ♅
20 34 ☽ ☌ ♃
23 11 ☽ ± ☉
23 25 ☽ ⊼ ♇

17 Wednesday
06 12 ☽ ± ☿
06 53 ☽ □ ♆
10 24 ☽ ⊼ ☉
10 34 ☽ ∠ ♄
18 03 ☽ ⊼ ☿
21 36 ☽ ⚹ ♂

18 Thursday
01 49 ☽ ⊥ ♄
03 50 ☉ ∥ ☿
11 49 ☽ ⊻ ♄
12 35 ♂ ∠ ♅
15 07 ♂ ∠ ♃
15 23 ♀ ⊥ ♇
16 21 ☽ ⊼ ♀
18 21 ☽ Q ♂
20 45 ☽ # ♀
23 06 ☽ ⊻ ♅
23 10 ☽ ⊻ ♃

19 Friday
00 35 ☿ ⅊ ♅
00 58 ☿ ⅊ ♃
02 13 ☽ ☍ ♇
02 46 ☽ ± ♀
04 40 ☿ □ ♂
08 56 ☽ ⊥ ♅
08 57 ☽ ⊥ ♃
09 15 ☽ △ ♆
16 09 ☽ ☍ ☉
16 55 ♃ ☌ ♅
18 13 ☽ ⅊ ♀
23 30 ☽ ∠ ♃
23 31 ☽ ∠ ♅

20 Saturday
00 41 ☽ □ ♂
02 51 ☽ ☍ ☿
09 34 ☽ ⅊ ♆
09 59 ☽ # ☉
12 30 ☽ ☌ ♄
15 16 ☽ # ☿
19 27 ☽ ∥ ♆
19 40 ☽ △ ♀
21 32 ☽ ∥ ♄
23 29 ☽ ⚹ ♃
23 35 ☽ ⚹ ♅

21 Sunday
00 02 ☽ ± ♆
02 48 ☽ ⊼ ♇
09 39 ☽ ⊼ ♆
12 26 ☽ ± ♇
17 49 ☿ # ♆
18 37 ☽ Q ♃
18 45 ☽ Q ♅
19 51 ☽ ⊼ ☉

22 Monday
02 20 ☽ △ ♂
02 53 ☽ ⅊ ♇
06 10 ☽ ± ☉
06 14 ☿ # ♄
09 04 ☿ ∠ ♆
09 51 ☽ ⊼ ☿
12 22 ☽ ⊻ ♄
20 56 ☽ ± ☿
21 44 ☽ ⅊ ☉
22 00 ☽ ⊥ ♄
22 28 ☽ □ ♀
23 27 ☽ □ ♃
23 41 ☽ □ ♅

23 Tuesday
03 08 ☽ △ ♇
03 19 ☽ ⅊ ♂
05 08 ☿ ☍ ♄
10 06 ☽ ☍ ♆
12 34 ☽ ∠ ♄
13 42 ☽ ⅊ ☿
15 40 ♀ ☍ ♃
16 56 ☽ ∥ ☿
18 30 ☽ ± ♂
20 26 ♀ ☍ ♅
21 37 ☉ ⅊ ♃
22 47 ☽ # ♄

24 Wednesday
00 04 ☽ △ ☉
00 39 ☽ # ♆
01 39 ☉ ⅊ ♅
03 37 ☽ ∥ ☉
04 45 ☽ ⊼ ♂
05 40 ☽ Q ♀
13 15 ☽ ⚹ ♄
18 16 ☽ △ ☿
23 59 ☉ ∠ ♀

25 Thursday
00 51 ☽ △ ♃
01 13 ☽ △ ♅
03 02 ☽ ⚹ ♀
05 02 ☽ □ ♇
09 26 ☽ Q ♄
12 19 ☽ ⊼ ♆
13 49 ♅ # ♇
18 57 ☉ # ♆
22 41 ☽ ± ♆

26 Friday
00 54 ☽ ∥ ♀
02 30 ☽ ⅊ ♃
02 56 ☽ ⅊ ♅
06 30 ☽ ∠ ♀
06 56 ☽ □ ☉
09 41 ☽ ☍ ♂
09 48 ♃ # ♇
12 14 ☽ Q ♇
14 17 ♀ ⊻ ♇
14 27 ☽ ⅊ ♆
15 50 ☉ # ♄
16 32 ☽ □ ♄
18 15 ☽ ± ♃
18 43 ☽ ± ♅
19 35 ☿ △ ♃
23 05 ☿ △ ♅
23 26 ☽ ⊥ ♀

27 Saturday
04 53 ☽ ⊼ ♃
05 24 ☽ ⊼ ♅
06 24 ☽ □ ☿
07 05 ☿ ⊥ ♆
09 38 ☽ ⚹ ♇
10 54 ☽ ⊻ ♀
17 20 ☽ △ ♆
17 35 ☽ Q ☉

28 Sunday
05 32 ♃ ∥ ♅
06 35 ☿ ⊻ ♇
13 02 ☽ ∠ ♇
17 25 ☽ ⚹ ☉
17 51 ☽ ⊼ ♂
20 30 ☽ Q ☿
22 41 ☽ △ ♄

MARCH 2038

LONGITUDES

Date	Sidereal time h m s	Sun ☉ ° ' "	Moon ☽ ° ' "	Moon ☽ 24.00 ° ' "	Mercury ☿ ° '	Venus ♀ ° '	Mars ♂ ° '	Jupiter ♃ ° '	Saturn ♄ ° '	Uranus ♅ ° '	Neptune ♆ ° '	Pluto ♇ ° '
01	22 37 36	10 ♓ 59 26	20 ♑ 10 13	26 ♑ 31 33	25 ♓ 01	25 ♑ 17	10 ♊ 49	20 ♋ 03	13 ♍ 00	20 ♋ 24	27 ♈ 08	22 ♒ 52
02	22 41 32	11 59 40	02 ♒ 49 48	09 ♒ 05 14	26 45	26 09	11 18	20 R 01	12 R 55	20 R 22	27 09	22 54
03	22 45 29	12 59 52	15 ♒ 18 07	21 ♒ 28 41	28 ♓ 26	27 03	11 47	19 58	12 51	20 21	27 11	22 56
04	22 49 26	14 00 03	27 ♒ 37 05	03 ♓ 43 31	00 ♈ 04	27 57	12 17	19 56	12 46	20 20	27 13	22 57
05	22 53 22	15 00 12	09 ♓ 48 08	15 ♓ 51 03	01 38	28 51	12 46	19 54	12 41	20 19	27 15	22 59
06	22 57 19	16 00 19	21 ♓ 52 25	27 ♓ 52 21	03 08	29 ♑ 46	13 16	19 52	12 36	20 18	27 16	23 01
07	23 01 15	17 00 24	03 ♈ 51 00	09 ♈ 48 33	04 33	00 ♒ 42	13 46	19 50	12 31	20 16	27 18	23 02
08	23 05 12	18 00 27	15 ♈ 45 12	21 ♈ 41 09	05 52	01 38	14 16	19 48	12 27	20 15	27 20	23 04
09	23 09 08	19 00 28	27 ♈ 36 41	03 ♉ 32 07	07 05	02 35	14 46	19 47	12 22	20 14	27 22	23 06
10	23 13 05	20 00 27	09 ♉ 27 48	15 ♉ 24 08	08 11	03 32	15 16	19 46	12 17	20 13	27 24	23 07
11	23 17 01	21 00 24	21 ♉ 21 35	27 ♉ 20 39	09 11	04 30	15 47	19 45	12 12	20 12	27 26	23 09
12	23 20 58	22 00 19	03 ♊ 21 51	09 ♊ 25 47	10 03	05 28	16 18	19 44	12 08	20 12	27 27	23 10
13	23 24 54	23 00 12	15 ♊ 33 01	21 ♊ 44 13	10 47	06 26	16 48	19 44	12 03	20 11	27 29	23 12
14	23 28 51	24 00 03	27 ♊ 59 58	04 ♋ 20 56	11 23	07 25	17 19	19 43	11 58	20 10	27 31	23 14
15	23 32 48	24 59 51	10 ♋ 47 41	17 ♋ 20 46	11 50	08 25	17 50	19 43	11 54	20 09	27 33	23 15
16	23 36 44	25 59 37	24 ♋ 00 39	00 ♌ 47 43	12 09	09 24	18 21	19 D 43	11 49	20 09	27 35	23 17
17	23 40 41	26 59 21	07 ♌ 42 12	14 ♌ 44 10	12 20	10 24	18 52	19 44	11 45	20 08	27 37	23 18
18	23 44 37	27 59 03	21 ♌ 53 31	29 ♌ 09 54	12 R 22	11 25	19 24	19 44	11 40	20 08	27 39	23 20
19	23 48 34	28 58 42	06 ♍ 32 46	14 ♍ 01 18	12 16	12 26	19 55	19 45	11 36	20 07	27 41	23 21
20	23 52 30	29 ♓ 58 19	21 ♍ 34 29	29 ♍ 11 05	12 02	13 27	20 27	19 45	11 31	20 07	27 43	23 23
21	23 56 27	00 ♈ 57 55	06 ♎ 49 46	14 ♎ 29 05	11 40	14 28	20 58	19 46	11 27	20 06	27 45	23 24
22	00 00 23	01 57 28	22 ♎ 07 33	29 ♎ 43 46	11 12	15 30	21 30	19 48	11 22	20 06	27 48	23 26
23	00 04 20	02 56 59	07 ♏ 16 28	14 ♏ 44 31	10 37	16 32	22 02	19 49	11 18	20 06	27 50	23 27
24	00 08 17	03 56 29	22 ♏ 07 01	29 ♏ 23 18	09 57	17 35	22 34	19 51	11 14	20 05	27 52	23 29
25	00 12 13	04 55 56	06 ♐ 32 55	13 ♐ 35 39	09 13	18 37	23 06	19 52	11 10	20 05	27 54	23 30
26	00 16 10	05 55 23	20 ♐ 31 27	27 ♐ 20 27	08 25	19 40	23 38	19 54	11 05	20 05	27 56	23 31
27	00 20 06	06 54 47	04 ♑ 02 57	10 ♑ 39 18	07 35	20 43	24 10	19 57	11 01	20 05	27 58	23 33
28	00 24 03	07 54 10	17 ♑ 09 57	23 ♑ 35 24	06 43	21 47	24 42	19 59	10 57	20 D 05	28 00	23 34
29	00 27 59	08 53 30	29 ♑ 56 10	06 ♒ 12 45	05 51	22 51	25 15	20 01	10 53	20 05	28 02	23 35
30	00 31 56	09 52 50	12 ♒ 25 41	18 ♒ 35 26	05 01	23 55	25 47	20 04	10 50	20 05	28 05	23 37
31	00 35 52	10 ♈ 52 07	24 ♒ 42 26	00 ♓ 47 07	04 ♈ 11	24 ♒ 59	26 ♊ 20	20 ♋ 07	10 ♍ 46	20 ♋ 05	28 ♈ 07	23 ♒ 38

Date	Moon True ☊ ° '	Moon Mean ☊ ° '	Moon ☽ Latitude ° '
01	18 ♋ 30	16 ♋ 55	00 S 09
02	18 R 29	16 52	01 16
03	18 26	16 49	02 19
04	18 20	16 46	03 14
05	18 11	16 43	03 59
06	18 00	16 39	04 33
07	17 49	16 36	04 54
08	17 37	16 33	05 03
09	17 26	16 30	04 59
10	17 17	16 27	04 41
11	17 10	16 24	04 12
12	17 06	16 20	03 31
13	17 04	16 17	02 39
14	17 D 03	16 14	01 40
15	17 04	16 11	00 S 33
16	17 R 04	16 08	00 N 37
17	17 02	16 05	01 47
18	16 58	16 01	02 53
19	16 51	15 58	03 50
20	16 42	15 55	04 33
21	16 32	15 52	04 57
22	16 21	15 49	05 00
23	16 11	15 45	04 41
24	16 04	15 42	04 04
25	15 59	15 39	03 12
26	15 56	15 36	02 10
27	15 D 56	15 33	01 N 02
28	15 R 56	15 30	00 S 07
29	15 55	15 26	01 13
30	15 53	15 23	02 15
31	15 ♋ 48	15 ♋ 20	03 S 09

DECLINATIONS

Date	Sun ☉ ° '	Moon ☽ ° '	Mercury ☿ ° '	Venus ♀ ° '	Mars ♂ ° '	Jupiter ♃ ° '	Saturn ♄ ° '	Uranus ♅ ° '	Neptune ♆ ° '	Pluto ♇ ° '
01	07 S 27	22 S 04	02 S 07	17 S 21	24 N 03	22 N 23	08 N 36	22 N 23	08 N 53	22 S 20
02	07 04	20 45	01 15	17 18	24 07	22 24	08 38	22 23	08 53	22 20
03	06 41	18 27	00 S 23	17 14	24 10	22 24	08 40	22 23	08 54	22 19
04	06 18	15 19	00 N 28	17 10	24 14	22 24	08 42	22 23	08 55	22 19
05	05 54	11 35	01 18	17 06	24 18	22 25	08 44	22 23	08 55	22 18
06	05 31	07 24	02 06	17 01	24 21	22 25	08 46	22 24	08 56	22 18
07	05 08	02 S 58	02 52	16 55	24 24	22 25	08 48	22 24	08 57	22 17
08	04 44	01 N 32	03 36	16 50	24 28	22 26	08 50	22 24	08 58	22 17
09	04 21	05 58	04 17	16 44	24 31	22 26	08 52	22 24	08 58	22 17
10	03 57	10 12	04 56	16 37	24 34	22 26	08 53	22 24	08 59	22 16
11	03 34	14 03	05 32	16 30	24 37	22 26	08 55	22 24	09 00	22 16
12	03 10	17 23	06 04	16 22	24 40	22 26	08 57	22 24	09 00	22 16
13	02 47	20 01	06 33	16 14	24 43	22 26	08 59	22 24	09 01	22 15
14	02 23	21 46	06 57	16 06	24 45	22 27	09 01	22 25	09 02	22 15
15	01 59	22 27	07 18	15 57	24 48	22 27	09 03	22 25	09 03	22 14
16	01 36	21 54	07 34	15 47	24 51	22 27	09 04	22 25	09 03	22 14
17	01 12	20 04	07 46	15 37	24 53	22 27	09 06	22 25	09 04	22 14
18	00 48	16 57	07 54	15 27	24 55	22 27	09 08	22 25	09 05	22 13
19	00 24	12 40	07 57	15 16	24 57	22 26	09 09	22 25	09 05	22 13
20	00 S 01	07 31	07 55	15 05	24 59	22 26	09 11	22 25	09 06	22 13
21	00 N 23	01 N 50	07 49	14 53	25 01	22 26	09 13	22 25	09 07	22 12
22	00 47	03 S 59	07 39	14 41	25 03	22 26	09 14	22 25	09 08	22 12
23	01 10	09 30	07 25	14 28	25 05	22 26	09 16	22 25	09 08	22 12
24	01 34	14 22	07 06	14 15	25 06	22 26	09 18	22 25	09 09	22 11
25	01 58	18 14	06 45	14 01	25 08	22 26	09 19	22 25	09 10	22 11
26	02 21	20 56	06 20	13 47	25 09	22 25	09 21	22 25	09 11	22 11
27	02 45	22 20	05 53	13 33	25 10	22 25	09 22	22 25	09 12	22 11
28	03 08	22 26	05 23	13 18	25 11	22 25	09 24	22 25	09 12	22 10
29	03 31	21 21	04 52	13 02	25 12	22 24	09 25	22 25	09 13	22 10
30	03 55	19 14	04 21	12 47	25 13	22 24	09 27	22 25	09 14	22 10
31	04 N 18	16 S 15	03 N 49	12 S 31	25 N 13	22 N 24	09 N 28	22 N 25	09 N 15	22 S 10

ZODIAC SIGN ENTRIES

Date	h	m	Planets
02	06	36	☽ ♒
04	10	56	☿ ♈
04	16	40	☽ ♓
06	17	53	♀ ♒
07	04	16	☽ ♈
09	16	50	☽ ♉
12	05	18	☽ ♊
14	15	48	☽ ♋
16	22	36	☽ ♌
19	01	22	☽ ♍
20	12	40	☉ ♈
21	01	17	☽ ♎
23	00	26	☽ ♏
25	01	01	☽ ♐
27	04	44	☽ ♑
29	12	07	☽ ♒
31	22	27	☽ ♓

LATITUDES

Date	Mercury ☿ ° '	Venus ♀ ° '	Mars ♂ ° '	Jupiter ♃ ° '	Saturn ♄ ° '	Uranus ♅ ° '	Neptune ♆ ° '	Pluto ♇ ° '
01	00 S 09	03 N 48	02 N 00	00 N 27	02 N 05	00 N 30	01 S 41	08 S 58
04	00 N 29	03 28	01 59	00 27	02 06	00 30	01 41	08 58
07	01 09	03 09	01 58	00 27	02 06	00 30	01 40	08 58
10	01 50	02 49	01 58	00 28	02 06	00 30	01 40	08 59
13	02 29	02 30	01 57	00 28	02 06	00 30	01 40	08 59
16	03 01	02 11	01 56	00 28	02 06	00 30	01 40	09 00
19	03 22	01 53	01 55	00 28	02 06	00 30	01 40	09 01
22	03 30	01 35	01 54	00 28	02 06	00 30	01 40	09 01
25	03 22	01 17	01 53	00 28	02 06	00 30	01 40	09 02
28	02 58	01 00	01 51	00 28	02 06	00 29	01 40	09 02
31	02 N 21	00 N 43	01 N 50	00 N 28	02 N 06	00 N 29	01 S 40	09 S 03

DATA

Julian Date	2465484
Delta T	+77 seconds
Ayanamsa	24° 23' 09"
Synetic vernal point	04° ♓ 43' 49"
True obliquity of ecliptic	23° 26' 01"

MOON'S PHASES, APSIDES AND POSITIONS ☽

Date	h	m	Phase	Longitude ° '	Eclipse Indicator
05	23	15	●	15 ♓ 28	
14	03	42	☽	23 ♊ 39	
21	02	10	○	00 ♎ 33	
27	17	36	☾	07 ♑ 09	

Day	h	m		
09	00	46	Apogee	
21	17	20	Perigee	
08	03	50	0N	
15	13	50	Max dec	22° N 27'
21	19	33	0S	
28	02	01	Max dec	22° S 33'

LONGITUDES

Date	Chiron ⚷ ° '	Ceres ⚳ ° '	Pallas ⚴ ° '	Juno ⚵ ° '	Vesta ⚶ ° '	Black Moon Lilith ⚸ ° '
01	20 ♊ 24	18 ♑ 58	19 ♐ 00	26 ♏ 46	12 ♉ 57	16 ♈ 12
11	20 ♊ 31	22 ♑ 19	21 ♐ 22	27 ♏ 13	16 ♉ 33	17 ♈ 19
21	20 ♊ 46	25 ♑ 28	23 ♐ 15	27 ♏ 08	20 ♉ 20	18 ♈ 26
31	21 ♊ 09	28 ♑ 22	24 ♐ 35	26 ♏ 29	24 ♉ 15	19 ♈ 34

ASPECTARIAN

h m Aspects

01 Monday
03 56 ☉ □ ♂
05 28 ☽ ± ♂
05 48 ☽ ⊥ ♇
11 47 ☽ ☍ ♃
12 25 ☽ ☍ ♅
17 06 ☽ ⊻ ♇
19 12 ☿ ⚹ ♀
22 21 ☽ ☌ ♀
22 36 ☽ ⚹ ☿
23 05 ☽ bQ ♂
23 56 ☽ ∠ ☉

02 Tuesday
01 10 ☽ □ ♆
02 43 ☽ bQ ♄
15 57 ☉ ∠ ♆
17 47 ☿ ⊻ ♆
18 35 ☽ ⊥ ☉
19 47 ☽ ± ♄
20 39 ♀ bQ ♂

03 Wednesday
04 56 ☽ △ ♂
07 09 ☽ ⊻ ☉
07 16 ☽ ⊼ ♄
07 50 ☽ ∠ ☿
08 33 ☉ ☍ ♄
11 46 ☽ Q ♆
15 49 ♀ □ ♆
19 15 ☿ ⊥ ♇
21 02 ☽ ⊼ ♃
21 47 ☽ ⊼ ♅
22 13 ☽ ∥ ♀

04 Thursday
02 52 ☽ ☌ ♇
04 01 ☽ ⊥ ☿
07 32 ♀ bQ ♄
08 34 ♂ ∠ ♆
08 42 ☽ ± ♃
09 29 ☽ ± ♅
11 12 ☽ ⚹ ♆
12 42 ☽ ⊻ ♀
16 30 ☿ Q ♂
17 32 ☽ ⊻ ☿

05 Friday
01 26 ☽ ⊥ ♀
02 20 ☽ bQ ♃
03 08 ☽ bQ ♅
08 15 ♂ □ ♄
16 51 ☽ ∠ ♆
17 40 ☽ ☍ ♄
18 08 ☽ □ ♂
20 42 ☽ ∠ ♀
23 15 ☽ ☌ ☉

06 Saturday
03 25 ☽ # ♆
04 24 ☽ # ♄
08 00 ☽ △ ♃
08 51 ☽ △ ♅
10 48 ☽ ⊥ ♆
14 17 ☽ ⊻ ♇
22 49 ☽ ⊻ ♆
23 18 ☽ ∥ ☉

07 Sunday
02 19 ☽ ⊥ ♇
05 08 ☽ ⚹ ♀
07 38 ☽ Q ♂
12 30 ☽ # ☿
13 35 ☽ ☌ ☿
15 04 ♂ ⊥ ♃
20 27 ☽ ∠ ♇

08 Monday
05 21 ☽ ⊼ ♄
07 21 ☽ Q ♀
08 52 ☽ ⚹ ♂
11 31 ♂ ⊥ ♅
16 59 ☽ ⊻ ☉
17 24 ☽ ± ♄
20 11 ☽ □ ♃
21 06 ☽ □ ♅

09 Tuesday
01 08 ☽ ∥ ☿
02 50 ☽ ⚹ ♇
03 51 ☽ # ☉
06 14 ☽ ⊥ ☉
11 30 ☽ ☌ ♆
11 30 ☽ bQ ♄
11 56 ♄ bQ ♆
13 21 ☉ # ☿
16 34 ☽ ∠ ♂
22 57 ☽ □ ♀

10 Wednesday
02 09 ☽ ∠ ☉
03 12 ☽ Q ♇
04 21 ☽ ∥ ♄
04 55 ☽ ∥ ♆
06 19 ☉ △ ♃
08 34 ☽ Q ♃
09 10 ☽ ⊻ ☿
09 30 ☽ Q ♅
10 23 ☿ ∠ ♇
11 36 ☽ ⊥ ♂
17 05 ☉ △ ♅
17 40 ☽ △ ♄
22 27 ☽ ⊥ ☿

11 Thursday
00 16 ☽ ⊻ ♂
08 46 ☽ ⚹ ♃
09 41 ☽ ⚹ ♅
11 14 ☽ ⚹ ☉
15 36 ☽ □ ♇
18 08 ☽ ∠ ☿
22 24 ☉ ⊥ ♆

12 Friday
00 12 ☽ ⊻ ♆
04 33 ☽ # ♀
12 11 ☽ ⊥ ♆
13 23 ☽ Q ☉
14 44 ☽ ∠ ♃
15 37 ☽ ∠ ♅
16 32 ☽ △ ♀

13 Saturday
02 06 ☽ ⚹ ☿
03 09 ♀ ± ♄
05 12 ☽ □ ♄
06 00 ☽ ∠ ♆
08 27 ☽ ⊥ ♃
09 20 ☽ ⊥ ♅
14 33 ☽ ☌ ♂
16 53 ☉ ⊻ ♇
20 07 ☽ ⊻ ♃
20 59 ☽ ⊻ ♅

14 Sunday
00 24 ☽ bQ ♀
02 45 ☽ Q ☿
02 51 ☽ △ ♇
03 42 ☽ □ ☉
11 05 ☽ ⚹ ♆
15 43 ☽ Q ♄
19 02 ☽ ± ♀
23 58 ☽ # ♇

15 Monday
07 13 ☽ ⊼ ♀
07 17 ☽ bQ ♇
08 11 ☽ ∥ ♅
09 42 ☽ Q ♆
11 24 ♄ ∥ ♆
12 18 ☽ ∥ ♃
13 59 ☽ □ ☿
14 01 ☽ ⚹ ♄
15 00 ☿ ⊼ ♄
15 23 ☽ ∥ ♃
17 52 ♃ St D
19 26 ☽ ∥ ♅
23 51 ☽ ± ♇

16 Tuesday
01 25 ☽ ⊻ ♂
03 42 ☽ # ♇
04 18 ☽ ☌ ♃
05 04 ☽ ☌ ♅
10 41 ☽ ⊼ ♇
12 38 ☽ ⊥ ♂
15 49 ☽ △ ☉
16 58 ☽ ∠ ♄
18 22 ☽ □ ♆

17 Wednesday
05 07 ☽ ∠ ♂
08 38 ☽ ⊥ ♄
17 00 ☽ ☍ ♀
18 53 ☽ ⊻ ♄
19 54 ☽ bQ ☉
19 58 ☽ △ ☿

18 Thursday
03 47 ☉ ⊻ ♆
06 08 ☿ St R
07 41 ☽ ⚹ ♂
08 24 ☽ ⊻ ♃
09 04 ☽ ⊻ ♅
12 10 ☽ ± ☉
14 23 ☽ ☍ ♇
17 37 ♀ ⊼ ♄
18 22 ☽ ⊥ ♃
19 00 ☽ ⊥ ♅
21 01 ☽ bQ ☿
21 30 ☽ # ♀
21 33 ☽ △ ♆
22 48 ☽ ⊼ ☉

19 Friday
03 55 ♂ ⊻ ♃
04 13 ☽ Q ♂
08 43 ☿ ⚹ ♀
09 05 ☽ ∠ ♃
09 41 ☽ ∠ ♅
11 33 ☽ ± ☿
20 05 ☽ ☌ ♄
21 03 ♂ ⊻ ♅
21 05 ☽ ⊼ ☿
21 19 ☉ ⊥ ♇
21 54 ☽ bQ ♆
22 08 ☽ ⊼ ♀

20 Saturday
04 38 ☽ ∥ ♄
04 59 ☽ ∥ ♆
08 23 ☽ ± ♀
09 07 ☽ ⚹ ♃
09 41 ☽ ⚹ ♅
10 09 ☽ □ ♂
10 13 ☽ ∥ ☿
12 14 ☽ ± ♆
14 51 ☽ ⊼ ♇
21 43 ☽ ⊼ ♆
23 37 ☽ bQ ♀

21 Sunday
00 20 ☽ ± ♇
02 10 ☽ ☍ ☉
04 04 ☽ Q ♃
04 35 ☽ Q ♅
14 28 ☽ bQ ♇
17 36 ☽ ∥ ☉
19 12 ☽ ⊻ ♄
19 23 ☽ ☍ ☿
21 47 ☽ # ☉

22 Monday
00 50 ☽ △ ♀
02 26 ☿ ⊼ ♄
04 34 ☽ ⊥ ♄
08 20 ☽ □ ♃
08 49 ☽ □ ♅
10 59 ☽ △ ♂
14 03 ☽ △ ♇
18 40 ☽ ∠ ♄
19 01 ♀ Q ♆
20 57 ☽ ☍ ♆

23 Tuesday
03 05 ☽ # ☿
04 37 ☽ ⊼ ☉
10 21 ☽ # ♆
10 56 ☽ # ♄
11 36 ☽ bQ ♂
14 52 ☽ ± ☉
17 09 ☽ ⊼ ☿
18 25 ☽ ⚹ ♄

24 Wednesday
00 07 ☿ Q ♂
02 25 ☽ ± ☿
02 36 ☽ ± ♂
04 02 ☽ □ ♀
06 26 ☽ bQ ☉
08 17 ☽ △ ♃
08 41 ☽ △ ♅
11 25 ☽ ∥ ♀
12 45 ☽ ⊼ ♂
13 49 ☽ Q ♄
14 14 ☽ □ ♇
16 27 ☽ bQ ☿
21 29 ☽ ⊼ ♆

25 Thursday
07 31 ☽ ± ♆
09 04 ☽ △ ☉
09 10 ☽ bQ ♃
09 32 ☽ bQ ♅
12 08 ☽ Q ♀
16 17 ☽ △ ☿
19 48 ☽ □ ♄
20 25 ☽ Q ♇
22 50 ☽ bQ ♆

26 Friday
00 31 ☽ ± ♃
00 51 ☽ ± ♅
07 02 ♂ △ ♇
09 02 ☿ ∠ ♇
10 23 ☽ ⚹ ♀
10 55 ☽ ⊼ ♃
11 14 ☽ ⊼ ♅
17 16 ☽ ⚹ ♇
17 34 ♀ ⊼ ♃
17 40 ☽ ☍ ♂
21 27 ♀ ⊼ ♅
23 49 ♃ ∥ ♅

27 Saturday
01 05 ☽ △ ♆
07 44 ☽ ∥ ♇
15 01 ☽ # ♃
15 09 ☽ # ♅
15 17 ☽ ∠ ♀
17 36 ☽ □ ☉
18 00 ☽ □ ☿
20 10 ☽ ∠ ♇
20 41 ☉ ☌ ☿

28 Sunday
00 37 ☽ △ ♄
04 29 ♅ St D
09 13 ☽ ⊥ ♀
11 14 ☿ ∠ ♀
12 45 ☽ ⊥ ♇
13 00 ☽ # ♅
13 19 ☽ # ♃
17 15 ☽ ☍ ♃
17 26 ☽ ☍ ♅
20 50 ☽ ∥ ♇
21 23 ☽ ⊻ ♀
23 59 ☽ ⊻ ♇

29 Monday
01 14 ☽ Q ☿
02 43 ☽ ⊼ ♂
04 22 ☽ bQ ♄
04 31 ☉ ∠ ♇
05 44 ☽ Q ☉
08 23 ☽ □ ♆
14 36 ☽ ± ♂
21 25 ☽ ± ♄
22 36 ☽ ⚹ ☿

30 Tuesday
05 06 ♀ ☌ ♇
06 39 ☽ ⚹ ☉
08 40 ☽ bQ ♂
08 55 ☽ ⊼ ♄
19 07 ☽ Q ♆
21 41 ♃ ☌ ♅
23 18 ☉ ∥ ☿

31 Wednesday
01 50 ☽ ∠ ☿
02 56 ☽ ⊼ ♅
02 57 ☽ ⊼ ♃
09 33 ☉ ⊼ ♄
09 53 ☽ ☌ ♇
12 36 ☽ ☌ ♀
14 29 ☽ ∠ ☉
14 43 ☽ ± ♅
14 47 ☽ ± ♃
15 20 ☽ △ ♂
18 27 ☽ ⊥ ☿
18 44 ☽ ⚹ ♆

All ephemeris data is given at 12.00 UT and the Moon's longitude is additionally given for 24.00 UT

APRIL 2038

LONGITUDES

Date	Sidereal time h m s	Sun ☉ ° ' "	Moon ☽ ° ' "	Moon ☽ 24.00 ° ' "	Mercury ☿ ° '	Venus ♀ ° '	Mars ♂ ° '	Jupiter ♃ ° '	Saturn ♄ ° '	Uranus ♅ ° '	Neptune ♆ ° '	Pluto ♇ ° '
01	00 39 49	11 ♈ 51 22	06 ♓ 49 50	12 ♓ 50 56	03 ♈ 25	26 ♒ 03	26 ♊ 52	20 ♋ 10	10 ♍ 42	20 ♋ 06	28 ♈ 09	23 ♒ 39
02	00 43 46	12 50 35	18 ♓ 50 40	24 ♓ 49 18	02 R 41	27 08	27 25	20 13	10 R 38	20 06	28 11	23 41
03	00 47 42	13 49 47	00 ♈ 47 02	06 ♈ 44 05	02 02	28 13	27 58	20 17	10 35	20 06	28 13	23 42
04	00 51 39	14 48 56	12 ♈ 40 35	18 ♈ 36 43	01 28	29 ♒ 18	28 31	20 20	10 31	20 06	28 16	23 43
05	00 55 35	15 48 04	24 ♈ 32 38	00 ♉ 28 31	00 58	00 ♓ 23	29 04	20 24	10 28	20 07	28 18	23 44
06	00 59 32	16 47 09	06 ♉ 24 31	12 ♉ 20 52	00 33	01 29	29 ♊ 37	20 28	10 24	20 07	28 20	23 45
07	01 03 28	17 46 13	18 ♉ 17 48	24 ♉ 15 35	00 14	02 35	00 ♋ 10	20 32	10 21	20 08	28 22	23 47
08	01 07 25	18 45 14	00 ♊ 14 32	06 ♊ 15 02	00 ♈ 01	03 40	00 43	20 36	10 18	20 08	28 24	23 48
09	01 11 21	19 44 13	12 ♊ 17 27	18 ♊ 22 15	29 ♓ 52	04 46	01 16	20 41	10 14	20 09	28 27	23 49
10	01 15 18	20 43 10	24 ♊ 29 56	00 ♋ 41 01	29 50	05 53	01 50	20 46	10 11	20 10	28 29	23 50
11	01 19 15	21 42 04	06 ♋ 56 04	13 ♋ 15 38	29 ♓ 52	06 59	02 23	20 50	10 08	20 11	28 31	23 51
12	01 23 11	22 40 57	19 ♋ 40 18	26 ♋ 10 37	00 ♈ 00	08 05	02 56	20 55	10 05	20 11	28 33	23 52
13	01 27 08	23 39 47	02 ♌ 47 07	09 ♌ 30 15	00 D 13	09 12	03 30	21 00	10 03	20 12	28 36	23 53
14	01 31 04	24 38 34	16 ♌ 20 22	23 ♌ 17 42	00 31	10 19	04 03	21 06	10 00	20 13	28 38	23 54
15	01 35 01	25 37 20	00 ♍ 22 20	07 ♍ 34 10	00 53	11 26	04 37	21 11	09 57	20 14	28 40	23 55
16	01 38 57	26 36 03	14 ♍ 52 52	22 ♍ 17 53	01 20	12 33	05 11	21 17	09 55	20 15	28 42	23 56
17	01 42 54	27 34 44	29 ♍ 48 23	07 ♎ 23 23	01 51	13 40	05 45	21 22	09 52	20 16	28 45	23 57
18	01 46 50	28 33 23	15 ♎ 01 38	22 ♎ 41 46	02 27	14 47	06 18	21 28	09 50	20 17	28 47	23 58
19	01 50 47	29 ♈ 31 59	00 ♏ 22 19	08 ♏ 01 47	03 06	15 55	06 52	21 34	09 48	20 18	28 49	23 59
20	01 54 44	00 ♉ 30 34	15 ♏ 38 45	23 ♏ 11 52	03 49	17 02	07 26	21 41	09 46	20 20	28 52	24 00
21	01 58 40	01 29 08	00 ♐ 40 01	08 ♐ 02 15	04 36	18 10	08 00	21 47	09 43	20 21	28 54	24 01
22	02 02 37	02 27 39	15 ♐ 17 52	22 ♐ 26 25	05 26	19 18	08 34	21 53	09 41	20 22	28 56	24 02
23	02 06 33	03 26 09	29 ♐ 27 37	06 ♑ 21 28	06 19	20 26	09 08	22 00	09 40	20 23	28 58	24 02
24	02 10 30	04 24 37	13 ♑ 08 06	19 ♑ 47 47	07 15	21 34	09 42	22 07	09 38	20 25	29 01	24 03
25	02 14 26	05 23 04	26 ♑ 20 55	02 ♒ 47 59	08 14	22 42	10 17	22 14	09 36	20 26	29 03	24 04
26	02 18 23	06 21 29	09 ♒ 09 32	15 ♒ 26 07	09 17	23 50	10 51	22 21	09 35	20 28	29 05	24 05
27	02 22 19	07 19 52	21 ♒ 38 19	27 ♒ 46 44	10 21	24 59	11 25	22 28	09 33	20 29	29 07	24 05
28	02 26 16	08 18 14	03 ♓ 51 56	09 ♓ 54 25	11 29	26 07	12 00	22 35	09 32	20 31	29 10	24 06
29	02 30 13	09 16 34	15 ♓ 54 44	21 ♓ 53 21	12 39	27 16	12 34	22 43	09 30	20 33	29 12	24 07
30	02 34 09	10 ♉ 14 53	27 ♓ 50 40	03 ♈ 47 05	13 ♈ 52	28 ♓ 25	13 ♋ 08	22 ♋ 50	09 ♍ 29	20 ♋ 34	29 ♈ 14	24 ♒ 07

Date	Moon True ☊ ° '	Moon Mean ☊ ° '	Moon ☽ Latitude ° '
01	15 ♋ 40	15 ♋ 17	03 S 54
02	15 R 30	15 14	04 28
03	15 17	15 10	04 50
04	15 02	15 07	04 59
05	14 47	15 04	04 55
06	14 34	15 01	04 38
07	14 22	14 58	04 09
08	14 13	14 55	03 29
09	14 07	14 51	02 39
10	14 04	14 48	01 41
11	14 02	14 45	00 S 38
12	14 D 02	14 42	00 N 30
13	14 R 02	14 39	01 37
14	14 01	14 36	02 42
15	13 57	14 32	03 39
16	13 51	14 29	04 24
17	13 42	14 26	04 52
18	13 32	14 23	05 01
19	13 22	14 20	04 48
20	13 12	14 16	04 15
21	13 04	14 13	03 24
22	12 59	14 10	02 21
23	12 57	14 07	01 N 11
24	12 D 56	14 04	00 S 01
25	12 56	14 01	01 11
26	12 R 56	13 57	02 15
27	12 55	13 54	03 11
28	12 52	13 51	03 56
29	12 47	13 48	04 31
30	12 ♋ 39	13 ♋ 45	04 S 53

DECLINATIONS

Date	Sun ☉ ° '	Moon ☽ ° '	Mercury ☿ ° '	Venus ♀ ° '	Mars ♂ ° '	Jupiter ♃ ° '	Saturn ♄ ° '	Uranus ♅ ° '	Neptune ♆ ° '	Pluto ♇ ° '
01	04 N 41	12 S 37	03 N 17	12 S 14	25 N 14	22 N 23	09 N 30	22 N 25	09 N 16	22 S 09
02	05 04	08 31	02 46	11 57	25 14	22 23	09 31	22 25	09 16	22 09
03	05 27	04 S 07	02 16	11 40	25 14	22 22	09 32	22 25	09 17	22 09
04	05 50	00 N 25	01 48	11 22	25 14	22 22	09 33	22 25	09 18	22 09
05	06 13	04 56	01 21	11 04	25 14	22 21	09 35	22 25	09 19	22 09
06	06 36	09 16	00 56	10 45	25 14	22 20	09 36	22 25	09 19	22 09
07	06 58	13 17	00 34	10 26	25 14	22 20	09 37	22 24	09 20	22 08
08	07 21	16 47	00 N 14	10 07	25 13	22 19	09 38	22 24	09 21	22 08
09	07 43	19 38	00 S 03	09 47	25 13	22 19	09 39	22 24	09 22	22 08
10	08 05	21 38	00 18	09 27	25 12	22 18	09 40	22 24	09 23	22 08
11	08 27	22 38	00 30	09 07	25 11	22 17	09 42	22 24	09 23	22 08
12	08 49	22 29	00 39	08 47	25 10	22 16	09 43	22 24	09 24	22 08
13	09 11	21 07	00 46	08 26	25 09	22 16	09 44	22 24	09 25	22 08
14	09 33	18 31	00 50	08 05	25 07	22 15	09 44	22 24	09 26	22 07
15	09 54	14 45	00 52	07 43	25 06	22 14	09 45	22 23	09 27	22 07
16	10 15	10 01	00 52	07 21	25 04	22 13	09 46	22 23	09 28	22 07
17	10 37	04 N 33	00 49	06 59	25 03	22 12	09 47	22 23	09 28	22 07
18	10 58	01 S 17	00 44	06 37	25 01	22 11	09 48	22 23	09 29	22 07
19	11 18	07 06	00 36	06 14	24 58	22 11	09 49	22 23	09 30	22 07
20	11 39	12 27	00 27	05 52	24 56	22 10	09 49	22 23	09 31	22 07
21	11 59	16 57	00 15	05 28	24 54	22 09	09 50	22 22	09 31	22 07
22	12 19	20 17	00 S 02	05 05	24 51	22 08	09 51	22 22	09 32	22 07
23	12 39	22 15	00 N 13	04 42	24 49	22 06	09 51	22 22	09 33	22 07
24	12 59	22 48	00 31	04 18	24 46	22 05	09 52	22 22	09 34	22 07
25	13 19	22 02	00 49	03 54	24 43	22 04	09 52	22 21	09 35	22 07
26	13 38	20 08	01 10	03 30	24 40	22 03	09 53	22 21	09 35	22 07
27	13 57	17 18	01 32	03 06	24 36	22 02	09 53	22 21	09 36	22 07
28	14 16	13 45	01 56	02 41	24 33	22 01	09 54	22 21	09 37	22 07
29	14 35	09 43	02 21	02 17	24 29	22 00	09 54	22 20	09 38	22 07
30	14 N 53	05 S 20	02 N 48	01 S 52	24 N 26	21 N 58	09 N 54	22 N 20	09 N 39	22 S 07

ZODIAC SIGN ENTRIES

Date	h m	Planets
03	10 25	☽ ♈
05	03 23	♀ ♓
05	23 02	☽ ♉
07	04 57	♂ ♋
08	11 31	☽ ♊
08	13 08	☿ ♓
10	22 41	☽ ♋
12	11 56	☿ ♈
13	06 58	☽ ♌
15	11 22	☽ ♍
17	12 18	☽ ♎
19	11 25	☽ ♏
19	23 28	☉ ♉
21	10 55	☽ ♐
23	12 56	☽ ♑
25	18 46	☽ ♒
28	04 22	☽ ♓
30	16 21	☽ ♈

LATITUDES

Date	Mercury ☿ ° '	Venus ♀ ° '	Mars ♂ ° '	Jupiter ♃ ° '	Saturn ♄ ° '	Uranus ♅ ° '	Neptune ♆ ° '	Pluto ♇ ° '
01	02 N 06	00 N 38	01 N 50	00 N 28	02 N 06	00 N 29	01 S 40	09 S 03
04	01 19	00 22	01 49	00 28	02 06	00 29	01 40	09 04
07	00 N 31	00 N 08	01 48	00 28	02 05	00 29	01 40	09 05
10	00 S 15	00 S 07	01 47	00 28	02 05	00 29	01 40	09 06
13	00 56	00 20	01 46	00 28	02 05	00 29	01 40	09 06
16	01 31	00 33	01 44	00 28	02 04	00 29	01 40	09 07
19	02 00	00 45	01 43	00 29	02 04	00 29	01 40	09 08
22	02 23	00 55	01 42	00 29	02 04	00 29	01 40	09 09
25	02 39	01 06	01 41	00 29	02 03	00 29	01 40	09 10
28	02 50	01 15	01 40	00 29	02 03	00 29	01 40	09 10
31	02 S 55	01 S 23	01 N 39	00 N 29	02 N 03	00 N 29	01 S 40	09 S 11

DATA

Julian Date	2465515
Delta T	+77 seconds
Ayanamsa	24° 23' 12"
Synetic vernal point	04° ♓ 43' 47"
True obliquity of ecliptic	23° 26' 02"

MOON'S PHASES, APSIDES AND POSITIONS ☽

Date	h m	Phase	Longitude ° '	Eclipse Indicator
04	16 43	●	15 ♈ 01	
12	18 02	☽	22 ♋ 56	
19	10 36	○	29 ♎ 29	
26	06 15	☾	06 ♒ 07	

Day	h m		
05	05 47	Apogee	
19	04 36	Perigee	
04	09 48	0N	
11	21 08	Max dec	22° N 43'
18	06 45	0S	
24	09 39	Max dec	22° S 49'

LONGITUDES

Date	Chiron ⚷ ° '	Ceres ⚳ ° '	Pallas ⚴ ° '	Juno ⚵ ° '	Vesta ⚶ ° '	Black Moon Lilith ⚸ ° '
01	21 ♊ 12	28 ♑ 38	24 ♐ 41	26 ♏ 23	24 ♉ 38	19 ♈ 40
11	21 ♊ 42	01 ♒ 13	25 ♐ 17	25 ♏ 08	28 ♉ 40	20 ♈ 48
21	22 ♊ 18	03 ♒ 27	25 ♐ 10	23 ♏ 24	02 ♊ 47	21 ♈ 55
31	22 ♊ 59	05 ♒ 19	24 ♐ 16	21 ♏ 20	06 ♊ 58	23 ♈ 02

All ephemeris data is given at 12.00 UT and the Moon's longitude is additionally given for 24.00 UT

ASPECTARIAN

h m	Aspects
01 Thursday	
05 36	☽ ⊻ ☿
08 33	☽ ⅊ ♅
08 41	☽ ⅊ ♃
09 53	☽ ⊥ ☉
12 45	♀ ± ♅
14 33	♀ ± ♃
14 35	☽ ∥ ♀
19 40	☽ ☍ ♄
22 55	☽ ⊻ ☉
02 Friday	
00 38	☽ ∠ ♆
05 59	☿ ⊥ ♀
06 25	☽ # ♄
07 47	☽ # ♆
14 31	☽ △ ♅
14 46	☽ △ ♃
18 43	☽ ⊥ ♆
21 43	☽ ⊻ ♇
03 Saturday	
00 32	♀ △ ♂
05 24	☽ # ☉
06 03	☽ □ ♂
06 19	☽ ⊻ ♀
06 49	☽ ⊻ ♆
09 48	☽ ⊥ ♇
12 05	♀ ✶ ♆
14 24	☽ ☌ ☿
19 37	☽ ⊥ ♀
23 01	☽ # ☿
04 Sunday	
00 11	♂ ✶ ♆
03 59	☽ ∠ ♇
07 40	☽ ⊼ ♄
12 17	♂ Q ♄
15 37	☽ ∠ ♀
16 43	☽ ☌ ☉
18 36	☽ ∥ ☿
19 44	☽ ± ♄
20 08	☽ Q ♂
05 Monday	
03 02	☽ □ ♅
03 34	☽ □ ♃
10 22	☽ ✶ ♇
13 51	☽ ⅊ ♄
19 37	☽ ☌ ♆
19 37	☽ ∥ ☉
21 01	☿ ⊻ ♀
21 35	☽ ✶ ♂
06 Tuesday	
00 32	☽ ⊻ ☿
01 02	☽ ✶ ♀
03 10	☉ ± ♄
10 41	☽ Q ♇
12 17	☽ ⊥ ☿
12 18	☽ ∥ ♆
13 53	☽ ∥ ♄
15 28	☽ Q ♅
16 11	☽ Q ♃
19 59	☽ # ♀
20 02	☽ △ ♄
07 Wednesday	
03 44	☽ Q ♀
05 22	☽ ∠ ♂
05 58	☽ ∠ ☿
10 51	☽ ⊻ ☉
14 10	☿ □ ♂
15 42	☽ ✶ ♅
16 32	☽ ✶ ♃
23 03	☽ □ ♇
08 Thursday	
00 00	☽ ⊥ ☉
00 23	☽ ⊥ ♂
08 19	☽ ⊻ ♆
09 51	☉ Q ♂
11 32	☽ ✶ ☿
13 00	☽ ⊻ ♂
19 33	☽ □ ♀
19 39	☽ ∠ ☉
20 21	☽ ⊥ ♆
21 48	☽ ∠ ♅
22 47	☽ ∠ ♃
09 Friday	
04 31	☉ ∠ ♀
07 57	☽ □ ♄
11 11	☽ Q ☿
14 17	☽ ∠ ♆
15 41	☽ ⊥ ♅
16 45	☽ ⊥ ♃
20 19	♀ ⅊ ♅
21 11	♀ # ♄
22 15	☉ □ ♅
10 Saturday	
03 31	☽ ⊻ ♅
03 58	☽ ✶ ☉
04 38	☽ ⊻ ♃
06 28	☿ ⊥ ♇
09 14	♀ ⅊ ♃
10 42	☽ △ ♇
12 12	☿ St D
13 02	☉ □ ♃
17 29	♀ # ♆
19 09	☽ Q ♄
19 46	☽ ✶ ♆
20 59	☽ # ♇
22 22	☽ □ ☿
11 Sunday	
00 40	☽ ∥ ♃
02 52	☽ ☌ ♂
03 33	☽ ∥ ♅
03 46	☿ ⊥ ♇
05 17	☽ Q ☉
12 06	☽ △ ♀
15 39	☽ ⅊ ♇
18 05	☽ ✶ ♄
18 50	☽ Q ♆
12 Monday	
08 38	☽ ± ♇
10 30	☉ # ♀
12 58	☽ ☌ ♅
14 20	☽ ☌ ♃
14 31	☽ ∥ ♅
17 41	☽ ∥ ♃
18 02	☽ □ ☉
18 55	☽ ⅊ ♀
19 46	☽ ⊼ ♇
20 46	☽ # ♇
21 58	☽ ∠ ♄
13 Tuesday	
04 23	☽ □ ♆
07 16	☽ △ ☿
12 49	☽ ± ♀
13 20	☽ ⊻ ♂
14 15	☽ ⊥ ♄
17 34	☉ ✶ ♇
14 Wednesday	
00 29	☽ ⊼ ♀
00 30	☽ ⊥ ♂
00 55	☽ ⊻ ♄
04 08	☉ ∥ ♆
05 30	♀ ☍ ♄
10 31	☽ ⅊ ☿
16 55	☽ ∠ ♂
18 44	☽ ⊻ ♅
20 17	☽ ⊻ ♃
20 21	☉ ⅊ ♄
15 Thursday	
01 03	☽ ☍ ♇
01 41	☉ ∥ ♄
02 27	☽ ± ☿
03 22	☽ △ ☉
05 00	☽ ⊥ ♅
06 35	☽ ⊥ ♃
09 07	☽ △ ♆
12 53	☽ ⊼ ☿
19 24	☽ ✶ ♂
20 08	☽ ∠ ♅
21 46	☽ ∠ ♃
16 Friday	
03 53	☽ ☌ ♄
06 15	☽ ⅊ ☉
07 52	☽ ☍ ♀
10 05	☽ ⅊ ♆
10 55	☽ ∥ ☉
13 06	☽ ∥ ♄
14 32	☽ ∥ ♆
15 53	☽ Q ♂
20 43	☽ ✶ ♅
21 55	☽ ± ☉
22 26	☽ ✶ ♃
17 Saturday	
00 41	☽ ± ♆
00 50	☽ # ♀
02 39	☽ ⊼ ♇
08 12	☽ ⊼ ☉
10 18	☽ ⊼ ♆
12 14	☽ ± ♇
13 45	♀ ∠ ♆
15 23	☽ ☍ ☿
15 55	☽ Q ♅
17 42	☽ Q ♃
21 46	☽ □ ♂
18 Sunday	
02 29	☽ ⅊ ♇
03 39	☽ # ☿
03 52	☽ ⊻ ♄
09 46	☽ ∥ ☿
11 36	☽ ⊼ ♀
13 16	☽ ⊥ ♄
17 48	☉ ☌ ♆
20 15	☽ □ ♅
21 44	☽ ± ♀
22 09	☽ □ ♃
19 Monday	
02 00	☽ △ ♇
03 18	☽ ∠ ♄
08 36	☽ ∥ ♀
09 34	☽ ☍ ♆
10 36	☽ ☍ ☉
12 55	☽ ⅊ ♀
16 28	☽ ⊼ ☿
22 25	☽ # ♆
22 34	☽ △ ♂
23 49	☽ # ♄
20 Tuesday	
02 21	☽ ± ☿
02 44	☽ ✶ ♄
07 54	☽ # ☉
14 23	☽ △ ♀
17 17	☽ ⅊ ☿
19 26	☽ △ ♅
21 38	☽ △ ♃
21 41	☽ Q ♄
23 12	☽ ⅊ ♂
21 Wednesday	
01 17	☽ □ ♇
09 08	☽ ⊼ ♆
13 25	☽ ⊼ ☉
14 15	☽ ± ♂
18 44	☽ △ ☿
18 53	☽ ± ♆
19 37	☽ ⅊ ♅
22 01	☽ ⅊ ♃
23 53	☽ ± ☉
22 Thursday	
00 26	☽ ⊼ ♂
02 44	☽ □ ♄
06 34	☽ Q ♇
09 44	☽ ⅊ ♆
10 27	☽ ± ♅
13 00	☽ ± ♃
15 53	☽ ⅊ ☉
19 16	☽ □ ♀
20 31	☽ ⊼ ♅
23 09	☽ ⊼ ♃
23 Friday	
01 02	♃ # ♇
02 42	☽ ✶ ♇
07 43	♂ ⅊ ♇
09 28	☽ # ♃
09 34	☽ ∥ ♇
11 09	♀ △ ♅
11 09	☽ △ ♆
14 18	☽ # ♅
19 25	☽ △ ☉
24 Saturday	
00 47	☽ □ ☿
04 44	☽ ∠ ♇
05 04	☽ Q ♀
05 38	☽ ☍ ♂
05 47	☽ △ ♄
08 54	♂ ✶ ♄
20 51	☽ ⊥ ♇
25 Sunday	
00 55	♀ △ ♃
01 09	☽ ☍ ♅
04 22	☽ ☍ ♃
04 40	☽ ✶ ♀
05 39	☽ # ♅
07 48	☽ ⊻ ♇
08 47	☽ ⅊ ♄
10 33	☽ ∥ ♇
11 21	☽ # ♃
11 47	☽ Q ☿
17 01	☽ □ ♆
19 32	♀ ⊥ ♆
26 Monday	
01 29	☽ ± ♄
05 00	☉ Q ♇
06 15	☽ □ ☉
07 26	☿ ∠ ♇
11 20	☽ ∠ ♀
12 15	☽ ✶ ☿
12 47	☽ ⊼ ♄
15 22	☽ ⊼ ♂
17 04	♀ ⊻ ♇
18 36	☿ ⊼ ♄
22 39	♂ Q ♆
27 Tuesday	
03 13	☽ Q ♆
03 26	☽ ± ♂
06 19	☽ ⊥ ♀
09 46	☽ ⊼ ♅
13 38	☽ ⊼ ♃
16 47	☽ ☌ ♇
19 11	☽ ⊻ ♀
19 49	☽ Q ☉
19 59	☽ ∠ ☿
21 30	☽ ± ♅
21 47	☽ ⅊ ♂
28 Wednesday	
01 29	☽ ± ♃
02 41	☽ ✶ ♆
09 00	☽ # ☉
15 17	☽ ⅊ ♅
15 33	☽ ⊥ ☿
17 25	☉ Q ♅
19 28	☽ ⅊ ♃
21 35	☽ ✶ ☉
23 13	☽ ☍ ♄
29 Thursday	
04 46	☽ ⊻ ☿
04 58	☽ △ ♂
08 33	☽ ∠ ♆
08 37	☿ □ ♂
09 52	☿ # ♀
10 57	☽ # ♄
12 29	☽ # ♆
17 34	☉ △ ♄
21 19	☽ △ ♅
30 Friday	
01 48	☽ △ ♃
02 41	☽ ⊥ ♆
04 29	☽ ⊻ ♇
06 18	☽ ∠ ☉
13 16	☽ ☌ ♀
14 49	☽ ⊻ ♆
16 36	☽ ⊥ ♇

MAY 2038

Date	Sidereal time h m s	LONGITUDES Sun ☉ ° ' "	Moon ☽ ° ' "	Moon ☽ 24.00 ° ' "	Mercury ☿ ° '	Venus ♀ ° '	Mars ♂ ° '	Jupiter ♃ ° '	Saturn ♄ ° '	Uranus ♅ ° '	Neptune ♆ ° '	Pluto ♇ ° '
01	02 38 06	11 ♉ 13 09	09 ♈ 42 56	15 ♈ 38 33	15 ♈ 07	29 ♓ 33	13 ♋ 43	22 ♋ 58	09 ♍ 28	20 ♋ 36	29 ♈ 16	24 ♒ 08
02	02 42 02	12 11 25	21 ♈ 34 10	27 ♈ 30 03	16 24	00 ♈ 42	14 17	23 06	09 R 27	20 38	29 19	24 09
03	02 45 59	13 09 39	03 ♉ 26 22	09 ♉ 23 21	17 43	01 51	14 52	23 14	09 26	20 40	29 21	24 09
04	02 49 55	14 07 51	15 ♉ 21 10	21 ♉ 19 59	19 05	03 00	15 27	23 22	09 26	20 42	29 23	24 10
05	02 53 52	15 06 01	27 ♉ 19 59	03 ♊ 21 22	20 29	04 09	16 01	23 30	09 25	20 44	29 25	24 10
06	02 57 48	16 04 10	09 ♊ 24 19	15 ♊ 29 06	21 55	05 19	16 36	23 39	09 24	20 46	29 27	24 11
07	03 01 45	17 02 17	21 ♊ 35 57	27 ♊ 45 11	23 23	06 28	17 11	23 47	09 24	20 48	29 30	24 11
08	03 05 42	18 00 22	03 ♋ 57 06	10 ♋ 12 04	24 53	07 37	17 46	23 56	09 24	20 50	29 32	24 12
09	03 09 38	18 58 25	16 ♋ 30 29	22 ♋ 52 44	26 25	08 47	18 20	24 05	09 23	20 52	29 34	24 12
10	03 13 35	19 56 26	29 ♋ 19 15	05 ♌ 50 27	27 59	09 56	18 55	24 13	09 23	20 54	29 36	24 12
11	03 17 31	20 54 26	12 ♌ 26 44	19 ♌ 08 28	29 ♈ 36	11 06	19 30	24 22	09 D 23	20 56	29 38	24 13
12	03 21 28	21 52 24	25 ♌ 55 58	02 ♍ 49 28	01 ♉ 14	12 15	20 05	24 31	09 23	20 58	29 40	24 13
13	03 25 24	22 50 19	09 ♍ 49 05	16 ♍ 54 49	02 54	13 25	20 40	24 41	09 23	21 01	29 42	24 13
14	03 29 21	23 48 13	24 ♍ 06 28	01 ♎ 23 44	04 36	14 35	21 15	24 50	09 24	21 03	29 45	24 14
15	03 33 17	24 46 05	08 ♎ 46 03	16 ♎ 12 42	06 21	15 44	21 50	24 59	09 24	21 05	29 47	24 14
16	03 37 14	25 43 56	23 ♎ 42 48	01 ♏ 15 16	08 07	16 54	22 25	25 09	09 25	21 08	29 49	24 14
17	03 41 11	26 41 45	08 ♏ 48 57	16 ♏ 22 36	09 56	18 04	23 01	25 18	09 25	21 10	29 51	24 14
18	03 45 07	27 39 32	23 ♏ 54 57	01 ♐ 24 47	11 46	19 14	23 36	25 28	09 26	21 13	29 53	24 15
19	03 49 04	28 37 18	08 ♐ 50 59	16 ♐ 12 33	13 38	20 24	24 11	25 38	09 27	21 15	29 55	24 15
20	03 53 00	29 ♉ 35 03	23 ♐ 28 39	00 ♑ 38 40	15 33	21 34	24 46	25 48	09 28	21 18	29 57	24 15
21	03 56 57	00 ♊ 32 46	07 ♑ 42 07	14 ♑ 38 47	17 29	22 44	25 21	25 58	09 29	21 20	29 ♈ 59	24 15
22	04 00 53	01 30 28	21 ♑ 28 33	28 ♑ 11 31	19 28	23 55	25 57	26 08	09 30	21 23	00 ♉ 01	24 15
23	04 04 50	02 28 09	04 ♒ 47 53	11 ♒ 17 59	21 28	25 05	26 32	26 18	09 31	21 25	00 03	24 15
24	04 08 46	03 25 49	17 ♒ 42 15	24 ♒ 01 09	23 30	26 15	27 08	26 28	09 32	21 28	00 05	24 15
25	04 12 43	04 23 29	00 ♓ 15 13	06 ♓ 25 01	25 34	27 26	27 43	26 38	09 34	21 31	00 07	24 R 15
26	04 16 40	05 21 07	12 ♓ 31 08	18 ♓ 34 09	27 40	28 36	28 18	26 49	09 35	21 34	00 09	24 15
27	04 20 36	06 18 44	24 ♓ 34 37	00 ♈ 33 08	29 ♉ 47	29 ♈ 47	28 54	26 59	09 37	21 36	00 11	24 15
28	04 24 33	07 16 20	06 ♈ 30 11	12 ♈ 26 19	01 ♊ 55	00 ♉ 57	29 ♋ 29	27 10	09 39	21 39	00 13	24 15
29	04 28 29	08 13 55	18 ♈ 21 58	24 ♈ 17 36	04 05	02 08	00 ♌ 05	27 21	09 40	21 42	00 15	24 15
30	04 32 26	09 11 29	00 ♉ 13 35	06 ♉ 10 18	06 15	03 18	00 41	27 31	09 42	21 45	00 16	24 15
31	04 36 22	10 ♊ 09 03	12 ♉ 08 04	18 ♉ 07 09	08 ♊ 27	04 ♉ 29	01 ♌ 16	27 ♋ 42	09 ♍ 44	21 ♋ 48	00 ♉ 18	24 ♒ 15

Date	Moon True ☊ ° '	Moon Mean ☊ ° '	Moon ☽ Latitude ° '	DECLINATIONS Sun ☉ ° '	Moon ☽ ° '	Mercury ☿ ° '	Venus ♀ ° '	Mars ♂ ° '	Jupiter ♃ ° '	Saturn ♄ ° '	Uranus ♅ ° '	Neptune ♆ ° '	Pluto ♇ ° '
01	12 ♋ 29	13 ♋ 42	05 S 03	15 N 12	00 S 48	03 N 15	01 S 27	24 N 22	21 N 57	09 N 55	22 N 20	09 N 39	22 S 07
02	12 R 17	13 38	04 59	15 29	03 N 47	03 45	01 02	24 18	21 56	09 55	22 20	09 40	22 07
03	12 06	13 35	04 43	15 47	08 13	04 15	00 37	24 14	21 54	09 55	22 19	09 41	22 07
04	11 55	13 32	04 14	16 05	12 23	04 47	00 S 11	24 09	21 53	09 55	22 19	09 42	22 07
05	11 46	13 29	03 34	16 22	16 06	05 19	00 N 14	24 05	21 52	09 55	22 19	09 42	22 08
06	11 39	13 26	02 43	16 39	19 10	05 53	00 39	24 00	21 50	09 55	22 18	09 43	22 08
07	11 35	13 22	01 45	16 55	21 25	06 28	01 05	23 55	21 49	09 55	22 18	09 44	22 08
08	11 32	13 19	00 S 41	17 11	22 42	07 04	01 30	23 50	21 47	09 55	22 18	09 45	22 08
09	11 D 32	13 16	00 N 27	17 28	22 51	07 40	01 56	23 45	21 46	09 55	22 17	09 45	22 08
10	11 33	13 13	01 34	17 43	21 49	08 18	02 22	23 40	21 44	09 55	22 17	09 46	22 08
11	11 34	13 10	02 38	17 59	19 36	08 56	02 47	23 35	21 43	09 55	22 17	09 47	22 08
12	11 R 34	13 07	03 36	18 14	16 15	09 35	03 13	23 29	21 41	09 55	22 17	09 48	22 08
13	11 33	13 03	04 22	18 29	11 56	10 15	03 39	23 23	21 40	09 55	22 16	09 49	22 09
14	11 30	13 00	04 54	18 43	06 10	10 55	04 04	23 18	21 38	09 54	22 16	09 49	22 09
15	11 25	12 57	05 08	18 57	01 N 14	11 36	04 30	23 12	21 36	09 54	22 16	09 50	22 09
16	11 19	12 54	05 01	19 11	04 S 32	12 17	04 56	23 06	21 35	09 54	22 15	09 50	22 09
17	11 12	12 51	04 33	19 25	10 07	12 59	05 21	22 59	21 33	09 54	22 15	09 51	22 09
18	11 06	12 48	03 47	19 38	15 05	13 41	05 47	22 53	21 31	09 53	22 14	09 52	22 10
19	11 01	12 44	02 45	19 51	19 03	14 23	06 12	22 46	21 29	09 53	22 14	09 52	22 10
20	10 58	12 41	01 33	20 03	21 43	15 06	06 38	22 40	21 28	09 53	22 14	09 53	22 10
21	10 57	12 38	00 N 18	20 16	22 55	15 47	07 03	22 33	21 26	09 52	22 13	09 54	22 10
22	10 D 57	12 35	00 S 57	20 27	22 39	16 28	07 28	22 26	21 24	09 51	22 13	09 55	22 11
23	10 58	12 32	02 06	20 39	21 06	17 10	07 53	22 19	21 22	09 51	22 12	09 55	22 11
24	11 00	12 28	03 06	20 50	18 29	17 50	08 18	22 11	21 20	09 50	22 11	09 56	22 11
25	11 01	12 25	03 56	21 00	15 04	18 31	08 43	22 04	21 18	09 49	22 11	09 56	22 11
26	11 R 00	12 22	04 34	21 11	11 05	19 10	09 08	21 57	21 16	09 48	22 11	09 57	22 11
27	10 59	12 19	04 59	21 21	06 43	19 48	09 33	21 49	21 14	09 48	22 10	09 58	22 12
28	10 56	12 16	05 10	21 31	02 S 10	20 25	09 57	21 41	21 12	09 47	22 10	09 58	22 12
29	10 51	12 13	05 09	21 40	02 N 26	21 00	10 22	21 33	21 10	09 46	22 09	09 59	22 12
30	10 46	12 09	04 54	21 49	06 58	21 34	10 46	21 25	21 08	09 45	22 09	10 00	22 13
31	10 ♋ 41	12 ♋ 06	04 S 26	21 N 58	11 N 15	22 N 06	11 N 10	21 N 17	21 N 06	09 N 44	22 N 08	10 N 00	22 S 13

ZODIAC SIGN ENTRIES

Date	h	m	Planets
01	21	17	♀ ♈
03	05	03	☽ ♉
05	17	19	☽ ♊
08	04	22	☽ ♋
10	13	15	☽ ♌
11	18	01	☿ ♉
12	19	06	☽ ♍
14	21	43	☽ ♎
16	22	00	☽ ♏
18	21	44	☽ ♐
20	22	23	☉ ♊
20	22	55	☽ ♑
22	00	17	♆ ♉
23	03	16	☽ ♒
25	11	31	☽ ♓
27	14	31	☿ ♊
27	16	34	♀ ♉
27	22	53	☽ ♈
29	08	38	♂ ♌
30	11	33	☽ ♉

LATITUDES

Date	Mercury ☿ ° '	Venus ♀ ° '	Mars ♂ ° '	Jupiter ♃ ° '	Saturn ♄ ° '	Uranus ♅ ° '	Neptune ♆ ° '	Pluto ♇ ° '
01	02 S 55	01 S 23	01 N 39	00 N 29	02 N 03	00 N 29	01 S 40	09 S 11
04	02 54	01 31	01 37	00 29	02 02	00 29	01 40	09 12
07	02 49	01 37	01 36	00 29	02 02	00 29	01 40	09 13
10	02 38	01 43	01 35	00 29	02 01	00 29	01 40	09 14
13	02 22	01 47	01 34	00 29	02 01	00 29	01 40	09 15
16	02 02	01 51	01 33	00 29	02 00	00 29	01 40	09 16
19	01 38	01 54	01 32	00 29	02 00	00 29	01 40	09 17
22	01 10	01 56	01 30	00 29	01 59	00 29	01 40	09 18
25	00 39	01 58	01 29	00 29	01 59	00 29	01 40	09 19
28	00 S 08	01 58	01 28	00 29	01 59	00 29	01 40	09 19
31	00 N 24	01 S 58	01 N 27	00 N 30	01 N 58	00 N 28	01 S 41	09 S 20

DATA

Julian Date	2465545
Delta T	+77 seconds
Ayanamsa	24° 23' 15"
Synetic vernal point	04° ♓ 43' 44"
True obliquity of ecliptic	23° 26' 02"

MOON'S PHASES, APSIDES AND POSITIONS ☽

Date	h	m	Phase	Longitude ° '	Eclipse Indicator
04	09	20	●	14 ♉ 01	
12	04	18	☽	21 ♌ 34	
18	18	23	○	27 ♏ 55	
25	20	43	☾	04 ♓ 44	

Day	h	m		
02	08	50	Apogee	
17	13	40	Perigee	
29	20	25	Apogee	
01	16	09	ON	
09	03	17	Max dec	22° N 56'
15	17	09	OS	
21	19	31	Max dec	22° S 59'
28	23	18	ON	

LONGITUDES

Date	Chiron ⚷ ° '	Ceres ⚳ ° '	Pallas ⚴ ° '	Juno ⚵ ° '	Vesta ⚶ ° '	Black Moon Lilith ⚸ ° '
01	22 ♊ 59	05 ♒ 19	24 ♐ 16	21 ♏ 20	06 ♊ 58	23 ♈ 02
11	23 ♊ 44	06 ♒ 43	22 ♐ 36	19 ♏ 40	11 ♊ 12	24 ♈ 09
21	24 ♊ 33	07 ♒ 38	20 ♐ 15	16 ♏ 53	15 ♊ 29	25 ♈ 16
31	25 ♊ 25	07 ♒ 59	17 ♐ 27	14 ♏ 54	19 ♊ 47	26 ♈ 24

ASPECTARIAN

h m	Aspects
01 Saturday	
00 16	☽ # ☿
02 05	☽ ⊥ ☉
04 54	☉ Q ♃
05 51	♀ ⊻ ♆
08 14	☽ ∥ ♀
10 49	☽ ∠ ♇
11 30	☽ ⊼ ♄
15 19	☽ ⊻ ☉
18 42	☿ ± ♄
20 30	☽ □ ♂
22 46	☽ # ♀
23 38	☽ ± ♄
02 Sunday	
00 11	♀ ⊥ ♇
00 14	☽ ☌ ☿
10 06	☽ □ ♅
11 48	☽ ∥ ☿
15 08	☽ □ ♃
17 13	☽ ⚹ ♇
17 50	☽ bQ ♄
03 Monday	
03 42	☽ ☌ ♆
08 27	☽ ⊻ ♀
10 47	☽ Q ♂
17 29	☽ Q ♇
20 12	☽ ∥ ♆
21 32	☽ ∥ ♄
21 51	☽ ⊥ ♀
22 34	☽ Q ♅
04 Tuesday	
00 05	☽ △ ♄
03 54	☽ Q ♃
09 20	☽ ☌ ☉
12 11	☽ ⚹ ♂
17 53	☽ ∠ ♀
20 28	☽ ⊻ ☿
22 45	☽ ⚹ ♅
05 Wednesday	
04 16	☽ ⚹ ♃
05 41	☽ □ ♇
10 04	☽ ⊥ ☿
14 05	☽ ∥ ☉
16 10	☽ ⊻ ♆
16 17	☿ □ ♅
19 43	☽ ∠ ♂
06 Thursday	
03 01	☽ ⚹ ♀
04 09	☽ ⊥ ♆
04 45	☽ ∠ ♅
06 23	☽ ∠ ☿
10 29	☽ ∠ ♃
12 00	☽ □ ♄
14 29	☽ ⊥ ♂
22 00	☽ ∠ ♆
22 36	☽ ⊥ ♅
07 Friday	
02 17	☽ ⊻ ☉
02 54	☽ ⊻ ♂
04 26	☽ ⊥ ♃
05 13	☽ Q ♀
10 25	☽ ⊻ ♅
15 03	☽ ⊥ ☉
15 58	☽ ⚹ ☿
16 20	☽ ⊻ ♃
17 03	☽ △ ♇
17 31	☽ ∥ ♃
19 16	☿ □ ♃
20 43	☉ ⚹ ♂
22 51	☽ # ♇
23 18	☽ Q ♄
08 Saturday	
01 01	☿ ⚹ ♇
02 01	☽ ∥ ♅
03 25	☽ ⚹ ♆
04 15	☿ bQ ♄
10 01	☽ ∠ ☉
18 25	☽ Q ☿
19 46	☽ □ ♀
22 05	☽ bQ ♇
22 27	☽ ⚹ ♄
09 Sunday	
02 35	☽ Q ♆
06 11	♂ ± ♇
15 12	☽ ± ♇
15 38	☽ ☌ ♂
17 03	☽ ⚹ ☉
20 14	☽ ☌ ♅
20 52	♀ ∠ ♇
10 Monday	
00 40	♀ ⊼ ♄
02 24	☽ ☌ ♃
02 29	☽ ⊼ ♇
02 49	☽ ∠ ♄
04 19	☽ ∥ ♅
07 02	☽ # ♇
09 10	♃ ⊼ ♇
09 11	☽ □ ☿
12 31	☽ □ ♆
13 14	☽ ∥ ♃
17 14	☽ Q ☉
19 30	☽ ⊥ ♄
11 Tuesday	
06 23	♄ St D
06 28	☽ ⊻ ♄
09 19	☽ △ ♀
12 39	☿ ☌ ♆
12 42	☉ ⚹ ♅
13 55	♃ ∠ ♄
23 41	☽ ∥ ☉
12 Wednesday	
01 13	☽ ⊻ ♂
03 14	☽ ⊻ ♅
04 18	☽ □ ☉
08 59	☽ ☍ ♇
09 30	☽ ⊻ ♃
12 17	☽ ⊥ ♂
13 50	☽ ⊥ ♅
14 32	☽ bQ ♀
18 33	☽ △ ♆
19 37	☿ ∥ ♆
20 06	☽ ⊥ ♃
22 30	☽ △ ☿
23 52	☿ ∥ ♄
13 Thursday	
04 36	☽ ∠ ♂
05 28	☽ ∠ ♅
07 31	☽ ± ♀
11 16	☽ ☌ ♄
11 45	☽ ∠ ♃
18 39	☽ ⊼ ♀
19 18	☽ ∥ ☿
20 18	☽ bQ ♆
21 54	☽ ∥ ♄
22 22	☽ ∥ ♆
14 Friday	
02 56	♂ ☌ ♅
03 30	☽ bQ ☿
06 54	☽ ⚹ ♅
07 04	☽ ⚹ ♂
11 24	☽ ± ♆
11 28	☽ △ ☉
12 12	☽ ⊼ ♇
13 13	☽ ⚹ ♃
20 25	☽ ± ☿
21 19	☽ ⊼ ♆
22 06	☽ ± ♇
22 38	☉ □ ♇
23 12	☽ ∥ ♀
15 Saturday	
02 44	☽ Q ♅
03 40	☽ Q ♂
04 59	♀ ± ♄
07 33	☽ ⊼ ☿
09 05	☽ Q ♃
10 27	☿ Q ♇
12 45	☽ bQ ♇
13 02	☽ ⊻ ♄
13 44	☽ bQ ☉
18 32	☉ ⚹ ♃
22 42	☽ ⊥ ♄
16 Sunday	
00 11	☽ ☍ ♀
05 12	☽ ± ☉
07 52	☽ □ ♅
09 52	☽ □ ♂
12 50	☽ △ ♇
13 07	☽ ∠ ♄
13 45	☽ # ♀
14 18	☽ □ ♃
15 26	☽ ⊼ ☉
21 44	☽ ☍ ♆
17 Monday	
01 45	☿ Q ♅
05 16	☿ △ ♄
10 49	☽ # ♆
11 00	☽ # ♄
12 57	☽ ⚹ ♄
14 00	☽ ☍ ☿
18 Tuesday	
03 33	☽ # ☿
03 55	☽ ⊼ ♀
07 40	☽ △ ♅
08 02	☽ Q ♄
08 47	☿ Q ♂
11 28	☽ △ ♂
12 31	☽ □ ♇
14 17	☽ ± ♀
14 30	☽ △ ♃
18 23	☽ ☍ ☉
21 34	☽ ⊼ ♆
19 Wednesday	
05 57	☽ bQ ♀
07 14	☽ ± ♆
07 47	☽ bQ ♅
11 52	☿ Q ♃
12 34	☽ bQ ♂
12 58	☽ □ ♄
14 11	♄ ∥ ♆
14 37	♂ ⊼ ♇
14 55	☽ bQ ♃
17 31	☽ Q ♇
18 24	☽ # ☉
20 57	☽ ⊼ ☿
21 54	☽ bQ ♆
22 28	☽ ± ♅
22 57	♂ ∠ ♄
20 Thursday	
03 53	☽ ± ♂
05 50	☽ ± ♃
06 05	♀ □ ♅
08 19	☽ ± ☿
08 22	☽ ⊼ ♅
08 34	☽ △ ♀
08 58	☽ # ♃
13 17	☽ ⚹ ♇
14 15	☽ ⊼ ♂
15 54	☽ ⊼ ♃
18 10	☽ ∥ ♇
18 54	☽ # ♅
21 26	☉ ⊻ ♆
22 51	☽ △ ♆
22 57	☽ ⊼ ☉
21 Friday	
01 42	☽ bQ ☿
01 54	☽ # ♂
09 53	☽ ± ☉
14 40	☽ ∠ ♇
15 03	☽ △ ♄
22 Saturday	
02 35	☽ bQ ☉
06 19	☽ ⊥ ♇
07 50	☽ △ ☿
11 50	☽ ☍ ♅
16 44	☽ □ ♀
16 56	☽ ⊻ ♇
17 23	☽ bQ ♄
17 33	☽ # ♂
19 00	♀ ⚹ ♇
20 20	☽ ☍ ♂
20 24	☽ ☍ ♃
21 10	☽ # ♅
21 34	☽ ∥ ♇
22 23	♂ ☌ ♃
23 Sunday	
00 11	♀ bQ ♄
03 20	☽ □ ♆
07 25	☽ △ ☉
08 50	☽ # ♃
09 39	☽ ± ♄
11 30	☿ ⚹ ♅
16 27	☽ # ☉
20 42	☽ ⊼ ♄
24 Monday	
04 52	☽ Q ♀
12 01	♂ ∥ ♅
12 43	☽ Q ♆
12 53	♂ # ♇
16 02	☽ # ☿
17 09	♀ □ ♃
19 10	☽ ⊼ ♅
20 47	☿ □ ♇
21 12	♅ # ♇
25 Tuesday	
00 27	☽ ☌ ♇
01 10	☽ □ ☿
02 19	♇ St R
04 56	☽ ⊼ ♃
05 58	☽ ⚹ ♀
06 41	☽ ± ♅
06 51	☽ ⊼ ♂
11 44	☽ ⚹ ♆
16 42	☽ ± ♃
19 04	☽ ± ♂
20 43	☽ □ ☉
23 53	♀ □ ♂
26 Wednesday	
00 14	☽ bQ ♅
01 29	☿ ⚹ ♃
06 12	☽ ☍ ♄
10 35	☽ bQ ♃
13 38	☽ bQ ♂
14 22	☽ ∠ ♀
17 13	☽ ∠ ♆
18 20	☽ # ♆
19 11	☽ # ♄
19 32	☽ Q ☿
21 25	☉ ∥ ♃
21 57	☽ # ♀
22 15	☿ ⚹ ♂
27 Thursday	
06 02	☽ △ ♅
08 34	☉ ⊥ ♆
10 13	☽ ⊥ ♀
11 12	☽ ⊥ ♆
11 21	☽ ⊻ ♇
11 25	☽ Q ☉
12 00	☿ ⊻ ♀
16 36	☿ ⊻ ♆
16 55	☽ △ ♃
19 45	☉ ∠ ♅
20 27	♀ ☌ ♆
21 08	☽ △ ♂
23 17	☽ ⊻ ♆
23 24	☽ ⊥ ♇
23 35	☽ ⊻ ♀
28 Friday	
00 43	☽ ⚹ ☿
02 09	♀ ∥ ♄
13 15	♀ ∥ ♆
13 41	☽ ⚹ ☉
17 33	☽ ∠ ♇
18 22	☽ ⊼ ♄
29 Saturday	
01 45	☉ ∥ ♂
06 32	☽ ± ♄
13 46	☽ ∠ ☿
18 18	☿ ∥ ♃
18 47	☽ □ ♅
18 48	♂ □ ♆
22 43	☽ ∠ ☉
23 24	☉ ⊥ ♀
23 55	☽ ⚹ ♇
30 Sunday	
00 48	☽ bQ ♄
06 27	☽ □ ♃
06 37	☿ ∥ ♂
12 04	☽ ⊥ ☿
12 06	☽ ☌ ♆
12 13	☿ ⊥ ♆
12 57	☽ □ ♂
17 34	☿ ∠ ♅
18 31	☽ ⊥ ☉
18 54	☽ ☌ ♀
31 Monday	
00 09	☽ Q ♇
01 18	☉ □ ♄
02 54	☽ ⊻ ☿
03 23	☽ ∥ ♄
03 23	☉ ∥ ☿
04 50	☽ ∥ ♆
07 10	☽ △ ♄
07 17	☽ Q ♅
07 40	☽ ⊻ ☉
11 26	☽ ∥ ♀
13 37	☿ ∥ ♅
17 42	☿ # ♇
19 17	☽ Q ♃

All ephemeris data is given at 12.00 UT and the Moon's longitude is additionally given for 24.00 UT

JUNE 2038

LONGITUDES

Date	Sidereal time h m s	Sun ☉ ° ' "	Moon ☽ ° ' "	Moon ☽ 24.00 ° ' "	Mercury ☿ ° '	Venus ♀ ° '	Mars ♂ ° '	Jupiter ♃ ° '	Saturn ♄ ° '	Uranus ♅ ° '	Neptune ♆ ° '	Pluto ♇ ° '
01	04 40 19	11 ♊ 06 35	24 ♉ 07 50	00 ♊ 10 19	10 ♊ 38	05 ♉ 40	01 ♌ 52	27 ♋ 53	09 ♍ 46	21 ♋ 51	00 ♉ 20	24 ♒ 14
02	04 44 15	12 04 07	06 ♊ 14 49	12 ♊ 21 30	12 51	06 51	02 28	28 04	09 49	21 54	00 22	24 R 14
03	04 48 12	13 01 37	18 ♊ 30 31	24 ♊ 42 03	15 03	08 01	03 03	28 15	09 51	21 57	00 24	24 14
04	04 52 09	13 59 07	00 ♋ 56 14	07 ♋ 13 13	17 14	09 12	03 39	28 27	09 53	22 00	00 25	24 14
05	04 56 05	14 56 35	13 ♋ 33 09	19 ♋ 56 11	19 26	10 23	04 15	28 38	09 56	22 03	00 27	24 14
06	05 00 02	15 54 03	26 ♋ 22 29	02 ♌ 52 14	21 36	11 34	04 51	28 49	09 59	22 06	00 29	24 13
07	05 03 58	16 51 29	09 ♌ 25 36	16 ♌ 02 46	23 45	12 45	05 27	29 00	10 01	22 09	00 31	24 13
08	05 07 55	17 48 54	22 ♌ 43 52	29 ♌ 29 04	25 53	13 56	06 02	29 12	10 04	22 12	00 32	24 13
09	05 11 51	18 46 18	06 ♍ 18 29	13 ♍ 12 10	27 ♊ 59	15 07	06 38	29 23	10 07	22 16	00 34	24 12
10	05 15 48	19 43 41	20 ♍ 10 08	27 ♍ 12 20	00 ♋ 04	16 18	07 14	29 35	10 10	22 19	00 36	24 12
11	05 19 44	20 41 03	04 ♎ 18 36	11 ♎ 28 41	02 07	17 29	07 50	29 47	10 13	22 22	00 37	24 11
12	05 23 41	21 38 24	18 ♎ 42 13	25 ♎ 58 44	04 08	18 41	08 26	29 ♋ 58	10 16	22 25	00 39	24 11
13	05 27 38	22 35 44	03 ♏ 17 39	10 ♏ 38 17	06 07	19 52	09 02	00 ♌ 10	10 19	22 28	00 40	24 10
14	05 31 34	23 33 02	17 ♏ 59 52	25 ♏ 21 34	08 03	21 03	09 38	00 22	10 23	22 32	00 42	24 10
15	05 35 31	24 30 20	02 ♐ 42 30	10 ♐ 01 48	09 58	22 14	10 14	00 34	10 26	22 35	00 43	24 09
16	05 39 27	25 27 38	17 ♐ 18 36	24 ♐ 32 08	11 50	23 26	10 51	00 46	10 30	22 38	00 45	24 09
17	05 43 24	26 24 54	01 ♑ 41 40	08 ♑ 46 35	13 40	24 37	11 27	00 58	10 33	22 42	00 46	24 08
18	05 47 20	27 22 11	15 ♑ 46 23	22 ♑ 40 43	15 27	25 48	12 03	01 10	10 37	22 45	00 48	24 08
19	05 51 17	28 19 26	29 ♑ 29 18	06 ♒ 12 04	17 12	27 00	12 39	01 22	10 41	22 49	00 49	24 07
20	05 55 13	29 ♊ 16 42	12 ♒ 49 01	19 ♒ 20 15	18 55	28 11	13 15	01 34	10 45	22 52	00 51	24 07
21	05 59 10	00 ♋ 13 57	25 ♒ 46 01	02 ♓ 06 38	20 35	29 ♉ 23	13 51	01 46	10 49	22 55	00 52	24 06
22	06 03 07	01 11 11	08 ♓ 22 27	14 ♓ 33 56	22 13	00 ♊ 34	14 28	01 58	10 53	22 59	00 53	24 05
23	06 07 03	02 08 26	20 ♓ 41 34	26 ♓ 45 54	23 49	01 46	15 04	02 11	10 57	23 02	00 55	24 04
24	06 11 00	03 05 40	02 ♈ 47 27	08 ♈ 46 48	25 22	02 58	15 40	02 23	11 01	23 06	00 56	24 04
25	06 14 56	04 02 55	14 ♈ 44 32	20 ♈ 41 12	26 52	04 09	16 17	02 36	11 05	23 09	00 57	24 03
26	06 18 53	05 00 09	26 ♈ 37 22	02 ♉ 33 34	28 20	05 21	16 53	02 48	11 10	23 13	00 58	24 02
27	06 22 49	05 57 23	08 ♉ 30 19	14 ♉ 28 08	29 ♋ 46	06 33	17 30	03 00	11 14	23 16	01 00	24 01
28	06 26 46	06 54 37	20 ♉ 27 27	26 ♉ 28 42	01 ♌ 09	07 45	18 06	03 13	11 19	23 20	01 01	24 01
29	06 30 42	07 51 51	02 ♊ 32 15	08 ♊ 38 28	02 30	08 56	18 43	03 26	11 23	23 23	01 02	24 00
30	06 34 39	08 ♋ 49 06	14 ♊ 47 38	21 ♊ 00 00	03 ♌ 48	10 ♊ 08	19 ♌ 19	03 ♌ 38	11 ♍ 28	23 ♋ 27	01 ♉ 03	23 ♒ 59

Date	Moon True ☊ ° '	Moon Mean ☊ ° '	Moon ☽ Latitude ° '
01	10 ♋ 35	12 ♋ 03	03 S 47
02	10 R 31	12 00	02 56
03	10 28	11 57	01 57
04	10 27	11 54	00 S 52
05	10 D 26	11 50	00 N 17
06	10 27	11 47	01 26
07	10 28	11 44	02 33
08	10 30	11 41	03 32
09	10 31	11 38	04 21
10	10 R 31	11 34	04 55
11	10 31	11 31	05 13
12	10 30	11 28	05 12
13	10 28	11 25	04 51
14	10 25	11 22	04 10
15	10 24	11 19	03 13
16	10 22	11 15	02 04
17	10 21	11 12	00 N 48
18	10 D 21	11 09	00 S 30
19	10 22	11 06	01 44
20	10 23	11 03	02 50
21	10 24	10 59	03 46
22	10 24	10 56	04 29
23	10 25	10 53	04 58
24	10 25	10 50	05 14
25	10 R 25	10 47	05 16
26	10 25	10 44	05 04
27	10 24	10 40	04 40
28	10 24	10 37	04 03
29	10 23	10 34	03 15
30	10 ♋ 23	10 ♋ 31	02 S 17

DECLINATIONS

Date	Sun ☉ ° '	Moon ☽ ° '	Mercury ☿ ° '	Venus ♀ ° '	Mars ♂ ° '	Jupiter ♃ ° '	Saturn ♄ ° '	Uranus ♅ ° '	Neptune ♆ ° '	Pluto ♇ ° '
01	22 N 06	15 N 08	22 N 36	11 N 33	21 N 08	21 N 04	09 N 43	22 N 08	10 N 01	22 S 14
02	22 14	18 27	23 04	11 57	21 00	21 02	09 42	22 07	10 02	22 14
03	22 21	20 59	23 29	12 20	20 51	20 59	09 41	22 07	10 02	22 14
04	22 28	22 34	23 52	12 43	20 43	20 57	09 40	22 06	10 03	22 15
05	22 35	23 02	24 12	13 06	20 34	20 55	09 39	22 06	10 03	22 15
06	22 41	22 17	24 29	13 28	20 25	20 53	09 38	22 05	10 04	22 16
07	22 47	20 21	24 43	13 51	20 15	20 50	09 37	22 05	10 04	22 16
08	22 53	17 16	24 55	14 13	20 06	20 48	09 36	22 04	10 05	22 16
09	22 58	13 14	25 04	14 34	19 57	20 46	09 34	22 04	10 05	22 17
10	23 02	08 25	25 10	14 56	19 47	20 43	09 33	22 03	10 06	22 17
11	23 06	03 N 05	25 14	15 17	19 38	20 41	09 32	22 03	10 06	22 18
12	23 10	02 S 31	25 15	15 37	19 28	20 38	09 31	22 02	10 07	22 18
13	23 14	08 03	25 13	15 58	19 18	20 36	09 29	22 02	10 07	22 18
14	23 17	13 11	25 09	16 18	19 08	20 33	09 28	22 01	10 08	22 19
15	23 19	17 32	25 03	16 37	18 58	20 31	09 26	22 01	10 08	22 19
16	23 21	20 46	24 54	16 57	18 47	20 28	09 25	22 00	10 09	22 20
17	23 22	22 38	24 44	17 15	18 37	20 26	09 23	21 59	10 09	22 20
18	23 24	23 00	24 31	17 34	18 27	20 23	09 22	21 59	10 09	22 21
19	23 25	21 57	24 17	17 52	18 16	20 21	09 20	21 58	10 10	22 21
20	23 26	19 41	24 01	18 10	18 05	20 18	09 19	21 58	10 11	22 22
21	23 26	16 28	23 44	18 27	17 54	20 15	09 17	21 57	10 11	22 22
22	23 26	12 35	23 25	18 44	17 43	20 13	09 15	21 57	10 12	22 22
23	23 25	08 15	23 05	19 00	17 32	20 10	09 14	21 56	10 12	22 23
24	23 24	03 S 41	22 43	19 16	17 21	20 07	09 12	21 55	10 12	22 23
25	23 22	00 N 57	22 21	19 31	17 10	20 04	09 10	21 55	10 13	22 24
26	23 20	05 32	21 57	19 46	16 58	20 01	09 08	21 54	10 13	22 24
27	23 18	09 55	21 33	20 00	16 47	19 59	09 07	21 54	10 14	22 25
28	23 15	13 57	21 08	20 14	16 35	19 56	09 05	21 53	10 14	22 25
29	23 12	17 29	20 43	20 27	16 24	19 53	09 03	21 52	10 14	22 26
30	23 N 08	20 N 18	20 N 16	20 N 40	16 N 12	19 N 50	09 N 01	21 N 52	10 N 15	22 S 26

ZODIAC SIGN ENTRIES

Date	h	m	Planets
01	23	40	☽ ♊
04	10	12	☽ ♋
06	18	43	☽ ♌
09	00	55	☽ ♍
10	11	13	☿ ♋
11	04	44	☽ ♎
12	15	25	♃ ♌
13	06	36	☽ ♏
15	07	34	☽ ♐
17	09	09	☽ ♑
19	12	55	☽ ♒
21	06	09	☉ ♋
21	19	59	☽ ♓
22	00	28	♀ ♊
24	06	26	☽ ♈
26	18	50	☽ ♉
27	15	58	☿ ♌
29	06	59	☽ ♊

LATITUDES

Date	Mercury ☿ ° '	Venus ♀ ° '	Mars ♂ ° '	Jupiter ♃ ° '	Saturn ♄ ° '	Uranus ♅ ° '	Neptune ♆ ° '	Pluto ♇ ° '
01	00 N 34	01 S 58	01 N 26	00 N 30	01 N 58	00 N 28	01 S 41	09 S 21
04	01 02	01 56	01 25	00 30	01 57	00 28	01 41	09 22
07	01 26	01 54	01 24	00 30	01 57	00 28	01 41	09 22
10	01 44	01 52	01 22	00 30	01 57	00 28	01 41	09 23
13	01 56	01 48	01 21	00 30	01 56	00 28	01 41	09 24
16	02 01	01 44	01 20	00 30	01 56	00 28	01 41	09 25
19	01 59	01 40	01 19	00 30	01 55	00 28	01 41	09 26
22	01 50	01 34	01 17	00 30	01 55	00 28	01 42	09 27
25	01 36	01 29	01 16	00 30	01 55	00 28	01 42	09 27
28	01 16	01 23	01 15	00 31	01 54	00 28	01 42	09 28
31	00 N 51	01 S 16	01 N 14	00 N 31	01 N 54	00 N 28	01 S 42	09 S 29

DATA

Julian Date	2465576
Delta T	+77 seconds
Ayanamsa	24° 23' 20"
Synetic vernal point	04° ♓ 43' 39"
True obliquity of ecliptic	23° 26' 01"

MOON'S PHASES, APSIDES AND POSITIONS ☽

Date	h	m	Phase	Longitude ° '	Eclipse Indicator
03	00	24	●	12 ♊ 34	
10	11	11	☽	19 ♍ 42	
17	02	30	○	26 ♐ 02	
24	12	40	☾	03 ♈ 07	

Day	h	m		
14	15	35	Perigee	
26	13	00	Apogee	
05	09	24	Max dec	23° N 02'
12	01	16	0S	
18	05	58	Max dec	23° S 03'
25	07	05	0N	

LONGITUDES

Date	Chiron ⚷ ° '	Ceres ⚳ ° '	Pallas ⚴ ° '	Juno ⚵ ° '	Vesta ⚶ ° '	Black Moon Lilith ⚸ ° '
01	25 ♊ 30	07 ♒ 59	17 ♐ 09	14 ♏ 43	20 ♊ 13	26 ♈ 30
11	26 ♊ 24	07 ♒ 41	14 ♐ 12	13 ♏ 10	24 ♊ 33	27 ♈ 38
21	27 ♊ 18	06 ♒ 47	11 ♐ 27	12 ♏ 06	28 ♊ 53	28 ♈ 45
31	28 ♊ 26	05 ♒ 21	09 ♐ 11	11 ♏ 34	03 ♋ 13	29 ♈ 52

All ephemeris data is given at 12.00 UT and the Moon's longitude is additionally given for 24.00 UT

ASPECTARIAN

h m	Aspects
01 Tuesday	
02 23	☿ □ ♄
03 03	☽ Q ♂
07 26	☽ ✶ ♅
12 13	☽ □ ♇
16 41	☉ ∥ ♅
19 35	☽ ✶ ♃
21 04	☉ ☌ ☿
23 44	♀ Q ♇
02 Wednesday	
00 21	☽ ⊻ ♆
04 08	☽ ✶ ♂
05 56	♂ ∥ ♃
12 01	☿ ⊥ ♀
12 14	☽ ⊥ ♆
12 18	☉ # ♇
13 17	☽ ∠ ♅
13 18	☽ ⊻ ♀
14 44	☿ ∠ ♃
19 02	☽ □ ♄
03 Thursday	
00 24	☽ ☌ ☉
01 36	☽ ∠ ♃
02 20	☽ ⊥ ♀
03 47	☽ ☌ ☿
05 55	☽ ∠ ♆
07 00	☽ ⊥ ♅
10 35	☽ ∥ ♂
11 04	☽ ∠ ♂
12 01	☽ ∥ ♃
15 53	☿ ∠ ♆
18 42	☽ ⊻ ♅
19 07	☉ ∠ ♃
19 23	☽ ⊥ ♃
21 41	☽ ∠ ♀
22 07	☿ ⊥ ♅
23 06	☽ △ ♇
04 Friday	
03 11	☽ ∥ ♅
05 22	☽ ⊥ ♂
05 32	☽ # ♇
06 08	☽ Q ♄
07 08	☽ ⊻ ♃
09 45	☽ ∥ ☉
11 01	☽ ✶ ♆
17 27	☽ ⊻ ♂
22 21	♂ ⊥ ♄
05 Saturday	
02 27	♀ △ ♄
03 49	☽ ₽ ♇
04 52	♀ Q ♅
05 08	☽ ✶ ♄
05 24	☽ ✶ ♀
09 18	☿ ∠ ♂
09 55	☽ Q ♆
14 50	☽ ⊻ ☉
20 47	☽ ± ♇
06 Sunday	
01 11	☉ ∠ ♆
01 18	☽ ⊻ ☿
03 00	☽ ⊥ ☉
04 01	☽ ☌ ♅
04 22	☽ ∥ ☉
06 15	☽ Q ♀
08 00	☽ ⊼ ♇
09 24	☽ ∠ ♄
12 31	☽ # ♇
14 43	☽ ⊥ ☿
15 21	☽ ∥ ♅
16 36	☽ ☌ ♃
17 19	☉ ⊥ ♅
17 44	☿ ⊻ ♅
19 37	☽ □ ♆
21 02	☽ ∠ ☉
07 Monday	
02 05	☽ ⊥ ♄
02 53	☿ ⊥ ♃
04 22	☽ ☌ ♂
06 59	☽ ∥ ♃
10 32	☽ ∠ ☿
12 51	☽ ∥ ♂
13 05	☽ ⊻ ♄
17 11	☿ △ ♇
18 38	☽ □ ♀
08 Tuesday	
02 30	☽ ✶ ☉
11 03	☽ ⊻ ♅
14 38	☽ ☍ ♇
18 40	☽ ✶ ☿
21 47	☽ ⊥ ♅
23 40	☽ ⊻ ♃
09 Wednesday	
01 32	☽ Q ☉
01 53	☽ △ ♆
05 10	☽ ∥ ♀
10 22	☽ ⊥ ♃
12 36	☽ ⊻ ♂
13 29	☿ Q ♄
13 40	☽ ∠ ♅
18 40	☽ ☌ ♄
19 34	☽ Q ☿
23 31	☽ ⊥ ♂
10 Thursday	
02 15	☽ ∠ ♃
03 59	☽ ∥ ♆
04 07	☽ ₽ ♆
04 44	☽ △ ♀
05 47	☿ ⊻ ♃
06 36	☽ ∥ ♄
11 11	☽ □ ☉
15 41	☽ ✶ ♅
15 42	☽ ∠ ♂
18 12	☿ ✶ ♆
18 53	☽ ⊼ ♇
19 34	☽ ± ♆
11 Friday	
04 15	☽ ✶ ♃
05 03	☽ ± ♇
05 46	☽ ⊼ ♆
07 21	☿ ⊥ ♂
07 41	☽ □ ☿
08 39	☽ ₽ ♀
12 06	☽ Q ♅
18 11	☽ ✶ ♂
18 58	♀ Q ♃
20 11	☽ ₽ ♇
21 56	☽ ⊻ ♄
22 43	☿ ∠ ♀
12 Saturday	
00 40	☽ Q ♃
01 06	☽ ± ♀
07 57	☽ ⊥ ♄
11 57	☽ ⊼ ♀
14 59	☽ Q ♂
17 12	☽ △ ☉
18 10	☽ □ ♅
21 02	☽ △ ♇
22 52	☽ ∠ ♄
13 Sunday	
06 49	☽ □ ♃
07 42	☽ ☍ ♆
08 47	☉ ⊻ ♅
17 19	☽ △ ☿
18 25	☽ # ♄
19 31	☽ ₽ ☉
21 23	☽ # ♆
21 47	☽ □ ♂
23 32	☽ ✶ ♄
14 Monday	
11 13	☽ ± ☉
17 25	☽ ☍ ♀
19 10	☽ Q ♄
19 25	☽ △ ♅
21 29	☽ ₽ ☿
21 41	☽ ⊼ ☉
22 03	☽ □ ♇
15 Tuesday	
01 53	☿ ₽ ♇
03 19	☉ △ ♇
06 00	☽ # ♀
08 27	☽ △ ♃
08 45	☽ ⊼ ♆
13 50	☉ ⊥ ♃
14 21	☽ ± ☿
17 14	☿ ⊻ ♂
18 17	☿ ✶ ♄
18 35	☽ ± ♆
19 21	♀ ✶ ♅
20 01	☽ ₽ ♅
20 43	♂ ⊻ ♄
21 00	☽ # ♂
16 Wednesday	
00 43	☽ □ ♄
00 53	☽ △ ♂
01 38	☽ ⊼ ☿
03 29	☽ Q ♇
09 20	☽ # ♃
09 24	☽ ₽ ♃
09 25	☽ ₽ ♆
10 08	♃ □ ♆
10 53	☽ ± ♅
20 53	☽ ⊼ ♅
23 04	☽ ⊼ ♀
23 21	☽ ✶ ♇
17 Thursday	
00 09	☿ Q ♆
00 33	☽ ± ♃
01 28	☽ # ♅
02 27	♀ □ ♇
02 30	☽ ☍ ☉
02 48	☽ ₽ ♂
06 31	☽ ∥ ♇
10 01	☽ ± ♀
10 27	☽ △ ♆
10 45	☽ ⊼ ♃
13 59	☉ ∠ ♂
18 37	☽ ± ♂
18 Friday	
00 37	☽ ∠ ♇
02 40	☽ ₽ ♀
03 06	☽ △ ♄
05 18	☽ ⊼ ♂
11 22	☽ ☍ ☿
16 04	☽ ⊥ ♇
19 Saturday	
00 11	☽ ☍ ♅
02 32	☽ ⊻ ♇
05 14	☽ ₽ ♄
05 43	☽ ∥ ♇
07 10	☽ △ ♀
09 47	☽ ⊼ ☉
11 38	☽ # ♅
14 22	☽ □ ♆
15 23	☽ ☍ ♃
21 17	☽ ± ☉
21 19	☽ ± ♄
21 38	☉ Q ♄
20 Sunday	
00 43	☿ ± ♇
06 21	☽ # ♃
08 12	☽ ⊼ ♄
08 14	♀ ∥ ♂
12 50	☽ ☍ ♂
14 53	☽ ₽ ☉
23 05	☽ # ♀
23 06	☽ Q ♆
21 Monday	
00 53	☽ ⊼ ☿
01 22	☽ # ♂
06 39	☽ ⊼ ♅
08 52	☽ ☌ ♇
13 46	☽ ± ☿
17 59	☽ ± ♅
19 31	☽ □ ♀
21 07	☽ △ ☉
21 39	☽ ✶ ♆
23 32	☽ ⊼ ♃
22 Tuesday	
04 20	☉ ✶ ♆
09 27	☽ ₽ ☿
11 01	☉ ∥ ☿
11 13	☽ ± ♃
11 14	☽ ₽ ♅
16 52	☽ ☍ ♄
18 29	♀ ⊻ ♆
23 52	☿ ☌ ♅
23 Wednesday	
00 25	☽ ⊼ ♂
01 26	☽ # ♆
02 36	☽ ∠ ♆
04 59	☽ ₽ ♃
06 42	☽ # ♄
09 59	☽ Q ♀
12 47	☽ ± ♂
13 16	☉ ⊻ ♃
16 02	☿ ⊼ ♇
16 39	☽ △ ♅
18 40	☽ ⊻ ♇
19 03	☽ △ ☿
20 20	☽ ⊥ ♆
22 03	♀ ✶ ♃
24 Thursday	
06 34	☽ ⊥ ♇
07 33	☽ ₽ ♂
08 17	☽ ⊻ ♆
11 11	☽ △ ♃
12 23	☽ ✶ ♀
12 40	☽ □ ☉
22 53	☿ ∠ ♄
25 Friday	
00 33	☽ ∠ ♇
01 23	☉ ⊻ ♀
04 36	☽ ⊼ ♄
08 52	☿ # ♇
15 16	☽ △ ♂
16 46	☽ ± ♄
17 04	♀ Q ♂
21 54	☽ ∠ ♀
26 Saturday	
04 02	☽ Q ☉
05 04	☽ □ ♅
06 47	☽ ✶ ♇
11 04	☽ ₽ ♄
15 20	☿ ∥ ♅
15 57	☽ □ ☿
18 08	☽ ⊥ ♀
20 49	☽ ☌ ♆
27 Sunday	
00 42	☽ □ ♃
06 25	☽ ✶ ☉
07 00	☽ Q ♇
07 30	☽ ∥ ♄
07 36	☽ ⊻ ♀
09 40	♀ ∥ ♃
13 45	☽ ∥ ♆
17 32	☽ △ ♄
17 36	☽ Q ♅
21 07	♀ ⊥ ♆
28 Monday	
07 02	☽ □ ♂
09 03	☽ Q ☿
09 30	☿ □ ♆
13 33	☽ Q ♃
15 09	☽ ∠ ☉
17 46	☽ ✶ ♅
19 05	☽ □ ♇
29 Tuesday	
00 24	♀ ∠ ♅
04 34	☽ ∥ ♂
09 01	☽ ⊻ ♆
10 33	☽ ⊥ ☉
11 55	☽ ✶ ☿
13 47	☽ ✶ ♃
20 38	☽ Q ♂
20 52	☽ ⊥ ♆
21 16	☿ ∥ ♀
23 22	☽ ⊻ ☉
23 34	☽ ∠ ♅
30 Wednesday	
01 57	☽ ☌ ♀
05 29	☽ □ ♄
07 37	☽ ∥ ♃
08 24	☿ ☌ ♃
11 48	☽ ∥ ☿
14 27	☽ ∠ ♆
16 06	☉ ₽ ♇
16 17	☽ ∥ ♀
17 10	☽ ⊥ ♅
19 34	☽ ∠ ♃
20 38	☽ ∠ ☿
21 13	☽ ✶ ♂

JULY 2038

LONGITUDES

Date	Sidereal time h m s	Sun ☉ ° ' "	Moon ☽ ° ' "	Moon ☽ 24.00 ° ' "	Mercury ☿ ° '	Venus ♀ ° '	Mars ♂ ° '	Jupiter ♃ ° '	Saturn ♄ ° '	Uranus ♅ ° '	Neptune ♆ ° '	Pluto ♇ ° '
01	06 38 36	09 ♋ 46 20	27 ♊ 15 45	03 ♋ 35 02	05 ♌ 03	11 ♊ 20	19 ♌ 56	03 ♌ 51	11 ♍ 33	23 ♋ 31	01 ♉ 04	23 ♒ 58
02	06 42 32	10 43 34	09 ♋ 57 59	16 ♋ 24 38	06 16	12 32	20 32	04 04	11 38	23 34	01 05	23 R 57
03	06 46 29	11 40 48	22 ♋ 55 00	29 ♋ 29 04	07 26	13 44	21 09	04 16	11 42	23 38	01 06	23 56
04	06 50 25	12 38 02	06 ♌ 06 45	12 ♌ 47 59	08 33	14 56	21 45	04 29	11 47	23 41	01 07	23 55
05	06 54 22	13 35 16	19 ♌ 32 37	26 ♌ 20 31	09 37	16 08	22 22	04 42	11 52	23 45	01 08	23 55
06	06 58 18	14 32 29	03 ♍ 11 32	10 ♍ 05 27	10 38	17 20	22 59	04 55	11 58	23 49	01 09	23 54
07	07 02 15	15 29 42	17 ♍ 02 05	24 ♍ 01 14	11 36	18 32	23 36	05 07	12 03	23 52	01 10	23 53
08	07 06 11	16 26 55	01 ♎ 02 38	08 ♎ 06 05	12 31	19 45	24 12	05 20	12 08	23 56	01 11	23 52
09	07 10 08	17 24 08	15 ♎ 11 18	22 ♎ 18 00	13 23	20 57	24 49	05 33	12 13	23 59	01 11	23 51
10	07 14 05	18 21 21	29 ♎ 25 53	06 ♏ 34 37	14 11	22 09	25 26	05 46	12 19	24 03	01 13	23 50
11	07 18 01	19 18 33	13 ♏ 43 51	20 ♏ 53 13	14 56	23 21	26 03	05 59	12 24	24 07	01 13	23 49
12	07 21 58	20 15 46	28 ♏ 02 17	05 ♐ 10 37	15 36	24 33	26 40	06 12	12 30	24 10	01 14	23 48
13	07 25 54	21 12 58	12 ♐ 17 48	19 ♐ 23 20	16 14	25 46	27 17	06 25	12 35	24 14	01 15	23 46
14	07 29 51	22 10 11	26 ♐ 26 47	03 ♑ 27 41	16 47	26 58	27 53	06 38	12 41	24 18	01 15	23 45
15	07 33 47	23 07 24	10 ♑ 25 06	17 ♑ 20 06	17 16	28 11	28 30	06 51	12 47	24 21	01 16	23 44
16	07 37 44	24 04 37	24 ♑ 10 51	00 ♒ 57 32	17 40	29 ♊ 23	29 07	07 04	12 52	24 25	01 17	23 43
17	07 41 40	25 01 50	07 ♒ 39 54	14 ♒ 17 46	18 01	00 ♋ 35	29 ♌ 44	07 17	12 58	24 29	01 17	23 42
18	07 45 37	25 59 04	20 ♒ 51 01	27 ♒ 19 39	18 16	01 48	00 ♍ 21	07 31	13 04	24 32	01 18	23 41
19	07 49 34	26 56 18	03 ♓ 43 43	10 ♓ 03 21	18 27	03 00	00 58	07 44	13 10	24 36	01 18	23 40
20	07 53 30	27 53 33	16 ♓ 18 46	22 ♓ 30 14	18 34	04 13	01 35	07 57	13 16	24 40	01 19	23 39
21	07 57 27	28 50 48	28 ♓ 38 07	04 ♈ 42 48	18 R 35	05 26	02 13	08 10	13 22	24 43	01 19	23 37
22	08 01 23	29 ♋ 48 05	10 ♈ 44 45	16 ♈ 44 27	18 32	06 38	02 50	08 23	13 28	24 47	01 20	23 36
23	08 05 20	00 ♌ 45 22	22 ♈ 42 28	28 ♈ 39 19	18 23	07 51	03 27	08 36	13 34	24 51	01 20	23 35
24	08 09 16	01 42 39	04 ♉ 35 37	10 ♉ 31 56	18 10	09 04	04 04	08 50	13 40	24 54	01 21	23 34
25	08 13 13	02 39 58	16 ♉ 28 53	22 ♉ 27 02	17 52	10 17	04 41	09 03	13 47	24 58	01 21	23 33
26	08 17 09	03 37 18	28 ♉ 26 59	04 ♊ 29 17	17 29	11 29	05 19	09 16	13 53	25 02	01 21	23 31
27	08 21 06	04 34 38	10 ♊ 34 28	16 ♊ 43 01	17 01	12 42	05 56	09 29	13 59	25 05	01 22	23 30
28	08 25 03	05 32 00	22 ♊ 55 23	29 ♊ 11 57	16 30	13 55	06 33	09 42	14 06	25 09	01 22	23 29
29	08 28 59	06 29 22	05 ♋ 33 03	11 ♋ 58 54	15 55	15 08	07 11	09 56	14 12	25 12	01 22	23 28
30	08 32 56	07 26 46	18 ♋ 29 40	25 ♋ 06 25	15 16	16 21	07 48	10 09	14 19	25 16	01 22	23 26
31	08 36 52	08 ♌ 24 10	01 ♌ 46 08	08 ♌ 31 40	14 ♌ 35	17 ♋ 34	08 ♍ 26	10 ♌ 22	14 ♍ 25	25 ♋ 20	01 ♉ 22	23 ♒ 25

Date	Moon True ☊ ° '	Moon Mean ☊ ° '	Moon ☽ Latitude ° '
01	10 ♋ 23	10 ♋ 28	01 S 12
02	10 D 23	10 25	00 S 02
03	10 R 23	10 21	01 N 09
04	10 23	10 18	02 18
05	10 23	10 15	03 20
06	10 22	10 12	04 13
07	10 22	10 09	04 51
08	10 21	10 05	05 13
09	10 21	10 02	05 16
10	10 D 21	09 59	04 59
11	10 22	09 56	04 25
12	10 22	09 53	03 34
13	10 23	09 50	02 29
14	10 24	09 46	01 N 16
15	10 24	09 43	00 00
16	10 R 24	09 40	01 S 15
17	10 23	09 37	02 25
18	10 21	09 34	03 25
19	10 19	09 31	04 13
20	10 16	09 27	04 48
21	10 14	09 24	05 09
22	10 12	09 21	05 15
23	10 11	09 18	05 07
24	10 D 11	09 15	04 47
25	10 11	09 11	04 14
26	10 13	09 08	03 30
27	10 14	09 05	02 36
28	10 16	09 02	01 34
29	10 17	08 59	00 S 26
30	10 R 17	08 56	00 N 45
31	10 ♋ 15	08 ♋ 52	01 N 55

DECLINATIONS

Date	Sun ☉	Moon ☽	Mercury ☿	Venus ♀	Mars ♂	Jupiter ♃	Saturn ♄	Uranus ♅	Neptune ♆	Pluto ♇
01	23 N 04	22 N 12	19 N 50	20 N 53	16 N 00	19 N 47	08 N 59	21 N 51	10 N 15	22 S 27
02	23 00	23 01	19 23	21 04	15 48	19 44	08 57	21 51	10 15	22 28
03	22 55	22 37	18 56	21 15	15 36	19 41	08 55	21 50	10 15	22 28
04	22 50	20 58	18 28	21 26	15 23	19 38	08 53	21 49	10 16	22 29
05	22 44	18 07	18 01	21 36	15 11	19 35	08 51	21 49	10 16	22 29
06	22 38	14 15	17 34	21 46	14 59	19 32	08 49	21 48	10 16	22 30
07	22 32	09 35	17 07	21 54	14 46	19 29	08 47	21 47	10 17	22 30
08	22 25	04 N 22	16 40	22 03	14 34	19 26	08 45	21 47	10 17	22 31
09	22 18	01 S 08	16 13	22 10	14 21	19 23	08 43	21 46	10 17	22 31
10	22 11	06 36	15 47	22 17	14 08	19 20	08 41	21 46	10 17	22 32
11	22 03	11 45	15 22	22 24	13 55	19 17	08 39	21 45	10 18	22 32
12	21 54	16 15	14 57	22 30	13 42	19 13	08 36	21 44	10 18	22 33
13	21 46	19 48	14 33	22 35	13 29	19 10	08 34	21 44	10 18	22 33
14	21 37	22 07	14 10	22 40	13 16	19 07	08 32	21 43	10 18	22 34
15	21 27	23 02	13 48	22 43	13 03	19 04	08 30	21 42	10 18	22 35
16	21 17	22 31	13 27	22 47	12 49	19 01	08 27	21 42	10 18	22 35
17	21 07	20 41	13 07	22 49	12 36	18 57	08 25	21 41	10 18	22 36
18	20 57	17 47	12 49	22 51	12 22	18 54	08 23	21 41	10 18	22 36
19	20 46	14 04	12 32	22 53	12 09	18 51	08 20	21 40	10 19	22 37
20	20 35	09 49	12 17	22 53	11 55	18 47	08 18	21 39	10 19	22 37
21	20 23	05 16	12 04	22 53	11 42	18 44	08 16	21 39	10 19	22 38
22	20 11	00 S 35	11 52	22 53	11 28	18 41	08 13	21 38	10 19	22 38
23	19 59	04 N 04	11 43	22 52	11 14	18 37	08 11	21 37	10 19	22 39
24	19 46	08 32	11 35	22 50	11 00	18 34	08 08	21 37	10 19	22 40
25	19 34	12 42	11 30	22 47	10 46	18 31	08 06	21 36	10 19	22 40
26	19 20	16 24	11 27	22 44	10 32	18 27	08 03	21 35	10 19	22 41
27	19 07	19 27	11 26	22 40	10 18	18 24	08 01	21 35	10 19	22 41
28	18 53	21 41	11 27	22 35	10 03	18 20	07 58	21 34	10 19	22 42
29	18 39	22 53	11 31	22 30	09 49	18 17	07 56	21 33	10 19	22 42
30	18 24	22 54	11 37	22 24	09 35	18 13	07 53	21 33	10 19	22 42
31	18 N 10	21 N 38	11 N 45	22 N 18	09 N 20	18 N 10	07 N 51	21 N 32	10 N 19	22 S 43

ZODIAC SIGN ENTRIES

Date	h	m	Planets
01	17	13	☽ ♋
04	00	56	☽ ♌
06	06	25	☽ ♍
08	10	13	☽ ♎
10	12	57	☽ ♏
12	15	18	☽ ♐
14	18	04	☽ ♑
16	22	18	☽ ♒
17	00	16	♀ ♋
17	22	07	♂ ♍
19	05	00	☽ ♓
21	14	41	☽ ♈
22	17	00	☉ ♌
24	02	43	☽ ♉
26	15	05	☽ ♊
29	01	31	☽ ♋
31	08	50	☽ ♌

LATITUDES

Date	Mercury ☿	Venus ♀	Mars ♂	Jupiter ♃	Saturn ♄	Uranus ♅	Neptune ♆	Pluto ♇
01	00 N 51	01 S 16	01 N 14	00 N 31	01 N 54	00 N 28	01 S 42	09 S 29
04	00 N 22	01 09	01 12	00 31	01 53	00 28	01 42	09 30
07	00 S 12	01 02	01 11	00 31	01 53	00 28	01 42	09 30
10	00 49	00 55	01 10	00 31	01 53	00 28	01 43	09 31
13	01 29	00 47	01 08	00 31	01 53	00 28	01 43	09 32
16	02 11	00 39	01 07	00 32	01 52	00 28	01 43	09 32
19	02 54	00 31	01 06	00 32	01 52	00 28	01 43	09 33
22	03 34	00 23	01 04	00 32	01 52	00 28	01 43	09 33
25	04 11	00 15	01 03	00 32	01 52	00 29	01 43	09 34
28	04 38	00 S 07	01 02	00 32	01 51	00 29	01 44	09 34
31	04 S 55	00 N 01	01 N 01	00 N 33	01 N 51	00 N 29	01 S 44	09 S 34

DATA

Julian Date	2465606
Delta T	+77 seconds
Ayanamsa	24° 23' 25"
Synetic vernal point	04° ♓ 43' 34"
True obliquity of ecliptic	23° 26' 01"

MOON'S PHASES, APSIDES AND POSITIONS ☽

Date	h	m	Phase	Longitude ° '	Eclipse Indicator
02	13	32	●	10 ♋ 47	Annular
09	16	01	☽	17 ♎ 34	
16	11	48	○	24 ♑ 04	
24	05	40	☾	01 ♉ 28	

Day	h	m		
11	19	42	Perigee	
24	07	18	Apogee	
02	16	19	Max dec	23° N 02'
09	07	07	0S	
15	15	11	Max dec	23° S 02'
22	14	58	0N	
30	09	59	Max dec	23° N 03'

LONGITUDES

Date	Chiron ⚷ ° '	Ceres ⚳ ° '	Pallas ⚴ ° '	Juno ⚵ ° '	Vesta ⚶ ° '	Black Moon Lilith ⚸ ° '
01	28 ♊ 12	05 ♒ 21	09 ♐ 11	11 ♏ 34	03 ♋ 13	29 ♈ 52
11	29 ♊ 05	03 ♒ 28	07 ♐ 34	11 ♏ 34	07 ♋ 32	00 ♉ 59
21	29 ♊ 56	01 ♒ 19	06 ♐ 42	12 ♏ 05	11 ♋ 51	02 ♉ 06
31	00 ♋ 45	29 ♑ 08	06 ♐ 34	13 ♏ 04	16 ♋ 08	03 ♉ 13

All ephemeris data is given at 12.00 UT and the Moon's longitude is additionally given for 24.00 UT

ASPECTARIAN

h m Aspects

01 Thursday
04 47 ☽ ⊻ ♅
05 43 ☽ △ ♇
06 29 ☽ ∥ ♅
13 08 ☽ ⊥ ♃
14 40 ☿ ∥ ♃
15 47 ☽ ⊥ ☿
16 22 ☽ Q ♄
16 28 ♀ □ ♄
16 38 ☽ # ♇
19 15 ☽ ✶ ♆
22 18 ☿ ⊥ ♄

02 Friday
00 42 ☽ ⊻ ♃
03 16 ☽ ∠ ♂
04 20 ☽ ⊻ ☿
10 06 ☽ ⅊ ♇
10 32 ☽ ∥ ☉
13 32 ☽ ☌ ☉
15 07 ☽ ✶ ♄
17 17 ☽ ⊻ ♀
17 50 ☽ Q ♆
20 57 ☽ ⊥ ♂

03 Saturday
01 08 ☽ ∥ ☉
02 50 ☽ ± ♇
05 33 ☽ ⊥ ♀
08 35 ☽ ⊻ ♂
12 44 ☉ ✶ ♄
13 19 ☽ ☌ ♅
13 52 ☽ ⊼ ♇
15 16 ☽ # ♇
18 59 ☽ ∠ ♄
23 42 ☽ ∠ ♀

04 Sunday
01 42 ☽ ∥ ♅
02 58 ☽ □ ♆
07 08 ☽ ∥ ♀
09 01 ☽ ☌ ♃
11 25 ☽ ⊥ ♄
16 46 ☽ ☌ ☿
22 15 ☽ ⊻ ♄

05 Monday
00 27 ☉ Q ♆
00 36 ☽ ⊻ ☉
00 37 ☽ ∥ ♃
05 22 ☽ ✶ ♀
12 00 ♀ ∠ ♆
12 05 ☽ ⊥ ☉
12 52 ☽ ∥ ☿
17 14 ☽ ☌ ♂
19 28 ☽ ⊻ ♅
19 42 ☽ ☍ ♇

06 Tuesday
04 37 ☽ Q ♀
05 08 ☽ ∠ ☉
06 04 ☽ ⊥ ♅
07 42 ☽ ∥ ♂
08 26 ☽ △ ♆
15 03 ☽ ⊻ ♃
18 14 ♀ ∥ ♅
21 49 ☽ ∠ ♅
21 55 ♀ ⊥ ♅

07 Wednesday
01 38 ☽ ⊥ ♃
01 56 ☽ ⊻ ☿
03 20 ☽ ☌ ♄
08 38 ☽ ∥ ♆
09 09 ☽ ✶ ☉
10 30 ☽ ⅊ ♆
13 03 ☽ ⊥ ☿
14 07 ♅ ⊼ ♇
14 50 ☽ □ ♀
15 50 ☽ ∥ ♄
17 24 ☽ ∠ ♃
18 20 ☉ # ♇
22 51 ♂ ☍ ♇
23 44 ☽ ⊼ ♇
23 47 ☽ ⊻ ♂
23 48 ☽ ✶ ♅

08 Thursday
00 04 ♂ ⊻ ♅
00 33 ☿ ⊻ ♄
01 59 ☽ ± ♆
05 35 ☽ ∠ ☿
07 15 ☽ Q ☉
09 59 ☽ ± ♇
10 30 ☽ ⊥ ♂
12 14 ☽ ⊼ ♆
19 25 ☽ ✶ ♃
20 21 ☽ Q ♅

09 Friday
01 16 ☽ ⅊ ♇
02 30 ☽ ∠ ♂
02 30 ♀ ∠ ♃
06 57 ☽ ⊻ ♄
08 45 ☽ ✶ ☿
16 01 ☽ □ ☉
16 03 ☽ Q ♃
17 09 ☽ ⊥ ♄
22 37 ☽ △ ♀
22 55 ☉ ± ♇

10 Saturday
00 36 ☉ ∥ ♀
02 35 ☽ △ ♇
02 55 ☽ □ ♅
04 58 ☽ ✶ ♂
06 13 ☽ Q ☿
08 25 ☽ ∠ ♄
14 59 ☽ ☍ ♆
21 23 ☽ # ♄
22 49 ☽ □ ♃

11 Sunday
02 02 ☽ Q ♂
02 09 ☽ ⅊ ♀
04 59 ☽ # ♆
09 45 ☽ ✶ ♄
14 06 ☽ □ ☿
18 38 ☽ ± ♀
20 58 ♀ △ ♇
22 01 ☽ △ ☉
22 34 ☽ # ♂

12 Monday
03 57 ♀ ⊻ ♅
04 53 ☽ □ ♇
05 14 ☽ # ☿
05 29 ☽ △ ♅
05 37 ☽ ⊼ ♀
06 01 ☽ Q ♄
09 35 ☽ □ ♂
17 22 ☽ ⊼ ♆

13 Tuesday
01 01 ☽ ⅊ ☉
01 56 ☽ △ ♃
03 29 ☽ ± ♆
03 51 ♀ # ♇
06 49 ☽ ⅊ ♅
07 15 ☽ # ♃
11 07 ☽ Q ♇
12 30 ☽ □ ♄
16 12 ☉ ⊥ ♂
17 17 ☽ ± ☉
17 39 ☉ ∥ ♅
18 41 ☽ ⅊ ♆
18 55 ☽ △ ☿
22 05 ☽ ± ♅

14 Wednesday
03 41 ☽ ⅊ ♃
04 12 ☽ ⊼ ☉
05 52 ☽ # ☉
06 44 ☽ # ♅
07 25 ☽ ✶ ♇
08 19 ☽ ⊼ ♅
12 58 ☽ ☍ ♀
14 35 ☽ △ ♂
19 16 ☽ ± ♃
19 50 ☽ ∥ ♇
20 14 ☽ △ ♆
20 31 ♃ ⊥ ♄
21 27 ☽ ⅊ ☿
22 28 ☽ # ♀

15 Thursday
05 44 ☽ ⊼ ♃
09 05 ☽ ∠ ♇
13 29 ☽ ± ☿
16 06 ☽ △ ♄
17 35 ☽ ⅊ ♂

16 Friday
00 15 ☽ ⊼ ☿
00 41 ☽ ⊥ ♇
01 28 ♀ ✶ ♂
03 11 ☉ ⊼ ♇
06 04 ☽ # ♀
10 03 ☽ ± ♂
10 27 ☽ ∥ ♇
11 11 ☽ ⊻ ♇
11 48 ☽ ☍ ☉
12 25 ☽ ☍ ♅
18 34 ☽ ⅊ ♄
21 09 ☽ ⊼ ♂
21 10 ☉ ☌ ♅
22 06 ☽ ⊼ ♀

17 Saturday
00 35 ☽ □ ♆
00 44 ☽ # ♅
07 04 ☽ # ☉
09 53 ☽ ± ♀
10 44 ☽ ± ♄
11 19 ☽ ☍ ♃
20 10 ♀ Q ♄
21 39 ☽ ⊼ ♄

18 Sunday
02 00 ♀ ✶ ♆
03 27 ☽ # ♃
03 49 ☽ ⅊ ♀
04 59 ♀ ⊥ ♃
07 11 ☽ ☍ ☿
09 09 ☽ Q ♆
17 13 ☽ ☌ ♇
18 51 ☽ ⊼ ♅
19 54 ♂ ⊥ ♅
22 15 ☽ ⊼ ☉

19 Monday
06 05 ☽ ± ♅
06 33 ☽ ☍ ♂
07 27 ☽ ✶ ♆
10 23 ☽ ± ☉
10 30 ☽ △ ♀
19 42 ☽ ⊼ ♃
21 32 ☽ # ☿
21 57 ☿ ∠ ♀
23 11 ☽ ⅊ ♅
23 46 ☽ # ♂

20 Tuesday
01 07 ♂ △ ♆
04 53 ☽ ⅊ ☉
06 05 ☽ ☍ ♄
07 22 ☽ ± ♃
09 18 ☽ # ♆
12 00 ☽ ∠ ♆
16 22 ☽ ⊼ ☿
20 12 ☽ # ♄
22 27 ☉ ∠ ♄

21 Wednesday
01 06 ☽ ⅊ ♃
02 12 ☽ ⊻ ♇
04 04 ☽ ± ☿
04 18 ☽ △ ♅
05 30 ☽ ⊥ ♆
06 58 ☿ St R
12 27 ☽ △ ☉
13 57 ☽ ⊥ ♇
17 18 ☽ ⊻ ♆
19 26 ☽ ⊼ ♂
21 44 ☽ ⅊ ☿

22 Thursday
02 55 ☽ □ ♀
07 12 ☽ △ ♃
07 44 ☽ ∠ ♇
07 58 ☽ ± ♂
17 29 ☽ ⊼ ♄

23 Friday
02 58 ☽ ⅊ ♂
03 25 ☽ △ ☿
05 38 ☽ ± ♄
13 46 ☽ ✶ ♇
16 20 ☽ □ ♅
19 03 ☽ Q ♀
23 56 ☽ ⅊ ♄

24 Saturday
02 18 ♀ ⅊ ♇
02 43 ☉ □ ♆
05 26 ☽ ☌ ♆
05 40 ☽ □ ☉
06 16 ♀ ⊻ ♃
09 48 ☽ ∥ ♄
10 53 ☽ △ ♂
13 58 ☽ Q ♇
20 43 ☽ □ ♃
22 01 ☽ ∥ ♆
22 03 ☽ ✶ ♀

25 Sunday
01 10 ☽ ∥ ♂
04 53 ☽ Q ♅
04 57 ☽ ∥ ☿
06 30 ☽ △ ♄
14 42 ☽ □ ☿
21 09 ☽ Q ☉

26 Monday
02 10 ☽ □ ♇
05 08 ☽ ✶ ♅
07 39 ☽ ∠ ♀
09 36 ☽ Q ♃
11 51 ☿ ⊥ ♀
17 47 ☽ ⊻ ♆
23 10 ☽ ✶ ☉

27 Tuesday
01 28 ☽ Q ☿
02 22 ☽ □ ♂
03 11 ☽ ∥ ♃
03 32 ☽ ⊥ ♀
04 45 ♀ # ♇
05 40 ☽ ⊥ ♆
08 57 ♂ ∥ ♆
09 14 ☽ ∥ ☉
09 49 ☽ ✶ ♃
11 02 ☽ ∠ ♅
16 38 ☽ ⊻ ♀
18 44 ☽ □ ♄
23 18 ☽ ∠ ♆

28 Wednesday
00 06 ☽ ✶ ☿
00 59 ♀ Q ♆
04 41 ☽ ⊥ ♅
07 01 ☽ ∠ ☉
10 28 ☽ ∥ ♅
13 04 ☽ △ ♇
15 18 ☽ Q ♂
15 29 ☽ ∠ ♃
15 49 ♀ ✶ ♄
16 17 ☽ ⊻ ♅

29 Thursday
01 40 ☽ ⊥ ☉
02 22 ☽ ∥ ♀
03 39 ☽ ∠ ☿
04 07 ☽ ✶ ♆
05 38 ☽ Q ♄
06 22 ☽ # ♇
08 54 ☽ ⊥ ♃
13 54 ☽ ⊻ ☉
15 13 ☽ ✶ ♂
17 27 ☽ ⅊ ♇
19 46 ☽ ⊥ ☿
20 20 ☽ ⊻ ♃
22 05 ☿ ⊻ ♀

30 Friday
02 34 ☽ Q ♆
04 15 ☽ ✶ ♄
06 21 ☽ ⊻ ☿
07 40 ☽ ☌ ♀
10 04 ☽ ± ♇
17 38 ☽ # ♇
20 15 ☽ ∠ ♂
21 00 ☽ ⊼ ♇

31 Saturday
00 23 ☽ ☌ ♅
01 33 ☽ ∥ ♀
07 46 ☽ ∠ ♄
09 09 ♀ ± ♇
11 17 ☽ □ ♆
11 37 ☉ ∥ ♃
12 29 ☉ ⊥ ♄
13 11 ☽ ∥ ♅
13 14 ☽ ⊥ ♂
13 50 ☉ ⊻ ♂
16 33 ☿ ⊻ ♄
23 54 ☽ ⊥ ♄

AUGUST 2038

LONGITUDES

Date	Sidereal time h m s	Sun ☉ ° ' "	Moon ☽ ° ' "	Moon ☽ 24.00 ° ' "	Mercury ☿ ° '	Venus ♀ ° '	Mars ♂ ° '	Jupiter ♃ ° '	Saturn ♄ ° '	Uranus ♅ ° '	Neptune ♆ ° '	Pluto ♇ ° '
01	08 40 49	09 ♌ 21 35	15 ♌ 21 47	22 ♌ 16 09	13 ♌ 51	18 ♋ 47	09 ♍ 03	10 ♌ 35	14 ♍ 32	25 ♋ 23	01 ♉ 22	23 ♒ 24
02	08 44 45	10 19 01	29 ♌ 14 22	06 ♍ 15 57	13 R 06	20 00	09 41	10 49	14 38	25 27	01 22	23 R 23
03	08 48 42	11 16 27	13 ♍ 20 22	20 ♍ 27 01	12 20	21 13	10 18	11 02	14 45	25 30	01 22	23 21
04	08 52 38	12 13 55	27 ♍ 35 21	04 ♎ 44 45	11 35	22 26	10 56	11 15	14 52	25 34	01 R 22	23 20
05	08 56 35	13 11 23	11 ♎ 54 39	19 ♎ 04 33	10 51	23 40	11 33	11 28	14 59	25 38	01 22	23 19
06	09 00 32	14 08 51	26 ♎ 13 57	03 ♏ 22 27	10 08	24 53	12 11	11 42	15 05	25 41	01 22	23 17
07	09 04 28	15 06 21	10 ♏ 29 43	17 ♏ 35 25	09 28	26 06	12 49	11 55	15 12	25 45	01 22	23 16
08	09 08 25	16 03 51	24 ♏ 39 22	01 ♐ 41 20	08 52	27 19	13 26	12 08	15 19	25 48	01 22	23 15
09	09 12 21	17 01 22	08 ♐ 41 12	15 ♐ 38 51	08 20	28 33	14 04	12 21	15 26	25 52	01 22	23 13
10	09 16 18	17 58 53	22 ♐ 34 09	29 ♐ 27 01	07 52	29 ♋ 46	14 42	12 35	15 33	25 55	01 22	23 12
11	09 20 14	18 56 26	06 ♑ 17 21	13 ♑ 05 04	07 31	00 ♌ 59	15 20	12 48	15 40	25 58	01 22	23 11
12	09 24 11	19 53 59	19 ♑ 50 02	26 ♑ 32 09	07 15	02 13	15 58	13 01	15 47	26 02	01 21	23 10
13	09 28 07	20 51 34	03 ♒ 11 16	09 ♒ 47 17	07 06	03 26	16 35	13 14	15 54	26 05	01 21	23 08
14	09 32 04	21 49 09	16 ♒ 20 04	22 ♒ 49 31	07 D 03	04 40	17 13	13 27	16 01	26 09	01 21	23 07
15	09 36 01	22 46 46	29 ♒ 15 32	05 ♓ 38 05	07 07	05 53	17 51	13 41	16 08	26 12	01 20	23 06
16	09 39 57	23 44 24	11 ♓ 57 10	18 ♓ 12 48	07 19	07 07	18 29	13 54	16 15	26 15	01 20	23 04
17	09 43 54	24 42 03	24 ♓ 25 04	00 ♈ 34 08	07 38	08 20	19 07	14 07	16 22	26 19	01 20	23 03
18	09 47 50	25 39 44	06 ♈ 40 11	12 ♈ 43 30	08 05	09 34	19 45	14 20	16 30	26 22	01 19	23 02
19	09 51 47	26 37 26	18 ♈ 44 25	24 ♈ 43 17	08 38	10 48	20 23	14 33	16 37	26 25	01 19	23 00
20	09 55 43	27 35 10	00 ♉ 40 35	06 ♉ 36 45	09 19	12 02	21 01	14 46	16 44	26 29	01 18	22 59
21	09 59 40	28 32 55	12 ♉ 32 21	18 ♉ 27 57	10 07	13 15	21 39	14 59	16 51	26 32	01 18	22 58
22	10 03 36	29 ♌ 30 42	24 ♉ 24 08	00 ♊ 21 32	11 03	14 29	22 18	15 12	16 59	26 35	01 17	22 56
23	10 07 33	00 ♍ 28 30	06 ♊ 20 48	12 ♊ 22 33	12 04	15 43	22 56	15 25	17 06	26 38	01 16	22 55
24	10 11 30	01 26 21	18 ♊ 27 27	24 ♊ 36 06	13 13	16 57	23 34	15 38	17 13	26 42	01 16	22 54
25	10 15 26	02 24 13	00 ♋ 49 07	07 ♋ 07 03	14 27	18 11	24 12	15 51	17 21	26 45	01 15	22 52
26	10 19 23	03 22 07	13 ♋ 30 22	19 ♋ 59 28	15 47	19 25	24 51	16 04	17 28	26 48	01 14	22 51
27	10 23 19	04 20 02	26 ♋ 34 40	03 ♌ 16 10	17 13	20 39	25 29	16 17	17 35	26 51	01 14	22 50
28	10 27 16	05 18 00	10 ♌ 03 59	16 ♌ 58 02	18 44	21 53	26 07	16 30	17 43	26 54	01 13	22 48
29	10 31 12	06 15 59	23 ♌ 58 03	01 ♍ 03 37	20 19	23 07	26 46	16 43	17 50	26 57	01 12	22 47
30	10 35 09	07 13 59	08 ♍ 14 06	15 ♍ 28 49	21 58	24 21	27 24	16 56	17 58	27 00	01 11	22 46
31	10 39 05	08 ♍ 12 01	22 ♍ 46 53	00 ♎ 07 21	23 ♌ 41	25 ♌ 35	28 ♍ 03	17 ♌ 08	18 ♍ 05	27 ♋ 03	01 ♉ 11	22 ♒ 45

Date	Moon True ☊ ° '	Moon Mean ☊ ° '	Moon ☽ Latitude ° '
01	10 ♋ 12	08 ♋ 49	03 N 00
02	10 R 08	08 46	03 56
03	10 04	08 43	04 39
04	09 59	08 40	05 04
05	09 56	08 37	05 11
06	09 53	08 33	04 59
07	09 52	08 30	04 28
08	09 D 52	08 27	03 41
09	09 53	08 24	02 42
10	09 54	08 21	01 33
11	09 55	08 17	00 N 20
12	09 R 55	08 14	00 S 54
13	09 53	08 11	02 03
14	09 49	08 08	03 04
15	09 43	08 05	03 55
16	09 35	08 02	04 33
17	09 27	07 58	04 57
18	09 19	07 55	05 07
19	09 12	07 52	05 03
20	09 07	07 49	04 46
21	09 04	07 46	04 17
22	09 02	07 43	03 37
23	09 D 02	07 39	02 47
24	09 03	07 36	01 49
25	09 04	07 33	00 S 44
26	09 R 04	07 30	00 N 24
27	09 03	07 27	01 32
28	08 59	07 23	02 38
29	08 53	07 20	03 36
30	08 45	07 17	04 22
31	08 ♋ 36	07 ♋ 14	04 N 53

DECLINATIONS

Date	Sun ☉ ° '	Moon ☽ ° '	Mercury ☿ ° '	Venus ♀ ° '	Mars ♂ ° '	Jupiter ♃ ° '	Saturn ♄ ° '	Uranus ♅ ° '	Neptune ♆ ° '	Pluto ♇ ° '
01	17 N 54	19 N 06	11 N 55	22 N 10	09 N 06	18 N 06	07 N 48	21 N 31	10 N 19	22 S 44
02	17 39	15 25	12 08	22 03	08 51	18 03	07 45	21 31	10 19	22 45
03	17 23	10 50	12 21	21 54	08 37	17 59	07 43	21 30	10 19	22 45
04	17 08	05 37	12 37	21 45	08 22	17 56	07 40	21 30	10 19	22 46
05	16 51	00 N 04	12 54	21 35	08 07	17 52	07 37	21 29	10 19	22 46
06	16 35	05 S 28	13 11	21 25	07 52	17 48	07 35	21 28	10 19	22 47
07	16 18	10 43	13 30	21 14	07 38	17 45	07 32	21 28	10 19	22 47
08	16 01	15 21	13 49	21 02	07 23	17 41	07 29	21 27	10 19	22 48
09	15 44	19 05	14 08	20 50	07 08	17 38	07 27	21 26	10 19	22 48
10	15 26	21 41	14 28	20 37	06 53	17 34	07 24	21 26	10 19	22 49
11	15 09	22 57	14 47	20 24	06 38	17 30	07 21	21 25	10 19	22 49
12	14 51	22 51	15 05	20 10	06 23	17 27	07 18	21 24	10 18	22 50
13	14 32	21 26	15 23	19 55	06 07	17 23	07 16	21 24	10 18	22 50
14	14 14	18 52	15 39	19 40	05 52	17 19	07 13	21 23	10 18	22 51
15	13 55	15 24	15 55	19 24	05 37	17 16	07 10	21 23	10 18	22 52
16	13 36	11 17	16 08	19 08	05 22	17 12	07 07	21 22	10 18	22 52
17	13 17	06 46	16 20	18 51	05 06	17 08	07 05	21 21	10 18	22 53
18	12 58	02 S 03	16 31	18 34	04 51	17 04	07 02	21 21	10 17	22 53
19	12 38	02 N 40	16 39	18 16	04 36	17 01	06 59	21 20	10 17	22 54
20	12 19	07 14	16 44	17 57	04 20	16 57	06 56	21 20	10 17	22 54
21	11 59	11 31	16 48	17 39	04 05	16 53	06 53	21 19	10 17	22 54
22	11 38	15 22	16 48	17 19	03 49	16 49	06 50	21 18	10 16	22 55
23	11 18	18 38	16 46	16 59	03 34	16 46	06 48	21 18	10 16	22 55
24	10 58	21 08	16 42	16 39	03 18	16 42	06 45	21 17	10 16	22 56
25	10 37	22 42	16 34	16 18	03 03	16 38	06 42	21 17	10 16	22 56
26	10 16	23 09	16 23	15 57	02 47	16 34	06 39	21 16	10 15	22 57
27	09 55	22 21	16 10	15 35	02 31	16 31	06 36	21 15	10 15	22 57
28	09 34	20 15	15 53	15 13	02 16	16 27	06 33	21 15	10 15	22 58
29	09 13	16 56	15 33	14 50	02 00	16 23	06 30	21 14	10 14	22 58
30	08 51	12 32	15 11	14 27	01 44	16 19	06 27	21 14	10 14	22 59
31	08 N 30	07 N 20	14 N 45	14 N 03	01 N 28	16 N 16	06 N 24	21 N 13	10 N 14	22 S 59

ZODIAC SIGN ENTRIES

Date	h m	Planets
02	13 18	☽ ♍
04	16 03	☽ ♎
06	18 20	☽ ♏
08	21 07	☽ ♐
10	16 34	♀ ♌
11	00 58	☽ ♑
13	06 14	☽ ♒
15	13 23	☽ ♓
17	22 53	☽ ♈
20	10 38	☽ ♉
22	23 17	☽ ♊
23	00 10	☉ ♍
25	10 26	☽ ♋
27	18 10	☽ ♌
29	22 13	☽ ♍
31	23 48	☽ ♎

LATITUDES

Date	Mercury ☿ ° '	Venus ♀ ° '	Mars ♂ ° '	Jupiter ♃ ° '	Saturn ♄ ° '	Uranus ♅ ° '	Neptune ♆ ° '	Pluto ♇ ° '
01	04 S 57	00 N 03	01 N 00	00 N 33	01 N 51	00 N 29	01 S 44	09 S 35
04	04 53	00 11	00 58	00 33	01 51	00 29	01 44	09 35
07	04 32	00 19	00 57	00 33	01 51	00 29	01 44	09 35
10	03 58	00 26	00 56	00 34	01 51	00 29	01 44	09 36
13	03 13	00 33	00 54	00 34	01 51	00 29	01 45	09 36
16	02 23	00 40	00 53	00 34	01 51	00 29	01 45	09 36
19	01 30	00 46	00 51	00 34	01 51	00 29	01 45	09 36
22	00 S 40	00 52	00 50	00 35	01 51	00 29	01 45	09 36
25	00 N 05	00 58	00 48	00 35	01 51	00 29	01 45	09 36
28	00 42	01 03	00 47	00 35	01 51	00 29	01 45	09 36
31	01 N 12	01 N 08	00 N 46	00 N 36	01 N 51	00 N 29	01 S 46	09 S 36

DATA

Julian Date	2465637
Delta T	+77 seconds
Ayanamsa	24° 23' 30"
Synetic vernal point	04° ♓ 43' 29"
True obliquity of ecliptic	23° 26' 02"

MOON'S PHASES, APSIDES AND POSITIONS ☽

Date	h m	Phase	Longitude ° '	Eclipse Indicator
01	00 40	●	08 ♌ 54	
07	20 21	☽	15 ♏ 26	
14	22 57	○	22 ♒ 15	
22	23 12	☾	29 ♉ 58	
30	10 13	●	07 ♍ 10	

Day	h m		
05	21 37	Perigee	
21	01 52	Apogee	
05	12 18	OS	
11	22 09	Max dec	23° S 05'
18	22 25	ON	
26	08 57	Max dec	23° N 09'

LONGITUDES

Date	Chiron ⚷ ° '	Ceres ⚳ ° '	Pallas ⚴ ° '	Juno ⚵ ° '	Vesta ⚶ ° '	Black Moon Lilith ⚸ ° '
01	00 ♋ 49	28 ♑ 56	06 ♐ 36	13 ♏ 11	16 ♋ 33	03 ♉ 20
11	01 ♋ 34	26 ♑ 58	07 ♐ 12	14 ♏ 37	20 ♋ 48	04 ♉ 27
21	02 ♋ 13	25 ♑ 24	08 ♐ 23	16 ♏ 24	24 ♋ 59	05 ♉ 34
31	02 ♋ 47	24 ♑ 21	10 ♐ 04	18 ♏ 30	29 ♋ 07	06 ♉ 41

All ephemeris data is given at 12.00 UT and the Moon's longitude is additionally given for 24.00 UT

ASPECTARIAN

h m	Aspects
01 Sunday	
00 24	☽ ⊻ ♂
00 40	☽ ☌ ☉
03 30	☽ ☌ ♃
09 29	☽ ☌ ☿
10 32	☽ ⊻ ♄
18 32	☽ ⊻ ♀
19 19	☽ ∥ ♃
21 12	☽ ∥ ☉
02 Monday	
01 56	☽ ☍ ♇
05 27	☽ ⊻ ♅
05 54	☽ ⊥ ♀
15 39	☽ △ ♆
15 48	☽ ⊥ ♅
22 47	☽ ∠ ♀
03 Tuesday	
04 07	☉ ☌ ♃
04 49	☽ ∥ ☿
06 37	☽ ☌ ♂
07 11	☽ ∠ ♅
08 02	☽ ⊻ ♃
08 15	☽ ⊻ ☉
10 24	☽ ⊻ ☿
14 24	☽ ☌ ♄
14 27	☽ ∥ ♆
17 08	☽ ⚼ ♆
18 20	☽ ⊥ ♃
19 08	☽ ⊥ ☉
20 01	☽ ⊥ ☿
20 37	♂ ∠ ♅
23 00	☽ ∥ ♂
04 Wednesday	
01 55	♆ St R
02 32	☽ ✶ ♀
02 44	☽ ∥ ♄
02 55	☉ ☌ ☿
04 52	☽ ⊼ ♇
08 17	☽ ± ♆
08 35	☽ ✶ ♅
09 43	☽ ∠ ♃
10 24	☽ ∠ ☿
11 21	☽ ∠ ☉
14 55	☽ ± ♇
18 21	☽ ⊼ ♆
20 12	☿ ☌ ♃
23 26	☿ ⊻ ♂
05 Thursday	
00 33	☽ Q ♀
04 48	☽ Q ♅
05 17	♀ ⊼ ♇
05 59	☽ ⚼ ♇
07 07	♂ ⊻ ♃
10 18	☽ ✶ ☿
11 15	☽ ✶ ♃
11 23	☽ ⊻ ♂
14 18	☽ ✶ ☉
17 10	☽ ⊻ ♄
21 53	☽ ⊥ ♂
06 Friday	
03 18	☽ ⊥ ♄
03 45	♀ ∥ ♅
05 27	☽ Q ☿
07 04	☽ △ ♇
07 41	☽ Q ♃
09 31	☽ □ ♀
11 05	☽ □ ♅
11 51	☽ Q ☉
13 40	☽ ∠ ♂
18 32	☽ ∠ ♄
20 38	☽ ☍ ♆
21 19	☽ # ♄
22 15	☽ # ♂
07 Saturday	
04 36	♀ ☌ ♅
10 06	☽ # ♆
10 21	☽ □ ☿
14 26	☽ □ ♃
14 46	☉ ⊻ ♄
16 06	☽ ✶ ♂
20 01	☽ ✶ ♄
20 21	☽ □ ☉
20 40	☿ ⊥ ♄
22 57	♂ ∥ ♄
08 Sunday	
02 59	☽ # ☿
09 36	☽ □ ♇
13 24	☽ Q ♂
13 57	☽ △ ♅
15 38	☽ # ☉
16 34	☽ Q ♄
16 58	☽ △ ♀
23 27	☽ ⊼ ♆
09 Monday	
01 59	☽ # ♃
09 44	☽ ± ♆
11 24	☽ △ ☿
15 45	☽ ⚼ ♅
16 22	☽ Q ♇
17 34	☿ ⊥ ♂
18 25	☽ △ ♃
21 10	☽ ⚼ ♀
21 43	☽ □ ♂
23 44	☽ □ ♄
10 Tuesday	
01 14	☽ ⚼ ♆
01 39	☽ # ♀
03 27	☽ △ ☉
07 22	☽ ± ♅
09 05	☽ # ♅
09 12	♀ ∠ ♂
12 31	☽ ⚼ ☿
13 06	☽ ✶ ♇
14 17	☽ ± ♀
17 51	☽ ⊼ ♅
20 52	☽ ⚼ ♃
11 Wednesday	
01 47	☽ ⊼ ♀
03 21	☽ △ ♆
03 48	☽ ± ☿
04 56	♀ ∠ ♄
07 32	☽ ∥ ♇
07 34	☽ ⚼ ☉
12 55	☽ ± ♃
14 06	☽ ⊼ ☿
15 20	☽ ∠ ♇
19 14	♀ □ ♆
23 41	☽ ⊼ ♃
12 Thursday	
00 38	☽ ± ☉
02 22	☉ ∥ ☿
03 44	♂ ☌ ♄
04 43	☽ △ ♄
04 46	☽ △ ♂
07 15	☽ ⊥ ♇
12 08	☽ ⊼ ☉
12 34	☽ ∥ ♇
17 56	☽ ⊻ ♇
23 09	☽ ☍ ♅
13 Friday	
03 01	♂ ⚼ ♆
07 49	☽ ⚼ ♄
08 41	☽ □ ♆
08 58	☽ ⚼ ♂
12 23	☽ # ♅
12 30	☽ ☍ ♀
19 03	☽ ☍ ☿
14 Saturday	
00 19	☽ ± ♄
02 09	☽ ± ♂
04 46	☽ # ♀
06 37	☽ ☍ ♃
08 49	☿ St D
11 25	☽ ⊼ ♄
13 43	☽ ⊼ ♂
17 33	☽ Q ♆
22 57	☽ ☍ ☉
23 33	☽ # ♃
15 Sunday	
00 31	☽ ☌ ♇
06 16	☽ ⊼ ♅
08 56	☽ # ☿
15 54	☽ ✶ ♆
17 33	☽ ± ♅
19 39	☉ ☍ ♇
21 43	☽ # ☉
16 Monday	
01 49	☽ ⊼ ♀
03 01	☽ ⊼ ☿
10 40	☽ ⚼ ♅
14 28	☽ ± ♀
14 40	☽ ± ☿
15 47	☽ ⊼ ♃
17 09	☿ ☌ ♀
17 22	☽ # ♆
20 19	☽ ☍ ♄
20 23	☽ ∠ ♆
17 Tuesday	
01 12	☽ ☍ ♂
03 13	♄ ⚼ ♆
03 31	☽ ± ♃
08 26	☽ ⚼ ☿
09 21	☽ ⊻ ♇
09 41	☽ ⚼ ♀
10 22	☽ # ♄
12 36	☽ ⊼ ☉
13 46	☽ ⊥ ♆
15 42	☽ △ ♅
20 59	☽ # ♂
21 00	☽ ⊥ ♇
21 19	☽ ⚼ ♃
18 Wednesday	
01 18	☽ ± ☉
01 29	☽ ⊻ ♆
14 41	☽ ∠ ♇
14 54	☽ △ ☿
18 23	☽ △ ♀
20 35	☽ ⚼ ☉
19 Thursday	
03 29	☽ △ ♃
06 42	☉ ⊻ ♅
07 42	☽ ⊼ ♄
08 01	♀ ⊥ ♄
15 29	☽ ⊼ ♂
19 50	☽ ± ♄
20 32	☽ ✶ ♇
21 29	♂ ⊥ ♃
21 30	☽ ∥ ♂
20 Friday	
03 30	☽ □ ♅
04 13	☽ ± ♂
05 13	☽ △ ☉
10 25	☽ ∥ ♄
13 16	☽ ☌ ♆
14 09	☽ ⚼ ♄
20 41	☽ Q ♇
23 25	☽ ⚼ ♂
21 Saturday	
04 52	☽ ∥ ♆
06 44	☽ □ ☿
13 37	☽ □ ♀
14 31	☽ ∥ ☉
16 03	☽ Q ♅
17 03	☽ □ ♃
20 50	☽ △ ♄
22 Sunday	
07 30	☽ △ ♂
09 03	☽ □ ♇
10 07	☿ ⊥ ♄
16 25	☽ ✶ ♅
21 57	☽ ∥ ☿
21 58	☽ ∥ ♃
22 12	☽ Q ☿
22 39	☿ ∥ ♃
23 12	☽ □ ☉
23 Monday	
00 30	☽ ∥ ♀
01 51	☽ ⊻ ♆
05 01	♀ ☌ ♃
06 02	☽ Q ♃
06 08	☽ Q ♀
11 33	♂ ⊼ ♇
13 51	☽ ⊥ ♆
22 35	☽ ∠ ♅
24 Tuesday	
00 33	☽ ✶ ☿
06 21	☽ ✶ ♃
07 20	☿ ∥ ♀
07 23	♀ ∥ ♃
07 40	☿ ∥ ♃
07 41	☉ △ ♆
07 41	☽ ∠ ♆
08 42	☽ ✶ ♀
09 33	☽ □ ♄
13 50	☽ ∥ ♅
14 05	☽ Q ☉
16 24	☽ ⊥ ♅
17 55	♀ ⊻ ♄
20 40	☽ △ ♇
22 32	☽ □ ♂
25 Wednesday	
04 07	☽ ⊻ ♅
09 04	☽ ∠ ☿
12 04	☽ ∠ ♃
12 50	☽ ✶ ♆
13 03	♀ ⊥ ♂
15 17	☽ ✶ ☉
17 00	☽ ∠ ♀
18 48	☽ # ♇
20 43	☽ Q ♄
20 58	☉ ⊥ ♅
26 Thursday	
01 24	☽ ⚼ ♇
04 12	☽ ⊥ ☿
05 27	☽ ⊥ ♃
10 42	☽ Q ♂
11 30	☽ Q ♆
11 48	☽ ⊥ ♀
12 53	☉ ∥ ♆
16 45	☽ ⊻ ☿
16 51	☽ ⊻ ♃
17 41	☿ ☌ ♃
18 12	☽ ± ♇
19 25	☽ ✶ ♄
21 44	☽ ∠ ☉
22 36	☽ # ♇
27 Friday	
00 04	☽ ⊻ ♀
02 43	♂ ± ♆
05 12	☽ ⊼ ♇
09 55	☽ ✶ ♂
12 29	☽ ☌ ♅
15 25	☽ ⊥ ☉
18 35	☿ ⊻ ♄
20 21	☽ □ ♆
22 53	☽ ∠ ♄
28 Saturday	
02 20	☽ ∥ ♅
02 58	☽ ⊻ ☉
13 56	☽ ∠ ♂
14 54	☽ ⊥ ♄
23 22	☽ ☌ ♃
29 Sunday	
01 24	☽ ⊻ ♄
04 56	☽ ☌ ☿
05 49	♀ ☍ ♇
06 16	☽ ⊥ ♂
09 59	☽ ☍ ♇
10 24	☽ ☌ ♀
15 19	☽ ∥ ♃
16 58	☽ ⊻ ♂
17 05	☽ ⊻ ♅
19 42	♂ ✶ ♅
20 50	☽ ∥ ☿
22 51	☿ ⊥ ♂
30 Monday	
00 14	☽ △ ♆
01 14	☽ ∥ ♀
03 14	☽ ⊥ ♅
10 13	☽ ☌ ☉
18 16	☽ ∠ ♅
23 00	☽ ∥ ♆
23 11	☿ ☍ ♇
31 Tuesday	
01 09	☽ ⚼ ♆
02 36	☽ ⊻ ♃
04 13	☽ ☌ ♄
06 32	☽ ∥ ☉
11 56	☽ ⊼ ♇
12 36	☽ ⊥ ♃
13 40	☽ ⊻ ☿
15 55	☽ ± ♆
16 05	☽ ∥ ♄
17 00	☽ ⊻ ♀
19 00	☽ ✶ ♅
21 00	☽ ☌ ♂
21 44	☽ ± ♇

SEPTEMBER 2038

LONGITUDES

Date	Sidereal time h	m	s	Sun ☉ °	'	"	Moon ☽ °	'	"	Moon ☽ 24.00 °	'	"	Mercury ☿ °	'	Venus ♀ °	'	Mars ♂ °	'	Jupiter ♃ °	'	Saturn ♄ °	'	Uranus ♅ °	'	Neptune ♆ °	'	Pluto ♇ °	'
01	10	43	02	09 ♍	10	05	07 ♎	29	15	14 ♎	51	34	25 ♌	26	26 ♌	49	28 ♍	41	17 ♌	21	18 ♍	12	27 ♋	06	01 ♉	10	22 ♒	43
02	10	46	59	10	08	10	22 ♎	13	22	29 ♎	33	44	27	15	28	03	29	20	17	34	18	20	27	09	01 R	09	22 R	42
03	10	50	55	11	06	16	06 ♏	51	54	14 ♏	07	13	29 ♌	06	29 ♌	17	29 ♍	58	17	46	18	27	27	12	01	08	22	41
04	10	54	52	12	04	24	21 ♏	19	11	28 ♏	27	25	00 ♍	58	00 ♍	31	00 ♎	37	17	59	18	35	27	15	01	07	22	40
05	10	58	48	13	02	34	05 ♐	31	41	12 ♐	31	53	02	52	01	46	01	16	18	12	18	42	27	17	01	06	22	38
06	11	02	45	14	00	44	19 ♐	27	58	26 ♐	20	01	04	46	03	00	01	54	18	24	18	50	27	20	01	05	22	37
07	11	06	41	14	58	57	03 ♑	08	10	09 ♑	52	33	06	41	04	14	02	33	18	37	18	57	27	23	01	04	22	36
08	11	10	38	15	57	10	16 ♑	33	23	23 ♑	10	49	08	37	05	28	03	12	18	49	19	05	27	26	01	03	22	35
09	11	14	34	16	55	25	29 ♑	45	02	06 ♒	16	12	10	33	06	43	03	51	19	02	19	12	27	28	01	02	22	34
10	11	18	31	17	53	42	12 ♒	44	28	19 ♒	09	54	12	28	07	57	04	29	19	14	19	20	27	31	01	01	22	32
11	11	22	28	18	52	00	25 ♒	32	37	01 ♓	52	39	14	23	09	12	05	08	19	26	19	27	27	34	01	00	22	31
12	11	26	24	19	50	20	08 ♓	10	04	14 ♓	24	53	16	18	10	26	05	47	19	39	19	35	27	36	00	58	22	30
13	11	30	21	20	48	42	20 ♓	37	09	26 ♓	46	54	18	12	11	40	06	26	19	51	19	42	27	39	00	57	22	29
14	11	34	17	21	47	06	02 ♈	54	12	08 ♈	59	08	20	06	12	55	07	05	20	03	19	50	27	41	00	56	22	28
15	11	38	14	22	45	31	15 ♈	01	50	21 ♈	02	28	21	58	14	09	07	44	20	15	19	57	27	44	00	55	22	27
16	11	42	10	23	43	58	27 ♈	01	16	02 ♉	58	28	23	50	15	24	08	23	20	27	20	05	27	46	00	54	22	26
17	11	46	07	24	42	28	08 ♉	54	24	14 ♉	49	27	25	41	16	38	09	02	20	39	20	12	27	49	00	52	22	24
18	11	50	03	25	40	59	20 ♉	44	02	26 ♉	38	39	27	30	17	53	09	42	20	51	20	20	27	51	00	51	22	23
19	11	54	00	26	39	33	02 ♊	33	49	08 ♊	30	06	29 ♍	19	19	08	10	21	21	03	20	27	27	53	00	50	22	22
20	11	57	57	27	38	09	14 ♊	28	08	20 ♊	28	32	01 ♎	07	20	22	11	00	21	15	20	35	27	55	00	48	22	21
21	12	01	53	28	36	47	26 ♊	31	59	02 ♋	39	09	02	54	21	37	11	39	21	27	20	42	27	58	00	47	22	20
22	12	05	50	29 ♍	35	27	08 ♋	50	43	15 ♋	07	19	04	40	22	52	12	19	21	38	20	50	28	00	00	46	22	19
23	12	09	46	00 ♎	34	10	21 ♋	29	36	27 ♋	58	05	06	25	24	06	12	58	21	50	20	57	28	02	00	44	22	18
24	12	13	43	01	32	55	04 ♌	33	16	11 ♌	15	30	08	09	25	21	13	37	22	02	21	05	28	04	00	43	22	17
25	12	17	39	02	31	42	18 ♌	05	01	25 ♌	01	53	09	51	26	36	14	17	22	13	21	12	28	06	00	41	22	16
26	12	21	36	03	30	31	02 ♍	05	57	09 ♍	16	53	11	33	27	51	14	56	22	25	21	20	28	08	00	40	22	15
27	12	25	32	04	29	22	16 ♍	34	08	23 ♍	56	55	13	14	29 ♍	05	15	36	22	36	21	27	28	10	00	39	22	14
28	12	29	29	05	28	15	01 ♎	24	16	08 ♎	55	03	14	54	00 ♎	20	16	15	22	47	21	34	28	12	00	37	22	14
29	12	33	26	06	27	10	16 ♎	28	01	24 ♎	01	50	16	34	01	35	16	55	22	58	21	42	28	14	00	36	22	13
30	12	37	22	07 ♎	26	08	01 ♏	35	13	09 ♏	06	55	18 ♎	12	02 ♎	50	17 ♎	35	23 ♌	10	21 ♍	49	28 ♋	16	00 ♉	34	22 ♒	12

Date	Moon True ☊ °	'	Moon Mean ☊ °	'	Moon ☽ Latitude °	'
01	08 ♋	27	07 ♋	11	05 N	04
02	08 R	19	07	08	04	55
03	08	12	07	04	04	27
04	08	08	07	01	03	42
05	08	07	06	58	02	44
06	08 D	07	06	55	01	38
07	08	07	06	52	00 N	26
08	08 R	07	06	48	00 S	45
09	08	05	06	45	01	52
10	08	00	06	42	02	53
11	07	53	06	39	03	43
12	07	43	06	36	04	22
13	07	31	06	33	04	48
14	07	18	06	29	05	00
15	07	05	06	26	04	58
16	06	54	06	23	04	43
17	06	44	06	20	04	15
18	06	36	06	17	03	37
19	06	32	06	14	02	49
20	06	30	06	10	01	54
21	06 D	29	06	07	00 S	52
22	06 R	29	06	04	00 N	12
23	06	29	06	01	01	19
24	06	26	05	58	02	23
25	06	22	05	54	03	21
26	06	14	05	51	04	09
27	06	05	05	48	04	43
28	05	54	05	45	04	59
29	05	42	05	42	04	55
30	05 ♋	32	05 ♋	39	04 N	30

DECLINATIONS

Date	Sun ☉ °	'	Moon ☽ °	'	Mercury ☿ °	'	Venus ♀ °	'	Mars ♂ °	'	Jupiter ♃ °	'	Saturn ♄ °	'	Uranus ♅ °	'	Neptune ♆ °	'	Pluto ♇ °	'
01	08 N	08	01 N	41	14 N	17	13 N	40	01 N	13	16 N	12	06 N	22	21 N	13	10 N	13	22 S	59
02	07	46	04 S	05	13	46	13	15	00	57	16	08	06	19	21	12	10	13	23	00
03	07	24	09	36	13	13	12	51	00	41	16	04	06	16	21	12	10	13	23	00
04	07	02	14	31	12	38	12	26	00	25	16	01	06	13	21	11	10	12	23	01
05	06	40	18	31	12	01	12	00	00 N	09	15	57	06	10	21	11	10	12	23	01
06	06	17	21	24	11	22	11	35	00 S	06	15	53	06	07	21	10	10	12	23	01
07	05	55	22	57	10	42	11	09	00	22	15	49	06	04	21	10	10	11	23	02
08	05	32	23	09	10	00	10	43	00	38	15	46	06	01	21	09	10	11	23	02
09	05	10	22	02	09	16	10	16	00	54	15	42	05	58	21	08	10	10	23	03
10	04	47	19	45	08	32	09	49	01	10	15	38	05	55	21	08	10	10	23	03
11	04	24	16	30	07	47	09	22	01	26	15	34	05	53	21	08	10	09	23	03
12	04	01	12	33	07	01	08	55	01	42	15	31	05	50	21	07	10	09	23	04
13	03	38	08	08	06	15	08	27	01	57	15	27	05	47	21	07	10	09	23	04
14	03	15	03 S	26	05	28	07	59	02	13	15	23	05	44	21	06	10	08	23	04
15	02	52	01 N	20	04	41	07	31	02	29	15	19	05	41	21	06	10	08	23	05
16	02	29	06	01	03	53	07	03	02	45	15	16	05	38	21	05	10	07	23	05
17	02	06	10	26	03	06	06	34	03	01	15	12	05	35	21	05	10	07	23	05
18	01	43	14	27	02	18	06	05	03	17	15	08	05	32	21	04	10	06	23	05
19	01	20	17	54	01	30	05	36	03	33	15	05	05	29	21	04	10	06	23	06
20	00	56	20	39	00 N	43	05	07	03	48	15	01	05	26	21	03	10	05	23	06
21	00	33	22	31	00 S	04	04	38	04	04	14	57	05	24	21	03	10	05	23	06
22	00 N	10	23	21	00	52	04	09	04	20	14	54	05	21	21	03	10	04	23	06
23	00 S	14	23	01	01	38	03	39	04	36	14	50	05	18	21	02	10	04	23	07
24	00	37	21	26	02	25	03	10	04	52	14	46	05	15	21	02	10	03	23	07
25	01	00	18	35	03	11	02	40	05	07	14	43	05	12	21	01	10	03	23	07
26	01	24	14	36	03	57	02	10	05	23	14	39	05	09	21	01	10	02	23	07
27	01	47	09	39	04	42	01	40	05	39	14	36	05	06	21	01	10	02	23	08
28	02	10	04 N	01	05	27	01	10	05	55	14	32	05	04	21	00	10	01	23	08
29	02	34	01 S	56	06	12	00	40	06	10	14	29	05	01	21	00	10	01	23	08
30	02 S	57	07 S	48	06 S	55	00 N	10	06 S	26	14 N	25	04 N	58	21 N	00	10 N	00	23 S	08

ZODIAC SIGN ENTRIES

Date	h	m	Planets
03	00	43	☽ ♏
03	13	05	♂ ♎
03	23	41	☿ ♍
04	01	51	♀ ♍
05	02	37	☽ ♐
07	06	27	☽ ♑
09	12	27	☽ ♒
11	20	26	☽ ♓
14	06	18	☽ ♈
16	18	00	☽ ♉
19	06	48	☽ ♊
19	21	03	☿ ♎
21	18	49	☽ ♋
22	22	02	☉ ♎
24	03	43	☽ ♌
26	08	27	☽ ♍
28	05	33	♀ ♎
28	09	45	☽ ♎
30	09	29	☽ ♏

LATITUDES

Date	Mercury ☿ °	'	Venus ♀ °	'	Mars ♂ °	'	Jupiter ♃ °	'	Saturn ♄ °	'	Uranus ♅ °	'	Neptune ♆ °	'	Pluto ♇ °	'
01	01 N	19	01 N	09	00 N	45	00 N	36	01 N	51	00 N	29	01 S	46	09 S	36
04	01	37	01	13	00	44	00	36	01	51	00	29	01	46	09	36
07	01	46	01	17	00	42	00	36	01	51	00	29	01	46	09	36
10	01	48	01	20	00	41	00	37	01	51	00	29	01	46	09	36
13	01	43	01	22	00	39	00	37	01	51	00	29	01	46	09	36
16	01	34	01	24	00	38	00	38	01	51	00	30	01	46	09	36
19	01	21	01	25	00	36	00	38	01	51	00	30	01	46	09	35
22	01	05	01	26	00	34	00	38	01	52	00	30	01	47	09	35
25	00	47	01	26	00	33	00	39	01	52	00	30	01	47	09	35
28	00	27	01	25	00	31	00	39	01	52	00	30	01	47	09	34
31	00 N	07	01 N	24	00 N	30	00 N	40	01 N	52	00 N	30	01 S	47	09 S	34

DATA

Julian Date	2465668
Delta T	+77 seconds
Ayanamsa	24° 23' 34"
Synetic vernal point	04° ♓ 43' 25"
True obliquity of ecliptic	23° 26' 03"

MOON'S PHASES, APSIDES AND POSITIONS ☽

Date	h	m	Phase	Longitude °	'	Eclipse Indicator
06	01	51	☽	13 ♐	36	
13	12	24	○	20 ♓	50	
21	16	27	☾	28 ♊	48	
28	18	58	●	05 ♎	45	

Day	h	m		
01	22	31	Perigee	
17	19	04	Apogee	
30	00	19	Perigee	
01	18	59	0S	
08	03	23	Max dec	23° S 14'
15	05	15	0N	
22	17	23	Max dec	23° N 23'
29	04	15	0S	

LONGITUDES

Date	Chiron ⚷ °	'	Ceres ⚳ °	'	Pallas ⚴ °	'	Juno ⚵ °	'	Vesta ⚶ °	'	Black Moon Lilith ⚸ °	'
01	02 ♋	50	24 ♑	17	10 ♐	16	18 ♏	44	29 ♋	32	06 ♉	48
11	03 ♋	18	23 ♑	53	12 ♐	25	21 ♏	07	03 ♌	34	07 ♉	55
21	03 ♋	38	24 ♑	06	14 ♐	54	23 ♏	45	07 ♌	30	09 ♉	02
31	03 ♋	51	24 ♑	52	17 ♐	41	26 ♏	34	11 ♌	19	10 ♉	09

All ephemeris data is given at 12.00 UT and the Moon's longitude is additionally given for 24.00 UT

ASPECTARIAN

h m	Aspects
01 Wednesday	
00 48	☽ ⊥ ☿
01 42	☽ ⊼ ♆
03 31	☽ ∠ ♃
03 41	☽ ⊥ ♀
12 23	☽ ⚼ ♇
13 20	♂ ± ♇
14 03	☽ ∥ ♂
14 38	☽ Q ♅
14 56	☽ ⊻ ☉
17 28	☽ ∠ ☿
17 46	♀ ⊻ ♅
19 41	☽ ∠ ♀
23 29	☽ # ♂
02 Thursday	
01 23	☽ ⊥ ☉
04 17	☽ ⚹ ♃
05 36	☽ ⊻ ♄
10 39	☿ ⊻ ♅
12 47	☽ △ ♇
15 28	☽ ⊥ ♄
17 06	☽ ∠ ☉
20 04	☽ □ ♅
21 23	☽ ⚹ ☿
21 24	☽ # ♄
22 24	☽ ⚹ ♀
03 Friday	
00 09	☽ ⊻ ♂
00 11	☽ Q ♃
02 35	☽ ☍ ♆
02 50	☽ # ☉
06 20	☽ ∠ ♄
10 28	☽ ⊥ ♂
14 50	☽ # ♆
19 27	☿ ☌ ♀
19 30	☽ ⚹ ☉
19 59	☽ Q ♀
20 01	☽ Q ☿
04 Saturday	
02 02	☽ ∠ ♂
02 13	☽ # ♀
03 28	☽ # ☿
05 14	☿ ⊻ ♂
06 21	☽ □ ♃
07 23	☽ ⚹ ♄
13 54	☿ △ ♆
14 14	☽ □ ♇
15 43	♀ ⊻ ♂
16 26	☉ ∠ ♅
16 57	☽ Q ☉
20 09	☽ # ♃
21 59	☽ △ ♅
23 21	♀ △ ♆
05 Sunday	
03 44	☽ Q ♄
04 24	☽ ⚹ ♂
04 29	☽ ⊼ ♆
04 59	☽ □ ♀
06 11	♂ ⊼ ♆
06 45	☽ □ ☿
13 19	☿ ∥ ♀
14 41	☽ ± ♆
17 35	☿ ⊥ ♅
20 44	☽ Q ♇
23 37	☽ ⚼ ♅
06 Monday	
01 51	☽ □ ☉
01 54	☽ Q ♂
06 08	☽ ⚼ ♆
09 38	☽ # ♅
10 08	☽ △ ♃
10 53	☽ □ ♄
15 16	☽ ± ♅
17 29	☽ ⚹ ♇
18 50	♀ ⊥ ♅
07 Tuesday	
00 38	☉ ∥ ♄
01 49	☽ ⊼ ♅
08 20	☽ △ ♆
10 55	☽ □ ♂
12 51	☽ ⚼ ♃
14 09	☽ △ ♀
14 13	☽ ∥ ♇
19 22	☽ △ ☿
19 55	☽ ∠ ♇
08 Wednesday	
05 10	☽ ± ♃
05 37	☿ ∥ ♆
10 50	☽ △ ☉
12 02	☽ ⊥ ♇
14 18	☉ ⚼ ♆
16 09	☽ ⊼ ♃
16 25	☽ ∥ ♇
16 36	☽ △ ♄
19 49	☽ ⚼ ♀
22 53	☽ ⊻ ♇
09 Thursday	
02 59	☽ ⚼ ☿
07 49	☽ ☍ ♅
13 57	☽ ± ♀
14 21	☽ □ ♆
16 19	☽ ⚼ ☉
17 06	♀ ∥ ♆
19 55	☽ △ ♂
20 16	☽ ⚼ ♄
21 59	♂ ∠ ♃
22 21	☽ ± ☿
22 44	☽ # ♅
10 Friday	
02 10	☽ ⊼ ♀
10 18	☽ ± ☉
11 24	☽ ⊼ ☿
12 36	☿ ∠ ♅
13 07	☽ ± ♄
22 25	☽ ⊼ ☉
23 42	☽ Q ♆
11 Saturday	
00 20	☽ ☍ ♃
00 26	☽ ⊼ ♄
01 17	☽ ⚼ ♂
06 19	☽ ☌ ♇
15 50	☽ ⊼ ♅
17 30	♃ ⊻ ♄
18 08	☽ # ♃
19 10	☽ ± ♂
22 18	☽ ⚹ ♆
12 Sunday	
03 15	☽ ± ♅
04 43	☉ ☌ ♄
05 52	☉ ⊻ ♃
07 12	☽ ⊼ ♂
07 55	☿ ⚼ ♆
16 49	☽ ☍ ♀
20 33	☽ ⚼ ♅
13 Monday	
01 18	☽ # ♆
02 59	☽ ∠ ♆
06 28	☽ ☍ ☿
10 07	☽ # ♀
10 13	☽ ☍ ♄
10 29	☽ ⊼ ♃
12 24	☽ ☍ ☉
15 37	☽ ⊻ ♇
20 25	☽ ⊥ ♆
22 21	☽ ± ♃
23 39	☽ # ☿
14 Tuesday	
00 14	☽ # ♄
01 44	☽ △ ♅
03 18	☽ ⊥ ♇
03 22	☿ ∥ ♄
07 28	♀ ∠ ♅
08 08	☽ ⊻ ♆
08 27	☿ ☌ ♄
11 23	☿ ⊻ ♃
12 57	☽ # ☉
16 18	☽ ⚼ ♃
17 46	☽ ∥ ♂
20 43	☽ ☍ ♂
20 58	☽ ∠ ♇
15 Wednesday	
01 28	♂ ⚼ ♇
04 23	☉ ⊼ ♇
10 04	☽ ⊼ ♀
18 03	☿ ⊼ ♇
18 10	☽ # ♂
19 12	☽ ∥ ☉
21 56	☽ ⊼ ♄
22 36	☽ △ ♃
23 25	☽ ± ♀
16 Thursday	
02 18	☉ # ♂
02 35	☽ ∥ ☿
02 47	☽ ⚹ ♇
04 25	☽ ⊼ ☿
04 49	☽ ⊼ ☉
09 22	☉ ☌ ☿
10 02	☽ ∥ ♄
10 06	☽ ± ♄
13 31	☽ □ ♅
16 56	☽ ∥ ♀
17 57	☽ ± ☉
18 42	☽ ± ☿
19 36	☽ ⚼ ♀
19 47	☽ ☌ ♆
21 23	♀ ⚼ ♆
17 Friday	
01 38	☿ ± ♆
02 55	☽ Q ♇
04 26	☽ ⚼ ♄
10 13	☽ ∥ ♆
12 17	☽ ⊼ ♂
13 46	☽ ⚼ ☉
13 48	☿ # ♂
15 57	☉ ± ♆
16 15	☽ ⚼ ☿
18 Saturday	
01 10	☽ ± ♂
02 03	☽ Q ♅
02 21	☿ ⊥ ♃
05 32	☽ △ ♀
11 11	☽ △ ♄
12 15	☽ □ ♃
15 21	☽ □ ♇
16 25	☽ ∥ ♃
16 37	☿ ⚹ ♅
18 04	♂ Q ♅
20 31	☽ ⚼ ♂
22 57	☽ △ ☉
23 33	☿ ± ♇
19 Sunday	
02 29	☽ ⚹ ♅
04 15	☽ △ ☿
08 29	☽ ⊻ ♆
18 31	♀ ∥ ♄
20 36	☽ ⊥ ♆
22 40	☉ ∥ ☿
20 Monday	
00 03	☉ ⊥ ♃
01 20	☽ Q ♃
04 38	☽ △ ♂
07 53	☿ ⊼ ♆
08 53	☽ ∠ ♅
14 40	☽ ∠ ♆
16 22	☽ ∥ ♅
16 32	♀ ☌ ♄
19 22	☉ ⚹ ♅
21 Tuesday	
00 20	☽ □ ♄
01 09	☽ □ ♀
01 46	☽ ⚹ ♃
02 56	☽ ⊥ ♅
03 43	☽ △ ♇
05 21	☉ ± ♇
08 05	♀ ⊻ ♃
14 49	☽ ⊻ ♅
16 27	☽ □ ☉
20 20	☽ ⚹ ♆
21 46	☉ # ☿
22 Wednesday	
01 04	☽ # ♇
01 45	♀ ⊼ ♇
02 34	☽ □ ☿
05 58	♀ # ♂
07 40	☽ ∠ ♃
09 04	☽ ⚼ ♇
11 58	☽ Q ♄
16 18	☽ Q ♀
19 01	☽ □ ♂
19 30	☽ Q ♆
23 Thursday	
01 11	☽ ⊥ ♃
02 15	☽ ± ♇
06 04	☽ Q ☉
09 16	☽ # ♇
10 59	☽ ⚹ ♄
12 39	☽ ⊻ ♃
13 31	☽ ⊼ ♇
16 02	☉ ⊼ ♆
17 23	☽ ⚹ ♀
18 16	☽ Q ☿
18 35	☿ ∠ ♃
23 59	♀ ± ♆
24 Friday	
00 09	☽ ☌ ♅
00 15	☿ ⚼ ♇
05 02	☽ □ ♆
06 07	☽ ⚹ ☉
06 24	☽ Q ♂
14 47	☽ ∠ ♄
16 05	☽ ∥ ♅
19 24	☽ ⚹ ☿
23 27	☽ ∠ ♀
25 Saturday	
02 07	☿ # ♀
05 00	☽ ⚹ ♂
06 55	☽ ⊥ ♄
10 57	☽ ∠ ☉
15 31	☿ Q ♅
16 48	☽ ⊥ ♀
17 28	☽ ⊻ ♄
17 59	♂ # ♄
18 11	♃ ☍ ♇
19 15	☽ ☍ ♇
19 16	☽ ☌ ♃
26 Sunday	
01 19	☽ ∠ ☿
03 39	☽ ⊥ ☉
04 06	☽ ⊻ ♀
05 17	☽ ⊻ ♅
08 10	☽ ∠ ♂
09 35	☽ △ ♆
11 42	☽ ∥ ♃
14 33	☽ ⊻ ☉
15 26	☽ ⊥ ♅
17 49	♀ ⚹ ♅
18 34	☽ ⊥ ☿
19 53	♀ ± ♇
23 59	☽ ⊥ ♂
27 Monday	
00 54	♀ ⊥ ♃
05 50	☽ ⊻ ☿
06 24	☽ ∠ ♅
08 53	☉ # ♀
10 17	☽ ∥ ♆
10 20	☽ ⊻ ♂
10 29	☽ ⚼ ♆
20 01	☽ ☌ ♄
21 14	☽ ⊼ ♇
21 57	☽ ⊻ ♃
28 Tuesday	
00 05	☿ # ♄
01 06	☽ ± ♆
04 30	☽ # ♂
06 45	☽ # ☿
06 51	☽ ⚹ ♅
06 54	☽ ± ♇
07 40	☽ ∥ ♄
07 45	☽ ⊥ ♃
10 08	☽ ☌ ♀
10 44	☽ ⊼ ♆
17 20	♀ ⊼ ♆
18 58	☽ ☌ ☉
19 03	☽ # ☉
21 18	☽ ⚼ ♇
22 20	☽ ∠ ♃
29 Wednesday	
00 37	☽ ∥ ♀
02 04	☽ Q ♅
07 20	☽ # ♀
10 51	☿ ∥ ♂
12 10	☽ ☌ ☿
12 45	☽ ☌ ♂
14 43	☽ ∥ ☉
20 22	☽ ⊻ ♄
20 41	☿ ☌ ♂
21 06	☽ △ ♇
22 27	☽ ⚹ ♃
30 Thursday	
00 22	☽ # ♄
05 58	☽ ⊥ ♄
06 02	☽ ∥ ♂
06 15	☉ ⚼ ♇
06 42	☽ □ ♅
07 49	☽ ∥ ☿
10 23	☽ ☍ ♆
14 09	☽ ⊻ ♀
17 46	☽ Q ♃
20 24	☽ ∠ ♄
21 28	☽ # ♆
21 58	☽ ⊻ ☉

OCTOBER 2038

LONGITUDES

Date	Sidereal time h m s	Sun ☉ ° ' "	Moon ☽ ° ' "	Moon ☽ 24.00 ° ' "	Mercury ☿ ° '	Venus ♀ ° '	Mars ♂ ° '	Jupiter ♃ ° '	Saturn ♄ ° '	Uranus ♅ ° '	Neptune ♆ ° '	Pluto ♇ ° '
01	12 41 19	08 ♎ 25 07	16 ♏ 35 47	24 ♏ 00 51	19 ♎ 49	04 ♎ 05	18 ♎ 14	23 ♌ 21	21 ♍ 57	28 ♋ 17	00 ♉ 33	22 ♒ 11
02	12 45 15	09 24 08	01 ♐ 21 21	08 ♐ 36 40	21 26	05 20	18 54	23 32	22 04	28 19	00 R 31	22 R 10
03	12 49 12	10 23 11	15 ♐ 46 26	22 ♐ 50 25	23 02	06 34	19 34	23 43	22 11	28 21	00 29	22 09
04	12 53 08	11 22 15	29 ♐ 48 35	06 ♑ 41 02	24 36	07 49	20 14	23 53	22 18	28 22	00 28	22 09
05	12 57 05	12 21 22	13 ♑ 27 57	20 ♑ 09 37	26 11	09 04	20 54	24 04	22 26	28 24	00 26	22 08
06	13 01 01	13 20 30	26 ♑ 46 22	03 ♒ 18 34	27 44	10 19	21 33	24 15	22 33	28 26	00 25	22 07
07	13 04 58	14 19 40	09 ♒ 46 36	16 ♒ 10 48	29 ♎ 16	11 34	22 13	24 25	22 40	28 27	00 23	22 06
08	13 08 55	15 18 51	22 ♒ 31 33	28 ♒ 49 09	00 ♏ 48	12 49	22 53	24 36	22 47	28 29	00 22	22 06
09	13 12 51	16 18 04	05 ♓ 03 55	11 ♓ 16 06	02 19	14 04	23 33	24 46	22 54	28 30	00 20	22 05
10	13 16 48	17 17 20	17 ♓ 25 56	23 ♓ 33 35	03 49	15 19	24 13	24 57	23 01	28 31	00 18	22 04
11	13 20 44	18 16 36	29 ♓ 39 15	05 ♈ 43 04	05 18	16 34	24 53	25 07	23 08	28 33	00 17	22 04
12	13 24 41	19 15 55	11 ♈ 45 10	17 ♈ 45 40	06 47	17 49	25 34	25 17	23 15	28 34	00 15	22 03
13	13 28 37	20 15 16	23 ♈ 44 44	29 ♈ 42 30	08 15	19 04	26 14	25 27	23 22	28 35	00 13	22 02
14	13 32 34	21 14 39	05 ♉ 39 07	11 ♉ 34 48	09 42	20 19	26 54	25 37	23 29	28 36	00 12	22 02
15	13 36 30	22 14 04	17 ♉ 29 46	23 ♉ 24 18	11 08	21 34	27 34	25 47	23 36	28 37	00 10	22 01
16	13 40 27	23 13 32	29 ♉ 18 42	05 ♊ 13 20	12 33	22 49	28 15	25 56	23 43	28 38	00 08	22 01
17	13 44 24	24 13 01	11 ♊ 08 38	17 ♊ 05 01	13 58	24 04	28 55	26 06	23 50	28 39	00 07	22 00
18	13 48 20	25 12 33	23 ♊ 03 00	29 ♊ 03 09	15 21	25 19	29 ♎ 35	26 16	23 57	28 40	00 05	22 00
19	13 52 17	26 12 07	05 ♋ 06 02	11 ♋ 12 15	16 44	26 34	00 ♏ 16	26 25	24 04	28 41	00 03	22 00
20	13 56 13	27 11 43	17 ♋ 22 27	23 ♋ 37 16	18 06	27 49	00 56	26 34	24 10	28 42	00 02	21 59
21	14 00 10	28 11 22	29 ♋ 57 20	06 ♌ 23 15	19 26	29 ♎ 05	01 37	26 44	24 17	28 42	00 ♉ 00	21 59
22	14 04 06	29 ♎ 11 03	12 ♌ 55 35	19 ♌ 34 48	20 46	00 ♏ 20	02 17	26 53	24 24	28 43	29 ♈ 58	21 58
23	14 08 03	00 ♏ 10 46	26 ♌ 21 16	03 ♍ 15 14	22 05	01 35	02 58	27 02	24 30	28 44	29 57	21 58
24	14 11 59	01 10 31	10 ♍ 16 47	17 ♍ 25 46	23 22	02 50	03 39	27 10	24 37	28 44	29 55	21 58
25	14 15 56	02 10 18	24 ♍ 41 51	02 ♎ 04 28	24 38	04 05	04 19	27 19	24 43	28 45	29 53	21 57
26	14 19 53	03 10 08	09 ♎ 32 49	17 ♎ 05 52	25 52	05 20	05 00	27 28	24 50	28 45	29 52	21 57
27	14 23 49	04 10 00	24 ♎ 42 24	02 ♏ 21 06	27 05	06 35	05 41	27 36	24 56	28 46	29 50	21 57
28	14 27 46	05 09 53	10 ♏ 00 31	17 ♏ 39 13	28 17	07 51	06 22	27 45	25 03	28 46	29 48	21 57
29	14 31 42	06 09 49	25 ♏ 15 51	02 ♐ 49 09	29 ♏ 26	09 06	07 03	27 53	25 09	28 47	29 47	21 57
30	14 35 39	07 09 47	10 ♐ 18 01	17 ♐ 41 34	00 ♐ 34	10 21	07 44	28 01	25 15	28 47	29 45	21 56
31	14 39 35	08 ♏ 09 46	24 ♐ 59 07	02 ♑ 10 12	01 ♐ 39	11 ♏ 36	08 ♏ 25	28 ♌ 09	25 ♍ 21	28 ♋ 47	29 ♈ 43	21 ♒ 56

Date	Moon True ☊ ° '	Moon Mean ☊ ° '	Moon ☽ Latitude ° '
01	05 ♋ 23	05 ♋ 35	03 N 47
02	05 R 18	05 32	02 49
03	05 15	05 29	01 41
04	05 D 14	05 26	00 N 29
05	05 R 14	05 23	00 S 43
06	05 14	05 20	01 51
07	05 11	05 16	02 52
08	05 07	05 13	03 42
09	04 59	05 10	04 21
10	04 49	05 07	04 47
11	04 37	05 04	04 59
12	04 23	05 00	04 58
13	04 09	04 57	04 43
14	03 57	04 54	04 16
15	03 46	04 51	03 39
16	03 39	04 48	02 51
17	03 34	04 45	01 56
18	03 31	04 41	00 S 55
19	03 D 31	04 38	00 N 08
20	03 31	04 35	01 13
21	03 R 31	04 32	02 16
22	03 30	04 29	03 14
23	03 28	04 26	04 03
24	03 22	04 22	04 40
25	03 15	04 19	05 01
26	03 06	04 16	05 02
27	02 57	04 13	04 43
28	02 48	04 10	04 03
29	02 41	04 06	03 06
30	02 37	04 03	01 57
31	02 ♋ 35	04 ♋ 00	00 N 41

DECLINATIONS

Date	Sun ☉ ° '	Moon ☽ ° '	Mercury ☿ ° '	Venus ♀ ° '	Mars ♂ ° '	Jupiter ♃ ° '	Saturn ♄ ° '	Uranus ♅ ° '	Neptune ♆ ° '	Pluto ♇ ° '
01	03 S 20	13 S 10	07 S 39	00 S 20	06 S 41	14 N 22	04 N 55	20 N 59	09 N 59	23 S 08
02	03 44	17 40	08 22	00 50	06 57	14 18	04 52	20 59	09 59	23 08
03	04 07	21 00	09 04	01 20	07 13	14 15	04 50	20 59	09 58	23 09
04	04 30	22 57	09 45	01 51	07 28	14 11	04 47	20 58	09 58	23 09
05	04 53	23 28	10 26	02 21	07 44	14 08	04 44	20 58	09 57	23 09
06	05 16	22 37	11 07	02 51	07 59	14 04	04 41	20 58	09 57	23 09
07	05 39	20 33	11 46	03 21	08 14	14 01	04 38	20 58	09 56	23 09
08	06 02	17 30	12 25	03 51	08 30	13 58	04 36	20 57	09 55	23 09
09	06 25	13 42	13 03	04 21	08 45	13 54	04 33	20 57	09 55	23 09
10	06 47	09 22	13 41	04 51	09 00	13 51	04 30	20 57	09 54	23 09
11	07 10	04 S 43	14 18	05 20	09 15	13 48	04 28	20 57	09 54	23 09
12	07 32	00 N 04	14 54	05 50	09 31	13 45	04 25	20 56	09 53	23 09
13	07 55	04 49	15 29	06 20	09 46	13 41	04 22	20 56	09 53	23 10
14	08 17	09 22	16 03	06 49	10 01	13 38	04 20	20 56	09 52	23 10
15	08 39	13 33	16 36	07 18	10 16	13 35	04 17	20 56	09 51	23 10
16	09 01	17 13	17 09	07 47	10 31	13 32	04 15	20 56	09 51	23 10
17	09 23	20 12	17 41	08 16	10 45	13 29	04 12	20 56	09 50	23 10
18	09 45	22 20	18 12	08 45	11 00	13 26	04 09	20 55	09 50	23 10
19	10 07	23 29	18 41	09 14	11 15	13 23	04 07	20 55	09 49	23 10
20	10 28	23 31	19 10	09 42	11 30	13 20	04 04	20 55	09 48	23 10
21	10 50	22 22	19 38	10 10	11 44	13 17	04 02	20 55	09 48	23 10
22	11 11	20 02	20 05	10 38	11 59	13 14	03 59	20 55	09 47	23 09
23	11 32	16 32	20 31	11 06	12 13	13 11	03 57	20 55	09 47	23 09
24	11 53	12 02	20 55	11 33	12 28	13 08	03 54	20 55	09 46	23 09
25	12 14	06 43	21 19	12 00	12 42	13 05	03 52	20 55	09 45	23 09
26	12 34	00 N 51	21 41	12 27	12 56	13 02	03 49	20 55	09 45	23 09
27	12 54	05 S 11	22 02	12 54	13 10	13 00	03 47	20 54	09 44	23 09
28	13 14	10 58	22 22	13 20	13 24	12 57	03 45	20 54	09 44	23 09
29	13 34	16 04	22 41	13 46	13 38	12 54	03 42	20 54	09 43	23 09
30	13 54	20 04	22 58	14 12	13 52	12 52	03 40	20 54	09 43	23 09
31	14 S 14	22 S 40	23 S 14	14 S 37	14 S 06	12 N 49	03 N 38	20 N 54	09 N 42	23 S 09

ZODIAC SIGN ENTRIES

Date	h	m	Planets
02	09	46	☽ ♐
04	12	20	☽ ♑
06	17	54	☽ ♒
07	23	25	☿ ♏
09	02	16	☽ ♓
11	12	41	☽ ♈
14	00	35	☽ ♉
16	13	24	☽ ♊
19	01	53	☽ ♋
19	02	36	♂ ♏
21	12	05	☽ ♌
21	12	20	♆ ♈
22	05	43	♀ ♏
23	07	41	☉ ♏
23	18	22	☽ ♍
25	20	39	☽ ♎
27	20	19	☽ ♏
29	19	31	☽ ♐
29	23	58	☿ ♐
31	20	21	☽ ♑

LATITUDES

Date	Mercury ☿ ° '	Venus ♀ ° '	Mars ♂ ° '	Jupiter ♃ ° '	Saturn ♄ ° '	Uranus ♅ ° '	Neptune ♆ ° '	Pluto ♇ ° '
01	00 N 07	01 N 24	00 N 30	00 N 40	01 N 52	00 N 30	01 S 47	09 S 34
04	00 S 15	01 22	00 28	00 40	01 53	00 30	01 47	09 34
07	00 36	01 20	00 27	00 41	01 53	00 30	01 47	09 33
10	00 57	01 17	00 25	00 41	01 53	00 30	01 47	09 33
13	01 18	01 14	00 23	00 42	01 54	00 30	01 47	09 32
16	01 38	01 10	00 22	00 42	01 54	00 31	01 47	09 32
19	01 56	01 06	00 20	00 43	01 55	00 31	01 47	09 31
22	02 13	01 01	00 18	00 43	01 55	00 31	01 47	09 31
25	02 28	00 56	00 17	00 44	01 56	00 31	01 47	09 30
28	02 40	00 50	00 15	00 44	01 56	00 31	01 47	09 30
31	02 S 48	00 N 44	00 N 13	00 N 45	01 N 57	00 N 31	01 S 47	09 S 29

DATA

Julian Date	2465698
Delta T	+77 seconds
Ayanamsa	24° 23' 36"
Synetic vernal point	04° ♓ 43' 22"
True obliquity of ecliptic	23° 26' 03"

MOON'S PHASES, APSIDES AND POSITIONS ☽

Date	h	m	Phase	Longitude ° '	Eclipse Indicator
05	09	52	☽	12 ♑ 16	
13	04	22	○	19 ♈ 56	
21	08	23	☾	28 ♋ 02	
28	03	53	●	04 ♏ 50	

Day	h	m		
15	06	57	Apogee	
28	10	14	Perigee	
05	08	44	Max dec	23° S 29'
12	11	38	ON	
20	00	52	Max dec	23° N 39'
26	15	24	OS	

LONGITUDES

Date	Chiron ⚷ ° '	Ceres ⚳ ° '	Pallas ⚴ ° '	Juno ⚵ ° '	Vesta ⚶ ° '	Black Moon Lilith ⚸ ° '
01	03 ♋ 51	24 ♑ 52	17 ♐ 41	26 ♏ 34	11 ♌ 19	10 ♉ 09
11	03 ♋ 56	26 ♑ 09	20 ♐ 42	29 ♏ 33	14 ♌ 58	11 ♉ 16
21	03 ♋ 53	27 ♑ 53	23 ♐ 55	02 ♐ 40	18 ♌ 26	12 ♉ 23
31	03 ♋ 42	00 ♒ 02	27 ♐ 17	05 ♐ 55	21 ♌ 39	13 ♉ 30

All ephemeris data is given at 12.00 UT and the Moon's longitude is additionally given for 24.00 UT

ASPECTARIAN

h m	Aspects
01 Friday	
00 35	☽ ⊥ ♀
08 15	☽ ⊥ ☉
09 48	☉ ∠ ♃
14 46	☽ ⊻ ♂
16 22	☽ ∠ ♀
17 46	☽ ♯ ♃
17 50	☽ ⊻ ☿
20 42	☽ ⚹ ♄
21 01	☽ □ ♇
23 03	☽ □ ♃
23 49	☽ ∠ ☉
02 Saturday	
00 57	☽ ⊥ ♂
04 47	☽ ⊥ ☿
07 01	☽ △ ♅
10 37	☽ ⊼ ♆
16 24	☽ ∠ ♂
16 30	☽ Q ♄
19 10	☽ ⚹ ♀
20 30	☽ ± ♆
21 25	☽ ∠ ☿
22 16	☿ ⊻ ♄
22 58	☿ △ ♇
03 Sunday	
02 18	☽ ⚹ ☉
02 35	☽ Q ♇
06 43	♄ ⊼ ♇
07 55	☽ bQ ♅
11 02	☉ Q ♅
11 31	☽ bQ ♆
11 48	☽ ♯ ♅
17 12	☽ Q ♀
18 44	☽ ⚹ ♂
22 49	☽ ⚹ ♇
22 59	☽ □ ♄
23 05	♀ bQ ♇
23 11	☽ ± ♅
23 41	☿ ⚹ ♃
04 Monday	
00 04	☽ Q ☉
01 40	☽ △ ♃
01 53	☽ ⚹ ☿
09 31	☽ ⊼ ♅
13 08	☽ △ ♆
16 06	☽ ∥ ♇
16 25	☽ Q ♂
19 05	☿ ♯ ♆
05 Tuesday	
00 48	☽ ∠ ♇
01 25	☽ Q ☿
03 25	☽ □ ♀
03 42	☉ ♯ ♄
04 06	☽ bQ ♃
09 52	☽ □ ☉
12 00	♀ ∠ ♃
16 45	☽ ⊥ ♇
20 21	☽ ± ♃
06 Wednesday	
01 34	☽ ∥ ♇
02 01	☽ □ ♂
03 33	☽ ⊻ ♇
04 15	☽ △ ♄
07 20	☽ ⊼ ♃
13 59	☽ □ ☿
14 07	♀ Q ♅
15 02	☽ ☍ ♅
18 39	☽ □ ♆
23 00	☿ □ ♅
07 Thursday	
01 46	☿ ⊥ ♄
07 52	♂ △ ♇
08 02	☽ bQ ♄
08 06	☽ ♯ ♅
15 42	☽ △ ♀
21 14	☽ △ ☉
08 Friday	
01 02	☽ ± ♄
04 08	☽ Q ♆
05 11	☿ ☍ ♆
07 30	♂ ⊻ ♄
11 11	☽ ☌ ♇
12 30	☽ ⊼ ♄
12 44	☽ △ ♂
16 00	☽ ☍ ♃
23 12	☽ bQ ♀
23 22	☽ ⊼ ♅
09 Saturday	
02 55	☽ ⚹ ♆
04 08	☽ bQ ☉
05 58	☽ △ ☿
10 46	☽ ♯ ♃
10 54	☽ ± ♅
15 13	☽ ∥ ☿
18 27	☽ ± ♀
19 08	☽ bQ ♂
21 02	♀ ♯ ♄
23 00	☽ ± ☉
10 Sunday	
04 22	☽ bQ ♅
07 24	☽ ⊼ ♀
07 52	☽ ∠ ♆
09 07	☽ ♯ ♆
11 42	☽ ⊼ ☉
13 38	☽ ± ♂
13 49	☽ ∥ ♂
15 05	☽ bQ ☿
18 02	☿ ♯ ♃
21 04	☽ ⊻ ♇
23 03	☽ ☍ ♄
11 Monday	
00 26	☽ ∥ ☉
01 26	☽ ⊥ ♆
02 04	☽ ⊼ ♂
02 56	☽ ⊼ ♃
08 52	☽ ⊥ ♇
09 08	☽ ∥ ♀
09 48	☽ △ ♅
11 13	☽ ± ☿
13 14	☽ ⊻ ♆
13 17	☽ ♯ ♄
14 56	☽ ± ♃
22 42	♂ ⚹ ♃
12 Tuesday	
00 44	☽ ⊼ ☿
02 39	☽ ∠ ♇
09 02	☽ bQ ♃
21 17	☿ Q ♃
13 Wednesday	
01 31	☽ ☍ ♀
04 22	☽ ☍ ☉
08 35	☽ ⚹ ♇
09 44	☽ ∥ ♄
11 15	☽ ⊼ ♄
14 22	☿ ∠ ♄
15 28	☽ △ ♃
17 18	☽ ☍ ♂
20 43	☽ ♯ ♀
21 45	☽ □ ♅
22 35	♂ ♯ ♆
23 26	☽ ± ♄
14 Thursday	
01 00	☽ ☌ ♆
05 37	☽ ♯ ☉
08 44	☽ Q ♇
14 44	☽ ∥ ♆
15 46	☽ ♯ ♂
17 48	☽ bQ ♄
21 18	☽ ☍ ☿
15 Friday	
06 56	☉ △ ♇
10 13	☽ Q ♅
12 11	☽ ∥ ♃
20 41	♀ △ ♇
21 11	☽ □ ♇
21 15	☽ ⊼ ♀
22 30	☽ ⊼ ☉
16 Saturday	
00 32	☽ △ ♄
05 04	☽ □ ♃
09 42	☽ ⊼ ♂
10 38	☽ ⚹ ♅
10 53	☽ ± ♀
11 28	☽ ♯ ☿
11 49	☽ ± ☉
13 41	☽ ⊻ ♆
22 37	☽ ± ♂
17 Sunday	
01 34	☉ ⊻ ♄
01 50	☽ ⊥ ♆
02 21	♂ □ ♅
07 04	♀ ⊻ ♄
07 18	☽ bQ ♀
07 45	☽ bQ ☉
17 05	☽ ∠ ♅
17 57	☽ bQ ♂
18 04	☽ Q ♃
18 27	☽ ⊼ ☿
19 09	☽ ∥ ♅
20 00	☽ ∠ ♆
18 Monday	
01 41	☉ ☌ ♀
08 09	☽ ± ☿
09 54	☽ △ ♇
11 14	☽ ⊥ ♅
13 49	☽ □ ♄
16 43	☽ △ ☉
16 49	☉ ♯ ♆
17 05	☽ △ ♀
18 31	☽ ⚹ ♃
23 15	☽ ⊻ ♅
19 Tuesday	
01 51	☽ △ ♂
02 01	☽ ⚹ ♆
02 45	☽ ♯ ♇
03 22	♂ ⊥ ♄
04 28	☽ bQ ☿
04 54	♂ ☍ ♆
08 37	♀ ⚹ ♃
11 02	♄ ± ♆
15 44	☽ bQ ♇
18 11	☉ ⚹ ♃
20 Wednesday	
00 35	☽ ∠ ♃
01 38	☽ Q ♆
01 48	☽ Q ♄
09 19	☽ ± ♇
13 34	☽ △ ☿
17 17	♀ ♯ ♆
18 14	☽ ⊥ ♃
20 52	☽ ⊼ ♇
22 37	☽ ♯ ♇
21 Thursday	
01 10	☽ ⚹ ♄
04 53	♀ □ ♅
05 50	☽ ⊻ ♃
08 23	☽ □ ☉
09 39	☽ ☌ ♅
10 10	☽ □ ♀
12 05	☽ □ ♆
15 17	☽ □ ♂
22 Friday	
00 40	☉ □ ♅
04 15	☽ ∥ ♅
05 20	♀ ☍ ♆
05 30	☽ ∠ ♄
11 38	☽ ♯ ☿
13 27	♀ ⊥ ♄
20 19	☽ Q ☉
21 58	☽ ⊥ ♄
22 46	☽ Q ♀
23 Saturday	
01 58	☽ Q ♂
03 39	☽ □ ☿
04 15	☽ ☍ ♇
06 29	☉ ☍ ♆
08 43	☽ ⊻ ♄
09 58	☿ □ ♇
13 12	☽ ☌ ♃
16 10	☽ ⊻ ♅
18 15	☽ △ ♆
19 12	☽ ⚹ ☉
20 52	☉ ⊥ ♄
22 01	☽ ⚹ ♀
24 Sunday	
00 06	☽ ⚹ ♂
02 33	☽ ⊥ ♅
06 30	☽ ∥ ♃
10 01	☽ ♯ ♂
11 26	☿ ♯ ♅
12 41	☽ ♯ ☉
14 01	☽ Q ☿
14 07	☽ ♯ ♀
17 50	☽ ∠ ♅
19 47	☽ bQ ♆
22 37	☽ ∥ ♆
22 39	☽ ∠ ☉
25 Monday	
01 52	☽ ∠ ♀
02 43	☽ ∠ ♂
07 30	☽ ⊼ ♇
10 41	☽ ± ♆
11 53	☽ ⚹ ☿
12 03	☽ ☌ ♄
13 56	☿ ⚹ ♄
14 35	☽ ⊥ ☉
16 20	☽ ⊻ ♃
17 19	☽ ± ♇
18 03	☽ ⊥ ♀
18 12	☽ ⊥ ♂
18 37	☽ ⚹ ♅
20 27	☽ ⊼ ♆
22 03	♀ ☌ ♂
23 56	☽ ∥ ♄
26 Tuesday	
01 02	☽ ⊻ ☉
02 09	☽ ⊥ ♃
04 22	☽ ⊻ ♂
04 38	☽ ⊻ ♀
07 51	☽ bQ ♇
13 56	☽ Q ♅
14 18	☽ ∠ ☿
16 42	☽ ∠ ♃
21 08	♂ ♯ ♃
27 Wednesday	
05 49	☽ ⊥ ☿
06 29	☽ ♯ ♄
07 40	☽ △ ♇
12 22	☽ ⊻ ♄
14 00	☉ ∥ ♀
16 04	☽ ⊻ ☿
16 36	☽ ⚹ ♃
16 51	♀ ♯ ♃
17 38	☉ ♯ ♃
18 23	☽ □ ♅
20 02	☽ ☍ ♆
21 51	☽ ⊥ ♄
23 41	☿ □ ♃
28 Thursday	
03 53	☽ ☌ ☉
06 02	☽ ☌ ♂
06 42	☽ ♯ ♆
08 18	☽ ☌ ♀
11 35	☽ Q ♃
12 03	☽ ∠ ♄
20 10	♀ ∥ ♂
20 47	☽ ♯ ♃
22 10	☿ △ ♅
22 54	☽ ∥ ☉
23 28	☽ ∥ ♂
23 37	☽ ∥ ♀
29 Friday	
06 45	☽ □ ♇
11 49	☽ ⚹ ♄
16 11	☽ □ ♃
17 34	☽ △ ♅
19 01	☿ ⊼ ♆
19 08	☽ ⊼ ♆
19 09	☽ ☌ ☿
30 Saturday	
03 25	☉ ∥ ♂
04 42	☽ ± ♆
04 52	♀ Q ♃
06 35	☽ ⊻ ☉
07 04	☽ Q ♄
07 40	☽ ⊻ ♂
10 01	♀ ∠ ♄
11 25	☽ Q ♇
12 05	☽ ⊻ ♀
16 58	☽ ⊥ ☉
17 38	☽ bQ ♅
17 49	☽ ⊥ ♂
18 25	☽ ♯ ♅
19 11	☽ bQ ♆
22 43	☽ ⊥ ♀
31 Sunday	
04 12	☿ ∥ ♇
06 58	☽ ⚹ ♇
08 22	☽ ± ♅
08 46	☽ ∠ ☉
09 16	☽ ∠ ♂
12 37	☽ □ ♄
14 57	☽ ∠ ♀
17 19	☽ △ ♃
18 19	☽ ⊼ ♅
19 16	☽ ∥ ♇
19 52	☽ △ ♆
23 09	☽ ∥ ☿

NOVEMBER 2038

LONGITUDES

Date	Sidereal time h m s	Sun ☉ ° ' "	Moon ☽ ° ' "	Moon ☽ 24.00 ° ' "	Mercury ☿ ° '	Venus ♀ ° '	Mars ♂ ° '	Jupiter ♃ ° '	Saturn ♄ ° '	Uranus ♅ ° '	Neptune ♆ ° '	Pluto ♇ ° '
01	14 43 32	09 ♏ 09 47	09 ♑ 14 34	16 ♑ 12 09	02 ♐ 41	12 ♏ 51	09 ♏ 06	28 ♌ 17	25 ♍ 28	28 ♋ 47	29 ♈ 42	21 ♒ 56
02	14 47 28	10 09 50	23 ♑ 03 02	29 ♑ 47 26	03 41	14 07	09 47	28 28	25 34	28 47	29 R 40	21 R 56
03	14 51 25	11 09 54	06 ♒ 25 41	12 ♒ 58 09	04 37	15 22	10 28	28 32	25 40	28 47	29 38	21 56
04	14 55 22	12 09 59	19 ♒ 25 17	25 ♒ 47 34	05 30	16 37	11 09	28 40	25 46	28 R 47	29 37	21 56
05	14 59 18	13 10 06	02 ♓ 05 28	08 ♓ 19 28	06 19	17 52	11 50	28 47	25 52	28 47	29 35	21 D 56
06	15 03 15	14 10 15	14 ♓ 30 03	20 ♓ 37 38	07 03	19 07	12 32	28 54	25 58	28 47	29 33	21 56
07	15 07 11	15 10 25	26 ♓ 42 37	02 ♈ 45 24	07 43	20 23	13 13	29 01	26 03	28 47	29 32	21 56
08	15 11 08	16 10 37	08 ♈ 46 19	14 ♈ 45 41	08 17	21 38	13 54	29 08	26 09	28 47	29 30	21 56
09	15 15 04	17 10 50	20 ♈ 43 47	26 ♈ 40 52	08 45	22 53	14 36	29 15	26 15	28 47	29 29	21 56
10	15 19 01	18 11 05	02 ♉ 37 09	08 ♉ 32 53	09 06	24 08	15 17	29 21	26 20	28 46	29 27	21 56
11	15 22 57	19 11 22	14 ♉ 28 15	20 ♉ 23 27	09 19	25 24	15 59	29 28	26 26	28 46	29 25	21 57
12	15 26 54	20 11 40	26 ♉ 18 42	02 ♊ 14 13	09 25	26 39	16 40	29 34	26 31	28 45	29 24	21 57
13	15 30 51	21 12 00	08 ♊ 10 14	14 ♊ 07 00	09 R 21	27 54	17 22	29 40	26 37	28 45	29 22	21 57
14	15 34 47	22 12 22	20 ♊ 04 49	26 ♊ 03 59	09 08	29 ♏ 09	18 03	29 46	26 42	28 44	29 21	21 57
15	15 38 44	23 12 46	02 ♋ 04 52	08 ♋ 07 50	08 46	00 ♐ 24	18 45	29 52	26 47	28 44	29 19	21 58
16	15 42 40	24 13 11	14 ♋ 13 17	20 ♋ 21 42	08 12	01 40	19 27	29 ♌ 58	26 52	28 43	29 18	21 58
17	15 46 37	25 13 39	26 ♋ 33 32	02 ♌ 49 17	07 29	02 55	20 08	00 ♍ 03	26 57	28 42	29 16	21 58
18	15 50 33	26 14 08	09 ♌ 09 26	15 ♌ 34 30	06 35	04 10	20 50	00 09	27 03	28 42	29 15	21 58
19	15 54 30	27 14 39	22 ♌ 04 58	28 ♌ 41 15	05 33	05 25	21 32	00 14	27 07	28 41	29 13	21 59
20	15 58 26	28 15 11	05 ♍ 23 44	12 ♍ 12 43	04 22	06 41	22 14	00 19	27 12	28 40	29 12	21 59
21	16 02 23	29 ♏ 15 46	19 ♍ 08 23	26 ♍ 10 48	03 06	07 56	22 56	00 24	27 17	28 39	29 11	22 00
22	16 06 20	00 ♐ 16 22	03 ♎ 19 50	10 ♎ 35 11	01 45	09 11	23 38	00 29	27 22	28 38	29 09	22 00
23	16 10 16	01 17 00	17 ♎ 56 23	25 ♎ 22 44	00 ♐ 23	10 26	24 20	00 33	27 27	28 37	29 08	22 01
24	16 14 13	02 17 39	02 ♏ 53 22	10 ♏ 27 14	29 ♏ 03	11 42	25 02	00 37	27 31	28 36	29 06	22 01
25	16 18 09	03 18 21	18 ♏ 03 10	25 ♏ 39 53	27 47	12 57	25 44	00 42	27 36	28 35	29 05	22 02
26	16 22 06	04 19 03	03 ♐ 16 06	10 ♐ 50 32	26 37	14 12	26 26	00 46	27 40	28 35	29 04	22 02
27	16 26 02	05 19 48	18 ♐ 21 58	25 ♐ 49 20	25 36	15 28	27 09	00 49	27 44	28 33	29 02	22 03
28	16 29 59	06 20 33	03 ♑ 11 41	10 ♑ 28 17	24 45	16 43	27 51	00 53	27 48	28 31	29 01	22 03
29	16 33 55	07 21 20	17 ♑ 38 33	24 ♑ 42 09	24 05	17 58	28 33	00 57	27 53	28 30	29 00	22 04
30	16 37 52	08 ♐ 22 07	01 ♒ 38 51	08 ♒ 28 40	23 ♏ 37	19 ♐ 13	29 ♏ 16	01 ♍ 00	27 ♍ 57	28 ♋ 29	28 ♈ 59	22 ♒ 05

Date	Moon True ☊ ° '	Moon Mean ☊ ° '	Moon ☽ Latitude ° '
01	02 ♋ 35	03 ♋ 57	00 S 36
02	02 D 36	03 54	01 48
03	02 36	03 51	02 52
04	02 R 36	03 47	03 45
05	02 34	03 44	04 25
06	02 30	03 41	04 53
07	02 24	03 38	05 06
08	02 16	03 35	05 06
09	02 07	03 32	04 52
10	01 58	03 28	04 25
11	01 50	03 25	03 47
12	01 43	03 22	03 00
13	01 39	03 19	02 04
14	01 36	03 16	01 S 02
15	01 D 35	03 12	00 N 03
16	01 36	03 09	01 08
17	01 38	03 06	02 12
18	01 39	03 03	03 11
19	01 40	03 00	04 01
20	01 R 40	02 57	04 41
21	01 39	02 53	05 05
22	01 36	02 50	05 13
23	01 32	02 47	05 00
24	01 27	02 44	04 27
25	01 23	02 41	03 35
26	01 20	02 37	02 28
27	01 18	02 34	01 N 11
28	01 D 18	02 31	00 S 10
29	01 18	02 28	01 29
30	01 ♋ 20	02 ♋ 25	02 S 40

DECLINATIONS

Date	Sun ☉ ° '	Moon ☽ ° '	Mercury ☿ ° '	Venus ♀ ° '	Mars ♂ ° '	Jupiter ♃ ° '	Saturn ♄ ° '	Uranus ♅ ° '	Neptune ♆ ° '	Pluto ♇ ° '
01	14 S 33	23 S 42	23 S 28	15 S 02	14 S 19	12 N 47	03 N 35	20 N 54	09 N 41	23 S 09
02	14 52	23 14	23 41	15 26	14 33	12 44	03 33	20 54	09 41	23 08
03	15 11	21 26	23 52	15 50	14 46	12 42	03 31	20 54	09 40	23 08
04	15 29	18 33	24 02	16 14	15 00	12 39	03 29	20 54	09 40	23 08
05	15 47	14 51	24 09	16 37	15 13	12 37	03 27	20 54	09 39	23 08
06	16 05	10 36	24 15	17 00	15 26	12 35	03 24	20 55	09 39	23 08
07	16 23	05 59	24 19	17 22	15 39	12 32	03 22	20 55	09 38	23 07
08	16 40	01 S 12	24 21	17 44	15 52	12 30	03 20	20 55	09 38	23 07
09	16 58	03 N 35	24 21	18 06	16 05	12 28	03 18	20 55	09 37	23 07
10	17 15	08 13	24 19	18 27	16 18	12 26	03 16	20 55	09 37	23 07
11	17 31	12 33	24 13	18 47	16 30	12 24	03 14	20 55	09 36	23 07
12	17 47	16 25	24 06	19 07	16 43	12 22	03 12	20 55	09 35	23 06
13	18 03	19 38	23 55	19 27	16 55	12 20	03 10	20 55	09 35	23 06
14	18 19	22 02	23 42	19 45	17 07	12 18	03 08	20 55	09 34	23 06
15	18 34	23 28	23 25	20 04	17 19	12 16	03 06	20 56	09 34	23 06
16	18 49	23 49	23 05	20 21	17 31	12 14	03 05	20 56	09 33	23 05
17	19 04	23 00	22 42	20 39	17 43	12 13	03 03	20 56	09 33	23 05
18	19 18	21 02	22 16	20 55	17 55	12 11	03 01	20 56	09 32	23 05
19	19 32	17 57	21 46	21 12	18 06	12 09	02 59	20 56	09 32	23 04
20	19 46	13 53	21 14	21 27	18 18	12 08	02 57	20 57	09 32	23 04
21	19 59	08 59	20 39	21 42	18 29	12 06	02 56	20 57	09 31	23 04
22	20 12	03 N 28	20 03	21 56	18 40	12 05	02 54	20 57	09 30	23 03
23	20 25	02 S 24	19 27	22 09	18 51	12 03	02 52	20 57	09 30	23 03
24	20 37	08 17	18 51	22 22	19 02	12 02	02 51	20 57	09 30	23 03
25	20 49	13 45	18 16	22 35	19 13	12 01	02 49	20 58	09 29	23 02
26	21 00	18 23	17 45	22 46	19 23	12 00	02 48	20 58	09 29	23 02
27	21 11	21 45	17 17	22 57	19 33	11 59	02 46	20 58	09 29	23 02
28	21 22	23 34	16 53	23 07	19 44	11 58	02 45	20 59	09 28	23 01
29	21 32	23 44	16 33	23 17	19 54	11 57	02 43	20 59	09 27	23 01
30	21 S 42	22 S 23	16 S 19	23 S 26	20 S 03	11 N 56	02 N 42	20 N 59	09 N 27	23 S 00

ZODIAC SIGN ENTRIES

Date	h m	Planets
03	00 23	☽ ♒
05	08 00	☽ ♓
07	18 31	☽ ♈
10	06 42	☽ ♉
12	19 28	☽ ♊
15	04 12	♀ ♐
15	07 51	☽ ♋
16	21 20	♃ ♍
17	18 37	☽ ♌
20	02 22	☽ ♍
22	05 31	☉ ♐
22	06 26	☽ ♎
23	18 54	☿ ♏
24	07 24	☽ ♏
26	06 50	☽ ♐
28	06 47	☽ ♑
30	09 08	☽ ♒

LATITUDES

Date	Mercury ☿ ° '	Venus ♀ ° '	Mars ♂ ° '	Jupiter ♃ ° '	Saturn ♄ ° '	Uranus ♅ ° '	Neptune ♆ ° '	Pluto ♇ ° '
01	02 S 50	00 N 42	00 N 13	00 N 45	01 N 57	00 N 31	01 S 47	09 S 29
04	02 51	00 36	00 11	00 46	01 57	00 31	01 47	09 28
07	02 46	00 29	00 09	00 47	01 58	00 31	01 47	09 28
10	02 32	00 22	00 08	00 47	01 59	00 31	01 47	09 27
13	02 06	00 15	00 06	00 48	01 59	00 32	01 47	09 27
16	01 26	00 08	00 04	00 49	02 00	00 32	01 47	09 26
19	00 S 33	00 N 01	00 N 02	00 49	02 00	00 32	01 47	09 26
22	00 N 28	00 S 07	00 N 01	00 50	02 01	00 32	01 47	09 25
25	01 25	00 14	00 S 01	00 51	02 02	00 32	01 47	09 24
28	02 08	00 21	00 03	00 51	02 02	00 32	01 46	09 24
31	02 N 32	00 S 28	00 S 05	00 N 52	02 N 03	00 N 32	01 S 46	09 S 23

DATA

Julian Date	2465729
Delta T	+77 seconds
Ayanamsa	24° 23' 40"
Synetic vernal point	04° ♓ 43' 19"
True obliquity of ecliptic	23° 26' 03"

MOON'S PHASES, APSIDES AND POSITIONS ☽

Date	h m	Phase	Longitude ° '	Eclipse Indicator
03	21 24	☽	11 ♒ 33	
11	22 27	○	19 ♉ 38	
19	22 10	☾	27 ♌ 40	
26	13 47	●	04 ♐ 24	

Day	h m		
11	08 42	Apogee	
25	22 33	Perigee	
01	16 16	Max dec	23° S 44'
08	18 01	0N	
16	07 18	Max dec	23° N 50'
23	02 14	0S	
29	02 27	Max dec	23° S 52'

LONGITUDES

Date	Chiron ⚷ ° '	Ceres ⚳ ° '	Pallas ⚴ ° '	Juno ⚵ ° '	Vesta ⚶ ° '	Black Moon Lilith ⚸ ° '
01	03 ♋ 40	00 ♒ 16	27 ♐ 38	06 ♐ 15	21 ♌ 58	13 ♉ 36
11	03 ♋ 21	02 ♒ 47	01 ♑ 50	09 ♐ 35	24 ♌ 52	14 ♉ 43
21	02 ♋ 55	05 ♒ 34	04 ♑ 44	13 ♐ 01	27 ♌ 25	15 ♉ 50
31	02 ♋ 23	08 ♒ 37	08 ♑ 25	16 ♐ 29	29 ♌ 32	16 ♉ 57

ASPECTARIAN

h m	Aspects
01 Monday	
00 00	☽ ⊻ ☿
07 00	☉ ☌ ♂
08 04	☽ ∠ ♇
10 58	☽ ⊥ ☿
11 44	☽ ⚹ ♂
11 51	☽ ⚹ ☉
18 49	☽ ⚹ ♀
19 00	☽ bQ ♃
23 32	☽ ⊥ ♇
02 Tuesday	
02 24	☽ ∥ ☿
03 43	☽ ∠ ☿
09 39	☽ Q ♂
10 02	☽ ⊻ ♇
10 19	☽ Q ☉
10 52	☽ ± ♃
13 55	☽ ∥ ♇
16 29	☽ △ ♄
17 59	☽ Q ♀
18 47	☉ Q ♃
21 37	☽ ⊼ ♃
22 12	☽ ☍ ♅
22 38	☉ ∠ ♄
23 45	☽ □ ♆
03 Wednesday	
08 28	☽ ⚹ ☿
14 59	♂ Q ♃
17 03	☽ # ♅
19 48	☽ □ ♂
19 49	☽ bQ ♄
20 01	♂ ∠ ♄
21 24	☽ □ ☉
22 02	♅ St R
04 Thursday	
06 12	☽ □ ♀
08 09	☽ Q ☿
08 37	☽ Q ♆
12 39	☽ ± ♄
15 05	☿ ± ♆
16 43	☽ ☌ ♇
20 18	♇ St D
05 Friday	
00 02	☽ ⊼ ♄
02 06	☽ ∥ ♀
05 37	☽ ☍ ♃
05 42	☽ ⊼ ♅
06 44	☽ ∥ ☉
07 13	☽ ⚹ ♆
09 56	☽ ∥ ♂
13 14	♃ ⊻ ♅
17 11	☽ ± ♅
20 39	☽ □ ☿
06 Saturday	
01 03	☽ # ♃
07 56	☽ △ ♂
10 37	☽ bQ ♅
11 18	☽ △ ☉
12 07	☽ ∠ ♆
17 06	☽ # ♆
22 05	☽ △ ♀
07 Sunday	
02 34	☽ ⊻ ♇
05 44	☽ ⊥ ♆
10 42	☽ ☍ ♄
14 25	☽ ⊥ ♇
15 10	☽ bQ ♂
16 06	☽ △ ♅
16 37	☽ ⊼ ♃
17 35	☽ ⊻ ♆
19 29	☽ bQ ☉
08 Monday	
01 17	☽ # ♄
04 40	☽ ± ♃
04 49	☿ Q ♄
07 14	☽ bQ ♀
08 20	☽ ∠ ♇
10 10	☽ ± ♂
10 58	☽ △ ☿
15 04	☽ ± ☉
17 50	♀ □ ♇
22 51	☽ bQ ♃
22 55	☽ ⊼ ♂
09 Tuesday	
03 21	☽ ± ♀
04 12	☽ ⊼ ☉
10 36	☽ ∥ ♄
14 26	☽ ⚹ ♇
16 51	☽ ⊼ ♀
18 17	☽ bQ ☿
23 12	☽ ⊼ ♄
10 Wednesday	
04 13	☽ □ ♅
05 21	☽ △ ♃
05 36	☽ ☌ ♆
11 26	☽ ± ♄
13 00	☽ ± ☿
14 40	☽ Q ♇
19 26	☽ ∥ ♆
11 Thursday	
01 24	☽ ⊼ ☿
04 45	♃ △ ♆
05 47	☽ bQ ♄
11 06	☽ ∥ ♃
15 15	☽ ☍ ♂
16 39	☽ Q ♅
22 27	☽ ☍ ☉
12 Friday	
03 09	☽ □ ♇
09 25	♀ ⚹ ♄
12 26	☽ △ ♄
12 45	☽ ☍ ♀
14 10	☽ # ♂
14 55	☿ St R
16 57	☽ ⚹ ♅
18 14	☽ ⊻ ♆
18 39	☽ □ ♃
22 29	☽ # ☉
13 Saturday	
06 21	☽ ⊥ ♆
10 16	☽ # ♀
14 22	☽ ☍ ☿
23 15	☽ ∠ ♅
23 49	☽ ∥ ♅
14 Sunday	
00 29	☽ ∠ ♆
04 08	♀ △ ♅
05 57	☉ □ ♇
07 19	☽ Q ♃
07 41	☽ ⊼ ♂
15 38	♀ ⊼ ♆
15 46	☽ △ ♇
16 40	☽ ⊼ ☉
17 20	☽ ⊥ ♅
20 28	☽ ± ♂
15 Monday	
00 50	♀ □ ♃
01 22	☽ □ ♄
03 52	☽ # ♇
05 20	☽ ⊻ ♅
05 46	☽ ± ☉
06 31	☽ ⚹ ♆
07 34	☽ ⚹ ♃
08 17	☽ ⊼ ♀
10 48	☿ Q ♄
11 10	☽ # ☿
15 31	☽ bQ ♂
21 35	☽ ± ♀
21 41	☽ bQ ♇
16 Tuesday	
00 42	☽ ⊼ ☿
01 16	☽ bQ ☉
06 16	☽ Q ♆
11 58	☽ ± ☿
12 04	☿ ∥ ♇
13 17	☽ Q ♄
13 28	☽ ∠ ♃
15 25	☽ ± ♇
17 20	☽ bQ ♀
22 50	☽ △ ♂
17 Wednesday	
03 07	☽ ⊼ ♇
04 35	☽ bQ ☿
07 08	☽ ⊥ ♃
09 12	☽ △ ☉
10 31	☽ # ♇
12 46	☽ ⚹ ♄
16 08	☽ ☌ ♅
17 12	☽ □ ♆
18 19	☽ # ☿
18 46	☽ ⊻ ♃
18 Thursday	
01 32	☽ △ ♀
07 30	☽ △ ☿
12 51	☽ ∥ ♅
12 52	☽ # ♀
12 56	♀ # ♅
17 27	☽ ∠ ♄
19 Friday	
01 25	☽ # ☉
08 15	♀ ± ♆
08 55	☉ ⚹ ♄
10 14	☽ ⊥ ♄
10 56	☽ □ ♂
11 00	☽ # ♂
11 49	☽ ☍ ♇
13 14	☿ ☌ ♀
18 58	☿ ± ♆
21 14	☽ ⊻ ♄
22 10	☽ □ ☉
23 59	☽ ⊻ ♅
20 Saturday	
00 57	☽ △ ♆
02 52	☽ ☌ ♃
03 28	♂ □ ♇
05 25	☿ ∥ ♀
10 20	☽ □ ☿
10 42	☽ ⊥ ♅
14 30	☽ □ ♀
21 00	☽ Q ♂
21 02	☽ ∥ ♃
21 41	☉ △ ♅
23 59	☿ # ♅
21 Sunday	
02 31	☽ ∠ ♅
03 26	☽ bQ ♆
08 31	☽ Q ☉
09 32	☽ ∥ ♆
10 00	☉ ⊼ ♆
15 04	☽ Q ☿
16 54	☽ ⊼ ♇
18 50	☽ ⚹ ♂
18 54	☽ ± ♆
22 Monday	
00 42	☽ Q ♀
01 57	☽ ☌ ♄
03 04	☽ ± ♇
04 09	☽ ⚹ ♅
05 01	☽ ⊼ ♆
06 30	☽ ⚹ ☉
07 12	☽ ⊻ ♃
07 34	☉ ∥ ☿
09 36	☽ ⚹ ☿
14 20	☽ ∥ ♄
15 37	♀ Q ♄
17 13	☉ □ ♃
17 15	☽ ⊥ ♃
18 06	☽ bQ ♇
21 14	☽ ∠ ♂
22 37	☽ ⚹ ♀
23 Tuesday	
00 04	☽ Q ♅
02 57	☉ ☌ ☿
03 42	♀ Q ♇
08 06	☽ ∠ ♃
08 12	☽ ∠ ☿
09 07	☽ ∠ ☉
09 18	☿ □ ♃
12 40	☽ ⊥ ♂
13 53	☽ # ♄
18 35	☽ △ ♇
21 33	☽ ⊥ ☿
22 50	☽ ⊻ ♂
24 Wednesday	
00 42	☽ ⊥ ☉
01 12	☽ ∠ ♀
03 23	☽ ⊻ ♄
05 10	☽ □ ♅
05 59	☽ ☍ ♆
06 13	☿ ∥ ♂
06 22	☽ ⊻ ☿
08 22	☽ ⚹ ♃
10 56	☿ ⊼ ♆
10 59	☽ ⊻ ☉
13 00	☽ ⊥ ♄
16 52	☽ ⊥ ♀
17 06	☽ # ♆
20 25	☿ △ ♅
25 Thursday	
03 13	☽ ⊻ ♀
03 20	☽ ∠ ♄
03 30	☽ Q ♃
04 05	☽ # ♃
15 29	☿ ⚹ ♄
18 16	☽ □ ♇
23 54	♀ bQ ♅
26 Friday	
00 42	☽ ☌ ♂
02 14	☽ ☌ ☿
03 07	☽ ⚹ ♄
04 35	☽ △ ♅
05 22	☽ ⊼ ♆
07 16	☉ # ♅
08 01	☽ □ ♃
08 43	☽ ∥ ☿
09 20	♀ bQ ♆
13 47	☽ ☌ ☉
14 23	☿ ☌ ♂
14 50	☽ ± ♆
18 28	☽ ∥ ♂
22 11	☽ Q ♄
22 44	☽ Q ♇
27 Saturday	
04 18	☽ bQ ♅
05 06	☽ bQ ♆
05 19	☉ ± ♆
05 22	☽ # ♅
06 47	☽ ∥ ☉
06 56	☽ ☌ ♀
17 55	☽ ⚹ ♇
18 42	☽ ± ♅
21 57	♀ ∥ ♇
22 59	☽ ⊻ ☿
28 Sunday	
02 08	☽ ∥ ♇
02 38	☽ ∥ ♀
02 51	☽ ⊻ ♂
03 11	☽ □ ♄
04 23	☽ ⊼ ♅
05 12	☽ △ ♆
08 12	☽ ⊥ ☿
08 13	☽ △ ♃
10 26	♂ ⚹ ♄
13 08	☽ ⊥ ♂
17 33	☽ ⊻ ☉
18 21	☽ ∠ ♇
22 18	☽ ∠ ☿
29 Monday	
04 16	☽ ⊥ ☉
04 47	☽ ∠ ♂
09 08	☽ bQ ♃
09 21	☽ ⊥ ♇
10 11	♂ △ ♅
12 36	☽ ⊻ ♀
19 30	☽ ⊻ ♇
20 37	☽ ∠ ☉
22 03	☽ ∥ ♀
22 33	☽ ⚹ ☿
23 48	☽ ⊥ ♀
30 Tuesday	
00 28	☽ ± ♃
02 44	♂ ⊼ ♆
03 42	☽ ∥ ♇
05 33	☽ △ ♄
06 30	☽ ☍ ♅
07 22	☽ □ ♆
07 38	☽ ⚹ ♂
10 52	☽ ⊼ ♃
16 57	☽ ∠ ♀
18 46	☽ Q ☿
18 54	☽ ∥ ☉

All ephemeris data is given at 12.00 UT and the Moon's longitude is additionally given for 24.00 UT

DECEMBER 2038

Date	Sidereal time h m s	LONGITUDES Sun ☉ ° ' "	Moon ☽ ° ' "	Moon ☽ 24.00 ° ' "	Mercury ☿ ° '	Venus ♀ ° '	Mars ♂ ° '	Jupiter ♃ ° '	Saturn ♄ ° '	Uranus ♅ ° '	Neptune ♆ ° '	Pluto ⯓ ° '
01	16 41 49	09 ♐ 22 56	15 ♒ 11 42	21 ♒ 48 13	23 ♏ 20	20 ♐ 29	29 ♏ 58	01 ♍ 03	28 ♍ 01	28 ♋ 27	28 ♈ 58	22 ♒ 06
02	16 45 45	10 23 45	28 ♒ 18 33	04 ♓ 43 07	23 R 14	21 44	00 ♐ 41	01 06	28 04	28 R 26	28 R 56	22 06
03	16 49 42	11 24 35	11 ♓ 02 25	17 ♓ 16 56	23 D 19	22 59	01 23	01 09	28 08	28 24	28 55	22 07
04	16 53 38	12 25 26	23 ♓ 27 14	29 ♓ 33 50	23 34	24 14	02 06	01 11	28 12	28 23	28 54	22 08
05	16 57 35	13 26 18	05 ♈ 37 18	11 ♈ 38 09	23 58	25 30	02 48	01 14	28 16	28 21	28 53	22 09
06	17 01 31	14 27 11	17 ♈ 36 54	23 ♈ 34 01	24 30	26 45	03 31	01 16	28 19	28 20	28 52	22 09
07	17 05 28	15 28 04	29 ♈ 29 59	05 ♉ 25 13	25 09	28 00	04 13	01 18	28 22	28 18	28 51	22 10
08	17 09 24	16 28 59	11 ♉ 20 06	17 ♉ 14 59	25 55	29 ♐ 15	04 56	01 20	28 26	28 16	28 50	22 11
09	17 13 21	17 29 54	23 ♉ 10 14	29 ♉ 06 07	26 47	00 ♑ 31	05 39	01 21	28 29	28 14	28 49	22 12
10	17 17 18	18 30 50	05 ♊ 02 56	11 ♊ 00 56	27 44	01 46	06 22	01 23	28 32	28 13	28 48	22 13
11	17 21 14	19 31 47	17 ♊ 00 20	23 ♊ 01 22	28 45	03 01	07 05	01 24	28 35	28 11	28 47	22 14
12	17 25 11	20 32 45	29 ♊ 04 15	05 ♋ 09 10	29 ♏ 50	04 16	07 48	01 25	28 38	28 09	28 46	22 15
13	17 29 07	21 33 43	11 ♋ 16 21	17 ♋ 25 59	00 ♐ 58	05 32	08 30	01 26	28 41	28 07	28 45	22 16
14	17 33 04	22 34 43	23 ♋ 38 18	29 ♋ 53 31	02 09	06 47	09 13	01 26	28 44	28 05	28 44	22 17
15	17 37 00	23 35 43	06 ♌ 11 52	12 ♌ 33 36	03 23	08 02	09 57	01 27	28 46	28 03	28 44	22 18
16	17 40 57	24 36 45	18 ♌ 58 56	25 ♌ 28 10	04 39	09 17	10 40	01 27	28 49	28 01	28 43	22 19
17	17 44 53	25 37 47	02 ♍ 01 30	08 ♍ 39 12	05 57	10 32	11 23	01 R 27	28 51	27 59	28 42	22 20
18	17 48 50	26 38 50	15 ♍ 20 50	22 ♍ 08 27	07 16	11 48	12 06	01 27	28 53	27 57	28 41	22 21
19	17 52 47	27 39 54	29 ♍ 00 19	05 ♎ 57 05	08 37	13 03	12 49	01 27	28 56	27 55	28 41	22 22
20	17 56 43	28 40 59	12 ♎ 58 45	20 ♎ 05 11	10 00	14 18	13 32	01 26	28 58	27 53	28 40	22 23
21	18 00 40	29 ♐ 42 05	27 ♎ 16 09	04 ♏ 31 18	11 23	15 33	14 16	01 26	29 00	27 50	28 39	22 24
22	18 04 36	00 ♑ 43 12	11 ♏ 50 09	19 ♏ 12 04	12 48	16 48	14 59	01 25	29 02	27 48	28 39	22 26
23	18 08 33	01 44 19	26 ♏ 36 22	04 ♐ 02 11	14 13	18 04	15 43	01 23	29 03	27 46	28 38	22 27
24	18 12 29	02 45 27	11 ♐ 28 37	18 ♐ 54 42	15 39	19 19	16 26	01 22	29 05	27 44	28 38	22 28
25	18 16 26	03 46 36	26 ♐ 19 27	03 ♑ 41 55	17 06	20 34	17 10	01 21	29 07	27 41	28 37	22 29
26	18 20 22	04 47 45	11 ♑ 01 09	18 ♑ 16 20	18 33	21 49	17 53	01 19	29 08	27 39	28 37	22 31
27	18 24 19	05 48 55	25 ♑ 26 43	02 ♒ 31 44	20 01	23 04	18 37	01 17	29 10	27 37	28 36	22 32
28	18 28 16	06 50 04	09 ♒ 30 53	16 ♒ 23 53	21 30	24 20	19 20	01 15	29 11	27 34	28 36	22 33
29	18 32 12	07 51 14	23 ♒ 10 35	29 ♒ 50 55	22 59	25 35	20 04	01 13	29 12	27 32	28 35	22 35
30	18 36 09	08 52 24	06 ♓ 25 02	12 ♓ 53 09	24 28	26 50	20 48	01 10	29 13	27 30	28 35	22 36
31	18 40 05	09 ♑ 53 33	19 ♓ 15 34	25 ♓ 32 43	25 ♐ 58	28 ♑ 05	21 ♐ 32	01 ♍ 08	29 ♍ 14	27 ♋ 27	28 ♈ 35	22 ♒ 37

Date	Moon True ☊ ° '	Moon Mean ☊ ° '	Moon ☽ Latitude ° '	DECLINATIONS Sun ☉ ° '	Moon ☽ ° '	Mercury ☿ ° '	Venus ♀ ° '	Mars ♂ ° '	Jupiter ♃ ° '	Saturn ♄ ° '	Uranus ♅ ° '	Neptune ♆ ° '	Pluto ⯓ ° '
01	01 ♋ 21	02 ♋ 22	03 S 39	21 S 51	19 S 46	16 S 09	23 S 34	20 S 13	11 N 55	02 N 41	20 N 59	09 N 27	23 S 00
02	01 D 22	02 18	04 25	22 00	16 12	16 03	23 41	20 23	11 54	02 39	21 00	09 26	23 00
03	01 23	02 15	04 56	22 09	11 59	16 03	23 48	20 32	11 53	02 38	21 00	09 26	22 59
04	01 R 22	02 12	05 13	22 17	07 23	16 06	23 54	20 41	11 52	02 37	21 00	09 26	22 58
05	01 21	02 09	05 15	22 24	02 S 35	16 12	23 59	20 50	11 52	02 36	21 01	09 25	22 58
06	01 19	02 06	05 03	22 32	02 N 14	16 22	24 04	20 59	11 51	02 34	21 01	09 25	22 58
07	01 17	02 03	04 38	22 39	06 57	16 34	24 07	21 08	11 51	02 33	21 02	09 25	22 57
08	01 14	01 59	04 02	22 45	11 23	16 48	24 10	21 16	11 50	02 32	21 02	09 24	22 57
09	01 12	01 56	03 15	22 51	15 25	17 05	24 13	21 24	11 50	02 31	21 02	09 24	22 57
10	01 10	01 53	02 20	22 56	18 51	17 23	24 14	21 32	11 50	02 30	21 03	09 24	22 56
11	01 09	01 50	01 18	23 01	21 31	17 42	24 15	21 40	11 50	02 29	21 03	09 23	22 56
12	01 08	01 47	00 S 11	23 06	23 14	18 02	24 15	21 48	11 49	02 28	21 03	09 23	22 55
13	01 D 09	01 43	00 N 56	23 10	23 53	18 22	24 14	21 55	11 49	02 28	21 04	09 23	22 55
14	01 09	01 40	02 02	23 14	23 22	18 43	24 13	22 02	11 49	02 27	21 04	09 23	22 54
15	01 10	01 37	03 03	23 17	21 40	19 05	24 10	22 10	11 50	02 26	21 05	09 23	22 54
16	01 11	01 34	03 56	23 19	18 51	19 26	24 07	22 16	11 50	02 25	21 05	09 22	22 53
17	01 11	01 31	04 38	23 22	15 04	19 47	24 04	22 23	11 50	02 24	21 06	09 22	22 53
18	01 12	01 28	05 06	23 23	10 28	20 08	23 59	22 29	11 50	02 24	21 06	09 22	22 52
19	01 R 12	01 24	05 17	23 25	05 N 15	20 29	23 54	22 36	11 51	02 23	21 06	09 21	22 52
20	01 12	01 21	05 11	23 26	00 S 21	20 49	23 48	22 42	11 51	02 23	21 07	09 21	22 51
21	01 12	01 18	04 45	23 26	06 03	21 08	23 41	22 47	11 51	02 22	21 07	09 21	22 51
22	01 11	01 15	04 01	23 26	11 33	21 27	23 34	22 53	11 52	02 21	21 08	09 21	22 50
23	01 D 11	01 12	03 01	23 25	16 28	21 45	23 26	22 58	11 53	02 21	21 08	09 21	22 50
24	01 12	01 09	01 47	23 24	20 23	22 03	23 17	23 04	11 53	02 21	21 09	09 21	22 49
25	01 12	01 05	00 N 27	23 23	22 56	22 19	23 08	23 09	11 54	02 20	21 09	09 20	22 48
26	01 R 12	01 02	00 S 54	23 21	23 53	22 35	22 57	23 13	11 55	02 20	21 10	09 20	22 48
27	01 11	00 59	02 11	23 18	23 11	22 49	22 46	23 18	11 56	02 20	21 10	09 20	22 47
28	01 11	00 56	03 17	23 15	21 02	23 03	22 35	23 22	11 57	02 19	21 11	09 20	22 47
29	01 10	00 53	04 10	23 12	17 44	23 15	22 22	23 26	11 58	02 19	21 11	09 20	22 46
30	01 09	00 49	04 48	23 08	13 37	23 27	22 09	23 30	11 59	02 19	21 12	09 20	22 46
31	01 ♋ 08	00 ♋ 46	05 S 10	23 S 04	09 S 00	23 S 37	21 S 56	23 S 33	12 N 00	02 N 19	21 N 12	09 N 20	22 S 45

ZODIAC SIGN ENTRIES

Date	h	m	Planets
01	13	06	♂ ♐
02	15	09	☽ ♓
05	00	52	☽ ♈
07	13	01	☽ ♉
09	02	12	♀ ♑
10	01	49	☽ ♊
12	13	50	☽ ♋
12	15	41	☿ ♐
15	00	12	☽ ♌
17	08	18	☽ ♍
19	13	44	☽ ♎
21	16	32	☽ ♏
21	19	02	☉ ♑
23	17	29	☽ ♐
25	17	58	☽ ♑
27	19	42	☽ ♒
30	00	16	☽ ♓

LATITUDES

Date	Mercury ☿ ° '	Venus ♀ ° '	Mars ♂ ° '	Jupiter ♃ ° '	Saturn ♄ ° '	Uranus ♅ ° '	Neptune ♆ ° '	Pluto ⯓ ° '
01	02 N 32	00 S 28	00 S 05	00 N 52	02 N 03	00 N 32	01 S 46	09 S 23
04	02 39	00 35	00 07	00 53	02 04	00 32	01 46	09 23
07	02 34	00 42	00 09	00 54	02 05	00 32	01 46	09 22
10	02 20	00 49	00 10	00 54	02 06	00 32	01 46	09 22
13	02 01	00 55	00 12	00 55	02 06	00 33	01 46	09 21
16	01 40	01 01	00 14	00 56	02 07	00 33	01 46	09 21
19	01 16	01 07	00 16	00 57	02 08	00 33	01 46	09 20
22	00 53	01 12	00 18	00 58	02 09	00 33	01 45	09 20
25	00 30	01 17	00 20	00 58	02 10	00 33	01 45	09 20
28	00 N 07	01 21	00 22	00 59	02 11	00 33	01 45	09 19
31	00 S 15	01 S 25	00 S 23	01 N 00	02 N 11	00 N 33	01 S 45	09 S 19

DATA

Julian Date	2465759
Delta T	+77 seconds
Ayanamsa	24° 23' 45"
Synetic vernal point	04° ♓ 43' 14"
True obliquity of ecliptic	23° 26' 02"

MOON'S PHASES, APSIDES AND POSITIONS ☽

Date	h	m	Phase	Longitude ° '	Eclipse Indicator
03	12	46	☽	11 ♓ 27	
11	17	30	○	19 ♊ 46	
19	09	29	☾	27 ♍ 33	
26	01	02	●	04 ♑ 20	Total

Day	h	m	
08	13	27	Apogee
24	08	16	Perigee

06	00	50	0N	
13	13	26	Max dec	23° N 53'
20	10	32	0S	
26	13	40	Max dec	23° S 53'

LONGITUDES

Date	Chiron ⚷ ° '	Ceres ⚳ ° '	Pallas ⚴ ° '	Juno ⚵ ° '	Vesta ⚶ ° '	Black Moon Lilith ⚸ ° '
01	02 ♋ 23	08 ♒ 37	08 ♑ 25	16 ♐ 29	29 ♌ 32	16 ♉ 57
11	01 ♋ 47	11 ♒ 52	12 ♑ 08	20 ♐ 00	01 ♍ 08	18 ♉ 04
21	01 ♋ 09	15 ♒ 17	15 ♑ 53	23 ♐ 33	02 ♍ 07	19 ♉ 11
31	00 ♋ 31	18 ♒ 51	19 ♑ 39	27 ♐ 05	02 ♍ 24	20 ♉ 18

ASPECTARIAN

h m Aspects

01 Wednesday
00 46 ☽ ⚹ ☉
01 58 ☽ # ♅
05 53 ☽ Q ♂
08 03 ☽ ♇ ♄
08 36 ☽ ∥ ♂
15 11 ☽ Q ♆
22 35 ☽ ⚹ ♀

02 Thursday
00 10 ☽ Q ☉
00 26 ☽ ± ♄
00 32 ☽ ☌ ⯓
02 39 ☽ □ ☿
03 53 ☉ Q ♄
05 00 ☉ Q ⯓
11 34 ☽ ⊼ ♄
12 13 ☽ ⊼ ♅
12 31 ☿ St D
12 51 ☽ ∥ ☿
13 10 ☽ ⚹ ♆
16 40 ☽ □ ♂
17 13 ☽ ☍ ♃
19 10 ♀ ⚹ ⯓
23 15 ☽ Q ♀
23 26 ☽ ± ♅

03 Friday
01 05 ♀ ± ♅
01 44 ♄ ± ⯓
03 20 ♂ □ ♃
12 32 ☽ # ♃
12 46 ☽ □ ☉
16 31 ☽ ♇ ♅
17 31 ☽ ∠ ♆
19 31 ☿ ⊻ ♀

04 Saturday
01 31 ☽ # ♆
09 25 ☽ ⊻ ⯓
10 56 ☽ ⊥ ♆
12 14 ☽ △ ☿
13 43 ☽ □ ♀
21 11 ☽ ⊥ ⯓
21 21 ☽ ☍ ♄
21 39 ☽ △ ♅
22 41 ☽ ⊻ ♆

05 Sunday
03 15 ☽ ⊼ ♃
06 03 ☽ △ ♂
10 01 ☉ ♇ ♅
11 57 ☽ # ♄
15 02 ☽ ∠ ⯓
15 12 ☽ ± ♃
18 57 ☽ ♇ ☿
22 22 ☉ ♇ ♆

06 Monday
05 03 ☽ △ ☉
09 16 ☽ ♇ ♃
13 40 ☽ ∥ ♄
13 53 ☽ ± ☿
13 55 ☽ ♇ ♂
14 20 ♄ ⚹ ♅
18 29 ♂ # ♅
21 10 ☽ ⚹ ⯓

07 Tuesday
02 40 ☽ ⊼ ☿
08 37 ☽ △ ♀
09 15 ☽ ± ♂
09 34 ☽ □ ♅
09 43 ☽ ⊼ ♄
10 41 ☽ ☌ ♆
14 09 ☽ ♇ ☉
15 39 ☽ △ ♃
17 30 ♀ ⊼ ♅
19 25 ♀ □ ♄
21 29 ☽ Q ⯓
21 55 ☽ ± ♄
22 11 ☽ ⊼ ♂

08 Wednesday
01 03 ☽ ∥ ♆
03 58 ♀ △ ♆
08 34 ♂ ± ♆
10 07 ☽ ± ☉
14 33 ☽ ∥ ♃
16 16 ☽ ♇ ♄
18 38 ☽ ♇ ♀
21 59 ☽ Q ♅
23 25 ☽ ⊼ ☉

09 Thursday
10 02 ☽ □ ⯓
15 02 ☽ ± ♀
19 56 ☽ ☍ ☿
22 14 ☽ ⚹ ♅
22 48 ☽ △ ♄
23 24 ☽ ⊻ ♆

10 Friday
00 05 ☽ # ☿
04 25 ♀ △ ♃
04 35 ☽ □ ♃
04 36 ☽ ⊼ ♀
11 13 ☉ ∥ ⯓
11 30 ☽ ⊥ ♆
14 49 ☽ ☍ ♂
23 14 ☿ △ ♅

11 Saturday
04 22 ☽ ∠ ♅
05 34 ☽ ∠ ♆
07 12 ☽ ∥ ♅
08 07 ☿ ⚹ ♄
12 51 ☿ ⊼ ♆
13 49 ☽ # ♂
16 47 ☽ Q ♃
17 30 ☽ ☍ ☉
22 18 ☽ ⊥ ♅
22 26 ☽ △ ⯓

12 Sunday
06 16 ☽ # ⯓
09 09 ☽ # ☉
10 11 ☽ ⊻ ♅
11 08 ☽ □ ♄
11 24 ☽ ⚹ ♆
13 39 ☽ ⊼ ☿
16 38 ☽ ⚹ ♃
23 27 ☽ ☍ ♀

13 Monday
02 41 ☽ ± ☿
04 08 ☽ ♇ ⯓
06 15 ☽ ⊼ ♂
10 59 ☽ Q ♆
18 42 ☽ ± ♂
21 33 ☿ □ ♃
21 44 ☽ ± ⯓
22 04 ☽ ∠ ♃
22 07 ☽ ♇ ☿
22 35 ☽ Q ♄

14 Tuesday
00 41 ☉ ± ♅
04 50 ☉ ⚹ ⯓
09 23 ☽ ⊼ ⯓
09 46 ☽ ⊼ ☉
13 12 ☽ ♇ ♂
14 40 ☽ # ☉
15 28 ☽ ⊥ ♃
17 48 ♄ ⊼ ♆
20 25 ☽ # ⯓
20 31 ☽ ☌ ♅
21 41 ♀ ∠ ⯓
21 47 ☽ □ ♆
21 48 ☽ ⚹ ♄
22 19 ☽ ± ☉

15 Wednesday
02 58 ☽ ⊻ ♃
06 04 ☽ △ ☿
06 47 ☽ # ♂
15 51 ☽ ⊼ ♀
16 56 ☽ ♇ ☉
17 51 ☽ ∥ ♅
19 30 ☽ △ ♂

16 Thursday
00 10 ♂ Q ⯓
02 19 ☽ ∠ ♄
04 22 ☽ ± ♀
08 10 ☽ # ☿
13 12 ☿ ± ♆
17 22 ♂ Q ♄
18 11 ☽ ☍ ⯓
19 07 ☽ ⊥ ♄
22 52 ☽ ♇ ♀
23 18 ☽ △ ☉

17 Friday
04 38 ☽ ⊻ ♅
05 56 ☽ △ ♆
06 11 ☽ ⊻ ♄
08 39 ♃ St R
10 57 ☽ ☌ ♃
15 33 ☽ ⊥ ♅
19 54 ☽ □ ☿

18 Saturday
04 59 ☽ △ ♀
05 13 ☽ ∥ ♃
05 51 ☽ □ ♂
07 43 ☽ ∠ ♅
09 02 ☽ ♇ ♆
17 17 ☽ ∥ ♆

19 Sunday
00 23 ☽ ⊼ ⯓
00 57 ☽ ± ♆
01 42 ♀ ⊻ ♂
07 24 ☽ Q ☿
09 29 ☽ □ ☉
10 06 ☽ ⚹ ♅
10 54 ☽ ± ⯓
11 26 ☽ ⊼ ♆
11 52 ☽ ☌ ♄
14 58 ♂ ♇ ♅
15 19 ☽ Q ♂
16 14 ☽ ⊻ ♃
17 39 ☉ ⊼ ♅

20 Monday
00 26 ☽ ∥ ♄
02 27 ☽ ♇ ⯓
02 33 ☽ ⊥ ♃
06 22 ☽ ⚹ ☿
06 44 ☽ Q ♅
11 36 ☉ △ ♆
13 00 ☽ ⚹ ♂
14 28 ☽ □ ♀
16 06 ♂ ♇ ♆
17 51 ☽ ∠ ♃
18 45 ☽ Q ☉
18 50 ☉ □ ♄
18 55 ☿ Q ⯓
20 30 ☽ # ♄

21 Tuesday
03 53 ☽ △ ⯓
05 09 ☿ Q ♄
10 22 ☽ ∠ ☿
10 54 ☿ # ♅
12 57 ☽ □ ♅
14 18 ☽ ☍ ♆
14 52 ☽ ⊻ ♄
15 29 ☽ ∠ ♂
16 21 ☽ ⚹ ☉
18 53 ☽ ⚹ ♃
23 23 ☽ Q ♀

22 Wednesday
00 07 ♂ ∥ ⯓
00 49 ☽ ⊥ ♄
02 10 ☽ # ♆
02 51 ☽ ⊥ ☿
04 28 ♀ ♇ ♃
04 36 ♀ ⊥ ⯓
07 06 ☽ ⊥ ♂
12 11 ☿ ♇ ♅
13 27 ☽ # ♃
13 44 ☽ ⊻ ☿
14 34 ☽ Q ♃
15 35 ☽ ∠ ♄
17 24 ☽ ⊻ ♂
18 48 ☽ ∠ ☉
20 52 ☽ ⚹ ♀

23 Thursday
02 19 ☿ ♇ ♆
03 58 ☉ △ ♃
05 16 ☽ □ ⯓
10 30 ☽ ⊥ ☉
13 45 ☉ ∥ ♀
13 52 ☽ △ ♅
15 17 ☽ ⊼ ♆
15 58 ☽ ⚹ ♄
19 43 ☽ □ ♃
20 54 ☽ ⊻ ☉
23 23 ☽ ∠ ♀

24 Friday
00 58 ☽ ± ♆
10 22 ☽ Q ⯓
11 22 ☽ Q ♄
14 01 ☽ ♇ ♅
15 14 ☽ ⊥ ♀
15 28 ☽ ♇ ♆
17 55 ☽ # ♅
19 27 ☽ ☌ ☿
20 24 ☽ ☌ ♂

25 Saturday
01 49 ☽ ⊻ ♀
03 43 ☽ ∥ ☿
04 31 ☽ ± ♅
05 47 ☽ ⚹ ⯓
10 22 ☽ ∥ ⯓
10 24 ♀ ∥ ♂
14 06 ☿ ☌ ♂
14 13 ☽ ⊼ ♅
14 30 ☽ ∥ ♀
15 10 ☽ ∥ ♂
15 43 ☽ △ ♆
16 32 ☽ □ ♄
18 50 ☽ ∥ ☉
20 09 ☽ △ ♃

26 Sunday
01 02 ☽ ☌ ☉
06 14 ☽ ∠ ⯓
20 44 ☽ ♇ ♃
21 05 ☽ ⊥ ⯓
23 57 ☽ ⊻ ♂

27 Monday
01 27 ♀ ⊻ ⯓
01 53 ☽ ⊻ ☿
07 06 ☽ ⊻ ⯓
07 38 ☽ ☌ ♀
08 53 ☿ ∥ ⯓
09 13 ☿ ∥ ♀
09 40 ♀ ∥ ⯓
09 54 ☽ ∥ ☉
10 16 ☽ ∥ ♂
10 31 ☽ ⊥ ♂
11 44 ☽ ± ♃
13 05 ☽ ⊥ ☿
14 27 ☉ ∥ ♂
15 39 ☽ ☍ ♅
16 46 ☽ ∥ ☿
17 20 ☽ □ ♆
17 50 ☽ ∥ ⯓
18 17 ☽ △ ♄
18 44 ☽ ∥ ♀
21 51 ☽ ⊼ ♃

28 Tuesday
02 36 ☽ ∠ ♂
06 10 ☽ ∠ ☿
07 01 ☽ ⊻ ☉
10 44 ☽ # ♅
13 14 ☿ ± ♅
18 14 ☽ ⊥ ☉
20 08 ☽ ♇ ♄

29 Wednesday
00 21 ☽ Q ♆
05 11 ♀ ± ♃
05 18 ☉ ∠ ⯓
05 24 ☿ ⚹ ⯓
06 09 ☽ ⚹ ♂
06 50 ☉ ∥ ☿
10 56 ☽ ☌ ⯓
11 23 ☽ ∠ ☉
11 36 ☽ ⚹ ☿
12 03 ☽ ± ♄
16 45 ☽ ⊻ ♀
19 47 ☽ ⊼ ♅
21 43 ☽ ⚹ ♆
22 50 ☽ ⊼ ♄

30 Thursday
02 26 ☽ ☍ ♃
04 44 ☽ ⊥ ♀
04 58 ☽ Q ♂
06 39 ☽ ± ♅
10 24 ♀ ⊥ ♂
12 06 ☽ Q ☿
16 55 ☽ ⚹ ☉
20 41 ☽ # ♃
21 11 ☿ ∥ ♂
23 07 ☽ ∠ ♀
23 14 ☽ ♇ ♅

31 Friday
00 17 ♀ ☍ ♅
01 18 ☽ ∠ ♆
09 43 ♂ ± ♅
10 22 ☽ # ♆
16 34 ☽ □ ♂
17 27 ☽ Q ☉
18 19 ☽ ⊥ ♆
18 24 ☽ ⊻ ⯓
21 30 ♀ □ ♆

All ephemeris data is given at 12.00 UT and the Moon's longitude is additionally given for 24.00 UT

JANUARY 2039

LONGITUDES

Date	Sidereal time h m s	Sun ☉ ° ' "	Moon ☽ ° ' "	Moon ☽ 24.00 ° ' "	Mercury ☿ ° '	Venus ♀ ° '	Mars ♂ ° '	Jupiter ♃ ° '	Saturn ♄ ° '	Uranus ♅ ° '	Neptune ♆ ° '	Pluto ♇ ° '
01	18 44 02	10 ♑ 54 43	01 ♈ 45 03	07 ♈ 53 07	27 ♐ 28	29 ♑ 20	22 ♐ 15	01 ♍ 05	29 ♍ 15	27 ♋ 25	28 ♈ 35	22 ♒ 39
02	18 47 58	11 55 52	13 ♈ 57 28	19 ♈ 58 42	28 ♐ 59	00 ♒ 35	22 59	01 R 02	29 15	27 R 22	28 R 34	22 40
03	18 51 55	12 57 01	25 ♈ 57 24	01 ♉ 54 10	00 ♑ 30	01 50	23 43	00 58	29 16	27 20	28 34	22 41
04	18 55 51	13 58 09	07 ♉ 49 37	13 ♉ 44 19	02 02	03 05	24 27	00 55	29 17	27 17	28 34	22 43
05	18 59 48	14 59 18	19 ♉ 38 49	25 ♉ 33 40	03 33	04 20	25 11	00 51	29 17	27 15	28 34	22 44
06	19 03 45	16 00 26	01 ♊ 29 22	07 ♊ 26 22	05 06	05 35	25 55	00 48	29 17	27 12	28 34	22 46
07	19 07 41	17 01 34	13 ♊ 25 06	19 ♊ 25 57	06 38	06 50	26 39	00 44	29 17	27 10	28 34	22 47
08	19 11 38	18 02 42	25 ♊ 29 15	01 ♋ 35 17	08 11	08 05	27 23	00 40	29 R 17	27 07	28 34	22 49
09	19 15 34	19 03 50	07 ♋ 44 18	13 ♋ 56 28	09 45	09 20	28 07	00 35	29 17	27 04	28 D 34	22 50
10	19 19 31	20 04 57	20 ♋ 11 57	26 ♋ 30 50	11 19	10 35	28 51	00 31	29 17	27 02	28 34	22 52
11	19 23 27	21 06 04	02 ♌ 53 11	09 ♌ 19 00	12 53	11 50	29 ♐ 36	00 26	29 17	26 59	28 34	22 53
12	19 27 24	22 07 11	15 ♌ 48 17	22 ♌ 20 59	14 28	13 05	00 ♑ 20	00 21	29 17	26 57	28 34	22 55
13	19 31 20	23 08 17	28 ♌ 57 02	05 ♍ 36 21	16 03	14 20	01 04	00 17	29 16	26 54	28 34	22 56
14	19 35 17	24 09 24	12 ♍ 18 50	19 ♍ 04 24	17 38	15 35	01 49	00 11	29 15	26 51	28 34	22 58
15	19 39 14	25 10 30	25 ♍ 52 57	02 ♎ 44 21	19 14	16 50	02 33	00 06	29 15	26 49	28 34	22 59
16	19 43 10	26 11 36	09 ♎ 38 32	16 ♎ 35 22	20 51	18 05	03 17	00 ♍ 01	29 14	26 46	28 35	23 01
17	19 47 07	27 12 42	23 ♎ 34 45	00 ♏ 36 32	22 28	19 20	04 02	29 ♌ 55	29 13	26 44	28 35	23 03
18	19 51 03	28 13 47	07 ♏ 40 35	14 ♏ 46 42	24 06	20 35	04 46	29 49	29 12	26 41	28 35	23 04
19	19 55 00	29 ♑ 14 53	21 ♏ 54 41	29 ♏ 04 13	25 44	21 49	05 31	29 43	29 11	26 38	28 36	23 06
20	19 58 56	00 ♒ 15 58	06 ♐ 15 01	13 ♐ 26 40	27 23	23 04	06 16	29 37	29 10	26 36	28 36	23 08
21	20 02 53	01 17 03	20 ♐ 38 44	27 ♐ 50 42	29 ♑ 02	24 19	07 00	29 31	29 08	26 33	28 36	23 09
22	20 06 49	02 18 08	05 ♑ 01 59	12 ♑ 12 01	00 ♒ 42	25 34	07 45	29 25	29 07	26 31	28 37	23 11
23	20 10 46	03 19 12	19 ♑ 20 09	26 ♑ 25 46	02 22	26 48	08 30	29 19	29 05	26 28	28 37	23 13
24	20 14 43	04 20 15	03 ♒ 28 15	10 ♒ 27 00	04 03	28 03	09 14	29 12	29 04	26 25	28 38	23 14
25	20 18 39	05 21 18	17 ♒ 21 33	24 ♒ 11 25	05 45	29 ♒ 18	09 59	29 05	29 02	26 23	28 39	23 16
26	20 22 36	06 22 19	00 ♓ 56 17	07 ♓ 35 55	07 27	00 ♓ 33	10 44	28 59	29 00	26 20	28 39	23 18
27	20 26 32	07 23 20	14 ♓ 10 11	20 ♓ 39 05	09 10	01 47	11 29	28 52	28 58	26 18	28 40	23 19
28	20 30 29	08 24 20	27 ♓ 02 43	03 ♈ 21 17	10 53	03 02	12 14	28 45	28 56	26 15	28 40	23 21
29	20 34 25	09 25 18	09 ♈ 35 06	15 ♈ 44 34	12 37	04 16	12 59	28 38	28 54	26 13	28 41	23 23
30	20 38 22	10 26 16	21 ♈ 50 06	27 ♈ 52 16	14 22	05 31	13 44	28 31	28 51	26 10	28 42	23 24
31	20 42 18	11 ♒ 27 12	03 ♉ 51 37	09 ♉ 48 47	16 ♒ 07	06 ♓ 45	14 ♑ 29	28 ♌ 23	28 ♍ 49	26 ♋ 08	28 ♈ 43	23 ♒ 26

Date	Moon True ☊ ° '	Moon Mean ☊ ° '	Moon ☽ Latitude ° '
01	01 ♋ 08	00 ♋ 43	05 S 17
02	01 D 07	00 40	05 09
03	01 08	00 37	04 48
04	01 09	00 34	04 15
05	01 10	00 30	03 31
06	01 11	00 27	02 37
07	01 13	00 24	01 37
08	01 14	00 21	00 S 32
09	01 R 13	00 18	00 N 36
10	01 12	00 15	01 43
11	01 10	00 11	02 46
12	01 08	00 08	03 42
13	01 04	00 05	04 27
14	01 01	00 ♋ 02	04 58
15	00 57	29 ♊ 59	05 13
16	00 55	29 55	05 11
17	00 54	29 52	04 50
18	00 D 55	29 49	04 12
19	00 56	29 46	03 18
20	00 57	29 43	02 12
21	00 59	29 40	00 N 56
22	00 R 59	29 36	00 S 22
23	00 58	29 33	01 39
24	00 55	29 30	02 48
25	00 50	29 27	03 46
26	00 44	29 24	04 30
27	00 38	29 20	04 58
28	00 31	29 17	05 10
29	00 26	29 14	05 07
30	00 22	29 11	04 49
31	00 ♋ 20	29 ♊ 08	04 S 19

DECLINATIONS

Date	Sun ☉ ° '	Moon ☽ ° '	Mercury ☿ ° '	Venus ♀ ° '	Mars ♂ ° '	Jupiter ♃ ° '	Saturn ♄ ° '	Uranus ♅ ° '	Neptune ♆ ° '	Pluto ♇ ° '
01	22 S 59	04 S 09	23 S 47	21 S 41	23 S 37	12 N 01	02 N 19	21 N 13	09 N 20	22 S 45
02	22 54	00 N 45	23 55	21 26	23 40	12 03	02 19	21 13	09 20	22 44
03	22 48	05 33	24 02	21 11	23 42	12 04	02 19	21 14	09 20	22 43
04	22 42	10 06	24 07	20 55	23 45	12 06	02 19	21 14	09 20	22 43
05	22 36	14 16	24 12	20 38	23 47	12 07	02 19	21 15	09 20	22 42
06	22 28	17 53	24 15	20 20	23 50	12 09	02 19	21 15	09 20	22 42
07	22 21	20 48	24 17	20 02	23 51	12 10	02 19	21 16	09 20	22 41
08	22 13	22 50	24 17	19 44	23 53	12 12	02 20	21 16	09 20	22 41
09	22 05	23 48	24 16	19 25	23 54	12 14	02 20	21 17	09 20	22 40
10	21 56	23 37	24 14	19 05	23 56	12 16	02 20	21 17	09 20	22 39
11	21 47	22 12	24 11	18 45	23 56	12 17	02 21	21 18	09 20	22 39
12	21 37	19 37	24 06	18 24	23 57	12 19	02 21	21 18	09 20	22 38
13	21 27	16 00	23 59	18 03	23 57	12 21	02 21	21 19	09 20	22 38
14	21 17	11 31	23 52	17 41	23 58	12 23	02 22	21 19	09 21	22 37
15	21 06	06 25	23 43	17 19	23 57	12 25	02 22	21 20	09 21	22 36
16	20 54	00 N 57	23 32	16 56	23 57	12 27	02 23	21 20	09 21	22 36
17	20 43	04 S 39	23 20	16 33	23 57	12 30	02 24	21 21	09 21	22 35
18	20 31	10 05	23 07	16 09	23 56	12 32	02 24	21 21	09 21	22 35
19	20 18	15 03	22 52	15 45	23 55	12 34	02 25	21 22	09 21	22 34
20	20 05	19 11	22 35	15 20	23 53	12 36	02 26	21 22	09 21	22 33
21	19 52	22 10	22 17	14 55	23 52	12 39	02 27	21 23	09 22	22 33
22	19 39	23 42	21 58	14 30	23 50	12 41	02 27	21 23	09 22	22 32
23	19 25	23 40	21 37	14 04	23 48	12 44	02 28	21 24	09 22	22 32
24	19 10	22 06	21 15	13 38	23 45	12 46	02 29	21 24	09 22	22 31
25	18 56	19 13	20 51	13 12	23 43	12 49	02 30	21 25	09 23	22 30
26	18 41	15 21	20 25	12 45	23 40	12 51	02 31	21 25	09 23	22 30
27	18 25	10 49	19 58	12 17	23 37	12 54	02 32	21 26	09 23	22 29
28	18 10	05 55	19 30	11 50	23 33	12 56	02 33	21 26	09 23	22 29
29	17 54	00 S 54	19 00	11 22	23 30	12 59	02 34	21 27	09 24	22 28
30	17 37	04 N 02	18 28	10 54	23 26	13 02	02 35	21 27	09 24	22 28
31	17 S 21	08 N 44	17 S 55	10 S 26	23 S 22	13 N 04	02 N 37	21 N 28	09 N 24	22 S 27

ZODIAC SIGN ENTRIES

Date	h	m	Planets
01	08	36	☽ ♈
02	00	45	♀ ♒
03	04	05	☿ ♑
03	20	09	☽ ♉
06	08	59	☽ ♊
08	20	53	☽ ♋
11	06	35	☽ ♌
12	01	12	♂ ♑
13	13	54	☽ ♍
15	19	13	☽ ♎
16	14	56	♃ ♌
17	22	58	☽ ♏
20	01	33	☽ ♐
20	05	44	☉ ♒
22	02	00	☿ ♒
22	03	36	☽ ♑
24	06	04	☽ ♒
26	01	32	♀ ♓
26	10	19	☽ ♓
28	17	36	☽ ♈
31	04	15	☽ ♉

LATITUDES

Date	Mercury ☿ ° '	Venus ♀ ° '	Mars ♂ ° '	Jupiter ♃ ° '	Saturn ♄ ° '	Uranus ♅ ° '	Neptune ♆ ° '	Pluto ♇ ° '
01	00 S 22	01 S 26	00 S 24	01 N 00	02 N 12	00 N 33	01 S 45	09 S 19
04	00 42	01 29	00 26	01 01	02 13	00 33	01 45	09 18
07	01 01	01 32	00 28	01 02	02 13	00 33	01 44	09 18
10	01 17	01 34	00 30	01 02	02 14	00 33	01 44	09 18
13	01 32	01 35	00 32	01 03	02 15	00 33	01 44	09 18
16	01 44	01 36	00 34	01 04	02 16	00 33	01 44	09 18
19	01 54	01 36	00 35	01 04	02 17	00 33	01 44	09 17
22	02 01	01 36	00 37	01 05	02 18	00 33	01 44	09 17
25	02 05	01 35	00 39	01 05	02 18	00 33	01 43	09 17
28	02 05	01 33	00 41	01 06	02 19	00 33	01 43	09 17
31	02 S 01	01 S 30	00 S 43	01 N 07	02 N 20	00 N 33	01 S 43	09 S 17

DATA

Julian Date	2465790
Delta T	+77 seconds
Ayanamsa	24° 23' 50"
Synetic vernal point	04° ♓ 43' 08"
True obliquity of ecliptic	23° 26' 03"

MOON'S PHASES, APSIDES AND POSITIONS ☽

Date	h	m	Phase	Longitude ° '	Eclipse Indicator
02	07	37	☽	11 ♈ 45	
10	11	45	○	20 ♋ 04	
17	18	42	☾	27 ♎ 30	
24	13	36	●	04 ♒ 24	

Day	h	m		
05	05	07	Apogee	
21	05	20	Perigee	
02	08	19	0N	
09	20	10	Max dec	23° N 53'
16	16	03	0S	
22	23	24	Max dec	23° S 53'
29	16	21	0N	

LONGITUDES

Date	Chiron ⚷ ° '	Ceres ⚳ ° '	Pallas ⚴ ° '	Juno ⚵ ° '	Vesta ⚶ ° '	Black Moon Lilith ⚸ ° '
01	00 ♋ 27	19 ♒ 13	20 ♑ 01	27 ♐ 26	02 ♍ 24	20 ♉ 24
11	29 ♊ 50	22 ♒ 54	23 ♑ 46	00 ♑ 57	01 ♍ 51	21 ♉ 31
21	29 ♊ 17	26 ♒ 40	27 ♑ 30	04 ♑ 27	00 ♍ 32	22 ♉ 38
31	28 ♊ 49	00 ♓ 31	01 ♒ 11	07 ♑ 52	28 ♌ 32	23 ♉ 45

All ephemeris data is given at 12.00 UT and the Moon's longitude is additionally given for 24.00 UT

ASPECTARIAN

h m Aspects

01 Saturday
02 34 ☽ □ ☿
03 37 ☽ △ ♅
05 51 ☽ ⊻ ♆
05 58 ☽ ⊥ ♇
06 47 ☽ ⚹ ♀
07 08 ☽ ☍ ♄
10 17 ♀ △ ♄
10 42 ☽ ⊼ ♃
11 03 ☿ ⊼ ♅
20 59 ☽ ⋕ ♄
22 22 ☽ ± ♃
23 33 ☽ ∠ ♇

02 Sunday
01 10 ♂ ⚹ ♇
05 29 ☿ △ ♆
07 37 ☽ □ ☉
08 58 ☽ Q ♀
16 06 ☽ ꝑ ♃
16 23 ☿ □ ♄
19 44 ☽ ∥ ♄
20 08 ♀ ⊼ ♃

03 Monday
05 25 ☽ ⚹ ♇
07 12 ☽ △ ♂
07 54 ♀ ⋕ ♅
14 45 ☽ □ ♅
17 16 ☽ ☌ ♆
18 41 ☽ ⊼ ♄
19 11 ☿ △ ♃
22 05 ☽ △ ♃
22 31 ☽ △ ☿

04 Tuesday
01 16 ☽ □ ♀
05 41 ☽ Q ♇
06 50 ☽ ± ♄
07 50 ☽ ∥ ♆
08 52 ☉ ∥ ♇
15 31 ☽ ꝑ ♂
23 14 ☽ ∥ ♃

05 Wednesday
01 06 ☽ ꝑ ♄
01 39 ☽ △ ☉
03 05 ☽ Q ♅
09 27 ☽ ꝑ ☿
11 00 ☽ ± ♂
18 17 ☽ □ ♇
23 58 ☽ ⊼ ♂

06 Thursday
03 21 ☽ ⚹ ♅
06 05 ☽ ⊻ ♆
06 26 ☽ ± ☿
07 18 ☉ ꝑ ♃
07 33 ☽ △ ♄
10 36 ☽ □ ♃
10 56 ☽ ꝑ ☉
18 12 ☽ ⊥ ♆
20 22 ☽ ⊼ ☿
21 15 ☽ △ ♀

07 Friday
05 45 ☽ ⋕ ♀
06 13 ☉ ⊥ ♇
06 46 ☽ ± ☉
09 29 ☽ ∠ ♅
12 17 ☽ ∠ ♆
16 36 ☽ ∥ ♅
19 52 ☽ ⊼ ☉
22 33 ☽ Q ♃

08 Saturday
03 23 ☽ ⊥ ♅
03 43 ♂ ⊼ ♅
04 01 ☽ ⋕ ☉
04 13 ☿ ⊻ ♀
06 04 ☿ ∠ ♇
06 42 ☽ △ ♇
06 43 ☽ ꝑ ♀
09 41 ☽ ⋕ ♇
11 55 ♄ St R
15 12 ☽ ⊻ ♅
15 59 ☽ ☍ ♂
17 01 ♆ St D
18 03 ☽ ⚹ ♆
19 30 ☽ □ ♄
22 07 ☽ ⚹ ♃

09 Sunday
02 28 ☽ ± ♀
12 11 ☽ ꝑ ♇
15 28 ☽ ⊼ ♀
16 27 ☽ ☍ ☿
17 28 ☽ Q ♆

10 Monday
02 21 ♂ △ ♆
03 05 ☽ ∠ ♃
05 36 ☽ ± ♇
06 26 ☽ Q ♄
11 45 ☽ ☍ ☉
17 05 ☽ ⊼ ♇
20 10 ☽ ⊥ ♃

11 Tuesday
00 56 ☽ ☌ ♅
01 56 ♂ □ ♄
03 52 ☽ □ ♆
05 14 ☽ ⚹ ♄
05 26 ☽ ⊼ ♂
06 15 ☽ ⋕ ♇
07 26 ☽ ⊻ ♃
17 07 ☽ ⋕ ☉
17 22 ☽ ± ♂
21 38 ☽ ∥ ♅

12 Wednesday
06 27 ☽ ☍ ♀
09 11 ☽ ⊼ ☿
09 11 ☽ ∠ ♄
11 05 ☽ ꝑ ♂
12 46 ♂ △ ♃
21 43 ☽ ± ☿
21 48 ☽ ⋕ ♀

13 Thursday
00 33 ☽ ⊼ ☉
00 56 ☿ ꝑ ♃
01 03 ☽ ☍ ♇
01 41 ☽ ⊥ ♄
07 11 ☉ ⊻ ♇
08 18 ☽ ⊻ ♅
10 41 ♀ ꝑ ♄
11 18 ☽ △ ♆
12 22 ☽ ± ☉
12 34 ☽ ⊻ ♄
14 23 ☽ ☌ ♃
16 03 ☽ △ ♂
16 18 ☽ ꝑ ☿
18 23 ☿ ∥ ♂
19 07 ☽ ⊥ ♅

14 Friday
01 41 ☿ ⊥ ♇
05 54 ☽ ꝑ ☉
06 05 ☉ ⋕ ♅
07 41 ☽ ∥ ♃
11 11 ☽ ∠ ♅
12 43 ☉ ± ♃
14 14 ☽ ꝑ ♆
18 25 ☽ ⊼ ♀
22 33 ☽ ∥ ♆
22 44 ☽ △ ☿

15 Saturday
06 06 ☽ ± ♀
06 11 ☽ ± ♆
06 54 ☽ ⊼ ♇
06 59 ♀ Q ♆
10 39 ☽ △ ☉
13 38 ☽ ⚹ ♅
16 43 ☽ ⊼ ♆
17 28 ☽ ± ♇
17 54 ☽ ☌ ♄
19 21 ☽ ⊻ ♃
23 28 ☽ ꝑ ♀

16 Sunday
00 20 ☽ □ ♂
05 44 ☽ ⊥ ♃
05 48 ☽ ∥ ♄
09 11 ☽ ꝑ ♇
10 30 ☽ Q ♅
21 13 ☽ ∠ ♃
21 51 ♀ ∠ ♂

17 Monday
01 03 ☉ ☍ ♅
02 17 ☽ ⋕ ♄
04 00 ☽ △ ♀
09 12 ☽ Q ♂
09 51 ☽ □ ☿
11 05 ☽ △ ♇
17 22 ☽ □ ♅
18 42 ☽ □ ☉
20 33 ☽ ☍ ♆
20 39 ☿ ⊻ ♇
21 37 ☽ ⊻ ♄
22 45 ☽ ⚹ ♃

18 Tuesday
06 48 ☽ ⚹ ♂
07 48 ☽ ⊥ ♄
08 13 ☿ ± ♃
08 38 ☽ ⋕ ♆
18 58 ☽ Q ♃
20 27 ☽ Q ☿
20 29 ☉ □ ♆
23 01 ☽ ∠ ♄
23 31 ☽ ⋕ ♃

19 Wednesday
03 33 ☽ Q ☉
09 32 ☽ ∠ ♂
10 27 ☉ △ ♄
11 50 ☽ □ ♀
14 00 ☽ □ ♇
15 23 ☽ ∥ ♀
19 14 ☽ ⚹ ☿
19 54 ☽ △ ♅
22 14 ☉ ⊼ ♃
23 12 ☽ ⊼ ♆

20 Thursday
00 10 ☽ ⚹ ♄
00 56 ☿ ☍ ♅
01 00 ☽ □ ♃
01 14 ☽ ⚹ ☉
01 26 ☽ ⊥ ♂
09 15 ☽ ± ♆
12 01 ☽ ⊻ ♂
13 06 ♀ ☌ ♇
13 42 ♀ ± ♄
14 34 ☿ ∥ ♇
17 52 ☽ ∥ ☉
20 09 ☽ Q ♇
20 11 ☽ Q ♄
20 48 ☽ Q ♀
20 54 ☽ ꝑ ♅
23 33 ☽ ∠ ☿

21 Friday
00 16 ☽ ꝑ ♆
04 11 ☽ ∠ ☉
04 30 ☽ ⋕ ♅
04 35 ♄ ± ♇
05 51 ☿ □ ♆
11 51 ☽ ± ♅
13 08 ☽ ∥ ☿
13 32 ☿ △ ♄
16 11 ☽ ⚹ ♇
16 12 ☽ ∥ ♇
16 30 ☽ ⊥ ☿
18 41 ☿ ⊼ ♃
18 42 ☽ ⚹ ♀
20 19 ☽ ⊥ ☉
21 49 ☽ ⊼ ♅

22 Saturday
01 17 ☽ △ ♆
02 08 ☽ □ ♄
02 41 ☽ △ ♃
03 48 ☽ ⊻ ☿
07 05 ☽ ⊻ ☉
16 30 ☽ ∥ ♂
16 47 ☽ ☌ ♂
17 17 ☽ ∠ ♇
22 08 ☽ ∠ ♀

23 Sunday
02 30 ♂ ∠ ♇
03 36 ☽ ꝑ ♃
05 39 ♀ ⊼ ♅
07 23 ☽ ∥ ♂
08 25 ☽ ⊥ ♇
14 43 ☽ ⊥ ♀
18 33 ☽ ⊻ ♇
18 40 ☽ ± ♃

24 Monday
00 02 ☽ ☍ ♅
01 52 ☽ ⊻ ♀
01 57 ☿ ⋕ ♅
03 44 ☽ □ ♆
04 29 ☽ △ ♄
04 46 ☽ ⊼ ♃
07 18 ☽ ∥ ♇
13 08 ☽ ☌ ☿
13 36 ☽ ☌ ☉
18 46 ☽ ⋕ ♅
21 37 ☽ ∥ ☿
22 09 ☉ ☌ ☿
22 28 ☽ ⊻ ♂
23 15 ♀ ⚹ ♆

25 Tuesday
06 13 ☽ ꝑ ♄
06 58 ♀ ⊼ ♄
08 19 ♀ ☍ ♃
09 28 ☽ ⊥ ♂
10 45 ☽ Q ♆
14 10 ☽ ∥ ☉
21 56 ☽ ± ♄
22 23 ☽ ☌ ♇

26 Wednesday
02 11 ☽ ∠ ♂
03 50 ☽ ⊼ ♅
05 36 ♃ ⊻ ♄
06 43 ♀ ⋕ ♃
07 55 ☽ ⚹ ♆
08 32 ☽ ☍ ♃
08 33 ☽ ⊼ ♄
11 13 ☽ ☌ ♀
14 30 ☽ ± ♅
22 35 ☽ ⊻ ☉

27 Thursday
01 23 ☽ ⋕ ♃
01 29 ☽ ⊻ ☿
03 36 ☽ ∥ ♀
06 45 ☽ ꝑ ♅
06 46 ☽ ⚹ ♂
10 27 ☽ ⊥ ☉
11 04 ☽ ∠ ♆
14 07 ☽ ⊥ ☿
19 06 ☽ ⋕ ♆
21 30 ♀ ± ♅

28 Friday
03 46 ☽ ⊥ ♆
04 34 ☽ ∠ ☉
05 02 ☽ ⊻ ♇
06 22 ☽ Q ♂
09 28 ☽ ∠ ☿
10 31 ☽ △ ♅
15 05 ☽ ⊻ ♆
15 11 ☽ ⊼ ♃
15 34 ☽ ☍ ♄
16 22 ☽ ⊥ ♇

29 Saturday
00 38 ☽ ⊻ ♀
01 30 ♃ △ ♆
02 32 ☽ ± ♃
04 03 ☽ ⋕ ♄
09 40 ☽ ∠ ♇
11 39 ☽ ⚹ ☉
13 29 ☽ ⊥ ♀
18 52 ☽ ⚹ ☿
19 01 ☽ □ ♂
19 47 ☽ ꝑ ♃
20 38 ☿ ⊻ ♂

30 Sunday
04 53 ☽ ∥ ♄
05 11 ☿ ꝑ ♄
05 58 ♂ ꝑ ♃
09 06 ☽ ∠ ♀
13 18 ☽ Q ☉
15 07 ☽ ⚹ ♇
20 35 ☽ □ ♅
22 31 ☽ Q ☿

31 Monday
01 08 ☽ △ ♃
01 40 ☽ ☌ ♆
01 56 ☽ ⊼ ♄
13 55 ☽ ± ♄
15 11 ☽ Q ♇
15 36 ☽ ∥ ♆
18 31 ☽ ⚹ ♀
20 10 ☿ Q ♆
20 14 ☽ ⋕ ♀

FEBRUARY 2039

Date	Sidereal time h m s	Sun ☉ ° ' "	Moon ☽ ° ' "	Moon ☽ 24.00 ° ' "	Mercury ☿ ° '	Venus ♀ ° '	Mars ♂ ° '	Jupiter ♃ ° '	Saturn ♄ ° '	Uranus ♅ ° '	Neptune ♆ ° '	Pluto ♇ ° '
01	20 46 15	12 ♒ 28 07	15 ♉ 44 22	21 ♉ 39 03	17 ♒ 53	08 ♓ 00	15 ♑ 14	28 ♌ 16	28 ♍ 47	26 ♋ 05	28 ♈ 43	23 ♒ 28
02	20 50 12	13 29 00	27 ♉ 33 29	03 ♊ 28 19	19 39	09 14	15 59	28 R 08	28 R 44	26 R 03	28 44	23 30
03	20 54 08	14 29 53	09 ♊ 24 12	15 ♊ 21 45	21 25	10 29	16 44	28 01	28 41	26 00	28 45	23 31
04	20 58 05	15 30 44	21 ♊ 21 34	27 ♊ 24 12	23 11	11 43	17 29	27 53	28 39	25 58	28 46	23 33
05	21 02 01	16 31 33	03 ♋ 30 08	09 ♋ 39 50	24 59	12 57	18 14	27 46	28 36	25 55	28 47	23 35
06	21 05 58	17 32 22	15 ♋ 53 39	22 ♋ 11 53	26 46	14 12	18 59	27 38	28 33	25 53	28 48	23 37
07	21 09 54	18 33 09	28 ♋ 34 45	05 ♌ 02 22	28 ♒ 33	15 26	19 45	27 30	28 30	25 51	28 49	23 38
08	21 13 51	19 33 55	11 ♌ 34 44	18 ♌ 11 46	00 ♓ 20	16 40	20 30	27 23	28 27	25 48	28 50	23 40
09	21 17 47	20 34 39	24 ♌ 53 19	01 ♍ 39 05	02 07	17 54	21 15	27 15	28 24	25 46	28 51	23 42
10	21 21 44	21 35 22	08 ♍ 28 45	15 ♍ 21 53	03 53	19 08	22 01	27 07	28 21	25 44	28 52	23 44
11	21 25 41	22 36 04	22 ♍ 18 02	29 ♍ 16 43	05 38	20 23	22 46	26 59	28 17	25 41	28 53	23 45
12	21 29 37	23 36 45	06 ♎ 17 26	13 ♎ 19 42	07 22	21 37	23 31	26 51	28 14	25 39	28 54	23 47
13	21 33 34	24 37 24	20 ♎ 23 03	27 ♎ 27 06	09 04	22 51	24 17	26 43	28 10	25 37	28 56	23 49
14	21 37 30	25 38 03	04 ♏ 31 28	11 ♏ 35 52	10 44	24 05	25 02	26 35	28 07	25 35	28 57	23 51
15	21 41 27	26 38 40	18 ♏ 40 03	25 ♏ 43 49	12 22	25 19	25 48	26 27	28 03	25 33	28 58	23 52
16	21 45 23	27 39 16	02 ♐ 47 02	09 ♐ 49 35	13 56	26 32	26 33	26 19	27 59	25 31	28 59	23 54
17	21 49 20	28 39 51	16 ♐ 51 20	23 ♐ 52 12	15 27	27 46	27 19	26 11	27 56	25 29	29 01	23 56
18	21 53 16	29 40 25	00 ♑ 52 05	07 ♑ 50 49	16 54	29 ♓ 00	28 04	26 04	27 52	25 27	29 02	23 58
19	21 57 13	00 ♓ 40 58	14 ♑ 48 15	21 ♑ 44 09	18 16	00 ♈ 14	28 50	25 56	27 48	25 25	29 03	23 59
20	22 01 10	01 41 29	28 ♑ 38 18	05 ♒ 30 23	19 33	01 27	29 ♑ 36	25 48	27 44	25 23	29 05	24 01
21	22 05 06	02 41 59	12 ♒ 20 06	19 ♒ 07 08	20 43	02 41	00 ♒ 21	25 40	27 40	25 21	29 06	24 03
22	22 09 03	03 42 28	25 ♒ 51 07	02 ♓ 31 44	21 47	03 55	01 07	25 32	27 36	25 19	29 07	24 05
23	22 12 59	04 42 55	09 ♓ 08 43	15 ♓ 41 48	22 43	05 08	01 53	25 24	27 32	25 17	29 09	24 06
24	22 16 56	05 43 20	22 ♓ 10 47	28 ♓ 35 35	23 31	06 22	02 39	25 17	27 28	25 15	29 10	24 08
25	22 20 52	06 43 43	04 ♈ 56 08	11 ♈ 12 29	24 11	07 35	03 24	25 09	27 23	25 13	29 12	24 10
26	22 24 49	07 44 05	17 ♈ 24 48	23 ♈ 33 17	24 41	08 49	04 10	25 01	27 19	25 12	29 14	24 12
27	22 28 45	08 44 24	29 ♈ 38 15	05 ♉ 40 04	25 02	10 02	04 56	24 54	27 15	25 10	29 15	24 13
28	22 32 42	09 ♓ 44 42	11 ♉ 39 13	17 ♉ 36 11	25 ♓ 14	11 ♈ 15	05 ♒ 42	24 ♌ 46	27 ♍ 10	25 ♋ 08	29 ♈ 17	24 ♒ 15

Date	Moon True ☊ ° '	Moon Mean ☊ ° '	Moon ☽ Latitude ° '
01	00 ♋ 20	29 ♊ 05	03 S 39
02	00 D 21	29 01	02 49
03	00 23	28 58	01 51
04	00 24	28 55	00 S 49
05	00 R 24	28 52	00 N 17
06	00 23	28 49	01 23
07	00 19	28 46	02 26
08	00 14	28 42	03 23
09	00 06	28 39	04 11
10	29 ♊ 57	28 36	04 45
11	29 48	28 33	05 03
12	29 39	28 30	05 04
13	29 33	28 26	04 46
14	29 28	28 23	04 11
15	29 26	28 20	03 20
16	29 D 26	28 17	02 18
17	29 27	28 14	01 N 07
18	29 R 27	28 11	00 S 08
19	29 26	28 07	01 21
20	29 23	28 04	02 29
21	29 17	28 01	03 28
22	29 08	27 58	04 14
23	28 57	27 55	04 45
24	28 45	27 52	05 00
25	28 32	27 48	05 00
26	28 22	27 45	04 46
27	28 13	27 42	04 19
28	28 ♊ 07	27 ♊ 39	03 S 40

DECLINATIONS

Date	Sun ☉ ° '	Moon ☽ ° '	Mercury ☿ ° '	Venus ♀ ° '	Mars ♂ ° '	Jupiter ♃ ° '	Saturn ♄ ° '	Uranus ♅ ° '	Neptune ♆ ° '	Pluto ♇ ° '
01	17 S 04	13 N 03	17 S 21	09 S 57	23 S 17	13 N 07	02 N 38	21 N 28	09 N 25	22 S 26
02	16 46	16 52	16 45	09 28	23 12	13 10	02 39	21 29	09 25	22 26
03	16 29	20 01	16 08	08 59	23 08	13 13	02 40	21 29	09 25	22 25
04	16 11	22 20	15 29	08 29	23 02	13 15	02 42	21 30	09 26	22 25
05	15 53	23 40	14 49	08 00	22 57	13 18	02 43	21 30	09 26	22 24
06	15 34	23 51	14 07	07 30	22 51	13 21	02 44	21 30	09 27	22 23
07	15 16	22 49	13 25	07 00	22 46	13 24	02 46	21 31	09 27	22 23
08	14 57	20 34	12 41	06 30	22 39	13 26	02 47	21 31	09 27	22 22
09	14 38	17 09	11 56	05 59	22 33	13 29	02 49	21 32	09 28	22 22
10	14 18	12 47	11 11	05 29	22 27	13 32	02 50	21 32	09 28	22 21
11	13 59	07 42	10 24	04 58	22 20	13 35	02 52	21 33	09 29	22 21
12	13 39	02 N 09	09 37	04 27	22 13	13 38	02 53	21 33	09 29	22 20
13	13 19	03 S 32	08 49	03 57	22 05	13 40	02 55	21 33	09 30	22 20
14	12 58	09 05	08 02	03 26	21 58	13 43	02 56	21 34	09 30	22 19
15	12 38	14 10	07 14	02 54	21 50	13 46	02 58	21 34	09 31	22 18
16	12 17	18 28	06 26	02 23	21 42	13 49	03 00	21 35	09 31	22 18
17	11 56	21 40	05 39	01 52	21 34	13 52	03 01	21 35	09 32	22 17
18	11 35	23 33	04 53	01 21	21 25	13 54	03 03	21 35	09 32	22 17
19	11 14	23 57	04 08	00 49	21 16	13 57	03 05	21 36	09 33	22 16
20	10 52	22 52	03 25	00 S 18	21 07	14 00	03 06	21 36	09 33	22 16
21	10 31	20 25	02 44	00 N 13	20 58	14 03	03 08	21 36	09 34	22 15
22	10 09	16 52	02 05	00 45	20 49	14 05	03 10	21 37	09 34	22 15
23	09 47	12 32	01 29	01 16	20 39	14 08	03 12	21 37	09 35	22 14
24	09 25	07 42	00 56	01 48	20 30	14 11	03 14	21 37	09 35	22 14
25	09 02	02 S 38	00 27	02 19	20 19	14 13	03 15	21 38	09 36	22 13
26	08 40	02 N 26	00 S 02	02 50	20 09	14 16	03 17	21 38	09 37	22 12
27	08 17	07 18	00 N 20	03 22	19 59	14 18	03 19	21 38	09 37	22 12
28	07 S 55	11 N 50	00 N 36	03 N 53	19 S 48	14 N 21	03 N 21	21 N 39	09 N 38	22 S 12

ZODIAC SIGN ENTRIES

Date	h	m	Planets
02	16	57	☽ ♊
05	05	07	☽ ♋
07	14	39	☽ ♌
08	07	27	☿ ♓
09	21	05	☽ ♍
12	01	14	☽ ♎
14	04	19	☽ ♏
16	07	16	☽ ♐
18	10	31	☽ ♑
18	19	46	☉ ♓
19	07	31	♀ ♈
20	14	22	☽ ♒
21	00	46	♂ ♒
22	19	27	☽ ♓
25	02	39	☽ ♈
27	12	43	☽ ♉

LATITUDES

Date	Mercury ☿ ° '	Venus ♀ ° '	Mars ♂ ° '	Jupiter ♃ ° '	Saturn ♄ ° '	Uranus ♅ ° '	Neptune ♆ ° '	Pluto ♇ ° '
01	01 S 58	01 S 29	00 S 44	01 N 07	02 N 20	00 N 33	01 S 43	09 S 17
04	01 48	01 26	00 45	01 07	02 21	00 33	01 43	09 17
07	01 32	01 22	00 47	01 08	02 22	00 33	01 43	09 17
10	01 10	01 17	00 49	01 08	02 22	00 33	01 43	09 17
13	00 42	01 12	00 51	01 08	02 23	00 33	01 42	09 18
16	00 S 08	01 06	00 53	01 09	02 23	00 33	01 42	09 18
19	00 N 32	01 00	00 54	01 09	02 24	00 33	01 42	09 18
22	01 16	00 53	00 56	01 09	02 25	00 33	01 42	09 18
25	02 02	00 45	00 58	01 09	02 25	00 33	01 42	09 19
28	02 44	00 37	00 59	01 09	02 25	00 33	01 42	09 19
31	03 N 17	00 S 29	01 S 01	01 N 09	02 N 26	00 N 33	01 S 42	09 S 19

DATA

Julian Date	2465821
Delta T	+77 seconds
Ayanamsa	24° 23' 55"
Synetic vernal point	04° ♓ 43' 04"
True obliquity of ecliptic	23° 26' 03"

MOON'S PHASES, APSIDES AND POSITIONS ☽

Date	h	m	Phase	Longitude ° '	Eclipse Indicator
01	04	45	☽	12 ♉ 10	
09	03	39	○	20 ♌ 14	
16	02	36	☾	27 ♏ 16	
23	03	17	●	04 ♓ 21	

Day	h	m		
02	01	37	Apogee	
15	16	57	Perigee	
06	03	52	Max dec	23° N 56'
12	21	05	0S	
19	06	18	Max dec	24° S 00'
26	00	26	0N	

LONGITUDES

Date	Chiron ⚷ ° '	Ceres ⚳ ° '	Pallas ⚴ ° '	Juno ⚵ ° '	Vesta ⚶ ° '	Black Moon Lilith ⚸ ° '
01	28 ♊ 47	00 ♓ 54	01 ♒ 33	08 ♑ 13	28 ♌ 18	23 ♉ 51
11	28 ♊ 26	04 ♓ 48	05 ♒ 09	11 ♑ 33	25 ♌ 41	24 ♉ 58
21	28 ♊ 13	08 ♓ 44	08 ♒ 41	14 ♑ 47	23 ♌ 10	26 ♉ 05
31	28 ♊ 08	12 ♓ 40	12 ♒ 06	17 ♑ 54	20 ♌ 44	27 ♉ 11

ASPECTARIAN

h m Aspects

01 Tuesday
04 45 ☽ □ ☉
08 02 ☽ Ᵽ ♄
08 40 ☽ Q ♅
10 54 ☽ △ ♂
12 21 ☽ ∥ ♃
17 06 ☽ □ ☿
21 39 ☽ Q ♀

02 Wednesday
03 43 ☽ □ ♇
08 56 ☽ ⚹ ♅
10 15 ☉ ∥ ☿
10 56 ♄ ⊼ ♆
11 17 ☽ # ☿
11 22 ☽ # ☉
13 10 ☽ □ ♃
14 23 ♀ # ♆
14 23 ☽ △ ♄
14 24 ☽ ⊻ ♆
17 42 ☉ Ᵽ ♄
19 25 ☽ Ᵽ ♂

03 Thursday
02 35 ☽ ⊥ ♆
14 25 ☽ □ ♀
14 52 ☽ ± ♂
15 13 ☽ ∠ ♅
20 46 ☽ ∠ ♆
21 50 ♀ Ᵽ ♅
23 13 ☽ △ ☉

04 Friday
01 10 ☽ Q ♃
02 03 ☽ ∥ ♅
03 44 ☽ ⊼ ♂
04 43 ☿ ± ♄
09 13 ☽ ⊥ ♅
12 55 ☽ # ♇
14 15 ♂ ⊥ ♇
16 18 ☽ △ ☿
16 22 ☽ △ ♇
16 49 ☿ ☌ ♇
18 36 ☿ ⊥ ♂
21 07 ☽ ⊻ ♅
21 42 ☽ # ♂

05 Saturday
00 50 ☽ ⚹ ♃
02 24 ☽ □ ♄
02 43 ☽ ⚹ ♆
07 46 ☽ Ᵽ ☉
18 10 ☉ Q ♆
21 55 ☽ Ᵽ ♇

06 Sunday
00 21 ☿ ⊼ ♅
02 11 ☽ Q ♆
02 44 ☽ Ᵽ ☿
02 53 ☽ ± ☉
04 12 ♀ ∠ ♆
05 48 ☽ ∠ ♃
08 23 ☽ △ ♀
13 15 ☽ Q ♄
15 17 ☽ ± ♇
15 25 ☽ ⊼ ☉
18 17 ☽ ☍ ♂
22 49 ☽ ± ☿
22 49 ☽ ⊥ ♃
22 51 ☿ ☍ ♃

07 Monday
02 42 ☽ ⊼ ♇
06 53 ☽ ☌ ♅
10 01 ☽ ⊻ ♃
11 16 ☿ ⊼ ♄
11 51 ☽ ⚹ ♄
11 57 ☽ ⊼ ☿
12 26 ☽ □ ♆
12 37 ☿ # ♃
12 58 ☽ # ♂
15 32 ☿ ⚹ ♆
15 50 ☽ Ᵽ ♀
17 54 ☽ # ♇

08 Tuesday
03 17 ☽ ∥ ♅
10 10 ☽ ± ♀
15 24 ☽ ∠ ♄
22 12 ☽ ⊼ ♀

09 Wednesday
03 39 ☽ ☍ ☉
05 07 ☽ ⊼ ♂
07 24 ☿ ± ♅
07 34 ☽ ⊥ ♄
09 52 ☽ ☍ ♇
11 49 ♂ ± ♃
13 34 ☽ ⊻ ♅
16 09 ☽ ☌ ♃
16 28 ☽ ± ♂
18 13 ☽ ⊻ ♄
19 03 ☽ △ ♆

10 Thursday
00 10 ☽ ⊥ ♅
02 44 ☽ ☍ ☿
03 34 ☽ # ☉
08 15 ☽ ∥ ♃
09 16 ☽ Ᵽ ♂
15 55 ☽ ∠ ♅
21 21 ☽ # ☿
21 25 ☽ Ᵽ ♆

11 Friday
03 55 ☽ ∥ ♆
04 56 ☉ ± ♄
08 10 ♂ ∥ ♇
08 21 ☽ ☍ ♀
12 34 ☽ ⊼ ☉
12 51 ☽ △ ♂
13 01 ☽ ± ♆
14 31 ☽ ⊼ ♇
17 49 ☽ ⚹ ♅
19 59 ☽ ⊻ ♃
22 15 ☽ ☌ ♄
23 21 ☽ ⊼ ♆
23 41 ☽ ± ☉

12 Saturday
00 51 ☽ ± ♇
01 11 ☽ # ♀
03 26 ☉ ⊻ ♂
06 11 ☽ ⊥ ♃
08 55 ☽ ∥ ♄
13 11 ☉ # ♃
14 05 ☽ ⊼ ☿
14 19 ☽ Q ♅
15 55 ☿ # ♆
16 13 ☉ ☌ ♇
16 16 ☽ Ᵽ ♇
16 16 ☽ Ᵽ ☉
20 41 ♂ ⊻ ♇
21 24 ☽ ∠ ♃

13 Sunday
01 43 ☽ ± ☿
09 20 ☽ # ♄
13 33 ☽ ∥ ♀
13 37 ♀ ⊥ ♆
16 35 ☽ ⊼ ♀
17 38 ☿ ∠ ♂
17 50 ☽ △ ♇
18 59 ☽ □ ♂
19 05 ☽ Ᵽ ☿
19 45 ☽ △ ☉
20 52 ☽ □ ♅
22 40 ☽ ⚹ ♃

14 Monday
01 10 ☽ ⊻ ♄
02 31 ☽ ☍ ♆
03 44 ☽ ± ♀
07 21 ♀ ⊻ ♇
07 54 ☽ ∥ ☿
09 53 ☿ Ᵽ ♅
10 48 ☉ ⊼ ♅
11 18 ☽ ⊥ ♄
13 53 ☽ # ♆
18 50 ☽ Q ♃
20 28 ☽ Ᵽ ♀
23 55 ☽ △ ☿

15 Tuesday
02 31 ☽ ∠ ♄
03 16 ☽ Q ♂
04 28 ♂ ☍ ♅
04 55 ☽ ∥ ☉
08 02 ☉ ☍ ♃
09 26 ♀ # ♄
10 00 ☽ # ♃
16 30 ♀ △ ♅
20 52 ☽ □ ♇
23 39 ☽ △ ♅

16 Wednesday
00 22 ☽ △ ♀
00 48 ☽ ⚹ ♂
01 07 ☽ □ ♃
02 36 ☽ □ ☉
03 53 ☽ ⚹ ♄
05 32 ☽ ⊼ ♆
05 47 ♂ ⊼ ♃
08 12 ♀ ⊼ ♃
12 44 ♀ ⚹ ♂
12 48 ☿ ∠ ♆
15 46 ☽ ± ♆
19 31 ☉ ⊼ ♄

17 Thursday
00 14 ☽ Q ♄
01 08 ☽ Ᵽ ♅
03 34 ☽ Q ♇
03 48 ☽ ∠ ♂
07 08 ☽ Ᵽ ♆
08 22 ♂ # ♅
09 19 ☽ □ ☿
11 00 ☽ ∥ ♂
11 09 ☽ # ♅
11 39 ☽ Q ☉
14 55 ♀ ☍ ♄
16 28 ☽ ± ♅
18 06 ☽ ∥ ♇
20 04 ☽ ⊥ ♂
20 23 ☉ ⚹ ♆

18 Friday
00 08 ☽ ⚹ ♇
02 43 ☽ ⊼ ♅
03 50 ☽ △ ♃
05 54 ♂ △ ♄
06 52 ☽ □ ♄
06 56 ☽ ⊻ ♂
08 29 ☽ □ ♀
08 51 ☽ △ ♆
09 47 ☽ ⚹ ☉
12 38 ♀ ⊻ ♆
19 42 ☽ Q ☿

19 Saturday
01 57 ☽ ∠ ♇
05 22 ☽ Ᵽ ♃
07 13 ♀ ⊥ ♇
13 38 ☽ ∠ ☉
17 31 ☽ ⊥ ♇
18 30 ☽ Q ♀
18 38 ☽ ⚹ ☿
19 09 ♂ □ ♆
20 47 ☽ ± ♃

20 Sunday
03 57 ☽ ⊻ ♇
04 44 ☉ ± ♅
06 20 ☽ ☍ ♅
06 28 ☽ ⊥ ☉
07 06 ☽ ⊼ ♃
10 26 ☽ △ ♄
12 46 ☽ □ ♆
13 46 ☽ ☌ ♂
17 24 ☽ ⚹ ♀
17 45 ☽ ⊻ ☉
17 59 ♀ ± ♃
19 11 ☽ ∥ ♇
22 22 ☿ # ♄
23 19 ☽ ∠ ☿

21 Monday
01 49 ☽ # ♅
07 12 ☽ ∥ ♂
12 35 ☽ Ᵽ ♄
13 38 ☉ ⊻ ♀
16 36 ☽ ⊥ ☿
20 26 ☽ Q ♆
22 24 ☽ ∠ ♀

22 Tuesday
04 09 ☽ ⊻ ☿
04 27 ☽ ± ♄
08 49 ☽ ☌ ♇
11 02 ☽ ⊼ ♅
11 26 ☽ ☍ ♃
15 07 ☽ ⊼ ♄
16 04 ☽ ⊥ ♀
17 53 ☽ ⚹ ♆
21 47 ☽ ± ♅
22 02 ☽ ⊻ ♂

23 Wednesday
03 17 ☽ ☌ ☉
03 38 ☽ # ♃
03 58 ☽ ⊻ ♀
09 33 ☽ ⊥ ♂
14 04 ☽ Ᵽ ♅
16 44 ☿ # ♀
21 10 ☽ ∠ ♆

24 Thursday
00 40 ☿ ⊥ ♆
00 42 ☉ # ♆
02 50 ☽ # ♆
03 01 ☽ ∥ ☉
03 04 ☽ ∠ ♂
13 51 ☽ ⊥ ♆
14 39 ☽ ☌ ☿
15 39 ☽ ⊻ ♇
17 43 ☽ △ ♅
17 43 ☽ ⊼ ♃
18 15 ♃ ⊻ ♅
21 49 ☽ ☍ ♄

25 Friday
01 07 ☽ ⊻ ♆
02 56 ☽ ⊥ ♇
04 53 ☽ ± ♃
08 55 ☽ ⚹ ♂
09 06 ☽ # ♄
11 13 ☿ ⊻ ♇
13 22 ☽ # ♀
15 43 ☽ ⊻ ☉
17 36 ☽ ☌ ♀
20 06 ☽ ∠ ♇
21 52 ☽ Ᵽ ♃
23 17 ☽ ∥ ☿

26 Saturday
01 25 ☽ # ☿
04 15 ☽ ⊥ ☉
09 26 ☽ Q ♂
14 12 ☽ ∥ ♀
16 10 ☽ ∥ ♄
19 42 ♀ ∠ ♇
23 19 ☽ ∠ ☉

27 Sunday
01 17 ☽ ⚹ ♇
02 41 ☽ ⊻ ☿
02 44 ☽ △ ♃
03 11 ☽ □ ♅
03 54 ☿ ⊼ ♃
07 18 ☽ ⊼ ♄
09 35 ♀ Ᵽ ♃
09 59 ♀ ∥ ♄
11 14 ☽ ☌ ♆
14 50 ☽ ⊥ ☿
16 40 ☽ # ☉
19 07 ☽ ± ♄
23 15 ☽ □ ♂
23 36 ☿ △ ♅

28 Monday
00 01 ☽ ∥ ♆
01 08 ☽ Q ♇
07 49 ☽ ⚹ ☉
09 06 ☽ ∠ ☿
11 06 ☽ ⊻ ♀
13 02 ☽ Ᵽ ♄
14 59 ☽ Q ♅
21 07 ☉ Ᵽ ♅

All ephemeris data is given at 12.00 UT and the Moon's longitude is additionally given for 24.00 UT

MARCH 2039

LONGITUDES

Date	Sidereal time h m s	Sun ☉ ° ' "	Moon ☽ ° ' "	Moon ☽ 24.00 ° ' "	Mercury ☿ ° '	Venus ♀ ° '	Mars ♂ ° '	Jupiter ♃ ° '	Saturn ♄ ° '	Uranus ♅ ° '	Neptune ♆ ° '	Pluto ♇ ° '
01	22 36 39	10 ♓ 44 58	23 ♉ 31 33	29 ♉ 25 57	25 ♓ 15	12 ♈ 28	06 ♒ 28	24 ♌ 39	27 ♍ 06	25 ♋ 07	29 ♈ 18	24 ♒ 17
02	22 40 35	11 45 12	05 ♊ 20 00	11 ♊ 14 25	25 R 08	13 42	07 14	24 R 32	27 R 01	25 R 05	29 20	24 18
03	22 44 32	12 45 24	17 ♊ 09 51	23 ♊ 07 01	24 50	14 55	08 00	24 24	26 57	25 04	29 22	24 20
04	22 48 28	13 45 33	29 ♊ 06 37	05 ♋ 09 19	24 24	16 08	08 46	24 17	26 52	25 02	29 23	24 22
05	22 52 25	14 45 41	11 ♋ 15 46	17 ♋ 26 33	23 50	17 21	09 31	24 10	26 48	25 01	29 25	24 23
06	22 56 21	15 45 47	23 ♋ 42 14	00 ♌ 03 15	23 09	18 33	10 17	24 03	26 43	24 59	29 27	24 25
07	23 00 18	16 45 50	06 ♌ 29 59	13 ♌ 02 42	22 21	19 46	11 03	23 56	26 39	24 58	29 28	24 27
08	23 04 14	17 45 52	19 ♌ 41 30	26 ♌ 26 23	21 28	20 59	11 49	23 50	26 34	24 57	29 30	24 28
09	23 08 11	18 45 51	03 ♍ 17 11	10 ♍ 13 34	20 32	22 12	12 36	23 43	26 29	24 55	29 32	24 30
10	23 12 08	19 45 49	17 ♍ 15 05	24 ♍ 21 07	19 33	23 24	13 22	23 36	26 25	24 54	29 34	24 32
11	23 16 04	20 45 44	01 ♎ 30 57	08 ♎ 43 46	18 33	24 37	14 08	23 30	26 20	24 53	29 36	24 33
12	23 20 01	21 45 38	15 ♎ 58 42	23 ♎ 14 55	17 34	25 49	14 54	23 24	26 15	24 52	29 38	24 35
13	23 23 57	22 45 30	00 ♏ 31 31	07 ♏ 47 45	16 37	27 02	15 40	23 18	26 10	24 51	29 39	24 36
14	23 27 54	23 45 20	15 ♏ 02 53	22 ♏ 16 21	15 42	28 14	16 26	23 12	26 06	24 50	29 41	24 38
15	23 31 50	24 45 08	29 ♏ 27 37	06 ♐ 36 22	14 51	29 ♈ 26	17 12	23 06	26 01	24 49	29 43	24 40
16	23 35 47	25 44 55	13 ♐ 42 20	20 ♐ 45 22	14 04	00 ♉ 38	17 58	23 00	25 56	24 48	29 45	24 41
17	23 39 43	26 44 40	27 ♐ 45 23	04 ♑ 42 24	13 23	01 50	18 44	22 55	25 52	24 47	29 47	24 43
18	23 43 40	27 44 24	11 ♑ 36 28	18 ♑ 27 38	12 48	03 02	19 31	22 49	25 47	24 46	29 49	24 44
19	23 47 37	28 44 06	25 ♑ 15 59	02 ♒ 01 35	12 18	04 14	20 17	22 44	25 42	24 46	29 51	24 46
20	23 51 33	29 ♓ 43 46	08 ♒ 44 30	15 ♒ 24 44	11 55	05 26	21 03	22 39	25 37	24 45	29 53	24 47
21	23 55 30	00 ♈ 43 24	22 ♒ 02 18	28 ♒ 37 10	11 38	06 38	21 49	22 34	25 33	24 44	29 55	24 49
22	23 59 26	01 43 01	05 ♓ 09 16	11 ♓ 38 31	11 27	07 49	22 36	22 29	25 28	24 44	29 57	24 50
23	00 03 23	02 42 35	18 ♓ 04 50	24 ♓ 28 08	11 22	09 01	23 22	22 24	25 23	24 43	29 ♈ 59	24 52
24	00 07 19	03 42 08	00 ♈ 48 21	07 ♈ 05 25	11 D 23	10 13	24 08	22 20	25 18	24 43	00 ♉ 01	24 53
25	00 11 16	04 41 38	13 ♈ 19 20	19 ♈ 30 07	11 30	11 24	24 54	22 15	25 14	24 42	00 03	24 54
26	00 15 12	05 41 07	25 ♈ 37 50	01 ♉ 42 38	11 43	12 35	25 41	22 11	25 09	24 42	00 05	24 56
27	00 19 09	06 40 33	07 ♉ 44 43	13 ♉ 44 21	12 00	13 46	26 27	22 07	25 05	24 42	00 08	24 57
28	00 23 06	07 39 57	19 ♉ 41 51	25 ♉ 37 37	12 23	14 58	27 13	22 04	25 00	24 41	00 10	24 59
29	00 27 02	08 39 19	01 ♊ 32 05	07 ♊ 25 46	12 50	16 09	28 00	22 00	24 55	24 41	00 12	25 00
30	00 30 59	09 38 39	13 ♊ 19 12	19 ♊ 13 01	13 21	17 20	28 46	21 56	24 51	24 41	00 14	25 01
31	00 34 55	10 ♈ 37 57	25 ♊ 07 49	01 ♋ 04 16	13 ♓ 57	18 ♉ 30	29 ♒ 32	21 ♌ 53	24 ♍ 46	24 ♋ 41	00 ♉ 16	25 ♒ 03

MOON NODES AND LATITUDE

Date	Moon True ☊ ° '	Moon Mean ☊ ° '	Moon ☽ Latitude ° '
01	28 ♊ 03	27 ♊ 36	02 S 52
02	28 R 01	27 32	01 58
03	28 D 01	27 29	00 S 57
04	28 R 02	27 26	00 N 06
05	28 01	27 23	01 10
06	27 58	27 20	02 12
07	27 53	27 17	03 09
08	27 46	27 13	03 57
09	27 35	27 10	04 34
10	27 23	27 07	04 56
11	27 11	27 04	05 00
12	27 00	27 01	04 44
13	26 51	26 58	04 11
14	26 44	26 54	03 21
15	26 40	26 51	02 18
16	26 D 39	26 48	01 N 08
17	26 R 39	26 45	00 S 06
18	26 38	26 42	01 18
19	26 37	26 38	02 25
20	26 33	26 35	03 23
21	26 25	26 32	04 08
22	26 15	26 29	04 40
23	26 03	26 26	04 57
24	25 49	26 23	04 59
25	25 36	26 19	04 46
26	25 23	26 16	04 20
27	25 13	26 13	03 42
28	25 05	26 10	02 55
29	25 00	26 07	02 01
30	24 58	26 04	01 01
31	24 ♊ 57	26 ♊ 00	00 N 01

DECLINATIONS

Date	Sun ☉ ° '	Moon ☽ ° '	Mercury ☿ ° '	Venus ♀ ° '	Mars ♂ ° '	Jupiter ♃ ° '	Saturn ♄ ° '	Uranus ♅ ° '	Neptune ♆ ° '	Pluto ♇ ° '
01	07 S 32	15 N 52	00 N 48	04 N 24	19 S 37	14 N 24	03 N 23	21 N 39	09 N 38	22 S 11
02	07 09	19 16	00 56	04 55	19 26	14 26	03 25	21 39	09 39	22 11
03	06 46	21 52	00 58	05 26	19 15	14 28	03 27	21 39	09 40	22 11
04	06 23	23 32	00 55	05 57	19 03	14 31	03 29	21 40	09 40	22 10
05	06 00	24 07	00 48	06 27	18 52	14 33	03 30	21 40	09 41	22 10
06	05 37	23 31	00 36	06 58	18 40	14 35	03 32	21 40	09 42	22 09
07	05 13	21 41	00 N 20	07 28	18 28	14 38	03 34	21 40	09 42	22 09
08	04 50	18 40	00 S 01	07 58	18 15	14 40	03 36	21 41	09 43	22 08
09	04 27	14 33	00 24	08 28	18 03	14 42	03 38	21 41	09 43	22 08
10	04 03	09 34	00 50	08 58	17 50	14 44	03 40	21 41	09 44	22 07
11	03 40	03 N 59	01 18	09 28	17 38	14 46	03 42	21 41	09 45	22 07
12	03 16	01 S 55	01 48	09 57	17 25	14 48	03 44	21 41	09 46	22 07
13	02 52	07 44	02 19	10 26	17 11	14 50	03 46	21 41	09 46	22 06
14	02 29	13 09	02 50	10 55	16 58	14 52	03 48	21 42	09 47	22 06
15	02 05	17 47	03 20	11 24	16 45	14 54	03 50	21 42	09 48	22 05
16	01 41	21 19	03 50	11 53	16 31	14 56	03 52	21 42	09 48	22 05
17	01 18	23 31	04 18	12 21	16 17	14 58	03 54	21 42	09 49	22 05
18	00 54	24 14	04 45	12 49	16 03	15 00	03 56	21 42	09 50	22 04
19	00 30	23 27	05 10	13 16	15 49	15 01	03 58	21 42	09 50	22 04
20	00 S 06	21 20	05 33	13 44	15 34	15 03	03 59	21 42	09 51	22 04
21	00 N 17	18 04	05 53	14 11	15 20	15 04	04 01	21 43	09 52	22 03
22	00 41	13 58	06 11	14 37	15 05	15 06	04 03	21 43	09 53	22 03
23	01 05	09 16	06 27	15 04	14 51	15 07	04 05	21 43	09 53	22 03
24	01 28	04 S 15	06 40	15 30	14 36	15 09	04 07	21 43	09 54	22 02
25	01 52	00 N 52	06 51	15 55	14 21	15 10	04 09	21 43	09 55	22 02
26	02 15	05 52	06 59	16 20	14 05	15 11	04 11	21 43	09 56	22 02
27	02 39	10 35	07 05	16 45	13 50	15 13	04 13	21 43	09 56	22 02
28	03 02	14 50	07 08	17 10	13 35	15 14	04 14	21 43	09 57	22 01
29	03 26	18 29	07 10	17 34	13 19	15 15	04 16	21 43	09 58	22 01
30	03 49	21 23	07 09	17 57	13 03	15 16	04 18	21 43	09 59	22 01
31	04 N 12	23 N 22	07 S 05	18 N 20	12 S 47	15 N 17	04 N 20	21 N 43	09 N 59	22 S 01

ZODIAC SIGN ENTRIES

Date	h	m	Planets
02	01	09	☽ ♊
04	13	46	☽ ♋
06	23	54	☽ ♌
09	06	16	☽ ♍
11	09	28	☽ ♎
13	11	08	☽ ♏
15	12	54	☽ ♐
15	23	17	♀ ♉
17	15	52	☽ ♑
19	20	24	☽ ♒
20	18	32	☉ ♈
22	02	32	☽ ♓
23	20	41	♆ ♉
24	10	28	☽ ♈
26	20	37	☽ ♉
29	08	53	☽ ♊
31	21	50	☽ ♋

LATITUDES

Date	Mercury ☿ ° '	Venus ♀ ° '	Mars ♂ ° '	Jupiter ♃ ° '	Saturn ♄ ° '	Uranus ♅ ° '	Neptune ♆ ° '	Pluto ♇ ° '
01	02 N 56	00 S 35	01 S 00	01 N 09	02 N 26	00 N 33	01 S 42	09 S 19
04	03 25	00 26	01 01	01 09	02 26	00 33	01 42	09 19
07	03 39	00 17	01 03	01 09	02 26	00 33	01 41	09 20
10	03 35	00 S 08	01 05	01 09	02 27	00 33	01 41	09 20
13	03 13	00 N 02	01 06	01 09	02 27	00 33	01 41	09 21
16	02 38	00 12	01 08	01 09	02 27	00 33	01 41	09 21
19	01 55	00 22	01 09	01 09	02 27	00 33	01 41	09 22
22	01 10	00 32	01 10	01 09	02 27	00 33	01 41	09 22
25	00 N 26	00 42	01 12	01 09	02 27	00 33	01 41	09 23
28	00 S 15	00 53	01 13	01 08	02 27	00 32	01 41	09 24
31	00 S 50	01 N 03	01 S 14	01 N 08	02 N 27	00 N 32	01 S 41	09 S 24

DATA

Julian Date	2465849
Delta T	+77 seconds
Ayanamsa	24° 23' 59"
Synetic vernal point	04° ♓ 43' 00"
True obliquity of ecliptic	23° 26' 04"

MOON'S PHASES, APSIDES AND POSITIONS ☽

Date	h	m	Phase	Longitude ° '	Eclipse Indicator
03	02	15	☽	12 ♊ 21	
10	16	35	○	19 ♍ 57	
17	10	08	☾	26 ♐ 40	
24	17	59	●	03 ♈ 57	

Day	h	m		
01	22	42	Apogee	
13	18	51	Perigee	
29	17	31	Apogee	
05	12	09	Max dec	24° N 07'
12	04	16	0S	
18	11	22	Max dec	24° S 14'
25	07	56	0N	

LONGITUDES

Date	Chiron ⚷ ° '	Ceres ⚳ ° '	Pallas ⚴ ° '	Juno ⚵ ° '	Vesta ⚶ ° '	Black Moon Lilith ⚸ ° '
01	28 ♊ 09	11 ♓ 53	11 ♒ 26	17 ♑ 17	21 ♌ 11	26 ♉ 58
11	28 ♊ 11	15 ♓ 49	14 ♒ 45	20 ♑ 45	19 ♌ 10	28 ♉ 05
21	28 ♊ 21	19 ♓ 45	17 ♒ 56	23 ♑ 04	17 ♌ 49	29 ♉ 11
31	28 ♊ 39	23 ♓ 38	20 ♒ 57	25 ♑ 38	17 ♌ 16	00 ♊ 18

ASPECTARIAN

h m Aspects

01 Tuesday
00 35 ☽ ⊥ ♀
02 43 ☽ ∥ ♃
04 12 ☿ St R
10 17 ☽ Q ☉
13 32 ☽ □ ♇
14 15 ☽ □ ♃
15 13 ☽ ✶ ♅
15 30 ☽ ✶ ☿
19 12 ☽ △ ♄
20 56 ☽ ∠ ♀
23 46 ☽ ⚺ ♆

02 Wednesday
12 00 ☽ ⊥ ♆
13 19 ☽ # ♂
15 35 ☽ Q ☿
16 07 ☽ △ ♂
16 58 ☿ △ ♅
21 38 ☽ ∠ ♅

03 Thursday
02 15 ☽ □ ☉
02 28 ☽ Q ♃
06 19 ☽ ∠ ♆
06 55 ☽ ✶ ♀
09 44 ☽ ∥ ♅
15 34 ☽ # ♇
15 49 ☽ ⊥ ♅
23 41 ♃ ☍ ♇

04 Friday
00 33 ☽ ⚼ ♂
02 27 ☽ ✶ ♃
02 29 ☽ △ ♇
02 57 ☽ □ ☿
03 52 ☽ ⚺ ♅
07 34 ☽ □ ♄
09 49 ☽ Q ♀
12 33 ☽ ✶ ♆
13 55 ☿ ⚺ ♇
19 00 ☿ ⚻ ♃
19 45 ☽ ± ♂
23 51 ☉ # ♀
23 54 ☿ ∠ ♂

05 Saturday
03 30 ☉ ∠ ♆
07 56 ☽ ∠ ♃
08 20 ☽ ⚼ ♇
08 22 ☽ ⚻ ♂
12 18 ☽ Q ♆
18 50 ☽ Q ♄
19 25 ☽ △ ☉

06 Sunday
01 05 ☽ □ ♀
01 17 ☽ ⊥ ♃
01 51 ☽ ± ♇
02 24 ☿ ⊥ ♆
11 00 ☽ △ ☿
12 40 ☽ ⚺ ♃
13 22 ☽ ⚻ ♇
14 26 ☽ ☌ ♅
17 41 ☽ ✶ ♄
22 53 ☽ □ ♆

07 Monday
02 27 ☽ ⚼ ☉
07 16 ☽ # ♇
12 10 ☽ ∥ ♅
13 28 ☽ ⚼ ☿
20 29 ☽ ± ☉
20 54 ☽ ☍ ♂
21 23 ☽ ∠ ♄

08 Tuesday
04 39 ♂ ⚼ ♄
04 53 ☽ ± ☿
08 16 ☽ ⚻ ☉
13 33 ☽ ⊥ ♄
14 32 ☽ △ ♀
14 49 ☽ # ♂
14 59 ☽ ⚻ ☿
17 30 ☿ ⚺ ♀
18 11 ☉ ⊥ ♂
19 19 ☽ ☌ ♃
20 32 ☽ ☍ ♇
21 20 ☽ ⚺ ♅

09 Wednesday
00 09 ☽ ⚺ ♄
05 26 ☽ △ ♆
07 53 ☽ ⊥ ♅
11 14 ☽ ∥ ♃
19 25 ☽ ⚼ ♀
23 28 ☽ ∠ ♅

10 Thursday
04 59 ☽ ⚻ ♂
07 25 ☽ ⚼ ♆
09 27 ☉ ☌ ☿
11 16 ☽ ∥ ♆
12 17 ☽ ± ♀
14 28 ☽ ∥ ♀
14 38 ☿ ⊥ ♂
15 39 ☽ ☍ ☿
15 43 ♀ △ ♃
15 47 ☽ ± ♂
16 35 ☽ ☍ ☉
22 40 ☽ ⚺ ♃
22 42 ☽ ± ♆
23 22 ☽ ⚻ ♀

11 Friday
00 19 ☽ ⚻ ♇
00 55 ☽ ✶ ♅
03 22 ☽ ☌ ♄
07 47 ☽ ⚼ ♂
08 40 ☽ ⊥ ♃
08 47 ☽ ⚻ ♆
09 42 ☉ # ♄
10 23 ☽ ± ♇
10 48 ♀ ✶ ♇
11 24 ☿ ⊥ ♀
13 08 ☽ ∥ ♄
13 24 ☽ # ☉
17 18 ♀ □ ♅
20 55 ☽ Q ♅
22 06 ☽ # ☿
23 32 ☽ ∠ ♃

12 Saturday
01 23 ☽ ⚼ ♇
02 09 ♀ ∥ ♆
10 07 ☽ △ ♂
11 32 ☽ ∥ ☿
14 28 ☽ ⚻ ☿
17 10 ☽ ∥ ☉
19 28 ☽ # ♄
20 05 ♀ ⚻ ♄
22 15 ☽ ⚻ ☉
23 45 ☽ ± ☿

13 Sunday
00 10 ☽ ✶ ♃
02 13 ☽ △ ♇
02 39 ☽ □ ♅
04 52 ☽ ⚺ ♄
05 43 ☽ ☍ ♀
08 52 ☽ ± ☉
10 34 ☽ ☍ ♆
13 41 ☽ ⚼ ☿
14 42 ☽ ⊥ ♄
19 49 ☽ Q ♃
20 44 ☽ # ♆
23 45 ☉ ⚻ ♃

14 Monday
00 49 ☽ # ♀
00 49 ☽ ⚼ ☉
01 24 ☿ ⚺ ♂
02 44 ☉ ∥ ☿
05 29 ☽ ∠ ♄
10 22 ☉ ⊥ ♆
13 01 ☽ △ ☿
14 25 ☽ □ ♂
20 27 ☽ # ♃
23 08 ♀ Q ♂

15 Tuesday
01 27 ☽ □ ♃
03 33 ☽ △ ☉
03 58 ☽ □ ♇
04 15 ☽ △ ♅
06 16 ☽ ✶ ♄
06 27 ☽ ∥ ♂
09 42 ☉ ⚺ ♇
11 57 ☽ ⚻ ♀
12 26 ☽ ⚻ ♆
13 29 ☉ △ ♅
15 35 ☿ ∠ ♆
16 56 ☿ ∠ ♀
17 53 ♀ ☌ ♆
22 11 ☽ Q ♂
22 32 ☽ ± ♆
22 57 ☽ ± ♀

16 Wednesday
02 18 ☽ Q ♄
04 55 ♂ Q ♆
05 24 ☽ ⚼ ♅
10 16 ☽ Q ♇
12 36 ☽ □ ☿
13 31 ☿ # ♄
13 47 ☽ ⚼ ♆
15 16 ☽ # ♅
15 35 ☽ ⚼ ♀
16 13 ☉ ☍ ♄
18 47 ☽ ∥ ♇
19 40 ☽ ✶ ♂
20 39 ☽ ± ♅

17 Thursday
03 44 ☽ △ ♃
06 46 ☽ ✶ ♇
06 54 ☽ ⚻ ♅
08 45 ☽ □ ♄
10 08 ☽ □ ☉
12 23 ♀ ± ♄
15 30 ☽ △ ♆
17 59 ☽ Q ☿
19 42 ☽ △ ♀
22 56 ☽ ∠ ♂

18 Friday
05 27 ☽ ⚼ ♃
08 44 ☽ ∠ ♇
14 00 ☽ ✶ ☿
15 31 ☽ ⊥ ♂
19 34 ♂ ± ♄
19 48 ☽ Q ☉
21 04 ☽ ± ♃

19 Saturday
00 30 ☽ ⊥ ♇
02 41 ☽ ⚺ ♂
07 33 ☽ ⚻ ♃
11 06 ☽ ⚺ ♇
11 06 ☽ ☍ ♅
11 08 ♅ ⚻ ♇
11 55 ☉ ± ♃
12 46 ☽ △ ♄
15 30 ☽ ∠ ☿
18 38 ☽ ✶ ☉
20 09 ☽ □ ♆

20 Sunday
05 09 ☽ ∥ ♇
05 30 ☽ □ ♀
07 04 ☽ ⊥ ☿
08 35 ☽ # ♅
15 21 ☽ ⚼ ♄
15 54 ☉ ⚺ ♆
17 34 ☽ ⚺ ☿
23 38 ☽ ∠ ☉

21 Monday
04 31 ☽ Q ♆
07 30 ☽ ± ♄
11 35 ☽ ☌ ♂
12 57 ☽ ☍ ♃
14 10 ☉ ⊥ ♇
15 44 ♀ Q ♇
16 54 ☽ ⚻ ♅
17 03 ☽ ☌ ♇
17 11 ☽ Q ♀
17 17 ☽ ⊥ ☉
18 21 ☽ ⚻ ♄
23 12 ☉ ⊥ ♀

22 Tuesday
02 25 ☽ ✶ ♆
03 52 ☽ ± ♅
05 09 ☽ ⚺ ☉
05 26 ☽ ∥ ♂
05 47 ☽ # ♃
08 44 ☽ # ♀
08 51 ♂ ☍ ♃
11 03 ♂ # ♃
17 26 ☽ ✶ ♀
20 27 ☽ ⚼ ♅
23 33 ☽ ☌ ☿

23 Wednesday
04 22 ♀ # ♂
06 13 ☽ ∠ ♆
08 57 ☽ # ♆
15 43 ♀ ∥ ♃
19 20 ☿ St D
20 04 ☽ ⚻ ♃
22 33 ☽ ⚺ ♂
23 07 ☽ ⊥ ♆

24 Thursday
00 18 ☽ ∠ ♀
00 28 ☽ △ ♅
00 46 ☽ ⚺ ♇
01 00 ☽ ∥ ☿
01 39 ☽ ☍ ♄
07 20 ☽ ± ♃
10 30 ☽ ⚺ ♆
10 39 ☽ ⊥ ♂
12 09 ☽ ⊥ ♇
12 38 ☽ # ♄
13 53 ♀ ⚼ ♄
17 59 ☽ ☌ ☉
19 10 ☽ ⊥ ♀

25 Friday
00 06 ☽ # ☉
00 23 ☽ ⚼ ♃
04 59 ☽ ∠ ♂
05 24 ☽ ∠ ♇
05 43 ♂ ⚻ ♅
07 54 ☽ ⚺ ♀
08 27 ☽ ⚺ ☿
12 01 ♂ ☌ ♇
14 28 ☿ ✶ ♀
17 08 ☽ ∥ ☉
20 14 ☽ ⊥ ☿
21 08 ♂ ⚻ ♄

26 Saturday
03 44 ☽ ∥ ♄
05 17 ☽ △ ♃
10 10 ☽ □ ♅
10 37 ☽ ✶ ♇
11 04 ☽ ⚻ ♄
12 06 ☽ ✶ ♂
14 10 ☽ ∠ ☿
14 13 ♀ Q ♅
17 40 ☽ # ☿
20 49 ☽ ☌ ♆
22 50 ☽ ± ♄

27 Sunday
08 36 ☽ ∥ ♆
09 41 ☽ ⚺ ☉
10 25 ☽ Q ♇
13 30 ☽ Q ♂
16 38 ☽ ⚼ ♄
20 46 ☽ ✶ ☿
21 54 ☽ Q ♅
22 10 ☉ ⚼ ♃
22 45 ☽ ⊥ ☉

28 Monday
01 24 ☽ ☌ ♀
04 59 ☽ # ♂
14 23 ☽ ∥ ♃
16 45 ☽ □ ♃
17 29 ♄ ⚻ ♇
18 33 ☽ ∠ ☉
21 49 ☽ Q ☿
22 05 ☽ ✶ ♅
22 40 ☽ △ ♄
22 42 ☽ □ ♇

29 Tuesday
04 18 ☽ □ ♂
04 36 ☽ ∥ ♀
09 16 ☽ ⚺ ♆
21 31 ☽ ⊥ ♆

30 Wednesday
03 50 ☽ ✶ ☉
04 35 ☽ ∠ ♅
05 09 ☽ Q ♃
12 05 ☽ □ ☿
15 24 ☽ ∥ ♅
15 54 ☽ ∠ ♆
18 31 ☽ # ♇
21 04 ☽ ⚺ ♀
21 24 ☉ ∠ ♇
22 55 ☽ ⊥ ♅

31 Thursday
05 27 ☽ ✶ ♃
06 29 ☽ Q ☉
10 36 ☽ ⊥ ♀
11 05 ☽ ⚺ ♅
11 17 ☽ □ ♄
11 50 ☽ △ ♇
19 59 ☉ ∥ ♄
21 32 ☽ △ ♂
22 25 ☽ ✶ ♆

All ephemeris data is given at 12.00 UT and the Moon's longitude is additionally given for 24.00 UT

APRIL 2039

LONGITUDES

Date	Sidereal time h m s	Sun ☉ ° ' "	Moon ☽ ° ' "	Moon ☽ 24.00 ° ' "	Mercury ☿ ° '	Venus ♀ ° '	Mars ♂ ° '	Jupiter ♃ ° '	Saturn ♄ ° '	Uranus ♅ ° '	Neptune ♆ ° '	Pluto ♇ ° '
01	00 38 52	11 ♈ 37 12	07 ♋ 03 05	13 ♋ 04 55	14 ♓ 37	19 ♉ 41	00 ♓ 19	21 ♌ 50	24 ♍ 42	24 ♋ 41	00 ♉ 18	25 ♒ 04
02	00 42 48	12 36 25	19 ♋ 10 29	25 ♋ 20 27	15 21	20 52	01 05	21 R 47	24 R 38	24 D 41	00 20	25 05
03	00 46 45	13 35 35	01 ♌ 35 28	07 ♌ 56 06	16 08	22 02	01 51	21 45	24 33	24 41	00 23	25 07
04	00 50 41	14 34 44	14 ♌ 22 52	20 ♌ 56 12	16 58	23 13	02 38	21 42	24 29	24 41	00 25	25 08
05	00 54 38	15 33 50	27 ♌ 36 24	04 ♍ 23 36	17 52	24 23	03 24	21 40	24 25	24 41	00 27	25 09
06	00 58 35	16 32 53	11 ♍ 17 49	18 ♍ 18 51	18 49	25 33	04 10	21 37	24 20	24 41	00 29	25 10
07	01 02 31	17 31 55	25 ♍ 26 19	02 ♎ 39 39	19 48	26 43	04 56	21 35	24 16	24 42	00 31	25 11
08	01 06 28	18 30 54	09 ♎ 58 07	17 ♎ 20 46	20 50	27 53	05 43	21 34	24 12	24 42	00 34	25 13
09	01 10 24	19 29 51	24 ♎ 46 36	02 ♏ 14 29	21 55	29 ♉ 03	06 29	21 32	24 08	24 42	00 36	25 14
10	01 14 21	20 28 46	09 ♏ 43 17	17 ♏ 11 52	23 03	00 ♊ 12	07 15	21 31	24 04	24 43	00 38	25 15
11	01 18 17	21 27 40	24 ♏ 39 11	02 ♐ 04 17	24 12	01 22	08 02	21 29	24 00	24 43	00 40	25 16
12	01 22 14	22 26 31	09 ♐ 26 20	16 ♐ 44 40	25 24	02 31	08 48	21 28	23 56	24 44	00 43	25 17
13	01 26 10	23 25 21	23 ♐ 58 48	01 ♑ 08 22	26 39	03 41	09 34	21 27	23 52	24 44	00 45	25 18
14	01 30 07	24 24 10	08 ♑ 13 09	15 ♑ 13 05	27 55	04 50	10 21	21 27	23 49	24 45	00 47	25 19
15	01 34 04	25 22 56	22 ♑ 08 11	28 ♑ 58 33	29 ♓ 13	05 59	11 07	21 26	23 45	24 46	00 49	25 20
16	01 38 00	26 21 41	05 ♒ 44 21	12 ♒ 25 47	00 ♈ 34	07 08	11 53	21 26	23 41	24 46	00 52	25 21
17	01 41 57	27 20 24	19 ♒ 03 05	25 ♒ 36 27	01 56	08 16	12 40	21 26	23 38	24 47	00 54	25 22
18	01 45 53	28 19 05	02 ♓ 06 09	08 ♓ 32 22	03 20	09 25	13 26	21 D 26	23 34	24 48	00 56	25 23
19	01 49 50	29 ♈ 17 45	14 ♓ 55 18	21 ♓ 15 09	04 46	10 33	14 12	21 26	23 31	24 49	00 58	25 24
20	01 53 46	00 ♉ 16 23	27 ♓ 32 03	03 ♈ 46 10	06 14	11 42	14 58	21 27	23 28	24 50	01 01	25 25
21	01 57 43	01 14 59	09 ♈ 57 36	16 ♈ 06 29	07 44	12 50	15 45	21 27	23 24	24 51	01 03	25 26
22	02 01 39	02 13 33	22 ♈ 12 56	28 ♈ 17 05	09 16	13 58	16 31	21 28	23 21	24 52	01 05	25 27
23	02 05 36	03 12 05	04 ♉ 19 05	10 ♉ 19 04	10 49	15 06	17 17	21 29	23 18	24 53	01 07	25 28
24	02 09 33	04 10 36	16 ♉ 17 15	22 ♉ 13 51	12 24	16 13	18 03	21 30	23 15	24 54	01 10	25 29
25	02 13 29	05 09 05	28 ♉ 09 06	04 ♊ 03 19	14 01	17 21	18 50	21 32	23 12	24 55	01 12	25 29
26	02 17 26	06 07 31	09 ♊ 56 50	15 ♊ 50 03	15 40	18 28	19 36	21 33	23 09	24 57	01 14	25 30
27	02 21 22	07 05 56	21 ♊ 43 22	27 ♊ 37 17	17 20	19 36	20 22	21 35	23 07	24 58	01 17	25 31
28	02 25 19	08 04 19	03 ♋ 32 17	09 ♋ 28 56	19 02	20 43	21 08	21 37	23 04	24 59	01 19	25 32
29	02 29 15	09 02 40	15 ♋ 27 48	21 ♋ 29 30	20 46	21 49	21 54	21 39	23 01	25 01	01 21	25 32
30	02 33 12	10 ♉ 00 58	27 ♋ 34 38	03 ♌ 43 51	22 ♈ 31	22 ♊ 56	22 ♓ 40	21 ♌ 41	22 ♍ 59	25 ♋ 02	01 ♉ 23	25 ♒ 33

Date	Moon True ☊ ° '	Moon Mean ☊ ° '	Moon ☽ Latitude ° '
01	24 ♊ 58	25 ♊ 57	01 N 04
02	24 R 58	25 54	02 05
03	24 56	25 51	03 02
04	24 53	25 48	03 51
05	24 47	25 44	04 30
06	24 38	25 41	04 55
07	24 28	25 38	05 03
08	24 18	25 35	04 52
09	24 08	25 32	04 21
10	24 00	25 29	03 33
11	23 55	25 25	02 29
12	23 52	25 22	01 N 16
13	23 D 51	25 19	00 S 01
14	23 51	25 16	01 16
15	23 52	25 13	02 25
16	23 R 52	25 09	03 25
17	23 50	25 06	04 12
18	23 45	25 03	04 44
19	23 38	25 00	05 02
20	23 30	24 57	05 05
21	23 20	24 54	04 53
22	23 10	24 50	04 28
23	23 01	24 47	03 51
24	22 54	24 44	03 04
25	22 49	24 41	02 09
26	22 46	24 38	01 09
27	22 45	24 35	00 S 06
28	22 D 45	24 31	00 N 58
29	22 46	24 28	02 00
30	22 ♊ 48	24 ♊ 25	02 N 58

DECLINATIONS

Date	Sun ☉ ° '	Moon ☽ ° '	Mercury ☿ ° '	Venus ♀ ° '	Mars ♂ ° '	Jupiter ♃ ° '	Saturn ♄ ° '	Uranus ♅ ° '	Neptune ♆ ° '	Pluto ♇ ° '
01	04 N 36	24 N 19	07 S 00	18 N 43	12 S 31	15 N 18	04 N 21	21 N 43	10 00	22 S 00
02	04 59	24 08	06 53	19 05	12 15	15 19	04 23	21 43	10 01	22 00
03	05 22	22 45	06 43	19 27	11 59	15 20	04 25	21 43	10 02	22 00
04	05 45	20 12	06 32	19 49	11 43	15 20	04 26	21 43	10 03	22 00
05	06 08	16 31	06 19	20 09	11 26	15 21	04 28	21 43	10 03	21 59
06	06 30	11 52	06 04	20 30	11 10	15 21	04 30	21 43	10 04	21 59
07	06 53	06 27	05 48	20 50	10 53	15 22	04 31	21 43	10 05	21 59
08	07 15	00 N 32	05 29	21 09	10 37	15 23	04 33	21 43	10 06	21 59
09	07 38	05 S 32	05 09	21 28	10 20	15 23	04 34	21 42	10 06	21 59
10	08 00	11 22	04 48	21 46	10 03	15 23	04 36	21 42	10 07	21 59
11	08 22	16 31	04 25	22 03	09 46	15 24	04 37	21 42	10 08	21 58
12	08 44	20 36	04 01	22 21	09 29	15 24	04 39	21 42	10 09	21 58
13	09 06	23 19	03 35	22 37	09 12	15 24	04 40	21 42	10 10	21 58
14	09 27	24 27	03 08	22 53	08 55	15 24	04 42	21 42	10 10	21 58
15	09 49	24 00	02 39	23 09	08 37	15 24	04 43	21 42	10 11	21 58
16	10 10	22 08	02 10	23 23	08 20	15 24	04 44	21 42	10 12	21 58
17	10 31	19 05	01 39	23 38	08 02	15 24	04 46	21 41	10 13	21 58
18	10 52	15 09	01 06	23 51	07 45	15 24	04 47	21 41	10 14	21 58
19	11 13	10 35	00 S 33	24 04	07 27	15 24	04 48	21 41	10 14	21 58
20	11 34	05 39	00 N 02	24 17	07 10	15 23	04 49	21 41	10 15	21 58
21	11 54	00 S 33	00 37	24 28	06 52	15 23	04 51	21 41	10 16	21 57
22	12 15	04 N 30	01 14	24 40	06 34	15 23	04 52	21 40	10 17	21 57
23	12 35	09 20	01 52	24 50	06 16	15 22	04 53	21 40	10 17	21 57
24	12 55	13 46	02 31	25 00	05 59	15 22	04 54	21 40	10 18	21 57
25	13 14	17 39	03 10	25 09	05 41	15 21	04 55	21 40	10 19	21 57
26	13 34	20 48	03 51	25 18	05 23	15 21	04 56	21 40	10 20	21 57
27	13 53	23 05	04 33	25 26	05 05	15 20	04 57	21 39	10 21	21 57
28	14 12	24 21	05 15	25 33	04 47	15 19	04 58	21 39	10 21	21 57
29	14 31	24 32	05 58	25 40	04 29	15 18	04 59	21 39	10 22	21 57
30	14 N 49	23 N 33	06 N 42	25 N 46	04 S 11	15 N 18	05 N 00	21 N 38	10 N 23	21 S 57

ZODIAC SIGN ENTRIES

Date	h	m	Planets
01	02	23	♂ ♓
03	08	58	☽ ♌
05	16	15	☽ ♍
07	19	36	☽ ♎
09	20	24	☽ ♏
10	07	45	♀ ♊
11	20	39	☽ ♐
13	22	05	☽ ♑
16	01	49	☽ ♒
16	02	00	☿ ♈
18	08	06	☽ ♓
20	05	18	☉ ♉
20	16	44	☽ ♈
23	03	24	☽ ♉
25	15	45	☽ ♊
28	04	50	☽ ♋
30	16	44	☽ ♌

LATITUDES

Date	Mercury ☿ ° '	Venus ♀ ° '	Mars ♂ ° '	Jupiter ♃ ° '	Saturn ♄ ° '	Uranus ♅ ° '	Neptune ♆ ° '	Pluto ♇ ° '
01	01 S 01	01 N 07	01 S 15	01 N 08	02 N 27	00 N 32	01 S 41	09 S 25
04	01 30	01 17	01 16	01 08	02 27	00 32	01 41	09 25
07	01 54	01 27	01 17	01 07	02 27	00 32	01 41	09 26
10	02 13	01 37	01 18	01 07	02 27	00 32	01 41	09 27
13	02 27	01 46	01 19	01 07	02 26	00 32	01 41	09 28
16	02 36	01 55	01 20	01 06	02 26	00 32	01 41	09 28
19	02 40	02 04	01 21	01 06	02 26	00 32	01 41	09 29
22	02 39	02 12	01 21	01 06	02 25	00 32	01 41	09 30
25	02 33	02 20	01 22	01 05	02 25	00 32	01 41	09 31
28	02 23	02 27	01 23	01 05	02 25	00 32	01 41	09 32
31	02 S 08	02 N 33	01 S 23	01 N 04	02 N 24	00 N 32	01 S 41	09 S 33

DATA

Julian Date	2465880
Delta T	+77 seconds
Ayanamsa	24° 24' 02"
Synetic vernal point	04° ♓ 42' 57"
True obliquity of ecliptic	23° 26' 04"

MOON'S PHASES, APSIDES AND POSITIONS ☽

Date	h	m	Phase	Longitude ° '	Eclipse Indicator
01	21	55	☽	12 ♋ 02	
09	02	53	○	19 ♎ 07	
15	18	07	☾	25 ♑ 38	
23	09	35	●	03 ♉ 06	

Day	h	m		
10	17	41	Perigee	
26	07	01	Apogee	
01	20	16	Max dec	24° N 23'
08	14	05	0S	
14	17	07	Max dec	24° S 29'
21	14	36	0N	
29	03	37	Max dec	24° N 36'

LONGITUDES

Date	Chiron ⚷ ° '	Ceres ⚳ ° '	Pallas ⚴ ° '	Juno ⚵ ° '	Vesta ⚶ ° '	Black Moon Lilith ⚸ ° '
01	28 ♊ 42	24 ♓ 02	21 ♒ 14	25 ♑ 53	17 ♌ 15	00 ♊ 24
11	29 ♊ 08	27 ♓ 52	24 ♒ 02	28 ♑ 10	17 ♌ 34	01 ♊ 31
21	29 ♊ 42	01 ♈ 40	26 ♒ 36	00 ♒ 07	18 ♌ 36	02 ♊ 38
31	00 ♋ 21	05 ♈ 22	28 ♒ 53	01 ♒ 42	20 ♌ 17	03 ♊ 44

ASPECTARIAN

01 Friday
06 45 ☽ ∠ ♀
11 34 ☽ ∠ ♃
11 51 ♂ ✶ ♆
18 01 ☽ Ᵽ ♇
19 11 ♄ ✶ ♅
21 55 ☽ □ ☉
22 29 ☽ Q ♆
23 11 ☽ Q ♄
23 28 ♂ ± ♅
23 33 ♅ St D

02 Saturday
03 59 ☽ △ ☿
05 22 ☽ ⊥ ♃
05 31 ☽ Ᵽ ♂
11 50 ☽ ± ♇
11 54 ☿ ∠ ♆
15 39 ☽ ✶ ♀
17 05 ☽ ⊻ ♃
22 33 ☽ ✶ ♄
22 43 ☽ ☌ ♅
23 32 ☽ ⊼ ♇

03 Sunday
00 16 ☽ ± ♂
06 13 ♀ □ ♃
09 41 ☽ □ ♆
11 04 ☽ Ᵽ ☿
12 32 ☽ ⊼ ♂
17 07 ☽ Q ♀
20 22 ☽ # ♇
23 07 ☽ ∥ ♅

04 Monday
02 57 ☽ ∠ ♄
05 13 ☽ ± ☿
12 24 ☽ △ ☉
14 38 ☽ ∥ ♀
17 06 ☽ ⊼ ☿
19 29 ☽ ⊥ ♄

05 Tuesday
01 21 ☽ ☌ ♃
03 13 ♄ ± ♆
05 40 ☽ □ ♀
06 18 ☽ ⊻ ♄
06 46 ☽ ⊻ ♅
07 36 ☽ ☍ ♇
12 36 ♀ △ ♄
17 04 ☽ △ ♆
17 28 ☽ ⊥ ♅
17 40 ☽ Ᵽ ☉
18 15 ♀ ✶ ♅
18 30 ☽ ∥ ♃
19 23 ☉ # ☿
22 52 ☽ ☍ ♂

06 Wednesday
04 05 ♀ □ ♇
09 13 ☽ ∠ ♅
10 37 ☽ ± ☉
15 28 ☽ # ♂
19 12 ☽ Ᵽ ♆
20 17 ☽ ∥ ♆
21 40 ☽ ⊼ ☉

07 Thursday
01 48 ☽ ☍ ☿
05 33 ☽ ⊻ ♃
10 03 ☽ ☌ ♄
10 17 ☽ ∥ ☉
10 28 ☽ ± ♆
10 45 ☽ ✶ ♅
11 35 ☽ ⊼ ♇
14 19 ☽ △ ♀
14 52 ☽ # ☿
15 35 ☽ ⊥ ♃
19 55 ☽ ∥ ♄
20 29 ☽ ⊼ ♆
21 35 ☽ ± ♇

08 Friday
04 38 ☽ ⊼ ♂
06 26 ☽ ∠ ♃
06 38 ☽ Q ♅
12 24 ☽ Ᵽ ♇
15 00 ☽ ± ♂
17 09 ☽ Ᵽ ♀

09 Saturday
02 53 ☽ ☍ ☉
03 43 ☿ ⊼ ♃
06 24 ☽ Ᵽ ♂
06 47 ☽ ✶ ♃
07 02 ☽ ⊼ ☿
08 09 ☽ # ♄
08 58 ☽ ± ♀
10 34 ☽ ∥ ☿
10 58 ☽ ⊻ ♄
11 53 ☽ □ ♅
12 44 ☽ △ ♇
17 28 ☽ ± ☿
19 27 ☽ ⊼ ♀
20 35 ☽ ⊥ ♄
20 59 ☽ # ☉
21 23 ☽ ☍ ♆

10 Sunday
02 03 ☽ Q ♃
06 12 ♂ # ♆
06 41 ☽ ∥ ♂
06 42 ☽ # ♆
07 29 ♀ ∥ ♅
07 50 ☽ △ ♂
09 05 ☽ Ᵽ ☿
10 57 ☽ ∠ ♄
21 11 ♀ ⊻ ♆

11 Monday
00 10 ☿ # ♄
05 05 ♀ # ♇
06 23 ☽ # ♃
06 30 ☽ ⊼ ☉
06 55 ☽ □ ♃
08 05 ☿ ☍ ♄
10 57 ☽ ✶ ♄
11 13 ☽ △ ☿
12 06 ☽ △ ♅
12 42 ☉ △ ♃
12 59 ☽ □ ♇
16 51 ☽ ± ☉
21 44 ☿ ⊥ ♆
21 45 ☽ ⊼ ♆
22 25 ☿ △ ♅
23 46 ☽ ☍ ♀

12 Tuesday
06 19 ☽ Q ♄
07 32 ☽ ± ♆
08 30 ☽ Ᵽ ☉
09 35 ☿ ⊻ ♇
10 54 ☽ □ ♂
12 28 ☽ Ᵽ ♅
18 19 ☽ Q ♇
20 14 ☽ # ♅
22 19 ☽ Ᵽ ♆
22 28 ☽ ∥ ♇

13 Wednesday
03 17 ☽ ± ♅
03 28 ☽ # ♀
07 48 ☽ △ ♃
11 00 ☽ △ ☉
11 50 ☽ □ ♄
13 16 ☽ ⊼ ♅
14 13 ☽ ✶ ♇
15 41 ☉ # ♂
16 53 ☽ □ ☿
17 12 ♂ Ᵽ ♅
18 21 ☽ Q ♂
22 24 ☉ ⊼ ♄
23 22 ☽ △ ♆

14 Thursday
03 19 ☿ ± ♃
05 44 ☽ ⊼ ♀
08 59 ☽ Ᵽ ♃
15 36 ☽ ∠ ♇
15 50 ☽ ✶ ♂
16 51 ☽ ± ♀
20 34 ☉ □ ♅

15 Friday
00 23 ☽ ± ♃
02 34 ☽ Q ☿
07 07 ☽ ⊥ ♇
09 48 ☽ Ᵽ ♀
10 47 ☽ ⊼ ♃
10 53 ☉ ✶ ♇
14 48 ☽ △ ♄
16 35 ☽ ☍ ♅
17 36 ☽ ⊻ ♇
18 07 ☽ □ ☉
19 23 ☽ ∠ ♂
23 40 ☽ # ♀

16 Saturday
01 47 ☽ ✶ ☿
03 18 ☽ □ ♆
06 14 ♀ ⊥ ♆
12 17 ☽ ⊥ ♂
13 39 ☽ ∥ ♇
13 58 ☉ ∥ ♆
14 43 ☽ △ ♀
16 07 ☽ # ♅
17 15 ☽ Ᵽ ♄
17 25 ☿ ⊻ ♆
23 42 ☽ ⊻ ♂

17 Sunday
02 05 ☿ ⊥ ♇
04 43 ☽ Q ☉
07 42 ☽ ∠ ☿
09 26 ☽ ± ♄
11 43 ☽ Q ♆
14 51 ♃ St D
16 21 ☽ ☍ ♃
20 20 ☽ ⊼ ♄
22 30 ☽ ⊼ ♅
23 35 ☽ ☌ ♇

18 Monday
02 07 ☽ ⊥ ☿
04 25 ☽ ✶ ☉
09 35 ☽ ± ♅
09 50 ☽ ✶ ♆
10 35 ☽ # ♃
12 21 ♀ Q ♃
14 35 ☽ ⊻ ☿
20 11 ♀ ∠ ♅

19 Tuesday
01 08 ☉ ∠ ♂
02 23 ☽ Ᵽ ♅
02 59 ☽ □ ♀
09 00 ☽ # ☉
10 33 ☽ ☌ ♂
10 43 ☽ ∠ ☉
13 42 ☽ # ♆
14 00 ☽ ∠ ♆
17 08 ☉ ± ♄

20 Wednesday
00 21 ☽ ⊼ ♃
04 15 ☽ ☍ ♄
04 18 ☽ ∥ ♂
05 14 ☽ ⊥ ☉
06 49 ☽ △ ♅
07 09 ☽ ⊥ ♆
07 57 ☽ ⊻ ♇
11 50 ☽ ± ♃
15 19 ☿ Ᵽ ♃
15 52 ☽ # ♄
16 34 ☽ Q ♀
17 42 ☽ ⊻ ☉
18 42 ☽ ⊻ ♆
19 28 ☽ ⊥ ♇

21 Thursday
05 11 ☽ Ᵽ ♃
06 52 ☉ ☌ ♆
07 05 ☽ ☌ ☿
11 42 ☽ # ☿
12 55 ☽ ∠ ♇
18 10 ☽ ✶ ♀
18 16 ☽ ∥ ☿
21 58 ♂ ∠ ♆

22 Friday
00 03 ☽ ⊻ ♂
10 32 ☽ △ ♃
12 38 ☽ ⊥ ♂
13 45 ☽ ∥ ♄
14 14 ☽ ⊼ ♄
17 14 ☽ □ ♅
18 23 ☽ ✶ ♇
21 31 ☽ # ♂

23 Saturday
02 04 ☽ ± ♄
02 44 ☽ ∠ ♀
05 37 ☽ ☌ ♆
06 29 ☿ ∠ ♇
07 40 ☽ ∠ ♂
09 35 ☽ ☌ ☉
16 59 ☽ ∥ ♆
18 17 ☽ Q ♇
19 56 ☽ Ᵽ ♄
22 33 ☽ ⊥ ♀

24 Sunday
02 59 ☽ ⊻ ☿
05 11 ☽ Q ♅
06 43 ☽ ∥ ☉
10 37 ♀ ∠ ♆
11 52 ☽ ⊻ ♀
15 49 ☽ ✶ ♂
16 56 ☽ ⊥ ☿
21 22 ☽ ∥ ♃
22 33 ☽ □ ♃

25 Monday
02 01 ☽ △ ♄
05 27 ☽ ✶ ♅
06 35 ☽ □ ♇
14 02 ☽ ∠ ☿
17 49 ☽ Q ♂
18 13 ☽ ⊻ ♆

26 Tuesday
03 31 ☽ ⊻ ☉
06 28 ☽ ⊥ ♆
11 12 ☽ Q ♃
12 00 ☽ ∠ ♅
16 50 ☽ ⊥ ☉
19 53 ☽ ∥ ♅
22 19 ♀ ⊥ ♅
22 54 ☽ # ♇

27 Wednesday
00 52 ☽ ∠ ♆
01 34 ☽ ✶ ☿
06 23 ☽ ⊥ ♅
07 12 ☽ ☌ ♀
09 02 ☽ □ ♂
11 43 ☽ ✶ ♃
12 50 ☽ ∠ ☉
14 49 ☽ □ ♄
18 37 ☽ ⊻ ♅
19 43 ☽ △ ♇
21 54 ♂ # ♄
22 22 ☉ Q ♇

28 Thursday
00 48 ☿ # ♂
02 08 ☿ ∥ ♄
06 04 ☽ Q ☿
07 29 ☽ ✶ ♆
11 52 ☉ Ᵽ ♄
18 14 ☽ ∠ ♃
21 58 ☽ ✶ ☉

29 Friday
02 07 ☽ Ᵽ ♇
03 08 ☽ Q ♄
03 43 ♂ ⊼ ♃
07 45 ☽ Q ♆
08 08 ♀ ✶ ♃
12 22 ☽ ⊥ ♃
17 33 ♀ □ ♂
20 08 ☽ ± ♇

30 Saturday
00 05 ☽ Q ☉
00 21 ☽ □ ☿
00 21 ☽ ⊻ ♃
00 24 ☿ △ ♃
01 41 ☽ △ ♂
01 56 ☽ ⊻ ♀
02 59 ☽ ✶ ♄
07 00 ☽ ☌ ♅
08 01 ☽ ⊼ ♇
12 59 ♀ □ ♄
14 55 ☽ ⊥ ♀
15 32 ☿ ⊻ ♂
18 03 ☿ ⊼ ♄
19 28 ☽ □ ♆
21 13 ♂ ☍ ♄

All ephemeris data is given at 12.00 UT and the Moon's longitude is additionally given for 24.00 UT

MAY 2039

LONGITUDES

Date	Sidereal time h m s	Sun ☉ ° ' "	Moon ☽ ° ' "	Moon ☽ 24.00 ° ' "	Mercury ☿ ° '	Venus ♀ ° '	Mars ♂ ° '	Jupiter ♃ ° '	Saturn ♄ ° '	Uranus ♅ ° '	Neptune ♆ ° '	Pluto ♇ ° '
01	02 37 08	10 ♉ 59 15	09 ♌ 57 45	16 ♌ 16 57	24 ♈ 19	24 ♊ 03	23 ♓ 26	21 ♌ 44	22 ♍ 57	25 ♋ 04	01 ♉ 26	25 ♒ 34
02	02 41 05	11 57 29	22 ♌ 41 59	29 ♌ 13 21	26 08	25 09	24 12	21 46	22 R 54	25 05	01 28	25 34
03	02 45 02	12 55 42	05 ♍ 51 28	12 ♍ 36 38	27 59	26 15	24 58	21 49	22 52	25 07	01 30	25 35
04	02 48 58	13 53 52	19 ♍ 29 01	26 ♍ 28 38	29 ♈ 51	27 21	25 44	21 52	22 50	25 09	01 32	25 35
05	02 52 55	14 52 00	03 ♎ 35 18	10 ♎ 48 40	01 ♉ 46	28 26	26 30	21 55	22 48	25 10	01 34	25 36
06	02 56 51	15 50 07	18 ♎ 08 10	25 ♎ 33 03	03 42	29 ♊ 32	27 16	21 59	22 46	25 12	01 37	25 37
07	03 00 48	16 48 12	03 ♏ 02 23	10 ♏ 35 05	05 40	00 ♋ 37	28 02	22 02	22 44	25 14	01 39	25 37
08	03 04 44	17 46 15	18 ♏ 09 59	25 ♏ 45 48	07 40	01 42	28 48	22 06	22 42	25 16	01 41	25 38
09	03 08 41	18 44 16	03 ♐ 21 20	10 ♐ 55 20	09 41	02 46	29 ♓ 34	22 10	22 41	25 18	01 43	25 38
10	03 12 37	19 42 16	18 ♐ 26 43	25 ♐ 54 29	11 44	03 51	00 ♈ 20	22 14	22 39	25 20	01 45	25 38
11	03 16 34	20 40 15	03 ♑ 17 48	10 ♑ 36 01	13 49	04 55	01 06	22 18	22 38	25 22	01 48	25 39
12	03 20 31	21 38 12	17 ♑ 48 39	24 ♑ 55 22	15 55	05 59	01 52	22 22	22 36	25 24	01 50	25 39
13	03 24 27	22 36 08	01 ♒ 56 00	08 ♒ 50 33	18 02	07 03	02 38	22 27	22 35	25 26	01 52	25 40
14	03 28 24	23 34 02	15 ♒ 39 04	22 ♒ 21 45	20 10	08 06	03 23	22 31	22 34	25 28	01 54	25 40
15	03 32 20	24 31 55	28 ♒ 58 51	05 ♓ 30 42	22 20	09 09	04 09	22 36	22 33	25 30	01 56	25 40
16	03 36 17	25 29 47	11 ♓ 57 37	18 ♓ 19 59	24 30	10 12	04 55	22 41	22 32	25 32	01 58	25 40
17	03 40 13	26 27 38	24 ♓ 38 10	00 ♈ 52 34	26 41	11 15	05 40	22 46	22 31	25 34	02 00	25 41
18	03 44 10	27 25 28	07 ♈ 03 31	13 ♈ 11 25	28 ♉ 52	12 17	06 26	22 51	22 30	25 37	02 02	25 41
19	03 48 06	28 23 16	19 ♈ 16 34	25 ♈ 19 18	01 ♊ 04	13 19	07 12	22 57	22 30	25 39	02 05	25 41
20	03 52 03	29 ♉ 21 03	01 ♉ 19 56	07 ♉ 18 45	03 15	14 21	07 57	23 02	22 29	25 41	02 07	25 41
21	03 56 00	00 ♊ 18 49	13 ♉ 16 01	19 ♉ 12 00	05 26	15 22	08 43	23 08	22 29	25 44	02 09	25 41
22	03 59 56	01 16 34	25 ♉ 06 59	01 ♊ 01 13	07 36	16 23	09 28	23 14	22 29	25 46	02 11	25 42
23	04 03 53	02 14 17	06 ♊ 54 57	12 ♊ 48 28	09 44	17 24	10 14	23 20	22 28	25 49	02 13	25 42
24	04 07 49	03 12 00	18 ♊ 42 04	24 ♊ 36 02	11 52	18 24	10 59	23 26	22 28	25 51	02 15	25 42
25	04 11 46	04 09 40	00 ♋ 30 42	06 ♋ 26 24	13 58	19 24	11 45	23 32	22 D 28	25 54	02 17	25 42
26	04 15 42	05 07 20	12 ♋ 23 31	18 ♋ 22 27	16 03	20 24	12 30	23 39	22 28	25 56	02 19	25 42
27	04 19 39	06 04 58	24 ♋ 23 38	00 ♌ 27 28	18 05	21 23	13 15	23 45	22 28	25 59	02 21	25 R 42
28	04 23 35	07 02 35	06 ♌ 34 28	12 ♌ 45 05	20 06	22 22	14 00	23 52	22 29	26 02	02 23	25 42
29	04 27 32	08 00 10	18 ♌ 59 48	25 ♌ 19 07	22 04	23 20	14 46	23 59	22 29	26 04	02 25	25 42
30	04 31 29	08 57 44	01 ♍ 43 30	08 ♍ 13 23	23 59	24 18	15 31	24 06	22 30	26 07	02 27	25 42
31	04 35 25	09 ♊ 55 16	14 ♍ 49 10	21 ♍ 31 11	25 ♊ 52	25 ♋ 16	16 ♈ 16	24 ♌ 13	22 ♍ 30	26 ♋ 10	02 ♉ 28	25 ♒ 42

Moon Nodes and Latitude / DECLINATIONS

Date	Moon True ☊ ° '	Moon Mean ☊ ° '	Moon ☽ Latitude ° '	Sun ☉ ° '	Moon ☽ ° '	Mercury ☿ ° '	Venus ♀ ° '	Mars ♂ ° '	Jupiter ♃ ° '	Saturn ♄ ° '	Uranus ♅ ° '	Neptune ♆ ° '	Pluto ♇ ° '
01	22 ♊ 49	24 ♊ 22	03 N 48	15 N 07	21 N 25	07 N 27	25 N 51	03 S 53	15 N 17	05 N 00	21 N 38	10 N 24	21 S 58
02	22 R 48	24 19	04 29	15 25	18 11	08 12	25 56	03 34	15 16	05 01	21 38	10 24	21 58
03	22 46	24 15	04 58	15 43	13 58	08 58	26 00	03 16	15 15	05 02	21 38	10 25	21 58
04	22 42	24 12	05 11	16 00	08 55	09 44	26 03	02 58	15 14	05 03	21 37	10 26	21 58
05	22 37	24 09	05 05	16 18	03 N 15	10 31	26 06	02 40	15 12	05 03	21 37	10 27	21 58
06	22 31	24 06	04 41	16 35	02 S 47	11 18	26 08	02 22	15 11	05 04	21 37	10 27	21 58
07	22 26	24 03	03 57	16 51	08 49	12 05	26 10	02 04	15 10	05 04	21 36	10 28	21 58
08	22 24	24 00	02 56	17 08	14 25	12 52	26 11	01 46	15 09	05 05	21 36	10 29	21 58
09	22 19	23 56	01 42	17 24	19 09	13 40	26 10	01 27	15 07	05 05	21 35	10 30	21 58
10	22 18	23 53	00 N 21	17 39	22 35	14 27	26 10	01 09	15 06	05 06	21 35	10 30	21 58
11	22 D 18	23 50	01 S 00	17 55	24 23	15 14	26 09	00 51	15 04	05 06	21 35	10 31	21 59
12	22 20	23 47	02 15	18 10	24 29	16 00	26 07	00 33	15 03	05 07	21 34	10 32	21 59
13	22 21	23 44	03 20	18 25	22 59	16 46	26 05	00 S 15	15 01	05 07	21 34	10 33	21 59
14	22 22	23 41	04 12	18 40	20 09	17 31	26 02	00 N 03	15 00	05 07	21 34	10 33	21 59
15	22 R 22	23 37	04 48	18 54	16 20	18 15	25 58	00 21	14 58	05 07	21 33	10 34	21 59
16	22 21	23 34	05 09	19 08	11 50	18 58	25 54	00 39	14 56	05 08	21 33	10 35	21 59
17	22 19	23 31	05 14	19 22	06 56	19 39	25 49	00 58	14 54	05 08	21 32	10 35	22 00
18	22 16	23 28	05 04	19 35	01 S 51	20 19	25 44	01 16	14 53	05 08	21 32	10 36	22 00
19	22 12	23 25	04 40	19 48	03 N 13	20 57	25 38	01 34	14 51	05 08	21 31	10 37	22 00
20	22 08	23 21	04 04	20 00	08 07	21 33	25 31	01 52	14 49	05 08	21 31	10 37	22 00
21	22 05	23 18	03 18	20 13	12 41	22 07	25 24	02 09	14 47	05 08	21 30	10 38	22 01
22	22 02	23 15	02 23	20 25	16 44	22 38	25 17	02 27	14 45	05 08	21 30	10 39	22 01
23	22 00	23 12	01 22	20 36	20 06	23 07	25 08	02 45	14 43	05 08	21 30	10 39	22 01
24	21 59	23 09	00 S 18	20 48	22 39	23 34	25 00	03 03	14 41	05 08	21 29	10 40	22 01
25	22 D 00	23 06	00 N 47	20 58	24 13	23 57	24 50	03 20	14 38	05 08	21 29	10 41	22 02
26	22 01	23 03	01 51	21 09	24 41	24 18	24 41	03 39	14 36	05 07	21 28	10 41	22 02
27	22 02	22 59	02 50	21 19	24 01	24 37	24 30	03 56	14 34	05 07	21 28	10 42	22 02
28	22 03	22 56	03 43	21 29	22 13	24 53	24 20	04 14	14 32	05 07	21 27	10 42	22 02
29	22 04	22 53	04 26	21 38	19 20	25 06	24 08	04 31	14 29	05 06	21 26	10 43	22 03
30	22 05	22 50	04 57	21 47	15 29	25 16	23 57	04 49	14 27	05 06	21 26	10 44	22 03
31	22 ♊ 05	22 ♊ 47	05 N 15	21 N 56	10 N 49	25 N 24	23 N 44	05 N 06	14 N 24	05 N 06	21 N 25	10 N 45	22 S 03

ZODIAC SIGN ENTRIES

Date	h	m	Planets
03	01	25	☽ ♍
04	13	49	☿ ♉
05	05	58	☽ ♎
06	22	26	♀ ♋
07	07	09	☽ ♏
09	06	42	☽ ♐
10	01	29	♂ ♈
11	06	38	☽ ♑
13	08	40	☽ ♒
15	13	52	☽ ♓
17	22	18	☽ ♈
19	00	22	☿ ♊
20	09	20	☽ ♉
21	04	11	☉ ♊
22	21	55	☽ ♊
25	10	58	☽ ♋
27	23	06	☽ ♌
30	08	47	☽ ♍

LATITUDES

Date	Mercury ☿ ° '	Venus ♀ ° '	Mars ♂ ° '	Jupiter ♃ ° '	Saturn ♄ ° '	Uranus ♅ ° '	Neptune ♆ ° '	Pluto ♇ ° '
01	02 S 08	02 N 33	01 S 23	01 N 04	02 N 24	00 N 32	01 S 41	09 S 33
04	01 48	02 39	01 24	01 04	02 24	00 32	01 41	09 34
07	01 24	02 44	01 24	01 04	02 23	00 32	01 41	09 34
10	00 57	02 47	01 24	01 03	02 23	00 32	01 41	09 35
13	00 S 27	02 50	01 25	01 03	02 22	00 31	01 41	09 36
16	00 N 04	02 52	01 25	01 03	02 21	00 31	01 41	09 37
19	00 36	02 53	01 25	01 02	02 21	00 31	01 41	09 38
22	01 05	02 52	01 25	01 02	02 20	00 31	01 41	09 39
25	01 30	02 50	01 24	01 02	02 20	00 31	01 41	09 40
28	01 49	02 47	01 24	01 01	02 19	00 31	01 41	09 41
31	02 N 02	02 N 42	01 S 24	01 N 01	02 N 19	00 N 31	01 S 41	09 S 42

DATA

Julian Date	2465910
Delta T	+77 seconds
Ayanamsa	24° 24' 05"
Synetic vernal point	04° ♓ 42' 53"
True obliquity of ecliptic	23° 26' 04"

MOON'S PHASES, APSIDES AND POSITIONS ☽

Date	h	m	Phase	Longitude ° '	Eclipse Indicator
01	14	07	☽	11 ♌ 04	
08	11	20	○	17 ♏ 45	
15	03	17	☾	24 ♒ 11	
23	01	38	●	01 ♊ 49	
31	02	24	☽	09 ♍ 32	

Day	h	m	
09	01	54	Perigee
23	12	20	Apogee
06	01	01	0S
12	01	16	Max dec 24° S 39'
18	20	42	0N
26	10	03	Max dec 24° N 42'

LONGITUDES

Date	Chiron ⚷ ° '	Ceres ⚳ ° '	Pallas ⚴ ° '	Juno ⚵ ° '	Vesta ⚶ ° '	Black Moon Lilith ⚸ ° '
01	00 ♋ 21	05 ♈ 22	28 ♒ 53	01 ♒ 42	20 ♌ 17	03 ♊ 44
11	01 ♋ 06	08 ♈ 59	00 ♓ 52	02 ♒ 51	22 ♌ 31	04 ♊ 51
21	01 ♋ 56	12 ♈ 29	02 ♓ 29	03 ♒ 30	25 ♌ 14	05 ♊ 57
31	02 ♋ 49	15 ♈ 51	03 ♓ 41	03 ♒ 35	28 ♌ 21	07 ♊ 04

ASPECTARIAN

h m Aspects

01 Sunday
02 37 ☿ ∗ ♀
06 55 ☽ # ♇
08 08 ☽ ∠ ♄
08 53 ☽ ꝑ ♂
09 58 ☽ ∥ ♅
10 04 ☽ ∠ ♀
14 07 ☽ □ ☉
22 06 ☿ □ ♅
23 53 ☉ ∥ ♃

02 Monday
01 12 ☽ ⊥ ♄
03 05 ☽ ± ♂
04 36 ☿ ∗ ♇
10 16 ☽ ☌ ♃
10 44 ♀ ⊻ ♅
12 23 ☽ ⊻ ♄
14 58 ☽ ⊼ ♂
16 26 ☽ ⊻ ♅
16 57 ☽ ∗ ♀
17 19 ☽ ☍ ♇
19 23 ☽ △ ☿
21 20 ♀ △ ♇

03 Tuesday
03 16 ☽ ∥ ☉
03 26 ☽ ⊥ ♅
04 07 ☽ △ ♆
05 14 ☽ ∥ ♃
16 36 ♂ △ ♅
16 39 ☽ Q ♀
16 47 ☉ Q ♅
19 36 ☽ ∠ ♅
23 11 ☿ ± ♄

04 Wednesday
01 31 ☽ △ ☉
02 39 ☽ ꝑ ☿
05 11 ☽ ∥ ♆
05 16 ♂ ⊥ ♆
06 52 ☽ ꝑ ♆
07 13 ♂ ⊻ ♇
08 49 ☽ ∥ ☿
16 08 ☽ ⊻ ♃
17 45 ☽ ☌ ♄
20 42 ☽ ± ☿
21 45 ☽ ∗ ♅
22 26 ☽ ± ♆
22 30 ☽ ⊼ ♇
23 22 ☽ ☍ ♂

05 Thursday
02 25 ☽ ⊥ ♃
02 36 ☽ □ ♀
04 35 ☽ ∥ ♄
05 17 ☽ ꝑ ☉
08 28 ☽ ⊼ ☿
08 37 ☽ ⊼ ♆
08 40 ☽ ± ♇
09 34 ☿ ☌ ♆
09 54 ☿ ∥ ♆
14 28 ☽ # ♂
17 35 ☽ ∠ ♃
17 59 ☽ Q ♅
21 25 ☽ ± ☉
23 40 ☽ ꝑ ♇

06 Friday
03 19 ☿ ⊥ ♂
07 59 ☽ ⊼ ☉
10 26 ☽ ∥ ♂
18 15 ☽ ∗ ♃
19 30 ☽ ⊻ ♄
21 01 ☽ # ♄
23 28 ☽ □ ♅

07 Saturday
00 06 ☽ △ ♇
03 34 ☽ ⊼ ♂
05 08 ☽ ⊥ ♄
07 49 ☽ △ ♀
09 46 ☽ ☍ ♆
11 52 ♂ ± ♃
13 36 ☽ Q ♃
13 41 ☽ ± ♂
16 49 ☽ ☍ ☿
18 50 ☽ # ♆
19 28 ☽ ∠ ♄

08 Sunday
03 54 ☽ # ☿
04 44 ☽ ꝑ ♂
09 30 ☽ ꝑ ♀
11 20 ☽ ☍ ☉
11 33 ☿ Q ♇
11 43 ♀ ∗ ♆
12 31 ☿ ꝑ ♄
15 20 ☽ # ♃
18 14 ☽ □ ♃
19 10 ☽ ∗ ♄
23 14 ☽ △ ♅
23 47 ☽ □ ♇

09 Monday
00 48 ☽ ± ♀
01 48 ☽ # ☉
05 42 ☽ △ ♂
09 24 ☽ ⊼ ♆
11 01 ☽ ⊼ ♀
14 05 ☽ Q ♄
18 56 ☽ ± ♆
23 01 ☽ ꝑ ♅
23 36 ☽ ⊼ ☿

10 Tuesday
03 51 ☽ # ♅
04 19 ☽ Q ♇
06 50 ☽ ∥ ♇
09 18 ☽ ꝑ ♆
10 41 ☽ ± ☿
13 25 ☽ ± ♅
14 09 ☽ ⊼ ☉
18 06 ☽ △ ♃
18 45 ☽ □ ♄
23 05 ☽ ⊼ ♅
23 34 ☽ ∗ ♇

11 Wednesday
00 29 ☽ ± ☉
03 30 ☽ ꝑ ☿
06 44 ☿ Q ♅
07 25 ☿ ∥ ♃
08 13 ☽ □ ♂
09 32 ☽ △ ♆
14 52 ☽ ☍ ♀
16 10 ☽ ꝑ ☉
18 35 ☽ ꝑ ♃

12 Thursday
00 05 ☽ ∠ ♇
05 21 ♂ ⊥ ♇
08 16 ☽ △ ☿
09 34 ☽ ± ♃
10 53 ♂ ⊻ ♆
15 06 ☽ ⊥ ♇
15 38 ☽ Q ♂
18 54 ☽ △ ☉
19 43 ☽ ⊼ ♃
20 04 ☽ △ ♄

13 Friday
00 50 ☽ ☍ ♅
01 15 ☽ ⊻ ♇
04 54 ☿ ∠ ♂
07 43 ☉ □ ♃
11 37 ☉ △ ♄
11 53 ☽ □ ♆
13 16 ☽ ∗ ♂
21 36 ☽ ⊼ ♀
21 39 ☽ ∥ ♇
21 42 ♀ ∠ ♃
21 47 ☽ ꝑ ♄

14 Saturday
01 13 ☽ # ♅
09 02 ☽ ± ♀
13 38 ☽ ± ♄
17 10 ☽ ∠ ♂
19 36 ☽ Q ♆
21 22 ☽ # ☉
21 37 ☽ □ ☿
23 34 ♃ ⊻ ♄

15 Sunday
00 21 ☽ ⊼ ♄
00 21 ☽ ☍ ♃
02 22 ☽ # ☿
02 29 ☽ ꝑ ♀
03 17 ☽ □ ☉
05 39 ☽ ⊼ ♅
05 58 ☽ ☌ ♇
10 24 ☽ ⊥ ♂
14 25 ☿ △ ♄
15 06 ☿ □ ♃
16 37 ☽ ± ♅
17 26 ☽ ∗ ♆
19 37 ☽ # ♃
22 05 ☽ ⊻ ♂

16 Monday
08 26 ☽ △ ♀
09 20 ☽ ꝑ ♅
13 00 ☉ ∗ ♅
13 13 ☽ Q ☿
15 07 ☽ Q ☉
16 27 ☉ □ ♇
18 15 ☽ # ♆
19 29 ♀ Q ♄
20 39 ☉ ∥ ☿
21 27 ☽ ∠ ♆
22 51 ♀ ꝑ ♇
23 34 ☿ ∗ ♅

17 Tuesday
00 55 ☿ □ ♇
02 49 ☿ ∠ ♀
07 36 ☉ ☌ ☿
07 58 ☽ ☍ ♄
08 24 ☽ ⊼ ♃
13 48 ☽ △ ♅
14 00 ☽ ⊻ ♇
14 38 ☽ ⊥ ♆
15 47 ☽ ∗ ☉
16 46 ☽ ∗ ☿
19 59 ☽ ± ♃
20 32 ☽ # ♄

18 Wednesday
01 33 ☽ ⊥ ♇
02 14 ☽ ⊻ ♆
10 42 ☽ ☌ ♂
13 34 ☽ ꝑ ♃
14 37 ☽ # ♂
19 05 ☽ ∠ ♇
23 10 ☽ □ ♀
23 24 ☽ ∠ ☉

19 Thursday
03 34 ☽ ∥ ♂
04 16 ☽ ∠ ☿
18 23 ☽ ⊼ ♄
18 42 ☽ ⊥ ☉
19 20 ☽ △ ♃
21 13 ☽ ∥ ♄
23 19 ☿ ⊻ ♆

20 Friday
00 42 ☽ □ ♅
00 44 ☽ ∗ ♇
02 02 ☽ ⊥ ☿
06 16 ♀ Q ♆
06 19 ☽ ± ♄
07 42 ☽ ⊻ ☉
10 41 ☿ ∥ ♅
11 42 ♅ ⊼ ♇
13 20 ☉ ∠ ♀
13 34 ☽ ☌ ♆
14 13 ☽ Q ♀
15 02 ♂ ꝑ ♃
16 42 ☽ ⊻ ☿

21 Saturday
00 21 ☽ ꝑ ♄
00 46 ☽ Q ♇
00 57 ☽ ∥ ♆
02 12 ☽ ⊻ ♂
07 29 ☿ # ♇
12 56 ☽ Q ♅
15 07 ☽ ⊥ ♂
16 39 ☽ ∗ ♀
23 56 ☽ ∥ ♃

22 Sunday
06 39 ☽ △ ♄
08 09 ☽ □ ♃
10 36 ☽ ∠ ♂
13 10 ☽ □ ♇
13 20 ☽ ∗ ♅
18 38 ☿ ⊥ ♆

23 Monday
01 38 ☽ ☌ ☉
01 57 ☽ ∠ ♀
02 24 ☽ ⊻ ♆
10 13 ♀ ⊥ ♃
11 20 ☉ ⊻ ♆
14 39 ☽ ⊥ ♆
16 23 ☽ ∥ ☉
19 02 ☽ ☌ ☿
19 13 ☽ ∗ ♂
19 58 ☽ ∠ ♅
20 30 ☿ ∗ ♂
21 04 ☽ Q ♃
21 59 ☽ ⊥ ♀
23 57 ☽ ∥ ♅

24 Tuesday
00 16 ☿ ∠ ♅
02 45 ♂ ∠ ♇
05 09 ☽ # ♇
06 49 ☿ Q ♃
09 02 ☽ ∠ ♆
11 21 ☽ ⊻ ♀
14 21 ☽ ⊥ ♅
19 40 ☽ □ ♄
21 19 ☽ Q ♂
21 43 ☽ ∗ ♃
23 37 ☿ ⊥ ♀

25 Wednesday
02 14 ☽ △ ♇
02 36 ☽ ⊻ ♅
03 06 ♄ St D
04 24 ☽ ∥ ☿
15 36 ☽ ∗ ♆
19 01 ♀ ± ♇
20 03 ☽ ⊻ ☉

26 Thursday
04 23 ☽ ∠ ♃
08 08 ☽ Q ♄
08 35 ☽ ꝑ ♇
09 13 ☽ ⊥ ☉
09 17 ☽ ∥ ♀
12 14 ☽ □ ♂
15 52 ☽ Q ♆
17 50 ☽ ∥ ♀
19 19 ♇ St R
20 51 ☽ ⊻ ☿
22 38 ☽ ⊥ ♃

27 Friday
00 49 ☽ ∥ ☿
02 39 ☽ ± ♇
03 06 ☿ ∠ ♆
04 50 ☽ ∠ ☉
05 29 ☽ ☌ ♀
06 17 ☿ ∥ ♀
08 11 ☽ ∗ ♄
10 43 ☽ ⊻ ♃
11 16 ☽ ⊥ ☿
14 35 ☽ ⊼ ♇
15 10 ☽ ☌ ♅

28 Saturday
03 45 ☽ □ ♆
07 32 ☉ ∥ ♅
08 32 ☽ ∠ ☿
11 12 ☿ ⊥ ♅
13 00 ☽ ∗ ☉
13 45 ☽ # ♇
13 46 ☽ ∠ ♄
14 48 ♀ ∗ ♄
18 36 ☽ ∥ ☉
19 18 ☽ ∥ ♅

29 Sunday
03 21 ☽ △ ♂
07 11 ☽ ⊥ ♄
14 05 ☽ Q ☉
17 17 ☿ □ ♄
18 39 ☽ ⊻ ♄
18 54 ☽ ∗ ☿
20 57 ☽ ⊻ ♀
21 34 ☽ ☌ ♃
22 32 ☉ ⊥ ♆

30 Monday
00 43 ☽ ☍ ♇
01 28 ☽ ⊻ ♅
06 07 ♀ ⊻ ♃
09 09 ☽ ⊥ ♀
09 36 ☽ ꝑ ♂
12 44 ☽ ⊥ ♅
13 20 ☽ △ ♆
13 29 ☿ ∗ ♃
17 44 ☽ ∥ ♃
20 08 ☿ ⊻ ♀
21 14 ☽ Q ☿

31 Tuesday
02 24 ☽ □ ☉
03 05 ☽ ∠ ♀
03 14 ☽ ± ♂
05 21 ☽ ∠ ♅
09 39 ☿ △ ♇
11 17 ♂ ∥ ♄
12 19 ☽ ∥ ♆
14 46 ☽ ⊼ ♂
15 49 ☿ ⊻ ♅
16 47 ☽ ꝑ ♆
22 42 ♀ ⊼ ♇

All ephemeris data is given at 12.00 UT and the Moon's longitude is additionally given for 24.00 UT

JUNE 2039

Date	Sidereal time h m s	Sun ☉ ° ' "	Moon ☽ ° ' "	Moon ☽ 24.00 ° ' "	Mercury ☿ ° '	Venus ♀ ° '	Mars ♂ ° '	Jupiter ♃ ° '	Saturn ♄ ° '	Uranus ♅ ° '	Neptune ♆ ° '	Pluto ♇ ° '
01	04 39 22	10 ♊ 52 47	28 ♍ 19 40	05 ♎ 14 47	27 ♊ 43	26 ♋ 13	17 ♈ 01	24 ♌ 20	22 ♍ 31	26 ♋ 13	02 ♉ 30	25 ♒ 41
02	04 43 18	11 50 17	12 ♎ 16 31	19 ♎ 24 45	29 ♊ 31	27 10	17 46	24 28	22 32	26 15	02 32	25 R 41
03	04 47 15	12 47 45	26 ♎ 39 10	03 ♏ 59 19	01 ♋ 16	28 06	18 31	24 35	22 33	26 18	02 34	25 41
04	04 51 11	13 45 12	11 ♏ 24 33	18 ♏ 54 04	02 59	29 01	19 16	24 43	22 34	26 21	02 36	25 41
05	04 55 08	14 42 38	26 ♏ 26 53	04 ♐ 01 56	04 39	29 ♋ 56	20 01	24 51	22 35	26 24	02 38	25 41
06	04 59 04	15 40 03	11 ♐ 38 04	19 ♐ 14 03	06 16	00 ♌ 51	20 45	24 58	22 36	26 27	02 39	25 40
07	05 03 01	16 37 28	26 ♐ 48 43	04 ♑ 20 53	07 50	01 45	21 30	25 06	22 37	26 30	02 41	25 40
08	05 06 58	17 34 51	11 ♑ 49 30	19 ♑ 13 39	09 21	02 38	22 15	25 15	22 39	26 33	02 43	25 40
09	05 10 54	18 32 14	26 ♑ 32 32	03 ♒ 45 33	10 50	03 31	22 59	25 23	22 40	26 36	02 45	25 39
10	05 14 51	19 29 36	10 ♒ 52 13	17 ♒ 52 18	12 16	04 23	23 44	25 31	22 42	26 39	02 46	25 39
11	05 18 47	20 26 57	24 ♒ 45 39	01 ♓ 32 19	13 38	05 15	24 29	25 40	22 44	26 42	02 48	25 39
12	05 22 44	21 24 18	08 ♓ 12 27	14 ♓ 46 18	14 58	06 06	25 13	25 48	22 45	26 46	02 50	25 38
13	05 26 40	22 21 38	21 ♓ 14 13	27 ♓ 36 36	16 15	06 56	25 58	25 57	22 47	26 49	02 51	25 38
14	05 30 37	23 18 58	03 ♈ 53 54	10 ♈ 06 38	17 28	07 45	26 42	26 06	22 49	26 52	02 53	25 37
15	05 34 33	24 16 18	16 ♈ 15 18	22 ♈ 20 23	18 39	08 34	27 26	26 14	22 51	26 55	02 54	25 37
16	05 38 30	25 13 37	28 ♈ 22 26	04 ♉ 21 55	19 46	09 22	28 11	26 23	22 54	26 58	02 56	25 36
17	05 42 27	26 10 55	10 ♉ 19 20	16 ♉ 15 07	20 51	10 09	28 55	26 33	22 56	27 02	02 57	25 36
18	05 46 23	27 08 14	22 ♉ 09 44	28 ♉ 03 35	21 52	10 55	29 ♈ 39	26 42	22 58	27 05	02 59	25 35
19	05 50 20	28 05 32	03 ♊ 57 03	09 ♊ 50 30	22 49	11 41	00 ♉ 23	26 51	23 01	27 08	03 00	25 35
20	05 54 16	29 ♊ 02 49	15 ♊ 44 15	21 ♊ 38 38	23 43	12 26	01 07	27 01	23 03	27 11	03 02	25 34
21	05 58 13	00 ♋ 00 07	27 ♊ 33 56	03 ♋ 30 26	24 34	13 09	01 51	27 10	23 06	27 15	03 03	25 34
22	06 02 09	00 57 23	09 ♋ 28 25	15 ♋ 28 07	25 20	13 52	02 35	27 20	23 09	27 18	03 05	25 33
23	06 06 06	01 54 40	21 ♋ 29 48	27 ♋ 33 44	26 03	14 34	03 19	27 29	23 12	27 22	03 06	25 32
24	06 10 02	02 51 55	03 ♌ 40 09	09 ♌ 49 19	26 43	15 14	04 02	27 39	23 15	27 25	03 07	25 32
25	06 13 59	03 49 11	16 ♌ 01 31	22 ♌ 17 00	27 18	15 54	04 46	27 49	23 18	27 28	03 09	25 31
26	06 17 56	04 46 25	28 ♌ 36 03	04 ♍ 58 58	27 49	16 33	05 29	27 59	23 21	27 32	03 10	25 30
27	06 21 52	05 43 40	11 ♍ 26 00	17 ♍ 57 27	28 15	17 10	06 13	28 09	23 24	27 35	03 11	25 29
28	06 25 49	06 40 53	24 ♍ 33 34	01 ♎ 14 34	28 38	17 46	06 56	28 19	23 27	27 39	03 13	25 28
29	06 29 45	07 38 06	08 ♎ 00 40	14 ♎ 52 01	28 56	18 21	07 40	28 29	23 31	27 42	03 14	25 28
30	06 33 42	08 ♋ 35 19	21 ♎ 48 40	28 ♎ 50 37	29 ♋ 09	18 ♌ 55	08 ♉ 23	28 ♌ 40	23 ♍ 34	27 ♋ 46	03 ♉ 15	25 ♒ 27

Date	Moon True ☊ ° '	Moon Mean ☊ ° '	Moon ☽ Latitude ° '
01	22 ♊ 04	22 ♊ 43	05 N 15
02	22 R 04	22 40	04 58
03	22 03	22 37	04 21
04	22 02	22 34	03 27
05	22 01	22 31	02 17
06	22 00	22 27	00 N 57
07	22 D 00	22 24	00 S 27
08	22 01	22 21	01 48
09	22 01	22 18	03 00
10	22 01	22 15	04 00
11	22 01	22 12	04 43
12	22 01	22 08	05 09
13	22 R 01	22 05	05 18
14	22 D 01	22 02	05 11
15	22 02	21 59	04 50
16	22 02	21 56	04 16
17	22 02	21 53	03 32
18	22 03	21 49	02 39
19	22 03	21 46	01 39
20	22 04	21 43	00 S 35
21	22 R 04	21 40	00 N 31
22	22 03	21 37	01 35
23	22 02	21 33	02 36
24	22 01	21 30	03 31
25	21 59	21 27	04 17
26	21 58	21 24	04 51
27	21 56	21 21	05 11
28	21 55	21 18	05 16
29	21 D 55	21 14	05 04
30	21 ♊ 55	21 ♊ 11	04 N 35

DECLINATIONS

Date	Sun ☉	Moon ☽	Mercury ☿	Venus ♀	Mars ♂	Jupiter ♃	Saturn ♄	Uranus ♅	Neptune ♆	Pluto ♇
01	22 N 04	05 N 29	25 N 30	23 N 32	05 N 24	14 N 22	05 N 05	21 N 25	10 N 45	22 S 04
02	22 12	00 S 17	25 33	23 19	05 41	14 19	05 05	21 24	10 46	22 04
03	22 20	06 13	25 34	23 05	05 58	14 17	05 04	21 24	10 46	22 04
04	22 27	11 58	25 33	22 52	06 15	14 14	05 04	21 23	10 47	22 05
05	22 34	17 08	25 29	22 37	06 32	14 11	05 03	21 23	10 47	22 05
06	22 40	21 14	25 24	22 23	06 49	14 09	05 02	21 22	10 48	22 05
07	22 46	23 50	25 17	22 08	07 06	14 06	05 02	21 22	10 49	22 06
08	22 51	24 42	25 09	21 53	07 23	14 03	05 01	21 21	10 49	22 06
09	22 56	23 48	24 58	21 37	07 40	14 00	05 00	21 20	10 50	22 07
10	23 01	21 20	24 47	21 21	07 56	13 57	04 59	21 20	10 50	22 07
11	23 05	17 42	24 34	21 05	08 13	13 54	04 58	21 19	10 51	22 07
12	23 09	13 15	24 19	20 49	08 29	13 51	04 57	21 19	10 51	22 08
13	23 13	08 20	24 04	20 32	08 46	13 48	04 57	21 18	10 52	22 08
14	23 16	03 S 13	23 48	20 15	09 02	13 45	04 56	21 17	10 52	22 09
15	23 19	01 N 56	23 31	19 58	09 18	13 42	04 55	21 17	10 53	22 09
16	23 21	06 54	23 13	19 40	09 34	13 39	04 55	21 16	10 53	22 10
17	23 23	11 34	22 54	19 23	09 50	13 36	04 52	21 15	10 54	22 10
18	23 24	15 45	22 35	19 05	10 06	13 33	04 51	21 15	10 54	22 10
19	23 25	19 19	22 15	18 47	10 21	13 29	04 50	21 14	10 55	22 11
20	23 26	22 05	21 55	18 29	10 37	13 26	04 49	21 13	10 55	22 11
21	23 26	23 55	21 35	18 11	10 52	13 23	04 48	21 13	10 56	22 12
22	23 26	24 41	21 15	17 53	11 08	13 19	04 46	21 12	10 56	22 12
23	23 25	24 17	20 54	17 34	11 23	13 16	04 45	21 11	10 56	22 13
24	23 24	22 45	20 34	17 16	11 38	13 12	04 44	21 11	10 57	22 13
25	23 23	20 06	20 13	16 58	11 53	13 09	04 42	21 10	10 57	22 14
26	23 21	16 30	19 53	16 39	12 08	13 05	04 41	21 09	10 58	22 14
27	23 19	12 04	19 34	16 21	12 23	13 02	04 40	21 09	10 58	22 15
28	23 16	07 00	19 15	16 02	12 37	12 58	04 38	21 08	10 58	22 15
29	23 13	01 29	18 56	15 44	12 52	12 55	04 37	21 07	10 59	22 16
30	23 N 09	04 S 15	18 N 38	15 N 25	13 N 06	12 N 51	04 N 35	21 N 07	10 N 59	22 S 16

ZODIAC SIGN ENTRIES

Date	h m	Planets
01	14 55	☽ ♎
02	18 31	☿ ♋
03	17 30	☽ ♏
05	13 35	♀ ♌
05	17 38	☽ ♐
07	17 04	☽ ♑
09	17 44	☽ ♒
11	21 16	☽ ♓
14	04 33	☽ ♈
16	15 15	☽ ♉
18	23 31	♂ ♉
19	03 57	☽ ♊
21	11 57	☉ ♋
21	16 55	☽ ♋
24	04 48	☽ ♌
26	14 38	☽ ♍
28	21 47	☽ ♎

LATITUDES

Date	Mercury ☿	Venus ♀	Mars ♂	Jupiter ♃	Saturn ♄	Uranus ♅	Neptune ♆	Pluto ♇
01	02 N 05	02 N 41	01 S 24	01 N 01	02 N 18	00 N 31	01 S 41	09 S 42
04	02 09	02 34	01 23	01 01	02 18	00 31	01 41	09 43
07	02 05	02 26	01 23	01 00	02 17	00 31	01 42	09 44
10	01 55	02 16	01 22	01 00	02 17	00 31	01 42	09 45
13	01 38	02 04	01 22	01 00	02 16	00 31	01 42	09 46
16	01 15	01 50	01 21	00 59	02 15	00 31	01 42	09 47
19	00 46	01 34	01 20	00 59	02 15	00 31	01 42	09 47
22	00 N 11	01 16	01 19	00 59	02 14	00 31	01 42	09 48
25	00 S 29	00 56	01 18	00 59	02 14	00 31	01 42	09 49
28	01 13	00 34	01 17	00 59	02 13	00 31	01 43	09 50
31	01 S 59	00 N 09	01 S 15	00 N 58	02 N 13	00 N 31	01 S 43	09 S 51

DATA

Julian Date	2465941
Delta T	+77 seconds
Ayanamsa	24° 24' 10"
Synetic vernal point	04° ♓ 42' 49"
True obliquity of ecliptic	23° 26' 04"

MOON'S PHASES, APSIDES AND POSITIONS ☽

Date	h m	Phase	Longitude ° '	Eclipse Indicator
06	18 48	○	15 ♐ 56	partial
13	14 16	☾	22 ♓ 27	
21	17 21	●	00 ♋ 13	Annular
29	11 17	☽	07 ♎ 36	

Day	h m		
06	12 05	Perigee	
19	16 11	Apogee	
02	10 52	0S	
08	11 26	Max dec	24° S 42'
15	02 57	0N	
22	15 57	Max dec	24° N 42'
29	18 15	0S	

LONGITUDES

Date	Chiron ⚷ ° '	Ceres ⚳ ° '	Pallas ⚴ ° '	Juno ⚵ ° '	Vesta ⚶ ° '	Black Moon Lilith ⚸ ° '
01	02 ♋ 54	16 ♈ 11	03 ♓ 46	03 ♒ 34	28 ♌ 41	07 ♊ 10
11	03 ♋ 50	19 ♈ 22	04 ♓ 27	03 ♒ 00	02 ♍ 10	08 ♊ 17
21	04 ♋ 47	22 ♈ 22	04 ♓ 37	01 ♒ 51	05 ♍ 56	09 ♊ 23
31	05 ♋ 45	25 ♈ 07	04 ♓ 12	00 ♒ 09	09 ♍ 57	10 ♊ 30

ASPECTARIAN

h m Aspects

01 Wednesday
01 46 ☽ ☌ ♄
04 56 ☽ ⊻ ♃
07 22 ☽ ⊼ ♇
08 01 ☽ ✶ ♀
08 17 ☽ ✶ ♅
08 48 ☽ ± ♆
10 07 ☉ ⋕ ♇
10 46 ☽ □ ☿
11 50 ♀ ☌ ♅
12 23 ☽ ∥ ♂
13 42 ☽ ∥ ♄
15 32 ☽ ⊥ ♃
17 51 ☽ ± ♇
19 17 ☽ ⊼ ♆
20 43 ☉ ∠ ♅

02 Thursday
05 08 ☽ Q ♅
06 20 ☽ Q ♀
07 11 ☽ ∠ ♃
09 18 ☽ ᵽ ♇
11 12 ☽ △ ☉
17 46 ☿ Q ♂
21 45 ☽ ☍ ♂

03 Friday
05 12 ☽ ⊻ ♄
05 57 ☉ Q ♃
07 23 ☽ ⋕ ♄
08 34 ☽ ✶ ♃
10 24 ☽ △ ♇
10 57 ☽ ⋕ ♂
11 26 ☽ □ ♅
14 01 ☽ ᵽ ☉
14 32 ☽ □ ♀
15 07 ☽ ⊥ ♄
20 35 ☽ △ ☿
21 42 ☽ ☍ ♆

04 Saturday
04 21 ☽ Q ♃
05 42 ☽ ± ☉
05 47 ☽ ∠ ♄
06 24 ☿ ✶ ♆
06 54 ☽ ⋕ ♆
16 01 ☽ ⊼ ☉
21 59 ☽ ⋕ ♃
23 51 ☽ ᵽ ☿

05 Sunday
01 14 ☽ ⊼ ♂
05 51 ☽ ✶ ♄
09 26 ☽ □ ♃
10 47 ☽ □ ♇
11 16 ☽ ± ♂
11 56 ☽ △ ♅
15 54 ☽ ± ☿
16 32 ☉ ∥ ♀
17 53 ☽ △ ♀
21 48 ☽ ⊼ ♆

06 Monday
00 53 ☽ Q ♄
02 15 ☽ ᵽ ♂
02 32 ☽ ⊼ ☿
07 17 ☽ ± ♆
11 43 ☽ ᵽ ♅
12 59 ☽ ⋕ ♅
15 13 ☽ Q ♇
18 28 ☽ ∥ ♇
18 48 ☽ ☍ ☉
19 04 ☽ ᵽ ♀
20 08 ☽ ⋕ ♀
21 31 ☽ ᵽ ♆
23 44 ☽ ⋕ ☉

07 Tuesday
01 58 ☽ ± ♅
03 09 ☽ △ ♂
05 21 ☽ □ ♄
09 16 ☽ △ ♃
10 11 ☽ ✶ ♇
10 12 ☽ ± ♀
11 30 ☽ ⊼ ♅
15 19 ♀ ⋕ ♇
20 21 ☽ ⊼ ♀
21 22 ☽ △ ♆

08 Wednesday
07 35 ☽ ☍ ☿
09 26 ☽ ᵽ ♃
10 08 ☽ ∠ ♇
14 09 ♀ □ ♆
15 27 ☉ ∠ ♆
21 58 ☽ ⊼ ☉

09 Thursday
00 08 ☽ ± ♃
00 42 ☽ ⊥ ♇
01 19 ♂ ⊼ ♄
03 46 ☿ ∠ ♃
05 37 ☽ △ ♄
05 51 ☽ □ ♂
05 58 ☉ ∠ ♀
08 28 ☽ ± ☉
09 06 ☿ ᵽ ♇
09 18 ☿ Q ♄
10 04 ☽ ⊼ ♃
10 32 ☽ ⊻ ♇
12 06 ☽ ☍ ♅
21 40 ☽ ⋕ ☉
22 19 ☽ □ ♆

10 Friday
00 21 ☽ ☍ ♀
00 28 ☽ ᵽ ☉
05 42 ☽ ∥ ♇
06 37 ☽ ᵽ ♄
11 52 ☽ ⋕ ♀
12 05 ☽ ⋕ ♅
13 33 ☽ Q ♂
14 25 ♀ ∥ ♅
14 38 ☽ ⊼ ☿
22 00 ☽ ± ♄

11 Saturday
02 05 ☽ ± ☿
03 54 ☽ △ ☉
05 04 ☽ Q ♆
08 26 ☽ ⊼ ♄
09 40 ♃ ☍ ♇
11 28 ☽ ✶ ♂
13 33 ☽ ☌ ♇
13 36 ☽ ☍ ♃
15 26 ☽ ⊼ ♅
18 50 ☉ ⊥ ♅
19 36 ☽ ᵽ ☿

12 Sunday
02 09 ☽ ± ♅
02 17 ☽ ✶ ♆
07 55 ☽ ⊼ ♀
08 55 ☽ ⋕ ♃
09 22 ☿ Q ♆
15 53 ☽ ∠ ♂
18 30 ☽ ᵽ ♅
19 34 ☽ ± ♀
23 54 ☽ ⋕ ♆

13 Monday
01 27 ♂ ✶ ♇
01 43 ☽ △ ☿
05 41 ☽ ∠ ♆
09 28 ☽ ⊥ ♂
10 06 ☽ ⋕ ♂
11 29 ♂ △ ♃
13 23 ☽ ᵽ ♀
14 16 ☽ □ ☉
14 55 ☽ ☍ ♄
20 15 ☽ ⊻ ♇
20 57 ☽ ⊼ ♃
21 26 ☽ ⊻ ♂
22 32 ☽ △ ♅
22 35 ☽ ⊥ ♆
23 09 ☉ □ ♄

14 Tuesday
03 59 ☽ ⋕ ♄
07 39 ☽ ⊥ ♇
08 30 ☽ ± ♃
10 03 ☽ ⊻ ♆
14 09 ♀ ∠ ♄
17 48 ♂ □ ♅
19 57 ☽ △ ♀

15 Wednesday
00 59 ☽ ∠ ♇
02 05 ☽ ᵽ ♃
03 33 ☽ Q ☉
17 12 ☽ □ ☿

16 Thursday
01 04 ☽ ⊼ ♄
02 12 ☽ ∥ ♄
02 24 ☉ ∥ ☿
05 11 ☽ ✶ ☉
06 29 ☽ ✶ ♇
08 00 ☽ △ ♃
08 23 ☿ ± ♇
09 11 ☽ □ ♅
11 35 ☽ ☌ ♂
13 03 ☽ ± ♄
21 09 ☽ ☌ ♆
21 28 ☉ △ ♇

17 Friday
02 18 ☽ ∥ ♂
03 57 ☿ ⊥ ♃
06 31 ☽ Q ♇
07 10 ☽ ᵽ ♄
08 27 ☽ ∥ ♆
08 44 ☽ Q ☿
11 38 ☽ □ ♀
12 44 ♂ ± ♄
13 53 ☽ ∠ ☉
21 34 ☽ Q ♅
22 48 ☉ ✶ ♃
23 08 ☽ ∥ ♃

18 Saturday
09 44 ☽ ⊥ ☉
10 30 ☉ ⊻ ♅
11 20 ☽ ✶ ☿
13 39 ☽ △ ♄
18 58 ☽ □ ♇
21 21 ☽ □ ♃
22 03 ☽ ✶ ♅
23 01 ☽ ⊻ ☉

19 Sunday
02 42 ☽ Q ♀
04 15 ☽ ⊻ ♂
08 27 ☽ ∥ ♀
10 04 ☽ ⊻ ♆
17 10 ☿ ⋕ ♇
17 17 ☽ ⊥ ♂
17 21 ☿ ✶ ♄
20 33 ☽ ∠ ☿
22 19 ☽ ⊥ ♆

20 Monday
03 41 ☽ ∥ ♅
04 45 ☽ ∠ ♅
04 49 ☽ ✶ ♀
10 30 ☽ ∥ ☿
10 30 ☽ Q ♃
12 49 ☽ ∠ ♂
13 02 ☽ ⋕ ♇
16 21 ☽ ⊥ ☿
16 40 ☽ ∠ ♆
23 08 ☽ ⊥ ♅

21 Tuesday
02 55 ☽ □ ♄
04 09 ☽ ∥ ☉
05 28 ☽ ⊻ ☿
07 57 ☽ △ ♇
11 11 ☽ ✶ ♃
11 21 ☽ ⊻ ♅
13 16 ☽ ∠ ♀
17 00 ♂ ∥ ♆
17 21 ☽ ☌ ☉
21 13 ☽ ✶ ♂
23 07 ☽ ✶ ♆

22 Wednesday
06 30 ♃ ⊻ ♅
08 34 ☽ ⊥ ♀
14 09 ☽ ᵽ ♇
15 02 ☿ ∥ ♅
15 22 ☽ Q ♄
17 48 ☽ ∠ ♃
18 42 ☿ ⊼ ♇
21 21 ☽ ⊻ ♀
22 53 ☽ Q ♂
23 15 ☽ Q ♆

23 Thursday
05 00 ♂ ☌ ♆
08 07 ☽ ± ♇
11 59 ☽ ⊥ ♃
15 23 ☽ ✶ ♄
20 00 ☽ ⊼ ♇
21 33 ☽ ☌ ☿
23 39 ☽ ☌ ♅

24 Friday
00 01 ☽ ⊻ ♃
03 38 ☽ ∥ ☉
10 18 ☽ ⊻ ☉
10 56 ☽ □ ♆
12 46 ☽ □ ♂
17 35 ☽ ⋕ ♇
18 41 ☉ ✶ ♆
20 58 ☽ ∠ ♄
22 59 ☽ ⊥ ☉

25 Saturday
03 23 ☽ ∥ ♅
10 59 ☽ ∥ ☿
11 45 ☽ ☌ ♀
14 27 ☽ ⊥ ♄
17 49 ☽ ∠ ☉
20 59 ☿ ☌ ♅

26 Sunday
01 59 ☽ ⊻ ♄
06 08 ☽ ☍ ♇
09 58 ☽ ⊻ ♅
10 27 ☽ ⊻ ☿
10 49 ☽ ☌ ♃
11 00 ☽ ∥ ♀
20 37 ☽ △ ♆
21 19 ☽ ⊥ ♅
22 11 ☽ ⊥ ☿

27 Monday
00 32 ☽ ✶ ☉
01 44 ☽ △ ♂
02 12 ☿ ⊻ ♃
07 02 ☽ ∥ ♃
10 30 ☽ ∥ ♂
14 09 ☽ ∠ ♅
15 29 ☽ ∠ ☿
17 26 ☽ ∥ ♆
22 07 ♀ ⊥ ♄
23 04 ☽ ⊻ ♀

28 Tuesday
00 27 ☽ ᵽ ♆
00 29 ☽ Q ☉
06 59 ☽ ᵽ ♂
10 00 ☽ ☌ ♄
10 31 ☽ ⊥ ♀
13 39 ☽ ⊼ ♇
16 47 ☽ ± ♆
17 35 ☽ ✶ ♅
18 52 ☽ ⊻ ♃
19 31 ☽ ✶ ☿
22 31 ☽ ∥ ♄

29 Wednesday
00 07 ☽ ± ♂
00 24 ☽ ± ♇
03 24 ☽ ∠ ♀
03 32 ☽ ⊼ ♆
05 34 ♂ Q ♇
05 42 ☽ ⊥ ♃
11 17 ☽ □ ☉
11 21 ☽ ⊼ ♂
14 49 ☉ ✶ ♂
15 00 ☽ Q ♅
15 53 ♂ ∥ ♃
16 19 ☽ ᵽ ♇
17 13 ☽ Q ☿
21 44 ☽ ∠ ♃

30 Thursday
06 48 ☽ ✶ ♀
13 23 ☽ ⋕ ♄
15 02 ☽ ⊻ ♄
18 14 ☽ △ ♇
18 51 ♂ ᵽ ♄
22 13 ☽ □ ♅
23 50 ☽ ✶ ♃

All ephemeris data is given at 12.00 UT and the Moon's longitude is additionally given for 24.00 UT

JULY 2039

LONGITUDES

Date	Sidereal time	Sun ☉	Moon ☽	Moon ☽ 24.00	Mercury ☿	Venus ♀	Mars ♂	Jupiter ♃	Saturn ♄	Uranus ♅	Neptune ♆	Pluto ⯓
	h m s	° ' "	° ' "	° ' "	° '	° '	° '	° '	° '	° '	° '	° '
01	06 37 38	09 ♋ 32 31	05 ♏ 57 47	13 ♏ 09 55	29 ♋ 18	19 ♌ 27	09 ♉ 06	28 ♌ 50	23 ♍ 38	27 ♋ 49	03 ♉ 16	25 ♒ 26
02	06 41 35	10 29 43	20 ♏ 26 41	27 ♏ 47 37	29 22	19 58	09 49	29 01	23 42	27 53	03 17	25 R 25
03	06 45 31	11 26 54	05 ♐ 12 06	12 ♐ 39 23	29 R 22	20 27	10 32	29 11	23 45	27 56	03 18	25 25
04	06 49 28	12 24 05	20 ♐ 08 36	27 ♐ 38 47	29 17	20 55	11 15	29 22	23 49	28 00	03 19	25 24
05	06 53 25	13 21 17	05 ♑ 08 56	12 ♑ 37 59	29 07	21 22	11 58	29 33	23 53	28 04	03 21	25 23
06	06 57 21	14 18 28	20 ♑ 04 52	27 ♑ 28 35	28 53	21 46	12 41	29 43	23 57	28 07	03 22	25 22
07	07 01 18	15 15 39	04 ♒ 48 12	12 ♒ 02 53	28 35	22 09	13 24	29 ♌ 54	24 01	28 11	03 23	25 21
08	07 05 14	16 12 50	19 ♒ 11 58	26 ♒ 14 56	28 12	22 31	14 06	00 ♍ 05	24 05	28 14	03 23	25 20
09	07 09 11	17 10 01	03 ♓ 11 26	10 ♓ 01 14	27 46	22 50	14 49	00 16	24 10	28 18	03 24	25 19
10	07 13 07	18 07 13	16 ♓ 44 20	23 ♓ 20 49	27 16	23 08	15 31	00 27	24 14	28 22	03 25	25 18
11	07 17 04	19 04 25	29 ♓ 50 54	06 ♈ 14 57	26 43	23 23	16 14	00 39	24 18	28 25	03 26	25 17
12	07 21 00	20 01 38	12 ♈ 33 23	18 ♈ 46 42	26 08	23 37	16 56	00 50	24 23	28 29	03 27	25 16
13	07 24 57	20 58 51	24 ♈ 55 26	01 ♉ 00 11	25 31	23 49	17 38	01 01	24 27	28 33	03 28	25 15
14	07 28 54	21 56 04	07 ♉ 01 32	13 ♉ 00 06	24 52	23 58	18 21	01 12	24 32	28 36	03 29	25 14
15	07 32 50	22 53 18	18 ♉ 56 31	24 ♉ 51 22	24 13	24 06	19 03	01 24	24 37	28 40	03 29	25 13
16	07 36 47	23 50 33	00 ♊ 45 13	06 ♊ 38 38	23 33	24 11	19 45	01 35	24 41	28 44	03 30	25 12
17	07 40 43	24 47 48	12 ♊ 32 09	18 ♊ 26 14	22 54	24 14	20 26	01 47	24 46	28 47	03 31	25 11
18	07 44 40	25 45 04	24 ♊ 21 20	00 ♋ 17 53	22 17	24 R 15	21 08	01 59	24 51	28 51	03 31	25 10
19	07 48 36	26 42 20	06 ♋ 16 13	12 ♋ 16 41	21 41	24 14	21 50	02 10	24 56	28 55	03 32	25 09
20	07 52 33	27 39 37	18 ♋ 19 31	24 ♋ 25 00	21 08	24 10	22 31	02 22	25 01	28 58	03 33	25 07
21	07 56 29	28 36 55	00 ♌ 33 17	06 ♌ 44 32	20 38	24 04	23 13	02 34	25 06	29 02	03 33	25 06
22	08 00 26	29 ♋ 34 13	12 ♌ 58 52	19 ♌ 16 22	20 13	23 55	23 54	02 46	25 11	29 06	03 34	25 05
23	08 04 23	00 ♌ 31 31	25 ♌ 37 06	02 ♍ 01 08	19 51	23 44	24 36	02 57	25 17	29 09	03 34	25 04
24	08 08 19	01 28 50	08 ♍ 28 28	14 ♍ 59 08	19 34	23 31	25 17	03 09	25 22	29 13	03 35	25 03
25	08 12 16	02 26 09	21 ♍ 33 09	28 ♍ 10 34	19 22	23 15	25 58	03 21	25 27	29 17	03 35	25 02
26	08 16 12	03 23 28	04 ♎ 51 24	11 ♎ 35 40	19 16	22 57	26 39	03 33	25 33	29 20	03 35	25 00
27	08 20 09	04 20 48	18 ♎ 23 24	25 ♎ 14 38	19 D 15	22 37	27 20	03 45	25 38	29 24	03 36	24 59
28	08 24 05	05 18 09	02 ♏ 09 24	09 ♏ 07 39	19 20	22 14	28 00	03 58	25 44	29 28	03 36	24 58
29	08 28 02	06 15 30	16 ♏ 09 23	23 ♏ 14 29	19 31	21 50	28 41	04 10	25 50	29 31	03 36	24 57
30	08 31 58	07 12 51	00 ♐ 22 50	07 ♐ 34 10	19 48	21 23	29 ♉ 21	04 22	25 55	29 35	03 37	24 55
31	08 35 55	08 ♌ 10 13	14 ♐ 48 12	22 ♐ 04 31	20 ♋ 12	20 ♌ 54	00 ♊ 02	04 ♍ 34	26 ♍ 01	29 ♋ 39	03 ♉ 37	24 ♒ 54

DECLINATIONS

Date	Moon True ☊	Moon Mean ☊	Moon ☽ Latitude	Sun ☉	Moon ☽	Mercury ☿	Venus ♀	Mars ♂	Jupiter ♃	Saturn ♄	Uranus ♅	Neptune ♆	Pluto ⯓
	° '	° '	° '	° '	° '	° '	° '	° '	° '	° '	° '	° '	° '
01	21 ♊ 56	21 ♊ 08	03 N 48	23 N 05	09 S 55	18 N 21	15 N 07	13 N 20	12 N 47	04 N 33	21 N 06	10 N 59	22 S 17
02	21 D 58	21 05	02 46	23 01	15 12	18 05	14 49	13 34	12 44	04 32	21 05	11 00	22 17
03	21 59	21 01	01 31	22 56	19 40	17 50	14 31	13 48	12 40	04 30	21 05	11 00	22 18
04	21 59	20 58	00 N 10	22 51	22 54	17 35	14 13	14 02	12 36	04 28	21 04	11 00	22 18
05	21 R 59	20 55	01 S 12	22 46	24 32	17 22	13 55	14 15	12 32	04 27	21 03	11 01	22 19
06	21 57	20 52	02 29	22 40	24 23	17 11	13 38	14 29	12 28	04 25	21 02	11 01	22 19
07	21 55	20 49	03 35	22 34	22 32	17 00	13 21	14 42	12 24	04 23	21 02	11 01	22 20
08	21 52	20 46	04 25	22 27	19 15	16 51	13 04	14 55	12 21	04 21	21 01	11 02	22 21
09	21 48	20 43	04 58	22 20	14 58	16 43	12 47	15 08	12 17	04 20	21 00	11 02	22 21
10	21 45	20 39	05 13	22 13	10 02	16 37	12 31	15 21	12 13	04 18	20 59	11 02	22 22
11	21 43	20 36	05 11	22 05	04 S 49	16 33	12 15	15 34	12 09	04 16	20 59	11 02	22 22
12	21 41	20 33	04 54	21 56	00 N 27	16 30	11 59	15 46	12 05	04 14	20 58	11 03	22 23
13	21 D 41	20 30	04 23	21 48	05 34	16 28	11 44	15 59	12 01	04 12	20 57	11 03	22 23
14	21 42	20 27	03 41	21 39	10 22	16 28	11 29	16 11	11 57	04 10	20 57	11 03	22 24
15	21 43	20 24	02 50	21 30	14 43	16 30	11 15	16 23	11 52	04 08	20 56	11 03	22 24
16	21 45	20 21	01 53	21 20	18 28	16 33	11 01	16 35	11 48	04 06	20 55	11 03	22 25
17	21 47	20 17	00 S 50	21 10	21 28	16 37	10 48	16 47	11 44	04 04	20 54	11 03	22 26
18	21 R 47	20 14	00 N 14	20 59	23 33	16 43	10 35	16 58	11 40	04 02	20 54	11 04	22 26
19	21 46	20 11	01 18	20 49	24 35	16 49	10 23	17 10	11 36	04 00	20 53	11 04	22 27
20	21 43	20 08	02 20	20 38	24 29	16 57	10 12	17 21	11 31	03 58	20 52	11 04	22 27
21	21 39	20 04	03 16	20 26	23 13	17 06	10 01	17 32	11 27	03 55	20 51	11 04	22 28
22	21 33	20 01	04 03	20 14	20 48	17 16	09 50	17 43	11 23	03 53	20 51	11 04	22 28
23	21 26	19 58	04 39	20 02	17 21	17 27	09 41	17 53	11 19	03 51	20 50	11 04	22 29
24	21 20	19 55	05 02	19 50	13 03	17 38	09 32	18 04	11 14	03 49	20 49	11 04	22 30
25	21 14	19 52	05 10	19 37	08 05	17 50	09 24	18 14	11 10	03 47	20 48	11 04	22 30
26	21 09	19 49	05 01	19 24	02 N 41	18 02	09 17	18 25	11 05	03 44	20 48	11 05	22 31
27	21 06	19 45	04 35	19 10	02 S 58	18 14	09 10	18 35	11 01	03 42	20 47	11 05	22 31
28	21 05	19 42	03 54	18 56	08 33	18 26	09 05	18 44	10 57	03 40	20 46	11 05	22 32
29	21 D 05	19 39	02 58	18 42	13 50	18 38	09 00	18 54	10 52	03 37	20 45	11 05	22 32
30	21 06	19 36	01 50	18 28	18 26	18 50	08 56	19 04	10 48	03 35	20 44	11 05	22 33
31	21 ♊ 07	19 ♊ 33	00 N 34	18 N 13	22 S 00	19 N 02	08 N 53	19 N 13	10 N 43	03 N 32	20 N 44	11 N 05	22 S 33

ZODIAC SIGN ENTRIES

Date	h	m	Planets
01	01	58	☽ ♏
03	03	35	☽ ♐
05	03	46	☽ ♑
07	04	07	☽ ♒
08	00	24	♃ ♍
09	06	28	☽ ♓
11	12	17	☽ ♈
13	22	01	☽ ♉
16	10	28	☽ ♊
18	23	24	☽ ♋
21	10	55	☽ ♌
22	22	48	☉ ♌
23	20	14	☽ ♍
26	03	17	☽ ♎
28	08	16	☽ ♏
30	11	22	☽ ♐
31	10	58	♂ ♊

LATITUDES

Date	Mercury ☿	Venus ♀	Mars ♂	Jupiter ♃	Saturn ♄	Uranus ♅	Neptune ♆	Pluto ⯓
	° '	° '	° '	° '	° '	° '	° '	° '
01	01 S 59	00 N 09	01 S 15	00 N 58	02 N 13	00 N 31	01 S 43	09 S 51
04	02 46	00 S 19	01 14	00 58	02 12	00 31	01 43	09 51
07	03 31	00 49	01 13	00 58	02 11	00 31	01 43	09 52
10	04 09	01 22	01 11	00 58	02 11	00 31	01 43	09 53
13	04 38	01 57	01 09	00 58	02 11	00 31	01 43	09 53
16	04 54	02 35	01 08	00 58	02 10	00 31	01 44	09 54
19	04 55	03 14	01 06	00 58	02 10	00 31	01 44	09 54
22	04 42	03 55	01 04	00 57	02 09	00 31	01 44	09 55
25	04 15	04 37	01 02	00 57	02 09	00 31	01 44	09 56
28	03 38	05 18	01 00	00 57	02 08	00 31	01 44	09 56
31	02 S 55	05 S 57	00 S 58	00 N 57	02 N 08	00 N 31	01 S 45	09 S 56

DATA

Julian Date	2465971
Delta T	+77 seconds
Ayanamsa	24° 24' 15"
Synetic vernal point	04° ♓ 42' 44"
True obliquity of ecliptic	23° 26' 04"

MOON'S PHASES, APSIDES AND POSITIONS ☽

Date	h	m	Phase	Longitude ° '	Eclipse Indicator
06	02	03	○	13 ♑ 55	
13	03	38	☾	20 ♈ 39	
21	07	54	●	28 ♋ 27	
28	17	50	☽	05 ♏ 32	

Day	h	m		
04	20	34	Perigee	
17	03	22	Apogee	
05	22	00	Max dec	24° S 41'
12	09	56	0N	
19	21	56	Max dec	24° N 41'
26	23	27	0S	

LONGITUDES

Date	Chiron ⚷	Ceres ⚳	Pallas ⚴	Juno ⚵	Vesta ⚶	Black Moon Lilith ⚸
	° '	° '	° '	° '	° '	° '
01	05 ♋ 45	25 ♈ 07	04 ♓ 12	00 ♒ 09	09 ♍ 57	10 ♊ 30
11	06 ♋ 43	27 ♈ 36	03 ♓ 11	28 ♑ 01	14 ♍ 10	11 ♊ 36
21	07 ♋ 39	29 ♈ 46	01 ♓ 37	25 ♑ 40	18 ♍ 34	12 ♊ 43
31	08 ♋ 34	01 ♉ 34	29 ♒ 33	23 ♑ 19	23 ♍ 07	13 ♊ 49

All ephemeris data is given at 12.00 UT and the Moon's longitude is additionally given for 24.00 UT

ASPECTARIAN

h m	Aspects
01 Friday	
00 40	☽ □ ☿
01 17	☽ ⊥ ♄
04 07	☽ Q ♀
07 28	☽ ☍ ♆
16 29	☽ ∠ ♄
16 41	☽ # ♆
17 31	☽ ☍ ♂
18 24	☽ △ ☉
20 14	☽ Q ♃
02 Saturday	
00 34	☽ # ♃
03 57	☽ # ♂
10 14	☉ ♇ ⯓
10 17	☽ # ♀
11 11	☽ □ ♀
17 20	☽ ⚹ ♄
20 08	☽ □ ⯓
20 50	☽ ♇ ☉
21 48	☿ St R
03 Sunday	
00 11	☽ △ ♅
02 01	☽ # ☿
02 09	☽ □ ♃
02 34	☽ △ ☿
08 56	☽ ⊼ ♆
12 26	☽ ± ☉
12 54	☽ Q ♄
18 37	☽ ± ♆
20 17	☉ Q ♄
21 02	☽ ⊼ ♂
21 11	☽ # ♅
22 45	☽ ⊼ ☉
04 Monday	
00 30	☽ ♇ ♅
01 12	☽ Q ⯓
02 40	☽ ♇ ☿
04 56	☿ ⊻ ♃
06 39	☽ ∥ ⯓
07 09	☽ ± ♂
09 05	☽ ♇ ♆
11 36	☽ # ☉
13 17	☽ △ ♀
14 59	☽ ± ♅
16 59	☽ ± ☿
17 54	☽ □ ♄
20 24	☽ ⚹ ⯓
20 38	♀ ∥ ♂
22 16	☽ ♇ ♂
05 Tuesday	
00 37	☽ ⊼ ♅
02 28	☽ ⊼ ☿
02 56	☽ △ ♃
09 06	☽ △ ♆
14 00	☽ ♇ ♀
20 22	☽ ∠ ⯓
23 29	☽ △ ♂
06 Wednesday	
02 03	☽ ☍ ☉
03 15	☽ ♇ ♃
04 51	☽ ± ♀
10 51	☽ ⊥ ⯓
14 49	☽ ⊼ ♀
17 58	☽ ± ♃
18 18	☽ △ ♄
20 33	☽ ⊻ ⯓
07 Thursday	
00 58	☉ ∠ ♃
01 06	☽ ☍ ♅
02 01	☽ ☍ ☿
03 52	☽ ⊼ ♃
09 39	☽ □ ♆
11 41	☽ # ☉
13 44	☽ ∥ ⯓
14 57	☉ Q ♆
19 00	☽ ♇ ♄
08 Friday	
00 10	☽ # ♅
03 00	☽ □ ♂
06 37	☽ ⊼ ☉
10 07	☽ ± ♄
10 08	☿ ☌ ♅
15 43	☽ Q ♆
17 29	☽ ± ☉
17 45	☽ ☍ ♀
20 21	☽ ⊼ ♄
22 25	☽ ☌ ⯓
23 24	☉ ⊥ ♀
09 Saturday	
02 24	☽ # ☿
02 54	☽ ⊼ ☿
03 29	☽ ⊼ ♅
06 52	☽ ☍ ♃
08 26	☉ # ⯓
10 05	☽ ♇ ☉
11 07	☽ # ♂
11 19	☽ Q ♂
12 23	☽ ⚹ ♆
12 58	☽ ± ☿
13 57	☽ ± ♅
23 30	☽ # ♀
10 Sunday	
01 31	☽ # ♃
04 17	☽ ♇ ☿
05 55	☽ ♇ ♅
07 00	☿ Q ♂
07 18	☽ # ♆
09 42	☽ ⚹ ♂
14 41	☽ △ ☉
15 02	☽ ∠ ♆
23 50	☽ ⊼ ♀
11 Monday	
01 42	☽ ☍ ♄
03 34	☽ ⊻ ⯓
06 27	☽ △ ☿
07 31	☽ ⊥ ♆
09 20	☽ △ ♅
11 08	☽ ± ♀
13 30	☽ ⊼ ♃
14 32	☽ # ♄
14 40	☽ ⊥ ⯓
14 44	☽ ∠ ♂
17 13	☉ ± ⯓
18 43	☽ ⊻ ♆
19 03	♂ Q ♅
12 Tuesday	
00 48	♀ ∥ ♃
00 56	☽ ± ♃
04 21	☽ ♇ ♀
07 38	☽ ∠ ⯓
08 43	☽ ⊥ ♂
18 23	☽ ♇ ♃
20 56	☽ ⊻ ♂
13 Wednesday	
03 38	☽ □ ☉
05 32	☽ ∥ ♄
09 47	☽ △ ♀
11 04	☽ ⊼ ♄
12 38	☽ ⚹ ⯓
13 06	☽ □ ☿
19 10	☽ □ ♅
22 13	☿ ⊼ ⯓
22 59	☽ ± ♄
14 Thursday	
00 13	☽ △ ♃
02 22	☿ ⊥ ♃
04 54	☽ ☌ ♆
12 25	☽ Q ⯓
15 35	☽ ∥ ♆
17 03	☽ ♇ ♄
17 37	☽ ∥ ♀
18 20	☽ Q ☉
20 14	☽ ∥ ♃
23 07	☽ Q ☿
23 07	☿ ⚹ ♄
15 Friday	
07 22	☽ Q ♅
12 13	☽ ☌ ♂
15 40	☿ ⊻ ♀
20 42	☽ ⚹ ☉
22 08	☽ ⚹ ☿
22 33	☽ □ ♀
22 42	☽ ∥ ♂
22 58	☽ ∥ ☿
23 35	☽ △ ♄
16 Saturday	
00 43	☽ □ ⯓
05 23	☿ ∥ ♂
07 43	☉ ☌ ☿
07 51	☽ ⚹ ♅
08 19	♀ ∥ ♆
13 44	☽ □ ♃
17 36	☽ ⊻ ♆
21 14	☉ ⊻ ♀
17 Sunday	
03 04	☽ ∠ ☿
05 50	☽ ⊥ ♆
05 56	☽ ∠ ☉
07 00	☽ ∥ ♅
09 26	☽ ∥ ☉
11 15	☉ ⚹ ♄
11 23	☽ Q ♀
14 33	☽ ∠ ♅
20 26	☽ ⊥ ☿
21 26	☉ ⊼ ⯓
21 47	☽ # ⯓
18 Monday	
00 10	☽ ∠ ♆
01 51	☽ ⊥ ☉
02 58	☽ Q ♃
05 04	☽ ⊻ ♂
08 00	☽ ⊻ ☿
08 36	♀ St R
08 56	☽ ⊥ ♅
11 47	☽ ⚹ ♀
13 01	☽ □ ♄
13 38	☽ △ ⯓
15 04	☽ ⊻ ☉
17 58	☽ ⊥ ♂
19 04	☉ ⊥ ♃
21 07	☽ ⊻ ♅
19 Tuesday	
02 09	☉ ∥ ♅
03 38	☽ ⚹ ♃
06 30	☽ ⚹ ♆
09 17	☿ ⚹ ♂
13 12	☽ ∠ ♂
17 53	☽ ∠ ♀
19 44	☽ ♇ ⯓
20 Wednesday	
01 24	☽ Q ♄
06 29	☽ Q ♆
10 04	☽ ∠ ♃
11 41	☽ ⊥ ♀
13 34	☽ ± ⯓
17 19	☽ ☌ ☿
20 47	☽ ⚹ ♂
23 25	☽ ⊻ ♀
21 Thursday	
01 16	☽ ⚹ ♄
01 22	☽ ⊼ ⯓
04 05	☽ ⊥ ♃
07 54	☽ ☌ ☉
09 01	☽ ☌ ♅
12 03	♄ ⊼ ⯓
15 58	☽ ⊻ ♃
17 50	☽ □ ♆
20 38	☽ # ⯓
21 35	☽ Q ♂
23 11	☉ ☌ ♅
22 Friday	
06 36	☽ ∠ ♄
11 36	☽ ∥ ♅
12 20	♀ □ ♂
16 41	☽ ∥ ☉
23 56	☽ ⊥ ♄
23 Saturday	
01 23	☽ ⊻ ☿
08 30	☽ ☌ ♀
08 47	☽ ∥ ♂
09 57	☽ □ ♂
10 58	☽ ☍ ⯓
11 21	☽ ⊻ ♄
11 27	☽ ∥ ☿
12 25	☽ ⊥ ☿
18 40	☽ ⊻ ♅
21 57	☽ ⊻ ☉
24 Sunday	
01 58	☽ ☌ ♃
02 54	☽ △ ♆
04 04	♂ □ ⯓
04 53	☽ ∠ ☿
05 56	☽ ⊥ ♅
10 01	☽ ⊥ ☉
15 33	♂ △ ♄
21 15	☽ ∥ ♃
21 54	☽ ∥ ♆
22 38	☽ ∠ ♅
25 Monday	
03 54	☽ ∠ ☉
05 45	☽ ∥ ♀
06 35	☽ ♇ ♆
08 04	☽ ⚹ ☿
15 01	☽ ⊻ ♀
18 18	☽ ⊼ ⯓
19 08	☽ ☌ ♄
20 26	☽ △ ♂
22 56	☽ ± ♆
26 Tuesday	
01 38	☽ ⊥ ♀
02 03	☽ ⚹ ♅
05 06	☽ ± ⯓
05 35	☽ Q ☿
07 22	☽ ∥ ♄
09 11	☽ ⚹ ☉
09 38	☽ ⊻ ♃
09 44	☽ ⊼ ♆
16 00	♃ △ ♆
16 59	♃ ∥ ♆
16 59	☉ □ ♆
17 15	☉ ⊻ ♃
17 24	☽ ∠ ♀
20 30	☽ ⊥ ♃
21 10	☽ ♇ ⯓
23 36	☽ Q ♅
27 Wednesday	
00 44	☽ ♇ ♂
03 06	☿ St D
08 08	☽ Q ☉
12 39	☽ ∠ ♃
13 31	☽ □ ☿
15 07	☽ # ♄
17 25	☽ ± ♂
19 13	☽ ⚹ ♀
23 32	☽ △ ⯓
28 Thursday	
00 46	☽ ⊻ ♄
04 26	☽ ⊼ ♂
07 19	☽ □ ♅
11 16	☽ ⊥ ♄
14 15	☽ # ♀
14 30	☽ ☍ ♆
15 10	☽ ⚹ ♃
15 29	☽ Q ♀
17 50	☽ □ ☉
22 27	☽ # ♃
23 12	☽ # ♆
29 Friday	
00 09	☉ ∥ ♂
02 51	☽ ∠ ♄
12 01	☽ Q ♃
15 33	☉ ∥ ☿
17 49	☽ △ ☿
21 19	☽ □ ♀
30 Saturday	
02 51	☽ □ ⯓
04 28	☽ ⚹ ♄
10 12	☽ ☍ ♂
10 39	☽ △ ♅
12 09	☽ # ☉
14 28	☽ # ☿
15 49	☽ # ♂
17 24	☽ ⊼ ♆
18 46	☽ □ ♃
19 35	☽ ♇ ☿
20 50	♂ ⚹ ♅
31 Sunday	
00 13	☽ △ ☉
00 40	☽ Q ♄
02 27	☽ # ♅
03 24	☽ ± ♆
08 52	☽ Q ⯓
10 58	☽ ± ☿
11 44	☽ ♇ ♅
16 48	☽ ∥ ⯓
18 18	☽ ♇ ♆
21 12	☽ ⊼ ☿
21 44	☽ △ ♀

AUGUST 2039

Date	Sidereal time h m s	Sun ☉	Moon ☽	Moon ☽ 24.00	Mercury ☿	Venus ♀	Mars ♂	Jupiter ♃	Saturn ♄	Uranus ♅	Neptune ♆	Pluto ♇
01	08 39 52	09 ♌ 07 35	29 ♐ 22 36	06 ♑ 41 51	20 ♋ 42	20 ♌ 24	00 ♊ 42	04 ♍ 47	26 ♍ 07	29 ♋ 42	03 ♉ 37	24 ♒ 53
02	08 43 48	10 04 58	14 ♑ 01 34	21 ♑ 20 58	21 18	19 R 52	01 22	04 59	26 13	29 46	03 37	24 R 52
03	08 47 45	11 02 22	28 ♑ 39 13	05 ♒ 55 29	22 00	19 19	02 02	05 11	26 19	29 49	03 37	24 50
04	08 51 41	11 59 47	13 ♒ 08 55	20 ♒ 18 42	22 48	18 44	02 42	05 24	26 25	29 53	03 37	24 49
05	08 55 38	12 57 12	27 ♒ 24 07	04 ♓ 24 32	23 42	18 08	03 22	05 36	26 31	29 ♋ 57	03 37	24 48
06	08 59 34	13 54 39	11 ♓ 19 26	18 ♓ 08 29	24 43	17 32	04 02	05 49	26 37	00 ♌ 00	03 37	24 47
07	09 03 31	14 52 06	24 ♓ 51 26	01 ♈ 28 13	25 49	16 55	04 41	06 01	26 43	00 04	03 R 37	24 45
08	09 07 27	15 49 35	07 ♈ 58 54	14 ♈ 23 41	27 00	16 17	05 21	06 14	26 49	00 07	03 37	24 44
09	09 11 24	16 47 05	20 ♈ 42 52	26 ♈ 56 52	28 18	15 40	06 00	06 26	26 55	00 11	03 37	24 43
10	09 15 21	17 44 36	03 ♉ 06 10	09 ♉ 11 19	29 ♋ 40	15 03	06 39	06 39	27 02	00 15	03 37	24 41
11	09 19 17	18 42 09	15 ♉ 12 56	21 ♉ 11 39	01 ♌ 07	14 26	07 18	06 52	27 08	00 18	03 37	24 40
12	09 23 14	19 39 43	27 ♉ 08 08	03 ♊ 03 03	02 39	13 50	07 57	07 04	27 14	00 22	03 37	24 39
13	09 27 10	20 37 18	08 ♊ 57 05	14 ♊ 50 52	04 16	13 14	08 36	07 17	27 21	00 25	03 37	24 37
14	09 31 07	21 34 55	20 ♊ 45 04	26 ♊ 40 17	05 57	12 40	09 15	07 30	27 27	00 29	03 36	24 36
15	09 35 03	22 32 34	02 ♋ 37 06	08 ♋ 36 01	07 41	12 07	09 53	07 43	27 34	00 32	03 36	24 35
16	09 39 00	23 30 14	14 ♋ 37 32	20 ♋ 42 04	09 29	11 35	10 32	07 55	27 40	00 36	03 36	24 33
17	09 42 56	24 27 55	26 ♋ 49 58	03 ♌ 01 30	11 19	11 05	11 10	08 08	27 47	00 39	03 35	24 32
18	09 46 53	25 25 38	09 ♌ 16 53	15 ♌ 36 14	13 12	10 37	11 48	08 21	27 54	00 43	03 35	24 31
19	09 50 50	26 23 22	21 ♌ 59 37	28 ♌ 27 00	15 07	10 11	12 26	08 34	28 00	00 46	03 35	24 29
20	09 54 46	27 21 07	04 ♍ 58 17	11 ♍ 33 20	17 04	09 47	13 04	08 47	28 07	00 49	03 34	24 28
21	09 58 43	28 18 54	18 ♍ 11 55	24 ♍ 53 49	19 03	09 25	13 42	09 00	28 14	00 53	03 34	24 27
22	10 02 39	29 ♌ 16 42	01 ♎ 38 47	08 ♎ 26 30	21 02	09 05	14 19	09 13	28 21	00 56	03 33	24 25
23	10 06 36	00 ♍ 14 31	15 ♎ 16 44	22 ♎ 09 13	23 02	08 47	14 57	09 25	28 27	00 59	03 33	24 24
24	10 10 32	01 12 21	29 ♎ 03 44	06 ♏ 00 04	25 02	08 32	15 34	09 38	28 34	01 03	03 32	24 23
25	10 14 29	02 10 13	12 ♏ 58 05	19 ♏ 57 38	27 02	08 19	16 11	09 51	28 41	01 06	03 32	24 22
26	10 18 25	03 08 06	26 ♏ 58 37	04 ♐ 00 57	29 ♌ 02	08 09	16 48	10 04	28 48	01 09	03 31	24 20
27	10 22 22	04 06 00	11 ♐ 04 31	18 ♐ 09 13	01 ♍ 02	08 00	17 24	10 17	28 55	01 12	03 30	24 19
28	10 26 19	05 03 56	25 ♐ 14 55	02 ♑ 21 27	03 01	07 55	18 01	10 30	29 02	01 16	03 30	24 18
29	10 30 15	06 01 52	09 ♑ 28 34	16 ♑ 35 58	04 59	07 51	18 37	10 43	29 09	01 19	03 29	24 16
30	10 34 12	06 59 50	23 ♑ 43 17	00 ♒ 50 05	06 56	07 D 50	19 14	10 56	29 16	01 22	03 28	24 15
31	10 38 08	07 ♍ 57 49	07 ♒ 55 51	15 ♒ 00 03	08 ♍ 53	07 ♌ 52	19 ♊ 50	11 ♍ 09	29 ♍ 23	01 ♌ 25	03 ♉ 28	24 ♒ 14

DECLINATIONS

Date	Moon True ☊	Moon Mean ☊	Moon ☽ Latitude	Sun ☉	Moon ☽	Mercury ☿	Venus ♀	Mars ♂	Jupiter ♃	Saturn ♄	Uranus ♅	Neptune ♆	Pluto ♇
01	21 ♊ 07	19 ♊ 30	00 S 44	17 N 58	24 S 10	19 N 12	08 N 51	19 N 22	10 N 39	03 N 30	20 N 43	11 N 05	22 S 34
02	21 R 05	19 26	02 00	17 43	24 41	19 23	08 49	19 31	10 34	03 28	20 42	11 05	22 35
03	21 01	19 23	03 08	17 27	23 30	19 32	08 49	19 40	10 30	03 25	20 41	11 05	22 35
04	20 55	19 20	04 03	17 11	20 45	19 39	08 49	19 48	10 25	03 23	20 41	11 05	22 36
05	20 47	19 17	04 42	16 55	16 46	19 46	08 50	19 57	10 20	03 20	20 40	11 05	22 36
06	20 38	19 14	05 02	16 39	11 58	19 51	08 52	20 05	10 16	03 18	20 39	11 05	22 37
07	20 30	19 10	05 05	16 22	06 43	19 54	08 54	20 13	10 11	03 15	20 38	11 05	22 37
08	20 23	19 07	04 52	16 05	01 S 19	19 56	08 57	20 21	10 07	03 13	20 38	11 04	22 38
09	20 18	19 04	04 25	15 48	04 N 00	19 55	09 01	20 29	10 02	03 10	20 37	11 04	22 38
10	20 15	19 01	03 45	15 31	09 01	19 52	09 06	20 36	09 57	03 07	20 36	11 04	22 39
11	20 13	18 58	02 56	15 13	13 35	19 47	09 11	20 43	09 53	03 05	20 35	11 04	22 40
12	20 D 14	18 55	02 01	14 55	17 33	19 39	09 16	20 51	09 48	03 02	20 35	11 04	22 40
13	20 14	18 51	01 S 00	14 37	20 48	19 29	09 22	20 58	09 43	03 00	20 34	11 04	22 41
14	20 R 15	18 48	00 N 03	14 18	23 09	19 16	09 29	21 04	09 38	02 57	20 33	11 04	22 41
15	20 14	18 45	01 06	14 00	24 30	19 01	09 36	21 11	09 34	02 54	20 33	11 04	22 42
16	20 11	18 42	02 06	13 41	24 44	18 42	09 43	21 17	09 29	02 52	20 32	11 04	22 42
17	20 06	18 39	03 02	13 22	23 46	18 21	09 50	21 24	09 24	02 49	20 31	11 03	22 43
18	19 58	18 36	03 50	13 03	21 38	17 58	09 58	21 30	09 19	02 46	20 30	11 03	22 43
19	19 48	18 32	04 28	12 43	18 24	17 31	10 06	21 36	09 14	02 43	20 30	11 03	22 44
20	19 37	18 29	04 53	12 23	14 13	17 03	10 14	21 42	09 10	02 41	20 29	11 03	22 44
21	19 26	18 26	05 02	12 03	09 18	16 32	10 21	21 47	09 05	02 38	20 28	11 03	22 45
22	19 15	18 23	04 55	11 43	03 N 52	15 58	10 29	21 53	09 00	02 35	20 27	11 02	22 45
23	19 07	18 20	04 32	11 23	01 S 50	15 23	10 37	21 58	08 55	02 32	20 27	11 02	22 46
24	19 00	18 16	03 52	11 03	07 31	14 46	10 45	22 03	08 50	02 30	20 26	11 02	22 46
25	18 57	18 13	02 58	10 42	12 53	14 07	10 53	22 08	08 45	02 27	20 25	11 02	22 47
26	18 56	18 10	01 54	10 21	17 38	13 27	11 01	22 12	08 41	02 24	20 25	11 01	22 47
27	18 D 56	18 07	00 N 42	10 00	21 25	12 46	11 08	22 17	08 36	02 21	20 24	11 01	22 48
28	18 R 56	18 04	00 S 33	09 39	23 54	12 03	11 16	22 21	08 31	02 18	20 23	11 01	22 48
29	18 54	18 01	01 47	09 18	24 52	11 19	11 23	22 26	08 26	02 15	20 23	11 01	22 48
30	18 51	17 57	02 53	08 56	24 12	10 35	11 30	22 30	08 21	02 13	20 22	11 00	22 49
31	18 ♊ 45	17 ♊ 54	03 S 49	08 N 35	21 S 58	09 N 49	11 N 36	22 N 34	08 N 16	02 N 10	20 N 21	11 N 00	22 S 49

ZODIAC SIGN ENTRIES

Date	h	m	Planets
01	13	01	☽ ♑
03	14	13	☽ ♒
05	16	26	☽ ♓
06	10	00	♅ ♌
07	21	19	☽ ♈
10	05	56	☽ ♉
10	17	38	☿ ♌
12	17	48	☽ ♊
15	06	43	☽ ♋
17	18	09	☽ ♌
20	02	52	☽ ♍
22	09	05	☽ ♎
23	05	59	☉ ♍
24	13	37	☽ ♏
26	17	09	☽ ♐
26	23	37	☿ ♍
28	20	01	☽ ♑
30	22	35	☽ ♒

LATITUDES

Date	Mercury ☿	Venus ♀	Mars ♂	Jupiter ♃	Saturn ♄	Uranus ♅	Neptune ♆	Pluto ♇
01	02 S 40	06 S 09	00 S 57	00 N 57	02 N 08	00 N 31	01 S 45	09 S 57
04	01 53	06 43	00 55	00 57	02 08	00 31	01 45	09 57
07	01 06	07 11	00 52	00 57	02 07	00 31	01 45	09 57
10	00 S 21	07 33	00 50	00 57	02 07	00 31	01 45	09 58
13	00 N 18	07 47	00 47	00 57	02 07	00 31	01 45	09 58
16	00 51	07 53	00 44	00 57	02 06	00 31	01 46	09 58
19	01 17	07 53	00 41	00 57	02 06	00 31	01 46	09 58
22	01 34	07 45	00 39	00 57	02 06	00 31	01 46	09 58
25	01 44	07 33	00 36	00 57	02 06	00 32	01 46	09 58
28	01 46	07 16	00 32	00 57	02 06	00 32	01 46	09 58
31	01 N 42	06 S 55	00 S 29	00 N 58	02 N 06	00 N 32	01 S 46	09 S 58

DATA

Julian Date	2466002
Delta T	+77 seconds
Ayanamsa	24° 24' 21"
Synetic vernal point	04° ♓ 42' 38"
True obliquity of ecliptic	23° 26' 04"

MOON'S PHASES, APSIDES AND POSITIONS ☽

Date	h	m	Phase	Longitude	Eclipse Indicator
04	09	57	○	11 ♒ 55	
11	19	36	☾	19 ♉ 00	
19	20	50	●	26 ♌ 45	
26	23	16	☽	03 ♐ 35	

Day	h	m		
01	22	38	Perigee	
13	19	40	Apogee	
29	03	13	Perigee	
02	07	11	Max dec	24° S 44'
08	17	51	0N	
16	04	36	Max dec	24° N 47'
23	04	19	0S	
29	14	07	Max dec	24° S 53'

LONGITUDES

Date	Chiron ⚷	Ceres ⚳	Pallas ⚴	Juno ⚵	Vesta ⚶	Black Moon Lilith ⚸
01	08 ♋ 39	01 ♉ 43	29 ♒ 19	23 ♑ 05	23 ♍ 35	13 ♊ 56
11	09 ♋ 30	03 ♉ 02	26 ♒ 53	21 ♑ 02	28 ♍ 16	15 ♊ 02
21	10 ♋ 18	03 ♉ 50	24 ♒ 18	19 ♑ 27	03 ♎ 05	16 ♊ 09
31	11 ♋ 00	04 ♉ 03	21 ♒ 50	18 ♑ 28	08 ♎ 01	17 ♊ 15

ASPECTARIAN

h m Aspects

01 Monday
02 38 ☽ ± ♅
02 46 ☽ ⚼ ☉
04 38 ☽ ⚹ ♇
05 13 ☿ ⊻ ♀
06 36 ☽ □ ♄
12 32 ☽ ⊼ ♅
14 17 ☽ ⊼ ♂
18 35 ☽ ± ☉
18 57 ☽ △ ♆
20 59 ☽ △ ♃
21 32 ☽ ⚼ ♀
22 58 ♀ ⊥ ♄

02 Tuesday
00 35 ☽ ± ♂
05 06 ☽ ⊼ ☉
05 11 ☽ ∠ ♇
11 45 ☽ ± ♀
16 01 ☽ ⚼ ♂
19 55 ☽ ⊥ ♇
21 13 ☽ ⊼ ♀
21 54 ☽ ⚼ ♃

03 Wednesday
00 29 ☽ ☍ ☿
05 44 ☽ ⊻ ♇
08 07 ☽ △ ♄
12 54 ☽ ± ♃
13 56 ☽ ☍ ♅
17 51 ☽ △ ♂
19 35 ☉ ∠ ♄
20 11 ☽ □ ♆
21 23 ☽ ∥ ♇
22 56 ☽ ⊼ ♃

04 Thursday
09 05 ☽ ⚼ ♄
09 57 ☽ ☍ ☉
12 28 ☽ # ♅
18 05 ☽ # ♂
19 02 ☽ # ☿
20 58 ☽ ☍ ♀

05 Friday
00 15 ☽ ± ♄
02 13 ☽ Q ♆
05 17 ☽ ⊼ ☿
07 35 ☽ ☌ ♇
10 28 ☽ ⊼ ♄
11 09 ☽ # ☉
16 13 ☽ ± ☿
16 21 ☽ ⊼ ♅
21 17 ♂ ⊻ ♆
22 39 ☽ ⚹ ♆
22 43 ☽ □ ♂

06 Saturday
02 16 ☽ ☍ ♃
02 43 ☽ ± ♅
08 57 ☽ ⚼ ☿
13 27 ☿ ⊼ ♇
15 56 ♆ St R
16 12 ☽ # ♆
16 52 ☽ ⊼ ☉
18 29 ☽ ⚼ ♅
20 04 ☽ # ♃
22 27 ☽ ⊼ ♀

07 Sunday
00 51 ☽ ∠ ♆
02 15 ☽ # ♀
04 18 ☽ ± ☉
07 54 ☽ Q ♂
08 40 ☽ ± ♀
11 49 ☽ ⊻ ♇
13 53 ☽ △ ☿
15 23 ☽ ☍ ♄
16 13 ♀ Q ♂
17 00 ☽ ⊥ ♆
19 10 ♃ ⊥ ♅
21 29 ☽ △ ♅
21 47 ☽ ⚼ ☉
22 40 ☽ ⊥ ♇

08 Monday
00 14 ☽ ⚼ ♀
03 30 ☽ # ♄
03 57 ☽ ⊻ ♆
06 52 ☽ ⚹ ♂
07 59 ☿ ⚹ ♄
08 42 ☽ ⊼ ♃
15 15 ☽ ∠ ♇
19 02 ☉ ☌ ♀
20 04 ☽ ± ♃

09 Tuesday
02 51 ☽ △ ♀
03 54 ☽ △ ☉
08 13 ☽ ∥ ♄
12 35 ☽ ∠ ♂
13 25 ☽ ⚼ ♃
19 40 ☽ ⚹ ♇

10 Wednesday
00 03 ☽ ⊼ ♄
04 25 ☽ □ ☿
06 23 ☽ □ ♅
06 57 ☽ ⊥ ♂
11 44 ♂ □ ♃
11 51 ☽ ± ♄
12 25 ♂ ∥ ♅
12 25 ☽ ∥ ♀
13 01 ☽ ☌ ♆
16 39 ☽ ∥ ♃
19 03 ☽ Q ♇
19 06 ☽ △ ♃
19 23 ☽ ⊻ ♂
22 05 ☿ ☌ ♅
22 28 ☽ ∥ ♆

11 Thursday
05 48 ☽ ⚼ ♄
07 06 ☿ ⊥ ♃
10 30 ☽ □ ♀
18 13 ☽ Q ♅
19 36 ☽ □ ☉
20 45 ☽ ∥ ☉
20 58 ☽ Q ☿

12 Friday
06 58 ☽ □ ♇
12 13 ☽ △ ♄
18 34 ☽ ⚹ ♅
21 03 ☽ Q ♀

13 Saturday
00 57 ☽ ⚹ ☿
01 08 ☽ ⊻ ♆
02 07 ☽ ∥ ☿
02 22 ☿ □ ♆
08 33 ☽ □ ♃
10 05 ☽ ∥ ♅
10 33 ☉ Q ♂
11 15 ☽ ☌ ♂
11 16 ☽ Q ☉
13 20 ☽ ⊥ ♆
13 28 ☽ ∥ ♂
20 19 ☽ ⚹ ♀

14 Sunday
01 14 ☽ ∠ ♅
06 18 ☽ # ♇
07 39 ☽ ∠ ♆
08 22 ☉ ⊥ ♄
12 27 ☽ ∠ ☿
13 50 ☽ ⚹ ☉
19 36 ☽ ⊥ ♅
19 38 ♀ ∠ ♄
19 48 ☽ △ ♇
21 48 ☽ Q ♃

15 Monday
01 18 ♂ ⊥ ♆
01 23 ☽ ∠ ♀
01 42 ☽ □ ♄
07 45 ♀ ∥ ♃
07 47 ☽ ⊻ ♅
09 47 ☽ ⊥ ☿
12 26 ☿ ⊻ ♃
13 59 ☽ ⚹ ♆
18 43 ☽ ⊥ ♀
20 26 ♄ ± ♆
22 24 ☽ ⚹ ♃
22 45 ☽ ∠ ☉
23 56 ☽ ⊻ ☿

16 Tuesday
01 56 ☽ ⚼ ♇
03 24 ☽ ⊻ ♂
06 13 ☽ ⊻ ♀
13 55 ☽ Q ♆
14 05 ☽ Q ♄
15 59 ☽ ⊥ ♂
18 11 ☽ ⊥ ☉
19 46 ☽ ± ♇

17 Wednesday
04 39 ☽ ∠ ♃
06 59 ☽ ⊻ ☉
07 31 ☽ ⊼ ♇
09 03 ☿ ⚹ ♂
09 40 ☿ ☌ ♀
10 19 ♀ ⚹ ♂
10 38 ☽ ∠ ♂
13 42 ☉ ☍ ♇
13 52 ☽ ⚹ ♄
19 27 ☽ ☌ ♅
22 28 ☽ ⊥ ♃

18 Thursday
01 05 ☽ □ ♆
01 22 ☽ # ♇
07 49 ☿ ∠ ♄
10 11 ☽ ⊻ ♃
13 06 ☽ ∥ ♂
14 28 ☽ ☌ ♀
17 03 ☽ ⚹ ♂
18 56 ☽ ∠ ♄
20 47 ☽ ☌ ☿
21 17 ☽ ∥ ♅

19 Friday
12 01 ☽ ⊥ ♄
16 39 ☽ ☍ ♇
16 47 ☽ Q ♂
18 12 ☽ ∥ ☿
20 50 ☽ ☌ ☉
23 16 ☽ ⊻ ♄

20 Saturday
04 21 ☽ ⊻ ♅
09 26 ☽ △ ♆
15 24 ☽ ⊥ ♅
19 04 ☽ ☌ ♃
20 32 ☽ ⊻ ♀
21 59 ☽ ∥ ☉

21 Sunday
03 28 ☽ □ ♂
03 50 ☽ ∥ ♆
07 07 ☽ ⊥ ♀
07 12 ☽ ∥ ♀
07 48 ☽ ∠ ♅
09 34 ☉ ⊻ ♄
12 39 ☽ ⚼ ♆
13 01 ☽ ∥ ♃
13 47 ☽ ⊻ ☿
22 52 ☽ ∠ ♀
23 11 ☽ ⊼ ♇

22 Monday
02 23 ☽ ⊥ ☿
04 44 ☽ ± ♆
06 05 ☽ ☌ ♄
06 13 ♀ ⊻ ♃
07 29 ☽ ⊻ ☉
09 50 ☽ ± ♇
10 44 ☽ ⚹ ♅
15 23 ☽ ⊼ ♆
17 28 ☽ ∥ ♄
18 55 ☽ ⊥ ☉
21 05 ☽ ∠ ☿

23 Tuesday
00 50 ☽ ⚹ ♀
01 34 ☽ ⊻ ♃
01 42 ☽ ⚼ ♇
04 43 ☿ ⊥ ♄
07 58 ☽ Q ♅
11 23 ☽ △ ♂
11 56 ☽ ∠ ☉
12 16 ☽ ⊥ ♃
14 55 ☽ # ♄
21 26 ☽ Q ♀

24 Wednesday
03 49 ☽ ⚹ ☿
03 53 ☽ △ ♇
04 12 ☽ ∠ ♃
04 18 ☿ ☍ ♇
07 45 ☉ ⊻ ♅
11 08 ☽ ⊻ ♄
12 53 ☉ ∥ ♆
14 43 ☽ ⚼ ♂
15 27 ☽ □ ♅
15 59 ☽ ⚹ ☉
17 38 ☽ # ♃
19 44 ☽ ☍ ♆
21 37 ☽ ⊥ ♄

25 Thursday
02 34 ☽ # ♀
02 34 ☽ # ☉
02 36 ☉ ∥ ♀
03 26 ☽ # ♆
04 06 ☽ Q ☿
04 07 ☽ □ ♀
06 34 ☽ ⚹ ♃
06 59 ☽ ± ♂
08 31 ♂ ∠ ♅
13 15 ☽ ∠ ♄
14 13 ☽ Q ☉
17 11 ☽ # ☿
17 46 ☽ ⊼ ♂

26 Friday
03 29 ☽ Q ♃
07 30 ☽ □ ♇
07 52 ☿ Q ♂
09 02 ☿ ⊻ ♄
13 37 ♀ ∥ ♆
15 08 ☽ ⚹ ♄
16 05 ☽ □ ☿
19 09 ☽ △ ♅
21 24 ☉ △ ♆
23 09 ☽ ⊼ ♆
23 16 ☽ □ ☉

27 Saturday
04 49 ☽ # ♅
06 50 ☽ △ ♀
09 20 ☽ ± ♆
10 39 ☽ □ ♃
11 44 ☽ Q ♄
14 06 ☽ Q ♇
14 15 ☿ ⊻ ♅
19 10 ☽ # ♂
20 44 ☽ ⚼ ♅
23 13 ☽ ☍ ♂
23 36 ☽ ∥ ♇

28 Sunday
00 35 ☽ ⚼ ♆
08 04 ☽ ⚼ ♀
10 23 ☽ ⚹ ♇
12 01 ☽ ± ♅
17 51 ☿ △ ♆
18 27 ☽ □ ♄
22 11 ☽ ⊼ ♅
23 12 ☽ ± ♀

29 Monday
01 54 ☽ △ ♆
03 13 ☽ △ ☿
05 46 ☽ △ ☉
06 34 ♂ ∠ ♆
09 17 ☽ ⊼ ♀
10 15 ☿ ∥ ♀
11 39 ☽ ∠ ♇
14 08 ☽ △ ♃
22 07 ☿ ∥ ♆

30 Tuesday
02 48 ☽ ⊥ ♇
04 06 ☽ ⊼ ♂
08 31 ☽ ⚼ ☿
08 53 ☽ ⚼ ☉
10 14 ♀ St D
12 54 ☽ ⊻ ♇
13 26 ☉ ☌ ☿
14 39 ☽ ± ♂
15 48 ☽ ⚼ ♃
17 26 ☿ ⊥ ♅
21 26 ☽ △ ♄
21 43 ☉ ⊥ ♅
23 11 ☿ ⊻ ♀

31 Wednesday
00 57 ☽ ☍ ♅
01 10 ☽ ± ☉
02 06 ☽ ± ☿
04 26 ☽ ∥ ♇
04 26 ☽ □ ♆
06 31 ☽ ⚼ ♂
06 58 ☽ # ♂
07 14 ☽ ± ♃
09 22 ☉ ⊻ ♀
11 53 ☽ ☍ ♀
12 04 ☽ ⊼ ☉
13 52 ☽ ⊼ ☿
17 33 ☽ ⊼ ♃
23 03 ☽ ⚼ ♄
23 51 ☽ # ♅

All ephemeris data is given at 12.00 UT and the Moon's longitude is additionally given for 24.00 UT

SEPTEMBER 2039

Date	Sidereal time h m s	Sun ☉ ° ' "	Moon ☽ ° ' "	Moon ☽ 24.00 ° ' "	Mercury ☿ ° '	Venus ♀ ° '	Mars ♂ ° '	Jupiter ♃ ° '	Saturn ♄ ° '	Uranus ♅ ° '	Neptune ♆ ° '	Pluto ♇ ° '
01	10 42 05	08 ♍ 55 50	22 ♒ 02 06	29 ♒ 01 25	10 ♍ 48	07 ♌ 55	20 ♊ 26	11 ♍ 22	29 ♍ 30	01 ♌ 28	03 ♉ 27	24 ♒ 13
02	10 46 01	09 53 52	05 ♓ 57 24	12 ♓ 49 33	12 43	08 01	21 01	11 35	29 37	01 31	03 R 26	24 R 11
03	10 49 58	10 51 56	19 ♓ 37 23	26 ♓ 20 31	14 36	08 09	21 37	11 48	29 45	01 34	03 25	24 10
04	10 53 54	11 50 02	02 ♈ 58 40	09 ♈ 31 40	16 28	08 20	22 12	12 01	29 52	01 37	03 24	24 09
05	10 57 51	12 48 09	15 ♈ 59 28	22 ♈ 22 07	18 19	08 32	22 47	12 14	29 ♍ 59	01 40	03 23	24 07
06	11 01 48	13 46 18	28 ♈ 39 47	04 ♉ 52 46	20 09	08 46	23 22	12 27	00 ♎ 06	01 43	03 22	24 06
07	11 05 44	14 44 29	11 ♉ 01 24	17 ♉ 06 09	21 57	09 03	23 57	12 40	00 14	01 46	03 21	24 05
08	11 09 41	15 42 42	23 ♉ 07 32	29 ♉ 06 08	23 44	09 21	24 32	12 53	00 21	01 49	03 20	24 04
09	11 13 37	16 40 57	05 ♊ 02 35	10 ♊ 57 31	25 31	09 42	25 06	13 06	00 28	01 52	03 19	24 03
10	11 17 34	17 39 14	16 ♊ 51 38	22 ♊ 45 36	27 16	10 04	25 40	13 20	00 35	01 55	03 18	24 01
11	11 21 30	18 37 34	28 ♊ 40 09	04 ♋ 35 57	29 ♍ 00	10 28	26 14	13 33	00 43	01 58	03 17	24 00
12	11 25 27	19 35 55	10 ♋ 33 39	16 ♋ 33 55	00 ♎ 42	10 53	26 48	13 46	00 50	02 01	03 16	23 59
13	11 29 23	20 34 18	22 ♋ 37 19	28 ♋ 44 25	02 24	11 20	27 22	13 59	00 57	02 03	03 15	23 58
14	11 33 20	21 32 44	04 ♌ 55 41	11 ♌ 11 30	04 04	11 49	27 55	14 11	01 05	02 06	03 14	23 57
15	11 37 17	22 31 11	17 ♌ 32 10	23 ♌ 57 55	05 44	12 19	28 28	14 24	01 12	02 09	03 13	23 56
16	11 41 13	23 29 41	00 ♍ 28 50	07 ♍ 04 53	07 22	12 51	29 01	14 37	01 20	02 11	03 12	23 55
17	11 45 10	24 28 12	13 ♍ 45 58	20 ♍ 31 48	09 00	13 24	29 ♊ 34	14 50	01 27	02 14	03 10	23 53
18	11 49 06	25 26 46	27 ♍ 22 03	04 ♎ 16 16	10 36	13 59	00 ♋ 06	15 03	01 34	02 16	03 09	23 52
19	11 53 03	26 25 21	11 ♎ 13 58	18 ♎ 14 34	12 11	14 35	00 38	15 16	01 42	02 19	03 08	23 51
20	11 56 59	27 23 58	25 ♎ 17 30	02 ♏ 22 11	13 45	15 12	01 10	15 29	01 49	02 21	03 07	23 50
21	12 00 56	28 22 37	09 ♏ 28 04	16 ♏ 34 39	15 18	15 50	01 42	15 42	01 57	02 24	03 05	23 49
22	12 04 52	29 ♍ 21 18	23 ♏ 41 28	00 ♐ 48 09	16 51	16 29	02 13	15 55	02 04	02 26	03 04	23 48
23	12 08 49	00 ♎ 20 00	07 ♐ 54 24	14 ♐ 59 56	18 22	17 10	02 45	16 08	02 11	02 29	03 03	23 47
24	12 12 46	01 18 44	22 ♐ 04 37	29 ♐ 08 16	19 52	17 52	03 15	16 20	02 19	02 31	03 01	23 46
25	12 16 42	02 17 30	06 ♑ 10 48	13 ♑ 12 08	21 21	18 34	03 46	16 33	02 26	02 33	03 00	23 45
26	12 20 39	03 16 18	20 ♑ 12 09	27 ♑ 10 47	22 49	19 18	04 17	16 46	02 34	02 35	02 58	23 44
27	12 24 35	04 15 07	04 ♒ 07 53	11 ♒ 03 19	24 17	20 03	04 47	16 59	02 41	02 38	02 57	23 43
28	12 28 32	05 13 58	17 ♒ 56 53	24 ♒ 48 21	25 43	20 48	05 17	17 11	02 49	02 40	02 56	23 42
29	12 32 28	06 12 50	01 ♓ 37 29	08 ♓ 23 59	27 08	21 35	05 46	17 24	02 56	02 42	02 55	23 41
30	12 36 25	07 ♎ 11 45	15 ♓ 07 34	21 ♓ 47 58	28 ♎ 32	22 ♌ 22	06 ♋ 15	17 ♍ 37	03 ♎ 04	02 ♌ 44	02 ♉ 53	23 ♒ 40

Date	Moon True ☊ ° '	Moon Mean ☊ ° '	Moon ☽ Latitude ° '
01	18 ♊ 36	17 ♊ 51	04 S 29
02	18 R 25	17 48	04 54
03	18 12	17 45	05 01
04	18 00	17 42	04 51
05	17 50	17 38	04 26
06	17 41	17 35	03 48
07	17 35	17 32	03 00
08	17 32	17 29	02 05
09	17 31	17 26	01 06
10	17 D 30	17 22	00 S 03
11	17 R 30	17 19	00 N 59
12	17 29	17 16	01 59
13	17 26	17 13	02 52
14	17 20	17 10	03 43
15	17 12	17 07	04 22
16	17 01	17 03	04 48
17	16 49	17 00	05 00
18	16 37	16 57	04 55
19	16 25	16 54	04 33
20	16 16	16 51	03 54
21	16 09	16 47	03 01
22	16 05	16 44	01 56
23	16 03	16 41	00 N 43
24	16 D 03	16 38	00 S 32
25	16 R 03	16 35	01 45
26	16 02	16 32	02 51
27	15 59	16 28	03 46
28	15 53	16 25	04 28
29	15 45	16 22	04 53
30	15 ♊ 34	16 ♊ 19	05 S 02

DECLINATIONS

Date	Sun ☉ ° '	Moon ☽ ° '	Mercury ☿ ° '	Venus ♀ ° '	Mars ♂ ° '	Jupiter ♃ ° '	Saturn ♄ ° '	Uranus ♅ ° '	Neptune ♆ ° '	Pluto ♇ ° '
01	08 N 13	18 S 24	09 N 03	11 N 43	22 N 37	08 N 11	02 N 07	20 N 21	11 N 00	22 S 50
02	07 51	13 52	08 17	11 48	22 41	08 06	02 04	20 20	10 59	22 50
03	07 29	08 43	07 30	11 54	22 44	08 01	02 01	20 19	10 59	22 51
04	07 07	03 S 16	06 43	11 59	22 48	07 56	01 58	20 19	10 59	22 51
05	06 45	02 N 12	05 56	12 04	22 51	07 51	01 55	20 18	10 58	22 51
06	06 23	07 27	05 09	12 09	22 54	07 47	01 53	20 17	10 58	22 52
07	06 00	12 16	04 22	12 13	22 57	07 42	01 50	20 17	10 58	22 52
08	05 38	16 32	03 34	12 16	22 59	07 37	01 47	20 16	10 57	22 53
09	05 15	20 04	02 47	12 19	23 02	07 32	01 44	20 15	10 57	22 53
10	04 53	22 44	02 00	12 22	23 04	07 27	01 41	20 15	10 56	22 53
11	04 30	24 25	01 13	12 24	23 07	07 22	01 38	20 14	10 56	22 54
12	04 07	24 59	00 N 27	12 26	23 09	07 17	01 35	20 14	10 56	22 54
13	03 44	24 24	00 S 20	12 27	23 11	07 12	01 32	20 13	10 55	22 54
14	03 21	22 38	01 06	12 28	23 13	07 07	01 29	20 13	10 55	22 55
15	02 58	19 44	01 52	12 29	23 14	07 02	01 26	20 12	10 54	22 55
16	02 35	15 48	02 37	12 28	23 16	06 57	01 23	20 11	10 54	22 55
17	02 12	11 00	03 22	12 28	23 17	06 52	01 20	20 11	10 53	22 56
18	01 49	05 N 34	04 06	12 26	23 19	06 47	01 17	20 10	10 53	22 56
19	01 25	00 S 15	04 50	12 25	23 20	06 42	01 15	20 10	10 52	22 56
20	01 02	06 08	05 33	12 22	23 21	06 37	01 12	20 09	10 52	22 57
21	00 39	11 47	06 16	12 20	23 22	06 32	01 09	20 09	10 52	22 57
22	00 N 15	16 50	06 59	12 16	23 23	06 27	01 06	20 08	10 51	22 57
23	00 S 08	20 55	07 41	12 13	23 24	06 23	01 03	20 08	10 51	22 57
24	00 31	23 44	08 22	12 08	23 25	06 18	01 00	20 07	10 50	22 58
25	00 55	25 02	09 02	12 04	23 25	06 13	00 57	20 07	10 50	22 58
26	01 18	24 44	09 42	11 58	23 26	06 08	00 54	20 06	10 49	22 58
27	01 41	22 53	10 21	11 52	23 26	06 03	00 51	20 06	10 49	22 58
28	02 05	19 41	11 00	11 46	23 27	05 58	00 48	20 05	10 48	22 58
29	02 28	15 27	11 38	11 39	23 27	05 53	00 45	20 05	10 47	22 59
30	02 S 51	10 S 30	12 S 15	11 N 31	23 N 27	05 N 48	00 N 42	20 N 04	10 N 47	22 S 59

ZODIAC SIGN ENTRIES

Date	h	m	Planets
02	01	41	☽ ♓
04	06	36	☽ ♈
05	15	15	♄ ♎
06	14	34	☽ ♉
09	01	49	☽ ♊
11	14	42	☽ ♋
12	02	05	☿ ♎
14	02	27	☽ ♌
16	11	07	☽ ♍
18	07	28	♂ ♋
18	16	35	☽ ♎
20	19	59	☽ ♏
22	22	39	☽ ♐
23	03	49	☉ ♎
25	01	28	☽ ♑
27	04	52	☽ ♒
29	09	08	☽ ♓

LATITUDES

Date	Mercury ☿ ° '	Venus ♀ ° '	Mars ♂ ° '	Jupiter ♃ ° '	Saturn ♄ ° '	Uranus ♅ ° '	Neptune ♆ ° '	Pluto ♇ ° '
01	01 N 40	06 S 48	00 S 28	00 N 58	02 N 05	00 N 32	01 S 46	09 S 58
04	01 30	06 24	00 25	00 58	02 05	00 32	01 47	09 58
07	01 16	05 59	00 21	00 58	02 05	00 32	01 47	09 58
10	01 00	05 33	00 18	00 58	02 05	00 32	01 47	09 58
13	00 41	05 07	00 14	00 58	02 05	00 32	01 47	09 58
16	00 N 20	04 40	00 10	00 58	02 05	00 32	01 47	09 58
19	00 S 01	04 13	00 06	00 59	02 05	00 32	01 47	09 58
22	00 24	03 47	00 S 02	00 59	02 05	00 32	01 47	09 57
25	00 46	03 22	00 N 03	00 59	02 05	00 32	01 47	09 57
28	01 09	02 57	00 07	00 59	02 06	00 32	01 48	09 57
31	01 S 31	02 S 32	00 N 12	01 N 00	02 N 06	00 N 33	01 S 48	09 S 56

DATA

Julian Date	2466033
Delta T	+77 seconds
Ayanamsa	24° 24' 25"
Synetic vernal point	04° ♓ 42' 34"
True obliquity of ecliptic	23° 26' 05"

MOON'S PHASES, APSIDES AND POSITIONS ☽

Date	h	m	Phase	Longitude ° '	Eclipse Indicator
02	19	23	○	10 ♓ 12	
10	13	46	☾	17 ♊ 44	
18	08	23	●	25 ♍ 18	
25	04	52	☽	02 ♑ 00	

Day	h	m		
10	14	24	Apogee	
23	02	11	Perigee	
05	02	16	ON	
12	12	07	Max dec	24° N 59'
19	10	59	0S	
25	19	26	Max dec	25° S 07'

LONGITUDES

Date	Chiron ⚷ ° '	Ceres ⚳ ° '	Pallas ⚴ ° '	Juno ⚵ ° '	Vesta ⚶ ° '	Black Moon Lilith ⚸ ° '
01	11 ♋ 04	04 ♉ 03	21 ♒ 36	18 ♑ 24	08 ♒ 30	17 ♊ 22
11	11 ♋ 40	03 ♉ 37	19 ♒ 28	18 ♑ 08	13 ♒ 31	18 ♊ 28
21	12 ♋ 09	02 ♉ 33	17 ♒ 48	18 ♑ 31	18 ♒ 37	19 ♊ 35
31	12 ♋ 31	00 ♉ 57	16 ♒ 42	19 ♑ 32	23 ♒ 47	20 ♊ 41

All ephemeris data is given at 12.00 UT and the Moon's longitude is additionally given for 24.00 UT

ASPECTARIAN

h m Aspects

01 Thursday
09 08 ☽ △ ♂
11 00 ☽ Q ♆
14 32 ☽ ± ♄
14 55 ☉ ∥ ♃
15 43 ☽ ☌ ♇
20 03 ☿ ☌ ♃
02 Friday
00 57 ☽ ⊼ ♄
04 17 ☽ ⊼ ♅
07 37 ☽ ✶ ♆
14 44 ☽ ± ♅
15 38 ☽ ⊼ ♀
18 13 ☿ ∥ ♃
19 23 ☽ ☍ ☉
21 41 ☽ # ♀
21 59 ☽ ☍ ♃
03 Saturday
01 41 ☽ # ♆
01 42 ☽ ☍ ☿
02 14 ☽ ± ♀
05 53 ☿ ⊥ ♀
06 35 ☽ ⚼ ♅
09 52 ☽ ∠ ♆
12 51 ☉ ∥ ☿
15 08 ☽ # ♃
15 42 ☽ □ ♂
17 50 ☽ # ☉
18 17 ☽ # ☿
18 22 ☽ ⚼ ♀
20 05 ☽ ⊻ ♇
04 Sunday
01 55 ☽ ⊥ ♆
06 18 ☽ ☍ ♄
06 52 ☽ ⊥ ♇
09 32 ☽ △ ♅
12 46 ☽ ⊻ ♆
14 06 ☿ ∠ ♅
17 41 ☽ # ♄
18 03 ☉ ☌ ♃
21 56 ☽ △ ♀
23 17 ☽ ∠ ♇
05 Monday
01 52 ☽ Q ♂
04 54 ☽ ⊼ ♃
05 35 ☽ ⊼ ☉
10 47 ☽ ∥ ♄
12 57 ☿ ⚼ ♆
16 17 ☽ ± ♃
17 05 ☽ ⊼ ☿
17 13 ♂ # ♇
17 42 ☽ ± ☉
06 Tuesday
01 25 ☽ ✶ ♂
02 43 ☽ ∥ ☿
03 18 ☽ ✶ ♇
06 22 ☽ ± ☿
07 21 ☽ ∥ ☉
09 39 ☽ ⚼ ♃
12 01 ♄ ± ♇
12 14 ☽ ⚼ ☉
13 33 ☽ ∥ ♃
14 48 ☽ ⊼ ♄
17 55 ☽ □ ♅
21 04 ☽ ☌ ♆
07 Wednesday
02 21 ☽ Q ♇
02 31 ☽ ± ♄
02 40 ☽ ⚼ ☿
05 13 ☽ ∥ ♆
07 44 ☽ ∠ ♂
08 02 ☽ □ ♀
11 40 ☽ ∥ ♀
15 18 ☽ △ ♃
17 20 ♂ △ ♇
19 58 ☽ △ ☉
20 20 ♀ ∠ ♂
20 22 ☽ ⚼ ♄
22 58 ☉ ⊥ ♀
08 Thursday
02 22 ☽ ⊥ ♂
05 23 ☽ Q ♅
13 27 ☽ △ ☿
13 52 ☽ □ ♇
14 57 ☽ ⊻ ♂
16 18 ☿ ⊼ ♇
20 43 ☽ Q ♀
22 11 ☿ ∠ ♀
09 Friday
02 39 ☽ △ ♄
03 44 ☿ □ ♂
05 33 ☽ ✶ ♅
08 31 ☽ ⊻ ♆
13 32 ☽ ∥ ♅
16 49 ☉ ∠ ♅
20 40 ☽ ⊥ ♆
21 44 ☽ ✶ ♀
10 Saturday
04 41 ☽ □ ♃
12 07 ☽ ∠ ♅
12 36 ☿ ± ♆
13 46 ☽ □ ☉
13 46 ☽ # ♇
14 56 ☽ ∠ ♆
15 58 ☽ ∥ ♂
22 26 ☿ ∥ ♄
23 17 ♂ ⊥ ♅
11 Sunday
02 33 ☽ △ ♇
03 48 ☉ ⚼ ♆
05 15 ☽ ∠ ♀
06 29 ☽ ⊥ ♅
06 49 ☽ ☌ ♂
12 46 ☽ □ ☿
16 11 ☽ □ ♄
17 56 ☽ Q ♃
18 42 ☽ ⊻ ♅
21 20 ☽ ✶ ♆
12 Monday
00 09 ☽ ⊥ ♀
01 59 ☿ ± ♇
05 31 ☽ Q ☉
08 50 ☽ ⚼ ♇
12 40 ☽ ⊻ ♀
13 57 ☿ ☌ ♄
18 31 ☽ ✶ ♃
21 24 ☽ Q ♆
13 Tuesday
02 48 ☽ ± ♇
04 41 ☽ Q ♄
06 54 ☽ Q ☿
06 57 ☿ ✶ ♅
07 36 ☽ ✶ ☉
14 38 ☽ ⊼ ♇
21 45 ☽ ⊻ ♂
14 Wednesday
00 02 ☿ ⊼ ♆
00 41 ☽ ∠ ♃
04 29 ☽ ✶ ♄
05 50 ☽ ∥ ♂
06 31 ☽ ☌ ♅
08 44 ☽ □ ♆
09 07 ☽ # ♇
09 57 ☽ ⊥ ♂
10 06 ☽ ✶ ☿
15 23 ☽ ∠ ☉
18 23 ☽ ⊥ ♃
23 30 ☿ # ♄
15 Thursday
01 44 ☽ ☌ ♀
03 58 ☽ ∠ ♂
06 00 ☽ ⊻ ♃
08 38 ☽ ∥ ♅
09 28 ☽ ∠ ♄
09 56 ☽ ⊥ ☉
18 52 ☽ ∠ ☿
22 05 ☽ ⊻ ☉
23 55 ☽ ☍ ♇
16 Friday
02 26 ☽ ⊥ ♄
09 12 ☽ ✶ ♂
11 21 ☉ # ☿
13 34 ☽ ⊻ ♄
13 51 ☽ ⊥ ☿
15 08 ☽ ⊻ ♅
16 56 ☽ △ ♆
22 00 ☉ ⊼ ♇
17 Saturday
02 02 ☽ ⊥ ♅
02 16 ☽ ⊻ ☿
05 02 ☽ ∥ ♀
07 53 ☽ Q ♂
10 30 ☿ ⚼ ♇
11 19 ☽ ⊻ ♀
12 30 ☽ ∥ ♆
13 57 ☽ ☌ ♃
18 11 ☽ ∠ ♅
19 49 ☽ ⚼ ♆
22 27 ☽ ⊥ ♀
18 Sunday
05 53 ☽ ⊼ ♇
06 43 ☽ ∥ ♃
08 23 ☽ ☌ ☉
11 37 ☽ ± ♆
14 56 ☽ ∠ ♀
16 22 ☽ ± ♇
16 58 ☽ □ ♂
17 27 ☽ # ☿
19 23 ☽ ☌ ♄
20 34 ☽ ✶ ♅
22 03 ☽ ⊼ ♆
19 Monday
04 43 ☽ ∥ ☉
05 53 ☽ ∥ ♄
07 55 ☽ ⚼ ♇
13 50 ☽ ☌ ☿
16 00 ☽ # ♄
16 28 ☽ # ☉
17 18 ☽ Q ♅
18 00 ☽ ✶ ♀
19 02 ☽ ⊻ ♃
20 Tuesday
00 46 ☉ ∥ ♄
05 02 ☉ ± ♆
05 26 ☽ ⊥ ♃
09 16 ☽ ∥ ☿
09 32 ☽ △ ♇
13 57 ☽ # ♃
15 23 ☽ Q ♀
15 51 ☽ ⊻ ☉
20 57 ☽ ∠ ♃
21 32 ☿ Q ♅
22 21 ☽ △ ♂
23 10 ☽ ⊻ ♄
21 Wednesday
00 01 ☽ □ ♅
01 14 ☽ ☍ ♆
02 46 ☽ ⊥ ☉
04 37 ♀ ⊻ ♃
07 56 ☽ # ♆
09 24 ☽ ⊥ ♄
14 25 ☽ # ♀
19 05 ☽ ∠ ☉
19 06 ☿ ⊻ ♃
20 06 ☿ # ♃
22 41 ☽ ✶ ♃
23 03 ☽ ⊻ ☿
23 15 ☽ □ ♀
22 Thursday
00 40 ☽ ⚼ ♂
00 44 ☽ ∠ ♄
02 13 ☿ ✶ ♀
02 37 ♂ □ ♄
10 24 ☽ ⊥ ☿
12 11 ☽ □ ♇
16 26 ☽ ± ♂
19 14 ☽ Q ♃
22 16 ☽ ✶ ☉
22 38 ♂ ⊻ ♅
22 46 ☉ ± ♇
23 Friday
02 16 ☽ ✶ ♄
02 48 ☽ △ ♅
02 57 ☽ ⊼ ♂
03 24 ☽ ∠ ☿
03 48 ☽ ⊼ ♆
06 49 ☽ # ♅
13 05 ♀ ∠ ♄
13 55 ☽ ± ♆
18 33 ☽ Q ♇
20 03 ☽ Q ☉
22 44 ☽ Q ♄
24 Saturday
01 23 ♂ ✶ ♆
02 07 ☽ □ ♃
04 05 ☽ ∥ ♇
04 14 ☽ ⚼ ♅
04 29 ☽ △ ♀
05 08 ☽ ⚼ ♆
07 48 ☽ ✶ ☿
08 32 ☽ # ♂
14 52 ☽ ✶ ♇
19 33 ☽ ± ♅
25 Sunday
04 52 ☽ □ ☉
05 34 ☽ □ ♄
05 48 ☽ ⊼ ♅
06 35 ☽ △ ♆
06 37 ☽ Q ☿
07 19 ☽ ⚼ ♀
07 44 ☽ ☍ ♂
14 03 ☉ # ♄
16 08 ☉ ☌ ♄
16 23 ☽ ∠ ♇
18 38 ☉ ✶ ♅
23 31 ☽ ± ♀
26 Monday
04 53 ☉ ⊼ ♆
06 01 ☽ △ ♃
07 46 ☽ ⊥ ♇
09 27 ♀ ∠ ♂
10 22 ☽ ⊼ ♀
10 55 ☿ ⊥ ♃
17 02 ☽ □ ☿
18 04 ☽ ⊻ ♇
19 09 ♄ ✶ ♅
27 Tuesday
02 53 ☿ △ ♇
06 24 ☽ # ♂
08 13 ☽ ⚼ ♃
09 23 ☽ ☍ ♅
09 29 ☽ △ ♄
09 58 ☽ □ ♆
11 08 ☽ ∥ ♇
12 13 ☽ △ ☉
13 10 ☽ ⊼ ♂
23 57 ☽ ± ♂
28 Wednesday
00 03 ☽ ± ♃
04 34 ☿ # ♆
04 49 ♂ Q ♃
09 25 ☽ # ♅
10 39 ☽ ⊼ ♃
11 46 ☽ ⚼ ♄
14 06 ☉ □ ♂
16 13 ☽ ⚼ ♂
16 18 ☽ ⚼ ☉
17 12 ☽ Q ♆
17 17 ☽ ☍ ♀
22 03 ☽ ☌ ♇
29 Thursday
03 10 ☽ △ ☿
03 40 ☽ ± ♄
06 47 ♄ ⊼ ♆
09 19 ☽ ± ☉
12 34 ☿ # ♀
13 54 ☽ ⊼ ♅
14 15 ☽ ✶ ♆
14 20 ☽ ⊼ ♄
19 36 ☽ △ ♂
20 45 ☽ ⊼ ☉
30 Friday
00 34 ☽ ± ♅
04 43 ☽ ∥ ☿
07 08 ☽ # ♀
08 49 ☽ ⚼ ☿
10 42 ☽ # ♆
16 32 ☽ ☍ ♃
16 41 ☽ ⚼ ♅
16 56 ☽ ∠ ♆

OCTOBER 2039

Date	Sidereal time h m s	Sun ☉ ° ' "	Moon ☽ ° ' "	Moon ☽ 24.00 ° ' "	Mercury ☿ ° '	Venus ♀ ° '	Mars ♂ ° '	Jupiter ♃ ° '	Saturn ♄ ° '	Uranus ♅ ° '	Neptune ♆ ° '	Pluto ♇ ° '
01	12 40 21	08 ♎ 10 41	28 ♓ 24 54	04 ♈ 58 09	29 ♎ 55	23 ♌ 11	06 ♋ 44	17 ♍ 49	03 ♎ 11	02 ♌ 46	02 ♉ 51	23 ♒ 40
02	12 44 18	09 09 39	11 ♈ 27 32	17 ♈ 52 54	01 ♏ 17	24 00	07 13	18 02	03 18	02 48	02 R 50	23 R 39
03	12 48 15	10 08 39	24 ♈ 14 12	00 ♉ 31 28	02 37	24 50	07 41	18 14	03 26	02 50	02 48	23 38
04	12 52 11	11 07 42	06 ♉ 44 46	12 ♉ 54 16	03 57	25 40	08 09	18 27	03 33	02 52	02 47	23 37
05	12 56 08	12 06 46	19 ♉ 00 15	25 ♉ 03 00	05 15	26 32	08 37	18 39	03 41	02 53	02 45	23 36
06	13 00 04	13 05 53	01 ♊ 02 56	07 ♊ 00 29	06 32	27 24	09 05	18 52	03 48	02 55	02 44	23 35
07	13 04 01	14 05 02	12 ♊ 56 12	18 ♊ 50 36	07 48	28 17	09 32	19 04	03 55	02 57	02 42	23 35
08	13 07 57	15 04 14	24 ♊ 44 18	00 ♋ 37 57	09 02	29 ♌ 10	09 59	19 17	04 03	02 59	02 40	23 34
09	13 11 54	16 03 27	06 ♋ 32 12	12 ♋ 27 42	10 14	00 ♍ 04	10 25	19 29	04 10	03 00	02 39	23 33
10	13 15 50	17 02 43	18 ♋ 25 10	24 ♋ 25 16	11 25	00 59	10 51	19 41	04 17	03 02	02 37	23 33
11	13 19 47	18 02 02	00 ♌ 28 39	06 ♌ 35 56	12 34	01 54	11 17	19 53	04 25	03 03	02 36	23 32
12	13 23 44	19 01 22	12 ♌ 47 45	19 ♌ 04 35	13 41	02 50	11 42	20 06	04 32	03 05	02 34	23 31
13	13 27 40	20 00 45	25 ♌ 26 54	01 ♍ 55 03	14 46	03 47	12 07	20 18	04 39	03 06	02 32	23 31
14	13 31 37	21 00 10	08 ♍ 29 18	15 ♍ 09 45	15 49	04 44	12 32	20 30	04 46	03 08	02 31	23 30
15	13 35 33	21 59 37	21 ♍ 56 23	28 ♍ 49 03	16 49	05 41	12 56	20 42	04 54	03 09	02 29	23 30
16	13 39 30	22 59 07	05 ♎ 47 25	12 ♎ 51 02	17 46	06 39	13 20	20 54	05 01	03 10	02 27	23 29
17	13 43 26	23 58 38	19 ♎ 59 18	27 ♎ 11 31	18 41	07 38	13 43	21 06	05 08	03 12	02 26	23 28
18	13 47 23	24 58 12	04 ♏ 26 53	11 ♏ 44 33	19 32	08 37	14 06	21 18	05 15	03 13	02 24	23 28
19	13 51 19	25 57 47	19 ♏ 03 38	26 ♏ 23 18	20 19	09 36	14 28	21 29	05 22	03 14	02 22	23 27
20	13 55 16	26 57 25	03 ♐ 42 44	11 ♐ 01 11	21 02	10 36	14 50	21 41	05 30	03 15	02 21	23 27
21	13 59 13	27 57 04	18 ♐ 18 01	25 ♐ 32 40	21 41	11 37	15 12	21 53	05 37	03 16	02 19	23 27
22	14 03 09	28 56 46	02 ♑ 44 43	09 ♑ 53 49	22 15	12 37	15 33	22 04	05 44	03 17	02 17	23 26
23	14 07 06	29 ♎ 56 29	16 ♑ 59 43	24 ♑ 02 15	22 44	13 39	15 54	22 16	05 51	03 18	02 16	23 26
24	14 11 02	00 ♏ 56 13	01 ♒ 01 21	07 ♒ 56 57	23 06	14 40	16 14	22 28	05 58	03 19	02 14	23 25
25	14 14 59	01 55 59	14 ♒ 49 04	21 ♒ 37 44	23 22	15 42	16 34	22 39	06 05	03 20	02 12	23 25
26	14 18 55	02 55 47	28 ♒ 22 58	05 ♓ 04 49	23 31	16 44	16 53	22 50	06 12	03 20	02 11	23 25
27	14 22 52	03 55 37	11 ♓ 43 20	18 ♓ 18 32	23 R 33	17 47	17 12	23 02	06 19	03 21	02 09	23 25
28	14 26 48	04 55 28	24 ♓ 50 28	01 ♈ 19 09	23 26	18 50	17 30	23 13	06 25	03 22	02 07	23 24
29	14 30 45	05 55 21	07 ♈ 44 37	14 ♈ 06 51	23 10	19 53	17 48	23 24	06 32	03 22	02 06	23 24
30	14 34 42	06 55 15	20 ♈ 25 55	26 ♈ 41 51	22 45	20 57	18 05	23 35	06 39	03 23	02 04	23 24
31	14 38 38	07 ♏ 55 12	02 ♉ 54 43	09 ♉ 04 35	22 ♏ 11	22 ♍ 01	18 ♋ 21	23 ♍ 46	06 ♎ 46	03 ♌ 23	02 ♉ 02	23 ♒ 24

Date	Moon True ☊ ° '	Moon Mean ☊ ° '	Moon ☽ Latitude ° '
01	15 ♊ 23	16 ♊ 16	04 S 55
02	15 R 12	16 13	04 32
03	15 01	16 09	03 55
04	14 53	16 06	03 08
05	14 47	16 03	02 13
06	14 44	16 00	01 13
07	14 43	15 57	00 S 10
08	14 D 44	15 53	00 N 54
09	14 44	15 50	01 54
10	14 R 45	15 47	02 51
11	14 44	15 44	03 40
12	14 41	15 41	04 21
13	14 36	15 38	04 50
14	14 29	15 34	05 05
15	14 20	15 31	05 04
16	14 12	15 28	04 45
17	14 03	15 25	04 09
18	13 56	15 22	03 16
19	13 52	15 19	02 10
20	13 50	15 15	00 N 54
21	13 D 49	15 12	00 S 24
22	13 50	15 09	01 41
23	13 51	15 06	02 50
24	13 R 52	15 03	03 48
25	13 51	14 59	04 32
26	13 48	14 56	04 59
27	13 44	14 53	05 10
28	13 38	14 50	05 04
29	13 31	14 47	04 43
30	13 25	14 44	04 08
31	13 ♊ 19	14 ♊ 40	03 S 22

DECLINATIONS

Date	Sun ☉ ° '	Moon ☽ ° '	Mercury ☿ ° '	Venus ♀ ° '	Mars ♂ ° '	Jupiter ♃ ° '	Saturn ♄ ° '	Uranus ♅ ° '	Neptune ♆ ° '	Pluto ♇ ° '
01	03 S 15	05 S 08	12 S 51	11 N 23	23 N 27	05 N 44	00 N 39	20 N 04	10 N 46	22 S 59
02	03 38	00 N 22	13 27	11 15	23 28	05 39	00 37	20 04	10 46	22 59
03	04 01	05 45	14 01	11 06	23 27	05 34	00 34	20 03	10 45	22 59
04	04 24	10 48	14 35	10 56	23 27	05 29	00 31	20 03	10 45	22 59
05	04 47	15 20	15 08	10 46	23 27	05 24	00 28	20 02	10 44	23 00
06	05 10	19 11	15 40	10 36	23 27	05 20	00 25	20 02	10 44	23 00
07	05 33	22 11	16 11	10 25	23 27	05 15	00 22	20 02	10 43	23 00
08	05 56	24 13	16 41	10 13	23 27	05 10	00 19	20 01	10 43	23 00
09	06 19	25 11	17 10	10 01	23 26	05 05	00 16	20 01	10 42	23 00
10	06 42	24 59	17 38	09 49	23 26	05 01	00 14	20 01	10 41	23 00
11	07 04	23 38	18 05	09 36	23 26	04 56	00 11	20 00	10 41	23 00
12	07 27	21 08	18 30	09 22	23 25	04 51	00 08	20 00	10 40	23 00
13	07 49	17 35	18 54	09 08	23 25	04 47	00 05	20 00	10 40	23 00
14	08 12	13 06	19 17	08 54	23 24	04 42	00 N 02	19 59	10 39	23 00
15	08 34	07 51	19 39	08 39	23 24	04 37	00 00	19 59	10 39	23 00
16	08 56	02 N 04	19 59	08 24	23 23	04 33	00 S 03	19 59	10 38	23 01
17	09 18	03 S 58	20 17	08 08	23 23	04 28	00 06	19 59	10 37	23 01
18	09 40	09 55	20 34	07 52	23 22	04 24	00 09	19 58	10 37	23 01
19	10 02	15 24	20 49	07 36	23 22	04 19	00 11	19 58	10 36	23 01
20	10 23	20 00	21 02	07 19	23 21	04 15	00 14	19 58	10 36	23 01
21	10 45	23 19	21 13	07 01	23 21	04 10	00 17	19 58	10 35	23 01
22	11 06	25 05	21 22	06 43	23 20	04 06	00 20	19 58	10 35	23 01
23	11 27	25 10	21 28	06 25	23 20	04 01	00 22	19 57	10 34	23 00
24	11 48	23 38	21 32	06 07	23 20	03 57	00 25	19 57	10 33	23 00
25	12 09	20 43	21 34	05 48	23 19	03 52	00 28	19 57	10 33	23 00
26	12 29	16 42	21 32	05 29	23 19	03 48	00 30	19 57	10 32	23 00
27	12 50	11 56	21 27	05 09	23 19	03 44	00 33	19 57	10 32	23 00
28	13 10	06 42	21 18	04 49	23 19	03 39	00 35	19 57	10 31	23 00
29	13 30	01 S 16	21 06	04 29	23 18	03 35	00 38	19 57	10 31	23 00
30	13 49	04 N 09	20 50	04 09	23 18	03 31	00 40	19 56	10 30	23 00
31	14 S 09	09 N 19	20 S 29	03 N 48	23 N 18	03 N 27	00 S 43	19 N 56	10 N 29	23 S 00

ZODIAC SIGN ENTRIES

Date	h	m	Planets
01	13	30	☿ ♏
01	14	53	☽ ♈
03	23	00	☽ ♉
06	09	54	☽ ♊
08	22	43	☽ ♋
09	10	10	♀ ♍
11	11	04	☽ ♌
13	20	28	☽ ♍
16	02	03	☽ ♎
18	04	39	☽ ♏
20	05	55	☽ ♐
22	07	25	☽ ♑
23	13	25	☉ ♏
24	10	14	☽ ♒
26	14	53	☽ ♓
28	21	33	☽ ♈
31	06	22	☽ ♉

LATITUDES

Date	Mercury ☿ ° '	Venus ♀ ° '	Mars ♂ ° '	Jupiter ♃ ° '	Saturn ♄ ° '	Uranus ♅ ° '	Neptune ♆ ° '	Pluto ♇ ° '
01	01 S 31	02 S 32	00 N 12	01 N 00	02 N 06	00 N 33	01 S 48	09 S 56
04	01 52	02 09	00 16	01 00	02 06	00 33	01 48	09 56
07	02 12	01 46	00 21	01 00	02 06	00 33	01 48	09 55
10	02 30	01 24	00 27	01 01	02 06	00 33	01 48	09 55
13	02 46	01 03	00 32	01 01	02 06	00 33	01 48	09 54
16	02 58	00 43	00 37	01 01	02 07	00 33	01 48	09 54
19	03 07	00 24	00 43	01 02	02 07	00 33	01 48	09 53
22	03 09	00 S 06	00 49	01 02	02 07	00 33	01 48	09 53
25	03 06	00 N 11	00 55	01 02	02 08	00 33	01 48	09 52
28	02 46	00 26	01 02	01 03	02 08	00 33	01 48	09 52
31	02 S 15	00 N 41	01 N 08	01 N 03	02 N 08	00 N 34	01 S 48	09 S 51

DATA

Julian Date	2466063
Delta T	+77 seconds
Ayanamsa	24° 24' 28"
Synetic vernal point	04° ♓ 42' 31"
True obliquity of ecliptic	23° 26' 06"

MOON'S PHASES, APSIDES AND POSITIONS ☽

Date	h	m	Phase	Longitude ° '	Eclipse Indicator
02	07	23	○	08 ♈ 58	
10	08	59	☾	16 ♋ 55	
17	19	09	●	24 ♎ 16	
24	11	50	☽	00 ♒ 56	
31	22	36	○	08 ♉ 22	

Day	h	m		
08	10	01	Apogee	
20	05	04	Perigee	
02	10	25	0N	
09	20	05	Max dec	25° N 15'
16	20	17	0S	
23	01	06	Max dec	25° S 20'
29	17	34	0N	

LONGITUDES

Date	Chiron ⚷ ° '	Ceres ⚳ ° '	Pallas ⚴ ° '	Juno ⚵ ° '	Vesta ⚶ ° '	Black Moon Lilith ⚸ ° '
01	12 ♋ 31	00 ♉ 57	16 ♒ 42	19 ♑ 32	23 ♎ 47	20 ♊ 41
11	12 ♋ 45	28 ♈ 55	16 ♒ 11	21 ♑ 06	29 ♎ 00	21 ♊ 48
21	12 ♋ 50	26 ♈ 39	16 ♒ 15	23 ♑ 10	04 ♏ 16	22 ♊ 54
31	12 ♋ 47	24 ♈ 25	16 ♒ 50	25 ♑ 40	09 ♏ 35	24 ♊ 00

ASPECTARIAN

h m	Aspects
01 Saturday	
01 52	☽ ⊼ ♀
02 53	☽ ± ☿
03 22	☽ ⊻ ♇
04 05	♃ ∠ ♅
09 10	☽ ⊥ ♆
09 22	☽ # ♃
11 47	☉ ∠ ♀
13 29	☽ ± ♀
14 16	☽ ⊥ ♇
15 03	☽ ⊼ ☿
15 12	♃ ⊡ ♆
19 44	☽ ∥ ☉
19 58	☽ △ ♅
20 06	☽ ⊻ ♆
20 48	☽ ☍ ♄
23 35	☉ ⊡ ♇
02 Sunday	
01 58	♀ ☍ ♇
03 51	☽ □ ♂
06 47	☽ ∠ ♇
07 07	☽ ⊡ ♀
07 23	☽ ☍ ☉
07 43	☽ # ♄
13 04	☽ ∥ ♄
03 Monday	
00 29	☽ ⊼ ♃
01 22	♅ □ ♆
03 34	☽ # ☉
10 51	☽ ⚹ ♇
11 11	☽ ∥ ♃
12 00	☽ ± ♃
13 12	☽ △ ♀
14 52	☽ Q ♂
15 10	☿ ☍ ♆
15 47	☿ □ ♅
04 Tuesday	
01 13	☿ ∠ ♃
04 03	☿ ⊻ ♄
04 21	☽ ☌ ♆
04 28	☽ □ ♅
05 31	☽ ⊡ ♃
05 46	☽ ⊼ ♄
05 57	☽ ☍ ☿
09 49	☽ Q ♇
11 44	☽ ∥ ♆
12 41	☽ ∥ ♀
14 51	☽ ⚹ ♂
17 31	☽ ± ♄
21 16	☽ ⊼ ☉
05 Wednesday	
10 05	☽ ± ☉
10 42	☽ # ☿
11 11	♂ ⊡ ♇
11 18	☽ △ ♃
11 21	☽ ⊡ ♄
15 45	☽ Q ♅
17 20	♀ ∥ ♆
21 07	☽ □ ♇
21 31	☽ ∠ ♂
06 Thursday	
04 07	☽ □ ♀
05 33	☽ ⊡ ☉
15 22	☽ ⊻ ♆
15 46	☽ ⚹ ♅
16 14	☽ ⊥ ♂
17 35	☽ △ ♄
18 06	☽ ∥ ♅
20 05	☉ # ♃
07 Friday	
00 22	☽ ⊼ ☿
00 49	♀ ⊥ ♄
03 26	☽ ⊥ ♆
04 49	☽ ⊻ ♂
13 57	☽ ± ☿
14 32	☽ △ ☉
19 20	☽ Q ♀
20 12	☽ # ♇
21 39	☽ ∠ ♆
22 12	☽ ∠ ♅
08 Saturday	
00 41	☽ □ ♃
01 20	☽ ∥ ♂
09 37	☽ △ ♇
09 39	☉ Q ♅
10 23	☽ ⊡ ☿
16 34	☽ ⊥ ♅
21 46	☽ ⚹ ♀
09 Sunday	
04 07	☽ ⚹ ♆
04 48	☽ ⊻ ♅
07 08	☽ □ ♄
10 26	☿ ⊥ ♄
13 57	☽ Q ♃
16 05	☽ ⊡ ♇
17 42	☿ △ ♂
20 10	☽ ☌ ♂
20 20	☽ △ ☿
10 Monday	
04 22	☽ Q ♆
06 41	☽ ∠ ♀
08 59	☽ □ ☉
10 15	☽ ± ♇
14 35	☽ ⚹ ♃
19 50	☽ Q ♄
22 15	☽ ⊼ ♇
11 Tuesday	
02 12	☽ ⊥ ♀
14 27	☽ ∥ ♂
15 02	☽ ⊻ ♀
16 09	☽ □ ♆
17 05	☽ ☌ ♅
19 07	☽ # ♇
19 48	☽ ⚹ ♄
20 49	☽ ∠ ♃
23 51	☽ Q ☉
12 Wednesday	
05 16	♀ △ ♆
09 49	☽ ⊻ ♂
13 52	☽ □ ☿
14 32	☽ ⊥ ♃
18 25	♀ ⊻ ♅
20 27	☽ ∥ ♅
21 43	☽ ⊥ ♂
13 Thursday	
00 54	☽ ⚹ ☉
00 59	☽ ∠ ♄
02 09	☽ ⊻ ♃
04 35	☽ # ☿
08 22	☽ ☍ ♇
15 13	☽ ∠ ♂
18 02	☽ ⊥ ♄
20 36	☉ ⊻ ♃
14 Friday	
01 07	☽ △ ♆
02 12	☽ ⊻ ♅
02 45	☽ Q ☿
04 37	☽ ☌ ♀
05 11	☽ ⊻ ♄
07 07	☽ ∠ ☉
13 10	☽ ⊥ ♅
13 19	♀ ⊻ ♄
19 31	☽ ⚹ ♂
23 34	☽ ∥ ♆
15 Saturday	
00 39	☽ ⊥ ☉
02 13	☽ ⚹ ☿
04 09	☽ ⊡ ♆
05 18	☽ ∠ ♅
08 20	☽ ∥ ♀
09 05	☽ # ☉
09 47	☽ ☌ ♃
12 06	☽ ⊻ ☉
14 43	☽ ⊼ ♇
15 24	☉ # ♀
17 24	☽ Q ♂
19 56	☽ ± ♆
16 Sunday	
01 10	☽ ± ♇
01 47	☽ ∥ ♃
06 18	☽ ⊼ ♆
06 27	☽ ∠ ☿
07 30	☽ ⚹ ♅
10 40	☽ ☌ ♄
12 04	☿ # ♅
13 35	☽ ⊻ ♀
16 35	☽ ⊡ ♇
20 00	☽ # ♄
20 33	☽ ∥ ♄
22 53	☽ ⊥ ☿
23 56	☉ △ ♇
17 Monday	
00 32	☽ ⊥ ♀
01 10	☽ □ ♂
03 56	☽ Q ♅
09 39	☽ ⊻ ☿
13 53	☽ ⊻ ♃
13 58	☽ # ♃
16 44	☽ ∠ ♀
17 49	☽ △ ♇
19 09	☽ ☌ ☉
18 Tuesday	
00 00	☽ ⊥ ♃
03 57	☽ # ♀
08 38	☽ ☍ ♆
09 57	☽ □ ♅
10 52	☽ ∥ ☉
13 20	☽ ⊻ ♄
14 53	☽ # ♆
15 05	☽ ∠ ♃
19 22	☽ ⚹ ♀
23 17	☽ ⊥ ♄
19 Wednesday	
02 45	♀ ⊥ ♅
04 17	☽ △ ♂
06 07	♂ Q ♆
14 10	☽ ☌ ☿
14 10	☽ ∠ ♄
14 12	☿ ∠ ♄
16 02	☽ ⚹ ♃
16 28	☽ Q ♀
19 12	☽ □ ♇
20 Thursday	
00 08	☽ ⊻ ☉
05 29	☽ ⊡ ♂
09 46	☽ ⊼ ♆
10 40	☽ ⊥ ☉
11 14	☽ △ ♅
11 48	☽ # ♅
11 57	☽ Q ♃
14 57	☽ ⚹ ♄
18 51	☽ ∥ ☿
19 35	☽ ± ♆
20 38	☽ ± ♂
21 Friday	
00 09	☽ □ ♀
00 42	☽ Q ♇
01 40	☉ # ♆
02 32	☽ ∠ ☉
06 45	☽ ⊼ ♂
09 11	☽ ∥ ♇
09 52	☉ ⊥ ♃
10 23	☽ ⊡ ♆
10 51	☽ Q ♄
11 57	☽ ⊡ ♅
12 14	☽ # ♂
17 50	☽ ⊻ ☿
18 00	☽ □ ♃
20 30	☽ ⚹ ♇
23 53	☿ ⚹ ♃
22 Saturday	
02 53	☽ ± ♅
04 13	☽ ⊥ ☿
05 11	☽ ⚹ ☉
11 14	☽ △ ♆
12 54	☽ ⊼ ♅
17 02	☽ □ ♄
19 50	☽ ∠ ☿
21 32	☽ ∠ ♇
23 Sunday	
02 48	☽ Q ☉
05 53	☽ △ ♀
10 05	☽ ☍ ♂
12 44	☽ ⊥ ♇
21 06	☽ △ ♃
22 03	☽ ⚹ ☿
22 57	☽ ⊻ ♇
24 Monday	
09 29	☽ ⊡ ♀
11 50	☽ □ ☉
14 05	☽ □ ♆
15 06	☽ # ♂
15 58	☽ ☍ ♅
18 08	☽ ∥ ♇
19 13	☽ Q ☿
20 37	☽ △ ♄
23 18	☽ ⊡ ♃
25 Tuesday	
02 19	☽ ± ♀
05 58	☽ ∥ ☿
13 41	☽ ⊼ ♀
15 08	☽ ⊼ ♂
15 16	☽ ± ♃
17 01	☽ # ♅
17 36	☿ □ ♇
18 23	☉ ☍ ♆
21 28	☽ Q ♆
23 07	☽ ⊡ ♄
26 Wednesday	
01 59	☽ ± ♂
02 00	☽ ⊼ ♃
03 10	☽ ☌ ♇
03 17	☽ □ ☿
15 16	☽ ± ♄
16 47	♀ ⚹ ♂
18 25	☽ ⊡ ♂
18 46	☽ ⚹ ♆
20 47	☽ △ ☉
20 53	☽ ⊼ ♅
21 51	♀ ⊡ ♆
21 57	☉ □ ♅
27 Thursday	
02 08	☽ ⊼ ♄
03 53	☿ St R
07 42	☽ ± ♅
08 01	☽ ∥ ☉
18 38	☽ # ♆
21 52	☽ ∠ ♆
22 12	☽ △ ♂
28 Friday	
00 00	☽ ☍ ♀
00 05	☽ ⊡ ♅
01 07	♀ ∠ ♅
02 13	☽ ⊡ ☉
04 33	♂ ± ♇
08 57	☽ ☍ ♃
09 21	☽ ⊻ ♇
09 26	☽ △ ☿
14 22	☽ ⊥ ♆
14 42	☿ □ ♇
20 11	☽ ± ☉
20 26	☽ ⊥ ♇
20 54	☽ # ♀
29 Saturday	
00 23	☿ ⚹ ♃
01 28	☽ ⊻ ♆
01 41	☽ # ♃
03 49	☽ △ ♅
08 18	☽ ⊼ ☉
09 43	☽ ☍ ♄
12 09	♃ ⊼ ♇
12 46	☽ ⊡ ☿
13 14	☽ ∠ ♇
14 46	☽ ∥ ♄
20 26	☽ # ♄
30 Sunday	
04 39	☉ ⊻ ♄
05 15	☽ ± ☿
07 25	☽ □ ♂
09 13	☽ ∥ ♃
12 00	☽ ∥ ♀
13 05	☽ ⊼ ♀
16 15	☽ ⊼ ☿
17 40	☽ ⚹ ♇
18 07	☽ ⊼ ♃
31 Monday	
01 39	☽ ± ♀
05 49	☽ ± ♃
10 19	☽ ☌ ♆
12 55	☽ □ ♅
14 16	☿ ⚹ ♀
16 49	☽ Q ♇
17 43	☽ ∥ ♆
18 51	☽ Q ♂
19 33	☽ ⊼ ♄
20 44	☽ ⊡ ♀
22 36	☽ ☍ ☉
23 34	☽ ⊡ ♃

All ephemeris data is given at 12.00 UT and the Moon's longitude is additionally given for 24.00 UT

LONGITUDES

Date	Sidereal time h m s	Sun ☉ ° ' "	Moon ☽ ° ' "	Moon ☽ 24.00 ° ' "	Mercury ☿ ° '	Venus ♀ ° '	Mars ♂ ° '	Jupiter ♃ ° '	Saturn ♄ ° '	Uranus ♅ ° '	Neptune ♆ ° '	Pluto ♇ ° '
01	14 42 35	08 ♏ 55 10	15 ♉ 11 36	21 ♉ 15 54	21 ♏ 27	23 ♍ 05	18 ♋ 38	23 ♍ 57	06 ♎ 52	03 ♌ 24	02 ♉ 01	23 ♒ 24
02	14 46 31	09 55 10	27 ♉ 17 41	03 ♊ 17 12	20 R 35	24 10	18 53	24 08	06 59	03 24	01 R 59	23 R 23
03	14 50 28	10 55 13	09 ♊ 14 43	15 ♊ 10 35	19 34	25 15	19 08	24 19	07 06	03 24	01 57	23 23
04	14 54 24	11 55 17	21 ♊ 05 10	26 ♊ 58 52	18 25	26 20	19 22	24 29	07 12	03 25	01 56	23 23
05	14 58 21	12 55 23	02 ♋ 52 10	08 ♋ 45 34	17 12	27 26	19 36	24 40	07 19	03 25	01 54	23 23
06	15 02 17	13 55 32	14 ♋ 39 35	20 ♋ 34 48	15 54	28 31	19 49	24 50	07 25	03 25	01 52	23 23
07	15 06 14	14 55 42	26 ♋ 31 49	02 ♌ 31 14	14 35	29 ♍ 37	20 02	25 01	07 32	03 25	01 51	23 D 23
08	15 10 10	15 55 55	08 ♌ 33 00	14 ♌ 39 44	13 17	00 ♎ 44	20 13	25 11	07 38	03 25	01 49	23 23
09	15 14 07	16 56 09	20 ♌ 50 04	27 ♌ 05 13	12 03	01 50	20 25	25 21	07 44	03 R 25	01 47	23 23
10	15 18 04	17 56 26	03 ♍ 25 46	09 ♍ 52 11	10 54	02 57	20 35	25 32	07 51	03 25	01 46	23 23
11	15 22 00	18 56 44	16 ♍ 24 54	23 ♍ 04 11	09 54	04 04	20 45	25 42	07 57	03 25	01 44	23 23
12	15 25 57	19 57 04	29 ♍ 50 17	06 ♎ 43 13	09 04	05 11	20 54	25 52	08 03	03 25	01 43	23 24
13	15 29 53	20 57 27	13 ♎ 42 54	20 ♎ 49 05	08 24	06 19	21 03	26 01	08 09	03 25	01 41	23 24
14	15 33 50	21 57 51	28 ♎ 01 19	05 ♏ 19 00	07 55	07 26	21 10	26 11	08 15	03 24	01 39	23 24
15	15 37 46	22 58 17	12 ♏ 41 23	20 ♏ 07 33	07 39	08 34	21 17	26 21	08 21	03 24	01 38	23 24
16	15 41 43	23 58 44	27 ♏ 36 31	05 ♐ 07 10	07 D 34	09 42	21 23	26 31	08 27	03 24	01 36	23 24
17	15 45 39	24 59 14	12 ♐ 38 26	20 ♐ 09 12	07 40	10 51	21 29	26 40	08 33	03 23	01 35	23 25
18	15 49 36	25 59 45	27 ♐ 38 22	05 ♑ 05 10	07 56	11 59	21 34	26 49	08 39	03 23	01 33	23 25
19	15 53 33	27 00 17	12 ♑ 28 34	19 ♑ 47 56	08 23	13 08	21 38	26 59	08 45	03 22	01 32	23 25
20	15 57 29	28 00 50	27 ♑ 02 42	04 ♒ 12 27	08 57	14 17	21 41	27 08	08 51	03 21	01 30	23 26
21	16 01 26	29 ♏ 01 25	11 ♒ 16 55	18 ♒ 15 56	09 40	15 26	21 43	27 17	08 56	03 21	01 29	23 26
22	16 05 22	00 ♐ 02 01	25 ♒ 09 28	01 ♓ 57 36	10 29	16 35	21 44	27 26	09 02	03 20	01 28	23 27
23	16 09 19	01 02 38	08 ♓ 40 28	15 ♓ 18 15	11 25	17 44	21 45	27 34	09 07	03 19	01 26	23 27
24	16 13 15	02 03 16	21 ♓ 51 13	28 ♓ 19 38	12 26	18 54	21 R 45	27 43	09 13	03 19	01 25	23 27
25	16 17 12	03 03 55	04 ♈ 43 49	11 ♈ 04 03	13 32	20 04	21 44	27 52	09 18	03 18	01 23	23 28
26	16 21 08	04 04 36	17 ♈ 20 39	23 ♈ 33 55	14 41	21 13	21 42	28 00	09 24	03 17	01 22	23 28
27	16 25 05	05 05 17	29 ♈ 44 08	05 ♉ 51 35	15 54	22 23	21 40	28 08	09 29	03 16	01 21	23 29
28	16 29 02	06 06 00	11 ♉ 56 32	17 ♉ 59 14	17 10	23 33	21 36	28 17	09 34	03 15	01 19	23 30
29	16 32 58	07 06 43	23 ♉ 59 57	29 ♉ 58 54	18 29	24 44	21 32	28 25	09 39	03 14	01 18	23 30
30	16 36 55	08 ♐ 07 30	05 ♊ 56 20	11 ♊ 52 31	19 ♏ 50	25 ♎ 54	21 ♋ 27	28 ♍ 33	09 ♎ 44	03 ♌ 13	01 ♉ 17	23 ♒ 31

Date	Moon True ☊ ° '	Moon Mean ☊ ° '	Moon ☽ Latitude ° '
01	13 ♊ 14	14 ♊ 37	02 S 27
02	13 R 11	14 34	01 26
03	13 10	14 31	00 S 21
04	13 D 10	14 28	00 N 43
05	13 11	14 25	01 46
06	13 13	14 21	02 44
07	13 15	14 18	03 36
08	13 16	14 15	04 19
09	13 R 16	14 12	04 51
10	13 15	14 09	05 10
11	13 14	14 05	05 14
12	13 10	14 02	05 02
13	13 07	13 59	04 31
14	13 04	13 56	03 43
15	13 01	13 53	02 40
16	13 00	13 50	01 24
17	12 59	13 46	00 N 02
18	13 D 00	13 43	01 S 20
19	13 01	13 40	02 36
20	13 02	13 37	03 41
21	13 03	13 34	04 30
22	13 03	13 31	05 02
23	13 R 03	13 27	05 16
24	13 02	13 24	05 13
25	13 02	13 21	04 55
26	13 00	13 18	04 22
27	12 59	13 15	03 38
28	12 58	13 11	02 44
29	12 58	13 08	01 43
30	12 ♊ 57	13 ♊ 05	00 S 39

DECLINATIONS

Date	Sun ☉ ° '	Moon ☽ ° '	Mercury ☿ ° '	Venus ♀ ° '	Mars ♂ ° '	Jupiter ♃ ° '	Saturn ♄ ° '	Uranus ♅ ° '	Neptune ♆ ° '	Pluto ♇ ° '
01	14 S 28	14 N 03	20 S 05	03 N 27	23 N 18	03 N 22	00 S 46	19 N 56	10 N 29	23 S 00
02	14 47	18 10	19 36	03 05	23 18	03 18	00 48	19 56	10 28	22 59
03	15 06	21 29	19 03	02 44	23 19	03 14	00 50	19 56	10 28	22 59
04	15 25	23 51	18 27	02 22	23 19	03 10	00 53	19 56	10 27	22 59
05	15 43	25 10	17 47	01 59	23 19	03 06	00 55	19 56	10 27	22 59
06	16 01	25 21	17 06	01 37	23 20	03 02	00 58	19 56	10 26	22 59
07	16 19	24 23	16 23	01 14	23 20	02 58	01 00	19 56	10 25	22 59
08	16 36	22 17	15 40	00 51	23 21	02 54	01 03	19 56	10 25	22 58
09	16 53	19 09	14 58	00 28	23 22	02 50	01 05	19 56	10 24	22 58
10	17 10	15 04	14 19	00 N 05	23 23	02 46	01 07	19 56	10 24	22 58
11	17 27	10 11	13 43	00 S 18	23 24	02 43	01 10	19 56	10 23	22 58
12	17 43	04 N 41	13 12	00 42	23 25	02 39	01 12	19 56	10 23	22 57
13	18 00	01 S 14	12 46	01 06	23 26	02 35	01 14	19 57	10 22	22 57
14	18 15	07 17	12 25	01 30	23 27	02 31	01 16	19 57	10 22	22 57
15	18 31	13 06	12 10	01 54	23 29	02 28	01 19	19 57	10 21	22 57
16	18 46	18 15	12 00	02 18	23 31	02 24	01 21	19 57	10 21	22 56
17	19 01	22 17	11 56	02 42	23 32	02 20	01 23	19 57	10 20	22 56
18	19 15	24 45	11 57	03 06	23 34	02 17	01 25	19 57	10 20	22 56
19	19 29	25 26	12 02	03 31	23 37	02 13	01 27	19 57	10 19	22 56
20	19 43	24 21	12 11	03 55	23 39	02 10	01 29	19 58	10 19	22 55
21	19 56	21 43	12 24	04 20	23 41	02 06	01 31	19 58	10 18	22 55
22	20 09	17 52	12 40	04 45	23 44	02 03	01 33	19 58	10 18	22 55
23	20 22	13 12	12 59	05 09	23 46	02 00	01 35	19 58	10 17	22 54
24	20 34	08 02	13 20	05 34	23 49	01 57	01 37	19 58	10 17	22 54
25	20 46	02 S 38	13 43	05 58	23 52	01 53	01 39	19 59	10 16	22 54
26	20 58	02 N 46	14 07	06 23	23 55	01 50	01 41	19 59	10 16	22 53
27	21 09	07 59	14 32	06 48	23 59	01 47	01 43	19 59	10 16	22 53
28	21 19	12 49	14 59	07 12	24 02	01 44	01 45	19 59	10 15	22 52
29	21 30	17 06	15 26	07 37	24 06	01 41	01 46	20 00	10 15	22 52
30	21 S 39	20 N 39	15 S 54	08 S 01	24 N 09	01 N 38	01 S 48	20 N 00	10 N 14	22 S 52

ZODIAC SIGN ENTRIES

Date	h	m	Planets
02	17	25	☽ ♊
05	06	09	☽ ♋
07	18	58	☽ ♌
07	20	13	♀ ♎
10	05	32	☽ ♍
12	12	17	☽ ♎
14	15	16	☽ ♏
16	15	49	☽ ♐
18	15	48	☽ ♑
20	16	56	☽ ♒
22	11	12	☉ ♐
22	20	32	☽ ♓
25	03	07	☽ ♈
27	12	31	☽ ♉
30	00	02	☽ ♊

LATITUDES

Date	Mercury ☿ ° '	Venus ♀ ° '	Mars ♂ ° '	Jupiter ♃ ° '	Saturn ♄ ° '	Uranus ♅ ° '	Neptune ♆ ° '	Pluto ♇ ° '
01	02 S 02	00 N 46	01 N 11	01 N 03	02 N 09	00 N 34	01 S 48	09 S 51
04	01 11	00 59	01 18	01 04	02 09	00 34	01 48	09 50
07	00 S 11	01 11	01 25	01 04	02 10	00 34	01 48	09 50
10	00 N 49	01 22	01 32	01 05	02 10	00 34	01 48	09 49
13	01 37	01 32	01 40	01 06	02 11	00 34	01 48	09 49
16	02 09	01 41	01 48	01 06	02 11	00 34	01 48	09 48
19	02 23	01 49	01 56	01 07	02 12	00 34	01 48	09 47
22	02 25	01 55	02 04	01 07	02 12	00 34	01 48	09 47
25	02 17	02 01	02 13	01 08	02 13	00 35	01 47	09 46
28	02 04	02 05	02 22	01 09	02 13	00 35	01 47	09 46
31	01 N 46	02 N 09	02 N 31	01 N 09	02 N 14	00 N 35	01 S 47	09 S 45

DATA

Julian Date	2466094
Delta T	+77 seconds
Ayanamsa	24° 24' 31"
Synetic vernal point	04° ♓ 42' 28"
True obliquity of ecliptic	23° 26' 05"

MOON'S PHASES, APSIDES AND POSITIONS ☽

Date	h	m	Phase	Longitude ° '	Eclipse Indicator
09	03	46	☾	16 ♌ 35	
16	05	46	●	23 ♏ 43	
22	21	17	☽	00 ♓ 25	
30	16	50	○	08 ♊ 20	partial

Day	h	m		
05	04	18	Apogee	
17	09	06	Perigee	
06	03	47	Max dec	25° N 25'
13	07	04	0S	
19	09	04	Max dec	25° S 27'
25	23	38	0N	

LONGITUDES

Date	Chiron ⚷ ° '	Ceres ⚳ ° '	Pallas ⚴ ° '	Juno ⚵ ° '	Vesta ⚶ ° '	Black Moon Lilith ⚸ ° '
01	12 ♋ 47	24 ♈ 12	16 ♒ 56	25 ♑ 56	10 ♏ 07	24 ♊ 07
11	12 ♋ 34	22 ♈ 16	18 ♒ 03	28 ♑ 52	15 ♏ 27	25 ♊ 14
21	12 ♋ 14	20 ♈ 49	19 ♒ 35	02 ♒ 07	20 ♏ 48	26 ♊ 20
31	11 ♋ 46	19 ♈ 59	21 ♒ 29	05 ♒ 40	26 ♏ 10	27 ♊ 27

All ephemeris data is given at 12.00 UT and the Moon's longitude is additionally given for 24.00 UT

ASPECTARIAN

h m Aspects

01 Tuesday
00 31 ☿ ∠ ♄
07 24 ☽ ± ♄
12 52 ☉ ∠ ♃
14 28 ☽ # ☉
17 49 ♀ ∥ ♃
18 46 ♀ ⊼ ♇
18 55 ☽ ✶ ♂
19 23 ☿ # ♅
23 34 ☽ ☍ ☿

02 Wednesday
00 16 ☽ Q ♅
01 20 ☽ ♇ ♄
04 13 ☽ □ ♇
05 09 ☽ △ ♀
05 36 ☽ △ ♃
11 04 ♀ ☌ ♃
20 23 ☽ # ☿
21 22 ☽ ⊻ ♆

03 Thursday
00 04 ☽ ∥ ♅
00 14 ☽ ✶ ♅
01 29 ☽ ∠ ♂
07 37 ☽ △ ♄
09 24 ☽ ⊥ ♆
15 42 ☽ ⊼ ☉
19 40 ☿ △ ♂
20 01 ☽ ⊥ ♂

04 Friday
01 59 ☽ # ♇
03 15 ♀ ± ♆
03 34 ☽ ∠ ♆
04 58 ☽ ± ☉
05 30 ☽ ∥ ♂
06 34 ☽ ∠ ♅
07 05 ☽ ⊼ ☿
08 27 ☽ ⊻ ♂
16 41 ☽ △ ♇
18 10 ☽ ± ☿
19 02 ☽ □ ♃
23 46 ☽ □ ♀

05 Saturday
00 53 ☽ ⊥ ♅
00 59 ☽ ♇ ☉
10 02 ☽ ✶ ♆
10 45 ☽ ♇ ☿
13 07 ☽ ⊻ ♅
21 08 ☽ □ ♄
22 29 ☉ ⊥ ♄
23 14 ☽ ♇ ♇

06 Sunday
08 15 ☽ Q ♃
10 22 ☽ △ ☉
10 24 ☽ Q ♆
13 53 ♇ St D
14 16 ☽ △ ☿
16 10 ☽ Q ♀
17 32 ☽ ± ♇
22 39 ☽ ☌ ♂

07 Monday
05 40 ☽ ⊼ ♇
06 51 ♀ ± ♇
08 26 ☉ ☌ ☿
08 54 ☽ ✶ ♃
09 58 ☽ Q ♄
11 37 ☿ ∠ ♀
13 40 ☉ ∥ ☿
18 50 ☽ ✶ ♀
22 38 ☽ □ ♆

08 Tuesday
01 24 ☽ ∥ ♂
01 26 ♀ # ♄
01 48 ☽ ☌ ♅
05 21 ☽ # ♇
05 56 ☿ ⊥ ♄
10 09 ☽ ✶ ♄
15 15 ☽ ∠ ♃
20 26 ☽ □ ☿
21 06 ♅ St R

09 Wednesday
03 29 ☽ ∠ ♀
03 46 ☽ □ ☉
06 35 ☽ ∥ ♅
09 06 ☽ ⊥ ♃
11 02 ♀ ⊼ ♆
11 10 ☽ ⊻ ♂
15 43 ☽ ∠ ♄
16 55 ☽ ☍ ♇
20 49 ☽ ⊻ ♃
22 33 ☽ ⊥ ♀
22 52 ☽ ⊥ ♂

10 Thursday
00 59 ☽ # ☉
02 45 ♀ Q ♂
04 09 ☽ Q ☿
08 52 ☽ △ ♆
09 00 ☽ ⊥ ♄
09 57 ☉ ∠ ♀
11 01 ☽ ⊻ ♀
11 59 ☽ ⊻ ♅
16 06 ☽ ∠ ♂
16 31 ☽ # ☿
17 06 ☽ Q ☉
19 28 ☿ ∠ ♃
20 19 ☽ ⊻ ♄
22 04 ♀ ✶ ♅
23 10 ☽ ⊥ ♅

11 Friday
00 53 ☽ ✶ ☿
10 04 ☿ ⊥ ♀
11 04 ☽ ∥ ♆
12 35 ☽ ♇ ♆
15 38 ☽ ∠ ♅
16 58 ☽ ✶ ☉
17 22 ♃ ± ♆
19 56 ☽ ✶ ♂

12 Saturday
00 35 ☽ ⊼ ♇
02 20 ☽ ∠ ☿
04 43 ☽ ± ♆
04 53 ☽ ☌ ♃
11 13 ☽ ± ♇
15 17 ☽ ⊼ ♆
17 21 ☽ ⊥ ☿
17 26 ☽ Q ♂
18 15 ☽ ✶ ♅
20 29 ☽ ∥ ♃
21 38 ☽ ∠ ☉
22 10 ☽ ☌ ♀

13 Sunday
02 10 ☽ # ♄
02 24 ☽ ☌ ♄
02 53 ☽ ♇ ♇
03 15 ☽ # ♀
03 16 ☽ ⊻ ☿
11 23 ☽ ∥ ♀
11 59 ☽ ∥ ♄
14 16 ☽ ⊥ ☉
14 21 ☉ △ ♂
14 53 ☽ Q ♅
17 17 ☽ # ♃
21 05 ☿ ⊻ ♄
21 14 ♀ ∥ ♄

14 Monday
00 30 ☽ □ ♂
01 10 ☽ ⊻ ☉
04 19 ☽ △ ♇
08 56 ☽ ⊻ ♃
17 59 ☽ ☍ ♆
18 57 ☽ ⊥ ♃
19 52 ☿ ⊻ ♀
20 52 ☽ □ ♅

15 Tuesday
00 26 ☽ # ♆
03 55 ☽ ☌ ☿
04 45 ☽ ⊻ ♀
04 55 ☽ ⊻ ♄
07 00 ♀ ☌ ♄
08 07 ☽ ∥ ☿
08 24 ♀ ♇ ♇
09 48 ☽ ∠ ♃
14 43 ☽ ⊥ ♄
15 18 ☽ ⊥ ♀
22 12 ☉ ∠ ♄
22 20 ☉ □ ♇
23 35 ♄ ♇ ♇

16 Wednesday
01 58 ☽ △ ♂
05 16 ☽ □ ♇
05 18 ☽ ∠ ♄
05 46 ☽ ☌ ☉
06 58 ☽ ∠ ♀
10 13 ☽ ✶ ♃
10 39 ☿ St D
14 48 ☽ ∥ ☉
17 15 ♀ # ♃
18 23 ☽ ⊼ ♆
21 09 ☽ # ♅
21 14 ☽ △ ♅

17 Thursday
02 07 ☽ ♇ ♂
03 57 ☽ ± ♆
03 59 ☽ ⊻ ☿
05 26 ☽ ✶ ♄
05 36 ☽ Q ♃
08 54 ☽ ✶ ♀
10 02 ☽ Q ♇
13 39 ☽ ⊥ ☿
16 34 ☽ ± ♂
17 02 ☽ ∥ ♇
18 17 ☽ ♇ ♆
21 10 ☽ ♇ ♅
22 22 ☽ # ♂

18 Friday
00 44 ☽ Q ♄
02 12 ☽ ⊼ ♂
04 17 ☽ ∠ ☿
05 13 ☽ ✶ ♇
05 39 ☽ Q ♀
09 10 ☽ ⊻ ☉
10 40 ☽ □ ♃
11 35 ☽ ± ♅
18 17 ☽ △ ♆
19 31 ☽ ⊥ ☉
21 14 ☽ ⊼ ♅

19 Saturday
05 06 ☽ ✶ ☿
05 24 ☽ ∠ ♇
05 54 ☽ □ ♄
11 10 ☽ ∠ ☉
11 12 ☉ ✶ ♃
13 10 ☽ □ ♀
20 06 ☽ ⊥ ♇

20 Sunday
01 27 ☽ Q ☿
03 04 ☽ ☍ ♂
06 00 ☽ ⊻ ♇
07 06 ☿ ⊻ ♄
12 08 ☽ △ ♃
13 44 ☽ ✶ ☉
19 27 ☽ □ ♆
19 41 ☽ # ♂
22 34 ☽ ☍ ♅

21 Monday
02 31 ☽ ∥ ♇
07 59 ☽ △ ♄
09 05 ☽ □ ☿
10 16 ♀ Q ♅
11 32 ☽ Q ☉
13 43 ☽ ♇ ♃
14 48 ☉ # ♅
19 45 ☽ △ ♀
23 13 ☽ ∥ ☉
23 41 ☽ # ♅

22 Tuesday
02 05 ☽ Q ♆
05 25 ☽ ± ♃
06 01 ☽ ⊼ ♂
09 00 ☽ ☌ ♇
10 01 ☽ ♇ ♄
16 02 ☽ ⊼ ♃
16 33 ☽ ± ♂
21 17 ☽ □ ☉
23 05 ☽ ✶ ♆

23 Wednesday
00 23 ☽ ♇ ♀
02 00 ☽ ± ♄
02 26 ☽ ⊼ ♅
08 33 ☽ ♇ ♂
12 49 ☽ ⊼ ♄
12 59 ☽ ∥ ☿
13 10 ☽ ± ♅
17 21 ☽ △ ☿
18 03 ☽ ± ♀
20 47 ♂ St R
21 05 ☉ ⊼ ♆

24 Thursday
01 43 ☽ # ♆
02 02 ☽ ∠ ♆
05 30 ☽ ♇ ♅
06 02 ☽ ⊼ ♀
11 49 ☽ △ ♂
14 58 ☽ ⊻ ♇
18 34 ☽ ⊥ ♆
19 24 ☿ ∠ ♃
22 14 ☽ ∥ ♀
22 59 ☽ ☍ ♃
23 17 ☽ ♇ ☿

25 Friday
02 07 ☽ ⊥ ♇
05 44 ☽ ⊻ ♆
08 36 ☽ △ ☉
09 18 ☽ △ ♅
15 19 ☽ # ♃
16 18 ☽ ∥ ♄
17 21 ☉ △ ♅
17 48 ☽ ± ☿
19 04 ☽ ∠ ♇
20 43 ☽ ☍ ♄

26 Saturday
06 23 ☽ ⊼ ☿
07 06 ☽ # ♄
07 51 ☽ ∥ ♃
15 38 ☽ ♇ ☉
20 15 ☽ ☍ ♀
20 23 ☽ □ ♂
21 38 ♀ □ ♂
23 50 ☽ ✶ ♇

27 Sunday
03 08 ☿ ⊥ ♄
05 57 ☽ # ♀
08 51 ☽ ⊼ ♃
10 37 ☽ ± ☉
15 08 ☽ ☌ ♆
18 53 ☽ □ ♅
20 43 ☽ ± ♃
23 02 ☽ ∥ ♆
23 16 ☽ Q ♇
23 26 ☽ ⊼ ☉

28 Monday
07 16 ☽ ⊼ ♄
07 24 ☽ Q ♂
08 51 ♃ # ♄
10 40 ♀ △ ♇
14 41 ☽ ♇ ♃
19 14 ☽ ± ♄
23 14 ☉ ♇ ♂
23 38 ☽ ☍ ☿

29 Tuesday
01 05 ☽ # ☿
06 28 ☽ Q ♅
07 06 ☽ ✶ ♂
11 00 ☽ □ ♇
13 19 ☽ ♇ ♄
13 37 ☽ ⊼ ♀
16 22 ☉ ± ♆
20 57 ☽ △ ♃

30 Wednesday
02 38 ☽ ⊻ ♆
02 58 ☽ ± ♀
06 30 ☽ ✶ ♅
07 07 ☽ ∥ ♅
13 01 ☽ ∠ ♂
14 42 ☽ ⊥ ♆
16 50 ☽ ☍ ☉
19 44 ☽ △ ♄
20 33 ☽ # ☉
23 08 ☽ ♇ ♀

DECEMBER 2039

LONGITUDES

Date	Sidereal time h m s	Sun ☉ ° ′ ″	Moon ☽ ° ′ ″	Moon ☽ 24.00 ° ′ ″	Mercury ☿ ° ′	Venus ♀ ° ′	Mars ♂ ° ′	Jupiter ♃ ° ′	Saturn ♄ ° ′	Uranus ♅ ° ′	Neptune ♆ ° ′	Pluto ♇ ° ′
01	16 40 51	09 ♐ 08 15	17 ♊ 47 40	23 ♊ 42 04	21 ♏ 12	27 ♎ 05	21 ♋ 21	28 ♍ 40	09 ♎ 49	03 ♌ 11	01 ♉ 15	23 ♒ 31
02	16 44 48	10 09 03	29 ♊ 36 00	05 ♋ 29 44	22 36	28 15	21 R 14	28 48	09 54	03 R 10	01 R 14	23 32
03	16 48 44	11 09 51	11 ♋ 23 35	17 ♋ 17 56	24 02	29 ♎ 26	21 06	28 56	09 59	03 09	01 13	23 33
04	16 52 41	12 10 42	23 ♋ 13 07	29 ♋ 09 32	25 28	00 ♏ 37	20 58	29 03	10 04	03 08	01 12	23 34
05	16 56 37	13 11 33	05 ♌ 07 37	11 ♌ 07 49	26 56	01 48	20 48	29 10	10 08	03 06	01 11	23 34
06	17 00 34	14 12 26	17 ♌ 10 37	23 ♌ 16 29	28 24	02 59	20 38	29 18	10 13	03 05	01 10	23 35
07	17 04 31	15 13 20	29 ♌ 25 57	05 ♍ 39 32	29 ♏ 53	04 11	20 27	29 25	10 17	03 03	01 08	23 36
08	17 08 27	16 14 15	11 ♍ 57 44	18 ♍ 21 03	01 ♐ 23	05 22	20 15	29 31	10 22	03 02	01 07	23 37
09	17 12 24	17 15 11	24 ♍ 49 58	01 ♎ 24 53	02 53	06 34	20 02	29 38	10 26	03 00	01 06	23 38
10	17 16 20	18 16 09	08 ♎ 06 10	14 ♎ 54 05	04 23	07 45	19 49	29 45	10 30	02 59	01 05	23 38
11	17 20 17	19 17 07	21 ♎ 48 48	28 ♎ 50 20	05 54	08 57	19 34	29 51	10 34	02 57	01 04	23 39
12	17 24 13	20 18 07	05 ♏ 58 35	13 ♏ 13 14	07 25	10 09	19 19	29 ♍ 57	10 38	02 55	01 03	23 40
13	17 28 10	21 19 09	20 ♏ 33 50	27 ♏ 59 43	08 57	11 21	19 03	00 ♎ 04	10 42	02 54	01 02	23 41
14	17 32 06	22 20 11	05 ♐ 30 05	13 ♐ 03 54	10 28	12 33	18 46	00 10	10 46	02 52	01 01	23 42
15	17 36 03	23 21 14	20 ♐ 40 05	28 ♐ 17 24	12 00	13 45	18 29	00 15	10 50	02 50	01 00	23 43
16	17 40 00	24 22 18	05 ♑ 54 35	13 ♑ 30 22	13 32	14 57	18 11	00 21	10 54	02 48	01 00	23 44
17	17 43 56	25 23 22	21 ♑ 03 33	28 ♑ 33 01	15 04	16 09	17 52	00 26	10 58	02 46	00 59	23 45
18	17 47 53	26 24 27	05 ♒ 57 48	13 ♒ 17 04	16 37	17 22	17 33	00 32	11 01	02 44	00 58	23 46
19	17 51 49	27 25 33	20 ♒ 30 33	27 ♒ 36 48	18 09	18 34	17 13	00 37	11 04	02 42	00 57	23 47
20	17 55 46	28 26 38	04 ♓ 36 34	11 ♓ 29 27	19 42	19 46	16 52	00 42	11 08	02 40	00 56	23 48
21	17 59 42	29 ♐ 27 44	18 ♓ 15 30	24 ♓ 54 56	21 15	20 59	16 31	00 47	11 11	02 38	00 56	23 50
22	18 03 39	00 ♑ 28 50	01 ♈ 28 03	07 ♈ 55 16	22 48	22 12	16 09	00 51	11 14	02 36	00 55	23 51
23	18 07 35	01 29 56	14 ♈ 17 00	20 ♈ 33 47	24 22	23 24	15 47	00 56	11 17	02 34	00 54	23 52
24	18 11 32	02 31 03	26 ♈ 46 06	02 ♉ 54 30	25 55	24 37	15 25	01 00	11 20	02 32	00 54	23 53
25	18 15 29	03 32 09	08 ♉ 59 29	15 ♉ 01 34	27 29	25 50	15 02	01 05	11 23	02 30	00 53	23 54
26	18 19 25	04 33 16	21 ♉ 01 14	26 ♉ 58 58	29 ♐ 03	27 03	14 39	01 08	11 26	02 28	00 53	23 55
27	18 23 22	05 34 23	02 ♊ 55 10	08 ♊ 50 15	00 ♑ 37	28 16	14 16	01 12	11 29	02 26	00 52	23 57
28	18 27 18	06 35 30	14 ♊ 44 35	20 ♊ 38 31	02 11	29 ♏ 29	13 52	01 16	11 31	02 23	00 52	23 58
29	18 31 15	07 36 38	26 ♊ 33 40	02 ♋ 26 21	03 46	00 ♐ 42	13 29	01 19	11 34	02 21	00 51	23 59
30	18 35 11	08 37 45	08 ♋ 20 48	14 ♋ 15 56	05 21	01 55	13 05	01 23	11 36	02 19	00 51	24 00
31	18 39 08	09 ♑ 38 53	20 ♋ 11 59	26 ♋ 09 11	06 ♑ 56	03 ♐ 08	12 ♋ 41	01 ♎ 26	11 ♎ 38	02 ♌ 16	00 ♉ 50	24 ♒ 02

MOON NODES, LATITUDE AND DECLINATIONS

Date	Moon True ☊ ° ′	Moon Mean ☊ ° ′	Moon ☽ Latitude ° ′	Sun ☉ Dec ° ′	Moon ☽ Dec ° ′	Mercury ☿ Dec ° ′	Venus ♀ Dec ° ′	Mars ♂ Dec ° ′	Jupiter ♃ Dec ° ′	Saturn ♄ Dec ° ′	Uranus ♅ Dec ° ′	Neptune ♆ Dec ° ′	Pluto ♇ Dec ° ′
01	12 ♊ 57	13 ♊ 02	00 N 27	21 S 49	23 N 19	16 S 21	08 S 26	24 N 13	01 N 35	01 S 50	20 00	10 N 14	22 S 51
02	12 D 57	12 59	01 31	21 58	24 57	16 49	08 50	24 17	01 32	01 52	20 01	10 14	22 51
03	12 58	12 56	02 32	22 07	25 28	17 17	09 14	24 21	01 29	01 53	20 01	10 13	22 50
04	12 58	12 52	03 26	22 15	24 49	17 44	09 38	24 26	01 27	01 55	20 01	10 13	22 50
05	12 R 58	12 49	04 11	22 23	23 03	18 12	10 02	24 30	01 24	01 57	20 02	10 12	22 50
06	12 58	12 46	04 46	22 30	20 14	18 38	10 26	24 34	01 21	01 58	20 02	10 12	22 49
07	12 58	12 43	05 09	22 37	16 29	19 05	10 50	24 39	01 19	02 00	20 02	10 12	22 49
08	12 57	12 40	05 18	22 43	11 58	19 30	11 13	24 44	01 16	02 01	20 03	10 11	22 48
09	12 D 57	12 36	05 11	22 49	06 49	19 55	11 36	24 48	01 14	02 03	20 03	10 11	22 48
10	12 58	12 33	04 48	22 55	01 N 12	20 20	12 00	24 53	01 12	02 04	20 03	10 11	22 47
11	12 58	12 30	04 08	23 00	04 S 40	20 43	12 23	24 58	01 09	02 06	20 04	10 11	22 47
12	12 59	12 27	03 12	23 05	10 30	21 06	12 45	25 03	01 07	02 07	20 04	10 10	22 46
13	13 00	12 24	02 02	23 09	15 56	21 27	13 08	25 08	01 05	02 08	20 05	10 10	22 46
14	13 00	12 21	00 N 42	23 13	20 32	21 48	13 30	25 13	01 03	02 09	20 05	10 10	22 45
15	13 R 00	12 17	00 S 42	23 16	23 49	22 08	13 52	25 18	01 00	02 11	20 06	10 09	22 45
16	12 59	12 14	02 04	23 19	25 27	22 27	14 14	25 24	00 58	02 12	20 06	10 09	22 44
17	12 58	12 11	03 16	23 21	25 01	22 44	14 35	25 29	00 57	02 13	20 07	10 09	22 44
18	12 57	12 08	04 13	23 23	22 52	23 01	14 56	25 34	00 55	02 14	20 07	10 09	22 43
19	12 55	12 05	04 53	23 25	19 17	23 16	15 17	25 39	00 53	02 15	20 07	10 08	22 43
20	12 53	12 02	05 13	23 26	14 41	23 31	15 37	25 44	00 51	02 16	20 08	10 08	22 42
21	12 52	11 58	05 15	23 26	09 29	23 44	15 57	25 49	00 49	02 17	20 08	10 08	22 42
22	12 D 52	11 55	05 00	23 26	04 S 00	23 56	16 17	25 54	00 48	02 18	20 09	10 08	22 41
23	12 52	11 52	04 30	23 26	01 N 28	24 07	16 36	25 59	00 46	02 19	20 09	10 08	22 41
24	12 53	11 49	03 49	23 25	06 46	24 16	16 55	26 03	00 45	02 20	20 10	10 07	22 40
25	12 55	11 46	02 57	23 23	11 42	24 24	17 14	26 08	00 43	02 21	20 10	10 07	22 40
26	12 57	11 42	01 58	23 21	16 06	24 31	17 32	26 13	00 42	02 22	20 11	10 07	22 39
27	12 58	11 39	00 S 55	23 19	19 50	24 37	17 50	26 17	00 41	02 23	20 12	10 07	22 39
28	12 R 58	11 36	00 N 10	23 16	22 43	24 41	18 07	26 21	00 39	02 24	20 12	10 07	22 38
29	12 57	11 33	01 14	23 13	24 37	24 44	18 24	26 26	00 38	02 24	20 13	10 07	22 37
30	12 55	11 30	02 15	23 09	25 25	24 46	18 40	26 30	00 37	02 25	20 13	10 07	22 37
31	12 ♊ 52	11 ♊ 27	03 N 11	23 S 05	25 N 03	24 S 46	18 S 56	26 N 33	00 N 36	02 S 26	20 N 14	10 N 07	22 S 36

ZODIAC SIGN ENTRIES

Date	h	m	Planets
02	12	49	☽ ♋
03	23	27	♀ ♏
05	01	42	☽ ♌
07	13	06	☽ ♍
07	13	52	☿ ♐
09	21	26	☽ ♎
12	01	58	☽ ♏
12	22	04	♃ ♎
14	03	13	☽ ♐
16	02	41	☽ ♑
18	02	20	☽ ♒
20	04	04	☽ ♓
22	00	40	☉ ♑
22	09	18	☽ ♈
24	18	18	☽ ♉
27	02	34	☿ ♑
27	06	06	☽ ♊
28	22	18	♀ ♐
29	19	02	☽ ♋

LATITUDES

Date	Mercury ☿ ° ′	Venus ♀ ° ′	Mars ♂ ° ′	Jupiter ♃ ° ′	Saturn ♄ ° ′	Uranus ♅ ° ′	Neptune ♆ ° ′	Pluto ♇ ° ′
01	01 N 46	02 N 09	02 N 31	01 N 09	02 N 14	00 N 35	01 S 47	09 S 45
04	01 26	02 11	02 39	01 10	02 15	00 35	01 47	09 44
07	01 04	02 13	02 48	01 11	02 16	00 35	01 47	09 44
10	00 42	02 13	02 57	01 11	02 16	00 35	01 47	09 43
13	00 N 20	02 13	03 05	01 12	02 17	00 35	01 47	09 43
16	00 S 01	02 11	03 13	01 13	02 18	00 35	01 47	09 42
19	00 22	02 09	03 21	01 14	02 19	00 35	01 46	09 42
22	00 42	02 06	03 28	01 14	02 19	00 35	01 46	09 41
25	01 00	02 02	03 34	01 15	02 20	00 35	01 46	09 41
28	01 16	01 58	03 40	01 16	02 21	00 36	01 46	09 41
31	01 S 31	01 N 53	03 N 45	01 N 17	02 N 22	00 N 36	01 S 46	09 S 40

DATA

Julian Date	2466124
Delta T	+77 seconds
Ayanamsa	24° 24′ 36″
Synetic vernal point	04° ♓ 42′ 23″
True obliquity of ecliptic	23° 26′ 05″

MOON'S PHASES, APSIDES AND POSITIONS ☽

Date	h	m	Phase	Longitude ° ′	Eclipse Indicator
08	20	45	☾	16 ♍ 36	
15	16	32	●	23 ♐ 33	Total
22	10	02	☽	00 ♈ 24	
30	12	38	○	08 ♋ 39	

Day	h	m		
02	15	58	Apogee	
15	20	49	Perigee	
29	16	17	Apogee	
03	10	35	Max dec	25° N 28′
10	16	57	0S	
16	19	31	Max dec	25° S 28′
23	05	30	0N	
30	16	31	Max dec	25° N 26′

LONGITUDES

Date	Chiron ⚷ ° ′	Ceres ⚳ ° ′	Pallas ⚴ ° ′	Juno ⚵ ° ′	Vesta ⚶ ° ′	Black Moon Lilith ⚸ ° ′
01	11 ♋ 46	19 ♈ 59	21 ♒ 29	05 ♒ 40	26 ♏ 10	27 ♊ 27
11	11 ♋ 13	19 ♈ 48	23 ♒ 41	09 ♒ 27	01 ♐ 31	28 ♊ 33
21	10 ♋ 35	20 ♈ 15	26 ♒ 10	13 ♒ 28	06 ♐ 52	29 ♊ 40
31	09 ♋ 55	21 ♈ 18	28 ♒ 51	17 ♒ 40	12 ♐ 10	00 ♋ 46

All ephemeris data is given at 12.00 UT and the Moon's longitude is additionally given for 24.00 UT

ASPECTARIAN

h m	Aspects
01 Thursday	
07 05	☽ ⊥ ♂
07 06	☽ # ♇
08 53	☽ ∠ ♆
12 48	☽ ∠ ♅
14 20	☿ △ ♂
19 09	☽ ⊻ ♂
19 51	☽ ⊼ ☿
23 39	☽ △ ♇
23 40	☽ ∥ ♂
02 Friday	
05 34	☉ ⚹ ♄
07 04	☽ ⊥ ♅
08 58	☽ △ ♀
09 42	☽ ± ☿
10 22	☽ □ ♃
15 20	☽ ⚹ ♆
19 15	☽ ⊻ ♅
03 Saturday	
00 28	♀ ⊻ ♃
03 52	☿ □ ♇
05 38	☉ Q ♃
06 12	☽ bQ ♇
06 31	☽ bQ ☿
09 06	☽ □ ♄
11 29	☽ ⊼ ☉
15 42	☽ Q ♆
21 10	☉ Q ♇
23 22	☽ Q ♃
04 Sunday	
00 31	☽ ± ♇
00 50	☽ ± ☉
04 48	☿ ∠ ♄
07 29	☽ ☌ ♂
12 41	☽ ⊼ ♇
17 11	☽ △ ☿
18 30	☽ ∥ ♂
20 45	☽ bQ ☉
21 51	☽ Q ♄
23 33	♀ ☍ ♆
23 54	☽ ⚹ ♃
05 Monday	
04 05	☽ □ ♆
04 35	☽ □ ♀
07 57	☽ ☌ ♅
14 13	☽ # ♇
18 13	☽ # ☉
22 05	☽ ⚹ ♄
22 10	♀ # ♆
06 Tuesday	
05 35	☽ △ ☉
06 14	☽ ∠ ♃
13 25	☽ ∥ ♅
13 48	♀ □ ♅
18 43	☽ ⊻ ♂
20 19	☽ Q ♀
20 34	☉ ± ♂
21 42	☽ # ☿
07 Wednesday	
00 09	☽ ⊥ ♃
00 37	☽ ☍ ♇
03 42	☿ ⚹ ♃
03 53	☽ ∠ ♄
06 17	☽ ⊥ ♂
11 57	☽ ⊻ ♃
13 00	☽ □ ☿
15 18	☽ △ ♆
18 59	☽ ⊻ ♅
21 26	☽ ⊥ ♄
22 07	☽ ⚹ ♀
23 25	☽ ∠ ♂
08 Thursday	
06 27	☽ ⊥ ♅
07 58	☿ ⊼ ♆
08 57	☽ ⊻ ♄
09 19	☉ bQ ♆
15 23	☽ # ♀
15 29	♀ ⊥ ♃
19 49	☽ bQ ♆
20 35	☽ ∥ ♆
20 45	☽ □ ☉
23 23	☽ ∠ ♅
09 Friday	
03 17	☽ ⚹ ♂
03 44	☽ Q ☿
05 21	☽ ∠ ♀
05 48	☉ ∥ ♇
09 27	♃ ± ♇
09 47	☽ ⊼ ♇
12 30	☽ ± ♆
13 59	☿ △ ♅
19 39	☿ # ♅
20 46	☽ ± ♇
20 51	☽ ☌ ♃
23 25	☽ ⊼ ♆
23 29	☽ ⊥ ♀
10 Saturday	
00 54	☽ Q ♂
02 50	☽ ⚹ ♅
04 30	☽ ⚹ ☿
05 18	☉ bQ ♅
08 23	☽ # ♄
08 28	☽ Q ☉
11 19	☽ ⊻ ♀
12 01	☽ ∥ ♃
12 57	☽ bQ ♇
16 17	☽ ☌ ♄
17 50	☿ bQ ♂
21 47	☽ # ♃
11 Sunday	
00 07	☽ Q ♅
01 29	☽ ∥ ♄
07 17	☽ ⚹ ☉
08 12	☽ □ ♂
10 14	☽ ∠ ☿
15 10	☽ △ ♇
17 23	☉ ⊼ ♂
12 Monday	
01 49	☽ ⊻ ♃
03 27	☽ ⊥ ☿
03 45	☽ ☍ ♆
06 19	☿ ± ♆
06 54	☽ □ ♅
10 38	☽ # ♆
10 47	☽ ∠ ☉
11 58	☽ ⊥ ♃
14 41	☽ ⊻ ☿
19 33	☽ ☌ ♀
19 47	☽ ⊻ ♄
22 22	☽ ∥ ♀
22 31	♀ ⊻ ♄
13 Tuesday	
02 48	☽ ⊥ ☉
02 57	☽ ∠ ♃
05 41	☽ ⊥ ♄
09 35	☽ △ ♂
13 19	☽ ⊻ ☉
17 04	☽ □ ♇
20 21	☽ ∠ ♄
14 Wednesday	
03 25	☽ ⚹ ♃
04 52	☽ ⊼ ♆
07 48	☽ △ ♅
09 18	☽ bQ ♂
09 22	☽ # ♅
14 25	☽ ± ♆
16 58	☿ ⚹ ♄
20 24	☽ ⚹ ♄
20 47	☽ ☌ ☿
20 52	☽ ∥ ☿
21 51	☽ Q ♇
22 38	☽ Q ♃
22 59	☉ Q ♄
23 19	☽ ± ♂
15 Thursday	
00 08	☽ ⊻ ♀
02 53	☽ ∥ ♇
04 39	☽ bQ ♆
06 58	☽ ∥ ☉
07 32	☿ Q ♇
07 32	☽ bQ ♅
08 37	☽ ⊼ ♂
10 25	☽ ⊥ ♀
15 26	☽ Q ♄
16 15	☿ Q ♃
16 32	☽ ☌ ☉
16 49	☽ ⚹ ♇
18 17	☿ ± ♂
20 45	☉ ⚹ ♇
21 41	☽ ± ♅
16 Friday	
01 44	♂ Q ♃
01 48	☽ ∠ ♀
03 11	☽ □ ♃
04 16	☽ △ ♆
07 07	☽ ⊼ ♅
13 28	☽ # ♂
16 28	☽ ∠ ♇
19 54	☽ □ ♄
20 39	♀ ∠ ♃
23 27	☽ # ♂
17 Saturday	
01 24	☽ ⊻ ☿
03 31	☽ ⚹ ♀
06 44	☽ ⊥ ♇
07 01	☽ ☍ ♂
11 26	☿ ∥ ♇
12 02	☽ ⊥ ☿
16 18	☽ ⊻ ♇
19 26	☽ ⊻ ☉
20 03	♂ ± ♇
18 Sunday	
00 22	☽ Q ♀
02 00	☿ bQ ♆
03 08	☽ △ ♃
03 54	☽ □ ♆
04 08	☽ ∠ ☿
04 52	♀ ⊥ ♄
05 48	☽ ⊥ ☉
06 47	☽ ☍ ♅
07 36	☽ ∥ ☉
10 52	☽ ∥ ☿
13 10	☽ ∥ ♇
14 52	♀ △ ♂
19 36	☉ ± ♅
20 18	☽ △ ♄
21 35	☽ ∠ ☉
23 53	☿ ⊼ ♂
19 Monday	
03 49	☽ bQ ♃
05 09	☿ bQ ♅
06 38	☽ ⊼ ♂
07 00	☽ # ♅
07 37	☽ ⚹ ☿
08 28	☽ □ ♀
09 25	☽ Q ♆
16 26	☽ ± ♂
17 32	☽ ☌ ♇
18 58	☽ ± ♃
21 25	☽ bQ ♄
20 Tuesday	
00 35	☽ ⚹ ☉
02 27	☉ ∥ ☿
05 14	☽ ⊼ ♃
05 41	☽ ⚹ ♆
06 22	☽ Q ☿
07 23	☽ bQ ♂
07 41	☽ ∥ ♀
08 40	☽ ⊼ ♅
12 54	☽ ± ♄
16 55	☿ ⊻ ♀
19 03	☽ ± ♅
22 59	☽ Q ☉
23 25	☽ ⊼ ♄
21 Wednesday	
07 51	☽ ∠ ♆
08 58	☽ △ ♂
09 03	☽ # ♆
10 54	☽ bQ ♅
17 23	☽ △ ♀
18 05	☽ □ ☿
22 02	☽ ⊻ ♇
22 Thursday	
00 01	☽ ⊥ ♆
09 00	☽ ⊥ ♇
10 02	☽ □ ☉
10 52	☽ ☍ ♃
10 59	☽ ⊻ ♆
14 06	☽ △ ♅
18 53	☿ Q ♄
19 23	☽ ∥ ♄
21 37	☉ □ ♃
22 12	☉ △ ♆
23 45	☽ bQ ♀
23 Friday	
01 45	☽ ∠ ♇
02 05	☽ # ♃
04 14	☿ ⚹ ♇
04 37	♃ ⊼ ♆
06 18	☽ ☍ ♄
08 54	☽ ∥ ♃
14 47	☽ □ ♂
15 47	☽ # ♄
18 35	☽ ± ♀
21 15	♀ □ ♇
24 Saturday	
06 24	☽ ⚹ ♇
07 22	☽ ⊼ ♀
10 07	☽ △ ☿
12 24	☉ ⊼ ♅
20 03	☽ ☌ ♆
20 19	☽ ⊼ ♃
21 15	☿ ± ♅
23 14	☽ □ ♅
25 Sunday	
00 15	☽ △ ☉
00 37	☽ Q ♂
04 06	☽ ∥ ♆
05 53	☽ Q ♇
08 11	☽ ± ♃
16 46	☽ ⊼ ♄
19 58	☽ bQ ☿
23 24	♀ ∠ ♄
23 39	☽ ⚹ ♂
26 Monday	
02 10	☽ bQ ♃
04 47	☽ ± ♄
08 47	☽ bQ ☉
10 53	☽ Q ♅
16 41	☽ ± ☿
17 51	☽ □ ♇
21 21	☽ # ♀
22 56	☽ bQ ♄
27 Tuesday	
01 30	☽ ☍ ♀
04 36	☽ ± ☉
04 51	☽ ∠ ♂
06 38	☽ ⊼ ☿
07 51	☽ ⊻ ♆
08 31	☽ △ ♃
11 00	☽ ⚹ ♅
14 36	☽ ∥ ♅
15 50	☿ △ ♆
17 53	☽ ⊼ ☉
20 00	☽ ⊥ ♆
21 20	☿ □ ♃
22 30	☽ ⊥ ♂
28 Wednesday	
03 00	♀ bQ ♂
05 25	☽ △ ♄
10 17	☽ ⊻ ♂
11 07	☽ # ♇
14 16	☽ ∠ ♆
14 56	☿ ⊼ ♅
17 22	☽ ∠ ♅
17 35	☽ # ☉
29 Thursday	
06 48	☽ △ ♇
11 37	☽ ⊥ ♅
14 11	☽ # ☿
15 07	♀ ⊼ ♆
20 46	☽ ⚹ ♆
21 26	☽ ⊼ ♀
21 47	☽ □ ♃
23 47	☽ ⊻ ♅
30 Friday	
00 59	♀ ⚹ ♃
04 59	☽ ☍ ☿
11 01	☽ ± ♀
12 38	☽ ☍ ☉
13 21	☽ bQ ♇
18 37	☽ □ ♄
19 36	♀ △ ♅
21 07	☽ Q ♆
21 08	☉ ∠ ♇
21 18	☽ ☌ ♂
31 Saturday	
02 30	♂ Q ♆
07 21	☽ bQ ♀
07 37	☽ ± ♇
10 26	☽ Q ♃
18 18	☽ # ☿
19 44	☽ ⊼ ♇

JANUARY 2040

Date	Sidereal time h m s	LONGITUDES Sun ☉ ° ' "	Moon ☽ ° ' "	Moon ☽ 24.00 ° ' "	Mercury ☿ ° '	Venus ♀ ° '	Mars ♂ ° '	Jupiter ♃ ° '	Saturn ♄ ° '	Uranus ♅ ° '	Neptune ♆ ° '	Pluto ♇ ° '
01	18 43 04	10 ♑ 40 01	02 ♌ 07 44	08 ♌ 07 54	08 ♑ 32	04 ♐ 21	12 ♋ 17	01 ♎ 29	11 ♎ 40	02 ♌ 14	00 ♉ 50	24 ♒ 03
02	18 47 01	11 41 09	14 ♌ 09 53	20 ♌ 13 57	10 08	05 34	11 R 53	01 32	11 42	02 R 12	00 R 50	24 05
03	18 50 58	12 42 17	26 ♌ 20 21	02 ♍ 29 24	11 44	06 48	11 30	01 34	11 44	02 09	00 49	24 06
04	18 54 54	13 43 25	08 ♍ 41 24	14 ♍ 56 40	13 21	08 01	11 06	01 36	11 46	02 07	00 49	24 07
05	18 58 51	14 44 34	21 ♍ 15 34	27 ♍ 38 28	14 57	09 15	10 42	01 39	11 48	02 04	00 49	24 09
06	19 02 47	15 45 43	04 ♎ 05 43	10 ♎ 37 43	16 35	10 28	10 19	01 41	11 50	02 02	00 49	24 10
07	19 06 44	16 46 52	17 ♎ 14 49	23 ♎ 57 20	18 12	11 42	09 56	01 43	11 51	01 59	00 49	24 12
08	19 10 40	17 48 01	00 ♏ 45 33	07 ♏ 39 42	19 51	12 55	09 33	01 44	11 53	01 57	00 49	24 13
09	19 14 37	18 49 10	14 ♏ 39 54	21 ♏ 46 09	21 29	14 09	09 11	01 46	11 54	01 54	00 48	24 14
10	19 18 33	19 50 20	28 ♏ 58 19	06 ♐ 16 08	23 08	15 22	08 49	01 47	11 55	01 52	00 48	24 16
11	19 22 30	20 51 30	13 ♐ 39 07	21 ♐ 06 39	24 47	16 36	08 27	01 48	11 56	01 49	00 D 48	24 17
12	19 26 27	21 52 39	28 ♐ 37 54	06 ♑ 11 51	26 27	17 50	08 06	01 49	11 57	01 47	00 48	24 19
13	19 30 23	22 53 49	13 ♑ 47 24	21 ♑ 23 18	28 07	19 04	07 46	01 50	11 58	01 44	00 49	24 20
14	19 34 20	23 54 58	28 ♑ 58 15	06 ♒ 30 59	29 ♑ 48	20 17	07 26	01 50	11 59	01 42	00 49	24 22
15	19 38 16	24 56 07	14 ♒ 00 15	21 ♒ 24 57	01 ♒ 28	21 31	07 06	01 50	12 00	01 39	00 49	24 24
16	19 42 13	25 57 15	28 ♒ 44 06	05 ♓ 56 57	03 10	22 45	06 47	01 50	12 00	01 36	00 49	24 25
17	19 46 09	26 58 22	13 ♓ 02 55	20 ♓ 01 38	04 51	23 59	06 29	01 R 50	12 01	01 34	00 49	24 27
18	19 50 06	27 59 29	26 ♓ 52 57	03 ♈ 36 54	06 33	25 13	06 12	01 50	12 01	01 31	00 49	24 28
19	19 54 02	29 ♑ 00 35	10 ♈ 13 40	16 ♈ 43 36	08 15	26 27	05 55	01 50	12 01	01 29	00 50	24 30
20	19 57 59	00 ♒ 01 39	23 ♈ 07 08	29 ♈ 24 48	09 57	27 40	05 39	01 49	12 01	01 26	00 50	24 32
21	20 01 56	01 02 44	05 ♉ 37 12	11 ♉ 44 59	11 39	28 ♐ 54	05 24	01 48	12 R 01	01 23	00 50	24 33
22	20 05 52	02 03 47	17 ♉ 48 46	23 ♉ 49 14	13 21	00 ♑ 08	05 09	01 47	12 01	01 21	00 51	24 35
23	20 09 49	03 04 49	29 ♉ 47 02	05 ♊ 42 47	15 02	01 22	04 55	01 46	12 01	01 18	00 51	24 36
24	20 13 45	04 05 51	11 ♊ 37 07	17 ♊ 30 35	16 44	02 36	04 42	01 44	12 01	01 16	00 51	24 38
25	20 17 42	05 06 51	23 ♊ 23 43	29 ♊ 17 00	18 24	03 50	04 30	01 43	12 00	01 13	00 52	24 40
26	20 21 38	06 07 51	05 ♋ 10 51	11 ♋ 05 42	20 04	05 04	04 18	01 41	12 00	01 10	00 52	24 41
27	20 25 35	07 08 49	17 ♋ 01 50	22 ♋ 59 35	21 43	06 19	04 08	01 39	11 59	01 08	00 53	24 43
28	20 29 31	08 09 47	28 ♋ 59 10	05 ♌ 00 47	23 20	07 33	03 58	01 37	11 58	01 05	00 54	24 45
29	20 33 28	09 10 43	11 ♌ 04 35	17 ♌ 10 43	24 56	08 47	03 49	01 34	11 58	01 03	00 54	24 46
30	20 37 25	10 11 39	23 ♌ 19 15	29 ♌ 30 17	26 29	10 01	03 41	01 32	11 57	01 00	00 55	24 48
31	20 41 21	11 ♒ 12 34	05 ♍ 43 54	12 ♍ 00 08	28 ♒ 00	11 ♑ 15	03 ♋ 33	01 ♎ 29	11 ♎ 56	00 ♌ 57	00 ♉ 56	24 ♒ 50

Date	Moon True ☊ ° '	Moon Mean ☊ ° '	Moon ☽ Latitude ° '	DECLINATIONS Sun ☉ ° '	Moon ☽ ° '	Mercury ☿ ° '	Venus ♀ ° '	Mars ♂ ° '	Jupiter ♃ ° '	Saturn ♄ ° '	Uranus ♅ ° '	Neptune ♆ ° '	Pluto ♇ ° '
01	12 ♊ 47	11 ♊ 23	03 N 58	23 S 00	23 N 33	24 S 45	19 S 12	26 N 37	00 N 35	02 S 26	20 N 14	10 N 07	22 S 36
02	12 R 42	11 20	04 35	22 55	20 58	24 42	19 27	26 41	00 35	02 27	20 15	10 07	22 35
03	12 36	11 17	05 00	22 50	17 26	24 38	19 41	26 44	00 34	02 27	20 15	10 06	22 35
04	12 32	11 14	05 12	22 44	13 07	24 33	19 55	26 47	00 33	02 28	20 16	10 06	22 34
05	12 28	11 11	05 09	22 37	08 12	24 26	20 08	26 50	00 32	02 28	20 16	10 06	22 33
06	12 26	11 08	04 51	22 30	02 N 49	24 18	20 21	26 53	00 32	02 28	20 17	10 06	22 33
07	12 D 25	11 04	04 17	22 23	02 S 49	24 08	20 33	26 55	00 31	02 29	20 18	10 06	22 32
08	12 26	11 01	03 28	22 15	08 29	23 56	20 45	26 58	00 31	02 29	20 18	10 06	22 32
09	12 27	10 58	02 26	22 07	13 55	23 43	20 56	27 00	00 31	02 29	20 19	10 06	22 31
10	12 28	10 55	01 N 13	21 58	18 44	23 29	21 07	27 02	00 30	02 30	20 19	10 06	22 31
11	12 R 29	10 52	00 S 06	21 49	22 32	23 13	21 17	27 04	00 30	02 30	20 20	10 07	22 30
12	12 28	10 48	01 27	21 39	24 52	22 55	21 26	27 05	00 30	02 30	20 21	10 07	22 29
13	12 25	10 45	02 42	21 30	25 24	22 36	21 35	27 07	00 30	02 30	20 21	10 07	22 29
14	12 20	10 42	03 46	21 19	24 02	22 15	21 43	27 08	00 30	02 30	20 22	10 07	22 28
15	12 13	10 39	04 33	21 08	20 58	21 53	21 51	27 09	00 30	02 30	20 22	10 07	22 28
16	12 06	10 36	05 01	20 57	16 36	21 29	21 58	27 10	00 30	02 30	20 23	10 07	22 27
17	11 59	10 33	05 09	20 46	11 05	21 04	22 04	27 11	00 31	02 30	20 23	10 07	22 26
18	11 54	10 29	04 59	20 34	05 48	20 37	22 10	27 12	00 31	02 30	20 24	10 07	22 26
19	11 50	10 26	04 32	20 21	00 S 07	20 09	22 15	27 12	00 32	02 29	20 25	10 07	22 25
20	11 48	10 23	03 53	20 08	05 N 23	19 39	22 19	27 12	00 32	02 29	20 25	10 08	22 24
21	11 D 48	10 20	03 03	19 55	10 31	19 08	22 23	27 12	00 33	02 29	20 26	10 08	22 24
22	11 49	10 17	02 06	19 42	15 07	18 35	22 26	27 12	00 33	02 29	20 26	10 08	22 23
23	11 50	10 14	01 05	19 28	19 03	18 01	22 29	27 12	00 34	02 28	20 27	10 08	22 23
24	11 51	10 10	00 S 01	19 14	22 09	17 25	22 30	27 12	00 35	02 28	20 28	10 08	22 22
25	11 R 50	10 07	01 N 02	18 59	24 18	16 49	22 31	27 12	00 36	02 28	20 28	10 09	22 21
26	11 47	10 04	02 02	18 44	25 22	16 11	22 32	27 11	00 37	02 27	20 29	10 09	22 21
27	11 41	10 01	02 57	18 29	25 17	15 32	22 32	27 11	00 38	02 27	20 29	10 09	22 20
28	11 33	09 58	03 45	18 13	24 01	14 53	22 31	27 10	00 39	02 26	20 30	10 09	22 20
29	11 23	09 54	04 23	17 57	21 39	14 13	22 29	27 10	00 40	02 26	20 30	10 10	22 19
30	11 12	09 51	04 49	17 41	18 17	13 32	22 27	27 09	00 41	02 25	20 31	10 10	22 18
31	11 ♊ 00	09 ♊ 48	05 N 02	17 S 25	14 N 05	12 S 51	22 S 24	27 N 08	00 N 43	02 S 24	20 N 32	10 N 10	22 S 18

ZODIAC SIGN ENTRIES

Date	h m	Planets
01	07 44	☽ ♌
03	19 09	☽ ♍
06	04 24	☽ ♎
08	10 40	☽ ♏
10	13 42	☽ ♐
12	14 10	☽ ♑
14	13 38	☽ ♒
14	14 58	☿ ♒
16	14 05	☽ ♓
18	17 32	☽ ♈
20	11 21	☉ ♒
21	01 08	☽ ♉
22	09 16	♀ ♑
23	12 26	☽ ♊
26	01 28	☽ ♋
28	14 01	☽ ♌
31	00 57	☽ ♍

LATITUDES

Date	Mercury ☿ ° '	Venus ♀ ° '	Mars ♂ ° '	Jupiter ♃ ° '	Saturn ♄ ° '	Uranus ♅ ° '	Neptune ♆ ° '	Pluto ♇ ° '
01	01 S 36	01 N 51	03 N 46	01 N 17	02 N 22	00 N 36	01 S 46	09 S 40
04	01 47	01 45	03 50	01 18	02 23	00 36	01 46	09 40
07	01 57	01 38	03 52	01 19	02 24	00 36	01 45	09 39
10	02 03	01 31	03 54	01 20	02 25	00 36	01 45	09 39
13	02 06	01 24	03 55	01 20	02 25	00 36	01 45	09 39
16	02 06	01 16	03 55	01 21	02 26	00 36	01 45	09 39
19	02 01	01 08	03 54	01 22	02 27	00 36	01 45	09 38
22	01 51	01 00	03 53	01 23	02 28	00 36	01 45	09 38
25	01 35	00 51	03 51	01 24	02 29	00 36	01 44	09 38
28	01 00	00 43	03 48	01 24	02 30	00 36	01 44	09 38
31	00 S 43	00 N 34	03 N 45	01 N 25	02 N 31	00 N 36	01 S 44	09 S 38

DATA

Julian Date	2466155
Delta T	+77 seconds
Ayanamsa	24° 24' 42"
Synetic vernal point	04° ♓ 42' 17"
True obliquity of ecliptic	23° 26' 05"

MOON'S PHASES, APSIDES AND POSITIONS ☽

Date	h m	Phase	Longitude ° '	Eclipse Indicator
07	11 05	☾	16 ♎ 45	
14	03 25	●	23 ♑ 33	
21	02 21	☽	00 ♉ 38	
29	07 55	○	09 ♌ 00	

Day	h m		
13	09 50	Perigee	
25	22 34	Apogee	
07	00 06	0S	
13	06 40	Max dec	25° S 27'
19	12 32	0N	
26	22 14	Max dec	25° N 28'

LONGITUDES

Date	Chiron ⚷ ° '	Ceres ⚳ ° '	Pallas ⚴ ° '	Juno ⚵ ° '	Vesta ⚶ ° '	Black Moon Lilith ⚸ ° '
01	09 ♋ 51	21 ♈ 26	29 ♒ 07	18 ♒ 06	12 ♐ 42	00 ♋ 53
11	09 ♋ 11	23 ♈ 03	02 ♓ 01	22 ♒ 30	17 ♐ 57	01 ♋ 59
21	08 ♋ 33	25 ♈ 07	05 ♓ 12	27 ♒ 03	23 ♐ 05	03 ♋ 06
31	07 ♋ 59	27 ♈ 35	08 ♓ 12	01 ♓ 43	28 ♐ 16	04 ♋ 12

ASPECTARIAN

h m	Aspects
01 Sunday	
07 04	☽ Q ♄
09 24	☽ □ ♆
10 42	☽ ⚹ ♃
12 13	☽ ☌ ♅
16 57	☽ △ ♀
18 04	☽ # ☉
19 57	☿ ∠ ♇
21 58	☽ # ♇
02 Monday	
02 45	☽ ⊼ ☿
06 37	☽ ⊼ ☉
07 06	☽ ⚹ ♄
07 38	☽ ⊻ ♂
12 31	☉ □ ♄
15 27	☉ ☍ ♂
16 29	☽ ± ☿
16 40	♀ ± ♂
16 42	☽ ∠ ♃
17 21	☽ ‖ ♅
19 09	☽ ⊥ ♂
19 37	☽ ± ☉
22 13	♂ □ ♄
22 21	☽ # ♀
03 Tuesday	
01 10	☉ ⊥ ♀
07 36	☽ ☍ ♇
09 07	☿ ☍ ♂
10 29	☽ ⊥ ♃
12 06	☿ □ ♄
12 18	☽ ∠ ♂
12 33	♀ ± ♆
12 47	☽ ∠ ♄
12 53	☽ ⚼ ☿
14 55	☽ ⚼ ☉
20 45	☽ △ ♆
22 14	☽ ⊻ ♃
23 19	☽ ⊻ ♅
04 Wednesday	
06 21	☽ ⊥ ♄
10 34	☽ □ ♀
10 54	☽ ⊥ ♅
16 30	☽ ⚹ ♂
17 56	☽ ⊻ ♄
22 15	☽ △ ☿
22 31	☽ △ ☉
05 Thursday	
01 40	☽ ⚼ ♆
02 59	☽ ‖ ♆
03 23	☉ ☌ ☿
04 05	☽ ∠ ♅
14 39	☽ Q ♂
17 27	☽ ⊼ ♇
18 42	☽ ± ♆
06 Friday	
00 27	☽ Q ♀
02 18	☉ ‖ ♇
03 55	♀ # ♅
04 42	☽ ± ♇
05 17	☿ ⊥ ♀
05 55	☽ ⊼ ♆
07 31	☽ ☌ ♃
08 11	☽ ⚹ ♅
09 47	♀ ⊼ ♂
13 29	☽ # ♄
21 21	☽ ⚼ ♇
21 50	☽ ‖ ♃
23 07	☽ □ ♂
07 Saturday	
00 54	☽ ⚹ ♀
02 13	☽ ☌ ♄
02 20	☽ # ♃
06 08	☽ Q ♅
10 35	☽ ‖ ♄
11 05	☽ □ ☉
11 47	☿ ⊥ ♇
13 58	☽ □ ☿
15 11	♀ ⚹ ♄
21 59	♀ Q ♇
08 Sunday	
00 27	☽ △ ♇
06 31	☽ ∠ ♀
12 05	☽ ☍ ♆
13 43	☽ ⊻ ♃
14 04	☽ □ ♅
19 01	☽ # ♆
21 29	☽ Q ☉
22 03	☉ ⊥ ♇
23 45	☽ ⊥ ♀
09 Monday	
00 09	☽ ⊥ ♃
01 57	☽ Q ☿
02 52	☽ △ ♂
04 20	♀ Q ♃
07 16	☽ ⊻ ♄
08 27	♂ ⚼ ♇
11 02	☽ ⊻ ♀
15 34	☽ ∠ ♃
17 29	☽ ⊥ ♄
19 35	☽ ⚹ ☉
10 Tuesday	
01 01	☽ ⚹ ☿
03 38	☽ ⚼ ♂
04 10	☽ □ ♇
08 35	☽ ∠ ♄
15 02	☽ ⊼ ♆
16 39	☽ ⚹ ♃
16 46	☽ △ ♅
18 11	☽ ± ♂
20 29	♀ ⚼ ♆
21 07	☽ # ♅
22 23	☽ ∠ ☉
11 Wednesday	
00 53	☽ ± ♆
02 43	☽ ‖ ♀
03 46	☽ ⊼ ♂
04 42	☿ ⊻ ♇
04 57	☽ ∠ ☿
05 07	♆ St D
06 58	☽ ‖ ☉
09 13	☽ ⚹ ♄
09 48	☽ Q ♇
11 41	☽ ‖ ♇
12 14	☽ Q ♃
14 05	☽ ⊥ ☉
15 29	☽ ⚼ ♆
16 10	♀ ⚼ ♅
16 55	☽ ‖ ☿
17 06	☽ ⚼ ♅
17 11	☽ ☌ ♀
21 10	♃ ⚹ ♅
21 18	☽ ⊥ ☿
12 Thursday	
00 26	☽ ⊻ ☉
04 32	☽ Q ♄
05 07	☽ ⚹ ♇
07 28	☽ ± ♅
08 06	☽ ⊻ ☿
15 27	☽ △ ♆
16 59	☽ ⊼ ♅
17 04	☽ □ ♃
13 Friday	
02 41	☽ ☍ ♂
04 40	☉ ‖ ♀
04 58	☽ ∠ ♇
09 07	☽ □ ♄
19 12	☽ ⊥ ♇
20 48	☿ ‖ ♇
21 03	☽ ⊻ ♀
14 Saturday	
03 25	☽ ☌ ☉
04 42	☽ ⊻ ♇
07 22	☽ ⊥ ♀
13 28	☽ ☌ ☿
14 55	☽ □ ♆
16 19	☽ ☍ ♅
16 33	☽ △ ♃
22 54	☉ ⊻ ♇
22 56	☽ ∠ ♀
15 Sunday	
01 10	☽ ⊼ ♂
01 46	☽ ‖ ♇
02 34	☿ □ ♆
05 12	☽ ‖ ☿
06 23	☽ ‖ ♀
08 46	☽ △ ♄
10 35	☽ ± ♂
10 52	☽ ‖ ☉
13 32	☿ ‖ ♀
14 28	☿ ☍ ♅
15 41	☽ # ♅
16 34	☽ ⚼ ♃
17 13	☿ △ ♃
19 46	☽ Q ♆
16 Monday	
00 51	☽ ⚼ ♂
01 17	☽ ⚹ ♀
04 54	☽ ☌ ♇
07 05	☽ ⊻ ☉
07 14	☽ ± ♃
09 08	☽ ⚼ ♄
13 18	♃ St R
15 27	☽ ⚹ ♆
16 44	☽ ⊼ ♅
17 09	☽ ⊼ ♃
17 44	☽ ⊥ ☉
20 19	☽ ⊻ ☿
22 56	☽ Q ♀
17 Tuesday	
00 06	☽ ± ♄
01 08	☽ △ ♂
02 44	☽ ± ♅
07 45	☽ ⊥ ☿
10 02	☽ ∠ ☉
10 14	☽ ⊼ ♄
12 35	♀ Q ♄
16 44	☽ ∠ ♆
17 38	☽ # ♆
18 00	☽ ⚼ ♅
21 17	♀ ⚹ ♇
18 Wednesday	
01 20	☽ ∠ ☿
07 45	☽ ⊻ ♇
07 46	☿ ⊼ ♂
08 22	☽ ⊥ ♆
08 46	☽ □ ♀
14 07	☽ ⚹ ☉
17 16	☉ ‖ ☿
17 49	♀ ± ♅
18 23	☽ ⊥ ♇
19 00	☽ ⊻ ♆
20 13	☽ △ ♅
20 48	☽ ☍ ♃
22 52	☿ # ♅
19 Thursday	
01 58	☽ ‖ ♄
04 19	☽ □ ♂
05 42	☉ # ♅
07 51	☽ ⚹ ☿
10 17	☽ # ♃
10 40	☽ ∠ ♇
13 33	☽ Q ☉
14 48	☽ ‖ ♃
15 17	☽ ☍ ♄
23 15	☽ # ♄
20 Friday	
09 26	☽ Q ☿
12 59	☽ Q ♂
14 40	☽ ⚹ ♇
21 37	☽ △ ♀
21 Saturday	
02 21	☽ □ ☉
02 44	☽ ☌ ♆
03 50	☽ □ ♅
04 12	♄ St R
04 36	☽ ⊼ ♃
07 04	☉ □ ♆
08 53	☿ ± ♂
10 07	☽ ‖ ♆
11 34	☽ ⚹ ♂
13 49	☽ Q ♇
16 15	☽ ± ♃
16 44	♀ ‖ ♇
17 17	☿ △ ♄
19 46	☉ ☍ ♅
22 Sunday	
00 32	☽ ⊼ ♄
01 43	☽ □ ☿
05 32	☉ △ ♃
06 06	☽ ⚼ ♀
09 58	☽ ⚼ ♃
12 25	☽ ± ♄
15 02	☽ Q ♅
16 34	☽ ∠ ♂
23 Monday	
01 33	☽ □ ♇
01 46	♀ △ ♆
02 06	☽ ± ♀
06 06	☽ # ☿
06 25	☽ ⚼ ♄
10 17	☽ ⊥ ♂
10 40	♀ ⊼ ♅
14 09	☽ ⊻ ♆
14 41	☽ # ☉
15 03	☽ ⚹ ♅
15 35	☽ ⊼ ♀
16 00	☽ △ ♃
19 17	☽ △ ☉
19 27	♀ □ ♃
22 02	☽ ‖ ♅
22 12	☽ ⊻ ♂
24 Tuesday	
02 19	☽ ⊥ ♆
12 09	☿ ⚼ ♃
12 48	☽ △ ♄
13 58	☽ # ♇
15 18	☽ # ♀
20 38	☽ ∠ ♆
21 25	☽ ∠ ♅
23 51	☉ ⊼ ♂
25 Wednesday	
00 09	☽ △ ☿
04 41	☽ ⚼ ☉
14 35	☽ △ ♇
15 42	☽ ⊥ ♅
18 37	☿ Q ♆
23 02	♀ ☍ ♂
26 Thursday	
00 45	☽ ± ☉
02 03	☿ ⚼ ♂
03 14	☽ ⚹ ♆
03 52	☽ ⊻ ♅
04 54	☽ □ ♃
10 15	☽ ☌ ♂
11 45	☽ ⚼ ☿
11 46	☽ ☍ ♀
12 09	☿ ∠ ♀
14 07	☽ ⊼ ☉
21 10	☽ ⚼ ♇
27 Friday	
01 49	☽ □ ♄
03 37	☽ Q ♆
08 56	☽ ± ☿
15 25	☽ ± ♇
17 16	☽ Q ♃
22 56	☽ ⊼ ☿
28 Saturday	
03 30	☽ ⊼ ♇
13 58	☽ Q ♄
15 49	☽ □ ♆
16 10	☽ ☌ ♅
17 13	☽ ⚹ ♃
21 47	☽ ⊻ ♂
29 Sunday	
04 40	☽ # ♀
06 15	☽ # ♇
06 57	☽ ⊼ ♀
07 55	☽ ☍ ☉
09 32	☽ ⊥ ♂
09 33	☿ ☌ ♇
13 44	☽ ⚹ ♄
20 07	☽ ± ♀
20 56	☽ ‖ ♅
21 33	☿ ± ♃
22 47	☽ ∠ ♃
30 Monday	
01 10	☉ ± ♂
03 02	☽ ∠ ♂
07 47	♀ ∠ ♇
14 53	☽ ☍ ♇
15 40	☽ ⚼ ♀
16 02	☽ # ☉
16 17	☽ ⊥ ♃
19 01	☽ ☍ ☿
19 02	☽ ∠ ♄
19 05	☿ ⚼ ♄
31 Tuesday	
02 44	☽ △ ♆
02 50	☽ ⊻ ♅
03 51	☽ ⊻ ♃
07 32	☉ ⊻ ♀
07 51	☽ ⚹ ♂
12 22	☽ ⊥ ♄
14 21	☽ ⊥ ♅
19 31	☽ # ☿
23 24	☽ ⊼ ☉
23 43	☽ △ ♀
23 50	☽ ⊻ ♄

All ephemeris data is given at 12.00 UT and the Moon's longitude is additionally given for 24.00 UT

FEBRUARY 2040

Date	Sidereal time (h m s)	Sun ☉ (° ' ")	Moon ☽ (° ' ")	Moon ☽ 24.00 (° ' ")	Mercury ☿ (° ')	Venus ♀ (° ')	Mars ♂ (° ')	Jupiter ♃ (° ')	Saturn ♄ (° ')	Uranus ♅ (° ')	Neptune ♆ (° ')	Pluto ♇ (° ')
01	20 45 18	12 ♒ 13 28	18 ♍ 19 05	24 ♍ 40 51	29 ♒ 27	12 ♑ 29	03 ♋ 26	01 ♎ 26	11 ♎ 54	00 ♌ 55	00 ♉ 56	24 ♒ 52
02	20 49 14	13 14 21	01 ♎ 05 31	07 ♎ 33 14	00 ♓ 51	13 43	03 R 21	01 R 23	11 R 53	00 R 52	00 57	24 53
03	20 53 11	14 15 13	14 ♎ 04 09	20 ♎ 38 27	02 10	14 58	03 16	01 20	11 52	00 50	00 58	24 55
04	20 57 07	15 16 04	27 ♎ 16 21	03 ♏ 58 03	03 24	16 12	03 11	01 16	11 50	00 47	00 59	24 57
05	21 01 04	16 16 55	10 ♏ 43 47	17 ♏ 33 45	04 32	17 26	03 08	01 13	11 49	00 45	00 59	24 58
06	21 05 00	17 17 44	24 ♏ 28 07	01 ♐ 27 02	05 33	18 40	03 05	01 09	11 47	00 42	01 00	25 00
07	21 08 57	18 18 33	08 ♐ 30 31	15 ♐ 38 33	06 26	19 54	03 03	01 05	11 45	00 40	01 01	25 02
08	21 12 54	19 19 21	22 ♐ 50 57	00 ♑ 07 26	07 12	21 09	03 02	01 01	11 43	00 37	01 02	25 04
09	21 16 50	20 20 09	07 ♑ 27 31	14 ♑ 50 37	07 48	22 23	03 D 02	00 56	11 42	00 35	01 03	25 05
10	21 20 47	21 20 55	22 ♑ 15 56	29 ♑ 42 33	08 15	23 37	03 02	00 52	11 39	00 32	01 04	25 07
11	21 24 43	22 21 40	07 ♒ 09 28	14 ♒ 35 35	08 31	24 52	03 03	00 47	11 37	00 30	01 05	25 09
12	21 28 40	23 22 23	21 ♒ 59 45	29 ♒ 20 52	08 37	26 06	03 05	00 42	11 35	00 28	01 06	25 11
13	21 32 36	24 23 05	06 ♓ 37 56	13 ♓ 50 00	08 R 33	27 20	03 08	00 37	11 33	00 25	01 07	25 12
14	21 36 33	25 23 46	20 ♓ 56 18	27 ♓ 56 17	08 18	28 35	03 11	00 32	11 30	00 23	01 09	25 14
15	21 40 29	26 24 25	04 ♈ 49 33	11 ♈ 35 53	07 52	29 ♑ 49	03 15	00 27	11 28	00 21	01 10	25 16
16	21 44 26	27 25 03	18 ♈ 15 15	24 ♈ 47 50	07 17	01 ♒ 03	03 20	00 22	11 25	00 19	01 11	25 18
17	21 48 23	28 25 38	01 ♉ 13 54	07 ♉ 33 52	06 33	02 18	03 25	00 16	11 22	00 16	01 12	25 19
18	21 52 19	29 ♒ 26 13	13 ♉ 48 16	19 ♉ 57 39	05 41	03 32	03 31	00 10	11 20	00 14	01 13	25 21
19	21 56 16	00 ♓ 26 45	26 ♉ 02 42	02 ♊ 04 05	04 43	04 46	03 38	00 ♎ 04	11 17	00 12	01 15	25 23
20	22 00 12	01 27 15	08 ♊ 02 30	13 ♊ 58 39	03 40	06 01	03 45	29 ♍ 58	11 14	00 10	01 16	25 25
21	22 04 09	02 27 44	19 ♊ 53 14	25 ♊ 46 56	02 34	07 15	03 53	29 52	11 11	00 08	01 17	25 26
22	22 08 05	03 28 11	01 ♋ 40 25	07 ♋ 34 17	01 27	08 29	04 02	29 46	11 08	00 06	01 19	25 28
23	22 12 02	04 28 36	13 ♋ 29 06	19 ♋ 25 24	00 ♓ 21	09 44	04 11	29 40	11 04	00 04	01 20	25 30
24	22 15 58	05 29 00	25 ♋ 23 38	01 ♌ 24 13	29 ♒ 16	10 58	04 21	29 33	11 01	00 02	01 22	25 32
25	22 19 55	06 29 21	07 ♌ 27 29	13 ♌ 33 42	28 15	12 12	04 31	29 27	10 58	00 ♌ 00	01 23	25 33
26	22 23 52	07 29 41	19 ♌ 43 04	25 ♌ 55 44	27 19	13 27	04 42	29 20	10 54	29 ♋ 58	01 25	25 35
27	22 27 48	08 29 58	02 ♍ 11 44	08 ♍ 31 07	26 28	14 41	04 54	29 13	10 51	29 56	01 26	25 37
28	22 31 45	09 30 14	14 ♍ 53 48	21 ♍ 19 44	25 43	15 55	05 06	29 06	10 47	29 54	01 28	25 38
29	22 35 41	10 ♓ 30 28	27 ♍ 48 47	04 ♎ 20 49	25 ♒ 06	17 ♒ 10	05 ♋ 19	28 ♍ 59	10 ♎ 43	29 ♋ 52	01 ♉ 29	25 ♒ 40

Date	Moon True ☊ (° ')	Moon Mean ☊ (° ')	Moon ☽ Latitude (° ')
01	10 ♊ 49	09 ♊ 45	05 N 01
02	10 R 40	09 42	04 45
03	10 33	09 39	04 14
04	10 29	09 35	03 29
05	10 28	09 32	02 31
06	10 D 28	09 29	01 24
07	10 28	09 26	00 N 10
08	10 R 28	09 23	01 S 05
09	10 25	09 20	02 18
10	10 20	09 16	03 23
11	10 12	09 13	04 13
12	10 01	09 10	04 47
13	09 50	09 07	05 01
14	09 38	09 04	04 56
15	09 28	09 00	04 33
16	09 20	08 57	03 55
17	09 15	08 54	03 07
18	09 12	08 51	02 10
19	09 11	08 48	01 09
20	09 D 11	08 45	00 S 06
21	09 R 11	08 41	00 N 56
22	09 09	08 38	01 56
23	09 05	08 35	02 51
24	08 58	08 32	03 38
25	08 48	08 29	04 17
26	08 36	08 25	04 44
27	08 23	08 22	04 58
28	08 09	08 19	04 58
29	07 ♊ 55	08 ♊ 16	04 N 42

DECLINATIONS

Date	Sun ☉ (° ')	Moon ☽ (° ')	Mercury ☿ (° ')	Venus ♀ (° ')	Mars ♂ (° ')	Jupiter ♃ (° ')	Saturn ♄ (° ')	Uranus ♅ (° ')	Neptune ♆ (° ')	Pluto ♇ (° ')
01	17 S 08	09 N 14	12 S 10	22 S 20	27 N 07	00 N 44	02 S 24	20 N 32	10 N 10	22 S 17
02	16 51	03 N 55	11 29	22 16	27 06	00 46	02 23	20 33	10 11	22 17
03	16 33	01 S 39	10 48	22 11	27 05	00 47	02 22	20 33	10 11	22 16
04	16 15	07 15	10 09	22 05	27 04	00 49	02 21	20 34	10 11	22 16
05	15 57	12 38	09 31	21 59	27 03	00 51	02 20	20 34	10 12	22 15
06	15 39	17 31	08 54	21 52	27 02	00 52	02 19	20 35	10 12	22 14
07	15 20	21 33	08 20	21 44	27 00	00 54	02 19	20 35	10 12	22 14
08	15 01	24 20	07 48	21 36	26 59	00 56	02 18	20 36	10 13	22 13
09	14 42	25 31	07 20	21 27	26 58	00 58	02 17	20 36	10 13	22 13
10	14 23	24 56	06 54	21 18	26 57	01 00	02 16	20 37	10 14	22 12
11	14 03	22 34	06 33	21 07	26 55	01 02	02 14	20 37	10 14	22 11
12	13 44	18 42	06 16	20 56	26 54	01 04	02 13	20 38	10 14	22 11
13	13 23	13 44	06 03	20 45	26 52	01 06	02 12	20 38	10 15	22 10
14	13 03	08 07	05 55	20 33	26 51	01 08	02 11	20 39	10 15	22 10
15	12 43	02 S 15	05 52	20 20	26 49	01 11	02 10	20 39	10 16	22 09
16	12 22	03 N 32	05 53	20 07	26 48	01 13	02 09	20 40	10 16	22 09
17	12 01	08 59	05 59	19 53	26 46	01 16	02 07	20 40	10 17	22 08
18	11 40	13 54	06 10	19 39	26 45	01 18	02 06	20 41	10 17	22 08
19	11 19	18 09	06 25	19 24	26 43	01 21	02 05	20 41	10 18	22 07
20	10 57	21 33	06 43	19 08	26 42	01 23	02 03	20 42	10 18	22 06
21	10 36	23 59	07 05	18 52	26 40	01 26	02 02	20 42	10 19	22 06
22	10 14	25 22	07 29	18 35	26 38	01 28	02 00	20 43	10 19	22 05
23	09 52	25 35	07 54	18 18	26 36	01 31	01 59	20 43	10 20	22 05
24	09 30	24 38	08 21	18 00	26 35	01 34	01 58	20 43	10 20	22 04
25	09 08	22 32	08 48	17 42	26 33	01 37	01 56	20 44	10 21	22 04
26	08 45	19 23	09 15	17 23	26 31	01 39	01 55	20 44	10 21	22 03
27	08 23	15 19	09 41	17 04	26 29	01 42	01 53	20 45	10 22	22 03
28	08 00	10 31	10 06	16 44	26 27	01 45	01 51	20 45	10 23	22 02
29	07 S 38	05 N 11	10 S 30	16 S 24	26 N 25	01 N 48	01 S 50	20 N 45	10 N 23	22 S 02

ZODIAC SIGN ENTRIES

Date	h m	Planets
01	21 16	☿ ♓
02	09 58	☽ ♎
04	16 54	☽ ♏
06	21 31	☽ ♐
08	23 48	☽ ♑
11	00 28	☽ ♒
13	01 04	☽ ♓
15	03 34	☽ ♈
15	15 35	♀ ♒
17	09 41	☽ ♉
19	01 24	☉ ♓
19	19 52	☽ ♊
20	05 36	♃ ♍
22	08 35	☽ ♋
23	19 37	☿ ♒
24	21 12	☽ ♌
25	07 39	♅ ♋
27	07 48	☽ ♍
29	16 02	☽ ♎

LATITUDES

Date	Mercury ☿ (° ')	Venus ♀ (° ')	Mars ♂ (° ')	Jupiter ♃ (° ')	Saturn ♄ (° ')	Uranus ♅ (° ')	Neptune ♆ (° ')	Pluto ♇ (° ')
01	00 S 32	00 N 31	03 N 44	01 N 26	02 N 31	00 N 36	01 S 44	09 S 38
04	00 N 07	00 22	03 40	01 26	02 32	00 36	01 44	09 38
07	00 52	00 13	03 37	01 27	02 32	00 36	01 44	09 38
10	01 42	00 N 05	03 33	01 28	02 33	00 36	01 43	09 38
13	02 30	00 S 04	03 29	01 28	02 34	00 36	01 43	09 38
16	03 11	00 12	03 24	01 29	02 35	00 36	01 43	09 39
19	03 37	00 20	03 20	01 30	02 35	00 36	01 43	09 39
22	03 43	00 28	03 16	01 30	02 36	00 36	01 43	09 39
25	03 30	00 36	03 12	01 31	02 37	00 36	01 43	09 39
28	03 01	00 43	03 07	01 31	02 37	00 36	01 43	09 40
31	02 N 23	00 S 49	03 N 03	01 N 32	02 N 38	00 N 36	01 S 42	09 S 40

DATA

Julian Date	2466186
Delta T	+77 seconds
Ayanamsa	24° 24' 47"
Synetic vernal point	04° ♓ 42' 12"
True obliquity of ecliptic	23° 26' 06"

MOON'S PHASES, APSIDES AND POSITIONS ☽

Date	h m	Phase	Longitude (° ')	Eclipse Indicator
05	22 32	☾	16 ♏ 44	
12	14 24	●	23 ♒ 28	
19	21 34	☽	00 ♊ 51	
28	00 59	○	09 ♍ 03	

Day	h m		
10	18 37	Perigee	
22	15 13	Apogee	
03	04 58	0S	
09	16 08	Max dec	25° S 33'
15	21 16	0N	
23	04 37	Max dec	25° N 38'

LONGITUDES

Date	Chiron ⚷ (° ')	Ceres ⚳ (° ')	Pallas ⚴ (° ')	Juno ⚵ (° ')	Vesta ⚶ (° ')	Black Moon Lilith ⚸ (° ')
01	07 ♋ 56	27 ♈ 51	08 ♓ 32	02 ♓ 11	28 ♐ 47	04 ♋ 19
11	07 ♋ 29	00 ♉ 40	11 ♓ 48	06 ♓ 59	03 ♑ 48	05 ♋ 26
21	07 ♋ 08	03 ♉ 46	15 ♓ 08	11 ♓ 53	08 ♑ 41	06 ♋ 32
31	06 ♋ 56	07 ♉ 04	18 ♓ 32	16 ♓ 53	13 ♑ 26	07 ♋ 39

ASPECTARIAN

h m	Aspects
01 Wednesday	
00 56	♀ □ ♄
01 05	♅ □ ♆
04 39	☉ △ ♄
06 35	☽ Q ♂
07 27	☽ ∠ ♅
07 29	☽ ⚼ ♆
07 34	☽ ∥ ♆
11 48	☽ ± ☉
02 Thursday	
00 22	☽ ⊼ ♇
00 30	☽ ± ♆
06 13	☽ ⚼ ☉
06 27	♀ ∥ ♇
11 29	☽ ⊼ ☿
11 35	☽ ⚹ ♅
11 37	☽ ± ♇
11 44	☽ ⊼ ♆
12 25	☿ ⊼ ♅
12 33	☽ ☌ ♃
13 52	☿ ⚹ ♆
16 10	☽ □ ♂
18 44	☽ # ♄
21 15	☿ ⊼ ♃
23 55	☽ ± ☿
03 Friday	
01 38	☽ ∥ ♃
04 21	☽ ⚼ ♇
07 57	☽ ☌ ♄
08 19	☽ # ♃
09 44	☽ Q ♅
12 22	☽ △ ☉
13 48	☽ □ ♀
15 04	☽ ∥ ♄
18 16	☽ ⚼ ☿
04 Saturday	
07 48	☽ △ ♇
08 04	☿ △ ♂
10 31	☿ # ♆
18 18	☽ □ ♅
18 40	☽ ☍ ♆
19 09	☽ ⊻ ♃
22 34	☽ △ ♂
23 22	☽ ∥ ☿
05 Sunday	
00 01	☽ △ ☿
00 54	☽ # ♆
01 40	☽ Q ♀
05 48	☽ ⊥ ♃
10 26	☉ ⚼ ♃
13 54	☽ ⊻ ♄
21 36	☽ ∠ ♃
22 32	☽ □ ☉
06 Monday	
00 25	☽ ⊥ ♄
00 56	☽ ⚹ ♀
00 57	☽ ⚼ ♂
02 57	☽ ∥ ☉
12 55	☽ □ ♇
15 59	☽ ∠ ♄
16 30	☽ ± ♂
17 58	☿ ± ♄
18 36	♀ ⊥ ♇
22 41	☽ △ ♅
23 15	☽ ⊼ ♆
23 26	☽ ⚹ ♃
07 Tuesday	
02 45	☽ ⊼ ♂
05 18	☽ ∠ ♀
05 39	☽ # ♅
06 06	☉ ⚼ ♂
07 59	☽ Q ☉
08 16	☽ □ ☿
09 29	☽ ± ♆
13 17	☽ ∥ ♀
16 53	☽ ∥ ♇
17 28	☽ ⚹ ♄
18 20	☿ ± ♅
19 38	☽ Q ♇
19 40	☽ Q ♃
21 57	☽ ⊥ ♀
08 Wednesday	
00 00	☽ ⚼ ♅
00 11	♃ ± ♇
00 39	☽ ⚼ ♆
04 55	♃ ⊼ ♆
05 08	☉ Q ♆
05 42	☽ ⚹ ☉
08 55	☽ ⊻ ♀
13 27	☽ Q ♄
14 56	☽ ± ♅
15 40	☽ ⚹ ♇
16 04	☽ Q ☿
09 Thursday	
00 47	☽ ⊼ ♅
01 24	☽ □ ♃
01 31	☽ △ ♆
04 46	☽ ☍ ♂
08 17	☽ ∠ ☉
11 48	♂ St D
12 35	☽ ⚹ ☿
16 18	☽ ∠ ♇
18 52	☽ □ ♄
10 Friday	
00 00	☽ ⊥ ☉
01 14	☿ ∠ ♀
06 54	☽ ⊥ ♇
10 25	☽ ⊻ ☉
13 37	☽ ∠ ☿
14 23	☽ ☌ ♀
16 37	☽ ⊻ ♇
11 Saturday	
01 18	☽ ☍ ♅
01 47	☽ △ ♃
02 13	☽ □ ♆
04 25	☽ ⊥ ☿
05 23	☽ ⊼ ♂
14 14	☽ ⊻ ☿
14 44	☽ ∥ ♇
15 04	☽ ± ♂
17 42	♀ ⊻ ♇
19 11	☽ △ ♄
22 25	☽ ∥ ♀
12 Sunday	
01 00	☽ # ♅
01 51	☽ ⚼ ♃
05 38	☽ ⚼ ♂
07 18	☽ Q ♆
13 38	☿ St R
14 24	☽ ☌ ☉
16 23	☽ ± ♃
17 11	☽ ☌ ♇
19 18	☽ ⊻ ♀
19 27	☽ ⚼ ♄
13 Monday	
01 47	☽ ⊼ ♅
02 09	☽ ⊼ ♃
02 54	☽ ⚹ ♆
06 03	☽ ⊥ ♀
06 12	☽ △ ♂
10 12	☽ ± ♄
11 39	☽ ± ♅
13 37	☽ ∥ ☉
15 08	☽ ☌ ☿
17 14	☉ ± ♃
20 09	☽ ⊼ ♄
22 24	☽ ∠ ♀
14 Tuesday	
00 40	♀ # ♅
02 38	☽ ⚼ ♅
03 05	☽ # ♆
03 52	☽ ∠ ♆
08 04	☉ ☌ ♇
19 12	☽ ⊥ ♆
19 22	☽ ⊻ ♇
20 13	☽ ⊻ ☉
21 12	☽ ∥ ☿
15 Wednesday	
02 23	☽ ⚹ ♀
04 12	☽ △ ♅
04 24	☽ ☍ ♃
05 35	☽ ⊻ ♆
05 45	☽ ⊥ ♇
07 26	☽ ⊥ ☉
09 13	☽ □ ♂
12 23	☽ ∥ ♄
13 15	☉ ⚼ ♄
16 23	☽ # ♃
17 10	☽ ⊻ ☿
21 38	☽ ∠ ♇
21 59	♀ ☍ ♅
23 28	♀ △ ♃
23 43	☽ ☍ ♄
16 Thursday	
00 36	☽ ∠ ☉
01 39	☽ Q ♀
02 15	☽ ∥ ♃
03 25	☽ ⊥ ☿
06 11	☽ # ♄
14 32	♀ □ ♆
14 48	☿ ⊥ ♀
17 39	☽ Q ♂
18 59	☽ ∠ ☿
22 18	☽ # ☿
17 Friday	
00 57	☽ ⚹ ♇
06 18	☽ ⚹ ☉
09 35	♃ ⚹ ♅
10 12	☽ ⊼ ♃
10 12	☽ □ ♅
11 57	☽ ☌ ♆
14 13	☽ □ ♀
16 09	☽ ⚹ ♂
18 04	☽ ∥ ♆
20 14	☿ ± ♅
21 26	☽ ⚹ ☿
21 27	☽ ± ♃
23 34	☽ Q ♇
18 Saturday	
01 28	☽ # ☉
07 01	☽ Q ☉
07 14	☽ ⊼ ♄
11 46	♀ ⊼ ♂
14 38	☽ ⚼ ♃
18 49	☽ ± ♄
19 00	☽ Q ☿
20 36	☽ Q ♅
21 16	☽ ∠ ♂
21 29	☿ ± ♄
19 Sunday	
03 55	☉ ⊼ ♃
06 20	☉ ⊼ ♅
10 41	☽ □ ♇
11 22	☿ ⊻ ♀
12 28	☽ ⚼ ♄
15 11	☽ ⊥ ♂
19 35	☽ # ♀
19 57	☽ △ ♃
20 14	☽ ⚹ ♅
21 34	☽ □ ☉
22 22	☽ ⊻ ♆
20 Monday	
03 17	☽ ⊻ ♂
03 55	☽ □ ☿
05 21	☽ ∥ ♅
07 26	☽ △ ♀
07 27	☉ ⚹ ♆
10 11	☿ △ ♂
10 26	☽ ⊥ ♆
16 44	☽ # ♇
18 24	☽ △ ♄
21 Tuesday	
02 22	☽ ∠ ♅
04 41	☽ ∠ ♆
13 13	☉ ☌ ☿
17 22	☽ ⚼ ♀
20 36	☽ ⊥ ♅
23 20	☽ △ ♇
22 Wednesday	
06 19	☉ # ♆
08 09	☽ □ ♃
08 48	☽ ⊻ ♅
11 16	☽ ⚹ ♆
11 35	☽ △ ☿
13 51	☽ ± ♀
14 59	☿ ⚹ ♆
16 00	☽ △ ☉
16 52	☽ ☌ ♂
23 Thursday	
03 29	☽ ⊼ ♀
03 49	☉ △ ♂
05 56	☽ ⚼ ♇
07 08	☽ □ ♄
11 42	☽ Q ♆
15 27	☽ ⚼ ☿
18 29	☿ ⊼ ♅
20 22	☽ Q ♃
22 16	♀ ± ♂
24 Friday	
00 11	☽ ± ♇
01 14	☽ ⚼ ☉
01 27	☉ ± ♄
04 53	☿ ⊼ ♃
08 05	☽ ± ☿
12 16	☽ ⊼ ♇
12 57	♀ △ ♄
19 08	☽ ⊼ ☿
19 13	☽ Q ♄
20 14	☽ ⚹ ♃
20 55	☽ ± ☉
21 14	☽ ☌ ♅
23 56	☽ □ ♆
25 Saturday	
00 33	☉ ± ♅
06 07	☽ ⊻ ♂
09 55	☽ ⊼ ☉
16 10	☽ # ♇
18 08	☽ ⊥ ♂
18 52	☽ ⚹ ♄
21 41	☉ ∥ ☿
22 24	☽ ☍ ♀
26 Sunday	
01 36	☽ ∠ ♃
02 37	☽ ∥ ♅
11 59	☽ ∠ ♂
18 56	☽ ⊥ ♃
23 22	☽ ☍ ♇
23 54	☽ ∠ ♄
27 Monday	
01 27	☽ # ♀
01 43	☽ ☍ ☿
03 46	♀ ⚼ ♃
06 22	☽ ⊻ ♃
07 41	☽ ⊻ ♅
10 33	☽ △ ♆
17 01	☽ ⊥ ♄
17 14	☽ ⚹ ♂
19 05	☽ ⊥ ♅
28 Tuesday	
00 59	☽ ☍ ☉
04 19	☽ ⊻ ♄
09 44	☿ ⚼ ♄
12 00	☽ ∠ ♅
12 41	☽ ∥ ♆
13 49	☽ # ☿
14 07	☽ ⊼ ♀
14 53	☿ ☌ ♇
14 56	☽ ⚼ ♆
16 11	☽ Q ♂
29 Wednesday	
00 26	☽ # ☉
02 29	☽ ± ♀
04 46	☿ # ♆
07 12	☽ ⊼ ☿
07 42	☽ ± ♆
08 02	☽ ⊼ ♇
14 09	☽ ☌ ♃
15 47	☽ ⚹ ♅
16 51	☉ ⊼ ♄
17 47	☽ ± ☿
18 46	☽ ⊼ ♆
19 06	☽ ± ♇
20 50	☽ ⚼ ♀
21 38	♃ # ♄

All ephemeris data is given at 12.00 UT and the Moon's longitude is additionally given for 24.00 UT

MARCH 2040

LONGITUDES

Date	Sidereal time h m s	Sun ☉ ° ' "	Moon ☽ ° ' "	Moon ☽ 24.00 ° ' "	Mercury ☿ ° '	Venus ♀ ° '	Mars ♂ ° '	Jupiter ♃ ° '	Saturn ♄ ° '	Uranus ♅ ° '	Neptune ♆ ° '	Pluto ♇ ° '
01	22 39 38	11 ♓ 30 41	10 ♎ 55 42	17 ♎ 33 17	24 ♒ 35	18 ♒ 24	05 ♋ 32	28 ♍ 52	10 ♎ 40	29 ♋ 50	01 ♉ 31	25 ♒ 42
02	22 43 34	12 30 52	24 ♎ 13 29	00 ♏ 56 12	24 R 11	19 38	05 45	28 R 45	10 R 36	29 R 49	01 32	25 43
03	22 47 31	13 31 01	07 ♏ 41 22	14 ♏ 28 58	23 54	20 53	05 59	28 38	10 32	29 47	01 34	25 45
04	22 51 27	14 31 09	21 ♏ 18 59	28 ♏ 11 28	23 45	22 07	06 14	28 30	10 28	29 45	01 36	25 47
05	22 55 24	15 31 15	05 ♐ 06 27	12 ♐ 03 57	23 D 41	23 21	06 29	28 23	10 24	29 44	01 37	25 48
06	22 59 21	16 31 20	19 ♐ 04 01	26 ♐ 06 38	23 45	24 36	06 45	28 16	10 20	29 42	01 39	25 50
07	23 03 17	17 31 24	03 ♑ 11 43	10 ♑ 19 08	23 54	25 50	07 00	28 08	10 16	29 41	01 41	25 52
08	23 07 14	18 31 25	17 ♑ 28 40	24 ♑ 39 58	24 09	27 04	07 17	28 00	10 12	29 39	01 42	25 53
09	23 11 10	19 31 26	01 ♒ 52 36	09 ♒ 06 03	24 30	28 19	07 34	27 53	10 08	29 38	01 44	25 55
10	23 15 07	20 31 24	16 ♒ 19 39	23 ♒ 32 42	24 56	29 ♒ 33	07 51	27 45	10 03	29 36	01 46	25 57
11	23 19 03	21 31 21	00 ♓ 44 24	07 ♓ 53 59	25 26	00 ♓ 48	08 09	27 38	09 59	29 35	01 48	25 58
12	23 23 00	22 31 16	15 ♓ 00 39	22 ♓ 03 40	26 01	02 02	08 27	27 30	09 55	29 34	01 50	26 00
13	23 26 56	23 31 08	29 ♓ 02 21	05 ♈ 56 08	26 41	03 16	08 45	27 22	09 50	29 32	01 51	26 01
14	23 30 53	24 30 59	12 ♈ 44 36	19 ♈ 27 26	27 24	04 31	09 04	27 14	09 46	29 31	01 53	26 03
15	23 34 50	25 30 48	26 ♈ 04 28	02 ♉ 35 41	28 10	05 45	09 23	27 06	09 42	29 30	01 55	26 05
16	23 38 46	26 30 35	09 ♉ 01 12	15 ♉ 21 15	29 00	06 59	09 43	26 59	09 37	29 29	01 57	26 06
17	23 42 43	27 30 19	21 ♉ 36 11	27 ♉ 46 26	29 ♒ 54	08 14	10 03	26 51	09 33	29 28	01 59	26 08
18	23 46 39	28 30 02	03 ♊ 52 31	09 ♊ 55 00	00 ♓ 50	09 28	10 23	26 43	09 28	29 27	02 01	26 09
19	23 50 36	29 ♓ 29 42	15 ♊ 54 31	21 ♊ 51 43	01 49	10 42	10 44	26 35	09 24	29 26	02 03	26 11
20	23 54 32	00 ♈ 29 20	27 ♊ 47 18	03 ♋ 41 55	02 51	11 57	11 05	26 27	09 19	29 25	02 05	26 12
21	23 58 29	01 28 56	09 ♋ 36 16	15 ♋ 31 01	03 55	13 11	11 27	26 20	09 14	29 24	02 07	26 14
22	00 02 25	02 28 29	21 ♋ 26 50	27 ♋ 24 19	05 02	14 25	11 48	26 12	09 10	29 24	02 09	26 15
23	00 06 22	03 28 00	03 ♌ 24 02	09 ♌ 26 33	06 11	15 39	12 10	26 04	09 05	29 23	02 11	26 17
24	00 10 19	04 27 29	15 ♌ 32 17	21 ♌ 41 41	07 22	16 54	12 32	25 57	09 00	29 22	02 13	26 18
25	00 14 15	05 26 56	27 ♌ 55 02	04 ♍ 12 36	08 35	18 08	12 55	25 49	08 56	29 21	02 15	26 20
26	00 18 12	06 26 20	10 ♍ 34 32	17 ♍ 00 55	09 51	19 22	13 18	25 41	08 51	29 21	02 17	26 21
27	00 22 08	07 25 42	23 ♍ 31 42	00 ♎ 06 48	11 08	20 36	13 41	25 34	08 46	29 20	02 19	26 22
28	00 26 05	08 25 02	06 ♎ 46 01	13 ♎ 29 06	12 27	21 51	14 05	25 27	08 42	29 20	02 21	26 24
29	00 30 01	09 24 20	20 ♎ 15 45	27 ♎ 05 37	13 47	23 05	14 28	25 19	08 37	29 19	02 23	26 25
30	00 33 58	10 23 36	03 ♏ 58 21	10 ♏ 53 32	15 10	24 19	14 52	25 12	08 32	29 19	02 26	26 26
31	00 37 54	11 ♈ 22 50	17 ♏ 50 50	24 ♏ 49 54	16 ♓ 34	25 ♓ 33	15 ♋ 16	25 ♍ 05	08 ♎ 28	29 ♋ 19	02 ♉ 28	26 ♒ 28

Date	Moon True ☊ ° '	Moon Mean ☊ ° '	Moon ☽ Latitude ° '
01	07 ♊ 44	08 ♊ 13	04 N 12
02	07 R 36	08 10	03 27
03	07 31	08 06	02 30
04	07 28	08 03	01 24
05	07 D 27	08 00	00 N 12
06	07 R 27	07 57	01 S 01
07	07 27	07 54	02 12
08	07 24	07 51	03 15
09	07 19	07 47	04 06
10	07 11	07 44	04 42
11	07 01	07 41	04 59
12	06 49	07 38	04 58
13	06 37	07 35	04 39
14	06 27	07 31	04 04
15	06 18	07 28	03 16
16	06 12	07 25	02 19
17	06 09	07 22	01 17
18	06 08	07 19	00 S 12
19	06 D 08	07 16	00 N 52
20	06 09	07 12	01 53
21	06 R 09	07 09	02 49
22	06 07	07 06	03 37
23	06 02	07 03	04 17
24	05 56	07 00	04 45
25	05 47	06 57	05 01
26	05 37	06 53	05 03
27	05 26	06 50	04 49
28	05 16	06 47	04 20
29	05 08	06 44	03 36
30	05 02	06 41	02 38
31	04 ♊ 58	06 ♊ 37	01 N 31

DECLINATIONS

Date	Sun ☉ ° '	Moon ☽ ° '	Mercury ☿ ° '	Venus ♀ ° '	Mars ♂ ° '	Jupiter ♃ ° '	Saturn ♄ ° '	Uranus ♅ ° '	Neptune ♆ ° '	Pluto ♇ ° '
01	07 S 15	00 S 28	10 S 52	16 S 04	26 N 24	01 N 51	01 S 48	20 N 46	10 N 24	22 S 01
02	06 52	06 11	11 12	15 42	26 22	01 54	01 47	20 46	10 24	22 01
03	06 29	11 42	11 31	15 21	26 20	01 57	01 45	20 46	10 25	22 00
04	06 06	16 44	11 47	14 59	26 17	02 00	01 43	20 47	10 26	22 00
05	05 42	20 57	12 01	14 36	26 15	02 03	01 41	20 47	10 26	22 00
06	05 19	24 00	12 13	14 14	26 13	02 06	01 40	20 47	10 27	21 59
07	04 56	25 35	12 23	13 50	26 11	02 09	01 38	20 48	10 27	21 59
08	04 32	25 31	12 31	13 27	26 09	02 12	01 36	20 48	10 28	21 58
09	04 09	23 44	12 37	13 03	26 06	02 15	01 35	20 48	10 29	21 58
10	03 45	20 25	12 41	12 39	26 04	02 18	01 33	20 49	10 29	21 57
11	03 22	15 52	12 42	12 14	26 02	02 22	01 31	20 49	10 30	21 57
12	02 58	10 29	12 42	11 49	25 59	02 25	01 29	20 49	10 31	21 57
13	02 34	04 S 39	12 40	11 24	25 57	02 28	01 27	20 49	10 31	21 56
14	02 11	01 N 18	12 36	10 58	25 54	02 31	01 25	20 50	10 32	21 56
15	01 47	07 02	12 30	10 32	25 51	02 34	01 24	20 50	10 33	21 55
16	01 23	12 18	12 23	10 06	25 49	02 37	01 22	20 50	10 33	21 55
17	01 00	16 56	12 13	09 39	25 46	02 40	01 20	20 50	10 34	21 55
18	00 36	20 43	12 02	09 13	25 43	02 44	01 18	20 50	10 35	21 54
19	00 S 12	23 33	11 50	08 46	25 40	02 47	01 16	20 51	10 35	21 54
20	00 N 12	25 18	11 36	08 18	25 37	02 50	01 14	20 51	10 36	21 54
21	00 35	25 53	11 20	07 51	25 34	02 53	01 12	20 51	10 37	21 53
22	00 59	25 18	11 03	07 23	25 31	02 56	01 11	20 51	10 38	21 53
23	01 23	23 33	10 44	06 55	25 28	02 59	01 09	20 51	10 38	21 53
24	01 46	20 43	10 24	06 27	25 24	03 02	01 07	20 51	10 39	21 52
25	02 10	16 54	10 03	05 59	25 21	03 05	01 05	20 51	10 40	21 52
26	02 33	12 16	09 40	05 31	25 17	03 08	01 03	20 52	10 40	21 52
27	02 57	07 00	09 15	05 02	25 14	03 11	01 01	20 52	10 41	21 51
28	03 20	01 N 18	08 50	04 34	25 10	03 14	00 59	20 52	10 42	21 51
29	03 44	04 S 35	08 23	04 05	25 07	03 17	00 57	20 52	10 43	21 51
30	04 07	10 21	07 54	03 36	25 03	03 20	00 56	20 52	10 43	21 51
31	04 N 30	15 S 42	07 S 25	03 S 07	24 N 59	03 N 22	00 S 54	20 N 52	10 N 44	21 S 50

ZODIAC SIGN ENTRIES

Date	h	m	Planets
02	22	20	☽ ♏
05	03	09	☽ ♐
07	06	36	☽ ♑
09	08	53	☽ ♒
10	20	39	♀ ♓
11	10	46	☽ ♓
13	13	40	☽ ♈
15	19	12	☽ ♉
17	14	43	☿ ♓
18	04	22	☽ ♊
20	00	12	☉ ♈
20	16	29	☽ ♋
23	05	12	☽ ♌
25	15	59	☽ ♍
27	23	48	☽ ♎
30	05	05	☽ ♏

LATITUDES

Date	Mercury ☿ ° '	Venus ♀ ° '	Mars ♂ ° '	Jupiter ♃ ° '	Saturn ♄ ° '	Uranus ♅ ° '	Neptune ♆ ° '	Pluto ♇ ° '
01	02 N 36	00 S 47	03 N 05	01 N 31	02 N 38	00 N 36	01 S 42	09 S 40
04	01 56	00 54	03 00	01 32	02 38	00 36	01 42	09 40
07	01 14	01 00	02 56	01 32	02 39	00 36	01 42	09 41
10	00 N 34	01 05	02 52	01 32	02 39	00 36	01 42	09 41
13	00 S 03	01 10	02 48	01 33	02 39	00 36	01 42	09 42
16	00 36	01 15	02 44	01 33	02 40	00 35	01 42	09 42
19	01 05	01 18	02 41	01 33	02 40	00 35	01 42	09 43
22	01 29	01 22	02 37	01 33	02 40	00 35	01 42	09 43
25	01 50	01 24	02 33	01 33	02 40	00 35	01 42	09 44
28	02 06	01 27	02 30	01 33	02 40	00 35	01 42	09 45
31	02 S 18	01 S 28	02 N 26	01 N 33	02 N 41	00 N 35	01 S 41	09 S 45

DATA

Julian Date	2466215
Delta T	+77 seconds
Ayanamsa	24° 24' 51"
Synetic vernal point	04° ♓ 42' 08"
True obliquity of ecliptic	23° 26' 07"

MOON'S PHASES, APSIDES AND POSITIONS ☽

Date	h	m	Phase	Longitude ° '	Eclipse Indicator
06	07	19	☾	16 ♐ 20	
13	01	46	●	23 ♓ 06	
20	17	59	☽	00 ♋ 44	
28	15	12	○	08 ♎ 33	

Day	h	m		
09	12	15	Perigee	
21	11	20	Apogee	
01	10	03	OS	
07	22	55	Max dec	25° S 46'
14	06	44	ON	
21	12	03	Max dec	25° N 53'
28	17	19	OS	

LONGITUDES

Date	Chiron ⚷ ° '	Ceres ⚳ ° '	Pallas ⚴ ° '	Juno ⚵ ° '	Vesta ⚶ ° '	Black Moon Lilith ⚸ ° '
01	06 ♋ 57	06 ♉ 44	18 ♓ 11	16 ♓ 23	12 ♑ 58	07 ♋ 32
11	06 ♋ 53	10 ♉ 13	21 ♓ 37	21 ♓ 27	17 ♑ 34	08 ♋ 39
21	06 ♋ 58	13 ♉ 52	25 ♓ 05	26 ♓ 36	21 ♑ 59	09 ♋ 46
31	07 ♋ 11	17 ♉ 38	28 ♓ 32	01 ♈ 48	26 ♑ 09	10 ♋ 52

All ephemeris data is given at 12.00 UT and the Moon's longitude is additionally given for 24.00 UT

ASPECTARIAN

h m	Aspects
01 Thursday	
01 59	☽ □ ♂
02 21	☽ ∥ ♃
02 25	☽ # ♄
02 49	♄ bQ ♇
09 38	☽ bQ ☿
11 31	☽ ☌ ♄
11 35	☽ bQ ♇
13 09	☽ ⚻ ☉
13 39	☽ Q ♅
17 35	☽ ∥ ♄
17 51	☽ # ♃
02 Friday	
00 54	☽ ± ☉
02 55	☽ △ ♀
10 00	♀ Q ♆
11 56	☽ △ ☿
14 42	☽ △ ♇
14 43	☽ ∥ ☉
18 22	☽ bQ ☉
20 02	☽ ⚺ ♃
21 58	☽ □ ♅
03 Saturday	
01 06	☽ ☍ ♆
06 15	☽ # ♆
06 37	☽ ⊥ ♃
08 56	☽ △ ♂
11 06	☽ ∥ ☿
14 41	♀ bQ ♂
17 01	☽ ⚺ ♄
22 24	☽ ∠ ♃
23 07	☽ △ ☉
04 Sunday	
03 32	☽ ⊥ ♄
03 53	☽ ∥ ♀
11 51	☽ bQ ♂
13 32	☽ □ ♀
16 13	☽ □ ☿
17 31	☉ bQ ♅
18 52	♀ ± ♃
19 13	☽ ∠ ♄
19 49	☽ □ ♇
05 Monday	
00 26	☽ ⚹ ♃
02 42	☽ △ ♅
03 50	☽ ± ♂
05 57	☽ ⚻ ♆
10 59	☽ # ♅
11 20	☿ St D
14 25	☽ ⚻ ♂
16 21	☽ ± ♆
18 34	☿ ☌ ♀
19 09	☽ ∥ ♇
21 02	☽ Q ♃
21 06	☽ ⚹ ♄
23 23	☽ Q ☿
23 50	☽ Q ♀
06 Tuesday	
03 01	☽ Q ♇
04 33	☽ bQ ♅
07 19	☽ □ ☉
07 51	☽ bQ ♆
15 08	☉ ∠ ♆
17 33	☽ Q ♄
19 53	☽ ± ♅
20 03	☽ ⚹ ☿
22 20	☽ ⚹ ♀
23 33	☽ ⚹ ♇
07 Wednesday	
01 34	♀ bQ ♄
03 31	☽ □ ♃
06 04	☽ ⚻ ♅
09 26	☽ △ ♆
12 33	♀ ☌ ♇
16 14	☽ Q ☉
18 33	☽ ☍ ♂
21 47	☽ ∠ ☿
23 51	☽ □ ♄
08 Thursday	
00 56	☽ ∠ ♇
02 05	☽ ∠ ♀
13 10	☽ ⊥ ☿
13 53	☽ ⚹ ☉
16 02	☽ ⊥ ♇
18 35	☽ ⊥ ♀
23 24	☽ ⚺ ☿
09 Friday	
02 04	☽ ⚺ ♇
04 24	♀ ⚻ ♃
05 25	☽ △ ♃
05 31	☽ ⚺ ♀
08 16	☽ ☍ ♅
11 46	☽ □ ♆
16 44	☽ ∠ ☉
21 38	☽ ⚻ ♂
10 Saturday	
01 38	☽ △ ♄
02 03	☽ ∥ ♇
06 07	☽ bQ ♃
07 48	☽ ± ♂
08 47	☽ ⊥ ☉
09 35	☽ # ♅
10 16	☿ ∥ ♀
13 00	♀ ⚻ ♅
17 32	☿ bQ ♄
17 44	☽ Q ♆
19 29	☽ ⚺ ☉
20 56	☽ ± ♃
23 04	☽ bQ ♂
11 Sunday	
02 27	☽ bQ ♄
02 49	☽ ☌ ☿
04 02	☽ ☌ ♇
06 51	☽ ⚻ ♃
10 04	☽ ⚻ ♅
12 06	☽ ☌ ♀
13 46	☽ ⚹ ♆
17 24	☽ ± ♄
20 06	☽ ± ♅
12 Monday	
00 41	☽ △ ♂
02 26	☽ ∥ ☿
03 26	☽ ⚻ ♄
05 50	☽ ∥ ♀
07 56	♀ ⚹ ♆
11 00	☿ ☌ ♇
11 14	☽ bQ ♅
11 53	☽ # ♆
15 05	☽ ∠ ♆
13 Tuesday	
01 46	☽ ☌ ☉
06 30	☽ ⊥ ♆
06 47	☽ ⚺ ♇
07 43	☽ ⚺ ☿
09 08	☽ ☍ ♃
12 52	☽ △ ♅
16 54	☽ ⚺ ♆
17 11	☽ ⊥ ♇
17 52	☉ # ♃
18 39	☽ ⊥ ☿
20 04	☽ ⚺ ♀
20 43	☽ # ♃
20 56	☽ ∥ ☉
22 26	♀ ± ♄
14 Wednesday	
00 55	☽ ∥ ♄
05 21	☽ □ ♂
06 46	☽ ☍ ♄
07 39	☽ ⊥ ♀
07 43	☿ ⚻ ♃
09 00	☽ ∠ ♇
11 21	☽ ∠ ☿
12 32	☽ # ♄
15 24	☽ # ☉
17 04	☽ ∥ ♃
15 Thursday	
01 20	☽ ∠ ♀
07 18	♀ ± ♅
10 54	☽ ⚺ ☉
11 28	♀ # ♆
12 00	☽ ⚹ ♇
13 52	☽ ⚻ ♃
14 28	☽ Q ♂
16 06	☽ ⚹ ☿
18 17	☽ □ ♅
22 07	☉ ⊥ ♆
22 47	☽ ☌ ♆
22 50	☽ ⊥ ☉
16 Friday	
00 49	☽ ± ♃
01 56	☉ ⚺ ♇
02 27	☽ # ♀
03 44	☽ ∥ ♆
06 09	♂ □ ♄
07 47	☽ ⚹ ♀
10 16	☽ Q ♇
12 20	☽ # ☿
13 07	☽ ⚻ ♄
13 21	☽ ⚹ ♂
13 38	☉ ∥ ♄
16 01	☽ Q ☿
17 06	☽ ∠ ☉
17 32	☽ bQ ♃
21 58	☉ ☍ ♃
17 Saturday	
00 26	☽ ± ♄
00 48	☿ ⚻ ♅
04 03	☽ Q ♅
09 03	☽ Q ♀
17 40	☽ bQ ♄
18 53	☽ ∠ ♂
20 48	☽ □ ♇
22 05	☽ △ ♃
18 Sunday	
00 29	☽ ⚹ ☉
03 18	☽ ⚹ ♅
05 30	☽ □ ☿
08 19	☽ ⚺ ♆
12 03	♀ ⚻ ♄
12 51	☽ ∥ ♅
13 03	☽ ⊥ ♂
20 14	☽ ⊥ ♆
20 59	☽ # ♇
23 02	☽ △ ♄
19 Monday	
00 22	☽ □ ♀
01 20	☽ ⚺ ♂
02 21	☽ Q ☉
09 03	☽ ∠ ♅
10 34	☉ △ ♅
12 51	♀ △ ♂
14 18	☽ ∠ ♆
17 36	☿ ⚹ ♆
20 Tuesday	
03 10	☽ ⊥ ♅
08 47	☽ △ ♇
09 20	☽ □ ♃
15 18	☽ ⚺ ♅
17 59	☽ □ ☉
19 20	☽ ∥ ♂
20 37	♂ bQ ♇
20 44	☽ ⚹ ♆
21 48	☿ ± ♄
23 18	☽ △ ☿
21 Wednesday	
11 16	☽ □ ♄
15 18	☽ bQ ♇
15 51	☽ ☌ ♂
20 07	☽ △ ♀
21 11	☽ Q ♆
21 29	☽ Q ♃
22 Thursday	
03 41	♃ ⚻ ♇
03 51	☉ ⚺ ♆
06 31	☉ ⊥ ♇
06 52	☽ ∥ ♂
08 51	☽ bQ ☿
09 35	☽ ± ♇
11 30	♀ bQ ♅
19 28	☿ ± ♅
19 35	♃ ± ♆
21 29	☽ ⚹ ♃
21 42	☽ ⚻ ♇
22 49	☉ # ♄
23 26	☽ Q ♄
23 Friday	
03 58	☽ ☌ ♅
04 53	☽ ± ☿
05 53	☽ bQ ♀
09 34	☽ □ ♆
12 09	☽ △ ☉
18 08	☽ ⚻ ☿
19 00	☿ # ♆
23 13	☽ ⚹ ♄
24 Saturday	
01 50	☽ ± ♀
03 04	☽ ∠ ♃
03 13	☽ # ♇
05 56	☽ ⚺ ♂
10 59	☽ ∥ ♅
14 57	☽ ⚻ ♀
18 03	☽ ⊥ ♂
18 26	♀ ∠ ♆
20 19	☽ bQ ☉
20 31	☽ ⊥ ♃
25 Sunday	
04 22	☽ ∠ ♄
08 00	☽ ⚺ ♃
08 56	☽ ☍ ♇
12 00	☽ ∠ ♂
14 45	☽ ⚺ ♅
15 11	☽ ± ☉
18 09	☿ ⚻ ♄
20 18	☽ △ ♆
21 31	☽ ⊥ ♄
26 Monday	
02 10	☽ ⊥ ♅
03 34	☽ ⚻ ☉
08 47	☽ ⚺ ♄
10 29	☽ ☍ ☿
17 15	☽ ⚹ ♂
19 03	☽ ∠ ♅
19 31	☽ ∥ ♆
27 Tuesday	
00 32	☽ bQ ♆
01 11	☽ # ☿
06 04	☽ ☍ ♀
06 25	♂ Q ♃
15 42	☽ ☌ ♃
16 04	☽ Q ♂
17 07	☽ ± ♆
17 12	☽ ⚻ ♇
21 10	☽ # ♀
22 35	☽ ⚹ ♅
28 Wednesday	
04 02	☽ ⚻ ♆
04 04	☽ ∥ ♃
04 05	☽ ∥ ☉
04 07	☽ ± ♇
04 29	☉ ∥ ♃
13 16	☽ # ♄
15 12	☽ ☍ ☉
15 26	☽ ☌ ♄
18 15	☉ ☍ ♄
20 10	☽ Q ♅
20 18	☽ bQ ♇
21 18	☽ ∥ ♄
23 16	☽ ⚻ ☿
29 Thursday	
01 27	☽ □ ♂
06 38	☽ # ♃
06 38	♂ Q ♆
08 15	☽ # ☉
10 06	☽ ∥ ♀
11 04	☽ ± ☿
17 27	☽ ⚻ ♀
20 49	☽ ⚺ ♃
21 22	☿ bQ ♅
21 46	☉ # ♀
22 50	☽ △ ♇
30 Friday	
02 28	☽ ∥ ☿
03 54	☽ □ ♅
04 37	☽ bQ ☿
04 51	☿ △ ♂
05 00	☽ ± ♀
07 13	☽ ⊥ ♃
09 18	☽ ☍ ♆
13 34	☽ # ♆
19 53	☽ ⚺ ♄
22 11	☽ bQ ♀
22 42	☽ ∠ ♃
23 59	☽ ⚻ ☉
31 Saturday	
00 17	♀ # ♃
03 32	♀ ☍ ♃
06 12	☽ ⊥ ♄
07 26	☽ △ ♂
09 33	☽ △ ☿
11 08	☽ ± ☉
14 02	☉ ∠ ♇
21 36	☽ ∠ ♄

LONGITUDES

Date	Sidereal time h m s	Sun ☉ ° ' "	Moon ☽ ° ' "	Moon ☽ 24.00 ° ' "	Mercury ☿ ° '	Venus ♀ ° '	Mars ♂ ° '	Jupiter ♃ ° '	Saturn ♄ ° '	Uranus ♅ ° '	Neptune ♆ ° '	Pluto ♇ ° '
01	00 41 51	12 ♈ 22 02	01 ♐ 50 24	08 ♐ 52 04	19 ♓ 00	26 ♓ 48	15 ♋ 41	24 ♍ 57	08 ♎ 23	29 ♋ 19	02 ♉ 30	26 ♒ 29
02	00 45 48	13 21 13	15 ♐ 54 40	22 ♐ 57 57	19 27	28 02	16 06	24 R 50	08 R 18	29 R 18	02 32	26 30
03	00 49 44	14 20 22	00 ♑ 01 46	07 ♑ 05 56	20 56	29 ♓ 16	16 31	24 43	08 14	29 18	02 34	26 32
04	00 53 41	15 19 29	14 ♑ 10 18	21 ♑ 14 42	22 27	00 ♈ 30	16 56	24 36	08 09	29 18	02 36	26 33
05	00 57 37	16 18 34	28 ♑ 18 57	05 ♒ 22 51	23 59	01 44	17 21	24 29	08 04	29 18	02 39	26 34
06	01 01 34	17 17 38	12 ♒ 26 09	19 ♒ 28 35	25 32	02 58	17 47	24 23	08 00	29 D 18	02 41	26 35
07	01 05 30	18 16 40	26 ♒ 29 50	03 ♓ 29 31	27 08	04 13	18 13	24 16	07 55	29 18	02 43	26 37
08	01 09 27	19 15 40	10 ♓ 27 17	17 ♓ 22 42	28 ♓ 44	05 27	18 39	24 10	07 51	29 18	02 45	26 38
09	01 13 23	20 14 38	24 ♓ 15 24	01 ♈ 04 57	00 ♈ 22	06 41	19 05	24 03	07 46	29 18	02 47	26 39
10	01 17 20	21 13 34	07 ♈ 51 00	14 ♈ 33 14	02 02	07 55	19 32	23 57	07 42	29 19	02 50	26 40
11	01 21 17	22 12 28	21 ♈ 11 22	27 ♈ 45 12	03 43	09 09	19 58	23 51	07 37	29 19	02 52	26 41
12	01 25 13	23 11 21	04 ♉ 14 38	10 ♉ 39 36	05 26	10 23	20 25	23 45	07 33	29 19	02 54	26 42
13	01 29 10	24 10 11	17 ♉ 00 09	23 ♉ 16 25	07 10	11 38	20 52	23 39	07 28	29 20	02 56	26 43
14	01 33 06	25 08 59	29 ♉ 28 37	05 ♊ 37 02	08 56	12 52	21 20	23 33	07 24	29 20	02 58	26 44
15	01 37 03	26 07 45	11 ♊ 42 00	17 ♊ 43 58	10 44	14 06	21 47	23 27	07 20	29 20	03 01	26 45
16	01 40 59	27 06 29	23 ♊ 43 24	29 ♊ 40 48	12 33	15 20	22 15	23 22	07 15	29 21	03 03	26 46
17	01 44 56	28 05 11	05 ♋ 36 46	11 ♋ 31 51	14 23	16 34	22 43	23 17	07 11	29 22	03 05	26 47
18	01 48 52	29 ♈ 03 51	17 ♋ 26 41	23 ♋ 21 54	16 15	17 48	23 11	23 11	07 07	29 22	03 08	26 48
19	01 52 49	00 ♉ 02 28	29 ♋ 18 08	05 ♌ 15 59	18 09	19 02	23 39	23 06	07 03	29 23	03 10	26 49
20	01 56 46	01 01 03	11 ♌ 16 06	17 ♌ 19 04	20 04	20 16	24 07	23 01	06 59	29 24	03 12	26 50
21	02 00 42	01 59 36	23 ♌ 25 27	29 ♌ 35 45	22 01	21 30	24 36	22 57	06 55	29 25	03 14	26 51
22	02 04 39	02 58 07	05 ♍ 50 27	12 ♍ 09 55	23 59	22 44	25 05	22 52	06 51	29 25	03 17	26 52
23	02 08 35	03 56 35	18 ♍ 34 30	25 ♍ 04 23	25 59	23 58	25 34	22 48	06 47	29 26	03 19	26 53
24	02 12 32	04 55 02	01 ♎ 39 43	08 ♎ 20 30	28 ♈ 01	25 12	26 03	22 43	06 43	29 27	03 21	26 54
25	02 16 28	05 53 26	15 ♎ 06 39	21 ♎ 57 57	00 ♉ 04	26 26	26 32	22 39	06 39	29 28	03 23	26 55
26	02 20 25	06 51 49	28 ♎ 54 06	05 ♏ 54 40	02 08	27 40	27 01	22 35	06 36	29 29	03 26	26 55
27	02 24 21	07 50 10	12 ♏ 59 09	20 ♏ 06 59	04 13	28 ♈ 54	27 31	22 31	06 32	29 31	03 28	26 56
28	02 28 18	08 48 28	27 ♏ 17 32	04 ♐ 30 09	06 20	00 ♉ 08	28 00	22 28	06 28	29 32	03 30	26 57
29	02 32 15	09 46 46	11 ♐ 44 10	18 ♐ 58 56	08 27	01 22	28 30	22 24	06 25	29 33	03 32	26 58
30	02 36 11	10 ♉ 45 02	26 ♐ 13 49	03 ♑ 28 14	10 ♉ 36	02 ♉ 36	29 ♋ 00	22 ♍ 21	06 ♎ 21	29 ♋ 34	03 ♉ 35	26 ♒ 58

Date	Moon True ☊ ° '	Moon Mean ☊ ° '	Moon ☽ Latitude ° '
01	04 ♊ 57	06 ♊ 34	00 N 17
02	04 D 57	06 31	00 S 59
03	04 58	06 28	02 11
04	04 59	06 25	03 15
05	04 R 58	06 22	04 07
06	04 56	06 18	04 45
07	04 51	06 15	05 05
08	04 45	06 12	05 06
09	04 37	06 09	04 50
10	04 30	06 06	04 18
11	04 23	06 03	03 32
12	04 18	05 59	02 35
13	04 14	05 56	01 32
14	04 13	05 53	00 S 26
15	04 D 13	05 50	00 N 41
16	04 14	05 47	01 44
17	04 16	05 43	02 43
18	04 18	05 40	03 34
19	04 18	05 37	04 16
20	04 R 18	05 34	04 48
21	04 16	05 31	05 07
22	04 12	05 28	05 12
23	04 08	05 24	05 03
24	04 03	05 21	04 37
25	03 59	05 18	03 56
26	03 55	05 15	03 00
27	03 53	05 12	01 52
28	03 52	05 09	00 N 36
29	03 D 52	05 05	00 S 43
30	03 ♊ 53	05 ♊ 02	02 S 00

DECLINATIONS

Date	Sun ☉ ° '	Moon ☽ ° '	Mercury ☿ ° '	Venus ♀ ° '	Mars ♂ ° '	Jupiter ♃ ° '	Saturn ♄ ° '	Uranus ♅ ° '	Neptune ♆ ° '	Pluto ♇ ° '
01	04 N 53	20 S 15	06 S 54	02 S 38	24 N 55	03 N 25	00 S 52	20 N 52	10 N 45	21 S 50
02	05 16	23 40	06 22	02 09	24 51	03 28	00 50	20 52	10 46	21 50
03	05 39	25 37	05 49	01 39	24 47	03 31	00 48	20 52	10 46	21 50
04	06 02	25 55	05 15	01 10	24 42	03 33	00 46	20 52	10 47	21 49
05	06 25	24 32	04 39	00 41	24 38	03 36	00 45	20 52	10 48	21 49
06	06 47	21 37	04 02	00 S 11	24 33	03 39	00 43	20 52	10 49	21 49
07	07 10	17 27	03 25	00 N 18	24 29	03 41	00 41	20 52	10 49	21 49
08	07 32	12 22	02 46	00 48	24 24	03 44	00 39	20 52	10 50	21 49
09	07 55	06 43	02 06	01 17	24 19	03 46	00 37	20 52	10 51	21 48
10	08 17	00 S 50	01 25	01 47	24 14	03 48	00 36	20 52	10 52	21 48
11	08 39	04 N 59	00 S 43	02 16	24 09	03 51	00 34	20 52	10 53	21 48
12	09 01	10 30	00 00	02 46	24 04	03 53	00 32	20 51	10 53	21 48
13	09 22	15 26	00 N 44	03 15	23 59	03 55	00 31	20 51	10 54	21 48
14	09 44	19 37	01 29	03 44	23 54	03 57	00 29	20 51	10 55	21 48
15	10 05	22 51	02 14	04 13	23 48	04 00	00 27	20 51	10 56	21 48
16	10 26	25 01	03 01	04 43	23 43	04 02	00 26	20 51	10 56	21 47
17	10 48	26 02	03 48	05 12	23 37	04 04	00 24	20 51	10 57	21 47
18	11 08	25 50	04 37	05 41	23 32	04 05	00 23	20 51	10 58	21 47
19	11 29	24 28	05 25	06 09	23 26	04 07	00 21	20 50	10 59	21 47
20	11 50	22 00	06 15	06 38	23 20	04 09	00 20	20 50	10 59	21 47
21	12 10	18 32	07 05	07 07	23 14	04 11	00 18	20 50	11 00	21 47
22	12 30	14 12	07 56	07 35	23 08	04 13	00 17	20 50	11 01	21 47
23	12 50	09 10	08 47	08 03	23 01	04 14	00 15	20 50	11 02	21 47
24	13 10	03 N 35	09 38	08 31	22 55	04 16	00 14	20 49	11 03	21 47
25	13 29	02 S 19	10 29	08 59	22 48	04 17	00 12	20 49	11 03	21 47
26	13 48	08 16	11 21	09 27	22 42	04 19	00 11	20 49	11 04	21 47
27	14 07	13 57	12 13	09 54	22 35	04 20	00 10	20 49	11 05	21 47
28	14 26	18 58	13 04	10 21	22 28	04 21	00 08	20 48	11 06	21 47
29	14 45	22 54	13 55	10 48	22 21	04 22	00 07	20 48	11 06	21 47
30	15 N 03	25 S 23	14 N 45	11 N 15	22 N 14	04 N 23	00 S 06	20 N 48	11 N 07	21 S 47

ZODIAC SIGN ENTRIES

Date	h	m	Planets
01	08	51	☽ ♐
03	11	57	☽ ♑
04	02	16	♀ ♈
05	14	52	☽ ♒
07	18	00	☽ ♓
09	06	33	☿ ♈
09	22	05	☽ ♈
12	04	08	☽ ♉
14	13	01	☽ ♊
17	00	39	☽ ♋
19	10	59	☉ ♉
19	13	24	☽ ♌
22	00	47	☽ ♍
24	08	59	☽ ♎
25	11	19	☿ ♉
26	13	53	☽ ♏
28	09	22	♀ ♉
28	16	31	☽ ♐
30	18	15	☽ ♑

LATITUDES

Date	Mercury ☿ ° '	Venus ♀ ° '	Mars ♂ ° '	Jupiter ♃ ° '	Saturn ♄ ° '	Uranus ♅ ° '	Neptune ♆ ° '	Pluto ♇ ° '
01	02 S 21	01 S 29	02 N 25	01 N 33	02 N 41	00 N 35	01 S 41	09 S 45
04	02 27	01 29	02 22	01 32	02 41	00 35	01 41	09 46
07	02 28	01 29	02 18	01 32	02 41	00 35	01 41	09 47
10	02 25	01 29	02 15	01 32	02 40	00 35	01 41	09 48
13	02 18	01 28	02 12	01 31	02 40	00 35	01 41	09 48
16	02 06	01 26	02 09	01 31	02 40	00 35	01 41	09 49
19	01 50	01 24	02 06	01 31	02 40	00 35	01 41	09 50
22	01 29	01 21	02 03	01 30	02 39	00 35	01 41	09 51
25	01 04	01 18	02 00	01 29	02 39	00 35	01 41	09 52
28	00 36	01 14	01 57	01 29	02 39	00 34	01 41	09 53
31	00 S 05	01 S 10	01 N 54	01 N 28	02 N 38	00 N 34	01 S 41	09 S 54

DATA

Julian Date	2466246
Delta T	+77 seconds
Ayanamsa	24° 24' 54"
Synetic vernal point	04° ♓ 42' 05"
True obliquity of ecliptic	23° 26' 07"

MOON'S PHASES, APSIDES AND POSITIONS ☽

Date	h	m	Phase	Longitude ° '	Eclipse Indicator
04	14	06	☾	15 ♑ 25	
11	14	00	●	22 ♈ 17	
19	13	37	☽	00 ♌ 06	
27	02	38	○	07 ♏ 27	

Day	h	m		
03	20	46	Perigee	
18	07	17	Apogee	
30	04	30	Perigee	
04	04	13	Max dec	26° S 00'
10	15	24	0N	
17	20	08	Max dec	26° N 06'
25	02	39	0S	

LONGITUDES

Date	Chiron ⚷ ° '	Ceres ⚳ ° '	Pallas ⚴ ° '	Juno ⚵ ° '	Vesta ⚶ ° '	Black Moon Lilith ☾ ° '
01	07 ♋ 13	18 ♉ 02	28 ♓ 53	02 ♈ 19	26 ♑ 33	10 ♋ 59
11	07 ♋ 36	21 ♉ 55	02 ♈ 21	07 ♈ 36	00 ♒ 25	12 ♋ 06
21	08 ♋ 06	25 ♉ 54	05 ♈ 47	12 ♈ 55	03 ♒ 57	13 ♋ 12
31	08 ♋ 44	29 ♉ 57	09 ♈ 11	18 ♈ 17	07 ♒ 05	14 ♋ 19

ASPECTARIAN

h m	Aspects
01 Sunday	
00 19	☽ ✱ ♃
02 31	☽ △ ♀
02 49	☽ □ ♇
03 28	☿ ∠ ♆
03 46	☽ Ᵽ ☉
05 56	♀ ⊻ ♇
06 07	♀ ⊥ ♆
07 40	☽ △ ♅
09 58	☽ Ᵽ ♂
13 08	☽ ⊼ ♆
15 43	☽ # ♅
20 40	☽ Q ♃
22 00	☽ ∥ ♇
23 07	☽ ✱ ♄
23 24	☽ ± ♆
02 Monday	
01 47	☽ ± ♂
07 19	☽ △ ☉
09 16	☽ Ᵽ ♅
09 36	☽ Q ♇
12 19	☽ ⊼ ♂
14 46	☽ Ᵽ ♆
18 44	☽ □ ☿
19 26	☽ Q ♄
23 41	☽ # ♂
03 Tuesday	
00 34	☽ ± ♅
03 03	☽ □ ♃
06 03	☽ ✱ ♇
10 35	☽ □ ♀
10 46	☽ ⊼ ♅
12 43	♀ △ ♅
16 09	☉ # ☿
16 19	☽ △ ♆
04 Wednesday	
01 50	☽ □ ♄
04 55	☽ Q ☿
07 33	☽ ∠ ♇
14 06	☽ □ ☉
16 49	☽ ☍ ♂
20 03	☽ Q ♀
22 50	☽ ⊥ ♇
05 Thursday	
03 44	☽ ✱ ☿
05 34	☽ △ ♃
08 28	♀ ∥ ♄
09 02	☽ ⊻ ♇
10 52	☽ # ♂
13 40	☽ ☍ ♅
18 22	☽ ✱ ♀
19 22	☽ □ ♆
19 23	☿ ☍ ♃
21 45	♅ St D
22 57	☽ Q ☉
06 Friday	
04 23	♀ ⊥ ♇
04 29	☽ △ ♄
06 04	♀ ⊻ ♆
06 50	☽ Ᵽ ♃
08 22	☽ ∠ ☿
10 40	☽ ∥ ♇
16 52	☽ # ♅
20 54	☽ ✱ ☉
21 24	☽ ⊼ ♂
22 03	☽ ± ♃
22 21	☽ ∠ ♀
07 Saturday	
01 38	☽ ⊥ ☿
02 05	☽ Q ♆
02 17	☿ # ♃
04 08	☿ ⊻ ♇
05 41	☿ ⊥ ♆
05 55	☽ Ᵽ ♄
07 58	☽ ± ♂
08 13	☽ ⊼ ♃
09 09	☉ □ ♂
12 12	☽ ☌ ♇
13 13	☽ ⊻ ☿
15 13	☽ ⊥ ♀
16 48	☽ ⊼ ♅
21 15	☽ ± ♄
22 42	☽ ✱ ♆
23 53	☽ Ᵽ ♂
08 Sunday	
00 31	☽ ∠ ☉
02 31	☽ ⊻ ♀
03 07	☽ ± ♅
05 23	♀ # ♄
07 31	☽ ⊼ ♄
17 14	☽ ⊥ ☉
18 40	☽ Ᵽ ♅
18 40	☽ # ♆
20 21	☿ △ ♅
09 Monday	
00 41	☽ ∠ ♆
02 40	☽ △ ♂
04 27	☽ ⊻ ☉
07 22	☽ # ☉
11 39	☽ ☍ ♃
16 12	☽ ⊻ ♇
16 27	☽ ⊥ ♆
20 52	☽ △ ♅
10 Tuesday	
00 00	☽ # ♃
00 14	☽ ☌ ☿
02 47	☽ ⊥ ♇
03 03	☽ ⊻ ♆
04 36	☿ # ♀
07 53	♀ ☍ ♄
08 28	☽ # ♀
09 20	☽ ∥ ☿
11 43	☽ ☍ ♄
12 08	☽ ☌ ♀
12 59	☽ ∥ ♄
17 48	☽ # ♄
18 50	☽ ∠ ♇
20 13	☽ # ☿
21 08	☿ ⊥ ♇
23 32	☿ ⊻ ♆
23 39	☽ ∥ ♀
11 Wednesday	
07 11	☽ ∥ ♃
09 43	☽ □ ♂
14 00	☽ ☌ ☉
16 48	☽ ⊼ ♃
17 15	☿ ∥ ♄
22 03	☽ ✱ ♇
12 Thursday	
02 53	☽ □ ♅
03 44	☽ ± ♃
04 51	☽ ∥ ☉
09 30	☽ ☌ ♆
13 49	☽ ∥ ♆
14 34	☽ ⊻ ☿
18 07	☽ ⊼ ♄
20 05	☽ Q ♂
20 20	☽ Q ♇
20 20	☽ Ᵽ ♃
13 Friday	
00 22	☉ ⊼ ♃
00 44	☽ ⊻ ♀
03 35	☽ ⊥ ☿
03 38	♂ ± ♇
05 05	☿ # ♄
05 21	☽ ± ♄
12 37	☽ Q ♅
13 19	☽ ⊥ ♀
13 54	♀ ∠ ♇
15 56	☿ ☍ ♄
19 40	☽ ✱ ♂
22 24	☽ Ᵽ ♄
23 30	☽ ∠ ☿
14 Saturday	
00 37	☽ △ ♃
02 54	☽ ⊻ ☉
06 41	☽ □ ♇
08 31	☽ ∠ ♀
11 43	☽ ✱ ♅
15 32	☽ ⊥ ☉
18 51	☽ ⊻ ♆
20 19	☽ ∥ ♅
23 40	♀ ∥ ♃
15 Sunday	
01 55	☽ ∠ ♂
03 15	☽ # ♇
03 25	☽ △ ♄
06 40	☽ ⊥ ♆
09 44	☽ ✱ ☿
10 46	☽ ∠ ☉
17 15	☽ ∠ ♅
17 18	☽ ✱ ♀
20 27	☽ ⊥ ♂
20 44	☽ ∥ ♂
16 Monday	
00 36	☽ ∠ ♆
01 46	☿ ∠ ♇
03 40	☉ ✱ ♇
08 55	☽ ⊻ ♂
11 15	☽ ⊥ ♅
11 17	☽ □ ♃
13 57	☽ Q ☿
18 09	☽ △ ♇
19 25	☽ ✱ ☉
20 06	☽ Q ♀
23 21	☽ ⊻ ♅
17 Tuesday	
06 52	☽ ✱ ♆
15 10	☽ □ ♄
19 55	☿ ∥ ♃
21 53	☽ Q ☉
23 24	☽ Q ♃
23 28	☉ ∥ ♆
18 Wednesday	
00 33	☽ Ᵽ ♇
07 17	☽ Q ♆
09 08	☽ □ ☿
12 19	♂ ✱ ♃
12 48	☽ □ ♀
14 49	☉ ± ♃
18 50	☽ ± ♇
19 38	☉ □ ♅
23 34	☽ ✱ ♃
19 Thursday	
00 06	☽ ☌ ♂
03 27	☽ Q ♄
06 59	☽ ⊼ ♇
12 10	☽ ☌ ♅
13 37	☽ □ ☉
19 48	☽ □ ♆
23 52	☽ ∥ ♂
20 Friday	
03 29	☽ ✱ ♄
05 34	☽ ∠ ♃
13 42	☽ # ♇
18 46	☿ ☌ ♀
20 49	☽ ∥ ♅
23 21	☽ ⊥ ♃
21 Saturday	
07 49	☽ △ ♀
08 43	☽ △ ☿
09 03	☽ ∠ ♄
11 04	☽ ⊻ ♃
13 48	☿ ∥ ♀
14 23	☽ ⊻ ♂
18 42	☽ ☍ ♇
22 52	☿ ⊼ ♃
23 39	☽ ⊻ ♅
22 Sunday	
01 47	♂ Q ♄
02 28	☽ ⊥ ♄
02 30	☽ ⊥ ♂
06 02	☽ △ ☉
07 04	☽ △ ♆
11 12	☽ ⊥ ♅
13 55	☽ ⊻ ♄
14 23	♀ ⊼ ♃
16 00	☽ Ᵽ ♀
19 07	☽ Ᵽ ☿
19 52	☉ ☌ ♆
19 55	☽ ∥ ☉
20 23	☽ ∠ ♂
23 Monday	
03 28	☽ ∥ ♆
04 16	☽ ∠ ♅
05 15	☿ □ ♂
10 45	☽ ± ♀
11 31	☽ Ᵽ ♆
12 44	☽ Ᵽ ☉
13 29	☽ ∥ ☿
15 07	☽ ± ☿
16 31	☽ ∥ ♀
19 46	☽ ☌ ♃
22 42	☿ ✱ ♇
23 01	☽ ⊼ ♀
24 Tuesday	
01 23	☽ ✱ ♂
03 20	☽ ⊼ ♇
04 09	☽ ± ♆
04 10	☽ ⊼ ☿
06 37	☽ ± ☉
08 00	☽ ✱ ♅
09 10	☽ ∥ ♃
14 14	☽ ± ♇
15 04	☽ ⊼ ♆
18 20	☽ ⊼ ☉
20 05	☿ ± ♃
21 03	☽ ☌ ♄
23 54	☽ Q ♂
25 Wednesday	
01 47	☽ # ♄
03 31	☽ ∥ ♄
05 05	☿ □ ♅
05 34	☽ Q ♅
06 20	☽ Ᵽ ♇
14 59	♀ □ ♂
19 55	☽ # ♃
21 18	♀ ✱ ♇
26 Thursday	
01 08	☽ ⊻ ♃
03 59	☿ ∥ ♆
05 43	☉ ⊼ ♄
07 11	♂ ⊼ ♇
08 35	☽ △ ♇
08 38	☽ □ ♂
09 40	☽ ☍ ♀
11 28	☽ ⊥ ♃
13 01	☽ □ ♅
17 14	☽ # ♀
18 30	☽ ☍ ☿
19 47	☽ ☍ ♆
23 36	☽ # ♆
27 Friday	
01 06	☽ ⊻ ♄
02 38	☽ ☍ ☉
02 47	☽ ∠ ♃
03 03	☽ # ☿
03 12	☿ ☌ ♆
04 46	☉ Ᵽ ♃
04 59	♀ ± ♃
11 14	☽ ⊥ ♄
12 48	☽ # ☉
23 59	♀ □ ♅
28 Saturday	
02 19	☽ ∠ ♄
03 58	☽ ✱ ♃
11 25	☽ □ ♇
13 14	☽ △ ♂
13 35	☿ ⊼ ♄
15 30	☉ Q ♇
15 44	☽ △ ♅
17 11	☽ ⊼ ♀
22 16	☽ # ♅
22 22	☽ ⊼ ♆
23 53	☽ Q ♃
29 Sunday	
00 30	☿ Ᵽ ♃
03 12	☽ ✱ ♄
04 05	☽ ± ♀
04 18	☽ ∥ ♇
05 37	☽ ⊼ ☿
08 13	☽ # ♂
08 21	☽ ± ♆
08 31	☽ ⊼ ☉
15 01	☽ Ᵽ ♂
16 40	☽ Ᵽ ♅
17 17	☽ ± ☿
17 21	☽ Q ♇
17 43	☿ Q ♇
19 11	☽ ± ☉
20 23	☽ Ᵽ ♀
23 01	☽ Q ♄
23 18	☽ Ᵽ ♆
30 Monday	
04 29	♀ ∥ ♆
05 36	☽ □ ♃
06 27	☽ ± ♂
07 35	☽ ± ♅
10 46	☽ Ᵽ ☿
11 09	☽ Ᵽ ☉
13 14	☽ ✱ ♇
15 10	☉ ☌ ☿
16 45	☽ ⊼ ♂
17 32	☽ ⊼ ♅
23 32	☽ △ ♀

All ephemeris data is given at 12.00 UT and the Moon's longitude is additionally given for 24.00 UT

MAY 2040

LONGITUDES

Date	Sidereal time h m s	Sun ☉ ° ' "	Moon ☽ ° ' "	Moon ☽ 24.00 ° ' "	Mercury ☿ ° '	Venus ♀ ° '	Mars ♂ ° '	Jupiter ♃ ° '	Saturn ♄ ° '	Uranus ♅ ° '	Neptune ♆ ° '	Pluto ♇ ° '
01	02 40 08	11 ♉ 43 16	10 ♑ 41 40	17 ♑ 53 38	12 ♉ 45	03 ♉ 50	29 ♋ 30	22 ♍ 18	06 ♎ 18	29 ♋ 36	03 ♉ 37	26 ♒ 59
02	02 44 04	12 41 28	25 ♑ 03 45	02 ♒ 11 40	14 54	05 04	00 ♌ 00	22 R 15	06 R 15	29 37	03 39	27 00
03	02 48 01	13 39 40	09 ♒ 17 06	16 ♒ 19 49	17 04	06 18	00 30	22 12	06 11	29 38	03 41	27 00
04	02 51 57	14 37 49	23 ♒ 19 39	00 ♓ 16 25	19 14	07 32	01 00	22 10	06 08	29 40	03 44	27 01
05	02 55 54	15 35 58	07 ♓ 10 01	14 ♓ 00 23	21 23	08 46	01 31	22 07	06 05	29 41	03 46	27 01
06	02 59 50	16 34 05	20 ♓ 47 24	27 ♓ 31 03	23 31	10 00	02 02	22 05	06 02	29 43	03 48	27 02
07	03 03 47	17 32 10	04 ♈ 11 16	10 ♈ 48 01	25 39	11 14	02 32	22 03	05 59	29 44	03 50	27 02
08	03 07 44	18 30 14	17 ♈ 21 17	23 ♈ 51 03	27 46	12 27	03 03	22 01	05 56	29 46	03 52	27 03
09	03 11 40	19 28 16	00 ♉ 17 20	06 ♉ 40 11	29 ♉ 50	13 41	03 34	22 00	05 54	29 48	03 55	27 03
10	03 15 37	20 26 17	12 ♉ 59 36	19 ♉ 15 43	01 ♊ 54	14 55	04 05	21 58	05 51	29 50	03 57	27 04
11	03 19 33	21 24 17	25 ♉ 28 36	01 ♊ 38 25	03 55	16 09	04 37	21 57	05 48	29 51	03 59	27 04
12	03 23 30	22 22 15	07 ♊ 45 19	13 ♊ 49 33	05 53	17 23	05 08	21 56	05 46	29 53	04 01	27 05
13	03 27 26	23 20 11	19 ♊ 51 22	25 ♊ 51 03	07 50	18 37	05 39	21 55	05 43	29 55	04 03	27 05
14	03 31 23	24 18 06	01 ♋ 48 56	07 ♋ 45 23	09 43	19 51	06 11	21 54	05 41	29 57	04 06	27 05
15	03 35 19	25 15 59	13 ♋ 40 51	19 ♋ 35 45	11 34	21 05	06 43	21 53	05 39	29 ♋ 59	04 08	27 06
16	03 39 16	26 13 51	25 ♋ 30 34	01 ♌ 25 49	13 22	22 18	07 14	21 53	05 37	00 ♌ 01	04 10	27 06
17	03 43 13	27 11 40	07 ♌ 22 03	13 ♌ 19 47	15 07	23 32	07 46	21 53	05 35	00 03	04 12	27 06
18	03 47 09	28 09 28	19 ♌ 19 36	25 ♌ 22 05	16 49	24 46	08 18	21 D 53	05 33	00 05	04 14	27 07
19	03 51 06	29 ♉ 07 15	01 ♍ 27 48	07 ♍ 37 17	18 28	26 00	08 50	21 53	05 31	00 08	04 16	27 07
20	03 55 02	00 ♊ 04 59	13 ♍ 51 07	20 ♍ 09 46	20 03	27 14	09 23	21 53	05 29	00 10	04 18	27 07
21	03 58 59	01 02 42	26 ♍ 33 42	03 ♎ 03 18	21 35	28 27	09 55	21 54	05 28	00 12	04 20	27 07
22	04 02 55	02 00 24	09 ♎ 38 52	16 ♎ 20 39	23 04	29 ♉ 41	10 27	21 54	05 26	00 14	04 22	27 07
23	04 06 52	02 58 03	23 ♎ 08 43	00 ♏ 03 04	24 29	00 ♊ 55	11 00	21 55	05 25	00 17	04 24	27 07
24	04 10 48	03 55 42	07 ♏ 03 31	14 ♏ 09 48	25 51	02 09	11 32	21 56	05 23	00 19	04 26	27 07
25	04 14 45	04 53 19	21 ♏ 21 27	28 ♏ 37 53	27 09	03 23	12 05	21 58	05 22	00 21	04 28	27 07
26	04 18 42	05 50 55	05 ♐ 58 22	13 ♐ 22 05	28 24	04 36	12 38	21 59	05 21	00 24	04 30	27 08
27	04 22 38	06 48 30	20 ♐ 48 07	28 ♐ 15 28	29 ♊ 35	05 50	13 10	22 01	05 20	00 26	04 32	27 08
28	04 26 35	07 46 03	05 ♑ 43 10	13 ♑ 10 13	00 ♋ 43	07 04	13 43	22 02	05 19	00 29	04 34	27 R 08
29	04 30 31	08 43 36	20 ♑ 35 41	27 ♑ 58 42	01 47	08 18	14 16	22 04	05 18	00 31	04 36	27 07
30	04 34 28	09 41 08	05 ♒ 18 31	12 ♒ 34 28	02 47	09 31	14 49	22 06	05 17	00 34	04 38	27 07
31	04 38 24	10 ♊ 38 38	19 ♒ 46 04	26 ♒ 52 55	03 ♋ 43	10 ♊ 45	15 ♌ 22	22 ♍ 09	05 ♎ 16	00 ♌ 37	04 ♉ 40	27 ♒ 07

Date	Moon True ☊ ° '	Moon Mean ☊ ° '	Moon ☽ Latitude ° '
01	03 ♊ 54	04 ♊ 59	03 S 09
02	03 D 55	04 56	04 06
03	03 56	04 53	04 47
04	03 R 56	04 49	05 10
05	03 55	04 46	05 15
06	03 53	04 43	05 02
07	03 51	04 40	04 32
08	03 49	04 37	03 49
09	03 47	04 34	02 55
10	03 45	04 30	01 52
11	03 44	04 27	00 S 46
12	03 D 44	04 24	00 N 22
13	03 45	04 21	01 28
14	03 46	04 18	02 29
15	03 46	04 14	03 24
16	03 47	04 11	04 09
17	03 48	04 08	04 44
18	03 48	04 05	05 07
19	03 R 48	04 02	05 17
20	03 48	03 59	05 12
21	03 48	03 55	04 53
22	03 48	03 52	04 17
23	03 48	03 49	03 27
24	03 D 48	03 46	02 23
25	03 48	03 43	01 N 09
26	03 R 48	03 40	00 S 12
27	03 48	03 36	01 33
28	03 47	03 33	02 48
29	03 47	03 30	03 52
30	03 46	03 27	04 39
31	03 ♊ 46	03 ♊ 24	05 S 08

DECLINATIONS

Date	Sun ☉ ° '	Moon ☽ ° '	Mercury ☿ ° '	Venus ♀ ° '	Mars ♂ ° '	Jupiter ♃ ° '	Saturn ♄ ° '	Uranus ♅ ° '	Neptune ♆ ° '	Pluto ♇ ° '
01	15 N 21	26 S 09	15 N 35	11 N 42	22 N 07	04 N 24	00 S 05	20 N 48	11 N 08	21 S 47
02	15 39	25 09	16 24	12 08	22 00	04 25	00 03	20 47	11 09	21 47
03	15 56	22 32	17 11	12 34	21 52	04 26	00 02	20 47	11 09	21 47
04	16 13	18 37	17 58	12 59	21 45	04 27	00 S 01	20 47	11 10	21 47
05	16 30	13 45	18 42	13 25	21 37	04 28	00 00	20 46	11 11	21 47
06	16 47	08 16	19 25	13 50	21 29	04 29	00 N 01	20 46	11 12	21 47
07	17 04	02 S 30	20 06	14 14	21 21	04 29	00 02	20 46	11 12	21 48
08	17 20	03 N 17	20 45	14 39	21 13	04 30	00 03	20 45	11 13	21 48
09	17 36	08 51	21 22	15 03	21 05	04 30	00 04	20 45	11 14	21 48
10	17 51	13 57	21 56	15 26	20 57	04 31	00 05	20 44	11 14	21 48
11	18 07	18 23	22 28	15 49	20 49	04 31	00 06	20 44	11 15	21 48
12	18 22	21 58	22 57	16 12	20 40	04 31	00 06	20 44	11 16	21 48
13	18 36	24 31	23 24	16 35	20 32	04 31	00 07	20 43	11 17	21 48
14	18 51	25 55	23 48	16 56	20 23	04 31	00 08	20 43	11 17	21 49
15	19 05	26 07	24 10	17 18	20 14	04 31	00 09	20 42	11 18	21 49
16	19 18	25 07	24 29	17 39	20 05	04 31	00 09	20 42	11 19	21 49
17	19 32	23 00	24 46	18 00	19 56	04 31	00 10	20 41	11 19	21 49
18	19 45	19 53	25 00	18 20	19 47	04 31	00 10	20 41	11 20	21 49
19	19 58	15 53	25 12	18 40	19 38	04 31	00 11	20 40	11 21	21 50
20	20 10	11 09	25 21	18 59	19 28	04 30	00 12	20 40	11 21	21 50
21	20 22	05 51	25 29	19 17	19 19	04 30	00 12	20 39	11 22	21 50
22	20 34	00 N 07	25 34	19 36	19 09	04 30	00 12	20 39	11 23	21 50
23	20 45	05 S 47	25 37	19 53	19 00	04 29	00 13	20 38	11 23	21 51
24	20 56	11 37	25 38	20 10	18 50	04 28	00 13	20 38	11 24	21 51
25	21 06	17 00	25 38	20 27	18 40	04 28	00 13	20 37	11 25	21 51
26	21 17	21 30	25 36	20 43	18 30	04 27	00 14	20 37	11 25	21 51
27	21 27	24 40	25 32	20 58	18 20	04 26	00 14	20 36	11 26	21 52
28	21 36	26 07	25 27	21 13	18 09	04 25	00 14	20 36	11 27	21 52
29	21 45	25 40	25 20	21 28	17 59	04 24	00 14	20 35	11 27	21 52
30	21 54	23 27	25 13	21 41	17 49	04 23	00 14	20 35	11 28	21 53
31	22 N 02	19 S 45	25 N 03	21 N 54	17 N 38	04 N 22	00 N 14	20 N 34	11 N 28	21 S 53

ZODIAC SIGN ENTRIES

Date	h	m	Planets
02	12	07	♂ ♌
02	20	18	☽ ♒
04	23	32	☽ ♓
07	04	27	☽ ♈
09	11	28	☽ ♉
09	13	51	☿ ♊
11	20	48	☽ ♊
14	08	21	☽ ♋
15	22	13	♅ ♌
16	21	06	☽ ♌
19	09	08	☽ ♍
20	09	56	☉ ♊
21	18	23	☽ ♎
22	18	06	♀ ♊
23	23	55	☽ ♏
26	02	15	☽ ♐
27	20	42	☿ ♋
28	02	48	☽ ♑
30	03	18	☽ ♒

LATITUDES

Date	Mercury ☿ ° '	Venus ♀ ° '	Mars ♂ ° '	Jupiter ♃ ° '	Saturn ♄ ° '	Uranus ♅ ° '	Neptune ♆ ° '	Pluto ♇ ° '
01	00 S 05	01 S 10	01 N 54	01 N 28	02 N 38	00 N 34	01 S 41	09 S 54
04	00 N 27	01 05	01 51	01 28	02 38	00 34	01 41	09 54
07	00 58	01 00	01 49	01 27	02 37	00 34	01 41	09 55
10	01 26	00 55	01 46	01 26	02 37	00 34	01 41	09 56
13	01 49	00 49	01 44	01 26	02 36	00 34	01 41	09 57
16	02 06	00 43	01 41	01 25	02 36	00 34	01 42	09 58
19	02 16	00 37	01 38	01 24	02 35	00 34	01 42	09 59
22	02 19	00 30	01 36	01 24	02 35	00 34	01 42	10 00
25	02 14	00 23	01 33	01 23	02 34	00 34	01 42	10 01
28	02 01	00 16	01 31	01 22	02 33	00 34	01 42	10 02
31	01 N 41	00 S 09	01 N 29	01 N 22	02 N 33	00 N 34	01 S 42	10 S 03

DATA

Julian Date	2466276
Delta T	+77 seconds
Ayanamsa	24° 24' 57"
Synetic vernal point	04° ♓ 42' 01"
True obliquity of ecliptic	23° 26' 06"

MOON'S PHASES, APSIDES AND POSITIONS ☽

Date	h	m	Phase	Longitude ° '	Eclipse Indicator
03	20	00	☾	13 ♒ 59	
11	03	28	●	21 ♉ 04	Partial
19	07	00	☽	28 ♌ 55	
26	11	47	○	05 ♐ 50	total

Day	h	m		
16	01	04	Apogee	
28	02	33	Perigee	
01	10	17	Max dec	26° S 09'
07	22	20	ON	
15	03	55	Max dec	26° N 11'
22	12	31	OS	
28	18	24	Max dec	26° S 11'

LONGITUDES

Date	Chiron ⚷ ° '	Ceres ⚳ ° '	Pallas ⚴ ° '	Juno ⚵ ° '	Vesta ⚶ ° '	Black Moon Lilith ⚸ ° '
01	08 ♋ 44	29 ♉ 57	09 ♈ 11	18 ♈ 17	07 ♒ 05	14 ♋ 19
11	09 ♋ 28	04 ♊ 02	12 ♈ 32	23 ♈ 41	09 ♒ 45	15 ♋ 26
21	10 ♋ 17	08 ♊ 11	15 ♈ 49	29 ♈ 07	11 ♒ 53	16 ♋ 33
31	11 ♋ 11	12 ♊ 21	19 ♈ 01	04 ♉ 34	13 ♒ 23	17 ♋ 40

All ephemeris data is given at 12.00 UT and the Moon's longitude is additionally given for 24.00 UT

ASPECTARIAN

h m Aspects

01 Tuesday
00 12 ☽ △ ♆
01 13 ☉ ∥ ☿
04 43 ☽ □ ♄
07 09 ☿ ± ♄
07 37 ♀ ☌ ♆
13 50 ☽ △ ☉
14 09 ☽ ∠ ♇
16 01 ☽ △ ☿
16 50 ♂ ☌ ♅

02 Wednesday
01 31 ☉ ± ♄
05 11 ☽ ⊥ ♇
07 18 ☽ △ ♃
15 15 ☽ ⊻ ♇
19 40 ☽ ☍ ♅
20 36 ☽ ☍ ♂

03 Thursday
02 30 ☽ □ ♆
06 27 ☽ □ ♀
06 46 ☽ △ ♄
08 29 ☽ ꝑ ♃
09 59 ♀ ⊼ ♄
16 48 ☽ # ♂
17 10 ☽ ∥ ♇
18 26 ☿ Q ♅
20 00 ☽ □ ☉
23 36 ☽ # ♅
23 45 ☽ ± ♃

04 Friday
03 27 ♂ # ♇
03 40 ☽ □ ☿
05 05 ♀ ꝑ ♃
08 15 ☽ ꝑ ♄
08 50 ☿ Q ♂
09 14 ☽ Q ♆
10 00 ☽ ⊼ ♃
14 59 ☽ # ☿
16 10 ☽ Q ♀
18 22 ☽ ☌ ♇
22 58 ☽ ⊼ ♅
23 35 ☽ # ☉
23 43 ☽ ± ♄

05 Saturday
01 47 ☽ ⊼ ♂
05 18 ☽ Q ☉
06 03 ☽ ✶ ♆
08 47 ☿ ꝑ ♄
09 25 ☽ ± ♅
10 07 ☽ ⊼ ♄
12 38 ☽ ± ♂
13 25 ☽ # ♀
15 04 ☽ ✶ ♀
16 36 ☽ Q ☿
17 08 ♀ Q ♇
20 09 ☿ △ ♃
23 28 ☽ # ♆

06 Sunday
01 14 ☽ ꝑ ♅
03 57 ☽ ✶ ☉
05 04 ☽ ꝑ ♂
08 28 ☽ ∠ ♆
14 18 ☽ ☍ ♃
17 47 ☽ ✶ ☿
20 15 ☽ ∠ ♀
23 08 ☽ ⊻ ♇

07 Monday
00 33 ☽ ⊥ ♆
03 50 ☽ # ♃
03 58 ☽ △ ♅
08 47 ☽ ∠ ☉
08 54 ☽ △ ♂
09 56 ☽ ⊥ ♇
11 22 ☽ ⊻ ♆
14 04 ☽ ⊥ ♀
15 15 ☽ ☍ ♄
17 15 ☉ Q ♅
22 10 ☽ # ♄
22 30 ☽ ∥ ♄

08 Tuesday
01 58 ☽ ∠ ☿
02 06 ☽ ⊻ ♀
02 16 ☽ ∠ ♇
02 17 ♀ ± ♄
02 24 ☽ ⊥ ☉
03 51 ☿ □ ♇
12 00 ☿ ∥ ♅
14 17 ☽ ⊻ ☉
17 07 ☽ ∥ ♃
20 35 ☽ ⊼ ♃
21 41 ☽ ⊥ ☿

09 Wednesday
02 54 ☿ ∥ ♂
05 58 ☽ ✶ ♇
07 43 ☽ ± ♃
11 00 ☽ ⊻ ☿
11 05 ☽ □ ♅
11 30 ☿ ✶ ♅
18 25 ☽ □ ♂
18 49 ☽ ☌ ♆
22 30 ☽ ⊼ ♄
22 54 ☽ ∥ ♆

10 Thursday
00 35 ☽ ꝑ ♃
04 32 ☽ Q ♇
04 56 ♂ □ ♆
06 03 ☿ # ♇
09 50 ☽ ± ♄
16 05 ☽ ☌ ♀
20 17 ☽ ∥ ♀
21 16 ☽ Q ♅
21 46 ☉ ꝑ ♄

11 Friday
03 00 ☽ ꝑ ♄
03 28 ☽ ☌ ☉
05 11 ☽ △ ♃
06 13 ☽ Q ♂
10 14 ☽ ∥ ☉
12 54 ☿ ⊻ ♆
15 06 ☽ □ ♇
20 32 ☽ ✶ ♅
23 27 ☿ ✶ ♂

12 Saturday
01 10 ☉ △ ♃
01 38 ♂ ∥ ♅
02 54 ☽ ∥ ♂
02 57 ☽ ∥ ♅
04 38 ☽ ⊻ ♆
06 37 ☽ ✶ ♂
07 38 ☽ ☌ ☿
08 06 ☽ △ ♄
10 30 ☿ △ ♄
10 45 ☽ # ♇
16 29 ☽ ⊥ ♆
21 52 ☽ ∥ ☿
22 06 ♀ Q ♅

13 Sunday
02 09 ☽ ∠ ♅
09 14 ☽ ⊻ ♀
10 24 ☽ ∠ ♆
13 40 ☽ ∠ ♂
14 52 ♂ ✶ ♄
16 06 ☽ □ ♃
19 34 ☽ ⊻ ☉
20 09 ☽ ⊥ ♅
22 37 ☽ ⊥ ♀

14 Monday
02 29 ☽ △ ♇
05 31 ☉ Q ♂
08 14 ☽ ⊻ ♅
08 34 ☽ ⊥ ♂
08 41 ☽ ⊥ ☉
16 37 ☽ ✶ ♆
16 51 ☿ ⊥ ♆
18 49 ☽ ∠ ♀
19 47 ☽ □ ♄
21 14 ☽ ⊻ ♂

15 Tuesday
03 55 ♀ ꝑ ♄
04 19 ☽ Q ♃
04 28 ☽ ∠ ☉
06 57 ☽ ⊻ ☿
08 47 ☽ ꝑ ♇
16 59 ☽ Q ♆
19 55 ♂ ∠ ♃
21 19 ☽ ⊥ ☿

16 Wednesday
03 03 ☽ ± ♇
03 45 ♀ △ ♃
04 39 ☽ ✶ ♃
04 45 ☽ ✶ ♀
08 10 ☽ Q ♄
13 36 ☽ ✶ ☉
15 14 ☽ ⊼ ♇
18 49 ☽ ∠ ☿
19 28 ☽ ∥ ☿
21 10 ☽ ☌ ♅

17 Thursday
05 35 ☽ □ ♆
07 53 ☽ Q ♀
08 24 ☽ ✶ ♄
09 45 ☉ □ ♇
11 01 ☽ ∠ ♃
11 05 ☿ ∠ ♅
12 51 ☽ ☌ ♂
16 00 ☽ Q ☉
21 59 ☽ # ♇

18 Friday
03 58 ♃ St D
05 06 ☽ ⊥ ♃
06 10 ☽ ✶ ☿
06 28 ☽ ∥ ♅
12 41 ☽ ∥ ♂
12 51 ☽ ∥ ☉
14 21 ☉ ∥ ♂
14 26 ☽ ∠ ♄
17 05 ☽ ⊻ ♃
21 07 ☽ ∥ ♀

19 Saturday
00 02 ☽ □ ♀
03 27 ☽ ☍ ♇
07 00 ☽ □ ☉
08 11 ☽ ⊥ ♄
09 22 ☽ ⊻ ♅
09 44 ☽ Q ☿
17 30 ☽ △ ♆
19 54 ☽ ⊻ ♄
21 07 ☽ ⊥ ♅

20 Sunday
00 24 ☿ ∠ ♆
03 00 ☽ ⊻ ♂
09 48 ♀ □ ♇
11 02 ☽ ∥ ♆
14 03 ☉ ✶ ♅
14 31 ☽ ∠ ♅
15 03 ☽ ⊥ ♂
17 08 ♀ Q ♂
22 24 ☽ ꝑ ♆

21 Monday
01 25 ☽ □ ☿
03 15 ☽ ☌ ♃
08 48 ☽ ∠ ♂
13 02 ☽ ⊼ ♇
13 13 ♀ ∥ ♂
15 19 ☽ ± ♆
15 54 ☽ △ ♀
17 01 ☿ □ ♃
17 46 ☽ ∥ ♃
18 46 ☽ ✶ ♅
20 58 ☽ △ ☉

22 Tuesday
00 07 ☽ ± ♇
02 23 ☽ ⊼ ♆
04 22 ☽ ☌ ♄
11 40 ☽ ∥ ♄
13 21 ☽ # ♄
13 31 ☽ ✶ ♂
16 27 ☽ ꝑ ♇
16 41 ☽ Q ♅
21 58 ☽ ꝑ ♀
22 45 ☉ ∥ ♅
23 05 ♀ ✶ ♅

23 Wednesday
02 11 ☽ ꝑ ☉
06 44 ☽ # ♃
08 23 ☿ ⊥ ♅
09 51 ☽ ⊻ ♃
11 43 ☽ Q ♂
14 36 ☽ △ ☿
15 24 ☽ ± ♀
18 56 ☽ △ ♇
19 10 ☽ ± ☉
20 20 ☽ ⊥ ♃

24 Thursday
00 25 ☽ □ ♅
02 48 ☽ ⊼ ♀
06 16 ☽ ⊼ ☉
07 32 ☽ ☍ ♆
09 09 ☽ ⊻ ♄
11 06 ☽ # ♆
11 48 ☽ ∠ ♃
19 05 ☽ ꝑ ☿
19 19 ☽ ⊥ ♄
19 53 ☽ □ ♂

25 Friday
01 17 ☉ ⊻ ♆
09 41 ☿ ∠ ♂
10 21 ☽ ∠ ♄
11 30 ☿ △ ♇
11 37 ☽ ± ☿
13 00 ☽ ✶ ♃
19 58 ☽ # ♂
21 31 ☽ □ ♇
22 28 ☽ ⊼ ☿
23 39 ☉ △ ♄

26 Saturday
02 52 ☽ △ ♅
02 59 ♀ ∥ ♅
06 46 ☽ # ♅
07 02 ☽ # ♀
08 45 ☽ Q ♃
09 34 ☽ ☍ ♀
09 37 ☽ ⊼ ♆
10 03 ♀ ⊻ ♆
10 36 ☽ # ☉
10 59 ☽ ✶ ♄
11 47 ☽ ☍ ☉
14 14 ☽ ∥ ♇
19 23 ☽ ± ♆
23 13 ☽ △ ♂

27 Sunday
02 13 ♀ △ ♄
02 50 ☽ Q ♇
03 19 ☽ ꝑ ♅
06 24 ☽ Q ♄
09 58 ☽ ꝑ ♆
13 20 ♇ St R
13 57 ☽ □ ♃
17 52 ☽ ± ♅
22 11 ☽ # ☿
22 11 ☽ ✶ ♇

28 Monday
00 19 ☽ ꝑ ♂
03 18 ☽ ☍ ☿
03 33 ☽ ⊼ ♅
06 47 ☿ ⊻ ♅
10 09 ☽ △ ♆
11 21 ☽ □ ♄
14 22 ☽ ⊼ ♀
15 21 ☽ ± ♂
15 31 ☽ ⊼ ☉
22 19 ☽ ∠ ♇

29 Tuesday
00 54 ☽ ± ♀
01 23 ☽ ⊼ ♂
01 51 ☽ ± ☉
12 52 ☽ ⊥ ♇
14 24 ☽ △ ♃
16 46 ☽ ꝑ ♀
17 26 ☽ ꝑ ☉
17 28 ☽ # ☿
22 36 ☽ ⊻ ♇

30 Wednesday
04 12 ☽ ☍ ♅
07 33 ☽ ⊼ ☿
07 57 ☉ # ♇
10 54 ☽ □ ♆
11 57 ☽ △ ♄
14 58 ☽ ꝑ ♃
18 07 ☽ ± ☿
19 35 ☽ △ ♀
19 44 ☽ △ ☉
22 39 ☽ # ☉
23 12 ☽ ∥ ♇
23 44 ☽ # ♀

31 Thursday
02 25 ☉ ☌ ♀
04 22 ☽ ☍ ♂
05 55 ☽ ± ♃
07 18 ☽ # ♅
09 17 ♀ # ♇
10 07 ☽ ꝑ ☿
10 23 ♀ ⊥ ♆
12 42 ☉ ⊥ ♆
12 51 ☽ ꝑ ♄
16 00 ☽ ⊼ ♃
16 54 ☽ Q ♆
23 37 ☽ # ♂

JUNE 2040

LONGITUDES

Date	Sidereal time h m s	Sun ☉ ° ' "	Moon ☽ ° ' "	Moon ☽ 24.00 ° ' "	Mercury ☿ ° '	Venus ♀ ° '	Mars ♂ ° '	Jupiter ♃ ° '	Saturn ♄ ° '	Uranus ♅ ° '	Neptune ♆ ° '	Pluto ♇ ° '
01	04 42 21	11 ♊ 36 08	03 ♓ 54 46	10 ♓ 51 29	04 ♋ 35	11 ♊ 59	15 ♌ 56	22 ♍ 11	05 ♎ 16	00 ♌ 39	04 ♉ 42	27 ♒ 07
02	04 46 17	12 33 38	17 ♓ 43 01	24 ♓ 29 27	05 24	13 13	16 29	22 14	05 R 15	00 42	04 44	27 R 07
03	04 50 14	13 31 06	01 ♈ 10 53	07 ♈ 47 32	06 08	14 26	17 02	22 17	05 15	00 45	04 46	27 07
04	04 54 11	14 28 34	14 ♈ 19 36	20 ♈ 47 22	06 48	15 40	17 36	22 19	05 15	00 48	04 48	27 07
05	04 58 07	15 26 01	27 ♈ 11 05	03 ♉ 31 04	07 24	16 54	18 09	22 23	05 14	00 50	04 50	27 07
06	05 02 04	16 23 27	09 ♉ 47 34	16 ♉ 00 53	07 56	18 08	18 43	22 26	05 14	00 53	04 51	27 06
07	05 06 00	17 20 53	22 ♉ 11 16	28 ♉ 19 00	08 23	19 21	19 17	22 29	05 D 14	00 56	04 53	27 06
08	05 09 57	18 18 18	04 ♊ 24 20	10 ♊ 27 29	08 46	20 35	19 50	22 33	05 14	00 59	04 55	27 06
09	05 13 53	19 15 42	16 ♊ 28 43	22 ♊ 28 15	09 04	21 49	20 24	22 37	05 15	01 02	04 57	27 05
10	05 17 50	20 13 06	28 ♊ 26 20	04 ♋ 23 13	09 18	23 03	20 58	22 41	05 15	01 05	04 58	27 05
11	05 21 46	21 10 29	10 ♋ 19 09	16 ♋ 14 23	09 27	24 16	21 32	22 45	05 15	01 08	05 00	27 05
12	05 25 43	22 07 50	22 ♋ 09 15	28 ♋ 04 01	09 32	25 30	22 06	22 49	05 16	01 11	05 02	27 04
13	05 29 40	23 05 11	03 ♌ 59 03	09 ♌ 54 43	09 R 32	26 44	22 40	22 54	05 17	01 14	05 04	27 04
14	05 33 36	24 02 32	15 ♌ 51 23	21 ♌ 49 29	09 27	27 57	23 15	22 58	05 17	01 17	05 05	27 03
15	05 37 33	24 59 51	27 ♌ 49 29	03 ♍ 51 50	09 19	29 ♊ 11	23 49	23 03	05 18	01 20	05 07	27 03
16	05 41 29	25 57 09	09 ♍ 57 02	16 ♍ 05 36	09 06	00 ♋ 25	24 23	23 08	05 19	01 23	05 08	27 03
17	05 45 26	26 54 27	22 ♍ 18 03	28 ♍ 34 54	08 49	01 39	24 58	23 13	05 20	01 27	05 10	27 02
18	05 49 22	27 51 43	04 ♎ 56 40	11 ♎ 23 50	08 28	02 52	25 32	23 18	05 21	01 30	05 11	27 01
19	05 53 19	28 48 59	17 ♎ 56 50	24 ♎ 36 03	08 04	04 06	26 07	23 24	05 23	01 33	05 13	27 01
20	05 57 15	29 ♊ 46 14	01 ♏ 21 47	08 ♏ 14 14	07 36	05 20	26 42	23 29	05 24	01 36	05 14	27 00
21	06 01 12	00 ♋ 43 29	15 ♏ 13 28	22 ♏ 19 26	07 07	06 33	27 16	23 35	05 25	01 39	05 16	27 00
22	06 05 09	01 40 42	29 ♏ 31 52	06 ♐ 50 22	06 35	07 47	27 51	23 41	05 27	01 43	05 17	26 59
23	06 09 05	02 37 56	14 ♐ 14 20	21 ♐ 42 57	06 01	09 01	28 26	23 46	05 29	01 46	05 19	26 59
24	06 13 02	03 35 09	29 ♐ 15 18	06 ♑ 50 14	05 26	10 15	29 01	23 53	05 30	01 49	05 20	26 58
25	06 16 58	04 32 21	14 ♑ 26 35	22 ♑ 03 03	04 51	11 28	29 ♌ 36	23 59	05 32	01 53	05 22	26 57
26	06 20 55	05 29 33	29 ♑ 38 22	07 ♒ 11 18	04 16	12 42	00 ♍ 11	24 05	05 34	01 56	05 23	26 56
27	06 24 51	06 26 45	14 ♒ 40 42	22 ♒ 05 36	03 42	13 56	00 46	24 12	05 36	02 00	05 24	26 56
28	06 28 48	07 23 57	29 ♒ 25 10	06 ♓ 38 46	03 09	15 09	01 21	24 18	05 38	02 03	05 26	26 55
29	06 32 44	08 21 09	13 ♓ 45 57	20 ♓ 46 28	02 38	16 23	01 56	24 25	05 40	02 06	05 27	26 54
30	06 36 41	09 ♋ 18 21	27 ♓ 40 15	04 ♈ 27 23	02 ♋ 09	17 ♋ 37	02 ♍ 31	24 ♍ 32	05 ♎ 43	02 ♌ 10	05 ♉ 28	26 ♒ 53

Date	Moon True ☊ ° '	Moon Mean ☊ ° '	Moon ☽ Latitude ° '
01	03 ♊ 45	03 ♊ 20	05 S 17
02	03 D 45	03 17	05 08
03	03 46	03 14	04 42
04	03 46	03 11	04 01
05	03 47	03 08	03 09
06	03 48	03 05	02 09
07	03 49	03 01	01 S 04
08	03 R 50	02 58	00 N 03
09	03 49	02 55	01 09
10	03 48	02 52	02 12
11	03 46	02 49	03 08
12	03 43	02 46	03 56
13	03 40	02 42	04 34
14	03 37	02 39	05 00
15	03 35	02 36	05 13
16	03 33	02 33	05 13
17	03 32	02 30	04 58
18	03 D 32	02 26	04 29
19	03 33	02 23	03 45
20	03 34	02 20	02 48
21	03 36	02 17	01 39
22	03 36	02 14	00 N 22
23	03 R 36	02 11	00 S 58
24	03 34	02 07	02 16
25	03 31	02 04	03 25
26	03 27	02 01	04 20
27	03 23	01 58	04 56
28	03 19	01 55	05 11
29	03 15	01 52	05 07
30	03 ♊ 13	01 ♊ 48	04 S 44

DECLINATIONS

Date	Sun ☉ ° '	Moon ☽ ° '	Mercury ☿ ° '	Venus ♀ ° '	Mars ♂ ° '	Jupiter ♃ ° '	Saturn ♄ ° '	Uranus ♅ ° '	Neptune ♆ ° '	Pluto ♇ ° '
01	22 N 10	14 S 59	24 N 53	22 N 07	17 N 27	04 N 21	00 N 14	20 N 33	11 N 29	21 S 53
02	22 18	09 34	24 42	22 19	17 17	04 20	00 14	20 33	11 30	21 54
03	22 25	03 S 50	24 30	22 30	17 06	04 18	00 14	20 32	11 30	21 54
04	22 32	01 N 56	24 17	22 40	16 55	04 17	00 14	20 32	11 31	21 54
05	22 38	07 31	24 04	22 50	16 44	04 16	00 14	20 31	11 31	21 55
06	22 44	12 42	23 49	22 59	16 33	04 14	00 14	20 30	11 32	21 55
07	22 50	17 17	23 35	23 08	16 21	04 12	00 14	20 30	11 33	21 55
08	22 55	21 04	23 19	23 16	16 10	04 11	00 13	20 29	11 33	21 56
09	23 00	23 54	23 04	23 23	15 59	04 09	00 13	20 28	11 34	21 56
10	23 04	25 37	22 48	23 30	15 47	04 07	00 13	20 28	11 34	21 57
11	23 08	26 10	22 31	23 36	15 35	04 05	00 12	20 27	11 35	21 57
12	23 12	25 30	22 15	23 41	15 24	04 04	00 12	20 26	11 35	21 57
13	23 15	23 41	21 59	23 45	15 12	04 02	00 11	20 26	11 36	21 58
14	23 18	20 51	21 42	23 49	15 00	04 00	00 11	20 25	11 36	21 58
15	23 20	17 07	21 26	23 52	14 48	03 58	00 10	20 24	11 37	21 59
16	23 22	12 39	21 10	23 55	14 36	03 55	00 10	20 24	11 37	21 59
17	23 24	07 37	20 55	23 56	14 23	03 53	00 09	20 23	11 38	22 00
18	23 25	02 N 09	20 39	23 57	14 11	03 51	00 09	20 22	11 38	22 00
19	23 26	03 S 34	20 25	23 58	13 59	03 49	00 08	20 21	11 39	22 00
20	23 26	09 19	20 11	23 57	13 46	03 46	00 07	20 21	11 39	22 01
21	23 26	14 49	19 57	23 56	13 33	03 44	00 06	20 20	11 40	22 01
22	23 25	19 41	19 45	23 54	13 21	03 41	00 05	20 19	11 40	22 02
23	23 25	23 28	19 33	23 52	13 08	03 39	00 05	20 18	11 40	22 02
24	23 23	25 42	19 22	23 48	12 55	03 36	00 04	20 18	11 41	22 03
25	23 21	26 03	19 13	23 44	12 42	03 34	00 03	20 17	11 41	22 03
26	23 19	24 27	19 04	23 40	12 29	03 31	00 02	20 16	11 42	22 04
27	23 17	21 08	18 57	23 34	12 16	03 28	00 N 01	20 15	11 42	22 04
28	23 14	16 31	18 51	23 28	12 03	03 25	00 00	20 15	11 42	22 05
29	23 10	11 06	18 47	23 21	11 50	03 22	00 S 01	20 14	11 43	22 05
30	23 N 07	05 S 16	18 N 43	23 N 14	11 N 36	03 N 20	00 S 02	20 N 13	11 N 43	22 S 06

ZODIAC SIGN ENTRIES

Date	h m	Planets
01	05 18	☽ ♓
03	09 52	☽ ♈
05	17 19	☽ ♉
08	03 19	☽ ♊
10	15 09	☽ ♋
13	03 55	☽ ♌
15	16 20	☽ ♍
16	03 53	♀ ♋
18	02 41	☽ ♎
20	09 36	☽ ♏
20	17 46	☉ ♋
22	12 46	☽ ♐
24	13 11	☽ ♑
26	04 42	♂ ♍
26	12 34	☽ ♒
28	12 57	☽ ♓
30	16 06	☽ ♈

LATITUDES

Date	Mercury ☿ ° '	Venus ♀ ° '	Mars ♂ ° '	Jupiter ♃ ° '	Saturn ♄ ° '	Uranus ♅ ° '	Neptune ♆ ° '	Pluto ♇ ° '
01	01 N 32	00 S 07	01 N 28	01 N 22	02 N 32	00 N 34	01 S 42	10 S 03
04	01 02	00 00	01 26	01 21	02 32	00 34	01 42	10 04
07	00 N 24	00 N 08	01 23	01 20	02 31	00 34	01 42	10 05
10	00 S 19	00 15	01 21	01 20	02 30	00 34	01 42	10 06
13	01 07	00 22	01 19	01 19	02 30	00 34	01 42	10 07
16	01 57	00 29	01 16	01 18	02 29	00 34	01 43	10 08
19	02 47	00 35	01 14	01 18	02 28	00 34	01 43	10 09
22	03 32	00 42	01 12	01 17	02 27	00 33	01 43	10 09
25	04 08	00 48	01 10	01 16	02 27	00 33	01 43	10 10
28	04 33	00 54	01 08	01 16	02 26	00 33	01 43	10 11
31	04 S 44	01 N 00	01 N 06	01 N 15	02 N 25	00 N 33	01 S 43	10 S 12

DATA

Julian Date	2466307
Delta T	+77 seconds
Ayanamsa	24° 25' 03"
Synetic vernal point	04° ♓ 41' 56"
True obliquity of ecliptic	23° 26' 06"

MOON'S PHASES, APSIDES AND POSITIONS ☽

Date	h m	Phase	Longitude ° '	Eclipse Indicator
02	02 17	☾	12 ♓ 10	
09	18 03	●	19 ♊ 30	
17	21 32	☽	27 ♍ 17	
24	19 19	○	03 ♑ 53	

Day	h m		
12	14 25	Apogee	
25	09 41	Perigee	
04	03 54	ON	
11	10 38	Max dec	26° N 10'
18	21 06	OS	
25	04 16	Max dec	26° S 09'

LONGITUDES

Date	Chiron ⚷ ° '	Ceres ⚳ ° '	Pallas ⚴ ° '	Juno ⚵ ° '	Vesta ⚶ ° '	Black Moon Lilith ⚸ ° '
01	11 ♋ 16	12 ♊ 46	19 ♈ 20	05 ♉ 07	13 ♒ 30	17 ♋ 46
11	12 ♋ 14	16 ♊ 57	22 ♈ 24	10 ♉ 35	14 ♒ 12	18 ♋ 53
21	13 ♋ 14	21 ♊ 08	25 ♈ 21	16 ♉ 02	14 ♒ 06	20 ♋ 00
31	14 ♋ 15	25 ♊ 19	28 ♈ 06	21 ♉ 29	13 ♒ 13	21 ♋ 07

All ephemeris data is given at 12.00 UT and the Moon's longitude is additionally given for 24.00 UT

ASPECTARIAN

h m Aspects

01 Friday
00 24 ☽ ☌ ♇
04 03 ☽ ± ♄
06 24 ☽ ⚻ ♅
13 14 ☽ △ ☿
13 22 ☽ ⚹ ♆
14 19 ☽ ⚻ ♄
15 26 ☿ ⚹ ♆
16 44 ☽ ± ♅

02 Saturday
00 06 ♂ ⊥ ♃
02 17 ☽ □ ☉
03 19 ☽ □ ♀
03 44 ☽ # ♆
07 42 ☿ □ ♄
08 06 ☉ ∥ ♀
08 27 ☽ BQ ♅
09 44 ☽ ⚻ ♂
15 34 ☽ ∠ ♆
20 00 ☽ ☍ ♃
20 47 ☽ ± ♂

03 Sunday
04 41 ☽ ⊻ ♇
07 38 ☽ ⊥ ♆
10 03 ☽ # ♃
11 13 ☽ △ ♅
12 39 ☽ Q ☉
13 37 ☽ BQ ♂
14 30 ☽ Q ♀
15 30 ☽ ⊥ ♇
18 30 ☽ ⊻ ♆
19 22 ☽ ☍ ♄
21 28 ☽ □ ☿

04 Monday
02 54 ☽ # ♄
04 53 ☽ ∥ ♄
07 55 ☽ ∠ ♇
12 18 ☽ ⚹ ☉
14 31 ♀ ∠ ♅
14 44 ☽ ⚹ ♀
18 20 ☽ △ ♂
21 52 ☽ ∥ ♃

05 Tuesday
02 56 ☽ ⚻ ♃
08 29 ☽ Q ☿
11 51 ☽ ⚹ ♇
14 16 ☽ ± ♃
18 39 ☽ ∠ ☉
18 56 ☽ □ ♅
21 53 ☽ ∠ ♀
22 42 ☉ ∠ ♅

06 Wednesday
02 32 ☽ ☌ ♆
03 17 ☽ ⚻ ♄
06 21 ☽ ∥ ♆
07 27 ☽ BQ ♃
08 17 ☽ ⚹ ☿
10 41 ☽ Q ♇
13 15 ☽ ⊥ ☉
14 47 ☽ ± ♄
16 59 ☽ ⊥ ♀
17 55 ♄ St D

07 Thursday
01 48 ☽ ⊻ ☉
05 38 ☽ Q ♅
05 52 ☽ ⊻ ♀
06 04 ☽ □ ♂
07 03 ☽ ∥ ♂
08 12 ☽ BQ ♄
09 10 ♀ ⚹ ♂
12 36 ☽ △ ♃
14 25 ☽ ∠ ☿
21 36 ☽ □ ♇
22 38 ♀ ∠ ♆

08 Friday
05 13 ☽ ⚹ ♅
07 53 ☽ ∥ ♅
08 40 ☽ ⊥ ☿
13 01 ☽ ⊻ ♆
13 39 ☽ △ ♄
15 20 ☿ ∥ ♀
18 26 ☽ # ♇
19 08 ☽ Q ♂
20 53 ☽ ⊻ ☿

09 Saturday
00 57 ☽ ⊥ ♆
03 05 ☽ ∥ ☉
04 36 ☽ ∥ ☿
05 10 ♂ ∠ ♄
06 40 ☽ ∥ ♀
11 06 ☽ ∠ ♅
16 06 ☉ ∥ ☿
18 03 ☽ ☌ ☉
18 57 ☽ ∠ ♆
20 15 ☽ ⚹ ♂
23 54 ☽ ☌ ♀

10 Sunday
00 21 ☽ □ ♃
04 30 ♀ □ ♃
05 13 ☽ ⊥ ♅
05 41 ☉ ∠ ♆
09 16 ☽ △ ♇
17 21 ☽ ⊻ ♅

11 Monday
01 13 ☽ ⚹ ♆
01 45 ☽ □ ♄
03 58 ☽ ∠ ♂
10 14 ☽ ☌ ☿
12 52 ☽ Q ♃
15 34 ☽ BQ ♇
23 06 ☽ ⊥ ♂

12 Tuesday
01 35 ☽ Q ♆
05 32 ♀ ⊥ ♅
09 48 ☽ ± ♇
10 25 ☉ ⚹ ♂
11 54 ☽ ⊻ ♂
11 57 ☽ ⊻ ☉
13 22 ☽ ⚹ ♃
14 16 ☽ Q ♄
19 35 ☽ ⊻ ♀
21 58 ☽ ⚻ ♇

13 Wednesday
00 19 ☿ St R
01 12 ☽ ⊥ ☉
06 24 ☽ ☌ ♅
06 45 ☉ □ ♃
09 10 ☽ ⊥ ♀
11 16 ☽ ∥ ♀
13 21 ☿ # ♇
14 11 ☽ □ ♆
14 37 ☽ ⚹ ♄
16 09 ☽ ∥ ☉
18 31 ♀ △ ♇
19 58 ☽ ∠ ♃
21 02 ☽ ∠ ☉
22 41 ♂ ⊻ ♃
23 11 ☽ ⊻ ☿

14 Thursday
03 26 ☽ # ♇
04 52 ☽ ∥ ☿
05 29 ☽ ∠ ♀
11 12 ☽ ⊥ ☿
14 16 ☽ ⊥ ♃
15 04 ☽ ∥ ♅
20 56 ☽ ∠ ♄

15 Friday
02 24 ☽ ⊻ ♃
03 35 ☽ ☌ ♂
05 05 ☽ ∠ ☿
05 52 ☽ ⚹ ☉
10 27 ☽ ☍ ♇
14 57 ☽ ⊥ ♄
15 01 ☽ ⚹ ♀
19 01 ☽ ⊻ ♅
21 02 ☉ ⊥ ♅

16 Saturday
01 33 ☽ ∥ ♂
02 30 ☽ △ ♆
02 52 ☽ ⊻ ♄
03 14 ☿ ∠ ♂
06 56 ☽ ⊥ ♅
07 44 ☽ Q ☉
10 21 ☽ ⚹ ☿
17 08 ☽ ∥ ♆
17 22 ☽ Q ♀

17 Sunday
00 38 ☽ ∠ ♅
07 53 ☽ BQ ♆
07 53 ♀ ⊻ ♅
09 12 ☽ Q ☿
13 46 ☽ ☌ ♃
15 09 ☉ △ ♇
17 21 ☽ ⊻ ♂
21 03 ☽ ⚻ ♇
21 32 ☽ □ ☉

18 Monday
01 08 ☽ ± ♆
04 38 ☽ ∥ ♃
05 18 ☽ ⊥ ♂
05 29 ☽ ⚹ ♅
07 42 ☽ □ ♀
08 24 ☽ ± ♇
12 28 ☽ ⚻ ♆
12 46 ☽ ☌ ♄
18 23 ☽ □ ☿
20 31 ☽ ∥ ♄
21 40 ☽ # ♄
22 54 ☽ ∠ ♂

19 Tuesday
01 09 ☽ BQ ♇
03 56 ☽ Q ♅
13 00 ☽ # ♃
18 00 ☿ ∥ ♅
21 54 ☽ ⊻ ♃

20 Wednesday
03 22 ☽ ⚹ ♂
04 18 ☽ △ ♇
08 40 ☽ ⊥ ♃
08 59 ☽ △ ☉
10 15 ♀ ⚹ ♆
12 25 ☽ □ ♅
13 23 ♀ □ ♄
18 48 ☽ ☍ ♆
19 05 ☽ ⊻ ♄
19 38 ☽ △ ♀
21 58 ☽ # ♆
22 32 ☽ △ ☿

21 Thursday
00 31 ☽ ∠ ♃
00 50 ♂ ☍ ♇
01 21 ☽ Q ♂
05 29 ☽ ⊥ ♄
06 32 ☽ # ♂
12 55 ☽ BQ ☉
19 35 ☿ ☌ ♀
20 49 ☽ ∠ ♄
23 14 ☽ BQ ☿
23 43 ☽ BQ ♀

22 Friday
02 12 ☽ ⚹ ♃
05 09 ☽ ± ☉
07 47 ☽ □ ♇
09 06 ☽ □ ♂
12 20 ☽ # ☿
12 55 ☉ ⊻ ♅
13 40 ☽ ± ☿
15 31 ☽ # ♅
15 37 ☽ △ ♅
15 47 ☽ ⚻ ☉
16 04 ☽ ± ♀
21 29 ☽ ⚻ ♆
21 45 ☽ ⚹ ♄
22 10 ☽ Q ♃
23 09 ☽ ⚻ ☿

23 Saturday
01 54 ☽ ∥ ♇
02 46 ☽ ⚻ ♀
07 16 ☽ ± ♆
11 34 ☽ # ☉
13 11 ☽ Q ♇
15 08 ☽ # ♀
16 05 ☽ BQ ♅
17 13 ☽ Q ♄
21 46 ☽ BQ ♆

24 Sunday
03 23 ☽ □ ♃
06 32 ☽ ± ♅
08 22 ☽ ⚹ ♇
09 26 ☿ □ ♄
11 36 ☽ △ ♂
16 02 ☿ ⚹ ♆
16 05 ☽ ⚻ ♅
19 19 ☽ ☍ ☉
21 26 ☽ ☍ ☿
21 39 ☽ △ ♆
21 55 ☽ □ ♄

25 Monday
06 54 ☽ ☍ ♀
08 05 ☽ ∠ ♇
09 26 ♂ ⊥ ♄
12 15 ☽ BQ ♂
16 53 ☉ ☌ ☿
21 20 ♀ BQ ♇
22 16 ☽ ⊥ ♇
22 53 ♀ Q ♃

26 Tuesday
03 01 ☽ ± ♂
03 09 ☽ △ ♃
07 44 ☽ ⊻ ♇
09 09 ☉ ⚹ ♆
12 53 ☽ ⚻ ♂
13 57 ☉ □ ♄
15 39 ☽ ☍ ♅
19 04 ☽ # ♀
19 05 ☽ ⚻ ☿
21 08 ☽ □ ♆
21 26 ☽ △ ♄
21 37 ☽ # ☉
21 56 ☽ ⚻ ☉

27 Wednesday
03 08 ☽ BQ ♃
04 18 ☽ ± ☿
06 12 ☽ ∥ ♇
08 10 ☽ ± ☉
10 41 ☽ ⚻ ♀
17 03 ☽ # ♅
17 43 ☽ ± ♃
18 15 ☽ BQ ☿
21 15 ☽ ± ♀
21 36 ☽ BQ ♄
23 42 ☽ BQ ☉

28 Thursday
00 19 ☽ # ☿
02 09 ☽ Q ♆
03 32 ☽ ⚻ ♃
07 53 ☽ ☌ ♇
12 21 ☽ ± ♄
13 20 ☽ BQ ♀
15 19 ☽ ☍ ♂
16 22 ☽ ⚻ ♅
17 57 ☽ △ ☿
21 59 ☽ ⚹ ♆
22 20 ☽ ⚻ ♄

29 Friday
02 12 ☽ △ ☉
02 25 ☽ ± ♅
08 47 ☽ # ♂
09 24 ☽ # ♆
16 53 ☽ △ ♀
17 43 ☽ BQ ♅
19 48 ♂ ⊻ ♅
23 27 ☽ ∠ ♆

30 Saturday
00 12 ♂ ∥ ♆
03 29 ☿ ⚹ ♂
06 28 ☽ ☍ ♃
08 39 ♀ ∠ ♂
09 07 ♀ Q ♆
10 38 ☽ ⊻ ♇
11 18 ☿ ⊻ ♅
15 10 ☽ ⊥ ♆
19 39 ☽ □ ☿
19 57 ☽ # ♃
19 58 ☽ △ ♅
20 57 ☽ ⚻ ♂
21 12 ☽ ⊥ ♇

JULY 2040

LONGITUDES

Date	Sidereal time h m s	Sun ☉ ° ' "	Moon ☽ ° ' "	Moon ☽ 24.00 ° ' "	Mercury ☿ ° '	Venus ♀ ° '	Mars ♂ ° '	Jupiter ♃ ° '	Saturn ♄ ° '	Uranus ♅ ° '	Neptune ♆ ° '	Pluto ♇ ° '
01	06 40 38	10 ♋ 15 33	11 ♈ 08 03	17 ♈ 42 36	01 ♋ 43	18 ♋ 50	03 ♍ 07	24 ♍ 39	05 ♎ 45	02 ♌ 13	05 ♉ 29	26 ♒ 53
02	06 44 34	11 12 46	24 ♈ 11 23	00 ♉ 34 52	01 R 21	20 04	03 42	24 46	05 48	02 17	05 30	26 R 52
03	06 48 31	12 09 58	06 ♉ 53 32	13 ♉ 07 53	01 02	21 18	04 17	24 53	05 50	02 20	05 32	26 51
04	06 52 27	13 07 11	19 ♉ 18 24	25 ♉ 25 35	00 48	22 31	04 53	25 01	05 53	02 24	05 33	26 50
05	06 56 24	14 04 24	01 ♊ 29 55	07 ♊ 31 49	00 38	23 45	05 29	25 08	05 56	02 27	05 34	26 49
06	07 00 20	15 01 38	13 ♊ 31 42	19 ♊ 29 59	00 32	24 59	06 04	25 16	05 58	02 31	05 35	26 48
07	07 04 17	15 58 51	25 ♊ 26 59	01 ♋ 23 02	00 D 32	26 13	06 40	25 24	06 01	02 34	05 36	26 47
08	07 08 13	16 56 05	07 ♋ 18 25	13 ♋ 13 24	00 36	27 26	07 16	25 32	06 04	02 38	05 37	26 47
09	07 12 10	17 53 19	19 ♋ 08 13	25 ♋ 03 07	00 46	28 40	07 52	25 40	06 08	02 42	05 38	26 46
10	07 16 07	18 50 33	00 ♌ 58 19	06 ♌ 54 01	01 01	29 ♋ 54	08 27	25 48	06 11	02 45	05 39	26 45
11	07 20 03	19 47 47	12 ♌ 50 28	18 ♌ 47 53	01 21	01 ♌ 07	09 03	25 56	06 14	02 49	05 40	26 44
12	07 24 00	20 45 01	24 ♌ 46 32	00 ♍ 46 40	01 46	02 21	09 39	26 05	06 17	02 52	05 41	26 43
13	07 27 56	21 42 15	06 ♍ 48 36	12 ♍ 52 39	02 17	03 35	10 15	26 13	06 21	02 56	05 41	26 42
14	07 31 53	22 39 29	18 ♍ 59 11	25 ♍ 08 36	02 53	04 49	10 52	26 22	06 25	03 00	05 42	26 41
15	07 35 49	23 36 43	01 ♎ 21 18	07 ♎ 37 43	03 34	06 02	11 28	26 30	06 28	03 03	05 43	26 40
16	07 39 46	24 33 58	13 ♎ 58 20	20 ♎ 23 36	04 20	07 16	12 04	26 39	06 32	03 07	05 44	26 38
17	07 43 42	25 31 12	26 ♎ 53 59	03 ♏ 29 56	05 11	08 30	12 40	26 48	06 36	03 11	05 44	26 37
18	07 47 39	26 28 27	10 ♏ 11 50	17 ♏ 00 02	06 08	09 43	13 17	26 57	06 40	03 14	05 45	26 36
19	07 51 36	27 25 42	23 ♏ 54 46	00 ♐ 56 12	07 09	10 57	13 53	27 06	06 44	03 18	05 46	26 35
20	07 55 32	28 22 57	08 ♐ 04 19	15 ♐ 18 56	08 16	12 11	14 29	27 15	06 48	03 22	05 46	26 34
21	07 59 29	29 ♋ 20 12	22 ♐ 39 42	00 ♑ 06 01	09 27	13 24	15 06	27 25	06 52	03 25	05 47	26 33
22	08 03 25	00 ♌ 17 28	07 ♑ 37 06	15 ♑ 11 59	10 43	14 38	15 43	27 34	06 56	03 29	05 48	26 32
23	08 07 22	01 14 44	22 ♑ 49 30	00 ♒ 28 20	12 03	15 52	16 19	27 44	07 00	03 33	05 48	26 31
24	08 11 18	02 12 01	08 ♒ 07 06	15 ♒ 44 24	13 28	17 05	16 56	27 53	07 05	03 36	05 49	26 29
25	08 15 15	03 09 18	23 ♒ 18 52	00 ♓ 49 14	14 58	18 19	17 33	28 03	07 09	03 40	05 49	26 28
26	08 19 11	04 06 36	08 ♓ 14 26	15 ♓ 33 32	16 32	19 33	18 09	28 13	07 14	03 44	05 50	26 27
27	08 23 08	05 03 55	22 ♓ 45 51	29 ♓ 50 58	18 09	20 46	18 46	28 23	07 18	03 48	05 50	26 26
28	08 27 05	06 01 15	06 ♈ 48 38	13 ♈ 38 50	19 51	22 00	19 23	28 33	07 23	03 51	05 50	26 25
29	08 31 01	06 58 36	20 ♈ 21 42	26 ♈ 57 32	21 36	23 14	20 00	28 43	07 28	03 55	05 51	26 23
30	08 34 58	07 55 58	03 ♉ 26 47	09 ♉ 49 55	23 25	24 27	20 37	28 53	07 32	03 59	05 51	26 22
31	08 38 54	08 ♌ 53 21	16 ♉ 07 31	22 ♉ 20 13	25 ♋ 17	25 ♌ 41	21 ♍ 14	29 ♍ 03	07 ♎ 37	04 ♌ 02	05 ♉ 51	26 ♒ 21

Moon True ☊ / Mean ☊ / Latitude and DECLINATIONS

Date	Moon True ☊ ° '	Moon Mean ☊ ° '	Moon ☽ Latitude ° '	Sun ☉ ° '	Moon ☽ ° '	Mercury ☿ ° '	Venus ♀ ° '	Mars ♂ ° '	Jupiter ♃ ° '	Saturn ♄ ° '	Uranus ♅ ° '	Neptune ♆ ° '	Pluto ♇ ° '
01	03 ♊ 13	01 ♊ 45	04 S 06	23 N 02	00 N 38	18 N 42	23 N 06	11 N 23	03 N 17	00 S 04	20 N 12	11 N 43	22 S 06
02	03 D 13	01 42	03 17	22 58	06 20	18 41	22 57	11 09	03 14	00 05	20 11	11 44	22 07
03	03 15	01 39	02 19	22 53	11 38	18 42	22 47	10 56	03 10	00 06	20 11	11 44	22 07
04	03 16	01 36	01 15	22 47	16 21	18 45	22 37	10 42	03 07	00 07	20 10	11 44	22 08
05	03 17	01 32	00 S 10	22 42	20 18	18 48	22 26	10 29	03 04	00 09	20 09	11 45	22 09
06	03 R 16	01 29	00 N 55	22 35	23 20	18 53	22 15	10 15	03 01	00 10	20 08	11 45	22 09
07	03 14	01 26	01 57	22 29	25 19	18 59	22 03	10 01	02 58	00 11	20 07	11 45	22 10
08	03 10	01 23	02 54	22 22	26 07	19 07	21 50	09 47	02 54	00 13	20 07	11 46	22 10
09	03 03	01 20	03 42	22 14	25 44	19 15	21 36	09 33	02 51	00 14	20 06	11 46	22 11
10	02 55	01 17	04 21	22 07	24 11	19 24	21 22	09 19	02 48	00 16	20 05	11 46	22 11
11	02 47	01 13	04 49	21 59	21 34	19 34	21 08	09 05	02 44	00 17	20 04	11 46	22 12
12	02 38	01 10	05 04	21 50	18 02	19 45	20 53	08 51	02 41	00 19	20 03	11 47	22 12
13	02 30	01 07	05 06	21 41	13 45	19 56	20 37	08 36	02 37	00 20	20 02	11 47	22 13
14	02 23	01 04	04 54	21 32	08 52	20 07	20 20	08 22	02 34	00 22	20 02	11 47	22 13
15	02 19	01 01	04 29	21 22	03 N 34	20 19	20 03	08 08	02 30	00 23	20 01	11 47	22 14
16	02 16	00 58	03 50	21 12	01 S 59	20 31	19 46	07 53	02 26	00 25	20 00	11 47	22 15
17	02 D 16	00 54	02 58	21 02	07 36	20 42	19 28	07 39	02 23	00 27	19 59	11 48	22 15
18	02 16	00 51	01 56	20 51	13 03	20 54	19 09	07 24	02 19	00 29	19 58	11 48	22 16
19	02 17	00 48	00 N 45	20 40	18 01	21 05	18 50	07 09	02 15	00 30	19 57	11 48	22 16
20	02 R 17	00 45	00 S 31	20 29	22 10	21 15	18 30	06 55	02 11	00 32	19 57	11 48	22 17
21	02 16	00 42	01 47	20 17	25 01	21 25	18 10	06 40	02 08	00 34	19 56	11 48	22 17
22	02 12	00 38	02 57	20 05	26 10	21 33	17 49	06 25	02 04	00 36	19 55	11 48	22 18
23	02 06	00 35	03 56	19 53	25 23	21 41	17 28	06 10	02 00	00 38	19 54	11 48	22 19
24	01 58	00 32	04 38	19 40	22 43	21 47	17 06	05 55	01 56	00 40	19 53	11 49	22 19
25	01 49	00 29	05 01	19 27	18 28	21 51	16 44	05 41	01 52	00 41	19 52	11 49	22 20
26	01 40	00 26	05 02	19 14	13 09	21 54	16 21	05 26	01 48	00 43	19 51	11 49	22 20
27	01 33	00 23	04 44	19 00	07 13	21 54	15 58	05 10	01 44	00 45	19 51	11 49	22 21
28	01 27	00 19	04 08	18 46	01 S 06	21 53	15 35	04 55	01 40	00 47	19 50	11 49	22 21
29	01 24	00 16	03 20	18 32	04 N 52	21 49	15 11	04 40	01 36	00 49	19 49	11 49	22 22
30	01 23	00 13	02 23	18 17	10 25	21 43	14 47	04 25	01 31	00 51	19 48	11 49	22 23
31	01 ♊ 23	00 ♊ 10	01 S 20	18 N 02	15 N 23	21 N 34	14 N 22	04 N 10	01 N 27	00 S 54	19 N 47	11 N 49	22 S 23

ZODIAC SIGN ENTRIES

Date	h	m	Planets
02	22	54	☽ ♉
05	09	02	☽ ♊
07	21	12	☽ ♋
10	10	02	☽ ♌
10	14	02	♀ ♌
12	22	27	☽ ♍
15	09	24	☽ ♎
17	17	40	☽ ♏
19	22	25	☽ ♐
21	23	50	☽ ♑
22	04	41	☉ ♌
23	23	16	☽ ♒
25	22	41	☽ ♓
28	00	15	☽ ♈
30	05	36	☽ ♉

LATITUDES

Date	Mercury ☿ ° '	Venus ♀ ° '	Mars ♂ ° '	Jupiter ♃ ° '	Saturn ♄ ° '	Uranus ♅ ° '	Neptune ♆ ° '	Pluto ♇ ° '
01	04 S 44	01 N 00	01 N 06	01 N 15	02 N 25	00 N 33	01 S 43	10 S 12
04	04 41	01 05	01 04	01 15	02 25	00 33	01 44	10 13
07	04 27	01 10	01 01	01 14	02 24	00 33	01 44	10 13
10	04 02	01 14	00 59	01 14	02 23	00 33	01 44	10 14
13	03 29	01 18	00 57	01 13	02 23	00 33	01 44	10 15
16	02 51	01 21	00 55	01 13	02 22	00 33	01 44	10 15
19	02 10	01 24	00 53	01 12	02 21	00 33	01 44	10 16
22	01 27	01 26	00 51	01 12	02 21	00 33	01 45	10 16
25	00 45	01 28	00 49	01 11	02 20	00 33	01 45	10 17
28	00 S 05	01 29	00 47	01 11	02 20	00 33	01 45	10 17
31	00 N 30	01 N 30	00 N 45	01 N 10	02 N 19	00 N 33	01 S 45	10 S 18

DATA

Julian Date	2466337
Delta T	+77 seconds
Ayanamsa	24° 25' 08"
Synetic vernal point	04° ♓ 41' 51"
True obliquity of ecliptic	23° 26' 06"

MOON'S PHASES, APSIDES AND POSITIONS ☽

Date	h	m	Phase	Longitude ° '	Eclipse Indicator
01	10	18	☾	10 ♈ 11	
09	09	15	●	17 ♋ 47	
17	09	16	☽	25 ♎ 25	
24	02	06	○	01 ♒ 48	
30	21	06	☾	08 ♉ 18	

Day	h	m		
09	20	28	Apogee	
23	19	18	Perigee	
01	09	26	0N	
08	16	14	Max dec	26° N 08'
16	03	31	0S	
22	14	29	Max dec	26° S 11'
28	16	21	0N	

LONGITUDES

Date	Chiron ⚷ ° '	Ceres ⚳ ° '	Pallas ⚴ ° '	Juno ⚵ ° '	Vesta ⚶ ° '	Black Moon Lilith ⚸ ° '
01	14 ♋ 15	25 ♊ 19	28 ♈ 06	21 ♉ 29	13 ♒ 13	21 ♋ 07
11	15 ♋ 18	29 ♊ 29	00 ♉ 40	26 ♉ 54	11 ♒ 35	22 ♋ 14
21	16 ♋ 20	03 ♋ 38	02 ♉ 58	02 ♊ 17	09 ♒ 24	23 ♋ 21
31	17 ♋ 20	07 ♋ 44	04 ♉ 56	07 ♊ 35	06 ♒ 58	24 ♋ 28

All ephemeris data is given at 12.00 UT and the Moon's longitude is additionally given for 24.00 UT

ASPECTARIAN

h m Aspects

01 Sunday
01 49 ☽ ⊻ ♆
02 17 ☽ ☍ ♄
08 10 ☽ ± ♂
09 12 ☽ ∥ ♄
09 40 ☽ ⋕ ♄
10 18 ☽ □ ☉
13 21 ☽ ∠ ♇
22 54 ☽ ∥ ♃

02 Monday
01 20 ☽ ⊡ ♂
03 15 ☽ Q ☿
03 33 ☽ □ ♀
07 05 ☉ ∥ ♀
13 05 ☽ ⚻ ♃
17 00 ☽ ✶ ♇
22 11 ☽ Q ☉

03 Tuesday
00 28 ☽ ± ♃
01 07 ☽ ✶ ☿
03 17 ☽ □ ♅
03 22 ♀ ± ♇
04 09 ☉ ⊡ ♇
06 48 ☽ △ ♂
08 51 ☽ ∥ ♂
09 23 ☽ ☌ ♆
09 58 ☽ ⚻ ♄
12 30 ☽ ∥ ♆
15 45 ☽ Q ♇
17 07 ☽ Q ♀
17 48 ☽ ⊡ ♃
21 32 ☽ ± ♄
22 59 ☽ ✶ ☉

04 Wednesday
05 17 ☽ ∠ ☿
08 56 ☉ Q ♃
14 08 ☽ Q ♅
15 05 ☽ ⊡ ♄
19 00 ☽ ✶ ♀
22 36 ☽ ⊥ ☿
23 18 ☽ △ ♃

05 Thursday
02 09 ☽ ∥ ☿
02 46 ☽ □ ♇
06 47 ☽ ∠ ☉
10 18 ☽ ⊻ ☿
11 00 ☽ ∥ ♅
13 55 ☽ ✶ ♅
15 31 ♀ Q ♄
15 34 ♂ △ ♆
20 05 ☽ ⊻ ♆
20 19 ☽ □ ♂
20 50 ☽ △ ♄

06 Friday
01 38 ☽ ⋕ ♇
02 13 ☽ ⊥ ☉
03 02 ☽ ∥ ♀
04 05 ☽ ∠ ♀
05 32 ☽ ∥ ☉
07 46 ♂ ⊻ ♄
08 05 ☽ ⊥ ♆
15 16 ☽ ⊻ ☉
18 13 ♀ ✶ ♃
20 03 ☽ ∠ ♅
23 03 ♀ ⋕ ♇

07 Saturday
00 13 ☽ ⊥ ♀
02 12 ☽ ∠ ♆
02 20 ☿ St D
10 20 ☽ Q ♂
11 54 ☽ □ ♃
13 43 ☽ ⊻ ♀
14 17 ☽ ⊥ ♅
14 42 ☽ △ ♇
22 19 ☽ ☌ ☿
23 12 ♀ ⚻ ♇

08 Sunday
02 29 ☽ ⊻ ♅
08 34 ☽ ✶ ♆
09 29 ☽ □ ♄
11 54 ☽ ✶ ♂
21 03 ☽ ⊡ ♇

09 Monday
00 46 ☽ Q ♃
05 23 ☉ Q ♆
08 56 ☽ Q ♆
09 15 ☽ ☌ ☉
15 17 ☽ ± ♇
19 57 ☽ ∠ ♂
22 10 ☽ Q ♄
22 54 ☉ ⋕ ♇

10 Tuesday
01 23 ☽ ✶ ♃
03 26 ☽ ⚻ ♇
09 34 ☽ ☌ ♀
12 06 ☽ ⊻ ☿
15 10 ☽ ⊥ ♂
15 38 ☽ ☌ ♅
21 28 ☽ □ ♆
22 36 ☽ ✶ ♄

11 Wednesday
00 34 ☽ ⊥ ☿
01 14 ♂ ⊥ ♅
03 57 ☽ ⊻ ♂
06 59 ☽ ⋕ ♇
08 07 ☽ ∠ ♃
08 36 ☽ ∥ ☉
15 35 ☽ ∥ ♀
18 31 ☿ ⊻ ♀
19 19 ☽ ∠ ☿
22 55 ☽ ∥ ♅

12 Thursday
01 36 ☽ ∥ ☿
02 27 ☽ ⊥ ♃
03 13 ☽ ⊻ ☉
04 59 ☽ ∠ ♄
11 01 ☉ ± ♇
14 38 ☽ ⊻ ♃
15 52 ☽ ☍ ♇
16 18 ☽ ⊥ ☉
22 44 ♀ ☌ ♅
23 05 ☽ ⊥ ♄

13 Friday
02 35 ☽ ✶ ☿
04 16 ☽ ⊻ ♅
04 51 ☽ ⊻ ♀
09 46 ☽ △ ♆
11 05 ☽ ⊻ ♄
11 46 ☽ ∠ ☉
16 14 ☽ ⊥ ♅
18 06 ☽ ⊥ ♀
19 11 ☽ ☌ ♂
21 57 ☽ ∥ ♆

14 Saturday
01 16 ☿ ∥ ♅
03 31 ☽ Q ☿
10 03 ☽ ∠ ♅
13 47 ☽ ∠ ♀
13 56 ♀ ⊥ ♂
14 27 ☽ ∥ ♂
15 22 ☽ ⊡ ♆
16 41 ☿ ⊻ ♅
19 46 ☽ ✶ ☉
23 02 ☿ ∥ ♀

15 Sunday
02 32 ☽ ☌ ♃
02 57 ☽ ⚻ ♇
05 40 ♀ □ ♆
08 51 ☽ ± ♆
14 30 ☽ ± ♇
15 17 ☽ ✶ ♅
15 38 ♀ ∥ ♅
16 30 ☽ □ ☿
16 44 ☽ ∥ ♃
20 22 ☽ ⚻ ♆
20 50 ☽ Q ☉
20 54 ♀ ✶ ♄
21 50 ☽ ☌ ♄
21 56 ☽ ✶ ♀

16 Monday
01 46 ☽ ⋕ ♄
05 17 ☽ ∥ ♄
07 37 ☽ ⊡ ♇
08 14 ☽ ⊻ ♂
10 25 ♃ ⚻ ♇
11 05 ☉ Q ♄
13 57 ☽ ⋕ ♃
14 10 ☽ Q ♅
20 03 ☽ ⊥ ♂
22 57 ☽ Q ♀

17 Tuesday
09 16 ☽ □ ☉
11 30 ☽ △ ♇
11 49 ☽ ⊻ ♃
12 12 ☽ ⋕ ♂
13 29 ☽ ∠ ♂
22 52 ☽ ⊥ ♃
23 28 ☽ □ ♅

18 Wednesday
02 29 ☿ ✶ ♆
04 03 ☽ ☍ ♆
04 10 ☽ △ ☿
05 39 ☽ ⊻ ♄
06 23 ☽ ⋕ ♆
09 21 ☉ ∥ ☿
11 04 ☽ □ ♀
15 09 ☽ ∠ ♃
15 13 ☉ ⚻ ♇
16 23 ☽ ⊥ ♄
17 43 ☽ ✶ ♂

19 Thursday
01 32 ☿ □ ♄
02 15 ☉ ✶ ♃
08 12 ☽ ∠ ♄
08 43 ☽ ⊡ ☿
15 32 ☽ Q ♂
15 58 ☽ ⋕ ♀
16 35 ☽ □ ♇
17 32 ☽ ✶ ♃
18 28 ☽ △ ☉
22 26 ☽ ⋕ ♅

20 Friday
01 24 ☽ ± ☿
01 58 ☽ ⋕ ☉
04 04 ☽ △ ♅
05 53 ☽ ⋕ ☿
08 09 ☽ ⚻ ♆
09 51 ☽ ✶ ♄
12 20 ☽ ⚻ ☿
12 50 ☽ ∥ ♇
13 46 ♀ ∠ ♃
14 00 ☽ Q ♃
18 10 ☽ ± ♆
19 27 ☽ △ ♀
21 26 ☽ ⊡ ☉
22 45 ☽ Q ♇
23 06 ☽ □ ♂

21 Saturday
05 04 ☽ ⊡ ♅
05 47 ☽ Q ♄
08 57 ☽ ⊡ ♆
13 10 ☽ ± ☉
18 17 ☽ ✶ ♇
19 44 ☽ ± ♅
19 46 ☽ □ ♃
22 07 ☽ ⊡ ♀
23 31 ☽ ⚻ ☉

22 Sunday
05 23 ☽ ⚻ ♅
09 06 ☽ △ ♆
10 54 ☽ □ ♄
13 45 ☽ ± ♀
17 22 ☽ ☍ ☿
18 12 ☽ ∠ ♇

23 Monday
00 05 ☽ ⚻ ♀
01 20 ☽ △ ♂
02 37 ☿ ⊡ ♇
08 22 ☽ ⊥ ♇
09 20 ☉ ∥ ♅
17 12 ☉ ∠ ♂
17 47 ☽ ⊻ ♇
19 47 ☽ △ ♃

24 Tuesday
01 53 ☽ ⊡ ♂
02 06 ☽ ☍ ☉
04 53 ☽ ☍ ♅
05 53 ♀ ⊻ ♂
08 22 ☽ □ ♆
10 22 ☽ △ ♄
14 36 ☽ ∥ ♇
16 36 ☽ ± ♂
17 55 ☽ ⋕ ☿
19 35 ☽ ⊡ ♃
21 19 ☽ ⚻ ☿

25 Wednesday
02 28 ☽ ⚻ ♂
03 22 ☽ ☍ ♀
04 50 ☽ ⋕ ♅
06 49 ☽ ⋕ ☉
07 51 ☽ ± ☿
09 58 ☽ ± ♃
10 09 ☽ ⊡ ♄
12 48 ☽ Q ♆
17 02 ☽ ☌ ♇
19 38 ☽ ⚻ ♃
20 54 ☽ ⋕ ♀
23 51 ☽ ⊡ ☿

26 Thursday
00 36 ☽ ± ♄
01 49 ☉ ☌ ♅
04 39 ☽ ⚻ ♅
04 50 ☽ ⚻ ☉
06 45 ☿ Q ♃
08 05 ☽ ✶ ♆
10 21 ☽ ⚻ ♄
14 26 ☽ ± ♅
15 16 ☽ ± ☉
17 32 ☽ ⋕ ♆

27 Friday
03 19 ☽ △ ☿
05 02 ☽ ☍ ♂
05 20 ☽ ⊡ ♅
07 10 ☽ ⊡ ☉
07 18 ☿ Q ♆
08 21 ☽ ⚻ ♀
08 46 ☽ ∠ ♆
12 56 ♂ ∠ ♅
18 11 ☽ ⊻ ♇
19 24 ☽ ± ♀
20 21 ☽ ⋕ ♂
21 36 ☽ ☍ ♃
23 59 ☽ ⊥ ♆

28 Saturday
01 51 ☿ ✶ ♂
04 24 ☽ ⊥ ♇
06 51 ☽ △ ♅
07 24 ☉ □ ♆
09 46 ☽ ⋕ ♃
10 19 ☽ ⊻ ♆
10 32 ☽ △ ☉
12 21 ☽ ⊡ ♀
13 00 ☽ ☍ ♄
13 13 ☽ ∥ ♄
19 32 ☽ ⋕ ♄
19 41 ☿ ± ♇
20 01 ♀ ∠ ♄
20 02 ☽ ∠ ♇
22 51 ☽ ∥ ♃

29 Sunday
00 22 ♀ ⊥ ♃
11 15 ☽ ∥ ♂
11 19 ☽ ⚻ ♂
14 36 ☽ □ ☿
17 43 ☽ △ ♀
22 45 ☽ ± ♂
22 57 ☽ ✶ ♇

30 Monday
01 16 ☉ ✶ ♄
03 25 ☽ ⚻ ♃
13 00 ☽ □ ♅
14 43 ☽ ± ♃
16 16 ☽ ⊡ ♂
16 30 ☽ ☌ ♆
18 27 ☽ ∥ ♆
19 43 ☽ ⚻ ♄
21 03 ♂ ⊡ ♆
21 06 ☽ □ ☉
21 13 ☽ Q ♇

31 Tuesday
05 36 ☽ Q ☿
07 11 ☽ ± ♄
07 15 ☽ ∥ ♀
07 58 ☽ ⊡ ♃
16 34 ☿ Q ♄
22 23 ☽ △ ♂
23 29 ☽ Q ♅

AUGUST 2040

Date	Sidereal time h m s	Sun ☉ ° ' "	Moon ☽ ° ' "	Moon ☽ 24.00 ° ' "	Mercury ☿ ° '	Venus ♀ ° '	Mars ♂ ° '	Jupiter ♃ ° '	Saturn ♄ ° '	Uranus ♅ ° '	Neptune ♆ ° '	Pluto ♇ ° '
01	08 42 51	09 ♌ 50 45	28 ♉ 28 36	04 ♊ 33 19	27 ♋ 11	26 ♌ 54	21 ♍ 51	29 ♍ 13	07 ♎ 42	04 ♌ 06	05 ♉ 51	26 ♒ 20
02	08 46 47	10 48 10	10 ♊ 34 59	16 ♊ 34 11	29 ♋ 08	28 08	22 28	29 24	07 47	04 10	05 52	26 R 19
03	08 50 44	11 45 36	22 ♊ 31 28	28 ♊ 27 22	01 ♌ 07	29 ♌ 22	23 06	29 34	07 52	04 13	05 52	26 17
04	08 54 40	12 43 04	04 ♋ 22 21	10 ♋ 16 52	03 08	00 ♍ 35	23 43	29 45	07 58	04 17	05 52	26 16
05	08 58 37	13 40 33	16 ♋ 11 16	22 ♋ 05 55	05 10	01 49	24 20	29 ♍ 56	08 03	04 21	05 52	26 15
06	09 02 34	14 38 02	28 ♋ 01 06	03 ♌ 57 05	07 13	03 03	24 58	00 ♎ 06	08 08	04 24	05 52	26 13
07	09 06 30	15 35 33	09 ♌ 54 04	15 ♌ 52 15	09 17	04 16	25 35	00 17	08 13	04 28	05 52	26 12
08	09 10 27	16 33 05	21 ♌ 51 47	27 ♌ 52 51	11 21	05 30	26 13	00 28	08 19	04 32	05 R 52	26 11
09	09 14 23	17 30 38	03 ♍ 55 34	10 ♍ 00 05	13 25	06 43	26 50	00 39	08 24	04 35	05 52	26 10
10	09 18 20	18 28 12	16 ♍ 06 33	22 ♍ 15 09	15 29	07 57	27 28	00 50	08 30	04 39	05 52	26 08
11	09 22 16	19 25 46	28 ♍ 26 04	04 ♎ 39 31	17 32	09 10	28 06	01 01	08 35	04 42	05 52	26 07
12	09 26 13	20 23 22	10 ♎ 55 47	17 ♎ 15 07	19 35	10 24	28 43	01 12	08 41	04 46	05 52	26 06
13	09 30 09	21 20 59	23 ♎ 37 52	00 ♏ 04 22	21 37	11 38	29 21	01 23	08 47	04 50	05 52	26 04
14	09 34 06	22 18 37	06 ♏ 34 58	13 ♏ 10 04	23 39	12 51	29 ♍ 59	01 35	08 53	04 53	05 52	26 03
15	09 38 03	23 16 15	19 ♏ 50 00	26 ♏ 35 08	25 39	14 05	00 ♎ 37	01 46	08 58	04 57	05 51	26 02
16	09 41 59	24 13 55	03 ♐ 25 45	10 ♐ 22 04	27 38	15 18	01 15	01 57	09 04	05 00	05 51	26 00
17	09 45 56	25 11 36	17 ♐ 24 13	24 ♐ 32 11	29 ♌ 35	16 32	01 53	02 09	09 10	05 04	05 51	25 59
18	09 49 52	26 09 18	01 ♑ 45 30	09 ♑ 04 49	01 ♍ 32	17 45	02 31	02 20	09 16	05 07	05 50	25 58
19	09 53 49	27 07 01	16 ♑ 28 36	23 ♑ 56 28	03 27	18 59	03 09	02 32	09 22	05 11	05 50	25 56
20	09 57 45	28 04 45	01 ♒ 27 31	09 ♒ 00 39	05 21	20 12	03 47	02 44	09 28	05 14	05 50	25 55
21	10 01 42	29 ♌ 02 30	16 ♒ 34 40	24 ♒ 08 16	07 14	21 26	04 26	02 55	09 35	05 18	05 49	25 54
22	10 05 38	00 ♍ 00 16	01 ♓ 40 09	09 ♓ 09 02	09 05	22 39	05 04	03 07	09 41	05 21	05 49	25 52
23	10 09 35	00 58 04	16 ♓ 33 44	23 ♓ 53 14	10 54	23 53	05 42	03 19	09 47	05 25	05 48	25 51
24	10 13 32	01 55 54	01 ♈ 06 42	08 ♈ 13 29	12 43	25 06	06 21	03 31	09 53	05 28	05 48	25 50
25	10 17 28	02 53 45	15 ♈ 13 10	22 ♈ 05 34	14 30	26 19	06 59	03 43	10 00	05 31	05 47	25 49
26	10 21 25	03 51 37	28 ♈ 50 39	05 ♉ 28 36	16 16	27 33	07 37	03 55	10 06	05 35	05 47	25 47
27	10 25 21	04 49 32	11 ♉ 59 44	18 ♉ 24 28	18 00	28 ♍ 46	08 16	04 07	10 12	05 38	05 46	25 46
28	10 29 18	05 47 28	24 ♉ 43 20	01 ♊ 56 55	19 43	00 ♎ 00	08 55	04 19	10 19	05 41	05 46	25 45
29	10 33 14	06 45 26	07 ♊ 05 52	13 ♊ 10 51	21 25	01 13	09 33	04 31	10 25	05 45	05 45	25 43
30	10 37 11	07 43 26	19 ♊ 12 31	25 ♊ 11 32	23 05	02 26	10 12	04 43	10 32	05 48	05 44	25 42
31	10 41 07	08 ♍ 41 28	01 ♋ 08 34	07 ♋ 04 13	24 ♍ 44	03 ♎ 40	10 ♎ 51	04 ♎ 55	10 ♎ 39	05 ♌ 51	05 ♉ 44	25 ♒ 41

Date	Moon True ☊ ° '	Moon Mean ☊ ° '	Moon ☽ Latitude ° '
01	01 ♊ 23	00 ♊ 07	00 S 16
02	01 R 23	00 03	00 N 49
03	01 21	00 ♊ 00	01 50
04	01 17	29 ♉ 57	02 46
05	01 10	29 54	03 34
06	01 00	29 51	04 13
07	00 48	29 48	04 41
08	00 35	29 44	04 57
09	00 21	29 41	05 00
10	00 09	29 38	04 49
11	29 ♉ 59	29 35	04 25
12	29 51	29 32	03 47
13	29 46	29 29	02 58
14	29 44	29 25	01 59
15	29 D 43	29 22	00 N 52
16	29 R 43	29 19	00 S 20
17	29 42	29 16	01 32
18	29 40	29 13	02 41
19	29 35	29 09	03 40
20	29 28	29 06	04 26
21	29 18	29 03	04 53
22	29 07	29 00	05 00
23	28 56	28 57	04 46
24	28 46	28 54	04 14
25	28 39	28 50	03 27
26	28 34	28 47	02 30
27	28 31	28 44	01 26
28	28 D 31	28 41	00 S 20
29	28 R 31	28 38	00 N 45
30	28 30	28 35	01 47
31	28 ♉ 28	28 ♉ 31	02 N 43

DECLINATIONS

Date	Sun ☉ ° '	Moon ☽ ° '	Mercury ☿ ° '	Venus ♀ ° '	Mars ♂ ° '	Jupiter ♃ ° '	Saturn ♄ ° '	Uranus ♅ ° '	Neptune ♆ ° '	Pluto ♇ ° '
01	17 N 47	19 N 34	21 N 23	13 N 57	03 N 55	01 N 23	00 S 56	19 N 46	11 N 49	22 S 24
02	17 31	22 50	21 09	13 32	03 39	01 19	00 58	19 45	11 49	22 24
03	17 15	25 03	20 52	13 06	03 24	01 14	01 00	19 45	11 49	22 25
04	16 59	26 07	20 33	12 40	03 09	01 10	01 02	19 44	11 49	22 25
05	16 43	26 00	20 11	12 13	02 53	01 06	01 04	19 43	11 49	22 26
06	16 26	24 41	19 47	11 46	02 38	01 01	01 07	19 42	11 49	22 27
07	16 10	22 16	19 20	11 19	02 22	00 57	01 09	19 41	11 49	22 27
08	15 52	18 54	18 51	10 52	02 07	00 53	01 11	19 40	11 49	22 28
09	15 35	14 43	18 20	10 24	01 51	00 48	01 13	19 39	11 49	22 28
10	15 17	09 55	17 47	09 56	01 36	00 44	01 16	19 39	11 49	22 29
11	14 59	04 N 40	17 12	09 28	01 20	00 39	01 18	19 38	11 49	22 29
12	14 41	00 S 50	16 36	08 59	01 05	00 35	01 20	19 37	11 49	22 30
13	14 23	06 25	15 57	08 31	00 49	00 30	01 23	19 36	11 48	22 30
14	14 04	11 50	15 18	08 02	00 34	00 25	01 25	19 35	11 48	22 31
15	13 46	16 51	14 37	07 33	00 18	00 21	01 28	19 34	11 48	22 31
16	13 27	21 13	13 56	07 03	00 N 02	00 16	01 30	19 33	11 48	22 32
17	13 07	24 22	13 13	06 34	00 S 14	00 11	01 33	19 33	11 48	22 32
18	12 48	26 06	12 30	06 04	00 29	00 07	01 35	19 32	11 48	22 33
19	12 28	26 04	11 46	05 34	00 45	00 N 02	01 37	19 31	11 48	22 33
20	12 08	24 09	11 01	05 04	01 01	00 S 03	01 40	19 30	11 47	22 34
21	11 48	20 31	10 16	04 34	01 17	00 07	01 43	19 29	11 47	22 34
22	11 28	15 33	09 30	04 04	01 32	00 12	01 45	19 28	11 47	22 35
23	11 08	09 42	08 45	03 33	01 48	00 17	01 48	19 28	11 47	22 35
24	10 47	03 S 27	07 59	03 02	02 04	00 22	01 50	19 27	11 47	22 36
25	10 26	02 N 48	07 13	02 32	02 20	00 26	01 53	19 26	11 46	22 36
26	10 05	08 43	06 27	02 02	02 36	00 31	01 55	19 25	11 46	22 37
27	09 44	14 04	05 41	01 31	02 51	00 36	01 58	19 24	11 46	22 37
28	09 23	18 37	04 54	01 00	03 07	00 41	02 01	19 24	11 46	22 38
29	09 02	22 14	04 09	00 N 29	03 23	00 46	02 03	19 23	11 45	22 38
30	08 40	24 47	03 23	00 S 02	03 39	00 51	02 06	19 22	11 45	22 39
31	08 N 19	26 N 09	02 N 37	00 S 33	03 S 55	00 S 56	02 S 09	19 N 21	11 N 45	22 S 39

ZODIAC SIGN ENTRIES

Date	h	m	Planets
01	15	00	☽ ♊
02	22	31	☿ ♌
04	00	30	♀ ♍
04	03	08	☽ ♋
05	22	03	♃ ♎
06	16	01	☽ ♌
09	04	13	☽ ♍
11	15	02	☽ ♎
13	23	52	☽ ♏
14	12	36	♂ ♎
16	06	00	☽ ♐
17	17	02	☿ ♍
18	09	05	☽ ♑
20	09	41	☽ ♒
22	09	20	☽ ♓
22	11	53	☉ ♍
24	10	09	☽ ♈
26	14	05	☽ ♉
28	12	08	♀ ♎
28	22	10	☽ ♊
31	09	41	☽ ♋

LATITUDES

Date	Mercury ☿ ° '	Venus ♀ ° '	Mars ♂ ° '	Jupiter ♃ ° '	Saturn ♄ ° '	Uranus ♅ ° '	Neptune ♆ ° '	Pluto ♇ ° '
01	00 N 41	01 N 30	00 N 44	01 N 10	02 N 19	00 N 34	01 S 45	10 S 18
04	01 08	01 30	00 42	01 10	02 18	00 34	01 45	10 18
07	01 28	01 29	00 41	01 09	02 18	00 34	01 46	10 19
10	01 40	01 27	00 39	01 09	02 18	00 34	01 46	10 19
13	01 45	01 25	00 37	01 09	02 17	00 34	01 46	10 19
16	01 44	01 22	00 35	01 08	02 17	00 34	01 46	10 19
19	01 38	01 19	00 33	01 08	02 16	00 34	01 46	10 20
22	01 27	01 15	00 31	01 08	02 16	00 34	01 46	10 20
25	01 12	01 10	00 29	01 08	02 16	00 34	01 47	10 20
28	00 55	01 05	00 27	01 07	02 15	00 34	01 47	10 20
31	00 N 35	01 N 00	00 N 25	01 N 07	02 N 15	00 N 34	01 S 47	10 S 20

DATA

Julian Date	2466368
Delta T	+77 seconds
Ayanamsa	24° 25' 13"
Synetic vernal point	04° ♓ 41' 46"
True obliquity of ecliptic	23° 26' 07"

MOON'S PHASES, APSIDES AND POSITIONS ☽

Date	h	m	Phase	Longitude ° '	Eclipse Indicator
08	00	26	●	16 ♌ 05	
15	18	36	☽	23 ♏ 32	
22	09	10	○	29 ♒ 53	
29	11	16	☾	06 ♊ 44	

Day	h	m		
05	23	58	Apogee	
21	03	58	Perigee	
04	21	25	Max dec	26° N 13'
12	08	24	0S	
18	23	32	Max dec	26° S 19'
25	01	09	0N	

LONGITUDES

Date	Chiron ⚷ ° '	Ceres ⚳ ° '	Pallas ⚴ ° '	Juno ⚵ ° '	Vesta ⚶ ° '	Black Moon Lilith ⚸ ° '
01	17 ♋ 26	08 ♋ 08	05 ♉ 07	08 ♊ 06	06 ♒ 44	24 ♋ 34
11	18 ♋ 25	12 ♋ 11	06 ♉ 40	13 ♊ 18	04 ♒ 25	25 ♋ 41
21	19 ♋ 20	16 ♋ 09	07 ♉ 45	18 ♊ 21	02 ♒ 32	26 ♋ 48
31	20 ♋ 11	20 ♋ 01	08 ♉ 15	23 ♊ 14	01 ♒ 19	27 ♋ 55

ASPECTARIAN

h m	Aspects
01 Wednesday	
00 38	☽ ꝑ ♄
00 54	♀ ☍ ♇
01 24	☿ ⊼ ♇
01 48	☽ ∥ ☉
02 20	☿ ⊻ ♀
07 48	☽ □ ♇
08 35	☽ □ ♀
09 00	☽ ✶ ☿
10 39	☽ Q ☉
13 19	☽ ∥ ♅
13 29	☽ △ ♃
23 09	☽ ✶ ♅
23 39	☽ ∥ ☿
02 Thursday	
02 35	☽ ⊻ ♆
06 23	☽ △ ♄
08 22	☽ # ♇
12 29	☽ ✶ ☉
14 33	☽ ⊥ ♆
15 30	☿ ✶ ♃
20 31	☽ ∠ ☿
03 Friday	
00 24	☽ Q ♀
05 18	☽ ∠ ♅
08 39	☽ ∠ ♆
13 13	☽ □ ♂
16 48	♀ ⊻ ♃
18 18	☽ ⊥ ☿
19 36	☽ △ ♇
21 19	☽ ∠ ☉
23 35	☽ ⊥ ♅
04 Saturday	
02 29	☽ □ ♃
03 26	☽ ✶ ♀
08 57	☽ ⊻ ☿
11 49	☽ ⊻ ♅
15 02	☽ ✶ ♆
17 11	☽ ⊥ ☉
19 20	☽ □ ♄
05 Sunday	
01 59	☽ ꝑ ♇
02 04	☿ ☌ ♅
03 45	☽ Q ♂
04 14	♃ ± ♆
06 27	☽ ⊻ ☉
13 25	☽ ∠ ♀
15 25	☽ Q ♆
15 35	☽ Q ♃
16 52	♀ ⊥ ♄
16 59	♃ # ♄
20 13	☽ ± ♇
20 17	☿ □ ♆
06 Monday	
05 28	☽ ✶ ♂
08 09	☽ Q ♄
08 22	☽ ⊼ ♇
09 37	♀ ∥ ♆
09 48	☽ ⊥ ♀
16 17	☽ ✶ ♃
17 01	☿ ∥ ♅
23 13	☿ ✶ ♄
23 20	☽ ⊻ ♀
07 Tuesday	
00 59	☽ ☌ ♅
02 29	☉ ∠ ♃
03 52	☽ □ ♆
08 36	☽ ✶ ♄
10 29	☽ ☌ ☿
10 31	☽ # ♇
13 27	☽ ∠ ♂
16 03	♀ ⊻ ♅
22 59	☽ ∠ ♃
08 Wednesday	
00 26	☽ ☌ ☉
05 24	♆ St R
06 59	☽ ∥ ♅
08 31	☽ ⊥ ♂
09 49	☿ ∠ ♂
10 49	♂ ⊼ ♇
12 16	☽ ∥ ☿
14 55	☽ ∠ ♄
17 17	☽ ⊥ ♃
19 20	♀ △ ♆
20 36	☽ ☍ ♇
21 09	☽ ⊻ ♂
09 Thursday	
05 24	☽ ⊻ ♃
07 00	☽ ∥ ☉
08 58	☽ ⊥ ♄
13 19	☽ ⊻ ♅
15 51	☽ △ ♆
18 09	☽ ☌ ♀
20 55	☽ ⊻ ♄
10 Friday	
01 13	☽ ⊥ ♅
02 49	☽ ∥ ♆
10 31	☽ ⊻ ☿
11 53	☽ ∥ ♀
16 30	☿ ∠ ♃
17 01	☽ ⊻ ☉
18 57	☽ ∠ ♅
21 18	☽ ꝑ ♆
23 38	♀ ⊻ ♄
11 Saturday	
00 36	☽ ⊥ ☿
05 41	☽ ⊥ ☉
07 31	☽ ⊼ ♇
11 18	☽ ☌ ♂
14 46	☽ ± ♆
14 52	♂ # ♄
17 04	☽ ☌ ♃
19 06	☽ ± ♇
21 29	☽ ∠ ☿
12 Sunday	
00 09	☽ ✶ ♅
00 31	☽ ∠ ☉
02 19	☽ ⊼ ♆
02 40	☽ # ♄
03 20	☽ ∥ ♂
05 50	☽ ∥ ♃
07 41	☽ ☌ ♄
10 53	☽ ⊻ ♀
10 53	☽ # ♃
12 19	☽ ꝑ ♇
12 59	☽ # ♂
14 11	☽ ∥ ♄
19 31	♀ ⊥ ♅
23 08	☽ Q ♅
23 30	☽ ⊥ ♀
13 Monday	
05 53	☉ ☌ ☿
07 22	☽ ✶ ☉
07 31	☽ ✶ ☿
16 33	☽ △ ♇
18 11	☽ ∠ ♀
20 25	☽ # ♀
23 13	☽ ⊻ ♂
14 Tuesday	
02 39	☽ ⊻ ♃
07 17	♂ ± ♆
07 30	☽ Q ☉
08 52	☽ □ ♅
09 58	☽ Q ☿
10 40	☽ ☍ ♆
10 51	☽ ⊥ ♂
11 52	☽ # ♆
13 51	☽ ⊥ ♃
14 57	☿ ∠ ♄
16 14	☽ ⊻ ♄
17 57	☿ ⊥ ♂
21 46	☽ # ☉
15 Wednesday	
00 36	☽ ✶ ♀
02 19	☽ # ☿
03 12	☽ ⊥ ♄
04 03	☽ ∠ ♂
05 17	♂ ∥ ♃
06 25	☽ ∠ ♃
13 38	☿ ⊥ ♃
16 34	☿ ☍ ♇
18 36	☽ □ ☉
19 26	☽ ∠ ♄
23 00	☽ □ ♇
16 Thursday	
00 07	☽ □ ☿
00 12	☽ Q ♀
02 29	☽ # ♅
07 33	☉ ∠ ♄
08 01	☽ ✶ ♂
09 24	☽ ✶ ♃
14 45	☽ △ ♅
16 12	☽ ⊼ ♆
17 30	♃ ± ♇
21 11	☽ ∥ ♇
21 51	☽ ✶ ♄
17 Friday	
02 32	☽ ± ♆
05 44	☽ Q ♂
06 12	☽ Q ♇
06 24	☽ Q ♃
09 19	♂ # ♃
10 23	☽ □ ♀
15 44	♂ ± ♇
16 31	☽ ꝑ ♅
17 46	☉ ∥ ☿
17 49	☽ ꝑ ♆
18 24	☽ Q ♄
18 Saturday	
02 02	☽ △ ☉
02 24	☽ ✶ ♇
02 28	♂ ☌ ♃
07 18	☉ ☍ ♇
07 37	☽ ± ♅
11 34	☽ △ ☿
12 58	☽ □ ♃
13 18	☽ □ ♂
17 33	☽ ⊼ ♅
17 49	☉ ⊥ ♃
18 42	☽ △ ♆
23 12	☿ ⊻ ♃
19 Sunday	
00 24	☽ □ ♄
03 03	☽ ∠ ♇
04 27	☽ ꝑ ☉
06 20	☿ ⊻ ♂
10 52	☿ ∥ ♆
10 55	☿ ⊥ ♄
14 36	☉ ⊥ ♂
15 39	☽ ꝑ ☿
16 24	☽ △ ♀
17 34	☽ ⊥ ♇
19 59	☽ ± ☉
20 Monday	
03 11	☽ ⊻ ♇
06 15	☽ ⊼ ☉
08 10	☽ ± ☿
10 31	☿ ⊻ ♅
12 42	♀ ∠ ♅
14 03	☽ △ ♃
15 52	☽ △ ♂
18 02	☽ ☍ ♅
18 05	☿ △ ♆
18 29	☽ ꝑ ♀
18 57	☽ □ ♆
19 04	☽ ⊼ ☿
23 42	☽ ∥ ♇
21 Tuesday	
00 12	♀ ꝑ ♆
00 49	☽ △ ♄
10 01	☽ ± ♀
13 19	☉ ∥ ♆
14 10	☽ ꝑ ♃
16 43	☽ ꝑ ♂
17 28	☽ # ♅
20 22	☽ ⊼ ♀
23 30	☽ Q ♆
22 Wednesday	
00 47	☽ ꝑ ♄
02 47	☽ ☌ ♇
04 39	☽ ± ♃
07 39	☽ ± ♂
09 10	☽ ☍ ☉
14 21	☽ ⊼ ♃
15 14	☽ ± ♄
17 40	☽ ⊼ ♂
17 55	☽ ⊼ ♅
18 38	☽ ✶ ♆
20 20	☿ ⊻ ♄
23 57	♂ ✶ ♅
23 Thursday	
00 57	☽ ⊼ ♄
01 33	☽ ☍ ☿
03 37	☽ ± ♅
03 43	☽ # ♆
06 00	☽ # ☉
10 52	♂ ∥ ♄
15 53	♂ ⊼ ♆
16 14	☽ # ☿
18 19	☽ ꝑ ♅
18 52	☿ ⊥ ♅
18 56	☽ ∠ ♆
24 Friday	
01 05	☽ ☍ ♀
03 13	☽ ⊻ ♇
09 48	☽ ⊥ ♆
13 12	☽ ⊥ ♇
13 28	☽ ⊼ ☉
13 39	☽ # ♀
16 05	☽ ☍ ♃
17 02	☽ ∥ ♂
18 05	☽ ∥ ♄
19 21	☽ △ ♅
19 53	☽ ⊻ ♆
21 13	☽ ☍ ♂
23 37	☽ ∥ ♃
25 Saturday	
00 21	☽ ± ☉
02 06	♀ ⊼ ♇
02 43	☽ # ♃
02 57	☽ ☍ ♄
04 25	☽ ∠ ♇
08 23	☽ # ♄
10 04	☽ # ♂
10 34	☽ ⊼ ☿
11 02	☽ ∥ ♀
17 00	☽ ꝑ ☉
18 25	♀ # ♂
22 34	☽ ± ☿
26 Sunday	
03 37	☽ ∥ ☿
06 33	☽ ✶ ♇
09 27	☽ ⊼ ♀
13 41	☉ ⊻ ♃
16 27	♀ # ♄
17 00	☽ ꝑ ☿
17 31	☽ ∥ ☉
18 42	☉ ⊥ ♄
21 18	☽ ⊼ ♃
21 21	☽ ± ♀
21 46	☽ △ ☉
27 Monday	
00 14	☽ □ ♅
00 33	☽ ☌ ♆
01 17	☽ ∥ ♆
04 12	☽ Q ♇
04 45	☽ ⊼ ♂
08 28	☽ ± ♃
08 40	☽ ⊼ ♄
15 39	☽ ꝑ ♀
16 27	☽ ± ♂
19 56	☽ ± ♄
28 Tuesday	
00 59	☽ △ ☿
01 33	☽ ꝑ ♃
07 28	♀ ± ♆
09 21	☉ ⊻ ♅
10 01	☽ Q ♅
10 22	☽ ꝑ ♂
11 13	☉ △ ♆
13 09	☽ ꝑ ♄
13 57	☽ □ ♇
16 36	☽ ∥ ♅
23 16	☽ △ ♀
29 Wednesday	
00 45	♀ # ♃
02 13	☿ ∠ ♅
02 38	☿ ꝑ ♆
06 52	☽ △ ♃
09 20	☽ ✶ ♅
09 21	☽ ⊻ ♆
11 16	☽ □ ☉
13 08	♅ □ ♆
15 12	☽ # ♇
17 06	☽ △ ♂
18 36	☽ △ ♄
21 09	☽ ⊥ ♆
21 47	♀ ± ♇
30 Thursday	
05 45	☿ # ♂
15 03	☽ ∠ ♆
15 12	☽ ∠ ♅
21 01	☽ □ ☿
31 Friday	
01 00	☽ △ ♇
02 13	☽ Q ☉
02 54	♂ ☌ ♄
06 02	♂ ꝑ ♇
09 23	☽ ⊥ ♅
17 41	☽ □ ♀
18 47	♄ ꝑ ♇
19 47	☽ □ ♃
21 16	☽ ✶ ♆
21 35	☽ ⊻ ♅

All ephemeris data is given at 12.00 UT and the Moon's longitude is additionally given for 24.00 UT

SEPTEMBER 2040

LONGITUDES

Date	Sidereal time h m s	Sun ☉ ° ' "	Moon ☽ ° ' "	Moon ☽ 24.00 ° ' "	Mercury ☿ ° '	Venus ♀ ° '	Mars ♂ ° '	Jupiter ♃ ° '	Saturn ♄ ° '	Uranus ♅ ° '	Neptune ♆ ° '	Pluto ♇ ° '
01	10 45 04	09 ♍ 39 32	12 ♋ 59 05	18 ♋ 53 42	26 ♍ 22	04 ♎ 53	11 ♎ 30	05 ♎ 08	10 ♎ 45	05 ♌ 54	05 ♉ 43	25 ♒ 40
02	10 49 01	10 37 38	24 ♋ 48 33	00 ♌ 44 07	27 59	06 06	12 09	05 20	10 52	05 58	05 R 42	25 R 38
03	10 52 57	11 35 46	06 ♌ 40 45	12 ♌ 38 48	29 ♍ 34	07 20	12 47	05 32	10 59	06 01	05 41	25 37
04	10 56 54	12 33 55	18 ♌ 38 34	24 ♌ 40 17	01 ♎ 09	08 33	13 26	05 45	11 05	06 04	05 40	25 36
05	11 00 50	13 32 06	00 ♍ 44 06	06 ♍ 50 12	02 42	09 46	14 06	05 57	11 12	06 07	05 39	25 35
06	11 04 47	14 30 19	12 ♍ 58 39	19 ♍ 09 33	04 13	10 59	14 45	06 10	11 19	06 10	05 39	25 33
07	11 08 43	15 28 34	25 ♍ 22 57	01 ♎ 38 53	05 44	12 13	15 24	06 22	11 26	06 13	05 38	25 32
08	11 12 40	16 26 50	07 ♎ 57 23	14 ♎ 18 32	07 13	13 26	16 03	06 35	11 33	06 16	05 37	25 31
09	11 16 36	17 25 08	20 ♎ 42 21	27 ♎ 08 57	08 41	14 39	16 42	06 47	11 40	06 19	05 36	25 30
10	11 20 33	18 23 28	03 ♏ 38 26	10 ♏ 10 55	10 08	15 52	17 22	07 00	11 47	06 22	05 35	25 28
11	11 24 30	19 21 49	16 ♏ 46 36	23 ♏ 25 38	11 34	17 06	18 01	07 12	11 54	06 25	05 34	25 27
12	11 28 26	20 20 12	00 ♐ 08 13	06 ♐ 54 33	12 58	18 19	18 41	07 25	12 01	06 28	05 33	25 26
13	11 32 23	21 18 36	13 ♐ 44 50	20 ♐ 39 11	14 21	19 32	19 20	07 38	12 08	06 31	05 32	25 25
14	11 36 19	22 17 03	27 ♐ 37 43	04 ♑ 40 27	15 43	20 45	20 00	07 51	12 15	06 34	05 30	25 24
15	11 40 16	23 15 30	11 ♑ 47 18	18 ♑ 58 05	17 03	21 58	20 39	08 03	12 22	06 37	05 29	25 23
16	11 44 12	24 14 00	26 ♑ 12 28	03 ♒ 30 00	18 22	23 11	21 19	08 16	12 29	06 40	05 28	25 21
17	11 48 09	25 12 30	10 ♒ 50 05	18 ♒ 11 56	19 40	24 24	21 59	08 29	12 36	06 42	05 27	25 20
18	11 52 05	26 11 03	25 ♒ 34 44	02 ♓ 57 30	20 56	25 37	22 38	08 42	12 43	06 45	05 26	25 19
19	11 56 02	27 09 37	10 ♓ 19 14	17 ♓ 38 57	22 10	26 50	23 18	08 54	12 50	06 48	05 25	25 18
20	11 59 59	28 08 13	24 ♓ 55 38	02 ♈ 08 24	23 23	28 03	23 58	09 07	12 57	06 50	05 23	25 17
21	12 03 55	29 ♍ 06 51	09 ♈ 16 26	16 ♈ 19 06	24 34	29 ♎ 16	24 38	09 20	13 05	06 53	05 22	25 16
22	12 07 52	00 ♎ 05 31	23 ♈ 15 54	00 ♉ 06 30	25 43	00 ♏ 29	25 18	09 33	13 12	06 55	05 21	25 15
23	12 11 48	01 04 13	06 ♉ 50 44	13 ♉ 28 35	26 50	01 42	25 58	09 46	13 19	06 58	05 19	25 14
24	12 15 45	02 02 57	20 ♉ 00 36	26 ♉ 25 51	27 56	02 55	26 38	09 59	13 26	07 00	05 18	25 13
25	12 19 41	03 01 43	02 ♊ 45 53	09 ♊ 00 47	28 59	04 08	27 18	10 12	13 34	07 03	05 17	25 12
26	12 23 38	04 00 32	15 ♊ 11 04	21 ♊ 17 19	29 ♎ 59	05 21	27 59	10 25	13 41	07 05	05 15	25 11
27	12 27 34	04 59 23	27 ♊ 20 10	03 ♋ 20 14	00 ♏ 56	06 34	28 39	10 38	13 48	07 08	05 14	25 10
28	12 31 31	05 58 16	09 ♋ 18 11	15 ♋ 14 40	01 52	07 47	29 ♎ 19	10 51	13 56	07 10	05 13	25 09
29	12 35 28	06 57 12	21 ♋ 10 18	27 ♋ 05 42	02 44	09 00	00 ♏ 00	11 04	14 03	07 12	05 11	25 08
30	12 39 24	07 ♎ 56 09	03 ♌ 01 28	08 ♌ 58 08	03 ♏ 33	10 ♏ 12	00 ♏ 40	11 ♎ 17	14 ♎ 10	07 ♌ 14	05 ♉ 10	25 ♒ 07

Moon nodes and latitude

Date	Moon True ☊ ° '	Moon Mean ☊ ° '	Moon ☽ Latitude ° '
01	28 ♉ 24	28 ♉ 28	03 N 32
02	28 R 17	28 25	04 12
03	28 08	28 22	04 40
04	27 56	28 19	04 57
05	27 43	28 15	05 00
06	27 30	28 12	04 50
07	27 18	28 09	04 26
08	27 08	28 06	03 49
09	27 00	28 03	02 59
10	26 55	28 00	02 00
11	26 53	27 56	00 N 53
12	26 D 53	27 53	00 S 17
13	26 53	27 50	01 29
14	26 R 53	27 47	02 36
15	26 52	27 44	03 35
16	26 48	27 41	04 22
17	26 42	27 37	04 53
18	26 34	27 34	05 04
19	26 25	27 31	04 55
20	26 16	27 28	04 27
21	26 08	27 25	03 43
22	26 02	27 21	02 46
23	25 58	27 18	01 41
24	25 56	27 15	00 S 32
25	25 D 56	27 12	00 N 37
26	25 57	27 09	01 42
27	25 58	27 06	02 41
28	25 R 58	27 02	03 32
29	25 57	26 59	04 13
30	25 ♉ 54	26 ♉ 56	04 N 44

DECLINATIONS

Date	Sun ☉	Moon ☽	Mercury ☿	Venus ♀	Mars ♂	Jupiter ♃	Saturn ♄	Uranus ♅	Neptune ♆	Pluto ♇
01	07 N 57	26 N 19	01 N 52	01 S 04	04 S 10	01 S 01	02 S 11	19 N 20	11 N 44	22 S 40
02	07 35	25 17	01 07	01 35	04 26	01 06	02 14	19 20	11 44	22 40
03	07 13	23 07	00 N 22	02 05	04 42	01 10	02 17	19 19	11 44	22 40
04	06 51	19 56	00 S 22	02 36	04 58	01 15	02 19	19 19	11 44	22 41
05	06 28	15 53	01 06	03 07	05 13	01 20	02 22	19 19	11 43	22 41
06	06 06	11 09	01 50	03 38	05 29	01 25	02 25	19 17	11 43	22 41
07	05 43	05 54	02 33	04 09	05 45	01 30	02 28	19 16	11 43	22 42
08	05 21	00 N 21	03 16	04 39	06 00	01 35	02 30	19 15	11 42	22 42
09	04 58	05 S 19	03 58	05 10	06 16	01 40	02 33	19 14	11 42	22 43
10	04 35	10 51	04 39	05 41	06 32	01 45	02 36	19 14	11 41	22 43
11	04 13	16 00	05 21	06 11	06 47	01 50	02 39	19 13	11 41	22 44
12	03 50	20 28	06 01	06 41	07 03	01 56	02 41	19 12	11 41	22 44
13	03 27	23 55	06 41	07 11	07 19	02 01	02 44	19 12	11 40	22 44
14	03 04	26 01	07 20	07 41	07 34	02 06	02 47	19 11	11 40	22 45
15	02 41	26 29	07 58	08 11	07 50	02 11	02 50	19 10	11 39	22 45
16	02 17	25 12	08 36	08 41	08 05	02 16	02 53	19 10	11 39	22 45
17	01 54	22 12	09 13	09 10	08 20	02 21	02 55	19 09	11 39	22 46
18	01 31	17 45	09 49	09 40	08 36	02 26	02 58	19 08	11 38	22 46
19	01 08	12 15	10 24	10 09	08 51	02 31	03 01	19 08	11 38	22 46
20	00 44	06 S 06	10 58	10 38	09 06	02 36	03 04	19 07	11 37	22 46
21	00 N 21	00 N 16	11 32	11 06	09 22	02 41	03 07	19 06	11 37	22 47
22	00 S 02	06 28	12 04	11 35	09 37	02 46	03 10	19 06	11 36	22 47
23	00 26	12 13	12 35	12 03	09 52	02 51	03 12	19 05	11 36	22 47
24	00 49	17 14	13 06	12 31	10 07	02 56	03 15	19 05	11 35	22 48
25	01 12	21 19	13 35	12 58	10 22	03 01	03 18	19 04	11 35	22 48
26	01 36	24 18	14 02	13 26	10 37	03 06	03 21	19 03	11 34	22 48
27	01 59	26 05	14 29	13 53	10 52	03 12	03 24	19 03	11 34	22 48
28	02 22	26 38	14 54	14 19	11 07	03 17	03 27	19 02	11 33	22 48
29	02 46	25 56	15 17	14 46	11 22	03 22	03 29	19 02	11 33	22 49
30	03 S 09	24 N 05	15 S 39	15 S 12	11 S 37	03 S 27	03 S 32	19 N 01	11 N 32	22 S 49

ZODIAC SIGN ENTRIES

Date	h	m	Planets
02	22	31	☽ ♌
03	18	29	♀ ♎
05	10	33	☽ ♍
07	20	51	☽ ♎
10	05	17	☽ ♏
12	11	45	☽ ♐
14	16	03	☽ ♑
16	18	15	☽ ♒
18	19	11	☽ ♓
20	20	26	☽ ♈
22	02	21	♀ ♏
22	09	45	☉ ♎
22	23	49	☽ ♉
25	06	44	☽ ♊
26	12	23	☿ ♏
27	17	19	☽ ♋
29	12	10	♂ ♏
30	05	53	☽ ♌

LATITUDES

Date	Mercury ☿	Venus ♀	Mars ♂	Jupiter ♃	Saturn ♄	Uranus ♅	Neptune ♆	Pluto ♇
01	00 N 28	00 N 58	00 N 24	01 N 07	02 N 15	00 N 34	01 S 47	10 S 20
04	00 N 05	00 51	00 22	01 07	02 15	00 34	01 47	10 20
07	00 S 18	00 44	00 21	01 07	02 14	00 34	01 47	10 20
10	00 42	00 37	00 19	01 07	02 14	00 34	01 47	10 20
13	01 06	00 29	00 17	01 06	02 14	00 34	01 48	10 19
16	01 31	00 21	00 15	01 06	02 14	00 34	01 48	10 19
19	01 54	00 13	00 13	01 06	02 14	00 34	01 48	10 19
22	02 17	00 N 04	00 11	01 06	02 14	00 35	01 48	10 19
25	02 39	00 S 05	00 09	01 06	02 13	00 35	01 48	10 18
28	02 58	00 14	00 07	01 06	02 13	00 35	01 48	10 18
31	03 S 13	00 S 23	00 N 05	01 N 06	02 N 13	00 N 35	01 S 48	10 S 18

DATA

Julian Date	2466399
Delta T	+77 seconds
Ayanamsa	24° 25' 17"
Synetic vernal point	04° ♓ 41' 42"
True obliquity of ecliptic	23° 26' 08"

MOON'S PHASES, APSIDES AND POSITIONS ☽

Date	h	m	Phase	Longitude ° '	Eclipse Indicator
06	15	14	●	14 ♍ 38	
14	02	07	☽	21 ♐ 53	
20	17	43	○	28 ♓ 22	
28	04	41	☾	05 ♋ 40	

Day	h	m	
02	10	24	Apogee
18	06	49	Perigee
30	02	53	Apogee

Day	h	m		
01	03	16	Max dec	26° N 24'
08	13	28	OS	
15	06	32	Max dec	26° S 32'
21	11	01	ON	
28	10	26	Max dec	26° N 38'

LONGITUDES

Date	Chiron ⚷ ° '	Ceres ⚳ ° '	Pallas ⚴ ° '	Juno ⚵ ° '	Vesta ⚶ ° '	Black Moon Lilith ⚸ ° '
01	20 ♋ 15	20 ♋ 24	08 ♉ 16	23 ♊ 43	01 ♒ 15	28 ♋ 02
11	21 ♋ 00	24 ♋ 09	08 ♉ 03	28 ♊ 21	00 ♒ 53	29 ♋ 09
21	21 ♋ 39	27 ♋ 46	07 ♉ 04	02 ♋ 42	01 ♒ 18	00 ♌ 16
31	22 ♋ 11	01 ♌ 12	05 ♉ 19	06 ♋ 41	02 ♒ 25	01 ♌ 23

All ephemeris data is given at 12.00 UT and the Moon's longitude is additionally given for 24.00 UT

ASPECTARIAN

h m	Aspects
01 Saturday	
01 37	☿ ⊼ ♇
02 16	☿ ⋕ ♄
04 39	☽ ⚹ ☉
07 17	☽ ⚼ ♇
07 26	☽ □ ♄
08 48	☽ □ ♂
09 14	♀ ∥ ♃
15 16	☽ Q ☿
17 48	♀ ☌ ♃
21 35	☽ Q ♆
02 Sunday	
01 32	☽ ± ♇
03 13	☿ ⋕ ♀
04 07	♀ ⊼ ♆
08 57	☽ Q ♃
09 03	♀ ⚹ ♅
10 24	☽ Q ♀
12 35	☿ ⋕ ♃
13 41	☽ ⊼ ♇
13 48	☽ ∠ ☉
18 39	☉ ⊻ ♄
19 26	☽ ⚹ ☿
20 18	☽ Q ♄
23 25	☽ Q ♂
03 Monday	
09 37	☽ ⊥ ☉
09 40	☽ ⚹ ♃
10 00	☽ □ ♆
10 39	☽ ☌ ♅
13 27	☽ ⚹ ♀
13 41	☿ ± ♆
15 49	☽ ⋕ ♇
20 44	☽ ⚹ ♄
21 33	♀ ∥ ♄
22 46	☽ ⊻ ☉
22 58	☉ ⊥ ♅
04 Tuesday	
01 00	☽ ⚹ ♂
03 59	♃ ⊼ ♆
06 16	☽ ∠ ☿
16 05	☽ ∥ ♅
16 16	☽ ∠ ♃
18 52	☿ ± ♇
22 52	☽ ∠ ♀
05 Wednesday	
01 49	☽ ☍ ♇
02 51	☽ ⊥ ☿
02 57	☽ ∠ ♄
08 34	☽ ∠ ♂
10 26	☽ ⊥ ♃
16 25	☽ ⊻ ☿
18 39	☽ ⊥ ♀
20 42	☿ ∥ ♃
20 53	☽ ⊥ ♄
21 41	☽ △ ♆
22 27	☽ ⊻ ♃
22 38	☽ ⊻ ♅
06 Thursday	
03 16	☽ ⊥ ♂
03 34	♀ ⚼ ♇
07 42	☽ ⊻ ♀
08 44	☽ ⊻ ♄
09 16	☽ ∥ ♆
10 25	☽ ⊥ ♅
13 32	♃ ⚹ ♅
15 14	☽ ☌ ☉
15 38	☽ ⊻ ♂
19 02	♀ ☌ ♄
07 Friday	
02 51	☽ ⚼ ♆
03 57	☽ ∠ ♅
06 03	☉ ⊻ ♂
08 43	☿ ∥ ♄
10 19	☿ ⊼ ♆
11 11	☉ ⋕ ♂
12 17	☽ ⊼ ♇
12 39	☽ ⋕ ♂
12 51	☽ ∥ ☉
19 02	☽ ⋕ ♀
20 06	☿ ⚹ ♅
20 08	☽ ± ♆
23 46	☽ ± ♇
23 52	☿ ☌ ♃
08 Saturday	
00 56	☽ ⋕ ☿
02 51	☽ ⋕ ♄
06 47	☽ ⋕ ♃
07 33	☽ ⊼ ♆
08 47	☽ ⚹ ♅
09 20	☽ ☌ ♃
10 25	☽ ☌ ☿
16 50	☽ ⚼ ♇
18 51	☽ ☌ ♄
20 21	☽ ∥ ♃
23 27	☽ ☌ ♀
09 Sunday	
00 12	☽ ∥ ♄
04 06	☽ ☌ ♂
05 20	☽ ⊻ ☉
05 28	☽ ∥ ☿
06 39	☉ ⋕ ♀
07 31	☽ Q ♅
10 38	☽ ⋕ ☉
11 20	☽ ∥ ♀
16 17	☽ ∥ ♂
17 29	☽ ⊥ ☉
20 55	☽ △ ♇
10 Monday	
10 29	☉ ⋕ ☿
11 30	☽ ∠ ☉
15 34	☽ ☍ ♆
15 47	☽ ⋕ ♆
17 02	☽ □ ♅
17 31	☿ ⚼ ♇
18 16	☽ ⊻ ♃
11 Tuesday	
01 23	☽ ⊻ ☿
03 03	☽ ⊻ ♄
05 25	☽ ⊥ ♃
12 38	☽ ⊻ ♀
13 36	☽ ⊥ ☿
14 02	☽ ⊥ ♄
14 22	☽ ⊻ ♂
17 03	☽ ⚹ ☉
18 02	☿ ☌ ♄
21 58	☽ ∠ ♃
12 Wednesday	
00 33	☽ ⊥ ♀
01 44	☽ ⊥ ♂
03 37	☽ □ ♇
03 52	♂ Q ♅
04 48	☽ ⋕ ♅
06 22	☽ ∠ ♄
07 42	☽ ∠ ☿
15 13	♀ Q ♅
16 13	☽ Q ☉
17 01	☉ ⚼ ♆
18 12	☽ ∠ ♀
18 36	☽ ∠ ♂
21 35	☽ ⊼ ♆
23 16	☽ △ ♅
13 Thursday	
01 06	☽ ⚹ ♃
02 51	☽ ∥ ♇
03 32	♀ ☌ ♂
08 07	☽ ± ♆
09 08	☽ ⚹ ♄
11 25	☽ Q ♇
13 11	☽ ⚹ ☿
17 23	☉ ∠ ♅
22 12	☽ ⚹ ♂
22 23	☽ Q ♃
23 02	☽ ⚹ ♀
23 46	☽ ⚼ ♆
23 49	♀ ∥ ♂
14 Friday	
01 32	☽ ⚼ ♅
02 07	☽ □ ☉
06 09	☽ Q ♄
08 11	☽ ⚹ ♇
12 10	☽ Q ☿
17 02	☽ ± ♅
19 49	☽ Q ♂
21 34	☽ Q ♀
15 Saturday	
01 24	☽ △ ♆
02 49	☿ ∥ ♂
03 15	☽ ⊼ ♅
03 25	☉ ⋕ ♄
05 37	☽ □ ♃
09 38	☽ ∠ ♇
12 58	☽ □ ♄
21 43	☽ □ ☿
16 Sunday	
00 40	☽ ⊥ ♇
03 31	☽ □ ♂
06 33	☽ □ ♀
08 30	☽ △ ☉
10 36	☽ ⊻ ♇
13 27	☉ ⋕ ♃
17 26	☿ Q ♅
17 Monday	
03 12	☽ □ ♆
04 00	☿ ∥ ♀
05 14	☽ ☍ ♅
08 06	☽ △ ♃
08 19	☽ ∥ ♇
10 54	☽ ⚼ ☉
14 54	☽ △ ♄
15 10	☉ ⊼ ♇
18 Tuesday	
02 37	☽ ± ☉
03 45	☽ △ ☿
05 11	☽ ⋕ ♅
06 08	♀ △ ♇
07 00	☽ △ ♂
08 31	☽ Q ♆
08 53	☽ ⚼ ♃
11 35	☽ ☌ ♇
12 05	☽ △ ♀
13 03	☽ ⊼ ☉
15 30	☽ ⚼ ♄
23 44	☽ ± ♃
19 Wednesday	
04 00	☽ ⚹ ♆
06 14	☽ ⊼ ♅
06 16	☽ ± ♄
06 24	☽ ⚼ ☿
08 33	☽ ⚼ ♂
09 39	☽ ⊼ ♃
14 30	☽ ⋕ ♆
14 42	☽ ⚼ ♀
16 03	☽ ± ♅
16 09	☽ ⊼ ♄
18 45	☽ ∥ ☿
19 47	☽ ∥ ♀
22 27	☽ ± ☿
23 59	☽ ± ♂
20 Thursday	
00 55	☽ ∥ ♂
04 31	☽ ∠ ♆
06 49	☽ ± ♀
06 53	☽ ⚼ ♅
09 13	☽ ⊼ ☿
10 20	☽ ⊼ ♂
12 36	☽ ⊻ ♇
17 40	☽ ⊼ ♀
17 43	☽ ☍ ☉
19 24	☽ ⊥ ♆
20 01	☉ ⊻ ♀
22 34	☽ ⊥ ♇
23 23	☽ ∥ ♄
21 Friday	
01 03	☽ ∥ ♃
05 25	☽ ⊻ ♆
07 57	☽ △ ♅
09 32	☽ ⋕ ☉
12 06	☽ ☍ ♃
12 20	☽ ∥ ☉
13 41	☽ ∠ ♇
15 13	☿ ☌ ♂
15 36	☿ ⋕ ♆
18 05	☉ ± ♆
18 31	☽ ☍ ♄
21 21	☽ ⋕ ♃
22 57	☽ ⋕ ♄
22 Saturday	
02 16	☿ △ ♇
10 12	♂ △ ♇
13 20	♀ ⋕ ♆
15 28	☽ ⚹ ♇
15 44	☽ ☍ ♂
16 40	☽ ☍ ☿
23 Sunday	
00 54	☽ ⊼ ☉
01 25	☽ ⋕ ♂
01 55	☽ ☍ ♀
09 17	☽ ☌ ♆
09 18	☽ ∥ ♆
11 12	☽ ⋕ ♀
12 13	☽ □ ♅
12 26	☽ ± ☉
12 42	☽ Q ♇
13 52	☽ ⋕ ☿
15 56	☉ ± ♇
17 21	☽ ⊼ ♃
23 49	☽ ⊼ ♄
24 Monday	
04 27	☽ ± ♃
06 06	☽ ⚼ ☉
10 57	☽ ± ♄
21 21	☽ Q ♅
21 26	☽ ⚼ ♃
21 42	☽ □ ♇
22 07	☽ ∥ ♅
25 Tuesday	
01 05	☽ ⊼ ♂
03 56	☽ ⚼ ♄
04 10	☽ ⊼ ☿
12 30	♃ ⚼ ♇
12 33	☽ △ ☉
13 06	☽ ± ♂
14 54	☽ ⊼ ♀
16 36	☽ ± ☿
16 48	☽ ⊻ ♆
20 14	☽ ⚹ ♅
22 49	☽ ⋕ ♇
26 Wednesday	
02 32	☽ △ ♃
03 43	☽ ± ♀
04 22	☽ ⊥ ♆
07 27	☽ ⚼ ♂
09 02	☽ △ ♄
10 12	♀ ☍ ♆
11 34	☽ ⚼ ☿
21 57	☽ ∠ ♆
23 16	☽ ⚼ ♀
27 Thursday	
01 37	☽ ∠ ♅
07 42	☽ △ ♇
14 46	☽ △ ♂
17 50	☉ ⊼ ♆
19 36	☽ ⊥ ♅
19 50	☽ △ ☿
23 30	♀ □ ♅
28 Friday	
03 47	☽ ⚹ ♆
04 41	☽ □ ☉
07 41	☽ ⊻ ♅
08 35	☽ △ ♀
13 43	☽ ⚼ ♇
15 10	☽ □ ♃
21 26	☽ □ ♄
29 Saturday	
03 57	☽ Q ♆
07 53	☽ ± ♇
18 22	☉ ⚹ ♅
20 01	☽ ⊼ ♇
20 21	☽ Q ☉
30 Sunday	
04 16	☽ Q ♃
05 23	♂ ⋕ ♆
06 57	☽ □ ♂
10 15	☽ Q ♄
13 09	☽ □ ☿
16 19	☽ □ ♆
20 33	☽ ☌ ♅
22 49	☽ ⚹ ☉
23 25	☽ ⋕ ♇

OCTOBER 2040

Date	Sidereal time h m s	Sun ☉	Moon ☽	Moon ☽ 24.00	Mercury ☿	Venus ♀	Mars ♂	Jupiter ♃	Saturn ♄	Uranus ♅	Neptune ♆	Pluto ♇
01	12 43 21	08 ♎ 55 09	14 ♌ 56 14	20 ♌ 56 11	04 ♍ 19	11 ♏ 25	01 ♏ 21	11 ♎ 30	14 ♎ 18	07 ♌ 17	05 ♉ 08	25 ♒ 06
02	12 47 17	09 54 12	26 ♌ 58 26	03 ♍ 03 17	05 00	12 38	02 01	11 43	14 25	07 19	05 R 07	25 R 06
03	12 51 14	10 53 16	09 ♍ 11 04	15 ♍ 21 58	05 38	13 51	02 42	11 56	14 32	07 21	05 05	25 05
04	12 55 10	11 52 22	21 ♍ 36 10	27 ♍ 53 47	06 11	15 03	03 23	12 09	14 40	07 23	05 04	25 04
05	12 59 07	12 51 31	04 ♎ 14 51	10 ♎ 39 22	06 39	16 16	04 03	12 22	14 47	07 25	05 02	25 03
06	13 03 03	13 50 42	17 ♎ 07 17	23 ♎ 38 31	07 01	17 29	04 44	12 35	14 54	07 27	05 01	25 02
07	13 07 00	14 49 54	00 ♏ 12 58	06 ♏ 50 32	07 18	18 41	05 25	12 48	15 02	07 29	04 59	25 02
08	13 10 57	15 49 09	13 ♏ 31 03	20 ♏ 14 24	07 28	19 54	06 06	13 01	15 09	07 31	04 58	25 01
09	13 14 53	16 48 26	27 ♏ 00 28	03 ♐ 49 09	07 32	21 07	06 47	13 14	15 16	07 33	04 56	25 00
10	13 18 50	17 47 44	10 ♐ 40 19	17 ♐ 33 55	07 R 28	22 19	07 28	13 27	15 24	07 34	04 54	24 59
11	13 22 46	18 47 05	24 ♐ 29 51	01 ♑ 28 02	07 17	23 32	08 09	13 39	15 31	07 36	04 53	24 59
12	13 26 43	19 46 27	08 ♑ 28 22	15 ♑ 30 44	06 57	24 44	08 50	13 52	15 38	07 38	04 51	24 58
13	13 30 39	20 45 51	22 ♑ 35 00	29 ♑ 40 57	06 30	25 57	09 31	14 05	15 46	07 39	04 50	24 57
14	13 34 36	21 45 17	06 ♒ 48 22	13 ♒ 56 56	05 53	27 09	10 13	14 18	15 53	07 41	04 48	24 57
15	13 38 32	22 44 44	21 ♒ 06 15	28 ♒ 15 55	05 09	28 21	10 54	14 31	16 00	07 43	04 46	24 56
16	13 42 29	23 44 13	05 ♓ 25 26	12 ♓ 34 15	04 17	29 ♏ 34	11 35	14 44	16 08	07 44	04 45	24 56
17	13 46 26	24 43 44	19 ♓ 41 46	26 ♓ 47 25	03 17	00 ♐ 46	12 17	14 57	16 15	07 46	04 43	24 55
18	13 50 22	25 43 16	03 ♈ 50 36	10 ♈ 50 44	02 12	01 59	12 58	15 10	16 22	07 47	04 41	24 55
19	13 54 19	26 42 51	17 ♈ 47 18	24 ♈ 39 50	01 ♏ 01	03 11	13 40	15 23	16 30	07 48	04 40	24 54
20	13 58 15	27 42 27	01 ♉ 27 57	08 ♉ 11 22	29 ♎ 47	04 23	14 22	15 36	16 37	07 50	04 38	24 54
21	14 02 12	28 42 06	14 ♉ 49 52	21 ♉ 23 22	28 32	05 35	15 03	15 49	16 44	07 51	04 36	24 53
22	14 06 08	29 ♎ 41 46	27 ♉ 51 52	04 ♊ 15 28	27 19	06 47	15 45	16 02	16 51	07 52	04 35	24 53
23	14 10 05	00 ♏ 41 29	10 ♊ 34 22	16 ♊ 48 50	26 08	07 59	16 27	16 14	16 59	07 53	04 33	24 52
24	14 14 01	01 41 14	22 ♊ 59 14	29 ♊ 05 58	25 03	09 12	17 08	16 27	17 06	07 54	04 31	24 52
25	14 17 58	02 41 01	05 ♋ 09 31	11 ♋ 10 23	24 06	10 24	17 50	16 40	17 13	07 55	04 30	24 52
26	14 21 55	03 40 51	17 ♋ 09 08	23 ♋ 06 20	23 17	11 36	18 32	16 53	17 20	07 56	04 28	24 51
27	14 25 51	04 40 42	29 ♋ 02 36	04 ♌ 58 30	22 39	12 48	19 14	17 05	17 27	07 57	04 26	24 51
28	14 29 48	05 40 36	10 ♌ 54 41	16 ♌ 51 43	22 11	14 00	19 56	17 18	17 35	07 58	04 25	24 51
29	14 33 44	06 40 32	22 ♌ 50 12	28 ♌ 50 42	21 56	15 11	20 38	17 31	17 42	07 59	04 23	24 51
30	14 37 41	07 40 30	04 ♍ 53 44	10 ♍ 59 49	21 D 51	16 23	21 21	17 43	17 49	08 00	04 21	24 50
31	14 41 37	08 ♏ 40 30	17 ♍ 09 21	23 ♍ 22 46	21 ♎ 58	17 ♐ 35	22 ♏ 03	17 ♎ 56	17 ♎ 56	08 ♌ 00	04 ♉ 20	24 ♒ 50

Date	Moon True ☊	Moon Mean ☊	Moon ☽ Latitude	Decl. Sun ☉	Moon ☽	Mercury ☿	Venus ♀	Mars ♂	Jupiter ♃	Saturn ♄	Uranus ♅	Neptune ♆	Pluto ♇
01	25 ♉ 49	26 ♉ 53	05 N 02	03 S 32	21 N 10	15 S 59	15 S 37	11 S 51	03 S 32	03 S 35	19 N 01	11 N 32	22 S 49
02	25 R 42	26 50	05 08	03 55	17 20	16 18	16 03	12 06	03 37	03 38	19 00	11 31	22 49
03	25 34	26 47	04 59	04 18	12 44	16 34	16 28	12 20	03 42	03 41	19 00	11 31	22 49
04	25 25	26 43	04 37	04 42	07 34	16 48	16 52	12 35	03 47	03 44	18 59	11 30	22 50
05	25 18	26 40	04 00	05 05	01 N 59	16 59	17 16	12 49	03 52	03 46	18 59	11 30	22 50
06	25 11	26 37	03 11	05 28	03 S 47	17 08	17 40	13 04	03 57	03 49	18 58	11 29	22 50
07	25 07	26 34	02 11	05 51	09 30	17 14	18 03	13 18	04 02	03 52	18 58	11 29	22 50
08	25 04	26 31	01 N 02	06 13	14 54	17 18	18 26	13 32	04 07	03 55	18 57	11 28	22 50
09	25 D 04	26 27	00 S 11	06 36	19 39	17 17	18 49	13 46	04 12	03 58	18 57	11 28	22 50
10	25 05	26 24	01 24	06 59	23 26	17 14	19 11	14 00	04 17	04 00	18 56	11 27	22 50
11	25 06	26 21	02 33	07 21	25 53	17 06	19 32	14 14	04 22	04 03	18 56	11 27	22 50
12	25 07	26 18	03 34	07 44	26 44	16 54	19 53	14 28	04 27	04 06	18 56	11 26	22 50
13	25 R 08	26 15	04 23	08 06	25 52	16 39	20 13	14 42	04 32	04 09	18 55	11 25	22 50
14	25 07	26 12	04 56	08 29	23 21	16 18	20 33	14 55	04 37	04 12	18 55	11 25	22 51
15	25 04	26 08	05 11	08 51	19 22	15 53	20 53	15 09	04 42	04 14	18 55	11 24	22 51
16	25 01	26 05	05 07	09 13	14 16	15 24	21 12	15 22	04 47	04 17	18 54	11 24	22 51
17	24 56	26 02	04 44	09 35	08 25	14 50	21 30	15 35	04 52	04 20	18 54	11 23	22 51
18	24 52	25 59	04 03	09 56	02 S 11	14 12	21 48	15 49	04 57	04 23	18 54	11 23	22 51
19	24 48	25 56	03 09	10 18	04 N 04	13 31	22 05	16 02	05 02	04 25	18 53	11 22	22 51
20	24 45	25 52	02 04	10 39	10 02	12 48	22 21	16 15	05 07	04 28	18 53	11 22	22 51
21	24 43	25 49	00 S 54	11 01	15 25	12 02	22 37	16 27	05 12	04 31	18 53	11 21	22 51
22	24 D 43	25 46	00 N 17	11 22	19 58	11 17	22 53	16 40	05 17	04 34	18 52	11 20	22 51
23	24 44	25 43	01 26	11 43	23 27	10 32	23 07	16 53	05 22	04 36	18 52	11 20	22 51
24	24 45	25 40	02 29	12 04	25 44	09 49	23 22	17 05	05 27	04 39	18 52	11 19	22 50
25	24 47	25 37	03 25	12 24	26 44	09 10	23 35	17 18	05 32	04 42	18 52	11 19	22 50
26	24 48	25 33	04 10	12 45	26 28	08 34	23 48	17 30	05 36	04 44	18 51	11 18	22 50
27	24 49	25 30	04 44	13 05	24 59	08 04	24 00	17 42	05 41	04 47	18 51	11 18	22 50
28	24 R 49	25 27	05 06	13 25	22 24	07 38	24 12	17 54	05 46	04 50	18 51	11 17	22 50
29	24 48	25 24	05 15	13 45	18 52	07 19	24 23	18 06	05 51	04 52	18 51	11 16	22 50
30	24 47	25 21	05 10	14 04	14 32	07 06	24 33	18 18	05 56	04 55	18 51	11 16	22 50
31	24 ♉ 44	25 ♉ 18	04 N 52	14 S 23	09 N 33	06 S 58	24 S 42	18 S 29	06 S 00	04 S 57	18 N 51	11 N 15	22 S 50

ZODIAC SIGN ENTRIES

Date	h m	Planets
02	17 59	☽ ♍
05	03 59	☽ ♎
07	11 36	☽ ♏
09	17 17	☽ ♐
11	21 29	☽ ♑
14	00 32	☽ ♒
16	02 54	☽ ♓
16	20 40	♀ ♐
18	05 27	☽ ♈
20	07 53	☿ ♎
20	09 24	☽ ♉
22	16 00	☽ ♊
22	19 20	☉ ♏
25	01 47	☽ ♋
27	13 56	☽ ♌
30	02 18	☽ ♍

LATITUDES

Date	Mercury ☿	Venus ♀	Mars ♂	Jupiter ♃	Saturn ♄	Uranus ♅	Neptune ♆	Pluto ♇
01	03 S 13	00 S 23	00 N 05	01 N 06	02 N 13	00 N 35	01 S 48	10 S 18
04	03 25	00 33	00 04	01 06	02 13	00 35	01 48	10 17
07	03 29	00 42	00 N 02	01 06	02 13	00 35	01 48	10 17
10	03 25	00 51	00 00	01 06	02 13	00 35	01 48	10 16
13	03 08	01 01	00 S 02	01 06	02 14	00 35	01 49	10 16
16	02 36	01 10	00 04	01 06	02 14	00 35	01 49	10 15
19	01 49	01 19	00 06	01 06	02 14	00 36	01 49	10 15
22	00 S 49	01 28	00 07	01 06	02 14	00 36	01 49	10 14
25	00 N 12	01 36	00 09	01 06	02 14	00 36	01 49	10 14
28	01 04	01 44	00 11	01 07	02 14	00 36	01 49	10 13
31	01 N 42	01 S 51	00 S 13	01 N 07	02 N 15	00 N 36	01 S 49	10 S 12

DATA

Julian Date	2466429
Delta T	+77 seconds
Ayanamsa	24° 25' 20"
Synetic vernal point	04° ♓ 41' 38"
True obliquity of ecliptic	23° 26' 08"

MOON'S PHASES, APSIDES AND POSITIONS ☽

Date	h m	Phase	Longitude	Eclipse Indicator
06	05 26	●	13 ♎ 34	
13	08 41	☽	20 ♑ 38	
20	04 50	○	27 ♈ 25	
28	00 27	☾	05 ♌ 12	

Day	h m		
15	10 55	Perigee	
27	22 33	Apogee	
05	20 18	OS	
12	12 03	Max dec	26° S 44'
18	20 21	ON	
25	18 45	Max dec	26° N 47'

LONGITUDES

Date	Chiron ⚷	Ceres ⚳	Pallas ⚴	Juno ⚵	Vesta ⚶	Black Moon Lilith ⚸
01	22 ♋ 11	01 ♌ 12	05 ♉ 19	06 ♋ 41	02 ♒ 25	01 ♌ 23
11	22 ♋ 35	04 ♌ 24	02 ♉ 52	10 ♋ 13	04 ♒ 11	02 ♌ 30
21	22 ♋ 50	07 ♌ 21	29 ♈ 53	13 ♋ 12	06 ♒ 30	03 ♌ 37
31	22 ♋ 57	09 ♌ 58	26 ♈ 43	15 ♋ 32	09 ♒ 17	04 ♌ 45

All ephemeris data is given at 12.00 UT and the Moon's longitude is additionally given for 24.00 UT

ASPECTARIAN

h m	Aspects
01 Monday	
04 08	☽ □ ♀
04 57	☽ ⚹ ♃
10 42	☽ ⚹ ♄
11 45	☉ ∥ ♃
13 47	♀ ⊻ ♃
15 31	☉ ∥ ♄
21 21	☽ Q ♂
02 Tuesday	
02 08	☽ ∥ ♅
03 40	☽ Q ☿
07 32	☽ ∠ ☉
08 16	☽ ☍ ♇
11 28	☽ ∠ ♃
15 47	☿ ☍ ♆
16 34	☉ ⚼ ♇
16 53	☽ ∠ ♄
17 25	☽ ⋕ ☿
18 29	☽ ⋕ ♀
20 02	☽ Q ♀
22 04	♃ ∥ ♄
22 33	☽ ⚹ ♂
03 Wednesday	
02 52	☽ ⊥ ☉
04 01	☽ △ ♆
04 42	☽ ⚹ ☿
05 31	☽ ⊥ ♃
08 24	☽ ⊻ ♅
10 43	☽ ⊥ ♄
13 51	☽ ⋕ ♂
15 36	☽ ⊻ ☉
17 26	☽ ⊻ ♃
17 55	☽ ∥ ♆
20 07	☽ ⊥ ♅
22 02	☽ ⚹ ♀
22 30	☽ ⊻ ♄
04 Thursday	
02 13	☿ ∥ ♀
03 17	♀ ⊻ ♄
05 27	☽ ∠ ♂
09 03	☽ ⚼ ♆
11 09	☽ ∠ ☿
13 30	☽ ∠ ♅
18 36	☽ ⊼ ♇
20 24	☉ ☌ ♃
23 38	☽ ⊥ ♂
23 42	☽ ⋕ ☉
05 Friday	
02 11	☽ ± ♆
04 09	☽ ⋕ ♃
04 30	☽ ⋕ ♄
04 58	☽ ⊥ ☿
05 47	☽ ∠ ♀
05 59	☽ ± ♇
11 37	☽ ⊻ ♂
13 29	☽ ⊼ ♆
16 39	☽ ⊻ ☿
17 58	☽ ⚹ ♅
22 52	☽ ⚼ ♇
06 Saturday	
00 27	☽ ⊥ ♀
03 26	☽ ☌ ♃
05 26	☽ ☌ ☉
07 52	☽ ☌ ♄
12 09	☽ ∥ ♄
12 43	☽ ∥ ♃
12 44	☽ ⊻ ♀
16 18	☽ Q ♅
19 28	☽ ∥ ☉
21 21	♂ ☍ ♆
07 Sunday	
02 33	☽ △ ♇
14 30	♀ ⊥ ♃
17 26	☉ ☌ ♄
20 32	☽ ⋕ ♆
20 38	☽ ☍ ♆
21 56	☽ ☌ ♂
08 Monday	
01 01	☽ ☌ ☿
01 11	☽ □ ♅
05 25	☽ ∥ ♂
11 04	☽ ⊻ ♃
14 57	☽ ⊻ ♄
16 27	☽ ⊻ ☉
21 58	☽ ⊥ ♃
23 35	☽ ∥ ☿
09 Tuesday	
00 31	☽ ☌ ♀
01 45	☽ ⊥ ♄
03 58	☽ ⊥ ☉
06 59	☽ ∥ ♀
08 08	☽ ⋕ ♅
08 27	☽ □ ♇
12 04	☿ St R
14 11	☽ ∠ ♃
15 37	♀ ⊥ ♄
17 49	☽ ∠ ♄
20 51	♀ ⋕ ♅
21 07	☽ ∠ ☉
10 Wednesday	
01 56	☽ ⊼ ♆
06 06	☽ ⊻ ♂
06 26	☽ ⊻ ☿
06 34	☽ △ ♅
07 41	☽ ∥ ♇
11 58	☿ ☌ ♂
12 25	☽ ± ♆
15 54	♂ □ ♅
16 02	☽ Q ♇
16 49	☽ ⊥ ☿
16 54	☽ ⚹ ♃
17 08	☽ ⊥ ♂
20 18	☽ ⚹ ♄
11 Thursday	
01 21	☽ ⚹ ☉
04 02	☽ ⚼ ♆
08 14	☽ ∠ ☿
08 43	☽ ⚼ ♅
09 33	☽ ∠ ♂
10 10	☽ ⊻ ♀
12 50	☽ ⚹ ♇
14 02	☽ Q ♃
17 15	☽ Q ♄
21 29	☽ ⊥ ♀
23 39	☽ Q ☉
12 Friday	
00 15	☽ ± ♅
05 49	☽ △ ♆
08 25	☉ Q ♅
09 28	☽ ⚹ ☿
10 33	☽ ⊼ ♅
12 39	☽ ⚹ ♂
14 22	☽ ∠ ♀
14 33	☽ ∠ ♇
16 34	♀ □ ♇
21 21	☽ □ ♃
13 Saturday	
00 20	☽ □ ♄
05 18	☽ Q ☿
05 51	☽ ⊥ ♇
08 41	☽ □ ☉
10 07	☽ Q ♂
16 01	☽ ⊻ ♇
18 13	☽ ⚹ ♀
14 Sunday	
08 38	☽ □ ♆
10 32	☽ □ ☿
13 29	☽ ☍ ♅
15 32	☽ ∥ ♇
16 18	☽ Q ♀
18 01	☽ □ ♂
15 Monday	
00 48	☽ △ ♃
03 23	☽ △ ♄
04 21	☽ ∥ ♀
14 23	☽ ⋕ ♅
14 47	☽ Q ♆
14 57	☽ △ ☉
18 25	☽ ☌ ♇
23 13	☿ ☍ ♆
16 Tuesday	
01 16	☽ □ ♀
02 19	☽ ⚼ ♃
04 44	☽ ⚼ ♄
06 32	☽ ∥ ☿
07 22	☽ ∥ ♂
10 12	☽ △ ☿
10 52	☽ ⚹ ♆
12 52	☿ ∥ ♂
15 53	☽ ⊼ ♅
16 12	♀ ∠ ♃
17 39	☽ ± ♃
17 58	☽ ⚼ ☉
19 58	☽ ± ♄
22 53	☽ △ ♂
17 Wednesday	
00 07	☽ ⋕ ♆
01 59	☽ ± ♅
03 53	☽ ⊼ ♃
06 09	☽ ⊼ ♄
07 22	☉ ⊥ ♀
07 42	☽ ∥ ☉
09 47	☽ ⚼ ☿
10 15	☽ ± ☉
12 02	☽ ∠ ♆
16 33	☉ △ ♇
17 11	☽ ⚼ ♅
20 49	☽ ⊻ ♇
21 09	☽ ⊼ ☉
22 40	♀ ∠ ♄
23 56	☽ ± ☿
18 Thursday	
01 30	☽ ⚼ ♂
01 35	☽ ∥ ♃
03 15	☽ ⊥ ♆
03 43	☽ ∥ ♄
07 00	☽ ⊥ ♇
08 31	☽ △ ♀
09 24	☽ ⊼ ☿
13 27	☽ ⊻ ♆
14 13	☿ ⊻ ♀
17 38	☽ ± ♂
18 45	☽ △ ♅
22 23	☽ ∠ ♇
19 Friday	
04 29	☽ ⊼ ♂
07 46	☽ ☍ ♃
09 44	☽ ☍ ♄
12 45	☽ ⚼ ♀
13 23	☽ ⋕ ♄
15 49	☽ ⋕ ♃
20 Saturday	
00 25	☽ ⚹ ♇
04 50	☽ ☍ ☉
06 01	☽ ± ♀
09 16	☽ ☍ ☿
14 46	☽ ⋕ ☉
16 54	♀ ⊼ ♆
17 35	☽ ∥ ♆
17 38	☽ ☌ ♆
17 42	☽ ⊼ ♀
21 40	☽ Q ♇
22 26	☽ ⋕ ☿
23 22	☽ □ ♅
21 Sunday	
01 45	☿ ⊥ ♀
10 16	☉ ☌ ☿
12 26	☽ ☍ ♂
13 49	☽ ⊼ ♃
15 30	☽ ⊼ ♄
17 18	☽ ⋕ ♂
22 Monday	
00 59	☽ ± ♃
02 37	☽ ± ♄
05 45	☽ ∥ ♅
06 27	☽ □ ♇
08 17	☽ Q ♅
08 37	♀ ∥ ♇
10 04	☿ ⋕ ♆
10 11	☉ ∥ ☿
10 26	☉ ⋕ ♆
11 03	☽ ⊼ ☿
15 43	☽ ⊼ ☉
18 01	☽ ⚼ ♃
19 33	☽ ⚼ ♄
21 20	☽ ± ☿
23 Tuesday	
00 35	☽ ⊻ ♆
01 49	♂ ⊻ ♃
03 59	☽ ± ☉
06 34	☽ ☍ ♀
06 52	☽ ⚹ ♅
07 11	☽ ⋕ ♇
09 09	☽ ⋕ ♀
09 51	♀ △ ♅
11 58	☽ ⊥ ♆
12 59	☽ ⚼ ☿
22 41	☽ ⚼ ☉
23 05	☽ △ ♃
23 57	☽ ⊼ ♂
24 Wednesday	
00 26	☽ △ ♄
05 16	☽ ∠ ♆
10 14	♂ ⊻ ♄
11 50	☽ ∠ ♅
12 19	☽ ± ♂
15 41	☽ △ ♇
15 44	☽ △ ☿
16 25	☿ △ ♇
21 22	☿ ∠ ♀
25 Thursday	
05 34	☽ ⊥ ♅
06 39	☽ △ ☉
07 07	☽ ⚼ ♂
10 41	☽ ⚹ ♆
13 59	♀ ± ♆
17 30	☽ ⊻ ♅
21 22	☽ ⚼ ♇
23 36	☽ ⊼ ♀
26 Friday	
10 37	☽ Q ♆
11 26	☽ □ ♃
12 23	☽ □ ♄
12 59	☽ ± ♀
14 58	☽ △ ♂
15 26	☽ ± ♇
23 41	☽ □ ☿
27 Saturday	
03 32	☽ ⊼ ♇
06 24	☉ ☍ ♆
09 11	☽ ⚼ ♀
13 08	♀ Q ♇
21 29	☽ ⋕ ♀
22 53	☽ □ ♆
28 Sunday	
00 27	☽ □ ☉
00 27	☽ Q ♃
01 06	☽ Q ♄
06 03	☽ ☌ ♅
08 29	☽ ⋕ ♇
10 35	☽ Q ☿
18 55	☽ △ ♀
29 Monday	
01 07	☽ ⚹ ♃
01 34	☽ ⚹ ♄
07 19	☽ □ ♂
10 12	☽ ⚹ ☿
12 04	☽ ∥ ♅
16 01	☽ ☍ ♇
16 01	☽ Q ☉
16 20	☽ ⋕ ♂
30 Tuesday	
07 37	☽ ∠ ♃
07 50	☽ ∠ ♄
09 06	☿ St D
10 56	☽ △ ♆
14 11	☽ ⋕ ☉
15 53	☽ ∠ ☿
17 58	☽ ⚹ ☉
18 07	☽ ⊻ ♅
19 43	☉ □ ♅
21 18	☽ Q ♂
31 Wednesday	
01 39	☽ ⊥ ♃
01 44	☽ ⊥ ♄
04 03	☽ ∥ ♆
05 52	☽ ⊥ ♅
08 19	☿ ⊻ ♂
09 40	☽ ⊥ ☿
11 47	♃ ☌ ♄
12 55	☽ □ ♀
13 31	☽ ⊻ ♄
13 32	☽ ⊻ ♃
16 11	☽ ⚼ ♆
19 42	♀ ⚹ ♄
20 26	♀ ⚹ ♃
21 29	☽ ⊻ ☿
22 01	☽ ⚹ ♂
23 17	☽ ∠ ♅
23 39	☽ ⋕ ☿

NOVEMBER 2040

LONGITUDES

Date	Sidereal time h m s	Sun ☉ ° ' "	Moon ☽ ° ' "	Moon ☽ 24.00 ° ' "	Mercury ☿ ° '	Venus ♀ ° '	Mars ♂ ° '	Jupiter ♃ ° '	Saturn ♄ ° '	Uranus ♅ ° '	Neptune ♆ ° '	Pluto ♇ ° '
01	14 45 34	09 ♏ 40 32	29 ♍ 40 22	06 ♎ 02 24	22 ♎ 16	18 ♐ 47	22 ♏ 45	18 ♎ 08	18 ♎ 03	08 ♌ 01	04 ♉ 18	24 ♒ 50
02	14 49 30	10 40 37	12 ♎ 29 02	19 ♎ 00 23	22 43	19 59	23 28	18 21	18 10	08 02	04 R 16	24 R 50
03	14 53 27	11 40 43	25 ♎ 36 26	02 ♏ 17 06	23 20	21 10	24 10	18 33	18 17	08 02	04 15	24 50
04	14 57 24	12 40 51	09 ♏ 02 15	15 ♏ 51 37	24 05	22 22	24 52	18 46	18 24	08 03	04 13	24 50
05	15 01 20	13 41 01	22 ♏ 44 53	29 ♏ 41 43	24 57	23 33	25 35	18 58	18 31	08 03	04 11	24 50
06	15 05 17	14 41 13	06 ♐ 41 39	13 ♐ 44 16	25 56	24 45	26 18	19 11	18 38	08 03	04 10	24 49
07	15 09 13	15 41 27	20 ♐ 49 04	27 ♐ 55 34	27 01	25 57	27 00	19 23	18 44	08 04	04 08	24 D 49
08	15 13 10	16 41 42	05 ♑ 03 17	12 ♑ 11 46	28 11	27 08	27 43	19 35	18 51	08 04	04 06	24 49
09	15 17 06	17 41 59	19 ♑ 20 33	26 ♑ 29 14	29 ♎ 25	28 19	28 26	19 48	18 58	08 04	04 05	24 50
10	15 21 03	18 42 17	03 ♒ 37 25	10 ♒ 44 47	00 ♏ 43	29 ♐ 31	29 08	20 00	19 05	08 04	04 03	24 50
11	15 24 59	19 42 37	17 ♒ 51 01	24 ♒ 55 49	02 03	00 ♑ 42	29 ♏ 51	20 12	19 12	08 04	04 01	24 50
12	15 28 56	20 42 57	01 ♓ 58 57	09 ♓ 00 10	03 27	01 53	00 ♐ 34	20 24	19 18	08 04	04 00	24 50
13	15 32 53	21 43 20	15 ♓ 59 16	22 ♓ 56 02	04 52	03 04	01 17	20 36	19 25	08 R 04	03 58	24 50
14	15 36 49	22 43 43	29 ♓ 50 19	06 ♈ 41 54	06 20	04 15	02 00	20 48	19 31	08 04	03 57	24 50
15	15 40 46	23 44 08	13 ♈ 30 37	20 ♈ 16 20	07 49	05 26	02 43	21 00	19 38	08 04	03 55	24 50
16	15 44 42	24 44 34	26 ♈ 58 53	03 ♉ 38 08	09 19	06 37	03 26	21 12	19 44	08 04	03 54	24 51
17	15 48 39	25 45 02	10 ♉ 13 59	16 ♉ 46 18	10 50	07 48	04 10	21 24	19 51	08 04	03 52	24 51
18	15 52 35	26 45 32	23 ♉ 15 04	29 ♉ 40 13	12 22	08 59	04 53	21 35	19 57	08 04	03 50	24 51
19	15 56 32	27 46 03	06 ♊ 01 47	12 ♊ 19 48	13 55	10 09	05 36	21 47	20 04	08 03	03 49	24 51
20	16 00 28	28 46 35	18 ♊ 34 23	24 ♊ 45 39	15 28	11 20	06 19	21 59	20 10	08 03	03 47	24 52
21	16 04 25	29 ♏ 47 09	00 ♋ 53 48	06 ♋ 59 05	17 01	12 30	07 03	22 10	20 16	08 02	03 46	24 52
22	16 08 22	00 ♐ 47 45	13 ♋ 01 48	19 ♋ 02 16	18 35	13 41	07 46	22 22	20 22	08 02	03 44	24 53
23	16 12 18	01 48 22	25 ♋ 00 54	00 ♌ 58 06	20 09	14 51	08 30	22 33	20 29	08 01	03 43	24 53
24	16 16 15	02 49 01	06 ♌ 54 21	12 ♌ 50 09	21 43	16 01	09 13	22 45	20 35	08 01	03 42	24 53
25	16 20 11	03 49 42	18 ♌ 46 02	24 ♌ 42 34	23 18	17 11	09 57	22 56	20 41	08 00	03 40	24 54
26	16 24 08	04 50 24	00 ♍ 40 19	06 ♍ 39 52	24 52	18 21	10 41	23 07	20 47	07 59	03 39	24 54
27	16 28 04	05 51 07	12 ♍ 41 50	18 ♍ 46 46	26 26	19 31	11 24	23 18	20 53	07 59	03 37	24 55
28	16 32 01	06 51 52	24 ♍ 55 16	01 ♎ 07 53	28 01	20 41	12 08	23 30	20 58	07 58	03 36	24 55
29	16 35 57	07 52 39	07 ♎ 25 06	13 ♎ 47 23	29 ♏ 35	21 51	12 52	23 41	21 04	07 57	03 35	24 56
30	16 39 54	08 ♐ 53 27	20 ♎ 15 08	26 ♎ 48 39	01 ♐ 09	23 ♑ 00	13 ♐ 36	23 ♎ 51	21 ♎ 10	07 ♌ 56	03 ♉ 33	24 ♒ 57

Date	Moon True ☊ ° '	Moon Mean ☊ ° '	Moon ☽ Latitude ° '
01	24 ♉ 42	25 ♉ 14	04 N 19
02	24 R 40	25 11	03 32
03	24 38	25 08	02 34
04	24 37	25 05	01 25
05	24 37	25 02	00 N 10
06	24 D 37	24 58	01 S 06
07	24 38	24 55	02 20
08	24 38	24 52	03 26
09	24 39	24 49	04 19
10	24 39	24 46	04 56
11	24 40	24 43	05 15
12	24 R 40	24 39	05 14
13	24 39	24 36	04 55
14	24 39	24 33	04 19
15	24 39	24 30	03 29
16	24 D 39	24 27	02 28
17	24 39	24 24	01 19
18	24 39	24 20	00 S 08
19	24 R 39	24 17	01 N 03
20	24 39	24 14	02 09
21	24 38	24 11	03 08
22	24 37	24 08	03 57
23	24 36	24 04	04 35
24	24 35	24 01	05 02
25	24 35	23 58	05 15
26	24 34	23 55	05 15
27	24 D 34	23 52	05 01
28	24 35	23 49	04 33
29	24 36	23 45	03 52
30	24 ♉ 37	23 ♉ 42	02 N 59

DECLINATIONS

Sun ☉ ° '	Moon ☽ ° '	Mercury ☿ ° '	Venus ♀ ° '	Mars ♂ ° '	Jupiter ♃ ° '	Saturn ♄ ° '	Uranus ♅ ° '	Neptune ♆ ° '	Pluto ♇ ° '
14 S 43	04 N 05	06 S 56	24 S 51	18 S 40	06 S 05	05 S 00	18 N 50	11 N 15	22 S 50
15 01	01 S 40	07 00	24 59	18 52	06 10	05 03	18 50	11 14	22 50
15 20	07 31	07 08	25 07	19 03	06 14	05 05	18 50	11 14	22 49
15 38	13 10	07 20	25 13	19 14	06 19	05 08	18 50	11 13	22 49
15 57	18 17	07 37	25 19	19 25	06 24	05 10	18 50	11 13	22 49
16 14	22 31	07 56	25 25	19 35	06 28	05 13	18 50	11 12	22 49
16 32	25 27	08 19	25 29	19 46	06 33	05 15	18 50	11 11	22 49
16 49	26 46	08 44	25 33	19 56	06 37	05 18	18 50	11 11	22 48
17 06	26 19	09 12	25 36	20 06	06 42	05 20	18 50	11 10	22 48
17 23	24 08	09 41	25 39	20 16	06 47	05 23	18 50	11 10	22 48
17 40	20 28	10 11	25 40	20 26	06 51	05 25	18 50	11 09	22 48
17 56	15 39	10 43	25 41	20 36	06 55	05 27	18 50	11 09	22 48
18 12	10 04	11 16	25 42	20 45	07 00	05 30	18 50	11 08	22 47
18 27	04 S 02	11 49	25 41	20 54	07 04	05 32	18 50	11 08	22 47
18 42	02 N 07	12 22	25 40	21 03	07 09	05 34	18 50	11 07	22 47
18 57	08 06	12 56	25 38	21 12	07 13	05 37	18 50	11 07	22 46
19 12	13 38	13 30	25 35	21 21	07 17	05 39	18 50	11 06	22 46
19 26	18 27	14 04	25 32	21 29	07 22	05 41	18 50	11 06	22 46
19 40	22 20	14 38	25 28	21 38	07 26	05 44	18 51	11 05	22 46
19 53	25 05	15 11	25 23	21 46	07 30	05 46	18 51	11 05	22 45
20 06	26 34	15 44	25 17	21 54	07 34	05 48	18 51	11 04	22 45
20 19	26 44	16 17	25 11	22 02	07 39	05 50	18 51	11 04	22 45
20 31	25 38	16 49	25 04	22 09	07 43	05 52	18 51	11 03	22 44
20 43	23 24	17 20	24 56	22 16	07 47	05 54	18 51	11 03	22 44
20 55	20 11	17 51	24 48	22 23	07 51	05 57	18 52	11 03	22 44
21 06	16 08	18 20	24 39	22 30	07 55	05 59	18 52	11 02	22 43
21 17	11 25	18 50	24 29	22 37	07 59	06 01	18 52	11 02	22 43
21 27	06 12	19 18	24 19	22 43	08 03	06 03	18 52	11 01	22 42
21 37	00 37	19 45	24 08	22 50	08 07	06 05	18 53	11 01	22 42
21 S 47	05 S 09	20 S 12	23 S 56	22 S 56	08 S 11	06 S 07	18 N 53	11 N 00	22 S 42

ZODIAC SIGN ENTRIES

Date	h	m	Planets	
01	12	37	☽	♎
03	19	55	☽	♏
06	00	31	☽	♐
08	03	30	☽	♑
09	22	57	☿	♏
10	05	54	☽	♒
10	21	53	♀	♑
11	16	52	♂	♐
12	08	37	☽	♓
14	12	17	☽	♈
16	17	26	☽	♉
19	00	37	☽	♊
21	10	14	☽	♋
21	17	05	☉	♐
23	22	03	☽	♌
26	10	39	☽	♍
28	21	49	☽	♎
29	18	20	☿	♐

LATITUDES

Date	Mercury ☿ ° '	Venus ♀ ° '	Mars ♂ ° '	Jupiter ♃ ° '	Saturn ♄ ° '	Uranus ♅ ° '	Neptune ♆ ° '	Pluto ♇ ° '
01	01 N 52	01 S 54	00 S 13	01 N 07	02 N 15	00 N 36	01 S 49	10 S 12
04	02 09	02 01	00 15	01 07	02 15	00 36	01 49	10 12
07	02 14	02 07	00 17	01 07	02 15	00 36	01 49	10 11
10	02 10	02 13	00 19	01 07	02 16	00 36	01 49	10 10
13	02 00	02 18	00 21	01 08	02 16	00 37	01 48	10 10
16	01 45	02 22	00 22	01 08	02 16	00 37	01 48	10 09
19	01 27	02 25	00 24	01 08	02 17	00 37	01 48	10 08
22	01 07	02 28	00 26	01 09	02 17	00 37	01 48	10 08
25	00 47	02 29	00 27	01 09	02 18	00 37	01 48	10 07
28	00 26	02 30	00 29	01 09	02 18	00 37	01 48	10 06
31	00 N 05	02 S 30	00 S 31	01 N 10	02 N 19	00 N 37	01 S 48	10 S 06

DATA

Julian Date	2466460
Delta T	+77 seconds
Ayanamsa	24° 25' 24"
Synetic vernal point	04° ♓ 41' 35"
True obliquity of ecliptic	23° 26' 08"

MOON'S PHASES, APSIDES AND POSITIONS ☽

Date	h	m	Phase	Longitude ° '	Eclipse Indicator
04	18	56	●	12 ♏ 58	Partial
11	15	23	☽	19 ♒ 51	
18	19	06	○	27 ♉ 03	total
26	21	07	☾	05 ♍ 13	

Day	h	m		
09	06	00	Perigee	
24	19	06	Apogee	
02	05	07	0S	
08	17	53	Max dec	26° S 49'
15	03	42	0N	
22	03	07	Max dec	26° N 49'
29	14	35	0S	

LONGITUDES

Date	Chiron ⚷ ° '	Ceres ⚳ ° '	Pallas ⚴ ° '	Juno ⚵ ° '	Vesta ⚶ ° '	Black Moon Lilith ⚸ ° '
01	22 ♋ 57	10 ♌ 13	26 ♈ 24	15 ♋ 43	09 ♒ 35	04 ♌ 51
11	22 ♋ 53	12 ♌ 24	23 ♈ 27	17 ♋ 10	12 ♒ 47	05 ♌ 58
21	22 ♋ 40	14 ♌ 07	21 ♈ 05	17 ♋ 42	16 ♒ 18	07 ♌ 06
31	22 ♋ 19	15 ♌ 16	19 ♈ 32	17 ♋ 18	20 ♒ 05	08 ♌ 13

All ephemeris data is given at 12.00 UT and the Moon's longitude is additionally given for 24.00 UT

ASPECTARIAN

h m	Aspects
01 Thursday	
01 39	☽ ∠ ☉
02 47	☽ ⚻ ♇
03 32	☽ # ♃
08 07	☽ # ♄
09 24	☽ ± ♆
14 12	☽ ± ♇
20 12	☽ ⊥ ☉
20 43	☽ ⚻ ♆
22 09	♀ ⚼ ♆
02 Friday	
02 46	☽ Q ♀
03 42	☽ ✶ ♅
04 05	☽ ∠ ♂
07 05	☽ ⚼ ♇
08 22	☽ ⚺ ☉
09 03	♂ # ♅
21 41	☽ ⊥ ♂
22 33	☽ ☌ ♄
22 58	☽ ☌ ♃
03 Saturday	
01 52	☽ Q ♅
01 56	☽ ∥ ♄
03 09	☽ ✶ ♀
06 41	☽ ∥ ♃
07 40	☽ ☌ ☿
09 15	☽ ⚺ ♂
10 22	☽ ∥ ☿
10 35	☽ △ ♇
16 41	♂ ⊥ ♄
04 Sunday	
03 28	☽ ☍ ♆
03 35	☽ # ♆
06 49	♂ ⊥ ♃
08 45	☽ ∠ ♀
10 14	☽ □ ♅
10 24	♂ □ ♇
18 56	☽ ☌ ☉
23 54	☽ ∥ ☉
05 Monday	
01 42	♀ ⚼ ♅
02 07	☽ ⊥ ♀
04 34	☽ ⚺ ♄
05 20	☽ ⚺ ♃
08 33	☿ △ ♇
13 32	☽ ⚺ ♀
14 47	☽ # ♅
15 05	☽ ⊥ ♄
15 36	☽ □ ♇
15 55	☽ ⊥ ♃
16 06	☽ ⚺ ☿
17 10	☽ ☌ ♂
18 03	☽ ∥ ♂
06 Tuesday	
03 13	☽ ⊥ ☿
06 42	☽ ∠ ♄
07 38	☽ ∠ ♃
07 40	☽ ⚻ ♆
13 29	♀ ✶ ♇
14 00	☽ ∥ ♇
14 20	☽ △ ♅
17 54	☽ ± ♆
19 50	☽ ∠ ☿
22 27	☽ Q ♇
07 Wednesday	
02 39	☽ ⚺ ☉
06 00	♇ St D
08 27	☽ ✶ ♄
09 09	☽ ⚼ ♆
09 32	☽ ✶ ♃
10 58	☿ ⚺ ♂
12 29	☽ ∥ ♀
13 35	☽ ⊥ ☉
15 48	☽ ⚼ ♅
18 46	☽ ✶ ♇
21 27	☽ ☌ ♀
23 00	☽ ⚺ ♂
23 23	☽ ✶ ☿
08 Thursday	
04 53	☽ Q ♄
05 55	☽ ∠ ☉
06 05	☽ Q ♃
06 58	☽ ± ♅
09 38	☽ ⊥ ♂
10 24	☽ △ ♆
17 04	☽ ⚻ ♅
20 01	☽ ∠ ♇
21 25	☽ Q ☿
09 Friday	
01 33	☽ ∠ ♂
09 02	☽ ✶ ☉
11 08	☽ ⊥ ♇
11 22	☽ □ ♄
12 46	☽ □ ♃
17 18	♀ ⚺ ♂
21 13	☽ ⚺ ♇
21 44	☽ ∥ ♀
10 Saturday	
04 04	☽ ✶ ♂
04 27	☽ ⚺ ♀
06 36	☽ □ ☿
06 43	☽ Q ☉
12 43	☽ □ ♆
15 28	☽ ⊥ ♀
19 29	☽ ☍ ♅
21 02	♃ Q ♅
21 47	☽ ∥ ♇
22 05	☉ ⚺ ♄
11 Sunday	
01 20	☽ Q ♂
08 02	☽ ∠ ♀
12 11	☽ ∥ ♂
14 17	☽ △ ♄
15 23	☽ □ ☉
16 02	☽ △ ♃
19 03	☽ Q ♆
20 43	☽ # ♅
23 02	♀ Q ♄
23 50	☽ ☌ ♇
12 Monday	
01 48	☽ ∥ ☉
02 34	☉ ⚺ ♃
09 28	☽ □ ♂
11 49	☽ ✶ ♀
14 47	☽ △ ☿
15 26	☽ ✶ ♆
15 50	♀ ± ♅
16 00	☽ ⚼ ♄
17 54	♅ St R
17 55	☽ ⚼ ♃
21 08	☿ ☍ ♆
22 25	☽ ⚻ ♅
13 Tuesday	
00 34	♀ Q ♃
06 43	☿ # ♆
07 29	☽ ∥ ☿
07 32	☽ ± ♄
07 33	☽ # ♆
08 42	☽ ± ♅
09 35	☽ ± ♃
10 16	☽ Q ♀
17 08	☽ ∠ ♆
17 57	☽ ⚻ ♄
19 29	☽ ⚼ ☿
20 05	☽ ⚻ ♃
22 41	☽ △ ☉
14 Wednesday	
00 10	☽ ∥ ♃
00 14	☽ ⚼ ♅
03 18	☽ ⚺ ♇
05 52	♀ △ ♆
06 08	☽ ∥ ♄
08 42	☽ ⊥ ♆
12 58	☽ ± ☿
13 44	☽ ⊥ ♇
15 59	☽ △ ♂
19 09	☽ ⚺ ♆
20 27	☽ □ ♀
15 Thursday	
00 44	☽ ⚻ ☿
02 25	☽ △ ♅
02 55	☽ ⚼ ☉
05 31	☽ ∠ ♇
16 08	☿ □ ♅
19 53	☽ ⚼ ♂
20 05	☽ ± ☉
22 57	☽ ☍ ♄
16 Friday	
00 46	☉ # ♅
01 30	☽ ☍ ♃
01 47	☽ # ♄
07 40	☽ ⚻ ☉
08 10	☽ ✶ ♇
08 19	☽ # ♃
12 52	☽ ± ♂
14 24	☉ □ ♇
17 Saturday	
00 19	☽ ⚻ ♂
00 26	☽ ☌ ♆
00 44	☽ ∥ ♆
02 33	♂ ⚻ ♆
05 50	☽ Q ♇
07 07	☽ △ ♀
08 03	☽ □ ♅
11 20	☽ # ☿
13 15	☽ ☍ ☿
14 35	☉ ⊥ ♄
17 24	♀ ⚻ ♅
18 Sunday	
05 50	☽ ⚻ ♄
08 52	☽ ⚻ ♃
13 29	☽ ⚼ ♀
14 07	☽ ∥ ♅
14 54	♂ ∠ ♄
14 59	☽ □ ♇
17 05	☽ ± ♄
17 14	☽ Q ♅
17 46	☽ # ☉
19 06	☽ ☍ ☉
20 13	☽ ± ♃
19 Monday	
05 56	♀ ∠ ♇
06 52	☽ # ♂
07 49	☽ ⚺ ♆
08 05	☽ ± ♀
10 09	☽ ⚼ ♄
10 27	♄ Q ♅
11 08	☽ ☍ ♂
12 32	☉ ⊥ ♃
13 27	☽ ⚼ ♃
15 05	☽ # ♇
15 50	☽ ✶ ♅
19 11	☽ ⊥ ♆
20 39	☽ ⚻ ♀
20 Tuesday	
05 09	☽ ⚻ ☿
12 25	☽ ∠ ♆
15 06	☽ △ ♄
15 22	☽ # ♀
18 24	☽ ± ☿
18 42	☽ △ ♃
20 40	☽ ∠ ♅
21 Wednesday	
00 12	☽ △ ♇
09 37	☽ ⚻ ☉
14 15	☽ ⊥ ♅
14 32	☽ ⚼ ☿
17 38	♂ ∠ ♃
17 38	☽ ✶ ♆
22 30	☽ ± ☉
22 Thursday	
00 54	☽ ⚻ ♂
02 05	☽ ⚺ ♅
05 44	☽ ⚼ ♇
13 26	☽ ☍ ♀
13 34	☽ ± ♂
17 06	♀ ⊥ ♂
17 24	☽ Q ♆
18 01	☽ ⚼ ☉
20 28	♂ △ ♅
23 41	☽ ± ♇
23 Friday	
00 46	☽ △ ☿
02 48	☽ □ ♄
06 58	☽ □ ♃
08 45	☽ ⚼ ♂
11 44	☽ ⚻ ♇
17 17	☿ ⚺ ♄
19 51	☽ # ♀
24 Saturday	
02 58	☽ △ ☉
05 31	☽ □ ♆
14 14	☽ ☌ ♅
15 25	☽ Q ♄
17 00	☽ △ ♂
17 37	☽ # ♇
19 54	☽ Q ♃
20 53	☽ # ♂
25 Sunday	
03 02	♂ ± ♆
05 45	☿ ⚺ ♃
07 21	☽ # ☉
08 19	☉ ⚻ ♆
08 27	☽ ⚻ ♀
15 54	☽ ✶ ♄
20 20	☽ ∥ ♅
20 33	☽ ✶ ♃
21 54	☽ ± ♀
22 32	☽ □ ☿
26 Monday	
00 23	☽ ☍ ♇
00 49	☽ # ☿
12 37	☿ □ ♇
17 57	☽ △ ♆
17 57	☽ ⚼ ♀
21 07	☽ □ ☉
22 19	☽ ∠ ♄
23 29	♀ ⊥ ♇
27 Tuesday	
02 38	☽ ⚺ ♅
03 08	☽ ∠ ♃
09 17	☽ □ ♂
12 39	☉ ∠ ♄
13 54	☽ ∥ ♆
14 04	☿ # ♅
14 32	☽ ⊥ ♅
15 58	☽ Q ☿
16 21	☽ ⊥ ♄
19 07	☿ ⊥ ♄
21 15	☽ ⊥ ♃
23 40	☽ ⚼ ♆
28 Wednesday	
02 52	☽ △ ♀
03 49	☽ # ♃
04 15	☽ ⚺ ♄
08 11	☽ ∠ ♅
08 30	♂ ∥ ♇
09 11	☽ ⚺ ♃
11 53	☽ Q ☉
12 00	☽ ⚻ ♇
12 40	☽ # ♄
17 11	☽ ± ♆
18 37	♀ □ ♄
18 52	☽ ✶ ☿
22 43	☽ Q ♂
23 37	☽ ± ♇
29 Thursday	
04 42	☽ ⚻ ♆
12 57	☽ ✶ ☉
13 00	☽ ✶ ♅
13 34	☿ ⊥ ♃
13 37	☉ △ ♅
14 11	♂ Q ♇
16 46	☽ ⚼ ♇
22 54	☽ ✶ ♂
30 Friday	
03 22	☽ ∠ ☿
11 03	☉ ∠ ♃
11 25	☽ Q ♅
13 42	☽ ☌ ♄
16 01	☽ ∥ ♄
17 33	☽ □ ♀
18 43	☽ ☌ ♃
19 14	☽ ∠ ☉
20 37	☽ △ ♇
22 13	☽ ⊥ ☿

DECEMBER 2040

Date	Sidereal time h m s	Sun ☉ ° ' "	Moon ☽ ° ' "	Moon ☽ 24.00 ° ' "	Mercury ☿ ° '	Venus ♀ ° '	Mars ♂ ° '	Jupiter ♃ ° '	Saturn ♄ ° '	Uranus ♅ ° '	Neptune ♆ ° '	Pluto ♇ ° '
01	16 43 51	09 ♐ 54 17	03 ♏ 28 08	10 ♏ 13 40	02 ♐ 44	24 ♑ 10	14 ♐ 20	24 ♎ 02	21 ♎ 16	07 ♌ 55	03 ♉ 32	24 ♒ 57
02	16 47 47	10 55 08	17 ♏ 05 15	24 ♏ 02 43	04 18	25 19	15 04	24 13	21 21	07 R 54	03 R 31	24 58
03	16 51 44	11 56 00	01 ♐ 05 43	08 ♐ 13 50	05 52	26 28	15 48	24 24	21 27	07 53	03 30	24 59
04	16 55 40	12 56 54	15 ♐ 26 29	22 ♐ 42 55	07 26	27 37	16 32	24 34	21 32	07 52	03 28	24 59
05	16 59 37	13 57 48	00 ♑ 06 21	07 ♑ 23 53	09 01	28 46	17 16	24 45	21 38	07 50	03 27	25 00
06	17 03 33	14 58 44	14 ♑ 46 35	22 ♑ 09 31	10 35	29 ♑ 55	18 01	24 55	21 43	07 49	03 26	25 01
07	17 07 30	15 59 41	29 ♑ 31 44	06 ♒ 52 26	12 09	01 ♒ 04	18 45	25 06	21 49	07 48	03 25	25 02
08	17 11 26	17 00 38	14 ♒ 10 48	21 ♒ 26 13	13 43	02 12	19 29	25 16	21 54	07 47	03 24	25 02
09	17 15 23	18 01 36	28 ♒ 38 07	05 ♓ 46 07	15 17	03 21	20 13	25 26	21 59	07 45	03 23	25 03
10	17 19 20	19 02 34	12 ♓ 49 54	19 ♓ 49 20	16 51	04 29	20 58	25 36	22 04	07 44	03 22	25 04
11	17 23 16	20 03 33	26 ♓ 44 19	03 ♈ 34 52	18 25	05 37	21 42	25 46	22 09	07 42	03 21	25 05
12	17 27 13	21 04 32	10 ♈ 21 06	17 ♈ 03 09	20 00	06 45	22 27	25 56	22 14	07 41	03 19	25 06
13	17 31 09	22 05 32	23 ♈ 41 11	00 ♉ 15 24	21 34	07 53	23 11	26 06	22 19	07 39	03 18	25 07
14	17 35 06	23 06 33	06 ♉ 46 00	13 ♉ 13 13	23 08	09 00	23 56	26 16	22 24	07 38	03 18	25 08
15	17 39 02	24 07 34	19 ♉ 37 13	25 ♉ 58 12	24 43	10 07	24 41	26 25	22 28	07 36	03 17	25 09
16	17 42 59	25 08 36	02 ♊ 16 21	08 ♊ 31 48	26 18	11 14	25 25	26 35	22 33	07 34	03 16	25 10
17	17 46 55	26 09 38	14 ♊ 44 41	20 ♊ 55 10	27 52	12 21	26 10	26 44	22 38	07 33	03 15	25 11
18	17 50 52	27 10 41	27 ♊ 03 21	03 ♋ 09 23	29 ♐ 27	13 28	26 55	26 53	22 42	07 31	03 14	25 12
19	17 54 49	28 11 44	09 ♋ 13 23	15 ♋ 15 31	01 ♑ 02	14 35	27 40	27 02	22 46	07 29	03 13	25 13
20	17 58 45	29 ♐ 12 48	21 ♋ 15 56	27 ♋ 14 51	02 38	15 41	28 25	27 11	22 51	07 27	03 12	25 14
21	18 02 42	00 ♑ 13 53	03 ♌ 12 30	09 ♌ 09 08	04 13	16 47	29 10	27 20	22 55	07 25	03 12	25 15
22	18 06 38	01 14 58	15 ♌ 05 04	21 ♌ 00 38	05 49	17 53	29 ♐ 55	27 29	22 59	07 23	03 11	25 16
23	18 10 35	02 16 04	26 ♌ 56 15	02 ♍ 52 19	07 24	18 58	00 ♑ 40	27 38	23 03	07 21	03 10	25 17
24	18 14 31	03 17 11	08 ♍ 49 19	14 ♍ 47 46	09 00	20 03	01 25	27 46	23 07	07 19	03 10	25 18
25	18 18 28	04 18 18	20 ♍ 48 13	26 ♍ 51 14	10 37	21 08	02 10	27 55	23 11	07 17	03 09	25 19
26	18 22 24	05 19 25	02 ♎ 57 24	09 ♎ 07 21	12 13	22 13	02 55	28 03	23 15	07 15	03 08	25 21
27	18 26 21	06 20 34	15 ♎ 21 40	21 ♎ 40 58	13 50	23 17	03 40	28 11	23 19	07 13	03 08	25 22
28	18 30 18	07 21 43	28 ♎ 05 48	04 ♏ 36 41	15 26	24 22	04 26	28 19	23 22	07 11	03 07	25 23
29	18 34 14	08 22 52	11 ♏ 14 04	17 ♏ 58 18	17 03	25 25	05 11	28 27	23 26	07 09	03 07	25 24
30	18 38 11	09 24 02	24 ♏ 49 37	01 ♐ 48 04	18 40	26 29	05 56	28 35	23 29	07 07	03 06	25 26
31	18 42 07	10 ♑ 25 12	08 ♐ 53 36	16 ♐ 05 54	20 ♑ 18	27 ♒ 32	06 ♑ 42	28 ♎ 43	23 ♎ 33	07 ♌ 04	03 ♉ 06	25 ♒ 27

Date	Moon True ☊ ° '	Moon Mean ☊ ° '	Moon ☽ Latitude ° '
01	24 ♉ 38	23 ♉ 39	01 N 54
02	24 D 39	23 36	00 N 42
03	24 R 39	23 33	00 S 36
04	24 38	23 30	01 52
05	24 36	23 26	03 03
06	24 34	23 23	04 02
07	24 31	23 20	04 45
08	24 28	23 17	05 09
09	24 26	23 14	05 13
10	24 25	23 10	04 58
11	24 D 25	23 07	04 26
12	24 26	23 04	03 39
13	24 27	23 01	02 41
14	24 29	22 58	01 36
15	24 30	22 55	00 S 27
16	24 R 30	22 51	00 N 42
17	24 28	22 48	01 48
18	24 24	22 45	02 48
19	24 19	22 42	03 40
20	24 13	22 39	04 21
21	24 06	22 36	04 50
22	23 59	22 32	05 06
23	23 53	22 29	05 09
24	23 48	22 26	04 59
25	23 45	22 23	04 36
26	23 44	22 20	04 00
27	23 D 44	22 16	03 13
28	23 46	22 13	02 14
29	23 47	22 10	01 N 07
30	23 R 48	22 07	00 S 06
31	23 ♉ 46	22 ♉ 04	01 S 21

DECLINATIONS

Date	Sun ☉ ° '	Moon ☽ ° '	Mercury ☿ ° '	Venus ♀ ° '	Mars ♂ ° '	Jupiter ♃ ° '	Saturn ♄ ° '	Uranus ♅ ° '	Neptune ♆ ° '	Pluto ♇ ° '
01	21 S 56	10 S 53	20 S 38	23 S 44	23 S 02	08 S 15	06 S 09	18 N 53	11 N 00	22 S 41
02	22 05	16 16	21 02	23 31	23 07	08 19	06 11	18 53	11 00	22 41
03	22 13	20 57	21 26	23 17	23 12	08 22	06 13	18 54	10 59	22 40
04	22 21	24 30	21 48	23 03	23 18	08 26	06 14	18 54	10 59	22 40
05	22 28	26 29	22 10	22 49	23 22	08 30	06 16	18 54	10 58	22 40
06	22 35	26 37	22 30	22 33	23 27	08 34	06 18	18 55	10 58	22 39
07	22 42	24 53	22 50	22 17	23 31	08 37	06 20	18 55	10 58	22 39
08	22 48	21 30	23 08	22 01	23 36	08 41	06 22	18 56	10 57	22 38
09	22 54	16 50	23 25	21 44	23 40	08 44	06 23	18 56	10 57	22 38
10	22 59	11 20	23 41	21 26	23 43	08 48	06 25	18 56	10 57	22 37
11	23 04	05 S 22	23 56	21 08	23 47	08 51	06 27	18 57	10 56	22 37
12	23 08	00 N 44	24 09	20 49	23 50	08 55	06 28	18 57	10 56	22 36
13	23 12	06 42	24 21	20 30	23 53	08 58	06 30	18 58	10 56	22 36
14	23 15	12 16	24 32	20 11	23 56	09 02	06 32	18 58	10 56	22 35
15	23 18	17 12	24 42	19 51	23 58	09 05	06 33	18 58	10 55	22 35
16	23 21	21 18	24 50	19 30	24 00	09 08	06 35	18 59	10 55	22 34
17	23 23	24 21	24 57	19 09	24 02	09 11	06 36	18 59	10 55	22 34
18	23 24	26 12	25 02	18 48	24 04	09 15	06 37	19 00	10 55	22 33
19	23 25	26 46	25 06	18 26	24 05	09 18	06 39	19 00	10 54	22 33
20	23 26	26 03	25 09	18 04	24 06	09 21	06 40	19 01	10 54	22 32
21	23 26	24 08	25 10	17 41	24 07	09 24	06 42	19 01	10 54	22 32
22	23 26	21 11	25 10	17 18	24 08	09 27	06 43	19 02	10 54	22 31
23	23 25	17 22	25 09	16 55	24 08	09 30	06 44	19 02	10 54	22 31
24	23 24	12 53	25 06	16 31	24 09	09 33	06 45	19 03	10 53	22 30
25	23 22	07 52	25 01	16 07	24 08	09 35	06 47	19 04	10 53	22 30
26	23 20	02 N 30	24 55	15 42	24 08	09 38	06 48	19 04	10 53	22 29
27	23 17	03 S 05	24 48	15 18	24 07	09 41	06 49	19 05	10 53	22 28
28	23 14	08 42	24 39	14 53	24 06	09 44	06 50	19 05	10 53	22 28
29	23 10	14 08	24 28	14 27	24 05	09 46	06 51	19 06	10 53	22 27
30	23 06	19 06	24 16	14 02	24 04	09 49	06 52	19 06	10 53	22 27
31	23 S 02	23 S 06	24 S 02	13 S 36	24 S 02	09 S 51	06 S 53	19 N 07	10 N 52	22 S 26

ZODIAC SIGN ENTRIES

Date	h	m	Planets
01	05	46	☽ ♏
03	10	09	☽ ♐
05	11	56	☽ ♑
06	13	43	♀ ♒
07	12	46	☽ ♒
09	14	17	☽ ♓
11	17	42	☽ ♈
13	23	32	☽ ♉
16	07	40	☽ ♊
18	17	47	☽ ♋
18	20	17	☿ ♑
21	05	32	☽ ♌
21	06	33	☉ ♑
22	14	50	♂ ♑
23	18	12	☽ ♍
26	06	12	☽ ♎
28	15	32	☽ ♏
30	20	55	☽ ♐

LATITUDES

Date	Mercury ☿ ° '	Venus ♀ ° '	Mars ♂ ° '	Jupiter ♃ ° '	Saturn ♄ ° '	Uranus ♅ ° '	Neptune ♆ ° '	Pluto ♇ ° '
01	00 N 05	02 S 30	00 S 31	01 N 10	02 N 19	00 N 37	01 S 48	10 S 06
04	00 S 16	02 28	00 32	01 10	02 19	00 37	01 48	10 05
07	00 35	02 26	00 34	01 10	02 20	00 37	01 48	10 05
10	00 54	02 22	00 36	01 11	02 21	00 38	01 48	10 04
13	01 11	02 17	00 37	01 11	02 21	00 38	01 47	10 04
16	01 27	02 11	00 39	01 12	02 22	00 38	01 47	10 03
19	01 41	02 04	00 40	01 12	02 23	00 38	01 47	10 03
22	01 52	01 55	00 42	01 13	02 23	00 38	01 47	10 02
25	02 01	01 45	00 43	01 13	02 24	00 38	01 47	10 02
28	02 07	01 34	00 45	01 14	02 25	00 38	01 47	10 01
31	02 S 10	01 S 21	00 S 46	01 N 14	02 N 26	00 N 38	01 S 47	10 S 01

DATA

Julian Date	2466490
Delta T	+77 seconds
Ayanamsa	24° 25' 29"
Synetic vernal point	04° ♓ 41' 30"
True obliquity of ecliptic	23° 26' 07"

MOON'S PHASES, APSIDES AND POSITIONS ☽

Date	h	m	Phase	Longitude ° '	Eclipse Indicator
04	07	33	●	12 ♐ 46	
10	23	30	☽	19 ♓ 32	
18	12	16	○	27 ♊ 11	
26	17	02	☾	05 ♎ 32	

Day	h	m		
06	13	15	Perigee	
22	13	18	Apogee	
06	01	47	Max dec	26° S 48'
12	09	04	0N	
19	10	20	Max dec	26° N 46'
26	22	50	0S	

LONGITUDES

Date	Chiron ⚷ ° '	Ceres ⚳ ° '	Pallas ⚴ ° '	Juno ⚵ ° '	Vesta ⚶ ° '	Black Moon Lilith ⚸ ° '
01	22 ♋ 19	15 ♌ 16	19 ♈ 32	17 ♋ 18	20 ♒ 05	08 ♌ 13
11	21 ♋ 50	15 ♌ 47	18 ♈ 55	15 ♋ 58	24 ♒ 04	09 ♌ 20
21	21 ♋ 15	15 ♌ 37	19 ♈ 14	13 ♋ 55	28 ♒ 14	10 ♌ 27
31	20 ♋ 35	14 ♌ 44	20 ♈ 25	11 ♋ 30	02 ♓ 33	11 ♌ 34

All ephemeris data is given at 12.00 UT and the Moon's longitude is additionally given for 24.00 UT

ASPECTARIAN

h m Aspects

01 Saturday
00 45 ☽ ∥ ♃
03 26 ☉ ± ♆
04 08 ☽ ∠ ♂
09 01 ♀ □ ♃
10 30 ☽ ⊻ ☿
12 07 ☽ ☍ ♆
12 32 ☽ # ♆
12 51 ☽ ⊥ ☉
19 54 ☽ □ ♅
21 09 ☽ ⊥ ♂

02 Sunday
00 10 ☿ ⚻ ♆
00 21 ☽ ⊻ ☉
04 38 ♀ ⊻ ♇
04 49 ☽ Q ♀
08 17 ☽ ⊻ ♂
19 26 ☽ ⊻ ♄

03 Monday
00 27 ☽ ⊻ ♃
00 51 ☽ # ♅
01 35 ☽ □ ♇
03 27 ☽ ⚹ ♀
05 46 ☽ ⊥ ♄
10 48 ☽ ⊥ ♃
15 04 ☽ ∥ ☿
16 03 ☽ ⚻ ♆
18 22 ♀ ∥ ♂
19 51 ☽ ∥ ☉
21 02 ☽ ☌ ☿
21 05 ☽ ∠ ♄
21 25 ☿ ∠ ♄
22 30 ☽ ∥ ♇
23 24 ☽ △ ♅

04 Tuesday
01 47 ☽ ∥ ♀
02 05 ☽ ± ♆
02 07 ☽ ∠ ♃
02 29 ☽ ∥ ♂
06 55 ☽ ∠ ♀
07 33 ☽ ☌ ☉
07 56 ☽ Q ♇
12 57 ☉ Q ♇
13 54 ☽ ☌ ♂
17 00 ☽ bQ ♆
18 22 ☿ △ ♅
22 08 ☽ ⚹ ♄
23 04 ☽ ⊥ ♀

05 Wednesday
00 13 ☽ bQ ♅
03 14 ☽ ⚹ ♃
03 45 ☽ ⚹ ♇
09 45 ☽ ⊻ ♀
14 56 ☽ ± ♅
17 34 ☽ △ ♆
17 54 ☽ Q ♄
18 42 ☿ ± ♆
23 05 ☽ Q ♃

06 Thursday
00 42 ☽ ⚻ ♅
00 45 ☿ ∠ ♃
02 35 ♀ ∥ ♇
04 15 ☽ ∠ ♇
04 22 ☽ ⊻ ☿
09 42 ☉ ∥ ♀
12 21 ☽ ⊻ ☉
13 52 ☿ ∥ ♀
15 17 ☽ ⊥ ☿
17 32 ☽ ⊻ ♂
18 53 ☽ ⊥ ♇
21 06 ☉ ∥ ☿
22 22 ☿ ∥ ♇
22 50 ☽ ⊥ ☉
23 14 ☉ ∠ ♀
23 21 ☽ □ ♄

07 Friday
00 39 ☉ ∥ ♇
01 21 ♃ △ ♇
01 28 ♂ bQ ♆
03 48 ☽ ⊥ ♂
04 39 ☽ ⊻ ♇
04 42 ☽ □ ♃
07 39 ☽ ∠ ☿
14 34 ☽ ∠ ☉
14 43 ☽ ☌ ♀
18 20 ☽ □ ♆
19 15 ☽ ∠ ♂
22 45 ☽ ∥ ♂

08 Saturday
01 30 ☽ ☍ ♅
01 34 ☿ Q ♇
02 31 ☽ ∥ ☿
04 09 ☽ ∥ ☉
05 00 ☽ ∥ ♇
08 42 ☽ ∥ ♀
11 09 ☽ ⚹ ☿
17 01 ☽ ⚹ ☉
21 14 ☽ ⚹ ♂
23 55 ☽ Q ♆

09 Sunday
00 50 ☽ △ ♄
01 54 ☽ # ♅
06 01 ☽ ☌ ♇
06 35 ☽ △ ♃
09 28 ☽ Q ☿
12 40 ♀ □ ♆
14 31 ☽ Q ☉
18 21 ☽ Q ♂
19 57 ☽ ⚹ ♆
20 08 ☉ bQ ♆
20 36 ☽ ⊻ ♀

10 Monday
02 08 ☽ bQ ♄
03 20 ☽ ⚻ ♅
07 39 ☽ ⊥ ♀
08 09 ☽ bQ ♃
13 32 ☽ ± ♅
13 35 ☽ # ♆
16 53 ☿ ∥ ♂
17 34 ☽ ± ♄
19 46 ☽ □ ☿
21 28 ☽ ∠ ♆
22 14 ☽ ∥ ♃
23 30 ☽ □ ☉
23 46 ☽ ± ♃

11 Tuesday
00 26 ☽ ∠ ♀
02 46 ☽ □ ♂
03 59 ☽ ⚻ ♄
05 00 ☽ bQ ♅
07 44 ☽ ∥ ♄
09 07 ☽ ⊻ ♇
10 18 ☽ ⚻ ♃
10 46 ☿ bQ ♆
13 03 ☽ ⊥ ♆
19 37 ☽ ⊥ ♇
23 34 ☽ ⊻ ♆

12 Wednesday
04 13 ♂ ⚹ ♄
05 01 ☽ ⚹ ♀
07 16 ☽ △ ♅
11 33 ☽ ∠ ♇
19 16 ♂ bQ ♅

13 Thursday
04 27 ☽ Q ♀
07 24 ♀ ☍ ♅
07 38 ☽ △ ☿
08 52 ☽ △ ☉
09 30 ☽ ☍ ♄
11 03 ☽ △ ♂
11 10 ☽ # ♄
14 36 ☽ ⚹ ♇
16 27 ☽ ☍ ♃
17 42 ☉ ⚹ ♄
21 38 ☽ # ♃

14 Friday
00 02 ☿ ⚹ ♄
00 57 ☉ bQ ♅
04 19 ☿ bQ ♅
05 35 ☽ ☌ ♆
06 01 ☽ ∥ ♆
07 48 ♀ ∠ ♂
10 39 ☉ ☌ ☿
12 40 ☽ Q ♇
13 36 ☽ □ ♅
14 42 ☽ bQ ☉
14 54 ☽ bQ ☿
16 16 ☽ bQ ♂
16 32 ☽ □ ♀

15 Saturday
08 57 ☽ ± ☉
10 03 ☽ ± ☿
10 07 ☽ ± ♂
10 56 ☿ ☌ ♂
17 25 ☽ ⚻ ♄
18 35 ☿ ⚹ ♇
21 15 ☽ ⚻ ☉
21 42 ☽ ∥ ♅
22 09 ☽ ⚻ ♂
22 27 ☽ □ ♇
22 59 ☽ ⚻ ☿
23 16 ☽ Q ♅

16 Sunday
01 01 ☽ ⚻ ♃
01 37 ☽ # ♀
03 19 ♂ ⚹ ♇
04 52 ☽ ± ♄
09 20 ☿ ∠ ♀
12 24 ☉ ⚹ ♇
12 35 ☽ ± ♃
13 53 ☽ ⊻ ♆
16 48 ☿ ⚹ ♃
20 54 ☽ # ♇
22 08 ☽ ⚹ ♅
22 11 ☽ bQ ♄

17 Monday
01 24 ☽ ⊥ ♆
03 10 ☽ # ☉
06 06 ☽ bQ ♃
06 55 ☽ △ ♀
08 55 ☽ # ♂
12 49 ☉ ☌ ♂
18 22 ☽ # ☿
18 47 ☽ ∠ ♆
22 42 ♀ # ♅

18 Tuesday
03 08 ☽ ∠ ♅
03 25 ☽ △ ♄
03 55 ☉ ⚹ ♃
08 21 ☽ △ ♇
10 48 ♂ ⚹ ♃
11 40 ☽ △ ♃
11 43 ☽ ☍ ♂
12 16 ☽ ☍ ☉
15 03 ☽ bQ ♀
17 25 ☽ ☍ ☿
20 44 ☽ ⊥ ♅

19 Wednesday
00 08 ☽ ⚹ ♆
08 34 ☽ ⊻ ♅
10 35 ☽ ± ♀
13 58 ☽ bQ ♇
18 37 ☿ ± ♅
23 43 ☽ ⚻ ♀
23 55 ☽ Q ♆

20 Thursday
07 55 ☽ ± ♇
15 11 ☽ □ ♄
19 58 ☽ ⚻ ♇
20 42 ☿ △ ♆

21 Friday
00 02 ☽ □ ♃
00 45 ☽ # ☿
03 18 ☽ ⚻ ♂
05 27 ☽ ⚻ ☉
11 58 ☽ □ ♆
12 05 ☽ # ♂
14 21 ☽ ⚻ ☿
16 12 ☽ ± ♂
18 26 ☽ # ☉
18 40 ☽ ± ☉
20 29 ☽ ☌ ♅
23 04 ☿ Q ♄

22 Saturday
02 00 ☽ # ♇
03 40 ☽ Q ♄
04 21 ☽ ± ☿
11 38 ☽ bQ ♂
12 49 ☽ Q ♃
14 35 ☽ bQ ☉
15 12 ☉ ± ♅
18 14 ☽ ☍ ♀

23 Sunday
01 24 ☽ bQ ☿
02 05 ☽ ∥ ♅
04 06 ☽ ⚹ ♄
08 39 ☽ ☍ ♇
11 17 ☿ ⚻ ♅
13 25 ☽ ⚹ ♃
14 54 ☽ # ♀
20 03 ☽ △ ♂
23 47 ☽ △ ☉

24 Monday
00 35 ☽ △ ♆
08 59 ☽ ⊻ ♅
09 02 ☉ △ ♆
09 14 ♂ ± ♅
10 35 ☽ ∠ ♄
12 26 ☽ △ ☿
20 02 ☽ ∠ ♃
21 01 ☽ ⊥ ♅
21 49 ☽ ∥ ♆

25 Tuesday
00 36 ☿ Q ♃
04 05 ☽ # ♃
04 45 ☽ ⊥ ♄
06 43 ☽ bQ ♆
07 42 ☿ ∠ ♇
12 12 ♀ Q ♆
12 44 ☽ ⚻ ♀
14 14 ☽ ⊥ ♃
14 57 ☽ ∠ ♅
16 46 ☽ ⊻ ♄
17 00 ☽ # ♄
20 59 ☽ ⚻ ♇

26 Wednesday
00 34 ☽ ± ♆
01 47 ☽ ± ♀
02 15 ☽ ⊻ ♃
08 50 ☽ ± ♇
10 07 ☉ Q ♄
11 55 ☽ □ ♂
12 21 ☽ ⚻ ♆
17 02 ☽ □ ☉
18 53 ♂ △ ♆
20 21 ☽ ⚹ ♅
21 06 ☽ bQ ♀

27 Thursday
02 23 ☽ bQ ♇
08 38 ☽ □ ☿
12 27 ♀ △ ♄
19 19 ☽ Q ♅

28 Friday
00 44 ☽ Q ♂
03 08 ☽ ☌ ♄
03 58 ☽ ∥ ♄
04 24 ☽ △ ♀
06 28 ☽ Q ☉
06 57 ☽ △ ♇
07 57 ☉ ⚻ ♅
12 26 ☽ ☌ ♃
16 29 ☽ ∥ ♃
21 16 ☽ ☍ ♆
21 27 ☽ # ♆
23 14 ☽ Q ☿

29 Saturday
00 23 ☽ ⚹ ♂
04 38 ☽ □ ♅
06 26 ☽ ⚹ ☉
11 36 ♀ ☌ ♇
13 23 ☽ ∥ ♀
20 28 ♂ Q ♄
23 47 ☽ ⚹ ☿

30 Sunday
04 49 ☽ ∠ ♂
09 40 ☽ ⊻ ♄
11 12 ☽ ∠ ☉
12 14 ☽ # ♅
13 03 ☽ □ ♇
15 06 ☽ □ ♀
18 33 ☽ ⊻ ♃
20 04 ☽ ⊥ ♄
21 19 ☽ ⊥ ♂
23 21 ☿ ⊥ ♇

31 Monday
02 13 ☽ ⚻ ♆
03 52 ☽ ⊥ ☉
04 53 ☽ ⊥ ♃
05 09 ☽ ∠ ☿
07 31 ☽ ∥ ♇
08 06 ☽ ⊻ ♂
08 57 ☽ △ ♅
11 25 ☽ ∠ ♄
11 28 ☽ ∥ ☉
12 20 ☽ ± ♆
12 27 ☿ ∥ ♂
12 43 ☉ ∠ ♇
14 45 ☽ ⊻ ☉
18 23 ☽ ∥ ☿
18 47 ☽ ∥ ♂
19 37 ☽ Q ♇
19 54 ☉ Q ♃
20 07 ☽ ∠ ♃
22 09 ☽ ⊥ ☿
23 23 ♂ ⚻ ♅
23 56 ☽ Q ♀

JANUARY 2041

LONGITUDES

Date	Sidereal time h m s	Sun ☉ ° ' "	Moon ☽ ° ' "	Moon ☽ 24.00 ° ' "	Mercury ☿ ° '	Venus ♀ ° '	Mars ♂ ° '	Jupiter ♃ ° '	Saturn ♄ ° '	Uranus ♅ ° '	Neptune ♆ ° '	Pluto ♇ ° '
01	18 46 04	11 ♑ 26 23	23 ♐ 24 28	00 ♑ 48 36	21 ♑ 55	28 ♒ 35	07 ♑ 27	28 ♎ 50	23 ♎ 36	07 ♌ 02	03 ♉ 05	25 ♒ 28
02	18 50 00	12 27 34	08 ♑ 17 23	15 ♑ 49 45	23 32	29 ♒ 38	08 13	28 58	23 39	07 R 00	03 R 05	25 30
03	18 53 57	13 28 45	23 ♑ 24 27	01 ♒ 00 10	25 09	00 ♓ 40	08 58	29 05	23 42	06 58	03 05	25 31
04	18 57 53	14 29 55	08 ♒ 35 34	16 ♒ 09 19	26 47	01 42	09 44	29 12	23 45	06 55	03 04	25 32
05	19 01 50	15 31 06	23 ♒ 40 13	01 ♓ 07 11	28 ♑ 23	02 43	10 30	29 19	23 48	06 53	03 04	25 34
06	19 05 47	16 32 16	08 ♓ 29 17	15 ♓ 45 50	00 ♒ 00	03 44	11 15	29 26	23 51	06 50	03 04	25 35
07	19 09 43	17 33 26	22 ♓ 56 21	00 ♈ 00 31	01 36	04 45	12 01	29 33	23 53	06 48	03 04	25 37
08	19 13 40	18 34 35	06 ♈ 58 16	13 ♈ 49 38	03 12	05 45	12 47	29 39	23 56	06 46	03 03	25 38
09	19 17 36	19 35 44	20 ♈ 34 48	27 ♈ 14 06	04 46	06 45	13 32	29 46	23 58	06 43	03 03	25 39
10	19 21 33	20 36 52	03 ♉ 47 54	10 ♉ 16 37	06 20	07 44	14 18	29 52	24 01	06 41	03 03	25 41
11	19 25 29	21 38 00	16 ♉ 40 42	23 ♉ 00 39	07 52	08 42	15 04	29 ♎ 58	24 03	06 38	03 03	25 42
12	19 29 26	22 39 07	29 ♉ 16 53	05 ♊ 29 52	09 22	09 41	15 50	00 ♏ 04	24 05	06 36	03 03	25 44
13	19 33 22	23 40 14	11 ♊ 40 00	17 ♊ 47 40	10 51	10 38	16 36	00 10	24 07	06 33	03 D 03	25 45
14	19 37 19	24 41 20	23 ♊ 53 12	29 ♊ 56 53	12 17	11 36	17 22	00 16	24 09	06 31	03 03	25 47
15	19 41 16	25 42 26	05 ♋ 58 59	11 ♋ 59 44	13 40	12 32	18 08	00 21	24 11	06 28	03 03	25 48
16	19 45 12	26 43 31	17 ♋ 59 20	23 ♋ 57 57	14 59	13 28	18 54	00 26	24 13	06 26	03 03	25 50
17	19 49 09	27 44 36	29 ♋ 55 44	05 ♌ 52 51	16 14	14 24	19 40	00 32	24 15	06 23	03 04	25 52
18	19 53 05	28 45 40	11 ♌ 49 27	17 ♌ 45 41	17 25	15 18	20 26	00 37	24 16	06 20	03 04	25 53
19	19 57 02	29 ♑ 46 43	23 ♌ 41 45	29 ♌ 37 50	18 30	16 12	21 12	00 41	24 18	06 18	03 04	25 55
20	20 00 58	00 ♒ 47 46	05 ♍ 34 12	11 ♍ 31 07	19 28	17 06	21 59	00 46	24 19	06 15	03 04	25 56
21	20 04 55	01 48 49	17 ♍ 28 54	23 ♍ 27 55	20 19	17 58	22 45	00 51	24 20	06 13	03 05	25 58
22	20 08 51	02 49 51	29 ♍ 28 36	05 ♎ 31 22	21 02	18 50	23 31	00 55	24 21	06 10	03 05	26 00
23	20 12 48	03 50 52	11 ♎ 36 45	17 ♎ 45 17	21 37	19 42	24 17	00 59	24 22	06 07	03 05	26 01
24	20 16 45	04 51 53	23 ♎ 57 32	00 ♏ 14 05	22 01	20 32	25 04	01 03	24 23	06 05	03 06	26 03
25	20 20 41	05 52 54	06 ♏ 35 32	13 ♏ 02 29	22 15	21 22	25 50	01 07	24 24	06 02	03 06	26 04
26	20 24 38	06 53 54	19 ♏ 35 28	26 ♏ 15 01	22 R 18	22 10	26 36	01 11	24 25	05 59	03 06	26 06
27	20 28 34	07 54 53	03 ♐ 01 32	09 ♐ 55 20	22 09	22 58	27 23	01 14	24 25	05 57	03 07	26 08
28	20 32 31	08 55 52	16 ♐ 56 36	24 ♐ 05 17	21 50	23 45	28 09	01 17	24 26	05 54	03 08	26 09
29	20 36 27	09 56 51	01 ♑ 21 10	08 ♑ 43 48	21 19	24 31	28 56	01 21	24 26	05 52	03 08	26 11
30	20 40 24	10 57 48	16 ♑ 12 28	23 ♑ 46 13	20 37	25 16	29 ♑ 42	01 23	24 27	05 49	03 09	26 13
31	20 44 20	11 ♒ 58 45	01 ♒ 23 53	09 ♒ 04 07	19 ♒ 46	26 ♓ 00	00 ♒ 29	01 ♏ 26	24 ♎ 27	05 ♌ 46	03 ♉ 09	26 ♒ 14

Moon

Date	Moon True ☊ ° '	Moon Mean ☊ ° '	Moon ☽ Latitude ° '
01	23 ♉ 43	22 ♉ 01	02 S 33
02	23 R 38	21 57	03 36
03	23 31	21 54	04 25
04	23 22	21 51	04 56
05	23 14	21 48	05 06
06	23 07	21 45	04 55
07	23 02	21 41	04 26
08	23 00	21 38	03 41
09	22 D 59	21 35	02 45
10	23 00	21 32	01 41
11	23 00	21 29	00 S 34
12	23 R 00	21 26	00 N 33
13	22 58	21 22	01 38
14	22 54	21 19	02 37
15	22 47	21 16	03 28
16	22 37	21 13	04 09
17	22 25	21 10	04 39
18	22 11	21 07	04 57
19	21 58	21 03	05 01
20	21 46	21 00	04 53
21	21 35	20 57	04 32
22	21 28	20 54	03 58
23	21 23	20 51	03 14
24	21 21	20 47	02 20
25	21 D 20	20 44	01 17
26	21 20	20 41	00 N 09
27	21 R 20	20 38	01 S 01
28	21 18	20 35	02 11
29	21 13	20 32	03 15
30	21 05	20 28	04 07
31	20 ♉ 55	20 ♉ 25	04 S 43

DECLINATIONS

Date	Sun ☉ ° '	Moon ☽ ° '	Mercury ☿ ° '	Venus ♀ ° '	Mars ♂ ° '	Jupiter ♃ ° '	Saturn ♄ ° '	Uranus ♅ ° '	Neptune ♆ ° '	Pluto ♇ ° '
01	22 S 57	25 S 49	23 S 47	13 S 10	24 S 00	09 S 54	06 S 54	19 N 08	10 N 52	22 S 26
02	22 51	26 46	23 30	12 44	23 58	09 56	06 55	19 08	10 52	22 25
03	22 45	25 46	23 12	12 17	23 55	09 59	06 56	19 09	10 52	22 24
04	22 39	22 52	22 52	11 50	23 53	10 01	06 57	19 09	10 52	22 24
05	22 32	18 26	22 31	11 23	23 50	10 03	06 58	19 10	10 52	22 23
06	22 25	12 57	22 08	10 56	23 46	10 06	06 58	19 11	10 52	22 23
07	22 17	06 52	21 44	10 29	23 43	10 08	06 59	19 11	10 52	22 22
08	22 09	00 S 37	21 18	10 02	23 39	10 10	07 00	19 12	10 52	22 21
09	22 00	05 N 30	20 51	09 34	23 35	10 12	07 00	19 13	10 52	22 21
10	21 51	11 12	20 23	09 07	23 31	10 14	07 01	19 13	10 52	22 20
11	21 42	16 17	19 54	08 39	23 26	10 16	07 02	19 14	10 52	22 20
12	21 32	20 32	19 23	08 12	23 21	10 18	07 02	19 14	10 52	22 19
13	21 22	23 48	18 52	07 44	23 16	10 20	07 03	19 15	10 52	22 18
14	21 11	25 54	18 20	07 16	23 11	10 21	07 03	19 16	10 52	22 18
15	21 00	26 46	17 47	06 48	23 05	10 23	07 04	19 16	10 52	22 17
16	20 48	26 20	17 14	06 20	22 59	10 25	07 04	19 17	10 53	22 16
17	20 37	24 42	16 41	05 52	22 53	10 26	07 04	19 18	10 53	22 16
18	20 24	21 59	16 08	05 25	22 46	10 28	07 05	19 18	10 53	22 15
19	20 12	18 21	15 36	04 57	22 40	10 29	07 05	19 19	10 53	22 15
20	19 59	14 00	15 04	04 29	22 33	10 31	07 05	19 20	10 53	22 14
21	19 45	09 07	14 34	04 01	22 26	10 32	07 06	19 20	10 53	22 14
22	19 31	03 N 51	14 05	03 34	22 18	10 34	07 06	19 21	10 53	22 13
23	19 17	01 S 37	13 38	03 06	22 10	10 35	07 06	19 22	10 54	22 12
24	19 03	07 08	13 14	02 39	22 03	10 36	07 06	19 22	10 54	22 12
25	18 48	12 30	12 53	02 12	21 54	10 37	07 06	19 23	10 54	22 11
26	18 33	17 29	12 35	01 45	21 46	10 38	07 06	19 24	10 54	22 10
27	18 17	21 46	12 21	01 18	21 37	10 39	07 06	19 24	10 54	22 10
28	18 01	24 58	12 11	00 51	21 28	10 40	07 06	19 25	10 55	22 09
29	17 45	26 40	12 05	00 S 24	21 19	10 41	07 06	19 26	10 55	22 09
30	17 29	26 32	12 03	00 N 02	21 10	10 41	07 05	19 26	10 55	22 08
31	17 S 12	24 S 27	12 S 05	00 N 28	21 S 00	10 S 42	07 S 05	19 N 27	10 N 55	22 S 07

ZODIAC SIGN ENTRIES

Date	h m	Planets
01	22 42	☽ ♑
02	20 37	♀ ♓
03	22 25	☽ ♒
05	22 11	☽ ♓
06	11 59	☿ ♒
07	23 59	☽ ♈
10	05 02	☽ ♉
11	19 33	♃ ♏
12	13 23	☽ ♊
15	00 06	☽ ♋
17	12 09	☽ ♌
19	17 13	☉ ♒
20	00 45	☽ ♍
22	13 03	☽ ♎
24	23 33	☽ ♏
27	06 40	☽ ♐
29	09 47	☽ ♑
30	21 08	♂ ♒
31	09 48	☽ ♒

LATITUDES

Date	Mercury ☿ ° '	Venus ♀ ° '	Mars ♂ ° '	Jupiter ♃ ° '	Saturn ♄ ° '	Uranus ♅ ° '	Neptune ♆ ° '	Pluto ♇ ° '
01	02 S 10	01 S 17	00 S 47	01 N 14	02 N 26	00 N 38	01 S 46	10 S 01
04	02 07	01 02	00 48	01 15	02 27	00 38	01 46	10 00
07	01 59	00 46	00 49	01 16	02 27	00 38	01 46	10 00
10	01 45	00 29	00 51	01 16	02 28	00 38	01 46	10 00
13	01 25	00 S 10	00 52	01 17	02 29	00 38	01 46	09 59
16	00 56	00 N 10	00 53	01 18	02 30	00 38	01 46	09 59
19	00 S 20	00 32	00 55	01 18	02 31	00 38	01 45	09 59
22	00 N 25	00 55	00 56	01 19	02 31	00 38	01 45	09 59
25	01 17	01 20	00 57	01 19	02 32	00 39	01 45	09 59
28	02 10	01 47	00 58	01 20	02 33	00 39	01 45	09 59
31	02 N 57	02 N 14	00 S 59	01 N 21	02 N 34	00 N 39	01 S 45	09 S 59

DATA

Julian Date	2466521
Delta T	+78 seconds
Ayanamsa	24° 25' 35"
Synetic vernal point	04° ♓ 41' 24"
True obliquity of ecliptic	23° 26' 07"

MOON'S PHASES, APSIDES AND POSITIONS ☽

Date	h m	Phase	Longitude ° '	Eclipse Indicator
02	19 08	●	12 ♑ 46	
09	10 06	☽	19 ♈ 31	
17	07 11	○	27 ♋ 32	
25	10 33	☾	05 ♏ 49	

Day	h m		
03	19 26	Perigee	
18	22 58	Apogee	
02	11 51	Max dec	26° S 46'
08	14 22	0N	
15	16 01	Max dec	26° N 47'
23	04 57	0S	
29	22 24	Max dec	26° S 51'

LONGITUDES

Date	Chiron ⚷ ° '	Ceres ⚳ ° '	Pallas ⚴ ° '	Juno ⚵ ° '	Vesta ⚶ ° '	Black Moon Lilith ⚸ ° '
01	20 ♋ 31	14 ♌ 36	20 ♈ 35	11 ♋ 16	02 ♓ 59	11 ♌ 41
11	19 ♋ 49	12 ♌ 59	22 ♈ 37	08 ♋ 55	07 ♓ 25	13 ♌ 48
21	19 ♋ 08	10 ♌ 53	25 ♈ 18	07 ♋ 05	11 ♓ 56	13 ♌ 56
31	18 ♋ 28	08 ♌ 33	28 ♈ 33	05 ♋ 59	16 ♓ 31	15 ♌ 03

All ephemeris data is given at 12.00 UT and the Moon's longitude is additionally given for 24.00 UT

ASPECTARIAN

h m	Aspects
01 Tuesday	
03 17	☽ bQ ♆
09 16	☽ ⊻ ☿
09 46	☽ bQ ♅
12 19	☽ ⚹ ♄
15 22	☽ ⚹ ♇
18 39	♀ △ ♃
20 54	☽ ⚹ ♃
21 03	☽ ⚹ ♀
02 Wednesday	
00 20	☽ ± ♅
03 40	☽ △ ♆
07 46	☽ Q ♄
09 56	☽ ⊼ ♅
11 52	☽ ☌ ♂
13 44	☿ □ ♄
15 31	☽ ∠ ♇
15 45	☿ ⊥ ♀
16 18	☽ Q ♃
19 08	☽ ☌ ☉
22 50	☽ ∠ ♀
03 Thursday	
05 50	☽ ⊥ ♇
12 28	☽ □ ♄
14 08	☽ ⊥ ♀
15 06	☽ ☌ ☿
15 20	☽ ⊻ ♇
17 24	☿ ⊻ ♇
21 03	☽ □ ♃
04 Friday	
00 18	☽ ⊻ ♀
03 16	☽ □ ♆
04 59	☽ ∥ ♂
09 22	☽ ☍ ♅
11 59	☽ ∥ ☿
13 27	☽ ∥ ☉
13 54	☽ ⊻ ♂
14 58	☽ ∥ ♇
22 03	☽ ⊻ ☉
23 56	☽ ⊥ ♂
05 Saturday	
07 50	☽ Q ♆
08 18	☽ ⊥ ☉
08 29	☽ # ♅
10 19	☉ ∥ ☿
12 12	☽ △ ♄
14 17	♂ ∠ ♇
15 03	☽ ☌ ♇
15 05	☽ ∠ ♂
20 14	♀ ⚹ ♆
20 31	☽ ⊻ ☿
20 32	☿ ∥ ♇
21 10	☽ △ ♃
23 50	☽ ∠ ☉
06 Sunday	
02 54	☿ □ ♃
03 09	☽ ⚹ ♆
03 40	☽ ☌ ♀
07 26	☽ ⊥ ☿
09 19	☽ ⊼ ♅
12 35	☽ bQ ♄
15 50	♀ # ♆
16 48	☽ ⚹ ♂
18 41	♂ Q ♃
19 08	☽ ± ♅
19 15	☉ ∥ ♇
20 22	☽ # ♆
20 44	☽ ∥ ♀
21 52	☽ bQ ♃
23 22	☽ ∥ ♃
07 Monday	
00 04	☽ ∠ ☿
02 18	☽ ⚹ ☉
03 31	☽ ± ♄
03 49	☽ ∠ ♆
10 05	☽ bQ ♅
11 33	☽ ∥ ♄
13 02	☽ ± ♃
13 36	☽ ⊼ ♄
13 55	☽ Q ♂
16 31	☽ ⊻ ♇
18 58	☽ ⊥ ♆
23 18	☽ ⊼ ♃
08 Tuesday	
00 06	☽ Q ☉
02 46	☽ ⊥ ♇
04 38	☽ ⚹ ☿
05 14	☽ ⊻ ♆
05 27	♀ ∥ ♃
09 43	☽ ⊻ ♀
09 58	☿ □ ♆
11 38	☽ △ ♅
18 24	☽ ∠ ♇
21 00	☽ ⊥ ♀
22 45	☽ □ ♂
09 Wednesday	
04 19	☽ Q ☿
10 06	☽ □ ☉
11 28	♀ ⊼ ♅
13 30	☉ ⊥ ♇
14 15	☽ ∠ ♀
18 07	☽ ☍ ♄
18 11	☽ # ♄
21 09	☽ ⚹ ♇
10 Thursday	
03 39	☽ # ♀
04 44	☽ ☍ ♃
07 45	☽ # ♃
10 33	☽ ∥ ♆
10 38	☽ ☌ ♆
17 18	☽ □ ☿
17 18	☽ □ ♅
17 18	☿ ☍ ♅
19 11	☽ Q ♇
19 52	☽ ⚹ ♀
11 Friday	
08 47	☽ △ ♂
20 15	☽ Q ♀
20 47	♀ bQ ♄
22 12	☽ △ ☉
12 Saturday	
02 01	☽ ⊼ ♄
03 03	☽ Q ♅
04 06	☽ ∥ ♅
05 11	☽ □ ♇
05 46	☽ # ☿
13 31	☽ ⊼ ♃
13 33	☽ ± ♄
15 11	☽ bQ ♂
16 36	♆ St D
18 15	☽ # ☉
18 42	☿ # ♅
19 16	☽ ⊻ ♆
13 Sunday	
00 04	☽ # ♇
01 13	☽ ± ♃
02 05	☽ ⚹ ♅
02 16	☿ ⊻ ♀
05 38	☽ bQ ☉
06 54	☽ ⊥ ♆
07 02	☽ bQ ♄
07 36	☽ # ♂
09 47	☽ ± ♂
09 50	☽ □ ♀
10 11	☽ △ ☿
18 54	☽ bQ ♃
22 18	☽ ⊼ ♂
22 59	☉ □ ♄
14 Monday	
00 31	☽ ∠ ♆
00 50	☽ ± ☉
07 20	☽ ∠ ♅
12 32	☽ △ ♄
13 44	☽ ⊼ ☉
15 45	☽ △ ♇
19 35	☽ bQ ☿
22 52	♀ ∥ ♄
15 Tuesday	
00 43	☽ △ ♃
01 04	☽ ⊥ ♅
06 10	☽ ⚹ ♆
10 20	♀ ± ♅
12 58	☽ ⊻ ♅
14 26	☉ ⊻ ♇
15 46	☽ ± ☿
21 39	☽ bQ ♇
16 Wednesday	
02 11	☽ △ ♀
05 15	☽ ⊼ ☿
06 07	☽ Q ♆
13 57	☽ ☍ ♂
15 43	☽ ± ♇
17 Thursday	
00 32	☽ □ ♄
03 47	☽ ⊼ ♇
07 11	☽ ☍ ☉
10 50	☽ bQ ♀
13 13	☽ □ ♃
18 11	♂ ⊥ ♇
18 19	☽ □ ♆
18 Friday	
00 58	☽ ☌ ♅
05 37	☽ # ♂
06 29	☽ ± ♀
09 55	☽ # ♇
12 54	☽ Q ♄
19 37	☽ ⊼ ♀
20 52	♀ bQ ♃
23 47	☽ # ☉
19 Saturday	
00 27	☽ ☍ ☿
01 48	☽ Q ♃
06 08	☽ ∥ ♅
06 37	☽ ⊼ ♂
13 13	☽ ⚹ ♄
16 30	☽ ☍ ♇
19 35	☽ ± ♂
20 Sunday	
01 27	☽ ⊼ ☉
02 14	☽ ⚹ ♃
05 42	☽ # ☿
06 57	☽ △ ♆
11 16	☉ □ ♃
13 22	☽ ⊻ ♅
14 42	☽ ± ☉
15 02	☽ bQ ♂
19 34	☽ ∠ ♄
21 Monday	
01 26	☽ ⊥ ♅
03 33	☽ ∥ ♆
05 15	☽ # ♃
08 41	☽ ∠ ♃
10 32	☽ bQ ☉
13 04	☽ ☍ ♀
13 12	☽ bQ ♆
13 43	☽ ⊥ ♄
14 48	♀ ∠ ♆
18 06	☽ ⊼ ☿
19 27	☽ ∠ ♅
21 23	☽ # ♄
22 14	♀ ± ♄
23 17	☽ △ ♂
22 Tuesday	
01 46	☽ ⊻ ♄
02 51	☽ ⊥ ♃
05 02	☽ ⊼ ♇
06 52	☽ ± ☿
07 13	☽ ± ♆
13 24	☽ # ♀
13 31	☿ Q ♆
14 53	☽ ⊻ ♃
17 01	☽ ± ♇
17 56	☉ □ ♆
19 10	☽ ⊼ ♆
19 17	☽ △ ☉
23 Wednesday	
01 13	☽ ⚹ ♅
01 42	☽ bQ ☿
04 56	☉ # ♅
05 56	♂ ∥ ♇
10 50	☽ bQ ♇
14 40	♂ □ ♄
17 58	☽ ∥ ♀
24 Thursday	
00 40	☽ Q ♅
04 55	☽ ⊼ ♀
08 09	☽ △ ☿
11 51	☽ ∥ ♄
12 50	☽ ☌ ♄
14 15	☽ □ ♂
16 01	☽ △ ♇
17 18	☽ ± ♀
25 Friday	
01 37	☽ ☌ ♃
02 58	♀ bQ ♅
03 26	☽ ∥ ♃
04 43	☽ # ♆
05 26	☽ ☍ ♆
10 33	☽ □ ☉
10 57	☽ □ ♅
11 32	☽ bQ ♀
13 40	☽ ∥ ☿
15 28	☉ ☍ ♅
19 44	♂ ⊻ ♇
26 Saturday	
02 19	☽ Q ♂
06 14	☿ St R
15 28	☿ ⊻ ♀
16 53	☽ □ ☿
16 59	☽ △ ♀
17 15	☽ ∥ ☉
20 43	☽ ⊻ ♄
22 09	☽ # ♅
22 22	☽ Q ☉
23 45	☽ □ ♇
27 Sunday	
01 24	☽ ⚹ ♂
07 25	☽ ⊥ ♄
08 50	☽ ⊻ ♃
11 08	☽ ∥ ♂
12 10	☽ ⊼ ♆
14 33	☽ ∥ ♇
17 06	☽ △ ♅
17 54	☉ ∠ ♀
19 23	☽ ⊥ ♃
21 12	☽ ⚹ ☉
22 37	☽ ± ♆
23 09	☽ ∠ ♄
28 Monday	
00 09	☽ Q ☿
05 10	☽ ∠ ♂
07 15	☽ Q ♇
10 53	☽ ∠ ♃
14 00	☽ bQ ♆
18 39	☽ bQ ♅
19 58	☽ ⚹ ☿
21 16	☽ ⊥ ♂
29 Tuesday	
00 05	☽ □ ♀
00 35	☽ ⚹ ♄
00 38	☽ ∠ ☉
03 28	☽ ⚹ ♇
07 48	☽ ⊻ ♂
09 25	♀ ⊼ ♄
09 33	☽ ± ♅
11 59	☽ ⚹ ♃
14 55	☽ △ ♆
16 33	☽ ⊥ ☉
18 30	☿ Q ♆
19 20	☽ ⊼ ♅
19 45	☽ ∠ ☿
20 18	☽ Q ♄
30 Wednesday	
02 59	☽ ⊻ ☉
03 59	☽ ∠ ♇
07 04	☽ Q ♀
07 29	☽ Q ♃
09 35	☽ ⊥ ☿
16 11	♀ ± ♃
18 23	☽ ⊥ ♇
18 39	☽ ⊻ ☿
31 Thursday	
01 04	☽ □ ♄
03 06	☽ ⚹ ♀
03 53	☽ ⊻ ♇
08 29	☿ ⊥ ♀
10 29	☽ ☌ ♂
12 04	☽ □ ♃
14 45	☽ □ ♆
18 50	☽ ☍ ♅
20 13	♀ ⊻ ♇

FEBRUARY 2041

LONGITUDES

Date	Sidereal time h m s	Sun ☉ ° ' "	Moon ☽ ° ' "	Moon ☽ 24.00 ° ' "	Mercury ☿ ° '	Venus ♀ ° '	Mars ♂ ° '	Jupiter ♃ ° '	Saturn ♄ ° '	Uranus ♅ ° '	Neptune ♆ ° '	Pluto ♇ ° '
01	20 48 17	12 ♒ 59 41	16 ♒ 45 26	24 ♒ 26 18	18 ♒ 46	26 ♓ 43	01 ♒ 15	01 ♏ 29	24 ♎ 27	05 ♌ 44	03 ♉ 10	26 ♒ 16
02	20 52 14	14 00 35	02 ♓ 05 12	09 ♓ 40 44	17 R 41	27 25	02 02	01 31	24 R 27	05 R 41	03 11	26 18
03	20 56 10	15 01 29	17 ♓ 11 38	24 ♓ 36 51	16 30	28 06	02 49	01 33	24 27	05 39	03 11	26 20
04	21 00 07	16 02 21	01 ♈ 55 35	09 ♈ 07 15	15 18	28 45	03 35	01 35	24 26	05 36	03 12	26 21
05	21 04 03	17 03 11	16 ♈ 11 32	23 ♈ 08 21	14 04	29 ♓ 23	04 22	01 37	24 26	05 33	03 13	26 23
06	21 08 00	18 04 01	29 ♈ 57 47	06 ♉ 40 06	12 53	00 ♈ 00	05 09	01 39	24 26	05 31	03 14	26 25
07	21 11 56	19 04 49	13 ♉ 15 41	19 ♉ 45 00	11 45	00 36	05 55	01 40	24 25	05 28	03 15	26 26
08	21 15 53	20 05 35	26 ♉ 08 37	02 ♊ 27 06	10 42	01 10	06 42	01 42	24 25	05 26	03 16	26 28
09	21 19 49	21 06 20	08 ♊ 41 03	14 ♊ 51 04	09 45	01 42	07 29	01 43	24 24	05 23	03 17	26 30
10	21 23 46	22 07 03	20 ♊ 57 44	27 ♊ 01 35	08 55	02 13	08 16	01 44	24 23	05 21	03 18	26 32
11	21 27 43	23 07 45	03 ♋ 03 09	09 ♋ 02 53	08 13	02 43	09 02	01 45	24 22	05 18	03 19	26 33
12	21 31 39	24 08 25	15 ♋ 01 13	20 ♋ 58 32	07 38	03 10	09 49	01 45	24 21	05 16	03 20	26 35
13	21 35 36	25 09 03	26 ♋ 55 09	02 ♌ 51 21	07 12	03 36	10 36	01 45	24 19	05 14	03 21	26 37
14	21 39 32	26 09 40	08 ♌ 47 24	14 ♌ 43 29	06 54	04 01	11 23	01 46	24 18	05 11	03 22	26 39
15	21 43 29	27 10 16	20 ♌ 39 48	26 ♌ 36 30	06 43	04 23	12 10	01 R 46	24 17	05 09	03 23	26 40
16	21 47 25	28 10 50	02 ♍ 33 44	08 ♍ 31 39	06 D 40	04 43	12 57	01 45	24 15	05 06	03 24	26 42
17	21 51 22	29 ♒ 11 22	14 ♍ 30 24	20 ♍ 30 09	06 44	05 02	13 44	01 45	24 14	05 04	03 25	26 44
18	21 55 18	00 ♓ 11 53	26 ♍ 31 06	02 ♎ 33 27	06 55	05 18	14 30	01 44	24 12	05 02	03 26	26 45
19	21 59 15	01 12 23	08 ♎ 37 28	14 ♎ 43 27	07 11	05 33	15 17	01 44	24 10	04 59	03 28	26 47
20	22 03 12	02 12 51	20 ♎ 51 43	27 ♎ 02 39	07 34	05 45	16 04	01 43	24 08	04 57	03 29	26 49
21	22 07 08	03 13 18	03 ♏ 16 41	09 ♏ 34 15	08 02	05 55	16 51	01 41	24 06	04 55	03 30	26 51
22	22 11 05	04 13 43	15 ♏ 55 51	22 ♏ 21 57	08 34	06 03	17 38	01 40	24 04	04 53	03 31	26 52
23	22 15 01	05 14 07	28 ♏ 53 03	05 ♐ 29 37	09 12	06 08	18 25	01 38	24 02	04 51	03 33	26 54
24	22 18 58	06 14 30	12 ♐ 12 06	19 ♐ 00 51	09 53	06 11	19 12	01 37	24 00	04 48	03 34	26 56
25	22 22 54	07 14 52	25 ♐ 56 09	02 ♑ 58 07	10 39	06 R 12	19 59	01 35	23 57	04 46	03 36	26 58
26	22 26 51	08 15 12	10 ♑ 06 45	17 ♑ 21 49	11 28	06 10	20 46	01 33	23 55	04 44	03 37	26 59
27	22 30 47	09 15 31	24 ♑ 42 56	02 ♒ 09 27	12 20	06 06	21 33	01 30	23 52	04 42	03 39	27 01
28	22 34 44	10 ♓ 15 48	09 ♒ 40 30	17 ♒ 15 02	13 ♒ 16	05 ♈ 59	22 ♒ 21	01 ♏ 28	23 ♎ 50	04 ♌ 40	03 ♉ 40	27 ♒ 03

Date	Moon True ☊ ° '	Moon Mean ☊ ° '	Moon ☽ Latitude ° '
01	20 ♉ 44	20 ♉ 22	04 S 59
02	20 R 32	20 19	04 54
03	20 22	20 16	04 28
04	20 14	20 13	03 45
05	20 09	20 09	02 49
06	20 06	20 06	01 44
07	20 D 06	20 03	00 S 36
08	20 R 06	20 00	00 N 32
09	20 05	19 57	01 36
10	20 02	19 53	02 35
11	19 57	19 50	03 26
12	19 49	19 47	04 07
13	19 38	19 44	04 36
14	19 25	19 41	04 54
15	19 10	19 38	04 59
16	18 55	19 34	04 51
17	18 42	19 31	04 30
18	18 30	19 28	03 57
19	18 21	19 25	03 13
20	18 15	19 22	02 19
21	18 12	19 19	01 18
22	18 11	19 15	00 N 12
23	18 D 12	19 12	00 S 56
24	18 R 12	19 09	02 04
25	18 10	19 06	03 06
26	18 07	19 03	03 59
27	18 00	18 59	04 38
28	17 ♉ 52	18 ♉ 56	04 S 59

DECLINATIONS

Date	Sun ☉ ° '	Moon ☽ ° '	Mercury ☿ ° '	Venus ♀ ° '	Mars ♂ ° '	Jupiter ♃ ° '	Saturn ♄ ° '	Uranus ♅ ° '	Neptune ♆ ° '	Pluto ♇ ° '
01	16 S 55	20 S 34	12 S 11	00 N 54	20 S 50	10 S 43	07 S 05	19 N 28	10 N 56	22 S 07
02	16 37	15 18	12 21	01 19	20 40	10 44	07 05	19 28	10 56	22 06
03	16 20	09 10	12 34	01 44	20 30	10 44	07 04	19 29	10 56	22 06
04	16 02	02 S 40	12 49	02 09	20 20	10 45	07 04	19 29	10 57	22 05
05	15 43	03 N 46	13 06	02 34	20 09	10 45	07 04	19 30	10 57	22 04
06	15 25	09 50	13 25	02 58	19 58	10 46	07 03	19 31	10 57	22 04
07	15 06	15 15	13 45	03 21	19 47	10 46	07 03	19 31	10 58	22 03
08	14 47	19 48	14 05	03 45	19 35	10 46	07 02	19 32	10 58	22 03
09	14 28	23 20	14 26	04 07	19 24	10 46	07 02	19 33	10 58	22 02
10	14 08	25 42	14 45	04 30	19 12	10 46	07 01	19 33	10 59	22 01
11	13 48	26 50	15 04	04 51	19 00	10 46	07 01	19 34	10 59	22 01
12	13 28	26 40	15 22	05 13	18 47	10 46	07 00	19 34	10 59	22 00
13	13 08	25 17	15 39	05 33	18 35	10 46	06 59	19 35	11 00	22 00
14	12 48	22 47	15 55	05 53	18 22	10 46	06 59	19 35	11 00	21 59
15	12 27	19 19	16 09	06 12	18 10	10 46	06 58	19 36	11 01	21 59
16	12 06	15 05	16 21	06 31	17 57	10 46	06 57	19 37	11 01	21 58
17	11 45	10 15	16 33	06 49	17 43	10 45	06 56	19 37	11 02	21 57
18	11 24	05 N 00	16 42	07 06	17 30	10 45	06 55	19 38	11 02	21 57
19	11 03	00 S 28	16 50	07 22	17 16	10 44	06 55	19 38	11 03	21 56
20	10 41	05 59	16 57	07 37	17 03	10 44	06 54	19 39	11 03	21 56
21	10 19	11 23	17 02	07 51	16 49	10 43	06 53	19 39	11 03	21 55
22	09 57	16 25	17 05	08 05	16 35	10 43	06 52	19 40	11 04	21 55
23	09 35	20 49	17 07	08 17	16 20	10 42	06 51	19 40	11 04	21 54
24	09 13	24 18	17 07	08 28	16 06	10 41	06 50	19 41	11 05	21 54
25	08 51	26 28	17 06	08 39	15 51	10 40	06 49	19 41	11 06	21 53
26	08 28	27 01	17 03	08 48	15 36	10 39	06 47	19 42	11 06	21 53
27	08 06	25 44	16 59	08 55	15 21	10 38	06 46	19 42	11 07	21 52
28	07 S 43	22 S 38	16 S 54	09 N 02	15 S 06	10 S 37	06 S 45	19 N 43	11 N 07	21 S 52

ZODIAC SIGN ENTRIES

Date	h m	Planets
02	08 43	☽ ♓
04	08 49	☽ ♈
06	11 59	♀ ♈
06	12 04	☽ ♉
08	19 19	☽ ♊
11	05 55	☽ ♋
13	18 14	☽ ♌
16	06 50	☽ ♍
18	07 17	☉ ♓
18	18 55	☽ ♎
21	05 42	☽ ♏
23	14 02	☽ ♐
25	18 57	☽ ♑
27	20 32	☽ ♒

LATITUDES

Date	Mercury ☿ ° '	Venus ♀ ° '	Mars ♂ ° '	Jupiter ♃ ° '	Saturn ♄ ° '	Uranus ♅ ° '	Neptune ♆ ° '	Pluto ♇ ° '
01	03 N 10	02 N 24	00 S 59	01 N 21	02 N 34	00 N 39	01 S 45	09 S 58
04	03 35	02 53	01 00	01 22	02 35	00 39	01 44	09 58
07	03 39	03 24	01 01	01 22	02 36	00 39	01 44	09 59
10	03 23	03 56	01 02	01 23	02 37	00 39	01 44	09 59
13	02 55	04 29	01 03	01 24	02 37	00 39	01 44	09 59
16	02 19	05 03	01 04	01 25	02 38	00 39	01 44	09 59
19	01 41	05 37	01 04	01 25	02 39	00 38	01 44	09 59
22	01 04	06 11	01 05	01 26	02 40	00 38	01 43	09 59
25	00 N 29	06 44	01 06	01 26	02 40	00 38	01 43	10 00
28	00 S 04	07 16	01 06	01 27	02 41	00 38	01 43	10 00
31	00 S 33	07 N 44	01 S 07	01 N 28	02 N 42	00 N 38	01 S 43	10 S 00

DATA

Julian Date	2466552
Delta T	+78 seconds
Ayanamsa	24° 25' 41"
Synetic vernal point	04° ♓ 41' 18"
True obliquity of ecliptic	23° 26' 08"

MOON'S PHASES, APSIDES AND POSITIONS ☽

Date	h m	Phase	Longitude ° '	Eclipse Indicator
01	05 43	●	12 ♒ 44	
07	23 40	☽	19 ♉ 34	
16	02 21	○	27 ♌ 46	
24	00 29	☾	05 ♐ 46	

Day	h m	
01	07 41	Perigee
14	22 47	Apogee

Day	h m		
04	21 51	ON	
11	21 03	Max dec	26° N 55'
19	09 59	0S	
26	07 25	Max dec	27° S 03'

LONGITUDES

Date	Chiron ⚷ ° '	Ceres ⚳ ° '	Pallas ⚴ ° '	Juno ⚵ ° '	Vesta ⚶ ° '	Black Moon Lilith ⚸ ° '
01	18 ♋ 25	08 ♌ 19	28 ♈ 55	05 ♋ 55	16 ♓ 59	15 ♌ 10
11	17 ♋ 51	06 ♌ 04	02 ♉ 42	05 ♋ 54	21 ♓ 38	16 ♌ 17
21	17 ♋ 23	04 ♌ 12	06 ♉ 54	06 ♋ 23	26 ♓ 18	17 ♌ 24
31	17 ♋ 03	02 ♌ 55	11 ♉ 27	07 ♋ 45	01 ♈ 00	18 ♌ 31

ASPECTARIAN

h m	Aspects
01 Friday	
03 31	☽ ∥ ♇
03 45	☽ ∠ ♀
05 43	☽ ☌ ☉
10 30	☽ ∥ ♂
14 57	☽ ☌ ☿
16 01	♄ St R
17 30	☽ # ♅
18 29	☽ ⊥ ♀
18 53	☽ Q ♆
19 19	♂ □ ♃
02 Saturday	
00 01	☽ △ ♄
02 54	☽ ☌ ♇
03 36	♀ ⊥ ♆
04 19	☽ ⊻ ♀
06 06	☽ ∥ ☉
11 06	☽ △ ♃
11 55	☽ ⊻ ♂
13 43	☽ ⚹ ♆
17 40	☽ ⊼ ♅
21 54	☽ ⊥ ♂
23 29	☽ ∥ ☿
23 38	☽ ⚼ ♄
03 Sunday	
03 09	☽ ± ♅
05 19	☽ # ♆
06 05	☽ ∥ ♃
08 16	☽ ⊻ ☉
10 59	☽ ⚼ ♃
10 59	☽ ⊻ ☿
13 03	☽ ∠ ♂
13 36	☽ ∠ ♆
14 01	☽ ± ♄
17 32	☽ ⚼ ♅
18 38	☽ ⊥ ☉
19 49	☽ ∥ ♄
19 56	☽ ⊥ ☿
23 43	☽ ⊼ ♄
23 56	♂ □ ♆
04 Monday	
01 34	☽ ± ♃
02 49	☽ ⊻ ♇
03 59	☉ ☌ ☿
04 13	☽ ⊥ ♆
06 31	☽ ☌ ♀
09 31	☽ ∠ ☿
10 25	☽ ∠ ☉
11 27	☽ ⊼ ♃
12 43	☽ ⊥ ♇
13 48	☽ # ♀
14 07	☽ ⊻ ♆
14 55	☽ ⚹ ♂
18 05	☽ △ ♅
05 Tuesday	
03 48	☽ ∠ ♇
07 06	☽ ∥ ♀
07 56	☿ ∠ ♀
08 40	☽ ⚹ ☿
12 19	☽ Q ♂
13 35	☽ ⚹ ☉
06 Wednesday	
00 45	☽ # ♄
02 15	☽ ☍ ♄
03 45	☽ Q ☿
05 43	☽ ⚹ ♇
12 04	☽ ⊻ ♀
12 12	☽ Q ☉
15 00	☽ ☍ ♃
15 54	☽ # ♃
16 43	☽ ∥ ♆
17 50	☽ ☌ ♆
21 50	☽ □ ♂
21 53	☽ □ ♅
22 51	♂ ☍ ♅
23 18	☽ ⊥ ♀
07 Thursday	
03 11	☽ Q ♇
04 34	☽ # ☿
09 27	☽ □ ☿
11 21	☽ # ☉
16 29	☽ ∠ ♀
23 40	☽ □ ☉
08 Friday	
06 54	☽ Q ♅
08 44	☽ ⊼ ♄
10 25	☽ ∥ ♅
10 47	☽ # ♂
12 37	☽ □ ♇
18 27	♂ # ♅
20 05	☽ ± ♄
21 58	☽ ⚹ ♀
22 34	☽ ⊼ ♃
09 Saturday	
01 34	☽ ⊻ ♆
02 17	☽ # ♇
05 40	☽ ⚹ ♅
09 31	☽ △ ♂
10 07	☽ ± ♃
12 34	♀ ⊼ ♃
13 09	☽ ⊥ ♆
13 19	☉ ∥ ☿
13 22	☽ ⚼ ♄
13 55	☽ △ ☿
16 06	☉ Q ♆
22 11	☽ Q ♀
10 Sunday	
03 40	☽ ⚼ ♃
06 44	☽ ∠ ♆
10 48	☽ ∠ ♅
14 29	☽ △ ☉
16 51	☽ ⚼ ♂
17 29	☽ ⚼ ☿
18 45	☽ △ ♄
22 20	☿ ☌ ♂
23 02	☽ △ ♇
11 Monday	
03 50	♀ ⊥ ♇
04 33	☽ ⊥ ♅
09 23	☽ △ ♃
10 24	☽ ± ☿
11 17	☽ □ ♀
11 58	☽ ± ♂
12 31	☽ ⚹ ♆
16 29	☽ ⊻ ♅
21 49	☽ ⊼ ☿
23 06	☽ ⚼ ☉
12 Tuesday	
00 49	☽ ⊼ ♂
05 05	☽ ⚼ ♇
12 37	☽ Q ♆
16 45	☉ △ ♄
18 52	☽ ± ☉
20 38	♀ ⊻ ♆
23 14	☽ ± ♇
13 Wednesday	
06 46	☽ □ ♄
08 06	☽ ⊼ ☉
11 23	☽ ⊼ ♇
21 47	☽ □ ♃
14 Thursday	
01 00	☽ □ ♆
02 00	☽ △ ♀
04 44	☽ ☌ ♅
08 15	☽ ☍ ☿
17 37	☽ ☍ ♂
18 08	☽ # ♇
19 05	☽ Q ♄
20 21	♃ St R
23 47	☉ ☌ ♇
15 Friday	
09 20	☽ ⚼ ♀
10 10	☽ Q ♃
10 16	☽ ∥ ♅
19 17	☽ ⚹ ♄
19 22	☽ # ♂
16 Saturday	
00 09	☽ ☍ ♇
02 21	☽ ☍ ☉
04 03	☽ ± ♀
05 28	☽ # ☿
10 23	☽ ⚹ ♃
10 28	☿ St D
13 41	☽ △ ♆
16 28	☽ ⊼ ♀
17 06	☽ ⊻ ♅
20 17	☽ ⊼ ☿
17 Sunday	
01 26	☽ ∠ ♄
04 13	☽ # ☉
05 07	☽ ⊥ ♅
06 44	☉ ⊥ ♀
08 17	☽ ∥ ♆
08 25	☽ ± ☿
09 35	☽ # ♃
10 20	☽ ⊼ ♂
14 30	♀ △ ♅
16 29	☽ ∠ ♃
19 26	☽ ⊥ ♄
19 51	☽ ⚼ ♆
22 08	♀ # ♄
23 06	☽ ∠ ♅
23 11	☽ ± ♂
18 Monday	
02 39	☽ ⚼ ☿
03 05	☽ ∥ ♀
03 22	☽ # ♄
07 24	☽ ⊻ ♄
10 27	☽ ⊥ ♃
12 29	☽ ⊼ ♇
13 50	☽ ± ♆
18 22	☽ ⚼ ♂
19 59	☽ ⊼ ☉
22 22	☽ ⊻ ♃
19 Tuesday	
00 26	☽ ± ♇
01 46	☽ ⊼ ♆
04 51	☽ ⚹ ♅
05 48	☽ ☍ ♀
08 57	☽ ± ☉
09 05	☽ △ ☿
12 07	☉ # ♆
18 15	☽ ⚼ ♇
20 Wednesday	
00 10	☉ △ ♃
02 00	☽ △ ♂
04 14	☽ ⚼ ☉
04 23	☽ Q ♅
08 52	☉ ∥ ♃
15 56	☽ ∥ ♄
18 21	☽ ☌ ♄
19 19	☿ ∥ ♂
19 28	☽ # ♀
23 35	☽ △ ♇
21 Thursday	
07 30	☽ ∥ ☉
08 57	☽ ☌ ♃
09 00	☽ ∥ ♃
10 32	☽ # ♆
11 53	☽ △ ☉
12 26	☽ ☍ ♆
15 08	☽ □ ♅
17 06	☽ ⊼ ♀
18 50	☉ ⚹ ♆
21 27	☽ □ ☿
22 Friday	
04 37	☽ ± ♀
12 47	☽ ∥ ♂
15 24	☽ □ ♂
15 26	☽ ∥ ☿
21 37	☽ ⚼ ♀
23 Saturday	
02 58	☉ ⊼ ♅
03 07	☽ ⊻ ♄
05 17	☽ # ♅
08 22	☽ □ ♇
08 45	☽ Q ☿
14 06	☽ ⊥ ♄
17 01	☽ ⊻ ♃
18 41	☽ ∥ ♇
20 30	☽ ⊼ ♆
22 48	☽ △ ♅
24 Sunday	
00 29	☽ □ ☉
01 13	☽ △ ♀
02 31	☽ Q ♂
03 50	☽ ⊥ ♃
06 18	☽ ∠ ♄
07 18	☽ ± ♆
07 39	☽ ⚹ ☿
10 36	☉ ⊻ ♀
16 50	☽ Q ♇
19 46	☽ ∠ ♃
23 15	☽ ⚼ ♆
25 Monday	
01 05	☽ ⚹ ♂
01 21	☽ ⚼ ♅
06 07	♀ St R
08 36	☽ ⚹ ♄
10 43	☽ Q ☉
11 28	☽ ∠ ☿
13 46	☽ ⚹ ♇
16 51	☽ ± ♅
21 17	☉ # ♀
21 37	☽ ⚹ ♃
26 Tuesday	
01 05	☽ △ ♆
03 01	☽ ⊼ ♅
03 44	☽ ⊥ ☿
04 19	☽ ∠ ♂
05 00	☽ Q ♄
05 25	☽ □ ♀
08 40	☽ ⚹ ☉
14 23	☽ ⊻ ☿
15 08	☽ ∠ ♇
17 41	☽ Q ♃
20 10	☽ ⊥ ♂
23 08	♀ ∠ ♂
27 Wednesday	
03 12	☉ ⚼ ♄
05 58	☽ ⊥ ♇
06 35	☽ ⊻ ♂
10 38	☽ □ ♄
11 00	☽ Q ♀
11 12	☽ ∠ ☉
14 40	♂ Q ♆
15 44	☽ ⊻ ♇
22 56	☽ □ ♃
28 Thursday	
02 24	☽ □ ♆
02 45	☽ ⊥ ☉
04 03	☽ ☍ ♅
06 10	☽ ⚹ ♀
13 00	☽ ⊻ ☉
16 29	☽ ∥ ♇
18 05	☽ ☌ ☿
21 26	☉ ± ♅

All ephemeris data is given at 12.00 UT and the Moon's longitude is additionally given for 24.00 UT

MARCH 2041

LONGITUDES

Date	Sidereal time h m s	Sun ☉ ° ′ ″	Moon ☽ ° ′ ″	Moon ☽ 24.00 ° ′ ″	Mercury ☿ ° ′	Venus ♀ ° ′	Mars ♂ ° ′	Jupiter ♃ ° ′	Saturn ♄ ° ′	Uranus ♅ ° ′	Neptune ♆ ° ′	Pluto ♇ ° ′
01	22 38 41	11 ♓ 16 04	24 ♒ 51 48	02 ♓ 29 27	14 ♒ 14	05 ♈ 50	23 ♒ 08	01 ♏ 25	23 ♎ 47	04 ♌ 38	03 ♉ 42	27 ♒ 04
02	22 42 37	12 16 18	10 ♓ 06 35	17 ♓ 41 47	15 15	05 R 38	23 55	01 R 23	23 R 44	04 R 36	03 43	27 06
03	22 46 34	13 16 30	25 ♓ 13 43	02 ♈ 41 12	16 19	05 24	24 42	01 20	23 41	04 34	03 45	27 08
04	22 50 30	14 16 40	10 ♈ 03 11	17 ♈ 18 54	17 25	05 07	25 29	01 16	23 38	04 33	03 46	27 09
05	22 54 27	15 16 48	24 ♈ 27 44	01 ♉ 29 21	18 33	04 48	26 16	01 13	23 35	04 31	03 48	27 11
06	22 58 23	16 16 54	08 ♉ 23 37	15 ♉ 10 33	19 43	04 26	27 03	01 09	23 32	04 29	03 49	27 13
07	23 02 20	17 16 58	21 ♉ 50 24	28 ♉ 23 28	20 56	04 02	27 50	01 06	23 29	04 27	03 51	27 14
08	23 06 16	18 17 00	04 ♊ 50 15	11 ♊ 11 14	22 10	03 37	28 37	01 02	23 26	04 26	03 53	27 16
09	23 10 13	19 17 00	17 ♊ 27 01	23 ♊ 38 13	23 26	03 09	29 ♒ 24	00 58	23 22	04 24	03 55	27 18
10	23 14 10	20 16 58	29 ♊ 45 27	05 ♋ 49 20	24 43	02 39	00 ♓ 11	00 54	23 19	04 22	03 56	27 19
11	23 18 06	21 16 53	11 ♋ 50 30	17 ♋ 49 30	26 02	02 07	00 59	00 49	23 15	04 21	03 58	27 21
12	23 22 03	22 16 47	23 ♋ 46 54	29 ♋ 43 12	27 23	01 34	01 46	00 45	23 12	04 19	04 00	27 23
13	23 25 59	23 16 38	05 ♌ 38 54	11 ♌ 34 23	28 ♒ 46	00 59	02 33	00 40	23 08	04 18	04 02	27 24
14	23 29 56	24 16 27	17 ♌ 30 05	23 ♌ 26 17	00 ♓ 09	00 ♈ 24	03 20	00 35	23 05	04 16	04 03	27 26
15	23 33 52	25 16 14	29 ♌ 23 19	05 ♍ 21 24	01 35	29 ♓ 47	04 07	00 30	23 01	04 15	04 05	27 27
16	23 37 49	26 15 59	11 ♍ 20 47	17 ♍ 21 38	03 01	29 10	04 54	00 25	22 57	04 14	04 07	27 29
17	23 41 45	27 15 41	23 ♍ 24 06	29 ♍ 28 21	04 29	28 32	05 41	00 20	22 53	04 12	04 09	27 30
18	23 45 42	28 15 22	05 ♎ 34 30	11 ♎ 42 42	05 59	27 54	06 28	00 14	22 49	04 11	04 11	27 32
19	23 49 39	29 ♓ 15 01	17 ♎ 53 03	24 ♎ 05 44	07 29	27 16	07 15	00 09	22 45	04 10	04 13	27 33
20	23 53 35	00 ♈ 14 38	00 ♏ 20 54	06 ♏ 38 44	09 01	26 39	08 02	00 ♏ 03	22 41	04 09	04 15	27 35
21	23 57 32	01 14 13	12 ♏ 59 27	19 ♏ 23 17	10 35	26 02	08 49	29 ♎ 57	22 37	04 08	04 17	27 36
22	00 01 28	02 13 46	25 ♏ 50 29	02 ♐ 21 19	12 10	25 26	09 37	29 51	22 33	04 07	04 19	27 38
23	00 05 25	03 13 17	08 ♐ 56 04	15 ♐ 35 01	13 46	24 51	10 24	29 45	22 29	04 06	04 21	27 39
24	00 09 21	04 12 47	22 ♐ 18 25	29 ♐ 06 30	15 23	24 18	11 11	29 39	22 24	04 05	04 23	27 41
25	00 13 18	05 12 15	05 ♑ 59 26	12 ♑ 57 19	17 02	23 46	11 58	29 32	22 20	04 04	04 25	27 42
26	00 17 14	06 11 42	20 ♑ 00 09	27 ♑ 07 48	18 42	23 16	12 45	29 26	22 16	04 03	04 27	27 44
27	00 21 11	07 11 06	04 ♒ 20 03	11 ♒ 36 30	20 23	22 47	13 32	29 19	22 11	04 02	04 29	27 45
28	00 25 08	08 10 29	18 ♒ 56 35	26 ♒ 19 37	22 06	22 21	14 19	29 13	22 07	04 02	04 31	27 47
29	00 29 04	09 09 50	03 ♓ 44 46	11 ♓ 11 05	23 50	21 57	15 06	29 06	22 03	04 01	04 33	27 48
30	00 33 01	10 09 09	18 ♓ 37 34	26 ♓ 03 07	25 35	21 34	15 53	28 59	21 58	04 00	04 35	27 49
31	00 36 57	11 ♈ 08 26	03 ♈ 26 43	10 ♈ 47 19	27 ♓ 22	21 ♓ 15	16 ♓ 40	28 ♎ 52	21 ♎ 54	04 ♌ 00	04 ♉ 37	27 ♒ 51

MOON NODES AND LATITUDE

Date	Moon True ☊ ° ′	Moon Mean ☊ ° ′	Moon ☽ Latitude ° ′
01	17 ♉ 42	18 ♉ 53	05 S 00
02	17 R 32	18 50	04 39
03	17 22	18 47	03 59
04	17 15	18 44	03 04
05	17 11	18 40	01 58
06	17 08	18 37	00 S 46
07	17 D 08	18 34	00 N 25
08	17 09	18 31	01 33
09	17 09	18 28	02 35
10	17 R 09	18 25	03 27
11	17 07	18 21	04 10
12	17 02	18 18	04 41
13	16 55	18 15	04 59
14	16 46	18 12	05 05
15	16 36	18 09	04 58
16	16 26	18 05	04 37
17	16 17	18 02	04 04
18	16 09	17 59	03 20
19	16 03	17 56	02 25
20	15 59	17 53	01 23
21	15 58	17 50	00 N 16
22	15 D 59	17 46	00 S 53
23	16 00	17 43	02 01
24	16 01	17 40	03 03
25	16 R 02	17 37	03 58
26	16 01	17 34	04 39
27	15 58	17 30	05 04
28	15 54	17 27	05 10
29	15 49	17 24	04 55
30	15 43	17 21	04 21
31	15 ♉ 38	17 ♉ 18	03 S 29

DECLINATIONS

Date	Sun ☉	Moon ☽	Mercury ☿	Venus ♀	Mars ♂	Jupiter ♃	Saturn ♄	Uranus ♅	Neptune ♆	Pluto ♇
01	07 S 20	17 S 56	16 S 47	09 N 07	14 S 51	10 S 36	06 S 44	19 N 43	11 N 08	21 S 51
02	06 57	12 05	16 38	09 11	14 36	10 35	06 43	19 44	11 08	21 51
03	06 34	05 S 33	16 28	09 14	14 20	10 34	06 41	19 44	11 09	21 50
04	06 11	01 N 10	16 17	09 15	14 04	10 32	06 40	19 45	11 09	21 50
05	05 48	07 39	16 04	09 15	13 49	10 31	06 39	19 45	11 10	21 49
06	05 25	13 34	15 50	09 13	13 33	10 30	06 38	19 45	11 10	21 49
07	05 01	18 38	15 35	09 10	13 16	10 28	06 36	19 46	11 11	21 48
08	04 38	22 38	15 18	09 05	13 00	10 27	06 35	19 46	11 12	21 48
09	04 14	25 25	15 00	08 59	12 44	10 25	06 34	19 47	11 12	21 47
10	03 51	26 54	14 40	08 52	12 27	10 24	06 32	19 47	11 13	21 47
11	03 27	27 03	14 19	08 42	12 11	10 22	06 30	19 47	11 14	21 47
12	03 04	25 57	13 57	08 32	11 54	10 20	06 29	19 48	11 14	21 46
13	02 40	23 41	13 34	08 20	11 37	10 18	06 28	19 48	11 15	21 46
14	02 16	20 26	13 09	08 07	11 20	10 16	06 26	19 48	11 16	21 45
15	01 53	16 20	12 43	07 53	11 03	10 15	06 24	19 49	11 16	21 45
16	01 29	11 35	12 16	07 37	10 46	10 13	06 23	19 49	11 17	21 44
17	01 05	06 22	11 47	07 20	10 29	10 11	06 21	19 49	11 18	21 44
18	00 42	00 N 51	11 18	07 03	10 11	10 09	06 20	19 49	11 18	21 44
19	00 S 18	04 S 46	10 47	06 45	09 54	10 07	06 18	19 50	11 19	21 43
20	00 N 06	10 17	10 15	06 26	09 36	10 04	06 17	19 50	11 20	21 43
21	00 30	15 29	09 41	06 06	09 19	10 02	06 15	19 50	11 20	21 43
22	00 53	20 05	09 07	05 46	09 01	10 00	06 13	19 50	11 21	21 42
23	01 17	23 47	08 31	05 25	08 43	09 58	06 12	19 51	11 22	21 42
24	01 40	26 16	07 54	05 05	08 25	09 56	06 10	19 51	11 22	21 42
25	02 04	27 16	07 16	04 44	08 07	09 53	06 08	19 51	11 23	21 41
26	02 28	26 33	06 37	04 24	07 49	09 51	06 07	19 51	11 24	21 41
27	02 51	24 05	05 56	04 03	07 31	09 48	06 05	19 51	11 24	21 41
28	03 15	20 03	05 15	03 43	07 13	09 46	06 03	19 51	11 25	21 40
29	03 38	14 43	04 32	03 23	06 55	09 44	06 01	19 52	11 26	21 40
30	04 01	08 30	03 49	03 04	06 37	09 41	06 00	19 52	11 26	21 40
31	04 N 24	01 S 50	03 S 04	02 N 45	06 S 18	09 S 39	05 S 58	19 N 52	11 N 27	21 S 40

ZODIAC SIGN ENTRIES

Date	h	m	Planets
01	20	05	☽ ♓
03	19	40	☽ ♈
05	21	26	☽ ♉
08	02	59	☽ ♊
10	06	09	♂ ♓
10	12	29	☽ ♋
13	00	34	☽ ♌
14	09	21	☿ ♓
15	03	31	♀ ♓
15	13	14	☽ ♍
18	01	02	☽ ♎
20	06	07	☉ ♈
20	11	20	☽ ♏
21	00	02	♃ ♎
22	19	40	☽ ♐
25	01	34	☽ ♑
27	04	48	☽ ♒
29	05	57	☽ ♓
31	06	24	☽ ♈

LATITUDES

Date	Mercury ☿	Venus ♀	Mars ♂	Jupiter ♃	Saturn ♄	Uranus ♅	Neptune ♆	Pluto ♇
01	00 S 14	07 N 25	01 S 06	01 N 27	02 N 41	00 N 38	01 S 43	10 S 00
04	00 42	07 52	01 07	01 28	02 42	00 38	01 43	10 00
07	01 07	08 15	01 07	01 28	02 42	00 38	01 43	10 01
10	01 28	08 31	01 07	01 29	02 43	00 38	01 43	10 01
13	01 46	08 40	01 08	01 30	02 43	00 38	01 43	10 02
16	02 00	08 40	01 08	01 30	02 43	00 38	01 43	10 02
19	02 11	08 32	01 08	01 30	02 44	00 38	01 42	10 03
22	02 17	08 15	01 08	01 31	02 44	00 38	01 42	10 03
25	02 20	07 51	01 08	01 31	02 45	00 38	01 42	10 04
28	02 18	07 21	01 08	01 31	02 45	00 38	01 42	10 04
31	02 S 12	06 N 46	01 S 08	01 N 32	02 N 46	00 N 38	01 S 42	10 S 05

DATA

Julian Date	2466580
Delta T	+78 seconds
Ayanamsa	24° 25′ 44″
Synetic vernal point	04° ♓ 41′ 15″
True obliquity of ecliptic	23° 26′ 09″

MOON'S PHASES, APSIDES AND POSITIONS ☽

Date	h	m	Phase	Longitude ° ′	Eclipse Indicator
02	15	39	●	12 ♓ 25	
09	15	51	☽	19 ♊ 27	
17	20	19	○	27 ♍ 36	
25	10	32	☾	05 ♑ 09	

Day	h	m		
01	19	47	Perigee	
14	06	58	Apogee	
30	02	24	Perigee	
04	07	50	ON	
11	02	59	Max dec	27° N 09′
18	15	37	OS	
25	14	10	Max dec	27° S 16′
31	18	30	ON	

LONGITUDES

Date	Chiron ⚷ ° ′	Ceres ⚳ ° ′	Pallas ⚴ ° ′	Juno ⚵ ° ′	Vesta ⚶ ° ′	Black Moon Lilith ⚸ ° ′
01	17 ♋ 06	03 ♌ 07	10 ♉ 31	07 ♋ 25	00 ♈ 03	18 ♌ 18
11	16 ♋ 54	02 ♌ 24	15 ♉ 18	09 ♋ 18	04 ♈ 45	19 ♌ 25
21	16 ♋ 50	02 ♌ 24	20 ♉ 22	11 ♋ 42	09 ♈ 26	20 ♌ 32
31	16 ♋ 56	03 ♌ 05	25 ♉ 38	14 ♋ 32	14 ♈ 07	21 ♌ 40

ASPECTARIAN

01 Friday
03 41 ☽ # ♅
05 43 ☽ ∠ ♀
07 00 ☽ Q ♆
09 07 ☽ ☌ ♂
10 18 ☽ △ ♄
15 29 ☽ ☌ ♇
17 13 ☽ ∥ ☿
19 43 ☽ ⊥ ♀
22 17 ☽ △ ♃

02 Saturday
01 43 ☽ ∥ ♂
01 55 ☽ ✶ ♆
03 21 ☽ ⚻ ♅
05 02 ☽ ⚺ ♀
07 01 ♂ △ ♄
09 51 ☽ ⚼ ♄
12 47 ☽ ± ♅
15 35 ☽ # ♆
15 39 ☽ ☌ ☉
17 42 ☽ ∥ ♃
20 44 ☽ ⚺ ☿
21 52 ☽ ⚼ ♃
22 46 ☽ # ♀

03 Sunday
00 02 ☽ ± ♄
01 39 ☽ ∠ ♆
03 00 ☽ ⚼ ♅
04 11 ☉ ∥ ♄
06 59 ☽ ⊥ ☿
07 55 ☽ ∥ ♄
08 08 ☽ ∥ ☉
09 33 ☽ ⚻ ♄
11 06 ☽ ⚺ ♂
12 09 ☽ ± ♃
15 03 ☽ ⚺ ♇
16 02 ☽ ⊥ ♆
21 16 ☽ ⊥ ♂
21 46 ☽ ⚻ ♃
22 33 ☽ ∠ ☿

04 Monday
00 44 ☽ ⊥ ♇
01 44 ☽ ⚺ ♆
03 02 ☽ △ ♅
04 06 ☽ ☌ ♀
12 45 ☽ ∠ ♂
15 28 ☽ ∠ ♇
19 29 ☽ ⚺ ☉

05 Tuesday
01 12 ☽ ✶ ☿
05 23 ☽ # ☉
06 14 ☽ ⊥ ☉
08 11 ☽ # ♄
10 32 ☽ ☍ ♄
15 14 ☽ ✶ ♂
16 38 ☽ ✶ ♇
18 11 ☽ ∥ ♀
22 41 ☽ ∠ ☉
23 13 ☽ # ♃
23 20 ☽ Q ☿
23 29 ☽ ☍ ♃

06 Wednesday
01 55 ☽ ∥ ♆
04 01 ☽ ☌ ♆
05 12 ☽ □ ♅
05 17 ☽ ⚺ ♀
07 36 ☿ ∠ ♀
08 41 ♀ △ ♅
09 10 ☉ ⚼ ♃
11 54 ☽ # ♂
13 13 ☽ Q ♂
13 26 ☽ Q ♇
15 30 ☽ ⊥ ♀
17 07 ♂ ☌ ♇
21 43 ☽ # ☿

07 Thursday
03 06 ☽ ✶ ☉
07 05 ☽ ∠ ♀
10 11 ☽ □ ☿
13 07 ☽ Q ♅
14 59 ☽ ⚻ ♄
16 05 ♀ ⊥ ♂
16 35 ☉ ± ♄
18 11 ☽ ∥ ♅
21 54 ☽ □ ♇
22 02 ♀ ⚺ ♆
23 41 ☽ □ ♂

08 Friday
01 58 ☽ ± ♄
02 47 ☽ Q ☉
04 56 ☽ ⚻ ♃
06 25 ☽ # ♇
06 28 ☿ Q ♆
09 47 ☽ ✶ ♀
10 12 ☽ ⚺ ♆
11 14 ☽ ✶ ♅
16 06 ☽ ± ♃
18 44 ☽ ⚼ ♄
21 32 ☽ ⊥ ♆

09 Saturday
02 46 ☉ ∠ ♆
04 45 ♀ ⊥ ♇
07 43 ☽ Q ♀
09 09 ☽ ⚼ ♃
11 03 ☿ △ ♄
14 44 ☉ ⚼ ♅
14 49 ☽ ∠ ♆
15 46 ☽ ∠ ♅
15 51 ☽ □ ☉
23 26 ☽ △ ♄

10 Sunday
00 57 ☽ △ ☿
07 12 ☽ △ ♇
09 17 ☽ ⊥ ♅
12 55 ☽ △ ♂
14 13 ☽ △ ♃
17 28 ☽ □ ♀
20 17 ☽ ✶ ♆
21 06 ☽ ⚺ ♅

11 Monday
07 37 ♂ △ ♃
10 12 ☽ ⚼ ☿
12 59 ☿ ⊥ ♀
13 01 ☽ ⚼ ♇
20 17 ☽ Q ♆
20 52 ☽ ⚼ ♂

12 Tuesday
06 33 ☽ ± ☿
07 08 ☽ ± ♇
08 29 ♀ ⚺ ♂
08 42 ☽ △ ☉
10 50 ☽ □ ♄
11 48 ☿ ☌ ♇
16 17 ☽ ± ♂
19 17 ☽ ⚻ ♇
20 14 ☽ ⚻ ☿

13 Wednesday
01 59 ☽ □ ♃
03 00 ☽ △ ♀
05 16 ☽ ⚻ ♂
08 43 ☽ □ ♆
08 50 ☉ ⚻ ♄
09 16 ☽ ☌ ♅
17 49 ☽ ⚼ ☉
23 04 ☽ Q ♄

14 Thursday
03 00 ☽ # ♇
03 08 ♀ ⚻ ♃
07 56 ☽ ⚼ ♀
13 42 ☽ ± ☉
14 11 ☽ Q ♃
14 50 ☿ ⚺ ♀
15 56 ☽ ∥ ♅
18 16 ♂ # ♆
18 53 ☿ △ ♃
18 54 ☉ ± ♃
23 13 ☽ ✶ ♄

15 Friday
01 15 ☽ ± ♀
02 56 ☽ ⚻ ☉
08 06 ☽ ☍ ♇
11 07 ♂ ✶ ♆
12 45 ☽ ⚻ ♀
14 13 ☽ ✶ ♃
16 00 ♂ ⚻ ♅
17 00 ☽ ☍ ☿
21 29 ☽ △ ♆
21 46 ☽ ⚺ ♅
22 11 ☽ ☍ ♂

16 Saturday
05 14 ☽ ∠ ♄
08 22 ☽ # ☿
09 46 ☽ ⊥ ♅
13 25 ☽ ∥ ♆
16 06 ☽ # ♂
18 31 ☽ # ♃
20 04 ☽ ∠ ♃
23 07 ☽ ⊥ ♄

17 Sunday
03 32 ☽ ⚼ ♆
03 41 ☽ ∠ ♅
06 24 ☿ ✶ ♆
07 22 ☽ ∥ ♀
07 31 ☿ ⚻ ♅
10 59 ☽ ⚺ ♄
12 01 ☽ # ♄
13 49 ☽ ⊥ ♃
18 05 ☉ ⚺ ♇
20 08 ☽ ⚻ ♇
20 19 ☽ ☍ ☉
21 25 ☽ ± ♆
21 39 ☽ ☍ ♀

18 Monday
01 35 ☽ ⚺ ♃
01 52 ♀ ⊥ ♆
06 46 ☉ ☌ ♀
07 59 ☽ ± ♇
09 16 ☽ ⚻ ♆
09 17 ☽ ✶ ♅
10 09 ☉ ⊥ ♆
11 34 ☿ # ♆
12 42 ☽ # ☉
12 54 ☽ ⚻ ☿
13 53 ☽ ⚻ ♂
14 11 ♅ □ ♆
16 19 ♂ ∥ ♃
18 09 ☽ ∥ ☉

19 Tuesday
01 32 ♀ ⚺ ♇
01 38 ☽ ⚼ ♇
02 16 ☽ ± ☿
02 24 ☽ ± ♂
04 20 ☿ ☌ ♂
08 40 ☽ Q ♅
15 57 ☿ ⚼ ♄
18 32 ☽ ∥ ♄
20 01 ☽ # ♀
21 01 ☽ ⚼ ♂
21 22 ☽ ☌ ♄
22 09 ☽ ⚼ ☿

20 Wednesday
02 00 ♂ ⚼ ♄
05 15 ☽ ⚻ ♀
06 42 ☽ △ ♇
07 42 ☉ ⚻ ♃
09 07 ☽ ∥ ♂
11 02 ☽ ∥ ♃
11 26 ☽ ☌ ♃
11 47 ☽ ⚻ ☉
11 48 ☽ ∥ ☿
16 11 ☽ ± ♀
16 39 ☽ # ♆
19 14 ☽ □ ♅
19 27 ☽ ☍ ♆
19 54 ☿ ∥ ♃

21 Thursday
00 02 ♀ # ♄
00 12 ☽ ± ☉
03 37 ☽ △ ♂
05 09 ☿ ± ♅
06 49 ☽ △ ☿
08 29 ☽ ⚼ ♀
18 37 ☽ ⚼ ☉

22 Friday
05 55 ☽ ⚺ ♄
10 40 ☽ # ♅
11 17 ☽ △ ♀
15 19 ☽ □ ♇
16 58 ☽ ⊥ ♄
19 21 ☽ ⚺ ♃
19 35 ☿ ∥ ♂
21 44 ☽ ∥ ♇

23 Saturday
00 44 ☽ △ ☉
03 05 ♂ ± ♅
03 12 ☽ △ ♅
03 37 ☽ ⚻ ♆
06 15 ☽ ⊥ ♃
09 22 ☽ ∠ ♄
14 34 ☽ ± ♆
14 49 ☽ □ ♂
21 55 ☽ □ ☿
22 25 ☽ ∠ ♃
22 49 ☉ ⊥ ♇

24 Sunday
00 09 ☽ Q ♇
01 48 ☿ ⚼ ♃
06 16 ☽ ⚼ ♅
06 47 ☽ ⚼ ♆
08 50 ☉ △ ♅
12 11 ☽ ✶ ♄
15 24 ☽ □ ♀
16 08 ☉ ⚺ ♆
21 31 ☽ ✶ ♇
22 11 ☽ ± ♅

25 Monday
00 51 ☽ ✶ ♃
00 51 ☽ Q ♂
02 22 ☿ ± ♄
08 40 ☽ ⚻ ♅
09 09 ☽ Q ♄
09 15 ☽ △ ♆
10 06 ☽ Q ☿
10 32 ☽ □ ☉
21 30 ☽ Q ♃
21 37 ☽ Q ♀
22 55 ☽ ✶ ♂
23 36 ☽ ∠ ♇

26 Tuesday
01 24 ♀ ± ♃
09 29 ☽ ✶ ☿
14 56 ☽ ⊥ ♇
15 48 ☽ □ ♄
17 05 ☿ ⚼ ♅
17 19 ☽ ✶ ♀
19 36 ☽ Q ☉
22 57 ☿ ∠ ♆

27 Wednesday
01 02 ☽ ⚺ ♇
01 47 ☽ ∠ ♂
03 44 ☽ □ ♃
06 41 ☿ ∥ ♄
11 31 ☽ ☍ ♅
12 15 ☽ □ ♆
13 58 ☽ ∠ ☿
17 04 ☽ ✶ ☉
17 32 ☽ ∠ ♀
17 35 ☽ ⊥ ♂

28 Thursday
03 18 ☽ ∥ ♇
04 01 ☽ ⚺ ♂
06 44 ☽ ⊥ ☿
07 53 ☽ ⊥ ♀
09 19 ♂ ⚼ ♃
12 20 ☿ ⚻ ♄
12 56 ☽ # ♅
14 51 ☿ ☌ ♀
17 09 ☽ △ ♄
17 23 ☽ ⚺ ♀
17 48 ☽ ⚺ ☿
17 50 ☽ Q ♆
19 23 ☽ ∠ ☉

29 Friday
02 22 ☽ ☌ ♇
02 37 ☿ ± ♃
03 47 ☉ ∥ ♀
04 21 ♀ ⚻ ♄
04 33 ☽ △ ♃
11 00 ☽ ⊥ ☉
12 26 ☽ ⚻ ♅
13 18 ☽ ✶ ♆
17 18 ☽ ⚼ ♄
21 22 ☽ ⚺ ☉
22 06 ☽ ± ♅

30 Saturday
00 58 ☽ # ♆
04 34 ☽ ⚼ ♃
07 20 ☽ ☌ ♂
07 29 ☉ # ☿
07 34 ☽ ∥ ♃
07 44 ☽ ± ♄
12 37 ☽ ⚼ ♅
13 33 ☽ ∠ ♆
14 29 ♂ ± ♄
16 39 ☽ ☌ ♀
17 22 ☽ ⚻ ♄
18 59 ☽ ± ♃
19 13 ☽ ∥ ♂
21 09 ☽ ∥ ♄

31 Sunday
00 46 ☽ ☌ ☿
02 53 ☽ ⚺ ♇
03 19 ☽ # ☉
04 08 ☽ ⊥ ♆
04 38 ☽ ⚻ ♃
07 03 ☽ ∥ ☿
08 34 ☽ # ♀
12 39 ☽ ⊥ ♇
12 54 ☽ △ ♅
13 55 ☽ ⚺ ♆
18 34 ☿ ⚺ ♇

All ephemeris data is given at 12.00 UT and the Moon's longitude is additionally given for 24.00 UT

APRIL 2041

LONGITUDES

Date	Sidereal time h m s	Sun ☉ ° ' "	Moon ☽ ° ' "	Moon ☽ 24.00 ° ' "	Mercury ☿ ° '	Venus ♀ ° '	Mars ♂ ° '	Jupiter ♃ ° '	Saturn ♄ ° '	Uranus ♅ ° '	Neptune ♆ ° '	Pluto ♇ ° '
01	00 40 54	12 ♈ 07 41	18 ♈ 04 01	25 ♈ 16 00	29 ♓ 10	20 ♓ 57	17 ♓ 27	28 ♎ 45	21 ♎ 49	03 ♌ 59	04 ♉ 39	27 ♒ 52
02	00 44 50	13 06 54	02 ♉ 22 37	09 ♉ 23 23	00 ♈ 59	20 R 42	18 14	28 R 38	21 R 45	03 R 59	04 41	27 53
03	00 48 47	14 06 05	16 ♉ 17 56	23 ♉ 06 06	02 50	20 29	19 01	28 31	21 40	03 58	04 43	27 55
04	00 52 43	15 05 14	29 ♉ 47 53	06 ♊ 23 24	04 42	20 19	19 48	28 24	21 35	03 58	04 46	27 56
05	00 56 40	16 04 21	12 ♊ 52 52	19 ♊ 16 38	06 36	20 11	20 34	28 16	21 31	03 58	04 48	27 57
06	01 00 37	17 03 25	25 ♊ 35 08	01 ♋ 48 50	08 31	20 06	21 21	28 09	21 26	03 57	04 50	27 58
07	01 04 33	18 02 27	07 ♋ 58 17	14 ♋ 04 04	10 28	20 03	22 08	28 02	21 22	03 57	04 52	28 00
08	01 08 30	19 01 27	20 ♋ 06 44	26 ♋ 06 54	12 25	20 D 03	22 55	27 54	21 17	03 57	04 54	28 01
09	01 12 26	20 00 25	02 ♌ 05 10	08 ♌ 02 06	14 25	20 05	23 42	27 47	21 12	03 57	04 57	28 02
10	01 16 23	20 59 20	13 ♌ 58 17	19 ♌ 54 14	16 25	20 09	24 29	27 39	21 08	03 57	04 59	28 03
11	01 20 19	21 58 13	25 ♌ 50 27	01 ♍ 47 26	18 27	20 15	25 15	27 31	21 03	03 D 57	05 01	28 04
12	01 24 16	22 57 04	07 ♍ 45 37	13 ♍ 45 22	20 30	20 24	26 02	27 24	20 59	03 57	05 03	28 05
13	01 28 12	23 55 52	19 ♍ 47 03	25 ♍ 50 59	22 34	20 35	26 49	27 16	20 54	03 57	05 05	28 07
14	01 32 09	24 54 38	01 ♎ 57 24	08 ♎ 06 32	24 39	20 48	27 36	27 08	20 49	03 57	05 08	28 08
15	01 36 06	25 53 23	14 ♎ 18 33	20 ♎ 33 37	26 45	21 03	28 22	27 01	20 45	03 57	05 10	28 09
16	01 40 02	26 52 05	26 ♎ 51 48	03 ♏ 13 12	28 ♈ 51	21 20	29 09	26 53	20 40	03 58	05 12	28 10
17	01 43 59	27 50 45	09 ♏ 37 51	16 ♏ 05 47	00 ♉ 58	21 38	29 ♓ 56	26 45	20 36	03 58	05 14	28 11
18	01 47 55	28 49 24	22 ♏ 36 59	29 ♏ 11 29	03 05	21 59	00 ♈ 42	26 38	20 31	03 58	05 17	28 12
19	01 51 52	29 ♈ 48 00	05 ♐ 49 14	12 ♐ 30 15	05 12	22 22	01 29	26 30	20 26	03 59	05 19	28 13
20	01 55 48	00 ♉ 46 35	19 ♐ 14 29	26 ♐ 01 54	07 19	22 46	02 15	26 22	20 22	03 59	05 21	28 14
21	01 59 45	01 45 09	02 ♑ 52 27	09 ♑ 46 05	09 25	23 12	03 02	26 15	20 17	04 00	05 23	28 15
22	02 03 41	02 43 40	16 ♑ 42 42	23 ♑ 42 13	11 31	23 40	03 48	26 07	20 13	04 01	05 26	28 16
23	02 07 38	03 42 10	00 ♒ 44 28	07 ♒ 49 15	13 35	24 09	04 35	26 00	20 09	04 01	05 28	28 16
24	02 11 35	04 40 39	14 ♒ 56 21	22 ♒ 05 26	15 37	24 40	05 21	25 52	20 04	04 02	05 30	28 17
25	02 15 31	05 39 06	29 ♒ 16 10	06 ♓ 28 06	17 38	25 12	06 08	25 44	20 00	04 03	05 32	28 18
26	02 19 28	06 37 31	13 ♓ 40 45	20 ♓ 53 36	19 37	25 45	06 54	25 37	19 55	04 03	05 35	28 19
27	02 23 24	07 35 55	28 ♓ 06 02	05 ♈ 17 29	21 33	26 20	07 40	25 30	19 51	04 04	05 37	28 20
28	02 27 21	08 34 17	12 ♈ 27 17	19 ♈ 34 50	23 27	26 56	08 27	25 22	19 47	04 05	05 39	28 21
29	02 31 17	09 32 37	26 ♈ 39 33	03 ♉ 40 52	25 17	27 33	09 13	25 15	19 43	04 06	05 41	28 21
30	02 35 14	10 ♉ 30 56	10 ♉ 38 18	17 ♉ 31 26	27 ♉ 05	28 ♓ 12	09 ♈ 59	25 ♎ 08	19 ♎ 38	04 ♌ 07	05 ♉ 44	28 ♒ 22

Date	Moon True ☊ ° '	Moon Mean ☊ ° '	Moon ☽ Latitude ° '
01	15 ♉ 35	17 ♉ 15	02 S 24
02	15 R 32	17 11	01 S 11
03	15 D 32	17 08	00 N 04
04	15 32	17 05	01 17
05	15 34	17 02	02 24
06	15 35	16 59	03 22
07	15 37	16 56	04 09
08	15 R 37	16 52	04 43
09	15 36	16 49	05 05
10	15 35	16 46	05 13
11	15 32	16 43	05 09
12	15 28	16 40	04 50
13	15 25	16 36	04 19
14	15 21	16 33	03 36
15	15 19	16 30	02 42
16	15 17	16 27	01 40
17	15 16	16 24	00 N 31
18	15 D 16	16 21	00 S 40
19	15 17	16 17	01 51
20	15 18	16 14	02 57
21	15 19	16 11	03 54
22	15 20	16 08	04 38
23	15 20	16 05	05 06
24	15 R 20	16 02	05 16
25	15 19	15 58	05 07
26	15 19	15 55	04 38
27	15 18	15 52	03 52
28	15 17	15 49	02 52
29	15 16	15 46	01 41
30	15 ♉ 16	15 ♉ 42	00 S 26

DECLINATIONS

Date	Sun ☉ ° '	Moon ☽ ° '	Mercury ☿ ° '	Venus ♀ ° '	Mars ♂ ° '	Jupiter ♃ ° '	Saturn ♄ ° '	Uranus ♅ ° '	Neptune ♆ ° '	Pluto ♇ ° '
01	04 N 48	04 N 52	02 S 18	02 N 27	06 S 00	09 S 36	05 S 56	19 N 52	11 N 28	21 S 39
02	05 11	11 11	01 31	02 10	05 42	09 33	05 55	19 52	11 29	21 39
03	05 34	16 47	00 S 44	01 53	05 23	09 31	05 53	19 52	11 30	21 39
04	05 57	21 22	00 N 05	01 37	05 05	09 28	05 51	19 52	11 30	21 39
05	06 19	24 43	00 55	01 22	04 46	09 26	05 49	19 52	11 31	21 38
06	06 42	26 43	01 45	01 08	04 28	09 23	05 48	19 52	11 32	21 38
07	07 05	27 20	02 36	00 54	04 09	09 20	05 46	19 52	11 32	21 38
08	07 27	26 36	03 28	00 42	03 50	09 18	05 44	19 52	11 33	21 38
09	07 49	24 38	04 21	00 31	03 32	09 15	05 42	19 52	11 34	21 38
10	08 11	21 38	05 14	00 20	03 13	09 12	05 41	19 52	11 35	21 37
11	08 33	17 44	06 08	00 11	02 54	09 09	05 39	19 52	11 35	21 37
12	08 55	13 09	07 02	00 N 02	02 36	09 07	05 37	19 52	11 36	21 37
13	09 17	08 01	07 57	00 S 05	02 17	09 04	05 35	19 52	11 37	21 37
14	09 39	02 N 32	08 52	00 12	01 58	09 01	05 34	19 52	11 38	21 37
15	10 00	03 S 09	09 46	00 18	01 40	08 58	05 32	19 52	11 38	21 37
16	10 21	08 48	10 41	00 22	01 21	08 56	05 30	19 52	11 39	21 36
17	10 42	14 12	11 35	00 26	01 02	08 53	05 29	19 52	11 40	21 36
18	11 03	19 05	12 29	00 29	00 43	08 50	05 27	19 52	11 41	21 36
19	11 24	23 06	13 22	00 31	00 25	08 48	05 25	19 52	11 41	21 36
20	11 45	25 56	14 14	00 32	00 S 06	08 45	05 24	19 51	11 42	21 36
21	12 05	27 18	15 05	00 32	00 N 13	08 42	05 22	19 51	11 43	21 36
22	12 25	26 59	15 55	00 31	00 31	08 39	05 20	19 51	11 44	21 36
23	12 45	24 58	16 43	00 29	00 50	08 37	05 19	19 51	11 44	21 36
24	13 05	21 23	17 29	00 27	01 09	08 34	05 17	19 51	11 45	21 36
25	13 24	16 31	18 14	00 24	01 27	08 31	05 16	19 50	11 46	21 36
26	13 44	10 42	18 56	00 20	01 46	08 29	05 14	19 50	11 47	21 36
27	14 03	04 S 18	19 37	00 15	02 04	08 26	05 13	19 50	11 47	21 36
28	14 21	02 N 17	20 14	00 10	02 23	08 24	05 11	19 50	11 48	21 36
29	14 40	08 42	20 50	00 03	02 41	08 21	05 10	19 50	11 49	21 36
30	14 N 58	14 N 36	21 N 23	00 N 03	03 N 00	08 S 19	05 S 08	19 N 49	11 N 50	21 S 36

ZODIAC SIGN ENTRIES

Date	h	m	Planets
01	23	04	☿ ♈
02	07	58	☽ ♉
04	12	22	☽ ♊
06	20	29	☽ ♋
09	07	48	☽ ♌
11	20	24	☽ ♍
14	08	10	☽ ♎
16	17	56	☽ ♏
17	01	04	☿ ♉
17	14	18	♂ ♈
19	01	28	☽ ♐
19	16	55	☉ ♉
21	06	58	☽ ♑
23	10	44	☽ ♒
25	13	13	☽ ♓
27	15	10	☽ ♈
29	17	42	☽ ♉

LATITUDES

Date	Mercury ☿ ° '	Venus ♀ ° '	Mars ♂ ° '	Jupiter ♃ ° '	Saturn ♄ ° '	Uranus ♅ ° '	Neptune ♆ ° '	Pluto ♇ ° '
01	02 S 09	06 N 34	01 S 08	01 N 32	02 N 46	00 N 38	01 S 42	10 S 05
04	01 57	05 55	01 08	01 32	02 46	00 38	01 42	10 06
07	01 40	05 16	01 07	01 32	02 46	00 38	01 42	10 07
10	01 19	04 37	01 07	01 32	02 46	00 37	01 42	10 08
13	00 54	03 58	01 07	01 32	02 46	00 37	01 42	10 08
16	00 S 25	03 20	01 06	01 32	02 46	00 37	01 42	10 09
19	00 N 07	02 45	01 05	01 32	02 46	00 37	01 42	10 10
22	00 40	02 11	01 05	01 32	02 46	00 37	01 42	10 11
25	01 11	01 39	01 04	01 32	02 46	00 37	01 42	10 12
28	01 40	01 09	01 03	01 31	02 45	00 37	01 42	10 13
31	02 N 03	00 N 42	01 S 02	01 N 31	02 N 45	00 N 37	01 S 42	10 S 14

DATA

Julian Date	2466611
Delta T	+78 seconds
Ayanamsa	24° 25' 47"
Synetic vernal point	04° ♓ 41' 12"
True obliquity of ecliptic	23° 26' 09"

MOON'S PHASES, APSIDES AND POSITIONS ☽

Date	h	m	Phase	Longitude ° '	Eclipse Indicator
01	01	29	●	11 ♈ 42	
08	09	38	☽	18 ♋ 56	
16	12	01	○	26 ♎ 52	
23	17	24	☾	03 ♒ 55	
30	11	46	●	10 ♉ 30	Total

Day	h	m		
10	23	37	Apogee	
26	16	38	Perigee	
07	10	37	Max dec	27° N 20'
14	22	44	0S	
21	19	38	Max dec	27° S 23'
28	03	41	0N	

LONGITUDES

Date	Chiron ⚷ ° '	Ceres ⚳ ° '	Pallas ⚴ ° '	Juno ⚵ ° '	Vesta ⚶ ° '	Black Moon Lilith ⚸ ° '
01	16 ♋ 57	03 ♌ 12	26 ♉ 10	14 ♋ 50	14 ♈ 35	21 ♌ 46
11	17 ♋ 13	04 ♌ 35	01 ♊ 39	18 ♋ 02	19 ♈ 14	22 ♌ 54
21	17 ♋ 38	06 ♌ 30	07 ♊ 17	21 ♋ 29	23 ♈ 50	24 ♌ 01
31	18 ♋ 11	08 ♌ 54	13 ♊ 01	25 ♋ 09	28 ♈ 24	25 ♌ 08

All ephemeris data is given at 12.00 UT and the Moon's longitude is additionally given for 24.00 UT

ASPECTARIAN

h m	Aspects
01 Monday	
01 29	☽ ☌ ☉
03 24	☽ ∠ ♇
03 38	☽ ∥ ♀
03 41	☽ ⋕ ☿
04 26	☿ ⋕ ♀
05 08	☿ ⊥ ♆
06 55	☿ ⚻ ♃
10 55	☽ ⚺ ♂
11 44	☽ ∥ ☉
15 56	☽ ⋕ ♄
15 59	☽ ⋕ ♂
16 42	☽ ⚺ ♀
17 27	♂ ∥ ♄
18 12	☽ ☍ ♄
21 28	☽ ⊥ ♂
02 Tuesday	
02 34	☽ ⊥ ♀
04 24	☽ ⚹ ♇
05 39	☽ ⋕ ♃
05 43	☽ ☍ ♃
06 23	☉ ∠ ♇
09 17	☽ ⚺ ☿
13 12	☽ ∥ ♆
13 32	☽ ∠ ♂
14 43	☽ □ ♅
15 57	☽ ☌ ♆
17 35	☽ ∠ ♀
21 04	☽ ⊥ ☿
03 Wednesday	
00 53	☽ Q ♇
05 57	☉ ⋕ ♂
07 52	☽ ⚺ ☉
10 46	♂ ⚼ ♅
15 07	☽ ∠ ☿
17 03	☽ ⚹ ♂
19 12	☽ ⊥ ☉
19 16	☽ ⚹ ♀
21 24	☽ ⚻ ♄
21 59	☽ Q ♅
04 Thursday	
02 00	☿ ⊥ ♇
02 33	☿ △ ♅
03 30	☽ ∥ ♅
06 39	☉ ⋕ ♄
08 03	☽ ± ♄
08 38	☽ □ ♇
09 29	☽ ⚻ ♃
10 57	♂ ∠ ♆
12 34	☽ ∠ ☉
12 42	☿ ⚺ ♆
13 42	☽ ⋕ ♇
15 50	☽ Q ♂
16 31	☽ Q ♀
19 34	☽ ⚹ ♅
20 16	☽ ± ♃
21 02	☽ ⚺ ♆
22 25	☽ ⚹ ☿
05 Friday	
00 18	☽ ⚼ ♄
01 44	♀ ☌ ♂
08 07	☽ ⊥ ♆
12 43	☽ ⚼ ♃
18 28	☽ ⚹ ☉
22 05	☿ ∥ ♀
23 24	☽ ∠ ♅
06 Saturday	
00 37	☽ Q ☿
01 01	☽ ∠ ♆
01 37	☽ □ ♀
03 24	☽ □ ♂
04 08	☽ △ ♄
14 17	♂ ⚻ ♄
16 33	☽ ⊥ ♅
16 36	☽ △ ♇
16 52	☽ △ ♃
19 15	☽ Q ☉
07 Sunday	
04 09	☽ ⚺ ♅
05 55	☽ ⚹ ♆
09 02	♂ ± ♃
17 09	♃ △ ♇
17 49	☽ □ ☿
21 54	☽ ⚼ ♇
08 Monday	
05 08	♀ St D
05 36	☽ Q ♆
09 38	☽ □ ☉
11 52	☽ △ ♀
14 19	☽ □ ♄
15 48	☽ ± ♇
17 59	☽ △ ♂
19 13	☿ ∠ ♇
19 24	☿ ⋕ ♂
09 Tuesday	
03 25	☽ □ ♃
03 50	☽ ⚻ ♇
13 47	☉ ⚺ ♀
15 45	☽ ☌ ♅
17 46	☽ □ ♆
18 03	☽ ⚼ ♀
10 Wednesday	
02 17	☽ Q ♄
02 17	☽ ⚼ ♂
12 03	☽ ⋕ ♇
12 21	☽ ± ♀
15 12	☉ ☍ ♄
15 21	☽ Q ♃
17 58	☽ △ ☿
19 05	♅ St D
21 45	☽ ± ♂
23 20	☿ ⋕ ♄
23 27	☽ ∥ ♅
11 Thursday	
00 36	☽ ⚻ ♀
02 23	☽ ⚹ ♄
03 28	☽ △ ☉
10 44	☽ ⚻ ♂
15 22	☽ ⚹ ♃
16 31	☽ ☍ ♇
12 Friday	
04 20	☽ ⚺ ♅
06 31	☽ ⚼ ☿
06 33	☽ △ ♆
08 26	☽ ∠ ♄
10 44	☿ ⚺ ♀
12 25	☽ ⚼ ☉
16 23	☽ ⊥ ♅
17 23	☿ ☍ ♄
19 26	☽ ∥ ♆
21 11	☽ ∠ ♃
23 06	☉ ⋕ ♃
13 Saturday	
02 21	☽ ⊥ ♄
04 16	☽ ± ☿
06 38	☽ ∥ ☉
07 13	☽ ⋕ ♃
08 00	☽ ± ☉
10 21	☽ ∠ ♅
12 16	☽ ∥ ☿
12 36	☽ ⚼ ♆
13 36	☽ ☍ ♀
14 12	☽ ⚺ ♄
14 55	☽ ⊥ ♃
18 39	☽ ⚻ ☿
20 56	☽ ⚻ ☉
22 49	☽ ⋕ ♄
14 Sunday	
00 00	♂ ⚻ ♃
02 38	☽ ⚺ ♃
02 51	☽ ☍ ♂
04 29	☽ ⚻ ♇
06 26	☽ ± ♆
14 16	♀ ⚻ ♄
14 31	☽ ⋕ ♂
15 54	☽ ⚹ ♅
15 58	☿ ⋕ ♃
16 15	☽ ± ♇
17 42	☉ ☌ ☿
18 13	☽ ⚻ ♆
21 44	☽ ⋕ ♀
23 47	☽ ∥ ♀
15 Monday	
04 53	♂ ⚺ ♇
06 04	☽ ∥ ♂
09 45	☽ ⚼ ♇
10 09	♀ ± ♃
14 54	☿ ☍ ♃
15 10	☽ Q ♅
15 55	☿ ⊥ ♀
21 50	☉ ∥ ☿
22 01	☽ ∥ ♄
16 Tuesday	
00 17	☽ ☌ ♄
01 13	☽ ⚻ ♀
04 07	☿ ⚹ ♇
12 01	☽ ☍ ☉
12 02	☽ ☌ ♃
12 21	☉ ☍ ♃
12 32	☽ ∥ ♃
12 54	☽ ± ♀
13 43	♂ ⊥ ♆
14 28	☽ △ ♇
16 31	☽ ☍ ☿
16 36	☽ ⚻ ♂
17 22	☿ ⚺ ♂
19 12	☽ ⋕ ☉
21 47	☽ ⋕ ☿
17 Wednesday	
00 29	☽ ⋕ ♆
01 24	☽ □ ♅
03 46	☽ ☍ ♆
04 26	☉ ⊥ ♀
04 38	☽ ± ♂
06 16	☽ ⚼ ♀
14 09	☿ ∥ ♆
20 21	☉ ⚹ ♇
22 28	☽ ⚼ ♂
18 Thursday	
08 10	☽ ⚺ ♄
10 49	☽ △ ♀
16 16	☽ ⋕ ♅
19 05	☽ ⊥ ♄
19 16	☽ ⚺ ♃
22 06	☿ □ ♅
22 12	☽ □ ♇
19 Friday	
00 14	☽ ⚻ ☉
02 21	☽ ∥ ♇
03 40	☽ △ ♂
05 06	♀ ∥ ♂
06 04	☽ ⊥ ♃
08 41	☽ △ ♅
10 40	☽ ⚻ ☿
11 05	☽ ⚻ ♆
11 19	☽ ∠ ♄
11 58	☽ ± ☉
13 17	☿ ☌ ♆
21 54	☽ ± ♆
22 06	☽ ∠ ♃
23 29	☽ ± ☿
20 Saturday	
05 22	☽ ⚼ ☉
06 38	☽ Q ♇
09 10	☉ ∥ ♆
11 33	☽ ⚼ ♅
13 58	☽ ⚼ ♆
13 59	☽ ⚹ ♄
18 26	☿ ∠ ♀
18 27	☽ □ ♀
18 27	☽ ⚼ ☿
21 Sunday	
00 29	☽ ⚹ ♃
03 27	☽ ± ♅
03 53	☽ ⚹ ♇
04 54	☿ ⊥ ♂
09 53	☽ △ ☉
10 59	☽ Q ♄
12 17	☽ □ ♂
13 58	☽ ⚻ ♅
16 24	☽ △ ♆
21 16	☽ Q ♃
21 30	☿ Q ♇
22 Monday	
01 26	☽ △ ☿
02 59	☽ Q ♀
06 02	☽ ∠ ♇
11 39	♀ ⋕ ♂
18 00	☽ □ ♄
18 25	♂ △ ♅
21 16	☽ Q ♂
21 32	☽ ⊥ ♇
23 Tuesday	
00 21	☽ ⚹ ♀
02 24	♂ ⊥ ♇
03 59	☽ □ ♃
07 48	☽ ⚺ ♇
17 24	☽ □ ☉
17 34	☽ ☍ ♅
18 53	☽ ⚹ ♂
19 53	☉ □ ♅
20 02	☽ □ ♆
24 Wednesday	
02 46	☽ ∠ ♀
10 47	☽ ∥ ♇
13 21	☽ □ ☿
16 53	♂ ⚺ ♆
18 29	☽ ⊥ ♀
20 13	☽ ⋕ ♅
20 34	☽ △ ♄
21 36	☽ ∠ ♂
25 Thursday	
01 56	☽ Q ☉
02 24	☽ Q ♆
04 56	☽ ⚺ ♀
05 07	☽ ⋕ ☿
06 09	☽ △ ♃
09 08	☉ ☌ ♆
10 23	☽ ☌ ♇
13 31	☽ ⊥ ♂
19 58	☽ ⚻ ♅
21 30	☽ ⚼ ♄
22 29	☽ ⚹ ♆
23 25	☽ ⚹ ☉
26 Friday	
00 05	☽ ⚺ ♂
00 19	☽ Q ☿
00 32	☽ ⋕ ☉
05 58	☽ ± ♅
06 57	☽ ⚼ ♃
07 16	♀ ⚻ ♃
07 46	☽ ⋕ ♆
12 24	☽ ± ♄
15 39	☿ ⚻ ♄
20 34	☽ ∥ ♃
20 57	☽ ⚼ ♅
21 48	☽ ± ♃
22 20	☽ ⚻ ♄
23 25	☽ ⚹ ☿
23 30	☽ ∠ ♆
27 Saturday	
02 10	☽ ∠ ☉
07 42	☽ ⚻ ♃
08 41	☽ ∥ ♄
08 56	☽ ☌ ♀
12 23	☽ ⚺ ♇
14 32	☽ ⊥ ♆
18 15	☽ ⊥ ☉
18 35	☿ Q ♅
19 48	☽ ⋕ ♂
20 19	☿ ∥ ♅
20 41	☉ ⚺ ♂
21 58	☽ △ ♅
22 24	☽ ⊥ ♇
28 Sunday	
00 34	☽ ⚺ ♆
02 59	☽ ∥ ♀
04 16	☽ ∠ ☿
04 23	☽ ⋕ ♀
04 53	☽ ☌ ♂
05 01	☽ ⚺ ☉
11 59	☿ ∠ ♂
12 22	☽ ∥ ♂
13 30	☽ ∠ ♇
21 39	☽ ⊥ ☿
22 39	☽ ⋕ ♄
29 Monday	
00 17	☽ ☍ ♄
09 19	☽ ⚺ ☿
09 37	☽ ☍ ♃
10 39	☽ ⋕ ♃
11 32	☿ ⚻ ♃
13 36	☽ ⚺ ♀
14 54	☽ ⚹ ♇
17 24	☿ ± ♄
30 Tuesday	
00 19	☽ ∥ ♆
00 21	☽ ⊥ ♀
00 44	☽ □ ♅
03 30	☽ ☌ ♆
08 19	☉ Q ♇
10 48	☽ ⚺ ♂
11 32	☽ Q ♇
11 46	☽ ☌ ☉
13 42	☽ ∥ ☉
16 40	☽ ∠ ♀
18 24	♀ ⚺ ♇
21 43	☿ ⋕ ♇
21 52	☽ ⊥ ♂

MAY 2041

LONGITUDES

Date	Sidereal time h m s	Sun ☉ ° ' "	Moon ☽ ° ' "	Moon ☽ 24.00 ° ' "	Mercury ☿ ° '	Venus ♀ ° '	Mars ♂ ° '	Jupiter ♃ ° '	Saturn ♄ ° '	Uranus ♅ ° '	Neptune ♆ ° '	Pluto ♇ ° '
01	02 39 10	11 ♉ 29 13	24 ♉ 19 54	01 ♊ 03 30	28 ♉ 49	28 ♓ 51	10 ♈ 45	25 ♎ 01	19 ♎ 34	04 ♌ 08	05 ♉ 46	28 ♒ 23
02	02 43 07	12 27 28	07 ♊ 42 05	14 ♊ 15 35	00 ♊ 29	29 ♓ 32	11 31	24 R 53	19 R 30	04 09	05 48	28 23
03	02 47 04	13 25 41	20 ♊ 44 03	27 ♊ 07 38	02 06	00 ♈ 14	12 18	24 46	19 26	04 11	05 50	28 24
04	02 51 00	14 23 53	03 ♋ 26 33	09 ♋ 41 06	03 39	00 56	13 04	24 40	19 22	04 12	05 53	28 25
05	02 54 57	15 22 02	15 ♋ 51 39	21 ♋ 58 37	05 08	01 40	13 50	24 33	19 18	04 13	05 55	28 25
06	02 58 53	16 20 10	28 ♋ 02 29	04 ♌ 03 45	06 33	02 25	14 36	24 26	19 14	04 14	05 57	28 26
07	03 02 50	17 18 15	10 ♌ 02 58	16 ♌ 00 40	07 54	03 10	15 22	24 19	19 11	04 16	05 59	28 27
08	03 06 46	18 16 19	21 ♌ 57 27	27 ♌ 53 54	09 11	03 56	16 08	24 13	19 07	04 17	06 02	28 27
09	03 10 43	19 14 21	03 ♍ 50 35	09 ♍ 48 03	10 24	04 43	16 53	24 06	19 03	04 19	06 04	28 28
10	03 14 39	20 12 21	15 ♍ 46 53	21 ♍ 47 36	11 33	05 31	17 39	24 00	19 00	04 20	06 06	28 28
11	03 18 36	21 10 19	27 ♍ 50 41	03 ♎ 56 36	12 37	06 20	18 25	23 54	18 56	04 22	06 08	28 29
12	03 22 33	22 08 15	10 ♎ 05 46	16 ♎ 18 34	13 36	07 09	19 11	23 48	18 53	04 23	06 10	28 29
13	03 26 29	23 06 09	22 ♎ 35 17	28 ♎ 56 10	14 32	07 59	19 57	23 42	18 49	04 25	06 13	28 29
14	03 30 26	24 04 02	05 ♏ 21 23	11 ♏ 51 04	15 22	08 50	20 42	23 36	18 46	04 27	06 15	28 30
15	03 34 22	25 01 53	18 ♏ 25 13	25 ♏ 03 48	16 08	09 41	21 28	23 30	18 42	04 29	06 17	28 30
16	03 38 19	25 59 43	01 ♐ 46 41	08 ♐ 33 41	16 50	10 34	22 13	23 25	18 39	04 30	06 19	28 30
17	03 42 15	26 57 32	15 ♐ 24 33	22 ♐ 18 57	17 27	11 26	22 59	23 20	18 36	04 32	06 21	28 31
18	03 46 12	27 55 19	29 ♐ 16 31	06 ♑ 16 51	17 58	12 19	23 44	23 14	18 33	04 34	06 23	28 31
19	03 50 08	28 53 05	13 ♑ 19 31	20 ♑ 24 04	18 26	13 13	24 30	23 09	18 30	04 36	06 26	28 31
20	03 54 05	29 ♉ 50 50	27 ♑ 30 04	04 ♒ 37 05	18 48	14 07	25 15	23 04	18 27	04 38	06 28	28 32
21	03 58 02	00 ♊ 48 34	11 ♒ 44 40	18 ♒ 52 27	19 05	15 02	26 01	22 59	18 24	04 40	06 30	28 32
22	04 01 58	01 46 16	26 ♒ 00 04	03 ♓ 07 10	19 18	15 58	26 46	22 55	18 22	04 42	06 32	28 32
23	04 05 55	02 43 58	10 ♓ 13 26	17 ♓ 18 36	19 26	16 54	27 31	22 50	18 19	04 44	06 34	28 32
24	04 09 51	03 41 38	24 ♓ 22 25	01 ♈ 24 37	19 29	17 50	28 16	22 46	18 16	04 47	06 36	28 32
25	04 13 48	04 39 18	08 ♈ 25 00	15 ♈ 23 19	19 R 27	18 47	29 02	22 41	18 14	04 49	06 38	28 32
26	04 17 44	05 36 56	22 ♈ 19 22	29 ♈ 12 56	19 21	19 44	29 ♈ 47	22 37	18 12	04 51	06 40	28 32
27	04 21 41	06 34 34	06 ♉ 03 49	12 ♉ 51 47	19 11	20 41	00 ♉ 32	22 33	18 09	04 53	06 42	28 32
28	04 25 37	07 32 10	19 ♉ 36 40	26 ♉ 18 16	18 56	21 39	01 17	22 30	18 07	04 56	06 44	28 32
29	04 29 34	08 29 46	02 ♊ 56 25	09 ♊ 30 59	18 38	22 38	02 02	22 26	18 05	04 58	06 46	28 R 32
30	04 33 31	09 27 20	16 ♊ 01 50	22 ♊ 28 56	18 16	23 37	02 47	22 23	18 03	05 00	06 48	28 32
31	04 37 27	10 ♊ 24 54	28 ♊ 52 15	05 ♋ 11 49	17 ♊ 51	24 ♈ 36	03 ♉ 32	22 ♎ 19	18 ♎ 01	05 ♌ 03	06 ♉ 50	28 ♒ 32

Date	Moon True ☊ ° '	Moon Mean ☊ ° '	Moon ☽ Latitude ° '
01	15 ♉ 16	15 ♉ 39	00 N 50
02	15 D 16	15 36	02 01
03	15 17	15 33	03 04
04	15 17	15 30	03 56
05	15 R 17	15 27	04 36
06	15 17	15 23	05 03
07	15 17	15 20	05 15
08	15 D 17	15 17	05 15
09	15 17	15 14	05 00
10	15 17	15 11	04 33
11	15 18	15 08	03 54
12	15 18	15 04	03 03
13	15 19	15 01	02 03
14	15 19	14 58	00 N 55
15	15 R 20	14 55	00 S 17
16	15 19	14 52	01 30
17	15 18	14 48	02 39
18	15 17	14 45	03 40
19	15 15	14 42	04 28
20	15 14	14 39	05 01
21	15 12	14 36	05 15
22	15 12	14 33	05 10
23	15 D 12	14 29	04 46
24	15 12	14 26	04 05
25	15 14	14 23	03 09
26	15 15	14 20	02 03
27	15 16	14 17	00 S 50
28	15 R 16	14 14	00 N 24
29	15 15	14 10	01 36
30	15 13	14 07	02 41
31	15 ♉ 10	14 ♉ 04	03 N 37

DECLINATIONS

Date	Sun ☉ ° '	Moon ☽ ° '	Mercury ☿ ° '	Venus ♀ ° '	Mars ♂ ° '	Jupiter ♃ ° '	Saturn ♄ ° '	Uranus ♅ ° '	Neptune ♆ ° '	Pluto ♇ ° '
01	15 N 17	19 N 40	21 N 54	00 N 11	03 N 18	08 S 16	05 S 07	19 N 49	11 N 50	21 S 36
02	15 34	23 35	22 22	00 19	03 36	08 14	05 05	19 49	11 51	21 36
03	15 52	26 10	22 47	00 28	03 55	08 11	05 04	19 48	11 52	21 36
04	16 09	27 20	23 11	00 37	04 13	08 09	05 02	19 48	11 53	21 36
05	16 26	27 04	23 31	00 47	04 31	08 06	05 01	19 48	11 53	21 36
06	16 43	25 29	23 50	00 58	04 49	08 04	05 00	19 47	11 54	21 36
07	17 00	22 47	24 05	01 09	05 07	08 02	04 58	19 47	11 55	21 36
08	17 16	19 09	24 19	01 21	05 25	08 00	04 57	19 47	11 56	21 36
09	17 32	14 45	24 30	01 33	05 43	07 57	04 56	19 46	11 56	21 36
10	17 48	09 48	24 40	01 45	06 01	07 55	04 55	19 46	11 57	21 37
11	18 03	04 N 26	24 47	01 58	06 18	07 53	04 53	19 46	11 58	21 37
12	18 18	01 S 12	24 52	02 12	06 36	07 51	04 52	19 45	11 58	21 37
13	18 33	06 53	24 55	02 26	06 54	07 49	04 51	19 45	11 59	21 37
14	18 47	12 27	24 57	02 40	07 11	07 47	04 50	19 44	12 00	21 37
15	19 01	17 35	24 56	02 55	07 29	07 45	04 49	19 44	12 00	21 37
16	19 15	21 59	24 54	03 10	07 46	07 43	04 48	19 43	12 01	21 38
17	19 28	25 16	24 50	03 26	08 03	07 42	04 47	19 43	12 02	21 38
18	19 42	27 06	24 45	03 42	08 20	07 40	04 46	19 42	12 02	21 38
19	19 54	27 13	24 37	03 58	08 38	07 38	04 45	19 42	12 03	21 38
20	20 07	25 34	24 29	04 14	08 54	07 36	04 44	19 41	12 04	21 38
21	20 19	22 18	24 19	04 31	09 11	07 35	04 43	19 41	12 05	21 39
22	20 31	17 42	24 08	04 48	09 28	07 33	04 42	19 40	12 05	21 39
23	20 42	12 09	23 55	05 05	09 45	07 32	04 41	19 40	12 06	21 39
24	20 53	05 S 59	23 41	05 23	10 01	07 30	04 41	19 39	12 06	21 39
25	21 04	00 N 26	23 26	05 41	10 18	07 29	04 40	19 39	12 07	21 40
26	21 14	06 47	23 10	05 59	10 34	07 28	04 39	19 38	12 08	21 40
27	21 24	12 45	22 53	06 17	10 50	07 27	04 39	19 38	12 08	21 40
28	21 34	18 01	22 35	06 36	11 06	07 25	04 38	19 37	12 09	21 40
29	21 43	22 18	22 17	06 54	11 22	07 24	04 37	19 36	12 10	21 41
30	21 52	25 22	21 58	07 13	11 38	07 23	04 37	19 36	12 10	21 41
31	22 N 00	27 N 03	21 N 38	07 N 32	11 N 54	07 S 22	04 S 36	19 N 35	12 N 11	21 S 41

ZODIAC SIGN ENTRIES

Date	h	m	Planets
01	22	06	☽ ♊
02	04	57	☿ ♊
03	04	08	♀ ♈
04	05	26	☽ ♋
06	15	54	☽ ♌
09	04	15	☽ ♍
11	16	15	☽ ♎
14	02	00	☽ ♏
16	08	50	☽ ♐
18	13	15	☽ ♑
20	15	49	☉ ♊
20	16	13	☽ ♒
22	18	44	☽ ♓
24	21	35	☽ ♈
26	19	05	♂ ♉
27	01	22	☽ ♉
29	06	40	☽ ♊
31	14	08	☽ ♋

LATITUDES

Date	Mercury ☿ ° '	Venus ♀ ° '	Mars ♂ ° '	Jupiter ♃ ° '	Saturn ♄ ° '	Uranus ♅ ° '	Neptune ♆ ° '	Pluto ♇ ° '
01	02 N 03	00 N 42	01 S 02	01 N 31	02 N 45	00 N 37	01 S 42	10 S 14
04	02 20	00 N 16	01 02	01 31	02 45	00 37	01 42	10 14
07	02 30	00 S 07	01 01	01 30	02 44	00 37	01 42	10 15
10	02 31	00 29	00 59	01 30	02 44	00 37	01 42	10 16
13	02 24	00 48	00 58	01 29	02 43	00 37	01 42	10 17
16	02 08	01 06	00 57	01 29	02 43	00 36	01 42	10 18
19	01 42	01 22	00 56	01 28	02 42	00 36	01 42	10 19
22	01 08	01 36	00 55	01 27	02 42	00 36	01 42	10 20
25	00 N 25	01 49	00 53	01 27	02 41	00 36	01 42	10 21
28	00 S 23	02 00	00 52	01 26	02 40	00 36	01 42	10 22
31	01 S 15	02 S 09	00 S 50	01 N 25	02 N 40	00 N 36	01 S 42	10 S 23

DATA

Julian Date	2466641
Delta T	+78 seconds
Ayanamsa	24° 25' 51"
Synetic vernal point	04° ♓ 41' 08"
True obliquity of ecliptic	23° 26' 09"

MOON'S PHASES, APSIDES AND POSITIONS ☽

Date	h	m	Phase	Longitude ° '	Eclipse Indicator
08	03	54	☽	17 ♌ 57	
16	00	52	○	25 ♏ 33	partial
22	22	26	☾	02 ♓ 11	
29	22	56	●	08 ♊ 56	

Day	h	m		
08	18	42	Apogee	
22	01	23	Perigee	
04	19	21	Max dec	27° N 24'
12	06	58	0S	
19	01	38	Max dec	27° S 23'
25	10	22	0N	

LONGITUDES

Date	Chiron ⚷ ° '	Ceres ⚳ ° '	Pallas ⚴ ° '	Juno ⚵ ° '	Vesta ⚶ ° '	Black Moon Lilith ⚸ ° '
01	18 ♋ 11	08 ♌ 54	13 ♊ 01	25 ♋ 09	28 ♈ 24	25 ♌ 08
11	18 ♋ 52	11 ♌ 40	18 ♊ 51	28 ♋ 58	02 ♉ 55	26 ♌ 15
21	19 ♋ 39	14 ♌ 47	24 ♊ 45	02 ♌ 55	07 ♉ 22	27 ♌ 22
31	20 ♋ 32	18 ♌ 10	00 ♋ 41	06 ♌ 56	11 ♉ 44	28 ♌ 30

All ephemeris data is given at 12.00 UT and the Moon's longitude is additionally given for 24.00 UT

ASPECTARIAN

01 Wednesday
03 38 ☽ ⊼ ♄
05 56 ☿ □ ♇
08 07 ☽ Q ♅
12 50 ☽ ∥ ♅
13 05 ☿ ⚹ ♀
13 11 ☽ ⊼ ♃
14 11 ☽ ± ♄
14 41 ☽ ∠ ♂
19 13 ☽ □ ♇
20 29 ☽ ⚹ ♀
21 07 ☽ ☌ ☿
22 55 ☽ # ♇
23 48 ☽ ± ♃

02 Thursday
02 32 ☽ ∥ ☿
05 34 ☽ ⚹ ♅
06 14 ☽ bQ ♄
08 33 ☽ ⊻ ♆
15 57 ☽ bQ ♃
17 33 ☿ ± ♃
19 23 ☽ Q ♀
19 25 ☽ ⚹ ♂
19 31 ☽ ⊥ ♆
21 23 ☽ ⊻ ☉
21 53 ♀ ⊥ ♆

03 Friday
09 06 ☽ ∠ ♅
09 22 ☽ ⊥ ☉
09 36 ☽ △ ♄
12 12 ☽ ∠ ♆
19 05 ☽ Q ♂
19 30 ☽ △ ♃

04 Saturday
02 00 ☽ ⊥ ♅
02 25 ☽ △ ♇
03 40 ☽ ∠ ☉
06 57 ☽ □ ♀
12 27 ☽ ⊻ ☿
13 27 ☽ ⊻ ♅
16 41 ☽ ⚹ ♆
20 50 ☿ ⚹ ♅
23 00 ☿ bQ ♄
23 08 ♂ ∠ ♇

05 Sunday
01 34 ☽ ⊥ ☿
07 15 ☽ bQ ♇
07 46 ☽ □ ♂
10 57 ☽ ⚹ ☉
16 02 ☽ Q ♆
18 42 ☽ □ ♄
21 30 ☽ ∠ ☿

06 Monday
00 53 ☽ ± ♇
01 23 ☿ ⊻ ♆
04 55 ☽ □ ♃
12 38 ☽ Q ☉
12 47 ☽ ⊼ ♇
21 17 ☽ △ ♀

07 Tuesday
00 23 ☽ ☌ ♅
01 16 ♂ # ♄
02 24 ☽ ∥ ☿
03 50 ☽ □ ♆
06 16 ☽ Q ♄
07 10 ☽ ⚹ ☿
16 32 ☽ Q ♃
20 24 ☽ # ♇
23 25 ☽ △ ♂

08 Wednesday
03 54 ☽ □ ☉
05 29 ☽ bQ ♀
06 17 ☽ ⚹ ♄
08 10 ☽ ∥ ♅
10 16 ☽ Q ☿
12 26 ☿ bQ ♃
16 31 ☽ ⚹ ♃
22 06 ☽ ∥ ☉
23 03 ♀ △ ♅

09 Thursday
00 56 ☽ ± ♀
01 08 ☽ ☍ ♇
03 52 ♀ ⊥ ♇
07 39 ☉ ⊼ ♄
07 48 ☽ bQ ♂
12 25 ☽ ∠ ♄
12 57 ☽ ⊻ ♅
13 54 ☽ ⊼ ♀
16 30 ☽ △ ♆
22 31 ☽ ∠ ♃

10 Friday
01 03 ☽ ⊥ ♅
01 54 ☽ ∥ ♆
02 37 ☽ □ ☿
03 10 ☽ ± ♂
06 26 ☽ ⊥ ♄
16 00 ☽ ⊼ ♂
16 24 ☽ ⊥ ♃
18 23 ☽ ⊻ ♄
19 07 ☽ ∠ ♅
20 38 ☽ # ♃
21 37 ☽ △ ☉
22 39 ☽ bQ ♆

11 Saturday
00 46 ☿ ⊥ ♆
04 12 ☽ ∥ ♂
04 15 ☽ ⊻ ♃
06 00 ♀ ⊻ ♆
09 59 ☽ # ♄
13 15 ☽ ⊼ ♇
16 32 ☽ ± ♆
22 10 ☽ ∥ ♀

12 Sunday
00 51 ☽ ⚹ ♅
01 03 ☽ ± ♇
03 03 ♂ ☍ ♄
04 20 ☽ ⊼ ♆
05 45 ☽ bQ ☉
05 52 ☽ ☍ ♀
15 06 ☉ ∠ ♀
16 25 ☽ # ♀
18 30 ☉ Q ♅
18 34 ☽ bQ ♇
19 21 ☽ △ ☿

13 Monday
00 11 ☽ Q ♅
00 39 ☽ ± ☉
03 25 ☽ ∥ ♄
04 51 ☽ ☌ ♄
06 38 ☽ ☍ ♂
12 02 ☽ # ♂
13 04 ☽ ⊼ ☉
14 06 ☽ ☌ ♃
15 55 ☽ ∥ ♃
23 10 ☽ △ ♇

14 Tuesday
01 29 ☉ ⊼ ♃
02 03 ☽ bQ ☿
10 01 ☽ # ♆
10 18 ☽ □ ♅
13 40 ☽ ☍ ♆
18 54 ☽ ⊼ ♀
19 55 ☽ ± ☿

15 Wednesday
04 22 ☉ ± ♄
06 41 ☽ ± ♀
07 36 ☽ ⊼ ☿
12 31 ☽ ⊻ ♄
17 51 ☽ ⊼ ♂
19 43 ☽ # ☉
21 08 ☽ ⊻ ♃
23 05 ☽ # ♅
23 19 ☽ ⊥ ♄

16 Thursday
00 07 ☽ bQ ♀
00 52 ☽ ☍ ☉
05 17 ☽ ± ♂
06 10 ☽ □ ♇
07 49 ☽ ⊥ ♃
08 40 ♂ # ♃
09 51 ☽ ∥ ♇
15 19 ☽ ∠ ♄
16 51 ☽ △ ♅
20 04 ☽ ⊼ ♆
22 13 ☽ bQ ♂
23 40 ☽ ∠ ♃

17 Friday
04 34 ☽ △ ♀
06 39 ☽ ± ♆
08 17 ☽ # ☿
13 55 ☽ Q ♇
15 42 ☽ ☍ ☿
17 32 ☽ ⚹ ♄
19 12 ☽ bQ ♅
21 42 ♂ ☍ ♃
22 22 ☽ bQ ♆

18 Saturday
01 40 ☽ ⚹ ♃
01 55 ☽ △ ♂
09 30 ☽ ⊼ ☉
10 42 ☽ ⚹ ♇
10 47 ☽ ± ♅
13 21 ☉ ∥ ♅
14 11 ☽ Q ♄
20 33 ☽ ± ☉
21 06 ☽ ⊼ ♅
22 09 ☽ Q ♃

19 Sunday
00 13 ☽ △ ♆
02 54 ☉ □ ♇
11 48 ☽ □ ♀
12 20 ☽ ∠ ♇
13 01 ☽ bQ ☉
15 59 ☿ △ ♄
18 07 ☉ ± ♃
20 04 ♀ ∠ ♇
20 45 ☽ □ ♄
20 54 ☽ ⊼ ☿

20 Monday
03 35 ☽ ⊥ ♇
04 33 ☽ □ ♃
07 19 ☽ ± ☿
08 00 ☽ □ ♂
13 44 ☽ ⊻ ♇
16 15 ☽ △ ☉
20 20 ☽ Q ♀
21 46 ☽ # ☿
22 51 ☽ bQ ☿

21 Tuesday
00 04 ☽ ☍ ♅
03 09 ☽ □ ♆
15 50 ☽ ∥ ♇
16 02 ☽ Q ♂
17 56 ☽ ⚹ ♀
22 38 ☽ # ☉
23 11 ☽ △ ♄

22 Wednesday
00 34 ☽ △ ☿
02 25 ☽ # ♅
04 14 ♀ # ♄
06 49 ☽ △ ♃
09 31 ☽ Q ♆
13 22 ☽ ⚹ ♂
16 16 ☽ ☌ ♇
20 57 ☽ ∠ ♀
22 26 ☽ □ ☉

23 Thursday
00 22 ☽ bQ ♄
02 43 ☽ ⊼ ♅
05 48 ☽ ⚹ ♆
07 59 ☽ bQ ♃
12 11 ☽ # ♆
12 53 ☽ ± ♅
13 13 ☽ ⊥ ♀
15 32 ☽ ± ♄
16 06 ☽ ∠ ♂
21 07 ☽ # ♂
23 08 ☽ ± ♃

24 Friday
00 06 ☽ ⊻ ♀
01 40 ☽ ⊼ ♄
01 58 ☉ bQ ♄
03 40 ☽ □ ☿
04 10 ☽ bQ ♅
06 10 ☽ ∥ ♃
07 07 ☽ Q ☉
07 16 ☽ ∠ ♆
08 14 ☽ ⊥ ♂
09 16 ☽ ⊼ ♃
14 09 ☽ # ♀
15 38 ☿ St R
16 55 ☽ ∥ ♄
19 01 ☽ ⊻ ♂
19 06 ☽ ⊻ ♇
20 23 ♂ ⚹ ♇
22 39 ☽ ⊥ ♆
22 47 ♀ ☍ ♄

25 Saturday
05 05 ☽ ⚹ ☉
05 21 ☽ ⊥ ♇
05 48 ☽ △ ♅
08 56 ☽ ⊻ ♆
10 21 ☽ Q ☿
16 07 ☉ ⚹ ♅
20 49 ☽ ∠ ♇

26 Sunday
03 35 ☿ ⚹ ♀
03 51 ☽ # ♄
04 52 ☽ ☍ ♄
06 54 ☽ ⚹ ☿
07 10 ☽ ☌ ♀
08 45 ☽ ∥ ♀
08 49 ☽ ∠ ☉
12 31 ☽ ☍ ♃
14 37 ☽ # ♃
22 49 ☽ ⚹ ♇

27 Monday
01 39 ☽ ⊥ ☉
01 44 ☽ ☌ ♂
03 42 ☽ ∥ ♂
06 41 ☉ ⊥ ♂
08 44 ☽ ∠ ☿
09 27 ☽ ∥ ♆
09 56 ☽ □ ♅
12 58 ☽ ⊻ ☉
13 08 ☽ ☌ ♆
15 18 ☉ ⊻ ♆
19 53 ☽ Q ♇

28 Tuesday
00 21 ☽ ⊥ ☿
09 21 ☽ ⊼ ♄
10 49 ☽ ⊻ ☿
11 00 ☉ bQ ♃
15 57 ☽ ⊻ ♀
17 08 ☽ ⊼ ♃
17 57 ☽ Q ♅
20 03 ☽ ± ♄
20 14 ☽ ∥ ♅

29 Wednesday
03 27 ♇ St R
03 35 ☽ ⊥ ♀
03 53 ☽ ± ♃
04 02 ☽ □ ♇
05 26 ☉ # ♇
07 26 ♀ ☍ ♃
08 02 ☽ # ♇
08 08 ☽ ∥ ☉
10 15 ☽ ⊻ ♂
11 50 ☽ ∥ ☿
12 15 ☽ bQ ♄
15 42 ☽ ⚹ ♅
19 00 ☽ ⊻ ♆
20 09 ☽ bQ ♃
21 14 ☽ ∠ ♀
21 50 ☽ ⊥ ♂
22 56 ☽ ☌ ☉

30 Thursday
06 01 ☽ ⊥ ♆
15 26 ☽ ∠ ♂
15 44 ☽ △ ♄
16 01 ☽ ☌ ☿
16 52 ☉ ∥ ☿
19 24 ☽ ∠ ♅
22 05 ☿ ∠ ♂
22 46 ☽ ∠ ♆
23 45 ☽ △ ♃

31 Friday
00 39 ♀ # ♃
01 42 ☿ △ ♄
03 17 ☽ ⚹ ♀
08 03 ☿ # ♇
11 22 ☽ △ ♇
12 20 ☽ ⊥ ♅
21 22 ☽ ⚹ ♂
23 45 ☽ ⊻ ♅

JUNE 2041

LONGITUDES

Date	Sidereal time h m s	Sun ☉ ° ' "	Moon ☽ ° ' "	Moon ☽ 24.00 ° ' "	Mercury ☿ ° '	Venus ♀ ° '	Mars ♂ ° '	Jupiter ♃ ° '	Saturn ♄ ° '	Uranus ♅ ° '	Neptune ♆ ° '	Pluto ♇ ° '
01	04 41 24	11 ♊ 22 26	11 ♋ 27 43	17 ♋ 40 06	17 ♊ 23	25 ♈ 35	04 ♉ 16	22 ♎ 16	17 ♎ 59	05 ♌ 05	06 ♉ 52	28 ♒ 32
02	04 45 20	12 19 57	23 ♋ 49 10	29 ♋ 55 12	16 R 53	26 35	05 01	22 R 13	17 R 57	05 08	06 54	28 R 32
03	04 49 17	13 17 27	05 ♌ 58 29	11 ♌ 59 25	16 21	27 35	05 46	22 11	17 56	05 11	06 56	28 32
04	04 53 13	14 14 56	17 ♌ 58 26	23 ♌ 55 59	15 48	28 36	06 30	22 08	17 54	05 13	06 58	28 32
05	04 57 10	15 12 23	29 ♌ 52 34	05 ♍ 48 46	15 15	29 ♈ 36	07 15	22 06	17 53	05 16	07 00	28 32
06	05 01 06	16 09 49	11 ♍ 45 07	17 ♍ 42 13	14 41	00 ♉ 38	07 59	22 04	17 51	05 18	07 02	28 32
07	05 05 03	17 07 14	23 ♍ 40 40	29 ♍ 41 05	14 08	01 39	08 44	22 02	17 50	05 21	07 03	28 31
08	05 09 00	18 04 38	05 ♎ 44 05	11 ♎ 50 13	13 37	02 40	09 28	22 00	17 49	05 24	07 05	28 31
09	05 12 56	19 02 01	18 ♎ 00 05	24 ♎ 14 11	13 06	03 42	10 13	21 58	17 48	05 27	07 07	28 31
10	05 16 53	19 59 22	00 ♏ 33 00	06 ♏ 56 57	12 38	04 45	10 57	21 57	17 47	05 30	07 09	28 31
11	05 20 49	20 56 43	13 ♏ 26 22	20 ♏ 01 28	12 13	05 47	11 41	21 55	17 46	05 32	07 11	28 30
12	05 24 46	21 54 03	26 ♏ 42 24	03 ♐ 29 09	11 51	06 50	12 25	21 54	17 45	05 35	07 12	28 30
13	05 28 42	22 51 22	10 ♐ 21 37	17 ♐ 19 32	11 32	07 52	13 09	21 53	17 44	05 38	07 14	28 29
14	05 32 39	23 48 41	24 ♐ 22 28	01 ♑ 29 55	11 17	08 56	13 53	21 52	17 44	05 41	07 16	28 29
15	05 36 35	24 45 58	08 ♑ 41 14	15 ♑ 55 39	11 05	09 59	14 37	21 52	17 43	05 44	07 17	28 29
16	05 40 32	25 43 16	23 ♑ 12 21	00 ♒ 30 29	10 59	11 03	15 21	21 51	17 43	05 47	07 19	28 28
17	05 44 29	26 40 32	07 ♒ 49 12	15 ♒ 07 39	10 56	12 06	16 05	21 51	17 43	05 50	07 21	28 28
18	05 48 25	27 37 49	22 ♒ 25 03	29 ♒ 40 44	10 D 58	13 10	16 49	21 D 51	17 42	05 53	07 22	28 27
19	05 52 22	28 35 04	06 ♓ 54 06	14 ♓ 04 40	11 04	14 15	17 33	21 51	17 42	05 56	07 24	28 27
20	05 56 18	29 ♊ 32 20	21 ♓ 12 05	28 ♓ 16 05	11 16	15 19	18 16	21 52	17 D 42	05 59	07 25	28 26
21	06 00 15	00 ♋ 29 36	05 ♈ 16 31	12 ♈ 13 18	11 32	16 24	19 00	21 52	17 42	06 03	07 27	28 26
22	06 04 11	01 26 51	19 ♈ 06 27	25 ♈ 56 01	11 52	17 29	19 44	21 53	17 42	06 06	07 28	28 25
23	06 08 08	02 24 06	02 ♉ 42 05	09 ♉ 24 45	12 17	18 34	20 27	21 54	17 43	06 09	07 30	28 25
24	06 12 04	03 21 22	16 ♉ 04 08	22 ♉ 40 21	12 47	19 39	21 11	21 55	17 43	06 12	07 31	28 24
25	06 16 01	04 18 37	29 ♉ 13 28	05 ♊ 43 36	13 21	20 44	21 54	21 56	17 44	06 15	07 33	28 23
26	06 19 57	05 15 52	12 ♊ 10 47	18 ♊ 35 06	14 00	21 50	22 37	21 57	17 44	06 19	07 34	28 23
27	06 23 54	06 13 07	24 ♊ 56 34	01 ♋ 15 13	14 43	22 56	23 21	21 59	17 45	06 22	07 36	28 22
28	06 27 51	07 10 21	07 ♋ 31 06	13 ♋ 44 15	15 31	24 01	24 04	22 01	17 46	06 25	07 37	28 21
29	06 31 47	08 07 36	19 ♋ 54 44	26 ♋ 02 38	16 23	25 07	24 47	22 02	17 47	06 29	07 38	28 20
30	06 35 44	09 ♋ 04 50	02 ♌ 08 04	08 ♌ 11 12	17 ♊ 19	26 ♉ 14	25 ♉ 30	22 ♎ 05	17 ♎ 48	06 ♌ 32	07 ♉ 40	28 ♒ 20

MOON NODES AND LATITUDE

Date	Moon True ☊ ° '	Moon Mean ☊ ° '	Moon ☽ Latitude ° '
01	15 ♉ 06	14 ♉ 01	04 N 21
02	15 R 01	13 58	04 52
03	14 57	13 54	05 09
04	14 54	13 51	05 12
05	14 51	13 48	05 02
06	14 50	13 45	04 39
07	14 D 50	13 42	04 04
08	14 51	13 39	03 17
09	14 53	13 35	02 21
10	14 54	13 32	01 18
11	14 55	13 29	00 N 08
12	14 R 54	13 26	01 S 04
13	14 52	13 23	02 14
14	14 48	13 19	03 18
15	14 43	13 16	04 10
16	14 37	13 13	04 48
17	14 31	13 10	05 07
18	14 26	13 07	05 06
19	14 22	13 04	04 45
20	14 20	13 00	04 07
21	14 D 20	12 57	03 15
22	14 21	12 54	02 12
23	14 22	12 51	01 S 03
24	14 R 22	12 48	00 N 09
25	14 21	12 45	01 19
26	14 18	12 41	02 24
27	14 13	12 38	03 20
28	14 05	12 35	04 05
29	13 56	12 32	04 39
30	13 ♉ 46	12 ♉ 29	04 N 58

DECLINATIONS

Date	Sun ☉ ° '	Moon ☽ ° '	Mercury ☿ ° '	Venus ♀ ° '	Mars ♂ ° '	Jupiter ♃ ° '	Saturn ♄ ° '	Uranus ♅ ° '	Neptune ♆ ° '	Pluto ♇ ° '
01	22 N 08	27 N 16	21 N 18	07 N 51	12 N 10	07 S 21	04 S 36	19 N 35	12 N 11	21 S 42
02	22 16	26 07	20 58	08 10	12 25	07 21	04 35	19 34	12 12	21 42
03	22 23	23 45	20 38	08 29	12 40	07 20	04 35	19 33	12 13	21 42
04	22 30	20 23	20 18	08 48	12 56	07 19	04 34	19 33	12 13	21 43
05	22 37	16 13	19 59	09 07	13 11	07 18	04 34	19 32	12 14	21 43
06	22 43	11 27	19 40	09 27	13 26	07 18	04 34	19 31	12 14	21 43
07	22 49	06 14	19 23	09 46	13 40	07 17	04 34	19 31	12 15	21 44
08	22 54	00 N 45	19 06	10 06	13 55	07 17	04 33	19 30	12 16	21 44
09	22 59	04 S 53	18 51	10 25	14 09	07 17	04 33	19 29	12 16	21 44
10	23 03	10 27	18 37	10 44	14 24	07 16	04 33	19 29	12 17	21 45
11	23 08	15 45	18 24	11 04	14 38	07 16	04 33	19 28	12 17	21 45
12	23 11	20 27	18 13	11 23	14 52	07 16	04 33	19 27	12 18	21 46
13	23 15	24 12	18 04	11 42	15 06	07 16	04 33	19 26	12 18	21 46
14	23 17	26 36	17 57	12 02	15 20	07 16	04 33	19 26	12 19	21 46
15	23 20	27 19	17 51	12 21	15 33	07 16	04 33	19 25	12 19	21 47
16	23 22	26 10	17 47	12 40	15 47	07 16	04 33	19 24	12 20	21 47
17	23 24	23 15	17 46	12 59	16 00	07 16	04 33	19 23	12 20	21 48
18	23 25	18 51	17 46	13 18	16 13	07 16	04 33	19 23	12 21	21 48
19	23 26	13 23	17 48	13 36	16 26	07 16	04 33	19 22	12 21	21 49
20	23 26	07 17	17 51	13 55	16 39	07 17	04 34	19 21	12 22	21 49
21	23 26	00 S 53	17 56	14 13	16 51	07 17	04 34	19 20	12 22	21 50
22	23 26	05 N 27	18 03	14 31	17 04	07 18	04 34	19 19	12 22	21 50
23	23 25	11 26	18 12	14 50	17 16	07 18	04 35	19 19	12 23	21 51
24	23 24	16 47	18 21	15 07	17 28	07 19	04 35	19 18	12 23	21 51
25	23 22	21 16	18 32	15 25	17 40	07 20	04 35	19 17	12 24	21 52
26	23 20	24 37	18 45	15 42	17 52	07 20	04 36	19 16	12 24	21 52
27	23 17	26 40	18 58	16 00	18 03	07 21	04 36	19 15	12 25	21 53
28	23 14	27 18	19 12	16 17	18 14	07 22	04 37	19 14	12 25	21 53
29	23 11	26 33	19 27	16 33	18 26	07 23	04 37	19 14	12 25	21 54
30	23 N 07	24 N 31	19 N 43	16 N 50	18 N 37	07 S 24	04 S 38	19 N 13	12 N 26	21 S 54

ZODIAC SIGN ENTRIES

Date	h m	Planets
03	00 09	☽ ♌
05	12 15	☽ ♍
05	21 16	♀ ♉
08	00 38	☽ ♎
10	10 58	☽ ♏
12	17 51	☽ ♐
14	21 29	☽ ♑
16	23 10	☽ ♒
19	00 32	☽ ♓
20	23 36	☉ ♋
21	02 57	☽ ♈
23	07 12	☽ ♉
25	13 26	☽ ♊
27	21 37	☽ ♋
30	07 47	☽ ♌

LATITUDES

Date	Mercury ☿ ° '	Venus ♀ ° '	Mars ♂ ° '	Jupiter ♃ ° '	Saturn ♄ ° '	Uranus ♅ ° '	Neptune ♆ ° '	Pluto ♇ ° '
01	01 S 33	02 S 12	00 S 50	01 N 25	02 N 40	00 N 36	01 S 42	10 S 23
04	02 23	02 19	00 48	01 24	02 39	00 36	01 43	10 24
07	03 08	02 25	00 47	01 23	02 38	00 36	01 43	10 25
10	03 44	02 30	00 45	01 23	02 37	00 36	01 43	10 26
13	04 08	02 34	00 43	01 22	02 37	00 36	01 43	10 27
16	04 20	02 36	00 41	01 21	02 36	00 36	01 43	10 28
19	04 21	02 37	00 40	01 20	02 35	00 36	01 43	10 29
22	04 11	02 37	00 38	01 19	02 34	00 36	01 43	10 30
25	03 53	02 37	00 36	01 19	02 34	00 36	01 43	10 31
28	03 20	02 35	00 34	01 18	02 33	00 36	01 44	10 31
31	02 S 57	02 S 32	00 S 32	01 N 17	02 N 32	00 N 36	01 S 44	10 S 32

DATA

Julian Date	2466672
Delta T	+78 seconds
Ayanamsa	24° 25' 56"
Synetic vernal point	04° ♓ 41' 03"
True obliquity of ecliptic	23° 26' 08"

MOON'S PHASES, APSIDES AND POSITIONS ☽

Date	h m	Phase	Longitude ° '	Eclipse Indicator
06	21 41	☽	16 ♍ 33	
14	10 59	○	23 ♐ 46	
21	03 12	☾	00 ♈ 09	
28	11 17	●	07 ♋ 09	

Day	h m		
05	13 35	Apogee	
17	11 57	Perigee	
01	03 47	Max dec	27° N 21'
08	15 12	0S	
15	09 16	Max dec	27° S 20'
21	15 20	0N	
28	10 48	Max dec	27° N 19'

LONGITUDES

Date	Chiron ⚷ ° '	Ceres ⚳ ° '	Pallas ⚴ ° '	Juno ⚵ ° '	Vesta ⚶ ° '	Black Moon Lilith ☾ ° '
01	20 ♋ 37	18 ♌ 31	01 ♋ 16	07 ♌ 21	12 ♉ 10	28 ♌ 36
11	21 ♋ 35	22 ♌ 09	07 ♋ 13	11 ♌ 27	16 ♉ 27	29 ♌ 44
21	22 ♋ 37	25 ♌ 58	13 ♋ 09	15 ♌ 36	20 ♉ 37	00 ♍ 51
31	23 ♋ 42	29 ♌ 56	19 ♋ 04	19 ♌ 47	24 ♉ 41	01 ♍ 58

All ephemeris data is given at 12.00 UT and the Moon's longitude is additionally given for 24.00 UT

ASPECTARIAN

h m	Aspects
01 Saturday	
03 10	☽ ⚹ ♆
03 56	☽ Q ♀
11 49	☽ ⊻ ☉
15 01	♂ ∥ ♆
16 00	☽ bQ ♇
21 53	☽ Q ♂
23 00	☽ ⊻ ☿
02 Sunday	
00 23	☽ ⊥ ☉
00 35	☽ □ ♄
02 22	☽ Q ♆
08 53	☽ □ ♃
09 29	☽ ± ♇
10 14	☽ ⊥ ☿
15 55	♂ □ ♅
17 55	☽ □ ♀
19 29	☽ ∠ ☉
21 16	☽ ⚻ ♇
03 Monday	
02 43	☉ ⊥ ♆
03 13	☽ ∠ ☿
10 24	☽ ☌ ♅
11 33	☽ □ ♂
11 54	☽ Q ♄
13 55	☽ □ ♆
20 21	☽ Q ♃
22 09	☽ ∥ ☉
04 Tuesday	
03 19	☽ # ♇
03 52	☽ ⚹ ☉
07 50	☽ ⚹ ☿
10 31	♀ ⚹ ♇
11 51	☽ ⚹ ♄
12 33	☽ ∥ ☿
17 13	☽ ∥ ♅
20 21	☽ ⚹ ♃
05 Wednesday	
03 26	♂ ☌ ♆
06 08	☽ Q ☉
06 56	☽ Q ☿
09 17	☽ ☍ ♇
11 24	☽ △ ♀
12 39	☉ ☌ ☿
18 03	☽ ∠ ♄
21 45	☿ ∠ ♀
22 56	☽ ⊻ ♅
06 Thursday	
02 26	☽ △ ♆
02 33	☽ ∠ ♃
02 50	☽ ∥ ♂
03 54	☽ △ ♂
08 11	☽ ∥ ♆
11 06	☽ ⊥ ♅
12 12	☽ ⊥ ♄
17 40	☽ □ ☿
20 33	☽ bQ ♀
20 40	☽ ⊥ ♃
20 52	☽ ∥ ♀
21 41	☽ □ ☉
07 Friday	
00 17	☽ ⊻ ♄
00 46	☿ ∥ ♅
00 58	☿ ⊥ ♂
05 18	☽ ∠ ♅
07 17	☽ # ♃
08 42	☽ ⊻ ♃
08 44	☽ bQ ♆
12 07	☽ bQ ♂
16 19	☽ ± ♀
19 27	☽ # ♄
21 41	☽ ⚻ ♇
08 Saturday	
02 46	☽ ± ♆
05 23	☽ ⚻ ♀
05 30	☉ △ ♄
07 14	☽ ± ♂
09 36	☽ ± ♇
11 20	☽ ⚹ ♅
14 41	☽ ⚻ ♆
19 50	☽ ⚻ ♂
09 Sunday	
02 51	☽ △ ☿
03 17	☽ bQ ♇
10 36	☽ ∥ ♄
10 55	☽ Q ♅
11 32	☿ ⊥ ♆
11 36	☽ ☌ ♄
14 10	☽ △ ☉
19 38	☽ ☌ ♃
21 47	♂ Q ♇
22 14	☽ ∥ ♃
10 Monday	
06 41	☽ bQ ☿
08 08	☽ △ ♇
13 20	☽ # ♀
20 05	☽ # ♆
20 34	☽ ☍ ♀
21 01	☽ bQ ☉
21 19	☽ □ ♅
23 03	☽ ± ☿
11 Tuesday	
00 24	☽ ☍ ♆
01 17	☉ ∠ ♅
06 09	♀ □ ♅
06 33	☽ # ♂
08 35	☽ ☍ ♂
09 49	☽ ⚻ ☿
14 59	☽ ± ☉
19 17	☿ ⊥ ♀
19 53	☽ ⊻ ♄
23 22	☿ ⊻ ♂
12 Wednesday	
00 34	☽ # ☿
02 43	☽ ⚻ ☉
03 24	☽ ⊻ ♃
06 34	☽ # ♅
06 43	☽ ⊥ ♄
12 02	☉ △ ♃
14 08	☽ ⊥ ♃
15 11	☽ □ ♇
19 37	☽ ∥ ♇
19 53	☉ ∠ ♆
20 55	♀ ☌ ♆
22 42	☽ ∠ ♄
13 Thursday	
03 45	☽ △ ♅
04 58	☽ # ☉
05 58	☽ ∠ ♃
06 33	☽ ⚻ ♆
07 19	☉ ∠ ♀
07 19	☽ ⚻ ♀
13 59	☽ ☍ ☿
16 59	☽ ± ♆
17 06	☽ ⚻ ♂
18 34	☽ ± ♀
22 34	☽ Q ♇
14 Friday	
00 42	☽ ⚹ ♄
03 58	☽ ± ♂
05 43	☽ bQ ♅
07 46	☽ ⚹ ♃
08 25	☽ bQ ♆
10 59	☽ ☍ ☉
11 11	☽ bQ ♀
18 56	☽ ⚹ ♇
20 02	☽ bQ ♂
20 59	☽ ± ♅
21 01	☽ Q ♄
15 Saturday	
03 58	☽ Q ♃
05 42	♄ Q ♅
07 04	☽ ⚻ ♅
09 40	☽ △ ♆
10 03	♀ ∥ ♆
14 19	☽ △ ♀
15 57	☽ ⚻ ☿
19 57	☽ ∠ ♇
22 22	☽ △ ♂
23 08	♀ Q ♇
16 Sunday	
01 48	☽ ± ☿
02 57	☽ □ ♄
09 47	☽ □ ♃
10 34	☿ ⊻ ♀
10 47	☽ ⊥ ♇
16 26	☽ ⚻ ☉
16 32	☽ bQ ☿
20 39	☽ ⊻ ♇
17 Monday	
02 58	☽ ± ☉
08 44	☽ ☍ ♅
11 03	☽ # ☉
11 13	☽ □ ♆
13 41	☿ St D
17 07	☽ △ ☿
18 46	☽ bQ ☉
19 36	☽ □ ♀
20 43	☽ ∥ ♇
18 Tuesday	
02 18	☽ □ ♂
04 14	☽ △ ♄
06 14	♃ St D
09 27	☽ # ♅
11 04	☽ △ ♃
16 53	☽ Q ♆
17 06	☽ # ☿
21 13	☽ △ ☉
21 58	☽ ☌ ♇
23 36	☽ # ♂
19 Wednesday	
03 38	☽ Q ♀
05 01	☽ bQ ♄
08 34	☉ △ ♇
09 37	☽ Q ♂
10 23	☽ ⚻ ♅
11 10	☽ # ♀
11 55	☽ bQ ♃
12 50	☽ ⚹ ♆
16 13	☽ # ♆
17 09	♂ ⚻ ♄
19 03	☽ □ ☿
20 01	☽ ± ♄
20 27	☽ ± ♅
20 Thursday	
01 11	♄ St D
01 17	☽ ⚹ ♀
02 59	☽ ± ♃
06 06	☽ ⚻ ♄
06 48	☽ ⚹ ♂
11 39	☽ bQ ♅
11 59	☽ ∥ ♃
13 07	☽ ⚻ ♃
14 04	☽ ∠ ♆
22 15	☽ ∥ ♄
21 Friday	
00 01	☉ ⊥ ♅
00 17	☽ ⊻ ♇
01 56	☽ Q ☿
03 12	☽ □ ☉
04 47	☽ ∠ ♀
05 25	☽ ⊥ ♆
09 42	☽ ∠ ♂
10 33	☽ ⊥ ♇
13 20	☽ △ ♅
15 45	☽ ⊻ ♆
21 35	☽ ⊥ ♀
23 03	☽ ⚹ ☿
22 Saturday	
02 05	☽ ∠ ♇
02 06	☽ ⊥ ♂
08 38	☽ # ♄
08 54	☽ ⊻ ♀
09 33	☽ ☍ ♄
12 38	☽ Q ☉
13 09	☽ ⊻ ♂
16 52	☽ ☍ ♃
17 08	♀ ⚻ ♄
19 14	☽ # ♃
23 Sunday	
02 04	☽ ∠ ☿
04 23	☽ ⚹ ♇
11 26	☽ ⚹ ☉
16 03	☽ ∥ ♆
18 11	☽ □ ♅
18 38	☽ ⊥ ☿
20 35	☽ ☌ ♆
24 Monday	
01 47	☽ Q ♇
03 41	☽ ∥ ♀
05 49	☽ ⊻ ☿
15 00	☽ ⚻ ♄
15 27	☽ ∥ ♂
16 28	☽ ∠ ☉
19 05	☽ ☌ ♀
20 07	☽ ∥ ☿
21 49	☽ ☌ ♂
22 38	☽ ⚻ ♃
25 Tuesday	
00 43	☽ ∥ ♅
01 55	☽ ± ♄
02 51	☽ Q ♅
09 37	☽ ± ♃
10 11	☽ ⊥ ☉
10 28	☽ □ ♇
13 04	♂ ⚻ ♃
15 41	☽ # ♇
18 28	☽ bQ ♄
19 45	☿ ⊥ ♆
22 07	☽ ⊻ ☉
26 Wednesday	
01 02	☽ ⚹ ♅
01 47	☽ ∥ ☉
02 15	☽ bQ ♃
03 24	☽ ⊻ ♆
14 36	☽ ⊥ ♆
14 47	♀ ⚻ ♃
15 36	☽ ☌ ☿
22 25	☽ △ ♄
27 Thursday	
05 13	☽ ∠ ♅
06 23	☽ △ ♃
07 33	☽ ∠ ♆
07 49	☽ ⊻ ♀
08 47	☽ ⊻ ♂
15 58	☉ ⊻ ♅
18 29	☽ △ ♇
20 17	☽ ⊥ ♀
20 52	☽ ⊥ ♂
22 21	☽ ⊥ ♅
28 Friday	
01 48	♂ ± ♄
06 15	♀ ± ♄
09 53	☽ ⊻ ♅
11 17	☽ ☌ ☉
12 11	☽ ⚹ ♆
14 30	♀ ☌ ♂
15 09	☽ ∠ ♂
15 11	☽ ∠ ♀
15 49	☿ ∥ ♅
21 11	♀ Q ♅
23 15	☽ bQ ♇
23 26	☉ ⚹ ♆
29 Saturday	
01 02	♂ Q ♅
04 36	☽ ⊻ ☿
07 50	☽ □ ♄
11 28	☽ Q ♆
16 10	☽ □ ♃
16 44	☽ ± ♇
17 12	☽ ⊥ ☿
22 07	☽ ⚹ ♂
23 12	☽ ⚹ ♀
30 Sunday	
04 30	☽ ⚻ ♇
12 23	☽ ∠ ☿
19 16	☽ Q ♄
20 46	☽ ☌ ♅
22 58	☽ □ ♆
23 18	☽ Q ♂
23 49	☿ △ ♄

JULY 2041

LONGITUDES

Date	Sidereal time (h m s)	Sun ☉ (° ' ")	Moon ☽ (° ' ")	Moon ☽ 24.00 (° ' ")	Mercury ☿ (° ')	Venus ♀ (° ')	Mars ♂ (° ')	Jupiter ♃ (° ')	Saturn ♄ (° ')	Uranus ♅ (° ')	Neptune ♆ (° ')	Pluto ♇ (° ')
01	06 39 40	10 ♋ 02 04	14 ♌ 12 13	20 ♌ 11 24	18 ♊ 19	27 ♉ 20	26 ♉ 13	22 ♎ 07	17 ♎ 49	06 ♌ 35	07 ♉ 41	28 ♒ 19
02	06 43 37	10 59 17	26 ♌ 09 01	02 ♍ 05 27	19 24	28 26	26 56	22 09	17 50	06 39	07 42	28 R 18
03	06 47 33	11 56 31	08 ♍ 01 05	13 ♍ 56 23	20 32	29 ♉ 33	27 39	22 12	17 51	06 42	07 43	28 17
04	06 51 30	12 53 44	19 ♍ 51 50	25 ♍ 47 59	21 45	00 ♊ 40	28 21	22 15	17 53	06 46	07 45	28 17
05	06 55 26	13 50 56	01 ♎ 45 24	07 ♎ 44 43	23 02	01 47	29 04	22 18	17 54	06 49	07 46	28 16
06	06 59 23	14 48 09	13 ♎ 46 33	19 ♎ 51 32	24 22	02 54	29 ♉ 47	22 21	17 56	06 53	07 47	28 15
07	07 03 20	15 45 21	26 ♎ 00 20	02 ♏ 13 34	25 47	04 01	00 ♊ 29	22 24	17 57	06 56	07 48	28 14
08	07 07 16	16 42 33	08 ♏ 31 52	14 ♏ 55 45	27 15	05 08	01 12	22 27	17 59	07 00	07 49	28 13
09	07 11 13	17 39 45	21 ♏ 25 45	28 ♏ 02 17	28 ♊ 47	06 15	01 54	22 31	18 01	07 03	07 50	28 12
10	07 15 09	18 36 57	04 ♐ 45 36	11 ♐ 35 55	00 ♋ 22	07 23	02 36	22 35	18 03	07 07	07 51	28 11
11	07 19 06	19 34 09	18 ♐ 33 11	25 ♐ 37 14	02 01	08 31	03 19	22 39	18 05	07 10	07 52	28 10
12	07 23 02	20 31 21	02 ♑ 47 42	10 ♑ 04 01	03 44	09 38	04 01	22 43	18 07	07 14	07 53	28 09
13	07 26 59	21 28 33	17 ♑ 25 23	24 ♑ 50 52	05 30	10 46	04 43	22 47	18 09	07 18	07 54	28 08
14	07 30 55	22 25 45	02 ♒ 19 23	09 ♒ 49 46	07 19	11 54	05 25	22 52	18 12	07 21	07 55	28 07
15	07 34 52	23 22 58	17 ♒ 20 46	24 ♒ 51 11	09 11	13 02	06 07	22 56	18 14	07 25	07 56	28 06
16	07 38 49	24 20 11	02 ♓ 19 53	09 ♓ 45 48	11 06	14 11	06 49	23 01	18 17	07 28	07 57	28 05
17	07 42 45	25 17 25	17 ♓ 08 04	24 ♓ 25 58	13 04	15 19	07 31	23 06	18 19	07 32	07 57	28 04
18	07 46 42	26 14 39	01 ♈ 38 58	08 ♈ 46 44	15 04	16 27	08 13	23 11	18 22	07 36	07 58	28 03
19	07 50 38	27 11 54	15 ♈ 49 04	22 ♈ 45 56	17 06	17 36	08 54	23 16	18 25	07 39	07 59	28 02
20	07 54 35	28 09 09	29 ♈ 37 26	06 ♉ 23 45	19 09	18 45	09 36	23 21	18 28	07 43	08 00	28 01
21	07 58 31	29 ♋ 06 26	13 ♉ 05 08	19 ♉ 41 53	21 14	19 54	10 17	23 27	18 31	07 47	08 00	28 00
22	08 02 28	00 ♌ 03 43	26 ♉ 14 21	02 ♊ 42 51	23 20	21 02	10 59	23 32	18 34	07 50	08 01	27 59
23	08 06 24	01 01 01	09 ♊ 07 44	15 ♊ 29 18	25 27	22 11	11 40	23 38	18 37	07 54	08 02	27 58
24	08 10 21	01 58 20	21 ♊ 47 49	28 ♊ 03 34	27 34	23 21	12 22	23 44	18 40	07 58	08 02	27 56
25	08 14 18	02 55 40	04 ♋ 16 46	10 ♋ 27 35	29 ♋ 42	24 30	13 03	23 50	18 44	08 01	08 03	27 55
26	08 18 14	03 53 01	16 ♋ 36 13	22 ♋ 42 47	01 ♌ 49	25 39	13 44	23 56	18 47	08 05	08 03	27 54
27	08 22 11	04 50 22	28 ♋ 47 25	04 ♌ 50 15	03 56	26 48	14 25	24 03	18 51	08 09	08 04	27 53
28	08 26 07	05 47 44	10 ♌ 51 24	16 ♌ 51 00	06 02	27 58	15 06	24 09	18 54	08 12	08 04	27 52
29	08 30 04	06 45 07	22 ♌ 49 12	28 ♌ 46 12	08 08	29 ♊ 08	15 47	24 16	18 58	08 16	08 05	27 51
30	08 34 00	07 42 30	04 ♍ 42 11	10 ♍ 37 26	10 12	00 ♋ 17	16 28	24 23	19 02	08 20	08 05	27 49
31	08 37 57	08 ♌ 39 54	16 ♍ 32 13	22 ♍ 26 55	12 ♌ 16	01 ♋ 27	17 ♊ 08	24 ♎ 30	19 ♎ 05	08 ♌ 24	08 ♉ 05	27 ♒ 48

Date	Moon True ☊ (° ')	Moon Mean ☊ (° ')	Moon ☽ Latitude (° ')
01	13 ♉ 36	12 ♉ 25	05 N 04
02	13 R 27	12 22	04 57
03	13 20	12 19	04 37
04	13 15	12 16	04 05
05	13 12	12 13	03 22
06	13 11	12 10	02 30
07	13 D 11	12 06	01 30
08	13 12	12 03	00 N 25
09	13 R 12	12 00	00 S 44
10	13 10	11 57	01 52
11	13 06	11 54	02 56
12	12 59	11 51	03 51
13	12 51	11 47	04 33
14	12 41	11 44	04 57
15	12 31	11 41	05 01
16	12 22	11 38	04 44
17	12 15	11 35	04 08
18	12 11	11 31	03 17
19	12 09	11 28	02 15
20	12 D 09	11 25	01 S 06
21	12 R 09	11 22	00 N 05
22	12 08	11 19	01 14
23	12 06	11 16	02 17
24	12 01	11 12	03 13
25	11 53	11 09	03 58
26	11 42	11 06	04 32
27	11 30	11 03	04 52
28	11 16	11 00	05 00
29	11 03	10 57	04 53
30	10 50	10 53	04 34
31	10 ♉ 40	10 ♉ 50	04 N 04

DECLINATIONS

Date	Sun ☉	Moon ☽	Mercury ☿	Venus ♀	Mars ♂	Jupiter ♃	Saturn ♄	Uranus ♅	Neptune ♆	Pluto ♇
01	23 N 03	21 N 25	19 N 59	17 N 06	18 N 47	07 S 25	04 S 39	19 N 12	12 N 26	21 S 55
02	22 59	17 27	20 15	17 22	18 58	07 26	04 39	19 11	12 26	21 55
03	22 54	12 50	20 32	17 37	19 08	07 28	04 40	19 10	12 27	21 56
04	22 49	07 46	20 48	17 52	19 19	07 29	04 41	19 09	12 27	21 56
05	22 43	02 N 24	21 05	18 07	19 29	07 30	04 42	19 08	12 27	21 57
06	22 37	03 S 08	21 21	18 22	19 38	07 32	04 43	19 08	12 28	21 57
07	22 30	08 38	21 37	18 36	19 48	07 33	04 43	19 07	12 28	21 58
08	22 23	13 57	21 52	18 50	19 58	07 35	04 44	19 06	12 28	21 58
09	22 16	18 49	22 06	19 03	20 07	07 36	04 45	19 05	12 29	21 59
10	22 09	22 55	22 20	19 16	20 16	07 38	04 46	19 04	12 29	21 59
11	22 01	25 52	22 32	19 28	20 25	07 40	04 47	19 03	12 29	22 00
12	21 52	27 16	22 42	19 41	20 33	07 41	04 48	19 02	12 29	22 01
13	21 43	26 49	22 51	19 52	20 42	07 43	04 49	19 01	12 30	22 01
14	21 34	24 27	22 58	20 04	20 50	07 45	04 51	19 00	12 30	22 02
15	21 25	20 24	23 03	20 15	20 58	07 47	04 52	18 59	12 30	22 02
16	21 15	15 03	23 06	20 25	21 06	07 49	04 53	18 58	12 31	22 03
17	21 05	08 53	23 07	20 35	21 14	07 51	04 54	18 58	12 31	22 03
18	20 54	02 S 21	23 05	20 44	21 21	07 53	04 55	18 57	12 31	22 04
19	20 43	04 N 09	23 01	20 53	21 29	07 55	04 57	18 56	12 31	22 05
20	20 32	10 19	22 53	21 02	21 36	07 57	04 58	18 55	12 31	22 05
21	20 20	15 51	22 44	21 10	21 43	08 00	04 59	18 54	12 31	22 06
22	20 08	20 30	22 31	21 17	21 49	08 02	05 01	18 53	12 31	22 06
23	19 56	24 05	22 16	21 24	21 56	08 04	05 02	18 52	12 32	22 07
24	19 43	26 24	21 58	21 31	22 02	08 07	05 04	18 51	12 32	22 07
25	19 30	27 20	21 38	21 37	22 08	08 09	05 05	18 50	12 32	22 08
26	19 17	26 54	21 15	21 42	22 14	08 12	05 07	18 49	12 32	22 09
27	19 03	25 10	20 50	21 47	22 20	08 14	05 08	18 48	12 32	22 09
28	18 49	22 18	20 22	21 51	22 25	08 17	05 10	18 47	12 32	22 10
29	18 35	18 31	19 53	21 55	22 30	08 19	05 11	18 46	12 32	22 10
30	18 20	14 02	19 22	21 58	22 35	08 22	05 13	18 45	12 32	22 11
31	18 N 06	09 N 03	18 N 49	22 N 00	22 N 40	08 S 25	05 S 14	18 N 44	12 N 32	22 S 11

ZODIAC SIGN ENTRIES

Date	h	m	Planets
02	19	46	☽ ♍
03	21	43	♀ ♊
05	08	28	☽ ♎
06	19	30	♂ ♊
07	19	43	☽ ♏
10	03	31	☽ ♐
10	06	29	☿ ♋
12	07	21	☽ ♑
14	08	17	☽ ♒
16	08	15	☽ ♓
18	09	15	☽ ♈
20	12	40	☽ ♉
22	10	27	☉ ♌
22	18	57	☽ ♊
25	03	44	☽ ♋
25	15	27	☿ ♌
27	14	24	☽ ♌
30	02	29	☽ ♍
30	06	04	♀ ♋

LATITUDES

Date	Mercury ☿	Venus ♀	Mars ♂	Jupiter ♃	Saturn ♄	Uranus ♅	Neptune ♆	Pluto ♇
01	02 S 57	02 S 32	00 S 32	01 N 17	02 N 32	00 N 36	01 S 44	10 S 32
04	02 23	02 28	00 30	01 16	02 31	00 36	01 44	10 33
07	01 45	02 24	00 28	01 15	02 30	00 36	01 44	10 34
10	01 06	02 18	00 25	01 15	02 30	00 36	01 44	10 34
13	00 S 28	02 12	00 23	01 14	02 29	00 36	01 44	10 35
16	00 N 08	02 06	00 21	01 13	02 28	00 36	01 45	10 36
19	00 40	01 59	00 18	01 12	02 27	00 36	01 45	10 36
22	01 07	01 51	00 16	01 11	02 27	00 36	01 45	10 37
25	01 27	01 43	00 14	01 11	02 26	00 36	01 45	10 37
28	01 40	01 34	00 11	01 10	02 25	00 36	01 45	10 38
31	01 N 46	01 S 25	00 S 09	01 N 09	02 N 25	00 N 36	01 S 45	10 S 38

DATA

Julian Date	2466702
Delta T	+78 seconds
Ayanamsa	24° 26' 02"
Synetic vernal point	04° ♓ 40' 57"
True obliquity of ecliptic	23° 26' 08"

MOON'S PHASES, APSIDES AND POSITIONS ☽

Date	h	m	Phase	Longitude	Eclipse Indicator
06	14	12	☽	14 ♎ 53	
13	19	01	○	21 ♑ 45	
20	09	13	☾	28 ♈ 03	
28	01	02	●	05 ♌ 22	

Day	h	m		
03	06	50	Apogee	
15	09	15	Perigee	
30	20	16	Apogee	
05	22	27	0S	
12	18	24	Max dec	27° S 20'
18	20	37	0N	
25	16	16	Max dec	27° N 22'

LONGITUDES

Date	Chiron ⚷	Ceres ⚳	Pallas ⚴	Juno ⚵	Vesta ⚶	Black Moon Lilith ⚸
01	23 ♋ 42	29 ♌ 56	19 ♋ 04	19 ♌ 47	24 ♉ 41	01 ♍ 58
11	24 ♋ 48	04 ♍ 03	24 ♋ 56	24 ♌ 00	28 ♉ 36	03 ♍ 05
21	25 ♋ 55	08 ♍ 15	00 ♌ 44	28 ♌ 22	02 ♊ 22	04 ♍ 12
31	27 ♋ 02	12 ♍ 34	06 ♌ 29	02 ♍ 24	05 ♊ 56	05 ♍ 19

ASPECTARIAN

h m Aspects

01 Monday
00 00 ☽ ∥ ☉
01 18 ☽ Q ♀
02 58 ☽ ⊻ ☉
03 49 ☽ Q ♃
08 38 ☽ # ♇
15 59 ☽ ⊥ ☉
19 14 ☽ ⚹ ♄
20 39 ☽ ∥ ☿
21 03 ☽ ⚹ ☿

02 Tuesday
02 02 ☽ ∥ ♅
03 43 ☽ ∥ ♂
03 55 ☽ ⚹ ♃
05 35 ♀ ± ♃
09 05 ♀ □ ♇
11 39 ☽ ∠ ☉
12 29 ☽ ∥ ♀
13 40 ☽ □ ♂
16 20 ☽ ☍ ♇
17 06 ☽ □ ♀
23 43 ☽ Q ☿

03 Wednesday
01 31 ☽ ∠ ♄
09 20 ☽ ⊻ ♅
10 20 ☽ ∠ ♃
11 24 ☽ △ ♆
13 56 ☽ ∥ ♆
15 54 ♂ ∥ ♅
19 47 ☽ ⊥ ♄
20 39 ☽ ⚹ ☉
21 33 ☽ ⊥ ♅

04 Thursday
04 38 ☽ ⊥ ♃
07 57 ♂ ± ♃
07 58 ☽ ⊻ ♄
09 21 ♂ □ ♇
12 14 ☿ ∠ ♅
13 18 ☽ # ♃
15 52 ☽ ∠ ♅
16 16 ☽ □ ☿
16 50 ☽ ⊻ ♃
17 50 ☽ ⚼ ♆
21 26 ☉ ⚼ ♇
21 48 ☿ △ ♃
23 04 ☽ Q ☉

05 Friday
01 52 ☽ # ♄
04 58 ☽ ⊼ ♇
06 15 ☽ △ ♂
07 02 ☿ ∠ ♆
12 01 ☽ ± ♆
12 03 ☽ △ ♀
17 01 ☽ ± ♇
22 12 ☽ ⚹ ♅

06 Saturday
00 03 ☽ ⊼ ♆
09 36 ☉ ∠ ♂
10 57 ☽ ⚼ ♇
12 46 ♀ ⚼ ♄
14 06 ☽ ⚼ ♂
14 12 ☽ □ ☉
18 52 ☽ ∥ ♄
20 13 ☽ ☌ ♄
20 57 ☽ ⚼ ♀
22 07 ☽ Q ♅

07 Sunday
04 57 ☽ ☌ ♃
07 13 ☽ ∥ ♃
08 52 ☽ ± ♂
11 30 ☽ △ ☿
16 16 ☽ ± ♀
16 19 ☽ △ ♇
21 11 ☽ ⊼ ♂

08 Monday
04 55 ☽ ⊼ ♀
05 10 ☽ # ♆
09 05 ☽ □ ♅
10 39 ☽ ☍ ♆
19 55 ☽ ⚼ ☿
22 42 ☿ # ♇

09 Tuesday
03 11 ☿ △ ♇
04 31 ☽ △ ☉
05 43 ☽ ⊻ ♄
13 16 ☽ # ♀
13 23 ☽ # ♅
14 00 ☽ ⊻ ♃
14 48 ☽ ± ☿
15 22 ♀ ∥ ♅
16 44 ☽ ⊥ ♄
19 16 ☽ # ♂
21 15 ☉ □ ♄
23 08 ☉ ∥ ☿

10 Wednesday
00 17 ☽ □ ♇
00 55 ☽ ⊥ ♃
03 07 ☽ ⊼ ☿
05 57 ♀ ⚹ ♅
06 00 ☽ ∥ ♇
07 07 ☽ # ☉
07 54 ☽ # ☿
07 58 ☽ ☍ ♂
08 58 ☽ ∠ ♄
09 49 ☽ ⚼ ☉
16 10 ☽ △ ♅
16 29 ♀ ⚼ ♃
17 00 ☽ ∠ ♃
17 02 ☽ ☍ ♀
17 27 ☽ ⊼ ♆
22 08 ♀ ⊻ ♆
23 17 ☿ ⊥ ♅

11 Thursday
02 47 ☽ ± ☉
03 51 ♂ ⚼ ♄
03 56 ☽ ± ♆
07 55 ☽ Q ♇
11 12 ☽ ⚹ ♄
13 21 ☉ # ♇
13 52 ☽ ⊼ ☉
18 12 ☽ ⚼ ♅
19 00 ☽ ⚹ ♃
19 21 ☽ ⚼ ♆
19 39 ☉ Q ♆

12 Friday
04 16 ☽ ⚹ ♇
07 32 ☽ Q ♄
09 23 ☽ ± ♅
13 46 ☽ ☍ ☿
14 08 ☽ ⊼ ♂
15 12 ☽ Q ♃
18 26 ☿ ⊻ ♂
19 22 ☽ ⊼ ♅
20 25 ☽ △ ♆

13 Saturday
00 15 ☽ ⊼ ♀
00 31 ☽ ± ♂
05 02 ☽ ∠ ♇
10 51 ☽ ± ♀
13 12 ☽ □ ♄
15 54 ☽ ⚼ ♂
19 01 ☽ ☍ ☉
19 37 ☽ ⊥ ♇
20 43 ☽ □ ♃

14 Sunday
02 36 ☽ ⚼ ♀
04 24 ☉ ± ♇
05 16 ☽ ⊻ ♇
12 25 ☿ ⊻ ♅
17 12 ☽ △ ♂
19 44 ☿ ⚹ ♆
20 05 ☽ ☍ ♅
20 57 ☽ □ ♆
21 07 ☽ ⊼ ☿
21 39 ☽ # ☿
23 45 ☉ □ ♃

15 Monday
03 22 ☽ ∥ ♇
04 34 ☽ △ ♀
06 35 ☽ # ☉
08 04 ☽ ± ☿
09 11 ☽ # ♂
12 46 ☽ # ♀
13 26 ☽ △ ♄
18 52 ☽ # ♅
20 59 ☽ △ ♃
22 18 ☽ ⊼ ☉

16 Tuesday
00 32 ☽ ⚼ ☿
01 44 ☽ Q ♆
05 11 ☽ ☌ ♇
07 00 ♀ ⊥ ♆
08 34 ☽ ± ☉
13 32 ☽ ⚼ ♄
19 35 ☽ □ ♂
20 20 ☽ ⊼ ♅
21 04 ☽ ⚹ ♆
21 13 ☽ ⚼ ♃
22 12 ☽ # ♆
23 32 ☉ ∥ ♂

17 Wednesday
00 05 ☽ ⚼ ☉
04 08 ☽ ± ♄
04 20 ☽ △ ☿
06 06 ☽ ± ♅
08 47 ☽ □ ♀
11 56 ☽ ± ♃
12 03 ☿ ⚼ ♇
12 45 ♂ ⚹ ♅
13 57 ☽ ⊼ ♄
15 51 ☽ ∥ ♃
20 19 ☿ ⊥ ♂
20 54 ☽ ⚼ ♅
21 34 ☽ ∠ ♆
21 51 ☽ ⊼ ♃

18 Thursday
02 22 ☽ △ ☉
02 29 ☽ Q ♂
02 40 ☽ ∥ ♄
03 33 ♂ ⊻ ♆
06 01 ☽ ⊻ ♇
10 50 ♂ ⚼ ♃
12 32 ☽ ⊥ ♆
16 01 ☽ ⊥ ♇
17 07 ☽ Q ♀
22 02 ☽ △ ♅
22 38 ☽ ⊻ ♆
23 30 ☉ ∥ ♀
23 36 ☽ ⚹ ♂

19 Friday
07 15 ☽ ∠ ♇
14 34 ☽ □ ☿
15 00 ☽ # ♄
15 20 ☽ ⚹ ♀
16 29 ☽ ☍ ♄

20 Saturday
00 57 ☽ ☍ ♃
01 22 ☿ ⊻ ♀
02 32 ☽ # ♃
02 43 ☽ ∠ ♂
03 48 ☿ □ ♄
05 50 ♀ △ ♄
08 38 ☉ ⊼ ♇
09 11 ☽ ⚹ ♇
09 13 ☽ □ ☉
19 24 ☽ ⊥ ♂
19 58 ☽ ∠ ♀
21 12 ☽ ∥ ♆
21 45 ☿ Q ♆

21 Sunday
02 25 ☽ □ ♅
02 52 ☽ ☌ ♆
03 48 ☽ Q ☿
06 27 ☽ Q ♇
06 42 ☽ ⊻ ♂
13 36 ☽ ⊥ ♀
19 51 ☽ Q ☉
20 39 ☿ ± ♇
21 53 ☽ ⊼ ♄

22 Monday
01 32 ☽ ⊻ ♀
03 04 ☽ ∥ ♅
05 38 ☽ ⚹ ☿
07 00 ☽ ⊼ ♃
08 54 ☽ ± ♄
09 59 ☽ ∥ ☉
11 15 ☽ Q ♅
14 25 ☿ □ ♃
15 13 ☽ □ ♇
16 48 ☽ ∥ ♀
18 09 ☽ ± ♃
19 38 ☽ ⚹ ☉
20 12 ☽ ∥ ♂
21 48 ☽ # ♇
23 44 ☽ ∥ ☿

23 Tuesday
01 38 ☽ ⚼ ♄
09 41 ☽ ⚹ ♅
09 56 ☽ ⊻ ♆
11 04 ☽ ⚼ ♃
14 59 ☽ ∠ ☿
17 03 ☽ ☌ ♂
21 14 ☽ ⊥ ♆

24 Wednesday
00 21 ☿ # ♇
02 04 ☽ ∠ ☉
03 36 ♀ ∠ ♅
05 33 ♀ ∠ ♆
06 01 ☽ △ ♄
08 20 ☿ ∥ ♂
08 27 ☿ ∠ ♂
11 29 ☽ ⊥ ☿
14 14 ☽ ∠ ♅
14 22 ☽ ∠ ♆
15 15 ☽ ☌ ♀
15 44 ☽ △ ♃
16 09 ☿ ⊼ ♇
20 39 ☽ ⊥ ☉
20 57 ♀ △ ♃
23 45 ☽ △ ♇

25 Thursday
01 20 ☽ ⊻ ☿
07 37 ☽ ⊥ ♅
09 10 ☽ ⊻ ☉
11 43 ♂ # ♇
13 06 ☿ ∥ ♀
19 18 ☽ ⊻ ♅
19 18 ☽ ⚹ ♆
22 11 ♅ □ ♆

26 Friday
04 47 ☽ ⚼ ♇
06 03 ☽ ⊻ ♂
07 54 ☿ ⊥ ♀
16 18 ☽ □ ♄
18 30 ☽ ⊥ ♂
18 46 ☽ Q ♆
22 23 ☽ ± ♇
23 22 ♂ ⊥ ♆

27 Saturday
02 33 ☽ □ ♃
07 40 ☽ ⊻ ♀
10 12 ☽ ⊼ ♇
13 19 ☽ ∠ ♂
20 49 ☽ ⊥ ♀

28 Sunday
00 21 ☽ ☌ ☿
01 02 ☽ ☌ ☉
04 04 ☽ Q ♄
06 26 ☽ □ ♆
06 41 ☽ ☌ ♅
06 58 ☉ ☌ ☿
09 55 ♀ △ ♇
11 12 ☽ ∥ ♂
13 00 ☽ # ♇
14 37 ☽ Q ♃
15 08 ☽ ∥ ♀
15 49 ☉ ∥ ♅
16 40 ☽ ∠ ♀
21 00 ☽ ⚹ ♂
22 13 ☿ Q ♄

29 Monday
02 44 ☽ ∥ ☿
04 12 ☽ ⚹ ♄
10 34 ☽ ∥ ♅
11 24 ☿ □ ♆
11 38 ☽ ∥ ☉
13 41 ☿ ☌ ♅
14 56 ☽ ⚹ ♃
17 39 ☉ Q ♄
22 07 ☽ ☍ ♇
22 37 ☽ Q ♂

30 Tuesday
02 06 ☽ ⚹ ♀
10 37 ☽ ∠ ♄
18 37 ☽ ⊻ ☉
18 51 ☽ △ ♆
19 23 ☽ ⊻ ♅
19 27 ☽ ∥ ♆
21 26 ☉ □ ♆
21 34 ☽ ∠ ♃

31 Wednesday
01 30 ☽ ⊻ ☿
04 41 ☉ ☌ ♅
04 58 ☽ ⊥ ♄
05 03 ☽ Q ♀
07 37 ☽ ⊥ ♅
07 52 ☽ ⊥ ☉
13 18 ☽ □ ♂
14 53 ☿ Q ♃
14 55 ☽ # ♃
15 19 ☿ ∥ ♅
16 00 ☽ ⊥ ♃
16 14 ☽ ⊥ ☿
17 13 ☽ ⊻ ♄

All ephemeris data is given at 12.00 UT and the Moon's longitude is additionally given for 24.00 UT

AUGUST 2041

LONGITUDES

Date	Sidereal time h m s	Sun ☉ ° ' "	Moon ☽ ° ' "	Moon ☽ 24.00 ° ' "	Mercury ☿ ° '	Venus ♀ ° '	Mars ♂ ° '	Jupiter ♃ ° '	Saturn ♄ ° '	Uranus ♅ ° '	Neptune ♆ ° '	Pluto ♇ ° '
01	08 41 53	09 ♌ 37 19	28 ♍ 21 54	04 ♎ 17 37	14 ♌ 18	02 ♋ 37	17 ♊ 49	24 ♎ 37	19 ♎ 09	08 ♌ 27	08 ♉ 06	27 ♒ 47
02	08 45 50	10 34 44	10 ♎ 14 34	16 ♎ 13 16	16 19	03 47	18 30	24 44	19 13	08 31	08 06	27 R 46
03	08 49 47	11 32 10	22 ♎ 14 18	28 ♎ 18 16	18 18	04 57	19 10	24 51	19 17	08 35	08 06	27 44
04	08 53 43	12 29 37	04 ♏ 25 48	10 ♏ 37 34	20 16	06 07	19 51	24 58	19 21	08 38	08 06	27 43
05	08 57 40	13 27 04	16 ♏ 54 11	23 ♏ 16 17	22 13	07 17	20 31	25 06	19 26	08 42	08 07	27 42
06	09 01 36	14 24 32	29 ♏ 44 27	06 ♐ 19 14	24 08	08 27	21 11	25 14	19 30	08 46	08 07	27 41
07	09 05 33	15 22 01	13 ♐ 01 03	19 ♐ 50 13	26 01	09 38	21 51	25 21	19 34	08 49	08 07	27 39
08	09 09 29	16 19 31	26 ♐ 46 56	03 ♑ 51 10	27 53	10 48	22 31	25 29	19 39	08 53	08 07	27 38
09	09 13 26	17 17 01	11 ♑ 02 42	18 ♑ 21 05	29 ♌ 44	11 59	23 11	25 37	19 43	08 57	08 07	27 37
10	09 17 22	18 14 32	25 ♑ 45 41	03 ♒ 15 33	01 ♍ 32	13 09	23 51	25 46	19 48	09 00	08 07	27 36
11	09 21 19	19 12 05	10 ♒ 49 36	18 ♒ 26 34	03 20	14 20	24 31	25 54	19 53	09 04	08 R 07	27 34
12	09 25 16	20 09 38	26 ♒ 05 03	03 ♓ 43 37	05 05	15 31	25 11	26 02	19 57	09 08	08 07	27 33
13	09 29 12	21 07 13	11 ♓ 20 52	18 ♓ 55 28	06 50	16 42	25 50	26 11	20 02	09 11	08 07	27 32
14	09 33 09	22 04 48	26 ♓ 26 15	03 ♈ 52 14	08 32	17 53	26 30	26 19	20 07	09 15	08 07	27 30
15	09 37 05	23 02 25	11 ♈ 12 36	18 ♈ 26 50	10 14	19 04	27 09	26 28	20 12	09 19	08 07	27 29
16	09 41 02	24 00 04	25 ♈ 34 33	02 ♉ 35 37	11 53	20 15	27 49	26 37	20 17	09 22	08 06	27 28
17	09 44 58	24 57 44	09 ♉ 30 03	16 ♉ 18 02	13 32	21 26	28 28	26 46	20 22	09 26	08 06	27 26
18	09 48 55	25 55 26	22 ♉ 59 50	29 ♉ 35 49	15 09	22 37	29 07	26 55	20 27	09 29	08 06	27 25
19	09 52 51	26 53 10	06 ♊ 06 24	12 ♊ 32 03	16 44	23 49	29 ♊ 46	27 04	20 33	09 33	08 06	27 24
20	09 56 48	27 50 55	18 ♊ 53 13	25 ♊ 10 23	18 18	25 00	00 ♋ 25	27 13	20 38	09 37	08 05	27 23
21	10 00 45	28 48 42	01 ♋ 24 00	07 ♋ 34 29	19 51	26 11	01 04	27 23	20 43	09 40	08 05	27 21
22	10 04 41	29 ♌ 46 30	13 ♋ 42 14	19 ♋ 47 36	21 22	27 23	01 43	27 32	20 49	09 44	08 05	27 20
23	10 08 38	00 ♍ 44 20	25 ♋ 50 55	01 ♌ 52 28	22 51	28 35	02 22	27 42	20 54	09 47	08 04	27 19
24	10 12 34	01 42 12	07 ♌ 52 30	13 ♌ 51 15	24 19	29 ♋ 46	03 00	27 51	21 00	09 51	08 04	27 17
25	10 16 31	02 40 05	19 ♌ 48 56	25 ♌ 45 43	25 46	00 ♌ 58	03 39	28 01	21 05	09 54	08 03	27 16
26	10 20 27	03 38 00	01 ♍ 41 47	07 ♍ 37 20	27 11	02 10	04 17	28 11	21 11	09 58	08 03	27 15
27	10 24 24	04 35 56	13 ♍ 32 33	19 ♍ 27 38	28 35	03 22	04 56	28 21	21 17	10 01	08 02	27 13
28	10 28 20	05 33 53	25 ♍ 22 48	01 ♎ 18 20	29 ♍ 57	04 34	05 34	28 31	21 22	10 05	08 02	27 12
29	10 32 17	06 31 52	07 ♎ 14 29	13 ♎ 11 37	01 ♎ 18	05 46	06 12	28 41	21 28	10 08	08 01	27 11
30	10 36 14	07 29 53	19 ♎ 10 04	25 ♎ 10 15	02 37	06 58	06 50	28 51	21 34	10 11	08 01	27 10
31	10 40 10	08 ♍ 27 55	01 ♏ 12 38	07 ♏ 17 42	03 ♎ 54	08 ♌ 10	07 ♋ 28	29 ♎ 02	21 ♎ 40	10 ♌ 15	08 ♉ 00	27 ♒ 08

Date	Moon True ☊ ° '	Moon Mean ☊ ° '	Moon ☽ Latitude ° '
01	10 ♉ 32	10 ♉ 47	03 N 22
02	10 R 27	10 44	02 32
03	10 25	10 41	01 34
04	10 D 24	10 37	00 N 32
05	10 R 24	10 34	00 S 34
06	10 23	10 31	01 40
07	10 21	10 28	02 43
08	10 17	10 25	03 39
09	10 10	10 22	04 23
10	10 01	10 18	04 51
11	09 51	10 15	05 01
12	09 40	10 12	04 49
13	09 31	10 09	04 16
14	09 23	10 06	03 26
15	09 18	10 03	02 23
16	09 16	09 59	01 S 12
17	09 D 15	09 56	00 N 01
18	09 16	09 53	01 12
19	09 R 15	09 50	02 17
20	09 14	09 47	03 14
21	09 09	09 43	04 00
22	09 03	09 40	04 33
23	08 53	09 37	04 54
24	08 42	09 34	05 02
25	08 30	09 31	04 56
26	08 17	09 28	04 38
27	08 06	09 24	04 07
28	07 57	09 21	03 26
29	07 50	09 18	02 35
30	07 46	09 15	01 38
31	07 ♉ 44	09 ♉ 12	00 N 35

DECLINATIONS

Date	Sun ☉ ° '	Moon ☽ ° '	Mercury ☿ ° '	Venus ♀ ° '	Mars ♂ ° '	Jupiter ♃ ° '	Saturn ♄ ° '	Uranus ♅ ° '	Neptune ♆ ° '	Pluto ♇ ° '
01	17 N 50	03 N 45	18 N 14	22 N 02	22 N 45	08 S 28	05 S 16	18 N 43	12 N 32	22 S 12
02	17 35	01 S 43	17 38	22 03	22 49	08 30	05 18	18 42	12 32	22 13
03	17 19	07 12	17 01	22 04	22 53	08 33	05 20	18 41	12 32	22 13
04	17 03	12 30	16 22	22 04	22 57	08 36	05 21	18 40	12 32	22 14
05	16 47	17 26	15 43	22 04	23 01	08 39	05 23	18 39	12 32	22 14
06	16 30	21 44	15 03	22 03	23 05	08 42	05 25	18 38	12 32	22 15
07	16 14	25 03	14 21	22 01	23 08	08 45	05 27	18 38	12 32	22 15
08	15 56	27 03	13 40	21 59	23 12	08 48	05 29	18 37	12 32	22 16
09	15 39	27 21	12 57	21 56	23 15	08 51	05 31	18 36	12 32	22 17
10	15 22	25 46	12 14	21 53	23 17	08 55	05 33	18 35	12 32	22 17
11	15 04	22 20	11 31	21 49	23 20	08 58	05 34	18 34	12 32	22 18
12	14 46	17 20	10 48	21 44	23 22	09 01	05 36	18 33	12 32	22 18
13	14 27	11 15	10 04	21 39	23 25	09 04	05 38	18 32	12 32	22 19
14	14 09	04 S 34	09 20	21 33	23 27	09 08	05 40	18 31	12 32	22 19
15	13 50	02 N 15	08 36	21 27	23 29	09 11	05 42	18 30	12 32	22 20
16	13 31	08 46	07 52	21 20	23 30	09 14	05 45	18 29	12 32	22 20
17	13 12	14 40	07 08	21 12	23 32	09 18	05 47	18 28	12 32	22 21
18	12 53	19 41	06 24	21 04	23 33	09 21	05 49	18 27	12 31	22 21
19	12 33	23 35	05 40	20 55	23 34	09 25	05 51	18 26	12 31	22 22
20	12 13	26 11	04 56	20 46	23 35	09 28	05 53	18 25	12 31	22 22
21	11 53	27 25	04 13	20 36	23 36	09 32	05 55	18 24	12 31	22 23
22	11 33	27 16	03 30	20 25	23 36	09 35	05 57	18 23	12 31	22 23
23	11 13	25 47	02 47	20 14	23 37	09 39	05 59	18 22	12 31	22 24
24	10 52	23 09	02 04	20 02	23 37	09 42	06 02	18 21	12 30	22 24
25	10 31	19 33	01 22	19 50	23 37	09 46	06 04	18 20	12 30	22 25
26	10 10	15 11	00 N 41	19 37	23 37	09 50	06 06	18 20	12 30	22 25
27	09 49	10 16	00 S 01	19 23	23 36	09 53	06 08	18 19	12 30	22 26
28	09 28	04 N 59	00 41	19 09	23 36	09 57	06 11	18 18	12 30	22 26
29	09 07	00 S 30	01 21	18 55	23 35	10 01	06 13	18 17	12 29	22 27
30	08 45	06 00	02 01	18 40	23 34	10 05	06 15	18 16	12 29	22 27
31	08 N 24	11 S 21	02 S 40	18 N 24	23 N 33	10 S 09	06 S 18	18 N 15	12 N 29	22 S 28

ZODIAC SIGN ENTRIES

Date	h	m	Planets
01	15	19	☽ ♎
04	03	20	☽ ♏
06	12	29	☽ ♐
08	17	29	☽ ♑
09	15	37	☿ ♍
10	18	48	☽ ♒
12	18	09	☽ ♓
14	17	44	☽ ♈
16	19	33	☽ ♉
19	00	44	☽ ♊
19	20	27	♂ ♋
21	09	18	☽ ♋
22	17	36	☉ ♍
23	20	16	☽ ♌
24	16	33	♀ ♌
26	08	34	☽ ♍
28	12	52	☿ ♎
28	21	21	☽ ♎
31	09	36	☽ ♏

LATITUDES

Date	Mercury ☿ ° '	Venus ♀ ° '	Mars ♂ ° '	Jupiter ♃ ° '	Saturn ♄ ° '	Uranus ♅ ° '	Neptune ♆ ° '	Pluto ♇ ° '
01	01 N 47	01 S 22	00 S 08	01 N 09	02 N 24	00 N 36	01 S 46	10 S 39
04	01 44	01 13	00 05	01 08	02 24	00 36	01 46	10 39
07	01 37	01 04	00 S 03	01 08	02 23	00 36	01 46	10 39
10	01 24	00 54	00 00	01 07	02 23	00 36	01 46	10 40
13	01 09	00 45	00 N 02	01 07	02 22	00 36	01 46	10 40
16	00 50	00 35	00 05	01 06	02 21	00 36	01 46	10 40
19	00 28	00 26	00 08	01 05	02 21	00 36	01 47	10 40
22	00 N 05	00 16	00 11	01 05	02 20	00 36	01 47	10 41
25	00 S 20	00 S 07	00 14	01 04	02 20	00 36	01 47	10 41
28	00 46	00 N 02	00 16	01 04	02 19	00 36	01 47	10 41
31	01 S 13	00 N 11	00 N 19	01 N 03	02 N 19	00 N 36	01 S 47	10 S 41

DATA

Julian Date	2466733
Delta T	+78 seconds
Ayanamsa	24° 26' 07"
Synetic vernal point	04° ♓ 40' 52"
True obliquity of ecliptic	23° 26' 09"

MOON'S PHASES, APSIDES AND POSITIONS ☽

Date	h	m	Phase	Longitude ° '	Eclipse Indicator
05	04	53	☽	13 ♏ 10	
12	02	05	○	19 ♒ 46	
18	17	43	☾	26 ♉ 09	
26	16	16	●	03 ♍ 48	

Day	h	m		
12	16	08	Perigee	
27	02	17	Apogee	
02	04	28	0S	
09	04	01	Max dec	27° S 27'
15	04	02	0N	
21	21	11	Max dec	27° N 31'
29	09	50	0S	

LONGITUDES

Date	Chiron ⚷ ° '	Ceres ⚳ ° '	Pallas ⚴ ° '	Juno ⚵ ° '	Vesta ⚶ ° '	Black Moon Lilith ⚸ ° '
01	27 ♋ 09	13 ♍ 00	07 ♌ 03	02 ♍ 50	06 ♊ 17	05 ♍ 26
11	28 ♋ 14	17 ♍ 23	12 ♌ 42	07 ♍ 01	09 ♊ 37	06 ♍ 33
21	29 ♋ 18	21 ♍ 49	18 ♌ 15	11 ♍ 11	12 ♊ 42	07 ♍ 40
31	00 ♌ 18	26 ♍ 18	23 ♌ 42	15 ♍ 20	15 ♊ 28	08 ♍ 47

ASPECTARIAN

h m Aspects

01 Thursday
01 18 ☽ ⚼ ♆
01 59 ☽ ∠ ♅
03 45 ☽ ∠ ☉
04 18 ☽ ⊻ ♃
05 15 ☽ # ♄
08 31 ♀ ⊥ ♅
10 49 ☽ ⊼ ♇
14 17 ☽ ∠ ☿
19 33 ☽ ± ♆
21 32 ☽ □ ♀
22 57 ☽ ± ♇

02 Friday
07 41 ☽ ⊼ ♆
08 30 ☽ ✶ ♅
12 44 ☽ ✶ ☉
15 42 ☉ ∥ ☿
17 03 ☽ ⚼ ♇

03 Saturday
02 37 ☽ ✶ ☿
03 43 ☽ ∥ ♄
05 32 ☽ △ ♂
06 06 ☽ ☌ ♄
08 41 ☽ Q ♅
14 48 ☽ Q ☉
16 42 ♂ △ ♄
17 14 ☽ ☌ ♃
18 06 ☽ ∥ ♃
22 52 ☽ △ ♇

04 Sunday
00 26 ☿ ✶ ♄
04 03 ☿ ✶ ♂
06 59 ☽ Q ☿
12 11 ☽ # ♆
12 51 ☽ ⚼ ♂
15 37 ☽ △ ♀
19 08 ☽ ☍ ♆
20 12 ☽ □ ♅

05 Monday
01 51 ☉ Q ♃
04 24 ☽ # ☿
04 53 ☽ □ ☉
07 12 ☽ ± ♂
08 52 ☽ # ☉
14 19 ☿ ∠ ♀
16 48 ☽ ⊻ ♄
18 24 ☽ # ♅
19 13 ☽ ⊼ ♂
23 10 ☽ ⚼ ♀
23 47 ☽ □ ☿

06 Tuesday
03 34 ☽ ⊻ ♃
04 07 ☽ ⊥ ♄
04 56 ♀ ✶ ♆
06 41 ☉ ⊥ ♀
08 12 ☽ □ ♇
14 01 ☽ # ♀
14 45 ☽ ⊥ ♃
15 18 ☽ ∥ ♇
17 28 ☽ ± ♀
18 36 ♀ ⊻ ♅
20 45 ☽ ∠ ♄
20 58 ☽ # ♂

07 Wednesday
02 56 ☿ ✶ ♃
03 14 ☽ ⊼ ♆
04 29 ☽ △ ♅
05 22 ☽ ⊼ ♀
07 13 ☽ ∠ ♃
13 57 ☽ ± ♆
16 28 ☽ △ ☉
16 40 ☽ Q ♇
23 36 ☽ ✶ ♄

08 Thursday
04 18 ☽ ☍ ♂
05 41 ☽ ⚼ ♆
07 00 ☽ ⚼ ♅
08 49 ☿ ☍ ♇
09 46 ☽ ✶ ♃
13 28 ☽ ✶ ♇
14 10 ☽ △ ☿
20 18 ☽ ⚼ ☉
20 19 ☽ Q ♄
22 25 ☽ ± ♅

09 Friday
06 16 ☽ Q ♃
07 08 ☽ △ ♆
08 30 ☽ ⊼ ♅
09 23 ♂ ∠ ♆
12 25 ☽ ± ☉
13 41 ☽ ☍ ♀
14 35 ☽ ∠ ♇
18 56 ☽ ⚼ ☿
22 58 ☽ ⊼ ☉

10 Saturday
00 45 ♀ ⚼ ♇
02 03 ☿ ∥ ♆
02 18 ☽ □ ♄
05 17 ☽ ⊥ ♇
08 47 ☽ ⊼ ♂
11 36 ☽ ± ☿
12 00 ☽ □ ♃
14 56 ☽ ⊻ ♇
18 07 ♂ ∠ ♅
18 52 ☽ ± ♂
19 53 ♆ St R
22 31 ☽ ⊼ ☿

11 Sunday
06 09 ☽ # ♂
07 43 ☽ □ ♆
09 12 ☽ ☍ ♅
09 50 ☽ ⚼ ♂
12 12 ☽ ∥ ♇
14 52 ☽ # ♀
18 00 ☽ ⊼ ♀

12 Monday
02 05 ☽ ☍ ☉
02 20 ☽ △ ♄
04 13 ☽ ± ♀
06 26 ☉ ✶ ♄
06 44 ☽ # ♅
10 05 ☿ ∠ ♄
10 31 ☽ △ ♂
11 56 ☽ △ ♃
12 03 ☽ Q ♆
14 18 ☽ ☌ ♇
19 32 ☽ ⚼ ♀
23 09 ☽ # ☉

13 Tuesday
02 00 ☽ ⚼ ♄
03 58 ☽ ☍ ☿
06 54 ☽ ✶ ♆
07 12 ☽ # ♆
08 35 ☽ ⊼ ♅
11 44 ☽ ⚼ ♃
16 16 ☽ ± ♄
16 54 ☽ # ☿
18 06 ☽ ± ♅
19 54 ☽ ∥ ♃
21 10 ☽ △ ♀

14 Wednesday
01 51 ☽ ⊼ ♄
02 08 ☽ ± ♃
03 48 ♂ △ ♃
04 33 ☽ ⊼ ☉
05 59 ☿ △ ♆
06 41 ☽ ∠ ♆
08 06 ☽ ∥ ♄
08 29 ☽ ⚼ ♅
11 01 ☿ Q ♂
11 49 ☽ ⊼ ♃
12 06 ☽ □ ♂
13 43 ☽ ⊻ ♇
14 49 ☽ ± ☉
18 18 ☿ # ♃
21 09 ☽ ⊥ ♆
22 25 ☿ ⊻ ♅
23 24 ☽ ⊥ ♇

15 Thursday
06 26 ☽ ⚼ ☉
06 55 ☽ ⊻ ♆
08 52 ☽ △ ♅
10 11 ☽ ⊼ ☿
14 06 ☽ ∠ ♇
18 50 ☽ Q ♂
21 23 ☽ ± ☿
23 40 ♂ △ ♇

16 Friday
00 35 ☽ # ♄
02 12 ☽ □ ♀
03 01 ☽ ☍ ♄
07 36 ☿ ∠ ♃
08 55 ☽ ∥ ☿
09 08 ☽ △ ☉
09 14 ♀ Q ♆
12 51 ♀ □ ♄
13 47 ☽ ☍ ♃
13 49 ☽ # ♃
14 32 ☽ ⚼ ☿
15 12 ☽ ✶ ♇
15 59 ☽ ✶ ♂

17 Saturday
02 54 ☽ ∥ ♆
06 02 ☽ ∥ ☉
09 34 ☽ ☌ ♆
11 52 ☽ Q ♀
11 53 ☽ □ ♅
11 54 ☽ Q ♇
12 12 ♀ ± ♇
19 20 ☽ ∠ ♂
20 03 ☽ △ ☿

18 Sunday
01 08 ☿ ⊥ ♄
05 38 ☽ ∥ ♅
07 24 ☽ ⊼ ♄
11 15 ☽ ✶ ♀
12 14 ☽ ⊥ ♂
17 24 ☿ ⊥ ♅
17 43 ☽ □ ☉
18 18 ☽ ± ♄
19 11 ☽ ⊼ ♃
19 27 ☽ ∥ ♀
20 01 ☽ □ ♇
20 11 ☽ Q ♅
23 42 ☽ ⊻ ♂

19 Monday
03 40 ☽ # ♇
06 19 ☽ ± ♃
06 25 ☿ # ♄
10 57 ☽ ⚼ ♄
11 56 ☽ ∥ ♂
13 59 ☉ ∥ ♆
15 42 ☽ ⊻ ♆
17 23 ☉ ✶ ♃
17 32 ☽ ∠ ♀
18 27 ☽ ✶ ♅
23 16 ☽ ⚼ ♃

20 Tuesday
00 29 ☉ ☍ ♇
02 56 ☽ ⊥ ♆
05 46 ☽ Q ☉
10 44 ☽ □ ☿
12 14 ☽ ⊥ ♀
15 20 ☽ △ ♄
20 01 ☽ ∠ ♆
22 58 ☽ ∠ ♅

21 Wednesday
00 54 ☽ ⊻ ♀
04 08 ☽ △ ♃
04 12 ☽ △ ♇
06 35 ☽ ✶ ☉
08 49 ♃ △ ♇
11 19 ☽ ☌ ♂
16 25 ☽ ⊥ ♅

22 Thursday
00 59 ☽ ✶ ♆
02 18 ☽ Q ☿
02 41 ☿ ⊻ ♄
04 10 ☽ ⊻ ♅
09 19 ☽ ⚼ ♇
10 59 ♀ ⊼ ♇
14 17 ☽ ∠ ☉
15 08 ☿ ⊥ ♃
15 32 ♀ □ ♃

23 Friday
00 34 ☽ Q ♆
02 07 ☽ □ ♄
03 01 ☽ ± ♇
05 14 ☽ ✶ ☿
09 36 ☽ ⊥ ☉
14 54 ☽ ⊼ ♇
15 32 ☿ ⚼ ♆
15 43 ☽ □ ♃
18 02 ☽ ☌ ♀
22 35 ☽ ⊻ ☉

24 Saturday
01 42 ☽ ⊻ ♂
08 26 ☽ ∥ ♂
12 23 ☽ □ ♆
14 15 ☽ Q ♄
14 24 ☽ ⊥ ♂
15 18 ☽ ∠ ☿
15 58 ☽ ☌ ♅
17 29 ☽ # ♇
21 00 ☿ ∠ ♅

25 Sunday
04 15 ☽ Q ♃
09 31 ☽ ∠ ♂
10 13 ☽ ∥ ♀
11 53 ☽ ⊥ ☿
14 35 ☽ ✶ ♄
19 03 ☽ ∥ ♅
22 37 ♂ ⊥ ♅

26 Monday
01 39 ☽ ⊻ ☿
03 01 ☽ ☍ ♇
04 48 ☽ ✶ ♃
12 58 ☿ ⊼ ♇
13 04 ☽ ⊻ ♀
16 16 ☽ ☌ ☉
17 33 ☽ ✶ ♂
21 09 ☽ ∠ ♄

27 Tuesday
00 51 ☽ △ ♆
01 23 ☽ ∥ ♆
02 35 ☽ ⊥ ♀
04 49 ☽ ⊻ ♅
07 24 ☿ ⊻ ♃
08 05 ☉ # ♃
11 36 ☽ ∠ ♃
13 44 ☽ # ♃
14 14 ☽ ∥ ☉
15 32 ☽ ⊥ ♄
17 03 ☽ ⊥ ♅
19 15 ☽ Q ♂
22 53 ☽ ∠ ♀

28 Wednesday
03 48 ☽ ⊻ ♄
06 07 ☽ ⊥ ♃
06 42 ☽ # ♄
07 15 ☽ ⚼ ♆
11 23 ☽ ∠ ♅
11 48 ☉ ✶ ♂
15 41 ☽ ⊼ ♇
18 27 ☽ ⊻ ♃
22 27 ☽ ☌ ☿

29 Thursday
01 28 ☽ ± ♆
03 48 ☽ ± ♇
04 48 ☽ # ☿
08 41 ☽ ✶ ♀
09 46 ☽ □ ♂
10 15 ☉ ∠ ♄
10 26 ☽ ⊻ ☉
13 34 ☽ ⊼ ♆
16 15 ☽ ∥ ☿
17 52 ☽ ✶ ♅
21 57 ☽ ⚼ ♇
23 36 ☽ ⊥ ☉

30 Friday
01 08 ☿ ± ♆
06 06 ♀ ⊻ ♂
11 34 ☽ Q ♀
13 09 ☽ ∥ ♄
16 50 ☽ ☌ ♄
18 05 ☽ Q ♅
19 15 ☽ ∠ ☉
22 01 ☿ ± ♇
23 28 ☽ # ☉

31 Saturday
00 37 ☉ △ ♆
03 56 ☽ △ ♇
06 25 ☽ ∥ ♃
07 37 ☽ ☌ ♃
08 35 ♀ □ ♆
17 15 ☽ # ♆
17 56 ☽ ⊻ ☿

All ephemeris data is given at 12.00 UT and the Moon's longitude is additionally given for 24.00 UT

SEPTEMBER 2041

LONGITUDES

Date	Sidereal time h m s	Sun ☉ ° ' "	Moon ☽ ° ' "	Moon ☽ 24.00 ° ' "	Mercury ☿ ° '	Venus ♀ ° '	Mars ♂ ° '	Jupiter ♃ ° '	Saturn ♄ ° '	Uranus ♅ ° '	Neptune ♆ ° '	Pluto ♇ ° '
01	10 44 07	09 ♍ 25 58	13 ♏ 25 57	19 ♏ 37 57	05 ♎ 09	09 ♌ 23	08 ♋ 05	29 ♎ 12	21 ♎ 46	10 ♌ 18	07 ♉ 59	27 ♒ 07
02	10 48 03	10 24 03	25 ♏ 54 15	02 ♐ 15 25	06 23	10 35	08 43	29 23	21 52	10 22	07 R 59	27 R 06
03	10 52 00	11 22 10	08 ♐ 42 01	15 ♐ 14 34	07 35	11 47	09 21	29 33	21 58	10 25	07 58	27 04
04	10 55 56	12 20 17	21 ♐ 53 31	28 ♐ 39 16	08 45	13 00	09 58	29 44	22 04	10 28	07 57	27 03
05	10 59 53	13 18 26	05 ♑ 32 05	12 ♑ 32 06	09 53	14 12	10 36	29 ♎ 55	22 10	10 31	07 56	27 02
06	11 03 49	14 16 37	19 ♑ 39 17	26 ♑ 53 24	10 59	15 25	11 13	00 ♏ 05	22 17	10 35	07 56	27 01
07	11 07 46	15 14 49	04 ♒ 14 01	11 ♒ 40 26	12 02	16 38	11 50	00 16	22 23	10 38	07 55	26 59
08	11 11 43	16 13 03	19 ♒ 11 47	26 ♒ 46 59	13 04	17 50	12 27	00 27	22 29	10 41	07 54	26 58
09	11 15 39	17 11 18	04 ♓ 24 47	12 ♓ 03 50	14 02	19 03	13 04	00 38	22 35	10 44	07 53	26 57
10	11 19 36	18 09 34	19 ♓ 42 45	27 ♓ 20 07	14 58	20 16	13 41	00 49	22 42	10 47	07 52	26 56
11	11 23 32	19 07 53	04 ♈ 54 39	12 ♈ 25 09	15 51	21 29	14 17	01 01	22 48	10 50	07 51	26 55
12	11 27 29	20 06 13	19 ♈ 50 37	27 ♈ 10 16	16 41	22 42	14 54	01 12	22 55	10 54	07 50	26 53
13	11 31 25	21 04 36	04 ♉ 23 28	11 ♉ 29 52	17 28	23 55	15 30	01 23	23 01	10 57	07 49	26 52
14	11 35 22	22 03 00	18 ♉ 29 16	25 ♉ 21 39	18 11	25 08	16 07	01 34	23 08	11 00	07 48	26 51
15	11 39 18	23 01 27	02 ♊ 07 10	08 ♊ 46 06	18 51	26 21	16 43	01 46	23 15	11 03	07 47	26 50
16	11 43 15	23 59 56	15 ♊ 18 48	21 ♊ 45 42	19 26	27 34	17 19	01 57	23 21	11 06	07 46	26 49
17	11 47 12	24 58 27	28 ♊ 07 17	04 ♋ 24 05	19 57	28 ♌ 48	17 55	02 09	23 28	11 08	07 45	26 48
18	11 51 08	25 57 00	10 ♋ 36 38	16 ♋ 45 26	20 24	00 ♍ 01	18 31	02 21	23 35	11 11	07 44	26 47
19	11 55 05	26 55 36	22 ♋ 51 02	28 ♋ 53 55	20 45	01 14	19 06	02 32	23 41	11 14	07 42	26 45
20	11 59 01	27 54 14	04 ♌ 54 33	10 ♌ 53 23	21 01	02 28	19 42	02 44	23 48	11 17	07 41	26 44
21	12 02 58	28 52 53	16 ♌ 50 49	22 ♌ 47 14	21 11	03 41	20 17	02 56	23 55	11 20	07 40	26 43
22	12 06 54	29 ♍ 51 35	28 ♌ 42 57	04 ♍ 38 18	21 15	04 55	20 53	03 08	24 02	11 23	07 39	26 42
23	12 10 51	00 ♎ 50 19	10 ♍ 33 33	16 ♍ 28 58	21 R 12	06 09	21 28	03 20	24 09	11 25	07 38	26 41
24	12 14 47	01 49 05	22 ♍ 24 47	28 ♍ 21 14	21 03	07 22	22 03	03 32	24 16	11 28	07 36	26 40
25	12 18 44	02 47 53	04 ♎ 18 30	10 ♎ 16 51	20 46	08 36	22 38	03 44	24 22	11 31	07 35	26 39
26	12 22 41	03 46 42	16 ♎ 16 28	22 ♎ 17 35	20 22	09 50	23 12	03 56	24 29	11 33	07 34	26 38
27	12 26 37	04 45 34	28 ♎ 20 28	04 ♏ 25 22	19 51	11 03	23 47	04 08	24 36	11 36	07 32	26 37
28	12 30 34	05 44 28	10 ♏ 32 34	16 ♏ 42 24	19 13	12 17	24 21	04 20	24 43	11 38	07 31	26 36
29	12 34 30	06 43 24	22 ♏ 55 11	29 ♏ 11 16	18 27	13 31	24 56	04 33	24 51	11 41	07 30	26 35
30	12 38 27	07 ♎ 42 21	05 ♐ 31 03	11 ♐ 54 54	17 ♎ 34	14 ♍ 45	25 ♋ 30	04 ♏ 45	24 ♎ 58	11 ♌ 43	07 ♉ 28	26 ♒ 34

DECLINATIONS

Date	Moon True ☊ ° '	Moon Mean ☊ ° '	Moon ☽ Latitude ° '	Sun ☉ ° '	Moon ☽ ° '	Mercury ☿ ° '	Venus ♀ ° '	Mars ♂ ° '	Jupiter ♃ ° '	Saturn ♄ ° '	Uranus ♅ ° '	Neptune ♆ ° '	Pluto ♇ ° '
01	07 ♉ 44	09 ♉ 08	00 S 31	08 N 02	16 S 21	03 S 18	18 N 08	23 N 32	10 S 12	06 S 20	18 N 14	12 N 29	22 S 28
02	07 D 45	09 05	01 36	07 40	20 47	03 56	17 51	23 30	10 16	06 23	18 13	12 28	22 29
03	07 45	09 02	02 38	07 18	24 21	04 32	17 34	23 29	10 20	06 25	18 12	12 28	22 29
04	07 R 45	08 59	03 34	06 56	26 45	05 08	17 17	23 27	10 24	06 27	18 12	12 28	22 30
05	07 43	08 56	04 20	06 34	27 39	05 43	16 58	23 25	10 28	06 30	18 11	12 27	22 30
06	07 39	08 53	04 52	06 11	26 48	06 17	16 40	23 23	10 32	06 32	18 10	12 27	22 30
07	07 32	08 49	05 06	05 49	24 09	06 50	16 21	23 21	10 36	06 35	18 09	12 27	22 31
08	07 25	08 46	05 00	05 26	19 49	07 22	16 01	23 19	10 40	06 37	18 08	12 26	22 31
09	07 17	08 43	04 33	05 04	14 08	07 53	15 41	23 16	10 44	06 40	18 07	12 26	22 32
10	07 11	08 40	03 47	04 41	07 33	08 22	15 20	23 13	10 48	06 42	18 06	12 26	22 32
11	07 05	08 37	02 44	04 18	00 S 34	08 50	15 00	23 11	10 52	06 44	18 06	12 25	22 32
12	07 02	08 34	01 31	03 55	06 N 21	09 17	14 38	23 08	10 56	06 47	18 05	12 25	22 33
13	07 01	08 30	00 S 14	03 32	12 46	09 42	14 16	23 05	11 00	06 50	18 04	12 25	22 33
14	07 D 01	08 27	01 N 02	03 09	18 19	10 05	13 54	23 01	11 04	06 52	18 03	12 24	22 33
15	07 03	08 24	02 12	02 46	22 44	10 26	13 32	22 58	11 08	06 55	18 02	12 24	22 34
16	07 04	08 21	03 12	02 23	25 49	10 46	13 09	22 55	11 12	06 57	18 02	12 24	22 34
17	07 R 04	08 18	04 01	02 00	27 27	11 03	12 45	22 51	11 16	07 00	18 01	12 23	22 34
18	07 03	08 14	04 38	01 37	27 38	11 18	12 22	22 47	11 20	07 02	18 00	12 23	22 35
19	07 00	08 11	05 01	01 13	26 26	11 31	11 58	22 43	11 24	07 05	17 59	12 22	22 35
20	06 55	08 08	05 10	00 50	24 02	11 41	11 33	22 39	11 29	07 07	17 59	12 22	22 35
21	06 49	08 05	05 05	00 27	20 38	11 47	11 08	22 35	11 33	07 10	17 58	12 21	22 36
22	06 42	08 02	04 48	00 N 03	16 24	11 51	10 43	22 31	11 37	07 13	17 57	12 21	22 36
23	06 35	07 59	04 18	00 S 20	11 35	11 51	10 18	22 27	11 41	07 15	17 56	12 21	22 36
24	06 29	07 55	03 37	00 43	06 19	11 48	09 52	22 22	11 45	07 18	17 56	12 20	22 36
25	06 24	07 52	02 46	01 07	00 N 49	11 40	09 26	22 18	11 49	07 20	17 55	12 20	22 37
26	06 21	07 49	01 47	01 30	04 S 45	11 29	09 00	22 13	11 54	07 23	17 54	12 19	22 37
27	06 19	07 46	00 N 43	01 53	10 12	11 13	08 34	22 08	11 58	07 26	17 54	12 19	22 37
28	06 D 19	07 43	00 S 23	02 17	15 21	10 53	08 07	22 03	12 02	07 28	17 53	12 18	22 37
29	06 20	07 40	01 30	02 40	19 57	10 28	07 40	21 58	12 06	07 31	17 52	12 18	22 38
30	06 ♉ 21	07 ♉ 36	02 S 33	03 S 03	23 S 44	09 S 59	07 N 13	21 N 53	12 S 11	07 S 33	17 N 52	12 N 17	22 S 38

ZODIAC SIGN ENTRIES

Date	h	m	Planets
02	19	45	☽ ♐
05	02	22	☽ ♑
06	00	12	♃ ♏
07	05	06	☽ ♒
09	05	04	☽ ♓
11	04	13	☽ ♈
13	04	41	☽ ♉
15	08	13	☽ ♊
17	15	34	☽ ♋
18	11	41	♀ ♍
20	02	12	☽ ♌
22	14	36	☽ ♍
22	15	26	☉ ♎
25	03	19	☽ ♎
27	15	17	☽ ♏
30	01	33	☽ ♐

LATITUDES

Date	Mercury ☿ ° '	Venus ♀ ° '	Mars ♂ ° '	Jupiter ♃ ° '	Saturn ♄ ° '	Uranus ♅ ° '	Neptune ♆ ° '	Pluto ♇ ° '
01	01 S 22	00 N 14	00 N 20	01 N 03	02 N 19	00 N 36	01 S 47	10 S 41
04	01 49	00 23	00 23	01 03	02 18	00 36	01 48	10 41
07	02 15	00 31	00 26	01 02	02 18	00 36	01 48	10 41
10	02 41	00 39	00 30	01 02	02 17	00 36	01 48	10 40
13	03 04	00 46	00 33	01 01	02 17	00 36	01 48	10 40
16	03 25	00 53	00 36	01 01	02 17	00 36	01 48	10 40
19	03 41	01 00	00 39	01 00	02 16	00 37	01 48	10 40
22	03 51	01 05	00 42	01 00	02 16	00 37	01 48	10 40
25	03 51	01 11	00 46	01 00	02 16	00 37	01 48	10 39
28	03 38	01 16	00 49	00 59	02 16	00 37	01 49	10 39
31	03 S 09	01 N 20	00 N 53	00 N 59	02 N 16	00 N 37	01 S 49	10 S 38

DATA

Julian Date	2466764
Delta T	+78 seconds
Ayanamsa	24° 26' 11"
Synetic vernal point	04° ♓ 40' 48"
True obliquity of ecliptic	23° 26' 09"

MOON'S PHASES, APSIDES AND POSITIONS ☽

Date	h	m	Phase	Longitude ° '	Eclipse Indicator
03	17	19	☽	11 ♐ 35	
10	09	24	○	18 ♓ 03	
17	05	33	☾	24 ♊ 43	
25	08	41	●	02 ♎ 40	

Day	h	m		
10	02	20	Perigee	
23	05	14	Apogee	
05	12	46	Max dec	27° S 39'
11	13	55	0N	
18	02	59	Max dec	27° N 43'
25	15	33	0S	

LONGITUDES

Date	Chiron ⚷ ° '	Ceres ⚳ ° '	Pallas ⚴ ° '	Juno ⚵ ° '	Vesta ⚶ ° '	Black Moon Lilith ⚸ ° '
01	00 ♌ 24	26 ♍ 45	24 ♌ 14	15 ♍ 44	15 ♊ 43	08 ♍ 54
11	01 ♌ 19	01 ♎ 16	29 ♌ 34	19 ♍ 50	18 ♊ 04	10 ♍ 01
21	02 ♌ 10	05 ♎ 48	04 ♍ 46	23 ♍ 53	19 ♊ 59	11 ♍ 08
31	02 ♌ 54	10 ♎ 21	09 ♍ 51	27 ♍ 52	21 ♊ 22	12 ♍ 15

ASPECTARIAN

h m	Aspects
01 Sunday	
01 00	☽ △ ♂
01 22	☽ ☍ ♆
02 25	♀ ∥ ♅
03 13	☽ □ ♀
03 31	☽ ⚹ ☉
05 52	☽ □ ♅
07 04	☽ ⊥ ☿
08 12	♂ ⚹ ♆
17 31	☉ ⊻ ♀
20 23	♀ Q ♄
20 40	☽ # ♀
21 43	☽ # ♅
02 Monday	
02 26	☽ ∠ ☿
04 14	☽ ⊻ ♄
04 45	☽ Q ☉
07 19	♀ ☌ ♅
07 37	☽ ⊞ ♂
10 54	☉ ⊻ ♅
14 15	☽ □ ♇
15 45	☽ ⊥ ♄
18 40	☽ ⊻ ♃
22 35	☽ ∥ ♇
03 Tuesday	
01 32	☽ ± ♂
05 29	☽ # ♂
06 05	☽ ⊥ ♃
08 46	☽ ∠ ♄
09 44	☽ ⚹ ☿
10 39	☽ ⊼ ♆
13 15	☽ ⊼ ♂
15 11	☽ △ ♅
17 19	☽ □ ☉
18 16	☽ △ ♀
19 41	☿ ⊼ ♆
21 40	☽ ± ♆
22 53	☽ ∠ ♃
23 40	☽ Q ♇
04 Wednesday	
09 46	☽ Q ☿
12 19	☽ ⚹ ♄
13 54	☽ ⊞ ♆
18 24	☽ ⊞ ♅
21 10	☽ ⚹ ♇
23 54	☽ ⊞ ♀
05 Thursday	
02 04	☽ ⚹ ♃
09 06	♂ ⊻ ♅
09 37	☽ Q ♄
10 15	☽ ± ♅
15 45	☉ # ♄
16 09	☽ △ ♆
17 03	☽ ± ♀
20 07	☽ □ ☿
20 36	☽ ⊼ ♅
21 05	☽ ☍ ♂
23 05	☽ Q ♃
23 08	☽ ∠ ♇
06 Friday	
02 17	☽ △ ☉
02 38	☿ ⚹ ♅
04 13	☽ ⊼ ♀
09 27	☉ # ☿
14 16	☽ ⊥ ♇
16 24	☽ □ ♄
23 41	☿ ∥ ♄
07 Saturday	
00 11	☽ ⊻ ♇
00 24	☿ □ ♂
05 03	☽ ⊞ ☉
05 28	☽ □ ♃
10 54	☿ ⊞ ♇
12 43	☉ ∠ ♃
17 12	☽ # ♂
17 57	☽ □ ♆
18 01	♂ ⊞ ♇
20 40	☽ ± ☉
22 05	☽ ∥ ♇
22 22	☽ ☍ ♅
08 Sunday	
00 47	☽ ⊼ ♂
01 31	☽ △ ☿
06 56	☽ ⊼ ☉
09 39	☽ ☍ ♀
10 46	☽ ± ♂
17 15	☽ △ ♄
19 26	☉ ⊥ ♄
19 38	☽ # ♅
22 36	☽ Q ♆
09 Monday	
00 13	☉ ⊥ ♅
00 17	☽ ☌ ♇
01 36	☽ ⊞ ♂
02 18	♀ Q ♃
02 59	☽ ⊞ ☿
05 30	☽ # ♀
06 00	☽ △ ♃
12 24	♀ ⊥ ♂
17 01	☽ ⊞ ♄
17 26	☽ ⚹ ♆
18 04	☽ ± ☿
18 25	☽ # ♆
21 57	☽ ⊼ ♅
10 Tuesday	
00 34	☽ ∥ ♃
02 08	☽ △ ♂
04 05	☽ ⊼ ☿
05 49	☽ ⊞ ♃
07 14	☽ ± ♄
07 24	☽ ± ♅
09 18	☽ ∥ ☿
09 24	☽ ☍ ☉
12 57	☽ ⊼ ♀
14 56	☽ ∥ ♄
16 44	☽ ⊼ ♄
16 57	☽ ∠ ♆
20 08	☽ ± ♃
21 36	☽ ⊞ ♅
22 29	☽ # ☉
23 13	☽ ± ♀
23 21	☽ ⊻ ♇
11 Wednesday	
05 44	☽ ⊼ ♃
07 09	☽ ⊥ ♆
08 50	☽ ⊥ ♇
14 43	☽ ⊞ ♀
16 41	☽ ⊻ ♆
21 30	☽ △ ♅
23 10	☽ ∠ ♇
12 Thursday	
03 05	♄ Q ♅
03 39	☽ □ ♂
03 53	☽ ∥ ☉
06 35	☽ ☍ ☿
12 27	☽ ⊼ ☉
13 34	☽ # ♄
16 40	♀ ⚹ ♄
17 03	☽ ☍ ♄
17 05	☽ △ ♀
22 58	☽ ± ☉
23 24	☽ # ☿
23 31	☽ ⚹ ♇
13 Friday	
05 03	☽ # ♃
06 55	☽ ☍ ♃
10 27	☽ Q ♂
10 37	☽ ∥ ♆
15 02	☽ ⊞ ☉
17 45	☽ ☌ ♆
17 46	☽ ∥ ♀
19 32	☽ Q ♇
23 06	☽ □ ♅
14 Saturday	
07 42	☽ ⚹ ♂
10 46	☽ ∥ ♅
11 27	☽ ⊼ ☿
18 40	☽ △ ☉
20 09	☽ ⊼ ♄
22 27	☽ ± ☿
15 Sunday	
00 44	☽ □ ♀
02 36	☽ □ ♇
06 09	☉ ⊞ ♆
06 30	☽ Q ♅
06 49	☽ ± ♄
10 56	☽ # ♇
11 14	☽ ∠ ♂
11 21	☽ ⊼ ♃
13 29	☽ ∥ ♂
15 15	☽ ⊞ ☿
18 03	☉ ⊻ ♄
21 17	♀ ☍ ♇
22 12	☽ ⊻ ♆
22 20	☽ ± ♃
23 08	☽ ⊞ ♄
16 Monday	
04 13	☽ ⚹ ♅
04 17	☽ ⊥ ♂
09 09	☽ ⊥ ♆
12 32	☽ Q ♀
15 05	☽ ⊞ ♃
15 53	☽ ⊻ ♂
19 59	☽ △ ☿
17 Tuesday	
01 52	☽ ∠ ♆
03 07	☽ △ ♄
05 33	☽ □ ☉
08 14	☽ ∠ ♅
09 29	☽ △ ♇
13 25	☽ ⚹ ♀
19 48	☽ △ ♃
18 Wednesday	
01 29	☽ ⊥ ♅
06 25	☽ ⚹ ♆
10 53	♀ ∥ ♆
13 08	☽ ⊻ ♅
14 16	☽ ⊞ ♇
17 02	☿ ∥ ♃
18 12	☉ ∠ ♅
19 04	☽ Q ☉
21 33	☽ ∠ ♀
19 Thursday	
00 08	☉ ⊥ ♃
04 14	☽ ☌ ♂
05 48	☽ Q ♆
07 44	☽ □ ☿
07 52	☽ ± ♇
07 55	☉ ⊼ ♇
13 40	☽ □ ♄
17 16	☽ ⊥ ♀
19 44	☽ ⊼ ♇
20 47	☽ ⚹ ☉
20 Friday	
06 33	☽ ⊻ ♀
06 37	☿ # ♀
07 35	☽ □ ♃
11 35	♂ Q ♆
15 49	♀ # ♃
17 33	☽ □ ♆
18 22	♀ ⚹ ♃
20 22	☽ Q ☿
22 45	☽ ∥ ♂
22 59	☽ # ♇
21 Saturday	
00 51	☽ ☌ ♅
01 58	☽ Q ♄
05 29	☽ ∠ ☉
09 24	♂ # ♇
19 19	☽ ⊻ ♂
20 24	☽ Q ♃
20 50	☽ ⚹ ☿
22 Sunday	
01 17	☽ ⊥ ☉
02 25	☽ ⚹ ♄
03 39	☽ ∥ ♅
05 07	♂ ± ♇
07 56	☽ ☍ ♇
08 05	☽ ⊥ ♂
14 32	☽ ⊻ ☉
14 43	☿ St R
21 06	☽ ⚹ ♃
23 Monday	
02 01	☽ ☌ ♀
02 36	☿ □ ♂
03 14	☽ ∠ ☿
03 16	☽ ∠ ♂
06 04	☽ △ ♆
08 21	☽ ∥ ♆
09 06	☽ ∠ ♄
10 41	☽ # ☿
11 30	☽ # ♃
13 07	☿ ∠ ♀
13 45	☽ ⊻ ♅
18 32	☽ ∥ ♀
21 19	☽ ⊥ ☿
23 59	♀ ∠ ♂
24 Tuesday	
01 57	☽ ⊥ ♅
03 31	☽ ⊥ ♄
04 01	☽ ∠ ♃
06 52	☉ ± ♆
07 42	☽ # ♄
09 17	☽ ⊻ ☿
11 13	☽ ⚹ ♂
12 23	☽ ⊞ ♆
15 46	☽ ⊻ ♄
16 30	♀ △ ♆
17 48	☿ ∥ ♃
20 13	☽ ∠ ♅
20 35	☽ ⊼ ♇
22 31	☽ ⊥ ♃
25 Wednesday	
06 31	☽ ± ♆
08 29	☉ ± ♇
08 40	☽ ± ♇
08 41	☽ ☌ ☉
10 49	☽ ⊻ ♃
10 50	☽ # ☉
12 41	☽ Q ♂
18 34	☽ ⊼ ♆
20 58	☽ ∥ ☉
21 37	☽ ⊻ ♀
26 Thursday	
02 31	☽ ⚹ ♅
02 44	☽ ⊞ ♇
04 45	♀ ∠ ♄
11 00	☽ ⊥ ♀
16 49	☉ ⊻ ♃
19 52	☽ ☌ ☿
23 34	☽ ∥ ♄
27 Friday	
02 31	☽ □ ♂
02 34	☽ Q ♅
04 32	☽ ☌ ♄
05 13	☽ # ♀
06 58	☽ ∠ ♀
08 36	☽ △ ♇
16 19	☽ ∥ ☿
20 06	☽ ∥ ♃
21 35	☽ # ♆
22 54	♀ ⊻ ♅
23 38	☽ ☌ ♃
28 Saturday	
01 46	☽ ⊻ ☉
06 05	☽ ☍ ♆
14 09	☽ □ ♅
14 33	☽ ⊥ ☉
15 47	☽ ⚹ ♀
23 15	☿ ⊥ ♀
29 Sunday	
00 45	☽ # ♅
03 54	☽ ⊻ ☿
07 21	♂ □ ♄
09 30	☽ ∠ ☉
14 45	☽ ⊥ ☿
15 44	☽ ⊻ ♄
16 03	☽ △ ♂
17 32	☽ Q ♀
19 02	☽ □ ♇
19 13	♀ # ♄
23 47	☽ # ♂
30 Monday	
00 00	☉ Q ♂
03 17	☽ ⊥ ♄
04 19	☽ ∥ ♇
06 21	☉ ⊼ ♆
06 49	☽ ∠ ☿
10 32	☽ ⊻ ♃
15 40	☽ ⊼ ♆
16 28	☽ ⚹ ☉
20 25	☽ ∠ ♄
21 47	☽ ⊞ ♂
21 59	☽ ⊥ ♃
23 41	☽ △ ♅

All ephemeris data is given at 12.00 UT and the Moon's longitude is additionally given for 24.00 UT

OCTOBER 2041

Date	Sidereal time h m s	Sun ☉ ° ' "	Moon ☽ ° ' "	Moon ☽ 24.00 ° ' "	Mercury ☿ ° '	Venus ♀ ° '	Mars ♂ ° '	Jupiter ♃ ° '	Saturn ♄ ° '	Uranus ♅ ° '	Neptune ♆ ° '	Pluto ♇ ° '
01	12 42 23	08 ♎ 41 21	18 ♐ 23 11	24 ♐ 56 18	16 ♎ 36	15 ♍ 59	26 ♋ 04	04 ♏ 57	25 ♎ 05	11 ♌ 46	07 ♉ 27	26 ♒ 33
02	12 46 20	09 40 22	01 ♑ 34 35	08 ♑ 18 19	15 R 33	17 13	26 38	05 10	25 12	11 48	07 R 25	26 R 32
03	12 50 16	10 39 24	15 ♑ 07 44	22 ♑ 03 00	14 26	18 27	27 11	05 22	25 19	11 50	07 24	26 32
04	12 54 13	11 38 29	29 ♑ 04 09	06 ♒ 11 05	13 17	19 41	27 45	05 35	25 26	11 53	07 22	26 31
05	12 58 10	12 37 35	13 ♒ 23 34	20 ♒ 41 12	12 08	20 56	28 18	05 47	25 33	11 55	07 21	26 30
06	13 02 06	13 36 43	28 ♒ 03 25	05 ♓ 29 29	11 00	22 10	28 52	06 00	25 40	11 57	07 19	26 29
07	13 06 03	14 35 53	12 ♓ 58 32	20 ♓ 29 34	09 55	23 24	29 25	06 12	25 48	11 59	07 18	26 28
08	13 09 59	15 35 04	28 ♓ 01 28	05 ♈ 33 07	08 56	24 38	29 ♋ 57	06 25	25 55	12 01	07 16	26 28
09	13 13 56	16 34 18	13 ♈ 03 22	20 ♈ 31 05	08 03	25 53	00 ♌ 30	06 38	26 02	12 04	07 15	26 27
10	13 17 52	17 33 33	27 ♈ 55 15	05 ♉ 14 57	07 19	27 07	01 03	06 50	26 09	12 06	07 13	26 26
11	13 21 49	18 32 51	12 ♉ 29 27	19 ♉ 38 06	06 44	28 21	01 35	07 03	26 16	12 08	07 12	26 26
12	13 25 45	19 32 11	26 ♉ 40 30	03 ♊ 36 23	06 20	29 ♍ 36	02 07	07 16	26 24	12 09	07 10	26 25
13	13 29 42	20 31 33	10 ♊ 25 38	17 ♊ 08 17	06 06	00 ♎ 50	02 39	07 29	26 31	12 11	07 08	26 24
14	13 33 39	21 30 57	23 ♊ 44 31	00 ♋ 14 37	06 D 03	02 05	03 11	07 42	26 38	12 13	07 07	26 23
15	13 37 35	22 30 24	06 ♋ 38 56	12 ♋ 57 55	06 11	03 19	03 43	07 55	26 45	12 15	07 05	26 23
16	13 41 32	23 29 53	19 ♋ 12 03	25 ♋ 21 52	06 30	04 34	04 14	08 07	26 53	12 17	07 04	26 22
17	13 45 28	24 29 25	01 ♌ 27 56	07 ♌ 30 49	06 59	05 49	04 45	08 20	27 00	12 18	07 02	26 22
18	13 49 25	25 28 58	13 ♌ 31 03	19 ♌ 29 14	07 37	07 03	05 16	08 33	27 07	12 20	07 00	26 21
19	13 53 21	26 28 34	25 ♌ 25 52	01 ♍ 21 30	08 24	08 18	05 47	08 46	27 15	12 22	06 59	26 20
20	13 57 18	27 28 12	07 ♍ 16 37	13 ♍ 11 40	09 19	09 33	06 18	08 59	27 22	12 23	06 57	26 20
21	14 01 14	28 27 52	19 ♍ 07 06	25 ♍ 03 18	10 20	10 48	06 48	09 12	27 29	12 25	06 55	26 19
22	14 05 11	29 ♎ 27 34	01 ♎ 00 39	06 ♎ 59 26	11 28	12 02	07 19	09 25	27 36	12 26	06 54	26 19
23	14 09 08	00 ♏ 27 19	12 ♎ 59 59	19 ♎ 02 32	12 41	13 17	07 49	09 38	27 44	12 28	06 52	26 18
24	14 13 04	01 27 05	25 ♎ 07 19	01 ♏ 14 32	13 59	14 32	08 18	09 51	27 51	12 29	06 50	26 18
25	14 17 01	02 26 54	07 ♏ 24 22	13 ♏ 36 57	15 21	15 47	08 48	10 05	27 58	12 30	06 49	26 18
26	14 20 57	03 26 44	19 ♏ 52 28	26 ♏ 11 00	16 46	17 02	09 17	10 18	28 06	12 31	06 47	26 17
27	14 24 54	04 26 36	02 ♐ 32 41	08 ♐ 57 37	18 14	18 17	09 46	10 31	28 13	12 33	06 45	26 17
28	14 28 50	05 26 31	15 ♐ 25 55	21 ♐ 57 42	19 45	19 32	10 15	10 44	28 20	12 34	06 44	26 17
29	14 32 47	06 26 27	28 ♐ 33 02	05 ♑ 12 01	21 17	20 47	10 44	10 57	28 27	12 35	06 42	26 16
30	14 36 43	07 26 25	11 ♑ 54 45	18 ♑ 41 16	22 51	22 02	11 12	11 10	28 35	12 36	06 40	26 16
31	14 40 40	08 ♏ 26 24	25 ♑ 31 37	02 ♒ 25 48	24 ♎ 26	23 ♎ 17	11 ♌ 40	11 ♏ 23	28 ♎ 42	12 ♌ 37	06 ♉ 39	26 ♒ 16

Date	Moon True ☊ ° '	Moon Mean ☊ ° '	Moon ☽ Latitude ° '
01	06 ♉ 23	07 ♉ 33	03 S 31
02	06 D 24	07 30	04 18
03	06 R 24	07 27	04 53
04	06 23	07 24	05 12
05	06 21	07 20	05 12
06	06 18	07 17	04 52
07	06 15	07 14	04 12
08	06 13	07 11	03 15
09	06 11	07 08	02 04
10	06 10	07 05	00 S 45
11	06 D 09	07 01	00 N 35
12	06 10	06 58	01 51
13	06 11	06 55	02 58
14	06 13	06 52	03 54
15	06 14	06 49	04 35
16	06 14	06 46	05 03
17	06 R 14	06 42	05 15
18	06 13	06 39	05 14
19	06 12	06 36	04 59
20	06 11	06 33	04 31
21	06 10	06 30	03 52
22	06 09	06 26	03 03
23	06 08	06 23	02 05
24	06 08	06 20	01 N 01
25	06 D 08	06 17	00 S 07
26	06 08	06 14	01 16
27	06 08	06 11	02 21
28	06 08	06 07	03 21
29	06 R 08	06 04	04 12
30	06 08	06 01	04 49
31	06 ♉ 08	05 ♉ 58	05 S 12

DECLINATIONS

Date	Sun ☉ ° '	Moon ☽ ° '	Mercury ☿ ° '	Venus ♀ ° '	Mars ♂ ° '	Jupiter ♃ ° '	Saturn ♄ ° '	Uranus ♅ ° '	Neptune ♆ ° '	Pluto ♇ ° '
01	03 S 27	26 S 26	09 S 26	06 N 45	21 N 48	12 S 15	07 S 36	17 N 51	12 N 17	22 S 38
02	03 50	27 44	08 49	06 17	21 43	12 19	07 39	17 50	12 16	22 38
03	04 13	27 26	08 09	05 50	21 37	12 23	07 41	17 50	12 16	22 38
04	04 36	25 25	07 27	05 21	21 32	12 27	07 44	17 49	12 15	22 39
05	04 59	21 47	06 43	04 53	21 26	12 32	07 47	17 48	12 15	22 39
06	05 22	16 43	05 59	04 25	21 21	12 36	07 49	17 48	12 14	22 39
07	05 45	10 34	05 15	03 56	21 15	12 40	07 52	17 47	12 14	22 39
08	06 08	03 S 46	04 32	03 28	21 09	12 44	07 55	17 47	12 13	22 39
09	06 31	03 N 15	03 53	02 59	21 04	12 49	07 57	17 46	12 13	22 39
10	06 53	10 02	03 17	02 30	20 58	12 53	08 00	17 46	12 12	22 39
11	07 16	16 08	02 45	02 01	20 52	12 57	08 02	17 45	12 12	22 39
12	07 39	21 13	02 18	01 32	20 46	13 01	08 05	17 45	12 11	22 40
13	08 01	24 57	01 57	01 03	20 40	13 06	08 08	17 44	12 11	22 40
14	08 23	27 11	01 41	00 33	20 34	13 10	08 10	17 44	12 10	22 40
15	08 45	27 51	01 31	00 N 04	20 28	13 14	08 13	17 43	12 10	22 40
16	09 07	27 03	01 27	00 S 25	20 21	13 18	08 16	17 43	12 09	22 40
17	09 29	24 57	01 28	00 55	20 15	13 23	08 18	17 42	12 08	22 40
18	09 51	21 46	01 34	01 24	20 09	13 27	08 21	17 42	12 08	22 40
19	10 13	17 44	01 45	01 53	20 03	13 31	08 23	17 42	12 07	22 40
20	10 34	13 02	02 00	02 23	19 56	13 35	08 26	17 41	12 07	22 40
21	10 56	07 52	02 19	02 52	19 50	13 39	08 29	17 41	12 06	22 40
22	11 17	02 N 23	02 42	03 21	19 44	13 44	08 31	17 40	12 06	22 40
23	11 38	03 S 13	03 07	03 51	19 37	13 48	08 34	17 40	12 05	22 40
24	11 59	08 47	03 36	04 20	19 31	13 52	08 36	17 40	12 05	22 40
25	12 19	14 06	04 07	04 49	19 24	13 56	08 39	17 39	12 04	22 40
26	12 40	18 55	04 40	05 18	19 18	14 00	08 42	17 39	12 04	22 40
27	13 00	22 59	05 14	05 47	19 12	14 04	08 44	17 39	12 03	22 40
28	13 20	25 58	05 50	06 16	19 05	14 09	08 47	17 38	12 02	22 39
29	13 40	27 37	06 27	06 45	18 59	14 13	08 49	17 38	12 02	22 39
30	14 00	27 42	07 05	07 13	18 52	14 17	08 52	17 38	12 01	22 39
31	14 S 19	26 S 08	07 S 43	07 S 42	18 N 46	14 S 21	08 S 54	17 N 38	12 N 01	22 S 39

ZODIAC SIGN ENTRIES

Date	h	m	Planets
02	09	10	☽ ♑
04	13	35	☽ ♒
06	15	09	☽ ♓
08	13	53	♂ ♌
08	15	09	☽ ♈
10	15	23	☽ ♉
12	17	44	☽ ♊
12	19	46	♀ ♎
14	23	33	☽ ♋
17	09	06	☽ ♌
19	21	15	☽ ♍
22	09	58	☽ ♎
23	02	00	☉ ♏
24	21	34	☽ ♏
27	07	13	☽ ♐
29	14	38	☽ ♑
31	19	47	☽ ♒

LATITUDES

Date	Mercury ☿ ° '	Venus ♀ ° '	Mars ♂ ° '	Jupiter ♃ ° '	Saturn ♄ ° '	Uranus ♅ ° '	Neptune ♆ ° '	Pluto ♇ ° '
01	03 S 09	01 N 20	00 N 53	00 N 59	02 N 16	00 N 37	01 S 49	10 S 38
04	02 24	01 23	00 56	00 59	02 15	00 37	01 49	10 38
07	01 26	01 26	01 00	00 58	02 15	00 37	01 49	10 38
10	00 S 25	01 29	01 04	00 58	02 15	00 37	01 49	10 37
13	00 N 31	01 30	01 08	00 58	02 15	00 37	01 49	10 37
16	01 14	01 31	01 12	00 58	02 15	00 37	01 49	10 36
19	01 44	01 32	01 16	00 57	02 15	00 38	01 49	10 36
22	02 00	01 31	01 20	00 57	02 15	00 38	01 49	10 35
25	02 05	01 30	01 24	00 57	02 15	00 38	01 49	10 34
28	02 02	01 29	01 28	00 57	02 15	00 38	01 49	10 34
31	01 N 53	01 N 27	01 N 33	00 N 57	02 N 15	00 N 38	01 S 49	10 S 33

DATA

Julian Date	2466794
Delta T	+78 seconds
Ayanamsa	24° 26' 14"
Synetic vernal point	04° ♓ 40' 45"
True obliquity of ecliptic	23° 26' 10"

MOON'S PHASES, APSIDES AND POSITIONS ☽

Date	h	m	Phase	Longitude ° '	Eclipse Indicator
03	03	33	☽	10 ♑ 19	
09	18	03	○	16 ♈ 49	
16	21	05	☾	23 ♋ 52	
25	01	30	●	02 ♏ 01	Annular

Day	h	m		
08	12	09	Perigee	
20	16	17	Apogee	
02	19	44	Max dec	27° S 49'
09	00	53	0N	
15	10	39	Max dec	27° N 51'
22	22	16	0S	
30	01	16	Max dec	27° S 52'

LONGITUDES

Date	Chiron ⚷ ° '	Ceres ⚳ ° '	Pallas ⚴ ° '	Juno ⚵ ° '	Vesta ⚶ ° '	Black Moon Lilith ⚸ ° '
01	02 ♌ 54	10 ♎ 21	09 ♍ 51	27 ♍ 52	21 ♊ 22	12 ♍ 15
11	03 ♌ 30	14 ♎ 53	14 ♍ 47	01 ♎ 47	22 ♊ 09	13 ♍ 22
21	03 ♌ 58	19 ♎ 25	19 ♍ 34	05 ♎ 36	22 ♊ 15	14 ♍ 29
31	04 ♌ 18	23 ♎ 55	24 ♍ 11	09 ♎ 20	21 ♊ 38	15 ♍ 36

ASPECTARIAN

h m	Aspects
01 Tuesday	
02 52	☽ ± ♆
04 55	☽ Q ♇
07 06	☽ □ ♀
08 57	☽ ⚹ ☿
14 56	☽ ∠ ♃
15 14	☽ ± ♂
16 35	☽ Q ☉
18 32	☿ ⊻ ♀
19 26	☽ bQ ♆
22 37	♃ # ♆
02 Wednesday	
00 22	☽ ⚹ ♄
02 40	☽ ⊼ ♂
02 55	☽ ⚹ ♇
03 22	☽ bQ ♅
05 17	☽ Q ☿
08 21	♂ ⊼ ♇
18 31	☽ ⚹ ♃
19 35	☽ ± ♅
22 07	☽ Q ♄
22 25	☽ △ ♆
23 40	♀ ⊥ ♅
03 Thursday	
03 33	☽ □ ☉
05 42	☽ ∠ ♇
06 13	☽ ⊼ ♅
10 53	☽ □ ☿
15 58	☽ Q ♃
18 21	☽ △ ♀
21 22	☽ ⊥ ♇
04 Friday	
03 02	☿ ∥ ♄
05 45	☽ □ ♄
06 29	♀ ⊥ ♄
07 39	☽ ⊻ ♇
08 54	☉ bQ ♇
09 40	☽ ☍ ♂
18 01	☉ ⚹ ♅
22 24	☽ bQ ♀
23 09	☽ □ ♃
05 Saturday	
01 58	☽ □ ♆
06 24	☉ ☌ ☿
07 06	☽ ∥ ♇
08 46	♀ ∠ ♃
09 11	☉ # ♀
09 33	☽ ☍ ♅
10 04	☽ △ ☿
10 38	☽ △ ☉
13 51	☽ # ♂
14 46	☽ ± ♀
16 17	☿ ⚹ ♅
06 Sunday	
01 24	☿ bQ ♇
01 32	☽ ⊼ ♀
07 18	☽ # ♅
07 34	☽ Q ♆
08 06	☽ △ ♄
08 53	☽ bQ ☿
09 27	☽ ☌ ♇
12 58	☽ bQ ☉
13 21	☽ ⊼ ♂
13 59	☿ Q ♂
15 02	♀ bQ ♆
22 26	☽ ± ☿
23 24	☽ ± ♂
07 Monday	
01 00	☽ △ ♃
01 01	☉ ∥ ☿
02 55	☽ ⚹ ♆
04 15	☽ ∥ ♃
04 30	☽ ± ☉
05 49	☽ # ♆
07 26	☽ ⊼ ☿
08 29	☽ bQ ♄
10 25	☽ ⊼ ♅
14 23	☽ bQ ♂
14 47	☽ ⊼ ☉
20 02	☽ ± ♅
21 41	☽ ∥ ♄
22 58	☽ ± ♄
08 Tuesday	
01 20	☽ bQ ♃
02 51	☽ ∠ ♆
04 15	☽ ∥ ☉
06 07	☽ ☍ ♀
08 37	☽ ⊼ ♄
09 03	☽ ∥ ☿
09 30	☽ ⊻ ♇
10 24	☽ bQ ♅
13 08	☽ # ♀
15 12	☽ △ ♂
15 52	☽ ± ♃
17 10	☽ ⊥ ♆
19 04	☽ ⊥ ♇
09 Wednesday	
01 34	☽ ⊼ ♃
02 43	☽ ⊻ ♆
04 25	☽ ☍ ☿
09 25	☽ ∠ ♇
10 24	☽ △ ♅
11 08	☽ ∥ ♀
14 00	☽ # ☿
15 19	♀ ⊻ ♄
18 03	☽ ☍ ☉
22 54	♀ ⊼ ♇
10 Thursday	
00 01	☽ # ☉
04 35	☽ # ♄
09 06	☽ ☍ ♄
09 35	☽ ⚹ ♇
10 34	☽ ⊼ ♀
11 30	♀ ∠ ♅
15 36	☿ ⊼ ♆
17 18	☽ □ ♂
20 10	☽ ∥ ♆
21 17	☽ ± ♀
22 55	☽ # ♃
11 Friday	
01 41	☿ ⊻ ♃
02 47	☽ ⊼ ☿
02 51	☽ ☍ ♃
03 13	☽ ☌ ♆
05 15	☽ Q ♇
11 23	☽ □ ♅
12 24	☽ ± ☿
13 35	☽ bQ ♀
19 03	☽ ∥ ♅
22 55	☽ ⊼ ☉
12 Saturday	
00 23	☽ Q ♂
02 03	♃ ☍ ♆
03 05	☽ bQ ☿
09 41	☽ ∥ ♂
09 54	☽ ± ☉
11 31	☽ ⊼ ♄
11 33	☽ □ ♇
14 54	♄ △ ♇
17 32	☽ △ ♀
18 01	☽ Q ♅
20 16	☽ # ♇
21 47	☽ ⚹ ♂
21 59	☽ ± ♄
13 Sunday	
02 41	☽ bQ ☉
04 27	☽ △ ☿
06 12	☽ ⊻ ♆
06 43	☽ ⊼ ♃
13 57	☽ bQ ♄
15 08	☽ ⚹ ♅
16 49	☽ ⊥ ♆
17 31	☽ ± ♃
17 42	♀ ± ♆
20 12	☉ ∥ ♄
14 Monday	
01 28	☽ ∠ ♂
02 58	♀ ⊥ ♃
05 52	☿ St D
07 36	☽ △ ☉
09 02	☽ ∠ ♆
10 03	☽ bQ ♃
16 52	☽ △ ♇
17 22	☽ △ ♄
17 53	♀ ± ♇
18 25	☽ ∠ ♅
18 36	☽ ⊥ ♂
15 Tuesday	
05 05	☽ □ ♀
06 14	☽ ⊻ ♂
11 07	☽ □ ☿
11 15	☽ ⊥ ♅
12 50	☽ ⚹ ♆
14 25	☽ △ ♃
20 58	☽ bQ ♇
22 39	☽ ⊻ ♅
16 Wednesday	
00 55	♀ ⚹ ♂
11 44	☽ Q ♆
14 16	☽ ± ♇
19 16	☽ Q ♀
21 05	☽ □ ☉
22 42	☽ Q ☿
17 Thursday	
01 57	☽ ⊼ ♇
03 07	☽ □ ♄
07 27	☉ Q ♅
14 02	☿ ⊼ ♆
18 49	☽ ☌ ♂
21 36	☽ ⚹ ♀
23 01	☽ □ ♆
23 31	☽ ⚹ ☿
18 Friday	
01 54	☽ □ ♃
05 57	☽ # ♇
09 37	☽ ☌ ♅
11 03	♀ ⊼ ♆
11 55	☽ Q ☉
15 15	☽ Q ♄
22 27	☽ ∥ ♂
23 54	☿ ∥ ♀
19 Saturday	
07 12	☽ ∠ ♀
07 35	☽ ∠ ☿
08 45	☉ △ ♇
12 12	☽ ∥ ♅
13 50	☽ ☍ ♇
14 18	☽ ⚹ ☉
14 46	☽ Q ♃
15 42	☽ ⚹ ♄
18 10	☿ ☌ ♀
22 57	♀ ⊻ ♃
20 Sunday	
01 23	☿ ⊻ ♃
03 15	☽ ⊥ ☿
03 33	☽ ⊥ ♀
09 07	☉ ☌ ♄
09 21	☽ # ♃
09 56	☽ ⊻ ♂
11 20	☽ △ ♆
15 32	☽ ⚹ ♃
16 25	☽ ∥ ♆
16 30	☽ ⊻ ☿
17 09	☽ ⊻ ♀
22 23	☽ ⊻ ♅
22 26	☽ ∠ ♄
22 38	☽ ⊥ ♂
22 54	☽ # ☉
23 30	☽ ∠ ☉
21 Monday	
09 16	☽ # ♄
10 34	☽ ⊥ ♅
16 50	☽ ⊥ ♄
17 16	♂ □ ♆
17 40	☽ bQ ♆
17 41	☽ ∠ ♂
19 23	☽ ⊥ ☉
22 09	♀ bQ ♇
22 28	☽ ∠ ♃
22 Tuesday	
02 33	☽ ⊼ ♇
04 47	☽ ∠ ♅
05 05	☽ ⊻ ♄
08 10	☽ # ♀
08 36	☽ ⊻ ☉
08 55	☿ bQ ♇
10 47	☽ # ☿
11 46	☽ ± ♆
14 37	☽ ± ♇
16 56	☽ ⊥ ♃
19 46	♀ ⚹ ♅
23 47	☽ ⊼ ♆
23 Wednesday	
01 11	☽ ⚹ ♂
05 10	☽ ⊻ ♃
07 33	☿ ⚹ ♅
08 38	☽ bQ ♇
10 55	☽ ⚹ ♅
11 18	☽ ☌ ☿
11 35	☽ ∥ ☿
12 38	☽ ☌ ♀
14 57	☽ ∥ ♀
24 Thursday	
02 06	☽ Q ♂
10 44	☽ Q ♅
11 14	☽ ∥ ♄
14 19	☽ △ ♇
17 25	☽ ☌ ♄
18 48	☉ # ♆
25 Friday	
01 30	☽ ☌ ☉
02 40	☽ # ♆
03 14	☽ ∥ ☉
10 51	☽ ☍ ♆
11 15	☽ ∥ ♃
14 49	☽ □ ♂
17 16	☽ ☌ ♃
21 52	☽ □ ♅
26 Saturday	
05 17	☽ ⊻ ☿
05 23	☽ # ♅
05 58	☽ ⊻ ♀
14 01	☽ # ♂
18 14	☽ ⊥ ☿
18 41	☽ ⊥ ♀
27 Sunday	
00 12	☽ □ ♇
03 46	☽ ⊻ ♄
09 54	☽ ∥ ♇
13 29	☽ ∠ ☿
13 32	☽ ∠ ♀
15 10	☽ ⊥ ♄
15 52	☽ ⊻ ☉
16 11	☿ ☌ ♀
19 52	☽ ⊼ ♆
28 Monday	
02 02	☽ △ ♂
03 09	☽ ⊻ ♃
04 00	☽ ⊥ ☉
06 41	☽ △ ♅
07 00	☽ ± ♆
08 05	☽ ∠ ♄
09 52	☽ Q ♇
14 26	☽ ⊥ ♃
20 21	☽ ⚹ ♀
20 59	☽ ⚹ ☿
21 59	☽ ∠ ☉
23 33	☽ bQ ♆
29 Tuesday	
06 41	☽ bQ ♂
07 12	☽ ∠ ♃
07 52	☽ ⚹ ♇
10 14	☽ bQ ♅
11 50	☽ ⚹ ♄
18 02	☉ ☍ ♆
20 26	☽ Q ♀
21 41	☽ Q ☿
23 34	☽ ± ♂
30 Wednesday	
02 30	☽ ± ♅
02 39	☽ △ ♆
03 22	☽ ⚹ ☉
08 52	♂ □ ♃
09 36	☽ Q ♄
10 39	☽ ⚹ ♃
10 41	☽ ⊼ ♂
10 51	☽ ∠ ♇
13 13	☽ ⊼ ♅
19 37	☿ Q ♂
31 Thursday	
02 23	☽ Q ☉
02 47	☽ ⊥ ♇
07 41	☽ □ ♀
08 12	☽ Q ♃
09 29	☿ ∥ ♀
09 51	☽ □ ☿
13 17	☽ ⊻ ♇
14 43	☿ Q ♅
15 04	☉ ∥ ♃
17 34	☽ □ ♄
23 53	♀ Q ♂

All ephemeris data is given at 12.00 UT and the Moon's longitude is additionally given for 24.00 UT

NOVEMBER 2041

LONGITUDES

Date	Sidereal time h m s	Sun ☉ ° ' "	Moon ☽ ° ' "	Moon ☽ 24.00 ° ' "	Mercury ☿ ° '	Venus ♀ ° '	Mars ♂ ° '	Jupiter ♃ ° '	Saturn ♄ ° '	Uranus ♅ ° '	Neptune ♆ ° '	Pluto ♇ ° '
01	14 44 37	09 ♏ 26 25	09 ♒ 23 47	16 ♒ 25 28	26 ♎ 02	24 ♎ 32	12 ♌ 08	11 ♏ 36	28 ♎ 49	12 ♌ 38	06 ♉ 37	26 ♒ 16
02	14 48 33	10 26 27	23 ♒ 30 43	00 ♓ 39 16	27 39	25 47	12 36	11 50	28 56	12 39	06 R 35	26 R 16
03	14 52 30	11 26 31	07 ♓ 50 51	15 ♓ 05 02	29 ♎ 16	27 02	13 03	12 03	29 03	12 39	06 34	26 15
04	14 56 26	12 26 37	22 ♓ 21 21	29 ♓ 39 13	00 ♏ 54	28 17	13 30	12 16	29 11	12 40	06 32	26 15
05	15 00 23	13 26 44	06 ♈ 58 01	14 ♈ 17 01	02 32	29 ♎ 32	13 57	12 29	29 18	12 41	06 30	26 15
06	15 04 19	14 26 52	21 ♈ 35 28	28 ♈ 52 36	04 10	00 ♏ 47	14 23	12 42	29 25	12 42	06 28	26 15
07	15 08 16	15 27 02	06 ♉ 07 38	13 ♉ 19 49	05 48	02 03	14 49	12 55	29 32	12 42	06 27	26 15
08	15 12 12	16 27 15	20 ♉ 28 26	27 ♉ 32 53	07 26	03 18	15 15	13 09	29 39	12 43	06 25	26 15
09	15 16 09	17 27 28	04 ♊ 32 36	11 ♊ 27 09	09 04	04 33	15 41	13 22	29 46	12 43	06 24	26 D 15
10	15 20 06	18 27 44	18 ♊ 16 14	24 ♊ 59 38	10 42	05 48	16 06	13 35	29 ♎ 53	12 44	06 22	26 15
11	15 24 02	19 28 02	01 ♋ 37 18	08 ♋ 09 15	12 20	07 03	16 31	13 48	00 ♏ 00	12 44	06 20	26 15
12	15 27 59	20 28 22	14 ♋ 35 39	20 ♋ 56 46	13 57	08 19	16 56	14 01	00 07	12 44	06 19	26 15
13	15 31 55	21 28 43	27 ♋ 12 54	03 ♌ 24 29	15 34	09 34	17 20	14 14	00 14	12 45	06 17	26 15
14	15 35 52	22 29 06	09 ♌ 32 00	15 ♌ 35 59	17 11	10 49	17 44	14 27	00 21	12 45	06 15	26 15
15	15 39 48	23 29 32	21 ♌ 36 58	27 ♌ 35 33	18 48	12 05	18 08	14 41	00 28	12 45	06 14	26 16
16	15 43 45	24 29 59	03 ♍ 32 21	09 ♍ 27 58	20 24	13 20	18 32	14 54	00 35	12 45	06 12	26 16
17	15 47 41	25 30 27	15 ♍ 23 01	21 ♍ 18 06	22 01	14 35	18 55	15 07	00 42	12 45	06 11	26 16
18	15 51 38	26 30 58	27 ♍ 13 48	03 ♎ 10 40	23 36	15 51	19 17	15 20	00 49	12 R 45	06 09	26 16
19	15 55 35	27 31 30	09 ♎ 09 14	15 ♎ 09 59	25 12	17 06	19 39	15 33	00 56	12 45	06 07	26 16
20	15 59 31	28 32 05	21 ♎ 13 22	27 ♎ 19 46	26 48	18 21	20 01	15 46	01 02	12 45	06 06	26 17
21	16 03 28	29 ♏ 32 40	03 ♏ 29 31	09 ♏ 42 54	28 23	19 37	20 23	15 59	01 09	12 45	06 04	26 17
22	16 07 24	00 ♐ 33 18	16 ♏ 00 08	22 ♏ 21 20	29 ♏ 58	20 52	20 44	16 12	01 16	12 44	06 03	26 17
23	16 11 21	01 33 57	28 ♏ 46 36	05 ♐ 15 56	01 ♐ 33	22 07	21 05	16 25	01 22	12 44	06 01	26 18
24	16 15 17	02 34 37	11 ♐ 49 16	18 ♐ 26 30	03 07	23 23	21 25	16 38	01 29	12 44	06 00	26 18
25	16 19 14	03 35 19	25 ♐ 07 26	01 ♑ 51 51	04 42	24 38	21 45	16 51	01 36	12 43	05 59	26 19
26	16 23 10	04 36 02	08 ♑ 39 31	15 ♑ 30 08	06 16	25 54	22 04	17 04	01 42	12 43	05 57	26 19
27	16 27 07	05 36 47	22 ♑ 23 24	29 ♑ 19 03	07 51	27 09	22 23	17 17	01 49	12 42	05 56	26 20
28	16 31 04	06 37 32	06 ♒ 16 45	13 ♒ 16 14	09 25	28 25	22 42	17 29	01 55	12 41	05 54	26 20
29	16 35 00	07 38 18	20 ♒ 17 15	27 ♒ 19 32	10 59	29 ♏ 40	23 00	17 42	02 02	12 41	05 53	26 21
30	16 38 57	08 ♐ 39 06	04 ♓ 22 52	11 ♓ 27 02	12 ♐ 33	00 ♐ 56	23 ♌ 17	17 ♏ 55	02 ♏ 08	12 ♌ 41	05 ♉ 51	26 ♒ 21

Moon Nodes, Latitude and DECLINATIONS

Date	Moon True ☊ ° '	Moon Mean ☊ ° '	Moon ☽ Latitude ° '	Sun ☉ ° '	Moon ☽ ° '	Mercury ☿ ° '	Venus ♀ ° '	Mars ♂ ° '	Jupiter ♃ ° '	Saturn ♄ ° '	Uranus ♅ ° '	Neptune ♆ ° '	Pluto ♇ ° '
01	06 ♉ 08	05 ♉ 55	05 S 17	14 S 38	22 S 59	08 S 22	08 S 10	18 N 40	14 S 25	08 S 57	17 N 37	12 N 00	22 S 39
02	06 D 08	05 52	05 03	14 57	18 26	09 01	08 38	18 33	14 29	08 59	17 37	12 00	22 39
03	06 08	05 48	04 30	15 16	12 48	09 40	09 06	18 27	14 33	09 02	17 37	11 59	22 39
04	06 09	05 45	03 40	15 34	06 S 24	10 19	09 34	18 21	14 37	09 04	17 37	11 59	22 39
05	06 09	05 42	02 35	15 52	00 N 24	10 58	10 02	18 15	14 41	09 07	17 37	11 58	22 38
06	06 10	05 39	01 S 20	16 10	07 11	11 36	10 29	18 08	14 45	09 09	17 37	11 58	22 38
07	06 10	05 36	00 00	16 28	13 33	12 15	10 56	18 02	14 49	09 12	17 36	11 57	22 38
08	06 R 10	05 32	01 N 18	16 45	19 08	12 53	11 23	17 56	14 53	09 14	17 36	11 56	22 38
09	06 09	05 29	02 31	17 02	23 31	13 30	11 49	17 50	14 57	09 16	17 36	11 56	22 38
10	06 08	05 26	03 33	17 19	26 27	14 07	12 16	17 44	15 01	09 19	17 36	11 55	22 38
11	06 06	05 23	04 21	17 36	27 46	14 43	12 42	17 38	15 05	09 21	17 36	11 55	22 37
12	06 04	05 20	04 54	17 52	27 31	15 19	13 07	17 32	15 09	09 24	17 36	11 54	22 37
13	06 02	05 17	05 12	18 08	25 49	15 54	13 33	17 26	15 13	09 26	17 36	11 54	22 37
14	06 01	05 13	05 15	18 23	22 55	16 28	13 58	17 20	15 17	09 28	17 36	11 53	22 37
15	06 00	05 10	05 04	18 39	19 06	17 01	14 22	17 15	15 21	09 31	17 36	11 53	22 36
16	06 D 01	05 07	04 40	18 54	14 34	17 34	14 47	17 09	15 25	09 33	17 36	11 52	22 36
17	06 02	05 04	04 05	19 08	09 31	18 06	15 11	17 04	15 28	09 35	17 36	11 52	22 36
18	06 03	05 01	03 18	19 22	04 N 08	18 37	15 34	16 58	15 32	09 38	17 36	11 51	22 35
19	06 05	04 57	02 23	19 36	01 S 26	19 07	15 57	16 53	15 36	09 40	17 36	11 51	22 35
20	06 06	04 54	01 21	19 50	07 01	19 36	16 20	16 48	15 40	09 42	17 36	11 50	22 35
21	06 07	04 51	00 N 14	20 03	12 27	20 04	16 42	16 42	15 44	09 44	17 36	11 50	22 35
22	06 R 07	04 48	00 S 54	20 16	17 29	20 31	17 04	16 37	15 47	09 46	17 36	11 49	22 34
23	06 05	04 45	02 01	20 28	21 51	20 57	17 26	16 33	15 51	09 49	17 37	11 49	22 34
24	06 02	04 42	03 01	20 40	25 14	21 22	17 47	16 28	15 55	09 51	17 37	11 48	22 34
25	05 58	04 38	03 57	20 52	27 18	21 46	18 07	16 23	15 58	09 53	17 37	11 48	22 33
26	05 53	04 35	04 38	21 03	27 47	22 09	18 27	16 19	16 02	09 55	17 37	11 48	22 33
27	05 49	04 32	05 04	21 14	26 34	22 31	18 47	16 14	16 06	09 57	17 37	11 47	22 33
28	05 45	04 29	05 12	21 25	23 44	22 51	19 06	16 10	16 09	09 59	17 37	11 47	22 32
29	05 42	04 26	05 02	21 35	19 30	23 11	19 24	16 06	16 13	10 01	17 38	11 46	22 32
30	05 ♉ 41	04 ♉ 23	04 S 34	21 S 45	14 S 09	23 S 29	19 S 42	16 N 02	16 S 16	10 S 03	17 N 38	11 N 46	22 S 31

ZODIAC SIGN ENTRIES

Date	h	m	Planets
02	22	54	☽ ♓
03	22	43	☿ ♏
05	00	34	☽ ♈
05	20	51	♀ ♏
07	01	51	☽ ♉
09	04	11	☽ ♊
11	09	03	☽ ♋
11	10	58	♄ ♏
13	17	23	☽ ♌
16	04	51	☽ ♍
18	17	36	☽ ♎
21	05	13	☽ ♏
21	22	49	☉ ♐
22	12	31	☿ ♐
23	14	16	☽ ♐
25	20	41	☽ ♑
28	01	11	☽ ♒
29	18	20	♀ ♐
30	04	33	☽ ♓

LATITUDES

Date	Mercury ☿ ° '	Venus ♀ ° '	Mars ♂ ° '	Jupiter ♃ ° '	Saturn ♄ ° '	Uranus ♅ ° '	Neptune ♆ ° '	Pluto ♇ ° '
01	01 N 49	01 N 26	01 N 34	00 N 57	02 N 15	00 N 38	01 S 49	10 S 33
04	01 34	01 23	01 39	00 56	02 15	00 38	01 49	10 32
07	01 17	01 20	01 44	00 56	02 15	00 38	01 49	10 32
10	00 58	01 16	01 48	00 56	02 16	00 38	01 49	10 31
13	00 38	01 12	01 53	00 56	02 16	00 39	01 49	10 30
16	00 N 17	01 07	01 59	00 56	02 16	00 39	01 49	10 30
19	00 S 03	01 02	02 04	00 56	02 16	00 39	01 49	10 29
22	00 23	00 56	02 09	00 56	02 17	00 39	01 49	10 28
25	00 42	00 50	02 15	00 56	02 17	00 39	01 49	10 28
28	01 00	00 44	02 21	00 56	02 17	00 39	01 49	10 27
31	01 S 18	00 N 37	02 N 27	00 N 56	02 N 18	00 N 39	01 S 49	10 S 26

DATA

Julian Date	2466825
Delta T	+78 seconds
Ayanamsa	24° 26' 18"
Synetic vernal point	04° ♓ 40' 40"
True obliquity of ecliptic	23° 26' 09"

MOON'S PHASES, APSIDES AND POSITIONS ☽

Date	h	m	Phase	Longitude ° '	Eclipse Indicator
01	12	05	☽	09 ♒ 27	
08	04	43	○	16 ♉ 09	partial
15	16	06	☾	23 ♌ 40	
23	17	37	●	01 ♐ 48	
30	19	49	☽	08 ♓ 59	

Day	h	m		
05	15	54	Perigee	
17	10	12	Apogee	
05	10	38	ON	
11	19	51	Max dec	27° N 51'
19	05	52	OS	
26	06	58	Max dec	27° S 49'

LONGITUDES

Date	Chiron ⚷ ° '	Ceres ⚳ ° '	Pallas ⚴ ° '	Juno ⚵ ° '	Vesta ⚶ ° '	Black Moon Lilith ⚸ ° '
01	04 ♌ 19	24 ♎ 22	24 ♍ 38	09 ♎ 42	21 ♊ 32	15 ♍ 42
11	04 ♌ 28	28 ♎ 49	29 ♍ 02	13 ♎ 17	20 ♊ 06	16 ♍ 49
21	04 ♌ 27	03 ♏ 12	03 ♎ 14	16 ♎ 44	18 ♊ 03	17 ♍ 56
31	04 ♌ 16	07 ♏ 31	07 ♎ 10	20 ♎ 00	15 ♊ 34	19 ♍ 03

ASPECTARIAN

h m Aspects

01 Friday
07 14 ☽ □ ♆
12 05 ☽ □ ☉
13 53 ♀ Q ♅
14 01 ☽ ∥ ♇
15 21 ☿ △ ♇
15 51 ☽ □ ♃
16 51 ☽ ☍ ♂
17 32 ☽ ☍ ♅

02 Saturday
11 03 ☿ ∥ ♄
11 27 ☽ # ♂
13 48 ☽ Q ♆
14 42 ♂ ☌ ♅
15 46 ☽ # ♅
16 12 ☽ △ ♀
16 38 ☽ ☌ ♇
19 51 ☽ △ ☿
21 06 ♀ △ ♇
21 12 ☽ △ ♄

03 Sunday
02 31 ☽ ∥ ☉
05 01 ☽ ∥ ♃
07 42 ♀ ∥ ♄
08 34 ☿ ☌ ♄
09 52 ☽ ✶ ♆
15 11 ☽ # ♆
18 25 ☽ △ ☉
19 05 ☽ △ ♃
19 37 ☽ ⚼ ♀
19 59 ☽ ⚻ ♅
20 55 ☽ ⚻ ♂
22 23 ☽ ⚼ ♄
22 55 ☽ ∥ ☿

04 Monday
00 00 ☽ ⚼ ☿
01 10 ☽ ∥ ♀
02 18 ☽ ∥ ♄
05 55 ☽ ± ♅
06 33 ☉ ☌ ♃
07 09 ☽ ± ♂
10 39 ☽ ∠ ♆
11 52 ☽ ± ♀
13 22 ☽ ± ♄
16 43 ☽ ± ☿
17 30 ☉ □ ♅
18 25 ☽ ⊻ ♇
20 12 ☽ ⚼ ♃
20 45 ☽ ⚼ ♅
20 59 ☽ ⚼ ☉
22 25 ☽ ⚼ ♂
22 40 ☽ ⚻ ♀
23 19 ☽ ⚻ ♄

05 Tuesday
01 25 ☽ ⊥ ♆
03 49 ☽ ⚻ ☿
04 16 ☽ ⊥ ♇
06 53 ♀ ☌ ♄
11 12 ☽ ± ♃
11 14 ☽ ⊻ ♆
12 51 ☽ ± ☉
19 02 ☽ ∠ ♇
21 11 ☽ ⚻ ♃
21 23 ☽ △ ♅
23 24 ☽ ⚻ ☉
23 48 ☽ △ ♂

06 Wednesday
09 26 ☉ □ ♂
10 35 ♃ □ ♅
19 14 ☽ # ♄
19 40 ☽ ✶ ♇

07 Thursday
01 00 ☿ # ♆
01 00 ☽ ☍ ♄
01 04 ☽ # ♀
04 36 ☽ ☍ ♀
05 43 ☽ ∥ ♆
06 16 ☽ # ☿
11 24 ☽ ☍ ☿
12 32 ☽ ☌ ♆
15 32 ☽ Q ♇
17 09 ☽ # ♃
21 18 ☿ ☍ ♆
22 57 ☽ □ ♅
23 30 ☽ ☍ ♃

08 Friday
00 39 ☽ # ☉
02 57 ☽ □ ♂
04 43 ☽ ☍ ☉
05 00 ☽ ∥ ♅
06 34 ☽ ∥ ♂
21 47 ☽ □ ♇

09 Saturday
00 13 ♇ St D
03 44 ☽ ⚻ ♄
05 25 ☽ Q ♅
06 30 ☽ # ♇
10 28 ☽ Q ♂
12 01 ☽ ⚻ ♀
14 08 ☽ ± ♄
15 11 ☽ ⊻ ♆
17 56 ♀ # ♆
20 54 ☽ ⚻ ☿
23 28 ☽ ± ♀

10 Sunday
01 37 ☽ ⊥ ♆
02 14 ☽ ✶ ♅
03 36 ☽ ⚻ ♃
05 58 ☽ ⚼ ♄
08 03 ☽ ✶ ♂
08 51 ☽ ± ☿
12 22 ☽ ⚻ ☉
14 22 ☽ ± ♃
16 58 ☽ ⚼ ♀
17 29 ☽ ∠ ♆
22 31 ♀ ☍ ♆
23 57 ☽ ± ☉

11 Monday
02 16 ☽ △ ♇
03 07 ☽ ⚼ ☿
04 56 ☽ ∠ ♅
06 47 ☽ ⚼ ♃
09 02 ☽ △ ♄
11 49 ☽ ∠ ♂
12 38 ☉ # ♅
14 27 ☉ # ♂
17 38 ☽ ⚼ ☉
18 00 ☿ □ ♅
19 31 ♂ ∥ ♅
20 38 ☽ ✶ ♆
21 23 ☽ ⊥ ♅
23 02 ☽ △ ♀

12 Tuesday
04 27 ☿ ∥ ♃
04 56 ☽ ⊥ ♂
05 45 ☽ ⚼ ♇
08 31 ☽ ⊻ ♅
10 37 ☽ △ ☿
10 54 ☽ △ ♃
13 11 ☿ ☌ ♃
16 33 ☽ ⊻ ♂
18 59 ☽ Q ♆
22 41 ☽ ± ♇

13 Wednesday
00 04 ☽ △ ☉
10 09 ☽ ⚻ ♇
17 54 ☽ □ ♄

14 Thursday
05 35 ☽ □ ♆
14 12 ☽ # ♇
14 50 ☽ □ ♀
18 21 ☽ ☌ ♅
21 55 ☽ □ ♃
22 56 ☿ □ ♂

15 Friday
04 49 ☽ ☌ ♂
05 30 ☽ □ ☿
05 39 ☽ Q ♄
14 24 ☽ # ☉
16 06 ☽ □ ☉
20 15 ☿ # ♂
20 16 ☽ ∥ ♅
21 19 ☽ ☍ ♇
22 07 ☽ # ☿
22 23 ☽ ∥ ♂

16 Saturday
00 53 ♀ □ ♅
05 58 ☽ ✶ ♄
07 01 ☽ Q ♀
07 47 ☽ # ♃
10 40 ☽ Q ♃
11 00 ☽ # ♀
13 26 ☿ # ♅
17 14 ♂ Q ♄
17 23 ☽ △ ♆
23 24 ☽ Q ☿

17 Sunday
01 03 ☽ ∥ ♆
06 40 ☽ ⊻ ♅
07 50 ☽ Q ☉
10 11 ☽ ✶ ♀
11 26 ☽ ✶ ♃
11 41 ☽ # ♄
12 39 ☽ ∠ ♄
17 17 ♅ St R
18 50 ☽ ⊥ ♅
19 23 ☽ ⊻ ♂
23 43 ☽ ⚼ ♆

18 Monday
00 11 ♀ ☌ ♃
03 32 ☽ ✶ ☿
06 06 ☉ □ ♇
07 04 ☽ ⊥ ♄
07 56 ☽ ⊥ ♂
09 43 ♀ ∥ ♃
10 03 ☽ ⚻ ♇
10 25 ☽ ✶ ☉
13 03 ☽ ∠ ♅
17 53 ☽ ± ♆
18 23 ☽ ∠ ♃
19 18 ☽ ⊻ ♄
20 09 ☽ ∠ ♀
22 10 ☽ ± ♇

19 Tuesday
02 41 ☽ ∠ ♂
05 56 ☽ ⚻ ♆
12 48 ☽ ⊥ ♃
14 25 ☽ ∠ ☿
16 14 ☽ ⚼ ♇
16 21 ☽ ⊥ ♀
19 11 ☽ ✶ ♅
19 21 ☽ ∠ ☉

20 Wednesday
01 00 ☽ ⊻ ♃
04 12 ☿ □ ♇
05 40 ☽ ⊻ ♀
09 33 ☽ ✶ ♂
11 02 ☽ ⊥ ☿
14 49 ☽ ⊥ ☉
18 56 ☽ Q ♅
21 57 ☽ △ ♇
23 46 ☽ ∥ ♄

21 Thursday
00 35 ☽ ⊻ ☿
03 38 ☽ ⊻ ☉
07 25 ☽ ☌ ♄
09 12 ☽ # ♆
09 47 ☽ Q ♂
10 40 ☉ ∥ ☿
12 05 ♀ # ♂
16 59 ☽ ☍ ♆

22 Friday
03 30 ☽ ∥ ♃
05 48 ☽ □ ♅
07 46 ☽ # ♂
08 25 ♀ □ ♂
09 44 ☽ ∥ ♀
12 23 ☽ ☌ ♃
12 36 ☽ # ♅
21 12 ☽ □ ♂
22 12 ☽ ☌ ♀

23 Saturday
03 28 ☽ ∥ ☉
06 00 ☽ ∥ ☿
06 52 ☉ ⊻ ♄
07 23 ☽ □ ♇
09 10 ☿ ⊻ ♄
12 49 ☉ ☌ ☿
16 26 ☽ ∥ ♇
16 51 ☽ ⊻ ♄
17 37 ☽ ☌ ☉
17 51 ☽ ☌ ☿

24 Sunday
00 24 ♀ # ♅
01 22 ☽ ⚻ ♆
04 00 ☽ ⊥ ♄
12 19 ☽ ± ♆
13 39 ☽ △ ♅
16 31 ☽ Q ♇
20 32 ☽ ∠ ♄
20 52 ☽ ⊻ ♃

25 Monday
04 35 ☽ ⚼ ♆
05 48 ☽ △ ♂
07 51 ☽ ⊥ ♃
11 03 ☽ ⊻ ♀
14 07 ☽ ✶ ♇
16 38 ☽ ⚼ ♅
22 50 ☽ ⊥ ♀
23 37 ☽ ✶ ♄

26 Tuesday
00 10 ☽ ∠ ♃
04 16 ☽ ⊻ ☉
07 10 ☿ ⚻ ♆
07 14 ☽ △ ♆
07 15 ☽ ⊻ ☿
08 35 ☽ ± ♅
09 08 ☽ ⚼ ♂
15 41 ☽ ⊥ ☉
16 20 ☽ ∠ ♀
16 40 ☽ ∠ ♇
19 07 ☽ ⚻ ♅
19 10 ☽ ⊥ ☿
20 06 ♀ □ ♇
20 55 ☽ Q ♄

27 Wednesday
01 18 ☽ ± ♂
02 57 ☽ ✶ ♃
08 24 ☽ ⊥ ♇
08 40 ☽ ∠ ☉
11 28 ☿ ⊥ ♄
12 00 ☽ ⚻ ♂
12 53 ☽ ∠ ☿
13 53 ☿ ∥ ♇
18 50 ☽ ⊻ ♇
19 17 ☉ ⚻ ♆
21 05 ☽ ✶ ♀

28 Thursday
00 07 ☽ Q ♃
04 26 ☽ □ ♄
11 21 ☽ □ ♆
12 39 ☽ ✶ ☉
14 04 ♂ # ♃
17 11 ☽ ∥ ☿
18 04 ☽ ✶ ☿
19 36 ☽ ∥ ♇
19 48 ☽ Q ♀
23 01 ☽ ☍ ♅

29 Friday
01 29 ☽ ∥ ☉
07 31 ☽ □ ♃
10 48 ☽ Q ☉
12 26 ☽ ∥ ♀
16 43 ☽ ☍ ♂
17 10 ☽ Q ☿
18 07 ☽ Q ♆
20 53 ☽ # ♅
22 13 ☉ ⊥ ♄
22 20 ☽ ☌ ♇

30 Saturday
01 39 ☿ ± ♆
03 03 ☽ ∥ ♃
03 55 ☽ # ♂
05 33 ☽ □ ♀
08 09 ☽ △ ♄
14 02 ☿ △ ♅
14 30 ☽ ✶ ♆
19 49 ☽ □ ☉
21 42 ☽ # ♆

All ephemeris data is given at 12.00 UT and the Moon's longitude is additionally given for 24.00 UT

LONGITUDES

Date	Sidereal time h m s	Sun ☉ ° ' "	Moon ☽ ° ' "	Moon ☽ 24.00 ° ' "	Mercury ☿ ° '	Venus ♀ ° '
01	16 42 53	09 ♐ 39 54	18 ♓ 31 51	25 ♓ 37 05	14 ♐ 06	02 ♐ 11
02	16 46 50	10 40 43	02 ♈ 42 33	09 ♈ 48 01	15 40	03 26
03	16 50 46	11 41 32	16 ♈ 53 15	23 ♈ 57 56	17 14	04 42
04	16 54 43	12 42 23	01 ♉ 01 46	08 ♉ 04 25	18 48	05 57
05	16 58 39	13 43 14	15 ♉ 05 29	22 ♉ 04 33	20 21	07 13
06	17 02 36	14 44 07	29 ♉ 01 13	05 ♊ 55 03	21 55	08 28
07	17 06 33	15 45 01	12 ♊ 45 38	19 ♊ 32 35	23 28	09 44
08	17 10 29	16 45 55	26 ♊ 15 33	02 ♋ 54 15	25 02	10 59
09	17 14 26	17 46 51	09 ♋ 28 30	15 ♋ 58 09	26 36	12 15
10	17 18 22	18 47 47	22 ♋ 23 10	28 ♋ 43 36	28 09	13 30
11	17 22 19	19 48 45	04 ♌ 59 36	11 ♌ 11 24	29 ♐ 43	14 46
12	17 26 15	20 49 43	17 ♌ 19 18	23 ♌ 23 42	01 ♑ 16	16 01
13	17 30 12	21 50 43	29 ♌ 25 04	05 ♍ 23 53	02 50	17 17
14	17 34 08	22 51 43	11 ♍ 20 45	17 ♍ 16 16	04 23	18 32
15	17 38 05	23 52 45	23 ♍ 11 03	29 ♍ 05 45	05 57	19 48
16	17 42 02	24 53 48	05 ♎ 01 04	10 ♎ 57 39	07 30	21 03
17	17 45 58	25 54 51	16 ♎ 56 10	22 ♎ 57 15	09 03	22 19
18	17 49 55	26 55 55	29 ♎ 01 31	05 ♏ 09 32	10 36	23 34
19	17 53 51	27 57 01	11 ♏ 21 49	17 ♏ 38 49	12 08	24 50
20	17 57 48	28 58 07	24 ♏ 00 53	00 ♐ 28 17	13 41	26 05
21	18 01 44	29 ♐ 59 14	07 ♐ 01 10	13 ♐ 39 35	15 12	27 21
22	18 05 41	01 ♑ 00 21	20 ♐ 23 27	27 ♐ 12 30	16 43	28 36
23	18 09 37	02 01 29	04 ♑ 06 26	11 ♑ 04 46	18 14	29 ♐ 52
24	18 13 34	03 02 38	18 ♑ 06 56	25 ♑ 12 17	19 43	01 ♑ 08
25	18 17 31	04 03 47	02 ♒ 20 07	09 ♒ 29 43	21 11	02 23
26	18 21 27	05 04 56	16 ♒ 40 21	23 ♒ 51 22	22 38	03 39
27	18 25 24	06 06 05	01 ♓ 02 07	08 ♓ 12 05	24 03	04 54
28	18 29 20	07 07 14	15 ♓ 20 48	22 ♓ 27 55	25 26	06 10
29	18 33 17	08 08 22	29 ♓ 33 10	06 ♈ 36 22	26 47	07 25
30	18 37 13	09 09 31	13 ♈ 37 26	20 ♈ 36 18	28 05	08 41
31	18 41 10	10 ♑ 10 40	27 ♈ 32 57	04 ♉ 27 24	29 ♑ 20	09 ♑ 56

Date	Mars ♂ ° '	Jupiter ♃ ° '	Saturn ♄ ° '	Uranus ♅ ° '	Neptune ♆ ° '	Pluto ⯓ ° '
01	23 ♌ 35	18 ♏ 08	02 ♏ 14	12 ♌ 40	05 ♉ 50	26 ♒ 22
02	23 51	18 20	02 20	12 R 39	05 R 49	26 22
03	24 07	18 33	02 27	12 38	05 48	26 23
04	24 23	18 46	02 33	12 37	05 46	26 24
05	24 38	18 58	02 39	12 36	05 45	26 24
06	24 53	19 11	02 45	12 35	05 44	26 25
07	25 07	19 23	02 51	12 34	05 43	26 26
08	25 21	19 36	02 57	12 33	05 41	26 26
09	25 34	19 48	03 03	12 32	05 40	26 27
10	25 46	20 01	03 09	12 31	05 39	26 28
11	25 58	20 13	03 14	12 30	05 38	26 29
12	26 09	20 25	03 20	12 28	05 37	26 30
13	26 20	20 37	03 26	12 27	05 36	26 31
14	26 30	20 50	03 31	12 26	05 35	26 31
15	26 39	21 02	03 37	12 24	05 34	26 32
16	26 48	21 14	03 42	12 23	05 33	26 33
17	26 56	21 26	03 48	12 21	05 32	26 34
18	27 04	21 38	03 53	12 20	05 31	26 35
19	27 11	21 50	03 58	12 18	05 30	26 36
20	27 17	22 01	04 03	12 16	05 29	26 37
21	27 22	22 13	04 09	12 15	05 28	26 38
22	27 27	22 25	04 14	12 13	05 28	26 39
23	27 31	22 37	04 19	12 11	05 27	26 41
24	27 34	22 48	04 24	12 09	05 26	26 42
25	27 36	23 00	04 28	12 08	05 25	26 43
26	27 38	23 11	04 33	12 06	05 25	26 44
27	27 39	23 22	04 38	12 04	05 24	26 45
28	27 R 39	23 34	04 42	12 02	05 23	26 46
29	27 39	23 45	04 47	12 00	05 23	26 48
30	27 37	23 56	04 51	11 58	05 22	26 49
31	27 ♌ 35	24 ♏ 07	04 ♏ 56	11 ♌ 56	05 ♉ 22	26 ♒ 50

MOON

Date	Moon True ☊ ° '	Moon Mean ☊ ° '	Moon ☽ Latitude ° '
01	05 ♉ 41	04 ♉ 19	03 S 49
02	05 D 42	04 16	02 50
03	05 44	04 13	01 41
04	05 45	04 10	00 S 26
05	05 R 45	04 07	00 N 51
06	05 43	04 03	02 03
07	05 38	04 00	03 07
08	05 32	03 57	04 00
09	05 25	03 54	04 38
10	05 17	03 51	05 01
11	05 09	03 48	05 08
12	05 03	03 44	05 01
13	04 58	03 41	04 41
14	04 55	03 38	04 09
15	04 54	03 35	03 26
16	04 D 54	03 32	02 34
17	04 56	03 29	01 36
18	04 57	03 25	00 N 32
19	04 R 57	03 22	00 S 34
20	04 55	03 19	01 41
21	04 50	03 16	02 43
22	04 44	03 13	03 38
23	04 35	03 09	04 23
24	04 25	03 06	04 52
25	04 14	03 03	05 04
26	04 05	03 00	04 57
27	03 58	02 57	04 31
28	03 54	02 54	03 48
29	03 52	02 50	02 52
30	03 D 52	02 47	01 46
31	03 ♉ 52	02 ♉ 44	00 S 34

DECLINATIONS

Date	Sun ☉ ° '	Moon ☽ ° '	Mercury ☿ ° '	Venus ♀ ° '	Mars ♂ ° '	Jupiter ♃ ° '	Saturn ♄ ° '	Uranus ♅ ° '	Neptune ♆ ° '	Pluto ⯓ ° '
01	21 S 54	08 S 03	23 S 46	19 S 59	15 N 58	16 S 20	10 S 06	17 N 38	11 N 45	22 S 31
02	22 03	01 S 32	24 02	20 16	15 54	16 23	10 08	17 38	11 45	22 30
03	22 11	05 N 05	24 17	20 32	15 51	16 27	10 10	17 39	11 45	22 30
04	22 19	11 26	24 31	20 48	15 48	16 30	10 11	17 39	11 44	22 30
05	22 27	17 10	24 43	21 03	15 45	16 34	10 13	17 39	11 44	22 29
06	22 34	21 56	24 54	21 18	15 42	16 37	10 15	17 39	11 44	22 29
07	22 40	25 25	25 03	21 31	15 39	16 41	10 17	17 40	11 43	22 28
08	22 47	27 22	25 11	21 45	15 36	16 44	10 19	17 40	11 43	22 28
09	22 52	27 43	25 18	21 57	15 34	16 47	10 21	17 41	11 42	22 27
10	22 58	26 31	25 23	22 09	15 32	16 50	10 23	17 41	11 42	22 27
11	23 03	24 00	25 27	22 20	15 30	16 54	10 25	17 41	11 42	22 26
12	23 07	20 25	25 30	22 31	15 28	16 57	10 26	17 42	11 42	22 26
13	23 11	16 03	25 31	22 41	15 26	17 00	10 28	17 42	11 41	22 25
14	23 15	11 08	25 31	22 50	15 25	17 03	10 30	17 43	11 41	22 25
15	23 18	05 51	25 29	22 58	15 24	17 07	10 32	17 43	11 41	22 24
16	23 20	00 N 22	25 26	23 06	15 23	17 10	10 33	17 43	11 40	22 24
17	23 22	05 S 11	25 21	23 13	15 22	17 13	10 35	17 44	11 40	22 23
18	23 24	10 38	25 15	23 20	15 22	17 16	10 37	17 44	11 40	22 23
19	23 25	15 47	25 07	23 25	15 22	17 19	10 38	17 45	11 40	22 22
20	23 26	20 24	24 58	23 30	15 22	17 22	10 40	17 45	11 39	22 22
21	23 26	24 10	24 47	23 35	15 22	17 25	10 41	17 46	11 39	22 21
22	23 26	26 43	24 35	23 38	15 23	17 28	10 43	17 46	11 39	22 21
23	23 25	27 45	24 21	23 41	15 24	17 31	10 44	17 47	11 39	22 20
24	23 24	27 02	24 06	23 43	15 25	17 34	10 46	17 47	11 38	22 20
25	23 22	24 34	23 49	23 44	15 26	17 36	10 47	17 48	11 38	22 19
26	23 20	20 33	23 31	23 45	15 28	17 39	10 49	17 49	11 38	22 19
27	23 18	15 19	23 12	23 45	15 30	17 42	10 50	17 49	11 38	22 18
28	23 15	09 17	22 52	23 44	15 32	17 45	10 51	17 50	11 38	22 18
29	23 11	02 S 48	22 30	23 42	15 34	17 47	10 53	17 50	11 38	22 17
30	23 07	03 N 45	22 08	23 40	15 37	17 50	10 54	17 51	11 38	22 16
31	23 S 03	10 N 05	21 S 45	23 S 37	15 N 40	17 S 53	10 S 55	17 N 51	11 N 37	22 S 16

ZODIAC SIGN ENTRIES

Date	h	m	Planets	
02	07	25	☽	♈
04	10	15	☽	♉
06	13	42	☽	♊
08	18	44	☽	♋
11	02	26	☽	♌
11	16	24	☿	♑
13	13	10	☽	♍
16	01	50	☽	♎
18	13	55	☽	♏
20	23	08	☽	♐
21	12	18	☉	♑
23	04	52	☽	♑
23	14	33	♀	♑
25	08	05	☽	♒
27	10	16	☽	♓
29	12	46	☽	♈
31	16	15	☽	♉

LATITUDES

Date	Mercury ☿ ° '	Venus ♀ ° '	Mars ♂ ° '	Jupiter ♃ ° '	Saturn ♄ ° '	Uranus ♅ ° '	Neptune ♆ ° '	Pluto ⯓ ° '
01	01 S 18	00 N 37	02 N 27	00 N 56	02 N 18	00 N 39	01 S 49	10 S 26
04	01 33	00 30	02 33	00 56	02 18	00 39	01 48	10 26
07	01 47	00 23	02 39	00 56	02 19	00 40	01 48	10 25
10	01 58	00 16	02 46	00 56	02 19	00 40	01 48	10 25
13	02 07	00 09	02 52	00 56	02 19	00 40	01 48	10 24
16	02 13	00 N 02	02 59	00 56	02 20	00 40	01 48	10 23
19	02 15	00 S 05	03 06	00 56	02 21	00 40	01 48	10 23
22	02 12	00 13	03 13	00 56	02 21	00 40	01 48	10 22
25	02 05	00 20	03 20	00 57	02 21	00 40	01 47	10 22
28	01 51	00 27	03 28	00 57	02 22	00 40	01 47	10 21
31	01 S 29	00 S 33	03 N 35	00 N 57	02 N 23	00 N 40	01 S 47	10 S 21

DATA

Julian Date	2466855
Delta T	+78 seconds
Ayanamsa	24° 26' 23"
Synetic vernal point	04° ♓ 40' 35"
True obliquity of ecliptic	23° 26' 09"

MOON'S PHASES, APSIDES AND POSITIONS ☽

Date	h	m	Phase	Longitude ° '	Eclipse Indicator
07	17	42	○	15 ♊ 59	
15	13	33	☾	23 ♍ 57	
23	08	06	●	01 ♑ 52	
30	03	46	☽	08 ♈ 49	

Day	h	m		
02	16	16	Perigee	
15	07	08	Apogee	
27	09	47	Perigee	
02	17	32	0N	
09	05	05	Max dec	27° N 47'
16	13	35	0S	
23	14	22	Max dec	27° S 45'
29	22	14	0N	

LONGITUDES

Date	Chiron ⚷ ° '	Ceres ⚳ ° '	Pallas ⚴ ° '	Juno ⚵ ° '	Vesta ⚶ ° '	Black Moon Lilith ⚸ ° '
01	04 ♌ 16	07 ♏ 31	07 ♎ 10	20 ♎ 00	15 ♊ 34	19 ♍ 03
11	03 ♌ 56	11 ♏ 43	10 ♎ 50	23 ♎ 04	12 ♊ 57	20 ♍ 10
21	03 ♌ 27	15 ♏ 49	14 ♎ 10	25 ♎ 54	12 ♊ 30	21 ♍ 17
31	02 ♌ 52	19 ♏ 46	17 ♎ 07	28 ♎ 27	08 ♊ 31	22 ♍ 24

All ephemeris data is given at 12.00 UT and the Moon's longitude is additionally given for 24.00 UT

ASPECTARIAN

h m	Aspects
01 Sunday	
02 04	☽ ⊼ ♅
03 34	☽ □ ☿
04 16	☽ ∥ ♄
09 47	☽ ♇ ♄
11 19	☽ △ ♃
12 14	☽ ± ♅
13 05	♀ ⊻ ♄
15 54	☽ ∠ ♆
15 56	☿ Q ⯓
20 43	☽ ⊼ ♂
02 Monday	
01 08	☽ ± ♄
01 16	☽ ⊻ ⯓
03 27	☽ ♇ ♅
07 04	☽ ± ♂
07 06	☽ ⊥ ♆
11 22	☽ ⊼ ♄
11 26	☽ ⊥ ⯓
13 05	☽ ♇ ♃
13 21	☽ △ ♀
17 15	☽ ⊻ ♆
22 36	☽ ♇ ♂
03 Tuesday	
02 31	☽ △ ☉
02 40	☽ ∠ ⯓
04 33	☽ ± ♃
04 49	☽ △ ♅
12 39	☽ △ ☿
14 18	☉ ± ♆
14 52	☽ ⊼ ♃
15 29	☿ ∠ ♄
17 14	☽ ♇ ♀
04 Wednesday	
00 30	☽ △ ♂
04 07	☽ ⚹ ⯓
05 55	☽ ♇ ☉
07 09	☽ # ♄
08 31	♀ ⊼ ♆
10 00	☽ ± ♀
10 02	☉ △ ♅
11 28	☿ ⊻ ♃
13 13	☽ ∥ ♆
14 36	☽ ☍ ♄
17 17	☽ ♇ ☿
20 04	☽ ☌ ♆
21 13	☽ ⊼ ♀
22 25	☽ ± ☉
05 Thursday	
00 33	☽ Q ⯓
05 46	☽ ∥ ♂
07 45	☽ □ ♅
09 16	☽ # ♃
09 28	☽ ⊼ ☉
10 35	☽ ± ☿
14 13	☽ ∥ ♅
18 01	☿ ♇ ♆
18 46	☽ ☍ ♃
20 03	☉ ∥ ⯓
22 10	☽ ⊼ ☿
06 Friday	
04 21	☉ Q ⯓
04 43	☽ □ ♂
07 29	☽ □ ⯓
08 10	☽ # ♀
14 43	☽ Q ♅
15 10	☽ # ⯓
15 47	☽ # ☉
17 44	♀ ⊥ ♄
18 31	☽ ⊼ ♄
23 39	☽ ⊻ ♆
07 Saturday	
05 05	☽ ± ♄
06 08	☽ ☍ ♀
08 44	☽ # ☿
10 09	☽ ⊥ ♆
11 40	☽ ⚹ ♅
12 38	☽ Q ♂
17 42	☽ ☍ ☉
21 03	☽ ♇ ♄
23 55	☽ ⊼ ♃
08 Sunday	
02 03	☽ ∠ ♆
09 31	☽ ☍ ☿
10 20	☽ ⚹ ♂
10 48	☽ ± ♃
12 20	☽ △ ⯓
14 20	☽ ∠ ♅
17 33	☿ △ ♂
21 59	☿ ⊥ ♃
09 Monday	
00 10	☽ △ ♄
01 11	♀ ± ♆
03 19	☽ ♇ ♃
05 03	☽ ⚹ ♆
06 37	☽ ⊥ ♅
09 49	☿ ⚹ ⯓
14 02	☽ ∠ ♂
15 39	☽ ♇ ⯓
17 27	♀ △ ♅
17 38	☽ ⊻ ♅
17 39	☽ ⊼ ♀
18 56	☉ ∠ ♄
10 Tuesday	
02 18	☿ ♇ ♅
03 09	☽ Q ♆
04 41	☽ ⊼ ☉
06 00	☽ ± ♀
07 01	☽ ⊥ ♂
07 28	☽ △ ♃
08 24	☽ ± ⯓
16 56	☽ ± ☉
18 29	☽ ⊻ ♂
19 43	☽ ⊼ ⯓
23 49	☽ # ☿
11 Wednesday	
00 27	☽ ⊼ ☿
00 52	☽ ♇ ♀
06 34	♀ Q ⯓
08 36	☽ □ ♄
11 37	☽ ♇ ☉
13 14	☽ □ ♆
13 35	☽ ± ☿
18 53	☽ # ☉
23 14	☽ # ⯓
23 21	☽ # ♀
23 56	☉ ⊻ ♃
12 Thursday	
01 26	♀ ∥ ⯓
02 31	☽ ☌ ♅
07 02	☉ ♇ ♆
09 09	☽ △ ♀
09 38	☽ ♇ ☿
18 13	☽ □ ♃
19 33	☽ △ ☉
19 59	☽ Q ♄
13 Friday	
03 23	☽ ∥ ♅
05 45	☽ ☌ ♂
06 11	☽ ☍ ⯓
07 08	☽ # ♃
15 10	☽ ∥ ♂
19 52	☽ △ ☿
20 06	☽ ⚹ ♄
21 46	☿ ⚹ ♄
14 Saturday	
00 23	☽ △ ♆
06 49	☽ Q ♃
09 27	☽ ∥ ♆
11 41	♀ ∠ ♄
14 11	☽ ⊻ ♅
14 58	☽ # ♄
16 02	♂ ☍ ⯓
15 Sunday	
02 19	☽ ⊥ ♅
02 39	☽ ∠ ♄
04 18	☽ □ ♀
06 09	☿ △ ♆
06 41	☽ ♇ ♆
07 33	☽ ⚹ ♃
13 28	☿ ∠ ♃
13 33	☽ □ ☉
18 49	☽ ⊼ ⯓
18 59	☿ ± ♅
19 09	☽ ⊻ ♂
20 33	☽ ∠ ♅
21 04	☽ ⊥ ♄
16 Monday	
00 56	☽ ± ♆
02 26	♀ ♇ ♆
07 00	☽ ± ⯓
07 28	☽ ⊥ ♂
09 19	☽ ⊻ ♄
13 04	☽ ⊼ ♆
14 30	☽ ∠ ♃
15 59	♀ ⊻ ♃
17 46	☽ □ ☿
21 07	☽ Q ♀
17 Tuesday	
01 13	☽ ♇ ⯓
01 51	☽ ∠ ♂
02 50	☽ ⚹ ♅
05 23	☽ Q ☉
08 56	☽ ⊥ ♃
21 07	☽ ⊻ ♃
23 58	☽ ⚹ ♀
18 Wednesday	
02 45	☽ Q ♅
03 46	☉ ⚹ ⯓
07 11	☽ △ ⯓
07 30	☽ ⚹ ☉
08 06	☽ ⚹ ♂
11 02	☽ Q ☿
11 54	☽ ∥ ♄
15 30	☉ △ ♂
16 40	☽ # ♆
21 07	☉ ♇ ♅
21 35	☽ ☌ ♄
19 Thursday	
00 41	☽ ☍ ♆
03 34	☿ ∠ ⯓
07 45	☽ Q ♂
08 24	☉ ⊥ ♃
08 43	☽ ∠ ♀
09 58	☽ # ♂
10 30	☉ ∥ ♀
12 37	☿ ♇ ♂
13 42	☽ ⚹ ☿
13 48	☽ □ ♅
14 29	☿ ⊼ ♅
15 19	☽ ∠ ☉
19 40	☽ ∥ ♃
21 48	☽ # ♅
20 Friday	
03 49	☽ ⊥ ♀
08 12	☽ ☌ ♃
09 52	☽ ⊥ ☉
16 18	☽ ⊻ ♀
16 52	☽ □ ⯓
18 08	☽ □ ♂
21 50	☽ ∠ ☿
22 00	☽ ⊻ ☉
22 18	♀ ⚹ ⯓
23 43	☽ ∥ ⯓
21 Saturday	
06 43	☽ ⊻ ♄
06 49	☽ ∥ ☉
07 44	☽ ∥ ♀
09 11	☽ ⊼ ♆
10 06	♀ ♇ ♅
12 24	♀ △ ♂
16 29	☽ ⊥ ☿
16 32	☽ ∥ ☿
17 42	☽ ⊥ ♄
20 03	☽ ± ♆
21 26	☽ △ ♅
22 Sunday	
01 47	☽ Q ⯓
03 44	☿ Q ♄
04 39	☽ ⊻ ☿
07 40	♀ ⊥ ♃
09 56	☽ ∠ ♄
12 07	☽ ♇ ♆
15 38	☽ ⊻ ♃
23 03	☽ ⚹ ⯓
23 59	☽ ♇ ♅
23 Monday	
00 29	☽ △ ♂
02 19	☽ ⊥ ♃
03 54	☽ ☌ ♀
08 06	☽ ☌ ☉
12 21	☽ ⚹ ♄
14 19	☽ △ ♆
15 35	☽ ± ♅
18 08	☽ ∠ ♃
24 Tuesday	
01 02	☽ ∠ ⯓
01 52	☽ ⊼ ♅
02 31	☽ ♇ ♂
09 03	☽ Q ♄
15 02	☽ ☌ ☿
16 23	☽ ⊥ ⯓
17 52	☽ ± ♂
20 03	☽ ⚹ ♃
25 Wednesday	
02 32	☽ ⊻ ⯓
04 02	☽ ⊼ ♂
04 10	☿ ⊥ ⯓
12 05	☽ ⊻ ♀
15 07	☽ ⊻ ☉
15 36	☽ □ ♄
16 31	☽ Q ♃
17 10	☽ □ ♆
17 36	☽ ∥ ☿
17 39	☽ ∥ ♀
18 20	☿ ∥ ♀
19 06	☿ ± ♂
20 07	☽ ∥ ☉
22 29	☉ ⚹ ♄
23 07	☽ ⊥ ♀
26 Thursday	
01 56	☽ ⊥ ☉
02 26	☽ ∥ ⯓
04 22	☽ ☍ ♅
15 36	☽ ∠ ♀
18 08	☽ ∠ ☉
19 39	☉ △ ♆
22 40	☿ ⚹ ♃
23 01	☽ □ ♃
23 03	☽ ⊻ ☿
23 15	☽ Q ♆
27 Friday	
01 08	☽ # ♅
01 44	☽ ∥ ♃
04 11	☉ ∥ ☿
04 50	☽ ☌ ⯓
06 20	☽ ☍ ♂
06 30	♀ ⚹ ♄
10 11	☽ ⊥ ☿
11 07	☉ ± ♅
11 17	☽ # ♂
18 03	☽ △ ♄
19 06	☽ ⚹ ♀
19 18	☽ ⚹ ♆
21 08	☽ ⚹ ☉
21 25	♀ △ ♆
28 Saturday	
02 53	☽ ∠ ☿
02 57	☽ # ♆
05 39	♂ St R
06 00	☽ ∥ ♄
06 26	☽ ⊼ ♅
09 34	♀ ± ♅
16 30	☽ ± ♅
17 12	☽ Q ♀
18 51	☽ Q ☉
19 23	☽ ♇ ♄
20 29	☽ ∠ ♆
29 Sunday	
02 02	☽ △ ♃
06 50	☽ ⚹ ☿
07 19	☽ ⊻ ⯓
07 40	☽ ♇ ♅
08 46	☽ ⊼ ♂
10 41	☽ ± ♄
11 42	☽ ⊥ ♆
12 07	☿ ⊻ ⯓
17 31	☽ ⊥ ⯓
18 57	☽ ± ♂
20 56	☽ ⊼ ♄
21 54	☽ ⊻ ♆
30 Monday	
02 42	☽ □ ♀
02 56	☿ ∥ ⯓
03 27	☿ ⊼ ♂
03 46	☽ □ ☉
03 52	☽ ♇ ♃
05 20	☽ Q ☿
05 28	☉ ∠ ♃
08 54	☽ ∠ ⯓
09 10	☽ △ ♅
10 17	☽ ♇ ♂
17 41	♀ ∠ ♃
19 30	☽ ± ♃
20 03	♃ # ♅
31 Tuesday	
05 59	☽ ⊼ ♃
10 46	☽ ⚹ ⯓
12 04	☽ △ ♂
15 20	☽ # ♄
15 24	☽ □ ☿
18 09	☽ ∥ ♆

JANUARY 2042

LONGITUDES

Date	Sidereal time h m s	Sun ☉ ° ' "	Moon ☽ ° ' "	Moon ☽ 24.00 ° ' "	Mercury ☿ ° '	Venus ♀ ° '	Mars ♂ ° '	Jupiter ♃ ° '	Saturn ♄ ° '	Uranus ♅ ° '	Neptune ♆ ° '	Pluto ♇ ° '
01	18 45 06	11 ♑ 11 48	11 ♉ 19 39	18 ♉ 09 43	00 ♒ 31	11 ♑ 12	27 ♌ 32	24 ♏ 18	05 ♏ 00	11 ♌ 53	05 ♉ 21	26 ♒ 51
02	18 49 03	12 12 56	24 ♉ 57 34	01 ♊ 43 09	01 37	12 27	27 R 28	24 29	05 04	11 R 51	05 R 21	26 53
03	18 53 00	13 14 05	08 ♊ 26 25	15 ♊ 07 13	02 39	13 43	27 24	24 40	05 08	11 49	05 20	26 54
04	18 56 56	14 15 13	21 ♊ 45 26	28 ♊ 20 54	03 34	14 58	27 18	24 50	05 12	11 47	05 20	26 55
05	19 00 53	15 16 21	04 ♋ 53 27	11 ♋ 22 54	04 23	16 14	27 12	25 01	05 16	11 45	05 20	26 57
06	19 04 49	16 17 28	17 ♋ 49 07	24 ♋ 11 58	05 03	17 29	27 05	25 11	05 20	11 42	05 19	26 58
07	19 08 46	17 18 36	00 ♌ 31 22	06 ♌ 47 17	05 35	18 45	26 58	25 22	05 24	11 40	05 19	26 59
08	19 12 42	18 19 44	12 ♌ 59 45	19 ♌ 08 51	05 58	20 00	26 49	25 32	05 28	11 38	05 19	27 01
09	19 16 39	19 20 51	25 ♌ 14 46	01 ♍ 17 44	06 11	21 15	26 40	25 42	05 31	11 35	05 19	27 02
10	19 20 35	20 21 59	07 ♍ 18 03	13 ♍ 16 05	06 R 12	22 31	26 29	25 52	05 35	11 33	05 18	27 04
11	19 24 32	21 23 06	19 ♍ 12 18	25 ♍ 07 12	06 02	23 46	26 18	26 03	05 38	11 31	05 18	27 05
12	19 28 29	22 24 13	01 ♎ 01 19	06 ♎ 55 16	05 39	25 02	26 06	26 12	05 42	11 28	05 18	27 07
13	19 32 25	23 25 20	12 ♎ 49 42	18 ♎ 45 16	05 06	26 17	25 54	26 22	05 45	11 26	05 18	27 08
14	19 36 22	24 26 27	24 ♎ 42 41	00 ♏ 42 39	04 21	27 33	25 40	26 32	05 48	11 23	05 18	27 10
15	19 40 18	25 27 34	06 ♏ 45 52	12 ♏ 53 02	03 25	28 ♑ 48	25 26	26 42	05 51	11 21	05 D 18	27 11
16	19 44 15	26 28 41	19 ♏ 04 47	25 ♏ 21 45	02 21	00 ♒ 04	25 11	26 51	05 54	11 18	05 18	27 13
17	19 48 11	27 29 48	01 ♐ 44 28	08 ♐ 13 24	01 ♒ 11	01 19	24 56	27 01	05 57	11 16	05 18	27 14
18	19 52 08	28 30 54	14 ♐ 48 53	21 ♐ 31 07	29 ♑ 55	02 34	24 39	27 10	06 00	11 13	05 18	27 16
19	19 56 04	29 ♑ 32 00	28 ♐ 20 11	05 ♑ 15 56	28 37	03 50	24 22	27 19	06 02	11 11	05 18	27 17
20	20 00 01	00 ♒ 33 06	12 ♑ 18 05	19 ♑ 26 07	27 19	05 05	24 04	27 28	06 05	11 08	05 18	27 19
21	20 03 58	01 34 11	26 ♑ 39 24	03 ♒ 57 04	26 04	06 21	23 46	27 37	06 07	11 06	05 19	27 20
22	20 07 54	02 35 15	11 ♒ 18 09	18 ♒ 41 37	24 53	07 36	23 27	27 46	06 10	11 03	05 19	27 22
23	20 11 51	03 36 19	26 ♒ 06 22	03 ♓ 31 18	23 48	08 52	23 07	27 55	06 12	11 01	05 19	27 24
24	20 15 47	04 37 22	10 ♓ 55 24	18 ♓ 17 43	22 51	10 07	22 47	28 03	06 14	10 58	05 19	27 25
25	20 19 44	05 38 23	25 ♓ 37 27	02 ♈ 53 57	22 02	11 22	22 26	28 12	06 16	10 55	05 20	27 27
26	20 23 40	06 39 24	10 ♈ 06 43	17 ♈ 15 24	21 22	12 38	22 05	28 20	06 18	10 53	05 20	27 28
27	20 27 37	07 40 24	24 ♈ 19 48	01 ♉ 19 51	20 51	13 53	21 43	28 28	06 20	10 50	05 21	27 30
28	20 31 33	08 41 22	08 ♉ 15 34	15 ♉ 07 05	20 28	15 08	21 21	28 37	06 22	10 47	05 21	27 32
29	20 35 30	09 42 19	21 ♉ 54 34	28 ♉ 38 11	20 15	16 24	20 58	28 45	06 23	10 45	05 22	27 33
30	20 39 27	10 43 15	05 ♊ 18 11	11 ♊ 54 46	20 10	17 39	20 35	28 52	06 25	10 42	05 22	27 35
31	20 43 23	11 ♒ 44 10	18 ♊ 28 07	24 ♊ 58 25	20 ♑ 12	18 ♒ 54	20 ♌ 12	29 ♏ 00	06 ♏ 26	10 ♌ 40	05 ♉ 23	27 ♒ 37

MOON NODES AND LATITUDE

Date	Moon True ☊ ° '	Moon Mean ☊ ° '	Moon ☽ Latitude ° '
01	03 ♉ 52	02 ♉ 41	00 N 40
02	03 R 50	02 38	01 50
03	03 46	02 35	02 53
04	03 38	02 31	03 45
05	03 28	02 28	04 25
06	03 15	02 25	04 50
07	03 02	02 22	05 00
08	02 49	02 19	04 56
09	02 37	02 15	04 38
10	02 27	02 12	04 08
11	02 20	02 09	03 27
12	02 16	02 06	02 37
13	02 14	02 03	01 41
14	02 D 13	02 00	00 N 40
15	02 R 13	01 56	00 S 24
16	02 13	01 53	01 28
17	02 10	01 50	02 29
18	02 05	01 47	03 25
19	01 56	01 44	04 11
20	01 46	01 41	04 43
21	01 33	01 37	04 59
22	01 21	01 34	04 56
23	01 10	01 31	04 32
24	01 00	01 28	03 51
25	00 54	01 25	02 54
26	00 51	01 21	01 47
27	00 50	01 18	00 S 34
28	00 D 50	01 15	00 N 39
29	00 R 50	01 12	01 49
30	00 47	01 09	02 51
31	00 ♉ 43	01 ♉ 06	03 N 43

DECLINATIONS

Date	Sun ☉ ° '	Moon ☽ ° '	Mercury ☿ ° '	Venus ♀ ° '	Mars ♂ ° '	Jupiter ♃ ° '	Saturn ♄ ° '	Uranus ♅ ° '	Neptune ♆ ° '	Pluto ♇ ° '
01	22 S 58	15 N 51	21 S 21	23 S 33	15 N 43	17 S 55	10 S 56	17 N 52	11 N 37	22 S 15
02	22 52	20 47	20 56	23 29	15 47	17 58	10 58	17 53	11 37	22 15
03	22 47	24 33	20 32	23 23	15 51	18 01	10 59	17 53	11 37	22 14
04	22 40	26 55	20 07	23 17	15 55	18 03	11 00	17 54	11 37	22 13
05	22 34	27 45	19 43	23 11	15 59	18 06	11 01	17 55	11 37	22 13
06	22 27	27 02	19 19	23 03	16 04	18 08	11 02	17 55	11 37	22 12
07	22 19	24 55	18 56	22 55	16 09	18 10	11 03	17 56	11 37	22 12
08	22 11	21 38	18 35	22 46	16 14	18 13	11 04	17 57	11 37	22 11
09	22 02	17 28	18 15	22 37	16 19	18 15	11 05	17 57	11 37	22 10
10	21 54	12 39	17 57	22 26	16 25	18 18	11 06	17 58	11 37	22 10
11	21 44	07 26	17 41	22 16	16 31	18 20	11 07	17 59	11 37	22 09
12	21 34	02 N 00	17 28	22 04	16 37	18 22	11 08	17 59	11 37	22 09
13	21 24	03 S 31	17 18	21 52	16 43	18 24	11 09	18 00	11 37	22 08
14	21 14	08 57	17 11	21 39	16 50	18 26	11 09	18 01	11 37	22 07
15	21 03	14 09	17 07	21 25	16 56	18 29	11 10	18 01	11 37	22 07
16	20 51	18 54	17 05	21 11	17 03	18 31	11 11	18 02	11 37	22 06
17	20 40	22 57	17 06	20 56	17 11	18 33	11 12	18 03	11 37	22 05
18	20 27	25 58	17 10	20 41	17 18	18 35	11 12	18 04	11 37	22 05
19	20 15	27 36	17 16	20 25	17 25	18 37	11 13	18 04	11 37	22 04
20	20 02	27 34	17 23	20 08	17 33	18 39	11 14	18 05	11 37	22 04
21	19 48	25 42	17 32	19 51	17 41	18 41	11 14	18 06	11 38	22 03
22	19 35	22 07	17 42	19 33	17 49	18 43	11 15	18 06	11 38	22 02
23	19 21	17 05	17 52	19 14	17 57	18 45	11 15	18 07	11 38	22 02
24	19 06	11 02	18 04	18 56	18 05	18 46	11 16	18 08	11 38	22 01
25	18 52	04 S 24	18 15	18 36	18 13	18 48	11 16	18 09	11 38	22 00
26	18 36	02 N 22	18 27	18 16	18 21	18 50	11 17	18 09	11 38	22 00
27	18 21	08 54	18 38	17 55	18 29	18 52	11 17	18 10	11 39	21 59
28	18 05	14 53	18 49	17 34	18 38	18 53	11 17	18 11	11 39	21 59
29	17 49	20 00	19 00	17 13	18 46	18 55	11 18	18 12	11 39	21 58
30	17 33	24 00	19 11	16 51	18 54	18 57	11 18	18 12	11 39	21 57
31	17 S 16	26 N 39	19 S 20	16 S 28	19 N 02	18 S 58	11 S 18	18 N 13	11 N 39	21 S 57

ZODIAC SIGN ENTRIES

Date	h	m	Planets
01	01	20	☽ ♉
02	20	56	☽ ♊
05	03	01	☽ ♋
07	11	00	☽ ♌
09	21	25	☽ ♍
12	09	55	☽ ♎
14	22	35	☽ ♏
16	10	51	♀ ♒
17	08	45	☽ ♐
18	10	25	☿ ♑
19	14	54	☽ ♑
19	23	00	☉ ♒
21	17	31	☽ ♒
23	18	18	☽ ♓
25	19	12	☽ ♈
27	21	43	☽ ♉
30	02	27	☽ ♊

LATITUDES

Date	Mercury ☿ ° '	Venus ♀ ° '	Mars ♂ ° '	Jupiter ♃ ° '	Saturn ♄ ° '	Uranus ♅ ° '	Neptune ♆ ° '	Pluto ♇ ° '
01	01 S 20	00 S 35	03 N 37	00 N 57	02 N 23	00 N 40	01 S 47	10 S 21
04	00 47	00 42	03 44	00 57	02 24	00 40	01 47	10 20
07	00 S 04	00 48	03 51	00 57	02 24	00 41	01 47	10 20
10	00 N 48	00 54	03 58	00 57	02 25	00 41	01 47	10 20
13	01 44	00 59	04 05	00 58	02 26	00 41	01 46	10 19
16	02 37	01 04	04 11	00 58	02 26	00 41	01 46	10 19
19	03 14	01 09	04 16	00 58	02 27	00 41	01 46	10 19
22	03 31	01 13	04 21	00 58	02 28	00 41	01 46	10 19
25	03 26	01 17	04 26	00 59	02 29	00 41	01 46	10 18
28	03 05	01 20	04 29	00 59	02 29	00 41	01 45	10 18
31	02 N 36	01 S 23	04 N 32	00 N 59	02 N 30	00 N 41	01 S 45	10 S 18

DATA

Julian Date	2466886
Delta T	+78 seconds
Ayanamsa	24° 26' 29"
Synetic vernal point	04° ♓ 40' 30"
True obliquity of ecliptic	23° 26' 09"

MOON'S PHASES, APSIDES AND POSITIONS ☽

Date	h	m	Phase	Longitude ° '	Eclipse Indicator
06	08	54	○	16 ♋ 10	
14	11	24	☾	24 ♎ 25	
21	20	42	●	01 ♒ 56	
28	12	48	☽	08 ♉ 43	

Day	h	m		
12	03	59	Apogee	
23	22	27	Perigee	
05	12	43	Max dec	27° N 45'
12	20	42	0S	
19	23	34	Max dec	27° S 48'
26	03	36	0N	

LONGITUDES

Date	Chiron ⚷ ° '	Ceres ⚳ ° '	Pallas ⚴ ° '	Juno ⚵ ° '	Vesta ⚶ ° '	Black Moon Lilith ⚸ ° '
01	02 ♌ 48	20 ♏ 10	17 ♎ 23	28 ♎ 42	08 ♊ 22	22 ♍ 30
11	02 ♌ 07	23 ♏ 55	19 ♎ 50	00 ♏ 53	07 ♊ 06	23 ♍ 37
21	01 ♌ 24	27 ♏ 28	21 ♎ 44	02 ♏ 42	06 ♊ 35	24 ♍ 44
31	00 ♌ 40	00 ♐ 46	23 ♎ 00	04 ♏ 04	06 ♊ 48	25 ♍ 51

All ephemeris data is given at 12.00 UT and the Moon's longitude is additionally given for 24.00 UT

ASPECTARIAN

h m	Aspects
01 Wednesday	
00 53	☽ ☍ ♄
01 34	☽ ☌ ♆
07 40	☽ Q ♇
11 25	☽ ∥ ♂
11 45	☽ △ ♀
11 45	☽ △ ☉
12 17	☉ ☌ ♀
12 59	☽ □ ♅
21 16	☽ ∥ ♅
21 36	☽ ⋕ ♃
02 Thursday	
00 52	♀ ∠ ♇
00 56	♀ ⊼ ♅
03 49	☉ ⊼ ♅
03 53	☉ ∠ ♇
11 08	☽ ☍ ♃
12 23	♀ ⚼ ♂
12 49	☽ ⋕ ☿
15 24	☽ □ ♇
16 19	☽ ⚼ ☉
16 26	☽ □ ♂
16 52	☽ ⚼ ♀
17 41	☉ ⚼ ♂
20 24	☽ ⋕ ♇
20 40	☽ Q ♅
03 Friday	
00 02	☽ ⋕ ☉
00 49	☽ △ ☿
03 54	☽ ⋕ ♀
06 04	☽ ⊼ ♄
06 27	☽ ⊻ ♆
09 40	☽ ± ☉
10 33	☽ ± ♀
16 52	☽ ± ♄
17 12	☽ ⊥ ♆
18 03	☽ ⚹ ♅
21 19	☽ ⊼ ☉
22 27	☽ ⊼ ♀
04 Saturday	
00 25	☽ Q ♂
05 49	☽ ⚼ ☿
09 11	☽ ⚼ ♄
09 25	☽ ∠ ♆
17 40	☽ ⊼ ♃
21 07	☽ ∠ ♅
21 25	☽ △ ♇
22 02	☽ ⚹ ♂
23 18	☽ ± ☿
05 Sunday	
04 47	☽ ± ♃
11 00	☽ ⊼ ☿
12 42	☽ △ ♄
12 48	☽ ⚹ ♆
13 34	☽ ⊥ ♅
21 36	☽ ⚼ ♃
06 Monday	
00 38	☽ ⊻ ♅
01 04	☽ ⚼ ♇
01 25	☽ ∠ ♂
06 25	♄ ☍ ♆
08 54	☽ ☍ ☉
09 03	♀ Q ♄
11 04	☽ Q ♆
11 18	☽ ☍ ♀
17 55	☽ ± ♇
18 05	☽ ⊥ ♂
22 59	☿ □ ♆
07 Tuesday	
01 23	☿ □ ♄
02 04	☽ △ ♃
05 16	☽ ⊼ ♇
05 18	☽ ⊻ ♂
07 11	♂ ☍ ♇
14 16	☉ Q ♄
21 10	☽ □ ♆
21 23	☽ □ ♄
22 03	☽ ☍ ☿
08 Wednesday	
04 10	☽ ⋕ ♀
08 20	☽ ⋕ ☉
08 27	☽ ⋕ ♇
09 22	☽ ☌ ♅
11 40	☉ ∥ ♇
23 20	☽ ⊼ ☉
09 Thursday	
01 52	♀ ± ♂
03 14	☽ ⊼ ♀
07 29	☽ ⋕ ☿
07 43	♀ ⊥ ♇
07 48	☽ ⋕ ♃
08 35	☽ Q ♄
09 23	☽ ∥ ♅
11 23	☿ ∥ ♃
12 13	☽ ± ☉
12 55	☽ □ ♃
14 45	☽ ☌ ♂
15 33	☽ ☍ ♇
16 26	☽ ± ♀
17 50	☽ ∥ ♂
10 Friday	
02 47	☿ St R
07 46	☽ ⚼ ☉
08 00	☽ △ ♆
08 32	☽ ⚹ ♄
09 48	☽ ⊼ ☿
10 29	☿ ⋕ ♅
12 29	☽ ⚼ ♀
14 27	☉ ± ♂
16 55	☽ ∥ ♆
19 18	☽ ⋕ ♄
20 31	☽ ⊻ ♅
21 45	☽ ± ☿
11 Saturday	
01 25	☽ Q ♃
04 47	☉ ⊥ ♇
08 35	☽ ⊥ ♅
14 14	☽ ⚼ ♆
14 55	☽ ∠ ♄
15 36	☽ ⚼ ☿
16 50	☽ △ ☉
22 22	☽ △ ♀
12 Sunday	
02 04	☽ ⚹ ♃
02 10	☽ ⊻ ♂
02 12	♀ ∥ ♇
02 47	☽ ∠ ♅
04 02	☽ ⊼ ♇
05 25	♂ □ ♃
08 30	☽ ± ♆
09 17	☽ ⊥ ♄
10 21	☿ □ ♄
14 10	☽ ⊥ ♂
16 15	☽ ± ♇
20 42	☽ ⊼ ♆
21 03	☽ △ ☿
21 33	☽ ⊻ ♄
13 Monday	
04 07	☿ □ ♆
05 37	♀ ⊼ ♂
08 09	☽ ∠ ♂
09 00	☽ ∠ ♃
09 10	☽ ⚹ ♅
10 35	☽ ⚼ ♇
13 50	♀ ⚹ ♃
14 Tuesday	
03 29	☽ ⊥ ♃
04 29	♀ ⊻ ♇
09 21	☽ Q ♅
11 24	☽ □ ☉
13 54	☽ ⚹ ♂
15 42	☽ ⊻ ♃
16 55	☽ △ ♇
18 21	☽ □ ♀
22 01	☽ ∥ ♄
15 Wednesday	
00 06	☽ ⋕ ♆
05 53	☽ □ ☿
06 15	♆ St D
09 06	☽ ☍ ♆
10 11	☽ ☌ ♄
11 33	☉ ⊼ ♂
13 18	☽ Q ♂
20 58	☽ □ ♅
16 Thursday	
02 06	☽ ⋕ ♂
02 18	☽ Q ☉
02 31	☽ ∥ ☿
07 24	☽ ⋕ ♅
09 49	☽ Q ♀
09 54	☽ ∥ ♃
14 15	☽ Q ☿
18 44	☿ ⋕ ♂
22 28	☉ ⚹ ♃
22 29	☽ ∥ ☉
23 26	☽ □ ♂
17 Friday	
00 11	☽ ∥ ♀
03 00	☽ ☌ ♃
03 21	☽ ⚹ ☉
03 32	☽ □ ♇
05 41	☉ ⊻ ♇
06 29	☽ ∥ ♇
10 38	☿ ☌ ♀
11 02	☽ ⚹ ☿
11 07	☽ ⚹ ♀
18 37	☽ ⊼ ♆
19 50	☽ ⊻ ♄
18 Saturday	
05 30	☽ △ ♅
05 38	☽ ± ♆
06 52	☽ ⊥ ♄
09 27	☽ ∠ ☉
12 10	☽ ∠ ☿
12 48	☽ Q ♇
17 29	☽ ∠ ♀
21 50	☽ ⚼ ♆
23 06	☽ ∠ ♄
19 Sunday	
02 32	☉ ☌ ☿
02 50	☽ ⊥ ☿
02 53	☽ ⊥ ☉
05 11	☽ △ ♂
05 55	♃ □ ♇
08 14	☽ ⚼ ♅
10 10	☽ ⚹ ♇
10 12	☽ ⊻ ♃
11 02	☽ ⊥ ♀
12 27	☽ ⊻ ☿
14 15	☽ ⊻ ☉
20 45	☽ ⊥ ♃
22 29	☽ ⊻ ♀
23 49	☽ ± ♅
20 Monday	
00 04	☽ △ ♆
01 22	☽ ⚹ ♄
06 37	☽ ⚼ ♂
09 30	☿ ⚹ ♃
10 02	☽ ⊼ ♅
11 13	♂ Q ♄
12 01	☽ ∠ ♇
12 08	☿ ⊻ ♇
12 17	☽ ∠ ♃
16 11	♀ □ ♆
21 31	☽ ± ♂
21 46	☽ Q ♄
21 Tuesday	
03 10	☽ ⊥ ♇
07 18	☽ ⊼ ♂
07 35	♀ □ ♄
11 06	☽ ☌ ☿
13 08	☽ ⊻ ♇
13 36	☽ ⚹ ♃
20 42	☽ ☌ ☉
22 Wednesday	
01 26	☉ ∥ ♀
02 14	☽ □ ♆
03 36	☽ □ ♄
05 24	☽ ☌ ♀
09 29	☽ Q ♃
11 36	☽ ☍ ♅
12 25	☽ ∥ ♇
23 Thursday	
01 26	☽ ∥ ☉
01 47	☽ ∥ ♀
04 44	☽ ∥ ♃
07 16	☽ ☍ ♂
07 29	☽ Q ♆
07 30	☽ ⋕ ♅
08 21	☽ ⋕ ♂
08 31	☽ ⊻ ☿
08 41	☽ ∥ ☿
14 05	☽ ☌ ♇
14 57	☽ □ ♃
17 36	☽ ⊥ ☿
24 Friday	
01 02	☽ ⊻ ☉
02 55	☽ ⚹ ♆
04 23	☽ △ ♄
07 18	☽ ∠ ☿
09 43	☽ ⋕ ♆
10 34	☽ ⊻ ♀
10 44	♀ Q ♃
11 08	☽ ∥ ♄
11 29	☽ ⊥ ☉
12 04	☽ ⊼ ♅
15 08	☿ ⊼ ♂
19 41	☿ ⋕ ♂
21 14	☽ ⊥ ♀
21 30	☿ ⋕ ♅
21 48	☽ ± ♅
22 19	♂ ∥ ♅
22 23	♀ ∥ ♃
25 Saturday	
03 13	☽ ∠ ☉
03 19	☽ ∠ ♆
03 42	♀ ☍ ♅
04 39	☉ □ ♆
04 51	☽ ⚼ ♄
06 24	☽ ⚹ ☿
06 53	☽ ⊼ ♂
12 29	☽ ⚼ ♅
13 21	☽ ∠ ♀
15 00	☽ ⊻ ♇
16 17	☽ △ ♃
16 31	☽ ± ♂
16 46	☉ ∥ ♃
18 06	☽ ⊥ ♆
19 40	☽ ± ♄
26 Sunday	
00 56	☽ ⊥ ♇
01 14	☽ Q ☿
03 23	☉ □ ♄
03 56	☿ ∥ ♀
04 02	☽ ⊻ ♆
05 38	☽ ⊼ ♄
05 48	☽ ⚹ ☉
07 04	☽ ⚼ ♂
07 43	♀ ⋕ ♂
07 46	☿ ⊥ ♇
13 17	☽ △ ♅
15 58	☽ ∠ ♇
16 37	☽ ⚹ ♀
17 27	☽ ⚼ ♃
19 33	♀ ⋕ ♅
20 41	☉ ∥ ☿
27 Monday	
03 28	☽ Q ☉
03 35	☉ ⋕ ♂
06 15	☽ □ ☿
07 40	☽ △ ♂
08 49	☽ ± ♃
14 55	☽ Q ♀
17 26	☽ ⚹ ♇
19 09	☽ ⊼ ♃
21 15	☽ ⋕ ♄
22 40	☽ ∥ ♆
28 Tuesday	
03 51	☉ ⋕ ♅
06 57	☽ ☌ ♆
08 42	☽ ☍ ♄
12 48	☽ □ ☉
14 13	☽ Q ♇
16 24	☽ □ ♅
22 13	☿ ∥ ♃
23 16	☽ ⋕ ♀
29 Wednesday	
01 15	☽ □ ♀
01 45	☽ ⋕ ☉
02 58	☽ ∥ ♅
05 35	☽ ∥ ♂
06 29	☽ ⋕ ♃
06 46	☽ ⋕ ☿
09 05	☽ △ ☿
10 23	☽ □ ♂
22 05	☽ □ ♇
22 53	☽ ⋕ ♇
30 Thursday	
00 09	☽ Q ♅
00 19	☽ ☍ ♃
11 36	☉ ☍ ♅
11 45	☽ ⚼ ☿
12 07	☽ ⊻ ♆
14 01	☽ ⊼ ♄
15 33	☿ St D
16 07	☉ Q ♃
17 47	☽ Q ♂
21 25	♂ ⋕ ♃
21 46	☽ ⚹ ♅
22 39	☽ △ ☉
23 01	☽ ⊥ ♆
31 Friday	
00 57	☽ ± ♄
04 08	☽ ± ☿
11 48	☿ ⊼ ♂
12 53	☽ △ ♀
15 06	☽ ⚹ ♂
15 14	☽ ⊼ ☿
15 31	☽ ∠ ♆
17 29	☽ ⚼ ♄

FEBRUARY 2042

LONGITUDES

Date	Sidereal time h m s	Sun ☉ ° ' "	Moon ☽ ° ' "	Moon ☽ 24.00 ° ' "	Mercury ☿ ° '	Venus ♀ ° '	Mars ♂ ° '	Jupiter ♃ ° '	Saturn ♄ ° '	Uranus ♅ ° '	Neptune ♆ ° '	Pluto ♇ ° '
01	20 47 20	12 ♒ 45 04	01 ♋ 25 49	07 ♋ 50 26	20 ♑ 22	20 ♒ 09	19 ♌ 49	29 ♏ 08	06 ♏ 28	10 ♌ 37	05 ♉ 23	27 ♒ 38
02	20 51 16	13 45 56	14 ♋ 12 20	20 ♋ 31 36	20 D 39	21 25	19 R 25	29 15	06 29	10 R 34	05 24	27 40
03	20 55 13	14 46 47	26 ♋ 48 16	03 ♌ 02 21	21 02	22 40	19 01	29 22	06 30	10 32	05 24	27 42
04	20 59 09	15 47 37	09 ♌ 13 53	15 ♌ 22 54	21 31	23 55	18 37	29 30	06 31	10 29	05 25	27 43
05	21 03 06	16 48 26	21 ♌ 29 29	27 ♌ 33 41	22 05	25 10	18 14	29 37	06 32	10 26	05 26	27 45
06	21 07 02	17 49 13	03 ♍ 35 38	09 ♍ 35 29	22 44	26 26	17 50	29 43	06 33	10 24	05 27	27 47
07	21 10 59	18 50 00	15 ♍ 33 28	21 ♍ 29 48	23 27	27 41	17 26	29 50	06 34	10 21	05 27	27 48
08	21 14 56	19 50 45	27 ♍ 24 51	03 ♎ 18 56	24 15	28 ♒ 56	17 02	29 ♏ 57	06 34	10 19	05 28	27 50
09	21 18 52	20 51 29	09 ♎ 12 31	15 ♎ 06 03	25 06	00 ♓ 11	16 38	00 ♐ 03	06 35	10 16	05 29	27 52
10	21 22 49	21 52 12	21 ♎ 00 04	26 ♎ 55 08	26 00	01 26	16 14	00 10	06 35	10 14	05 30	27 54
11	21 26 45	22 52 54	02 ♏ 51 53	08 ♏ 50 56	26 57	02 41	15 51	00 16	06 36	10 11	05 31	27 55
12	21 30 42	23 53 35	14 ♏ 52 58	20 ♏ 58 39	27 58	03 57	15 27	00 22	06 36	10 08	05 32	27 57
13	21 34 38	24 54 15	27 ♏ 08 41	03 ♐ 23 42	29 ♑ 01	05 12	15 05	00 27	06 36	10 06	05 33	27 59
14	21 38 35	25 54 54	09 ♐ 44 22	16 ♐ 11 14	00 ♒ 06	06 27	14 42	00 33	06 R 36	10 03	05 34	28 01
15	21 42 31	26 55 32	22 ♐ 44 49	29 ♐ 25 30	01 13	07 42	14 20	00 39	06 36	10 01	05 35	28 02
16	21 46 28	27 56 08	06 ♑ 13 31	13 ♑ 09 00	02 23	08 57	13 58	00 44	06 36	09 58	05 36	28 04
17	21 50 25	28 56 44	20 ♑ 11 49	27 ♑ 21 43	03 34	10 12	13 36	00 49	06 35	09 56	05 37	28 06
18	21 54 21	29 ♒ 57 18	04 ♒ 38 09	12 ♒ 00 25	04 48	11 27	13 16	00 54	06 35	09 54	05 38	28 07
19	21 58 18	00 ♓ 57 50	19 ♒ 27 35	26 ♒ 58 34	06 03	12 42	12 55	00 59	06 34	09 51	05 39	28 09
20	22 02 14	01 58 21	04 ♓ 32 07	12 ♓ 06 58	07 19	13 57	12 35	01 04	06 34	09 49	05 41	28 11
21	22 06 11	02 58 51	19 ♓ 41 49	27 ♓ 15 23	08 38	15 12	12 16	01 08	06 33	09 46	05 42	28 13
22	22 10 07	03 59 18	04 ♈ 46 31	12 ♈ 14 11	09 57	16 27	11 57	01 13	06 32	09 44	05 43	28 14
23	22 14 04	04 59 44	19 ♈ 37 33	26 ♈ 55 57	11 18	17 42	11 39	01 17	06 31	09 42	05 44	28 16
24	22 18 00	06 00 08	04 ♉ 08 54	11 ♉ 16 06	12 41	18 57	11 22	01 21	06 30	09 40	05 46	28 18
25	22 21 57	07 00 30	18 ♉ 17 26	25 ♉ 12 54	14 04	20 11	11 05	01 25	06 29	09 37	05 47	28 19
26	22 25 54	08 00 51	02 ♊ 02 37	08 ♊ 46 49	15 29	21 26	10 49	01 29	06 28	09 35	05 48	28 21
27	22 29 50	09 01 09	15 ♊ 25 45	21 ♊ 59 47	16 55	22 41	10 34	01 32	06 26	09 33	05 50	28 23
28	22 33 47	10 ♓ 01 25	28 ♊ 29 13	04 ♋ 54 26	18 ♒ 22	23 ♓ 56	10 ♌ 20	01 ♐ 35	06 ♏ 25	09 ♌ 31	05 ♉ 51	28 ♒ 25

Date	Moon True ☊ ° '	Moon Mean ☊ ° '	Moon ☽ Latitude ° '
01	00 ♉ 35	01 ♉ 02	04 N 23
02	00 R 25	00 59	04 49
03	00 12	00 56	05 00
04	29 ♈ 59	00 53	04 57
05	29 45	00 50	04 40
06	29 33	00 46	04 10
07	29 22	00 43	03 30
08	29 15	00 40	02 40
09	29 10	00 37	01 44
10	29 08	00 34	00 N 43
11	29 D 08	00 31	00 S 20
12	29 09	00 27	01 23
13	29 R 09	00 24	02 24
14	29 08	00 21	03 19
15	29 05	00 18	04 06
16	28 59	00 15	04 41
17	28 52	00 12	05 01
18	28 42	00 08	05 03
19	28 33	00 05	04 45
20	28 24	00 ♉ 02	04 07
21	28 17	29 ♈ 59	03 11
22	28 12	29 56	02 02
23	28 10	29 52	00 S 46
24	28 D 10	29 49	00 N 32
25	28 11	29 46	01 46
26	28 12	29 43	02 52
27	28 R 12	29 40	03 47
28	28 ♈ 10	29 ♈ 37	04 N 28

DECLINATIONS

Date	Sun ☉ ° '	Moon ☽ ° '	Mercury ☿ ° '	Venus ♀ ° '	Mars ♂ ° '	Jupiter ♃ ° '	Saturn ♄ ° '	Uranus ♅ ° '	Neptune ♆ ° '	Pluto ♇ ° '
01	16 S 59	27 N 49	19 S 30	16 S 05	19 N 10	19 S 00	11 S 18	18 N 14	11 N 40	21 S 56
02	16 42	27 28	19 38	15 42	19 18	19 01	11 18	18 14	11 40	21 56
03	16 24	25 42	19 46	15 18	19 26	19 03	11 19	18 15	11 40	21 55
04	16 06	22 42	19 53	14 54	19 34	19 04	11 19	18 16	11 41	21 54
05	15 48	18 45	19 58	14 29	19 42	19 06	11 19	18 17	11 41	21 54
06	15 29	14 04	20 03	14 04	19 50	19 07	11 19	18 17	11 41	21 53
07	15 11	08 55	20 07	13 38	19 57	19 08	11 19	18 18	11 41	21 53
08	14 52	03 N 29	20 10	13 13	20 04	19 09	11 19	18 19	11 42	21 52
09	14 32	02 S 03	20 12	12 46	20 11	19 11	11 19	18 19	11 42	21 51
10	14 13	07 32	20 12	12 20	20 18	19 12	11 19	18 20	11 42	21 51
11	13 53	12 47	20 12	11 53	20 25	19 13	11 18	18 21	11 43	21 50
12	13 33	17 37	20 10	11 26	20 31	19 14	11 18	18 21	11 43	21 50
13	13 13	21 51	20 07	10 58	20 38	19 15	11 18	18 22	11 44	21 49
14	12 53	25 11	20 03	10 31	20 44	19 16	11 18	18 23	11 44	21 48
15	12 32	27 20	19 57	10 03	20 50	19 17	11 17	18 23	11 44	21 48
16	12 11	27 58	19 51	09 34	20 55	19 18	11 17	18 24	11 45	21 47
17	11 50	26 53	19 43	09 06	21 00	19 19	11 17	18 25	11 45	21 47
18	11 29	24 00	19 34	08 37	21 05	19 20	11 16	18 25	11 46	21 46
19	11 08	19 29	19 24	08 08	21 10	19 21	11 16	18 26	11 46	21 46
20	10 46	13 40	19 12	07 39	21 15	19 22	11 16	18 27	11 46	21 45
21	10 25	07 S 00	18 59	07 09	21 19	19 23	11 15	18 27	11 47	21 44
22	10 03	00 N 02	18 45	06 40	21 23	19 24	11 15	18 28	11 47	21 44
23	09 41	06 58	18 30	06 10	21 26	19 24	11 14	18 29	11 48	21 43
24	09 19	13 24	18 13	05 40	21 30	19 25	11 14	18 29	11 48	21 43
25	08 56	18 58	17 55	05 10	21 33	19 26	11 13	18 30	11 49	21 42
26	08 34	23 23	17 36	04 40	21 36	19 26	11 12	18 30	11 49	21 42
27	08 11	26 24	17 15	04 09	21 38	19 27	11 12	18 31	11 50	21 41
28	07 S 49	27 N 54	16 S 53	03 S 39	21 N 40	19 S 27	11 S 11	18 N 31	11 N 50	21 S 41

ZODIAC SIGN ENTRIES

Date	h	m	Planets
01	09	20	☽ ♋
03	18	08	☽ ♌
06	04	51	☽ ♍
08	17	15	☽ ♎
08	23	52	♃ ♐
09	08	27	♀ ♓
11	06	14	☽ ♏
13	17	30	☽ ♐
14	09	52	☿ ♒
16	01	01	☽ ♑
18	04	22	☽ ♒
18	13	04	☉ ♓
20	04	48	☽ ♓
22	04	22	☽ ♈
24	05	05	☽ ♉
26	08	23	☽ ♊
28	14	49	☽ ♋

LATITUDES

Date	Mercury ☿ ° '	Venus ♀ ° '	Mars ♂ ° '	Jupiter ♃ ° '	Saturn ♄ ° '	Uranus ♅ ° '	Neptune ♆ ° '	Pluto ♇ ° '
01	02 N 25	01 S 24	04 N 32	00 N 59	02 N 30	00 N 41	01 S 45	10 S 18
04	01 52	01 26	04 33	01 00	02 31	00 41	01 45	10 18
07	01 18	01 27	04 34	01 00	02 32	00 41	01 45	10 18
10	00 45	01 28	04 33	01 00	02 33	00 41	01 45	10 18
13	00 N 15	01 28	04 31	01 01	02 33	00 41	01 45	10 18
16	00 S 14	01 28	04 29	01 01	02 34	00 41	01 44	10 18
19	00 39	01 27	04 25	01 01	02 35	00 41	01 44	10 19
22	01 02	01 25	04 21	01 02	02 36	00 41	01 44	10 19
25	01 22	01 23	04 16	01 02	02 36	00 41	01 44	10 19
28	01 39	01 21	04 11	01 02	02 37	00 41	01 44	10 19
31	01 S 53	01 S 17	04 N 05	01 N 03	02 N 38	00 N 41	01 S 44	10 S 20

DATA

Julian Date	2466917
Delta T	+78 seconds
Ayanamsa	24° 26' 35"
Synetic vernal point	04° ♓ 40' 24"
True obliquity of ecliptic	23° 26' 09"

MOON'S PHASES, APSIDES AND POSITIONS ☽

Date	h	m	Phase	Longitude ° '	Eclipse Indicator
05	01	58	○	16 ♌ 23	
13	07	16	☾	24 ♏ 42	
20	07	39	●	01 ♓ 47	
26	23	29	☽	08 ♊ 30	

Day	h	m		
08	21	03	Apogee	
21	05	50	Perigee	
01	18	23	Max dec	27° N 52'
09	03	05	0S	
16	09	13	Max dec	27° S 59'
22	11	54	0N	
28	23	22	Max dec	28° N 04'

LONGITUDES

Date	Chiron ⚷ ° '	Ceres ⚳ ° '	Pallas ⚴ ° '	Juno ⚵ ° '	Vesta ⚶ ° '	Black Moon Lilith ⚸ ° '
01	00 ♌ 36	01 ♐ 05	23 ♎ 06	04 ♏ 10	06 ♊ 52	25 ♍ 57
11	29 ♋ 55	04 ♐ 04	23 ♎ 33	04 ♏ 59	07 ♊ 49	27 ♍ 04
21	29 ♋ 19	06 ♐ 42	23 ♎ 12	05 ♏ 15	09 ♊ 22	28 ♍ 11
31	28 ♋ 49	08 ♐ 56	21 ♎ 59	04 ♏ 55	11 ♊ 26	29 ♍ 17

All ephemeris data is given at 12.00 UT and the Moon's longitude is additionally given for 24.00 UT

ASPECTARIAN

01 Saturday
01 14 ☽ ∠ ♅
04 34 ☽ bQ ☉
04 56 ☽ △ ♇
06 58 ♀ ☍ ♂
07 40 ☽ ⊼ ♃
17 05 ☿ ⊻ ♀
17 56 ☽ ⊥ ♅
18 08 ☽ ∠ ♂
18 59 ☽ ± ♃
19 24 ☽ ✶ ♆
19 43 ☽ bQ ♀
21 26 ☽ △ ♄
22 49 ☽ ± ☉

02 Sunday
05 10 ☽ ⊻ ♅
09 05 ☽ bQ ♇
10 33 ☽ ⊥ ♂
11 06 ☽ ⊼ ☉
12 05 ☽ bQ ♃
14 32 ☽ ± ♀
18 03 ☽ Q ♆
21 35 ☽ ⊻ ♂

03 Monday
00 36 ☽ ☍ ☿
02 12 ☽ ± ♇
03 12 ☽ ⊼ ♀
13 43 ☽ ⊼ ♇
16 59 ☽ △ ♃

04 Tuesday
02 19 ♀ Q ♆
04 36 ☽ □ ♆
06 44 ☽ □ ♄
14 26 ☽ ☌ ♅
17 19 ☽ # ♇
21 32 ☿ ⊥ ♇

05 Wednesday
01 58 ☽ ☍ ☉
05 13 ☽ # ☿
05 47 ☽ ☌ ♂
06 48 ☽ ∥ ♂
10 05 ☽ # ♃
13 14 ☽ ⊼ ☿
14 33 ☽ ∥ ♅
18 01 ☽ Q ♄
20 07 ☽ ☍ ♀

06 Thursday
00 24 ☽ ☍ ♇
01 46 ☽ ± ☿
04 13 ☽ □ ♃
04 32 ☽ # ☉
12 01 ☽ # ♀
12 05 ☉ ☍ ♂
15 42 ☽ △ ♆
17 55 ☽ ✶ ♄
20 48 ☽ bQ ☿
23 18 ☽ ∥ ♆

07 Friday
01 02 ☽ # ♄
01 34 ☽ ⊻ ♅
13 36 ☽ ⊥ ♅
14 31 ♀ ☌ ♇
15 39 ☽ ⊻ ♂
16 39 ☽ Q ♃
19 14 ☽ ⊼ ☉
21 54 ☽ bQ ♆

08 Saturday
00 09 ☽ ∠ ♄
03 23 ☽ ⊥ ♂
05 06 ☽ △ ☿
07 45 ☽ ∠ ♅
08 31 ☽ ± ☉
12 52 ☽ ⊼ ♇
15 27 ☽ ⊼ ♀
16 11 ☽ ± ♆
17 12 ☽ ✶ ♃
18 26 ☽ ⊥ ♄
21 04 ☽ ∠ ♂

09 Sunday
01 06 ☽ ± ♇
04 24 ☽ ⊼ ♆
04 32 ☽ bQ ☉
05 07 ☽ ± ♀
06 39 ☽ ⊻ ♄
09 15 ♀ □ ♃
12 53 ☿ # ♂
14 09 ☽ ✶ ♅
19 28 ☽ bQ ♇

10 Monday
00 01 ☽ ∠ ♃
01 37 ☽ bQ ♀
02 37 ☽ ✶ ♂
13 56 ☽ △ ☉
14 29 ☽ Q ♅
18 28 ☽ ⊥ ♃
23 01 ☽ □ ☿

11 Tuesday
02 00 ☽ △ ♇
02 12 ☽ Q ♂
05 08 ☽ ∥ ♄
06 43 ☽ ⊻ ♃
07 01 ☽ # ♆
08 09 ☽ ∥ ♀
11 36 ☽ △ ♀
16 58 ☽ ∥ ☉
17 20 ☽ ☍ ♆
19 30 ☽ ☌ ♄
21 00 ♀ # ♆

12 Wednesday
02 37 ☽ □ ♅
03 15 ☉ Q ♆
11 43 ☿ ⊻ ♇
13 06 ☽ □ ♂
14 14 ☿ ⊥ ♀
14 20 ☽ Q ☿
15 54 ☽ # ♅
18 49 ♀ ∥ ♄
20 44 ☽ ∥ ♃

13 Thursday
01 46 ☽ ∥ ☿
04 25 ☽ # ♂
07 16 ☽ □ ☉
11 48 ☽ ∥ ♇
13 37 ☽ □ ♇
15 56 ☽ ✶ ☿
18 26 ☽ ☌ ♃
18 52 ♀ ✶ ♆
22 50 ♄ St R

14 Friday
04 07 ☽ ⊼ ♆
05 07 ☽ □ ♀
06 05 ☽ ⊻ ♄
12 36 ☽ △ ♅
14 58 ♀ △ ♄
15 25 ☽ ± ♆
17 21 ☽ ⊥ ♄
20 27 ☽ Q ☉
20 59 ☽ △ ♂
22 40 ☿ ✶ ♃
22 56 ☽ ∠ ☿
23 42 ☽ Q ♇

15 Saturday
08 03 ☽ bQ ♆
09 55 ☽ ∠ ♄
16 05 ☽ bQ ♅
16 54 ☽ ⊥ ☿
17 53 ☽ Q ♀
20 09 ☽ ✶ ☉
21 33 ☽ ✶ ♇
23 31 ☽ bQ ♂

16 Sunday
02 16 ☽ ⊻ ♃
04 37 ☽ ⊻ ☿
08 04 ☽ ± ♅
10 54 ☽ △ ♆
12 39 ☽ ✶ ♄
12 54 ☽ ⊥ ♃
14 57 ☽ ± ♂
15 12 ☉ ☌ ♇
17 13 ☽ ✶ ♀
18 30 ☽ ⊼ ♅
23 53 ☽ ∠ ♇

17 Monday
00 32 ☽ ∠ ☉
01 04 ☽ ⊼ ♂
04 32 ☽ ∠ ♃
07 06 ♀ ⊼ ♅
09 17 ☽ Q ♄
15 12 ☽ ⊥ ♇
16 59 ☽ ⊥ ☉
17 48 ☉ # ♆
21 12 ☽ ∠ ♀

18 Tuesday
01 15 ☽ ⊻ ♇
03 43 ☽ ⊻ ☉
05 50 ☽ ✶ ♃
12 17 ☽ ☌ ☿
13 27 ☽ ⊥ ♀
13 38 ☽ □ ♆
15 11 ☽ □ ♄
20 33 ☽ ☍ ♅

19 Wednesday
00 07 ☽ ⊻ ♀
00 53 ☽ ∥ ♇
01 32 ☽ Q ♃
01 42 ☽ ☍ ♂
02 34 ☉ ∥ ♄
04 03 ☽ # ♂
04 25 ☿ □ ♆
12 26 ☽ ∥ ☿
12 34 ☉ □ ♃
12 36 ☽ ∥ ♃
15 21 ♀ ⊼ ♂
16 42 ☽ # ♅
16 52 ☿ ∥ ♃
18 43 ☽ Q ♆
21 51 ☿ □ ♄

20 Thursday
01 54 ☽ ☌ ♇
06 28 ☽ □ ♃
07 39 ☽ ☌ ☉
13 49 ☽ ✶ ♆
15 12 ☽ △ ♄
16 50 ☽ ⊻ ☿
19 04 ☽ # ♆
20 20 ☽ ⊼ ♅
20 58 ☽ ∥ ♄
23 20 ☽ ∥ ☉

21 Friday
00 29 ☽ ⊼ ♂
03 13 ☽ ⊥ ☿
04 14 ☽ ☌ ♀
05 48 ☽ ± ♅
09 47 ☽ ± ♂
11 27 ☽ ∥ ♀
13 35 ☽ ∠ ♆
14 56 ☽ bQ ♄
18 50 ☽ ∠ ☿
20 02 ☽ bQ ♅
22 45 ♀ ± ♅
23 46 ☽ bQ ♂

22 Saturday
01 33 ☽ ⊻ ♇
03 54 ☽ ⊥ ♆
05 14 ☽ ± ♄
06 17 ☽ △ ♃
08 11 ☿ ☍ ♅
10 39 ☽ ⊻ ☉
11 08 ☽ ⊥ ♇
13 31 ☽ ⊻ ♆
14 49 ☽ ⊼ ♄
19 57 ☽ △ ♅
20 59 ☽ ⊥ ☉
21 09 ☽ ✶ ☿
23 19 ☽ △ ♂

23 Sunday
01 39 ☽ ∠ ♇
06 32 ☽ bQ ♃
08 34 ☽ ⊻ ♀
09 20 ☽ # ♀
11 24 ♀ ± ♂
12 39 ☽ ∠ ☉
13 33 ☿ # ♅
17 05 ☿ ☍ ♂
18 39 ☽ Q ☿
19 17 ☽ ⊥ ♀
21 16 ☽ # ☉
21 19 ☽ ± ♃

24 Monday
02 14 ☽ ✶ ♇
03 35 ☽ # ♄
05 46 ☽ ∥ ♆
06 06 ☉ ✶ ♆
07 18 ☽ ⊼ ♃
11 37 ☽ ∠ ♀
14 42 ☽ ☌ ♆
15 21 ☽ ✶ ☉
15 57 ☽ ☍ ♄
21 15 ☽ □ ♅
22 22 ☽ Q ♇
23 42 ☉ △ ♄
23 56 ☽ □ ♂

25 Tuesday
00 13 ☿ Q ♃
03 58 ☽ □ ☿
07 21 ☽ # ☿
09 45 ☽ ∥ ♅
13 20 ☽ Q ☉
14 12 ☽ # ♃
15 36 ☽ ✶ ♀
23 36 ♀ ∠ ♆

26 Wednesday
01 18 ☽ ∥ ♂
02 00 ☽ # ♇
04 10 ☽ Q ♅
05 29 ☽ □ ♇
05 40 ☿ ⊥ ♀
06 26 ☽ Q ♂
10 59 ☽ ☍ ♃
12 28 ♀ bQ ♄
14 43 ☽ Q ♀
18 42 ☽ ⊻ ♆
19 50 ☽ ⊼ ♄
23 29 ☽ □ ☉

27 Thursday
01 24 ☽ ✶ ♅
03 22 ☽ ✶ ♂
05 28 ☽ ⊥ ♆
06 36 ☽ ± ♄
15 03 ☽ △ ☿
21 52 ☽ ∠ ♆
22 57 ☽ bQ ♄

28 Friday
00 11 ☉ ⊼ ♅
02 40 ☽ □ ♀
04 39 ☽ ∠ ♅
06 15 ☽ ∠ ♂
11 51 ☽ △ ♇
15 59 ♂ # ♇
17 48 ☽ ⊼ ♃
17 53 ☉ ⊼ ♂
21 21 ☽ ⊥ ♅
22 18 ☽ bQ ☿
22 43 ☽ ⊥ ♂
22 52 ♀ bQ ♅

MARCH 2042

LONGITUDES

Date	Sidereal time h m s	Sun ☉ ° ′ ″	Moon ☽ ° ′ ″	Moon ☽ 24.00 ° ′ ″	Mercury ☿ ° ′	Venus ♀ ° ′	Mars ♂ ° ′	Jupiter ♃ ° ′	Saturn ♄ ° ′	Uranus ♅ ° ′	Neptune ♆ ° ′	Pluto ⯓ ° ′
01	22 37 43	11 ♓ 01 39	11 ♋ 15 47	17 ♋ 33 36	19 ♒ 51	25 ♓ 11	10 ♌ 06	01 ♐ 39	06 ♏ 23	09 ♌ 29	05 ♉ 53	28 ♒ 26
02	22 41 40	12 01 52	23 ♋ 48 12	29 ♋ 59 53	21 20	26 25	09 R 53	01 41	06 R 22	09 R 26	05 54	28 28
03	22 45 36	13 02 02	06 ♌ 08 56	12 ♌ 15 34	22 51	27 40	09 41	01 44	06 20	09 24	05 56	28 30
04	22 49 33	14 02 10	18 ♌ 20 02	24 ♌ 22 31	24 23	28 ♓ 55	09 29	01 47	06 18	09 22	05 57	28 31
05	22 53 29	15 02 16	00 ♍ 23 14	06 ♍ 22 22	25 56	00 ♈ 09	09 19	01 49	06 16	09 20	05 59	28 33
06	22 57 26	16 02 20	12 ♍ 20 05	18 ♍ 16 36	27 30	01 24	09 09	01 52	06 14	09 18	06 00	28 35
07	23 01 23	17 02 23	24 ♍ 12 07	00 ♎ 06 51	29 ♒ 05	02 39	09 00	01 54	06 12	09 16	06 02	28 36
08	23 05 19	18 02 23	06 ♎ 01 03	11 ♎ 55 00	00 ♓ 41	03 53	08 51	01 56	06 10	09 15	06 03	28 38
09	23 09 16	19 02 22	17 ♎ 49 01	23 ♎ 43 26	02 19	05 08	08 44	01 57	06 08	09 13	06 05	28 40
10	23 13 12	20 02 19	29 ♎ 38 39	05 ♏ 35 04	03 57	06 22	08 37	01 59	06 05	09 11	06 07	28 41
11	23 17 09	21 02 14	11 ♏ 33 10	17 ♏ 33 26	05 37	07 37	08 31	02 00	06 03	09 09	06 08	28 43
12	23 21 05	22 02 08	23 ♏ 36 23	29 ♏ 42 35	07 17	08 51	08 26	02 01	06 00	09 07	06 10	28 44
13	23 25 02	23 02 00	05 ♐ 52 35	12 ♐ 06 58	08 59	10 06	08 21	02 02	05 58	09 06	06 12	28 46
14	23 28 58	24 01 50	18 ♐ 26 17	24 ♐ 51 07	10 42	11 20	08 18	02 03	05 55	09 04	06 14	28 48
15	23 32 55	25 01 39	01 ♑ 21 55	07 ♑ 59 10	12 27	12 34	08 15	02 04	05 52	09 03	06 16	28 49
16	23 36 52	26 01 26	14 ♑ 43 12	21 ♑ 34 15	14 12	13 49	08 13	02 04	05 49	09 01	06 17	28 51
17	23 40 48	27 01 12	28 ♑ 32 27	05 ♒ 37 43	15 59	15 03	08 11	02 04	05 46	08 59	06 19	28 52
18	23 44 45	28 00 55	12 ♒ 49 49	20 ♒ 08 18	17 47	16 17	08 11	02 R 04	05 43	08 58	06 21	28 54
19	23 48 41	29 ♓ 00 37	27 ♒ 32 33	05 ♓ 01 43	19 36	17 32	08 D 11	02 04	05 40	08 57	06 23	28 55
20	23 52 38	00 ♈ 00 17	12 ♓ 34 48	20 ♓ 10 39	21 26	18 46	08 12	02 04	05 37	08 55	06 25	28 57
21	23 56 34	00 59 55	27 ♓ 48 00	05 ♈ 25 35	23 18	20 00	08 13	02 03	05 33	08 54	06 27	28 58
22	00 00 31	01 59 31	13 ♈ 02 04	20 ♈ 36 16	25 11	21 14	08 16	02 02	05 30	08 53	06 29	29 00
23	00 04 27	02 59 05	28 ♈ 07 02	05 ♉ 33 23	27 05	22 28	08 19	02 01	05 27	08 51	06 31	29 01
24	00 08 24	03 58 37	12 ♉ 54 31	20 ♉ 09 49	29 ♓ 00	23 42	08 22	02 00	05 23	08 50	06 32	29 03
25	00 12 21	04 58 06	27 ♉ 18 50	04 ♊ 21 18	00 ♈ 57	24 56	08 27	01 59	05 20	08 49	06 34	29 04
26	00 16 17	05 57 34	11 ♊ 17 07	18 ♊ 06 20	02 54	26 11	08 32	01 58	05 16	08 48	06 36	29 06
27	00 20 14	06 56 59	24 ♊ 49 07	01 ♋ 25 44	04 53	27 25	08 38	01 56	05 12	08 47	06 38	29 07
28	00 24 10	07 56 22	07 ♋ 56 31	14 ♋ 21 53	06 53	28 38	08 44	01 54	05 08	08 46	06 40	29 09
29	00 28 07	08 55 42	20 ♋ 42 16	26 ♋ 58 07	08 54	29 ♈ 52	08 51	01 52	05 05	08 45	06 43	29 10
30	00 32 03	09 55 01	03 ♌ 09 54	09 ♌ 18 07	10 56	01 ♉ 06	08 59	01 50	05 01	08 44	06 45	29 11
31	00 36 00	10 ♈ 54 16	15 ♌ 23 12	21 ♌ 25 36	12 ♈ 58	02 ♉ 20	09 ♌ 07	01 ♐ 47	04 ♏ 57	08 ♌ 43	06 ♉ 47	29 ♒ 13

DECLINATIONS

Date	Moon True ☊ ° ′	Moon Mean ☊ ° ′	Moon ☽ Latitude ° ′	Sun ☉ ° ′	Moon ☽ ° ′	Mercury ☿ ° ′	Venus ♀ ° ′	Mars ♂ ° ′	Jupiter ♃ ° ′	Saturn ♄ ° ′	Uranus ♅ ° ′	Neptune ♆ ° ′	Pluto ⯓ ° ′
01	28 ♈ 06	29 ♈ 33	04 N 55	07 S 26	27 N 51	16 S 30	03 S 08	21 N 42	19 S 28	11 S 10	18 N 32	11 N 51	21 S 40
02	28 R 00	29 30	05 07	07 03	26 23	16 06	02 37	21 44	19 28	11 10	18 33	11 51	21 40
03	27 53	29 27	05 05	06 40	23 39	15 40	02 07	21 45	19 29	11 09	18 33	11 52	21 39
04	27 44	29 24	04 49	06 17	19 54	15 14	01 36	21 46	19 29	11 08	18 34	11 52	21 39
05	27 36	29 21	04 20	05 54	15 23	14 45	01 05	21 47	19 30	11 07	18 34	11 53	21 38
06	27 28	29 18	03 39	05 30	10 19	14 16	00 34	21 48	19 30	11 06	18 35	11 54	21 38
07	27 22	29 14	02 50	05 07	04 N 54	13 45	00 S 03	21 48	19 30	11 05	18 35	11 54	21 37
08	27 18	29 11	01 53	04 44	00 S 40	13 13	00 N 28	21 49	19 31	11 04	18 36	11 55	21 37
09	27 15	29 08	00 N 51	04 20	06 12	12 40	00 59	21 48	19 31	11 03	18 36	11 55	21 36
10	27 D 15	29 05	00 S 13	03 57	11 33	12 06	01 30	21 48	19 31	11 02	18 37	11 56	21 36
11	27 16	29 02	01 17	03 33	16 31	11 30	02 01	21 48	19 31	11 01	18 37	11 57	21 35
12	27 17	28 58	02 19	03 10	20 55	10 53	02 31	21 47	19 31	11 00	18 37	11 57	21 35
13	27 19	28 55	03 16	02 46	24 30	10 15	03 02	21 46	19 31	10 59	18 38	11 58	21 35
14	27 20	28 52	04 04	02 22	26 59	09 36	03 33	21 45	19 31	10 58	18 38	11 58	21 34
15	27 R 20	28 49	04 42	01 59	28 08	08 55	04 04	21 43	19 31	10 57	18 39	11 59	21 34
16	27 19	28 46	05 06	01 35	27 41	08 14	04 34	21 42	19 31	10 56	18 39	12 00	21 33
17	27 16	28 43	05 13	01 11	25 33	07 31	05 05	21 40	19 31	10 55	18 39	12 00	21 33
18	27 12	28 39	05 02	00 47	21 47	06 47	05 35	21 38	19 31	10 54	18 40	12 01	21 32
19	27 08	28 36	04 30	00 S 24	16 33	06 01	06 05	21 35	19 31	10 53	18 40	12 02	21 32
20	27 04	28 33	03 40	00 00	10 14	05 15	06 36	21 33	19 31	10 51	18 40	12 02	21 32
21	27 01	28 30	02 34	00 N 24	03 S 13	04 28	07 06	21 30	19 31	10 50	18 41	12 03	21 31
22	26 59	28 27	01 S 16	00 48	03 N 59	03 39	07 35	21 28	19 30	10 49	18 41	12 03	21 31
23	26 D 59	28 24	00 N 06	01 11	10 54	02 49	08 05	21 25	19 30	10 48	18 41	12 04	21 31
24	26 59	28 20	01 27	01 35	17 05	01 59	08 35	21 22	19 30	10 46	18 42	12 05	21 30
25	27 01	28 17	02 40	01 58	22 09	01 07	09 04	21 18	19 29	10 45	18 42	12 05	21 30
26	27 02	28 14	03 41	02 22	25 46	00 S 15	09 33	21 15	19 29	10 44	18 42	12 06	21 30
27	27 03	28 11	04 27	02 46	27 47	00 N 39	10 02	21 11	19 29	10 42	18 42	12 07	21 29
28	27 R 03	28 08	04 58	03 09	28 10	01 33	10 31	21 07	19 28	10 41	18 43	12 08	21 29
29	27 03	28 04	05 14	03 32	27 01	02 28	10 59	21 03	19 28	10 40	18 43	12 08	21 29
30	27 02	28 01	05 14	03 56	24 32	03 23	11 27	20 59	19 27	10 38	18 43	12 09	21 28
31	27 ♈ 00	27 ♈ 58	05 N 00	04 N 19	20 N 59	04 N 19	11 N 55	20 N 55	19 S 26	10 S 37	18 N 43	12 N 10	21 S 28

ZODIAC SIGN ENTRIES

Date	h	m	Planets
03	00	00	☽ ♌
05	08	58	♀ ♈
05	11	14	☽ ♍
07	23	46	☽ ♎
08	01	47	☿ ♓
10	12	43	☽ ♏
13	00	34	☽ ♐
15	09	30	☽ ♑
17	14	29	☽ ♒
19	15	57	☽ ♓
20	11	53	☉ ♈
21	15	28	☽ ♈
23	15	02	☽ ♉
25	00	21	☿ ♈
25	16	33	☽ ♊
27	21	23	☽ ♋
29	14	29	♀ ♉
30	05	51	☽ ♌

LATITUDES

Date	Mercury ☿ ° ′	Venus ♀ ° ′	Mars ♂ ° ′	Jupiter ♃ ° ′	Saturn ♄ ° ′	Uranus ♅ ° ′	Neptune ♆ ° ′	Pluto ⯓ ° ′
01	01 S 44	01 S 20	04 N 09	01 N 03	02 N 37	00 N 41	01 S 44	10 S 19
04	01 57	01 16	04 03	01 03	02 38	00 41	01 43	10 20
07	02 06	01 12	03 57	01 03	02 39	00 41	01 43	10 20
10	02 11	01 08	03 50	01 04	02 39	00 41	01 43	10 21
13	02 13	01 03	03 43	01 04	02 40	00 41	01 43	10 21
16	02 11	00 57	03 36	01 04	02 40	00 40	01 43	10 22
19	02 04	00 51	03 30	01 05	02 41	00 40	01 43	10 22
22	01 53	00 45	03 23	01 05	02 41	00 40	01 43	10 23
25	01 38	00 38	03 16	01 05	02 42	00 40	01 43	10 23
28	01 18	00 31	03 10	01 06	02 42	00 40	01 43	10 24
31	00 S 53	00 S 23	03 N 03	01 N 06	02 N 43	00 N 40	01 S 43	10 S 25

DATA

Julian Date	2466945
Delta T	+78 seconds
Ayanamsa	24° 26′ 39″
Synetic vernal point	04° ♓ 40′ 20″
True obliquity of ecliptic	23° 26′ 10″

MOON'S PHASES, APSIDES AND POSITIONS ☽

Date	h	m	Phase	Longitude ° ′	Eclipse Indicator
06	20	10	○	16 ♍ 23	
14	23	21	☾	24 ♐ 30	
21	17	23	●	01 ♈ 13	
28	12	00	☽	07 ♋ 56	

Day	h	m		
08	04	46	Apogee	
21	17	33	Perigee	
08	09	09	0S	
15	17	40	Max dec	28° S 10′
21	22	44	0N	
28	05	37	Max dec	28° N 13′

LONGITUDES

Date	Chiron ⚷ ° ′	Ceres ⚳ ° ′	Pallas ⚴ ° ′	Juno ⚵ ° ′	Vesta ⚶ ° ′	Black Moon Lilith ⚸ ° ′
01	28 ♋ 54	08 ♐ 31	22 ♎ 18	05 ♏ 02	10 ♊ 59	29 ♍ 04
11	28 ♋ 31	10 ♐ 22	20 ♎ 25	04 ♏ 13	13 ♊ 24	00 ♎ 11
21	28 ♋ 17	11 ♐ 42	17 ♎ 50	02 ♏ 50	16 ♊ 11	01 ♎ 17
31	28 ♋ 13	12 ♐ 27	14 ♎ 49	00 ♏ 58	19 ♊ 16	02 ♎ 24

All ephemeris data is given at 12.00 UT and the Moon's longitude is additionally given for 24.00 UT

ASPECTARIAN

h m	Aspects
01 Saturday	
01 48	☽ ⚹ ♆
02 48	☽ △ ♄
05 07	☽ ± ♃
08 37	☽ ⊻ ♅
09 50	☽ ⊻ ♂
10 42	♀ ♇ ♂
11 31	☽ △ ☉
16 08	☽ ♇ ⯓
17 34	☽ ± ☿
22 17	☽ ♇ ♃
02 Sunday	
00 38	☽ Q ♆
06 36	☽ ⊼ ☿
09 25	☽ ± ⯓
17 38	☽ △ ♀
18 47	☽ ♇ ☉
21 03	☽ ⊼ ⯓
03 Monday	
03 21	☽ △ ♃
11 34	☽ □ ♆
12 22	☽ □ ♄
13 53	☽ ± ☉
18 22	☽ ☌ ♅
18 49	☽ ☌ ♂
04 Tuesday	
00 46	☽ ∥ ♂
01 31	☽ # ⯓
02 16	☽ ♇ ♀
02 44	☽ ⊼ ☉
04 17	♀ ⊻ ⯓
05 12	☿ Q ♆
14 21	☽ # ♃
19 29	☽ ∥ ♅
22 08	☽ ± ♀
23 49	☽ Q ♄
05 Wednesday	
01 47	☽ ☍ ☿
06 55	♂ ☌ ♅
08 19	☽ ☍ ⯓
08 28	♀ ⊥ ♆
11 29	☽ ⊼ ♀
14 08	♀ ± ♄
14 53	☽ □ ♃
15 24	☽ # ☿
17 34	☉ ± ♂
19 00	☉ ± ♅
23 14	☽ △ ♆
23 46	☽ ⚹ ♄
06 Thursday	
04 43	☽ ∥ ♆
05 40	☽ ⊻ ♂
05 55	☽ ⊻ ♅
08 21	☽ # ♄
17 36	☽ ⊥ ♂
17 59	☽ ⊥ ♅
20 10	☽ ☍ ☉
21 07	♀ △ ♃
07 Friday	
03 15	☽ Q ♃
04 42	☿ ☌ ⯓
05 34	☽ ♇ ♆
05 56	☽ ∠ ♄
10 59	☽ # ☉
11 35	☽ ∠ ♂
12 09	☽ ∠ ♅
20 57	☽ ⊼ ⯓
23 27	☽ ⊼ ☿
23 51	☽ ± ♆
08 Saturday	
00 08	☽ ⊥ ♄
03 40	☽ ⚹ ♃
07 09	☽ ☍ ♀
07 35	☽ ∥ ♀
09 11	☽ ± ⯓
11 04	☽ # ♀
12 05	☽ ⊼ ♆
12 18	☽ ⊻ ♄
13 34	☽ ± ☿
17 42	☽ ⚹ ♂
18 33	☽ ⚹ ♅
09 Sunday	
02 44	♀ ⊥ ⯓
03 32	☽ ♇ ⯓
04 24	☽ ∥ ☉
06 42	☿ □ ♃
10 14	☽ ∠ ♃
10 48	☽ ♇ ☿
14 43	☽ ⊼ ☉
17 52	☽ Q ♂
18 53	☽ Q ♅
10 Monday	
02 43	♄ ☍ ♆
04 01	☽ ± ☉
04 34	☽ ⊥ ♃
06 41	♀ ⊼ ♄
06 54	♀ ⊻ ♆
09 40	☽ ∥ ♄
10 03	☽ △ ⯓
13 47	☽ # ♆
14 17	☽ ∥ ☿
16 44	☽ ⊻ ♃
18 40	☿ # ♆
22 07	☽ △ ☿
23 54	☽ ♇ ☉
11 Tuesday	
00 58	☽ ☌ ♄
01 06	☽ ☍ ♆
03 10	☽ ⊼ ♀
05 57	☽ □ ♂
07 12	☽ □ ♅
12 11	☉ ♇ ♄
14 35	☉ ∠ ♆
16 36	☽ ± ♀
18 05	☿ △ ♄
19 45	☿ ⚹ ♆
22 59	☽ # ♅
12 Wednesday	
03 57	☽ ∥ ♃
04 21	♀ △ ♂
07 18	☿ ∥ ♄
08 37	☽ △ ☉
12 33	☽ ♇ ♀
16 02	☽ ∥ ⯓
17 07	♀ △ ♅
17 15	☽ # ♂
22 08	☽ □ ⯓
13 Thursday	
03 27	☿ ⊼ ♂
04 32	☽ ☌ ♃
04 49	☉ # ♀
12 10	☽ ⊻ ♄
12 38	☽ ⊼ ♆
13 28	☿ ⊼ ♅
16 46	☽ △ ♂
18 12	☽ △ ♅
18 58	☽ □ ☿
19 20	☉ ♇ ♂
21 01	☽ △ ♀
23 39	☽ ⊥ ♄
14 Friday	
00 11	☽ ± ♆
08 53	☽ Q ⯓
12 53	☉ ♇ ♅
16 38	☽ ∠ ♄
17 15	☽ ♇ ♆
21 04	☽ ♇ ♂
22 31	☽ ♇ ♅
23 21	☽ □ ☉
15 Saturday	
07 20	☽ ⚹ ⯓
10 04	☽ Q ☿
13 16	☽ ⊻ ♃
13 37	☽ ± ♂
15 03	☽ ± ♅
17 58	☿ ⊻ ♀
20 09	☽ ⚹ ♄
20 54	☽ △ ♆
16 Sunday	
00 08	☽ ⊥ ♃
00 26	☽ ⊼ ♂
01 52	☽ ⊼ ♅
10 14	☽ □ ♀
10 27	☽ ∠ ⯓
10 40	☽ Q ☉
10 57	☽ ⚹ ☿
12 08	☿ ± ♂
12 41	♀ ∠ ⯓
16 08	☽ ∠ ♃
17 26	☽ Q ♄
22 50	☿ ± ♅
17 Monday	
02 14	☽ ⊥ ⯓
09 12	☽ ⚹ ☉
12 34	☽ ⊻ ⯓
16 45	☽ ∠ ☿
18 00	☽ ⚹ ♃
20 23	☽ Q ♀
18 Tuesday	
00 11	☽ □ ♄
01 11	☽ □ ♆
02 47	♃ St R
04 17	☽ ☍ ♂
05 36	☽ ☍ ♅
10 01	☽ ⊥ ☿
12 20	☽ ∠ ☉
12 47	☽ # ♂
13 14	☽ ∥ ⯓
14 03	☽ Q ♃
18 14	☽ ⚹ ♀
19 51	♂ St D
21 18	☽ ⊻ ☿
23 07	☽ ∥ ♃
19 Wednesday	
02 59	☽ # ♅
03 07	♀ ♇ ♃
04 09	☽ ⊥ ☉
06 53	☽ Q ♆
09 52	☉ ⊻ ⯓
10 44	☿ # ♀
14 14	☽ ☌ ⯓
14 32	☽ ⊻ ☉
19 16	☽ □ ♃
20 43	☽ ∠ ♀
19 34	☿ ♇ ♅
20 55	☽ # ☉
23 18	☽ ⊥ ⯓
20 Thursday	
00 58	☽ △ ♄
01 32	☿ ♇ ♄
02 11	☽ ⚹ ♆
02 59	☉ ± ♄
05 02	☽ ⊼ ♂
05 28	☽ # ♆
06 12	☽ ⊼ ♅
09 44	☽ ∥ ♄
11 39	☿ ∠ ♆
12 19	☽ ⊥ ♀
14 34	☽ ± ♂
15 42	☽ ± ♅
22 10	☉ ⊥ ♆
22 38	☽ ⊻ ♀
23 48	☽ # ♀
21 Friday	
00 38	☽ ♇ ♄
01 59	☽ ∠ ♆
02 38	♂ # ⯓
03 56	☽ ☌ ☿
04 47	☽ ♇ ♂
05 52	☽ ♇ ♅
07 20	☽ ∥ ☿
11 01	☿ ♇ ♂
13 51	☽ ⊻ ⯓
14 45	☽ ± ♄
16 10	☽ ⊥ ♆
17 23	☽ ☌ ☉
18 41	☽ △ ♃
22 Saturday	
00 10	☽ ⊼ ♄
00 45	☽ ∥ ☉
01 38	☽ ⊻ ♆
04 27	☽ △ ♂
05 27	☽ △ ♅
11 01	☽ # ☿
13 08	☉ △ ♃
13 32	☽ ∠ ⯓
18 20	☽ ♇ ♃
23 Sunday	
01 15	☽ ∥ ♀
02 10	☽ ☌ ♀
08 39	☽ ± ♃
10 06	☽ ⊻ ☿
11 37	☽ # ♄
13 28	☽ ⚹ ⯓
16 17	☽ ∥ ♆
18 17	☽ ⊼ ♃
20 24	☽ ⊻ ☉
21 11	☽ ⊥ ☿
23 46	☽ ☍ ♄
24 Monday	
01 35	☽ ☌ ♆
04 33	☽ □ ♂
05 21	☽ □ ♅
06 51	☽ ⊥ ☉
08 57	☽ Q ⯓
12 34	☿ ⊻ ⯓
14 05	☽ ∠ ☿
16 35	☿ ± ♄
19 00	☽ ∥ ♅
19 41	☉ # ☿
22 39	☽ # ♃
22 46	☽ ∠ ☉
25 Tuesday	
07 21	☿ ⊥ ♆
07 35	☽ ∥ ♂
07 37	☽ ⊻ ♀
08 33	☽ # ⯓
10 32	☽ Q ♂
11 10	☽ Q ♅
14 37	☉ ⊥ ⯓
14 59	☽ □ ⯓
18 45	☽ ⊥ ♀
19 10	☽ ⚹ ☿
19 55	☽ ☍ ♃
20 08	☉ ⊼ ♄
26 Wednesday	
00 35	☿ △ ♃
01 37	☽ ⊼ ♄
02 03	☽ ⚹ ☉
03 52	☽ ⊻ ♆
07 11	☽ ⚹ ♂
07 41	☽ ⚹ ♅
07 53	♀ ± ♃
11 47	☽ ∠ ♀
11 58	☽ ± ♄
14 19	☽ ⊥ ♆
19 25	☽ Q ☿
27 Thursday	
00 40	☽ Q ☉
03 46	☽ ♇ ♄
04 15	☉ ⊻ ♆
06 17	☽ ∠ ♆
09 50	☽ ∠ ♂
10 08	☽ ∠ ♅
14 52	☿ ⊥ ⯓
15 42	☿ ⊼ ♄
17 10	☽ ⚹ ♀
19 48	☽ △ ⯓
28 Friday	
00 53	☽ ⊼ ♃
02 19	☽ ⊥ ♂
02 28	☽ ⊥ ♅
06 51	☽ △ ♄
09 27	☿ ⊻ ♆
09 39	☽ ⚹ ♆
09 41	☽ □ ☿
11 55	☽ ± ♃
12 00	☽ □ ☉
13 29	☽ ⊻ ♂
13 32	☽ ⊻ ♅
17 33	☽ Q ♀
17 50	♂ ☌ ♅
20 22	♀ # ♄
22 01	♀ ⚹ ⯓
23 37	☽ ♇ ⯓
29 Saturday	
04 44	☽ ♇ ♃
07 43	☉ △ ♅
08 12	☽ Q ♆
09 51	☉ △ ♂
10 15	☿ △ ♅
11 24	☿ △ ♂
12 42	☉ ☌ ☿
16 43	☽ ± ⯓
30 Sunday	
04 17	☽ ⊼ ⯓
07 33	☽ □ ♀
09 25	☽ △ ♃
15 35	☽ □ ♄
19 00	☽ □ ♆
22 52	☽ ☌ ♅
23 29	☽ ☌ ♂
31 Monday	
01 43	♀ ⊼ ♃
02 22	☽ △ ☉
06 15	☽ △ ☿
09 05	☽ # ⯓
12 06	☉ ∥ ☿
12 27	☽ ∥ ♂
20 59	☽ # ♃

APRIL 2042

LONGITUDES

Date	Sidereal time h m s	Sun ☉ ° ' "	Moon ☽ ° ' "	Moon ☽ 24.00 ° ' "	Mercury ☿ ° '	Venus ♀ ° '	Mars ♂ ° '	Jupiter ♃ ° '	Saturn ♄ ° '	Uranus ♅ ° '	Neptune ♆ ° '	Pluto ♇ ° '
01	00 39 56	11 ♈ 53 30	27 ♌ 25 44	03 ♍ 24 01	15 ♈ 01	03 ♉ 34	09 ♌ 16	01 ♐ 45	04 ♏ 53	08 ♌ 42	06 ♉ 49	29 ♒ 14
02	00 43 53	12 52 41	09 ♍ 20 49	15 ♍ 16 28	17 05	04 48	09 25	01 R 42	04 R 49	08 R 42	06 51	29 16
03	00 47 50	13 51 50	21 ♍ 11 19	27 ♍ 05 41	19 09	06 01	09 35	01 39	04 45	08 41	06 53	29 17
04	00 51 46	14 50 57	02 ♎ 59 50	08 ♎ 54 04	21 12	07 15	09 46	01 36	04 41	08 40	06 55	29 18
05	00 55 43	15 50 02	14 ♎ 48 39	20 ♎ 43 50	23 16	08 29	09 57	01 33	04 36	08 40	06 57	29 19
06	00 59 39	16 49 05	26 ♎ 39 53	02 ♏ 37 05	25 19	09 42	10 09	01 29	04 32	08 39	06 59	29 21
07	01 03 36	17 48 06	08 ♏ 35 43	14 ♏ 36 02	27 21	10 56	10 21	01 26	04 28	08 39	07 01	29 22
08	01 07 32	18 47 05	20 ♏ 38 22	26 ♏ 43 01	29 ♈ 22	12 09	10 33	01 22	04 24	08 38	07 04	29 23
09	01 11 29	19 46 02	02 ♐ 50 20	09 ♐ 00 39	01 ♉ 21	13 23	10 47	01 18	04 19	08 38	07 06	29 24
10	01 15 25	20 44 57	15 ♐ 14 21	21 ♐ 31 48	03 19	14 36	11 00	01 14	04 15	08 38	07 08	29 26
11	01 19 22	21 43 51	27 ♐ 52 32	04 ♑ 19 25	05 14	15 50	11 14	01 09	04 11	08 37	07 10	29 27
12	01 23 19	22 42 43	10 ♑ 50 21	17 ♑ 26 28	07 07	17 03	11 29	01 05	04 06	08 37	07 12	29 28
13	01 27 15	23 41 33	24 ♑ 08 03	00 ♒ 55 22	08 56	18 16	11 44	01 00	04 02	08 37	07 15	29 29
14	01 31 12	24 40 21	07 ♒ 48 32	14 ♒ 47 37	10 43	19 29	12 00	00 56	03 57	08 37	07 17	29 30
15	01 35 08	25 39 08	21 ♒ 52 34	29 ♒ 03 11	12 26	20 43	12 16	00 51	03 53	08 37	07 19	29 31
16	01 39 05	26 37 53	06 ♓ 19 09	13 ♓ 39 58	14 04	21 56	12 32	00 46	03 48	08 D 37	07 21	29 32
17	01 43 01	27 36 36	21 ♓ 04 59	28 ♓ 33 25	15 39	23 09	12 49	00 40	03 44	08 37	07 23	29 33
18	01 46 58	28 35 18	06 ♈ 04 22	13 ♈ 36 48	17 10	24 22	13 07	00 35	03 39	08 37	07 26	29 34
19	01 50 54	29 ♈ 33 57	21 ♈ 09 37	28 ♈ 41 42	18 36	25 35	13 24	00 30	03 35	08 37	07 28	29 35
20	01 54 51	00 ♉ 32 35	06 ♉ 11 56	13 ♉ 39 14	19 57	26 48	13 42	00 24	03 30	08 38	07 30	29 36
21	01 58 48	01 31 11	21 ♉ 02 38	28 ♉ 21 16	21 13	28 01	14 01	00 18	03 26	08 38	07 32	29 37
22	02 02 44	02 29 45	05 ♊ 34 26	12 ♊ 41 34	22 25	29 ♉ 14	14 20	00 12	03 21	08 38	07 35	29 38
23	02 06 41	03 28 17	19 ♊ 42 16	26 ♊ 36 20	23 31	00 ♊ 27	14 39	00 06	03 16	08 39	07 37	29 39
24	02 10 37	04 26 47	03 ♋ 23 41	10 ♋ 04 23	24 32	01 40	14 59	00 ♐ 00	03 12	08 39	07 39	29 40
25	02 14 34	05 25 15	16 ♋ 38 38	23 ♋ 06 45	25 28	02 53	15 19	29 ♏ 54	03 07	08 40	07 41	29 41
26	02 18 30	06 23 40	29 ♋ 29 06	05 ♌ 46 10	26 18	04 05	15 39	29 48	03 03	08 40	07 44	29 42
27	02 22 27	07 22 03	11 ♌ 58 27	18 ♌ 06 28	27 03	05 18	16 00	29 41	02 58	08 41	07 46	29 43
28	02 26 23	08 20 25	24 ♌ 10 48	00 ♍ 12 01	27 43	06 31	16 21	29 35	02 54	08 41	07 48	29 44
29	02 30 20	09 18 44	06 ♍ 10 41	12 ♍ 07 21	28 17	07 43	16 43	29 28	02 49	08 42	07 51	29 44
30	02 34 17	10 ♉ 17 01	18 ♍ 02 34	23 ♍ 56 50	28 ♉ 45	08 ♊ 56	17 ♌ 04	29 ♏ 21	02 ♏ 45	08 ♌ 43	07 ♉ 53	29 ♒ 45

Date	Moon True ☊ ° '	Moon Mean ☊ ° '	Moon ☽ Latitude ° '
01	26 ♈ 58	27 ♈ 55	04 N 33
02	26 R 55	27 52	03 54
03	26 53	27 49	03 05
04	26 52	27 45	02 08
05	26 51	27 42	01 06
06	26 51	27 39	00 N 01
07	26 D 51	27 36	01 S 05
08	26 51	27 33	02 08
09	26 52	27 30	03 07
10	26 53	27 26	03 58
11	26 53	27 23	04 38
12	26 54	27 20	05 05
13	26 R 54	27 17	05 17
14	26 54	27 14	05 12
15	26 54	27 10	04 48
16	26 53	27 07	04 06
17	26 D 54	27 04	03 06
18	26 54	27 01	01 53
19	26 54	26 58	00 S 32
20	26 R 54	26 55	00 N 51
21	26 53	26 51	02 10
22	26 53	26 48	03 19
23	26 52	26 45	04 13
24	26 51	26 42	04 51
25	26 51	26 39	05 12
26	26 50	26 35	05 17
27	26 D 50	26 32	05 06
28	26 51	26 29	04 42
29	26 52	26 26	04 06
30	26 ♈ 53	26 ♈ 23	03 N 19

DECLINATIONS

Date	Sun ☉ ° '	Moon ☽ ° '	Mercury ☿ ° '	Venus ♀ ° '	Mars ♂ ° '	Jupiter ♃ ° '	Saturn ♄ ° '	Uranus ♅ ° '	Neptune ♆ ° '	Pluto ♇ ° '
01	04 N 42	16 N 37	05 N 15	12 N 22	20 N 50	19 S 26	10 S 35	18 N 44	12 N 10	21 S 28
02	05 05	11 40	06 11	12 50	20 46	19 25	10 34	18 44	12 11	21 28
03	05 28	06 19	07 08	13 17	20 41	19 25	10 32	18 44	12 12	21 27
04	05 51	00 N 46	08 04	13 43	20 36	19 24	10 31	18 44	12 12	21 27
05	06 14	04 S 49	09 00	14 10	20 31	19 23	10 30	18 44	12 13	21 27
06	06 36	10 16	09 56	14 36	20 26	19 22	10 28	18 44	12 14	21 27
07	06 59	15 23	10 50	15 01	20 21	19 22	10 27	18 44	12 15	21 26
08	07 21	19 58	11 44	15 27	20 15	19 21	10 25	18 44	12 15	21 26
09	07 44	23 47	12 37	15 51	20 10	19 20	10 24	18 44	12 16	21 26
10	08 06	26 33	13 28	16 16	20 04	19 19	10 22	18 45	12 17	21 26
11	08 28	28 03	14 17	16 40	19 58	19 18	10 20	18 45	12 17	21 26
12	08 50	28 04	15 05	17 04	19 52	19 17	10 19	18 45	12 18	21 25
13	09 12	26 29	15 51	17 27	19 46	19 16	10 17	18 45	12 19	21 25
14	09 33	23 20	16 34	17 50	19 40	19 15	10 16	18 45	12 20	21 25
15	09 55	18 45	17 16	18 12	19 34	19 14	10 14	18 45	12 20	21 25
16	10 16	13 00	17 55	18 34	19 27	19 13	10 13	18 45	12 21	21 25
17	10 37	06 S 23	18 31	18 55	19 20	19 12	10 11	18 44	12 22	21 25
18	10 58	00 N 41	19 05	19 16	19 14	19 11	10 10	18 44	12 22	21 25
19	11 19	07 46	19 37	19 36	19 07	19 09	10 08	18 44	12 23	21 24
20	11 40	14 24	20 05	19 56	19 00	19 08	10 07	18 44	12 24	21 24
21	12 00	20 07	20 31	20 16	18 53	19 07	10 05	18 44	12 25	21 24
22	12 20	24 29	20 55	20 35	18 46	19 06	10 04	18 44	12 25	21 24
23	12 40	27 15	21 16	20 53	18 38	19 04	10 02	18 44	12 26	21 24
24	13 00	28 15	21 34	21 11	18 31	19 03	10 00	18 44	12 27	21 24
25	13 20	27 34	21 49	21 28	18 23	19 02	09 59	18 44	12 28	21 24
26	13 39	25 25	22 02	21 44	18 15	19 00	09 57	18 43	12 28	21 24
27	13 58	22 06	22 12	22 00	18 08	18 59	09 56	18 43	12 29	21 24
28	14 17	17 53	22 20	22 16	18 00	18 58	09 54	18 43	12 30	21 24
29	14 36	13 03	22 25	22 30	17 52	18 56	09 53	18 43	12 31	21 24
30	14 N 54	07 N 47	22 N 28	22 N 44	17 N 43	18 S 55	09 S 51	18 N 43	12 N 31	21 S 24

ZODIAC SIGN ENTRIES

Date	h	m	Planets	
01	17	10	☽	♍
04	05	54	☽	♎
06	18	44	☽	♏
08	19	35	☿	♉
09	06	27	☽	♐
11	15	57	☽	♑
13	22	23	☽	♒
16	01	34	☽	♓
18	02	18	☽	♈
19	22	40	☉	♉
20	02	05	☽	♉
22	02	43	☽	♊
23	03	05	♀	♊
24	05	59	☽	♋
24	12	41	♃	♏
26	12	59	☽	♌
28	23	36	☽	♍

LATITUDES

Date	Mercury ☿ ° '	Venus ♀ ° '	Mars ♂ ° '	Jupiter ♃ ° '	Saturn ♄ ° '	Uranus ♅ ° '	Neptune ♆ ° '	Pluto ♇ ° '
01	00 S 43	00 S 21	03 N 01	01 N 06	02 N 43	00 N 40	01 S 43	10 S 25
04	00 S 13	00 13	02 55	01 06	02 43	00 40	01 42	10 25
07	00 N 20	00 S 05	02 48	01 06	02 43	00 40	01 42	10 26
10	00 54	00 N 03	02 42	01 07	02 44	00 40	01 42	10 27
13	01 27	00 11	02 36	01 07	02 44	00 40	01 42	10 28
16	01 56	00 20	02 31	01 07	02 44	00 40	01 42	10 29
19	02 21	00 28	02 25	01 07	02 44	00 40	01 42	10 29
22	02 38	00 36	02 20	01 07	02 44	00 39	01 42	10 30
25	02 47	00 45	02 15	01 07	02 44	00 39	01 42	10 31
28	02 46	00 53	02 10	01 07	02 44	00 39	01 42	10 32
31	02 N 34	01 N 01	02 N 05	01 N 07	02 N 44	00 N 39	01 S 42	10 S 33

DATA

Julian Date	2466976
Delta T	+78 seconds
Ayanamsa	24° 26' 42"
Synetic vernal point	04° ♓ 40' 16"
True obliquity of ecliptic	23° 26' 10"

MOON'S PHASES, APSIDES AND POSITIONS ☽

Date	h	m	Phase	Longitude ° '	Eclipse Indicator
05	14	16	○	15 ♎ 56	
13	11	09	☾	23 ♑ 39	
20	02	19	●	00 ♉ 09	Total
27	02	19	☽	06 ♌ 59	

Day	h	m		
04	05	37	Apogee	
19	04	16	Perigee	
04	15	18	0S	
12	00	15	Max dec	28° S 15'
18	09	43	0N	
24	13	56	Max dec	28° N 15'

LONGITUDES

Date	Chiron ⚷ ° '	Ceres ⚳ ° '	Pallas ⚴ ° '	Juno ⚵ ° '	Vesta ⚶ ° '	Black Moon Lilith ⚸ ° '
01	28 ♋ 13	12 ♐ 30	14 ♎ 30	00 ♏ 46	19 ♊ 36	02 ♎ 30
11	28 ♋ 20	12 ♐ 33	11 ♎ 24	28 ♎ 33	22 ♊ 58	03 ♎ 37
21	28 ♋ 38	11 ♐ 56	08 ♎ 37	26 ♎ 14	26 ♊ 34	04 ♎ 44
31	29 ♋ 03	10 ♐ 42	06 ♎ 26	24 ♎ 02	00 ♋ 20	05 ♎ 50

All ephemeris data is given at 12.00 UT and the Moon's longitude is additionally given for 24.00 UT

ASPECTARIAN

h m	Aspects
01 Tuesday	
00 56	☽ ∥ ♅
01 07	♀ ∥ ♆
02 45	☿ ∠ ♇
02 57	☽ Q ♄
10 50	☽ ⚼ ☉
15 38	☽ ☍ ♇
18 17	☽ ⚼ ☿
20 38	☽ □ ♃
02 Wednesday	
01 45	☽ △ ♀
02 54	☽ ✶ ♄
06 34	☽ ± ☉
06 56	☽ △ ♆
07 01	☽ ∥ ♀
07 41	☿ ⚼ ♃
09 37	☽ ∥ ♆
10 41	☽ ⊻ ♅
12 09	☽ ⊻ ♂
12 22	♀ ☍ ♄
16 14	☽ ± ☿
17 06	☽ # ♄
19 48	☽ ⊼ ☉
22 49	☽ ⊥ ♅
03 Thursday	
00 28	☽ ⊥ ♂
06 58	☽ ⊼ ☿
08 54	☽ Q ♃
08 59	☽ ∥ ☿
09 05	☽ ∠ ♄
11 37	☽ ⚼ ♀
13 25	☽ ⚼ ♆
15 30	☽ ∥ ☉
17 04	☽ ∠ ♅
19 00	☽ ∠ ♂
22 23	☉ ∠ ♇
04 Friday	
03 16	☽ ⊥ ♄
04 28	☽ ⊼ ♇
05 19	♀ ☌ ♆
07 46	☽ ± ♆
08 02	☽ ± ♀
09 10	☽ ✶ ♃
15 24	☽ ⊻ ♄
16 42	☽ ± ♇
20 00	☽ ⊼ ♆
21 39	☽ ⊼ ♀
23 32	☽ ✶ ♅
05 Saturday	
01 58	☽ ✶ ♂
11 01	☽ ⚼ ♇
14 16	☽ ☍ ☉
15 30	☽ ∠ ♃
15 38	♀ □ ♅
18 35	☽ # ☉
23 51	☽ Q ♅
06 Sunday	
02 42	☽ Q ♂
04 23	☉ ⚼ ♃
08 42	☽ ☍ ☿
09 38	☽ ⊥ ♃
10 09	☽ # ☿
12 54	☽ ∥ ♄
13 58	☿ ± ♃
17 25	☽ △ ♇
21 00	☽ # ♆
21 41	☽ ⊻ ♃
22 16	♀ □ ♂
07 Monday	
01 49	☿ # ♄
03 46	☽ ☌ ♄
08 51	☽ ☍ ♆
10 02	☽ # ♀
12 06	☽ □ ♅
15 34	☽ □ ♂
17 12	☽ ☍ ♀
20 42	♀ Q ♇
08 Tuesday	
05 12	☽ # ♅
08 00	☽ ⊼ ☉
08 31	☽ ∥ ♃
12 14	☿ ✶ ♇
13 35	☽ # ♂
20 35	☽ ∥ ♇
20 55	☽ ± ☉
09 Wednesday	
02 25	☿ ∥ ♆
05 16	☽ □ ♇
08 33	☽ ⊼ ☿
09 00	☽ ☌ ♃
11 18	☿ ⊼ ♃
14 53	☽ ⊻ ♄
16 05	☽ ⚼ ☉
20 19	☽ ⊼ ♆
22 28	☽ ± ☿
23 16	☽ △ ♅
10 Thursday	
02 27	☽ ⊥ ♄
03 42	☽ △ ♂
07 57	☽ ± ♆
10 39	☽ ⊼ ♀
16 12	☽ Q ♇
18 57	☽ ⚼ ☿
19 37	☽ ∠ ♄
23 11	☿ ☍ ♄
23 20	☽ ± ♀
23 24	☽ △ ☉
11 Friday	
01 11	☽ ⚼ ♆
03 58	☽ ⚼ ♅
08 51	☽ ⚼ ♂
14 55	☽ ✶ ♇
18 04	☽ ⚼ ♀
18 05	☽ ⊻ ♃
20 51	☽ ± ♅
23 40	☽ ✶ ♄
12 Saturday	
01 58	☽ ± ♂
04 01	☽ △ ☿
05 08	☽ ⊥ ♃
05 19	☽ △ ♆
07 56	☽ ⊼ ♅
13 12	☽ ⊼ ♂
13 14	☿ ☌ ♆
18 37	☽ ∠ ♇
21 29	☽ ∠ ♃
21 32	☽ Q ♄
13 Sunday	
00 26	☽ △ ♀
07 44	☿ □ ♅
10 50	☽ ⊥ ♇
11 09	☽ □ ☉
21 29	☽ ⊻ ♇
14 Monday	
00 05	☽ ✶ ♃
05 20	☽ □ ♄
11 05	☽ □ ♆
13 24	☽ ☍ ♅
17 43	☽ □ ☿
17 46	☉ ± ♃
19 21	☽ ☍ ♂
20 46	☽ Q ♃
21 00	☽ Q ☉
22 54	☽ ∥ ♇
23 05	☿ Q ♇
15 Tuesday	
08 08	☽ # ♂
09 16	☿ □ ♂
09 46	☽ ∥ ♃
09 51	☽ □ ♀
12 03	☽ # ♅
14 21	☽ # ♀
17 47	☽ Q ♆
17 56	☽ # ☿
18 48	☽ ✶ ☉
18 56	♅ St D
16 Wednesday	
00 48	☽ ☌ ♇
02 53	☽ □ ♃
04 07	☽ Q ☿
07 53	☽ △ ♄
08 27	☉ # ♄
13 42	☽ ✶ ♆
14 27	☽ # ♆
15 46	☽ ⊼ ♅
18 27	☽ Q ♀
21 18	☽ ∠ ☉
21 40	☽ # ☉
22 22	☽ ⊼ ♂
22 26	☽ ∥ ♄
23 57	♀ ∥ ♅
17 Thursday	
01 33	☽ ± ♅
02 12	☽ ✶ ☿
08 13	☽ ⚼ ♄
08 17	☽ ± ♂
12 55	☽ ⊥ ☉
14 07	☽ ∠ ♆
15 37	☽ ✶ ♀
16 05	☽ ⚼ ♅
21 08	☿ ∥ ♅
22 37	☽ ± ♄
23 02	☽ ⚼ ♂
23 13	☽ ⊻ ☉
18 Friday	
01 37	☽ ⊻ ♇
03 18	☽ △ ♃
04 34	☽ ⊥ ♆
05 05	☽ ∠ ☿
05 56	♀ # ♃
08 10	☽ ⊼ ♄
09 56	♀ ∥ ♂
11 12	☽ ⊥ ♇
14 10	☽ ⊻ ♆
15 50	☿ # ♃
16 03	☽ △ ♅
17 13	☿ ∥ ♂
17 43	☽ ∠ ♀
20 58	☽ ⊥ ☿
23 25	☽ △ ♂
19 Saturday	
01 25	♂ # ♃
01 33	☽ ∠ ♇
03 03	☽ ⚼ ♃
07 30	☽ ⊻ ☿
09 17	☽ ⊥ ♀
11 50	☿ ∥ ♀
12 37	☉ ✶ ♇
17 16	☽ ± ♃
19 40	☽ ⊻ ♀
20 17	☽ # ♄
20 Sunday	
01 14	☽ ∥ ☉
01 27	☽ ✶ ♇
02 19	☽ ☌ ☉
02 47	☽ ⊼ ♃
04 29	☽ ∥ ♆
07 42	☽ ☍ ♄
08 29	♀ Q ♅
08 47	☉ ⊼ ♃
14 06	☽ ☌ ♆
15 54	☽ □ ♅
20 42	☽ Q ♇
21 Monday	
00 21	☽ □ ♂
05 46	☽ ∥ ♅
06 31	☽ ∥ ♂
07 28	☽ # ♃
12 19	☽ ☌ ☿
12 46	☽ ∥ ♀
14 08	☽ ∥ ☿
18 18	☽ # ♇
21 10	☽ Q ♅
22 Tuesday	
00 30	☽ ☌ ♀
02 07	☽ □ ♇
03 07	☽ ☍ ♃
06 28	☽ Q ♂
06 29	☽ ⊻ ☉
08 18	☽ ⊼ ♄
15 22	☽ ⊻ ♆
16 56	♂ ∥ ♅
17 09	☽ ✶ ♅
17 16	☽ ⊥ ☉
18 18	☽ ± ♄
18 31	☉ ∥ ♆
20 03	♀ □ ♇
23 Wednesday	
01 32	☽ ⊥ ♆
03 08	☽ ✶ ♂
05 41	♀ ☍ ♃
07 28	☉ ☍ ♄
09 33	☽ ⚼ ♄
09 43	☽ ∠ ☉
17 03	☽ ∠ ♆
18 50	☽ ∠ ♅
19 09	☽ ⊻ ☿
22 51	☿ # ♇
24 Thursday	
05 23	☽ △ ♇
05 48	☽ ∠ ♂
06 02	☽ ⊼ ♃
06 31	☽ ⊥ ☿
08 37	☽ ⊻ ♀
10 41	☽ ⊥ ♅
11 39	☽ △ ♄
14 01	☽ ✶ ☉
16 38	☽ ± ♃
19 39	☽ ✶ ♆
20 25	☽ ⊥ ♀
21 26	☽ ⊻ ♅
22 17	☽ ⊥ ♂
23 52	☽ ∠ ☿
25 Friday	
06 49	♀ # ♇
08 24	☽ ⚼ ♇
08 49	☽ ⚼ ♃
09 30	☽ ⊻ ♂
13 33	☽ Q ☉
14 31	☽ ∠ ♀
16 31	♀ ⊼ ♄
17 39	☽ Q ♆
26 Saturday	
00 03	♀ Q ♂
01 05	☽ ± ♇
05 35	☽ ✶ ☿
12 24	☽ ⊼ ♇
12 35	☽ △ ♃
18 44	☽ □ ♄
21 43	☽ ✶ ♀
23 27	☿ Q ♅
27 Sunday	
02 19	☽ □ ☉
03 49	☽ □ ♆
05 36	☽ ☌ ♅
05 59	☽ Q ☿
06 40	♃ □ ♇
11 19	☽ ∥ ☿
12 33	☽ ∥ ♀
16 18	☽ # ♇
20 06	☽ ☌ ♂
22 15	☉ ☌ ♆
23 34	☽ Q ♀
28 Monday	
05 32	☽ Q ♄
06 14	☽ # ♃
07 35	☽ ∥ ♅
11 26	☽ ∥ ♂
19 24	☽ □ ☿
20 44	☉ □ ♅
22 39	☽ □ ♃
23 04	☽ ☍ ♇
29 Tuesday	
00 23	☿ ∥ ♀
05 02	☽ ∥ ☉
05 17	☽ ✶ ♄
14 27	♀ ⊻ ♆
14 31	☽ ∥ ♆
15 22	☽ △ ♆
15 28	☽ □ ♀
17 06	☽ ⊻ ♅
18 53	☽ △ ☉
30 Wednesday	
02 40	☽ # ♄
05 14	☽ ⊥ ♅
07 39	♀ ✶ ♅
08 27	♀ ± ♄
09 58	☽ ⊻ ♂
10 37	☽ Q ♃
11 24	☽ ∠ ♄
21 52	☽ ⚼ ♆
22 33	☽ ⊥ ♂
23 32	☽ ∠ ♅

MAY 2042

LONGITUDES

Date	Sidereal time h m s	Sun ☉ ° ' "	Moon ☽ ° ' "	Moon ☽ 24.00 ° ' "	Mercury ☿ ° '	Venus ♀ ° '	Mars ♂ ° '	Jupiter ♃ ° '	Saturn ♄ ° '	Uranus ♅ ° '	Neptune ♆ ° '	Pluto ♇ ° '
01	02 38 13	11 ♉ 15 16	29 ♍ 50 39	05 ♎ 44 29	29 ♉ 08	10 ♊ 08	17 ♌ 27	29 ♏ 14	02 ♏ 40	08 ♌ 44	07 ♉ 55	29 ♒ 46
02	02 42 10	12 13 29	11 ♎ 38 45	17 ♎ 33 50	29 26	11 21	17 49	29 R 07	02 R 36	08 45	07 57	29 47
03	02 46 06	13 11 40	23 ♎ 30 07	29 ♎ 27 54	29 37	12 33	18 12	29 00	02 31	08 45	08 00	29 47
04	02 50 03	14 09 50	05 ♏ 27 31	11 ♏ 29 11	29 44	13 45	18 35	28 53	02 27	08 46	08 02	29 48
05	02 53 59	15 07 58	17 ♏ 33 09	23 ♏ 39 39	29 R 45	14 58	18 58	28 46	02 22	08 47	08 04	29 49
06	02 57 56	16 06 04	29 ♏ 48 49	06 ♐ 00 52	29 41	16 10	19 21	28 39	02 18	08 49	08 06	29 49
07	03 01 52	17 04 09	12 ♐ 15 55	18 ♐ 34 07	29 32	17 22	19 45	28 31	02 13	08 50	08 09	29 50
08	03 05 49	18 02 12	24 ♐ 55 36	01 ♑ 20 29	29 18	18 34	20 09	28 24	02 09	08 51	08 11	29 50
09	03 09 46	19 00 13	07 ♑ 48 53	14 ♑ 20 55	29 00	19 46	20 34	28 17	02 05	08 52	08 13	29 51
10	03 13 42	19 58 14	20 ♑ 56 43	27 ♑ 36 22	28 39	20 58	20 58	28 09	02 01	08 53	08 15	29 51
11	03 17 39	20 56 12	04 ♒ 19 58	11 ♒ 07 36	28 13	22 10	21 23	28 02	01 56	08 55	08 17	29 52
12	03 21 35	21 54 10	17 ♒ 59 19	24 ♒ 55 09	27 44	23 22	21 48	27 54	01 52	08 56	08 20	29 52
13	03 25 32	22 52 06	01 ♓ 55 03	08 ♓ 58 58	27 13	24 34	22 13	27 47	01 48	08 58	08 22	29 53
14	03 29 28	23 50 01	16 ♓ 06 42	23 ♓ 18 03	26 40	25 46	22 39	27 39	01 44	08 59	08 24	29 53
15	03 33 25	24 47 55	00 ♈ 32 41	07 ♈ 50 08	26 06	26 57	23 05	27 31	01 40	09 01	08 26	29 54
16	03 37 21	25 45 48	15 ♈ 09 55	22 ♈ 31 22	25 30	28 09	23 31	27 24	01 36	09 02	08 28	29 54
17	03 41 18	26 43 39	29 ♈ 53 46	07 ♉ 16 19	24 55	29 ♊ 21	23 57	27 16	01 32	09 04	08 31	29 54
18	03 45 15	27 41 29	14 ♉ 38 10	21 ♉ 58 26	24 20	00 ♋ 32	24 24	27 08	01 28	09 05	08 33	29 55
19	03 49 11	28 39 18	29 ♉ 16 13	06 ♊ 30 42	23 46	01 44	24 51	27 01	01 24	09 07	08 35	29 55
20	03 53 08	29 ♉ 37 05	13 ♊ 41 06	20 ♊ 46 43	23 14	02 55	25 18	26 53	01 21	09 09	08 37	29 55
21	03 57 04	00 ♊ 34 51	27 ♊ 46 59	04 ♋ 41 30	22 44	04 06	25 45	26 46	01 17	09 11	08 39	29 56
22	04 01 01	01 32 36	11 ♋ 29 56	18 ♋ 12 10	22 17	05 18	26 12	26 38	01 13	09 13	08 41	29 56
23	04 04 57	02 30 19	24 ♋ 48 11	01 ♌ 18 07	21 52	06 29	26 40	26 30	01 10	09 15	08 43	29 56
24	04 08 54	03 28 00	07 ♌ 42 12	14 ♌ 00 47	21 31	07 40	27 08	26 23	01 06	09 16	08 46	29 56
25	04 12 50	04 25 40	20 ♌ 14 19	26 ♌ 23 16	21 14	08 51	27 36	26 15	01 03	09 18	08 48	29 56
26	04 16 47	05 23 19	02 ♍ 28 13	08 ♍ 29 46	21 01	10 02	28 04	26 08	00 59	09 21	08 50	29 56
27	04 20 44	06 20 55	14 ♍ 28 32	20 ♍ 25 09	20 52	11 13	28 32	26 00	00 56	09 23	08 52	29 56
28	04 24 40	07 18 31	26 ♍ 20 16	02 ♎ 14 30	20 47	12 24	29 01	25 53	00 53	09 25	08 54	29 57
29	04 28 37	08 16 05	08 ♎ 08 28	14 ♎ 02 47	20 D 47	13 35	29 30	25 46	00 50	09 27	08 56	29 57
30	04 32 33	09 13 37	19 ♎ 58 00	25 ♎ 54 39	20 51	14 45	29 ♌ 59	25 38	00 47	09 29	08 58	29 57
31	04 36 30	10 ♊ 11 09	01 ♏ 53 13	07 ♏ 54 08	21 ♉ 00	15 ♋ 56	00 ♍ 28	25 ♏ 31	00 ♏ 44	09 ♌ 31	09 ♉ 00	29 ♒ 57

Date	Moon True ☊ ° '	Moon Mean ☊ ° '	Moon ☽ Latitude ° '
01	26 ♈ 54	26 ♈ 20	02 N 24
02	26 D 55	26 16	01 24
03	26 56	26 13	00 N 19
04	26 R 55	26 10	00 S 47
05	26 54	26 07	01 52
06	26 52	26 04	02 52
07	26 50	26 01	03 45
08	26 47	25 57	04 28
09	26 43	25 54	04 58
10	26 41	25 51	05 13
11	26 39	25 48	05 12
12	26 38	25 45	04 53
13	26 D 38	25 41	04 17
14	26 39	25 38	03 25
15	26 40	25 35	02 19
16	26 41	25 32	01 S 03
17	26 R 42	25 29	00 N 18
18	26 41	25 26	01 37
19	26 38	25 22	02 49
20	26 34	25 19	03 49
21	26 29	25 16	04 34
22	26 24	25 13	05 01
23	26 19	25 10	05 11
24	26 15	25 07	05 05
25	26 12	25 03	04 45
26	26 11	25 00	04 11
27	26 D 11	24 57	03 28
28	26 12	24 54	02 35
29	26 14	24 51	01 37
30	26 15	24 47	00 34
31	26 ♈ 15	24 ♈ 44	00 S 31

DECLINATIONS

Date	Sun ☉ ° '	Moon ☽ ° '	Mercury ☿ ° '	Venus ♀ ° '	Mars ♂ ° '	Jupiter ♃ ° '	Saturn ♄ ° '	Uranus ♅ ° '	Neptune ♆ ° '	Pluto ♇ ° '
01	15 N 12	02 N 16	22 N 28	22 N 58	17 N 35	18 S 53	09 S 50	18 N 42	12 N 32	21 S 24
02	15 30	03 S 19	22 26	23 11	17 27	18 52	09 48	18 42	12 33	21 24
03	15 48	08 50	22 22	23 23	17 18	18 50	09 47	18 42	12 34	21 24
04	16 05	14 05	22 15	23 35	17 09	18 49	09 46	18 42	12 34	21 24
05	16 22	18 51	22 05	23 46	17 01	18 47	09 44	18 41	12 35	21 24
06	16 39	22 54	21 54	23 56	16 52	18 46	09 43	18 41	12 36	21 24
07	16 56	25 58	21 40	24 06	16 43	18 44	09 41	18 41	12 36	21 24
08	17 12	27 48	21 24	24 15	16 34	18 43	09 40	18 40	12 37	21 24
09	17 28	28 09	21 07	24 23	16 25	18 41	09 39	18 40	12 38	21 24
10	17 44	26 57	20 47	24 30	16 15	18 39	09 37	18 40	12 38	21 25
11	17 59	24 13	20 26	24 37	16 06	18 38	09 36	18 39	12 39	21 25
12	18 14	20 05	20 04	24 43	15 56	18 36	09 35	18 39	12 40	21 25
13	18 29	14 47	19 40	24 49	15 47	18 35	09 33	18 38	12 41	21 25
14	18 44	08 37	19 16	24 54	15 37	18 33	09 32	18 38	12 41	21 25
15	18 58	01 S 54	18 51	24 58	15 27	18 31	09 31	18 38	12 42	21 25
16	19 12	05 N 00	18 25	25 01	15 17	18 30	09 30	18 37	12 43	21 25
17	19 25	11 42	18 00	25 04	15 07	18 28	09 28	18 37	12 43	21 26
18	19 39	17 46	17 35	25 06	14 57	18 26	09 27	18 36	12 44	21 26
19	19 51	22 44	17 10	25 07	14 47	18 25	09 26	18 36	12 45	21 26
20	20 04	26 14	16 47	25 07	14 36	18 23	09 25	18 35	12 45	21 26
21	20 16	27 59	16 24	25 07	14 26	18 21	09 24	18 35	12 46	21 26
22	20 28	27 56	16 03	25 06	14 15	18 20	09 23	18 34	12 47	21 27
23	20 40	26 16	15 44	25 05	14 05	18 18	09 21	18 34	12 47	21 27
24	20 51	23 15	15 26	25 02	13 54	18 16	09 20	18 33	12 48	21 27
25	21 01	19 14	15 10	24 59	13 43	18 15	09 19	18 32	12 49	21 27
26	21 12	14 30	14 57	24 56	13 32	18 13	09 18	18 32	12 49	21 28
27	21 22	09 18	14 45	24 51	13 21	18 12	09 17	18 31	12 50	21 28
28	21 32	03 N 50	14 36	24 46	13 10	18 10	09 16	18 31	12 50	21 28
29	21 41	01 S 45	14 29	24 40	12 59	18 08	09 16	18 30	12 51	21 28
30	21 50	07 17	14 25	24 34	12 47	18 07	09 15	18 30	12 52	21 29
31	21 N 58	12 S 36	14 N 22	24 N 27	12 N 36	18 S 05	09 S 14	18 N 29	12 N 52	21 S 29

ZODIAC SIGN ENTRIES

Date	h	m	Planets
01	12	19	☽ ♎
04	01	04	☽ ♏
06	12	22	☽ ♐
08	21	30	☽ ♑
11	04	17	☽ ♒
13	08	43	☽ ♓
15	11	06	☽ ♈
17	12	10	☽ ♉
18	01	13	♀ ♋
19	13	12	☽ ♊
20	21	31	☉ ♊
21	15	50	☽ ♋
23	21	35	☽ ♌
26	07	07	☽ ♍
28	19	27	☽ ♎
30	13	08	♂ ♍
31	08	13	☽ ♏

LATITUDES

Date	Mercury ☿ ° '	Venus ♀ ° '	Mars ♂ ° '	Jupiter ♃ ° '	Saturn ♄ ° '	Uranus ♅ ° '	Neptune ♆ ° '	Pluto ♇ ° '
01	02 N 34	01 N 01	02 N 05	01 N 07	02 N 44	00 N 39	01 S 42	10 S 33
04	02 12	01 08	02 00	01 07	02 43	00 39	01 42	10 34
07	01 40	01 16	01 55	01 07	02 43	00 39	01 42	10 35
10	00 57	01 23	01 51	01 07	02 43	00 39	01 42	10 36
13	00 N 08	01 30	01 46	01 07	02 43	00 39	01 42	10 37
16	00 S 44	01 36	01 42	01 07	02 42	00 39	01 42	10 38
19	01 35	01 41	01 38	01 06	02 42	00 39	01 42	10 39
22	02 22	01 47	01 34	01 06	02 41	00 39	01 42	10 40
25	03 00	01 51	01 30	01 06	02 41	00 38	01 43	10 41
28	03 28	01 55	01 26	01 05	02 40	00 38	01 43	10 42
31	03 S 46	01 N 59	01 N 23	01 N 05	02 N 40	00 N 38	01 S 43	10 S 43

DATA

Julian Date	2467006
Delta T	+78 seconds
Ayanamsa	24° 26' 46"
Synetic vernal point	04° ♓ 40' 13"
True obliquity of ecliptic	23° 26' 10"

MOON'S PHASES, APSIDES AND POSITIONS ☽

Date	h	m	Phase	Longitude ° '	Eclipse Indicator
05	06	48	○	14 ♏ 55	
12	19	18	☾	22 ♒ 12	
19	10	55	●	28 ♉ 37	
26	18	18	☽	05 ♍ 38	

Day	h	m		
01	15	00	Apogee	
17	09	14	Perigee	
29	07	08	Apogee	
01	21	45	0S	
09	05	41	Max dec	28° S 13'
15	18	37	0N	
21	23	24	Max dec	28° N 11'
29	04	30	0S	

LONGITUDES

Date	Chiron ⚷ ° '	Ceres ⚳ ° '	Pallas ⚴ ° '	Juno ⚵ ° '	Vesta ⚶ ° '	Black Moon Lilith ⚸ ° '
01	29 ♋ 03	10 ♐ 42	06 ♎ 26	24 ♎ 02	00 ♋ 20	05 ♎ 50
11	29 ♋ 38	08 ♐ 57	05 ♎ 01	22 ♎ 09	04 ♋ 15	06 ♎ 57
21	00 ♌ 20	06 ♐ 50	04 ♎ 24	20 ♎ 42	08 ♋ 18	08 ♎ 03
31	01 ♌ 10	04 ♐ 36	04 ♎ 34	19 ♎ 48	12 ♋ 26	09 ♎ 10

All ephemeris data is given at 12.00 UT and the Moon's longitude is additionally given for 24.00 UT

ASPECTARIAN

h m	Aspects
01 Thursday	
04 02	☽ ⚼ ☉
05 34	☽ ⊥ ♄
10 31	☽ △ ☿
10 47	☽ ✶ ♃
11 50	☽ ⚻ ♇
16 14	☽ ± ♆
17 28	☽ ∠ ♂
17 30	☿ ☍ ♃
17 42	☽ ⚺ ♄
02 Friday	
00 00	☽ ± ☉
00 03	☽ ± ♇
00 45	☉ Q ♇
04 29	☽ ⚻ ♆
06 06	☽ ✶ ♅
11 19	☽ △ ♀
13 17	☽ ⚻ ☉
16 58	☽ ∠ ♃
17 45	☽ ⚼ ☿
18 21	☽ ⚼ ♇
03 Saturday	
00 55	☽ ✶ ♂
06 27	☽ Q ♅
11 00	☽ ⊥ ♃
12 15	☽ ± ☿
16 13	☽ ∥ ♄
21 04	☽ ⚼ ♀
22 58	☽ ⚺ ♃
04 Sunday	
00 27	☽ ⚻ ☿
00 39	☽ △ ♇
01 54	☽ Q ♂
04 54	☽ # ♆
06 01	☽ ☌ ♄
17 06	☽ ± ♀
17 09	☽ ☍ ♆
17 36	♀ ⊥ ♆
18 37	☽ □ ♅
22 22	☽ # ☉
05 Monday	
02 41	☽ # ♂
05 13	☿ St R
06 19	☽ ⚻ ♀
06 48	☽ ☍ ☉
11 08	☽ # ♅
11 40	☽ ∥ ♃
14 52	☽ □ ♂
06 Tuesday	
02 30	☽ ∥ ♇
05 24	☉ ⚺ ♀
05 52	☽ # ☿
09 45	☽ ☌ ♃
11 45	☽ ☍ ☿
12 01	☽ □ ♇
16 47	☽ ⚺ ♄
19 31	☽ # ♀
23 49	☉ ∥ ♂
07 Wednesday	
04 04	☽ ⚻ ♆
04 18	☽ ⊥ ♄
05 24	☽ △ ♅
09 18	♀ ⚼ ♄
15 36	☽ ± ♆
21 24	☽ ∠ ♄
21 55	☽ ⚻ ☉
22 36	☽ Q ♇
22 45	☽ ☍ ♀
08 Thursday	
02 42	☽ △ ♂
08 42	☽ ⚼ ♆
09 58	☽ ⚼ ♅
10 11	☽ ± ☉
11 58	♂ Q ♄
12 09	☿ # ♇
18 27	☽ ⚺ ♃
20 02	☽ ⚻ ☿
21 12	☽ ✶ ♇
09 Friday	
01 26	☽ ✶ ♄
02 50	☽ ± ♅
04 23	☽ ⚼ ☉
05 31	☽ ⊥ ♃
06 56	☽ ± ☿
07 42	☽ ⚼ ♂
12 45	☽ △ ♆
13 57	☽ ⚻ ♅
21 57	☽ ∠ ♃
23 05	☽ ⚼ ☿
23 27	☽ Q ♄
10 Saturday	
00 47	☽ ± ♂
00 55	☽ ∠ ♇
09 38	♃ # ♅
10 06	☽ △ ☉
11 58	♀ ✶ ♂
12 03	☽ ⚻ ♂
12 03	☽ ⚻ ♀
17 16	☽ ⊥ ♇
23 56	☽ ± ♀
11 Sunday	
00 52	☽ ✶ ♃
01 27	☽ △ ☿
04 02	☽ ⚺ ♇
07 46	☽ □ ♄
09 13	☽ # ♀
17 31	☽ ⚼ ♀
19 02	☽ □ ♆
20 07	☽ ☍ ♅
21 59	☽ Q ♃
12 Monday	
01 16	☿ ☍ ♃
05 02	☽ ∥ ♇
07 35	☉ □ ♂
11 11	♀ ∠ ♆
12 06	☽ # ☿
18 50	☽ ☍ ♂
19 00	☽ # ♅
19 14	☽ ∥ ♃
19 18	☽ □ ☉
20 29	☽ # ☉
22 12	☽ △ ♀
23 37	♀ ∠ ♅
13 Tuesday	
02 28	☽ Q ♆
04 15	☽ □ ☿
04 59	☽ □ ♃
07 42	☽ # ♂
08 31	☽ ☌ ♇
11 48	☽ △ ♄
19 55	☉ # ♃
20 32	☽ # ♆
22 59	☽ ✶ ♆
23 06	☿ Q ♅
23 59	☽ ⚻ ♅
14 Wednesday	
02 39	☉ ∥ ♅
04 17	☽ Q ☉
08 36	☽ ∥ ♄
09 40	☽ Q ☿
10 06	☽ ± ♅
13 02	☽ ⚼ ♄
23 15	☽ ⚻ ♂
15 Thursday	
00 12	☽ ∠ ♆
00 22	☿ ⚺ ♀
01 09	☽ ⚼ ♅
01 48	☽ ✶ ☉
03 58	☽ ± ♄
04 55	☽ ✶ ☿
05 32	☽ □ ♀
07 03	☽ △ ♃
07 35	☉ ∥ ☿
09 31	☽ ± ♂
10 56	☽ ⚺ ♇
13 51	☽ ⚻ ♄
15 08	☽ ⊥ ♆
20 49	☽ ⊥ ♇
22 19	♀ ⚻ ♃
16 Friday	
00 41	☿ ∥ ♅
00 47	☽ ⚼ ♂
01 01	☽ ⚺ ♆
01 57	☽ △ ♅
04 18	☽ ∠ ☉
04 41	☽ ∠ ☿
07 31	☽ ⚼ ♃
07 36	☿ # ♃
08 04	☉ ☌ ☿
11 34	☽ ∠ ♇
13 45	☽ Q ♀
18 49	☽ ⊥ ☿
20 02	☽ ⊥ ☉
22 05	☽ ± ♃
17 Saturday	
02 02	☽ △ ♂
03 50	☽ # ♄
04 13	☽ ⚺ ☿
06 29	☽ ⚺ ☉
07 46	☽ ⚻ ♃
11 01	☽ ✶ ♀
12 01	☽ ✶ ♇
14 39	☽ ☍ ♄
15 49	☽ ∥ ♆
20 36	☉ Q ♅
23 23	♀ △ ♇
23 55	☉ ☍ ♃
18 Sunday	
00 44	☽ ∥ ♂
02 03	☽ ☌ ♆
02 57	☽ □ ♅
07 33	☽ Q ♇
09 18	☿ ⊥ ♀
10 33	☿ □ ♂
11 15	☽ ∥ ☿
13 36	☽ ∠ ♀
14 54	☽ # ♃
15 38	☽ ∥ ♅
20 43	☽ ∥ ☉
19 Monday	
03 17	☽ ☌ ☿
04 29	☽ □ ♂
05 02	☽ # ♇
05 39	☽ ⊥ ♀
05 54	♀ △ ♄
08 19	☽ ☍ ♃
08 27	☽ Q ♅
10 55	☽ ☌ ☉
13 04	☽ □ ♇
15 31	☽ ⚻ ♄
16 25	☽ ⚺ ♀
20 Tuesday	
01 26	☽ ± ♄
03 05	☽ ∥ ♀
03 29	☽ ⚺ ♆
04 23	☽ ✶ ♅
11 19	☽ Q ♂
11 27	♀ ± ♃
13 35	☽ ⊥ ♆
16 28	☽ ⚼ ♄
16 48	♀ ⊥ ♅
19 35	☉ □ ♇
21 Wednesday	
03 38	☽ ⚺ ☿
04 53	☽ ∠ ♆
05 48	☽ ∠ ♅
08 23	☽ ✶ ♂
10 15	☽ ⚻ ♃
13 35	☽ ⊥ ☿
15 42	☽ △ ♇
17 12	☽ ⚺ ☉
18 02	☽ △ ♄
20 33	☽ ± ♃
21 23	☽ ⊥ ♅
22 Thursday	
00 01	☽ ☌ ♀
04 28	☉ ⚻ ♄
04 28	☽ ⊥ ☉
04 46	☽ ∠ ☿
07 01	☽ ✶ ♆
07 56	☽ ⚺ ♅
11 27	☽ ∠ ♂
12 14	☽ ⚼ ♃
18 07	☽ ⚼ ♇
21 43	☽ ∠ ☉
23 Friday	
04 11	☽ ⊥ ♂
04 33	☽ Q ♆
05 33	♂ □ ♃
06 48	☽ ✶ ☿
10 24	☽ ± ♇
15 06	☽ △ ♃
15 33	☽ ⚺ ♂
18 03	☿ ∠ ♀
21 27	☽ ⚻ ♇
22 44	☽ ∥ ♀
23 41	☽ □ ♄
24 Saturday	
03 24	☽ ✶ ☉
04 21	☽ Q ☿
11 55	☽ ⚺ ♀
14 00	☽ □ ♆
14 59	☽ ☌ ♅
23 23	☽ # ♇
25 Sunday	
00 31	☽ ⊥ ♀
02 20	☽ ∥ ☉
04 01	☽ Q ☉
09 42	☽ Q ♄
10 49	♀ ✶ ♆
13 54	☽ □ ☿
15 43	☽ ∥ ♅
17 17	☽ # ♃
19 47	☽ ∠ ♀
21 32	♀ ⚺ ♅
23 37	☽ □ ♃
26 Monday	
02 57	☽ ☌ ♂
06 59	☽ ☍ ♇
09 05	☽ ✶ ♄
09 45	☽ ∥ ☿
16 46	☽ ∥ ♂
18 18	☽ □ ☉
19 56	☽ ∥ ♆
27 Tuesday	
00 42	☽ △ ♆
01 44	☽ ⚺ ♅
04 44	☽ ✶ ♀
08 07	♀ ⚼ ♃
11 04	☽ Q ♃
12 03	☽ # ♄
13 49	☽ ⊥ ♅
14 56	☽ ∠ ♄
28 Wednesday	
00 48	☽ △ ☿
01 54	☉ ± ♄
02 50	☉ # ♇
07 02	☽ ⚼ ♆
07 38	☽ Q ♀
08 05	☽ ∠ ♅
09 04	☽ ⊥ ♄
11 05	☽ ✶ ♃
17 40	☽ ⚺ ♂
19 19	☽ ⚻ ♇
21 12	☽ ⚺ ♄
29 Thursday	
01 22	☽ ± ♆
01 31	☿ St D
06 23	☽ ⊥ ♂
07 12	☽ ⚼ ☿
07 32	☽ ± ♇
12 17	☽ △ ☉
13 37	☽ ⚻ ♆
14 40	☽ ✶ ♅
17 16	☽ ∠ ♃
30 Friday	
00 16	☽ □ ♀
01 27	☽ ∠ ♂
01 34	☽ ± ☿
01 49	☽ ⚼ ♇
02 57	♂ ∥ ♆
05 13	☉ ⚺ ♆
10 19	♂ ☍ ♇
11 21	☽ ⊥ ♃
13 49	☽ ⚻ ☿
15 05	☽ Q ♅
15 48	♀ ⚼ ♇
16 36	♇ St R
18 44	☉ ✶ ♅
19 38	♀ ∠ ♂
20 41	☽ ∥ ♄
21 22	☽ ⚼ ☉
23 20	☽ ⚺ ♃
31 Saturday	
08 06	☽ △ ♇
09 02	☽ ✶ ♂
09 41	☽ ☌ ♄
11 57	☽ # ♂
13 15	☽ # ♆
17 00	☽ ± ☉
20 17	☽ # ☿
23 52	♂ ✶ ♄

JUNE 2042

LONGITUDES

Date	Sidereal time h m s	Sun ☉	Moon ☽	Moon ☽ 24.00	Mercury ☿	Venus ♀	Mars ♂	Jupiter ♃	Saturn ♄	Uranus ♅	Neptune ♆	Pluto ♇
01	04 40 26	11 ♊ 08 39	13 ♏ 57 48	20 ♏ 04 32	21 ♉ 13	17 ♋ 07	00 ♍ 57	25 ♏ 24	00 ♏ 41	09 ♌ 34	09 ♉ 02	29 ♒ 57
02	04 44 23	12 06 08	26 ♏ 14 36	02 ♐ 28 13	21 31	18 17	01 26	25 R 17	00 R 38	09 36	09 04	29 R 56
03	04 48 19	13 03 36	08 ♐ 45 30	15 ♐ 06 33	21 53	19 27	01 56	25 10	00 35	09 38	09 06	29 56
04	04 52 16	14 01 03	21 ♐ 31 22	27 ♐ 59 56	22 19	20 38	02 26	25 03	00 32	09 41	09 08	29 56
05	04 56 13	14 58 29	04 ♑ 32 09	11 ♑ 07 53	22 50	21 48	02 56	24 56	00 30	09 43	09 10	29 56
06	05 00 09	15 55 54	17 ♑ 47 00	24 ♑ 29 18	23 25	22 58	03 26	24 49	00 27	09 46	09 12	29 56
07	05 04 06	16 53 19	01 ♒ 14 37	08 ♒ 02 44	24 03	24 08	03 56	24 43	00 25	09 48	09 13	29 56
08	05 08 02	17 50 42	14 ♒ 53 29	21 ♒ 46 41	24 46	25 18	04 27	24 36	00 22	09 51	09 15	29 56
09	05 11 59	18 48 05	28 ♒ 42 12	05 ♓ 39 53	25 33	26 28	04 57	24 30	00 20	09 54	09 17	29 55
10	05 15 55	19 45 28	12 ♓ 39 37	19 ♓ 41 16	26 24	27 38	05 28	24 24	00 18	09 56	09 19	29 55
11	05 19 52	20 42 50	26 ♓ 44 43	03 ♈ 49 50	27 18	28 47	05 59	24 17	00 16	09 59	09 21	29 55
12	05 23 48	21 40 11	10 ♈ 56 28	18 ♈ 04 25	28 16	29 ♋ 57	06 30	24 11	00 14	10 02	09 23	29 55
13	05 27 45	22 37 33	25 ♈ 13 26	02 ♉ 23 11	29 ♉ 17	01 ♌ 06	07 01	24 05	00 12	10 04	09 24	29 54
14	05 31 42	23 34 53	09 ♉ 33 20	16 ♉ 43 25	00 ♊ 22	02 16	07 32	23 59	00 10	10 07	09 26	29 54
15	05 35 38	24 32 14	23 ♉ 52 55	01 ♊ 01 17	01 31	03 25	08 04	23 54	00 08	10 10	09 28	29 54
16	05 39 35	25 29 34	08 ♊ 07 55	15 ♊ 12 12	02 42	04 34	08 35	23 48	00 07	10 13	09 30	29 53
17	05 43 31	26 26 53	22 ♊ 13 30	29 ♊ 11 14	03 58	05 44	09 07	23 43	00 05	10 16	09 31	29 53
18	05 47 28	27 24 12	06 ♋ 04 52	12 ♋ 53 57	05 16	06 53	09 39	23 37	00 04	10 19	09 33	29 52
19	05 51 24	28 21 30	19 ♋ 38 06	26 ♋ 17 04	06 38	08 02	10 11	23 32	00 02	10 22	09 35	29 52
20	05 55 21	29 ♊ 18 48	02 ♌ 50 42	09 ♌ 18 59	08 03	09 10	10 43	23 27	00 01	10 25	09 36	29 51
21	05 59 17	00 ♋ 16 05	15 ♌ 42 00	21 ♌ 59 58	09 31	10 19	11 15	23 22	00 ♏ 00	10 28	09 38	29 51
22	06 03 14	01 13 21	28 ♌ 13 10	04 ♍ 22 00	11 02	11 28	11 48	23 17	29 ♎ 59	10 31	09 39	29 50
23	06 07 11	02 10 37	10 ♍ 26 57	16 ♍ 28 53	12 37	12 36	12 20	23 13	29 58	10 34	09 41	29 50
24	06 11 07	03 07 52	22 ♍ 27 22	28 ♍ 24 03	14 14	13 44	12 53	23 08	29 57	10 37	09 42	29 49
25	06 15 04	04 05 06	04 ♎ 19 15	10 ♎ 13 38	15 55	14 53	13 26	23 04	29 56	10 40	09 44	29 49
26	06 19 00	05 02 20	16 ♎ 07 53	22 ♎ 02 41	17 38	16 01	13 58	23 00	29 56	10 43	09 45	29 48
27	06 22 57	05 59 33	27 ♎ 58 41	03 ♏ 56 31	19 25	17 09	14 32	22 56	29 55	10 46	09 47	29 47
28	06 26 53	06 56 45	09 ♏ 56 49	16 ♏ 00 08	21 14	18 17	15 05	22 52	29 54	10 49	09 48	29 47
29	06 30 50	07 53 58	22 ♏ 06 58	28 ♏ 17 45	23 06	19 24	15 38	22 48	29 54	10 53	09 50	29 46
30	06 34 46	08 ♋ 51 10	04 ♐ 32 53	10 ♐ 52 36	25 ♊ 01	20 ♌ 32	16 ♍ 11	22 ♏ 45	29 ♎ 54	10 ♌ 56	09 ♉ 51	29 ♒ 45

MOON NODES AND LATITUDE

Date	Moon True ☊	Moon Mean ☊	Moon ☽ Latitude
01	26 ♈ 14	24 ♈ 41	01 S 34
02	26 R 10	24 38	02 35
03	26 04	24 35	03 29
04	25 57	24 32	04 14
05	25 49	24 28	04 46
06	25 40	24 25	05 04
07	25 33	24 22	05 05
08	25 27	24 19	04 49
09	25 23	24 16	04 17
10	25 21	24 13	03 29
11	25 D 21	24 09	02 27
12	25 22	24 06	01 17
13	25 23	24 03	00 S 01
14	25 R 22	24 00	01 N 15
15	25 19	23 57	02 26
16	25 13	23 53	03 28
17	25 06	23 50	04 16
18	24 56	23 47	04 48
19	24 46	23 44	05 02
20	24 36	23 41	05 00
21	24 27	23 38	04 43
22	24 20	23 34	04 13
23	24 16	23 31	03 31
24	24 13	23 28	02 41
25	24 D 13	23 25	01 44
26	24 13	23 22	00 N 43
27	24 R 13	23 19	00 S 20
28	24 12	23 15	01 23
29	24 09	23 12	02 22
30	24 ♈ 04	23 ♈ 09	03 S 17

DECLINATIONS

Date	Sun ☉	Moon ☽	Mercury ☿	Venus ♀	Mars ♂	Jupiter ♃	Saturn ♄	Uranus ♅	Neptune ♆	Pluto ♇
01	22 N 07	17 S 32	14 N 22	24 N 19	12 N 24	18 S 04	09 S 13	18 N 28	12 N 53	21 S 29
02	22 14	21 49	14 24	24 11	12 13	18 02	09 12	18 28	12 54	21 30
03	22 22	25 12	14 28	24 01	12 01	18 01	09 11	18 27	12 54	21 30
04	22 29	27 23	14 34	23 52	11 49	17 59	09 11	18 26	12 55	21 30
05	22 35	28 07	14 42	23 41	11 37	17 58	09 10	18 26	12 55	21 31
06	22 42	27 16	14 51	23 30	11 25	17 56	09 09	18 25	12 56	21 31
07	22 47	24 50	15 03	23 18	11 13	17 55	09 09	18 24	12 56	21 31
08	22 53	20 58	15 16	23 06	11 01	17 53	09 08	18 24	12 57	21 32
09	22 58	15 56	15 31	22 53	10 49	17 52	09 08	18 23	12 57	21 32
10	23 02	10 01	15 47	22 40	10 36	17 51	09 07	18 22	12 58	21 32
11	23 07	03 S 33	16 04	22 26	10 24	17 49	09 06	18 22	12 59	21 33
12	23 10	03 N 09	16 23	22 11	10 11	17 48	09 06	18 21	12 59	21 33
13	23 14	09 45	16 42	21 56	09 59	17 47	09 06	18 20	13 00	21 34
14	23 17	15 52	17 03	21 40	09 46	17 46	09 05	18 19	13 00	21 34
15	23 19	21 06	17 24	21 23	09 33	17 44	09 05	18 18	13 01	21 34
16	23 22	25 05	17 47	21 06	09 20	17 43	09 04	18 17	13 01	21 35
17	23 23	27 28	18 09	20 49	09 08	17 42	09 04	18 17	13 02	21 35
18	23 25	28 05	18 33	20 31	08 55	17 41	09 04	18 16	13 02	21 36
19	23 26	26 59	18 56	20 13	08 41	17 40	09 04	18 15	13 03	21 36
20	23 26	24 23	19 20	19 54	08 28	17 39	09 03	18 14	13 03	21 37
21	23 26	20 38	19 44	19 34	08 15	17 38	09 03	18 13	13 04	21 37
22	23 26	16 02	20 08	19 14	08 02	17 37	09 03	18 13	13 04	21 38
23	23 25	10 54	20 31	18 54	07 48	17 36	09 03	18 12	13 04	21 38
24	23 24	05 N 27	20 55	18 33	07 35	17 35	09 03	18 11	13 05	21 39
25	23 22	00 S 08	21 17	18 12	07 21	17 34	09 03	18 10	13 05	21 39
26	23 20	05 41	21 39	17 50	07 08	17 33	09 03	18 09	13 06	21 40
27	23 18	11 04	22 00	17 28	06 54	17 33	09 03	18 08	13 06	21 40
28	23 15	16 06	22 20	17 06	06 40	17 32	09 03	18 07	13 06	21 41
29	23 12	20 35	22 38	16 43	06 27	17 31	09 03	18 06	13 07	21 41
30	23 N 08	24 S 16	22 N 55	16 N 20	06 N 13	17 S 30	09 S 03	18 N 05	13 N 07	21 S 42

ZODIAC SIGN ENTRIES

Date	h	m	Planets
02	19	15	☽ ♐
05	03	41	☽ ♑
07	09	48	☽ ♒
09	14	14	☽ ♓
11	17	31	☽ ♈
12	13	03	♀ ♌
13	20	00	☽ ♉
14	03	57	☿ ♊
15	22	17	☽ ♊
18	01	25	☽ ♋
20	06	47	☽ ♌
21	05	16	☉ ♋
21	10	26	♄ ♎
22	15	28	☽ ♍
25	03	14	☽ ♎
27	16	05	☽ ♏
30	03	17	☽ ♐

LATITUDES

Date	Mercury ☿	Venus ♀	Mars ♂	Jupiter ♃	Saturn ♄	Uranus ♅	Neptune ♆	Pluto ♇
01	03 S 50	02 N 00	01 N 21	01 N 05	02 N 39	00 N 38	01 S 43	10 S 43
04	03 55	02 02	01 18	01 04	02 39	00 38	01 43	10 44
07	03 51	02 03	01 14	01 04	02 38	00 38	01 43	10 45
10	03 40	02 04	01 11	01 03	02 37	00 38	01 43	10 46
13	03 22	02 04	01 07	01 03	02 37	00 38	01 43	10 47
16	02 59	02 03	01 04	01 02	02 36	00 38	01 43	10 48
19	02 31	02 01	01 01	01 02	02 35	00 38	01 43	10 48
22	01 59	01 58	00 58	01 01	02 35	00 38	01 44	10 49
25	01 25	01 55	00 55	01 00	02 34	00 38	01 44	10 50
28	00 49	01 50	00 52	00 59	02 33	00 38	01 44	10 51
31	00 S 14	01 N 44	00 N 49	00 N 59	02 N 32	00 N 38	01 S 44	10 S 52

DATA

Julian Date	2467037
Delta T	+78 seconds
Ayanamsa	24° 26' 51"
Synetic vernal point	04° ♓ 40' 08"
True obliquity of ecliptic	23° 26' 09"

MOON'S PHASES, APSIDES AND POSITIONS ☽

Date	h	m	Phase	Longitude	Eclipse Indicator
03	20	48	○	13 ♐ 25	
11	01	00	☾	20 ♓ 17	
17	19	48	●	26 ♊ 46	
25	11	29	☽	04 ♎ 04	

Day	h	m		
13	22	03	Perigee	
26	01	29	Apogee	
05	11	20	Max dec	28° S 07'
12	00	44	0N	
18	08	24	Max dec	28° N 06'
25	11	27	0S	

LONGITUDES

Date	Chiron ⚷	Ceres ⚳	Pallas ⚴	Juno ⚵	Vesta ⚶	Black Moon Lilith ⚸
01	01 ♌ 16	04 ♐ 23	04 ♎ 38	19 ♎ 44	12 ♋ 52	09 ♎ 16
11	02 ♌ 12	02 ♐ 20	05 ♎ 34	19 ♎ 26	17 ♋ 06	10 ♎ 23
21	03 ♌ 14	00 ♐ 38	07 ♎ 06	19 ♎ 40	21 ♋ 25	11 ♎ 29
31	04 ♌ 20	29 ♏ 29	09 ♎ 08	20 ♎ 24	25 ♋ 48	12 ♎ 36

ASPECTARIAN

01 Sunday
02 13 ☽ ☍ ♆
03 16 ☽ □ ♅
05 57 ☽ ⚻ ☉
09 55 ☽ Q ♂
14 45 ☽ ∥ ♃
15 52 ☉ ⊥ ♀
16 55 ☽ ⋕ ♅
18 51 ☽ △ ♀

02 Monday
02 34 ☽ ☍ ☿
10 00 ☽ ∥ ♇
10 09 ☽ ☌ ♃
14 43 ☽ ⋕ ☉
19 08 ☽ □ ♇
20 26 ☽ ⚺ ♄
22 26 ☽ □ ♂

03 Tuesday
02 58 ☽ ⚼ ♀
03 10 ☽ ⋕ ♀
07 53 ☽ ⊥ ♄
12 39 ☽ ⚻ ♆
13 41 ☽ △ ♅
20 48 ☽ ☍ ☉
21 47 ☽ ± ♀

04 Wednesday
00 00 ☽ ± ♆
00 51 ☽ ∠ ♄
05 19 ☽ Q ♇
10 10 ☽ ⚻ ♀
13 32 ☽ ⚻ ☿
16 51 ☽ ⚼ ♆
17 53 ☽ ⚼ ♅
18 30 ☽ ⚺ ♃
22 34 ♀ Q ♆

05 Thursday
01 05 ☽ ± ☿
03 34 ☽ ✶ ♇
04 37 ☽ ✶ ♄
05 28 ☽ ⊥ ♃
08 57 ☽ △ ♂
10 30 ☽ ± ♅
16 51 ☉ ⊥ ♆
18 16 ☽ ⚼ ☿
20 27 ☽ △ ♆
21 28 ☽ ⚻ ♅
21 45 ☽ ∠ ♃

06 Friday
00 32 ☉ ⚼ ♄
02 25 ☽ Q ♄
06 52 ☽ ∠ ♇
08 25 ☽ ⚻ ☉
13 13 ☽ ⚼ ♂
20 00 ☽ ± ☉
22 10 ☽ ☍ ♀
22 34 ☽ △ ☿
23 01 ☽ ⊥ ♇

07 Saturday
00 30 ☽ ✶ ♃
05 55 ☽ ± ♂
07 49 ♀ ± ♇
08 12 ☿ ✶ ♀
09 41 ☽ ⚺ ♇
10 32 ☽ □ ♄
13 14 ☽ ⚼ ☉
16 57 ☽ ⚻ ♂
21 35 ☽ Q ♃
22 54 ♀ △ ♃
23 07 ☽ ⋕ ♀

08 Sunday
01 24 ☽ ⋕ ☉
02 06 ☽ □ ♆
03 08 ☽ ☍ ♅
07 18 ☿ ☍ ♃
09 00 ☽ ∥ ♇
17 32 ☽ △ ☉

09 Monday
00 56 ☽ ⋕ ♅
03 18 ☽ ∥ ♃
04 47 ☽ □ ♃
05 48 ☉ ∥ ♀
06 12 ☽ □ ☿
07 46 ☽ ⚻ ♀
09 33 ☽ Q ♆
13 44 ☽ ⋕ ☿
14 06 ☽ ☌ ♇
14 49 ☽ △ ♄
19 05 ☽ ± ♀
23 11 ☽ ☍ ♂

10 Tuesday
00 25 ☽ ⋕ ♆
06 16 ☽ ✶ ♆
07 19 ☽ ⚻ ♅
09 40 ☽ ⋕ ♂
11 56 ☽ ⚼ ♀
15 10 ☽ Q ☿
15 27 ☽ ∥ ♄
16 30 ☽ ⚼ ♄
17 37 ☽ ± ♅

11 Wednesday
01 00 ☽ □ ☉
07 48 ☽ ± ♄
07 51 ☽ △ ♃
07 55 ☽ ∠ ♆
09 00 ☽ ⚼ ♅
13 00 ☽ ✶ ☿
15 47 ☽ △ ♀
17 22 ☽ ⚺ ♇
17 57 ☽ ⚻ ♄
23 12 ☽ ⊥ ♆

12 Thursday
03 31 ☽ ⊥ ♇
04 13 ☽ ⚻ ♂
06 00 ☿ Q ♅
09 04 ☽ ⚼ ♃
09 22 ☽ ⚺ ♆
09 42 ☽ Q ☉
10 27 ☽ △ ♅
11 11 ♀ ⚻ ♇
14 43 ☽ ± ♂
16 12 ☽ ∠ ☿
17 41 ♀ □ ♄
18 41 ☽ ∠ ♇

13 Friday
00 06 ☽ ± ♃
06 25 ☽ ⚼ ♂
07 20 ☽ ✶ ☉
08 30 ☽ ⊥ ☿
08 32 ♀ ⊥ ♂
09 34 ☽ ⋕ ♄
10 06 ☽ ⚻ ♃
12 51 ☽ ∥ ♂
19 21 ☽ ⚺ ☿
19 50 ☽ ✶ ♇
20 19 ☽ ☍ ♄
22 44 ☽ □ ♀

14 Saturday
00 26 ☽ ∥ ♆
01 46 ☿ □ ♇
07 45 ☿ ⚻ ♄
08 30 ☽ △ ♂
10 15 ☽ ∠ ☉
11 48 ☽ ☌ ♆
12 57 ☽ □ ♅
15 55 ☽ Q ♇
17 23 ☽ ∥ ☿
20 07 ☽ ⋕ ♃
20 15 ♀ ⋕ ♇
21 18 ☉ ⚻ ♃
22 37 ☽ ∥ ♅

15 Sunday
02 24 ☽ ⊥ ☉
07 30 ☽ Q ♀
10 08 ☉ ∠ ♆
12 01 ☽ ☍ ♃
13 11 ☽ ⚺ ☉
13 24 ☽ ∥ ♀
14 28 ☽ ⋕ ♇
19 13 ☽ Q ♅
22 06 ☽ □ ♇
22 30 ☽ ⚻ ♄

16 Monday
00 26 ☽ ∥ ☉
01 59 ☽ ☌ ☿
04 37 ☉ ∠ ♅
05 27 ☽ ✶ ♀
08 36 ☽ ± ♄
08 36 ☿ ⋕ ♃
12 48 ☽ □ ♂
14 19 ☽ ⚺ ♆
15 32 ☽ ✶ ♅
23 49 ☽ ⚼ ♄

17 Tuesday
00 31 ☽ ⊥ ♆
09 12 ☽ ∠ ♀
14 32 ☽ ⚻ ♃
15 57 ☽ ∠ ♆
17 14 ☽ ∠ ♅
18 30 ♂ ⋕ ♄
19 22 ☿ ∥ ♅
19 48 ☽ ☌ ☉
20 45 ☽ Q ♂

18 Wednesday
00 49 ☽ ± ♃
01 11 ☽ △ ♇
01 32 ☽ △ ♄
02 06 ☽ ⊥ ♀
07 17 ♂ △ ♆
08 54 ☽ ⊥ ♅
10 25 ☽ ⚺ ☿
13 31 ☽ ⚺ ♀
16 25 ☽ ⚼ ♃
18 06 ☽ ✶ ♆
18 31 ☽ ✶ ♂
19 27 ☽ ⚺ ♅
22 07 ☽ ⊥ ☿

19 Thursday
01 53 ☿ ± ♄
01 56 ☉ Q ♂
03 29 ☽ ⚼ ♇
15 30 ☽ Q ♆
16 00 ☽ ∠ ☿
18 58 ☽ △ ♃
19 36 ☽ ± ♇
20 48 ♂ ⚺ ♅
22 25 ☽ ∠ ♂

20 Friday
05 01 ☽ ⚺ ☉
06 31 ☽ ⚻ ♇
06 49 ☽ □ ♄
15 09 ☉ ± ♃
15 36 ☽ ⊥ ♂
16 55 ☽ ⊥ ☉
18 45 ☽ ∥ ☉
21 16 ♀ □ ♆
22 51 ☽ ✶ ☿

21 Saturday
00 34 ☽ □ ♆
00 53 ☽ ☌ ♀
01 30 ☉ △ ♇
02 06 ☽ ☌ ♅
03 16 ☽ ⚺ ♂
05 22 ☉ △ ♄
06 16 ☽ ⋕ ♇
06 38 ☿ ∥ ♀
11 07 ☽ ∠ ☉
13 54 ☿ ⚺ ♆
15 08 ♀ ☌ ♅
16 21 ☽ Q ♄
16 22 ♂ Q ♃
16 35 ☽ ∥ ☿
18 21 ☽ ∥ ♀

22 Sunday
00 35 ☽ Q ☿
01 04 ☽ ∥ ♅
02 32 ☽ □ ♃
03 32 ☿ ✶ ♅
04 07 ☽ ⋕ ♃
15 09 ☽ ☍ ♇
15 25 ☽ ✶ ♄
18 20 ☽ ✶ ☉

23 Monday
01 24 ♀ ⚺ ♂
02 06 ☽ ∥ ♆
05 46 ☿ □ ♂
10 29 ☽ △ ♆
11 34 ☿ ✶ ♀
12 13 ☽ ⚺ ♅
13 30 ☽ Q ♃
15 55 ☽ ☌ ♂
16 43 ☽ ⚺ ♀
16 57 ☽ □ ☿
20 03 ☽ Q ☉
20 17 ☽ ⋕ ♄
20 58 ☽ ∠ ♄

24 Tuesday
00 13 ☽ ⊥ ♅
02 20 ☽ ∥ ♂
05 58 ☽ ⊥ ♀
13 22 ☽ ✶ ♃
15 00 ☽ ⊥ ♄
16 33 ☽ ⚼ ♆
18 24 ☽ ∠ ♅
22 15 ☿ ⚼ ♄

25 Wednesday
02 02 ☽ ∠ ♀
02 51 ☽ ⚻ ♇
03 07 ☽ ⚺ ♄
09 26 ☿ ⊥ ♆
10 48 ☽ ± ♆
11 29 ☽ □ ☉
14 15 ♀ ∥ ♅
15 01 ☽ ± ♇
19 34 ☽ ∠ ♃
23 01 ☽ ⚻ ♆

26 Thursday
00 57 ☽ ✶ ♅
03 28 ☉ ⊥ ♅
07 24 ☽ ⚺ ♂
09 18 ☽ ⚼ ♇
11 44 ☽ ✶ ♀
12 33 ☿ ⋕ ♇
13 45 ☽ ⊥ ♃
15 36 ☽ △ ☿
18 06 ☽ ⋕ ♂
20 11 ☽ ⊥ ♂

27 Friday
01 25 ☽ Q ♅
01 51 ☽ ⚺ ♃
02 52 ☽ ∥ ♄
07 05 ♀ ⋕ ♃
14 36 ☽ Q ♀
15 16 ☽ ∠ ♂
15 39 ☽ △ ♇
15 54 ☽ ☌ ♄
21 28 ☽ ⋕ ♆

28 Saturday
03 15 ☽ ⚼ ☿
04 14 ♀ Q ♄
04 43 ♂ ∠ ♄
05 30 ☽ △ ☉
11 43 ☽ ☍ ♆
13 45 ☽ □ ♅
16 39 ☽ ⋕ ♀
19 15 ☽ ∥ ♃
22 20 ☽ ⋕ ♅
22 40 ☽ ✶ ♂

29 Sunday
00 23 ☽ ± ☿
06 09 ☽ □ ♀
08 19 ☿ ⚻ ♃
09 49 ☉ ⚼ ♃
13 21 ☽ ☌ ♃
13 39 ☽ ⚼ ☉
14 17 ☽ ⚻ ☿
18 34 ☽ ∥ ♇
23 13 ☽ Q ♂

30 Monday
01 41 ☽ ⋕ ☿
02 49 ☽ □ ♇
03 05 ☽ ⚺ ♄
04 05 ☽ ⋕ ☉
08 30 ☽ ± ☉
09 52 ☿ ∠ ♆
14 34 ☽ ⊥ ♄
20 51 ☽ ⚻ ☉
22 05 ☽ ⚻ ♆
23 34 ☿ ∠ ♅

All ephemeris data is given at 12.00 UT and the Moon's longitude is additionally given for 24.00 UT

JULY 2042

LONGITUDES

Date	Sidereal time h m s	Sun ☉ ° ' "	Moon ☽ ° ' "	Moon ☽ 24.00 ° ' "	Mercury ☿ ° '	Venus ♀ ° '	Mars ♂ ° '	Jupiter ♃ ° '	Saturn ♄ ° '	Uranus ♅ ° '	Neptune ♆ ° '	Pluto ♇ ° '
01	06 38 43	09 ♋ 48 21	17 ♐ 17 08	23 ♐ 46 32	26 ♊ 59	21 ♌ 39	16 ♍ 45	22 ♏ 42	29 ♎ 54	10 ♌ 59	09 ♉ 52	29 ♒ 44
02	06 42 40	10 45 33	00 ♑ 20 48	06 ♑ 59 49	28 ♊ 58	22 46	17 18	22 R 39	29 R 53	11 03	09 54	29 R 44
03	06 46 36	11 42 44	13 ♑ 43 20	20 ♑ 31 04	01 ♋ 00	23 54	17 52	22 36	29 D 53	11 06	09 55	29 43
04	06 50 33	12 39 55	27 ♑ 22 36	04 ♒ 17 31	03 04	25 01	18 26	22 33	29 54	11 09	09 56	29 42
05	06 54 29	13 37 07	11 ♒ 15 17	18 ♒ 15 25	05 10	26 07	19 00	22 30	29 54	11 13	09 58	29 41
06	06 58 26	14 34 18	25 ♒ 17 24	02 ♓ 20 46	07 17	27 14	19 34	22 28	29 54	11 16	09 59	29 41
07	07 02 22	15 31 29	09 ♓ 25 03	16 ♓ 29 53	09 25	28 20	20 08	22 25	29 54	11 19	10 00	29 40
08	07 06 19	16 28 41	23 ♓ 34 55	00 ♈ 39 54	11 34	29 ♌ 27	20 42	22 23	29 55	11 23	10 01	29 39
09	07 10 15	17 25 53	07 ♈ 44 36	14 ♈ 48 51	13 43	00 ♍ 33	21 16	22 21	29 55	11 26	10 02	29 38
10	07 14 12	18 23 06	21 ♈ 52 32	28 ♈ 55 32	15 53	01 39	21 51	22 20	29 56	11 30	10 03	29 37
11	07 18 09	19 20 19	05 ♉ 57 44	12 ♉ 59 02	18 03	02 45	22 25	22 18	29 57	11 33	10 04	29 36
12	07 22 05	20 17 33	19 ♉ 59 17	26 ♉ 58 21	20 12	03 50	23 00	22 17	29 58	11 37	10 05	29 35
13	07 26 02	21 14 47	03 ♊ 55 59	10 ♊ 51 59	22 21	04 56	23 35	22 16	29 ♎ 59	11 40	10 06	29 34
14	07 29 58	22 12 01	17 ♊ 46 01	24 ♊ 37 48	24 30	06 01	24 09	22 15	00 ♏ 00	11 44	10 07	29 33
15	07 33 55	23 09 17	01 ♋ 26 59	08 ♋ 13 12	26 37	07 06	24 44	22 14	00 01	11 47	10 08	29 32
16	07 37 51	24 06 32	14 ♋ 56 08	21 ♋ 35 27	28 ♋ 43	08 11	25 19	22 13	00 02	11 51	10 09	29 31
17	07 41 48	25 03 48	28 ♋ 10 53	04 ♌ 42 14	00 ♌ 48	09 16	25 54	22 13	00 04	11 55	10 10	29 30
18	07 45 44	26 01 04	11 ♌ 09 20	17 ♌ 32 08	02 51	10 20	26 30	22 12	00 05	11 58	10 11	29 29
19	07 49 41	26 58 21	23 ♌ 50 38	00 ♍ 04 57	04 53	11 25	27 05	22 12	00 07	12 02	10 12	29 28
20	07 53 38	27 55 38	06 ♍ 15 17	12 ♍ 21 53	06 54	12 29	27 41	22 D 12	00 08	12 05	10 13	29 27
21	07 57 34	28 52 55	18 ♍ 25 08	24 ♍ 25 27	08 52	13 33	28 16	22 13	00 10	12 09	10 13	29 26
22	08 01 31	29 ♋ 50 12	00 ♎ 23 19	06 ♎ 19 17	10 50	14 36	28 52	22 13	00 12	12 13	10 14	29 25
23	08 05 27	00 ♌ 47 30	12 ♎ 13 57	18 ♎ 07 57	12 45	15 40	29 ♍ 27	22 14	00 14	12 16	10 15	29 24
24	08 09 24	01 44 48	24 ♎ 01 55	29 ♎ 56 33	14 39	16 43	00 ♎ 03	22 15	00 16	12 20	10 15	29 23
25	08 13 20	02 42 06	05 ♏ 52 33	11 ♏ 50 34	16 31	17 45	00 39	22 16	00 18	12 24	10 16	29 22
26	08 17 17	03 39 25	17 ♏ 51 19	23 ♏ 55 24	18 21	18 48	01 15	22 17	00 20	12 27	10 17	29 20
27	08 21 13	04 36 45	00 ♐ 03 28	06 ♐ 16 04	20 09	19 50	01 51	22 18	00 23	12 31	10 17	29 19
28	08 25 10	05 34 04	12 ♐ 33 41	18 ♐ 56 44	21 56	20 52	02 27	22 20	00 25	12 35	10 18	29 18
29	08 29 07	06 31 25	25 ♐ 25 31	02 ♑ 00 13	23 41	21 54	03 04	22 21	00 28	12 38	10 18	29 17
30	08 33 03	07 28 45	08 ♑ 40 54	15 ♑ 27 30	25 25	22 56	03 40	22 23	00 30	12 42	10 19	29 16
31	08 37 00	08 ♌ 26 07	22 ♑ 19 46	29 ♑ 17 22	27 ♌ 06	23 ♍ 57	04 ♎ 16	22 ♏ 25	00 ♏ 33	12 ♌ 46	10 ♉ 19	29 ♒ 15

Date	Moon True ☊ ° '	Moon Mean ☊ ° '	Moon ☽ Latitude ° '
01	23 ♈ 56	23 ♈ 06	04 S 02
02	23 R 45	23 03	04 37
03	23 34	22 59	04 57
04	23 22	22 56	05 00
05	23 11	22 53	04 46
06	23 03	22 50	04 15
07	22 57	22 47	03 28
08	22 53	22 44	02 28
09	22 52	22 40	01 19
10	22 D 52	22 37	00 S 05
11	22 R 52	22 34	01 N 09
12	22 50	22 31	02 18
13	22 46	22 28	03 19
14	22 39	22 24	04 07
15	22 30	22 21	04 41
16	22 18	22 18	04 58
17	22 06	22 15	04 59
18	21 53	22 12	04 44
19	21 42	22 09	04 15
20	21 33	22 05	03 35
21	21 27	22 02	02 45
22	21 24	21 59	01 49
23	21 22	21 56	00 N 48
24	21 D 22	21 53	00 S 14
25	21 R 22	21 50	01 16
26	21 21	21 46	02 16
27	21 19	21 43	03 10
28	21 14	21 40	03 57
29	21 07	21 37	04 33
30	20 57	21 34	04 55
31	20 ♈ 47	21 ♈ 30	05 S 02

DECLINATIONS

Date	Sun ☉ ° '	Moon ☽ ° '	Mercury ☿ ° '	Venus ♀ ° '	Mars ♂ ° '	Jupiter ♃ ° '	Saturn ♄ ° '	Uranus ♅ ° '	Neptune ♆ ° '	Pluto ♇ ° '
01	23 N 04	26 S 51	23 N 10	15 N 56	05 N 59	17 S 30	09 S 03	18 N 05	13 N 08	21 S 42
02	23 00	28 03	23 24	15 32	05 45	17 29	09 04	18 04	13 08	21 43
03	22 55	27 39	23 35	15 08	05 31	17 29	09 04	18 03	13 08	21 43
04	22 50	25 35	23 44	14 43	05 17	17 28	09 04	18 02	13 09	21 44
05	22 44	21 59	23 51	14 18	05 03	17 28	09 04	18 01	13 09	21 44
06	22 38	17 05	23 54	13 53	04 48	17 27	09 05	18 00	13 09	21 45
07	22 32	11 15	23 56	13 27	04 34	17 27	09 05	17 59	13 10	21 45
08	22 25	04 S 49	23 54	13 02	04 20	17 27	09 06	17 58	13 10	21 46
09	22 18	01 N 52	23 50	12 35	04 05	17 27	09 06	17 57	13 10	21 46
10	22 10	08 26	23 43	12 09	03 51	17 26	09 07	17 56	13 11	21 47
11	22 03	14 35	23 34	11 43	03 36	17 26	09 07	17 55	13 11	21 47
12	21 54	19 57	23 21	11 16	03 22	17 26	09 08	17 54	13 11	21 48
13	21 45	24 11	23 06	10 49	03 07	17 26	09 08	17 53	13 11	21 49
14	21 36	26 58	22 49	10 21	02 53	17 26	09 09	17 52	13 12	21 49
15	21 27	28 06	22 29	09 54	02 38	17 26	09 10	17 51	13 12	21 50
16	21 17	27 32	22 07	09 26	02 23	17 26	09 10	17 50	13 12	21 50
17	21 07	25 24	21 42	08 58	02 09	17 26	09 11	17 49	13 12	21 51
18	20 57	21 58	21 16	08 30	01 54	17 26	09 12	17 48	13 13	21 51
19	20 46	17 35	20 48	08 02	01 39	17 27	09 13	17 47	13 13	21 52
20	20 34	12 32	20 18	07 34	01 24	17 27	09 13	17 46	13 13	21 53
21	20 23	07 07	19 46	07 06	01 09	17 27	09 14	17 45	13 13	21 53
22	20 11	01 N 31	19 14	06 37	00 54	17 28	09 15	17 44	13 13	21 54
23	19 59	04 S 06	18 40	06 08	00 39	17 28	09 16	17 43	13 13	21 54
24	19 46	09 32	18 04	05 40	00 24	17 28	09 17	17 42	13 14	21 55
25	19 33	14 41	17 28	05 11	00 N 09	17 29	09 18	17 41	13 14	21 55
26	19 20	19 19	16 51	04 42	00 S 06	17 29	09 19	17 40	13 14	21 56
27	19 06	23 15	16 13	04 13	00 21	17 30	09 20	17 39	13 14	21 57
28	18 53	26 13	15 35	03 44	00 36	17 31	09 21	17 38	13 14	21 57
29	18 38	27 54	14 55	03 15	00 52	17 31	09 22	17 37	13 14	21 58
30	18 24	28 04	14 16	02 46	01 07	17 32	09 23	17 36	13 14	21 58
31	18 N 09	26 S 33	13 N 36	02 N 17	01 S 22	17 S 33	09 S 25	17 N 35	13 N 14	21 S 59

ZODIAC SIGN ENTRIES

Date	h	m	Planets
02	11	22	☽ ♑
03	00	11	☿ ♋
04	16	34	☽ ♒
06	20	01	☽ ♓
08	22	52	☽ ♈
09	00	03	♀ ♍
11	01	50	☽ ♉
13	05	13	☽ ♊
14	13	59	♄ ♏
15	09	27	☽ ♋
17	02	47	☿ ♌
17	15	20	☽ ♌
19	23	50	☽ ♍
22	11	13	☽ ♎
22	16	06	☉ ♌
24	09	51	♂ ♎
25	00	07	☽ ♏
27	11	53	☽ ♐
29	20	22	☽ ♑

LATITUDES

Date	Mercury ☿ ° '	Venus ♀ ° '	Mars ♂ ° '	Jupiter ♃ ° '	Saturn ♄ ° '	Uranus ♅ ° '	Neptune ♆ ° '	Pluto ♇ ° '
01	00 S 14	01 N 44	00 N 49	00 N 59	02 N 32	00 N 38	01 S 44	10 S 52
04	00 N 20	01 38	00 46	00 58	02 31	00 38	01 44	10 53
07	00 50	01 30	00 43	00 57	02 31	00 38	01 44	10 53
10	01 14	01 21	00 40	00 56	02 30	00 38	01 44	10 54
13	01 33	01 12	00 37	00 56	02 29	00 38	01 45	10 55
16	01 44	01 01	00 35	00 55	02 28	00 38	01 45	10 56
19	01 48	00 49	00 32	00 54	02 27	00 38	01 45	10 56
22	01 47	00 36	00 29	00 53	02 27	00 38	01 45	10 57
25	01 40	00 22	00 27	00 53	02 26	00 38	01 45	10 57
28	01 28	00 N 08	00 24	00 52	02 25	00 38	01 45	10 58
31	01 N 12	00 S 08	00 N 22	00 N 51	02 N 24	00 N 38	01 S 46	10 S 58

DATA

Julian Date	2467067
Delta T	+78 seconds
Ayanamsa	24° 26' 57"
Synetic vernal point	04° ♓ 40' 02"
True obliquity of ecliptic	23° 26' 09"

MOON'S PHASES, APSIDES AND POSITIONS ☽

Date	h	m	Phase	Longitude ° '	Eclipse Indicator
03	08	09	○	11 ♑ 34	
10	05	38	☾	18 ♈ 08	
17	05	52	●	24 ♋ 49	
25	05	02	☽	02 ♏ 25	

Day	h	m		
09	07	31	Perigee	
23	20	11	Apogee	
02	18	13	Max dec	28° S 06'
09	05	20	ON	
15	15	48	Max dec	28° N 08'
22	18	26	OS	
30	02	28	Max dec	28° S 12'

LONGITUDES

Date	Chiron ⚷ ° '	Ceres ⚳ ° '	Pallas ⚴ ° '	Juno ⚵ ° '	Vesta ⚶ ° '	Black Moon Lilith ⚸ ° '
01	04 ♌ 20	29 ♏ 29	09 ♎ 08	20 ♎ 24	25 ♋ 48	12 ♎ 36
11	05 ♌ 29	28 ♏ 55	11 ♎ 35	21 ♎ 34	00 ♌ 14	13 ♎ 42
21	06 ♌ 40	28 ♏ 59	14 ♎ 24	23 ♎ 07	04 ♌ 43	14 ♎ 49
31	07 ♌ 52	29 ♏ 38	17 ♎ 31	25 ♎ 01	09 ♌ 14	15 ♎ 55

All ephemeris data is given at 12.00 UT and the Moon's longitude is additionally given for 24.00 UT

ASPECTARIAN

(h m — Aspects)

01 Tuesday
00 09 ☽ △ ♅
04 14 ☉ ∥ ☿
07 32 ☽ ∠ ♄
09 22 ☽ ± ♆
10 57 ☽ □ ♂
12 51 ☽ Q ♇
13 44 ☉ ⚹ ♆
20 51 ☽ △ ♀
21 58 ☽ ⊻ ♃
23 36 ♂ ⊥ ♅

02 Wednesday
02 02 ☽ ⚼ ♆
04 07 ☽ ⚼ ♅
08 09 ☿ ± ♃
08 55 ☽ ⊥ ♃
09 03 ☽ ☍ ☿
09 18 ♀ □ ♃
10 53 ☽ ⚹ ♇
11 10 ☽ ⚹ ♄
19 35 ☉ ⊻ ♅
20 31 ☽ ± ♅
20 55 ☿ △ ♇
22 53 ☿ △ ♄
23 47 ♄ St D

03 Thursday
01 07 ☽ ∠ ♃
02 37 ☽ ⚼ ♀
05 13 ☽ △ ♆
07 19 ☽ ⊼ ♅
08 09 ☽ ☍ ☉
08 45 ☽ Q ♄
13 46 ☽ ∠ ♇
19 39 ☽ △ ♂
20 02 ☽ ± ♀

04 Friday
03 35 ☽ ⚹ ♃
05 35 ☽ ⊥ ♇
07 30 ☽ ⊼ ♀
16 03 ☽ ⊻ ♇
16 23 ☽ □ ♄
22 57 ☽ ⚼ ♂
23 38 ☽ ⊼ ☿

05 Saturday
00 24 ☽ Q ♃
01 05 ☽ # ☿
07 32 ☽ # ☉
09 46 ☽ □ ♆
11 49 ☽ ± ☿
11 55 ☽ ☍ ♅
12 35 ☿ ⊥ ♅
13 21 ☽ ∥ ♇
15 07 ☽ ± ♂
16 21 ☽ ⊼ ☉

06 Sunday
01 49 ☽ ⊼ ♂
03 22 ☽ ± ☉
05 57 ☽ ⚼ ☿
07 12 ☽ □ ♃
07 54 ☽ # ♅
10 22 ☽ ∥ ♃
14 03 ☿ ⚼ ♃
14 34 ☉ ⚼ ♇
15 35 ☽ ☍ ♀
16 21 ☿ Q ♂
16 35 ☽ Q ♆
19 27 ☽ ☌ ♇
19 49 ☽ ⚼ ☉
19 51 ☽ △ ♄

07 Monday
02 37 ☽ # ♀
04 29 ☽ # ♆
11 59 ☽ △ ☿
12 59 ☽ ⚹ ♆
15 15 ☽ ⊼ ♅
18 37 ☿ ⚹ ♆
20 15 ☽ ∥ ♄
21 19 ☽ ⚼ ♄
23 06 ☽ △ ☉

08 Tuesday
01 27 ☽ ± ♅
04 13 ♀ ∥ ♆
06 55 ☽ ☍ ♂
09 56 ☿ ⊻ ♅
09 59 ☽ △ ♃
12 34 ☽ ± ♄
13 50 ☽ # ♂
14 26 ☽ ∠ ♆
16 18 ♀ ☍ ♇
16 46 ☽ ⚼ ♅
22 16 ☽ ⊻ ♇
22 17 ♀ ⚹ ♄
22 44 ☽ ⊼ ♄
22 46 ☽ ⊼ ♀

09 Wednesday
05 42 ☽ ⊥ ♆
08 25 ☽ ⊥ ♇
09 48 ☽ ± ♀
11 21 ☽ ⚼ ♃
15 54 ☽ ⊻ ♆
18 18 ☽ △ ♅
19 45 ☽ ∥ ♂
22 02 ☿ ⚼ ♇
23 41 ☽ ∠ ♇
23 59 ☽ □ ☿

10 Thursday
02 22 ☽ ⚼ ♀
02 35 ☽ ± ♃
05 38 ☽ □ ☉
11 57 ☽ ⊼ ♂
12 46 ☽ ⊼ ♃
14 31 ☽ # ♄
22 35 ☽ ± ♂

11 Friday
01 10 ☽ ⚹ ♇
01 14 ☽ ∥ ♀
01 44 ☽ ☍ ♄
05 12 ☿ ∠ ♀
06 03 ☽ △ ♀
06 17 ☽ ∥ ♆
07 21 ♂ ⚹ ♃
12 10 ☽ Q ☿
14 31 ☽ Q ☉
14 36 ☽ ⚼ ♂
19 02 ☽ ☌ ♆
21 36 ☽ □ ♅
21 37 ☽ Q ♇

12 Saturday
00 12 ☽ # ♃
02 20 ☽ ∥ ♅
12 27 ☽ ⚹ ☿
12 34 ☽ ⚹ ☉
13 43 ☉ ☌ ☿
15 56 ☽ ☍ ♃
17 23 ☽ △ ♂
21 37 ☽ # ♇
21 51 ☽ ∥ ☉

13 Sunday
04 06 ☉ # ♇
04 29 ☽ □ ♇
04 37 ☽ Q ♅
05 10 ☽ ⊼ ♄
05 32 ☽ ∥ ☿
09 11 ☿ Q ♆
10 56 ☿ △ ♃
13 52 ☽ □ ♀
15 33 ☽ ± ♄
16 18 ☽ ∠ ☉
19 00 ☽ ∠ ☿
22 42 ☽ ⊻ ♆

14 Monday
01 27 ☽ ⚹ ♅
01 31 ☿ ± ♇
05 17 ♂ ⊥ ♄
06 47 ☿ ⚹ ♂
07 10 ☽ ⚼ ♄
09 04 ☽ ⊥ ☉
09 08 ☽ ⊥ ♆
10 00 ☉ Q ♆
13 04 ☉ △ ♃
13 30 ☽ ⊥ ☿
19 49 ☽ ⊼ ♃
20 19 ☽ ⊻ ☉
23 40 ☽ □ ♂
23 52 ☽ Q ♀

15 Tuesday
00 53 ☽ ∠ ♆
01 55 ☽ ⊻ ☿
03 45 ☽ ∠ ♅
06 20 ☽ ± ♃
08 38 ☽ △ ♇
09 28 ☽ △ ♄
19 43 ☽ ⊥ ♅
21 28 ☉ ± ♇
22 14 ☽ ⚼ ♃
22 53 ☽ ⚹ ♀

16 Wednesday
03 26 ☽ ⚹ ♆
04 52 ♂ ⚼ ♆
06 27 ☽ ⊻ ♅
08 59 ☽ Q ♂
11 15 ☽ ⚼ ♇
21 11 ☿ ⊼ ♇

17 Thursday
01 02 ☽ Q ♆
01 08 ☽ △ ♃
01 31 ♀ # ♄
03 24 ☿ □ ♄
03 29 ☽ ± ♇
03 51 ☿ # ♇
04 13 ☽ ∠ ♀
05 52 ☽ ☌ ☉
07 39 ☽ ⚹ ♂
14 25 ☽ ⊼ ♇
15 27 ☽ □ ♄
17 42 ☽ ☌ ☿
22 11 ☽ ⊥ ♀

18 Friday
08 27 ♀ △ ♆
09 02 ♀ Q ♃
10 11 ☽ □ ♆
10 20 ☽ ⊻ ♀
12 40 ☽ ∠ ♂
12 40 ☽ # ♇
13 32 ☽ ☌ ♅
16 42 ☽ ∥ ☿
18 18 ☽ ∥ ☉

19 Saturday
01 04 ☽ Q ♄
06 29 ☽ ⊥ ♂
08 52 ☽ □ ♃
09 32 ♂ ∠ ♅
10 57 ☽ ∥ ♅
12 41 ☽ # ♃
12 48 ♃ St D
14 43 ☉ ∥ ☿
18 30 ☽ ⊻ ☉
18 32 ☽ ⊻ ♂
19 23 ☉ ⚹ ♂
22 48 ☽ ☍ ♇

20 Sunday
00 05 ☽ ⚹ ♄
01 17 ☿ ⊥ ♀
02 45 ♀ ⊻ ♅
07 05 ☽ ⊥ ☉
08 54 ☽ ∥ ♆
13 30 ☽ ⊻ ☿
19 45 ☽ Q ♃
19 46 ☽ △ ♆
23 31 ☽ ⊻ ♅

21 Monday
01 24 ☽ ☌ ♀
02 14 ☽ ∠ ☉
02 46 ☽ # ♄
03 36 ☽ ⊥ ☿
05 32 ☽ ∠ ♄
11 28 ☽ ⊥ ♅
12 05 ☽ ∥ ♀
19 34 ☽ ⚹ ♃
23 31 ☽ ⊥ ♄

22 Tuesday
01 02 ☽ ∠ ☿
01 36 ☉ ⊼ ♇
01 37 ☽ ⚼ ♆
04 38 ☿ □ ♆
05 34 ☽ ∠ ♅
08 45 ☽ ☌ ♂
10 02 ☽ ⊼ ♇
10 47 ☽ ⚹ ☉
11 37 ☽ ⊻ ♄
14 43 ☽ ∥ ♂
19 47 ☽ ± ♆
21 28 ☉ □ ♄
21 51 ☽ # ♂
22 09 ☽ ± ♇

23 Wednesday
01 50 ☽ ∠ ♃
02 01 ♀ ∠ ♄
05 48 ☿ ☌ ♅
07 57 ☽ ⊼ ♆
09 40 ♂ ⊼ ♇
12 05 ☽ ⚹ ♅
13 14 ☽ Q ☉
13 15 ☽ ⚹ ☿
16 24 ☽ ⚼ ♇
19 39 ☽ ⊻ ♀
20 08 ☽ ⊥ ♃
20 11 ☽ # ♀

24 Thursday
08 21 ☽ ⊻ ♃
09 03 ☽ ⊥ ♀
10 50 ☽ ∥ ♄
12 37 ☽ Q ♅
18 19 ☽ Q ☿
19 40 ☿ ∠ ♂
21 07 ♂ ⊻ ♄
21 19 ☉ ∠ ♀
22 50 ☽ △ ♇

25 Friday
00 42 ☽ ☌ ♄
00 52 ☽ ⊻ ♂
03 20 ☿ ∥ ♅
05 02 ☽ □ ☉
05 02 ☽ # ♆
05 06 ☽ ∠ ♀
11 31 ☿ # ♃
13 39 ☽ ⊥ ♂
20 51 ☽ ☍ ♆

26 Saturday
00 24 ☽ # ☿
01 10 ☽ □ ♅
02 09 ☽ ∥ ♃
03 07 ☽ # ♅
03 33 ♀ ⊥ ♅
08 39 ☽ ∠ ♂
11 53 ☿ Q ♄
12 03 ☽ # ☉
13 09 ☽ □ ☿
14 03 ☽ ⚹ ♀
20 46 ☽ ☌ ♃

27 Sunday
02 00 ☿ ⊻ ♀
03 23 ☽ ∥ ♇
10 34 ☽ □ ♇
12 38 ☽ ⊻ ♄
15 40 ☽ ⚹ ♂
15 46 ☽ Q ♀
21 33 ☽ △ ☉

28 Monday
00 15 ☽ ⊥ ♄
07 42 ☽ ⊼ ♆
12 02 ☽ △ ♅
15 45 ☽ Q ♂
17 23 ☿ □ ♃
17 25 ☽ ∠ ♄
19 03 ☽ ± ♆
20 55 ☽ Q ♇

29 Tuesday
04 13 ☽ ⚼ ☉
04 56 ☽ □ ♀
06 19 ☽ ⊻ ♃
08 18 ☽ △ ☿
11 47 ☽ ⚼ ♆
16 05 ☽ ⚼ ♅
17 23 ☽ ⊥ ♃
19 03 ☽ ⚹ ♇
21 14 ☽ ⚹ ♄
22 03 ☽ ± ☉
22 55 ♀ ⚹ ♃

30 Wednesday
02 34 ☽ □ ♂
08 27 ☽ ± ♅
09 41 ☽ ∠ ♃
09 41 ☽ ⊼ ☉
14 54 ☽ △ ♆
15 31 ☽ ⚼ ☿
18 49 ☽ Q ♄
19 10 ☽ ⊼ ♅
21 53 ☽ ∠ ♇

31 Thursday
09 35 ☽ ± ☿
12 10 ☽ ⚹ ♃
13 35 ☽ ⊥ ♇
13 49 ♂ ± ♆
15 01 ☽ △ ♀
21 22 ☽ ⊼ ☿
23 54 ☽ ⊻ ♇

AUGUST 2042

LONGITUDES

Date	Sidereal time h m s	Sun ☉ ° ' "	Moon ☽ ° ' "	Moon ☽ 24.00 ° ' "	Mercury ☿ ° '	Venus ♀ ° '	Mars ♂ ° '	Jupiter ♃ ° '	Saturn ♄ ° '	Uranus ♅ ° '	Neptune ♆ ° '	Pluto ♇ ° '
01	08 40 56	09 ♌ 23 29	06 ♒ 19 47	13 ♒ 26 25	28 ♌ 46	24 ♍ 58	04 ♎ 53	22 ♏ 28	00 ♏ 36	12 ♌ 50	10 ♉ 20	29 ♒ 13
02	08 44 53	10 20 52	20 ♒ 36 33	27 ♒ 49 24	00 ♍ 24	25 58	05 30	22 30	00 39	12 53	10 20	29 R 12
03	08 48 49	11 18 16	05 ♓ 04 10	12 ♓ 20 03	02 01	26 58	06 06	22 33	00 42	12 57	10 20	29 11
04	08 52 46	12 15 41	19 ♓ 36 16	26 ♓ 52 07	03 36	27 58	06 43	22 35	00 45	13 01	10 21	29 10
05	08 56 42	13 13 07	04 ♈ 06 58	11 ♈ 20 17	05 09	28 58	07 20	22 38	00 48	13 04	10 21	29 08
06	09 00 39	14 10 34	18 ♈ 31 38	25 ♈ 40 40	06 41	29 ♍ 57	07 57	22 41	00 51	13 08	10 21	29 07
07	09 04 36	15 08 02	02 ♉ 47 09	09 ♉ 50 56	08 10	00 ♎ 55	08 34	22 45	00 54	13 12	10 21	29 06
08	09 08 32	16 05 32	16 ♉ 51 53	23 ♉ 49 58	09 38	01 54	09 11	22 48	00 58	13 16	10 21	29 05
09	09 12 29	17 03 03	00 ♊ 45 10	07 ♊ 37 28	11 05	02 52	09 48	22 52	01 01	13 19	10 22	29 03
10	09 16 25	18 00 36	14 ♊ 26 53	21 ♊ 13 24	12 29	03 49	10 25	22 55	01 05	13 23	10 22	29 02
11	09 20 22	18 58 10	27 ♊ 57 01	04 ♋ 37 40	13 52	04 47	11 03	22 59	01 08	13 27	10 22	29 01
12	09 24 18	19 55 45	11 ♋ 15 19	17 ♋ 49 55	15 13	05 43	11 40	23 04	01 12	13 30	10 22	29 00
13	09 28 15	20 53 22	24 ♋ 21 23	00 ♌ 49 38	16 33	06 39	12 18	23 08	01 16	13 34	10 R 22	28 58
14	09 32 11	21 51 00	07 ♌ 14 37	13 ♌ 36 16	17 50	07 35	12 55	23 12	01 20	13 38	10 22	28 57
15	09 36 08	22 48 39	19 ♌ 54 35	26 ♌ 09 33	19 05	08 31	13 33	23 17	01 24	13 41	10 22	28 56
16	09 40 05	23 46 20	02 ♍ 21 15	08 ♍ 29 45	20 19	09 25	14 11	23 22	01 28	13 45	10 22	28 54
17	09 44 01	24 44 01	14 ♍ 35 14	20 ♍ 37 53	21 30	10 20	14 49	23 27	01 32	13 49	10 22	28 53
18	09 47 58	25 41 44	26 ♍ 37 59	02 ♎ 35 52	22 39	11 14	15 27	23 32	01 36	13 52	10 21	28 52
19	09 51 54	26 39 28	08 ♎ 31 52	14 ♎ 26 28	23 46	12 07	16 05	23 37	01 40	13 56	10 21	28 51
20	09 55 51	27 37 13	20 ♎ 20 07	26 ♎ 13 21	24 50	12 59	16 43	23 42	01 45	14 00	10 21	28 49
21	09 59 47	28 35 00	02 ♏ 06 44	08 ♏ 00 52	25 53	13 51	17 21	23 48	01 49	14 03	10 21	28 48
22	10 03 44	29 ♌ 32 47	13 ♏ 56 23	19 ♏ 53 55	26 52	14 43	18 00	23 54	01 53	14 07	10 20	28 47
23	10 07 40	00 ♍ 30 36	25 ♏ 54 08	01 ♐ 57 41	27 49	15 34	18 38	23 59	01 58	14 11	10 20	28 45
24	10 11 37	01 28 26	08 ♐ 05 13	14 ♐ 17 21	28 43	16 24	19 16	24 05	02 03	14 14	10 20	28 44
25	10 15 34	02 26 17	20 ♐ 34 40	26 ♐ 57 41	29 ♍ 34	17 13	19 55	24 12	02 07	14 18	10 19	28 43
26	10 19 30	03 24 09	03 ♑ 26 49	10 ♑ 02 25	00 ♎ 22	18 02	20 33	24 18	02 12	14 21	10 19	28 41
27	10 23 27	04 22 03	16 ♑ 44 43	23 ♑ 33 46	01 06	18 50	21 12	24 24	02 17	14 25	10 19	28 40
28	10 27 23	05 19 58	00 ♒ 29 32	07 ♒ 31 44	01 47	19 37	21 51	24 31	02 22	14 28	10 18	28 39
29	10 31 20	06 17 54	14 ♒ 39 58	21 ♒ 53 40	02 24	20 23	22 30	24 38	02 27	14 32	10 18	28 38
30	10 35 16	07 15 52	29 ♒ 12 05	06 ♓ 34 21	02 57	21 09	23 09	24 44	02 32	14 35	10 17	28 36
31	10 39 13	08 ♍ 13 51	13 ♓ 59 31	21 ♓ 26 33	03 ♎ 25	21 ♎ 53	23 ♎ 48	24 ♏ 51	02 ♏ 37	14 ♌ 39	10 ♉ 16	28 ♒ 35

MOON NODES AND LATITUDE; DECLINATIONS

Date	Moon True ☊ ° '	Moon Mean ☊ ° '	Moon Latitude ° '	Sun ☉ ° '	Moon ☽ ° '	Mercury ☿ ° '	Venus ♀ ° '	Mars ♂ ° '	Jupiter ♃ ° '	Saturn ♄ ° '	Uranus ♅ ° '	Neptune ♆ ° '	Pluto ♇ ° '
01	20 ♈ 36	21 ♈ 27	04 S 51	17 N 54	23 S 23	12 N 55	01 N 47	01 S 37	17 S 34	09 S 26	17 N 34	13 N 14	22 S 00
02	20 R 26	21 24	04 22	17 39	18 45	12 15	01 18	01 53	17 34	09 27	17 33	13 15	22 00
03	20 18	21 21	03 35	17 23	12 59	11 34	00 49	02 08	17 35	09 28	17 32	13 15	22 01
04	20 12	21 18	02 35	17 07	06 S 29	10 53	00 N 20	02 23	17 36	09 30	17 31	13 15	22 01
05	20 09	21 15	01 24	16 51	00 N 21	10 12	00 S 09	02 38	17 37	09 31	17 29	13 15	22 02
06	20 08	21 11	00 S 09	16 34	07 08	09 31	00 38	02 54	17 38	09 32	17 28	13 15	22 02
07	20 D 08	21 08	01 N 07	16 18	13 29	08 50	01 07	03 09	17 39	09 34	17 27	13 15	22 03
08	20 R 09	21 05	02 18	16 01	19 04	08 09	01 36	03 24	17 40	09 35	17 26	13 15	22 04
09	20 08	21 02	03 19	15 43	23 33	07 28	02 05	03 40	17 42	09 36	17 25	13 15	22 04
10	20 05	20 59	04 08	15 26	26 38	06 48	02 33	03 55	17 43	09 38	17 24	13 15	22 04
11	20 00	20 56	04 42	15 08	28 08	06 08	03 02	04 11	17 44	09 39	17 23	13 15	22 05
12	19 52	20 52	05 01	14 50	27 57	05 28	03 31	04 26	17 45	09 41	17 22	13 15	22 06
13	19 43	20 49	05 03	14 32	26 13	04 49	03 59	04 41	17 47	09 42	17 21	13 15	22 06
14	19 33	20 46	04 50	14 13	23 08	04 10	04 27	04 57	17 48	09 44	17 21	13 14	22 07
15	19 23	20 43	04 23	13 55	19 00	03 31	04 55	05 12	17 49	09 45	17 20	13 14	22 07
16	19 14	20 40	03 43	13 36	14 06	02 54	05 23	05 27	17 51	09 47	17 19	13 14	22 08
17	19 07	20 36	02 54	13 17	08 44	02 16	05 51	05 43	17 52	09 49	17 17	13 14	22 08
18	19 02	20 33	01 57	12 57	03 N 08	01 40	06 19	05 58	17 54	09 50	17 16	13 14	22 09
19	18 59	20 30	00 N 56	12 38	02 S 31	01 04	06 47	06 13	17 55	09 52	17 15	13 14	22 10
20	18 D 59	20 27	00 S 07	12 18	08 03	00 N 30	07 14	06 29	17 57	09 54	17 14	13 14	22 10
21	18 59	20 24	01 10	11 58	13 18	00 S 04	07 41	06 44	17 58	09 55	17 13	13 14	22 10
22	19 00	20 21	02 11	11 38	18 06	00 37	08 08	06 59	18 00	09 57	17 12	13 13	22 11
23	19 01	20 17	03 06	11 17	22 15	01 09	08 35	07 15	18 02	09 59	17 11	13 13	22 12
24	19 R 01	20 14	03 54	10 57	25 30	01 40	09 01	07 30	18 03	10 01	17 10	13 13	22 12
25	19 00	20 11	04 32	10 36	27 38	02 09	09 27	07 45	18 05	10 03	17 09	13 13	22 13
26	18 56	20 08	04 58	10 15	28 22	02 37	09 53	08 01	18 07	10 04	17 08	13 13	22 13
27	18 51	20 05	05 09	09 54	27 30	03 04	10 19	08 16	18 08	10 06	17 07	13 13	22 14
28	18 44	20 02	05 03	09 33	24 58	03 29	10 44	08 31	18 10	10 08	17 06	13 12	22 14
29	18 38	19 58	04 38	09 12	20 52	03 52	11 09	08 46	18 12	10 10	17 04	13 12	22 15
30	18 31	19 55	03 55	08 51	15 25	04 13	11 33	09 01	18 14	10 12	17 03	13 12	22 15
31	18 ♈ 26	19 ♈ 52	02 S 56	08 N 29	09 S 00	04 S 32	11 S 58	09 S 16	18 S 16	10 S 14	17 N 02	13 N 12	22 S 16

ZODIAC SIGN ENTRIES

Date	h	m	Planets
01	01	13	☽ ♒
02	05	59	☿ ♍
03	03	37	☽ ♓
05	05	11	☽ ♈
06	13	20	♀ ♎
07	07	17	☽ ♉
09	10	41	☽ ♊
11	15	40	☽ ♋
13	22	28	☽ ♌
16	07	26	☽ ♍
18	18	46	☽ ♎
21	07	42	☽ ♏
22	23	18	☉ ♍
23	20	08	☽ ♐
26	00	54	☿ ♎
26	05	39	☽ ♑
28	11	09	☽ ♒
30	13	18	☽ ♓

LATITUDES

Date	Mercury ☿ ° '	Venus ♀ ° '	Mars ♂ ° '	Jupiter ♃ ° '	Saturn ♄ ° '	Uranus ♅ ° '	Neptune ♆ ° '	Pluto ♇ ° '
01	01 N 05	00 S 14	00 N 21	00 N 51	02 N 24	00 N 38	01 S 46	10 S 59
04	00 45	00 31	00 18	00 50	02 23	00 38	01 46	10 59
07	00 N 21	00 49	00 16	00 50	02 23	00 38	01 46	10 59
10	00 S 05	01 08	00 14	00 49	02 22	00 38	01 46	11 00
13	00 33	01 28	00 11	00 48	02 21	00 38	01 46	11 00
16	01 02	01 48	00 09	00 47	02 20	00 38	01 47	11 00
19	01 32	02 10	00 07	00 47	02 20	00 38	01 47	11 00
22	02 02	02 32	00 04	00 46	02 19	00 38	01 47	11 01
25	02 32	02 55	00 N 02	00 45	02 18	00 38	01 47	11 01
28	03 01	03 18	00 00	00 45	02 18	00 38	01 47	11 01
31	03 S 28	03 S 42	00 S 02	00 N 44	02 N 17	00 N 38	01 S 47	11 S 01

DATA

Julian Date	2467098
Delta T	+78 seconds
Ayanamsa	24° 27' 02"
Synetic vernal point	04° ♓ 39' 56"
True obliquity of ecliptic	23° 26' 10"

MOON'S PHASES, APSIDES AND POSITIONS ☽

Date	h	m	Phase	Longitude ° '	Eclipse Indicator
01	17	33	○	09 ♒ 37	
08	10	35	☾	16 ♉ 02	
15	18	01	●	23 ♌ 03	
23	21	55	☽	00 ♐ 55	
31	02	02	○	07 ♓ 50	

Day	h	m		
04	18	17	Perigee	
20	14	03	Apogee	
05	10	47	0N	
11	21	30	Max dec	28° N 15'
19	01	16	0S	
26	11	21	Max dec	28° S 22'

LONGITUDES

Date	Chiron ⚷ ° '	Ceres ⚳ ° '	Pallas ⚴ ° '	Juno ⚵ ° '	Vesta ⚶ ° '	Black Moon Lilith ⚸ ° '
01	08 ♌ 00	29 ♏ 44	17 ♎ 51	25 ♎ 13	09 ♌ 41	16 ♎ 02
11	09 ♌ 12	00 ♐ 58	21 ♎ 13	27 ♎ 26	14 ♌ 14	17 ♎ 08
21	10 ♌ 24	02 ♐ 40	24 ♎ 49	29 ♎ 53	18 ♌ 47	18 ♎ 15
31	11 ♌ 33	04 ♐ 46	28 ♎ 34	02 ♏ 32	23 ♌ 22	19 ♎ 21

All ephemeris data is given at 12.00 UT and the Moon's longitude is additionally given for 24.00 UT

ASPECTARIAN

(h m — Aspects)

01 Friday
00 37 ☿ ∥ ♆
02 12 ☽ □ ♄
02 55 ♀ ⊥ ♄
08 49 ☽ Q ♃
09 26 ☽ △ ♂
12 39 ♃ # ♅
14 34 ☿ ⊥ ♂
17 32 ♀ # ♂
17 33 ☽ ☍ ☉
18 30 ☿ ☍ ♇
18 37 ☽ ⚼ ♀
18 46 ☽ □ ♆
19 56 ☽ ∥ ♇
20 44 ♀ ⚼ ♆
23 01 ☽ ☍ ♅

02 Saturday
00 58 ♂ ± ♇
10 51 ☽ ± ♀
11 36 ☉ □ ♆
11 48 ☽ ⚼ ♂
15 10 ☽ □ ♃
15 35 ☿ ✶ ♄
17 11 ☽ # ☉
17 14 ☽ ∥ ♃
17 24 ☽ # ♅
18 09 ☉ # ♃
21 35 ☽ ⊼ ♀
22 01 ☉ ∥ ♅

03 Sunday
00 51 ☽ Q ♆
02 16 ☽ ☌ ♇
03 25 ☽ ± ♂
04 44 ☽ △ ♄
06 19 ☽ ☍ ☿
11 01 ☽ # ♆
13 47 ☽ ⊼ ♂
18 07 ☽ # ☿
20 42 ☽ ✶ ♆
23 02 ☽ ⊼ ☉

04 Monday
01 04 ☽ ⊼ ♅
01 11 ☽ ∥ ♄
05 36 ☽ ⚼ ♄
09 38 ☽ ± ☉
11 01 ☽ ± ♅
13 06 ♀ ∠ ♅
16 57 ☽ △ ♃
20 31 ☽ ± ♄
21 29 ☽ ∠ ♆

05 Tuesday
01 33 ☽ ⚼ ☉
01 56 ☽ ∥ ♂
01 57 ☽ ⚼ ♅
02 50 ☽ ☍ ♀
03 46 ☽ ⊻ ♇
06 29 ☽ ⊼ ♄
08 06 ☉ ☌ ♅
10 23 ☽ ∥ ♀
11 15 ☽ # ♀
12 23 ☽ ⊥ ♆
13 42 ☽ ⊥ ♇
13 55 ☽ ⊼ ☿
16 18 ♀ ⊼ ♇
17 34 ☽ ☍ ♂
17 52 ☽ ⚼ ♃
20 20 ☽ # ♂
22 21 ☽ ⊻ ♆

06 Wednesday
01 02 ♂ ∠ ♃
01 05 ☽ ± ☿
02 57 ☽ △ ♅
04 13 ☽ △ ☉
04 39 ☽ ∠ ♇
08 55 ☽ ± ♃
11 09 ☿ # ♄
17 54 ☽ ⚼ ☿
19 00 ☽ ⊼ ♃
19 54 ☽ ∥ ☿
20 52 ☽ # ♄

07 Thursday
05 47 ☽ ✶ ♇
08 37 ☽ ⊼ ♀
08 48 ☽ ☍ ♄
11 02 ☽ ∥ ♆
11 27 ♀ ⊻ ♄
19 33 ☽ ± ♀
22 13 ☽ △ ☿
22 16 ☽ ⊼ ♂
22 56 ☿ ⊻ ♂
23 01 ☽ ∥ ☉

08 Friday
00 52 ☽ ☌ ♆
02 07 ☽ Q ♇
04 34 ☽ ∥ ♅
05 35 ☽ # ♃
05 48 ☽ □ ♅
08 59 ☽ ± ♂
10 35 ☽ □ ☉
12 04 ☽ ⚼ ♀
22 16 ☽ ☍ ♃
23 54 ☿ △ ♆

09 Saturday
01 12 ☽ ⚼ ♂
03 16 ☽ # ♇
08 09 ☿ Q ♃
09 03 ☽ □ ♇
12 28 ☽ ⊼ ♄
13 00 ☽ Q ♅
15 57 ☽ △ ♀
20 04 ☽ Q ☉
22 59 ☽ ± ♄

10 Sunday
04 34 ☽ △ ♂
04 48 ☽ ⊻ ♆
08 09 ☽ □ ☿
09 36 ♂ ⊼ ♆
10 07 ☽ ✶ ♅
14 53 ☽ ⚼ ♄
15 23 ☽ ⊥ ♆
18 47 ☽ ✶ ☉

11 Monday
01 35 ♀ ± ♆
03 06 ☽ ⊼ ♃
04 10 ☿ ⊻ ♅
07 22 ☽ ∠ ♆
12 53 ☽ ∠ ♅
13 52 ☽ ± ♃
13 54 ☽ △ ♇
16 29 ☉ Q ♄
17 45 ☽ △ ♄
17 56 ♀ ± ♇
19 50 ☽ Q ☿
23 39 ☽ ∠ ☉

12 Tuesday
01 12 ☽ □ ♀
05 10 ☽ ⊥ ♅
06 10 ☽ ⚼ ♃
10 23 ☽ ✶ ♆
12 48 ☽ □ ♂
16 07 ☽ ⊻ ♅
16 59 ☽ ⚼ ♇
17 15 ☽ ⊥ ☉
20 02 ☽ ✶ ☿

13 Wednesday
05 06 ☽ ⊻ ☉
06 36 ☿ ∠ ♄
08 20 ☽ Q ♆
09 09 ♆ St R
09 27 ☽ ± ♇
09 44 ☽ △ ♃
12 36 ☽ Q ♀
15 15 ☿ # ♂
20 32 ☽ ⊼ ♇
23 35 ☽ Q ♂

14 Thursday
00 52 ☽ □ ♄
02 49 ☽ ∠ ☿
05 43 ☿ # ♀
12 42 ☽ ✶ ♀
17 52 ☽ □ ♆
18 25 ☽ # ♇
21 36 ☽ ⊥ ☿
23 16 ☽ ✶ ♂

15 Friday
00 06 ☽ ☌ ♅
05 28 ♀ ∠ ♃
10 15 ☽ ⊻ ☿
11 00 ☽ Q ♄
17 48 ♂ ✶ ♅
18 01 ☽ ☌ ☉
18 01 ☽ # ♃
18 30 ☽ □ ♃
19 27 ☽ ∠ ♀
20 37 ☽ ∥ ♅

16 Saturday
00 23 ☿ ⊥ ♅
00 48 ☉ □ ♃
01 50 ♂ ⚼ ♇
05 19 ☽ ☍ ♇
05 31 ☽ ∠ ♂
10 15 ☽ ✶ ♄
14 15 ☽ ⊥ ♀
14 31 ☽ ∥ ☉
16 00 ☽ ∥ ♆
19 30 ♀ ∥ ♂

17 Sunday
00 00 ☽ ⊥ ♂
02 56 ☽ ⊻ ♀
03 40 ☽ △ ♆
05 45 ☽ Q ♃
07 21 ☽ # ♄
10 28 ☽ ⊻ ♅
12 28 ☽ ⊻ ♂
12 48 ♀ ⊼ ♆
14 59 ☉ ∥ ♆
15 52 ☽ ∠ ♄
22 25 ☽ ⊥ ♅
23 29 ☽ # ♀

18 Monday
00 28 ☽ # ♂
03 11 ☽ ☌ ☿
05 44 ☽ ✶ ♃
09 27 ☽ ⚼ ♆
09 55 ☽ ⊥ ♄
09 57 ☽ ⊻ ☉
16 28 ☽ ⊼ ♇
16 31 ☽ ∠ ♅
18 56 ☽ ∥ ☿
22 03 ☽ ⊻ ♄
23 05 ☽ ⊥ ☉

19 Tuesday
03 33 ☽ ± ♆
04 33 ☽ ± ♇
06 24 ☽ # ☿
08 26 ☿ ✶ ♃
12 10 ☽ ∠ ♃
15 42 ☽ ⊼ ♆
18 54 ☽ ∠ ☉
19 51 ☽ ☌ ♀
22 46 ☽ ⚼ ♇
23 02 ☽ ✶ ♅

20 Wednesday
04 13 ☽ ☌ ♂
04 43 ☽ ∥ ♂
06 36 ☽ ⊥ ♃
08 01 ☽ ∥ ♀
18 55 ☽ ⊻ ♃
20 16 ☽ ∥ ♄
22 05 ☽ ⊻ ☿
23 36 ☽ Q ♅
23 38 ☽ ⚼ ♆

21 Thursday
04 10 ☽ ✶ ☉
05 16 ☽ △ ♇
06 05 ☽ # ☉
10 24 ♀ ⚼ ♇
10 28 ☿ ⊥ ♄
11 24 ☽ ☌ ♄
11 28 ☽ ⊥ ☿
11 38 ☽ # ♆
17 15 ☉ ☍ ♇
17 57 ♀ ✶ ♅

22 Friday
04 43 ☽ ☍ ♆
06 44 ☽ Q ☉
07 16 ☽ # ♅
07 26 ☽ ∠ ☿
07 34 ♂ ⊥ ♃
11 27 ☽ ∥ ♃
12 22 ☽ □ ♅
13 41 ☽ ⊻ ♀
20 38 ☽ ⊻ ♂

23 Saturday
02 41 ☽ ⊥ ♀
08 09 ☽ ☌ ♃
09 20 ☽ ⊥ ♂
11 42 ☽ ∥ ♇
16 07 ☽ ✶ ☿
17 39 ☽ □ ♇
21 41 ☉ ∠ ♀
21 55 ☽ ∠ ♀
21 55 ☽ □ ☉

24 Sunday
00 05 ☽ ⊻ ♄
04 08 ☽ ∠ ♂
11 55 ☽ ⊥ ♄
12 28 ☿ ⊼ ♇
16 21 ☽ ⊼ ♆
17 29 ☽ Q ☿
23 57 ☽ △ ♅

25 Monday
03 28 ☉ ✶ ♄
03 39 ☿ ∠ ♅
03 54 ☽ ± ♆
04 39 ☽ Q ♇
05 10 ☽ ✶ ♀
05 23 ☽ ∠ ♄
10 41 ☽ ✶ ♂
18 52 ☽ ⊻ ♃
20 56 ☽ ⚼ ♆

26 Tuesday
03 14 ☽ ✶ ♇
04 25 ☽ ⚼ ♅
05 17 ☽ Q ♀
05 57 ☽ □ ☿
06 09 ☽ ⊥ ♃
09 42 ☽ ✶ ♄
10 17 ☽ Q ♂
11 55 ☽ △ ☉
21 00 ☽ ± ♅
21 13 ♀ ⊥ ♃
22 45 ☽ ∠ ♃
23 32 ♀ ∥ ♄
23 35 ☉ # ♀
23 39 ☉ # ♄

27 Wednesday
00 30 ☽ △ ♆
06 32 ☽ ∠ ♇
07 35 ☽ Q ♄
07 50 ☽ ⊼ ♅
15 55 ☽ □ ♀
17 00 ☽ ⚼ ☉
20 15 ☽ □ ♂
22 25 ☽ ⊥ ♇

28 Thursday
01 35 ☽ ✶ ♃
08 50 ☽ ⊻ ♇
09 51 ☽ ± ☉
14 19 ☽ △ ☿
15 14 ☽ □ ♄
20 53 ☽ ⊼ ☉
22 22 ☽ Q ♃

29 Friday
04 40 ☽ □ ♆
04 50 ☽ ∥ ♇
11 47 ☽ ☍ ♅
14 28 ☿ ⊻ ♄
16 44 ☽ ⚼ ☿
22 02 ☽ △ ♀

30 Saturday
00 19 ☽ ∥ ♃
01 36 ☽ △ ♂
04 38 ☽ □ ♃
05 01 ☉ # ♂
05 16 ☽ # ♅
08 10 ☽ ± ☿
10 30 ☽ Q ♆
11 01 ☽ ☌ ♇
17 28 ☽ △ ♄
18 19 ☽ ⊼ ☿
20 39 ☽ # ♆
23 54 ☽ ⚼ ♀

31 Sunday
01 55 ☽ ∥ ♀
02 02 ☽ ☍ ☉
03 13 ☽ ⚼ ♂
06 00 ☽ ✶ ♆
07 38 ☽ ∥ ♄
11 05 ☽ ∥ ♂
13 04 ☽ ⊼ ♅
13 56 ☽ # ☉
15 13 ☽ ± ♀
17 53 ☽ ⚼ ♄
18 24 ☽ ± ♂
22 46 ☽ ± ♅

SEPTEMBER 2042

LONGITUDES

Date	Sidereal time h m s	Sun ☉ ° ' "	Moon ☽ ° ' "	Moon ☽ 24.00 ° ' "	Mercury ☿ ° '	Venus ♀ ° '	Mars ♂ ° '	Jupiter ♃ ° '	Saturn ♄ ° '	Uranus ♅ ° '	Neptune ♆ ° '	Pluto ♇ ° '
01	10 43 09	09 ♍ 11 51	28 ♓ 54 23	06 ♈ 22 01	03 ♎ 49	22 ♎ 37	24 ♎ 27	24 ♏ 59	02 ♏ 42	14 ♌ 42	10 ♉ 16	28 ♒ 34
02	10 47 06	10 09 54	13 ♈ 48 28	21 ♈ 12 52	04 08	23 20	25 06	25 06	02 47	14 46	10 R 15	28 R 32
03	10 51 03	11 07 58	28 ♈ 34 26	05 ♉ 52 33	04 22	24 01	25 45	25 13	02 53	14 49	10 15	28 31
04	10 54 59	12 06 04	13 ♉ 06 42	20 ♉ 16 31	04 31	24 42	26 24	25 21	02 58	14 53	10 14	28 30
05	10 58 56	13 04 13	27 ♉ 21 44	04 ♊ 22 15	04 R 34	25 22	27 04	25 28	03 04	14 56	10 13	28 29
06	11 02 52	14 02 23	11 ♊ 17 59	18 ♊ 08 59	04 30	26 00	27 43	25 36	03 09	14 59	10 12	28 27
07	11 06 49	15 00 36	24 ♊ 55 21	01 ♋ 37 13	04 21	26 38	28 23	25 44	03 15	15 03	10 12	28 26
08	11 10 45	15 58 50	08 ♋ 14 44	14 ♋ 48 07	04 05	27 14	29 02	25 52	03 20	15 06	10 11	28 25
09	11 14 42	16 57 07	21 ♋ 17 33	27 ♋ 43 14	03 43	27 49	29 ♎ 42	26 00	03 26	15 09	10 10	28 24
10	11 18 38	17 55 25	04 ♌ 05 22	10 ♌ 24 07	03 14	28 23	00 ♏ 22	26 09	03 32	15 13	10 09	28 22
11	11 22 35	18 53 46	16 ♌ 39 41	22 ♌ 52 15	02 39	28 55	01 02	26 17	03 37	15 16	10 08	28 21
12	11 26 32	19 52 08	29 ♌ 01 58	05 ♍ 09 02	01 57	29 26	01 41	26 26	03 43	15 19	10 07	28 20
13	11 30 28	20 50 32	11 ♍ 13 37	17 ♍ 15 53	01 10	29 ♎ 56	02 21	26 34	03 49	15 22	10 06	28 19
14	11 34 25	21 48 58	23 ♍ 16 04	29 ♍ 14 22	00 ♎ 18	00 ♏ 24	03 02	26 43	03 55	15 26	10 05	28 18
15	11 38 21	22 47 26	05 ♎ 11 01	11 ♎ 06 18	29 ♍ 21	00 50	03 42	26 52	04 01	15 29	10 04	28 16
16	11 42 18	23 45 56	17 ♎ 00 30	22 ♎ 53 57	28 21	01 15	04 22	27 01	04 07	15 32	10 03	28 15
17	11 46 14	24 44 27	28 ♎ 47 00	04 ♏ 40 04	27 19	01 39	05 02	27 10	04 13	15 35	10 02	28 14
18	11 50 11	25 43 01	10 ♏ 33 34	16 ♏ 27 58	26 15	02 00	05 43	27 19	04 19	15 38	10 01	28 13
19	11 54 07	26 41 36	22 ♏ 23 45	28 ♏ 21 28	25 13	02 20	06 23	27 29	04 26	15 41	10 00	28 12
20	11 58 04	27 40 13	04 ♐ 21 39	10 ♐ 24 51	24 12	02 38	07 04	27 38	04 32	15 44	09 59	28 11
21	12 02 01	28 38 51	16 ♐ 31 40	22 ♐ 42 40	23 16	02 54	07 44	27 48	04 38	15 47	09 58	28 10
22	12 05 57	29 ♍ 37 32	28 ♐ 58 25	05 ♑ 19 27	22 24	03 08	08 25	27 57	04 44	15 50	09 57	28 09
23	12 09 54	00 ♎ 36 14	11 ♑ 46 17	18 ♑ 19 19	21 39	03 20	09 06	28 07	04 51	15 53	09 55	28 08
24	12 13 50	01 34 57	24 ♑ 58 56	01 ♒ 45 24	21 02	03 30	09 46	28 17	04 57	15 56	09 54	28 06
25	12 17 47	02 33 42	08 ♒ 38 49	15 ♒ 39 11	20 33	03 37	10 27	28 27	05 04	15 59	09 53	28 05
26	12 21 43	03 32 29	22 ♒ 46 19	29 ♒ 59 53	20 14	03 43	11 08	28 37	05 10	16 01	09 52	28 04
27	12 25 40	04 31 18	07 ♓ 19 21	14 ♓ 44 00	20 04	03 46	11 49	28 47	05 17	16 04	09 50	28 03
28	12 29 36	05 30 09	22 ♓ 12 58	29 ♓ 45 15	20 D 05	03 R 47	12 30	28 57	05 23	16 07	09 49	28 02
29	12 33 33	06 29 01	07 ♈ 19 44	14 ♈ 55 14	20 16	03 45	13 11	29 08	05 30	16 10	09 48	28 01
30	12 37 30	07 ♎ 27 55	22 ♈ 30 34	00 ♉ 04 32	20 ♍ 37	03 ♏ 41	13 ♏ 53	29 ♏ 18	05 ♏ 36	16 ♌ 12	09 ♉ 46	28 ♒ 01

Date	Moon True ☊ ° '	Moon Mean ☊ ° '	Moon ☽ Latitude ° '
01	18 ♈ 23	19 ♈ 49	01 S 44
02	18 R 22	19 46	00 S 25
03	18 D 22	19 42	00 N 55
04	18 23	19 39	02 11
05	18 25	19 36	03 17
06	18 26	19 33	04 10
07	18 R 25	19 30	04 47
08	18 23	19 27	05 08
09	18 20	19 23	05 12
10	18 16	19 20	05 01
11	18 11	19 17	04 35
12	18 06	19 14	03 57
13	18 02	19 11	03 08
14	17 58	19 08	02 12
15	17 56	19 04	01 10
16	17 56	19 01	00 N 05
17	17 D 56	18 58	01 S 00
18	17 57	18 55	02 02
19	17 59	18 52	02 59
20	18 00	18 48	03 49
21	18 02	18 45	04 30
22	18 02	18 42	04 59
23	18 R 02	18 39	05 15
24	18 01	18 36	05 14
25	17 59	18 33	04 56
26	17 57	18 29	04 20
27	17 55	18 26	03 26
28	17 54	18 23	02 18
29	17 53	18 20	00 58
30	17 ♈ 53	18 ♈ 17	00 N 26

DECLINATIONS

Date	Sun ☉ ° '	Moon ☽ ° '	Mercury ☿ ° '	Venus ♀ ° '	Mars ♂ ° '	Jupiter ♃ ° '	Saturn ♄ ° '	Uranus ♅ ° '	Neptune ♆ ° '	Pluto ♇ ° '
01	08 N 07	02 S 01	04 S 49	12 S 22	09 S 31	18 S 18	10 S 16	17 N 01	13 N 12	22 S 16
02	07 45	05 N 04	05 04	12 45	09 46	18 20	10 18	17 01	13 11	22 16
03	07 23	11 50	05 16	13 08	10 01	18 22	10 20	17 00	13 11	22 17
04	07 01	17 51	05 25	13 31	10 16	18 24	10 22	16 59	13 11	22 17
05	06 39	22 46	05 32	13 53	10 31	18 26	10 24	16 58	13 11	22 18
06	06 17	26 15	05 35	14 15	10 46	18 28	10 26	16 57	13 10	22 18
07	05 54	28 07	05 35	14 37	11 01	18 30	10 28	16 56	13 10	22 19
08	05 32	28 18	05 31	14 58	11 15	18 32	10 30	16 55	13 10	22 19
09	05 09	26 53	05 23	15 18	11 30	18 34	10 32	16 54	13 09	22 19
10	04 46	24 06	05 12	15 38	11 45	18 36	10 34	16 53	13 09	22 20
11	04 24	20 13	04 56	15 58	11 59	18 38	10 36	16 52	13 09	22 20
12	04 01	15 30	04 36	16 17	12 14	18 40	10 38	16 51	13 08	22 21
13	03 38	10 15	04 13	16 35	12 28	18 43	10 40	16 50	13 08	22 21
14	03 15	04 N 41	03 46	16 53	12 43	18 45	10 42	16 49	13 08	22 21
15	02 52	01 S 00	03 15	17 10	12 57	18 47	10 45	16 48	13 07	22 22
16	02 29	06 36	02 41	17 26	13 11	18 49	10 47	16 47	13 07	22 22
17	02 05	11 58	02 05	17 42	13 25	18 52	10 49	16 46	13 07	22 22
18	01 42	16 55	01 26	17 57	13 39	18 54	10 51	16 45	13 06	22 23
19	01 19	21 15	00 47	18 11	13 53	18 56	10 53	16 45	13 06	22 23
20	00 56	24 46	00 S 07	18 25	14 07	18 58	10 56	16 44	13 05	22 23
21	00 32	27 14	00 N 33	18 37	14 21	19 01	10 58	16 43	13 05	22 24
22	00 N 09	28 25	01 11	18 49	14 35	19 03	11 00	16 42	13 05	22 24
23	00 S 14	28 08	01 47	19 00	14 49	19 05	11 02	16 41	13 04	22 24
24	00 38	26 17	02 20	19 10	15 02	19 08	11 05	16 40	13 04	22 25
25	01 01	22 51	02 50	19 19	15 16	19 10	11 07	16 40	13 03	22 25
26	01 24	18 01	03 16	19 26	15 29	19 13	11 09	16 39	13 03	22 25
27	01 48	12 01	03 37	19 33	15 42	19 15	11 11	16 38	13 02	22 25
28	02 11	05 S 12	03 53	19 39	15 56	19 17	11 14	16 37	13 02	22 26
29	02 34	02 N 01	04 05	19 43	16 09	19 20	11 16	16 36	13 02	22 26
30	02 S 58	09 N 09	04 N 11	19 S 46	16 S 22	19 S 22	11 S 18	16 N 36	13 N 01	22 S 26

ZODIAC SIGN ENTRIES

Date	h m	Planets
01	13 45	☽ ♈
03	14 20	☽ ♉
05	16 30	☽ ♊
07	21 05	☽ ♋
09	22 54	♂ ♏
10	04 17	☽ ♌
12	13 53	☽ ♍
13	15 31	♀ ♏
14	19 45	☿ ♍
15	01 32	☽ ♎
17	14 29	☽ ♏
20	03 18	☽ ♐
22	13 57	☽ ♑
22	21 11	☉ ♎
24	20 54	☽ ♒
27	00 00	☽ ♓
29	00 23	☽ ♈
30	23 53	☽ ♉

LATITUDES

Date	Mercury ☿ ° '	Venus ♀ ° '	Mars ♂ ° '	Jupiter ♃ ° '	Saturn ♄ ° '	Uranus ♅ ° '	Neptune ♆ ° '	Pluto ♇ ° '
01	03 S 36	03 S 50	00 S 03	00 N 44	02 N 17	00 N 38	01 S 48	11 S 01
04	03 57	04 15	00 05	00 43	02 16	00 38	01 48	11 01
07	04 12	04 40	00 07	00 43	02 16	00 38	01 48	11 01
10	04 15	05 05	00 09	00 42	02 15	00 38	01 48	11 01
13	04 05	05 30	00 11	00 42	02 15	00 38	01 48	11 00
16	03 39	05 54	00 14	00 41	02 14	00 38	01 48	11 00
19	02 55	06 18	00 16	00 41	02 14	00 38	01 48	11 00
22	01 59	06 40	00 17	00 40	02 14	00 38	01 49	11 00
25	00 59	07 01	00 19	00 40	02 13	00 39	01 49	10 59
28	00 S 03	07 19	00 21	00 39	02 13	00 39	01 49	10 59
31	00 N 45	07 S 34	00 S 23	00 N 39	02 N 13	00 N 39	01 S 49	10 S 59

DATA

Julian Date	2467129
Delta T	+78 seconds
Ayanamsa	24° 27' 06"
Synetic vernal point	04° ♓ 39' 52"
True obliquity of ecliptic	23° 26' 11"

MOON'S PHASES, APSIDES AND POSITIONS ☽

Date	h m	Phase	Longitude ° '	Eclipse Indicator
06	17 09	☾	14 ♊ 15	
14	08 50	●	21 ♍ 41	
22	13 21	☽	29 ♐ 41	
29	10 34	○	06 ♈ 26	partial

Day	h m		
01	15 52	Perigee	
17	04 29	Apogee	
29	23 52	Perigee	
01	18 50	0N	
08	02 35	Max dec	28° N 25'
15	07 48	0S	
22	19 38	Max dec	28° S 30'
29	05 20	0N	

LONGITUDES

Date	Chiron ⚷ ° '	Ceres ⚳ ° '	Pallas ⚴ ° '	Juno ⚵ ° '	Vesta ⚶ ° '	Black Moon Lilith ⚸ ° '
01	11 ♌ 40	05 ♐ 00	28 ♎ 57	02 ♏ 49	23 ♌ 49	19 ♎ 28
11	12 ♌ 46	07 ♐ 29	02 ♏ 52	05 ♏ 39	28 ♌ 24	20 ♎ 34
21	13 ♌ 48	10 ♐ 15	06 ♏ 54	08 ♏ 39	02 ♍ 58	21 ♎ 41
31	14 ♌ 45	13 ♐ 17	11 ♏ 01	11 ♏ 45	07 ♍ 31	22 ♎ 47

All ephemeris data is given at 12.00 UT and the Moon's longitude is additionally given for 24.00 UT

ASPECTARIAN

h m	Aspects
01 Monday	
01 22	☽ ⊼ ♀
02 51	☽ ∥ ☿
04 30	☽ ⊼ ♂
05 38	☽ △ ♃
06 09	☽ ∠ ♆
08 26	☽ ± ♄
11 27	☽ ⊻ ♇
13 18	☽ ⚼ ♅
18 09	☽ ⊼ ♄
20 05	☽ ☍ ☿
20 37	☽ ⊥ ♆
21 05	☽ ⊥ ♇
02 Tuesday	
05 43	☽ ⊼ ☉
05 58	☽ ⚼ ♃
06 16	☽ ⊻ ♆
06 51	☉ ∠ ♂
11 34	☽ ∠ ♇
11 59	♂ ⊻ ♃
12 01	☽ # ☿
13 33	☽ △ ♅
14 11	☉ △ ♆
16 05	☽ ± ☉
20 38	☽ ± ♃
20 51	☽ ∥ ☉
22 11	☿ ± ♆
03 Wednesday	
04 12	☽ ☍ ♀
05 07	☽ # ♂
06 28	☽ # ♄
06 28	☽ ⊼ ♃
07 10	☽ ☍ ♂
07 44	☽ ⚼ ☉
11 55	☽ ✱ ♇
14 49	♀ # ♆
17 06	☽ ∥ ♆
17 15	☽ # ♀
19 06	☽ ☍ ♄
21 38	☽ ⊼ ☿
04 Thursday	
07 13	☽ ☌ ♆
07 39	☽ ± ☿
07 39	☽ Q ♇
08 16	☽ ∥ ♅
08 39	☿ ± ♇
10 12	☽ △ ☉
14 24	☽ # ♃
14 58	☽ □ ♅
22 15	♂ ∥ ♄
22 46	☽ ⚼ ☿
05 Friday	
06 57	♂ Q ♅
08 26	☽ ⊼ ♀
08 46	☽ ☍ ♃
09 25	☽ # ♇
11 20	☿ St R
11 28	☽ ⊼ ♂
13 54	☽ □ ♇
17 09	♀ ⊻ ♃
19 10	☽ ± ♀
21 34	☽ Q ♅
21 49	☽ ⊼ ♄
22 14	☽ ± ♂
23 32	☉ Q ♃
06 Saturday	
00 18	☽ △ ☿
08 14	☽ ± ♄
10 06	☽ ⊻ ♆
11 28	☽ ⚼ ♀
14 36	☽ ⚼ ♂
17 09	☽ □ ☉
18 29	☽ ✱ ♅
20 35	☽ ⊥ ♆
23 30	☿ ± ♇
07 Sunday	
00 05	☽ ⚼ ♄
12 29	☽ ∠ ♆
12 59	☉ ⊻ ♅
13 28	☽ ⊼ ♃
14 00	♂ △ ♇
15 11	☽ △ ♀
18 16	☽ △ ♇
18 30	☽ △ ♂
21 13	☽ ∠ ♅
08 Monday	
00 20	☽ ± ♃
03 02	☽ △ ♄
03 39	☽ Q ☉
04 10	☿ ± ♆
04 37	☽ □ ☿
06 15	♀ Q ♅
13 31	☉ # ☿
13 34	☽ ⊥ ♅
15 32	☽ ✱ ♆
16 50	☽ ⚼ ♃
21 26	☽ ⚼ ♇
09 Tuesday	
00 36	☽ ⊻ ♅
03 19	☽ ✱ ☉
12 46	☽ Q ☿
13 37	☽ Q ♆
14 03	☽ ± ♇
20 53	☽ △ ♃
10 Wednesday	
00 25	☿ ⊻ ♄
00 45	☽ □ ♀
01 15	☽ ⊼ ♇
04 35	☽ □ ♂
09 37	☽ ∠ ☉
10 27	☽ ✱ ☿
10 56	☽ □ ♄
11 46	♀ △ ♇
23 31	☽ □ ♆
23 37	☽ # ♇
11 Thursday	
04 10	☽ ⊥ ☉
04 32	☉ ∠ ♄
09 18	☽ ☌ ♅
12 31	☽ Q ♀
13 48	☽ ∠ ☿
16 40	☽ ⊻ ☉
16 49	☽ Q ♂
20 21	☽ # ♃
21 39	☽ Q ♄
12 Friday	
05 27	☽ ∥ ♅
06 20	☽ ⊥ ☿
06 51	☽ □ ♃
08 29	☽ # ♀
10 38	☽ ☍ ♇
12 49	☽ ✱ ♀
16 32	☿ ⊻ ♂
17 24	☽ ⊻ ☿
17 30	☽ ✱ ♂
21 16	☽ ✱ ♄
23 04	☽ ∥ ♆
13 Saturday	
02 32	☽ # ♂
09 47	☽ △ ♆
10 10	☽ # ♄
18 43	☽ Q ♃
19 39	☽ ∠ ♀
20 16	☽ ⊻ ♅
14 Sunday	
00 54	☽ ∠ ♂
01 52	☉ ⊥ ♅
03 14	☽ ∠ ♄
07 12	♀ # ♅
08 18	☽ ⊥ ♅
08 47	☿ ∠ ♅
08 50	☽ ☌ ☉
10 15	☿ ⊻ ♀
14 21	☽ ⊥ ♀
15 39	☽ ⚼ ♆
16 17	☽ # ☿
18 32	☽ ∥ ☉
19 01	☽ ✱ ♃
20 00	☽ ⊥ ♂
21 25	☽ ⊥ ♄
22 05	☽ ⊼ ♇
15 Monday	
01 07	☽ ☌ ☿
02 27	☽ ∠ ♅
02 54	☽ ⊻ ♀
08 49	☽ ⊻ ♂
09 37	☽ ⊻ ♄
09 45	☽ ± ♆
10 10	☽ ± ♇
19 25	☽ # ☉
20 43	☽ ∥ ☿
21 54	☽ ⊼ ♆
16 Tuesday	
01 37	♂ ☌ ♄
01 43	☽ ∠ ♃
04 23	☽ ⚼ ♇
05 08	♂ # ♆
08 59	☽ ✱ ♅
11 47	☿ ⊥ ♂
14 15	☿ ⊼ ♇
16 54	☿ ⊥ ♄
20 16	☽ ⊥ ♃
17 Wednesday	
03 01	☽ ⊻ ☉
06 41	☽ ∥ ♄
08 40	☽ ⊻ ♃
09 14	☽ ⊻ ☿
09 32	☽ Q ♅
10 49	☉ # ☿
10 53	☽ △ ♇
14 49	☿ ✱ ♃
16 21	☽ ⊥ ☉
17 20	☽ # ♆
18 02	☽ ☌ ♀
19 08	☽ ∥ ♂
19 12	☉ ⚼ ♆
20 28	☽ ⊥ ☿
23 11	☽ ☌ ♄
18 Thursday	
01 31	☽ ☌ ♂
10 54	☽ ☍ ♆
11 12	☽ # ♅
12 21	☽ ∠ ☉
13 18	☽ ∠ ☿
16 20	☿ ⊥ ♀
17 42	☽ ∥ ♀
18 23	☉ ☌ ☿
22 21	☽ □ ♅
22 35	☽ ∥ ♃
22 49	☉ ⊥ ♀
19 Friday	
17 00	☿ ⚼ ♆
17 14	☽ ✱ ☿
19 06	☽ ∥ ♇
21 26	☽ ✱ ☉
22 22	☽ ☌ ♃
23 40	☽ □ ♇
20 Saturday	
08 28	☽ ⊻ ♀
10 59	☉ ✱ ♃
12 20	☽ ⊻ ♄
15 24	☽ Q ☿
17 41	☽ ⊻ ♂
20 41	☽ ⊥ ♀
23 08	☽ ⊼ ♆
23 27	☽ Q ☉
21 Sunday	
00 17	☉ ⊼ ♇
00 20	☽ ⊥ ♄
06 13	☽ ⊥ ♂
10 32	☽ △ ♅
10 54	☽ ± ♆
11 17	☽ Q ♇
11 37	☉ ⊥ ♄
11 45	☉ ∥ ☿
14 44	☽ ∠ ♀
18 06	☽ ∠ ♄
20 01	☿ ∠ ♂
22 Monday	
00 11	☽ □ ☿
00 44	☽ ∠ ♂
04 19	☽ ⚼ ♆
10 02	☽ ⊻ ♃
10 25	☽ ✱ ♇
13 21	☽ □ ☉
15 33	☽ ⚼ ♅
20 01	☽ ✱ ♀
21 33	☽ ⊥ ♃
23 00	☽ ✱ ♄
23 Tuesday	
04 36	☿ ⊥ ♅
06 46	☽ ✱ ♂
08 29	☽ ± ♅
08 35	☽ △ ♆
12 56	♃ □ ♇
14 30	☽ ∠ ♇
14 31	☽ ∠ ♃
18 38	☽ Q ♀
19 10	☉ ∠ ♅
19 35	☽ ⊼ ♅
21 23	☽ Q ♄
24 Wednesday	
05 11	☽ △ ☿
05 44	♀ ∥ ♃
05 56	☽ Q ♂
06 51	☽ ⊥ ♇
16 32	♂ ☍ ♆
17 33	☽ ⊻ ♇
17 57	☽ ✱ ♃
25 Thursday	
00 36	☽ △ ☉
03 12	☽ □ ♀
05 44	☽ □ ♄
06 47	☽ ⚼ ☿
14 08	☽ □ ♆
14 31	☽ ∥ ♇
15 09	☽ Q ♃
15 16	☽ □ ♂
21 52	☽ ± ☿
26 Friday	
00 36	☽ ☍ ♅
04 22	☽ ⚼ ☉
05 43	☽ ∥ ♀
06 41	☽ ∥ ♃
07 48	☽ ⊼ ☿
16 10	☿ ∠ ♄
16 31	☉ ⊻ ♀
17 53	☽ # ♅
19 41	☉ ± ♆
20 28	☽ Q ♆
20 31	☽ ± ☉
20 49	☽ ☌ ♇
21 50	☽ □ ♃
22 14	☽ ∥ ♂
27 Saturday	
00 49	☉ ± ♇
06 11	☽ △ ♀
07 06	☽ ⊼ ☉
08 09	☽ # ♆
08 38	☽ △ ♄
15 01	☽ ∥ ♄
16 05	☽ ✱ ♆
19 39	☽ △ ♂
22 18	☿ St D
28 Sunday	
02 12	☽ ⊼ ♅
06 30	☽ ⚼ ♀
08 34	☽ ☍ ☿
08 41	♀ St R
08 48	☉ ⊻ ♄
09 03	☽ ⚼ ♄
11 50	☽ ± ♅
16 09	☽ ∠ ♆
16 16	☽ # ☿
20 50	☽ ⚼ ♂
20 51	☽ ± ♀
21 16	☽ ⊻ ♇
21 34	☽ ∥ ☉
22 52	☽ △ ♃
23 30	☽ ± ♄
29 Monday	
02 12	☽ ⚼ ♅
06 21	☽ ⊼ ♀
06 25	☽ ⊥ ♆
06 47	☽ ⊥ ♇
09 05	☽ ⊼ ♄
10 34	☽ ☍ ☉
11 46	☽ ± ♂
13 57	☽ # ☉
15 54	☽ ⊻ ♆
18 58	☽ ∥ ☿
21 00	☽ ∠ ♇
21 42	☽ ⊼ ♂
22 52	☽ ⚼ ♃
30 Tuesday	
02 00	☽ △ ♅
08 55	☽ ⊼ ☿
11 40	☿ ∠ ♄
13 16	☽ ± ♃
18 42	☽ ± ☿
19 35	☽ # ♄
20 42	☽ ✱ ♇
22 54	☽ ⊼ ♃

OCTOBER 2042

LONGITUDES

Date	Sidereal time h m s	Sun ☉ ° ' "	Moon ☽ ° ' "	Moon ☽ 24.00 ° ' "	Mercury ☿ ° '	Venus ♀ ° '	Mars ♂ ° '	Jupiter ♃ ° '	Saturn ♄ ° '	Uranus ♅ ° '	Neptune ♆ ° '	Pluto ♇ ° '
01	12 41 26	08 ♎ 26 52	07 ♉ 36 03	15 ♉ 04 06	21 ♍ 07	03 ♏ 35	14 ♏ 34	29 ♏ 29	05 ♏ 43	16 ♌ 15	09 ♉ 45	28 ♒ 00
02	12 45 23	09 25 51	22 ♉ 27 50	29 ♉ 46 31	21 47	03 R 26	15 15	29 39	05 50	16 17	09 R 44	27 R 59
03	12 49 19	10 24 52	06 ♊ 59 38	14 ♊ 06 46	22 35	03 15	15 57	29 ♏ 50	05 57	16 20	09 42	27 58
04	12 53 16	11 23 56	21 ♊ 07 43	28 ♊ 02 24	23 32	03 02	16 38	00 ♐ 01	06 03	16 23	09 41	27 57
05	12 57 12	12 23 01	04 ♋ 50 51	11 ♋ 33 13	24 35	02 46	17 20	00 12	06 10	16 25	09 39	27 56
06	13 01 09	13 22 10	18 ♋ 09 45	24 ♋ 40 46	25 46	02 28	18 02	00 23	06 17	16 27	09 38	27 55
07	13 05 05	14 21 20	01 ♌ 06 36	07 ♌ 27 39	27 01	02 08	18 43	00 34	06 24	16 30	09 36	27 54
08	13 09 02	15 20 33	13 ♌ 44 20	19 ♌ 57 04	28 23	01 45	19 25	00 45	06 31	16 32	09 35	27 54
09	13 12 59	16 19 48	26 ♌ 06 16	02 ♍ 12 20	29 ♍ 48	01 20	20 07	00 56	06 38	16 34	09 33	27 53
10	13 16 55	17 19 05	08 ♍ 15 40	14 ♍ 16 38	01 ♎ 17	00 54	20 49	01 07	06 45	16 37	09 32	27 52
11	13 20 52	18 18 24	20 ♍ 15 36	26 ♍ 12 53	02 49	00 ♏ 25	21 31	01 19	06 51	16 39	09 30	27 51
12	13 24 48	19 17 46	02 ♎ 08 50	08 ♎ 03 44	04 24	29 ♎ 55	22 13	01 30	06 58	16 41	09 29	27 51
13	13 28 45	20 17 09	13 ♎ 57 52	19 ♎ 51 32	06 01	29 23	22 55	01 42	07 05	16 43	09 27	27 50
14	13 32 41	21 16 35	25 ♎ 45 00	01 ♏ 38 32	07 40	28 50	23 37	01 53	07 13	16 45	09 26	27 49
15	13 36 38	22 16 03	07 ♏ 32 25	13 ♏ 26 55	09 20	28 16	24 20	02 05	07 20	16 47	09 24	27 48
16	13 40 34	23 15 33	19 ♏ 22 22	25 ♏ 19 03	11 01	27 41	25 02	02 17	07 27	16 49	09 23	27 48
17	13 44 31	24 15 04	01 ♐ 17 19	07 ♐ 17 29	12 43	27 05	25 45	02 29	07 34	16 51	09 21	27 47
18	13 48 28	25 14 38	13 ♐ 19 57	19 ♐ 25 05	14 26	26 29	26 27	02 41	07 41	16 53	09 19	27 47
19	13 52 24	26 14 13	25 ♐ 33 19	01 ♑ 45 04	16 08	25 52	27 10	02 52	07 48	16 55	09 18	27 46
20	13 56 21	27 13 51	08 ♑ 00 46	14 ♑ 20 51	17 51	25 15	27 52	03 04	07 55	16 57	09 16	27 45
21	14 00 17	28 13 30	20 ♑ 45 46	27 ♑ 15 55	19 34	24 39	28 35	03 17	08 02	16 59	09 14	27 45
22	14 04 14	29 ♎ 13 10	03 ♒ 51 41	10 ♒ 33 24	21 17	24 03	29 ♏ 18	03 29	08 09	17 00	09 13	27 44
23	14 08 10	00 ♏ 12 53	17 ♒ 21 21	24 ♒ 15 41	22 59	23 28	00 ♐ 01	03 41	08 17	17 02	09 11	27 44
24	14 12 07	01 12 37	01 ♓ 16 28	08 ♓ 23 40	24 41	22 54	00 44	03 53	08 24	17 03	09 09	27 44
25	14 16 03	02 12 22	15 ♓ 37 02	22 ♓ 56 13	26 23	22 21	01 27	04 05	08 31	17 05	09 08	27 43
26	14 20 00	03 12 10	00 ♈ 20 38	07 ♈ 49 36	28 05	21 49	02 10	04 18	08 38	17 07	09 06	27 43
27	14 23 57	04 11 59	15 ♈ 22 12	22 ♈ 57 25	29 ♎ 46	21 19	02 53	04 30	08 45	17 08	09 04	27 42
28	14 27 53	05 11 50	00 ♉ 34 04	08 ♉ 10 58	01 ♏ 26	20 50	03 36	04 43	08 53	17 09	09 03	27 42
29	14 31 50	06 11 43	15 ♉ 46 50	23 ♉ 20 27	03 06	20 24	04 19	04 55	09 00	17 11	09 01	27 42
30	14 35 46	07 11 38	00 ♊ 50 39	08 ♊ 16 23	04 46	19 59	05 02	05 08	09 07	17 12	08 59	27 41
31	14 39 43	08 ♏ 11 35	15 ♊ 36 45	22 ♊ 51 03	06 ♏ 25	19 ♎ 37	05 ♐ 46	05 ♐ 20	09 ♏ 14	17 ♌ 13	08 ♉ 58	27 ♒ 41

Date	Moon True ☊ ° '	Moon Mean ☊ ° '	Moon ☽ Latitude ° '
01	17 ♈ 53	18 ♈ 13	01 N 47
02	17 D 54	18 10	03 01
03	17 55	18 07	04 00
04	17 55	18 04	04 44
05	17 56	18 01	05 10
06	17 R 56	17 58	05 18
07	17 56	17 54	05 09
08	17 55	17 51	04 46
09	17 55	17 48	04 10
10	17 55	17 45	03 23
11	17 D 55	17 42	02 28
12	17 55	17 39	01 27
13	17 55	17 35	00 N 22
14	17 R 55	17 32	00 S 43
15	17 55	17 29	01 47
16	17 54	17 26	02 46
17	17 54	17 23	03 38
18	17 53	17 19	04 22
19	17 52	17 16	04 54
20	17 51	17 13	05 13
21	17 50	17 10	05 17
22	17 D 50	17 07	05 05
23	17 51	17 04	04 36
24	17 51	17 00	03 51
25	17 53	16 57	02 50
26	17 54	16 54	01 36
27	17 54	16 51	00 S 14
28	17 R 54	16 48	01 N 10
29	17 52	16 45	02 28
30	17 50	16 41	03 36
31	17 ♈ 48	16 ♈ 38	04 N 27

DECLINATIONS

Date	Sun ☉ ° '	Moon ☽ ° '	Mercury ☿ ° '	Venus ♀ ° '	Mars ♂ ° '	Jupiter ♃ ° '	Saturn ♄ ° '	Uranus ♅ ° '	Neptune ♆ ° '	Pluto ♇ ° '
01	03 S 21	15 N 44	04 N 12	19 S 48	16 S 35	19 S 25	11 S 21	16 N 35	13 N 01	22 S 26
02	03 44	21 17	04 09	19 49	16 47	19 27	11 23	16 34	13 00	22 26
03	04 07	25 25	04 00	19 48	17 00	19 29	11 25	16 33	13 00	22 27
04	04 30	27 51	03 48	19 46	17 12	19 32	11 27	16 33	12 59	22 27
05	04 54	28 30	03 31	19 43	17 25	19 34	11 30	16 32	12 59	22 27
06	05 17	27 27	03 10	19 38	17 37	19 37	11 32	16 31	12 58	22 27
07	05 40	24 56	02 46	19 32	17 49	19 39	11 34	16 31	12 58	22 27
08	06 02	21 16	02 18	19 24	18 01	19 42	11 37	16 30	12 57	22 27
09	06 25	16 44	01 48	19 15	18 13	19 44	11 39	16 29	12 57	22 28
10	06 48	11 37	01 15	19 04	18 25	19 47	11 41	16 29	12 56	22 28
11	07 11	06 07	00 40	18 52	18 37	19 49	11 44	16 28	12 56	22 28
12	07 33	00 N 28	00 N 03	18 38	18 48	19 51	11 46	16 27	12 55	22 28
13	07 56	05 S 10	00 S 35	18 23	18 59	19 54	11 48	16 27	12 55	22 28
14	08 18	10 37	01 15	18 07	19 11	19 56	11 51	16 26	12 54	22 28
15	08 40	15 43	01 55	17 49	19 22	19 59	11 53	16 26	12 54	22 28
16	09 02	20 14	02 37	17 30	19 33	20 01	11 56	16 25	12 53	22 28
17	09 24	23 59	03 19	17 11	19 43	20 04	11 58	16 24	12 53	22 28
18	09 46	26 43	04 02	16 50	19 54	20 06	12 00	16 24	12 52	22 28
19	10 08	28 15	04 45	16 28	20 04	20 09	12 03	16 23	12 52	22 28
20	10 29	28 24	05 28	16 05	20 14	20 11	12 05	16 23	12 51	22 28
21	10 51	27 03	06 11	15 42	20 25	20 14	12 07	16 22	12 51	22 28
22	11 12	24 13	06 54	15 18	20 34	20 16	12 10	16 22	12 50	22 28
23	11 33	20 01	07 37	14 54	20 44	20 18	12 12	16 21	12 49	22 28
24	11 54	14 37	08 19	14 30	20 54	20 21	12 14	16 21	12 49	22 28
25	12 14	08 16	09 02	14 05	21 03	20 23	12 17	16 21	12 48	22 28
26	12 35	01 S 20	09 44	13 41	21 12	20 26	12 19	16 20	12 48	22 28
27	12 55	05 N 50	10 25	13 17	21 21	20 28	12 21	16 20	12 47	22 28
28	13 15	12 45	11 06	12 53	21 30	20 30	12 24	16 19	12 47	22 28
29	13 35	18 55	11 46	12 29	21 39	20 33	12 26	16 19	12 46	22 28
30	13 55	23 50	12 26	12 07	21 47	20 35	12 28	16 19	12 46	22 28
31	14 S 14	27 N 05	13 S 05	11 S 44	21 S 55	20 S 38	12 S 30	16 N 18	12 N 45	22 S 28

ZODIAC SIGN ENTRIES

Date	h m	Planets
03	00 22	☽ ♊
04	09 59	♃ ♐
05	03 26	☽ ♋
07	09 55	☽ ♌
09	15 18	☿ ♎
09	19 39	☽ ♍
12	07 39	☽ ♎
12	08 15	♀ ♎
14	20 39	☽ ♏
17	09 25	☽ ♐
19	20 37	☽ ♑
22	05 00	☽ ♒
23	06 49	☉ ♏
23	11 37	♂ ♐
24	09 50	☽ ♓
26	11 27	☽ ♈
27	15 27	☿ ♏
28	11 06	☽ ♉
30	10 39	☽ ♊

LATITUDES

Date	Mercury ☿ ° '	Venus ♀ ° '	Mars ♂ ° '	Jupiter ♃ ° '	Saturn ♄ ° '	Uranus ♅ ° '	Neptune ♆ ° '	Pluto ♇ ° '
01	00 N 45	07 S 34	00 S 23	00 N 39	02 N 13	00 N 39	01 S 49	10 S 59
04	01 20	07 44	00 25	00 38	02 12	00 39	01 49	10 58
07	01 43	07 49	00 27	00 38	02 12	00 39	01 49	10 58
10	01 55	07 47	00 29	00 37	02 12	00 39	01 49	10 57
13	01 58	07 38	00 30	00 37	02 11	00 39	01 49	10 57
16	01 54	07 22	00 32	00 36	02 11	00 39	01 49	10 56
19	01 44	06 57	00 34	00 36	02 11	00 39	01 49	10 56
22	01 31	06 26	00 35	00 35	02 11	00 40	01 49	10 55
25	01 14	05 49	00 37	00 35	02 11	00 40	01 49	10 55
28	00 56	05 08	00 39	00 35	02 11	00 40	01 49	10 54
31	00 N 36	04 S 24	00 S 40	00 N 34	02 N 11	00 N 40	01 S 49	10 S 53

DATA

Julian Date	2467159
Delta T	+78 seconds
Ayanamsa	24° 27' 10"
Synetic vernal point	04° ♓ 39' 49"
True obliquity of ecliptic	23° 26' 11"

MOON'S PHASES, APSIDES AND POSITIONS ☽

Date	h m	Phase	Longitude ° '	Eclipse Indicator
06	02 35	☾	12 ♋ 59	
14	02 03	●	20 ♎ 52	Annular
22	02 53	☽	28 ♑ 51	
28	19 48	○	05 ♉ 31	

Day	h m		
14	10 17	Apogee	
28	11 38	Perigee	
05	08 46	Max dec	28° N 31'
12	13 59	OS	
20	02 22	Max dec	28° S 31'
26	16 28	ON	

LONGITUDES

Date	Chiron ⚷ ° '	Ceres ⚳ ° '	Pallas ⚴ ° '	Juno ⚵ ° '	Vesta ⚶ ° '	Black Moon Lilith ⚸ ° '
01	14 ♌ 45	13 ♐ 17	11 ♏ 01	11 ♏ 45	07 ♍ 31	22 ♎ 47
11	15 ♌ 35	16 ♐ 31	15 ♏ 13	14 ♏ 57	12 ♍ 02	23 ♎ 54
21	16 ♌ 18	19 ♐ 56	19 ♏ 29	18 ♏ 14	16 ♍ 30	25 ♎ 00
31	16 ♌ 53	23 ♐ 29	23 ♏ 47	21 ♏ 35	20 ♍ 54	26 ♎ 07

All ephemeris data is given at 12.00 UT and the Moon's longitude is additionally given for 24.00 UT

ASPECTARIAN

h m	Aspects
01 Wednesday	
01 43	☽ ∥ ♆
05 38	☽ ☍ ♀
08 58	☽ ☍ ♄
09 32	☽ ⚼ ☿
12 36	♂ # ♅
13 27	☽ ⊼ ☉
15 20	☽ ∥ ♅
15 26	☽ ☌ ♆
15 27	☽ # ♂
15 50	☽ Q ♇
21 37	☉ ⊥ ♂
23 44	☽ ☍ ♂
23 47	☽ ± ☉
02 Thursday	
01 57	☽ □ ♅
03 21	☽ # ♃
05 04	☽ # ♀
10 50	☽ △ ☿
15 27	☽ ⚼ ☉
17 51	☽ # ♇
19 05	☉ ⊼ ♆
21 02	☽ □ ♇
23 57	☽ ☍ ♃
03 Friday	
04 25	☿ ⊥ ♅
05 51	☽ ⊼ ♀
07 02	☉ # ☿
07 33	☽ Q ♅
10 14	☽ ⊼ ♄
15 44	☽ ± ♀
16 33	☽ ⊻ ♆
18 10	☽ △ ☉
20 23	☽ ± ♄
04 Saturday	
02 19	♂ □ ♅
02 41	☽ ⊥ ♆
03 50	☽ ✶ ♅
03 54	☽ ⊼ ♂
06 46	☽ ⚼ ♀
11 52	☽ ⚼ ♄
14 44	☽ ± ♂
16 29	☽ □ ☿
18 08	☽ ∠ ♆
23 50	☽ △ ♇
05 Sunday	
03 40	☽ ⊼ ♃
05 55	☽ ∠ ♅
07 18	☽ ⚼ ♂
08 23	☽ △ ♀
13 24	☿ ⚼ ♆
14 22	☽ △ ♄
14 26	☽ ± ♃
20 34	☽ ✶ ♆
21 59	☽ ⊥ ♅
06 Monday	
01 12	☉ ⚼ ♇
02 29	☽ ⚼ ♇
02 35	☽ □ ☉
03 11	☽ Q ☿
06 51	☽ ⚼ ♃
08 53	☽ ⊻ ♅
11 44	☽ △ ♂
15 54	♀ ∥ ♃
18 21	☽ Q ♆
18 53	☽ ± ♇
22 49	☿ ⊥ ♀
07 Tuesday	
03 30	☽ ✶ ☿
06 00	☽ ⊼ ♇
10 58	☽ △ ♃
13 52	☽ □ ♀
14 32	☽ Q ☉
22 04	☽ □ ♄
08 Wednesday	
03 39	☿ ⊼ ♇
04 03	☽ □ ♆
04 52	☽ # ♇
11 13	☽ ∠ ☿
15 21	☽ ✶ ☉
17 24	☽ ☌ ♅
20 41	☽ # ♃
22 42	☽ # ♀
23 15	☽ Q ♀
23 37	☽ □ ♂
09 Thursday	
00 12	☉ ∠ ♃
04 49	☽ # ♂
06 53	☽ ⊥ ☿
09 05	☽ Q ♄
13 12	☽ ∥ ♅
15 29	☽ ☍ ♇
18 10	☉ ✶ ♅
20 15	☽ ⊻ ☿
21 39	☽ □ ♃
21 56	☽ ✶ ♀
23 11	☽ ∠ ☉
10 Friday	
02 37	☿ ⊥ ♄
03 29	♀ ⊻ ♃
05 58	☽ ∥ ♆
07 15	☿ ⊻ ♀
08 57	☽ ✶ ♄
09 05	☿ ✶ ♃
11 38	☽ # ♄
13 10	☽ Q ♂
14 31	☽ △ ♆
17 17	☿ ∠ ♅
18 38	☽ ⊥ ☉
11 Saturday	
02 40	☽ ∠ ♀
04 44	☽ ⊻ ♅
07 43	☽ ⊻ ☉
07 46	☽ # ☉
10 04	☽ Q ♃
14 41	☽ ✶ ♂
15 15	☽ ∠ ♄
16 49	☽ ⊥ ♅
20 03	☽ ⊥ ♀
20 32	☽ ⚼ ♆
22 16	☿ ± ♆
12 Sunday	
02 40	♀ ∥ ♂
03 18	☽ ⊼ ♇
03 34	☿ ± ♇
07 41	☽ ⊻ ♀
09 36	☽ ⊥ ♄
10 41	☽ ✶ ♃
11 04	☽ ∠ ♅
13 58	☽ ∥ ☿
14 01	☽ # ☿
14 42	☽ ± ♆
15 26	☽ ± ♇
17 18	☽ ☌ ☿
21 53	☽ ⊻ ♄
22 56	☽ ∠ ♂
13 Monday	
02 51	☽ ⊼ ♆
09 42	☽ ⚼ ♇
17 37	☽ ✶ ♅
17 39	☽ ∠ ♃
18 24	☽ ⊥ ♂
14 Tuesday	
00 51	☽ ∥ ☉
02 03	☽ ☌ ☉
04 48	☿ ⊻ ♄
07 24	☽ ⊻ ♂
12 17	☽ ⊥ ♃
15 27	♀ Q ♅
16 13	☽ △ ♇
17 38	☽ ∥ ♄
18 00	☽ ☌ ♀
18 08	☽ Q ♅
22 29	☽ # ♆
15 Wednesday	
00 43	☽ ⊻ ♃
11 34	☽ ☌ ♄
11 45	☿ ∠ ♂
12 53	☿ ⊼ ♆
15 35	☽ # ♅
15 47	☽ ☍ ♆
16 16	☽ ⊻ ☿
22 05	☽ ∥ ♀
16 Thursday	
06 27	☽ ⊥ ☿
06 49	☽ □ ♅
07 20	♀ △ ♇
07 55	☽ ∥ ♂
10 47	☽ ∥ ♃
20 34	☽ ⊻ ☉
17 Friday	
00 09	☽ ☌ ♂
01 39	☽ ∥ ♇
03 39	☽ ∠ ☿
03 58	☽ ⊻ ♀
04 59	☽ □ ♇
09 44	☽ ⊥ ☉
12 54	☿ ⚼ ♇
14 25	☽ ☌ ♃
15 26	☽ ⊥ ♀
18 Saturday	
00 40	☽ ⊻ ♄
04 04	☽ ⊼ ♆
05 20	☽ ∠ ☉
06 10	♀ ⊥ ♃
08 30	☽ ∠ ♀
12 30	♀ ⊻ ♂
12 42	☽ ⊥ ♄
14 31	☽ ✶ ☿
15 55	☽ ± ♆
16 50	☽ Q ♇
19 02	☽ △ ♅
19 Sunday	
06 29	☉ ☌ ♀
06 34	☽ ∠ ♄
09 33	☽ ⚼ ♆
12 35	☽ ✶ ♀
13 27	☽ ✶ ☉
15 19	☽ ⊻ ♂
16 18	☽ ✶ ♇
16 50	♀ # ♅
17 50	☽ Q ☿
23 05	☿ ✶ ♅
20 Monday	
00 21	☽ ⚼ ♅
01 35	♂ ∥ ♃
02 24	☽ ⊻ ♃
03 36	☽ ⊥ ♂
07 17	☉ ⊥ ♃
08 13	♂ □ ♇
10 38	☽ Q ♀
11 49	☽ ✶ ♄
14 03	☽ ⊥ ♃
14 23	☽ △ ♆
14 31	☽ Q ☉
15 31	☿ ∠ ♃
17 35	☽ ± ♅
21 00	☽ ∠ ♇
21 46	☽ ∠ ♂
21 Tuesday	
00 38	☉ △ ♇
04 55	☽ ⊼ ♅
07 18	☽ ∠ ♃
09 26	☽ □ ☿
10 38	☽ Q ♄
13 50	☽ ⊥ ♇
18 53	☽ □ ♀
22 Wednesday	
00 53	☽ ⊻ ♇
02 53	☽ □ ☉
03 14	☽ ✶ ♂
06 38	☉ Q ♅
11 18	☽ ✶ ♃
18 38	☉ ⊻ ♂
19 47	☽ □ ♄
21 35	☽ □ ♆
22 54	☽ ∥ ♇
23 Thursday	
01 53	♀ ⊥ ♂
02 03	☽ Q ♂
08 28	☽ ∥ ♂
09 01	☽ Q ♃
10 32	☽ ∥ ♃
11 26	☽ ☍ ♅
17 03	☿ ☌ ♀
22 12	☽ △ ♀
23 10	☽ △ ☿
24 Friday	
04 45	☽ # ♅
04 59	☽ Q ♆
05 57	☽ ☌ ♇
11 01	☽ □ ♂
11 53	☽ △ ☉
12 30	☽ ∥ ♀
12 58	☿ ⊥ ♂
16 29	☽ □ ♃
19 07	☽ # ♆
21 17	☽ ∥ ♄
22 06	☽ ∥ ☉
22 45	☽ ⚼ ♀
25 Saturday	
00 06	☽ △ ♄
01 15	☽ ✶ ♆
04 03	☽ ⚼ ☿
09 33	☽ ∥ ☿
13 09	☽ ± ♀
14 25	☽ ⊼ ♅
14 49	☽ ⚼ ☉
14 59	☉ ∥ ♄
20 51	☽ ± ☿
22 39	☽ ⊼ ♀
26 Sunday	
00 16	☽ ± ♅
01 03	☽ ⚼ ♄
01 55	☽ ∠ ♆
06 34	☽ ± ☉
06 52	☿ △ ♇
07 45	☽ ⊻ ♇
07 52	☽ ⊼ ☿
14 51	☽ ⚼ ♅
15 04	☽ △ ♂
15 35	☿ ⊥ ♃
15 43	☽ ± ♄
16 26	☽ ⊥ ♆
16 56	☽ ⊼ ☉
17 25	☽ ⊥ ♇
18 26	☽ △ ♃
27 Monday	
01 24	☽ ⊼ ♄
02 00	☽ ⊻ ♆
02 56	☿ Q ♅
03 05	☉ # ♆
07 46	☽ ∠ ♇
14 48	☽ △ ♅
16 10	☽ ⚼ ♂
18 38	☽ ⚼ ♃
21 07	☽ ☍ ♀
21 13	☉ ⊻ ♃
23 46	☉ ∥ ♀
28 Tuesday	
05 25	☽ # ☿
07 05	☽ ± ♂
07 29	☽ ✶ ♇
09 02	☽ ± ♃
10 41	☽ # ♄
12 05	☽ ∥ ♆
12 26	☽ # ♀
13 32	☽ ☍ ☿
13 54	☽ # ☉
17 01	☽ ⊼ ♂
18 15	♀ # ♆
18 37	☽ ⊼ ♃
19 48	☽ ☍ ☉
29 Wednesday	
01 12	☽ ☍ ♄
01 20	☽ ☌ ♆
01 23	☽ ∥ ♅
02 23	☽ Q ♇
14 13	☽ □ ♅
15 09	♄ ☍ ♆
15 22	♀ ∥ ♄
19 07	☽ ⊼ ♀
19 14	☽ # ♃
30 Thursday	
00 45	☽ # ♂
04 25	☽ ± ♀
04 30	☿ ∥ ♀
04 30	☽ # ♇
06 26	♀ ∠ ♃
06 57	☽ □ ♇
10 53	♀ ∠ ♂
13 25	☿ ∥ ♄
16 08	♂ ☌ ♃
18 03	☿ ⊻ ♃
18 31	☽ ⚼ ♀
19 00	☽ ☍ ♃
19 02	☽ Q ♅
19 06	☿ ⊻ ♂
19 06	☽ ☍ ♂
19 07	☽ ⊼ ☿
22 59	☽ ⊼ ☉
23 56	☿ # ♆
31 Friday	
01 08	☽ ⊻ ♆
01 29	☽ ⊼ ♄
06 06	☽ ± ☿
09 30	☽ ± ☉
10 56	☽ ⊥ ♆
11 23	☽ ± ♄
14 39	☽ ✶ ♅
18 28	☽ △ ♀
22 51	☽ ⚼ ☿

NOVEMBER 2042

Date	Sidereal time h m s	Sun ☉ ° ' "	Moon ☽ ° ' "	Moon ☽ 24.00 ° ' "	Mercury ☿ ° '	Venus ♀ ° '	Mars ♂ ° '	Jupiter ♃ ° '	Saturn ♄ ° '	Uranus ♅ ° '	Neptune ♆ ° '	Pluto ⯓ ° '
01	14 43 39	09 ♏ 11 34	29 ♊ 58 43	06 ♋ 59 25	08 ♏ 04	19 ♎ 17	06 ♐ 29	05 ♐ 33	09 ♏ 22	17 ♌ 14	08 ♉ 56	27 ♒ 41
02	14 47 36	10 11 36	13 ♋ 53 00	20 ♋ 39 29	09 42	18 R 59	07 13	05 46	09 29	17 16	08 R 54	27 R 41
03	14 51 32	11 11 39	27 ♋ 19 01	03 ♌ 51 53	11 20	18 44	07 56	05 59	09 36	17 17	08 53	27 40
04	14 55 29	12 11 45	10 ♌ 18 28	16 ♌ 39 15	12 57	18 31	08 40	06 11	09 43	17 18	08 51	27 40
05	14 59 26	13 11 52	22 ♌ 54 45	29 ♌ 05 32	14 34	18 21	09 24	06 24	09 50	17 19	08 49	27 40
06	15 03 22	14 12 02	05 ♍ 12 10	11 ♍ 15 14	16 11	18 13	10 07	06 37	09 58	17 20	08 48	27 40
07	15 07 19	15 12 14	17 ♍ 15 20	23 ♍ 13 01	17 47	18 08	10 51	06 50	10 05	17 20	08 46	27 40
08	15 11 15	16 12 27	29 ♍ 08 51	05 ♎ 03 19	19 23	18 05	11 35	07 03	10 12	17 21	08 44	27 40
09	15 15 12	17 12 43	10 ♎ 56 55	16 ♎ 50 05	20 58	18 D 05	12 19	07 16	10 19	17 22	08 42	27 40
10	15 19 08	18 13 01	22 ♎ 43 15	28 ♎ 36 45	22 33	18 07	13 03	07 29	10 26	17 23	08 41	27 40
11	15 23 05	19 13 20	04 ♏ 30 55	10 ♏ 26 05	24 07	18 11	13 47	07 42	10 34	17 23	08 39	27 D 40
12	15 27 01	20 13 42	16 ♏ 22 29	22 ♏ 20 23	25 42	18 18	14 31	07 55	10 41	17 24	08 38	27 40
13	15 30 58	21 14 05	28 ♏ 19 57	04 ♐ 21 25	27 16	18 27	15 16	08 08	10 48	17 25	08 36	27 40
14	15 34 55	22 14 29	10 ♐ 24 57	16 ♐ 30 44	28 ♏ 50	18 38	16 00	08 21	10 55	17 25	08 34	27 40
15	15 38 51	23 14 56	22 ♐ 38 55	28 ♐ 49 42	00 ♐ 23	18 51	16 44	08 35	11 02	17 25	08 33	27 40
16	15 42 48	24 15 24	05 ♑ 03 15	11 ♑ 19 46	01 56	19 07	17 29	08 48	11 09	17 26	08 31	27 40
17	15 46 44	25 15 53	17 ♑ 39 29	24 ♑ 02 36	03 29	19 24	18 13	09 01	11 16	17 26	08 29	27 40
18	15 50 41	26 16 23	00 ♒ 29 23	07 ♒ 00 06	05 02	19 44	18 58	09 14	11 24	17 26	08 28	27 41
19	15 54 37	27 16 55	13 ♒ 34 59	20 ♒ 14 19	06 34	20 05	19 42	09 28	11 31	17 26	08 26	27 41
20	15 58 34	28 17 28	26 ♒ 58 20	03 ♓ 47 17	08 06	20 28	20 27	09 41	11 38	17 27	08 25	27 41
21	16 02 30	29 ♏ 18 03	10 ♓ 41 18	17 ♓ 40 30	09 38	20 54	21 11	09 54	11 45	17 27	08 23	27 41
22	16 06 27	00 ♐ 18 38	24 ♓ 44 55	01 ♈ 54 29	11 10	21 20	21 56	10 08	11 52	17 27	08 22	27 42
23	16 10 24	01 19 14	09 ♈ 08 58	16 ♈ 28 00	12 42	21 49	22 41	10 21	11 59	17 R 27	08 20	27 42
24	16 14 20	02 19 52	23 ♈ 51 05	01 ♉ 17 32	14 13	22 19	23 26	10 35	12 06	17 27	08 18	27 42
25	16 18 17	03 20 31	08 ♉ 46 32	16 ♉ 17 07	15 44	22 51	24 11	10 48	12 12	17 27	08 17	27 43
26	16 22 13	04 21 11	23 ♉ 48 13	01 ♊ 18 39	17 15	23 24	24 56	11 01	12 19	17 26	08 16	27 43
27	16 26 10	05 21 53	08 ♊ 47 16	16 ♊ 12 55	18 45	23 58	25 41	11 15	12 26	17 26	08 14	27 43
28	16 30 06	06 22 36	23 ♊ 34 32	00 ♋ 51 08	20 16	24 34	26 26	11 28	12 33	17 26	08 13	27 44
29	16 34 03	07 23 20	08 ♋ 01 56	15 ♋ 06 18	21 46	25 12	27 11	11 42	12 40	17 26	08 11	27 44
30	16 37 59	08 ♐ 24 06	22 ♋ 03 49	28 ♋ 54 13	23 ♐ 15	25 ♎ 50	27 ♐ 56	11 ♐ 55	12 ♏ 47	17 ♌ 25	08 ♉ 10	27 ♒ 45

DECLINATIONS

Date	Moon True ☊ ° '	Moon Mean ☊ ° '	Moon ☽ Latitude ° '	Sun ☉ ° '	Moon ☽ ° '	Mercury ☿ ° '	Venus ♀ ° '	Mars ♂ ° '	Jupiter ♃ ° '	Saturn ♄ ° '	Uranus ♅ ° '	Neptune ♆ ° '	Pluto ⯓ ° '
01	17 ♈ 45	16 ♈ 35	05 N 00	14 S 33	28 N 27	13 S 43	11 S 23	22 S 03	20 S 40	12 S 33	16 N 18	12 N 45	22 S 28
02	17 R 42	16 32	05 15	14 52	27 56	14 21	11 02	22 11	20 42	12 35	16 18	12 44	22 28
03	17 40	16 29	05 11	15 11	25 46	14 58	10 43	22 19	20 45	12 37	16 17	12 44	22 27
04	17 40	16 25	04 51	15 30	22 19	15 34	10 24	22 26	20 47	12 40	16 17	12 43	22 27
05	17 D 40	16 22	04 18	15 48	17 56	16 10	10 06	22 34	20 49	12 42	16 17	12 43	22 27
06	17 41	16 19	03 33	16 06	12 54	16 44	09 50	22 41	20 51	12 44	16 17	12 42	22 27
07	17 43	16 16	02 40	16 24	07 29	17 18	09 34	22 47	20 54	12 46	16 17	12 41	22 27
08	17 45	16 13	01 41	16 41	01 N 53	17 51	09 20	22 54	20 56	12 49	16 16	12 41	22 27
09	17 46	16 10	00 N 37	16 58	03 S 46	18 23	09 07	23 00	20 58	12 51	16 16	12 40	22 26
10	17 R 46	16 06	00 S 27	17 15	09 15	18 54	08 55	23 06	21 01	12 53	16 16	12 40	22 26
11	17 45	16 03	01 30	17 32	14 27	19 24	08 44	23 12	21 03	12 55	16 16	12 39	22 26
12	17 42	16 00	02 30	17 48	19 08	19 53	08 34	23 18	21 05	12 57	16 16	12 39	22 26
13	17 37	15 57	03 24	18 04	23 06	20 21	08 26	23 23	21 07	13 00	16 16	12 38	22 26
14	17 31	15 54	04 08	18 20	26 06	20 48	08 18	23 29	21 09	13 02	16 15	12 38	22 25
15	17 24	15 51	04 42	18 35	27 56	21 14	08 12	23 34	21 12	13 04	16 15	12 37	22 25
16	17 18	15 47	05 04	18 50	28 24	21 39	08 06	23 38	21 14	13 06	16 15	12 37	22 25
17	17 12	15 44	05 10	19 05	27 24	22 03	08 02	23 43	21 16	13 08	16 15	12 36	22 24
18	17 07	15 41	05 02	19 19	24 57	22 26	07 59	23 47	21 18	13 10	16 15	12 36	22 24
19	17 04	15 38	04 38	19 33	21 11	22 48	07 57	23 51	21 20	13 13	16 15	12 35	22 24
20	17 03	15 35	03 59	19 47	16 15	23 09	07 56	23 55	21 22	13 15	16 15	12 35	22 24
21	17 D 04	15 31	03 05	20 00	10 24	23 28	07 56	23 58	21 24	13 17	16 15	12 34	22 23
22	17 05	15 28	01 58	20 13	03 S 54	23 47	07 56	24 02	21 26	13 19	16 15	12 34	22 23
23	17 06	15 25	00 S 43	20 25	02 N 58	24 04	07 58	24 05	21 28	13 21	16 15	12 34	22 23
24	17 R 06	15 22	00 N 36	20 37	09 49	24 20	08 01	24 07	21 30	13 23	16 15	12 33	22 22
25	17 04	15 19	01 54	20 49	16 14	24 34	08 04	24 10	21 32	13 25	16 15	12 33	22 22
26	17 00	15 16	03 05	21 01	21 42	24 48	08 08	24 12	21 34	13 27	16 15	12 32	22 22
27	16 54	15 12	04 02	21 12	25 45	25 00	08 13	24 14	21 36	13 29	16 16	12 32	22 21
28	16 46	15 09	04 42	21 22	27 59	25 10	08 19	24 16	21 38	13 31	16 16	12 31	22 21
29	16 38	15 06	05 04	21 32	28 15	25 20	08 25	24 17	21 40	13 33	16 16	12 31	22 20
30	16 ♈ 30	15 ♈ 03	05 N 06	21 S 42	26 N 39	25 S 28	08 S 33	24 S 19	21 S 42	13 S 35	16 N 16	12 N 30	22 S 20

ZODIAC SIGN ENTRIES

Date	h	m	Planets
01	12	02	☽ ♋
03	16	54	☽ ♌
06	01	46	☽ ♍
08	13	44	☽ ♎
11	02	49	☽ ♏
13	15	20	☽ ♐
15	06	05	☿ ♐
16	02	16	☽ ♑
18	11	06	☽ ♒
20	17	21	☽ ♓
22	04	37	☉ ♐
22	20	49	☽ ♈
24	21	55	☽ ♉
26	21	54	☽ ♊
28	22	35	☽ ♋

LATITUDES

Date	Mercury ☿ ° '	Venus ♀ ° '	Mars ♂ ° '	Jupiter ♃ ° '	Saturn ♄ ° '	Uranus ♅ ° '	Neptune ♆ ° '	Pluto ⯓ ° '
01	00 N 30	04 S 09	00 S 41	00 N 34	02 N 11	00 N 40	01 S 49	10 S 53
04	00 N 10	03 24	00 42	00 34	02 10	00 40	01 49	10 52
07	00 S 11	02 40	00 44	00 34	02 10	00 40	01 49	10 52
10	00 30	01 57	00 45	00 33	02 11	00 40	01 49	10 51
13	00 50	01 18	00 46	00 33	02 11	00 40	01 49	10 50
16	01 08	00 40	00 48	00 33	02 11	00 41	01 49	10 50
19	01 25	00 S 06	00 49	00 32	02 11	00 41	01 49	10 49
22	01 41	00 N 25	00 50	00 32	02 11	00 41	01 49	10 48
25	01 55	00 53	00 52	00 32	02 11	00 41	01 49	10 48
28	02 06	01 18	00 53	00 31	02 11	00 41	01 49	10 47
31	02 S 15	01 N 40	00 S 54	00 N 31	02 N 11	00 N 41	01 S 49	10 S 46

DATA

Julian Date	2467190
Delta T	+78 seconds
Ayanamsa	24° 27' 14"
Synetic vernal point	04° ♓ 39' 45"
True obliquity of ecliptic	23° 26' 10"

MOON'S PHASES, APSIDES AND POSITIONS ☽

Date	h	m	Phase	Longitude ° '	Eclipse Indicator
04	15	51	☾	12 ♌ 21	
12	20	28	●	20 ♏ 35	
20	14	31	☽	28 ♒ 24	
27	06	06	○	05 ♊ 07	

Day	h	m		
10	12	30	Apogee	
25	22	44	Perigee	
01	17	06	Max dec	28° N 29'
08	19	58	0S	
16	07	46	Max dec	28° S 25'
23	01	42	0N	
29	03	08	Max dec	28° N 22'

LONGITUDES

Date	Chiron ⚷ ° '	Ceres ⚳ ° '	Pallas ⚴ ° '	Juno ⚵ ° '	Vesta ⚶ ° '	Black Moon Lilith ⚸ ° '
01	16 ♌ 56	23 ♐ 51	24 ♏ 13	21 ♏ 55	21 ♍ 20	26 ♎ 13
11	17 ♌ 20	27 ♐ 33	28 ♏ 33	25 ♏ 19	25 ♍ 38	27 ♎ 20
21	17 ♌ 34	01 ♑ 20	02 ♐ 54	28 ♏ 43	29 ♍ 50	28 ♎ 26
31	17 ♌ 38	05 ♑ 12	07 ♐ 14	02 ♐ 08	03 ♎ 53	29 ♎ 33

All ephemeris data is given at 12.00 UT and the Moon's longitude is additionally given for 24.00 UT

ASPECTARIAN

h m	Aspects
01 Saturday	
01 31	☽ ♇ ☉
01 50	☽ ∠ ♆
02 27	☽ ♇ ♄
05 55	☉ ☍ ♆
08 07	☽ △ ⯓
15 51	☽ ∠ ♅
16 31	☉ ☌ ♄
21 40	☽ ⊼ ♃
23 45	☽ ⊼ ♂
02 Sunday	
00 30	☿ ☍ ♆
03 20	☽ ⚹ ♆
03 43	☽ △ ☿
04 14	☽ △ ♄
05 03	☽ △ ☉
07 24	☽ ⊥ ♅
08 14	☽ ± ♃
08 29	☿ ☌ ♄
09 53	☽ ♇ ⯓
10 45	☽ ± ♂
17 58	☽ ⊻ ♅
20 51	☽ □ ♀
03 Monday	
00 23	☽ ♇ ♃
00 25	☽ Q ♆
01 49	☽ ± ⯓
03 38	☽ ♇ ♂
06 45	☉ ☌ ☿
12 39	☽ ⊼ ⯓
04 Tuesday	
04 11	☽ △ ♃
05 02	☽ Q ♀
05 35	☉ ∥ ☿
08 44	☽ △ ♂
09 16	☽ □ ♆
10 53	☽ □ ♄
11 12	☽ # ⯓
11 19	☽ # ♂
14 57	♂ ∥ ⯓
15 51	☽ □ ☉
17 43	☽ □ ☿
17 47	♂ ⊼ ♆
20 51	☽ # ♃
05 Wednesday	
01 14	☽ ☌ ♅
03 21	☽ ⚹ ♀
16 53	☿ # ♅
17 57	♄ # ♆
19 49	☽ # ☿
20 10	☽ ∥ ♅
21 13	☽ ☍ ⯓
21 39	☽ Q ♄
21 54	☽ # ☉
06 Thursday	
05 34	☽ Q ☉
05 35	♂ ⊻ ♄
08 08	☽ ∠ ♀
09 40	☽ Q ☿
12 47	☽ # ♄
12 57	☽ ∥ ♆
14 51	☽ □ ♃
19 05	☽ △ ♆
21 31	☽ ⚹ ♄
22 23	☽ □ ♂
07 Friday	
01 49	☽ ⊥ ♀
02 18	☉ # ♅
02 29	☽ # ♀
05 22	☿ □ ♅
07 31	☽ ⚹ ☉
12 10	☽ ⊻ ♅
13 13	☽ ⚹ ☿
13 45	☽ ⊻ ♀
17 07	☿ ⊻ ♀
08 Saturday	
00 16	☽ ⊥ ♅
01 04	☽ ♇ ♆
03 33	☽ Q ♃
03 56	☽ ∠ ♄
08 59	☽ ⊼ ⯓
12 57	☽ Q ♂
16 34	☽ ∠ ☉
18 31	☽ ∠ ♅
19 16	☽ ± ♆
21 10	☽ ± ⯓
22 22	☽ ⊥ ♄
09 Sunday	
00 16	☽ ∠ ☿
04 05	♀ St D
04 21	☽ ⚹ ♃
07 27	☽ ⊼ ♆
10 42	☽ ⊻ ♄
12 35	☽ ⊥ ☉
14 59	☽ ⚹ ♂
15 30	☽ ♇ ⯓
15 46	☉ □ ♅
21 28	☽ ⊥ ☿
10 Monday	
01 06	☽ ⚹ ♅
01 58	☽ ⊻ ☉
02 34	☽ ☌ ♀
09 23	☉ ⊻ ♀
10 31	☽ ∥ ♀
11 30	☽ ∠ ♃
11 36	☽ ⊻ ☿
19 23	⯓ St D
22 04	☽ △ ⯓
23 35	☽ ∠ ♂
11 Tuesday	
01 34	☽ Q ♅
03 32	☽ # ♆
04 42	☽ ∥ ♄
06 10	☽ ⊥ ♃
12 59	☿ ⊥ ♀
18 35	☽ ⊻ ♃
19 05	☽ ⊥ ♂
20 22	☽ ☍ ♆
20 57	☽ # ♅
12 Wednesday	
00 23	☽ ☌ ♄
04 25	☽ ∥ ☉
08 01	☽ ⊻ ♂
14 04	☽ □ ♅
15 13	♂ ± ♆
15 55	☽ ⊻ ♀
16 45	☽ ∥ ☿
20 28	☽ ☌ ☉
23 19	☽ ∥ ♃
13 Thursday	
04 07	☽ ⊥ ♀
07 34	☽ ∥ ⯓
09 33	☽ ☌ ☿
10 40	☽ □ ⯓
14 06	☽ ∥ ♂
18 08	☿ □ ⯓
22 21	☽ ∠ ♀
14 Friday	
01 10	♂ Q ⯓
07 51	☽ ☌ ♃
08 22	☽ ⊼ ♆
13 00	☽ ⊻ ♄
20 10	☽ ± ♆
22 20	☽ Q ⯓
23 42	☽ ☌ ♂
15 Saturday	
00 55	☽ ⊥ ♄
01 47	☽ △ ♅
04 27	☽ ⚹ ♀
08 38	♃ ⊼ ♆
09 03	☿ ∥ ♃
13 16	☽ ⊻ ☉
13 44	☽ ♇ ♆
18 39	☽ ∠ ♄
21 45	☽ ⚹ ⯓
23 38	♂ ⊥ ♄
16 Sunday	
01 57	☽ ⊥ ☉
04 15	☽ Q ♀
05 09	☽ ⊻ ☿
06 57	☽ ♇ ♅
10 30	♂ △ ♅
18 18	☽ ⊥ ☿
18 37	☽ △ ♆
19 18	☽ ⊻ ♃
20 45	☽ ∠ ☉
23 47	☽ ⚹ ♄
17 Monday	
00 12	☽ ± ♅
02 33	☽ ∠ ⯓
06 55	☽ ⊥ ♃
11 35	☽ ⊼ ♅
13 07	☽ ⊻ ♂
13 47	☽ ∠ ☿
15 23	☽ □ ♀
16 51	☉ ⊥ ♀
19 33	☽ ⊥ ⯓
22 40	☽ Q ♄
18 Tuesday	
00 10	☽ ∠ ♃
01 05	☽ ⊥ ♂
03 30	☽ ⚹ ☉
06 03	☿ ∠ ♀
06 46	☽ ⊻ ⯓
09 42	☿ ∥ ⯓
18 48	☽ ∠ ♂
20 12	☽ ∥ ♂
21 30	☽ ⚹ ☿
19 Wednesday	
02 39	☽ □ ♆
03 25	☽ ∥ ☿
03 31	☽ Q ☉
04 22	☽ ⚹ ♃
05 04	☽ ∥ ⯓
08 12	☽ □ ♄
11 10	☽ ∥ ♃
18 59	☽ ☍ ♅
20 08	☽ ∥ ☉
21 28	☉ □ ⯓
22 11	☽ Q ☿
23 41	☽ ⚹ ♂
20 Thursday	
00 04	☽ △ ♀
02 26	☽ Q ♃
11 00	☽ Q ♆
12 01	☽ # ♅
13 15	☽ ☌ ⯓
14 03	♀ ⚹ ♂
14 31	☽ □ ☉
16 41	☿ ⊼ ♆
22 12	☽ Q ♂
21 Friday	
00 40	☽ ∥ ♄
03 25	☽ ♇ ♀
03 27	☽ # ♆
08 01	☽ ⚹ ♆
09 58	☽ □ ☿
10 38	☽ □ ♃
13 50	☽ △ ♄
16 55	☿ ☌ ♃
19 28	☽ ± ♀
21 21	☽ ∥ ♀
23 37	☽ ⊼ ♅
22 Saturday	
06 02	☽ ⊼ ♀
06 59	☽ □ ♂
09 39	☽ ∠ ♆
09 48	☽ ± ♅
15 16	♅ St R
15 35	☽ ♇ ♄
16 57	☽ ⊻ ⯓
22 02	☽ △ ☉
23 47	☿ ⊻ ♄
23 Sunday	
00 44	☽ ⊥ ♆
00 54	☽ ♇ ♅
02 58	☽ ⊥ ⯓
06 43	☽ ± ♄
10 39	☽ ⊻ ♆
13 46	☿ ∥ ♂
14 01	☽ △ ♃
16 41	☽ ⊼ ♄
17 50	☽ ∠ ⯓
18 30	☽ △ ☿
24 Monday	
00 38	☽ ♇ ☉
01 36	☽ △ ♅
05 32	☽ # ♀
08 13	♂ ♇ ♆
09 25	☽ ☍ ♀
11 17	☽ △ ♂
13 26	☿ ± ♆
14 50	☽ ♇ ♃
16 18	☽ ± ☉
18 13	☽ ⚹ ⯓
21 39	☽ ♇ ☿
21 55	☽ ∥ ♆
25 Tuesday	
01 06	☽ # ♄
02 40	☽ ⊼ ☉
05 32	☽ ± ♃
11 13	☽ ☌ ♆
11 37	☿ Q ⯓
12 06	☽ ∥ ♅
12 41	☽ ♇ ♂
13 30	☽ Q ⯓
13 42	☽ ± ☿
15 17	☽ ⊼ ♃
17 32	☽ ☍ ♄
26 Wednesday	
00 22	☽ ⊼ ☿
01 51	☽ □ ♅
03 49	☽ ± ♂
08 30	☽ # ☉
11 19	☽ ⊼ ♀
11 20	☽ # ♃
13 53	☽ ⊼ ♂
15 04	☿ △ ♅
15 19	☽ # ⯓
18 15	☽ □ ⯓
21 17	☽ ± ♀
27 Thursday	
01 44	☽ # ♂
06 06	☽ ☍ ☉
06 20	☽ # ☿
06 29	☿ ⊥ ♄
06 37	☽ Q ♅
11 07	☽ ⊻ ♆
12 19	☽ ♇ ♀
16 02	☽ ☍ ♃
17 56	☽ ⊼ ♄
20 47	☽ ⊥ ♆
28 Friday	
01 59	☽ ⚹ ♅
03 44	☽ ± ♄
05 58	☽ ☍ ☿
11 24	☽ ∠ ♆
13 42	☽ △ ♀
16 57	☽ ☍ ♂
18 35	☽ ♇ ♄
18 50	☽ △ ⯓
29 Saturday	
02 37	☽ ∠ ♅
10 50	☽ ⊼ ☉
12 16	☽ ⚹ ♆
17 44	☽ ⊥ ♅
18 18	☽ ⊼ ♃
19 54	☽ △ ♄
19 58	☽ ♇ ⯓
21 46	☽ ± ☉
30 Sunday	
03 58	☽ ⊻ ♅
04 43	☽ ± ♃
06 01	♂ ⚹ ⯓
06 10	♂ ∠ ♄
06 28	☉ ⊼ ♆
08 42	☽ Q ♆
10 30	☿ ♇ ♆
11 00	☉ ∥ ♃
11 27	☽ ± ⯓
14 20	☽ ⊼ ☿
14 31	☽ ♇ ☉
18 56	☽ □ ♀
20 39	☽ ♇ ♃
21 58	☽ ⊼ ⯓
22 03	☽ # ☿
22 53	☽ ⊼ ♂

DECEMBER 2042

LONGITUDES

Date	Sidereal time h m s	Sun ☉ ° ' "	Moon ☽ ° ' "	Moon ☽ 24.00 ° ' "	Mercury ☿ ° '	Venus ♀ ° '	Mars ♂ ° '	Jupiter ♃ ° '	Saturn ♄ ° '	Uranus ♅ ° '	Neptune ♆ ° '	Pluto ♇ ° '
01	16 41 56	09 ♐ 24 53	05 ♌ 37 29	12 ♌ 13 42	24 ♐ 45	26 ♎ 30	28 ♐ 41	12 ♐ 09	12 ♏ 53	17 ♌ 25	08 ♉ 08	27 ♒ 45
02	16 45 53	10 25 42	18 ♌ 43 09	25 ♌ 06 13	26 14	27 12	29 ♐ 27	12 22	13 00	17 R 24	08 R 07	27 46
03	16 49 49	11 26 32	01 ♍ 23 24	07 ♍ 35 16	27 42	27 54	00 ♑ 12	12 36	13 07	17 24	08 06	27 47
04	16 53 46	12 27 23	13 ♍ 42 25	19 ♍ 45 33	29 ♐ 10	28 37	00 57	12 49	13 13	17 23	08 04	27 47
05	16 57 42	13 28 15	25 ♍ 45 19	01 ♎ 42 25	00 ♑ 37	29 ♎ 22	01 43	13 03	13 20	17 22	08 03	27 48
06	17 01 39	14 29 09	07 ♎ 37 31	13 ♎ 31 18	02 03	00 ♏ 07	02 28	13 16	13 27	17 22	08 02	27 48
07	17 05 35	15 30 05	19 ♎ 24 22	25 ♎ 17 21	03 28	00 54	03 14	13 30	13 33	17 21	08 00	27 49
08	17 09 32	16 31 01	01 ♏ 10 46	07 ♏ 05 10	04 53	01 41	03 59	13 43	13 40	17 20	07 59	27 50
09	17 13 28	17 31 59	13 ♏ 00 58	18 ♏ 58 35	06 16	02 30	04 45	13 57	13 46	17 19	07 58	27 51
10	17 17 25	18 32 58	24 ♏ 58 23	01 ♐ 00 37	07 37	03 19	05 31	14 10	13 52	17 18	07 57	27 51
11	17 21 22	19 33 58	07 ♐ 05 31	13 ♐ 13 15	08 57	04 09	06 16	14 24	13 59	17 17	07 56	27 52
12	17 25 18	20 34 59	19 ♐ 23 56	25 ♐ 37 37	10 15	05 00	07 02	14 37	14 05	17 16	07 54	27 53
13	17 29 15	21 36 00	01 ♑ 54 21	08 ♑ 14 06	11 31	05 51	07 48	14 51	14 11	17 15	07 53	27 54
14	17 33 11	22 37 03	14 ♑ 36 50	21 ♑ 02 30	12 44	06 44	08 34	15 04	14 18	17 14	07 52	27 55
15	17 37 08	23 38 06	27 ♑ 31 04	04 ♒ 02 29	13 54	07 37	09 20	15 18	14 24	17 13	07 51	27 56
16	17 41 04	24 39 10	10 ♒ 36 43	17 ♒ 13 47	15 00	08 31	10 06	15 31	14 30	17 11	07 50	27 56
17	17 45 01	25 40 14	23 ♒ 53 40	00 ♓ 36 27	16 02	09 25	10 52	15 45	14 36	17 10	07 49	27 57
18	17 48 57	26 41 19	07 ♓ 22 12	14 ♓ 11 00	16 59	10 20	11 38	15 58	14 42	17 09	07 48	27 58
19	17 52 54	27 42 23	21 ♓ 02 58	27 ♓ 58 11	17 50	11 16	12 24	16 11	14 48	17 07	07 47	27 59
20	17 56 51	28 43 28	04 ♈ 56 43	11 ♈ 58 37	18 35	12 12	13 10	16 25	14 54	17 06	07 46	28 00
21	18 00 47	29 ♐ 44 34	19 ♈ 03 51	26 ♈ 12 17	19 13	13 09	13 57	16 38	15 00	17 05	07 45	28 01
22	18 04 44	00 ♑ 45 39	03 ♉ 23 44	10 ♉ 37 51	19 43	14 06	14 43	16 51	15 05	17 03	07 44	28 02
23	18 08 40	01 46 45	17 ♉ 54 12	25 ♉ 12 11	20 04	15 04	15 29	17 05	15 11	17 01	07 44	28 03
24	18 12 37	02 47 51	02 ♊ 31 07	09 ♊ 50 11	20 16	16 03	16 15	17 18	15 17	17 00	07 43	28 04
25	18 16 33	03 48 57	17 ♊ 08 30	24 ♊ 25 10	20 R 15	17 02	17 02	17 31	15 22	16 58	07 42	28 06
26	18 20 30	04 50 04	01 ♋ 39 13	08 ♋ 49 46	20 04	18 01	17 48	17 45	15 28	16 56	07 41	28 07
27	18 24 26	05 51 11	15 ♋ 56 01	22 ♋ 57 13	19 41	19 01	18 35	17 58	15 33	16 55	07 40	28 08
28	18 28 23	06 52 18	29 ♋ 52 50	06 ♌ 42 25	19 07	20 01	19 21	18 11	15 39	16 53	07 40	28 09
29	18 32 20	07 53 26	13 ♌ 25 44	20 ♌ 02 41	18 21	21 02	20 08	18 24	15 44	16 51	07 39	28 10
30	18 36 16	08 54 34	26 ♌ 33 20	02 ♍ 57 53	17 24	22 03	20 54	18 37	15 49	16 49	07 38	28 11
31	18 40 13	09 ♑ 55 42	09 ♍ 16 40	15 ♍ 30 07	16 ♑ 18	23 ♏ 05	21 ♑ 41	18 ♐ 50	15 ♏ 55	16 ♌ 47	07 ♉ 38	28 ♒ 13

Date	Moon True ☊ ° '	Moon Mean ☊ ° '	Moon ☽ Latitude ° '
01	16 ♈ 23	15 ♈ 00	04 N 50
02	16 R 19	14 57	04 20
03	16 16	14 53	03 37
04	16 D 15	14 50	02 46
05	16 16	14 47	01 48
06	16 17	14 44	00 N 46
07	16 R 17	14 41	00 S 17
08	16 16	14 37	01 19
09	16 12	14 34	02 18
10	16 06	14 31	03 11
11	15 57	14 28	03 56
12	15 46	14 25	04 31
13	15 33	14 22	04 54
14	15 20	14 18	05 02
15	15 09	14 15	04 55
16	14 59	14 12	04 33
17	14 53	14 09	03 56
18	14 49	14 06	03 05
19	14 47	14 02	02 02
20	14 D 47	13 59	00 S 52
21	14 R 47	13 56	00 N 23
22	14 46	13 53	01 37
23	14 43	13 50	02 46
24	14 36	13 47	03 44
25	14 27	13 43	04 27
26	14 16	13 40	04 53
27	14 03	13 37	05 00
28	13 51	13 34	04 49
29	13 40	13 31	04 21
30	13 32	13 28	03 41
31	13 ♈ 26	13 ♈ 24	02 N 50

DECLINATIONS

Date	Sun ☉ ° '	Moon ☽ ° '	Mercury ☿ ° '	Venus ♀ ° '	Mars ♂ ° '	Jupiter ♃ ° '	Saturn ♄ ° '	Uranus ♅ ° '	Neptune ♆ ° '	Pluto ♇ ° '
01	21 S 52	23 N 33	25 S 34	08 S 40	24 S 20	21 S 44	13 S 37	16 N 16	12 N 30	22 S 20
02	22 00	19 19	25 40	08 49	24 21	21 46	13 39	16 16	12 30	22 19
03	22 09	14 22	25 44	08 58	24 21	21 47	13 41	16 16	12 29	22 19
04	22 17	08 57	25 46	09 07	24 21	21 49	13 43	16 17	12 29	22 18
05	22 25	03 N 20	25 47	09 18	24 21	21 51	13 45	16 17	12 28	22 18
06	22 32	02 S 19	25 47	09 28	24 20	21 53	13 47	16 17	12 28	22 18
07	22 39	07 51	25 45	09 39	24 20	21 54	13 48	16 17	12 28	22 17
08	22 45	13 07	25 42	09 51	24 19	21 56	13 50	16 18	12 27	22 17
09	22 51	17 56	25 37	10 03	24 18	21 58	13 52	16 18	12 27	22 16
10	22 57	22 06	25 31	10 15	24 18	22 00	13 54	16 18	12 27	22 16
11	23 02	25 22	25 23	10 28	24 14	22 01	13 56	16 19	12 26	22 15
12	23 06	27 31	25 14	10 41	24 12	22 03	13 57	16 19	12 26	22 15
13	23 10	28 19	25 04	10 54	24 10	22 04	13 59	16 19	12 26	22 14
14	23 14	27 38	24 53	11 08	24 08	22 06	14 01	16 20	12 25	22 14
15	23 17	25 29	24 40	11 22	24 05	22 07	14 03	16 20	12 25	22 13
16	23 20	21 57	24 27	11 36	24 02	22 09	14 04	16 20	12 25	22 13
17	23 22	17 15	24 12	11 51	23 58	22 11	14 06	16 21	12 24	22 12
18	23 24	11 39	23 56	12 05	23 55	22 12	14 08	16 21	12 24	22 12
19	23 25	05 S 25	23 40	12 20	23 51	22 13	14 09	16 22	12 24	22 11
20	23 26	01 N 10	23 23	12 35	23 47	22 15	14 11	16 22	12 24	22 11
21	23 26	07 49	23 06	12 50	23 42	22 16	14 12	16 23	12 23	22 10
22	23 26	14 10	22 48	13 05	23 37	22 18	14 14	16 23	12 23	22 10
23	23 25	19 49	22 31	13 20	23 32	22 19	14 15	16 24	12 23	22 09
24	23 24	24 19	22 13	13 35	23 27	22 20	14 17	16 24	12 23	22 09
25	23 23	27 15	21 57	13 51	23 22	22 22	14 18	16 25	12 23	22 08
26	23 21	28 19	21 41	14 06	23 16	22 23	14 20	16 26	12 22	22 08
27	23 18	27 27	21 25	14 21	23 10	22 24	14 21	16 26	12 22	22 07
28	23 15	24 53	21 11	14 37	23 03	22 25	14 23	16 27	12 22	22 06
29	23 12	20 58	20 58	14 52	22 57	22 27	14 24	16 27	12 22	22 06
30	23 08	16 07	20 46	15 07	22 50	22 28	14 25	16 28	12 22	22 05
31	23 S 04	10 N 43	20 S 35	15 S 22	22 S 43	22 S 29	14 S 27	16 N 28	12 N 22	22 S 05

ZODIAC SIGN ENTRIES

Date	h	m	Planets
01	01	57	☽ ♌
03	05	43	♂ ♑
03	09	20	☽ ♍
05	01	51	☿ ♑
05	20	33	☽ ♎
06	08	11	♀ ♏
08	09	36	☽ ♏
10	22	00	☽ ♐
13	08	22	☽ ♑
15	16	35	☽ ♒
17	22	55	☽ ♓
20	03	30	☽ ♈
21	18	04	☉ ♑
22	06	21	☽ ♉
24	07	52	☽ ♊
26	09	15	☽ ♋
28	12	13	☽ ♌
30	18	26	☽ ♍

LATITUDES

Date	Mercury ☿ ° '	Venus ♀ ° '	Mars ♂ ° '	Jupiter ♃ ° '	Saturn ♄ ° '	Uranus ♅ ° '	Neptune ♆ ° '	Pluto ♇ ° '
01	02 S 15	01 N 40	00 S 54	00 N 31	02 N 11	00 N 41	01 S 49	10 S 46
04	02 20	01 59	00 55	00 31	02 12	00 41	01 49	10 46
07	02 22	02 16	00 56	00 31	02 12	00 41	01 49	10 45
10	02 18	02 31	00 57	00 31	02 12	00 42	01 49	10 44
13	02 09	02 43	00 58	00 30	02 12	00 42	01 48	10 44
16	01 52	02 53	00 59	00 30	02 13	00 42	01 48	10 43
19	01 26	03 01	01 00	00 30	02 13	00 42	01 48	10 43
22	00 49	03 08	01 00	00 30	02 14	00 42	01 48	10 42
25	00 S 02	03 12	01 01	00 30	02 14	00 42	01 48	10 41
28	00 N 54	03 15	01 02	00 29	02 14	00 42	01 48	10 41
31	01 N 52	03 N 16	01 S 02	00 N 29	02 N 15	00 N 42	01 S 48	10 S 40

DATA

Julian Date	2467220
Delta T	+78 seconds
Ayanamsa	24° 27' 19"
Synetic vernal point	04° ♓ 39' 40"
True obliquity of ecliptic	23° 26' 10"

MOON'S PHASES, APSIDES AND POSITIONS ☽

Date	h	m	Phase	Longitude ° '	Eclipse Indicator
04	09	19	☾	12 ♍ 21	
12	14	30	●	20 ♐ 41	
20	00	27	☽	28 ♓ 14	
26	17	43	○	05 ♋ 05	

Day	h	m		
08	00	46	Apogee	
24	02	20	Perigee	
06	02	09	0S	
13	13	08	Max dec	28° S 19'
20	07	47	0N	
26	13	05	Max dec	28° N 19'

LONGITUDES

Date	Chiron ⚷ ° '	Ceres ⚳ ° '	Pallas ⚴ ° '	Juno ⚵ ° '	Vesta ⚶ ° '	Black Moon Lilith ⚸ ° '
01	17 ♌ 38	05 ♑ 12	07 ♐ 14	02 ♐ 08	03 ♎ 53	29 ♎ 33
11	17 ♌ 31	09 ♑ 08	11 ♐ 33	05 ♐ 32	07 ♎ 45	00 ♏ 39
21	17 ♌ 14	13 ♑ 07	15 ♐ 50	08 ♐ 54	11 ♎ 23	01 ♏ 46
31	16 ♌ 48	17 ♑ 06	20 ♐ 04	12 ♐ 13	14 ♎ 44	02 ♏ 52

ASPECTARIAN

h m	Aspects
01 Monday	
02 11	☽ ± ☿
06 49	☽ # ♂
10 13	☽ ± ♂
16 32	☽ □ ♆
19 26	☽ △ ☉
19 30	☽ # ♇
20 25	☽ Ᵽ ☿
21 52	☽ # ☉
22 53	☽ # ♃
02 Tuesday	
00 03	☽ △ ♃
01 20	☽ □ ♄
03 35	☽ Ᵽ ♂
05 05	☽ Q ♀
09 33	☽ ☌ ♅
21 02	♀ ∠ ♃
03 Wednesday	
03 04	☽ ‖ ♅
04 00	☽ △ ☿
04 55	☽ ✶ ♀
05 04	☽ ☍ ♇
07 50	♀ △ ♇
09 34	☽ △ ♂
11 28	☽ Q ♄
13 17	☿ ✶ ♇
15 05	☽ # ♄
18 24	☿ ✶ ♀
19 21	☿ ∠ ♄
20 30	☽ ‖ ♆
04 Thursday	
00 58	☽ △ ♆
09 19	☽ □ ☉
10 13	☽ □ ♃
11 02	☽ ✶ ♄
11 18	☽ # ♀
11 49	☽ ∠ ♀
15 54	☉ ‖ ♇
19 16	☽ ⊻ ♅
23 05	☉ ☌ ♃
05 Friday	
06 35	☽ Ᵽ ♆
06 53	☽ ⊥ ♀
07 13	☽ ⊥ ♅
08 22	☉ ⊻ ♄
12 19	♀ Q ♅
16 07	☽ ⊼ ♇
17 14	☽ ∠ ♄
19 45	☽ ⊻ ♀
22 52	☽ Q ♃
23 08	☽ □ ☿
06 Saturday	
00 36	☽ Q ☉
00 40	☽ ± ♆
00 50	☽ □ ♂
01 20	☽ ∠ ♅
01 24	☉ ± ♆
04 15	☽ ± ♇
08 36	♂ Ᵽ ♅
11 38	☽ ⊥ ♄
12 49	☽ ⊼ ♆
17 10	☿ Ᵽ ♅
22 33	☽ Ᵽ ♇
23 43	☽ ✶ ♃
23 57	☽ ⊻ ♄
07 Sunday	
03 06	☿ ☌ ♂
03 17	☽ ✶ ☉
07 48	☽ ✶ ♅
15 59	☽ Q ♂
16 48	☽ Q ☿
19 36	☉ Q ♇
20 21	☽ ‖ ♀
23 27	♃ ⊻ ♄
08 Monday	
05 10	☽ △ ♇
06 54	☽ ∠ ♃
08 15	☽ Q ♅
08 54	☽ # ♆
12 45	☽ ∠ ☉
13 07	☽ ☌ ♀
15 28	☽ ‖ ♄
18 06	☽ ✶ ♂
20 31	☽ ✶ ☿
09 Tuesday	
01 33	☽ ⊥ ♃
01 48	☽ ☍ ♆
03 33	☽ # ♅
07 01	☉ △ ♅
07 21	☉ ∠ ♀
08 43	☽ ⊥ ☉
13 32	☽ ☌ ♄
13 54	♃ ± ♆
13 55	☽ ⊻ ♃
20 39	☽ □ ♅
21 57	☽ ⊻ ☉
10 Wednesday	
02 28	☽ ∠ ♂
06 43	☽ ∠ ☿
11 20	☽ ‖ ♃
13 04	☽ ‖ ♇
17 40	☿ △ ♆
17 41	☽ ‖ ☉
17 45	☽ □ ♇
21 38	☽ ⊥ ♂
11 Thursday	
02 51	☽ ⊥ ☿
02 54	☽ ‖ ♂
05 46	☽ ⊻ ♀
10 17	☽ ⊻ ♂
12 07	☽ ‖ ☿
13 38	☽ ⊼ ♆
16 06	☽ ⊻ ☿
18 26	☽ ⊥ ♀
22 53	☉ ⊥ ♄
12 Friday	
01 21	☽ ± ♆
01 36	☽ ⊻ ♄
02 33	☽ ☌ ♃
05 10	☽ Q ♇
07 53	☽ △ ♅
13 14	☽ ∠ ♀
13 20	☽ ⊥ ♄
14 30	☽ ☌ ☉
18 46	☽ Ᵽ ♆
13 Saturday	
04 20	☽ ✶ ♇
06 47	☽ ∠ ♄
06 59	☿ ± ♅
12 39	☽ Ᵽ ♅
14 38	♂ △ ♆
20 03	☽ ✶ ♀
23 20	☽ △ ♆
23 54	☽ ☌ ♂
14 Sunday	
05 40	☽ ± ♅
08 06	☽ ☌ ☿
08 48	☽ ∠ ♇
11 24	☽ ✶ ♄
12 52	☽ ⊻ ♃
15 42	☿ ∠ ♇
16 53	☽ ⊼ ♅
17 51	☉ Ᵽ ♆
20 16	☽ Q ♀
15 Monday	
00 16	☽ ⊥ ♃
01 38	☽ ⊥ ♇
04 12	☽ ⊻ ☉
09 55	☽ Q ♄
12 45	☽ ⊻ ♇
16 14	☽ ⊥ ☉
17 12	☽ ∠ ♃
18 15	♀ ☍ ♆
18 47	☽ ‖ ☿
22 43	☽ ‖ ♂
23 53	☿ ✶ ♄
16 Tuesday	
03 40	☽ ‖ ☉
06 57	☽ □ ♆
07 53	☽ □ ♀
10 06	☽ ∠ ☉
10 26	☽ ‖ ♇
10 49	☽ ‖ ♃
11 01	☽ ⊻ ♂
19 07	☽ □ ♄
20 39	☽ ⊻ ☿
21 04	☽ ✶ ♃
22 34	☽ ⊥ ♂
23 55	☽ ☍ ♅
17 Wednesday	
03 21	☿ ⊻ ♃
08 22	☽ ⊥ ☿
15 27	☽ Q ♆
15 27	☽ ✶ ☉
15 45	☽ ∠ ♂
16 07	☽ # ♅
19 00	☽ Q ♃
19 17	☽ ☌ ♇
21 12	♂ ± ♅
23 15	♀ ⊥ ♃
18 Thursday	
01 44	☽ ∠ ☿
01 52	☽ ‖ ♄
08 59	☽ # ♆
09 59	♃ ‖ ♇
10 20	☽ ‖ ♀
12 34	♃ Q ♇
12 46	☽ ✶ ♆
14 31	☽ Q ☉
15 11	☿ ‖ ♂
16 28	☿ ⊼ ♅
17 37	☽ △ ♀
19 58	☽ ✶ ♂
19 Friday	
01 00	☽ △ ♄
03 23	☽ □ ♃
05 10	☽ ⊼ ♅
06 02	☽ ✶ ☿
15 01	☽ ∠ ♆
15 36	☽ ± ♅
18 10	☽ Q ♂
18 24	♀ # ♆
18 44	☉ ✶ ♇
21 42	☽ Ᵽ ♀
20 Saturday	
00 03	☽ ⊻ ♇
00 27	☽ □ ☉
03 16	☽ Ᵽ ♄
04 06	☽ Q ☿
06 33	☽ ⊥ ♆
06 38	♂ ∠ ♇
07 08	☽ Ᵽ ♅
08 14	☉ ‖ ☿
10 23	☽ ⊥ ♇
14 18	☽ ± ♀
16 50	☽ ⊻ ♆
18 48	☽ ± ♄
21 Sunday	
01 16	☽ ⊼ ♀
01 46	☽ ∠ ♇
02 50	☽ □ ♂
05 04	☽ ⊼ ♄
07 50	☽ △ ♃
08 39	☽ △ ♅
12 16	☽ □ ☿
18 31	☉ ∠ ♄
22 Monday	
03 03	☽ ✶ ♇
05 05	☽ ‖ ♆
07 17	☽ △ ☉
07 35	☽ # ♀
09 24	☽ Ᵽ ♃
12 16	☽ # ♄
19 12	☽ ☌ ♆
21 02	☽ ‖ ♅
23 02	☽ Q ♇
23 Tuesday	
00 34	☽ ± ♃
01 21	♂ ✶ ♄
06 34	♃ △ ♅
07 00	☽ ☍ ♀
07 29	☽ ☍ ♄
07 47	☽ △ ♂
10 01	☽ Ᵽ ☉
10 33	☽ □ ♅
10 37	☽ ⊼ ♃
15 05	♀ ☌ ♄
15 37	☽ △ ☿
17 37	☉ Ᵽ ♅
23 31	☽ # ♇
24 Wednesday	
00 29	☽ # ♃
00 39	☽ # ☿
01 55	☽ ± ☉
03 05	☿ ‖ ♃
04 42	☽ □ ♇
06 30	☽ # ☉
06 51	☽ # ♂
09 49	☽ Ᵽ ♂
12 30	☽ ⊼ ☉
16 03	☽ Q ♅
16 30	☽ Ᵽ ☿
19 06	☿ ‖ ♇
20 30	☽ ⊻ ♆
25 Thursday	
00 43	☿ St R
01 24	☽ ± ♂
04 22	☉ ‖ ♂
06 21	☽ ⊥ ♆
07 16	☽ ± ☿
09 04	☽ ⊼ ♄
10 12	♂ ⊼ ♅
10 36	♀ □ ♅
11 43	☽ ✶ ♅
11 48	☽ ⊼ ♀
11 48	☽ ⊼ ♂
12 04	♀ ✶ ♂
12 38	☽ ☍ ♃
17 05	☽ ⊼ ☿
19 01	☽ ± ♄
21 09	☽ ∠ ♆
22 25	☽ ± ♀
26 Friday	
03 26	♀ ⊻ ♃
06 06	☽ △ ♇
09 26	♂ ⊻ ♃
10 01	☽ Ᵽ ♄
12 29	☽ ∠ ♅
14 27	☽ Ᵽ ♀
17 43	☽ ☍ ☉
22 04	☽ ✶ ♆
27 Saturday	
03 31	☽ ⊥ ♅
07 15	☽ Ᵽ ♇
11 21	☽ △ ♄
11 47	♀ ‖ ♄
13 40	☽ ⊻ ♅
15 31	☽ ⊼ ♃
16 46	☽ ☍ ♂
17 39	☽ △ ♀
18 11	☽ ☍ ☿
18 22	☽ Q ♆
22 33	☿ ✶ ♀
22 36	☽ ± ♇
28 Sunday	
01 58	☽ ± ♃
07 59	☿ ☌ ♂
08 59	☽ ⊼ ♇
17 53	☽ Ᵽ ♃
22 56	☽ # ☉
29 Monday	
00 21	☽ # ♂
01 18	☽ ⊼ ☉
01 41	☽ □ ♆
03 43	☽ # ♃
05 40	☽ # ♇
06 26	☉ △ ♆
10 40	☿ ⊻ ♃
11 59	☽ # ☿
12 54	☽ ± ☉
16 11	☽ □ ♄
18 10	☽ ☌ ♅
20 20	☽ ⊼ ☿
21 09	☽ △ ♃
30 Tuesday	
00 55	☽ ⊼ ♂
02 59	☽ □ ♀
06 35	☽ ± ☿
06 41	☽ Ᵽ ☉
10 23	☽ ‖ ♅
12 41	☽ ± ♂
15 03	☽ ☍ ♇
16 22	☽ # ♀
19 42	☽ # ♄
22 05	☽ Ᵽ ☿
31 Wednesday	
01 20	☿ ⊼ ♅
01 43	☽ Q ♄
04 52	☽ ‖ ♆
06 43	☽ Ᵽ ♂
08 51	☽ △ ♆
13 21	☽ △ ☉
15 46	☽ Q ♀
19 17	☿ ✶ ♄

All ephemeris data is given at 12.00 UT and the Moon's longitude is additionally given for 24.00 UT

JANUARY 2043

LONGITUDES

Date	Sidereal time h m s	Sun ☉ ° ' "	Moon ☽ ° ' "	Moon ☽ 24.00 ° ' "	Mercury ☿ ° '	Venus ♀ ° '	Mars ♂ ° '	Jupiter ♃ ° '	Saturn ♄ ° '	Uranus ♅ ° '	Neptune ♆ ° '	Pluto ♇ ° '
01	18 44 09	10 ♑ 56 51	21 ♍ 38 45	27 ♍ 43 12	15 ♑ 04	24 ♏ 07	22 ♑ 27	19 ♐ 03	16 ♏ 00	16 ♌ 45	07 ♉ 37	28 ♒ 14
02	18 48 06	11 58 00	03 ♎ 44 05	09 ♎ 42 05	13 R 46	25 09	23 14	19 16	16 05	16 R 43	07 R 37	28 15
03	18 52 02	12 59 09	15 ♎ 37 54	21 ♎ 32 16	12 24	26 12	24 01	19 29	16 10	16 41	07 36	28 16
04	18 55 59	14 00 19	27 ♎ 25 53	03 ♏ 19 26	11 03	27 15	24 47	19 42	16 15	16 39	07 36	28 18
05	18 59 55	15 01 29	09 ♏ 13 36	15 ♏ 09 01	09 45	28 18	25 34	19 55	16 20	16 37	07 35	28 19
06	19 03 52	16 02 39	21 ♏ 06 16	27 ♏ 05 53	08 32	29 ♏ 22	26 21	20 08	16 24	16 35	07 35	28 20
07	19 07 49	17 03 49	03 ♐ 08 22	09 ♐ 14 08	07 25	00 ♐ 26	27 08	20 21	16 29	16 33	07 35	28 22
08	19 11 45	18 04 59	15 ♐ 23 29	21 ♐ 36 42	06 27	01 31	27 55	20 34	16 34	16 31	07 34	28 23
09	19 15 42	19 06 10	27 ♐ 53 56	04 ♑ 15 17	05 39	02 35	28 41	20 47	16 38	16 29	07 34	28 24
10	19 19 38	20 07 20	10 ♑ 40 44	17 ♑ 10 12	05 00	03 40	29 ♑ 28	20 59	16 43	16 26	07 34	28 26
11	19 23 35	21 08 30	23 ♑ 43 33	00 ♒ 20 32	04 31	04 45	00 ♒ 15	21 12	16 47	16 24	07 33	28 27
12	19 27 31	22 09 40	07 ♒ 00 53	13 ♒ 44 20	04 12	05 51	01 02	21 24	16 52	16 22	07 33	28 29
13	19 31 28	23 10 49	20 ♒ 30 33	27 ♒ 19 14	04 02	06 56	01 49	21 37	16 56	16 19	07 33	28 30
14	19 35 24	24 11 58	04 ♓ 10 06	11 ♓ 02 51	04 D 02	08 02	02 36	21 49	17 00	16 17	07 33	28 32
15	19 39 21	25 13 06	17 ♓ 57 18	24 ♓ 53 14	04 09	09 08	03 23	22 02	17 04	16 15	07 33	28 33
16	19 43 18	26 14 14	01 ♈ 50 33	08 ♈ 49 07	04 25	10 15	04 10	22 14	17 08	16 12	07 33	28 34
17	19 47 14	27 15 21	15 ♈ 48 53	22 ♈ 49 47	04 47	11 21	04 58	22 27	17 12	16 10	07 33	28 36
18	19 51 11	28 16 26	29 ♈ 51 46	06 ♉ 54 46	05 16	12 28	05 45	22 39	17 16	16 07	07 D 33	28 37
19	19 55 07	29 ♑ 17 32	13 ♉ 58 42	21 ♉ 03 24	05 50	13 35	06 32	22 51	17 20	16 05	07 33	28 39
20	19 59 04	00 ♒ 18 36	28 ♉ 08 39	05 ♊ 14 13	06 30	14 42	07 19	23 03	17 23	16 02	07 33	28 41
21	20 03 00	01 19 39	12 ♊ 19 42	19 ♊ 24 43	07 14	15 50	08 06	23 15	17 27	16 00	07 33	28 42
22	20 06 57	02 20 42	26 ♊ 28 46	03 ♋ 31 18	08 03	16 57	08 53	23 27	17 30	15 57	07 33	28 44
23	20 10 53	03 21 44	10 ♋ 31 46	17 ♋ 29 35	08 56	18 05	09 41	23 39	17 34	15 55	07 33	28 45
24	20 14 50	04 22 44	24 ♋ 24 11	01 ♌ 15 04	09 52	19 13	10 28	23 51	17 37	15 52	07 34	28 47
25	20 18 47	05 23 44	08 ♌ 01 45	14 ♌ 43 52	10 51	20 21	11 15	24 03	17 40	15 50	07 34	28 48
26	20 22 43	06 24 43	21 ♌ 21 09	27 ♌ 53 25	11 53	21 29	12 02	24 14	17 43	15 47	07 34	28 50
27	20 26 40	07 25 42	04 ♍ 20 37	10 ♍ 42 48	12 58	22 38	12 50	24 26	17 46	15 45	07 34	28 52
28	20 30 36	08 26 39	17 ♍ 00 08	23 ♍ 12 53	14 05	23 46	13 37	24 38	17 49	15 42	07 35	28 53
29	20 34 33	09 27 36	29 ♍ 21 23	05 ♎ 26 06	15 14	24 55	14 24	24 49	17 52	15 39	07 35	28 55
30	20 38 29	10 28 32	11 ♎ 27 31	17 ♎ 26 13	16 25	26 04	15 12	25 00	17 55	15 37	07 36	28 57
31	20 42 26	11 ♒ 29 28	23 ♎ 22 46	29 ♎ 17 51	17 ♑ 38	27 ♐ 13	15 ♒ 59	25 ♐ 12	17 ♏ 57	15 ♌ 34	07 ♉ 36	28 ♒ 58

Date	Moon True ☊ ° '	Moon Mean ☊ ° '	Moon ☽ Latitude ° '
01	13 ♈ 23	13 ♈ 21	01 N 52
02	13 D 22	13 18	00 N 51
03	13 R 22	13 15	00 S 12
04	13 22	13 12	01 14
05	13 20	13 08	02 12
06	13 16	13 05	03 05
07	13 09	13 02	03 51
08	12 59	12 59	04 27
09	12 47	12 56	04 50
10	12 33	12 53	05 00
11	12 20	12 49	04 54
12	12 07	12 46	04 33
13	11 56	12 43	03 56
14	11 48	12 40	03 05
15	11 44	12 37	02 02
16	11 42	12 34	00 S 52
17	11 D 42	12 30	00 N 22
18	11 R 42	12 27	01 35
19	11 41	12 24	02 42
20	11 38	12 21	03 40
21	11 33	12 18	04 24
22	11 24	12 14	04 52
23	11 14	12 11	05 02
24	11 02	12 08	04 54
25	10 51	12 05	04 29
26	10 41	12 02	03 50
27	10 33	11 59	03 00
28	10 28	11 55	02 02
29	10 25	11 52	00 N 59
30	10 D 24	11 49	00 S 06
31	10 ♈ 25	11 ♈ 46	01 S 09

DECLINATIONS

Date	Sun ☉ ° '	Moon ☽ ° '	Mercury ☿ ° '	Venus ♀ ° '	Mars ♂ ° '	Jupiter ♃ ° '	Saturn ♄ ° '	Uranus ♅ ° '	Neptune ♆ ° '	Pluto ♇ ° '
01	22 S 59	05 N 02	20 S 26	15 S 37	22 S 36	22 S 30	14 S 28	16 N 29	12 N 21	22 S 04
02	22 54	00 S 42	20 18	15 52	22 28	22 31	14 29	16 30	12 21	22 03
03	22 48	06 20	20 11	16 07	22 20	22 32	14 31	16 30	12 21	22 03
04	22 42	11 42	20 06	16 22	22 12	22 33	14 32	16 31	12 21	22 02
05	22 35	16 39	20 02	16 36	22 04	22 34	14 33	16 32	12 21	22 02
06	22 28	21 01	20 00	16 51	21 55	22 35	14 34	16 32	12 21	22 01
07	22 21	24 33	19 59	17 05	21 46	22 36	14 35	16 33	12 21	22 00
08	22 13	27 03	19 59	17 19	21 37	22 37	14 37	16 34	12 21	22 00
09	22 05	28 15	20 01	17 32	21 28	22 38	14 38	16 34	12 21	21 59
10	21 56	27 59	20 04	17 46	21 18	22 39	14 39	16 35	12 21	21 59
11	21 46	26 11	20 09	17 59	21 08	22 40	14 40	16 36	12 21	21 58
12	21 37	22 55	20 14	18 12	20 58	22 41	14 41	16 36	12 21	21 57
13	21 27	18 22	20 21	18 24	20 48	22 42	14 42	16 37	12 21	21 57
14	21 16	12 51	20 28	18 36	20 37	22 43	14 43	16 38	12 21	21 56
15	21 05	06 38	20 36	18 48	20 26	22 43	14 44	16 39	12 21	21 55
16	20 54	00 S 04	20 44	19 00	20 15	22 44	14 45	16 39	12 21	21 55
17	20 42	06 N 34	20 52	19 11	20 04	22 45	14 46	16 40	12 21	21 54
18	20 30	12 54	21 01	19 22	19 53	22 46	14 47	16 41	12 21	21 54
19	20 18	18 37	21 09	19 32	19 41	22 46	14 47	16 42	12 21	21 53
20	20 05	23 19	21 17	19 42	19 29	22 47	14 48	16 42	12 21	21 52
21	19 52	26 38	21 25	19 52	19 17	22 48	14 49	16 43	12 21	21 52
22	19 38	28 15	21 32	20 01	19 05	22 48	14 50	16 44	12 21	21 51
23	19 24	28 02	21 39	20 10	18 52	22 49	14 51	16 45	12 21	21 50
24	19 10	26 03	21 45	20 18	18 39	22 50	14 51	16 46	12 21	21 50
25	18 55	22 35	21 51	20 26	18 26	22 50	14 52	16 46	12 22	21 49
26	18 40	18 01	21 55	20 33	18 13	22 51	14 53	16 47	12 22	21 49
27	18 25	12 42	21 59	20 40	18 00	22 51	14 53	16 48	12 22	21 48
28	18 09	07 00	22 02	20 46	17 46	22 52	14 54	16 49	12 22	21 47
29	17 53	01 N 09	22 04	20 52	17 33	22 52	14 55	16 49	12 22	21 47
30	17 37	04 S 37	22 04	20 58	17 19	22 53	14 55	16 50	12 23	21 46
31	17 S 20	10 S 09	22 S 04	21 S 02	17 S 05	22 S 53	14 S 56	16 N 51	12 N 23	21 S 45

ZODIAC SIGN ENTRIES

Date	h m	Planets
02	04 32	☽ ♎
04	17 14	☽ ♏
07	02 12	♀ ♐
07	05 47	☽ ♐
09	15 59	☽ ♑
11	04 09	♂ ♒
11	23 23	☽ ♒
14	04 42	☽ ♓
16	08 49	☽ ♈
18	12 14	☽ ♉
20	04 41	☉ ♒
20	15 08	☽ ♊
22	18 00	☽ ♋
24	21 48	☽ ♌
27	03 54	☽ ♍
29	13 16	☽ ♎

LATITUDES

Date	Mercury ☿ ° '	Venus ♀ ° '	Mars ♂ ° '	Jupiter ♃ ° '	Saturn ♄ ° '	Uranus ♅ ° '	Neptune ♆ ° '	Pluto ♇ ° '
01	02 N 10	03 N 16	01 S 02	00 N 29	02 N 15	00 N 42	01 S 47	10 S 40
04	02 53	03 16	01 03	00 29	02 15	00 42	01 47	10 40
07	03 15	03 14	01 03	00 29	02 16	00 42	01 47	10 39
10	03 16	03 11	01 04	00 29	02 17	00 43	01 47	10 39
13	03 02	03 06	01 04	00 29	02 17	00 43	01 47	10 39
16	02 38	03 01	01 04	00 28	02 18	00 43	01 47	10 38
19	02 10	02 55	01 05	00 28	02 18	00 43	01 46	10 38
22	01 39	02 48	01 05	00 28	02 19	00 43	01 46	10 38
25	01 09	02 40	01 05	00 28	02 19	00 43	01 46	10 38
28	00 40	02 31	01 05	00 28	02 20	00 43	01 46	10 38
31	00 N 12	02 N 22	01 S 05	00 N 28	02 N 21	00 N 43	01 S 46	10 S 37

DATA

Julian Date	2467251
Delta T	+78 seconds
Ayanamsa	24° 27' 25"
Synetic vernal point	04° ♓ 39' 34"
True obliquity of ecliptic	23° 26' 10"

MOON'S PHASES, APSIDES AND POSITIONS ☽

Date	h m	Phase	Longitude ° '	Eclipse Indicator
03	06 08	☾	12 ♎ 44	
11	06 53	●	20 ♑ 55	
18	09 05	☽	28 ♈ 09	
25	06 57	○	05 ♌ 11	

Day	h m		
04	20 00	Apogee	
19	20 20	Perigee	
02	09 02	0S	
09	19 50	Max dec	28° S 20'
16	12 13	0N	
22	21 08	Max dec	28° N 23'
29	16 45	0S	

LONGITUDES

Date	Chiron ⚷ ° '	Ceres ⚳ ° '	Pallas ⚴ ° '	Juno ⚵ ° '	Vesta ⚶ ° '	Black Moon Lilith ⚸ ° '
01	16 ♌ 45	17 ♑ 30	20 ♐ 29	12 ♐ 33	15 ♎ 03	02 ♏ 59
11	16 ♌ 10	21 ♑ 31	24 ♐ 37	15 ♐ 46	18 ♎ 02	04 ♏ 05
21	15 ♌ 29	25 ♑ 31	28 ♐ 40	18 ♐ 54	20 ♎ 35	05 ♏ 12
31	14 ♌ 45	29 ♑ 29	02 ♑ 35	21 ♐ 54	22 ♎ 37	06 ♏ 19

ASPECTARIAN

h m	Aspects
01 Thursday	
00 21	☽ △ ☿
00 53	☽ ⚹ ♄
02 28	☽ ⊻ ♅
04 53	♂ ⊥ ♇
06 50	☽ □ ♃
07 38	☉ ± ♅
13 42	☽ △ ♂
13 55	☽ ⊡ ♆
14 11	☽ ⊥ ♅
17 19	☽ ⚹ ♀
02 Friday	
01 02	☽ ⊼ ♇
03 07	♂ ∥ ♃
06 39	☽ ∠ ♄
07 45	☽ ± ♆
07 59	☽ ∠ ♅
13 02	☽ ± ♇
19 14	☽ Q ♃
19 47	☽ ⊼ ♆
20 56	☿ ∠ ♇
03 Saturday	
00 51	☽ ⊥ ♄
02 09	☽ ∠ ♀
06 08	☽ □ ☉
06 08	☽ □ ☿
06 10	☉ ☌ ☿
07 13	☽ ⊡ ♇
13 05	☽ ⊻ ♄
14 08	☽ ⚹ ♅
18 54	☉ ∠ ♇
19 59	☽ ⚹ ♃
22 11	☽ ⊥ ♀
04 Sunday	
00 03	☿ ∠ ♀
06 14	☽ □ ♂
11 36	☽ ⊻ ♀
13 46	☽ △ ♇
14 29	☽ Q ♅
14 58	☽ Q ☿
15 00	☽ # ♆
19 27	☿ ± ♅
22 12	☽ Q ☉
05 Monday	
01 27	☽ ∥ ♄
03 05	☽ ∠ ♃
03 52	♀ # ♅
08 41	☽ ☍ ♆
11 20	☽ # ♅
11 43	☽ ∥ ♀
12 13	♀ □ ♇
12 58	☽ ⚹ ☿
15 04	☉ ∥ ♃
17 46	♂ ∥ ♇
21 25	☽ Q ♂
21 41	☽ ⊥ ♃
06 Tuesday	
00 51	☽ ⚹ ☉
02 28	☽ ☌ ♄
02 55	☽ □ ♅
02 59	♂ ⊥ ♃
06 04	☽ ∥ ☿
10 01	☽ ⊻ ♃
16 26	☽ ∠ ☿
17 22	☽ ∥ ♂
18 12	☽ ∥ ♇
20 51	☽ ∥ ☉
21 16	☉ ⚹ ♄
21 58	☽ ∥ ♃
23 14	☽ ⚹ ♂
07 Wednesday	
00 18	☉ ⊼ ♅
02 30	☽ □ ♇
06 07	☽ ☌ ♀
08 27	☿ △ ♆
08 52	☽ ⊥ ☿
09 40	☽ ∠ ☉
19 48	☽ ⊻ ☿
20 44	☽ ⊼ ♆
23 23	☿ ⊥ ♀
08 Thursday	
01 19	♄ □ ♅
04 59	☽ ⊥ ☉
06 51	☽ ∠ ♂
08 28	☽ ± ♆
13 56	☽ Q ♇
14 10	☽ △ ♅
14 17	☽ ⊻ ♄
17 40	☽ ⊻ ☉
22 10	☽ ☌ ♃
09 Friday	
01 25	☽ ⊥ ♂
01 50	☽ ⊡ ♆
01 55	☽ ⊥ ♄
02 59	♂ ⊻ ♇
10 13	♂ Q ♄
12 58	☽ ⚹ ♇
13 36	☽ ⊻ ♂
18 45	☽ ⊡ ♅
19 07	☽ ∠ ♄
21 41	☽ ⊻ ♀
10 Saturday	
01 52	☽ ☌ ☿
03 40	☉ ∥ ♇
06 12	☽ △ ♆
09 57	☽ ⊥ ♀
11 33	☽ ± ♅
17 07	☽ ∠ ♇
22 37	☽ ⊼ ♅
23 13	☽ ⚹ ♄
11 Sunday	
04 05	☽ ∠ ♀
06 53	☽ ☌ ☉
07 19	☽ ⊻ ♃
08 12	☿ ⊻ ♀
09 41	☽ ⊥ ♇
13 39	☉ ⊻ ♃
18 25	☽ ⊥ ♃
20 36	☽ ⊻ ♇
21 14	☽ Q ♄
12 Monday	
00 35	☽ ☌ ♂
07 02	☽ ⊻ ☿
09 43	☽ ⚹ ♀
10 54	☽ ∠ ♃
12 58	☽ □ ♆
13 22	☽ ∥ ♃
17 36	☽ ⊥ ☿
17 36	☽ ∥ ♇
19 37	☉ ⊥ ♇
19 48	☽ ∥ ☉
23 27	☽ ∥ ♂
13 Tuesday	
02 30	☽ ∥ ☿
04 37	☽ ☍ ♅
05 38	☽ □ ♄
08 59	☽ Q ♀
09 25	☽ ∠ ☿
11 52	☽ ∥ ♀
13 59	☽ ⚹ ♃
17 06	☽ ⊻ ☉
20 00	☽ # ♅
20 53	☽ Q ♆
14 Wednesday	
01 23	♀ ⊼ ♆
01 51	☿ St D
02 06	☽ ☌ ♇
04 17	☽ ∥ ♄
04 30	☽ ⊥ ☉
09 06	☽ ⊻ ♂
11 23	☽ Q ♃
11 45	☽ ⚹ ☿
14 01	☽ # ♆
17 54	☽ ⚹ ♆
19 20	☽ □ ♀
20 13	☽ ⊥ ♂
21 29	☽ ∠ ☉
23 51	☿ ∥ ♂
15 Thursday	
08 50	☽ Q ☿
09 02	☽ ⊼ ♅
10 27	☽ △ ♄
12 48	☽ ∠ ♂
19 10	☽ □ ♃
19 24	☽ ± ♅
19 57	☽ ∠ ♆
16 Friday	
01 34	☽ ⚹ ☉
06 21	☽ ⊻ ♇
10 54	☽ ⊡ ♅
11 29	☽ ⊥ ♆
12 30	☽ ⊡ ♄
16 15	☽ ⚹ ♂
16 32	☽ □ ☿
16 43	☽ ⊥ ♇
21 49	☽ ⊻ ♆
23 52	☽ Q ☉
17 Saturday	
00 10	☉ ∥ ☿
00 50	☿ ⊻ ♂
03 42	☽ △ ♀
04 03	☽ ± ♄
08 12	☽ ∠ ♇
12 36	☽ △ ♅
14 05	☽ Q ♂
14 23	☽ ⊼ ♄
17 02	♆ St D
23 30	☽ △ ♃
18 Sunday	
00 17	♃ ⊡ ♆
07 34	☽ ⊡ ♀
09 05	☽ □ ☉
09 50	☽ ∥ ♆
09 53	☽ ⚹ ♇
19 33	☽ # ♄
20 29	☉ ⊻ ♇
21 34	☽ △ ☿
22 36	☽ □ ♂
22 58	☉ ⊥ ♃
19 Monday	
00 12	☽ ± ♀
01 05	☽ ☌ ♆
01 26	☽ ⊡ ♃
03 31	☽ ∥ ♅
06 20	☽ Q ♇
11 14	♀ ± ♆
11 16	☽ ⊼ ♀
12 50	☉ Q ♄
15 33	☽ □ ♅
16 25	☽ # ♀
16 47	☽ # ♂
16 56	☽ ± ♃
17 42	☽ ☍ ♄
19 36	☽ # ☉
21 42	♀ ∥ ♂
20 Tuesday	
00 11	☽ ⊡ ☿
00 33	☽ # ☿
03 15	☽ ⊼ ♃
03 57	☽ # ♇
08 55	☽ # ♃
12 54	☽ □ ♇
15 57	☽ △ ☉
16 12	☽ ± ☿
19 06	♂ □ ♆
21 57	☽ Q ♅
21 Wednesday	
02 54	☽ ⊼ ☿
03 55	☽ ⊻ ♆
04 26	☽ △ ♂
11 59	☉ ∥ ♀
14 04	☽ ⊥ ♆
15 32	♀ △ ♅
18 06	♂ ∠ ♃
18 12	☽ ⚹ ♅
18 26	☽ ☍ ♀
19 18	☽ ⊡ ☉
20 42	☽ ⊼ ♄
21 27	☿ △ ♆
22 Thursday	
05 20	☽ ∠ ♆
06 47	☽ ☍ ♃
06 55	☽ ± ♄
07 07	♀ Q ♇
07 20	☽ ⊡ ♂
11 45	☽ ± ☉
15 50	☽ △ ♇
19 36	☽ ∠ ♅
20 38	♃ ⊥ ♄
22 18	☽ ⊡ ♄
22 46	☽ ⊼ ☉
23 34	☽ ± ♂
23 Friday	
00 21	♀ ⊻ ♄
06 54	☽ ⚹ ♆
09 04	☽ ☍ ☿
10 27	☽ ⊼ ♂
10 57	☽ ⊥ ♅
17 33	☽ ⊡ ♇
21 15	☽ ⊻ ♅
24 Saturday	
00 10	☽ △ ♄
02 10	☽ ⊼ ♀
03 35	☽ Q ♆
09 10	☽ ± ♇
11 01	☽ ⊼ ♃
12 10	☿ ± ♅
13 32	☽ ± ♀
19 40	☽ ⊼ ♇
21 40	☽ ± ♃
25 Sunday	
06 11	☿ ∥ ♇
06 48	☽ ⊡ ♀
06 57	☽ ☍ ☉
10 33	☽ # ♃
11 10	☽ □ ♆
13 50	☽ ⊡ ♃
16 12	☽ # ☿
16 26	☽ # ♇
17 27	☽ ⊼ ☿
18 07	☽ ☍ ♂
21 38	☉ ∠ ♀
23 37	☽ # ♀
26 Monday	
01 56	☽ ☌ ♅
05 09	☽ ± ☿
05 22	☽ □ ♄
08 40	☽ # ☉
10 57	☽ # ♂
12 16	☽ △ ♀
17 22	☽ △ ♃
17 47	☽ ∥ ♅
23 03	☽ ⊡ ☿
27 Tuesday	
01 01	☿ ⊻ ♂
01 46	☽ ☍ ♇
02 26	☽ # ♄
10 56	♀ ⊡ ♆
13 28	☽ ∥ ♆
14 41	☽ Q ♄
15 29	☉ □ ♆
18 04	☽ △ ♆
18 18	☽ ⊼ ☉
28 Wednesday	
05 05	☽ ⊼ ♂
05 51	☽ △ ☿
06 40	☽ ± ☉
07 47	☿ ∠ ♇
09 31	☽ ⊻ ♅
13 07	♀ ⊥ ♄
13 35	☽ ⚹ ♄
17 22	☽ ± ♂
21 02	☽ ⊥ ♅
22 46	☽ ⊡ ♆
29 Thursday	
01 34	☽ ⊡ ☉
02 25	☽ □ ♀
02 59	☽ □ ♃
09 37	♀ ☌ ♃
11 08	☽ ⊼ ♇
12 06	☽ ⊡ ♂
14 33	☽ ∠ ♅
16 24	☽ ± ♆
18 56	☽ ∠ ♄
20 17	☿ ⊼ ♅
22 25	☉ ∠ ♃
23 00	☽ ± ♇
30 Friday	
04 17	☽ ⊼ ♆
09 51	☽ △ ☉
12 55	☽ ⊥ ♄
15 09	☽ Q ♃
16 59	☽ ⊡ ♇
17 46	☽ Q ♀
20 01	☽ △ ♂
20 18	☽ ⚹ ♅
23 05	☽ □ ☿
31 Saturday	
00 07	♂ ☍ ♅
01 00	☽ ⊻ ♄
15 44	☽ ⚹ ♃
18 25	☿ ⚹ ♄
20 28	☽ Q ♅
20 36	☽ ⚹ ♀
22 10	☽ # ♆
23 22	☽ △ ♇

All ephemeris data is given at 12.00 UT and the Moon's longitude is additionally given for 24.00 UT

FEBRUARY 2043

LONGITUDES

Date	Sidereal time h m s	Sun ☉ ° ' "	Moon ☽ ° ' "	Moon ☽ 24.00 ° ' "	Mercury ☿ ° '	Venus ♀ ° '	Mars ♂ ° '	Jupiter ♃ ° '	Saturn ♄ ° '	Uranus ♅ ° '	Neptune ♆ ° '	Pluto ♇ ° '
01	20 46 22	12 ♒ 30 22	05 ♏ 12 07	11 ♏ 06 14	18 ♑ 53	28 ♐ 22	16 ♒ 46	25 ♐ 23	18 ♏ 00	15 ♌ 31	07 ♉ 37	29 ♒ 00
02	20 50 19	13 31 16	17 ♏ 00 54	22 ♏ 56 48	20 09	29 ♐ 31	17 34	25 34	18 02	15 R 29	07 37	29 01
03	20 54 16	14 32 09	28 ♏ 54 36	04 ♐ 54 54	21 27	00 ♑ 41	18 21	25 45	18 05	15 26	07 38	29 03
04	20 58 12	15 33 02	10 ♐ 58 20	17 ♐ 05 26	22 46	01 50	19 08	25 56	18 07	15 24	07 38	29 05
05	21 02 09	16 33 53	23 ♐ 16 41	29 ♐ 32 29	24 06	03 00	19 56	26 07	18 09	15 21	07 39	29 06
06	21 06 05	17 34 44	05 ♑ 53 10	12 ♑ 18 58	25 28	04 09	20 43	26 18	18 11	15 18	07 40	29 08
07	21 10 02	18 35 33	18 ♑ 50 00	25 ♑ 26 15	26 50	05 19	21 31	26 29	18 13	15 16	07 40	29 10
08	21 13 58	19 36 22	02 ♒ 07 39	08 ♒ 53 58	28 14	06 29	22 18	26 39	18 15	15 13	07 41	29 12
09	21 17 55	20 37 09	15 ♒ 44 53	22 ♒ 40 00	29 ♑ 39	07 39	23 05	26 50	18 17	15 10	07 42	29 13
10	21 21 51	21 37 55	29 ♒ 38 47	06 ♓ 40 44	01 ♒ 05	08 50	23 53	27 00	18 18	15 08	07 43	29 15
11	21 25 48	22 38 40	13 ♓ 45 15	20 ♓ 51 45	02 32	10 00	24 40	27 11	18 20	15 05	07 43	29 17
12	21 29 44	23 39 23	27 ♓ 59 39	05 ♈ 08 23	04 00	11 10	25 28	27 21	18 21	15 03	07 44	29 18
13	21 33 41	24 40 04	12 ♈ 17 29	19 ♈ 26 29	05 28	12 21	26 15	27 31	18 23	15 00	07 45	29 20
14	21 37 38	25 40 44	26 ♈ 34 59	03 ♉ 42 41	06 58	13 31	27 03	27 41	18 24	14 57	07 46	29 22
15	21 41 34	26 41 23	10 ♉ 49 18	17 ♉ 54 37	08 29	14 42	27 50	27 51	18 25	14 55	07 47	29 24
16	21 45 31	27 41 59	24 ♉ 58 26	02 ♊ 00 38	10 00	15 52	28 38	28 01	18 26	14 52	07 48	29 25
17	21 49 27	28 42 34	09 ♊ 01 03	15 ♊ 59 34	11 33	17 03	29 ♒ 25	28 11	18 27	14 50	07 49	29 27
18	21 53 24	29 ♒ 43 07	22 ♊ 56 02	29 ♊ 50 20	13 06	18 14	00 ♓ 12	28 20	18 28	14 47	07 50	29 29
19	21 57 20	00 ♓ 43 39	06 ♋ 42 16	13 ♋ 31 42	14 40	19 25	01 00	28 30	18 29	14 45	07 51	29 30
20	22 01 17	01 44 08	20 ♋ 18 26	27 ♋ 02 17	16 15	20 36	01 47	28 39	18 29	14 42	07 52	29 32
21	22 05 13	02 44 36	03 ♌ 43 04	10 ♌ 20 35	17 52	21 47	02 35	28 48	18 30	14 40	07 54	29 34
22	22 09 10	03 45 02	16 ♌ 54 40	23 ♌ 25 12	19 29	22 58	03 22	28 58	18 30	14 37	07 55	29 36
23	22 13 07	04 45 26	29 ♌ 52 04	06 ♍ 15 13	21 07	24 09	04 10	29 07	18 30	14 35	07 56	29 37
24	22 17 03	05 45 48	12 ♍ 34 39	18 ♍ 50 24	22 45	25 21	04 57	29 16	18 31	14 33	07 57	29 39
25	22 21 00	06 46 09	25 ♍ 02 34	01 ♎ 11 21	24 25	26 32	05 44	29 25	18 31	14 30	07 59	29 41
26	22 24 56	07 46 28	07 ♎ 16 57	13 ♎ 19 40	26 06	27 43	06 32	29 33	18 R 31	14 28	08 00	29 42
27	22 28 53	08 46 45	19 ♎ 19 51	25 ♎ 17 54	27 48	28 ♑ 55	07 19	29 42	18 31	14 25	08 01	29 44
28	22 32 49	09 ♓ 47 02	01 ♏ 14 15	07 ♏ 09 24	29 ♒ 31	00 ♒ 06	08 ♓ 06	29 ♐ 50	18 ♏ 30	14 ♌ 23	08 ♉ 02	29 ♒ 46

MOON NODES AND LATITUDE; DECLINATIONS

Date	Moon True ☊ ° '	Moon Mean ☊ ° '	Moon ☽ Latitude ° '	Sun ☉ ° '	Moon ☽ ° '	Mercury ☿ ° '	Venus ♀ ° '	Mars ♂ ° '	Jupiter ♃ ° '	Saturn ♄ ° '	Uranus ♅ ° '	Neptune ♆ ° '	Pluto ♇ ° '
01	10 ♈ 26	11 ♈ 43	02 S 09	17 S 03	15 S 17	22 S 03	21 S 07	16 S 50	22 S 54	14 S 56	16 N 52	12 N 23	21 S 45
02	10 R 26	11 40	03 03	16 46	19 51	22 01	21 11	16 36	22 54	14 57	16 53	12 23	21 44
03	10 24	11 36	03 50	16 28	23 39	21 57	21 14	16 21	22 54	14 57	16 53	12 23	21 44
04	10 21	11 33	04 27	16 10	26 30	21 52	21 17	16 07	22 55	14 57	16 54	12 24	21 43
05	10 15	11 30	04 53	15 52	28 09	21 46	21 19	15 52	22 55	14 58	16 55	12 24	21 42
06	10 08	11 27	05 05	15 34	28 23	21 39	21 20	15 36	22 55	14 58	16 56	12 24	21 42
07	09 59	11 24	05 03	15 15	27 06	21 31	21 21	15 21	22 56	14 58	16 56	12 24	21 41
08	09 49	11 20	04 44	14 56	24 17	21 21	21 22	15 06	22 56	14 59	16 57	12 25	21 40
09	09 41	11 17	04 08	14 37	20 04	21 11	21 22	14 50	22 56	14 59	16 58	12 25	21 40
10	09 33	11 14	03 18	14 18	14 41	20 59	21 21	14 34	22 57	14 59	16 59	12 25	21 39
11	09 28	11 11	02 14	13 58	08 27	20 45	21 20	14 18	22 57	15 00	17 00	12 25	21 39
12	09 26	11 08	01 S 01	13 38	01 S 44	20 31	21 18	14 02	22 57	15 00	17 00	12 26	21 38
13	09 D 25	11 05	00 N 16	13 18	05 N 06	20 15	21 15	13 46	22 57	15 00	17 01	12 26	21 38
14	09 26	11 01	01 32	12 58	11 41	19 57	21 12	13 30	22 58	15 00	17 02	12 27	21 37
15	09 27	10 58	02 42	12 37	17 38	19 39	21 09	13 13	22 58	15 00	17 03	12 27	21 36
16	09 28	10 55	03 42	12 16	22 35	19 19	21 04	12 57	22 58	15 00	17 03	12 27	21 36
17	09 R 28	10 52	04 28	11 55	26 12	18 58	21 00	12 40	22 58	15 00	17 04	12 28	21 35
18	09 26	10 49	04 58	11 34	28 12	18 36	20 54	12 23	22 58	15 00	17 05	12 28	21 35
19	09 22	10 46	05 10	11 13	28 25	18 12	20 48	12 06	22 58	15 00	17 06	12 29	21 34
20	09 16	10 42	05 05	10 51	26 55	17 47	20 42	11 49	22 59	15 00	17 06	12 29	21 33
21	09 10	10 39	04 43	10 30	23 54	17 20	20 34	11 32	22 59	15 00	17 07	12 29	21 33
22	09 04	10 36	04 06	10 08	19 40	16 53	20 27	11 15	22 59	15 00	17 08	12 30	21 32
23	08 58	10 33	03 17	09 46	14 35	16 24	20 18	10 57	22 59	15 00	17 08	12 30	21 32
24	08 54	10 30	02 19	09 24	08 58	15 53	20 09	10 40	22 59	15 00	17 09	12 31	21 31
25	08 51	10 26	01 15	09 02	03 N 07	15 22	20 00	10 22	22 59	14 59	17 10	12 31	21 31
26	08 50	10 23	00 N 09	08 39	02 S 46	14 49	19 50	10 05	22 59	14 59	17 10	12 32	21 30
27	08 D 51	10 20	00 S 57	08 17	08 27	14 15	19 39	09 47	22 59	14 59	17 11	12 32	21 30
28	08 ♈ 52	10 ♈ 17	02 S 00	07 S 54	13 S 47	13 S 39	19 S 28	09 S 29	22 S 59	14 S 59	17 N 12	12 N 33	21 S 29

ZODIAC SIGN ENTRIES

Date	h m	Planets
01	01 26	☽ ♏
02	22 00	♀ ♑
03	14 11	☽ ♐
06	00 52	☽ ♑
08	08 12	☽ ♒
09	17 53	☿ ♒
10	12 36	☽ ♓
12	15 22	☽ ♈
14	17 45	☽ ♉
16	20 34	☽ ♊
18	05 43	♂ ♓
18	18 42	☉ ♓
19	00 17	☽ ♋
21	05 19	☽ ♌
23	12 15	☽ ♍
25	21 40	☽ ♎
28	09 30	☽ ♏
28	09 54	♀ ♒
28	18 44	☿ ♓

LATITUDES

Date	Mercury ☿ ° '	Venus ♀ ° '	Mars ♂ ° '	Jupiter ♃ ° '	Saturn ♄ ° '	Uranus ♅ ° '	Neptune ♆ ° '	Pluto ♇ ° '
01	00 N 03	02 N 19	01 S 05	00 N 28	02 N 21	00 N 43	01 S 46	10 S 37
04	00 S 22	02 09	01 05	00 28	02 22	00 43	01 45	10 37
07	00 45	01 58	01 05	00 28	02 22	00 43	01 45	10 37
10	01 05	01 48	01 05	00 27	02 23	00 43	01 45	10 37
13	01 23	01 37	01 04	00 27	02 23	00 43	01 45	10 37
16	01 39	01 26	01 04	00 27	02 24	00 43	01 45	10 38
19	01 51	01 14	01 03	00 27	02 25	00 43	01 45	10 38
22	02 00	01 03	01 03	00 27	02 25	00 43	01 44	10 38
25	02 06	00 52	01 02	00 27	02 26	00 43	01 44	10 38
28	02 09	00 40	01 02	00 27	02 27	00 43	01 44	10 38
31	02 S 07	00 N 29	01 S 01	00 N 27	02 N 27	00 N 43	01 S 44	10 S 39

DATA

Julian Date	2467282
Delta T	+78 seconds
Ayanamsa	24° 27' 30"
Synetic vernal point	04° ♓ 39' 29"
True obliquity of ecliptic	23° 26' 10"

LONGITUDES

Date	Chiron ⚷ ° '	Ceres ⚳ ° '	Pallas ⚴ ° '	Juno ⚵ ° '	Vesta ⚶ ° '	Black Moon Lilith ⚸ ° '
01	14 ♌ 40	29 ♑ 53	02 ♑ 58	22 ♐ 12	22 ♎ 47	06 ♏ 25
11	13 ♌ 55	03 ♒ 49	06 ♑ 44	25 ♐ 01	24 ♎ 09	07 ♏ 32
21	13 ♌ 11	07 ♒ 41	10 ♑ 19	27 ♐ 39	24 ♎ 47	08 ♏ 38
31	12 ♌ 32	11 ♒ 29	13 ♑ 42	00 ♑ 03	24 ♎ 38	09 ♏ 45

MOON'S PHASES, APSIDES AND POSITIONS ☽

Date	h m	Phase	Longitude ° '	Eclipse Indicator
02	04 15	☾	13 ♏ 12	
09	21 08	●	21 ♒ 00	
16	17 00	☽	27 ♉ 55	
23	21 58	○	05 ♍ 10	

Day	h m		
01	17 21	Apogee	
13	17 33	Perigee	
06	04 01	Max dec	28° S 29'
12	18 05	0N	
19	03 00	Max dec	28° N 33'
26	00 41	0S	

ASPECTARIAN

h m Aspects

01 Sunday
09 53 ♂ # ♅
10 19 ☽ ∥ ♄
15 49 ☽ Q ☿
16 54 ☽ ☍ ♆
19 28 ☽ ∥ ♂
19 59 ☽ # ♅
20 24 ☽ ∥ ☉
22 42 ☽ ∠ ♃

02 Monday
01 29 ♀ ⚹ ♇
02 58 ☉ # ♅
04 15 ☽ □ ☉
06 23 ☽ ∠ ♀
08 54 ☽ □ ♅
13 11 ☽ □ ♂
14 05 ☽ ☌ ♄
17 16 ☽ ⊥ ♃
19 08 ☽ ⚹ ☿
19 56 ☽ ∥ ♀
23 14 ☽ ∥ ♇

03 Tuesday
00 48 ☽ ∥ ☿
02 34 ☽ ⊥ ♀
03 19 ♂ □ ♄
05 34 ☽ ⊻ ♃
06 49 ☽ ∥ ♃
07 15 ♀ ⚼ ♅
12 17 ☽ □ ♇
15 55 ☽ ⊻ ♀
19 56 ☽ Q ☉

04 Wednesday
03 53 ☽ Q ♂
04 53 ☽ ∠ ☿
05 24 ☽ ⊼ ♆
08 26 ☉ ☍ ♅
17 15 ☽ ± ♆
17 49 ☿ ⊥ ♇
20 39 ☽ △ ♅
21 48 ☽ ⚹ ☉

05 Thursday
00 00 ☽ Q ♇
00 45 ☽ ⊥ ☿
02 02 ☽ ⊻ ♄
05 05 ☽ ⚹ ♂
10 47 ☽ ⚼ ♆
13 41 ☽ ⊥ ♄
13 47 ☽ ⊻ ☿
15 20 ♀ ∠ ♄
16 17 ☉ ∥ ♂
17 33 ☽ ☌ ♃
23 12 ☽ ⚹ ♇

06 Friday
01 29 ☽ ⚼ ♅
03 42 ☿ ∥ ♇
05 13 ☽ ∠ ☉
06 54 ☽ ∠ ♄
08 25 ☽ ☌ ♀
11 40 ☽ ∠ ♂
15 20 ☽ △ ♆
18 23 ☽ ± ♅
23 32 ☽ ⊥ ☉

07 Saturday
02 51 ☉ □ ♄
03 24 ☽ ∠ ♇
04 50 ☿ ⊻ ♃
05 28 ☽ ⊼ ♅
05 30 ☽ ⊥ ♂
10 52 ☽ ⚹ ♄
11 31 ☽ ⊻ ☉
17 12 ☽ ⊻ ♂
19 54 ☽ ⊥ ♇

08 Sunday
02 04 ☽ ⊻ ♃
04 13 ☽ ☌ ☿
06 45 ☽ ⊻ ♇
08 38 ☽ Q ♄
08 45 ☉ ∥ ♄
11 10 ☿ ∥ ♀
12 57 ☽ ⊥ ♃
20 28 ☽ ⊻ ♀
20 32 ☽ ∥ ♃
21 52 ☽ □ ♆
22 33 ♂ ∥ ♄

09 Monday
03 42 ☽ ∥ ♇
04 36 ☿ ⊻ ♇
05 04 ☽ ∠ ♃
05 21 ☽ ∥ ♀
06 05 ☽ ∥ ☿
08 01 ☽ ⊥ ♀
11 00 ☽ ☍ ♅
12 51 ♀ △ ♆
16 25 ☽ □ ♄
21 08 ☽ ☌ ☉
22 47 ☿ Q ♄

10 Tuesday
01 05 ☽ ∠ ♀
01 30 ☽ ☌ ♂
02 18 ☽ # ♅
05 14 ☽ Q ♆
07 25 ☽ ⚹ ♃
10 43 ☽ ∥ ♄
11 19 ☽ ☌ ♇
12 27 ☽ ∥ ♂
13 40 ☽ ∥ ☉
14 44 ☽ ⊻ ☿
15 29 ♀ ∠ ♂
18 02 ♀ ± ♅
20 59 ☽ # ♆

11 Wednesday
01 46 ☽ ⚹ ♆
02 08 ☽ ⊥ ☿
04 09 ☽ Q ♃
05 04 ☽ ⚹ ♀
14 15 ☽ ⊼ ♅
19 06 ☽ ∠ ☿
19 45 ☽ △ ♄

12 Thursday
00 07 ☿ ⊥ ♃
00 21 ☽ ± ♅
03 09 ☽ ∠ ♆
03 10 ☽ Q ♀
04 09 ☽ ⊻ ☉
07 30 ☽ ⊻ ♂
10 54 ☽ □ ♃
14 13 ☽ ⊻ ♇
15 00 ☽ ⊥ ☉
15 26 ☽ ⚼ ♅
18 10 ☽ ⊥ ♂
18 18 ☽ ⊥ ♆
20 33 ♂ Q ♆
21 01 ☽ ⚼ ♄
23 14 ☽ ⚹ ☿

13 Friday
00 18 ☽ ⊥ ♇
04 23 ☽ ⊻ ♆
07 16 ☽ ∠ ☉
10 09 ☽ ∠ ♂
12 06 ☽ □ ♀
12 09 ☽ ± ♄
15 26 ☽ ∠ ♇
16 32 ☽ △ ♅
21 42 ☽ Q ☿
22 14 ☽ ⊼ ♄

14 Saturday
10 22 ☽ ⚹ ☉
12 49 ☽ ⚹ ♂
13 53 ☽ △ ♃
14 11 ☉ Q ♆
14 56 ☽ ∥ ♆
16 40 ☽ # ☉
16 41 ☽ ⚹ ♇
18 43 ☽ # ♂

15 Sunday
00 55 ☿ □ ♆
00 59 ☽ # ♄
05 38 ♀ ∠ ♇
06 52 ☽ ☌ ♆
07 34 ☽ □ ☿
08 07 ☽ Q ☉
09 28 ☽ ∥ ♅
10 14 ☽ Q ♂
12 39 ♂ ⚹ ♃
12 58 ☽ Q ♇
15 28 ☽ ⚼ ♃
16 18 ♀ ⊼ ♅
18 54 ☽ □ ♅
19 09 ☽ △ ♀
20 34 ☽ # ☿
23 15 ☉ # ♆

16 Monday
00 53 ☽ ☍ ♄
04 08 ☽ # ♀
06 41 ☽ # ♇
06 55 ☽ ± ♃
14 06 ☽ # ♃
17 00 ☽ □ ☉
17 15 ☽ ⊼ ♃
18 36 ☽ □ ♂
19 36 ☽ □ ♇
20 57 ☉ ⚹ ♃
22 59 ☽ ⚼ ♀

17 Tuesday
01 26 ☽ Q ♅
09 57 ☽ ⊻ ♆
13 02 ♂ ☌ ♇
15 49 ☽ ± ♀
16 53 ☽ △ ☿
20 16 ☽ ⊥ ♆
21 58 ☽ ⚹ ♅

18 Wednesday
03 07 ☽ ⊼ ♀
04 16 ☽ ⊼ ♄
05 12 ♂ # ♆
06 06 ☉ ☌ ♇
11 50 ☽ ∠ ♆
14 39 ☽ ± ♄
16 04 ☿ ∠ ♃
16 43 ♀ ⚹ ♄
21 30 ☽ ☍ ♃
22 07 ☽ ⚼ ☿
23 24 ☽ △ ♇
23 53 ☽ ∠ ♅

19 Thursday
00 43 ☽ △ ☉
01 25 ☽ △ ♂
06 21 ☽ ⚼ ♄
13 07 ☿ ☍ ♅
14 01 ☽ ⚹ ♆
15 34 ☽ ⊥ ♅
15 54 ☽ ± ☿

20 Friday
01 45 ☽ ⚼ ♇
02 06 ☽ ⊻ ♅
03 52 ☽ ⊼ ☿
05 10 ☽ ⚼ ☉
05 22 ☽ ⚼ ♂
08 46 ☽ △ ♄
11 14 ☽ Q ♆
12 34 ☽ ☍ ♀
17 45 ☽ ± ♇
17 46 ☉ ☌ ♂
22 22 ☽ ± ♂
22 27 ☽ ± ☉

21 Saturday
03 04 ☽ ⊼ ♃
04 31 ☽ ⊼ ♇
09 49 ☽ ⊼ ♂
10 06 ☽ ⊼ ☉
13 59 ☽ ± ♃
17 44 ☽ # ♃
19 34 ☽ □ ♆
21 31 ☿ □ ♄
23 32 ☿ # ♅

22 Sunday
02 03 ☽ # ♇
06 32 ☽ ⚼ ♃
07 49 ☽ ☌ ♅
07 50 ☽ # ♀
14 55 ☽ □ ♄
17 23 ☽ ☍ ☿

23 Monday
00 17 ☽ ⊼ ♀
00 21 ☽ ∥ ♅
00 55 ♀ ⊥ ♇
02 56 ☽ # ☿
10 09 ☽ # ♄
10 34 ☽ △ ♃
11 32 ☽ ☍ ♇
12 36 ☽ ± ♀
20 35 ☽ ☍ ♂
21 04 ☽ ∥ ♆
21 58 ☽ ☍ ☉

24 Tuesday
00 29 ☽ Q ♄
03 12 ☽ △ ♆
04 32 ☽ # ♂
07 18 ☽ ⚼ ♀
10 07 ☽ # ☉
15 45 ☽ ⊻ ♅
23 22 ☽ ⚹ ♄

25 Wednesday
03 14 ☽ ⊥ ♅
07 59 ☽ ⚼ ♆
10 36 ☽ ⊼ ☿
15 12 ☽ △ ♀
20 36 ☽ □ ♃
20 40 ☽ ∠ ♅
21 04 ☽ ⊼ ♇
23 39 ♄ St R

26 Thursday
00 10 ☽ ± ☿
00 44 ♃ ⚼ ♅
01 34 ☽ ± ♆
04 26 ☿ ∥ ♄
04 34 ☽ ∠ ♄
08 53 ☽ ± ♇
10 24 ☽ ⊼ ♂
10 28 ☿ Q ♆
13 04 ☽ ⊼ ☉
13 25 ☽ ⊼ ♆
17 25 ☉ ⚹ ♆
20 49 ☽ ⚼ ☿
22 23 ☽ ⊥ ♄
23 08 ☽ ± ♂

27 Friday
02 04 ☽ ± ☉
02 13 ☽ ⚹ ♅
02 47 ☽ ⚼ ♇
08 41 ☽ Q ♃
10 21 ☽ ⊻ ♄
11 18 ☽ ∥ ☉
17 32 ☽ ∥ ♂
18 25 ☽ ⚼ ♂
20 52 ♃ ⚹ ♇
21 46 ☽ ⚼ ☉

28 Saturday
02 14 ☽ Q ♅
04 57 ♀ ⊻ ♇
05 49 ♀ ⊻ ♃
06 15 ☽ # ♆
07 55 ☽ △ ☿
09 00 ☽ △ ♇
09 07 ☽ ⚹ ♃
09 27 ☽ □ ♀
09 56 ♂ ⚹ ♆
11 28 ☽ ∥ ☿
15 29 ☿ ☌ ♇
16 48 ☿ ⚹ ♃
17 45 ☽ ∥ ♄
20 05 ♀ Q ♄

All ephemeris data is given at 12.00 UT and the Moon's longitude is additionally given for 24.00 UT

MARCH 2043

LONGITUDES

Date	Sidereal time h m s	Sun ☉ ° ' "	Moon ☽ ° ' "	Moon ☽ 24.00 ° ' "	Mercury ☿ ° '	Venus ♀ ° '	Mars ♂ ° '	Jupiter ♃ ° '	Saturn ♄ ° '	Uranus ♅ ° '	Neptune ♆ ° '	Pluto ♇ ° '
01	22 36 46	10 ♓ 47 16	13 ♏ 03 53	18 ♏ 58 16	01 ♓ 15	01 ♒ 18	08 ♓ 54	29 ♐ 58	18 ♏ 30	14 ♌ 21	08 ♉ 04	29 ♒ 47
02	22 40 42	11 47 29	24 ♏ 53 08	00 ♐ 49 05	03 00	02 29	09 41	00 ♑ 06	18 R 30	14 R 19	08 05	29 49
03	22 44 39	12 47 41	06 ♐ 46 45	12 ♐ 46 46	04 46	03 41	10 28	00 14	18 29	14 16	08 07	29 51
04	22 48 36	13 47 51	18 ♐ 49 44	24 ♐ 56 16	06 33	04 53	11 16	00 22	18 29	14 14	08 08	29 52
05	22 52 32	14 47 59	01 ♑ 06 55	07 ♑ 22 15	08 21	06 05	12 03	00 30	18 28	14 12	08 10	29 54
06	22 56 29	15 48 06	13 ♑ 42 42	20 ♑ 08 42	10 10	07 16	12 50	00 38	18 27	14 10	08 11	29 56
07	23 00 25	16 48 12	26 ♑ 40 32	03 ♒ 18 27	12 00	08 28	13 38	00 45	18 26	14 08	08 13	29 57
08	23 04 22	17 48 15	10 ♒ 02 31	16 ♒ 52 42	13 52	09 40	14 25	00 53	18 25	14 06	08 14	29 ♒ 59
09	23 08 18	18 48 17	23 ♒ 48 51	00 ♓ 50 38	15 44	10 52	15 12	01 00	18 24	14 04	08 16	00 ♓ 01
10	23 12 15	19 48 18	07 ♓ 57 37	15 ♓ 09 14	17 37	12 04	16 00	01 07	18 23	14 02	08 17	00 02
11	23 16 11	20 48 16	22 ♓ 24 46	29 ♓ 43 27	19 32	13 16	16 47	01 14	18 21	14 00	08 19	00 04
12	23 20 08	21 48 12	07 ♈ 04 27	14 ♈ 26 51	21 27	14 28	17 34	01 21	18 20	13 58	08 21	00 06
13	23 24 05	22 48 06	21 ♈ 49 49	29 ♈ 12 27	23 24	15 40	18 21	01 27	18 19	13 56	08 22	00 07
14	23 28 01	23 47 59	06 ♉ 33 57	13 ♉ 53 36	25 21	16 53	19 08	01 34	18 17	13 54	08 24	00 09
15	23 31 58	24 47 49	21 ♉ 10 44	28 ♉ 24 50	27 19	18 05	19 55	01 40	18 15	13 52	08 26	00 10
16	23 35 54	25 47 37	05 ♊ 35 26	12 ♊ 42 24	29 ♓ 17	19 17	20 43	01 46	18 13	13 51	08 28	00 12
17	23 39 51	26 47 23	19 ♊ 44 54	26 ♊ 43 24	01 ♈ 17	20 29	21 30	01 52	18 11	13 49	08 29	00 14
18	23 43 47	27 47 06	03 ♋ 37 36	10 ♋ 27 30	03 16	21 41	22 17	01 58	18 09	13 47	08 31	00 15
19	23 47 44	28 46 47	17 ♋ 13 08	23 ♋ 54 37	05 16	22 54	23 04	02 04	18 07	13 45	08 33	00 17
20	23 51 40	29 ♓ 46 26	00 ♌ 32 02	07 ♌ 05 31	07 16	24 06	23 51	02 09	18 05	13 44	08 35	00 18
21	23 55 37	00 ♈ 46 03	13 ♌ 35 14	20 ♌ 01 20	09 15	25 18	24 38	02 15	18 03	13 42	08 37	00 20
22	23 59 34	01 45 37	26 ♌ 23 59	02 ♍ 43 19	11 14	26 31	25 25	02 20	18 01	13 41	08 39	00 21
23	00 03 30	02 45 09	08 ♍ 59 30	15 ♍ 12 44	13 12	27 43	26 12	02 25	17 58	13 39	08 41	00 23
24	00 07 27	03 44 39	21 ♍ 23 08	27 ♍ 30 55	15 09	28 ♒ 56	26 59	02 30	17 56	13 38	08 42	00 24
25	00 11 23	04 44 06	03 ♎ 36 14	09 ♎ 39 18	17 04	00 ♓ 08	27 46	02 35	17 53	13 37	08 44	00 26
26	00 15 20	05 43 32	15 ♎ 40 19	21 ♎ 39 32	18 58	01 20	28 32	02 39	17 50	13 35	08 46	00 27
27	00 19 16	06 42 56	27 ♎ 37 12	03 ♏ 33 35	20 48	02 33	29 ♓ 19	02 44	17 48	13 34	08 48	00 29
28	00 23 13	07 42 18	09 ♏ 29 00	15 ♏ 23 49	22 36	03 45	00 ♈ 06	02 48	17 45	13 33	08 50	00 30
29	00 27 09	08 41 38	21 ♏ 18 22	27 ♏ 13 05	24 21	04 58	00 53	02 52	17 42	13 32	08 52	00 31
30	00 31 06	09 40 56	03 ♐ 08 24	09 ♐ 04 46	26 02	06 11	01 40	02 56	17 39	13 30	08 54	00 33
31	00 35 03	10 ♈ 40 12	15 ♐ 02 40	21 ♐ 02 39	27 ♈ 39	07 ♓ 23	02 ♈ 26	02 ♑ 59	17 ♏ 36	13 ♌ 29	08 ♉ 56	00 ♓ 34

Date	Moon True ☊ ° '	Moon Mean ☊ ° '	Moon ☽ Latitude ° '
01	08 ♈ 54	10 ♈ 14	02 S 57
02	08 D 56	10 11	03 46
03	08 57	10 07	04 27
04	08 R 57	10 04	04 56
05	08 56	10 01	05 12
06	08 54	09 58	05 14
07	08 51	09 55	05 00
08	08 48	09 51	04 29
09	08 45	09 48	03 43
10	08 42	09 45	02 42
11	08 41	09 42	01 29
12	08 40	09 39	00 S 09
13	08 D 40	09 36	01 N 12
14	08 41	09 32	02 28
15	08 42	09 29	03 34
16	08 43	09 26	04 26
17	08 44	09 23	05 00
18	08 R 44	09 20	05 16
19	08 43	09 17	05 13
20	08 42	09 13	04 54
21	08 41	09 10	04 20
22	08 40	09 07	03 34
23	08 40	09 04	02 37
24	08 39	09 01	01 34
25	08 39	08 57	00 N 28
26	08 D 39	08 54	00 S 39
27	08 39	08 51	01 43
28	08 39	08 48	02 43
29	08 R 39	08 45	03 35
30	08 39	08 42	04 19
31	08 ♈ 39	08 ♈ 38	04 S 51

DECLINATIONS

Date	Sun ☉ ° '	Moon ☽ ° '	Mercury ☿ ° '	Venus ♀ ° '	Mars ♂ ° '	Jupiter ♃ ° '	Saturn ♄ ° '	Uranus ♅ ° '	Neptune ♆ ° '	Pluto ♇ ° '
01	07 S 31	18 S 34	13 S 02	19 S 16	09 S 11	22 S 59	14 S 58	17 N 12	12 N 33	21 S 28
02	07 08	22 39	12 24	19 04	08 53	22 59	14 58	17 13	12 34	21 28
03	06 45	25 49	11 44	18 51	08 35	22 59	14 58	17 14	12 34	21 27
04	06 22	27 53	11 03	18 38	08 17	22 59	14 57	17 14	12 35	21 27
05	05 59	28 38	10 21	18 24	07 58	22 59	14 57	17 15	12 35	21 26
06	05 36	27 56	09 38	18 09	07 40	22 59	14 57	17 16	12 36	21 26
07	05 13	25 43	08 53	17 54	07 22	22 59	14 56	17 16	12 36	21 25
08	04 49	22 03	08 07	17 39	07 03	22 59	14 56	17 17	12 37	21 25
09	04 26	17 05	07 20	17 23	06 45	22 59	14 55	17 17	12 37	21 25
10	04 02	11 05	06 32	17 06	06 26	22 59	14 55	17 18	12 38	21 24
11	03 39	04 S 22	05 43	16 50	06 08	22 59	14 54	17 18	12 39	21 24
12	03 15	02 N 40	04 52	16 32	05 49	22 59	14 54	17 19	12 39	21 23
13	02 51	09 37	04 01	16 14	05 30	22 59	14 53	17 19	12 40	21 23
14	02 28	16 02	03 08	15 56	05 11	22 59	14 52	17 20	12 40	21 22
15	02 04	21 30	02 15	15 37	04 53	22 59	14 52	17 21	12 41	21 22
16	01 40	25 35	01 20	15 18	04 34	22 59	14 51	17 21	12 41	21 21
17	01 17	28 01	00 S 26	14 58	04 15	22 59	14 50	17 21	12 42	21 21
18	00 53	28 39	00 N 30	14 38	03 56	22 59	14 50	17 22	12 43	21 21
19	00 29	27 30	01 26	14 17	03 37	22 59	14 49	17 22	12 43	21 20
20	00 S 05	24 49	02 22	13 56	03 18	22 59	14 48	17 23	12 44	21 20
21	00 N 18	20 54	03 19	13 35	02 59	22 59	14 47	17 23	12 45	21 19
22	00 42	16 04	04 16	13 13	02 40	22 58	14 46	17 24	12 45	21 19
23	01 06	10 38	05 12	12 51	02 21	22 58	14 46	17 24	12 46	21 19
24	01 29	04 N 52	06 08	12 29	02 02	22 58	14 45	17 24	12 46	21 18
25	01 53	01 S 00	07 03	12 06	01 43	22 58	14 44	17 25	12 47	21 18
26	02 16	06 46	07 57	11 43	01 25	22 58	14 43	17 25	12 48	21 18
27	02 40	12 14	08 50	11 19	01 06	22 58	14 42	17 25	12 48	21 17
28	03 03	17 13	09 42	10 56	00 47	22 58	14 41	17 26	12 49	21 17
29	03 27	21 33	10 32	10 32	00 28	22 58	14 40	17 26	12 50	21 17
30	03 50	25 01	11 20	10 07	00 09	22 58	14 39	17 26	12 50	21 16
31	04 N 13	27 S 25	12 N 06	09 S 43	00 N 10	22 S 58	14 S 38	17 N 27	12 N 51	21 S 16

ZODIAC SIGN ENTRIES

Date	h m	Planets
01	17 06	♃ ♑
02	22 21	☽ ♐
05	09 51	☽ ♑
07	18 02	☽ ♒
09	00 45	♇ ♓
09	22 34	☽ ♓
12	00 27	☽ ♈
14	01 17	☽ ♉
16	02 39	☽ ♊
16	20 35	☿ ♈
18	05 41	☽ ♋
20	11 02	☽ ♌
20	17 28	☉ ♈
22	18 49	☽ ♍
25	04 53	☽ ♎
25	09 22	♀ ♓
27	16 48	☽ ♏
28	08 55	♂ ♈
30	05 38	☽ ♐

LATITUDES

Date	Mercury ☿ ° '	Venus ♀ ° '	Mars ♂ ° '	Jupiter ♃ ° '	Saturn ♄ ° '	Uranus ♅ ° '	Neptune ♆ ° '	Pluto ♇ ° '
01	02 S 09	00 N 37	01 S 02	00 N 27	02 N 27	00 N 43	01 S 44	10 S 38
04	02 06	00 26	01 01	00 27	02 28	00 43	01 44	10 39
07	01 59	00 15	01 00	00 27	02 28	00 43	01 44	10 39
10	01 47	00 N 04	00 59	00 27	02 29	00 43	01 44	10 39
13	01 31	00 S 06	00 59	00 27	02 30	00 43	01 44	10 40
16	01 09	00 16	00 58	00 27	02 30	00 43	01 43	10 40
19	00 43	00 26	00 57	00 27	02 31	00 43	01 43	10 41
22	00 S 12	00 35	00 56	00 27	02 31	00 42	01 43	10 41
25	00 N 22	00 43	00 55	00 26	02 32	00 42	01 43	10 42
28	00 58	00 51	00 53	00 26	02 32	00 42	01 43	10 43
31	01 N 34	00 S 59	00 S 52	00 N 26	02 N 33	00 N 42	01 S 43	10 S 43

DATA

Julian Date	2467310
Delta T	+78 seconds
Ayanamsa	24° 27' 34"
Synetic vernal point	04° ♓ 39' 25"
True obliquity of ecliptic	23° 26' 11"

MOON'S PHASES, APSIDES AND POSITIONS ☽

Date	h m	Phase	Longitude ° '	Eclipse Indicator
04	01 07	☾	13 ♐ 21	
11	09 09	●	20 ♓ 41	
18	01 03	☽	27 ♊ 20	
25	14 26	○	04 ♎ 50	total

Day	h m		
01	13 32	Apogee	
13	09 08	Perigee	
29	05 17	Apogee	
05	12 43	Max dec	28° S 38'
12	02 56	0N	
18	08 15	Max dec	28° N 40'
25	07 53	0S	

LONGITUDES

Date	Chiron ⚷ ° '	Ceres ⚳ ° '	Pallas ⚴ ° '	Juno ⚵ ° '	Vesta ⚶ ° '	Black Moon Lilith ⚸ ° '
01	12 ♌ 39	10 ♒ 44	13 ♑ 02	29 ♐ 35	24 ♎ 44	09 ♏ 32
11	12 ♌ 05	14 ♒ 27	16 ♑ 13	01 ♑ 46	23 ♎ 56	10 ♏ 38
21	11 ♌ 40	18 ♒ 04	19 ♑ 06	03 ♑ 38	22 ♎ 21	11 ♏ 45
31	11 ♌ 23	21 ♒ 33	21 ♑ 40	05 ♑ 08	20 ♎ 11	12 ♏ 52

ASPECTARIAN

h m	Aspects
01 Sunday	
01 49	☽ ☍ ♆
02 56	☽ △ ♂
04 50	☽ ⫲ ♅
06 57	☽ △ ☉
14 11	☿ ⊻ ♀
14 36	☽ □ ♅
15 40	☽ ∥ ♀
15 55	☽ ∠ ♃
23 02	☽ ☌ ♄
02 Monday	
02 05	☽ Q ♀
04 33	☽ ∥ ♇
05 56	☿ ⫲ ♆
10 24	☽ ⊥ ♃
14 16	☽ ∥ ♃
20 43	☉ Q ♃
22 00	☽ □ ♇
22 41	☽ ⊻ ♃
03 Tuesday	
05 05	☽ ⚹ ♀
07 14	☽ □ ☿
14 41	☽ ⊼ ♆
19 55	☽ □ ♂
04 Wednesday	
01 07	☽ □ ☉
02 41	☽ ± ♆
02 56	☽ △ ♅
10 07	☽ Q ♇
11 18	☽ ⊻ ♄
14 18	☽ ∠ ♀
20 29	☽ ꝑ ♆
22 09	☉ ⊼ ♅
23 05	☽ ⊥ ♄
05 Thursday	
01 10	☽ Q ☿
08 18	☽ ꝑ ♅
09 30	☿ ⚹ ♆
09 39	☽ ⚹ ♇
09 47	☽ ⊥ ♀
09 48	☽ Q ♂
10 32	♀ ⊥ ♂
10 48	☽ ☌ ♃
15 32	☽ Q ☉
16 31	☽ ∠ ♄
21 33	♀ ⊥ ♃
22 32	☽ ⊻ ♀
06 Friday	
01 32	☽ △ ♆
01 33	☽ ± ♅
04 11	☽ ⚹ ☿
04 21	♂ Q ♃
10 15	☽ ⚹ ♂
12 51	☽ ⊼ ♅
14 17	☽ ∠ ♇
16 15	☽ ⚹ ☉
20 51	☽ ⚹ ♄
07 Saturday	
06 40	♀ □ ♆
07 01	☽ ⊥ ♇
12 42	☽ ∠ ☿
15 47	☽ ∠ ♂
17 58	☽ ⊻ ♇
18 49	☽ Q ♄
19 28	☽ ⊻ ♃
22 03	☽ ∠ ☉
22 26	☿ Q ♃
08 Sunday	
02 38	♂ ⊼ ♅
06 20	☽ ⊥ ♃
06 39	☽ ∥ ♃
07 31	☽ ⊥ ☿
08 48	☽ □ ♆
08 57	☽ ⊥ ♂
11 17	☽ ☌ ♀
14 59	☿ ⊼ ♅
15 21	☽ ⊥ ☉
15 25	☽ ∥ ♇
19 07	☽ ☍ ♅
19 47	☽ ⊻ ☿
20 10	☽ ⊻ ♂
22 21	☽ ∠ ♃
09 Monday	
00 23	☿ ☌ ♂
02 28	☉ △ ♄
02 40	☽ □ ♄
02 41	☽ ⊻ ☉
10 38	☽ ∥ ♀
11 07	☽ ⫲ ♅
16 13	☽ Q ♆
20 02	♀ ⫲ ♅
21 05	☽ ∥ ♄
22 37	☽ ☌ ♇
10 Tuesday	
00 22	☽ ⚹ ♃
06 07	☽ ⫲ ♆
12 33	☽ ⚹ ♆
16 40	☿ ∥ ♂
17 12	☉ ± ♅
19 30	☽ ⊻ ♀
20 41	☽ Q ♃
21 27	☿ △ ♄
22 06	☽ ⊼ ♅
11 Wednesday	
02 10	☽ ☌ ♂
03 16	☿ ⊥ ♀
05 19	☽ △ ♄
05 35	☽ ∥ ♂
06 21	☽ ⊥ ♀
06 31	☽ ☌ ☿
06 41	☽ ∥ ☿
08 01	☽ ± ♅
09 09	☽ ☌ ☉
13 30	☽ ∠ ♆
14 39	☽ ∥ ☉
17 44	☿ ± ♅
22 29	☽ ∠ ♀
22 47	☽ ꝑ ♅
12 Thursday	
00 35	☽ ⊻ ♇
02 07	♀ ☍ ♅
02 35	☽ □ ♃
04 16	☽ ⊥ ♆
05 55	☽ ꝑ ♄
10 24	☽ ⊥ ♇
13 52	☽ ⫲ ☉
14 05	☽ ⊻ ♆
18 41	☽ ⫲ ☿
20 33	☽ ± ♄
20 58	☉ ☌ ☿
22 16	☽ ⫲ ♂
23 11	☽ △ ♅
13 Friday	
01 04	☽ ∠ ♇
01 06	☽ ⚹ ♀
06 02	☽ ⊻ ♂
06 17	☽ ⊼ ♄
10 42	♂ △ ♄
11 47	☿ ∠ ♆
13 42	☽ ⊻ ☉
14 56	☽ ⊻ ☿
16 20	☽ ⊥ ♂
22 21	☽ Q ♀
23 02	☽ ∥ ♆
14 Saturday	
00 10	☽ ⊥ ☉
01 31	☽ ⚹ ♇
02 11	☽ ⊥ ☿
02 11	☉ ∠ ♆
03 46	☽ △ ♃
05 07	♀ ∠ ♃
07 24	☽ ⫲ ♄
07 49	☽ ∠ ♂
11 35	☽ ⫲ ♀
15 01	☽ ☌ ♆
15 55	☽ ∠ ☉
17 16	☽ ∥ ♅
19 08	☽ ∠ ☿
21 09	☽ Q ♇
23 59	☽ □ ♅
15 Sunday	
04 30	☽ ꝑ ♃
06 25	☽ □ ♀
07 11	☽ ☍ ♄
09 49	☽ ⚹ ♂
10 27	♂ ± ♅
11 21	☽ ⫲ ♇
15 24	♀ □ ♄
18 26	☽ ⚹ ☉
19 29	☽ ± ♃
19 45	☽ ⫲ ♃
20 30	☉ ∥ ☿
23 47	☽ ⚹ ☿
16 Monday	
02 58	☽ □ ♇
05 33	☽ ⊼ ♃
05 44	☽ Q ♅
06 39	☿ ꝑ ♅
06 53	☽ Q ♂
15 59	☽ Q ☉
16 50	☽ ⊻ ♆
23 09	☿ ⊻ ♇
23 10	☽ Q ☿
17 Tuesday	
01 54	☽ ⚹ ♅
03 01	☽ ⊥ ♆
09 21	☽ ⊼ ♄
13 23	☽ △ ♀
15 10	☽ □ ♂
18 26	☽ ∠ ♆
19 31	☿ □ ♃
19 37	☽ ± ♄
21 35	♀ ∥ ♄
18 Wednesday	
01 03	☽ □ ☉
02 51	☿ ⊥ ♆
03 35	☽ ∠ ♅
06 07	☽ △ ♇
09 05	☽ ☍ ♃
10 41	☿ ꝑ ♄
11 11	☽ ꝑ ♄
11 16	☽ □ ☿
17 53	☽ ꝑ ♀
18 52	☉ ⫲ ☿
19 17	☽ ⊥ ♅
20 36	☽ ⚹ ♆
22 52	☉ ⊥ ♀
19 Thursday	
05 51	☽ ⊻ ♅
08 32	☽ ꝑ ♇
11 22	☽ ± ♀
11 30	☉ ꝑ ♅
13 37	☽ △ ♄
17 58	☽ Q ♆
21 36	♀ ⊻ ♂
23 08	☽ △ ♂
23 11	☽ ⊼ ♀
20 Friday	
00 20	☿ ⊥ ♇
00 41	☽ ± ♇
03 33	♂ ∠ ♆
10 30	☽ △ ☉
11 35	☽ ⊼ ♇
14 59	☽ ⊼ ♃
21 Saturday	
00 05	☽ ⫲ ♃
01 09	☉ ⊻ ♇
02 03	☽ ± ♃
02 32	☽ △ ☿
02 47	☽ □ ♆
04 09	☿ ⊻ ♆
04 12	☽ ꝑ ♂
05 45	☿ ⫲ ♂
09 39	☽ ⫲ ♇
12 13	☽ ☌ ♅
16 23	☽ ꝑ ☉
18 51	☽ ꝑ ♃
20 17	☽ □ ♄
22 00	☽ ± ♂
22 Sunday	
05 44	☽ ∥ ♅
10 01	☽ ⊼ ♂
10 41	☽ ± ☉
11 38	☽ ꝑ ☿
12 14	☽ ☍ ♀
14 43	♀ Q ♆
17 54	☽ ⫲ ♄
19 31	☽ ☍ ♇
20 37	☿ ∠ ♀
21 14	☿ ± ♄
23 02	☽ ⊼ ☉
23 20	☽ △ ♃
23 Monday	
01 42	☽ ⫲ ♀
02 49	☽ ∥ ♆
03 07	☉ □ ♃
06 13	☽ Q ♄
07 56	☽ ± ☿
10 05	☉ ⊥ ♆
11 23	☽ △ ♆
17 03	☉ ꝑ ♄
17 29	☿ △ ♅
17 44	♀ ⫲ ♆
20 58	☽ ⊻ ♅
21 38	☽ ⊼ ☿
24 Tuesday	
05 17	☽ ⚹ ♄
07 31	☽ ∥ ☿
08 35	☽ ⊥ ♅
15 11	☿ ∠ ♇
16 33	☽ ꝑ ♆
23 41	☽ ☍ ♂
25 Wednesday	
00 12	☽ ⫲ ♂
00 56	☽ ∥ ☉
02 10	☽ ∠ ♅
04 24	☽ ⊼ ♀
05 43	☽ ⊼ ♇
06 42	☉ ⫲ ♂
09 57	☽ □ ♃
10 17	☽ ± ♆
10 35	☽ ∠ ♄
14 26	☽ ☍ ☉
14 48	☽ ∥ ♂
15 52	☽ ⫲ ☉
17 34	☽ ± ♀
17 36	☽ ± ♇
18 01	♀ ☌ ♇
22 02	☿ ⊼ ♄
22 12	☽ ⊼ ♆
26 Thursday	
04 23	☽ ⊥ ♄
07 51	☽ ⚹ ♅
11 34	☽ ꝑ ♇
13 26	♂ ꝑ ♅
13 29	☽ ꝑ ♀
16 19	☽ ⊻ ♄
18 01	☽ ⫲ ☿
19 48	☽ ☍ ☿
22 02	☽ Q ♃
27 Friday	
06 05	☉ ⊥ ♇
07 52	☽ Q ♅
08 10	☽ ∥ ♀
14 38	☽ ⫲ ♆
15 40	☽ ⊼ ♂
15 44	♀ ⚹ ♃
17 47	☽ △ ♇
22 22	☽ ⚹ ♃
23 05	☽ △ ♀
23 29	☽ ∥ ♄
28 Saturday	
01 49	♃ ∠ ♄
04 40	☽ ± ♂
08 04	☽ ⊼ ☉
10 41	☽ ☍ ♆
13 04	☽ ⫲ ♅
20 14	☽ □ ♅
21 21	☽ ± ☉
29 Sunday	
00 12	☽ ꝑ ♂
00 44	♂ ⊻ ♇
04 42	☽ ☌ ♄
04 58	☽ ∠ ♃
10 21	☽ ∥ ♇
11 58	☿ ⫲ ♀
16 28	☉ ⊻ ♆
17 17	☽ ꝑ ☉
19 14	☽ ⊼ ☿
21 04	☽ ∥ ♃
23 21	☽ ⊥ ♃
30 Monday	
06 44	☽ □ ♇
08 47	☽ △ ♂
09 25	☽ ± ☿
11 34	☽ ⊻ ♃
18 50	☽ □ ♀
23 41	☽ ⊼ ♆
31 Tuesday	
02 25	☽ △ ☉
06 28	☽ ꝑ ☿
08 53	☽ △ ♅
11 47	☽ ± ♆
16 35	♂ ꝑ ♄
17 05	☽ ⊻ ♄
19 05	☽ Q ♇

All ephemeris data is given at 12.00 UT and the Moon's longitude is additionally given for 24.00 UT

APRIL 2043

Date	Sidereal time h m s	LONGITUDES Sun ☉ ° ′ ″	Moon ☽ ° ′ ″	Moon ☽ 24.00 ° ′ ″	Mercury ☿ ° ′	Venus ♀ ° ′	Mars ♂ ° ′	Jupiter ♃ ° ′	Saturn ♄ ° ′	Uranus ♅ ° ′	Neptune ♆ ° ′	Pluto ♇ ° ′
01	00 38 59	11 ♈ 39 27	27 ♐ 05 14	03 ♑ 10 57	29 ♈ 12	08 ♓ 36	03 ♈ 13	03 ♑ 03	17 ♏ 33	13 ♌ 28	08 ♉ 58	00 ♓ 36
02	00 42 56	12 38 40	09 ♑ 20 21	15 ♑ 33 59	00 ♉ 40	09 48	04 00	03 06	17 R 29	13 R 27	09 00	00 37
03	00 46 52	13 37 51	21 ♑ 52 23	28 ♑ 16 03	02 03	11 01	04 46	03 09	17 26	13 26	09 03	00 38
04	00 50 49	14 37 00	04 ♒ 45 24	11 ♒ 20 51	03 20	12 14	05 33	03 12	17 23	13 25	09 05	00 40
05	00 54 45	15 36 07	18 ♒ 02 42	24 ♒ 51 09	04 32	13 26	06 19	03 15	17 19	13 25	09 07	00 41
06	00 58 42	16 35 13	01 ♓ 46 17	08 ♓ 48 02	05 38	14 39	07 06	03 18	17 16	13 24	09 09	00 42
07	01 02 38	17 34 17	15 ♓ 56 11	23 ♓ 10 22	06 37	15 52	07 52	03 20	17 12	13 23	09 11	00 44
08	01 06 35	18 33 19	00 ♈ 30 02	07 ♈ 54 27	07 31	17 05	08 39	03 23	17 08	13 22	09 13	00 45
09	01 10 32	19 32 18	15 ♈ 22 45	22 ♈ 53 57	08 18	18 17	09 25	03 25	17 05	13 22	09 15	00 46
10	01 14 28	20 31 16	00 ♉ 26 56	08 ♉ 00 33	08 59	19 30	10 12	03 27	17 01	13 21	09 17	00 47
11	01 18 25	21 30 12	15 ♉ 33 39	23 ♉ 05 03	09 33	20 43	10 58	03 28	16 57	13 21	09 20	00 48
12	01 22 21	22 29 06	00 ♊ 33 43	07 ♊ 58 42	10 00	21 56	11 44	03 30	16 53	13 20	09 22	00 50
13	01 26 18	23 27 58	15 ♊ 19 10	22 ♊ 34 28	10 21	23 08	12 30	03 31	16 49	13 20	09 24	00 51
14	01 30 14	24 26 48	29 ♊ 44 07	06 ♋ 47 47	10 36	24 21	13 17	03 32	16 45	13 19	09 26	00 52
15	01 34 11	25 25 35	13 ♋ 45 19	20 ♋ 36 41	10 44	25 34	14 03	03 33	16 41	13 19	09 28	00 53
16	01 38 07	26 24 20	27 ♋ 21 59	04 ♌ 01 26	10 R 45	26 47	14 49	03 34	16 37	13 19	09 31	00 54
17	01 42 04	27 23 03	10 ♌ 35 18	17 ♌ 03 56	10 41	28 00	15 35	03 35	16 33	13 19	09 33	00 55
18	01 46 01	28 21 43	23 ♌ 27 42	29 ♌ 47 03	10 30	29 ♓ 13	16 21	03 35	16 29	13 19	09 35	00 56
19	01 49 57	29 ♈ 20 21	06 ♍ 02 21	12 ♍ 14 03	10 15	00 ♈ 25	17 07	03 35	16 25	13 18	09 37	00 57
20	01 53 54	00 ♉ 18 57	18 ♍ 22 34	24 ♍ 28 16	09 54	01 38	17 53	03 R 35	16 21	13 18	09 39	00 58
21	01 57 50	01 17 31	00 ♎ 31 33	06 ♎ 32 45	09 28	02 51	18 39	03 35	16 16	13 D 18	09 42	00 59
22	02 01 47	02 16 02	12 ♎ 32 13	18 ♎ 30 14	08 58	04 04	19 25	03 35	16 12	13 19	09 44	01 00
23	02 05 43	03 14 32	24 ♎ 27 06	00 ♏ 23 05	08 25	05 17	20 11	03 34	16 08	13 19	09 46	01 01
24	02 09 40	04 13 00	06 ♏ 18 26	12 ♏ 13 25	07 49	06 30	20 57	03 34	16 03	13 19	09 48	01 02
25	02 13 36	05 11 26	18 ♏ 08 16	24 ♏ 03 14	07 10	07 43	21 42	03 33	15 59	13 19	09 51	01 03
26	02 17 33	06 09 50	29 ♏ 58 34	05 ♐ 54 33	06 31	08 56	22 28	03 32	15 55	13 19	09 53	01 04
27	02 21 30	07 08 13	11 ♐ 51 27	17 ♐ 49 34	05 50	10 09	23 14	03 30	15 50	13 20	09 55	01 05
28	02 25 26	08 06 34	23 ♐ 49 15	29 ♐ 50 51	05 10	11 21	23 59	03 29	15 46	13 20	09 57	01 06
29	02 29 23	09 04 53	05 ♑ 54 43	12 ♑ 01 18	04 30	12 34	24 45	03 27	15 41	13 21	10 00	01 06
30	02 33 19	10 ♉ 03 11	18 ♑ 10 59	24 ♑ 24 16	03 ♉ 52	13 ♈ 47	25 ♈ 30	03 ♑ 25	15 ♏ 37	13 ♌ 21	10 ♉ 02	01 ♓ 07

Date	Moon True ☊ ° ′	Moon Mean ☊ ° ′	Moon ☽ Latitude ° ′
01	08 ♈ 39	08 ♈ 35	05 S 11
02	08 R 39	08 32	05 17
03	08 D 39	08 29	05 09
04	08 39	08 26	04 45
05	08 40	08 23	04 06
06	08 40	08 19	03 11
07	08 41	08 16	02 03
08	08 41	08 13	00 S 45
09	08 R 41	08 10	00 N 37
10	08 41	08 07	01 58
11	08 40	08 03	03 10
12	08 38	08 00	04 09
13	08 36	07 57	04 51
14	08 35	07 54	05 12
15	08 34	07 51	05 15
16	08 D 33	07 48	04 59
17	08 34	07 44	04 28
18	08 35	07 41	03 44
19	08 37	07 38	02 50
20	08 38	07 35	01 50
21	08 39	07 32	00 N 45
22	08 R 39	07 29	00 S 21
23	08 38	07 25	01 26
24	08 36	07 22	02 26
25	08 32	07 19	03 20
26	08 28	07 16	04 05
27	08 23	07 13	04 40
28	08 18	07 09	05 03
29	08 14	07 06	05 12
30	08 ♈ 10	07 ♈ 03	05 S 08

DECLINATIONS

Date	Sun ☉ ° ′	Moon ☽ ° ′	Mercury ☿ ° ′	Venus ♀ ° ′	Mars ♂ ° ′	Jupiter ♃ ° ′	Saturn ♄ ° ′	Uranus ♅ ° ′	Neptune ♆ ° ′	Pluto ♇ ° ′
01	04 N 37	28 S 35	12 N 50	09 S 18	00 N 29	22 S 58	14 S 37	17 N 27	12 N 52	21 S 16
02	05 00	28 23	13 31	08 52	00 48	22 58	14 36	17 27	12 52	21 15
03	05 23	26 45	14 10	08 27	01 07	22 58	14 35	17 27	12 53	21 15
04	05 46	23 41	14 46	08 01	01 26	22 58	14 34	17 28	12 54	21 15
05	06 08	19 19	15 19	07 35	01 45	22 58	14 33	17 28	12 54	21 15
06	06 31	13 49	15 49	07 09	02 03	22 58	14 32	17 28	12 55	21 14
07	06 54	07 26	16 16	06 43	02 22	22 58	14 31	17 28	12 56	21 14
08	07 16	00 S 30	16 40	06 16	02 41	22 58	14 30	17 28	12 57	21 14
09	07 39	06 N 37	17 01	05 49	03 00	22 58	14 28	17 29	12 57	21 14
10	08 01	13 28	17 18	05 22	03 18	22 58	14 27	17 29	12 58	21 13
11	08 23	19 32	17 32	04 55	03 37	22 58	14 26	17 29	12 59	21 13
12	08 45	24 20	17 43	04 28	03 55	22 58	14 25	17 29	12 59	21 13
13	09 07	27 26	17 51	04 01	04 14	22 58	14 24	17 29	13 00	21 13
14	09 28	28 39	17 55	03 33	04 32	22 58	14 23	17 29	13 01	21 13
15	09 50	27 57	17 56	03 06	04 50	22 58	14 21	17 29	13 01	21 13
16	10 11	25 34	17 53	02 38	05 09	22 58	14 20	17 29	13 02	21 12
17	10 32	21 53	17 48	02 10	05 27	22 58	14 19	17 29	13 03	21 12
18	10 53	17 13	17 39	01 42	05 45	22 58	14 18	17 29	13 04	21 12
19	11 14	11 56	17 26	01 14	06 03	22 58	14 17	17 29	13 04	21 12
20	11 35	06 17	17 11	00 46	06 21	22 58	14 15	17 29	13 05	21 12
21	11 55	00 N 28	16 53	00 S 18	06 39	22 58	14 14	17 29	13 06	21 12
22	12 16	05 S 17	16 33	00 N 10	06 57	22 58	14 13	17 29	13 06	21 12
23	12 36	10 48	16 10	00 38	07 15	22 58	14 11	17 29	13 07	21 12
24	12 55	15 55	15 45	01 07	07 32	22 58	14 10	17 29	13 08	21 12
25	13 15	20 26	15 19	01 35	07 50	22 58	14 09	17 29	13 09	21 11
26	13 34	24 08	14 51	02 03	08 07	22 58	14 08	17 29	13 09	21 11
27	13 54	26 50	14 22	02 31	08 25	22 58	14 06	17 29	13 10	21 11
28	14 13	28 20	13 53	02 59	08 42	22 58	14 05	17 29	13 11	21 11
29	14 31	28 30	13 23	03 28	08 59	22 58	14 04	17 28	13 11	21 11
30	14 N 50	27 S 17	12 N 54	03 N 56	09 N 17	22 S 58	14 S 03	17 N 28	13 N 12	21 S 11

ZODIAC SIGN ENTRIES

Date	h	m	Planets
01	17	45	☽ ♑
02	00	55	☿ ♉
04	03	13	☽ ♒
06	08	57	☽ ♓
08	11	11	☽ ♈
10	11	17	☽ ♉
12	11	06	☽ ♊
14	12	27	☽ ♋
16	16	44	☽ ♌
19	00	25	☽ ♍
19	03	37	♀ ♈
20	04	14	☉ ♉
21	10	57	☽ ♎
23	23	13	☽ ♏
26	12	03	☽ ♐
29	00	18	☽ ♑

LATITUDES

Date	Mercury ☿ ° ′	Venus ♀ ° ′	Mars ♂ ° ′	Jupiter ♃ ° ′	Saturn ♄ ° ′	Uranus ♅ ° ′	Neptune ♆ ° ′	Pluto ♇ ° ′
01	01 N 45	01 S 01	00 S 52	00 N 26	02 N 33	00 N 42	01 S 43	10 S 43
04	02 16	01 08	00 51	00 26	02 33	00 42	01 43	10 44
07	02 42	01 14	00 49	00 26	02 33	00 42	01 43	10 45
10	02 58	01 19	00 48	00 26	02 34	00 42	01 43	10 46
13	03 05	01 24	00 46	00 26	02 34	00 42	01 43	10 46
16	02 59	01 28	00 45	00 26	02 34	00 42	01 43	10 47
19	02 41	01 32	00 44	00 26	02 35	00 42	01 42	10 48
22	02 11	01 34	00 42	00 25	02 35	00 42	01 42	10 49
25	01 29	01 37	00 40	00 25	02 35	00 41	01 42	10 50
28	00 N 40	01 38	00 39	00 25	02 35	00 41	01 42	10 51
31	00 S 11	01 S 39	00 S 37	00 N 25	02 N 35	00 N 41	01 S 42	10 S 52

DATA

Julian Date	2467341
Delta T	+78 seconds
Ayanamsa	24° 27′ 38″
Synetic vernal point	04° ♓ 39′ 21″
True obliquity of ecliptic	23° 26′ 11″

MOON'S PHASES, APSIDES AND POSITIONS ☽

Date	h	m	Phase	Longitude ° ′	Eclipse Indicator
02	18	56	☾	12 ♑ 56	
09	19	07	●	19 ♈ 50	Total
16	10	09	☽	26 ♋ 20	
24	07	23	○	04 ♏ 02	

Day	h	m		
10	16	06	Perigee	
25	12	33	Apogee	
01	20	36	Max dec	28° S 41′
08	13	40	0N	
14	14	54	Max dec	28° N 39′
21	13	58	0S	
29	03	01	Max dec	28° S 36′

LONGITUDES

Date	Chiron ⚷ ° ′	Ceres ⚳ ° ′	Pallas ⚴ ° ′	Juno ⚵ ° ′	Vesta ⚶ ° ′	Black Moon Lilith ⚸ ° ′
01	11 ♌ 22	21 ♒ 53	21 ♑ 54	05 ♑ 16	19 ♎ 56	12 ♏ 58
11	11 ♌ 16	25 ♒ 12	24 ♑ 02	06 ♑ 18	17 ♎ 25	14 ♏ 05
21	11 ♌ 21	28 ♒ 19	25 ♑ 42	06 ♑ 51	14 ♎ 59	15 ♏ 12
31	11 ♌ 37	01 ♓ 13	26 ♑ 50	06 ♑ 53	12 ♎ 58	16 ♏ 19

All ephemeris data is given at 12.00 UT and the Moon's longitude is additionally given for 24.00 UT

ASPECTARIAN

h m Aspects

01 Wednesday
04 11 ♂ ⊥ ♆
05 00 ☽ ⊥ ♄
05 49 ☽ ⚼ ♆
06 27 ♂ □ ♃
09 21 ☉ ± ♄
10 55 ☽ Q ♀
13 11 ☿ ∥ ♆
14 44 ☽ ⚼ ♅
16 46 ☽ △ ☿
18 56 ☽ ✶ ♇
19 42 ♀ ✶ ♆
22 42 ☽ ∠ ♄
23 48 ☽ ☌ ♃

02 Thursday
00 53 ☽ □ ♂
08 21 ☽ ± ♅
11 09 ☿ ✶ ♇
11 21 ☽ △ ♆
13 00 ☽ ✶ ♀
18 56 ☽ □ ☉
19 56 ☽ ⚻ ♅

03 Friday
00 07 ☽ ∠ ♇
03 36 ☽ ✶ ♄
07 25 ☉ △ ♅
13 48 ☽ Q ♂
17 13 ☽ ⊥ ♇
20 37 ☽ ∠ ♀

04 Saturday
02 07 ☽ Q ♄
04 17 ☿ ⋕ ♄
04 26 ☽ ⚺ ♇
07 44 ☽ Q ☉
09 07 ☽ □ ☿
09 08 ☽ ⚺ ♃
09 27 ☿ △ ♃
13 32 ☽ ✶ ♂
14 58 ☽ ⊥ ♀
16 29 ☽ ∥ ♃
19 54 ☽ □ ♆
20 09 ☽ ⊥ ♃

05 Sunday
02 12 ☽ ∥ ♇
02 57 ☽ ⚺ ♀
03 44 ☽ ☍ ♅
07 18 ☽ ✶ ☉
10 43 ☽ □ ♄
11 26 ♀ ⚻ ♅
12 22 ☽ ∠ ♃
14 01 ☉ ∠ ♇
18 09 ☽ ∠ ♂
20 34 ☽ ⋕ ♅
20 38 ☽ Q ☿
23 29 ♂ ⊥ ♇

06 Monday
03 59 ☽ Q ♆
04 24 ☽ ⋕ ☿
09 05 ☽ ∥ ♄
10 10 ☽ ☌ ♇
10 46 ☽ ⊥ ♂
11 40 ☽ ∠ ☉
14 38 ☽ ✶ ♃
15 32 ☽ ⋕ ♆
19 08 ☽ ✶ ☿
21 38 ☽ ⚺ ♂

07 Tuesday
00 37 ☽ ✶ ♆
01 16 ♀ Q ♃
03 31 ☉ ⚻ ♄
04 08 ☽ ⊥ ☉
06 34 ☉ ⋕ ♀
07 44 ☽ ⚻ ♅
11 00 ☽ Q ♃
11 52 ☽ ☌ ♀
13 49 ☽ ⋕ ☉
14 06 ☽ △ ♄
14 45 ☽ ∥ ♀
14 55 ☽ ⚺ ☉
17 44 ☽ ± ♅
22 05 ☽ ∠ ☿

08 Wednesday
01 42 ☽ ∠ ♆
04 53 ☽ ⋕ ♂
08 32 ☽ ⚼ ♅
12 24 ☽ ⚺ ♇
13 14 ♀ △ ♄
13 45 ☽ ⊥ ☿
14 39 ☽ ⚼ ♄
16 26 ☽ ⊥ ♆
16 41 ☽ □ ♃
22 08 ☽ ⊥ ♇
23 11 ☽ ∥ ♂

09 Thursday
00 01 ☽ ⚺ ☿
01 55 ☽ ☌ ♂
02 09 ☽ ⚺ ♆
05 08 ☽ ± ♄
06 35 ♂ ⚺ ♆
08 46 ☽ △ ♅
09 26 ☽ ⋕ ♀
12 37 ☽ ∠ ♇
14 43 ☽ ⚻ ♄
15 40 ☽ ∥ ☉
17 04 ☽ ⚺ ♀
19 07 ☽ ☌ ☉

10 Friday
03 27 ☽ ⊥ ♀
09 06 ♀ ± ♅
10 10 ☽ ∥ ♆
12 32 ☽ ✶ ♇
15 40 ☽ ⋕ ♄
16 46 ☽ △ ♃
19 00 ☽ ∠ ♀

11 Saturday
01 25 ☿ ☌ ♆
02 04 ☽ ☌ ♆
02 05 ☽ ☌ ☿
03 19 ☽ ∥ ☿
03 25 ☽ ∥ ♅
04 18 ☽ ⚺ ♂
05 33 ☿ ∥ ♅
07 37 ☽ Q ♇
08 29 ☽ □ ♅
11 40 ♂ ± ♄
14 12 ☽ ☍ ♄
14 21 ☽ ⊥ ♂
16 39 ☽ ⚼ ♃
19 39 ☽ ⋕ ♇
20 56 ☽ ✶ ♀
22 08 ☽ ⚺ ☉

12 Sunday
04 19 ☽ ⋕ ♃
05 31 ☽ ∠ ♂
07 04 ☽ ± ♃
08 25 ☽ ⊥ ☉
12 26 ☽ □ ♇
13 15 ☽ Q ♅
16 45 ☽ ⚻ ♃
17 55 ☽ Q ♀

13 Monday
00 00 ☽ ∠ ☉
02 17 ☽ ⚺ ♆
03 42 ☽ ⚺ ☿
05 11 ♀ ⋕ ♂
07 08 ☽ ✶ ♂
08 44 ☽ ✶ ♅
12 08 ☽ ⊥ ♆
13 44 ☽ ⊥ ☿
14 28 ☽ ⚻ ♄

14 Tuesday
00 22 ☽ ± ♄
02 08 ☽ □ ♀
02 28 ☽ ✶ ☉
03 05 ☽ ∠ ♆
04 05 ☽ Q ♂
04 57 ☽ ∠ ☿
09 37 ☽ ∠ ♅
13 26 ♂ △ ♅
13 38 ♀ ∠ ♆
13 55 ☽ △ ♇
15 24 ☽ ⚼ ♄
18 27 ☽ ☍ ♃
21 27 ☉ ⚺ ♀

15 Wednesday
00 15 ☽ Q ☉
00 54 ☽ ⊥ ♅
04 34 ☽ ✶ ♆
06 43 ☽ ✶ ☿
11 15 ☽ ⚺ ♅
12 32 ☽ □ ♂
15 22 ☿ ∠ ♀
15 43 ☽ ⚼ ♇
17 05 ☽ △ ♄

16 Thursday
01 33 ☽ Q ♆
03 48 ☽ Q ☿
06 18 ☿ St R
07 36 ☽ ± ♇
10 09 ☽ □ ☉
10 51 ☽ △ ♀
18 22 ☽ ⚻ ♇
23 11 ☽ ⚻ ♃

17 Friday
05 41 ☽ ⋕ ♃
10 05 ☽ □ ♆
10 09 ☽ ± ♃
12 10 ☽ □ ☿
15 46 ☽ ⋕ ♇
16 54 ☽ ⚼ ♀
17 02 ☽ ☌ ♅
18 14 ♀ ⚼ ♅
21 50 ☽ △ ♂
22 43 ♂ ∠ ♇
22 59 ☽ □ ♄

18 Saturday
02 50 ☽ ⚼ ♃
09 55 ☽ ∥ ☿
10 44 ☽ ∥ ♅
11 28 ☽ ± ♀
15 48 ♂ ⚻ ♄
22 04 ☽ △ ☉

19 Sunday
00 04 ☽ ⚻ ♀
01 36 ☽ ⋕ ♄
02 13 ☽ ☍ ♇
03 58 ☽ ⚼ ♂
06 54 ☿ ∥ ♅
07 01 ☽ ∥ ♆
07 17 ☽ △ ♃
08 53 ☽ Q ♄
14 51 ☽ ∥ ☉
18 57 ☽ △ ♆
19 56 ☽ △ ☿
22 29 ☽ ± ♂
22 39 ♀ ⚺ ♇

20 Monday
02 05 ☽ ⚺ ♅
04 02 ♃ St R
05 29 ☽ ⚼ ☉
06 31 ♀ ⚼ ♄
08 03 ☽ ✶ ♄
10 58 ☽ ⚻ ♂
11 42 ☽ ∥ ♂
13 50 ☽ ⊥ ♅
17 50 ♅ St D

21 Tuesday
00 24 ☽ ⚼ ♆
00 25 ☽ ⚼ ☿
00 41 ☿ ☌ ♆
00 43 ☽ ± ☉
04 25 ☉ ✶ ♇
07 36 ☽ ∠ ♅
12 48 ☽ ⋕ ♀
12 55 ☽ ⚻ ♇
13 29 ☽ ∠ ♄
13 39 ☽ ⚻ ☉
14 57 ☽ ∥ ♀
17 09 ☽ ☍ ♀
17 38 ☽ ± ☿
18 06 ☽ □ ♃
18 20 ☽ ± ♆
20 42 ☿ ⊥ ♀

22 Wednesday
00 54 ☽ ± ♇
02 28 ♀ □ ♃
05 09 ☽ ⚻ ☿
05 10 ♀ ⊥ ♆
06 21 ☽ ⚻ ♆
07 21 ☽ ⊥ ♄
13 33 ☽ ✶ ♅
18 59 ☽ ⚼ ♇
19 19 ☽ ⚺ ♄
19 30 ☽ ⋕ ♂

23 Thursday
02 47 ☽ ☍ ♂
06 12 ☽ Q ♃
13 44 ☽ Q ♅
20 03 ☉ △ ♃
20 41 ☽ ⋕ ☉
22 35 ☽ ⋕ ♆

24 Friday
01 18 ☽ △ ♇
03 33 ☽ ∥ ♄
06 26 ☽ ✶ ♃
07 23 ☽ ☍ ☉
11 14 ☽ ⋕ ☿
12 26 ☽ ⚻ ♀
14 54 ☽ ☍ ☿
19 07 ☽ ☍ ♆
19 55 ☽ ⋕ ♅
22 48 ♀ ⊥ ♇

25 Saturday
01 59 ☽ ± ♀
02 13 ☽ □ ♅
03 44 ☉ ∥ ♆
05 04 ☿ ⚺ ♀
07 39 ☽ ☌ ♄
12 50 ☽ ∠ ♃
16 30 ☽ ∥ ♇
19 44 ☽ ⚻ ♂
22 20 ☽ ⚼ ♀

26 Sunday
03 46 ☽ ∥ ♃
07 03 ☽ ⊥ ♃
08 44 ☽ ± ♂
14 13 ☽ □ ♇
17 04 ☉ ☌ ☿
19 10 ☽ ⚺ ♃

27 Monday
00 31 ☽ ⚻ ☿
01 38 ☽ ⚻ ☉
04 11 ☽ ⚼ ♂
06 19 ♀ ± ♄
07 27 ♀ ⚺ ♆
08 05 ☽ ⚻ ♆
08 09 ☽ △ ♀
11 58 ☽ ± ☿
14 48 ☽ ± ☉
14 58 ☽ △ ♅
19 57 ☽ ⚺ ♄
20 12 ☽ ± ♆

28 Tuesday
01 25 ☿ ⋕ ♄
02 08 ☉ ∥ ☿
02 32 ☽ Q ♇
03 07 ☉ ⋕ ♄
05 05 ☽ ⚼ ☿
07 55 ☽ ⊥ ♄
10 27 ☽ ⚼ ☉
12 21 ☽ △ ♂
14 16 ☽ ⚼ ♆
21 00 ☽ ⚼ ♅

29 Wednesday
01 44 ☽ ∠ ♄
02 29 ☽ ✶ ♇
07 10 ☽ ☌ ♃
09 22 ☽ △ ☿
14 49 ☽ ± ♅
18 47 ☽ △ ☉
20 03 ☽ △ ♆
21 33 ☿ ∥ ♆

30 Thursday
02 31 ☽ □ ♀
02 35 ☽ ⚻ ♅
03 17 ♀ △ ♅
07 02 ☽ ✶ ♄
08 00 ☽ ∠ ♇
11 28 ☉ ☌ ♆

MAY 2043

LONGITUDES

Date	Sidereal time h m s	Sun ☉ ° ' "	Moon ☽ ° ' "	Moon ☽ 24.00 ° ' "	Mercury ☿ ° '	Venus ♀ ° '	Mars ♂ ° '	Jupiter ♃ ° '	Saturn ♄ ° '	Uranus ♅ ° '	Neptune ♆ ° '	Pluto ♇ ° '
01	02 37 16	11 ♉ 01 27	00 ♒ 41 34	07 ♒ 03 24	03 ♉ 16	15 ♈ 00	26 ♈ 16	03 ♑ 23	15 ♏ 32	13 ♌ 22	10 ♉ 04	01 ♓ 08
02	02 41 12	11 59 41	13 ♒ 30 11	20 ♒ 02 23	02 R 43	16 13	27 01	03 R 21	15 R 28	13 22	10 06	01 09
03	02 45 09	12 57 54	26 ♒ 40 23	03 ♓ 24 34	02 13	17 26	27 47	03 19	15 23	13 23	10 09	01 10
04	02 49 05	13 56 06	10 ♓ 15 12	17 ♓ 12 25	01 46	18 39	28 32	03 16	15 19	13 24	10 11	01 10
05	02 53 02	14 54 16	24 ♓ 16 18	01 ♈ 26 44	01 24	19 52	29 ♈ 17	03 14	15 14	13 24	10 13	01 11
06	02 56 59	15 52 25	08 ♈ 43 26	16 ♈ 05 55	01 05	21 05	00 ♉ 02	03 11	15 10	13 25	10 15	01 12
07	03 00 55	16 50 32	23 ♈ 33 33	01 ♉ 05 28	00 51	22 18	00 48	03 08	15 05	13 26	10 18	01 12
08	03 04 52	17 48 38	08 ♉ 40 37	16 ♉ 17 51	00 42	23 31	01 33	03 04	15 01	13 27	10 20	01 13
09	03 08 48	18 46 42	23 ♉ 55 51	01 ♊ 33 17	00 37	24 44	02 18	03 01	14 56	13 28	10 22	01 13
10	03 12 45	19 44 45	09 ♊ 08 50	16 ♊ 41 13	00 D 37	25 57	03 03	02 57	14 52	13 29	10 24	01 14
11	03 16 41	20 42 46	24 ♊ 09 17	01 ♋ 32 03	00 42	27 10	03 48	02 53	14 47	13 30	10 27	01 15
12	03 20 38	21 40 45	08 ♋ 48 42	15 ♋ 58 41	00 51	28 23	04 33	02 49	14 43	13 31	10 29	01 15
13	03 24 34	22 38 42	23 ♋ 01 36	29 ♋ 57 18	01 05	29 ♈ 36	05 18	02 45	14 38	13 32	10 31	01 16
14	03 28 31	23 36 38	06 ♌ 45 46	13 ♌ 27 12	01 23	00 ♉ 49	06 03	02 41	14 34	13 34	10 33	01 16
15	03 32 28	24 34 31	20 ♌ 01 54	26 ♌ 30 16	01 46	02 03	06 47	02 37	14 29	13 35	10 36	01 16
16	03 36 24	25 32 23	02 ♍ 52 47	09 ♍ 10 02	02 13	03 16	07 32	02 32	14 25	13 36	10 38	01 17
17	03 40 21	26 30 13	15 ♍ 22 32	21 ♍ 30 55	02 44	04 29	08 17	02 27	14 20	13 38	10 40	01 17
18	03 44 17	27 28 02	27 ♍ 35 45	03 ♎ 37 37	03 19	05 42	09 01	02 22	14 16	13 39	10 42	01 18
19	03 48 14	28 25 48	09 ♎ 37 03	15 ♎ 34 35	03 58	06 55	09 46	02 17	14 12	13 41	10 44	01 18
20	03 52 10	29 ♉ 23 33	21 ♎ 30 42	27 ♎ 25 50	04 41	08 08	10 31	02 12	14 08	13 42	10 47	01 18
21	03 56 07	00 ♊ 21 17	03 ♏ 20 24	09 ♏ 14 45	05 28	09 21	11 15	02 07	14 03	13 44	10 49	01 18
22	04 00 03	01 18 59	15 ♏ 09 13	21 ♏ 04 05	06 18	10 34	11 59	02 01	13 59	13 45	10 51	01 19
23	04 04 00	02 16 40	26 ♏ 59 37	02 ♐ 56 02	07 11	11 47	12 44	01 56	13 55	13 47	10 53	01 19
24	04 07 57	03 14 20	08 ♐ 53 32	14 ♐ 52 18	08 08	13 00	13 28	01 50	13 51	13 49	10 55	01 19
25	04 11 53	04 11 58	20 ♐ 52 32	26 ♐ 54 24	09 08	14 13	14 12	01 44	13 47	13 50	10 57	01 19
26	04 15 50	05 09 35	02 ♑ 58 04	09 ♑ 03 45	10 11	15 26	14 56	01 38	13 43	13 52	10 59	01 20
27	04 19 46	06 07 11	15 ♑ 11 38	21 ♑ 21 59	11 17	16 39	15 41	01 32	13 39	13 54	11 01	01 20
28	04 23 43	07 04 46	27 ♑ 35 01	03 ♒ 51 04	12 26	17 52	16 25	01 26	13 35	13 56	11 04	01 20
29	04 27 39	08 02 20	10 ♒ 10 25	16 ♒ 33 26	13 39	19 06	17 09	01 19	13 31	13 58	11 06	01 20
30	04 31 36	08 59 53	23 ♒ 00 27	29 ♒ 31 51	14 53	20 19	17 53	01 13	13 27	14 00	11 08	01 20
31	04 35 32	09 ♊ 57 25	06 ♓ 08 00	12 ♓ 49 16	16 ♉ 11	21 ♉ 32	18 ♉ 37	01 ♑ 06	13 ♏ 23	14 ♌ 02	11 ♉ 10	01 ♓ 20

Date	Moon True ☊ ° '	Moon Mean ☊ ° '	Moon ☽ Latitude ° '
01	08 ♈ 08	07 ♈ 00	04 S 49
02	08 D 08	06 57	04 15
03	08 09	06 54	03 27
04	08 10	06 50	02 26
05	08 11	06 47	01 S 15
06	08 R 12	06 44	00 N 03
07	08 11	06 41	01 23
08	08 08	06 38	02 38
09	08 04	06 34	03 43
10	07 58	06 31	04 31
11	07 52	06 28	05 00
12	07 46	06 25	05 09
13	07 41	06 22	04 58
14	07 38	06 19	04 30
15	07 36	06 15	03 49
16	07 D 36	06 12	02 57
17	07 38	06 09	01 57
18	07 39	06 06	00 N 54
19	07 R 39	06 03	00 S 11
20	07 38	06 00	01 14
21	07 35	05 56	02 14
22	07 29	05 53	03 07
23	07 21	05 50	03 53
24	07 12	05 47	04 28
25	07 01	05 44	04 52
26	06 51	05 40	05 03
27	06 41	05 37	05 01
28	06 33	05 34	04 44
29	06 27	05 31	04 14
30	06 24	05 28	03 30
31	06 ♈ 23	05 ♈ 25	02 S 34

DECLINATIONS

Date	Sun ☉ ° '	Moon ☽ ° '	Mercury ☿ ° '	Venus ♀ ° '	Mars ♂ ° '	Jupiter ♃ ° '	Saturn ♄ ° '	Uranus ♅ ° '	Neptune ♆ ° '	Pluto ♇ ° '
01	15 N 08	24 S 42	12 N 26	04 N 24	09 N 34	22 S 59	14 S 01	17 N 28	13 N 13	21 S 11
02	15 26	20 50	11 58	04 52	09 50	22 59	14 00	17 28	13 14	21 11
03	15 44	15 52	11 32	05 20	10 07	22 59	13 59	17 28	13 14	21 11
04	16 01	09 59	11 08	05 47	10 24	22 59	13 57	17 27	13 15	21 11
05	16 18	03 S 26	10 46	06 15	10 40	22 59	13 56	17 27	13 16	21 11
06	16 35	03 N 30	10 25	06 43	10 57	22 59	13 55	17 27	13 16	21 11
07	16 52	10 26	10 07	07 10	11 13	22 59	13 54	17 27	13 17	21 12
08	17 08	16 53	09 51	07 37	11 29	22 59	13 52	17 26	13 18	21 12
09	17 24	22 21	09 38	08 04	11 46	23 00	13 51	17 26	13 18	21 12
10	17 40	26 17	09 27	08 31	12 02	23 00	13 50	17 26	13 19	21 12
11	17 56	28 19	09 19	08 58	12 17	23 00	13 48	17 25	13 20	21 12
12	18 11	28 17	09 13	09 25	12 33	23 00	13 47	17 25	13 20	21 12
13	18 26	26 22	09 10	09 51	12 49	23 00	13 46	17 25	13 21	21 12
14	18 40	22 56	09 09	10 17	13 04	23 00	13 45	17 24	13 22	21 12
15	18 55	18 25	09 10	10 43	13 19	23 00	13 44	17 24	13 22	21 12
16	19 09	13 11	09 14	11 09	13 34	23 01	13 42	17 23	13 23	21 13
17	19 22	07 34	09 20	11 35	13 49	23 01	13 41	17 23	13 24	21 13
18	19 36	01 N 47	09 28	12 00	14 04	23 01	13 40	17 22	13 24	21 13
19	19 48	03 S 58	09 38	12 25	14 19	23 01	13 39	17 22	13 25	21 13
20	20 01	09 32	09 50	12 49	14 34	23 01	13 38	17 22	13 26	21 13
21	20 13	14 43	10 04	13 14	14 48	23 01	13 36	17 21	13 26	21 13
22	20 25	19 22	10 20	13 38	15 02	23 02	13 35	17 21	13 27	21 14
23	20 37	23 15	10 37	14 02	15 16	23 02	13 34	17 20	13 28	21 14
24	20 48	26 12	10 56	14 25	15 30	23 02	13 33	17 20	13 28	21 14
25	20 59	27 59	11 17	14 48	15 44	23 02	13 32	17 19	13 29	21 14
26	21 09	28 28	11 39	15 11	15 58	23 02	13 31	17 18	13 30	21 15
27	21 20	27 33	12 02	15 33	16 11	23 02	13 30	17 18	13 30	21 15
28	21 29	25 17	12 26	15 55	16 24	23 03	13 29	17 17	13 31	21 15
29	21 39	21 45	12 52	16 17	16 38	23 03	13 28	17 17	13 31	21 15
30	21 48	17 09	13 18	16 38	16 51	23 03	13 27	17 16	13 32	21 16
31	21 N 56	11 S 39	13 N 46	16 N 59	17 N 03	23 S 03	13 S 26	17 N 15	13 N 33	21 S 16

ZODIAC SIGN ENTRIES

Date	h	m	Planets
01	10	41	☽ ♒
03	17	57	☽ ♓
05	21	36	☽ ♈
06	10	41	♂ ♉
07	22	16	☽ ♉
09	21	33	☽ ♊
11	21	30	☽ ♋
13	19	45	♀ ♉
14	00	05	☽ ♌
16	06	33	☽ ♍
18	16	46	☽ ♎
21	03	09	☉ ♊
21	05	13	☽ ♏
23	18	05	☽ ♐
26	06	08	☽ ♑
28	16	38	☽ ♒
31	00	51	☽ ♓

LATITUDES

Date	Mercury ☿ ° '	Venus ♀ ° '	Mars ♂ ° '	Jupiter ♃ ° '	Saturn ♄ ° '	Uranus ♅ ° '	Neptune ♆ ° '	Pluto ♇ ° '
01	00 S 11	01 S 39	00 S 37	00 N 25	02 N 35	00 N 41	01 S 42	10 S 52
04	01 01	01 39	00 35	00 25	02 35	00 41	01 42	10 53
07	01 46	01 38	00 34	00 25	02 35	00 41	01 42	10 53
10	02 23	01 37	00 32	00 25	02 35	00 41	01 42	10 54
13	02 52	01 35	00 30	00 24	02 34	00 41	01 43	10 55
16	03 12	01 32	00 28	00 24	02 34	00 41	01 43	10 56
19	03 24	01 29	00 27	00 24	02 34	00 41	01 43	10 57
22	03 28	01 26	00 25	00 24	02 34	00 41	01 43	10 58
25	03 26	01 22	00 23	00 23	02 33	00 40	01 43	10 59
28	03 17	01 17	00 21	00 23	02 33	00 40	01 43	11 00
31	03 S 02	01 S 12	00 S 19	00 N 23	02 N 32	00 N 40	01 S 43	11 S 01

DATA

Julian Date	2467371
Delta T	+78 seconds
Ayanamsa	24° 27' 42"
Synetic vernal point	04° ♓ 39' 17"
True obliquity of ecliptic	23° 26' 10"

MOON'S PHASES, APSIDES AND POSITIONS ☽

Date	h	m	Phase	Longitude ° '	Eclipse Indicator
02	08	59	☾	11 ♒ 52	
09	03	21	●	18 ♉ 26	
15	21	05	☽	24 ♌ 56	
23	23	37	○	02 ♐ 45	
31	19	25	☾	10 ♓ 15	

Day	h	m		
09	02	29	Perigee	
22	14	31	Apogee	
05	23	57	ON	
11	23	37	Max dec	28° N 33'
18	19	23	OS	
26	08	21	Max dec	28° S 29'

LONGITUDES

Date	Chiron ⚷ ° '	Ceres ⚳ ° '	Pallas ⚴ ° '	Juno ⚵ ° '	Vesta ⚶ ° '	Black Moon Lilith ⚸ ° '
01	11 ♌ 37	01 ♓ 13	26 ♑ 50	06 ♑ 53	12 ♎ 58	16 ♏ 19
11	12 ♌ 02	03 ♓ 52	27 ♑ 22	06 ♑ 22	11 ♎ 38	17 ♏ 25
21	12 ♌ 37	06 ♓ 13	27 ♑ 14	05 ♑ 17	11 ♎ 07	18 ♏ 32
31	13 ♌ 21	08 ♓ 13	26 ♑ 25	03 ♑ 41	11 ♎ 25	19 ♏ 39

All ephemeris data is given at 12.00 UT and the Moon's longitude is additionally given for 24.00 UT

ASPECTARIAN

h m Aspects

01 Friday
01 23 ☽ ⊥ ♇
03 01 ☽ □ ♂
06 02 ☽ Q ♄
06 50 ☿ △ ♃
12 50 ☽ ⊻ ♇
16 40 ☽ □ ☿
16 50 ☽ Q ♀
17 05 ☽ ⊻ ♃
21 54 ♀ ⊼ ♄
23 37 ☽ ∥ ♃

02 Saturday
04 19 ☽ ⊥ ♃
05 41 ☽ □ ♆
08 59 ☽ □ ☉
10 05 ☽ ∥ ♇
10 31 ♀ ∠ ♇
11 45 ☽ ☍ ♅
14 58 ☽ Q ♂
15 36 ☽ □ ♄
17 32 ☽ ⚹ ♀
20 54 ☽ ∠ ♃

03 Sunday
00 44 ☽ Q ☿
04 49 ☽ # ♅
12 34 ☽ # ☉
14 06 ☽ ⚹ ♂
14 39 ☽ Q ♆
16 52 ☉ Q ♇
20 01 ☽ ☌ ♇
20 06 ☽ ∥ ♄
20 16 ☽ Q ☉
21 34 ☽ ⚹ ☿
22 25 ☉ □ ♅
23 07 ☽ # ♆
23 18 ☽ ∠ ♀
23 48 ☽ ⚹ ♃

04 Monday
07 14 ☽ # ☿
10 28 ☽ # ♂
11 53 ☽ ⚹ ♆
16 34 ☽ ⊥ ♀
17 27 ☽ ⊼ ♅
18 01 ☽ ∠ ♂
18 51 ☽ ⚹ ☉
20 39 ☽ Q ♃
20 42 ☽ △ ♄
22 57 ☽ ∠ ☿

05 Tuesday
02 33 ☽ # ♀
03 45 ☽ ± ♅
03 51 ☽ ⊻ ♀
10 15 ☽ ⊥ ♂
13 36 ☽ ∠ ♆
13 51 ☽ ⊥ ☿
15 19 ☿ ∥ ♂
18 57 ☽ ⚼ ♅
19 38 ☉ ☍ ♄
20 52 ☽ ⊻ ♂
21 56 ☽ ⚼ ♄
22 07 ☽ ∠ ☉
23 34 ☽ ⊻ ♇
23 39 ☽ ⊻ ☿

06 Wednesday
02 54 ☽ □ ♃
03 28 ☿ ⚹ ♇
04 38 ☽ ⊥ ♆
09 29 ☽ ⊥ ♇
12 43 ☽ ± ♄
14 01 ☽ ⊥ ☉
14 31 ☽ ⊻ ♆
19 40 ☽ △ ♅
22 26 ☽ ⊼ ♄
23 48 ☽ ∥ ♀

07 Thursday
00 10 ☽ ∠ ♇
00 27 ☽ ⊻ ☉
09 49 ☽ ☌ ♀
10 58 ☽ ∥ ☿
13 33 ☿ ☌ ♂
14 57 ☽ ∥ ♂
22 19 ☽ ∥ ♆
23 29 ☽ ☌ ☿

08 Friday
00 08 ☽ ☌ ♂
00 11 ☽ ⚹ ♇
00 30 ☽ # ♄
01 14 ♂ ⚹ ♇
03 10 ☽ △ ♃
13 04 ☽ ∥ ☉
14 13 ☽ ∥ ♅
14 37 ☽ ☌ ♆
18 06 ☉ ⚼ ♃
19 09 ☽ Q ♇
19 32 ☽ □ ♅
21 56 ☽ ☍ ♄

09 Saturday
02 44 ☽ ⚼ ♃
03 21 ☽ ☌ ☉
06 25 ☽ # ♇
13 23 ☽ ⊻ ♀
14 15 ☉ ∥ ♅
15 19 ☽ # ♃
16 50 ☽ ± ♃
22 31 ☽ ⊻ ☿
23 29 ☽ □ ♇
23 39 ☽ ⊥ ♀
23 52 ☽ Q ♅

10 Sunday
00 20 ☿ St D
01 51 ☽ ⊻ ♂
02 15 ☽ ⊼ ♃
07 59 ☽ ⊥ ☿
09 09 ♂ △ ♃
11 50 ☽ ⊥ ♂
14 00 ☽ ⊻ ♆
15 07 ☽ ∠ ♀
18 54 ☽ ⚹ ♅
21 02 ☽ ⊼ ♄
22 20 ☽ ∠ ☿
23 35 ☽ ⊥ ♆

11 Monday
02 56 ☽ ∠ ♂
06 04 ☽ ⊻ ☉
06 36 ☽ ± ♄
14 06 ☽ ∠ ♆
16 26 ☽ ⊥ ☉
17 20 ☽ ⚹ ♀
19 03 ☽ ∠ ♅
21 06 ☽ ⚼ ♄
22 44 ☽ ⚹ ☿
23 32 ☽ △ ♇

12 Tuesday
02 10 ☽ ☍ ♃
03 17 ☿ ∥ ♀
04 34 ☽ ⚹ ♂
08 13 ☽ ∠ ☉
09 51 ☽ ⊥ ♅
14 47 ☽ ⚹ ♆
14 52 ☽ Q ♀
18 50 ☽ Q ☿
19 52 ☽ ⊻ ♅
21 49 ☽ △ ♄

13 Wednesday
00 28 ☽ ⚼ ♇
01 41 ☽ Q ♂
11 08 ☽ Q ♆
11 18 ☽ ⚹ ☉
15 51 ☽ ± ♇

14 Thursday
00 30 ☽ □ ♀
02 16 ☽ □ ☿
02 17 ☽ ⊼ ♇
03 02 ☿ ⚹ ♇
04 49 ☽ ⊼ ♃
09 48 ☽ Q ☉
10 39 ☽ □ ♂
11 36 ☽ # ♃
15 24 ☽ ± ♃
18 48 ☽ □ ♆
20 46 ♀ ⚹ ♇
21 46 ☽ # ♇

15 Friday
00 13 ☽ ☌ ♅
01 56 ☽ □ ♄
03 51 ☿ ☌ ♀
07 35 ☽ ⚼ ♃
09 41 ☽ ∥ ☉
16 54 ☽ ∥ ♅
17 11 ♂ ∥ ♆
21 05 ☽ □ ☉
22 32 ♀ △ ♃

16 Saturday
08 58 ☽ ☍ ♇
09 43 ☽ # ♄
10 23 ☽ ∥ ♂
10 41 ☽ △ ☿
11 07 ☽ Q ♄
11 08 ☽ ∥ ♆
11 21 ☽ △ ♃
12 48 ☽ △ ♀
20 13 ☽ ∥ ♀
21 26 ☽ △ ♂
23 37 ♂ # ♄

17 Sunday
01 07 ☿ △ ♃
02 51 ☽ △ ♆
04 43 ☽ ∥ ☿
08 36 ☽ ⊻ ♅
10 00 ☽ ⚹ ♄
10 51 ☉ ± ♃
16 49 ☽ ⚼ ☿
20 19 ☽ ⊥ ♅
20 53 ☽ ⚼ ♀

18 Monday
04 29 ☽ ⚼ ♂
08 14 ☽ ⚼ ♆
11 25 ☽ ± ☿
11 43 ☽ △ ☉
14 06 ☽ ∠ ♅
15 18 ☽ ∠ ♄
16 38 ☽ ± ♀
19 21 ☽ ⊼ ♇
21 26 ☽ □ ♃
23 30 ☽ ± ♂

19 Tuesday
00 01 ☽ ⊼ ☿
02 12 ☽ ± ♆
05 57 ☽ ⊼ ♀
07 21 ☽ ± ♇
09 10 ☽ ⊥ ♄
12 19 ☽ ⊼ ♂
14 16 ☽ ⊼ ♆
20 11 ☽ ⚹ ♅
20 21 ☽ ⚼ ☉
21 10 ☽ ⊻ ♄

20 Wednesday
01 28 ☽ ⚼ ♇
09 22 ☽ Q ♃
13 26 ☽ # ☿
16 09 ☽ ± ☉
20 31 ☽ Q ♅
21 05 ♂ ☌ ♆

21 Thursday
04 17 ☽ # ♀
05 24 ☽ ⊼ ☉
05 52 ☽ # ♆
06 42 ☽ ∥ ♄
07 52 ☽ △ ♇
09 31 ☽ ⚹ ♃
12 25 ☽ # ♂
16 37 ☽ ☍ ☿

22 Friday
00 54 ♀ ∥ ♆
01 10 ☽ # ♅
01 37 ☽ ☍ ♀
03 14 ☽ ☍ ♆
05 09 ☽ ☍ ♂
09 09 ☽ □ ♅
09 34 ♀ # ♄
09 38 ☽ ☌ ♄
11 54 ☉ □ ♇
15 46 ☽ ∠ ♃
17 44 ♀ ☌ ♆
18 21 ☽ # ☉
22 51 ☽ ∥ ♇
23 15 ☉ Q ♅

23 Saturday
04 01 ☉ ⊼ ♃
09 51 ☽ ⊥ ♃
10 25 ☽ ∥ ♃
20 44 ☽ □ ♇
21 53 ☽ ⊻ ♃
23 37 ☽ ☍ ☉

24 Sunday
07 12 ♂ Q ♇
10 20 ☽ ⊼ ☿
16 05 ☽ ⊼ ♆
18 19 ♀ Q ♇
20 16 ♄ □ ♅
21 11 ☽ ⊼ ♀
21 47 ☽ ⊼ ♂
21 53 ☽ ⊻ ♄
21 54 ☽ △ ♅
23 15 ♂ ☍ ♄
23 28 ☽ ± ☿
23 39 ♂ □ ♅

25 Monday
03 45 ♀ ☍ ♄
04 09 ☽ ± ♆
04 22 ♀ □ ♅
08 54 ☽ Q ♇
09 49 ☽ ⊥ ♄
10 33 ☽ ± ♀
10 34 ☽ ± ♂
11 17 ♀ ☌ ♂
19 06 ☽ ⚼ ☿
22 08 ☽ ⚼ ♆

26 Tuesday
03 37 ☽ ∠ ♄
03 53 ☽ ⚼ ♅
05 38 ☽ ⚼ ♂
06 26 ☽ ⚼ ♀
08 45 ☽ ⚹ ♇
09 23 ☽ ☌ ♃
16 42 ☽ ⊼ ☉
21 41 ☽ ± ♅

27 Wednesday
00 23 ☉ # ♇
03 36 ☽ △ ☿
03 49 ☽ △ ♆
05 29 ☽ ± ☉
06 11 ☿ ☌ ♆
06 49 ♄ # ♆
08 59 ☽ ⚹ ♄
09 28 ☽ ⊼ ♅
09 44 ♀ ⚼ ♃
13 00 ☽ △ ♂
14 13 ☽ ∠ ♇
15 10 ☽ △ ♀

28 Thursday
00 29 ☽ ⚼ ☉
07 40 ☽ ⊥ ♇
08 09 ☽ Q ♄
12 26 ♂ ⚼ ♃
19 11 ☽ ⊻ ♇
19 19 ☽ ⊻ ♃

29 Friday
04 07 ☽ ∥ ♃
05 53 ☿ Q ♇
06 39 ☽ ⊥ ♃
07 38 ☽ △ ☉
09 33 ☿ ☍ ♄
09 35 ♃ ⚹ ♇
12 38 ☽ # ☉
13 44 ☽ □ ♆
14 53 ☽ ∥ ♇
18 15 ☽ □ ♄
18 30 ☿ □ ♅
19 10 ☽ ☍ ♅
19 14 ☽ □ ☿
23 28 ☽ ∠ ♃

30 Saturday
01 54 ☽ □ ♂
06 29 ☽ □ ♀
11 25 ☽ # ♅
13 22 ☽ # ♂
14 14 ☽ # ♀
19 10 ☿ # ♄
23 17 ☽ Q ♆

31 Sunday
00 18 ☿ ∥ ♆
02 57 ☽ ⚹ ♃
03 17 ☽ ☌ ♇
03 49 ☽ # ☿
04 06 ☽ # ♆
04 33 ☽ ∥ ♄
08 05 ☽ Q ☿
10 35 ☿ ⚼ ♃
12 55 ☽ Q ♂
18 44 ☽ Q ♀
19 25 ☽ □ ☉
21 04 ☽ ⚹ ♆

JUNE 2043

Date	Sidereal time h m s	Sun ☉ ° ' "	Moon ☽ ° ' "	Moon ☽ 24.00 ° ' "	Mercury ☿ ° '	Venus ♀ ° '	Mars ♂ ° '	Jupiter ♃ ° '	Saturn ♄ ° '	Uranus ♅ ° '	Neptune ♆ ° '	Pluto ♇ ° '
01	04 39 29	10 ♊ 54 57	19 ♓ 35 58	26 ♓ 28 21	17 ♉ 32	22 ♉ 45	19 ♉ 21	01 ♑ 00	13 ♏ 19	14 ♌ 04	11 ♉ 12	01 ♓ 20
02	04 43 26	11 52 27	03 ♈ 26 38	10 ♈ 30 51	18 55	23 58	20 04	00 R 53	13 R 15	14 06	11 14	01 R 20
03	04 47 22	12 49 57	17 ♈ 40 58	24 ♈ 56 47	20 21	25 11	20 48	00 46	13 12	14 08	11 16	01 20
04	04 51 19	13 47 26	02 ♉ 17 53	09 ♉ 43 41	21 50	26 25	21 32	00 39	13 08	14 11	11 18	01 20
05	04 55 15	14 44 55	17 ♉ 13 24	24 ♉ 46 04	23 21	27 38	22 15	00 32	13 05	14 13	11 20	01 20
06	04 59 12	15 42 23	02 ♊ 20 34	09 ♊ 55 39	24 55	28 ♉ 51	22 59	00 25	13 01	14 15	11 22	01 20
07	05 03 08	16 39 50	17 ♊ 30 00	25 ♊ 02 19	26 31	00 ♊ 04	23 43	00 18	12 58	14 18	11 24	01 20
08	05 07 05	17 37 16	02 ♋ 31 20	09 ♋ 55 53	28 11	01 17	24 26	00 10	12 55	14 20	11 25	01 19
09	05 11 01	18 34 41	17 ♋ 14 59	24 ♋ 27 49	29 ♉ 52	02 31	25 10	00 ♑ 03	12 51	14 22	11 27	01 19
10	05 14 58	19 32 05	01 ♌ 33 49	08 ♌ 32 35	01 ♊ 37	03 44	25 53	29 ♐ 56	12 48	14 25	11 29	01 19
11	05 18 55	20 29 28	15 ♌ 24 00	22 ♌ 08 04	03 24	04 57	26 36	29 48	12 45	14 27	11 31	01 19
12	05 22 51	21 26 50	28 ♌ 44 59	05 ♍ 15 06	05 13	06 10	27 19	29 41	12 42	14 30	11 33	01 18
13	05 26 48	22 24 11	11 ♍ 38 53	17 ♍ 56 52	07 05	07 24	28 03	29 33	12 39	14 32	11 35	01 18
14	05 30 44	23 21 31	24 ♍ 09 39	00 ♎ 17 53	08 59	08 37	28 46	29 26	12 36	14 35	11 37	01 18
15	05 34 41	24 18 50	06 ♎ 22 13	12 ♎ 23 20	10 56	09 50	29 ♉ 29	29 18	12 33	14 38	11 38	01 18
16	05 38 37	25 16 08	18 ♎ 21 52	24 ♎ 18 28	12 55	11 03	00 ♊ 12	29 10	12 31	14 40	11 40	01 17
17	05 42 34	26 13 25	00 ♏ 13 43	06 ♏ 08 12	14 56	12 17	00 55	29 03	12 28	14 43	11 42	01 17
18	05 46 30	27 10 42	12 ♏ 02 26	17 ♏ 56 54	16 59	13 30	01 38	28 55	12 25	14 46	11 44	01 16
19	05 50 27	28 07 58	23 ♏ 52 01	29 ♏ 48 09	19 04	14 43	02 21	28 47	12 23	14 49	11 45	01 16
20	05 54 24	29 ♊ 05 13	05 ♐ 45 38	11 ♐ 44 44	21 11	15 57	03 03	28 40	12 21	14 52	11 47	01 16
21	05 58 20	00 ♋ 02 27	17 ♐ 45 40	23 ♐ 48 37	23 19	17 10	03 46	28 32	12 18	14 54	11 49	01 15
22	06 02 17	00 59 41	29 ♐ 53 44	06 ♑ 01 07	25 28	18 23	04 29	28 24	12 16	14 57	11 50	01 15
23	06 06 13	01 56 55	12 ♑ 10 52	18 ♑ 23 02	27 38	19 37	05 11	28 17	12 14	15 00	11 52	01 14
24	06 10 10	02 54 08	24 ♑ 37 42	00 ♒ 54 56	29 ♊ 49	20 50	05 54	28 09	12 12	15 03	11 53	01 13
25	06 14 06	03 51 21	07 ♒ 14 47	13 ♒ 37 23	02 ♋ 00	22 04	06 36	28 02	12 10	15 06	11 55	01 13
26	06 18 03	04 48 34	20 ♒ 02 51	26 ♒ 31 19	04 12	23 17	07 19	27 54	12 08	15 09	11 56	01 12
27	06 21 59	05 45 47	03 ♓ 02 57	09 ♓ 37 59	06 23	24 30	08 01	27 46	12 06	15 12	11 58	01 12
28	06 25 56	06 42 59	16 ♓ 16 37	22 ♓ 59 04	08 34	25 44	08 43	27 39	12 05	15 15	11 59	01 11
29	06 29 53	07 40 12	29 ♓ 45 34	06 ♈ 36 21	10 44	26 57	09 26	27 32	12 03	15 18	12 01	01 10
30	06 33 49	08 ♋ 37 25	13 ♈ 31 32	20 ♈ 31 16	12 ♋ 53	28 ♊ 11	10 ♊ 08	27 ♐ 24	12 ♏ 01	15 ♌ 22	12 ♉ 02	01 ♓ 10

Date	Moon True ☊ ° '	Moon Mean ☊ ° '	Moon ☽ Latitude ° '	DECLINATIONS Sun ☉ ° '	Moon ☽ ° '	Mercury ☿ ° '	Venus ♀ ° '	Mars ♂ ° '	Jupiter ♃ ° '	Saturn ♄ ° '	Uranus ♅ ° '	Neptune ♆ ° '	Pluto ♇ ° '
01	06 ♈ 23	05 ♈ 21	01 S 28	22 N 05	05 S 28	14 N 14	17 N 19	17 N 16	23 S 03	13 S 25	17 N 15	13 N 33	21 S 16
02	06 D 23	05 18	00 S 16	22 13	01 N 08	14 43	17 39	17 28	23 03	13 24	17 14	13 34	21 17
03	06 R 23	05 15	01 N 00	22 20	07 52	15 13	17 59	17 41	23 04	13 23	17 14	13 34	21 17
04	06 21	05 12	02 13	22 27	14 21	15 44	18 18	17 53	23 04	13 22	17 13	13 35	21 17
05	06 16	05 09	03 19	22 34	20 09	16 14	18 37	18 05	23 04	13 21	17 12	13 36	21 18
06	06 09	05 06	04 12	22 40	24 44	16 46	18 55	18 16	23 04	13 20	17 11	13 36	21 18
07	05 59	05 02	04 47	22 46	27 36	17 17	19 12	18 28	23 04	13 20	17 11	13 37	21 18
08	05 49	04 59	05 01	22 52	28 26	17 48	19 29	18 39	23 04	13 19	17 10	13 37	21 19
09	05 39	04 56	04 56	22 57	27 13	18 20	19 46	18 50	23 04	13 18	17 09	13 38	21 19
10	05 30	04 53	04 32	23 01	24 13	18 51	20 02	19 01	23 04	13 17	17 09	13 38	21 19
11	05 24	04 50	03 52	23 06	19 55	19 22	20 18	19 12	23 05	13 16	17 08	13 39	21 20
12	05 20	04 46	03 01	23 10	14 44	19 53	20 33	19 23	23 05	13 16	17 07	13 39	21 20
13	05 18	04 43	02 02	23 13	09 04	20 23	20 47	19 33	23 05	13 15	17 06	13 40	21 21
14	05 D 18	04 40	00 N 59	23 16	03 N 13	20 52	21 01	19 43	23 05	13 14	17 06	13 41	21 21
15	05 R 18	04 37	00 S 06	23 19	02 S 37	21 20	21 14	19 53	23 05	13 14	17 05	13 41	21 21
16	05 17	04 34	01 09	23 21	08 15	21 47	21 27	20 03	23 05	13 13	17 04	13 42	21 22
17	05 15	04 31	02 08	23 23	13 33	22 12	21 39	20 13	23 05	13 12	17 03	13 42	21 22
18	05 10	04 27	03 01	23 24	18 19	22 36	21 51	20 22	23 05	13 12	17 02	13 43	21 23
19	05 03	04 24	03 47	23 25	22 24	22 58	22 02	20 31	23 05	13 11	17 01	13 43	21 23
20	04 53	04 21	04 23	23 26	25 34	23 19	22 12	20 40	23 05	13 11	17 01	13 43	21 24
21	04 41	04 18	04 47	23 26	27 38	23 37	22 22	20 49	23 05	13 10	17 00	13 44	21 24
22	04 28	04 15	04 59	23 26	28 25	23 52	22 31	20 58	23 05	13 10	16 59	13 44	21 24
23	04 15	04 12	04 57	23 25	27 48	24 06	22 39	21 06	23 05	13 10	16 58	13 45	21 25
24	04 02	04 08	04 41	23 24	25 48	24 16	22 47	21 14	23 05	13 09	16 57	13 45	21 25
25	03 52	04 05	04 11	23 23	22 30	24 24	22 54	21 22	23 05	13 09	16 56	13 46	21 26
26	03 44	04 02	03 28	23 21	18 05	24 30	23 01	21 30	23 06	13 08	16 55	13 46	21 26
27	03 39	03 59	02 34	23 19	12 46	24 32	23 06	21 38	23 06	13 08	16 54	13 47	21 27
28	03 37	03 56	01 30	23 16	06 48	24 31	23 12	21 45	23 06	13 08	16 53	13 47	21 27
29	03 D 36	03 52	00 20	23 13	00 24	24 28	23 16	21 52	23 06	13 08	16 52	13 47	21 28
30	03 ♈ 36	03 ♈ 49	00 N 52	23 N 09	06 N 08	24 N 22	23 N 20	21 N 59	23 S 06	13 S 07	16 N 52	13 N 48	21 S 28

ZODIAC SIGN ENTRIES

Date	h	m	Planets
02	06	06	☽ ♈
04	08	16	☽ ♉
06	08	18	☽ ♊
07	10	37	♀ ♊
08	07	57	☽ ♋
09	13	47	☿ ♊
09	21	43	♃ ♐
10	09	20	☽ ♌
12	14	18	☽ ♍
14	23	25	☽ ♎
16	05	23	♂ ♊
17	11	32	☽ ♏
20	00	24	☽ ♐
21	10	58	☉ ♋
22	12	12	☽ ♑
24	14	00	☿ ♋
24	22	15	☽ ♒
27	06	24	☽ ♓
29	12	25	☽ ♈

LATITUDES

Date	Mercury ☿ ° '	Venus ♀ ° '	Mars ♂ ° '	Jupiter ♃ ° '	Saturn ♄ ° '	Uranus ♅ ° '	Neptune ♆ ° '	Pluto ♇ ° '
01	02 S 56	01 S 10	00 S 18	00 N 23	02 N 32	00 N 40	01 S 43	11 S 02
04	02 35	01 05	00 16	00 22	02 32	00 40	01 43	11 03
07	02 09	00 59	00 14	00 22	02 31	00 40	01 43	11 04
10	01 40	00 52	00 12	00 22	02 31	00 40	01 43	11 05
13	01 08	00 46	00 10	00 21	02 30	00 40	01 43	11 06
16	00 34	00 39	00 08	00 21	02 29	00 40	01 43	11 06
19	00 S 01	00 32	00 06	00 21	02 29	00 40	01 44	11 07
22	00 N 31	00 25	00 04	00 20	02 28	00 40	01 44	11 08
25	00 59	00 18	00 S 02	00 20	02 27	00 40	01 44	11 09
28	01 22	00 10	00 00	00 19	02 27	00 40	01 44	11 10
31	01 N 39	00 S 03	00 N 02	00 N 19	02 N 26	00 N 40	01 S 44	11 S 11

DATA

Julian Date	2467402
Delta T	+78 seconds
Ayanamsa	24° 27' 47"
Synetic vernal point	04° ♓ 39' 12"
True obliquity of ecliptic	23° 26' 10"

MOON'S PHASES, APSIDES AND POSITIONS ☽

Date	h	m	Phase	Longitude ° '	Eclipse Indicator
07	10	35	●	16 ♊ 36	
14	10	19	☽	23 ♍ 17	
22	14	21	○	01 ♑ 05	
30	02	53	☾	08 ♈ 16	

Day	h	m		
06	11	49	Perigee	
18	23	23	Apogee	
02	07	58	0N	
08	09	29	Max dec	28° N 27'
15	01	11	0S	
22	13	36	Max dec	28° S 25'
29	13	30	0N	

LONGITUDES

Date	Chiron ⚷ ° '	Ceres ⚳ ° '	Pallas ⚴ ° '	Juno ⚵ ° '	Vesta ⚶ ° '	Black Moon Lilith ⚸ ° '
01	13 ♌ 26	08 ♓ 23	26 ♑ 18	03 ♑ 30	11 ♎ 29	19 ♏ 46
11	14 ♌ 18	09 ♓ 57	24 ♑ 43	01 ♑ 28	12 ♎ 38	20 ♏ 53
21	15 ♌ 16	11 ♓ 02	22 ♑ 32	29 ♐ 13	14 ♎ 27	21 ♏ 59
31	16 ♌ 21	11 ♓ 36	19 ♑ 55	26 ♐ 57	16 ♎ 52	23 ♏ 06

ASPECTARIAN

h m	Aspects
01 Monday	
00 24	☽ Q ♃
00 56	☽ △ ♄
01 07	♀ ∥ ♂
02 12	☽ ⊼ ♅
06 41	♀ ∥ ♅
07 57	☽ ⚹ ☿
08 37	♇ St R
09 57	♂ ∥ ♅
11 31	☽ ⚹ ♂
12 50	☽ ± ♅
18 04	☽ ⚹ ♀
19 15	☉ ⊻ ♆
23 33	☽ ∠ ♆
02 Tuesday	
03 08	☽ ⊑ ♄
04 32	☽ ⊑ ♅
05 25	☽ Q ☉
07 38	☽ □ ♃
08 23	☽ ⊻ ♇
12 54	☽ ∠ ☿
14 56	☽ ∠ ♂
15 03	☽ ⊥ ♆
18 28	☽ ± ♄
18 37	☽ ⊥ ♇
22 17	☽ ∠ ♀
03 Wednesday	
01 14	☽ ⊻ ♆
03 19	☽ ⚹ ☉
04 21	♀ ± ♃
04 33	☽ ⊼ ♄
05 49	☽ ⊥ ☿
06 05	☽ △ ♅
06 56	☽ ⊥ ♂
09 45	☽ ∠ ♇
14 44	☽ ⊥ ♀
16 55	☽ ⊻ ☿
17 27	☽ ⊻ ♂
20 36	☉ ⊼ ♄
04 Thursday	
01 32	☽ ⊻ ♀
02 39	☿ ☌ ♂
05 54	☽ ∠ ☉
08 15	☽ ⋕ ♄
09 03	☽ ∥ ♆
09 20	☽ △ ♃
10 26	☽ ⚹ ♇
17 51	☽ ∥ ☿
21 30	☽ ⊥ ☉
22 05	☉ ⚹ ♅
23 20	☽ ∥ ♅
05 Friday	
02 33	☽ ☌ ♆
02 37	☽ ∥ ♂
04 49	☽ ∥ ♀
05 24	☽ ☍ ♄
05 47	☽ Q ♇
07 11	☽ □ ♅
07 47	☽ ⊻ ☉
09 19	☽ ⊑ ♃
17 20	☽ ⋕ ♇
20 25	☽ ☌ ♂
22 52	☽ ☌ ☿
23 32	☽ ± ♃
23 56	☽ ∥ ☉
06 Saturday	
02 18	☽ ⋕ ♃
04 55	☿ ± ♃
05 59	☽ ☌ ♀
08 58	☽ ⊼ ♃
10 24	☽ □ ♇
11 52	☽ Q ♅
07 Sunday	
02 18	☽ ⊻ ♆
04 50	☽ ⊼ ♄
06 54	☽ ⚹ ♅
07 24	☿ ∥ ♅
10 35	☽ ☌ ☉
11 50	☽ ⊥ ♆
14 19	☽ ± ♄
15 59	♀ ⊼ ♃
22 23	☽ ⊻ ♂
08 Monday	
02 12	☽ ∠ ♆
04 09	☽ ⊻ ☿
04 32	♂ ± ♃
04 37	☽ ⊑ ♄
06 52	☽ ∠ ♅
06 54	☉ ⊥ ♆
08 15	☽ ☍ ♃
08 28	☽ ⊥ ♂
09 50	☽ ⊻ ♀
10 04	☽ △ ♇
12 38	♀ □ ♇
15 00	☽ ⊥ ☿
20 24	☽ ⊥ ♀
21 25	☽ ⊥ ♅
23 46	☽ ∠ ♂
09 Tuesday	
02 28	☽ ⚹ ♆
04 48	☽ △ ♄
07 15	☽ ⊻ ♅
07 34	☽ ∠ ☿
09 11	♀ Q ♅
10 28	☽ ⊑ ♇
12 28	☽ ∠ ♀
14 18	☿ ⊼ ♃
14 21	☽ ⊻ ☉
18 36	☉ ± ♄
22 20	☽ Q ♆
10 Wednesday	
01 04	☽ ⊥ ☉
01 26	☽ ± ♇
01 52	☽ ⚹ ♂
07 58	☿ □ ♇
09 14	☽ ⊼ ♃
11 35	☽ ⊼ ♇
12 06	☽ ⚹ ☿
16 04	☽ ⚹ ♀
17 27	☽ ∠ ☉
19 01	☽ ⋕ ♃
19 11	☽ ∥ ☉
19 25	☽ ± ♃
23 08	☿ Q ♅
23 27	☽ Q ♂
11 Thursday	
00 04	☿ ∥ ♂
04 44	☽ ⋕ ♇
04 59	☉ ⋕ ♃
05 10	☽ □ ♆
07 21	☽ □ ♄
10 09	☽ ∥ ♀
10 20	☽ ☌ ♅
10 57	☽ ⊑ ♃
11 59	☽ Q ☿
14 24	☽ ∥ ☿
15 01	☽ Q ♀
15 21	☽ ∥ ♂
21 45	☽ ⚹ ☉
12 Friday	
01 17	☽ ∥ ♅
09 15	☽ □ ♂
13 41	☽ △ ♃
15 34	☽ Q ♄
16 40	☽ ∥ ♆
16 42	☽ ☍ ♇
18 24	☽ ⋕ ♄
21 20	☽ Q ☉
13 Saturday	
01 57	☽ □ ☿
03 09	☽ □ ♀
11 52	☽ △ ♆
13 39	♃ ⊑ ♅
13 53	☽ ⚹ ♄
17 30	☽ ⊻ ♅
23 09	☿ ☌ ♀
14 Sunday	
05 03	☽ ⊥ ♅
10 19	☽ □ ☉
16 47	☽ ⊑ ♆
18 41	☽ ∠ ♄
21 33	☽ △ ♂
22 11	☽ □ ♃
22 38	☽ ∠ ♅
15 Monday	
01 58	☽ ⊼ ♇
03 05	☿ ∥ ♀
06 52	♂ ⊼ ♃
10 33	☽ ± ♆
12 22	☽ ⊥ ♄
13 20	☿ ⋕ ♇
13 50	☽ ± ♇
19 41	☽ △ ♀
20 45	☿ ⊻ ♆
22 32	☽ ⊼ ♆
22 52	☽ △ ☿
16 Tuesday	
00 17	☽ ⊻ ♄
01 24	♀ ⋕ ♇
04 33	☽ ⚹ ♅
05 14	☽ ⊑ ♂
07 16	☿ ⊼ ♄
07 49	☽ ⊑ ♇
09 38	☽ Q ♃
17 Wednesday	
00 17	♀ ⊻ ♆
00 32	☽ ± ♂
03 10	☽ △ ☉
04 51	☽ Q ♅
05 20	☽ ⊑ ♀
09 26	☿ ⚹ ♅
09 38	☽ ⚹ ♃
10 24	☽ ∥ ♄
11 16	☽ ⊑ ☿
12 44	☽ ⋕ ♆
13 29	☽ ⊼ ♂
14 08	☽ △ ♇
15 33	♀ ⊼ ♄
18 Thursday	
00 11	♂ □ ♇
00 16	☉ ∠ ♆
01 42	☽ ± ♀
05 16	☽ ⋕ ♅
09 24	☽ ± ☿
11 22	☽ ☍ ♆
12 18	☽ ⊑ ☉
12 47	☽ ☌ ♄
15 19	☽ ⊼ ♀
15 47	☽ ∠ ♃
17 33	☽ □ ♅
17 55	☉ ⊑ ♄
20 43	☿ ⊥ ♆
23 56	☽ ⋕ ♂
19 Friday	
00 11	☽ ⊼ ☿
04 19	☿ ± ♄
05 34	☽ ∥ ♇
08 11	☽ ± ☉
09 30	☽ ⋕ ♀
09 51	☽ ⊥ ♃
13 49	♀ ⚹ ♅
16 18	☽ ⋕ ☿
16 39	☽ ∥ ♃
19 01	☽ ⋕ ☉
19 46	☿ ⋕ ♃
21 23	☽ ⊼ ☉
21 51	☽ ⊻ ♃
20 Saturday	
02 36	☉ ☍ ♃
02 57	☽ □ ♇
04 54	♂ Q ♅
06 13	☽ ☍ ♂
21 28	☉ ∥ ☿
21 Sunday	
00 06	☽ ⊼ ♆
01 09	☽ ⊻ ♄
06 17	☽ △ ♅
08 26	☉ ∠ ♅
10 41	☽ ☍ ♀
12 06	☽ ± ♆
13 05	☽ ⊥ ♄
14 58	☽ Q ♇
22 Monday	
00 54	♀ ⊥ ♆
01 23	☽ ☍ ☿
05 58	☽ ⊑ ♆
06 50	☽ ∠ ♄
09 06	☽ ☌ ♃
09 40	♀ ± ♄
12 07	☽ ⊑ ♅
14 21	☽ ☍ ☉
14 39	☽ ⚹ ♇
18 11	☉ △ ♇
21 32	☽ ⊼ ♂
23 Tuesday	
03 22	☿ ∠ ♆
05 48	☽ ± ♅
07 37	☿ ⊑ ♄
09 57	☽ ± ♂
11 23	☽ △ ♆
12 06	☽ ⚹ ♄
17 30	☽ ⊼ ♅
18 43	☿ ☍ ♃
19 51	☽ ∠ ♇
24 Wednesday	
03 56	☽ ⊼ ♀
04 25	☽ ⊑ ♂
11 11	☽ Q ♄
13 08	☽ ⊥ ♇
14 39	☿ ∠ ♅
16 41	☽ ± ♀
18 40	☽ ⊻ ♃
23 42	☽ ⋕ ☿
23 59	☽ ⊼ ☿
25 Thursday	
00 35	☽ ⊻ ♇
03 22	☿ △ ♇
05 04	☽ ⊼ ☉
05 58	☽ ⊥ ♃
06 21	☽ ⋕ ☉
08 17	☽ ∥ ♃
09 34	☽ ⋕ ♀
10 43	☽ △ ♂
11 36	☽ ⊑ ♀
13 44	☽ ± ☿
17 19	☽ ± ☉
18 21	☽ ∥ ♇
18 30	☽ ⋕ ♂
20 49	☽ □ ♆
21 14	☽ □ ♄
22 47	☽ ∠ ♃
23 19	♂ ⋕ ♇
26 Friday	
02 50	☽ ☍ ♅
10 05	☽ ⊑ ☿
11 31	☽ ⊑ ☉
17 37	☽ ⋕ ♅
18 38	☽ △ ♀
23 56	☉ ☌ ☿
27 Saturday	
02 24	☽ ⚹ ♃
06 20	☽ Q ♆
07 44	☽ ⋕ ♆
08 10	♀ ⋕ ♃
08 36	☽ ☌ ♇
10 28	☽ ∥ ♄
17 21	☽ △ ☉
19 18	☽ △ ☿
21 35	☽ □ ♂
28 Sunday	
00 09	☽ Q ♃
04 16	☽ ⚹ ♆
04 26	☽ △ ♄
10 09	☽ ⊼ ♅
14 38	☿ ⊻ ♂
19 50	☿ ⊥ ♅
20 57	☽ ± ♅
29 Monday	
01 47	☉ ∥ ♀
06 33	☽ □ ♀
07 09	☽ ∠ ♆
07 13	☽ ⊑ ♄
07 40	☽ Q ♂
08 06	☽ □ ♃
12 58	☽ ⊑ ♅
13 15	♀ ∠ ♆
13 49	♀ ⊑ ♄
14 29	☽ ⊻ ♇
22 10	♀ ☍ ♃
22 59	☽ ⊥ ♆
23 00	☽ ± ♄
30 Tuesday	
00 59	☽ ⊥ ♇
02 27	☿ ⚹ ♆
02 28	☿ △ ♄
02 53	☽ □ ☉
03 38	♄ ☍ ♆
05 49	☽ ⚹ ♂
09 25	☽ ⊼ ♄
09 26	☽ ⊻ ♆
10 42	☽ □ ☿
15 10	☽ △ ♅
16 32	☽ ∠ ♇
17 00	☽ Q ♀

All ephemeris data is given at 12.00 UT and the Moon's longitude is additionally given for 24.00 UT

JULY 2043

LONGITUDES

Date	Sidereal time h m s	Sun ☉ ° ' "	Moon ☽ ° ' "	Moon ☽ 24.00 ° ' "	Mercury ☿ ° '	Venus ♀ ° '	Mars ♂ ° '	Jupiter ♃ ° '	Saturn ♄ ° '	Uranus ♅ ° '	Neptune ♆ ° '	Pluto ♇ ° '
01	06 37 46	09 ♋ 34 37	27 ♈ 35 33	04 ♉ 44 18	15 ♋ 01	29 ♊ 24	10 ♊ 50	27 ♐ 17	12 ♏ 00	15 ♌ 25	12 ♉ 04	01 ♓ 09
02	06 41 42	10 31 50	11 ♉ 57 19	19 ♉ 14 13	17 08	00 ♋ 38	11 32	27 R 09	11 R 59	15 28	12 05	01 R 08
03	06 45 39	11 29 04	26 ♉ 34 29	03 ♊ 57 27	19 14	01 51	12 14	27 02	11 57	15 31	12 07	01 08
04	06 49 35	12 26 17	11 ♊ 22 18	18 ♊ 48 05	21 17	03 05	12 56	26 55	11 56	15 34	12 08	01 07
05	06 53 32	13 23 31	26 ♊ 13 45	03 ♋ 38 12	23 20	04 18	13 38	26 48	11 55	15 38	12 09	01 06
06	06 57 28	14 20 45	11 ♋ 00 22	18 ♋ 19 11	25 20	05 32	14 20	26 41	11 54	15 41	12 10	01 05
07	07 01 25	15 17 59	25 ♋ 33 41	02 ♌ 43 04	27 19	06 45	15 02	26 34	11 53	15 44	12 12	01 05
08	07 05 22	16 15 12	09 ♌ 46 39	16 ♌ 43 58	29 ♋ 16	07 59	15 44	26 27	11 53	15 48	12 13	01 04
09	07 09 18	17 12 26	23 ♌ 34 41	00 ♍ 18 42	01 ♌ 10	09 13	16 25	26 20	11 52	15 51	12 14	01 03
10	07 13 15	18 09 39	06 ♍ 56 02	13 ♍ 26 55	03 03	10 26	17 07	26 14	11 51	15 54	12 15	01 02
11	07 17 11	19 06 53	19 ♍ 51 39	26 ♍ 10 40	04 54	11 40	17 48	26 07	11 51	15 58	12 16	01 01
12	07 21 08	20 04 06	02 ♎ 24 29	08 ♎ 33 41	06 44	12 54	18 30	26 01	11 51	16 01	12 18	01 00
13	07 25 04	21 01 19	14 ♎ 38 54	20 ♎ 40 46	08 31	14 07	19 11	25 54	11 50	16 05	12 19	00 59
14	07 29 01	21 58 32	26 ♎ 39 57	02 ♏ 37 09	10 16	15 21	19 53	25 48	11 50	16 08	12 20	00 58
15	07 32 57	22 55 46	08 ♏ 32 59	14 ♏ 28 07	11 59	16 35	20 34	25 42	11 50	16 12	12 21	00 57
16	07 36 54	23 52 59	20 ♏ 23 08	26 ♏ 18 37	13 40	17 48	21 15	25 36	11 D 50	16 15	12 22	00 56
17	07 40 51	24 50 13	02 ♐ 15 06	08 ♐ 13 02	15 19	19 02	21 56	25 30	11 50	16 19	12 23	00 55
18	07 44 47	25 47 26	14 ♐ 12 51	20 ♐ 14 56	16 57	20 16	22 38	25 24	11 50	16 22	12 24	00 54
19	07 48 44	26 44 40	26 ♐ 19 34	02 ♑ 27 00	18 33	21 29	23 19	25 19	11 51	16 26	12 24	00 53
20	07 52 40	27 41 55	08 ♑ 37 25	14 ♑ 50 57	20 06	22 43	24 00	25 13	11 51	16 29	12 25	00 52
21	07 56 37	28 39 09	21 ♑ 07 40	27 ♑ 27 36	21 38	23 57	24 41	25 08	11 52	16 33	12 26	00 51
22	08 00 33	29 ♋ 36 24	03 ♒ 50 43	10 ♒ 16 59	23 08	25 11	25 21	25 03	11 52	16 36	12 27	00 50
23	08 04 30	00 ♌ 33 40	16 ♒ 46 18	23 ♒ 18 37	24 35	26 25	26 02	24 58	11 53	16 40	12 28	00 49
24	08 08 26	01 30 56	29 ♒ 53 51	06 ♓ 31 54	26 01	27 38	26 43	24 53	11 54	16 44	12 29	00 48
25	08 12 23	02 28 13	13 ♓ 12 44	19 ♓ 56 17	27 25	28 ♋ 52	27 24	24 48	11 55	16 47	12 29	00 47
26	08 16 20	03 25 31	26 ♓ 42 31	03 ♈ 31 28	28 ♌ 46	00 ♌ 06	28 04	24 44	11 56	16 51	12 30	00 46
27	08 20 16	04 22 49	10 ♈ 23 05	17 ♈ 17 26	00 ♍ 06	01 20	28 45	24 39	11 57	16 55	12 31	00 45
28	08 24 13	05 20 09	24 ♈ 14 28	01 ♉ 14 13	01 23	02 34	29 ♊ 25	24 35	11 58	16 58	12 31	00 44
29	08 28 09	06 17 30	08 ♉ 16 35	15 ♉ 21 29	02 38	03 48	00 ♋ 06	24 31	11 59	17 02	12 32	00 42
30	08 32 06	07 14 51	22 ♉ 28 45	29 ♉ 38 07	03 51	05 02	00 46	24 27	12 01	17 06	12 32	00 41
31	08 36 02	08 ♌ 12 14	06 ♊ 49 15	14 ♊ 01 43	05 ♍ 01	06 ♌ 16	01 ♋ 27	24 ♐ 23	12 ♏ 02	17 ♌ 09	12 ♉ 33	00 ♓ 40

DECLINATIONS

Date	Moon True ☊ ° '	Moon Mean ☊ ° '	Moon ☽ Latitude ° '	Sun ☉ ° '	Moon ☽ ° '	Mercury ☿ ° '	Venus ♀ ° '	Mars ♂ ° '	Jupiter ♃ ° '	Saturn ♄ ° '	Uranus ♅ ° '	Neptune ♆ ° '	Pluto ♇ ° '
01	03 ♈ 36	03 ♈ 46	02 N 03	23 N 05	12 N 32	24 N 13	23 N 23	22 N 06	23 S 06	13 S 07	16 N 51	13 N 48	21 S 29
02	03 R 33	03 43	03 08	23 01	18 24	24 02	23 25	22 12	23 06	13 07	16 50	13 49	21 29
03	03 28	03 40	04 01	22 56	23 18	23 48	23 27	22 19	23 06	13 07	16 49	13 49	21 30
04	03 21	03 37	04 39	22 51	26 45	23 32	23 28	22 25	23 06	13 07	16 48	13 49	21 30
05	03 11	03 33	04 59	22 46	28 21	23 14	23 28	22 31	23 06	13 07	16 47	13 50	21 31
06	03 00	03 30	04 58	22 40	27 56	22 54	23 28	22 36	23 05	13 07	16 46	13 50	21 31
07	02 49	03 27	04 38	22 34	25 34	22 32	23 27	22 42	23 05	13 07	16 45	13 50	21 32
08	02 40	03 24	04 01	22 27	21 40	22 07	23 25	22 47	23 05	13 07	16 44	13 51	21 33
09	02 32	03 21	03 10	22 20	16 39	21 42	23 23	22 52	23 05	13 07	16 43	13 51	21 33
10	02 28	03 18	02 11	22 12	10 59	21 15	23 20	22 57	23 05	13 07	16 42	13 51	21 34
11	02 25	03 14	01 N 06	22 04	05 N 02	20 46	23 16	23 02	23 05	13 07	16 41	13 51	21 34
12	02 D 25	03 11	00 00	21 56	00 S 57	20 16	23 12	23 06	23 05	13 07	16 40	13 52	21 35
13	02 25	03 08	01 S 05	21 48	06 46	19 45	23 06	23 10	23 05	13 07	16 39	13 52	21 35
14	02 R 25	03 05	02 05	21 39	12 14	19 13	23 01	23 14	23 05	13 07	16 38	13 52	21 36
15	02 24	03 02	03 00	21 29	17 11	18 40	22 54	23 18	23 05	13 07	16 36	13 53	21 36
16	02 21	02 58	03 46	21 20	21 28	18 06	22 47	23 21	23 05	13 08	16 35	13 53	21 37
17	02 15	02 55	04 23	21 09	24 54	17 32	22 39	23 25	23 05	13 08	16 34	13 53	21 38
18	02 07	02 52	04 48	20 59	27 16	16 57	22 30	23 28	23 05	13 08	16 33	13 53	21 38
19	01 57	02 49	05 01	20 48	28 24	16 22	22 21	23 31	23 05	13 09	16 32	13 53	21 39
20	01 46	02 46	05 01	20 37	28 09	15 46	22 11	23 33	23 05	13 09	16 31	13 54	21 39
21	01 35	02 43	04 46	20 26	26 29	15 09	22 01	23 35	23 05	13 10	16 30	13 54	21 40
22	01 25	02 39	04 16	20 14	23 26	14 33	21 50	23 38	23 05	13 10	16 29	13 54	21 40
23	01 16	02 36	03 33	20 02	19 12	13 56	21 38	23 40	23 05	13 11	16 28	13 54	21 41
24	01 10	02 33	02 38	19 49	13 59	13 19	21 26	23 42	23 05	13 11	16 27	13 54	21 42
25	01 06	02 30	01 34	19 36	08 03	12 43	21 13	23 44	23 05	13 12	16 26	13 55	21 42
26	01 05	02 27	00 S 23	19 23	01 S 40	12 06	20 59	23 45	23 05	13 12	16 25	13 55	21 43
27	01 D 05	02 23	00 N 50	19 10	04 N 52	11 29	20 45	23 46	23 05	13 13	16 23	13 55	21 43
28	01 06	02 20	02 01	18 56	11 16	10 53	20 30	23 47	23 05	13 13	16 22	13 55	21 44
29	01 R 06	02 17	03 06	18 42	17 11	10 17	20 14	23 48	23 05	13 14	16 21	13 55	21 45
30	01 05	02 14	04 00	18 27	22 15	09 41	19 58	23 49	23 05	13 15	16 20	13 55	21 45
31	01 ♈ 02	02 ♈ 11	04 N 40	18 N 13	26 N 02	09 N 05	19 N 42	23 N 49	23 S 05	13 S 15	16 N 19	13 N 55	21 S 46

ZODIAC SIGN ENTRIES

Date	h m	Planets
01	16 03	☽ ♉
01	23 42	♀ ♋
03	17 35	☽ ♊
05	18 06	☽ ♋
07	19 25	☽ ♌
08	21 15	☿ ♌
09	23 26	☽ ♍
12	07 21	☽ ♎
14	18 43	☽ ♏
17	07 27	☽ ♐
19	19 13	☽ ♑
22	04 47	☽ ♒
22	21 53	☉ ♌
24	12 11	☽ ♓
26	10 00	♀ ♌
26	17 48	☽ ♈
27	10 18	☿ ♍
28	21 53	☽ ♉
29	08 31	♂ ♋
31	00 37	☽ ♊

LATITUDES

Date	Mercury ☿ ° '	Venus ♀ ° '	Mars ♂ ° '	Jupiter ♃ ° '	Saturn ♄ ° '	Uranus ♅ ° '	Neptune ♆ ° '	Pluto ♇ ° '
01	01 N 39	00 S 03	00 N 02	00 N 19	02 N 26	00 N 40	01 S 44	11 S 11
04	01 49	00 N 04	00 04	00 18	02 25	00 40	01 44	11 12
07	01 52	00 11	00 06	00 18	02 24	00 40	01 44	11 13
10	01 49	00 18	00 08	00 18	02 24	00 39	01 45	11 13
13	01 41	00 25	00 10	00 17	02 23	00 39	01 45	11 14
16	01 27	00 32	00 13	00 17	02 22	00 39	01 45	11 15
19	01 09	00 39	00 15	00 16	02 21	00 39	01 45	11 15
22	00 47	00 45	00 17	00 16	02 20	00 39	01 45	11 16
25	00 N 22	00 51	00 19	00 15	02 20	00 39	01 45	11 17
28	00 S 07	00 56	00 21	00 15	02 19	00 39	01 46	11 17
31	00 S 37	01 N 01	00 N 23	00 N 14	02 N 18	00 N 39	01 S 46	11 S 18

DATA

Julian Date	2467432
Delta T	+78 seconds
Ayanamsa	24° 27' 52"
Synetic vernal point	04° ♓ 39' 07"
True obliquity of ecliptic	23° 26' 10"

MOON'S PHASES, APSIDES AND POSITIONS ☽

Date	h m	Phase	Longitude ° '	Eclipse Indicator
06	17 51	●	14 ♋ 35	
14	01 47	☽	21 ♎ 34	
22	03 24	○	29 ♑ 16	
29	08 23	☾	06 ♉ 09	

Day	h m		
04	16 00	Perigee	
16	14 31	Apogee	
05	18 56	Max dec	28° N 27'
12	08 08	0S	
19	19 48	Max dec	28° S 29'
26	18 08	0N	

LONGITUDES

Date	Chiron ⚷ ° '	Ceres ⚳ ° '	Pallas ⚴ ° '	Juno ⚵ ° '	Vesta ⚶ ° '	Black Moon Lilith ⚸ ° '
01	16 ♌ 21	11 ♓ 36	19 ♑ 55	26 ♐ 57	16 ♎ 52	23 ♏ 06
11	17 ♌ 30	11 ♓ 36	17 ♑ 08	24 ♐ 53	19 ♎ 46	24 ♏ 13
21	18 ♌ 43	11 ♓ 00	14 ♑ 27	23 ♐ 11	23 ♎ 06	25 ♏ 20
31	19 ♌ 59	09 ♓ 50	12 ♑ 07	22 ♐ 00	26 ♎ 46	26 ♏ 27

All ephemeris data is given at 12.00 UT and the Moon's longitude is additionally given for 24.00 UT

ASPECTARIAN

h m	Aspects
01 Wednesday	
07 36	☉ ⊥ ♅
08 53	☽ ∠ ♂
11 11	☉ ⋕ ♃
11 28	☽ △ ♃
11 58	☽ Q ☉
14 17	☽ ⋕ ♄
15 20	☽ ✱ ♀
16 31	☿ ⊻ ♅
16 59	☽ ∥ ♆
17 59	☽ ✱ ♇
22 43	☽ Q ☿
02 Thursday	
00 43	☿ ⚼ ♇
00 47	☽ ⊥ ♂
05 17	☽ ∥ ♅
08 40	♀ ∠ ♅
09 28	☽ ✱ ☉
11 16	☽ ⊻ ♂
12 02	☽ ☍ ♄
12 13	☽ ☌ ♆
12 20	☽ ⚼ ♃
13 57	☽ Q ♇
17 49	☽ □ ♅
18 37	☽ ∠ ♀
18 53	☿ ⊥ ♂
21 56	♀ △ ♇
21 59	☽ ✱ ☿
23 43	♃ ⚼ ♆
03 Friday	
02 28	☽ ⋕ ♇
02 38	♂ ⊼ ♄
03 01	☽ ± ♃
06 32	☽ ∥ ♂
07 27	♂ ⊻ ♆
10 04	☽ ∥ ☉
10 43	☽ ⊥ ♀
10 52	☽ ⋕ ♃
11 51	☽ ∠ ☉
12 45	☽ ⊼ ♃
12 54	☽ ∥ ♀
14 47	☽ ∥ ☿
19 24	☽ □ ♇
21 22	☽ ⊻ ♀
23 20	☽ Q ♅
23 38	☉ △ ♄
04 Saturday	
02 27	☽ ∠ ☿
03 28	☽ ⊥ ☉
04 06	☉ ✱ ♆
06 58	♃ ∠ ♄
12 55	☽ ⊼ ♄
13 14	☽ ⊻ ♆
13 50	☽ ⊻ ☉
14 39	☽ ☌ ♂
17 41	☿ ∥ ♀
18 49	☽ ✱ ♅
19 20	☽ ⊥ ☿
22 36	☽ ± ♄
22 56	☽ ⊥ ♆
05 Sunday	
06 34	☽ ⊻ ☿
12 55	☽ ☍ ♃
13 07	☽ ⚼ ♄
13 30	☽ ∠ ♆
19 09	☽ ∠ ♅
19 53	☽ △ ♇
21 56	☿ Q ♆
22 29	☿ ⋕ ♃
06 Monday	
02 16	☽ ☌ ♀
09 04	☿ ± ♇
09 50	☽ ⊥ ♅
10 51	☉ ⊻ ♂
13 28	☽ △ ♄
13 55	☽ ✱ ♆
17 43	☽ ⊻ ♂
17 51	☽ ☌ ☉
18 58	☉ ∥ ♂
19 41	☽ ⊻ ♅
20 19	☽ ⚼ ♇
07 Tuesday	
03 11	☿ ∥ ♂
03 24	☿ ⊼ ♃
04 06	☽ ⊥ ♂
09 06	☉ ∥ ☿
09 43	☽ Q ♆
11 11	☽ ± ♇
13 40	☽ ⊼ ♃
15 23	☽ ☌ ☿
19 51	☽ ∠ ♂
21 13	☽ ⊼ ♇
23 39	☽ ± ♃
23 42	☉ ⊻ ♅
08 Wednesday	
02 05	☽ ∥ ♀
04 06	☽ ⋕ ♃
05 58	☽ ∥ ♂
07 15	☉ ⚼ ♇
07 38	☽ ∥ ☉
08 39	☽ ⊻ ♀
09 15	☽ ∥ ☿
12 37	☽ ⋕ ♇
14 32	♂ ✱ ♅
14 51	☽ ⚼ ♃
15 36	☽ □ ♄
16 12	☽ □ ♆
19 56	☽ ⊥ ♀
22 25	☽ ☌ ♅
22 48	☽ ✱ ♂
09 Thursday	
00 00	☽ ⊻ ☉
10 25	☿ ⊼ ♇
11 18	☽ ⊥ ☉
11 44	☽ ∥ ♅
13 14	☽ ∠ ♀
16 51	☽ △ ♃
16 56	☿ ∠ ♂
19 36	☿ ⋕ ♇
21 05	☽ Q ♂
23 12	☽ Q ♄
10 Friday	
00 07	☽ ∥ ♆
01 06	♀ ⊥ ♅
01 19	☽ ☍ ♇
01 58	☿ ± ♃
03 13	☽ ⋕ ♄
03 47	☽ ⊻ ☿
04 36	☽ ∠ ☉
16 32	☽ ⊥ ☿
19 06	☽ ✱ ♀
21 03	☽ ✱ ♄
21 48	☽ △ ♆
11 Saturday	
04 39	☽ ⊻ ♅
07 55	☽ □ ♂
10 29	☽ ✱ ☉
12 06	☽ ∠ ☿
13 28	♂ ± ♄
15 35	♀ △ ♄
15 59	☽ ⊥ ♅
19 59	☽ Q ♀
23 47	☽ □ ♃
12 Sunday	
00 06	♀ ✱ ♆
01 17	☽ ∠ ♄
02 07	☽ ⚼ ♆
04 40	♂ ⊥ ♆
08 16	♂ ⋕ ♃
09 17	☽ ⊼ ♇
09 18	☽ ∠ ♅
11 17	☽ Q ☉
18 41	☽ ⊥ ♄
19 34	☽ ± ♆
20 56	☽ ± ♇
21 51	☽ ✱ ☿
13 Monday	
02 40	♀ ∥ ♂
06 27	☽ ⊻ ♄
07 22	☽ ⊼ ♆
10 32	☽ Q ♃
10 50	☽ □ ♀
14 39	☽ ⚼ ♇
14 51	☽ ✱ ♅
17 10	♀ ⋕ ♃
21 34	☽ △ ♂
14 Tuesday	
01 40	☽ Q ☿
01 47	☽ □ ☉
10 17	☽ ✱ ♃
14 58	☽ Q ♅
16 09	☽ ∥ ♄
18 37	☉ ⋕ ♇
19 00	☿ ⚼ ♃
19 40	☽ ⋕ ♆
20 40	☽ △ ♇
15 Wednesday	
00 02	♀ ⚼ ♇
04 08	♀ ⊻ ♅
05 35	☽ ⚼ ♂
09 04	☽ ⋕ ♅
09 52	☿ □ ♄
16 19	☽ ∠ ♃
17 07	☿ □ ♆
18 05	♄ St D
18 39	☽ ☌ ♄
18 59	☽ ⋕ ☿
19 42	☽ ☍ ♆
20 08	☽ □ ☿
16 Thursday	
00 57	☽ ± ♂
03 34	☽ □ ♅
06 10	☽ △ ♀
10 25	☽ ⊥ ♃
11 09	☽ ⋕ ☉
12 55	☽ ∥ ♇
13 52	☽ ⊼ ♂
19 42	☽ △ ☉
20 09	☽ ⋕ ♀
22 28	☽ ⊻ ♃
22 30	☽ ∥ ♃
17 Friday	
00 15	☉ Q ♆
00 35	☽ ⋕ ♂
09 20	☽ □ ♇
14 09	☉ ± ♇
16 00	☽ ⚼ ♀
18 Saturday	
02 57	☿ ☌ ♅
03 11	☉ ⊼ ♃
04 34	☽ ⚼ ☉
07 15	☽ ⊻ ♄
08 22	☽ ⊼ ♆
12 06	☽ ± ♀
16 19	☽ △ ♅
18 18	☽ △ ☿
19 13	☽ ⊥ ♄
20 20	☽ ± ♆
21 20	☽ Q ♇
19 Sunday	
00 02	☽ ± ☉
01 23	☽ ⊼ ♀
04 36	☿ ∥ ♅
05 42	☽ ☍ ♂
10 01	☽ ☌ ♃
12 54	☽ ⊼ ☉
13 01	☽ ∠ ♄
14 08	☽ ⚼ ♆
20 56	☽ ✱ ♇
22 03	☽ ⚼ ♅
20 Monday	
04 12	☽ ⚼ ☿
15 37	☽ ± ♅
18 14	☽ ✱ ♄
19 21	☽ △ ♆
21 Tuesday	
00 03	☽ ± ☿
01 57	☽ ∠ ♇
03 13	☽ ⊼ ♅
13 05	☽ ⊼ ☿
17 12	☽ Q ♄
17 56	☽ ☍ ♀
19 04	☽ ⊥ ♇
19 07	☽ ⊼ ♂
19 33	☽ ⊻ ♃
21 36	♀ Q ♆
22 Wednesday	
02 18	♂ ☍ ♃
03 24	☽ ☍ ☉
05 25	♀ ± ♇
06 22	☽ ⊻ ♇
06 47	☽ ⊥ ♃
07 04	☽ ± ♂
09 34	♀ ⊼ ♃
10 42	☽ ⋕ ♂
14 17	☽ ∥ ♃
19 42	♀ ⊻ ♂
22 20	☽ ⋕ ♀
22 43	☽ ∥ ♇
23 29	☽ ∠ ♃
23 Thursday	
00 49	☽ ⚼ ♂
02 58	☽ □ ♄
04 02	☽ □ ♆
06 09	♀ ⋕ ♇
07 32	☽ ⋕ ☉
11 48	☽ ☍ ♅
13 15	☿ ∥ ♆
17 55	☿ △ ♃
18 23	☉ ⊼ ♇
21 18	☉ ± ♃
24 Friday	
01 05	☽ ⋕ ♅
02 56	☽ ✱ ♃
04 05	☽ ☍ ☿
05 54	☽ △ ♂
07 29	☽ ⊼ ♀
12 18	☽ ⋕ ♆
13 03	☽ Q ♆
13 38	☽ ☌ ♇
15 06	☽ ⋕ ☿
15 10	☽ ⊼ ☉
15 21	☽ ∥ ♄
17 24	☿ ⋕ ♄
18 32	♂ ⚼ ♄
19 28	☽ ± ♀
25 Saturday	
00 33	☽ Q ♃
02 50	☽ ± ☉
09 40	☽ △ ♄
10 42	☽ ✱ ♆
11 30	☿ ✱ ♂
13 18	☽ ⚼ ♀
15 21	♂ ∠ ♆
18 25	☽ ⊼ ♅
20 11	☽ ⚼ ☉
26 Sunday	
05 08	☽ ± ♅
08 31	☽ □ ♃
12 23	☽ ⚼ ♄
13 24	☽ ∠ ♆
14 32	☽ □ ♂
16 02	☽ ⊼ ☿
18 35	☽ △ ♀
19 08	☽ ⊻ ♇
21 06	☽ ⚼ ♅
23 28	♀ ± ♃
27 Monday	
00 43	☽ △ ☉
00 43	♀ ⊼ ♇
03 42	☽ ± ☿
04 14	☽ ± ♄
05 14	☽ ⊥ ♆
05 39	☽ ⊥ ♇
09 17	☿ Q ♄
14 43	☽ ⊼ ♄
15 42	☽ ⊻ ♆
21 02	☽ ⚼ ☿
21 19	☽ ∠ ♇
23 23	☽ △ ♅
23 38	☽ Q ♂
23 54	☿ ☍ ♇
28 Tuesday	
10 38	☽ ∥ ☿
12 35	☽ △ ♃
19 30	☉ ⊥ ♂
19 39	☽ ⋕ ♄
21 21	☽ ✱ ♂
22 24	☽ ∥ ♆
23 07	☽ ✱ ♇
29 Wednesday	
01 27	☽ △ ☿
03 38	☽ □ ♀
08 23	☽ □ ☉
08 26	☽ ∥ ♅
14 05	☽ ⚼ ♃
18 18	☽ ☍ ♄
18 22	☽ ∥ ☉
19 13	☽ ☌ ♆
19 30	☽ Q ♇
30 Thursday	
00 08	☽ ∠ ♂
01 08	☽ ∥ ♀
02 53	☽ □ ♅
05 15	☽ ± ♃
09 07	♂ △ ♇
09 22	☽ ⋕ ♇
13 01	☽ Q ♀
15 17	☽ ⊼ ♃
16 02	☽ ⊥ ♂
16 33	☽ ⋕ ♃
16 59	☽ Q ☉
20 50	☽ ∥ ♂
31 Friday	
01 45	☽ □ ♇
02 35	☽ ⊻ ♂
08 44	☽ □ ☿
09 12	☽ Q ♅
10 59	☽ ✱ ♀
14 28	☽ ✱ ☉
20 42	☽ ⊼ ♄
21 33	☽ ⊻ ♆

AUGUST 2043

LONGITUDES

Date	Sidereal time h m s	Sun ☉ ° ' "	Moon ☽ ° ' "	Moon ☽ 24.00 ° ' "	Mercury ☿ ° '	Venus ♀ ° '	Mars ♂ ° '	Jupiter ♃ ° '	Saturn ♄ ° '	Uranus ♅ ° '	Neptune ♆ ° '	Pluto ♇ ° '
01	08 39 59	09 ♌ 09 38	21 ♊ 14 59	28 ♊ 28 27	06 ♍ 10	07 ♌ 30	02 ♋ 07	24 ♐ 19	12 ♏ 04	17 ♌ 13	12 ♉ 33	00 ♓ 39
02	08 43 55	10 07 03	05 ♋ 41 26	12 ♋ 53 12	07 15	08 44	02 47	24 R 16	12 05	17 17	12 34	00 R 38
03	08 47 52	11 04 29	20 ♋ 03 02	27 ♋ 10 11	08 18	09 58	03 27	24 13	12 07	17 20	12 34	00 37
04	08 51 49	12 01 57	04 ♌ 13 57	11 ♌ 13 43	09 19	11 12	04 07	24 10	12 09	17 24	12 35	00 35
05	08 55 45	12 59 25	18 ♌ 08 56	24 ♌ 59 10	10 17	12 26	04 47	24 07	12 11	17 28	12 35	00 35
06	08 59 42	13 56 53	01 ♍ 44 08	08 ♍ 23 37	11 11	13 40	05 27	24 04	12 13	17 31	12 35	00 34
07	09 03 38	14 54 23	14 ♍ 57 34	21 ♍ 26 03	12 03	14 54	06 07	24 02	12 15	17 35	12 35	00 33
08	09 07 35	15 51 54	27 ♍ 49 15	04 ♎ 07 25	12 51	16 08	06 47	23 59	12 17	17 39	12 36	00 32
09	09 11 31	16 49 25	10 ♎ 20 56	16 ♎ 30 14	13 36	17 22	07 27	23 57	12 20	17 43	12 36	00 30
10	09 15 28	17 46 57	22 ♎ 35 50	28 ♎ 38 16	14 18	18 36	08 07	23 55	12 22	17 46	12 36	00 29
11	09 19 24	18 44 30	04 ♏ 38 07	10 ♏ 36 00	14 56	19 50	08 46	23 53	12 25	17 50	12 36	00 28
12	09 23 21	19 42 04	16 ♏ 32 33	22 ♏ 28 24	15 30	21 04	09 26	23 51	12 27	17 54	12 37	00 27
13	09 27 18	20 39 39	28 ♏ 24 09	04 ♐ 20 27	15 59	22 19	10 05	23 50	12 30	17 58	12 37	00 25
14	09 31 14	21 37 15	10 ♐ 17 52	16 ♐ 16 58	16 25	23 33	10 45	23 49	12 33	18 01	12 37	00 24
15	09 35 11	22 34 52	22 ♐ 18 16	28 ♐ 22 16	16 46	24 47	11 24	23 48	12 36	18 05	12 37	00 23
16	09 39 07	23 32 30	04 ♑ 29 22	10 ♑ 39 58	17 02	26 01	12 04	23 47	12 39	18 09	12 R 37	00 22
17	09 43 04	24 30 09	16 ♑ 54 20	23 ♑ 12 43	17 13	27 15	12 43	23 46	12 42	18 12	12 37	00 20
18	09 47 00	25 27 49	29 ♑ 35 16	06 ♒ 02 04	17 18	28 30	13 22	23 46	12 45	18 16	12 37	00 19
19	09 50 57	26 25 30	12 ♒ 33 07	19 ♒ 08 22	17 R 19	29 ♌ 44	14 01	23 45	12 48	18 20	12 37	00 18
20	09 54 53	27 23 13	25 ♒ 47 41	02 ♓ 30 51	17 13	00 ♍ 58	14 40	23 45	12 51	18 24	12 36	00 16
21	09 58 50	28 20 56	09 ♓ 17 40	16 ♓ 07 49	17 02	02 12	15 19	23 D 45	12 55	18 27	12 36	00 15
22	10 02 47	29 ♌ 18 41	23 ♓ 01 01	29 ♓ 56 55	16 46	03 26	15 58	23 45	12 58	18 31	12 36	00 14
23	10 06 43	00 ♍ 16 28	06 ♈ 55 12	13 ♈ 55 32	16 23	04 41	16 37	23 46	13 02	18 35	12 36	00 12
24	10 10 40	01 14 16	20 ♈ 57 35	28 ♈ 01 03	15 55	05 55	17 16	23 46	13 05	18 38	12 36	00 11
25	10 14 36	02 12 06	05 ♉ 05 39	12 ♉ 11 05	15 21	07 09	17 55	23 47	13 09	18 42	12 35	00 10
26	10 18 33	03 09 58	19 ♉ 17 05	26 ♉ 23 23	14 42	08 24	18 33	23 48	13 13	18 46	12 35	00 09
27	10 22 29	04 07 52	03 ♊ 29 44	10 ♊ 35 51	13 59	09 38	19 12	23 49	13 17	18 49	12 35	00 07
28	10 26 26	05 05 47	17 ♊ 41 28	24 ♊ 46 18	13 11	10 52	19 50	23 51	13 21	18 53	12 34	00 06
29	10 30 22	06 03 45	01 ♋ 50 03	08 ♋ 52 22	12 19	12 07	20 29	23 52	13 25	18 56	12 34	00 05
30	10 34 19	07 01 44	15 ♋ 52 56	22 ♋ 51 25	11 25	13 21	21 07	23 54	13 29	19 00	12 33	00 03
31	10 38 16	07 ♍ 59 45	29 ♋ 47 27	06 ♌ 40 42	10 ♍ 29	14 ♍ 35	21 ♋ 46	23 ♐ 56	13 ♏ 33	19 ♌ 04	12 ♉ 33	00 ♓ 01

Date	Moon True ☊ ° '	Moon Mean ☊ ° '	Moon ☽ Latitude ° '
01	00 ♈ 57	02 ♈ 08	05 N 02
02	00 R 50	02 04	05 05
03	00 43	02 01	04 49
04	00 35	01 58	04 16
05	00 28	01 55	03 27
06	00 23	01 52	02 28
07	00 20	01 49	01 22
08	00 19	01 45	00 N 13
09	00 D 19	01 42	00 S 54
10	00 20	01 39	01 58
11	00 22	01 36	02 55
12	00 23	01 33	03 44
13	00 R 22	01 29	04 24
14	00 21	01 26	04 52
15	00 17	01 23	05 08
16	00 12	01 20	05 10
17	00 07	01 17	04 57
18	00 01	01 14	04 31
19	29 ♓ 55	01 10	03 49
20	29 51	01 07	02 55
21	29 48	01 04	01 50
22	29 46	01 01	00 S 37
23	29 D 46	00 58	00 N 39
24	29 47	00 55	01 54
25	29 49	00 51	03 02
26	29 50	00 48	03 59
27	29 51	00 45	04 42
28	29 R 50	00 42	05 07
29	29 49	00 39	05 14
30	29 46	00 35	05 02
31	29 ♓ 43	00 ♈ 32	04 N 32

DECLINATIONS

Date	Sun ☉ ° '	Moon ☽ ° '	Mercury ☿ ° '	Venus ♀ ° '	Mars ♂ ° '	Jupiter ♃ ° '	Saturn ♄ ° '	Uranus ♅ ° '	Neptune ♆ ° '	Pluto ♇ ° '
01	17 N 58	28 N 10	08 N 30	19 N 24	23 N 49	23 S 05	13 S 16	16 N 18	13 N 55	21 S 46
02	17 42	28 24	07 56	19 07	23 49	23 05	13 17	16 17	13 56	21 47
03	17 27	26 42	07 22	18 48	23 49	23 05	13 18	16 16	13 56	21 47
04	17 11	23 20	06 49	18 30	23 49	23 05	13 18	16 15	13 56	21 48
05	16 55	18 40	06 17	18 10	23 48	23 05	13 19	16 13	13 56	21 49
06	16 38	13 10	05 46	17 51	23 47	23 05	13 20	16 12	13 56	21 49
07	16 22	07 11	05 16	17 30	23 46	23 05	13 21	16 11	13 56	21 50
08	16 05	01 N 04	04 47	17 09	23 45	23 05	13 22	16 10	13 56	21 50
09	15 48	04 S 56	04 19	16 48	23 44	23 05	13 23	16 09	13 56	21 51
10	15 30	10 37	03 52	16 26	23 42	23 05	13 24	16 08	13 56	21 51
11	15 12	15 49	03 27	16 04	23 40	23 05	13 25	16 07	13 56	21 52
12	14 54	20 22	03 04	15 41	23 38	23 05	13 26	16 05	13 56	21 53
13	14 36	24 05	02 42	15 18	23 36	23 05	13 27	16 04	13 56	21 53
14	14 18	26 48	02 22	14 55	23 34	23 06	13 28	16 03	13 56	21 54
15	13 59	28 20	02 04	14 31	23 31	23 06	13 29	16 02	13 56	21 54
16	13 40	28 31	01 48	14 07	23 28	23 06	13 30	16 01	13 56	21 55
17	13 21	27 17	01 35	13 42	23 25	23 06	13 31	16 00	13 56	21 55
18	13 02	24 38	01 24	13 17	23 22	23 06	13 33	15 59	13 55	21 56
19	12 42	20 42	01 16	12 51	23 19	23 06	13 34	15 57	13 55	21 56
20	12 23	15 40	01 10	12 26	23 16	23 06	13 35	15 56	13 55	21 57
21	12 03	09 47	01 08	11 59	23 12	23 07	13 36	15 55	13 55	21 58
22	11 43	03 S 20	01 09	11 33	23 08	23 07	13 38	15 54	13 55	21 58
23	11 22	03 N 21	01 13	11 06	23 04	23 07	13 39	15 53	13 55	21 59
24	11 02	09 56	01 21	10 39	23 00	23 07	13 40	15 52	13 55	21 59
25	10 41	16 04	01 31	10 12	22 56	23 07	13 41	15 51	13 55	22 00
26	10 21	21 22	01 46	09 44	22 51	23 07	13 43	15 50	13 54	22 00
27	10 00	25 27	02 03	09 16	22 47	23 08	13 44	15 48	13 54	22 01
28	09 38	27 58	02 24	08 48	22 42	23 08	13 46	15 47	13 54	22 01
29	09 17	28 39	02 48	08 20	22 37	23 08	13 47	15 46	13 54	22 02
30	08 56	27 29	03 15	07 51	22 32	23 08	13 48	15 45	13 54	22 02
31	08 N 34	24 N 37	03 N 44	07 N 22	22 N 26	23 S 09	13 S 50	15 N 44	13 N 54	22 S 03

ZODIAC SIGN ENTRIES

Date	h m	Planets
02	02 32	☽ ♋
04	04 48	☽ ♌
06	08 54	☽ ♍
08	16 08	☽ ♎
11	02 43	☽ ♏
13	15 14	☽ ♐
16	03 12	☽ ♑
18	12 46	☽ ♒
19	17 16	♀ ♍
20	19 31	☽ ♓
23	00 05	☽ ♈
23	05 10	☉ ♍
25	03 22	☽ ♉
27	06 06	☽ ♊
29	08 53	☽ ♋
31	12 22	☽ ♌

LATITUDES

Date	Mercury ☿ ° '	Venus ♀ ° '	Mars ♂ ° '	Jupiter ♃ ° '	Saturn ♄ ° '	Uranus ♅ ° '	Neptune ♆ ° '	Pluto ♇ ° '
01	00 S 48	01 N 03	00 N 24	00 N 14	02 N 18	00 N 39	01 S 46	11 S 18
04	01 21	01 07	00 26	00 14	02 17	00 39	01 46	11 18
07	01 55	01 12	00 29	00 13	02 16	00 39	01 46	11 19
10	02 30	01 15	00 31	00 13	02 15	00 39	01 46	11 19
13	03 04	01 18	00 33	00 12	02 15	00 39	01 46	11 19
16	03 36	01 21	00 35	00 12	02 14	00 39	01 47	11 20
19	04 04	01 23	00 37	00 11	02 13	00 39	01 47	11 20
22	04 25	01 24	00 40	00 11	02 12	00 39	01 47	11 20
25	04 36	01 25	00 42	00 10	02 12	00 40	01 47	11 20
28	04 33	01 25	00 44	00 10	02 11	00 40	01 47	11 20
31	04 S 13	01 N 25	00 N 46	00 N 09	02 N 10	00 N 40	01 S 48	11 S 20

DATA

Julian Date	2467463
Delta T	+78 seconds
Ayanamsa	24° 27' 58"
Synetic vernal point	04° ♓ 39' 01"
True obliquity of ecliptic	23° 26' 10"

MOON'S PHASES, APSIDES AND POSITIONS ☽

Date	h m	Phase	Longitude ° '	Eclipse Indicator
05	02 23	●	12 ♌ 36	
12	18 57	☽	19 ♏ 59	
20	15 04	○	27 ♒ 31	
27	13 09	☾	04 ♊ 11	

Day	h m		
01	05 01	Perigee	
13	08 31	Apogee	
26	13 42	Perigee	
02	02 47	Max dec	28° N 33'
08	16 14	0S	
16	03 16	Max dec	28° S 37'
23	00 01	0N	
29	08 50	Max dec	28° N 40'

LONGITUDES

Date	Chiron ⚷ ° '	Ceres ⚳ ° '	Pallas ⚴ ° '	Juno ⚵ ° '	Vesta ⚶ ° '	Black Moon Lilith ⚸ ° '
01	20 ♌ 06	09 ♓ 41	11 ♑ 54	21 ♐ 55	27 ♎ 10	26 ♏ 34
11	21 ♌ 24	07 ♓ 58	10 ♑ 10	21 ♐ 21	01 ♏ 10	27 ♏ 41
21	22 ♌ 42	05 ♓ 54	09 ♑ 04	21 ♐ 22	05 ♏ 25	28 ♏ 48
31	24 ♌ 00	03 ♓ 41	08 ♑ 37	21 ♐ 57	09 ♏ 52	29 ♏ 55

ASPECTARIAN

h m	Aspects
01 Saturday	
05 16	☽ ✶ ♅
06 42	☽ ± ♄
07 31	☽ ⊥ ♆
14 16	☽ ∠ ♀
15 52	☉ ⚼ ♃
15 56	♂ ∠ ♅
17 05	☽ ☍ ♃
17 11	☽ ∠ ☉
17 14	☽ Q ☿
21 40	☽ ⚼ ♄
22 29	☽ ∠ ♆
02 Sunday	
03 36	☽ △ ♇
06 18	☽ ∠ ♅
06 37	☽ ⊥ ♀
06 56	☽ ☌ ♂
09 12	☽ ⊥ ☉
14 26	♀ ⊥ ♂
14 49	☽ ✶ ☿
17 32	☽ ⊻ ♀
19 54	☽ ⊻ ☉
21 21	☽ ⊥ ♅
22 02	♀ ⚼ ♃
22 41	☽ △ ♄
23 28	☽ ✶ ♆
03 Monday	
04 34	☽ ⚼ ♇
07 26	☽ ⊻ ♅
17 54	☽ ∠ ☿
18 59	☽ ⊼ ♃
19 37	☽ Q ♆
19 40	☽ ± ♇
04 Tuesday	
05 06	☽ ± ♃
05 48	☽ ⊼ ♇
09 09	☽ ∥ ♂
10 19	☽ ⊥ ☿
11 48	☽ ⊻ ♂
13 28	☽ ⋕ ♃
15 04	☉ □ ♄
20 25	☽ ⚼ ♃
20 31	☽ ⋕ ♇
21 22	☽ ⊻ ☿
22 36	☽ ⊥ ♂
22 55	♆ Q ♇
05 Wednesday	
01 06	☽ ☌ ♀
01 37	☽ □ ♄
01 44	☉ □ ♆
02 20	☽ □ ♆
02 23	☽ ☌ ☉
07 04	♀ □ ♄
10 48	☽ ☌ ♅
14 28	☽ ∥ ♀
14 59	♀ □ ♆
15 01	☽ ∠ ♂
20 24	☽ ∥ ☉
22 25	☽ △ ♃
23 03	☽ ∥ ♅
06 Thursday	
08 48	☽ ∥ ♆
09 17	☽ Q ♄
09 53	☽ ☍ ♇
11 16	☽ ⋕ ♄
19 02	☽ ✶ ♂
07 Friday	
06 18	☽ ☌ ☿
07 01	☽ ✶ ♄
07 39	☽ △ ♆
11 53	☽ ⊻ ♀
11 54	☽ ⊻ ☉
12 40	☉ ☌ ♀
16 52	☽ ⊻ ♅
18 09	☽ Q ♂
18 12	☿ ✶ ♄
20 13	☽ ∥ ☿
23 54	☽ ⊥ ☉
08 Saturday	
00 10	☽ ⊥ ♀
04 04	☉ ∥ ♅
04 05	☿ △ ♆
04 06	☽ ⊥ ♅
04 48	☽ □ ♃
11 00	☽ ∠ ♄
11 35	☽ ⚼ ♆
17 05	☽ ⊼ ♇
18 15	☽ ∠ ☉
18 58	☽ ∠ ♀
21 13	☽ ∠ ♅
09 Sunday	
04 13	☽ ⊥ ♄
04 33	☽ ± ♇
04 45	☽ ± ♆
06 05	☽ □ ♂
09 40	☽ ⋕ ☿
15 06	☽ Q ♃
15 52	☽ ⊻ ♄
16 23	☽ ⊼ ♆
18 44	☽ ⊻ ☿
19 00	♀ ☌ ♅
21 59	☽ ⚼ ♇
10 Monday	
01 42	☽ ✶ ☉
02 26	☽ ✶ ♅
03 14	☽ ✶ ♀
07 12	☽ ⊥ ☿
11 44	☉ ☌ ♅
14 36	☽ ✶ ♃
11 Tuesday	
00 34	☽ ∥ ♄
02 04	☽ ∠ ☿
02 20	☽ Q ♅
03 00	☽ ⋕ ♆
03 31	☽ Q ☉
03 37	☽ △ ♇
05 45	☽ Q ♀
09 11	♀ ∥ ♅
09 13	☽ ⋕ ☉
13 10	☽ ⋕ ♀
13 27	☽ ⋕ ♅
20 31	☽ ∠ ♃
20 48	☽ △ ♂
12 Wednesday	
03 43	☽ ☌ ♄
04 03	☽ ☍ ♆
09 47	☽ ✶ ☿
14 39	☽ ⊥ ♃
14 45	☽ □ ♅
18 57	☽ □ ☉
21 06	☽ ∥ ♇
22 14	☽ □ ♀
13 Thursday	
02 46	☽ ⊻ ♃
04 54	☽ ⚼ ♂
04 58	☽ ∥ ♃
08 32	☽ ⋕ ♂
11 08	☽ Q ☿
16 02	☽ □ ♇
14 Friday	
00 10	☽ ± ♂
12 57	☽ ⊼ ♂
16 32	☽ ⊻ ♄
16 39	☽ ⊼ ♆
17 06	♀ △ ♃
15 Saturday	
00 39	☽ □ ☿
03 33	☽ △ ♅
04 10	☽ Q ♇
04 35	☽ ⊥ ♄
04 39	☽ ± ♆
12 36	☽ △ ☉
14 57	☽ ☌ ♃
16 25	☉ ∥ ♆
17 28	☽ △ ♀
21 19	♄ ☍ ♆
21 56	♆ St R
22 30	☽ ⚼ ♆
22 31	☽ ∠ ♄
16 Sunday	
03 53	☽ ✶ ♇
09 22	☽ ⚼ ♅
15 28	♂ ⊥ ♅
17 53	☉ △ ♃
20 33	☽ ⚼ ☉
22 44	♀ ∥ ♆
23 53	☉ ⋕ ♄
17 Monday	
02 05	☽ ⚼ ♀
02 56	☽ ± ♅
03 30	☽ ☍ ♂
03 45	☽ △ ♆
03 53	☽ ✶ ♄
08 16	♂ ✶ ♆
08 58	☽ ∠ ♇
11 12	♂ △ ♄
12 35	☽ △ ☿
14 30	☽ ⊼ ♅
15 18	☽ ± ☉
21 11	☽ ± ♀
21 38	♀ ⋕ ♄
18 Tuesday	
01 03	☽ ⊻ ♃
02 04	☽ ⊥ ♇
02 52	☽ Q ♄
03 37	☽ ⊼ ☉
06 52	♀ ∠ ♂
09 44	☽ ⊼ ♀
12 19	☽ ⊥ ♃
13 19	☽ ⊻ ♇
17 05	☽ ⚼ ☿
20 42	☽ ⋕ ♂
22 17	☽ ∥ ♃
19 Wednesday	
01 33	☿ St R
05 02	☽ ∠ ♃
05 12	☽ ∥ ♇
09 44	☽ ± ☿
12 06	☽ □ ♆
12 27	☽ □ ♄
14 49	☽ ⊼ ♂
20 39	☽ ⊼ ☿
22 22	♀ ☍ ♇
22 35	☽ ☍ ♅
20 Thursday	
02 18	☽ ± ♂
08 20	☽ ✶ ♃
09 44	♀ Q ♄
10 47	☽ ⋕ ♅
15 04	☽ ☍ ☉
18 35	♃ St D
19 17	☽ ⚼ ♂
19 25	☽ ⋕ ♆
19 58	☽ ☌ ♇
20 36	☽ Q ♆
20 47	☽ ∥ ♄
22 11	☽ ☍ ♀
23 23	☉ ∥ ♀
21 Friday	
02 33	☽ ⋕ ☉
02 36	☽ ⋕ ♀
05 45	☽ Q ♃
08 46	♂ ⚼ ♇
17 49	☽ ✶ ♆
18 23	☽ △ ♄
23 07	☽ △ ♂
22 Saturday	
01 21	☽ ☍ ☿
04 08	☽ ⊼ ♅
13 17	☽ □ ♃
14 37	☽ ± ♅
19 50	☽ ⋕ ☿
19 56	☽ ∠ ♆
20 37	☽ ⚼ ♄
20 59	♂ ⋕ ♃
23 43	☽ ⊼ ☉
23 Sunday	
00 26	☽ ⊻ ♇
04 17	☽ ∥ ☿
06 14	☽ ⚼ ♅
06 48	☿ ✶ ♂
07 46	☽ ⊼ ♀
09 51	☉ ☍ ♇
10 45	☽ ⊥ ♇
10 49	☽ ± ☉
11 27	☽ ⊥ ♆
12 11	☽ ± ♄
19 04	☽ ± ♀
21 43	☽ ⊻ ♆
22 31	☽ ⊼ ♄
24 Monday	
02 08	☽ ∠ ♇
03 21	☽ ⚼ ☉
03 42	☽ ⊼ ☿
05 24	☽ □ ♂
08 01	☉ Q ♄
08 02	☽ △ ♅
11 55	☽ ⚼ ♀
13 34	☽ ± ☿
14 32	☽ ∥ ♀
15 56	☽ ∥ ☉
16 48	☽ △ ♃
25 Tuesday	
02 21	☽ ⋕ ♄
03 16	☽ ∥ ♆
03 37	☽ ✶ ♇
04 17	☽ ⚼ ☿
06 44	☽ △ ☉
11 05	☽ ∥ ♅
13 27	☽ Q ♂
15 49	☽ △ ♀
18 15	☽ ⚼ ♃
23 55	☽ Q ♇
26 Wednesday	
00 41	☽ ☌ ♆
01 42	☽ ☍ ♄
04 37	☽ △ ☿
09 30	☽ ± ♃
10 42	☽ ✶ ♂
11 06	☽ □ ♅
15 14	☽ ⋕ ♇
19 38	☽ ⊼ ♃
19 41	☽ ∥ ♂
20 25	♂ ⊻ ♅
21 20	☽ ⋕ ♃
27 Thursday	
06 16	☽ □ ♇
13 09	☽ □ ☉
13 15	☽ ∠ ♂
16 59	☉ ∠ ♂
17 38	☽ Q ♅
23 21	☽ □ ♀
28 Friday	
03 20	☽ ⊻ ♆
04 37	☽ ⊼ ♄
04 47	☽ □ ☿
05 10	☽ ⊥ ♂
07 29	☿ ✶ ♄
13 29	☽ ⊥ ♆
14 01	☽ ✶ ♅
14 49	☽ ± ♄
15 49	☽ ⊻ ♂
21 50	☽ Q ☉
22 27	☽ ☍ ♃
29 Saturday	
04 44	☽ ∠ ♆
05 14	☿ △ ♆
06 09	☽ ⚼ ♄
08 47	☽ Q ♀
08 59	☽ △ ♇
09 34	☽ Q ☿
14 21	☿ ☌ ♀
15 36	☽ ∠ ♅
19 44	☽ ✶ ☉
20 43	♀ △ ♆
30 Sunday	
04 49	☽ ✶ ☿
06 18	☽ ✶ ♆
07 02	☽ ⊥ ♅
07 14	☽ ✶ ♀
07 52	☽ △ ♄
10 33	☽ ⚼ ♇
14 42	♀ ✶ ♄
17 23	☽ ⊻ ♅
21 27	☽ ☌ ♂
23 22	☽ ∠ ☉
31 Monday	
01 50	☽ ⊼ ♃
02 01	☽ ± ♇
02 55	☽ Q ♆
05 00	☽ ∠ ☿
11 37	☽ ∠ ♀
12 15	☽ ± ♃
12 23	☽ ⊼ ♇
16 07	☽ ⊥ ☉
19 39	☽ ⊥ ☿
21 10	☽ ⋕ ♃

All ephemeris data is given at 12.00 UT and the Moon's longitude is additionally given for 24.00 UT

SEPTEMBER 2043

Date	Sidereal time h m s	LONGITUDES Sun ☉ ° ' "	Moon ☽ ° ' "	Moon ☽ 24.00 ° '	Mercury ☿ ° '	Venus ♀ ° '	Mars ♂ ° '	Jupiter ♃ ° '	Saturn ♄ ° '	Uranus ♅ ° '	Neptune ♆ ° '	Pluto ♇ ° '
01	10 42 12	08 ♍ 57 48	13 ♌ 30 50	20 ♌ 17 34	09 ♍ 32	15 ♍ 50	22 ♋ 24	23 ♐ 58	13 ♏ 37	19 ♌ 07	12 ♉ 32	00 ♓ 00
02	10 46 09	09 55 52	27 ♌ 00 38	03 ♍ 39 49	08 R 37	17 04	23 02	24 00	13 42	19 11	12 R 32	29 ♒ 58
03	10 50 05	10 53 58	10 ♍ 14 56	16 ♍ 45 53	07 43	18 18	23 40	24 03	13 46	19 14	12 31	29 R 57
04	10 54 02	11 52 06	23 ♍ 12 39	29 ♍ 35 14	06 51	19 33	24 18	24 05	13 50	19 18	12 31	29 56
05	10 57 58	12 50 16	05 ♎ 53 44	12 ♎ 08 18	06 05	20 47	24 56	24 08	13 55	19 21	12 30	29 54
06	11 01 55	13 48 27	18 ♎ 19 11	24 ♎ 26 39	05 23	22 02	25 34	24 11	14 00	19 25	12 29	29 53
07	11 05 51	14 46 39	00 ♏ 31 04	06 ♏ 32 49	04 48	23 16	26 12	24 14	14 04	19 28	12 28	29 52
08	11 09 48	15 44 53	12 ♏ 32 20	18 ♏ 30 07	04 20	24 30	26 50	24 18	14 09	19 32	12 28	29 51
09	11 13 45	16 43 09	24 ♏ 26 40	00 ♐ 22 34	04 00	25 45	27 28	24 21	14 14	19 35	12 27	29 49
10	11 17 41	17 41 27	06 ♐ 18 20	12 ♐ 14 35	03 49	26 59	28 05	24 25	14 19	19 39	12 26	29 48
11	11 21 38	18 39 46	18 ♐ 11 54	24 ♐ 10 52	03 D 46	28 14	28 43	24 29	14 24	19 42	12 25	29 47
12	11 25 34	19 38 06	00 ♑ 12 04	06 ♑ 16 03	03 53	29 ♍ 28	29 20	24 33	14 29	19 45	12 25	29 46
13	11 29 31	20 36 28	12 ♑ 23 22	18 ♑ 34 31	04 08	00 ♎ 43	29 ♋ 58	24 37	14 34	19 49	12 24	29 45
14	11 33 27	21 34 52	24 ♑ 49 57	01 ♒ 10 03	04 33	01 57	00 ♌ 35	24 41	14 39	19 52	12 23	29 43
15	11 37 24	22 33 17	07 ♒ 35 10	14 ♒ 05 31	05 06	03 12	01 12	24 46	14 44	19 55	12 22	29 42
16	11 41 20	23 31 44	20 ♒ 41 17	27 ♒ 22 30	05 49	04 26	01 50	24 51	14 50	19 58	12 21	29 41
17	11 45 17	24 30 12	04 ♓ 09 07	11 ♓ 01 00	06 39	05 40	02 27	24 56	14 55	20 02	12 20	29 40
18	11 49 14	25 28 43	17 ♓ 57 52	24 ♓ 59 19	07 37	06 55	03 04	25 01	15 01	20 05	12 19	29 39
19	11 53 10	26 27 15	02 ♈ 04 55	09 ♈ 14 05	08 43	08 09	03 41	25 06	15 06	20 08	12 18	29 38
20	11 57 07	27 25 49	16 ♈ 26 11	23 ♈ 40 33	09 55	09 24	04 18	25 11	15 12	20 11	12 17	29 36
21	12 01 03	28 24 25	00 ♉ 56 28	08 ♉ 13 12	11 13	10 38	04 54	25 17	15 17	20 14	12 16	29 35
22	12 05 00	29 ♍ 23 03	15 ♉ 30 04	22 ♉ 46 22	12 36	11 53	05 31	25 22	15 23	20 17	12 15	29 34
23	12 08 56	00 ♎ 21 44	00 ♊ 01 31	07 ♊ 14 55	14 03	13 07	06 08	25 28	15 28	20 21	12 13	29 33
24	12 12 53	01 20 27	14 ♊ 26 06	21 ♊ 34 40	15 36	14 22	06 44	25 34	15 34	20 24	12 12	29 32
25	12 16 49	02 19 12	28 ♊ 40 18	05 ♋ 42 43	17 12	15 36	07 21	25 40	15 40	20 27	12 11	29 31
26	12 20 46	03 17 59	12 ♋ 41 45	19 ♋ 37 15	18 51	16 51	07 57	25 47	15 46	20 30	12 10	29 30
27	12 24 43	04 16 49	26 ♋ 29 11	03 ♌ 17 29	20 32	18 05	08 34	25 53	15 52	20 33	12 09	29 29
28	12 28 39	05 15 41	10 ♌ 02 10	16 ♌ 43 16	22 16	19 20	09 10	26 00	15 58	20 35	12 07	29 28
29	12 32 36	06 14 35	23 ♌ 20 48	29 ♌ 54 50	24 00	20 34	09 46	26 07	16 04	20 38	12 06	29 27
30	12 36 32	07 ♎ 13 32	06 ♍ 25 25	12 ♍ 52 39	25 ♍ 46	21 ♎ 49	10 ♌ 22	26 ♐ 14	16 ♏ 10	20 ♌ 41	12 ♉ 05	29 ♒ 26

Date	Moon True ☊ ° '	Moon Mean ☊ ° '	Moon ☽ Latitude ° '	DECLINATIONS Sun ☉ ° '	Moon ☽ ° '	Mercury ☿ ° '	Venus ♀ ° '	Mars ♂ ° '	Jupiter ♃ ° '	Saturn ♄ ° '	Uranus ♅ ° '	Neptune ♆ ° '	Pluto ♇ ° '
01	29 ♓ 40	00 ♈ 29	03 N 47	08 N 12	20 N 23	04 N 14	06 N 53	22 N 21	23 S 09	13 S 51	15 N 43	13 N 53	22 S 03
02	29 R 37	00 26	02 50	07 51	15 10	04 47	06 24	22 15	23 09	13 53	15 42	13 53	22 04
03	29 36	00 23	01 45	07 29	09 20	05 19	05 55	22 10	23 09	13 54	15 41	13 53	22 04
04	29 35	00 20	00 N 35	07 07	03 N 14	05 52	05 25	22 04	23 10	13 56	15 40	13 53	22 04
05	29 D 35	00 16	00 S 35	06 44	02 S 52	06 25	04 55	21 58	23 10	13 57	15 39	13 52	22 05
06	29 35	00 13	01 42	06 22	08 45	06 57	04 26	21 52	23 10	13 59	15 38	13 52	22 05
07	29 37	00 10	02 43	06 00	14 12	07 27	03 56	21 45	23 10	14 01	15 36	13 52	22 06
08	29 38	00 07	03 36	05 37	19 01	07 55	03 26	21 39	23 11	14 02	15 35	13 52	22 06
09	29 39	00 04	04 19	05 15	23 04	08 20	02 55	21 32	23 11	14 04	15 34	13 51	22 07
10	29 40	00 ♈ 01	04 51	04 52	26 08	08 42	02 25	21 25	23 11	14 05	15 33	13 51	22 07
11	29 40	29 ♓ 57	05 10	04 29	28 04	09 01	01 55	21 19	23 12	14 07	15 32	13 51	22 07
12	29 R 40	29 54	05 17	04 06	28 43	09 16	01 24	21 11	23 12	14 09	15 31	13 50	22 08
13	29 39	29 51	05 09	03 43	27 59	09 27	00 54	21 04	23 12	14 10	15 30	13 50	22 08
14	29 38	29 48	04 47	03 20	25 51	09 35	00 N 23	20 57	23 12	14 12	15 29	13 50	22 09
15	29 37	29 45	04 10	02 57	22 24	09 38	00 S 08	20 50	23 13	14 14	15 28	13 50	22 09
16	29 37	29 41	03 19	02 34	17 44	09 37	00 38	20 42	23 13	14 16	15 27	13 49	22 09
17	29 36	29 38	02 16	02 11	12 06	09 32	01 09	20 35	23 13	14 17	15 26	13 49	22 10
18	29 35	29 35	01 S 04	01 48	05 S 44	09 22	01 39	20 27	23 14	14 19	15 25	13 48	22 10
19	29 D 36	29 32	00 N 14	01 25	01 N 02	09 09	02 10	20 19	23 14	14 21	15 24	13 48	22 10
20	29 36	29 29	01 32	01 01	07 53	08 52	02 41	20 11	23 14	14 23	15 23	13 48	22 11
21	29 36	29 26	02 45	00 38	14 23	08 32	03 11	20 03	23 15	14 24	15 22	13 47	22 11
22	29 36	29 22	03 48	00 N 15	20 07	08 08	03 42	19 55	23 15	14 26	15 21	13 47	22 11
23	29 R 36	29 19	04 36	00 S 09	24 39	07 40	04 12	19 46	23 15	14 28	15 20	13 47	22 12
24	29 36	29 16	05 06	00 32	27 36	07 10	04 43	19 38	23 16	14 30	15 19	13 46	22 12
25	29 36	29 13	05 17	00 55	28 43	06 38	05 13	19 29	23 16	14 32	15 18	13 46	22 12
26	29 D 36	29 10	05 09	01 19	27 57	06 03	05 43	19 21	23 16	14 33	15 17	13 45	22 12
27	29 36	29 07	04 43	01 42	25 29	05 26	06 13	19 12	23 17	14 35	15 17	13 45	22 13
28	29 36	29 03	04 02	02 05	21 36	04 47	06 43	19 03	23 17	14 37	15 16	13 45	22 13
29	29 37	29 00	03 08	02 29	16 41	04 06	07 13	18 54	23 17	14 39	15 15	13 44	22 13
30	29 ♓ 38	28 ♓ 57	02 N 05	02 S 52	11 N 05	03 N 24	07 S 42	18 N 45	23 S 18	14 S 41	15 N 14	13 N 44	22 S 13

ZODIAC SIGN ENTRIES

Date	h	m	Planets
01	03	35	♇ ♒
02	17	23	☽ ♍
05	00	47	☽ ♎
07	10	58	☽ ♏
09	23	14	☽ ♐
12	11	36	☽ ♑
12	22	15	♀ ♎
13	13	26	♂ ♌
14	21	48	☽ ♒
17	04	40	☽ ♓
19	08	29	☽ ♈
21	10	27	☽ ♉
23	03	07	☉ ♎
23	11	57	☽ ♊
25	14	15	☽ ♋
27	18	11	☽ ♌
30	00	09	☽ ♍

LATITUDES

Date	Mercury ☿ ° '	Venus ♀ ° '	Mars ♂ ° '	Jupiter ♃ ° '	Saturn ♄ ° '	Uranus ♅ ° '	Neptune ♆ ° '	Pluto ♇ ° '
01	04 S 03	01 N 25	00 N 47	00 N 09	02 N 10	00 N 40	01 S 48	11 S 20
04	03 21	01 24	00 49	00 09	02 09	00 40	01 48	11 20
07	02 28	01 22	00 52	00 08	02 09	00 40	01 48	11 20
10	01 31	01 20	00 54	00 08	02 08	00 40	01 48	11 20
13	00 S 34	01 17	00 56	00 07	02 08	00 40	01 48	11 20
16	00 N 15	01 14	00 59	00 07	02 07	00 40	01 48	11 20
19	00 55	01 10	01 01	00 07	02 06	00 40	01 49	11 20
22	01 24	01 05	01 03	00 06	02 06	00 40	01 49	11 19
25	01 43	01 00	01 06	00 06	02 05	00 40	01 49	11 19
28	01 52	00 55	01 08	00 05	02 05	00 40	01 49	11 19
31	01 N 53	00 N 49	01 N 11	00 N 05	02 N 05	00 N 40	01 S 49	11 S 18

DATA

Julian Date	2467494
Delta T	+78 seconds
Ayanamsa	24° 28' 03"
Synetic vernal point	04° ♓ 38' 56"
True obliquity of ecliptic	23° 26' 11"

MOON'S PHASES, APSIDES AND POSITIONS ☽

Date	h	m	Phase	Longitude ° '	Eclipse Indicator
03	13	17	●	10 ♍ 57	
11	13	01	☽	18 ♐ 42	
19	01	47	○	26 ♓ 02	total
25	18	40	☾	02 ♋ 36	

Day	h	m		
10	03	41	Apogee	
22	00	35	Perigee	
05	00	38	OS	
12	11	29	Max dec	28° S 43'
19	08	22	ON	
25	14	09	Max dec	28° N 43'

LONGITUDES

Date	Chiron ⚷ ° '	Ceres ⚳ ° '	Pallas ⚴ ° '	Juno ⚵ ° '	Vesta ⚶ ° '	Black Moon Lilith ⚸ ° '
01	24 ♌ 07	03 ♓ 28	08 ♑ 37	22 ♐ 02	10 ♏ 19	00 ♐ 02
11	25 ♌ 23	01 ♓ 22	08 ♑ 52	23 ♐ 11	14 ♏ 58	01 ♐ 09
21	26 ♌ 36	29 ♒ 35	09 ♑ 40	24 ♐ 48	19 ♏ 46	02 ♐ 16
31	27 ♌ 46	28 ♒ 18	10 ♑ 58	26 ♐ 48	24 ♏ 41	03 ♐ 23

ASPECTARIAN

h m	Aspects
01 Tuesday	
01 28	☽ ∥ ♂
03 21	☽ # ♇
03 24	☽ ⊻ ☉
03 59	☽ bQ ♃
04 52	☽ ⊥ ♀
05 28	☽ ⊻ ☿
10 17	☽ □ ♆
12 11	☽ □ ♄
16 30	☽ ⊻ ♀
19 16	☉ ☌ ☿
21 58	☽ ☌ ♅
02 Wednesday	
04 32	☽ ⊻ ♂
06 36	☽ △ ♃
09 42	☽ ∥ ♅
15 49	☽ ⊥ ♂
16 25	♄ # ♆
17 19	☽ ☍ ♇
17 26	☽ # ♄
17 27	☽ ∥ ♆
20 29	☽ Q ♄
20 52	☿ ∠ ♂
03 Thursday	
07 39	☽ ☌ ☿
08 58	☽ ∠ ♂
13 17	☽ ☌ ☉
16 10	☽ △ ♆
18 30	☽ ✶ ♄
19 51	☽ ∥ ☉
22 09	♂ ± ♇
04 Friday	
01 33	☿ ∥ ♀
02 31	☽ ∥ ☿
02 42	☽ ∥ ♀
03 04	♂ ⊼ ♃
04 26	☽ ☌ ♀
04 40	☽ ⊻ ♅
06 55	♀ ⊻ ♅
09 31	♂ # ♇
13 39	☽ □ ♃
14 10	☽ ✶ ♂
15 56	☽ ⊥ ♅
19 30	♂ Q ♆
20 04	☽ bQ ♆
22 39	☽ ∠ ♄
05 Saturday	
00 38	☽ ⊼ ♇
03 41	☉ △ ♆
09 03	☽ ∠ ♅
12 01	☽ ± ♇
12 20	☽ ⊻ ☿
13 09	☽ ± ♆
14 06	☽ Q ♂
15 54	☽ ⊥ ♄
19 35	☽ # ♀
20 30	☉ ∥ ☿
23 14	☽ ⊥ ☿
06 Sunday	
00 02	☽ Q ♃
00 41	☽ ⊼ ♆
02 29	☽ ⊻ ☉
02 42	☽ # ☉
03 32	☽ ⊻ ♄
03 45	☽ # ☿
05 20	☽ bQ ♇
14 09	☽ ✶ ♅
15 09	☽ ⊥ ☉
15 50	☽ ∠ ☿
17 01	☉ ✶ ♄
20 04	☽ ⊻ ♀
23 32	☽ ✶ ♃
07 Monday	
03 00	☽ □ ♂
09 14	☽ ⊥ ♀
10 24	☽ ∠ ☉
10 29	☽ # ♆
10 43	☽ △ ♇
11 09	☽ ∥ ♄
13 54	☽ Q ♅
18 40	☽ # ♅
20 11	☽ ✶ ☿
08 Tuesday	
05 13	☽ ∠ ♀
05 28	☽ ∠ ♃
07 39	♀ □ ♃
11 51	☽ ☍ ♆
15 16	☽ ☌ ♄
19 01	☽ ✶ ☉
19 24	☽ Q ☿
23 38	☽ ⊥ ♃
09 Wednesday	
02 08	☽ □ ♅
02 35	☽ # ♂
05 51	☽ ∥ ♇
08 41	♀ ⊥ ♅
11 49	☽ ⊻ ♃
12 50	☽ ∥ ♃
14 57	☽ ✶ ♀
18 27	☽ △ ♂
21 25	☽ Q ☉
22 52	☽ □ ♇
10 Thursday	
03 24	☿ ⊥ ♂
07 01	☽ □ ☿
18 04	☽ Q ♀
20 34	♀ bQ ♆
11 Friday	
00 23	☽ ⊼ ♆
02 28	☽ bQ ♂
04 18	☽ ⊻ ♄
07 04	☿ St D
11 10	☽ Q ♇
12 27	☽ ± ♆
13 01	☽ □ ☉
15 02	☽ △ ♅
16 27	☽ ⊥ ♄
21 34	☽ ± ♂
12 Saturday	
00 40	☽ ☌ ♃
06 27	☽ bQ ♆
06 55	♀ ✶ ♂
10 12	☽ ⊼ ♂
10 23	☽ □ ♀
10 34	☽ ∠ ♄
11 08	☽ ✶ ♇
12 15	♀ ∠ ♄
15 07	☉ ⊻ ♅
17 35	♀ ⊼ ♇
19 24	☽ △ ☿
21 03	☽ bQ ♅
13 Sunday	
03 49	♂ ⊼ ♇
12 01	☽ △ ♆
14 47	☽ ± ♅
16 16	☽ ✶ ♄
16 34	☽ ∠ ♇
14 Monday	
01 29	☽ bQ ☿
02 27	☽ ⊼ ♅
05 15	☽ △ ☉
09 53	☽ ⊥ ♇
11 43	☽ ⊻ ♃
15 29	☽ Q ♄
16 34	♂ ± ♃
19 21	☽ ± ☿
21 16	☽ ⊻ ♇
23 10	☽ ⊥ ♃
23 28	☽ ☍ ♂
15 Tuesday	
02 55	☽ △ ♀
07 03	☽ ∥ ♃
07 09	☽ ⊼ ☿
11 56	☽ bQ ☉
13 26	☽ ∥ ♇
16 04	☽ ∠ ♃
20 49	☽ □ ♆
20 59	☽ # ♂
16 Wednesday	
01 17	☽ □ ♄
05 49	☽ ± ☉
09 30	☽ bQ ♀
10 42	☽ ☍ ♅
17 31	☽ ⊼ ☉
19 31	☽ ✶ ♃
22 14	☽ # ♅
22 57	♀ ∠ ♅
17 Thursday	
03 10	☽ ∥ ♄
03 17	☽ ± ♀
04 05	☽ ☌ ♇
05 05	☽ # ♆
05 16	☽ Q ♆
08 51	☽ ⊼ ♂
11 50	♀ ± ♇
14 56	☽ ⊼ ♀
16 42	☽ ☍ ☿
16 54	☽ Q ♃
19 52	☽ ± ♂
22 12	☽ # ☿
23 22	☉ □ ♃
18 Friday	
00 32	♀ ± ♆
02 16	☽ ✶ ♆
06 53	☽ △ ♄
12 10	☽ bQ ♂
15 39	☽ ⊼ ♅
15 43	☉ # ♀
19 Saturday	
00 07	☽ □ ♃
01 32	☽ ∥ ♀
01 47	☽ ☍ ☉
01 55	☽ ± ♅
02 55	☽ # ☉
03 42	☉ ⊥ ♅
03 56	☽ ∠ ♆
07 52	☽ ⊻ ♇
08 38	☽ bQ ♄
13 13	☽ ∥ ☉
14 48	☽ △ ♂
16 15	☽ # ♀
17 09	☽ bQ ♅
17 57	☽ ⊥ ♇
19 04	☽ ⊥ ♆
23 10	☽ ☍ ♀
23 51	☽ ± ♄
20 Sunday	
00 07	☽ ⊼ ☿
05 05	☽ ⊻ ♆
07 43	♀ ⊥ ♄
08 21	☉ bQ ♆
08 58	☽ ∠ ♇
09 55	☽ ⊼ ♄
11 02	☽ ± ☿
15 23	☽ ∥ ☿
18 15	☽ △ ♅
21 Monday	
01 47	☿ ⊥ ♂
02 35	☽ △ ♃
03 24	☽ bQ ☿
07 31	☽ ⊼ ☉
09 44	☽ ∥ ♆
09 46	☽ ✶ ♇
12 06	☽ # ♄
15 52	☽ ∥ ♅
18 08	☽ ± ☉
18 49	☽ □ ♂
22 Tuesday	
03 30	☽ bQ ♃
05 28	☽ ⊼ ♀
05 32	☽ Q ♇
06 02	☿ △ ♆
06 38	☽ ☌ ♆
06 42	☽ △ ☿
10 02	☽ bQ ☉
11 06	☽ ∥ ♂
11 48	☽ ☍ ♄
16 17	☽ ± ♀
16 31	☉ ⊼ ♇
18 26	☽ ± ♃
18 55	♀ ⊼ ♆
19 56	☽ □ ♅
22 06	☽ # ♇
23 Wednesday	
01 49	☽ Q ♂
03 48	☽ # ♃
04 25	☽ ⊼ ♃
08 33	☽ bQ ♀
11 13	☽ □ ♇
12 36	☽ △ ☉
15 03	☉ ∠ ♄
19 23	♀ Q ♃
22 35	☽ ✶ ♂
24 Thursday	
01 52	☽ Q ♅
08 16	☽ ⊻ ♆
11 24	☿ ✶ ♄
11 52	☽ △ ♀
13 55	☽ ⊼ ♄
14 12	☽ □ ☿
15 19	♀ bQ ♇
18 19	☽ ⊥ ♆
22 02	☽ ✶ ♅
25 Friday	
00 04	☽ ± ♄
00 49	☽ ∠ ♂
06 53	☽ ☍ ♃
09 29	☽ ∠ ♆
13 22	♀ ⊻ ♄
13 26	☽ △ ♇
15 25	☽ bQ ♄
16 45	☽ ⊥ ♂
18 40	☽ □ ☉
23 35	☽ ∠ ♅
26 Saturday	
00 36	☽ Q ☿
03 28	☽ ⊻ ♂
11 05	☽ ✶ ♆
15 07	☽ bQ ♇
15 07	☽ ⊥ ♅
17 21	☽ △ ♄
19 07	☿ # ♀
19 53	☽ □ ♀
27 Sunday	
00 08	☽ ✶ ☿
01 34	☽ ⊻ ♅
04 04	☽ Q ☉
06 45	☽ ± ♇
07 54	☽ Q ♆
10 56	☽ ⊼ ♃
12 03	☿ ⊻ ♅
17 16	☽ ⊼ ♇
21 36	☽ ± ♃
28 Monday	
02 30	☽ # ♃
02 50	☽ ✶ ☉
06 19	☽ ∠ ☿
06 41	☽ Q ♀
08 39	☽ # ♇
10 22	☽ ☌ ♂
13 44	☽ bQ ♃
15 44	☽ □ ♆
16 56	☉ ± ♇
20 29	☉ ∠ ♅
22 43	☽ □ ♄
29 Tuesday	
00 51	☽ ⊥ ☿
01 21	☽ ∥ ♂
06 26	☽ ✶ ♀
07 04	☽ ☌ ♅
07 52	☽ ∠ ☉
08 36	☉ ± ♆
13 23	☽ ⊻ ☿
13 26	♀ ✶ ♅
17 05	☽ △ ♃
18 26	☽ ∥ ♅
20 57	☽ # ♄
23 08	☽ ☍ ♇
30 Wednesday	
00 57	☽ ∥ ♆
01 38	☽ ⊥ ☉
03 40	☿ ∠ ♂
07 48	☽ Q ♄
12 48	☽ ∠ ♀
13 36	☽ ⊻ ☉
18 32	☿ □ ♃
19 41	☽ ⊻ ♂
22 30	☽ △ ♆
23 48	☉ # ☿

All ephemeris data is given at 12.00 UT and the Moon's longitude is additionally given for 24.00 UT

OCTOBER 2043

LONGITUDES

Date	Sidereal time h m s	Sun ☉ ° ' "	Moon ☽ ° ' "	Moon ☽ 24.00 ° ' "	Mercury ☿ ° '	Venus ♀ ° '	Mars ♂ ° '	Jupiter ♃ ° '	Saturn ♄ ° '	Uranus ♅ ° '	Neptune ♆ ° '	Pluto ♇ ° '
01	12 40 29	08 ♎ 12 30	19 ♍ 16 34	25 ♍ 37 17	27 ♍ 33	23 ♎ 03	10 ♌ 58	26 ♐ 21	16 ♏ 16	20 ♌ 44	12 ♉ 03	29 ♒ 25
02	12 44 25	09 11 31	01 ♎ 54 53	08 ♎ 09 28	29 ♍ 21	24 18	11 34	26 28	16 22	20 47	12 R 02	29 R 24
03	12 48 22	10 10 33	14 ♎ 21 09	20 ♎ 30 04	01 ♎ 09	25 32	12 10	26 35	16 28	20 50	12 01	29 23
04	12 52 18	11 09 38	26 ♎ 36 24	02 ♏ 40 19	02 56	26 47	12 46	26 43	16 35	20 52	11 59	29 22
05	12 56 15	12 08 44	08 ♏ 42 04	14 ♏ 41 53	04 44	28 01	13 21	26 50	16 41	20 55	11 58	29 21
06	13 00 12	13 07 53	20 ♏ 40 04	26 ♏ 36 58	06 32	29 ♎ 16	13 57	26 58	16 47	20 58	11 57	29 20
07	13 04 08	14 07 04	02 ♐ 32 55	08 ♐ 28 20	08 19	00 ♏ 30	14 32	27 06	16 54	21 00	11 55	29 20
08	13 08 05	15 06 16	14 ♐ 23 40	20 ♐ 19 23	10 06	01 45	15 07	27 14	17 00	21 03	11 54	29 19
09	13 12 01	16 05 30	26 ♐ 15 59	02 ♑ 14 01	11 52	02 59	15 43	27 22	17 07	21 05	11 52	29 18
10	13 15 58	17 04 46	08 ♑ 14 01	14 ♑ 16 33	13 38	04 14	16 18	27 31	17 13	21 08	11 51	29 17
11	13 19 54	18 04 04	20 ♑ 22 13	26 ♑ 31 34	15 23	05 28	16 53	27 39	17 20	21 10	11 49	29 16
12	13 23 51	19 03 24	02 ♒ 45 11	09 ♒ 03 35	17 08	06 43	17 28	27 48	17 26	21 13	11 48	29 16
13	13 27 47	20 02 45	15 ♒ 27 17	21 ♒ 56 44	18 52	07 57	18 03	27 56	17 33	21 15	11 46	29 15
14	13 31 44	21 02 08	28 ♒ 32 17	05 ♓ 14 14	20 35	09 12	18 38	28 05	17 40	21 17	11 45	29 14
15	13 35 41	22 01 33	12 ♓ 02 44	18 ♓ 57 49	22 17	10 26	19 12	28 14	17 46	21 19	11 43	29 14
16	13 39 37	23 01 00	25 ♓ 59 24	03 ♈ 07 11	23 59	11 40	19 47	28 23	17 53	21 22	11 41	29 13
17	13 43 34	24 00 28	10 ♈ 20 44	17 ♈ 39 26	25 40	12 55	20 21	28 32	18 00	21 24	11 40	29 12
18	13 47 30	24 59 59	25 ♈ 02 31	02 ♉ 29 03	27 21	14 09	20 56	28 42	18 06	21 26	11 38	29 12
19	13 51 27	25 59 31	09 ♉ 58 02	17 ♉ 28 20	29 ♎ 00	15 24	21 30	28 51	18 13	21 28	11 37	29 11
20	13 55 23	26 59 06	24 ♉ 58 49	02 ♊ 28 22	00 ♏ 40	16 38	22 04	29 01	18 20	21 30	11 35	29 10
21	13 59 20	27 58 43	09 ♊ 55 54	17 ♊ 20 29	02 18	17 53	22 38	29 10	18 27	21 32	11 33	29 10
22	14 03 16	28 58 22	24 ♊ 41 14	01 ♋ 57 29	03 56	19 07	23 12	29 20	18 34	21 34	11 32	29 09
23	14 07 13	29 ♎ 58 04	09 ♋ 08 44	16 ♋ 14 36	05 33	20 22	23 46	29 30	18 41	21 36	11 30	29 09
24	14 11 10	00 ♏ 57 48	23 ♋ 14 53	00 ♌ 09 33	07 10	21 36	24 20	29 40	18 48	21 38	11 28	29 08
25	14 15 06	01 57 34	06 ♌ 58 38	13 ♌ 42 19	08 46	22 51	24 53	29 ♐ 50	18 55	21 40	11 27	29 08
26	14 19 03	02 57 22	20 ♌ 20 50	26 ♌ 54 28	10 22	24 05	25 27	00 ♑ 00	19 02	21 41	11 25	29 07
27	14 22 59	03 57 12	03 ♍ 23 34	09 ♍ 48 29	11 57	25 19	26 00	00 11	19 09	21 43	11 23	29 07
28	14 26 56	04 57 05	16 ♍ 09 34	22 ♍ 27 11	13 32	26 34	26 33	00 21	19 16	21 45	11 22	29 07
29	14 30 52	05 57 00	28 ♍ 41 38	04 ♎ 53 15	15 06	27 48	27 06	00 31	19 23	21 46	11 20	29 06
30	14 34 49	06 56 57	11 ♎ 02 20	17 ♎ 09 08	16 39	29 ♏ 03	27 39	00 42	19 30	21 48	11 18	29 06
31	14 38 45	07 ♏ 56 55	23 ♎ 13 53	29 ♎ 16 48	18 ♏ 12	00 ♐ 17	28 ♌ 12	00 ♑ 53	19 ♏ 37	21 ♌ 49	11 ♉ 17	29 ♒ 06

Moon nodes and latitude

Date	Moon True ☊ ° '	Moon Mean ☊ ° '	Moon ☽ Latitude ° '
01	29 ♓ 38	28 ♓ 54	00 N 57
02	29 R 38	28 51	00 S 13
03	29 38	28 47	01 21
04	29 37	28 44	02 24
05	29 35	28 41	03 20
06	29 33	28 38	04 06
07	29 31	28 35	04 41
08	29 29	28 32	05 05
09	29 27	28 28	05 15
10	29 26	28 25	05 12
11	29 D 26	28 22	04 55
12	29 27	28 19	04 24
13	29 28	28 16	03 39
14	29 29	28 12	02 42
15	29 31	28 09	01 35
16	29 32	28 06	00 S 19
17	29 R 31	28 03	00 N 59
18	29 30	28 00	02 16
19	29 27	27 57	03 24
20	29 23	27 53	04 18
21	29 19	27 50	04 55
22	29 15	27 47	05 12
23	29 13	27 44	05 08
24	29 11	27 41	04 46
25	29 D 11	27 38	04 07
26	29 12	27 34	03 16
27	29 13	27 31	02 16
28	29 15	27 28	01 11
29	29 16	27 25	00 N 03
30	29 R 15	27 22	01 S 04
31	29 ♓ 12	27 ♓ 18	02 S 07

DECLINATIONS

Date	Sun ☉ ° '	Moon ☽ ° '	Mercury ☿ ° '	Venus ♀ ° '	Mars ♂ ° '	Jupiter ♃ ° '	Saturn ♄ ° '	Uranus ♅ ° '	Neptune ♆ ° '	Pluto ♇ ° '
01	03 S 15	05 N 07	02 N 42	08 S 12	18 N 36	23 S 18	14 S 43	15 N 13	13 N 43	22 S 14
02	03 39	00 S 57	01 58	08 41	18 27	23 18	14 45	15 12	13 43	22 14
03	04 02	06 54	01 13	09 10	18 18	23 19	14 47	15 11	13 42	22 14
04	04 25	12 30	00 N 28	09 39	18 09	23 19	14 48	15 10	13 42	22 14
05	04 48	17 33	00 S 17	10 08	17 59	23 19	14 50	15 10	13 42	22 14
06	05 11	21 52	01 03	10 36	17 50	23 20	14 52	15 09	13 41	22 15
07	05 34	25 16	01 49	11 05	17 40	23 20	14 54	15 08	13 41	22 15
08	05 57	27 34	02 35	11 33	17 31	23 20	14 56	15 07	13 40	22 15
09	06 20	28 38	03 20	12 00	17 21	23 20	14 58	15 06	13 40	22 15
10	06 43	28 22	04 06	12 28	17 11	23 21	15 00	15 06	13 39	22 15
11	07 05	26 45	04 52	12 55	17 01	23 21	15 02	15 05	13 39	22 15
12	07 28	23 49	05 37	13 22	16 52	23 21	15 04	15 04	13 38	22 15
13	07 50	19 41	06 21	13 48	16 42	23 22	15 06	15 04	13 38	22 16
14	08 12	14 31	07 06	14 14	16 32	23 22	15 08	15 03	13 37	22 16
15	08 35	08 30	07 50	14 40	16 22	23 22	15 10	15 02	13 37	22 16
16	08 57	01 S 53	08 33	15 06	16 12	23 22	15 12	15 01	13 36	22 16
17	09 19	05 N 00	09 16	15 31	16 01	23 22	15 14	15 01	13 36	22 16
18	09 41	11 48	09 58	15 56	15 51	23 23	15 16	15 00	13 35	22 16
19	10 02	18 01	10 40	16 20	15 41	23 23	15 18	15 00	13 35	22 16
20	10 24	23 11	11 21	16 44	15 31	23 23	15 20	14 59	13 34	22 16
21	10 45	26 48	12 02	17 07	15 21	23 23	15 22	14 58	13 34	22 16
22	11 06	28 31	12 41	17 30	15 10	23 23	15 24	14 58	13 33	22 16
23	11 28	28 14	13 20	17 53	15 00	23 24	15 26	14 57	13 33	22 16
24	11 48	26 07	13 59	18 15	14 49	23 24	15 28	14 57	13 32	22 16
25	12 09	22 31	14 36	18 37	14 39	23 24	15 30	14 56	13 32	22 16
26	12 30	17 48	15 13	18 58	14 29	23 24	15 31	14 55	13 31	22 16
27	12 50	12 23	15 49	19 19	14 18	23 24	15 33	14 55	13 31	22 16
28	13 10	06 33	16 24	19 39	14 08	23 24	15 35	14 54	13 30	22 16
29	13 30	00 N 34	16 58	19 58	13 57	23 24	15 37	14 54	13 30	22 16
30	13 50	05 S 21	17 32	20 18	13 46	23 24	15 39	14 54	13 29	22 16
31	14 S 09	10 S 59	18 S 04	20 S 36	13 N 36	23 S 24	15 S 41	14 N 53	13 N 28	22 S 16

ZODIAC SIGN ENTRIES

Date	h m	Planets
02	08 20	☽ ♎
02	20 45	☿ ♎
04	18 42	☽ ♏
07	02 18	♀ ♏
07	06 51	☽ ♐
09	19 31	☽ ♑
12	06 43	☽ ♒
14	14 38	☽ ♓
16	18 46	☽ ♈
18	20 00	☽ ♉
20	02 23	☿ ♏
20	20 02	☽ ♊
22	20 45	☽ ♋
23	12 47	☉ ♏
24	23 43	☽ ♌
26	11 30	♃ ♑
27	05 42	☽ ♍
29	14 31	☽ ♎
31	06 28	♀ ♐

LATITUDES

Date	Mercury ☿ ° '	Venus ♀ ° '	Mars ♂ ° '	Jupiter ♃ ° '	Saturn ♄ ° '	Uranus ♅ ° '	Neptune ♆ ° '	Pluto ♇ ° '
01	01 N 53	00 N 49	01 N 11	00 N 05	02 N 05	00 N 40	01 S 49	11 S 18
04	01 47	00 43	01 13	00 05	02 04	00 40	01 49	11 18
07	01 37	00 36	01 15	00 04	02 04	00 41	01 49	11 17
10	01 23	00 30	01 18	00 04	02 03	00 41	01 49	11 17
13	01 07	00 22	01 20	00 04	02 03	00 41	01 49	11 16
16	00 48	00 15	01 23	00 03	02 03	00 41	01 49	11 16
19	00 29	00 N 07	01 26	00 03	02 02	00 41	01 49	11 15
22	00 N 09	00 00	01 28	00 03	02 02	00 41	01 50	11 15
25	00 S 12	00 S 08	01 31	00 02	02 02	00 41	01 50	11 14
28	00 32	00 16	01 33	00 02	02 01	00 41	01 50	11 13
31	00 S 52	00 S 24	01 N 36	00 N 02	02 N 01	00 N 41	01 S 50	11 S 13

DATA

Julian Date	2467524
Delta T	+78 seconds
Ayanamsa	24° 28' 06"
Synetic vernal point	04° ♓ 38' 53"
True obliquity of ecliptic	23° 26' 11"

MOON'S PHASES, APSIDES AND POSITIONS ☽

Date	h m	Phase	Longitude ° '	Eclipse Indicator
03	03 12	●	09 ♎ 49	Annular
11	07 05	☽	17 ♑ 52	
18	11 56	○	25 ♈ 00	
25	02 27	☾	01 ♌ 34	

Day	h m		
07	22 28	Apogee	
19	23 38	Perigee	
02	08 13	0S	
09	19 19	Max dec	28° S 42'
16	18 38	0N	
22	20 34	Max dec	28° N 39'
29	14 17	0S	

LONGITUDES

Date	Chiron ⚷ ° '	Ceres ⚳ ° '	Pallas ⚴ ° '	Juno ⚵ ° '	Vesta ⚶ ° '	Black Moon Lilith ⚸ ° '
01	27 ♌ 46	28 ♒ 18	10 ♑ 58	26 ♐ 48	24 ♏ 41	03 ♐ 23
11	28 ♌ 50	27 ♒ 35	12 ♑ 41	29 ♐ 10	29 ♏ 43	04 ♐ 30
21	29 ♌ 48	27 ♒ 29	14 ♑ 45	01 ♑ 51	04 ♐ 50	05 ♐ 37
31	00 ♍ 39	27 ♒ 59	17 ♑ 07	04 ♑ 47	10 ♐ 01	06 ♐ 44

All ephemeris data is given at 12.00 UT and the Moon's longitude is additionally given for 24.00 UT

ASPECTARIAN

h m	Aspects
01 Thursday	
00 39	☿ ⊥ ♅
00 40	☽ # ♀
05 23	☿ ⚼ ♆
06 18	☽ ✶ ♄
07 22	☽ ⊥ ♀
07 27	☽ ⊥ ♂
08 51	♀ Q ♂
14 45	☽ ⊻ ♅
18 55	☽ # ☉
19 54	☽ ⊻ ♀
22 53	☽ ∥ ☿
02 Friday	
01 17	☽ ∠ ♂
01 30	☽ □ ♃
02 10	☽ ⊥ ♅
02 42	☽ ⚼ ♆
06 16	☽ ☌ ☿
06 52	♂ ⚼ ♃
07 12	☽ ⊼ ♇
10 57	☽ ∠ ♄
12 44	☿ ⊼ ♇
15 35	☽ # ☿
18 41	☽ ± ♇
19 27	☽ ∠ ♅
19 54	☽ ± ♆
23 29	☽ ∥ ☉
03 Saturday	
03 12	☽ ☌ ☉
04 25	☽ ⊥ ♄
06 08	♂ □ ♆
07 28	☽ ⊼ ♆
07 32	☽ ✶ ♂
12 04	☽ ⚼ ♇
12 28	☽ Q ♃
16 10	☽ ⊻ ♄
16 42	☿ ∠ ♄
20 07	☉ ⊥ ♄
22 26	☽ ∥ ♀
04 Sunday	
00 41	☽ ✶ ♅
08 10	☽ Q ♂
10 36	♀ ✶ ♃
12 13	☽ ✶ ♃
12 22	☽ ☌ ♀
17 27	☽ △ ♇
17 28	☽ # ♆
22 41	☽ ∥ ♄
05 Monday	
00 19	☽ # ♅
00 26	☽ Q ♅
02 43	☽ ⊻ ☿
07 43	☉ ⊼ ♆
14 11	☽ # ♂
16 47	☽ ⊥ ☿
18 20	☽ ∠ ♃
18 31	☽ ☍ ♆
19 30	☽ ⊻ ☉
20 13	☿ ± ♇
21 47	☽ □ ♂
06 Tuesday	
04 08	☽ ☌ ♄
04 10	☿ ∠ ♅
04 14	☿ ± ♆
07 49	♃ ⚼ ♆
08 38	☽ ⊥ ☉
12 35	☽ □ ♅
12 37	☽ ⊥ ♃
13 33	♀ △ ♇
14 03	☽ ∠ ☿
14 22	☽ ∥ ♇
21 29	☽ ∥ ♃
07 Wednesday	
00 51	☽ ⊻ ♃
04 26	☽ ∠ ☉
05 29	☽ □ ♇
07 22	☽ ⊻ ♀
17 02	☉ ⚼ ♇
20 57	☽ ⊥ ♀
08 Thursday	
01 45	☽ ✶ ☿
06 57	☽ ⊼ ♆
13 14	☉ ✶ ♂
13 33	☽ △ ♂
13 34	☽ ✶ ☉
15 41	☉ Q ♃
17 19	☽ ∠ ♀
17 20	☽ ⊻ ♄
17 54	☽ Q ♇
19 04	☽ ± ♆
09 Friday	
01 02	☿ ⊥ ♄
01 30	☽ △ ♅
05 34	☽ ⊥ ♄
06 19	☽ Q ☿
11 59	☿ ⊼ ♆
13 13	☽ ⚼ ♆
14 02	♀ Q ♅
14 15	☽ ☌ ♃
16 01	☽ Q ☉
18 06	☽ ✶ ♇
21 25	☽ ⚼ ♂
23 52	☽ ∠ ♄
10 Saturday	
03 04	☽ ✶ ♀
07 47	☽ ⚼ ♅
15 48	☉ ⊻ ♄
16 19	☽ ± ♂
19 10	☽ △ ♆
20 52	☿ ⚼ ♇
11 Sunday	
00 01	☽ ∠ ♇
00 33	☽ □ ☿
01 44	☽ ± ♅
04 48	☽ ⊼ ♂
05 39	☽ Q ♀
05 58	☽ ✶ ♄
07 05	☽ □ ☉
13 34	☽ ⊼ ♅
15 58	☿ Q ♃
17 40	☽ ⊥ ♇
12 Monday	
02 21	☽ ⊻ ♃
05 18	☽ ⊻ ♇
05 34	☽ Q ♄
10 35	♂ □ ♄
14 01	☽ ⊥ ♃
14 34	♄ # ♅
15 04	☽ ∥ ♃
16 35	☿ ⊻ ♄
19 02	☿ ✶ ♂
20 22	☽ □ ♀
21 52	☽ ∥ ♇
13 Tuesday	
02 32	♀ # ♆
05 07	☽ □ ♆
07 15	☽ ∠ ♃
15 55	☽ □ ♄
17 02	☽ ☍ ♂
19 17	☽ △ ☿
21 12	☽ △ ☉
22 45	☽ ☍ ♅
14 Wednesday	
02 54	☽ # ♂
09 23	☽ ∥ ♄
09 43	☽ # ♅
11 10	☽ ✶ ♃
13 05	☽ ∥ ♀
13 16	☽ ☌ ♇
14 10	☽ Q ♆
15 47	☽ # ♆
18 19	☉ ✶ ♅
22 09	☿ ✶ ♅
15 Thursday	
02 26	☽ ⚼ ☿
02 29	☽ ⚼ ☉
03 13	☉ ☌ ☿
08 48	☽ Q ♃
08 54	☽ △ ♀
11 26	☽ ✶ ♆
11 44	☽ ∥ ☉
14 16	☽ ∥ ☿
19 28	☽ ± ☉
20 25	☽ ± ☿
22 01	☽ △ ♄
16 Friday	
00 57	☽ ⊼ ♂
04 06	☽ ⊼ ♅
06 34	☽ ⊼ ☉
07 59	♀ # ♅
08 08	☽ ⊼ ☿
11 38	☽ ± ♂
12 18	♀ ☍ ♆
13 11	☽ ∠ ♆
13 16	☽ ⚼ ♀
14 20	☽ ± ♅
16 06	☽ □ ♃
17 27	☽ ⊻ ♇
18 11	♀ ∥ ♄
23 42	☽ ⚼ ♄
17 Saturday	
03 23	☽ ⚼ ♂
03 29	☽ ⊥ ♇
04 15	☽ ⊥ ♆
05 27	☽ ⚼ ♅
05 47	☽ ± ♀
14 10	☽ ⊻ ♆
14 44	☽ ± ♄
15 05	☉ ∥ ☿
16 38	☽ ⊼ ♀
18 21	☽ ∠ ♇
18 Sunday	
00 39	☽ ⊼ ♄
01 47	♀ ∠ ♃
03 57	☽ # ☉
04 41	☽ # ☿
05 03	☽ △ ♂
06 08	☽ △ ♅
09 01	♀ # ♂
11 56	☽ ☍ ☉
16 12	☽ ☍ ☿
17 58	☽ △ ♃
18 36	☽ ∥ ♆
18 42	☽ ✶ ♇
23 57	☽ ∥ ♅
19 Monday	
01 03	☽ # ♄
02 52	☽ ∥ ♂
04 40	☽ # ♀
09 30	☿ ✶ ♃
10 38	♂ ☌ ♅
13 57	☽ Q ♇
14 31	☿ △ ♇
14 38	☽ ☌ ♆
18 17	☽ ⚼ ♃
21 28	☽ ☍ ♀
20 Tuesday	
01 18	☽ ☍ ♄
06 26	☽ □ ♅
07 10	☽ □ ♂
07 13	☽ # ♇
08 49	☽ ± ♃
13 05	☽ # ♃
15 26	☽ ⊼ ☉
18 31	☽ ⊼ ♃
18 42	☽ □ ♇
22 13	☽ ⊼ ☿
21 Wednesday	
01 44	☽ ± ☉
09 03	☽ ± ☿
09 50	♂ # ♄
10 55	♃ ✶ ♇
11 21	☽ Q ♅
13 11	☽ Q ♂
14 37	☽ ⊻ ♆
17 17	☽ ⚼ ☉
22 Thursday	
00 08	♀ ☌ ♄
00 20	☽ ⊥ ♆
01 25	☽ ⚼ ☿
01 55	☽ ⊼ ♄
02 03	☽ ⊼ ♀
06 28	☿ Q ♅
06 53	☽ ✶ ♅
09 28	☽ ✶ ♂
11 48	☽ ± ♄
12 47	☽ ± ♀
15 01	☽ ∠ ♆
16 21	☉ △ ♇
19 21	☽ △ ♇
19 35	☽ △ ☉
19 44	☽ ☍ ♃
22 27	☉ ✶ ♃
23 Friday	
02 47	☽ ⚼ ♄
05 04	☽ ⚼ ♀
05 14	☽ △ ☿
07 43	☽ ∠ ♅
11 20	☽ ∠ ♂
15 58	☽ ✶ ♆
16 43	☿ Q ♂
18 38	♂ ∥ ♅
19 30	☿ # ♆
20 26	☽ ⚼ ♇
22 56	☽ ⊥ ♅
24 Saturday	
03 12	☽ ⊥ ♂
04 17	☽ △ ♄
08 53	☽ △ ♀
09 12	☽ ⊻ ♅
11 49	☽ ± ♇
12 23	☽ Q ♆
12 34	♀ □ ♅
13 56	☽ ⊻ ♂
22 13	☽ ⊼ ♇
23 16	☽ ⊼ ♃
25 Sunday	
02 27	☽ □ ☉
06 49	☽ # ♃
09 57	☽ ± ♃
13 22	☿ # ♂
13 23	☽ # ♇
15 37	☽ □ ☿
19 56	☽ □ ♆
26 Monday	
00 38	☿ # ♅
02 13	☽ ⚼ ♃
06 50	☽ # ♀
09 35	☽ □ ♄
10 13	♀ ⊥ ♃
13 12	☽ Q ☉
14 27	☽ ☌ ♅
19 32	☽ □ ♀
21 43	☽ ☌ ♂
22 20	☽ # ♄
22 35	☽ # ☿
27 Tuesday	
00 58	☿ ∥ ♄
01 04	☽ ∥ ♅
03 30	☽ ∥ ♂
03 41	☿ ☍ ♆
04 05	☽ ☍ ♇
04 43	☽ Q ☿
05 57	☽ △ ♃
06 10	☉ Q ♅
07 11	☽ ∥ ♆
10 11	☽ # ☉
13 08	☽ ✶ ☉
19 04	☽ Q ♄
28 Wednesday	
02 57	☽ △ ♆
06 18	☽ ✶ ☿
08 39	☽ Q ♀
11 39	♀ □ ♂
17 57	☽ ✶ ♄
19 50	☽ ∠ ☉
22 40	☽ ⊻ ♅
29 Thursday	
07 28	☽ ⚼ ♆
08 48	☽ ⊻ ♂
10 06	☽ ✶ ♀
10 13	☽ ⊥ ♅
11 14	☉ # ♆
12 47	☽ ⊼ ♇
14 38	☽ ⊥ ☉
15 05	☽ ∠ ☿
15 35	☽ □ ♃
19 28	☿ ∠ ♃
20 56	☽ ⊥ ♂
23 07	☽ ∠ ♄
30 Friday	
00 25	☽ ± ♇
00 51	☽ ± ♆
03 18	☽ ⊻ ☉
03 42	☽ ∠ ♅
09 16	☉ # ♂
11 08	☽ ⊥ ☿
12 31	☽ ⊼ ♆
13 00	♀ □ ♇
15 19	☽ ∠ ♂
16 52	☽ ⊥ ♄
18 00	☽ ⚼ ♇
18 34	☽ ∠ ♀
31 Saturday	
00 37	☽ ⊻ ☿
03 16	☽ Q ♃
04 47	☽ ⊻ ♄
09 13	☽ ✶ ♅
14 20	☽ ⊥ ♀
22 20	☽ ✶ ♂
23 13	☽ # ♆
23 25	☽ # ♂
23 37	☽ △ ♇

NOVEMBER 2043

Date	Sidereal time h m s	Sun ☉ ° ′ ″	Moon ☽ ° ′ ″	Moon ☽ 24.00 ° ′ ″	Mercury ☿ ° ′	Venus ♀ ° ′	Mars ♂ ° ′	Jupiter ♃ ° ′	Saturn ♄ ° ′	Uranus ♅ ° ′	Neptune ♆ ° ′	Pluto ♇ ° ′
01	14 42 42	08 ♏ 56 56	05 ♏ 18 06	11 ♏ 17 57	19 ♏ 45	01 ♐ 32	28 ♌ 45	01 ♑ 04	19 ♏ 44	21 ♌ 51	11 ♉ 15	29 ♒ 05
02	14 46 39	09 56 59	17 ♏ 16 33	23 ♏ 14 04	21 17	02 46	29 18	01 14	19 51	21 52	11 R 13	29 R 05
03	14 50 35	10 57 04	29 ♏ 10 43	05 ♐ 06 41	22 49	04 00	29 ♌ 50	01 25	19 58	21 54	11 12	29 05
04	14 54 32	11 57 10	11 ♐ 02 12	16 ♐ 57 31	24 20	05 15	00 ♍ 22	01 37	20 05	21 55	11 10	29 05
05	14 58 28	12 57 19	22 ♐ 52 56	28 ♐ 48 46	25 51	06 29	00 55	01 48	20 12	21 56	11 08	29 04
06	15 02 25	13 57 29	04 ♑ 45 24	10 ♑ 43 12	27 21	07 44	01 27	01 59	20 20	21 57	11 07	29 04
07	15 06 21	14 57 40	16 ♑ 42 37	22 ♑ 44 09	28 ♏ 51	08 58	01 59	02 10	20 27	21 59	11 05	29 04
08	15 10 18	15 57 53	28 ♑ 48 17	04 ♒ 55 36	00 ♐ 21	10 12	02 30	02 22	20 34	22 00	11 03	29 04
09	15 14 14	16 58 08	11 ♒ 06 38	17 ♒ 21 58	01 50	11 27	03 02	02 33	20 41	22 01	11 02	29 04
10	15 18 11	17 58 24	23 ♒ 42 12	00 ♓ 07 53	03 19	12 41	03 33	02 45	20 48	22 02	11 00	29 04
11	15 22 08	18 58 41	06 ♓ 39 33	13 ♓ 17 40	04 47	13 56	04 05	02 57	20 55	22 03	10 58	29 04
12	15 26 04	19 59 00	20 ♓ 02 39	26 ♓ 54 46	06 14	15 10	04 36	03 08	21 03	22 03	10 57	29 D 04
13	15 30 01	20 59 21	03 ♈ 54 10	11 ♈ 00 51	07 41	16 24	05 07	03 20	21 10	22 04	10 55	29 04
14	15 33 57	21 59 42	18 ♈ 14 36	25 ♈ 35 00	09 08	17 39	05 38	03 32	21 17	22 05	10 53	29 04
15	15 37 54	23 00 05	03 ♉ 01 23	10 ♉ 32 53	10 34	18 53	06 08	03 44	21 24	22 06	10 52	29 04
16	15 41 50	24 00 30	18 ♉ 08 25	25 ♉ 46 44	11 59	20 07	06 39	03 56	21 31	22 06	10 50	29 04
17	15 45 47	25 00 57	03 ♊ 26 26	11 ♊ 06 05	13 24	21 22	07 09	04 08	21 38	22 07	10 48	29 04
18	15 49 43	26 01 25	18 ♊ 44 14	26 ♊ 19 30	14 48	22 36	07 40	04 21	21 46	22 07	10 47	29 04
19	15 53 40	27 01 55	03 ♋ 50 40	11 ♋ 16 39	16 11	23 50	08 10	04 33	21 53	22 08	10 45	29 04
20	15 57 37	28 02 26	18 ♋ 36 36	25 ♋ 49 56	17 33	25 04	08 39	04 45	22 00	22 08	10 44	29 05
21	16 01 33	29 ♏ 03 00	02 ♌ 56 15	09 ♌ 55 23	18 53	26 19	09 09	04 58	22 07	22 09	10 42	29 05
22	16 05 30	00 ♐ 03 35	16 ♌ 47 22	23 ♌ 32 23	20 13	27 33	09 39	05 10	22 14	22 09	10 40	29 05
23	16 09 26	01 04 12	00 ♍ 10 45	06 ♍ 42 54	21 31	28 ♐ 47	10 08	05 23	22 21	22 09	10 39	29 05
24	16 13 23	02 04 50	13 ♍ 09 18	19 ♍ 30 31	22 48	00 ♑ 01	10 37	05 35	22 28	22 10	10 37	29 06
25	16 17 19	03 05 30	25 ♍ 47 06	01 ♎ 59 36	24 02	01 16	11 06	05 48	22 36	22 10	10 36	29 06
26	16 21 16	04 06 12	08 ♎ 08 34	14 ♎ 14 31	25 15	02 30	11 35	06 01	22 43	22 10	10 34	29 06
27	16 25 12	05 06 55	20 ♎ 17 57	26 ♎ 19 18	26 25	03 44	12 03	06 13	22 50	22 10	10 33	29 07
28	16 29 09	06 07 40	02 ♏ 18 58	08 ♏ 17 20	27 33	04 58	12 31	06 26	22 57	22 R 10	10 31	29 07
29	16 33 06	07 08 26	14 ♏ 14 42	20 ♏ 11 21	28 37	06 12	13 00	06 39	23 04	22 10	10 30	29 08
30	16 37 02	08 ♐ 09 14	26 ♏ 07 30	02 ♐ 03 22	29 ♐ 38	07 ♑ 27	13 ♍ 27	06 ♑ 52	23 ♏ 11	22 ♌ 10	10 ♉ 28	29 ♒ 08

Date	Moon True ☊ ° ′	Moon Mean ☊ ° ′	Moon ☽ Latitude ° ′
01	29 ♓ 08	27 ♓ 15	03 S 03
02	29 R 01	27 12	03 51
03	28 53	27 09	04 28
04	28 45	27 06	04 53
05	28 36	27 03	05 06
06	28 29	26 59	05 06
07	28 23	26 56	04 52
08	28 19	26 53	04 26
09	28 17	26 50	03 46
10	28 D 17	26 47	02 55
11	28 18	26 44	01 54
12	28 19	26 40	00 S 44
13	28 R 19	26 37	00 N 30
14	28 18	26 34	01 45
15	28 14	26 31	02 55
16	28 07	26 28	03 54
17	27 59	26 24	04 37
18	27 50	26 21	05 00
19	27 41	26 18	05 03
20	27 34	26 15	04 44
21	27 28	26 12	04 08
22	27 26	26 09	03 19
23	27 25	26 05	02 20
24	27 D 25	26 02	01 15
25	27 26	25 59	00 N 09
26	27 R 25	25 56	00 S 57
27	27 22	25 53	01 58
28	27 17	25 50	02 54
29	27 09	25 46	03 41
30	26 ♓ 58	25 ♓ 43	04 S 19

DECLINATIONS

Date	Sun ☉ ° ′	Moon ☽ ° ′	Mercury ☿ ° ′	Venus ♀ ° ′	Mars ♂ ° ′	Jupiter ♃ ° ′	Saturn ♄ ° ′	Uranus ♅ ° ′	Neptune ♆ ° ′	Pluto ♇ ° ′
01	14 S 29	16 S 10	18 S 36	20 S 54	13 N 25	23 S 24	15 S 43	14 N 53	13 N 28	22 S 16
02	14 48	20 40	19 07	21 11	13 15	23 24	15 45	14 52	13 27	22 15
03	15 07	24 20	19 37	21 28	13 04	23 24	15 47	14 52	13 27	22 15
04	15 25	26 56	20 06	21 44	12 54	23 24	15 49	14 51	13 26	22 15
05	15 44	28 20	20 33	22 00	12 43	23 24	15 51	14 51	13 26	22 15
06	16 02	28 27	21 00	22 15	12 33	23 24	15 53	14 51	13 25	22 15
07	16 19	27 13	21 26	22 29	12 22	23 24	15 55	14 50	13 25	22 15
08	16 37	24 44	21 51	22 43	12 11	23 24	15 57	14 50	13 24	22 15
09	16 54	21 04	22 14	22 56	12 01	23 24	15 59	14 50	13 24	22 14
10	17 11	16 22	22 37	23 09	11 50	23 24	16 01	14 50	13 23	22 14
11	17 28	10 50	22 59	23 20	11 40	23 24	16 03	14 49	13 23	22 14
12	17 44	04 S 37	23 19	23 31	11 29	23 23	16 05	14 49	13 22	22 14
13	18 00	02 N 01	23 38	23 42	11 19	23 23	16 06	14 49	13 22	22 13
14	18 16	08 46	23 56	23 51	11 08	23 23	16 08	14 49	13 21	22 13
15	18 31	15 15	24 13	24 00	10 58	23 23	16 10	14 48	13 21	22 13
16	18 46	20 58	24 28	24 08	10 48	23 23	16 12	14 48	13 20	22 13
17	19 01	25 22	24 43	24 16	10 37	23 22	16 14	14 48	13 20	22 12
18	19 15	27 57	24 56	24 23	10 27	23 22	16 16	14 48	13 19	22 12
19	19 30	28 25	25 07	24 29	10 17	23 22	16 18	14 48	13 19	22 12
20	19 43	26 50	25 18	24 34	10 06	23 21	16 19	14 48	13 18	22 12
21	19 57	23 32	25 26	24 39	09 56	23 21	16 21	14 48	13 18	22 11
22	20 10	18 58	25 34	24 43	09 46	23 21	16 23	14 48	13 17	22 11
23	20 22	13 35	25 40	24 46	09 36	23 20	16 25	14 48	13 17	22 11
24	20 34	07 47	25 45	24 48	09 26	23 20	16 27	14 48	13 17	22 10
25	20 46	01 N 49	25 49	24 50	09 16	23 19	16 29	14 47	13 16	22 10
26	20 58	04 S 06	25 51	24 51	09 06	23 19	16 30	14 48	13 16	22 10
27	21 09	09 45	25 51	24 51	08 56	23 18	16 32	14 48	13 15	22 09
28	21 20	15 00	25 50	24 50	08 46	23 18	16 34	14 48	13 15	22 09
29	21 30	19 38	25 48	24 49	08 36	23 17	16 36	14 48	13 14	22 09
30	21 S 40	23 S 28	25 S 45	24 S 47	08 N 26	23 S 17	16 S 37	14 N 48	13 N 14	22 S 08

ZODIAC SIGN ENTRIES

Date	h m	Planets
01	01 26	☽ ♏
03	13 40	☽ ♐
03	19 22	♂ ♍
06	02 24	☽ ♑
08	06 25	☿ ♐
08	14 21	☽ ♒
10	23 45	☽ ♓
13	05 20	☽ ♈
15	07 09	☽ ♉
17	06 37	☽ ♊
19	05 51	☽ ♋
21	07 01	☽ ♌
22	10 35	☉ ♐
23	11 40	☽ ♍
24	11 34	♀ ♑
25	20 08	☽ ♎
28	07 21	☽ ♏
30	19 50	☽ ♐
30	20 54	☿ ♑

LATITUDES

Date	Mercury ☿ ° ′	Venus ♀ ° ′	Mars ♂ ° ′	Jupiter ♃ ° ′	Saturn ♄ ° ′	Uranus ♅ ° ′	Neptune ♆ ° ′	Pluto ♇ ° ′
01	00 S 58	00 S 27	01 N 37	00 N 02	02 N 01	00 N 41	01 S 50	11 S 13
04	01 17	00 35	01 40	00 01	02 01	00 42	01 50	11 12
07	01 34	00 42	01 43	00 01	02 01	00 42	01 50	11 11
10	01 50	00 50	01 45	00 N 01	02 01	00 42	01 50	11 11
13	02 04	00 57	01 48	00 00	02 01	00 42	01 50	11 10
16	02 16	01 05	01 51	00 00	02 00	00 42	01 49	11 09
19	02 25	01 11	01 54	00 00	02 00	00 42	01 49	11 08
22	02 30	01 18	01 57	00 S 01	02 00	00 42	01 49	11 08
25	02 31	01 24	02 01	00 01	02 00	00 43	01 49	11 07
28	02 26	01 30	02 04	00 01	02 00	00 43	01 49	11 06
31	02 S 13	01 S 35	02 N 07	00 S 01	02 N 00	00 N 43	01 S 49	11 S 06

DATA

Julian Date	2467555
Delta T	+78 seconds
Ayanamsa	24° 28′ 09″
Synetic vernal point	04° ♓ 38′ 50″
True obliquity of ecliptic	23° 26′ 10″

MOON'S PHASES, APSIDES AND POSITIONS ☽

Date	h m	Phase	Longitude ° ′	Eclipse Indicator
01	19 57	●	09 ♏ 17	
10	00 13	☽	17 ♒ 29	
16	21 52	○	24 ♉ 25	
23	13 46	☾	01 ♍ 09	

Day	h m	
04	13 13	Apogee
17	09 15	Perigee

Day	h m		
06	01 54	Max dec	28° S 34′
13	04 50	ON	
19	05 16	Max dec	28° N 30′
25	19 17	OS	

LONGITUDES

Date	Chiron ⚷ ° ′	Ceres ⚳ ° ′	Pallas ⚴ ° ′	Juno ⚵ ° ′	Vesta ⚶ ° ′	Black Moon Lilith ⚸ ° ′
01	00 ♍ 44	28 ♒ 04	17 ♑ 22	05 ♑ 06	10 ♐ 32	06 ♐ 51
11	01 ♍ 25	29 ♒ 11	20 ♑ 01	08 ♑ 18	15 ♐ 48	07 ♐ 58
21	01 ♍ 58	00 ♓ 46	22 ♑ 52	11 ♑ 42	21 ♐ 06	09 ♐ 05
31	02 ♍ 20	02 ♓ 48	25 ♑ 52	15 ♑ 16	26 ♐ 26	10 ♐ 12

All ephemeris data is given at 12.00 UT and the Moon's longitude is additionally given for 24.00 UT

ASPECTARIAN

h m	Aspects
01 Sunday	
01 25	♀ ⊻ ♃
03 22	☽ ∥ ☉
03 25	☽ ⚹ ♃
03 36	☽ ⊻ ♀
05 48	☽ # ♅
05 52	♂ ∥ ♆
09 06	☽ Q ♅
09 51	☽ ∥ ♄
11 44	☿ ☌ ♄
19 57	☽ ☌ ☉
23 25	☽ Q ♂
23 53	☽ ☍ ♆
02 Monday	
02 07	☽ ∥ ☿
02 46	♂ ☍ ♇
09 53	☽ ∠ ♃
15 18	☽ ∥ ♀
17 14	☽ ☌ ♄
17 29	☉ # ♅
21 15	☽ ☌ ☿
21 16	☽ □ ♅
21 21	☿ □ ♅
21 39	☽ ∥ ♇
03 Tuesday	
04 18	☽ ⊥ ♃
05 19	☽ ∥ ♃
11 48	☽ □ ♇
13 23	☽ □ ♂
16 37	☽ ⊻ ♃
17 41	☉ ☍ ♆
22 54	☽ ☌ ♀
04 Wednesday	
12 16	☽ ⊼ ♆
14 02	☽ ⊻ ☉
05 Thursday	
00 14	☽ Q ♇
00 24	☽ ± ♆
03 19	☽ ⊥ ☉
06 31	☽ ⊻ ♄
09 41	☉ Q ♂
10 05	☽ △ ♅
11 04	☿ ⊥ ♃
18 35	☽ bQ ♆
18 48	☽ ⊥ ♄
18 53	☽ ⊻ ☿
23 12	☉ ∥ ♄
23 13	☽ ∠ ☉
06 Friday	
00 31	☽ ⚹ ♇
05 00	☽ △ ♂
06 19	☽ ☌ ♃
08 46	☽ ⊥ ☿
11 44	♀ ∥ ♇
13 09	☽ ∠ ♄
16 26	☽ bQ ♅
18 41	☽ ⊻ ♀
07 Saturday	
00 45	☽ △ ♆
05 28	☽ ∠ ☿
06 43	☽ ∠ ♇
08 07	☽ ⊥ ♀
08 11	☽ ⚹ ☉
10 32	☽ ± ♅
12 33	☽ bQ ♂
15 24	☿ □ ♇
19 31	☽ ⚹ ♄
22 30	☽ ⊼ ♅
08 Sunday	
00 39	☽ ⊥ ♇
01 54	♂ △ ♃
02 33	♆ Q ♇
04 05	☽ ∠ ♀
07 16	☽ ± ♂
10 12	☽ Q ☉
12 31	☽ ⊻ ♇
15 27	☽ ⚹ ☿
19 06	☽ ⊻ ♃
19 27	☽ Q ♄
19 36	☽ ⊼ ♂
21 33	☽ ∥ ♃
09 Monday	
01 14	☽ ∥ ♀
04 02	♀ ⊼ ♆
05 01	☽ ∥ ♇
05 38	☽ ∥ ☿
06 59	☽ ⊥ ♃
11 50	☿ ∥ ♇
11 50	☽ □ ♆
12 43	☽ ⚹ ♀
17 56	☽ Q ☿
10 Tuesday	
00 13	☽ □ ☉
00 33	☽ ∠ ♃
01 31	☿ ⊻ ♃
05 22	☉ ∠ ♃
06 29	☽ □ ♄
08 22	☽ ∥ ☉
08 50	☽ ☍ ♅
13 39	☽ ∥ ♄
14 03	☽ Q ♀
18 16	☿ □ ♂
19 03	☽ # ♅
21 53	☽ Q ♆
22 01	☽ ☌ ♇
11 Wednesday	
01 20	☽ # ♆
05 05	☽ ⚹ ♃
07 05	☽ ☍ ♂
08 08	☽ □ ☿
08 28	☽ # ♂
19 12	♀ ∥ ♃
19 48	☽ ⚹ ♆
12 Thursday	
02 28	☽ □ ♀
03 10	☽ Q ♃
11 27	♇ St D
11 53	☽ △ ☉
13 46	☽ △ ♄
15 32	☽ ⊼ ♅
17 35	☿ ∥ ♃
22 18	☽ ∠ ♆
13 Friday	
02 00	☽ ± ♅
03 43	☽ ⊻ ♇
11 01	☽ □ ♃
13 43	☽ ⊥ ♆
13 58	☽ ⊥ ♇
14 08	☽ ⊼ ♂
15 49	☽ bQ ☉
15 52	☽ bQ ♄
16 40	☉ ☌ ♄
17 22	☽ bQ ♅
19 09	☽ △ ☿
21 40	♀ ± ♆
22 09	☿ ∥ ♀
23 49	☽ ⊻ ♆
14 Saturday	
00 37	☽ ± ♂
00 44	♀ Q ♇
05 05	☽ ∠ ♇
07 04	☽ ± ♄
08 01	☽ ± ☉
10 55	☽ △ ♀
14 07	☉ □ ♅
16 04	☽ bQ ♂
17 02	☽ ⊼ ♄
18 18	☽ △ ♅
18 36	☽ ⊼ ☉
20 22	☽ ∥ ♂
22 41	☽ bQ ☿
15 Sunday	
04 45	☽ ∥ ♆
05 38	☽ ⚹ ♇
10 17	☽ ∥ ♅
13 09	☽ △ ♃
13 30	☽ bQ ♀
14 44	☽ ± ☿
15 36	☽ # ♄
16 51	☿ ⊼ ♆
17 10	☽ △ ♂
16 Monday	
00 28	☽ ☌ ♆
00 49	☽ Q ♇
01 17	☽ ⊼ ☿
01 47	☽ # ☉
05 06	☽ ± ♀
13 16	☽ bQ ♃
15 24	☽ ⊼ ♀
17 21	☽ ☍ ♄
17 58	☽ # ♇
18 14	☽ □ ♅
21 52	☽ ☍ ☉
17 Tuesday	
00 01	☽ # ♃
03 35	☽ ± ♃
04 51	☽ # ♀
05 09	☽ □ ♇
07 28	☽ # ☿
13 07	☽ ⊼ ♃
18 01	☽ □ ♂
18 01	♀ ⊻ ♄
22 28	☽ Q ♅
23 31	☽ ⊻ ♆
18 Wednesday	
02 46	♀ △ ♅
05 11	☽ ☍ ☿
08 55	☽ ⊥ ♆
16 48	☽ ⊼ ♄
17 21	☽ ⚹ ♅
18 38	☽ ☍ ♀
23 07	☽ ∠ ♆
23 19	☽ Q ♂
19 Thursday	
00 21	☽ ⊼ ☉
02 24	☽ ± ♄
04 22	☽ △ ♇
10 36	☽ ± ☉
13 09	☽ ☍ ♃
16 55	☽ bQ ♄
17 18	☽ ∠ ♅
19 11	☽ ⚹ ♂
21 53	☿ ± ♆
23 08	☽ ⚹ ♆
20 Friday	
02 12	☽ bQ ☉
03 45	☿ Q ♇
04 33	☽ bQ ♇
07 56	☽ ⊥ ♅
10 04	☽ ⊼ ☿
17 39	☽ △ ♄
17 51	☽ ⊻ ♅
18 48	☽ Q ♆
19 24	☽ ± ♇
20 40	☽ ∠ ♂
21 01	☽ ± ☿
23 45	☽ ⊼ ♀
23 59	☽ # ☿
21 Saturday	
00 23	♀ bQ ♆
04 54	☽ △ ☉
05 06	☽ # ♀
05 27	☽ ⊼ ♇
09 19	☉ ⊥ ♃
10 50	☽ ± ♀
12 23	☽ ⊥ ♂
12 43	☉ □ ♇
13 02	☽ # ♃
13 48	☽ bQ ☿
15 30	☽ ⊼ ♃
18 14	♄ □ ♅
19 38	☽ # ♇
23 03	☽ ⊻ ♂
22 Sunday	
01 19	☽ □ ♆
02 01	☽ ± ♃
03 50	☽ bQ ♀
06 25	☽ # ☉
18 05	☽ bQ ♃
18 43	☽ △ ☿
21 31	☽ ☌ ♅
21 45	☽ □ ♄
23 46	☽ # ♄
23 Monday	
02 45	♀ ⊥ ♄
06 50	☽ ∥ ♅
09 12	☽ △ ♀
10 01	☽ ☍ ♇
13 18	☽ ∥ ♆
13 46	☽ □ ☉
17 54	♀ ⚹ ♇
21 41	☽ △ ♃
23 57	☿ △ ♅
24 Tuesday	
02 35	♂ Q ♄
05 05	☽ ∥ ♂
05 16	☿ ⊻ ♄
06 56	☽ Q ♄
07 04	☽ ☌ ♂
07 16	☽ △ ♆
12 11	♂ △ ♆
25 Wednesday	
02 14	☽ Q ☉
05 03	☽ ⊻ ♅
05 49	☽ ⚹ ♄
08 17	☽ □ ☿
11 38	☽ bQ ♆
16 35	☽ ⊥ ♅
18 24	☽ ⊼ ♇
23 45	☽ □ ♀
26 Thursday	
03 24	☽ ⚹ ☉
05 02	☽ ± ♆
06 03	☽ ± ♇
07 45	☽ □ ♃
10 05	☽ ∠ ♅
11 09	☽ ∠ ♄
16 45	☽ ⊼ ♆
18 17	☿ bQ ♆
19 01	☽ ⊻ ♂
23 08	☽ Q ☿
23 44	☽ bQ ♇
27 Friday	
05 03	☽ ⊥ ♄
07 22	☽ ⊥ ♂
08 30	☽ # ♂
11 36	☽ ∠ ☉
15 11	☽ Q ♀
15 38	♅ St R
15 43	☽ ⚹ ♅
17 05	☽ ⊻ ♄
19 57	☽ Q ♃
28 Saturday	
01 29	☽ ⚹ ☿
02 01	☽ ∠ ♂
03 43	☽ # ♆
05 35	☽ △ ♇
07 13	☽ ⊥ ☉
11 01	☽ # ♅
15 42	☽ Q ♅
17 57	☽ ⚹ ♀
19 47	☽ ∥ ♄
20 22	☽ ⊻ ☉
20 26	☽ ⚹ ♃
21 19	☉ ⊻ ♃
29 Sunday	
04 27	☽ ☍ ♆
09 22	☽ ⚹ ♂
10 38	☽ ∠ ☿
22 29	♀ ☌ ♃
23 30	☽ ∥ ☉
23 30	☿ ⊥ ♄
23 44	☿ ⚹ ♇
30 Monday	
03 02	☽ ∥ ♇
03 14	☽ ∠ ♃
03 41	☽ ∠ ♀
03 59	☽ □ ♅
05 59	☽ ☌ ♄
06 31	☽ ⊥ ☿
06 32	♀ bQ ♅
10 36	☽ Q ♂
10 41	☽ ∥ ♃
18 05	☽ □ ♇
19 45	☽ ⊻ ☿
21 46	☽ ⊥ ♃
21 48	☽ ∥ ♀

DECEMBER 2043

Date	Sidereal time h m s	Sun ☉ ° ′ ″	Moon ☽ ° ′ ″	Moon ☽ 24.00 ° ′ ″	Mercury ☿ ° ′	Venus ♀ ° ′	Mars ♂ ° ′	Jupiter ♃ ° ′	Saturn ♄ ° ′	Uranus ♅ ° ′	Neptune ♆ ° ′	Pluto ♇ ° ′
01	16 40 59	09 ♐ 10 03	07 ♐ 59 09	13 ♐ 55 00	00 ♑ 35	08 ♑ 41	13 ♍ 55	07 ♑ 05	23 ♏ 18	22 ♌ 09	10 ♉ 27	29 ♒ 08
02	16 44 55	10 10 54	19 ♐ 51 04	25 ♐ 47 31	01 28	09 55	14 23	07 18	23 25	22 R 09	10 R 25	29 09
03	16 48 52	11 11 45	01 ♑ 44 30	07 ♑ 42 14	02 15	11 09	14 50	07 31	23 32	22 09	10 24	29 10
04	16 52 48	12 12 38	13 ♑ 40 55	19 ♑ 40 47	02 56	12 23	15 17	07 44	23 39	22 09	10 23	29 10
05	16 56 45	13 13 31	25 ♑ 42 07	01 ♒ 45 15	03 30	13 37	15 43	07 58	23 46	22 08	10 21	29 11
06	17 00 41	14 14 26	07 ♒ 50 32	13 ♒ 58 23	03 57	14 51	16 10	08 11	23 53	22 08	10 20	29 11
07	17 04 38	15 15 21	20 ♒ 09 16	26 ♒ 23 40	04 15	16 05	16 36	08 24	23 59	22 07	10 19	29 12
08	17 08 35	16 16 17	02 ♓ 42 06	09 ♓ 05 05	04 24	17 19	17 02	08 37	24 06	22 07	10 17	29 13
09	17 12 31	17 17 13	15 ♓ 33 10	22 ♓ 06 54	04 R 23	18 33	17 28	08 51	24 13	22 06	10 16	29 13
10	17 16 28	18 18 11	28 ♓ 46 44	05 ♈ 33 06	04 12	19 47	17 53	09 04	24 20	22 05	10 15	29 14
11	17 20 24	19 19 08	12 ♈ 26 21	19 ♈ 26 41	03 49	21 01	18 19	09 18	24 27	22 05	10 13	29 15
12	17 24 21	20 20 07	26 ♈ 34 08	03 ♉ 48 35	03 14	22 15	18 44	09 31	24 33	22 04	10 12	29 16
13	17 28 17	21 21 06	11 ♉ 09 40	18 ♉ 36 46	02 29	23 29	19 08	09 45	24 40	22 03	10 11	29 16
14	17 32 14	22 22 06	26 ♉ 09 04	03 ♊ 45 30	01 32	24 43	19 33	09 58	24 47	22 02	10 10	29 17
15	17 36 10	23 23 06	11 ♊ 24 48	19 ♊ 05 34	00 ♑ 26	25 57	19 57	10 12	24 53	22 01	10 09	29 18
16	17 40 07	24 24 07	26 ♊ 46 16	04 ♋ 25 24	29 ♐ 12	27 11	20 20	10 25	25 00	22 00	10 08	29 19
17	17 44 04	25 25 09	12 ♋ 01 30	19 ♋ 33 15	27 53	28 24	20 44	10 39	25 06	21 59	10 07	29 20
18	17 48 00	26 26 12	26 ♋ 59 30	04 ♌ 19 21	26 30	29 ♑ 38	21 07	10 53	25 13	21 58	10 06	29 21
19	17 51 57	27 27 15	11 ♌ 32 07	18 ♌ 37 23	25 07	00 ♒ 52	21 30	11 06	25 19	21 57	10 04	29 22
20	17 55 53	28 28 20	25 ♌ 34 57	02 ♍ 24 51	23 47	02 05	21 53	11 20	25 26	21 56	10 03	29 22
21	17 59 50	29 ♐ 29 24	09 ♍ 07 18	15 ♍ 42 38	22 32	03 19	22 15	11 34	25 32	21 54	10 03	29 23
22	18 03 46	00 ♑ 30 30	22 ♍ 11 19	28 ♍ 33 55	21 24	04 33	22 37	11 48	25 39	21 53	10 02	29 24
23	18 07 43	01 31 37	04 ♎ 51 03	11 ♎ 03 20	20 25	05 46	22 58	12 01	25 45	21 52	10 01	29 25
24	18 11 39	02 32 44	17 ♎ 11 25	23 ♎ 15 58	19 36	07 00	23 19	12 15	25 51	21 50	10 00	29 26
25	18 15 36	03 33 52	29 ♎ 17 36	05 ♏ 16 54	18 58	08 13	23 40	12 29	25 57	21 49	09 59	29 28
26	18 19 33	04 35 00	11 ♏ 14 26	17 ♏ 10 43	18 30	09 27	24 01	12 43	26 03	21 47	09 58	29 29
27	18 23 29	05 36 09	23 ♏ 06 13	29 ♏ 01 20	18 13	10 40	24 21	12 57	26 10	21 46	09 57	29 30
28	18 27 26	06 37 19	04 ♐ 56 26	10 ♐ 51 51	18 07	11 54	24 40	13 10	26 16	21 44	09 56	29 31
29	18 31 22	07 38 29	16 ♐ 47 49	22 ♐ 44 35	18 D 09	13 07	25 00	13 24	26 22	21 43	09 56	29 32
30	18 35 19	08 39 40	28 ♐ 42 20	04 ♑ 41 13	18 21	14 20	25 19	13 38	26 28	21 41	09 55	29 33
31	18 39 15	09 ♑ 40 50	10 ♑ 41 22	16 ♑ 42 54	18 ♐ 41	15 ♒ 34	25 ♍ 37	13 ♑ 52	26 ♏ 33	21 ♌ 39	09 ♉ 54	29 ♒ 34

Date	Moon True ☊ ° ′	Moon Mean ☊ ° ′	Moon ☽ Latitude ° ′
01	26 ♓ 45	25 ♓ 40	04 S 44
02	26 R 30	25 37	04 58
03	26 16	25 34	04 59
04	26 03	25 30	04 46
05	25 52	25 27	04 21
06	25 44	25 24	03 44
07	25 39	25 21	02 55
08	25 37	25 18	01 58
09	25 D 36	25 15	00 S 53
10	25 R 36	25 11	00 N 17
11	25 35	25 08	01 28
12	25 33	25 05	02 36
13	25 28	25 02	03 35
14	25 19	24 59	04 22
15	25 09	24 56	04 51
16	24 57	24 52	05 00
17	24 46	24 49	04 46
18	24 36	24 46	04 14
19	24 28	24 43	03 25
20	24 23	24 40	02 26
21	24 21	24 36	01 19
22	24 D 20	24 33	00 N 11
23	24 R 20	24 30	00 S 55
24	24 20	24 27	01 58
25	24 17	24 24	02 53
26	24 12	24 21	03 41
27	24 04	24 17	04 18
28	23 53	24 14	04 44
29	23 39	24 11	04 58
30	23 25	24 08	04 59
31	23 ♓ 11	24 ♓ 05	04 S 46

DECLINATIONS

Date	Sun ☉ ° ′	Moon ☽ ° ′	Mercury ☿ ° ′	Venus ♀ ° ′	Mars ♂ ° ′	Jupiter ♃ ° ′	Saturn ♄ ° ′	Uranus ♅ ° ′	Neptune ♆ ° ′	Pluto ♇ ° ′
01	21 S 49	26 S 19	25 S 40	24 S 44	08 N 17	23 S 16	16 S 39	14 N 48	13 N 13	22 S 08
02	21 58	28 00	25 33	24 40	08 07	23 16	16 41	14 48	13 13	22 07
03	22 07	28 24	25 25	24 36	07 58	23 15	16 42	14 48	13 13	22 07
04	22 15	27 29	25 16	24 31	07 48	23 14	16 44	14 48	13 12	22 07
05	22 23	25 16	25 06	24 25	07 39	23 14	16 46	14 48	13 12	22 06
06	22 30	21 54	24 55	24 18	07 30	23 13	16 47	14 49	13 11	22 06
07	22 37	17 32	24 42	24 11	07 21	23 12	16 49	14 49	13 11	22 05
08	22 44	12 21	24 29	24 03	07 12	23 11	16 51	14 49	13 11	22 05
09	22 50	06 30	24 14	23 54	07 03	23 11	16 52	14 49	13 10	22 04
10	22 55	00 S 14	23 58	23 44	06 54	23 10	16 54	14 50	13 10	22 04
11	23 00	06 N 16	23 41	23 34	06 45	23 09	16 56	14 50	13 10	22 03
12	23 05	12 40	23 24	23 23	06 36	23 08	16 57	14 50	13 09	22 03
13	23 09	18 35	23 05	23 12	06 28	23 07	16 59	14 50	13 09	22 03
14	23 13	23 32	22 46	23 00	06 19	23 06	17 00	14 51	13 09	22 02
15	23 16	26 57	22 26	22 47	06 11	23 05	17 02	14 51	13 08	22 02
16	23 19	28 23	22 06	22 33	06 03	23 04	17 03	14 51	13 08	22 01
17	23 21	27 39	21 46	22 19	05 55	23 03	17 05	14 52	13 08	22 01
18	23 23	24 54	21 27	22 04	05 47	23 02	17 06	14 52	13 07	22 00
19	23 25	20 36	21 08	21 48	05 39	23 01	17 08	14 53	13 07	21 59
20	23 26	15 16	20 51	21 32	05 31	23 00	17 09	14 53	13 07	21 59
21	23 26	09 23	20 35	21 15	05 23	22 59	17 11	14 54	13 07	21 58
22	23 26	03 N 16	20 22	20 58	05 16	22 58	17 12	14 54	13 06	21 58
23	23 26	02 S 46	20 11	20 40	05 09	22 57	17 13	14 54	13 06	21 57
24	23 25	08 34	20 02	20 22	05 01	22 56	17 15	14 55	13 06	21 57
25	23 23	13 55	19 57	20 03	04 54	22 55	17 16	14 55	13 06	21 56
26	23 21	18 42	19 54	19 43	04 47	22 53	17 18	14 56	13 06	21 56
27	23 19	22 42	19 54	19 23	04 41	22 52	17 19	14 56	13 05	21 55
28	23 16	25 46	19 56	19 02	04 34	22 51	17 20	14 57	13 05	21 55
29	23 13	27 43	20 00	18 41	04 28	22 50	17 21	14 58	13 05	21 54
30	23 09	28 24	20 06	18 19	04 22	22 48	17 23	14 58	13 05	21 53
31	23 S 05	27 S 46	20 S 14	17 S 57	04 N 15	22 S 47	17 S 24	14 N 59	13 N 05	21 S 53

ZODIAC SIGN ENTRIES

Date	h	m	Planets
03	08	29	☽ ♑
05	20	32	☽ ♒
08	06	53	☽ ♓
10	14	11	☽ ♈
12	17	43	☽ ♉
14	18	05	☽ ♊
15	20	44	☿ ♐
16	17	03	☽ ♋
18	16	54	☽ ♌
18	19	09	♀ ♒
20	19	44	☽ ♍
22	00	01	☉ ♑
23	02	43	☽ ♎
25	13	25	☽ ♏
28	01	59	☽ ♐
30	14	36	☽ ♑

LATITUDES

Date	Mercury ☿ ° ′	Venus ♀ ° ′	Mars ♂ ° ′	Jupiter ♃ ° ′	Saturn ♄ ° ′	Uranus ♅ ° ′	Neptune ♆ ° ′	Pluto ♇ ° ′
01	02 S 13	01 S 35	02 N 07	00 S 01	02 N 00	00 N 43	01 S 49	11 S 06
04	01 52	01 39	02 11	00 02	02 01	00 43	01 49	11 05
07	01 20	01 44	02 14	00 02	02 01	00 43	01 49	11 04
10	00 S 36	01 47	02 17	00 02	02 01	00 43	01 49	11 04
13	00 N 19	01 50	02 21	00 03	02 01	00 43	01 49	11 03
16	01 20	01 52	02 25	00 03	02 01	00 43	01 49	11 02
19	02 13	01 53	02 29	00 03	02 01	00 44	01 48	11 02
22	02 48	01 54	02 32	00 03	02 01	00 44	01 48	11 01
25	03 02	01 54	02 36	00 04	02 02	00 44	01 48	11 01
28	02 59	01 53	02 40	00 04	02 02	00 44	01 48	11 00
31	02 N 44	01 S 51	02 N 45	00 S 04	02 N 02	00 N 44	01 S 48	10 S 59

DATA

Julian Date	2467585
Delta T	+78 seconds
Ayanamsa	24° 28' 14"
Synetic vernal point	04° ♓ 38' 44"
True obliquity of ecliptic	23° 26' 09"

MOON'S PHASES, APSIDES AND POSITIONS ☽

Date	h	m	Phase	Longitude ° ′	Eclipse Indicator
01	14	37	●	09 ♐ 17	
09	15	27	☽	17 ♓ 26	
16	08	02	○	24 ♊ 14	
23	05	04	☾	01 ♎ 14	
31	09	48	●	09 ♑ 35	

Day	h	m		
01	17	25	Apogee	
15	22	08	Perigee	
28	19	47	Apogee	
03	07	17	Max dec	28° S 26'
10	12	51	0N	
16	15	47	Max dec	28° N 25'
23	00	54	0S	
30	12	28	Max dec	28° S 24'

LONGITUDES

Date	Chiron ⚷ ° ′	Ceres ⚳ ° ′	Pallas ⚴ ° ′	Juno ⚵ ° ′	Vesta ⚶ ° ′	Black Moon Lilith ⚸ ° ′
01	02 ♍ 20	02 ♓ 48	25 ♑ 52	15 ♑ 16	26 ♐ 26	10 ♐ 12
11	02 ♍ 31	05 ♓ 11	29 ♑ 01	19 ♑ 00	01 ♑ 47	11 ♐ 19
21	02 ♍ 30	07 ♓ 54	02 ♒ 15	22 ♑ 51	07 ♑ 09	12 ♐ 27
31	02 ♍ 19	10 ♓ 52	05 ♒ 35	26 ♑ 48	12 ♑ 32	13 ♐ 34

ASPECTARIAN

h m	Aspects
01 Tuesday	
00 01	☽ ⊥ ♀
03 50	♀ ∠ ♄
05 47	☽ ∥ ☿
10 09	☽ ⊻ ♃
13 34	☽ ⊻ ♀
14 37	☽ ☌ ☉
16 58	☽ ⊼ ♆
19 51	♃ ₽ ♅
02 Wednesday	
00 29	☽ □ ♂
05 05	☽ ± ♆
06 32	☽ Q ♇
16 39	☽ △ ♅
17 35	☉ ⊼ ♆
19 16	☽ ⊻ ♄
21 42	♀ △ ♆
23 14	☽ ₽ ♆
03 Thursday	
06 47	☽ ⚹ ♇
07 30	☽ ⊥ ♄
12 00	☉ ∥ ♇
13 04	☽ ☌ ☿
17 04	☉ ⊻ ♀
22 53	☽ ₽ ♅
23 51	☽ ☌ ♃
04 Friday	
01 48	☽ ∠ ♄
05 23	☽ △ ♆
08 47	☽ ⊻ ☉
09 06	☽ ☌ ♀
12 59	☽ ∠ ♇
15 19	☽ △ ♂
16 56	☽ ± ♅
21 54	☽ ⊥ ☉
05 Saturday	
04 54	☽ ⊼ ♅
06 58	☽ ⊥ ♇
08 06	☽ ⚹ ♄
13 31	☽ ∥ ☿
17 28	☽ ∠ ☉
18 54	☽ ⊻ ♇
19 11	☽ ∥ ♀
22 20	☽ ₽ ♂
22 57	♀ ∠ ♇
06 Sunday	
03 31	☽ ∥ ♃
04 04	☽ ⊻ ☿
08 06	☽ Q ♄
08 21	☽ ∥ ☉
10 49	☽ ∥ ♇
12 41	☽ ⊻ ♃
16 15	☽ ⊥ ☿
16 44	☽ ± ♂
16 52	☽ □ ♆
07 Monday	
00 38	☽ ⊥ ♃
01 39	☽ ⚹ ☉
03 14	☽ ⊻ ♀
04 52	☽ ⊼ ♂
10 13	☽ ∠ ☿
12 36	♀ ± ♅
15 31	☽ ∥ ♄
15 47	☽ ☍ ♅
16 08	☽ ⊥ ♀
18 22	☽ ∠ ♃
19 28	☽ □ ♄
08 Tuesday	
00 59	☽ # ♅
02 51	☽ Q ☉
03 28	♀ △ ♂
03 38	☽ Q ♆
05 22	☽ ☌ ♇
08 21	☽ # ♆
11 12	☽ ∠ ♀
12 22	☉ ± ♆
15 14	☽ ⚹ ☿
21 56	☿ St R
23 20	☽ ⚹ ♃
09 Wednesday	
02 13	☽ ⚹ ♆
09 50	☽ # ♂
10 26	☉ Q ♇
13 31	☽ Q ☿
15 27	☽ □ ☉
15 38	☽ ☍ ♂
18 05	☽ ⚹ ♀
19 15	☉ □ ♂
21 52	☽ Q ♃
23 58	☽ ⊼ ♅
10 Thursday	
03 57	☽ △ ♄
05 40	☽ ∠ ♆
10 46	☽ ± ♅
12 49	☽ ⊻ ♇
17 54	☽ Q ♀
21 23	☽ □ ☿
21 41	☽ ⊥ ♆
23 27	☽ ⊥ ♇
11 Friday	
02 41	☽ ₽ ♅
06 27	☽ □ ♃
06 46	☽ ₽ ♄
08 10	☽ ⊻ ♆
13 46	☽ ∥ ♂
15 07	☽ ∠ ♇
22 23	☽ ⊼ ♂
22 23	☽ ± ♄
12 Saturday	
00 42	☽ △ ☉
04 04	☽ □ ♀
04 26	☽ △ ♅
08 20	♀ ⊼ ♅
08 36	☽ ⊼ ♄
08 49	☽ ± ♂
13 06	☿ ∥ ♀
13 54	☽ ∥ ♆
16 29	☽ ⚹ ♇
19 53	♃ ∠ ♄
20 29	☽ ∥ ♅
22 33	☽ △ ☿
13 Sunday	
00 12	☽ ₽ ♂
02 23	☉ ∥ ♃
03 35	☽ ₽ ☉
05 09	☽ # ♄
07 49	♀ ⊥ ♇
07 52	☉ ∥ ☿
09 24	☿ ∥ ♃
09 40	☽ △ ♃
10 25	☽ ☌ ♆
12 11	☽ Q ♇
15 57	☉ ∥ ♀
19 16	☽ ± ☉
21 36	☽ ₽ ☿
21 54	♀ ∥ ♃
14 Monday	
01 12	☽ △ ♂
04 03	☽ # ♇
04 12	☉ △ ♅
05 28	☽ □ ♅
05 34	☽ ⊼ ☉
08 07	☽ # ☿
09 10	☽ # ♀
09 31	☽ △ ♀
09 39	☽ # ♃
09 48	☽ ☍ ♄
10 06	☽ ₽ ♃
10 13	☽ # ☉
11 05	☽ ± ☿
13 21	♀ ⚹ ♄
16 58	☽ □ ♇
19 57	☽ ⊼ ☿
15 Tuesday	
00 31	☽ ± ♃
03 20	☿ ⊥ ♄
07 11	♃ △ ♆
09 49	☽ Q ♅
10 01	☽ ⊻ ♆
10 04	☽ ⊼ ♃
11 12	☽ ₽ ♀
19 23	☽ ⊥ ♆
16 Wednesday	
01 41	☽ □ ♂
02 30	☽ ± ♀
04 33	☽ ⚹ ♅
08 02	☽ ☍ ☉
09 12	☽ ⊼ ♄
09 26	☽ ∠ ♆
09 58	☿ ⚹ ♇
12 41	☽ ⊼ ♀
15 31	☽ ☍ ☿
15 59	☽ △ ♇
18 40	☽ ± ♄
18 43	☿ ∥ ♇
17 Thursday	
03 45	☉ ⊻ ♄
04 02	☽ ∠ ♅
04 49	☉ ₽ ♆
06 39	☽ Q ♂
07 07	☿ ⊻ ♀
08 56	☽ ₽ ♄
08 58	☽ ⚹ ♆
09 47	☽ ☍ ♃
15 40	☽ ₽ ♇
18 17	☽ ⊥ ♅
18 Friday	
02 16	☽ ⚹ ♂
03 53	☽ ⊻ ♅
04 05	☽ Q ♆
06 05	☽ ± ♇
06 15	♀ ⊻ ♇
09 06	☽ △ ♄
11 02	☽ ⊼ ☉
11 17	☽ ⊼ ☿
12 40	☉ ☌ ☿
15 50	☽ ⊼ ♇
16 42	☽ ☍ ♀
18 23	♀ ∥ ♇
20 14	☽ ± ☿
21 17	☽ # ☉
21 34	☽ ± ☉
23 21	☽ # ♃
19 Saturday	
03 22	☽ ∠ ♂
04 58	☽ # ♇
05 35	☽ # ♀
08 46	☿ ⊻ ♄
09 12	☽ # ☿
09 33	☽ □ ♆
09 51	☽ ₽ ☿
11 16	☽ ⊼ ♃
12 53	☿ ₽ ♆
13 40	☽ ₽ ☉
14 25	☿ ⊥ ♀
18 52	☽ ⊥ ♂
21 34	☽ ± ♃
20 Sunday	
03 57	☽ # ♄
05 24	☽ ⊻ ♂
05 41	☽ ☌ ♅
09 09	☽ △ ☿
11 44	☽ □ ♄
13 20	☽ ₽ ♃
13 38	☽ ∥ ♅
15 05	♂ ⊻ ♅
17 27	☽ △ ☉
18 38	☽ ☍ ♇
20 58	☽ ∥ ♆
21 Monday	
00 34	☽ ⊼ ♀
09 37	☉ ⚹ ♇
12 23	☽ ± ♀
13 40	☽ △ ♆
16 27	☿ □ ♂
16 30	☽ △ ♃
20 05	☽ Q ♄
22 Tuesday	
01 14	☿ △ ♅
04 01	☽ ∥ ♂
06 34	☽ ₽ ♀
10 39	☽ □ ☿
11 26	☽ ⊻ ♅
12 49	☽ ☌ ♂
17 19	☽ ₽ ♆
18 32	☽ ⚹ ♄
22 41	☽ ⊥ ♅
23 Wednesday	
01 37	☽ ⊼ ♇
05 04	☽ □ ☉
08 04	☿ ∠ ♀
10 23	☽ ± ♆
13 06	☽ ± ♇
13 58	☽ △ ♀
15 52	☽ ∠ ♅
17 47	☉ ⊥ ♄
18 26	☽ Q ☿
21 28	☽ # ♂
21 57	☽ ⊼ ♆
23 30	☽ ∠ ♄
24 Thursday	
02 09	☽ □ ♃
06 36	☽ ₽ ♇
16 29	☽ ⚹ ☿
17 17	☽ ⊥ ♄
19 13	☽ Q ☉
21 09	☽ ⚹ ♅
25 Friday	
00 28	☽ ⊻ ♂
05 17	☽ ⊻ ♄
06 17	♀ Q ♄
08 09	☽ # ♆
12 20	☽ △ ♇
12 47	☽ ⊥ ♂
14 25	☽ Q ♃
16 47	☽ # ♅
20 39	☿ ∥ ♀
20 58	☽ ∠ ☿
21 02	☽ Q ♅
21 21	☽ ⚹ ☉
26 Saturday	
00 13	♀ ₽ ♂
04 33	☽ ∥ ♄
07 22	☽ ∠ ♂
07 58	☽ □ ♀
09 26	☽ ☍ ♆
14 29	☽ ⊥ ☿
15 02	☽ ⚹ ♃
17 15	☽ ∥ ♀
18 41	☽ ∥ ☿
22 06	♀ □ ♆
27 Sunday	
02 18	☽ ⊻ ☿
06 27	☽ ∠ ☉
06 52	☽ ∥ ♇
09 17	☽ □ ♅
13 07	☽ ∥ ♃
14 36	☽ ⚹ ♂
16 10	☽ ∥ ☉
18 15	☽ ☌ ♄
22 00	☽ ∠ ♃
28 Monday	
00 35	☽ Q ♀
00 59	☽ □ ♇
02 25	☽ ⊥ ☉
14 39	☉ ₽ ♅
15 37	☽ Q ♂
15 44	☽ ⊻ ☉
16 37	☽ ⊥ ♃
16 46	☿ St D
22 07	☽ ⊼ ♆
29 Tuesday	
03 42	☽ ⚹ ♀
05 00	☽ ⊻ ♃
07 05	♂ ₽ ♆
10 15	☽ ± ♆
13 29	☽ Q ♇
14 46	☽ ☌ ☿
19 01	♀ ⊻ ♃
21 54	☽ △ ♅
30 Wednesday	
04 23	☽ ₽ ♆
04 59	☽ □ ♂
07 27	☽ ⊻ ♄
13 25	☽ ∠ ♀
13 42	☽ ⚹ ♇
19 36	☽ ⊥ ♄
31 Thursday	
03 57	☽ ₽ ♅
09 29	☽ ⊥ ♀
09 48	☽ ☌ ☉
10 26	☽ △ ♆
13 45	☽ ∠ ♄
17 15	☉ △ ♆
18 28	☽ ☌ ♃
19 45	☽ ∠ ♇
21 52	☽ ± ♅
22 48	☽ ⊻ ♀

All ephemeris data is given at 12.00 UT and the Moon's longitude is additionally given for 24.00 UT

JANUARY 2044

LONGITUDES

Date	Sidereal time h m s	Sun ☉ ° ' "	Moon ☽ ° ' "	Moon ☽ 24.00 ° ' "	Mercury ☿ ° '	Venus ♀ ° '	Mars ♂ ° '	Jupiter ♃ ° '	Saturn ♄ ° '	Uranus ♅ ° '	Neptune ♆ ° '	Pluto ♇ ° '
01	18 43 12	10 ♑ 42 01	22 ♑ 45 57	28 ♑ 50 38	19 ♐ 08	16 ♒ 47	25 ♍ 55	14 ♑ 06	26 ♏ 39	21 ♌ 37	09 ♉ 54	29 ♒ 36
02	18 47 08	11 43 12	04 ♒ 57 05	11 ♒ 05 27	19 42	18 00	26 13	14 20	26 45	21 R 36	09 R 53	29 37
03	18 51 05	12 44 23	17 ♒ 15 56	23 ♒ 28 45	20 21	19 13	26 30	14 34	26 51	21 34	09 53	29 38
04	18 55 02	13 45 33	29 ♒ 44 11	06 ♓ 02 29	21 06	20 26	26 47	14 48	26 56	21 32	09 52	29 39
05	18 58 58	14 46 44	12 ♓ 24 01	18 ♓ 49 08	21 56	21 39	27 03	15 02	27 02	21 30	09 51	29 41
06	19 02 55	15 47 54	25 ♓ 18 14	01 ♈ 51 41	22 50	22 52	27 19	15 16	27 08	21 28	09 51	29 42
07	19 06 51	16 49 04	08 ♈ 29 54	15 ♈ 13 14	23 48	24 05	27 34	15 30	27 13	21 26	09 51	29 43
08	19 10 48	17 50 13	22 ♈ 02 01	28 ♈ 56 28	24 49	25 18	27 49	15 44	27 19	21 24	09 50	29 44
09	19 14 44	18 51 22	05 ♉ 56 45	13 ♉ 02 53	25 53	26 30	28 04	15 57	27 24	21 22	09 50	29 46
10	19 18 41	19 52 30	20 ♉ 14 43	27 ♉ 31 56	26 59	27 43	28 17	16 11	27 29	21 20	09 49	29 47
11	19 22 37	20 53 38	04 ♊ 54 02	12 ♊ 20 18	28 08	28 ♒ 56	28 31	16 25	27 34	21 18	09 49	29 49
12	19 26 34	21 54 46	19 ♊ 49 51	27 ♊ 21 38	29 ♐ 20	00 ♓ 08	28 44	16 39	27 40	21 16	09 49	29 50
13	19 30 31	22 55 53	04 ♋ 54 27	12 ♋ 27 01	00 ♑ 33	01 21	28 56	16 53	27 45	21 13	09 49	29 51
14	19 34 27	23 56 59	19 ♋ 58 05	27 ♋ 26 23	01 48	02 33	29 08	17 07	27 50	21 11	09 48	29 53
15	19 38 24	24 58 06	04 ♌ 50 45	12 ♌ 10 11	03 04	03 45	29 19	17 21	27 55	21 09	09 48	29 54
16	19 42 20	25 59 11	19 ♌ 23 50	26 ♌ 31 04	04 22	04 58	29 29	17 35	27 59	21 07	09 48	29 56
17	19 46 17	27 00 17	03 ♍ 31 28	10 ♍ 24 48	05 41	06 10	29 40	17 49	28 04	21 04	09 48	29 57
18	19 50 13	28 01 22	17 ♍ 11 01	23 ♍ 50 14	07 01	07 22	29 49	18 03	28 09	21 02	09 48	29 ♒ 59
19	19 54 10	29 ♑ 02 26	00 ♎ 22 45	06 ♎ 48 57	08 23	08 34	29 ♍ 58	18 16	28 14	21 00	09 48	00 ♓ 00
20	19 58 06	00 ♒ 03 30	13 ♎ 09 19	19 ♎ 24 24	09 46	09 46	00 ♎ 06	18 30	28 18	20 57	09 D 48	00 02
21	20 02 03	01 04 34	25 ♎ 34 48	01 ♏ 41 09	11 09	10 58	00 14	18 44	28 23	20 55	09 48	00 03
22	20 06 00	02 05 38	07 ♏ 44 06	13 ♏ 44 18	12 34	12 09	00 21	18 58	28 27	20 52	09 48	00 05
23	20 09 56	03 06 41	19 ♏ 42 21	25 ♏ 38 54	13 59	13 21	00 27	19 12	28 32	20 50	09 48	00 06
24	20 13 53	04 07 44	01 ♐ 34 31	07 ♐ 29 44	15 25	14 32	00 33	19 25	28 36	20 48	09 48	00 08
25	20 17 49	05 08 47	13 ♐ 25 03	19 ♐ 20 56	16 52	15 44	00 38	19 39	28 40	20 45	09 48	00 09
26	20 21 46	06 09 49	25 ♐ 17 47	01 ♑ 15 59	18 20	16 55	00 42	19 53	28 44	20 43	09 48	00 11
27	20 25 42	07 10 50	07 ♑ 15 49	13 ♑ 17 33	19 49	18 06	00 46	20 06	28 48	20 40	09 49	00 13
28	20 29 39	08 11 50	19 ♑ 21 23	25 ♑ 27 31	21 18	19 17	00 49	20 20	28 52	20 37	09 49	00 14
29	20 33 35	09 12 50	01 ♒ 36 04	07 ♒ 47 07	22 48	20 28	00 51	20 34	28 56	20 35	09 49	00 16
30	20 37 32	10 13 49	14 ♒ 00 46	20 ♒ 17 05	24 19	21 39	00 52	20 47	29 00	20 32	09 50	00 17
31	20 41 29	11 ♒ 14 46	26 ♒ 36 05	02 ♓ 57 51	25 ♑ 50	22 ♓ 50	00 ♎ 53	21 ♑ 01	29 ♏ 03	20 ♌ 30	09 ♉ 50	00 ♓ 19

Date	Moon True ☊ ° '	Moon Mean ☊ ° '	Moon ☽ Latitude ° '
01	22 ♓ 57	24 ♓ 01	04 S 21
02	22 R 46	23 58	03 44
03	22 38	23 55	02 55
04	22 33	23 52	01 58
05	22 30	23 49	00 S 53
06	22 D 30	23 46	00 N 15
07	22 30	23 42	01 24
08	22 R 31	23 39	02 30
09	22 29	23 36	03 30
10	22 26	23 33	04 18
11	22 20	23 30	04 50
12	22 12	23 27	05 03
13	22 02	23 23	04 56
14	21 53	23 20	04 28
15	21 44	23 17	03 43
16	21 38	23 14	02 43
17	21 34	23 11	01 35
18	21 32	23 07	00 N 23
19	21 D 33	23 04	00 S 47
20	21 34	23 01	01 53
21	21 35	22 58	02 52
22	21 R 35	22 55	03 42
23	21 33	22 52	04 22
24	21 29	22 48	04 49
25	21 23	22 45	05 05
26	21 16	22 42	05 07
27	21 07	22 39	04 56
28	20 58	22 36	04 31
29	20 50	22 33	03 54
30	20 44	22 29	03 05
31	20 ♓ 39	22 ♓ 26	02 S 06

DECLINATIONS

Date	Sun ☉ ° '	Moon ☽ ° '	Mercury ☿ ° '	Venus ♀ ° '	Mars ♂ ° '	Jupiter ♃ ° '	Saturn ♄ ° '	Uranus ♅ ° '	Neptune ♆ ° '	Pluto ♇ ° '
01	23 S 00	25 S 48	20 S 23	17 S 34	04 N 10	22 S 46	17 S 25	14 N 59	13 N 04	21 S 52
02	22 55	22 38	20 33	17 11	04 04	22 44	17 26	15 00	13 04	21 52
03	22 50	18 26	20 44	16 47	03 58	22 43	17 28	15 01	13 04	21 51
04	22 44	13 24	20 55	16 23	03 53	22 41	17 29	15 01	13 04	21 50
05	22 37	07 44	21 07	15 58	03 48	22 40	17 30	15 02	13 04	21 50
06	22 30	01 S 38	21 18	15 33	03 43	22 39	17 31	15 02	13 04	21 49
07	22 23	04 N 39	21 30	15 08	03 38	22 37	17 32	15 03	13 04	21 49
08	22 15	10 54	21 42	14 42	03 34	22 36	17 33	15 04	13 04	21 48
09	22 07	16 48	21 53	14 16	03 29	22 34	17 35	15 05	13 04	21 47
10	21 58	21 56	22 04	13 50	03 25	22 32	17 36	15 05	13 04	21 47
11	21 49	25 52	22 14	13 23	03 21	22 31	17 37	15 06	13 04	21 46
12	21 39	28 05	22 24	12 56	03 17	22 29	17 38	15 07	13 04	21 45
13	21 29	28 17	22 33	12 28	03 14	22 28	17 39	15 07	13 03	21 45
14	21 19	26 23	22 41	12 01	03 11	22 26	17 40	15 08	13 03	21 44
15	21 08	22 39	22 48	11 32	03 07	22 24	17 41	15 09	13 03	21 44
16	20 57	17 35	22 55	11 04	03 05	22 23	17 42	15 10	13 04	21 43
17	20 45	11 41	23 00	10 35	03 02	22 21	17 43	15 10	13 04	21 42
18	20 33	05 N 25	23 05	10 07	03 00	22 19	17 43	15 11	13 04	21 42
19	20 21	00 S 52	23 08	09 37	02 58	22 17	17 44	15 12	13 04	21 41
20	20 08	06 56	23 10	09 08	02 56	22 16	17 45	15 13	13 04	21 40
21	19 55	12 34	23 12	08 39	02 54	22 14	17 46	15 14	13 04	21 40
22	19 41	17 35	23 12	08 09	02 53	22 12	17 47	15 14	13 04	21 39
23	19 28	21 51	23 11	07 39	02 52	22 10	17 48	15 15	13 04	21 39
24	19 13	25 12	23 08	07 09	02 51	22 08	17 48	15 16	13 04	21 38
25	18 59	27 27	23 05	06 38	02 50	22 06	17 49	15 17	13 04	21 37
26	18 44	28 28	23 00	06 08	02 50	22 05	17 50	15 18	13 04	21 37
27	18 28	28 09	22 54	05 37	02 50	22 03	17 51	15 18	13 04	21 36
28	18 13	26 31	22 47	05 06	02 50	22 01	17 51	15 19	13 05	21 35
29	17 57	23 36	22 38	04 36	02 51	21 59	17 52	15 20	13 05	21 35
30	17 41	19 34	22 28	04 05	02 51	21 57	17 53	15 21	13 05	21 34
31	17 S 24	14 S 38	22 S 17	03 S 33	02 N 53	21 S 55	17 S 53	15 N 22	13 N 05	21 S 33

ZODIAC SIGN ENTRIES

Date	h	m	Planets
02	02	17	☽ ♒
04	12	30	☽ ♓
06	20	36	☽ ♈
09	01	49	☽ ♉
11	04	02	☽ ♊
12	09	15	♀ ♓
13	01	22	☿ ♑
13	04	12	☽ ♋
15	04	08	☽ ♌
17	05	56	☽ ♍
19	09	30	♇ ♓
19	11	18	☽ ♎
19	17	59	♂ ♎
20	10	37	☉ ♒
21	20	40	☽ ♏
24	08	49	☽ ♐
26	21	28	☽ ♑
29	08	53	☽ ♒
31	18	25	☽ ♓

LATITUDES

Date	Mercury ☿ ° '	Venus ♀ ° '	Mars ♂ ° '	Jupiter ♃ ° '	Saturn ♄ ° '	Uranus ♅ ° '	Neptune ♆ ° '	Pluto ♇ ° '
01	02 N 37	01 S 51	02 N 46	00 S 04	02 N 02	00 N 44	01 S 48	10 S 59
04	02 14	01 48	02 50	00 05	02 03	00 44	01 48	10 59
07	01 47	01 44	02 55	00 05	02 03	00 44	01 47	10 58
10	01 20	01 40	02 59	00 05	02 03	00 44	01 47	10 58
13	00 53	01 35	03 04	00 05	02 04	00 44	01 47	10 58
16	00 27	01 29	03 08	00 06	02 04	00 44	01 47	10 57
19	00 N 02	01 22	03 13	00 06	02 05	00 44	01 47	10 57
22	00 S 21	01 14	03 17	00 06	02 05	00 45	01 47	10 57
25	00 43	01 06	03 22	00 07	02 05	00 45	01 46	10 56
28	01 02	00 57	03 26	00 07	02 06	00 45	01 46	10 56
31	01 S 20	00 S 47	03 N 31	00 S 07	02 N 06	00 N 45	01 S 46	10 S 56

DATA

Julian Date	2467616
Delta T	+79 seconds
Ayanamsa	24° 28' 21"
Synetic vernal point	04° ♓ 38' 38"
True obliquity of ecliptic	23° 26' 09"

MOON'S PHASES, APSIDES AND POSITIONS ☽

Date	h	m	Phase	Longitude ° '	Eclipse Indicator
08	04	02	☽	17 ♈ 30	
14	18	51	○	24 ♋ 14	
21	23	47	☾	01 ♏ 35	
30	04	04	●	09 ♒ 54	

Day	h	m		
13	09	14	Perigee	
25	09	40	Apogee	
06	18	17	0N	
13	02	09	Max dec	28° N 28'
19	08	38	0S	
26	18	33	Max dec	28° S 31'

LONGITUDES

Date	Chiron ⚷ ° '	Ceres ⚳ ° '	Pallas ⚴ ° '	Juno ⚵ ° '	Vesta ⚶ ° '	Black Moon Lilith ⚸ ° '
01	02 ♍ 17	11 ♓ 11	05 ♒ 55	27 ♑ 12	13 ♑ 04	13 ♐ 41
11	01 ♍ 55	14 ♓ 24	09 ♒ 18	01 ♒ 16	18 ♑ 25	14 ♐ 48
21	01 ♍ 23	17 ♓ 48	12 ♒ 42	05 ♒ 24	23 ♑ 45	15 ♐ 55
31	00 ♍ 44	21 ♓ 21	16 ♒ 08	09 ♒ 35	29 ♑ 03	17 ♐ 02

All ephemeris data is given at 12.00 UT and the Moon's longitude is additionally given for 24.00 UT

ASPECTARIAN

h m Aspects

01 Friday
02 35 ☉ ⊥ ♀
04 29 ☽ ⊻ ☿
09 45 ☽ ⊼ ♅
13 38 ☽ ⊥ ♇
16 53 ☽ ⊥ ☿
18 24 ☽ △ ♂
19 45 ☽ ✱ ♄
20 30 ♀ ‖ ♄

02 Saturday
01 30 ☽ ⊻ ♇
10 08 ☽ ‖ ☉
11 21 ☽ ‖ ♃
11 28 ☽ ∠ ☿
12 48 ☉ ∠ ♄
16 55 ☽ ‖ ♇
19 30 ☽ Q ♄
21 38 ☽ □ ♆

03 Sunday
00 07 ☽ ‖ ☿
00 32 ☽ bQ ♂
02 25 ☽ ⊻ ☉
06 40 ☽ ⊻ ♃
15 07 ☽ ⊥ ☉
16 11 ☽ ☌ ♀
16 56 ☽ ‖ ♄
18 20 ☽ ✱ ☿
18 24 ☽ ± ♂
18 30 ☽ ⊥ ♃
19 47 ♃ ∠ ♇
20 17 ☽ ☍ ♅
21 01 ☽ ‖ ♀

04 Monday
04 40 ☽ # ♅
06 13 ☽ ⊼ ♂
06 37 ☽ □ ♄
08 25 ☽ Q ♆
09 58 ☽ ∠ ☉
11 51 ☽ ☌ ♇
12 07 ☽ ∠ ♃
13 28 ☽ # ♆
18 51 ☽ Q ☿
20 48 ♀ ± ♂
20 51 ♀ ⊥ ♃
21 52 ☉ ‖ ♃

05 Tuesday
00 14 ☿ △ ♅
07 13 ☽ ✱ ♆
09 07 ♀ ☍ ♅
09 32 ☉ ∠ ♇
09 35 ♂ ✱ ♄
16 51 ☽ ✱ ☉
17 01 ☽ ✱ ♃
19 38 ☉ ☌ ♃

06 Wednesday
03 51 ☽ # ♂
04 29 ☉ ± ♅
04 56 ☽ ⊼ ♅
07 02 ☽ ⊻ ♀
07 05 ☽ □ ☿
09 00 ☿ ✱ ♀
11 10 ☽ ∠ ♆
15 22 ☽ △ ♄
15 40 ☽ Q ♃
15 46 ☽ ☍ ♂
15 58 ☽ ± ♅
16 58 ☽ Q ☉
19 12 ☽ ⊥ ♀
20 04 ☽ ⊻ ♇

07 Thursday
00 36 ♂ ⊥ ♅
03 36 ☽ ⊥ ♆
06 42 ♃ ± ♅
06 59 ☽ ⊥ ♇
08 10 ☽ ‖ ♂
08 18 ☽ bQ ♅
13 09 ☽ ∠ ♀
14 25 ☽ ⊻ ♆
16 29 ♀ # ♅
18 42 ☽ bQ ♄
23 08 ☽ ∠ ♇

08 Friday
00 42 ☽ □ ♃
04 02 ☽ □ ☉
10 43 ☽ ± ♄
10 54 ☽ △ ♅
12 35 ☿ bQ ♆
17 14 ☽ △ ☿
18 14 ☽ ✱ ♀
20 33 ☽ ‖ ♆
21 14 ☽ ⊼ ♄
22 14 ☽ ⊼ ♂

09 Saturday
00 32 ☿ ‖ ♇
01 24 ☽ ✱ ♇
02 10 ☽ # ♀
04 44 ☽ ‖ ♅
08 44 ☽ ± ♂
15 24 ☽ # ♄
16 45 ☽ Q ♀
18 35 ☽ ☌ ♆
21 03 ☽ bQ ☿
21 51 ☽ Q ♇

10 Sunday
00 13 ☽ bQ ♂
04 40 ☉ ‖ ☿
05 09 ☽ △ ♃
07 01 ♀ □ ♄
11 11 ☽ # ♇
11 20 ☽ △ ☉
12 08 ☽ # ☉
12 41 ☽ # ☿
13 20 ☽ ± ☿
13 48 ☽ □ ♅
14 03 ♀ Q ♆
15 11 ☽ # ♃
23 21 ☿ ⊻ ♄

11 Monday
00 00 ☽ ☍ ♄
00 03 ☽ ⊼ ☿
01 25 ☽ □ ♀
01 27 ☽ △ ♂
01 53 ♀ ⊼ ♂
03 43 ☽ □ ♇
06 16 ☽ bQ ♃
13 44 ☽ bQ ☉
19 05 ☽ Q ♅
19 11 ☉ ‖ ♇
19 56 ☽ ⊻ ♆
21 04 ☽ ± ♃
21 10 ☉ ⊼ ♅
21 21 ☿ □ ♂

12 Tuesday
05 11 ♀ # ♆
05 17 ☽ ± ☉
05 35 ☽ ⊥ ♆
05 49 ♀ ☌ ♇
06 50 ☽ ⊼ ♃
14 17 ☽ ✱ ♅
15 34 ☽ ⊼ ☉
19 57 ☽ ∠ ♆
22 16 ☿ ✱ ♇

13 Wednesday
00 01 ☿ ‖ ♃
00 33 ☽ ⊼ ♄
02 22 ☽ □ ♂
03 57 ☽ △ ♇
04 27 ☽ ☍ ☿
05 51 ☽ △ ♀
10 08 ☽ ± ♄
14 05 ☽ ∠ ♅
19 48 ☽ ✱ ♆

14 Thursday
00 32 ☽ bQ ♄
01 18 ♀ ∠ ♃
03 52 ☽ bQ ♇
04 23 ☽ ⊥ ♅
07 22 ☽ ☍ ♃
07 24 ☽ Q ♂
07 48 ☽ bQ ♀
10 19 ☉ ⊥ ♇
13 57 ☽ ⊻ ♅
14 56 ☽ Q ♆
18 17 ☽ ± ♇
18 51 ☽ ☍ ☉
23 30 ☽ ± ♀

15 Friday
00 42 ☽ △ ♄
02 55 ☽ ✱ ♂
03 58 ☽ ⊼ ♇
08 50 ☽ ⊼ ☿
10 04 ☽ ⊼ ♀
11 11 ☽ # ☿
13 19 ☽ # ♃
16 49 ☽ # ♇
19 34 ☽ ± ☿
20 04 ☽ # ☉
20 06 ☽ □ ♆

16 Saturday
03 44 ☽ ∠ ♂
04 42 ☿ ⊥ ♄
08 55 ☽ ⊼ ♃
11 31 ☽ # ♄
11 56 ☽ bQ ☿
14 52 ☽ ☌ ♅
18 58 ☽ ⊥ ♂
19 09 ☽ ± ♃
22 08 ☽ ‖ ♅
23 58 ☽ ⊼ ☉

17 Sunday
02 35 ☽ □ ♄
05 17 ☽ ⊻ ♂
05 51 ☽ ☍ ♇
06 37 ☽ ‖ ♆
10 45 ☽ bQ ♃
11 02 ☽ ± ☉
16 08 ☽ △ ☿
16 37 ☽ # ♀
17 01 ☽ ☍ ♀
18 48 ☿ bQ ♅
22 55 ☽ △ ♆

18 Monday
04 00 ☽ bQ ☉
10 09 ☽ Q ♄
13 34 ☽ △ ♃
15 16 ☉ ✱ ♄
18 54 ☽ ⊻ ♅
21 15 ☽ ‖ ♂

19 Tuesday
01 45 ☽ bQ ♆
05 47 ☽ ⊥ ♅
08 01 ☽ ✱ ♄
09 19 ☽ △ ☉
11 14 ☽ ☌ ♂
11 18 ☽ ⊼ ♇
18 21 ☽ ± ♆
19 49 ♂ ⊼ ♇
20 04 ☽ # ♂
22 26 ☽ ∠ ♅
22 30 ☽ ± ♇

20 Wednesday
04 46 ☽ □ ☿
04 53 ☽ ⊼ ♀
05 37 ☽ ⊼ ♆
05 45 ♆ St D
11 15 ☉ ⊻ ♇
12 12 ☿ ✱ ♀
12 17 ☽ ∠ ♄
12 37 ☿ △ ♆
12 41 ♀ ✱ ♆
13 11 ☉ △ ♂
15 35 ☽ bQ ♇
17 31 ☽ ± ♀
20 24 ☽ ‖ ♀
22 27 ☽ □ ♃

21 Thursday
02 57 ☽ ✱ ♅
05 43 ☽ ⊥ ♄
12 49 ☽ bQ ♀
14 15 ☽ # ♆
17 31 ☽ ⊻ ♄
19 55 ☽ Q ☿
20 48 ☽ △ ♇
21 13 ☽ ⊻ ♂
23 47 ☽ □ ☉

22 Friday
00 20 ☽ # ♅
02 23 ☽ Q ♅
09 13 ☽ ⊥ ♂
10 26 ☽ Q ♃
12 59 ☽ ‖ ♄
16 07 ☽ ☍ ♆
21 48 ☽ △ ♀
22 40 ☽ ‖ ☉
22 56 ☽ ✱ ☿

23 Saturday
03 22 ☽ ∠ ♂
03 53 ☿ ∠ ♄
10 41 ☽ ‖ ♇
10 57 ☽ ✱ ♃
13 59 ☽ ‖ ♃
14 16 ☽ □ ♅
15 06 ☽ Q ☉
20 36 ☽ ‖ ☿

24 Sunday
01 49 ☿ ± ♅
05 56 ☽ ☌ ♄
07 04 ☿ ∠ ♇
09 04 ☽ □ ♇
09 20 ☽ ∠ ☿
09 54 ☽ ✱ ♂
17 40 ☽ ✱ ☉
17 53 ☽ ∠ ♃

25 Monday
04 41 ☽ ⊼ ♆
06 07 ☽ ⊥ ☿
10 24 ☽ Q ♂
12 29 ☽ ⊥ ♃
16 50 ☽ ± ♆
17 12 ☽ □ ♀
19 58 ☽ ⊻ ☿
21 37 ☽ Q ♇

26 Tuesday
00 51 ☽ ⊻ ♃
02 47 ☽ △ ♅
02 53 ☽ ∠ ☉
11 01 ☽ bQ ♆
18 58 ☽ ⊻ ♄
21 51 ☽ ✱ ♇
22 42 ☽ ⊥ ☉
22 55 ☽ □ ♂

27 Wednesday
07 03 ☽ ⊥ ♄
08 49 ☽ bQ ♅
09 26 ☽ Q ♀
11 49 ☽ ⊻ ☉
17 05 ☽ △ ♆
17 38 ☿ ☌ ♃

28 Thursday
01 05 ☽ ∠ ♄
01 26 ☿ ⊼ ♅
02 41 ☽ ± ♅
03 50 ☽ ∠ ♇
11 51 ☽ ✱ ♀
13 58 ☽ ☌ ♃
14 29 ☽ ⊼ ♅
16 22 ☽ ☌ ☿
21 37 ☽ ⊥ ♇

29 Friday
06 46 ☽ ✱ ♄
09 23 ☽ ⊻ ♇
10 32 ☽ △ ♂
13 55 ♃ ⊼ ♅
14 06 ♀ ⊼ ♅
14 10 ♀ ✱ ♃
18 38 ☽ ‖ ☿
18 49 ☉ ‖ ♄
20 19 ☽ ∠ ♀
22 29 ☽ ‖ ♃

30 Saturday
00 50 ☽ ‖ ♇
02 25 ☉ □ ♆
03 56 ☽ □ ♆
04 04 ☽ ☌ ☉
06 10 ☽ Q ♄
11 38 ☿ ⊥ ♇
15 29 ☽ ⊥ ♀
15 34 ☽ bQ ♂
20 39 ☽ ‖ ♄
22 15 ☽ ‖ ☉

31 Sunday
00 27 ☽ ☍ ♅
01 12 ☽ ⊻ ♃
04 07 ☽ ⊻ ♀
07 16 ☉ Q ♄
08 40 ☽ # ♅
08 44 ☽ ± ♂
10 21 ☽ ⊻ ☿
12 48 ☽ ⊥ ♃
14 20 ☽ Q ♆
16 40 ☽ □ ♄
18 50 ☽ # ♆
19 02 ☽ ☌ ♇
20 05 ☽ ⊼ ♂
23 10 ♂ St R
23 14 ☽ ⊥ ☿

FEBRUARY 2044

LONGITUDES

Date	Sidereal time h m s	Sun ☉ ° ' "	Moon ☽ ° ' "	Moon ☽ 24.00 ° ' "	Mercury ☿ ° '	Venus ♀ ° '	Mars ♂ ° '	Jupiter ♃ ° '	Saturn ♄ ° '	Uranus ♅ ° '	Neptune ♆ ° '	Pluto ♇ ° '
01	20 45 25	12 ♒ 15 43	09 ♓ 22 25	15 ♓ 49 51	27 ♑ 22	24 ♓ 01	00 ♎ 53	21 ♑ 14	29 ♏ 07	20 ♌ 27	09 ♉ 50	00 ♓ 21
02	20 49 22	13 16 38	22 ♓ 20 14	28 ♓ 53 40	28 ♑ 55	25 11	00 R 52	21 28	29 11	20 R 25	09 51	00 22
03	20 53 18	14 17 33	05 ♈ 30 15	12 ♈ 10 07	00 ♒ 29	26 22	00 51	21 41	29 14	20 22	09 51	00 24
04	20 57 15	15 18 25	18 ♈ 53 23	25 ♈ 40 12	02 03	27 32	00 48	21 55	29 17	20 19	09 52	00 26
05	21 01 11	16 19 17	02 ♉ 30 38	09 ♉ 24 48	03 38	28 42	00 45	22 08	29 21	20 17	09 52	00 27
06	21 05 08	17 20 07	16 ♉ 22 42	23 ♉ 24 19	05 14	29 ♓ 52	00 41	22 21	29 24	20 14	09 53	00 29
07	21 09 04	18 20 56	00 ♊ 29 32	07 ♊ 38 09	06 50	01 ♈ 02	00 37	22 35	29 27	20 11	09 54	00 31
08	21 13 01	19 21 43	14 ♊ 49 51	22 ♊ 04 13	08 27	02 12	00 31	22 48	29 30	20 09	09 54	00 32
09	21 16 58	20 22 29	29 ♊ 20 43	06 ♋ 38 42	10 06	03 21	00 25	23 01	29 33	20 06	09 55	00 34
10	21 20 54	21 23 13	13 ♋ 57 27	21 ♋ 16 09	11 44	04 31	00 18	23 14	29 35	20 04	09 56	00 36
11	21 24 51	22 23 55	28 ♋ 33 58	05 ♌ 50 02	13 24	05 40	00 11	23 27	29 38	20 01	09 56	00 37
12	21 28 47	23 24 37	13 ♌ 03 31	20 ♌ 13 38	15 04	06 49	00 ♎ 02	23 40	29 41	19 58	09 57	00 39
13	21 32 44	24 25 16	27 ♌ 19 40	04 ♍ 21 01	16 46	07 58	29 ♍ 53	23 53	29 43	19 56	09 58	00 41
14	21 36 40	25 25 54	11 ♍ 17 13	18 ♍ 07 54	18 28	09 07	29 43	24 06	29 46	19 53	09 59	00 42
15	21 40 37	26 26 31	24 ♍ 52 52	01 ♎ 32 01	20 11	10 15	29 32	24 19	29 48	19 50	10 00	00 44
16	21 44 33	27 27 07	08 ♎ 05 23	14 ♎ 33 09	21 54	11 23	29 20	24 32	29 50	19 48	10 01	00 46
17	21 48 30	28 27 41	20 ♎ 55 34	27 ♎ 12 59	23 39	12 32	29 08	24 45	29 52	19 45	10 02	00 47
18	21 52 27	29 ♒ 28 14	03 ♏ 25 48	09 ♏ 34 31	25 25	13 40	28 55	24 57	29 54	19 43	10 03	00 49
19	21 56 23	00 ♓ 28 46	15 ♏ 39 41	21 ♏ 41 49	27 11	14 47	28 41	25 10	29 56	19 40	10 04	00 51
20	22 00 20	01 29 16	27 ♏ 41 33	03 ♐ 39 27	28 ♒ 58	15 55	28 26	25 23	29 ♏ 58	19 38	10 05	00 53
21	22 04 16	02 29 46	09 ♐ 36 08	15 ♐ 32 12	00 ♓ 47	17 02	28 11	25 35	00 ♐ 00	19 35	10 06	00 54
22	22 08 13	03 30 14	21 ♐ 28 14	27 ♐ 24 48	02 36	18 10	27 55	25 48	00 01	19 32	10 07	00 56
23	22 12 09	04 30 40	03 ♑ 22 25	09 ♑ 21 37	04 26	19 17	27 38	26 00	00 03	19 30	10 08	00 58
24	22 16 06	05 31 05	15 ♑ 22 52	21 ♑ 26 34	06 17	20 23	27 21	26 13	00 04	19 27	10 09	00 59
25	22 20 02	06 31 29	27 ♑ 33 06	03 ♒ 42 46	08 09	21 30	27 03	26 25	00 06	19 25	10 10	01 01
26	22 23 59	07 31 51	09 ♒ 55 52	16 ♒ 12 34	10 01	22 36	26 44	26 37	00 07	19 22	10 12	01 03
27	22 27 56	08 32 12	22 ♒ 33 02	28 ♒ 57 20	11 54	23 42	26 25	26 49	00 08	19 20	10 13	01 05
28	22 31 52	09 32 31	05 ♓ 25 30	11 ♓ 57 31	13 48	24 48	26 05	27 01	00 09	19 18	10 14	01 06
29	22 35 49	10 ♓ 32 48	18 ♓ 33 18	25 ♓ 12 44	15 ♓ 42	25 ♈ 54	25 ♍ 44	27 ♑ 13	00 ♐ 10	19 ♌ 15	10 ♉ 15	01 ♓ 08

Moon Nodes and Latitude

Date	Moon True ☊ ° '	Moon Mean ☊ ° '	Moon ☽ Latitude ° '
01	20 ♓ 36	22 ♓ 23	01 S 01
02	20 D 36	22 20	00 N 09
03	20 37	22 17	01 20
04	20 38	22 13	02 28
05	20 40	22 10	03 29
06	20 40	22 07	04 18
07	20 R 40	22 04	04 53
08	20 38	22 01	05 11
09	20 34	21 58	05 09
10	20 30	21 54	04 47
11	20 26	21 51	04 06
12	20 22	21 48	03 10
13	20 19	21 45	02 03
14	20 17	21 42	00 N 49
15	20 D 17	21 39	00 S 25
16	20 18	21 35	01 36
17	20 20	21 32	02 41
18	20 21	21 29	03 36
19	20 23	21 26	04 19
20	20 23	21 23	04 51
21	20 R 23	21 19	05 10
22	20 22	21 16	05 15
23	20 21	21 13	05 07
24	20 18	21 10	04 46
25	20 16	21 07	04 11
26	20 14	21 04	03 25
27	20 13	21 00	02 27
28	20 12	20 57	01 21
29	20 ♓ 11	20 ♓ 54	00 S 09

DECLINATIONS

Date	Sun ☉ ° '	Moon ☽ ° '	Mercury ☿ ° '	Venus ♀ ° '	Mars ♂ ° '	Jupiter ♃ ° '	Saturn ♄ ° '	Uranus ♅ ° '	Neptune ♆ ° '	Pluto ♇ ° '
01	17 S 07	08 S 59	22 S 04	03 S 02	02 N 54	21 S 53	17 S 54	15 N 23	13 N 05	21 S 33
02	16 50	02 S 54	21 50	02 31	02 56	21 51	17 55	15 23	13 05	21 32
03	16 32	03 N 25	21 35	02 00	02 58	21 49	17 55	15 24	13 06	21 32
04	16 15	09 41	21 19	01 28	03 00	21 47	17 56	15 25	13 06	21 31
05	15 57	15 36	21 01	00 57	03 02	21 45	17 56	15 26	13 06	21 30
06	15 38	20 51	20 41	00 S 26	03 05	21 43	17 57	15 27	13 06	21 30
07	15 20	25 02	20 21	00 N 06	03 08	21 41	17 57	15 28	13 07	21 29
08	15 01	27 43	19 58	00 37	03 12	21 39	17 58	15 28	13 07	21 29
09	14 42	28 35	19 35	01 08	03 16	21 36	17 58	15 29	13 07	21 28
10	14 22	27 27	19 10	01 40	03 19	21 34	17 59	15 30	13 07	21 27
11	14 03	24 28	18 44	02 11	03 23	21 32	17 59	15 31	13 08	21 27
12	13 43	19 56	18 16	02 42	03 28	21 30	17 59	15 32	13 08	21 26
13	13 23	14 19	17 47	03 14	03 33	21 28	18 00	15 33	13 08	21 25
14	13 03	08 05	17 17	03 45	03 38	21 26	18 00	15 33	13 09	21 25
15	12 42	01 N 39	16 45	04 16	03 43	21 24	18 00	15 34	13 09	21 24
16	12 21	04 S 41	16 12	04 47	03 49	21 22	18 01	15 35	13 09	21 24
17	12 00	10 39	15 37	05 17	03 54	21 19	18 01	15 36	13 10	21 23
18	11 39	16 02	15 01	05 48	04 00	21 17	18 01	15 37	13 10	21 22
19	11 18	20 40	14 23	06 19	04 07	21 15	18 01	15 37	13 10	21 22
20	10 57	24 22	13 44	06 49	04 13	21 13	18 02	15 38	13 11	21 21
21	10 35	26 59	13 04	07 20	04 20	21 11	18 02	15 39	13 11	21 21
22	10 13	28 24	12 23	07 50	04 27	21 08	18 02	15 40	13 12	21 20
23	09 51	28 31	11 40	08 20	04 34	21 06	18 02	15 41	13 12	21 19
24	09 29	27 17	10 55	08 49	04 42	21 04	18 02	15 41	13 12	21 19
25	09 07	24 45	10 10	09 19	04 49	21 02	18 02	15 42	13 13	21 18
26	08 45	21 02	09 23	09 48	04 57	21 00	18 02	15 43	13 13	21 18
27	08 22	16 19	08 35	10 18	05 05	20 57	18 02	15 44	13 14	21 17
28	07 59	10 46	07 46	10 47	05 13	20 55	18 02	15 45	13 14	21 17
29	07 S 37	04 S 40	06 S 56	11 N 15	05 N 21	20 S 53	18 S 02	15 N 45	13 N 15	21 S 16

ZODIAC SIGN ENTRIES

Date	h	m	Planets	
03	02	01	☽	♈
03	04	39	☿	♒
05	07	37	☽	♉
06	14	41	♀	♈
07	11	10	☽	♊
09	13	05	☽	♋
11	14	22	☽	♌
12	17	26	♂	♍
13	16	33	☽	♍
15	21	13	☽	♎
18	05	22	☽	♏
19	00	36	☉	♓
20	16	38	☽	♐
21	01	40	☿	♓
21	14	20	♄	♐
23	05	13	☽	♑
25	16	47	☽	♒
28	01	57	☽	♓

LATITUDES

Date	Mercury ☿ ° '	Venus ♀ ° '	Mars ♂ ° '	Jupiter ♃ ° '	Saturn ♄ ° '	Uranus ♅ ° '	Neptune ♆ ° '	Pluto ♇ ° '
01	01 S 25	00 S 43	03 N 32	00 S 07	02 N 06	00 N 45	01 S 46	10 S 56
04	01 39	00 32	03 37	00 08	02 07	00 45	01 46	10 56
07	01 51	00 21	03 41	00 08	02 07	00 45	01 46	10 56
10	01 59	00 S 08	03 45	00 08	02 08	00 45	01 45	10 56
13	02 04	00 N 04	03 49	00 09	02 09	00 45	01 45	10 56
16	02 06	00 18	03 52	00 09	02 09	00 45	01 45	10 56
19	02 04	00 32	03 55	00 09	02 10	00 45	01 45	10 56
22	01 58	00 46	03 57	00 10	02 10	00 45	01 45	10 56
25	01 47	01 01	03 58	00 10	02 11	00 45	01 44	10 56
28	01 31	01 16	03 59	00 10	02 11	00 45	01 44	10 57
31	01 S 10	01 N 31	04 N 00	00 S 11	02 N 12	00 N 45	01 S 44	10 S 57

DATA

Julian Date	2467647
Delta T	+79 seconds
Ayanamsa	24° 28' 26"
Synetic vernal point	04° ♓ 38' 32"
True obliquity of ecliptic	23° 26' 10"

MOON'S PHASES, APSIDES AND POSITIONS ☽

Date	h	m	Phase	Longitude ° '	Eclipse Indicator
06	13	46	☽	17 ♉ 25	
13	06	42	○	24 ♌ 12	
20	20	20	☾	01 ♐ 50	
28	20	12	●	09 ♓ 53	Annular

Day	h	m		
10	10	42	Perigee	
22	05	21	Apogee	
02	23	03	0N	
09	10	29	Max dec	28° N 35'
15	18	10	0S	
23	01	57	Max dec	28° S 38'

LONGITUDES

Date	Chiron ⚷ ° '	Ceres ⚳ ° '	Pallas ⚴ ° '	Juno ⚵ ° '	Vesta ⚶ ° '	Black Moon Lilith ⚸ ° '
01	00 ♍ 39	21 ♓ 43	16 ♒ 29	10 ♒ 01	29 ♑ 35	17 ♐ 09
11	29 ♌ 55	25 ♓ 24	19 ♒ 54	14 ♒ 15	04 ♒ 49	18 ♐ 16
21	29 ♌ 08	29 ♓ 10	23 ♒ 17	18 ♒ 32	10 ♒ 01	19 ♐ 24
31	28 ♌ 22	03 ♈ 01	26 ♒ 38	22 ♒ 50	15 ♒ 08	20 ♐ 31

All ephemeris data is given at 12.00 UT and the Moon's longitude is additionally given for 24.00 UT

ASPECTARIAN

h m	Aspects
01 Monday	
06 02	☽ ∠ ♃
12 52	☽ ⚹ ♆
17 50	☽ ⊻ ☉
18 09	♀ # ♂
18 20	☽ ∠ ☿
02 Tuesday	
04 58	♀ ∠ ♆
05 53	☽ ⊥ ☉
08 28	☽ ⊼ ♅
10 22	☽ ⚹ ♃
11 07	☿ ∥ ♃
11 53	☽ # ♂
13 35	☽ ∥ ♀
16 07	☿ ⚹ ♄
16 36	☽ ∠ ♆
17 45	☽ ☌ ♀
19 26	☽ ± ♅
23 47	☽ ∠ ☉
03 Wednesday	
00 34	☽ △ ♄
01 40	☽ ⚹ ☿
02 43	☽ ⊻ ♇
03 34	☽ ☍ ♂
07 02	☽ # ♀
08 39	☽ Q ♃
09 01	☽ ⊥ ♆
10 16	☽ ∥ ♂
10 45	☿ ⊻ ♇
11 45	☽ bQ ♅
12 04	♀ ± ♅
13 37	☽ ⊥ ♇
17 28	☿ △ ♂
17 38	☿ ∥ ♇
19 51	☽ ⊻ ♆
04 Thursday	
02 13	☽ Q ☿
03 46	☽ bQ ♄
05 06	☽ ⚹ ☉
05 49	☽ ∠ ♇
14 32	☽ △ ♅
17 27	☽ □ ♃
19 50	☽ ± ♄
23 15	☉ bQ ♂
05 Friday	
01 36	☽ ∥ ♆
04 05	☽ Q ☉
04 43	☽ ⊻ ♀
06 26	☽ ⊼ ♄
08 24	☽ ⚹ ♇
08 56	☽ ⊼ ♂
11 16	☽ ∥ ♅
13 22	☽ # ☉
14 13	☽ □ ☿
16 11	☽ ⊥ ♀
19 21	☽ ± ♂
22 12	☽ # ♄
06 Saturday	
00 48	☽ ☌ ♆
01 48	♀ △ ♄
05 17	☽ Q ♇
09 10	☽ ∠ ♀
10 49	☽ bQ ♂
11 14	☽ # ☿
13 46	☽ □ ☉
15 16	☽ # ♇
16 23	☽ # ♃
18 35	☽ □ ♅
22 23	☽ △ ♃
07 Sunday	
00 55	♀ ⊻ ♇
02 14	☉ # ♅
03 53	♀ ☍ ♂
10 14	☽ ☍ ♄
12 02	☽ □ ♇
12 12	☽ △ ♂
13 00	☽ ⚹ ♀
08 Monday	
00 01	☽ △ ☿
00 05	☽ bQ ♃
00 53	☽ Q ♅
03 47	☽ ⊻ ♆
09 18	♂ ⊼ ♇
10 51	☽ Q ♀
13 47	☽ ⊥ ♆
15 19	☽ ± ♃
20 05	☽ △ ☉
20 48	☽ ⚹ ♅
09 Tuesday	
01 24	☽ ⊼ ♃
04 06	☽ bQ ☿
04 42	☽ ∠ ♆
05 50	☉ ☍ ♅
09 24	☿ □ ♆
12 20	☽ ⊼ ♄
13 45	☽ □ ♂
14 01	☽ △ ♇
19 10	☽ □ ♀
20 48	☽ ± ☿
21 26	☽ ∠ ♅
22 14	☽ ± ♄
22 39	☽ bQ ☉
23 45	♀ ⊥ ♆
10 Wednesday	
05 23	☽ ⚹ ♆
07 54	☽ ⊼ ☿
09 48	☿ Q ♄
12 10	☽ ⊥ ♅
13 03	☽ bQ ♄
14 31	☽ ± ☉
14 41	☽ bQ ♇
19 04	☽ Q ♂
21 59	☽ ⊻ ♅
22 58	♀ bQ ♅
11 Thursday	
01 06	☽ Q ♆
01 06	☽ ⊼ ☉
03 28	☽ ☍ ♃
05 30	☽ ± ♇
06 34	♀ Q ♃
13 46	☽ △ ♄
14 38	☽ ⚹ ♂
15 24	☽ ⊼ ♇
12 Friday	
00 44	☽ △ ♀
04 20	☽ # ♃
04 44	☽ # ♇
06 50	☽ □ ♆
08 26	♀ ⊥ ♇
11 30	☿ bQ ♂
15 16	☽ ∠ ♂
15 48	☽ ☍ ☿
19 54	☉ ⊻ ♃
20 12	☽ # ☿
20 43	☽ # ♄
23 32	☽ ☌ ♅
13 Saturday	
00 52	☉ ± ♂
01 53	☿ ∥ ♄
03 58	☽ bQ ♀
06 05	☽ ⊼ ♃
06 13	☽ ⊥ ♂
06 42	☽ ☍ ☉
07 03	☽ ∥ ♅
15 57	☽ # ☉
16 05	☽ □ ♄
16 18	☽ ⊻ ♂
16 26	☽ ± ♃
16 41	☽ ∥ ♆
17 43	☽ ☍ ♇
20 37	☽ ± ♀
14 Sunday	
04 56	☉ # ♆
05 29	♀ ∥ ♂
06 18	♂ ⚹ ♄
07 53	☽ ⊼ ♀
08 09	☽ bQ ♃
09 44	☽ △ ♆
23 23	☽ Q ♄
15 Monday	
02 24	☽ ⊼ ☿
03 00	☽ ∥ ♀
03 03	☽ ⊻ ♅
04 25	☽ ∥ ♂
05 33	♃ ∥ ♇
06 32	♀ ⊻ ♆
07 27	☿ ☍ ♅
10 59	☽ △ ♃
12 12	☽ bQ ♆
13 43	☽ ⊥ ♅
14 40	☽ ± ☿
15 02	☽ ⊼ ☉
20 15	☽ ☌ ♂
20 53	☽ ⚹ ♄
22 34	☽ ⊼ ♇
16 Tuesday	
02 47	☽ ± ☉
04 30	☽ ± ♆
05 58	☽ ∠ ♅
08 34	☽ # ♂
09 29	☽ bQ ☿
09 33	☽ ± ♇
12 24	☽ # ♀
15 33	☽ ⊼ ♆
18 42	☽ ☍ ♀
20 46	☽ bQ ☉
17 Wednesday	
00 34	☽ ∠ ♄
01 30	☉ Q ♆
02 18	☽ bQ ♇
05 38	☿ ± ♂
09 47	☽ ⚹ ♅
12 39	☿ # ♅
17 27	☽ ∥ ☉
17 37	☽ ⊥ ♄
17 57	♃ ⊥ ♇
18 01	☽ △ ☿
19 24	☽ □ ♃
22 51	☽ # ♆
18 Thursday	
01 06	☉ ⊼ ♂
03 25	☽ ⊻ ♂
03 40	☽ △ ☉
05 01	☿ ⊻ ♃
05 10	☽ ⊻ ♄
06 56	☽ △ ♇
07 43	☽ ∥ ☿
08 41	☽ Q ♅
10 00	☽ # ♅
14 50	☽ ⊥ ♂
21 47	☽ ∥ ♄
22 42	☉ □ ♄
19 Friday	
00 56	☽ ☍ ♆
06 59	☽ Q ♃
08 09	☽ ∠ ♂
10 06	☽ ⊼ ♀
15 14	♀ bQ ♄
15 24	☽ ∥ ♃
16 06	☽ ∥ ♇
19 56	☽ □ ♅
21 02	☉ ☌ ♇
23 15	☽ ± ♀
23 53	☿ Q ♆
20 Saturday	
05 41	☿ ⊼ ♂
07 17	☽ ⚹ ♃
08 44	☉ ⊥ ♃
11 07	♀ ∠ ♇
13 28	☽ ⚹ ♂
15 02	☽ □ ☿
16 35	☽ ☌ ♄
18 25	☽ □ ♇
19 09	☽ bQ ♀
20 20	☽ □ ☉
21 Sunday	
01 28	☿ □ ♄
07 59	☿ # ♆
13 00	☽ ⊼ ♆
13 09	☽ Q ♂
13 42	☿ ☌ ♇
14 02	☽ ∠ ♃
22 Monday	
00 05	☿ ⊥ ♃
01 09	☽ ± ♆
04 37	☽ △ ♀
06 51	☽ Q ♇
08 07	☽ △ ♅
08 33	☽ ⊥ ♃
09 55	☽ Q ☿
12 04	☽ Q ☉
19 22	☽ bQ ♆
20 54	☽ ⊻ ♃
23 Tuesday	
00 43	☽ □ ♂
05 18	☽ ⊻ ♄
06 49	☿ ∠ ♀
07 08	☽ ⚹ ♇
14 15	☽ bQ ♅
14 16	☉ ☌ ☿
14 30	☽ ⚹ ☉
14 31	☽ ⚹ ☿
16 38	♀ △ ♅
17 23	☽ ⊥ ♄
24 Wednesday	
01 34	☽ △ ♆
08 11	☽ ± ♅
11 23	☽ ∠ ♄
13 13	☽ ∠ ♇
20 03	☽ ⊼ ♅
22 55	☽ □ ♀
23 06	☽ ∠ ☉
25 Thursday	
01 47	☽ ∠ ☿
06 24	☉ # ♀
07 01	☽ ⊥ ♇
09 44	☽ ☌ ♃
11 02	☽ △ ♂
16 59	☽ ⚹ ♄
18 19	☽ ⊥ ☉
18 44	☉ ∠ ♀
18 47	☽ ⊻ ♇
22 33	☽ ⊥ ☿
26 Friday	
04 04	☿ # ♀
06 59	☽ ⊻ ☉
10 32	☽ ∥ ♇
12 12	☽ ⊻ ☿
12 15	☽ ∥ ♃
12 30	☽ □ ♆
13 25	☽ Q ♀
14 16	☿ ⚹ ♆
15 22	☽ bQ ♂
16 12	☽ Q ♄
17 14	♂ △ ♃
27 Saturday	
03 44	☽ ∥ ♄
05 57	☽ ☍ ♅
08 04	☽ ± ♂
10 49	☿ ∠ ♃
14 22	☽ ⚹ ♀
14 40	☽ # ♅
19 04	☽ ⊼ ♂
20 08	☽ ⊻ ♃
21 36	♀ ± ♄
22 38	☽ Q ♆
23 59	☉ ∥ ☿
28 Sunday	
01 43	☽ # ♆
02 13	☽ □ ♄
03 59	☽ ☌ ♇
07 29	☽ ⊥ ♃
11 59	☽ # ♀
20 12	☽ ☌ ☉
20 47	☽ ∠ ♀
20 52	☽ ⚹ ♆
23 53	☽ ∥ ☉
29 Monday	
00 18	☽ ∠ ♃
01 55	☽ ∥ ☿
04 55	☉ ⚹ ♆
05 58	☽ ☌ ☿
09 25	☽ # ♂
09 28	♀ ⊼ ♂
13 15	☽ ⊼ ♅
14 38	☽ ⊥ ♀

MARCH 2044

LONGITUDES

Date	Sidereal time h m s	Sun ☉ ° ' "	Moon ☽ ° ' "	Moon ☽ 24.00 ° ' "	Mercury ☿ ° '	Venus ♀ ° '	Mars ♂ ° '	Jupiter ♃ ° '	Saturn ♄ ° '	Uranus ♅ ° '	Neptune ♆ ° '	Pluto ♇ ° '
01	22 39 45	11 ♓ 33 03	01 ♈ 55 39	08 ♈ 41 52	17 ♓ 37	26 ♈ 59	25 ♍ 24	27 ♑ 25	00 ♐ 11	19 ♌ 13	10 ♉ 17	01 ♓ 10
02	22 43 42	12 33 17	15 ♈ 31 11	22 ♈ 23 20	19 32	28 04	25 R 02	27 37	00 12	19 R 10	10 18	01 11
03	22 47 38	13 33 28	29 ♈ 18 07	06 ♉ 15 15	21 28	29 ♈ 09	24 41	27 49	00 12	19 08	10 20	01 13
04	22 51 35	14 33 38	13 ♉ 14 30	20 ♉ 15 37	23 23	00 ♉ 13	24 18	28 01	00 13	19 06	10 21	01 15
05	22 55 31	15 33 45	27 ♉ 18 20	04 ♊ 22 24	25 17	01 17	23 56	28 12	00 13	19 03	10 22	01 16
06	22 59 28	16 33 51	11 ♊ 27 33	18 ♊ 33 31	27 12	02 21	23 33	28 24	00 13	19 01	10 24	01 18
07	23 03 25	17 33 54	25 ♊ 40 00	02 ♋ 46 43	29 ♓ 05	03 25	23 10	28 35	00 14	18 59	10 25	01 20
08	23 07 21	18 33 55	09 ♋ 53 21	16 ♋ 59 33	00 ♈ 57	04 28	22 47	28 46	00 14	18 57	10 27	01 21
09	23 11 18	19 33 54	24 ♋ 04 59	01 ♌ 09 16	02 47	05 31	22 24	28 58	00 R 14	18 55	10 28	01 23
10	23 15 14	20 33 51	08 ♌ 12 01	15 ♌ 12 50	04 35	06 33	22 00	29 09	00 14	18 53	10 30	01 25
11	23 19 11	21 33 45	22 ♌ 11 20	29 ♌ 07 09	06 21	07 35	21 37	29 20	00 13	18 50	10 32	01 26
12	23 23 07	22 33 37	05 ♍ 59 53	12 ♍ 49 14	08 03	08 37	21 13	29 31	00 13	18 48	10 33	01 28
13	23 27 04	23 33 28	19 ♍ 34 53	26 ♍ 16 36	09 42	09 39	20 49	29 42	00 13	18 46	10 35	01 29
14	23 31 00	24 33 16	02 ♎ 54 11	09 ♎ 27 31	11 17	10 40	20 26	29 ♑ 53	00 12	18 44	10 37	01 31
15	23 34 57	25 33 02	15 ♎ 56 33	22 ♎ 21 16	12 48	11 40	20 02	00 ♒ 03	00 11	18 42	10 38	01 33
16	23 38 54	26 32 47	28 ♎ 41 46	04 ♏ 58 11	14 13	12 41	19 39	00 14	00 11	18 40	10 40	01 34
17	23 42 50	27 32 29	11 ♏ 10 47	17 ♏ 19 48	15 33	13 40	19 16	00 24	00 10	18 39	10 42	01 36
18	23 46 47	28 32 10	23 ♏ 25 37	29 ♏ 28 37	16 47	14 40	18 53	00 35	00 09	18 37	10 43	01 37
19	23 50 43	29 ♓ 31 49	05 ♐ 29 15	11 ♐ 28 00	17 55	15 39	18 30	00 45	00 08	18 35	10 45	01 39
20	23 54 40	00 ♈ 31 27	17 ♐ 25 24	23 ♐ 22 00	18 56	16 37	18 08	00 55	00 07	18 33	10 47	01 40
21	23 58 36	01 31 03	29 ♐ 18 22	05 ♑ 15 06	19 49	17 35	17 46	01 05	00 06	18 31	10 49	01 42
22	00 02 33	02 30 37	11 ♑ 12 46	17 ♑ 11 59	20 36	18 32	17 24	01 15	00 04	18 30	10 51	01 43
23	00 06 29	03 30 09	23 ♑ 13 18	29 ♑ 17 18	21 15	19 29	17 03	01 25	00 03	18 28	10 53	01 45
24	00 10 26	04 29 39	05 ♒ 24 31	11 ♒ 35 26	21 46	20 26	16 42	01 35	00 01	18 26	10 55	01 46
25	00 14 23	05 29 08	17 ♒ 50 30	24 ♒ 10 06	22 10	21 22	16 22	01 45	00 ♐ 00	18 25	10 56	01 48
26	00 18 19	06 28 34	00 ♓ 34 33	07 ♓ 04 05	22 26	22 17	16 02	01 54	29 ♏ 58	18 23	10 58	01 49
27	00 22 16	07 27 59	13 ♓ 38 51	20 ♓ 18 55	22 34	23 12	15 43	02 04	29 56	18 22	11 00	01 51
28	00 26 12	08 27 22	27 ♓ 04 11	03 ♈ 54 30	22 R 34	24 06	15 24	02 13	29 55	18 21	11 02	01 52
29	00 30 09	09 26 43	10 ♈ 49 35	17 ♈ 49 03	22 27	24 59	15 06	02 22	29 53	18 19	11 04	01 54
30	00 34 05	10 26 01	24 ♈ 52 25	01 ♉ 59 07	22 13	25 52	14 49	02 31	29 50	18 18	11 06	01 55
31	00 38 02	11 ♈ 25 18	09 ♉ 08 33	16 ♉ 20 01	21 ♈ 53	26 ♉ 44	14 ♍ 32	02 ♒ 40	29 ♏ 48	18 ♌ 17	11 ♉ 08	01 ♓ 56

Date	Moon True ☊ ° '	Moon Mean ☊ ° '	Moon ☽ Latitude ° '
01	20 ♓ 11	20 ♓ 51	01 N 05
02	20 D 12	20 48	02 16
03	20 13	20 45	03 20
04	20 13	20 41	04 13
05	20 14	20 38	04 52
06	20 14	20 35	05 13
07	20 R 14	20 32	05 15
08	20 14	20 29	04 58
09	20 14	20 25	04 23
10	20 D 14	20 22	03 32
11	20 14	20 19	02 29
12	20 14	20 16	01 18
13	20 14	20 13	00 N 04
14	20 R 14	20 10	01 S 10
15	20 13	20 06	02 18
16	20 13	20 03	03 17
17	20 12	20 00	04 06
18	20 11	19 57	04 43
19	20 10	19 54	05 06
20	20 09	19 50	05 16
21	20 09	19 47	05 12
22	20 D 09	19 44	04 55
23	20 09	19 41	04 25
24	20 11	19 38	03 43
25	20 12	19 35	02 49
26	20 13	19 31	01 46
27	20 14	19 28	00 S 36
28	20 R 14	19 25	00 N 38
29	20 13	19 22	01 51
30	20 11	19 19	02 59
31	20 ♓ 08	19 ♓ 16	03 N 57

DECLINATIONS

Date	Sun ☉ ° '	Moon ☽ ° '	Mercury ☿ ° '	Venus ♀ ° '	Mars ♂ ° '	Jupiter ♃ ° '	Saturn ♄ ° '	Uranus ♅ ° '	Neptune ♆ ° '	Pluto ♇ ° '
01	07 S 14	01 N 45	06 S 05	11 N 44	05 N 30	20 S 51	18 S 02	15 N 46	13 N 15	21 S 16
02	06 51	08 12	05 12	12 12	05 38	20 49	18 02	15 47	13 15	21 15
03	06 28	14 20	04 20	12 40	05 47	20 46	18 02	15 47	13 16	21 15
04	06 05	19 50	03 26	13 08	05 55	20 44	18 02	15 48	13 16	21 14
05	05 41	24 17	02 32	13 35	06 04	20 42	18 02	15 49	13 17	21 14
06	05 18	27 19	01 37	14 02	06 12	20 40	18 02	15 50	13 17	21 13
07	04 55	28 37	00 S 42	14 29	06 21	20 38	18 02	15 50	13 18	21 13
08	04 31	28 02	00 N 12	14 55	06 30	20 35	18 02	15 51	13 19	21 12
09	04 08	25 37	01 07	15 22	06 38	20 33	18 02	15 52	13 19	21 12
10	03 44	21 38	02 01	15 47	06 47	20 31	18 01	15 52	13 20	21 11
11	03 21	16 28	02 54	16 13	06 55	20 29	18 01	15 53	13 20	21 11
12	02 57	10 31	03 46	16 38	07 04	20 27	18 01	15 53	13 21	21 10
13	02 33	04 N 11	04 37	17 03	07 12	20 25	18 01	15 54	13 21	21 10
14	02 10	02 S 13	05 26	17 27	07 20	20 22	18 00	15 55	13 22	21 09
15	01 46	08 23	06 14	17 51	07 28	20 20	18 00	15 55	13 22	21 09
16	01 22	14 05	06 59	18 15	07 36	20 18	18 00	15 56	13 23	21 08
17	00 59	19 05	07 42	18 38	07 44	20 16	18 00	15 56	13 23	21 08
18	00 35	23 11	08 22	19 01	07 52	20 14	17 59	15 57	13 24	21 08
19	00 S 11	26 14	08 59	19 23	07 59	20 12	17 59	15 57	13 25	21 07
20	00 N 13	28 05	09 33	19 45	08 06	20 10	17 58	15 58	13 25	21 07
21	00 36	28 38	10 04	20 06	08 13	20 08	17 58	15 59	13 26	21 06
22	01 00	27 52	10 31	20 27	08 20	20 06	17 58	15 59	13 26	21 06
23	01 24	25 48	10 54	20 48	08 27	20 04	17 57	15 59	13 27	21 06
24	01 47	22 31	11 14	21 08	08 33	20 01	17 57	16 00	13 28	21 05
25	02 11	18 10	11 30	21 28	08 39	19 59	17 56	16 00	13 28	21 05
26	02 34	12 56	11 41	21 47	08 45	19 57	17 56	16 01	13 29	21 04
27	02 58	06 59	11 49	22 06	08 50	19 56	17 55	16 01	13 30	21 04
28	03 21	00 S 35	11 52	22 25	08 56	19 54	17 55	16 02	13 30	21 04
29	03 45	05 N 59	11 51	22 42	09 01	19 52	17 54	16 02	13 31	21 03
30	04 08	12 25	11 46	23 00	09 05	19 50	17 53	16 02	13 31	21 03
31	04 N 31	18 N 17	11 N 37	23 N 17	09 N 09	19 S 48	17 S 53	16 N 03	13 N 32	21 S 03

ZODIAC SIGN ENTRIES

Date	h	m	Planets
01	08	34	☽ ♈
03	13	12	☽ ♉
04	07	08	♀ ♉
05	16	35	☽ ♊
07	19	19	☽ ♋
07	23	49	☿ ♈
09	22	02	☽ ♌
12	01	32	☽ ♍
14	06	44	☽ ♎
15	04	27	♃ ♒
16	14	29	☽ ♏
19	01	02	☽ ♐
19	23	20	☉ ♈
21	13	24	☽ ♑
24	01	24	☽ ♒
25	10	02	♄ ♏
26	10	56	☽ ♓
28	17	10	☽ ♈
30	20	40	☽ ♉

LATITUDES

Date	Mercury ☿ ° '	Venus ♀ ° '	Mars ♂ ° '	Jupiter ♃ ° '	Saturn ♄ ° '	Uranus ♅ ° '	Neptune ♆ ° '	Pluto ♇ ° '
01	01 S 17	01 N 26	04 N 00	00 S 11	02 N 12	00 N 45	01 S 44	10 S 57
04	00 52	01 41	03 59	00 11	02 12	00 45	01 44	10 57
07	00 S 22	01 57	03 58	00 11	02 13	00 45	01 44	10 58
10	00 N 12	02 13	03 56	00 12	02 13	00 45	01 44	10 58
13	00 50	02 28	03 53	00 12	02 14	00 44	01 44	10 58
16	01 29	02 44	03 49	00 12	02 14	00 44	01 44	10 59
19	02 07	02 59	03 44	00 13	02 15	00 44	01 43	10 59
22	02 40	03 15	03 39	00 13	02 15	00 44	01 43	11 00
25	03 05	03 30	03 33	00 14	02 16	00 44	01 43	11 00
28	03 19	03 44	03 27	00 14	02 16	00 44	01 43	11 01
31	03 N 20	03 N 58	03 N 20	00 S 15	02 N 17	00 N 44	01 S 43	11 S 02

DATA

Julian Date	2467676
Delta T	+79 seconds
Ayanamsa	24° 28' 30"
Synetic vernal point	04° ♓ 38' 29"
True obliquity of ecliptic	23° 26' 10"

MOON'S PHASES, APSIDES AND POSITIONS ☽

Date	h	m	Phase	Longitude ° '	Eclipse Indicator
06	21	17	☽	16 ♊ 57	
13	19	41	○	23 ♍ 53	total
21	16	52	☾	01 ♑ 43	
29	09	26	●	09 ♈ 20	

Day	h	m		
07	20	34	Perigee	
21	01	55	Apogee	
01	05	30	0N	
07	16	32	Max dec	28° N 39'
14	03	38	0S	
21	10	03	Max dec	28° S 39'
28	14	10	0N	

LONGITUDES

Date	Chiron ⚷ ° '	Ceres ⚳ ° '	Pallas ⚴ ° '	Juno ⚵ ° '	Vesta ⚶ ° '	Black Moon Lilith ⚸ ° '
01	28 ♌ 27	02 ♈ 38	26 ♒ 18	22 ♒ 24	14 ♒ 38	20 ♐ 24
11	27 ♌ 44	06 ♈ 32	29 ♒ 35	26 ♒ 42	19 ♒ 40	21 ♐ 31
21	27 ♌ 06	10 ♈ 29	02 ♓ 47	01 ♓ 00	24 ♒ 37	22 ♐ 39
31	26 ♌ 37	14 ♈ 27	05 ♓ 54	05 ♓ 18	29 ♒ 28	23 ♐ 46

All ephemeris data is given at 12.00 UT and the Moon's longitude is additionally given for 24.00 UT

ASPECTARIAN

01 Tuesday
00 02 ☽ ± ♅
00 06 ☽ ∠ ♆
00 37 ☽ ☍ ♂
02 23 ☽ ⊻ ♀
03 50 ☽ ✶ ♃
08 53 ☽ △ ♄
10 38 ☽ ⊻ ♇
16 03 ☽ ₽ ♅
16 11 ☽ ⊥ ♆
19 19 ♂ ₽ ♆
21 18 ☽ ⊥ ♇
23 55 ♀ □ ♃

02 Wednesday
01 28 ☽ Q ♃
01 49 ♂ ⊥ ♅
01 53 ☿ # ♂
02 08 ☽ # ☿
02 11 ☽ ∥ ♂
02 49 ☽ ⊻ ♆
06 23 ☽ ⊻ ☉
07 13 ☽ # ☉
07 31 ☿ ⊼ ♅
11 26 ☽ ₽ ♄
13 10 ☽ ∠ ♇
13 53 ☉ ∠ ♃
17 44 ☽ ⊥ ☉
18 22 ☽ △ ♅
20 10 ☽ ⊻ ☿

03 Thursday
03 09 ☽ ± ♄
04 11 ☽ ⊼ ♂
04 42 ☽ ∥ ♀
07 39 ☽ ∥ ♆
08 18 ☽ ⊥ ☿
09 23 ☽ □ ♃
10 37 ☽ ∠ ☉
11 42 ☽ ☌ ♀
13 34 ☽ ⊼ ♄
14 19 ☽ ± ♂
15 19 ☽ ✶ ♇
18 01 ☽ ∥ ♅

04 Friday
02 19 ☽ ∠ ☿
03 44 ☽ # ♄
05 26 ☽ ₽ ♂
07 02 ☽ ☌ ♆
11 51 ♀ ⊼ ♄
12 00 ☽ Q ♇
13 30 ♀ ± ♂
14 26 ☽ ✶ ☉
16 21 ☽ # ♃
18 52 ☽ # ♇
19 09 ☉ # ♂
19 49 ♀ ∥ ♆
21 46 ☿ ☍ ♂
21 59 ☽ □ ♅

05 Saturday
06 25 ☽ △ ♂
08 02 ☽ ✶ ☿
09 08 ☿ ± ♅
11 38 ♀ ✶ ♇
11 55 ☿ ⊥ ♀
12 28 ☽ Q ☉
13 04 ☿ ∠ ♆
13 33 ☽ △ ♃
16 57 ☽ ☍ ♄
18 45 ☽ □ ♇
19 19 ☽ ⊻ ♀

06 Sunday
04 30 ☽ Q ♅
06 19 ☽ ⊥ ♀
07 34 ☽ Q ☿
10 12 ☽ ⊻ ♆
15 19 ☽ ₽ ♃
20 22 ☽ ⊥ ♆
21 17 ☽ □ ☉
22 46 ☽ ∠ ♀

07 Monday
00 45 ☽ ✶ ♅
04 59 ☿ ✶ ♃
06 44 ☽ ± ♃
07 54 ☽ □ ♂
11 35 ☽ ∠ ♆
16 59 ☽ ⊼ ♃
18 38 ☽ □ ☿
19 42 ☽ ⊼ ♄
21 34 ☽ △ ♇

08 Tuesday
02 00 ☽ ∠ ♅
02 07 ☽ ✶ ♀
02 44 ☿ △ ♄
05 49 ☽ ± ♄
12 57 ☽ ✶ ♆
13 28 ☽ Q ♂
17 09 ☽ ⊥ ♅
17 24 ☿ ⊻ ♇
20 50 ☉ ⊼ ♅
21 01 ☽ ₽ ♄
22 26 ♄ St R
22 56 ☽ ₽ ♇

09 Wednesday
00 00 ☽ Q ♀
03 16 ☽ ⊻ ♅
03 46 ☽ △ ☉
09 13 ☽ ✶ ♂
09 16 ☽ Q ♆
14 12 ☽ ± ♇
20 23 ☽ ☍ ♃
22 25 ☽ △ ♄

10 Thursday
00 25 ☽ ⊼ ♇
02 40 ☿ ₽ ♅
04 56 ☽ △ ☿
07 10 ☽ ₽ ☉
08 58 ☽ □ ♀
10 01 ☽ ∠ ♂
10 51 ☿ ⊥ ♆
14 17 ☽ # ♇
15 56 ☽ □ ♆
16 38 ♀ ∥ ♅
17 39 ☽ # ♃
19 33 ♀ ₽ ♂
23 43 ☽ ± ☉

11 Friday
00 59 ☽ ⊥ ♂
05 13 ☽ # ♄
06 15 ☽ ☌ ♅
10 20 ☽ ₽ ☿
10 50 ☽ ⊼ ☉
11 02 ☽ ⊻ ♂
12 50 ☉ ☍ ♂
13 00 ☽ ∥ ♀
14 28 ☽ ∥ ♅
20 28 ☉ # ☿

12 Saturday
00 32 ☽ ⊼ ♃
00 54 ☽ ∥ ♆
01 55 ☽ □ ♄
03 28 ☿ ⊥ ♇
04 04 ☽ ☍ ♇
04 09 ☽ ± ☿
11 09 ☽ ± ♃
16 07 ☽ ⊼ ☿
16 58 ☽ △ ♀
20 01 ☽ △ ♆

13 Sunday
00 54 ☽ ∥ ♂
03 12 ☽ ₽ ♃
09 33 ☽ Q ♄
09 35 ☿ ⊻ ♀
10 33 ☽ ∥ ☿
10 34 ☽ ⊻ ♅
14 09 ☽ ☌ ♂
18 27 ☽ # ☉
19 41 ☽ ☍ ☉
21 16 ☽ ⊥ ♅
21 49 ☽ ₽ ♀
22 46 ☽ ₽ ♆

14 Monday
01 24 ☿ ⊻ ♆
06 26 ☽ △ ♃
07 06 ☽ ✶ ♄
09 28 ☽ ⊼ ♇
10 44 ♀ ☌ ♆
11 48 ☽ ∥ ☉
13 31 ☽ ∠ ♅
15 07 ☽ ± ♆
15 28 ☽ ± ♀
16 18 ☉ ± ♅
20 27 ☽ ± ♇
22 28 ☿ Q ♃

15 Tuesday
02 09 ☽ ⊼ ♆
02 10 ☽ # ☿
03 25 ☽ ⊼ ♀
05 24 ☽ ☍ ☿
08 16 ☽ # ♂
10 36 ☽ ∠ ♄
13 07 ☽ ₽ ♇
14 10 ☉ ∠ ♆
17 09 ☽ ✶ ♅
19 26 ☽ ⊻ ♂
21 10 ♀ # ♄

16 Wednesday
03 27 ☽ ⊥ ♄
05 07 ♃ ✶ ♄
06 24 ☽ ⊥ ♂
07 34 ☽ ⊼ ☉
08 54 ☽ # ♆
14 49 ☽ ⊻ ♄
14 58 ☽ □ ♃
15 46 ☽ Q ♅
17 29 ☽ △ ♇
19 59 ☽ ± ☉
20 28 ☽ # ♅
23 03 ☽ ∠ ♂

17 Thursday
04 56 ☿ ₽ ♄
06 27 ☽ ∥ ♄
09 28 ☽ # ♀
10 05 ♀ Q ♇
11 03 ☽ ☍ ♆
13 48 ☿ ∥ ♂
14 53 ☽ ₽ ☉
17 16 ☽ ☍ ♀
18 21 ☽ ∥ ♃
21 30 ☽ ⊼ ☿
23 17 ☽ ∥ ♇

18 Friday
02 19 ☽ Q ♃
02 32 ☽ □ ♅
03 19 ☽ ✶ ♂
08 37 ☿ ∠ ♇
10 36 ☽ ± ☿
23 02 ☽ △ ☉

19 Saturday
01 19 ☽ ☌ ♄
02 21 ☽ Q ♂
02 24 ☽ ✶ ♃
04 18 ☽ □ ♇
06 20 ☽ ₽ ☿
06 39 ♂ ⊻ ♅
22 00 ☿ ⊼ ♂
22 36 ☽ ⊼ ♆

20 Sunday
02 17 ☉ △ ♄
03 05 ☿ △ ♅
08 56 ☽ ∠ ♃
10 14 ☽ ⊼ ♀
10 43 ☽ ± ♆
13 08 ♂ Q ♄
13 23 ☽ □ ♂
14 16 ☽ △ ♅
15 18 ☽ △ ☿
16 33 ☽ Q ♇
23 25 ☽ ± ♀
23 37 ☉ ✶ ♃

21 Monday
03 22 ☽ ⊥ ♃
04 56 ☽ ₽ ♆
13 15 ♀ # ♃
13 35 ☽ ⊻ ♄
15 17 ♀ △ ♂
15 39 ☽ ⊻ ♃
16 28 ☉ ⊻ ♇
16 50 ☽ ✶ ♇
16 52 ☽ □ ☉
19 12 ☽ ₽ ♀
20 30 ☽ ₽ ♅

22 Tuesday
01 40 ☽ ⊥ ♄
10 54 ♀ □ ♅
11 16 ☽ △ ♆
14 34 ☽ ± ♅
19 44 ☽ ∠ ♄
23 04 ☽ ∠ ♇

23 Wednesday
00 03 ☽ △ ♂
02 33 ☽ ⊼ ♅
03 56 ☽ △ ♀
07 53 ☽ □ ☿
08 17 ☽ Q ☉
11 11 ☉ ₽ ♅
17 01 ☽ ⊥ ♇

24 Thursday
01 28 ☽ ✶ ♄
04 25 ☽ ☌ ♃
04 52 ☽ ⊻ ♇
04 57 ☽ ₽ ♂
08 19 ♀ # ♇
10 03 ☽ ✶ ☉
17 33 ♂ ₽ ♃
19 37 ☽ # ♀
20 32 ☽ ∥ ♇
20 48 ☽ Q ☿
22 00 ☽ ± ♂
22 22 ☉ ⊥ ♆
22 43 ☽ □ ♆

25 Friday
00 49 ☽ Q ♄
02 31 ☽ ∥ ♃
09 15 ☽ ⊼ ♂
13 06 ☽ ☍ ♅
13 10 ☽ ∥ ♄
17 28 ☽ ∠ ☉
19 13 ☽ □ ♀
19 25 ☿ ± ♂
20 25 ☽ ✶ ☿
20 57 ♃ ⊻ ♇
22 20 ☽ # ♅

26 Saturday
09 00 ☽ Q ♆
09 37 ☽ # ♆
10 52 ☽ □ ♄
11 48 ☽ ⊥ ☉
14 19 ☽ ☌ ♇
14 30 ☽ ⊻ ♃
16 47 ☿ ⊻ ♀
17 05 ☽ # ☿
23 48 ☽ ⊻ ☉

27 Sunday
00 49 ☽ ∠ ☿
01 42 ☽ ⊥ ♃
04 52 ☽ # ♂
07 11 ☽ ✶ ♆
07 13 ☽ Q ♀
15 39 ☽ ☍ ♂
17 17 ☽ ⊥ ☿
18 14 ☽ ∠ ♃
20 30 ☽ ⊼ ♅
21 25 ☉ ⊥ ♇

28 Monday
01 31 ☿ St R
02 22 ☽ # ☉
04 03 ☽ ⊻ ☿
06 22 ☽ ✶ ♀
07 11 ☽ ± ♅
10 10 ☽ ∠ ♆
16 59 ☽ △ ♄
20 27 ☽ ⊻ ♇
21 09 ☽ ✶ ♃
23 00 ☽ ₽ ♅

29 Tuesday
02 00 ☽ ⊥ ♆
03 18 ☽ ∥ ☉
06 55 ☽ ⊥ ♇
09 26 ☽ ☌ ☉
10 27 ☽ ∠ ♀
12 25 ☽ ⊻ ♆
18 10 ☽ Q ♃
18 57 ☽ ₽ ♄
19 12 ☽ ⊼ ♂
22 26 ☽ ∠ ♇
23 16 ☽ ∥ ♂

30 Wednesday
00 50 ☽ △ ♅
02 56 ☽ ⊥ ♀
05 15 ☽ ± ♂
07 36 ☽ ☌ ☿
09 35 ☽ ∥ ☿
10 15 ☽ ± ♄
13 48 ☽ ⊻ ♀
16 21 ☽ ∥ ♆
20 11 ☽ ₽ ♂
20 22 ☽ ⊼ ♄
23 54 ☽ ✶ ♇

31 Thursday
01 02 ☽ □ ♃
02 28 ☽ ∥ ♅
04 50 ☉ ⊻ ♆
10 14 ☽ # ♄
15 20 ☽ ☌ ♆
16 05 ☽ ⊻ ☉
18 46 ☽ # ♃
20 01 ☽ Q ♇
20 50 ☽ △ ♂

APRIL 2044

Date	Sidereal time h m s	Sun ☉ ° ' "	Moon ☽ ° ' "	Moon ☽ 24.00 ° ' "	Mercury ☿ ° '	Venus ♀ ° '	Mars ♂ ° '	Jupiter ♃ ° '	Saturn ♄ ° '	Uranus ♅ ° '	Neptune ♆ ° '	Pluto ♇ ° '	Date
01	00 41 58	12 ♈ 24 33	23 ♉ 32 50	00 ♊ 46 19	21 ♈ 26	27 ♉ 36	14 ♍ 16	02 ♒ 49	29 ♏ 46	18 ♌ 15	11 ♉ 10	01 ♓ 58	01
02	00 45 55	13 23 45	07 ♊ 59 49	15 ♊ 12 43	20 R 54	28 27	14 R 01	02 58	29 R 44	18 R 14	11 12	01 59	02
03	00 49 52	14 22 55	22 ♊ 24 27	29 ♊ 34 34	20 18	29 ♉ 17	13 46	03 06	29 41	18 13	11 14	02 00	03
04	00 53 48	15 22 03	06 ♋ 42 38	13 ♋ 48 21	19 37	00 ♊ 06	13 32	03 15	29 39	18 12	11 16	02 02	04
05	00 57 45	16 21 08	20 ♋ 51 28	27 ♋ 51 48	18 53	00 54	13 19	03 23	29 36	18 11	11 18	02 03	05
06	01 01 41	17 20 11	04 ♌ 49 14	11 ♌ 43 41	18 07	01 42	13 07	03 31	29 34	18 10	11 21	02 04	06
07	01 05 38	18 19 12	18 ♌ 35 07	25 ♌ 23 32	17 20	02 28	12 55	03 39	29 31	18 09	11 23	02 06	07
08	01 09 34	19 18 10	02 ♍ 08 53	08 ♍ 51 13	16 33	03 14	12 44	03 47	29 28	18 08	11 25	02 07	08
09	01 13 31	20 17 06	15 ♍ 30 30	22 ♍ 06 46	15 46	03 59	12 34	03 55	29 25	18 07	11 27	02 08	09
10	01 17 27	21 16 00	28 ♍ 39 59	05 ♎ 10 09	15 01	04 43	12 25	04 02	29 22	18 06	11 29	02 09	10
11	01 21 24	22 14 52	11 ♎ 37 15	18 ♎ 01 17	14 18	05 25	12 17	04 10	29 19	18 06	11 31	02 11	11
12	01 25 21	23 13 41	24 ♎ 22 16	00 ♏ 40 12	13 38	06 07	12 09	04 17	29 16	18 05	11 33	02 12	12
13	01 29 17	24 12 29	06 ♏ 55 08	13 ♏ 07 08	13 01	06 48	12 02	04 24	29 13	18 04	11 36	02 13	13
14	01 33 14	25 11 15	19 ♏ 16 19	25 ♏ 22 51	12 29	07 27	11 56	04 31	29 10	18 04	11 38	02 14	14
15	01 37 10	26 09 59	01 ♐ 26 54	07 ♐ 28 43	12 01	08 05	11 51	04 38	29 06	18 03	11 40	02 15	15
16	01 41 07	27 08 41	13 ♐ 28 37	19 ♐ 26 54	11 37	08 42	11 46	04 45	29 03	18 03	11 42	02 16	16
17	01 45 03	28 07 21	25 ♐ 24 00	01 ♑ 20 19	11 19	09 18	11 42	04 52	28 59	18 02	11 44	02 17	17
18	01 49 00	29 ♈ 06 00	07 ♑ 16 22	13 ♑ 12 38	11 06	09 53	11 39	04 58	28 56	18 02	11 47	02 18	18
19	01 52 56	00 ♉ 04 37	19 ♑ 09 42	25 ♑ 08 08	10 58	10 26	11 37	05 04	28 52	18 02	11 49	02 19	19
20	01 56 53	01 03 12	01 ♒ 08 32	07 ♒ 11 32	10 55	10 57	11 36	05 11	28 49	18 02	11 51	02 20	20
21	02 00 50	02 01 46	13 ♒ 17 45	19 ♒ 27 47	10 D 57	11 27	11 35	05 17	28 45	18 01	11 53	02 21	21
22	02 04 46	03 00 18	25 ♒ 42 14	02 ♓ 01 40	11 04	11 56	11 D 35	05 22	28 41	18 01	11 55	02 22	22
23	02 08 43	03 58 48	08 ♓ 26 34	14 ♓ 57 22	11 16	12 23	11 36	05 28	28 37	18 01	11 58	02 23	23
24	02 12 39	04 57 17	21 ♓ 34 25	28 ♓ 17 56	11 33	12 48	11 37	05 34	28 33	18 01	12 00	02 24	24
25	02 16 36	05 55 43	05 ♈ 08 01	12 ♈ 04 36	11 55	13 12	11 39	05 39	28 30	18 D 01	12 02	02 25	25
26	02 20 32	06 54 09	19 ♈ 07 28	26 ♈ 16 13	12 21	13 34	11 42	05 44	28 26	18 01	12 04	02 26	26
27	02 24 29	07 52 32	03 ♉ 30 16	10 ♉ 48 53	12 51	13 54	11 46	05 49	28 22	18 01	12 07	02 27	27
28	02 28 25	08 50 54	18 ♉ 11 10	25 ♉ 36 07	13 25	14 12	11 50	05 54	28 17	18 01	12 09	02 28	28
29	02 32 22	09 49 14	03 ♊ 02 39	10 ♊ 29 40	14 03	14 29	11 55	05 59	28 13	18 02	12 11	02 29	29
30	02 36 19	10 ♉ 47 32	17 ♊ 56 05	25 ♊ 20 51	14 ♈ 45	14 ♊ 43	12 ♍ 01	06 ♒ 03	28 ♏ 09	18 ♌ 02	12 ♉ 13	02 ♓ 30	30

Date	Moon True ☊ ° '	Moon Mean ☊ ° '	Moon ☽ Latitude ° '
01	20 ♓ 05	19 ♓ 12	04 N 41
02	20 R 02	19 09	05 07
03	20 00	19 06	05 14
04	19 59	19 03	05 01
05	19 D 58	19 00	04 30
06	19 59	18 56	03 43
07	20 01	18 53	02 44
08	20 02	18 50	01 36
09	20 03	18 47	00 N 25
10	20 R 03	18 44	00 S 47
11	20 01	18 41	01 55
12	19 58	18 37	02 56
13	19 53	18 34	03 48
14	19 47	18 31	04 28
15	19 40	18 28	04 55
16	19 34	18 25	05 08
17	19 28	18 22	05 08
18	19 24	18 18	04 55
19	19 22	18 15	04 29
20	19 D 21	18 12	03 51
21	19 22	18 09	03 03
22	19 23	18 06	02 05
23	19 24	18 02	00 S 59
24	19 R 25	17 59	00 N 12
25	19 24	17 56	01 24
26	19 20	17 53	02 33
27	19 15	17 50	03 35
28	19 08	17 47	04 23
29	19 00	17 43	04 54
30	18 ♓ 53	17 ♓ 40	05 N 06

DECLINATIONS

Date	Sun ☉ ° '	Moon ☽ ° '	Mercury ☿ ° '	Venus ♀ ° '	Mars ♂ ° '	Jupiter ♃ ° '	Saturn ♄ ° '	Uranus ♅ ° '	Neptune ♆ ° '	Pluto ♇ ° '
01	04 N 54	23 N 11	11 N 24	23 N 33	09 N 13	19 S 46	17 S 52	16 N 03	13 N 33	21 S 03
02	05 17	26 41	11 08	23 49	09 17	19 44	17 52	16 04	13 33	21 02
03	05 40	28 26	10 48	24 04	09 20	19 42	17 51	16 04	13 34	21 02
04	06 03	28 16	10 25	24 19	09 23	19 40	17 50	16 04	13 35	21 02
05	06 26	26 15	10 00	24 34	09 26	19 39	17 50	16 04	13 35	21 01
06	06 48	22 40	09 32	24 48	09 29	19 37	17 49	16 05	13 36	21 01
07	07 11	17 51	09 02	25 01	09 31	19 35	17 48	16 05	13 37	21 01
08	07 33	12 13	08 32	25 14	09 32	19 33	17 48	16 05	13 37	21 01
09	07 56	06 N 06	08 00	25 26	09 34	19 32	17 47	16 05	13 38	21 00
10	08 18	00 S 11	07 29	25 38	09 35	19 30	17 46	16 06	13 39	21 00
11	08 40	06 22	06 57	25 49	09 36	19 29	17 45	16 06	13 39	21 00
12	09 01	12 11	06 27	26 00	09 36	19 27	17 45	16 06	13 40	21 00
13	09 23	17 24	05 57	26 10	09 37	19 25	17 44	16 06	13 41	21 00
14	09 45	21 50	05 29	26 20	09 37	19 24	17 43	16 06	13 41	21 00
15	10 06	25 15	05 03	26 29	09 36	19 22	17 42	16 06	13 42	20 59
16	10 27	27 31	04 39	26 37	09 35	19 21	17 41	16 07	13 43	20 59
17	10 48	28 29	04 17	26 45	09 34	19 20	17 41	16 07	13 44	20 59
18	11 09	28 09	03 58	26 53	09 33	19 18	17 40	16 07	13 44	20 59
19	11 30	26 30	03 41	27 00	09 32	19 17	17 39	16 07	13 45	20 59
20	11 50	23 40	03 27	27 06	09 30	19 15	17 38	16 07	13 46	20 59
21	12 11	19 45	03 15	27 12	09 28	19 14	17 37	16 07	13 46	20 59
22	12 31	14 54	03 06	27 17	09 25	19 13	17 36	16 07	13 47	20 58
23	12 51	09 19	03 00	27 22	09 23	19 12	17 35	16 07	13 48	20 58
24	13 10	03 S 10	02 56	27 26	09 20	19 10	17 34	16 07	13 48	20 58
25	13 30	03 N 19	02 54	27 30	09 17	19 09	17 33	16 07	13 49	20 58
26	13 49	09 51	02 55	27 33	09 13	19 08	17 33	16 07	13 50	20 58
27	14 08	16 02	02 59	27 35	09 10	19 07	17 32	16 07	13 50	20 58
28	14 27	21 27	03 04	27 37	09 06	19 06	17 31	16 07	13 51	20 58
29	14 45	25 35	03 12	27 38	09 02	19 05	17 30	16 07	13 52	20 58
30	15 N 04	27 N 58	03 N 22	27 N 38	08 N 57	19 S 04	17 S 29	16 N 06	13 N 53	20 S 58

ZODIAC SIGN ENTRIES

Date	h m	Planets
01	22 43	☽ ♊
04	00 43	☽ ♋
04	09 09	♀ ♊
06	03 41	☽ ♌
08	08 10	☽ ♍
10	14 27	☽ ♎
12	22 43	☽ ♏
15	09 08	☽ ♐
17	21 18	☽ ♑
19	10 07	☉ ♉
20	09 43	☽ ♒
22	20 10	☽ ♓
25	03 00	☽ ♈
27	06 12	☽ ♉
29	07 06	☽ ♊

LATITUDES

Date	Mercury ☿ ° '	Venus ♀ ° '	Mars ♂ ° '	Jupiter ♃ ° '	Saturn ♄ ° '	Uranus ♅ ° '	Neptune ♆ ° '	Pluto ♇ ° '
01	03 N 17	04 N 02	03 N 17	00 S 15	02 N 17	00 N 44	01 S 43	11 S 02
04	02 58	04 15	03 10	00 15	02 17	00 44	01 43	11 03
07	02 25	04 27	03 02	00 16	02 18	00 44	01 43	11 03
10	01 42	04 38	02 54	00 16	02 18	00 44	01 43	11 04
13	00 53	04 48	02 46	00 17	02 18	00 44	01 43	11 05
16	00 N 04	04 56	02 39	00 17	02 19	00 44	01 43	11 06
19	00 S 43	05 03	02 31	00 18	02 19	00 43	01 43	11 06
22	01 23	05 07	02 23	00 18	02 19	00 43	01 43	11 07
25	01 57	05 09	02 16	00 19	02 19	00 43	01 43	11 08
28	02 25	05 09	02 08	00 19	02 19	00 43	01 43	11 09
31	02 S 45	05 N 05	02 N 01	00 S 20	02 N 19	00 N 43	01 S 43	11 S 10

DATA

Julian Date	2467707
Delta T	+79 seconds
Ayanamsa	24° 28' 33"
Synetic vernal point	04° ♓ 38' 25"
True obliquity of ecliptic	23° 26' 10"

LONGITUDES

Date	Chiron ⚷ ° '	Ceres ⚳ ° '	Pallas ⚴ ° '	Juno ⚵ ° '	Vesta ⚶ ° '	Black Moon Lilith ⚸ ° '
01	26 ♌ 34	14 ♈ 51	06 ♓ 12	05 ♓ 43	29 ♒ 56	23 ♐ 52
11	26 ♌ 15	18 ♈ 49	09 ♓ 11	09 ♓ 59	04 ♓ 38	25 ♐ 00
21	26 ♌ 06	22 ♈ 47	12 ♓ 00	14 ♓ 12	09 ♓ 12	26 ♐ 07
31	26 ♌ 08	26 ♈ 44	14 ♓ 39	18 ♓ 21	13 ♓ 35	27 ♐ 14

MOON'S PHASES, APSIDES AND POSITIONS ☽

Date	h m	Phase	Longitude ° '	Eclipse Indicator
05	03 45	☽	16 ♋ 01	
12	09 39	○	23 ♎ 08	
20	11 48	☾	01 ♒ 03	
27	19 42	●	08 ♉ 11	

Day	h m		
02	02 07	Perigee	
17	20 39	Apogee	
29	18 22	Perigee	
03	21 55	Max dec	28° N 36'
10	11 17	0S	
17	17 46	Max dec	28° S 32'
24	23 48	0N	

ASPECTARIAN

h m	Aspects
01 Friday	
00 46	☽ # ♇
02 50	☽ ⊥ ☉
03 13	☽ □ ♅
08 37	☽ ⊻ ☿
09 11	☿ ⊥ ♀
14 13	☽ ∥ ♀
18 15	☽ ⊥ ☿
18 53	☽ ∠ ☉
19 09	☽ ☌ ♀
22 18	☽ ☍ ♄
02 Saturday	
02 00	☽ □ ♇
03 33	☽ △ ♃
08 40	☽ ∠ ☿
09 05	☽ Q ♅
17 21	☽ ⊻ ♆
19 38	☉ ∠ ♀
21 38	☽ ⚹ ☉
21 50	☽ □ ♂
23 59	☉ ⊼ ♂
03 Sunday	
03 21	☽ ⊥ ♆
04 45	☽ ⚼ ♃
05 01	☽ ⚹ ♅
08 37	☽ ⚹ ☿
18 25	☽ ∠ ♆
19 08	☽ Q ☉
19 11	☉ ⚼ ♄
19 56	☽ ± ♃
23 28	♀ ☍ ♄
04 Monday	
00 09	☽ ⊼ ♄
00 12	☽ ⊻ ♀
03 26	☽ Q ♂
03 49	☽ Q ☿
04 06	☽ △ ♇
06 06	☽ ∠ ♅
06 06	☽ ⊼ ♃
08 31	☉ Q ♃
10 13	☽ ± ♄
10 54	☽ ⊥ ♀
15 53	☿ ± ♂
19 44	☽ ⚹ ♆
21 16	☽ ⊥ ♅
23 22	☽ ⚹ ♂
05 Tuesday	
01 23	☽ ⚼ ♄
03 03	☽ ∠ ♀
03 45	☽ □ ☉
05 30	☽ ⚼ ♇
07 27	☽ ⊻ ♅
08 48	☽ □ ☿
16 12	☽ Q ♆
20 54	☽ ± ♇
23 43	☽ ∥ ♀
06 Wednesday	
00 36	☽ ∠ ♂
02 57	☽ △ ♄
05 25	☉ ∠ ♇
06 17	☽ ⚹ ♀
07 15	☽ ⊼ ♇
09 44	☽ ☍ ♃
10 37	☿ △ ♅
14 36	☿ ∥ ♂
15 55	☽ ⊥ ♂
20 47	☽ # ♇
22 40	☉ ☌ ☿
23 21	☽ □ ♆
23 54	♀ □ ♇
07 Thursday	
02 13	☽ ⊻ ♂
03 51	☽ # ♃
04 22	☽ Q ♀
07 54	☉ △ ♅
09 54	☿ ∠ ♀
09 56	☽ △ ☿
11 14	☽ ☌ ♅
11 30	☽ △ ☉
12 12	☽ # ♄
19 10	☿ ∠ ♇
19 49	☽ ∥ ♅
08 Friday	
00 20	☉ ± ♂
06 14	☽ ∥ ♆
07 15	☽ □ ♄
10 59	☽ ⚼ ☿
11 56	☽ ☍ ♇
14 03	☽ □ ♀
14 57	☽ ⊼ ♃
16 09	☽ ⚼ ☉
22 35	☽ ∥ ♂
09 Saturday	
01 48	☽ ± ♃
02 13	☽ ± ☿
03 56	☽ ∥ ☿
04 39	☽ △ ♆
05 20	☽ ∥ ☉
06 46	☽ ☌ ♂
08 04	☿ Q ♃
09 20	♀ △ ♃
09 37	☽ ± ☉
12 26	☽ ⊼ ☿
14 09	☉ ∥ ☿
15 27	☽ Q ♄
16 44	☽ ⊻ ♅
18 14	☽ ⚼ ♃
21 22	☽ ⊼ ☉
10 Sunday	
03 39	☽ ⊥ ♅
07 59	☽ ⚼ ♆
13 17	☽ ⚹ ♄
18 26	☽ ⊼ ♇
20 11	☽ ∠ ♅
22 00	☽ △ ♃
23 48	☽ △ ♀
11 Monday	
00 37	☽ ± ♆
05 34	☽ ± ♇
11 03	☿ ⚼ ♄
11 49	☽ ⊼ ♆
13 13	☽ ⊻ ♂
14 12	☽ # ☿
16 45	☽ ☍ ☿
17 02	☽ ∠ ♄
21 53	☽ # ☉
22 25	☽ ⚼ ♇
12 Tuesday	
00 08	☽ ⚹ ♅
00 21	☽ ⊥ ♂
01 08	☽ # ♂
05 29	☽ ⚼ ♀
09 39	☽ ☍ ☉
09 55	☽ ⊥ ♄
10 48	♀ Q ♅
12 55	☉ ± ♄
17 14	☽ ∠ ♂
18 33	☽ # ♆
21 17	☽ ⊻ ♄
22 52	☽ Q ♅
23 34	☽ ± ♀
13 Wednesday	
02 57	☽ △ ♇
05 43	☽ # ♅
07 07	☽ □ ♃
11 45	☽ ⊼ ♀
13 37	☽ ∥ ♄
21 04	☽ ☍ ♆
21 49	☽ ⚹ ♂
22 18	☽ ∥ ♃
23 17	☽ ⊼ ☿
14 Thursday	
02 56	☉ ∥ ♂
07 03	☽ ∥ ♇
09 38	☽ □ ♅
10 30	☽ ± ☿
18 26	☽ Q ♃
21 05	☽ Q ♂
15 Friday	
00 38	☽ ⊼ ☉
03 31	☽ ⚼ ☿
07 23	☽ ☌ ♄
13 33	☽ ± ☉
13 36	☽ □ ♇
18 24	☽ ⚹ ♃
23 58	☿ ⊼ ♂
16 Saturday	
00 12	☽ # ♀
01 57	☽ ☍ ♀
03 24	☉ ⚼ ♂
07 03	☿ ⊻ ♆
08 23	☽ △ ☿
08 26	☽ ⊼ ♆
08 36	☽ □ ♂
09 06	☽ ⚼ ☉
20 31	☽ ± ♆
21 11	☽ △ ♅
17 Sunday	
00 44	☽ ∠ ♃
01 41	☽ Q ♇
03 48	♂ △ ♆
14 43	☽ ⚼ ♆
18 00	☽ △ ☉
19 03	☽ ⊥ ♃
19 13	☽ ⊻ ♄
18 Monday	
01 57	☽ ⚹ ♇
03 26	☽ ⚼ ♅
07 17	☽ ⊥ ♄
07 18	☽ ⊻ ♃
08 05	☉ ⊼ ♄
17 32	☽ ⊼ ♀
19 37	☽ □ ☿
20 50	☽ △ ♂
21 08	☽ △ ♆
21 37	☽ ± ♅
19 Tuesday	
01 23	☽ ∠ ♄
06 14	☽ ± ♀
06 46	☽ # ♀
08 18	☽ ∠ ♇
09 43	☽ ⊼ ♅
20 Wednesday	
01 10	☽ ⚼ ♀
02 24	☽ ⊥ ♇
02 56	☽ ⚼ ♂
07 22	☽ ⚹ ♄
07 34	☽ Q ☿
10 05	☿ ⚹ ♀
11 48	☽ □ ☉
13 34	☿ St D
14 23	☽ ⊻ ♇
20 05	☽ ☌ ♃
20 50	☽ ± ♂
21 Thursday	
05 05	☽ ∥ ♇
07 02	☽ Q ♄
07 22	☽ ⚹ ☿
08 15	☽ △ ♀
08 39	☽ ⊼ ♂
09 14	☽ □ ♆
14 46	☽ ∥ ♃
18 04	♀ □ ♂
20 13	☉ ⚹ ♇
21 12	☽ ☍ ♅
23 06	☽ ∥ ♄
23 36	♂ St D
22 Friday	
02 13	☽ Q ☉
06 22	☽ # ♅
11 25	♀ ⊻ ♆
12 43	☽ ∠ ☿
17 03	☽ # ♆
17 39	☽ □ ♄
20 03	☽ Q ♆
22 02	☽ # ☉
23 Saturday	
00 40	☽ ☌ ♇
02 59	☽ ⚹ ☉
05 59	☽ ⊥ ☿
06 25	☽ ⊻ ♃
11 44	☽ # ♂
17 20	☽ ⊻ ☿
17 38	☽ ⊥ ♃
17 50	☽ ☍ ♂
18 31	☽ ⚹ ♆
19 32	☽ □ ♀
24 Sunday	
05 35	☽ ⊼ ♅
08 51	☽ ∠ ☉
10 10	☽ ∠ ♃
12 53	☽ # ☿
16 23	☽ ± ♅
17 16	☿ ⊼ ♂
18 51	♅ St D
21 43	☽ ∠ ♆
25 Monday	
00 24	☽ △ ♄
02 11	☽ ⊥ ☉
04 24	☉ □ ♃
04 55	☽ Q ♀
07 15	☽ ⊻ ♇
08 18	☽ ⚼ ♅
10 29	☽ ∥ ☿
12 54	☽ ⚹ ♃
13 29	☽ ⊻ ☉
13 34	☽ ⊥ ♆
15 53	♂ ± ♃
17 43	☽ ⊥ ♇
20 03	☿ ⊻ ♆
23 19	☽ ⊼ ♂
23 58	☽ ⊻ ♆
26 Tuesday	
00 04	☽ ☌ ☿
02 18	☽ ⚹ ♀
02 21	☽ ⚼ ♄
09 08	☽ ∠ ♇
09 35	☽ ± ♂
09 38	☽ Q ♃
09 42	☽ ∥ ♂
10 08	☽ △ ♅
12 59	☉ ∥ ♆
17 32	☽ ± ♄
27 Wednesday	
00 46	☽ ⚼ ♂
03 12	☽ ∥ ♆
03 31	☽ ⊼ ♄
03 57	☽ ∥ ☉
04 12	☽ ∠ ♀
10 16	☽ ⚹ ♇
12 17	☽ ∥ ♅
12 57	☉ ⊥ ♀
15 50	☽ □ ♃
18 09	☽ # ♄
19 23	☽ ⊥ ♀
19 42	☽ ☌ ☉
28 Thursday	
01 03	☽ # ♃
01 37	☽ △ ♂
02 09	☽ ☌ ♆
03 55	☽ ⊻ ☿
05 24	☽ ⊻ ♀
05 57	☽ Q ♇
07 25	☿ ⚼ ♄
09 39	☽ # ♇
11 44	☽ □ ♅
14 05	☽ ⊥ ☿
29 Friday	
04 16	☽ ☍ ♄
05 16	☽ ∠ ☿
11 05	☽ □ ♇
16 45	☽ △ ♃
16 48	☽ Q ♅
23 41	☽ ⊻ ☉
30 Saturday	
02 23	☽ □ ♂
02 46	☽ ⊻ ♆
06 36	☽ ⚹ ☿
06 43	☽ ☌ ♀
07 08	☽ ∥ ♀
10 01	☿ ⚹ ♀
10 02	☽ ⊥ ☉
12 09	☽ ⚹ ♅
12 28	☽ ⊥ ♆
17 04	☽ ⚼ ♃

All ephemeris data is given at 12.00 UT and the Moon's longitude is additionally given for 24.00 UT

MAY 2044

LONGITUDES

Date	Sidereal time h m s	Sun ☉ ° ' "	Moon ☽ ° ' "	Moon ☽ 24.00 ° ' "	Mercury ☿ ° '	Venus ♀ ° '	Mars ♂ ° '	Jupiter ♃ ° '	Saturn ♄ ° '	Uranus ♅ ° '	Neptune ♆ ° '	Pluto ♇ ° '
01	02 40 15	11 ♉ 45 48	02 ♋ 43 04	10 ♋ 01 57	15 ♈ 31	14 ♊ 55	12 ♍ 07	06 ♒ 07	28 ♏ 05	18 ♌ 02	12 ♉ 16	02 ♓ 30
02	02 44 12	12 44 02	17 ♋ 16 51	24 ♋ 27 18	16 20	15 05	12 14	06 12	28 R 01	18 03	12 18	02 31
03	02 48 08	13 42 14	01 ♌ 32 59	08 ♌ 33 45	17 12	15 13	12 22	06 16	27 57	18 03	12 20	02 32
04	02 52 05	14 40 24	15 ♌ 29 33	22 ♌ 20 27	18 08	15 18	12 30	06 19	27 52	18 04	12 22	02 33
05	02 56 01	15 38 32	29 ♌ 06 37	05 ♍ 48 15	19 06	15 22	12 39	06 23	27 48	18 04	12 25	02 33
06	02 59 58	16 36 38	12 ♍ 25 38	18 ♍ 59 02	20 08	15 R 22	12 49	06 26	27 44	18 05	12 27	02 34
07	03 03 54	17 34 42	25 ♍ 28 43	01 ♎ 54 59	21 12	15 21	12 59	06 30	27 39	18 05	12 29	02 35
08	03 07 51	18 32 44	08 ♎ 18 06	14 ♎ 38 17	22 19	15 17	13 10	06 33	27 35	18 06	12 31	02 35
09	03 11 48	19 30 44	20 ♎ 55 44	27 ♎ 10 40	23 29	15 11	13 21	06 36	27 30	18 07	12 34	02 36
10	03 15 44	20 28 43	03 ♏ 23 13	09 ♏ 33 31	24 42	15 02	13 33	06 38	27 26	18 08	12 36	02 36
11	03 19 41	21 26 40	15 ♏ 41 42	21 ♏ 47 53	25 57	14 51	13 45	06 41	27 22	18 09	12 38	02 37
12	03 23 37	22 24 35	27 ♏ 52 09	03 ♐ 54 39	27 14	14 37	13 58	06 43	27 17	18 10	12 40	02 37
13	03 27 34	23 22 29	09 ♐ 55 30	15 ♐ 54 52	28 34	14 21	14 12	06 46	27 13	18 11	12 43	02 38
14	03 31 30	24 20 21	21 ♐ 52 57	27 ♐ 49 57	29 ♈ 56	14 02	14 25	06 48	27 08	18 12	12 45	02 38
15	03 35 27	25 18 12	03 ♑ 46 10	09 ♑ 41 53	01 ♉ 21	13 41	14 40	06 49	27 04	18 13	12 47	02 39
16	03 39 23	26 16 02	15 ♑ 37 30	21 ♑ 33 23	02 47	13 18	14 55	06 51	26 59	18 14	12 49	02 39
17	03 43 20	27 13 51	27 ♑ 30 01	03 ♒ 27 53	04 16	12 53	15 10	06 52	26 55	18 15	12 52	02 40
18	03 47 17	28 11 38	09 ♒ 27 33	15 ♒ 29 35	05 48	12 26	15 26	06 54	26 50	18 16	12 54	02 40
19	03 51 13	29 ♉ 09 24	21 ♒ 34 36	27 ♒ 43 13	07 21	11 57	15 43	06 55	26 46	18 17	12 56	02 40
20	03 55 10	00 ♊ 07 09	03 ♓ 56 04	10 ♓ 13 47	08 57	11 26	16 00	06 56	26 41	18 19	12 58	02 41
21	03 59 06	01 04 53	16 ♓ 36 59	23 ♓ 06 14	10 35	10 54	16 17	06 56	26 37	18 20	13 00	02 41
22	04 03 03	02 02 35	29 ♓ 42 00	06 ♈ 24 43	12 15	10 20	16 35	06 57	26 32	18 22	13 02	02 41
23	04 06 59	03 00 17	13 ♈ 14 40	20 ♈ 11 57	13 58	09 45	16 53	06 57	26 28	18 23	13 05	02 42
24	04 10 56	03 57 58	27 ♈ 16 32	04 ♉ 28 09	15 42	09 09	17 12	06 57	26 23	18 25	13 07	02 42
25	04 14 52	04 55 37	11 ♉ 46 21	19 ♉ 10 24	17 29	08 32	17 31	06 R 57	26 19	18 26	13 09	02 42
26	04 18 49	05 53 16	26 ♉ 39 24	04 ♊ 12 14	19 18	07 55	17 50	06 57	26 15	18 28	13 11	02 42
27	04 22 46	06 50 54	11 ♊ 47 38	19 ♊ 24 16	21 10	07 18	18 10	06 56	26 10	18 30	13 13	02 42
28	04 26 42	07 48 30	27 ♊ 00 42	04 ♋ 35 37	23 03	06 40	18 30	06 56	26 06	18 31	13 15	02 42
29	04 30 39	08 46 05	12 ♋ 07 44	19 ♋ 35 57	24 59	06 02	18 51	06 55	26 02	18 33	13 17	02 43
30	04 34 35	09 43 39	26 ♋ 59 19	04 ♌ 17 09	26 56	05 25	19 12	06 54	25 57	18 35	13 19	02 43
31	04 38 32	10 ♊ 41 12	11 ♌ 28 55	18 ♌ 34 22	28 ♉ 56	04 ♊ 49	19 ♍ 33	06 ♒ 53	25 ♏ 53	18 ♌ 37	13 ♉ 21	02 ♓ 43

Date	Moon True ☊ ° '	Moon Mean ☊ ° '	Moon ☽ Latitude ° '
01	18 ♓ 46	17 ♓ 37	04 N 57
02	18 R 42	17 34	04 29
03	18 39	17 31	03 45
04	18 D 38	17 28	02 48
05	18 39	17 24	01 43
06	18 40	17 21	00 N 33
07	18 R 40	17 18	00 S 36
08	18 38	17 15	01 43
09	18 33	17 12	02 43
10	18 27	17 08	03 35
11	18 17	17 05	04 15
12	18 06	17 02	04 44
13	17 54	16 59	04 59
14	17 42	16 56	05 01
15	17 32	16 53	04 50
16	17 23	16 49	04 27
17	17 16	16 46	03 51
18	17 13	16 43	03 06
19	17 11	16 40	02 11
20	17 D 11	16 37	01 10
21	17 11	16 34	00 S 03
22	17 R 11	16 30	01 N 06
23	17 08	16 27	02 13
24	17 04	16 24	03 15
25	16 56	16 21	04 06
26	16 47	16 18	04 42
27	16 36	16 14	04 59
28	16 26	16 11	04 55
29	16 16	16 08	04 31
30	16 09	16 05	03 48
31	16 ♓ 05	16 ♓ 02	02 N 52

DECLINATIONS

Date	Sun ☉ ° '	Moon ☽ ° '	Mercury ☿ ° '	Venus ♀ ° '	Mars ♂ ° '	Jupiter ♃ ° '	Saturn ♄ ° '	Uranus ♅ ° '	Neptune ♆ ° '	Pluto ♇ ° '
01	15 N 22	28 N 22	03 N 34	27 N 38	08 N 53	19 S 03	17 S 28	16 N 06	13 N 53	20 S 58
02	15 39	26 46	03 48	27 38	08 48	19 02	17 27	16 06	13 54	20 58
03	15 57	23 28	04 04	27 36	08 43	19 02	17 26	16 06	13 55	20 58
04	16 14	18 52	04 22	27 34	08 37	19 01	17 25	16 06	13 55	20 58
05	16 31	13 23	04 41	27 31	08 32	19 00	17 24	16 06	13 56	20 58
06	16 48	07 25	05 02	27 27	08 26	18 59	17 23	16 05	13 57	20 58
07	17 04	01 N 14	05 25	27 23	08 20	18 59	17 22	16 05	13 57	20 58
08	17 21	04 S 52	05 49	27 18	08 14	18 58	17 21	16 05	13 58	20 58
09	17 36	10 41	06 14	27 12	08 07	18 58	17 20	16 05	13 59	20 58
10	17 52	16 00	06 41	27 05	08 01	18 57	17 19	16 04	13 59	20 58
11	18 07	20 36	07 09	26 58	07 54	18 57	17 18	16 04	14 00	20 59
12	18 22	24 17	07 38	26 49	07 47	18 56	17 17	16 04	14 01	20 59
13	18 37	26 52	08 08	26 40	07 40	18 56	17 16	16 03	14 01	20 59
14	18 51	28 12	08 40	26 30	07 33	18 56	17 15	16 03	14 02	20 59
15	19 05	28 13	09 12	26 19	07 25	18 55	17 14	16 03	14 03	20 59
16	19 19	26 56	09 46	26 07	07 18	18 55	17 13	16 02	14 03	20 59
17	19 32	24 26	10 20	25 54	07 10	18 55	17 12	16 02	14 04	20 59
18	19 45	20 52	10 55	25 40	07 02	18 55	17 11	16 01	14 05	21 00
19	19 58	16 23	11 31	25 26	06 53	18 55	17 10	16 01	14 05	21 00
20	20 10	11 09	12 07	25 10	06 45	18 55	17 09	16 01	14 06	21 00
21	20 22	05 S 20	12 44	24 54	06 37	18 55	17 08	16 00	14 07	21 00
22	20 34	00 N 53	13 22	24 37	06 28	18 55	17 07	16 00	14 07	21 00
23	20 45	07 16	13 59	24 20	06 19	18 55	17 06	15 59	14 08	21 01
24	20 56	13 32	14 38	24 01	06 10	18 55	17 06	15 59	14 08	21 01
25	21 07	19 16	15 16	23 42	06 01	18 55	17 05	15 58	14 09	21 01
26	21 17	23 59	15 54	23 23	05 52	18 55	17 04	15 58	14 10	21 01
27	21 27	27 09	16 33	23 03	05 42	18 56	17 03	15 57	14 10	21 01
28	21 37	28 19	17 11	22 43	05 33	18 56	17 02	15 56	14 11	21 02
29	21 46	27 23	17 49	22 22	05 23	18 57	17 01	15 56	14 11	21 02
30	21 54	24 29	18 26	22 02	05 13	18 57	17 00	15 55	14 12	21 02
31	22 N 03	20 N 05	19 N 03	21 N 41	05 N 03	18 S 58	16 S 59	15 N 55	14 N 13	21 S 03

ZODIAC SIGN ENTRIES

Date	h	m	Planets
01	07	34	☽ ♋
03	09	22	☽ ♌
05	13	35	☽ ♍
07	20	25	☽ ♎
10	05	27	☽ ♏
12	16	14	☽ ♐
14	13	08	☿ ♉
15	04	23	☽ ♑
17	17	02	☽ ♒
20	04	25	☽ ♓
20	09	02	☉ ♊
22	12	32	☽ ♈
24	16	34	☽ ♉
26	17	20	☽ ♊
28	16	43	☽ ♋
30	16	56	☽ ♌

LATITUDES

Date	Mercury ☿ ° '	Venus ♀ ° '	Mars ♂ ° '	Jupiter ♃ ° '	Saturn ♄ ° '	Uranus ♅ ° '	Neptune ♆ ° '	Pluto ♇ ° '
01	02 S 45	05 N 05	02 N 01	00 S 20	02 N 19	00 N 43	01 S 43	11 S 10
04	02 58	04 58	01 54	00 20	02 20	00 43	01 43	11 11
07	03 05	04 47	01 47	00 21	02 20	00 43	01 43	11 12
10	03 06	04 31	01 40	00 21	02 20	00 43	01 43	11 13
13	03 01	04 11	01 34	00 22	02 19	00 43	01 43	11 14
16	02 51	03 45	01 28	00 22	02 19	00 42	01 43	11 15
19	02 35	03 15	01 22	00 23	02 19	00 42	01 43	11 16
22	02 16	02 39	01 16	00 23	02 19	00 42	01 43	11 17
25	01 51	02 00	01 11	00 24	02 19	00 42	01 43	11 18
28	01 24	01 19	01 05	00 25	02 19	00 42	01 43	11 19
31	00 S 53	00 N 36	01 N 00	00 S 25	02 N 18	00 N 42	01 S 43	11 S 20

DATA

Julian Date	2467737
Delta T	+79 seconds
Ayanamsa	24° 28' 38"
Synetic vernal point	04° ♓ 38' 21"
True obliquity of ecliptic	23° 26' 10"

MOON'S PHASES, APSIDES AND POSITIONS ☽

Date	h	m	Phase	Longitude ° '	Eclipse Indicator
04	10	28	☽	14 ♌ 37	
12	00	17	○	21 ♏ 56	
20	04	02	☾	29 ♒ 48	
27	03	40	●	06 ♊ 31	

Day	h	m		
15	11	08	Apogee	
28	00	05	Perigee	
01	04	35	Max dec	28° N 27'
07	16	49	OS	
15	00	20	Max dec	28° S 22'
22	08	38	ON	
28	13	16	Max dec	28° N 20'

LONGITUDES

Date	Chiron ⚷ ° '	Ceres ⚳ ° '	Pallas ⚴ ° '	Juno ⚵ ° '	Vesta ⚶ ° '	Black Moon Lilith ⚸ ° '
01	26 ♌ 08	26 ♈ 44	14 ♓ 39	18 ♓ 21	13 ♓ 35	27 ♐ 14
11	26 ♌ 21	00 ♉ 39	17 ♓ 05	22 ♓ 27	17 ♓ 47	28 ♐ 21
21	26 ♌ 45	04 ♉ 32	19 ♓ 17	26 ♓ 26	21 ♓ 45	29 ♐ 29
31	27 ♌ 19	08 ♉ 21	21 ♓ 13	00 ♈ 19	25 ♓ 28	00 ♑ 36

ASPECTARIAN

h m Aspects

01 Sunday
01 37 ☽ ∠ ☉
03 03 ☽ Q ☿
03 05 ☽ ∠ ♆
04 29 ☽ ⊼ ♄
07 44 ☽ Q ♂
07 45 ☽ ± ♃
11 39 ☽ △ ♇
12 31 ☽ ∠ ♅
13 02 ♂ ± ♃
14 13 ☽ ± ♄
17 36 ☽ ⊼ ♃
22 02 ☉ △ ♂

02 Monday
00 49 ☉ ☌ ♆
02 22 ☽ ∥ ♀
03 19 ☽ ⊥ ♅
03 34 ☽ ⚹ ♂
03 43 ☽ ⚹ ♆
03 55 ☽ ⚹ ☉
04 57 ☽ ᖷ ♄
08 19 ☽ ⊻ ♀
10 19 ☽ □ ☿
12 24 ☽ ᖷ ♇
13 16 ☽ ⊻ ♅
18 24 ☽ ⊥ ♀
23 46 ☽ Q ♆

03 Tuesday
01 23 ☽ Q ☉
03 29 ☽ ± ♇
04 21 ♂ △ ♆
04 51 ☽ ∠ ♂
05 55 ☽ △ ♄
09 43 ☽ ∠ ♀
13 40 ☽ ⊼ ♇
20 05 ☽ ☍ ♃
20 18 ☽ ⊥ ♂
20 44 ☿ ∠ ♇

04 Wednesday
00 14 ☉ ∥ ♅
01 42 ☽ ⋕ ♇
06 34 ☽ □ ♆
06 45 ☽ ⊻ ♂
08 43 ☉ Q ♇
10 14 ☿ △ ♅
10 28 ☽ □ ☉
11 17 ☽ ⋕ ♃
11 40 ☽ ⚹ ♀
16 29 ☽ ☌ ♅
16 57 ☽ △ ☿
17 10 ☿ Q ♃
18 40 ☽ ⋕ ♄
23 02 ☿ ± ♂
23 17 ☽ ∥ ☉

05 Thursday
00 30 ☽ ∥ ♅
04 40 ☉ ⊻ ♀
08 52 ☽ Q ♀
09 41 ☽ □ ♄
09 44 ☽ ∥ ♆
18 10 ☽ ☍ ♇
21 41 ☽ ᖷ ☿

06 Friday
01 06 ☽ ⊼ ♃
07 55 ☽ ∥ ♂
09 01 ♀ St R
12 01 ☽ ± ♃
12 02 ☽ △ ♆
12 43 ☽ ☌ ♂
15 23 ☽ ± ☿
17 22 ☽ □ ♀
17 59 ☽ Q ♄
20 15 ☽ △ ☉
20 45 ☽ ∥ ☿
22 21 ☽ ⊻ ♅

07 Saturday
03 22 ☽ ⊼ ☿
04 36 ☽ ᖷ ♃
09 25 ☽ ⊥ ♅
15 45 ☽ ᖷ ♆
16 01 ☽ ⚹ ♄
21 08 ☿ ± ♄

08 Sunday
00 52 ☉ □ ♅
01 15 ☽ ⊼ ♇
02 13 ☽ ∠ ♅
02 19 ☽ ᖷ ☉
08 38 ☽ ± ♆
08 41 ☽ △ ♃
12 32 ☽ ± ♇
12 44 ☉ ⋕ ♄
16 04 ☽ ⋕ ☿
20 01 ☽ ⊼ ♆
20 03 ☽ ∠ ♄
20 42 ☽ ± ☉
21 20 ☽ ⊻ ♂

09 Monday
01 08 ☽ △ ♀
01 25 ☽ ⋕ ♂
05 38 ☽ ᖷ ♇
06 37 ☽ ⚹ ♅
08 56 ☽ ⊥ ♂
09 04 ☽ ⊼ ☉
13 06 ☽ ⊥ ♄
17 25 ☽ ☍ ☿

10 Tuesday
00 34 ☽ ⊻ ♄
02 29 ☽ ∠ ♂
02 33 ☽ ⋕ ♆
05 35 ☽ ᖷ ♀
05 41 ☽ Q ♅
10 29 ☽ △ ♇
12 20 ☽ ⋕ ♅
18 20 ☽ □ ♃
18 26 ☽ ∥ ♄
21 45 ☽ ⋕ ☉
22 49 ☽ ± ♀

11 Wednesday
02 51 ☽ ∥ ♃
06 00 ☽ ☍ ♆
08 08 ☽ ⚹ ♂
10 21 ☽ ⊼ ♀
14 10 ☽ ∥ ♇
16 49 ☽ □ ♅

12 Thursday
00 17 ☽ ☍ ☉
05 45 ☽ Q ♃
08 10 ☽ Q ♂
10 35 ☽ ⊼ ☿
10 51 ☽ ☌ ♄
12 53 ☿ ⊼ ♄
17 50 ☿ ∥ ♂
21 27 ☽ □ ♇
23 58 ☽ ± ☿

13 Friday
05 39 ☽ ⚹ ♃
09 52 ☽ ⋕ ♀
17 36 ☽ ⊼ ♆
19 03 ♀ □ ♂
20 13 ☽ ᖷ ☿
20 38 ☽ ☍ ♀
20 43 ☽ □ ♂
23 19 ☿ ∠ ♀

14 Saturday
01 20 ☿ ᖷ ♂
04 34 ☽ △ ♅
05 40 ☽ ± ♆
09 30 ☽ Q ♇
11 49 ☽ ∠ ♃
17 23 ☽ ⊼ ☉
19 21 ☉ ⋕ ♃
22 32 ☽ ⊻ ♄
23 52 ☽ ᖷ ♆

15 Sunday
06 02 ☽ ⊥ ♃
06 25 ☽ △ ☿
06 34 ☽ ± ☉
09 44 ☽ ⚹ ♇
10 35 ☽ ⊥ ♄
10 52 ☽ ᖷ ♅
18 12 ☽ ⊻ ♃

16 Monday
02 24 ☽ ᖷ ☉
04 41 ☽ ∠ ♄
05 07 ☽ ± ♅
06 19 ☽ △ ♆
07 28 ☽ ⊼ ♀
09 46 ☿ ⚹ ♇
10 32 ☽ △ ♂
16 07 ☽ ∠ ♇
17 10 ♂ Q ♄
17 17 ☽ ⊼ ♅
19 12 ☽ ± ♀
22 03 ☽ ⋕ ♀

17 Tuesday
04 37 ☉ ☍ ♄
10 18 ☽ ⊥ ♇
10 49 ☽ ⚹ ♄
11 25 ☽ △ ☉
12 45 ☽ ᖷ ♀
13 28 ♀ ⊻ ♆
17 30 ☽ ᖷ ♂
22 23 ☽ ⊻ ♇

18 Wednesday
03 36 ☽ □ ☿
06 52 ☽ ☌ ♃
10 46 ☽ Q ♄
11 16 ☽ ∥ ♇
11 58 ☽ ± ♂
17 42 ☽ △ ♀
18 06 ☽ ⋕ ☉
18 52 ☽ □ ♆
19 35 ☿ ⊥ ♀
23 01 ☽ ∥ ♃

19 Thursday
00 10 ☽ ⊼ ♂
05 10 ☿ □ ♃
05 31 ☽ ☍ ♅
08 04 ☽ ∥ ♄
13 47 ☽ ⋕ ♅
20 29 ☽ Q ☿
22 05 ☽ □ ♄
22 53 ☽ ⋕ ♆

20 Friday
04 02 ☽ □ ☉
06 16 ☽ Q ♆
08 11 ☽ ⋕ ☿
09 35 ☽ ☌ ♇
17 44 ☽ ⊻ ♃
22 59 ☽ ⚹ ☿

21 Saturday
01 42 ☽ □ ♀
05 06 ☽ ⊥ ♃
05 13 ☽ ⚹ ♆
06 45 ☽ ⋕ ♂
11 22 ☽ ☍ ♂
15 12 ☽ ⊼ ♅
15 23 ☿ ⊻ ♀
16 57 ☽ Q ☉
21 52 ☽ ∠ ♃

22 Sunday
02 17 ☽ ± ♅
06 18 ☽ △ ♄
06 56 ☽ ∠ ☿
09 00 ☽ ∠ ♆
09 38 ☽ Q ♀
16 32 ☽ ⚹ ☉
17 22 ☽ ⊻ ♇
18 35 ☽ ᖷ ♅
23 20 ☿ ☌ ♆

23 Monday
00 57 ☽ ⚹ ♃
01 09 ☽ ⊥ ♆
01 24 ☽ ⊥ ☿
04 02 ☽ ⊥ ♇
04 10 ☉ □ ♇
06 09 ☽ ⚹ ♀
08 30 ☽ ∥ ♂
08 55 ☽ ᖷ ♄
11 42 ☽ ⊻ ♆
13 26 ☽ ⊻ ☿
17 18 ☿ ∥ ♆
18 27 ☽ ⊼ ♂
19 42 ☽ ∠ ♇
20 51 ☽ ∠ ☉
20 54 ☽ △ ♅
21 52 ☽ Q ♃
22 06 ☿ Q ♇

24 Tuesday
00 23 ☽ ± ♄
04 57 ☽ ± ♂
06 57 ☽ ∠ ♀
10 31 ☽ ⊼ ♄
13 15 ☽ ⊥ ☉
14 03 ♃ St R
14 25 ☽ ∥ ♆
16 53 ☽ ∥ ☿
20 24 ☽ ᖷ ♂
21 04 ☽ ⚹ ♇
21 25 ☽ ⊥ ♀
21 50 ☽ ∥ ♅
22 07 ☉ ⋕ ♇
23 58 ☽ ⊻ ☉

25 Wednesday
02 28 ☽ ⋕ ♄
04 06 ☽ □ ♃
06 55 ☽ ⊻ ♀
10 27 ☽ ⋕ ♃
12 20 ☿ △ ♂
14 15 ☽ ☌ ♆
16 46 ☽ Q ♇
20 12 ☽ ⋕ ♇
20 59 ☽ ∥ ☉
21 31 ☽ △ ♂
22 34 ☽ ☌ ☿
22 50 ☽ □ ♅

26 Thursday
00 46 ☿ □ ♅
08 46 ☽ ∥ ♀
11 21 ☽ ☍ ♄
13 55 ☿ ∥ ♅
21 37 ☽ □ ♇

27 Friday
02 50 ☉ Q ♅
03 36 ☽ Q ♅
03 40 ☽ ☌ ☉
04 20 ☽ △ ♃
05 10 ☽ ☌ ♀
14 15 ☽ ⊻ ♆
14 16 ☉ △ ♃
18 42 ☉ ☌ ♀
22 16 ☽ □ ♂
22 35 ☽ ⚹ ♅
23 44 ☽ ⊥ ♆

28 Saturday
01 41 ♀ △ ♃
03 59 ☽ ᖷ ♃
04 51 ☽ ⊻ ☿
06 20 ☿ ⋕ ♄
10 34 ☽ ⊼ ♄
13 21 ♂ ⊻ ♅
13 58 ☽ ∠ ♆
15 41 ☽ ⊥ ☿
17 16 ♀ Q ♅
18 11 ☽ ± ♃
20 00 ☽ ± ♄
21 01 ☽ △ ♇
22 19 ☽ ∠ ♅

29 Sunday
02 41 ☽ ⊻ ♀
03 23 ☽ Q ♂
03 42 ☽ ⊼ ♃
06 16 ☽ ⊻ ☉
08 03 ☽ ∠ ☿
10 15 ☽ ᖷ ♄
11 52 ☽ ⊥ ♀
12 41 ☽ ⊥ ♅
13 52 ☽ ⚹ ♆
16 31 ☽ ⊥ ☉
20 57 ☽ ᖷ ♇
22 20 ☽ ⊻ ♅
23 03 ☽ ⚹ ♂

30 Monday
00 26 ☿ ☍ ♄
01 45 ☽ ∠ ♀
08 03 ☽ ∠ ☉
09 17 ☽ Q ♆
10 19 ☽ △ ♄
11 33 ☽ ± ♇
11 54 ☽ ⚹ ☿
18 03 ☉ ∥ ♀
21 24 ☽ ⊼ ♇

31 Tuesday
00 09 ☽ ∠ ♂
01 19 ☽ ⚹ ♀
02 25 ☽ ∥ ☉
03 22 ☽ ∥ ♀
04 19 ☽ ☍ ♃
07 17 ☽ ⋕ ♇
08 09 ☿ ⋕ ♃
10 34 ☽ ⚹ ☉
10 56 ☽ Q ☿
15 10 ☽ □ ♆
15 35 ☽ ⊥ ♂
16 20 ☽ ∥ ☿
17 17 ☽ ⋕ ♃
20 39 ☽ Q ♀

All ephemeris data is given at 12.00 UT and the Moon's longitude is additionally given for 24.00 UT

JUNE 2044

Date	Sidereal time h m s	Sun ☉	Moon ☽	Moon ☽ 24.00	Mercury ☿	Venus ♀	Mars ♂	Jupiter ♃	Saturn ♄	Uranus ♅	Neptune ♆	Pluto ♇
01	04 42 28	11 ♊ 38 43	25 ♌ 33 21	02 ♍ 25 57	00 ♊ 58	04 ♊ 14	19 ♍ 55	06 ♒ 51	25 ♏ 49	18 ♌ 39	13 ♉ 24	02 ♓ 43
02	04 46 25	12 36 12	09 ♍ 12 22	15 ♍ 52 52	03 01	03 R 39	20 17	06 R 50	25 R 44	18 41	13 26	02 R 43
03	04 50 21	13 33 41	22 ♍ 27 52	28 ♍ 57 46	05 06	03 06	20 40	06 48	25 40	18 43	13 28	02 43
04	04 54 18	14 31 08	05 ♎ 23 01	11 ♎ 44 04	07 13	02 34	21 03	06 46	25 36	18 45	13 30	02 43
05	04 58 15	15 28 34	18 ♎ 01 23	24 ♎ 15 23	09 21	02 04	21 26	06 44	25 32	18 47	13 32	02 43
06	05 02 11	16 25 58	00 ♏ 26 27	06 ♏ 34 57	11 31	01 36	21 49	06 41	25 28	18 49	13 34	02 42
07	05 06 08	17 23 22	12 ♏ 41 14	18 ♏ 45 34	13 41	01 10	22 13	06 39	25 24	18 51	13 36	02 42
08	05 10 04	18 20 45	24 ♏ 48 13	00 ♐ 49 23	15 53	00 46	22 37	06 36	25 20	18 53	13 37	02 42
09	05 14 01	19 18 06	06 ♐ 49 17	12 ♐ 48 05	18 05	00 24	23 02	06 33	25 16	18 56	13 39	02 42
10	05 17 57	20 15 27	18 ♐ 45 57	24 ♐ 43 03	20 17	00 ♊ 04	23 26	06 30	25 12	18 58	13 41	02 42
11	05 21 54	21 12 48	00 ♑ 39 33	06 ♑ 35 36	22 29	29 ♉ 47	23 51	06 27	25 09	19 00	13 43	02 42
12	05 25 50	22 10 07	12 ♑ 31 26	18 ♑ 27 16	24 40	29 32	24 17	06 24	25 05	19 03	13 45	02 41
13	05 29 47	23 07 26	24 ♑ 23 22	00 ♒ 20 00	26 52	29 19	24 42	06 20	25 01	19 05	13 47	02 41
14	05 33 44	24 04 44	06 ♒ 17 33	12 ♒ 16 23	29 ♊ 02	29 09	25 08	06 16	24 57	19 07	13 49	02 41
15	05 37 40	25 02 02	18 ♒ 16 56	24 ♒ 19 41	01 ♋ 11	29 01	25 34	06 12	24 54	19 10	13 51	02 40
16	05 41 37	25 59 19	00 ♓ 25 07	06 ♓ 33 49	03 19	28 55	26 01	06 08	24 50	19 12	13 52	02 40
17	05 45 33	26 56 36	12 ♓ 46 20	19 ♓ 03 15	05 26	28 52	26 27	06 04	24 47	19 15	13 54	02 40
18	05 49 30	27 53 53	25 ♓ 25 10	01 ♈ 52 39	07 31	28 D 52	26 54	05 59	24 43	19 18	13 56	02 39
19	05 53 26	28 51 09	08 ♈ 26 15	15 ♈ 06 25	09 34	28 53	27 22	05 55	24 40	19 20	13 58	02 39
20	05 57 23	29 ♊ 48 26	21 ♈ 53 32	28 ♈ 47 53	11 35	28 57	27 49	05 50	24 37	19 23	13 59	02 38
21	06 01 19	00 ♋ 45 42	05 ♉ 49 33	12 ♉ 58 28	13 35	29 04	28 17	05 45	24 34	19 26	14 01	02 38
22	06 05 16	01 42 58	20 ♉ 14 21	27 ♉ 36 41	15 32	29 12	28 45	05 40	24 31	19 29	14 03	02 38
23	06 09 13	02 40 14	05 ♊ 04 44	12 ♊ 37 32	17 27	29 23	29 13	05 35	24 28	19 31	14 04	02 37
24	06 13 09	03 37 29	20 ♊ 13 55	27 ♊ 52 33	19 21	29 35	29 ♍ 41	05 30	24 25	19 34	14 06	02 37
25	06 17 06	04 34 45	05 ♋ 32 02	13 ♋ 10 55	21 12	29 ♉ 50	00 ♎ 10	05 24	24 22	19 37	14 08	02 36
26	06 21 02	05 32 00	20 ♋ 47 46	28 ♋ 21 19	23 01	00 ♊ 06	00 39	05 18	24 19	19 40	14 09	02 35
27	06 24 59	06 29 15	05 ♌ 50 22	13 ♌ 14 01	24 47	00 25	01 08	05 13	24 16	19 43	14 11	02 35
28	06 28 55	07 26 29	20 ♌ 31 29	27 ♌ 42 18	26 32	00 45	01 38	05 07	24 13	19 46	14 12	02 34
29	06 32 52	08 23 43	04 ♍ 46 10	11 ♍ 42 58	28 14	01 07	02 07	05 01	24 11	19 49	14 14	02 34
30	06 36 48	09 ♋ 20 56	18 ♍ 32 48	25 ♍ 15 54	29 ♋ 54	01 ♊ 30	02 ♎ 37	04 ♒ 54	24 ♏ 08	19 ♌ 52	14 ♉ 15	02 ♓ 33

Date	Moon True ☊	Moon Mean ☊	Moon ☽ Latitude
01	16 ♓ 03	15 ♓ 59	01 N 46
02	16 D 02	15 55	00 N 36
03	16 R 02	15 52	00 S 34
04	16 01	15 49	01 40
05	15 59	15 46	02 40
06	15 53	15 43	03 31
07	15 45	15 39	04 12
08	15 35	15 36	04 40
09	15 22	15 33	04 56
10	15 08	15 30	04 59
11	14 55	15 27	04 48
12	14 42	15 24	04 26
13	14 32	15 20	03 51
14	14 24	15 17	03 06
15	14 19	15 14	02 13
16	14 17	15 11	01 13
17	14 16	15 08	00 S 08
18	14 D 16	15 05	00 N 59
19	14 R 16	15 01	02 04
20	14 14	14 58	03 05
21	14 11	14 55	03 57
22	14 04	14 52	04 37
23	13 56	14 49	04 58
24	13 46	14 45	05 00
25	13 36	14 42	04 40
26	13 28	14 39	04 01
27	13 21	14 36	03 05
28	13 17	14 33	01 58
29	13 15	14 30	00 45
30	13 ♓ 15	14 ♓ 26	00 S 28

DECLINATIONS

Date	Sun ☉	Moon ☽	Mercury ☿	Venus ♀	Mars ♂	Jupiter ♃	Saturn ♄	Uranus ♅	Neptune ♆	Pluto ♇
01	22 N 11	14 N 40	19 N 39	21 N 21	04 N 53	18 S 58	16 S 58	15 N 54	14 N 13	21 S 03
02	22 18	08 40	20 14	21 00	04 43	18 59	16 57	15 53	14 14	21 03
03	22 25	02 N 28	20 48	20 40	04 33	18 59	16 57	15 53	14 14	21 04
04	22 32	03 S 40	21 21	20 21	04 22	19 00	16 56	15 52	14 15	21 04
05	22 39	09 32	21 52	20 02	04 11	19 01	16 55	15 51	14 16	21 04
06	22 45	14 55	22 21	19 43	04 01	19 02	16 54	15 51	14 16	21 05
07	22 50	19 38	22 48	19 25	03 50	19 02	16 53	15 50	14 17	21 05
08	22 55	23 30	23 13	19 08	03 39	19 03	16 52	15 49	14 17	21 05
09	23 00	26 19	23 36	18 51	03 28	19 04	16 52	15 49	14 18	21 06
10	23 05	27 55	23 56	18 36	03 17	19 05	16 51	15 48	14 18	21 06
11	23 09	28 14	24 13	18 21	03 05	19 06	16 50	15 47	14 19	21 06
12	23 12	27 15	24 28	18 07	02 54	19 07	16 49	15 46	14 19	21 07
13	23 15	25 02	24 40	17 54	02 42	19 08	16 49	15 46	14 20	21 07
14	23 18	21 42	24 50	17 43	02 31	19 09	16 48	15 45	14 20	21 08
15	23 21	17 27	24 56	17 32	02 19	19 11	16 47	15 44	14 21	21 08
16	23 22	12 27	25 00	17 22	02 07	19 12	16 47	15 43	14 21	21 08
17	23 24	06 53	25 01	17 13	01 55	19 13	16 46	15 42	14 22	21 09
18	23 25	00 S 55	24 59	17 05	01 44	19 14	16 45	15 41	14 22	21 09
19	23 26	05 N 15	24 55	16 58	01 31	19 16	16 45	15 41	14 23	21 10
20	23 26	11 23	24 48	16 51	01 19	19 17	16 44	15 40	14 23	21 10
21	23 26	17 12	24 39	16 46	01 07	19 18	16 44	15 39	14 24	21 11
22	23 25	22 14	24 27	16 42	00 55	19 20	16 43	15 38	14 24	21 11
23	23 25	26 02	24 14	16 38	00 42	19 21	16 42	15 37	14 25	21 12
24	23 23	28 04	23 58	16 36	00 30	19 23	16 42	15 36	14 25	21 12
25	23 21	27 59	23 41	16 34	00 17	19 24	16 41	15 35	14 25	21 12
26	23 19	25 48	23 22	16 32	00 N 04	19 26	16 41	15 34	14 26	21 13
27	23 17	21 48	23 01	16 32	00 S 08	19 27	16 40	15 33	14 26	21 13
28	23 14	16 30	22 39	16 32	00 21	19 29	16 40	15 32	14 27	21 14
29	23 10	10 28	22 15	16 33	00 34	19 31	16 40	15 31	14 27	21 14
30	23 N 06	04 N 06	21 N 50	16 N 35	00 S 47	19 S 32	16 S 39	15 N 30	14 N 27	21 S 15

ZODIAC SIGN ENTRIES

Date	h	m	Planets
01	00	42	☿ ♊
01	19	44	☽ ♍
04	01	56	☽ ♎
06	11	09	☽ ♏
08	22	21	☽ ♐
10	17	15	♀ ♉
11	10	40	☽ ♑
13	23	20	☽ ♒
14	22	45	☿ ♋
16	11	11	☽ ♓
18	20	32	☽ ♈
20	16	51	☉ ♋
21	02	04	☽ ♉
23	03	51	☽ ♊
25	03	20	☽ ♋
25	03	35	♂ ♎
26	03	21	♀ ♊
27	02	38	☽ ♌
29	03	53	☽ ♍
30	13	29	☿ ♌

LATITUDES

Date	Mercury ☿	Venus ♀	Mars ♂	Jupiter ♃	Saturn ♄	Uranus ♅	Neptune ♆	Pluto ♇
01	00 S 42	00 N 22	00 N 58	00 S 25	02 N 18	00 N 42	01 S 43	11 S 20
04	00 S 10	00 S 20	00 53	00 26	02 18	00 42	01 43	11 21
07	00 N 22	01 00	00 49	00 27	02 17	00 42	01 43	11 22
10	00 51	01 36	00 44	00 27	02 17	00 42	01 43	11 23
13	01 17	02 09	00 39	00 28	02 16	00 42	01 43	11 24
16	01 36	02 37	00 35	00 28	02 16	00 42	01 43	11 25
19	01 50	03 01	00 31	00 29	02 15	00 41	01 43	11 26
22	01 56	03 21	00 27	00 30	02 15	00 41	01 44	11 27
25	01 56	03 38	00 23	00 30	02 14	00 41	01 44	11 28
28	01 50	03 51	00 19	00 31	02 14	00 41	01 44	11 29
31	01 N 38	04 S 01	00 N 16	00 S 31	02 N 13	00 N 41	01 S 44	11 S 30

DATA

Julian Date	2467768
Delta T	+79 seconds
Ayanamsa	24° 28' 43"
Synetic vernal point	04° ♓ 38' 16"
True obliquity of ecliptic	23° 26' 09"

MOON'S PHASES, APSIDES AND POSITIONS ☽

Date	h	m	Phase	Longitude	Eclipse Indicator
02	18	33	☽	12 ♍ 52	
10	15	16	○	20 ♐ 23	
18	17	00	☾	28 ♓ 06	
25	10	24	●	04 ♋ 31	

Day	h	m	
11	18	16	Apogee
25	09	20	Perigee

Day	h	m		
03	21	34	0S	
11	05	48	Max dec	28° S 17'
18	15	37	0N	
24	23	12	Max dec	28° N 18'

LONGITUDES

Date	Chiron ⚷	Ceres ⚳	Pallas ⚴	Juno ⚵	Vesta ⚶	Black Moon Lilith ⚸
01	27 ♌ 23	08 ♉ 43	21 ♓ 23	00 ♈ 42	25 ♓ 49	00 ♑ 43
11	28 ♌ 07	12 ♉ 27	22 ♓ 56	04 ♈ 25	29 ♓ 12	01 ♑ 20
21	28 ♌ 59	16 ♉ 05	24 ♓ 06	07 ♈ 57	02 ♈ 14	02 ♑ 57
31	29 ♌ 59	19 ♉ 37	24 ♓ 50	11 ♈ 15	04 ♈ 50	04 ♑ 04

ASPECTARIAN

h m Aspects

01 Wednesday
00 06 ☽ ☌ ♅
02 02 ☽ ⚺ ♂
02 09 ☽ # ♄
06 47 ☽ ∥ ♅
08 27 ☽ Q ☉
12 26 ☽ □ ♄
13 50 ☽ ∥ ♆
23 04 ☽ □ ☿
23 13 ♇ St R

02 Thursday
00 29 ☽ ☍ ♇
02 32 ☽ □ ♀
07 46 ☽ ⚻ ♃
08 27 ☿ □ ♇
08 38 ♀ # ♇
17 46 ☿ ☌ ♀
18 28 ☽ ± ♃
18 33 ☽ □ ☉
19 35 ☽ △ ♆
20 06 ☽ Q ♄

03 Friday
03 46 ☽ ∥ ♂
05 07 ☽ ⚺ ♅
08 25 ☿ ∥ ♀
08 36 ☽ ☌ ♂
09 22 ☉ ⚺ ♆
10 47 ☽ ⚼ ♃
16 08 ☽ ⊥ ♅
17 52 ☽ ⚹ ♄
23 06 ☽ ⚼ ♆
23 15 ☿ # ♇

04 Saturday
05 36 ♀ □ ♇
06 33 ☿ Q ♅
06 56 ☽ △ ♀
06 57 ☿ △ ♃
06 59 ☽ ⚻ ♇
08 55 ☽ ∠ ♅
14 35 ☽ △ ♃
14 41 ☽ # ♂
15 59 ☽ ± ♆
16 09 ☽ △ ☿
18 16 ☽ ± ♇
21 48 ☽ ∠ ♄

05 Sunday
03 23 ☽ ⚻ ♆
06 43 ☽ △ ☉
10 15 ☽ ⚼ ♀
11 24 ☽ ⚼ ♇
13 27 ☽ ⚹ ♅
14 53 ☽ ⊥ ♄
18 45 ☽ ⚺ ♂

06 Monday
02 24 ☽ ⚺ ♄
02 45 ☽ ⚼ ☿
02 56 ☽ ± ♀
04 45 ♂ ⚼ ♃
06 44 ☽ ⊥ ♂
08 56 ☽ # ♆
12 44 ☽ Q ♅
14 06 ☽ ⚼ ☉
14 11 ☽ ⚻ ♀
16 25 ☽ # ♅
16 25 ☽ △ ♇
21 35 ☽ ∥ ♄

07 Tuesday
00 03 ☽ ± ☿
00 10 ☽ □ ♃
00 53 ☽ ∠ ♂
08 42 ☽ ∥ ♃
09 14 ☽ ± ☉
10 51 ☽ # ♀
10 54 ☿ ⚺ ♆
13 47 ☽ ☍ ♆
14 25 ☽ ⚻ ☿
14 41 ☉ ∥ ☿
20 20 ☽ ∥ ♇
22 05 ☽ ⚻ ☉

08 Wednesday
00 13 ☽ □ ♅
07 30 ☽ ⚹ ♂
07 56 ☽ # ☉
09 47 ☽ # ☿
11 36 ☽ Q ♃
13 03 ☽ ☌ ♄
18 17 ♀ # ♃
23 31 ☽ ☍ ♀

09 Thursday
02 09 ☉ ⚹ ♅
03 45 ☽ □ ♇
08 17 ☽ Q ♂
11 28 ☽ ⚹ ♃
21 12 ☉ ⊥ ♆
21 26 ☿ ⚹ ♅

10 Friday
01 45 ☽ ⚻ ♆
05 29 ☿ ⊥ ♆
11 38 ☉ ☌ ☿
12 24 ☽ △ ♅
13 52 ☽ ± ♆
15 16 ☽ ☍ ☉
15 44 ☽ ☍ ☿
15 53 ☽ Q ♇
17 30 ☽ ∠ ♃
21 46 ☽ □ ♂

11 Saturday
00 55 ☽ ⚺ ♄
01 03 ☿ ⚼ ♃
08 04 ☽ ⚼ ♆
10 15 ☽ ⚻ ♀
11 35 ☽ ⊥ ♃
12 58 ☽ ⊥ ♄
16 07 ☽ ⚹ ♇
17 36 ☉ ⚼ ♃
18 47 ☽ ⚼ ♅
22 07 ☽ ± ♀
23 39 ☽ ⚺ ♃

12 Sunday
06 38 ☿ □ ♂
07 05 ☽ ∠ ♄
13 03 ☽ ± ♅
14 29 ☽ △ ♆
15 58 ☽ ⚼ ♀
16 20 ☿ ⚻ ♄
22 27 ☽ ∠ ♇

13 Monday
01 14 ☽ ⚻ ♅
09 13 ☽ ⚻ ☉
12 40 ☽ △ ♂
13 16 ☽ ⚹ ♄
14 47 ☽ # ☿
16 38 ☽ ⊥ ♇
18 07 ☽ ⚻ ☿
21 48 ☽ △ ♀
22 24 ☽ ± ☉

14 Tuesday
01 29 ☽ # ☉
03 19 ♂ ⚹ ♄
04 44 ☽ ⚺ ♇
08 54 ☽ ± ☿
09 31 ☿ ∠ ♆
11 16 ♂ ⊥ ♅
11 57 ☽ ☌ ♃
13 10 ☿ ⚺ ♀
13 20 ☽ Q ♄
15 34 ☽ ∥ ♇
18 05 ☽ ⚼ ☉
20 01 ☽ ⚼ ♂

15 Wednesday
01 21 ☿ ± ♃
02 52 ☽ ∥ ♃
03 07 ☽ □ ♆
06 55 ☽ ⚼ ☿
08 47 ☉ ⚻ ♄
08 51 ☿ ± ♄
11 36 ☽ # ♀
13 46 ☽ ☍ ♅
14 40 ☽ ± ♂
15 24 ☽ ∥ ♄
20 40 ☽ # ♅

16 Thursday
01 04 ☽ □ ♄
02 32 ☽ △ ☉
03 00 ☽ ⚻ ♂
03 15 ☽ # ♆
04 39 ☿ △ ♇
09 05 ☽ □ ♀
13 06 ☉ □ ♂
14 51 ☽ Q ♆
16 24 ☽ ☌ ♇
18 52 ☽ △ ☿
22 16 ☿ ∠ ♅
23 06 ☽ ⚺ ♃

17 Friday
05 44 ☿ ⊥ ♀
10 39 ☽ ⊥ ♃
14 10 ☽ ⚹ ♆
19 01 ☿ ⚻ ♃
19 50 ☽ Q ♀

18 Saturday
00 25 ☽ ⚻ ♅
03 43 ☽ ∠ ♃
06 37 ♀ St D
08 43 ☽ # ♂
10 42 ☽ △ ♄
11 46 ☽ ± ♅
14 53 ☽ ☍ ♂
17 00 ☽ □ ☉
18 26 ☽ ⚹ ♀
18 34 ☽ ∠ ♆
22 01 ☽ ∥ ♂

19 Sunday
01 26 ☽ ⚺ ♇
04 30 ☽ ⚼ ♅
07 26 ☽ ⚹ ♃
11 08 ☽ ⊥ ♆
12 23 ☽ ⊥ ♇
13 01 ☉ ⚺ ♀
13 10 ☿ ⚼ ♄
14 13 ☽ ⚼ ♄
14 25 ☽ □ ☿
14 46 ☉ ∠ ♆
21 52 ☽ ∠ ♀
21 58 ☽ ⚺ ♆

20 Monday
04 15 ☽ Q ☉
04 31 ☽ ∠ ♇
04 53 ☽ Q ♃
06 15 ☽ ± ♄
07 34 ☽ △ ♅
12 39 ☉ ± ♃
13 53 ☽ ⊥ ♀
16 44 ☽ ⚻ ♄
22 40 ☽ ⚻ ♂

21 Tuesday
00 08 ☽ ∥ ♆
00 21 ☽ ⚺ ♀
02 45 ☽ ⚹ ☉
03 35 ☽ Q ☿
05 24 ☽ ∥ ♅
06 35 ☽ ⚹ ♇
07 15 ☉ ± ♄
09 17 ☽ ± ♂
09 59 ☽ # ♄
10 07 ☿ ⊥ ♅
10 12 ☽ ∥ ♀
11 53 ☽ □ ♃
17 23 ☿ ⚹ ♆
18 15 ☿ ∠ ♀
21 33 ☽ # ♃

22 Wednesday
00 55 ☽ ⚼ ♂
01 46 ☽ ☌ ♆
02 45 ☽ Q ♇
03 03 ☽ ⚹ ☿
04 33 ♀ # ♄
05 48 ☽ ∠ ☉
06 32 ☽ # ♇
10 45 ☽ □ ♅
18 33 ☽ ∥ ☉
18 57 ☽ ☍ ♄
21 33 ☽ ⊥ ☉

23 Thursday
00 07 ☽ ∥ ☿
02 17 ☽ △ ♂
02 44 ☽ ☌ ♀
04 12 ♂ ⚼ ♆
07 11 ☽ ∠ ☿
07 53 ☽ ⚺ ☉
07 57 ☿ Q ♂
08 04 ☽ □ ♇
10 41 ☉ △ ♇
12 48 ☽ △ ♃
14 00 ☿ ⚼ ♇
15 55 ☽ Q ♅
23 36 ☽ ⊥ ☿

24 Friday
02 09 ♀ △ ♂
02 19 ☽ ⚺ ♆
10 25 ☽ ⚺ ☿
10 57 ☽ ⚹ ♅
11 47 ☽ ⊥ ♆
12 24 ☽ ⚼ ♃
14 58 ☿ ⚺ ♅
18 33 ☽ ⚻ ♄

25 Saturday
01 56 ☽ ∠ ♆
02 27 ☽ ± ♃
02 55 ☽ ⚺ ♀
03 19 ☽ □ ♂
03 55 ☽ ± ♄
07 24 ☽ △ ♇
10 24 ☽ ☌ ☉
10 33 ☽ ∠ ♅
11 48 ☽ ⚻ ♃
12 28 ☽ ⊥ ♀
13 00 ☉ ∠ ♅
17 59 ☽ ⚼ ♄

26 Sunday
00 43 ☽ ⊥ ♅
01 30 ☽ ⚹ ♆
02 51 ☽ ∠ ♀
06 48 ☉ ⚻ ♃
06 57 ☽ ⚼ ♇
08 30 ☽ Q ♂
10 12 ☽ ⚺ ♅
15 15 ☉ ∥ ☿
15 58 ☽ ☌ ☿
17 33 ☽ △ ♄
20 30 ☽ Q ♆
21 11 ☽ ± ♇

27 Monday
03 06 ☽ ⚹ ♀
04 02 ☽ ∥ ☉
04 11 ☽ ⚹ ♂
05 00 ☽ ∥ ☿
05 08 ☿ △ ♄
06 46 ☽ ⚻ ♇
09 02 ☉ ⊥ ♀
10 59 ☽ ☍ ♃
13 07 ☽ ⚺ ☉
14 50 ☽ # ♇
22 54 ☽ Q ♀
23 08 ☽ # ♃
23 32 ☽ ⊥ ☉

28 Tuesday
01 34 ☽ □ ♆
05 20 ☽ ∠ ♂
07 25 ☿ Q ♆
10 44 ☽ ☌ ♅
11 20 ☽ # ♄
11 52 ☽ ∥ ♀
12 36 ☿ ± ♇
15 25 ☽ ∠ ☉
16 00 ☽ ∥ ♅
18 08 ☽ □ ♄
20 25 ☽ ∥ ♆
20 48 ☽ ⊥ ♂
23 23 ☽ ⚺ ☿

29 Wednesday
05 35 ☽ □ ♀
07 19 ☽ ⚺ ♂
08 14 ☽ ☍ ♇
10 57 ☽ ⊥ ☿
12 25 ☽ ⚻ ♃
18 42 ☽ ⚹ ☉
22 42 ☽ ± ♃

30 Thursday
00 46 ☽ Q ♄
04 25 ☽ △ ♆
04 41 ☽ ∠ ☿
06 56 ☉ ⚼ ♄
08 51 ♂ ⚻ ♇
14 21 ☽ ⚺ ♅
14 24 ☽ ⚼ ♃
17 22 ☽ Q ☉
21 57 ☽ ⚹ ♄

All ephemeris data is given at 12.00 UT and the Moon's longitude is additionally given for 24.00 UT

JULY 2044

Date	Sidereal time h m s	Sun ☉ ° ' "	Moon ☽ ° ' "	Moon ☽ 24.00 ° ' "	Mercury ☿ ° '	Venus ♀ ° '	Mars ♂ ° '	Jupiter ♃ ° '	Saturn ♄ ° '	Uranus ♅ ° '	Neptune ♆ ° '	Pluto ♇ ° '
01	06 40 45	10 ♋ 18 09	01 ♎ 52 35	08 ♎ 23 16	01 ♌ 32	01 ♊ 56	03 ♎ 07	04 ♒ 48	24 ♏ 06	19 ♌ 55	14 ♉ 17	02 ♓ 32
02	06 44 42	11 15 21	14 ♎ 48 28	21 ♎ 08 40	03 07	02 22	03 37	04 R 42	24 R 04	19 58	14 18	02 R 32
03	06 48 38	12 12 33	27 ♎ 24 25	03 ♏ 36 14	04 41	02 51	04 08	04 35	24 01	20 01	14 19	02 31
04	06 52 35	13 09 45	09 ♏ 44 39	15 ♏ 50 09	06 12	03 21	04 38	04 29	23 59	20 04	14 21	02 30
05	06 56 31	14 06 57	21 ♏ 53 13	27 ♏ 54 15	07 41	03 52	05 09	04 22	23 57	20 07	14 22	02 29
06	07 00 28	15 04 08	03 ♐ 53 41	09 ♐ 51 51	09 07	04 24	05 40	04 15	23 55	20 11	14 23	02 29
07	07 04 24	16 01 20	15 ♐ 49 05	21 ♐ 45 40	10 31	04 58	06 11	04 08	23 53	20 14	14 25	02 28
08	07 08 21	16 58 31	27 ♐ 41 51	03 ♑ 37 52	11 53	05 34	06 43	04 01	23 51	20 17	14 26	02 27
09	07 12 17	17 55 42	09 ♑ 33 57	15 ♑ 30 17	13 13	06 10	07 14	03 54	23 50	20 20	14 27	02 26
10	07 16 14	18 52 54	21 ♑ 27 04	27 ♑ 24 29	14 30	06 48	07 46	03 47	23 48	20 24	14 28	02 25
11	07 20 11	19 50 06	03 ♒ 22 46	09 ♒ 22 08	15 44	07 26	08 18	03 40	23 46	20 27	14 30	02 24
12	07 24 07	20 47 17	15 ♒ 22 49	21 ♒ 25 06	16 56	08 06	08 50	03 32	23 45	20 30	14 31	02 24
13	07 28 04	21 44 30	27 ♒ 29 16	03 ♓ 35 40	18 05	08 47	09 22	03 25	23 44	20 34	14 32	02 23
14	07 32 00	22 41 42	09 ♓ 44 39	15 ♓ 56 36	19 12	09 29	09 55	03 17	23 42	20 37	14 33	02 22
15	07 35 57	23 38 55	22 ♓ 11 59	28 ♓ 31 12	20 16	10 12	10 28	03 10	23 41	20 40	14 34	02 21
16	07 39 53	24 36 09	04 ♈ 54 42	11 ♈ 22 58	21 17	10 56	11 00	03 02	23 40	20 44	14 35	02 20
17	07 43 50	25 33 23	17 ♈ 56 25	24 ♈ 35 26	22 15	11 40	11 33	02 55	23 39	20 47	14 36	02 19
18	07 47 46	26 30 38	01 ♉ 20 24	08 ♉ 11 32	23 10	12 26	12 06	02 47	23 38	20 51	14 37	02 18
19	07 51 43	27 27 54	15 ♉ 09 01	22 ♉ 12 51	24 02	13 12	12 40	02 39	23 37	20 54	14 38	02 17
20	07 55 40	28 25 11	29 ♉ 22 54	06 ♊ 38 52	24 51	14 00	13 13	02 32	23 37	20 58	14 39	02 16
21	07 59 36	29 ♋ 22 28	14 ♊ 00 14	21 ♊ 26 19	25 36	14 48	13 47	02 24	23 36	21 01	14 40	02 15
22	08 03 33	00 ♌ 19 46	28 ♊ 56 13	06 ♋ 28 55	26 17	15 36	14 21	02 16	23 36	21 05	14 41	02 14
23	08 07 29	01 17 05	14 ♋ 03 14	21 ♋ 37 55	26 55	16 26	14 55	02 08	23 35	21 08	14 42	02 13
24	08 11 26	02 14 25	29 ♋ 11 43	06 ♌ 43 23	27 29	17 16	15 29	02 01	23 35	21 12	14 42	02 12
25	08 15 22	03 11 45	14 ♌ 11 45	21 ♌ 35 48	27 59	18 07	16 03	01 53	23 35	21 16	14 43	02 11
26	08 19 19	04 09 06	28 ♌ 54 39	06 ♍ 07 37	28 24	18 58	16 37	01 45	23 34	21 19	14 44	02 09
27	08 23 15	05 06 27	13 ♍ 14 12	20 ♍ 14 07	28 45	19 50	17 12	01 37	23 D 34	21 23	14 45	02 08
28	08 27 12	06 03 49	27 ♍ 07 13	03 ♎ 53 31	29 02	20 43	17 47	01 30	23 34	21 26	14 45	02 07
29	08 31 09	07 01 11	10 ♎ 33 14	17 ♎ 06 36	29 13	21 36	18 22	01 22	23 35	21 30	14 46	02 06
30	08 35 05	07 58 34	23 ♎ 34 02	29 ♎ 55 59	29 20	22 30	18 57	01 14	23 35	21 34	14 46	02 05
31	08 39 02	08 ♌ 55 57	06 ♏ 12 56	12 ♏ 25 25	29 ♌ 22	23 ♊ 24	19 ♎ 32	01 ♒ 07	23 ♏ 35	21 ♌ 37	14 ♉ 47	02 ♓ 04

Date	Moon True ☊ ° '	Moon Mean ☊ ° '	Moon ☽ Latitude ° '
01	13 ♓ 15	14 ♓ 23	01 S 38
02	13 R 15	14 20	02 40
03	13 14	14 17	03 33
04	13 11	14 14	04 15
05	13 05	14 11	04 44
06	12 58	14 07	05 01
07	12 48	14 04	05 04
08	12 38	14 01	04 54
09	12 27	13 58	04 31
10	12 18	13 55	03 57
11	12 10	13 51	03 12
12	12 05	13 48	02 18
13	12 02	13 45	01 17
14	12 00	13 42	00 S 12
15	12 D 01	13 39	00 N 55
16	12 02	13 36	02 00
17	12 03	13 32	03 02
18	12 R 03	13 29	03 55
19	12 02	13 26	04 36
20	11 59	13 23	05 02
21	11 54	13 20	05 09
22	11 48	13 17	04 56
23	11 42	13 13	04 22
24	11 37	13 10	03 29
25	11 33	13 07	02 23
26	11 31	13 04	01 N 08
27	11 D 30	13 01	00 S 09
28	11 31	12 57	01 24
29	11 32	12 54	02 32
30	11 34	12 51	03 29
31	11 ♓ 34	12 ♓ 48	04 S 15

DECLINATIONS

Date	Sun ☉ ° '	Moon ☽ ° '	Mercury ☿ ° '	Venus ♀ ° '	Mars ♂ ° '	Jupiter ♃ ° '	Saturn ♄ ° '	Uranus ♅ ° '	Neptune ♆ ° '	Pluto ♇ ° '
01	23 N 02	02 S 14	21 N 25	16 N 37	01 S 00	19 S 34	16 S 39	15 N 29	14 N 28	21 S 15
02	22 58	08 18	20 58	16 39	01 13	19 36	16 38	15 28	14 28	21 16
03	22 53	13 51	20 30	16 42	01 26	19 38	16 38	15 27	14 29	21 17
04	22 47	18 45	20 01	16 46	01 40	19 39	16 38	15 26	14 29	21 17
05	22 41	22 48	19 32	16 50	01 53	19 41	16 37	15 25	14 29	21 18
06	22 35	25 51	19 03	16 54	02 06	19 43	16 37	15 24	14 30	21 18
07	22 28	27 43	18 32	16 59	02 20	19 45	16 37	15 23	14 30	21 19
08	22 21	28 19	18 02	17 04	02 33	19 47	16 37	15 22	14 30	21 19
09	22 14	27 36	17 31	17 10	02 47	19 48	16 37	15 21	14 31	21 20
10	22 06	25 38	17 00	17 16	03 00	19 50	16 36	15 20	14 31	21 20
11	21 58	22 30	16 29	17 22	03 14	19 52	16 36	15 19	14 31	21 21
12	21 50	18 25	15 58	17 28	03 28	19 54	16 36	15 18	14 31	21 21
13	21 41	13 33	15 26	17 34	03 41	19 56	16 36	15 17	14 32	21 22
14	21 32	08 06	14 56	17 41	03 55	19 58	16 36	15 16	14 32	21 23
15	21 22	02 S 15	14 25	17 48	04 09	20 00	16 36	15 15	14 32	21 23
16	21 12	03 N 48	13 55	17 54	04 23	20 02	16 36	15 14	14 33	21 24
17	21 02	09 50	13 25	18 01	04 36	20 03	16 36	15 13	14 33	21 24
18	20 51	15 36	12 55	18 08	04 50	20 05	16 36	15 11	14 33	21 25
19	20 40	20 46	12 27	18 15	05 04	20 07	16 36	15 10	14 33	21 25
20	20 28	24 55	11 59	18 22	05 18	20 09	16 36	15 09	14 33	21 26
21	20 17	27 35	11 32	18 29	05 32	20 11	16 36	15 08	14 34	21 27
22	20 05	28 21	11 05	18 36	05 46	20 13	16 36	15 07	14 34	21 27
23	19 52	27 02	10 40	18 43	06 00	20 15	16 36	15 06	14 34	21 28
24	19 39	23 44	10 16	18 50	06 14	20 17	16 36	15 05	14 34	21 28
25	19 26	18 51	09 54	18 57	06 28	20 19	16 37	15 03	14 34	21 29
26	19 13	12 55	09 33	19 03	06 43	20 20	16 37	15 02	14 35	21 29
27	18 59	06 N 26	09 13	19 10	06 57	20 22	16 37	15 01	14 35	21 30
28	18 45	00 S 09	08 55	19 16	07 11	20 24	16 37	15 00	14 35	21 31
29	18 31	06 30	08 39	19 22	07 25	20 26	16 38	14 59	14 35	21 31
30	18 16	12 23	08 25	19 28	07 39	20 28	16 38	14 58	14 35	21 32
31	18 N 01	17 S 36	08 N 13	19 N 34	07 S 53	20 S 30	16 S 38	14 N 56	14 N 35	21 S 32

ZODIAC SIGN ENTRIES

Date	h	m	Planets
01	08	35	☽ ♎
03	17	00	☽ ♏
06	04	12	☽ ♐
08	16	39	☽ ♑
11	05	13	☽ ♒
13	16	57	☽ ♓
16	02	47	☽ ♈
18	09	38	☽ ♉
20	13	02	☽ ♊
22	03	43	☉ ♌
22	13	42	☽ ♋
24	13	17	☽ ♌
26	13	48	☽ ♍
28	17	05	☽ ♎
31	00	08	☽ ♏

LATITUDES

Date	Mercury ☿ ° '	Venus ♀ ° '	Mars ♂ ° '	Jupiter ♃ ° '	Saturn ♄ ° '	Uranus ♅ ° '	Neptune ♆ ° '	Pluto ♇ ° '
01	01 N 38	04 S 01	00 N 16	00 S 31	02 N 13	00 N 41	01 S 44	11 S 30
04	01 21	04 08	00 12	00 32	02 12	00 41	01 44	11 30
07	00 59	04 12	00 08	00 32	02 12	00 41	01 44	11 31
10	00 32	04 14	00 05	00 33	02 11	00 41	01 44	11 32
13	00 N 02	04 14	00 N 02	00 33	02 10	00 41	01 45	11 33
16	00 S 31	04 13	00 S 01	00 34	02 09	00 41	01 45	11 34
19	01 08	04 09	00 04	00 34	02 09	00 41	01 45	11 34
22	01 46	04 04	00 07	00 35	02 08	00 41	01 45	11 35
25	02 26	03 58	00 10	00 35	02 07	00 41	01 45	11 35
28	03 05	03 51	00 13	00 36	02 06	00 41	01 45	11 36
31	03 S 42	03 S 43	00 S 16	00 S 36	02 N 06	00 N 41	01 S 46	11 S 37

DATA

Julian Date	2467798
Delta T	+79 seconds
Ayanamsa	24° 28' 49"
Synetic vernal point	04° ♓ 38' 10"
True obliquity of ecliptic	23° 26' 09"

MOON'S PHASES, APSIDES AND POSITIONS ☽

Date	h	m	Phase	Longitude ° '	Eclipse Indicator
02	04	48	☽	10 ♎ 58	
10	06	22	○	18 ♑ 39	
18	02	47	☾	26 ♈ 09	
24	17	10	●	02 ♌ 27	
31	17	40	☽	09 ♏ 10	

Day	h	m		
08	20	44	Apogee	
23	18	16	Perigee	
01	03	26	0S	
08	10	57	Max dec	28° S 19'
15	21	00	0N	
22	08	55	Max dec	28° N 22'
28	11	29	0S	

LONGITUDES

Date	Chiron ⚷ ° '	Ceres ⚳ ° '	Pallas ⚴ ° '	Juno ⚵ ° '	Vesta ⚶ ° '	Black Moon Lilith ⚸ ° '
01	29 ♌ 59	19 ♉ 37	24 ♓ 50	11 ♈ 15	04 ♈ 50	04 ♑ 04
11	01 ♍ 05	23 ♉ 00	25 ♓ 03	14 ♈ 15	06 ♈ 56	05 ♑ 11
21	02 ♍ 16	26 ♉ 13	24 ♓ 44	16 ♈ 56	08 ♈ 28	06 ♑ 18
31	03 ♍ 32	29 ♉ 14	23 ♓ 50	19 ♈ 10	09 ♈ 21	07 ♑ 25

All ephemeris data is given at 12.00 UT and the Moon's longitude is additionally given for 24.00 UT

ASPECTARIAN

h m	Aspects
01 Friday	
00 03	☽ # ♂
01 08	☽ ⊥ ♅
07 05	☽ ∥ ♂
07 15	☽ Ꝑ ♆
11 17	☽ ⚹ ☿
12 06	☽ △ ♀
13 13	☽ ⚻ ♇
14 22	☽ ☌ ♂
17 20	☽ △ ♃
17 36	☽ ∠ ♅
20 04	☿ # ♇
20 10	☿ ⚹ ♀
23 49	☽ ± ♆
02 Saturday	
00 16	☽ ± ♇
01 17	☽ ∠ ♄
03 02	☿ ⚻ ♇
04 48	☽ □ ☉
05 59	♀ # ♄
11 03	☽ ⚻ ♆
12 40	☽ Q ☿
17 01	☽ Ꝑ ♀
17 07	☽ Ꝑ ♇
18 07	☽ ⊥ ♄
19 43	♀ □ ♇
21 48	☽ ⚹ ♅
23 18	☿ ⚹ ♂
03 Sunday	
05 31	☽ ⚺ ♄
10 41	☿ ☍ ♃
10 53	☽ ± ♀
14 51	☽ # ♆
19 26	☽ # ♅
20 57	☽ Q ♅
21 52	☽ △ ♇
22 58	☽ ⚻ ♀
04 Monday	
01 09	☽ ∥ ♄
01 35	☽ ⚺ ♂
01 41	☽ # ♀
01 47	☽ □ ♃
04 05	☽ □ ☿
05 46	♂ △ ♃
13 50	☽ ⊥ ♂
16 57	☽ ∥ ♃
18 16	☽ # ☿
19 18	☽ △ ☉
21 04	☽ ☍ ♆
05 Tuesday	
02 22	☽ ∥ ♇
05 15	☿ # ♃
08 24	☽ ∠ ♂
08 29	☽ □ ♅
10 32	♂ ∠ ♅
11 15	☽ # ☉
12 12	☉ ⊥ ♅
12 56	☽ Q ♃
16 06	☽ ☌ ♄
18 31	☉ ⚹ ♆
06 Wednesday	
03 40	☽ Ꝑ ☉
06 21	♀ △ ♃
09 09	☽ □ ♇
12 42	☽ ⚹ ♃
13 05	☽ ☍ ♀
15 43	☽ ⚹ ♂
23 18	☽ ± ☉
23 54	☽ △ ☿
07 Thursday	
09 10	☽ ⚻ ♆
12 27	☽ ⚻ ☉
17 00	☽ Q ♂
18 38	☽ ∠ ♃
20 57	☽ △ ♅
21 18	☽ ± ♆
21 22	☽ Q ♇
08 Friday	
04 15	☽ ⚺ ♄
10 09	☽ Ꝑ ☿
12 38	☽ ⊥ ♃
15 31	☽ Ꝑ ♆
16 21	☽ ⊥ ♄
21 36	☽ ⚹ ♇
22 59	♆ Q ♇
23 46	☉ Ꝑ ♇
09 Saturday	
00 39	☽ ⚺ ♃
03 25	☽ Ꝑ ♅
04 45	☽ ⚻ ♀
06 39	☽ ± ☿
07 05	☽ □ ♂
10 31	☽ ∠ ♄
17 33	☽ ± ♀
20 16	☽ ⚻ ☿
21 41	☽ ± ♅
21 54	☽ △ ♆
10 Sunday	
01 47	☿ ∥ ♀
03 53	☽ ∠ ♇
06 22	☽ ☍ ☉
09 51	☽ ⚻ ♅
11 39	☿ □ ♆
12 44	☽ Ꝑ ♀
16 43	☽ ⚹ ♄
22 00	☽ ⊥ ♇
11 Monday	
06 11	☿ # ♄
10 03	☽ ⚺ ♇
12 33	☽ ☌ ♃
15 37	☽ # ☉
16 39	♂ ± ♇
16 48	☽ Q ♄
19 21	☽ ∥ ♇
20 36	☽ △ ♀
21 01	♂ ± ♆
22 19	☽ △ ♂
12 Tuesday	
03 57	☽ ∥ ♃
04 25	☉ ⚺ ♅
08 18	♂ ∠ ♄
10 16	☽ □ ♆
15 26	☽ ☍ ☿
16 53	☽ # ♀
21 23	☽ ∥ ♄
22 14	☽ ☍ ♅
23 40	☽ ⚻ ☉
13 Wednesday	
02 02	☽ # ☿
03 33	♀ Q ♅
03 49	☽ # ♅
04 35	☽ □ ♄
05 34	☽ Ꝑ ♂
07 27	☽ # ♆
12 33	☽ ± ☉
19 37	☿ ∥ ♅
21 36	☽ ☌ ♇
21 56	☽ Q ♆
23 32	☽ ⚺ ♃
14 Thursday	
00 06	☽ ± ♂
07 41	☽ Ꝑ ☉
11 07	☽ ⊥ ♃
11 27	☽ □ ♀
12 21	☽ ⚻ ♂
21 20	☽ ⚹ ♆
15 Friday	
04 21	☽ ∠ ♃
04 40	☽ ∥ ♂
06 11	☿ ∥ ♆
07 58	☽ ⚻ ☿
09 04	☽ ⚻ ♅
09 21	☉ # ♇
12 55	☉ △ ♄
14 50	☽ △ ♄
14 59	☽ △ ☉
20 26	☽ ± ☿
20 33	☽ ± ♅
22 00	☿ ☌ ♅
16 Saturday	
00 05	☽ Q ♀
02 00	☽ ∠ ♆
07 11	☽ ⚺ ♇
08 32	☽ ⚹ ♃
13 32	☽ Ꝑ ♅
14 23	☽ # ♂
14 46	☽ Ꝑ ☿
18 21	☽ ⊥ ♇
18 50	☽ ⊥ ♆
18 58	☽ Ꝑ ♄
21 59	♀ △ ♂
23 48	☽ ☍ ♂
23 50	☽ ⚹ ♀
17 Sunday	
05 54	☽ ⚺ ♆
06 32	☽ Q ♃
10 52	☽ ∠ ♇
11 28	☽ ± ♄
17 11	☽ △ ♅
20 24	☽ △ ☿
22 18	☽ ⚻ ♄
18 Monday	
01 31	☽ ∥ ☿
02 47	☽ □ ☉
04 39	☽ ∠ ♀
06 44	☉ ± ♇
07 28	☽ ∥ ♆
10 13	☽ ∥ ♅
13 41	☽ ⚹ ♇
14 31	☽ □ ♃
14 45	☉ Q ♆
16 22	☽ # ♄
21 28	☽ ⊥ ♀
23 35	☽ ∥ ♀
19 Tuesday	
00 30	☿ □ ♄
07 33	☽ ⚻ ♂
08 28	☽ ⚺ ♀
08 44	☽ # ♃
10 31	☽ Q ♇
11 07	☽ ☌ ♆
11 29	☽ ∥ ☉
12 35	☽ Q ☉
15 22	☽ # ♇
18 14	☽ ± ♂
21 50	☽ □ ♅
20 Wednesday	
02 22	☽ ☍ ♄
03 59	☽ □ ☿
09 59	☽ Ꝑ ♂
10 17	☽ ⚹ ☉
16 46	☽ □ ♇
17 10	☽ △ ♃
21 Thursday	
03 51	☽ Q ♅
08 07	♀ ⚺ ♆
11 18	☽ Q ☿
11 37	☽ △ ♂
12 39	☽ ∠ ☉
13 04	☽ ⚺ ♆
13 21	☽ ☌ ♀
17 27	☽ Ꝑ ♃
21 49	☉ # ♃
22 46	☽ ⊥ ♆
23 23	☽ ⚹ ♅
22 Friday	
03 28	☽ ⚻ ♄
04 08	☽ ⊥ ☉
07 35	☽ ⚹ ☿
07 47	☽ ± ♃
13 03	☽ ± ♄
13 11	☽ ∠ ♆
14 22	☽ ⚺ ☉
17 14	☽ △ ♇
17 16	☽ ⚻ ♃
20 36	♃ ⚺ ♇
23 25	☽ ∠ ♅
23 Saturday	
02 32	♂ ⚻ ♆
03 20	☽ Ꝑ ♄
08 29	☽ ∠ ☿
13 01	☽ ⚹ ♆
13 25	☽ □ ♂
13 44	☽ ⊥ ♅
15 59	☽ ⚺ ♀
17 00	☽ Ꝑ ♇
23 16	☽ ⚺ ♅
23 18	☽ ⊥ ☿
24 Sunday	
02 02	☽ ⊥ ♀
03 05	☽ △ ♄
05 41	♀ Ꝑ ♃
06 55	☉ ☍ ♃
07 14	☽ ± ♇
08 03	☽ Q ♆
09 11	☽ ⚺ ☿
10 51	☉ ⚻ ♇
16 26	☽ ☍ ♃
16 46	☽ ⚻ ♇
17 10	☽ ☌ ☉
17 10	☽ ∠ ♀
17 19	☉ ∠ ♀
19 05	☽ Q ♂
23 53	☽ # ♇
25 Monday	
05 28	☽ # ♃
09 16	☽ ∥ ☉
11 34	☽ ∥ ♀
12 51	☽ □ ♆
15 07	☽ ⚹ ♂
18 43	☽ ⚹ ♀
21 25	☽ # ♄
23 30	☽ ☌ ♅
26 Tuesday	
03 14	☽ □ ♄
03 44	☽ ∥ ♅
05 34	☽ ∥ ♆
11 08	☽ ☌ ☿
15 37	☽ Q ♀
16 40	☽ ⚻ ♃
16 41	☽ ∠ ♂
17 22	☽ ☍ ♇
21 19	☽ ⚺ ☉
23 25	☉ ∥ ♀
27 Wednesday	
01 19	☽ ∥ ☿
02 36	☽ ± ♃
07 27	♄ St D
08 08	☽ ⊥ ☉
08 24	☽ ⊥ ♂
09 11	☽ Q ♄
09 31	♂ Ꝑ ♇
10 14	☽ # ♂
14 34	☽ △ ♆
17 44	☽ Ꝑ ♃
17 53	☉ Q ♂
19 04	☽ ⚺ ♂
28 Thursday	
00 04	☽ □ ♀
00 39	☽ ∠ ☉
02 03	☽ ⚺ ♅
03 31	♂ ⊥ ♄
05 48	☽ ⚹ ♄
12 34	☽ ⊥ ♅
13 03	♀ ⊥ ♆
15 25	☽ ⚺ ☿
16 39	☽ Ꝑ ♆
19 39	☽ △ ♃
20 50	☽ ⚻ ♇
29 Friday	
02 15	☽ ⊥ ☿
04 39	☽ ∠ ♅
05 07	☽ ⚹ ☉
07 34	☽ ± ♇
08 25	☽ ∠ ♄
08 45	☽ ± ♆
09 04	♀ ⚹ ♅
15 43	☽ ∥ ♂
18 46	☽ ∠ ☿
19 41	☽ ⚻ ♆
20 10	☽ # ☿
23 58	☽ Ꝑ ♇
30 Saturday	
00 52	☽ ⊥ ♄
02 59	☽ ☌ ♂
04 46	☽ Q ☉
08 14	☽ ⚹ ♅
09 51	☽ △ ♀
12 02	☽ ⚺ ♄
21 42	☽ # ♆
22 55	☽ ⚹ ☿
23 21	☽ # ♅
31 Sunday	
02 20	☽ □ ♃
04 04	☽ △ ♇
07 00	☽ Q ♅
07 17	☽ ∥ ♄
08 21	☿ St R
14 01	☽ # ☉
16 33	☽ Ꝑ ♀
16 52	♀ ⚻ ♄
17 40	☽ □ ☉
21 55	☽ Q ☿
22 25	☽ # ♀

AUGUST 2044

LONGITUDES

Date	Sidereal time h m s	Sun ☉ ° ' "	Moon ☽ ° ' "	Moon ☽ 24.00 ° ' "	Mercury ☿ ° '	Venus ♀ ° '	Mars ♂ ° '	Jupiter ♃ ° '	Saturn ♄ ° '	Uranus ♅ ° '	Neptune ♆ ° '	Pluto ♇ ° '
01	08 42 58	09 ♌ 53 21	18 ♏ 34 00	24 ♏ 39 13	29 ♌ 19	24 ♊ 19	20 ♎ 07	00 ♒ 59	23 ♏ 36	21 ♌ 41	14 ♉ 48	02 ♓ 03
02	08 46 55	10 50 45	00 ♐ 41 38	06 ♐ 41 47	29 R 10	25 14	20 42	00 R 51	23 36	21 45	14 48	02 R 01
03	08 50 51	11 48 10	12 ♐ 37 16	18 ♐ 37 16	28 56	26 10	21 18	00 44	23 37	21 48	14 49	02 00
04	08 54 48	12 45 36	24 ♐ 33 33	00 ♑ 29 25	28 37	27 07	21 54	00 36	23 38	21 52	14 49	01 59
05	08 58 44	13 43 03	06 ♑ 25 17	12 ♑ 21 29	28 13	28 03	22 30	00 29	23 39	21 56	14 49	01 58
06	09 02 41	14 40 30	18 ♑ 18 21	24 ♑ 16 09	27 44	29 01	23 06	00 22	23 39	22 00	14 50	01 57
07	09 06 38	15 37 59	00 ♒ 15 10	06 ♒ 15 38	27 11	29 ♊ 58	23 42	00 14	23 41	22 03	14 50	01 55
08	09 10 34	16 35 28	12 ♒ 17 46	18 ♒ 21 47	26 33	00 ♋ 56	24 18	00 07	23 42	22 07	14 50	01 54
09	09 14 31	17 32 58	24 ♒ 27 50	00 ♓ 36 09	25 52	01 55	24 54	00 ♒ 00	23 43	22 11	14 51	01 53
10	09 18 27	18 30 30	06 ♓ 46 54	13 ♓ 00 16	25 07	02 54	25 31	29 ♑ 53	23 44	22 14	14 51	01 52
11	09 22 24	19 28 02	19 ♓ 16 27	25 ♓ 35 40	24 20	03 53	26 07	29 46	23 46	22 18	14 51	01 50
12	09 26 20	20 25 36	01 ♈ 58 06	08 ♈ 23 59	23 32	04 53	26 44	29 39	23 47	22 22	14 51	01 49
13	09 30 17	21 23 11	14 ♈ 53 33	21 ♈ 27 00	22 42	05 53	27 21	29 33	23 49	22 26	14 51	01 48
14	09 34 13	22 20 48	28 ♈ 04 33	04 ♉ 46 24	21 53	06 53	27 58	29 26	23 50	22 29	14 52	01 47
15	09 38 10	23 18 26	11 ♉ 32 43	18 ♉ 23 36	21 04	07 54	28 35	29 20	23 52	22 33	14 52	01 45
16	09 42 07	24 16 05	25 ♉ 19 07	02 ♊ 19 15	20 18	08 55	29 12	29 13	23 54	22 37	14 52	01 44
17	09 46 03	25 13 47	09 ♊ 23 54	16 ♊ 32 51	19 35	09 56	29 ♎ 50	29 07	23 56	22 41	14 52	01 43
18	09 50 00	26 11 30	23 ♊ 45 46	01 ♋ 02 14	18 55	10 58	00 ♏ 27	29 01	23 58	22 44	14 R 52	01 41
19	09 53 56	27 09 15	08 ♋ 21 39	15 ♋ 43 21	18 21	12 00	01 05	28 55	24 00	22 48	14 52	01 40
20	09 57 53	28 07 01	23 ♋ 06 35	00 ♌ 30 27	17 52	13 03	01 42	28 49	24 03	22 52	14 52	01 39
21	10 01 49	29 ♌ 04 49	07 ♌ 54 05	15 ♌ 16 33	17 29	14 05	02 20	28 43	24 05	22 55	14 51	01 38
22	10 05 46	00 ♍ 02 38	22 ♌ 36 55	29 ♌ 54 19	17 13	15 08	02 58	28 37	24 08	22 59	14 51	01 36
23	10 09 42	01 00 28	07 ♍ 07 59	14 ♍ 17 11	17 04	16 11	03 36	28 32	24 10	23 03	14 51	01 35
24	10 13 39	01 58 20	21 ♍ 21 21	28 ♍ 20 03	17 D 02	17 15	04 14	28 27	24 13	23 07	14 51	01 34
25	10 17 36	02 56 14	05 ♎ 12 57	11 ♎ 59 53	17 09	18 18	04 53	28 21	24 15	23 10	14 51	01 32
26	10 21 32	03 54 08	18 ♎ 40 49	25 ♎ 15 49	17 23	19 22	05 31	28 16	24 18	23 14	14 50	01 31
27	10 25 29	04 52 04	01 ♏ 45 04	08 ♏ 08 52	17 46	20 27	06 10	28 12	24 21	23 18	14 50	01 30
28	10 29 25	05 50 02	14 ♏ 27 34	20 ♏ 41 35	18 16	21 31	06 48	28 07	24 24	23 21	14 50	01 29
29	10 33 22	06 48 00	26 ♏ 51 26	02 ♐ 57 36	18 55	22 36	07 27	28 02	24 27	23 25	14 49	01 27
30	10 37 18	07 46 00	09 ♐ 00 40	15 ♐ 01 10	19 41	23 41	08 06	27 58	24 30	23 29	14 49	01 26
31	10 41 15	08 ♍ 44 02	20 ♐ 59 41	26 ♐ 56 46	20 ♌ 34	24 ♋ 46	08 ♏ 45	27 ♑ 54	24 ♏ 34	23 ♌ 32	14 ♉ 49	01 ♓ 25

Date	Moon True ☊ ° '	Moon Mean ☊ ° '	Moon ☽ Latitude ° '
01	11 ♓ 34	12 ♓ 45	04 S 48
02	11 R 32	12 42	05 07
03	11 29	12 38	05 12
04	11 25	12 35	05 04
05	11 21	12 32	04 43
06	11 16	12 29	04 10
07	11 12	12 26	03 26
08	11 09	12 23	02 32
09	11 06	12 19	01 30
10	11 06	12 16	00 S 24
11	11 D 06	12 13	00 N 45
12	11 07	12 10	01 53
13	11 08	12 07	02 56
14	11 09	12 03	03 51
15	11 10	12 00	04 35
16	11 R 11	11 57	05 04
17	11 10	11 54	05 16
18	11 09	11 51	05 08
19	11 08	11 48	04 41
20	11 06	11 44	03 55
21	11 05	11 41	02 53
22	11 04	11 38	01 40
23	11 03	11 35	00 N 22
24	11 D 03	11 32	00 S 57
25	11 04	11 28	02 10
26	11 04	11 25	03 14
27	11 05	11 22	04 05
28	11 06	11 19	04 43
29	11 06	11 16	05 07
30	11 R 06	11 13	05 17
31	11 ♓ 06	11 ♓ 09	05 S 12

DECLINATIONS

Date	Sun ☉ ° '	Moon ☽ ° '	Mercury ☿ ° '	Venus ♀ ° '	Mars ♂ ° '	Jupiter ♃ ° '	Saturn ♄ ° '	Uranus ♅ ° '	Neptune ♆ ° '	Pluto ♇ ° '
01	17 N 46	21 S 57	08 N 04	19 N 39	08 S 07	20 S 31	16 S 39	14 N 55	14 N 35	21 S 33
02	17 31	25 17	07 57	19 45	08 22	20 33	16 39	14 54	14 35	21 34
03	17 15	27 28	07 52	19 49	08 36	20 35	16 39	14 53	14 35	21 34
04	16 59	28 23	07 50	19 54	08 50	20 37	16 40	14 52	14 36	21 35
05	16 42	28 00	07 50	19 58	09 04	20 38	16 40	14 50	14 36	21 35
06	16 26	26 19	07 53	20 03	09 18	20 40	16 41	14 49	14 36	21 36
07	16 09	23 26	07 59	20 06	09 32	20 42	16 41	14 48	14 36	21 36
08	15 52	19 32	08 08	20 10	09 47	20 43	16 42	14 47	14 36	21 37
09	15 34	14 47	08 19	20 13	10 01	20 45	16 42	14 46	14 36	21 38
10	15 17	09 23	08 33	20 15	10 15	20 46	16 43	14 44	14 36	21 38
11	14 59	03 S 33	08 49	20 18	10 29	20 48	16 44	14 43	14 36	21 39
12	14 41	02 N 30	09 07	20 20	10 43	20 49	16 44	14 42	14 36	21 39
13	14 22	08 34	09 28	20 21	10 57	20 51	16 45	14 41	14 36	21 40
14	14 04	14 23	09 49	20 22	11 11	20 52	16 45	14 40	14 36	21 40
15	13 45	19 39	10 12	20 23	11 25	20 54	16 46	14 38	14 36	21 41
16	13 26	24 00	10 36	20 23	11 39	20 55	16 47	14 37	14 36	21 42
17	13 07	27 03	11 01	20 23	11 53	20 57	16 48	14 36	14 36	21 42
18	12 47	28 25	11 25	20 23	12 07	20 58	16 48	14 35	14 36	21 43
19	12 27	27 51	11 50	20 22	12 21	20 59	16 49	14 33	14 35	21 43
20	12 08	25 19	12 13	20 20	12 34	21 01	16 50	14 32	14 35	21 44
21	11 48	21 05	12 36	20 18	12 48	21 02	16 51	14 31	14 35	21 44
22	11 27	15 33	12 57	20 16	13 02	21 03	16 52	14 30	14 35	21 45
23	11 07	09 14	13 17	20 13	13 16	21 04	16 52	14 29	14 35	21 45
24	10 46	02 N 33	13 34	20 09	13 29	21 05	16 53	14 27	14 35	21 46
25	10 26	04 S 04	13 50	20 05	13 43	21 06	16 54	14 26	14 35	21 46
26	10 05	10 18	14 03	20 01	13 56	21 07	16 55	14 25	14 35	21 47
27	09 43	15 55	14 13	19 56	14 10	21 09	16 56	14 24	14 35	21 47
28	09 22	20 41	14 21	19 51	14 23	21 10	16 57	14 23	14 34	21 48
29	09 01	24 26	14 25	19 45	14 37	21 10	16 58	14 21	14 34	21 48
30	08 39	27 01	14 27	19 38	14 50	21 11	16 59	14 20	14 34	21 49
31	08 N 18	28 S 19	14 N 25	19 N 31	15 S 03	21 S 12	17 S 00	14 N 19	14 N 34	21 S 49

ZODIAC SIGN ENTRIES

Date	h	m	Planets
02	10	37	☽ ♐
04	23	00	☽ ♑
07	11	30	☽ ♒
07	12	42	♀ ♋
09	12	43	♃ ♑
09	22	50	☽ ♓
12	08	18	☽ ♈
14	15	28	☽ ♉
16	20	02	☽ ♊
17	18	43	♂ ♏
18	22	18	☽ ♋
20	23	11	☽ ♌
22	10	55	☉ ♍
23	00	09	☽ ♍
25	02	53	☽ ♎
27	08	45	☽ ♏
29	18	10	☽ ♐

LATITUDES

Date	Mercury ☿ ° '	Venus ♀ ° '	Mars ♂ ° '	Jupiter ♃ ° '	Saturn ♄ ° '	Uranus ♅ ° '	Neptune ♆ ° '	Pluto ♇ ° '
01	03 S 53	03 S 40	00 S 17	00 S 36	02 N 05	00 N 41	01 S 46	11 S 37
04	04 24	03 30	00 20	00 36	02 05	00 41	01 46	11 37
07	04 45	03 20	00 22	00 37	02 04	00 41	01 46	11 38
10	04 52	03 09	00 25	00 37	02 03	00 41	01 46	11 38
13	04 44	02 57	00 27	00 37	02 02	00 41	01 46	11 38
16	04 20	02 45	00 30	00 37	02 02	00 41	01 47	11 39
19	03 41	02 33	00 32	00 38	02 01	00 41	01 47	11 39
22	02 51	02 20	00 34	00 38	02 00	00 41	01 47	11 39
25	01 57	02 07	00 36	00 38	01 59	00 41	01 47	11 39
28	01 04	01 54	00 38	00 38	01 59	00 41	01 47	11 39
31	00 S 14	01 S 40	00 S 40	00 S 38	01 N 58	00 N 41	01 S 47	11 S 39

DATA

Julian Date	2467829
Delta T	+79 seconds
Ayanamsa	24° 28' 54"
Synetic vernal point	04° ♓ 38' 05"
True obliquity of ecliptic	23° 26' 09"

MOON'S PHASES, APSIDES AND POSITIONS ☽

Date	h	m	Phase	Longitude ° '	Eclipse Indicator
08	21	14	○	16 ♒ 58	
16	10	03	☾	24 ♉ 11	
23	01	06	●	00 ♍ 34	Total
30	09	19	☽	07 ♐ 40	

Day	h	m		
05	05	16	Apogee	
20	22	52	Perigee	
04	16	47	Max dec	28° S 25'
12	02	08	ON	
18	17	06	Max dec	28° N 28'
24	21	11	OS	
31	23	47	Max dec	28° S 29'

LONGITUDES

Date	Chiron ⚷ ° '	Ceres ⚳ ° '	Pallas ⚴ ° '	Juno ⚵ ° '	Vesta ⚶ ° '	Black Moon Lilith ⚸ ° '
01	03 ♍ 40	29 ♉ 32	23 ♓ 43	19 ♈ 22	09 ♈ 24	07 ♑ 32
11	05 ♍ 00	02 ♊ 17	22 ♓ 11	21 ♈ 02	09 ♈ 29	08 ♑ 39
21	06 ♍ 21	04 ♊ 46	20 ♓ 09	22 ♈ 04	08 ♈ 47	09 ♑ 46
31	07 ♍ 44	06 ♊ 53	17 ♓ 45	22 ♈ 23	07 ♈ 20	10 ♑ 53

All ephemeris data is given at 12.00 UT and the Moon's longitude is additionally given for 24.00 UT

ASPECTARIAN

h m Aspects

01 Monday
03 28 ☽ ∥ ♃
04 37 ☽ ☍ ♆
08 11 ☿ # ♂
09 32 ☽ ∥ ♇
11 28 ☽ ± ♀
12 48 ☽ Q ♃
15 12 ☽ ⚺ ♂
18 10 ☽ □ ♅
21 55 ☽ ☌ ♄

02 Tuesday
00 16 ☽ ⚻ ♀
03 13 ♀ ± ♃
03 40 ☽ ⊥ ♂
09 00 ☽ □ ☿
12 19 ☽ ⚹ ♃
14 39 ☽ □ ♇
22 32 ☽ ∠ ♂

03 Wednesday
10 06 ☽ △ ☉
16 19 ☽ ⚻ ♆
18 06 ☽ ∠ ♃

04 Thursday
02 46 ☽ Q ♇
04 26 ☽ ± ♆
06 20 ☽ ⚹ ♂
06 32 ☽ △ ♅
10 07 ☽ ⚺ ♄
10 46 ♂ ⚹ ♅
12 06 ☽ ⊥ ♃
17 36 ☽ ☍ ♀
19 03 ☽ bQ ☉
19 58 ☽ △ ☿
22 16 ☽ ⊥ ♄
22 39 ☽ bQ ♆

05 Friday
00 07 ☽ ⚺ ♃
03 00 ☽ ⚹ ♇
07 54 ☽ Q ♂
13 02 ☽ bQ ♅
14 47 ☉ # ♄
14 51 ☿ ⚹ ♀
14 51 ☽ ± ☉
16 30 ☽ ∠ ♄

06 Saturday
01 14 ☽ bQ ☿
04 02 ☽ ⚻ ☉
04 59 ☽ △ ♆
07 19 ☽ ± ♅
09 15 ☽ ∠ ♇
15 54 ☉ □ ♆
18 37 ☽ ± ☿
19 28 ☽ ⚻ ♅
22 09 ☽ □ ♂
22 47 ☽ ⚹ ♄

07 Sunday
03 20 ☽ ⊥ ♇
04 29 ♀ ± ♄
06 08 ☽ ⚻ ☿
08 35 ♀ ∠ ♆
11 14 ♂ ⚺ ♄
11 23 ☽ ⚻ ♀
11 59 ☽ ☌ ♃
15 20 ☽ ⚺ ♇
17 58 ♀ ⚻ ♃
22 51 ☽ Q ♄
23 57 ☽ ∥ ♇

08 Monday
00 25 ☽ ± ♀
05 19 ☽ ∥ ♃
08 32 ☽ # ♀
17 03 ☽ □ ♆
19 51 ☽ bQ ♀
21 14 ☽ ☍ ☉

09 Tuesday
02 44 ☽ ∥ ♄
07 29 ☽ ☍ ♅
08 01 ☽ # ☉
10 32 ☽ □ ♄
11 11 ♀ △ ♇
12 06 ☽ # ♅
12 52 ☽ # ♆
12 54 ☽ △ ♂
14 36 ☽ ☍ ☿
22 44 ☽ ⚺ ♃

10 Wednesday
02 28 ☽ ☌ ♇
03 49 ☽ △ ♀
04 22 ☽ Q ♆
05 10 ☿ ⚹ ♂
08 29 ☽ ∥ ♂
10 17 ☽ ⊥ ♃
15 23 ☽ # ☿
19 34 ☽ bQ ♂

11 Thursday
03 28 ☽ ∠ ♃
03 33 ☽ ⚹ ♆
12 24 ☽ ⚻ ☉
13 42 ☽ ± ♂
17 47 ☽ ⚻ ♅
20 33 ☽ △ ♄
21 03 ☽ ⚻ ☿

12 Friday
00 43 ☽ ± ☉
01 39 ☽ ⚻ ♂
04 37 ☿ □ ♄
05 12 ☽ ± ♅
07 42 ☽ ± ☿
07 42 ☽ ⚹ ♃
08 02 ☽ ∠ ♆
10 09 ☉ ∥ ♅
11 43 ☽ ⚺ ♇
17 54 ☽ □ ♀
18 23 ☉ ∥ ♆
18 59 ☽ bQ ☉
22 07 ☽ bQ ♅
22 54 ☽ ⊥ ♇
23 30 ☽ bQ ☿

13 Saturday
00 45 ☽ bQ ♄
00 51 ☽ ⊥ ♆
05 53 ☽ Q ♃
11 56 ☽ ⚺ ♆
15 30 ☽ ∠ ♇
15 50 ☽ ∥ ☿
17 22 ☽ ± ♄
19 25 ☿ ☌ ♅
22 03 ☽ # ♂

14 Sunday
00 49 ☽ △ ☉
01 26 ☽ △ ☿
01 51 ☽ △ ♅
04 20 ☽ ⚻ ♄
05 41 ☉ ☌ ☿
05 46 ☽ Q ♀
10 43 ☽ ∥ ☉
11 47 ☽ ☍ ♂
11 53 ☿ ∠ ♀
12 56 ☽ ∥ ♆
13 12 ☽ ∥ ♅
14 26 ☽ □ ♃
15 48 ☉ ☌ ♅
18 38 ☽ ⚹ ♇
22 28 ☽ # ♄

15 Monday
03 14 ♀ ∠ ♅
05 02 ☽ ⚹ ♀
15 43 ☽ ∥ ♀
15 53 ☽ Q ♇
17 50 ☽ ☌ ♆
18 22 ☽ # ♃
22 31 ☽ # ♇

16 Tuesday
02 34 ☉ □ ♄
03 46 ☽ □ ☿
07 18 ☽ □ ♅
09 23 ☽ ∠ ♀
09 33 ☽ ☍ ♄
10 03 ☽ □ ☉
11 40 ♀ bQ ♄
12 34 ♂ □ ♃
18 39 ☽ △ ♃
18 59 ☽ ⚻ ♂
22 59 ☽ □ ♇

17 Wednesday
02 02 ☽ ⊥ ♀
05 41 ☽ ± ♂
09 04 ☽ Q ☿
12 38 ♆ St R
12 59 ☽ ⚺ ♀
14 10 ☽ Q ♅
17 33 ♅ ∥ ♆
18 55 ☽ Q ☉
19 52 ☽ bQ ♃
21 11 ☽ ⚺ ♆
21 32 ☽ bQ ♂

18 Thursday
04 18 ☽ ⚹ ☿
07 11 ☽ ⊥ ♆
10 18 ☽ ⚹ ♅
10 46 ☽ ± ♃
12 21 ☽ ⚻ ♄
16 18 ☽ ⚹ ☉
20 37 ☽ ⚻ ♃
21 15 ☿ Q ♂
22 04 ☽ ∠ ♆
22 16 ☽ ± ♄
23 32 ☽ △ ♂

19 Friday
01 03 ☽ △ ♇
04 05 ☽ ∠ ☿
11 05 ☽ ∠ ♅
13 04 ☽ bQ ♄
16 52 ☉ # ♂
17 17 ☿ ⊥ ♀
18 17 ☽ ⊥ ☿
18 24 ☽ ☌ ♀
18 37 ☽ ∠ ☉
22 36 ☽ ⚹ ♆

20 Saturday
01 31 ☽ bQ ♇
01 49 ☽ ⊥ ♅
03 44 ☽ ⚺ ☿
08 55 ☉ ∥ ☿
09 53 ♂ △ ♇
10 17 ☽ ⊥ ☉
11 36 ☽ ⚺ ♅
13 31 ☽ △ ♄
16 07 ☽ ± ♇
18 05 ☽ Q ♆
20 41 ☽ ⚺ ☉
21 12 ☽ ☍ ♃

21 Sunday
01 50 ☽ ⚻ ♇
02 34 ☽ □ ♂
03 48 ☉ ⚻ ♃
08 46 ☽ # ♇
10 11 ☉ ∠ ♀
12 14 ☽ # ♃
15 43 ☽ ∥ ♀
22 50 ☽ ⚺ ♀
23 19 ☽ □ ♆

22 Monday
03 18 ☽ ☌ ☿
05 38 ♀ ⚹ ♆
06 43 ☽ # ♄
09 11 ☽ Q ♂
09 23 ☽ ⊥ ♀
12 37 ☽ ☌ ♅
14 29 ☽ □ ♄
15 50 ☽ ∥ ♆
16 11 ☽ ∥ ♅
21 29 ☽ # ♂
21 38 ☽ ∥ ☿
21 49 ☽ ⚻ ♃

23 Tuesday
01 06 ☽ ☌ ☉
01 21 ☽ ∠ ♀
02 47 ☽ ☍ ♇
04 39 ☽ ∥ ☉
05 51 ☽ ⚹ ♂
07 42 ☽ ± ♃
07 54 ☿ # ♂
20 27 ☽ Q ♄
20 48 ♀ bQ ♇
22 40 ☽ bQ ♃

24 Wednesday
00 57 ☽ △ ♆
02 00 ☉ ☍ ♇
04 12 ☿ St D
04 26 ☽ ⚹ ♀
04 39 ☽ ⚺ ☿
07 15 ☿ ⚺ ♀
08 13 ☽ ∠ ♂
08 45 ♀ ⊥ ♅
14 54 ☽ ⊥ ☿
15 01 ☽ ⚺ ♅
16 54 ☽ ⚹ ♄

25 Thursday
00 07 ☽ △ ♃
00 24 ☽ ⊥ ♂
01 24 ☽ ⊥ ♅
02 37 ☽ bQ ♆
02 42 ☽ Q ♀
05 35 ☽ ⚻ ♇
06 34 ☽ ∠ ☿
07 42 ☽ ⚺ ☉
11 23 ☽ ⚺ ♂
16 05 ☽ ± ♇
17 14 ☽ ∠ ♅
18 24 ☽ ± ♆
19 04 ☽ ⊥ ☉
19 09 ☽ ∠ ♄

26 Friday
00 14 ♂ Q ♅
05 05 ☽ ⚻ ♆
08 06 ☽ bQ ♇
09 37 ☽ ⚹ ☿
11 08 ☽ # ☉
11 19 ☽ ⊥ ♄
12 26 ☽ ∠ ☉
13 22 ☽ □ ♀
20 19 ☽ ⚹ ♅
20 31 ☉ ± ♃
22 17 ☽ ⚺ ♄

27 Saturday
03 50 ☽ ∥ ♂
04 10 ☽ # ☿
05 12 ☽ # ♅
05 26 ☽ □ ♃
05 58 ☽ # ♆
08 11 ☽ Q ☿
11 32 ☽ △ ♇
16 47 ☽ ∥ ♄
18 18 ☽ ⚹ ☉
18 40 ☽ Q ♅
20 41 ☽ ☌ ♂

28 Sunday
02 18 ☿ # ♂
07 30 ☽ # ♀
10 47 ♂ # ♅
12 42 ☽ ☍ ♆
14 43 ☽ ∥ ♃
15 09 ☽ Q ♃
18 28 ☽ ∥ ♇
18 55 ☿ ∥ ♅
19 01 ☽ Q ☉
19 41 ☽ □ ☿

29 Monday
02 54 ☽ △ ♀
05 15 ☽ □ ♅
07 17 ☽ ☌ ♄
07 43 ♂ # ♆
14 18 ☽ ⚹ ♃
21 01 ☽ □ ♇

30 Tuesday
07 20 ♀ ⚺ ♅
09 19 ☽ □ ☉
10 05 ☽ ⚺ ♂
11 16 ☽ bQ ♀
19 51 ☽ ∠ ♃
22 44 ☽ ⊥ ♂
23 35 ☽ ⚻ ♆

31 Wednesday
07 04 ☽ ± ♀
07 17 ♀ △ ♄
08 49 ☽ Q ♇
11 04 ☽ △ ☿
11 38 ☽ ± ♆
12 58 ☉ ⚹ ♂
13 48 ☽ ⊥ ♃
17 09 ☽ △ ♅
17 52 ☽ ∠ ♂
19 13 ☽ ⚺ ♄
20 22 ☽ ⚻ ♀

SEPTEMBER 2044

LONGITUDES

Date	Sidereal time h m s	Sun ☉ ° ' "	Moon ☽ ° ' "	Moon ☽ 24.00 ° ' "	Mercury ☿ ° '	Venus ♀ ° '	Mars ♂ ° '	Jupiter ♃ ° '	Saturn ♄ ° '	Uranus ♅ ° '	Neptune ♆ ° '	Pluto ♇ ° '
01	10 45 11	09 ♍ 42 05	02 ♑ 53 01	08 ♑ 48 56	21 ♌ 35	25 ♋ 51	09 ♏ 24	27 ♑ 50	24 ♏ 37	23 ♌ 36	14 ♉ 48	01 ♓ 23
02	10 49 08	10 40 09	14 ♑ 45 10	20 ♑ 41 53	22 42	26 57	10 03	27 R 46	24 40	23 40	14 R 48	01 R 22
03	10 53 05	11 38 14	26 ♑ 39 52	02 ♒ 39 26	23 56	28 03	10 42	27 42	24 44	23 43	14 47	01 21
04	10 57 01	12 36 21	08 ♒ 40 59	14 ♒ 44 52	25 16	29 ♋ 09	11 22	27 39	24 47	23 47	14 46	01 20
05	11 00 58	13 34 30	20 ♒ 51 22	27 ♒ 00 46	26 41	00 ♌ 15	12 01	27 35	24 51	23 50	14 46	01 18
06	11 04 54	14 32 40	03 ♓ 13 17	09 ♓ 29 05	28 11	01 21	12 41	27 32	24 55	23 54	14 45	01 17
07	11 08 51	15 30 52	15 ♓ 48 17	22 ♓ 11 00	29 ♌ 46	02 28	13 20	27 29	24 59	23 58	14 45	01 16
08	11 12 47	16 29 05	28 ♓ 37 15	05 ♈ 07 05	01 ♍ 24	03 35	14 00	27 26	25 03	24 01	14 44	01 15
09	11 16 44	17 27 21	11 ♈ 40 27	18 ♈ 17 18	03 06	04 42	14 40	27 24	25 07	24 05	14 43	01 13
10	11 20 40	18 25 38	24 ♈ 57 35	01 ♉ 41 12	04 51	05 49	15 20	27 21	25 11	24 08	14 43	01 12
11	11 24 37	19 23 57	08 ♉ 28 03	15 ♉ 17 59	06 38	06 56	16 00	27 19	25 15	24 12	14 42	01 11
12	11 28 34	20 22 19	22 ♉ 10 52	29 ♉ 06 33	08 27	08 04	16 40	27 17	25 19	24 15	14 41	01 10
13	11 32 30	21 20 42	06 ♊ 04 53	13 ♊ 05 40	10 17	09 12	17 21	27 15	25 23	24 18	14 40	01 08
14	11 36 27	22 19 08	20 ♊ 08 41	27 ♊ 13 44	12 09	10 19	18 01	27 13	25 28	24 22	14 39	01 07
15	11 40 23	23 17 36	04 ♋ 20 33	11 ♋ 28 51	14 01	11 28	18 42	27 12	25 32	24 25	14 38	01 06
16	11 44 20	24 16 06	18 ♋ 38 18	25 ♋ 48 31	15 53	12 36	19 22	27 11	25 37	24 29	14 37	01 05
17	11 48 16	25 14 38	02 ♌ 59 08	10 ♌ 09 41	17 46	13 44	20 03	27 10	25 41	24 32	14 36	01 04
18	11 52 13	26 13 13	17 ♌ 19 42	24 ♌ 28 39	19 39	14 53	20 44	27 09	25 46	24 35	14 35	01 03
19	11 56 09	27 11 49	01 ♍ 36 03	08 ♍ 41 19	21 32	16 02	21 25	27 08	25 51	24 39	14 34	01 01
20	12 00 06	28 10 28	15 ♍ 43 58	22 ♍ 43 29	23 24	17 10	22 05	27 07	25 55	24 42	14 33	01 00
21	12 04 03	29 ♍ 09 08	29 ♍ 39 23	06 ♎ 31 16	25 16	18 20	22 47	27 07	26 00	24 45	14 32	00 59
22	12 07 59	00 ♎ 07 50	13 ♎ 18 48	20 ♎ 01 42	27 07	19 29	23 28	27 07	26 05	24 48	14 31	00 58
23	12 11 56	01 06 35	26 ♎ 39 46	03 ♏ 12 56	28 ♍ 58	20 38	24 09	27 D 07	26 10	24 52	14 30	00 57
24	12 15 52	02 05 21	09 ♏ 41 11	16 ♏ 04 36	00 ♎ 47	21 48	24 50	27 07	26 15	24 55	14 29	00 56
25	12 19 49	03 04 09	22 ♏ 23 21	28 ♏ 37 44	02 36	22 57	25 32	27 08	26 20	24 58	14 28	00 55
26	12 23 45	04 02 59	04 ♐ 48 03	10 ♐ 54 42	04 24	24 07	26 13	27 08	26 25	25 01	14 27	00 54
27	12 27 42	05 01 50	16 ♐ 58 10	22 ♐ 58 56	06 12	25 17	26 55	27 09	26 31	25 04	14 25	00 53
28	12 31 38	06 00 44	28 ♐ 57 33	04 ♑ 54 36	07 58	26 27	27 37	27 10	26 36	25 07	14 24	00 52
29	12 35 35	06 59 39	10 ♑ 50 41	16 ♑ 46 24	09 43	27 37	28 19	27 12	26 41	25 10	14 23	00 51
30	12 39 32	07 ♎ 58 36	22 ♑ 42 23	28 ♑ 39 13	11 ♎ 28	28 ♌ 47	29 ♏ 00	27 ♑ 13	26 ♏ 47	25 ♌ 13	14 ♉ 22	00 ♓ 50

Date	Moon True ☊ ° '	Moon Mean ☊ ° '	Moon ☽ Latitude ° '
01	11 ♓ 06	11 ♓ 06	04 S 54
02	11 R 06	11 03	04 24
03	11 D 06	11 00	03 42
04	11 06	10 57	02 50
05	11 06	10 54	01 50
06	11 06	10 50	00 S 44
07	11 R 06	10 47	00 N 26
08	11 06	10 44	01 36
09	11 05	10 41	02 41
10	11 05	10 38	03 40
11	11 04	10 34	04 27
12	11 03	10 31	04 59
13	11 02	10 28	05 15
14	11 02	10 25	05 12
15	11 D 02	10 22	04 51
16	11 02	10 19	04 11
17	11 04	10 15	03 15
18	11 05	10 12	02 08
19	11 05	10 09	00 N 52
20	11 R 05	10 06	00 S 26
21	11 05	10 03	01 41
22	11 03	10 00	02 48
23	11 00	09 56	03 45
24	10 57	09 53	04 29
25	10 54	09 50	04 58
26	10 51	09 47	05 12
27	10 49	09 44	05 12
28	10 47	09 40	04 59
29	10 D 47	09 37	04 32
30	10 ♓ 48	09 ♓ 34	03 S 54

DECLINATIONS

Date	Sun ☉ ° '	Moon ☽ ° '	Mercury ☿ ° '	Venus ♀ ° '	Mars ♂ ° '	Jupiter ♃ ° '	Saturn ♄ ° '	Uranus ♅ ° '	Neptune ♆ ° '	Pluto ♇ ° '
01	07 N 56	28 S 19	14 N 20	19 N 24	15 S 16	21 S 13	17 S 01	14 N 18	14 N 34	21 S 50
02	07 34	27 00	14 11	19 16	15 29	21 14	17 02	14 17	14 34	21 50
03	07 12	24 27	14 00	19 08	15 42	21 15	17 03	14 15	14 33	21 51
04	06 50	20 49	13 45	18 59	15 55	21 15	17 04	14 14	14 33	21 51
05	06 27	16 17	13 26	18 49	16 08	21 16	17 05	14 13	14 33	21 52
06	06 05	11 00	13 05	18 39	16 21	21 17	17 06	14 12	14 33	21 52
07	05 43	05 S 12	12 41	18 29	16 33	21 17	17 08	14 11	14 32	21 53
08	05 20	00 N 55	12 13	18 18	16 46	21 18	17 09	14 10	14 32	21 53
09	04 57	07 05	11 43	18 06	16 58	21 18	17 09	14 08	14 32	21 53
10	04 35	13 04	11 11	17 54	17 11	21 19	17 11	14 07	14 32	21 54
11	04 12	18 32	10 36	17 42	17 23	21 19	17 12	14 06	14 31	21 54
12	03 49	23 08	09 59	17 29	17 35	21 20	17 13	14 05	14 31	21 55
13	03 26	26 29	09 20	17 15	17 47	21 20	17 15	14 04	14 31	21 55
14	03 03	28 16	08 40	17 01	17 59	21 21	17 16	14 03	14 30	21 55
15	02 40	28 12	07 58	16 47	18 11	21 21	17 17	14 02	14 30	21 56
16	02 17	26 17	07 15	16 32	18 23	21 21	17 18	14 01	14 30	21 56
17	01 53	22 39	06 30	16 16	18 34	21 21	17 20	13 59	14 30	21 56
18	01 30	17 40	05 45	16 00	18 46	21 21	17 21	13 58	14 29	21 57
19	01 07	11 43	04 59	15 44	18 57	21 22	17 22	13 57	14 29	21 57
20	00 44	05 N 14	04 13	15 27	19 08	21 22	17 24	13 56	14 29	21 58
21	00 N 20	01 S 24	03 26	15 10	19 20	21 22	17 25	13 55	14 28	21 58
22	00 S 03	07 50	02 39	14 52	19 30	21 22	17 26	13 54	14 28	21 58
23	00 26	13 47	01 52	14 34	19 41	21 22	17 28	13 53	14 27	21 59
24	00 50	18 58	01 05	14 16	19 51	21 22	17 29	13 52	14 27	21 59
25	01 13	23 10	00 N 17	13 57	20 02	21 22	17 30	13 51	14 27	21 59
26	01 37	26 13	00 S 30	13 37	20 12	21 22	17 32	13 50	14 26	21 59
27	02 00	27 59	01 17	13 17	20 22	21 22	17 33	13 49	14 26	22 00
28	02 23	28 24	02 04	12 57	20 33	21 21	17 34	13 48	14 25	22 00
29	02 47	27 31	02 50	12 36	20 42	21 21	17 36	13 47	14 25	22 00
30	03 S 10	25 S 22	03 S 37	12 N 15	20 S 52	21 S 21	17 S 37	13 N 46	14 N 25	22 S 00

ZODIAC SIGN ENTRIES

Date	h	m	Planets
01	06	10	☽ ♑
03	18	41	☽ ♒
05	06	40	♀ ♌
06	05	47	☽ ♓
07	15	30	☿ ♍
08	14	33	☽ ♈
10	21	00	☽ ♉
13	01	32	☽ ♊
15	04	41	☽ ♋
17	07	01	☽ ♌
19	09	18	☽ ♍
21	12	36	☽ ♎
22	08	48	☉ ♎
23	18	06	☽ ♏
24	01	38	☿ ♎
26	02	39	☽ ♐
28	14	06	☽ ♑

LATITUDES

Date	Mercury ☿ ° '	Venus ♀ ° '	Mars ♂ ° '	Jupiter ♃ ° '	Saturn ♄ ° '	Uranus ♅ ° '	Neptune ♆ ° '	Pluto ♇ ° '
01	00 N 01	01 S 36	00 S 41	00 S 38	01 N 58	00 N 41	01 S 48	11 S 39
04	00 41	01 23	00 43	00 38	01 57	00 41	01 48	11 39
07	01 12	01 10	00 45	00 38	01 56	00 41	01 48	11 39
10	01 33	00 56	00 47	00 38	01 56	00 41	01 48	11 39
13	01 45	00 44	00 49	00 38	01 55	00 41	01 48	11 39
16	01 49	00 31	00 50	00 39	01 55	00 41	01 48	11 39
19	01 47	00 19	00 52	00 39	01 54	00 41	01 48	11 39
22	01 39	00 S 07	00 54	00 39	01 53	00 41	01 49	11 38
25	01 27	00 N 05	00 55	00 39	01 53	00 41	01 49	11 38
28	01 12	00 16	00 57	00 39	01 52	00 42	01 49	11 38
31	00 N 54	00 N 27	00 S 58	00 S 39	01 N 52	00 N 42	01 S 49	11 S 37

DATA

Julian Date	2467860
Delta T	+79 seconds
Ayanamsa	24° 28' 58"
Synetic vernal point	04° ♓ 38' 00"
True obliquity of ecliptic	23° 26' 09"

MOON'S PHASES, APSIDES AND POSITIONS ☽

Date	h	m	Phase	Longitude ° '	Eclipse Indicator
07	11	24	○	15 ♓ 29	total
14	15	58	☾	22 ♊ 29	
21	11	03	●	29 ♍ 07	
29	03	31	☽	06 ♑ 39	

Day	h	m		
01	20	23	Apogee	
17	12	18	Perigee	
29	15	05	Apogee	
08	08	27	0N	
14	23	19	Max dec	28° N 28'
21	06	54	0S	
28	07	41	Max dec	28° S 26'

LONGITUDES

Date	Chiron ⚷ ° '	Ceres ⚳ ° '	Pallas ⚴ ° '	Juno ⚵ ° '	Vesta ⚶ ° '	Black Moon Lilith ⚸ ° '
01	07 ♍ 52	07 ♊ 05	17 ♓ 29	22 ♈ 22	07 ♈ 10	11 ♑ 00
11	09 ♍ 15	08 ♊ 46	14 ♓ 54	21 ♈ 49	05 ♈ 12	12 ♑ 07
21	10 ♍ 37	09 ♊ 58	12 ♓ 22	20 ♈ 30	02 ♈ 33	13 ♑ 14
31	11 ♍ 56	10 ♊ 37	10 ♓ 06	18 ♈ 32	00 ♈ 01	14 ♑ 21

ASPECTARIAN

h m Aspects

01 Thursday
01 50 ☽ ⊻ ♃
02 00 ♀ ± ♇
05 46 ☽ bQ ♆
07 24 ☽ ⊥ ♄
08 59 ☽ ✶ ♇
20 02 ☿ ∥ ♅
20 14 ☽ bQ ☿
23 37 ☽ bQ ♅

02 Friday
01 41 ☽ ∠ ♄
01 57 ☽ ✶ ♂
03 01 ☽ △ ☉
08 40 ♀ Q ♆
12 05 ☽ △ ♆
15 16 ☽ ∠ ♇
16 23 ☽ ± ☿
17 54 ☽ ± ♅

03 Saturday
03 35 ☽ Q ♂
04 57 ♀ ☍ ♃
05 52 ☽ ⊼ ☿
06 03 ☽ ⊼ ♅
07 44 ☿ ☌ ♅
08 06 ☽ ✶ ♄
09 22 ☽ ⊥ ♇
11 56 ☽ bQ ☉
14 04 ☽ ☌ ♃
15 03 ☽ ☍ ♀
21 22 ☽ ⊻ ♇

04 Sunday
03 14 ☿ □ ♄
05 50 ☽ ∥ ♇
07 31 ☽ ± ☉
08 13 ☽ Q ♄
09 28 ☽ ∥ ♃
12 52 ☉ bQ ♃
16 53 ☉ Q ♄
17 37 ☽ □ ♂
20 27 ☽ ⊼ ☉
22 39 ☽ # ♀

05 Monday
00 03 ☽ □ ♆
08 02 ☽ ∥ ♄
12 40 ☽ ∥ ♂
17 51 ☽ ☍ ♅
19 50 ☽ □ ♄
20 12 ☽ # ♆
21 45 ☽ # ♅

06 Tuesday
00 55 ☽ ☍ ☿
01 03 ☽ ⊻ ♃
02 03 ☿ ⊼ ♃
02 10 ☽ # ☿
08 03 ☽ ⊼ ♀
08 16 ☽ ☌ ♇
10 33 ♀ ⊼ ♇
11 06 ☽ Q ♆
12 36 ☽ ⊥ ♃
17 07 ☉ △ ♆
20 42 ☽ ± ♀

07 Wednesday
05 44 ☽ ∠ ♃
07 05 ☽ △ ♂
09 48 ☽ # ☉
10 00 ☽ ✶ ♆
11 24 ☽ ☍ ☉
15 26 ☽ bQ ♀

08 Thursday
03 23 ☽ ⊼ ♅
05 19 ☽ △ ♄
09 41 ☿ ☍ ♇
09 49 ☽ ✶ ♃
12 45 ☽ bQ ♂
14 03 ☽ ∠ ♆
14 36 ☽ ± ♅
16 51 ☽ ⊻ ♇
17 55 ☽ ⊼ ☿
22 01 ☽ △ ♀

09 Friday
02 04 ☿ Q ♂
03 53 ☽ ⊥ ♇
04 09 ☽ ∥ ☉
06 13 ☽ ± ♂
06 36 ☽ ± ☿
06 37 ☽ ⊥ ♆
07 14 ☽ bQ ♅
07 51 ☽ Q ♃
09 08 ☽ bQ ♄
13 46 ♂ ☍ ♆
15 58 ☿ ± ♃
17 32 ☽ ⊻ ♆
17 44 ☽ ⊼ ♂
20 15 ☽ ∠ ♇
23 19 ☽ ⊼ ☉

10 Saturday
01 25 ☽ bQ ☿
01 33 ☽ ± ♄
04 55 ☽ ∥ ☿
10 31 ☽ △ ♅
10 58 ☽ ± ☉
12 23 ☽ ⊼ ♄
12 37 ♂ ∥ ♄
12 38 ♂ Q ♃
16 16 ☽ □ ♃
16 25 ☽ ∥ ♅
18 08 ☽ ∥ ♆
23 07 ☽ ✶ ♇

11 Sunday
04 15 ☽ bQ ☉
05 51 ☽ # ♄
06 29 ☽ # ♂
08 16 ☽ △ ☿
08 16 ☽ ∥ ♀
09 03 ☽ □ ♀
20 17 ☽ Q ♇
22 48 ☿ ⊻ ♀
22 56 ☽ ☌ ♆

12 Monday
01 55 ☽ ☍ ♂
01 55 ☽ # ♃
05 03 ☽ # ♇
06 00 ♀ # ♂
08 37 ☽ △ ☉
15 36 ☽ □ ♅
17 28 ☽ ☍ ♄
19 20 ☽ Q ♀
20 49 ☽ △ ♃

13 Tuesday
03 31 ☽ □ ♇
13 04 ♀ # ♄
17 48 ☽ ✶ ♀
20 18 ☽ □ ☿
22 32 ☽ bQ ♃
22 42 ☽ Q ♅

14 Wednesday
02 40 ☽ ⊻ ♆
08 12 ☽ ⊼ ♂
12 52 ☽ ⊥ ♆
13 02 ☿ bQ ♃
13 50 ☽ ± ♃
15 58 ☽ □ ☉
18 54 ☽ ± ♂
19 11 ☽ ✶ ♅
21 03 ☽ ⊼ ♄
21 32 ☽ ∠ ♀
23 58 ☽ ⊼ ♃

15 Thursday
04 05 ☽ ∠ ♆
05 37 ☿ Q ♄
06 33 ☽ △ ♇
07 15 ☽ ± ♄
07 29 ☽ Q ☿
10 51 ☽ bQ ♂
14 03 ☽ ⊥ ♀
19 56 ☿ △ ♆
20 34 ☽ ∠ ♅
22 28 ☽ bQ ♄

16 Friday
00 32 ☽ Q ☉
01 00 ☽ ⊻ ♀
05 17 ☽ ✶ ♆
06 42 ☽ ✶ ☿
07 43 ☽ bQ ♇
11 44 ☽ ⊥ ♅
13 17 ☽ △ ♂
17 27 ☉ ⊻ ♅
21 49 ☽ ⊻ ♅
22 07 ☽ ✶ ☉
22 46 ☽ ± ♇
23 44 ☽ △ ♄

17 Saturday
01 21 ☽ Q ♆
02 16 ☽ ☍ ♃
08 47 ☽ ⊼ ♇
11 35 ☽ ∠ ☿
15 49 ☽ # ♇
18 50 ☽ # ♃
23 47 ☉ ✶ ♄

18 Sunday
01 01 ☽ ∠ ☉
04 55 ☽ ⊥ ☿
06 00 ♀ □ ♆
07 20 ☽ # ♂
07 25 ☽ □ ♆
07 33 ☽ ☌ ♀
13 21 ☽ # ♄
16 29 ☽ ⊻ ☿
17 12 ☽ ⊥ ☉
17 59 ☽ □ ♂
19 21 ☽ ∥ ♀

19 Monday
00 14 ☽ ☌ ♅
01 12 ☽ ∥ ♆
02 14 ☽ □ ♄
03 17 ☽ ∥ ♅
04 02 ☽ ⊻ ☉
04 28 ☽ ⊼ ♃
09 32 ☿ ✶ ♂
10 25 ☉ △ ♃
11 02 ☽ ☍ ♇
14 35 ☽ ± ♃

20 Tuesday
01 54 ☽ Q ♂
04 24 ☿ ⊥ ♀
05 51 ☽ bQ ♃
08 53 ☽ Q ♄
10 00 ☽ △ ♆
14 41 ☽ ⊻ ♀
16 09 ☽ ∥ ☿
23 28 ☽ ✶ ♂

21 Wednesday
01 56 ☽ ⊥ ♀
03 13 ☽ ☌ ☿
03 28 ☽ ⊻ ♅
05 11 ☿ ⊻ ♅
05 18 ☽ ∥ ☉
05 37 ☽ ✶ ♄
07 36 ☽ △ ♃
08 21 ☽ # ☉
11 03 ☽ ☌ ☉
11 48 ☽ bQ ♆
13 55 ☽ ⊥ ♅
14 19 ☽ ⊼ ♇
18 40 ☽ # ☿
18 59 ☽ ∠ ♀
21 18 ☉ bQ ♆
21 58 ☿ ✶ ♄

22 Thursday
00 48 ☽ ± ♇
02 58 ☽ ∠ ♂
03 32 ☽ ± ♆
05 46 ☽ ∠ ♅
08 01 ☽ ∠ ♄
11 58 ☿ △ ♃
14 09 ☽ ⊼ ♆
14 53 ♃ St D
16 43 ☽ bQ ♇
19 48 ☽ ⊥ ♂

23 Friday
00 03 ☽ ✶ ♀
00 11 ☽ ⊥ ♄
05 31 ☉ ⊥ ♅
07 11 ☽ ⊻ ♂
08 10 ☉ ⊼ ♇
08 43 ☽ ✶ ♅
11 06 ☽ ⊻ ♄
12 27 ☽ # ♅
12 50 ☽ □ ♃
14 56 ☽ # ♆
15 14 ☽ # ♀
16 52 ☽ ⊻ ☿
19 02 ☿ bQ ♆
19 49 ☽ △ ♇
20 47 ☽ ⊻ ☉
20 55 ♀ ∥ ♆

24 Saturday
00 00 ☽ Q ♀
04 41 ☽ ∥ ♄
05 43 ☽ ⊥ ☿
06 49 ☽ Q ♅
08 47 ☽ ⊥ ☉
13 41 ☿ ⊥ ♅
13 52 ☿ ⊼ ♇
14 45 ♂ □ ♅
16 51 ☽ ∥ ♂
17 03 ☉ # ☿
20 59 ☽ ☍ ♆
22 12 ☽ Q ♃

25 Sunday
00 57 ☽ ∥ ♃
01 22 ☽ ∠ ☿
03 05 ☽ ∠ ☉
04 37 ☽ ∥ ♇
13 11 ☽ □ ♀
16 57 ☽ □ ♅
18 23 ☽ ☌ ♂
19 16 ♀ ∥ ♅
19 38 ☽ ☌ ♄
21 07 ☽ ✶ ♃

26 Monday
01 32 ☉ ☌ ☿
04 24 ☽ □ ♇
10 24 ☽ ✶ ☉
11 06 ☽ ✶ ☿
19 59 ♂ ☌ ♄

27 Tuesday
02 27 ☽ ∠ ♃
06 57 ☽ ⊼ ♆
07 28 ♀ ☌ ♅
12 08 ☽ Q ☉
14 51 ☽ Q ☿
15 48 ☽ Q ♇
18 52 ☽ ± ♆
20 21 ☽ ⊥ ♃
20 26 ♂ ✶ ♃
21 11 ☿ ± ♇

28 Wednesday
04 15 ☽ △ ♅
06 24 ☽ △ ♀
07 13 ☽ ⊻ ♄
08 24 ☽ ⊻ ♃
09 07 ☽ ⊻ ♂
12 54 ☽ bQ ♆
15 27 ♀ □ ♄
15 50 ☽ ✶ ♇
17 54 ☿ ± ♆
19 24 ☽ ⊥ ♄
21 58 ☽ ⊥ ♂

29 Thursday
03 12 ♀ ⊼ ♃
03 31 ☽ □ ☉
07 59 ☉ ∥ ☿
08 27 ☉ ± ♇
09 20 ☽ □ ☿
10 38 ☽ bQ ♅
13 43 ☽ ∠ ♄
15 58 ☽ bQ ♀
17 18 ☽ ∠ ♂
18 18 ☿ ∠ ♅
19 09 ☽ △ ♆
22 07 ☽ ∠ ♇

30 Friday
04 55 ☽ ± ♅
12 11 ☽ ± ♀
16 17 ☽ ⊥ ♇
16 36 ☿ ∠ ♄
17 06 ☽ ⊼ ♅
20 17 ☽ ✶ ♄
21 08 ☽ ☌ ♃
21 13 ☉ ± ♆
23 16 ♀ □ ♂

All ephemeris data is given at 12.00 UT and the Moon's longitude is additionally given for 24.00 UT

OCTOBER 2044

LONGITUDES

Date	Sidereal time h m s	Sun ☉ ° ′ ″	Moon ☽ ° ′ ″	Moon ☽ 24.00 ° ′ ″	Mercury ☿ ° ′	Venus ♀ ° ′	Mars ♂ ° ′	Jupiter ♃ ° ′	Saturn ♄ ° ′	Uranus ♅ ° ′	Neptune ♆ ° ′	Pluto ♇ ° ′
01	12 43 28	08 ♎ 57 34	04 ♒ 37 30	10 ♒ 37 50	13 ♎ 12	29 ♌ 58	29 ♏ 42	27 ♑ 15	26 ♏ 52	25 ♌ 16	14 ♉ 20	00 ♓ 49
02	12 47 25	09 56 35	16 ♒ 40 44	22 ♒ 46 44	14 54	01 ♍ 08	00 ♐ 25	27 17	26 58	25 19	14 R 19	00 R 48
03	12 51 21	10 55 37	28 ♒ 56 16	05 ♓ 09 46	16 36	02 19	01 07	27 19	27 04	25 22	14 18	00 47
04	12 55 18	11 54 41	11 ♓ 27 33	17 ♓ 49 53	18 18	03 29	01 49	27 21	27 09	25 25	14 16	00 46
05	12 59 14	12 53 46	24 ♓ 16 56	00 ♈ 48 49	19 58	04 40	02 31	27 23	27 15	25 28	14 15	00 45
06	13 03 11	13 52 54	07 ♈ 25 30	14 ♈ 06 54	21 37	05 51	03 14	27 26	27 21	25 30	14 14	00 44
07	13 07 07	14 52 04	20 ♈ 52 48	27 ♈ 42 57	23 16	07 02	03 56	27 29	27 27	25 33	14 12	00 43
08	13 11 04	15 51 16	04 ♉ 36 57	11 ♉ 34 23	24 54	08 14	04 39	27 32	27 32	25 36	14 11	00 43
09	13 15 00	16 50 30	18 ♉ 34 45	25 ♉ 37 32	26 31	09 25	05 21	27 35	27 38	25 39	14 09	00 42
10	13 18 57	17 49 46	02 ♊ 42 12	09 ♊ 48 14	28 07	10 36	06 04	27 38	27 44	25 41	14 08	00 41
11	13 22 54	18 49 04	16 ♊ 55 06	24 ♊ 02 21	29 ♎ 43	11 48	06 47	27 42	27 50	25 44	14 06	00 40
12	13 26 50	19 48 25	01 ♋ 09 33	08 ♋ 16 20	01 ♏ 18	12 59	07 30	27 45	27 56	25 46	14 05	00 39
13	13 30 47	20 47 49	15 ♋ 22 24	22 ♋ 27 29	02 52	14 11	08 13	27 49	28 03	25 49	14 03	00 39
14	13 34 43	21 47 14	29 ♋ 31 22	06 ♌ 33 54	04 25	15 23	08 56	27 53	28 09	25 51	14 02	00 38
15	13 38 40	22 46 42	13 ♌ 34 56	20 ♌ 34 20	05 58	16 35	09 39	27 57	28 15	25 54	14 00	00 37
16	13 42 36	23 46 12	27 ♌ 31 59	04 ♍ 27 46	07 30	17 47	10 22	28 02	28 21	25 56	13 59	00 36
17	13 46 33	24 45 45	11 ♍ 21 32	18 ♍ 13 08	09 02	18 59	11 06	28 07	28 28	25 59	13 57	00 36
18	13 50 29	25 45 19	25 ♍ 02 23	01 ♎ 49 06	10 33	20 11	11 49	28 11	28 34	26 01	13 56	00 35
19	13 54 26	26 44 56	08 ♎ 33 04	15 ♎ 14 04	12 03	21 24	12 32	28 16	28 40	26 03	13 54	00 35
20	13 58 23	27 44 35	21 ♎ 51 54	28 ♎ 26 20	13 32	22 36	13 16	28 21	28 47	26 05	13 52	00 34
21	14 02 19	28 44 16	04 ♏ 57 14	11 ♏ 24 27	15 01	23 49	14 00	28 27	28 53	26 08	13 51	00 33
22	14 06 16	29 ♎ 43 59	17 ♏ 47 53	24 ♏ 07 31	16 30	25 01	14 43	28 32	29 00	26 10	13 49	00 33
23	14 10 12	00 ♏ 43 44	00 ♐ 23 23	06 ♐ 35 36	17 57	26 14	15 27	28 38	29 06	26 12	13 48	00 32
24	14 14 09	01 43 31	12 ♐ 44 19	18 ♐ 49 48	19 24	27 27	16 11	28 44	29 13	26 14	13 46	00 32
25	14 18 05	02 43 19	24 ♐ 52 22	00 ♑ 52 23	20 50	28 39	16 55	28 50	29 19	26 16	13 44	00 31
26	14 22 02	03 43 09	06 ♑ 50 19	12 ♑ 46 39	22 16	29 ♍ 52	17 39	28 56	29 26	26 18	13 43	00 31
27	14 25 58	04 43 02	18 ♑ 41 56	24 ♑ 36 46	23 41	01 ♎ 05	18 23	29 02	29 33	26 20	13 41	00 30
28	14 29 55	05 42 55	00 ♒ 31 45	06 ♒ 27 34	25 04	02 18	19 07	29 09	29 39	26 22	13 39	00 30
29	14 33 52	06 42 51	12 ♒ 24 52	18 ♒ 24 18	26 28	03 31	19 51	29 15	29 46	26 23	13 38	00 30
30	14 37 48	07 42 47	24 ♒ 26 34	00 ♓ 32 19	27 50	04 45	20 36	29 22	29 ♏ 53	26 25	13 36	00 29
31	14 41 45	08 ♏ 42 46	06 ♓ 42 09	12 ♓ 56 39	29 ♏ 11	05 ♎ 58	21 ♐ 20	29 ♑ 29	00 ♐ 00	26 ♌ 27	13 ♉ 34	00 ♓ 29

Date	Moon True ☊ ° ′	Moon Mean ☊ ° ′	Moon ☽ Latitude ° ′
01	10 ♓ 50	09 ♓ 31	03 S 06
02	10 D 51	09 28	02 09
03	10 53	09 25	01 S 05
04	10 R 53	09 21	00 N 03
05	10 53	09 18	01 13
06	10 50	09 15	02 20
07	10 46	09 12	03 21
08	10 41	09 09	04 11
09	10 35	09 06	04 48
10	10 30	09 02	05 07
11	10 25	08 59	05 08
12	10 22	08 56	04 50
13	10 21	08 53	04 14
14	10 D 21	08 50	03 23
15	10 22	08 46	02 20
16	10 23	08 43	01 N 09
17	10 R 24	08 40	00 S 05
18	10 23	08 37	01 18
19	10 19	08 34	02 26
20	10 14	08 31	03 24
21	10 06	08 27	04 11
22	09 57	08 24	04 44
23	09 48	08 21	05 02
24	09 39	08 18	05 05
25	09 31	08 15	04 55
26	09 25	08 12	04 32
27	09 21	08 08	03 57
28	09 19	08 05	03 12
29	09 D 19	08 02	02 19
30	09 20	07 59	01 19
31	09 ♓ 21	07 ♓ 56	00 S 14

DECLINATIONS

Date	Sun ☉ ° ′	Moon ☽ ° ′	Mercury ☿ ° ′	Venus ♀ ° ′	Mars ♂ ° ′	Jupiter ♃ ° ′	Saturn ♄ ° ′	Uranus ♅ ° ′	Neptune ♆ ° ′	Pluto ♇ ° ′
01	03 S 33	22 S 06	04 S 23	11 N 54	21 S 02	21 S 20	17 S 39	13 N 45	14 N 24	22 S 01
02	03 56	17 53	05 08	11 32	21 11	21 20	17 40	13 44	14 24	22 01
03	04 19	12 51	05 53	11 10	21 20	21 19	17 42	13 43	14 23	22 01
04	04 43	07 13	06 38	10 48	21 29	21 19	17 43	13 42	14 23	22 01
05	05 06	01 S 10	07 22	10 25	21 38	21 19	17 44	13 41	14 23	22 01
06	05 29	05 N 05	08 05	10 02	21 47	21 18	17 46	13 40	14 22	22 02
07	05 51	11 15	08 48	09 38	21 56	21 17	17 47	13 39	14 22	22 02
08	06 14	17 00	09 31	09 15	22 04	21 17	17 49	13 39	14 21	22 02
09	06 37	21 57	10 12	08 51	22 12	21 16	17 50	13 38	14 21	22 02
10	07 00	25 43	10 54	08 26	22 20	21 16	17 52	13 37	14 20	22 02
11	07 22	27 54	11 34	08 02	22 28	21 15	17 53	13 36	14 20	22 02
12	07 45	28 16	12 14	07 37	22 35	21 14	17 55	13 35	14 19	22 02
13	08 07	26 46	12 53	07 12	22 43	21 13	17 56	13 34	14 19	22 03
14	08 29	23 33	13 31	06 46	22 50	21 12	17 58	13 34	14 18	22 03
15	08 51	18 59	14 09	06 20	22 57	21 12	17 59	13 33	14 18	22 03
16	09 13	13 25	14 46	05 55	23 03	21 11	18 01	13 32	14 17	22 03
17	09 35	07 13	15 22	05 29	23 10	21 10	18 02	13 31	14 17	22 03
18	09 57	00 N 46	15 57	05 02	23 16	21 09	18 04	13 30	14 17	22 03
19	10 19	05 S 38	16 31	04 36	23 22	21 08	18 05	13 30	14 16	22 03
20	10 40	11 41	17 05	04 09	23 28	21 07	18 07	13 29	14 16	22 03
21	11 01	17 06	17 38	03 42	23 34	21 06	18 08	13 28	14 15	22 03
22	11 23	21 40	18 09	03 15	23 39	21 05	18 10	13 28	14 15	22 03
23	11 44	25 09	18 40	02 48	23 45	21 03	18 11	13 27	14 14	22 03
24	12 04	27 22	19 10	02 21	23 50	21 02	18 13	13 26	14 14	22 03
25	12 25	28 15	19 39	01 53	23 54	21 01	18 15	13 26	14 13	22 03
26	12 45	27 47	20 07	01 26	23 59	21 00	18 16	13 25	14 13	22 03
27	13 06	26 03	20 34	00 58	24 03	20 58	18 18	13 24	14 12	22 03
28	13 26	23 10	21 00	00 31	24 07	20 57	18 19	13 24	14 12	22 03
29	13 45	19 18	21 25	00 N 03	24 11	20 56	18 21	13 23	14 11	22 03
30	14 05	14 37	21 49	00 S 25	24 14	20 54	18 22	13 23	14 11	22 03
31	14 S 24	09 S 16	22 S 12	00 S 53	24 S 18	20 S 53	18 S 24	13 N 22	14 N 10	22 S 03

ZODIAC SIGN ENTRIES

Date	h	m	Planets
01	02	43	☽ ♒
01	12	51	♀ ♍
01	22	01	♂ ♐
03	14	03	☽ ♓
05	22	31	☽ ♈
08	03	59	☽ ♉
10	07	25	☽ ♊
11	16	21	☿ ♏
12	10	03	☽ ♋
14	12	49	☽ ♌
16	16	16	☽ ♍
18	20	46	☽ ♎
21	02	52	☽ ♏
22	18	26	☉ ♏
23	11	15	☽ ♐
25	22	15	☽ ♑
26	14	31	♀ ♎
28	10	56	☽ ♒
30	22	57	☽ ♓
31	12	52	♄ ♐

LATITUDES

Date	Mercury ☿ ° ′	Venus ♀ ° ′	Mars ♂ ° ′	Jupiter ♃ ° ′	Saturn ♄ ° ′	Uranus ♅ ° ′	Neptune ♆ ° ′	Pluto ♇ ° ′
01	00 N 54	00 N 27	00 S 58	00 S 39	01 N 52	00 N 42	01 S 49	11 S 37
04	00 35	00 37	00 59	00 39	01 51	00 42	01 49	11 37
07	00 N 15	00 46	01 01	00 38	01 51	00 42	01 49	11 36
10	00 S 06	00 55	01 02	00 38	01 50	00 42	01 49	11 36
13	00 27	01 03	01 03	00 38	01 50	00 42	01 49	11 35
16	00 47	01 11	01 04	00 38	01 49	00 42	01 49	11 35
19	01 08	01 17	01 05	00 38	01 49	00 42	01 49	11 34
22	01 27	01 23	01 06	00 38	01 49	00 42	01 49	11 33
25	01 45	01 29	01 07	00 38	01 48	00 42	01 50	11 33
28	02 02	01 33	01 08	00 38	01 48	00 43	01 50	11 32
31	02 S 17	01 N 37	01 S 09	00 S 38	01 N 48	00 N 43	01 S 50	11 S 32

DATA

Julian Date	2467890
Delta T	+79 seconds
Ayanamsa	24° 29′ 02″
Synetic vernal point	04° ♓ 37′ 57″
True obliquity of ecliptic	23° 26′ 10″

MOON'S PHASES, APSIDES AND POSITIONS ☽

Date	h	m	Phase	Longitude ° ′	Eclipse Indicator
07	00	30	○	14 ♈ 24	
13	21	52	☾	21 ♋ 12	
20	23	36	●	28 ♎ 13	
28	23	28	☽	06 ♒ 12	

Day	h	m		
12	15	52	Perigee	
27	11	20	Apogee	
05	16	29	0N	
12	04	37	Max dec	28° N 21′
18	14	52	0S	
25	15	35	Max dec	28° S 16′

LONGITUDES

Date	Chiron ⚷ ° ′	Ceres ⚳ ° ′	Pallas ⚴ ° ′	Juno ⚵ ° ′	Vesta ⚶ ° ′	Black Moon Lilith ⚸ ° ′
01	11 ♍ 56	10 ♊ 37	10 ♓ 06	18 ♈ 32	00 ♈ 01	14 ♑ 21
11	13 ♍ 12	10 ♊ 39	08 ♓ 19	16 ♈ 13	27 ♓ 46	15 ♑ 28
21	14 ♍ 24	10 ♊ 02	07 ♓ 06	13 ♈ 53	26 ♓ 03	16 ♑ 35
31	15 ♍ 30	08 ♊ 47	06 ♓ 30	11 ♈ 58	25 ♓ 03	17 ♑ 42

ASPECTARIAN

h m	Aspects
01 Saturday	
01 30	☽ ✶ ♂
01 36	☽ ⊼ ♀
04 21	☽ ⊻ ♇
12 37	☽ ‖ ♇
16 47	☽ ‖ ♃
18 23	☽ ‖ ♂
20 34	☽ Q ♄
21 26	☽ △ ☉
02 Sunday	
03 01	☽ Q ♂
03 50	☿ ⊼ ♆
05 14	♀ ☍ ♇
07 20	☽ □ ♆
07 55	☽ △ ☿
13 05	☽ ‖ ♄
21 37	☉ ∠ ♅
03 Monday	
00 00	☿ ∠ ♂
00 26	☿ Ᵽ ♇
01 00	♂ □ ♇
05 00	☽ # ♆
05 02	☽ ☍ ♅
05 38	☽ Ᵽ ☉
08 06	☽ # ♅
08 19	☽ □ ♄
08 50	☽ ⊻ ♃
09 39	♂ ‖ ♃
15 34	☽ ☌ ♇
16 27	☽ □ ♂
17 58	☽ Ᵽ ☿
18 29	☽ Q ♆
19 12	☽ ☍ ♀
19 58	☽ # ♀
20 28	☽ ⊥ ♃
04 Tuesday	
00 32	☽ ± ☉
08 59	♀ ± ♃
12 56	☽ ⊼ ☉
13 41	☽ ∠ ♃
13 49	☽ ± ☿
14 07	☽ ‖ ☿
17 19	☽ ✶ ♆
18 32	☉ ∠ ♄
21 29	☽ ‖ ☉
21 34	☿ ∠ ♀
05 Wednesday	
02 47	☽ ⊼ ☿
14 11	☽ ⊼ ♅
17 31	☽ △ ♄
17 44	☽ ✶ ♃
21 08	☽ ∠ ♆
23 52	☽ ⊻ ♇
06 Thursday	
01 14	☽ ± ♅
03 58	☽ △ ♂
07 45	☿ ⊥ ♄
08 53	☽ ⊼ ♀
10 46	☽ ⊥ ♇
13 27	☽ ⊥ ♆
13 36	☽ # ☉
15 37	☽ Q ♃
17 34	☽ Ᵽ ♅
20 14	☉ ⊼ ♆
20 44	☽ ± ♀
20 54	☽ Ᵽ ♄
07 Friday	
00 11	☽ ⊻ ♆
00 30	☽ ☍ ☉
01 07	☽ # ☿
02 52	☽ ∠ ♇
06 01	☽ ‖ ♀
08 22	☽ Ᵽ ♂
13 00	☽ ± ♄
14 14	☽ Ᵽ ♀
16 46	☽ ☍ ☿
20 15	☽ △ ♅
21 45	☽ ‖ ♅
23 36	☽ ⊼ ♄
23 37	☽ □ ♃
08 Saturday	
00 40	☽ ‖ ♆
01 03	☽ ± ♂
04 46	♃ ✶ ♄
05 13	☽ ✶ ♇
06 08	☿ # ♀
06 18	♂ ‖ ♇
08 30	☉ Ᵽ ♇
12 03	☽ ⊼ ♂
15 40	☽ # ♄
18 49	☽ △ ♀
22 42	☿ ✶ ♅
09 Sunday	
01 56	☽ Q ♇
04 27	☽ ☌ ♆
08 22	☽ # ♃
08 48	☽ ⊼ ☉
12 25	☽ # ♇
13 22	☽ # ♂
19 49	☽ ± ☉
10 Monday	
00 04	☽ □ ♅
03 14	☽ ⊼ ☿
03 22	☽ △ ♃
03 32	☽ ☍ ♄
04 28	☿ □ ♃
05 55	☿ ⊻ ♄
08 35	☽ □ ♇
12 14	☽ Ᵽ ☉
14 42	☽ ± ☿
17 59	☽ ☍ ♂
11 Tuesday	
02 34	☽ □ ♀
04 51	☽ Ᵽ ♃
06 37	☽ Q ♅
07 16	☽ ⊻ ♆
07 49	☽ Ᵽ ☿
15 26	☽ △ ☉
17 22	☽ ⊥ ♆
20 05	☽ ± ♃
12 Wednesday	
02 22	☿ △ ♇
02 54	☽ ✶ ♅
06 14	☽ ⊼ ♃
06 32	☽ ⊼ ♄
06 59	♀ Ᵽ ♃
07 56	☉ # ♀
08 30	☽ ∠ ♆
11 09	☽ △ ♇
11 41	☽ Q ♀
12 15	☽ △ ☿
16 43	☽ ± ♄
17 42	☿ ⊥ ♂
23 16	☽ ⊼ ♂
13 Thursday	
04 16	☽ ∠ ♅
08 02	☽ Ᵽ ♄
09 28	♀ △ ♆
09 47	☽ ✶ ♆
09 48	☽ ✶ ♀
09 56	☽ ± ♂
12 27	☽ Ᵽ ♇
19 33	☽ ⊥ ♅
21 52	☽ □ ☉
14 Friday	
01 59	☽ Ᵽ ♂
03 42	☽ ± ♇
05 45	☽ ⊻ ♅
06 05	☽ Q ♆
09 12	☽ ☍ ♃
09 38	☽ △ ♄
13 28	☿ # ♅
13 36	☽ ∠ ♀
13 53	☽ ⊼ ♇
16 10	☽ # ♂
20 39	☽ # ♇
21 23	☽ □ ☿
21 41	☉ ⊥ ♄
15 Saturday	
01 07	☽ # ♃
04 41	♀ Q ♄
04 54	☽ △ ♂
06 23	☽ ⊥ ♀
06 50	☽ Q ☉
12 44	☽ □ ♆
16 32	☽ # ♄
17 38	☽ ⊻ ♀
17 54	☿ # ♆
16 Sunday	
05 01	☽ ✶ ☉
07 00	☽ # ☿
08 04	☽ Q ☿
08 24	☽ ‖ ♆
09 14	☽ ☌ ♅
10 23	☉ ⊥ ♀
11 30	☽ ‖ ♅
12 52	☽ ⊼ ♃
13 26	☽ □ ♄
17 19	☽ ☍ ♇
18 59	☿ Q ♅
23 19	☽ ± ♃
17 Monday	
03 30	☽ # ☉
07 26	☽ ✶ ☿
09 00	☽ ∠ ☉
11 31	☽ □ ♂
15 04	☽ Ᵽ ♃
16 31	☽ △ ♆
19 02	☽ ‖ ♀
20 59	☽ Q ♄
18 Tuesday	
01 58	☽ ⊥ ☉
02 38	☽ ☌ ♀
13 00	☽ ∠ ☿
13 22	☽ ⊻ ☉
13 44	☽ ⊻ ♅
17 36	☽ △ ♃
18 17	☽ ✶ ♄
18 32	☉ ✶ ♅
18 52	☽ Ᵽ ♆
20 55	☽ Q ♂
21 48	☽ ⊼ ♇
19 Wednesday	
00 23	☽ ⊥ ♅
06 58	☽ ⊥ ☿
08 20	☽ # ♀
08 28	☽ ± ♇
10 50	☽ ± ♆
16 30	☽ ∠ ♅
19 03	☽ ⊻ ☿
19 34	☽ ✶ ♂
21 16	☽ ∠ ♄
21 35	☽ ⊼ ♆
20 Thursday	
00 36	☽ Ᵽ ♇
03 24	☿ ⊻ ♂
07 36	☽ ‖ ☉
13 29	☽ ⊻ ♀
13 40	☽ ⊥ ♄
15 23	♂ ∠ ♃
17 19	☿ ☍ ♆
19 37	☽ # ♅
19 43	☽ ✶ ♅
22 59	☽ # ♆
23 36	☽ ☌ ☉
23 56	☽ □ ♃
21 Friday	
00 22	☽ ∠ ♂
00 44	☽ ⊻ ♄
01 33	☽ ⊥ ♀
03 54	☽ △ ♇
04 14	☉ □ ♃
07 22	♂ ⊼ ♆
14 48	☽ ‖ ☿
16 00	☉ ⊻ ♄
17 02	☽ ‖ ♄
17 54	☽ Q ♅
17 58	☽ ⊥ ♂
19 54	☽ ∠ ♀
22 Saturday	
04 32	☽ ☍ ♆
05 52	☽ ⊻ ♂
08 34	☽ ‖ ♃
09 13	☽ ☌ ☿
09 36	☽ Q ♃
11 00	☉ ∠ ♂
12 29	☿ ‖ ♄
12 46	☿ Q ♃
14 17	☽ ‖ ♇
23 Sunday	
01 02	☽ ‖ ♂
03 10	☽ ✶ ♀
03 56	☽ □ ♅
07 27	☉ △ ♇
08 35	☽ ✶ ♃
09 30	☽ ☌ ♄
11 18	♀ ⊻ ♅
12 17	☽ □ ♇
12 43	☽ ⊻ ☉
24 Monday	
01 20	☽ ⊥ ☉
04 51	☽ Q ♀
13 57	☽ ∠ ♃
14 01	☽ ⊼ ♆
19 12	☽ ☌ ♂
20 33	☽ ∠ ☉
23 24	☽ Q ♇
25 Tuesday	
01 49	☽ ± ♆
02 54	☽ ⊻ ☿
07 54	☽ ⊥ ♃
13 34	♀ Ᵽ ♆
14 47	☽ △ ♅
15 39	♀ △ ♃
16 27	☽ ⊥ ☿
19 42	☽ Ᵽ ♆
19 58	☽ ⊻ ♃
20 25	☽ □ ♀
20 59	☽ ⊻ ♄
23 17	☽ ✶ ♇
26 Wednesday	
02 28	♀ ✶ ♄
05 09	☽ ✶ ☉
09 09	☽ ⊥ ♄
12 58	☽ ∠ ☿
21 02	☽ Ᵽ ♅
27 Thursday	
00 36	♀ ⊼ ♇
01 52	☽ △ ♆
03 30	☽ ∠ ♄
05 32	☽ ∠ ♇
07 37	☽ Q ☉
11 19	☽ ⊻ ♂
15 19	☽ ± ♅
16 01	♂ Q ♇
23 27	☽ ✶ ☿
23 47	☽ ⊥ ♇
28 Friday	
00 18	☽ ⊥ ♂
03 31	☽ ⊼ ♅
05 07	☽ ‖ ♂
09 03	☿ ‖ ♃
09 10	☽ ☌ ♃
09 57	☉ # ♅
10 13	☽ ✶ ♄
11 57	☽ ⊻ ♇
13 06	♀ ⊥ ♅
16 01	☽ △ ♀
19 33	☽ ‖ ♇
19 45	☽ ∠ ♂
23 28	☽ □ ☉
29 Saturday	
00 44	☽ ‖ ☿
02 30	☽ ‖ ♃
03 00	☽ Q ☿
04 52	♂ ± ♆
10 42	☽ Q ♄
10 46	☿ □ ♅
14 26	☽ □ ♆
17 11	☽ ‖ ♄
30 Sunday	
01 37	☽ Ᵽ ♀
03 52	☽ ✶ ♂
14 04	☽ # ♆
14 22	☽ ‖ ☉
15 55	☽ ☍ ♅
17 48	☽ # ♅
18 52	☉ # ♆
19 31	☽ □ ☿
21 25	☽ ± ♀
21 48	☽ ⊻ ♃
22 49	☽ □ ♄
23 54	☽ ☌ ♇
31 Monday	
02 03	☽ Q ♆
02 16	☿ ‖ ♇
05 02	☽ Q ♂
05 28	☉ Q ♅
09 37	☽ ⊥ ♃
10 25	☽ ⊼ ♀
16 13	☽ △ ☉
17 49	☿ ✶ ♃
22 08	♀ ± ♇

All ephemeris data is given at 12.00 UT and the Moon's longitude is additionally given for 24.00 UT

NOVEMBER 2044

LONGITUDES

Date	Sidereal time h m s	Sun ☉ ° ' "	Moon ☽ ° ' "	Moon ☽ 24.00 ° ' "	Mercury ☿ ° '	Venus ♀ ° '	Mars ♂ ° '	Jupiter ♃ ° '	Saturn ♄ ° '	Uranus ♅ ° '	Neptune ♆ ° '	Pluto ♇ ° '
01	14 45 41	09 ♏ 42 46	19 ♓ 16 20	25 ♓ 41 39	00 ♐ 31	07 ♎ 11	22 ♐ 04	29 ♑ 36	00 ♐ 07	26 ♌ 29	13 ♉ 33	00 ♓ 29
02	14 49 38	10 42 48	02 ♈ 12 55	08 ♈ 50 23	01 51	08 25	22 49	29 43	00 13	26 30	13 R 31	00 R 28
03	14 53 34	11 42 51	15 ♈ 34 07	22 ♈ 24 04	03 08	09 38	23 33	29 51	00 20	26 32	13 29	00 28
04	14 57 31	12 42 57	29 ♈ 20 02	06 ♉ 21 37	04 25	10 51	24 18	29 ♑ 58	00 27	26 33	13 28	00 28
05	15 01 27	13 43 04	13 ♉ 28 17	20 ♉ 39 22	05 40	12 05	25 03	00 ♒ 06	00 34	26 35	13 26	00 28
06	15 05 24	14 43 13	27 ♉ 54 02	05 ♊ 11 25	06 54	13 19	25 47	00 14	00 41	26 36	13 24	00 27
07	15 09 21	15 43 23	12 ♊ 30 34	19 ♊ 50 31	08 06	14 32	26 32	00 22	00 48	26 37	13 23	00 27
08	15 13 17	16 43 36	27 ♊ 10 20	04 ♋ 29 10	09 15	15 46	27 17	00 30	00 55	26 39	13 21	00 27
09	15 17 14	17 43 51	11 ♋ 46 15	19 ♋ 00 57	10 23	17 00	28 02	00 38	01 02	26 40	13 19	00 27
10	15 21 10	18 44 08	26 ♋ 12 45	03 ♌ 21 17	11 28	18 14	28 47	00 47	01 09	26 41	13 17	00 27
11	15 25 07	19 44 26	10 ♌ 26 17	17 ♌ 27 39	12 30	19 28	29 ♐ 32	00 55	01 16	26 42	13 16	00 27
12	15 29 03	20 44 47	24 ♌ 25 21	01 ♍ 19 24	13 29	20 42	00 ♑ 17	01 04	01 23	26 43	13 14	00 27
13	15 33 00	21 45 10	08 ♍ 09 56	14 ♍ 57 04	14 25	21 56	01 02	01 13	01 30	26 45	13 12	00 D 27
14	15 36 56	22 45 34	21 ♍ 40 57	28 ♍ 21 44	15 17	23 10	01 48	01 22	01 37	26 46	13 11	00 27
15	15 40 53	23 46 01	04 ♎ 59 34	11 ♎ 34 32	16 04	24 24	02 33	01 31	01 44	26 46	13 09	00 27
16	15 44 50	24 46 29	18 ♎ 06 45	24 ♎ 36 15	16 46	25 38	03 18	01 40	01 51	26 47	13 07	00 27
17	15 48 46	25 46 59	01 ♏ 03 04	07 ♏ 27 11	17 22	26 52	04 04	01 49	01 59	26 48	13 06	00 27
18	15 52 43	26 47 31	13 ♏ 48 36	20 ♏ 07 16	17 52	28 07	04 49	01 59	02 06	26 49	13 04	00 27
19	15 56 39	27 48 04	26 ♏ 23 11	02 ♐ 36 19	18 15	29 ♎ 21	05 35	02 08	02 13	26 50	13 03	00 27
20	16 00 36	28 48 39	08 ♐ 46 42	14 ♐ 54 22	18 30	00 ♏ 35	06 21	02 18	02 20	26 50	13 01	00 28
21	16 04 32	29 ♏ 49 15	20 ♐ 59 24	27 ♐ 01 56	18 36	01 50	07 06	02 28	02 27	26 51	12 59	00 28
22	16 08 29	00 ♐ 49 53	03 ♑ 02 11	09 ♑ 00 23	18 R 33	03 04	07 52	02 38	02 34	26 51	12 58	00 28
23	16 12 25	01 50 32	14 ♑ 56 52	20 ♑ 51 58	18 21	04 19	08 38	02 48	02 41	26 52	12 56	00 28
24	16 16 22	02 51 13	26 ♑ 46 09	02 ♒ 39 53	17 57	05 33	09 24	02 58	02 48	26 52	12 55	00 29
25	16 20 19	03 51 54	08 ♒ 33 44	14 ♒ 28 16	17 23	06 48	10 09	03 08	02 56	26 53	12 53	00 29
26	16 24 15	04 52 37	20 ♒ 24 06	26 ♒ 21 56	16 38	08 02	10 55	03 19	03 03	26 53	12 51	00 29
27	16 28 12	05 53 20	02 ♓ 22 25	08 ♓ 26 17	15 43	09 17	11 41	03 29	03 10	26 53	12 50	00 30
28	16 32 08	06 54 05	14 ♓ 34 12	20 ♓ 46 52	14 38	10 31	12 27	03 40	03 17	26 53	12 48	00 30
29	16 36 05	07 54 51	27 ♓ 04 56	03 ♈ 29 01	13 26	11 46	13 13	03 51	03 24	26 54	12 47	00 30
30	16 40 01	08 ♐ 55 37	09 ♈ 59 37	16 ♈ 37 09	12 ♐ 07	13 ♏ 01	14 ♑ 00	04 ♒ 02	03 ♐ 31	26 ♌ 54	12 ♉ 45	00 ♓ 31

Moon nodes and latitude

Date	Moon True ☊ ° '	Moon Mean ☊ ° '	Moon ☽ Latitude ° '
01	09 ♓ 21	07 ♓ 52	00 N 53
02	09 R 18	07 49	01 59
03	09 13	07 46	03 00
04	09 06	07 43	03 53
05	08 56	07 40	04 33
06	08 46	07 37	04 57
07	08 35	07 33	05 02
08	08 26	07 30	04 47
09	08 19	07 27	04 13
10	08 15	07 24	03 23
11	08 13	07 21	02 22
12	08 D 13	07 17	01 N 12
13	08 13	07 14	00 00
14	08 R 12	07 11	01 S 11
15	08 10	07 08	02 17
16	08 04	07 05	03 14
17	07 56	07 02	04 01
18	07 44	06 58	04 34
19	07 31	06 55	04 54
20	07 16	06 52	05 00
21	07 02	06 49	04 51
22	06 50	06 46	04 30
23	06 40	06 43	03 56
24	06 32	06 39	03 13
25	06 28	06 36	02 22
26	06 26	06 33	01 24
27	06 D 25	06 30	00 S 21
28	06 R 25	06 27	00 N 43
29	06 24	06 23	01 47
30	06 ♓ 22	06 ♓ 20	02 N 47

DECLINATIONS

Date	Sun ☉ ° '	Moon ☽ ° '	Mercury ☿ ° '	Venus ♀ ° '	Mars ♂ ° '	Jupiter ♃ ° '	Saturn ♄ ° '	Uranus ♅ ° '	Neptune ♆ ° '	Pluto ♇ ° '
01	14 S 43	03 S 26	22 S 33	01 S 21	24 S 21	20 S 51	18 S 25	13 N 22	14 N 10	22 S 03
02	15 02	02 N 42	22 54	01 49	24 24	20 50	18 27	13 21	14 09	22 02
03	15 21	08 54	23 13	02 17	24 26	20 48	18 28	13 21	14 09	22 02
04	15 39	14 52	23 31	02 45	24 28	20 46	18 30	13 20	14 08	22 02
05	15 57	20 13	23 48	03 13	24 30	20 45	18 31	13 20	14 08	22 02
06	16 15	24 31	24 03	03 41	24 32	20 43	18 33	13 19	14 07	22 02
07	16 33	27 17	24 17	04 09	24 34	20 41	18 34	13 19	14 07	22 02
08	16 50	28 11	24 30	04 37	24 35	20 40	18 36	13 18	14 06	22 02
09	17 07	27 07	24 41	05 05	24 36	20 38	18 37	13 18	14 06	22 01
10	17 24	24 14	24 51	05 32	24 37	20 36	18 39	13 18	14 05	22 01
11	17 40	19 53	24 59	06 00	24 37	20 34	18 40	13 17	14 05	22 01
12	17 56	14 31	25 06	06 28	24 37	20 32	18 42	13 17	14 04	22 01
13	18 12	08 31	25 11	06 55	24 37	20 30	18 43	13 17	14 04	22 01
14	18 28	02 N 13	25 15	07 23	24 37	20 28	18 45	13 16	14 03	22 00
15	18 43	04 S 04	25 16	07 50	24 36	20 26	18 46	13 16	14 03	22 00
16	18 58	10 06	25 16	08 17	24 35	20 24	18 48	13 16	14 02	22 00
17	19 12	15 36	25 15	08 44	24 34	20 22	18 49	13 16	14 02	22 00
18	19 26	20 21	25 11	09 11	24 33	20 20	18 51	13 15	14 01	21 59
19	19 40	24 06	25 05	09 37	24 31	20 18	18 52	13 15	14 01	21 59
20	19 53	26 42	24 58	10 04	24 29	20 16	18 54	13 15	14 00	21 59
21	20 07	27 58	24 48	10 30	24 27	20 13	18 55	13 15	14 00	21 58
22	20 19	27 53	24 35	10 56	24 24	20 11	18 56	13 15	13 59	21 58
23	20 32	26 31	24 21	11 22	24 22	20 09	18 58	13 14	13 59	21 58
24	20 44	23 57	24 04	11 47	24 18	20 06	18 59	13 14	13 58	21 57
25	20 55	20 24	23 44	12 12	24 15	20 04	19 01	13 14	13 58	21 57
26	21 06	16 00	23 22	12 37	24 12	20 02	19 02	13 14	13 58	21 57
27	21 17	10 58	22 57	13 02	24 08	19 59	19 03	13 14	13 57	21 56
28	21 28	05 S 25	22 31	13 26	24 04	19 57	19 05	13 14	13 57	21 56
29	21 38	00 N 29	22 02	13 50	23 59	19 54	19 06	13 14	13 56	21 56
30	21 S 47	06 N 31	21 S 32	14 S 14	23 S 54	19 S 52	19 S 08	13 N 14	13 N 56	21 S 55

ZODIAC SIGN ENTRIES

Date	h	m	Planets
01	02	35	☿ ♐
02	07	57	☽ ♈
04	13	09	☽ ♉
04	17	32	♃ ♒
06	15	28	☽ ♊
08	16	38	☽ ♋
10	18	21	☽ ♌
12	02	48	♂ ♑
12	21	41	☽ ♍
15	02	57	☽ ♎
17	10	02	☽ ♏
19	18	58	☽ ♐
20	00	36	♀ ♏
21	16	15	☉ ♐
22	05	55	☽ ♑
24	18	35	☽ ♒
27	07	16	☽ ♓
29	17	30	☽ ♈

LATITUDES

Date	Mercury ☿ ° '	Venus ♀ ° '	Mars ♂ ° '	Jupiter ♃ ° '	Saturn ♄ ° '	Uranus ♅ ° '	Neptune ♆ ° '	Pluto ♇ ° '
01	02 S 21	01 N 38	01 S 09	00 S 38	01 N 47	00 N 43	01 S 50	11 S 31
04	02 32	01 41	01 10	00 38	01 47	00 43	01 50	11 31
07	02 40	01 43	01 10	00 38	01 47	00 43	01 50	11 30
10	02 43	01 44	01 11	00 38	01 47	00 43	01 50	11 29
13	02 41	01 45	01 11	00 38	01 46	00 43	01 50	11 28
16	02 31	01 45	01 12	00 38	01 46	00 43	01 49	11 28
19	02 11	01 44	01 12	00 38	01 46	00 44	01 49	11 27
22	01 39	01 42	01 12	00 38	01 46	00 44	01 49	11 26
25	00 S 54	01 40	01 13	00 38	01 46	00 44	01 49	11 26
28	00 N 03	01 37	01 13	00 38	01 46	00 44	01 49	11 25
31	01 N 03	01 N 33	01 S 13	00 S 38	01 N 46	00 N 44	01 S 49	11 S 24

DATA

Julian Date	2467921
Delta T	+79 seconds
Ayanamsa	24° 29' 05"
Synetic vernal point	04° ♓ 37' 53"
True obliquity of ecliptic	23° 26' 09"

MOON'S PHASES, APSIDES AND POSITIONS ☽

Date	h	m	Phase	Longitude ° '	Eclipse Indicator
05	12	27	○	13 ♉ 44	
12	05	09	☾	20 ♌ 28	
19	14	58	●	27 ♏ 56	
27	19	36	☽	06 ♓ 13	

Day	h	m		
08	06	29	Perigee	
24	06	47	Apogee	
02	01	32	0N	
08	10	57	Max dec	28° N 11'
14	20	24	0S	
21	22	36	Max dec	28° S 06'
29	10	06	0N	

LONGITUDES

Date	Chiron ⚷ ° '	Ceres ⚳ ° '	Pallas ⚴ ° '	Juno ⚵ ° '	Vesta ⚶ ° '	Black Moon Lilith ⚸ ° '
01	15 ♍ 36	08 ♊ 38	06 ♓ 28	11 ♈ 49	24 ♓ 59	17 ♑ 49
11	16 ♍ 34	06 ♊ 47	06 ♓ 32	10 ♈ 41	24 ♓ 49	18 ♑ 56
21	17 ♍ 24	04 ♊ 34	07 ♓ 11	10 ♈ 27	25 ♓ 23	20 ♑ 02
31	18 ♍ 05	02 ♊ 14	08 ♓ 20	11 ♈ 08	26 ♓ 37	21 ♑ 09

All ephemeris data is given at 12.00 UT and the Moon's longitude is additionally given for 24.00 UT

ASPECTARIAN

h m	Aspects
01 Tuesday	
01 10	☽ ⚹ ♆
03 04	☽ ∠ ♃
03 51	☿ ☌ ♄
11 11	☿ □ ♇
17 35	☽ □ ♂
18 52	♀ ± ♆
19 40	☽ ∥ ♀
23 02	☽ bQ ☉
02 Wednesday	
01 29	☽ ⊼ ♅
05 14	☽ ∠ ♆
07 23	☽ ⚹ ♃
08 18	☽ # ♀
08 19	☽ △ ♄
08 49	☽ ⊻ ♇
11 15	☽ △ ☿
12 32	☽ ± ♅
16 55	☽ ± ☉
19 44	☽ ⊥ ♇
21 36	☽ ⊥ ♆
03 Thursday	
00 22	☽ ☍ ♀
04 36	☽ ⊼ ☉
04 48	☽ bQ ♅
05 19	☽ Q ♃
08 19	☽ ⊻ ♆
11 35	☽ bQ ♄
11 49	☽ ∠ ♇
17 01	☽ bQ ☿
23 10	♂ ⊥ ♃
04 Friday	
02 49	☽ △ ♂
03 30	☽ ± ♄
05 40	☽ ∥ ♅
07 12	☽ △ ♅
08 57	☽ ∥ ♆
10 16	☽ ± ☿
13 06	☽ □ ♃
13 56	☽ ⊼ ♄
13 56	☽ ⚹ ♇
14 14	♄ □ ♇
15 29	☽ # ☉
21 33	☽ ⊼ ☿
05 Saturday	
01 54	♀ ∠ ♅
03 55	☽ # ♄
05 20	☉ ☍ ♆
05 55	☽ bQ ♂
09 27	☽ ⊼ ♀
10 18	☽ Q ♇
11 56	☽ ☌ ♆
12 27	☽ ☍ ☉
14 35	☽ # ♃
20 26	☽ ± ♀
21 17	☽ # ♇
21 50	☽ ± ♂
06 Sunday	
08 20	☽ ⊼ ♂
08 49	☽ # ☿
09 51	☽ □ ♅
12 11	☽ # ♂
12 44	☽ bQ ♀
13 45	♀ ⊼ ♆
15 53	☽ △ ♃
16 13	☽ □ ♇
16 38	☽ ☍ ♄
07 Monday	
04 08	☽ ☍ ☿
11 56	♀ Q ♂
13 25	☽ ⊻ ♆
14 51	♂ △ ♅
15 28	☽ Q ♅
15 38	☽ △ ♀
16 43	☽ bQ ♃
17 39	☽ ⊼ ☉
23 13	☽ ⊥ ♆
08 Tuesday	
04 07	♃ ⊻ ♇
04 11	☽ ± ☉
05 51	♀ bQ ♇
07 35	☽ ± ♃
11 08	☽ ⚹ ♅
12 12	☽ ☍ ♂
13 55	☽ ∠ ♆
15 11	♀ ∠ ♄
17 23	☽ △ ♇
17 30	☽ ⊼ ♃
18 11	☽ ⊼ ♄
20 01	☽ bQ ☉
23 35	☿ ∥ ♂
09 Wednesday	
04 08	☽ ± ♄
09 31	☽ ⊼ ☿
11 50	☽ ∠ ♅
14 33	☽ ⚹ ♆
18 05	☽ bQ ♇
19 07	☽ bQ ♄
20 15	☽ ± ☿
20 46	♂ bQ ♆
21 27	☽ □ ♀
22 36	☽ △ ☉
10 Thursday	
02 46	☽ ⊥ ♅
08 00	☽ # ☿
09 03	☽ ± ♇
09 29	☽ # ♂
10 28	☽ Q ♆
12 28	☽ bQ ☿
12 48	☽ ⊻ ♅
13 08	☉ Q ♃
16 33	☽ ⊼ ♂
19 06	☽ ⊼ ♇
19 44	☽ ☍ ♃
20 21	☽ △ ♄
11 Friday	
01 05	☽ # ♇
03 13	☽ ± ♂
06 28	☽ Q ♀
08 38	☽ # ♃
15 48	☽ △ ☿
16 48	☽ □ ♆
17 46	☽ # ♄
19 23	☽ bQ ♂
21 53	☽ # ☉
12 Saturday	
04 56	☽ ⚹ ♀
05 09	☽ □ ☉
05 49	☿ ⊼ ♆
13 52	☽ ∥ ♆
16 00	☽ ☌ ♅
17 06	♂ ⚹ ♇
17 06	☽ ∥ ♅
17 23	☉ ⊻ ♀
22 28	☽ ☍ ♇
22 47	☽ △ ♂
23 40	☽ ⊼ ♃
13 Sunday	
00 13	☽ □ ♄
04 10	♇ St D
09 36	☽ ∠ ♀
10 18	☽ ± ♃
15 01	☽ Q ☉
17 42	☽ # ♀
18 42	♂ ⊻ ♃
20 53	☽ △ ♆
23 49	☽ □ ☿
14 Monday	
02 24	☽ bQ ♃
03 07	☽ ⊥ ♀
05 24	♂ ⊻ ♄
08 17	☽ Q ♄
14 05	☽ ⚹ ☉
14 55	☽ ⊻ ♀
21 07	☽ ⊻ ♅
23 39	☽ bQ ♆
15 Tuesday	
03 46	☽ ⊼ ♇
05 37	☽ △ ♃
06 03	☽ ⚹ ♄
07 18	☽ □ ♂
07 58	☽ ⊥ ♅
10 12	☽ Q ☿
14 39	☽ ± ♇
15 55	☽ ± ♆
18 21	☉ ∥ ♄
19 26	☽ ∠ ☉
16 Wednesday	
00 23	☽ ∠ ♅
02 51	☽ ⊼ ♆
04 00	☽ ∥ ♀
07 06	☽ bQ ♇
07 29	☿ ∠ ♃
09 23	☽ ⚹ ☿
09 40	☽ ∠ ♄
13 19	☽ ⊥ ☉
16 44	♀ ⊥ ♄
18 16	☽ Q ♂
17 Thursday	
01 21	☽ ⊻ ☉
01 28	☽ # ♅
02 27	☽ ⊥ ♄
03 23	☽ ☌ ♀
04 04	☽ ⚹ ♅
04 52	☽ # ♆
10 36	♀ ⚹ ♅
10 53	☽ △ ♇
13 27	☽ □ ♃
13 45	☽ ⊻ ♄
14 34	☽ ∠ ☿
18 00	☽ ⚹ ♂
18 Friday	
02 33	☽ Q ♅
03 53	☽ ∥ ♄
06 47	☽ ∥ ☉
08 11	☽ ⊥ ☿
10 36	☽ ☍ ♆
11 57	☽ ∥ ♃
12 33	☉ □ ♅
19 58	☽ ⊻ ☿
21 39	☽ ∥ ♇
23 53	☽ Q ♃
19 Saturday	
00 10	☽ ∠ ♂
12 51	☽ □ ♅
14 58	☽ ☌ ☉
15 08	☽ ∥ ♂
18 20	☽ ⊻ ♀
18 33	☽ ⊥ ♂
19 32	☽ ∥ ☿
19 51	☽ □ ♇
23 14	☽ ⚹ ♃
23 21	☽ ☌ ♄
20 Sunday	
06 57	☽ ⊻ ♂
07 15	☽ ⊥ ♀
07 20	☿ Q ♇
09 30	♀ △ ♇
20 16	☽ ⊼ ♆
21 Monday	
02 51	☽ ∠ ♀
04 56	♃ ⚹ ♄
04 56	☽ ∠ ♃
07 00	☽ Q ♇
07 16	☽ ☌ ☿
08 03	☽ ± ♆
16 50	☿ St R
22 48	☉ ∥ ♃
23 01	☽ ⊥ ♃
23 38	☽ △ ♅
22 Tuesday	
01 18	♀ ⊻ ♄
01 53	☽ bQ ♆
02 11	♀ □ ♃
03 19	☉ □ ♇
06 51	☽ ⚹ ♇
07 11	☽ ⊻ ☉
11 03	☽ ⊻ ♄
11 10	☽ ⊻ ♃
12 04	☽ ⚹ ♀
20 20	☽ ⊥ ☉
20 22	☿ ∠ ♀
22 22	☽ ☌ ♂
23 14	☽ ⊥ ♄
23 Wednesday	
00 10	☿ Q ♇
05 46	☽ bQ ♅
07 56	☽ △ ♆
10 42	☿ ∥ ♂
13 04	☽ ∠ ♇
14 14	♂ ⊥ ♄
15 05	☽ Q ♀
16 12	☽ ∠ ☉
17 37	☽ ∠ ♄
18 42	☽ ⊻ ☿
24 Thursday	
00 00	☽ ± ♅
06 29	☽ ⊥ ☿
07 20	☽ ⊥ ♇
09 10	☽ ∥ ♂
10 46	☉ ☌ ♄
11 04	☽ ∥ ☿
11 27	☿ ∠ ♃
12 13	☽ ⊼ ♅
15 15	☉ ⚹ ♃
19 33	☽ ⊻ ♇
25 Friday	
00 03	☽ ∠ ☿
00 25	☽ ⚹ ♄
00 48	☽ ☌ ♃
01 33	☽ ⚹ ☉
02 14	☽ ∥ ♇
07 59	☽ □ ♀
08 58	☽ ∥ ☉
13 58	☽ ∥ ♃
15 28	☽ ⊻ ♂
19 59	☽ ∥ ♄
20 46	☽ □ ♆
26 Saturday	
01 03	☽ Q ♄
04 13	☽ Q ☉
04 29	☽ ⊥ ♂
04 52	☽ ⚹ ☿
16 26	☉ ⊥ ♂
22 06	☽ # ♆
23 52	☽ ∠ ♂
27 Sunday	
01 03	☽ ☍ ♅
01 31	☽ # ♅
03 10	☽ ∥ ♀
03 24	☽ Q ☿
04 24	♀ Q ♅
08 15	☽ ☌ ♇
08 56	☽ Q ♆
13 35	☽ □ ♄
14 15	☽ ⊻ ♃
16 36	☿ ⊥ ♀
18 16	♂ bQ ♅
19 36	☽ □ ☉
23 51	♀ # ♅
28 Monday	
02 16	☽ ⊥ ♃
03 12	☽ △ ♀
07 36	☽ ⚹ ♂
08 34	☽ ⚹ ♆
12 07	☽ □ ☿
20 03	☽ ∠ ♃
22 38	♂ △ ♆
29 Tuesday	
08 15	☽ Q ♂
11 21	☽ bQ ♀
11 39	☽ ⊼ ♅
13 19	☽ ∠ ♆
14 25	☿ ⊻ ♂
16 59	☿ ∥ ♇
17 40	♀ # ♆
18 27	☽ ⊻ ♇
22 54	☽ ± ♅
23 57	☽ △ ♄
30 Wednesday	
00 17	☿ ⊼ ♆
00 51	☽ ⚹ ♃
02 39	☉ ∥ ☿
03 42	☿ ⊻ ♀
05 37	☽ ⊥ ♇
05 57	☽ ± ♀
06 04	☽ ⊥ ♆
07 10	♀ ☍ ♆
09 53	☽ △ ☉
15 28	☽ bQ ♅
15 31	☽ △ ☿
17 01	☽ ⊻ ♆
18 04	☽ ⊼ ♀
19 43	☽ □ ♂
22 01	☽ ∠ ♇
23 05	☽ Q ♃

DECEMBER 2044

Date	Sidereal time h m s	LONGITUDES Sun ☉ ° ′ ″	Moon ☽ ° ′ ″	Moon ☽ 24.00 ° ′ ″	Mercury ☿ ° ′	Venus ♀ ° ′	Mars ♂ ° ′	Jupiter ♃ ° ′	Saturn ♄ ° ′	Uranus ♅ ° ′	Neptune ♆ ° ′	Pluto ⯓ ° ′
01	16 43 58	09 ♐ 56 25	23 ♈ 21 56	00 ♉ 14 05	10 ♐ 45	14 ♏ 15	14 ♑ 46	04 ♒ 12	03 ♐ 38	26 ♌ 54	12 ♉ 44	00 ♓ 31
02	16 47 54	10 57 13	07 ♉ 13 33	14 ♉ 20 06	09 R 23	15 30	15 32	04 23	03 45	26 R 54	12 R 43	00 32
03	16 51 51	11 58 03	21 ♉ 51 15	28 ♉ 52 19	08 02	16 45	16 18	04 35	03 53	26 54	12 41	00 32
04	16 55 48	12 58 53	06 ♊ 16 27	13 ♊ 44 33	06 46	18 00	17 04	04 46	04 00	26 54	12 40	00 33
05	16 59 44	13 59 45	21 ♊ 15 28	28 ♊ 47 54	05 37	19 15	17 51	04 57	04 07	26 53	12 38	00 33
06	17 03 41	15 00 38	06 ♋ 20 35	13 ♋ 52 15	04 37	20 29	18 37	05 09	04 14	26 53	12 37	00 34
07	17 07 37	16 01 32	21 ♋ 21 44	28 ♋ 48 02	03 48	21 44	19 24	05 20	04 21	26 53	12 36	00 35
08	17 11 34	17 02 27	06 ♌ 10 19	13 ♌ 27 53	03 09	22 59	20 10	05 32	04 28	26 52	12 34	00 35
09	17 15 30	18 03 23	20 ♌ 40 19	27 ♌ 47 19	02 42	24 14	20 57	05 43	04 35	26 52	12 33	00 36
10	17 19 27	19 04 21	04 ♍ 48 45	11 ♍ 44 40	02 26	25 29	21 43	05 55	04 42	26 52	12 32	00 37
11	17 23 23	20 05 19	18 ♍ 35 12	25 ♍ 20 34	02 D 21	26 44	22 30	06 07	04 49	26 51	12 30	00 37
12	17 27 20	21 06 19	02 ♎ 01 03	08 ♎ 37 00	02 26	27 59	23 16	06 19	04 56	26 50	12 29	00 38
13	17 31 17	22 07 20	15 ♎ 08 43	21 ♎ 36 33	02 41	29 ♏ 14	24 03	06 31	05 03	26 50	12 28	00 39
14	17 35 13	23 08 22	28 ♎ 00 51	04 ♏ 21 53	03 04	00 ♐ 29	24 50	06 43	05 10	26 49	12 27	00 40
15	17 39 10	24 09 25	10 ♏ 39 55	16 ♏ 55 13	03 35	01 44	25 36	06 56	05 17	26 48	12 26	00 41
16	17 43 06	25 10 29	23 ♏ 07 58	29 ♏ 18 20	04 13	02 59	26 23	07 08	05 23	26 48	12 24	00 41
17	17 47 03	26 11 33	05 ♐ 26 29	11 ♐ 32 32	04 58	04 14	27 10	07 20	05 30	26 47	12 23	00 42
18	17 50 59	27 12 39	17 ♐ 36 36	23 ♐ 38 47	05 48	05 29	27 57	07 33	05 37	26 46	12 22	00 43
19	17 54 56	28 13 45	29 ♐ 39 14	05 ♑ 38 02	06 42	06 44	28 44	07 45	05 44	26 45	12 21	00 44
20	17 58 52	29 ♐ 14 52	11 ♑ 35 23	17 ♑ 31 26	07 41	08 00	29 ♑ 31	07 58	05 51	26 44	12 20	00 45
21	18 02 49	00 ♑ 15 59	23 ♑ 26 26	29 ♑ 20 37	08 44	09 15	00 ♒ 18	08 10	05 57	26 43	12 19	00 46
22	18 06 46	01 17 07	05 ♒ 14 18	11 ♒ 07 50	09 50	10 30	01 04	08 23	06 04	26 42	12 18	00 47
23	18 10 42	02 18 14	17 ♒ 01 38	22 ♒ 56 10	10 58	11 45	01 51	08 36	06 11	26 41	12 17	00 48
24	18 14 39	03 19 22	28 ♒ 51 54	04 ♓ 49 24	12 10	13 00	02 38	08 49	06 17	26 39	12 16	00 49
25	18 18 35	04 20 31	10 ♓ 49 35	16 ♓ 52 04	13 23	14 15	03 26	09 02	06 24	26 38	12 15	00 50
26	18 22 32	05 21 39	22 ♓ 58 29	29 ♓ 09 11	14 39	15 31	04 13	09 15	06 31	26 37	12 14	00 51
27	18 26 28	06 22 47	05 ♈ 24 47	11 ♈ 45 56	15 56	16 46	05 00	09 28	06 37	26 35	12 13	00 52
28	18 30 25	07 23 55	18 ♈ 13 13	24 ♈ 47 11	17 15	18 01	05 47	09 41	06 44	26 34	12 13	00 53
29	18 34 21	08 25 04	01 ♉ 28 15	08 ♉ 16 46	18 35	19 16	06 34	09 54	06 50	26 33	12 12	00 54
30	18 38 18	09 26 12	15 ♉ 12 53	22 ♉ 16 36	19 56	20 31	07 21	10 07	06 57	26 31	12 11	00 55
31	18 42 15	10 ♑ 27 20	29 ♉ 27 43	06 ♊ 45 47	21 ♐ 19	21 ♐ 47	08 ♒ 08	10 ♒ 21	07 ♐ 03	26 ♌ 30	12 ♉ 10	00 ♓ 57

Date	Moon True ☊ ° ′	Moon Mean ☊ ° ′	Moon ☽ Latitude ° ′
01	06 ♓ 16	06 ♓ 17	03 N 41
02	06 R 08	06 14	04 23
03	05 58	06 11	04 51
04	05 46	06 08	05 00
05	05 34	06 04	04 49
06	05 24	06 01	04 18
07	05 16	05 58	03 29
08	05 10	05 55	02 27
09	05 08	05 52	01 16
10	05 07	05 49	00 N 02
11	05 D 07	05 45	01 S 11
12	05 R 07	05 42	02 17
13	05 05	05 39	03 15
14	05 00	05 36	04 02
15	04 52	05 33	04 36
16	04 42	05 29	04 56
17	04 29	05 26	05 02
18	04 16	05 23	04 54
19	04 03	05 20	04 33
20	03 51	05 17	04 00
21	03 41	05 14	03 16
22	03 34	05 10	02 25
23	03 30	05 07	01 27
24	03 29	05 04	00 S 25
25	03 D 29	05 01	00 N 39
26	03 30	04 58	01 43
27	03 30	04 55	02 43
28	03 R 30	04 51	03 36
29	03 27	04 48	04 20
30	03 24	04 45	04 51
31	03 ♓ 15	04 ♓ 42	05 N 06

DECLINATIONS

Date	Sun ☉ ° ′	Moon ☽ ° ′	Mercury ☿ ° ′	Venus ♀ ° ′	Mars ♂ ° ′	Jupiter ♃ ° ′	Saturn ♄ ° ′	Uranus ♅ ° ′	Neptune ♆ ° ′	Pluto ⯓ ° ′
01	21 S 56	12 N 30	21 S 01	14 S 38	23 S 49	19 S 49	19 S 09	13 N 14	13 N 55	21 S 55
02	22 05	18 04	20 30	15 01	23 44	19 46	19 10	13 14	13 55	21 54
03	22 13	22 50	20 00	15 23	23 39	19 44	19 12	13 14	13 55	21 54
04	22 21	26 17	19 32	15 46	23 33	19 41	19 13	13 14	13 54	21 54
05	22 29	27 57	19 06	16 07	23 27	19 38	19 14	13 14	13 54	21 53
06	22 36	27 35	18 44	16 29	23 21	19 36	19 16	13 15	13 53	21 53
07	22 42	25 11	18 26	16 50	23 14	19 33	19 17	13 15	13 53	21 52
08	22 48	21 06	18 12	17 11	23 07	19 30	19 18	13 15	13 53	21 52
09	22 54	15 48	18 02	17 31	23 00	19 27	19 20	13 15	13 52	21 51
10	22 59	09 46	17 56	17 50	22 53	19 24	19 21	13 15	13 52	21 51
11	23 04	03 N 26	17 53	18 10	22 45	19 21	19 22	13 16	13 52	21 50
12	23 08	02 S 54	17 55	18 28	22 37	19 18	19 23	13 16	13 51	21 50
13	23 12	08 58	17 59	18 47	22 29	19 15	19 25	13 16	13 51	21 50
14	23 15	14 32	18 07	19 04	22 21	19 12	19 26	13 16	13 51	21 49
15	23 18	19 23	18 16	19 22	22 12	19 09	19 27	13 17	13 50	21 49
16	23 21	23 19	18 28	19 38	22 03	19 06	19 28	13 17	13 50	21 48
17	23 23	26 09	18 41	19 54	21 54	19 03	19 29	13 17	13 50	21 48
18	23 24	27 44	18 56	20 10	21 45	19 00	19 31	13 18	13 49	21 47
19	23 25	27 59	19 12	20 25	21 35	18 57	19 32	13 18	13 49	21 46
20	23 26	26 55	19 28	20 39	21 25	18 53	19 33	13 18	13 49	21 46
21	23 26	24 38	19 45	20 53	21 15	18 50	19 34	13 19	13 49	21 45
22	23 26	21 18	20 03	21 07	21 05	18 47	19 35	13 19	13 48	21 45
23	23 25	17 07	20 20	21 19	20 54	18 43	19 36	13 20	13 48	21 44
24	23 24	12 15	20 38	21 31	20 43	18 40	19 37	13 20	13 48	21 44
25	23 22	06 54	20 55	21 43	20 32	18 37	19 38	13 20	13 48	21 43
26	23 20	01 S 13	21 12	21 53	20 21	18 33	19 39	13 21	13 47	21 43
27	23 17	04 N 38	21 29	22 03	20 09	18 30	19 41	13 21	13 47	21 42
28	23 14	10 29	21 45	22 13	19 58	18 26	19 42	13 22	13 47	21 41
29	23 10	16 03	22 00	22 22	19 46	18 23	19 43	13 22	13 47	21 41
30	23 06	21 02	22 15	22 30	19 34	18 19	19 44	13 23	13 47	21 40
31	23 S 01	25 N 00	22 S 29	22 S 37	19 S 21	18 S 16	19 S 45	13 N 24	13 N 46	21 S 40

ZODIAC SIGN ENTRIES

Date	h	m	Planets
01	23	36	☽ ♉
04	01	50	☽ ♊
06	01	55	☽ ♋
08	01	57	☽ ♌
10	03	46	☽ ♍
12	08	21	☽ ♎
14	02	42	♀ ♐
14	15	45	☽ ♏
17	01	21	☽ ♐
19	12	42	☽ ♑
21	03	03	♂ ♒
21	05	43	☉ ♑
22	01	20	☽ ♒
24	14	17	☽ ♓
27	01	38	☽ ♈
29	09	23	☽ ♉
31	12	53	☽ ♊

LATITUDES

Date	Mercury ☿ ° ′	Venus ♀ ° ′	Mars ♂ ° ′	Jupiter ♃ ° ′	Saturn ♄ ° ′	Uranus ♅ ° ′	Neptune ♆ ° ′	Pluto ⯓ ° ′
01	01 N 03	01 N 33	01 S 13	00 S 38	01 N 46	00 N 44	01 S 49	11 S 24
04	01 56	01 29	01 13	00 38	01 45	00 44	01 49	11 23
07	02 31	01 25	01 13	00 38	01 45	00 44	01 49	11 23
10	02 46	01 20	01 13	00 38	01 45	00 44	01 49	11 22
13	02 45	01 14	01 12	00 38	01 45	00 45	01 49	11 21
16	02 34	01 08	01 12	00 38	01 45	00 45	01 49	11 21
19	02 16	01 02	01 12	00 38	01 45	00 45	01 48	11 20
22	01 54	00 55	01 11	00 38	01 45	00 45	01 48	11 19
25	01 30	00 48	01 11	00 38	01 46	00 45	01 48	11 19
28	01 05	00 41	01 10	00 38	01 46	00 45	01 48	11 18
31	00 N 40	00 N 34	01 S 10	00 S 38	01 N 46	00 N 45	01 S 48	11 S 18

DATA

Julian Date	2467951
Delta T	+79 seconds
Ayanamsa	24° 29′ 10″
Synetic vernal point	04° ♓ 37′ 49″
True obliquity of ecliptic	23° 26′ 08″

MOON'S PHASES, APSIDES AND POSITIONS ☽

Date	h	m	Phase	Longitude ° ′	Eclipse Indicator
04	23	34	○	13 ♊ 28	
11	14	52	☾	20 ♍ 13	
19	08	53	●	28 ♐ 06	
27	14	00	☽	06 ♈ 28	

Day	h	m		
06	08	17	Perigee	
21	20	39	Apogee	
05	19	42	Max dec	28° N 04′
12	00	55	0S	
19	04	26	Max dec	28° S 03′
26	17	01	0N	

LONGITUDES

Date	Chiron ⚷ ° ′	Ceres ⚳ ° ′	Pallas ⚴ ° ′	Juno ⚵ ° ′	Vesta ⚶ ° ′	Black Moon Lilith ⚸ ° ′
01	18 ♍ 05	02 ♊ 14	08 ♓ 20	11 ♈ 08	26 ♓ 37	21 ♑ 09
11	18 ♍ 35	00 ♊ 03	09 ♓ 57	12 ♈ 42	28 ♓ 26	22 ♑ 16
21	18 ♍ 54	28 ♉ 16	11 ♓ 58	15 ♈ 03	00 ♈ 45	23 ♑ 23
31	19 ♍ 01	27 ♉ 05	14 ♓ 20	18 ♈ 04	03 ♈ 29	24 ♑ 30

ASPECTARIAN

h m Aspects

01 Thursday
03 32 ☽ ♇ ♄
08 29 ☉ ‖ ⯓
14 48 ♅ St R
14 59 ☽ ♇ ☉
15 04 ☽ ‖ ♅
15 49 ☽ ♇ ☿
17 56 ☽ ‖ ♆
18 12 ☽ △ ♅
19 33 ☽ ± ♄
20 11 ☉ ☌ ☿
21 33 ☽ # ♀

02 Friday
00 30 ☽ ⚹ ⯓
06 01 ☽ ± ☿
06 01 ☽ ⊼ ♄
07 06 ☽ □ ♃
07 49 ☽ ± ☉
10 17 ☿ ⊥ ♂
11 58 ♂ ∠ ⯓
13 24 ♀ ⚹ ♂
15 20 ☽ ⊼ ☿
17 10 ☽ # ♄
18 48 ☽ ⊼ ☉
19 57 ☽ # ♃
20 59 ☽ Q ⯓
21 15 ☽ ☌ ♆
22 29 ☽ # ☿

03 Saturday
02 48 ☽ △ ♂
03 17 ☽ ☍ ♀
06 52 ☽ # ⯓
08 30 ☽ # ☉
16 44 ☽ # ♂
20 46 ☽ □ ♅

04 Sunday
02 43 ☽ □ ⯓
02 56 ☿ ‖ ♃
04 36 ☉ ⊼ ♆
04 50 ☽ ♇ ♂
08 17 ☽ ☍ ♄
09 32 ☽ △ ♃
12 44 ☽ ☍ ☿
20 09 ☽ ± ♂
22 15 ☽ ⊻ ♆
23 34 ☽ ☍ ☉

05 Monday
01 50 ☽ Q ♅
04 33 ☿ ‖ ♄
06 16 ☽ ⊼ ♂
07 50 ☽ ⊥ ♆
08 30 ☽ ⊼ ♀
09 54 ☽ ♇ ♃
18 55 ☽ ± ♀
20 58 ☽ ⚹ ♅
22 08 ☽ ∠ ♆

06 Tuesday
00 24 ☽ ± ♃
01 04 ☿ ⚹ ♃
02 48 ☽ △ ⯓
08 37 ☽ ⊼ ♄
09 25 ☽ ⊼ ☿
10 04 ☽ ⊼ ♃
10 31 ☽ ♇ ♀
18 14 ☽ ± ♄
18 26 ☽ ± ☿
20 50 ☽ ∠ ♅
21 27 ☿ ☌ ♄
21 59 ☽ ⚹ ♆

07 Wednesday
02 43 ☽ ♇ ⯓
02 44 ☿ ∠ ♂
02 49 ☽ ⊼ ☉
08 05 ☽ ♇ ☿
08 40 ☽ ☍ ♂
08 44 ☽ ♇ ♄
10 19 ♂ ∠ ♄
11 14 ☽ ⊥ ♅
12 40 ☽ △ ♀
13 09 ☽ ± ☉
17 11 ☽ ± ⯓
17 12 ☽ Q ♆
20 53 ☽ ⊻ ♅
22 41 ♆ Q ⯓

08 Thursday
00 52 ☽ # ♂
02 54 ☽ ⊼ ⯓
03 10 ☽ # ☉
04 46 ☽ ♇ ☉
07 15 ☽ △ ☿
08 02 ☽ # ⯓
09 11 ☽ △ ♄
10 56 ☽ ☍ ♃
19 47 ☽ # ♃
20 35 ☽ # ♄
22 30 ☽ □ ♆

09 Friday
00 22 ♀ Q ♃
02 08 ☽ # ☿
05 06 ☽ # ♀
07 18 ☽ △ ☉
09 42 ♂ ± ♅
12 29 ☽ ⊼ ♂
18 34 ☽ □ ♀
19 54 ☽ ‖ ♆
22 23 ☽ ‖ ♅
22 26 ☽ ☌ ♅
23 11 ☽ ± ♂
23 23 ☉ ± ♆
23 42 ☉ ‖ ♂

10 Saturday
00 59 ☉ Q ⯓
04 48 ☽ ☍ ⯓
07 59 ☽ □ ☿
11 48 ☽ □ ♄
13 56 ☽ ⊼ ♃
15 29 ☽ ♇ ♂
17 29 ☿ ‖ ♀

11 Sunday
00 29 ☽ ± ♃
01 21 ☽ △ ♆
04 33 ☽ Q ♀
07 18 ♃ ‖ ♄
11 32 ☿ St D
14 13 ♀ □ ♅
14 52 ☽ □ ☉
15 08 ☽ Q ☿
16 33 ☽ ♇ ♃
19 21 ☽ △ ♂
19 33 ☽ Q ♄

12 Monday
02 41 ☽ ⊻ ♅
03 51 ☽ ♇ ♆
03 59 ☽ ⚹ ♀
09 30 ☽ ⊼ ⯓
12 46 ☽ ⚹ ☿
13 29 ☽ ⊥ ♅
17 20 ☽ ⚹ ♄
18 17 ☉ ∠ ♃
19 56 ☽ △ ♃
20 06 ☽ ± ♆
20 24 ☽ ± ⯓

13 Tuesday
01 59 ☽ Q ☉
05 54 ☽ ∠ ♅
07 04 ☽ ⊼ ♆
10 08 ☽ ∠ ♀
12 56 ☽ ♇ ⯓
16 49 ☽ ∠ ☿
21 10 ☽ ∠ ♄

14 Wednesday
02 04 ☽ ⚹ ☉
04 39 ☽ ⊥ ♀
05 38 ☽ □ ♂
06 21 ☽ # ♅
06 49 ♂ ⊥ ⯓
08 55 ☽ # ♆
09 45 ☽ ⚹ ♅
10 10 ☽ ⊥ ☿
14 11 ☽ ⊥ ♄
15 26 ♀ □ ⯓
17 00 ☽ △ ⯓
17 10 ☽ ⊻ ♀
21 12 ♀ ‖ ♃
21 56 ☽ ⊻ ☿

15 Thursday
01 38 ☽ ⊻ ♄
04 45 ☽ □ ♃
05 55 ☽ ‖ ☿
08 27 ☽ Q ♅
08 52 ☽ ∠ ☉
10 48 ☽ ‖ ♃
11 54 ☽ ‖ ♀
12 23 ☽ ‖ ♄
15 22 ☽ ☍ ♆
18 00 ☽ Q ♂
20 12 ♀ ‖ ♄

16 Friday
02 02 ☽ ‖ ⯓
03 40 ☽ ⊥ ☉
03 55 ☽ ‖ ♂
12 14 ☽ ‖ ☉
15 56 ☽ Q ♃
16 19 ☽ ⊻ ☉
18 44 ☽ ⚹ ♂
19 06 ☽ □ ♅

17 Saturday
00 20 ♂ ⊼ ♅
02 43 ☽ □ ⯓
09 22 ☽ ☌ ♀
11 00 ☽ ☌ ☿
12 08 ☽ ☌ ♄
15 47 ☽ ⚹ ♃

18 Sunday
01 38 ☉ △ ♅
01 39 ☽ ⊼ ♆
02 08 ☽ ∠ ♂
05 51 ♂ ‖ ⯓
06 28 ☿ ☌ ♄
13 30 ☽ ± ♆
14 12 ☽ Q ⯓
14 44 ♀ ☌ ♄
15 40 ☉ ♇ ♆
16 48 ☿ ‖ ♃
21 13 ☽ ⊥ ♂
21 59 ☽ ∠ ♃

19 Monday
06 12 ☽ △ ♅
07 24 ☽ ♇ ♆
08 53 ☽ ☌ ☉
08 56 ☿ ☌ ♀
10 01 ☽ ⊻ ♂
14 10 ☽ ⚹ ⯓
16 17 ☽ ⊥ ♃

20 Tuesday
00 19 ☽ ⊻ ♄
03 24 ☽ ⊻ ☿
03 54 ☽ ⊻ ♀
04 33 ☽ ⊻ ♃
11 17 ♀ ⚹ ♃
12 17 ☽ ♇ ♅
12 31 ☽ ⊥ ♄
13 30 ☽ △ ♆
16 38 ☽ ⊥ ☿
17 26 ☽ ⊥ ♀
18 48 ☿ ‖ ♄
20 15 ☿ ⚹ ♃
20 25 ☽ ∠ ⯓

21 Wednesday
06 29 ☽ ± ♅
06 55 ☽ ∠ ♄
12 38 ☽ ∠ ☿
13 50 ☽ ∠ ♀
14 34 ☉ ⊻ ♂
14 42 ☽ ⊥ ⯓
18 39 ☽ ⊼ ♅
21 25 ☽ ‖ ☉
23 57 ☉ ⚹ ⯓

22 Thursday
02 50 ♂ ⊻ ⯓
02 55 ☽ ⊻ ⯓
02 55 ☽ ☌ ♂
03 11 ☽ ⊻ ☉
09 06 ☽ ‖ ⯓
10 05 ♀ ‖ ♂
13 07 ☽ ‖ ♀
13 26 ☽ ‖ ♂
13 43 ☽ ⚹ ♄
16 34 ☽ ⊥ ☉
18 32 ☽ ☌ ♃
19 07 ☽ ‖ ☿
22 18 ☽ ‖ ♄
22 20 ☽ ⚹ ☿
23 59 ☽ ⚹ ♀

23 Friday
02 22 ☽ □ ♆
03 08 ☽ ‖ ♃
12 37 ☽ ∠ ☉
14 22 ☽ Q ♄
20 48 ☉ ⊥ ♃
22 06 ♀ ⊼ ♆

24 Saturday
01 25 ☽ Q ☿
03 16 ☽ Q ♀
04 39 ☽ # ♆
06 54 ☽ # ♅
07 33 ☽ ☍ ♅
14 08 ☿ ⊼ ♆
14 50 ☽ Q ♆
15 56 ☽ ☌ ⯓
16 34 ☿ ‖ ♂
20 09 ☽ ⊻ ♂
21 49 ☽ ⚹ ☉

25 Sunday
03 05 ☽ □ ♄
08 22 ☽ ⊻ ♃
09 01 ☽ ⊥ ♂
13 06 ♀ ‖ ⯓
14 51 ☽ ⚹ ♆
17 41 ☽ □ ☿
19 37 ☽ □ ♀
20 31 ☽ ⊥ ♃
23 58 ☽ Q ☉

26 Monday
04 07 ☽ ∠ ♂
14 32 ☽ ∠ ♃
19 04 ☽ ⊼ ♅
20 17 ☽ ∠ ♆

27 Tuesday
03 18 ☽ ⊻ ⯓
06 37 ☽ ± ♅
11 09 ☽ ⚹ ♂
13 33 ☽ ⊥ ♆
14 00 ☽ □ ☉
14 19 ☽ △ ♄
14 46 ☽ ⊥ ⯓
18 19 ☉ ⊻ ♄
19 48 ☽ ⚹ ♃
23 39 ☽ ♇ ♅

28 Wednesday
00 51 ☽ ⊻ ♆
07 09 ☿ ‖ ⯓
07 41 ☽ ∠ ⯓
10 00 ☽ △ ☿
11 08 ☽ Q ♂
11 35 ☽ △ ♀
15 43 ♀ ± ♆
18 28 ☽ Q ♃
18 29 ☽ ♇ ♄

29 Thursday
00 15 ☽ ‖ ♅
02 01 ☽ ‖ ♆
03 12 ☽ △ ♅
04 56 ♀ Q ⯓
05 14 ☿ ± ♆
10 52 ☽ ± ♄
11 00 ☽ ⚹ ⯓
16 09 ☽ ♇ ☿
17 28 ☽ ♇ ♀
17 48 ♂ ‖ ♄
17 51 ☿ Q ⯓
21 32 ♂ ⚹ ♄
21 33 ☽ ⊼ ♄
21 33 ☽ □ ♂
22 38 ☽ # ♃

30 Friday
01 13 ☽ △ ☉
03 04 ☽ □ ♃
04 47 ☽ # ♂
05 17 ☽ # ♄
06 47 ☽ ☌ ♆
08 03 ☽ Q ⯓
09 34 ☽ ± ☿
10 42 ☽ ± ♀
15 24 ☽ # ⯓
18 58 ☽ # ☿
20 16 ☽ # ♀
20 55 ☽ ⊼ ☿
21 55 ☽ ⊼ ♀
23 19 ☽ # ☉

31 Saturday
04 49 ☽ ♇ ☉
07 05 ☽ □ ♅
08 37 ☉ ⊻ ♃
14 27 ☽ □ ⯓
20 51 ☽ ± ☉

All ephemeris data is given at 12.00 UT and the Moon's longitude is additionally given for 24.00 UT

JANUARY 2045

Date	Sidereal time h m s	LONGITUDES Sun ☉ ° ' "	Moon ☽ ° ' "	Moon ☽ 24.00 ° ' "	Mercury ☿ ° '	Venus ♀ ° '	Mars ♂ ° '	Jupiter ♃ ° '	Saturn ♄ ° '	Uranus ♅ ° '	Neptune ♆ ° '	Pluto ♇ ° '
01	18 46 11	11 ♑ 28 28	14 ♊ 10 08	21 ♊ 39 54	22 ♐ 42	23 ♐ 02	08 ♒ 55	10 ♒ 34	07 ♐ 09	26 ♌ 28	12 ♉ 10	00 ♓ 58
02	18 50 08	12 29 37	29 ♊ 14 00	06 ♋ 51 10	24 06	24 17	09 43	10 47	07 15	26 R 26	12 R 09	00 59
03	18 54 04	13 30 45	14 ♋ 30 03	22 ♋ 09 14	25 32	25 32	10 30	11 01	07 22	26 25	12 08	01 00
04	18 58 01	14 31 53	29 ♋ 47 20	07 ♌ 23 03	26 58	26 47	11 17	11 14	07 28	26 23	12 08	01 01
05	19 01 57	15 33 01	14 ♌ 55 09	22 ♌ 22 41	28 24	28 03	12 04	11 28	07 34	26 21	12 07	01 03
06	19 05 54	16 34 09	29 ♌ 44 48	07 ♍ 00 54	29 ♐ 51	29 ♐ 18	12 52	11 41	07 40	26 19	12 07	01 04
07	19 09 50	17 35 17	14 ♍ 10 36	21 ♍ 13 41	01 ♑ 19	00 ♑ 33	13 39	11 55	07 46	26 18	12 06	01 05
08	19 13 47	18 36 25	28 ♍ 10 07	05 ♎ 00 01	02 48	01 48	14 26	12 09	07 52	26 16	12 06	01 07
09	19 17 44	19 37 34	11 ♎ 43 35	18 ♎ 21 09	04 17	03 04	15 14	12 22	07 58	26 14	12 05	01 08
10	19 21 40	20 38 42	24 ♎ 53 05	01 ♏ 19 50	05 46	04 19	16 01	12 36	08 04	26 12	12 05	01 09
11	19 25 37	21 39 51	07 ♏ 41 50	13 ♏ 59 32	07 17	05 34	16 48	12 50	08 10	26 10	12 05	01 11
12	19 29 33	22 40 59	20 ♏ 13 24	26 ♏ 23 51	08 47	06 49	17 36	13 04	08 16	26 08	12 04	01 12
13	19 33 30	23 42 08	02 ♐ 31 18	08 ♐ 36 08	10 18	08 05	18 23	13 18	08 21	26 06	12 04	01 13
14	19 37 26	24 43 16	14 ♐ 38 43	20 ♐ 39 23	11 50	09 20	19 10	13 32	08 27	26 04	12 04	01 15
15	19 41 23	25 44 24	26 ♐ 38 24	02 ♑ 36 03	13 22	10 35	19 58	13 45	08 33	26 02	12 03	01 16
16	19 45 19	26 45 32	08 ♑ 32 36	14 ♑ 28 16	14 55	11 51	20 45	13 59	08 38	25 59	12 03	01 18
17	19 49 16	27 46 40	20 ♑ 23 17	26 ♑ 17 51	16 28	13 06	21 32	14 13	08 44	25 57	12 03	01 19
18	19 53 13	28 47 47	02 ♒ 12 11	08 ♒ 06 32	18 02	14 21	22 20	14 27	08 49	25 55	12 03	01 21
19	19 57 09	29 ♑ 48 54	14 ♒ 01 08	19 ♒ 56 13	19 36	15 36	23 07	14 42	08 55	25 53	12 03	01 22
20	20 01 06	00 ♒ 49 58	25 ♒ 52 06	01 ♓ 49 04	21 11	16 52	23 55	14 56	09 00	25 50	12 03	01 24
21	20 05 02	01 51 03	07 ♓ 47 29	13 ♓ 47 44	22 46	18 07	24 42	15 10	09 05	25 48	12 03	01 25
22	20 08 59	02 52 07	19 ♓ 50 11	25 ♓ 55 18	24 22	19 22	25 29	15 24	09 10	25 46	12 D 03	01 27
23	20 12 55	03 53 10	02 ♈ 03 33	08 ♈ 15 26	25 58	20 38	26 17	15 38	09 15	25 43	12 03	01 28
24	20 16 52	04 54 12	14 ♈ 31 25	20 ♈ 52 02	27 35	21 53	27 04	15 52	09 20	25 41	12 03	01 30
25	20 20 48	05 55 13	27 ♈ 17 45	03 ♉ 49 04	29 ♑ 13	23 08	27 52	16 06	09 25	25 39	12 03	01 31
26	20 24 45	06 56 13	10 ♉ 26 24	17 ♉ 10 05	00 ♒ 51	24 23	28 39	16 21	09 30	25 36	12 03	01 33
27	20 28 42	07 57 12	24 ♉ 00 24	00 ♊ 57 29	02 30	25 39	29 ♒ 27	16 35	09 35	25 34	12 03	01 34
28	20 32 38	08 58 10	08 ♊ 01 21	15 ♊ 11 50	04 09	26 54	00 ♓ 14	16 49	09 40	25 31	12 04	01 36
29	20 36 35	09 59 07	22 ♊ 28 35	29 ♊ 51 06	05 50	28 09	01 01	17 03	09 44	25 29	12 04	01 38
30	20 40 31	11 00 02	07 ♋ 18 38	14 ♋ 50 18	07 30	29 ♑ 24	01 49	17 17	09 49	25 26	12 04	01 39
31	20 44 28	12 ♒ 00 57	22 ♋ 25 02	00 ♌ 01 38	09 ♒ 12	00 ♒ 40	02 ♓ 36	17 ♒ 32	09 ♐ 54	25 ♌ 24	12 ♉ 04	01 ♓ 41

Date	Moon True ☊ ° '	Moon Mean ☊ ° '	Moon ☽ Latitude ° '
01	03 ♓ 07	04 ♓ 39	05 N 01
02	02 R 59	04 35	04 35
03	02 51	04 32	03 50
04	02 46	04 29	02 48
05	02 42	04 26	01 34
06	02 41	04 23	00 N 16
07	02 D 41	04 20	01 S 02
08	02 43	04 16	02 14
09	02 44	04 13	03 16
10	02 R 44	04 10	04 05
11	02 42	04 07	04 41
12	02 39	04 04	05 03
13	02 34	04 01	05 10
14	02 27	03 57	05 03
15	02 20	03 54	04 43
16	02 13	03 51	04 11
17	02 07	03 48	03 27
18	02 02	03 45	02 35
19	01 58	03 41	01 37
20	01 57	03 38	00 S 33
21	01 D 57	03 35	00 N 32
22	01 58	03 32	01 37
23	01 59	03 29	02 38
24	02 01	03 26	03 33
25	02 02	03 22	04 19
26	02 R 03	03 19	04 53
27	02 02	03 16	05 12
28	02 00	03 13	05 13
29	01 57	03 10	04 55
30	01 54	03 06	04 17
31	01 ♓ 51	03 ♓ 03	03 N 21

DECLINATIONS

Date	Sun ☉ ° '	Moon ☽ ° '	Mercury ☿ ° '	Venus ♀ ° '	Mars ♂ ° '	Jupiter ♃ ° '	Saturn ♄ ° '	Uranus ♅ ° '	Neptune ♆ ° '	Pluto ♇ ° '
01	22 S 56	27 N 29	22 S 42	22 S 44	19 S 09	18 S 12	19 S 46	13 N 24	13 N 46	21 S 39
02	22 51	28 01	22 54	22 50	18 56	18 09	19 47	13 25	13 46	21 39
03	22 45	26 27	23 05	22 56	18 43	18 05	19 48	13 25	13 46	21 38
04	22 39	22 56	23 15	23 00	18 30	18 01	19 49	13 26	13 46	21 37
05	22 32	17 52	23 24	23 04	18 16	17 58	19 49	13 27	13 46	21 37
06	22 24	11 48	23 33	23 08	18 03	17 54	19 50	13 27	13 46	21 36
07	22 17	05 N 16	23 40	23 10	17 49	17 50	19 51	13 28	13 46	21 35
08	22 09	01 S 49	23 45	23 12	17 35	17 46	19 52	13 29	13 46	21 35
09	22 00	07 38	23 50	23 13	17 21	17 42	19 53	13 29	13 45	21 34
10	21 51	13 26	23 53	23 14	17 06	17 39	19 54	13 30	13 45	21 34
11	21 42	18 30	23 56	23 14	16 52	17 35	19 55	13 31	13 45	21 33
12	21 32	22 39	23 57	23 12	16 37	17 31	19 56	13 31	13 45	21 32
13	21 21	25 43	23 56	23 11	16 22	17 27	19 57	13 32	13 45	21 32
14	21 11	27 34	23 55	23 08	16 07	17 23	19 57	13 33	13 45	21 31
15	21 00	28 06	23 52	23 05	15 52	17 19	19 58	13 34	13 45	21 30
16	20 48	27 20	23 48	23 01	15 37	17 15	19 59	13 34	13 45	21 30
17	20 36	25 18	23 42	22 57	15 21	17 11	20 00	13 35	13 45	21 29
18	20 24	22 11	23 35	22 52	15 06	17 07	20 00	13 36	13 45	21 29
19	20 11	18 10	23 27	22 46	14 50	17 03	20 01	13 37	13 45	21 28
20	19 58	13 25	23 17	22 39	14 34	16 59	20 02	13 37	13 45	21 27
21	19 45	08 09	23 06	22 32	14 18	16 55	20 03	13 38	13 45	21 27
22	19 31	02 S 32	22 54	22 24	14 01	16 51	20 03	13 39	13 45	21 26
23	19 17	03 N 14	22 40	22 15	13 45	16 47	20 04	13 40	13 46	21 25
24	19 02	09 00	22 24	22 05	13 28	16 42	20 05	13 41	13 46	21 25
25	18 47	14 32	22 08	21 55	13 12	16 38	20 05	13 42	13 46	21 24
26	18 32	19 35	21 49	21 45	12 55	16 34	20 06	13 42	13 46	21 23
27	18 17	23 48	21 30	21 33	12 38	16 30	20 07	13 43	13 46	21 23
28	18 01	26 48	21 09	21 21	12 21	16 26	20 07	13 44	13 46	21 22
29	17 45	28 08	20 46	21 09	12 04	16 21	20 08	13 45	13 46	21 21
30	17 28	27 31	20 22	20 55	11 46	16 17	20 08	13 46	13 46	21 21
31	17 S 11	24 N 52	19 S 56	20 S 41	11 S 29	16 S 13	20 S 09	13 N 47	13 N 46	21 S 20

ZODIAC SIGN ENTRIES

Date	h	m	Planets
02	13	13	☽ ♋
04	12	20	☽ ♌
06	12	25	☽ ♍
06	14	20	☿ ♑
07	01	26	♀ ♑
08	15	12	☽ ♎
10	21	31	☽ ♏
13	07	03	☽ ♐
15	18	46	☽ ♑
18	07	31	☽ ♒
19	16	22	☉ ♒
20	20	20	☽ ♓
23	07	59	☽ ♈
25	17	00	☽ ♉
25	23	34	☿ ♒
27	22	21	☽ ♊
28	04	58	♂ ♓
30	00	14	☽ ♋
30	23	22	♀ ♒
31	23	57	☽ ♌

LATITUDES

Date	Mercury ☿ ° '	Venus ♀ ° '	Mars ♂ ° '	Jupiter ♃ ° '	Saturn ♄ ° '	Uranus ♅ ° '	Neptune ♆ ° '	Pluto ♇ ° '
01	00 N 32	00 N 31	01 S 10	00 S 38	01 N 46	00 N 45	01 S 48	11 S 18
04	00 N 09	00 24	01 09	00 39	01 46	00 45	01 48	11 17
07	00 S 14	00 16	01 08	00 39	01 46	00 46	01 47	11 17
10	00 35	00 08	01 08	00 39	01 46	00 46	01 47	11 16
13	00 54	00 N 01	01 07	00 39	01 47	00 46	01 47	11 16
16	01 12	00 S 07	01 06	00 39	01 47	00 46	01 47	11 15
19	01 27	00 14	01 05	00 39	01 47	00 46	01 47	11 15
22	01 41	00 22	01 04	00 39	01 47	00 46	01 47	11 15
25	01 51	00 29	01 03	00 40	01 48	00 46	01 46	11 15
28	01 59	00 35	01 02	00 40	01 48	00 46	01 46	11 14
31	02 S 04	00 S 42	01 S 00	00 S 40	01 N 48	00 N 46	01 S 46	11 S 14

DATA

Julian Date	2467982
Delta T	+79 seconds
Ayanamsa	24° 29' 16"
Synetic vernal point	04° ♓ 37' 42"
True obliquity of ecliptic	23° 26' 08"

MOON'S PHASES, APSIDES AND POSITIONS ☽

Date	h	m	Phase	Longitude ° '	Eclipse Indicator
03	10	20	○	13 ♋ 27	
10	03	32	☾	20 ♎ 17	
18	04	25	●	28 ♑ 28	
26	05	09	☽	06 ♉ 39	

Day	h	m		
03	19	32	Perigee	
17	22	50	Apogee	
02	06	18	Max dec	28° N 05'
08	07	10	OS	
15	09	42	Max dec	28° S 07'
22	22	36	ON	
29	16	43	Max dec	28° N 10'

LONGITUDES

Date	Chiron ⚷ ° '	Ceres ⚳ ° '	Pallas ⚴ ° '	Juno ⚵ ° '	Vesta ⚶ ° '	Black Moon Lilith ⚸ ° '
01	19 ♍ 01	27 ♉ 00	14 ♓ 35	18 ♈ 24	03 ♈ 46	24 ♑ 36
11	18 ♍ 56	26 ♉ 32	17 ♓ 16	22 ♈ 01	06 ♈ 53	25 ♑ 43
21	18 ♍ 39	26 ♉ 46	20 ♓ 13	26 ♈ 05	10 ♈ 18	26 ♑ 50
31	18 ♍ 12	27 ♉ 38	23 ♓ 26	00 ♉ 32	13 ♈ 52	27 ♑ 57

ASPECTARIAN

h m Aspects

01 Sunday
00 33 ☽ ☍ ♄
03 02 ☽ △ ♂
06 06 ☽ △ ♃
07 20 ☽ ⊼ ☉
08 46 ☽ ⊻ ♆
11 50 ☉ bQ ♅
12 29 ☽ Q ♅
18 24 ☽ ⊥ ♆
21 11 ☿ ∥ ♀

02 Monday
03 03 ☽ ☍ ☿
03 27 ☽ ☍ ♀
04 01 ☉ △ ♆
04 27 ☽ bQ ♂
06 28 ☽ bQ ♃
07 35 ☽ ⚹ ♅
07 53 ☉ ∥ ☿
08 43 ☽ ∠ ♆
13 31 ☉ ∥ ♀
14 46 ☽ △ ♇
19 27 ☽ ± ♂
20 53 ☽ ± ♃

03 Tuesday
00 43 ☽ ⊼ ♄
05 23 ☽ ⊼ ♂
06 27 ☽ ⊼ ♃
07 10 ☽ ∠ ♅
08 04 ☉ ⊥ ♄
08 18 ☽ ⚹ ♆
10 01 ♀ ∠ ♂
10 12 ☽ ± ♄
10 20 ☽ ☍ ☉
10 53 ☿ ∠ ♂
13 13 ☿ ☌ ♀
14 22 ☽ bQ ♇
21 15 ☽ ⊥ ♅
21 40 ☿ ∠ ♃
23 06 ♀ ∠ ♃

04 Wednesday
00 25 ☽ bQ ♄
02 34 ☿ △ ♅
03 07 ☽ Q ♆
04 24 ♀ △ ♅
04 30 ☽ ± ♇
06 39 ☽ ⊻ ♅
06 52 ☽ ⊼ ♀
07 05 ☽ ⊼ ☿
09 58 ♂ ☌ ♃
10 14 ☽ # ☿
11 34 ☽ # ♀
13 34 ☽ # ☉
13 57 ☽ ⊼ ♇
14 50 ☿ bQ ♆
17 09 ☽ ± ♀
17 31 ☽ ± ☿
18 28 ♀ bQ ♆
18 50 ☽ # ♇

05 Thursday
00 13 ☽ △ ♄
03 24 ☽ # ♄
06 24 ☽ ☍ ♃
07 12 ☽ ☍ ♂
07 32 ☽ □ ♆
08 44 ☽ bQ ♀
09 19 ☽ bQ ☿
10 10 ☽ # ♂
11 35 ☽ # ♃
13 05 ☽ ⊼ ☉
13 28 ♂ □ ♆
23 27 ☽ ± ☉
23 55 ☉ ∠ ♇

06 Friday
04 32 ☽ ∥ ♆
05 45 ☽ ∥ ♅
06 25 ☽ ☌ ♅
11 12 ☽ △ ♀
12 12 ☽ △ ☿
14 10 ☽ ☍ ♇
15 13 ☽ bQ ☉

07 Saturday
01 11 ☽ □ ♄
08 07 ☿ ⚹ ♇
08 08 ☽ ⊼ ♃
08 31 ☽ △ ♆
09 26 ♂ ∥ ♃
11 03 ☽ ⊼ ♂
18 14 ☽ △ ☉
18 27 ☽ ± ♃
21 51 ☽ ± ♂
22 26 ♀ ⚹ ♇

08 Sunday
07 07 ♃ □ ♆
07 59 ☽ Q ♄
08 42 ☽ ⊻ ♅
10 08 ☽ bQ ♆
10 11 ☽ bQ ♃
14 21 ☽ bQ ♂
17 09 ☽ ⊼ ♇
19 01 ☽ □ ♀
19 09 ☽ ⊥ ♅
21 06 ☽ □ ☿

09 Monday
01 56 ☽ ± ♆
03 46 ☽ ± ♇
05 14 ☽ ⚹ ♄
11 07 ☽ ∠ ♅
12 39 ☽ ⊼ ♆
13 11 ☽ △ ♃
18 43 ☽ △ ♂
19 58 ☽ bQ ♇

10 Tuesday
01 48 ☉ ± ♅
03 32 ☽ □ ☉
06 45 ☽ Q ♀
08 37 ☽ ∠ ♄
09 41 ☽ Q ☿
12 17 ☽ # ♅
13 25 ☽ # ♆
14 26 ☽ ⚹ ♅
23 41 ☽ △ ♇

11 Wednesday
01 29 ☽ ⊥ ♄
03 39 ☿ ⊥ ♃
04 11 ☽ ∥ ♂
07 22 ☽ ∥ ♃
07 32 ☽ ⚹ ♀
11 06 ☽ ⚹ ☿
12 53 ☽ Q ♅
12 54 ☽ ⊻ ♄
16 04 ☽ Q ☉
19 34 ☽ ∥ ♄
20 20 ☽ ☍ ♆
21 58 ☽ □ ♃

12 Thursday
03 07 ☿ ⊻ ♄
04 57 ☽ ∥ ♇
05 10 ☽ ∥ ☉
06 35 ☽ □ ♂
10 19 ☉ ∥ ♇
15 27 ☽ ∠ ♀
15 42 ☽ ∥ ♀
17 12 ☽ ⚹ ☉
17 34 ♀ ⊥ ♃
19 53 ☽ ∠ ☿
21 01 ☽ ∥ ☿
23 27 ☽ □ ♅

13 Friday
03 02 ☉ ∠ ♄
09 27 ☽ □ ♇
09 32 ☽ Q ♃
11 02 ☽ ⊥ ♀
16 01 ☽ ⊥ ☿
17 45 ♀ ⊻ ♄
20 08 ☽ Q ♂
23 36 ☽ ☌ ♄

14 Saturday
00 11 ☿ bQ ♅
00 14 ☽ ⊻ ♀
01 19 ☽ ∠ ☉
05 36 ☽ ⊻ ☿
06 52 ☽ ⊼ ♆
09 44 ☽ ⚹ ♃
15 35 ☿ △ ♆
18 49 ☽ ± ♆
20 53 ☽ ⊥ ☉
21 12 ☽ Q ♇
21 40 ☽ ⚹ ♂

15 Sunday
00 41 ☉ ⊥ ♇
10 01 ☽ ⊻ ☉
10 46 ☽ △ ♅
12 50 ☽ bQ ♆
16 21 ☽ ∠ ♃
18 31 ☉ ⊼ ♅
19 09 ☿ ⊻ ♃
20 08 ♀ bQ ♅
21 20 ☽ ⚹ ♇

16 Monday
05 57 ☽ ∠ ♂
06 54 ☿ ⊥ ♂
07 29 ☿ ⊥ ♄
08 06 ♂ Q ♄
10 52 ☽ ⊥ ♃
12 12 ☽ ⊻ ♄
16 01 ♀ △ ♆
16 56 ☽ bQ ♅
19 06 ☽ △ ♆
19 28 ☽ ☌ ♀
23 15 ☽ ⊻ ♃

17 Tuesday
00 26 ☽ ⊥ ♄
01 28 ☽ ⊥ ♂
02 50 ☽ ☌ ☿
03 44 ☽ ∠ ♇
09 43 ☿ ∠ ♇
11 07 ☽ ± ♅
14 30 ☽ ⊻ ♂
18 50 ☽ ∠ ♄
22 02 ☽ ⊥ ♇
23 16 ☽ ⊼ ♅

18 Wednesday
01 48 ☽ ∥ ☿
04 25 ☽ ☌ ☉
07 18 ☽ ∥ ♀
10 15 ☽ ⊻ ♇
14 27 ♀ ⊻ ♃
16 39 ☽ ∥ ♇
21 36 ♀ ⊥ ♄
23 53 ☽ ∥ ☉

19 Thursday
01 33 ☽ ∥ ♄
01 33 ☽ ⚹ ♄
08 00 ☽ □ ♆
13 24 ☽ ☌ ♃
15 36 ☽ ⊻ ♀
16 12 ☿ ± ♅
18 00 ☽ ∥ ♃

20 Friday
01 03 ☽ ⊻ ☿
02 04 ☽ Q ♄
02 50 ♀ ∠ ♇
05 12 ☽ ⊥ ♀
05 33 ☉ ∥ ♄
06 09 ☽ ∥ ♂
07 46 ☽ ☌ ♂
10 23 ☽ # ♆
11 01 ☽ # ♅
11 57 ☽ ☍ ♅
15 03 ☽ ⊥ ☿
20 26 ☽ Q ♆
22 57 ☽ ⊻ ☉
23 10 ☽ ☌ ♇

21 Saturday
01 31 ☽ ∠ ♀
01 32 ☉ ⊻ ♇
11 56 ☽ ∠ ☿
12 08 ☽ ⊥ ☉
14 37 ☽ □ ♄
16 20 ♆ St D
20 31 ☽ ⚹ ♆

22 Sunday
03 01 ☽ ⊻ ♃
07 45 ☽ ∠ ☉
08 59 ☿ ∠ ♄
10 58 ☽ ⚹ ♀
15 09 ☽ ⊥ ♃
18 11 ♀ ⊥ ♂
19 15 ♀ ± ♅
19 51 ♂ ☍ ♅
22 17 ☽ ⚹ ☿
23 39 ☽ ⊼ ♅
23 56 ☽ ⊻ ♂

23 Monday
02 13 ☽ ∠ ♆
04 26 ☿ ⊥ ♇
08 27 ☿ ⊼ ♅
09 10 ☽ ∠ ♃
10 51 ☽ ⊻ ♇
11 09 ♂ # ♆
11 21 ☽ ± ♅
12 28 ☽ ⊥ ♂
13 14 ☽ Q ♀
15 52 ☽ ⚹ ☉
18 58 ♂ # ♅
19 44 ☽ ⊥ ♆
21 08 ☿ ⊻ ♂
22 30 ☽ ⊥ ♇

24 Tuesday
01 09 ☽ Q ☿
02 01 ☽ △ ♄
04 41 ☽ bQ ♅
07 00 ☽ ∠ ♂
07 17 ☽ ⊻ ♆
14 36 ☽ ⚹ ♃
15 45 ☽ ∠ ♇
16 55 ☽ Q ☉

25 Wednesday
03 25 ☽ □ ♀
06 20 ☽ # ♂
06 37 ☽ bQ ♄
08 14 ☽ ∥ ♅
08 33 ☽ ∥ ♆
08 56 ☽ △ ♅
13 07 ☽ ⚹ ♂
13 32 ☽ Q ♃
16 03 ☽ □ ☿
19 48 ☽ ⚹ ♇
21 30 ☽ # ♃
23 21 ☽ ± ♄

26 Thursday
05 09 ☽ □ ☉
07 01 ☽ # ☉
10 18 ☽ ⊼ ♄
12 24 ☽ Q ♂
14 21 ♀ ∠ ♄
14 40 ☽ # ♄
14 54 ☽ ☌ ♆
17 35 ☽ Q ♇
21 34 ☽ # ♇
22 20 ☿ ⊻ ♇
22 44 ☽ □ ♃
23 05 ☽ # ♀
23 11 ☽ # ☿

27 Friday
02 09 ☿ ∥ ♀
10 29 ♀ ⊼ ♅
10 36 ♀ ⊥ ♇
14 42 ☽ □ ♅
15 08 ☽ △ ♀
20 26 ☿ ∥ ♇
21 58 ☽ □ ♂

28 Saturday
01 05 ☽ □ ♇
04 35 ☽ △ ☿
06 46 ♂ Q ♆
10 12 ♀ ∥ ♇
13 43 ☽ △ ☉
14 47 ☽ ☍ ♄
18 47 ☽ ⊻ ♆
19 08 ☽ bQ ♀
21 11 ☽ Q ♅

29 Sunday
02 56 ☽ △ ♃
04 45 ☽ ⊥ ♆
05 46 ☉ ⚹ ♄
08 57 ☽ bQ ☿
11 25 ☽ ± ♀
16 24 ☽ bQ ☉
16 53 ☽ ⚹ ♅
19 29 ☽ ∠ ♆
22 06 ☽ ⊼ ♀

30 Monday
01 29 ☽ ± ☿
02 40 ☽ △ ♂
02 54 ☽ △ ♇
03 49 ☽ bQ ♃
07 00 ♂ ☌ ♇
08 02 ☽ ± ☉
12 21 ☽ ⊼ ☿
16 02 ☽ ⊼ ♄
16 59 ☽ ∠ ♅
18 19 ☽ ⊼ ☉
18 28 ☽ ± ♃
19 36 ☽ ⚹ ♆

31 Tuesday
00 49 ☿ ∥ ♄
01 37 ☽ ± ♄
02 54 ☽ bQ ♇
03 58 ☽ bQ ♂
04 09 ☽ ⊼ ♃
05 31 ♅ ∥ ♆
07 14 ☽ ⊥ ♅
13 25 ☉ □ ♆
14 37 ☽ Q ♆
15 56 ☽ bQ ♄
16 41 ☽ ⊻ ♅
17 09 ☽ ± ♇
18 58 ☽ ± ♂
22 16 ☿ ⚹ ♄

All ephemeris data is given at 12.00 UT and the Moon's longitude is additionally given for 24.00 UT

FEBRUARY 2045

LONGITUDES

Date	Sidereal time h m s	Sun ☉ ° ' "	Moon ☽ ° ' "	Moon ☽ 24.00 ° ' "	Mercury ☿ ° '	Venus ♀ ° '	Mars ♂ ° '	Jupiter ♃ ° '	Saturn ♄ ° '	Uranus ♅ ° '	Neptune ♆ ° '	Pluto ♇ ° '
01	20 48 24	13 ♒ 01 50	07 ♌ 38 52	15 ♌ 15 26	10 ♒ 54	01 ♒ 55	03 ♓ 23	17 ♒ 46	09 ♐ 58	25 ♌ 21	12 ♉ 05	01 ♓ 42
02	20 52 21	14 02 42	22 ♌ 50 06	00 ♍ 21 39	12 37	03 10	04 11	18 00	10 02	25 R 19	12 05	01 44
03	20 56 17	15 03 33	07 ♍ 49 04	15 ♍ 11 24	14 21	04 25	04 58	18 15	10 07	25 16	12 06	01 46
04	21 00 14	16 04 22	22 ♍ 27 57	29 ♍ 38 10	16 05	05 40	05 45	18 29	10 11	25 13	12 06	01 47
05	21 04 11	17 05 11	06 ♎ 41 41	13 ♎ 38 19	17 50	06 56	06 33	18 43	10 15	25 11	12 07	01 49
06	21 08 07	18 05 59	20 ♎ 28 02	27 ♎ 10 59	19 36	08 11	07 20	18 58	10 19	25 08	12 07	01 51
07	21 12 04	19 06 46	03 ♏ 47 23	10 ♏ 17 34	21 22	09 26	08 07	19 12	10 23	25 06	12 08	01 52
08	21 16 00	20 07 33	16 ♏ 41 58	23 ♏ 01 03	23 10	10 41	08 55	19 26	10 27	25 03	12 08	01 54
09	21 19 57	21 08 18	29 ♏ 15 19	05 ♐ 25 17	24 57	11 56	09 42	19 41	10 31	25 00	12 09	01 56
10	21 23 53	22 09 02	11 ♐ 31 30	17 ♐ 34 31	26 46	13 12	10 29	19 55	10 35	24 58	12 10	01 57
11	21 27 50	23 09 45	23 ♐ 34 51	29 ♐ 33 00	28 ♒ 35	14 27	11 16	20 09	10 38	24 55	12 10	01 59
12	21 31 46	24 10 27	05 ♑ 29 27	11 ♑ 24 41	00 ♓ 24	15 42	12 04	20 24	10 42	24 53	12 11	02 01
13	21 35 43	25 11 08	17 ♑ 19 06	23 ♑ 13 07	02 14	16 57	12 51	20 38	10 45	24 50	12 12	02 02
14	21 39 40	26 11 48	29 ♑ 07 06	05 ♒ 01 23	04 04	18 12	13 38	20 52	10 49	24 47	12 13	02 04
15	21 43 36	27 12 26	10 ♒ 56 16	16 ♒ 52 04	05 54	19 27	14 25	21 07	10 52	24 45	12 14	02 06
16	21 47 33	28 13 03	22 ♒ 49 03	28 ♒ 47 26	07 44	20 43	15 12	21 21	10 55	24 42	12 14	02 07
17	21 51 29	29 ♒ 13 38	04 ♓ 47 28	10 ♓ 49 24	09 34	21 58	15 59	21 35	10 58	24 39	12 15	02 09
18	21 55 26	00 ♓ 14 12	16 ♓ 53 25	22 ♓ 59 46	11 23	23 13	16 47	21 49	11 01	24 37	12 16	02 11
19	21 59 22	01 14 44	29 ♓ 08 39	05 ♈ 20 19	13 12	24 28	17 34	22 04	11 04	24 34	12 17	02 12
20	22 03 19	02 15 15	11 ♈ 35 00	17 ♈ 52 56	15 00	25 43	18 21	22 18	11 07	24 32	12 18	02 14
21	22 07 15	03 15 44	24 ♈ 14 22	00 ♉ 39 34	16 46	26 58	19 08	22 32	11 10	24 29	12 19	02 16
22	22 11 12	04 16 11	07 ♉ 08 47	13 ♉ 42 16	18 31	28 13	19 55	22 46	11 12	24 26	12 20	02 17
23	22 15 09	05 16 36	20 ♉ 20 15	27 ♉ 02 56	20 13	29 ♒ 28	20 42	23 01	11 15	24 24	12 21	02 19
24	22 19 05	06 16 59	03 ♊ 50 31	10 ♊ 43 05	21 52	00 ♓ 43	21 29	23 15	11 17	24 21	12 22	02 21
25	22 23 02	07 17 20	17 ♊ 40 41	24 ♊ 43 19	23 29	01 58	22 16	23 29	11 20	24 19	12 24	02 23
26	22 26 58	08 17 40	01 ♋ 50 48	09 ♋ 02 55	25 01	03 13	23 03	23 43	11 22	24 16	12 25	02 24
27	22 30 55	09 17 57	16 ♋ 19 18	23 ♋ 39 26	26 29	04 28	23 49	23 57	11 24	24 14	12 26	02 26
28	22 34 51	10 ♓ 18 12	01 ♌ 02 42	08 ♌ 28 22	27 ♓ 52	05 ♓ 43	24 ♓ 36	24 ♒ 11	11 ♐ 26	24 ♌ 11	12 ♉ 27	02 ♓ 28

Date	Moon True ☊ ° '	Moon Mean ☊ ° '	Moon ☽ Latitude ° '
01	01 ♓ 49	03 ♓ 00	02 N 10
02	01 R 48	02 57	00 N 49
03	01 D 48	02 54	00 S 33
04	01 49	02 51	01 52
05	01 50	02 47	03 01
06	01 51	02 44	03 58
07	01 52	02 41	04 40
08	01 53	02 38	05 06
09	01 R 52	02 35	05 17
10	01 52	02 32	05 13
11	01 51	02 28	04 55
12	01 50	02 25	04 24
13	01 49	02 22	03 43
14	01 48	02 19	02 52
15	01 47	02 16	01 53
16	01 47	02 12	00 S 50
17	01 D 47	02 09	00 N 17
18	01 47	02 06	01 23
19	01 47	02 03	02 26
20	01 R 47	02 00	03 24
21	01 47	01 57	04 12
22	01 47	01 53	04 49
23	01 46	01 50	05 11
24	01 46	01 47	05 17
25	01 D 46	01 44	05 06
26	01 47	01 41	04 35
27	01 47	01 38	03 47
28	01 ♓ 48	01 ♓ 34	02 N 43

DECLINATIONS

Date	Sun ☉ ° '	Moon ☽ ° '	Mercury ☿ ° '	Venus ♀ ° '	Mars ♂ ° '	Jupiter ♃ ° '	Saturn ♄ ° '	Uranus ♅ ° '	Neptune ♆ ° '	Pluto ♇ ° '
01	16 S 54	20 N 27	19 S 29	20 S 27	11 S 12	16 S 08	20 S 09	13 N 48	13 N 47	21 S 20
02	16 37	14 41	19 01	20 12	10 54	16 04	20 10	13 48	13 47	21 19
03	16 19	08 07	18 31	19 56	10 36	16 00	20 10	13 49	13 47	21 18
04	16 01	01 N 16	18 00	19 40	10 19	15 55	20 11	13 50	13 47	21 18
05	15 43	05 S 26	17 27	19 23	10 01	15 51	20 11	13 51	13 47	21 17
06	15 24	11 40	16 52	19 05	09 43	15 47	20 12	13 52	13 48	21 16
07	15 05	17 10	16 16	18 47	09 25	15 42	20 12	13 53	13 48	21 16
08	14 46	21 42	15 39	18 29	09 06	15 38	20 13	13 54	13 48	21 15
09	14 27	25 08	15 00	18 09	08 48	15 33	20 13	13 55	13 48	21 14
10	14 08	27 19	14 20	17 50	08 30	15 29	20 14	13 55	13 49	21 14
11	13 48	28 11	13 39	17 30	08 12	15 24	20 14	13 56	13 49	21 13
12	13 28	27 43	12 56	17 09	07 53	15 20	20 14	13 57	13 49	21 13
13	13 07	26 00	12 12	16 48	07 35	15 16	20 15	13 58	13 49	21 12
14	12 47	23 08	11 27	16 26	07 16	15 11	20 15	13 59	13 50	21 11
15	12 26	19 18	10 40	16 04	06 58	15 07	20 15	14 00	13 50	21 11
16	12 06	14 41	09 53	15 42	06 39	15 02	20 16	14 01	13 50	21 10
17	11 44	09 30	09 04	15 19	06 20	14 58	20 16	14 02	13 51	21 10
18	11 23	03 S 54	08 15	14 56	06 01	14 53	20 16	14 03	13 51	21 09
19	11 02	01 N 54	07 25	14 32	05 43	14 48	20 17	14 03	13 51	21 08
20	10 40	07 42	06 35	14 08	05 24	14 44	20 17	14 04	13 52	21 08
21	10 18	13 18	05 44	13 43	05 05	14 39	20 17	14 05	13 52	21 07
22	09 57	18 26	04 53	13 18	04 46	14 35	20 17	14 06	13 53	21 07
23	09 35	22 49	04 02	12 53	04 27	14 30	20 18	14 07	13 53	21 06
24	09 12	26 06	03 12	12 27	04 08	14 26	20 18	14 08	13 53	21 05
25	08 50	27 56	02 22	12 01	03 49	14 21	20 18	14 09	13 54	21 05
26	08 27	28 00	01 33	11 35	03 30	14 17	20 18	14 09	13 54	21 04
27	08 05	26 11	00 46	11 08	03 11	14 12	20 18	14 10	13 55	21 04
28	07 S 42	22 N 34	00 00	10 S 41	02 S 52	14 S 07	20 S 18	14 N 11	13 N 55	21 S 03

ZODIAC SIGN ENTRIES

Date	h m	Planets
02	23 25	☽ ♍
05	00 37	☽ ♎
07	05 06	☽ ♏
09	13 27	☽ ♐
12	00 54	☽ ♑
12	06 44	☿ ♓
14	13 48	☽ ♒
17	02 25	☽ ♓
18	06 22	☉ ♓
19	13 40	☽ ♈
21	22 46	☽ ♉
23	22 07	♀ ♓
24	05 14	☽ ♊
26	08 54	☽ ♋
28	10 18	☽ ♌

LATITUDES

Date	Mercury ☿ ° '	Venus ♀ ° '	Mars ♂ ° '	Jupiter ♃ ° '	Saturn ♄ ° '	Uranus ♅ ° '	Neptune ♆ ° '	Pluto ♇ ° '
01	02 S 05	00 S 44	01 S 00	00 S 40	01 N 48	00 N 46	01 S 46	11 S 14
04	02 05	00 50	00 59	00 40	01 49	00 46	01 46	11 14
07	02 00	00 56	00 57	00 40	01 49	00 46	01 46	11 14
10	01 52	01 01	00 56	00 41	01 49	00 46	01 45	11 14
13	01 38	01 06	00 55	00 41	01 50	00 46	01 45	11 14
16	01 18	01 10	00 53	00 41	01 50	00 46	01 45	11 14
19	00 53	01 14	00 52	00 41	01 50	00 46	01 45	11 14
22	00 S 22	01 17	00 50	00 42	01 51	00 46	01 45	11 14
25	00 N 15	01 20	00 49	00 42	01 51	00 46	01 45	11 14
28	00 56	01 23	00 47	00 42	01 52	00 46	01 44	11 15
31	01 N 38	01 S 24	00 S 45	00 S 43	01 N 52	00 N 46	01 S 44	11 S 15

DATA

Julian Date	2468013
Delta T	+79 seconds
Ayanamsa	24° 29' 22"
Synetic vernal point	04° ♓ 37' 36"
True obliquity of ecliptic	23° 26' 08"

MOON'S PHASES, APSIDES AND POSITIONS ☽

Date	h m	Phase	Longitude ° '	Eclipse Indicator
01	21 05	○	13 ♌ 25	
08	19 03	☾	20 ♏ 25	
16	23 51	●	28 ♒ 43	Annular
24	16 37	☽	06 ♊ 29	

Day	h m		
01	08 48	Perigee	
14	03 12	Apogee	
04	16 29	0S	
11	15 32	Max dec	28° S 12'
19	04 12	0N	
26	00 59	Max dec	28° N 12'

LONGITUDES

Date	Chiron ⚷ ° '	Ceres ⚳ ° '	Pallas ⚴ ° '	Juno ⚵ ° '	Vesta ⚶ ° '	Black Moon Lilith ⚸ ° '
01	18 ♍ 09	27 ♉ 45	23 ♓ 42	01 ♉ 00	14 ♈ 15	28 ♑ 03
11	17 ♍ 33	29 ♉ 15	27 ♓ 04	05 ♉ 46	18 ♈ 04	29 ♑ 10
21	16 ♍ 50	01 ♊ 15	00 ♈ 35	10 ♉ 47	22 ♈ 01	00 ♒ 17
31	16 ♍ 03	03 ♊ 40	04 ♈ 15	15 ♉ 58	26 ♈ 06	01 ♒ 23

All ephemeris data is given at 12.00 UT and the Moon's longitude is additionally given for 24.00 UT

ASPECTARIAN

h m Aspects

01 Wednesday
02 10 ☽ ☍ ♀
02 38 ☽ ⚻ ♇
04 56 ☽ ⚻ ♂
07 48 ☽ # ♇
07 56 ♀ ⚺ ♇
11 59 ☽ # ♀
13 19 ☽ # ♄
15 40 ☽ △ ♄
16 43 ☽ # ☿
17 47 ☽ ☍ ☿
18 59 ☽ □ ♆
21 05 ☽ ☍ ☉

02 Thursday
04 02 ☽ # ☉
04 14 ☽ ☍ ♃
04 34 ☿ □ ♆
06 32 ☽ # ♃
14 34 ♀ ∥ ♄
15 19 ☽ ∥ ♅
15 25 ☽ ∥ ♆
15 56 ☽ ☌ ♅

03 Friday
02 13 ☽ ☍ ♇
02 43 ☽ # ♂
06 01 ☽ ⚻ ♀
07 08 ☽ ☍ ♂
15 44 ☽ □ ♄
16 37 ☽ ± ♀
18 57 ☽ △ ♆

04 Saturday
00 03 ☽ ⚻ ☿
00 40 ☽ ⚻ ☉
05 18 ☽ ⚻ ♃
08 45 ☽ ⚼ ♀
11 17 ☽ ± ☿
11 18 ☽ ± ☉
11 36 ☉ ☌ ☿
15 25 ☽ ± ♃
16 15 ♀ ⚺ ♂
16 35 ☽ ⚺ ♅
19 45 ☽ ⚼ ♆
21 36 ☽ Q ♄
21 48 ☉ ∥ ♃

05 Sunday
02 38 ☽ ⊥ ♅
03 32 ☽ ⚼ ☉
03 40 ☽ ⚻ ♇
04 29 ☽ ⚼ ☿
06 50 ☽ ⚼ ♃
11 00 ☽ ± ♆
11 44 ☽ ⚻ ♂
12 26 ☽ △ ♀
13 56 ☽ ± ♇
17 59 ☽ ∠ ♅
18 09 ☽ ⚹ ♄
21 21 ☽ ⚻ ♆
22 43 ☽ ± ♂

06 Monday
01 59 ☿ ☌ ♃
04 34 ☽ ∥ ♂
05 36 ☽ ⚼ ♇
07 29 ☽ △ ☉
09 17 ☽ △ ♃
10 14 ☽ △ ☿
15 31 ☽ ⚼ ♂
20 18 ☽ ⚹ ♅
20 42 ☽ ∠ ♄
20 52 ☽ # ♆
21 11 ☽ # ♅

07 Tuesday
03 01 ☽ ∥ ☉
05 20 ☽ ∥ ♃
08 17 ☽ ∥ ☿
08 29 ☽ △ ♇
13 06 ☽ ⊥ ♄
14 43 ☉ ☌ ♃
18 03 ☽ Q ♅
19 29 ☽ ∥ ♀
20 29 ☽ △ ♂
23 31 ☽ □ ♀

08 Wednesday
00 14 ☽ ⚺ ♄
02 08 ☿ Q ♄
03 26 ☽ ☍ ♆
03 27 ☽ ∥ ♄
07 13 ♀ ⚹ ♄
09 19 ☽ ∥ ♇
12 58 ☿ ∥ ♃
17 17 ☽ □ ♃
19 03 ☽ □ ☉

09 Thursday
02 19 ☽ □ ☿
03 50 ☽ □ ♅
12 40 ☿ ☍ ♅
13 29 ☽ Q ♀
16 02 ♀ □ ♆
17 12 ☽ □ ♇

10 Friday
04 45 ☽ Q ♃
09 03 ☽ Q ☉
09 48 ☽ □ ♂
10 07 ☽ ☌ ♄
13 15 ☽ ⚻ ♆
15 01 ♂ □ ♄
15 41 ☽ ⚹ ♀
19 32 ☽ Q ☿
22 45 ☉ Q ♄

11 Saturday
01 11 ☽ ± ♆
01 59 ☉ # ♅
02 11 ☿ # ♅
02 22 ☉ ∥ ☿
04 47 ☽ Q ♇
05 00 ☽ ⚹ ♃
06 18 ☿ # ♆
10 35 ☉ # ♆
11 05 ☽ ⚹ ☉
14 41 ☽ △ ♅
19 13 ☽ ⚼ ♆
23 51 ☽ ⚹ ☿

12 Sunday
00 15 ☽ Q ♂
01 11 ☽ ∠ ♀
04 57 ☽ ⚹ ♇
09 09 ☿ Q ♆
11 48 ☽ ∠ ♃
15 53 ♂ ⚹ ♆
20 10 ☽ ∠ ☉
20 51 ☽ ⚼ ♅
21 32 ☽ ⊥ ♀
22 36 ☽ ⚺ ♄

13 Monday
01 35 ☽ △ ♆
02 16 ☽ ⚹ ♂
03 57 ☉ ☍ ♅
06 26 ☽ ⊥ ♃
09 26 ☿ ☌ ♇
10 51 ☽ ⊥ ♄
11 10 ☽ ⚺ ♀
11 26 ☽ ∠ ♇
11 47 ☽ ∠ ☿
15 04 ☽ ± ♅
16 09 ☽ ⊥ ☉
18 53 ☽ ⚺ ♃

14 Tuesday
03 13 ☽ ⚻ ♅
05 14 ☽ ∠ ♄
05 30 ☽ ⚺ ☉
05 47 ☽ ⊥ ♇
09 28 ☽ ⊥ ☿
10 57 ☽ ∠ ♂
18 00 ☽ ⚺ ♇
23 54 ☽ ⚺ ☿

15 Wednesday
00 52 ☽ ∥ ♇
06 29 ☽ ∥ ♄
06 32 ☽ ⊥ ♂
11 51 ☽ ⚹ ♄
14 37 ☽ □ ♆
19 33 ☽ ⚺ ♂

16 Thursday
06 37 ☽ ∥ ♀
07 15 ☽ ☌ ♀
08 59 ☽ ☌ ♃
10 17 ☽ ∥ ♃
12 12 ☽ Q ♄
15 15 ☽ # ♅
15 46 ☽ ☍ ♅
16 05 ☽ # ♆
23 51 ☽ ☌ ☉

17 Friday
01 10 ☽ ∥ ☉
02 55 ☽ Q ♆
03 05 ♀ ☌ ♃
06 43 ☽ ☌ ♇
14 11 ☽ ∥ ☿
23 12 ☽ ☌ ☿

18 Saturday
00 21 ☽ □ ♄
02 31 ☽ ∥ ♂
02 51 ☽ ⚹ ♆
06 59 ☿ □ ♄
08 08 ♀ Q ♄
11 46 ☽ ☌ ♂
12 48 ☉ Q ♆
15 09 ♀ ∥ ♃
21 54 ☽ ⚺ ♃
23 44 ☿ ⚹ ♆

19 Sunday
01 51 ☽ ⚺ ♀
03 07 ☽ ⚻ ♅
08 23 ☽ ∠ ♆
09 51 ☽ ⊥ ♃
13 53 ♀ ☍ ♅
14 45 ☽ ± ♅
14 52 ☽ ⊥ ♀
16 27 ☽ ⚺ ☉
17 57 ☽ ⚺ ♇

20 Monday
01 51 ☽ ⊥ ♆
02 54 ☽ # ♂
03 37 ☽ ∠ ♃
05 03 ☽ ⊥ ☉
05 34 ☽ ⊥ ♇
07 54 ☽ # ☿
08 04 ☽ ⚼ ♅
10 10 ☽ ∠ ♀
11 06 ☽ △ ♄
11 32 ☉ ☌ ♇
13 23 ☽ ⚺ ♆
15 07 ♀ # ♅
19 35 ☽ ⚺ ☿
22 48 ☽ ∠ ♇
23 45 ☽ ∠ ☉
23 47 ☽ # ☉

21 Tuesday
01 44 ☽ ⚺ ♂
03 16 ♀ # ♆
08 44 ☽ ⚹ ♃
08 47 ☽ ⊥ ☿
12 27 ☽ △ ♅
13 43 ☽ # ♀
13 47 ☽ ⊥ ♂
14 32 ☽ ∥ ♆
15 31 ☽ ∥ ♅
15 37 ☽ ⚼ ♄
17 40 ☽ ⚹ ♀
17 59 ☽ # ♃

22 Wednesday
03 01 ☽ ⚹ ♇
04 15 ☽ ∠ ☿
06 15 ☽ ⚹ ☉
07 33 ☽ Q ♃
07 37 ☽ ∠ ♂
08 25 ☽ ± ♄
17 16 ☿ ∥ ♂
18 14 ☽ Q ♀
19 28 ☽ ⚻ ♄
21 31 ☽ ☌ ♆
21 32 ☽ # ♄

23 Thursday
01 06 ☽ Q ♇
01 58 ☽ # ♇
06 02 ☽ Q ☉
11 45 ☽ ⚹ ☿
12 41 ☽ ⚹ ♂
16 53 ☽ □ ♃
19 15 ☽ □ ♅

24 Friday
01 02 ☿ ☌ ♂
05 12 ♀ Q ♆
05 57 ☽ □ ♀
09 22 ☽ □ ♇
11 19 ☽ Q ♂
12 04 ☽ Q ☿
16 37 ☽ □ ☉
17 36 ♃ Q ♄

25 Saturday
01 02 ☽ ☍ ♄
02 47 ☽ Q ♅
02 53 ☽ ⚺ ♆
12 03 ☿ ⚺ ♃
13 14 ☽ ⊥ ♆
19 54 ♀ ☌ ♇
20 17 ☽ □ ♂
22 04 ☽ △ ♃
23 07 ☽ □ ☿
23 16 ☽ ⚹ ♅

26 Sunday
00 29 ☿ ⚻ ♅
04 32 ☽ ∠ ♆
12 56 ☽ △ ♇
14 32 ☽ △ ♀
23 33 ☽ △ ☉
23 38 ☽ ⚼ ♃

27 Monday
00 20 ☽ ∠ ♅
03 53 ☽ ⚻ ♄
05 35 ☽ ⚹ ♆
13 47 ☽ ± ♄
13 50 ☽ ⚼ ♇
14 43 ☽ ± ♃
15 07 ☽ ⊥ ♅
17 32 ♂ ⚺ ♃
17 39 ☽ ⚼ ♀
19 55 ♃ # ♅
23 42 ♂ ⚻ ♅

28 Tuesday
00 41 ☽ ⚻ ♃
00 53 ☽ ⚺ ♅
00 58 ☽ △ ♂
01 17 ☽ Q ♆
02 00 ☽ ⚼ ☉
04 30 ☿ ∠ ♆
04 31 ☽ ⚼ ♄
04 33 ☽ ± ♇
06 20 ☽ △ ☿
09 40 ☽ ± ♀
11 56 ♃ ☍ ♅
14 18 ☽ ⚻ ♇
17 39 ☽ ± ☉
19 47 ☽ # ♇
20 16 ☽ ⚻ ♀
23 21 ☽ # ♄

MARCH 2045

LONGITUDES

Date	Sidereal time h m s	Sun ☉ ° ' "	Moon ☽ ° ' "	Moon ☽ 24.00 ° ' "	Mercury ☿ ° '	Venus ♀ ° '	Mars ♂ ° '	Jupiter ♃ ° '	Saturn ♄ ° '	Uranus ♅ ° '	Neptune ♆ ° '	Pluto ♇ ° '
01	22 38 48	11 ♓ 18 25	15 ♌ 55 34	23 ♌ 23 23	29 ♓ 10	06 ♓ 58	25 ♓ 23	24 ♒ 25	11 ♐ 28	24 ♌ 09	12 ♉ 29	02 ♓ 29
02	22 42 44	12 18 37	00 ♍ 50 49	08 ♍ 16 53	00 ♈ 21	08 13	26 10	24 39	11 30	24 R 06	12 30	02 31
03	22 46 41	13 18 46	15 ♍ 40 35	23 ♍ 00 58	01 25	09 28	26 57	24 53	11 32	24 04	12 31	02 33
04	22 50 38	14 18 54	00 ♎ 17 11	07 ♎ 28 29	02 22	10 43	27 43	25 07	11 34	24 01	12 32	02 34
05	22 54 34	15 18 59	14 ♎ 34 16	21 ♎ 34 03	03 11	11 58	28 30	25 21	11 35	23 59	12 34	02 36
06	22 58 31	16 19 04	28 ♎ 27 32	05 ♏ 14 33	03 52	13 13	29 ♓ 17	25 35	11 37	23 56	12 35	02 38
07	23 02 27	17 19 06	11 ♏ 55 04	18 ♏ 29 13	04 25	14 28	00 ♈ 03	25 49	11 38	23 54	12 37	02 39
08	23 06 24	18 19 07	24 ♏ 57 14	01 ♐ 19 27	04 48	15 43	00 50	26 03	11 39	23 52	12 38	02 41
09	23 10 20	19 19 06	07 ♐ 36 17	13 ♐ 48 14	05 03	16 58	01 37	26 16	11 41	23 49	12 40	02 43
10	23 14 17	20 19 04	19 ♐ 55 50	25 ♐ 59 39	05 08	18 13	02 23	26 30	11 42	23 47	12 41	02 44
11	23 18 13	21 19 00	02 ♑ 00 18	07 ♑ 58 23	05 R 05	19 28	03 10	26 44	11 43	23 45	12 43	02 46
12	23 22 10	22 18 55	13 ♑ 54 30	19 ♑ 49 17	04 53	20 42	03 56	26 57	11 44	23 43	12 44	02 47
13	23 26 07	23 18 48	25 ♑ 43 17	01 ♒ 37 05	04 32	21 57	04 43	27 11	11 44	23 40	12 46	02 49
14	23 30 03	24 18 39	07 ♒ 31 12	13 ♒ 26 09	04 03	23 12	05 29	27 25	11 45	23 38	12 47	02 51
15	23 34 00	25 18 28	19 ♒ 22 24	25 ♒ 20 22	03 29	24 27	06 15	27 38	11 46	23 36	12 49	02 52
16	23 37 56	26 18 15	01 ♓ 20 25	07 ♓ 22 54	02 48	25 42	07 02	27 52	11 46	23 34	12 51	02 54
17	23 41 53	27 18 01	13 ♓ 28 05	19 ♓ 36 12	02 03	26 56	07 48	28 05	11 47	23 32	12 52	02 55
18	23 45 49	28 17 44	25 ♓ 47 27	02 ♈ 01 58	01 12	28 11	08 34	28 19	11 47	23 30	12 54	02 57
19	23 49 46	29 ♓ 17 26	08 ♈ 19 51	14 ♈ 41 09	00 ♈ 19	29 ♓ 26	09 20	28 32	11 47	23 28	12 56	02 59
20	23 53 42	00 ♈ 17 05	21 ♈ 05 54	27 ♈ 34 06	29 ♓ 24	00 ♈ 41	10 07	28 45	11 47	23 26	12 58	03 00
21	23 57 39	01 16 43	04 ♉ 05 43	10 ♉ 40 41	28 29	01 55	10 53	28 58	11 R 47	23 24	12 59	03 02
22	00 01 36	02 16 18	17 ♉ 18 58	24 ♉ 00 30	27 35	03 10	11 39	29 12	11 47	23 22	13 01	03 03
23	00 05 32	03 15 51	00 ♊ 45 12	07 ♊ 33 01	26 42	04 25	12 25	29 25	11 47	23 20	13 03	03 05
24	00 09 29	04 15 22	14 ♊ 23 52	21 ♊ 17 42	25 52	05 39	13 11	29 38	11 46	23 18	13 05	03 06
25	00 13 25	05 14 51	28 ♊ 14 26	05 ♋ 13 58	25 06	06 54	13 57	29 ♒ 51	11 46	23 17	13 07	03 08
26	00 17 22	06 14 17	12 ♋ 16 13	19 ♋ 21 02	24 24	08 09	14 43	00 ♓ 04	11 45	23 15	13 09	03 09
27	00 21 18	07 13 41	26 ♋ 28 14	03 ♌ 37 36	23 46	09 23	15 29	00 17	11 45	23 13	13 11	03 11
28	00 25 15	08 13 02	10 ♌ 48 50	18 ♌ 01 34	23 14	10 38	16 14	00 29	11 44	23 12	13 12	03 12
29	00 29 11	09 12 21	25 ♌ 15 24	02 ♍ 29 47	22 48	11 52	17 00	00 42	11 43	23 10	13 14	03 13
30	00 33 08	10 11 38	09 ♍ 44 11	16 ♍ 57 58	22 27	13 07	17 46	00 55	11 42	23 09	13 16	03 15
31	00 37 05	11 ♈ 10 53	24 ♍ 10 26	01 ♎ 20 54	22 ♓ 12	14 ♈ 21	18 ♈ 32	01 ♓ 07	11 ♐ 41	23 ♌ 07	13 ♉ 18	03 ♓ 16

Date	Moon True ☊ ° '	Moon Mean ☊ ° '	Moon ☽ Latitude ° '
01	01 ♓ 49	01 ♓ 31	01 N 27
02	01 D 49	01 28	00 N 05
03	01 R 49	01 25	01 S 16
04	01 48	01 22	02 31
05	01 46	01 18	03 35
06	01 44	01 15	04 25
07	01 42	01 12	04 58
08	01 41	01 09	05 14
09	01 39	01 06	05 15
10	01 39	01 03	05 00
11	01 D 39	00 59	04 33
12	01 40	00 56	03 54
13	01 42	00 53	03 06
14	01 44	00 50	02 10
15	01 45	00 47	01 08
16	01 46	00 44	00 S 02
17	01 R 45	00 40	01 N 04
18	01 43	00 37	02 08
19	01 40	00 34	03 07
20	01 36	00 31	03 58
21	01 31	00 28	04 38
22	01 26	00 24	05 03
23	01 22	00 21	05 12
24	01 19	00 18	05 04
25	01 17	00 15	04 39
26	01 D 17	00 12	03 56
27	01 18	00 09	02 59
28	01 19	00 05	01 50
29	01 20	00 ♓ 02	00 N 33
30	01 R 20	29 ♒ 59	00 S 46
31	01 ♓ 18	29 ♒ 56	02 S 01

DECLINATIONS

Date	Sun ☉ ° '	Moon ☽ ° '	Mercury ☿ ° '	Venus ♀ ° '	Mars ♂ ° '	Jupiter ♃ ° '	Saturn ♄ ° '	Uranus ♅ ° '	Neptune ♆ ° '	Pluto ♇ ° '
01	07 S 19	17 N 27	00 N 44	10 S 14	02 S 33	14 S 03	20 S 19	14 N 12	13 N 55	21 S 03
02	06 56	11 15	01 25	09 47	02 14	13 58	20 19	14 13	13 56	21 02
03	06 33	04 N 29	02 04	09 19	01 55	13 54	20 19	14 14	13 56	21 02
04	06 10	02 S 26	02 40	08 51	01 36	13 49	20 19	14 14	13 57	21 01
05	05 47	09 03	03 12	08 23	01 16	13 45	20 19	14 15	13 57	21 01
06	05 24	15 03	03 41	07 55	00 57	13 40	20 19	14 16	13 58	21 00
07	05 01	20 07	04 05	07 26	00 38	13 35	20 19	14 17	13 58	21 00
08	04 37	24 05	04 26	06 57	00 S 19	13 31	20 19	14 17	13 59	20 59
09	04 14	26 45	04 42	06 28	00 00	13 26	20 19	14 18	13 59	20 59
10	03 50	28 03	04 53	05 59	00 N 19	13 22	20 19	14 19	14 00	20 58
11	03 27	27 58	05 00	05 30	00 38	13 17	20 19	14 20	14 00	20 58
12	03 03	26 36	05 02	05 00	00 57	13 13	20 19	14 20	14 01	20 57
13	02 39	24 03	04 59	04 30	01 16	13 08	20 19	14 21	14 01	20 57
14	02 16	20 29	04 52	04 01	01 35	13 04	20 19	14 22	14 02	20 56
15	01 52	16 05	04 40	03 31	01 54	12 59	20 19	14 22	14 02	20 56
16	01 28	11 02	04 24	03 01	02 12	12 54	20 19	14 23	14 02	20 55
17	01 04	05 S 31	04 04	02 31	02 31	12 50	20 19	14 24	14 03	20 55
18	00 41	00 N 17	03 41	02 01	02 50	12 45	20 19	14 24	14 04	20 54
19	00 S 17	06 10	03 15	01 31	03 09	12 41	20 19	14 25	14 04	20 54
20	00 N 07	11 55	02 47	01 00	03 27	12 36	20 18	14 26	14 05	20 54
21	00 31	17 14	02 17	00 S 30	03 46	12 32	20 18	14 26	14 06	20 53
22	00 54	21 50	01 45	00 00	04 05	12 27	20 18	14 27	14 06	20 53
23	01 18	25 24	01 14	00 N 31	04 23	12 23	20 18	14 28	14 07	20 52
24	01 41	27 34	00 42	01 01	04 42	12 19	20 18	14 28	14 07	20 52
25	02 05	28 04	00 N 11	01 31	05 00	12 14	20 18	14 29	14 08	20 52
26	02 29	26 47	00 S 19	02 02	05 18	12 10	20 17	14 29	14 09	20 51
27	02 52	23 47	00 48	02 32	05 37	12 05	20 17	14 30	14 09	20 51
28	03 15	19 17	01 15	03 02	05 55	12 01	20 17	14 30	14 10	20 50
29	03 39	13 37	01 40	03 32	06 13	11 56	20 17	14 31	14 10	20 50
30	04 02	07 13	02 03	04 02	06 31	11 52	20 17	14 31	14 11	20 50
31	04 N 25	00 N 28	02 S 23	04 N 32	06 N 49	11 S 48	20 S 16	14 N 32	14 N 12	20 S 49

ZODIAC SIGN ENTRIES

Date	h	m	Planets
02	04	45	☿ ♈
02	10	38	☽ ♍
04	11	31	☽ ♎
06	14	43	☽ ♏
07	10	14	♂ ♈
08	21	29	☽ ♐
11	07	59	☽ ♑
13	20	42	☽ ♒
16	09	20	☽ ♓
18	20	06	☽ ♈
19	20	22	☿ ♓
19	22	56	♀ ♈
20	05	07	☉ ♈
21	04	29	☽ ♉
23	10	40	☽ ♊
25	15	02	☽ ♋
26	05	08	♃ ♓
27	17	56	☽ ♌
29	19	52	☽ ♍
31	21	44	☽ ♎

LATITUDES

Date	Mercury ☿ ° '	Venus ♀ ° '	Mars ♂ ° '	Jupiter ♃ ° '	Saturn ♄ ° '	Uranus ♅ ° '	Neptune ♆ ° '	Pluto ♇ ° '
01	01 N 10	01 S 23	00 S 47	00 S 42	01 N 52	00 N 46	01 S 44	11 S 15
04	01 53	01 25	00 45	00 43	01 52	00 46	01 44	11 15
07	02 33	01 26	00 44	00 43	01 52	00 46	01 44	11 15
10	03 06	01 26	00 42	00 44	01 53	00 46	01 44	11 16
13	03 28	01 26	00 40	00 44	01 53	00 46	01 44	11 16
16	03 35	01 25	00 38	00 44	01 54	00 46	01 43	11 16
19	03 24	01 24	00 36	00 45	01 54	00 46	01 43	11 17
22	02 58	01 22	00 34	00 45	01 55	00 46	01 43	11 17
25	02 19	01 20	00 33	00 46	01 55	00 46	01 43	11 18
28	01 34	01 17	00 31	00 46	01 55	00 46	01 43	11 18
31	00 N 47	01 S 13	00 S 29	00 S 46	01 N 56	00 N 46	01 S 43	11 S 19

DATA

Julian Date	2468041
Delta T	+79 seconds
Ayanamsa	24° 29' 26"
Synetic vernal point	04° ♓ 37' 32"
True obliquity of ecliptic	23° 26' 09"

MOON'S PHASES, APSIDES AND POSITIONS ☽

Date	h	m	Phase	Longitude ° '	Eclipse Indicator
03	07	53	○	13 ♍ 08	partial
10	12	50	☾	20 ♐ 21	
18	17	15	●	28 ♓ 31	
26	00	56	☽	05 ♋ 47	

Day	h	m		
01	18	50	Perigee	
13	18	31	Apogee	
29	17	28	Perigee	
04	03	31	0S	
10	22	39	Max dec	28° S 11'
18	10	49	0N	
25	06	55	Max dec	28° N 06'
31	13	38	0S	

LONGITUDES

Date	Chiron ⚷ ° '	Ceres ⚳ ° '	Pallas ⚴ ° '	Juno ⚵ ° '	Vesta ⚶ ° '	Black Moon Lilith ⚸ ° '
01	16 ♍ 13	03 ♊ 09	03 ♈ 30	14 ♉ 55	25 ♈ 17	01 ♒ 10
11	15 ♍ 25	05 ♊ 52	07 ♈ 16	20 ♉ 13	29 ♈ 26	02 ♒ 17
21	14 ♍ 39	08 ♊ 54	11 ♈ 08	25 ♉ 38	03 ♉ 23	03 ♒ 23
31	13 ♍ 57	12 ♊ 11	15 ♈ 06	01 ♊ 07	07 ♉ 58	04 ♒ 30

ASPECTARIAN

h m	Aspects
01 Wednesday	
02 35	☽ bQ ♂
04 02	☽ ⊼ ☉
04 49	☽ △ ♄
06 26	☽ □ ♆
08 54	☽ bQ ☿
16 04	☉ □ ♄
17 52	☽ ± ♂
02 Thursday	
00 40	☽ ± ☿
00 58	☽ ∥ ♅
01 10	☽ ☌ ♅
01 44	☽ # ♃
01 52	☽ ☍ ♃
02 02	☽ ∥ ♆
04 03	☽ ⊼ ♂
07 02	☿ ± ♅
11 08	☽ ⊼ ☿
14 42	☽ ☍ ♇
16 34	☉ ⚹ ♆
17 44	☽ # ♀
20 22	☿ ⊥ ♃
23 54	♃ # ♆
03 Friday	
01 00	☽ ☍ ♀
04 19	☽ # ☉
05 15	☽ □ ♄
06 52	☽ △ ♆
07 53	☽ ☍ ☉
07 57	☿ # ♂
19 40	☽ ∥ ☿
21 19	☽ # ♂
04 Saturday	
01 41	☽ ⊻ ♅
03 19	☽ ⊼ ♃
06 13	♂ ∠ ♆
07 27	☽ bQ ♆
07 31	☽ ☍ ♂
09 12	☽ ∥ ♂
10 48	☽ Q ♄
11 34	☽ ⊥ ♅
12 54	☽ # ☿
13 24	☽ ± ♃
15 41	☽ ☍ ☿
15 48	☽ ⊼ ♇
17 46	☿ ⊻ ♇
22 27	☽ ± ♆
05 Sunday	
00 37	☽ ∥ ☉
01 52	☽ ± ♇
02 34	☽ ∠ ♅
04 28	♀ □ ♄
04 43	☽ bQ ♃
06 56	☽ ⚹ ♄
07 09	☽ ⊼ ♀
08 35	☽ ⊼ ♆
09 40	☽ ∥ ♀
13 22	☽ ⊼ ☉
17 11	☽ bQ ♇
18 23	☽ ± ♀
23 37	♀ ⚹ ♆
06 Monday	
00 28	☽ ± ☉
04 08	☽ ⚹ ♅
06 16	☽ ∥ ♃
06 53	☽ △ ♃
07 25	☽ # ♆
08 41	☽ # ♅
08 46	☽ ∠ ♄
11 32	☽ bQ ♀
13 32	☽ ⊼ ♂
17 26	☽ bQ ☉
19 22	☽ △ ♇
21 59	☽ ⊼ ☿
07 Tuesday	
00 41	☽ ⊥ ♄
00 48	☽ ± ♂
01 12	☽ Q ♅
07 25	♂ ± ♅
09 10	☽ ± ☿
11 29	☽ ⊻ ♄
13 01	☽ ∥ ♄
13 16	☽ ☍ ♆
16 41	☽ ∥ ♇
17 07	☽ △ ♀
18 04	☽ bQ ♂
22 40	☽ △ ☉
08 Wednesday	
02 10	☽ bQ ☿
09 58	☽ □ ♅
14 05	☽ □ ♃
18 37	☉ # ☿
23 48	☽ △ ♂
09 Thursday	
02 37	☽ □ ♇
07 01	☽ △ ☿
19 53	☽ ☌ ♄
21 48	☽ ⊼ ♆
10 Friday	
01 09	☽ Q ♃
08 14	☽ □ ♀
09 33	☽ ± ♆
12 50	☽ □ ☉
13 36	☽ Q ♇
14 42	☿ St R
17 06	♂ ⊥ ♃
19 36	☽ △ ♅
23 17	♂ ⊻ ♇
11 Saturday	
01 16	☽ ⚹ ♃
03 24	☽ bQ ♆
13 32	☽ ⚹ ♇
14 29	☽ □ ♂
18 06	☽ □ ☿
12 Sunday	
00 15	☽ Q ♀
01 31	☽ bQ ♅
04 04	☽ Q ☉
07 35	☽ ⊻ ♄
07 58	☽ ∠ ♃
09 37	☽ △ ♆
10 27	☿ # ♀
19 41	☽ ± ♅
19 45	☽ ⊥ ♄
19 54	☽ ∠ ♇
13 Monday	
02 35	☽ ⊥ ♃
03 26	☽ ⚹ ♀
05 26	☽ Q ♂
05 44	☽ Q ☿
06 39	☽ ⚹ ☉
07 51	☽ ⊼ ♅
08 29	☿ ☌ ♂
14 05	☽ ∠ ♄
14 14	☽ ⊥ ♇
15 02	☽ ⊻ ♃
20 23	☉ ⊼ ♅
14 Tuesday	
02 28	☽ ⊻ ♇
05 17	☽ ⚹ ☿
07 34	☽ ⚹ ♂
09 13	☽ ∥ ♇
12 58	☽ ∥ ♄
13 33	☽ ∠ ♀
15 58	☽ ∠ ☉
20 10	♀ ⊼ ♅
20 36	☽ ⚹ ♄
22 43	☽ □ ♆
15 Wednesday	
07 50	☿ ⊥ ♃
09 55	☽ ⊥ ♀
10 18	☽ ∠ ☿
11 03	☉ # ♂
11 51	☽ ⊥ ☉
16 03	☽ ∠ ♂
20 24	☽ # ♅
20 29	☽ ☍ ♅
20 50	☽ Q ♄
22 02	☽ # ♆
23 24	☽ ⊻ ♀
16 Thursday	
01 01	☽ ⊻ ☉
03 16	☽ ∥ ♃
03 27	☽ ⊥ ☿
04 55	☽ ☌ ♃
06 10	♂ ⊥ ♆
09 02	☿ ⊻ ♇
11 01	☽ Q ♆
11 20	☽ ⊥ ♂
14 45	☽ ⊻ ☿
15 07	☽ ☌ ♇
17 Friday	
00 04	☽ ⊻ ♂
08 40	☽ □ ♄
10 50	☽ ⚹ ♆
11 49	♀ # ♂
18 27	☽ # ☿
23 50	☽ # ♂
18 Saturday	
01 41	☽ ∥ ♀
02 15	☉ ∠ ♆
06 24	♀ ∠ ♆
07 35	☽ ⊼ ♅
07 46	☽ ∥ ☉
09 54	♂ bQ ♅
12 24	☉ ⊻ ♃
13 29	☽ # ☉
14 51	♀ ⊻ ♃
16 05	☽ ∠ ♆
16 56	☽ ⊻ ♃
17 08	☽ ☌ ♀
17 15	☽ ☌ ☉
18 28	☽ # ♀
19 07	☽ ± ♅
21 44	☽ ☌ ☿
22 23	☉ ☌ ♀
22 57	☽ ∥ ♂
19 Sunday	
00 16	♂ ⊥ ♇
00 54	☽ ∥ ☿
01 47	☽ ⊻ ♇
04 39	☽ ⊥ ♃
09 20	☽ ⊥ ♆
12 15	☽ bQ ♅
12 38	♀ ± ♅
13 14	☽ ⊥ ♇
14 02	☽ ☌ ♂
15 19	☿ ∥ ♂
16 06	☉ ± ♅
18 32	☽ △ ♄
20 43	☽ ⊻ ♆
21 51	☿ ☌ ♀
22 00	☽ ∠ ♃
20 Monday	
00 56	☉ ☌ ☿
06 12	☽ ∠ ♇
11 13	☿ ± ♅
14 59	☽ # ♃
16 20	☽ △ ♅
18 12	♄ St R
21 32	☽ ∥ ♆
22 33	☽ bQ ♄
23 04	☽ ∥ ♅
21 Tuesday	
01 43	☿ ⊻ ♃
02 22	☽ ⊻ ☿
02 26	☽ ⚹ ♃
06 25	☽ ⊻ ☉
07 36	☽ ⊻ ♀
10 02	☽ ⚹ ♇
11 46	☉ # ♀
12 40	☽ ⊥ ☿
15 05	☽ ± ♄
18 17	☽ ⊥ ☉
19 43	☽ ⊥ ♀
22 Wednesday	
00 36	☿ ∠ ♆
00 45	☽ Q ♃
01 08	☽ ⊻ ♂
02 00	☽ ⊼ ♄
03 28	☽ # ♄
03 59	☽ ∠ ☿
04 14	☽ ☌ ♆
06 36	☽ # ♇
07 55	☽ Q ♇
09 44	♀ ⊻ ♇
11 55	☽ ∠ ☉
12 38	☽ ⊥ ♂
13 41	☽ ∠ ♀
16 13	♂ △ ♄
22 50	☽ □ ♅
23 Thursday	
05 14	☽ ⚹ ☿
05 43	☽ ∠ ♂
07 23	☉ ⊻ ♇
09 35	☽ □ ♃
10 09	☉ ∥ ☿
16 07	☽ □ ♇
16 48	☽ ⚹ ☉
19 07	☽ ⚹ ♀
24 Friday	
01 12	☽ Q ☿
04 37	☿ ∥ ♀
06 37	☽ Q ♅
07 25	☽ ☍ ♄
08 45	♂ ⊻ ♆
09 42	☽ ⊻ ♆
09 45	☽ ⚹ ♂
11 21	♀ ⊥ ♃
15 30	☽ Q ☉
18 15	☽ Q ♀
20 11	☽ ⊥ ♆
25 Saturday	
03 27	☽ ⚹ ♅
06 51	☽ □ ☿
07 49	☽ Q ♂
11 47	☽ ∠ ♆
14 48	☽ △ ♃
16 12	♀ ⊥ ♆
20 25	☽ △ ♇
26 Sunday	
00 56	☽ □ ☉
04 18	☽ □ ♀
05 10	☽ ∠ ♅
06 32	☉ ⊥ ♃
11 08	☽ ⊼ ♄
13 29	☽ ⚹ ♆
14 01	♀ bQ ♅
16 23	☽ □ ♂
16 49	☽ bQ ♃
20 26	☽ ⊥ ♅
21 18	☽ ± ♄
21 59	☽ bQ ♇
27 Monday	
03 14	♂ ∠ ♃
06 33	☽ ⊻ ♅
07 38	☽ △ ☿
07 51	♀ ⊥ ♇
08 15	☽ ± ♃
09 49	☽ Q ♆
10 41	☉ ⊥ ♆
12 28	☽ bQ ♄
13 11	☽ ± ♇
18 29	☽ ⊼ ♃
23 16	☽ ⊼ ♇
28 Tuesday	
04 24	☽ # ♇
07 11	☽ # ♄
07 21	☽ △ ☉
07 51	☽ bQ ☿
11 30	☉ bQ ♅
11 40	☽ △ ♀
13 32	☽ △ ♄
14 07	☿ ⊼ ♅
16 00	☽ □ ♆
21 32	☽ △ ♂
22 21	☽ ± ☿
29 Wednesday	
08 02	☽ ⊼ ☿
08 28	☽ ∥ ♅
08 33	☽ ☌ ♅
09 06	♀ △ ♄
09 49	☽ ∥ ♆
10 08	☽ bQ ☉
12 28	☉ ⊥ ♇
14 56	☽ bQ ♀
18 35	☽ # ♃
21 10	☽ ☍ ♃
23 48	☽ bQ ♂
30 Thursday	
01 14	☽ ☍ ♇
02 08	☽ ± ☉
07 15	☽ ± ♀
12 04	☉ ∥ ♀
12 49	☽ ⊼ ☉
14 25	☽ ∥ ♂
15 07	♀ ⊻ ♆
15 16	☽ □ ♄
15 33	☽ ± ♂
17 53	☽ △ ♆
18 08	☽ ⊼ ♀
22 35	☽ ∥ ♀
22 45	☽ ∥ ☉
31 Friday	
02 04	☽ ⊼ ♂
03 39	♂ ∠ ♇
05 31	☽ # ☿
08 45	☽ ☍ ☿
10 15	☽ ⊻ ♅
18 55	☽ bQ ♆
20 15	☽ ⊥ ♅
21 12	☽ Q ♄
22 34	☽ ∥ ☿
23 48	☽ ⊼ ♃

All ephemeris data is given at 12.00 UT and the Moon's longitude is additionally given for 24.00 UT

APRIL 2045

LONGITUDES

Date	Sidereal time h m s	Sun ☉ ° ' "	Moon ☽ ° ' "	Moon ☽ 24.00 ° ' "	Mercury ☿ ° '	Venus ♀ ° '	Mars ♂ ° '	Jupiter ♃ ° '	Saturn ♄ ° '	Uranus ♅ ° '	Neptune ♆ ° '	Pluto ♇ ° '
01	00 41 01	12 ♈ 10 05	08 ♎ 28 41	15 ♎ 33 06	22 ♓ 02	15 ♈ 36	19 ♈ 17	01 ♓ 20	11 ♐ 40	23 ♌ 06	13 ♉ 20	03 ♓ 18
02	00 44 58	13 09 15	22 ♎ 33 32	29 ♎ 29 26	21 R 59	16 50	20 03	01 32	11 R 39	23 R 04	13 22	03 19
03	00 48 54	14 08 24	06 ♏ 20 22	13 ♏ 05 59	22 D 01	18 05	20 48	01 45	11 38	23 03	13 24	03 20
04	00 52 51	15 07 30	19 ♏ 46 04	26 ♏ 20 31	22 08	19 19	21 34	01 57	11 36	23 02	13 26	03 22
05	00 56 47	16 06 35	02 ♐ 49 22	09 ♐ 12 44	22 21	20 34	22 19	02 10	11 35	23 00	13 28	03 23
06	01 00 44	17 05 38	15 ♐ 30 55	21 ♐ 44 13	22 39	21 48	23 05	02 22	11 33	22 59	13 30	03 24
07	01 04 40	18 04 39	27 ♐ 53 05	03 ♑ 58 01	23 01	23 02	23 50	02 34	11 32	22 58	13 32	03 26
08	01 08 37	19 03 38	09 ♑ 59 35	15 ♑ 58 22	23 29	24 17	24 36	02 46	11 30	22 57	13 35	03 27
09	01 12 34	20 02 36	21 ♑ 55 00	27 ♑ 50 10	24 00	25 31	25 21	02 58	11 28	22 56	13 37	03 28
10	01 16 30	21 01 32	03 ♒ 44 30	09 ♒ 38 40	24 36	26 45	26 06	03 10	11 26	22 55	13 39	03 29
11	01 20 27	22 00 26	15 ♒ 33 20	21 ♒ 29 07	25 15	28 00	26 51	03 21	11 24	22 54	13 41	03 31
12	01 24 23	22 59 18	27 ♒ 26 38	03 ♓ 26 27	25 59	29 ♈ 14	27 36	03 33	11 22	22 53	13 43	03 32
13	01 28 20	23 58 08	09 ♓ 29 06	15 ♓ 35 02	26 45	00 ♉ 28	28 22	03 45	11 20	22 52	13 45	03 33
14	01 32 16	24 56 57	21 ♓ 44 40	27 ♓ 58 20	27 36	01 42	29 07	03 56	11 17	22 51	13 47	03 34
15	01 36 13	25 55 43	04 ♈ 16 16	10 ♈ 38 40	28 29	02 57	29 ♈ 52	04 08	11 15	22 50	13 49	03 35
16	01 40 09	26 54 28	17 ♈ 05 35	23 ♈ 37 01	29 ♓ 26	04 11	00 ♉ 37	04 19	11 12	22 49	13 52	03 37
17	01 44 06	27 53 11	00 ♉ 12 51	06 ♉ 52 56	00 ♈ 25	05 25	01 22	04 30	11 10	22 49	13 54	03 38
18	01 48 03	28 51 52	13 ♉ 36 58	20 ♉ 24 39	01 27	06 39	02 06	04 42	11 07	22 48	13 56	03 39
19	01 51 59	29 ♈ 50 31	27 ♉ 15 36	04 ♊ 09 25	02 32	07 53	02 51	04 53	11 04	22 48	13 58	03 40
20	01 55 56	00 ♉ 49 09	11 ♊ 05 41	18 ♊ 03 59	03 40	09 08	03 36	05 04	11 02	22 47	14 00	03 41
21	01 59 52	01 47 44	25 ♊ 03 56	02 ♋ 05 11	04 49	10 22	04 21	05 15	10 59	22 47	14 03	03 42
22	02 03 49	02 46 16	09 ♋ 07 23	16 ♋ 10 17	06 02	11 36	05 05	05 25	10 56	22 46	14 05	03 43
23	02 07 45	03 44 47	23 ♋ 13 40	00 ♌ 17 19	07 16	12 50	05 50	05 36	10 53	22 46	14 07	03 44
24	02 11 42	04 43 15	07 ♌ 21 07	14 ♌ 24 54	08 33	14 04	06 35	05 47	10 50	22 45	14 09	03 45
25	02 15 38	05 41 42	21 ♌ 28 34	28 ♌ 31 58	09 52	15 18	07 19	05 57	10 47	22 45	14 12	03 46
26	02 19 35	06 40 05	05 ♍ 34 58	12 ♍ 37 22	11 13	16 32	08 03	06 07	10 43	22 45	14 14	03 47
27	02 23 32	07 38 27	19 ♍ 38 56	26 ♍ 39 26	12 36	17 46	08 48	06 18	10 40	22 45	14 16	03 48
28	02 27 28	08 36 47	03 ♎ 38 30	10 ♎ 35 49	14 01	19 00	09 32	06 28	10 37	22 45	14 18	03 49
29	02 31 25	09 35 05	17 ♎ 30 58	24 ♎ 23 33	15 29	20 14	10 16	06 38	10 33	22 45	14 21	03 49
30	02 35 21	10 ♉ 33 20	01 ♏ 13 10	07 ♏ 59 23	16 ♈ 58	21 ♉ 28	11 ♉ 01	06 ♓ 48	10 ♐ 30	22 ♌ 45	14 ♉ 23	03 ♓ 50

Date	Moon True ☊ ° '	Moon Mean ☊ ° '	Moon ☽ Latitude ° '
01	01 ♓ 15	29 ♒ 53	03 S 08
02	01 R 09	29 50	04 02
03	01 02	29 46	04 41
04	00 54	29 43	05 03
05	00 47	29 40	05 09
06	00 41	29 37	04 59
07	00 37	29 34	04 35
08	00 35	29 30	03 59
09	00 D 34	29 27	03 13
10	00 35	29 24	02 20
11	00 36	29 21	01 20
12	00 37	29 18	00 S 17
13	00 R 37	29 15	00 N 48
14	00 35	29 11	01 51
15	00 30	29 08	02 50
16	00 23	29 05	03 42
17	00 14	29 02	04 24
18	00 04	28 59	04 52
19	29 ♒ 54	28 55	05 04
20	29 45	28 52	04 58
21	29 38	28 49	04 35
22	29 33	28 46	03 55
23	29 31	28 43	03 01
24	29 D 31	28 40	01 55
25	29 31	28 36	00 N 43
26	29 R 31	28 33	00 S 32
27	29 30	28 30	01 45
28	29 26	28 27	02 51
29	29 19	28 24	03 46
30	29 ♒ 10	28 ♒ 21	04 S 27

DECLINATIONS

Date	Sun ☉ ° '	Moon ☽ ° '	Mercury ☿ ° '	Venus ♀ ° '	Mars ♂ ° '	Jupiter ♃ ° '	Saturn ♄ ° '	Uranus ♅ ° '	Neptune ♆ ° '	Pluto ♇ ° '
01	04 N 49	06 S 14	02 S 41	05 N 02	07 N 07	11 S 43	20 S 16	14 N 32	14 N 12	20 S 49
02	05 12	12 31	02 56	05 32	07 25	11 39	20 16	14 33	14 13	20 49
03	05 35	18 03	03 09	06 01	07 42	11 35	20 15	14 33	14 13	20 49
04	05 57	22 32	03 19	06 31	08 00	11 31	20 15	14 33	14 14	20 48
05	06 20	25 46	03 26	07 00	08 17	11 26	20 15	14 34	14 15	20 48
06	06 43	27 36	03 31	07 30	08 35	11 22	20 15	14 34	14 15	20 48
07	07 05	28 00	03 34	07 59	08 52	11 18	20 14	14 34	14 16	20 47
08	07 28	27 02	03 34	08 28	09 09	11 14	20 14	14 35	14 16	20 47
09	07 50	24 50	03 32	08 56	09 27	11 10	20 13	14 35	14 17	20 47
10	08 12	21 35	03 27	09 25	09 44	11 05	20 13	14 35	14 18	20 47
11	08 34	17 27	03 20	09 53	10 00	11 01	20 13	14 36	14 19	20 47
12	08 56	12 38	03 12	10 21	10 17	10 57	20 12	14 36	14 19	20 46
13	09 18	07 17	03 01	10 49	10 34	10 53	20 12	14 36	14 19	20 46
14	09 39	01 S 34	02 48	11 16	10 51	10 49	20 12	14 37	14 20	20 46
15	10 01	04 N 18	02 34	11 44	11 07	10 45	20 11	14 37	14 21	20 46
16	10 22	10 08	02 18	12 11	11 23	10 41	20 11	14 37	14 21	20 46
17	10 43	15 40	01 59	12 37	11 40	10 37	20 10	14 37	14 22	20 46
18	11 04	20 34	01 40	13 04	11 56	10 33	20 10	14 37	14 23	20 45
19	11 25	24 28	01 18	13 30	12 12	10 30	20 09	14 38	14 23	20 45
20	11 45	27 01	00 55	13 56	12 28	10 26	20 09	14 38	14 24	20 45
21	12 06	27 55	00 31	14 21	12 43	10 22	20 08	14 38	14 25	20 45
22	12 26	27 02	00 S 05	14 46	12 59	10 18	20 08	14 38	14 25	20 45
23	12 46	24 24	00 N 22	15 11	13 14	10 14	20 07	14 38	14 26	20 45
24	13 06	20 17	00 50	15 35	13 30	10 11	20 07	14 38	14 27	20 45
25	13 25	15 01	01 20	15 59	13 45	10 07	20 06	14 38	14 28	20 44
26	13 44	08 58	01 51	16 23	14 00	10 03	20 06	14 38	14 29	20 44
27	14 03	02 N 29	02 24	16 46	14 15	10 00	20 05	14 38	14 29	20 44
28	14 22	04 S 04	02 57	17 08	14 29	09 56	20 05	14 38	14 30	20 44
29	14 41	10 21	03 31	17 31	14 44	09 53	20 04	14 38	14 31	20 44
30	14 N 59	16 S 04	04 N 07	17 N 52	14 N 58	09 S 49	20 S 04	14 N 38	14 N 31	20 S 44

ZODIAC SIGN ENTRIES

Date	h	m	Planets	
03	00	53	☽	♏
05	06	45	☽	♐
07	16	10	☽	♑
10	04	24	☽	♒
12	17	08	☽	♓
13	02	52	♀	♉
15	03	53	☽	♈
15	16	27	♂	♉
17	02	02	☿	♈
17	11	37	☽	♉
19	15	53	☉	♉
19	16	47	☽	♊
21	20	26	☽	♋
23	23	31	☽	♌
26	02	30	☽	♍
28	05	44	☽	♎
30	09	51	☽	♏

LATITUDES

Date	Mercury ☿ ° '	Venus ♀ ° '	Mars ♂ ° '	Jupiter ♃ ° '	Saturn ♄ ° '	Uranus ♅ ° '	Neptune ♆ ° '	Pluto ♇ ° '
01	00 N 31	01 S 12	00 S 28	00 S 47	01 N 56	00 N 46	01 S 43	11 S 19
04	00 S 13	01 08	00 26	00 47	01 56	00 45	01 43	11 20
07	00 52	01 03	00 25	00 48	01 57	00 45	01 43	11 21
10	01 25	00 58	00 23	00 48	01 57	00 45	01 43	11 22
13	01 53	00 53	00 21	00 49	01 57	00 45	01 43	11 22
16	02 15	00 47	00 19	00 49	01 58	00 45	01 43	11 23
19	02 31	00 41	00 17	00 50	01 58	00 45	01 43	11 24
22	02 42	00 34	00 15	00 50	01 58	00 45	01 42	11 25
25	02 48	00 27	00 13	00 51	01 58	00 45	01 42	11 26
28	02 48	00 20	00 11	00 52	01 58	00 45	01 42	11 27
31	02 S 43	00 S 13	00 S 09	00 S 52	01 N 59	00 N 45	01 S 42	11 S 28

DATA

Julian Date	2468072
Delta T	+79 seconds
Ayanamsa	24° 29' 29"
Synetic vernal point	04° ♓ 37' 29"
True obliquity of ecliptic	23° 26' 09"

MOON'S PHASES, APSIDES AND POSITIONS ☽

Date	h	m	Phase	Longitude ° '	Eclipse Indicator
01	18	43	○	12 ♎ 27	
09	07	52	☾	19 ♑ 52	
17	07	27	●	27 ♈ 42	
24	07	12	☽	04 ♌ 32	

Day	h	m		
10	13	51	Apogee	
24	22	32	Perigee	
07	06	51	Max dec	28° S 02'
14	18	27	0N	
21	12	08	Max dec	27° N 55'
27	21	05	0S	

LONGITUDES

Date	Chiron ⚷ ° '	Ceres ⚳ ° '	Pallas ⚴ ° '	Juno ⚵ ° '	Vesta ⚶ ° '	Black Moon Lilith ⚸ ° '
01	13 ♍ 54	12 ♊ 32	15 ♈ 30	01 ♊ 41	08 ♉ 24	04 ♒ 37
11	13 ♍ 19	16 ♊ 04	19 ♈ 33	07 ♊ 03	12 ♉ 44	05 ♒ 43
21	12 ♍ 54	19 ♊ 47	23 ♈ 41	12 ♊ 47	17 ♉ 06	06 ♒ 50
31	12 ♍ 38	23 ♊ 39	27 ♈ 53	18 ♊ 22	21 ♉ 28	07 ♒ 56

ASPECTARIAN

h m Aspects

01 Saturday
00 07 ☉ △ ♄
03 15 ☽ ⊼ ♇
06 29 ☽ # ☉
07 15 ☽ # ♀
10 02 ☽ ± ♃
10 04 ☽ ± ♆
11 21 ☽ ∠ ♅
12 35 ♃ Q ♆
13 23 ☽ ± ♇
15 22 ☽ # ♂
17 24 ☽ ⚹ ♄
18 43 ☽ ☍ ☉
20 15 ☽ ⊼ ♆

02 Sunday
01 15 ☽ ☍ ♀
01 32 ☽ ⚼ ♃
04 43 ☽ ⚼ ♇
05 06 ♀ ∠ ♃
07 26 ☽ ☍ ♂
08 35 ☽ ∥ ♃
11 01 ☽ ⊼ ☿
12 53 ☽ ⚹ ♅
15 10 ☿ St D
17 28 ☉ ⊻ ♆
18 59 ☽ # ♆
19 03 ☽ ∠ ♄
20 22 ☽ # ♅
21 23 ☽ ± ☿

03 Monday
03 49 ☽ △ ♃
06 43 ☽ △ ♇
09 44 ☽ Q ♅
10 45 ☽ ⊥ ♄
13 12 ☽ ⚼ ☿
17 08 ♀ ∠ ♇
21 22 ☽ ⊻ ♄
23 03 ☽ ∥ ♄

04 Tuesday
00 35 ☽ ☍ ♆
02 00 ☽ ∥ ♇
02 58 ☽ ⊼ ☉
11 06 ☽ ⊼ ♀
14 40 ☽ ± ☉
15 28 ☽ ⊼ ♂
16 22 ☽ △ ☿
17 55 ☽ □ ♅
23 11 ☽ ± ♀

05 Wednesday
03 08 ☽ ± ♂
08 33 ☽ ⚼ ☉
10 45 ☽ □ ♃
13 03 ☽ □ ♇
13 19 ☿ ⊻ ♂
17 40 ☽ ⚼ ♀
20 58 ☽ ⚼ ♂

06 Thursday
04 27 ☽ ☌ ♄
08 09 ☽ ⊼ ♆
08 58 ♂ △ ♅
15 17 ☽ △ ☉
19 42 ☽ ± ♆
20 13 ☉ ∠ ♃
21 29 ☽ Q ♃
23 23 ☽ Q ♇

07 Friday
01 28 ☽ △ ♀
02 11 ☽ □ ☿
02 24 ☽ △ ♅
03 34 ☽ △ ♂
08 40 ☿ ⊼ ♅
10 33 ♀ △ ♅
11 32 ☿ ⊻ ♀
13 18 ☽ ⚼ ♆
20 45 ☉ ∠ ♇
21 23 ☽ ⚹ ♃
22 57 ☽ ⚹ ♇

08 Saturday
07 55 ☽ ⚼ ♅
15 00 ☽ ⊻ ♄
15 06 ☽ Q ☿
19 12 ☽ △ ♆

09 Sunday
01 56 ☽ ± ♅
03 02 ☽ ⊥ ♄
03 36 ♀ ☌ ♂
03 53 ☽ ∠ ♃
05 01 ☽ ∠ ♇
07 52 ☽ □ ☉
14 02 ☽ ⊼ ♅
16 26 ☽ ⚹ ☿
19 26 ☽ □ ♂
20 09 ☽ □ ♀
21 12 ☽ ∠ ♄
22 24 ☽ ⊥ ♃
23 17 ☽ ⊥ ♇

10 Monday
05 53 ♀ ⚼ ♄
10 48 ☽ ⊻ ♃
11 29 ☽ ⊻ ♇
17 02 ☽ ∥ ♇
20 26 ☽ ∥ ♄
22 04 ♂ ⚼ ♄
23 43 ☽ Q ☉

11 Tuesday
00 35 ☽ ∠ ☿
03 35 ☽ ⚹ ♄
08 11 ☽ □ ♆
10 29 ☽ Q ♂
13 00 ☽ Q ♀
19 58 ☽ ⊥ ☿

12 Wednesday
02 14 ☽ ⚹ ☉
02 33 ☽ # ♅
02 49 ☽ ☍ ♅
03 49 ☽ Q ♄
03 55 ☽ # ♆
03 58 ♀ ∥ ♂
08 51 ☽ ⊻ ☿
09 00 ♃ ☌ ♇
09 20 ☉ △ ♅
12 21 ☽ ⚹ ♂
16 00 ☽ ⚹ ♀
19 49 ☽ ∥ ♃
20 35 ☽ Q ♆
21 37 ☽ # ♀
22 12 ☽ # ♂

13 Thursday
00 12 ☽ ☌ ♇
00 25 ☽ ☌ ♃
03 43 ☽ # ☉
10 53 ☽ ∠ ☉
15 20 ♀ # ♃
15 37 ☽ □ ♄
20 08 ☽ ∠ ♂
20 26 ☽ ⚹ ♆

14 Friday
01 06 ☽ ∠ ♀
06 06 ☽ ⊥ ☉
06 42 ☽ ∥ ☿
10 24 ♂ # ♃
14 08 ☽ ⊼ ♅
14 49 ☽ ⊥ ♂
18 43 ☽ ⊻ ☉
20 30 ☽ ⊥ ♀

15 Saturday
00 08 ☽ ☌ ☿
01 36 ☽ ∠ ♆
01 40 ☽ ± ♅
03 05 ☽ ⊻ ♂
05 13 ☽ # ☿
09 13 ☽ ⊻ ♀
10 42 ☽ ⊻ ♇
11 44 ☽ ⊻ ♃
18 43 ☽ ⚼ ♅
18 44 ☽ ⊥ ♆
19 31 ☉ ⚼ ♄
20 58 ☿ ± ♅
21 09 ☿ ∠ ♆
22 02 ☽ ⊥ ♇
23 12 ☽ ⊥ ♃

16 Sunday
00 42 ♀ ⚹ ♇
01 05 ☽ △ ♄
05 59 ☽ ⊻ ♆
13 02 ☽ ∥ ☉
14 17 ☽ # ♃
14 48 ☽ ∠ ♇
15 08 ♀ ⚹ ♃
16 10 ☽ ∠ ♃
17 33 ☽ ∥ ♂
21 24 ☽ ∥ ♀
22 32 ☽ △ ♅

17 Monday
04 40 ☽ ⚼ ♄
06 13 ☽ ∥ ♆
06 18 ☉ # ♃
07 13 ♀ ± ♄
07 18 ☽ ∥ ♅
07 27 ☽ ☌ ☉
12 24 ☽ ⊻ ☿
14 12 ☽ ☌ ♂
18 10 ☽ ⚹ ♇
19 51 ☽ ⚹ ♃
20 53 ☽ ± ♄
22 20 ☽ ☌ ♀

18 Tuesday
00 06 ☽ ⊥ ☿
07 35 ☽ ⊼ ♄
09 54 ☽ # ♄
12 34 ☽ ☌ ♆
13 03 ☽ # ♇
15 36 ☽ Q ♇
17 27 ☽ ∠ ☿
17 31 ☽ Q ♃

19 Wednesday
04 11 ☽ □ ♅
16 51 ☽ ⊻ ☉
21 59 ☽ ⚹ ☿
22 18 ☽ ⊻ ♂
23 09 ☽ □ ♇

20 Thursday
01 26 ☽ □ ♃
04 03 ☽ ⊥ ☉
08 16 ☽ ⊻ ♀
08 21 ☿ ⊻ ♂
09 16 ☽ ⊥ ♂
11 28 ☽ Q ♅
11 53 ☽ ☍ ♄
12 27 ☿ ⊻ ♇
14 41 ♂ ⚹ ♇
17 02 ☽ ⊻ ♆
19 37 ☽ ⊥ ♀
20 34 ☽ Q ☿
20 45 ☽ ∠ ☉

21 Friday
01 39 ☽ ∠ ♂
03 22 ☽ ⊥ ♆
08 05 ☽ ⚹ ♅
12 33 ☽ ∠ ♀
15 59 ♀ ∥ ♆
18 49 ☽ ∠ ♆
21 54 ☿ ⊻ ♃
23 34 ♀ ⊼ ♄

22 Saturday
00 22 ☽ ⚹ ☉
02 46 ☽ △ ♇
03 59 ♀ ∥ ♅
04 44 ☽ ⚹ ♂
05 37 ☽ △ ♃
06 14 ☽ □ ☿
07 13 ♂ ± ♄
09 42 ☽ ∠ ♅
15 04 ☽ ⊼ ♄
16 37 ☽ ⚹ ♀
20 28 ☽ ⚹ ♆
22 20 ☽ Q ☉

23 Sunday
01 01 ☽ ⊥ ♅
01 15 ☽ ± ♄
02 08 ♂ ⚹ ♃
02 19 ☽ Q ♂
04 21 ☽ ⚼ ♇
07 29 ☽ ⚼ ♃
11 13 ☽ ⊻ ♅
11 38 ☉ ⚹ ♇
14 59 ☽ Q ♀
16 30 ☽ ⚼ ♄
16 56 ☽ Q ♆
19 40 ☽ ± ♇
21 15 ☿ ⚼ ♅
22 58 ☽ ± ♃

24 Monday
04 25 ☿ ⊥ ♆
05 52 ☽ ⊼ ♇
07 12 ☽ □ ☉
09 17 ☽ ⊼ ♃
09 41 ☽ # ♇
10 36 ☽ □ ♂
12 52 ☽ # ♄
13 49 ♀ ☌ ♆
14 15 ☽ △ ☿
14 31 ☉ ± ♄
17 53 ☽ △ ♄
23 35 ☽ □ ♆

25 Tuesday
00 30 ☽ □ ♀
08 10 ☽ ∥ ♀
10 06 ☿ ⊥ ♇
13 36 ☽ ∥ ♅
14 10 ☽ ☌ ♅
14 18 ☽ ∥ ♆
17 02 ☽ ∥ ♂
18 15 ☽ ∥ ☉
18 22 ☽ ⚼ ☿
19 40 ☉ ⚹ ♃
21 10 ♀ Q ♇

26 Wednesday
03 35 ☿ △ ♄
07 47 ☽ # ♃
08 55 ☽ ☍ ♇
11 19 ☽ ± ☿
12 56 ☽ ☍ ♃
13 59 ☽ △ ☉
16 27 ☽ △ ♂
20 43 ☽ □ ♄
22 39 ☽ ⊼ ☿

27 Thursday
02 47 ☽ △ ♆
05 57 ☿ ⊥ ♃
08 28 ☽ △ ♀
12 19 ☽ ∥ ☿
17 18 ☽ ⊻ ♅
17 30 ☽ ⚼ ☉
19 30 ☽ ⚼ ♂
23 57 ♀ Q ♃

28 Friday
03 23 ☽ Q ♄
03 35 ☽ ⊥ ♅
04 31 ☽ ⚼ ♆
07 31 ☽ # ☿
10 06 ☽ ± ☉
11 49 ☽ ± ♂
12 17 ☽ ⊼ ♇
12 40 ☽ ⚼ ♀
12 43 ♂ ∥ ♆
16 50 ☿ ⊻ ♆
16 55 ☽ ⊼ ♃
19 04 ☽ ∠ ♅
20 04 ☽ ± ♆
21 13 ☽ ⊼ ☉
22 08 ☉ ∥ ♆
22 39 ☽ ± ♇
22 44 ☽ ⊼ ♂
23 58 ☽ ⚹ ♄

29 Saturday
02 13 ♂ ∥ ♅
03 25 ☽ ± ♃
05 44 ☽ ± ♀
06 28 ☽ ⊼ ♆
08 02 ☽ ☍ ☿
08 25 ☉ ∥ ♅
10 09 ☽ ∥ ♃
14 17 ☽ ⚼ ♇
17 11 ☽ ⊼ ♀
18 22 ♅ St D
19 15 ☽ ⚼ ♃
20 25 ♂ ⊼ ♄
21 07 ☽ ⚹ ♅

30 Sunday
01 58 ☽ ∠ ♄
05 11 ☽ # ♆
05 41 ☽ # ♅
06 57 ☽ # ☉
06 57 ☽ # ♂
07 21 ☉ ∥ ♂
10 35 ☉ ⊼ ♄
16 38 ☽ △ ♇
17 46 ☽ ⊥ ♄
18 14 ☽ Q ♅
21 04 ☽ # ♀
22 00 ☽ △ ♃

All ephemeris data is given at 12.00 UT and the Moon's longitude is additionally given for 24.00 UT

MAY 2045

LONGITUDES

Date	Sidereal time h m s	Sun ☉ ° ' "	Moon ☽ ° ' "	Moon ☽ 24.00 ° ' "	Mercury ☿ ° '	Venus ♀ ° '	Mars ♂ ° '	Jupiter ♃ ° '	Saturn ♄ ° '	Uranus ♅ ° '	Neptune ♆ ° '	Pluto ♇ ° '
01	02 39 18	11 ♉ 31 35	14 ♏ 41 52	21 ♏ 20 17	18 ♈ 29	22 ♉ 42	11 ♉ 45	06 ♓ 58	10 ♐ 26	22 ♌ 45	14 ♉ 25	03 ♓ 51
02	02 43 14	12 29 47	27 ♏ 54 24	04 ♐ 24 02	20 02	23 55	12 29	07 07	10 R 23	22 D 45	14 27	03 52
03	02 47 11	13 27 58	10 ♐ 49 07	17 ♐ 09 39	21 37	25 09	13 13	07 17	10 19	22 45	14 30	03 53
04	02 51 07	14 26 07	23 ♐ 25 46	29 ♐ 37 38	23 13	26 23	13 57	07 26	10 15	22 45	14 32	03 53
05	02 55 04	15 24 15	05 ♑ 45 35	11 ♑ 49 57	24 52	27 37	14 41	07 36	10 11	22 46	14 34	03 54
06	02 59 01	16 22 21	17 ♑ 51 13	23 ♑ 49 52	26 33	28 ♉ 51	15 25	07 45	10 07	22 46	14 36	03 55
07	03 02 57	17 20 25	29 ♑ 46 28	05 ♒ 41 38	28 ♈ 15	00 ♊ 04	16 09	07 54	10 04	22 46	14 39	03 55
08	03 06 54	18 18 29	11 ♒ 36 01	17 ♒ 30 16	00 ♉ 00	01 18	16 53	08 03	10 00	22 47	14 41	03 56
09	03 10 50	19 16 30	23 ♒ 25 06	29 ♒ 21 10	01 46	02 32	17 37	08 12	09 56	22 47	14 43	03 57
10	03 14 47	20 14 31	05 ♓ 19 11	11 ♓ 19 49	03 34	03 46	18 20	08 21	09 52	22 48	14 45	03 57
11	03 18 43	21 12 30	17 ♓ 23 41	23 ♓ 31 23	05 24	04 59	19 04	08 29	09 48	22 48	14 48	03 58
12	03 22 40	22 10 28	29 ♓ 43 29	06 ♈ 00 26	07 16	06 13	19 48	08 38	09 43	22 49	14 50	03 58
13	03 26 36	23 08 24	12 ♈ 22 36	18 ♈ 50 18	09 10	07 27	20 31	08 46	09 39	22 50	14 52	03 59
14	03 30 33	24 06 19	25 ♈ 23 41	02 ♉ 02 47	11 06	08 40	21 15	08 54	09 35	22 51	14 54	04 00
15	03 34 30	25 04 13	08 ♉ 47 31	15 ♉ 37 39	13 04	09 54	21 58	09 02	09 31	22 51	14 57	04 00
16	03 38 26	26 02 05	22 ♉ 32 49	29 ♉ 32 32	15 03	11 07	22 42	09 10	09 27	22 52	14 59	04 00
17	03 42 23	26 59 56	06 ♊ 36 12	13 ♊ 43 09	17 05	12 21	23 25	09 18	09 22	22 53	15 01	04 01
18	03 46 19	27 57 46	20 ♊ 52 39	28 ♊ 03 57	19 08	13 35	24 08	09 25	09 18	22 54	15 03	04 01
19	03 50 16	28 55 34	05 ♋ 16 18	12 ♋ 29 00	21 13	14 48	24 52	09 33	09 14	22 55	15 05	04 02
20	03 54 12	29 ♉ 53 21	19 ♋ 41 26	26 ♋ 53 02	23 19	16 02	25 35	09 40	09 09	22 56	15 08	04 02
21	03 58 09	00 ♊ 51 06	04 ♌ 03 20	11 ♌ 11 58	25 27	17 15	26 18	09 47	09 05	22 57	15 10	04 02
22	04 02 05	01 48 49	18 ♌ 18 41	25 ♌ 23 16	27 35	18 29	27 01	09 55	09 01	22 59	15 12	04 03
23	04 06 02	02 46 30	02 ♍ 25 37	09 ♍ 25 40	29 ♉ 45	19 42	27 44	10 01	08 56	23 00	15 14	04 03
24	04 09 59	03 44 10	16 ♍ 23 22	23 ♍ 18 43	01 ♊ 56	20 55	28 27	10 08	08 52	23 01	15 16	04 03
25	04 13 55	04 41 49	00 ♎ 11 43	07 ♎ 02 19	04 08	22 09	29 10	10 15	08 47	23 02	15 19	04 04
26	04 17 52	05 39 25	13 ♎ 50 30	20 ♎ 36 13	06 19	23 22	29 ♉ 53	10 21	08 43	23 04	15 21	04 04
27	04 21 48	06 37 01	27 ♎ 19 20	03 ♏ 59 47	08 31	24 36	00 ♊ 36	10 27	08 38	23 05	15 23	04 04
28	04 25 45	07 34 35	10 ♏ 37 23	17 ♏ 12 01	10 43	25 49	01 18	10 34	08 34	23 07	15 25	04 04
29	04 29 41	08 32 08	23 ♏ 43 32	00 ♐ 11 47	12 54	27 02	02 01	10 40	08 29	23 08	15 27	04 04
30	04 33 38	09 29 39	06 ♐ 36 39	12 ♐ 58 03	15 05	28 15	02 44	10 45	08 25	23 10	15 29	04 04
31	04 37 34	10 ♊ 27 10	19 ♐ 15 57	25 ♐ 30 21	17 ♊ 14	29 ♊ 29	03 ♊ 26	10 ♓ 51	08 ♐ 21	23 ♌ 11	15 ♉ 31	04 ♓ 04

Date	Moon True ☊ ° '	Moon Mean ☊ ° '	Moon ☽ Latitude ° '
01	28 ♒ 59	28 ♒ 17	04 S 52
02	28 R 47	28 14	05 01
03	28 35	28 11	04 55
04	28 25	28 08	04 34
05	28 17	28 05	04 00
06	28 11	28 01	03 16
07	28 08	27 58	02 24
08	28 07	27 55	01 26
09	28 D 07	27 52	00 S 25
10	28 R 07	27 49	00 N 38
11	28 05	27 46	01 40
12	28 02	27 42	02 39
13	27 57	27 39	03 31
14	27 48	27 36	04 14
15	27 38	27 33	04 45
16	27 26	27 30	04 59
17	27 14	27 27	04 56
18	27 03	27 23	04 35
19	26 54	27 20	03 56
20	26 48	27 17	03 02
21	26 45	27 14	01 57
22	26 44	27 11	00 N 44
23	26 D 44	27 07	00 S 30
24	26 R 44	27 04	01 42
25	26 42	27 01	02 47
26	26 38	26 58	03 42
27	26 31	26 55	04 24
28	26 21	26 52	04 50
29	26 10	26 48	05 01
30	25 57	26 45	04 56
31	25 ♒ 45	26 ♒ 42	04 S 36

DECLINATIONS

Date	Sun ☉ ° '	Moon ☽ ° '	Mercury ☿ ° '	Venus ♀ ° '	Mars ♂ ° '	Jupiter ♃ ° '	Saturn ♄ ° '	Uranus ♅ ° '	Neptune ♆ ° '	Pluto ♇ ° '
01	15 N 17	20 S 54	04 N 44	18 N 14	15 N 13	09 S 46	20 S 03	14 N 38	14 N 32	20 S 44
02	15 35	24 35	05 21	18 35	15 27	09 42	20 03	14 38	14 33	20 44
03	15 53	26 56	05 59	18 55	15 41	09 39	20 02	14 38	14 33	20 44
04	16 10	27 50	06 39	19 15	15 55	09 36	20 02	14 38	14 34	20 44
05	16 27	27 19	07 19	19 34	16 08	09 32	20 01	14 38	14 34	20 44
06	16 44	25 29	08 00	19 53	16 22	09 29	20 00	14 37	14 35	20 44
07	17 00	22 33	08 41	20 11	16 35	09 26	20 00	14 37	14 36	20 44
08	17 17	18 41	09 23	20 29	16 48	09 23	19 59	14 37	14 36	20 44
09	17 33	14 06	10 06	20 46	17 01	09 20	19 59	14 37	14 37	20 44
10	17 48	08 58	10 49	21 03	17 14	09 17	19 58	14 37	14 38	20 44
11	18 04	03 S 26	11 33	21 19	17 27	09 14	19 57	14 36	14 38	20 45
12	18 19	02 N 19	12 17	21 34	17 39	09 11	19 57	14 36	14 39	20 45
13	18 33	08 08	13 01	21 49	17 51	09 08	19 56	14 36	14 40	20 45
14	18 48	13 46	13 46	22 03	18 03	09 05	19 56	14 36	14 40	20 45
15	19 02	18 55	14 30	22 17	18 15	09 02	19 55	14 35	14 41	20 45
16	19 16	23 13	15 14	22 30	18 27	08 59	19 54	14 35	14 42	20 45
17	19 29	26 16	15 58	22 42	18 39	08 57	19 54	14 35	14 42	20 45
18	19 42	27 42	16 42	22 54	18 50	08 54	19 53	14 34	14 43	20 45
19	19 55	27 16	17 25	23 05	19 01	08 51	19 52	14 34	14 44	20 46
20	20 07	25 00	18 07	23 16	19 12	08 49	19 52	14 34	14 44	20 46
21	20 20	21 08	18 49	23 25	19 23	08 46	19 51	14 33	14 45	20 46
22	20 31	16 03	19 29	23 34	19 34	08 44	19 50	14 33	14 45	20 46
23	20 43	10 08	20 08	23 43	19 44	08 42	19 50	14 32	14 46	20 46
24	20 54	03 N 48	20 45	23 50	19 54	08 39	19 49	14 32	14 47	20 47
25	21 04	02 S 38	21 21	23 57	20 04	08 37	19 49	14 31	14 47	20 47
26	21 15	08 52	21 54	24 04	20 14	08 35	19 48	14 31	14 48	20 47
27	21 25	14 36	22 26	24 09	20 24	08 33	19 47	14 30	14 48	20 47
28	21 34	19 35	22 55	24 14	20 33	08 31	19 47	14 30	14 49	20 48
29	21 43	23 33	23 22	24 19	20 42	08 29	19 46	14 29	14 50	20 48
30	21 52	26 16	23 46	24 22	20 51	08 27	19 45	14 29	14 50	20 48
31	22 N 01	27 S 36	24 N 08	24 N 25	21 N 00	08 S 25	19 S 45	14 N 28	14 N 51	20 S 48

ZODIAC SIGN ENTRIES

Date	h	m	Planets
02	15	51	☽ ♐
05	00	44	☽ ♑
07	10	34	♀ ♊
07	12	27	☽ ♒
08	12	07	☿ ♉
10	01	18	☽ ♓
12	12	32	☽ ♈
14	20	20	☽ ♉
17	00	47	☽ ♊
19	03	13	☽ ♋
20	14	46	☉ ♊
21	05	13	☽ ♌
23	07	51	☽ ♍
23	14	42	☿ ♊
25	11	40	☽ ♎
26	16	01	♂ ♊
27	16	48	☽ ♏
29	23	38	☽ ♐
31	22	15	♀ ♋

LATITUDES

Date	Mercury ☿ ° '	Venus ♀ ° '	Mars ♂ ° '	Jupiter ♃ ° '	Saturn ♄ ° '	Uranus ♅ ° '	Neptune ♆ ° '	Pluto ♇ ° '
01	02 S 43	00 S 13	00 S 09	00 S 52	01 N 59	00 N 45	01 S 42	11 S 28
04	02 33	00 S 06	00 07	00 53	01 59	00 44	01 42	11 29
07	02 19	00 N 01	00 05	00 53	01 59	00 44	01 42	11 29
10	02 00	00 09	00 03	00 54	01 59	00 44	01 42	11 30
13	01 37	00 16	00 S 01	00 55	01 59	00 44	01 42	11 31
16	01 10	00 24	00 N 01	00 56	01 59	00 44	01 42	11 32
19	00 40	00 31	00 02	00 56	01 59	00 44	01 42	11 33
22	00 S 09	00 38	00 04	00 57	01 59	00 44	01 43	11 34
25	00 N 23	00 45	00 06	00 58	01 59	00 44	01 43	11 35
28	00 53	00 52	00 08	00 59	01 58	00 44	01 43	11 36
31	01 N 19	00 N 59	00 N 10	00 S 59	01 N 58	00 N 43	01 S 43	11 S 37

DATA

Julian Date	2468102
Delta T	+79 seconds
Ayanamsa	24° 29' 33"
Synetic vernal point	04° ♓ 37' 26"
True obliquity of ecliptic	23° 26' 08"

MOON'S PHASES, APSIDES AND POSITIONS ☽

Date	h	m	Phase	Longitude ° '	Eclipse Indicator
01	05	52	○	11 ♏ 17	
09	02	51	☾	18 ♒ 54	
16	18	27	●	26 ♉ 18	
23	12	38	☽	02 ♍ 48	
30	17	52	○	09 ♐ 44	

Day	h	m		
08	09	23	Apogee	
20	10	18	Perigee	
04	15	02	Max dec	27° S 50'
12	02	24	0N	
18	18	33	Max dec	27° N 46'
25	02	09	0S	
31	22	15	Max dec	27° S 43'

LONGITUDES

Date	Chiron ⚷ ° '	Ceres ⚳ ° '	Pallas ⚴ ° '	Juno ⚵ ° '	Vesta ⚶ ° '	Black Moon Lilith ⚸ ° '
01	12 ♍ 38	23 ♊ 39	27 ♈ 53	18 ♊ 22	21 ♉ 28	07 ♒ 56
11	12 ♍ 34	27 ♊ 38	02 ♉ 08	23 ♊ 56	25 ♉ 51	09 ♒ 03
21	12 ♍ 40	01 ♋ 44	06 ♉ 27	29 ♊ 28	00 ♊ 14	10 ♒ 09
31	12 ♍ 58	05 ♋ 56	10 ♉ 48	04 ♋ 59	04 ♊ 36	11 ♒ 16

ASPECTARIAN

h m	Aspects
01 Monday	
04 24	☽ ⊻ ♄
05 52	☽ ☍ ☉
06 24	☽ ☍ ♂
07 26	☽ ‖ ♄
11 07	☽ ‖ ♇
11 30	☽ ☍ ♆
13 04	♀ □ ♅
17 52	☿ ∠ ♇
19 43	☽ ⊼ ☿
02 Tuesday	
02 34	☽ □ ♅
03 57	☽ ☍ ♀
08 05	☽ ± ☿
10 42	☉ ☌ ♂
23 01	☽ □ ♇
03 Wednesday	
03 00	☽ bQ ☿
05 17	☽ □ ♃
11 03	☽ ☌ ♄
16 48	☽ ⊼ ♂
17 24	☽ ⊼ ☉
18 57	☽ ⊼ ♆
23 09	☿ ∠ ♃
04 Thursday	
04 55	☽ ± ♂
05 06	☿ △ ♅
05 46	☽ ± ☉
06 25	☽ ± ♆
09 02	☽ Q ♇
10 42	☽ △ ♅
11 32	☽ △ ☿
14 27	☉ ☌ ♆
15 55	☽ Q ♃
18 20	☽ ⊼ ♀
23 22	☽ bQ ♂
23 51	☽ bQ ♆
05 Friday	
00 37	☽ bQ ☉
07 19	☽ ± ♀
07 57	♂ ☌ ♆
08 21	☽ ✶ ♇
15 40	☽ ✶ ♃
15 57	☽ bQ ♅
16 29	☿ bQ ♄
20 42	☽ ⊻ ♄
06 Saturday	
00 28	☉ Q ♇
03 05	☽ bQ ♀
05 30	☽ △ ♆
06 49	☽ △ ♂
08 34	☽ ⊥ ♄
08 47	☽ △ ☉
09 49	☽ ± ♅
14 07	☽ ∠ ♇
21 22	♀ # ♄
21 52	☽ ⊼ ♅
21 57	☽ ∠ ♃
07 Sunday	
02 32	☽ ∠ ♄
04 31	♂ Q ♇
08 15	☽ ⊥ ♇
08 24	☽ □ ☿
12 40	☽ △ ♀
16 22	☽ ⊥ ♃
20 25	☽ ⊻ ♇
23 52	☽ ‖ ♇
08 Monday	
02 11	☽ # ♀
04 26	☽ ‖ ♄
04 41	☽ ⊻ ♃
08 45	☽ ✶ ♄
11 41	☿ # ♃
18 17	☽ □ ♆
19 19	☽ # ☉
21 48	☽ # ♂
23 26	☽ □ ♂
09 Tuesday	
02 51	☽ □ ☉
03 16	☽ Q ☿
05 02	♅ ‖ ♆
09 00	☽ Q ♄
09 27	☽ # ♆
09 28	☽ # ♅
09 34	♀ # ♇
10 43	☽ ☍ ♅
10 Wednesday	
04 36	☽ # ☿
06 50	☽ Q ♆
07 51	☽ ✶ ☿
08 31	☽ □ ♀
09 16	☽ ☌ ♇
10 36	☽ ‖ ♃
14 10	☽ Q ♂
14 55	☉ Q ♃
15 43	☿ ± ♄
15 52	♀ □ ♇
17 08	☿ ✶ ♇
18 07	☽ ☌ ♃
18 21	☽ Q ☉
19 42	☿ ⊻ ♀
21 01	☽ □ ♄
11 Thursday	
06 51	☽ ✶ ♆
15 30	☽ ✶ ♂
18 58	☽ ∠ ☿
20 07	☽ ✶ ☉
22 37	☽ ⊼ ♅
12 Friday	
00 10	☽ Q ♀
10 15	☽ ± ♅
12 12	☽ ∠ ♆
15 30	☽ ⊥ ☿
20 08	☽ ⊻ ♇
22 17	☽ ∠ ♂
13 Saturday	
01 44	☽ ✶ ♀
03 23	☽ ∠ ☉
03 27	☽ bQ ♅
04 11	☉ □ ♅
04 55	☽ ⊻ ☿
05 09	☽ ⊻ ♃
05 23	☽ ⊥ ♆
06 30	☿ ✶ ♃
06 55	☽ △ ♄
07 30	☽ ⊥ ♇
16 08	☽ # ♃
16 14	☽ ⊥ ♂
16 30	☽ ⊥ ♃
16 40	☽ ⊻ ♆
17 50	☿ ⊼ ♄
21 35	☽ ⊥ ☉
22 02	♂ Q ♃
14 Sunday	
00 17	☽ ∠ ♇
03 59	☽ ⊻ ♂
07 21	☽ △ ♅
08 32	☽ ∠ ♀
09 15	☽ ∠ ♃
09 28	☽ ⊻ ☉
10 32	☽ bQ ♄
11 59	☽ ‖ ☿
15 41	☽ ‖ ♅
16 03	☽ ‖ ♆
17 05	♀ □ ♃
15 Monday	
02 26	☽ ⊥ ♀
02 40	☽ ± ♄
03 29	☽ ✶ ♇
04 55	♀ ☍ ♄
08 36	☽ ‖ ♂
12 26	☽ ✶ ♃
12 36	☽ ‖ ☉
13 16	☽ ⊼ ♄
14 09	☽ ⊻ ♀
14 52	☿ ‖ ♅
17 05	☽ # ♄
18 03	☿ ‖ ♆
20 47	☽ ☌ ☿
21 32	☽ # ♇
22 50	☽ ☌ ♆
16 Tuesday	
00 39	☽ Q ♇
06 59	♀ Q ♅
07 15	☽ ‖ ♀
09 36	☽ Q ♃
11 03	☿ ☌ ♆
12 16	☽ ☌ ♂
12 34	☽ □ ♅
18 02	♂ □ ♅
18 27	☽ ☌ ☉
23 22	☿ Q ♇
17 Wednesday	
07 37	☽ □ ♇
16 36	☽ □ ♃
16 39	☽ ☍ ♄
19 14	☽ Q ♅
21 01	♃ □ ♄
22 37	☽ ☌ ♀
18 Thursday	
02 13	☽ ⊻ ♆
08 35	☽ ⊻ ☿
12 18	☽ ⊥ ♆
15 23	☽ ✶ ♅
17 44	☽ ⊻ ♂
20 18	☽ ⊥ ☿
19 Friday	
00 41	☽ ⊻ ☉
03 21	☽ ∠ ♆
04 16	☽ ⊥ ♂
07 18	☉ # ♄
09 56	☽ △ ♇
11 23	☽ ⊥ ☉
13 50	☽ ∠ ☿
16 08	☿ Q ♃
16 25	☽ ∠ ♅
17 50	♀ ⊻ ♆
18 33	☽ ⊼ ♄
19 11	☽ △ ♃
20 02	☽ ∠ ♂
20 Saturday	
03 26	☽ ∠ ☉
04 23	☽ ✶ ♆
04 29	☽ ± ♄
05 20	☽ ⊻ ♀
07 25	☽ ⊥ ♅
07 41	☿ □ ♅
10 54	☽ bQ ♇
16 15	☽ ⊥ ♀
17 25	☽ ⊻ ♅
19 05	☽ ✶ ☿
19 24	☽ bQ ♄
20 23	☽ bQ ♃
22 20	☽ ✶ ♂
23 24	☽ ‖ ♀
21 Sunday	
00 26	☽ Q ♆
01 56	☽ ± ♇
06 15	☽ ✶ ☉
08 42	☽ ∠ ♀
11 33	☽ ± ♃
11 58	☽ ⊼ ♇
13 54	☽ # ♇
15 59	☽ ‖ ☉
18 32	☽ # ♄
18 41	☽ Q ☿
19 30	☽ Q ♂
20 24	☽ △ ♄
20 30	☽ ‖ ♂
21 43	☽ ⊼ ♃
22 11	☽ ‖ ☿
22 Monday	
02 25	☿ ☌ ♂
03 51	☽ Q ☉
06 44	☽ □ ♆
12 18	☽ ✶ ♀
16 02	☿ ‖ ♂
17 27	☽ ‖ ♆
18 21	☽ ‖ ♅
19 55	☽ ☌ ♅
23 Tuesday	
01 06	☿ # ♄
03 34	☽ □ ♂
06 36	☽ □ ☿
10 38	☽ Q ♀
12 38	☽ □ ☉
14 47	☽ ☍ ♇
17 37	☽ # ♃
20 06	☉ # ♇
23 06	☽ □ ♄
24 Wednesday	
00 49	♂ # ♄
01 08	☽ ☍ ♃
10 04	☽ △ ♆
13 04	☿ # ♇
19 04	♀ ⊥ ♆
20 00	☉ □ ♇
20 16	☉ ‖ ☿
20 37	☽ □ ♀
23 30	☽ ⊻ ♅
25 Thursday	
06 05	☽ Q ♄
09 59	☽ ⊥ ♅
10 06	☽ △ ♂
11 16	☿ □ ♇
12 12	☽ bQ ♆
18 46	☽ ⊼ ♇
20 12	☽ △ ☿
20 29	☽ △ ☉
23 06	☉ ☌ ☿
26 Friday	
01 47	☽ ∠ ♅
03 00	☽ ✶ ♄
04 02	☽ ± ♆
05 19	☽ ± ♇
05 47	☽ ⊼ ♃
05 52	♀ ✶ ♅
10 53	☽ ‖ ♃
13 56	☽ bQ ♂
14 40	☽ ⊼ ♆
16 29	☽ ± ♃
21 16	☽ bQ ♇
27 Saturday	
01 01	☽ bQ ☉
03 52	☽ bQ ☿
04 25	☽ ✶ ♅
05 27	☽ ∠ ♄
06 38	☽ △ ♀
06 51	☽ ± ♂
08 38	☽ bQ ♃
11 33	☽ # ♅
12 53	☽ # ♆
13 16	☿ ☍ ♄
18 12	☽ ⊼ ♂
18 22	☽ ± ☉
21 30	☽ ⊥ ♄
23 11	☽ ± ☿
28 Sunday	
00 08	☽ △ ♇
02 00	☽ Q ♅
06 02	☽ ⊼ ☉
08 17	☽ ⊻ ♄
10 14	☿ □ ♃
11 53	☽ △ ♃
12 12	☽ ⊼ ☿
12 23	☽ bQ ♀
13 00	☽ ‖ ♄
16 24	☿ Q ♅
17 27	☽ # ♂
18 37	☽ ‖ ♇
20 46	☽ ☍ ♆
23 36	☽ # ☉
29 Monday	
06 32	☽ ± ♀
10 31	☽ # ☿
10 55	☽ □ ♅
10 58	☉ ☍ ♄
17 47	☽ # ♀
18 46	☽ ⊼ ♀
30 Tuesday	
02 38	♂ # ♇
04 18	☽ ☍ ♂
07 14	☽ □ ♇
15 23	☽ ☌ ♄
16 36	☿ ⊻ ♆
17 52	☽ ☍ ☉
19 52	☽ □ ♃
31 Wednesday	
04 50	☽ ⊼ ♆
07 19	☽ ☍ ☿
16 20	☽ ± ♆
17 23	☽ Q ♇
19 33	☽ △ ♅
23 01	☉ □ ♃

All ephemeris data is given at 12.00 UT and the Moon's longitude is additionally given for 24.00 UT

JUNE 2045

LONGITUDES

Date	Sidereal time h m s	Sun ☉ ° ′ ″	Moon ☽ ° ′ ″	Moon ☽ 24.00 ° ′ ″	Mercury ☿ ° ′	Venus ♀ ° ′	Mars ♂ ° ′	Jupiter ♃ ° ′	Saturn ♄ ° ′	Uranus ♅ ° ′	Neptune ♆ ° ′	Pluto ♇ ° ′
01	04 41 31	11 ♊ 24 40	01 ♑ 41 19	07 ♑ 49 00	19 ♊ 23	00 ♋ 42	04 ♊ 09	10 ♓ 57	08 ♐ 16	23 ♌ 13	15 ♉ 33	04 ♓ 05
02	04 45 28	12 22 08	13 ♑ 53 36	19 ♑ 55 22	21 29	01 55	04 52	11 02	08 R 12	23 15	15 35	04 05
03	04 49 24	13 19 36	25 ♑ 54 40	01 ♒ 51 52	23 35	03 08	05 34	11 07	08 07	23 17	15 38	04 05
04	04 53 21	14 17 03	07 ♒ 47 25	13 ♒ 41 51	25 38	04 22	06 16	11 12	08 03	23 18	15 40	04 R 05
05	04 57 17	15 14 29	19 ♒ 35 43	25 ♒ 29 36	27 39	05 35	06 59	11 17	07 58	23 20	15 42	04 05
06	05 01 14	16 11 54	01 ♓ 24 08	07 ♓ 19 58	29 ♊ 39	06 48	07 41	11 21	07 54	23 22	15 44	04 04
07	05 05 10	17 09 19	13 ♓ 17 48	19 ♓ 18 17	01 ♋ 36	08 01	08 23	11 26	07 50	23 24	15 46	04 04
08	05 09 07	18 06 43	25 ♓ 22 07	01 ♈ 29 56	03 31	09 14	09 06	11 30	07 45	23 26	15 48	04 04
09	05 13 03	19 04 06	07 ♈ 42 21	13 ♈ 59 58	05 23	10 27	09 48	11 34	07 41	23 28	15 50	04 04
10	05 17 00	20 01 29	20 ♈ 23 17	26 ♈ 52 42	07 13	11 40	10 30	11 38	07 37	23 30	15 51	04 04
11	05 20 57	20 58 52	03 ♉ 28 32	10 ♉ 10 58	09 00	12 53	11 12	11 42	07 32	23 32	15 53	04 04
12	05 24 53	21 56 14	17 ♉ 00 02	23 ♉ 55 35	10 45	14 06	11 54	11 46	07 28	23 35	15 55	04 04
13	05 28 50	22 53 35	00 ♊ 57 20	08 ♊ 04 49	12 28	15 19	12 36	11 49	07 24	23 37	15 57	04 03
14	05 32 46	23 50 56	15 ♊ 17 23	22 ♊ 34 17	14 08	16 32	13 18	11 52	07 20	23 39	15 59	04 03
15	05 36 43	24 48 17	29 ♊ 54 35	07 ♋ 17 21	15 45	17 45	14 00	11 55	07 15	23 41	16 01	04 03
16	05 40 39	25 45 37	14 ♋ 41 33	22 ♋ 06 12	17 20	18 58	14 41	11 58	07 11	23 44	16 03	04 03
17	05 44 36	26 42 56	29 ♋ 30 18	06 ♌ 53 01	18 52	20 11	15 23	12 01	07 07	23 46	16 05	04 02
18	05 48 32	27 40 14	14 ♌ 13 32	21 ♌ 31 14	20 22	21 24	16 05	12 03	07 03	23 49	16 06	04 02
19	05 52 29	28 37 31	28 ♌ 45 35	05 ♍ 56 12	21 49	22 37	16 47	12 06	06 59	23 51	16 08	04 02
20	05 56 26	29 ♊ 34 48	13 ♍ 02 49	20 ♍ 05 18	23 13	23 49	17 28	12 08	06 55	23 54	16 10	04 01
21	06 00 22	00 ♋ 32 03	27 ♍ 03 34	03 ♎ 57 39	24 35	25 02	18 10	12 10	06 51	23 56	16 12	04 01
22	06 04 19	01 29 18	10 ♎ 47 36	17 ♎ 33 32	25 54	26 15	18 51	12 11	06 48	23 59	16 13	04 00
23	06 08 15	02 26 33	24 ♎ 15 35	00 ♏ 53 54	27 10	27 28	19 33	12 13	06 44	24 01	16 15	04 00
24	06 12 12	03 23 46	07 ♏ 28 36	13 ♏ 59 50	28 23	28 40	20 14	12 14	06 40	24 04	16 17	03 59
25	06 16 08	04 21 00	20 ♏ 27 42	26 ♏ 52 21	29 ♋ 33	29 ♋ 53	20 55	12 15	06 36	24 07	16 18	03 59
26	06 20 05	05 18 12	03 ♐ 13 51	09 ♐ 32 18	00 ♌ 40	01 ♌ 06	21 37	12 16	06 33	24 09	16 20	03 58
27	06 24 01	06 15 25	15 ♐ 47 48	22 ♐ 00 26	01 45	02 18	22 18	12 17	06 29	24 12	16 22	03 58
28	06 27 58	07 12 37	28 ♐ 10 18	04 ♑ 17 31	02 46	03 31	22 59	12 18	06 26	24 15	16 23	03 57
29	06 31 55	08 09 48	10 ♑ 22 13	16 ♑ 24 33	03 44	04 43	23 40	12 18	06 22	24 18	16 25	03 57
30	06 35 51	09 ♋ 07 00	22 ♑ 24 43	28 ♑ 22 57	04 ♌ 39	05 ♌ 56	24 ♊ 21	12 ♓ 18	06 ♐ 19	24 ♌ 21	16 ♉ 26	03 ♓ 56

Date	Moon True ☊ ° ′	Moon Mean ☊ ° ′	Moon ☽ Latitude ° ′
01	25 ♒ 34	26 ♒ 39	04 S 04
02	25 R 25	26 36	03 21
03	25 19	26 33	02 29
04	25 16	26 29	01 31
05	25 14	26 26	00 S 30
06	25 D 14	26 23	00 N 33
07	25 15	26 20	01 35
08	25 R 15	26 17	02 33
09	25 13	26 13	03 26
10	25 09	26 10	04 10
11	25 03	26 07	04 43
12	24 55	26 04	05 01
13	24 46	26 01	05 02
14	24 36	25 58	04 45
15	24 28	25 54	04 08
16	24 21	25 51	03 15
17	24 16	25 48	02 08
18	24 14	25 45	00 N 53
19	24 D 13	25 42	00 S 24
20	24 14	25 39	01 39
21	24 15	25 35	02 47
22	24 R 15	25 32	03 44
23	24 12	25 29	04 27
24	24 08	25 26	04 54
25	24 02	25 23	05 07
26	23 54	25 19	05 03
27	23 46	25 16	04 45
28	23 38	25 13	04 14
29	23 30	25 10	03 31
30	23 ♒ 24	25 ♒ 07	02 S 39

DECLINATIONS

Date	Sun ☉ ° ′	Moon ☽ ° ′	Mercury ☿ ° ′	Venus ♀ ° ′	Mars ♂ ° ′	Jupiter ♃ ° ′	Saturn ♄ ° ′	Uranus ♅ ° ′	Neptune ♆ ° ′	Pluto ♇ ° ′
01	22 N 09	27 S 29	24 N 27	24 N 27	21 N 09	08 S 23	19 S 44	14 N 28	14 N 51	20 S 49
02	22 16	26 02	24 43	24 28	21 17	08 21	19 44	14 27	14 52	20 49
03	22 24	23 24	24 56	24 29	21 26	08 19	19 43	14 26	14 53	20 49
04	22 31	19 47	25 07	24 29	21 34	08 18	19 42	14 26	14 53	20 50
05	22 37	15 25	25 15	24 28	21 41	08 16	19 42	14 25	14 54	20 50
06	22 43	10 28	25 21	24 26	21 49	08 15	19 41	14 24	14 54	20 50
07	22 49	05 S 06	25 23	24 24	21 56	08 13	19 40	14 24	14 55	20 51
08	22 54	00 N 30	25 24	24 21	22 04	08 12	19 40	14 23	14 55	20 51
09	22 59	06 13	25 22	24 17	22 11	08 11	19 39	14 22	14 56	20 51
10	23 04	11 50	25 18	24 13	22 17	08 09	19 39	14 22	14 56	20 52
11	23 08	17 06	25 11	24 08	22 24	08 08	19 38	14 21	14 57	20 52
12	23 11	21 43	25 03	24 02	22 30	08 07	19 38	14 20	14 57	20 52
13	23 15	25 16	24 53	23 56	22 36	08 06	19 37	14 19	14 58	20 53
14	23 18	27 20	24 41	23 48	22 42	08 05	19 36	14 19	14 58	20 53
15	23 20	27 35	24 27	23 41	22 48	08 04	19 36	14 18	14 59	20 54
16	23 22	25 51	24 12	23 32	22 53	08 03	19 35	14 17	14 59	20 54
17	23 24	22 21	23 56	23 23	22 59	08 03	19 35	14 16	15 00	20 54
18	23 25	17 25	23 38	23 13	23 04	08 02	19 34	14 15	15 00	20 55
19	23 26	11 31	23 19	23 02	23 09	08 01	19 34	14 15	15 01	20 55
20	23 26	05 N 08	22 59	22 51	23 13	08 01	19 33	14 14	15 01	20 56
21	23 26	01 S 23	22 38	22 39	23 18	08 00	19 33	14 13	15 02	20 56
22	23 26	07 42	22 17	22 27	23 22	08 00	19 32	14 12	15 02	20 57
23	23 25	13 32	21 54	22 13	23 26	08 00	19 32	14 11	15 03	20 57
24	23 23	18 38	21 31	22 00	23 29	07 59	19 31	14 10	15 03	20 58
25	23 22	22 47	21 08	21 45	23 33	07 59	19 31	14 09	15 04	20 58
26	23 20	25 45	20 44	21 30	23 36	07 59	19 30	14 08	15 04	20 58
27	23 17	27 24	20 20	21 15	23 39	07 59	19 30	14 07	15 04	20 59
28	23 14	27 39	19 56	20 58	23 42	07 59	19 29	14 06	15 05	20 59
29	23 11	26 32	19 31	20 42	23 45	07 59	19 29	14 05	15 05	21 00
30	23 N 07	24 S 12	19 N 07	20 N 24	23 N 47	07 S 59	19 S 29	14 N 04	15 N 06	21 S 00

ZODIAC SIGN ENTRIES

Date	h	m	Planets
01	08	43	☽ ♑
03	20	14	☽ ♒
06	09	09	☽ ♓
06	16	20	☿ ♋
08	21	05	☽ ♈
11	05	42	☽ ♉
13	10	23	☽ ♊
15	12	09	☽ ♋
17	12	48	☽ ♌
19	14	04	☽ ♍
20	22	34	☉ ♋
21	17	06	☽ ♎
23	22	22	☽ ♏
25	14	20	♀ ♌
25	21	27	☿ ♌
26	05	53	☽ ♐
28	15	35	☽ ♑

LATITUDES

Date	Mercury ☿ ° ′	Venus ♀ ° ′	Mars ♂ ° ′	Jupiter ♃ ° ′	Saturn ♄ ° ′	Uranus ♅ ° ′	Neptune ♆ ° ′	Pluto ♇ ° ′
01	01 N 26	01 N 01	00 N 11	01 S 00	01 N 58	00 N 43	01 S 43	11 S 38
04	01 45	01 07	00 13	01 00	01 58	00 43	01 43	11 39
07	01 58	01 13	00 15	01 01	01 58	00 43	01 43	11 40
10	02 03	01 18	00 16	01 02	01 57	00 43	01 43	11 41
13	02 02	01 23	00 18	01 03	01 57	00 43	01 43	11 42
16	01 54	01 27	00 20	01 04	01 57	00 43	01 43	11 43
19	01 40	01 31	00 22	01 05	01 56	00 43	01 43	11 44
22	01 20	01 34	00 24	01 06	01 56	00 43	01 43	11 45
25	00 55	01 37	00 26	01 06	01 55	00 43	01 44	11 46
28	00 N 24	01 39	00 27	01 07	01 55	00 43	01 44	11 47
31	00 S 11	01 N 41	00 N 29	01 S 08	01 N 54	00 N 43	01 S 44	11 S 47

DATA

Julian Date	2468133
Delta T	+79 seconds
Ayanamsa	24° 29' 38"
Synetic vernal point	04° ♓ 37' 20"
True obliquity of ecliptic	23° 26' 07"

MOON'S PHASES, APSIDES AND POSITIONS ☽

Date	h	m	Phase	Longitude ° ′	Eclipse Indicator
07	20	23	☾	17 ♓ 29	
15	03	05	●	24 ♊ 27	
21	18	28	☽	00 ♎ 48	
29	07	16	○	07 ♑ 59	

Day	h	m		
05	03	20	Apogee	
17	02	03	Perigee	
08	09	51	0N	
15	02	56	Max dec	27° N 43'
21	06	52	0S	
28	04	16	Max dec	27° S 43'

LONGITUDES

Date	Chiron ⚷ ° ′	Ceres ⚳ ° ′	Pallas ⚴ ° ′	Juno ⚵ ° ′	Vesta ⚶ ° ′	Black Moon Lilith ⚸ ° ′
01	13 ♍ 01	06 ♋ 21	11 ♉ 14	05 ♋ 32	05 ♊ 02	11 ♒ 22
11	13 ♍ 31	10 ♋ 37	15 ♉ 39	10 ♋ 59	09 ♊ 22	12 ♒ 29
21	14 ♍ 10	14 ♋ 57	20 ♉ 05	16 ♋ 23	13 ♊ 40	13 ♒ 35
31	14 ♍ 59	19 ♋ 19	24 ♉ 33	21 ♋ 43	17 ♊ 56	14 ♒ 42

All ephemeris data is given at 12.00 UT and the Moon's longitude is additionally given for 24.00 UT

ASPECTARIAN

h m	Aspects
01 Thursday	
06 37	☽ Q ♃
07 03	☉ Q ♅
09 07	♀ ∠ ♆
09 28	♂ □ ♇
09 47	☽ ⚼ ♆
09 52	☽ ☍ ♀
12 06	☿ ∥ ♀
16 40	☽ ⚹ ♇
17 06	☽ ⊼ ♂
02 Friday	
00 49	☽ ⊻ ♄
00 49	☽ ⚼ ♅
05 37	☽ ± ♂
06 18	☽ ⚹ ♃
08 43	☽ ⊼ ☉
12 36	☽ ⊥ ♄
13 10	☿ ⊥ ♆
15 23	☽ △ ♆
18 41	☽ ± ♅
21 40	☽ ± ☉
22 19	☽ ∠ ♇
03 Saturday	
00 12	☽ # ☿
00 37	☽ ⚼ ♂
03 21	☽ # ♀
06 20	☽ ⊼ ☿
06 26	☽ ∠ ♄
06 42	☽ ⊼ ♅
08 28	☿ ⚹ ♅
12 25	☽ ∠ ♃
16 21	☽ ⊥ ♇
16 38	♇ St R
17 17	☽ ⚼ ☉
19 05	☽ # ☉
20 56	☽ ± ☿
04 Sunday	
01 21	☽ # ♂
04 15	☽ ⊼ ♀
04 28	☽ ⊻ ♇
05 41	☽ ∥ ♇
06 27	♀ △ ♇
06 43	☽ ⊥ ♃
08 44	☽ △ ♂
12 31	☽ ⚹ ♄
12 31	☽ ∥ ♄
17 49	☽ ± ♀
18 58	☽ ⊻ ♃
18 59	☽ ⚼ ☿
05 Monday	
02 21	☽ △ ☉
04 02	☽ □ ♆
12 46	☽ Q ♄
14 14	☽ ⚼ ♀
14 38	☽ # ♆
17 02	☽ # ♅
19 38	☽ ☍ ♅
23 44	☉ ⊻ ♆
06 Tuesday	
07 43	☽ △ ☿
16 43	☽ Q ♆
17 25	☽ ☌ ♇
18 38	♂ ☍ ♄
22 10	☽ ∥ ♃
07 Wednesday	
00 10	☽ △ ♀
01 04	☽ □ ♄
01 28	☿ ∠ ♆
01 31	☽ □ ♂
08 14	☽ ☌ ♃
08 30	♀ ⊼ ♄
16 57	☽ ⚹ ♆
19 51	♀ ∠ ♅
20 23	☽ □ ☉
08 Thursday	
05 28	♀ ⊻ ♂
08 11	☽ ⊼ ♅
15 36	☽ Q ♂
19 09	☿ △ ♇
20 00	☽ ± ♅
22 39	☽ ∠ ♆
09 Friday	
04 59	☽ ⊻ ♇
06 44	☽ □ ☿
10 40	☽ Q ☉
11 57	☽ △ ♄
13 28	☽ ⚼ ♅
16 04	☽ ⊥ ♆
16 14	☽ ⚹ ♂
16 32	☽ ⊥ ♇
17 49	☽ □ ♀
19 26	☽ ⊻ ♃
20 16	☽ # ♃
10 Saturday	
03 29	☽ ⊻ ♆
06 50	☽ ⊥ ♃
09 32	☽ ∠ ♇
11 16	☽ ⚹ ☉
11 22	♀ △ ♃
16 06	☽ ⚼ ♄
17 03	☿ ⊼ ♄
17 48	☽ △ ♅
22 00	☽ ∠ ♂
22 23	☽ Q ☿
23 14	☽ ∥ ♅
23 37	☽ ∠ ♃
11 Sunday	
01 54	☽ ∥ ♆
05 35	☿ ∠ ♅
06 50	☽ Q ♀
08 31	☽ ± ♄
13 04	☽ ⚹ ♇
15 16	☽ ⊥ ♂
16 51	☽ ∠ ☉
19 15	☽ ⊼ ♄
23 24	☽ ⚹ ☿
12 Monday	
00 11	♀ ± ♄
00 22	♂ Q ♅
00 35	☽ # ♄
02 33	☽ ⊻ ♂
02 45	☽ ⚹ ♃
06 25	☽ ⚹ ♀
06 50	♂ □ ♃
07 13	☽ # ♇
10 00	☽ ⊥ ☉
10 07	☽ ☌ ♆
10 21	☽ Q ♇
11 37	☉ ⊥ ♆
16 46	☽ ∥ ♂
21 04	☽ ∥ ☉
21 12	☽ ⊻ ☉
23 26	☽ □ ♅
23 46	☽ Q ♃
13 Tuesday	
02 10	☽ ∥ ♀
02 32	☿ △ ♃
05 15	☽ ∠ ☿
08 59	☽ ∥ ☿
10 49	☽ ∠ ♀
15 16	☿ ⊻ ♂
17 14	☽ □ ♇
22 32	☽ ⊥ ☿
22 48	☽ ☍ ♄
14 Wednesday	
00 49	☿ ± ♄
00 52	♀ ⚹ ♆
03 22	☽ ⊥ ♀
05 57	☽ Q ♅
06 18	☽ □ ♃
06 50	☉ ⚹ ♅
08 32	☽ ☌ ♂
09 50	☽ ⊻ ☿
13 09	☽ ⊻ ♆
14 15	☽ ⊻ ♀
23 04	☽ ⊥ ♆
15 Thursday	
01 48	☽ ⚹ ♅
03 05	☽ ☌ ☉
10 46	♀ ⊥ ♅
13 48	☽ ∠ ♆
16 02	☿ ⚹ ♆
18 44	☽ △ ♇
23 54	☽ ⊼ ♄
16 Friday	
02 19	☽ ∠ ♅
07 35	☽ △ ♃
09 01	♆ Q ♇
09 35	☽ ± ♄
12 00	☽ ⊻ ♂
13 31	♀ ⚼ ♇
14 12	☽ ⚹ ♆
16 47	☽ ☌ ☿
16 56	☽ ⊥ ♅
18 17	☿ ⊥ ♅
19 02	☽ ⚼ ♇
19 32	☽ ☌ ♀
22 12	☽ ⊥ ♂
17 Saturday	
00 05	☽ ⚼ ♄
01 43	☽ ∥ ☿
02 40	☽ ⊻ ♅
05 42	☽ ∥ ♀
05 54	☽ ∥ ☉
07 10	☽ ⊻ ☉
07 57	☽ ⚼ ♃
08 27	☽ ∥ ♂
09 37	☽ ± ♇
09 41	☽ Q ♆
10 02	☉ ∥ ♀
13 30	☽ ∠ ♂
14 38	☿ ⚼ ♇
17 35	☽ ⊥ ☉
19 22	☽ ⊼ ♇
19 38	☽ # ♇
22 37	☽ ± ♃
18 Sunday	
00 20	☽ △ ♄
02 10	☽ # ♄
08 26	☽ ⊼ ♃
09 16	☽ ∠ ☉
12 53	♂ ⊻ ♆
15 06	☽ □ ♆
15 12	☽ ⚹ ♂
22 09	☽ ∥ ♆
23 13	☽ ⊻ ☿
19 Monday	
00 21	♀ ⚼ ♄
00 52	☽ ⊻ ♀
01 15	☽ ∥ ♅
02 22	♀ ∥ ♂
03 50	☽ ☌ ♅
04 01	☉ ∥ ☿
10 15	☽ ⊥ ☿
11 44	☽ ⊥ ♀
11 46	☽ ⚹ ☉
12 02	☽ Q ♂
14 49	☿ ⚼ ♄
19 46	♀ ⊥ ♂
20 47	☽ ☍ ♇
22 27	☿ ∥ ♂
20 Tuesday	
01 18	☽ # ♃
01 42	☽ □ ♄
02 57	☽ ∠ ☿
04 11	☽ ∠ ♀
09 20	☽ Q ☉
10 26	☽ ☍ ♃
13 26	♀ ⊻ ♅
17 19	☽ △ ♆
19 55	☽ □ ♂
20 54	☿ ⊥ ♂
21 Wednesday	
00 11	☿ ⊻ ♅
06 35	☽ ⊻ ♅
07 15	☽ ⚹ ☿
08 10	☽ ⚹ ♀
08 13	☽ Q ♄
09 56	☿ ∥ ♀
17 00	☽ ⊥ ♅
18 28	☽ □ ☉
19 12	☽ ⚼ ♆
22 Thursday	
00 05	☽ ⊼ ♇
05 00	☽ ⚹ ♄
05 09	☉ ∠ ♆
06 21	☽ Q ☿
07 05	☽ Q ♀
08 47	☽ ∠ ♅
10 37	☽ ± ♇
11 00	☽ ± ♆
13 11	☽ ∥ ♃
14 28	☽ ⊼ ♃
21 39	☽ ⊼ ♆
23 Friday	
01 09	☽ ± ♃
02 34	☽ ⚼ ♇
03 05	☽ △ ♂
07 03	♀ ⚼ ♃
07 29	☽ ∠ ♄
07 36	☉ ∥ ♂
11 34	☽ ⚹ ♅
13 05	☿ ⚼ ♃
14 51	☽ # ♅
17 20	☽ ⚼ ♃
17 46	☽ □ ☿
18 21	☽ □ ♀
18 44	☽ # ♆
22 33	♀ ± ♇
23 38	☽ ⊥ ♄
24 Saturday	
03 58	☽ △ ☉
04 04	♀ Q ♆
04 15	☿ ± ♇
05 38	☽ △ ♇
07 40	☽ ⚼ ♂
09 24	☽ Q ♅
09 58	☿ Q ♆
10 31	☽ ⊻ ♄
16 38	☽ ∥ ♄
20 46	☽ △ ♃
25 Sunday	
00 42	☽ ∥ ♇
01 08	☽ ± ♂
02 36	☽ # ☿
02 46	☉ △ ♇
04 16	☽ ☍ ♆
05 49	☽ # ♀
09 46	☽ ⚼ ☉
12 55	☽ ⊼ ♂
15 58	☽ # ☉
17 25	☽ # ♂
18 51	☽ □ ♅
21 51	☿ # ♇
26 Monday	
03 59	☽ ± ☉
06 42	☽ △ ☿
07 32	☽ △ ♀
13 24	☽ □ ♇
16 15	☽ ⊼ ☉
18 16	☽ ☌ ♄
27 Tuesday	
05 15	☽ □ ♃
13 05	☽ ⊼ ♆
14 00	☽ ⚼ ☿
14 15	♂ ⊥ ♆
15 13	☽ ⚼ ♀
17 25	☉ ⊼ ♄
23 54	☽ Q ♇
28 Wednesday	
00 43	☽ ± ♆
01 18	☽ ☍ ♂
04 19	☽ △ ♅
09 01	☽ ± ☿
10 32	♀ # ♇
10 34	☽ ± ♀
16 09	☽ Q ♃
18 19	☽ ⚼ ♆
20 39	♀ ⊼ ♇
21 47	☽ ⊼ ☿
23 19	☽ ⚹ ♇
23 37	☽ ⊼ ♀
29 Thursday	
04 08	☽ ⊻ ♄
07 16	☽ ☍ ☉
09 52	☽ ⚼ ♅
14 13	☿ # ♄
15 50	☽ ⚹ ♃
15 57	☽ ⊥ ♄
17 18	☿ ⊼ ♇
30 Friday	
00 02	☽ △ ♆
03 50	☽ ± ♅
05 02	☽ ∠ ♇
09 48	☽ ∠ ♄
11 28	♂ ⚹ ♅
15 14	☽ # ♂
15 54	☽ ⊼ ♅
16 09	☽ ⊼ ♂
18 01	☉ ∠ ♅
19 15	♀ △ ♄
19 27	♀ ± ♃
20 32	☽ # ☉
21 50	☽ ∠ ♃
23 05	☽ ⊥ ♇

JULY 2045

LONGITUDES

Date	Sidereal time h m s	Sun ☉ ° ′ ″	Moon ☽ ° ′ ″	Moon ☽ 24.00 ° ′ ″	Mercury ☿ ° ′	Venus ♀ ° ′	Mars ♂ ° ′	Jupiter ♃ ° ′	Saturn ♄ ° ′	Uranus ♅ ° ′	Neptune ♆ ° ′	Pluto ♇ ° ′
01	06 39 48	10 ♋ 04 11	04 ♒ 19 30	10 ♒ 14 43	05 ♌ 30	07 ♌ 08	25 ♊ 03	12 ♓ 18	06 ♐ 15	24 ♌ 24	16 ♉ 28	03 ♓ 55
02	06 43 44	11 01 23	16 ♒ 08 55	22 ♒ 02 31	06 18	08 21	25 44	12 R 18	06 R 12	24 26	16 29	03 R 55
03	06 47 41	11 58 34	27 ♒ 55 57	03 ♓ 49 41	07 02	09 33	26 25	12 18	06 09	24 29	16 31	03 54
04	06 51 37	12 55 45	09 ♓ 44 15	15 ♓ 40 12	07 42	10 46	27 05	12 17	06 06	24 32	16 32	03 53
05	06 55 34	13 52 57	21 ♓ 38 06	27 ♓ 38 34	08 19	11 58	27 46	12 16	06 03	24 35	16 34	03 53
06	06 59 30	14 50 09	03 ♈ 40 59	09 ♈ 49 38	08 51	13 10	28 27	12 15	06 00	24 38	16 35	03 52
07	07 03 27	15 47 21	16 ♈ 01 28	22 ♈ 18 17	09 20	14 23	29 08	12 14	05 57	24 42	16 36	03 51
08	07 07 24	16 44 34	28 ♈ 40 38	05 ♉ 09 02	09 44	15 35	29 ♊ 49	12 13	05 54	24 45	16 38	03 50
09	07 11 20	17 41 47	11 ♉ 43 52	18 ♉ 25 28	10 04	16 47	00 ♋ 29	12 11	05 51	24 48	16 39	03 49
10	07 15 17	18 39 00	25 ♉ 14 00	02 ♊ 09 33	10 19	17 59	01 10	12 10	05 49	24 51	16 40	03 49
11	07 19 13	19 36 14	09 ♊ 11 58	16 ♊ 20 57	10 29	19 11	01 50	12 08	05 46	24 54	16 42	03 48
12	07 23 10	20 33 29	23 ♊ 36 03	00 ♋ 56 35	10 35	20 24	02 31	12 06	05 43	24 57	16 43	03 47
13	07 27 06	21 30 43	08 ♋ 21 45	15 ♋ 50 33	10 R 36	21 36	03 12	12 03	05 41	25 01	16 44	03 46
14	07 31 03	22 27 59	23 ♋ 21 55	00 ♌ 54 42	10 33	22 48	03 52	12 01	05 39	25 04	16 45	03 45
15	07 34 59	23 25 14	08 ♌ 27 45	15 ♌ 59 55	10 24	24 00	04 32	11 58	05 36	25 07	16 46	03 44
16	07 38 56	24 22 29	23 ♌ 30 08	00 ♍ 57 26	10 11	25 12	05 13	11 55	05 34	25 11	16 48	03 43
17	07 42 53	25 19 45	08 ♍ 20 59	15 ♍ 40 06	09 53	26 24	05 53	11 52	05 32	25 14	16 49	03 42
18	07 46 49	26 17 01	22 ♍ 54 14	00 ♎ 03 01	09 31	27 36	06 33	11 49	05 30	25 17	16 50	03 41
19	07 50 46	27 14 17	07 ♎ 06 12	14 ♎ 03 43	09 05	28 48	07 14	11 45	05 28	25 21	16 51	03 40
20	07 54 42	28 11 33	20 ♎ 55 33	27 ♎ 41 49	08 34	29 ♌ 59	07 54	11 42	05 26	25 24	16 52	03 39
21	07 58 39	29 ♋ 08 49	04 ♏ 22 41	10 ♏ 58 24	08 00	01 ♍ 11	08 34	11 38	05 25	25 28	16 53	03 38
22	08 02 35	00 ♌ 06 06	17 ♏ 29 15	23 ♏ 55 32	07 23	02 23	09 14	11 34	05 23	25 31	16 54	03 37
23	08 06 32	01 03 23	00 ♐ 17 35	06 ♐ 35 42	06 44	03 35	09 54	11 30	05 21	25 35	16 54	03 36
24	08 10 28	02 00 40	12 ♐ 50 13	19 ♐ 01 29	06 03	04 46	10 34	11 26	05 20	25 38	16 55	03 35
25	08 14 25	02 57 58	25 ♐ 09 46	01 ♑ 15 22	05 20	05 58	11 14	11 21	05 18	25 42	16 56	03 34
26	08 18 22	03 55 16	07 ♑ 18 35	13 ♑ 19 41	04 37	07 09	11 54	11 17	05 17	25 45	16 57	03 33
27	08 22 18	04 52 35	19 ♑ 18 56	25 ♑ 16 36	03 55	08 21	12 33	11 12	05 16	25 49	16 58	03 32
28	08 26 15	05 49 54	01 ♒ 12 55	07 ♒ 08 10	03 13	09 32	13 13	11 07	05 15	25 52	16 59	03 31
29	08 30 11	06 47 14	13 ♒ 02 37	18 ♒ 56 32	02 33	10 44	13 53	11 02	05 14	25 56	16 59	03 30
30	08 34 08	07 44 35	24 ♒ 50 13	00 ♓ 43 58	01 56	11 55	14 33	10 56	05 13	25 59	17 00	03 29
31	08 38 04	08 ♌ 41 57	06 ♓ 38 08	12 ♓ 33 04	01 ♌ 22	13 ♍ 07	15 ♋ 12	10 ♓ 51	05 ♐ 12	26 ♌ 03	17 ♉ 01	03 ♓ 28

Date	Moon True ☊ ° ′	Moon Mean ☊ ° ′	Moon ☽ Latitude ° ′
01	23 ♒ 21	25 ♒ 04	01 S 41
02	23 R 19	25 00	00 S 39
03	23 D 19	24 57	00 N 25
04	23 20	24 54	01 28
05	23 21	24 51	02 28
06	23 22	24 48	03 22
07	23 R 23	24 44	04 08
08	23 22	24 41	04 44
09	23 20	24 38	05 06
10	23 17	24 35	05 12
11	23 12	24 32	05 00
12	23 07	24 29	04 29
13	23 03	24 25	03 40
14	22 59	24 22	02 35
15	22 57	24 19	01 N 19
16	22 D 57	24 16	00 S 03
17	22 58	24 13	01 24
18	22 59	24 10	02 37
19	23 00	24 06	03 39
20	23 01	24 03	04 27
21	23 R 01	24 00	04 58
22	23 00	23 57	05 13
23	22 58	23 54	05 12
24	22 55	23 50	04 56
25	22 52	23 47	04 25
26	22 49	23 44	03 45
27	22 47	23 41	02 54
28	22 45	23 38	01 56
29	22 44	23 35	00 S 53
30	22 D 44	23 31	00 N 12
31	22 ♒ 44	23 ♒ 28	01 N 16

DECLINATIONS

Date	Sun ☉ ° ′	Moon ☽ ° ′	Mercury ☿ ° ′	Venus ♀ ° ′	Mars ♂ ° ′	Jupiter ♃ ° ′	Saturn ♄ ° ′	Uranus ♅ ° ′	Neptune ♆ ° ′	Pluto ♇ ° ′
01	23 N 03	20 S 49	18 N 43	20 N 06	23 N 50	08 S 00	19 S 28	14 N 03	15 N 06	21 S 01
02	22 59	16 37	18 19	19 48	23 52	08 00	19 28	14 02	15 06	21 01
03	22 54	11 48	17 55	19 29	23 53	08 00	19 27	14 01	15 07	21 02
04	22 48	06 33	17 32	19 10	23 55	08 01	19 27	14 00	15 07	21 03
05	22 43	01 S 03	17 09	18 50	23 56	08 01	19 27	13 59	15 07	21 03
06	22 37	04 N 34	16 47	18 29	23 58	08 02	19 26	13 58	15 08	21 04
07	22 30	10 07	16 25	18 08	23 59	08 03	19 26	13 57	15 08	21 04
08	22 23	15 25	16 05	17 47	23 59	08 04	19 26	13 56	15 08	21 05
09	22 16	20 12	15 45	17 25	24 00	08 04	19 26	13 55	15 09	21 05
10	22 08	24 07	15 27	17 02	24 00	08 05	19 25	13 54	15 09	21 06
11	22 00	26 46	15 10	16 39	24 00	08 06	19 25	13 53	15 09	21 06
12	21 52	27 46	14 54	16 16	24 00	08 07	19 25	13 52	15 10	21 07
13	21 43	26 50	14 40	15 53	24 00	08 09	19 25	13 51	15 10	21 07
14	21 34	23 58	14 27	15 28	23 59	08 10	19 24	13 50	15 10	21 08
15	21 24	19 25	14 16	15 04	23 59	08 11	19 24	13 49	15 11	21 09
16	21 14	13 38	14 06	14 39	23 58	08 12	19 24	13 47	15 11	21 09
17	21 04	07 09	13 59	14 14	23 57	08 14	19 24	13 46	15 11	21 10
18	20 53	00 N 24	13 53	13 48	23 55	08 15	19 24	13 45	15 11	21 10
19	20 43	06 S 11	13 49	13 22	23 54	08 17	19 24	13 44	15 12	21 11
20	20 31	12 17	13 47	12 56	23 52	08 19	19 23	13 43	15 12	21 11
21	20 20	17 39	13 48	12 29	23 50	08 20	19 23	13 42	15 12	21 12
22	20 08	22 03	13 50	12 03	23 48	08 22	19 23	13 40	15 12	21 13
23	19 55	25 18	13 54	11 35	23 46	08 24	19 23	13 39	15 12	21 13
24	19 43	27 14	13 59	11 08	23 43	08 26	19 23	13 38	15 13	21 14
25	19 30	27 47	14 07	10 40	23 40	08 28	19 23	13 37	15 13	21 14
26	19 16	26 59	14 16	10 12	23 37	08 30	19 23	13 36	15 13	21 15
27	19 03	24 55	14 27	09 44	23 34	08 32	19 23	13 34	15 13	21 16
28	18 49	21 46	14 39	09 15	23 31	08 34	19 23	13 33	15 13	21 16
29	18 34	17 45	14 52	08 47	23 27	08 36	19 23	13 32	15 13	21 17
30	18 20	13 03	15 06	08 18	23 24	08 38	19 23	13 31	15 14	21 17
31	18 N 05	07 S 54	15 N 21	07 N 49	23 N 20	08 S 41	19 S 23	13 N 30	15 N 14	21 S 18

ZODIAC SIGN ENTRIES

Date	h	m	Planets
01	03	16	☽ ♒
03	16	13	☽ ♓
06	04	41	☽ ♈
08	14	28	☽ ♉
08	18	44	♂ ♋
10	20	17	☽ ♊
12	22	28	☽ ♋
14	22	33	☽ ♌
16	22	27	☽ ♍
18	23	55	☽ ♎
20	12	12	♀ ♍
21	04	07	☽ ♏
22	09	27	☉ ♌
23	11	27	☽ ♐
25	21	31	☽ ♑
28	09	32	☽ ♒
30	22	31	☽ ♓

LATITUDES

Date	Mercury ☿ ° ′	Venus ♀ ° ′	Mars ♂ ° ′	Jupiter ♃ ° ′	Saturn ♄ ° ′	Uranus ♅ ° ′	Neptune ♆ ° ′	Pluto ♇ ° ′
01	00 S 11	01 N 41	00 N 29	01 S 08	01 N 54	00 N 43	01 S 44	11 S 47
04	00 51	01 41	00 31	01 09	01 54	00 42	01 44	11 48
07	01 33	01 42	00 33	01 10	01 53	00 42	01 44	11 49
10	02 17	01 41	00 34	01 11	01 53	00 42	01 44	11 50
13	03 01	01 40	00 36	01 12	01 52	00 42	01 44	11 51
16	03 43	01 38	00 38	01 13	01 52	00 42	01 45	11 51
19	04 19	01 35	00 39	01 13	01 51	00 42	01 45	11 52
22	04 45	01 31	00 41	01 14	01 50	00 42	01 45	11 53
25	04 58	01 27	00 43	01 15	01 50	00 42	01 45	11 53
28	04 55	01 22	00 44	01 16	01 49	00 42	01 45	11 54
31	04 S 37	01 N 16	00 N 46	01 S 16	01 N 48	00 N 42	01 S 45	11 S 55

DATA

Julian Date	2468163
Delta T	+79 seconds
Ayanamsa	24° 29′ 44″
Synetic vernal point	04° ♓ 37′ 15″
True obliquity of ecliptic	23° 26′ 07″

MOON'S PHASES, APSIDES AND POSITIONS ☽

Date	h	m	Phase	Longitude ° ′	Eclipse Indicator
07	11	31	☾	15 ♈ 46	
14	10	28	●	22 ♋ 24	
21	01	52	☽	28 ♎ 45	
28	22	11	○	06 ♒ 14	

Day	h	m		
02	17	52	Apogee	
15	07	07	Perigee	
30	01	45	Apogee	
05	16	30	0N	
12	12	43	Max dec	27° N 46′
18	13	27	0S	
25	09	40	Max dec	27° S 48′

LONGITUDES

Date	Chiron ⚷ ° ′	Ceres ⚳ ° ′	Pallas ⚴ ° ′	Juno ⚵ ° ′	Vesta ⚶ ° ′	Black Moon Lilith ⚸ ° ′
01	14 ♍ 59	19 ♋ 19	24 ♉ 33	21 ♋ 43	17 ♊ 56	14 ♒ 42
11	15 ♍ 57	23 ♋ 44	29 ♉ 03	26 ♋ 59	22 ♊ 08	15 ♒ 48
21	17 ♍ 01	28 ♋ 10	03 ♊ 32	02 ♌ 10	26 ♊ 17	16 ♒ 55
31	18 ♍ 12	02 ♌ 37	08 ♊ 02	07 ♌ 17	00 ♋ 21	18 ♒ 01

ASPECTARIAN

01 Saturday
00 37 ♃ St R
04 58 ☽ ± ♂
10 43 ☽ ∥ ♇
11 11 ☽ ⊻ ♇
14 33 ☽ ☍ ☿
15 54 ☽ ⚹ ♄
16 01 ☽ ⊥ ♃
16 42 ☽ # ♀
18 21 ☽ ☍ ♀
20 09 ☽ ∥ ♄

02 Sunday
00 18 ☽ ⚼ ♂
00 40 ☽ ⊼ ☉
01 50 ☽ # ☿
04 11 ☽ ⊻ ♃
09 14 ☿ △ ♄
12 09 ☿ ± ♃
12 42 ☽ □ ♆
13 56 ☽ ± ☉
16 10 ☽ Q ♄
19 47 ☽ # ♆

03 Monday
01 10 ☽ # ♅
04 58 ☽ ☍ ♅
08 42 ☽ △ ♂
09 53 ☽ ⚼ ☉
14 00 ♀ # ♄
16 08 ☉ ± ♄
19 59 ☉ △ ♃

04 Tuesday
00 08 ☽ ☌ ♇
01 25 ☽ Q ♆
04 39 ☽ □ ♄
05 28 ☽ ∥ ♃
07 38 ☽ ⊼ ☿
14 18 ☽ ⊼ ♀
17 09 ☽ ☌ ♃
19 02 ☽ △ ☉
20 29 ☽ ± ☿

05 Wednesday
01 47 ☽ ⚹ ♆
03 48 ☽ ± ♀
15 32 ☽ ⚼ ☿
17 56 ☽ ⊼ ♅
18 04 ♀ ⊼ ♃
23 50 ☽ ⚼ ♀

06 Thursday
00 59 ☽ □ ♂
05 56 ☽ ± ♅
07 49 ☽ ∠ ♆
12 19 ☽ ⊻ ♇
16 29 ☽ △ ♄
22 32 ☽ △ ☿
23 41 ☽ ⚼ ♅

07 Friday
00 03 ☽ ⊥ ♇
00 50 ♀ ∠ ♂
01 30 ☽ ⊥ ♆
02 56 ☽ # ♃
04 42 ☽ ⊻ ♃
08 29 ☽ △ ♀
11 31 ☽ □ ☉
13 07 ☽ ⊻ ♆
14 15 ☽ Q ♂
16 14 ☽ ⊥ ♃
17 25 ☽ ∠ ♇
21 23 ☽ ⚼ ♄

08 Saturday
04 35 ☽ △ ♅
05 08 ☽ ∥ ♅
09 07 ☉ ⚹ ♆
09 16 ☽ ∠ ♃
10 42 ☽ ∥ ♆
14 14 ☽ ⚹ ♂
14 16 ☽ ± ♄
14 58 ☽ ∥ ☿
21 34 ☽ ⚹ ♇
22 38 ☽ ∥ ♀

09 Sunday
00 08 ☽ Q ☉
01 20 ☽ ⊼ ♄
07 54 ☽ # ♄
08 54 ☽ □ ☿
09 20 ♀ □ ♆
12 50 ☽ ⚹ ♃
16 59 ☽ # ♇
19 07 ☽ ∠ ♂
19 21 ☽ Q ♇
20 51 ☽ ☌ ♆
21 58 ☽ □ ♀
23 31 ☽ ⚹ ☉
23 31 ☽ ∥ ☉

10 Monday
10 08 ☽ Q ♃
11 14 ☽ ∥ ♂
11 20 ☽ □ ♅
11 52 ☽ ⊥ ♂
15 59 ☉ ⚼ ♇
17 20 ☉ ⊥ ♅
17 26 ☽ Q ☿
22 49 ☽ ⊻ ♂

11 Tuesday
02 49 ☽ □ ♇
03 37 ☽ ∠ ☉
06 11 ☽ ☍ ♄
06 37 ♂ ∠ ♆
08 17 ☽ Q ♀
12 52 ☿ ∥ ♆
14 12 ☽ ⚹ ☿
16 56 ☽ □ ♃
18 16 ☽ Q ♅
19 57 ☽ ⊥ ☉

12 Wednesday
00 36 ☽ ⊻ ♆
06 14 ☽ ⚹ ♀
06 38 ☽ ⊻ ☉
10 32 ☽ ⊥ ♆
14 14 ☽ ⚹ ♅
15 16 ☽ ∠ ☿
16 01 ☉ ⚼ ♄

13 Thursday
01 16 ☽ ∠ ♆
03 15 ☽ ☌ ♂
04 01 ☉ ⊻ ♀
04 35 ☽ △ ♇
05 38 ☿ St R
05 56 ☽ ⊥ ☿
07 42 ☽ ⊼ ♄
08 54 ☽ ∠ ♀
14 40 ☽ ∠ ♅
15 36 ☽ ⊻ ☿
17 20 ☽ ± ♄
17 55 ☽ △ ♃

14 Friday
00 37 ☽ ⊥ ♀
01 27 ☽ ⚹ ♆
04 39 ☽ ⚼ ♇
05 07 ☽ ⊥ ♅
07 41 ☽ ⚼ ♄
08 03 ♂ △ ♇
10 28 ☽ ☌ ☉
11 01 ☽ ⊻ ♀
11 49 ☽ ∥ ♂
14 43 ☽ ⊻ ♅
17 47 ☽ ⚼ ♃
18 58 ☽ ± ♇
20 35 ☽ Q ♆

15 Saturday
02 04 ☽ ∥ ☉
03 45 ☽ # ♇
04 30 ☽ ⊼ ♇
05 28 ☽ ⊻ ♂
05 37 ♀ ∥ ♆
07 28 ☽ △ ♄
08 03 ☽ ± ♃
08 07 ☿ ⊥ ♂
12 02 ☽ # ♄
15 03 ☽ ☌ ☿
15 28 ☽ ⊥ ♂
17 34 ☽ ⊼ ♃

16 Sunday
01 15 ☽ □ ♆
05 57 ☽ ∥ ♆
06 29 ☽ ∠ ♂
07 45 ☽ ∥ ♀
10 08 ☽ ∥ ☿
11 24 ☽ ∥ ♅
11 35 ♀ ☌ ♅
13 30 ☽ ⊻ ☉
14 42 ☽ ☌ ♅
14 57 ☽ ☌ ♀
23 42 ☉ # ♇
23 49 ☽ ⊥ ☉

17 Monday
00 09 ♂ ⊼ ♄
04 27 ☽ ☍ ♇
07 26 ☽ □ ♄
07 48 ☽ ⚹ ♂
08 06 ☽ # ♃
09 25 ☉ ⊻ ♅
14 27 ☽ ⊻ ☿
15 27 ☽ ∠ ☉
17 44 ☽ ☍ ♃

18 Tuesday
00 04 ☽ ⊥ ☿
01 54 ☽ △ ♆
04 25 ☽ Q ♂
06 25 ☿ ∥ ♀
13 00 ☽ Q ♄
14 37 ☽ ∠ ☿
15 02 ♀ ∥ ♅
16 00 ☽ ⊻ ♅
18 04 ☽ ⚹ ☉
20 35 ☽ ⊻ ♀

19 Wednesday
00 36 ☉ ⚼ ♃
02 09 ☽ ⊥ ♅
03 02 ☽ ⚼ ♆
06 09 ☽ ⊼ ♇
07 41 ☽ ⊥ ♀
09 13 ☽ ⚹ ♄
12 13 ☽ □ ♂
15 16 ☽ ⚹ ☿
15 56 ☽ Q ☉
16 25 ☽ ± ♇
17 36 ☽ ∠ ♅
18 26 ☽ ± ♆
19 58 ☽ ⊼ ♃
20 03 ☽ ∥ ♃
22 47 ☉ ± ♇

20 Thursday
00 38 ☽ ∠ ♀
04 52 ☽ ⊼ ♆
06 22 ☽ ± ♃
08 01 ☽ ⚼ ♇
11 09 ☽ ∠ ♄
11 24 ☽ Q ☿
14 31 ☽ # ♀
18 00 ☽ # ♅
18 21 ☽ # ☿
19 57 ☽ ⚹ ♅
22 10 ☽ ⚼ ♃

21 Friday
00 32 ☽ # ♆
01 16 ☿ ⊻ ♂
01 52 ☽ □ ☉
03 05 ☽ ⊥ ♄
05 07 ☉ Q ♆
05 41 ☽ ⚹ ♀
10 40 ☽ △ ♇
13 52 ☽ ⊻ ♄
17 37 ☽ Q ♅
18 18 ☽ □ ☿
20 00 ☽ △ ♂
20 47 ☽ ∥ ♄

22 Saturday
01 09 ☽ △ ♃
01 15 ☽ # ☉
05 41 ☽ Q ♀
06 53 ☽ ∥ ♇
10 54 ☽ ☍ ♆
23 38 ☽ # ♂

23 Sunday
01 16 ☽ ⚼ ♂
03 03 ☽ □ ♅
12 35 ♀ ☍ ♇
13 34 ☽ △ ☉
18 17 ☽ □ ♇
18 54 ☽ □ ♀
19 14 ☽ ± ♂
21 37 ☽ ☌ ♄
23 38 ☽ △ ☿

24 Monday
07 22 ☽ ⊼ ♂
09 18 ☽ □ ♃
14 44 ♂ ∠ ♅
19 55 ☽ ⊼ ♆
20 46 ☽ ⚼ ☉
23 02 ♀ □ ♄

25 Tuesday
03 05 ☽ ⚼ ☿
04 06 ☿ ⊻ ♀
04 59 ☽ Q ♇
07 38 ☽ ± ♆
11 27 ☿ ± ♃
13 03 ☽ △ ♅
13 03 ☿ △ ♄
14 45 ♂ ± ♄
15 51 ☽ ± ☉
16 00 ♂ △ ♃
19 45 ☽ ± ☿
20 11 ☽ Q ♃
23 36 ☉ # ♄

26 Wednesday
01 22 ☽ ⚼ ♆
02 57 ☉ ⊼ ♇
04 33 ☽ ⚹ ♇
04 42 ☽ ⊼ ☉
06 58 ☽ ⊼ ☿
07 59 ☽ ⊻ ♄
11 40 ☽ △ ♀
18 53 ☽ ⚼ ♅
19 51 ☽ ⚹ ♃
19 54 ☽ ⊥ ♄
21 40 ☽ ☍ ♂
22 05 ☉ ☌ ☿

27 Thursday
07 16 ☽ △ ♆
10 26 ☽ ∠ ♇
13 00 ☽ ± ♅
13 54 ☽ ∠ ♄
19 23 ☉ ± ♃
21 01 ☽ ⚼ ♀
21 36 ☉ △ ♄
23 26 ☽ # ♂

28 Friday
01 08 ☽ ⊼ ♅
01 13 ☿ ⊼ ♇
01 46 ☽ ∠ ♃
04 32 ☽ ⊥ ♇
07 51 ☿ ⊥ ♀
15 18 ☽ ∥ ♇
15 50 ☽ ☍ ☿
16 39 ☽ ⊻ ♇
17 14 ☽ ± ♀
19 50 ☽ ⊥ ♃
20 09 ☽ ⚹ ♄
22 11 ☽ ☍ ☉

29 Saturday
02 47 ☽ ∥ ♄
06 46 ☽ ⊼ ♀
07 10 ☽ # ☉
07 56 ☽ ⊻ ♃
13 49 ☽ ⊼ ♂
17 34 ♀ ☍ ♃
20 02 ☽ □ ♆
20 12 ♀ # ♃
20 30 ☽ Q ♄

30 Sunday
01 16 ☽ # ♆
02 24 ☽ # ☿
02 45 ☽ ± ♂
09 46 ☽ # ♅
14 21 ☽ ☍ ♅
22 09 ☽ ⚼ ♂

31 Monday
00 46 ☿ ∥ ♆
01 46 ☽ ⊼ ☿
05 34 ☽ ☌ ♇
08 29 ☽ ∥ ♃
08 42 ☽ Q ♆
09 05 ☽ □ ♄
12 24 ☽ # ♀
13 25 ☽ ± ☿
16 33 ☽ ⊼ ☉
20 29 ☽ ☌ ♃

All ephemeris data is given at 12.00 UT and the Moon's longitude is additionally given for 24.00 UT

AUGUST 2045

LONGITUDES

Date	Sidereal time h m s	Sun ☉ ° ′ ″	Moon ☽ ° ′ ″	Moon ☽ 24.00 ° ′ ″	Mercury ☿ ° ′	Venus ♀ ° ′	Mars ♂ ° ′	Jupiter ♃ ° ′	Saturn ♄ ° ′	Uranus ♅ ° ′	Neptune ♆ ° ′	Pluto ♇ ° ′
01	08 42 01	09 ♌ 39 19	18 ♓ 29 09	24 ♓ 26 48	00 ♌ 52	14 ♍ 18	15 ♋ 52	10 ♓ 45	05 ♐ 11	26 ♌ 07	17 ♉ 01	03 ♓ 27
02	08 45 57	10 36 43	00 ♈ 26 25	06 ♈ 28 29	00 R 27	15 29	16 32	10 R 40	05 R 11	26 10	17 02	03 R 25
03	08 49 54	11 34 08	12 ♈ 57 27	18 ♈ 41 50	00 ♌ 06	16 40	17 11	10 34	05 10	26 14	17 02	03 24
04	08 53 51	12 31 33	24 ♈ 54 07	01 ♉ 10 47	29 ♋ 52	17 51	17 51	10 28	05 10	26 17	17 03	03 23
05	08 57 47	13 29 01	07 ♉ 32 20	13 ♉ 59 14	29 43	19 02	18 30	10 22	05 09	26 21	17 03	03 22
06	09 01 44	14 26 29	20 ♉ 31 54	27 ♉ 10 42	29 D 40	20 13	19 09	10 15	05 09	26 25	17 04	03 21
07	09 05 40	15 23 59	03 ♊ 55 53	10 ♊ 47 41	29 44	21 24	19 49	10 09	05 09	26 29	17 04	03 19
08	09 09 37	16 21 30	17 ♊ 46 08	24 ♊ 51 10	29 ♋ 55	22 35	20 28	10 02	05 09	26 32	17 05	03 18
09	09 13 33	17 19 02	02 ♋ 02 34	09 ♋ 19 55	00 ♌ 12	23 46	21 07	09 56	05 D 09	26 36	17 05	03 17
10	09 17 30	18 16 36	16 ♋ 42 40	24 ♋ 10 04	00 37	24 57	21 47	09 49	05 09	26 40	17 05	03 16
11	09 21 26	19 14 11	01 ♌ 41 13	09 ♌ 15 05	01 08	26 08	22 26	09 42	05 09	26 43	17 05	03 15
12	09 25 23	20 11 47	16 ♌ 50 34	24 ♌ 26 28	01 46	27 18	23 05	09 35	05 10	26 47	17 06	03 13
13	09 29 20	21 09 24	02 ♍ 01 34	09 ♍ 34 43	02 31	28 29	23 44	09 28	05 10	26 51	17 06	03 12
14	09 33 16	22 07 03	17 ♍ 04 49	24 ♍ 30 52	03 22	29 ♍ 40	24 23	09 21	05 11	26 55	17 06	03 11
15	09 37 13	23 04 42	01 ♎ 52 02	09 ♎ 07 39	04 20	00 ♎ 50	25 02	09 14	05 11	26 58	17 07	03 09
16	09 41 09	24 02 23	16 ♎ 17 10	23 ♎ 20 16	05 24	02 01	25 41	09 06	05 12	27 02	17 07	03 08
17	09 45 06	25 00 04	00 ♏ 16 45	07 ♏ 06 36	06 35	03 11	26 20	08 59	05 13	27 06	17 07	03 07
18	09 49 02	25 57 47	13 ♏ 49 56	20 ♏ 26 56	07 51	04 21	26 59	08 52	05 14	27 10	17 07	03 06
19	09 52 59	26 55 30	26 ♏ 57 55	03 ♐ 23 16	09 12	05 31	27 38	08 44	05 15	27 13	17 07	03 04
20	09 56 55	27 53 15	09 ♐ 43 25	15 ♐ 58 49	10 39	06 42	28 16	08 36	05 16	27 17	17 R 07	03 03
21	10 00 52	28 51 01	22 ♐ 09 59	28 ♐ 17 23	12 11	07 52	28 55	08 29	05 17	27 21	17 07	03 02
22	10 04 49	29 ♌ 48 48	04 ♑ 21 32	10 ♑ 22 54	13 47	09 02	29 ♋ 34	08 21	05 18	27 25	17 07	03 01
23	10 08 45	00 ♍ 46 36	16 ♑ 21 57	22 ♑ 19 10	15 27	10 12	00 ♌ 12	08 13	05 20	27 28	17 07	02 59
24	10 12 42	01 44 26	28 ♑ 14 56	04 ♒ 09 40	17 11	11 22	00 51	08 06	05 21	27 32	17 07	02 58
25	10 16 38	02 42 16	10 ♒ 03 45	15 ♒ 57 32	18 58	12 31	01 30	07 58	05 23	27 36	17 06	02 57
26	10 20 35	03 40 09	21 ♒ 51 20	27 ♒ 45 27	20 48	13 41	02 08	07 50	05 24	27 39	17 06	02 55
27	10 24 31	04 38 02	03 ♓ 40 13	09 ♓ 35 52	22 40	14 51	02 47	07 42	05 26	27 43	17 06	02 54
28	10 28 28	05 35 57	15 ♓ 32 41	21 ♓ 30 55	24 34	16 00	03 25	07 34	05 28	27 47	17 06	02 53
29	10 32 24	06 33 54	27 ♓ 30 30	03 ♈ 32 42	26 29	17 10	04 03	07 26	05 30	27 51	17 05	02 52
30	10 36 21	07 31 52	09 ♈ 36 46	15 ♈ 43 19	28 ♌ 26	18 19	04 42	07 18	05 32	27 54	17 05	02 50
31	10 40 18	08 ♍ 29 52	21 ♈ 52 37	28 ♈ 04 59	00 ♍ 23	19 ♎ 28	05 ♌ 20	07 ♓ 10	05 ♐ 34	27 ♌ 58	17 ♉ 05	02 ♓ 49

Date	Moon True ☊ ° ′	Moon Mean ☊ ° ′	Moon ☽ Latitude ° ′
01	22 ♒ 45	23 ♒ 25	02 N 18
02	22 D 46	23 22	03 14
03	22 47	23 19	04 03
04	22 48	23 16	04 41
05	22 48	23 12	05 06
06	22 R 48	23 09	05 17
07	22 48	23 06	05 12
08	22 48	23 03	04 48
09	22 47	23 00	04 06
10	22 46	22 56	03 07
11	22 46	22 53	01 55
12	22 46	22 50	00 N 33
13	22 D 46	22 47	00 S 51
14	22 46	22 44	02 11
15	22 R 46	22 41	03 21
16	22 46	22 37	04 16
17	22 46	22 34	04 54
18	22 46	22 31	05 14
19	22 D 45	22 28	05 17
20	22 46	22 25	05 04
21	22 46	22 22	04 37
22	22 47	22 18	03 58
23	22 47	22 15	03 09
24	22 48	22 12	02 12
25	22 49	22 09	01 10
26	22 49	22 06	00 S 05
27	22 R 49	22 02	01 N 00
28	22 48	21 59	02 03
29	22 47	21 56	03 01
30	22 44	21 53	03 51
31	22 ♒ 42	21 ♒ 50	04 N 31

DECLINATIONS

Date	Sun ☉ ° ′	Moon ☽ ° ′	Mercury ☿ ° ′	Venus ♀ ° ′	Mars ♂ ° ′	Jupiter ♃ ° ′	Saturn ♄ ° ′	Uranus ♅ ° ′	Neptune ♆ ° ′	Pluto ♇ ° ′
01	17 N 50	02 S 26	15 N 36	07 N 19	23 N 16	08 S 43	19 S 23	13 N 28	15 N 14	21 S 18
02	17 34	03 N 09	15 52	06 50	23 12	08 45	19 24	13 27	15 14	21 19
03	17 19	08 41	16 08	06 20	23 07	08 48	19 24	13 26	15 14	21 20
04	17 03	13 59	16 23	05 50	23 03	08 50	19 24	13 25	15 14	21 20
05	16 46	18 51	16 39	05 20	22 58	08 53	19 24	13 23	15 14	21 21
06	16 30	22 58	16 54	04 50	22 53	08 55	19 24	13 22	15 14	21 21
07	16 13	26 02	17 08	04 20	22 48	08 58	19 24	13 21	15 14	21 22
08	15 56	27 39	17 22	03 49	22 42	09 01	19 25	13 20	15 14	21 23
09	15 39	27 31	17 34	03 19	22 37	09 03	19 25	13 18	15 14	21 23
10	15 21	25 29	17 45	02 48	22 31	09 06	19 25	13 17	15 14	21 24
11	15 03	21 39	17 55	02 18	22 25	09 09	19 25	13 16	15 15	21 24
12	14 45	16 18	18 03	01 47	22 19	09 12	19 26	13 15	15 15	21 25
13	14 27	09 57	18 09	01 16	22 13	09 14	19 26	13 15	15 15	21 25
14	14 08	03 N 05	18 13	00 45	22 07	09 17	19 26	13 12	15 15	21 26
15	13 49	03 S 49	18 16	00 N 15	22 00	09 20	19 27	13 11	15 15	21 27
16	13 30	10 20	18 15	00 S 16	21 54	09 23	19 27	13 09	15 15	21 27
17	13 11	16 09	18 13	00 47	21 47	09 26	19 27	13 08	15 14	21 28
18	12 52	20 58	18 08	01 18	21 40	09 29	19 28	13 07	15 14	21 28
19	12 32	24 37	18 00	01 49	21 33	09 32	19 28	13 06	15 14	21 29
20	12 12	26 55	17 49	02 20	21 26	09 35	19 29	13 04	15 14	21 29
21	11 52	27 49	17 36	02 51	21 18	09 38	19 29	13 03	15 14	21 30
22	11 32	27 20	17 20	03 22	21 11	09 41	19 29	13 02	15 14	21 30
23	11 12	25 33	17 00	03 52	21 03	09 44	19 30	13 00	15 14	21 31
24	10 51	22 40	16 38	04 23	20 55	09 47	19 30	12 59	15 14	21 32
25	10 31	18 51	16 14	04 54	20 47	09 50	19 31	12 58	15 14	21 32
26	10 10	14 18	15 47	05 24	20 39	09 53	19 31	12 57	15 14	21 33
27	09 49	09 14	15 17	05 55	20 31	09 56	19 32	12 55	15 14	21 33
28	09 27	03 S 49	14 45	06 25	20 22	09 59	19 33	12 54	15 14	21 34
29	09 06	01 N 46	14 10	06 55	20 14	10 02	19 33	12 53	15 13	21 34
30	08 45	07 21	13 34	07 25	20 05	10 05	19 34	12 51	15 13	21 35
31	08 N 23	12 N 43	12 N 56	07 S 55	19 N 56	10 S 08	19 S 34	12 N 50	15 N 13	21 S 35

ZODIAC SIGN ENTRIES

Date	h	m	Planets
02	11	07	☽ ♈
03	21	25	☿ ♋
04	21	45	☽ ♉
07	05	02	☽ ♊
08	19	51	☿ ♌
09	08	36	☽ ♋
11	09	19	☽ ♌
13	08	47	☽ ♍
14	18	57	♀ ♎
15	08	56	☽ ♎
17	11	31	☽ ♏
19	17	39	☽ ♐
22	03	22	☽ ♑
22	16	39	☉ ♍
23	04	17	♂ ♌
24	15	33	☽ ♒
27	04	33	☽ ♓
29	16	57	☽ ♈
31	07	19	☿ ♍

LATITUDES

Date	Mercury ☿ ° ′	Venus ♀ ° ′	Mars ♂ ° ′	Jupiter ♃ ° ′	Saturn ♄ ° ′	Uranus ♅ ° ′	Neptune ♆ ° ′	Pluto ♇ ° ′
01	04 S 28	01 N 14	00 N 47	01 S 17	01 N 48	00 N 42	01 S 45	11 S 55
04	03 52	01 08	00 48	01 17	01 47	00 42	01 46	11 55
07	03 08	01 00	00 50	01 18	01 47	00 42	01 46	11 56
10	02 19	00 52	00 51	01 19	01 46	00 42	01 46	11 56
13	01 29	00 44	00 53	01 19	01 45	00 42	01 46	11 56
16	00 S 41	00 34	00 55	01 20	01 45	00 42	01 46	11 57
19	00 N 03	00 25	00 56	01 20	01 44	00 42	01 46	11 57
22	00 40	00 14	00 58	01 21	01 43	00 42	01 47	11 57
25	01 09	00 N 04	00 59	01 21	01 42	00 42	01 47	11 57
28	01 30	00 S 08	01 01	01 21	01 42	00 42	01 47	11 58
31	01 N 42	00 S 19	01 N 02	01 S 22	01 N 41	00 N 42	01 S 47	11 S 58

DATA

Julian Date	2468194
Delta T	+79 seconds
Ayanamsa	24° 29′ 49″
Synetic vernal point	04° ♓ 37′ 09″
True obliquity of ecliptic	23° 26′ 07″

MOON'S PHASES, APSIDES AND POSITIONS ☽

Date	h	m	Phase	Longitude ° ′	Eclipse Indicator
05	23	57	☾	13 ♉ 58	
12	17	39	●	20 ♌ 25	Total
19	11	55	☽	26 ♏ 55	
27	14	08	○	04 ♓ 43	

Day	h	m		
12	16	30	Perigee	
26	04	00	Apogee	
01	22	30	0N	
08	22	24	Max dec	27° N 49′
14	22	40	0S	
21	15	25	Max dec	27° S 50′
29	04	25	0N	

LONGITUDES

Date	Chiron ⚷ ° ′	Ceres ⚳ ° ′	Pallas ⚴ ° ′	Juno ⚵ ° ′	Vesta ⚶ ° ′	Black Moon Lilith ⚸ ° ′
01	18 ♍ 19	03 ♌ 03	08 ♊ 29	07 ♌ 47	00 ♋ 45	18 ♒ 08
11	19 ♍ 36	07 ♌ 31	12 ♊ 57	12 ♌ 48	04 ♋ 42	19 ♒ 14
21	20 ♍ 56	11 ♌ 58	17 ♊ 21	17 ♌ 43	08 ♋ 39	20 ♒ 21
31	22 ♍ 20	16 ♌ 24	21 ♊ 41	22 ♌ 32	12 ♋ 15	21 ♒ 27

ASPECTARIAN

h m Aspects

01 Tuesday
02 36 ☽ ☍ ♀
05 47 ☽ ± ☉
06 24 ☽ △ ♂
06 55 ☽ ⚼ ☿
09 02 ☽ ⚹ ♆

02 Wednesday
01 30 ☽ ⚼ ☉
03 25 ☽ ⊼ ♅
11 25 ☿ ∠ ♀
12 01 ☽ △ ☿
13 07 ☉ ⊼ ♃
15 10 ☽ ∠ ♆
15 28 ☽ ± ♅
17 56 ☽ ⊻ ♇
21 25 ☽ △ ♄

03 Thursday
02 36 ☽ ∥ ♀
05 48 ☽ ⊥ ♇
06 40 ♂ ⚹ ♆
08 07 ☽ ⊻ ♃
09 01 ☽ ⊥ ♆
09 23 ☽ ⚼ ♅
09 53 ☽ △ ☉
12 30 ☽ # ♃
19 33 ♀ △ ♆
19 47 ☽ ⊥ ♃
20 47 ☽ ⊻ ♆
20 55 ☽ ⊼ ♀
21 34 ☽ □ ♂
23 25 ☽ ∠ ♇

04 Friday
02 51 ☽ ⚼ ♄
09 19 ☽ ∥ ♅
09 46 ☽ ± ♀
11 24 ♀ ⚹ ♂
13 04 ☽ ∠ ♃
14 41 ☽ △ ♅
17 54 ☽ ∥ ♆
20 09 ☽ ± ♄
21 22 ☽ □ ☿

05 Saturday
00 09 ☽ ∥ ☿
01 58 ☽ ∥ ☉
04 09 ☽ ⚹ ♇
04 44 ☽ ⚼ ♀
07 11 ♂ ⚼ ♇
07 31 ☽ ⊼ ♄
09 57 ☽ Q ♂
14 58 ☽ # ♄
17 14 ☽ ⚹ ♃
17 42 ☉ ∥ ☿
23 57 ☽ □ ☉

06 Sunday
01 57 ☽ # ♇
02 31 ☽ Q ♇
05 40 ☽ ☌ ♆
06 47 ☽ Q ☿
09 22 ☽ ⚹ ♂
09 27 ☿ St D
11 23 ☽ △ ♀
11 25 ☽ ∥ ♂
15 06 ☽ Q ♃
22 41 ☽ □ ♅

07 Monday
04 31 ☽ ⚹ ☿
10 56 ☽ □ ♇
10 59 ☽ Q ☉
11 19 ☉ ⊥ ♀
13 38 ☽ ∠ ♂
14 09 ☽ ☍ ♄
22 48 ☽ □ ♃

08 Tuesday
00 19 ♂ ⚼ ♄
06 04 ☽ ⊥ ♂
06 26 ☽ Q ♅
07 02 ☽ ∠ ☿
09 25 ☽ ⚹ ☉
10 49 ☽ ⊻ ♆
14 49 ♂ ⊥ ♅
15 01 ♄ St D
16 49 ☽ ⊻ ♂
20 55 ☽ □ ♀
21 01 ☽ ⊥ ♆
22 37 ☽ ⊥ ☿
23 23 ♀ Q ♄

09 Wednesday
02 53 ☽ ⚹ ♅
06 08 ☉ □ ♆
08 53 ☽ ⊻ ☿
12 04 ☽ ∠ ♆
12 29 ☽ ∠ ☉
14 03 ☽ △ ♇
17 08 ☽ ⊼ ♄

10 Thursday
00 53 ☽ △ ♃
02 58 ☽ ± ♄
03 46 ☽ ∠ ♅
04 18 ☽ ⊥ ☉
05 22 ☽ Q ♀
12 37 ☽ ⚹ ♆
14 30 ☽ ⚼ ♇
14 42 ☽ ⊻ ☉
17 33 ☽ ⚼ ♄
18 24 ☽ ⊥ ♅
20 32 ☽ ☌ ♂
20 42 ☉ ∥ ♆

11 Friday
00 57 ☽ ⚼ ♃
02 23 ☽ ⚹ ♀
04 03 ☽ ⊻ ♅
04 56 ☽ ± ♇
07 46 ☽ ∥ ♂
07 52 ☽ Q ♆
11 05 ☽ ☌ ☿
13 12 ☽ # ♇
14 28 ☽ ⊼ ♇
15 11 ☽ ± ♃
17 31 ☽ △ ♄
22 41 ☽ # ♄

12 Saturday
00 37 ☽ ⊼ ♃
00 47 ♀ ⊻ ♅
04 14 ☽ ∠ ♀
04 55 ☽ ∥ ☿
12 24 ☽ □ ♆
16 15 ☽ ∥ ♆
17 39 ☽ ☌ ☉
18 30 ☽ ∥ ☉
19 38 ☽ ⊥ ♀
22 18 ☽ ⊻ ♂
23 59 ☽ ∥ ♅

13 Sunday
03 46 ☽ ☌ ♅
05 55 ☽ ⊻ ♀
08 12 ☽ ⊥ ♂
12 49 ☽ ⊻ ☿
13 52 ☽ ☍ ♇
14 32 ☽ # ♃
16 59 ☽ □ ♄
22 54 ☽ ⊥ ☿
23 08 ☽ ∠ ♂
23 44 ☽ ☍ ♃

14 Monday
07 05 ☿ ⊼ ♇
10 59 ♂ ⚼ ♃
11 36 ☿ ± ♃
12 03 ☽ △ ♆
14 12 ☽ ∠ ☿
20 41 ☽ ⊻ ☉
20 41 ☽ ∥ ♀
21 50 ☽ Q ♄

15 Tuesday
00 20 ☽ ⚹ ♂
00 22 ☽ # ♀
03 58 ☽ ⊻ ♅
07 07 ☽ ⊥ ☉
10 10 ☽ ☌ ♀
12 24 ☽ ⚼ ♆
13 49 ☽ ⊥ ♅
14 07 ☽ ⊼ ♇
16 22 ☽ ⚹ ☿
17 28 ☽ ⚹ ♄
20 55 ☽ Q ♂
22 59 ☽ ∠ ☉

16 Wednesday
00 02 ☽ ± ♇
00 04 ☽ ⊼ ♃
03 18 ☽ ± ♆
04 49 ☽ ∠ ♅
07 31 ☿ △ ♄
08 20 ☽ ∥ ♃
10 02 ☽ ± ♃
13 24 ☽ ⊼ ♆
14 03 ☽ Q ☿
14 06 ♀ ⚼ ♆
15 08 ☽ ⚼ ♇
18 38 ☽ ∠ ♄
23 10 ☽ # ♅
23 59 ☽ # ☉

17 Thursday
01 12 ☽ ⚼ ♃
02 11 ☽ ⚹ ☉
04 49 ☽ □ ♂
06 27 ☽ ⚹ ♅
08 00 ☽ # ♆
10 08 ☽ ⊥ ♄
10 08 ♀ ⊥ ♅
10 40 ♀ ⊼ ♇
16 02 ☉ ∥ ♅
16 57 ☽ △ ♇
17 33 ☽ ⊻ ♀
20 39 ☽ ⊻ ♄
21 29 ☽ # ☿

18 Friday
00 10 ☽ □ ☿
00 43 ☽ Q ☉
03 11 ☽ △ ♃
03 36 ☽ Q ♅
03 51 ☽ ∥ ♄
05 10 ☽ ⊥ ♀
14 51 ☽ ∥ ♇
15 53 ☽ # ♂
16 09 ♂ ± ♇
17 56 ☽ ☍ ♆
19 22 ♂ ⊻ ♅
22 59 ☽ ∠ ♀

19 Saturday
04 33 ☿ ⊼ ♃
06 09 ♀ ⚹ ♄
11 55 ☽ □ ☉
12 29 ☽ □ ♅
13 18 ☽ △ ♂
19 54 ☉ ☌ ♅
23 23 ☽ □ ♇

20 Sunday
00 20 ♆ St R
00 47 ♂ # ♇
03 31 ☽ ☌ ♄
05 39 ☽ ⚹ ♀
09 54 ☽ □ ♃
14 01 ☽ △ ☿
19 10 ☽ ⚼ ♂

21 Monday
02 11 ☽ ⊼ ♆
07 03 ☽ Q ♀
09 47 ☽ Q ♇
13 33 ☽ ± ♂
13 51 ☽ ± ♆
17 07 ☉ ⊻ ♂
19 17 ♂ Q ♆
20 21 ☽ Q ♃
22 12 ☽ △ ♅
23 17 ☽ ⚼ ☿
23 26 ♀ ⊼ ♃

22 Tuesday
01 59 ☽ ⊼ ♂
02 14 ☽ △ ☉
07 33 ☽ ⚼ ♆
09 20 ☽ ⚹ ♇
11 35 ♀ ± ♇
13 53 ☽ ⊻ ♄
19 52 ☽ ⚹ ♃
19 54 ☽ ± ☿
22 18 ☽ □ ♀

23 Wednesday
01 52 ☽ ⊥ ♄
04 08 ☽ ⚼ ♅
09 52 ☽ ⊼ ☿
10 43 ☽ ⚼ ☉
13 30 ☽ △ ♆
15 15 ☽ ∠ ♇
20 00 ☽ ∠ ♄
22 20 ☽ ± ♅

24 Thursday
01 40 ☽ ∠ ♃
06 28 ☽ ± ☉
06 50 ♀ ± ♆
09 24 ☽ ⊥ ♇
10 33 ☽ ⊼ ♅
10 55 ☿ □ ♆
17 35 ☽ ☍ ♂
19 43 ☽ ⊼ ☉
19 43 ☽ ⊥ ♃
19 44 ☽ ∥ ♇
21 33 ☽ ⊻ ♇

25 Friday
00 03 ☽ # ♂
02 27 ☽ ⚹ ♄
07 46 ☽ ⊻ ♃
08 09 ☽ ∥ ♄
13 35 ♀ ∠ ♅
17 33 ☽ △ ♀
17 51 ☉ ☍ ♇

26 Saturday
02 20 ☽ □ ♆
02 34 ♂ ± ♃
02 55 ☽ Q ♄
03 45 ☽ # ☿
07 22 ☽ # ♆
09 28 ☽ ☍ ☿
14 42 ♀ ± ♃
18 41 ☽ # ♅
23 52 ☽ ☍ ♅

27 Sunday
03 24 ☽ ⚼ ♀
04 29 ☉ # ♃
08 47 ♀ Q ♂
08 48 ☽ ∥ ♃
09 10 ☽ # ☉
10 05 ☽ ⊼ ♂
10 27 ☽ ☌ ♇
14 08 ☽ ☍ ☉
14 22 ☿ ∥ ♆
14 54 ☽ Q ♆
15 35 ☽ □ ♄
16 35 ♂ ⊼ ♇
20 04 ☽ ☌ ♃
22 56 ☽ ± ♂
23 37 ☽ ± ♀

28 Monday
01 35 ☽ ∥ ♀
08 36 ☉ □ ♄
13 02 ☽ ⊼ ♀
15 07 ☽ ⚹ ♆
18 06 ☽ ⚼ ♂

29 Tuesday
09 34 ☽ ⊼ ☿
10 31 ♀ ⊼ ♆
12 40 ☽ ⊼ ♅
21 06 ☽ ∠ ♆
22 37 ☽ ⊻ ♇
23 48 ☽ ± ☿

30 Wednesday
00 39 ☽ ± ♅
01 44 ☽ △ ♂
02 12 ♀ ⚼ ♇
03 55 ☽ △ ♄
05 20 ☿ ☌ ♅
07 02 ☉ ☍ ♃
07 30 ☽ ⊻ ♃
07 32 ☽ ⊼ ☉
10 28 ☽ ⊥ ♇
12 21 ☽ # ♀
14 54 ☽ ⊥ ♆
17 44 ☽ ∥ ☉
18 31 ☽ ⚼ ♅
19 11 ☽ ⊥ ♃
20 22 ☽ ± ☉
20 56 ☽ ⚼ ☿

31 Thursday
00 12 ☽ # ♃
02 39 ☽ ⊻ ♆
04 06 ☽ ∠ ♇
06 50 ☽ ☍ ♀
09 27 ☽ ⚼ ♄
12 33 ☽ ∥ ♅
12 34 ☽ ∠ ♃
12 52 ☽ ∥ ☿
15 25 ☽ ⚼ ☉
15 29 ☿ ∥ ♅
21 27 ♂ △ ♄
23 48 ☽ ∥ ♆
23 50 ☽ △ ♅

All ephemeris data is given at 12.00 UT and the Moon's longitude is additionally given for 24.00 UT

SEPTEMBER 2045

Date	Sidereal time h m s	Sun ☉	Moon ☽	Moon ☽ 24.00	Mercury ☿	Venus ♀	Mars ♂	Jupiter ♃	Saturn ♄	Uranus ♅	Neptune ♆	Pluto ♇
01	10 44 14	09 ♍ 27 54	04 ♉ 20 42	10 ♉ 40 05	02 ♍ 20	20 ♎ 38	05 ♌ 58	07 ♓ 02	05 ♐ 36	28 ♌ 02	17 ♉ 04	02 ♓ 48
02	10 48 11	10 25 57	17 ♉ 03 28	23 ♉ 31 08	04 18	21 47	06 36	06 R 55	05 39	28 05	17 R 04	02 R 46
03	10 52 07	11 24 03	00 ♊ 03 26	06 ♊ 40 37	06 16	22 56	07 15	06 47	05 41	28 09	17 03	02 45
04	10 56 04	12 22 11	13 ♊ 22 58	20 ♊ 10 42	08 13	24 05	07 53	06 39	05 44	28 13	17 03	02 44
05	11 00 00	13 20 21	27 ♊ 03 57	04 ♋ 02 49	10 10	25 13	08 31	06 31	05 46	28 16	17 02	02 43
06	11 03 57	14 18 32	11 ♋ 07 16	18 ♋ 17 08	12 06	26 22	09 09	06 23	05 49	28 20	17 02	02 41
07	11 07 53	15 16 46	25 ♋ 30 27	02 ♌ 52 00	14 01	27 31	09 47	06 15	05 52	28 24	17 01	02 40
08	11 11 50	16 15 02	10 ♌ 16 00	17 ♌ 43 28	15 56	28 39	10 25	06 08	05 55	28 27	17 01	02 39
09	11 15 47	17 13 20	25 ♌ 13 34	02 ♍ 45 19	17 49	29 ♎ 48	11 03	06 00	05 58	28 31	17 00	02 38
10	11 19 43	18 11 39	10 ♍ 17 40	17 ♍ 49 28	19 42	00 ♏ 56	11 41	05 53	06 01	28 35	16 59	02 36
11	11 23 40	19 10 00	25 ♍ 19 35	02 ♎ 46 55	21 34	02 04	12 19	05 45	06 04	28 38	16 59	02 35
12	11 27 36	20 08 23	10 ♎ 10 26	17 ♎ 29 12	23 24	03 13	12 56	05 38	06 07	28 42	16 58	02 34
13	11 31 33	21 06 48	24 ♎ 42 26	01 ♏ 49 32	25 14	04 21	13 34	05 30	06 10	28 45	16 57	02 33
14	11 35 29	22 05 15	08 ♏ 50 03	15 ♏ 43 43	27 02	05 29	14 12	05 23	06 14	28 49	16 56	02 31
15	11 39 26	23 03 43	22 ♏ 30 27	29 ♏ 10 20	28 ♍ 50	06 36	14 49	05 16	06 17	28 52	16 56	02 30
16	11 43 22	24 02 13	05 ♐ 43 33	12 ♐ 10 27	00 ♎ 36	07 44	15 27	05 09	06 21	28 56	16 55	02 29
17	11 47 19	25 00 44	18 ♐ 31 26	24 ♐ 47 00	02 21	08 52	16 05	05 02	06 24	28 59	16 54	02 28
18	11 51 16	25 59 17	00 ♑ 57 44	07 ♑ 04 12	04 05	09 59	16 42	04 55	06 28	29 03	16 53	02 27
19	11 55 12	26 57 52	13 ♑ 07 00	19 ♑ 06 47	05 48	11 06	17 19	04 48	06 32	29 06	16 52	02 26
20	11 59 09	27 56 29	25 ♑ 04 09	00 ♒ 59 42	07 30	12 13	17 57	04 41	06 36	29 10	16 51	02 24
21	12 03 05	28 55 07	06 ♒ 54 01	12 ♒ 47 39	09 11	13 20	18 34	04 35	06 40	29 13	16 50	02 23
22	12 07 02	29 ♍ 53 46	18 ♒ 41 08	24 ♒ 34 56	10 51	14 27	19 12	04 28	06 44	29 16	16 49	02 22
23	12 10 58	00 ♎ 52 28	00 ♓ 29 29	06 ♓ 25 12	12 31	15 34	19 49	04 22	06 48	29 20	16 48	02 21
24	12 14 55	01 51 11	12 ♓ 22 25	18 ♓ 21 28	14 09	16 41	20 26	04 16	06 52	29 23	16 47	02 20
25	12 18 51	02 49 56	24 ♓ 22 35	00 ♈ 26 01	15 46	17 47	21 03	04 10	06 56	29 26	16 46	02 19
26	12 22 48	03 48 43	06 ♈ 31 57	12 ♈ 40 31	17 22	18 53	21 40	04 04	07 01	29 29	16 45	02 18
27	12 26 45	04 47 33	18 ♈ 51 51	25 ♈ 06 03	18 57	19 59	22 17	03 58	07 05	29 33	16 44	02 17
28	12 30 41	05 46 24	01 ♉ 23 10	07 ♉ 43 19	20 32	21 05	22 55	03 52	07 09	29 36	16 42	02 16
29	12 34 38	06 45 17	14 ♉ 06 31	20 ♉ 32 52	22 05	22 11	23 32	03 47	07 14	29 39	16 41	02 15
30	12 38 34	07 ♎ 44 13	27 ♉ 02 25	03 ♊ 35 15	23 ♎ 38	23 ♏ 16	24 ♌ 08	03 ♓ 42	07 ♐ 19	29 ♌ 42	16 ♉ 40	02 ♓ 14

Date	Moon True ☊	Moon Mean ☊	Moon ☽ Latitude
01	22 ♒ 40	21 ♒ 47	05 N 00
02	22 R 38	21 43	05 14
03	22 37	21 40	05 13
04	22 D 36	21 37	04 55
05	22 37	21 34	04 21
06	22 38	21 31	03 30
07	22 39	21 28	02 24
08	22 41	21 24	01 N 08
09	22 R 41	21 21	00 S 14
10	22 40	21 18	01 35
11	22 38	21 15	02 50
12	22 35	21 12	03 52
13	22 31	21 08	04 37
14	22 26	21 05	05 05
15	22 23	21 02	05 13
16	22 20	20 59	05 05
17	22 18	20 56	04 41
18	22 D 18	20 53	04 05
19	22 19	20 49	03 18
20	22 21	20 46	02 24
21	22 23	20 43	01 24
22	22 24	20 40	00 S 20
23	22 R 23	20 37	00 N 44
24	22 21	20 33	01 47
25	22 17	20 30	02 45
26	22 12	20 27	03 36
27	22 05	20 24	04 19
28	21 57	20 21	04 49
29	21 49	20 18	05 05
30	21 ♒ 42	20 ♒ 14	05 N 07

DECLINATIONS

Date	Sun ☉	Moon ☽	Mercury ☿	Venus ♀	Mars ♂	Jupiter ♃	Saturn ♄	Uranus ♅	Neptune ♆	Pluto ♇
01	08 N 01	17 N 40	12 N 16	08 S 25	19 N 47	10 S 11	19 S 35	12 N 49	15 N 13	21 S 36
02	07 39	21 57	11 35	08 55	19 38	10 14	19 35	12 48	15 13	21 36
03	07 17	25 15	10 52	09 24	19 29	10 17	19 36	12 46	15 13	21 37
04	06 55	27 17	10 08	09 53	19 20	10 20	19 37	12 45	15 12	21 37
05	06 33	27 45	09 24	10 22	19 10	10 23	19 37	12 44	15 12	21 37
06	06 10	26 27	08 38	10 51	19 01	10 26	19 38	12 43	15 12	21 38
07	05 48	23 23	07 52	11 20	18 51	10 29	19 39	12 41	15 12	21 38
08	05 25	18 45	07 06	11 49	18 41	10 32	19 40	12 40	15 12	21 39
09	05 03	12 53	06 19	12 17	18 31	10 35	19 40	12 39	15 11	21 39
10	04 40	06 N 14	05 31	12 45	18 21	10 38	19 41	12 38	15 11	21 40
11	04 17	00 S 45	04 44	13 12	18 11	10 40	19 42	12 36	15 11	21 40
12	03 54	07 35	03 56	13 40	18 01	10 43	19 42	12 35	15 11	21 41
13	03 31	13 52	03 09	14 07	17 51	10 46	19 43	12 34	15 10	21 41
14	03 08	19 15	02 21	14 34	17 40	10 48	19 44	12 33	15 10	21 41
15	02 45	23 26	01 34	15 01	17 30	10 51	19 45	12 32	15 10	21 42
16	02 22	26 15	00 N 46	15 27	17 19	10 54	19 46	12 30	15 09	21 42
17	01 59	27 37	00 S 01	15 53	17 08	10 56	19 46	12 29	15 09	21 42
18	01 36	27 31	00 48	16 19	16 57	10 59	19 47	12 28	15 09	21 43
19	01 12	26 05	01 34	16 44	16 47	11 01	19 48	12 27	15 09	21 43
20	00 49	23 28	02 21	17 09	16 36	11 04	19 49	12 26	15 08	21 44
21	00 26	19 54	03 06	17 33	16 24	11 06	19 50	12 24	15 08	21 44
22	00 N 02	15 33	03 52	17 58	16 13	11 08	19 51	12 23	15 08	21 44
23	00 S 21	10 37	04 37	18 21	16 02	11 10	19 52	12 22	15 07	21 45
24	00 44	05 S 16	05 21	18 45	15 51	11 13	19 52	12 21	15 07	21 45
25	01 08	00 N 17	06 05	19 08	15 39	11 15	19 53	12 20	15 07	21 45
26	01 31	05 54	06 49	19 30	15 28	11 17	19 54	12 19	15 06	21 45
27	01 54	11 22	07 32	19 53	15 16	11 19	19 55	12 18	15 06	21 46
28	02 18	16 28	08 14	20 14	15 04	11 21	19 56	12 17	15 06	21 46
29	02 41	20 55	08 56	20 36	14 53	11 23	19 57	12 15	15 05	21 46
30	03 S 04	24 N 28	09 S 37	20 S 57	14 N 41	11 S 25	19 S 58	12 N 14	15 N 05	21 S 46

ZODIAC SIGN ENTRIES

Date	h m	Planets
01	03 41	☽ ♉
03	11 54	☽ ♊
05	17 04	☽ ♋
07	19 19	☽ ♌
09	16 15	♀ ♏
09	19 37	☽ ♍
11	19 31	☽ ♎
13	20 54	☽ ♏
16	01 30	☽ ♐
16	03 52	☿ ♎
18	10 07	☽ ♑
20	21 59	☽ ♒
22	14 33	☉ ♎
23	11 00	☽ ♓
25	23 09	☽ ♈
28	09 22	☽ ♉
30	17 26	☽ ♊

LATITUDES

Date	Mercury ☿	Venus ♀	Mars ♂	Jupiter ♃	Saturn ♄	Uranus ♅	Neptune ♆	Pluto ♇
01	01 N 44	00 S 23	01 N 03	01 S 22	01 N 41	00 N 42	01 S 47	11 S 58
04	01 47	00 36	01 04	01 22	01 40	00 42	01 47	11 58
07	01 43	00 48	01 05	01 22	01 40	00 42	01 48	11 58
10	01 34	01 01	01 07	01 22	01 39	00 42	01 48	11 57
13	01 22	01 14	01 09	01 22	01 38	00 42	01 48	11 57
16	01 06	01 26	01 10	01 22	01 38	00 42	01 48	11 57
19	00 48	01 39	01 12	01 22	01 37	00 42	01 48	11 57
22	00 28	01 52	01 13	01 22	01 36	00 42	01 48	11 57
25	00 N 07	02 05	01 14	01 21	01 36	00 43	01 48	11 56
28	00 S 14	02 18	01 16	01 21	01 35	00 43	01 49	11 56
31	00 S 36	02 S 30	01 N 17	01 S 21	01 N 35	00 N 43	01 S 49	11 S 56

DATA

Julian Date	2468225
Delta T	+79 seconds
Ayanamsa	24° 29' 53"
Synetic vernal point	04° ♓ 37' 06"
True obliquity of ecliptic	23° 26' 08"

MOON'S PHASES, APSIDES AND POSITIONS ☽

Date	h m	Phase	Longitude	Eclipse Indicator
04	10 03	☾	12 ♊ 17	
11	01 28	●	18 ♍ 44	
18	01 30	☽	25 ♐ 34	
26	06 11	○	03 ♈ 34	

Day	h m		
10	02 17	Perigee	
22	12 13	Apogee	
05	06 30	Max dec	27° N 47'
11	09 27	0S	
17	22 20	Max dec	27° S 44'
25	10 45	0N	

LONGITUDES

Date	Chiron ⚷	Ceres ⚳	Pallas ⚴	Juno ⚵	Vesta ⚶	Black Moon Lilith ⚸
01	22 ♍ 28	16 ♌ 51	22 ♊ 07	23 ♌ 00	12 ♋ 37	21 ♒ 34
11	23 ♍ 54	21 ♌ 15	26 ♊ 18	27 ♌ 42	16 ♋ 08	22 ♒ 40
21	25 ♍ 21	25 ♌ 37	00 ♋ 17	02 ♍ 16	19 ♋ 27	23 ♒ 47
31	26 ♍ 47	29 ♌ 54	04 ♋ 01	06 ♍ 42	22 ♋ 31	24 ♒ 53

All ephemeris data is given at 12.00 UT and the Moon's longitude is additionally given for 24.00 UT

ASPECTARIAN

h m	Aspects
01 Friday	
00 57	☉ # ♀
02 54	☽ ± ♄
07 28	☽ △ ☿
09 03	☽ ✶ ♇
11 35	♀ ∠ ♄
14 25	☽ ⊼ ♄
15 16	☽ □ ♂
17 05	☽ ✶ ♃
17 31	☿ ☍ ♇
22 13	☽ # ♄
22 32	☽ △ ☉
22 58	☽ ∥ ♂
02 Saturday	
07 44	☽ Q ♇
09 53	☽ # ♇
12 01	☽ ☌ ♆
14 27	♀ ₽ ♃
15 25	☽ Q ♃
18 53	♂ # ♄
21 25	♂ ⊼ ♃
21 38	☽ ⊼ ♀
03 Sunday	
02 44	☽ Q ♂
04 49	☿ □ ♄
08 30	☽ □ ♅
09 44	☽ ± ♀
16 54	☽ □ ♇
17 57	☿ ☍ ♃
22 15	☽ ☍ ♄
04 Monday	
00 04	☽ □ ♃
01 11	☽ □ ☿
01 40	☽ ✶ ♂
03 35	☽ ₽ ♀
05 54	☿ ⊻ ♂
05 57	☿ # ♃
10 03	☽ □ ☉
16 55	☿ # ♀
17 02	☽ Q ♅
18 29	☽ ⊻ ♆
05 Tuesday	
05 01	☽ ⊥ ♆
05 32	☽ ∠ ♂
08 31	☽ △ ♀
12 39	♀ ∥ ♃
13 56	☿ ∠ ♀
14 06	☽ ✶ ♅
14 12	☽ Q ☿
19 55	☽ Q ☉
20 34	☽ ∠ ♆
21 42	☽ △ ♇
21 49	☽ ⊥ ♂
06 Wednesday	
02 59	☽ ⊼ ♄
04 04	☽ △ ♃
08 31	☽ ⊻ ♂
13 10	☽ ± ♄
13 54	☽ ✶ ☿
15 44	☽ ∠ ♅
17 45	☽ ✶ ☉
21 54	☽ ✶ ♆
22 59	☽ ₽ ♇
07 Thursday	
04 15	☽ ₽ ♄
05 00	☽ ₽ ♃
06 48	☽ ⊥ ♅
13 52	☽ ± ♇
15 32	☽ □ ♀
16 43	☽ ⊻ ♅
17 43	☽ Q ♆
18 35	☽ ∠ ☿
19 41	☽ ± ♃
20 20	☽ ∠ ☉
21 57	☽ # ♇
23 40	☽ ⊼ ♇
08 Friday	
04 56	☽ △ ♄
05 22	☽ ⊼ ♃
07 32	♀ ✶ ♅
07 51	☽ # ♄
11 22	☽ ⊥ ☿
11 58	☽ ⊥ ☉
12 15	☽ ☌ ♂
12 19	☽ ∥ ♂
20 20	☉ ☌ ☿
21 13	☿ ⊥ ♂
22 18	☽ ⊻ ☉
22 27	☽ ⊻ ☿
22 51	☽ □ ♆
23 08	☽ Q ♀
23 38	☉ ⊥ ♂
09 Saturday	
01 37	☿ △ ♆
03 04	☽ ∥ ♆
06 36	☉ △ ♆
12 55	☽ ∥ ♅
14 09	☽ # ♀
15 33	♀ ⊥ ♄
17 16	☽ ☌ ♅
17 45	♃ □ ♄
19 53	☽ ✶ ♀
20 31	☽ # ♃
23 47	☽ ☍ ♇
10 Sunday	
05 01	☽ ☍ ♃
05 09	☽ □ ♄
06 09	♀ # ♅
14 18	☽ ⊻ ♂
14 47	☽ ∥ ☿
17 44	☽ ∥ ☉
21 44	☽ ∠ ♀
22 40	☽ △ ♆
11 Monday	
00 17	☽ ⊥ ♂
01 28	☽ ☌ ☉
05 07	☽ ☌ ☿
09 58	☽ Q ♄
13 18	☽ ⊥ ♀
15 19	☽ ∠ ♂
17 20	☽ ⊻ ♅
22 36	♀ △ ♇
22 42	☽ ₽ ♆
23 38	☽ # ☉
23 40	☽ ⊼ ♇
23 45	☽ ⊻ ♀
12 Tuesday	
00 24	☽ # ☿
03 04	☽ ⊥ ♅
04 40	☽ ⊼ ♃
05 22	☽ ✶ ♄
09 23	☽ ± ♇
13 18	☽ ± ♆
13 57	☉ ∥ ☿
14 21	☽ ± ♃
16 43	☽ ✶ ♂
17 47	☽ ∠ ♅
21 37	☿ Q ♄
23 08	☽ ⊼ ♆
23 42	☽ ∥ ♃
13 Wednesday	
00 07	☽ ₽ ♇
05 03	☽ ₽ ♃
05 34	☽ ⊻ ☉
06 05	☽ ∠ ♄
06 48	☽ # ♅
13 00	☽ ⊻ ☿
13 07	☽ ∥ ♀
13 30	☽ Q ♂
16 20	☽ ⊥ ☉
17 25	☽ # ♆
18 50	☽ ✶ ♅
21 14	☽ ⊥ ♄
14 Thursday	
00 36	☽ ⊥ ☿
01 12	☽ △ ♇
04 42	☽ # ♂
05 43	☽ ☌ ♀
06 07	☽ △ ♃
07 30	☽ ⊻ ♄
08 46	☽ ∠ ☉
10 11	♀ △ ♃
14 29	☽ ∥ ♄
15 26	☽ Q ♅
18 23	☽ ∠ ☿
21 46	☽ □ ♂
15 Friday	
01 06	☽ ∥ ♇
02 07	☽ ☍ ♆
04 48	♀ ⊻ ♄
12 38	☿ ⊻ ♅
13 04	☽ ✶ ☉
20 11	♀ # ♆
23 30	☽ □ ♅
16 Saturday	
01 08	☽ ✶ ☿
06 03	☽ □ ♇
08 53	☿ ∠ ♂
10 56	☽ □ ♃
12 37	☽ Q ☉
13 09	☽ ☌ ♄
16 04	☽ ⊻ ♀
20 04	☉ Q ♄
17 Sunday	
02 50	☽ Q ☿
04 23	☽ ⊥ ♀
05 49	☿ ₽ ♆
07 07	☽ △ ♂
08 55	☽ ⊼ ♆
13 33	☿ ⊼ ♇
15 42	☽ Q ♇
20 21	☽ ± ♆
20 32	☽ Q ♃
23 14	☽ ∠ ♀
18 Monday	
01 30	☽ □ ☉
07 55	☿ ⊥ ♀
08 15	☽ △ ♅
13 31	☽ ₽ ♂
13 48	☽ ₽ ♆
14 54	☽ ✶ ♇
18 49	♂ □ ♆
19 01	♃ Q ♆
19 08	☽ □ ☿
19 40	☽ ✶ ♃
22 46	☿ ⊼ ♃
22 52	☽ ⊻ ♄
19 Tuesday	
01 49	☿ ⊥ ♅
04 28	☉ # ☿
07 35	☽ ✶ ♀
08 14	☽ ± ♂
10 50	☽ ⊥ ♄
11 56	♀ Q ♅
13 45	♀ # ♂
13 58	☽ ₽ ♅
19 29	☽ △ ♆
20 36	☽ ∠ ♇
20 52	☽ ⊼ ♂
22 36	☿ ✶ ♄
20 Wednesday	
01 15	☽ ∠ ♃
04 57	☽ ∠ ♄
08 08	☽ ± ♅
10 07	☽ Q ♀
14 42	☽ ⊥ ♇
18 20	☽ △ ☉
19 15	☽ ⊥ ♃
20 19	☽ ⊼ ♅
21 Thursday	
00 25	☽ ∥ ♇
00 41	☿ ± ♇
02 51	☽ ⊻ ♇
07 19	☽ ⊻ ♃
11 31	☽ ✶ ♄
12 22	☽ ∥ ♄
17 26	☽ △ ☿
19 44	☉ ⊻ ♅
22 Friday	
00 13	☽ ∥ ♀
02 29	☽ □ ♀
03 35	☽ ₽ ☉
06 44	☿ ± ♃
08 12	☽ □ ♆
08 20	☽ # ♂
11 26	☿ ± ♆
12 05	☽ Q ♄
13 05	☽ ☍ ♂
14 08	☽ # ♆
23 34	☽ ± ☉
23 Saturday	
03 41	☽ # ♅
04 58	☽ ₽ ☿
09 23	☽ ∥ ♃
09 38	☽ ☍ ♅
12 51	☽ ⊼ ☉
15 46	☽ ☌ ♇
19 47	☽ ☌ ♃
20 43	☽ Q ♆
24 Sunday	
00 50	☽ □ ♄
02 08	☽ ± ☿
10 18	☉ ₽ ♆
11 42	☽ ∥ ☿
14 16	♀ ☍ ♆
15 39	☿ ∠ ♅
16 07	☽ ⊼ ☿
20 50	☽ ✶ ♆
21 31	☽ △ ♀
23 34	☉ ⊼ ♇
25 Monday	
05 02	☽ ⊼ ♂
06 20	☽ ∥ ☉
15 50	☽ # ☉
17 36	☽ ± ♂
21 35	☉ ∠ ♀
22 05	☽ ⊼ ♅
26 Tuesday	
02 36	☽ ∠ ♆
02 48	☿ ⊼ ♆
03 41	☽ ⊻ ♇
06 11	☽ ☍ ☉
06 17	☽ ₽ ♀
07 11	☽ ⊻ ♃
09 57	☽ ± ♅
11 00	☿ ₽ ♇
12 17	☽ ₽ ♂
12 56	☽ △ ♄
15 27	☽ ⊥ ♇
16 30	☽ # ☿
17 35	☉ ⊼ ♃
18 51	☽ ⊥ ♃
20 14	☽ ⊥ ♆
27 Wednesday	
01 38	☽ ± ♀
03 36	☽ ₽ ♅
07 52	☽ ⊻ ♆
08 57	☽ ∠ ♇
11 45	☽ # ♃
12 11	☿ ₽ ♃
12 12	☽ ∠ ♃
12 12	☽ ☍ ☿
14 23	☽ ⊼ ♀
14 55	♀ ∥ ♄
16 11	☽ ∥ ♅
18 15	☽ ₽ ♄
18 57	☽ △ ♂
28 Thursday	
05 18	☽ ∥ ♆
05 27	☽ ∥ ♂
07 29	☉ ⊥ ♅
08 35	☽ △ ♅
09 33	♂ ∥ ♆
11 34	☽ ± ♄
13 40	☽ ✶ ♇
16 41	☽ ✶ ♃
21 01	☽ ⊼ ☉
23 00	☽ ⊼ ♄
29 Friday	
06 21	☽ # ♄
09 15	☽ ± ☉
09 53	☽ # ♀
12 15	☽ Q ♇
14 26	☿ ∠ ♄
15 06	☽ Q ♃
16 49	☽ ☌ ♆
17 02	☿ ⊻ ♀
17 09	☽ # ♇
30 Saturday	
00 43	☉ ✶ ♄
03 24	☽ ₽ ☉
04 25	☽ ☍ ♀
04 52	☽ ⊼ ☿
06 23	☽ □ ♂
16 55	☽ □ ♅
17 23	☽ ± ☿
21 30	☽ □ ♇
23 50	☉ ± ♇

OCTOBER 2045

LONGITUDES

Date	Sidereal time h m s	Sun ☉ ° ' "	Moon ☽ ° ' "	Moon ☽ 24.00 ° ' "	Mercury ☿ ° '	Venus ♀ ° '	Mars ♂ ° '	Jupiter ♃ ° '	Saturn ♄ ° '	Uranus ♅ ° '	Neptune ♆ ° '	Pluto ♇ ° '
01	12 42 31	08 ♎ 43 11	10 ♊ 11 28	16 ♊ 51 10	25 ♎ 09	24 ♏ 22	24 ♌ 45	03 ♓ 36	07 ♐ 23	29 ♌ 45	16 ♉ 39	02 ♓ 13
02	12 46 27	09 42 11	23 ♊ 34 27	00 ♋ 21 26	26 40	25 27	25 22	03 R 31	07 28	29 49	16 R 38	02 R 12
03	12 50 24	10 41 13	07 ♋ 12 14	14 ♋ 06 55	28 10	26 32	25 59	03 27	07 33	29 52	16 36	02 11
04	12 54 20	11 40 18	21 ♋ 05 32	28 ♋ 08 06	29 ♎ 39	27 37	26 36	03 22	07 38	29 55	16 35	02 10
05	12 58 17	12 39 25	05 ♌ 14 32	12 ♌ 24 41	01 ♏ 08	28 41	27 13	03 17	07 43	29 ♌ 58	16 34	02 09
06	13 02 14	13 38 35	19 ♌ 38 18	26 ♌ 55 00	02 35	29 ♏ 46	27 49	03 13	07 48	00 ♍ 01	16 32	02 08
07	13 06 10	14 37 46	04 ♍ 14 18	11 ♍ 35 34	04 02	00 ♐ 50	28 26	03 09	07 53	00 04	16 31	02 07
08	13 10 07	15 37 00	18 ♍ 58 04	26 ♍ 20 58	05 27	01 54	29 02	03 05	07 58	00 06	16 30	02 06
09	13 14 03	16 36 16	03 ♎ 43 20	11 ♎ 04 13	06 52	02 57	29 ♌ 39	03 01	08 03	00 09	16 28	02 05
10	13 18 00	17 35 34	18 ♎ 22 39	25 ♎ 37 41	08 16	04 01	00 ♍ 15	02 58	08 09	00 12	16 27	02 05
11	13 21 56	18 34 54	02 ♏ 48 28	09 ♏ 54 14	09 38	05 04	00 52	02 54	08 14	00 15	16 25	02 04
12	13 25 53	19 34 17	16 ♏ 54 22	23 ♏ 48 25	11 00	06 07	01 28	02 51	08 19	00 18	16 24	02 03
13	13 29 49	20 33 41	00 ♐ 36 05	07 ♐ 17 13	12 21	07 10	02 04	02 48	08 25	00 20	16 22	02 02
14	13 33 46	21 33 07	13 ♐ 51 52	20 ♐ 20 11	13 40	08 12	02 41	02 45	08 31	00 23	16 21	02 01
15	13 37 43	22 32 35	26 ♐ 42 29	02 ♑ 59 10	14 59	09 14	03 17	02 42	08 36	00 26	16 19	02 01
16	13 41 39	23 32 04	09 ♑ 10 43	15 ♑ 17 45	16 16	10 16	03 53	02 40	08 42	00 28	16 18	02 00
17	13 45 36	24 31 36	21 ♑ 20 51	27 ♑ 20 42	17 32	11 18	04 29	02 38	08 47	00 31	16 16	01 59
18	13 49 32	25 31 09	03 ♒ 17 57	09 ♒ 13 20	18 46	12 19	05 05	02 36	08 53	00 34	16 15	01 59
19	13 53 29	26 30 44	15 ♒ 07 30	21 ♒ 01 08	19 59	13 20	05 41	02 34	08 59	00 36	16 13	01 58
20	13 57 25	27 30 20	26 ♒ 54 52	02 ♓ 49 20	21 10	14 20	06 17	02 32	09 05	00 38	16 12	01 57
21	14 01 22	28 29 58	08 ♓ 45 07	14 ♓ 42 43	22 19	15 20	06 53	02 31	09 11	00 41	16 10	01 57
22	14 05 18	29 ♎ 29 38	20 ♓ 42 36	26 ♓ 45 13	23 26	16 20	07 28	02 29	09 17	00 43	16 08	01 56
23	14 09 15	00 ♏ 29 20	02 ♈ 50 53	08 ♈ 59 53	24 31	17 19	08 04	02 28	09 23	00 46	16 07	01 56
24	14 13 12	01 29 04	15 ♈ 12 25	21 ♈ 28 37	25 34	18 18	08 40	02 28	09 29	00 48	16 05	01 55
25	14 17 08	02 28 50	27 ♈ 48 32	04 ♉ 12 10	26 34	19 17	09 15	02 27	09 35	00 50	16 04	01 55
26	14 21 05	03 28 37	10 ♉ 39 26	17 ♉ 10 13	27 31	20 15	09 51	02 26	09 41	00 52	16 02	01 54
27	14 25 01	04 28 27	23 ♉ 44 20	00 ♊ 21 37	28 24	21 13	10 27	02 26	09 47	00 54	16 00	01 54
28	14 28 58	05 28 18	07 ♊ 01 50	13 ♊ 44 46	29 ♏ 14	22 10	11 02	02 D 26	09 54	00 56	15 59	01 53
29	14 32 54	06 28 12	20 ♊ 30 13	27 ♊ 18 01	00 ♐ 01	23 07	11 37	02 26	10 00	00 59	15 57	01 53
30	14 36 51	07 28 08	04 ♋ 08 01	11 ♋ 00 05	00 42	24 03	12 13	02 27	10 06	01 01	15 55	01 52
31	14 40 47	08 ♏ 28 06	17 ♋ 54 08	24 ♋ 50 07	01 ♐ 19	24 ♐ 59	12 ♍ 48	02 ♓ 27	10 ♐ 13	01 ♍ 03	15 ♉ 54	01 ♓ 52

Date	Moon True ☊ ° '	Moon Mean ☊ ° '	Moon ☽ Latitude ° '
01	21 ♒ 37	20 ♒ 11	04 N 52
02	21 R 33	20 08	04 22
03	21 32	20 05	03 36
04	21 D 32	20 02	02 37
05	21 33	19 59	01 27
06	21 34	19 55	00 N 10
07	21 R 33	19 52	01 S 08
08	21 30	19 49	02 22
09	21 25	19 46	03 26
10	21 18	19 43	04 17
11	21 09	19 39	04 50
12	20 59	19 36	05 04
13	20 50	19 33	05 00
14	20 42	19 30	04 41
15	20 37	19 27	04 07
16	20 33	19 24	03 22
17	20 32	19 20	02 29
18	20 D 32	19 17	01 31
19	20 33	19 14	00 S 29
20	20 R 33	19 11	00 N 34
21	20 31	19 08	01 35
22	20 27	19 05	02 33
23	20 20	19 01	03 25
24	20 11	18 58	04 08
25	20 00	18 55	04 39
26	19 47	18 52	04 57
27	19 34	18 49	05 00
28	19 23	18 45	04 47
29	19 14	18 42	04 18
30	19 07	18 39	03 34
31	19 ♒ 04	18 ♒ 36	02 N 37

DECLINATIONS

Date	Sun ☉ ° '	Moon ☽ ° '	Mercury ☿ ° '	Venus ♀ ° '	Mars ♂ ° '	Jupiter ♃ ° '	Saturn ♄ ° '	Uranus ♅ ° '	Neptune ♆ ° '	Pluto ♇ ° '
01	03 S 27	26 N 48	10 S 17	21 S 17	14 N 29	11 S 26	19 S 59	12 N 13	15 N 04	21 S 47
02	03 51	27 38	10 57	21 37	14 17	11 28	20 00	12 12	15 04	21 47
03	04 14	26 50	11 36	21 56	14 05	11 30	20 01	12 11	15 04	21 47
04	04 37	24 22	12 15	22 15	13 53	11 31	20 02	12 10	15 03	21 47
05	05 00	20 21	12 53	22 34	13 41	11 33	20 03	12 09	15 03	21 48
06	05 23	15 05	13 30	22 52	13 28	11 34	20 04	12 08	15 02	21 48
07	05 46	08 54	14 06	23 09	13 16	11 36	20 05	12 07	15 02	21 48
08	06 09	02 N 11	14 41	23 26	13 04	11 37	20 06	12 06	15 02	21 48
09	06 32	04 S 38	15 16	23 42	12 51	11 38	20 07	12 05	15 01	21 48
10	06 54	11 09	15 50	23 58	12 39	11 39	20 08	12 04	15 01	21 48
11	07 17	16 58	16 22	24 13	12 27	11 41	20 09	12 03	15 00	21 49
12	07 39	21 44	16 54	24 28	12 14	11 42	20 10	12 02	15 00	21 49
13	08 02	25 10	17 26	24 42	12 01	11 43	20 11	12 01	14 59	21 49
14	08 24	27 06	17 56	24 56	11 49	11 43	20 12	12 00	14 59	21 49
15	08 46	27 31	18 25	25 09	11 36	11 44	20 13	11 59	14 58	21 49
16	09 08	26 29	18 53	25 22	11 24	11 45	20 14	11 59	14 58	21 49
17	09 30	24 12	19 20	25 34	11 11	11 46	20 15	11 58	14 58	21 49
18	09 52	20 53	19 46	25 45	10 58	11 46	20 16	11 57	14 57	21 49
19	10 14	16 45	20 11	25 56	10 45	11 47	20 17	11 56	14 57	21 49
20	10 35	12 00	20 34	26 06	10 32	11 47	20 18	11 55	14 56	21 49
21	10 56	06 49	20 56	26 16	10 20	11 48	20 19	11 54	14 56	21 49
22	11 18	01 S 20	21 18	26 25	10 07	11 48	20 20	11 54	14 55	21 49
23	11 38	04 N 16	21 37	26 33	09 54	11 48	20 21	11 53	14 55	21 49
24	11 59	09 48	21 56	26 41	09 41	11 48	20 22	11 52	14 54	21 49
25	12 20	15 02	22 12	26 48	09 28	11 48	20 23	11 51	14 54	21 49
26	12 40	19 43	22 28	26 55	09 15	11 48	20 24	11 50	14 53	21 49
27	13 01	23 33	22 41	27 01	09 02	11 48	20 25	11 50	14 53	21 49
28	13 21	26 12	22 53	27 06	08 49	11 48	20 26	11 49	14 52	21 49
29	13 40	27 23	23 03	27 11	08 36	11 48	20 27	11 48	14 52	21 49
30	14 00	26 56	23 11	27 15	08 23	11 48	20 28	11 48	14 51	21 49
31	14 S 19	24 N 50	23 S 17	27 S 19	08 N 10	11 S 47	20 S 29	11 N 47	14 N 51	21 S 49

ZODIAC SIGN ENTRIES

Date	h	m	Planets
02	23	22	☽ ♋
04	17	35	☿ ♏
05	03	10	☽ ♌
06	07	11	♅ ♍
06	17	25	♀ ♐
07	05	04	☽ ♍
09	05	56	☽ ♎
10	01	55	♂ ♍
11	07	18	☽ ♏
13	10	56	☽ ♐
15	18	16	☽ ♑
18	05	20	☽ ♒
20	18	16	☽ ♓
23	00	12	☉ ♏
23	06	24	☽ ♈
25	16	08	☽ ♉
27	23	21	☽ ♊
29	11	42	☿ ♐
30	04	45	☽ ♋

LATITUDES

Date	Mercury ☿ ° '	Venus ♀ ° '	Mars ♂ ° '	Jupiter ♃ ° '	Saturn ♄ ° '	Uranus ♅ ° '	Neptune ♆ ° '	Pluto ♇ ° '
01	00 S 36	02 S 30	01 N 17	01 S 21	01 N 35	00 N 43	01 S 49	11 S 56
04	00 58	02 42	01 19	01 21	01 34	00 43	01 49	11 55
07	01 19	02 54	01 20	01 20	01 34	00 43	01 49	11 55
10	01 39	03 05	01 21	01 20	01 33	00 43	01 49	11 54
13	01 59	03 15	01 23	01 19	01 33	00 43	01 49	11 54
16	02 17	03 25	01 24	01 19	01 32	00 43	01 49	11 53
19	02 32	03 34	01 26	01 18	01 32	00 43	01 49	11 53
22	02 45	03 42	01 27	01 18	01 31	00 43	01 49	11 52
25	02 54	03 49	01 28	01 17	01 31	00 43	01 49	11 51
28	02 58	03 54	01 30	01 17	01 30	00 44	01 49	11 51
31	02 S 56	03 S 59	01 N 31	01 S 16	01 N 30	00 N 44	01 S 49	11 S 50

DATA

Julian Date	2468255
Delta T	+79 seconds
Ayanamsa	24° 29' 57"
Synetic vernal point	04° ♓ 37' 02"
True obliquity of ecliptic	23° 26' 08"

MOON'S PHASES, APSIDES AND POSITIONS ☽

Date	h	m	Phase	Longitude ° '	Eclipse Indicator
03	18	32	☾	10 ♋ 57	
10	10	37	●	17 ♎ 32	
17	18	55	☽	24 ♑ 49	
25	21	31	○	02 ♉ 53	

Day	h	m		
08	08	06	Perigee	
20	04	03	Apogee	
02	12	31	Max dec	27° N 38'
08	19	41	0S	
15	06	30	Max dec	27° S 33'
22	17	44	0N	
29	17	33	Max dec	27° N 26'

LONGITUDES

Date	Chiron ⚷ ° '	Ceres ⚳ ° '	Pallas ⚴ ° '	Juno ⚵ ° '	Vesta ⚶ ° '	Black Moon Lilith ⚸ ° '
01	26 ♍ 47	29 ♌ 54	04 ♋ 01	06 ♍ 42	22 ♋ 31	24 ♒ 53
11	28 ♍ 11	04 ♍ 08	07 ♋ 23	11 ♍ 00	25 ♋ 18	26 ♒ 00
21	29 ♍ 33	08 ♍ 14	10 ♋ 15	15 ♍ 07	27 ♋ 42	27 ♒ 06
31	00 ♎ 51	12 ♍ 14	12 ♋ 29	19 ♍ 02	29 ♋ 41	28 ♒ 13

All ephemeris data is given at 12.00 UT and the Moon's longitude is additionally given for 24.00 UT

ASPECTARIAN

h m	Aspects
01 Sunday	
00 07	☽ □ ♃
01 25	☿ ⚹ ♂
06 53	☽ ☍ ♄
09 07	☽ △ ☉
11 56	☽ ⚼ ☿
16 51	☽ Q ♂
23 37	☽ ⊻ ♆
02 Monday	
01 40	☽ Q ♅
07 58	☉ ± ♃
08 03	♀ □ ♂
10 19	☽ ⊥ ♆
15 20	☽ ⚹ ♂
15 37	☽ ⊼ ♀
18 11	☽ △ ☿
23 04	☽ ⚹ ♅
03 Tuesday	
00 35	♀ ∥ ♇
02 13	☽ ∠ ♆
03 07	☽ ± ♀
03 13	☽ △ ♇
05 28	☽ △ ♃
07 45	☿ ∥ ♃
10 02	☉ ± ♆
12 36	☽ ⊼ ♄
18 32	☽ □ ☉
18 53	☽ ∠ ♂
20 10	☽ ⚼ ♀
23 05	☽ ± ♄
04 Wednesday	
01 20	☽ ∠ ♅
04 16	☽ ⚹ ♆
05 16	☽ ⚼ ♇
07 09	☉ ∠ ♂
07 21	☽ ⚼ ♃
09 08	☿ # ♅
11 07	☽ ⊥ ♂
14 39	☽ ⚼ ♄
16 16	☿ ⚹ ♅
16 50	☽ ⊥ ♅
20 39	☽ ± ♇
21 49	☽ ⊻ ♂
22 38	☽ ± ♃
05 Thursday	
00 01	☽ △ ♀
00 42	☽ # ♀
00 44	☽ Q ♆
03 04	☽ ⊻ ♅
03 42	☽ Q ☉
04 16	☽ □ ☿
04 19	☽ # ♇
06 48	☽ ⊼ ♇
08 44	☽ ⊼ ♃
13 34	☽ # ♄
16 10	☽ △ ♄
22 13	☿ ⊥ ♄
06 Friday	
01 19	☽ ⚹ ☉
04 38	☿ △ ♇
06 52	☽ □ ♆
11 28	☿ # ♂
12 13	☽ ∥ ♆
13 44	☽ Q ☿
17 53	☽ # ☿
17 54	♀ □ ♅
18 47	☽ ∥ ♂
22 02	☿ △ ♃
23 51	☽ ∥ ♅
07 Saturday	
01 56	☽ # ♃
02 04	☽ ☌ ♂
03 54	☽ ∠ ☉
05 08	☽ ☌ ♅
05 59	☽ □ ♀
08 32	☽ ☍ ♇
10 14	☽ ☍ ♃
11 37	☽ ⚹ ☿
17 59	☽ □ ♄
19 41	☽ ⊥ ☉
22 45	☽ # ☉
22 59	☉ ∠ ♅
08 Sunday	
06 10	☽ ⊻ ☉
07 59	☽ △ ♆
13 37	☽ Q ♀
14 40	☽ ∠ ☿
16 43	♀ □ ♇
23 27	☽ Q ♄
09 Monday	
01 51	☿ # ♆
05 05	☽ ⊻ ♂
06 10	☽ ⊻ ♅
06 51	☽ ⊥ ☿
08 20	☽ ⚼ ♆
08 48	☉ ⊼ ♆
09 21	☽ ⊼ ♇
10 39	☽ ⚹ ♀
10 52	☽ ⊼ ♃
13 24	♀ □ ♃
15 16	☽ ⊥ ♂
15 59	☽ ⊥ ♅
17 40	☽ ⊻ ☿
19 07	☽ ⚹ ♄
19 07	☽ ± ♇
19 12	☽ ∥ ☉
20 37	☽ ± ♃
23 00	☽ ± ♆
23 38	☉ ⚼ ♇
10 Tuesday	
06 38	☽ ∠ ♂
06 46	☽ ∠ ♅

h m	Aspects
08 49	☽ ⊼ ♆
09 44	♂ ☌ ♅
09 52	☽ ⚼ ♇
09 53	☿ ⊻ ♄
10 37	☽ ☌ ☉
11 19	☽ ⚼ ♃
13 08	☽ ∠ ♀
13 57	☽ ∥ ♃
15 32	☽ # ♅
17 39	☽ # ♂
19 56	☽ ∠ ♄
20 25	☉ ⚼ ♃
11 Wednesday	
03 27	☽ # ♆
05 14	☽ ⊥ ♀
07 42	☽ ⚹ ♅
08 35	☽ ⚹ ♂
09 03	☽ ∥ ☿
10 45	☽ △ ♇
11 02	☽ ⊥ ♄
12 10	☽ △ ♃
16 06	☽ ⊻ ♀
21 13	☽ ⊻ ♄
12 Thursday	
00 47	☽ ☌ ☿
03 20	☽ ∥ ♄
04 03	☽ Q ♅
05 49	☽ Q ♂
11 07	☽ ☍ ♆
12 26	☽ ∥ ♇
16 58	☽ ⊻ ☉
13 Friday	
04 17	☽ ⊥ ☉
07 45	☽ ∥ ♀
10 35	♂ ☍ ♇
11 32	☽ □ ♅
11 54	☿ Q ♅
12 24	♂ ∥ ♅
14 33	☽ □ ♇
14 45	☽ □ ♂
15 55	☽ □ ♃
21 36	☽ ∠ ☉
14 Saturday	
00 46	☽ ☌ ♀
02 09	☽ ☌ ♄
11 36	☽ ⊻ ☿
14 46	♂ ☍ ♃
16 34	☽ ⊼ ♆
19 50	♀ ☌ ♄
21 42	♂ # ♃
23 24	☽ Q ♇
23 59	☽ ⊥ ☿
15 Sunday	
00 44	☽ Q ♃
03 28	☽ ⚹ ☉
03 44	☽ ± ♆
18 57	☽ ∠ ☿
19 07	☽ △ ♅
20 47	☽ ⚼ ♆
22 07	☽ ⚹ ♇
22 32	☿ Q ♂
23 26	☽ ⚹ ♃
16 Monday	
01 12	☽ △ ♂
04 18	☽ Q ☉
11 03	☽ ⊻ ♄
12 39	☿ ☍ ♆
14 19	☽ ⊻ ♀
16 19	☉ ∠ ♄
22 54	☽ ⊥ ♄
17 Tuesday	
00 13	☽ ∥ ♀
00 24	☽ ⚼ ♅
01 57	☽ △ ♆
03 12	☽ ⊥ ♀
03 21	☽ ∠ ♇
03 32	☽ ⚹ ☿
04 38	☽ ∠ ♃
08 06	☽ ⚼ ♂
16 55	☽ ∠ ♄
18 21	☽ ± ♅
18 55	☽ □ ☉
21 16	☽ ⊥ ♇
22 32	☽ ⊥ ♃
22 49	☽ ∠ ♀
18 Wednesday	
03 03	☽ ± ♂
05 52	☽ ∥ ♇
06 18	☽ Q ☿
06 27	☽ ⊼ ♅
09 20	☽ ⊻ ♇
10 35	☽ ⊻ ♃
15 48	☽ ⊼ ♂
15 53	☽ ∥ ♄
18 18	☽ ∥ ☿
23 25	☽ ⚹ ♄
19 Thursday	
08 00	☽ ⚹ ♀
14 13	☽ □ ♆
18 42	☿ ∥ ♄
21 31	☽ # ♆
22 59	☽ □ ☿
20 Friday	
00 02	☽ Q ♄
10 09	☉ # ♂
10 43	☽ Q ♀
12 26	☽ # ♅
13 04	☽ ∥ ♃
13 19	☽ △ ☉
18 19	☽ ∥ ☉
19 16	☽ # ♂
19 36	☽ ☍ ♅

h m	Aspects
22 14	☽ ☌ ♇
23 24	☽ ☌ ♃
21 Saturday	
02 45	☽ Q ♆
08 01	☽ ☍ ♂
12 52	☽ □ ♄
22 26	☽ ⚼ ☉
22 Sunday	
02 28	☽ □ ♀
02 53	☽ ⚹ ♆
07 27	♀ ⊼ ♆
17 58	☽ △ ☿
18 02	☽ ± ☉
23 Monday	
06 57	☽ ⊼ ☉
07 53	☽ ⊼ ♅
08 36	☽ ∠ ♆
10 12	☽ ⊻ ♇
11 16	☽ ⊻ ♃
18 46	☉ ⚹ ♅
19 40	☽ ± ♅
21 55	☽ ⊥ ♇
22 43	☽ ⊼ ♂
22 58	☽ ⊥ ♃
23 08	☉ ∥ ♃
24 Tuesday	
00 51	☽ △ ♄
02 08	☽ ⊥ ♆
02 13	☽ ⚼ ☿
03 47	☿ ∥ ♇
03 49	☉ # ♅
10 54	☽ ± ♂
11 31	☽ ∥ ♂
13 08	☽ ⚼ ♅
13 41	☽ ⊻ ♆
15 17	☽ ∠ ♇
16 19	☽ ∠ ♃
18 27	☽ △ ♀
21 00	☽ # ♃
21 05	☽ ± ☿
21 15	☽ ∥ ♅
22 21	☉ △ ♇
22 31	☽ # ☉
25 Wednesday	
04 57	☽ ⚼ ♂
05 51	☽ ⚼ ♄
09 27	☽ ⊼ ☿
11 14	☉ △ ♃
11 21	☽ ∥ ♆
17 43	☽ △ ♅
19 42	☽ ⚹ ♇
20 43	☽ ⚹ ♃
21 31	☽ ☍ ☉
22 56	☽ ± ♄
26 Thursday	
01 09	☽ ⚼ ♀
03 18	♀ Q ♇
03 55	♂ □ ♄
10 11	☽ ⊼ ♄
10 26	☽ △ ♂
15 54	☽ # ♄
16 38	♀ Q ♃
17 36	☉ ⊥ ♄
17 59	☽ Q ♇
18 59	☽ Q ♃
19 10	☽ ± ♀
21 53	☽ ☌ ♆
27 Friday	
00 24	☽ # ♇
05 37	☽ # ☿
07 02	☽ ⊼ ♀
21 03	☽ ☍ ☿
28 Saturday	
01 01	☽ □ ♅
02 45	☽ □ ♇
03 45	☽ □ ♃
04 36	♃ St D
07 17	♀ ± ♆
08 59	☽ ⊼ ☉
17 10	☽ ☍ ♄
19 29	☽ □ ♂
20 35	☽ ± ☉
29 Sunday	
03 52	☽ # ♀
03 56	☽ ⊻ ♆
09 17	☽ Q ♅
13 51	☽ ⚼ ☉
14 33	☽ ⊥ ♆
16 58	☽ ☍ ♀
30 Monday	
04 49	☽ Q ♂
05 00	☽ # ♀
05 40	☽ ⊼ ☿
06 23	☽ ∠ ♆
06 30	☽ ⚹ ♅
08 02	☽ △ ♇
09 02	☽ △ ♃
16 43	☽ ± ☿
18 18	☽ △ ☉
20 19	♃ # ♅
22 31	☽ ⊼ ♄
31 Tuesday	
00 21	☿ □ ♅
02 44	☽ ⚹ ♂
08 31	☽ ⚹ ♆
08 46	☽ ∠ ♅
09 02	☽ ± ♄
09 07	☽ ⚼ ☿
10 12	☽ ⚼ ♇
11 13	☽ ⚼ ♃
23 08	☽ # ☿

NOVEMBER 2045

LONGITUDES

Date	Sidereal time h m s	Sun ☉	Moon ☽	Moon ☽ 24.00	Mercury ☿	Venus ♀	Mars ♂	Jupiter ♃	Saturn ♄	Uranus ♅	Neptune ♆	Pluto ♇
	h m s	° ′ ″	° ′ ″	° ′ ″	° ′	° ′	° ′	° ′	° ′	° ′	° ′	° ′
01	14 44 44	09 ♏ 28 07	01 ♌ 48 01	08 ♌ 47 47	01 ♐ 50	25 ♐ 54	13 ♍ 23	02 ♓ 28	10 ♐ 19	01 ♍ 04	15 ♉ 52	01 ♓ 52
02	14 48 41	10 28 09	15 ♌ 49 23	22 ♌ 52 48	02 15	26 49	13 58	02 29	10 25	01 06	15 R 50	01 R 51
03	14 52 37	11 28 13	29 ♌ 57 54	07 ♍ 04 35	02 33	27 43	14 33	02 30	10 32	01 08	15 49	01 51
04	14 56 34	12 28 20	14 ♍ 12 36	21 ♍ 21 40	02 44	28 37	15 08	02 32	10 38	01 10	15 47	01 51
05	15 00 30	13 28 29	28 ♍ 31 22	05 ♎ 41 16	02 R 47	29 ♐ 30	15 43	02 33	10 45	01 12	15 45	01 50
06	15 04 27	14 28 39	12 ♎ 50 47	19 ♎ 59 17	02 41	00 ♑ 22	16 18	02 35	10 52	01 13	15 44	01 50
07	15 08 23	15 28 52	27 ♎ 06 05	04 ♏ 10 31	02 26	01 13	16 53	02 37	10 58	01 15	15 42	01 50
08	15 12 20	16 29 07	11 ♏ 11 53	18 ♏ 09 31	02 01	02 04	17 27	02 39	11 05	01 16	15 40	01 50
09	15 16 16	17 29 23	25 ♏ 02 51	01 ♐ 51 24	01 27	02 55	18 02	02 41	11 12	01 18	15 39	01 50
10	15 20 13	18 29 41	08 ♐ 34 47	15 ♐ 12 45	00 ♐ 42	03 44	18 37	02 44	11 18	01 19	15 37	01 50
11	15 24 10	19 30 01	21 ♐ 45 10	28 ♐ 12 03	29 ♏ 48	04 33	19 11	02 47	11 25	01 21	15 35	01 49
12	15 28 06	20 30 23	04 ♑ 33 32	10 ♑ 49 52	28 45	05 21	19 46	02 50	11 32	01 22	15 33	01 49
13	15 32 03	21 30 46	17 ♑ 01 24	23 ♑ 08 36	27 35	06 08	20 20	02 53	11 39	01 24	15 32	01 49
14	15 35 59	22 31 10	29 ♑ 11 58	05 ♒ 12 05	26 19	06 54	20 54	02 56	11 45	01 25	15 30	01 49
15	15 39 56	23 31 35	11 ♒ 09 36	17 ♒ 05 11	24 59	07 39	21 28	03 00	11 52	01 26	15 28	01 D 49
16	15 43 52	24 32 02	22 ♒ 59 31	28 ♒ 53 19	23 38	08 23	22 02	03 04	11 59	01 27	15 27	01 49
17	15 47 49	25 32 31	04 ♓ 47 16	10 ♓ 42 05	22 19	09 06	22 36	03 08	12 06	01 28	15 25	01 49
18	15 51 45	26 33 00	16 ♓ 38 25	22 ♓ 36 57	21 04	09 48	23 10	03 12	12 13	01 29	15 23	01 49
19	15 55 42	27 33 31	28 ♓ 38 15	04 ♈ 42 52	19 56	10 29	23 44	03 16	12 20	01 30	15 22	01 50
20	15 59 39	28 34 03	10 ♈ 51 18	17 ♈ 03 57	18 56	11 09	24 18	03 21	12 27	01 31	15 20	01 50
21	16 03 35	29 ♏ 34 36	23 ♈ 21 09	29 ♈ 43 07	18 06	11 48	24 52	03 25	12 34	01 32	15 19	01 50
22	16 07 32	00 ♐ 35 11	06 ♉ 09 58	12 ♉ 41 44	17 27	12 25	25 25	03 30	12 41	01 33	15 17	01 50
23	16 11 28	01 35 47	19 ♉ 18 19	25 ♉ 59 31	17 00	13 01	25 59	03 35	12 48	01 34	15 15	01 50
24	16 15 25	02 36 25	02 ♊ 45 04	09 ♊ 34 34	16 44	13 36	26 32	03 40	12 55	01 34	15 14	01 51
25	16 19 21	03 37 04	16 ♊ 27 36	23 ♊ 23 39	16 D 40	14 09	27 05	03 46	13 02	01 35	15 12	01 51
26	16 23 18	04 37 44	00 ♋ 22 14	07 ♋ 22 50	16 47	14 41	27 39	03 51	13 09	01 36	15 11	01 51
27	16 27 14	05 38 26	14 ♋ 24 58	21 ♋ 28 09	17 04	15 11	28 12	03 57	13 16	01 36	15 09	01 51
28	16 31 11	06 39 09	28 ♋ 32 01	05 ♌ 36 13	17 30	15 40	28 45	04 03	13 23	01 37	15 07	01 52
29	16 35 08	07 39 54	12 ♌ 40 28	19 ♌ 44 32	18 04	16 07	29 18	04 09	13 30	01 37	15 06	01 52
30	16 39 04	08 ♐ 40 40	26 ♌ 48 16	03 ♍ 51 32	18 ♏ 46	16 ♑ 32	29 ♍ 51	04 ♓ 15	13 ♐ 37	01 ♍ 37	15 ♉ 04	01 ♓ 53

Moon Node and Latitude

Date	Moon True ☊	Moon Mean ☊	Moon ☽ Latitude
	° ′	° ′	° ′
01	19 ♒ 02	18 ♒ 33	01 N 30
02	19 D 02	18 30	00 N 17
03	19 R 02	18 26	00 S 58
04	19 00	18 23	02 09
05	18 56	18 20	03 12
06	18 49	18 17	04 04
07	18 39	18 14	04 40
08	18 27	18 11	04 58
09	18 14	18 07	04 58
10	18 01	18 04	04 41
11	17 50	18 01	04 10
12	17 42	17 58	03 26
13	17 36	17 55	02 34
14	17 33	17 51	01 36
15	17 32	17 48	00 S 34
16	17 D 32	17 45	00 N 29
17	17 R 32	17 42	01 30
18	17 31	17 39	02 28
19	17 27	17 36	03 19
20	17 20	17 32	04 03
21	17 11	17 29	04 36
22	17 00	17 26	04 56
23	16 48	17 23	05 01
24	16 35	17 20	04 49
25	16 23	17 17	04 21
26	16 14	17 13	03 37
27	16 08	17 10	02 40
28	16 04	17 07	01 32
29	16 03	17 04	00 18
30	16 ♒ 03	17 ♒ 01	00 S 57

DECLINATIONS

Date	Sun ☉	Moon ☽	Mercury ☿	Venus ♀	Mars ♂	Jupiter ♃	Saturn ♄	Uranus ♅	Neptune ♆	Pluto ♇
	° ′	° ′	° ′	° ′	° ′	° ′	° ′	° ′	° ′	° ′
01	14 S 39	21 N 13	23 S 21	27 S 22	07 N 56	11 S 47	20 S 31	11 N 46	14 N 50	21 S 49
02	14 58	16 22	23 25	27 25	07 43	11 46	20 32	11 46	14 50	21 49
03	15 16	10 35	23 22	27 27	07 30	11 46	20 33	11 45	14 50	21 49
04	15 35	04 N 14	23 18	27 28	07 17	11 45	20 34	11 45	14 49	21 49
05	15 53	02 S 21	23 11	27 29	07 04	11 44	20 35	11 44	14 48	21 49
06	16 11	08 49	23 01	27 30	06 51	11 43	20 36	11 43	14 48	21 48
07	16 28	14 47	22 48	27 29	06 38	11 43	20 37	11 43	14 48	21 48
08	16 46	19 54	22 31	27 29	06 25	11 42	20 38	11 42	14 47	21 48
09	17 03	23 50	22 10	27 27	06 12	11 41	20 39	11 42	14 47	21 48
10	17 20	26 22	21 46	27 26	05 59	11 39	20 40	11 42	14 46	21 48
11	17 36	27 20	21 18	27 23	05 45	11 38	20 41	11 41	14 46	21 48
12	17 52	26 48	20 47	27 21	05 32	11 37	20 42	11 40	14 45	21 47
13	18 08	24 54	20 12	27 17	05 19	11 36	20 43	11 40	14 45	21 47
14	18 24	21 52	19 35	27 14	05 06	11 34	20 44	11 40	14 44	21 47
15	18 39	17 58	18 56	27 09	04 53	11 33	20 45	11 39	14 44	21 47
16	18 54	13 24	18 16	27 05	04 40	11 31	20 46	11 39	14 43	21 47
17	19 09	08 21	17 37	27 00	04 27	11 30	20 47	11 38	14 43	21 47
18	19 23	03 S 00	16 59	26 54	04 14	11 28	20 48	11 38	14 43	21 46
19	19 37	02 N 31	16 24	26 48	04 01	11 26	20 49	11 38	14 42	21 46
20	19 50	08 01	15 53	26 42	03 48	11 24	20 50	11 38	14 41	21 45
21	20 03	13 20	15 26	26 35	03 35	11 23	20 51	11 37	14 41	21 45
22	20 16	18 13	15 04	26 28	03 22	11 21	20 52	11 37	14 40	21 45
23	20 29	22 22	14 47	26 21	03 10	11 19	20 53	11 37	14 40	21 45
24	20 41	25 26	14 35	26 13	02 57	11 17	20 54	11 37	14 40	21 44
25	20 52	27 05	14 29	26 05	02 44	11 15	20 55	11 36	14 39	21 44
26	21 04	27 03	14 27	25 56	02 31	11 12	20 56	11 36	14 39	21 44
27	21 15	25 18	14 29	25 48	02 18	11 10	20 57	11 36	14 38	21 43
28	21 25	21 57	14 36	25 39	02 06	11 08	20 58	11 36	14 38	21 43
29	21 35	17 17	14 46	25 30	01 53	11 05	20 58	11 36	14 37	21 42
30	21 S 45	11 N 41	14 S 59	25 S 20	01 N 40	11 S 03	20 S 59	11 N 36	14 N 37	21 S 42

ZODIAC SIGN ENTRIES

Date	h	m	Planets
01	08	54	☽ ♌
03	12	04	☽ ♍
05	14	28	☽ ♎
06	01	57	♀ ♑
07	16	54	☽ ♏
09	20	43	☽ ♐
11	07	01	☿ ♏
12	03	23	☽ ♑
14	13	36	☽ ♒
17	02	16	☽ ♓
19	14	42	☽ ♈
21	22	04	☉ ♐
22	00	32	☽ ♉
24	07	08	☽ ♊
26	11	22	☽ ♋
28	14	29	☽ ♌
30	17	26	☽ ♍
30	18	55	♂ ♎

LATITUDES

Date	Mercury ☿	Venus ♀	Mars ♂	Jupiter ♃	Saturn ♄	Uranus ♅	Neptune ♆	Pluto ♇
	° ′	° ′	° ′	° ′	° ′	° ′	° ′	° ′
01	02 S 53	04 S 00	01 N 32	01 S 16	01 N 30	00 N 44	01 S 49	11 S 50
04	02 39	04 03	01 33	01 16	01 30	00 44	01 49	11 49
07	02 12	04 04	01 35	01 15	01 29	00 44	01 49	11 48
10	01 30	04 03	01 36	01 15	01 29	00 44	01 49	11 48
13	00 S 36	04 00	01 37	01 14	01 29	00 44	01 49	11 47
16	00 N 26	03 55	01 39	01 13	01 28	00 44	01 49	11 46
19	01 22	03 48	01 40	01 13	01 28	00 45	01 49	11 45
22	02 03	03 38	01 40	01 12	01 28	00 45	01 49	11 45
25	02 28	03 25	01 43	01 12	01 28	00 45	01 49	11 44
28	02 33	03 09	01 44	01 11	01 27	00 45	01 49	11 43
31	02 N 28	02 S 50	01 N 46	01 S 11	01 N 27	00 N 45	01 S 49	11 S 42

DATA

Julian Date	2468286
Delta T	+79 seconds
Ayanamsa	24° 30′ 01″
Synetic vernal point	04° ♓ 36′ 58″
True obliquity of ecliptic	23° 26′ 07″

MOON'S PHASES, APSIDES AND POSITIONS ☽

Date	h	m	Phase	Longitude ° ′	Eclipse Indicator
02	02	10	☾	10 ♌ 04	
08	21	49	●	16 ♏ 54	
16	15	26	☽	24 ♒ 41	
24	11	43	○	02 ♊ 36	

Day	h	m		
04	21	32	Perigee	
16	23	59	Apogee	
29	18	23	Perigee	
05	03	26	0S	
11	15	11	Max dec	27° S 21′
19	01	07	0N	
25	23	43	Max dec	27° N 17′

LONGITUDES

Date	Chiron ⚷	Ceres ⚳	Pallas ⚴	Juno ⚵	Vesta ⚶	Black Moon Lilith ⚸
	° ′	° ′	° ′	° ′	° ′	° ′
01	00 ♎ 59	12 ♍ 37	12 ♋ 40	19 ♍ 25	29 ♋ 51	28 ♒ 19
11	02 ♎ 11	16 ♍ 25	14 ♋ 18	23 ♍ 06	01 ♌ 15	29 ♒ 26
21	03 ♎ 18	20 ♍ 02	14 ♋ 18	26 ♍ 32	02 ♌ 32	00 ♓ 33
31	04 ♎ 16	23 ♍ 23	13 ♋ 27	29 ♍ 39	02 ♌ 10	01 ♓ 39

ASPECTARIAN

h m	Aspects
01 Wednesday	
00 23	☽ ⊥ ♅
00 45	☽ ⚼ ♄
01 08	☽ ⊼ ♀
01 46	☽ ± ♇
02 48	☽ ± ♃
05 15	☽ Q ♆
05 52	☽ ∠ ♂
08 33	☽ # ♇
10 45	☽ ⊻ ♅
12 03	☽ △ ☿
12 06	☽ ⊼ ♇
12 12	☽ ± ♀
13 09	☽ ⊼ ♃
13 32	☿ □ ♇
15 52	☽ # ♄
22 00	☽ ⊥ ♂
02 Thursday	
02 10	☽ □ ☉
02 30	☉ # ♆
02 43	☽ △ ♄
04 41	☽ ⚼ ♀
08 42	☽ ⊻ ♂
10 47	☉ ⊻ ♄
12 02	☽ □ ♆
17 49	☽ # ☉
18 41	☽ ∥ ♆
03 Friday	
07 19	☿ □ ♃
07 20	☽ # ♃
07 22	☽ ∥ ♅
07 57	☽ △ ♀
11 06	☽ Q ☉
13 59	☽ ☌ ♅
15 11	☽ ☍ ♇
16 18	☽ ☍ ♃
16 26	☽ □ ☿
04 Saturday	
00 15	☽ ∥ ♂
05 57	☽ □ ♄
08 52	☽ ✶ ☉
13 37	☽ ☌ ♂
14 38	☽ △ ♆
23 00	☽ Q ☿
05 Sunday	
05 05	☉ Q ♅
08 06	☿ St R
11 55	☽ ∠ ☉
12 23	☽ Q ♄
13 24	♂ △ ♆
13 44	☽ □ ♀
15 44	☽ ⚼ ♆
16 29	☽ ⊻ ♅
17 33	☽ ⊼ ♇
18 46	☽ ⊼ ♃
19 06	☽ ✶ ☿
06 Monday	
02 33	☽ ⊥ ♅
03 36	☽ ± ♇
04 07	☽ ⊥ ☉
04 49	☽ # ♂
04 50	☽ ± ♃
06 47	☽ ± ♆
08 39	☽ ✶ ♄
11 37	♃ # ♅
14 57	☽ ⊻ ☉
16 50	☽ ⊼ ♆
17 41	☽ ∠ ♅
18 03	☽ ⊻ ♂
18 42	☽ ⚼ ♇
19 58	☽ ⚼ ♃
20 00	☽ ∠ ☿
21 46	♀ ⚼ ♆
21 52	☽ Q ♀
21 58	☿ □ ♃
23 21	☽ ∥ ♃
23 21	☽ # ♅
07 Tuesday	
04 34	☽ ⊥ ♂
10 04	☽ ∠ ♄
10 53	☽ ⊥ ☿
12 03	☽ # ♆
12 43	♀ △ ♅
17 03	☉ ☍ ♆
19 02	☽ ✶ ♅
19 26	☽ ✶ ♀
19 52	☽ ∥ ☉
20 01	☽ △ ♇
20 27	☽ ∠ ♂
20 48	☽ ⊻ ☿
21 22	☽ △ ♃
08 Wednesday	
01 28	☽ ⊥ ♄
05 08	♀ ✶ ♇
11 02	☿ ⊻ ♀
11 48	☽ ⊻ ♄
15 34	☽ Q ♅
15 56	☽ ∥ ♄
19 41	☽ ☍ ♆
20 41	☿ □ ♇
21 49	☽ ☌ ☉
22 36	☽ ∥ ♇
22 46	☽ ∠ ♀
23 15	☽ ✶ ♂
09 Thursday	
01 48	☽ ∥ ☿
05 23	♀ ✶ ♃
15 29	☽ ⊥ ♀
16 52	☿ □ ♅
21 10	☽ Q ♂
22 43	☽ ☌ ☿
23 02	☽ □ ♅
23 57	☽ □ ♇
10 Friday	
01 31	☽ □ ♃
02 46	☽ ⊻ ♀
10 25	☿ ∥ ♇
13 31	☿ Q ♂
16 57	☽ ☌ ♄
18 31	☉ ✶ ♂
11 Saturday	
00 42	☽ ⊼ ♆
07 03	☽ □ ♂
07 30	☽ ⊻ ☉
08 27	☽ Q ♇
10 12	☽ Q ♃
11 42	☽ ± ♆
17 14	☉ ∠ ♀
19 33	☽ ⊥ ☉
12 Sunday	
01 54	☽ ⊻ ☿
04 27	☽ ⚼ ♆
04 31	☿ ⊥ ♀
05 57	☽ △ ♅
06 49	☽ ✶ ♇
08 42	☽ ✶ ♃
12 20	☽ ⊥ ☿
13 35	☽ ☌ ♀
13 58	☽ ∠ ☉
15 15	☿ ∥ ♄
13 Monday	
01 28	☽ ⊻ ♄
04 09	☽ ∠ ☿
09 06	☽ △ ♆
10 46	☽ ⚼ ♅
11 36	☽ ∠ ♇
13 13	☽ ⊥ ♄
13 41	☽ ∠ ♃
18 47	☽ △ ♂
21 35	☽ ✶ ☉
14 Tuesday	
04 28	☽ ± ♅
05 18	☽ ⊥ ♇
06 49	☽ ✶ ☿
07 06	☽ ∠ ♄
07 29	☽ ⊥ ♃
12 37	☽ ∥ ♇
16 25	☽ ⊼ ♅
17 14	☽ ⊻ ♇
17 33	♇ St D
19 29	☽ ∥ ♄
19 30	☽ ⊻ ♃
23 37	☽ Q ☉
15 Wednesday	
02 05	☽ ⚼ ♂
04 26	☽ Q ☿
04 26	☽ ⊻ ♀
05 27	☽ ∥ ☿
08 19	☽ ∥ ☉
13 27	☽ ✶ ♄
17 22	☽ ⊥ ♀
19 16	☉ ∥ ☿
20 43	☽ □ ♆
21 10	☽ ± ♂
16 Thursday	
02 53	☉ ☌ ☿
05 19	☽ # ♆
09 58	☽ ⊼ ♂
12 51	☽ ∠ ♀
13 11	☽ □ ☿
14 03	☽ Q ♄
15 01	☿ ∠ ♀
15 26	☽ □ ☉
20 33	☽ # ♅
21 12	☽ ∥ ♃
17 Friday	
05 15	☽ ☍ ♅
05 58	☽ ☌ ♇
08 19	☿ ✶ ♂
08 36	☽ ☌ ♃
09 13	☽ Q ♆
21 19	☽ ✶ ♀
18 Saturday	
02 59	☽ □ ♄
06 20	☽ # ♂
09 29	☽ ✶ ♆
20 06	☽ △ ☿
23 00	☽ Q ♀
19 Sunday	
01 46	☽ ☍ ♂
09 40	☽ △ ☉
15 25	☽ ∠ ♆
17 41	☽ ⊼ ♅
18 18	☽ ∥ ♂
18 19	☽ ⊻ ♇
21 13	☽ ⊻ ♃
23 27	☽ ⚼ ☿
20 Monday	
05 29	☽ ± ♅
06 06	☽ ⊥ ♇
09 02	☽ ⊥ ♃
09 03	☽ ⊥ ♆
10 24	♃ Q ♆
12 36	☽ □ ♀
15 07	☽ △ ♄
15 45	☽ ± ☿
17 43	☽ ⚼ ☉
20 39	☽ ⊻ ♆
22 58	☽ ⚼ ♅
23 33	☽ ∠ ♇
21 Tuesday	
02 32	☽ ∠ ♃
02 34	☽ ⊼ ☿
03 02	☽ # ♃
04 06	☽ ∥ ♅
12 28	☽ ± ☉
14 59	☽ ⊼ ♂
18 20	☽ ∥ ♆
20 02	☽ ⚼ ♄
21 13	☽ # ☿
22 Wednesday	
00 44	☽ ⊼ ☉
02 47	☽ ± ♂
03 25	☽ △ ♅
03 57	☽ ✶ ♇
07 02	☽ ✶ ♃
12 58	☽ ± ♄
20 11	☽ ⚼ ♂
23 48	☽ # ☉
23 Thursday	
00 03	☽ △ ♀
00 05	☽ ⊼ ♄
01 04	♀ ⊻ ♄
02 05	☽ Q ♇
02 43	☽ # ♄
04 41	☽ ☌ ♆
05 13	☽ Q ♃
07 57	☽ ☍ ☿
08 00	☽ # ♇
11 09	☉ □ ♅
17 46	☉ □ ♇
24 Friday	
00 29	☽ △ ♂
01 44	☿ # ♆
04 19	☽ ⚼ ♀
09 55	☽ □ ♅
10 24	☽ □ ♇
11 43	☽ ☍ ☉
13 38	☽ □ ♃
20 12	☽ # ♀
20 54	☽ ± ♀
25 Saturday	
05 59	☽ ☍ ♄
07 49	☽ ⊼ ♀
08 55	☿ St D
09 49	☽ ⊻ ♆
12 22	☽ ⊼ ☿
15 49	☉ □ ♃
17 22	☉ ∥ ♄
17 25	☽ Q ♅
20 12	☽ ⊥ ♆
22 48	☽ ± ☿
26 Sunday	
07 07	☽ □ ♂
11 40	☽ ∠ ♆
14 06	☽ ✶ ♅
14 28	☽ ⚼ ☿
14 32	☽ △ ♇
18 01	☽ △ ♃
19 52	☽ ⊼ ☉
27 Monday	
06 44	☽ # ♀
06 54	☽ ± ☉
10 02	☽ ⊼ ♄
10 20	♀ △ ♆
13 15	☽ ✶ ♆
13 21	☽ ☍ ♀
15 09	☽ Q ♂
15 44	☽ ∠ ♅
16 09	☽ ⚼ ♇
16 38	☽ △ ☿
19 47	☽ ⚼ ♃
20 20	☽ ± ♄
23 25	☽ ⚼ ☉
28 Tuesday	
07 02	☽ ⊥ ♅
07 28	☽ ± ♇
09 37	☽ Q ♆
11 11	☽ ± ♃
11 45	☽ ⚼ ♄
12 22	☽ ✶ ♂
13 22	☽ # ♇
14 57	☽ # ☉
16 58	☿ # ♆
17 14	☽ ⊻ ♅
17 37	☽ # ♄
17 39	☽ ⊼ ♇
21 26	☽ ⊼ ♃
29 Wednesday	
02 51	☽ △ ☉
13 25	☽ △ ♄
14 52	☽ ∠ ♂
16 06	☽ □ ♆
18 02	☽ ⊼ ♀
21 36	☽ □ ☿
22 49	☽ # ☿
23 50	☽ ∥ ♆
30 Thursday	
04 32	☽ ± ♀
05 31	☉ ∥ ♇
06 46	☽ ⊥ ♂
12 21	☽ ∥ ♅
14 34	☽ # ♃
17 11	♀ ⚼ ♅
17 22	☽ ⊻ ♂
20 12	☽ ☌ ♅
20 17	☽ ⚼ ♀
20 38	☽ ☍ ♇
21 40	♂ ⚼ ♆

All ephemeris data is given at 12.00 UT and the Moon's longitude is additionally given for 24.00 UT

DECEMBER 2045

LONGITUDES

Date	Sidereal time h m s	Sun ☉ ° ' "	Moon ☽ ° ' "	Moon ☽ 24.00 ° ' "	Mercury ☿ ° '	Venus ♀ ° '	Mars ♂ ° '	Jupiter ♃ ° '	Saturn ♄ ° '	Uranus ♅ ° '	Neptune ♆ ° '	Pluto ♇ ° '
01	16 43 01	09 ♐ 41 28	10 ♍ 54 13	17 ♍ 56 14	19 ♏ 34	16 ♑ 56	00 ♎ 23	04 ♓ 22	13 ♐ 44	01 ♍ 38	15 ♉ 03	01 ♓ 53
02	16 46 57	10 42 17	24 ♍ 57 29	01 ♎ 57 52	20 28	17 18	00 56	04 29	13 52	01 38	15 R 01	01 53
03	16 50 54	11 43 08	08 ♎ 57 13	15 ♎ 55 22	21 27	17 38	01 28	04 35	13 59	01 38	15 00	01 54
04	16 54 50	12 44 00	22 ♎ 52 04	29 ♎ 47 04	22 31	17 56	02 01	04 42	14 06	01 38	14 58	01 54
05	16 58 47	13 44 54	06 ♏ 40 05	13 ♏ 30 44	23 38	18 12	02 33	04 49	14 13	01 39	14 57	01 55
06	17 02 43	14 45 48	20 ♏ 18 43	27 ♏ 03 39	24 49	18 25	03 05	04 57	14 20	01 39	14 56	01 55
07	17 06 40	15 46 44	03 ♐ 45 13	10 ♐ 23 07	26 02	18 37	03 38	05 04	14 27	01 R 39	14 54	01 56
08	17 10 37	16 47 42	16 ♐ 57 05	23 ♐ 26 55	27 19	18 47	04 10	05 12	14 34	01 38	14 53	01 57
09	17 14 33	17 48 40	29 ♐ 52 32	06 ♑ 13 51	28 37	18 54	04 41	05 19	14 41	01 38	14 51	01 57
10	17 18 30	18 49 39	12 ♑ 30 56	18 ♑ 43 54	29 ♏ 57	18 59	05 13	05 27	14 48	01 38	14 50	01 58
11	17 22 26	19 50 38	24 ♑ 52 59	00 ♒ 58 27	01 ♐ 18	19 01	05 45	05 35	14 56	01 38	14 49	01 59
12	17 26 23	20 51 39	07 ♒ 00 41	13 ♒ 00 07	02 41	19 R 02	06 16	05 43	15 03	01 38	14 47	01 59
13	17 30 19	21 52 40	18 ♒ 57 14	24 ♒ 52 36	04 06	18 59	06 48	05 52	15 10	01 37	14 46	02 00
14	17 34 16	22 53 41	00 ♓ 46 48	06 ♓ 40 27	05 31	18 54	07 19	06 00	15 17	01 37	14 45	02 01
15	17 38 12	23 54 43	12 ♓ 34 13	18 ♓ 28 46	06 57	18 47	07 50	06 09	15 24	01 36	14 44	02 02
16	17 42 09	24 55 46	24 ♓ 24 46	00 ♈ 22 55	08 24	18 37	08 21	06 18	15 31	01 36	14 42	02 02
17	17 46 06	25 56 48	06 ♈ 23 52	12 ♈ 28 15	09 51	18 25	08 52	06 26	15 38	01 35	14 41	02 03
18	17 50 02	26 57 52	18 ♈ 36 41	24 ♈ 49 41	11 20	18 10	09 22	06 35	15 45	01 35	14 40	02 04
19	17 53 59	27 58 55	01 ♉ 07 46	07 ♉ 31 19	12 48	17 53	09 53	06 45	15 52	01 34	14 39	02 05
20	17 57 55	28 ♐ 59 59	14 ♉ 00 37	20 ♉ 35 53	14 17	17 33	10 23	06 54	15 59	01 33	14 38	02 06
21	18 01 52	00 ♑ 01 04	27 ♉ 17 10	04 ♊ 04 24	15 47	17 11	10 54	07 03	16 06	01 32	14 37	02 07
22	18 05 48	01 02 08	10 ♊ 57 22	17 ♊ 55 42	17 17	16 47	11 24	07 13	16 13	01 32	14 36	02 08
23	18 09 45	02 03 14	24 ♊ 58 57	02 ♋ 06 30	18 47	16 21	11 54	07 23	16 20	01 31	14 35	02 09
24	18 13 41	03 04 19	09 ♋ 17 40	16 ♋ 31 42	20 18	15 53	12 24	07 32	16 27	01 30	14 34	02 10
25	18 17 38	04 05 25	23 ♋ 47 47	01 ♌ 05 07	21 49	15 23	12 53	07 42	16 34	01 29	14 33	02 11
26	18 21 35	05 06 32	08 ♌ 22 55	15 ♌ 40 28	23 20	14 51	13 23	07 52	16 41	01 28	14 32	02 12
27	18 25 31	06 07 39	22 ♌ 57 04	00 ♍ 12 08	24 52	14 18	13 52	08 03	16 47	01 27	14 31	02 13
28	18 29 28	07 08 46	07 ♍ 25 12	14 ♍ 35 51	26 24	13 44	14 21	08 13	16 54	01 25	14 30	02 14
29	18 33 24	08 09 54	21 ♍ 43 45	28 ♍ 48 43	27 56	13 09	14 50	08 23	17 01	01 24	14 29	02 15
30	18 37 21	09 11 02	05 ♎ 50 33	12 ♎ 49 09	29 ♐ 29	12 33	15 19	08 34	17 08	01 23	14 28	02 16
31	18 41 17	10 ♑ 12 11	19 ♎ 44 29	26 ♎ 36 31	01 ♑ 02	11 ♑ 56	15 ♎ 48	08 ♓ 45	17 ♐ 15	01 ♍ 22	14 ♉ 27	02 ♓ 17

Date	Moon True ☊ ° '	Moon Mean ☊ ° '	Moon ☽ Latitude ° '
01	16 ♒ 03	16 ♒ 57	02 S 08
02	16 R 03	16 54	03 12
03	16 00	16 51	04 03
04	15 54	16 48	04 40
05	15 46	16 45	05 00
06	15 36	16 42	05 03
07	15 25	16 38	04 49
08	15 14	16 35	04 19
09	15 05	16 32	03 37
10	14 57	16 29	02 44
11	14 53	16 26	01 45
12	14 50	16 22	00 S 42
13	14 D 50	16 19	00 N 22
14	14 51	16 16	01 25
15	14 52	16 13	02 24
16	14 53	16 10	03 17
17	14 R 52	16 07	04 02
18	14 50	16 03	04 37
19	14 45	16 00	05 00
20	14 38	15 57	05 09
21	14 31	15 54	05 01
22	14 23	15 51	04 36
23	14 16	15 48	03 55
24	14 10	15 44	02 58
25	14 07	15 41	01 48
26	14 05	15 38	00 N 31
27	14 D 05	15 35	00 S 48
28	14 06	15 32	02 04
29	14 08	15 28	03 11
30	14 09	15 25	04 05
31	14 ♒ 08	15 ♒ 22	04 S 45

DECLINATIONS

Date	Sun ☉ ° '	Moon ☽ ° '	Mercury ☿ ° '	Venus ♀ ° '	Mars ♂ ° '	Jupiter ♃ ° '	Saturn ♄ ° '	Uranus ♅ ° '	Neptune ♆ ° '	Pluto ♇ ° '
01	21 S 54	05 N 30	15 S 14	25 S 10	01 N 28	11 S 00	21 S 00	11 N 36	14 N 37	21 S 42
02	22 03	00 S 56	15 32	25 00	01 15	10 58	21 01	11 36	14 36	21 41
03	22 11	07 16	15 52	24 50	01 03	10 55	21 02	11 36	14 36	21 41
04	22 19	13 13	16 13	24 40	00 50	10 52	21 03	11 36	14 35	21 40
05	22 27	18 27	16 35	24 29	00 38	10 50	21 04	11 36	14 35	21 40
06	22 34	22 41	16 57	24 18	00 26	10 47	21 05	11 36	14 35	21 40
07	22 41	25 37	17 21	24 07	00 13	10 44	21 06	11 36	14 34	21 39
08	22 47	27 06	17 45	23 56	00 N 01	10 41	21 06	11 36	14 34	21 39
09	22 53	27 03	18 09	23 45	00 S 11	10 38	21 07	11 36	14 33	21 38
10	22 58	25 34	18 33	23 34	00 23	10 35	21 08	11 36	14 33	21 38
11	23 03	22 52	18 57	23 22	00 35	10 32	21 09	11 36	14 33	21 37
12	23 07	19 12	19 20	23 11	00 47	10 29	21 10	11 36	14 32	21 37
13	23 11	14 47	19 44	22 59	00 59	10 26	21 11	11 36	14 32	21 36
14	23 15	09 52	20 07	22 47	01 11	10 22	21 11	11 37	14 32	21 36
15	23 18	04 S 37	20 29	22 35	01 23	10 19	21 12	11 37	14 31	21 35
16	23 20	00 N 48	20 50	22 23	01 35	10 16	21 13	11 37	14 31	21 35
17	23 22	06 15	21 11	22 11	01 46	10 12	21 14	11 37	14 31	21 34
18	23 24	11 34	21 31	21 59	01 58	10 09	21 15	11 38	14 30	21 34
19	23 25	16 33	21 50	21 47	02 10	10 05	21 15	11 38	14 30	21 33
20	23 26	20 57	22 09	21 35	02 21	10 02	21 16	11 38	14 30	21 33
21	23 26	24 26	22 26	21 23	02 32	09 58	21 17	11 38	14 30	21 32
22	23 26	26 39	22 42	21 11	02 44	09 54	21 18	11 39	14 29	21 32
23	23 25	27 15	22 58	20 58	02 55	09 50	21 18	11 39	14 29	21 31
24	23 24	26 04	23 12	20 46	03 06	09 47	21 19	11 39	14 29	21 31
25	23 22	23 07	23 25	20 34	03 17	09 43	21 20	11 40	14 29	21 30
26	23 20	18 40	23 37	20 22	03 28	09 39	21 20	11 40	14 28	21 30
27	23 18	13 06	23 48	20 10	03 39	09 35	21 21	11 41	14 28	21 29
28	23 15	06 52	23 58	19 58	03 50	09 31	21 22	11 41	14 28	21 28
29	23 11	00 N 21	24 06	19 46	04 01	09 27	21 22	11 42	14 28	21 28
30	23 07	06 S 05	24 13	19 35	04 11	09 23	21 23	11 42	14 28	21 27
31	23 S 03	12 S 07	24 S 19	19 S 23	04 S 22	09 S 19	21 S 24	11 N 43	14 N 27	21 S 27

ZODIAC SIGN ENTRIES

Date	h	m	Planets
02	20	38	☽ ♎
05	00	22	☽ ♏
07	05	15	☽ ♐
09	12	14	☽ ♑
10	12	57	☿ ♐
11	22	04	☽ ♒
14	10	25	☽ ♓
16	23	14	☽ ♈
19	09	52	☽ ♉
21	11	35	☉ ♑
21	16	49	☽ ♊
23	20	28	☽ ♋
25	22	13	☽ ♌
27	23	40	☽ ♍
30	02	01	☽ ♎
30	20	07	☿ ♑

LATITUDES

Date	Mercury ☿ ° '	Venus ♀ ° '	Mars ♂ ° '	Jupiter ♃ ° '	Saturn ♄ ° '	Uranus ♅ ° '	Neptune ♆ ° '	Pluto ♇ ° '
01	02 N 28	02 S 50	01 N 46	01 S 11	01 N 27	00 N 45	01 S 49	11 S 42
04	02 16	02 27	01 47	01 10	01 27	00 45	01 49	11 42
07	01 58	02 00	01 49	01 10	01 27	00 45	01 49	11 41
10	01 37	01 29	01 50	01 09	01 27	00 45	01 49	11 40
13	01 15	00 54	01 52	01 09	01 26	00 46	01 49	11 39
16	00 52	00 S 15	01 53	01 08	01 26	00 46	01 49	11 39
19	00 30	00 N 27	01 54	01 08	01 26	00 46	01 48	11 38
22	00 N 07	01 13	01 56	01 07	01 26	00 46	01 48	11 37
25	00 S 14	02 00	01 57	01 07	01 26	00 46	01 48	11 37
28	00 34	02 46	01 59	01 07	01 26	00 46	01 48	11 36
31	00 S 53	03 N 32	02 N 00	01 S 06	01 N 26	00 N 46	01 S 48	11 S 36

DATA

Julian Date	2468316
Delta T	+79 seconds
Ayanamsa	24° 30' 06"
Synetic vernal point	04° ♓ 36' 53"
True obliquity of ecliptic	23° 26' 06"

MOON'S PHASES, APSIDES AND POSITIONS ☽

Date	h	m	Phase	Longitude ° '	Eclipse Indicator
01	09	46	☾	09 ♍ 36	
08	11	41	●	16 ♐ 47	
16	13	08	☽	24 ♓ 59	
24	00	49	○	02 ♋ 36	
30	18	11	☾	09 ♎ 27	

Day	h	m		
14	21	17	Apogee	
26	15	00	Perigee	
02	08	32	0S	
08	23	14	Max dec	27° S 15'
16	08	30	0N	
23	08	18	Max dec	27° N 16'
29	13	18	0S	

LONGITUDES

Date	Chiron ⚷ ° '	Ceres ⚳ ° '	Pallas ⚴ ° '	Juno ⚵ ° '	Vesta ⚶ ° '	Black Moon Lilith ⚸ ° '
01	04 ♎ 16	23 ♍ 23	13 ♋ 27	29 ♍ 39	02 ♌ 10	01 ♓ 39
11	05 ♎ 06	26 ♍ 27	11 ♋ 27	02 ♎ 26	01 ♌ 32	02 ♓ 45
21	05 ♎ 46	29 ♍ 08	08 ♋ 31	04 ♎ 50	00 ♌ 09	03 ♓ 52
31	06 ♎ 15	01 ♎ 24	05 ♋ 11	06 ♎ 45	28 ♋ 07	04 ♓ 58

All ephemeris data is given at 12.00 UT and the Moon's longitude is additionally given for 24.00 UT

ASPECTARIAN

h m Aspects

01 Friday
00 47 ☽ ☍ ♃
05 57 ☽ Q ☿
08 50 ♀ ∠ ♇
09 46 ☽ □ ☉
16 53 ☽ □ ♄
19 03 ☽ △ ♆
22 34 ☽ △ ♀

02 Saturday
03 36 ☽ ∥ ♂
03 46 ☽ ⚹ ☿
13 10 ☽ # ♂
18 55 ☽ Q ☉
20 39 ☽ ⅊ ♆
22 39 ☽ ☌ ♂
23 26 ☽ ⊻ ♅
23 53 ☽ ⊼ ♇
23 55 ☽ Q ♄

03 Sunday
04 26 ☽ ⊼ ♃
07 22 ☽ ∠ ☿
08 50 ☉ ⊥ ♀
09 44 ☽ ⊥ ♅
10 11 ☽ ± ♇
12 05 ☽ ± ♆
14 50 ☽ ± ♃
17 08 ☽ ⚹ ☉
19 17 ♂ ⊻ ♅
20 43 ☽ ⚹ ♄
22 23 ☽ ⊼ ♆

04 Monday
00 06 ☽ ⊥ ☿
01 14 ☽ ∠ ♅
01 41 ☽ ⅊ ♇
02 20 ☽ ∥ ♃
03 17 ☽ □ ♀
05 12 ☽ # ♅
06 29 ☽ ⅊ ♃
07 05 ♂ ⊼ ♇
11 20 ☽ ⊻ ☿
16 38 ♂ Q ♄
17 55 ☽ # ♆
21 06 ☽ ∠ ☉
22 54 ☽ ∠ ♄

05 Tuesday
02 13 ☽ ∥ ☿
03 14 ☽ ⚹ ♅
03 42 ☽ △ ♇
04 32 ☽ ⊻ ♂
08 45 ☽ △ ♃
11 09 ☽ Q ♀
14 02 ☽ ⊥ ☉
14 44 ☽ ⊥ ♄
15 26 ☽ ⊥ ♂

06 Wednesday
00 14 ☽ Q ♅
00 29 ☉ ☌ ♄
01 21 ☽ ⊻ ♄
01 25 ☽ ⊻ ☉
02 04 ☽ ∥ ♄
02 30 ☽ ☍ ♆
05 34 ☽ ∥ ♇
07 55 ☽ ∠ ♂
08 36 ☽ ⚹ ♀
11 14 ☽ ∥ ☉
15 45 ☉ ⊼ ♆
16 03 ♅ St R
20 47 ☽ ☌ ☿
23 11 ☽ ∥ ♀

07 Thursday
04 22 ☉ Q ♂
08 12 ☽ □ ♅
08 43 ☽ □ ♇
11 45 ☽ ∠ ♀
11 46 ☽ ⚹ ♂
14 23 ☽ □ ♃

08 Friday
04 16 ☽ ⊥ ♀
07 36 ☽ ☌ ♄
08 12 ☽ ⊼ ♆
10 29 ☽ Q ♂
11 41 ☽ ☌ ☉
15 24 ☽ ⊻ ♀
17 31 ☽ Q ♇
19 14 ☽ ± ♆
23 38 ☽ Q ♃

09 Saturday
09 22 ☽ ⊻ ☿
11 58 ☽ ⅊ ♆
15 19 ☽ △ ♅
15 55 ☽ ⚹ ♇
21 28 ☽ □ ♂
21 58 ☽ ⊥ ☿
22 23 ☽ ⚹ ♃

10 Sunday
15 50 ☉ ⊻ ♀
16 27 ☽ ⊻ ♄
16 27 ☽ △ ♆
16 35 ♄ ⊼ ♆
17 15 ☽ ∠ ☿
19 56 ☽ ⅊ ♅
20 35 ☽ ∠ ♇

11 Monday
00 32 ☽ ☌ ♀
01 17 ☽ ⊻ ☉
02 14 ♂ ⊼ ♃
03 31 ☽ ∠ ♃
04 11 ☽ ⊥ ♄
07 57 ☽ ∥ ♀
10 43 ☽ ∥ ☉
13 28 ☽ ± ♅
14 03 ☽ ⊥ ☉
14 09 ☽ ⊥ ♇
15 10 ☉ Q ♇
17 39 ☿ □ ♅
20 49 ☽ ∥ ♇
21 22 ☽ ⊥ ♃
22 01 ☽ ∠ ♄
23 46 ☿ □ ♇
23 53 ☽ ∥ ♄

12 Tuesday
01 18 ☽ ⊼ ♅
01 28 ♀ St R
02 00 ☽ ⊻ ♇
02 17 ☽ ⚹ ☿
09 24 ☽ ⊻ ♃
09 30 ☽ ∠ ☉
10 23 ☉ ± ♆
10 27 ☽ △ ♂
11 12 ☽ ∥ ☿
17 14 ☉ ∥ ♀

13 Wednesday
03 34 ☽ □ ♆
04 16 ☽ ⚹ ♄
05 27 ☽ Q ☿
10 15 ☿ ∠ ♀
12 04 ☽ ⊻ ♀
13 17 ☽ # ♆
18 01 ☽ ⅊ ♂
18 28 ☽ ⚹ ☉

14 Thursday
00 09 ☽ ⊥ ♀
03 44 ☽ # ♅
04 49 ☽ Q ♄
09 37 ☽ ∥ ♃
13 08 ☽ ± ♂
13 42 ☽ ☍ ♅
14 31 ☽ ☌ ♇
16 00 ☽ Q ♆
18 19 ☽ ∠ ♀
21 07 ☿ □ ♃
21 10 ☽ Q ☉
22 46 ☽ ☌ ♃
22 58 ☽ □ ☿

15 Friday
01 39 ♂ ⊥ ♅
01 55 ☽ ⊼ ♂
16 23 ☽ ⚹ ♆
17 48 ☽ □ ♄
18 28 ☉ Q ♃
21 12 ♂ ± ♇

16 Saturday
00 28 ☽ ⚹ ♀
01 54 ☽ ∥ ♂
10 46 ☿ ⚹ ♂
13 08 ☽ □ ☉
15 34 ☽ # ♂
22 38 ☽ ∠ ♆

17 Sunday
00 17 ☽ Q ♀
02 25 ☽ ⊼ ♅
03 20 ☽ ⊻ ♇
04 07 ♂ ± ♆
12 05 ☽ ⊻ ♃
14 22 ☽ ± ♅
15 17 ☽ ⊥ ♇
15 18 ☿ ∥ ♄
16 32 ☽ ⊥ ♆
17 06 ☽ ☍ ♂
19 47 ☽ △ ☿

18 Monday
00 05 ☽ ⊥ ♃
04 19 ☽ ⊻ ♆
05 34 ☽ # ♃
06 22 ☽ △ ♄
08 03 ☽ ⅊ ♅
08 59 ☽ ∠ ♇
11 10 ☽ □ ♀
12 18 ☽ ∥ ♅
15 19 ☿ ∥ ♇
17 51 ☽ ∠ ♃
23 34 ☿ ⊥ ♀

19 Tuesday
01 54 ☽ ∥ ♆
04 51 ☽ ⅊ ☿
05 30 ☽ △ ☉
09 34 ☿ ∥ ♀
11 30 ☽ ⅊ ♄
12 49 ☽ △ ♅
13 48 ☽ ⚹ ♇
22 41 ☽ ⚹ ♃

20 Wednesday
00 03 ☽ ± ☿
04 30 ☽ ± ♄
05 03 ☽ ⊼ ♂
11 59 ☽ ⅊ ☉
12 09 ☽ Q ♇
12 35 ☽ ⊼ ☿
13 08 ☽ ☌ ♆
13 58 ☽ # ♄
15 39 ☽ ⊼ ♄
15 40 ☽ # ♇
15 42 ☽ # ♀
16 28 ♀ ∥ ♇
16 32 ☽ ± ♂
17 25 ☿ ⊼ ♆
18 19 ☽ △ ♀
20 06 ☽ # ☿
21 02 ☽ Q ♃

21 Thursday
02 37 ☉ ⅊ ♆
04 18 ☽ # ☉
05 41 ☽ ± ☉
09 25 ☽ ⅊ ♂
16 46 ♀ ∠ ♇
17 15 ☽ ⊼ ☉
17 30 ☿ ☌ ♄
19 32 ☽ □ ♅
20 27 ☽ ⅊ ♀
20 33 ☽ □ ♇
23 09 ♀ ∥ ♄

22 Friday
05 25 ☽ □ ♃
05 49 ☿ ⊻ ♀
11 43 ☽ ± ♀
12 48 ☽ △ ♂
18 16 ☽ ⊻ ♆
21 09 ☽ ☍ ♄
21 45 ☽ ⊼ ♀
23 24 ☉ △ ♅

23 Saturday
00 12 ☽ ☍ ☿
02 43 ☽ Q ♅
03 10 ♀ ⅊ ♅
04 32 ☽ ⊥ ♆
12 50 ♀ ⊻ ♄
14 08 ☉ ⚹ ♇
19 45 ☽ ∠ ♆
22 59 ☽ ⚹ ♅

24 Sunday
00 04 ☽ △ ♇
00 49 ☽ ☍ ☉
09 03 ☽ △ ♃
09 44 ☿ Q ♇
16 06 ☿ ± ♆
17 20 ☽ □ ♂
20 44 ☽ ⚹ ♆
22 34 ☽ ☍ ♀
23 56 ☽ ∠ ♅
23 58 ☽ ⊼ ♄

25 Monday
01 03 ☽ ⅊ ♇
07 38 ☉ ∥ ☿
08 21 ☽ ⊼ ☿
09 57 ☽ ± ♄
10 09 ☽ # ☿
10 11 ☽ ⅊ ♃
10 19 ☽ # ☉
14 46 ☽ ⊥ ♅
15 55 ☽ ± ♇
16 31 ☽ Q ♆
19 24 ☽ ± ☿
21 35 ☽ # ♇
22 29 ☽ # ♄

26 Tuesday
00 05 ☽ Q ♂
00 38 ☽ ⊻ ♅
00 53 ☽ ⅊ ♄
01 10 ☽ ± ♃
01 49 ☽ ⊼ ♇
03 14 ☽ # ♀
06 13 ☽ ⊼ ☉
11 09 ☽ ⊼ ♃
11 55 ☽ ⅊ ☿
16 49 ☽ ± ☉
20 31 ☽ ⚹ ♂
22 06 ☽ □ ♆
22 16 ☽ ⊼ ♀

27 Wednesday
01 46 ☽ △ ♄
02 43 ♀ △ ♆
06 26 ☽ ∥ ♆
07 48 ☽ ± ♀
08 46 ☽ ⅊ ☉
15 32 ☽ △ ☿
17 39 ☽ ∥ ♅
21 54 ♀ □ ♂
22 07 ☽ ⅊ ♀
22 08 ☽ ∠ ♂

28 Thursday
01 12 ♂ ± ♃
01 55 ☽ # ♃
02 02 ☽ ☌ ♅
03 21 ☽ ☍ ♇
08 46 ☿ Q ♃
11 03 ☿ Q ♂
11 31 ☽ △ ☉
13 21 ☽ ☍ ♃
13 37 ☽ ⊥ ♂
18 51 ♂ ⊼ ♆
22 08 ☽ △ ♀
22 56 ☽ # ♂
23 49 ☽ △ ♆

29 Friday
00 00 ☽ ⊻ ♂
04 00 ☽ □ ♄
18 22 ☉ ⚹ ♃
23 48 ☽ □ ☿

30 Saturday
01 08 ☽ ⅊ ♆
04 23 ☽ ⊻ ♅
04 39 ☽ ∥ ♂
05 52 ☽ ⊼ ♇
10 46 ☽ Q ♄
11 54 ☿ ⅊ ♆
14 38 ☽ ⊥ ♅
16 10 ☽ ± ♇
16 30 ☽ ± ♆
16 44 ☽ ⊼ ♃
18 11 ☽ □ ☉
23 03 ☽ □ ♀

31 Sunday
00 45 ☽ ∥ ♃
02 50 ☽ ⊼ ♆
03 13 ☽ ± ♃
04 55 ☽ ☌ ♂
06 08 ☽ ∠ ♅
07 37 ☽ ⚹ ♄
07 43 ☽ ⅊ ♇
10 20 ☽ # ♅
10 36 ☽ Q ☿
17 07 ☿ △ ♅
19 04 ☽ ⅊ ♃
22 05 ☽ # ♆

JANUARY 2046

LONGITUDES

Date	Sidereal time h m s	Sun ☉ ° ' "	Moon ☽ ° ' "	Moon ☽ 24.00 ° ' "	Mercury ☿ ° '	Venus ♀ ° '	Mars ♂ ° '	Jupiter ♃ ° '	Saturn ♄ ° '	Uranus ♅ ° '	Neptune ♆ ° '	Pluto ♇ ° '
01	18 45 14	11 ♑ 13 20	03 ♏ 25 13	10 ♏ 10 38	02 ♑ 35	11 ♑ 20	16 ♎ 16	08 ♓ 55	17 ♐ 21	01 ♍ 20	14 ♉ 27	02 ♓ 18
02	18 49 10	12 14 30	16 ♏ 52 45	23 ♏ 31 35	04 08	10 R 43	16 45	09 06	17 28	01 R 19	14 R 26	02 19
03	18 53 07	13 15 40	00 ♐ 07 09	06 ♐ 39 26	05 42	10 07	17 13	09 17	17 35	01 17	14 25	02 20
04	18 57 04	14 16 51	13 ♐ 08 28	19 ♐ 34 13	07 16	09 31	17 41	09 28	17 41	01 16	14 25	02 22
05	19 01 00	15 18 01	25 ♐ 56 43	02 ♑ 15 59	08 51	08 56	18 09	09 39	17 48	01 14	14 24	02 23
06	19 04 57	16 19 12	08 ♑ 32 04	14 ♑ 45 01	10 25	08 22	18 36	09 51	17 55	01 13	14 23	02 24
07	19 08 53	17 20 22	20 ♑ 54 57	27 ♑ 01 58	12 01	07 49	19 03	10 02	18 01	01 11	14 23	02 25
08	19 12 50	18 21 32	03 ♒ 06 15	09 ♒ 08 01	13 36	07 18	19 31	10 14	18 08	01 10	14 22	02 27
09	19 16 46	19 22 42	15 ♒ 07 30	21 ♒ 05 01	15 12	06 48	19 57	10 25	18 14	01 08	14 22	02 28
10	19 20 43	20 23 52	27 ♒ 00 55	02 ♓ 55 33	16 49	06 20	20 24	10 37	18 20	01 06	14 21	02 29
11	19 24 39	21 25 01	08 ♓ 49 23	14 ♓ 42 53	18 26	05 54	20 51	10 49	18 27	01 04	14 21	02 31
12	19 28 36	22 26 10	20 ♓ 36 33	26 ♓ 30 56	20 03	05 30	21 17	11 00	18 33	01 02	14 20	02 32
13	19 32 33	23 27 18	02 ♈ 26 35	08 ♈ 24 07	21 41	05 09	21 43	11 12	18 39	01 00	14 20	02 33
14	19 36 29	24 28 26	14 ♈ 24 09	20 ♈ 27 15	23 19	04 49	22 09	11 24	18 46	00 59	14 20	02 35
15	19 40 26	25 29 33	26 ♈ 34 04	02 ♉ 45 12	24 58	04 32	22 34	11 37	18 52	00 57	14 19	02 36
16	19 44 22	26 30 39	09 ♉ 01 11	15 ♉ 22 33	26 37	04 17	22 59	11 49	18 58	00 55	14 19	02 37
17	19 48 19	27 31 44	21 ♉ 49 47	28 ♉ 23 15	28 17	04 05	23 24	12 01	19 04	00 53	14 19	02 39
18	19 52 15	28 32 49	05 ♊ 03 14	11 ♊ 49 55	29 ♑ 57	03 55	23 49	12 13	19 10	00 51	14 19	02 40
19	19 56 12	29 ♑ 33 53	18 ♊ 43 20	25 ♊ 43 21	01 ♒ 38	03 48	24 14	12 26	19 16	00 48	14 18	02 42
20	20 00 08	00 ♒ 34 57	02 ♋ 49 42	10 ♋ 01 58	03 19	03 43	24 38	12 38	19 22	00 46	14 18	02 43
21	20 04 05	01 35 59	17 ♋ 19 31	24 ♋ 41 36	05 01	03 41	25 02	12 51	19 28	00 44	14 18	02 45
22	20 08 02	02 37 01	02 ♌ 07 21	09 ♌ 35 47	06 43	03 D 41	25 25	13 04	19 34	00 42	14 18	02 46
23	20 11 58	03 38 02	17 ♌ 05 50	24 ♌ 36 26	08 26	03 43	25 49	13 17	19 40	00 40	14 18	02 48
24	20 15 55	04 39 03	02 ♍ 06 31	09 ♍ 35 02	10 09	03 48	26 12	13 29	19 45	00 37	14 D 18	02 49
25	20 19 51	05 40 02	17 ♍ 01 04	24 ♍ 23 48	11 53	03 55	26 35	13 42	19 51	00 35	14 18	02 51
26	20 23 48	06 41 01	01 ♎ 42 30	08 ♎ 56 39	13 37	04 05	26 57	13 55	19 57	00 33	14 18	02 52
27	20 27 44	07 42 00	16 ♎ 05 48	23 ♎ 09 41	15 22	04 17	27 19	14 08	20 02	00 30	14 18	02 54
28	20 31 41	08 42 58	00 ♏ 08 09	07 ♏ 01 10	17 06	04 31	27 41	14 21	20 08	00 28	14 18	02 55
29	20 35 37	09 43 55	13 ♏ 48 47	20 ♏ 31 09	18 51	04 47	28 03	14 34	20 13	00 26	14 18	02 57
30	20 39 34	10 44 52	27 ♏ 08 28	03 ♐ 40 59	20 36	05 05	28 24	14 48	20 18	00 23	14 19	02 59
31	20 43 31	11 ♒ 45 48	10 ♐ 09 00	16 ♐ 32 47	22 ♒ 21	05 ♑ 25	28 ♎ 45	15 ♓ 01	20 ♐ 24	00 ♍ 21	14 ♉ 19	03 ♓ 00

Moon True ☊, Mean ☊, Latitude

Date	Moon True ☊ ° '	Moon Mean ☊ ° '	Moon ☽ Latitude ° '
01	14 ♒ 07	15 ♒ 19	05 S 07
02	14 R 03	15 16	05 12
03	13 58	15 13	05 00
04	13 53	15 09	04 33
05	13 48	15 06	03 52
06	13 44	15 03	03 01
07	13 40	15 00	02 02
08	13 39	14 57	00 S 58
09	13 D 38	14 54	00 N 08
10	13 39	14 50	01 13
11	13 40	14 47	02 15
12	13 42	14 44	03 10
13	13 44	14 41	03 58
14	13 45	14 38	04 36
15	13 R 45	14 34	05 02
16	13 45	14 31	05 15
17	13 43	14 28	05 13
18	13 42	14 25	04 55
19	13 40	14 22	04 19
20	13 38	14 19	03 27
21	13 36	14 15	02 21
22	13 36	14 12	01 N 03
23	13 D 35	14 09	00 S 19
24	13 36	14 06	01 41
25	13 36	14 03	02 55
26	13 37	14 00	03 57
27	13 38	13 56	04 42
28	13 38	13 53	05 09
29	13 R 38	13 50	05 18
30	13 38	13 47	05 09
31	13 ♒ 38	13 ♒ 44	04 S 45

DECLINATIONS

Date	Sun ☉ ° '	Moon ☽ ° '	Mercury ☿ ° '	Venus ♀ ° '	Mars ♂ ° '	Jupiter ♃ ° '	Saturn ♄ ° '	Uranus ♅ ° '	Neptune ♆ ° '	Pluto ♇ ° '
01	22 S 58	17 S 27	24 S 24	19 S 12	04 S 32	09 S 15	21 S 24	11 N 43	14 N 27	21 S 26
02	22 52	21 51	24 27	19 01	04 43	09 11	21 25	11 44	14 27	21 25
03	22 46	25 03	24 29	18 50	04 53	09 06	21 26	11 44	14 27	21 25
04	22 40	26 53	24 30	18 40	05 03	09 02	21 26	11 45	14 27	21 24
05	22 33	27 15	24 29	18 30	05 13	08 58	21 27	11 45	14 27	21 24
06	22 26	26 10	24 27	18 21	05 23	08 53	21 27	11 46	14 27	21 23
07	22 19	23 49	24 24	18 12	05 33	08 49	21 28	11 47	14 26	21 22
08	22 11	20 24	24 19	18 03	05 43	08 44	21 28	11 47	14 26	21 22
09	22 02	16 10	24 13	17 55	05 52	08 40	21 29	11 48	14 26	21 21
10	21 53	11 22	24 05	17 48	06 02	08 35	21 30	11 49	14 26	21 21
11	21 44	06 11	23 56	17 41	06 11	08 31	21 30	11 49	14 26	21 20
12	21 34	00 S 48	23 45	17 34	06 21	08 26	21 31	11 50	14 26	21 19
13	21 24	04 N 37	23 33	17 29	06 30	08 22	21 31	11 51	14 26	21 19
14	21 13	09 55	23 20	17 23	06 39	08 17	21 32	11 51	14 26	21 18
15	21 02	14 56	23 05	17 19	06 48	08 12	21 32	11 52	14 26	21 17
16	20 51	19 29	22 48	17 15	06 57	08 08	21 33	11 53	14 26	21 17
17	20 39	23 16	22 30	17 11	07 06	08 03	21 33	11 54	14 26	21 16
18	20 27	25 58	22 10	17 08	07 14	07 58	21 34	11 54	14 26	21 15
19	20 14	27 16	21 49	17 06	07 23	07 53	21 34	11 55	14 26	21 15
20	20 01	26 52	21 27	17 04	07 31	07 48	21 34	11 56	14 26	21 14
21	19 48	24 38	21 02	17 03	07 39	07 43	21 35	11 57	14 26	21 13
22	19 34	20 43	20 37	17 02	07 47	07 38	21 35	11 57	14 26	21 13
23	19 20	15 24	20 09	17 01	07 55	07 33	21 36	11 58	14 26	21 12
24	19 06	09 09	19 41	17 01	08 03	07 28	21 36	11 59	14 26	21 12
25	18 51	02 N 26	19 10	17 02	08 11	07 23	21 37	12 00	14 26	21 11
26	18 36	04 S 18	18 38	17 02	08 19	07 18	21 37	12 01	14 26	21 10
27	18 20	10 40	18 05	17 03	08 26	07 13	21 37	12 02	14 26	21 10
28	18 05	16 20	17 30	17 05	08 34	07 08	21 38	12 02	14 26	21 09
29	17 49	21 02	16 54	17 06	08 41	07 03	21 38	12 03	14 26	21 08
30	17 32	24 32	16 17	17 08	08 48	06 58	21 38	12 04	14 27	21 08
31	17 S 15	26 S 39	15 S 38	17 S 10	08 S 55	06 S 53	21 S 39	12 N 05	14 N 27	21 S 07

ZODIAC SIGN ENTRIES

Date	h	m	Planets
01	05	58	☽ ♏
03	11	47	☽ ♐
05	19	41	☽ ♑
08	05	51	☽ ♒
10	18	03	☽ ♓
13	07	04	☽ ♈
15	18	41	☽ ♉
18	02	55	☽ ♊
18	12	42	☿ ♒
19	22	16	☉ ♒
20	07	15	☽ ♋
22	08	35	☽ ♌
24	08	37	☽ ♍
26	09	11	☽ ♎
28	11	46	☽ ♏
30	17	14	☽ ♐

LATITUDES

Date	Mercury ☿ ° '	Venus ♀ ° '	Mars ♂ ° '	Jupiter ♃ ° '	Saturn ♄ ° '	Uranus ♅ ° '	Neptune ♆ ° '	Pluto ♇ ° '
01	00 S 59	03 N 46	02 N 01	01 S 06	01 N 26	00 N 46	01 S 48	11 S 35
04	01 16	04 26	02 02	01 06	01 26	00 46	01 48	11 35
07	01 31	05 01	02 04	01 05	01 26	00 47	01 47	11 34
10	01 43	05 30	02 05	01 05	01 26	00 47	01 47	11 34
13	01 53	05 52	02 07	01 05	01 26	00 47	01 47	11 33
16	02 01	06 08	02 08	01 05	01 26	00 47	01 47	11 33
19	02 05	06 17	02 10	01 04	01 26	00 47	01 47	11 33
22	02 05	06 22	02 11	01 04	01 26	00 47	01 46	11 32
25	02 02	06 21	02 13	01 04	01 27	00 47	01 46	11 32
28	01 54	06 17	02 14	01 04	01 27	00 47	01 46	11 32
31	01 S 40	06 N 10	02 N 16	01 S 03	01 N 27	00 N 47	01 S 46	11 S 32

DATA

Julian Date	2468347
Delta T	+79 seconds
Ayanamsa	24° 30' 11"
Synetic vernal point	04° ♓ 36' 47"
True obliquity of ecliptic	23° 26' 06"

MOON'S PHASES, APSIDES AND POSITIONS ☽

Date	h	m	Phase	Longitude ° '	Eclipse Indicator
07	04	24	●	17 ♑ 01	
15	09	43	☽	25 ♈ 24	
22	12	51	○	02 ♌ 39	partial
29	04	11	☾	09 ♏ 24	

Day	h	m		
11	16	49	Apogee	
23	19	10	Perigee	
05	05	51	Max dec	27° S 17'
12	15	34	0N	
19	18	37	Max dec	27° N 20'
25	20	36	0S	

LONGITUDES

Date	Chiron ⚷ ° '	Ceres ⚳ ° '	Pallas ⚴ ° '	Juno ⚵ ° '	Vesta ⚶ ° '	Black Moon Lilith ⚸ ° '
01	06 ♎ 17	01 ♎ 36	04 ♋ 51	06 ♎ 55	27 ♋ 53	05 ♓ 05
11	06 ♎ 33	03 ♎ 17	01 ♋ 48	08 ♎ 16	25 ♋ 23	06 ♓ 12
21	06 ♎ 37	04 ♎ 22	29 ♊ 37	09 ♎ 00	22 ♋ 44	07 ♓ 18
31	06 ♎ 30	04 ♎ 47	28 ♊ 34	09 ♎ 05	20 ♋ 18	08 ♓ 25

All ephemeris data is given at 12.00 UT and the Moon's longitude is additionally given for 24.00 UT

ASPECTARIAN

h m Aspects

01 Monday
04 00 ☽ Q ☉
05 06 ☽ Q ♀
07 42 ☿ ⚹ ♇
08 20 ☽ ⚹ ♅
10 01 ☽ △ ♇
10 06 ☽ ∠ ♄
10 19 ☽ ⚹ ☿
13 36 ☉ ☌ ♀
15 04 ♂ ∠ ♅
20 29 ☽ ∥ ♀
21 54 ☽ △ ♃

02 Tuesday
01 27 ☽ ⚹ ♀
02 13 ☽ ⊥ ♄
03 00 ☽ ⚹ ☉
05 37 ☽ Q ♅
07 37 ☽ ☍ ♆
09 18 ☽ ∥ ♄
09 21 ☽ ∥ ♇
11 45 ☽ ⊻ ♂
13 04 ☽ ⊻ ♄
16 36 ☽ ∠ ☿
18 29 ☽ ∥ ☉
21 28 ♄ ∥ ♇
22 59 ☽ ⊥ ♂

03 Wednesday
03 17 ☽ ∠ ♀
06 51 ☽ ∥ ☿
08 19 ☽ ∠ ☉
11 08 ☽ ⊥ ☿
14 08 ☽ □ ♅
15 59 ☽ ∠ ♂
16 04 ☽ □ ♇
18 46 ♂ bQ ♇
19 00 ☽ ⊥ ♀
23 38 ☽ ⊻ ☿

04 Thursday
02 14 ☽ ⊥ ☉
05 06 ☽ □ ♃
05 35 ☽ ⊻ ♀
12 36 ♂ ⚹ ♄
13 31 ♀ ⚹ ♃
14 18 ☽ ⊻ ☉
14 21 ☽ ⊼ ♆
15 00 ☉ △ ♆
20 33 ☽ ☌ ♄
20 47 ☽ ⚹ ♂

05 Friday
01 30 ☽ Q ♇
01 34 ☽ ± ♆
13 02 ☿ ☌ ♀
15 17 ☽ Q ♃
18 32 ☽ bQ ♆
20 16 ☽ Q ♂
22 01 ☽ △ ♅

06 Saturday
00 14 ☽ ⚹ ♇
01 16 ♆ Q ♇
02 02 ☿ ⚹ ♃
09 34 ☉ bQ ♅
11 42 ☽ ☌ ♀
14 34 ☽ ⚹ ♃
16 10 ☽ ☌ ☿
23 17 ☽ △ ♆

07 Sunday
02 48 ☽ bQ ♅
04 24 ☽ ☌ ☉
05 11 ☽ ∠ ♇
06 18 ☽ ⊻ ♄
06 53 ☽ ∥ ☿
08 14 ☽ □ ♂
14 01 ☉ ∠ ♇
18 08 ☽ ⊥ ♄
20 12 ☽ ∠ ♃
20 21 ☽ ± ♅
22 49 ☽ ⊥ ♇
23 50 ☽ ∥ ☉

08 Monday
05 06 ☽ ∥ ♄
05 48 ☽ ∥ ♇
05 52 ☉ ⊻ ♄
08 09 ☽ ⊼ ♅
10 41 ☽ ⊻ ♇
12 03 ☽ ∠ ♄
14 16 ☽ ⊥ ♃
20 00 ☽ ⊻ ♀
23 26 ☿ △ ♆

09 Tuesday
02 13 ☽ ∥ ♀
02 25 ☽ ⊻ ♃
07 31 ☽ ⊥ ♀
10 28 ☽ □ ♆
12 11 ☽ ⊻ ☿
18 19 ☽ ⚹ ♄
20 55 ☽ # ♆
21 22 ☽ ⊻ ☉
22 06 ☽ △ ♂

10 Wednesday
00 56 ☽ ∠ ♀
01 35 ☿ bQ ♅
02 09 ☽ ⊥ ☿
09 51 ☽ # ♅
10 38 ☽ ⊥ ☉
12 10 ☉ □ ♂
18 48 ☽ Q ♄
20 16 ☽ ☍ ♅
22 12 ☿ ∠ ♇
22 50 ☽ Q ♆
23 08 ☽ ☌ ♇
23 17 ☽ ∠ ☿

11 Thursday
01 11 ☽ ∥ ♃
05 42 ☽ bQ ♂
06 16 ☽ ⚹ ♀
06 38 ☽ ∠ ☉
11 57 ☽ ∥ ♂
12 19 ☿ ⊻ ♄
16 07 ☽ ☌ ♃
23 14 ☽ ⚹ ♆

12 Friday
00 44 ☽ ± ♂
05 53 ☽ Q ♀
07 47 ☽ □ ♄
10 41 ☽ ⚹ ☿
13 25 ☽ ⊼ ♂
16 04 ☽ ⚹ ☉
19 48 ☉ ∥ ♄

13 Saturday
05 43 ☽ ∠ ♆
09 06 ☽ ⊼ ♅
12 14 ☽ ⊻ ♇
12 41 ☿ □ ♂
14 53 ☽ Q ☿
17 18 ☽ □ ♀
18 39 ☽ Q ☉
20 41 ☽ # ♂
21 11 ☽ ± ♅
23 51 ☽ ⊥ ♆

14 Sunday
00 20 ☽ ⊥ ♇
00 48 ☉ ∥ ♇
04 37 ☽ # ♃
05 55 ☽ ⊻ ♃
11 51 ☽ ⊻ ♆
15 07 ☽ bQ ♅
18 04 ☽ ⊥ ♃
18 19 ☽ ∠ ♇
19 31 ☉ ⊥ ♄
20 44 ☽ △ ♄
21 05 ☽ ∥ ♅
23 28 ☉ ± ♅

15 Monday
03 53 ☽ ☍ ♂
08 22 ☽ □ ☿
09 29 ☽ ∥ ♆
09 43 ☽ □ ☉
10 28 ☿ ⊥ ♄
10 45 ♀ Q ♂
11 44 ☿ ± ♅
12 05 ☽ ∠ ♃
20 29 ☽ △ ♅
23 44 ☽ ⚹ ♇
23 58 ☽ # ♀

16 Tuesday
02 15 ☽ bQ ♄
03 07 ☽ △ ♀
08 02 ☉ ☌ ☿
12 07 ☿ ⊥ ♇
14 45 ☉ ⊥ ♇
15 15 ☿ ∠ ♃
17 23 ☽ ⚹ ♃
19 32 ☽ ± ♄
19 42 ☽ # ☉
20 55 ☉ ∠ ♃
22 01 ☽ ☌ ♆
22 37 ☽ Q ♇
22 42 ☽ # ♇

17 Wednesday
00 26 ☽ # ♄
06 51 ☽ ⊼ ♄
06 59 ☽ bQ ♀
07 07 ☽ # ☿
15 00 ☽ ⊼ ♂
16 05 ☽ Q ♃
23 18 ☽ ± ♀
23 19 ☽ △ ☉

18 Thursday
01 30 ☽ △ ☿
02 17 ☽ ± ♂
04 28 ☽ □ ♅
07 44 ☽ □ ♇
09 59 ☽ ⊼ ♀
18 54 ☽ bQ ♂

19 Friday
00 29 ☿ ⊼ ♅
00 53 ☽ □ ♃
04 12 ☽ bQ ☉
04 20 ☽ ⊻ ♆
07 53 ☽ bQ ☿
12 09 ☽ Q ♅
12 57 ☽ ☍ ♄
14 44 ☽ ⊥ ♆
20 58 ☽ ± ☉
21 44 ☽ △ ♂

20 Saturday
01 27 ☽ ± ☿
03 23 ☿ ⊻ ♇
04 03 ☿ ∥ ♄
06 04 ☽ ∠ ♆
07 56 ☽ ⊼ ☉
08 33 ☽ ⚹ ♅
11 49 ☽ △ ♇
12 56 ☽ ⊼ ☿
13 29 ☽ ☍ ♀
16 18 ☉ ⊼ ♅
17 26 ☿ ⊻ ♀

21 Sunday
00 55 ☿ ∥ ♇
03 45 ☿ ∠ ♄
04 33 ☽ △ ♃
07 03 ☽ ⚹ ♆
09 24 ☽ ∠ ♅
12 41 ☽ bQ ♇
15 31 ☽ ⊼ ♄
19 17 ♂ ∥ ♃
22 58 ♀ St D

22 Monday
00 02 ☽ ⊥ ♅
00 53 ☽ □ ♂
01 20 ☽ ± ♄
02 36 ☽ Q ♆
03 21 ☽ ± ♇
05 22 ☽ bQ ♃
07 23 ☽ # ♄
09 22 ☽ # ♇
09 43 ☽ ⊻ ♅
12 34 ☽ # ☿
12 51 ☽ ☍ ☉
13 03 ☽ ⊼ ♇
14 30 ☽ ⊼ ♀
15 42 ☉ ⊻ ♇
15 57 ☽ bQ ♄
17 27 ☿ ⊥ ♃
17 55 ☽ # ☉
20 03 ☽ ± ♃
20 21 ☽ ☍ ☿

23 Tuesday
00 09 ☽ ± ♀
05 12 ☽ # ♀
05 48 ☽ ⊼ ♃
06 37 ☽ Q ♂
07 32 ☽ □ ♆
14 10 ☉ ⊻ ♀
14 36 ☽ bQ ♀
15 55 ☽ ∥ ♆
16 07 ☽ △ ♄

24 Wednesday
01 28 ☽ ∥ ♅
02 18 ☽ ⚹ ♂
03 35 ♆ St D
06 45 ☿ ⊥ ♀
09 38 ☽ ☌ ♅
13 09 ☽ ☍ ♇
14 43 ☉ ∠ ♄
14 44 ☽ △ ♀
15 54 ☽ # ♂
16 22 ☽ ⊼ ☉
18 09 ☽ # ♃

25 Thursday
02 37 ☽ ⊼ ☿
02 43 ☽ ± ☉
02 59 ☽ ∠ ♂
06 34 ☽ ☍ ♃
07 36 ☽ △ ♆
13 36 ☽ ± ☿
16 37 ☽ □ ♄
17 56 ☽ ⊥ ♂
18 22 ☽ bQ ☉

26 Friday
03 59 ☽ ⊻ ♂
06 14 ☽ bQ ☿
08 02 ☽ bQ ♆
10 05 ☽ ⊻ ♅
13 55 ☽ ⊼ ♇
15 31 ☉ ∥ ☿
15 58 ☽ □ ♀
16 43 ☿ ⊻ ♃
19 59 ☽ ⊥ ♅
20 52 ☽ △ ☉
21 24 ☿ □ ♆
22 24 ☽ Q ♄
22 56 ☽ ± ♆
22 56 ☽ ∥ ♃
23 54 ☽ ± ♇

27 Saturday
03 11 ☽ ∥ ♂
08 39 ☽ ⊼ ♃
08 59 ☽ ⊼ ♆
10 35 ☽ △ ☿
11 01 ☽ ∠ ♅
15 03 ☽ bQ ♇
17 28 ☽ # ♅
18 43 ☽ ⚹ ♄
18 57 ☽ ± ♃
22 40 ☽ Q ♀

28 Sunday
01 09 ☉ ⊥ ♃
03 34 ☽ # ♆
06 26 ♃ ⚹ ♆
07 40 ☽ ☌ ♂
10 38 ☽ bQ ♃
12 35 ☽ ⚹ ♅
15 29 ☽ ∥ ♀
16 51 ☽ △ ♇
16 56 ☽ ∥ ☿
19 45 ☽ ⚹ ♀
19 51 ☽ ∥ ☉
20 45 ☽ ∠ ♄

29 Monday
04 11 ☽ □ ☉
04 35 ☿ ∥ ♀
09 33 ☽ Q ♅
12 38 ☽ ∥ ♇
12 44 ☽ ⊥ ♄
12 53 ☽ ☍ ♆
13 23 ☽ △ ♃
15 35 ☽ ∥ ♄
22 22 ☽ □ ☿
22 54 ☽ ∠ ♀
23 32 ☽ ⊻ ♄

30 Tuesday
03 14 ☿ ∠ ♀
07 42 ☿ ⚹ ♄
13 05 ♄ ± ♆
14 22 ☽ ⊻ ♂
15 11 ☽ Q ☉
15 38 ☽ ⊥ ♀
17 55 ☽ □ ♅
22 43 ☽ □ ♇
23 33 ☉ ⊥ ♀

31 Wednesday
01 42 ☽ ⊥ ♂
02 58 ☽ ⊻ ♀
12 26 ☽ Q ☿
15 16 ☽ ⚹ ☉
18 47 ☉ ∥ ♀
18 55 ☽ ∠ ♂
19 48 ☽ ⊼ ♆
21 17 ☽ □ ♃

FEBRUARY 2046

LONGITUDES

Date	Sidereal time h m s	Sun ☉ ° ' "	Moon ☽ ° ' "	Moon ☽ 24.00 ° ' "	Mercury ☿ ° '	Venus ♀ ° '	Mars ♂ ° '	Jupiter ♃ ° '	Saturn ♄ ° '	Uranus ♅ ° '	Neptune ♆ ° '	Pluto ⯓ ° '
01	20 47 27	12 ♒ 46 44	22 ♐ 52 39	29 ♐ 08 56	24 ♒ 06	05 ♑ 47	29 ♎ 06	15 ♓ 14	20 ♐ 29	00 ♍ 18	14 ♉ 19	03 ♓ 02
02	20 51 24	13 47 39	05 ♑ 21 55	11 ♑ 31 53	25 50	06 11	29 26	15 28	20 34	00 R 16	14 20	03 03
03	20 55 20	14 48 33	17 ♑ 39 08	23 ♑ 43 55	27 34	06 36	29 ♎ 45	15 41	20 39	00 13	14 20	03 05
04	20 59 17	15 49 25	29 ♑ 46 31	05 ♒ 47 09	29 ♒ 17	07 03	00 ♏ 05	15 55	20 44	00 11	14 20	03 07
05	21 03 13	16 50 17	11 ♒ 46 05	17 ♒ 43 32	00 ♓ 58	07 32	00 24	16 08	20 49	00 08	14 21	03 08
06	21 07 10	17 51 08	23 ♒ 39 46	29 ♒ 35 01	02 38	08 03	00 43	16 22	20 54	00 06	14 21	03 10
07	21 11 06	18 51 57	05 ♓ 29 32	11 ♓ 23 35	04 16	08 35	01 01	16 36	20 59	00 03	14 22	03 11
08	21 15 03	19 52 45	17 ♓ 17 30	23 ♓ 11 33	05 52	09 08	01 19	16 49	21 04	00 ♍ 01	14 22	03 13
09	21 19 00	20 53 32	29 ♓ 06 06	05 ♈ 01 32	07 25	09 43	01 36	17 03	21 09	29 ♌ 58	14 23	03 15
10	21 22 56	21 54 17	10 ♈ 58 12	16 ♈ 56 34	08 54	10 19	01 53	17 17	21 13	29 56	14 23	03 16
11	21 26 53	22 55 00	22 ♈ 57 04	29 ♈ 00 11	10 19	10 57	02 09	17 31	21 18	29 53	14 24	03 18
12	21 30 49	23 55 43	05 ♉ 06 24	11 ♉ 16 15	11 39	11 35	02 26	17 45	21 22	29 50	14 25	03 20
13	21 34 46	24 56 23	17 ♉ 30 14	23 ♉ 48 52	12 54	12 15	02 41	17 59	21 27	29 48	14 25	03 21
14	21 38 42	25 57 02	00 ♊ 12 40	06 ♊ 42 04	14 03	12 56	02 56	18 13	21 31	29 45	14 26	03 23
15	21 42 39	26 57 39	13 ♊ 17 32	19 ♊ 59 23	15 05	13 38	03 11	18 27	21 35	29 42	14 27	03 25
16	21 46 35	27 58 15	26 ♊ 47 55	03 ♋ 43 16	15 59	14 21	03 25	18 41	21 39	29 40	14 27	03 26
17	21 50 32	28 58 49	10 ♋ 45 28	17 ♋ 54 24	16 45	15 06	03 39	18 55	21 43	29 37	14 28	03 28
18	21 54 29	29 ♒ 59 21	25 ♋ 09 46	02 ♌ 31 05	17 21	15 51	03 52	19 09	21 47	29 35	14 29	03 30
19	21 58 25	00 ♓ 59 51	09 ♌ 57 41	17 ♌ 28 43	17 49	16 37	04 05	19 23	21 51	29 32	14 30	03 31
20	22 02 22	02 00 20	25 ♌ 03 11	02 ♍ 39 55	18 07	17 24	04 17	19 37	21 55	29 29	14 31	03 33
21	22 06 18	03 00 47	10 ♍ 17 42	17 ♍ 55 14	18 14	18 12	04 29	19 51	21 59	29 27	14 32	03 35
22	22 10 15	04 01 12	25 ♍ 31 14	03 ♎ 04 28	18 R 12	19 00	04 40	20 06	22 03	29 24	14 33	03 37
23	22 14 11	05 01 36	10 ♎ 33 49	17 ♎ 58 18	17 59	19 50	04 50	20 20	22 06	29 21	14 34	03 38
24	22 18 08	06 01 58	25 ♎ 17 07	02 ♏ 29 40	17 37	20 40	05 00	20 34	22 10	29 19	14 35	03 40
25	22 22 04	07 02 19	09 ♏ 35 31	16 ♏ 34 28	17 06	21 31	05 10	20 48	22 13	29 16	14 36	03 42
26	22 26 01	08 02 39	23 ♏ 26 27	00 ♐ 11 34	16 26	22 23	05 19	21 03	22 17	29 14	14 37	03 43
27	22 29 58	09 02 58	06 ♐ 50 04	13 ♐ 22 15	15 39	23 15	05 27	21 17	22 20	29 11	14 38	03 45
28	22 33 54	10 ♓ 03 15	19 ♐ 48 35	26 ♐ 09 30	14 ♓ 46	24 ♑ 08	05 ♏ 34	21 ♓ 32	22 ♐ 23	29 ♌ 08	14 ♉ 39	03 ♓ 47

Date	Moon True ☊ ° '	Moon Mean ☊ ° '	Moon ☽ Latitude ° '
01	13 ♒ 37	13 ♒ 40	04 S 06
02	13 D 37	13 37	03 17
03	13 37	13 34	02 20
04	13 38	13 31	01 16
05	13 38	13 28	00 S 10
06	13 R 38	13 25	00 N 55
07	13 37	13 21	01 59
08	13 37	13 18	02 56
09	13 36	13 15	03 47
10	13 34	13 12	04 27
11	13 33	13 09	04 57
12	13 32	13 06	05 14
13	13 32	13 02	05 16
14	13 D 32	12 59	05 04
15	13 32	12 56	04 35
16	13 33	12 53	03 51
17	13 34	12 50	02 52
18	13 35	12 46	01 40
19	13 36	12 43	00 N 20
20	13 R 36	12 40	01 S 03
21	13 34	12 37	02 22
22	13 32	12 34	03 31
23	13 29	12 31	04 25
24	13 26	12 27	04 59
25	13 24	12 24	05 14
26	13 22	12 21	05 10
27	13 21	12 18	04 49
28	13 ♒ 21	12 ♒ 15	04 S 14

DECLINATIONS

Date	Sun ☉ ° '	Moon ☽ ° '	Mercury ☿ ° '	Venus ♀ ° '	Mars ♂ ° '	Jupiter ♃ ° '	Saturn ♄ ° '	Uranus ♅ ° '	Neptune ♆ ° '	Pluto ⯓ ° '
01	16 S 58	27 S 21	14 S 58	17 S 12	09 S 01	06 S 47	21 S 39	12 N 06	14 N 27	21 S 06
02	16 41	26 37	14 17	17 14	09 08	06 42	21 39	12 07	14 27	21 06
03	16 23	24 35	13 35	17 17	09 15	06 37	21 40	12 08	14 27	21 05
04	16 05	21 26	12 51	17 19	09 21	06 32	21 40	12 09	14 27	21 04
05	15 47	17 25	12 07	17 22	09 27	06 26	21 40	12 10	14 27	21 04
06	15 29	12 45	11 23	17 24	09 33	06 21	21 40	12 10	14 28	21 03
07	15 10	07 39	10 38	17 27	09 39	06 16	21 41	12 11	14 28	21 02
08	14 51	02 S 19	09 53	17 29	09 45	06 10	21 41	12 12	14 28	21 02
09	14 32	03 N 07	09 07	17 32	09 50	06 05	21 41	12 13	14 28	21 01
10	14 12	08 26	08 23	17 34	09 56	05 59	21 41	12 14	14 29	21 01
11	13 53	13 31	07 38	17 36	10 01	05 54	21 42	12 15	14 29	21 00
12	13 33	18 09	06 55	17 38	10 06	05 48	21 42	12 16	14 29	20 59
13	13 12	22 06	06 14	17 40	10 11	05 43	21 42	12 17	14 29	20 59
14	12 52	25 08	05 34	17 42	10 16	05 37	21 42	12 18	14 30	20 58
15	12 31	26 57	04 56	17 44	10 20	05 32	21 43	12 19	14 30	20 57
16	12 11	27 15	04 20	17 45	10 25	05 26	21 43	12 20	14 30	20 57
17	11 50	25 51	03 48	17 46	10 29	05 21	21 43	12 21	14 30	20 56
18	11 28	22 45	03 19	17 47	10 33	05 15	21 43	12 22	14 31	20 56
19	11 07	18 04	02 54	17 47	10 37	05 10	21 43	12 22	14 31	20 55
20	10 46	12 02	02 33	17 47	10 41	05 04	21 43	12 23	14 31	20 54
21	10 24	05 N 31	02 17	17 47	10 44	04 59	21 44	12 24	14 32	20 54
22	10 02	01 S 27	02 05	17 47	10 48	04 53	21 44	12 25	14 32	20 53
23	09 40	08 14	01 59	17 46	10 51	04 47	21 44	12 26	14 32	20 53
24	09 18	14 25	01 57	17 45	10 54	04 42	21 44	12 27	14 33	20 52
25	08 56	19 39	02 00	17 43	10 57	04 36	21 44	12 28	14 33	20 51
26	08 33	23 38	02 08	17 42	11 00	04 30	21 44	12 29	14 34	20 51
27	08 11	26 12	02 20	17 39	11 02	04 25	21 44	12 30	14 34	20 50
28	07 S 48	27 S 16	02 S 37	17 S 37	11 S 04	04 S 19	21 S 44	12 N 31	14 N 34	20 S 50

ZODIAC SIGN ENTRIES

Date	h m	Planets
02	01 38	☽ ♑
04	05 56	♂ ♏
04	12 27	☽ ♒
04	22 12	☿ ♓
07	00 51	☽ ♓
08	18 53	♅ ♌
09	13 49	☽ ♈
12	01 58	☽ ♉
14	11 36	☽ ♊
16	17 34	☽ ♋
18	12 16	☉ ♓
18	19 55	☽ ♌
20	19 48	☽ ♍
22	19 06	☽ ♎
24	19 50	☽ ♏
26	23 39	☽ ♐

LATITUDES

Date	Mercury ☿ ° '	Venus ♀ ° '	Mars ♂ ° '	Jupiter ♃ ° '	Saturn ♄ ° '	Uranus ♅ ° '	Neptune ♆ ° '	Pluto ⯓ ° '
01	01 S 34	06 N 07	02 N 16	01 S 03	01 N 27	00 N 47	01 S 46	11 S 32
04	01 13	05 56	02 18	01 03	01 27	00 47	01 46	11 31
07	00 44	05 43	02 19	01 03	01 27	00 47	01 45	11 31
10	00 S 09	05 29	02 20	01 03	01 27	00 47	01 45	11 31
13	00 N 32	05 13	02 22	01 03	01 28	00 47	01 45	11 31
16	01 17	04 56	02 23	01 03	01 28	00 47	01 45	11 31
19	02 04	04 39	02 24	01 03	01 28	00 47	01 45	11 31
22	02 48	04 21	02 25	01 03	01 28	00 47	01 45	11 31
25	03 22	04 02	02 26	01 02	01 28	00 47	01 44	11 32
28	03 40	03 44	02 27	01 02	01 29	00 47	01 44	11 32
31	03 N 40	03 N 25	02 N 27	01 S 02	01 N 29	00 N 47	01 S 44	11 S 32

DATA

Julian Date	2468378
Delta T	+79 seconds
Ayanamsa	24° 30' 17"
Synetic vernal point	04° ♓ 36' 42"
True obliquity of ecliptic	23° 26' 06"

MOON'S PHASES, APSIDES AND POSITIONS ☽

Date	h m	Phase	Longitude ° '	Eclipse Indicator
05	23 10	●	17 ♒ 19	Annular
14	03 20	☽	25 ♉ 35	
20	23 44	○	02 ♍ 30	
27	16 23	☾	09 ♐ 14	

Day	h m		
08	05 23	Apogee	
21	06 51	Perigee	
01	11 24	Max dec	27° S 21'
08	22 14	0N	
16	04 32	Max dec	27° N 20'
22	07 00	0S	
28	17 13	Max dec	27° S 18'

LONGITUDES

Date	Chiron ⚷ ° '	Ceres ⚳ ° '	Pallas ⚴ ° '	Juno ⚵ ° '	Vesta ⚶ ° '	Black Moon Lilith ⚸ ° '
01	06 ♎ 28	04 ♎ 47	28 ♊ 32	09 ♎ 04	20 ♋ 05	08 ♓ 31
11	06 ♎ 08	04 ♎ 23	28 ♊ 45	08 ♎ 23	18 ♋ 12	09 ♓ 38
21	05 ♎ 38	03 ♎ 17	00 ♋ 01	07 ♎ 03	17 ♋ 00	10 ♓ 45
31	05 ♎ 00	01 ♎ 34	02 ♋ 08	05 ♎ 10	16 ♋ 35	11 ♓ 51

ASPECTARIAN

h m	Aspects
01 Thursday	
07 08	☽ ± ♆
07 25	☽ ☌ ♄
08 29	☽ Q ⯓
14 42	☽ ⚹ ☿
22 12	☽ ∠ ☉
02 Friday	
00 13	☽ ⚹ ♂
00 20	☽ ♇ ♆
02 11	☽ △ ♅
06 10	☿ # ♆
07 31	☽ ⚹ ⯓
08 15	☽ Q ♃
13 38	☽ ☌ ♀
17 08	☽ ⊥ ☉
03 Saturday	
00 07	☽ Q ♂
00 23	☽ ∠ ☿
00 38	☉ □ ♆
05 29	☽ △ ♆
05 55	☽ ⊻ ☉
07 15	☽ ♇ ♅
08 04	☽ ⚹ ♃
12 51	☽ ∠ ⯓
17 58	☽ ⊻ ♄
20 59	☽ ⊥ ☿
04 Sunday	
00 56	☽ ± ♅
05 55	☽ ⊥ ♄
06 41	☽ ⊥ ⯓
10 29	☽ ∥ ♄
10 51	☽ ⊻ ☿
12 38	☽ □ ♂
12 48	☽ ⊼ ♅
14 19	☽ ∠ ♃
14 23	☽ ∥ ⯓
14 42	☉ ⊻ ♃
18 40	☽ ⊻ ⯓
18 43	♂ ⚹ ♅
23 59	☽ ∠ ♄
05 Monday	
00 29	☿ ☍ ♅
01 59	☿ △ ♂
03 09	☽ ⊻ ♀
08 40	☽ ⊥ ♃
10 53	☿ # ♅
12 18	☽ ∥ ♀
15 43	☽ ⊥ ♀
17 11	☽ □ ♆
20 58	☽ ⊻ ♃
21 22	☽ ∥ ☉
23 10	☽ ☌ ☉
06 Tuesday	
03 33	☽ # ♆
06 23	☽ ⚹ ♄
07 51	☿ Q ♆
10 42	☽ ∠ ♀
14 49	☽ # ♅
16 06	☿ Q ♄
19 48	☿ ☌ ⯓
19 48	☽ ∥ ☿
07 Wednesday	
01 00	☽ ☍ ♅
02 40	☽ △ ♂
02 58	☽ ∥ ♂
05 38	☽ Q ♆
06 52	☽ Q ♄
07 19	☽ ☌ ⯓
09 08	☽ ☌ ☿
18 28	☽ ∥ ♃
18 35	☽ ⚹ ♀
08 Thursday	
06 03	☽ ⚹ ♆
09 57	☽ ♇ ♂
11 02	☽ ☌ ♃
15 42	☿ ∥ ♂
17 45	☽ ⊻ ☉
19 44	☽ □ ♄
20 13	☽ Q ♀
09 Friday	
04 43	☽ ± ♂
07 06	☽ ⊥ ☉
12 34	☽ ∠ ♆
13 45	☽ ⊼ ♅
16 11	☉ # ♆
17 11	☽ ⊼ ♂
18 27	☉ ⚹ ♄
20 25	☽ ⊻ ⯓
10 Saturday	
01 03	☽ # ♃
01 51	☽ ± ♅
03 02	☽ ∠ ☉
06 47	☽ ⊥ ♆
07 14	☽ ⊻ ☿
08 34	☽ ⊥ ⯓
10 37	☽ □ ♀
11 45	☽ # ☿
18 53	☽ ⊻ ♆
19 01	☽ # ♂
19 56	☽ ♇ ♅
20 59	☽ ⊥ ☿
11 Sunday	
00 56	☽ ⊻ ♃
02 42	☽ ∠ ⯓
05 51	☽ ∥ ♅
08 41	☽ △ ♄
11 56	☽ ⚹ ☉
13 08	☽ ⊥ ♃
13 40	☽ # ☉
14 14	♄ Q ⯓
16 48	☽ ∥ ♆
17 18	☽ ∠ ☿
12 Monday	
01 41	☽ △ ♅
06 38	☽ ☍ ♂
07 17	☽ ∠ ♃
08 30	☽ ⚹ ⯓
09 13	☽ # ♀
09 33	☿ ⚹ ♀
13 45	☽ Q ☉
14 29	☽ ♇ ♄
13 Tuesday	
01 19	☽ △ ♀
02 12	☽ ⚹ ☿
04 40	☽ # ⯓
06 05	☽ ☌ ♆
07 52	☽ Q ⯓
08 02	☽ ± ♄
09 18	☽ # ♄
12 55	☽ ⚹ ♃
19 33	☽ ⊼ ♄
14 Wednesday	
03 20	☽ □ ☉
03 28	☽ Q ☿
07 31	☽ ♇ ♀
09 12	☿ ∥ ♃
11 09	☽ □ ♅
12 00	☽ Q ♃
17 10	☽ ⊼ ♂
17 54	☽ □ ⯓
20 46	☿ ⚹ ♆
15 Thursday	
01 08	☽ ± ♀
04 24	☽ ± ♂
12 40	☽ ⊼ ♀
14 05	☽ ⊻ ♆
15 28	☽ □ ☿
19 54	☽ Q ♅
20 57	☽ ♇ ♂
21 25	☽ □ ♃
16 Friday	
00 49	☽ ⊥ ♆
01 59	☉ # ♅
02 55	☽ ☍ ♄
14 12	☽ △ ☉
14 17	♂ △ ⯓
15 22	♀ △ ♆
16 38	☽ ∠ ♆
16 59	☽ ⚹ ♅
21 30	♀ ♇ ♅
23 32	☽ △ ⯓
23 41	☽ △ ♂
17 Saturday	
17 51	☽ ♇ ☉
18 16	☽ ⚹ ♆
18 29	☽ ∠ ♅
19 42	☽ ☍ ♀
22 32	☽ △ ☿
18 Sunday	
00 58	☽ ♇ ⯓
01 54	☽ △ ♃
02 35	☉ ☍ ♅
06 25	☽ ⊼ ♄
09 24	☽ ⊥ ♅
09 56	☽ ± ☉
13 03	♀ Q ♂
14 10	☽ Q ♆
15 50	☽ ± ⯓
16 20	☽ ± ♄
17 57	☽ # ♄
19 12	☽ ⊻ ♅
20 28	☽ ⊼ ☉
22 15	☽ # ⯓
19 Monday	
00 09	☽ ♇ ☿
01 37	☽ ⊼ ⯓
02 24	☽ □ ♂
02 52	☽ ♇ ♃
06 59	☽ ♇ ♄
13 16	☽ # ♀
15 03	☽ ± ☿
17 34	☽ ± ♃
19 16	☽ □ ♆
23 12	☽ ⊼ ♀
20 Tuesday	
00 49	☽ ⊼ ☿
02 56	☽ ∥ ♆
03 16	☽ ⊼ ♃
07 02	☽ △ ♄
07 34	☽ Q ♂
09 14	☽ ± ♀
11 12	☽ ∥ ♅
16 26	☉ ∥ ♂
17 32	☽ # ♂
17 36	☽ # ☉
18 59	☽ ☌ ♅
23 44	☽ ☍ ☉
21 Wednesday	
00 13	☽ ♇ ♀
00 21	☉ Q ♆
01 25	☽ ☍ ⯓
02 44	☽ ⚹ ♂
13 22	☿ ⚹ ♀
13 53	☽ # ♃
18 00	☿ St R
18 40	☽ △ ♆
23 30	☽ # ☿
23 52	♀ ∠ ⯓
22 Thursday	
00 30	☽ ☍ ☿
01 08	☽ △ ♀
01 55	☉ ☌ ⯓
02 38	☽ ∠ ♂
03 17	☽ ☍ ♃
06 29	☽ □ ♄
10 17	☉ ∠ ♀
12 38	☉ Q ♄
14 09	☽ ∥ ☿
17 03	☽ ⊥ ♂
18 08	☽ ⊻ ♅
18 24	☽ ♇ ♆
23 47	☽ ∥ ♃
23 Friday	
00 53	☽ ⊼ ⯓
02 29	☽ ⊼ ☉
02 43	☽ ⊻ ♂
03 40	☽ ⊥ ♅
06 38	☉ △ ♂
08 47	☽ ± ♆
10 30	☽ ± ⯓
11 15	☽ Q ♄
12 48	☽ ± ☉
17 00	☽ ∥ ☉
18 07	☽ ∠ ♅
18 28	☽ ⊼ ♆
21 51	☽ ∥ ♂
23 46	☽ ⊼ ☿
24 Saturday	
01 07	☽ ♇ ⯓
03 57	☽ □ ♀
03 59	☽ # ♅
04 07	☽ ⊼ ♃
04 29	☽ ♇ ☉
06 50	☽ ⚹ ♄
08 00	♀ ⚹ ♃
09 20	☽ ± ☿
12 31	☽ # ♆
14 10	☽ ± ♃
18 40	☽ ⚹ ♅
23 48	☽ ♇ ☿
25 Sunday	
02 00	☽ △ ⯓
02 36	☽ ∥ ♀
04 24	☽ ☌ ♂
05 28	☽ ♇ ♃
07 20	☽ △ ☉
07 57	☽ ∠ ♄
11 52	☽ Q ♀
14 51	☽ Q ♅
18 30	☽ ∥ ⯓
20 36	☽ ☍ ♆
23 26	☽ ⊥ ♄
23 35	☽ ∥ ♄
26 Monday	
00 20	☽ △ ☿
07 43	☽ △ ♃
08 50	♀ ⊻ ♄
09 56	☽ ⊻ ♄
10 01	☽ ⚹ ♀
22 14	☽ □ ♅
27 Tuesday	
06 23	☽ □ ⯓
09 27	☽ ⊻ ♂
10 07	♀ ± ♅
14 47	☽ ∠ ♀
16 23	☽ □ ☉
20 32	☽ ⊥ ♂
28 Wednesday	
02 22	☽ ⊼ ♆
03 12	☽ □ ☿
08 38	☽ ⊥ ♀
13 27	☽ ∠ ♂
13 36	☽ ± ♆
14 43	☿ ⚹ ♆
15 17	☽ □ ♃
15 43	☽ Q ⯓
16 52	☽ ☌ ♄
20 47	☽ ⊻ ♀

All ephemeris data is given at 12.00 UT and the Moon's longitude is additionally given for 24.00 UT

MARCH 2046

Date	Sidereal time h m s	Sun ☉ ° ′ ″	Moon ☽ ° ′ ″	Moon ☽ 24.00 ° ′ ″	Mercury ☿ ° ′
01	22 37 51	11 ♓ 03 30	02 ♑ 25 31	08 ♑ 37 11	13 ♓ 48
02	22 41 47	12 03 44	14 ♑ 45 01	20 ♑ 49 32	12 R 47
03	22 45 44	13 03 57	26 ♑ 51 15	02 ♒ 50 39	11 45
04	22 49 40	14 04 07	08 ♒ 48 11	14 ♒ 44 17	10 42
05	22 53 37	15 04 16	20 ♒ 39 18	26 ♒ 33 38	09 40
06	22 57 33	16 04 24	02 ♓ 27 35	08 ♓ 21 26	08 42
07	23 01 30	17 04 29	14 ♓ 15 28	20 ♓ 09 55	07 47
08	23 05 27	18 04 33	26 ♓ 05 02	02 ♈ 01 01	06 56
09	23 09 23	19 04 35	07 ♈ 58 06	13 ♈ 56 29	06 11
10	23 13 20	20 04 34	19 ♈ 56 26	25 ♈ 58 10	05 32
11	23 17 16	21 04 32	02 ♉ 01 56	08 ♉ 08 03	04 59
12	23 21 13	22 04 28	14 ♉ 16 48	20 ♉ 28 31	04 33
13	23 25 09	23 04 22	26 ♉ 43 35	03 ♊ 02 22	04 13
14	23 29 06	24 04 13	09 ♊ 25 15	15 ♊ 52 39	04 00
15	23 33 02	25 04 02	22 ♊ 24 58	29 ♊ 02 36	03 53
16	23 36 59	26 03 49	05 ♋ 45 52	12 ♋ 35 06	03 D 53
17	23 40 56	27 03 34	19 ♋ 30 30	26 ♋ 32 11	03 58
18	23 44 52	28 03 17	03 ♌ 40 09	10 ♌ 54 14	04 09
19	23 48 49	29 ♓ 02 57	18 ♌ 14 06	25 ♌ 39 14	04 26
20	23 52 45	00 ♈ 02 35	03 ♍ 08 54	10 ♍ 42 13	04 47
21	23 56 42	01 02 10	18 ♍ 18 06	25 ♍ 55 19	05 14
22	00 00 38	02 01 44	03 ♎ 32 35	11 ♎ 08 31	05 45
23	00 04 35	03 01 15	18 ♎ 41 50	26 ♎ 11 16	06 20
24	00 08 31	04 00 45	03 ♏ 35 43	10 ♏ 54 15	07 00
25	00 12 28	05 00 13	18 ♏ 06 09	25 ♏ 10 55	07 43
26	00 16 25	05 59 39	02 ♐ 08 15	08 ♐ 58 05	08 30
27	00 20 21	06 59 04	15 ♐ 40 30	22 ♐ 15 46	09 21
28	00 24 18	07 58 26	28 ♐ 44 16	05 ♑ 06 29	10 14
29	00 28 14	08 57 47	11 ♑ 23 00	17 ♑ 34 25	11 11
30	00 32 11	09 57 06	23 ♑ 41 23	29 ♑ 44 33	12 10
31	00 36 07	10 ♈ 56 23	05 ♒ 44 34	11 ♒ 42 05	13 ♓ 12

Date	Venus ♀ ° ′	Mars ♂ ° ′	Jupiter ♃ ° ′	Saturn ♄ ° ′	Uranus ♅ ° ′	Neptune ♆ ° ′	Pluto ♇ ° ′
01	25 ♑ 02	05 ♏ 41	21 ♓ 46	22 ♐ 26	29 ♌ 06	14 ♉ 41	03 ♓ 48
02	25 57	05 48	22 00	22 29	29 R 03	14 42	03 50
03	26 51	05 53	22 15	22 32	29 01	14 43	03 52
04	27 47	05 58	22 29	22 35	28 58	14 44	03 53
05	28 43	06 03	22 44	22 37	28 56	14 46	03 55
06	29 ♑ 39	06 06	22 58	22 40	28 53	14 47	03 57
07	00 ♒ 36	06 09	23 13	22 43	28 51	14 48	03 58
08	01 34	06 12	23 27	22 45	28 48	14 50	04 00
09	02 32	06 13	23 42	22 47	28 46	14 51	04 02
10	03 30	06 14	23 56	22 49	28 44	14 53	04 03
11	04 29	06 R 14	24 11	22 52	28 41	14 54	04 05
12	05 28	06 13	24 25	22 54	28 39	14 56	04 06
13	06 28	06 12	24 40	22 56	28 36	14 57	04 08
14	07 28	06 10	24 54	22 57	28 34	14 59	04 10
15	08 28	06 07	25 09	22 59	28 32	15 00	04 11
16	09 29	06 03	25 23	23 01	28 30	15 02	04 13
17	10 30	05 59	25 38	23 02	28 27	15 04	04 14
18	11 31	05 53	25 52	23 04	28 25	15 05	04 16
19	12 33	05 47	26 07	23 05	28 23	15 07	04 18
20	13 35	05 41	26 21	23 06	28 21	15 09	04 19
21	14 37	05 33	26 36	23 08	28 19	15 10	04 21
22	15 40	05 25	26 50	23 09	28 17	15 12	04 22
23	16 43	05 16	27 05	23 10	28 15	15 14	04 24
24	17 46	05 06	27 19	23 11	28 13	15 16	04 25
25	18 49	04 56	27 34	23 11	28 11	15 17	04 27
26	19 53	04 44	27 48	23 12	28 09	15 19	04 28
27	20 57	04 32	28 03	23 12	28 07	15 21	04 30
28	22 01	04 20	28 17	23 13	28 05	15 23	04 31
29	23 05	04 06	28 31	23 13	28 03	15 25	04 33
30	24 10	03 52	28 46	23 14	28 02	15 27	04 34
31	25 ♒ 15	03 ♏ 37	29 ♓ 00	23 ♐ 14	28 ♌ 00	15 ♉ 29	04 ♓ 35

Date	Moon True ☊ ° ′	Moon Mean ☊ ° ′	Moon ☽ Latitude ° ′
01	13 ♒ 23	12 ♒ 11	03 S 27
02	13 D 24	12 08	02 31
03	13 26	12 05	01 30
04	13 27	12 02	00 S 26
05	13 R 27	11 59	00 N 39
06	13 25	11 56	01 42
07	13 22	11 52	02 40
08	13 17	11 49	03 32
09	13 10	11 46	04 14
10	13 03	11 43	04 46
11	12 56	11 40	05 05
12	12 50	11 37	05 10
13	12 45	11 33	05 01
14	12 42	11 30	04 37
15	12 41	11 27	03 59
16	12 D 41	11 24	03 07
17	12 42	11 21	02 02
18	12 44	11 17	00 N 49
19	12 R 44	11 14	00 S 30
20	12 42	11 11	01 48
21	12 38	11 08	03 00
22	12 32	11 05	03 59
23	12 25	11 02	04 41
24	12 17	10 58	05 03
25	12 09	10 55	05 05
26	12 03	10 52	04 48
27	11 59	10 49	04 16
28	11 56	10 46	03 31
29	11 D 56	10 43	02 37
30	11 56	10 39	01 36
31	11 ♒ 57	10 ♒ 36	00 S 33

DECLINATIONS

Date	Sun ☉ ° ′	Moon ☽ ° ′	Mercury ☿ ° ′	Venus ♀ ° ′	Mars ♂ ° ′	Jupiter ♃ ° ′	Saturn ♄ ° ′	Uranus ♅ ° ′	Neptune ♆ ° ′	Pluto ♇ ° ′
01	07 S 25	26 S 51	02 S 57	17 S 33	11 S 06	04 S 13	21 S 45	12 N 32	14 N 35	20 S 49
02	07 02	25 07	03 20	17 30	11 08	04 08	21 45	12 32	14 35	20 49
03	06 39	22 15	03 46	17 26	11 10	04 02	21 45	12 33	14 36	20 48
04	06 16	18 28	04 14	17 22	11 12	03 56	21 45	12 34	14 36	20 48
05	05 53	13 59	04 43	17 17	11 13	03 51	21 45	12 35	14 36	20 47
06	05 30	09 00	05 13	17 11	11 14	03 45	21 45	12 36	14 37	20 47
07	05 06	03 S 44	05 42	17 06	11 15	03 39	21 45	12 37	14 37	20 46
08	04 43	01 N 41	06 11	17 00	11 16	03 33	21 45	12 38	14 38	20 45
09	04 19	07 03	06 39	16 53	11 16	03 28	21 45	12 38	14 38	20 45
10	03 56	12 12	07 05	16 46	11 16	03 22	21 45	12 39	14 39	20 44
11	03 32	16 56	07 30	16 38	11 16	03 16	21 45	12 40	14 39	20 44
12	03 09	21 03	07 53	16 30	11 16	03 10	21 45	12 41	14 40	20 43
13	02 45	24 18	08 14	16 22	11 16	03 05	21 45	12 42	14 40	20 43
14	02 21	26 26	08 32	16 13	11 15	02 59	21 45	12 43	14 41	20 43
15	01 58	27 12	08 48	16 03	11 14	02 53	21 45	12 43	14 41	20 42
16	01 34	26 25	09 02	15 53	11 14	02 47	21 45	12 44	14 42	20 42
17	01 10	24 02	09 13	15 43	11 12	02 42	21 45	12 45	14 42	20 41
18	00 46	20 07	09 22	15 32	11 11	02 36	21 45	12 46	14 43	20 41
19	00 S 23	14 53	09 29	15 21	11 09	02 30	21 45	12 46	14 43	20 40
20	00 N 01	08 40	09 33	15 09	11 07	02 25	21 45	12 47	14 44	20 40
21	00 25	01 N 52	09 36	14 57	11 05	02 19	21 45	12 48	14 44	20 39
22	00 48	05 S 04	09 36	14 44	11 03	02 13	21 45	12 48	14 45	20 39
23	01 12	11 39	09 34	14 31	11 01	02 07	21 45	12 49	14 45	20 39
24	01 36	17 28	09 30	14 18	10 58	02 02	21 45	12 50	14 46	20 38
25	01 59	22 06	09 24	14 04	10 55	01 56	21 45	12 50	14 46	20 38
26	02 23	25 18	09 16	13 49	10 52	01 50	21 45	12 51	14 47	20 37
27	02 46	26 54	09 06	13 34	10 49	01 45	21 45	12 52	14 48	20 37
28	03 10	26 56	08 54	13 19	10 45	01 39	21 45	12 52	14 48	20 37
29	03 33	25 33	08 41	13 03	10 41	01 33	21 45	12 53	14 49	20 36
30	03 56	22 56	08 26	12 47	10 37	01 27	21 45	12 53	14 49	20 36
31	04 N 20	19 S 22	08 S 09	12 S 30	10 S 33	01 S 22	21 S 44	12 N 54	14 N 50	20 S 36

ZODIAC SIGN ENTRIES

Date	h m	Planets
01	07 20	☽ ♑
03	18 18	☽ ♒
06	07 00	☽ ♓
06	20 42	♀ ♒
08	19 55	☽ ♈
11	07 59	☽ ♉
13	18 14	☽ ♊
16	01 43	☽ ♋
18	05 51	☽ ♌
20	06 58	☽ ♍
20	10 58	☉ ♈
22	06 25	☽ ♎
24	06 09	☽ ♏
26	08 17	☽ ♐
28	14 22	☽ ♑
31	00 31	☽ ♒

LATITUDES

Date	Mercury ☿ ° ′	Venus ♀ ° ′	Mars ♂ ° ′	Jupiter ♃ ° ′	Saturn ♄ ° ′	Uranus ♅ ° ′	Neptune ♆ ° ′	Pluto ♇ ° ′
01	03 N 42	03 N 37	02 N 27	01 S 02	01 N 29	00 N 47	01 S 44	11 S 32
04	03 35	03 18	02 27	01 02	01 29	00 47	01 44	11 32
07	03 11	02 59	02 28	01 02	01 29	00 47	01 44	11 32
10	02 34	02 41	02 28	01 03	01 30	00 47	01 44	11 33
13	01 52	02 22	02 27	01 03	01 30	00 47	01 44	11 33
16	01 08	02 04	02 27	01 03	01 30	00 47	01 43	11 34
19	00 N 26	01 45	02 26	01 03	01 30	00 47	01 43	11 34
22	00 S 13	01 28	02 25	01 03	01 31	00 47	01 43	11 35
25	00 47	01 11	02 23	01 03	01 31	00 47	01 43	11 35
28	01 16	00 54	02 21	01 03	01 31	00 47	01 43	11 36
31	01 S 41	00 N 38	02 N 18	01 S 03	01 N 31	00 N 47	01 S 43	11 S 36

DATA

Julian Date	2468406
Delta T	+79 seconds
Ayanamsa	24° 30′ 21″
Synetic vernal point	04° ♓ 36′ 38″
True obliquity of ecliptic	23° 26′ 06″

MOON'S PHASES, APSIDES AND POSITIONS ☽

Date	h m	Phase	Longitude ° ′	Eclipse Indicator
07	18 15	●	17 ♓ 20	
15	17 13	☽	25 ♊ 17	
22	09 27	○	01 ♎ 55	
29	06 57	☾	08 ♑ 45	

Day	h m		
07	07 09	Apogee	
21	19 05	Perigee	
08	04 33	ON	
15	12 16	Max dec	27° N 12′
21	18 26	OS	
28	00 31	Max dec	27° S 07′

LONGITUDES

Date	Chiron ⚷ ° ′	Ceres ⚳ ° ′	Pallas ⚴ ° ′	Juno ⚵ ° ′	Vesta ⚶ ° ′	Black Moon Lilith ⚸ ° ′
01	05 ♎ 08	01 ♎ 57	01 ♋ 39	05 ♎ 35	16 ♋ 36	11 ♓ 38
11	04 ♎ 25	29 ♍ 52	04 ♋ 20	03 ♎ 21	16 ♋ 47	12 ♓ 44
21	03 ♎ 38	27 ♍ 35	07 ♋ 33	00 ♎ 54	17 ♋ 40	13 ♓ 51
31	02 ♎ 51	25 ♍ 23	11 ♋ 11	28 ♍ 31	19 ♋ 11	14 ♓ 58

All ephemeris data is given at 12.00 UT and the Moon's longitude is additionally given for 24.00 UT

ASPECTARIAN

01 Thursday
04 58 ☽ Q ☉
05 38 ☽ △ ♅
06 43 ☽ bQ ♆
10 53 ☽ Q ☿
14 40 ☽ ✶ ♇
18 22 ☽ ✶ ♂

02 Friday
02 31 ☽ Q ♃
06 15 ☽ ✶ ☉
08 27 ☽ ✶ ☿
10 38 ☽ bQ ♅
11 54 ☽ △ ♆
18 03 ☽ Q ♂
20 04 ☽ ∠ ♇
20 31 ☉ ☌ ☿

03 Saturday
02 38 ☽ ✶ ♃
03 21 ☽ ⊻ ♄
04 22 ☽ ± ♅
10 37 ☿ ∠ ♀
11 48 ☽ ∠ ☿
12 00 ☽ ☌ ♀
14 01 ☽ ⊥ ♇
14 39 ☽ ∠ ☉
15 22 ☽ ⊥ ♄
15 33 ☽ ∥ ♄
16 18 ☽ ⊼ ♅
21 50 ☽ ∥ ♇
23 21 ☿ ∥ ♃

04 Sunday
02 05 ☽ ⊻ ♇
04 24 ☽ ⊥ ☿
06 15 ☽ □ ♂
09 17 ☽ ∠ ♃
09 31 ☽ ∠ ♄
10 23 ☽ ⊥ ☉
14 51 ♀ ⊥ ♇
15 31 ☽ ⊻ ☿
18 21 ☽ ∥ ♀
23 10 ♃ □ ♄
23 38 ☽ ⊻ ☉

05 Monday
00 02 ☽ □ ♆
03 52 ☽ ⊥ ♃
04 25 ☉ ✶ ♆
08 50 ☽ # ♆
09 32 ♀ ⊥ ♄
16 01 ☽ ✶ ♄
16 18 ☽ ⊻ ♃
17 15 ♀ ⊼ ♅
18 55 ☽ # ♅

06 Tuesday
01 32 ☽ ∥ ♂
04 46 ☽ ☍ ♅
05 48 ☽ ⊻ ♀
12 40 ☽ Q ♆
15 02 ☽ ☌ ♇
16 30 ☽ Q ♄
19 04 ☽ ⊥ ♀
19 27 ☽ △ ♂
19 42 ☉ ∥ ☿
23 45 ☽ ☌ ☿

07 Wednesday
03 53 ☽ ∥ ☿
05 20 ☽ ∥ ☉
12 20 ☽ ∥ ♃
13 07 ☽ ✶ ♆
14 59 ☽ ∠ ♀
18 15 ☽ ☌ ☉

08 Thursday
02 03 ☽ bQ ♂
03 29 ☿ ⊥ ♀
05 13 ☽ □ ♄
06 33 ☽ ☌ ♃
17 29 ☽ ⊼ ♅
19 36 ☽ ∠ ♆
20 10 ☽ # ♃
20 20 ☽ ± ♂

09 Friday
00 04 ☽ ✶ ♀
00 34 ☽ # ☉
04 02 ☽ ⊻ ♇
05 34 ☽ ± ♅
08 28 ☽ ⊼ ♂
08 36 ☽ ⊻ ☿
10 00 ☽ # ☿
10 54 ☿ △ ♂
13 47 ☽ ⊥ ♆
16 09 ☽ ⊥ ♇
20 01 ☽ ⊥ ☿
23 36 ☽ bQ ♅

10 Saturday
01 51 ☽ ⊻ ♆
02 21 ☽ Q ♀
07 34 ☽ # ♂
10 13 ☽ ∠ ♇
12 18 ☽ ⊻ ☉
13 07 ☽ ∠ ☿
14 13 ☽ ∥ ♅
17 46 ☽ △ ♄
20 07 ☽ ⊻ ♃

11 Sunday
00 05 ☽ ∥ ♆
01 19 ☽ ⊥ ☉
01 53 ♀ ⊻ ♇
02 11 ♂ St R
05 24 ☽ △ ♅
08 16 ☽ ⊥ ♃
10 27 ☽ # ♀
15 43 ☉ bQ ♂
16 03 ☽ ✶ ♇
17 15 ☽ □ ♀
17 36 ☽ ✶ ☿
17 49 ☿ Q ♄
20 15 ☿ ⊻ ♀
20 16 ☽ ☍ ♂
20 40 ☽ ∠ ☉
23 30 ☽ bQ ♄

12 Monday
02 20 ☽ ∠ ♃
09 35 ☉ ∥ ♃
09 56 ☽ # ♇
13 16 ☽ ☌ ♆
15 33 ☽ Q ♇
16 16 ☽ Q ☿
16 38 ☽ # ♄
17 05 ☽ ± ♄

13 Tuesday
04 23 ☽ ✶ ☉
04 42 ☽ ⊼ ♄
05 45 ♀ □ ♂
07 58 ☽ ✶ ♃
08 22 ☉ □ ♄
15 35 ☽ □ ♅
19 01 ☿ ☌ ♇

14 Wednesday
01 58 ☽ □ ☿
02 06 ☽ □ ♇
05 11 ☽ Q ☉
05 54 ☽ ⊼ ♂
07 12 ☽ Q ♃
08 01 ☽ △ ♀
17 06 ☽ ± ♂
22 22 ☽ ⊻ ♆

15 Thursday
00 11 ♀ ∠ ♄
01 14 ☽ Q ♅
09 25 ☽ ⊥ ♆
09 38 ☽ bQ ♂
13 02 ☽ ☍ ♄
14 04 ☽ bQ ♀
14 31 ☉ ☌ ♃
17 03 ☽ □ ♃
17 13 ☽ □ ☉
23 03 ☽ ✶ ♅

16 Friday
01 45 ☽ ∠ ♆
02 13 ☿ St D
07 37 ☽ ± ♀
08 38 ☽ △ ☿
09 15 ☽ △ ♇
12 30 ☽ △ ♂
19 05 ☽ ⊼ ♀

17 Saturday
01 33 ☽ ∠ ♅
04 18 ☽ ✶ ♆
11 03 ☽ bQ ☿
11 32 ☽ bQ ♇
16 06 ♀ ∠ ♃
17 03 ☽ ⊥ ♅
18 04 ☽ ⊼ ♄
22 39 ☽ △ ♃

18 Sunday
00 55 ☽ Q ♆
01 52 ☽ △ ☉
02 35 ☽ ± ☿
02 55 ☽ ± ♇
03 02 ☽ # ♄
03 12 ☽ ⊻ ♅
04 16 ☽ ± ♄
09 03 ☽ # ♇
12 49 ☽ ⊼ ☿
13 00 ☽ ⊼ ♇
15 41 ☽ □ ♂
19 19 ☽ bQ ♄
20 30 ☉ ⊼ ♅

19 Monday
00 09 ☽ bQ ♃
00 12 ☿ ☌ ♇
02 00 ☽ ☍ ♀
04 40 ☽ bQ ☉
06 54 ☽ □ ♆
10 01 ☽ # ♀
12 42 ☽ ∥ ♆
15 06 ☽ ± ♃
19 52 ☽ △ ♄
20 22 ☽ ± ☉
20 31 ☽ ∥ ♅
20 56 ☽ Q ♂

20 Tuesday
00 57 ☽ ⊼ ♃
02 50 ☽ # ♂
04 08 ☉ ± ♂
04 21 ☽ ☌ ♅
06 41 ☽ ⊼ ☉
08 46 ☽ # ☿
13 52 ☽ ☍ ♇
14 28 ☉ ∠ ♆
14 41 ☽ ☍ ☿
16 00 ☽ ✶ ♂

21 Wednesday
05 46 ☽ ⊼ ♀
06 29 ☿ Q ♄
07 03 ☽ △ ♆
10 26 ☽ # ♃
15 31 ☽ ∠ ♂
15 55 ☽ ± ♀
16 45 ☽ ∥ ☉
19 37 ☽ □ ♄
20 20 ☽ # ☉

22 Thursday
00 13 ☿ △ ♂
01 03 ♀ □ ♆
01 16 ☽ ☍ ♃
02 13 ☽ ∥ ♃
03 44 ☽ ⊻ ♅
05 34 ☽ ⊥ ♂
06 43 ☽ bQ ♆
07 08 ☽ bQ ♀
09 27 ☽ ☍ ☉
11 14 ♀ # ♆
13 09 ☽ ⊥ ♅
13 18 ☽ ⊼ ♇
14 56 ☽ ⊻ ♂
15 37 ☽ ⊼ ☿
20 57 ☽ ± ♆
22 48 ☽ ± ♇

23 Friday
00 01 ☽ Q ♄
01 29 ☽ ± ☿
03 21 ☽ ∠ ♅
04 11 ☽ ∥ ☿
06 28 ☽ ⊼ ♆
08 36 ☽ △ ♀
09 34 ☽ ∥ ♂
13 07 ☽ bQ ♇
16 25 ☽ bQ ☿
16 31 ☽ # ♅
19 09 ☽ ✶ ♄
22 55 ☽ ∥ ♀

24 Saturday
00 19 ☽ # ♆
01 40 ☽ ⊼ ♃
03 17 ☽ ✶ ♅
11 33 ☽ ± ♃
12 44 ☽ ⊼ ☉
13 21 ☽ △ ♇
14 26 ☽ ☌ ♂
16 40 ☉ ± ♅
17 51 ☽ △ ☿
19 30 ☽ ∠ ♄
22 08 ☉ ⊻ ♇
22 50 ☽ Q ♅
23 18 ☽ ± ☉

25 Sunday
02 35 ☽ bQ ♃
03 37 ☽ ∥ ♇
07 17 ☽ ☍ ♆
09 16 ☉ # ♃
09 54 ☽ ∥ ♄
10 27 ☉ ⊼ ♂
10 28 ☽ ⊥ ♄
13 18 ☽ □ ♀
15 27 ☽ bQ ☉
20 36 ☽ ⊻ ♄

26 Monday
04 22 ☽ △ ♃
05 06 ☽ □ ♅
16 05 ☽ □ ♇
16 29 ☽ ⊻ ♂
19 17 ☽ △ ☉
22 56 ☽ Q ♀
23 54 ☽ □ ☿

27 Tuesday
02 56 ☽ ⊥ ♂
11 25 ☽ ⊼ ♆
16 33 ♂ △ ♇
18 18 ♃ ⊼ ♅
18 54 ☽ ∠ ♂
22 21 ☽ ± ♆
22 26 ☽ ✶ ♀

28 Wednesday
00 27 ☽ Q ♇
01 45 ☽ ☌ ♄
10 47 ☽ △ ♅
10 59 ☽ Q ☿
11 08 ☽ □ ♃
14 08 ♂ ± ♃
15 05 ☽ bQ ♆
19 42 ♀ ⊥ ♃
22 20 ☽ ✶ ♂
22 54 ☽ ✶ ♇

29 Thursday
05 06 ☽ ∠ ♀
06 57 ☽ □ ☉
11 34 ☽ ✶ ☿
14 56 ♀ ✶ ♄
15 13 ☽ bQ ♅
19 49 ☽ △ ♆
20 58 ☽ Q ♂
22 09 ☽ Q ♃
23 16 ☉ ⊥ ♆

30 Friday
00 07 ☽ ⊥ ♀
02 44 ♀ # ♅
03 53 ☽ ∠ ♇
08 44 ☽ ± ♅
11 05 ☽ ⊻ ♄
13 02 ☽ ⊻ ♀
19 31 ☽ ∠ ☿
20 34 ☽ ⊼ ♅
20 44 ☽ ∥ ♄
21 11 ☽ Q ☉
21 41 ☽ ⊥ ♇
22 15 ☽ ✶ ♃
22 59 ☽ ⊥ ♄

31 Saturday
03 19 ☉ ⊥ ♇
04 22 ☽ ∥ ♇
07 49 ☽ □ ♂
09 41 ☽ ⊻ ♇
15 13 ☽ ⊥ ☿
17 00 ☽ ∠ ♄
23 25 ☽ ✶ ☉

APRIL 2046

LONGITUDES

Date	Sidereal time h m s	Sun ☉ ° ' "	Moon ☽ ° ' "	Moon ☽ 24.00 ° ' "	Mercury ☿ ° '	Venus ♀ ° '	Mars ♂ ° '	Jupiter ♃ ° '	Saturn ♄ ° '	Uranus ♅ ° '	Neptune ♆ ° '	Pluto ♇ ° '
01	00 40 04	11 ♈ 55 39	17 ♒ 37 43	23 ♒ 32 02	14 ♓ 17	26 ♒ 20	03 ♏ 21	29 ♓ 15	23 ♐ 14	27 ♌ 58	15 ♉ 31	04 ♓ 37
02	00 44 00	12 54 52	29 ♒ 25 34	05 ♓ 18 49	15 24	27 25	03 R 05	29 29	23 R 14	27 R 56	15 32	04 38
03	00 47 57	13 54 04	11 ♓ 12 15	17 ♓ 06 14	16 33	28 31	02 48	29 43	23 14	27 55	15 34	04 40
04	00 51 54	14 53 14	23 ♓ 01 08	28 ♓ 57 16	17 45	29 ♒ 36	02 30	29 ♓ 58	23 13	27 53	15 36	04 41
05	00 55 50	15 52 22	04 ♈ 54 52	10 ♈ 54 10	18 59	00 ♓ 42	02 12	00 ♈ 12	23 13	27 52	15 38	04 42
06	00 59 47	16 51 27	16 ♈ 55 20	22 ♈ 58 31	20 14	01 48	01 53	00 26	23 13	27 50	15 40	04 44
07	01 03 43	17 50 31	29 ♈ 03 51	05 ♉ 11 26	21 32	02 54	01 34	00 40	23 12	27 49	15 42	04 45
08	01 07 40	18 49 33	11 ♉ 21 21	17 ♉ 33 44	22 52	04 01	01 14	00 54	23 11	27 48	15 45	04 46
09	01 11 36	19 48 33	23 ♉ 48 41	00 ♊ 06 19	24 14	05 07	00 54	01 09	23 11	27 46	15 47	04 48
10	01 15 33	20 47 30	06 ♊ 26 47	12 ♊ 50 16	25 37	06 14	00 33	01 23	23 10	27 45	15 49	04 49
11	01 19 29	21 46 26	19 ♊ 16 58	25 ♊ 47 06	27 02	07 21	00 ♏ 12	01 37	23 09	27 44	15 51	04 50
12	01 23 26	22 45 19	02 ♋ 20 56	08 ♋ 58 43	28 29	08 28	29 ♎ 51	01 51	23 08	27 43	15 53	04 51
13	01 27 23	23 44 10	15 ♋ 40 44	22 ♋ 27 13	29 ♓ 58	09 35	29 29	02 05	23 07	27 41	15 55	04 52
14	01 31 19	24 42 58	29 ♋ 18 26	06 ♌ 14 33	01 ♈ 28	10 42	29 07	02 19	23 06	27 40	15 57	04 54
15	01 35 16	25 41 45	13 ♌ 15 41	20 ♌ 21 50	03 00	11 49	28 45	02 33	23 04	27 39	15 59	04 55
16	01 39 12	26 40 29	27 ♌ 32 54	04 ♍ 48 38	04 34	12 57	28 22	02 47	23 03	27 38	16 01	04 56
17	01 43 09	27 39 10	12 ♍ 08 37	19 ♍ 32 17	06 09	14 04	28 00	03 00	23 01	27 37	16 04	04 57
18	01 47 05	28 37 50	26 ♍ 58 52	04 ♎ 27 27	07 46	15 12	27 37	03 14	23 00	27 36	16 06	04 58
19	01 51 02	29 ♈ 36 27	11 ♎ 57 00	19 ♎ 26 23	09 25	16 20	27 14	03 28	22 58	27 36	16 08	04 59
20	01 54 58	00 ♉ 35 02	26 ♎ 54 24	04 ♏ 19 52	11 05	17 28	26 52	03 42	22 56	27 35	16 10	05 00
21	01 58 55	01 33 35	11 ♏ 41 41	18 ♏ 58 48	12 47	18 36	26 29	03 55	22 54	27 34	16 12	05 01
22	02 02 52	02 32 07	26 ♏ 10 23	03 ♐ 15 45	14 31	19 44	26 06	04 09	22 53	27 33	16 14	05 03
23	02 06 48	03 30 37	10 ♐ 14 24	17 ♐ 06 03	16 16	20 52	25 44	04 23	22 51	27 33	16 17	05 04
24	02 10 45	04 29 05	23 ♐ 50 36	00 ♑ 28 08	18 03	22 00	25 21	04 36	22 48	27 32	16 19	05 05
25	02 14 41	05 27 32	06 ♑ 58 54	13 ♑ 23 15	19 52	23 09	24 59	04 50	22 46	27 32	16 21	05 06
26	02 18 38	06 25 57	19 ♑ 41 41	25 ♑ 54 44	21 42	24 18	24 37	05 03	22 44	27 31	16 23	05 06
27	02 22 34	07 24 20	02 ♒ 03 02	08 ♒ 07 14	23 34	25 26	24 16	05 17	22 42	27 31	16 26	05 07
28	02 26 31	08 22 41	14 ♒ 08 03	20 ♒ 06 08	25 28	26 35	23 54	05 30	22 39	27 30	16 28	05 08
29	02 30 27	09 21 02	26 ♒ 02 11	01 ♓ 56 53	27 23	27 44	23 33	05 43	22 36	27 30	16 30	05 09
30	02 34 24	10 ♉ 19 20	07 ♓ 50 52	13 ♓ 44 43	29 ♈ 20	28 ♓ 53	23 ♎ 13	05 ♈ 56	22 ♐ 34	27 ♌ 30	16 ♉ 32	05 ♓ 10

Date	Moon True ☊ ° '	Moon Mean ☊ ° '	Moon ☽ Latitude ° '
01	11 ♒ 57	10 ♒ 33	00 N 30
02	11 R 56	10 30	01 32
03	11 51	10 27	02 30
04	11 45	10 23	03 21
05	11 35	10 20	04 03
06	11 24	10 17	04 35
07	11 12	10 14	04 55
08	10 59	10 11	05 02
09	10 48	10 08	04 54
10	10 38	10 04	04 32
11	10 32	10 01	03 56
12	10 28	09 58	03 07
13	10 26	09 55	02 07
14	10 D 26	09 52	00 N 59
15	10 R 26	09 49	00 S 15
16	10 25	09 45	01 29
17	10 22	09 42	02 39
18	10 17	09 39	03 40
19	10 08	09 36	04 26
20	09 58	09 33	04 53
21	09 47	09 29	05 01
22	09 36	09 26	04 48
23	09 26	09 23	04 19
24	09 19	09 20	03 35
25	09 14	09 17	02 42
26	09 12	09 14	01 41
27	09 D 11	09 10	00 S 38
28	09 R 11	09 07	00 N 26
29	09 11	09 04	01 28
30	09 ♒ 09	09 ♒ 01	02 N 25

DECLINATIONS

Date	Sun ☉ ° '	Moon ☽ ° '	Mercury ☿ ° '	Venus ♀ ° '	Mars ♂ ° '	Jupiter ♃ ° '	Saturn ♄ ° '	Uranus ♅ ° '	Neptune ♆ ° '	Pluto ♇ ° '
01	04 N 43	15 S 04	07 S 51	12 S 13	10 S 29	01 S 16	21 S 44	12 N 55	14 N 50	20 S 35
02	05 06	10 14	07 31	11 56	10 25	01 11	21 44	12 55	14 51	20 35
03	05 29	05 S 03	07 10	11 38	10 20	01 05	21 44	12 56	14 52	20 35
04	05 52	00 N 18	06 47	11 20	10 15	00 59	21 44	12 56	14 52	20 34
05	06 15	05 40	06 23	11 01	10 10	00 54	21 44	12 57	14 53	20 34
06	06 37	10 53	05 58	10 43	10 05	00 48	21 44	12 57	14 54	20 34
07	07 00	15 44	05 31	10 23	10 00	00 42	21 44	12 58	14 54	20 33
08	07 22	20 01	05 02	10 04	09 54	00 37	21 44	12 58	14 55	20 33
09	07 45	23 28	04 33	09 44	09 49	00 31	21 44	12 58	14 55	20 33
10	08 07	25 51	04 02	09 23	09 43	00 26	21 43	12 59	14 56	20 33
11	08 29	26 56	03 30	09 03	09 37	00 20	21 43	12 59	14 57	20 32
12	08 51	26 32	02 57	08 42	09 31	00 15	21 43	13 00	14 57	20 32
13	09 13	24 37	02 22	08 21	09 26	00 09	21 43	13 00	14 58	20 32
14	09 34	21 15	01 47	07 59	09 20	00 S 04	21 43	13 00	14 58	20 32
15	09 56	16 36	01 10	07 37	09 14	00 N 02	21 43	13 01	14 59	20 32
16	10 17	10 55	00 S 33	07 15	09 07	00 07	21 43	13 01	15 00	20 31
17	10 38	04 N 33	00 N 06	06 53	09 01	00 13	21 42	13 01	15 00	20 31
18	10 59	02 S 10	00 46	06 30	08 55	00 18	21 42	13 02	15 01	20 31
19	11 20	08 48	01 27	06 07	08 49	00 23	21 42	13 02	15 02	20 31
20	11 40	14 55	02 08	05 44	08 43	00 29	21 42	13 02	15 02	20 31
21	12 01	20 06	02 51	05 21	08 37	00 34	21 42	13 02	15 03	20 31
22	12 21	23 58	03 34	04 57	08 31	00 39	21 42	13 02	15 04	20 31
23	12 41	25 15	04 19	04 34	08 25	00 45	21 42	13 03	15 04	20 30
24	13 01	26 53	05 04	04 10	08 19	00 50	21 41	13 03	15 05	20 30
25	13 20	25 56	05 50	03 45	08 13	00 55	21 41	13 03	15 06	20 30
26	13 40	23 40	06 36	03 21	08 08	01 00	21 41	13 03	15 06	20 30
27	13 59	20 19	07 23	02 57	08 02	01 06	21 41	13 03	15 07	20 30
28	14 18	16 10	08 11	02 32	07 56	01 11	21 41	13 03	15 07	20 30
29	14 36	11 27	08 59	02 07	07 51	01 16	21 40	13 03	15 08	20 30
30	14 N 55	06 S 22	09 N 48	01 S 42	07 S 46	01 N 21	21 S 40	13 N 03	15 N 09	20 S 30

ZODIAC SIGN ENTRIES

Date	h	m	Planets
02	13	10	☽ ♓
04	16	11	♃ ♈
04	20	37	♀ ♓
05	02	07	☽ ♈
07	13	50	☽ ♉
09	23	48	☽ ♊
12	01	51	♂ ♎
12	07	43	☽ ♋
13	12	34	☿ ♈
14	13	12	☽ ♌
16	16	04	☽ ♍
18	16	51	☽ ♎
19	21	39	☉ ♉
20	16	59	☽ ♏
22	18	27	☽ ♐
24	23	09	☽ ♑
27	07	59	☽ ♒
29	20	03	☽ ♓
30	20	06	☿ ♉

LATITUDES

Date	Mercury ☿ ° '	Venus ♀ ° '	Mars ♂ ° '	Jupiter ♃ ° '	Saturn ♄ ° '	Uranus ♅ ° '	Neptune ♆ ° '	Pluto ♇ ° '
01	01 S 48	00 N 33	02 N 17	01 S 03	01 N 31	00 N 47	01 S 43	11 S 37
04	02 07	00 18	02 13	01 04	01 32	00 47	01 43	11 37
07	02 21	00 N 03	02 09	01 04	01 32	00 47	01 43	11 38
10	02 30	00 S 11	02 05	01 04	01 32	00 47	01 43	11 39
13	02 34	00 24	01 59	01 04	01 32	00 46	01 42	11 39
16	02 34	00 36	01 54	01 04	01 33	00 46	01 42	11 40
19	02 29	00 47	01 47	01 05	01 33	00 46	01 42	11 41
22	02 20	00 58	01 41	01 05	01 33	00 46	01 42	11 42
25	02 06	01 08	01 33	01 05	01 33	00 46	01 42	11 43
28	01 47	01 17	01 26	01 06	01 33	00 46	01 42	11 44
31	01 S 24	01 S 25	01 N 18	01 S 06	01 N 33	00 N 46	01 S 42	11 S 45

DATA

Julian Date	2468437
Delta T	+79 seconds
Ayanamsa	24° 30' 24"
Synetic vernal point	04° ♓ 36' 34"
True obliquity of ecliptic	23° 26' 06"

MOON'S PHASES, APSIDES AND POSITIONS ☽

Date	h	m	Phase	Longitude ° '	Eclipse Indicator
06	11	52	●	16 ♈ 51	
14	03	21	☽	24 ♋ 22	
20	18	21	○	00 ♏ 51	
27	23	30	☾	07 ♒ 52	

Day	h	m		
03	12	57	Apogee	
19	03	11	Perigee	
04	10	39	ON	
11	17	48	Max dec	26° N 58'
18	04	19	OS	
24	09	16	Max dec	26° S 53'

LONGITUDES

Date	Chiron ⚷ ° '	Ceres ⚳ ° '	Pallas ⚴ ° '	Juno ⚵ ° '	Vesta ⚶ ° '	Black Moon Lilith ⚸ ° '
01	02 ♎ 46	25 ♍ 10	11 ♋ 34	28 ♍ 18	19 ♋ 22	15 ♓ 04
11	02 ♎ 01	23 ♍ 23	15 ♋ 33	26 ♍ 15	21 ♋ 29	16 ♓ 11
21	01 ♎ 21	22 ♍ 10	19 ♋ 45	24 ♍ 41	24 ♋ 03	17 ♓ 18
31	00 ♎ 49	21 ♍ 38	24 ♋ 08	23 ♍ 43	27 ♋ 00	18 ♓ 25

All ephemeris data is given at 12.00 UT and the Moon's longitude is additionally given for 24.00 UT

ASPECTARIAN

h m Aspects

01 Sunday
04 31 ☽ ⚺ ☿
05 00 ☽ ∠ ♃
07 41 ☽ □ ♆
13 10 ☽ ⋕ ♆
14 46 ♄ St R
22 57 ☽ ⋕ ♅
23 23 ☽ ✶ ♄
23 39 ☽ ⊥ ♃

02 Monday
03 16 ☽ ∥ ♀
07 30 ☽ ☌ ♀
08 38 ☽ ∠ ☉
08 59 ☽ ☍ ♅
11 09 ☽ ∥ ♂
12 07 ☽ ⚺ ♃
12 38 ☉ ⚼ ♅
15 08 ☿ ✶ ♆
19 17 ☽ △ ♂
20 25 ☽ Q ♆
22 39 ☽ ☌ ♇
23 10 ♀ ☍ ♅
23 50 ☽ Q ♄

03 Tuesday
01 38 ☽ ∥ ☿
04 39 ☽ ⊥ ☉
10 11 ☽ ⋕ ☉
17 59 ☽ ⚺ ☉
20 55 ☽ ✶ ♆

04 Wednesday
00 06 ☽ ☌ ☿
01 05 ☽ ⚼ ♂
06 09 ☽ ∥ ♃
08 09 ☿ ⚼ ♂
12 25 ☽ □ ♄
15 00 ☽ ⋕ ♃
18 53 ☽ ± ♂
21 50 ☽ ⚻ ♅
21 51 ♀ ⚺ ♃

05 Thursday
02 19 ☽ ☌ ♃
02 40 ☽ ⚺ ♀
03 23 ☽ ∠ ♆
06 09 ☉ ⚺ ♆
06 41 ☽ ⚻ ♂
09 53 ☽ ± ♅
11 35 ☽ ⚺ ♇
14 47 ☽ ∥ ☉
14 59 ☽ ⋕ ☿
15 57 ☽ ⊥ ♀
16 13 ☉ ⋕ ☿
21 30 ☽ ⊥ ♆
23 38 ☽ ⊥ ♇

06 Friday
03 53 ☽ ⚼ ♅
08 17 ☽ ⋕ ♂
09 31 ☽ ⚺ ♆
11 13 ☽ ⋕ ♀
11 44 ☽ ∠ ♀
11 52 ☽ ☌ ☉
13 28 ♀ △ ♂
17 35 ☽ ∠ ♇
19 22 ☽ ⚺ ☿
21 58 ☽ ∥ ♅
23 04 ☉ ∠ ♀

07 Saturday
00 27 ☽ △ ♄
07 41 ☽ ∥ ♆
08 38 ☽ ⊥ ☿
09 33 ☽ △ ♅
15 13 ☽ ⚺ ♃
16 29 ♃ ∠ ♆
16 47 ☽ ☍ ♂
20 17 ☽ ✶ ♀
23 09 ☽ ✶ ♇

08 Sunday
03 11 ☽ ⊥ ♃
04 23 ☽ ∠ ☿
05 51 ☽ ⚼ ♄
05 58 ♀ Q ♆
15 21 ☽ ⋕ ♇
17 43 ☿ □ ♄
20 31 ☽ ☌ ♆
20 59 ☽ ∠ ♃
21 53 ☽ Q ♀
22 30 ☽ Q ♇
23 04 ☽ ⋕ ♄
23 16 ☽ ± ♄

09 Monday
01 58 ♂ ⚻ ♃
03 40 ☽ ⚺ ☉
03 46 ♀ ∥ ♂
04 46 ♀ ☌ ♇
10 47 ☽ ⚻ ♄
11 34 ☉ ∠ ♇
12 54 ☽ ✶ ☿
13 16 ♀ Q ♄
16 09 ☽ ⊥ ☉
19 33 ☽ □ ♅
21 24 ☿ ± ♂

10 Tuesday
01 09 ☽ ⚻ ♂
02 14 ☽ ✶ ♃
08 55 ☽ □ ♇
10 40 ☽ ∠ ☉
11 33 ☽ □ ♀
12 12 ☽ ± ♂
14 29 ☽ Q ☿

11 Wednesday
01 15 ☽ Q ♃
04 37 ☽ ⚼ ♂
05 24 ☽ Q ♅
05 36 ☽ ⚺ ♆
16 45 ☽ ⊥ ♆
16 59 ☽ ✶ ☉
19 08 ☽ ☍ ♄
23 21 ☿ ⚻ ♅

12 Thursday
03 32 ☽ ✶ ♅
04 04 ☽ □ ☿
06 56 ☉ ⋕ ♀
07 34 ☽ △ ♂
09 19 ☽ ∠ ♆
11 04 ☽ □ ♃
16 33 ☽ △ ♇
16 43 ☽ Q ☉
21 03 ☉ △ ♄

13 Friday
00 05 ☽ △ ♀
05 48 ☿ ⚻ ♂
06 40 ☽ ∠ ♅
12 25 ☽ ✶ ♆
19 28 ☽ ⚼ ♇
22 38 ☽ ⊥ ♅
23 08 ☉ ⋕ ♂

14 Saturday
01 09 ☽ ⚻ ♄
03 21 ☽ □ ☉
03 35 ☿ ∠ ♆
05 08 ☽ ⚼ ♀
09 09 ☽ ⚺ ♅
09 11 ☽ ⋕ ♄
09 38 ☽ Q ♆
11 17 ☽ ± ♇
11 38 ☽ ± ♄
11 41 ☽ □ ♂
16 09 ☽ ⋕ ♇
16 13 ☽ △ ☿
17 18 ☽ △ ♃
21 42 ☽ ⚻ ♇
22 09 ☽ ± ♀

15 Sunday
03 09 ☽ ⚼ ♄
03 35 ☿ ☌ ♃
09 20 ☽ ⚻ ♀
16 38 ☽ □ ♆
17 45 ☽ Q ♂
19 12 ☽ ∥ ♆
19 22 ☽ ⚼ ♃
21 01 ☽ ⚼ ☿
21 56 ☿ ± ♅

16 Monday
03 34 ☽ ∥ ♅
04 31 ☽ △ ♄
10 27 ☽ △ ☉
10 42 ☽ ± ♃
12 09 ☽ ☌ ♅
13 20 ☽ ✶ ♂
13 54 ☽ ± ☿
14 22 ☽ ∥ ☉
17 39 ☿ ⚺ ♇
18 54 ♀ ⚼ ♂
19 06 ☽ ⋕ ♂
20 47 ☽ ⚻ ♃

17 Tuesday
00 13 ☽ ☍ ♇
01 00 ☽ ⚻ ☿
01 55 ☿ ⋕ ♃
02 54 ☽ ⋕ ♀
11 16 ☉ △ ♅
12 53 ☽ ⚼ ☉
13 21 ☽ ∠ ♂
15 24 ☽ ☍ ♀
16 29 ☿ ∥ ♃
18 07 ☉ ☍ ♂
18 23 ☽ △ ♆

18 Wednesday
02 31 ☽ ∥ ☿
03 22 ☽ ∥ ♃
03 35 ☽ ⊥ ♂
04 30 ☽ ± ☉
05 17 ☽ ⋕ ♃
05 36 ☽ □ ♄
06 29 ☽ ⋕ ☿
12 49 ♂ ✶ ♅
13 00 ☽ ⚺ ♂
13 00 ☽ ⚺ ♅
14 50 ☽ ⚻ ☉
18 37 ☽ ⚼ ♆
22 12 ☽ ☍ ♃
22 38 ☽ ⊥ ♅

19 Thursday
00 50 ☽ ⚻ ♇
01 33 ♄ Q ♇
02 44 ☽ ∥ ♀
07 26 ☽ ☍ ☿
07 43 ♀ ✶ ♆
09 05 ☽ ± ♆
10 26 ☽ Q ♄
10 28 ☽ ± ♇
12 05 ☽ ∥ ♂
13 02 ☽ ∠ ♅
18 43 ☽ ⚻ ♆
19 35 ☽ ⚻ ♀
22 10 ☽ ⋕ ☉
22 33 ☿ ⊥ ♆

20 Friday
00 34 ♃ ± ♅
00 54 ☽ ⚼ ♇
04 17 ☽ ⋕ ♅
05 38 ☽ ✶ ♄
06 00 ☽ ± ♀
10 51 ☿ ⊥ ♇
11 56 ☽ ☌ ☉
12 30 ☽ ⋕ ♆
13 05 ☽ ✶ ♅
18 21 ☽ ☍ ☉
21 43 ☽ ⚼ ♀
23 08 ☽ ⚻ ♃

21 Saturday
01 07 ☽ △ ♇
05 50 ☽ ∠ ♄
08 32 ☽ Q ♅
08 58 ☿ ⚼ ♅
09 03 ☽ ± ♃
14 02 ☽ ⚻ ☿
14 12 ☽ ∥ ♇
19 26 ☽ ☍ ♆
20 33 ☽ ⊥ ♄
20 52 ☽ ∥ ♄

22 Sunday
00 06 ☽ ⚼ ♃
00 20 ☽ △ ♀
01 15 ☽ ± ☿
06 29 ☽ ⚺ ♄
11 53 ☽ ⚺ ♂
14 20 ☽ □ ♅
17 57 ♀ ± ♂
18 25 ☽ ⚼ ☿
21 46 ☽ ⊥ ♂
23 34 ☽ ⚻ ☉

23 Monday
01 44 ☽ △ ♃
03 04 ☽ □ ♇
10 38 ☽ ± ☉
12 09 ☿ ⚺ ♆
12 50 ☽ ∠ ♂
17 13 ☿ ⋕ ♀
22 35 ☽ ⚻ ♆

24 Tuesday
00 07 ☽ △ ☿
03 37 ☽ ⚼ ☉
08 24 ☽ □ ♀
09 15 ☽ ± ♆
10 09 ☽ ☌ ♄
10 37 ☽ Q ♇
14 30 ☉ ∥ ♅
14 39 ☽ ✶ ♂
15 47 ☉ ⚺ ♃
18 40 ☽ △ ♅

25 Wednesday
01 35 ☽ ⚼ ♆
02 48 ☉ ✶ ♇
04 17 ♀ □ ♄
07 56 ☽ □ ♃
08 30 ☽ ✶ ♇
08 57 ☽ △ ☉
12 01 ☽ Q ♂
15 04 ☿ ∠ ♇
20 33 ☽ Q ♀
22 22 ☽ ⚼ ♅

26 Thursday
05 40 ☽ △ ♆
12 48 ☽ ∠ ♇
15 30 ☽ ± ♅
16 31 ☽ □ ☿
17 14 ♀ ⚻ ♂
17 49 ☽ ⚺ ♄
18 26 ♃ ⚺ ♇
18 35 ☽ Q ♃
21 13 ☽ □ ♂
21 46 ☽ ✶ ♀

27 Friday
01 03 ☿ △ ♄
02 57 ☽ ∥ ♄
03 07 ☽ ⚻ ♅
05 26 ☽ ⊥ ♄
06 15 ☽ ⊥ ♇
10 48 ☽ ∥ ♇
18 04 ☽ ⚺ ♇
18 29 ☽ ✶ ♃
18 47 ☉ ⚼ ♄
19 26 ☿ ☍ ♂
23 07 ☽ ∠ ♄
23 30 ☽ □ ☉

28 Saturday
05 29 ☿ ⋕ ♂
06 22 ☽ ∠ ♀
10 24 ☽ Q ☿
16 41 ☽ □ ♆
17 31 ☽ ⋕ ♆
21 13 ☽ ⋕ ☉

29 Sunday
01 02 ☽ ∠ ♃
02 21 ☽ ⊥ ♀
04 07 ☽ ⋕ ♅
05 05 ☽ ✶ ♄
07 07 ☽ △ ♂
07 14 ♀ ⚻ ♅
13 28 ☿ △ ♅
14 54 ☽ Q ☉
14 58 ☽ ☍ ♅
15 16 ☽ ✶ ☿
15 48 ☽ ⚺ ♀
19 37 ☽ ⊥ ♃
22 13 ☽ ⋕ ☿
22 30 ☿ ⚺ ♀

30 Monday
05 15 ☽ Q ♆
05 21 ☽ Q ♄
05 27 ☽ ∥ ♂
06 32 ☽ ☌ ♇
08 03 ☽ ⚺ ♃
12 43 ☽ ⚼ ♂
17 29 ☽ ✶ ☉
19 38 ♄ ± ♆

MAY 2046

LONGITUDES

Date	Sidereal time h m s	Sun ☉ ° ' "	Moon ☽ ° ' "	Moon ☽ 24.00 ° ' "	Mercury ☿ ° '	Venus ♀ ° '	Mars ♂ ° '	Jupiter ♃ ° '	Saturn ♄ ° '	Uranus ♅ ° '	Neptune ♆ ° '	Pluto ♇ ° '
01	02 38 21	11 ♉ 17 37	19 ♓ 39 02	25 ♓ 34 19	01 ♉ 19	00 ♈ 02	22 ♎ 53	06 ♈ 10	22 ♐ 31	27 ♌ 30	16 ♉ 35	05 ♓ 11
02	02 42 17	12 15 52	01 ♈ 31 01	07 ♈ 29 34	03 19	01 11	22 R 33	06 23	22 R 28	27 R 30	16 37	05 12
03	02 46 14	13 14 06	13 ♈ 30 17	19 ♈ 33 29	05 21	02 20	22 14	06 36	22 26	27 29	16 39	05 13
04	02 50 10	14 12 18	25 ♈ 39 23	01 ♉ 48 08	07 25	03 30	21 55	06 49	22 23	27 29	16 41	05 13
05	02 54 07	15 10 29	07 ♉ 59 51	14 ♉ 14 35	09 29	04 39	21 37	07 02	22 20	27 D 29	16 44	05 14
06	02 58 03	16 08 38	20 ♉ 32 23	26 ♉ 53 11	11 36	05 48	21 20	07 14	22 16	27 29	16 46	05 15
07	03 02 00	17 06 45	03 ♊ 16 59	09 ♊ 43 43	13 43	06 58	21 03	07 27	22 13	27 30	16 48	05 16
08	03 05 56	18 04 51	16 ♊ 13 19	22 ♊ 45 44	15 52	08 07	20 47	07 40	22 10	27 30	16 50	05 16
09	03 09 53	19 02 55	29 ♊ 20 57	05 ♋ 58 56	18 01	09 17	20 32	07 53	22 07	27 30	16 53	05 17
10	03 13 50	20 00 57	12 ♋ 39 44	19 ♋ 23 21	20 11	10 27	20 17	08 05	22 03	27 30	16 55	05 18
11	03 17 46	20 58 57	26 ♋ 09 53	02 ♌ 59 24	22 22	11 37	20 03	08 18	22 00	27 31	16 57	05 18
12	03 21 43	21 56 55	09 ♌ 51 58	16 ♌ 47 41	24 32	12 46	19 50	08 30	21 57	27 31	16 59	05 19
13	03 25 39	22 54 52	23 ♌ 46 36	00 ♍ 48 42	26 43	13 56	19 37	08 42	21 53	27 31	17 02	05 19
14	03 29 36	23 52 46	07 ♍ 53 56	15 ♍ 02 10	28 ♉ 53	15 06	19 26	08 55	21 49	27 32	17 04	05 20
15	03 33 32	24 50 39	22 ♍ 13 10	29 ♍ 26 33	01 ♊ 03	16 16	19 15	09 07	21 46	27 32	17 06	05 20
16	03 37 29	25 48 30	06 ♎ 41 52	13 ♎ 58 31	03 12	17 26	19 05	09 19	21 42	27 33	17 08	05 21
17	03 41 25	26 46 19	21 ♎ 15 49	28 ♎ 32 59	05 20	18 36	18 56	09 31	21 38	27 34	17 11	05 21
18	03 45 22	27 44 06	05 ♏ 49 10	13 ♏ 03 30	07 26	19 46	18 47	09 43	21 34	27 34	17 13	05 22
19	03 49 19	28 41 52	20 ♏ 15 08	27 ♏ 23 13	09 30	20 57	18 39	09 55	21 31	27 35	17 15	05 22
20	03 53 15	29 ♉ 39 37	04 ♐ 27 02	11 ♐ 25 57	11 33	22 07	18 33	10 07	21 27	27 36	17 17	05 23
21	03 57 12	00 ♊ 37 21	18 ♐ 19 28	25 ♐ 07 14	13 34	23 17	18 26	10 19	21 23	27 37	17 19	05 23
22	04 01 08	01 35 03	01 ♑ 49 02	08 ♑ 24 49	15 32	24 28	18 21	10 30	21 19	27 38	17 22	05 23
23	04 05 05	02 32 44	14 ♑ 54 39	21 ♑ 18 46	17 27	25 38	18 17	10 42	21 15	27 39	17 24	05 24
24	04 09 01	03 30 24	27 ♑ 37 27	03 ♒ 51 10	19 21	26 49	18 13	10 53	21 11	27 40	17 26	05 24
25	04 12 58	04 28 02	10 ♒ 00 22	16 ♒ 05 37	21 11	27 59	18 10	11 05	21 06	27 41	17 28	05 24
26	04 16 54	05 25 40	22 ♒ 07 33	28 ♒ 06 46	22 59	29 ♈ 10	18 08	11 16	21 02	27 42	17 30	05 25
27	04 20 51	06 23 17	04 ♓ 03 56	09 ♓ 59 44	24 44	00 ♉ 20	18 07	11 27	20 58	27 43	17 33	05 25
28	04 24 47	07 20 52	15 ♓ 54 48	21 ♓ 49 48	26 26	01 31	18 06	11 38	20 54	27 44	17 35	05 25
29	04 28 44	08 18 27	27 ♓ 45 21	03 ♈ 42 03	28 05	02 42	18 D 07	11 49	20 50	27 46	17 37	05 25
30	04 32 41	09 16 01	09 ♈ 40 28	15 ♈ 41 06	29 ♊ 41	03 53	18 08	12 00	20 45	27 47	17 39	05 25
31	04 36 37	10 ♊ 13 34	21 ♈ 44 25	27 ♈ 50 49	01 ♋ 14	05 ♉ 03	18 ♎ 09	12 ♈ 11	20 ♐ 41	27 ♌ 48	17 ♉ 41	05 ♓ 25

Date	Moon True ☊ ° '	Moon Mean ☊ ° '	Moon ☽ Latitude ° '
01	09 ♒ 04	08 ♒ 58	03 N 16
02	08 R 57	08 55	03 59
03	08 47	08 51	04 32
04	08 35	08 48	04 53
05	08 21	08 45	05 00
06	08 08	08 42	04 53
07	07 56	08 39	04 32
08	07 46	08 35	03 56
09	07 38	08 32	03 07
10	07 34	08 29	02 08
11	07 32	08 26	01 N 00
12	07 D 31	08 23	00 S 12
13	07 R 31	08 20	01 25
14	07 31	08 16	02 34
15	07 28	08 13	03 33
16	07 23	08 10	04 20
17	07 16	08 07	04 51
18	07 06	08 04	05 02
19	06 56	08 00	04 54
20	06 46	07 57	04 28
21	06 37	07 54	03 47
22	06 30	07 51	02 53
23	06 26	07 48	01 52
24	06 24	07 45	00 S 47
25	06 D 23	07 41	00 N 19
26	06 24	07 38	01 23
27	06 25	07 35	02 22
28	06 R 24	07 32	03 15
29	06 22	07 29	03 59
30	06 18	07 26	04 33
31	06 ♒ 12	07 ♒ 22	04 N 56

DECLINATIONS

Date	Sun ☉ ° '	Moon ☽ ° '	Mercury ☿ ° '	Venus ♀ ° '	Mars ♂ ° '	Jupiter ♃ ° '	Saturn ♄ ° '	Uranus ♅ ° '	Neptune ♆ ° '	Pluto ♇ ° '
01	15 N 13	01 S 05	10 N 37	01 S 17	07 S 41	01 N 26	21 S 40	13 N 03	15 N 09	20 S 30
02	15 31	04 N 16	11 26	00 52	07 36	01 31	21 40	13 03	15 10	20 30
03	15 49	09 30	12 15	00 26	07 31	01 36	21 40	13 03	15 11	20 30
04	16 06	14 27	13 04	00 S 01	07 27	01 41	21 39	13 03	15 11	20 30
05	16 23	18 54	13 53	00 N 25	07 23	01 46	21 39	13 03	15 12	20 30
06	16 40	22 35	14 42	00 50	07 19	01 51	21 39	13 03	15 13	20 30
07	16 56	25 15	15 30	01 16	07 15	01 56	21 39	13 03	15 13	20 30
08	17 13	26 38	16 18	01 41	07 11	02 01	21 39	13 03	15 14	20 30
09	17 29	26 33	17 05	02 07	07 08	02 06	21 38	13 03	15 14	20 30
10	17 45	24 57	17 50	02 33	07 05	02 11	21 38	13 03	15 15	20 30
11	18 00	21 54	18 35	02 59	07 02	02 15	21 38	13 03	15 16	20 30
12	18 15	17 34	19 18	03 25	07 00	02 20	21 38	13 03	15 16	20 30
13	18 30	12 15	19 59	03 50	06 57	02 25	21 38	13 02	15 17	20 30
14	18 44	06 N 14	20 38	04 16	06 55	02 30	21 37	13 02	15 18	20 30
15	18 59	00 S 11	21 16	04 42	06 54	02 34	21 37	13 02	15 18	20 30
16	19 12	06 39	21 51	05 08	06 52	02 39	21 37	13 02	15 19	20 31
17	19 26	12 47	22 23	05 33	06 51	02 44	21 37	13 01	15 20	20 31
18	19 39	18 12	22 54	05 59	06 50	02 48	21 36	13 01	15 20	20 31
19	19 52	22 31	23 21	06 25	06 50	02 53	21 36	13 01	15 21	20 31
20	20 04	25 25	23 47	06 50	06 49	02 57	21 36	13 01	15 21	20 31
21	20 17	26 41	24 09	07 15	06 49	03 02	21 36	13 00	15 22	20 31
22	20 28	26 18	24 29	07 41	06 49	03 06	21 35	13 00	15 23	20 32
23	20 40	24 27	24 46	08 06	06 50	03 10	21 35	13 00	15 23	20 32
24	20 51	21 24	25 01	08 31	06 51	03 15	21 35	12 59	15 24	20 32
25	21 02	17 26	25 13	08 56	06 52	03 19	21 35	12 59	15 24	20 32
26	21 12	12 49	25 23	09 21	06 53	03 23	21 34	12 58	15 25	20 32
27	21 22	07 48	25 30	09 45	06 55	03 27	21 34	12 58	15 26	20 33
28	21 32	02 S 33	25 35	10 10	06 57	03 32	21 34	12 57	15 26	20 33
29	21 41	02 N 46	25 38	10 34	06 59	03 36	21 34	12 57	15 27	20 33
30	21 50	08 01	25 39	10 58	07 01	03 40	21 33	12 56	15 27	20 33
31	21 N 59	13 N 02	25 N 37	11 N 22	07 S 04	03 N 44	21 S 33	12 N 56	15 N 28	20 S 34

ZODIAC SIGN ENTRIES

Date	h	m	Planets	
01	11	21	♀	♈
02	08	57	☽	♈
04	20	29	☽	♉
07	05	51	☽	♊
09	13	11	☽	♋
11	18	45	☽	♌
13	22	37	☽	♍
15	00	18	☿	♊
16	00	55	☽	♎
18	02	23	☽	♏
20	04	25	☽	♐
20	20	28	☉	♊
22	08	44	☽	♑
24	16	34	☽	♒
27	03	48	☽	♓
27	05	04	♀	♉
29	16	32	☽	♈
30	16	51	☿	♋

LATITUDES

Date	Mercury ☿ ° '	Venus ♀ ° '	Mars ♂ ° '	Jupiter ♃ ° '	Saturn ♄ ° '	Uranus ♅ ° '	Neptune ♆ ° '	Pluto ♇ ° '
01	01 S 24	01 S 25	01 N 18	01 S 06	01 N 33	00 N 46	01 S 42	11 S 45
04	00 58	01 32	01 10	01 06	01 34	00 46	01 42	11 46
07	00 S 28	01 38	01 03	01 07	01 34	00 46	01 42	11 46
10	00 N 03	01 43	00 55	01 07	01 34	00 45	01 42	11 47
13	00 35	01 48	00 47	01 08	01 34	00 45	01 42	11 48
16	01 04	01 51	00 39	01 08	01 34	00 45	01 42	11 49
19	01 30	01 54	00 31	01 09	01 34	00 45	01 42	11 50
22	01 51	01 56	00 24	01 09	01 34	00 45	01 42	11 51
25	02 05	01 57	00 17	01 10	01 34	00 45	01 42	11 52
28	02 12	01 57	00 10	01 10	01 34	00 45	01 42	11 54
31	02 N 12	01 S 57	00 N 03	01 S 11	01 N 34	00 N 45	01 S 42	11 S 55

DATA

Julian Date	2468467
Delta T	+79 seconds
Ayanamsa	24° 30' 28"
Synetic vernal point	04° ♓ 36' 31"
True obliquity of ecliptic	23° 26' 06"

MOON'S PHASES, APSIDES AND POSITIONS ☽

Date	h	m	Phase	Longitude ° '	Eclipse Indicator
06	02	56	●	15 ♉ 47	
13	10	25	☽	22 ♌ 51	
20	03	15	○	29 ♏ 19	
27	17	07	☾	06 ♓ 36	

Day	h	m		
01	03	49	Apogee	
16	23	39	Perigee	
28	21	59	Apogee	
01	16	52	0N	
08	22	46	Max dec	26° N 47'
15	11	19	0S	
21	18	20	Max dec	26° S 44'
28	23	32	0N	

LONGITUDES

Date	Chiron ⚷ ° '	Ceres ⚳ ° '	Pallas ⚴ ° '	Juno ⚵ ° '	Vesta ⚶ ° '	Black Moon Lilith ⚸ ° '
01	00 ♎ 49	21 ♍ 38	24 ♋ 08	23 ♍ 43	27 ♋ 00	18 ♓ 25
11	00 ♎ 25	21 ♍ 48	28 ♋ 37	23 ♍ 21	00 ♌ 16	19 ♓ 31
21	00 ♎ 13	22 ♍ 36	03 ♌ 12	23 ♍ 34	03 ♌ 49	20 ♓ 38
31	00 ♎ 11	24 ♍ 00	07 ♌ 50	24 ♍ 19	07 ♌ 35	21 ♓ 45

All ephemeris data is given at 12.00 UT and the Moon's longitude is additionally given for 24.00 UT

ASPECTARIAN

h m	Aspects
01 Tuesday	
03 51	☽ ∠ ☿
04 43	♀ # ♃
05 44	☽ ∗ ♆
06 31	☽ ± ♂
07 04	☉ ∥ ♆
10 27	☽ # ♃
11 02	☽ ∥ ♀
17 48	☽ □ ♄
18 22	☽ ⊼ ♂
21 50	☽ # ♀
23 29	☽ ∥ ♃
02 Wednesday	
01 48	☽ ⊥ ☿
02 40	☽ ∠ ☉
03 53	☽ ⊼ ♅
11 15	☽ ☌ ♀
12 12	☽ ∠ ♆
15 35	☉ ⊥ ♃
15 58	☽ ± ♅
16 22	☽ ⊻ ☿
18 39	♂ ∗ ♄
19 24	☽ ⊻ ♇
21 14	♀ ∠ ♆
21 57	☽ ☌ ♃
22 23	☽ ⊥ ☉
03 Thursday	
02 58	☽ # ♂
06 18	☽ ⊥ ♆
07 25	☽ ⊥ ♇
09 59	☽ bQ ♅
10 19	☿ ∗ ♇
11 25	☽ ⊻ ☉
18 16	☽ ⊻ ♆
04 Friday	
01 18	☽ ∠ ♇
03 44	☽ ∥ ☿
04 14	☿ ⊻ ♃
04 50	☽ ☍ ♂
05 02	☽ ∥ ♅
05 35	☽ △ ♄
11 33	☿ ∥ ♅
11 38	☿ bQ ♄
11 56	♀ ± ♅
15 35	☽ △ ♅
15 46	☽ ∥ ♆
19 51	♅ St D
21 05	☽ ∥ ☉
05 Saturday	
04 51	☽ ⊻ ♀
06 39	☽ ∗ ♇
10 06	☽ ⊻ ♃
10 43	☽ bQ ♄
15 27	☽ ☌ ☿
17 38	☽ ⊥ ♀
21 43	☽ # ♇
21 50	☽ ⊥ ♃
06 Sunday	
00 18	♀ ⊻ ♇
02 56	☽ ☌ ☉
03 55	☽ ± ♄
04 48	☽ ☌ ♆
05 20	☽ # ♄
05 44	☽ Q ♇
12 33	☽ ∠ ♀
13 28	☽ ⊼ ♂
15 04	☉ ± ♄
15 17	☽ ⊼ ♄
15 17	☽ ∠ ♃
17 18	☿ ⊥ ♀
07 Monday	
00 34	☽ ± ♂
01 08	☽ □ ♅
03 18	☿ ∥ ♆
03 59	☉ ☌ ♆
08 42	☿ ⊥ ♃
15 40	☉ Q ♇
15 41	☽ □ ♇
17 04	☽ bQ ♂
19 33	☽ ∗ ♀
19 54	☽ ∗ ♃
08 Tuesday	
00 25	♀ ☌ ♃
10 40	☽ Q ♅
11 12	☽ ⊻ ☿
13 08	☽ ⊻ ♆
15 21	☿ ± ♄
15 42	☽ ⊻ ☉
18 26	☽ Q ♃
19 52	☽ Q ♀
20 13	☽ △ ♂
22 52	☽ ☍ ♄
23 07	☿ ☌ ♆
09 Wednesday	
00 11	☽ ⊥ ♆
00 23	☽ ⊥ ☿
03 33	☽ ⊥ ☉
03 48	☿ Q ♇
08 38	☽ ∗ ♅
10 24	♀ ∥ ♃
16 36	☽ ∠ ♆
19 56	☽ ∠ ☿
21 11	☽ ∠ ☉
22 45	☽ △ ♇
10 Thursday	
03 39	☽ □ ♃
07 16	☉ ∥ ☿
07 39	☽ □ ♀
08 38	☉ ☌ ☿
11 08	♂ bQ ♇
11 43	☽ ∠ ♅
12 59	☿ ⊼ ♂
17 20	☉ ⊼ ♂
19 37	☽ ∗ ♆
22 00	♀ ⊥ ♆
11 Friday	
01 21	☽ □ ♂
01 37	☽ bQ ♇
02 07	☽ ∗ ☉
03 46	☽ ⊥ ♅
03 59	☽ ∗ ☿
04 40	☽ ⊼ ♄
05 37	♀ ⊥ ♇
08 08	☿ ⊼ ♄
13 39	☽ # ♄
14 22	☽ ⊻ ♅
15 13	☽ ± ♄
16 55	☽ Q ♆
17 32	☽ ± ♇
20 27	☽ # ♇
23 23	☿ ∠ ♃
12 Saturday	
00 54	☽ Q ☉
04 03	☽ ⊼ ♇
04 22	☽ ∥ ☿
05 07	☽ Q ☿
06 41	♀ bQ ♅
06 56	☽ bQ ♄
08 31	☽ Q ♂
08 47	☽ ∥ ☉
09 35	☽ △ ♃
11 51	☉ ⊼ ♄
17 30	☽ △ ♀
22 50	☽ ∥ ♆
13 Sunday	
00 22	☽ □ ♆
00 58	☿ ± ♂
04 59	☽ ∗ ♂
08 40	☽ ∥ ♅
08 46	☽ △ ♄
10 25	☽ □ ☉
11 53	☽ bQ ♃
17 57	☽ □ ☿
18 24	☽ ☌ ♅
20 55	☿ □ ♅
21 36	☽ bQ ♀
14 Monday	
03 27	☽ ± ♃
06 13	☽ ∠ ♂
07 06	☿ # ♇
07 40	☽ ☍ ♇
09 19	☽ # ♂
13 03	☉ ∠ ♃
13 44	☽ ⊼ ♃
14 13	☽ ± ♀
18 58	☽ ∥ ♀
21 11	☽ ⊥ ♂
15 Tuesday	
01 11	☽ ⊼ ♀
01 54	☽ ∥ ♃
03 26	☽ △ ♆
07 06	☽ ⊻ ♂
11 14	☽ □ ♄
16 41	☽ △ ☉
17 15	☿ ∠ ♀
20 32	☉ ± ♂
20 51	☽ ⊻ ♅
20 56	☽ # ♃
16 Wednesday	
02 26	☿ # ♄
04 27	☽ bQ ♆
05 13	☽ △ ☿
05 41	♀ ⊻ ♆
05 55	☽ # ♀
06 48	☽ ⊥ ♅
09 46	☽ ⊼ ♇
12 51	☽ ∥ ♂
16 23	☽ ☍ ♃
16 56	☽ Q ♄
19 16	☽ bQ ☉
19 21	☽ ± ♆
19 40	☽ ± ♇
21 12	☿ bQ ♂
21 40	☽ ∠ ♅
17 Thursday	
05 15	☽ ⊼ ♆
07 15	☽ ☍ ♀
08 11	☽ ☌ ♂
10 12	☽ bQ ☿
10 30	☽ bQ ♇
11 08	☽ ± ☉
12 18	☿ □ ♇
12 37	☽ ∗ ♄
13 01	☽ # ♅
17 51	♀ ☍ ♂
21 43	☽ ⊼ ☉
22 23	☽ ∗ ♅
22 48	☽ # ♆
18 Friday	
03 32	☽ ± ☿
07 56	☉ □ ♅
11 15	☽ △ ♇
13 15	☽ ∠ ♄
15 07	☽ ⊼ ☿
18 13	☽ Q ♅
18 33	☽ ⊼ ♃
19 40	☽ # ☉
23 57	☽ ∥ ♇
19 Saturday	
00 09	♀ ∠ ♇
04 07	☽ ⊥ ♄
04 39	☽ ± ♃
06 13	☽ ∥ ♄
06 58	☽ ☍ ♆
09 21	☽ ⊻ ♂
13 16	☽ ⊼ ♀
14 06	☽ ⊻ ♄
17 17	☿ ∗ ♃
18 38	☽ # ☿
19 20	☽ ⊥ ♂
19 57	☽ bQ ♃
22 57	♀ △ ♄
20 Sunday	
00 16	☽ ± ♀
00 21	☽ □ ♅
03 15	☽ ☍ ☉
10 28	☽ ∠ ♂
11 15	♀ # ♂
13 35	☽ □ ♇
16 59	☽ bQ ♀
21 52	☽ △ ♃
21 Monday	
02 17	☽ ☍ ☿
10 15	☽ ⊼ ♆
12 12	☽ ∗ ♂
17 21	☽ ☌ ♄
20 50	☽ ± ♆
20 55	☽ Q ♇
21 35	☽ △ ♀
22 Tuesday	
04 28	☽ △ ♅
09 23	☽ Q ♂
11 33	☽ ⊼ ☉
12 59	☽ bQ ♆
13 15	☿ Q ♅
18 27	☉ # ♇
18 29	☽ ∗ ♇
23 19	☽ ± ☉
23 Wednesday	
04 05	☽ □ ♃
07 47	☽ bQ ♅
09 13	☽ # ☿
09 22	♆ Q ♇
11 14	☿ ⊻ ♆
16 39	☽ △ ♆
17 19	☽ bQ ☉
17 35	☽ ⊼ ☿
18 15	☽ □ ♂
22 01	☿ △ ♂
22 16	☽ ∠ ♇
23 48	☽ ⊻ ♄
24 Thursday	
00 39	☽ ± ♅
05 10	☉ bQ ♂
06 53	☽ ± ☿
10 17	☽ □ ♀
10 44	☽ ∥ ♄
11 09	☽ ⊥ ♄
12 04	☽ ⊼ ♅
14 27	☽ Q ♃
15 24	☽ ⊥ ♇
15 25	☽ # ☉
17 38	☽ ∥ ♇
25 Friday	
00 17	☽ △ ☉
03 01	☽ ⊻ ♇
03 14	☽ bQ ☿
04 25	☽ ∠ ♄
05 40	♀ △ ♅
11 00	☿ ☍ ♄
14 08	☽ ∗ ♃
22 50	☽ # ♆
26 Saturday	
01 03	☽ Q ♀
02 47	☽ □ ♆
04 04	☽ △ ♂
09 50	☽ ∗ ♄
11 15	☽ # ♅
11 31	☉ □ ♇
14 00	☽ △ ☿
16 17	☿ Q ♃
19 18	☿ ⊥ ♆
20 25	☽ ∠ ♃
23 11	☽ ☍ ♅
27 Sunday	
03 31	☽ # ♀
03 40	☽ ∗ ♀
07 02	♃ ⊥ ♇
09 48	☽ Q ♄
10 05	☽ bQ ♂
14 43	☽ ☌ ♇
14 51	☽ ⊥ ♃
15 00	☽ Q ♆
16 06	☽ ∥ ♂
17 07	☽ □ ☉
17 15	☉ ⊥ ♀
28 Monday	
03 01	♃ ⊥ ♆
03 11	☽ ⊻ ♃
04 17	☽ ± ♂
07 40	☽ # ♃
13 22	☽ ∠ ♀
15 23	☽ ∗ ♆
15 32	♂ St D
16 27	☽ ⊼ ♂
16 42	☉ # ♄
22 03	☽ □ ♄
29 Tuesday	
07 14	☿ ∗ ♅
08 49	☽ Q ☉
09 37	☽ ⊥ ♀
12 01	☽ ⊼ ♅
12 46	☽ □ ☿
15 47	☽ ∥ ♃
21 50	☽ ∠ ♆
23 05	☽ ⊻ ♀
30 Wednesday	
00 09	☽ ± ♅
03 28	☽ ⊻ ♇
07 20	☽ # ♂
11 07	☽ ∗ ☉
15 30	☽ ⊥ ♇
15 58	☽ ⊥ ♆
16 43	☽ ☌ ♃
18 14	☽ bQ ♅
31 Thursday	
03 07	☽ ∥ ♀
03 57	☽ ⊻ ♆
04 53	☽ ☍ ♂
06 20	☽ Q ☿
09 24	☽ ∠ ♇
09 55	☽ △ ♄
11 28	☽ ∥ ♅
19 26	♀ ∗ ♇
19 27	☽ ∠ ☉
23 57	☽ △ ♅
23 58	♀ bQ ♄

Date	Sidereal time h m s	Sun ☉	Moon ☽	Moon ☽ 24.00	Mercury ☿	Venus ♀
01	04 40 34	11 ♊ 11 07	04 ♉ 00 38	10 ♉ 14 07	02 ♋ 44	06 ♉ 14
02	04 44 30	12 08 38	16 ♉ 31 27	22 ♉ 52 45	04 11	07 25
03	04 48 27	13 06 08	29 ♉ 18 04	05 ♊ 47 21	05 35	08 36
04	04 52 23	14 03 38	12 ♊ 20 31	18 ♊ 57 24	06 56	09 47
05	04 56 20	15 01 07	25 ♊ 37 49	02 ♋ 21 31	08 13	10 58
06	05 00 16	15 58 34	09 ♋ 08 15	15 ♋ 57 47	09 27	12 09
07	05 04 13	16 56 01	22 ♋ 49 49	29 ♋ 44 08	10 38	13 20
08	05 08 10	17 53 27	06 ♌ 40 30	13 ♌ 38 42	11 46	14 32
09	05 12 06	18 50 51	20 ♌ 38 34	27 ♌ 39 55	12 50	15 43
10	05 16 03	19 48 14	04 ♍ 42 34	11 ♍ 46 24	13 51	16 54
11	05 19 59	20 45 36	18 ♍ 51 13	25 ♍ 56 51	14 48	18 05
12	05 23 56	21 42 58	03 ♎ 03 04	10 ♎ 09 36	15 42	19 17
13	05 27 52	22 40 18	17 ♎ 16 10	24 ♎ 22 24	16 32	20 28
14	05 31 49	23 37 37	01 ♏ 27 54	08 ♏ 32 15	17 18	21 39
15	05 35 45	24 34 55	15 ♏ 34 57	22 ♏ 35 33	18 00	22 51
16	05 39 42	25 32 12	29 ♏ 33 32	06 ♐ 28 26	18 38	24 02
17	05 43 39	26 29 29	13 ♐ 19 49	20 ♐ 07 19	19 12	25 14
18	05 47 35	27 26 45	26 ♐ 50 35	03 ♑ 29 23	19 41	26 25
19	05 51 32	28 24 00	10 ♑ 03 34	16 ♑ 33 03	20 07	27 37
20	05 55 28	29 ♊ 21 16	22 ♑ 57 51	29 ♑ 18 05	20 28	28 ♉ 48
21	05 59 25	00 ♋ 18 30	05 ♒ 33 58	11 ♒ 45 44	20 44	00 ♊ 00
22	06 03 21	01 15 44	17 ♒ 53 45	23 ♒ 58 25	20 56	01 11
23	06 07 18	02 12 58	00 ♓ 00 13	05 ♓ 59 38	21 04	02 23
24	06 11 14	03 10 12	11 ♓ 57 14	17 ♓ 53 35	21 06	03 35
25	06 15 11	04 07 26	23 ♓ 49 16	29 ♓ 44 55	21 R 05	04 47
26	06 19 08	05 04 40	05 ♈ 41 08	11 ♈ 38 32	20 59	05 58
27	06 23 04	06 01 53	17 ♈ 37 43	23 ♈ 39 14	20 48	07 10
28	06 27 01	06 59 07	29 ♈ 43 40	05 ♉ 51 29	20 33	08 22
29	06 30 57	07 56 20	12 ♉ 03 09	18 ♉ 19 05	20 14	09 34
30	06 34 54	08 ♋ 53 34	24 ♉ 39 35	01 ♊ 04 55	19 ♋ 51	10 ♊ 46

Date	Mars ♂	Jupiter ♃	Saturn ♄	Uranus ♅	Neptune ♆	Pluto ♇
01	18 ♎ 12	12 ♈ 21	20 ♐ 37	27 ♌ 50	17 ♉ 43	05 ♓ 26
02	18 15	12 32	20 R 32	27 51	17 45	05 26
03	18 19	12 42	20 28	27 53	17 47	05 26
04	18 24	12 53	20 23	27 54	17 50	05 26
05	18 30	13 03	20 19	27 56	17 52	05 R 26
06	18 36	13 13	20 15	27 58	17 54	05 26
07	18 43	13 23	20 10	27 59	17 56	05 26
08	18 51	13 33	20 06	28 01	17 58	05 26
09	18 59	13 43	20 01	28 03	18 00	05 25
10	19 08	13 53	19 57	28 05	18 02	05 25
11	19 18	14 02	19 52	28 07	18 04	05 25
12	19 28	14 12	19 48	28 09	18 06	05 25
13	19 39	14 21	19 44	28 11	18 08	05 25
14	19 51	14 30	19 39	28 13	18 10	05 25
15	20 03	14 39	19 35	28 15	18 11	05 24
16	20 16	14 48	19 30	28 17	18 13	05 24
17	20 30	14 57	19 26	28 19	18 15	05 24
18	20 44	15 06	19 22	28 21	18 17	05 24
19	20 58	15 14	19 17	28 24	18 19	05 23
20	21 13	15 23	19 13	28 26	18 21	05 23
21	21 29	15 31	19 09	28 28	18 22	05 22
22	21 45	15 39	19 04	28 31	18 24	05 22
23	22 02	15 47	19 00	28 33	18 26	05 22
24	22 19	15 55	18 56	28 36	18 28	05 21
25	22 37	16 03	18 52	28 38	18 29	05 21
26	22 55	16 11	18 47	28 41	18 31	05 20
27	23 14	16 18	18 43	28 43	18 33	05 20
28	23 34	16 25	18 39	28 46	18 34	05 19
29	23 53	16 33	18 35	28 48	18 36	05 19
30	24 ♎ 14	16 ♈ 40	18 ♐ 31	28 ♌ 51	18 ♉ 38	05 ♓ 18

Date	Moon True ☊	Moon Mean ☊	Moon ☽ Latitude
01	06 ♒ 03	07 ♒ 19	05 N 05
02	05 R 54	07 16	05 00
03	05 45	07 13	04 40
04	05 36	07 10	04 06
05	05 29	07 06	03 17
06	05 24	07 03	02 16
07	05 21	07 00	01 N 07
08	05 D 21	06 57	00 S 07
09	05 21	06 54	01 22
10	05 22	06 51	02 32
11	05 23	06 47	03 33
12	05 R 23	06 44	04 22
13	05 20	06 41	04 54
14	05 16	06 38	05 09
15	05 11	06 35	05 04
16	05 05	06 32	04 42
17	04 59	06 28	04 03
18	04 54	06 25	03 12
19	04 50	06 22	02 11
20	04 47	06 19	01 S 04
21	04 D 47	06 16	00 N 04
22	04 47	06 12	01 11
23	04 49	06 09	02 14
24	04 51	06 06	03 09
25	04 52	06 03	03 57
26	04 R 52	06 00	04 34
27	04 52	05 57	04 59
28	04 50	05 53	05 12
29	04 47	05 50	05 11
30	04 ♒ 43	05 ♒ 47	04 N 55

DECLINATIONS

Date	Sun ☉	Moon ☽	Mercury ☿	Venus ♀	Mars ♂	Jupiter ♃	Saturn ♄	Uranus ♅	Neptune ♆	Pluto ♇
01	22 N 07	17 N 38	25 N 34	11 N 46	07 S 07	03 N 48	21 S 33	12 N 55	15 N 28	20 S 34
02	22 15	21 34	25 30	12 09	07 10	03 52	21 33	12 55	15 29	20 34
03	22 22	24 33	25 23	12 32	07 14	03 56	21 32	12 54	15 30	20 34
04	22 29	26 20	25 15	12 55	07 17	03 59	21 32	12 54	15 30	20 35
05	22 36	26 39	25 06	13 18	07 21	04 03	21 32	12 53	15 31	20 35
06	22 42	25 23	24 55	13 41	07 26	04 07	21 31	12 52	15 31	20 35
07	22 48	22 36	24 43	14 03	07 30	04 11	21 31	12 52	15 32	20 36
08	22 53	18 29	24 31	14 25	07 35	04 14	21 31	12 51	15 32	20 36
09	22 58	13 19	24 17	14 46	07 40	04 18	21 31	12 51	15 33	20 36
10	23 03	07 25	24 02	15 07	07 45	04 22	21 30	12 50	15 33	20 37
11	23 07	01 N 08	23 46	15 28	07 50	04 25	21 30	12 49	15 34	20 37
12	23 11	05 S 13	23 30	15 49	07 56	04 29	21 30	12 48	15 34	20 37
13	23 14	11 18	23 14	16 09	08 02	04 32	21 30	12 48	15 35	20 38
14	23 17	16 48	22 56	16 29	08 08	04 35	21 29	12 47	15 35	20 38
15	23 19	21 21	22 39	16 48	08 14	04 39	21 29	12 46	15 36	20 39
16	23 22	24 38	22 21	17 08	08 20	04 42	21 29	12 45	15 36	20 39
17	23 23	26 25	22 03	17 26	08 27	04 45	21 29	12 45	15 37	20 39
18	23 25	26 35	21 45	17 45	08 34	04 48	21 28	12 44	15 37	20 40
19	23 25	25 13	21 27	18 02	08 41	04 51	21 28	12 43	15 38	20 40
20	23 26	22 32	21 09	18 20	08 48	04 54	21 28	12 42	15 38	20 41
21	23 26	18 48	20 52	18 37	08 55	04 57	21 28	12 41	15 39	20 41
22	23 26	14 20	20 34	18 53	09 02	05 00	21 27	12 41	15 39	20 42
23	23 25	09 23	20 17	19 09	09 10	05 03	21 27	12 40	15 40	20 42
24	23 24	04 S 09	20 01	19 25	09 18	05 06	21 27	12 39	15 40	20 43
25	23 22	01 N 10	19 45	19 40	09 26	05 09	21 27	12 38	15 40	20 43
26	23 20	06 27	19 30	19 55	09 34	05 11	21 26	12 37	15 41	20 43
27	23 18	11 31	19 16	20 09	09 42	05 14	21 26	12 36	15 41	20 44
28	23 15	16 14	19 02	20 22	09 51	05 17	21 26	12 35	15 42	20 44
29	23 12	20 22	18 50	20 36	09 59	05 19	21 26	12 34	15 42	20 45
30	23 N 08	23 N 41	18 N 38	20 N 48	10 S 08	05 N 22	21 S 26	12 N 33	15 N 42	20 S 45

ZODIAC SIGN ENTRIES

Date	h	m	Planets
01	04	12	☽ ♉
03	13	18	☽ ♊
05	19	48	☽ ♋
08	00	27	☽ ♌
10	03	59	☽ ♍
12	06	51	☽ ♎
14	09	31	☽ ♏
16	12	46	☽ ♐
18	17	41	☽ ♑
21	01	20	☽ ♒
21	04	15	☉ ♋
21	12	05	♀ ♊
23	12	00	☽ ♓
26	00	31	☽ ♈
28	12	32	☽ ♉
30	21	59	☽ ♊

LATITUDES

Date	Mercury ☿	Venus ♀	Mars ♂	Jupiter ♃	Saturn ♄	Uranus ♅	Neptune ♆	Pluto ♇
01	02 N 10	01 S 57	00 N 01	01 S 11	01 N 34	00 N 45	01 S 42	11 S 55
04	02 00	01 55	00 S 05	01 11	01 33	00 45	01 42	11 56
07	01 43	01 53	00 11	01 12	01 33	00 44	01 43	11 57
10	01 19	01 50	00 17	01 12	01 33	00 44	01 43	11 58
13	00 49	01 46	00 22	01 13	01 33	00 44	01 43	11 59
16	00 N 13	01 42	00 27	01 14	01 33	00 44	01 43	12 00
19	00 S 29	01 37	00 32	01 15	01 32	00 44	01 43	12 01
22	01 15	01 32	00 37	01 15	01 32	00 44	01 43	12 02
25	02 03	01 26	00 41	01 16	01 32	00 44	01 43	12 03
28	02 51	01 20	00 45	01 17	01 31	00 44	01 43	12 04
31	03 S 36	01 S 14	00 S 49	01 S 17	01 N 31	00 N 44	01 S 43	12 S 05

DATA

Julian Date	2468498
Delta T	+79 seconds
Ayanamsa	24° 30' 32"
Synetic vernal point	04° ♓ 36' 26"
True obliquity of ecliptic	23° 26' 05"

MOON'S PHASES, APSIDES AND POSITIONS ☽

Date	h	m	Phase	Longitude	Eclipse Indicator
04	15	22	●	14 ♊ 12	
11	15	28	☽	20 ♍ 54	
18	13	10	○	27 ♐ 30	
26	10	40	☾	05 ♈ 01	

Day	h	m	
12	04	20	Perigee
25	16	39	Apogee

Day	h	m		
05	04	56	Max dec	26° N 43'
11	16	17	0S	
18	02	32	Max dec	26° S 43'
25	06	44	0N	

LONGITUDES

Date	Chiron ⚷	Ceres ⚳	Pallas ⚴	Juno ⚵	Vesta ⚶	Black Moon Lilith ⚸
01	00 ♎ 11	24 ♍ 10	08 ♌ 18	24 ♍ 25	07 ♌ 58	21 ♓ 52
11	00 ♎ 22	26 ♍ 07	12 ♌ 58	26 ♍ 41	11 ♌ 58	22 ♓ 59
21	00 ♎ 44	28 ♍ 30	17 ♌ 40	27 ♍ 22	16 ♌ 07	24 ♓ 06
31	01 ♎ 16	01 ♎ 14	22 ♌ 23	29 ♍ 22	20 ♌ 26	24 ♓ 12

ASPECTARIAN

h m Aspects

01 Friday
00 20 ☽ ∥ ♆
09 11 ☽ ✶ ☿
11 47 ☿ ∠ ♆
14 28 ☽ ⊥ ☉
14 44 ☽ ✶ ♇
15 05 ☽ ⚼ ♄
16 46 ☽ ☌ ♀

02 Saturday
02 58 ☽ ⊻ ☉
04 18 ☽ ⊻ ♃
05 25 ☽ # ♇
08 15 ☽ ± ♄
11 51 ☽ # ♄
13 43 ☽ Q ♇
14 21 ☽ ☌ ♆
15 18 ☽ ⊼ ♂
15 52 ☽ ⊥ ♃
16 58 ☽ ∥ ☉
17 40 ☽ ∠ ☿
19 33 ☽ ⊼ ♄
23 56 ☉ ✶ ♃

03 Sunday
02 39 ☽ ± ♂
09 00 ☽ ∠ ♃
09 19 ☿ △ ♇
09 21 ☽ □ ♅
12 35 ☽ ⊥ ☿
19 30 ☽ ⚼ ♂
20 33 ☽ ∥ ☿
23 20 ☽ □ ♇

04 Monday
00 58 ☽ ⊻ ☿
06 52 ☽ ⊻ ♀
10 11 ♀ ∥ ♅
13 00 ☽ ✶ ♃
15 22 ☽ ☌ ☉
16 14 ♃ ⚼ ♅
18 30 ☽ Q ♅
18 53 ☽ ⊥ ♀
21 59 ☽ ⊻ ♆
23 05 ☽ △ ♂

05 Tuesday
02 31 ☽ ☍ ♄
07 25 ♇ St R
08 49 ☽ ⊥ ♆
10 57 ☽ Q ♃
12 40 ☽ ∠ ♀
16 08 ☽ ✶ ♅

06 Wednesday
00 55 ☽ ∠ ♆
05 27 ☽ △ ♇
11 37 ☉ Q ♅
12 37 ☽ ☌ ☿
17 31 ☽ ∥ ☿
17 49 ☽ ✶ ♀
18 45 ☽ ∠ ♅
19 17 ☽ □ ♃

07 Thursday
00 56 ☽ ⊻ ☉
03 25 ☽ ✶ ♆
04 46 ☽ □ ♂
07 23 ☽ ⊼ ♄
07 49 ☽ ⚼ ♇
10 32 ☽ ⊥ ♅
10 43 ☽ ∥ ☉
12 12 ☽ ⊥ ☉
13 07 ♀ ⊻ ♃
16 47 ☽ Q ♀
17 47 ☽ ± ♄
19 04 ☽ # ♄
21 00 ☽ ⊻ ♅
23 28 ☽ ± ♇

08 Friday
00 22 ☽ Q ♆
00 34 ☽ # ♇
03 48 ♀ ± ♄
04 59 ☽ ∠ ☉
09 17 ☽ ⚼ ♄
09 51 ☽ ⊼ ♇
12 18 ☽ Q ♂
13 51 ☉ ⊻ ♆
21 31 ☽ ⊻ ☿
23 59 ☽ △ ♃

09 Saturday
02 09 ☽ ∥ ♆
02 46 ☽ □ ♀
06 02 ☽ ∥ ♀
07 27 ☽ □ ♆
08 39 ☽ ⊥ ☿
08 42 ☽ ✶ ☉
09 08 ☽ ✶ ♂
10 57 ☽ △ ♄
14 02 ☽ ∥ ♅
16 06 ☉ △ ♂
17 05 ☿ ∠ ♅

10 Sunday
00 41 ☽ ☌ ♅
01 16 ☽ ∠ ☿
01 57 ☽ ⚼ ♃
06 42 ☽ Q ☉
10 45 ☽ # ♂
11 01 ☽ ∠ ♂
12 51 ☿ □ ♃
13 13 ☽ ☍ ♇
15 22 ☉ ☍ ♄
17 27 ☽ ± ♃
22 32 ♀ Q ♇
23 42 ☽ ∥ ♃

11 Monday
02 29 ☽ ⊥ ♂
03 45 ☽ ⊼ ♃
04 39 ☽ ✶ ☿
10 35 ☽ △ ♀
10 39 ☽ △ ♆
11 27 ♀ ☌ ♆
12 46 ☽ ⊻ ♂
13 43 ☽ □ ♄
15 28 ☽ □ ☉
18 31 ♀ ∥ ♆

12 Tuesday
02 22 ☽ Q ☿
03 42 ☽ ⊻ ♅
09 09 ☽ # ♃
12 04 ☽ ⚼ ♆
13 51 ☽ ⊥ ♅
14 15 ☽ ⚼ ♀
16 00 ☽ ⊼ ♇
16 40 ♀ ⊼ ♂
19 59 ☽ Q ♄
21 59 ♀ ⊼ ♄
22 39 ☽ ∥ ♂

13 Wednesday
02 07 ☽ ± ♇
03 18 ☽ ± ♆
05 05 ☽ ∠ ♅
06 50 ☽ ± ♀
07 01 ☽ ☍ ♃
09 23 ♀ ⊥ ♃
10 40 ☽ □ ☿
11 34 ☉ ∥ ☿
13 27 ☽ ⊼ ♆
16 05 ☽ ☌ ♂
16 08 ☽ ✶ ♄
17 19 ☽ ⚼ ♇
17 53 ☽ ⊼ ♀
18 11 ☽ # ♅
18 27 ♂ ✶ ♄
21 47 ☽ △ ☉

14 Thursday
06 23 ☽ # ♆
06 29 ☽ ✶ ♅
10 25 ☽ # ♀
17 23 ☽ ∠ ♄
18 41 ☽ △ ♇

15 Friday
01 02 ☽ ⚼ ☉
01 49 ☉ ⊥ ♆
02 53 ☽ Q ♅
07 50 ☽ ∥ ♇
08 36 ☽ ⊥ ♄
10 24 ☽ ⊼ ♃
12 50 ☽ ∥ ♄
16 19 ☽ △ ☿
16 28 ☽ ☍ ♆
17 30 ☽ ± ☉
18 48 ☽ ⊻ ♄
19 33 ☿ ✶ ♆
19 41 ☽ # ☿
19 46 ☽ ⊻ ♂
20 46 ☽ ± ♃

16 Saturday
01 17 ☽ # ☉
01 35 ☽ ☍ ♀
04 33 ☽ ⊼ ☉
06 14 ☽ ⊥ ♂
09 47 ☽ □ ♅
12 26 ☽ ⚼ ♃
19 22 ☽ ⚼ ☿
22 04 ☽ ∠ ♂
22 08 ☽ □ ♇

17 Sunday
02 07 ♂ ⚼ ♇
11 45 ☽ ± ☿
14 53 ☽ △ ♃
20 42 ☽ ⊼ ♆
21 45 ☿ ⊼ ♄
22 43 ☽ ☌ ♄
22 46 ☽ ⊼ ☿

18 Monday
00 53 ☽ ✶ ♂
01 41 ☉ Q ♃
05 50 ☽ Q ♇
07 24 ☽ ± ♆
11 10 ☽ ⊼ ♀
13 10 ☽ ☍ ☉
14 44 ☽ △ ♅
19 49 ♀ ± ♂
22 49 ☽ Q ♂
23 03 ☽ ± ♀
23 39 ☽ ⚼ ♆

19 Tuesday
03 27 ☽ ✶ ♇
10 31 ☿ # ♄
11 51 ☉ ✶ ♅
17 10 ☽ ⚼ ♀
18 10 ☽ ⚼ ♅
21 40 ☽ □ ♃

20 Wednesday
03 19 ☽ △ ♆
04 19 ♀ □ ♅
05 00 ☽ ⊻ ♄
05 03 ☽ # ☉
06 06 ☿ ⚼ ♇
07 09 ☽ ∠ ♇
07 11 ☽ ☍ ☿
08 40 ☽ □ ♂
11 00 ☽ ± ♅
16 13 ☽ ⊥ ♄
19 30 ☽ ∥ ♄
22 23 ☽ ⊼ ♅
22 25 ☽ # ☿

21 Thursday
00 09 ☽ ⊥ ♇
00 12 ☽ △ ♀
00 37 ☽ ∥ ♇
01 06 ☽ ⊼ ☉
08 01 ☽ Q ♃
09 17 ☽ ∠ ♄
11 38 ☽ ⊻ ♇
13 02 ☽ # ♀
13 33 ☽ ± ☉
23 54 ♀ ∠ ♃

22 Friday
01 57 ☿ # ♇
05 16 ☽ # ♆
07 33 ☽ ✶ ♃
08 31 ☽ ⚼ ☉
13 00 ☽ □ ♆
14 18 ☽ ✶ ♄
18 04 ☽ ⊼ ☿
19 12 ☉ ⊻ ♀
19 47 ☽ △ ♂
20 17 ☽ # ♅

23 Saturday
06 05 ☽ ± ☿
09 06 ☽ ☍ ♅
12 59 ☽ ∥ ♂
13 35 ☽ ∠ ♃
13 59 ☽ Q ♄
16 48 ☽ △ ☉
17 17 ☽ □ ♀
22 43 ☽ ☌ ♇

24 Sunday
00 12 ☽ ⚼ ☿
00 55 ☽ Q ♆
02 26 ☽ ⚼ ♂
07 46 ☽ # ♃
07 51 ☽ ⊥ ♃
14 58 ☿ St R
19 35 ☉ ∠ ♆
20 06 ☽ ⊻ ♃
21 03 ☽ ± ♂

25 Monday
01 11 ☽ ✶ ♆
02 01 ☽ □ ♄
06 29 ☽ △ ☿
09 30 ☽ ⊼ ♂
09 39 ☽ Q ♀
15 57 ☿ ∥ ♀
21 47 ☽ ⊼ ♅
23 20 ♀ □ ♇

26 Tuesday
06 11 ☽ ∥ ♃
07 37 ☽ ∠ ♆
09 57 ☽ ± ♅
10 40 ☽ □ ☉
11 18 ☽ ⊻ ♇
12 03 ☿ ∠ ♀
12 39 ☽ ✶ ♀
18 28 ☉ △ ♇
23 23 ☽ ⊥ ♇

27 Wednesday
01 48 ☽ ⊥ ♆
03 00 ☽ # ♂
04 09 ☽ ⚼ ♅
09 19 ☽ ☌ ♃
13 50 ☽ ⊻ ♆
14 10 ☽ △ ♄
17 16 ☽ ∥ ♅
17 23 ☽ ∠ ♇
18 13 ☽ □ ☿
22 03 ☽ ∠ ♀
23 29 ☽ ☍ ♂

28 Thursday
01 50 ☽ Q ☉
09 06 ☽ ∥ ♆
10 06 ☽ △ ♅
17 14 ♀ ⚼ ♂
17 45 ☽ ⊥ ♀
19 39 ☽ ⚼ ♄
22 57 ☽ ✶ ♇

29 Friday
03 00 ☽ ∥ ☿
03 23 ☽ ✶ ☉
04 49 ☽ Q ☿
06 33 ☿ ⚼ ♇
06 41 ☽ ⊻ ♀
07 26 ♄ ⊼ ♆
13 01 ☽ ± ♄
13 29 ☽ ∥ ♀
14 25 ☽ # ♇
18 56 ☽ # ♄
20 42 ☽ ⊻ ♃
22 04 ☽ Q ♇

30 Saturday
00 26 ☽ ⊼ ♄
00 34 ☽ ☌ ♆
03 12 ☽ ✶ ☿
06 34 ♀ # ♇
07 32 ☽ ∥ ☉
08 12 ☽ ⊥ ♃
10 26 ☽ ∠ ☉
11 10 ☽ ⊼ ♂
19 53 ☽ □ ♅
22 42 ☽ ± ♂

All ephemeris data is given at 12.00 UT and the Moon's longitude is additionally given for 24.00 UT

JULY 2046

LONGITUDES

Date	Sidereal time h m s	Sun ☉ ° ' "	Moon ☽ ° ' "	Moon ☽ 24.00 ° ' "	Mercury ☿ ° '	Venus ♀ ° '	Mars ♂ ° '	Jupiter ♃ ° '	Saturn ♄ ° '	Uranus ♅ ° '	Neptune ♆ ° '	Pluto ♇ ° '
01	06 38 50	09 ♋ 50 48	07 ♊ 35 14	14 ♊ 10 36	19 ♋ 25	11 ♊ 58	24 ♎ 34	16 ♈ 47	18 ♐ 27	28 ♌ 54	18 ♉ 39	05 ♓ 17
02	06 42 47	10 48 02	20 ♊ 50 59	27 ♊ 36 17	18 R 56	13 10	24 55	16 53	18 R 23	28 57	18 41	05 R 17
03	06 46 43	11 45 16	04 ♋ 26 15	11 ♋ 20 36	18 24	14 22	25 17	17 00	18 19	28 59	18 42	05 16
04	06 50 40	12 42 29	18 ♋ 18 57	25 ♋ 20 51	17 49	15 34	25 39	17 06	18 15	29 02	18 44	05 16
05	06 54 37	13 39 43	02 ♌ 25 47	09 ♌ 33 13	17 14	16 46	26 01	17 12	18 11	29 05	18 45	05 15
06	06 58 33	14 36 57	16 ♌ 42 35	23 ♌ 53 20	16 37	17 59	26 24	17 19	18 08	29 08	18 47	05 14
07	07 02 30	15 34 10	01 ♍ 04 56	08 ♍ 16 49	15 59	19 11	26 47	17 24	18 04	29 11	18 48	05 13
08	07 06 26	16 31 23	15 ♍ 28 31	22 ♍ 39 35	15 22	20 23	27 11	17 30	18 00	29 14	18 50	05 13
09	07 10 23	17 28 36	29 ♍ 49 36	06 ♎ 58 13	14 46	21 35	27 35	17 36	17 57	29 17	18 51	05 12
10	07 14 19	18 25 49	14 ♎ 05 04	21 ♎ 09 55	14 11	22 47	28 00	17 41	17 53	29 20	18 52	05 11
11	07 18 16	19 23 02	28 ♎ 12 30	05 ♏ 12 35	13 38	24 00	28 25	17 46	17 50	29 23	18 54	05 10
12	07 22 12	20 20 14	12 ♏ 10 00	19 ♏ 04 33	13 08	25 12	28 50	17 52	17 46	29 26	18 55	05 09
13	07 26 09	21 17 27	25 ♏ 56 07	02 ♐ 44 32	12 41	26 25	29 15	17 56	17 43	29 29	18 56	05 09
14	07 30 06	22 14 39	09 ♐ 29 42	16 ♐ 11 30	12 18	27 37	29 ♎ 41	18 01	17 40	29 32	18 57	05 08
15	07 34 02	23 11 52	22 ♐ 49 50	29 ♐ 24 40	11 59	28 ♊ 49	00 ♏ 07	18 06	17 37	29 36	18 59	05 07
16	07 37 59	24 09 05	05 ♑ 55 55	12 ♑ 23 34	11 45	00 ♋ 02	00 34	18 10	17 34	29 39	19 00	05 06
17	07 41 55	25 06 18	18 ♑ 47 39	25 ♑ 08 11	11 35	01 14	01 01	18 14	17 31	29 42	19 01	05 05
18	07 45 52	26 03 31	01 ♒ 25 16	07 ♒ 38 59	11 31	02 27	01 28	18 18	17 28	29 45	19 02	05 04
19	07 49 48	27 00 45	13 ♒ 49 30	19 ♒ 57 02	11 D 32	03 40	01 56	18 22	17 25	29 49	19 03	05 03
20	07 53 45	27 58 00	26 ♒ 01 48	02 ♓ 04 06	11 38	04 52	02 24	18 26	17 22	29 52	19 04	05 02
21	07 57 41	28 55 15	08 ♓ 04 15	14 ♓ 02 37	11 50	06 05	02 52	18 30	17 19	29 55	19 05	05 01
22	08 01 38	29 ♋ 52 30	19 ♓ 59 36	25 ♓ 55 38	12 08	07 18	03 20	18 33	17 16	29 ♌ 59	19 06	05 00
23	08 05 35	00 ♌ 49 46	01 ♈ 51 13	07 ♈ 46 50	12 32	08 30	03 49	18 36	17 14	00 ♍ 02	19 07	04 59
24	08 09 31	01 47 03	13 ♈ 43 01	19 ♈ 40 18	13 01	09 43	04 18	18 39	17 11	00 05	19 08	04 58
25	08 13 28	02 44 21	25 ♈ 39 16	01 ♉ 40 28	13 36	10 56	04 48	18 42	17 09	00 09	19 09	04 57
26	08 17 24	03 41 40	07 ♉ 44 28	13 ♉ 51 50	14 17	12 09	05 17	18 44	17 06	00 12	19 10	04 56
27	08 21 21	04 39 00	20 ♉ 03 06	26 ♉ 18 47	15 04	13 22	05 47	18 47	17 04	00 16	19 11	04 55
28	08 25 17	05 36 21	02 ♊ 39 20	09 ♊ 05 10	15 56	14 35	06 18	18 49	17 02	00 19	19 12	04 54
29	08 29 14	06 33 43	15 ♊ 36 38	22 ♊ 13 58	16 54	15 48	06 48	18 51	17 00	00 23	19 13	04 53
30	08 33 10	07 31 05	28 ♊ 57 20	05 ♋ 46 45	17 58	17 01	07 19	18 53	16 58	00 26	19 13	04 52
31	08 37 07	08 ♌ 28 29	12 ♋ 42 09	19 ♋ 43 18	19 ♋ 06	18 ♋ 14	07 ♏ 50	18 ♈ 54	16 ♐ 56	00 ♍ 30	19 ♉ 14	04 ♓ 51

Date	Moon True ☊ ° '	Moon Mean ☊ ° '	Moon ☽ Latitude ° '
01	04 ♒ 39	05 ♒ 44	04 N 23
02	04 R 36	05 41	03 37
03	04 33	05 38	02 38
04	04 31	05 34	01 28
05	04 31	05 31	00 N 11
06	04 D 31	05 28	01 S 07
07	04 32	05 25	02 21
08	04 33	05 22	03 27
09	04 34	05 18	04 20
10	04 35	05 15	04 56
11	04 R 35	05 12	05 14
12	04 34	05 09	05 13
13	04 33	05 06	04 54
14	04 31	05 03	04 19
15	04 29	04 59	03 30
16	04 28	04 56	02 31
17	04 27	04 53	01 26
18	04 27	04 50	00 S 17
19	04 D 27	04 47	00 N 52
20	04 28	04 44	01 57
21	04 28	04 40	02 56
22	04 29	04 37	03 47
23	04 29	04 34	04 27
24	04 29	04 31	04 57
25	04 30	04 28	05 13
26	04 R 30	04 24	05 17
27	04 D 30	04 21	05 06
28	04 30	04 18	04 40
29	04 30	04 15	03 59
30	04 30	04 12	03 05
31	04 ♒ 30	04 ♒ 09	01 N 58

DECLINATIONS

Date	Sun ☉ ° '	Moon ☽ ° '	Mercury ☿ ° '	Venus ♀ ° '	Mars ♂ ° '	Jupiter ♃ ° '	Saturn ♄ ° '	Uranus ♅ ° '	Neptune ♆ ° '	Pluto ♇ ° '
01	23 N 04	25 N 54	18 N 28	21 N 00	10 S 17	05 N 24	21 S 25	12 N 32	15 N 43	20 S 46
02	23 00	26 44	18 18	21 12	10 26	05 26	21 25	12 31	15 43	20 46
03	22 55	26 00	18 10	21 22	10 35	05 29	21 25	12 30	15 44	20 47
04	22 50	23 38	18 04	21 33	10 44	05 31	21 25	12 29	15 44	20 47
05	22 44	19 48	17 58	21 43	10 53	05 33	21 25	12 28	15 44	20 48
06	22 38	14 46	17 54	21 52	11 03	05 35	21 24	12 27	15 45	20 49
07	22 32	08 53	17 51	22 00	11 12	05 37	21 24	12 26	15 45	20 49
08	22 25	02 N 32	17 49	22 08	11 22	05 39	21 24	12 25	15 45	20 50
09	22 18	03 S 54	17 49	22 16	11 31	05 41	21 24	12 24	15 46	20 50
10	22 10	10 06	17 50	22 22	11 41	05 43	21 24	12 23	15 46	20 51
11	22 02	15 43	17 53	22 28	11 51	05 44	21 24	12 22	15 46	20 51
12	21 54	20 27	17 56	22 34	12 01	05 46	21 23	12 21	15 47	20 52
13	21 45	24 00	18 01	22 39	12 11	05 48	21 23	12 20	15 47	20 52
14	21 36	26 08	18 07	22 43	12 21	05 49	21 23	12 18	15 47	20 53
15	21 26	26 45	18 14	22 47	12 32	05 51	21 23	12 17	15 48	20 53
16	21 17	25 49	18 22	22 49	12 42	05 52	21 23	12 16	15 48	20 54
17	21 06	23 32	18 30	22 52	12 52	05 54	21 23	12 15	15 48	20 55
18	20 56	20 07	18 40	22 53	13 03	05 55	21 23	12 14	15 48	20 55
19	20 45	15 51	18 50	22 54	13 13	05 56	21 23	12 13	15 49	20 56
20	20 34	11 00	19 00	22 55	13 24	05 57	21 22	12 12	15 49	20 56
21	20 22	05 49	19 11	22 54	13 34	05 58	21 22	12 10	15 49	20 57
22	20 10	00 S 29	19 22	22 53	13 45	05 59	21 22	12 09	15 49	20 58
23	19 58	04 N 49	19 33	22 52	13 55	06 00	21 22	12 08	15 49	20 58
24	19 46	09 58	19 45	22 49	14 06	06 01	21 22	12 07	15 50	20 59
25	19 33	14 46	19 55	22 46	14 17	06 02	21 22	12 06	15 50	20 59
26	19 19	19 04	20 06	22 43	14 28	06 03	21 22	12 04	15 50	21 00
27	19 06	22 39	20 16	22 39	14 38	06 03	21 22	12 03	15 50	21 00
28	18 52	25 15	20 25	22 34	14 49	06 04	21 22	12 02	15 50	21 01
29	18 38	26 37	20 33	22 28	15 00	06 04	21 22	12 01	15 51	21 02
30	18 23	26 31	20 40	22 22	15 11	06 05	21 22	11 59	15 51	21 02
31	18 N 08	24 N 47	20 N 45	22 N 15	15 S 22	06 N 05	21 S 22	11 N 58	15 N 51	21 S 03

ZODIAC SIGN ENTRIES

Date	h	m	Planets
03	04	13	☽ ♋
05	07	54	☽ ♌
07	10	12	☽ ♍
09	12	17	☽ ♎
11	15	04	☽ ♏
13	19	09	☽ ♐
15	05	13	♂ ♏
16	01	05	☽ ♑
16	11	22	♀ ♋
18	09	17	☽ ♒
20	19	53	☽ ♓
22	15	09	☉ ♌
22	22	30	♅ ♍
23	08	15	☽ ♈
25	20	40	☽ ♉
28	07	00	☽ ♊
30	13	51	☽ ♋

LATITUDES

Date	Mercury ☿ ° '	Venus ♀ ° '	Mars ♂ ° '	Jupiter ♃ ° '	Saturn ♄ ° '	Uranus ♅ ° '	Neptune ♆ ° '	Pluto ♇ ° '
01	03 S 36	01 S 14	00 S 49	01 S 17	01 N 31	00 N 44	01 S 43	12 S 05
04	04 13	01 07	00 53	01 18	01 31	00 44	01 44	12 06
07	04 40	01 00	00 56	01 19	01 30	00 43	01 44	12 06
10	04 52	00 52	01 00	01 20	01 30	00 43	01 44	12 07
13	04 50	00 44	01 03	01 20	01 29	00 43	01 44	12 08
16	04 34	00 37	01 06	01 21	01 29	00 43	01 44	12 09
19	04 07	00 29	01 09	01 22	01 28	00 43	01 44	12 10
22	03 31	00 21	01 11	01 23	01 28	00 43	01 44	12 10
25	02 50	00 13	01 14	01 24	01 27	00 43	01 45	12 11
28	02 05	00 S 05	01 16	01 25	01 27	00 43	01 45	12 11
31	01 S 20	00 N 03	01 S 19	01 S 25	01 N 26	00 N 43	01 S 45	12 S 12

DATA

Julian Date	2468528
Delta T	+79 seconds
Ayanamsa	24° 30' 38"
Synetic vernal point	04° ♓ 36' 21"
True obliquity of ecliptic	23° 26' 04"

MOON'S PHASES, APSIDES AND POSITIONS ☽

Date	h	m	Phase	Longitude ° '	Eclipse Indicator
04	01	39	●	12 ♋ 18	
10	19	53	☽	18 ♎ 45	
18	00	55	○	25 ♑ 37	partial
26	03	19	☾	03 ♉ 21	

Day	h	m		
07	18	12	Perigee	
23	10	31	Apogee	
02	12	59	Max dec	26° N 44'
08	21	26	0S	
15	09	19	Max dec	26° S 45'
22	14	12	0N	
29	22	21	Max dec	26° N 46'

LONGITUDES

Date	Chiron ⚷ ° '	Ceres ⚳ ° '	Pallas ⚴ ° '	Juno ⚵ ° '	Vesta ⚶ ° '	Black Moon Lilith ⚸ ° '
01	01 ♎ 16	01 ♎ 14	22 ♌ 23	29 ♍ 22	20 ♌ 26	25 ♓ 12
11	01 ♎ 59	04 ♎ 16	27 ♌ 07	01 ♎ 40	24 ♌ 51	26 ♓ 19
21	02 ♎ 50	07 ♎ 33	01 ♍ 50	04 ♎ 13	29 ♌ 24	27 ♓ 26
31	03 ♎ 50	11 ♎ 03	06 ♍ 33	06 ♎ 57	04 ♍ 02	28 ♓ 33

ASPECTARIAN

01 Sunday
01 11 ☽ ∠ ♃
04 34 ☽ ⊥ ☉
06 22 ☽ ∠ ☿
07 47 ☽ □ ♇
15 44 ☽ ⚼ ♂
16 27 ☽ ⚺ ☉
20 48 ☽ ☌ ♀
22 15 ☽ ⊥ ☿

02 Monday
04 50 ☽ ⚹ ♃
04 58 ☽ Q ♅
07 36 ☽ ☍ ♄
08 06 ☽ ⚺ ♆
08 39 ☿ ⊥ ♀
08 41 ☽ ⚺ ☿
18 50 ☽ ⊥ ♆
19 27 ☽ △ ♂
22 52 ☿ ⚹ ♆

03 Tuesday
02 23 ☽ Q ♃
02 25 ☽ ⚹ ♅
10 43 ☽ ∠ ♆
13 27 ☽ △ ♇
15 41 ☿ ⊼ ♄
17 39 ♀ ⋕ ♄

04 Wednesday
01 39 ☽ ☌ ☉
04 38 ☽ ∠ ♅
06 51 ☽ ⚺ ♀
09 54 ☽ □ ♃
11 11 ☽ ☌ ☿
11 54 ☽ ⊼ ♄
12 43 ☽ ⚹ ♆
15 19 ☽ ⚼ ♇
18 01 ☽ ∥ ☉
18 05 ☽ ⊥ ♀
20 06 ☽ ⊥ ♅
22 06 ☽ ± ♄

05 Thursday
00 51 ☽ □ ♂
01 32 ☽ ∥ ♀
02 55 ☽ ⋕ ♄
06 20 ☽ ⚺ ♅
06 30 ☽ ⋕ ♇
06 37 ☽ ± ♇
09 10 ☽ Q ♆
10 47 ☽ ∠ ♀
12 38 ☿ □ ♃
13 17 ☽ ⚼ ♄
16 45 ☽ ⊼ ♇
18 01 ☿ ⚺ ♀
18 31 ♀ Q ♅
21 28 ☽ ∥ ☿
21 29 ♀ ⚹ ♃
23 14 ☉ ∠ ♅

06 Friday
07 41 ☽ ∥ ♆
08 02 ☽ Q ♂
08 15 ☽ ⚺ ☉
11 51 ☽ ⚺ ☿
13 01 ☽ △ ♃
14 19 ☽ ⚹ ♀
14 22 ☽ △ ♄
14 55 ♀ ☍ ♄
15 28 ☽ □ ♆
19 00 ☽ ⊥ ☉
21 27 ☽ ⊥ ☿
21 47 ☽ ∥ ♅

07 Saturday
03 05 ☽ ⋕ ♂
04 23 ♀ ⚺ ♆
04 39 ☽ ⚹ ♂
08 49 ☽ ☌ ♅
11 05 ☽ ∠ ☉
11 51 ☽ ∠ ☿
12 10 ☽ Q ♀
14 13 ☽ ⚼ ♃
18 23 ☉ ☌ ☿
18 54 ☽ ☍ ♇

08 Sunday
00 25 ☽ ∥ ♃
05 20 ☽ ± ♃
06 21 ☽ ∠ ♂
11 50 ☽ ⚹ ☿
13 52 ☽ ⚹ ☉
15 24 ☽ ⊼ ♃
16 12 ☽ □ ♄
17 36 ☽ △ ♆
20 57 ☽ □ ♀
21 49 ☽ ⊥ ♂

09 Monday
07 04 ☽ Q ☿
08 08 ☽ ⚺ ♂
11 05 ☽ ⚺ ♅
11 22 ☽ Q ☉
15 20 ☉ □ ♃
15 26 ☉ ∥ ♀
18 46 ☽ ⚼ ♆
18 46 ☽ ⋕ ♃
21 01 ☽ ⊼ ♇
21 12 ☽ ⊥ ♅
22 14 ☽ Q ♄
23 09 ☉ ⊼ ♄

10 Tuesday
06 10 ☿ ∠ ♅
07 06 ☽ ± ♇
09 57 ☽ ± ♆
12 10 ☽ □ ☿
12 25 ☽ ∠ ♅
18 08 ☽ ☍ ♃
18 25 ☽ ⚹ ♄
18 41 ☽ ∥ ♂
19 53 ☽ □ ☉
20 07 ☽ ⊼ ♆
21 21 ☽ ⋕ ♅
22 19 ☽ ⚼ ♇
23 24 ☉ ⚹ ♆

11 Wednesday
04 09 ☽ △ ♀
12 16 ☽ ⋕ ♆
12 21 ☽ ☌ ♂
14 01 ☽ ⚹ ♅
19 53 ☽ ∠ ♄
21 30 ♃ △ ♄
22 26 ☽ ⋕ ☿
23 55 ☽ △ ♇

12 Thursday
06 12 ♀ ⊥ ♆
07 33 ☉ ⚼ ♇
08 17 ☽ ⚼ ♀
10 44 ☽ Q ♅
11 19 ☽ ⊥ ♄
13 37 ☽ △ ☿
14 26 ☽ ∥ ♇
17 36 ☽ ∥ ♄
20 27 ☽ ⋕ ☉
21 42 ☽ ⚺ ♄
21 56 ☽ ⊼ ♃
23 44 ☽ ☍ ♆

13 Friday
01 24 ☽ ± ♀
01 33 ☽ ⋕ ♀
03 15 ☽ △ ☉
08 29 ☽ ± ♃
12 55 ☽ ⊼ ♀
14 59 ☽ ⚼ ☿
18 02 ☽ ⚺ ♂
18 16 ☽ □ ♅

14 Saturday
00 26 ☽ ⚼ ♃
02 47 ♂ ⚹ ♅
04 14 ☽ □ ♇
05 00 ☽ ⊥ ♂
05 46 ♂ ⋕ ♅
06 28 ☽ ± ☿
07 41 ☽ ⚼ ☉
16 54 ☽ ⊼ ☿
21 36 ☽ ∠ ♂

15 Sunday
01 02 ☽ ± ☉
02 36 ☽ ☌ ♄
03 23 ☽ △ ♃
05 01 ☽ ⊼ ♆
12 31 ☽ Q ♇
12 43 ☽ ⊼ ☉
15 55 ☽ ± ♆
20 42 ☉ ⋕ ♄
21 52 ☉ ± ♄
22 32 ☉ ⊥ ♅

16 Monday
00 02 ☽ ☍ ♀
00 23 ☽ △ ♅
01 46 ☽ ⚹ ♂
03 59 ♀ ⚹ ♅
08 25 ☽ ⚼ ♆
10 28 ☽ ⚹ ♇
14 55 ♀ Q ♃
22 39 ☽ ☍ ☿

17 Tuesday
00 46 ☽ Q ♂
04 17 ☽ ⚼ ♅
04 55 ♀ △ ♂
09 35 ☽ ⚺ ♄
10 57 ☽ □ ♃
12 25 ☽ △ ♆
14 26 ☽ ∠ ♇
17 15 ☽ ⋕ ♀
20 52 ☽ ⊥ ♄
21 19 ☽ ± ♅

18 Wednesday
00 55 ☽ ☍ ☉
03 53 ☽ ∥ ♄
06 33 ☽ ⋕ ☉
06 54 ☽ ∥ ♇
07 30 ☽ ⊥ ♇
08 48 ☽ ⊼ ♅
12 06 ☽ □ ♂
13 39 ☉ ⋕ ♇
13 59 ☽ ∠ ♄
14 11 ☽ ⊼ ♀
19 00 ☽ ⚺ ♇
19 50 ☿ St D
20 18 ☽ ⋕ ☿
21 27 ☽ Q ♃

19 Thursday
03 01 ☽ ± ♀
07 30 ☽ ⊼ ☿
12 12 ☽ ⋕ ♆
18 59 ☽ ⚹ ♄
19 17 ☽ ± ☿
19 56 ♀ ∠ ♆
20 57 ☽ ⚹ ♃
22 15 ☽ □ ♆
22 30 ☽ ⚼ ♀

20 Friday
00 51 ☽ ∥ ♂
06 18 ☽ ⋕ ♅
10 28 ♂ ∠ ♄
13 13 ☽ ⚼ ☿
15 16 ♀ △ ♇
16 10 ☽ ⊼ ☉
18 35 ☽ Q ♄
19 39 ☽ ☍ ♅

21 Saturday
01 10 ☽ △ ♂
02 48 ☽ ∠ ♃
05 09 ☽ ± ☉
05 54 ☽ ☌ ♇
07 34 ☽ △ ♀
10 02 ☽ Q ♆
11 19 ☽ ⋕ ♃
14 30 ☉ ± ♇
19 44 ☽ △ ☿
20 55 ☽ ⊥ ♃

22 Sunday
00 47 ☽ ⚼ ☉
06 32 ☽ □ ♄
08 31 ☽ ⚼ ♂
09 04 ☽ ⚺ ♃
10 12 ☽ ⚹ ♆
14 41 ☉ ⚺ ♅

23 Monday
03 29 ☽ ± ♂
08 18 ☽ ⊼ ♅
09 45 ☽ △ ☉
16 09 ☽ ⊼ ♂
16 36 ☽ ∠ ♆
17 26 ☽ ∥ ♃
18 20 ☽ ⚺ ♇
19 30 ☉ Q ♆
20 30 ☽ ± ♅

24 Tuesday
03 00 ☽ □ ♀
06 28 ☽ ⊥ ♇
10 32 ☽ □ ☿
10 50 ☽ ⊥ ♆
13 03 ☉ ∥ ☿
14 47 ☽ ⚼ ♅
18 58 ☽ △ ♄
21 41 ☉ ⚼ ♄
21 59 ☽ ☌ ♃
22 25 ☽ ∥ ♅
22 57 ☽ ⚺ ♆

25 Wednesday
00 35 ☽ ∠ ♇
09 20 ☽ ⋕ ♂
17 37 ☽ ∥ ♆
19 17 ☽ Q ♀
19 26 ♂ △ ♇
21 00 ☽ △ ♅

26 Thursday
00 33 ☽ Q ☿
00 54 ☽ ⚼ ♄
03 19 ☽ □ ☉
06 28 ☽ ⚹ ♇
06 58 ☽ ☍ ♂
13 26 ☽ ∥ ☉
18 35 ☽ ± ♄
18 37 ☽ ∥ ☿
21 36 ☽ ⚹ ♀

27 Friday
00 14 ☽ ⋕ ♇
01 40 ☽ ⚹ ☿
02 42 ☽ ⋕ ♄
05 57 ☽ Q ♇
06 15 ☽ ⊼ ♄
09 32 ☽ ⚺ ♃
10 19 ☽ ☌ ♆
11 55 ☽ ∥ ♀
17 25 ☽ Q ☉
17 57 ☿ ∠ ♅
18 37 ☉ ⊼ ♇
21 06 ☽ ⊥ ♃

28 Saturday
05 35 ☽ ∠ ♀
07 35 ☽ □ ♅
08 31 ☽ ∠ ☿
14 11 ☽ ∠ ♃
16 12 ☽ □ ♇
17 58 ☽ ⚹ ☉
19 05 ☽ ⊼ ♂

29 Sunday
00 12 ☽ ⊥ ♀
02 39 ☽ ⊥ ☿
03 19 ♀ ∠ ♅
06 39 ☽ ± ♂
12 22 ☽ ⚺ ♀
14 05 ☿ ⊼ ♄
14 31 ☽ ☍ ♄
14 33 ☽ ⚺ ☿
17 03 ☽ Q ♅
17 54 ☽ ⚹ ♃
18 33 ☽ ⚺ ♆
23 38 ☽ ∠ ☉
23 40 ☽ ⚼ ♂

30 Monday
01 01 ☉ □ ♂
05 21 ☽ ⊥ ♆
11 02 ♀ ⊼ ♄
14 38 ☽ ⚹ ♅
15 24 ☽ Q ♃
16 52 ☽ ⊥ ☉
21 17 ☽ ∠ ♆
22 23 ☽ △ ♇

31 Tuesday
03 15 ☽ △ ♂
04 09 ☽ ⚺ ☉
07 47 ☿ □ ♃
14 36 ☿ ⚹ ♆
16 49 ☽ ∠ ♅
19 14 ☽ ⊼ ♄
22 22 ☽ ☌ ♀
22 38 ☽ □ ♃
23 11 ☽ ⚹ ♆
23 59 ☽ ☌ ☿

All ephemeris data is given at 12.00 UT and the Moon's longitude is additionally given for 24.00 UT

AUGUST 2046

Date	Sidereal time h m s	Sun ☉ ° ' "	Moon ☽ ° ' "	Moon ☽ 24.00 ° ' "	Mercury ☿ ° '	Venus ♀ ° '	Mars ♂ ° '	Jupiter ♃ ° '	Saturn ♄ ° '	Uranus ♅ ° '	Neptune ♆ ° '	Pluto ♇ ° '
01	08 41 04	09 ♌ 25 54	26 ♋ 49 50	04 ♌ 01 17	20 ♋ 20	19 ♋ 27	08 ♏ 21	18 ♈ 56	16 ♐ 54	00 ♍ 33	19 ♉ 15	04 ♓ 50
02	08 45 00	10 23 20	11 ♌ 17 01	18 ♌ 36 17	21 40	20 40	08 53	18 57	16 R 52	00 37	19 16	04 R 48
03	08 48 57	11 20 46	25 ♌ 58 15	03 ♍ 22 04	23 04	21 53	09 25	18 58	16 50	00 40	19 16	04 47
04	08 52 53	12 18 13	10 ♍ 46 45	18 ♍ 11 25	24 32	23 06	09 57	18 59	16 49	00 44	19 17	04 46
05	08 56 50	13 15 41	25 ♍ 35 07	02 ♎ 57 03	26 06	24 20	10 29	18 59	16 47	00 48	19 17	04 45
06	09 00 46	14 13 10	10 ♎ 16 25	17 ♎ 32 36	27 43	25 33	11 02	19 00	16 46	00 51	19 18	04 44
07	09 04 43	15 10 39	24 ♎ 45 01	01 ♏ 53 17	29 ♋ 25	26 46	11 35	19 00	16 45	00 55	19 18	04 43
08	09 08 39	16 08 09	08 ♏ 57 04	15 ♏ 56 11	01 ♌ 10	28 00	12 08	19 R 00	16 43	00 58	19 19	04 41
09	09 12 36	17 05 40	22 ♏ 50 33	29 ♏ 40 10	02 58	29 ♋ 13	12 41	19 00	16 42	01 02	19 19	04 40
10	09 16 33	18 03 12	06 ♐ 25 06	13 ♐ 05 30	04 50	00 ♌ 26	13 14	19 00	16 41	01 06	19 20	04 39
11	09 20 29	19 00 45	19 ♐ 41 31	26 ♐ 13 21	06 44	01 40	13 48	18 59	16 40	01 10	19 20	04 38
12	09 24 26	19 58 18	02 ♑ 41 15	09 ♑ 05 26	08 40	02 53	14 22	18 58	16 40	01 13	19 20	04 37
13	09 28 22	20 55 53	15 ♑ 26 08	21 ♑ 43 35	10 38	04 07	14 56	18 57	16 39	01 17	19 21	04 35
14	09 32 19	21 53 28	27 ♑ 58 00	04 ♒ 09 37	12 38	05 20	15 31	18 56	16 38	01 21	19 21	04 34
15	09 36 15	22 51 05	10 ♒ 18 39	16 ♒ 25 17	14 39	06 34	16 05	18 55	16 38	01 24	19 21	04 33
16	09 40 12	23 48 42	22 ♒ 29 44	28 ♒ 32 12	16 40	07 47	16 40	18 53	16 37	01 28	19 22	04 32
17	09 44 08	24 46 21	04 ♓ 32 55	10 ♓ 32 04	18 42	09 01	17 15	18 52	16 37	01 32	19 22	04 30
18	09 48 05	25 44 01	16 ♓ 29 55	22 ♓ 26 42	20 44	10 15	17 50	18 50	16 36	01 36	19 22	04 29
19	09 52 02	26 41 43	28 ♓ 22 43	04 ♈ 18 15	22 46	11 28	18 25	18 47	16 36	01 39	19 22	04 28
20	09 55 58	27 39 26	10 ♈ 13 38	16 ♈ 09 13	24 47	12 42	19 01	18 45	16 36	01 43	19 22	04 27
21	09 59 55	28 37 10	22 ♈ 05 26	28 ♈ 02 40	26 48	13 56	19 37	18 43	16 D 36	01 47	19 22	04 25
22	10 03 51	29 ♌ 34 56	04 ♉ 01 24	10 ♉ 02 07	28 ♌ 48	15 10	20 13	18 40	16 36	01 51	19 22	04 24
23	10 07 48	00 ♍ 32 44	16 ♉ 05 19	22 ♉ 11 31	00 ♍ 47	16 24	20 49	18 37	16 37	01 54	19 R 22	04 23
24	10 11 44	01 30 34	28 ♉ 21 17	04 ♊ 35 09	02 46	17 38	21 25	18 34	16 37	01 58	19 22	04 21
25	10 15 41	02 28 25	10 ♊ 53 40	17 ♊ 17 20	04 43	18 51	22 01	18 31	16 37	02 02	19 22	04 20
26	10 19 37	03 26 19	23 ♊ 46 38	00 ♋ 22 00	06 39	20 05	22 38	18 27	16 38	02 06	19 22	04 19
27	10 23 34	04 24 14	07 ♋ 03 47	13 ♋ 52 14	08 34	21 19	23 15	18 24	16 38	02 09	19 22	04 18
28	10 27 31	05 22 10	20 ♋ 47 30	27 ♋ 49 34	10 28	22 33	23 52	18 20	16 39	02 13	19 22	04 16
29	10 31 27	06 20 09	04 ♌ 58 17	12 ♌ 13 16	12 21	23 47	24 29	18 16	16 40	02 17	19 21	04 15
30	10 35 24	07 18 09	19 ♌ 34 00	26 ♌ 59 45	14 12	25 02	25 06	18 11	16 41	02 21	19 21	04 14
31	10 39 20	08 ♍ 16 11	04 ♍ 29 36	12 ♍ 02 31	16 ♍ 02	26 ♌ 16	25 ♏ 44	18 ♈ 07	16 ♐ 42	02 ♍ 24	19 ♉ 21	04 ♓ 12

Date	Moon True ☊ ° '	Moon Mean ☊ ° '	Moon ☽ Latitude ° '
01	04 ♒ 31	04 ♒ 05	00 N 42
02	04 R 31	04 02	00 S 37
03	04 30	03 59	01 56
04	04 30	03 56	03 08
05	04 29	03 53	04 07
06	04 28	03 49	04 49
07	04 27	03 46	05 12
08	04 26	03 43	05 16
09	04 D 26	03 40	05 00
10	04 27	03 37	04 29
11	04 28	03 34	03 43
12	04 29	03 30	02 47
13	04 30	03 27	01 43
14	04 31	03 24	00 S 36
15	04 R 31	03 21	00 N 32
16	04 30	03 18	01 38
17	04 28	03 15	02 38
18	04 25	03 11	03 31
19	04 22	03 08	04 14
20	04 18	03 05	04 47
21	04 14	03 02	05 07
22	04 11	02 59	05 14
23	04 09	02 55	05 07
24	04 08	02 52	04 46
25	04 D 09	02 49	04 12
26	04 10	02 46	03 23
27	04 11	02 43	02 23
28	04 13	02 40	01 N 13
29	04 R 13	02 36	00 S 04
30	04 12	02 33	01 23
31	04 ♒ 10	02 ♒ 30	02 S 38

DECLINATIONS

Date	Sun ☉	Moon ☽	Mercury ☿	Venus ♀	Mars ♂	Jupiter ♃	Saturn ♄	Uranus ♅	Neptune ♆	Pluto ♇
01	17 N 53	21 N 29	20 N 49	22 N 07	15 S 32	06 N 05	21 S 22	11 N 57	15 N 51	21 S 03
02	17 38	16 47	20 52	21 59	15 43	06 06	21 22	11 55	15 51	21 04
03	17 22	11 02	20 52	21 50	15 54	06 06	21 22	11 54	15 51	21 05
04	17 06	04 N 38	20 51	21 41	16 05	06 06	21 22	11 53	15 51	21 05
05	16 50	02 S 01	20 47	21 31	16 16	06 06	21 22	11 52	15 51	21 06
06	16 34	08 30	20 41	21 20	16 27	06 06	21 22	11 50	15 52	21 06
07	16 17	14 25	20 32	21 08	16 38	06 06	21 22	11 49	15 52	21 07
08	16 00	19 27	20 21	20 56	16 48	06 05	21 22	11 48	15 52	21 08
09	15 43	23 19	20 08	20 44	16 59	06 05	21 22	11 46	15 52	21 08
10	15 25	25 47	19 51	20 31	17 10	06 05	21 23	11 45	15 52	21 09
11	15 07	26 45	19 32	20 17	17 21	06 04	21 23	11 44	15 52	21 09
12	14 49	26 11	19 10	20 03	17 31	06 04	21 23	11 42	15 52	21 10
13	14 31	24 15	18 46	19 48	17 42	06 03	21 23	11 41	15 52	21 10
14	14 12	21 09	18 19	19 32	17 53	06 02	21 23	11 40	15 52	21 11
15	13 54	17 08	17 50	19 16	18 03	06 02	21 23	11 38	15 52	21 11
16	13 35	12 28	17 19	18 59	18 14	06 01	21 23	11 37	15 52	21 12
17	13 16	07 23	16 45	18 42	18 24	06 00	21 23	11 36	15 52	21 12
18	12 56	02 S 05	16 10	18 25	18 35	05 59	21 24	11 35	15 52	21 13
19	12 37	03 N 15	15 33	18 06	18 45	05 58	21 24	11 33	15 52	21 14
20	12 17	08 26	14 54	17 48	18 55	05 57	21 24	11 32	15 52	21 14
21	11 57	13 20	14 14	17 28	19 06	05 55	21 24	11 30	15 52	21 15
22	11 37	17 46	13 32	17 09	19 16	05 54	21 24	11 29	15 52	21 15
23	11 17	21 32	12 50	16 49	19 26	05 53	21 25	11 28	15 52	21 16
24	10 56	24 26	12 07	16 28	19 36	05 51	21 25	11 26	15 52	21 16
25	10 35	26 13	11 22	16 07	19 46	05 50	21 25	11 25	15 52	21 17
26	10 15	26 41	10 37	15 45	19 56	05 48	21 25	11 24	15 51	21 18
27	09 54	25 38	09 52	15 23	20 06	05 47	21 26	11 22	15 51	21 18
28	09 32	23 02	09 06	15 01	20 15	05 45	21 26	11 21	15 51	21 19
29	09 11	18 57	08 20	14 38	20 25	05 43	21 26	11 20	15 51	21 19
30	08 50	13 38	07 33	14 14	20 34	05 42	21 26	11 18	15 51	21 20
31	08 N 28	07 N 25	06 N 46	13 N 51	20 S 44	05 N 40	21 S 27	11 N 17	15 N 51	21 S 20

ZODIAC SIGN ENTRIES

Date	h	m	Planets
01	17	18	☽ ♌
03	18	32	☽ ♍
05	19	11	☽ ♎
07	20	08	☿ ♌
07	20	49	☽ ♏
10	00	35	☽ ♐
10	03	24	♀ ♌
12	07	00	☽ ♑
14	15	56	☽ ♒
17	02	55	☽ ♓
19	15	17	☽ ♈
22	03	56	☽ ♉
22	22	24	☉ ♍
23	02	26	☿ ♍
24	15	11	☽ ♊
26	23	20	☽ ♋
29	03	40	☽ ♌
31	04	49	☽ ♍

LATITUDES

Date	Mercury ☿	Venus ♀	Mars ♂	Jupiter ♃	Saturn ♄	Uranus ♅	Neptune ♆	Pluto ♇
01	01 S 05	00 N 06	01 S 19	01 S 26	01 N 26	00 N 43	01 S 45	12 S 12
04	00 S 22	00 13	01 22	01 27	01 25	00 43	01 45	12 12
07	00 N 17	00 21	01 24	01 27	01 25	00 43	01 45	12 13
10	00 49	00 28	01 25	01 28	01 24	00 43	01 46	12 14
13	01 15	00 35	01 27	01 29	01 23	00 43	01 46	12 14
16	01 33	00 42	01 29	01 30	01 23	00 43	01 46	12 14
19	01 43	00 48	01 30	01 31	01 22	00 43	01 46	12 15
22	01 46	00 54	01 32	01 31	01 22	00 43	01 46	12 15
25	01 43	00 59	01 33	01 32	01 21	00 43	01 46	12 15
28	01 35	01 05	01 34	01 33	01 20	00 43	01 47	12 15
31	01 N 22	01 N 09	01 S 35	01 S 34	01 N 20	00 N 43	01 S 47	12 S 15

DATA

Julian Date	2468559
Delta T	+79 seconds
Ayanamsa	24° 30' 44"
Synetic vernal point	04° ♓ 36' 15"
True obliquity of ecliptic	23° 26' 05"

MOON'S PHASES, APSIDES AND POSITIONS ☽

Date	h	m	Phase	Longitude	Eclipse Indicator
02	10	25	●	10 ♌ 20	Total
09	01	16	☽	16 ♏ 40	
16	14	50	○	23 ♒ 56	
24	18	36	☾	01 ♊ 46	
31	18	25	●	08 ♍ 32	

Day	h	m		
04	09	15	Perigee	
20	01	40	Apogee	
05	04	42	0S	
11	15	02	Max dec	26° S 45'
18	21	22	0N	
26	07	32	Max dec	26° N 42'

LONGITUDES

Date	Chiron ⚷	Ceres ⚳	Pallas ⚴	Juno ⚵	Vesta ⚶	Black Moon Lilith ⚸
01	03 ♎ 56	11 ♎ 25	07 ♍ 01	07 ♎ 14	04 ♍ 30	28 ♓ 40
11	05 ♎ 04	15 ♎ 06	11 ♍ 44	10 ♎ 10	09 ♍ 14	29 ♓ 47
21	06 ♎ 17	18 ♎ 46	16 ♍ 25	13 ♎ 14	14 ♍ 02	00 ♈ 54
31	07 ♎ 36	22 ♎ 53	21 ♍ 06	16 ♎ 25	18 ♍ 54	02 ♈ 01

All ephemeris data is given at 12.00 UT and the Moon's longitude is additionally given for 24.00 UT

ASPECTARIAN

h m Aspects

01 Wednesday
00 12 ☽ ꝑ ♇
01 34 ♀ □ ♃
02 20 ☿ ꝑ ♇
05 24 ☽ ± ♄
07 58 ☽ ǁ ♀
08 00 ♀ ⚹ ♆
08 09 ☽ ⊥ ♅
12 41 ☽ # ♄
14 28 ☽ # ♇
15 20 ☽ ± ♇
15 45 ☽ ǁ ☿
18 15 ☽ ⚺ ♅
19 20 ♀ ꝑ ♇
19 24 ☽ Q ♆
20 27 ☽ ꝑ ♄

02 Thursday
01 19 ☽ ⊼ ♇
07 53 ☽ ǁ ☉
07 54 ☽ □ ♂
10 25 ☽ ☌ ☉
16 11 ☽ ǁ ♆
16 36 ☽ # ♂
21 09 ☽ △ ♄

03 Friday
00 35 ☽ △ ♃
01 05 ☽ □ ♆
04 45 ☽ ⚺ ♀
05 25 ♂ # ♆
06 45 ☽ ⚺ ☿
08 24 ☿ ± ♄
08 35 ☽ ǁ ♅
14 26 ☽ Q ♂
15 24 ☽ ⊥ ♀
17 34 ☽ ⊥ ☿
19 40 ☽ ☌ ♅

04 Saturday
00 59 ☽ ꝑ ♃
02 17 ☽ ☍ ♇
06 22 ♀ ± ♄
06 38 ☽ ǁ ♃
07 17 ☽ ∠ ♀
09 46 ☽ ∠ ☿
10 36 ☽ ⚹ ♂
14 38 ☽ ⚺ ☉
15 09 ☿ ⊥ ♅
15 34 ☽ ± ♃
21 45 ☽ □ ♄

05 Sunday
01 02 ☽ ⊥ ☉
01 17 ☽ ⊼ ♃
01 46 ☽ △ ♆
09 46 ☽ ⚹ ♀
11 50 ☽ ∠ ♂
12 56 ☽ ⚹ ☿
16 39 ☽ ∠ ☉
20 31 ☽ ⚺ ♅
21 37 ♀ ⊥ ♅

06 Monday
00 53 ♂ ⊥ ♄
02 12 ☽ ꝑ ♆
02 55 ☽ ⊼ ♇
02 57 ☽ # ♃
02 59 ☽ Q ♄
03 04 ☽ ⊥ ♂
06 22 ☽ ⊥ ♅
06 46 ♀ # ♄
07 07 ☽ Q ♀
10 59 ☽ Q ☿
12 45 ☽ ± ♇
13 17 ☽ ⚺ ♂
16 59 ☽ ± ♆
17 58 ☉ # ♂
18 58 ☽ ⚹ ☉
21 14 ☽ ∠ ♅
22 42 ☽ ⚹ ♄

07 Tuesday
01 08 ☽ # ♅
02 15 ☿ ± ♇
02 25 ☽ ☍ ♃
02 55 ☽ ⊼ ♆
03 37 ☽ ꝑ ♇
14 52 ♀ # ♇
15 42 ☽ □ ♀
16 22 ☽ Q ☉
18 24 ☽ # ♆
19 54 ☽ # ☉
20 55 ☽ □ ☿
22 18 ☽ ǁ ♂
22 24 ☽ ⚹ ♅
23 44 ☽ ∠ ♄

08 Wednesday
04 14 ♃ St R
04 45 ☽ △ ♇
09 20 ☿ ⚺ ♅
14 01 ☿ Q ♆
15 02 ☽ ⊥ ♄
16 43 ☽ # ☿
17 40 ☽ ☌ ♂
18 56 ☽ Q ♅
19 25 ☿ ꝑ ♄
19 56 ☽ # ♀
21 27 ☽ ǁ ♇
22 55 ☽ ǁ ♄
23 17 ☉ ǁ ♆

09 Thursday
01 16 ☽ □ ☉
01 21 ☽ ⚺ ♄
01 28 ♀ ± ♇
02 27 ☉ △ ♄
05 18 ☽ ⊼ ♃
05 52 ☽ ☍ ♆
15 46 ☽ ± ♃

10 Friday
00 18 ☽ △ ♀
02 29 ☽ □ ♅
05 07 ♂ Q ♅
07 40 ☽ ꝑ ♃
08 42 ☽ △ ☿
08 51 ☽ □ ♇
09 42 ☿ ⊼ ♇

11 Saturday
00 49 ☽ ⚺ ♂
01 36 ♀ ⚺ ♅
05 33 ♀ Q ♆
05 55 ☽ ꝑ ♀
06 30 ☽ ☌ ♄
10 40 ☽ △ ☉
10 42 ☽ △ ♃
11 18 ☉ △ ♃
11 21 ☽ ⊼ ♆
12 12 ♀ ꝑ ♄
12 13 ☽ ⊥ ♂
16 23 ☽ ꝑ ☿
17 22 ☽ Q ♇
20 08 ☉ □ ♆
22 22 ☽ ± ♆

12 Sunday
00 06 ☽ ± ♀
05 32 ☽ ∠ ♂
09 15 ☽ △ ♅
11 58 ☽ ± ☿
12 25 ☽ ⊼ ♀
15 05 ☽ ꝑ ♆
15 35 ☽ ⚹ ♇
16 37 ☽ ꝑ ☉

13 Monday
01 15 ☽ ⊼ ☿
10 58 ☽ ± ☉
11 01 ☽ ⚹ ♂
13 37 ☽ ꝑ ♅
14 18 ☽ ⚺ ♄
18 42 ☽ □ ♃
19 27 ☽ △ ♆
19 54 ☽ ∠ ♇
21 11 ♀ ⊼ ♇
23 21 ☽ ⊼ ☉

14 Tuesday
01 45 ☽ ⊥ ♄
06 55 ☽ ± ♅
10 28 ☽ ǁ ♄
11 05 ☽ Q ♂
11 48 ☽ ǁ ♇
13 10 ☽ ⊥ ♇
18 34 ☽ ⊼ ♅
19 06 ☽ ∠ ♄
22 59 ☽ # ♀

15 Wednesday
00 46 ☽ ⚺ ♇
03 52 ☽ ☍ ♀
04 09 ☿ # ♂
05 23 ☽ Q ♃
07 07 ☽ ǁ ♂
07 39 ☽ # ☿
18 51 ☽ # ♆
22 11 ☽ ☍ ☿
23 54 ☽ □ ♂

16 Thursday
00 24 ☽ ⚹ ♄
04 53 ☽ ⚹ ♃
05 48 ☽ □ ♆
06 09 ☽ # ☉
10 07 ♂ ⚺ ♄
11 27 ☿ △ ♄
11 59 ☿ □ ♂
14 50 ☽ ☍ ☉
16 10 ☽ # ♅

17 Friday
00 09 ☽ Q ♄
05 56 ☽ ☍ ♅
10 37 ☽ ∠ ♃
11 55 ☽ ☌ ♇
13 54 ☿ △ ♃
17 38 ☽ Q ♆
18 23 ☽ # ♃
19 53 ☿ □ ♆
21 59 ☽ ⊼ ♀

18 Saturday
03 23 ♀ # ♂
04 38 ☽ ⊥ ♃
11 26 ☽ ± ♀
12 13 ☽ □ ♄
14 50 ☽ △ ♂
16 41 ☽ ⚺ ♃
17 47 ☽ ⚹ ♆
22 17 ☽ ⊼ ☿
23 43 ☿ ǁ ♆

19 Sunday
07 42 ☽ ꝑ ♀
08 18 ☽ ⊼ ☉
12 56 ☽ ± ☿
18 40 ☽ ⊼ ♅
21 31 ☽ ± ☉
22 45 ☽ ꝑ ♂

20 Monday
00 08 ☽ ∠ ♆
00 18 ☽ ⚺ ♇
00 23 ☽ ǁ ♃
02 03 ♂ ⊼ ♃
06 53 ☽ ± ♅
10 55 ☽ ꝑ ☿
12 26 ☽ ⊥ ♇
17 21 ☽ ꝑ ☉
17 36 ☽ △ ♀
17 57 ☽ ± ♂
18 22 ☽ ⊥ ♆
19 11 ♄ St D

21 Tuesday
00 55 ☽ △ ♄
01 13 ☽ ꝑ ♅
02 15 ♂ ☍ ♆
02 51 ☽ ǁ ♅
05 12 ☽ ☌ ♃
05 28 ☽ ǁ ☉
06 30 ☽ ⚺ ♆
06 37 ☽ ∠ ♇
06 44 ☽ ⊼ ♂
16 01 ☽ ǁ ☿
23 25 ☽ △ ☿

22 Wednesday
01 18 ☽ ǁ ♆
02 19 ☽ △ ☉
07 09 ☽ ꝑ ♄
07 36 ☽ △ ♅
08 41 ☽ ǁ ♀
12 45 ☽ ⚹ ♇
13 26 ♆ St R
21 22 ☽ # ♂
21 57 ☉ ǁ ♅

23 Thursday
01 08 ☽ ± ♄
06 14 ☉ ☌ ☿
10 04 ☽ # ♇
11 05 ☽ # ♄
12 34 ☽ Q ♇
12 40 ☽ □ ♀
13 02 ☽ ⊼ ♄
16 13 ♀ △ ♄
16 58 ☽ ⚺ ♃
18 28 ☽ ☌ ♆
21 47 ☽ ☍ ♂

24 Friday
01 58 ☿ ☌ ♅
04 40 ☽ ⊥ ♃
18 36 ☽ □ ☉
19 01 ☽ □ ♅
21 33 ☿ ꝑ ♃
22 00 ☽ ∠ ♃
22 05 ☽ □ ☿
23 32 ☽ □ ♇

25 Saturday
00 11 ☉ ☌ ♅
03 30 ☽ Q ♀
05 32 ♀ △ ♃
07 19 ☿ ☍ ♇
10 26 ☿ ǁ ♅
21 54 ♀ □ ♆
22 46 ☽ ☍ ♄

26 Sunday
02 13 ☽ ⚹ ♃
03 52 ☽ ⚺ ♆
04 29 ☽ ⚹ ♀
04 52 ♀ ǁ ♆
05 10 ☽ Q ♅
07 21 ☽ Q ☉
09 48 ☽ ⊼ ♂
12 20 ☉ ꝑ ♃
13 54 ☽ Q ☿
14 55 ☽ ⊥ ♆
21 18 ☽ ± ♂

27 Monday
00 06 ☽ Q ♃
03 11 ☽ ⚹ ♅
06 53 ☽ ⚹ ☉
07 04 ☽ △ ♇
07 11 ☽ ∠ ♆
09 18 ☉ ☍ ♇
10 17 ☉ ǁ ☿
10 33 ☽ ∠ ♀
14 12 ☽ ꝑ ♂
15 07 ☽ ⚹ ☿

28 Tuesday
03 58 ☽ ⊥ ♀
04 50 ☽ ⊼ ♄
05 48 ☽ ∠ ♅
07 46 ☽ □ ♃
09 23 ☽ ꝑ ♇
09 32 ☽ ⚹ ♆
11 13 ☽ ∠ ☉
15 12 ☽ ± ♄
15 19 ☽ ⚺ ♀
17 31 ☽ △ ♂
21 14 ☽ ∠ ☿
21 19 ☽ ⊥ ♅
22 23 ☽ # ♄
23 06 ☽ # ♇

29 Wednesday
00 44 ☽ ± ♇
03 40 ☽ ⊥ ☉
04 12 ♂ ± ♃
04 33 ☽ # ♂
05 57 ☽ Q ♆
06 28 ☽ ꝑ ♄
07 29 ☽ ⚺ ♅
10 48 ☽ ⊼ ♇
10 55 ☿ ± ♃
14 26 ☽ ⚺ ☉
14 37 ☿ Q ♂
14 38 ☽ ⊥ ☿

30 Thursday
01 59 ☽ ⚺ ☿
02 35 ☽ ǁ ♆
07 18 ☽ △ ♄
09 19 ☽ ǁ ♀
09 46 ☽ △ ♃
11 39 ☽ □ ♆
15 11 ♀ □ ♂
21 21 ☽ ǁ ♅
21 21 ☽ □ ♂
21 38 ☽ ☌ ♀

31 Friday
07 53 ☽ ǁ ☉
08 39 ☽ ☌ ♅
09 49 ☽ ꝑ ♃
11 33 ☽ ☍ ♇
14 40 ☽ ǁ ☿
18 24 ☽ ǁ ♃
18 25 ☽ ☌ ☉
20 45 ☿ □ ♄

SEPTEMBER 2046

LONGITUDES

Date	Sidereal time h m s	Sun ☉ ° ' "	Moon ☽ ° ' "	Moon ☽ 24.00 ° ' "	Mercury ☿ ° '	Venus ♀ ° '	Mars ♂ ° '	Jupiter ♃ ° '	Saturn ♄ ° '	Uranus ♅ ° '	Neptune ♆ ° '	Pluto ♇ ° '
01	10 43 17	09 ♍ 14 14	19 ♍ 37 18	27 ♍ 12 41	17 ♍ 51	27 ♌ 30	26 ♏ 22	18 ♈ 03	16 ♐ 43	02 ♍ 28	19 ♉ 21	04 ♓ 11
02	10 47 13	10 12 19	04 ♎ 47 25	12 ♎ 20 15	19 39	28 44	27 00	17 R 58	16 44	02 32	19 R 20	04 R 10
03	10 51 10	11 10 25	19 ♎ 50 01	27 ♎ 15 42	21 25	29 ♌ 58	27 38	17 53	16 45	02 36	19 20	04 09
04	10 55 06	12 08 33	04 ♏ 36 26	11 ♏ 51 34	23 10	01 ♍ 12	28 16	17 48	16 47	02 39	19 19	04 07
05	10 59 03	13 06 43	19 ♏ 00 36	26 ♏ 03 15	24 54	02 27	28 54	17 43	16 48	02 43	19 19	04 06
06	11 03 00	14 04 54	02 ♐ 59 24	09 ♐ 49 06	26 37	03 41	29 ♏ 33	17 37	16 50	02 47	19 18	04 05
07	11 06 56	15 03 06	16 ♐ 32 32	23 ♐ 09 58	28 18	04 55	00 ♐ 11	17 32	16 51	02 50	19 18	04 04
08	11 10 53	16 01 20	29 ♐ 41 46	06 ♑ 08 20	29 ♍ 58	06 10	00 50	17 26	16 53	02 54	19 17	04 02
09	11 14 49	16 59 36	12 ♑ 30 09	18 ♑ 47 41	01 ♎ 37	07 24	01 29	17 20	16 55	02 58	19 17	04 01
10	11 18 46	17 57 52	25 ♑ 01 22	01 ♒ 11 41	03 15	08 38	02 08	17 14	16 57	03 01	19 16	04 00
11	11 22 42	18 56 11	07 ♒ 19 05	13 ♒ 23 56	04 52	09 53	02 47	17 08	16 59	03 05	19 16	03 59
12	11 26 39	19 54 31	19 ♒ 26 39	25 ♒ 27 35	06 28	11 07	03 27	17 02	17 01	03 09	19 15	03 57
13	11 30 35	20 52 52	01 ♓ 27 01	07 ♓ 25 15	08 02	12 22	04 06	16 55	17 03	03 12	19 14	03 56
14	11 34 32	21 51 16	13 ♓ 22 33	19 ♓ 19 08	09 36	13 36	04 46	16 49	17 06	03 16	19 14	03 55
15	11 38 29	22 49 41	25 ♓ 15 14	01 ♈ 11 01	11 08	14 51	05 26	16 42	17 08	03 20	19 13	03 54
16	11 42 25	23 48 08	07 ♈ 06 44	13 ♈ 02 33	12 39	16 05	06 05	16 35	17 11	03 23	19 12	03 52
17	11 46 22	24 46 37	18 ♈ 58 41	24 ♈ 55 23	14 09	17 20	06 46	16 28	17 13	03 27	19 11	03 51
18	11 50 18	25 45 08	00 ♉ 52 52	06 ♉ 51 27	15 39	18 34	07 26	16 21	17 16	03 30	19 10	03 50
19	11 54 15	26 43 41	12 ♉ 51 25	18 ♉ 53 08	17 07	19 49	08 06	16 14	17 19	03 34	19 10	03 49
20	11 58 11	27 42 16	24 ♉ 56 58	01 ♊ 03 20	18 33	21 04	08 46	16 07	17 22	03 37	19 09	03 48
21	12 02 08	28 40 54	07 ♊ 12 43	13 ♊ 25 33	19 59	22 18	09 27	16 00	17 25	03 41	19 08	03 47
22	12 06 04	29 ♍ 39 33	19 ♊ 42 23	26 ♊ 03 43	21 24	23 33	10 08	15 52	17 28	03 44	19 07	03 46
23	12 10 01	00 ♎ 38 15	02 ♋ 30 03	09 ♋ 01 55	22 47	24 48	10 48	15 45	17 31	03 48	19 06	03 44
24	12 13 58	01 37 00	15 ♋ 39 47	22 ♋ 24 02	24 10	26 02	11 29	15 37	17 34	03 51	19 05	03 43
25	12 17 54	02 35 46	29 ♋ 15 01	06 ♌ 12 57	25 31	27 17	12 10	15 30	17 37	03 55	19 04	03 42
26	12 21 51	03 34 35	13 ♌ 17 53	20 ♌ 29 43	26 50	28 32	12 52	15 22	17 41	03 58	19 03	03 41
27	12 25 47	04 33 25	27 ♌ 48 10	05 ♍ 12 42	28 09	29 ♍ 47	13 33	15 14	17 44	04 01	19 02	03 40
28	12 29 44	05 32 18	12 ♍ 42 35	20 ♍ 16 51	29 ♎ 26	01 ♎ 02	14 14	15 06	17 48	04 05	19 01	03 39
29	12 33 40	06 31 13	27 ♍ 54 21	05 ♎ 33 44	00 ♏ 42	02 16	14 56	14 59	17 51	04 08	18 59	03 38
30	12 37 37	07 ♎ 30 11	13 ♎ 13 37	20 ♎ 52 30	01 ♏ 56	03 ♎ 31	15 ♐ 38	14 ♈ 51	17 ♐ 55	04 ♍ 11	18 ♉ 58	03 ♓ 37

Date	Moon True ☊ ° '	Moon Mean ☊ ° '	Moon ☽ Latitude ° '
01	04 ♒ 06	02 ♒ 27	03 S 42
02	04 R 01	02 24	04 31
03	03 56	02 21	05 01
04	03 51	02 17	05 10
05	03 47	02 14	05 00
06	03 45	02 11	04 31
07	03 D 44	02 08	03 48
08	03 45	02 05	02 54
09	03 46	02 01	01 53
10	03 48	01 58	00 S 47
11	03 R 48	01 55	00 N 19
12	03 47	01 52	01 24
13	03 43	01 49	02 24
14	03 37	01 46	03 17
15	03 30	01 42	04 01
16	03 20	01 39	04 35
17	03 10	01 36	04 56
18	03 01	01 33	05 05
19	02 52	01 30	05 01
20	02 45	01 27	04 43
21	02 40	01 23	04 12
22	02 38	01 20	03 29
23	02 D 38	01 17	02 34
24	02 38	01 14	01 30
25	02 39	01 11	00 N 18
26	02 R 38	01 07	00 S 57
27	02 36	01 04	02 10
28	02 31	01 01	03 16
29	02 24	00 58	04 10
30	02 ♒ 15	00 ♒ 55	04 S 46

DECLINATIONS

Date	Sun ☉ ° '	Moon ☽ ° '	Mercury ☿ ° '	Venus ♀ ° '	Mars ♂ ° '	Jupiter ♃ ° '	Saturn ♄ ° '	Uranus ♅ ° '	Neptune ♆ ° '	Pluto ♇ ° '
01	08 N 06	00 N 42	05 N 59	13 N 26	20 S 53	05 N 38	21 S 27	11 N 16	15 N 51	21 S 21
02	07 44	06 S 03	05 12	13 02	21 02	05 36	21 27	11 14	15 51	21 21
03	07 22	12 24	04 25	12 37	21 11	05 34	21 28	11 13	15 50	21 22
04	07 00	17 55	03 39	12 12	21 20	05 32	21 28	11 12	15 50	21 22
05	06 38	22 16	02 52	11 46	21 29	05 29	21 28	11 10	15 50	21 23
06	06 16	25 11	02 05	11 21	21 38	05 27	21 29	11 09	15 50	21 23
07	05 53	26 32	01 19	10 54	21 46	05 25	21 29	11 08	15 50	21 24
08	05 31	26 20	00 N 33	10 28	21 55	05 23	21 29	11 06	15 50	21 24
09	05 08	24 43	00 S 13	10 01	22 03	05 20	21 30	11 05	15 49	21 24
10	04 45	21 54	00 58	09 34	22 11	05 18	21 30	11 04	15 49	21 25
11	04 23	18 08	01 43	09 07	22 19	05 15	21 31	11 02	15 49	21 25
12	04 00	13 40	02 28	08 39	22 27	05 13	21 31	11 01	15 49	21 26
13	03 37	08 43	03 12	08 12	22 35	05 10	21 31	11 00	15 49	21 26
14	03 14	03 S 30	03 55	07 44	22 42	05 08	21 32	10 59	15 48	21 27
15	02 51	01 N 48	04 39	07 15	22 50	05 05	21 32	10 57	15 48	21 27
16	02 28	07 01	05 21	06 47	22 57	05 02	21 33	10 56	15 48	21 27
17	02 04	12 00	06 03	06 18	23 04	04 59	21 33	10 55	15 47	21 28
18	01 41	16 32	06 45	05 49	23 11	04 57	21 34	10 53	15 47	21 28
19	01 18	20 28	07 26	05 20	23 18	04 54	21 34	10 52	15 47	21 29
20	00 55	23 34	08 06	04 51	23 24	04 51	21 34	10 51	15 47	21 29
21	00 31	25 39	08 46	04 22	23 31	04 48	21 35	10 50	15 46	21 29
22	00 N 08	26 30	09 25	03 52	23 37	04 45	21 35	10 48	15 46	21 30
23	00 S 15	25 59	10 04	03 23	23 43	04 42	21 36	10 47	15 46	21 30
24	00 39	24 00	10 41	02 53	23 49	04 39	21 36	10 46	15 45	21 30
25	01 02	20 36	11 18	02 23	23 54	04 36	21 37	10 45	15 45	21 31
26	01 25	15 55	11 54	01 53	24 00	04 33	21 37	10 44	15 45	21 31
27	01 49	10 12	12 29	01 23	24 05	04 30	21 38	10 42	15 44	21 31
28	02 12	03 N 46	13 04	00 53	24 10	04 27	21 38	10 41	15 44	21 31
29	02 35	03 S 00	13 37	00 23	24 15	04 24	21 39	10 40	15 44	21 32
30	02 S 59	09 S 37	14 S 10	00 S 07	24 S 19	04 N 21	21 S 39	10 N 39	15 N 43	21 S 32

ZODIAC SIGN ENTRIES

Date	h m	Planets
02	04 25	☽ ♎
03	12 34	♀ ♍
04	04 27	☽ ♏
06	06 48	☽ ♐
07	05 03	♂ ♐
08	12 25	☿ ♎
08	12 34	☽ ♑
10	21 40	☽ ♒
13	09 05	☽ ♓
15	21 36	☽ ♈
18	10 14	☽ ♉
20	21 56	☽ ♊
22	20 22	☉ ♎
23	07 22	☽ ♋
25	13 18	☽ ♌
27	15 35	☽ ♍
27	16 16	♀ ♎
28	22 43	☿ ♏
29	15 17	☽ ♎

LATITUDES

Date	Mercury ☿ ° '	Venus ♀ ° '	Mars ♂ ° '	Jupiter ♃ ° '	Saturn ♄ ° '	Uranus ♅ ° '	Neptune ♆ ° '	Pluto ♇ ° '
01	01 N 17	01 N 10	01 S 35	01 S 34	01 N 20	00 N 43	01 S 47	12 S 15
04	01 01	01 14	01 36	01 34	01 19	00 43	01 47	12 15
07	00 42	01 18	01 37	01 35	01 18	00 43	01 47	12 15
10	00 N 21	01 20	01 38	01 36	01 18	00 43	01 47	12 15
13	00 S 01	01 23	01 39	01 36	01 17	00 43	01 47	12 15
16	00 23	01 24	01 39	01 36	01 17	00 43	01 48	12 15
19	00 46	01 25	01 40	01 37	01 16	00 43	01 48	12 15
22	01 10	01 26	01 40	01 37	01 16	00 43	01 48	12 14
25	01 33	01 25	01 40	01 37	01 15	00 43	01 48	12 14
28	01 55	01 25	01 40	01 38	01 14	00 43	01 48	12 14
31	02 S 16	01 N 23	01 S 41	01 S 38	01 N 14	00 N 43	01 S 48	12 S 13

DATA

Julian Date	2468590
Delta T	+79 seconds
Ayanamsa	24° 30' 48"
Synetic vernal point	04° ♓ 36' 11"
True obliquity of ecliptic	23° 26' 05"

MOON'S PHASES, APSIDES AND POSITIONS ☽

Date	h m	Phase	Longitude ° '	Eclipse Indicator
07	09 07	☽	14 ♐ 56	
15	06 39	○	22 ♓ 37	
23	08 16	☾	00 ♋ 29	
30	02 25	●	07 ♎ 07	

Day	h m		
01	14 27	Perigee	
16	09 54	Apogee	
30	00 39	Perigee	
01	14 28	0S	
07	20 46	Max dec	26° S 38'
15	03 51	0N	
22	15 06	Max dec	26° N 31'
29	01 24	0S	

LONGITUDES

Date	Chiron ⚷ ° '	Ceres ⚳ ° '	Pallas ⚴ ° '	Juno ⚵ ° '	Vesta ⚶ ° '	Black Moon Lilith ⚸ ° '
01	07 ♎ 44	23 ♎ 17	21 ♍ 34	16 ♎ 44	19 ♍ 24	02 ♈ 08
11	09 ♎ 06	27 ♎ 21	26 ♍ 14	20 ♎ 01	24 ♍ 20	03 ♈ 15
21	10 ♎ 32	01 ♏ 29	00 ♎ 52	23 ♎ 22	29 ♍ 19	04 ♈ 22
31	11 ♎ 59	05 ♏ 41	05 ♎ 28	26 ♎ 46	04 ♎ 20	05 ♈ 29

All ephemeris data is given at 12.00 UT and the Moon's longitude is additionally given for 24.00 UT

ASPECTARIAN

01 Saturday
00 04 ☽ ± ♃
03 19 ☽ Q ♂
07 24 ☽ □ ♄
08 50 ☽ ☌ ☿
09 31 ☽ ⊼ ♃
11 33 ☽ △ ♆
14 25 ☿ ⊼ ♃
23 07 ☽ ⚹ ♂
23 27 ☿ ∥ ♃

02 Sunday
01 33 ☽ ⊻ ♀
07 51 ☿ △ ♆
08 24 ☽ ⊻ ♅
09 16 ☽ # ☿
10 22 ☽ # ♃
11 01 ☽ ⊼ ♇
11 17 ☽ bQ ♆
11 54 ☽ ⊥ ♀
11 55 ☽ Q ♄
17 50 ☽ # ☉
17 58 ☽ ⊥ ♅
20 32 ☽ ± ♇
21 12 ☽ ⊻ ☉
23 57 ☽ ∠ ♂

03 Monday
01 35 ☽ ± ♆
03 30 ☽ ∠ ♀
07 03 ☽ ⚹ ♄
07 21 ☽ # ♅
07 26 ☽ ⊥ ☉
08 23 ☽ ∠ ♅
08 53 ☽ ☍ ♃
10 54 ☽ bQ ♇
11 11 ☽ ⊼ ♆
12 50 ☽ # ♀
14 53 ☽ ⊻ ☿
15 01 ☽ ⊥ ♂
22 57 ☽ ∠ ☉

04 Tuesday
01 09 ☽ ⊻ ♂
01 54 ☽ ⊥ ☿
02 25 ☽ # ♆
04 08 ☉ ± ♃
05 55 ☽ ⚹ ♀
07 21 ☽ ∠ ♄
08 47 ☽ ⚹ ♅
11 12 ☽ △ ♇
17 54 ♂ ∥ ♇
18 40 ☽ ∠ ☿
22 13 ☽ ⊥ ♄

05 Wednesday
01 22 ☽ ⚹ ☉
03 36 ☽ Q ♀
04 45 ☽ Q ♅
06 25 ☽ ∥ ♇
06 52 ☽ ∥ ♂
06 59 ☽ ∥ ♄
08 16 ☽ ⊻ ♄
09 49 ☽ ⊼ ♃
10 24 ♂ ∥ ♄
10 59 ♃ bQ ♅
12 31 ☽ ☍ ♆
16 47 ♀ bQ ♃
17 31 ♀ ☌ ♅
19 56 ☽ ± ♃
23 09 ☽ Q ☉
23 25 ☽ ⚹ ☿

06 Thursday
05 43 ☽ ☌ ♂
11 22 ☽ bQ ♃
11 38 ☽ □ ♅
13 20 ☽ □ ♀
13 54 ☽ □ ♇
19 33 ♀ ☍ ♇
23 14 ♀ ∥ ♅
23 16 ☽ Q ☿

07 Friday
09 07 ☽ □ ☉
12 34 ☽ ☌ ♄
13 46 ☽ △ ♃
16 58 ☽ ⊼ ♆
21 58 ☽ Q ♇

08 Saturday
03 54 ☽ ± ♆
12 35 ☽ □ ☿
14 13 ☽ ⊻ ♂
17 59 ☽ △ ♅
20 03 ☽ ⚹ ♇
20 32 ☽ bQ ♆
21 39 ☉ ∥ ♃

09 Sunday
01 20 ☽ △ ♀
02 01 ☽ ⊥ ♂
08 37 ☿ ⚹ ♂
10 05 ☉ □ ♄
19 40 ☉ ⊼ ♃
19 59 ☽ ∠ ♂
20 26 ☽ ⊻ ♄
21 08 ☽ □ ♃
21 16 ☽ △ ☉
22 27 ☽ bQ ♅

10 Monday
00 24 ☽ ∠ ♇
00 55 ☽ △ ♆
07 59 ☽ ⊥ ♄
08 27 ☿ ⊻ ♅
09 02 ☽ bQ ♀
10 00 ☽ ∥ ♂
14 48 ☽ ∥ ♄
15 14 ♂ bQ ♃
15 24 ☽ ∥ ♇
15 54 ☽ ± ♅
17 46 ☽ ⊥ ♇
22 52 ☿ ⊼ ♇

11 Tuesday
01 31 ☽ ∠ ♄
02 37 ☽ ⚹ ♂
03 00 ☿ bQ ♆
03 39 ☽ ⊼ ♅
04 30 ☽ ± ♀
04 47 ☽ bQ ☉
05 27 ☽ ⊻ ♇
06 28 ☽ △ ☿
07 45 ☽ Q ♃
13 48 ☿ Q ♄
17 37 ☽ ⊼ ♀
19 56 ☉ △ ♆
23 58 ♂ □ ♅

12 Wednesday
00 03 ☽ ± ☉
00 50 ☽ # ♆
03 36 ☽ Q ♂
07 10 ☽ ⚹ ♄
07 14 ☽ ⚹ ♃
10 21 ♀ ± ♃
11 37 ☽ □ ♆
13 00 ☽ ⊼ ☉
13 20 ♃ △ ♄
16 38 ☽ bQ ☿

13 Thursday
01 07 ☽ # ♅
06 06 ♂ □ ♇
07 11 ☽ Q ♄
12 56 ☽ ∠ ♃
13 21 ☽ ± ☿
14 43 ☽ # ♀
15 33 ☽ ☍ ♅
16 59 ☽ ☌ ♇
17 38 ☽ □ ♂
21 00 ☉ # ☿
23 37 ☽ Q ♆

14 Friday
03 14 ☽ ⊼ ☿
04 34 ☽ # ♃
06 42 ☿ ⊥ ♅
06 53 ☽ ⊥ ♃
10 20 ☽ ∥ ☿
12 31 ☽ ☍ ♀
13 21 ☽ # ☉
16 54 ☿ ± ♇
18 52 ☽ ⊻ ♃
19 32 ☽ □ ♄
23 48 ☽ ⚹ ♆

15 Saturday
06 39 ☽ ☍ ☉
16 26 ☽ ∥ ☉

16 Sunday
01 49 ☿ # ♃
02 51 ☽ ∥ ♃
03 02 ☽ # ☿
04 25 ☽ ⊼ ♅
05 27 ☽ ⊻ ♇
06 07 ☽ ∠ ♆
09 49 ☽ △ ♂
10 58 ☽ ∥ ♀
16 37 ☽ ± ♅
17 35 ☽ ⊥ ♇
20 37 ☿ ± ♆
20 52 ♀ ⊼ ♃

17 Monday
00 18 ☽ ⊥ ♆
00 51 ☽ ☍ ☿
06 39 ☽ ∥ ♅
06 59 ☽ ☌ ♃
08 17 ☽ ⊼ ♀
08 26 ☽ △ ♄
09 50 ♀ □ ♄
10 34 ☉ Q ♂
10 55 ☽ bQ ♅
11 45 ☽ ∠ ♇
12 25 ☽ ⊻ ♆
16 59 ☿ # ♀
17 57 ☽ bQ ♂
21 49 ☽ ± ♀

18 Tuesday
00 45 ☽ ⊼ ☉
07 50 ☽ ∥ ♆
13 10 ☽ ± ♂
13 54 ☽ ± ☉
14 48 ☽ bQ ♄
17 18 ☽ △ ♅
17 56 ☽ ⚹ ♇
18 02 ☽ bQ ♀
22 48 ☿ ☍ ♃
23 28 ♀ △ ♆

19 Wednesday
01 55 ☽ ⊼ ♂
08 54 ☽ ± ♄
09 33 ☽ bQ ☉
15 29 ☿ ⚹ ♄
17 53 ☽ Q ♇
18 40 ☽ ⊻ ♃
19 04 ☽ # ♇
19 45 ☽ # ♄
20 55 ☽ ⊼ ♄
21 38 ☽ ⊼ ☿

20 Thursday
00 32 ☽ ☌ ♆
03 26 ☽ △ ♀
06 28 ☽ ⊥ ♃
10 23 ☽ # ♂
11 07 ☽ ± ☿
12 13 ♀ ∥ ♃
13 09 ☿ ∠ ♅
15 58 ☿ bQ ♇
17 54 ☽ △ ☉
21 45 ☿ ⊼ ♆

21 Friday
00 00 ☽ ∠ ♃
05 06 ☽ □ ♅
05 20 ☽ □ ♇
07 07 ☽ bQ ☿
16 35 ☽ ☍ ♂

22 Saturday
04 46 ☽ ⚹ ♃
07 42 ☽ ☍ ♄
10 53 ☽ ⊻ ♆
15 36 ☽ △ ☿
15 52 ☽ Q ♅
18 18 ♅ ☍ ♇
20 04 ☽ □ ♀
22 12 ☽ ⊥ ♆

23 Sunday
03 15 ☽ Q ♃
08 16 ☽ □ ☉
14 17 ☽ △ ♇
14 24 ☽ ⚹ ♅
14 57 ☽ ∠ ♆

24 Monday
04 04 ☽ ⊼ ♂
08 47 ☽ Q ♀
11 56 ☽ □ ♃
13 35 ☽ # ♂
15 01 ☿ # ♅
15 25 ☽ ⊼ ♄
15 27 ☽ ± ♂
17 28 ☽ bQ ♇
17 44 ☽ ∠ ♅
18 06 ☽ ⚹ ♆
19 37 ☽ Q ☉

25 Tuesday
02 07 ☽ ± ♄
04 46 ☽ □ ☿
05 46 ☽ # ♄
06 26 ☽ # ♇
08 12 ☽ bQ ♂
08 14 ☽ ⚹ ♀
09 19 ☽ ± ♇
09 40 ☽ ⊥ ♅
15 08 ☽ Q ♆
17 51 ☽ bQ ♄
18 14 ☽ ⚹ ☉
19 41 ☽ ⊼ ♇
20 05 ☽ ⊻ ♅

26 Wednesday
11 14 ☽ △ ♂
12 26 ☽ ∠ ♀
12 48 ☽ ∥ ♆
14 37 ☉ ⊼ ♇
14 51 ☽ Q ☿
15 26 ☽ △ ♃
19 21 ☽ △ ♄
21 28 ☽ ∠ ☉
21 35 ☽ □ ♆
22 10 ☉ ⊻ ♅
23 18 ☉ bQ ♆

27 Thursday
00 37 ☉ # ♀
03 40 ☽ # ☿
04 48 ☽ ⊥ ♀
10 01 ☽ ∥ ♅
12 37 ☽ ⚹ ☿
13 19 ☽ ⊥ ☉
15 31 ☽ ⊻ ♀
15 56 ☽ bQ ♃
21 30 ☽ ☍ ♇
22 07 ☽ ☌ ♅
23 43 ☽ ⊻ ☉

28 Friday
04 04 ☿ ∠ ♂
06 18 ☽ ± ♃
09 30 ☽ ∥ ♃
14 33 ☽ □ ♂
15 00 ☽ ∠ ☿
15 47 ☽ ⊼ ♃
17 18 ☽ # ☉
18 39 ☉ Q ♄
20 06 ☽ □ ♄
21 59 ☽ △ ♆
23 05 ☽ ∥ ♀

29 Saturday
03 25 ☽ # ♀
06 30 ☽ ⊥ ☿
10 29 ☽ ∥ ☉
13 16 ♂ △ ♃
16 46 ☽ ⊻ ☿
16 57 ☽ # ♃
19 27 ☽ ☌ ♀
20 15 ☽ Q ♂
20 58 ☽ ⊼ ♇
21 32 ☽ bQ ♆
21 48 ☽ ⊻ ♅

30 Sunday
00 30 ☽ Q ♄
02 25 ☽ ☌ ☉
06 21 ☽ ± ♇
07 14 ☽ ⊥ ♅
11 36 ☽ ± ♆
13 48 ♀ ⊼ ♇
14 31 ☽ ☍ ♃
15 53 ☽ # ♅
15 56 ☽ ⚹ ♂
16 36 ♀ Q ♂
19 23 ☽ ⚹ ♄
20 26 ☽ bQ ♇
20 34 ♀ bQ ♆
21 00 ☽ ⊼ ♆
21 23 ☽ ∠ ♅

OCTOBER 2046

Date	Sidereal time h m s	Sun ☉ ° ' "	Moon ☽ ° ' "	Moon ☽ 24.00 ° ' "	Mercury ☿ ° '	Venus ♀ ° '	Mars ♂ ° '	Jupiter ♃ ° '	Saturn ♄ ° '	Uranus ♅ ° '	Neptune ♆ ° '	Pluto ♇ ° '
01	12 41 33	08 ♎ 29 10	28 ♎ 28 59	06 ♏ 01 44	03 ♏ 08	04 ♎ 46	16 ♐ 19	14 ♈ 43	17 ♐ 59	04 ♍ 15	18 ♉ 57	03 ♓ 36
02	12 45 30	09 28 11	13 ♏ 29 34	20 ♏ 51 31	04 19	06 01	17 01	14 R 35	18 03	04 18	18 R 56	03 R 35
03	12 49 27	10 27 14	28 ♏ 06 50	05 ♐ 15 01	05 29	07 16	17 43	14 27	18 07	04 21	18 55	03 34
04	12 53 23	11 26 19	12 ♐ 15 47	19 ♐ 09 04	06 36	08 31	18 25	14 18	18 11	04 24	18 53	03 33
05	12 57 20	12 25 25	25 ♐ 55 00	02 ♑ 33 52	07 41	09 46	19 08	14 10	18 15	04 28	18 52	03 32
06	13 01 16	13 24 34	09 ♑ 06 03	15 ♑ 32 03	08 44	11 01	19 50	14 02	18 19	04 31	18 51	03 31
07	13 05 13	14 23 44	21 ♑ 52 26	28 ♑ 07 48	09 44	12 16	20 32	13 54	18 23	04 34	18 50	03 30
08	13 09 09	15 22 56	04 ♒ 18 45	10 ♒ 25 54	10 42	13 31	21 15	13 46	18 28	04 37	18 48	03 29
09	13 13 06	16 22 09	16 ♒ 29 51	22 ♒ 31 11	11 37	14 46	21 57	13 38	18 32	04 40	18 47	03 28
10	13 17 02	17 21 25	28 ♒ 30 26	04 ♓ 28 05	12 28	16 00	22 40	13 30	18 36	04 43	18 46	03 27
11	13 20 59	18 20 42	10 ♓ 24 37	16 ♓ 20 25	13 16	17 15	23 23	13 22	18 41	04 46	18 44	03 27
12	13 24 56	19 20 01	22 ♓ 15 51	28 ♓ 11 13	14 01	18 30	24 06	13 14	18 46	04 49	18 43	03 26
13	13 28 52	20 19 22	04 ♈ 06 48	10 ♈ 02 50	14 41	19 45	24 49	13 06	18 50	04 52	18 41	03 25
14	13 32 49	21 18 45	15 ♈ 59 31	21 ♈ 57 00	15 17	21 01	25 32	12 58	18 55	04 55	18 40	03 24
15	13 36 45	22 18 10	27 ♈ 55 28	03 ♉ 55 02	15 48	22 16	26 15	12 50	19 00	04 57	18 38	03 23
16	13 40 42	23 17 37	09 ♉ 55 52	15 ♉ 58 05	16 13	23 31	26 58	12 43	19 05	05 00	18 37	03 23
17	13 44 38	24 17 06	22 ♉ 02 01	28 ♉ 07 27	16 33	24 46	27 42	12 35	19 10	05 03	18 35	03 22
18	13 48 35	25 16 37	04 ♊ 15 00	10 ♊ 24 46	16 45	26 01	28 25	12 27	19 15	05 06	18 34	03 21
19	13 52 31	26 16 11	16 ♊ 37 03	22 ♊ 52 12	16 51	27 16	29 09	12 19	19 20	05 08	18 32	03 21
20	13 56 28	27 15 46	29 ♊ 10 34	05 ♋ 32 33	16 R 50	28 31	29 ♐ 52	12 12	19 25	05 11	18 31	03 20
21	14 00 25	28 15 24	11 ♋ 58 34	18 ♋ 29 04	16 40	29 ♎ 46	00 ♑ 36	12 04	19 30	05 14	18 29	03 19
22	14 04 21	29 ♎ 15 05	25 ♋ 04 29	01 ♌ 45 13	16 22	01 ♏ 01	01 20	11 57	19 36	05 16	18 28	03 19
23	14 08 18	00 ♏ 14 47	08 ♌ 31 40	15 ♌ 24 07	15 55	02 16	02 04	11 50	19 41	05 19	18 26	03 18
24	14 12 14	01 14 32	22 ♌ 22 47	29 ♌ 27 46	15 19	03 31	02 48	11 43	19 46	05 21	18 25	03 17
25	14 16 11	02 14 19	06 ♍ 38 58	13 ♍ 56 08	14 34	04 47	03 32	11 36	19 52	05 24	18 23	03 17
26	14 20 07	03 14 08	21 ♍ 18 48	28 ♍ 46 18	13 41	06 02	04 16	11 29	19 57	05 26	18 21	03 16
27	14 24 04	04 14 00	06 ♎ 17 44	13 ♎ 52 02	12 40	07 17	05 00	11 22	20 03	05 29	18 20	03 16
28	14 28 00	05 13 53	21 ♎ 27 56	29 ♎ 04 06	11 32	08 32	05 45	11 15	20 09	05 31	18 18	03 15
29	14 31 57	06 13 49	06 ♏ 39 10	14 ♏ 11 43	10 19	09 47	06 29	11 08	20 14	05 33	18 16	03 15
30	14 35 54	07 13 46	21 ♏ 40 30	29 ♏ 04 22	09 03	11 02	07 13	11 02	20 20	05 36	18 15	03 15
31	14 39 50	08 ♏ 13 46	06 ♐ 22 22	13 ♐ 33 46	07 ♏ 45	12 ♏ 18	07 ♑ 58	10 ♈ 55	20 ♐ 26	05 ♍ 38	18 ♉ 13	03 ♓ 14

Date	Moon True ☊ ° '	Moon Mean ☊ ° '	Moon ☽ Latitude ° '
01	02 ♒ 05	00 ♒ 52	05 S 02
02	01 R 55	00 48	04 57
03	01 47	00 45	04 32
04	01 42	00 42	03 51
05	01 39	00 39	02 58
06	01 37	00 36	01 57
07	01 D 38	00 33	00 S 52
08	01 R 38	00 29	00 N 14
09	01 37	00 26	01 18
10	01 34	00 23	02 17
11	01 28	00 20	03 10
12	01 20	00 17	03 54
13	01 08	00 13	04 28
14	00 55	00 10	04 50
15	00 41	00 07	04 59
16	00 27	00 04	04 56
17	00 14	00 ♒ 01	04 39
18	00 03	29 ♑ 58	04 09
19	29 ♑ 55	29 54	03 27
20	29 51	29 51	02 34
21	29 49	29 48	01 33
22	29 D 48	29 45	00 N 25
23	29 R 48	29 42	00 S 46
24	29 47	29 38	01 56
25	29 44	29 35	03 01
26	29 39	29 32	03 56
27	29 30	29 29	04 36
28	29 20	29 26	04 57
29	29 08	29 23	04 58
30	28 57	29 19	04 37
31	28 ♑ 47	29 ♑ 16	03 S 59

DECLINATIONS

Date	Sun ☉ ° '	Moon ☽ ° '	Mercury ☿ ° '	Venus ♀ ° '	Mars ♂ ° '	Jupiter ♃ ° '	Saturn ♄ ° '	Uranus ♅ ° '	Neptune ♆ ° '	Pluto ♇ ° '
01	03 S 22	15 S 38	14 S 41	00 S 37	24 S 24	04 N 18	21 S 40	10 N 38	15 N 43	21 S 32
02	03 45	20 36	15 12	01 07	24 28	04 15	21 40	10 36	15 43	21 33
03	04 08	24 09	15 41	01 37	24 32	04 11	21 41	10 35	15 42	21 33
04	04 31	26 04	16 09	02 08	24 36	04 08	21 41	10 34	15 42	21 33
05	04 55	26 20	16 36	02 38	24 40	04 05	21 42	10 33	15 42	21 33
06	05 18	25 04	17 02	03 08	24 43	04 02	21 42	10 32	15 41	21 33
07	05 40	22 30	17 27	03 38	24 46	03 59	21 43	10 31	15 41	21 34
08	06 03	18 57	17 50	04 08	24 49	03 56	21 44	10 30	15 40	21 34
09	06 26	14 39	18 11	04 38	24 52	03 53	21 44	10 29	15 40	21 34
10	06 49	09 51	18 31	05 08	24 54	03 50	21 45	10 27	15 40	21 34
11	07 11	04 S 44	18 49	05 37	24 56	03 47	21 45	10 26	15 39	21 34
12	07 34	00 N 31	19 06	06 07	24 58	03 44	21 46	10 25	15 39	21 34
13	07 56	05 44	19 20	06 36	25 00	03 41	21 46	10 24	15 38	21 35
14	08 19	10 45	19 32	07 06	25 01	03 38	21 47	10 23	15 38	21 35
15	08 41	15 23	19 42	07 35	25 03	03 35	21 47	10 22	15 38	21 35
16	09 03	19 27	19 50	08 04	25 04	03 32	21 48	10 21	15 37	21 35
17	09 25	22 45	19 55	08 33	25 04	03 29	21 49	10 20	15 37	21 35
18	09 47	25 04	19 56	09 01	25 05	03 26	21 49	10 19	15 36	21 35
19	10 08	26 11	19 55	09 30	25 05	03 23	21 50	10 18	15 36	21 35
20	10 30	26 00	19 50	09 58	25 05	03 20	21 50	10 17	15 35	21 35
21	10 51	24 26	19 42	10 26	25 05	03 17	21 51	10 16	15 35	21 35
22	11 12	21 31	19 29	10 54	25 04	03 15	21 51	10 16	15 34	21 35
23	11 33	17 23	19 13	11 22	25 04	03 12	21 52	10 15	15 34	21 35
24	11 54	12 13	18 52	11 49	25 03	03 09	21 52	10 14	15 34	21 35
25	12 15	06 N 16	18 27	12 16	25 01	03 07	21 53	10 13	15 33	21 35
26	12 35	00 S 11	17 57	12 43	25 00	03 04	21 54	10 12	15 33	21 35
27	12 56	06 44	17 23	13 09	24 58	03 01	21 54	10 11	15 32	21 35
28	13 16	12 58	16 46	13 35	24 56	02 59	21 55	10 10	15 32	21 35
29	13 36	18 25	16 05	14 01	24 54	02 56	21 55	10 10	15 31	21 35
30	13 55	22 38	15 22	14 26	24 51	02 54	21 56	10 09	15 31	21 35
31	14 S 15	25 S 17	14 S 37	14 S 51	24 S 48	02 N 52	21 S 56	10 N 08	15 N 30	21 S 35

ZODIAC SIGN ENTRIES

Date	h	m	Planets
01	14	24	☽ ♏
03	15	09	☽ ♐
05	19	21	☽ ♑
08	03	37	☽ ♒
10	15	00	☽ ♓
13	03	40	☽ ♈
15	16	10	☽ ♉
18	03	41	☽ ♊
20	13	34	☽ ♋
20	16	08	♂ ♑
21	16	29	♀ ♏
22	20	52	☽ ♌
23	06	03	☉ ♏
25	00	54	☽ ♍
27	01	58	☽ ♎
29	01	28	☽ ♏
31	01	31	☽ ♐

LATITUDES

Date	Mercury ☿ ° '	Venus ♀ ° '	Mars ♂ ° '	Jupiter ♃ ° '	Saturn ♄ ° '	Uranus ♅ ° '	Neptune ♆ ° '	Pluto ♇ ° '
01	02 S 16	01 N 23	01 S 41	01 S 38	01 N 14	00 N 43	01 S 48	12 S 13
04	02 35	01 22	01 41	01 38	01 13	00 44	01 48	12 13
07	02 52	01 19	01 40	01 38	01 13	00 44	01 48	12 13
10	03 06	01 16	01 40	01 38	01 12	00 44	01 49	12 12
13	03 15	01 13	01 40	01 37	01 12	00 44	01 49	12 12
16	03 17	01 09	01 40	01 37	01 11	00 44	01 49	12 11
19	03 11	01 04	01 39	01 37	01 11	00 44	01 49	12 10
22	02 53	00 59	01 39	01 37	01 10	00 44	01 49	12 10
25	02 21	00 54	01 38	01 36	01 10	00 44	01 49	12 09
28	01 33	00 48	01 37	01 36	01 10	00 44	01 49	12 08
31	00 S 34	00 N 42	01 S 37	01 S 35	01 N 09	00 N 44	01 S 49	12 S 08

DATA

Julian Date	2468620
Delta T	+79 seconds
Ayanamsa	24° 30' 51"
Synetic vernal point	04° ♓ 36' 08"
True obliquity of ecliptic	23° 26' 05"

MOON'S PHASES, APSIDES AND POSITIONS ☽

Date	h	m	Phase	Longitude ° '	Eclipse Indicator
06	20	41	☽	13 ♑ 46	
14	23	41	○	21 ♈ 48	
22	20	08	☾	29 ♋ 35	
29	11	17	●	06 ♏ 12	

Day	h	m		
13	11	10	Apogee	
28	11	33	Perigee	
05	03	49	Max dec	26° S 25'
12	09	40	0N	
19	20	42	Max dec	26° N 17'
26	11	21	0S	

LONGITUDES

Date	Chiron ⚷ ° '	Ceres ⚳ ° '	Pallas ⚴ ° '	Juno ⚵ ° '	Vesta ⚶ ° '	Black Moon Lilith ⚸ ° '
01	11 ♎ 59	05 ♏ 41	05 ♎ 28	26 ♎ 46	04 ♎ 20	05 ♈ 29
11	13 ♎ 26	09 ♏ 56	10 ♎ 03	00 ♏ 12	09 ♎ 24	06 ♈ 36
21	14 ♎ 53	14 ♏ 12	14 ♎ 35	03 ♏ 40	14 ♎ 28	07 ♈ 43
31	16 ♎ 18	18 ♏ 30	19 ♎ 04	07 ♏ 08	19 ♎ 34	08 ♈ 50

All ephemeris data is given at 12.00 UT and the Moon's longitude is additionally given for 24.00 UT

ASPECTARIAN

h m	Aspects
01 Monday	
01 29	♀ ⊻ ♅
07 35	☽ ∥ ☿
08 34	☿ ∠ ♄
12 22	☽ # ♆
16 43	☽ ∠ ♂
19 10	☽ ∠ ♄
20 02	☽ ☌ ☿
20 07	☽ △ ♇
21 04	☿ △ ♇
21 11	☽ ✶ ♅
22 53	☽ ⊻ ♀
02 Tuesday	
05 03	☽ ⊻ ☉
07 49	☽ ⊥ ♂
09 24	☽ ⊥ ♀
09 39	☽ ⊥ ♄
11 28	☿ ✶ ♅
12 34	♀ Q ♄
13 44	☽ ⊼ ♃
14 41	☉ ± ♇
15 26	☽ ⊥ ☉
16 34	☽ Q ♅
17 29	☽ ∥ ♇
18 01	☽ ⊻ ♂
18 18	☽ ∥ ♄
19 26	☽ ⊻ ♄
20 50	☽ ☍ ♆
23 26	☽ ± ♃
03 Wednesday	
01 24	☽ ∠ ♀
07 15	☽ ∠ ☉
09 23	☉ ⊥ ♅
13 02	☿ # ♆
14 12	☽ bQ ♃
14 52	☉ # ♃
15 38	☽ ∥ ♂
21 08	☽ □ ♇
22 31	☽ □ ♅
04 Thursday	
01 27	☽ ⊻ ☿
02 44	♂ ☌ ♄
04 56	☽ ✶ ♀
10 28	☽ ✶ ☉
12 37	☽ ⊥ ☿
15 30	☽ △ ♃
22 20	☽ ☌ ♄
23 18	☽ ☌ ♂
23 32	☽ ⊼ ♆
05 Friday	
03 32	♂ ⊼ ♆
03 51	☽ Q ♀
04 13	☽ Q ♇
05 44	☽ ∠ ☿
07 40	♀ ± ♇
09 07	☽ Q ☉
10 08	☽ ± ♆
22 39	☉ ± ♆
06 Saturday	
01 45	☽ ✶ ♇
01 59	♀ ⊥ ♅
02 21	☽ bQ ♆
03 31	☽ △ ♅
11 15	☽ ✶ ☿
15 55	☽ □ ♀
15 56	☽ ∥ ♂
20 41	☽ □ ☉
21 06	☽ □ ♃
07 Sunday	
01 29	☉ ☍ ♃
05 20	☽ ⊻ ♄
05 36	☽ ∠ ♇
06 13	☽ △ ♆
07 35	☽ bQ ♅
09 18	☽ ⊻ ♂
11 42	☽ Q ☿
16 50	☽ ⊥ ♄
17 52	☽ ∥ ♄
19 00	☽ ∥ ♇
21 28	☽ ⊥ ♂
22 43	♀ ± ♆
22 47	☽ ⊥ ♇
08 Monday	
00 53	☽ ± ♅
03 21	♀ # ♃
07 06	☽ Q ♃
10 20	☽ ∠ ♄
10 24	☽ ⊻ ♇
12 35	☽ ⊼ ♅
16 01	☽ ∠ ♂
16 32	♀ ☍ ♃
18 03	☽ ∥ ☿
19 57	♂ Q ♇
09 Tuesday	
01 33	☽ □ ☿
06 23	☽ ✶ ♃
06 36	☽ # ♆
08 09	☽ △ ♀
11 43	☽ △ ☉
16 04	☽ ✶ ♄
16 32	☽ □ ♆
23 34	☽ ✶ ♂
10 Wednesday	
09 01	☽ # ♅
11 59	☽ ∠ ♃
16 15	☽ Q ♄
16 23	☿ ⊥ ♄
17 37	☽ bQ ♀
20 27	☽ bQ ☉
21 57	☽ ☌ ♇
11 Thursday	
00 33	☽ ☍ ♅
01 12	☽ Q ♂
01 22	☽ ∥ ☉
04 35	☽ Q ♆
05 55	☽ ⊥ ♃
08 15	☽ ∥ ♀
13 55	☽ ± ♀
14 22	☉ bQ ♇
14 29	☿ ⊼ ♃
16 16	☽ ± ☉
16 26	☽ # ♃
17 55	☽ ⊻ ♃
18 12	☽ △ ☿
20 55	☉ ✶ ♄
21 17	☉ ⊼ ♆
12 Friday	
00 36	♄ ⊼ ♆
03 30	☽ ⊼ ♀
04 49	☽ ✶ ♆
04 51	☽ □ ♄
05 31	☽ ⊼ ☉
10 32	♀ bQ ♇
15 51	♀ ⊼ ♆
15 57	☽ □ ♂
17 10	♀ ✶ ♄
13 Saturday	
00 16	☉ ∠ ♅
02 31	☽ bQ ☿
02 35	☽ ∥ ♃
07 54	♂ ± ♆
10 35	☽ ⊻ ♇
11 09	☽ ∠ ♆
13 31	☽ ⊼ ♅
14 05	♀ ∠ ♅
16 33	☽ # ♀
21 46	☽ ± ☿
22 43	☽ ⊥ ♇
23 15	☽ # ☉
14 Sunday	
01 42	☽ ± ♅
05 18	☽ ⊥ ♆
05 58	☽ ☌ ♃
10 16	☽ ∥ ♅
10 30	☽ ⊼ ☿
16 51	☽ ∠ ♇
17 23	☽ ⊻ ♆
17 56	☽ △ ♄
19 56	☽ bQ ♅
23 17	☽ ☍ ♀
23 41	☽ ☍ ☉
15 Monday	
08 26	☽ △ ♂
13 19	☽ ∥ ♆
16 02	☉ ☌ ♀
22 56	☽ ✶ ♇
16 Tuesday	
00 15	☽ bQ ♄
02 08	☽ △ ♅
14 30	☽ # ☿
16 20	☽ bQ ♂
17 28	☽ ⊻ ♃
18 19	☽ ± ♄
22 49	☽ Q ♇
17 Wednesday	
00 52	☽ ☍ ☿
02 42	☽ # ♇
04 24	☽ # ♄
05 13	☽ ☌ ♆
05 15	☽ ⊥ ♃
06 18	☽ ⊼ ♄
09 25	♃ ⊥ ♆
11 18	☽ ± ♂
16 50	☽ ⊼ ☉
18 00	☽ ⊼ ♀
22 49	☽ ∠ ♃
23 52	☽ ⊼ ♂
18 Thursday	
05 40	☽ ± ☉
07 08	☽ ± ♀
10 15	☽ □ ♇
12 16	☽ # ♂
13 40	☽ □ ♅
19 Friday	
00 46	☽ bQ ☉
02 39	☽ bQ ♀
03 47	☽ ✶ ♃
12 28	☽ ⊼ ☿
15 42	☽ ⊻ ♆
17 16	☽ ☍ ♄
18 55	☿ St R
22 46	☉ # ♅
23 59	☽ ± ☿
20 Saturday	
00 34	☽ Q ♅
02 38	☽ Q ♃
03 09	☽ ⊥ ♆
08 04	☽ △ ☉
10 37	☽ △ ♀
13 24	☽ ☍ ♂
16 58	☽ bQ ☿
19 50	☽ △ ♇
20 10	☽ ∠ ♆
23 22	☽ ✶ ♅
21 Sunday	
03 54	♀ # ♅
04 21	☽ # ♂
12 11	☽ □ ♃
20 30	☽ △ ☿
23 42	☽ bQ ♇
23 59	☽ ✶ ♆
22 Monday	
01 58	☽ ⊼ ♄
03 15	☽ ∠ ♅
09 43	☽ # ♄
11 33	☽ # ♇
12 38	☿ ∠ ♂
12 57	☽ ± ♄
16 02	☽ ± ♇
19 35	☽ ⊥ ♅
20 08	☽ □ ☉
21 40	☽ Q ♆
23 47	☽ □ ♀
23 54	☽ ⊼ ♂
23 Tuesday	
01 27	☽ # ☿
02 31	♀ ✶ ♂
02 46	☽ ⊼ ♇
05 10	☽ bQ ♄
06 19	☽ ⊻ ♅
11 08	☽ ± ♂
17 44	☽ △ ♃
20 58	☽ ∥ ♆
24 Wednesday	
00 23	☽ □ ☿
03 42	☽ bQ ♂
05 12	☽ □ ♆
06 13	☽ Q ☉
07 31	☽ △ ♄
07 36	♀ △ ♇
10 24	☽ Q ♀
13 15	☽ # ☉
13 35	☽ # ♀
19 18	☽ bQ ♃
20 20	☽ ∥ ♅
25 Thursday	
03 57	♂ ✶ ♇
04 06	☽ ✶ ☉
05 34	☽ Q ☿
06 24	☽ ☍ ♇
06 32	☽ △ ♂
07 18	♂ bQ ♆
08 22	☉ ∥ ♀
08 35	☽ ✶ ♀
09 55	☽ ☌ ♅
10 16	☽ ± ♃
13 53	♀ ∠ ♄
20 06	☽ ⊼ ♃
23 59	☽ ∥ ♃
26 Friday	
00 19	☽ ✶ ☿
00 20	♀ ✶ ♅
05 31	☿ ⊥ ♄
06 39	☽ ∠ ☉
07 13	☽ △ ♆
09 48	☽ □ ♄
11 30	☽ ∠ ♀
12 54	☉ △ ♇
17 50	♃ ± ♅
22 13	☽ ⊥ ☉
22 29	☽ # ♃
23 07	☽ ∠ ☿
27 Saturday	
03 17	☽ ⊥ ♀
07 11	☽ ⊼ ♇
07 17	☽ bQ ♆
08 29	☽ ⊻ ☉
09 50	☽ □ ♂
10 42	☽ ⊻ ♅
12 33	☽ ⊥ ☿
13 42	☽ ⊻ ♀
14 48	☽ Q ♄
16 43	☽ ± ♇
19 59	☽ ☍ ♃
20 14	☽ ⊥ ♅
21 25	☽ ⊻ ☿
21 33	☽ ± ♆
28 Sunday	
01 02	☽ # ♅
04 15	♂ △ ♅
06 56	☽ bQ ♇
07 01	☽ ⊼ ♆
09 43	☉ ∠ ♄
09 54	☽ ✶ ♄
10 30	☽ ∠ ♅
13 19	☽ ∥ ☉
14 45	☽ ∥ ♀
15 47	☽ Q ♂
18 18	☿ ⊼ ♃
19 08	☉ ✶ ♅
22 45	☽ # ♆
29 Monday	
02 25	☽ ∥ ☿
06 37	☽ △ ♇
09 45	☽ ∠ ♄
10 15	☽ ✶ ♅
11 17	☽ ☌ ☉
11 43	☽ ✶ ♂
17 04	☿ ☌ ♀
17 22	☽ ☌ ☿
17 26	☽ ☌ ♀
19 05	☽ ⊼ ♃
30 Tuesday	
00 09	☽ ⊥ ♄
04 35	☽ ± ♃
05 11	☽ ∥ ♇
05 25	☽ Q ♅
06 30	☽ ☍ ♆
07 03	☿ # ♆
07 19	☽ ∥ ♄
09 50	☽ ⊻ ♄
11 24	☉ ✶ ♂
11 51	♀ ⊼ ♃
12 56	☽ ∠ ♂
19 00	☽ bQ ♃
31 Wednesday	
04 21	☽ ⊥ ♂
06 19	☽ ∥ ♂
06 49	☽ □ ♇
07 02	☉ ☌ ☿
07 15	☿ ∥ ♀
09 32	☿ ✶ ♂
10 46	☽ □ ♅
14 06	☽ ⊻ ☿
14 47	☽ ⊻ ♂
15 19	☽ ⊻ ☉
19 31	☽ △ ♃
20 31	☉ ∥ ☿
22 49	☽ ⊻ ♀
23 19	☽ ⊥ ☿

NOVEMBER 2046

LONGITUDES

Date	Sidereal time h m s	Sun ☉ ° ' "	Moon ☽ ° ' "	Moon ☽ 24.00 ° ' "	Mercury ☿ ° '	Venus ♀ ° '	Mars ♂ ° '	Jupiter ♃ ° '	Saturn ♄ ° '	Uranus ♅ ° '	Neptune ♆ ° '	Pluto ♇ ° '
01	14 43 47	09 ♏ 13 47	20 ♐ 38 04	27 ♐ 34 58	06 ♏ 29	13 ♏ 33	08 ♑ 43	10 ♈ 49	20 ♐ 32	05 ♍ 40	18 ♉ 11	03 ♓ 14
02	14 47 43	10 13 50	04 ♑ 24 25	11 ♑ 06 32	05 R 17	14 48	09 27	10 R 43	20 38	05 42	18 R 10	03 R 13
03	14 51 40	11 13 54	17 ♑ 41 35	24 ♑ 09 59	04 11	16 03	10 12	10 37	20 44	05 44	18 08	03 13
04	14 55 36	12 14 00	00 ♒ 32 14	06 ♒ 48 55	03 12	17 18	10 57	10 32	20 50	05 46	18 06	03 13
05	14 59 33	13 14 08	13 ♒ 00 40	19 ♒ 08 08	02 23	18 34	11 42	10 26	20 56	05 48	18 05	03 12
06	15 03 29	14 14 17	25 ♒ 12 00	01 ♓ 12 54	01 45	19 49	12 27	10 21	21 02	05 50	18 03	03 12
07	15 07 26	15 14 28	07 ♓ 11 30	13 ♓ 08 23	01 19	21 04	13 12	10 15	21 08	05 52	18 01	03 12
08	15 11 23	16 14 40	19 ♓ 04 10	24 ♓ 59 22	01 03	22 19	13 57	10 10	21 14	05 54	18 00	03 12
09	15 15 19	17 14 53	00 ♈ 54 28	06 ♈ 49 54	01 D 00	23 34	14 42	10 05	21 21	05 55	17 58	03 11
10	15 19 16	18 15 08	12 ♈ 46 05	18 ♈ 43 18	01 08	24 50	15 27	10 00	21 27	05 57	17 56	03 11
11	15 23 12	19 15 25	24 ♈ 41 53	00 ♉ 42 01	01 26	26 05	16 12	09 56	21 33	05 59	17 55	03 11
12	15 27 09	20 15 43	06 ♉ 43 55	12 ♉ 47 43	01 53	27 20	16 57	09 51	21 40	06 00	17 53	03 11
13	15 31 05	21 16 03	18 ♉ 53 32	25 ♉ 01 27	02 30	28 35	17 43	09 47	21 46	06 02	17 51	03 11
14	15 35 02	22 16 25	01 ♊ 11 34	07 ♊ 23 55	03 15	29 ♏ 50	18 28	09 43	21 52	06 04	17 50	03 11
15	15 38 58	23 16 49	13 ♊ 38 35	19 ♊ 55 38	04 06	01 ♐ 06	19 14	09 39	21 59	06 05	17 48	03 11
16	15 42 55	24 17 14	26 ♊ 15 12	02 ♋ 37 22	05 04	02 21	19 59	09 36	22 06	06 06	17 46	03 D 11
17	15 46 52	25 17 41	09 ♋ 02 18	15 ♋ 30 11	06 07	03 36	20 45	09 32	22 12	06 08	17 45	03 11
18	15 50 48	26 18 09	22 ♋ 01 13	28 ♋ 35 39	07 15	04 51	21 30	09 29	22 19	06 09	17 43	03 11
19	15 54 45	27 18 40	05 ♌ 13 43	11 ♌ 55 40	08 27	06 07	22 16	09 26	22 25	06 10	17 41	03 11
20	15 58 41	28 19 12	18 ♌ 41 45	25 ♌ 32 11	09 42	07 22	23 02	09 23	22 32	06 12	17 40	03 11
21	16 02 38	29 ♏ 19 46	02 ♍ 27 08	09 ♍ 26 42	11 01	08 37	23 48	09 20	22 39	06 13	17 38	03 11
22	16 06 34	00 ♐ 20 22	16 ♍ 30 52	23 ♍ 39 31	12 22	09 52	24 33	09 18	22 45	06 14	17 36	03 11
23	16 10 31	01 20 59	00 ♎ 52 24	08 ♎ 09 07	13 45	11 08	25 19	09 15	22 52	06 15	17 35	03 12
24	16 14 27	02 21 38	15 ♎ 29 06	22 ♎ 51 38	15 10	12 23	26 05	09 13	22 59	06 16	17 33	03 12
25	16 18 24	03 22 19	00 ♏ 15 54	07 ♏ 40 54	16 36	13 38	26 51	09 11	23 06	06 17	17 31	03 12
26	16 22 21	04 23 02	15 ♏ 05 36	22 ♏ 28 57	18 04	14 53	27 37	09 10	23 13	06 18	17 30	03 12
27	16 26 17	05 23 46	29 ♏ 49 51	07 ♐ 07 19	19 32	16 09	28 23	09 08	23 19	06 19	17 28	03 12
28	16 30 14	06 24 31	14 ♐ 20 27	21 ♐ 28 29	21 02	17 24	29 09	09 07	23 26	06 19	17 27	03 13
29	16 34 10	07 25 18	28 ♐ 30 47	05 ♑ 26 55	22 32	18 39	29 ♑ 55	09 06	23 33	06 20	17 25	03 13
30	16 38 07	08 ♐ 26 06	12 ♑ 16 38	18 ♑ 59 49	24 ♏ 03	19 ♐ 54	00 ♒ 42	09 ♈ 05	23 ♐ 40	06 ♍ 21	17 ♉ 23	03 ♓ 13

Date	Moon True ☊ ° '	Moon Mean ☊ ° '	Moon ☽ Latitude ° '
01	28 ♑ 40	29 ♑ 13	03 S 06
02	28 R 36	29 10	02 04
03	28 34	29 07	00 S 57
04	28 D 33	29 04	00 N 11
05	28 33	29 00	01 16
06	28 R 33	28 57	02 16
07	28 31	28 54	03 10
08	28 26	28 51	03 54
09	28 19	28 48	04 28
10	28 09	28 44	04 51
11	27 57	28 41	05 01
12	27 44	28 38	04 58
13	27 31	28 35	04 41
14	27 19	28 32	04 11
15	27 09	28 29	03 29
16	27 02	28 25	02 36
17	26 58	28 22	01 34
18	26 57	28 19	00 N 26
19	26 D 57	28 16	00 S 44
20	26 58	28 13	01 54
21	26 R 58	28 10	02 58
22	26 57	28 06	03 53
23	26 53	28 03	04 35
24	26 48	28 00	05 00
25	26 40	27 57	05 05
26	26 31	27 54	04 50
27	26 23	27 50	04 16
28	26 16	27 47	03 25
29	26 10	27 44	02 23
30	26 ♑ 07	27 ♑ 41	01 S 14

DECLINATIONS

Date	Sun ☉ ° '	Moon ☽ ° '	Mercury ☿ ° '	Venus ♀ ° '	Mars ♂ ° '	Jupiter ♃ ° '	Saturn ♄ ° '	Uranus ♅ ° '	Neptune ♆ ° '	Pluto ♇ ° '
01	14 S 34	26 S 12	13 S 53	15 S 16	24 S 45	02 N 50	21 S 57	10 N 07	15 N 30	21 S 35
02	14 53	25 26	13 10	15 40	24 42	02 47	21 57	10 07	15 29	21 35
03	15 12	23 13	12 29	16 04	24 38	02 45	21 58	10 06	15 29	21 35
04	15 30	19 52	11 51	16 28	24 34	02 43	21 59	10 05	15 28	21 35
05	15 49	15 42	11 18	16 51	24 30	02 41	21 59	10 04	15 28	21 35
06	16 06	10 59	10 51	17 13	24 26	02 39	22 00	10 04	15 28	21 35
07	16 24	05 56	10 28	17 35	24 21	02 37	22 00	10 03	15 27	21 34
08	16 42	00 S 44	10 12	17 57	24 16	02 36	22 01	10 02	15 27	21 34
09	16 59	04 N 28	10 01	18 18	24 11	02 34	22 01	10 02	15 26	21 34
10	17 16	09 30	09 56	18 39	24 05	02 32	22 02	10 01	15 26	21 34
11	17 32	14 14	09 56	18 59	23 59	02 31	22 02	10 01	15 25	21 34
12	17 48	18 27	10 01	19 19	23 54	02 29	22 03	10 00	15 25	21 34
13	18 04	21 56	10 11	19 38	23 47	02 28	22 03	10 00	15 24	21 33
14	18 20	24 29	10 24	19 56	23 41	02 26	22 04	09 59	15 24	21 33
15	18 35	25 53	10 41	20 15	23 34	02 25	22 04	09 59	15 23	21 33
16	18 50	25 59	11 00	20 32	23 27	02 24	22 05	09 58	15 23	21 33
17	19 05	24 41	11 22	20 49	23 20	02 23	22 05	09 58	15 22	21 32
18	19 19	22 04	11 47	21 05	23 12	02 22	22 06	09 57	15 22	21 32
19	19 33	18 15	12 13	21 21	23 04	02 21	22 06	09 57	15 22	21 32
20	19 47	13 25	12 40	21 36	22 56	02 20	22 07	09 56	15 21	21 32
21	20 00	07 50	13 08	21 51	22 48	02 19	22 07	09 56	15 21	21 31
22	20 13	01 N 44	13 38	22 04	22 40	02 18	22 08	09 56	15 20	21 31
23	20 26	04 S 33	14 08	22 18	22 31	02 18	22 08	09 55	15 20	21 31
24	20 38	10 42	14 38	22 30	22 22	02 17	22 09	09 55	15 19	21 30
25	20 50	16 19	15 08	22 42	22 13	02 17	22 09	09 55	15 19	21 30
26	21 01	20 59	15 39	22 53	22 03	02 16	22 09	09 54	15 18	21 30
27	21 12	24 16	16 09	23 04	21 53	02 16	22 10	09 54	15 18	21 29
28	21 22	25 55	16 39	23 14	21 43	02 16	22 10	09 54	15 18	21 29
29	21 33	25 49	17 09	23 23	21 33	02 16	22 11	09 54	15 17	21 29
30	21 S 42	24 S 06	17 S 38	23 S 31	21 S 23	02 N 16	22 S 11	09 N 53	15 N 17	21 S 28

ZODIAC SIGN ENTRIES

Date	h	m	Planets
02	04	14	☽ ♑
04	10	59	☽ ♒
06	21	34	☽ ♓
09	10	10	☽ ♈
11	22	36	☽ ♉
14	09	41	☽ ♊
14	15	03	♀ ♐
16	19	04	☽ ♋
19	02	33	☽ ♌
21	07	46	☽ ♍
22	03	56	☉ ♐
23	10	33	☽ ♎
25	11	34	☽ ♏
27	12	17	☽ ♐
29	14	25	♂ ♒
29	14	33	☽ ♑

LATITUDES

Date	Mercury ☿ ° '	Venus ♀ ° '	Mars ♂ ° '	Jupiter ♃ ° '	Saturn ♄ ° '	Uranus ♅ ° '	Neptune ♆ ° '	Pluto ♇ ° '
01	00 S 13	00 N 40	01 S 36	01 S 35	01 N 09	00 N 44	01 S 49	12 S 08
04	00 N 46	00 34	01 35	01 34	01 09	00 45	01 49	12 07
07	01 33	00 27	01 34	01 34	01 08	00 45	01 49	12 06
10	02 03	00 20	01 33	01 33	01 08	00 45	01 49	12 05
13	02 18	00 13	01 32	01 32	01 08	00 45	01 49	12 05
16	02 20	00 N 06	01 31	01 31	01 07	00 45	01 49	12 04
19	02 13	00 S 02	01 30	01 31	01 07	00 45	01 49	12 03
22	02 01	00 09	01 29	01 30	01 07	00 45	01 49	12 02
25	01 44	00 16	01 27	01 29	01 06	00 45	01 49	12 01
28	01 24	00 23	01 26	01 28	01 06	00 46	01 49	12 01
31	01 N 03	00 S 31	01 S 24	01 S 27	01 N 06	00 N 46	01 S 49	12 S 00

DATA

Julian Date	2468651
Delta T	+79 seconds
Ayanamsa	24° 30' 54"
Synetic vernal point	04° ♓ 36' 04"
True obliquity of ecliptic	23° 26' 04"

LONGITUDES

Date	Chiron ⚷ ° '	Ceres ⚳ ° '	Pallas ⚴ ° '	Juno ⚵ ° '	Vesta ⚶ ° '	Black Moon Lilith ☾ ° '
01	16 ♎ 27	18 ♏ 56	19 ♎ 31	07 ♏ 28	20 ♎ 04	08 ♈ 57
11	17 ♎ 48	23 ♏ 15	23 ♎ 56	10 ♏ 56	25 ♎ 10	10 ♈ 04
21	19 ♎ 06	27 ♏ 33	28 ♎ 17	14 ♏ 21	00 ♏ 15	11 ♈ 11
31	20 ♎ 18	01 ♐ 49	02 ♏ 32	17 ♏ 43	05 ♏ 19	12 ♈ 18

MOON'S PHASES, APSIDES AND POSITIONS ☽

Date	h	m	Phase	Longitude ° '	Eclipse Indicator
05	12	29	☽	13 ♒ 15	
13	17	04	○	21 ♉ 29	
21	06	10	☾	29 ♌ 05	
27	21	50	●	05 ♐ 49	

Day	h	m		
09	19	32	Apogee	
25	18	00	Perigee	
01	12	39	Max dec	26° S 12'
08	15	23	0N	
16	01	35	Max dec	26° N 06'
22	18	39	0S	
28	22	30	Max dec	26° S 05'

ASPECTARIAN

h m Aspects

01 Thursday
02 07 ☽ ⊥ ☉
07 50 ☽ ⊼ ♆
09 57 ☽ ⊥ ♀
11 49 ☽ ☌ ♄
13 01 ☽ Q ♇
13 21 ☽ ∠ ☿
18 06 ☽ ± ♆
18 40 ☽ ∠ ☉

02 Friday
01 28 ♀ # ♆
03 03 ☽ ∠ ♀
03 46 ☿ ✶ ♅
05 29 ☿ ∠ ♄
08 30 ♀ ⊥ ♄
09 48 ☽ bQ ♆
09 54 ☽ ✶ ♇
13 26 ☽ ✶ ☿
14 18 ☽ △ ♅
21 33 ☽ ☌ ♂
21 55 ☽ ∥ ♂
22 41 ☉ ⊼ ♃
23 13 ☽ □ ♃
23 16 ☽ ✶ ☉

03 Saturday
08 40 ☽ ✶ ♀
09 26 ☽ Q ☿
12 49 ☽ △ ♆
12 58 ☽ ∠ ♇
17 38 ☽ bQ ♅
17 39 ☽ ⊻ ♄
21 44 ☽ ∥ ♄
22 07 ♀ ± ♃
23 07 ☽ Q ☉

04 Sunday
00 02 ♂ □ ♃
00 33 ☽ ∥ ♇
04 56 ☽ ⊥ ♄
05 43 ☽ ⊥ ♇
08 13 ☽ Q ♃
09 25 ☽ Q ♀
09 40 ☉ # ♆
10 32 ☽ ± ♅
11 48 ☿ △ ♇
16 45 ☽ □ ☿
17 05 ☽ ⊻ ♇
21 05 ♀ Q ♅
22 01 ☽ ⊼ ♅
22 11 ☽ ∠ ♄

05 Monday
03 02 ♀ ☍ ♆
06 15 ☽ ∥ ♀
07 01 ☽ ✶ ♃
09 16 ☽ ⊻ ♂
11 25 ☽ ∥ ☉
12 29 ☽ □ ☉
13 13 ☽ # ♆
21 46 ☽ ⊥ ♂
21 54 ☽ □ ♆

06 Tuesday
00 07 ☽ □ ♀
03 40 ☽ ✶ ♄
12 17 ☽ ∠ ♃
12 43 ☽ ∥ ☿
16 28 ☽ # ♅
16 46 ☽ ∠ ♂

07 Wednesday
00 34 ☽ △ ☿
03 47 ☽ Q ♄
03 59 ☽ ☌ ♇
06 08 ☽ ⊥ ♃
09 12 ☉ ⊥ ♄
09 19 ☽ ☍ ♅
09 39 ☽ Q ♆
13 28 ♀ ⊻ ♄
14 33 ☿ Q ♂
18 08 ☽ ⊻ ♃

08 Thursday
00 55 ☽ ✶ ♂
01 59 ♄ Q ♇
03 24 ☽ # ♃
05 45 ☽ △ ☉
05 59 ☽ bQ ☿
09 50 ☽ ✶ ♆
10 20 ☉ ± ♃
16 26 ☽ □ ♄
19 22 ☽ △ ♀

09 Friday
00 02 ☽ ± ☿
02 53 ☽ Q ♂
03 14 ☽ ∥ ♃
07 27 ☿ St D
10 07 ☿ # ♅
12 11 ☽ ⊼ ☿
14 58 ☽ bQ ☉
16 10 ☽ ∠ ♆
16 38 ☽ ⊻ ♇
22 11 ☽ ⊼ ♅

10 Saturday
04 38 ☉ Q ♅
04 43 ☉ ☍ ♆
04 46 ☽ ⊥ ♇
05 21 ☽ bQ ♀
06 28 ☽ ☌ ♃
10 20 ☽ ⊥ ♆
10 21 ☽ ± ♅
10 52 ☽ ± ☉
14 06 ☽ # ☿
14 32 ☽ ∥ ♅
15 15 ♀ bQ ♃
17 46 ☽ □ ♂
22 24 ☽ ⊻ ♆
22 56 ☽ ∠ ♇

11 Sunday
00 04 ☽ ⊼ ☉
01 39 ☽ ± ♀
04 31 ☽ bQ ♅
05 38 ☽ △ ♄
15 05 ☽ ⊼ ♀
18 27 ☽ ∥ ♆

12 Monday
01 57 ☽ ☍ ☿
04 57 ☽ ✶ ♇
07 52 ☽ # ☉
08 23 ☿ # ♅
10 34 ☽ △ ♅
11 51 ☽ bQ ♄
18 00 ☽ # ♀
18 09 ☽ ⊻ ♃

13 Tuesday
04 43 ☽ Q ♇
05 48 ☽ ± ♄
05 56 ☽ ⊥ ♃
09 03 ☽ # ♇
09 32 ☽ △ ♂
09 58 ☽ ☌ ♆
12 55 ☽ # ♄
16 18 ♂ △ ♆
17 04 ☽ ☍ ☉
17 41 ☽ ⊼ ♄
23 28 ☽ ∠ ♃

14 Wednesday
01 21 ☉ ⊻ ♄
02 51 ♂ ∠ ♇
03 36 ☽ # ♂
09 05 ☽ ☍ ♀
10 12 ☿ △ ♇
15 51 ☽ □ ♇
16 15 ☽ ⊼ ☿
16 42 ☽ bQ ♂
21 26 ☽ □ ♅

15 Thursday
04 23 ☽ ✶ ♃
04 41 ☽ ± ☿
11 09 ☽ ± ♂
19 56 ☽ ⊻ ♆
23 16 ☽ bQ ☿
23 21 ☽ ⊼ ♂

16 Friday
03 13 ☽ Q ♃
04 03 ☽ ☍ ♄
07 19 ☽ ⊥ ♆
07 56 ☽ Q ♅
07 57 ☽ ⊼ ☉
08 48 ♇ St D
18 53 ☉ bQ ♃
20 16 ☽ ± ☉

17 Saturday
00 15 ☽ ∠ ♆
00 44 ☽ ⊼ ♀
01 03 ☽ △ ♇
03 57 ♀ □ ♇
06 03 ☽ △ ☿
06 34 ☽ ✶ ♅
12 15 ☿ ✶ ♅
12 55 ☽ □ ♃
13 10 ☽ ± ♀
14 32 ☽ bQ ☉

18 Sunday
00 29 ♂ bQ ♅
02 32 ☽ # ♂
04 06 ☽ ✶ ♆
04 57 ☽ bQ ♇
07 36 ☽ bQ ♀
10 24 ☽ ∠ ♅
11 00 ☽ ☍ ♂
11 47 ☽ # ♄
12 32 ☽ ⊼ ♄
13 21 ☿ ∠ ♄
15 49 ☽ # ♇
18 24 ☽ # ♀
20 29 ☽ △ ☉
21 26 ☽ ± ♇
23 35 ☽ ± ♄

19 Monday
02 00 ☽ Q ♆
02 51 ☽ ⊥ ♅
04 55 ☽ # ☉
08 19 ☽ ⊼ ♇
13 14 ♀ □ ♅
13 42 ☽ ⊻ ♅
13 45 ☽ △ ♀
15 58 ☽ bQ ♄
17 40 ♂ ⊻ ♄
18 23 ☽ □ ☿
19 31 ☽ △ ♃

20 Tuesday
02 53 ☽ ∥ ♆
05 00 ♀ ∥ ♇
06 04 ☿ ⊼ ♃
10 10 ☽ □ ♆
15 07 ☽ # ☿
18 48 ☽ △ ♄
20 04 ☽ ⊼ ♂
21 57 ☽ bQ ♃

21 Wednesday
03 16 ☽ ∥ ♅
05 25 ☽ Q ☿
06 10 ☽ □ ☉
07 08 ☽ ± ♂
13 16 ☽ ☍ ♇
13 31 ☽ ± ♃
18 29 ☽ ☌ ♅
20 33 ♀ ∠ ♂
23 31 ☽ bQ ♂
23 37 ☽ □ ♀
23 47 ☽ ⊼ ♃

22 Thursday
01 17 ♀ △ ♃
04 13 ☽ ✶ ☿
09 49 ☽ ∥ ♃
13 50 ☽ △ ♆
15 19 ☽ Q ☉
18 08 ♀ ∥ ♄
19 36 ☿ Q ♂
22 34 ☽ □ ♄

23 Friday
02 15 ☽ △ ♂
03 25 ☽ # ♃
08 06 ☽ ∠ ☿
08 50 ☽ Q ♀
12 51 ☽ ✶ ☉
14 49 ☽ bQ ♆
15 50 ☽ ⊼ ♇
20 53 ☽ ⊻ ♅

24 Saturday
00 33 ☽ ⊥ ☿
01 43 ☽ ± ♇
01 47 ☽ ☍ ♃
02 46 ♀ ∥ ♂
04 35 ☽ Q ♄
05 35 ☽ ± ♆
06 27 ☽ ✶ ♀
06 44 ☽ ⊥ ♅
08 51 ☽ # ♅
11 25 ☽ ⊻ ☿
12 59 ☿ ± ♃
15 17 ☽ ∠ ☉
15 22 ☽ ⊼ ♆
16 25 ☽ bQ ♇
21 25 ☽ ∠ ♅

25 Sunday
00 17 ☽ ✶ ♄
05 04 ♃ ⊥ ♇
06 07 ☽ ∥ ☿
06 10 ☽ □ ♂
06 58 ☽ ⊥ ☉
07 28 ☽ # ♆
07 53 ☉ □ ♇
09 07 ☽ ∠ ♀
16 45 ☽ △ ♇
17 24 ☽ ⊻ ☉
20 11 ☿ # ♆
20 39 ♂ ∥ ♄
20 53 ☿ ⊥ ♄
21 45 ☽ ✶ ♅
22 12 ♂ Q ♃
22 57 ♂ ⊥ ♇

26 Monday
00 46 ☽ ∠ ♄
01 01 ☽ ⊥ ♀
02 25 ☽ ⊼ ♃
02 57 ☿ ☍ ♆
11 38 ☽ ⊻ ♀
12 07 ☽ ± ♃
12 15 ☽ ∥ ☉
12 54 ☽ Q ♂
15 10 ☽ ∥ ♇
15 27 ☽ ⊥ ♄
15 53 ☽ ☍ ♆
15 54 ☿ Q ♅
17 12 ☽ Q ♅
17 20 ☽ ☌ ☿
18 26 ☽ ∥ ♂
19 27 ☽ ∥ ♄

27 Tuesday
01 14 ☽ ∥ ♀
01 17 ☽ ⊻ ♄
02 43 ☽ bQ ♃
09 30 ☽ ✶ ♂
17 33 ☽ □ ♇
21 50 ☽ ☌ ☉
22 40 ☽ □ ♅

28 Wednesday
03 19 ☽ △ ♃
09 55 ☉ □ ♅
11 40 ☽ ∠ ♂
12 52 ♀ ⊼ ♆
12 57 ♄ ± ♆
17 11 ☽ ⊼ ♆
17 37 ☽ ☌ ♀
22 26 ♂ ⊥ ♄
23 34 ☽ Q ♇

29 Thursday
00 34 ☽ ⊻ ☿
02 55 ☉ ∥ ♇
03 19 ☽ ± ♆
03 27 ☽ ☌ ♄
03 42 ☽ ⊥ ♂
12 02 ☽ ⊥ ☿
12 37 ☉ ∥ ♂
14 34 ☽ ⊻ ♂
18 43 ☽ bQ ♆
20 07 ☽ ✶ ♇
22 42 ♂ ∥ ♇

30 Friday
01 01 ♂ ± ♅
01 33 ☽ △ ♅
04 41 ☽ ⊻ ☉
05 35 ☿ ⊻ ♄
05 35 ☽ ∠ ☿
06 22 ☽ □ ♃
12 38 ☿ bQ ♃
16 09 ☽ ⊥ ☉
17 07 ☽ ∥ ♀
21 06 ☽ △ ♆
22 37 ☽ ∠ ♇

All ephemeris data is given at 12.00 UT and the Moon's longitude is additionally given for 24.00 UT

DECEMBER 2046

LONGITUDES

Date	Sidereal time h m s	Sun ☉ ° ' "	Moon ☽ ° ' "	Moon ☽ 24.00 ° ' "	Mercury ☿ ° '	Venus ♀ ° '	Mars ♂ ° '	Jupiter ♃ ° '	Saturn ♄ ° '	Uranus ♅ ° '	Neptune ♆ ° '	Pluto ⯓ ° '
01	16 42 03	09 ♐ 26 55	25 ♑ 36 32	02 ♒ 06 59	25 ♏ 34	21 ♐ 10	01 ♒ 28	09 ♈ 04	23 ♐ 47	06 ♍ 21	17 ♉ 22	03 ♓ 14
02	16 46 00	10 27 44	08 ♒ 31 29	14 ♒ 50 28	27 05	22 25	02 14	09 R 04	23 54	06 22	17 R 20	03 14
03	16 49 56	11 28 35	21 ♒ 04 24	27 ♒ 13 51	28 ♏ 37	23 40	03 00	09 04	24 01	06 22	17 19	03 15
04	16 53 53	12 29 26	03 ♓ 19 26	09 ♓ 21 46	00 ♐ 10	24 55	03 47	09 D 04	24 08	06 23	17 17	03 15
05	16 57 50	13 30 19	15 ♓ 21 30	21 ♓ 19 16	01 42	26 11	04 33	09 04	24 15	06 23	17 16	03 16
06	17 01 46	14 31 12	27 ♓ 15 42	03 ♈ 11 25	03 15	27 26	05 19	09 04	24 22	06 23	17 14	03 16
07	17 05 43	15 32 05	09 ♈ 07 01	15 ♈ 03 02	04 47	28 41	06 06	09 05	24 29	06 24	17 13	03 17
08	17 09 39	16 33 00	21 ♈ 00 00	26 ♈ 58 23	06 20	29 ♐ 56	06 52	09 06	24 36	06 24	17 12	03 17
09	17 13 36	17 33 55	02 ♉ 58 36	09 ♉ 01 02	07 53	01 ♑ 12	07 38	09 07	24 43	06 24	17 10	03 18
10	17 17 32	18 34 52	15 ♉ 05 59	21 ♉ 13 42	09 26	02 27	08 25	09 08	24 50	06 24	17 09	03 18
11	17 21 29	19 35 49	27 ♉ 24 03	03 ♊ 38 13	10 59	03 42	09 11	09 10	24 57	06 24	17 07	03 19
12	17 25 25	20 36 46	09 ♊ 55 15	16 ♊ 15 32	12 33	04 57	09 58	09 11	25 04	06 R 24	17 06	03 20
13	17 29 22	21 37 45	22 ♊ 39 05	29 ♊ 05 53	14 06	06 12	10 44	09 13	25 11	06 24	17 05	03 20
14	17 33 19	22 38 45	05 ♋ 35 51	12 ♋ 08 55	15 39	07 28	11 31	09 15	25 19	06 24	17 03	03 21
15	17 37 15	23 39 45	18 ♋ 45 01	25 ♋ 24 04	17 13	08 43	12 18	09 17	25 26	06 24	17 02	03 22
16	17 41 12	24 40 46	02 ♌ 05 57	08 ♌ 50 38	18 46	09 58	13 04	09 20	25 33	06 24	17 01	03 23
17	17 45 08	25 41 48	15 ♌ 38 02	22 ♌ 28 05	20 20	11 13	13 51	09 22	25 40	06 23	17 00	03 23
18	17 49 05	26 42 51	29 ♌ 20 44	06 ♍ 15 55	21 54	12 28	14 37	09 25	25 47	06 23	16 58	03 24
19	17 53 01	27 43 54	13 ♍ 13 35	20 ♍ 13 39	23 28	13 43	15 24	09 28	25 54	06 22	16 57	03 25
20	17 56 58	28 44 59	27 ♍ 15 59	04 ♎ 20 26	25 02	14 59	16 11	09 32	26 01	06 22	16 56	03 26
21	18 00 54	29 ♐ 46 04	11 ♎ 26 47	18 ♎ 34 47	26 36	16 14	16 57	09 35	26 08	06 21	16 55	03 27
22	18 04 51	00 ♑ 47 10	25 ♎ 44 04	02 ♏ 54 15	28 11	17 29	17 44	09 39	26 15	06 21	16 54	03 28
23	18 08 48	01 48 17	10 ♏ 04 51	17 ♏ 15 21	29 ♐ 46	18 44	18 31	09 43	26 22	06 20	16 53	03 28
24	18 12 44	02 49 25	24 ♏ 25 10	01 ♐ 33 41	01 ♑ 20	19 59	19 17	09 47	26 29	06 20	16 52	03 29
25	18 16 41	03 50 34	08 ♐ 40 17	15 ♐ 44 21	02 55	21 14	20 04	09 51	26 36	06 19	16 50	03 30
26	18 20 37	04 51 43	22 ♐ 45 18	29 ♐ 42 36	04 31	22 30	20 51	09 55	26 44	06 18	16 49	03 31
27	18 24 34	05 52 52	06 ♑ 35 47	13 ♑ 24 28	06 06	23 45	21 38	10 00	26 51	06 17	16 48	03 32
28	18 28 30	06 54 02	20 ♑ 08 21	26 ♑ 47 15	07 42	25 00	22 24	10 05	26 58	06 16	16 48	03 33
29	18 32 27	07 55 12	03 ♒ 21 05	09 ♒ 49 51	09 18	26 15	23 11	10 10	27 05	06 15	16 47	03 34
30	18 36 23	08 56 21	16 ♒ 13 41	22 ♒ 32 46	10 55	27 30	23 58	10 15	27 12	06 14	16 46	03 35
31	18 40 20	09 ♑ 57 31	28 ♒ 47 25	04 ♓ 57 56	12 ♑ 32	28 ♑ 45	24 ♒ 45	10 ♈ 20	27 ♐ 19	06 ♍ 13	16 ♉ 45	03 ♓ 37

DECLINATIONS

Date	Moon True ☊ ° '	Moon Mean ☊ ° '	Moon ☽ Latitude ° '	Sun ☉ ° '	Moon ☽ ° '	Mercury ☿ ° '	Venus ♀ ° '	Mars ♂ ° '	Jupiter ♃ ° '	Saturn ♄ ° '	Uranus ♅ ° '	Neptune ♆ ° '	Pluto ⯓ ° '
01	26 ♑ 06	27 ♑ 38	00 S 03	21 S 52	21 S 04	18 S 07	23 S 39	21 S 12	02 N 16	22 S 12	09 N 53	15 N 16	21 S 28
02	26 D 07	27 35	01 N 07	22 01	17 03	18 35	23 46	21 01	02 16	22 12	09 53	15 16	21 28
03	26 08	27 31	02 11	22 09	12 24	19 03	23 52	20 50	02 16	22 12	09 53	15 16	21 27
04	26 10	27 28	03 08	22 17	07 22	19 30	23 58	20 39	02 16	22 13	09 53	15 15	21 27
05	26 R 10	27 25	03 55	22 25	02 S 09	19 56	24 02	20 27	02 16	22 13	09 53	15 15	21 26
06	26 09	27 22	04 32	22 32	03 N 04	20 21	24 07	20 15	02 17	22 14	09 53	15 14	21 26
07	26 06	27 19	04 56	22 39	08 09	20 45	24 10	20 03	02 17	22 14	09 53	15 14	21 25
08	26 01	27 16	05 09	22 45	12 57	21 09	24 13	19 51	02 18	22 14	09 52	15 14	21 25
09	25 55	27 12	05 07	22 51	17 19	21 31	24 14	19 39	02 19	22 15	09 52	15 13	21 25
10	25 48	27 09	04 53	22 57	21 01	21 52	24 15	19 26	02 19	22 15	09 52	15 13	21 24
11	25 41	27 06	04 24	23 02	23 52	22 13	24 16	19 13	02 20	22 15	09 53	15 13	21 24
12	25 35	27 03	03 42	23 06	25 36	22 32	24 15	19 00	02 21	22 16	09 53	15 12	21 23
13	25 30	27 00	02 49	23 10	26 03	22 50	24 14	18 47	02 22	22 16	09 53	15 12	21 23
14	25 26	26 56	01 46	23 14	25 05	23 07	24 12	18 34	02 23	22 16	09 53	15 12	21 22
15	25 25	26 53	00 N 36	23 17	22 43	23 24	24 10	18 20	02 24	22 17	09 53	15 11	21 22
16	25 D 25	26 50	00 S 37	23 20	19 06	23 38	24 06	18 06	02 26	22 17	09 53	15 11	21 21
17	25 26	26 47	01 49	23 22	14 25	23 52	24 02	17 52	02 27	22 17	09 53	15 11	21 21
18	25 27	26 44	02 55	23 24	08 58	24 04	23 57	17 38	02 28	22 18	09 53	15 10	21 20
19	25 29	26 41	03 53	23 25	03 N 00	24 15	23 52	17 24	02 30	22 18	09 53	15 10	21 20
20	25 30	26 37	04 37	23 26	03 S 09	24 25	23 45	17 09	02 31	22 18	09 54	15 10	21 19
21	25 R 29	26 34	05 05	23 26	09 12	24 34	23 38	16 55	02 33	22 18	09 54	15 09	21 19
22	25 27	26 31	05 14	23 26	14 49	24 41	23 31	16 40	02 35	22 19	09 54	15 09	21 18
23	25 25	26 28	05 04	23 25	19 38	24 47	23 22	16 25	02 37	22 19	09 54	15 09	21 18
24	25 21	26 25	04 35	23 24	23 18	24 51	23 13	16 10	02 38	22 19	09 55	15 09	21 17
25	25 18	26 22	03 49	23 23	25 31	24 55	23 03	15 54	02 40	22 19	09 55	15 08	21 16
26	25 15	26 18	02 49	23 21	26 03	24 56	22 52	15 39	02 42	22 20	09 55	15 08	21 16
27	25 13	26 15	01 41	23 18	24 57	24 57	22 41	15 23	02 44	22 20	09 56	15 08	21 15
28	25 13	26 12	00 S 28	23 15	22 23	24 56	22 29	15 07	02 47	22 20	09 56	15 08	21 15
29	25 D 13	26 09	00 N 45	23 12	18 41	24 53	22 16	14 51	02 49	22 20	09 57	15 08	21 14
30	25 14	26 06	01 54	23 08	14 10	24 49	22 03	14 35	02 51	22 20	09 57	15 07	21 14
31	25 ♑ 15	26 ♑ 02	02 N 55	23 S 04	09 S 09	24 S 44	21 S 49	14 S 19	02 N 53	22 S 21	09 N 57	15 N 07	21 S 13

ZODIAC SIGN ENTRIES

Date	h	m	Planets
01	20	05	☽ ♒
04	05	26	☽ ♓
04	09	30	☿ ♐
06	17	32	☽ ♈
08	13	11	♀ ♑
09	06	04	☽ ♉
11	17	00	☽ ♊
14	01	40	☽ ♋
16	08	15	☽ ♌
18	13	08	☽ ♍
20	16	39	☽ ♎
21	17	28	☉ ♑
22	19	08	☽ ♏
23	15	40	☿ ♑
24	21	22	☽ ♐
27	00	30	☽ ♑
29	05	51	☽ ♒
31	14	20	☽ ♓

LATITUDES

Date	Mercury ☿ ° '	Venus ♀ ° '	Mars ♂ ° '	Jupiter ♃ ° '	Saturn ♄ ° '	Uranus ♅ ° '	Neptune ♆ ° '	Pluto ⯓ ° '
01	01 N 03	00 S 31	01 S 24	01 S 27	01 N 06	00 N 46	01 S 49	12 S 00
04	00 42	00 37	01 23	01 26	01 06	00 46	01 49	11 59
07	00 N 20	00 44	01 21	01 25	01 05	00 46	01 49	11 58
10	00 S 01	00 51	01 20	01 25	01 05	00 46	01 48	11 58
13	00 21	00 57	01 18	01 24	01 05	00 46	01 48	11 57
16	00 41	01 03	01 16	01 23	01 05	00 46	01 48	11 56
19	00 59	01 09	01 14	01 22	01 05	00 47	01 48	11 56
22	01 16	01 14	01 12	01 21	01 04	00 47	01 48	11 55
25	01 31	01 18	01 10	01 20	01 04	00 47	01 48	11 54
28	01 43	01 23	01 09	01 19	01 04	00 47	01 48	11 54
31	01 S 54	01 S 26	01 S 07	01 S 18	01 N 04	00 N 47	01 S 48	11 S 53

DATA

Julian Date	2468681
Delta T	+79 seconds
Ayanamsa	24° 31' 00"
Synetic vernal point	04° ♓ 35' 59"
True obliquity of ecliptic	23° 26' 03"

MOON'S PHASES, APSIDES AND POSITIONS ☽

Date	h	m	Phase	Longitude ° '	Eclipse Indicator
05	07	56	☽	13 ♓ 20	
13	09	55	○	21 ♊ 32	
20	14	43	☾	28 ♍ 52	
27	10	39	●	05 ♑ 49	

Day	h	m		
07	12	34	Apogee	
23	04	47	Perigee	
05	21	53	0N	
13	07	45	Max dec	26° N 04'
19	23	45	0S	
26	07	45	Max dec	26° S 05'

LONGITUDES

Date	Chiron ⚷ ° '	Ceres ⚳ ° '	Pallas ⚴ ° '	Juno ⚵ ° '	Vesta ⚶ ° '	Black Moon Lilith ⚸ ° '
01	20 ♎ 18	01 ♐ 49	02 ♏ 32	17 ♏ 43	05 ♏ 19	12 ♈ 18
11	21 ♎ 23	06 ♐ 04	06 ♏ 40	21 ♏ 01	10 ♏ 20	13 ♈ 26
21	22 ♎ 20	10 ♐ 16	10 ♏ 41	24 ♏ 13	15 ♏ 18	14 ♈ 33
31	23 ♎ 08	14 ♐ 23	14 ♏ 32	27 ♏ 19	20 ♏ 11	15 ♈ 40

All ephemeris data is given at 12.00 UT and the Moon's longitude is additionally given for 24.00 UT

ASPECTARIAN

h m Aspects

01 Saturday
03 04 ☽ ⊻ ♀
03 12 ☉ △ ♃
04 02 ☽ ∥ ♄
04 15 ☽ ♇ ♅
06 41 ☽ ∥ ☉
08 38 ☽ ⊻ ♄
09 14 ☽ ∥ ⯓
09 42 ☽ ∠ ☉
11 00 ☽ ∥ ♂
11 54 ☽ ✶ ☿
13 21 ♀ Q ⯓
14 41 ☽ Q ♃
14 58 ☽ ⊥ ⯓
15 09 ☽ ⊥ ♀
19 45 ☽ ⊥ ♄
20 44 ☽ ± ♅
23 28 ☽ ☌ ♂

02 Sunday
02 05 ☽ ⊻ ⯓
04 13 ☽ ∥ ☿
07 56 ☽ ⊼ ♅
09 41 ☽ ∠ ♀
12 43 ☽ ∠ ♄
13 01 ☽ ✶ ♃
13 13 ☽ Q ☿
15 59 ☽ ✶ ☉
21 33 ☽ # ♆

03 Monday
04 46 ☽ □ ♆
05 22 ♀ ± ♆
17 05 ☽ Q ☉
17 37 ☽ ✶ ♀
17 46 ☽ ✶ ♄
17 48 ☽ ∠ ♃
19 20 ♀ ☌ ♄
19 34 ♂ ⊻ ⯓
21 47 ☉ ∥ ♄

04 Tuesday
00 11 ☽ # ♅
01 08 ♃ St D
04 51 ☽ □ ☿
11 29 ☽ ⊥ ♃
11 51 ☽ ☌ ⯓
12 57 ☽ ⊻ ♂
15 53 ☽ Q ♆
17 37 ☽ Q ♄
18 04 ☽ ☍ ♅
19 58 ☽ Q ♀
23 24 ☽ ⊻ ♃

05 Wednesday
01 42 ☽ ⊥ ♂
07 56 ☽ □ ☉
11 28 ☽ # ♃
15 49 ☽ ✶ ♆
21 01 ☽ ∠ ♂

06 Thursday
06 05 ☽ □ ♄
08 22 ☿ ∥ ♂
08 22 ☽ ∥ ♃
12 23 ☽ □ ♀
12 23 ☿ □ ⯓
22 03 ☽ ∠ ♆

07 Friday
00 10 ☽ ⊻ ⯓
01 55 ☽ △ ☿
05 27 ☽ ✶ ♂
06 29 ☽ ⊼ ♅
11 56 ☽ ☌ ♃
12 20 ☽ ⊥ ⯓
16 14 ☽ ⊥ ♆
18 38 ☽ ± ♅
20 25 ☽ ∥ ♅
21 24 ♂ ⊼ ♅

08 Saturday
02 11 ☽ △ ☉
04 20 ☽ ⊻ ♆
06 32 ☽ ∠ ⯓
07 24 ☽ Q ♂
12 47 ☽ ♇ ☿
12 48 ☽ ♇ ♅
12 57 ☿ □ ♅
19 19 ☽ △ ♄

09 Sunday
00 08 ☽ ∥ ♆
02 50 ☉ ⊼ ♆
04 24 ☿ ✶ ♂
05 14 ☿ ∥ ⯓
08 01 ☽ △ ♀
09 30 ☽ ± ☿
11 06 ☽ ♇ ☉
12 38 ☽ ✶ ⯓
18 49 ☽ △ ♅
21 54 ☽ □ ♂
23 12 ☽ ⊼ ☿

10 Monday
00 13 ☽ ⊻ ♃
01 31 ☽ ♇ ♄
01 44 ☽ # ♂
06 22 ♀ ♇ ♆
06 35 ☽ ± ☉
07 14 ☿ △ ♃
10 31 ♀ ⊥ ♂
12 04 ☽ ⊥ ♃
12 24 ☽ Q ⯓
14 49 ☽ # ⯓
16 01 ☽ ☌ ♆
17 08 ☽ ♇ ♀
19 14 ☽ # ☿
19 24 ☽ ± ♄
19 27 ☽ ⊼ ☉
21 26 ☽ # ♄

11 Tuesday
03 47 ☽ # ☉
04 38 ♀ ✶ ⯓
05 41 ☽ ∠ ♃
07 12 ☽ ⊼ ♄
10 59 ♂ ✶ ♃
12 38 ☽ ± ♀
15 14 ☿ ∥ ♄
16 11 ♅ St R
16 22 ☽ # ♀
23 24 ☽ □ ⯓

12 Wednesday
01 28 ☽ ⊼ ♀
05 18 ☽ □ ♅
10 36 ☽ ✶ ♃
12 05 ☽ △ ♂
15 54 ♂ ∠ ♄
17 41 ☽ ☍ ☿

13 Thursday
01 34 ☽ ⊻ ♆
05 05 ☉ Q ⯓
09 19 ☽ Q ♃
09 55 ☽ ☍ ☉
12 48 ☽ ⊥ ♆
15 16 ☽ Q ♅
15 45 ♀ △ ♅
16 47 ☽ ☍ ♄
18 08 ☽ ♇ ♂

14 Friday
05 29 ☽ ∠ ♆
07 52 ☽ △ ⯓
11 51 ☽ ± ♂
13 28 ☽ ✶ ♅
15 47 ☽ ☍ ♀
18 43 ☽ □ ♃
21 29 ☉ ± ♆
22 48 ☽ # ♀
23 32 ☽ ⊼ ♂
23 54 ☉ ∥ ☿

15 Saturday
06 55 ☽ # ☿
07 27 ☽ # ☉
08 51 ☽ ⊼ ☿
08 54 ☽ ✶ ♆
09 16 ☿ ⊼ ♆
11 18 ☽ ♇ ⯓
15 25 ☽ # ♄
16 47 ☽ ∠ ♅
21 08 ☽ ± ☿
21 36 ☽ ⊼ ☉
21 58 ☽ # ⯓
23 25 ♀ □ ♃

16 Sunday
00 09 ☽ ⊼ ♄
03 32 ☽ ± ⯓
06 29 ☽ Q ♆
08 57 ☽ ⊥ ♅
09 15 ☽ ± ☉
11 00 ☽ ± ♄
14 17 ☽ ⊼ ⯓
15 23 ☽ ♇ ☿
17 49 ☽ # ♂
19 39 ☽ ⊻ ♅

17 Monday
00 54 ☽ △ ♃
02 34 ☽ ♇ ☉
03 09 ☽ ♇ ♄
03 25 ☽ ⊼ ♀
08 23 ☽ ∥ ♆
08 39 ☽ ☍ ♂
11 07 ☉ ☌ ♄
14 23 ☽ □ ♆
15 04 ☽ ± ♀
21 20 ☽ △ ☿
22 13 ♃ ⊥ ⯓

18 Tuesday
02 40 ☿ ∥ ♀
03 23 ☽ ♇ ♃
04 18 ☿ Q ⯓
05 44 ☽ △ ♄
07 03 ☽ △ ☉
08 06 ☽ ∥ ♅
08 25 ☽ ♇ ♀
19 03 ☽ ☍ ⯓
19 06 ☽ ± ♃

19 Wednesday
00 12 ☽ ☌ ♅
04 14 ☿ ± ♆
05 31 ☽ ⊼ ♃
12 56 ☽ △ ♀
14 00 ☽ ∥ ♃
15 57 ☽ ⊼ ♂
18 23 ☽ △ ♆

20 Thursday
02 49 ☽ ± ♂
07 44 ☽ □ ☿
09 34 ☽ # ♃
09 52 ☽ □ ♄
14 43 ☽ □ ☉
19 02 ☽ ♇ ♂
19 55 ☽ ♇ ♆
22 28 ☽ ⊼ ⯓

21 Friday
03 25 ☽ ⊻ ♅
04 14 ☿ ☌ ♄
08 37 ☽ ± ⯓
08 51 ☽ ☍ ♃
10 44 ♂ □ ♆
11 06 ☽ ± ♆
13 32 ☽ ⊥ ♅
14 53 ☽ # ♅
16 34 ☽ Q ♄
17 59 ☽ Q ☿
20 50 ☽ □ ♀
21 11 ☽ ⊼ ♆
21 48 ☽ △ ♂
23 27 ☽ Q ☉
23 47 ☽ ♇ ⯓

22 Saturday
00 54 ♀ △ ♆
04 39 ☽ ∠ ♅
12 53 ☽ ✶ ♄
13 34 ☽ # ♆
16 36 ☽ ✶ ☿
20 15 ☽ ∥ ♂
21 06 ☽ ✶ ☉

23 Sunday
00 42 ♀ ⊻ ♂
00 57 ☽ △ ⯓
02 15 ☉ ∥ ♀
05 45 ☽ ✶ ♅
05 52 ☽ Q ♀
06 56 ♀ ∠ ⯓
11 23 ☽ ⊼ ♃
13 40 ☉ ♇ ♆
14 11 ☽ ∠ ♄
20 47 ☽ ∠ ☿
21 27 ☽ ± ♃
21 47 ☽ ∥ ⯓
23 21 ☽ ☍ ♆

24 Monday
00 06 ☽ ∠ ☉
01 48 ☽ Q ♅
02 55 ☽ □ ♂
03 52 ☽ ✶ ♀
04 36 ☽ ∥ ♄
05 22 ☽ ⊥ ♄
11 18 ☽ ∥ ♀
12 36 ☽ ♇ ♃
12 47 ☽ ∥ ☉
13 44 ☽ ⊥ ☿
15 30 ☽ ⊻ ♄
16 21 ☽ ⊥ ☉
19 47 ☿ ♇ ♆

25 Tuesday
01 05 ☽ ⊻ ☿
03 13 ☽ ⊻ ☉
03 16 ☽ □ ⯓
03 16 ☽ ∥ ☿
03 57 ☉ ✶ ⯓
07 30 ☽ ∠ ♀
08 01 ☽ □ ♅
10 55 ☽ Q ♂
13 24 ♀ ♇ ♅
14 00 ☽ △ ♃
20 53 ☿ ✶ ⯓

26 Wednesday
00 14 ☽ ⊥ ♀
01 52 ☽ ⊼ ♆
08 32 ☽ ✶ ♂
09 53 ☽ Q ⯓
11 30 ☽ ⊻ ♀
12 07 ☽ ± ♆
18 54 ☽ ☌ ♄

27 Thursday
02 35 ☉ ☌ ☿
03 39 ☽ ♇ ♆
06 39 ☽ ✶ ⯓
10 39 ☽ ☌ ☉
11 02 ☽ ☌ ☿
11 28 ☽ △ ♅
12 01 ☽ ∥ ☿
12 04 ☽ ∠ ♂
14 41 ☿ △ ♅
18 01 ☽ □ ♃
21 25 ☉ △ ♅

28 Friday
03 18 ☿ ∠ ♂
04 55 ☽ ⊥ ♂
04 58 ☽ ∥ ☉
06 01 ☽ △ ♆
09 10 ☽ ∠ ⯓
11 13 ☽ ∥ ♀
11 16 ♂ # ♆
12 23 ☽ ∥ ♄
14 02 ☽ ♇ ♅
16 20 ☽ ⊻ ♂
20 06 ☽ ∥ ⯓
21 40 ☽ ☌ ♀

29 Saturday
00 26 ☽ ⊻ ♄
01 25 ☽ ⊥ ⯓
02 26 ☽ Q ♃
04 46 ♀ ∥ ♄
06 20 ☽ ± ♅
11 30 ☽ ⊥ ♄
12 25 ☽ ⊻ ⯓
17 21 ☽ ⊼ ♅
21 10 ☽ ⊻ ☉

30 Sunday
00 36 ☽ ⊻ ☿
00 42 ☽ ✶ ♃
01 30 ☿ □ ♃
04 21 ☽ ∠ ♄
05 31 ♀ ⊻ ♄
07 09 ☽ # ♆
09 22 ☽ ⊥ ☉
09 45 ☽ ∥ ♂
13 00 ☽ □ ♆
13 29 ☽ ⊥ ☿
13 45 ♀ ⊥ ⯓
14 41 ☉ ∠ ♂

31 Monday
03 23 ♀ Q ♃
03 42 ☽ ☌ ♂
03 58 ☽ ∠ ☉
05 18 ☽ ∠ ♃
08 16 ☽ # ♅
09 07 ☽ ✶ ♄
09 12 ☽ ∠ ☿
11 55 ☽ ⊻ ♀
21 22 ☽ ☌ ⯓
21 46 ☉ □ ♃
22 51 ☽ ⊥ ♃
23 34 ☽ Q ♆

JANUARY 2047

LONGITUDES

Date	Sidereal time h m s	Sun ☉ ° ' "	Moon ☽ ° ' "	Moon ☽ 24.00 ° ' "	Mercury ☿ ° '	Venus ♀ ° '	Mars ♂ ° '	Jupiter ♃ ° '	Saturn ♄ ° '	Uranus ♅ ° '	Neptune ♆ ° '	Pluto ♇ ° '
01	18 44 17	10 ♑ 58 41	11 ♓ 04 48	17 ♓ 08 28	14 ♑ 09	00 ♒ 00	25 ♒ 32	10 ♈ 26	27 ♐ 26	06 ♍ 12	16 ♉ 44	03 ♓ 38
02	18 48 13	11 59 51	23 ♓ 09 27	29 ♓ 08 19	15 46	01 15	26 18	10 31	27 33	06 R 11	16 R 43	03 39
03	18 52 10	13 01 00	05 ♈ 05 38	11 ♈ 02 00	17 23	02 30	27 05	10 37	27 40	06 10	16 42	03 40
04	18 56 06	14 02 09	16 ♈ 58 02	22 ♈ 54 19	19 01	03 45	27 52	10 43	27 46	06 09	16 42	03 41
05	19 00 03	15 03 18	28 ♈ 51 26	04 ♉ 50 00	20 40	05 00	28 39	10 50	27 53	06 07	16 41	03 42
06	19 03 59	16 04 27	10 ♉ 50 32	16 ♉ 53 34	22 18	06 15	29 ♒ 25	10 56	28 00	06 06	16 40	03 44
07	19 07 56	17 05 35	22 ♉ 59 35	29 ♉ 09 01	23 57	07 30	00 ♓ 12	11 03	28 07	06 04	16 40	03 45
08	19 11 52	18 06 43	05 ♊ 22 15	11 ♊ 39 35	25 36	08 45	00 59	11 09	28 14	06 03	16 39	03 46
09	19 15 49	19 07 51	18 ♊ 01 16	24 ♊ 27 29	27 16	10 00	01 46	11 16	28 21	06 02	16 38	03 47
10	19 19 46	20 08 58	00 ♋ 58 19	07 ♋ 33 46	28 ♑ 55	11 15	02 32	11 23	28 28	06 00	16 38	03 49
11	19 23 42	21 10 05	14 ♋ 13 47	20 ♋ 58 11	00 ♒ 35	12 30	03 19	11 30	28 34	05 58	16 37	03 50
12	19 27 39	22 11 12	27 ♋ 46 46	04 ♌ 39 13	02 15	13 45	04 06	11 38	28 41	05 57	16 37	03 51
13	19 31 35	23 12 18	11 ♌ 35 11	18 ♌ 34 16	03 55	15 00	04 53	11 45	28 48	05 55	16 36	03 52
14	19 35 32	24 13 25	25 ♌ 36 00	02 ♍ 39 56	05 35	16 15	05 39	11 53	28 54	05 53	16 36	03 54
15	19 39 28	25 14 30	09 ♍ 45 36	16 ♍ 52 30	07 15	17 30	06 26	12 01	29 01	05 52	16 35	03 55
16	19 43 25	26 15 36	24 ♍ 00 12	01 ♎ 08 14	08 54	18 45	07 13	12 09	29 08	05 50	16 35	03 57
17	19 47 21	27 16 41	08 ♎ 16 13	15 ♎ 23 47	10 33	19 59	07 59	12 17	29 14	05 48	16 35	03 58
18	19 51 18	28 17 47	22 ♎ 30 34	29 ♎ 36 17	12 11	21 14	08 46	12 25	29 21	05 46	16 34	03 59
19	19 55 15	29 ♑ 18 52	06 ♏ 40 40	13 ♏ 43 27	13 49	22 29	09 33	12 34	29 27	05 44	16 34	04 01
20	19 59 11	00 ♒ 19 57	20 ♏ 44 26	27 ♏ 43 24	15 25	23 44	10 19	12 42	29 34	05 42	16 34	04 02
21	20 03 08	01 21 01	04 ♐ 40 11	11 ♐ 34 34	17 00	24 58	11 06	12 51	29 40	05 40	16 34	04 04
22	20 07 04	02 22 05	18 ♐ 26 24	25 ♐ 15 31	18 33	26 13	11 52	13 00	29 46	05 38	16 34	04 05
23	20 11 01	03 23 09	02 ♑ 01 43	08 ♑ 44 53	20 04	27 28	12 39	13 09	29 53	05 36	16 33	04 07
24	20 14 57	04 24 13	15 ♑ 24 51	22 ♑ 01 29	21 33	28 43	13 26	13 18	29 ♐ 59	05 34	16 33	04 08
25	20 18 54	05 25 15	28 ♑ 34 41	05 ♒ 04 20	22 58	29 ♒ 57	14 12	13 27	00 ♑ 05	05 32	16 33	04 10
26	20 22 50	06 26 17	11 ♒ 30 25	17 ♒ 52 54	24 19	01 ♓ 12	14 59	13 37	00 11	05 30	16 33	04 11
27	20 26 47	07 27 18	24 ♒ 11 49	00 ♓ 27 14	25 36	02 26	15 45	13 46	00 18	05 28	16 D 33	04 13
28	20 30 44	08 28 18	06 ♓ 39 17	12 ♓ 48 09	26 48	03 41	16 32	13 56	00 24	05 25	16 33	04 14
29	20 34 40	09 29 16	18 ♓ 54 04	24 ♓ 57 19	27 54	04 55	17 18	14 06	00 30	05 23	16 33	04 16
30	20 38 37	10 30 14	00 ♈ 58 14	06 ♈ 57 13	28 54	06 10	18 05	14 16	00 36	05 21	16 34	04 17
31	20 42 33	11 ♒ 31 11	12 ♈ 54 41	18 ♈ 51 08	29 ♒ 45	07 ♓ 24	18 ♓ 51	14 ♈ 26	00 ♑ 42	05 ♍ 18	16 ♉ 34	04 ♓ 19

Date	Moon True ☊ ° '	Moon Mean ☊ ° '	Moon ☽ Latitude ° '
01	25 ♑ 16	25 ♑ 59	03 N 47
02	25 D 17	25 56	04 28
03	25 18	25 53	04 57
04	25 R 18	25 50	05 13
05	25 18	25 47	05 16
06	25 17	25 43	05 05
07	25 16	25 40	04 40
08	25 15	25 37	04 02
09	25 14	25 34	03 12
10	25 13	25 31	02 10
11	25 13	25 27	01 N 01
12	25 D 13	25 24	00 S 14
13	25 13	25 21	01 30
14	25 13	25 18	02 41
15	25 R 13	25 15	03 43
16	25 13	25 12	04 31
17	25 13	25 08	05 03
18	25 13	25 05	05 17
19	25 D 13	25 02	05 11
20	25 13	24 59	04 46
21	25 14	24 56	04 05
22	25 14	24 53	03 10
23	25 15	24 49	02 05
24	25 16	24 46	00 S 54
25	25 R 16	24 43	00 N 18
26	25 15	24 40	01 29
27	25 14	24 37	02 33
28	25 12	24 33	03 29
29	25 10	24 30	04 15
30	25 08	24 27	04 48
31	25 ♑ 06	24 ♑ 24	05 N 08

DECLINATIONS

Date	Sun ☉ ° '	Moon ☽ ° '	Mercury ☿ ° '	Venus ♀ ° '	Mars ♂ ° '	Jupiter ♃ ° '	Saturn ♄ ° '	Uranus ♅ ° '	Neptune ♆ ° '	Pluto ♇ ° '
01	22 S 59	03 S 54	24 S 37	21 S 34	14 S 03	02 N 56	22 S 21	09 N 58	15 N 07	21 S 12
02	22 54	01 N 23	24 29	21 19	13 46	02 58	22 21	09 58	15 07	21 12
03	22 48	06 34	24 19	21 03	13 29	03 01	22 21	09 59	15 07	21 11
04	22 42	11 29	24 08	20 46	13 13	03 03	22 21	09 59	15 06	21 11
05	22 35	15 59	23 55	20 29	12 56	03 06	22 21	10 00	15 06	21 10
06	22 28	19 54	23 40	20 12	12 39	03 09	22 22	10 00	15 06	21 09
07	22 20	23 02	23 24	19 53	12 22	03 12	22 22	10 01	15 06	21 09
08	22 13	25 10	23 07	19 34	12 05	03 15	22 22	10 01	15 06	21 08
09	22 04	26 05	22 47	19 15	11 47	03 18	22 22	10 02	15 06	21 08
10	21 55	25 36	22 27	18 55	11 30	03 21	22 22	10 03	15 06	21 07
11	21 46	23 41	22 04	18 34	11 12	03 24	22 22	10 03	15 06	21 06
12	21 36	20 22	21 41	18 13	10 55	03 27	22 22	10 04	15 05	21 06
13	21 26	15 52	21 15	17 52	10 37	03 30	22 22	10 04	15 05	21 05
14	21 16	10 28	20 49	17 30	10 19	03 33	22 22	10 05	15 05	21 04
15	21 05	04 N 28	20 20	17 07	10 01	03 36	22 22	10 06	15 05	21 04
16	20 54	01 S 46	19 51	16 44	09 43	03 40	22 23	10 07	15 05	21 03
17	20 42	07 55	19 20	16 21	09 25	03 43	22 23	10 07	15 05	21 02
18	20 30	13 39	18 48	15 57	09 07	03 47	22 23	10 08	15 05	21 02
19	20 17	18 38	18 14	15 32	08 49	03 50	22 23	10 09	15 05	21 01
20	20 05	22 32	17 39	15 07	08 30	03 54	22 23	10 09	15 05	21 01
21	19 51	25 05	17 04	14 42	08 12	03 57	22 23	10 10	15 05	21 00
22	19 38	26 05	16 27	14 17	07 54	04 01	22 23	10 11	15 05	20 59
23	19 24	25 30	15 50	13 51	07 35	04 05	22 23	10 12	15 05	20 59
24	19 09	23 27	15 13	13 24	07 17	04 08	22 23	10 13	15 05	20 58
25	18 55	20 08	14 35	12 58	06 58	04 12	22 23	10 13	15 05	20 57
26	18 40	15 54	13 57	12 31	06 39	04 16	22 23	10 14	15 05	20 57
27	18 24	11 02	13 19	12 03	06 21	04 20	22 23	10 15	15 05	20 56
28	18 08	05 49	12 42	11 36	06 02	04 24	22 23	10 16	15 05	20 55
29	17 52	00 S 29	12 06	11 08	05 43	04 28	22 23	10 17	15 06	20 55
30	17 36	04 N 47	11 32	10 39	05 24	04 32	22 23	10 18	15 06	20 54
31	17 S 19	09 N 50	10 S 59	10 S 11	05 S 06	04 N 36	22 S 23	10 N 18	15 N 06	20 S 53

ZODIAC SIGN ENTRIES

Date	h	m	Planets
01	11	55	♀ ♒
03	01	44	☽ ♈
05	14	18	☽ ♉
07	05	46	♂ ♓
08	01	39	☽ ♊
10	10	13	☽ ♋
11	03	33	☿ ♒
12	15	53	☽ ♌
14	19	29	☽ ♍
16	22	05	☽ ♎
19	00	40	☽ ♏
20	04	10	☉ ♒
21	03	56	☽ ♐
23	08	24	☽ ♑
24	15	40	♄ ♑
25	12	54	♀ ♓
25	14	37	☽ ♒
27	23	08	☽ ♓
30	10	04	☽ ♈
31	19	34	☿ ♓

LATITUDES

Date	Mercury ☿ ° '	Venus ♀ ° '	Mars ♂ ° '	Jupiter ♃ ° '	Saturn ♄ ° '	Uranus ♅ ° '	Neptune ♆ ° '	Pluto ♇ ° '
01	01 S 57	01 S 28	01 S 06	01 S 18	01 N 04	00 N 47	01 S 47	11 S 53
04	02 04	01 31	01 04	01 17	01 04	00 47	01 47	11 52
07	02 07	01 33	01 02	01 16	01 04	00 47	01 47	11 52
10	02 07	01 35	01 00	01 16	01 04	00 47	01 47	11 51
13	02 03	01 36	00 58	01 15	01 03	00 48	01 47	11 51
16	01 53	01 37	00 55	01 14	01 03	00 48	01 47	11 50
19	01 38	01 37	00 53	01 13	01 03	00 48	01 46	11 50
22	01 15	01 36	00 51	01 13	01 03	00 48	01 46	11 50
25	00 46	01 35	00 49	01 12	01 03	00 48	01 46	11 49
28	00 S 08	01 32	00 47	01 11	01 03	00 48	01 46	11 49
31	00 N 36	01 S 30	00 S 45	01 S 11	01 N 03	00 N 48	01 S 46	11 S 49

DATA

Julian Date	2468712
Delta T	+80 seconds
Ayanamsa	24° 31' 05"
Synetic vernal point	04° ♓ 35' 53"
True obliquity of ecliptic	23° 26' 03"

MOON'S PHASES, APSIDES AND POSITIONS ☽

Date	h	m	Phase	Longitude ° '	Eclipse Indicator
04	05	31	☽	13 ♈ 46	
12	01	21	○	21 ♋ 44	total
18	22	32	☾	28 ♎ 45	
26	01	44	●	06 ♒ 00	Partial

Day	h	m		
04	09	26	Apogee	
16	21	31	Perigee	
02	05	40	ON	
09	16	04	Max dec	26° N 06'
16	05	12	OS	
22	15	06	Max dec	26° S 06'
29	14	11	ON	

LONGITUDES

Date	Chiron ⚷ ° '	Ceres ⚳ ° '	Pallas ⚴ ° '	Juno ⚵ ° '	Vesta ⚶ ° '	Black Moon Lilith ⚸ ° '
01	23 ♎ 12	14 ♐ 48	14 ♏ 54	27 ♏ 37	20 ♏ 40	15 ♈ 47
11	23 ♎ 49	18 ♐ 50	18 ♏ 32	00 ♐ 33	25 ♏ 27	16 ♈ 54
21	24 ♎ 14	22 ♐ 45	21 ♏ 55	03 ♐ 18	00 ♐ 06	18 ♈ 01
31	24 ♎ 28	26 ♐ 33	25 ♏ 01	05 ♐ 50	04 ♐ 36	19 ♈ 08

All ephemeris data is given at 12.00 UT and the Moon's longitude is additionally given for 24.00 UT

ASPECTARIAN (h m, Aspects)

01 Tuesday
00 54 ☽ ⊥ ♀
02 26 ☽ ☍ ♅
08 37 ♂ ∠ ♃
08 43 ☽ Q ♄
10 42 ☽ ⊻ ♃
11 47 ☽ ✶ ☉
15 45 ♀ ± ♅
16 22 ☽ # ♃
18 59 ☽ ✶ ☿
20 39 ☽ ∠ ♀
23 11 ☽ ✶ ♆

02 Wednesday
13 50 ☽ Q ☉
18 45 ☽ ⊻ ♂
19 18 ☽ ∥ ♃
20 53 ☽ □ ♄
22 42 ☽ Q ☿
23 11 ♀ ∥ ♇

03 Thursday
01 59 ☿ △ ♆
05 11 ☽ ∠ ♆
06 10 ☽ ✶ ♀
07 40 ☽ ⊥ ♂
09 07 ☽ ⊻ ♇
14 09 ☽ ⊼ ♅
21 15 ☽ ⊥ ♇
23 16 ☽ ☌ ♃
23 20 ☽ ⊥ ♆

04 Friday
02 16 ☽ ± ♅
03 07 ☽ ∠ ♂
04 30 ☽ ∥ ♅
05 31 ☽ □ ☉
05 51 ♃ ⊥ ♆
06 58 ☿ ∠ ♇
08 46 ♂ ✶ ♄
09 16 ☽ Q ♀
10 35 ♀ ⊻ ♇
11 27 ☽ ⊻ ♆
12 22 ♀ ⊥ ♄
15 29 ☽ ∠ ♇
16 49 ☽ □ ☿
20 25 ☽ # ♂
20 25 ☽ ₽ ♅

05 Saturday
07 06 ☽ ∥ ♆
10 02 ☽ △ ♄
11 32 ☽ ✶ ♂
18 37 ☿ ₽ ♅
21 45 ☽ ✶ ♇

06 Sunday
01 47 ☽ □ ♀
02 33 ☽ △ ♅
08 59 ♀ ⊼ ♅
12 11 ☽ ⊻ ♃
13 14 ☽ Q ♂
13 50 ☽ # ♀
16 20 ☽ ₽ ♄
20 49 ☽ # ♇
21 42 ☽ Q ♇
23 20 ☽ △ ☉
23 33 ☽ ☌ ♆

07 Monday
00 11 ☽ ⊥ ♃
01 53 ☉ △ ♆
06 12 ☽ # ♄
06 18 ☽ # ☉
08 25 ☉ ∥ ♄
10 16 ☽ ± ♄
14 10 ☽ △ ☿
15 03 ☽ # ☿
18 01 ☽ ∠ ♃
18 52 ☿ ⊥ ♂
22 05 ☽ ⊼ ♄

08 Tuesday
02 59 ☽ □ ♂
07 16 ☽ ₽ ☉
08 55 ☽ □ ♇
13 18 ☽ □ ♅
19 12 ☽ △ ♀
23 09 ☽ ✶ ♃
23 30 ☽ ₽ ☿

09 Wednesday
01 59 ☽ ± ☉
03 44 ☉ ∠ ♇
09 24 ☽ ⊻ ♆
14 15 ☽ ⊼ ☉
18 57 ☽ ± ☿
19 41 ☿ ⊥ ♇
20 37 ☽ ⊥ ♆
21 53 ☽ Q ♃
23 11 ☽ Q ♅

10 Thursday
02 24 ☽ ₽ ♀
04 49 ☿ ⊻ ♄
07 21 ☽ ☍ ♄
07 41 ☽ ⊼ ☿
13 12 ☽ ∠ ♆
14 47 ♀ ✶ ♃
15 03 ☽ △ ♂
17 09 ☿ ∥ ♄
17 11 ☽ △ ♇
19 14 ☿ Q ♃
20 38 ☽ ± ♀
21 09 ☽ ✶ ♅

11 Friday
03 19 ☿ ± ♅
07 04 ☽ □ ♃
07 32 ☉ ₽ ♅
08 35 ☽ ⊼ ♀
16 16 ☽ ✶ ♆
19 45 ☽ ₽ ♂
20 13 ☽ ₽ ♇
22 40 ☽ # ♄
23 59 ☽ ∠ ♅

12 Saturday
01 21 ☽ ☍ ☉
02 28 ☽ # ☿
03 40 ☽ # ☉
04 12 ♂ ☌ ♇
07 27 ☽ # ♇
10 31 ♀ ∠ ♄
12 08 ☽ ± ♇
12 35 ☽ ± ♂
13 28 ☽ Q ♆
13 36 ☽ ⊼ ♄
15 47 ☽ ⊥ ♅
18 50 ☉ ∥ ☿
20 54 ☽ ☍ ☿
22 37 ☽ ⊼ ♇
23 42 ☽ ⊼ ♂

13 Sunday
00 09 ☽ ± ♄
01 12 ☽ # ♀
02 13 ☽ ⊻ ♅
03 42 ♂ Q ♆
11 23 ☿ ⊻ ♇
12 18 ☽ △ ♃
15 41 ☽ ∥ ♆
15 50 ☽ ₽ ♄
18 27 ☽ ☍ ♀
20 37 ☽ □ ♆
21 43 ☿ ∥ ♇

14 Monday
01 35 ☿ ⊥ ♄
09 28 ☽ ⊼ ☉
12 38 ☽ # ♂
12 49 ♃ ± ♅
13 33 ☽ ∥ ♅
14 00 ☿ ⊻ ♂
14 12 ☽ ₽ ♃
16 23 ☿ ⊼ ♅
17 40 ☽ △ ♄
18 39 ♀ □ ♆
19 00 ♂ ☍ ♅
20 28 ☽ ± ☉
20 31 ♂ ⊥ ♃

15 Tuesday
02 06 ☽ ☍ ♇
05 25 ☽ ☌ ♅
05 37 ☽ ± ♃
05 59 ♂ # ♅
06 03 ☽ ☍ ♂
07 11 ☽ ⊼ ☿
12 53 ☽ ₽ ☉
14 53 ☉ ∥ ♇
15 18 ☽ ∥ ♃
15 51 ☽ ⊼ ♃
18 39 ☽ ± ☿
23 31 ☽ △ ♆

16 Wednesday
02 18 ☽ ⊼ ♀
11 48 ☽ ₽ ☿
13 22 ☽ ± ♀
16 05 ☽ △ ☉
19 21 ☽ # ♃
20 41 ☽ □ ♄

17 Thursday
00 45 ☽ ₽ ♆
04 45 ☽ ⊼ ♇
05 57 ☽ ₽ ♀
07 51 ☽ ⊻ ♅
11 30 ☽ ⊼ ♂
14 51 ☽ ± ♇
15 53 ☽ ± ♆
16 20 ☽ △ ☿
17 45 ☽ ∥ ♂
17 56 ☽ ⊥ ♅
18 49 ☽ ☍ ♃
20 57 ☽ # ♅
22 11 ☽ ± ♂

18 Friday
01 59 ☽ ⊼ ♆
03 13 ☽ Q ♄
04 34 ☉ ⊥ ♇
06 03 ☽ ₽ ♇
09 04 ☽ ∠ ♅
09 39 ☽ △ ♀
14 15 ☽ ₽ ♂
15 45 ☿ ✶ ♃
18 30 ☽ # ♆
21 44 ☽ ∥ ♀
22 32 ☽ □ ☉
23 39 ☽ ✶ ♄

19 Saturday
07 28 ☽ △ ♇
10 09 ☽ ∥ ☿
10 24 ☽ ✶ ♅
15 39 ☉ ⊻ ♄
17 10 ☽ △ ♂
20 52 ☽ ∥ ☉
21 39 ☉ ± ♅
22 07 ☽ ⊼ ♃
22 11 ☿ ∠ ♄

20 Sunday
01 21 ☽ ∠ ♄
01 44 ☽ □ ☿
01 45 ☽ ∥ ♇
04 51 ☽ ☍ ♆
06 49 ☽ Q ♅
07 33 ☽ Q ☉
08 29 ☽ ± ♃
10 54 ☽ ∥ ♄
14 15 ♀ # ♆
16 53 ☽ ⊥ ♄
17 38 ☽ □ ♀
22 15 ☉ Q ♃

21 Monday
00 06 ☽ ₽ ♃
03 17 ☽ ⊻ ♄
05 17 ☿ □ ♆
05 48 ☽ ✶ ☉
10 57 ☽ □ ♇
12 39 ☽ Q ☿
13 44 ☽ □ ♅
23 50 ☽ □ ♂

22 Tuesday
02 23 ☽ △ ♃
03 53 ☽ Q ♀
08 23 ♂ Q ♄
08 42 ☽ ⊼ ♆
09 58 ☽ ∠ ☉
12 14 ☽ ✶ ☿
18 25 ☽ Q ♇
19 14 ☽ ± ♆

23 Wednesday
03 05 ☽ ✶ ♀
03 06 ☽ ⊥ ☉
08 09 ☽ ☌ ♄
09 24 ☽ Q ♂
11 10 ☽ ₽ ♆
14 37 ☽ ⊻ ☉
15 43 ☽ ✶ ♇
18 06 ☽ ∠ ☿
18 21 ☽ △ ♅

24 Thursday
03 01 ♀ ∠ ♃
05 28 ☉ ⊻ ♇
07 07 ♂ ⊻ ♃
08 08 ☽ □ ♃
08 11 ☽ ✶ ♂
08 36 ☽ ∠ ♀
12 16 ☽ ⊥ ☿
14 04 ☽ △ ♆
16 42 ☿ # ♆
18 45 ☽ ∠ ♇
20 39 ☽ ∥ ♄
21 19 ☽ ₽ ♅

25 Friday
00 29 ☽ ⊻ ☿
02 38 ☽ ⊥ ♀
06 44 ☽ ∥ ♇
11 14 ☽ ⊥ ♇
13 13 ☽ ∠ ♂
13 45 ☽ ± ♅
14 32 ☉ ⊼ ♅
14 48 ☽ ⊻ ♀
14 48 ☽ ⊻ ♄
14 50 ♀ ✶ ♄
17 22 ☽ Q ♃
19 56 ☽ ∥ ☉
22 19 ☽ ⊻ ♇

26 Saturday
00 49 ☽ ⊼ ♅
01 44 ☽ ☌ ☉
02 00 ☽ ⊥ ♄
05 32 ☉ ⊥ ♄
06 58 ☽ ⊥ ♂
14 53 ♆ St D
16 00 ☽ ✶ ♃
16 13 ☽ # ♆
18 57 ☽ ⊻ ♂
18 59 ☽ ∠ ♄
21 30 ☽ □ ♆
23 26 ☽ ∥ ☿

27 Sunday
06 42 ☽ ∥ ♀
15 00 ☽ ☌ ☿
15 42 ☽ # ♅
20 53 ☽ ∠ ♃
23 47 ☽ ✶ ♄

28 Monday
05 36 ☽ ☌ ♀
07 18 ☽ ☌ ♇
07 56 ☽ Q ♆
09 37 ☽ ☍ ♅
11 00 ☽ ∥ ♂
12 49 ♂ ✶ ♆
14 31 ☽ ⊥ ♃
15 51 ☽ ⊻ ☉
18 20 ☽ # ♃
22 53 ♀ ☌ ♇
23 18 ☽ Q ♄

29 Tuesday
02 24 ☽ ⊻ ♃
04 40 ☽ ⊥ ☉
04 53 ♀ Q ♆
07 23 ☽ ✶ ♆
08 38 ☽ ☌ ♂
20 37 ♀ ☍ ♅

30 Wednesday
00 05 ☽ ∠ ☉
07 30 ☽ ⊻ ☿
10 48 ☽ ∥ ♃
11 15 ☽ □ ♄
13 11 ☽ ∠ ♆
14 42 ☽ # ♂
18 39 ☽ ⊻ ♇
20 30 ☽ ⊥ ☿
20 45 ☽ ⊼ ♅
23 38 ☽ ⊻ ♀

31 Thursday
00 03 ☿ ∠ ♃
05 39 ♀ # ♅
06 45 ☽ ⊥ ♇
07 16 ☽ ⊥ ♆
08 47 ☽ ± ♅
08 56 ☽ ✶ ☉
13 07 ☽ ⊥ ♀
13 34 ☽ # ♀
14 23 ☽ ∥ ♅
15 06 ☽ ☌ ♃
15 59 ☽ ∠ ☿
17 12 ☽ # ☿
19 22 ☽ ⊻ ♆

FEBRUARY 2047

Date	Sidereal time h m s	Sun ☉	Moon ☽	Moon ☽ 24.00	Mercury ☿	Venus ♀	Mars ♂	Jupiter ♃	Saturn ♄	Uranus ♅	Neptune ♆	Pluto ♇
		° ' "	° ' "	° ' "	° '	° '	° '	° '	° '	° '	° '	° '
01	20 46 30	12 ♒ 32 06	24 ♈ 47 03	00 ♉ 43 00	00 ♓ 29	08 ♓ 39	19 ♓ 38	14 ♈ 36	00 ♑ 48	05 ♍ 16	16 ♉ 34	04 ♓ 20
02	20 50 26	13 33 00	06 ♉ 39 33	12 ♉ 37 17	01 03	09 53	20 24	14 46	00 54	05 R 14	16 34	04 22
03	20 54 23	14 33 53	18 ♉ 36 47	24 ♉ 38 41	01 28	11 08	21 10	14 56	00 59	05 11	16 34	04 23
04	20 58 19	15 34 44	00 ♊ 43 34	06 ♊ 52 01	01 42	12 22	21 57	15 07	01 05	05 09	16 35	04 25
05	21 02 16	16 35 35	13 ♊ 04 34	19 ♊ 21 46	01 R 45	13 36	22 43	15 18	01 11	05 06	16 35	04 27
06	21 06 13	17 36 24	25 ♊ 44 02	02 ♋ 11 48	01 37	14 50	23 29	15 28	01 16	05 04	16 35	04 28
07	21 10 09	18 37 11	08 ♋ 45 20	15 ♋ 24 51	01 18	16 05	24 15	15 39	01 22	05 02	16 36	04 30
08	21 14 06	19 37 57	22 ♋ 10 27	29 ♋ 02 04	00 49	17 19	25 02	15 50	01 27	04 59	16 36	04 32
09	21 18 02	20 38 42	05 ♌ 59 32	13 ♌ 02 31	00 ♓ 09	18 33	25 48	16 01	01 33	04 57	16 37	04 33
10	21 21 59	21 39 25	20 ♌ 10 33	27 ♌ 23 01	29 ♒ 21	19 47	26 34	16 12	01 38	04 54	16 37	04 35
11	21 25 55	22 40 06	04 ♍ 39 11	11 ♍ 58 13	28 24	21 01	27 20	16 23	01 44	04 51	16 38	04 36
12	21 29 52	23 40 47	19 ♍ 19 13	26 ♍ 41 14	27 22	22 15	28 06	16 35	01 49	04 49	16 38	04 38
13	21 33 48	24 41 26	04 ♎ 03 20	11 ♎ 24 37	26 15	23 29	28 52	16 46	01 54	04 46	16 39	04 40
14	21 37 45	25 42 04	18 ♎ 44 15	26 ♎ 01 30	25 06	24 43	29 ♓ 38	16 58	01 59	04 44	16 39	04 41
15	21 41 42	26 42 41	03 ♏ 15 44	10 ♏ 26 28	23 56	25 57	00 ♈ 24	17 09	02 04	04 41	16 40	04 43
16	21 45 38	27 43 17	17 ♏ 33 19	24 ♏ 36 03	22 48	27 10	01 10	17 21	02 09	04 39	16 41	04 45
17	21 49 35	28 43 52	01 ♐ 34 32	08 ♐ 28 44	21 42	28 24	01 56	17 33	02 14	04 36	16 42	04 46
18	21 53 31	29 ♒ 44 26	15 ♐ 18 41	22 ♐ 04 31	20 41	29 ♓ 38	02 42	17 45	02 19	04 33	16 42	04 48
19	21 57 28	00 ♓ 44 58	28 ♐ 46 23	05 ♑ 24 26	19 45	00 ♈ 52	03 28	17 57	02 24	04 31	16 43	04 50
20	22 01 24	01 45 30	11 ♑ 58 54	18 ♑ 29 57	18 55	02 05	04 14	18 09	02 29	04 28	16 44	04 51
21	22 05 21	02 45 59	24 ♑ 57 46	01 ♒ 22 32	18 12	03 19	04 59	18 21	02 33	04 25	16 45	04 53
22	22 09 17	03 46 28	07 ♒ 44 25	14 ♒ 03 32	17 37	04 32	05 45	18 33	02 38	04 23	16 46	04 55
23	22 13 14	04 46 55	20 ♒ 20 00	26 ♒ 33 56	17 09	05 46	06 31	18 46	02 42	04 20	16 47	04 56
24	22 17 11	05 47 20	02 ♓ 45 26	08 ♓ 54 34	16 48	06 59	07 17	18 58	02 47	04 18	16 48	04 58
25	22 21 07	06 47 44	15 ♓ 01 29	21 ♓ 06 15	16 35	08 12	08 02	19 10	02 51	04 15	16 49	05 00
26	22 25 04	07 48 05	27 ♓ 09 02	03 ♈ 09 58	16 29	09 26	08 48	19 23	02 56	04 12	16 50	05 01
27	22 29 00	08 48 25	09 ♈ 09 16	15 ♈ 07 10	16 D 30	10 39	09 33	19 36	03 00	04 10	16 51	05 03
28	22 32 57	09 ♓ 48 44	21 ♈ 03 57	26 ♈ 59 54	16 ♒ 37	11 ♈ 52	10 ♈ 19	19 ♈ 48	03 ♑ 04	04 ♍ 07	16 ♉ 52	05 ♓ 05

Date	Moon True ☊	Moon Mean ☊	Moon ☽ Latitude	DECLINATIONS Sun ☉	Moon ☽	Mercury ☿	Venus ♀	Mars ♂	Jupiter ♃	Saturn ♄	Uranus ♅	Neptune ♆	Pluto ♇
	° '	° '	° '	° '	° '	° '	° '	° '	° '	° '	° '	° '	° '
01	25 ♑ 04	24 ♑ 21	05 N 15	17 S 02	14 N 29	10 S 29	09 S 42	04 S 47	04 N 40	22 S 23	10 N 19	15 N 06	20 S 53
02	25 R 03	24 18	05 09	16 45	18 36	10 01	09 13	04 28	04 44	22 23	10 20	15 06	20 52
03	25 D 03	24 14	04 49	16 28	21 59	09 37	08 43	04 09	04 49	22 23	10 21	15 06	20 51
04	25 04	24 11	04 16	16 10	24 29	09 16	08 14	03 50	04 53	22 23	10 22	15 06	20 51
05	25 05	24 08	03 31	15 52	25 51	08 59	07 44	03 31	04 57	22 23	10 23	15 06	20 50
06	25 07	24 05	02 35	15 33	25 57	08 46	07 14	03 12	05 01	22 22	10 24	15 07	20 49
07	25 08	24 02	01 29	15 15	24 37	08 37	06 44	02 53	05 06	22 22	10 25	15 07	20 49
08	25 09	23 59	00 N 16	14 56	21 53	08 34	06 14	02 34	05 10	22 22	10 26	15 07	20 48
09	25 R 08	23 55	00 S 59	14 36	17 49	08 35	05 43	02 15	05 15	22 22	10 27	15 07	20 47
10	25 07	23 52	02 13	14 17	12 39	08 40	05 13	01 56	05 19	22 22	10 28	15 07	20 47
11	25 04	23 49	03 20	13 57	06 42	08 50	04 42	01 37	05 23	22 22	10 29	15 08	20 46
12	25 00	23 46	04 14	13 37	00 N 20	09 03	04 11	01 18	05 28	22 22	10 29	15 08	20 46
13	24 55	23 43	04 52	13 17	06 S 05	09 20	03 40	00 59	05 33	22 22	10 30	15 08	20 45
14	24 51	23 39	05 11	12 57	12 07	09 40	03 09	00 40	05 37	22 22	10 31	15 08	20 44
15	24 48	23 36	05 09	12 36	17 26	10 02	02 38	00 21	05 42	22 22	10 32	15 08	20 44
16	24 46	23 33	04 48	12 16	21 40	10 25	02 07	00 S 02	05 46	22 22	10 33	15 09	20 43
17	24 D 46	23 30	04 10	11 55	24 33	10 49	01 35	00 N 17	05 51	22 22	10 34	15 09	20 42
18	24 46	23 27	03 19	11 34	25 55	11 14	01 04	00 36	05 56	22 21	10 35	15 09	20 42
19	24 48	23 24	02 17	11 12	25 43	11 38	00 33	00 55	06 00	22 21	10 36	15 10	20 41
20	24 49	23 20	01 S 09	10 51	24 03	12 02	00 S 01	01 14	06 05	22 21	10 37	15 10	20 41
21	24 R 50	23 17	00 N 01	10 29	21 07	12 24	00 N 30	01 32	06 10	22 21	10 38	15 10	20 40
22	24 49	23 14	01 10	10 07	17 13	12 46	01 01	01 51	06 15	22 21	10 39	15 10	20 39
23	24 46	23 11	02 14	09 45	12 35	13 06	01 33	02 10	06 20	22 21	10 40	15 11	20 39
24	24 41	23 08	03 11	09 23	07 31	13 24	02 04	02 29	06 24	22 21	10 41	15 11	20 38
25	24 35	23 05	03 58	09 01	02 S 14	13 40	02 36	02 47	06 29	22 21	10 42	15 11	20 38
26	24 26	23 01	04 34	08 39	03 N 03	13 54	03 07	03 06	06 34	22 21	10 43	15 12	20 37
27	24 18	22 58	04 57	08 16	08 11	14 07	03 38	03 25	06 39	22 20	10 44	15 12	20 36
28	24 ♑ 09	22 ♑ 55	05 N 07	07 S 53	12 N 57	14 S 17	04 N 09	03 N 43	06 N 44	22 S 20	10 N 45	15 N 12	20 S 36

ZODIAC SIGN ENTRIES

Date	h	m	Planets
01	22	33	☽ ♉
04	10	34	☽ ♊
06	19	56	☽ ♋
09	01	41	☽ ♌
09	16	51	☿ ♒
11	04	20	☽ ♍
13	05	24	☽ ♎
14	23	22	♂ ♈
15	06	35	☽ ♏
17	09	17	☽ ♐
18	18	10	☉ ♓
18	19	12	♀ ♈
19	14	13	☽ ♑
21	21	25	☽ ♒
24	06	39	☽ ♓
26	17	41	☽ ♈

LATITUDES

Date	Mercury ☿	Venus ♀	Mars ♂	Jupiter ♃	Saturn ♄	Uranus ♅	Neptune ♆	Pluto ♇
	° '	° '	° '	° '	° '	° '	° '	° '
01	00 N 53	01 S 29	00 S 44	01 S 10	01 N 03	00 N 48	01 S 46	11 S 49
04	01 43	01 25	00 42	01 10	01 03	00 48	01 45	11 48
07	02 33	01 21	00 39	01 09	01 03	00 48	01 45	11 48
10	03 14	01 16	00 37	01 09	01 03	00 48	01 45	11 48
13	03 38	01 11	00 35	01 08	01 03	00 48	01 45	11 48
16	03 42	01 05	00 33	01 07	01 03	00 48	01 45	11 48
19	03 26	00 58	00 30	01 07	01 03	00 48	01 45	11 48
22	02 56	00 51	00 28	01 06	01 03	00 48	01 44	11 48
25	02 18	00 43	00 26	01 06	01 04	00 48	01 44	11 48
28	01 38	00 35	00 24	01 06	01 04	00 48	01 44	11 49
31	00 N 59	00 S 26	00 S 21	01 S 05	01 N 04	00 N 48	01 S 44	11 S 49

DATA

Julian Date	2468743
Delta T	+80 seconds
Ayanamsa	24° 31' 10"
Synetic vernal point	04° ♓ 35' 48"
True obliquity of ecliptic	23° 26' 03"

MOON'S PHASES, APSIDES AND POSITIONS ☽

Date	h	m	Phase	Longitude ° '	Eclipse Indicator
03	03	09	☽	14 ♉ 11	
10	14	40	○	21 ♌ 46	
17	06	42	☾	28 ♏ 31	
24	18	26	●	06 ♓ 04	

Day	h	m		
01	06	37	Apogee	
13	00	51	Perigee	
06	01	38	Max dec	26° N 04'
12	13	13	0S	
18	20	49	Max dec	26° S 01'
25	22	06	0N	

LONGITUDES

Date	Chiron ⚷	Ceres ⚳	Pallas ⚴	Juno ⚵	Vesta ⚶	Black Moon Lilith ⚸
	° '	° '	° '	° '	° '	° '
01	24 ♎ 28	26 ♐ 55	25 ♏ 19	06 ♐ 05	05 ♐ 03	19 ♈ 15
11	24 ♎ 29	00 ♑ 33	28 ♏ 02	08 ♐ 20	09 ♐ 20	20 ♈ 22
21	24 ♎ 18	04 ♑ 00	00 ♐ 20	10 ♐ 17	13 ♐ 23	21 ♈ 30
31	23 ♎ 56	07 ♑ 13	02 ♐ 08	11 ♐ 53	17 ♐ 08	22 ♈ 37

All ephemeris data is given at 12.00 UT and the Moon's longitude is additionally given for 24.00 UT

ASPECTARIAN

h m Aspects

01 Friday
00 50 ☽ ⊻ ♂
00 58 ☽ ∠ ♇
02 54 ☽ bQ ♅
09 26 ☽ ∠ ♀
10 51 ♀ ⊥ ♃
11 27 ☽ Q ☉
13 49 ☽ ⊥ ♂
15 23 ☽ ∥ ♆
18 46 ♂ # ♃
19 42 ☿ # ♅

02 Saturday
00 09 ☽ ✶ ☿
00 16 ☽ △ ♄
01 32 ☽ # ☉
03 00 ☿ ✶ ♄
07 22 ☽ ✶ ♇
09 07 ☽ △ ♅
09 17 ☽ ∠ ♂
19 15 ☽ ✶ ♀

03 Sunday
01 22 ☽ Q ☿
03 09 ☽ □ ☉
03 21 ☽ # ♇
04 33 ☽ ⊻ ♃
06 43 ☽ bQ ♄
07 33 ☽ Q ♇
07 55 ☽ ☌ ♆
15 12 ☽ # ♄
16 43 ☽ ⊥ ♃
17 27 ☽ ✶ ♂
22 01 ☽ Q ♀
22 44 ☉ ✶ ♃

04 Monday
00 47 ☽ ± ♄
10 47 ☽ ∠ ♃
12 43 ☽ ⊼ ♄
13 55 ☽ □ ☿
18 44 ☽ Q ♂
19 15 ☽ □ ♇
20 38 ☽ □ ♅

05 Tuesday
01 12 ☉ ∠ ♄
03 06 ♀ Q ♄
07 01 ☿ St R
11 46 ☉ □ ♆
13 07 ☽ □ ♀
16 19 ☽ ✶ ♃
18 43 ☽ ⊻ ♆
19 19 ☽ △ ☉

06 Wednesday
00 06 ☉ ⊥ ♂
06 06 ☽ ⊥ ♆
07 01 ☽ Q ♅
07 31 ☽ □ ♂
15 17 ☽ Q ♃
22 22 ☽ ☍ ♄
22 43 ☽ △ ☿
22 53 ☽ ∠ ♆

07 Thursday
01 50 ☽ bQ ☉
02 19 ♀ ⊻ ♃
04 13 ☽ △ ♇
05 13 ☽ ✶ ♅
08 51 ☿ ✶ ♄
19 34 ☽ ± ☉
21 50 ☉ # ♆
22 08 ♀ ✶ ♆

08 Friday
00 36 ☽ □ ♃
01 09 ☽ bQ ☿
02 07 ☽ ✶ ♆
02 31 ☽ △ ♀
07 09 ☽ ⊼ ☉
07 19 ☽ bQ ♇
08 09 ☽ ∠ ♅
08 27 ☽ # ♄
11 16 ☿ ∠ ♃
16 26 ☽ ± ☿
17 19 ☽ △ ♂
19 09 ☽ # ♇
23 08 ☽ ± ♇
23 15 ☽ Q ♆
23 53 ☽ ⊥ ♅

09 Saturday
02 26 ☽ ⊼ ☿
04 18 ☽ ⊼ ♄
07 24 ☽ bQ ♀
09 32 ☽ ⊼ ♇
10 12 ☽ ⊻ ♅
14 41 ☽ ± ♄
20 40 ☽ bQ ♂

10 Sunday
00 14 ☽ ± ♀
01 04 ☽ ∥ ♆
04 24 ☽ # ☉
05 15 ☽ △ ♃
06 02 ☽ bQ ♄
06 02 ☽ □ ♆
06 10 ♄ bQ ♆
07 46 ♀ # ♃
11 17 ☽ ⊼ ♀
12 41 ☽ ± ♂
14 40 ☽ ☍ ☉
21 07 ☽ ∥ ♅
23 14 ☽ ⊼ ♂

11 Monday
02 20 ☽ ☍ ☿
03 55 ☽ # ☿
06 33 ☽ bQ ♃
07 09 ☽ △ ♄
11 55 ☽ ☍ ♇
12 20 ☽ ☌ ♅
16 57 ☽ ∥ ♃
20 19 ☽ # ♀
21 32 ☽ ± ♃

12 Tuesday
02 22 ☿ ⊻ ♂
07 28 ☽ ⊼ ♃
07 37 ☽ △ ♆
08 12 ☽ # ♂
15 59 ♆ Q ♇
17 13 ☽ ☍ ♀
17 46 ☽ ∥ ♂
19 38 ☽ ⊼ ☉
19 44 ♃ ⊻ ♆

13 Wednesday
00 12 ☽ ⊼ ☿
03 05 ☽ ☍ ♂
03 35 ☽ ∥ ♀
06 07 ☽ ± ☉
08 04 ☽ bQ ♆
08 28 ☽ □ ♄
09 16 ☽ ± ☿
09 56 ☽ # ♃
12 59 ☽ ⊼ ♇
13 10 ☽ ⊻ ♅
21 52 ☽ bQ ☉
22 46 ☽ ± ♆
22 48 ☽ ± ♇
22 54 ☽ bQ ☿
22 56 ☽ ⊥ ♅

14 Thursday
01 23 ☽ ∥ ☿
05 22 ☉ ☌ ☿
05 24 ☽ # ♅
08 35 ☽ ⊼ ♆
09 03 ☽ ☍ ♃
13 34 ☽ bQ ♇
13 37 ☽ ∠ ♅
14 04 ☽ Q ♄
15 18 ☽ ∥ ☉
15 53 ☿ ⊻ ♀
21 42 ☽ △ ☿
22 45 ☽ ⊼ ♀

15 Friday
00 19 ☽ △ ☉
01 07 ☽ # ♆
01 26 ♅ ☍ ♇
06 11 ☿ ⊥ ♂
06 59 ☽ ⊼ ♂
09 36 ☽ ± ♀
10 01 ☽ ✶ ♄
14 22 ☽ ✶ ♅
14 26 ☽ △ ♇
17 32 ☽ ± ♂

16 Saturday
02 03 ☽ bQ ♀
05 58 ☽ ∥ ♇
09 31 ☽ bQ ♂
10 28 ☽ Q ♅
10 31 ☽ ☍ ♆
11 19 ☽ ∠ ♄
11 39 ☽ ⊼ ♃
16 48 ☽ ∥ ♄
20 15 ☽ □ ☿
20 20 ☿ # ♅
22 00 ☽ ± ♃

17 Sunday
02 45 ☽ ⊥ ♄
04 19 ♂ ∠ ♆
04 40 ☿ ⊥ ♀
06 00 ☽ △ ♀
06 42 ☽ □ ☉
12 39 ☽ △ ♂
13 09 ☽ ⊻ ♄
13 42 ☽ bQ ♃
17 13 ☽ □ ♅
17 33 ☽ □ ♇
22 40 ♂ □ ♄

18 Monday
01 08 ☽ Q ☿
14 28 ☽ ⊼ ♆
16 22 ☽ △ ♃
16 38 ☽ Q ☉
20 52 ☽ ✶ ☿
22 19 ☉ ∥ ☿
23 58 ☉ ⊻ ♀

19 Tuesday
01 08 ☽ ± ♆
01 19 ☽ Q ♇
01 29 ♀ # ♂
15 51 ☽ ✶ ☉
16 09 ☽ □ ♀
17 19 ☽ bQ ♆
18 35 ☽ ☌ ♄
20 59 ☽ □ ♂
22 08 ☽ ∠ ☿
22 20 ☽ △ ♅
22 58 ☽ ✶ ♇

20 Wednesday
05 01 ♀ ∠ ♆
07 09 ☿ ∠ ♂
13 37 ☽ ⊥ ☿
19 10 ♂ ⊼ ♅
20 12 ♀ □ ♄
20 45 ☽ △ ♆
21 31 ☽ ∠ ☉
23 32 ☽ □ ♃

21 Thursday
00 05 ☽ ⊻ ☿
01 45 ☽ bQ ♅
02 31 ☉ # ♅
02 33 ☽ ∠ ♇
02 59 ☽ ∥ ♄
04 30 ☽ Q ♀
06 35 ☉ ✶ ♄
07 56 ☿ ✶ ♃
08 06 ☽ Q ♂
08 33 ♂ ⊻ ♇
10 37 ☿ ∠ ♀
15 07 ☽ ∥ ♇
15 39 ☽ ⊥ ☉
18 27 ☽ ± ♅
19 21 ☽ ⊥ ♇

22 Friday
02 18 ☽ ⊻ ♄
03 52 ☽ ⊻ ☉
05 18 ☽ ✶ ♀
05 24 ☉ ∠ ♃
05 41 ☽ ⊼ ♅
06 39 ☽ ⊻ ♇
08 00 ☽ ✶ ♂
09 02 ♀ ⊼ ♅
09 43 ☽ Q ♃
11 15 ☿ ∠ ♄
13 42 ☽ ⊥ ♄
19 32 ♀ ⊻ ♇
22 56 ☽ # ♆

23 Saturday
01 50 ☉ ☍ ♅
05 11 ☽ □ ♆
06 06 ☽ ☌ ☿
06 56 ☽ ∠ ♄
08 56 ☽ ✶ ♃
09 40 ☽ ∥ ☿
11 55 ☉ Q ♆
12 55 ☽ ∠ ♀
14 25 ☽ ∠ ♂
15 53 ☉ ☌ ♇
21 15 ☽ # ♅

24 Sunday
02 39 ☽ ∥ ☉
08 11 ☽ ⊥ ♀
08 56 ☽ ⊥ ♂
12 03 ☽ ✶ ♄
13 05 ☿ □ ♆
14 23 ☽ ∠ ♃
14 59 ☽ ☍ ♅
15 58 ☽ Q ♆
16 19 ☽ ☌ ♇
17 02 ☽ # ♃
18 26 ☽ ☌ ☉
21 09 ☽ ⊻ ♀
21 23 ☽ ⊻ ♂

25 Monday
03 08 ♀ ☌ ♂
08 18 ☽ ⊥ ♃
09 38 ☽ # ♂
10 32 ☽ # ♀
11 40 ☽ Q ♄
15 03 ☽ ⊻ ☿
15 31 ☽ ✶ ♆
19 10 ♃ bQ ♅
20 19 ☽ ⊻ ♃

26 Tuesday
02 48 ☽ ⊥ ☿
10 30 ♀ ∥ ♂
12 14 ☽ ∥ ♂
12 19 ☽ ∥ ♀
20 38 ☽ ∠ ☿
21 05 ☿ St D
21 20 ☽ ∠ ♆
23 35 ☽ □ ♄

27 Wednesday
02 02 ☽ ⊼ ♅
02 44 ♀ ± ♅
03 45 ☽ ⊻ ♇
04 36 ☽ ∥ ♃
11 14 ☽ ⊻ ☉
12 24 ☽ # ☉
12 52 ☽ ☌ ♂
14 01 ☽ ± ♅
15 21 ☽ ☌ ♀
15 24 ☽ ⊥ ♆
15 49 ☽ ⊥ ♇
15 55 ♀ ⊥ ♆
20 06 ♀ ⊥ ♇

28 Thursday
00 25 ☽ ⊥ ☉
00 36 ☽ ∥ ♅
02 55 ☽ ✶ ☿
03 30 ☽ ⊻ ♆
06 05 ♂ ± ♅
08 05 ☽ bQ ♅
09 24 ☽ ☌ ♃
10 00 ☽ ∠ ♇
19 25 ☽ # ☿
20 17 ☽ ∠ ☉

MARCH 2047

LONGITUDES

Date	Sidereal time h m s	Sun ☉ ° ′ ″	Moon ☽ ° ′ ″	Moon ☽ 24.00 ° ′ ″	Mercury ☿ ° ′	Venus ♀ ° ′	Mars ♂ ° ′	Jupiter ♃ ° ′	Saturn ♄ ° ′	Uranus ♅ ° ′	Neptune ♆ ° ′	Pluto ♇ ° ′
01	22 36 53	10 ♓ 49 00	02 ♉ 55 25	08 ♉ 50 55	16 ♒ 51	13 ♈ 05	11 ♈ 04	20 ♈ 01	03 ♑ 08	04 ♍ 04	16 ♉ 53	05 ♓ 06
02	22 40 50	11 49 14	14 ♉ 46 50	20 ♉ 43 41	17 10	14 18	11 50	20 14	03 12	04 R 02	16 54	05 08
03	22 44 46	12 49 27	26 ♉ 42 00	02 ♊ 42 23	17 34	15 31	12 35	20 27	03 16	03 59	16 55	05 10
04	22 48 43	13 49 37	08 ♊ 45 24	14 ♊ 51 42	18 04	16 44	13 21	20 40	03 20	03 57	16 57	05 11
05	22 52 40	14 49 46	21 ♊ 01 54	27 ♊ 16 37	18 38	17 57	14 06	20 53	03 23	03 54	16 58	05 13
06	22 56 36	15 49 52	03 ♋ 36 28	10 ♋ 02 01	19 16	19 10	14 51	21 06	03 27	03 52	16 59	05 15
07	23 00 33	16 49 56	16 ♋ 33 46	23 ♋ 12 10	19 58	20 23	15 36	21 19	03 30	03 49	17 00	05 16
08	23 04 29	17 49 58	29 ♋ 57 30	06 ♌ 50 00	20 44	21 35	16 22	21 32	03 34	03 46	17 02	05 18
09	23 08 26	18 49 58	13 ♌ 49 40	20 ♌ 56 22	21 34	22 48	17 07	21 45	03 37	03 44	17 03	05 20
10	23 12 22	19 49 56	28 ♌ 09 44	05 ♍ 29 12	22 27	24 00	17 52	21 59	03 41	03 41	17 04	05 21
11	23 16 19	20 49 52	12 ♍ 53 59	20 ♍ 23 08	23 23	25 13	18 37	22 12	03 44	03 39	17 06	05 23
12	23 20 15	21 49 46	27 ♍ 55 29	05 ♎ 29 48	24 22	26 25	19 22	22 26	03 47	03 36	17 07	05 25
13	23 24 12	22 49 37	13 ♎ 04 45	20 ♎ 39 01	25 23	27 37	20 07	22 39	03 50	03 34	17 09	05 26
14	23 28 09	23 49 27	28 ♎ 11 19	05 ♏ 40 30	26 27	28 ♈ 49	20 52	22 53	03 53	03 32	17 10	05 28
15	23 32 05	24 49 16	13 ♏ 05 34	20 ♏ 25 40	27 33	00 ♉ 01	21 37	23 06	03 56	03 29	17 12	05 29
16	23 36 02	25 49 03	27 ♏ 40 14	04 ♐ 48 49	28 42	01 13	22 22	23 20	03 58	03 27	17 13	05 31
17	23 39 58	26 48 48	11 ♐ 51 15	18 ♐ 47 26	29 ♒ 53	02 25	23 06	23 33	04 01	03 24	17 15	05 33
18	23 43 55	27 48 31	25 ♐ 37 31	02 ♑ 21 43	01 ♓ 05	03 37	23 51	23 47	04 04	03 22	17 16	05 34
19	23 47 51	28 48 13	09 ♑ 00 21	15 ♑ 33 46	02 20	04 49	24 36	24 01	04 06	03 20	17 18	05 36
20	23 51 48	29 ♓ 47 53	22 ♑ 02 25	28 ♑ 26 44	03 37	06 01	25 20	24 15	04 09	03 18	17 20	05 37
21	23 55 44	00 ♈ 47 31	04 ♒ 47 08	11 ♒ 04 02	04 55	07 12	26 05	24 28	04 11	03 15	17 21	05 39
22	23 59 41	01 47 08	17 ♒ 17 50	23 ♒ 28 54	06 15	08 24	26 50	24 42	04 13	03 13	17 23	05 40
23	00 03 38	02 46 42	29 ♒ 37 33	05 ♓ 44 04	07 37	09 35	27 34	24 56	04 15	03 11	17 25	05 42
24	00 07 34	03 46 15	11 ♓ 48 43	17 ♓ 51 43	09 00	10 46	28 19	25 10	04 17	03 09	17 26	05 43
25	00 11 31	04 45 46	23 ♓ 53 14	29 ♓ 53 26	10 25	11 58	29 03	25 24	04 19	03 07	17 28	05 45
26	00 15 27	05 45 14	05 ♈ 52 30	11 ♈ 50 33	11 51	13 09	29 ♈ 47	25 38	04 21	03 04	17 30	05 46
27	00 19 24	06 44 41	17 ♈ 47 44	23 ♈ 44 13	13 19	14 20	00 ♉ 32	25 52	04 23	03 02	17 32	05 48
28	00 23 20	07 44 06	29 ♈ 40 11	05 ♉ 35 50	14 48	15 31	01 16	26 06	04 24	03 00	17 34	05 49
29	00 27 17	08 43 28	11 ♉ 31 25	17 ♉ 27 12	16 19	16 42	02 00	26 20	04 26	02 58	17 35	05 51
30	00 31 13	09 42 49	23 ♉ 23 30	29 ♉ 20 42	17 52	17 52	02 44	26 35	04 27	02 56	17 37	05 52
31	00 35 10	10 ♈ 42 07	05 ♊ 19 14	11 ♊ 19 32	19 ♓ 25	19 ♉ 03	03 ♉ 29	26 ♈ 49	04 ♑ 29	02 ♍ 54	17 ♉ 39	05 ♓ 54

Date	Moon True ☊ ° ′	Moon Mean ☊ ° ′	Moon ☽ Latitude ° ′
01	24 ♑ 02	22 ♑ 52	05 N 04
02	23 R 56	22 49	04 48
03	23 52	22 45	04 19
04	23 50	22 42	03 38
05	23 D 50	22 39	02 47
06	23 51	22 36	01 47
07	23 52	22 33	00 N 39
08	23 R 53	22 30	00 S 33
09	23 51	22 26	01 45
10	23 47	22 23	02 53
11	23 40	22 20	03 51
12	23 32	22 17	04 35
13	23 23	22 14	04 59
14	23 14	22 11	05 03
15	23 06	22 07	04 46
16	23 01	22 04	04 11
17	22 58	22 01	03 21
18	22 57	21 58	02 20
19	22 D 57	21 55	01 14
20	22 57	21 51	00 S 05
21	22 R 57	21 48	01 N 03
22	22 54	21 45	02 05
23	22 49	21 42	03 01
24	22 40	21 39	03 48
25	22 29	21 36	04 24
26	22 16	21 32	04 48
27	22 02	21 29	04 59
28	21 49	21 26	04 57
29	21 36	21 23	04 43
30	21 26	21 20	04 16
31	21 ♑ 18	21 ♑ 16	03 N 37

DECLINATIONS

Date	Sun ☉ ° ′	Moon ☽ ° ′	Mercury ☿ ° ′	Venus ♀ ° ′	Mars ♂ ° ′	Jupiter ♃ ° ′	Saturn ♄ ° ′	Uranus ♅ ° ′	Neptune ♆ ° ′	Pluto ♇ ° ′
01	07 S 31	17 N 14	14 S 26	04 N 40	04 N 02	06 N 49	22 S 20	10 N 46	15 N 13	20 S 35
02	07 08	20 51	14 33	05 11	04 20	06 54	22 20	10 47	15 13	20 35
03	06 45	23 36	14 38	05 42	04 39	06 59	22 20	10 48	15 14	20 34
04	06 22	25 21	14 41	06 13	04 57	07 04	22 20	10 49	15 14	20 34
05	05 58	25 55	14 42	06 44	05 15	07 09	22 20	10 50	15 14	20 33
06	05 35	25 10	14 42	07 14	05 33	07 14	22 20	10 50	15 15	20 33
07	05 12	23 03	14 40	07 44	05 51	07 19	22 19	10 51	15 15	20 32
08	04 48	19 37	14 36	08 14	06 10	07 24	22 19	10 52	15 16	20 31
09	04 25	15 00	14 30	08 44	06 28	07 29	22 19	10 53	15 16	20 31
10	04 02	09 24	14 23	09 14	06 45	07 34	22 19	10 54	15 17	20 30
11	03 38	03 N 09	14 15	09 44	07 03	07 39	22 19	10 55	15 17	20 30
12	03 14	03 S 22	14 04	10 13	07 21	07 44	22 19	10 56	15 17	20 29
13	02 51	09 45	13 52	10 42	07 39	07 49	22 19	10 57	15 18	20 29
14	02 27	15 32	13 39	11 11	07 56	07 54	22 19	10 58	15 18	20 28
15	02 03	20 18	13 24	11 40	08 14	08 00	22 19	10 58	15 19	20 28
16	01 40	23 42	13 08	12 08	08 31	08 05	22 18	10 59	15 19	20 27
17	01 16	25 31	12 50	12 36	08 48	08 10	22 18	11 00	15 20	20 27
18	00 52	25 42	12 31	13 04	09 05	08 15	22 18	11 01	15 20	20 27
19	00 29	24 21	12 10	13 31	09 23	08 20	22 18	11 02	15 21	20 26
20	00 S 05	21 43	11 48	13 58	09 40	08 25	22 18	11 03	15 21	20 26
21	00 N 19	18 03	11 25	14 25	09 56	08 31	22 18	11 03	15 22	20 25
22	00 43	13 39	11 01	14 52	10 13	08 36	22 18	11 04	15 22	20 25
23	01 06	08 46	10 35	15 18	10 30	08 41	22 18	11 05	15 23	20 24
24	01 30	03 S 37	10 08	15 44	10 46	08 46	22 18	11 06	15 23	20 24
25	01 54	01 N 37	09 39	16 09	11 03	08 51	22 17	11 06	15 24	20 24
26	02 17	06 45	09 09	16 34	11 19	08 56	22 17	11 07	15 24	20 23
27	02 41	11 35	08 39	16 59	11 35	09 02	22 17	11 08	15 25	20 23
28	03 04	15 59	08 06	17 23	11 51	09 07	22 17	11 09	15 25	20 22
29	03 28	19 46	07 33	17 47	12 07	09 12	22 17	11 09	15 26	20 22
30	03 51	22 44	06 59	18 10	12 23	09 17	22 17	11 10	15 27	20 22
31	04 N 14	24 N 45	06 S 23	18 N 33	12 N 39	09 N 22	22 S 17	11 N 11	15 N 27	20 S 21

ZODIAC SIGN ENTRIES

Date	h	m	Planets
01	06	05	☽ ♉
03	18	36	☽ ♊
06	05	11	☽ ♋
08	12	04	☽ ♌
10	15	02	☽ ♍
12	15	18	☽ ♎
14	14	54	☽ ♏
15	11	32	♀ ♉
16	15	54	☽ ♐
17	14	29	☿ ♓
18	19	46	☽ ♑
20	16	53	☉ ♈
21	02	56	☽ ♒
23	12	44	☽ ♓
26	00	13	☽ ♈
26	18	50	♂ ♉
28	12	40	☽ ♉
31	01	19	☽ ♊

LATITUDES

Date	Mercury ☿ ° ′	Venus ♀ ° ′	Mars ♂ ° ′	Jupiter ♃ ° ′	Saturn ♄ ° ′	Uranus ♅ ° ′	Neptune ♆ ° ′	Pluto ♇ ° ′
01	01 N 25	00 S 32	00 S 23	01 S 05	01 N 04	00 N 48	01 S 44	11 S 49
04	00 46	00 23	00 21	01 05	01 04	00 48	01 44	11 49
07	00 N 10	00 14	00 18	01 05	01 04	00 48	01 44	11 49
10	00 S 23	00 S 05	00 16	01 04	01 04	00 48	01 43	11 49
13	00 52	00 N 05	00 14	01 04	01 04	00 48	01 43	11 50
16	01 17	00 15	00 12	01 03	01 04	00 48	01 43	11 50
19	01 38	00 25	00 10	01 03	01 04	00 48	01 43	11 51
22	01 56	00 36	00 08	01 03	01 04	00 48	01 43	11 51
25	02 09	00 46	00 06	01 02	01 04	00 48	01 43	11 52
28	02 18	00 56	00 03	01 02	01 05	00 48	01 43	11 52
31	02 S 23	01 N 07	00 S 01	01 S 02	01 N 05	00 N 48	01 S 43	11 S 53

DATA

Julian Date	2468771
Delta T	+80 seconds
Ayanamsa	24° 31′ 14″
Synetic vernal point	04° ♓ 35′ 45″
True obliquity of ecliptic	23° 26′ 03″

MOON'S PHASES, APSIDES AND POSITIONS ☽

Date	h	m	Phase	Longitude ° ′	Eclipse Indicator
04	22	52	☽	14 ♊ 17	
12	01	37	○	21 ♍ 24	
18	16	11	☾	27 ♐ 59	
26	11	44	●	05 ♈ 45	

Day	h	m		
01	00	43	Apogee	
13	05	41	Perigee	
28	11	18	Apogee	
05	10	33	Max dec	25° N 55′
11	23	38	0S	
18	02	38	Max dec	25° S 49′
25	04	33	0N	

LONGITUDES

Date	Chiron ⚷ ° ′	Ceres ⚳ ° ′	Pallas ⚴ ° ′	Juno ⚵ ° ′	Vesta ⚶ ° ′	Black Moon Lilith ⚸ ° ′
01	24 ♎ 01	06 ♑ 36	01 ♐ 49	11 ♐ 36	16 ♐ 24	22 ♈ 23
11	23 ♎ 31	09 ♑ 37	03 ♐ 08	12 ♐ 53	19 ♐ 53	23 ♈ 31
21	22 ♎ 53	12 ♑ 22	03 ♐ 48	13 ♐ 43	22 ♐ 58	24 ♈ 38
31	22 ♎ 10	14 ♑ 46	03 ♐ 43	14 ♐ 03	25 ♐ 34	25 ♈ 45

All ephemeris data is given at 12.00 UT and the Moon's longitude is additionally given for 24.00 UT

ASPECTARIAN

01 Friday
00 14 ☽ ∥ ♆
03 34 ☽ Q ☿
05 48 ♂ ⊥ ♆
12 25 ☽ △ ♄
13 08 ♂ ⊥ ♇
14 19 ☽ △ ♅
15 25 ☿ □ ♆
16 26 ☽ ✶ ♇
23 44 ♃ ∠ ♇

02 Saturday
05 28 ☽ ✶ ☉
05 38 ☽ ⊻ ♂
10 03 ☽ # ♇
10 56 ☽ ⊻ ♀
13 00 ☉ ⊻ ♂
16 18 ☽ ☌ ♆
16 46 ☽ Q ♇
16 58 ☽ □ ☿
18 35 ☽ ⊥ ♂
18 56 ☽ ⚼ ♄
23 12 ☽ ⊻ ♃
23 51 ☉ # ♃
23 56 ☽ # ♄

03 Sunday
00 25 ☽ ⊥ ♀
07 54 ☽ Q ☉
11 29 ☽ ⊥ ♃
13 08 ☽ ± ♄
13 54 ☽ ∠ ♂
20 30 ☽ ∠ ♀

04 Monday
01 11 ☽ ⊼ ♄
02 30 ☽ □ ♅
04 55 ☽ □ ♇
05 46 ☽ ∠ ♃
15 51 ☉ # ♀
16 07 ♀ ⊻ ♆
21 37 ☽ ✶ ♂
22 52 ☽ □ ☉

05 Tuesday
01 04 ☿ ∠ ♄
04 05 ☽ ⊻ ♆
05 22 ☽ ✶ ♀
07 06 ☽ △ ☿
11 42 ☽ ✶ ♃
13 31 ☉ ⊥ ♃
13 40 ☽ Q ♅
15 44 ☽ ⊥ ♆
22 22 ☽ Q ♂

06 Wednesday
02 14 ☉ Q ♄
06 10 ♀ ⚼ ♅
06 54 ☽ Q ♀
08 56 ☽ ∠ ♆
11 01 ☽ Q ♃
11 42 ☽ ☍ ♄
11 51 ♀ ∥ ♃
12 28 ☽ ✶ ♅
13 03 ☉ # ♂
13 18 ☽ ⚼ ☿
15 05 ☽ △ ♇
16 37 ☿ ✶ ♀

07 Thursday
06 59 ☽ ± ☿
09 56 ♀ ∠ ♇
10 09 ☽ □ ♂
12 32 ☽ △ ☉
12 48 ☽ ✶ ♆
16 05 ☽ ∠ ♅
16 14 ☉ ✶ ♆
17 58 ☽ # ♄
18 33 ☽ ⊼ ☿
18 45 ☽ ⚼ ♇
19 37 ☽ □ ♀
20 45 ☽ □ ♃

08 Friday
06 28 ☽ # ♇
08 09 ☽ ⊥ ♅
10 21 ☽ Q ♆
10 47 ♀ ☌ ♃
10 50 ☽ ± ♇
17 26 ☽ ⚼ ☉
18 21 ☽ ⊼ ♄
18 40 ☽ ⊻ ♅
21 22 ☽ ⊼ ♇

09 Saturday
04 47 ☽ ± ♄
09 55 ♂ ⊻ ♆
10 11 ☽ ± ☉
10 44 ☽ ∥ ♆
14 19 ☽ # ☿
17 28 ☽ □ ♆
17 53 ☽ △ ♂
19 05 ☿ ✶ ♃
20 08 ☽ ⚼ ♄
21 06 ☽ ⊼ ☉

10 Sunday
01 35 ☽ △ ♃
01 54 ☽ ☍ ☿
04 29 ☽ △ ♀
05 55 ☽ ∥ ♅
12 38 ☽ ∥ ♀
15 27 ♄ △ ♅
19 10 ☽ ∥ ♃
20 08 ☽ ⚼ ♂
21 03 ☽ ☌ ♅
21 05 ☽ △ ♄
21 56 ☽ ∥ ♂
23 49 ☽ ☍ ♇

11 Monday
02 39 ☽ ⚼ ♃
07 17 ☽ ⚼ ♀
10 06 ☽ # ☉
11 31 ☽ ± ♂
13 00 ♂ ⚼ ♅
17 23 ☽ ± ♃
18 45 ☽ △ ♆
21 39 ☽ ⊼ ♂
23 00 ☽ ± ♀

12 Tuesday
01 37 ☽ ☍ ☉
03 07 ☽ ⊼ ♃
05 56 ☽ ⊼ ☿
09 24 ☽ ⊼ ♀
11 32 ☽ ∥ ☉
16 08 ☽ ± ☿
18 40 ☽ ⚼ ♆
20 59 ☽ ⊻ ♅
21 19 ☽ □ ♄
23 53 ☽ ⊼ ♇

13 Wednesday
03 30 ☽ # ♂
04 29 ☽ # ♃
06 27 ☽ ⊥ ♅
06 30 ☉ ⊻ ♃
07 25 ☽ ⚼ ☿
08 56 ☽ ± ♆
09 24 ☽ ± ♇
16 03 ☽ # ♀
16 43 ☽ # ♅
18 27 ☽ ⊼ ♆
20 40 ☽ ∠ ♅
22 42 ♂ ∠ ♇
23 41 ☽ ⚼ ♇
23 44 ☽ ☍ ♂

14 Thursday
00 33 ♀ ∥ ♅
01 55 ☽ Q ♄
03 24 ☽ ☍ ♃
04 06 ☽ ∥ ☿
04 33 ☽ ⊼ ☉
08 48 ♂ ∥ ♃
09 01 ☽ △ ☿
10 58 ☽ # ♆
13 06 ☽ ☍ ♀
14 48 ☽ ± ☉
20 31 ☽ ✶ ♅
21 09 ☽ ✶ ♄
23 41 ☽ △ ♇

15 Friday
06 19 ☽ ⚼ ☉
12 57 ☽ ∥ ♇
15 53 ☽ Q ♅
18 42 ☽ ☍ ♆
21 34 ☽ ∠ ♄

16 Saturday
00 50 ☽ ∥ ♄
02 42 ☽ ⊼ ♂
04 40 ☽ ⊼ ♃
08 41 ☽ △ ☉
12 31 ☽ ⊥ ♄
13 13 ☽ ± ♂
13 52 ☽ □ ☿
14 49 ☽ ± ♃
18 30 ☽ ⊼ ♀
21 40 ☽ □ ♅
22 37 ☽ ⊻ ♄

17 Sunday
01 13 ☽ □ ♇
05 14 ☽ ⚼ ♂
05 35 ☽ ± ♀
06 16 ☽ ⚼ ♃
19 18 ☿ # ♀
21 20 ☽ ⊼ ♆
22 32 ☽ ⚼ ♀
23 24 ☽ Q ☿

18 Monday
07 07 ♀ △ ♅
07 50 ☽ ± ♆
08 22 ☽ Q ♇
08 41 ☽ △ ♂
08 42 ☽ △ ♃
08 56 ♂ ☌ ♃
16 11 ☽ □ ☉
21 11 ♀ △ ♄
22 42 ☽ ✶ ☿
23 52 ☽ ⚼ ♆

19 Tuesday
01 46 ☽ △ ♅
03 06 ☽ ☌ ♄
03 40 ☽ △ ♀
05 49 ☽ ✶ ♇
10 33 ☉ ⊥ ♀

20 Wednesday
03 14 ☽ △ ♆
03 28 ☽ Q ☉
04 01 ♀ ✶ ♇
04 55 ☽ ∠ ☿
05 03 ☽ ⚼ ♅
06 15 ☿ ☍ ♅
07 27 ☽ ∥ ♄
09 21 ☽ ∠ ♇
16 11 ☽ □ ♃
18 33 ☽ □ ♂
21 08 ☽ ∥ ♇
21 48 ☽ ± ♅
22 11 ☿ ✶ ♄
23 36 ☽ ⊥ ☿

21 Thursday
02 15 ☽ ⊥ ♇
03 47 ☽ ✶ ☉
09 06 ☽ ⊼ ♅
10 51 ☽ ⊻ ♄
12 16 ☽ ⊻ ☿
13 39 ☽ ⊻ ♇
17 05 ☽ □ ♀
20 09 ☿ Q ♆
22 20 ☽ ⊥ ♄

22 Friday
01 32 ☿ ☌ ♇
02 59 ☽ Q ♃
03 02 ☽ # ♆
06 17 ☽ # ♀
06 56 ☽ Q ♂
08 40 ☿ # ♅
10 55 ☽ ∠ ☉
12 10 ☽ □ ♆
15 44 ☽ ∠ ♄

23 Saturday
00 53 ☽ # ♅
02 28 ☽ ∥ ☿
02 40 ☽ ✶ ♃
02 51 ☉ ∠ ♆
04 08 ☽ # ♂
05 56 ☽ ⊥ ☉
07 35 ☽ Q ♀
07 43 ☽ ✶ ♂
12 24 ☽ # ♃
14 42 ☿ # ♂
16 43 ♀ ∥ ♆
18 44 ☽ ⊻ ☉
18 57 ☽ ☍ ♅
21 07 ☽ ✶ ♄
21 23 ☉ ⊼ ♅
23 23 ☽ Q ♆
23 57 ☽ ☌ ♇

24 Sunday
05 43 ☽ ☌ ☿
08 41 ☽ ∠ ♃
09 43 ☽ ✶ ♀
15 10 ☽ ∠ ♂
20 54 ☽ Q ♄
21 00 ☽ # ☉
23 11 ☽ ✶ ♆

25 Monday
00 54 ☉ □ ♄
02 53 ☽ ⊥ ♃
10 13 ☽ ⊥ ♂
11 49 ☿ ∠ ♃
13 22 ☽ ∥ ☉
15 05 ☽ ⊻ ♃
17 37 ♂ ∥ ♅
18 49 ☽ ∠ ♀
23 00 ☽ ⊻ ♂

26 Tuesday
05 12 ☽ ∠ ♆
06 24 ☽ ⊼ ♅
08 56 ☽ □ ♄
11 44 ☽ ☌ ☉
11 48 ☽ ⊻ ♇
12 30 ☉ ⊻ ♇
14 50 ☽ ⊥ ♀
18 25 ☽ ± ♅
20 47 ☿ # ♃
22 37 ☽ # ☿
22 51 ☽ ∥ ♃
23 20 ☽ ⊥ ♆
23 53 ☽ ⊥ ♇

27 Wednesday
01 42 ☽ ⊻ ☿
04 15 ☽ ⊻ ♀
09 39 ☽ ∥ ♅
11 28 ☽ ⊻ ♆
12 00 ☽ ∥ ♂
12 29 ☽ ⚼ ♅
15 31 ☽ ⊥ ☿
18 05 ☽ ∠ ♇

28 Thursday
04 39 ☽ ☌ ♃
08 45 ☽ ∥ ♆
12 19 ☽ ∠ ☿
15 27 ☽ ☌ ♂
18 44 ☽ △ ♅
21 20 ☽ ∥ ♀
21 36 ☽ △ ♄

29 Friday
00 29 ☽ ✶ ♇
05 49 ☽ ⊻ ☉
13 43 ☿ Q ♄
16 21 ☽ # ♇
17 49 ☉ ± ♅
19 04 ☽ ⊥ ☉
23 09 ☽ ✶ ☿
23 37 ☽ ☌ ♀

30 Saturday
00 18 ☽ ☌ ♆
00 49 ☽ Q ♇
04 02 ☽ ⚼ ♄
06 42 ♀ ☌ ♆
07 49 ☽ # ♄
08 11 ☿ ✶ ♆
08 26 ☿ ∠ ♂
11 58 ♀ Q ♇
12 46 ☿ ✶ ♀
14 55 ☽ ∠ ☉
18 13 ♂ △ ♅
18 33 ☽ ⊻ ♃
22 13 ☽ ± ♄

31 Sunday
03 00 ☽ Q ☿
06 52 ☽ ⊥ ♃
07 10 ☽ □ ♅
08 04 ☽ ⊻ ♂
10 18 ☽ ⊼ ♄
13 09 ☽ □ ♇
20 50 ♀ ⚼ ♄
20 51 ☽ ⊥ ♂
23 43 ☽ ✶ ☉

APRIL 2047

LONGITUDES

Date	Sidereal time h m s	Sun ☉ ° ' "	Moon ☽ ° ' "	Moon ☽ 24.00 ° ' "	Mercury ☿ ° '	Venus ♀ ° '	Mars ♂ ° '	Jupiter ♃ ° '	Saturn ♄ ° '	Uranus ♅ ° '	Neptune ♆ ° '	Pluto ♇ ° '
01	00 39 07	11 ♈ 41 23	17 ♊ 22 07	23 ♊ 27 33	21 ♓ 01	20 ♉ 14	04 ♉ 13	27 ♈ 03	04 ♑ 30	02 ♍ 53	17 ♉ 41	05 ♓ 55
02	00 43 03	12 40 37	29 ♊ 36 23	05 ♋ 49 15	22 37	21 24	04 57	27 17	04 31	02 R 51	17 43	05 57
03	00 47 00	13 39 48	12 ♋ 06 44	18 ♋ 29 29	24 15	22 34	05 41	27 31	04 32	02 49	17 45	05 58
04	00 50 56	14 38 57	24 ♋ 58 04	01 ♌ 33 01	25 55	23 44	06 25	27 46	04 33	02 47	17 47	05 59
05	00 54 53	15 38 04	08 ♌ 14 49	15 ♌ 03 51	27 36	24 55	07 09	28 00	04 34	02 45	17 49	06 01
06	00 58 49	16 37 08	22 ♌ 00 21	29 ♌ 04 22	29 ♓ 18	26 04	07 52	28 14	04 34	02 44	17 51	06 02
07	01 02 46	17 36 10	06 ♍ 15 48	13 ♍ 34 17	01 ♈ 02	27 14	08 36	28 28	04 35	02 42	17 53	06 03
08	01 06 42	18 35 10	20 ♍ 59 14	28 ♍ 29 49	02 47	28 24	09 20	28 43	04 36	02 41	17 55	06 05
09	01 10 39	19 34 08	06 ♎ 04 59	13 ♎ 43 28	04 34	29 ♉ 33	10 04	28 57	04 36	02 39	17 57	06 06
10	01 14 36	20 33 03	21 ♎ 23 51	29 ♎ 04 40	06 22	00 ♊ 43	10 47	29 11	04 36	02 37	17 59	06 07
11	01 18 32	21 31 57	06 ♏ 44 24	14 ♏ 21 37	08 11	01 52	11 31	29 26	04 37	02 36	18 01	06 09
12	01 22 29	22 30 49	21 ♏ 54 59	29 ♏ 23 23	10 03	03 01	12 14	29 40	04 37	02 35	18 03	06 10
13	01 26 25	23 29 38	06 ♐ 45 55	14 ♐ 01 54	11 55	04 10	12 58	29 ♈ 55	04 R 37	02 33	18 05	06 11
14	01 30 22	24 28 27	21 ♐ 10 53	28 ♐ 12 40	13 50	05 19	13 41	00 ♉ 09	04 37	02 32	18 07	06 12
15	01 34 18	25 27 13	05 ♑ 07 15	11 ♑ 54 46	15 45	06 28	14 25	00 23	04 37	02 31	18 09	06 13
16	01 38 15	26 25 58	18 ♑ 35 31	25 ♑ 09 56	17 42	07 37	15 08	00 38	04 36	02 29	18 11	06 15
17	01 42 11	27 24 41	01 ♒ 38 27	08 ♒ 01 37	19 41	08 45	15 51	00 52	04 36	02 28	18 13	06 16
18	01 46 08	28 23 22	14 ♒ 19 59	20 ♒ 34 06	21 41	09 53	16 35	01 07	04 36	02 27	18 16	06 17
19	01 50 05	29 ♈ 22 02	26 ♒ 44 31	02 ♓ 51 45	23 43	11 01	17 18	01 21	04 35	02 26	18 18	06 18
20	01 54 01	00 ♉ 20 40	08 ♓ 56 18	14 ♓ 58 36	25 46	12 09	18 01	01 35	04 35	02 25	18 20	06 19
21	01 57 58	01 19 16	20 ♓ 59 04	26 ♓ 58 05	27 50	13 17	18 44	01 50	04 34	02 24	18 22	06 20
22	02 01 54	02 17 50	02 ♈ 55 58	08 ♈ 53 00	29 ♈ 55	14 25	19 27	02 04	04 33	02 23	18 24	06 21
23	02 05 51	03 16 23	14 ♈ 49 26	20 ♈ 45 29	02 ♉ 01	15 32	20 10	02 19	04 32	02 22	18 26	06 22
24	02 09 47	04 14 54	26 ♈ 41 21	02 ♉ 37 12	04 08	16 40	20 53	02 33	04 31	02 21	18 29	06 23
25	02 13 44	05 13 23	08 ♉ 33 14	14 ♉ 29 36	06 16	17 47	21 36	02 48	04 30	02 20	18 31	06 24
26	02 17 40	06 11 50	20 ♉ 26 28	26 ♉ 24 04	08 25	18 54	22 19	03 02	04 29	02 20	18 33	06 25
27	02 21 37	07 10 15	02 ♊ 22 35	08 ♊ 22 17	10 33	20 01	23 02	03 16	04 27	02 19	18 35	06 26
28	02 25 34	08 08 39	14 ♊ 23 27	20 ♊ 26 25	12 42	21 07	23 45	03 31	04 26	02 18	18 37	06 27
29	02 29 30	09 07 00	26 ♊ 31 32	02 ♋ 39 12	14 51	22 14	24 27	03 45	04 24	02 18	18 40	06 28
30	02 33 27	10 ♉ 05 20	08 ♋ 49 53	15 ♋ 04 02	16 ♉ 59	23 ♊ 20	25 ♉ 10	03 ♉ 59	04 ♑ 23	02 ♍ 17	18 ♉ 42	06 ♓ 29

Date	Moon True ☊ ° '	Moon Mean ☊ ° '	Moon ☽ Latitude ° '
01	21 ♑ 14	21 ♑ 13	02 N 49
02	21 R 12	21 10	01 51
03	21 D 11	21 07	00 N 48
04	21 R 11	21 04	00 S 20
05	21 10	21 01	01 29
06	21 08	20 57	02 35
07	21 03	20 54	03 34
08	20 55	20 51	04 21
09	20 45	20 48	04 50
10	20 34	20 45	05 00
11	20 23	20 42	04 48
12	20 14	20 38	04 16
13	20 06	20 35	03 27
14	20 02	20 32	02 26
15	20 00	20 29	01 18
16	19 D 59	20 26	00 S 07
17	19 R 59	20 22	01 N 01
18	19 59	20 19	02 05
19	19 56	20 16	03 01
20	19 51	20 13	03 48
21	19 43	20 10	04 24
22	19 32	20 07	04 48
23	19 19	20 03	04 59
24	19 06	20 00	04 58
25	18 52	19 57	04 43
26	18 40	19 54	04 16
27	18 30	19 51	03 38
28	18 22	19 48	02 49
29	18 17	19 44	01 53
30	18 ♑ 15	19 ♑ 41	00 N 50

DECLINATIONS

Date	Sun ☉ ° '	Moon ☽ ° '	Mercury ☿ ° '	Venus ♀ ° '	Mars ♂ ° '	Jupiter ♃ ° '	Saturn ♄ ° '	Uranus ♅ ° '	Neptune ♆ ° '	Pluto ♇ ° '
01	04 N 37	25 N 38	05 S 46	18 N 56	12 N 55	09 N 27	22 S 17	11 N 11	15 N 28	20 S 21
02	05 00	25 17	05 08	19 18	13 10	09 33	22 17	11 12	15 28	20 21
03	05 23	23 41	04 29	19 39	13 25	09 38	22 17	11 13	15 29	20 20
04	05 46	20 48	03 49	20 00	13 40	09 43	22 17	11 13	15 29	20 20
05	06 09	16 46	03 08	20 21	13 56	09 48	22 17	11 14	15 30	20 20
06	06 32	11 43	02 26	20 41	14 10	09 53	22 16	11 14	15 31	20 19
07	06 54	05 N 54	01 42	21 01	14 25	09 58	22 16	11 15	15 31	20 19
08	07 17	00 S 25	00 58	21 20	14 40	10 04	22 16	11 16	15 32	20 19
09	07 39	06 51	00 S 13	21 38	14 54	10 09	22 16	11 16	15 32	20 19
10	08 01	12 58	00 N 33	21 56	15 09	10 14	22 16	11 17	15 33	20 18
11	08 24	18 18	01 20	22 13	15 23	10 19	22 16	11 17	15 33	20 18
12	08 46	22 22	02 08	22 30	15 37	10 24	22 16	11 18	15 34	20 18
13	09 07	24 50	02 57	22 47	15 51	10 29	22 16	11 18	15 35	20 18
14	09 29	25 34	03 47	23 02	16 04	10 34	22 16	11 18	15 35	20 17
15	09 50	24 38	04 37	23 17	16 18	10 39	22 16	11 19	15 36	20 17
16	10 12	22 16	05 28	23 32	16 31	10 44	22 16	11 19	15 36	20 17
17	10 33	18 48	06 20	23 46	16 44	10 49	22 16	11 20	15 37	20 17
18	10 54	14 32	07 12	23 59	16 57	10 54	22 16	11 20	15 38	20 17
19	11 15	09 45	08 04	24 12	17 10	10 59	22 16	11 20	15 38	20 16
20	11 35	04 S 42	08 57	24 24	17 23	11 05	22 16	11 21	15 39	20 16
21	11 56	00 N 28	09 50	24 36	17 36	11 10	22 16	11 21	15 40	20 16
22	12 16	05 34	10 43	24 47	17 48	11 15	22 16	11 21	15 40	20 16
23	12 36	10 26	11 36	24 57	18 00	11 20	22 16	11 22	15 41	20 16
24	12 56	14 55	12 29	25 06	18 12	11 24	22 16	11 22	15 41	20 16
25	13 16	18 49	13 21	25 15	18 24	11 29	22 16	11 22	15 42	20 16
26	13 35	21 58	14 13	25 24	18 36	11 34	22 16	11 22	15 42	20 15
27	13 54	24 11	15 03	25 31	18 47	11 39	22 16	11 23	15 43	20 15
28	14 13	25 19	15 53	25 39	18 58	11 44	22 16	11 23	15 44	20 15
29	14 32	25 16	16 42	25 45	19 10	11 49	22 16	11 23	15 44	20 15
30	14 N 50	23 N 58	17 N 29	25 N 51	19 N 21	11 N 54	22 S 16	11 N 23	15 N 45	20 S 15

ZODIAC SIGN ENTRIES

Date	h m	Planets
02	12 46	☽ ♋
04	21 12	☽ ♌
06	21 46	☿ ♈
07	01 34	☽ ♍
09	02 23	☽ ♎
09	21 10	♀ ♊
11	01 27	☽ ♏
13	00 59	☽ ♐
13	21 03	♃ ♉
15	03 05	☽ ♑
17	08 57	☽ ♒
19	18 23	☽ ♓
20	03 32	☉ ♉
22	06 06	☽ ♈
22	12 59	☿ ♉
24	18 42	☽ ♉
27	07 14	☽ ♊
29	18 49	☽ ♋

LATITUDES

Date	Mercury ☿ ° '	Venus ♀ ° '	Mars ♂ ° '	Jupiter ♃ ° '	Saturn ♄ ° '	Uranus ♅ ° '	Neptune ♆ ° '	Pluto ♇ ° '
01	02 S 24	01 N 10	00 S 01	01 S 02	01 N 05	00 N 48	01 S 43	11 S 53
04	02 23	01 21	00 N 01	01 02	01 05	00 48	01 42	11 54
07	02 18	01 31	00 03	01 01	01 05	00 48	01 42	11 54
10	02 09	01 40	00 05	01 01	01 05	00 48	01 42	11 55
13	01 55	01 50	00 07	01 01	01 05	00 47	01 42	11 56
16	01 36	01 59	00 09	01 01	01 05	00 47	01 42	11 57
19	01 13	02 08	00 11	01 01	01 05	00 47	01 42	11 57
22	00 46	02 16	00 13	01 01	01 05	00 47	01 42	11 58
25	00 S 16	02 24	00 15	01 01	01 06	00 47	01 42	11 59
28	00 N 15	02 31	00 17	01 01	01 06	00 47	01 42	12 00
31	00 N 47	02 N 37	00 N 19	01 S 00	01 N 06	00 N 47	01 S 42	12 S 01

DATA

Julian Date	2468802
Delta T	+80 seconds
Ayanamsa	24° 31' 17"
Synetic vernal point	04° ♓ 35' 42"
True obliquity of ecliptic	23° 26' 03"

MOON'S PHASES, APSIDES AND POSITIONS ☽

Date	h m	Phase	Longitude ° '	Eclipse Indicator
03	15 11	☽	13 ♋ 48	
10	10 35	○	20 ♎ 30	
17	03 30	☾	27 ♑ 04	
25	04 40	●	04 ♉ 56	

Day	h m		
10	16 15	Perigee	
24	13 29	Apogee	
01	17 33	Max dec	25° N 40'
08	10 26	0S	
14	10 10	Max dec	25° S 34'
21	09 48	0N	
28	22 53	Max dec	25° N 27'

LONGITUDES

Date	Chiron ⚷ ° '	Ceres ⚳ ° '	Pallas ⚴ ° '	Juno ⚵ ° '	Vesta ⚶ ° '	Black Moon Lilith ⚸ ° '
01	22 ♎ 05	15 ♑ 00	03 ♐ 40	14 ♐ 03	25 ♐ 47	25 ♈ 52
11	21 ♎ 18	16 ♑ 58	02 ♐ 41	13 ♐ 47	27 ♐ 45	26 ♈ 59
21	20 ♎ 32	18 ♑ 31	00 ♐ 55	12 ♐ 59	29 ♐ 02	28 ♈ 06
31	19 ♎ 48	19 ♑ 32	28 ♏ 28	11 ♐ 38	29 ♐ 34	29 ♈ 14

All ephemeris data is given at 12.00 UT and the Moon's longitude is additionally given for 24.00 UT

ASPECTARIAN

h m	Aspects
01 Monday	
01 14	☽ ∠ ♃
11 49	☉ ⊥ ♆
12 37	☽ ⊻ ♆
12 41	☿ ⊥ ♃
15 53	☽ ∠ ♂
17 43	☉ ⊥ ♇
18 15	☽ ⊻ ♀
18 54	☽ Q ♅
20 16	☽ □ ☿
21 33	♂ △ ♄
02 Tuesday	
00 28	☽ ⊥ ♆
01 33	☽ Q ☉
07 15	☽ ⊥ ♀
07 24	☽ ⚹ ♃
15 02	☉ # ☿
18 02	☽ ∠ ♆
18 15	☽ ⚹ ♅
21 30	☽ ☍ ♄
22 58	☽ ⚹ ♂
03 Wednesday	
00 15	☽ △ ♇
02 28	☽ ∠ ♀
06 59	☽ Q ♃
15 11	☽ □ ☉
21 43	♂ ⚹ ♇
22 38	☽ ⚹ ♆
22 43	☽ ∠ ♅
23 07	☽ Q ♂
04 Thursday	
00 58	☽ # ♄
04 38	☽ ⚼ ♇
09 31	☽ ⚹ ♀
13 59	☽ △ ☿
15 12	☽ # ♇
15 20	☽ ⊥ ♅
16 53	☽ ∥ ♀
17 13	☽ □ ♃
20 49	☽ Q ♆
21 11	☽ ± ♇
05 Friday	
02 12	☽ ⊻ ♅
05 25	☽ ⊼ ♄
08 01	☽ ⊼ ♇
09 24	☽ Q ♀
09 55	☽ □ ♂
10 36	♀ # ♇
16 06	☽ ± ♄
18 28	☽ ∥ ♆
18 40	☿ ⊻ ♃
20 46	☽ ⚼ ☿
06 Saturday	
01 24	☽ ∥ ♂
02 00	☽ △ ☉
04 49	☽ □ ♆
07 49	☽ ⚼ ♄
14 06	☽ ∥ ♅
14 31	☽ ± ☿
19 33	☽ □ ♀
19 45	☽ ∥ ♃
22 46	☽ △ ♃
07 Sunday	
02 05	☽ ⊼ ☿
05 28	☽ ⚼ ☉
06 06	☽ ☌ ♅
08 15	☽ ∥ ☉
09 13	☽ △ ♄
11 39	☽ ☍ ♇
14 21	☉ ⚼ ♅
16 04	☽ △ ♂
18 57	☉ ⊻ ♆
21 25	☽ ± ☉
08 Monday	
00 02	☽ ⚼ ♃
06 08	☽ # ☿
07 02	☽ △ ♆
07 52	☽ ⊼ ☉
10 34	☿ ⊼ ♅
13 47	☿ ∠ ♆
13 50	☽ ∥ ☿
14 49	☽ ± ♃
16 03	♀ ± ♄
17 38	☽ ⚼ ♂
20 11	♀ ⊻ ♃
09 Tuesday	
00 32	☽ ⊼ ♃
00 39	☿ ⊥ ♂
00 50	☽ △ ♀
06 35	☽ ⊻ ♅
07 02	☽ ⚼ ♆
08 39	☽ ± ♂
09 17	☽ ☍ ☿
09 40	☽ □ ♄
12 02	☽ ⊼ ♇
12 33	☿ □ ♄
15 13	☽ # ☉
16 02	☽ ⊥ ♅
18 34	☽ ⊼ ♂
21 14	☽ ± ♆
21 28	☽ ± ♇
10 Wednesday	
00 48	☽ # ♃
02 24	☽ ⚼ ♀
05 06	☽ # ♅
06 07	☽ ∠ ♅
06 39	☽ ⊼ ♆
08 45	☿ ⊻ ♇
10 35	☽ ☍ ☉
11 34	☽ ⚼ ♇
13 53	☽ Q ♄
17 36	☽ ± ♀
21 40	☽ # ♂
23 04	☽ # ♆
11 Thursday	
00 22	☽ ☍ ♃
02 15	☉ ∠ ♇
03 45	☽ ⊼ ♀
05 31	☽ ⚹ ♅
08 40	☽ ⚹ ♄
11 04	☽ △ ♇
14 36	☽ ⊼ ☿
15 54	♀ # ♄
17 15	☿ ± ♅
19 53	☽ ☍ ♂
22 47	☽ ∥ ♇
12 Friday	
00 22	☽ Q ♅
01 22	☽ ± ☿
02 53	♀ □ ♅
05 50	☽ ☍ ♆
07 16	♂ ∥ ♆
08 20	☽ ∠ ♄
11 22	☽ ∥ ♄
13 01	☽ ⊼ ☉
13 11	☽ # ♀
17 43	☽ ⚼ ☿
22 45	☽ ⊥ ♄
23 20	☽ ± ☉
13 Saturday	
00 39	☽ ⊼ ♃
05 08	☽ □ ♅
07 24	☽ ☍ ♀
08 29	☽ ⊻ ♄
10 35	☽ ± ♃
11 03	☽ □ ♇
11 47	♄ St R
14 04	☿ ⊥ ♆
15 03	☽ ⚼ ☉
15 21	☿ ⊥ ♇
21 13	♀ ⊼ ♄
21 47	☽ △ ☿
22 46	☽ ⊼ ♂
14 Sunday	
01 41	☽ ⚼ ♃
06 49	☽ ⊼ ♆
09 15	☿ ⊻ ♂
09 21	☽ ± ♂
17 00	☽ ± ♆
17 08	☽ Q ♇
18 01	☽ △ ☉
15 Monday	
01 32	☽ ⚼ ♂
03 37	☽ △ ♃
06 49	♀ □ ♇
07 27	☽ △ ♅
08 33	☽ ⚼ ♆
09 57	♀ ⊥ ♃
11 06	☽ ☌ ♄
13 56	☽ ⚹ ♇
14 35	☽ ⊼ ♀
16 Tuesday	
01 44	☽ # ♀
02 12	☽ ± ♀
05 25	☽ △ ♂
09 22	☿ ⚼ ♅
10 01	☽ ⚼ ♅
10 07	☽ □ ☿
11 16	☽ △ ♆
11 59	☽ ∥ ♄
16 49	☽ ∠ ♇
17 58	☿ ⊻ ♆
20 01	☽ ⚼ ♀
17 Wednesday	
02 25	☽ ± ♅
02 31	☽ ∥ ♇
03 30	☽ □ ☉
09 26	☽ ⊥ ♇
10 32	☽ □ ♃
13 33	☽ ⊼ ♅
17 32	☽ ⊻ ♄
20 41	☽ ⊻ ♇
23 27	☽ # ♂
18 Thursday	
01 28	☽ Q ☿
02 41	☽ △ ♀
04 52	☽ ⊥ ♄
06 10	☽ # ♆
07 07	☿ ∠ ♇
12 43	☉ ∥ ♃
16 17	☽ Q ☉
16 34	☽ □ ♂
19 34	☽ □ ♆
21 22	☽ Q ♃
22 06	☽ ∠ ♄
19 Friday	
04 16	☽ # ♅
04 56	☽ ⚹ ☿
05 12	☽ # ☉
06 05	☽ # ♃
17 35	☽ ⚹ ☉
18 39	☉ ∥ ♅
18 55	☽ # ☿
19 19	♆ Q ♇
21 12	☽ ⚹ ♃
23 08	☽ ☍ ♅
20 Saturday	
03 23	☽ ⚹ ♄
05 52	☽ Q ♂
06 49	☽ ☌ ♇
06 50	☽ Q ♆
16 21	☽ ∠ ☿
19 03	☽ □ ♀
22 18	♂ Q ♇
23 00	♂ ☌ ♆
21 Sunday	
01 51	☽ ∠ ☉
03 10	☽ Q ♄
03 32	☽ ∠ ♃
06 45	☽ ⚹ ♆
07 13	☽ ⚹ ♂
14 02	☽ ⊥ ☿
21 28	☽ ⊥ ☉
21 55	☽ ⊥ ♃
23 37	☿ ∠ ♀
22 Monday	
04 37	☉ ☌ ♃
04 38	☽ ⊻ ☿
10 14	☽ ⊻ ♃
10 36	☽ ⊻ ☉
10 51	☽ Q ♀
10 54	☽ ⊼ ♅
12 57	☽ ∠ ♆
14 04	☉ △ ♅
15 02	♂ ⚼ ♄
15 15	☽ □ ♄
15 16	☽ ∠ ♂
18 54	☽ ⊻ ♇
22 59	☽ ± ♅
23 Tuesday	
03 48	☿ ∥ ♃
05 32	☿ ∥ ♅
07 02	☽ ⊥ ♇
07 10	☽ ⊥ ♆
10 36	☽ ⊥ ♂
13 36	☽ ⚹ ♀
15 45	☿ ☌ ♃
15 56	☿ △ ♅
16 40	☽ ∥ ♃
16 46	☽ ∥ ♅
17 08	☽ ⚼ ♅
17 20	♃ △ ♅
19 20	☽ ⊻ ♆
19 26	☽ ∥ ☿
23 10	♃ ∥ ♅
23 31	☽ ⊻ ♂
24 Wednesday	
00 12	☽ ∥ ☉
01 16	☽ ∠ ♇
14 17	☉ ☌ ☿
16 12	☿ △ ♄
16 31	☽ ∥ ♆
18 26	☉ △ ♄
23 07	☽ ∠ ♀
23 27	☽ △ ♅
25 Thursday	
00 06	☽ ☌ ♃
03 48	☽ △ ♄
04 40	☽ ☌ ☉
06 22	☽ ☌ ☿
07 39	☽ ⚹ ♇
08 07	☉ ∥ ☿
09 06	☽ ∥ ♂
12 14	♀ ∠ ♃
13 32	☿ ⚹ ♇
19 12	☽ ⊥ ♀
22 13	☽ # ♇
26 Friday	
04 13	♀ ⊻ ♆
07 55	☽ Q ♇
08 11	☽ ☌ ♆
08 34	☽ ⊻ ♀
10 03	☽ ⚼ ♄
14 44	☽ # ♄
16 01	☽ ☌ ♂
17 38	☉ ⚹ ♇
27 Saturday	
04 08	☽ ± ♄
11 53	☽ □ ♅
13 50	☽ ⊻ ♃
16 09	☽ ⊼ ♄
18 29	♀ Q ♅
20 09	☽ □ ♇
22 27	☽ ⊻ ☉
28 Sunday	
02 05	☽ ⊥ ♃
07 19	☿ ∥ ♆
07 54	☽ ⊻ ☿
11 28	☽ ⊥ ☉
20 21	☽ ∠ ♃
20 26	☽ ⊻ ♆
22 24	☽ ⊥ ☿
23 44	☽ Q ♅
29 Monday	
02 42	☽ ☌ ♀
06 51	☽ ∠ ☉
07 41	☽ ⊻ ♂
08 19	☽ ⊥ ♆
19 53	☽ ∠ ☿
20 11	☽ ⊥ ♂
23 18	☽ ⚹ ♅
30 Tuesday	
02 00	☽ ∠ ♆
02 25	☽ ⚹ ♃
03 23	☽ ☍ ♄
07 27	☽ △ ♇
14 38	☽ ⚹ ☉
14 44	☽ ∠ ♂
20 27	☿ ⊥ ♀

MAY 2047

LONGITUDES

Date	Sidereal time h m s	Sun ☉ ° ′ ″	Moon ☽ ° ′ ″	Moon ☽ 24.00 ° ′ ″	Mercury ☿ ° ′	Venus ♀ ° ′	Mars ♂ ° ′	Jupiter ♃ ° ′	Saturn ♄ ° ′	Uranus ♅ ° ′	Neptune ♆ ° ′	Pluto ⯓ ° ′
01	02 37 23	11 ♉ 03 37	21 ♋ 22 10	27 ♋ 44 48	19 ♉ 06	24 ♊ 26	25 ♉ 53	04 ♉ 14	04 ♑ 21	02 ♍ 17	18 ♉ 44	06 ♓ 30
02	02 41 20	12 01 52	04 ♌ 12 27	10 ♌ 45 36	21 12	25 32	26 35	04 28	04 R 19	02 R 16	18 46	06 31
03	02 45 16	13 00 06	17 ♌ 24 43	24 ♌ 10 12	23 16	26 38	27 18	04 43	04 17	02 16	18 49	06 32
04	02 49 13	13 58 17	01 ♍ 02 21	08 ♍ 01 21	25 19	27 43	28 00	04 57	04 15	02 16	18 51	06 32
05	02 53 09	14 56 26	15 ♍ 07 14	22 ♍ 19 50	27 20	28 48	28 43	05 11	04 13	02 15	18 53	06 33
06	02 57 06	15 54 33	29 ♍ 38 48	07 ♎ 03 34	29 ♉ 19	29 ♊ 53	29 ♉ 25	05 25	04 11	02 15	18 55	06 34
07	03 01 03	16 52 38	14 ♎ 33 20	22 ♎ 07 07	01 ♊ 15	00 ♋ 58	00 ♊ 07	05 40	04 09	02 15	18 58	06 35
08	03 04 59	17 50 42	29 ♎ 43 43	07 ♏ 21 49	03 08	02 02	00 50	05 54	04 07	02 15	19 00	06 35
09	03 08 56	18 48 43	15 ♏ 00 01	22 ♏ 36 57	04 58	03 06	01 32	06 08	04 04	02 15	19 02	06 36
10	03 12 52	19 46 43	00 ♐ 11 13	07 ♐ 41 37	06 46	04 10	02 14	06 22	04 02	02 D 15	19 04	06 37
11	03 16 49	20 44 42	15 ♐ 07 04	22 ♐ 26 41	08 30	05 14	02 56	06 36	03 59	02 15	19 07	06 37
12	03 20 45	21 42 39	29 ♐ 39 48	06 ♑ 45 58	10 11	06 17	03 38	06 51	03 57	02 15	19 09	06 38
13	03 24 42	22 40 35	13 ♑ 44 58	20 ♑ 36 46	11 48	07 20	04 20	07 05	03 54	02 15	19 11	06 39
14	03 28 38	23 38 30	27 ♑ 21 28	03 ♒ 59 22	13 22	08 23	05 03	07 19	03 51	02 16	19 13	06 39
15	03 32 35	24 36 23	10 ♒ 30 48	16 ♒ 56 17	14 53	09 26	05 44	07 33	03 48	02 16	19 16	06 40
16	03 36 32	25 34 15	23 ♒ 16 18	29 ♒ 31 25	16 20	10 28	06 26	07 47	03 46	02 16	19 18	06 40
17	03 40 28	26 32 05	05 ♓ 42 13	11 ♓ 49 18	17 43	11 30	07 08	08 01	03 43	02 17	19 20	06 41
18	03 44 25	27 29 55	17 ♓ 53 13	23 ♓ 54 32	19 02	12 31	07 50	08 15	03 39	02 17	19 22	06 41
19	03 48 21	28 27 43	29 ♓ 53 46	05 ♈ 51 26	20 18	13 33	08 32	08 29	03 36	02 17	19 25	06 42
20	03 52 18	29 ♉ 25 30	11 ♈ 47 58	17 ♈ 43 48	21 30	14 34	09 14	08 43	03 33	02 18	19 27	06 42
21	03 56 14	00 ♊ 23 16	23 ♈ 39 20	29 ♈ 34 52	22 38	15 34	09 55	08 57	03 30	02 19	19 29	06 43
22	04 00 11	01 21 01	05 ♉ 30 44	11 ♉ 27 12	23 42	16 35	10 37	09 11	03 26	02 19	19 31	06 43
23	04 04 07	02 18 45	17 ♉ 24 30	23 ♉ 22 52	24 42	17 34	11 19	09 24	03 23	02 20	19 34	06 43
24	04 08 04	03 16 27	29 ♉ 22 28	05 ♊ 23 31	25 38	18 34	12 00	09 38	03 20	02 21	19 36	06 44
25	04 12 01	04 14 08	11 ♊ 26 11	17 ♊ 30 39	26 30	19 33	12 42	09 52	03 16	02 22	19 38	06 44
26	04 15 57	05 11 48	23 ♊ 37 06	29 ♊ 45 44	27 17	20 32	13 23	10 06	03 12	02 22	19 40	06 44
27	04 19 54	06 09 27	05 ♋ 56 46	12 ♋ 10 27	28 00	21 30	14 05	10 19	03 09	02 23	19 42	06 45
28	04 23 50	07 07 04	18 ♋ 27 03	24 ♋ 46 50	28 39	22 28	14 46	10 33	03 05	02 24	19 45	06 45
29	04 27 47	08 04 40	01 ♌ 10 06	07 ♌ 37 11	29 14	23 25	15 27	10 46	03 01	02 25	19 47	06 45
30	04 31 43	09 02 15	14 ♌ 08 23	20 ♌ 44 01	29 ♊ 44	24 22	16 09	11 00	02 58	02 26	19 49	06 45
31	04 35 40	09 ♊ 59 48	27 ♌ 24 24	04 ♍ 09 47	00 ♋ 09	25 ♋ 19	16 ♊ 50	11 ♉ 13	02 ♑ 54	02 ♍ 27	19 ♉ 51	06 ♓ 45

Date	Moon True ☊ ° ′	Moon Mean ☊ ° ′	Moon ☽ Latitude ° ′
01	18 ♑ 15	19 ♑ 38	00 S 17
02	18 D 15	19 35	01 24
03	18 R 15	19 32	02 29
04	18 14	19 28	03 27
05	18 11	19 25	04 15
06	18 05	19 22	04 49
07	17 58	19 19	05 03
08	17 49	19 16	04 57
09	17 40	19 13	04 30
10	17 33	19 09	03 44
11	17 27	19 06	02 44
12	17 23	19 03	01 33
13	17 22	19 00	00 S 19
14	17 D 22	18 57	00 N 53
15	17 23	18 54	02 01
16	17 24	18 50	03 01
17	17 R 23	18 47	03 50
18	17 21	18 44	04 28
19	17 17	18 41	04 53
20	17 11	18 38	05 06
21	17 03	18 34	05 05
22	16 54	18 31	04 51
23	16 46	18 28	04 25
24	16 38	18 25	03 47
25	16 32	18 22	02 58
26	16 28	18 19	02 00
27	16 25	18 15	00 N 57
28	16 D 25	18 12	00 S 11
29	16 25	18 09	01 19
30	16 27	18 06	02 25
31	16 ♑ 28	18 ♑ 03	03 S 25

DECLINATIONS

Date	Sun ☉ ° ′	Moon ☽ ° ′	Mercury ☿ ° ′	Venus ♀ ° ′	Mars ♂ ° ′	Jupiter ♃ ° ′	Saturn ♄ ° ′	Uranus ♅ ° ′	Neptune ♆ ° ′	Pluto ⯓ ° ′
01	15 N 09	21 N 28	18 N 15	25 N 56	19 N 31	11 N 59	22 S 16	11 N 23	15 N 46	20 S 15
02	15 27	17 51	18 59	26 00	19 42	12 04	22 16	11 23	15 46	20 15
03	15 44	13 15	19 40	26 04	19 52	12 08	22 16	11 23	15 47	20 15
04	16 02	07 52	20 20	26 07	20 02	12 13	22 16	11 24	15 48	20 15
05	16 19	01 N 56	20 58	26 10	20 12	12 18	22 16	11 24	15 48	20 15
06	16 36	04 S 16	21 33	26 11	20 22	12 23	22 16	11 24	15 49	20 15
07	16 52	10 24	22 05	26 13	20 32	12 28	22 17	11 24	15 49	20 15
08	17 09	16 00	22 35	26 13	20 41	12 32	22 17	11 24	15 50	20 15
09	17 25	20 38	23 03	26 13	20 51	12 37	22 17	11 24	15 51	20 15
10	17 41	23 50	23 28	26 13	21 00	12 42	22 17	11 24	15 51	20 15
11	17 56	25 19	23 50	26 11	21 08	12 46	22 17	11 24	15 52	20 15
12	18 11	24 59	24 10	26 09	21 17	12 51	22 17	11 23	15 52	20 15
13	18 26	23 03	24 28	26 07	21 25	12 55	22 17	11 23	15 53	20 15
14	18 41	19 49	24 43	26 04	21 34	13 00	22 17	11 23	15 54	20 15
15	18 55	15 39	24 56	26 00	21 42	13 05	22 17	11 23	15 54	20 15
16	19 09	10 55	25 07	25 55	21 50	13 09	22 17	11 23	15 55	20 16
17	19 23	05 51	25 15	25 50	21 57	13 14	22 17	11 23	15 55	20 16
18	19 36	00 S 41	25 22	25 45	22 05	13 18	22 17	11 23	15 56	20 16
19	19 49	04 N 27	25 26	25 39	22 12	13 23	22 17	11 22	15 57	20 16
20	20 01	09 21	25 29	25 32	22 19	13 27	22 18	11 22	15 57	20 16
21	20 14	13 54	25 30	25 25	22 26	13 31	22 18	11 22	15 58	20 16
22	20 26	17 56	25 29	25 17	22 32	13 36	22 18	11 22	15 58	20 16
23	20 37	21 15	25 27	25 09	22 38	13 40	22 18	11 21	15 59	20 17
24	20 48	23 42	25 23	25 00	22 45	13 45	22 18	11 21	16 00	20 17
25	20 59	25 05	25 17	24 51	22 51	13 49	22 18	11 21	16 00	20 17
26	21 10	25 17	25 11	24 41	22 56	13 53	22 18	11 20	16 01	20 17
27	21 20	24 14	25 03	24 31	23 02	13 57	22 18	11 20	16 01	20 17
28	21 30	21 59	24 54	24 20	23 07	14 02	22 18	11 20	16 02	20 18
29	21 39	18 36	24 43	24 09	23 12	14 06	22 19	11 19	16 02	20 18
30	21 48	14 16	24 32	23 57	23 17	14 10	22 19	11 19	16 03	20 18
31	21 N 57	09 N 10	24 N 20	23 N 45	23 N 22	14 N 14	22 S 19	11 N 18	16 N 04	20 S 18

ZODIAC SIGN ENTRIES

Date	h	m	Planets
02	04	12	☽ ♌
04	10	12	☽ ♍
06	12	35	☽ ♎
06	14	34	♀ ♋
06	20	31	☿ ♊
07	07	45	♂ ♊
08	12	26	☽ ♏
10	11	42	☽ ♐
12	12	34	☽ ♑
14	16	45	☽ ♒
17	00	55	☽ ♓
19	12	13	☽ ♈
21	02	20	☉ ♊
22	00	51	☽ ♉
24	13	15	☽ ♊
27	00	28	☽ ♋
29	09	49	☽ ♌
31	02	38	☿ ♋
31	16	37	☽ ♍

LATITUDES

Date	Mercury ☿ ° ′	Venus ♀ ° ′	Mars ♂ ° ′	Jupiter ♃ ° ′	Saturn ♄ ° ′	Uranus ♅ ° ′	Neptune ♆ ° ′	Pluto ⯓ ° ′
01	00 N 47	02 N 37	00 N 19	01 S 00	01 N 06	00 N 47	01 S 42	12 S 01
04	01 17	02 42	00 20	01 00	01 06	00 47	01 42	12 02
07	01 43	02 47	00 22	01 00	01 06	00 47	01 42	12 03
10	02 04	02 51	00 24	01 00	01 06	00 46	01 42	12 04
13	02 17	02 53	00 25	01 00	01 06	00 46	01 42	12 05
16	02 24	02 55	00 27	01 00	01 06	00 46	01 42	12 06
19	02 22	02 55	00 29	01 00	01 06	00 46	01 42	12 07
22	02 12	02 54	00 30	01 00	01 06	00 46	01 42	12 08
25	01 54	02 52	00 32	01 00	01 06	00 46	01 42	12 09
28	01 28	02 48	00 34	01 00	01 05	00 46	01 42	12 10
31	00 N 54	02 N 43	00 N 35	01 S 00	01 N 05	00 N 46	01 S 42	12 S 11

DATA

Julian Date	2468832
Delta T	+80 seconds
Ayanamsa	24° 31′ 21″
Synetic vernal point	04° ♓ 35′ 38″
True obliquity of ecliptic	23° 26′ 02″

MOON'S PHASES, APSIDES AND POSITIONS ☽

Date	h	m	Phase	Longitude ° ′	Eclipse Indicator
03	03	26	☽	12 ♌ 39	
09	18	24	○	19 ♏ 04	
16	16	46	☾	25 ♒ 46	
24	20	27	●	03 ♊ 37	

Day	h	m		
09	02	49	Perigee	
21	20	14	Apogee	
05	19	32	0S	
11	19	32	Max dec	25° S 24′
18	15	09	0N	
26	03	56	Max dec	25° N 21′

LONGITUDES

Date	Chiron ⚷ ° ′	Ceres ⚳ ° ′	Pallas ⚴ ° ′	Juno ⚵ ° ′	Vesta ⚶ ° ′	Black Moon Lilith ⚸ ° ′
01	19 ♎ 48	19 ♑ 32	28 ♏ 28	11 ♐ 38	29 ♐ 34	29 ♈ 14
11	19 ♎ 09	20 ♑ 00	25 ♏ 33	09 ♐ 51	29 ♐ 16	00 ♉ 21
21	18 ♎ 39	19 ♑ 52	22 ♏ 32	07 ♐ 44	28 ♐ 11	01 ♉ 28
31	18 ♎ 17	19 ♑ 08	19 ♏ 44	05 ♐ 30	26 ♐ 23	02 ♉ 35

All ephemeris data is given at 12.00 UT and the Moon's longitude is additionally given for 24.00 UT

ASPECTARIAN

h m Aspects

01 Wednesday
02 02 ☽ Q ♃
04 14 ☽ ∠ ♅
05 11 ☿ Q ⯓
05 23 ☽ # ♄
06 49 ☽ ⚹ ☿
06 59 ☽ ⚹ ♆
07 51 ☿ ☌ ♆
12 15 ☽ ♇ ⯓
14 52 ☿ ♇ ♄
15 28 ☽ Q ☉
18 20 ☽ ⊻ ♀
18 47 ♀ ⊥ ♆
20 51 ☽ # ⯓
21 00 ☽ ⚹ ♂
21 15 ☽ ⊥ ♅
22 43 ♃ △ ♄

02 Thursday
01 05 ☽ ∥ ♂
05 09 ☽ ± ⯓
05 37 ☽ Q ♆
06 15 ☽ ∥ ☿
06 36 ☽ ⊥ ♀
08 26 ☽ ⊻ ♅
09 46 ☽ Q ☿
12 13 ☽ ⊼ ♄
12 30 ☽ □ ♃
16 15 ☽ ⊼ ⯓
20 30 ☽ Q ♂
23 11 ☽ ± ♄
23 22 ☽ ∥ ♆

03 Friday
00 19 ☽ ∥ ☉
00 38 ☽ ∠ ♀
03 26 ☽ □ ☉
14 30 ☽ □ ♆
15 21 ☽ ♇ ♄
15 48 ☉ ∥ ♆
17 08 ☽ ∥ ♃
20 39 ☽ ∥ ♅
21 24 ☿ ∥ ♂

04 Saturday
00 16 ☽ □ ☿
05 43 ☽ ⚹ ♀
06 26 ☽ □ ♂
08 47 ☿ # ⯓
14 07 ☽ ☌ ♅
17 32 ☽ △ ♄
18 51 ☽ △ ♃
20 09 ♂ ± ♄
21 29 ☽ ☍ ⯓

05 Sunday
04 07 ☽ Q ♀
06 19 ♀ ⊻ ♂
11 41 ☽ △ ☉
18 03 ♂ # ⯓
18 18 ☽ △ ♆
20 35 ☽ ♇ ♃
22 34 ☿ ± ♄

06 Monday
11 22 ☽ △ ☿
11 37 ☽ △ ♂
11 38 ☽ ± ♃
12 25 ☽ □ ♀
14 08 ☿ ☌ ♂
14 12 ☽ ♇ ☉
16 14 ☽ ⊻ ♅
18 58 ☽ ♇ ♆
19 21 ☽ □ ♄
21 31 ☽ ⊼ ♃
23 13 ☽ ⊼ ⯓

07 Tuesday
01 55 ☽ ⊥ ♅
04 00 ☿ ⊻ ♀
05 43 ☽ ± ☉
08 51 ☽ ± ⯓
09 27 ☽ ± ♆
12 57 ☽ ♇ ♂
15 04 ☽ ♇ ☿
15 57 ☽ ⊼ ☉
16 04 ☽ # ♅
16 17 ☽ ∠ ♅
19 01 ☽ ⊼ ♆
20 35 ☽ # ♃
20 48 ☿ # ♄
23 09 ☽ ♇ ⯓

08 Wednesday
00 01 ☽ Q ♄
00 44 ☿ □ ♅
03 54 ☽ ± ♂
07 20 ☽ ± ☿
11 12 ☽ # ♆
13 49 ☽ ⊼ ♂
15 54 ☽ △ ♀
15 58 ☽ ⚹ ♅
16 48 ♀ ⚹ ♅
17 41 ☽ # ☉
18 06 ☽ ⊼ ☿
18 53 ☽ ⚹ ♄
21 51 ☽ ☍ ♃
22 48 ☽ △ ⯓

09 Thursday
00 26 ☿ ⊼ ♄
06 43 ☉ Q ⯓
09 43 ☽ ∥ ⯓
10 49 ☽ Q ♅
13 18 ☽ # ♂
17 15 ☽ ♇ ♀
17 47 ☉ ☌ ♆
18 13 ☉ ♇ ♄
18 22 ☽ ☍ ♆
18 24 ☽ ∠ ♄
18 24 ☽ ☍ ☉
20 09 ♅ St D
22 48 ☽ ∥ ♄
23 23 ♄ ♇ ♆

10 Friday
05 53 ☿ ⊻ ♃
07 46 ☽ # ☿
08 33 ☽ ± ♀
08 35 ☽ ⊥ ♄
08 59 ♀ ☍ ♄
09 43 ♀ ∠ ♆
09 57 ☿ □ ⯓
12 26 ♂ □ ♅
15 17 ☽ □ ♅
15 25 ☽ ☍ ♂
18 07 ☽ ⊻ ♄
18 50 ☽ ⊼ ♀
22 02 ☽ ⊼ ♃
22 16 ☽ □ ⯓
23 54 ☽ ☍ ☿

11 Saturday
07 52 ☽ ± ♃
13 41 ♃ ⚹ ⯓
18 32 ☽ ⊼ ♆
21 51 ☽ ⊼ ☉
22 48 ☽ ♇ ♃

12 Sunday
03 37 ☽ Q ⯓
04 27 ☽ ± ♆
08 30 ☽ ± ☉
16 21 ☽ △ ♅
19 03 ☽ ⊼ ♂
19 11 ☽ ☌ ♄
19 34 ☽ ♇ ♆
19 58 ♀ △ ⯓
21 50 ♂ ⊼ ♄
22 40 ☽ # ☿
23 47 ☽ ⚹ ⯓

13 Monday
00 05 ☽ ☍ ♀
00 20 ☽ △ ♃
00 47 ☽ ♇ ☉
04 20 ♀ ⚹ ♃
05 48 ☽ ± ♂
08 12 ☽ ⊼ ☿
18 06 ☽ ♇ ♅
18 29 ☽ ∥ ♄
20 00 ☽ ± ☿
21 31 ☽ △ ♆
22 17 ☽ ♇ ♂

14 Tuesday
00 31 ☽ # ♂
01 50 ☽ ∠ ⯓
04 51 ☽ △ ☉
09 07 ☽ ∥ ⯓
10 02 ☽ ± ♅
10 55 ☿ ⊥ ♃
14 04 ☽ ♇ ☿
17 57 ☽ ⊥ ⯓
18 36 ☽ # ☉
20 51 ☽ ⊼ ♅
23 43 ☽ ⊻ ♄

15 Wednesday
02 42 ☽ △ ♂
04 53 ☽ ⊻ ⯓
06 25 ☽ □ ♃
09 49 ☽ ⊼ ♀
10 40 ☽ # ♆
10 42 ☽ ⊥ ♄
21 11 ☽ △ ☿
21 59 ☽ ± ♀

16 Thursday
01 07 ☽ # ♃
03 28 ☽ ∠ ♄
04 26 ☽ □ ♆
09 43 ☽ # ♅
16 34 ☽ ♇ ♀
16 46 ☽ □ ☉
16 54 ☽ Q ♃
20 07 ♂ □ ⯓

17 Friday
05 19 ☽ ☍ ♅
08 08 ☽ ⚹ ♄
13 55 ☽ ☌ ⯓
14 58 ☽ □ ♂
15 12 ☽ Q ♆
16 37 ☽ ⚹ ♃

18 Saturday
00 25 ☽ △ ♀
01 49 ☉ ∠ ♀
06 51 ☽ Q ☉
07 36 ☽ Q ♄
14 34 ☽ □ ☿
14 58 ☽ ⚹ ♆
15 45 ☉ ± ♄
18 27 ☿ ⊻ ♆
22 54 ☽ ∠ ♃

19 Sunday
04 50 ☽ Q ♂
08 52 ☽ ⚹ ☉
09 31 ♂ ⊻ ♃
11 49 ☿ Q ♅
16 49 ☽ ⊼ ♅
17 18 ☽ ⊥ ♃
19 26 ☽ □ ♄
21 07 ☽ ∠ ♆

20 Monday
01 42 ☽ ⊻ ⯓
04 55 ☽ ± ♅
05 39 ☽ ⊻ ♃
06 29 ☽ ⚹ ♂
06 50 ☽ Q ☿
07 51 ♂ # ♄
13 50 ☽ ⊥ ⯓
15 21 ☽ ⊥ ♆
17 47 ☽ ∠ ☉
18 07 ☽ □ ♀
20 50 ☿ ∥ ♀
22 21 ☽ ∥ ♅
23 08 ☽ ♇ ♅

21 Tuesday
03 32 ☽ ⊻ ♆
08 04 ☽ ∠ ⯓
09 43 ☽ ⚹ ☿
09 53 ☽ ∥ ♃
13 37 ☽ ⊥ ☉
14 43 ☽ ∠ ♂
17 14 ☉ # ⯓
23 52 ☽ ∥ ♆

22 Wednesday
02 50 ☽ ⊻ ☉
05 32 ☽ △ ♅
07 50 ☽ △ ♄
09 56 ☽ Q ♀
10 04 ☽ ⊥ ♂
14 26 ☽ ⚹ ⯓
15 04 ♀ ⊥ ♂
19 03 ☽ ∠ ☿
19 33 ☽ ☌ ♃
22 57 ☽ ⊻ ♂

23 Thursday
02 42 ☿ ∠ ♃
04 18 ☽ # ⯓
06 06 ♀ ∠ ♅
06 36 ☽ ∥ ☉
12 22 ☽ ⚹ ♀
12 30 ☉ □ ♅
13 57 ☽ ♇ ♄
14 39 ☽ Q ⯓
14 49 ☽ ⊥ ☿
16 20 ☽ ☌ ♆
21 08 ☽ # ♄

24 Friday
00 59 ☽ ∥ ♂
03 54 ☽ ⊻ ☿
07 55 ☽ ± ♄
11 04 ☿ ⊥ ♆
13 14 ☉ ⊼ ♄
17 56 ☽ □ ♅
19 51 ☽ ⊼ ♄
20 27 ☽ ☌ ☉
21 07 ☽ ∠ ♀

25 Saturday
02 40 ☽ □ ⯓
06 49 ☽ ∥ ♀
08 50 ☽ ⊻ ♃
14 04 ♀ ⚹ ♆
14 38 ☽ ☌ ♂
16 33 ☽ ⊥ ♀
18 38 ☽ ∥ ☿
20 55 ☽ ⊥ ♃

26 Sunday
04 14 ☽ ⊻ ♆
05 25 ☽ ⊻ ♀
05 37 ☽ Q ♅
14 57 ☽ ∠ ♃
16 02 ☽ ⊥ ♆
17 44 ☽ ∥ ☿
19 38 ☽ ☌ ☿

27 Monday
05 06 ☽ ⚹ ♅
06 36 ☽ ☍ ♄
07 16 ☽ ∥ ♀
09 36 ☽ ∠ ♆
12 27 ☽ ⊻ ☉
13 33 ☽ △ ⯓
18 02 ♀ ♇ ⯓
20 36 ☽ ⚹ ♃

28 Tuesday
00 58 ☽ ⊥ ☉
01 50 ☽ ∥ ♂
02 43 ☉ □ ⯓
04 33 ☽ ⊻ ♂
09 11 ☽ # ♄
10 00 ☽ ∠ ♅
14 28 ☽ ⚹ ♆
14 41 ♀ Q ♃
15 47 ☽ ∥ ☉
16 39 ☽ ⊥ ♂
18 16 ☽ ♇ ⯓
19 32 ☽ ∠ ☉
19 55 ☽ Q ♃
20 15 ☽ ☌ ♀

29 Wednesday
00 53 ☽ # ⯓
03 05 ☽ ⊥ ♅
08 13 ☽ ⊻ ☿
10 35 ☽ ∠ ♂
11 13 ☽ ± ⯓
13 09 ☽ Q ♆
14 21 ☽ ⊻ ♅
15 27 ☽ ⊼ ♄
19 54 ☽ ⊥ ☿
22 24 ☽ ⊼ ⯓

30 Thursday
01 52 ☽ ⚹ ☉
02 31 ☽ ± ♄
02 42 ☽ ∥ ♆
06 08 ☽ □ ♃
12 30 ☽ ∥ ♃
13 07 ☽ ∠ ☿
15 52 ☽ ⚹ ♂
18 56 ☽ ♇ ♄
22 22 ☽ □ ♆

31 Friday
01 32 ☽ Q ☉
02 17 ☽ ∥ ♅
07 58 ☽ ⊻ ♀
14 41 ☽ Q ♂
17 03 ☽ ⚹ ☿
19 29 ☽ ⊥ ♀
21 00 ☽ ☌ ♅
21 43 ☽ △ ♄

JUNE 2047

LONGITUDES

Date	Sidereal time h m s	Sun ☉ ° ' "	Moon ☽ ° ' "	Moon ☽ 24.00 ° ' "	Mercury ☿ ° '	Venus ♀ ° '	Mars ♂ ° '	Jupiter ♃ ° '	Saturn ♄ ° '	Uranus ♅ ° '	Neptune ♆ ° '	Pluto ♇ ° '
01	04 39 36	10 ♊ 57 19	11 ♍ 00 21	17 ♍ 56 15	00 ♋ 30	26 ♋ 15	17 ♊ 31	11 ♉ 27	02 ♑ 50	02 ♍ 29	19 ♉ 53	06 ♓ 46
02	04 43 33	11 54 49	24 ♍ 57 30	02 ♎ 04 02	00 46	27 10	18 12	11 40	02 R 46	02 30	19 55	06 46
03	04 47 30	12 52 18	09 ♎ 15 36	16 ♎ 31 51	00 58	28 05	18 53	11 54	02 42	02 31	19 57	06 46
04	04 51 26	13 49 46	23 ♎ 52 14	01 ♏ 16 04	01 05	29 00	19 34	12 07	02 38	02 32	20 00	06 46
05	04 55 23	14 47 12	08 ♏ 42 33	16 ♏ 10 43	01 07	29 ♋ 53	20 15	12 20	02 34	02 34	20 02	06 46
06	04 59 19	15 44 38	23 ♏ 39 32	01 ♐ 07 55	01 R 05	00 ♌ 47	20 56	12 33	02 30	02 35	20 04	06 46
07	05 03 16	16 42 02	08 ♐ 34 46	15 ♐ 59 02	00 58	01 39	21 37	12 46	02 25	02 37	20 06	06 R 46
08	05 07 12	17 39 26	23 ♐ 19 45	00 ♑ 36 02	00 47	02 31	22 18	12 59	02 21	02 38	20 08	06 46
09	05 11 09	18 36 48	07 ♑ 47 11	14 ♑ 52 36	00 32	03 23	22 59	13 12	02 17	02 40	20 10	06 46
10	05 15 05	19 34 10	21 ♑ 51 54	28 ♑ 44 51	00 ♋ 14	04 13	23 40	13 25	02 13	02 41	20 12	06 46
11	05 19 02	20 31 32	05 ♒ 31 21	12 ♒ 11 28	29 ♊ 51	05 03	24 21	13 38	02 09	02 43	20 14	06 46
12	05 22 59	21 28 52	18 ♒ 45 22	25 ♒ 13 20	29 26	05 52	25 01	13 51	02 04	02 45	20 16	06 46
13	05 26 55	22 26 12	01 ♓ 35 46	07 ♓ 53 05	28 58	06 41	25 42	14 04	02 00	02 47	20 18	06 45
14	05 30 52	23 23 32	14 ♓ 05 49	20 ♓ 14 28	28 27	07 29	26 23	14 16	01 56	02 48	20 20	06 45
15	05 34 48	24 20 51	26 ♓ 19 36	02 ♈ 21 48	27 55	08 16	27 03	14 29	01 51	02 50	20 22	06 45
16	05 38 45	25 18 10	08 ♈ 21 38	14 ♈ 19 40	27 22	09 02	27 44	14 42	01 47	02 52	20 24	06 45
17	05 42 41	26 15 28	20 ♈ 16 26	26 ♈ 12 28	26 48	09 47	28 24	14 54	01 42	02 54	20 26	06 45
18	05 46 38	27 12 46	02 ♉ 08 17	08 ♉ 04 20	26 14	10 31	29 05	15 06	01 38	02 56	20 28	06 44
19	05 50 34	28 10 04	14 ♉ 01 04	19 ♉ 58 52	25 41	11 14	29 ♊ 45	15 19	01 34	02 58	20 29	06 44
20	05 54 31	29 ♊ 07 21	25 ♉ 58 08	01 ♊ 59 09	25 08	11 57	00 ♋ 26	15 31	01 29	03 00	20 31	06 44
21	05 58 28	00 ♋ 04 38	08 ♊ 02 14	14 ♊ 07 38	24 37	12 38	01 06	15 43	01 25	03 02	20 33	06 43
22	06 02 24	01 01 55	20 ♊ 15 34	26 ♊ 26 12	24 09	13 19	01 46	15 55	01 20	03 05	20 35	06 43
23	06 06 21	01 59 12	02 ♋ 39 42	08 ♋ 56 12	23 43	13 58	02 27	16 07	01 16	03 07	20 37	06 43
24	06 10 17	02 56 28	15 ♋ 15 49	21 ♋ 38 38	23 20	14 36	03 07	16 19	01 11	03 09	20 39	06 42
25	06 14 14	03 53 43	28 ♋ 04 43	04 ♌ 34 09	23 01	15 13	03 47	16 31	01 07	03 11	20 40	06 42
26	06 18 10	04 50 58	11 ♌ 07 00	17 ♌ 43 17	22 45	15 49	04 27	16 43	01 03	03 14	20 42	06 41
27	06 22 07	05 48 13	24 ♌ 23 04	01 ♍ 06 22	22 34	16 23	05 08	16 55	00 58	03 16	20 44	06 41
28	06 26 03	06 45 27	07 ♍ 53 12	14 ♍ 43 34	22 27	16 57	05 48	17 06	00 54	03 19	20 46	06 40
29	06 30 00	07 42 40	21 ♍ 37 26	28 ♍ 34 44	22 25	17 29	06 28	17 18	00 49	03 21	20 47	06 40
30	06 33 57	08 ♋ 39 53	05 ♎ 35 21	12 ♎ 39 08	22 ♊ 27	17 ♌ 59	07 ♋ 08	17 ♉ 29	00 ♑ 45	03 ♍ 23	20 ♉ 49	06 ♓ 39

MOON NODES AND LATITUDE; DECLINATIONS

Date	Moon True ☊ ° '	Moon Mean ☊ ° '	Moon ☽ Latitude ° '	Sun ☉ ° '	Moon ☽ ° '	Mercury ☿ ° '	Venus ♀ ° '	Mars ♂ ° '	Jupiter ♃ ° '	Saturn ♄ ° '	Uranus ♅ ° '	Neptune ♆ ° '	Pluto ♇ ° '
01	16 ♑ 29	18 ♑ 00	04 S 14	22 N 05	03 N 31	24 N 07	23 N 32	23 N 26	14 N 18	22 S 19	11 N 18	16 N 04	20 S 19
02	16 R 28	17 56	04 50	22 13	02 S 26	23 53	23 19	23 31	14 22	22 19	11 17	16 05	20 19
03	16 26	17 53	05 09	22 20	08 24	23 38	23 06	23 35	14 26	22 19	11 17	16 05	20 19
04	16 23	17 50	05 09	22 27	14 03	23 22	22 52	23 38	14 30	22 19	11 16	16 06	20 19
05	16 18	17 47	04 48	22 34	18 57	23 07	22 38	23 42	14 34	22 19	11 16	16 06	20 20
06	16 14	17 44	04 08	22 40	22 41	22 51	22 24	23 45	14 38	22 19	11 16	16 07	20 20
07	16 10	17 40	03 11	22 46	24 52	22 35	22 09	23 49	14 42	22 20	11 15	16 07	20 20
08	16 08	17 37	02 02	22 52	25 17	22 18	21 54	23 52	14 46	22 20	11 14	16 08	20 21
09	16 06	17 34	00 S 45	22 57	23 58	22 01	21 38	23 54	14 50	22 20	11 14	16 08	20 21
10	16 D 06	17 31	00 N 32	23 01	21 08	21 43	21 23	23 57	14 54	22 20	11 13	16 09	20 21
11	16 07	17 28	01 45	23 06	17 12	21 26	21 07	23 59	14 58	22 20	11 12	16 09	20 22
12	16 08	17 25	02 50	23 10	12 30	21 09	20 51	24 01	15 01	22 20	11 12	16 10	20 22
13	16 10	17 21	03 44	23 13	07 24	20 53	20 34	24 03	15 05	22 20	11 11	16 10	20 23
14	16 11	17 18	04 27	23 16	02 S 09	20 36	20 17	24 05	15 09	22 20	11 10	16 11	20 23
15	16 R 11	17 15	04 56	23 19	03 N 04	20 21	20 01	24 06	15 12	22 21	11 10	16 11	20 23
16	16 11	17 12	05 12	23 21	08 05	20 06	19 43	24 08	15 16	22 21	11 09	16 12	20 24
17	16 09	17 09	05 14	23 23	12 46	19 51	19 26	24 09	15 20	22 21	11 08	16 12	20 24
18	16 07	17 05	05 03	23 24	16 56	19 38	19 09	24 09	15 23	22 21	11 07	16 13	20 24
19	16 04	17 02	04 38	23 25	20 28	19 26	18 51	24 10	15 27	22 21	11 07	16 13	20 25
20	16 01	16 59	04 02	23 26	23 09	19 15	18 34	24 11	15 30	22 21	11 06	16 14	20 25
21	15 59	16 56	03 14	23 26	24 50	19 05	18 16	24 11	15 34	22 21	11 05	16 14	20 26
22	15 57	16 53	02 17	23 26	25 21	18 57	17 58	24 11	15 37	22 21	11 04	16 15	20 26
23	15 56	16 50	01 13	23 25	24 37	18 50	17 40	24 11	15 40	22 22	11 03	16 15	20 27
24	15 56	16 46	00 N 04	23 24	22 37	18 45	17 22	24 10	15 44	22 22	11 03	16 16	20 27
25	15 D 56	16 43	01 S 07	23 23	19 27	18 41	17 04	24 09	15 47	22 22	11 02	16 16	20 28
26	15 57	16 40	02 15	23 21	15 16	18 39	16 46	24 09	15 50	22 22	11 01	16 16	20 28
27	15 58	16 37	03 17	23 18	10 17	18 38	16 29	24 08	15 54	22 22	11 00	16 17	20 28
28	15 58	16 34	04 10	23 16	04 N 45	18 39	16 11	24 06	15 57	22 22	10 59	16 17	20 29
29	15 59	16 31	04 49	23 13	01 S 06	18 42	15 53	24 05	16 00	22 22	10 58	16 18	20 29
30	15 ♑ 59	16 ♑ 27	05 S 12	23 N 09	06 S 59	18 N 46	15 N 35	24 N 03	16 N 03	22 S 22	10 N 57	16 N 18	20 S 30

ZODIAC SIGN ENTRIES

Date	h	m	Planets	
02	20	32	☽	♎
04	21	57	☽	♏
05	14	55	♀	♌
06	22	11	☽	♐
08	23	00	☽	♑
11	02	12	☽	♒
11	03	05	☿	♊
13	08	59	☽	♓
15	19	17	☽	♈
18	07	40	☽	♉
19	20	43	♂	♋
20	20	03	☽	♊
21	10	03	☉	♋
23	06	53	☽	♋
25	15	34	☽	♌
27	22	02	☽	♍
30	02	26	☽	♎

LATITUDES

Date	Mercury ☿ ° '	Venus ♀ ° '	Mars ♂ ° '	Jupiter ♃ ° '	Saturn ♄ ° '	Uranus ♅ ° '	Neptune ♆ ° '	Pluto ♇ ° '
01	00 N 41	02 N 41	00 N 36	01 S 01	01 N 05	00 N 46	01 S 42	12 S 11
04	00 S 03	02 34	00 37	01 01	01 05	00 45	01 42	12 12
07	00 51	02 25	00 39	01 01	01 05	00 45	01 42	12 13
10	01 43	02 15	00 40	01 01	01 05	00 45	01 42	12 14
13	02 33	02 02	00 41	01 01	01 05	00 45	01 42	12 15
16	03 19	01 48	00 43	01 01	01 05	00 45	01 42	12 16
19	03 56	01 31	00 44	01 01	01 04	00 45	01 42	12 17
22	04 22	01 12	00 45	01 02	01 04	00 45	01 43	12 18
25	04 34	00 51	00 47	01 01	01 04	00 45	01 43	12 19
28	04 34	00 27	00 48	01 01	01 04	00 45	01 43	12 20
31	04 S 23	00 00	00 N 49	01 S 02	01 N 03	00 N 44	01 S 43	12 S 21

DATA

Julian Date	2468863
Delta T	+80 seconds
Ayanamsa	24° 31' 26"
Synetic vernal point	04° ♓ 35' 33"
True obliquity of ecliptic	23° 26' 02"

MOON'S PHASES, APSIDES AND POSITIONS ☽

Date	h	m	Phase	Longitude ° '	Eclipse Indicator
01	11	54	☽	10 ♍ 57	
08	02	05	○	17 ♐ 16	
15	07	45	☾	24 ♓ 11	
23	10	36	●	01 ♋ 56	Partial
30	17	37	☽	08 ♎ 53	

Day	h	m		
06	09	26	Perigee	
18	10	29	Apogee	
02	02	15	0S	
08	05	33	Max dec	25° S 21'
14	21	49	0N	
22	10	04	Max dec	25° N 22'
29	07	32	0S	

LONGITUDES

Date	Chiron ⚷ ° '	Ceres ⚳ ° '	Pallas ⚴ ° '	Juno ⚵ ° '	Vesta ⚶ ° '	Black Moon Lilith ⚸ ° '
01	18 ♎ 15	19 ♑ 01	19 ♏ 29	05 ♐ 16	26 ♐ 10	02 ♉ 42
11	18 ♎ 05	17 ♑ 39	17 ♏ 16	03 ♐ 07	23 ♐ 53	03 ♉ 49
21	18 ♎ 06	15 ♑ 49	15 ♏ 47	01 ♐ 13	21 ♐ 29	04 ♉ 56
31	18 ♎ 18	13 ♑ 42	15 ♏ 04	29 ♏ 45	19 ♐ 21	06 ♉ 03

All ephemeris data is given at 12.00 UT and the Moon's longitude is additionally given for 24.00 UT

ASPECTARIAN

h m	Aspects
01 Saturday	
04 34	☽ ☍ ♇
08 28	♂ ⊥ ♃
11 54	☽ □ ☉
12 27	☽ ∠ ♀
12 47	☽ △ ♃
14 40	☽ Q ☿
18 05	♀ ⊥ ♅
20 17	♀ ‖ ♂
23 51	☽ □ ♂
02 Sunday	
03 23	☽ △ ♆
04 05	☉ ⊻ ♃
14 57	☽ ⚼ ♃
16 01	☽ ⚹ ♀
21 59	☽ □ ☿
03 Monday	
00 45	☽ ⊻ ♅
01 06	☽ □ ♄
04 49	☽ ⚼ ♆
06 19	☽ ± ♃
07 51	☽ ⊼ ♇
08 18	☉ # ♄
10 46	☽ ⊥ ♅
13 28	☽ Q ♀
16 26	☽ ⊼ ♃
16 50	☿ ‖ ♂
17 48	☽ ± ♇
18 24	☽ △ ☉
19 47	☽ ± ♆
23 57	☽ # ♅
04 Tuesday	
01 39	☽ ∠ ♅
04 38	☽ △ ♂
05 39	☽ ⊼ ♆
06 44	☽ Q ♄
08 34	☽ ⚼ ♇
14 07	☽ # ♃
20 37	☽ ⚼ ☉
20 52	☽ □ ♀
21 31	☽ # ♆
23 45	☽ △ ☿
05 Wednesday	
02 05	☽ ⚹ ♅
02 08	☽ ⚹ ♄
03 33	♂ ⊻ ♆
06 10	☽ ⚼ ♂
08 52	☽ △ ♇
11 12	♄ △ ♅
12 05	☿ St R
12 08	☽ ± ☉
16 44	☉ ‖ ♀
17 55	☽ ☍ ♃
19 55	☽ ‖ ♇
21 20	☽ ± ♂
21 25	☽ Q ♅
22 26	☽ ⊼ ☉
23 15	♂ Q ♅
23 53	☽ ⚼ ☿
06 Thursday	
02 09	☽ ∠ ♄
06 13	☽ ☍ ♆
07 26	☽ ⊼ ♂
09 12	☽ ‖ ♄
09 54	☽ # ♀
11 39	♀ ± ♇
11 53	☽ # ☉
13 14	☽ # ☿
14 16	☽ ± ☿
16 32	☽ ⊥ ♄
18 47	♀ # ♄
19 33	☿ ⊻ ♀
20 51	♇ St R
21 46	☽ # ♂
22 47	☉ ∠ ♀
23 39	☉ ‖ ☿
23 51	☽ ⊼ ☿
07 Friday	
00 09	☽ △ ♀
02 07	☽ ⊻ ♄
02 22	☽ □ ♅
09 04	☽ □ ♇
18 53	☽ ⊼ ♃
08 Saturday	
00 40	♀ Q ♆
01 55	☽ ⚼ ♀
02 05	☽ ☍ ☉
04 47	☽ ± ♃
06 45	☽ ⊼ ♆
07 41	♀ ⊼ ♄
09 13	☿ # ♄
10 14	☽ ☍ ♂
14 22	☽ Q ♇
15 21	♀ ⊻ ♅
16 37	☽ ± ♆
17 35	☽ ± ♀
19 48	☽ ⚼ ♃
09 Sunday	
00 07	☽ ☍ ☿
02 51	☽ ☌ ♄
03 25	☽ △ ♅
04 09	☽ ⊼ ♀
07 36	☽ ⚼ ♆
10 17	☽ ⚹ ♇
12 37	☽ # ♂
21 18	☽ △ ♃
21 43	☽ # ☉
10 Monday	
03 09	☽ ‖ ♄
04 48	☽ ⚼ ♅
07 12	☉ ⊥ ♃
07 24	☽ # ☿
07 45	☽ ⊼ ☉
09 07	☽ △ ♆
10 10	☽ # ♀
11 49	☽ ∠ ♇
15 17	☽ ⊼ ♂
17 15	☽ ‖ ♇
18 55	☽ ± ☉
20 25	☽ ± ♅
11 Tuesday	
02 14	☽ ⊼ ☿
02 20	☽ ± ♂
03 33	☽ ⊥ ♇
04 22	☉ ⊻ ♆
06 01	☽ ⊻ ♄
07 00	☽ ⊼ ♅
11 07	☽ ☍ ♀
12 00	☽ ⚼ ☉
12 35	☽ ± ☿
14 13	☽ ⊻ ♇
16 40	☽ ⊥ ♄
17 01	☉ Q ♅
17 34	☽ # ♆
19 13	☽ ⚼ ♂
23 39	☽ # ♃
12 Wednesday	
02 52	☽ □ ♃
03 33	☿ ⊥ ♀
04 20	☽ ⚼ ☿
08 55	☽ ∠ ♄
14 48	☽ □ ♆
17 26	☽ △ ☉
18 20	☽ # ♅
13 Thursday	
00 16	☽ △ ♂
07 12	☽ △ ☿
08 37	☿ ∠ ♃
12 45	☽ ⚹ ♄
12 54	☽ Q ♃
14 14	♀ ⊼ ♇
14 15	☽ ☍ ♅
21 50	☽ ☌ ♇
22 21	☽ ⊼ ♀
14 Friday	
00 50	☽ Q ♆
04 18	♀ # ♇
10 17	♂ ⊥ ♆
10 43	☽ ± ♀
11 40	☽ Q ♄
12 21	☽ ⚹ ♃
15 Saturday	
00 13	☽ ⚹ ♆
00 34	♀ ± ♄
05 31	☽ ⚼ ♀
07 45	☽ □ ☉
08 06	☿ # ♇
13 32	☽ □ ♂
15 02	☽ □ ☿
18 23	☽ ∠ ♃
22 55	☽ □ ♄
16 Sunday	
00 59	☽ ⊼ ♅
04 57	☿ ☌ ♂
06 03	☽ ∠ ♆
08 46	☽ ⊻ ♇
12 41	☽ ⊥ ♃
13 02	☽ ± ♅
13 26	☽ △ ♀
20 49	☽ ⊥ ♇
22 48	☽ Q ☉
17 Monday	
00 10	☽ ⊥ ♆
00 58	☽ ⊻ ♃
01 27	☽ Q ☿
03 26	☽ ‖ ♅
03 43	☽ Q ♂
07 12	☽ ⚼ ♅
12 19	☽ ⊻ ♆
14 58	☽ ∠ ♇
16 25	☉ ⊥ ♆
20 34	☉ ☌ ☿
18 Tuesday	
00 36	☽ ⚹ ☿
01 10	☽ ⚹ ☉
02 33	☽ ‖ ♃
02 57	☿ ⊥ ♆
05 26	☽ ⚹ ♂
07 34	☽ ‖ ♆
10 59	☽ △ ♄
13 37	☽ △ ♅
21 18	☽ ⚹ ♇
19 Wednesday	
01 18	☽ ‖ ♀
01 21	☿ ∠ ♀
04 50	☽ ‖ ☿
05 34	☽ ∠ ☿
06 02	☽ □ ♀
10 08	☽ ∠ ☉
11 37	☽ # ♇
13 35	☽ ∠ ♂
14 39	☽ ☌ ♃
17 05	☽ ⚼ ♄
21 29	☽ Q ♇
22 53	☽ ⊥ ☿
20 Thursday	
01 03	☽ ☌ ♆
03 54	☽ # ♄
05 49	☽ ⊥ ☉
08 44	☽ ⊥ ♂
10 24	☽ ⊻ ☿
11 02	☽ ± ♄
15 05	☽ ‖ ☉
16 33	♂ ∠ ♃
18 50	☽ ⊻ ☉
20 26	☽ Q ♀
21 26	☽ ⊻ ♂
22 56	☽ ⊼ ♄
21 Friday	
00 38	☽ ‖ ♂
02 04	☽ □ ♅
09 24	☽ □ ♇
21 36	☽ ⚹ ♀
21 58	♂ ☍ ♄
22 Saturday	
03 23	☽ ⊻ ♃
08 32	☉ ∠ ♃
12 38	☽ ⊻ ♆
13 36	☽ Q ♅
15 18	☽ ⊥ ♃
19 09	☉ ☍ ♄
19 18	☽ ☌ ☿
23 Sunday	
00 19	☽ ⊥ ♆
04 30	☽ ∠ ♀
09 00	☽ ∠ ♃
09 20	☽ ☍ ♄
10 36	☽ ☌ ☉
11 34	☽ ☌ ♂
12 52	☽ ⚹ ♅
17 40	☽ ∠ ♆
18 52	☽ ‖ ♂
19 45	☽ △ ♇
22 42	☽ ⊥ ♀
24 Monday	
00 19	♃ ⚼ ♄
04 10	☽ ‖ ☉
10 41	☽ ⊻ ♀
13 16	♂ ⚹ ♅
14 02	☽ ⚹ ♃
14 21	☽ # ♄
17 28	☽ ∠ ♅
17 30	☉ ⚹ ♅
22 09	☽ ⚹ ♆
25 Tuesday	
00 06	☽ ⚼ ♇
02 47	☽ ⊻ ☿
02 56	☉ ☌ ♂
05 13	☽ # ♇
10 21	☽ ⊥ ♅
12 50	☽ Q ♃
13 42	☽ ⊥ ☿
16 51	☽ ± ♇
16 54	☽ ‖ ☿
17 36	☽ ⊼ ♄
20 31	☽ Q ♆
21 29	☽ ⊻ ♅
23 08	☽ ⊻ ♂
23 37	☽ ⊻ ☉
26 Wednesday	
03 21	☽ ‖ ♀
03 54	☽ ⊼ ♇
04 35	☽ ± ♄
05 58	☽ ∠ ☿
06 42	☽ ‖ ♆
09 03	☽ ‖ ♃
10 44	☽ ⊥ ♂
11 28	☽ ⊥ ☉
14 09	☿ ⊥ ♃
20 18	♀ ⚼ ♄
20 55	☽ ⚼ ♄
20 57	☽ ☌ ♀
22 20	☽ □ ♃
27 Thursday	
03 56	☽ ∠ ♂
05 04	☽ ∠ ☉
05 25	☽ □ ♆
08 46	☽ ‖ ♅
08 46	☽ ⚹ ☿
10 06	☉ ∠ ♆
23 42	☽ △ ♄
28 Friday	
03 25	♀ ‖ ♆
03 53	☽ ☌ ♅
05 58	☽ Q ☿
08 07	☽ ⚹ ♂
09 52	☽ ⚹ ☉
09 52	☽ ☍ ♇
09 52	☉ △ ♇
10 36	♂ ∠ ♆
23 18	♀ □ ♃
29 Saturday	
03 54	♀ ‖ ♃
04 23	☽ △ ♃
04 31	☽ ⊻ ♀
06 14	☽ Q ♂
08 26	☽ Q ☉
10 33	☽ △ ♆
12 35	☿ St D
13 22	☽ □ ☿
15 20	☽ ⊥ ♀
19 03	♂ △ ♇
30 Sunday	
03 46	☽ □ ♄
06 38	☽ ⚼ ♃
07 23	☽ ∠ ♀
08 14	☽ ⊻ ♅
12 23	☽ ⚼ ♆
13 49	☽ ⊼ ♇
14 45	☽ □ ♂
17 37	☽ □ ☉
18 29	☽ ⊥ ♅
22 10	☽ ± ♃

JULY 2047

LONGITUDES

Date	Sidereal time h m s	Sun ☉ ° ' "	Moon ☽ ° ' "	Moon ☽ 24.00 ° ' "	Mercury ☿ ° '	Venus ♀ ° '	Mars ♂ ° '	Jupiter ♃ ° '	Saturn ♄ ° '	Uranus ♅ ° '	Neptune ♆ ° '	Pluto ♇ ° '
01	06 37 53	09 ♋ 37 06	19 ♎ 45 51	26 ♎ 55 13	22 ♊ 34	18 ♌ 28	07 ♋ 48	17 ♉ 41	00 ♑ 41	03 ♍ 26	20 ♉ 50	06 ♓ 39
02	06 41 50	10 34 18	04 ♏ 06 52	11 ♏ 20 23	22 D 46	18 55	08 28	17 52	00 R 36	03 29	20 52	06 R 38
03	06 45 46	11 31 29	18 ♏ 35 14	25 ♏ 50 53	23 03	19 21	09 08	18 03	00 32	03 31	20 54	06 37
04	06 49 43	12 28 41	03 ♐ 06 40	10 ♐ 21 58	23 25	19 45	09 48	18 14	00 28	03 34	20 55	06 37
05	06 53 39	13 25 52	17 ♐ 36 04	24 ♐ 48 18	23 51	20 08	10 28	18 25	00 23	03 37	20 57	06 36
06	06 57 36	14 23 03	01 ♑ 57 59	09 ♑ 04 29	24 23	20 28	11 07	18 36	00 19	03 39	20 58	06 36
07	07 01 32	15 20 14	16 ♑ 07 15	23 ♑ 05 45	25 00	20 47	11 47	18 47	00 15	03 42	21 00	06 35
08	07 05 29	16 17 25	29 ♑ 59 37	06 ♒ 48 30	25 41	21 04	12 27	18 58	00 11	03 45	21 01	06 34
09	07 09 26	17 14 36	13 ♒ 32 12	20 ♒ 10 37	26 27	21 19	13 07	19 08	00 07	03 48	21 03	06 33
10	07 13 22	18 11 48	26 ♒ 43 45	03 ♓ 11 40	27 18	21 32	13 46	19 19	00 ♑ 03	03 51	21 04	06 33
11	07 17 19	19 08 59	09 ♓ 34 34	15 ♓ 52 44	28 14	21 43	14 26	19 29	29 ♐ 58	03 53	21 06	06 32
12	07 21 15	20 06 11	22 ♓ 06 29	28 ♓ 16 14	29 ♊ 15	21 52	15 06	19 40	29 54	03 56	21 07	06 31
13	07 25 12	21 03 24	04 ♈ 22 26	10 ♈ 25 35	00 ♋ 20	21 59	15 45	19 50	29 50	03 59	21 08	06 30
14	07 29 08	22 00 37	16 ♈ 26 13	22 ♈ 24 54	01 29	22 03	16 25	20 00	29 47	04 02	21 10	06 29
15	07 33 05	22 57 50	28 ♈ 22 11	04 ♉ 18 39	02 43	22 06	17 04	20 10	29 43	04 05	21 11	06 29
16	07 37 01	23 55 04	10 ♉ 14 52	16 ♉ 11 26	04 01	22 R 06	17 44	20 20	29 39	04 08	21 12	06 28
17	07 40 58	24 52 19	22 ♉ 08 52	28 ♉ 07 43	05 24	22 03	18 23	20 29	29 35	04 12	21 13	06 27
18	07 44 55	25 49 34	04 ♊ 08 30	10 ♊ 11 40	06 51	21 59	19 03	20 39	29 31	04 15	21 15	06 26
19	07 48 51	26 46 50	16 ♊ 17 38	22 ♊ 26 49	08 22	21 51	19 42	20 49	29 28	04 18	21 16	06 25
20	07 52 48	27 44 07	28 ♊ 39 32	04 ♋ 56 03	09 57	21 42	20 21	20 58	29 24	04 21	21 17	06 24
21	07 56 44	28 41 24	11 ♋ 16 34	17 ♋ 41 16	11 36	21 30	21 01	21 07	29 20	04 24	21 18	06 23
22	08 00 41	29 ♋ 38 42	24 ♋ 10 11	00 ♌ 43 21	13 18	21 16	21 40	21 16	29 17	04 27	21 19	06 22
23	08 04 37	00 ♌ 36 00	07 ♌ 20 43	14 ♌ 02 08	15 04	21 00	22 19	21 25	29 13	04 31	21 20	06 21
24	08 08 34	01 33 19	20 ♌ 47 27	27 ♌ 36 24	16 54	20 41	22 58	21 34	29 10	04 34	21 21	06 20
25	08 12 30	02 30 38	04 ♍ 28 43	11 ♍ 24 05	18 46	20 20	23 38	21 43	29 07	04 37	21 22	06 19
26	08 16 27	03 27 58	18 ♍ 22 09	25 ♍ 22 32	20 41	19 57	24 17	21 52	29 03	04 41	21 23	06 18
27	08 20 24	04 25 18	02 ♎ 24 53	09 ♎ 28 50	22 39	19 31	24 56	22 00	29 00	04 44	21 24	06 17
28	08 24 20	05 22 38	16 ♎ 34 00	23 ♎ 40 03	24 38	19 04	25 35	22 09	28 57	04 47	21 25	06 16
29	08 28 17	06 19 59	00 ♏ 46 39	07 ♏ 53 29	26 40	18 35	26 14	22 17	28 54	04 51	21 26	06 15
30	08 32 13	07 17 20	15 ♏ 00 15	22 ♏ 06 40	28 ♋ 43	18 04	26 53	22 25	28 51	04 54	21 27	06 14
31	08 36 10	08 ♌ 14 42	29 ♏ 12 27	06 ♐ 17 19	00 ♌ 47	17 ♌ 31	27 ♋ 32	22 ♉ 33	28 ♐ 48	04 ♍ 58	21 ♉ 28	06 ♓ 13

Moon True ☊, Mean ☊, Latitude and DECLINATIONS

Date	Moon True ☊ ° '	Moon Mean ☊ ° '	Moon ☽ Latitude ° '	Sun ☉ ° '	Moon ☽ ° '	Mercury ☿ ° '	Venus ♀ ° '	Mars ♂ ° '	Jupiter ♃ ° '	Saturn ♄ ° '	Uranus ♅ ° '	Neptune ♆ ° '	Pluto ♇ ° '
01	15 ♑ 59	16 ♑ 24	05 S 16	23 N 05	12 S 36	18 N 51	15 N 18	24 N 01	16 N 06	22 S 23	10 N 56	16 N 18	20 S 30
02	15 R 59	16 21	05 01	23 01	17 36	18 58	15 00	23 59	16 09	22 23	10 55	16 19	20 31
03	15 58	16 18	04 27	22 56	21 38	19 06	14 43	23 57	16 12	22 23	10 54	16 19	20 31
04	15 58	16 15	03 36	22 51	24 18	19 15	14 26	23 55	16 15	22 23	10 53	16 20	20 32
05	15 58	16 11	02 31	22 45	25 22	19 25	14 09	23 52	16 18	22 23	10 52	16 20	20 32
06	15 58	16 08	01 S 17	22 39	24 42	19 36	13 53	23 49	16 21	22 23	10 51	16 20	20 33
07	15 D 58	16 05	00 N 01	22 33	22 27	19 48	13 37	23 46	16 24	22 23	10 50	16 21	20 34
08	15 R 58	16 02	01 17	22 26	18 54	20 00	13 21	23 43	16 27	22 23	10 49	16 21	20 34
09	15 58	15 59	02 27	22 19	14 24	20 13	13 05	23 39	16 30	22 23	10 48	16 21	20 35
10	15 57	15 56	03 27	22 12	09 21	20 26	12 50	23 36	16 32	22 24	10 47	16 22	20 35
11	15 57	15 52	04 15	22 04	04 S 02	20 39	12 35	23 32	16 35	22 24	10 46	16 22	20 36
12	15 57	15 49	04 50	21 56	01 N 19	20 53	12 20	23 28	16 38	22 24	10 45	16 22	20 36
13	15 56	15 46	05 10	21 47	06 29	21 06	12 06	23 24	16 40	22 24	10 44	16 23	20 37
14	15 56	15 43	05 17	21 38	11 20	21 19	11 53	23 19	16 43	22 24	10 43	16 23	20 37
15	15 D 56	15 40	05 09	21 29	15 42	21 31	11 40	23 15	16 46	22 24	10 42	16 23	20 38
16	15 56	15 37	04 49	21 19	19 27	21 43	11 27	23 10	16 48	22 24	10 40	16 23	20 39
17	15 57	15 33	04 16	21 09	22 25	21 54	11 15	23 05	16 51	22 24	10 39	16 24	20 39
18	15 58	15 30	03 31	20 59	24 26	22 03	11 04	23 00	16 53	22 25	10 38	16 24	20 40
19	15 59	15 27	02 37	20 48	25 20	22 12	10 53	22 55	16 55	22 25	10 37	16 24	20 40
20	16 00	15 24	01 35	20 37	25 00	22 18	10 43	22 49	16 58	22 25	10 36	16 24	20 41
21	16 00	15 21	00 N 26	20 25	23 23	22 24	10 34	22 44	17 00	22 25	10 35	16 25	20 41
22	16 R 00	15 17	00 S 45	20 13	20 32	22 27	10 25	22 38	17 02	22 25	10 33	16 25	20 42
23	15 59	15 14	01 55	20 01	16 34	22 28	10 17	22 32	17 05	22 25	10 32	16 25	20 43
24	15 58	15 11	03 00	19 49	11 42	22 27	10 10	22 26	17 07	22 25	10 31	16 25	20 43
25	15 56	15 08	03 56	19 36	06 12	22 24	10 03	22 19	17 09	22 25	10 30	16 26	20 44
26	15 54	15 05	04 39	19 23	00 N 19	22 19	09 57	22 13	17 11	22 25	10 28	16 26	20 44
27	15 51	15 02	05 06	19 09	05 S 39	22 10	09 52	22 06	17 13	22 26	10 27	16 26	20 45
28	15 50	14 58	05 15	18 55	11 21	21 59	09 48	21 59	17 15	22 26	10 26	16 26	20 45
29	15 49	14 55	05 04	18 41	16 29	21 46	09 45	21 52	17 17	22 26	10 25	16 26	20 46
30	15 D 49	14 52	04 35	18 27	20 43	21 29	09 42	21 45	17 19	22 26	10 23	16 26	20 47
31	15 ♑ 50	14 ♑ 49	03 S 49	18 N 12	23 S 42	21 N 11	09 N 41	21 N 38	17 N 21	22 S 26	10 N 22	16 N 27	20 S 47

ZODIAC SIGN ENTRIES

Date	h m	Planets
02	05 09	☽ ♏
04	06 52	☽ ♐
06	08 42	☽ ♑
08	12 01	☽ ♒
10	18 03	☽ ♓
11	02 59	♄ ♐
13	03 23	☽ ♈
13	04 57	☿ ♋
15	15 17	☽ ♉
18	03 45	☽ ♊
20	14 35	☽ ♋
22	20 55	☉ ♌
22	22 41	☽ ♌
25	04 11	☽ ♍
27	07 53	☽ ♎
29	10 41	☽ ♏
31	02 57	☿ ♌
31	13 20	☽ ♐

LATITUDES

Date	Mercury ☿ ° '	Venus ♀ ° '	Mars ♂ ° '	Jupiter ♃ ° '	Saturn ♄ ° '	Uranus ♅ ° '	Neptune ♆ ° '	Pluto ♇ ° '
01	04 S 23	00 00	00 N 49	01 S 02	01 N 03	00 N 44	01 S 43	12 S 21
04	04 02	00 S 29	00 50	01 03	01 03	00 44	01 43	12 22
07	03 33	01 00	00 52	01 03	01 03	00 44	01 43	12 23
10	02 58	01 35	00 53	01 03	01 02	00 44	01 43	12 24
13	02 20	02 11	00 54	01 03	01 02	00 44	01 43	12 25
16	01 40	02 50	00 55	01 04	01 02	00 44	01 44	12 26
19	00 59	03 31	00 56	01 04	01 01	00 44	01 44	12 26
22	00 S 19	04 13	00 57	01 04	01 01	00 44	01 44	12 27
25	00 N 17	04 54	00 58	01 05	01 01	00 44	01 44	12 28
28	00 49	05 34	00 59	01 05	01 00	00 44	01 44	12 28
31	01 N 14	06 S 11	01 N 00	01 S 06	01 N 00	00 N 44	01 S 44	12 S 29

DATA

Julian Date	2468893
Delta T	+80 seconds
Ayanamsa	24° 31' 31"
Synetic vernal point	04° ♓ 35' 28"
True obliquity of ecliptic	23° 26' 01"

MOON'S PHASES, APSIDES AND POSITIONS ☽

Date	h m	Phase	Longitude ° '	Eclipse Indicator
07	10 34	○	15 ♑ 17	total
15	00 09	☾	22 ♈ 30	
22	22 49	●	00 ♌ 05	Partial
29	22 03	☽	06 ♏ 44	

Day	h m		
04	05 01	Perigee	
16	04 00	Apogee	
30	09 03	Perigee	
05	14 43	Max dec	25° S 22'
12	06 04	0N	
19	17 46	Max dec	25° N 22'
26	13 15	0S	

LONGITUDES

Date	Chiron ⚷ ° '	Ceres ⚳ ° '	Pallas ⚴ ° '	Juno ⚵ ° '	Vesta ⚶ ° '	Black Moon Lilith ⚸ ° '
01	18 ♎ 18	13 ♑ 42	15 ♏ 04	29 ♏ 45	19 ♐ 21	06 ♉ 03
11	18 ♎ 41	11 ♑ 30	15 ♏ 08	28 ♏ 47	17 ♐ 47	07 ♉ 11
21	19 ♎ 14	09 ♑ 28	15 ♏ 52	28 ♏ 22	16 ♐ 57	08 ♉ 18
31	19 ♎ 57	07 ♑ 48	17 ♏ 13	28 ♏ 30	16 ♐ 55	09 ♉ 25

All ephemeris data is given at 12.00 UT and the Moon's longitude is additionally given for 24.00 UT

ASPECTARIAN

h m	Aspects
01 Monday	
00 00	☽ ± ♇
03 41	☽ ± ♆
04 42	☽ # ♅
08 27	☽ ⚻ ♃
09 44	☽ ⚹ ♀
09 45	☽ ∠ ♅
10 11	☽ Q ♄
13 49	☽ ⚻ ♆
15 09	☽ bQ ♇
16 46	☽ △ ☿
23 46	☽ # ♀
02 Tuesday	
04 33	☽ # ♃
05 24	☽ # ♆
06 11	☽ ⚹ ♄
06 31	☽ Q ♀
10 56	☽ ⚹ ♅
16 11	☽ △ ♇
18 10	☽ bQ ☿
19 35	☽ △ ♂
19 37	☽ # ☿
23 29	☽ △ ☉
03 Wednesday	
04 39	☽ ∥ ♇
06 55	☽ Q ♅
06 58	☽ ∠ ♄
09 23	☽ ± ☿
11 06	☽ ☍ ♃
13 18	☽ □ ♀
15 49	☽ ☍ ♆
17 35	☽ ∥ ♄
19 33	☽ ⚻ ☿
21 36	☽ bQ ♂
21 47	☽ ⊥ ♄
21 48	☽ # ☉
04 Thursday	
02 02	☽ bQ ☉
07 31	☽ # ♂
07 39	☽ ⚺ ♄
12 45	☽ □ ♅
13 11	☽ ± ♂
17 47	☽ □ ♇
17 57	☽ ± ☉
23 35	☽ ⚻ ♂
05 Friday	
04 35	☽ ⚻ ☉
13 23	☽ ⚻ ♃
16 19	☽ △ ♀
17 35	☽ ⚻ ♆
22 48	☽ ☍ ☿
23 30	☽ ± ♃
23 39	☽ Q ♇
06 Saturday	
03 37	☽ ± ♆
03 52	♃ ∥ ♆
09 15	☽ ☌ ♄
10 30	♃ Q ♇
14 47	☽ bQ ♃
14 51	☽ △ ♅
15 25	☉ ⊥ ♀
18 03	☽ bQ ♀
18 46	☽ bQ ♆
19 47	☽ ⚹ ♇
23 36	☽ # ♂
07 Sunday	
00 53	☿ ⊥ ♃
04 14	☽ ☍ ♂
09 40	☽ ± ♀
10 34	☽ ☍ ☉
11 06	☽ # ☉
12 28	☽ ∥ ♄
16 26	☽ bQ ♅
16 38	☽ △ ♃
20 11	☽ ⚻ ♀
20 23	☽ △ ♆
21 22	☽ ∠ ♇
08 Monday	
01 38	☽ ∥ ♇
04 05	☽ ⚻ ☿
05 36	☽ # ☿
07 14	♀ □ ♆
08 04	☽ ± ♅
12 19	☽ ⚺ ♄
13 00	☽ ⊥ ♇
15 08	☽ ± ☿
18 37	☽ ⚻ ♅
22 21	☉ # ♄
22 50	☽ ⊥ ♄
23 34	☽ ⚺ ♇
09 Tuesday	
01 26	☽ # ♃
02 04	☽ # ♆
08 02	☽ bQ ☿
11 12	☽ ⚻ ♂
14 49	☽ ∠ ♄
18 50	☽ # ♀
19 12	☽ ⚻ ☉
22 15	☽ □ ♃
22 35	☽ ± ♂
10 Wednesday	
01 36	☽ □ ♆
02 20	☽ ☍ ♀
05 20	☿ ⊥ ♆
05 22	☽ # ♅
06 59	☽ ± ☉
13 09	☽ △ ☿
15 59	☽ bQ ♂
18 06	☽ ⚹ ♄
11 Thursday	
00 58	☽ bQ ☉
01 16	☽ ☍ ♅
05 03	☿ # ♇
05 08	☉ ∠ ♅
06 16	☽ ☌ ♇
08 00	☽ Q ♃
11 05	☽ Q ♆
16 32	☽ Q ♄
21 45	☽ △ ♂
22 23	☉ ⚹ ♃
12 Friday	
07 12	☽ ⚹ ♃
07 48	☽ △ ☉
10 04	☽ ⚹ ♆
11 32	☽ ⚻ ♀
23 20	☽ ± ♀
13 Saturday	
02 04	☿ ☍ ♄
03 08	☽ □ ♄
03 14	☽ □ ☿
11 14	☽ ⚻ ♅
12 55	☽ ∠ ♃
14 05	☉ ⚹ ♆
15 30	☽ ∠ ♆
16 13	☽ ⚺ ♇
17 12	☽ bQ ♀
21 34	♀ ⊥ ♂
23 07	☉ bQ ♇
23 11	☽ ± ♅
14 Sunday	
04 07	☽ ⊥ ♇
07 03	☽ ⊥ ♃
08 49	☽ ∥ ♅
09 26	☽ ⊥ ♆
11 57	☽ □ ♂
13 13	☉ ⚺ ♀
14 44	☽ ∥ ♀
17 14	☽ bQ ♅
18 48	☽ Q ☿
19 14	☽ ⚺ ♃
21 30	☽ ⚺ ♆
22 08	☽ ∠ ♇
23 19	☽ △ ♀
15 Monday	
00 09	☽ □ ☉
09 22	☉ ∥ ☿
14 41	☽ △ ♄
16 04	☽ ∥ ♆
18 24	☽ ∥ ♃
21 51	☽ ⚹ ☿
23 36	☽ △ ♅
23 46	♀ St R
16 Tuesday	
02 20	☽ Q ♂
04 21	☽ ⚹ ♇
14 10	☿ ⚹ ♅
15 40	☽ Q ☉
20 50	☽ bQ ♄
20 50	☽ # ♇
17 Wednesday	
01 28	☽ ∥ ☉
03 59	☽ ⚹ ♂
04 34	☽ Q ♇
06 51	☽ ∥ ☿
08 01	☽ ∠ ☿
08 37	☽ ☌ ♃
10 08	☽ ☌ ♆
11 49	☽ □ ♀
11 54	☽ # ♄
13 42	☿ ∠ ♃
14 52	☽ ± ♄
17 57	☽ ⚹ ☉
18 30	☽ ∥ ♂
18 Thursday	
01 56	☿ ∠ ♆
02 50	☽ ⚻ ♄
04 31	☽ ⊥ ☿
05 17	☿ △ ♇
11 48	☽ ∠ ♂
12 12	☽ □ ♅
13 54	☿ ∠ ♀
16 33	☽ □ ♇
18 08	☽ ⚺ ☿
19 59	♂ ∠ ♅
23 28	☽ Q ♀
19 Friday	
02 23	☽ ∠ ☉
06 37	☽ ⊥ ♂
19 02	☽ ⚺ ♂
20 56	☽ ⚺ ♃
21 30	☽ ⊥ ☉
21 43	☽ ⚺ ♆
22 44	☽ ⚹ ♀
23 45	☽ Q ♅
20 Saturday	
03 27	☉ # ♇
08 42	☽ ⊥ ♃
09 21	☽ ⊥ ♆
10 05	☽ ⚺ ☉
13 25	☽ ☍ ♄
22 56	☽ ⚹ ♅
21 Sunday	
02 08	☽ ∠ ♃
02 35	☽ ∠ ♆
02 46	☽ △ ♇
03 08	☽ ∠ ♀
04 21	☉ ⊥ ♅
09 21	♀ ∥ ♅
12 42	☽ ☌ ☿
17 13	♂ ⚹ ♃
18 49	☽ ∥ ♂
19 14	☿ # ♄
19 48	☽ ⊥ ♀
21 20	☽ ∥ ☿
21 23	☽ # ♄
22 52	♂ ⚹ ♆
22 Monday	
01 25	♂ bQ ♇
01 25	♀ ⚺ ♂
03 15	☽ ∠ ♅
03 21	☉ ⚻ ♄
06 35	☽ ⚹ ♃
06 44	☽ ⚹ ♆
06 45	☽ ⚺ ♀
06 50	☽ bQ ♇
07 08	☽ ☌ ♂
07 37	♀ □ ♆
10 51	☽ # ♇
11 46	♀ □ ♃
14 18	☽ ∥ ☉
19 54	☽ ⊥ ♅
20 10	♃ ☌ ♆
21 20	☽ ⚻ ♄
22 49	☽ ☌ ☉
23 21	☽ ± ♇
23 Tuesday	
04 44	☽ Q ♆
04 50	☽ Q ♃
05 55	☉ ± ♇
06 51	☽ ⚺ ♅
08 11	☽ ± ♄
09 17	☽ ∥ ♃
10 13	☽ ⚻ ♇
11 05	☿ ⊥ ♀
12 50	☽ ∥ ♆
24 Wednesday	
00 17	☽ bQ ♄
02 53	☿ ∥ ♂
04 00	☽ ⚺ ☿
11 49	☽ ☌ ♀
13 00	☽ □ ♆
13 24	☽ □ ♃
13 50	♂ # ♄
16 03	☽ ⚺ ♂
16 18	☽ ⊥ ☿
17 25	☽ ∥ ♅
19 07	☽ ∥ ♀
25 Thursday	
02 40	☽ △ ♄
03 07	☽ ⊥ ♂
05 41	☿ # ♄
08 19	☽ ⚺ ☉
10 34	☽ ∠ ☿
12 15	☽ ☌ ♅
15 12	☽ ☍ ♇
19 31	☽ ⊥ ☉
19 34	☽ ∠ ♂
23 04	☿ ∠ ♅
26 Friday	
04 22	☿ ⚺ ♀
09 57	☉ Q ♆
12 11	☽ ∠ ☉
14 37	☽ ⚺ ♀
16 37	☽ ⚹ ☿
17 11	☽ △ ♆
18 03	☽ △ ♃
19 34	☿ bQ ♇
20 43	☿ ⚹ ♆
22 37	☽ ⚹ ♂
23 40	☉ Q ♃
27 Saturday	
00 36	☽ ⊥ ♀
03 36	☿ ⚹ ♃
06 13	☽ □ ♄
15 28	☽ ∠ ♀
15 40	☽ ⚹ ☉
15 57	☽ ⚺ ♅
16 25	☽ Q ☿
18 34	☽ ⚻ ♇
18 47	☽ bQ ♆
19 53	☽ bQ ♃
20 03	☽ Q ♂
20 17	☉ ⚺ ♅
28 Sunday	
01 49	☉ ± ♄
02 11	☽ ⊥ ♅
04 44	☽ ± ♇
05 23	☽ # ♀
08 01	☽ # ♅
10 03	☽ ± ♆
11 17	☽ ± ♃
11 41	☿ ∥ ♂
12 39	☽ Q ♄
13 28	☽ Q ☉
16 05	☽ ⚹ ♀
17 28	☽ ∠ ♅
19 56	☽ bQ ♇
20 12	☽ ⚻ ♆
21 31	☽ ⚻ ♃
29 Monday	
03 54	☽ □ ☿
03 58	☽ □ ♂
04 36	☿ ☌ ♂
08 50	☽ ⚹ ♄
09 58	☉ ⚻ ♇
11 41	☽ Q ♀
11 45	☽ # ♆
16 08	☽ # ♃
18 53	☽ ⚹ ♅
21 13	☽ △ ♇
22 03	☽ □ ☉
23 06	☽ # ☉
30 Tuesday	
10 03	☽ ∠ ♄
12 26	☽ ∥ ♇
13 32	☿ ⚻ ♄
14 16	☿ ⊥ ♅
15 13	☽ Q ♅
16 53	☽ # ☿
16 59	☽ □ ♀
18 57	☽ # ♂
22 53	☽ ☍ ♆
31 Wednesday	
00 26	☽ ∥ ♄
00 38	☽ ☍ ♃
01 12	☽ ⊥ ♄
05 29	☿ ± ♇
09 02	☽ △ ♂
11 19	☽ ⚺ ♄
15 07	☽ △ ☿
21 47	☽ □ ♅
23 52	☽ □ ♇

AUGUST 2047

LONGITUDES

Date	Sidereal time h m s	Sun ☉ ° ' "	Moon ☽ ° ' "	Moon ☽ 24.00 ° ' "	Mercury ☿ ° '	Venus ♀ ° '	Mars ♂ ° '	Jupiter ♃ ° '	Saturn ♄ ° '	Uranus ♅ ° '	Neptune ♆ ° '	Pluto ♇ ° '
01	08 40 06	09 ♌ 12 05	13 ♐ 21 01	20 ♐ 23 14	02 ♌ 52	16 ♌ 58	28 ♋ 11	22 ♉ 41	28 ♐ 45	05 ♍ 01	21 ♉ 28	06 ♓ 12
02	08 44 03	10 09 28	27 ♐ 23 44	04 ♑ 22 11	04 57	16 R 23	28 50	22 48	28 R 42	05 05	21 29	06 R 11
03	08 47 59	11 06 52	11 ♑ 18 18	18 ♑ 11 49	07 03	15 47	29 ♋ 29	22 56	28 40	05 08	21 30	06 10
04	08 51 56	12 04 16	25 ♑ 02 25	01 ♒ 49 49	09 08	15 10	00 ♌ 08	23 03	28 37	05 12	21 30	06 08
05	08 55 53	13 01 41	08 ♒ 33 47	15 ♒ 14 04	11 14	14 33	00 47	23 11	28 35	05 15	21 31	06 07
06	08 59 49	13 59 08	21 ♒ 50 29	28 ♒ 22 54	13 18	13 55	01 26	23 18	28 32	05 19	21 32	06 06
07	09 03 46	14 56 35	04 ♓ 51 14	11 ♓ 15 28	15 22	13 18	02 05	23 25	28 30	05 22	21 32	06 05
08	09 07 42	15 54 03	17 ♓ 35 37	23 ♓ 51 50	17 25	12 41	02 43	23 31	28 27	05 26	21 33	06 04
09	09 11 39	16 51 32	00 ♈ 04 15	06 ♈ 13 10	19 27	12 04	03 22	23 38	28 25	05 29	21 33	06 03
10	09 15 35	17 49 03	12 ♈ 18 51	18 ♈ 21 42	21 28	11 28	04 01	23 45	28 23	05 33	21 34	06 01
11	09 19 32	18 46 35	24 ♈ 22 08	00 ♉ 20 38	23 27	10 53	04 40	23 51	28 21	05 37	21 34	06 00
12	09 23 28	19 44 08	06 ♉ 17 43	12 ♉ 13 56	25 25	10 19	05 18	23 57	28 19	05 40	21 35	05 59
13	09 27 25	20 41 43	18 ♉ 09 52	24 ♉ 06 08	27 22	09 46	05 57	24 03	28 17	05 44	21 35	05 58
14	09 31 22	21 39 19	00 ♊ 03 20	06 ♊ 02 06	29 ♌ 18	09 15	06 36	24 09	28 15	05 48	21 36	05 56
15	09 35 18	22 36 57	12 ♊ 03 02	18 ♊ 06 44	01 ♍ 12	08 46	07 14	24 15	28 14	05 51	21 36	05 55
16	09 39 15	23 34 36	24 ♊ 13 47	00 ♋ 24 44	03 04	08 18	07 53	24 20	28 12	05 55	21 36	05 54
17	09 43 11	24 32 16	06 ♋ 40 04	13 ♋ 00 11	04 56	07 53	08 32	24 26	28 11	05 59	21 37	05 53
18	09 47 08	25 29 59	19 ♋ 25 28	25 ♋ 56 10	06 45	07 29	09 10	24 31	28 09	06 02	21 37	05 52
19	09 51 04	26 27 42	02 ♌ 32 25	09 ♌ 14 18	08 34	07 08	09 49	24 36	28 08	06 06	21 37	05 50
20	09 55 01	27 25 27	16 ♌ 01 42	22 ♌ 54 25	10 21	06 49	10 27	24 41	28 07	06 10	21 37	05 49
21	09 58 57	28 23 14	29 ♌ 52 07	06 ♍ 54 19	12 06	06 32	11 06	24 45	28 06	06 14	21 37	05 48
22	10 02 54	29 ♌ 21 02	14 ♍ 00 28	21 ♍ 09 54	13 50	06 17	11 44	24 50	28 05	06 17	21 37	05 46
23	10 06 51	00 ♍ 18 51	28 ♍ 21 52	05 ♎ 35 37	15 33	06 05	12 22	24 54	28 04	06 21	21 37	05 45
24	10 10 47	01 16 41	12 ♎ 50 21	20 ♎ 05 20	17 14	05 56	13 01	24 58	28 03	06 25	21 37	05 44
25	10 14 44	02 14 33	27 ♎ 19 51	04 ♏ 33 16	18 54	05 48	13 39	25 02	28 02	06 29	21 R 37	05 43
26	10 18 40	03 12 25	11 ♏ 45 03	18 ♏ 54 45	20 33	05 44	14 18	25 06	28 01	06 32	21 37	05 41
27	10 22 37	04 10 20	26 ♏ 02 00	03 ♐ 06 34	22 10	05 41	14 56	25 10	28 01	06 36	21 37	05 40
28	10 26 33	05 08 15	10 ♐ 08 17	17 ♐ 07 02	23 46	05 D 41	15 34	25 13	28 00	06 40	21 37	05 39
29	10 30 30	06 06 12	24 ♐ 02 48	00 ♑ 55 34	25 20	05 43	16 12	25 16	28 00	06 44	21 37	05 38
30	10 34 26	07 04 10	07 ♑ 45 23	14 ♑ 32 17	26 54	05 47	16 51	25 19	28 00	06 47	21 37	05 36
31	10 38 23	08 ♍ 02 09	21 ♑ 16 17	27 ♑ 57 27	28 ♍ 25	05 ♌ 54	17 ♌ 29	25 ♉ 22	28 ♐ 00	06 ♍ 51	21 ♉ 37	05 ♓ 35

Date	Moon True ☊ ° '	Moon Mean ☊ ° '	Moon ☽ Latitude ° '
01	15 ♑ 51	14 ♑ 46	02 S 49
02	15 D 53	14 43	01 40
03	15 53	14 39	00 S 25
04	15 R 53	14 36	00 N 50
05	15 52	14 33	02 01
06	15 49	14 30	03 04
07	15 45	14 27	03 56
08	15 40	14 23	04 35
09	15 35	14 20	05 01
10	15 30	14 17	05 11
11	15 27	14 14	05 08
12	15 24	14 11	04 51
13	15 23	14 08	04 22
14	15 D 23	14 04	03 42
15	15 25	14 01	02 52
16	15 26	13 58	01 53
17	15 28	13 55	00 N 48
18	15 R 28	13 52	00 S 22
19	15 26	13 49	01 31
20	15 23	13 45	02 38
21	15 18	13 42	03 37
22	15 12	13 39	04 24
23	15 05	13 36	04 55
24	14 58	13 33	05 08
25	14 52	13 29	05 01
26	14 49	13 26	04 35
27	14 47	13 23	03 52
28	14 D 47	13 20	02 55
29	14 47	13 17	01 49
30	14 48	13 14	00 38
31	14 ♑ 48	13 ♑ 10	00 N 35

DECLINATIONS

Date	Sun ☉ ° '	Moon ☽ ° '	Mercury ☿ ° '	Venus ♀ ° '	Mars ♂ ° '	Jupiter ♃ ° '	Saturn ♄ ° '	Uranus ♅ ° '	Neptune ♆ ° '	Pluto ♇ ° '
01	17 N 57	25 S 12	20 N 49	09 N 40	21 N 30	17 N 23	22 S 26	10 N 21	16 N 27	20 S 48
02	17 42	25 04	20 25	09 39	21 23	17 25	22 26	10 20	16 27	20 48
03	17 26	23 22	19 59	09 40	21 15	17 26	22 26	10 18	16 27	20 49
04	17 10	20 18	19 30	09 41	21 07	17 28	22 27	10 17	16 27	20 50
05	16 54	16 10	19 00	09 43	20 59	17 30	22 27	10 16	16 27	20 50
06	16 38	11 19	18 27	09 46	20 51	17 31	22 27	10 14	16 27	20 51
07	16 21	06 04	17 53	09 49	20 42	17 33	22 27	10 13	16 27	20 51
08	16 04	00 S 40	17 17	09 53	20 34	17 34	22 27	10 12	16 28	20 52
09	15 47	04 N 37	16 40	09 57	20 25	17 36	22 27	10 10	16 28	20 53
10	15 29	09 38	16 01	10 02	20 16	17 37	22 27	10 09	16 28	20 53
11	15 12	14 13	15 21	10 07	20 07	17 39	22 27	10 08	16 28	20 54
12	14 54	18 12	14 40	10 13	19 58	17 40	22 28	10 06	16 28	20 54
13	14 35	21 26	13 58	10 19	19 49	17 42	22 28	10 05	16 28	20 55
14	14 17	23 46	13 16	10 26	19 40	17 43	22 28	10 04	16 28	20 56
15	13 58	25 04	12 32	10 32	19 30	17 44	22 28	10 02	16 28	20 56
16	13 39	25 11	11 49	10 39	19 20	17 45	22 28	10 01	16 28	20 57
17	13 20	24 04	11 04	10 46	19 11	17 46	22 28	10 00	16 28	20 57
18	13 01	21 40	10 20	10 53	19 01	17 48	22 28	09 58	16 28	20 58
19	12 42	18 06	09 35	11 00	18 51	17 49	22 29	09 57	16 28	20 58
20	12 22	13 31	08 50	11 08	18 41	17 50	22 29	09 55	16 28	20 59
21	12 02	08 08	08 04	11 15	18 30	17 51	22 29	09 54	16 28	21 00
22	11 42	02 N 14	07 19	11 22	18 20	17 52	22 29	09 53	16 28	21 00
23	11 22	03 S 52	06 34	11 29	18 09	17 53	22 29	09 51	16 28	21 01
24	11 01	09 47	05 48	11 36	17 59	17 53	22 29	09 50	16 28	21 01
25	10 40	15 11	05 03	11 43	17 48	17 54	22 29	09 48	16 28	21 02
26	10 20	19 42	04 18	11 50	17 37	17 55	22 30	09 47	16 28	21 02
27	09 59	23 01	03 33	11 57	17 26	17 56	22 30	09 46	16 28	21 03
28	09 38	24 51	02 48	12 03	17 15	17 56	22 30	09 44	16 28	21 03
29	09 16	25 07	02 04	12 09	17 04	17 57	22 30	09 43	16 27	21 04
30	08 55	23 50	01 20	12 15	16 53	17 57	22 30	09 42	16 27	21 04
31	08 N 33	21 S 11	00 N 36	12 N 20	16 N 41	17 N 58	22 S 30	09 N 40	16 N 27	21 S 05

ZODIAC SIGN ENTRIES

Date	h	m	Planets
02	16	28	☽ ♑
04	06	59	♂ ♌
04	20	45	☽ ♒
07	02	59	☽ ♓
09	11	52	☽ ♈
11	23	18	☽ ♉
14	11	53	☽ ♊
14	20	51	☿ ♍
16	23	12	☽ ♋
19	07	24	☽ ♌
21	12	14	☽ ♍
23	04	11	☉ ♍
23	14	43	☽ ♎
25	16	26	☽ ♏
27	18	43	☽ ♐
29	22	23	☽ ♑

LATITUDES

Date	Mercury ☿ ° '	Venus ♀ ° '	Mars ♂ ° '	Jupiter ♃ ° '	Saturn ♄ ° '	Uranus ♅ ° '	Neptune ♆ ° '	Pluto ♇ ° '
01	01 N 20	06 S 23	01 N 01	01 S 06	00 N 59	00 N 44	01 S 44	12 S 29
04	01 36	06 54	01 01	01 06	00 59	00 44	01 45	12 30
07	01 44	07 18	01 02	01 06	00 59	00 44	01 45	12 30
10	01 46	07 35	01 03	01 07	00 58	00 44	01 45	12 31
13	01 41	07 45	01 04	01 07	00 58	00 44	01 45	12 31
16	01 32	07 47	01 05	01 08	00 57	00 44	01 45	12 31
19	01 19	07 43	01 06	01 08	00 57	00 44	01 45	12 32
22	01 02	07 33	01 07	01 09	00 56	00 44	01 46	12 32
25	00 43	07 19	01 08	01 09	00 56	00 44	01 46	12 32
28	00 N 22	07 00	01 08	01 10	00 55	00 44	01 46	12 32
31	00 S 02	06 S 39	01 N 09	01 S 10	00 N 55	00 N 44	01 S 46	12 S 32

DATA

Julian Date	2468924
Delta T	+80 seconds
Ayanamsa	24° 31' 36"
Synetic vernal point	04° ♓ 35' 22"
True obliquity of ecliptic	23° 26' 01"

MOON'S PHASES, APSIDES AND POSITIONS ☽

Date	h	m	Phase	Longitude ° '	Eclipse Indicator
05	20	38	○	13 ♒ 22	
13	17	34	☾	20 ♉ 55	
21	09	16	●	28 ♌ 17	
28	02	49	☽	04 ♐ 46	

Day	h	m		
12	22	37	Apogee	
24	23	25	Perigee	
01	22	08	Max dec	25° S 21'
08	15	00	0N	
16	02	25	Max dec	25° N 17'
22	20	48	0S	
29	03	58	Max dec	25° S 13'

LONGITUDES

Date	Chiron ⚷ ° '	Ceres ⚳ ° '	Pallas ⚴ ° '	Juno ⚵ ° '	Vesta ⚶ ° '	Black Moon Lilith ⚸ ° '
01	20 ♎ 02	07 ♑ 40	17 ♏ 23	28 ♏ 32	16 ♐ 58	09 ♉ 32
11	20 ♎ 55	06 ♑ 33	19 ♏ 18	29 ♏ 14	17 ♐ 47	10 ♉ 39
21	21 ♎ 56	06 ♑ 02	21 ♏ 37	00 ♐ 22	19 ♐ 19	11 ♉ 46
31	23 ♎ 03	06 ♑ 06	24 ♏ 18	01 ♐ 56	21 ♐ 28	12 ♉ 53

ASPECTARIAN

h m	Aspects
01 Thursday	
04 26	☽ △ ☉
11 43	☽ ⚼ ♂
13 21	☿ ⫲ ♇
17 55	☽ △ ♀
19 01	☿ Q ♆
21 02	☽ ⚼ ☿
22 13	♃ ± ♄
02 Friday	
01 52	☽ ⊼ ♆
03 49	☽ ± ♂
04 04	☽ ⊼ ♃
06 29	☽ Q ♇
07 27	♂ ⊼ ♄
07 53	☽ ⚼ ☉
09 13	☿ ± ♄
10 11	☿ Q ♃
12 09	☽ ± ♆
13 26	☿ ⊻ ♅
14 15	☽ ☌ ♄
14 27	☽ ± ♃
14 36	☽ ⊼ ♂
15 09	☽ ± ☿
18 34	☽ ⚼ ♀
21 38	♂ ⊥ ♅
03 Saturday	
00 30	☽ ± ☉
01 16	☽ △ ♅
01 54	☿ ⊼ ♇
03 06	☽ ⚹ ♇
03 19	☽ ⊼ ☿
03 40	☽ ⚼ ♆
06 06	☽ ⚼ ♃
09 28	☽ ± ♀
11 39	☽ ⊼ ☉
11 44	☉ ∥ ♃
19 27	☽ ⊼ ♀
20 22	☽ ∥ ♄
04 Sunday	
03 27	☽ ⚼ ♅
05 10	☽ ∠ ♇
05 47	☽ △ ♆
06 12	☽ ⫲ ♂
08 28	☽ ∥ ♇
08 29	☽ △ ♃
12 10	♂ ± ♇
17 47	☽ ⫲ ☿
18 17	☽ ⊻ ♄
19 22	☽ ± ♅
20 59	☽ ⊥ ♇
21 27	☽ ☍ ♂
05 Monday	
04 52	☽ ⫲ ♃
04 54	☽ ⊥ ♄
06 04	☽ ⊼ ♅
07 38	☽ ⊻ ♇
07 49	☽ ⫲ ☉
10 29	☽ ⫲ ♆
17 39	☽ ☍ ☿
20 38	☽ ☍ ☉
20 59	☽ ∠ ♄
22 17	☽ ☍ ♀
06 Tuesday	
01 10	☉ ⚼ ♄
11 04	☉ ☌ ♀
11 26	☽ □ ♆
11 48	♂ ⫲ ♇
14 39	☿ ⚼ ♄
14 41	☽ □ ♃
17 02	☽ ⫲ ♅
17 32	☿ ☌ ♀
19 09	☽ ⫲ ♀
07 Wednesday	
00 15	☽ ⚹ ♄
02 42	☉ ∥ ♆
02 46	☉ ☌ ☿
03 57	♀ ⚼ ♄
06 34	☽ ⊼ ♂
12 58	☽ ☍ ♅
14 17	☽ ☌ ♇
18 21	☽ ± ♂
20 46	☽ Q ♆
22 32	☽ Q ♄
08 Thursday	
00 24	☽ Q ♃
00 56	☿ ∥ ♃
03 07	☽ ⊼ ♀
08 31	☽ ⊼ ☉
11 36	☽ ⊼ ☿
12 16	☽ ⚼ ♂
13 58	☽ ± ♀
19 34	☽ ⚹ ♆
20 55	☽ ± ☉
23 27	☽ ⚹ ♃
09 Friday	
01 19	☽ ± ☿
06 27	☽ ⚼ ♀
08 48	☽ □ ♄
15 46	☽ ⚼ ☉
18 47	☽ △ ♂
19 02	♂ Q ♆
19 27	☿ ∥ ♆
22 13	☽ ⚼ ☿
22 37	☽ ⊼ ♅
23 38	☽ ⊻ ♇
10 Saturday	
00 40	☽ ∠ ♆
04 54	☽ ∠ ♃
10 24	☽ △ ♀
10 29	☽ ± ♅
11 25	☽ ⊥ ♇
13 15	☿ □ ♆
14 01	☽ ∥ ♀
14 33	☽ ∥ ♅
18 27	☽ ⊥ ♆
22 52	☽ ⊥ ♃
23 52	☽ △ ☉
11 Sunday	
01 01	♂ ± ♄
04 27	☽ ⚼ ♅
05 17	☽ ∠ ♇
06 24	☽ ⊻ ♆
09 48	☽ △ ☿
10 57	☽ ⊻ ♃
13 04	♀ ∥ ♅
17 03	☿ □ ♃
17 11	☽ ∥ ☉
17 33	☽ ∥ ☿
19 58	☽ △ ♄
21 51	☉ ∥ ☿
12 Monday	
01 04	☽ ∥ ♆
08 34	☽ ∥ ♃
09 53	☽ □ ♂
10 44	☽ △ ♅
11 22	☽ ⚹ ♇
19 46	☽ □ ♀
23 51	☽ ∥ ♂
13 Tuesday	
02 10	☽ ⚼ ♄
03 00	♂ ⊻ ♅
07 42	☽ ⫲ ♇
11 35	☽ Q ♇
12 24	♂ ⊼ ♇
16 25	♂ Q ♃
17 34	☽ □ ☉
18 55	☽ ☌ ♆
20 19	☽ ± ♄
21 23	☽ ⫲ ♄
23 12	☿ △ ♄
14 Wednesday	
00 00	☽ ☌ ♃
00 22	☽ Q ♂
06 35	☽ Q ♀
08 23	☽ ⊼ ♄
10 11	☽ □ ☿
10 27	☉ □ ♆
23 35	☽ □ ♅
23 48	☽ □ ♇
15 Thursday	
01 52	☽ ⚹ ♂
05 42	☽ ⚹ ♀
08 54	☽ Q ☉
16 Friday	
04 43	☽ Q ☿
06 52	☽ ⊻ ♆
07 19	♅ ☍ ♇
09 13	☽ ∠ ♂
10 15	☽ ∠ ♀
10 37	☽ ⚹ ☉
11 23	☽ Q ♅
12 13	☽ ⊻ ♃
18 34	☽ ⊥ ♆
19 43	☽ ☍ ♄
21 20	♀ ☌ ♂
23 56	☽ ⊥ ♃
17 Saturday	
03 07	☽ ⊥ ♀
03 38	☽ ⊥ ♂
08 06	☽ ⚹ ☿
08 55	☉ □ ♃
10 30	☽ △ ♇
10 41	☽ ⚹ ♅
11 53	☽ ∠ ♆
14 14	☽ ⊻ ♀
15 43	☽ ⊻ ♂
17 17	☽ ∠ ♃
17 54	☽ ∠ ☉
20 28	☿ ∥ ♀
18 Sunday	
00 18	☿ ☍ ♇
02 13	☿ ☌ ♅
05 13	☽ ⫲ ♄
12 09	☽ ⊥ ☉
14 39	☽ ⚼ ♇
15 00	☽ ∠ ♅
16 03	☽ ⚹ ♆
17 01	☽ ∠ ☿
17 25	☽ ⫲ ♇
20 00	☿ ⊻ ♀
21 27	☽ ⚹ ♃
23 51	☿ ∥ ♅
19 Monday	
00 05	☽ ⊻ ☉
04 01	☽ ⊼ ♄
07 07	☽ ± ♇
07 23	☽ ∥ ♂
07 34	☽ ⊥ ♅
12 03	☽ ⊥ ☿
13 42	☽ ∥ ♃
13 56	☽ Q ♆
14 52	☽ ± ♄
17 55	☽ ⊼ ♇
18 26	☽ ⊻ ♅
19 20	☽ Q ♃
20 02	☽ ☌ ♀
21 09	☽ ∥ ♆
20 Tuesday	
00 27	☽ ⊻ ☿
01 40	☽ ☌ ♂
06 52	☽ ⚼ ♄
14 18	☿ ⊻ ♂
17 46	☽ ∥ ☉
20 42	♀ Q ♃
21 46	☽ □ ♆
22 46	☽ ∥ ♀
21 Wednesday	
03 09	☽ □ ♃
04 23	☽ ∥ ♅
04 49	☉ △ ♄
08 57	☽ △ ♄
09 16	☽ ☌ ☉
12 17	☽ ∥ ☿
17 09	☿ ⊥ ♀
22 06	☽ ☍ ♇
22 54	☽ ☌ ♅
23 10	☽ ⊻ ♀
22 Thursday	
07 59	☽ ⊻ ♂
09 09	☽ ⊥ ♀
11 40	☽ ☌ ☿
12 09	♀ ⊻ ♅
18 33	☽ ⊥ ♂
23 Friday	
00 02	☽ ∠ ♀
00 46	☽ △ ♆
05 14	☉ ∥ ♀
06 12	☽ △ ♃
10 17	☽ ∠ ♂
11 30	☽ □ ♄
15 28	☽ ⊻ ☉
21 33	☽ ⫲ ☿
24 Saturday	
00 15	☽ ⊼ ♇
00 40	☽ ⚹ ♀
01 19	☽ ⊻ ♅
01 43	☽ ⚼ ♆
02 08	☽ ⊥ ☉
07 14	☽ ⚼ ♃
10 10	☽ ± ♇
11 18	☽ ⊥ ♅
12 11	☽ ⫲ ♅
12 18	☽ ⚹ ♂
13 11	♂ ⚼ ♄
16 37	☽ ± ♆
16 55	☽ ⫲ ☉
17 18	☽ Q ♄
18 06	☽ ∠ ☉
19 55	☽ ⫲ ♀
20 13	☽ ⊻ ☿
20 21	☽ Q ♀
22 12	☽ ± ♃
23 18	♂ ∥ ♃
25 Sunday	
01 03	☽ ⚼ ♇
01 18	♆ St R
02 15	☽ ∠ ♅
02 33	☽ ⊼ ♆
07 27	☽ ⊥ ☿
08 11	☽ ⊼ ♃
09 05	☽ Q ♂
13 10	☽ ⚹ ♄
18 14	☽ ⫲ ♆
20 44	☽ ⚹ ☉
26 Monday	
00 19	☽ ∠ ☿
00 40	☽ ⫲ ♂
01 46	☽ ⫲ ♃
01 54	☽ △ ♇
02 00	☽ □ ♀
03 16	☽ ⚹ ♅
05 57	☿ ⊥ ♂
14 07	☽ ∠ ♄
14 35	☿ ∠ ♀
16 27	☽ □ ♂
18 12	☽ Q ☉
20 34	☽ ∥ ♇
23 25	☽ Q ♅
27 Tuesday	
03 58	☿ △ ♆
04 34	☽ ☍ ♆
04 38	☽ ⚹ ☿
05 13	☽ ⊥ ♄
07 30	☽ ∥ ♄
10 31	☽ ☍ ♃
15 21	☽ ⊻ ♄
28 Wednesday	
02 08	♀ St D
02 49	☽ □ ☉
03 34	☽ Q ☿
03 47	☉ ∥ ♅
04 20	☽ □ ♇
04 22	☽ △ ♀
06 02	☽ □ ♅
21 47	☽ △ ♂
29 Thursday	
00 23	☉ ☍ ♇
01 51	☉ ⊻ ♀
06 12	☽ ⚼ ♀
07 47	☽ ⊼ ♆
10 54	☿ △ ♃
11 16	☽ Q ♇
14 08	☽ ⊼ ♃
14 32	☽ □ ☿
18 13	☽ ± ♆
18 53	☽ ☌ ♄
21 55	☽ ± ♀
30 Friday	
00 39	☽ ± ♃
01 06	☽ ⚼ ♂
04 35	☉ ☌ ♅
08 13	☽ ⚹ ♇
08 31	☽ ⊼ ♀
10 00	☽ ⚼ ♆
10 17	☽ △ ♅
10 42	☽ △ ☉
16 32	☽ ⚼ ♃
17 43	☽ ± ♂
31 Saturday	
01 25	☽ ∥ ♄
04 54	☽ ⊼ ♂
05 13	☿ □ ♄
10 46	☽ ∠ ♇
12 37	☽ △ ♆
12 44	☽ ∥ ♇
13 03	☽ ⚼ ♅
15 24	☽ ⚼ ☉
19 22	☽ △ ♃

All ephemeris data is given at 12.00 UT and the Moon's longitude is additionally given for 24.00 UT

SEPTEMBER 2047

LONGITUDES

Date	Sidereal time h m s	Sun ☉ ° ' "	Moon ☽ ° ' "	Moon ☽ 24.00 ° ' "	Mercury ☿ ° '	Venus ♀ ° '	Mars ♂ ° '	Jupiter ♃ ° '	Saturn ♄ ° '	Uranus ♅ ° '	Neptune ♆ ° '	Pluto ⯓ ° '
01	10 42 20	09 ♍ 00 10	04 ♒ 35 46	11 ♒ 11 14	29 ♍ 56	06 ♌ 03	18 ♌ 07	25 ♉ 25	27 ♐ 59	06 ♍ 55	21 ♉ 37	05 ♓ 34
02	10 46 16	09 58 12	17 ♒ 43 51	24 ♒ 13 34	01 ♎ 25	06 14	18 45	25 27	27 D 59	06 59	21 R 36	05 R 32
03	10 50 13	10 56 15	00 ♓ 40 19	07 ♓ 04 06	02 53	06 27	19 23	25 29	28 00	07 02	21 36	05 31
04	10 54 09	11 54 20	13 ♓ 24 51	19 ♓ 42 33	04 20	06 42	20 02	25 31	28 00	07 06	21 36	05 30
05	10 58 06	12 52 27	25 ♓ 57 13	02 ♈ 08 53	05 45	06 59	20 40	25 33	28 00	07 10	21 35	05 29
06	11 02 02	13 50 35	08 ♈ 17 38	14 ♈ 23 36	07 09	07 18	21 18	25 35	28 01	07 14	21 35	05 27
07	11 05 59	14 48 46	20 ♈ 26 59	26 ♈ 28 01	08 31	07 38	21 56	25 37	28 01	07 17	21 35	05 26
08	11 09 55	15 46 58	02 ♉ 27 00	08 ♉ 24 19	09 52	08 01	22 34	25 38	28 02	07 21	21 34	05 25
09	11 13 52	16 45 12	14 ♉ 20 21	20 ♉ 15 36	11 12	08 25	23 12	25 39	28 02	07 25	21 34	05 24
10	11 17 49	17 43 29	26 ♉ 10 33	02 ♊ 05 48	12 30	08 51	23 50	25 40	28 03	07 29	21 33	05 22
11	11 21 45	18 41 47	08 ♊ 01 56	13 ♊ 59 34	13 47	09 19	24 28	25 41	28 04	07 32	21 33	05 21
12	11 25 42	19 40 08	19 ♊ 59 22	26 ♊ 02 01	15 01	09 48	25 06	25 41	28 05	07 36	21 32	05 20
13	11 29 38	20 38 30	02 ♋ 08 09	08 ♋ 18 26	16 15	10 19	25 44	25 42	28 06	07 40	21 31	05 19
14	11 33 35	21 36 55	14 ♋ 33 30	20 ♋ 53 56	17 26	10 51	26 22	25 42	28 07	07 43	21 31	05 17
15	11 37 31	22 35 22	27 ♋ 20 14	03 ♌ 52 52	18 35	11 25	27 00	25 R 41	28 09	07 47	21 30	05 16
16	11 41 28	23 33 51	10 ♌ 32 08	17 ♌ 18 14	19 43	11 59	27 37	25 41	28 10	07 51	21 29	05 15
17	11 45 24	24 32 22	24 ♌ 11 11	01 ♍ 10 52	20 48	12 36	28 15	25 41	28 11	07 54	21 29	05 14
18	11 49 21	25 30 55	08 ♍ 16 56	15 ♍ 28 52	21 52	13 13	28 53	25 40	28 13	07 58	21 28	05 13
19	11 53 18	26 29 30	22 ♍ 45 59	00 ♎ 07 22	22 53	13 52	29 ♌ 31	25 39	28 15	08 02	21 27	05 11
20	11 57 14	27 28 07	07 ♎ 32 03	14 ♎ 58 53	23 51	14 32	00 ♍ 09	25 38	28 17	08 05	21 26	05 10
21	12 01 11	28 26 45	22 ♎ 26 43	29 ♎ 54 24	24 47	15 12	00 46	25 37	28 18	08 09	21 25	05 09
22	12 05 07	29 ♍ 25 26	07 ♏ 20 51	14 ♏ 45 04	25 39	15 54	01 24	25 35	28 20	08 13	21 25	05 08
23	12 09 04	00 ♎ 24 08	22 ♏ 06 12	29 ♏ 23 33	26 29	16 37	02 02	25 33	28 22	08 16	21 24	05 07
24	12 13 00	01 22 53	06 ♐ 36 35	13 ♐ 44 57	27 15	17 21	02 40	25 31	28 25	08 20	21 23	05 06
25	12 16 57	02 21 39	20 ♐ 48 27	27 ♐ 47 01	27 58	18 06	03 17	25 29	28 27	08 23	21 22	05 05
26	12 20 53	03 20 26	04 ♑ 40 44	11 ♑ 29 43	28 36	18 52	03 55	25 27	28 29	08 27	21 21	05 03
27	12 24 50	04 19 16	18 ♑ 14 11	24 ♑ 54 24	29 10	19 39	04 33	25 24	28 32	08 30	21 20	05 02
28	12 28 47	05 18 06	01 ♒ 30 38	08 ♒ 03 09	29 ♎ 40	20 27	05 10	25 22	28 34	08 34	21 19	05 01
29	12 32 43	06 16 59	14 ♒ 32 14	20 ♒ 58 07	00 ♏ 05	21 15	05 48	25 19	28 37	08 37	21 18	05 00
30	12 36 40	07 ♎ 15 53	27 ♒ 21 02	03 ♓ 41 11	00 ♏ 24	22 ♌ 04	06 ♍ 25	25 ♉ 16	28 ♐ 39	08 ♍ 41	21 ♉ 17	04 ♓ 59

Date	Moon True ☊ ° '	Moon Mean ☊ ° '	Moon ☽ Latitude ° '
01	14 ♑ 46	13 ♑ 07	01 N 44
02	14 R 42	13 04	02 47
03	14 36	13 01	03 40
04	14 27	12 58	04 21
05	14 17	12 54	04 49
06	14 06	12 51	05 02
07	13 55	12 48	05 02
08	13 46	12 45	04 48
09	13 39	12 42	04 22
10	13 34	12 39	03 44
11	13 32	12 35	02 57
12	13 D 31	12 32	02 02
13	13 32	12 29	01 N 00
14	13 R 32	12 26	00 S 05
15	13 31	12 23	01 13
16	13 28	12 20	02 18
17	13 23	12 16	03 18
18	13 15	12 13	04 08
19	13 05	12 10	04 43
20	12 54	12 07	05 00
21	12 44	12 04	04 57
22	12 34	12 00	04 34
23	12 28	11 57	03 53
24	12 23	11 54	02 57
25	12 22	11 51	01 51
26	12 D 21	11 48	00 S 41
27	12 R 21	11 45	00 N 31
28	12 20	11 41	01 39
29	12 17	11 38	02 41
30	12 ♑ 11	11 ♑ 35	03 N 34

DECLINATIONS

Date	Sun ☉ ° '	Moon ☽ ° '	Mercury ☿ ° '	Venus ♀ ° '	Mars ♂ ° '	Jupiter ♃ ° '	Saturn ♄ ° '	Uranus ♅ ° '	Neptune ♆ ° '	Pluto ⯓ ° '
01	08 N 12	17 S 25	00 S 07	12 N 26	16 N 30	17 N 58	22 S 31	09 N 39	16 N 27	21 S 05
02	07 50	12 52	00 50	12 31	16 18	17 59	22 31	09 37	16 27	21 06
03	07 28	07 48	01 33	12 35	16 06	17 59	22 31	09 36	16 27	21 06
04	07 06	02 S 30	02 15	12 39	15 55	18 00	22 31	09 35	16 27	21 07
05	06 44	02 N 48	02 56	12 43	15 43	18 00	22 31	09 33	16 27	21 07
06	06 21	07 55	03 37	12 46	15 31	18 00	22 31	09 32	16 26	21 08
07	05 59	12 38	04 17	12 49	15 19	18 00	22 32	09 30	16 26	21 08
08	05 36	16 49	04 57	12 52	15 06	18 01	22 32	09 29	16 26	21 09
09	05 14	20 18	05 36	12 54	14 54	18 01	22 32	09 28	16 26	21 09
10	04 51	22 55	06 14	12 56	14 42	18 01	22 32	09 26	16 26	21 10
11	04 28	24 33	06 52	12 57	14 29	18 01	22 32	09 25	16 26	21 10
12	04 05	25 05	07 28	12 58	14 17	18 01	22 32	09 23	16 25	21 11
13	03 42	24 25	08 04	12 58	14 04	18 01	22 33	09 22	16 25	21 11
14	03 19	22 33	08 40	12 58	13 52	18 01	22 33	09 21	16 25	21 11
15	02 56	19 30	09 14	12 57	13 39	18 00	22 33	09 19	16 25	21 12
16	02 33	15 22	09 47	12 56	13 26	18 00	22 33	09 18	16 25	21 12
17	02 10	10 21	10 19	12 54	13 13	18 00	22 33	09 17	16 24	21 13
18	01 47	04 N 38	10 50	12 52	13 00	18 00	22 34	09 15	16 24	21 13
19	01 24	01 S 28	11 20	12 50	12 47	17 59	22 34	09 14	16 24	21 13
20	01 00	07 35	11 48	12 47	12 34	17 59	22 34	09 13	16 24	21 14
21	00 37	13 19	12 16	12 43	12 21	17 58	22 34	09 11	16 23	21 14
22	00 N 14	18 16	12 41	12 39	12 07	17 58	22 34	09 10	16 23	21 14
23	00 S 10	22 02	13 05	12 34	11 54	17 57	22 34	09 09	16 23	21 15
24	00 33	24 19	13 28	12 29	11 41	17 57	22 35	09 07	16 22	21 15
25	00 56	24 58	13 48	12 23	11 27	17 56	22 35	09 06	16 22	21 15
26	01 20	24 02	14 07	12 17	11 14	17 56	22 35	09 05	16 22	21 16
27	01 43	21 41	14 24	12 10	11 00	17 55	22 35	09 03	16 22	21 16
28	02 06	18 12	14 38	12 03	10 47	17 54	22 35	09 02	16 21	21 16
29	02 30	13 54	14 50	11 55	10 33	17 53	22 36	09 01	16 21	21 17
30	02 S 53	09 S 03	14 S 59	11 N 47	10 N 19	17 N 52	22 S 36	09 N 00	16 N 21	21 S 17

ZODIAC SIGN ENTRIES

Date	h	m	Planets
01	03	41	☽ ♒
01	13	04	☿ ♎
03	10	45	☽ ♓
05	19	50	☽ ♈
08	07	05	☽ ♉
10	19	45	☽ ♊
13	07	49	☽ ♋
15	16	54	☽ ♌
17	21	59	☽ ♍
19	23	48	☽ ♎
20	06	29	♂ ♍
22	00	09	☽ ♏
23	02	08	☉ ♎
24	01	00	☽ ♐
26	03	51	☽ ♑
28	09	15	☽ ♒
29	07	04	☿ ♏
30	17	00	☽ ♓

LATITUDES

Date	Mercury ☿ ° '	Venus ♀ ° '	Mars ♂ ° '	Jupiter ♃ ° '	Saturn ♄ ° '	Uranus ♅ ° '	Neptune ♆ ° '	Pluto ⯓ ° '
01	00 S 10	06 S 32	01 N 09	01 S 10	00 N 55	00 N 44	01 S 46	12 S 32
04	00 34	06 08	01 10	01 11	00 54	00 44	01 46	12 32
07	00 59	05 43	01 11	01 11	00 54	00 44	01 46	12 32
10	01 25	05 17	01 11	01 12	00 53	00 44	01 47	12 32
13	01 50	04 52	01 12	01 12	00 53	00 44	01 47	12 32
16	02 14	04 26	01 12	01 12	00 52	00 44	01 47	12 32
19	02 37	04 00	01 13	01 13	00 52	00 44	01 47	12 32
22	02 59	03 35	01 14	01 13	00 51	00 44	01 47	12 32
25	03 17	03 10	01 15	01 13	00 51	00 44	01 47	12 31
28	03 31	02 46	01 15	01 14	00 50	00 44	01 48	12 31
31	03 S 38	02 S 22	01 N 16	01 S 14	00 N 50	00 N 44	01 S 48	12 S 31

DATA

Julian Date	2468955
Delta T	+80 seconds
Ayanamsa	24° 31' 41"
Synetic vernal point	04° ♓ 35' 18"
True obliquity of ecliptic	23° 26' 02"

MOON'S PHASES, APSIDES AND POSITIONS ☽

Date	h	m	Phase	Longitude ° '	Eclipse Indicator
04	08	54	○	11 ♓ 47	
12	11	18	☾	19 ♊ 38	
19	18	31	●	26 ♍ 45	
26	09	29	☽	03 ♑ 14	

Day	h	m		
09	17	00	Apogee	
21	15	23	Perigee	
04	23	15	0N	
12	10	48	Max dec	25° N 05'
19	06	19	0S	
25	09	34	Max dec	24° S 59'

LONGITUDES

Date	Chiron ⚷ ° '	Ceres ⚳ ° '	Pallas ⚴ ° '	Juno ⚵ ° '	Vesta ⚶ ° '	Black Moon Lilith ⚸ ° '
01	23 ♎ 10	06 ♑ 09	24 ♏ 36	02 ♐ 07	21 ♐ 42	12 ♉ 59
11	24 ♎ 24	06 ♑ 51	27 ♏ 36	04 ♐ 04	24 ♐ 25	14 ♉ 06
21	25 ♎ 43	08 ♑ 04	00 ♐ 50	06 ♐ 20	27 ♐ 33	15 ♉ 13
31	27 ♎ 05	09 ♑ 45	04 ♐ 16	08 ♐ 52	01 ♑ 03	16 ♉ 20

ASPECTARIAN

h m Aspects

01 Sunday
00 04 ☽ ⊻ ♄
02 29 ☽ △ ☿
02 54 ☽ ⊥ ⯓
05 18 ☽ ± ♅
08 49 ☽ # ♃
08 53 ☽ ± ☉
10 54 ☽ ⊥ ♄
13 45 ☽ ⊻ ⯓
14 40 ☽ ☍ ♀
16 14 ☽ ⊼ ♅
17 20 ♂ ∥ ♆
17 25 ☽ # ♆
17 26 ☽ # ♂
18 30 ♄ St D
20 39 ☽ ⊼ ☉

02 Monday
03 18 ☽ ∠ ♄
09 17 ☽ ♇ ☿
13 44 ☽ # ♀
13 59 ☽ ☍ ♂
19 09 ☽ □ ♆

03 Tuesday
02 19 ☽ □ ♃
03 38 ☽ # ♅
04 03 ☽ ± ☿
07 00 ☽ ✶ ♄
13 40 ☽ # ☉
16 41 ☽ ⊼ ☿
21 04 ☽ ☌ ⯓
23 02 ☽ ⊼ ♀

04 Wednesday
00 00 ☽ ☍ ♅
04 46 ☽ Q ♆
05 31 ☽ Q ♄
08 54 ☽ ☍ ☉
10 36 ☽ ± ♀
12 13 ☽ Q ♃
13 02 ☽ ∥ ☿

05 Thursday
01 17 ☽ ⊼ ♂
03 36 ☽ ✶ ♆
04 10 ☽ ♇ ♀
07 24 ☿ ⊼ ⯓
09 12 ☿ ∠ ♂
11 14 ☽ ✶ ♃
12 40 ☽ # ☿
13 26 ☽ ± ♂
15 42 ☉ ⊥ ♀
15 58 ☽ □ ♄

06 Friday
02 16 ☿ ♇ ♆
05 02 ☽ ∥ ☉
06 11 ♀ ⊻ ♅
06 27 ☽ ⊻ ⯓
07 52 ☽ ♇ ♂
08 39 ☽ ∠ ♆
09 28 ☽ ☍ ☿
09 54 ☽ ⊼ ♅
09 59 ☽ △ ♀
13 26 ☿ ⊻ ♅
15 16 ☿ ✶ ♀
16 30 ☽ ∠ ♃
18 12 ☽ ⊥ ⯓
19 55 ☽ ∥ ♅
21 45 ☽ ± ♅
22 45 ♂ □ ♆
23 52 ☽ ⊼ ☉

07 Saturday
02 20 ☽ ⊥ ♆
09 53 ♀ Q ♃
10 20 ☽ ⊥ ♃
11 58 ☽ ∠ ⯓
12 47 ☽ ± ☉
13 00 ☽ ∥ ♀
14 14 ☽ ⊻ ♆
15 06 ☽ △ ♂
15 41 ☽ ♇ ♅
22 18 ☽ ⊻ ♃

08 Sunday
02 12 ☽ ∥ ♂
03 07 ☽ △ ♄
08 21 ☽ ♇ ☉
09 37 ☽ ∥ ♆
17 57 ☽ ✶ ⯓
19 38 ☽ ∥ ♃
21 56 ☽ △ ♅
23 36 ☽ □ ♀

09 Monday
01 51 ☿ ♇ ♃
03 21 ☉ # ☿
04 51 ☽ ⊼ ☿
09 22 ☽ ♇ ♄
15 29 ☿ ± ⯓
17 20 ☽ △ ☉
18 11 ☽ Q ⯓
18 31 ☽ ± ☿
18 58 ☽ # ⯓

10 Tuesday
02 38 ☽ ☌ ♆
03 38 ☽ ± ♄
06 58 ☽ □ ♂
07 49 ☽ # ♄
10 58 ☽ ☌ ♃
13 26 ☽ Q ♀
15 01 ☽ ♇ ☿
15 49 ☽ ⊼ ♄

11 Wednesday
06 36 ☽ □ ⯓
07 17 ☿ ⊥ ♅
11 00 ☽ □ ♅
14 42 ☽ ✶ ♀
21 26 ☽ Q ♂

12 Thursday
00 55 ☽ △ ☿
11 18 ☽ □ ☉
15 04 ☽ ⊻ ♆
21 53 ☿ ± ♆
21 59 ☽ ∠ ♀
22 42 ☽ ✶ ♂
23 12 ☽ Q ♅
23 19 ☽ ⊻ ♃

13 Friday
02 57 ☽ ⊥ ♆
04 04 ☽ ☍ ♄
09 11 ☿ Q ♄
10 31 ♂ □ ♃
11 08 ☽ ⊥ ♃
16 26 ☽ ⊥ ♀
18 11 ☽ △ ⯓
20 32 ☽ ∠ ♆
22 48 ☽ ✶ ♅

14 Saturday
01 43 ☽ Q ☉
04 35 ☽ ⊻ ♀
04 36 ☽ ∠ ♃
05 33 ☽ ∠ ♂
09 29 ☉ △ ♆
12 01 ☽ # ♄
12 36 ♃ St R
18 01 ☽ □ ☿
22 50 ☽ ♇ ⯓
23 34 ☽ ⊥ ♂
23 47 ☽ # ⯓

15 Sunday
01 08 ☽ ✶ ♆
02 27 ☽ ✶ ☉
03 30 ☽ ∠ ♅
08 57 ☽ ✶ ♃
11 20 ☽ ⊻ ♂
13 30 ☽ ⊼ ♄
15 34 ☽ ± ⯓
15 57 ☿ # ♅
20 13 ☽ ⊥ ♅
21 23 ☽ ∥ ♃
23 18 ☽ Q ♆

16 Monday
00 30 ☽ ± ♄
02 30 ☽ ⊼ ⯓
06 29 ☽ Q ☿
06 29 ☽ ∥ ♆
06 53 ☽ Q ♃
07 09 ☽ ⊻ ♅
08 11 ☽ ∠ ☉
11 21 ☿ ± ♃
14 43 ☽ ☌ ♀
16 42 ☽ ♇ ♄
22 10 ☽ ∥ ♂
23 27 ☿ ♇ ⯓

17 Tuesday
00 10 ☽ ∥ ♀
01 25 ☽ ⊥ ☉
05 38 ☽ ✶ ☿
07 18 ☽ □ ♆
09 30 ♂ △ ♄
12 07 ☽ # ☿
12 39 ☽ ⊻ ☉
14 34 ☽ □ ♃
16 41 ☽ ∥ ♅
18 54 ☽ △ ♄
19 20 ☽ ☌ ♂

18 Wednesday
02 59 ☿ ⊼ ♆
06 50 ☽ ☍ ⯓
09 25 ☽ ∠ ☿
11 28 ☽ ☌ ♅
11 58 ♀ ♇ ♄
15 38 ☉ △ ♃
20 38 ☽ ⊻ ♀

19 Thursday
00 06 ☽ ∥ ☉
01 35 ☽ ⊥ ☿
05 54 ♀ ∥ ♂
07 01 ☽ ⊥ ♀
09 51 ☽ △ ♆
11 46 ☽ # ☉
12 12 ☽ ⊻ ☿
15 57 ☿ ∠ ♅
16 43 ☽ △ ♃
18 31 ☽ ☌ ☉
20 58 ☽ □ ♄
22 25 ☽ ∠ ♀
23 30 ☽ ⊻ ♂

20 Friday
08 11 ☽ ⊼ ⯓
09 39 ☽ ⊥ ♂
10 14 ☽ ♇ ♆
12 54 ☽ ⊻ ♅
16 59 ☽ ♇ ♃
17 51 ☽ ± ⯓
18 34 ☽ # ♅
22 36 ☽ ⊥ ♅
23 48 ☽ ✶ ♀

21 Saturday
00 43 ☽ ± ♆
00 48 ☽ ∠ ♂
02 07 ☽ Q ♄
06 56 ☽ ∥ ☿
07 27 ☽ ± ♃
07 54 ☽ # ♂
08 19 ☽ ♇ ⯓
08 27 ☉ □ ♄
09 22 ☽ # ♀
10 22 ☽ ⊼ ♆
13 08 ☽ ∠ ♅
15 16 ☿ # ♂
15 59 ☽ ☌ ☿
17 05 ☽ ⊼ ♃
20 02 ☽ Q ♀
21 27 ☽ ✶ ♄
22 20 ☽ ⊻ ☉

22 Sunday
01 59 ☽ ✶ ♂
02 16 ☽ # ♆
08 26 ☽ △ ⯓
08 40 ☽ ⊥ ☉
10 03 ☿ # ♀
10 05 ☿ ⊼ ♃
10 22 ☽ # ♃
13 24 ☽ ✶ ♅
21 44 ☽ ∠ ♄
22 15 ☽ Q ♂

23 Monday
00 17 ☽ ∠ ☉
02 35 ☽ □ ♀
06 10 ☽ ∥ ⯓
08 59 ☽ Q ♅
10 50 ☽ ☍ ♆
12 27 ☽ ⊥ ♄
16 27 ☽ ∥ ♄
17 39 ☽ ☍ ♃
19 36 ☽ ⊻ ☿
22 20 ☽ ⊻ ♄

24 Tuesday
02 40 ☽ ✶ ☉
05 07 ☽ □ ♂
06 06 ☽ ⊥ ☿
09 28 ☽ □ ⯓
14 53 ☽ □ ♅
21 59 ☽ ∠ ☿

25 Wednesday
00 13 ☽ Q ☉
07 08 ☽ △ ♀
12 57 ☽ ⊼ ♆
15 53 ☽ Q ⯓
20 01 ☽ ⊼ ♃
23 15 ☽ ± ♆

26 Thursday
00 55 ☽ ✶ ☿
01 11 ☽ ☌ ♄
06 23 ☽ ± ♃
07 04 ☿ ✶ ♄
09 29 ☽ □ ☉
10 30 ☽ ♇ ♀
10 36 ☽ △ ♂
12 40 ☽ ✶ ⯓
14 55 ☽ ♇ ♆
18 38 ☽ △ ♅
22 07 ☽ ♇ ♃
22 54 ☽ Q ☿

27 Friday
03 19 ☽ ± ♀
04 09 ☽ ∥ ♄
14 27 ☽ ♇ ♂
14 42 ☽ ⊼ ♀
15 14 ☽ ∠ ⯓
15 14 ☽ ∥ ⯓
17 32 ☽ △ ♆
21 30 ☽ ♇ ♅

28 Saturday
00 52 ☽ △ ♃
03 04 ☉ ⊻ ♂
05 16 ☉ ⊼ ⯓
06 29 ♂ ☍ ⯓
06 37 ☽ ⊻ ♄
07 28 ☽ ⊥ ⯓
07 31 ☽ ± ♂
08 31 ☽ □ ☿
13 50 ☽ # ♃
13 56 ☽ ± ♅
17 37 ☽ ⊥ ♄
18 25 ☽ ⊻ ⯓
19 02 ☽ ⊼ ♂
19 30 ☽ △ ☉
22 47 ☽ # ♆

29 Sunday
01 00 ☽ ⊼ ♅
07 16 ☽ ∥ ☿
07 48 ☉ ∠ ♀
10 16 ☽ ∠ ♄
12 15 ☉ ♇ ♆
13 09 ♀ □ ♆
22 18 ☽ # ♀

30 Monday
00 35 ☽ □ ♆
01 23 ☽ ☍ ♀
01 38 ☽ ♇ ☉
05 34 ☽ # ♂
08 05 ☽ □ ♃
12 14 ☽ # ♅
14 28 ☽ ✶ ♄
17 53 ☽ △ ☿
20 02 ☽ ± ☉

All ephemeris data is given at 12.00 UT and the Moon's longitude is additionally given for 24.00 UT

OCTOBER 2047

LONGITUDES

Date	Sidereal time h m s	Sun ☉	Moon ☽	Moon ☽ 24.00	Mercury ☿	Venus ♀	Mars ♂	Jupiter ♃	Saturn ♄	Uranus ♅	Neptune ♆	Pluto ♇
01	12 40 36	08 ♎ 14 50	09 ♓ 58 42	16 ♓ 13 43	00 ♏ 37	22 ♌ 55	07 ♍ 03	25 ♉ 13	28 ♐ 42	08 ♍ 44	21 ♉ 15	04 ♓ 58
02	12 44 33	09 13 48	22 ♓ 26 20	28 ♓ 36 40	00 44	23 45	07 40	25 R 09	28 45	08 47	21 R 14	04 R 57
03	12 48 29	10 12 48	04 ♈ 44 46	10 ♈ 50 43	00 R 44	24 37	08 18	25 05	28 48	08 51	21 13	04 56
04	12 52 26	11 11 50	16 ♈ 54 36	22 ♈ 56 32	00 37	25 29	08 55	25 02	28 51	08 54	21 12	04 55
05	12 56 22	12 10 54	28 ♈ 56 39	04 ♉ 55 06	00 23	26 22	09 33	24 57	28 54	08 57	21 11	04 54
06	13 00 19	13 10 00	10 ♉ 52 06	16 ♉ 47 55	00 ♏ 01	27 16	10 10	24 53	28 57	09 01	21 09	04 53
07	13 04 16	14 09 09	22 ♉ 42 49	28 ♉ 37 11	29 ♎ 30	28 10	10 48	24 49	29 01	09 04	21 08	04 52
08	13 08 12	15 08 19	04 ♊ 31 25	10 ♊ 25 59	28 52	29 ♌ 05	11 25	24 44	29 04	09 07	21 07	04 51
09	13 12 09	16 07 32	16 ♊ 21 22	22 ♊ 18 08	28 07	00 ♍ 00	12 02	24 40	29 08	09 10	21 06	04 50
10	13 16 05	17 06 48	28 ♊ 16 53	04 ♋ 18 15	27 13	00 56	12 40	24 35	29 11	09 13	21 04	04 50
11	13 20 02	18 06 05	10 ♋ 22 53	16 ♋ 31 26	26 14	01 53	13 17	24 30	29 15	09 17	21 03	04 49
12	13 23 58	19 05 25	22 ♋ 44 37	29 ♋ 03 03	25 09	02 50	13 54	24 24	29 19	09 20	21 02	04 48
13	13 27 55	20 04 47	05 ♌ 27 22	11 ♌ 58 08	23 59	03 48	14 32	24 19	29 22	09 23	21 00	04 47
14	13 31 51	21 04 12	18 ♌ 35 50	25 ♌ 20 50	22 47	04 46	15 09	24 13	29 26	09 26	20 59	04 46
15	13 35 48	22 03 38	02 ♍ 13 21	09 ♍ 13 26	21 35	05 44	15 46	24 08	29 30	09 29	20 57	04 45
16	13 39 45	23 03 07	16 ♍ 20 56	23 ♍ 35 29	20 23	06 43	16 23	24 02	29 34	09 32	20 56	04 45
17	13 43 41	24 02 38	00 ♎ 56 29	08 ♎ 23 07	19 16	07 43	17 01	23 56	29 38	09 35	20 55	04 44
18	13 47 38	25 02 12	15 ♎ 54 19	23 ♎ 28 55	18 14	08 43	17 38	23 49	29 43	09 38	20 53	04 43
19	13 51 34	26 01 47	01 ♏ 05 32	08 ♏ 42 48	17 19	09 43	18 15	23 43	29 47	09 41	20 52	04 42
20	13 55 31	27 01 25	16 ♏ 19 20	23 ♏ 53 47	16 33	10 44	18 52	23 37	29 51	09 44	20 50	04 42
21	13 59 27	28 01 04	01 ♐ 24 59	08 ♐ 51 55	15 57	11 45	19 29	23 30	29 ♐ 56	09 46	20 49	04 41
22	14 03 24	29 ♎ 00 46	16 ♐ 13 44	23 ♐ 29 52	15 32	12 47	20 06	23 23	00 ♑ 00	09 49	20 47	04 40
23	14 07 20	00 ♏ 00 29	00 ♑ 39 52	07 ♑ 43 35	15 18	13 49	20 43	23 16	00 05	09 52	20 45	04 40
24	14 11 17	01 00 14	14 ♑ 40 57	21 ♑ 32 05	15 D 16	14 51	21 20	23 09	00 09	09 55	20 44	04 39
25	14 15 14	02 00 00	28 ♑ 17 13	04 ♒ 56 40	15 24	15 54	21 57	23 02	00 14	09 57	20 42	04 39
26	14 19 10	02 59 48	11 ♒ 30 49	18 ♒ 00 04	15 43	16 57	22 34	22 55	00 19	10 00	20 41	04 38
27	14 23 07	03 59 38	24 ♒ 24 51	00 ♓ 45 35	16 13	18 01	23 11	22 48	00 24	10 03	20 39	04 37
28	14 27 03	04 59 29	07 ♓ 02 41	13 ♓ 16 32	16 51	19 04	23 48	22 40	00 29	10 05	20 38	04 37
29	14 31 00	05 59 22	19 ♓ 27 31	25 ♓ 35 56	17 38	20 08	24 25	22 33	00 34	10 08	20 36	04 36
30	14 34 56	06 59 17	01 ♈ 42 05	07 ♈ 46 14	18 32	21 13	25 02	22 25	00 39	10 10	20 34	04 36
31	14 38 53	07 ♏ 59 13	13 ♈ 48 35	19 ♈ 49 23	19 ♎ 33	22 ♍ 17	25 ♍ 39	22 ♉ 18	00 ♑ 44	10 ♍ 12	20 ♉ 33	04 ♓ 35

Moon Nodes and Latitude

Date	Moon True ☊	Moon Mean ☊	Moon ☽ Latitude
01	12 ♑ 02	11 ♑ 32	04 N 15
02	11 R 51	11 29	04 43
03	11 37	11 26	04 57
04	11 23	11 22	04 58
05	11 09	11 19	04 45
06	10 57	11 16	04 20
07	10 47	11 13	03 44
08	10 40	11 10	02 58
09	10 36	11 06	02 04
10	10 34	11 03	01 05
11	10 D 33	11 00	00 N 01
12	10 R 33	10 57	01 S 04
13	10 33	10 54	02 08
14	10 30	10 51	03 07
15	10 25	10 47	03 58
16	10 17	10 44	04 36
17	10 08	10 41	04 58
18	09 57	10 38	05 00
19	09 46	10 35	04 41
20	09 36	10 32	04 02
21	09 29	10 28	03 07
22	09 25	10 25	02 00
23	09 23	10 22	00 S 46
24	09 D 23	10 19	00 N 28
25	09 23	10 16	01 39
26	09 R 23	10 12	02 42
27	09 21	10 09	03 36
28	09 16	10 06	04 17
29	09 09	10 03	04 46
30	09 00	10 00	05 01
31	08 ♑ 49	09 ♑ 57	05 N 02

DECLINATIONS

Date	Sun ☉	Moon ☽	Mercury ☿	Venus ♀	Mars ♂	Jupiter ♃	Saturn ♄	Uranus ♅	Neptune ♆	Pluto ♇
01	03 S 16	03 S 54	15 S 05	11 N 38	10 N 06	17 N 52	22 S 36	08 N 58	16 N 20	21 S 17
02	03 39	01 N 20	15 08	11 29	09 52	17 51	22 36	08 57	16 20	21 18
03	04 03	06 26	15 08	11 19	09 38	17 50	22 36	08 56	16 20	21 18
04	04 26	11 14	15 04	11 09	09 24	17 49	22 36	08 55	16 19	21 18
05	04 49	15 32	14 56	10 58	09 10	17 48	22 37	08 53	16 19	21 18
06	05 12	19 12	14 44	10 47	08 56	17 46	22 37	08 52	16 19	21 18
07	05 35	22 03	14 28	10 35	08 42	17 45	22 37	08 51	16 18	21 19
08	05 58	23 57	14 07	10 23	08 28	17 44	22 37	08 50	16 18	21 19
09	06 20	24 48	13 41	10 10	08 14	17 43	22 37	08 49	16 17	21 19
10	06 43	24 30	13 12	09 57	08 00	17 42	22 37	08 47	16 17	21 19
11	07 06	23 03	12 37	09 43	07 45	17 40	22 38	08 46	16 17	21 19
12	07 28	20 28	12 00	09 29	07 31	17 39	22 38	08 45	16 16	21 20
13	07 51	16 50	11 18	09 15	07 17	17 38	22 38	08 44	16 16	21 20
14	08 13	12 17	10 35	09 00	07 03	17 36	22 38	08 43	16 16	21 20
15	08 35	06 59	09 50	08 44	06 48	17 35	22 38	08 42	16 15	21 20
16	08 58	01 N 09	09 04	08 28	06 34	17 33	22 38	08 41	16 15	21 20
17	09 19	04 S 55	08 19	08 12	06 20	17 32	22 39	08 40	16 14	21 20
18	09 41	10 52	07 37	07 55	06 05	17 30	22 39	08 39	16 14	21 20
19	10 03	16 14	06 57	07 38	05 51	17 28	22 39	08 37	16 13	21 20
20	10 25	20 35	06 21	07 21	05 37	17 27	22 39	08 36	16 13	21 20
21	10 46	23 29	05 51	07 03	05 22	17 25	22 39	08 35	16 13	21 21
22	11 07	24 42	05 25	06 45	05 08	17 23	22 39	08 34	16 12	21 21
23	11 28	24 12	05 05	06 26	04 53	17 22	22 39	08 33	16 12	21 21
24	11 49	22 09	04 51	06 07	04 39	17 20	22 40	08 32	16 11	21 21
25	12 10	18 53	04 43	05 48	04 24	17 18	22 40	08 31	16 11	21 21
26	12 30	14 44	04 41	05 28	04 10	17 16	22 40	08 30	16 10	21 21
27	12 51	09 59	04 44	05 08	03 55	17 15	22 40	08 29	16 10	21 21
28	13 11	04 S 56	04 52	04 48	03 41	17 13	22 40	08 29	16 10	21 21
29	13 31	00 N 13	05 04	04 27	03 26	17 11	22 40	08 28	16 09	21 21
30	13 51	05 17	05 20	04 06	03 12	17 09	22 40	08 27	16 09	21 21
31	14 S 10	10 N 05	05 S 41	03 N 45	02 N 57	17 N 07	22 S 40	08 N 26	16 N 08	21 S 21

ZODIAC SIGN ENTRIES

Date	h	m	Planets
03	02	43	☽ ♈
05	14	07	☽ ♉
06	12	31	☿ ♎
08	02	48	☽ ♊
09	11	56	♀ ♍
10	15	26	☽ ♋
13	01	47	☽ ♌
15	08	09	☽ ♍
17	10	28	☽ ♎
19	10	17	☽ ♏
21	09	44	☽ ♐
22	11	09	♄ ♑
23	10	53	☽ ♑
23	11	48	☉ ♏
25	15	04	☽ ♒
27	22	33	☽ ♓
30	08	39	☽ ♈

LATITUDES

Date	Mercury ☿	Venus ♀	Mars ♂	Jupiter ♃	Saturn ♄	Uranus ♅	Neptune ♆	Pluto ♇
01	03 S 38	02 S 22	01 N 16	01 S 14	00 N 50	00 N 44	01 S 48	12 S 31
04	03 37	01 59	01 16	01 15	00 49	00 44	01 48	12 30
07	03 23	01 37	01 17	01 15	00 49	00 44	01 48	12 30
10	02 54	01 16	01 17	01 15	00 48	00 44	01 48	12 29
13	02 09	00 56	01 18	01 15	00 48	00 44	01 48	12 29
16	01 11	00 37	01 18	01 16	00 48	00 44	01 48	12 28
19	00 S 10	00 18	01 18	01 16	00 47	00 44	01 48	12 28
22	00 N 45	00 S 01	01 19	01 16	00 47	00 44	01 48	12 27
25	01 27	00 N 15	01 19	01 16	00 46	00 45	01 48	12 26
28	01 54	00 31	01 20	01 16	00 46	00 45	01 48	12 26
31	02 N 08	00 N 45	01 N 20	01 S 16	00 N 46	00 N 45	01 S 48	12 S 25

DATA

Julian Date	2468985
Delta T	+80 seconds
Ayanamsa	24° 31' 43"
Synetic vernal point	04° ♓ 35' 15"
True obliquity of ecliptic	23° 26' 02"

MOON'S PHASES, APSIDES AND POSITIONS ☽

Date	h	m	Phase	Longitude	Eclipse Indicator
03	23	42	○	10 ♈ 42	
12	04	22	☾	18 ♋ 47	
19	03	28	●	25 ♎ 41	
25	19	13	☽	02 ♒ 18	

Day	h	m		
07	08	42	Apogee	
19	21	59	Perigee	
02	05	52	0N	
09	17	49	Max dec	24° N 50'
16	16	34	0S	
22	16	46	Max dec	24° S 44'
29	11	00	0N	

LONGITUDES

Date	Chiron ⚷	Ceres ⚳	Pallas ⚴	Juno ⚵	Vesta ⚶	Black Moon Lilith ⚸
01	27 ♎ 05	09 ♑ 45	04 ♐ 16	08 ♐ 52	01 ♑ 03	16 ♉ 20
11	28 ♎ 29	11 ♑ 50	07 ♐ 51	11 ♐ 38	04 ♑ 51	17 ♉ 27
21	29 ♎ 55	14 ♑ 15	11 ♐ 34	14 ♐ 37	08 ♑ 54	18 ♉ 34
31	01 ♏ 21	16 ♑ 59	15 ♐ 23	17 ♐ 45	13 ♑ 10	19 ♉ 41

ASPECTARIAN

h m	Aspects
01 Tuesday	
02 27	☽ ☌ ♇
06 06	☽ ☍ ♂
08 25	☽ ⊼ ☉
09 36	☽ ☍ ♅
10 37	☽ Q ♆
13 23	☽ Q ♄
14 40	☽ ∥ ☉
18 10	☽ Q ♃
22 57	☽ bQ ☿
02 Wednesday	
00 34	☉ ⊻ ♅
09 41	☽ ⚹ ♆
14 45	☽ ⊼ ♀
16 28	☽ ± ☿
17 14	☽ ⚹ ♃
23 42	☽ # ☉
03 Thursday	
00 19	☽ □ ♄
00 45	☿ St R
03 18	☽ ± ♀
04 10	☽ ⊼ ☿
09 10	☉ bQ ♃
12 22	☽ ⊻ ♇
14 53	☽ ∠ ♆
19 21	☽ ⊼ ♂
20 06	☽ ⊼ ♅
22 18	☽ bQ ♀
22 27	☽ ∠ ♃
23 42	☽ ☍ ☉
04 Friday	
00 10	☽ ⊥ ♇
00 12	☽ ∥ ♅
00 14	♀ □ ♃
03 03	☽ ∥ ♂
05 19	☉ ± ♇
07 51	☽ ± ♂
08 00	☽ ± ♅
08 37	☽ ⊥ ♆
11 04	♂ ☌ ♅
11 37	☽ ∥ ♀
16 11	☽ ⊥ ♃
17 58	☽ ∠ ♇
20 31	☽ ⊻ ♆
05 Saturday	
01 59	☽ bQ ♅
02 43	☽ bQ ♂
04 04	☽ ⊻ ♃
06 26	☽ △ ♀
08 35	☽ # ☿
11 55	☽ △ ♄
14 48	☽ ☍ ☿
16 42	☽ ∥ ♆
23 57	☽ ⚹ ♇
06 Sunday	
02 06	☽ ∥ ♃
08 14	☽ △ ♅
10 31	☽ △ ♂
17 04	☽ ⊼ ☉
18 16	☽ bQ ♄
19 09	♂ ∥ ♅
07 Monday	
00 10	☽ Q ♇
04 58	☽ # ♇
06 20	☽ ± ☉
08 48	☽ ☌ ♆
12 36	☽ ± ♄
16 14	☽ ☌ ♃
17 55	☽ # ♄
08 Tuesday	
00 00	☽ □ ♀
00 51	☽ ⊼ ♄
01 08	☽ ⊼ ☿
02 16	☽ bQ ☉
05 41	☿ ⚹ ♄
08 57	☿ ⚹ ♀
11 27	☉ ± ♆
11 29	☉ ⊥ ♅
11 43	♀ △ ♄
12 40	☽ ± ☿
12 40	☽ □ ♇
21 23	☽ □ ♅
09 Wednesday	
02 46	☽ □ ♂
05 51	☽ bQ ☿
11 30	☽ △ ☉
15 37	☽ Q ♀
21 33	☽ ⊻ ♆
10 Thursday	
04 38	☽ ⊻ ♃
05 13	☿ ∠ ♂
09 35	☽ ⊥ ♆
09 53	☽ Q ♅
10 02	☽ △ ☿
13 49	☽ ☍ ♄
13 52	☉ Q ♄
16 33	☽ ⊥ ♃
17 01	☽ Q ♂
17 45	☽ ⚹ ♀
11 Friday	
01 01	☽ △ ♇
03 28	☽ ∠ ♆
09 49	☽ ⚹ ♅
10 16	☽ ∠ ♃
16 40	☽ # ♄
18 00	☽ ⚹ ♂
20 44	☉ ± ♃
12 Saturday	
01 45	☽ ∠ ♀
04 22	☽ □ ☉
05 01	☽ # ♇
06 20	☽ bQ ♇
08 43	☽ ⚹ ♆
13 52	☉ # ♂
15 03	☽ ∠ ♅
15 09	☽ ⚹ ♃
16 12	☽ □ ☿
20 26	☽ ⊥ ♀
23 30	☽ ± ♇
13 Sunday	
00 20	☽ ∠ ♂
00 33	☽ ⊼ ♄
04 15	☿ ∠ ♅
04 42	☿ ⊼ ♃
04 54	☉ bQ ♇
07 15	☽ ∥ ♃
07 26	☽ Q ♆
08 07	☽ ⊥ ♅
08 39	☽ ⊻ ♀
10 45	☽ ⊼ ♇
11 51	☽ ± ♄
13 35	☽ Q ♃
15 17	☽ ∥ ♆
17 16	☽ Q ☉
17 58	☽ ⊥ ♂
19 17	☽ ⊻ ♅
23 02	☽ Q ☿
14 Monday	
04 27	☽ bQ ♄
05 29	☽ ⊻ ♂
09 53	☉ ⊼ ♆
12 12	♀ ☍ ♇
16 15	☽ □ ♆
16 46	☽ ⚹ ☉
17 11	☉ ⊥ ♂
18 51	☽ ⚹ ☿
21 24	☽ # ☿
21 57	☽ □ ♃
15 Tuesday	
03 59	☽ ∥ ♀
04 31	☽ ∥ ♅
05 27	☽ # ☉
06 42	☉ ☌ ☿
07 15	☽ △ ♄
09 26	☿ ⊥ ♂
12 47	☽ ∥ ♂
16 09	♀ ∥ ♅
16 22	☽ ☍ ♇
17 38	☉ # ♀
18 31	☽ ☌ ♀
18 34	☉ # ♅
18 54	☽ ∠ ☿
20 57	☽ ∠ ☉
21 13	☿ ∠ ♀
16 Wednesday	
00 29	☽ ☌ ♅
00 40	☿ ⊼ ♆
08 58	☽ ⊥ ☿
12 04	☽ ☌ ♂
13 16	☽ ⊥ ☉
14 20	☉ ∥ ☿
18 13	☽ ⊻ ☿
19 36	☽ △ ♆
23 55	☽ ⊻ ☉
17 Thursday	
00 38	☽ △ ♃
00 48	☿ # ♅
01 43	☿ bQ ♇
09 25	☉ ⊼ ♃
09 53	☽ □ ♄
17 22	☽ # ♂
18 07	☽ ⊼ ♇
18 28	☿ # ♀
20 01	☽ bQ ♆
23 42	☽ ⊻ ♀
18 Friday	
00 06	☽ ∥ ☿
00 30	☽ # ♀
00 47	☽ bQ ♃
01 40	☉ ∠ ♅
01 58	☽ ⊻ ♅
02 53	☽ # ♅
03 44	☽ ± ♇
06 49	☽ ∥ ☉
09 58	☽ ⊥ ♀
10 23	☽ ± ♆
11 34	☽ ⊥ ♅
14 51	☽ ⊻ ♂
14 53	☽ Q ♄
15 01	☽ ± ♃
15 28	☽ ☌ ☿
18 03	☽ bQ ♇
19 53	☽ ⊼ ♆
21 05	☿ ⊻ ♂
23 30	☿ ± ♃
19 Saturday	
00 13	☿ Q ♄
00 27	☽ ⊼ ♃
00 45	☽ ⊥ ♂
01 15	☽ ∠ ♀
01 51	☽ ∠ ♅
03 28	☽ ☌ ☉
03 29	♃ ± ♄
09 55	☽ ⚹ ♄
10 53	♀ ☌ ♅
11 57	☽ # ♆
15 33	☽ ∠ ♂
17 41	☽ △ ♇
18 09	☽ # ♃
20 Sunday	
01 34	☽ ⚹ ♅
02 34	☽ ⚹ ♀
09 22	☿ ⊥ ♀
09 40	☽ ∠ ♄
12 21	☽ ⊻ ☿
16 12	☽ ⚹ ♂
17 11	☽ ∥ ♇
19 08	☽ ☍ ♆
20 35	☽ Q ♅
21 27	☽ ⊥ ☿
22 54	☽ Q ♀
23 28	☽ ☍ ♃
23 59	☽ ⊥ ♄
21 Monday	
03 35	☽ ∥ ♄
06 11	☽ ⊻ ☉
09 36	☽ ⊻ ♄
11 17	☽ ∠ ☿
12 07	☽ Q ♂
16 28	☽ ⊥ ☉
17 15	☽ □ ♇
20 09	☿ ⊥ ♅
22 Tuesday	
01 31	☽ □ ♅
05 57	☽ □ ♀
08 07	☽ ∠ ☉
10 53	☽ ⚹ ☿
18 40	☽ □ ♂
19 29	☽ ⊼ ♆
22 37	☽ Q ♇
23 43	☽ ⊼ ♃
23 Wednesday	
05 27	☽ ± ♆
06 25	☽ Q ☿
09 40	☽ ± ♃
10 49	☽ ⚹ ☉
11 00	☽ ☌ ♄
13 15	♂ △ ♆
13 52	☉ ⚹ ♄
18 46	☽ ⚹ ♇
20 37	☽ bQ ♆
24 Thursday	
00 50	☽ bQ ♃
03 43	☽ △ ♅
05 21	☿ St D
07 22	☽ ∥ ♄
08 52	☽ Q ☉
12 20	☽ △ ♀
13 01	☽ □ ☿
18 46	☽ ∥ ♇
20 41	☽ ∠ ♇
22 12	☿ ⊻ ♀
22 34	☽ △ ♆
25 Friday	
00 13	☽ △ ♂
02 44	☽ △ ♃
06 02	☽ bQ ♅
12 38	☽ ⊥ ♇
15 31	☽ ⊻ ♄
17 06	☽ bQ ♀
19 13	☽ □ ☉
21 42	☽ # ♃
22 14	☽ ± ♅
23 27	☽ ⊻ ♇
26 Saturday	
02 26	☽ ⊥ ♄
04 04	☽ # ♆
04 26	☽ bQ ♂
09 13	☽ ⊼ ♅
10 53	☽ ± ♀
19 04	☽ ∠ ♄
20 02	☽ △ ☿
21 49	☽ ± ♂
22 44	☽ ∥ ☉
22 57	☽ ⊼ ♀
23 12	♂ △ ♃
27 Sunday	
03 52	☿ ⊥ ♅
04 58	☽ □ ♆
08 59	☽ □ ♃
09 35	☽ ⊼ ♂
19 15	☽ # ♅
23 23	☽ ⚹ ♄
28 Monday	
01 30	☽ bQ ☿
03 01	☉ △ ♇
06 52	☿ ± ♃
07 21	☽ ☌ ♇
07 44	☽ △ ☉
08 45	☿ # ♀
12 21	☽ ∥ ☿
12 44	☽ # ♀
15 02	☽ Q ♆
17 52	☽ ☍ ♅
18 10	☽ # ♂
18 54	☽ Q ♃
19 46	☽ ± ☿
22 32	☽ Q ♄
29 Tuesday	
08 11	☽ ⊼ ☿
13 27	☽ ☍ ♀
14 13	☽ ⚹ ♆
15 15	☽ bQ ☉
17 58	☽ ⚹ ♃
22 04	♀ △ ♆
22 12	☽ ☍ ♂
30 Wednesday	
02 29	☽ ∥ ♂
06 42	☽ ∥ ♀
09 54	☽ □ ♄
10 28	☽ ± ☉
12 20	☽ # ☿
15 08	☿ Q ♄
17 43	☽ ⊻ ♇
19 38	☽ ∠ ♆
23 12	☽ ∠ ♃
23 23	☽ ⊼ ☉
31 Thursday	
03 34	☽ ∥ ♅
04 49	☽ ⊼ ♅
05 36	☽ ⊥ ♇
12 08	♀ △ ♃
13 01	☿ bQ ♇
13 28	☽ ⊥ ♆
16 48	☽ ± ♅
16 54	☽ ⊥ ♃
23 32	☽ ∠ ♇

All ephemeris data is given at 12.00 UT and the Moon's longitude is additionally given for 24.00 UT

NOVEMBER 2047

LONGITUDES

Date	Sidereal time h m s	Sun ☉ ° ′ ″	Moon ☽ ° ′ ″	Moon ☽ 24.00 ° ′ ″	Mercury ☿ ° ′	Venus ♀ ° ′	Mars ♂ ° ′	Jupiter ♃ ° ′	Saturn ♄ ° ′	Uranus ♅ ° ′	Neptune ♆ ° ′	Pluto ♇ ° ′
01	14 42 49	08 ♏ 59 12	25 ♈ 48 46	01 ♉ 46 56	20 ♎ 39	23 ♍ 22	26 ♍ 16	22 ♉ 10	00 ♑ 49	10 ♍ 15	20 ♉ 31	04 ♓ 35
02	14 46 46	09 59 12	07 ♉ 44 03	13 ♉ 40 17	21 51	24 27	26 53	22 R 02	00 54	10 17	20 R 29	04 R 35
03	14 50 43	10 59 14	19 ♉ 35 49	25 ♉ 30 50	23 07	25 33	27 29	21 54	01 00	10 19	20 28	04 34
04	14 54 39	11 59 18	01 ♊ 25 35	07 ♊ 20 18	24 27	26 39	28 06	21 46	01 05	10 22	20 26	04 34
05	14 58 36	12 59 24	13 ♊ 15 17	19 ♊ 10 52	25 50	27 45	28 43	21 38	01 11	10 24	20 24	04 34
06	15 02 32	13 59 32	25 ♊ 07 26	01 ♋ 05 23	27 16	28 51	29 20	21 30	01 16	10 26	20 23	04 33
07	15 06 29	14 59 42	07 ♋ 05 11	13 ♋ 07 20	28 ♎ 44	29 ♍ 58	29 ♍ 56	21 22	01 22	10 28	20 21	04 33
08	15 10 25	15 59 54	19 ♋ 12 22	25 ♋ 20 50	00 ♏ 14	01 ♎ 04	00 ♎ 33	21 14	01 27	10 30	20 19	04 33
09	15 14 22	17 00 08	01 ♌ 33 20	07 ♌ 50 27	01 45	02 11	01 10	21 06	01 33	10 32	20 18	04 32
10	15 18 18	18 00 24	14 ♌ 12 45	20 ♌ 40 50	03 17	03 19	01 46	20 58	01 39	10 34	20 16	04 32
11	15 22 15	19 00 42	27 ♌ 15 10	03 ♍ 56 14	04 51	04 26	02 23	20 50	01 44	10 36	20 14	04 32
12	15 26 12	20 01 01	10 ♍ 44 20	17 ♍ 39 43	06 25	05 34	02 59	20 42	01 50	10 38	20 13	04 32
13	15 30 08	21 01 23	24 ♍ 42 26	01 ♎ 52 20	07 59	06 42	03 36	20 33	01 56	10 40	20 11	04 32
14	15 34 05	22 01 47	09 ♎ 09 05	16 ♎ 32 09	09 35	07 50	04 12	20 25	02 02	10 42	20 09	04 32
15	15 38 01	23 02 13	24 ♎ 00 44	01 ♏ 33 51	11 10	08 58	04 49	20 17	02 08	10 43	20 07	04 32
16	15 41 58	24 02 40	09 ♏ 10 20	16 ♏ 48 53	12 46	10 07	05 25	20 09	02 14	10 45	20 06	04 32
17	15 45 54	25 03 10	24 ♏ 28 06	02 ♐ 06 37	14 21	11 15	06 02	20 01	02 20	10 47	20 04	04 32
18	15 49 51	26 03 41	09 ♐ 43 04	17 ♐ 16 11	15 57	12 24	06 38	19 53	02 26	10 48	20 02	04 D 32
19	15 53 47	27 04 13	24 ♐ 44 54	02 ♑ 08 17	17 33	13 33	07 14	19 45	02 33	10 50	20 01	04 32
20	15 57 44	28 04 47	09 ♑ 25 38	16 ♑ 36 28	19 09	14 43	07 51	19 37	02 39	10 51	19 59	04 32
21	16 01 41	29 ♏ 05 22	23 ♑ 40 28	00 ♒ 37 33	20 44	15 52	08 27	19 29	02 45	10 53	19 57	04 32
22	16 05 37	00 ♐ 05 58	07 ♒ 27 44	14 ♒ 11 15	22 20	17 02	09 03	19 21	02 51	10 54	19 56	04 32
23	16 09 34	01 06 36	20 ♒ 48 22	27 ♒ 19 28	23 55	18 11	09 39	19 13	02 58	10 55	19 54	04 32
24	16 13 30	02 07 14	03 ♓ 45 02	10 ♓ 05 31	25 31	19 21	10 16	19 05	03 04	10 57	19 52	04 32
25	16 17 27	03 07 54	16 ♓ 21 26	22 ♓ 33 17	27 06	20 31	10 52	18 57	03 11	10 58	19 51	04 32
26	16 21 23	04 08 34	28 ♓ 41 35	04 ♈ 46 50	28 ♏ 41	21 41	11 28	18 50	03 17	10 59	19 49	04 32
27	16 25 20	05 09 16	10 ♈ 49 29	16 ♈ 49 59	00 ♐ 16	22 51	12 04	18 42	03 24	11 00	19 48	04 33
28	16 29 16	06 09 59	22 ♈ 48 44	28 ♈ 46 08	01 51	24 02	12 40	18 35	03 30	11 01	19 46	04 33
29	16 33 13	07 10 43	04 ♉ 42 30	10 ♉ 38 09	03 26	25 12	13 16	18 27	03 37	11 02	19 44	04 33
30	16 37 10	08 ♐ 11 28	16 ♉ 33 24	22 ♉ 28 30	05 ♐ 00	26 ♎ 23	13 ♎ 52	18 ♉ 20	03 ♑ 43	11 ♍ 03	19 ♉ 43	04 ♓ 34

Date	Moon True ☊ ° ′	Moon Mean ☊ ° ′	Moon ☽ Latitude ° ′
01	08 ♑ 37	09 ♑ 53	04 N 50
02	08 R 25	09 50	04 25
03	08 15	09 47	03 49
04	08 07	09 44	03 03
05	08 01	09 41	02 09
06	07 58	09 38	01 08
07	07 57	09 34	00 N 05
08	07 D 57	09 31	01 S 00
09	07 58	09 28	02 04
10	07 59	09 25	03 03
11	07 R 59	09 22	03 54
12	07 57	09 18	04 35
13	07 53	09 15	05 01
14	07 47	09 12	05 08
15	07 40	09 09	04 56
16	07 33	09 06	04 23
17	07 27	09 03	03 31
18	07 22	08 59	02 24
19	07 20	08 56	01 S 08
20	07 D 19	08 53	00 N 11
21	07 20	08 50	01 28
22	07 22	08 47	02 37
23	07 23	08 43	03 35
24	07 R 23	08 40	04 20
25	07 23	08 37	04 51
26	07 20	08 34	05 09
27	07 16	08 31	05 11
28	07 11	08 28	05 01
29	07 05	08 24	04 37
30	07 ♑ 00	08 ♑ 21	04 N 01

DECLINATIONS

Date	Sun ☉ ° ′	Moon ☽ ° ′	Mercury ☿ ° ′	Venus ♀ ° ′	Mars ♂ ° ′	Jupiter ♃ ° ′	Saturn ♄ ° ′	Uranus ♅ ° ′	Neptune ♆ ° ′	Pluto ♇ ° ′
01	14 S 29	14 N 28	06 S 04	03 N 23	02 N 43	17 N 05	22 S 40	08 N 25	16 N 08	21 S 20
02	14 48	18 16	06 30	03 01	02 28	17 03	22 41	08 24	16 07	21 20
03	15 07	21 18	06 58	02 39	02 14	17 01	22 41	08 23	16 07	21 20
04	15 26	23 25	07 28	02 17	01 59	16 59	22 41	08 22	16 07	21 20
05	15 44	24 31	08 00	01 54	01 45	16 57	22 41	08 22	16 06	21 20
06	16 02	24 29	08 34	01 32	01 30	16 55	22 41	08 21	16 06	21 20
07	16 20	23 19	09 08	01 09	01 16	16 53	22 41	08 20	16 05	21 20
08	16 37	21 04	09 43	00 45	01 01	16 51	22 41	08 19	16 05	21 20
09	16 55	17 48	10 19	00 N 22	00 47	16 49	22 41	08 19	16 04	21 20
10	17 11	13 39	10 55	00 S 02	00 32	16 47	22 41	08 18	16 04	21 19
11	17 28	08 45	11 31	00 25	00 18	16 45	22 41	08 17	16 03	21 19
12	17 44	03 N 18	12 07	00 49	00 N 03	16 43	22 41	08 17	16 03	21 19
13	18 01	02 S 30	12 43	01 13	00 S 11	16 41	22 41	08 16	16 02	21 19
14	18 16	08 21	13 19	01 38	00 26	16 39	22 41	08 15	16 02	21 19
15	18 32	13 54	13 55	02 02	00 40	16 37	22 41	08 15	16 02	21 18
16	18 47	18 42	14 30	02 26	00 54	16 35	22 41	08 14	16 01	21 18
17	19 01	22 18	15 05	02 51	01 09	16 33	22 41	08 13	16 01	21 18
18	19 16	24 17	15 39	03 15	01 23	16 31	22 41	08 13	16 00	21 18
19	19 30	24 28	16 12	03 40	01 37	16 29	22 41	08 12	16 00	21 18
20	19 44	22 54	16 45	04 05	01 52	16 27	22 41	08 12	15 59	21 17
21	19 57	19 55	17 18	04 29	02 06	16 25	22 41	08 11	15 59	21 17
22	20 10	15 52	17 49	04 54	02 20	16 23	22 41	08 11	15 59	21 17
23	20 23	11 09	18 20	05 19	02 34	16 21	22 41	08 10	15 58	21 16
24	20 35	06 05	18 50	05 44	02 48	16 19	22 41	08 10	15 58	21 16
25	20 47	00 S 54	19 18	06 09	03 02	16 17	22 41	08 10	15 57	21 16
26	20 58	04 N 12	19 47	06 33	03 16	16 15	22 41	08 09	15 57	21 15
27	21 09	09 03	20 14	06 58	03 30	16 13	22 41	08 09	15 56	21 15
28	21 20	13 31	20 40	07 23	03 44	16 11	22 41	08 08	15 56	21 15
29	21 30	17 26	21 05	07 47	03 58	16 09	22 41	08 08	15 56	21 14
30	21 S 40	20 N 38	21 S 29	08 S 12	04 S 12	16 N 07	22 S 41	08 N 08	15 N 55	21 S 14

ZODIAC SIGN ENTRIES

Date	h	m	Planets
01	20	25	☽ ♉
04	09	06	☽ ♊
06	21	49	☽ ♋
07	12	52	♀ ♎
07	14	26	♂ ♎
08	08	22	☿ ♏
09	09	00	☽ ♌
11	16	57	☽ ♍
13	20	53	☽ ♎
15	21	31	☽ ♏
17	20	41	☽ ♐
19	20	31	☽ ♑
21	22	55	☽ ♒
22	09	38	☉ ♐
24	04	59	☽ ♓
26	14	34	☽ ♈
27	07	57	☿ ♐
29	02	29	☽ ♉

LATITUDES

Date	Mercury ☿ ° ′	Venus ♀ ° ′	Mars ♂ ° ′	Jupiter ♃ ° ′	Saturn ♄ ° ′	Uranus ♅ ° ′	Neptune ♆ ° ′	Pluto ♇ ° ′
01	02 N 10	00 N 49	01 N 20	01 S 16	00 N 45	00 N 45	01 S 48	12 S 25
04	02 09	01 02	01 20	01 16	00 45	00 45	01 49	12 24
07	02 01	01 14	01 21	01 16	00 45	00 45	01 49	12 23
10	01 48	01 24	01 21	01 15	00 44	00 45	01 49	12 23
13	01 32	01 34	01 21	01 15	00 44	00 45	01 49	12 22
16	01 13	01 42	01 21	01 15	00 44	00 45	01 49	12 21
19	00 53	01 49	01 22	01 14	00 43	00 46	01 48	12 20
22	00 33	01 56	01 22	01 14	00 43	00 46	01 48	12 19
25	00 N 12	02 01	01 22	01 13	00 43	00 46	01 48	12 19
28	00 S 08	02 05	01 22	01 13	00 42	00 46	01 48	12 18
31	00 S 28	02 N 08	01 N 22	01 S 12	00 N 42	00 N 46	01 S 48	12 S 17

DATA

Julian Date	2469016
Delta T	+80 seconds
Ayanamsa	24° 31′ 47″
Synetic vernal point	04° ♓ 35′ 12″
True obliquity of ecliptic	23° 26′ 01″

MOON'S PHASES, APSIDES AND POSITIONS ☽

Date	h	m	Phase	Longitude ° ′	Eclipse Indicator
02	16	58	○	10 ♉ 12	
10	19	39	☾	18 ♌ 20	
17	12	59	●	25 ♏ 06	
24	08	41	☽	01 ♓ 59	

Day	h	m		
03	16	03	Apogee	
17	09	45	Perigee	
30	16	24	Apogee	
05	23	26	Max dec	24° N 38′
13	01	46	0S	
19	02	21	Max dec	24° S 36′
25	16	12	0N	

LONGITUDES

Date	Chiron ⚷ ° ′	Ceres ⚳ ° ′	Pallas ⚴ ° ′	Juno ⚵ ° ′	Vesta ⚶ ° ′	Black Moon Lilith ⚸ ° ′
01	01 ♏ 29	17 ♑ 16	15 ♐ 47	18 ♐ 04	13 ♑ 36	19 ♉ 48
11	02 ♏ 54	20 ♑ 16	19 ♐ 41	21 ♐ 23	18 ♑ 03	20 ♉ 55
21	04 ♏ 17	23 ♑ 28	23 ♐ 38	24 ♐ 48	22 ♑ 38	22 ♉ 02
31	05 ♏ 36	26 ♑ 51	27 ♐ 38	28 ♐ 19	27 ♑ 21	23 ♉ 09

ASPECTARIAN

h m	Aspects
01 Friday	
00 35	☽ ☍ ☿
01 25	☽ ⚺ ♆
04 46	☽ ⚺ ♃
06 37	☽ ⚻ ♀
09 07	☿ ⚻ ♆
10 52	☽ ⚼ ♅
12 07	☽ # ☉
12 57	☽ ⚻ ♂
19 52	☽ ± ♀
21 58	☽ ∥ ♆
22 08	☽ △ ♄
02 Saturday	
01 40	☽ ± ♂
03 55	☽ ∥ ♃
05 38	☽ ✶ ♇
15 10	☿ ⚻ ♃
15 50	☽ ⚼ ♀
16 58	☽ ☍ ☉
17 10	☽ △ ♅
19 29	☉ ✶ ♅
20 50	☽ ⚼ ♂
03 Sunday	
04 39	☽ ⚼ ♄
05 52	☽ Q ♇
12 22	☽ # ♇
13 45	☽ ☌ ♆
16 38	☽ ☌ ♃
20 03	☽ ⚻ ☿
23 02	☽ ± ♄
04 Monday	
01 18	☽ △ ♀
02 15	☽ # ♄
04 53	☽ △ ♂
09 46	☽ ± ☿
11 18	☽ ⚻ ♄
18 22	☽ □ ♇
05 Tuesday	
04 14	☿ ∠ ♅
06 11	☽ □ ♅
06 27	☽ ⚼ ☿
11 25	☽ ⚻ ☉
06 Wednesday	
00 41	☽ ± ☉
02 26	☽ ⚺ ♆
03 05	☿ # ♅
04 47	☽ ⚺ ♃
14 31	☽ ⊥ ♆
16 15	♀ ∥ ♂
16 39	☉ # ♆
16 45	☽ ⊥ ♃
16 55	☽ △ ☿
18 41	☽ Q ♅
20 16	☽ □ ♀
20 30	☽ ⚼ ☉
20 55	☽ □ ♂
07 Thursday	
00 27	☽ ☍ ♄
06 56	☽ △ ♇
08 30	☉ ∠ ♂
08 32	☽ ∠ ♆
10 35	☽ ∠ ♃
10 58	♀ ☌ ♂
18 45	☽ ✶ ♅
19 49	☉ ∠ ♀
20 03	☽ # ♄
08 Friday	
05 07	☽ △ ☉
09 39	☽ # ♇
10 38	☽ Q ♂
11 43	☽ Q ♀
12 40	☽ ⚼ ♇
14 11	☽ ✶ ♆
15 56	☽ ✶ ♃
20 32	☿ ⚺ ♂
20 56	♀ □ ♄
09 Saturday	
00 01	☉ ∠ ♄
00 20	☽ ∠ ♅
05 02	☉ # ♃
06 12	☽ ± ♇
08 39	☿ ✶ ♄
11 12	☽ ✶ ♂
11 59	☽ ⚻ ♄
12 25	☽ □ ☿
13 20	☽ ✶ ♀
13 25	☽ Q ♆
14 56	☽ Q ♃
17 10	☽ # ☉
17 43	☽ ⚻ ♇
17 44	☽ ⊥ ♅
18 10	☽ ∥ ♃
22 34	☽ ∥ ♆
23 32	☽ ± ♄
10 Sunday	
05 09	☽ ⚺ ♅
06 07	♂ □ ♄
13 18	☿ ⚺ ♀
16 34	☽ ⚼ ♄
17 00	☽ ∠ ♂
19 39	☽ □ ☉
20 21	☽ ∠ ♀
23 13	☽ □ ♆
11 Monday	
00 18	☽ # ☿
00 24	☽ □ ♃
02 53	☽ Q ☿
07 07	♀ # ♂
07 16	☿ △ ♇
10 20	☽ ⊥ ♂
14 06	♀ ⚻ ♇
14 08	☽ ∥ ♅
14 20	☽ ⊥ ♀
20 08	☽ △ ♄
21 40	☽ ⚺ ♂
12 Tuesday	
01 04	☽ ☍ ♇
02 04	☽ ⚺ ♀
03 24	☽ ✶ ☿
04 38	♀ ⚼ ♆
06 51	☽ Q ☉
11 49	☽ ☌ ♅
14 25	♀ ⚼ ♃
16 27	☉ ☍ ♆
21 42	☽ # ♀
13 Wednesday	
01 26	☽ # ♂
02 07	☽ ∥ ♂
02 11	☉ ☍ ♃
04 20	☽ △ ♆
05 01	☽ △ ♃
05 16	☽ ✶ ☉
06 24	☽ ∥ ♀
08 44	☽ ∠ ☿
14 Thursday	
00 11	☽ □ ♄
01 42	☽ ⊥ ☿
03 31	☽ ☌ ♂
04 24	☽ ⚻ ♇
05 27	☽ ⚼ ♆
05 56	☽ ⚼ ♃
08 16	☽ ∠ ☉
09 39	☽ ☌ ♀
11 36	☽ # ♅
12 47	☽ ⚺ ☿
14 15	☽ ± ♇
14 32	☽ ⚺ ♅
20 08	☽ ± ♆
20 30	☽ ± ♃
15 Friday	
00 00	☽ ⊥ ☉
00 17	☽ ⊥ ♅
00 43	♂ ⚻ ♇
03 25	☿ ⊥ ♂
04 20	☉ Q ♅
04 49	☽ ⚼ ♇
05 14	☿ ✶ ♅
05 45	☽ Q ♄
05 47	☽ ⚻ ♆
06 05	☽ ⚻ ♃
10 20	☽ ⚺ ☉
12 04	☽ ∥ ☿
14 44	☽ ∠ ♅
22 02	☽ # ♆
23 45	♂ ⚼ ♆
16 Saturday	
00 51	☽ # ♃
00 59	☽ ✶ ♄
03 11	♂ ⚼ ♃
04 41	☽ △ ♇
05 51	☽ ⚺ ♂
12 27	☽ ∥ ☉
13 36	☽ ⚺ ♀
14 29	☽ ✶ ♅
15 41	☽ ⊥ ♂
18 18	☽ ☌ ☿
20 40	♀ ± ♇
23 31	♃ ☌ ♆
23 47	☽ ⊥ ♀
17 Sunday	
00 45	☽ ∠ ♄
01 44	♀ ⚺ ♅
04 18	☽ ∥ ♇
05 04	☽ ☍ ♃
05 07	☽ ☍ ♆
06 23	☽ ∠ ♂
09 21	☽ Q ♅
12 59	☽ ☌ ☉
14 57	☽ ⊥ ♄
15 02	☽ ∠ ♀
15 28	☽ ∥ ♄
23 52	♇ St D
18 Monday	
00 26	☽ ⚺ ♄
03 48	☽ □ ♇
06 55	☽ ✶ ♂
13 44	☽ □ ♅
16 37	☽ ✶ ♀
22 00	☉ ⊥ ♄
23 04	☽ ⚺ ☿
19 Tuesday	
02 47	☽ Q ♂
03 03	☿ # ♆
04 02	☽ ⚻ ♃
04 24	☽ ⚻ ♆
08 25	☽ Q ♇
09 50	☽ ⊥ ☿
11 54	☿ ∠ ♄
13 25	☽ Q ♀
13 36	☽ ± ♃
14 02	☽ ± ♆
15 29	♀ ± ♃
16 02	☽ ⚺ ☉
21 15	♀ ± ♆
22 58	☿ # ♃
20 Wednesday	
00 45	☽ ☌ ♄
02 13	☽ ∠ ☿
02 32	☽ ⊥ ☉
03 55	☽ ✶ ♇
04 07	☽ ⚼ ♃
04 41	☽ ⚼ ♆
09 16	☽ □ ♂
14 12	☽ ∥ ♄
14 23	☽ △ ♅
18 28	☿ ☍ ♃
18 32	☽ ∠ ☉
21 35	☽ □ ♀
21 Thursday	
00 26	☿ ☍ ♆
02 13	☽ ∥ ♇
04 55	☽ △ ♃
04 56	☽ ∠ ♇
05 40	☽ △ ♆
06 21	☽ ✶ ☿
11 47	☽ ∥ ☉
15 47	☽ ⚼ ♅
20 22	☽ ⊥ ♇
22 04	☽ ✶ ☉
22 Friday	
02 18	☽ ∥ ☿
03 50	☽ ⚺ ♄
05 45	☽ Q ☿
06 50	☽ ⚺ ♇
07 29	☽ ± ♅
09 14	☽ # ♃
09 24	♀ ⊥ ♅
11 26	☽ # ♆
14 29	☽ ⊥ ♄
14 57	☽ △ ♂
18 07	☽ ⚻ ♅
20 43	☿ Q ♅
20 56	☽ Q ☉
23 Saturday	
06 46	☽ △ ♀
06 47	☽ ∠ ♄
09 07	☽ □ ♃
10 21	☽ □ ♆
18 31	☽ □ ☿
19 25	☽ ⚼ ♂
24 Sunday	
02 16	☽ # ♅
02 52	☿ ⊥ ♀
05 50	☿ ∠ ♂
07 03	♀ ⚻ ♃
08 41	☽ □ ☉
10 42	☽ ✶ ♄
13 00	☽ ± ♂
13 15	☽ ⚼ ♀
13 29	☽ ☌ ♇
13 32	☽ ∥ ♀
15 47	♀ ⚼ ♇
18 14	☽ Q ♃
19 46	☽ Q ♆
22 30	♀ ⚻ ♆
23 03	♂ ± ♇
25 Monday	
00 56	☽ ⚻ ♂
01 39	☽ ☍ ♅
02 32	☽ ∥ ♂
08 06	☽ ± ♀
09 42	☽ Q ♄
13 10	☉ ⚺ ♄
13 14	☿ ⊥ ♄
16 19	♂ ⚺ ♅
16 58	☽ ✶ ♃
18 44	☽ ✶ ♆
20 53	☽ ⚻ ♀
26 Tuesday	
02 53	♀ Q ♄
07 22	☽ # ♂
11 59	☽ △ ☿
21 07	☽ □ ♄
21 29	☉ □ ♇
22 01	☽ ∠ ♃
23 32	☽ ⚺ ♇
23 43	☽ △ ☉
27 Wednesday	
00 03	☽ ∠ ♆
00 31	☽ # ♀
07 23	☽ ∥ ♅
11 27	☽ ⊥ ♇
12 21	☽ ⚻ ♅
14 36	☽ ☍ ♂
15 42	☽ ⊥ ♃
17 54	☽ ⊥ ♆
22 13	☽ ⚼ ☿
28 Thursday	
00 22	☽ ± ♅
00 48	☉ ∥ ♇
03 35	☽ ⚺ ♃
05 27	☽ ∠ ♇
05 53	☽ ⚺ ♆
08 23	☽ ⚼ ☉
09 11	♂ ± ♃
14 43	☽ ☍ ♀
18 28	☽ ⚼ ♅
19 03	☽ ± ☿
20 01	♃ ⚼ ♄
29 Friday	
02 20	☽ ∥ ♆
03 48	☽ ∥ ♃
04 12	☽ ± ☉
09 01	☽ ⚻ ☿
09 46	☽ △ ♄
11 41	☽ ✶ ♇
15 02	☿ ⚺ ♄
17 28	☽ ⚻ ☉
21 22	☿ ∥ ♇
30 Saturday	
00 50	☽ △ ♅
05 09	♀ ∠ ♅
05 14	☿ □ ♇
06 13	♂ ± ♆
06 15	☽ ⚻ ♂
07 51	♀ # ♅
12 00	☽ Q ♇
15 34	☽ ☌ ♃
16 26	☽ ⚼ ♄
17 20	☽ # ♇
18 23	☽ ☌ ♆
19 04	☽ ± ♂
21 05	☽ # ☿
22 08	☽ # ☉

All ephemeris data is given at 12.00 UT and the Moon's longitude is additionally given for 24.00 UT

DECEMBER 2047

Date	Sidereal time h m s	LONGITUDES Sun ☉ ° ' "	Moon ☽ ° ' "	Moon ☽ 24.00 ° ' "	Mercury ☿ ° '	Venus ♀ ° '	Mars ♂ ° '	Jupiter ♃ ° '	Saturn ♄ ° '	Uranus ♅ ° '	Neptune ♆ ° '	Pluto ♇ ° '
01	16 41 06	09 ♐ 12 15	28 ♉ 23 42	04 ♊ 19 13	06 ♐ 35	27 ♎ 34	14 ♎ 28	18 ♉ 13	03 ♑ 50	11 ♍ 04	19 ♉ 41	04 ♓ 34
02	16 45 03	10 13 02	10 ♊ 15 18	16 ♊ 12 09	08 09	28 45	15 04	18 R 06	03 57	11 05	19 R 40	04 34
03	16 48 59	11 13 51	22 ♊ 10 01	28 ♊ 09 07	09 43	29 ♎ 56	15 39	17 59	04 03	11 06	19 38	04 35
04	16 52 56	12 14 41	04 ♋ 09 42	10 ♋ 12 03	11 18	01 ♏ 07	16 15	17 52	04 10	11 06	19 36	04 35
05	16 56 52	13 15 32	16 ♋ 16 26	22 ♋ 23 11	12 52	02 18	16 51	17 46	04 17	11 07	19 35	04 35
06	17 00 49	14 16 24	28 ♋ 32 38	04 ♌ 45 08	14 26	03 30	17 27	17 39	04 24	11 08	19 33	04 36
07	17 04 45	15 17 18	11 ♌ 01 04	17 ♌ 20 48	16 00	04 41	18 02	17 33	04 30	11 08	19 32	04 36
08	17 08 42	16 18 13	23 ♌ 44 46	00 ♍ 13 21	17 35	05 53	18 38	17 26	04 37	11 09	19 30	04 37
09	17 12 39	17 19 09	06 ♍ 46 56	13 ♍ 25 50	19 09	07 05	19 14	17 20	04 44	11 09	19 29	04 37
10	17 16 35	18 20 06	20 ♍ 10 21	27 ♍ 00 44	20 43	08 17	19 49	17 14	04 51	11 09	19 28	04 38
11	17 20 32	19 21 04	03 ♎ 57 05	10 ♎ 59 25	22 17	09 28	20 25	17 09	04 58	11 10	19 26	04 39
12	17 24 28	20 22 04	18 ♎ 07 38	25 ♎ 21 26	23 52	10 40	21 00	17 03	05 05	11 10	19 25	04 39
13	17 28 25	21 23 05	02 ♏ 40 24	10 ♏ 03 56	25 26	11 53	21 36	16 57	05 12	11 10	19 23	04 40
14	17 32 21	22 24 07	17 ♏ 31 15	25 ♏ 01 28	27 01	13 05	22 11	16 52	05 19	11 10	19 22	04 41
15	17 36 18	23 25 10	02 ♐ 33 31	10 ♐ 06 19	28 ♐ 35	14 17	22 46	16 47	05 26	11 11	19 21	04 41
16	17 40 14	24 26 13	17 ♐ 38 41	25 ♐ 09 28	00 ♑ 10	15 29	23 22	16 42	05 33	11 11	19 19	04 42
17	17 44 11	25 27 18	02 ♑ 37 33	10 ♑ 01 55	01 45	16 42	23 57	16 37	05 40	11 R 11	19 18	04 43
18	17 48 08	26 28 23	17 ♑ 21 38	24 ♑ 35 59	03 20	17 54	24 32	16 32	05 47	11 10	19 17	04 43
19	17 52 04	27 29 29	01 ♒ 44 21	08 ♒ 46 20	04 55	19 07	25 07	16 28	05 54	11 10	19 15	04 44
20	17 56 01	28 30 35	15 ♒ 41 40	22 ♒ 30 18	06 30	20 20	25 42	16 24	06 01	11 10	19 14	04 45
21	17 59 57	29 ♐ 31 41	29 ♒ 12 15	05 ♓ 47 44	08 05	21 32	26 17	16 20	06 08	11 10	19 13	04 46
22	18 03 54	00 ♑ 32 48	12 ♓ 17 03	18 ♓ 40 33	09 41	22 45	26 52	16 16	06 15	11 10	19 12	04 47
23	18 07 50	01 33 55	24 ♓ 58 44	01 ♈ 12 04	11 16	23 58	27 27	16 12	06 22	11 09	19 11	04 48
24	18 11 47	02 35 02	07 ♈ 21 07	13 ♈ 26 26	12 52	25 11	28 02	16 09	06 29	11 09	19 10	04 48
25	18 15 43	03 36 09	19 ♈ 28 35	25 ♈ 28 09	14 27	26 24	28 37	16 05	06 36	11 08	19 08	04 49
26	18 19 40	04 37 16	01 ♉ 25 42	07 ♉ 21 45	16 03	27 37	29 12	16 02	06 43	11 08	19 07	04 50
27	18 23 37	05 38 23	13 ♉ 16 50	19 ♉ 11 26	17 38	28 ♏ 50	29 ♎ 46	15 59	06 51	11 07	19 06	04 51
28	18 27 33	06 39 31	25 ♉ 06 01	01 ♊ 01 00	19 14	00 ♐ 03	00 ♏ 21	15 57	06 58	11 07	19 05	04 52
29	18 31 30	07 40 38	06 ♊ 56 47	12 ♊ 53 44	20 49	01 16	00 56	15 54	07 05	11 06	19 04	04 53
30	18 35 26	08 41 46	18 ♊ 52 08	24 ♊ 52 20	22 23	02 30	01 30	15 52	07 12	11 05	19 03	04 54
31	18 39 23	09 ♑ 42 54	00 ♋ 54 33	06 ♋ 58 59	23 ♑ 58	03 ♐ 43	02 ♏ 05	15 ♉ 50	07 ♑ 19	11 ♍ 04	19 ♉ 02	04 ♓ 55

Date	Moon True ☊ ° '	Moon Mean ☊ ° '	Moon ☽ Latitude ° '
01	06 ♑ 55	08 ♑ 18	03 N 15
02	06 R 51	08 15	02 21
03	06 49	08 12	01 20
04	06 48	08 09	00 N 15
05	06 D 48	08 05	00 S 52
06	06 49	08 02	01 57
07	06 51	07 59	02 58
08	06 52	07 56	03 51
09	06 54	07 53	04 34
10	06 R 54	07 49	05 03
11	06 53	07 46	05 16
12	06 52	07 43	05 10
13	06 50	07 40	04 45
14	06 48	07 37	04 00
15	06 46	07 34	02 58
16	06 45	07 30	01 44
17	06 45	07 27	00 S 23
18	06 D 45	07 24	00 N 59
19	06 45	07 21	02 14
20	06 46	07 18	03 20
21	06 47	07 15	04 12
22	06 48	07 11	04 49
23	06 48	07 08	05 11
24	06 R 48	07 05	05 18
25	06 48	07 02	05 10
26	06 48	06 59	04 49
27	06 47	06 55	04 16
28	06 47	06 52	03 32
29	06 D 47	06 49	02 38
30	06 47	06 46	01 38
31	06 ♑ 48	06 ♑ 43	00 N 33

DECLINATIONS

Date	Sun ☉ ° '	Moon ☽ ° '	Mercury ☿ ° '	Venus ♀ ° '	Mars ♂ ° '	Jupiter ♃ ° '	Saturn ♄ ° '	Uranus ♅ ° '	Neptune ♆ ° '	Pluto ♇ ° '
01	21 S 49	22 N 58	21 S 52	08 S 37	04 S 26	16 N 06	22 S 41	08 N 07	15 N 55	21 S 14
02	21 59	24 18	22 14	09 01	04 40	16 04	22 41	08 07	15 54	21 13
03	22 07	24 32	22 35	09 25	04 54	16 02	22 40	08 07	15 54	21 13
04	22 15	23 37	22 54	09 49	05 07	16 00	22 40	08 07	15 54	21 13
05	22 23	21 35	23 13	10 13	05 21	15 59	22 40	08 06	15 53	21 12
06	22 30	18 32	23 30	10 37	05 35	15 57	22 40	08 06	15 53	21 12
07	22 37	14 37	23 47	11 01	05 48	15 56	22 40	08 06	15 53	21 11
08	22 44	09 58	24 01	11 24	06 02	15 54	22 40	08 06	15 52	21 11
09	22 50	04 N 47	24 15	11 48	06 15	15 53	22 40	08 06	15 52	21 10
10	22 55	00 S 45	24 28	12 11	06 29	15 51	22 39	08 06	15 51	21 10
11	23 00	06 21	24 39	12 34	06 42	15 50	22 39	08 06	15 51	21 10
12	23 05	11 53	24 49	12 56	06 55	15 48	22 39	08 06	15 51	21 09
13	23 09	16 51	24 57	13 19	07 09	15 47	22 39	08 06	15 50	21 09
14	23 13	20 54	25 04	13 41	07 22	15 46	22 39	08 06	15 50	21 08
15	23 16	23 35	25 10	14 03	07 35	15 45	22 39	08 06	15 50	21 08
16	23 19	24 35	25 15	14 25	07 48	15 44	22 38	08 06	15 49	21 07
17	23 21	23 47	25 18	14 46	08 01	15 42	22 38	08 06	15 49	21 07
18	23 23	21 25	25 19	15 07	08 14	15 41	22 38	08 06	15 49	21 06
19	23 25	17 35	25 19	15 27	08 27	15 40	22 38	08 06	15 48	21 06
20	23 25	12 57	25 18	15 48	08 39	15 39	22 37	08 06	15 48	21 05
21	23 26	07 49	25 16	16 08	08 52	15 39	22 37	08 06	15 48	21 05
22	23 26	02 S 30	25 11	16 27	09 05	15 38	22 37	08 06	15 48	21 04
23	23 25	02 N 46	25 05	16 47	09 17	15 37	22 37	08 06	15 47	21 03
24	23 24	07 47	24 58	17 05	09 30	15 36	22 36	08 07	15 47	21 03
25	23 23	12 24	24 49	17 24	09 42	15 35	22 36	08 07	15 47	21 02
26	23 21	16 29	24 39	17 42	09 54	15 35	22 36	08 07	15 47	21 02
27	23 19	19 53	24 27	17 59	10 07	15 34	22 35	08 07	15 46	21 01
28	23 16	22 27	24 13	18 17	10 19	15 34	22 35	08 08	15 46	21 01
29	23 13	24 04	23 58	18 33	10 31	15 33	22 35	08 08	15 46	21 00
30	23 09	24 36	23 42	18 49	10 43	15 33	22 35	08 08	15 46	21 00
31	23 S 05	23 N 58	23 S 24	19 S 05	10 S 55	15 N 33	22 S 34	08 N 09	15 N 45	20 S 59

ZODIAC SIGN ENTRIES

Date	h	m	Planets
01	15	15	☽ ♊
03	13	23	♀ ♏
04	03	42	☽ ♋
06	14	49	☽ ♌
08	23	35	☽ ♍
11	05	11	☽ ♎
13	07	38	☽ ♏
15	07	56	☽ ♐
16	09	26	☿ ♑
17	07	46	☽ ♑
19	09	04	☽ ♒
21	13	26	☽ ♓
21	23	07	☉ ♑
23	21	40	☽ ♈
26	09	07	☽ ♉
27	21	26	♂ ♏
28	10	59	♀ ♐
28	21	56	☽ ♊
31	10	12	☽ ♋

LATITUDES

Date	Mercury ☿ ° '	Venus ♀ ° '	Mars ♂ ° '	Jupiter ♃ ° '	Saturn ♄ ° '	Uranus ♅ ° '	Neptune ♆ ° '	Pluto ♇ ° '
01	00 S 28	02 N 08	01 N 22	01 S 12	00 N 42	00 N 46	01 S 48	12 S 17
04	00 47	02 10	01 22	01 12	00 42	00 46	01 48	12 16
07	01 05	02 12	01 22	01 11	00 42	00 46	01 48	12 15
10	01 21	02 12	01 22	01 10	00 41	00 46	01 48	12 15
13	01 36	02 11	01 22	01 09	00 41	00 47	01 48	12 14
16	01 49	02 10	01 22	01 09	00 41	00 47	01 48	12 13
19	01 59	02 07	01 22	01 08	00 41	00 47	01 48	12 12
22	02 07	02 04	01 22	01 07	00 40	00 47	01 48	12 12
25	02 11	02 00	01 22	01 06	00 40	00 47	01 47	12 11
28	02 11	01 55	01 22	01 05	00 40	00 47	01 47	12 10
31	02 S 07	01 N 50	01 N 22	01 S 05	00 N 40	00 N 47	01 S 47	12 S 10

DATA

Julian Date	2469046
Delta T	+80 seconds
Ayanamsa	24° 31' 51"
Synetic vernal point	04° ♓ 35' 08"
True obliquity of ecliptic	23° 26' 00"

MOON'S PHASES, APSIDES AND POSITIONS ☽

Date	h	m	Phase	Longitude ° '	Eclipse Indicator
02	11	55	○	10 ♊ 13	
10	08	29	☾	18 ♍ 11	
16	23	38	●	24 ♐ 56	Partial
24	01	51	☽	02 ♈ 09	

Day	h	m		
15	21	45	Perigee	
28	02	45	Apogee	
03	04	46	Max dec	24° N 35'
10	08	47	0S	
16	13	20	Max dec	24° S 35'
22	23	18	0N	
30	11	09	Max dec	24° N 36'

LONGITUDES

Date	Chiron ⚷ ° '	Ceres ⚳ ° '	Pallas ⚴ ° '	Juno ⚵ ° '	Vesta ⚶ ° '	Black Moon Lilith ⚸ ° '
01	05 ♏ 36	26 ♑ 51	27 ♐ 38	28 ♐ 19	27 ♑ 21	23 ♉ 09
11	06 ♏ 50	00 ♒ 23	01 ♑ 38	01 ♑ 56	02 ♒ 08	24 ♉ 15
21	07 ♏ 59	04 ♒ 02	05 ♑ 39	05 ♑ 36	07 ♒ 00	25 ♉ 22
31	09 ♏ 00	07 ♒ 46	09 ♑ 38	09 ♑ 18	11 ♒ 55	26 ♉ 29

All ephemeris data is given at 12.00 UT and the Moon's longitude is additionally given for 24.00 UT

ASPECTARIAN

h m Aspects

01 Sunday
07 39 ☉ ∥ ☿
08 21 ☽ # ♄
10 08 ☽ ⊼ ♀
10 51 ☽ ± ♄
14 17 ☽ bQ ♂
23 07 ☽ ⊼ ♄
23 38 ☽ ± ♀

02 Monday
00 30 ☽ □ ♇
07 06 ☽ ☍ ☿
11 55 ☽ ☍ ☉
13 40 ☽ □ ♅
19 50 ☽ bQ ♀
22 12 ☽ △ ♂

03 Tuesday
03 40 ☽ ⊻ ♃
06 55 ☽ ⊻ ♆
08 43 ☉ □ ♅
15 37 ☽ ⊥ ♃
18 47 ☿ ∥ ♄
18 56 ☽ ⊥ ♆

04 Wednesday
01 54 ☽ Q ♅
05 16 ☽ △ ♀
09 06 ☿ □ ♅
09 27 ☽ ∠ ♃
12 01 ☽ ☍ ♄
12 51 ☽ △ ♇
12 53 ☽ ∠ ♆
20 27 ☽ # ☿

05 Thursday
00 42 ☽ # ♄
01 48 ☽ ⚹ ♅
04 16 ☽ ⊼ ☿
04 21 ☽ # ☉
05 31 ☽ ⊼ ☉
13 11 ☽ □ ♂
14 54 ☽ ⚹ ♃
15 28 ☽ # ♇
17 51 ☽ ± ☿
18 24 ☽ ± ☉
18 30 ☽ ⚹ ♆
18 32 ☽ bQ ♇
22 59 ♂ ⊥ ♅

06 Friday
04 59 ☉ ☌ ☿
07 18 ☽ ∠ ♅
12 06 ☽ ± ♇
13 33 ☽ bQ ☉
13 59 ☽ bQ ☿
14 08 ☽ Q ♃
17 50 ☽ Q ♆
19 03 ♂ ⊼ ♃
22 36 ☽ □ ♀
23 25 ☽ ⊼ ♄
23 43 ☽ ⊼ ♇

07 Saturday
00 44 ☽ ⊥ ♅
02 00 ☽ Q ♂
04 25 ☽ ∥ ♃
04 46 ☽ ∥ ♆
07 58 ♀ ⚹ ♄
10 21 ♀ △ ♇
11 01 ☽ ± ♄
12 14 ☽ ⊻ ♅
16 26 ♄ bQ ♆
20 49 ☽ △ ☉
21 15 ☉ ∥ ♄
22 48 ☽ △ ☿

08 Sunday
00 16 ☽ □ ♃
01 57 ☽ ⚹ ♂
04 05 ☽ □ ♆
04 13 ☽ bQ ♄
05 25 ☽ # ♀
10 03 ☿ ⊼ ♃
10 49 ♄ ⚹ ♇
12 17 ☽ Q ♀
20 53 ☽ ∥ ♅

09 Monday
05 38 ☽ # ♂
07 07 ☽ ∠ ♂
08 04 ☽ ☍ ♇
08 14 ☽ △ ♄
12 25 ☉ ⊼ ♃
12 35 ☽ ⚹ ♀
13 59 ☿ ⚹ ♂
17 04 ☿ ⊼ ♆
19 55 ☽ ☌ ♅
21 58 ♂ ⊼ ♆

10 Tuesday
00 11 ☽ ⊥ ♂
04 21 ♂ bQ ♇
06 50 ☽ △ ♃
07 34 ♃ ∥ ♆
08 29 ☽ □ ☉
10 44 ☽ △ ♆
11 21 ☽ ⊻ ♂
13 05 ☽ □ ☿
17 59 ☽ ∠ ♀

11 Wednesday
08 55 ☽ bQ ♃
11 06 ☽ ⊥ ♀
12 50 ☽ bQ ♆
13 11 ☽ ⊼ ♇
13 20 ☽ ∥ ♂
13 45 ☽ □ ♄
13 57 ☉ ⊼ ♆
17 27 ☿ Q ♇
18 16 ☽ Q ☉
19 17 ☽ # ♅
22 18 ☽ ⊻ ♀
23 25 ☽ ± ♇

12 Thursday
00 10 ☽ Q ☿
00 11 ☽ ± ♃
00 16 ☿ ± ♃
00 18 ☽ ⊻ ♅
04 06 ☽ ± ♆
10 12 ☽ ⊼ ♃
10 24 ☽ ⊥ ♅
14 08 ☽ ⊼ ♆
14 33 ☽ bQ ♇
16 01 ☽ ⚹ ☉
17 00 ☽ ☌ ♂
17 14 ☽ ∥ ♀
20 18 ☽ Q ♄
21 54 ♀ ⚹ ♅
22 41 ☽ ⚹ ☿

13 Friday
01 20 ☽ ∠ ♅
06 34 ☽ # ♃
06 49 ☽ # ♆
11 17 ☿ ± ♆
15 15 ☽ △ ♇
16 09 ☽ ⚹ ♄
18 29 ☽ ∠ ☉
23 47 ☉ ⚹ ♂

14 Saturday
01 47 ☽ ⚹ ♅
02 06 ☽ ∠ ☿
04 14 ☽ ☌ ♀
10 04 ☽ ⊥ ☉
10 57 ☽ ☍ ♃
13 43 ☽ ∥ ♇
14 57 ☽ ☍ ♆
16 30 ☽ ∠ ♄
18 15 ☽ ⊥ ☿
18 32 ☉ Q ♇
19 46 ☽ ⊻ ♂
20 23 ☽ ⊻ ☉
21 03 ☽ Q ♅
22 07 ☉ ± ♃

15 Sunday
01 56 ☽ ∥ ♄
04 57 ☽ ⊻ ☿
05 44 ☽ ⊥ ♂
06 59 ☽ ⊥ ♄
08 13 ☽ ∥ ☉
15 23 ☽ □ ♇
16 36 ☽ ⊻ ♄
20 38 ☽ ∠ ♂

16 Monday
01 42 ☽ □ ♅
08 16 ☽ ⊻ ♀
10 30 ☽ ⊼ ♃
11 53 ♃ Q ♇
14 40 ☽ ⊼ ♆
17 23 ♅ St R
18 40 ☽ ⊥ ♀
20 01 ☽ ± ♃
20 04 ☽ Q ♇
21 26 ♂ Q ♄
21 30 ☽ ⚹ ♂
23 38 ☽ ☌ ☉

17 Tuesday
00 15 ☽ ± ♆
08 26 ☉ ± ♆
08 42 ☿ ∠ ♀
10 06 ☿ bQ ♃
10 22 ☽ ∠ ♀
10 23 ☽ bQ ♃
10 25 ☽ ☌ ☿
10 31 ♀ ☍ ♃
14 42 ☽ bQ ♆
15 22 ☽ ⚹ ♇
16 57 ☽ ☌ ♄
17 30 ☽ ∥ ☉
17 36 ☽ Q ♂
20 52 ♂ # ♅

18 Wednesday
01 08 ☽ ∥ ♄
01 52 ☽ △ ♅
10 40 ☽ △ ♃
12 59 ☽ ⚹ ♀
13 46 ☽ ∥ ♇
15 10 ☽ △ ♆
15 54 ☽ ∠ ♇

19 Thursday
00 24 ☽ □ ♂
02 09 ☿ bQ ♆
02 38 ☽ bQ ♅
04 17 ☽ ⊻ ☉
06 56 ☽ ⊥ ♇
09 15 ☿ ⚹ ♇
10 51 ☽ Q ♀
14 46 ♀ ☍ ♆
15 12 ☽ ⊥ ☉
17 06 ☽ ⊻ ♇
17 50 ☽ ± ♅
18 05 ☽ ⊻ ☿
19 08 ☽ ⊻ ♄
21 37 ☽ # ♆
22 21 ☽ # ♃
22 39 ☽ ∥ ♀

20 Friday
02 29 ♀ # ♃
04 02 ☿ ☌ ♄
04 08 ☽ ⊼ ♅
05 33 ☽ ⊥ ♄
05 43 ☽ ⊥ ☿
07 53 ☽ ∠ ☉
12 28 ♀ # ♆
13 13 ☽ □ ♃
18 13 ☽ □ ♆
20 56 ☽ □ ♀
21 26 ☽ ∠ ♄
23 35 ☽ ∠ ☿

21 Saturday
03 06 ♀ ∠ ♄
06 31 ☽ △ ♂
06 56 ♂ ∠ ♅
07 21 ☽ ∥ ♂
10 40 ☽ # ♅
12 38 ☽ ⚹ ☉
16 44 ☿ Q ♂
21 16 ☽ Q ♃
22 07 ☽ ☌ ♇

22 Sunday
00 44 ☽ ⚹ ♄
02 35 ☽ Q ♆
06 29 ☽ ⚹ ☿
09 55 ☽ ☍ ♅
11 12 ☽ bQ ♂
12 32 ☽ Q ☉
19 25 ☽ ⚹ ♃
20 02 ♀ Q ♅
23 18 ☽ Q ♄

23 Monday
00 58 ☽ ⚹ ♆
03 56 ☉ bQ ♃
04 57 ☽ ± ♂
08 16 ☽ Q ☿
09 51 ☽ △ ♀
10 14 ☿ △ ♅
16 59 ☽ ⊼ ♂
23 57 ☽ ∠ ♃

24 Tuesday
01 51 ☽ □ ☉
05 46 ☽ ∠ ♆
07 01 ☽ ⊻ ♇
10 17 ☽ □ ♄
13 39 ☽ ∥ ♅
17 28 ☽ ⊥ ♃
18 11 ☽ bQ ♀
18 48 ☽ ⊥ ♇
19 28 ☽ ⊼ ♅
21 03 ☽ # ♂
23 26 ☽ ⊥ ♆

25 Wednesday
00 30 ☽ □ ☿
05 17 ☽ ⊻ ♃
07 21 ☽ ± ♅
11 20 ☽ ⊻ ♆
12 42 ☽ ∠ ♇
14 03 ☽ ± ♀

26 Thursday
00 28 ☉ bQ ♆
01 20 ☽ bQ ♅
03 26 ☽ ⊼ ♀
06 25 ☽ ∥ ♃
07 16 ☽ ☍ ♂
07 37 ☽ ∥ ♆
11 52 ☿ △ ♃
17 13 ☉ ⚹ ♇
18 54 ☽ ⚹ ♇
19 03 ☽ △ ☉
20 45 ☽ # ♀
22 49 ☽ △ ♄

27 Friday
07 37 ☽ △ ♅
17 29 ☽ ☌ ♃
19 16 ☽ Q ♇
21 36 ☽ # ♇
22 13 ☽ △ ☿
23 49 ☽ ☌ ♆

28 Saturday
04 21 ☽ bQ ☉
05 34 ☽ bQ ♄
09 57 ☿ △ ♆
13 30 ☽ # ♄
20 03 ☉ ☌ ♄
21 53 ☿ ∠ ♇
21 53 ☽ # ☉
23 08 ♀ ⊻ ♂
23 12 ☽ ⊼ ♂
23 12 ☽ ☍ ♀

29 Sunday
00 00 ☽ ± ♄
00 20 ☽ ± ☉
07 48 ♀ ⊥ ♄
07 50 ☽ □ ♇
09 21 ☽ bQ ☿
10 22 ☽ # ☿
11 57 ☽ ± ♂
12 16 ☽ ⊼ ♄
13 37 ☽ ⊼ ☉
20 22 ☽ □ ♅

30 Monday
05 59 ☽ ⊻ ♃
06 16 ☽ ± ☿
07 01 ☽ bQ ♂
12 22 ☽ ⊻ ♆
17 59 ☽ ⊥ ♃
20 06 ☽ ⊼ ☿

31 Tuesday
00 21 ☽ ⊥ ♆
08 22 ☽ Q ♅
11 51 ☽ ∠ ♃
12 06 ☉ ⊥ ♀
14 26 ☽ △ ♂
18 10 ☽ ⊼ ♀
18 11 ☽ ∠ ♆
19 57 ☽ △ ♇
23 52 ☽ # ☿

JANUARY 2048

Date	Sidereal time h m s	Sun ☉ ° ' "	Moon ☽ ° ' "	Moon ☽ 24.00 ° ' "	Mercury ☿ ° '	Venus ♀ ° '	Mars ♂ ° '	Jupiter ♃ ° '	Saturn ♄ ° '	Uranus ♅ ° '	Neptune ♆ ° '	Pluto ♇ ° '
01	18 43 19	10 ♑ 44 01	13 ♋ 05 52	19 ♋ 15 22	25 ♑ 32	04 ♐ 56	02 ♏ 39	15 ♉ 48	07 ♑ 26	11 ♍ 04	19 ♉ 01	04 ♓ 56
02	18 47 16	11 45 09	25 ♋ 27 37	01 ♌ 42 46	27 05	06 10	03 13	15 R 46	07 33	11 R 03	19 R 01	04 58
03	18 51 12	12 46 18	08 ♌ 00 57	14 ♌ 22 17	28 ♑ 37	07 23	03 48	15 45	07 40	11 02	19 00	04 59
04	18 55 09	13 47 26	20 ♌ 46 51	27 ♌ 14 47	00 ♒ 08	08 37	04 22	15 44	07 47	11 01	18 59	05 00
05	18 59 06	14 48 34	03 ♍ 46 11	10 ♍ 21 09	01 38	09 50	04 56	15 42	07 54	11 00	18 58	05 01
06	19 03 02	15 49 43	16 ♍ 59 46	23 ♍ 42 07	03 06	11 04	05 30	15 42	08 01	10 59	18 57	05 02
07	19 06 59	16 50 51	00 ♎ 28 18	07 ♎ 18 20	04 32	12 17	06 04	15 41	08 08	10 57	18 57	05 03
08	19 10 55	17 52 00	14 ♎ 12 16	21 ♎ 10 04	05 55	13 31	06 38	15 41	08 15	10 56	18 56	05 04
09	19 14 52	18 53 09	28 ♎ 11 40	05 ♏ 16 55	07 16	14 45	07 12	15 40	08 23	10 55	18 55	05 06
10	19 18 48	19 54 18	12 ♏ 25 37	19 ♏ 37 29	08 33	15 58	07 46	15 D 40	08 30	10 54	18 55	05 07
11	19 22 45	20 55 27	26 ♏ 52 07	04 ♐ 09 02	09 46	17 12	08 19	15 40	08 37	10 52	18 54	05 08
12	19 26 41	21 56 37	11 ♐ 27 42	18 ♐ 47 26	10 54	18 26	08 53	15 41	08 44	10 51	18 53	05 09
13	19 30 38	22 57 46	26 ♐ 07 32	03 ♑ 27 11	11 57	19 40	09 27	15 41	08 50	10 49	18 53	05 11
14	19 34 34	23 58 55	10 ♑ 45 37	18 ♑ 01 58	12 53	20 54	10 00	15 42	08 57	10 48	18 52	05 12
15	19 38 31	25 00 04	25 ♑ 15 28	02 ♒ 25 22	13 43	22 08	10 33	15 43	09 04	10 46	18 52	05 13
16	19 42 28	26 01 12	09 ♒ 30 58	16 ♒ 31 43	14 24	23 22	11 07	15 45	09 11	10 45	18 51	05 15
17	19 46 24	27 02 20	23 ♒ 27 08	00 ♓ 16 52	14 56	24 35	11 40	15 46	09 18	10 43	18 51	05 16
18	19 50 21	28 03 27	07 ♓ 00 44	13 ♓ 38 39	15 19	25 49	12 13	15 48	09 25	10 42	18 51	05 17
19	19 54 17	29 ♑ 04 33	20 ♓ 10 39	26 ♓ 36 56	15 31	27 03	12 46	15 49	09 32	10 40	18 50	05 19
20	19 58 14	00 ♒ 05 38	02 ♈ 57 44	09 ♈ 13 25	15 R 31	28 17	13 19	15 52	09 39	10 38	18 50	05 20
21	20 02 10	01 06 43	15 ♈ 24 28	21 ♈ 31 20	15 21	29 ♐ 31	13 52	15 54	09 46	10 36	18 50	05 22
22	20 06 07	02 07 47	27 ♈ 34 37	03 ♉ 34 52	14 58	00 ♑ 46	14 25	15 56	09 52	10 34	18 49	05 23
23	20 10 03	03 08 50	09 ♉ 32 44	15 ♉ 28 49	14 25	02 00	14 57	15 59	09 59	10 32	18 49	05 25
24	20 14 00	04 09 51	21 ♉ 23 47	27 ♉ 18 13	13 40	03 14	15 30	16 02	10 06	10 31	18 49	05 26
25	20 17 57	05 10 52	03 ♊ 12 45	09 ♊ 07 59	12 46	04 28	16 02	16 05	10 12	10 29	18 49	05 27
26	20 21 53	06 11 52	15 ♊ 04 27	21 ♊ 02 42	11 43	05 42	16 35	16 08	10 19	10 27	18 49	05 29
27	20 25 50	07 12 51	27 ♊ 03 13	03 ♋ 06 25	10 34	06 56	17 07	16 11	10 26	10 25	18 49	05 30
28	20 29 46	08 13 49	09 ♋ 12 42	15 ♋ 22 22	09 21	08 10	17 39	16 15	10 32	10 23	18 49	05 32
29	20 33 43	09 14 46	21 ♋ 35 41	27 ♋ 52 51	08 06	09 24	18 11	16 19	10 39	10 20	18 D 49	05 33
30	20 37 39	10 15 41	04 ♌ 13 58	10 ♌ 39 07	06 50	10 39	18 43	16 23	10 45	10 18	18 49	05 35
31	20 41 36	11 ♒ 16 36	17 ♌ 08 15	23 ♌ 41 18	05 ♒ 37	11 ♑ 53	19 ♏ 15	16 ♉ 27	10 ♑ 52	10 ♍ 16	18 ♉ 49	05 ♓ 37

Date	Moon True ☊ ° '	Moon Mean ☊ ° '	Moon ☽ Latitude ° '
01	06 ♑ 48	06 ♑ 40	00 S 35
02	06 R 47	06 36	01 42
03	06 47	06 33	02 45
04	06 46	06 30	03 41
05	06 45	06 27	04 26
06	06 44	06 24	04 59
07	06 43	06 21	05 15
08	06 43	06 17	05 14
09	06 D 43	06 14	04 55
10	06 43	06 11	04 18
11	06 44	06 08	03 23
12	06 46	06 05	02 16
13	06 46	06 01	00 S 59
14	06 R 47	05 58	00 N 22
15	06 46	05 55	01 40
16	06 44	05 52	02 51
17	06 41	05 49	03 50
18	06 38	05 46	04 34
19	06 35	05 42	05 02
20	06 32	05 39	05 14
21	06 30	05 36	05 11
22	06 29	05 33	04 54
23	06 D 29	05 30	04 24
24	06 30	05 26	03 43
25	06 32	05 23	02 53
26	06 33	05 20	01 55
27	06 35	05 17	00 N 52
28	06 R 35	05 14	00 S 14
29	06 34	05 11	01 21
30	06 31	05 07	02 25
31	06 ♑ 27	05 ♑ 04	03 S 23

DECLINATIONS

Date	Sun ☉ ° '	Moon ☽ ° '	Mercury ☿ ° '	Venus ♀ ° '	Mars ♂ ° '	Jupiter ♃ ° '	Saturn ♄ ° '	Uranus ♅ ° '	Neptune ♆ ° '	Pluto ♇ ° '
01	23 S 00	22 N 13	23 S 05	19 S 20	11 S 07	15 N 32	22 S 34	08 N 09	15 N 45	20 S 58
02	22 55	19 22	22 44	19 35	11 19	15 32	22 34	08 09	15 45	20 58
03	22 49	15 36	22 21	19 49	11 30	15 32	22 33	08 10	15 45	20 57
04	22 43	11 04	21 58	20 03	11 42	15 32	22 33	08 10	15 45	20 57
05	22 37	05 59	21 33	20 16	11 53	15 32	22 33	08 10	15 45	20 56
06	22 30	00 N 33	21 07	20 29	12 05	15 32	22 32	08 11	15 44	20 55
07	22 22	05 S 00	20 40	20 41	12 16	15 32	22 32	08 11	15 44	20 55
08	22 14	10 26	20 12	20 52	12 27	15 32	22 31	08 12	15 44	20 54
09	22 06	15 25	19 44	21 03	12 39	15 32	22 31	08 12	15 44	20 53
10	21 57	19 39	19 15	21 14	12 50	15 33	22 31	08 13	15 44	20 53
11	21 48	22 45	18 45	21 23	13 01	15 33	22 30	08 14	15 44	20 52
12	21 39	24 23	18 15	21 32	13 11	15 33	22 30	08 14	15 44	20 52
13	21 29	24 21	17 46	21 41	13 22	15 34	22 29	08 15	15 44	20 51
14	21 18	22 38	17 17	21 49	13 33	15 34	22 29	08 15	15 43	20 50
15	21 08	19 26	16 49	21 56	13 43	15 35	22 29	08 16	15 43	20 50
16	20 56	15 07	16 23	22 03	13 54	15 36	22 28	08 17	15 43	20 49
17	20 45	10 05	15 58	22 09	14 04	15 36	22 28	08 17	15 43	20 48
18	20 33	04 S 41	15 35	22 14	14 15	15 37	22 27	08 18	15 43	20 48
19	20 20	00 N 45	15 14	22 19	14 25	15 38	22 27	08 19	15 43	20 47
20	20 08	05 59	14 57	22 23	14 35	15 39	22 27	08 19	15 43	20 46
21	19 54	10 51	14 42	22 26	14 45	15 40	22 26	08 20	15 43	20 46
22	19 41	15 10	14 31	22 29	14 55	15 41	22 26	08 21	15 43	20 45
23	19 27	18 50	14 24	22 31	15 04	15 42	22 25	08 21	15 43	20 44
24	19 13	21 42	14 20	22 32	15 14	15 43	22 25	08 22	15 43	20 44
25	18 58	23 37	14 21	22 33	15 24	15 44	22 24	08 23	15 43	20 43
26	18 43	24 31	14 24	22 33	15 33	15 45	22 24	08 24	15 43	20 42
27	18 28	24 16	14 31	22 32	15 42	15 46	22 23	08 25	15 43	20 42
28	18 12	22 52	14 41	22 31	15 52	15 48	22 23	08 25	15 43	20 41
29	17 56	20 22	14 53	22 29	16 01	15 49	22 22	08 26	15 43	20 40
30	17 40	16 50	15 06	22 27	16 10	15 50	22 22	08 27	15 43	20 40
31	17 S 23	12 N 28	15 S 22	22 S 23	16 S 19	15 N 52	22 S 21	08 N 28	15 N 44	20 S 39

ZODIAC SIGN ENTRIES

Date	h	m	Planets
02	20	43	☽ ♌
04	09	54	☿ ♒
05	05	05	☽ ♍
07	11	10	☽ ♎
09	15	04	☽ ♏
11	17	10	☽ ♐
13	18	21	☽ ♑
15	19	56	☽ ♒
17	23	30	☽ ♓
20	06	23	☽ ♈
20	09	47	☉ ♒
21	21	15	♀ ♑
22	16	50	☽ ♉
25	05	29	☽ ♊
27	17	51	☽ ♋
30	04	01	☽ ♌

LATITUDES

Date	Mercury ☿ ° '	Venus ♀ ° '	Mars ♂ ° '	Jupiter ♃ ° '	Saturn ♄ ° '	Uranus ♅ ° '	Neptune ♆ ° '	Pluto ♇ ° '
01	02 S 05	01 N 48	01 N 21	01 S 04	00 N 40	00 N 47	01 S 47	12 S 10
04	01 53	01 42	01 21	01 03	00 39	00 48	01 47	12 09
07	01 35	01 36	01 21	01 02	00 39	00 48	01 47	12 08
10	01 10	01 29	01 20	01 02	00 39	00 48	01 47	12 08
13	00 S 35	01 21	01 20	01 01	00 39	00 48	01 46	12 07
16	00 N 08	01 13	01 19	01 00	00 39	00 48	01 46	12 07
19	00 59	01 05	01 19	00 59	00 39	00 48	01 46	12 06
22	01 54	00 57	01 18	00 58	00 38	00 48	01 46	12 06
25	02 45	00 48	01 18	00 57	00 38	00 48	01 46	12 06
28	03 21	00 40	01 17	00 56	00 38	00 48	01 46	12 05
31	03 N 36	00 N 31	01 N 16	00 S 56	00 N 38	00 N 48	01 S 45	12 S 05

DATA

Julian Date	2469077
Delta T	+80 seconds
Ayanamsa	24° 31' 57"
Synetic vernal point	04° ♓ 35' 01"
True obliquity of ecliptic	23° 26' 00"

MOON'S PHASES, APSIDES AND POSITIONS ☽

Date	h	m	Phase	Longitude ° '	Eclipse Indicator
01	06	57	○	10 ♋ 31	total
08	18	49	☾	18 ♎ 09	
15	11	32	●	24 ♑ 59	
22	21	56	☽	02 ♉ 33	
31	00	14	○	10 ♌ 47	

Day	h	m		
13	03	43	Perigee	
24	21	23	Apogee	
06	14	23	0S	
12	23	30	Max dec	24° S 35'
19	08	42	0N	
26	19	00	Max dec	24° N 33'

LONGITUDES

Date	Chiron ⚷ ° '	Ceres ⚳ ° '	Pallas ⚴ ° '	Juno ⚵ ° '	Vesta ⚶ ° '	Black Moon Lilith ⚸ ° '
01	09 ♏ 06	08 ♒ 09	10 ♑ 02	09 ♑ 41	12 ♒ 25	26 ♉ 36
11	09 ♏ 58	11 ♒ 58	13 ♑ 59	13 ♑ 25	17 ♒ 23	27 ♉ 43
21	10 ♏ 40	15 ♒ 51	17 ♑ 52	17 ♑ 11	22 ♒ 22	28 ♉ 49
31	11 ♏ 12	19 ♒ 46	21 ♑ 42	20 ♑ 56	27 ♒ 22	29 ♉ 56

All ephemeris data is given at 12.00 UT and the Moon's longitude is additionally given for 24.00 UT

ASPECTARIAN

h m	Aspects
01 Wednesday	
00 47	☽ ☍ ♄
02 38	☽ # ☉
06 57	☽ ☍ ☉
07 18	☽ ± ♀
08 01	☽ ✶ ♅
08 09	☽ # ♄
12 05	♀ □ ♇
17 16	☽ ✶ ♃
19 05	☉ ∥ ☿
19 34	☉ △ ♅
20 10	☿ Ꝑ ♅
23 32	☽ ✶ ♆
23 34	☽ # ♇
02 Thursday	
01 21	☽ Ꝑ ♇
02 47	☽ Ꝑ ♀
10 34	☽ # ♀
13 07	☽ ∠ ♅
15 33	☽ ☍ ☿
16 26	☽ Q ♃
18 44	☽ ± ♇
22 39	☽ Q ♆
23 13	☿ ∥ ♄
03 Friday	
03 35	☽ □ ♂
06 13	☽ ⊼ ♇
06 20	☽ ⊥ ♅
10 40	☽ △ ♀
11 10	☽ ∥ ♆
11 20	☽ ⊼ ♄
12 24	☽ ∥ ♃
17 42	☽ ⊻ ♅
17 47	☿ ⊥ ♇
18 12	♀ ⊻ ♄
21 46	☽ ⊼ ☉
22 47	☽ ± ♄
04 Saturday	
02 34	☽ □ ♃
08 39	☽ □ ♆
09 00	☽ # ♂
10 00	☽ ± ☉
15 05	☽ Q ♂
15 46	☽ Ꝑ ♄
05 Sunday	
01 56	☽ ∥ ♅
04 07	☽ Ꝑ ☉
07 34	☽ ⊼ ☿
14 13	☽ ✶ ♂
14 17	☽ ☍ ♇
15 38	♂ △ ♇
19 37	☽ △ ♄
19 56	☽ ± ☿
06 Monday	
00 11	☽ □ ♀
01 09	☽ ☌ ♅
03 05	☉ ∥ ♄
08 52	☉ △ ♃
09 40	☽ △ ♃
09 44	☽ △ ☉
10 23	♀ □ ♅
14 13	☽ Ꝑ ☿
15 31	☽ △ ♆
18 34	☽ ∠ ♂
22 54	☿ ∥ ♇
07 Tuesday	
03 59	♀ ⊥ ♂
11 15	☽ ⊥ ♂
11 39	☽ Q ♀
11 39	☿ ∥ ♀
12 22	☽ Ꝑ ♃
18 06	☽ Ꝑ ♆
19 15	☿ ± ♅
19 57	☽ △ ☿
20 04	☽ ⊼ ♇
21 09	☿ ⊻ ♇
22 15	☽ ⊻ ♂
08 Wednesday	
01 34	☽ □ ♄
01 58	☽ # ♅
04 08	☽ ± ♃
06 20	☽ ⊻ ♅
06 34	☽ ± ♇
09 48	☽ ± ♆
10 41	☽ ✶ ♀
14 33	☽ ⊼ ♃
15 43	♀ ∥ ♇
16 43	☽ ⊥ ♅
18 49	☽ □ ☉
20 09	☽ ⊼ ♆
21 48	☽ ∥ ♂
22 08	☽ Ꝑ ♇
09 Thursday	
08 08	☽ ∠ ♅
08 53	☽ Q ♄
09 59	☿ □ ♂
12 38	☽ # ♃
12 47	☉ △ ♆
13 38	☽ # ♆
14 53	☽ ∠ ♀
23 42	☽ △ ♇
10 Friday	
02 18	♃ St D
03 50	☽ Q ☉
03 51	☽ ☌ ♂
04 22	☉ Q ♂
04 52	☽ □ ☿
05 21	☽ ✶ ♄
06 07	♀ ⊼ ♃
07 30	☽ ⊥ ♀
09 26	☽ ✶ ♅
09 43	☽ ∥ ☿
10 53	☿ ⊻ ♄
17 03	☉ ∠ ♇
17 25	☽ ☍ ♃
18 28	☽ ⊻ ♀
20 22	☽ ∥ ♇
22 48	☽ ☍ ♆
23 33	☽ ∥ ♀
11 Saturday	
01 25	☽ ✶ ☉
03 57	☽ ∥ ☉
05 24	☽ Q ♅
06 34	☽ ∠ ♄
09 38	☽ ∥ ♄
13 36	☽ Q ☿
21 32	☽ ⊥ ♄
12 Sunday	
01 38	☽ □ ♇
03 30	♂ ✶ ♄
04 02	☽ ∠ ☉
07 29	☽ ⊻ ♄
07 36	☽ ⊻ ♂
10 54	☿ ⊼ ♅
11 00	☽ □ ♅
11 00	☽ ✶ ☿
17 50	☽ ⊥ ♂
18 55	☽ ⊼ ♃
19 53	☽ ⊥ ☉
20 16	☉ ∥ ♀
20 50	♀ ⊼ ♆
13 Monday	
00 09	☽ ⊼ ♆
00 28	☽ ☌ ♀
04 44	☽ ± ♃
06 26	☽ ⊻ ☉
07 10	☽ Q ♇
09 08	☽ ∠ ♂
09 58	☽ ± ♆
13 26	☽ ∠ ☿
19 29	☽ Ꝑ ♃
14 Tuesday	
00 42	☽ Ꝑ ♆
02 51	☽ ✶ ♇
05 13	☽ ⊥ ☿
09 01	☽ ☌ ♄
10 42	☽ ✶ ♂
12 04	☽ △ ♅
13 23	☽ ∥ ♄
15 44	☽ ⊻ ☿
18 54	☽ ∥ ♀
20 10	☽ △ ♃
23 51	☽ ∥ ☉
15 Wednesday	
01 23	☽ △ ♆
02 40	☽ ∥ ♇
03 37	☽ ∠ ♇
04 00	♀ ± ♃
06 18	☽ ⊻ ♀
07 19	☽ Q ♂
11 32	☽ ☌ ☉
12 52	☽ Ꝑ ♅
17 14	☽ ⊥ ♀
18 38	☽ ⊥ ♇
20 57	♂ ✶ ♅
16 Thursday	
03 56	☽ ± ♅
04 45	☽ ⊻ ♇
04 47	☽ ∥ ☿
05 44	☉ Ꝑ ♅
08 53	☽ # ♆
09 34	☽ # ♃
09 46	♀ Q ♇
09 51	☽ ∠ ♀
11 26	☽ ⊻ ♄
14 06	☽ ⊼ ♅
14 50	☽ □ ♂
17 50	☽ ∥ ♂
20 42	☽ ☌ ☿
21 47	☽ ⊥ ♄
22 40	☽ □ ♃
17 Friday	
04 00	☽ □ ♆
04 06	☉ ∥ ♇
13 30	☽ ∠ ♄
14 11	☽ ✶ ♀
17 00	♀ ± ♆
18 47	☽ ⊻ ☉
20 04	☽ # ♅
18 Saturday	
02 41	☿ # ♆
06 14	☽ Q ♃
06 17	☽ ⊥ ☉
08 55	☽ ☌ ♇
09 27	☿ # ♃
11 42	☽ Q ♆
13 36	☽ Q ♀
16 22	☽ ✶ ♄
18 37	☽ ☍ ♅
21 49	☽ △ ♂
23 51	☽ ∠ ☉
19 Sunday	
03 19	☽ ⊻ ☿
03 58	☽ ✶ ♃
09 31	☽ ✶ ♆
14 29	☽ ⊥ ☿
14 32	☽ Q ♄
17 45	☉ ⊥ ♇
20 Monday	
01 47	☿ St R
02 12	☽ □ ♀
02 48	☽ Ꝑ ♂
06 05	☽ ✶ ☉
07 24	☽ ∠ ☿
08 00	☽ ∠ ♃
12 54	♀ ∠ ♂
13 39	☽ ∠ ♆
16 33	☽ ⊻ ♇
20 43	☽ ± ♂
23 18	☽ ∥ ♅
21 Tuesday	
00 56	☽ □ ♄
01 16	☽ ⊥ ♃
02 41	☽ ⊼ ♅
04 07	☽ ⊥ ♇
06 59	☽ ⊥ ♆
07 08	☽ Q ☉
08 51	☽ ⊼ ♂
09 15	☿ ∥ ♂
11 32	♀ ∥ ♄
11 53	☽ ✶ ☿
12 57	☽ ⊻ ♃
14 20	☽ ± ♅
18 42	☽ ⊻ ♆
21 44	☽ ∠ ♇
22 Wednesday	
00 37	☿ ∠ ♀
08 02	☽ Ꝑ ♅
08 18	☽ # ☿
10 22	☽ # ♂
10 51	☽ Q ☿
15 03	☽ ∥ ♃
15 18	☽ ∥ ♆
15 35	♀ Ꝑ ♃
19 05	☽ △ ♀
21 56	☽ □ ☉
23 Thursday	
00 42	☿ □ ♂
03 39	☽ ✶ ♇
12 54	☽ △ ♄
14 00	☽ △ ♅
16 16	☽ # ☉
21 17	☽ □ ☿
23 28	☽ ☍ ♂
24 Friday	
01 04	☽ ☌ ♃
03 05	☽ # ♇
03 56	☽ Q ♇
04 49	☽ Ꝑ ♀
06 46	☽ ☌ ♆
19 34	☽ # ♄
19 35	☽ Ꝑ ♄
19 54	☉ ± ♅
21 07	☽ # ♀
21 32	♃ ∥ ♆
23 24	♀ Ꝑ ♆
25 Saturday	
01 14	☽ ± ♀
14 02	☽ ± ♄
14 02	♂ ☍ ♃
14 50	☽ ⊼ ♀
16 22	☽ △ ☉
16 34	☽ □ ♇
18 41	☉ ⊻ ♇
26 Sunday	
02 18	☽ ⊼ ♄
02 41	☽ □ ♅
05 48	☽ △ ☿
07 42	♀ ✶ ♇
12 13	☿ ⊥ ♀
14 09	☽ ⊻ ♃
15 10	☽ ⊼ ♂
19 31	☽ ⊻ ♆
27 Monday	
01 27	☽ Ꝑ ☉
02 15	☽ ⊥ ♃
03 46	☽ ± ♂
07 32	☽ ⊥ ♆
09 06	♄ △ ♅
09 19	☽ Ꝑ ☿
14 02	♂ # ♆
14 37	☿ ⊻ ♄
14 41	☽ Q ♅
15 18	☿ ⊼ ♅
20 15	☽ ∠ ♃
21 01	☽ ± ☉
22 30	☽ Ꝑ ♂
23 33	♂ # ♃
28 Tuesday	
01 23	☽ ∠ ♆
01 32	☽ ± ☿
04 46	☽ △ ♇
09 44	☽ ☍ ♀
09 54	☽ ⊼ ☉
12 15	☽ ⊼ ☿
14 16	☽ ✶ ♅
14 37	☽ ☍ ♄
16 11	☽ # ♀
17 40	☽ # ♄
18 28	☉ ⊻ ♀
23 21	☿ ⊻ ♀
23 50	☉ ☌ ☿
29 Wednesday	
01 25	♆ St D
01 46	☽ ✶ ♃
05 09	☽ △ ♂
06 39	☽ ✶ ♆
09 30	☽ # ♇
10 00	☽ Ꝑ ♇
19 09	☽ ∠ ♅
30 Thursday	
00 53	☽ Q ♃
03 13	☽ ± ♇
05 33	☽ Q ♆
05 37	♀ △ ♅
06 31	☽ # ☉
12 08	☽ ⊥ ♅
12 59	☉ ⊼ ♅
14 21	♀ ☌ ♄
14 32	☽ ⊼ ♇
15 52	☽ # ♂
16 14	♂ ☍ ♆
16 27	☽ ☍ ☿
17 51	☽ ∥ ♃
18 33	☽ ∥ ♆
21 28	☽ # ☿
23 19	☽ ⊻ ♅
31 Friday	
00 14	☽ ☍ ☉
00 18	☽ ⊼ ♄
01 02	☉ ⊻ ♄
01 15	☽ ⊼ ♀
10 44	☽ □ ♃
11 29	☽ ± ♄
12 15	☿ ⊻ ♇
13 31	☽ ± ♀
15 05	☽ □ ♆
16 03	☽ □ ♂

FEBRUARY 2048

Date	Sidereal time h m s	Sun ☉ ° ' "	Moon ☽ ° ' "	Moon ☽ 24.00 ° ' "	Mercury ☿ ° '	Venus ♀ ° '	Mars ♂ ° '	Jupiter ♃ ° '	Saturn ♄ ° '	Uranus ♅ ° '	Neptune ♆ ° '	Pluto ♇ ° '
01	20 45 32	12 ♒ 17 30	00 ♍ 18 08	06 ♍ 58 33	04 ♒ 28	13 ♑ 07	19 ♏ 47	16 ♉ 31	10 ♑ 58	10 ♍ 14	18 ♉ 49	05 ♓ 38
02	20 49 29	13 18 23	13 ♍ 42 21	20 ♍ 29 15	03 R 25	14 21	20 18	16 36	11 05	10 R 12	18 49	05 40
03	20 53 26	14 19 15	27 ♍ 19 00	04 ♎ 11 18	02 29	15 36	20 50	16 41	11 11	10 09	18 49	05 41
04	20 57 22	15 20 05	11 ♎ 05 54	18 ♎ 02 32	01 40	16 50	21 21	16 46	11 17	10 07	18 49	05 43
05	21 01 19	16 20 56	25 ♎ 00 57	02 ♏ 00 56	01 00	18 04	21 52	16 51	11 23	10 05	18 50	05 44
06	21 05 15	17 21 45	09 ♏ 02 19	16 ♏ 04 53	00 28	19 18	22 23	16 56	11 30	10 03	18 50	05 46
07	21 09 12	18 22 33	23 ♏ 08 29	00 ♐ 12 57	00 ♒ 05	20 33	22 54	17 01	11 36	10 00	18 50	05 48
08	21 13 08	19 23 21	07 ♐ 18 08	14 ♐ 23 49	29 ♑ 50	21 47	23 25	17 07	11 42	09 58	18 50	05 49
09	21 17 05	20 24 08	21 ♐ 29 49	28 ♐ 35 51	29 42	23 01	23 56	17 13	11 48	09 55	18 51	05 51
10	21 21 01	21 24 54	05 ♑ 41 38	12 ♑ 46 49	29 D 43	24 16	24 27	17 19	11 54	09 53	18 51	05 52
11	21 24 58	22 25 38	19 ♑ 51 00	26 ♑ 53 45	29 ♑ 50	25 30	24 57	17 25	12 00	09 50	18 52	05 54
12	21 28 55	23 26 22	03 ♒ 54 36	10 ♒ 53 04	00 ♒ 04	26 44	25 27	17 31	12 06	09 48	18 52	05 56
13	21 32 51	24 27 04	17 ♒ 48 37	24 ♒ 40 49	00 25	27 59	25 58	17 38	12 12	09 45	18 53	05 57
14	21 36 48	25 27 45	01 ♓ 29 13	08 ♓ 13 26	00 51	29 ♑ 13	26 28	17 44	12 18	09 43	18 53	05 59
15	21 40 44	26 28 24	14 ♓ 53 08	21 ♓ 28 07	01 22	00 ♒ 28	26 57	17 51	12 24	09 40	18 54	06 01
16	21 44 41	27 29 02	27 ♓ 58 15	04 ♈ 23 30	01 58	01 42	27 27	17 58	12 29	09 38	18 54	06 02
17	21 48 37	28 29 38	10 ♈ 43 56	16 ♈ 59 44	02 38	02 56	27 57	18 05	12 35	09 35	18 55	06 04
18	21 52 34	29 ♒ 30 13	23 ♈ 11 10	29 ♈ 18 34	03 23	04 11	28 26	18 13	12 40	09 33	18 56	06 06
19	21 56 30	00 ♓ 30 45	05 ♉ 22 24	11 ♉ 23 08	04 11	05 25	28 55	18 20	12 46	09 30	18 57	06 07
20	22 00 27	01 31 17	17 ♉ 21 20	23 ♉ 17 37	05 03	06 40	29 24	18 28	12 52	09 28	18 57	06 09
21	22 04 24	02 31 46	29 ♉ 12 36	05 ♊ 06 56	05 58	07 54	29 ♏ 53	18 35	12 57	09 25	18 58	06 10
22	22 08 20	03 32 13	11 ♊ 01 20	16 ♊ 56 27	06 56	09 08	00 ♐ 22	18 43	13 02	09 22	18 59	06 12
23	22 12 17	04 32 39	22 ♊ 52 58	28 ♊ 51 33	07 56	10 23	00 50	18 51	13 08	09 20	19 00	06 14
24	22 16 13	05 33 02	04 ♋ 52 50	10 ♋ 57 25	09 00	11 37	01 18	18 59	13 13	09 17	19 01	06 15
25	22 20 10	06 33 24	17 ♋ 05 50	23 ♋ 18 35	10 05	12 51	01 47	19 08	13 18	09 15	19 02	06 17
26	22 24 06	07 33 44	29 ♋ 36 04	05 ♌ 58 37	11 13	14 06	02 15	19 16	13 23	09 12	19 03	06 19
27	22 28 03	08 34 02	12 ♌ 26 26	18 ♌ 59 38	12 23	15 20	02 42	19 25	13 28	09 09	19 03	06 20
28	22 31 59	09 34 18	25 ♌ 38 13	02 ♍ 22 01	13 34	16 35	03 10	19 33	13 33	09 07	19 05	06 22
29	22 35 56	10 ♓ 34 33	09 ♍ 10 48	16 ♍ 04 11	14 ♒ 48	17 ♒ 49	03 ♐ 37	19 ♉ 42	13 ♑ 38	09 ♍ 04	19 ♉ 06	06 ♓ 24

Date	Moon True ☊ ° '	Moon Mean ☊ ° '	Moon ☽ Latitude ° '
01	06 ♑ 21	05 ♑ 01	04 S 12
02	06 R 15	04 58	04 47
03	06 08	04 55	05 07
04	06 03	04 52	05 09
05	05 59	04 48	04 53
06	05 57	04 45	04 20
07	05 D 57	04 42	03 31
08	05 58	04 39	02 29
09	05 59	04 36	01 18
10	06 00	04 32	00 S 02
11	05 R 59	04 29	01 N 14
12	05 56	04 26	02 25
13	05 51	04 23	03 26
14	05 43	04 20	04 14
15	05 34	04 17	04 47
16	05 24	04 13	05 04
17	05 15	04 10	05 05
18	05 08	04 07	04 51
19	05 02	04 04	04 25
20	04 59	04 01	03 47
21	04 57	03 58	03 00
22	04 D 58	03 54	02 05
23	04 58	03 51	01 04
24	04 59	03 48	00 N 01
25	04 R 58	03 45	01 S 04
26	04 55	03 42	02 08
27	04 50	03 38	03 06
28	04 42	03 35	03 56
29	04 ♑ 32	03 ♑ 32	04 S 34

DECLINATIONS

Date	Sun ☉ ° '	Moon ☽ ° '	Mercury ☿ ° '	Venus ♀ ° '	Mars ♂ ° '	Jupiter ♃ ° '	Saturn ♄ ° '	Uranus ♅ ° '	Neptune ♆ ° '	Pluto ♇ ° '
01	17 S 06	07 N 26	15 S 38	22 S 19	16 S 28	15 N 53	22 S 21	08 N 29	15 N 44	20 S 39
02	16 49	01 N 59	15 54	22 15	16 36	15 55	22 20	08 30	15 44	20 38
03	16 32	03 S 37	16 11	22 09	16 45	15 57	22 20	08 30	15 44	20 37
04	16 14	09 08	16 27	22 03	16 53	15 58	22 19	08 31	15 44	20 37
05	15 56	14 14	16 43	21 57	17 02	16 00	22 19	08 32	15 44	20 36
06	15 38	18 37	16 58	21 49	17 10	16 02	22 18	08 33	15 44	20 35
07	15 19	21 58	17 13	21 41	17 18	16 03	22 18	08 34	15 44	20 35
08	15 00	23 59	17 26	21 33	17 26	16 05	22 17	08 35	15 44	20 34
09	14 41	24 27	17 39	21 23	17 34	16 07	22 17	08 36	15 45	20 33
10	14 22	23 20	17 50	21 13	17 42	16 09	22 16	08 37	15 45	20 33
11	14 02	20 45	18 00	21 03	17 50	16 11	22 16	08 38	15 45	20 32
12	13 42	16 55	18 09	20 52	17 57	16 13	22 15	08 39	15 45	20 31
13	13 22	12 13	18 16	20 40	18 05	16 15	22 15	08 40	15 45	20 31
14	13 02	06 59	18 22	20 27	18 12	16 17	22 14	08 41	15 46	20 30
15	12 41	01 S 32	18 27	20 14	18 20	16 19	22 14	08 42	15 46	20 29
16	12 21	03 N 50	18 31	20 01	18 27	16 21	22 13	08 43	15 46	20 29
17	12 00	08 55	18 33	19 47	18 34	16 24	22 13	08 43	15 46	20 28
18	11 39	13 31	18 34	19 32	18 41	16 26	22 13	08 44	15 47	20 28
19	11 17	17 28	18 34	19 16	18 48	16 28	22 12	08 45	15 47	20 27
20	10 56	20 38	18 32	19 01	18 54	16 31	22 12	08 46	15 47	20 26
21	10 34	22 54	18 29	18 44	19 01	16 33	22 11	08 47	15 47	20 26
22	10 12	24 09	18 25	18 27	19 08	16 35	22 11	08 48	15 48	20 25
23	09 51	24 19	18 19	18 10	19 14	16 38	22 10	08 49	15 48	20 24
24	09 28	23 21	18 12	17 52	19 20	16 40	22 10	08 50	15 48	20 24
25	09 06	21 17	18 04	17 33	19 27	16 43	22 09	08 51	15 48	20 23
26	08 44	18 09	17 54	17 14	19 33	16 45	22 09	08 52	15 49	20 23
27	08 21	14 05	17 43	16 54	19 39	16 48	22 08	08 53	15 49	20 22
28	07 59	09 16	17 31	16 34	19 45	16 50	22 08	08 54	15 49	20 21
29	07 S 36	03 N 54	17 S 17	16 S 14	19 S 50	16 N 53	22 S 07	08 N 55	15 N 50	20 S 21

ZODIAC SIGN ENTRIES

Date	h	m	Planets
01	11	27	☽ ♍
03	16	42	☽ ♎
05	20	33	☽ ♏
07	18	21	☿ ♑
07	23	38	☽ ♐
10	02	22	☽ ♑
12	05	18	☽ ♒
12	05	34	☿ ♒
14	09	22	☽ ♓
15	03	06	♀ ♒
16	15	47	☽ ♈
18	23	48	☉ ♓
19	01	22	☽ ♉
21	13	36	☽ ♊
21	17	49	♂ ♐
24	02	17	☽ ♋
26	12	45	☽ ♌
28	19	48	☽ ♍

LATITUDES

Date	Mercury ☿ ° '	Venus ♀ ° '	Mars ♂ ° '	Jupiter ♃ ° '	Saturn ♄ ° '	Uranus ♅ ° '	Neptune ♆ ° '	Pluto ♇ ° '
01	03 N 37	00 N 28	01 N 16	00 S 55	00 N 38	00 N 48	01 S 45	12 S 05
04	03 25	00 19	01 15	00 55	00 38	00 49	01 45	12 05
07	02 59	00 11	01 14	00 54	00 38	00 49	01 45	12 05
10	02 26	00 N 02	01 13	00 53	00 38	00 49	01 45	12 05
13	01 50	00 S 06	01 11	00 52	00 38	00 49	01 45	12 05
16	01 14	00 15	01 10	00 51	00 37	00 49	01 44	12 04
19	00 40	00 23	01 09	00 51	00 37	00 49	01 44	12 04
22	00 N 07	00 30	01 07	00 50	00 37	00 49	01 44	12 05
25	00 S 22	00 38	01 05	00 49	00 37	00 49	01 44	12 05
28	00 48	00 45	01 04	00 49	00 37	00 49	01 44	12 05
31	01 S 11	00 S 51	01 N 02	00 S 48	00 N 37	00 N 49	01 S 43	12 S 05

DATA

Julian Date	2469108
Delta T	+80 seconds
Ayanamsa	24° 32' 03"
Synetic vernal point	04° ♓ 34' 56"
True obliquity of ecliptic	23° 26' 00"

MOON'S PHASES, APSIDES AND POSITIONS ☽

Date	h	m	Phase	Longitude ° '	Eclipse Indicator
07	03	17	☾	18 ♏ 00	
14	00	31	●	24 ♒ 59	
21	19	22	☽	02 ♊ 50	
29	14	38	○	10 ♍ 41	

Day	h	m		
09	09	04	Perigee	
21	18	29	Apogee	
02	20	32	0S	
09	07	11	Max dec	24° S 29'
15	18	48	0N	
23	03	33	Max dec	24° N 23'

LONGITUDES

Date	Chiron ⚷ ° '	Ceres ⚳ ° '	Pallas ⚴ ° '	Juno ⚵ ° '	Vesta ⚶ ° '	Black Moon Lilith ⚸ ° '
01	11 ♏ 15	20 ♒ 10	22 ♑ 04	21 ♑ 18	27 ♒ 52	00 ♊ 03
11	11 ♏ 35	24 ♒ 06	25 ♑ 47	25 ♑ 02	02 ♓ 52	01 ♊ 09
21	11 ♏ 43	28 ♒ 02	29 ♑ 23	28 ♑ 42	07 ♓ 51	02 ♊ 16
31	11 ♏ 40	01 ♓ 57	02 ♒ 51	02 ♒ 20	12 ♓ 49	03 ♊ 23

All ephemeris data is given at 12.00 UT and the Moon's longitude is additionally given for 24.00 UT

ASPECTARIAN

h m Aspects

01 Saturday
00 43 ♀ ∥ ♄
04 05 ☽ Ꝑ ♄
07 14 ☽ ∥ ♅
07 39 ☽ Ꝑ ♀
17 29 ☿ ± ♅
18 57 ☽ ⊼ ☿
20 38 ☿ # ♆
21 37 ☽ ☍ ♇

02 Sunday
01 59 ☽ Q ♂
04 54 ☽ ± ☿
05 46 ☽ ☌ ♅
07 17 ☽ △ ♄
11 14 ☽ ⊼ ☉
13 03 ☿ # ♃
13 16 ☽ △ ♀
17 10 ☽ △ ♃
19 47 ☽ Ꝑ ☿
21 03 ☽ △ ♆
22 43 ☽ ± ☉
23 55 ☉ ∥ ♂

03 Monday
00 09 ☽ ⚹ ♂
06 09 ☿ Q ♂
15 47 ☽ Ꝑ ☉
19 40 ☽ Ꝑ ♃
20 30 ☽ △ ☿
23 22 ☽ Ꝑ ♆

04 Tuesday
02 38 ☽ ⊼ ♇
02 42 ☉ ∥ ☿
03 26 ☽ ∠ ♂
09 17 ☽ # ♅
10 18 ☽ ⊻ ♅
10 33 ♀ △ ♃
11 25 ☽ ± ♃
12 20 ☽ □ ♄
13 04 ☽ ± ♇
14 59 ☽ ± ♆
19 39 ☽ ⊥ ♂
19 54 ☽ △ ☉
20 40 ☽ ⊥ ♅
21 51 ☽ ⊼ ♃
22 53 ☽ □ ♀

05 Wednesday
01 21 ☽ ⊼ ♆
04 38 ☽ Ꝑ ♇
06 23 ☽ ⊻ ♂
07 17 ☉ # ♃
12 07 ☽ ∠ ♅
19 34 ☽ Q ♄
19 44 ☽ # ♆
20 14 ☽ ∥ ☉
20 59 ♂ Q ♅
21 11 ☽ # ♃
21 52 ☽ □ ☿

06 Thursday
00 52 ☉ □ ♃
01 51 ☽ ∥ ☿
02 44 ♀ △ ♆
03 15 ☽ ∥ ♂
03 30 ☉ # ♆
06 24 ☽ △ ♇
08 46 ☽ Q ♀
13 42 ☽ ⚹ ♅
15 29 ☉ ⊥ ♄
16 13 ☽ ⚹ ♄

07 Friday
01 03 ☽ ∥ ♇
01 32 ☽ ☍ ♃
03 17 ☽ □ ☉
03 36 ☽ Q ☿
04 41 ☽ ☍ ♆
07 10 ☽ ⚹ ♀
09 46 ☽ ∥ ♀
10 04 ☽ Q ♅
11 35 ☽ ☌ ♂
15 05 ☽ ∥ ♄
16 53 ♀ ∠ ♇
17 54 ☽ ∠ ♄
22 57 ☉ □ ♆
23 32 ☽ ⚹ ☿

08 Saturday
09 16 ☽ ⊥ ♄
09 29 ☽ □ ♇
11 02 ☽ ∠ ♀
12 10 ☽ Q ☉
12 55 ☿ ∥ ♂
13 00 ☿ ⊥ ♇
16 29 ☽ □ ♅
19 30 ☽ ⊻ ♄

09 Sunday
00 36 ☽ ∠ ☿
03 43 ☽ ⊥ ♀
04 43 ☽ ⊼ ♃
07 31 ☽ ⊼ ♆
10 00 ☽ ⚹ ☉
14 50 ☽ ⊻ ♀
14 55 ☽ ± ♃
15 43 ☽ ⊥ ☿
15 59 ☽ Q ♇
16 16 ☽ ⊻ ♂
17 40 ☽ ± ♆
22 45 ☿ St D

10 Monday
01 51 ☽ ⊻ ☿
02 47 ☽ ⊥ ♂
06 15 ☽ Ꝑ ♃
08 53 ☽ Ꝑ ♆
12 18 ☽ ⚹ ♇
13 19 ☽ ∠ ☉
17 58 ♀ ⚹ ♂
18 35 ☽ ∠ ♂
19 04 ☽ △ ♅
22 35 ☽ ☌ ♄
23 27 ☽ ∥ ♄
23 37 ♀ Ꝑ ♅

11 Tuesday
05 44 ☽ ⊥ ☉
07 50 ☽ △ ♃
09 35 ☽ ∥ ♀
10 19 ☽ △ ♆
13 32 ☽ ∥ ♇
13 47 ☽ ∠ ♇
16 43 ☽ ⊻ ☉
20 28 ☽ Ꝑ ♅
20 31 ☿ ⊥ ♇
21 00 ☽ ⚹ ♂
22 33 ☽ ☌ ♀

12 Wednesday
05 10 ☽ ⊥ ♇
05 14 ☽ ∥ ☿
05 18 ☽ ☌ ☿
06 16 ☽ ∥ ♂
11 49 ☽ ± ♅
15 28 ☽ ⊻ ♇
15 50 ☽ # ♃
18 19 ☽ Q ♂
18 20 ☽ # ♆
22 06 ☽ ⊼ ♅

13 Thursday
02 12 ☽ ⊻ ♄
06 02 ☽ ∥ ☉
11 41 ☽ □ ♃
12 41 ☽ ⊥ ♄
13 52 ☽ □ ♆

14 Friday
00 31 ☽ ☌ ☉
02 47 ☽ □ ♂
04 26 ☽ # ♅
04 33 ☽ ∠ ♄
06 42 ♀ ∥ ♇
07 35 ☽ ⊻ ♀
10 49 ☽ ⊻ ☿
19 18 ☽ ⊥ ♀
19 37 ☽ Q ♃
20 01 ☽ ☌ ♇
21 37 ☽ Q ♆
21 54 ☽ ⊥ ☿

15 Saturday
02 38 ☽ ☍ ♅
03 05 ♀ ⊥ ♇
07 28 ☽ ⚹ ♄
13 09 ☽ ∠ ♀
14 48 ☽ ∠ ☿
17 26 ☽ ⚹ ♃
19 18 ☽ ⚹ ♆

16 Sunday
05 30 ☽ Q ♄
10 30 ☉ □ ♂
11 00 ☽ △ ♂
11 01 ☽ ⊻ ☉
12 05 ☉ ∠ ♄
14 08 ♂ ∠ ♄
19 42 ☽ ⚹ ♀
19 51 ☽ ⚹ ☿
21 25 ☽ ∠ ♃
22 51 ☿ ☌ ♀
23 06 ☽ ∠ ♆
23 11 ☽ ⊥ ☉

17 Monday
03 08 ☽ ⊻ ♇
05 47 ♃ Q ♇
09 12 ☿ ∥ ♂
09 50 ☽ ⊼ ♅
11 03 ☽ ∥ ♅
14 33 ☽ ⊥ ♇
14 37 ☽ ⊥ ♃
15 33 ☽ □ ♄
16 11 ☽ ⊥ ♆
16 24 ☽ Ꝑ ♂
17 44 ☽ ∠ ☉
19 55 ☽ Q ☿
20 56 ☽ Q ♀
21 16 ☽ ± ♅

18 Tuesday
00 10 ♀ ± ♅
02 15 ☽ ⊻ ♃
02 35 ☽ # ☉
03 44 ☽ ⊻ ♆
07 55 ☽ ∠ ♇
10 28 ☽ ± ♂
14 39 ☽ Ꝑ ♅
16 50 ☿ ± ♅
22 42 ☽ ⊼ ♂

19 Wednesday
01 12 ☽ ∥ ♆
01 30 ☽ ⚹ ☉
05 27 ☽ ∥ ♃
09 28 ☽ □ ☿
12 06 ☽ □ ♀
13 29 ☽ ⚹ ♇
19 36 ☽ # ☿
20 12 ☽ △ ♅
21 39 ☽ # ♂
23 54 ☽ # ♀

20 Thursday
01 53 ♀ ⊻ ♇
02 53 ☽ △ ♄
03 34 ☽ Q ☉
10 18 ☽ # ♇
13 36 ☽ Q ♇
14 15 ☽ ☌ ♃
15 14 ☽ ☌ ♆
18 22 ♀ ∥ ♂

21 Friday
03 16 ☽ # ♄
09 25 ☽ Ꝑ ♄
13 26 ☽ ☍ ♂
17 31 ☿ ⊻ ♇
19 22 ☽ □ ☉

22 Saturday
02 11 ☽ □ ♇
02 56 ☽ △ ☿
03 51 ☽ ± ♄
07 43 ☽ △ ♀
08 40 ☽ □ ♅
16 07 ☽ ⊼ ♄
16 22 ☿ ∥ ♀
16 24 ♀ ⊼ ♅

23 Sunday
03 46 ☽ ⊻ ♃
04 09 ☽ ⊻ ♆
12 08 ☽ Ꝑ ☿
16 01 ☽ ⊥ ♃
16 15 ☽ ⊥ ♆
17 36 ☽ Ꝑ ♀
20 54 ☽ Q ♅

24 Monday
04 36 ☽ ⊼ ♂
07 53 ☽ ± ☿
10 12 ☽ ∠ ♃
10 16 ☽ ∠ ♆
13 27 ☽ △ ☉
13 38 ☽ ± ♀
14 44 ☽ △ ♇
16 23 ♃ ☌ ♆
17 00 ☽ ± ♂
18 16 ☿ ⊼ ♅
20 41 ☽ ⚹ ♅
20 56 ☽ ⊼ ☿

25 Tuesday
02 48 ☽ ⊼ ♀
03 18 ☽ # ♄
04 33 ☽ ☍ ♄
05 21 ☉ ☌ ♇
11 21 ☽ Ꝑ ♂
15 45 ☽ ⚹ ♆
15 59 ☽ ⚹ ♃
19 41 ☽ # ♇
20 08 ☽ Ꝑ ♇
21 12 ♀ ⊻ ♄
21 23 ☽ Ꝑ ☉
23 23 ☉ Q ♆

26 Wednesday
01 44 ☽ ∠ ♅
02 31 ☽ # ♂
03 14 ☉ # ♅
03 48 ☉ Q ♃
13 21 ☽ ± ♇
13 43 ☽ # ☿
14 44 ☽ Q ♆
15 11 ☽ Q ♃
16 02 ☽ ± ☉
16 29 ♀ Q ♂
17 11 ☽ △ ♂
18 27 ☽ # ♀
18 46 ☽ ⊥ ♅
20 46 ☽ ∥ ♃

27 Thursday
00 39 ☽ ⊼ ♇
02 24 ☽ ∥ ♆
04 13 ☽ ⊼ ☉
05 56 ☽ ⊻ ♅
11 52 ☽ ☍ ☿
13 54 ☽ ⊼ ♄
17 53 ☽ ☍ ♀
19 11 ♀ # ♃

28 Friday
00 08 ☽ □ ♆
00 54 ☽ □ ♃
00 57 ☽ ± ♄
01 27 ☉ ☍ ♅
11 31 ☿ ⊻ ♄
13 41 ☽ ∥ ♅
17 15 ☽ Ꝑ ♄
18 25 ☽ # ☉

29 Saturday
01 53 ☽ □ ♂
07 06 ☽ ☍ ♇
11 48 ☽ ☌ ♅
14 38 ☽ ☍ ☉
19 49 ☽ △ ♄
22 46 ☽ ⊼ ☿

MARCH 2048

LONGITUDES

Date	Sidereal time h m s	Sun ☉ ° ' "	Moon ☽ ° ' "	Moon ☽ 24.00 ° ' "	Mercury ☿ ° '	Venus ♀ ° '	Mars ♂ ° '	Jupiter ♃ ° '	Saturn ♄ ° '	Uranus ♅ ° '	Neptune ♆ ° '	Pluto ♇ ° '
01	22 39 53	11 ♓ 34 45	23 ♍ 01 41	00 ♎ 02 44	16 ♒ 04	19 ♒ 03	04 ♐ 04	19 ♉ 51	13 ♑ 43	09 ♍ 01	19 ♉ 07	06 ♓ 25
02	22 43 49	12 34 56	07 ♎ 06 43	14 ♎ 12 58	17 21	20 18	04 31	20 00	13 48	08 R 59	19 08	06 27
03	22 47 46	13 35 05	21 ♎ 20 48	28 ♎ 29 35	18 39	21 32	04 58	20 09	13 52	08 56	19 09	06 29
04	22 51 42	14 35 12	05 ♏ 38 43	12 ♏ 47 39	19 59	22 47	05 24	20 19	13 57	08 54	19 10	06 30
05	22 55 39	15 35 18	19 ♏ 55 55	27 ♏ 03 10	21 21	24 01	05 51	20 28	14 02	08 51	19 11	06 32
06	22 59 35	16 35 23	04 ♐ 09 06	11 ♐ 13 30	22 44	25 15	06 17	20 38	14 06	08 48	19 12	06 34
07	23 03 32	17 35 26	18 ♐ 16 14	25 ♐ 17 14	24 09	26 30	06 42	20 48	14 10	08 46	19 14	06 35
08	23 07 28	18 35 27	02 ♑ 16 26	09 ♑ 13 48	25 34	27 44	07 08	20 57	14 15	08 43	19 15	06 37
09	23 11 25	19 35 27	16 ♑ 09 19	23 ♑ 02 58	27 01	28 ♒ 59	07 33	21 07	14 19	08 41	19 16	06 39
10	23 15 22	20 35 25	29 ♑ 54 40	06 ♒ 44 21	28 ♒ 30	00 ♓ 13	07 58	21 17	14 23	08 38	19 18	06 40
11	23 19 18	21 35 22	13 ♒ 31 52	20 ♒ 17 06	00 ♓ 00	01 27	08 23	21 28	14 27	08 35	19 19	06 42
12	23 23 15	22 35 17	26 ♒ 59 50	03 ♓ 39 51	01 30	02 42	08 47	21 38	14 31	08 33	19 20	06 44
13	23 27 11	23 35 09	10 ♓ 16 57	16 ♓ 50 54	03 03	03 56	09 12	21 48	14 35	08 30	19 22	06 45
14	23 31 08	24 35 00	23 ♓ 21 30	29 ♓ 48 35	04 36	05 10	09 35	21 59	14 39	08 28	19 23	06 47
15	23 35 04	25 34 49	06 ♈ 12 01	12 ♈ 31 42	06 11	06 25	09 59	22 09	14 43	08 25	19 25	06 48
16	23 39 01	26 34 36	18 ♈ 47 39	24 ♈ 59 55	07 46	07 39	10 22	22 20	14 47	08 23	19 26	06 50
17	23 42 57	27 34 21	01 ♉ 08 39	07 ♉ 14 03	09 23	08 54	10 45	22 31	14 50	08 20	19 28	06 52
18	23 46 54	28 34 04	13 ♉ 16 25	19 ♉ 16 05	11 02	10 08	11 08	22 42	14 54	08 18	19 29	06 53
19	23 50 51	29 ♓ 33 45	25 ♉ 13 31	01 ♊ 09 12	12 41	11 22	11 30	22 53	14 57	08 16	19 31	06 55
20	23 54 47	00 ♈ 33 23	07 ♊ 03 40	12 ♊ 57 31	14 22	12 37	11 52	23 04	15 00	08 13	19 32	06 56
21	23 58 44	01 33 00	18 ♊ 51 23	24 ♊ 45 56	16 04	13 51	12 14	23 15	15 04	08 11	19 34	06 58
22	00 02 40	02 32 34	00 ♋ 41 52	06 ♋ 39 52	17 47	15 05	12 35	23 26	15 07	08 09	19 36	06 59
23	00 06 37	03 32 05	12 ♋ 40 39	18 ♋ 44 53	19 32	16 20	12 56	23 38	15 10	08 06	19 37	07 01
24	00 10 33	04 31 35	24 ♋ 53 15	01 ♌ 06 22	21 18	17 34	13 17	23 49	15 13	08 04	19 39	07 02
25	00 14 30	05 31 02	07 ♌ 24 47	13 ♌ 49 00	23 05	18 48	13 37	24 01	15 16	08 02	19 41	07 04
26	00 18 26	06 30 27	20 ♌ 19 23	26 ♌ 56 13	24 53	20 02	13 57	24 12	15 19	08 00	19 43	07 05
27	00 22 23	07 29 49	03 ♍ 39 36	10 ♍ 29 32	26 43	21 17	14 17	24 24	15 21	07 57	19 44	07 07
28	00 26 20	08 29 09	17 ♍ 25 49	24 ♍ 28 04	28 ♓ 34	22 31	14 36	24 36	15 24	07 55	19 46	07 08
29	00 30 16	09 28 28	01 ♎ 35 44	08 ♎ 48 09	00 ♈ 26	23 45	14 55	24 48	15 27	07 53	19 48	07 10
30	00 34 13	10 27 44	16 ♎ 04 27	23 ♎ 23 45	02 20	24 59	15 13	25 00	15 29	07 51	19 50	07 11
31	00 38 09	11 ♈ 26 58	00 ♏ 45 02	08 ♏ 07 17	04 ♈ 15	26 ♓ 14	15 ♐ 31	25 ♉ 12	15 ♑ 31	07 ♍ 49	19 ♉ 52	07 ♓ 13

Date	Moon True ☊ ° '	Moon Mean ☊ ° '	Moon ☽ Latitude ° '
01	04 ♑ 21	03 ♑ 29	04 S 57
02	04 R 09	03 26	05 02
03	04 00	03 23	04 49
04	03 52	03 19	04 18
05	03 47	03 16	03 30
06	03 45	03 13	02 30
07	03 D 44	03 10	01 21
08	03 44	03 07	00 S 08
09	03 R 44	03 04	01 N 05
10	03 42	03 00	02 14
11	03 37	02 57	03 14
12	03 29	02 54	04 02
13	03 18	02 51	04 37
14	03 05	02 48	04 56
15	02 51	02 44	05 00
16	02 38	02 41	04 49
17	02 26	02 38	04 24
18	02 17	02 35	03 48
19	02 11	02 32	03 02
20	02 07	02 29	02 08
21	02 05	02 25	01 09
22	02 D 05	02 22	00 N 07
23	02 R 05	02 19	00 S 56
24	02 04	02 16	01 58
25	02 01	02 13	02 55
26	01 55	02 10	03 46
27	01 47	02 06	04 26
28	01 36	02 03	04 52
29	01 25	02 00	05 01
30	01 13	01 57	04 51
31	01 ♑ 02	01 ♑ 54	04 S 22

DECLINATIONS

Date	Sun ☉ ° '	Moon ☽ ° '	Mercury ☿ ° '	Venus ♀ ° '	Mars ♂ ° '	Jupiter ♃ ° '	Saturn ♄ ° '	Uranus ♅ ° '	Neptune ♆ ° '	Pluto ♇ ° '
01	07 S 13	01 S 46	17 S 02	15 S 53	19 S 56	16 N 55	22 S 07	08 N 56	15 N 50	20 S 20
02	06 50	07 27	16 46	15 32	20 02	16 58	22 06	08 57	15 50	20 20
03	06 27	12 47	16 28	15 10	20 07	17 01	22 06	08 58	15 51	20 19
04	06 04	17 27	16 09	14 47	20 13	17 03	22 05	08 59	15 51	20 19
05	05 41	21 06	15 49	14 25	20 18	17 06	22 05	09 00	15 52	20 18
06	05 18	23 26	15 27	14 02	20 23	17 09	22 04	09 01	15 52	20 18
07	04 54	24 16	15 05	13 38	20 28	17 12	22 04	09 02	15 52	20 17
08	04 31	23 33	14 41	13 14	20 33	17 14	22 04	09 03	15 53	20 16
09	04 07	21 22	14 15	12 50	20 38	17 17	22 03	09 04	15 53	20 16
10	03 44	17 59	13 49	12 26	20 43	17 20	22 03	09 05	15 54	20 15
11	03 20	13 39	13 21	12 01	20 48	17 23	22 02	09 06	15 54	20 15
12	02 56	08 43	12 52	11 36	20 52	17 26	22 02	09 07	15 54	20 14
13	02 33	03 S 26	12 21	11 10	20 57	17 29	22 01	09 08	15 55	20 14
14	02 09	01 N 54	11 50	10 44	21 01	17 32	22 01	09 09	15 55	20 13
15	01 45	07 03	11 17	10 18	21 06	17 34	22 01	09 10	15 56	20 13
16	01 22	11 48	10 43	09 52	21 10	17 37	22 00	09 11	15 56	20 12
17	00 58	15 59	10 08	09 25	21 14	17 40	22 00	09 12	15 57	20 12
18	00 34	19 26	09 31	08 58	21 19	17 43	22 00	09 12	15 57	20 11
19	00 S 10	22 00	08 54	08 31	21 23	17 46	21 59	09 13	15 58	20 11
20	00 N 13	23 36	08 15	08 04	21 27	17 49	21 59	09 14	15 58	20 11
21	00 37	24 07	07 35	07 36	21 31	17 52	21 58	09 15	15 59	20 10
22	01 01	23 33	06 54	07 08	21 34	17 55	21 58	09 16	15 59	20 10
23	01 24	21 54	06 12	06 40	21 38	17 58	21 58	09 17	16 00	20 09
24	01 48	19 13	05 29	06 12	21 42	18 01	21 57	09 18	16 00	20 09
25	02 11	15 35	04 44	05 44	21 46	18 04	21 57	09 18	16 00	20 08
26	02 35	11 08	03 59	05 15	21 49	18 07	21 57	09 19	16 01	20 08
27	02 58	06 02	03 12	04 47	21 53	18 10	21 57	09 20	16 02	20 08
28	03 22	00 N 29	02 24	04 18	21 56	18 13	21 56	09 21	16 02	20 07
29	03 45	05 S 14	01 36	03 49	22 00	18 16	21 56	09 22	16 03	20 07
30	04 08	10 48	00 46	03 20	22 03	18 19	21 56	09 22	16 03	20 07
31	04 N 32	15 S 49	00 N 04	02 S 51	22 S 06	18 N 22	21 S 55	09 N 23	16 N 04	20 S 06

ZODIAC SIGN ENTRIES

Date	h	m	Planets
01	23	55	☽ ♎
04	02	32	☽ ♏
06	04	59	☽ ♐
08	08	05	☽ ♑
10	07	50	♀ ♓
10	12	09	☽ ♒
11	12	08	☿ ♓
12	17	24	☽ ♓
15	00	21	☽ ♈
17	09	45	☽ ♉
19	21	40	☽ ♊
19	22	34	☉ ♈
22	10	36	☽ ♋
24	21	53	☽ ♌
27	05	29	☽ ♍
29	06	23	☿ ♈
29	09	20	☽ ♎
31	10	47	☽ ♏

LATITUDES

Date	Mercury ☿ ° '	Venus ♀ ° '	Mars ♂ ° '	Jupiter ♃ ° '	Saturn ♄ ° '	Uranus ♅ ° '	Neptune ♆ ° '	Pluto ♇ ° '
01	01 S 04	00 S 49	01 N 02	00 S 48	00 N 37	00 N 49	01 S 44	12 S 05
04	01 25	00 55	01 00	00 47	00 37	00 49	01 43	12 05
07	01 42	01 01	00 58	00 47	00 37	00 49	01 43	12 05
10	01 56	01 06	00 56	00 46	00 37	00 49	01 43	12 06
13	02 07	01 11	00 53	00 46	00 37	00 49	01 43	12 06
16	02 14	01 16	00 50	00 45	00 37	00 49	01 43	12 06
19	02 16	01 19	00 47	00 44	00 37	00 49	01 43	12 07
22	02 15	01 22	00 44	00 44	00 37	00 49	01 42	12 07
25	02 10	01 25	00 40	00 43	00 37	00 49	01 42	12 08
28	02 00	01 27	00 37	00 43	00 37	00 49	01 42	12 08
31	01 S 46	01 S 28	00 N 33	00 S 42	00 N 37	00 N 48	01 S 42	12 S 09

DATA

Julian Date	2469137
Delta T	+80 seconds
Ayanamsa	24° 32' 06"
Synetic vernal point	04° ♓ 34' 53"
True obliquity of ecliptic	23° 26' 00"

MOON'S PHASES, APSIDES AND POSITIONS ☽

Date	h	m	Phase	Longitude ° '	Eclipse Indicator
07	10	45	☾	17 ♐ 32	
14	14	28	●	24 ♓ 41	
22	16	03	☽	02 ♋ 43	
30	02	04	○	10 ♎ 03	

Day	h	m		
05	02	52	Perigee	
20	14	43	Apogee	
01	04	33	0S	
07	12	48	Max dec	24° S 16'
14	03	26	0N	
21	11	37	Max dec	24° N 07'
28	14	03	0S	

LONGITUDES

Date	Chiron ⚷ ° '	Ceres ⚳ ° '	Pallas ⚴ ° '	Juno ⚵ ° '	Vesta ⚶ ° '	Black Moon Lilith ⚸ ° '
01	11 ♏ 41	01 ♓ 33	02 ♒ 31	01 ♒ 58	12 ♓ 19	03 ♊ 16
11	11 ♏ 28	05 ♓ 27	05 ♒ 50	05 ♒ 31	17 ♓ 15	04 ♊ 23
21	11 ♏ 04	09 ♓ 18	08 ♒ 58	08 ♒ 58	22 ♓ 08	05 ♊ 29
31	10 ♏ 32	13 ♓ 05	11 ♒ 53	12 ♒ 17	26 ♓ 58	06 ♊ 36

All ephemeris data is given at 12.00 UT and the Moon's longitude is additionally given for 24.00 UT

ASPECTARIAN

h m	Aspects
01 Sunday	
04 30	☽ ⊼ ♀
05 15	☽ △ ♆
06 29	☽ △ ♃
10 10	☽ ± ☿
10 18	☽ Q ♂
12 21	☿ Q ♂
13 04	♀ □ ♆
15 07	♀ # ♆
15 49	☽ ± ♀
20 25	☿ # ♃
02 Monday	
01 37	♀ ⊥ ♄
03 06	☽ ⚼ ☿
05 32	♀ □ ♃
06 56	☽ ⚼ ♆
07 28	☽ ⚹ ♂
08 23	☽ ⚼ ♃
08 38	☽ ⚼ ♀
09 34	☽ ∥ ☉
10 53	☽ ⊼ ♇
15 09	☽ ⊻ ♅
18 36	☽ # ♅
21 03	☽ ± ♇
21 57	☽ ⊼ ☉
22 11	☽ ± ♆
23 21	☽ □ ♄
23 46	☽ ± ♃
03 Tuesday	
01 15	☽ ⊥ ♅
07 01	☽ △ ☿
08 18	☽ ⊼ ♆
08 49	☽ ± ☉
09 36	☽ ∠ ♂
09 59	☽ ⊼ ♃
12 13	☽ ⚼ ♇
12 21	☽ △ ♀
16 20	☽ ∠ ♅
19 28	☉ ⚹ ♄
21 03	☿ □ ♆
22 51	☽ ∥ ♀
04 Wednesday	
01 04	☽ ⚼ ☉
01 12	☽ ⊥ ♂
03 17	☽ # ♆
05 21	☽ ∥ ☿
05 46	☽ Q ♄
09 48	☽ # ♃
11 14	☿ ⊥ ♄
11 35	☽ ⊻ ♂
13 27	☽ △ ♇
17 26	☽ ⚹ ♅
18 29	☿ □ ♃
05 Thursday	
02 01	☽ ⚹ ♄
04 09	☽ △ ☉
05 53	☽ ∥ ♂
06 05	☽ ∥ ♇
08 54	☿ # ♆
10 45	☽ ☍ ♆
12 43	♂ ∥ ♇
12 55	☽ ☍ ♃
13 32	☽ Q ♅
14 39	☽ □ ☿
19 32	☽ □ ♀
20 28	☽ ∥ ♄
06 Friday	
03 25	☽ ∠ ♄
15 43	☽ ☌ ♂
16 06	☽ □ ♇
18 44	☽ ⊥ ♄
19 52	☽ □ ♅
07 Saturday	
00 24	☽ Q ☿
04 57	☽ Q ♀
04 59	☽ ⊻ ♄
05 04	♂ □ ♇
10 45	☽ □ ☉
13 38	☽ ⊼ ♆
16 21	☽ ⊼ ♃
22 50	☽ Q ♇
23 10	☽ ⚹ ☿
23 55	☽ ± ♆
08 Sunday	
02 45	☽ ± ♃
03 26	☽ ⚹ ♀
15 24	☽ ⚼ ♆
18 25	☽ ⚼ ♃
19 30	☽ ⚹ ♇
20 01	☽ Q ☉
20 38	☽ ⊻ ♂
23 05	☽ △ ♅
09 Monday	
04 00	☽ ∠ ☿
04 09	☉ ⚹ ♆
05 53	☽ ∥ ♄
07 21	☽ ⊥ ♂
07 51	☽ ∠ ♀
08 48	☽ ☌ ♄
17 26	☽ △ ♆
17 48	☽ ∥ ♂
18 26	☽ ⚹ ☉
19 00	♀ ∠ ♄
20 45	☽ △ ♃
20 46	☽ ∥ ♇
21 29	☽ ⊥ ☿
21 34	☽ ∠ ♇
23 29	☽ ∠ ♂
10 Tuesday	
01 03	☽ ⊥ ♀
01 03	☽ ⚼ ♅
09 13	☽ ⊻ ☿
12 35	☽ ⊻ ♀
13 20	☽ ⊥ ♇
15 51	☽ # ♃
16 46	☽ ± ♅
22 46	☽ ∠ ☉
23 54	☽ ⊻ ♇
11 Wednesday	
00 07	☽ # ♆
02 37	☽ ⚹ ♂
03 00	☿ ∠ ♄
03 18	☽ ⊼ ♅
08 13	☉ ⚹ ♃
13 39	☽ ⊻ ♄
13 46	☽ ∥ ☿
15 56	☽ ⊥ ☉
17 08	♂ ⊥ ♄
21 00	☽ ∥ ♀
22 18	☽ □ ♆
23 09	♂ □ ♅
12 Thursday	
00 22	☽ ⊥ ♄
00 33	☽ Q ♂
02 16	☽ □ ♃
03 29	☽ ⊻ ☉
10 07	☽ # ♅
16 33	☽ ∠ ♄
21 10	☽ ☌ ☿
23 18	☽ ☌ ♀
13 Friday	
05 35	☽ ☌ ♇
06 41	☽ Q ♆
08 47	☽ ☍ ♅
09 57	☽ □ ♂
11 07	☽ Q ♃
16 19	☽ ∥ ☉
19 54	☽ ⚹ ♄
14 Saturday	
04 39	☽ ⚹ ♆
09 25	☽ ⚹ ♃
13 05	☽ # ☉
14 28	☽ ☌ ☉
18 08	☽ Q ♄
15 Sunday	
08 37	☽ ∠ ♆
11 57	☽ ⊻ ☿
12 27	☽ ⊻ ♀
13 09	☽ ⊻ ♇
13 50	☽ ∠ ♃
16 11	☽ ⊼ ♅
19 23	☽ △ ♂
19 46	♀ ☌ ♇
21 40	☿ ☌ ♇
22 24	☽ ∥ ♅
16 Monday	
00 33	☽ ⊥ ♇
00 58	☽ ⊥ ☿
01 04	☽ ⊥ ♀
01 43	☽ ⊥ ♆
02 46	☽ # ♀
03 34	☽ ± ♅
04 06	☿ ☌ ♀
04 15	☽ □ ♄
06 54	☿ Q ♆
06 54	☽ # ☿
07 12	☽ ⊥ ♃
07 44	♀ Q ♆
13 14	☽ ⊻ ♆
17 06	☉ Q ♄
17 52	☽ ∠ ♇
18 56	☽ ⊻ ♃
20 17	☽ ∠ ♀
20 50	☽ ∠ ☿
20 50	☽ ⚼ ♅
20 51	☿ ☍ ♅
17 Tuesday	
01 08	☽ ⚼ ♂
01 40	♀ ☍ ♅
04 24	☽ ⊻ ☉
11 43	☽ ∥ ♆
17 12	☽ ⊥ ☉
19 20	☽ ± ♂
23 11	☽ ∥ ♃
23 17	☽ ⚹ ♇
23 49	♀ # ♅
18 Wednesday	
02 09	☽ △ ♅
05 02	☽ ⚹ ♀
06 32	☿ Q ♃
06 49	☽ ⚹ ☿
07 36	☽ ⊼ ♂
12 38	☽ ∠ ☉
13 57	☿ □ ♂
15 15	☽ △ ♄
18 13	☽ # ♇
23 15	☽ Q ♇
23 54	☿ # ♅
19 Thursday	
00 28	☽ ☌ ♆
00 48	♀ Q ♃
05 02	☽ # ♂
07 11	☽ ☌ ♃
07 49	☽ Q ♀
10 44	☽ Q ☿
11 46	☽ # ♄
15 42	♀ □ ♂
21 35	☽ ⚹ ☉
21 37	☽ ⚼ ♄
20 Friday	
11 45	☽ □ ♇
14 21	☽ □ ♅
15 59	☽ ± ♄
21 23	☿ ⚹ ♄
22 06	☽ ☍ ♂
21 Saturday	
00 13	☽ Q ☉
00 37	☽ □ ♀
04 15	☽ ⊼ ♄
05 22	☽ □ ☿
10 01	☿ ∥ ♀
13 27	☽ ⊻ ♆
21 04	☽ ⊻ ♃
22 Sunday	
01 39	☽ ⊥ ♆
02 49	☽ Q ♅
09 25	☽ ⊥ ♃
12 34	♀ ⚹ ♄
16 03	☽ □ ☉
19 52	☽ ∠ ♆
23 Monday	
00 41	☽ △ ♇
02 55	☽ ⚹ ♅
03 48	☽ ∠ ♃
11 20	☽ # ♄
12 32	☽ ⊼ ♂
13 18	☿ ⚹ ♆
14 46	☽ # ♂
16 57	☽ ☍ ♄
20 03	☽ △ ♀
24 Tuesday	
00 44	☽ ± ♂
01 45	☽ ⚹ ♆
03 49	☽ △ ☿
04 36	☽ # ♇
06 26	☽ ⚼ ♇
08 28	☽ ∠ ♅
09 53	☽ ⚹ ♃
15 07	☉ ∠ ♆
18 45	☽ ⚼ ♂
20 28	☽ ∥ ♃
23 54	☽ ± ♇
25 Wednesday	
01 04	☽ Q ♆
01 48	☽ ⊥ ♅
04 24	☽ ⚼ ♀
08 06	☽ △ ☉
09 18	☽ Q ♃
09 29	☽ ∥ ♆
11 20	☽ ⊼ ♇
13 10	☽ ⊻ ♅
13 28	☽ ⚼ ☿
23 11	☽ ± ♀
23 57	☽ △ ♂
26 Thursday	
01 54	☿ ⚹ ♃
02 44	☽ ⊼ ♄
05 26	♀ ⚹ ♆
08 56	☽ ± ☿
10 52	☽ □ ♆
11 26	☽ ⊼ ♀
13 49	☽ ± ♄
14 20	☽ ⚼ ☉
19 10	☽ □ ♃
20 49	☽ ∥ ♅
21 37	☽ ⊼ ☿
27 Friday	
02 28	☉ ⊻ ♇
06 07	☽ ⚼ ♄
07 51	☽ ± ☉
16 37	☉ # ☿
18 05	☽ # ♀
18 06	☽ ☍ ♇
19 17	☽ ⊼ ☉
19 33	☽ ☌ ♅
20 33	☿ Q ♄
22 44	☉ ⊼ ♅
28 Saturday	
00 31	☽ ∥ ☉
02 28	☽ # ☿
07 01	☽ □ ♂
08 30	☽ △ ♄
12 09	♂ ∥ ♄
16 01	☽ △ ♆
21 31	☽ ☍ ♀
22 39	☽ ∥ ☿
29 Sunday	
00 23	☽ △ ♃
05 20	☽ # ☉
06 31	☽ ∥ ♀
09 46	☽ ☍ ☿
13 48	☉ # ♀
14 15	☽ Q ♂
17 22	☽ ⚼ ♆
21 18	☽ ⊼ ♇
21 42	☉ ∠ ♃
22 27	☽ ⊻ ♅
30 Monday	
01 50	☽ ⚼ ♃
02 04	☽ ☍ ☉
05 42	☽ # ♅
07 14	☽ ± ♇
08 18	☽ ± ♆
08 21	☽ ⊥ ♅
10 34	☽ ⚹ ♂
11 02	☽ □ ♄
12 03	♀ ⚹ ♃
16 51	☽ ± ♃
18 10	☽ ⊼ ♆
22 02	☽ ⚼ ♇
23 05	☽ ∠ ♅
31 Tuesday	
02 49	☽ ⊼ ♃
03 57	☽ ⊼ ♀
11 37	☽ ∠ ♂
12 42	♂ ⊻ ♄
13 17	☽ # ♆
14 38	☽ ± ♀
16 32	☽ Q ♄
18 34	☽ ⊼ ☿
19 38	☿ ∠ ♆
22 32	☽ △ ♇
23 28	☽ ⚹ ♅

JUNE 2048

LONGITUDES

Date	Sidereal time h m s	Sun ☉ ° ' "	Moon ☽ ° ' "	Moon ☽ 24.00 ° ' "	Mercury ☿ ° '	Venus ♀ ° '	Mars ♂ ° '	Jupiter ♃ ° '	Saturn ♄ ° '	Uranus ♅ ° '	Neptune ♆ ° '	Pluto ♇ ° '
01	04 42 36	11 ♊ 40 31	16 ♒ 35 34	23 ♒ 29 20	04 ♊ 22	12 ♊ 40	14 ♐ 25	09 ♊ 06	14 ♑ 52	07 ♍ 11	22 ♉ 05	08 ♓ 05
02	04 46 32	12 38 01	00 ♓ 16 43	06 ♓ 57 54	03 R 55	13 54	14 R 05	09 20	14 R 49	07 12	22 07	08 05
03	04 50 29	13 35 29	13 ♓ 33 08	20 ♓ 02 46	03 32	15 08	13 45	09 34	14 46	07 13	22 09	08 05
04	04 54 25	14 32 57	26 ♓ 27 10	02 ♈ 46 46	03 12	16 21	13 24	09 48	14 42	07 14	22 12	08 05
05	04 58 22	15 30 23	09 ♈ 01 59	15 ♈ 13 15	02 55	17 35	13 04	10 02	14 39	07 15	22 14	08 05
06	05 02 18	16 27 50	21 ♈ 21 02	27 ♈ 25 44	02 43	18 49	12 44	10 16	14 35	07 16	22 16	08 05
07	05 06 15	17 25 15	03 ♉ 27 47	09 ♉ 27 35	02 35	20 03	12 24	10 30	14 31	07 17	22 18	08 R 05
08	05 10 12	18 22 40	15 ♉ 25 30	21 ♉ 21 53	02 31	21 16	12 03	10 44	14 28	07 19	22 20	08 05
09	05 14 08	19 20 05	27 ♉ 17 06	03 ♊ 11 27	02 D 31	22 30	11 43	10 58	14 24	07 20	22 22	08 05
10	05 18 05	20 17 28	09 ♊ 05 15	14 ♊ 58 47	02 36	23 44	11 23	11 12	14 20	07 22	22 24	08 05
11	05 22 01	21 14 51	20 ♊ 52 21	26 ♊ 46 14	02 45	24 58	11 03	11 26	14 16	07 23	22 26	08 05
12	05 25 58	22 12 14	02 ♋ 40 44	08 ♋ 36 06	02 59	26 11	10 44	11 40	14 12	07 25	22 28	08 05
13	05 29 54	23 09 35	14 ♋ 32 40	20 ♋ 30 43	03 18	27 25	10 24	11 54	14 08	07 26	22 30	08 05
14	05 33 51	24 06 56	26 ♋ 30 35	02 ♌ 32 37	03 41	28 39	10 05	12 08	14 04	07 28	22 32	08 05
15	05 37 47	25 04 16	08 ♌ 37 09	14 ♌ 44 34	04 08	29 ♊ 52	09 47	12 22	14 00	07 29	22 34	08 05
16	05 41 44	26 01 35	20 ♌ 55 15	27 ♌ 09 35	04 40	01 ♋ 06	09 29	12 36	13 56	07 31	22 36	08 04
17	05 45 41	26 58 53	03 ♍ 27 59	09 ♍ 50 51	05 16	02 20	09 11	12 49	13 52	07 33	22 38	08 04
18	05 49 37	27 56 10	16 ♍ 18 33	22 ♍ 51 28	05 56	03 34	08 54	13 03	13 48	07 35	22 40	08 04
19	05 53 34	28 53 27	29 ♍ 29 54	06 ♎ 14 09	06 41	04 47	08 37	13 17	13 44	07 37	22 42	08 04
20	05 57 30	29 ♊ 50 42	13 ♎ 04 24	20 ♎ 00 45	07 29	06 01	08 21	13 31	13 40	07 38	22 44	08 03
21	06 01 27	00 ♋ 47 57	27 ♎ 03 12	04 ♏ 11 37	08 22	07 15	08 05	13 44	13 35	07 40	22 46	08 03
22	06 05 23	01 45 11	11 ♏ 25 45	18 ♏ 45 09	09 18	08 28	07 51	13 58	13 31	07 42	22 47	08 03
23	06 09 20	02 42 25	26 ♏ 09 15	03 ♐ 37 18	10 19	09 42	07 36	14 12	13 27	07 44	22 49	08 02
24	06 13 16	03 39 38	11 ♐ 08 26	18 ♐ 41 38	11 23	10 56	07 23	14 25	13 22	07 46	22 51	08 02
25	06 17 13	04 36 51	26 ♐ 15 51	03 ♑ 49 55	12 31	12 10	07 10	14 39	13 18	07 49	22 53	08 01
26	06 21 10	05 34 03	11 ♑ 22 41	18 ♑ 53 02	13 43	13 23	06 58	14 52	13 14	07 51	22 55	08 01
27	06 25 06	06 31 16	26 ♑ 19 55	03 ♒ 42 24	14 59	14 37	06 46	15 06	13 09	07 53	22 56	08 00
28	06 29 03	07 28 27	10 ♒ 59 39	18 ♒ 11 03	16 18	15 51	06 36	15 19	13 05	07 55	22 58	08 00
29	06 32 59	08 25 39	25 ♒ 16 07	02 ♓ 14 32	17 41	17 04	06 26	15 33	13 01	07 58	23 00	07 59
30	06 36 56	09 ♋ 22 51	09 ♓ 06 10	15 ♓ 51 00	19 ♊ 07	18 ♋ 18	06 ♐ 17	15 ♊ 46	12 ♑ 56	08 ♍ 00	23 ♉ 02	07 ♓ 59

DECLINATIONS

Date	Moon True ☊ ° '	Moon Mean ☊ ° '	Moon ☽ Latitude ° '	Sun ☉ ° '	Moon ☽ ° '	Mercury ☿ ° '	Venus ♀ ° '	Mars ♂ ° '	Jupiter ♃ ° '	Saturn ♄ ° '	Uranus ♅ ° '	Neptune ♆ ° '	Pluto ♇ ° '
01	27 ♐ 42	28 ♐ 37	03 N 59	22 N 11	12 S 03	17 N 59	22 N 14	24 S 43	21 N 14	22 S 01	09 N 35	16 N 39	20 S 03
02	27 D 43	28 33	04 41	22 18	06 59	17 43	22 26	24 44	21 16	22 02	09 35	16 40	20 04
03	27 43	28 30	05 07	22 26	01 S 44	17 28	22 36	24 45	21 19	22 02	09 34	16 40	20 04
04	27 R 43	28 27	05 17	22 32	03 N 26	17 14	22 47	24 45	21 21	22 03	09 34	16 41	20 04
05	27 42	28 24	05 11	22 39	08 20	17 03	22 56	24 46	21 23	22 03	09 33	16 41	20 04
06	27 40	28 21	04 50	22 45	12 48	16 53	23 05	24 46	21 25	22 04	09 33	16 42	20 05
07	27 39	28 18	04 17	22 50	16 42	16 46	23 13	24 47	21 27	22 04	09 32	16 42	20 05
08	27 38	28 14	03 33	22 55	19 51	16 41	23 21	24 47	21 29	22 05	09 32	16 43	20 05
09	27 37	28 11	02 40	23 00	22 09	16 37	23 28	24 47	21 31	22 05	09 31	16 43	20 06
10	27 36	28 08	01 41	23 05	23 28	16 36	23 34	24 48	21 33	22 05	09 31	16 44	20 06
11	27 35	28 05	00 N 37	23 09	23 44	16 36	23 39	24 48	21 35	22 06	09 30	16 44	20 06
12	27 D 35	28 02	00 S 28	23 12	22 56	16 39	23 44	24 48	21 37	22 06	09 30	16 45	20 07
13	27 36	27 59	01 33	23 15	21 06	16 44	23 48	24 48	21 39	22 07	09 29	16 45	20 07
14	27 36	27 55	02 34	23 18	18 20	16 50	23 52	24 47	21 41	22 07	09 28	16 46	20 07
15	27 36	27 52	03 29	23 20	14 44	16 58	23 54	24 47	21 43	22 08	09 28	16 46	20 08
16	27 37	27 49	04 15	23 22	10 29	17 07	23 56	24 47	21 45	22 08	09 27	16 47	20 08
17	27 37	27 46	04 50	23 24	05 43	17 18	23 58	24 46	21 46	22 09	09 26	16 47	20 09
18	27 R 37	27 43	05 12	23 25	00 N 37	17 31	23 58	24 46	21 48	22 09	09 26	16 48	20 09
19	27 D 37	27 39	05 18	23 26	04 S 39	17 45	23 58	24 46	21 50	22 10	09 25	16 48	20 09
20	27 37	27 36	05 06	23 26	09 52	18 00	23 57	24 45	21 52	22 10	09 24	16 48	20 10
21	27 37	27 33	04 37	23 26	14 43	18 16	23 56	24 45	21 53	22 11	09 23	16 49	20 10
22	27 37	27 30	03 50	23 25	18 54	18 32	23 53	24 44	21 55	22 11	09 23	16 49	20 11
23	27 38	27 27	02 46	23 24	21 59	18 50	23 50	24 44	21 57	22 12	09 22	16 50	20 11
24	27 38	27 24	01 30	23 23	23 36	19 08	23 47	24 43	21 58	22 12	09 21	16 50	20 12
25	27 38	27 20	00 S 08	23 21	23 30	19 27	23 42	24 43	22 00	22 13	09 20	16 51	20 12
26	27 R 38	27 17	01 N 16	23 19	21 41	19 47	23 37	24 42	22 02	22 13	09 19	16 51	20 13
27	27 37	27 14	02 33	23 16	18 23	20 06	23 31	24 42	22 03	22 14	09 19	16 51	20 13
28	27 36	27 11	03 38	23 13	13 58	20 26	23 25	24 42	22 05	22 14	09 18	16 52	20 14
29	27 35	27 08	04 28	23 10	08 53	20 45	23 18	24 41	22 06	22 15	09 17	16 52	20 14
30	27 ♐ 34	27 ♐ 04	05 N 01	23 N 06	03 S 30	21 N 04	23 N 10	24 S 41	22 N 08	22 S 15	09 N 16	16 N 53	20 S 15

ZODIAC SIGN ENTRIES

Date	h	m	Planets
02	11	30	☽ ♓
04	18	43	☽ ♈
07	05	06	☽ ♉
09	17	31	☽ ♊
12	06	34	☽ ♋
14	18	57	☽ ♌
15	14	27	♀ ♋
17	05	25	☽ ♍
19	12	54	☽ ♎
20	15	54	☉ ♋
21	16	58	☽ ♏
23	18	12	☽ ♐
25	17	55	☽ ♑
27	17	57	☽ ♒
29	20	07	☽ ♓

LATITUDES

Date	Mercury ☿ ° '	Venus ♀ ° '	Mars ♂ ° '	Jupiter ♃ ° '	Saturn ♄ ° '	Uranus ♅ ° '	Neptune ♆ ° '	Pluto ♇ ° '
01	03 S 04	00 S 05	02 S 12	00 S 35	00 N 35	00 N 46	01 S 41	12 S 28
04	03 37	00 N 03	02 22	00 35	00 35	00 46	01 41	12 29
07	03 59	00 10	02 32	00 34	00 35	00 46	01 41	12 30
10	04 09	00 17	02 41	00 34	00 34	00 46	01 42	12 31
13	04 10	00 24	02 49	00 34	00 34	00 46	01 42	12 32
16	04 00	00 31	02 57	00 34	00 34	00 46	01 42	12 33
19	03 44	00 37	03 04	00 34	00 34	00 45	01 42	12 34
22	03 20	00 44	03 10	00 33	00 34	00 45	01 42	12 35
25	02 52	00 50	03 16	00 33	00 33	00 45	01 42	12 36
28	02 19	00 56	03 20	00 33	00 33	00 45	01 42	12 37
31	01 S 43	01 N 01	03 S 24	00 S 33	00 N 33	00 N 45	01 S 42	12 S 38

DATA

Julian Date	2469229
Delta T	+80 seconds
Ayanamsa	24° 32' 17"
Synetic vernal point	04° ♓ 34' 41"
True obliquity of ecliptic	23° 25' 58"

MOON'S PHASES, APSIDES AND POSITIONS ☽

Date	h	m	Phase	Longitude ° '	Eclipse Indicator
03	12	05	☾	13 ♓ 36	
11	12	50	●	21 ♊ 17	Annular
19	10	50	☽	28 ♍ 51	
26	02	08	○	05 ♑ 11	partial

Day	h	m		
10	20	37	Apogee	
25	09	53	Perigee	
03	19	58	0N	
11	06	00	Max dec	23° N 46'
18	14	48	0S	
24	22	46	Max dec	23° S 47'

LONGITUDES

Date	Chiron ⚷ ° '	Ceres ⚳ ° '	Pallas ⚴ ° '	Juno ⚵ ° '	Vesta ⚶ ° '	Black Moon Lilith ⚸ ° '
01	06 ♏ 12	04 ♈ 03	22 ♒ 28	27 ♒ 54	25 ♈ 05	13 ♊ 28
11	05 ♏ 45	06 ♈ 47	22 ♒ 27	29 ♒ 08	29 ♈ 12	14 ♊ 35
21	05 ♏ 28	09 ♈ 15	21 ♒ 50	29 ♒ 49	03 ♉ 08	15 ♊ 41
31	05 ♏ 20	11 ♈ 23	20 ♒ 35	29 ♒ 55	06 ♉ 53	16 ♊ 48

ASPECTARIAN

h m Aspects

01 Monday
02 56 ☽ △ ☉
04 36 ☽ △ ♀
08 20 ☽ ⚹ ♂
09 03 ☽ ⊻ ♄
19 24 ☽ ⊥ ♄
21 34 ☽ □ ♆
23 52 ☽ # ♅

02 Tuesday
04 44 ☽ Q ♂
11 11 ☽ ∠ ♄
14 47 ♀ ☍ ♂
18 19 ☽ □ ☿

03 Wednesday
00 26 ☽ ☍ ♅
02 02 ☽ ☌ ♇
04 36 ☽ □ ♃
05 08 ♀ ⊼ ♄
05 47 ☽ Q ♆
12 05 ☽ □ ☉
12 21 ☽ □ ♂
14 12 ☽ ⚹ ♄
14 51 ☉ ☍ ♂
15 12 ☽ □ ♀

04 Thursday
02 23 ☽ Q ☿
03 59 ☽ ⚹ ♆
12 28 ☽ Q ♄
14 36 ☽ Q ♃
15 36 ☉ ⊼ ♄

05 Friday
00 31 ☽ ⚹ ☿
00 31 ☽ Q ☉
04 39 ☽ Q ♀
08 31 ☽ ∠ ♆
08 34 ☽ ⊼ ♅
10 11 ☽ ⊻ ♇
13 59 ☽ ⚹ ♃
18 17 ☽ ∥ ♅
19 36 ☽ △ ♂
20 11 ☽ ± ♅
21 48 ☽ ⊥ ♇
22 49 ☽ □ ♄

06 Saturday
01 37 ☽ ⚹ ☉
02 00 ☽ ⊥ ♆
04 59 ☽ ∠ ☿
06 28 ☽ ⚹ ♀
13 48 ☽ ⊻ ♆
13 49 ☽ Ᵽ ♅
15 25 ☽ ∠ ♇
19 53 ☽ ∠ ♃
22 27 ☽ ⊥ ☿

07 Sunday
00 15 ☽ Ᵽ ♂
08 18 ♇ St R
09 45 ☽ ∠ ☉
10 15 ☽ ⊻ ☿
12 04 ☽ ∥ ♆
12 29 ☽ ∥ ☿
14 08 ☽ ⊥ ♃
15 31 ☽ ∠ ♀
17 42 ☽ ± ♂
19 40 ☽ △ ♅
21 15 ☽ ⚹ ♇

08 Monday
02 08 ☿ ∥ ♆
02 23 ☽ ⊻ ♃
05 20 ☽ ⊥ ☉
05 24 ☽ ⊼ ♂
10 04 ☽ △ ♄
11 39 ☽ ⊥ ♀
14 05 ☽ # ♇
18 29 ☽ ⊻ ☉
21 25 ☽ Q ♇
21 56 ☿ St D

09 Tuesday
01 11 ☽ ⊻ ♀
02 00 ☽ ☌ ♆
04 12 ☽ ∥ ♃
09 17 ♀ ⊻ ♆
11 07 ☽ # ♄
16 16 ☽ Ᵽ ♄
22 41 ☽ ☌ ☿

10 Wednesday
02 04 ☽ ∥ ☉
08 28 ☽ □ ♅
09 58 ☽ □ ♇
10 29 ☽ ± ♄
15 39 ☽ ∥ ♀
16 24 ☽ ☌ ♃
16 33 ☽ ☍ ♂
19 36 ♂ ☍ ♃
22 38 ☽ ⊼ ♄

11 Thursday
12 50 ☽ ☌ ☉
15 11 ☽ ⊻ ♆
16 20 ☽ ∥ ♀
20 28 ♀ Q ♅
21 12 ☽ Q ♅
21 17 ☽ ☌ ♀

12 Friday
03 26 ☽ ⊥ ♆
06 57 ☽ ∥ ☉
12 38 ☽ ⊻ ☿
18 53 ☉ ⊻ ♆
21 36 ☽ ⚹ ♅
21 44 ☽ ∠ ♆
22 57 ☽ △ ♇

13 Saturday
00 25 ☽ # ♄
01 06 ☽ ⊥ ☿
03 52 ☽ ⊼ ♂
06 07 ☽ ∥ ♃
06 34 ☽ ⊻ ♃
11 11 ☽ ☍ ♄
15 39 ☽ ± ♂
18 53 ☽ ⊥ ♃
19 47 ☽ ∠ ☿
19 47 ☿ ∥ ♆
21 23 ☽ # ♇

14 Sunday
03 53 ☽ ∠ ♅
04 02 ☽ ⚹ ♆
05 09 ☽ Ᵽ ♇
06 48 ☽ ⊻ ☉
09 14 ☽ Ᵽ ♂
09 46 ♀ ⊥ ♆
13 16 ☽ ∠ ♃
16 44 ☽ ⊻ ♀
19 48 ☽ ⊥ ☉
21 53 ☽ ⊥ ♅
22 16 ☽ ∥ ☿
23 04 ☽ ∥ ♆
23 04 ☽ ± ♇

15 Monday
02 47 ☽ ⚹ ☿
03 59 ☽ Q ♆
05 59 ☽ ⊥ ♀
09 46 ☽ ⊻ ♅
10 56 ☽ ⊼ ♇
14 14 ☽ △ ♂
15 06 ☽ ∠ ☉
19 29 ☽ ⚹ ♃
22 30 ☽ ⊼ ♄
22 51 ☉ Q ♅

16 Tuesday
01 37 ☽ ∠ ♀
03 21 ☽ Q ☿
10 07 ☽ ± ♄
15 15 ☽ □ ♆
17 25 ☽ ∥ ♅
19 13 ☽ Q ♃
22 38 ☽ ⚹ ☉

17 Wednesday
03 19 ☽ Ᵽ ♄
09 37 ☽ ⚹ ♀
15 34 ☽ □ ☿
19 43 ☽ ☌ ♅
20 40 ☽ ☍ ♇
22 31 ☽ □ ♂
23 13 ☽ Q ☉

18 Thursday
05 52 ☽ □ ♃
07 23 ☽ △ ♄
10 28 ☽ Q ♀
23 41 ☽ △ ♆

19 Friday
06 55 ☽ Q ♂
06 58 ☉ ⊥ ♆
10 50 ☽ □ ☉
22 23 ☽ □ ♀

20 Saturday
01 34 ☽ △ ☿
02 27 ☽ ⊻ ♅
02 37 ☽ Ᵽ ♆
03 13 ☽ ⊼ ♇
03 53 ☽ ⚹ ♂
09 51 ☽ # ♅
12 47 ☽ △ ♃
12 59 ☽ ⊥ ♅
13 01 ☽ □ ♄
13 42 ☽ ± ♇
16 33 ☿ ± ♄
16 33 ☿ □ ♅
18 21 ☽ ± ♆
23 51 ♃ ⊼ ♄

21 Sunday
03 39 ☿ □ ♇
04 32 ☽ ∠ ♅
04 41 ☽ ⊼ ♆
05 12 ☽ Ᵽ ♇
05 18 ☽ Ᵽ ☿
05 23 ☽ ∠ ♂
06 21 ☿ ☍ ♂
14 54 ☽ Ᵽ ♃
16 06 ♂ □ ♇
18 46 ☽ △ ☉
19 36 ☽ Q ♄
20 20 ☽ ⊥ ♂
20 35 ♀ ⚹ ♅
21 33 ☽ ± ☿
22 18 ♀ ∠ ♆
23 26 ☽ # ♆

22 Monday
01 43 ♀ ⊼ ♂
03 37 ♀ △ ♇
05 50 ☽ ⚹ ♅
06 10 ☽ ⊻ ♂
06 11 ☽ ± ♃
06 24 ☽ △ ♇
06 40 ☽ △ ♀
08 15 ☽ ⊼ ☿
09 33 ☽ # ☿
15 25 ☽ ⚹ ♄
16 15 ☽ ⊼ ♃
20 53 ☽ ∥ ♇
21 21 ☽ Ᵽ ☉
23 58 ♂ □ ♅

23 Tuesday
01 35 ☽ Q ♅
06 36 ☽ ☍ ♆
09 27 ☽ Ᵽ ♀
11 41 ☽ # ♃
12 57 ☽ ± ☉
14 15 ☽ ∥ ♄
15 41 ☽ ∠ ♄
23 15 ☽ ⊼ ☉

24 Wednesday
01 13 ☽ ± ♀
06 02 ☽ ⊥ ♄
06 06 ☽ ☌ ♂
06 37 ☽ □ ♅
06 48 ☽ # ☉
07 03 ☽ □ ♇
11 38 ☽ ⊼ ♀
12 25 ☽ ☍ ☿
12 56 ♂ ⊥ ♄
15 32 ☽ ⊻ ♄
17 18 ☽ ☍ ♃
19 20 ☽ # ♀

25 Thursday
04 11 ☽ # ♀
06 38 ☽ ⊼ ♆
11 37 ☽ Q ♇
15 29 ☽ # ☉
16 09 ☽ ± ♆

26 Friday
02 08 ☽ ☍ ☉
02 53 ☿ ⊼ ♄
04 54 ♀ ± ♂
05 04 ☽ ⊻ ♂
06 22 ☽ △ ♅
06 28 ☽ Ᵽ ♆
06 39 ☽ ⚹ ♇
06 49 ☽ ∥ ♄
08 49 ☽ # ♃
09 06 ♀ ☍ ♄
14 30 ☽ ⊥ ♂
14 56 ☽ ☌ ♄
15 29 ☽ ☍ ♀
16 04 ☽ ⊼ ☿
17 40 ☽ ⊼ ♃
23 53 ☽ ∥ ♇

27 Saturday
01 41 ☽ # ☿
02 33 ☽ ± ☿
03 26 ☽ ± ♃
04 44 ☽ ∠ ♂
06 25 ☽ Ᵽ ♅
06 31 ☽ △ ♆
06 38 ☽ ∠ ♇
14 43 ☿ ☌ ♃
17 22 ☉ ⊼ ♂
18 13 ☽ Ᵽ ♃
18 29 ☽ Ᵽ ☿
20 48 ☿ # ♇
20 50 ☽ # ♆
21 03 ☽ ± ♅
21 13 ☽ ⊥ ♇
23 37 ♀ ⊻ ♃

28 Sunday
04 50 ☽ ⚹ ♂
05 47 ☽ ⊼ ☉
06 54 ☽ ⊼ ♅
07 03 ☽ ⊻ ♇
15 27 ☽ ⊻ ♄
16 25 ☽ ± ☉
19 19 ☽ △ ♃
20 50 ☽ ⊼ ♀
21 46 ☽ △ ☿
23 43 ☉ ⚹ ♅

29 Monday
00 33 ☽ Q ♂
00 51 ☉ ∠ ♆
01 07 ☉ △ ♇
01 27 ☽ ⊥ ♄
07 54 ☽ ± ♀
08 08 ☽ □ ♆
08 38 ☽ Ᵽ ☉
10 11 ☽ # ♅
16 40 ☽ ∠ ♄

30 Tuesday
00 51 ☽ Ᵽ ♀
04 11 ♅ ☍ ♇
07 06 ☽ □ ♂
10 02 ☽ ☌ ♇
10 03 ☽ ☍ ♅
12 32 ☽ △ ☉
15 24 ☽ Q ♆
18 45 ☽ ⚹ ♄

All ephemeris data is given at 12.00 UT and the Moon's longitude is additionally given for 24.00 UT

JULY 2048

LONGITUDES

Date	Sidereal time h m s	Sun ☉ ° ' "	Moon ☽ ° ' "	Moon ☽ 24.00 ° ' "	Mercury ☿ ° '	Venus ♀ ° '	Mars ♂ ° '	Jupiter ♃ ° '	Saturn ♄ ° '	Uranus ♅ ° '	Neptune ♆ ° '	Pluto ♇ ° '
01	06 40 52	10 ♋ 20 03	22 ♓ 29 12	29 ♓ 01 00	20 ♊ 37	19 ♋ 32	06 ♐ 09	16 ♊ 00	12 ♑ 52	08 ♍ 02	23 ♉ 03	07 ♓ 58
02	06 44 49	11 17 15	05 ♈ 26 46	11 ♈ 46 56	22 11	20 45	06 R 01	16 13	12 R 47	08 05	23 05	07 R 58
03	06 48 45	12 14 27	18 ♈ 01 58	24 ♈ 12 25	23 47	21 59	05 54	16 26	12 43	08 07	23 07	07 57
04	06 52 42	13 11 40	00 ♉ 18 49	06 ♉ 21 44	25 28	23 13	05 49	16 40	12 38	08 10	23 08	07 57
05	06 56 39	14 08 52	12 ♉ 21 45	18 ♉ 19 23	27 11	24 26	05 44	16 53	12 34	08 12	23 10	07 56
06	07 00 35	15 06 05	24 ♉ 15 12	00 ♊ 09 42	28 ♊ 58	25 40	05 39	17 06	12 30	08 15	23 11	07 55
07	07 04 32	16 03 19	06 ♊ 03 23	11 ♊ 56 43	00 ♋ 47	26 54	05 36	17 19	12 25	08 17	23 13	07 55
08	07 08 28	17 00 32	17 ♊ 50 08	23 ♊ 44 01	02 40	28 07	05 34	17 32	12 21	08 20	23 14	07 54
09	07 12 25	17 57 46	29 ♊ 38 44	05 ♋ 34 38	04 35	29 ♋ 21	05 32	17 45	12 16	08 23	23 16	07 53
10	07 16 21	18 55 00	11 ♋ 32 00	17 ♋ 31 08	06 33	00 ♌ 35	05 31	17 58	12 12	08 25	23 17	07 52
11	07 20 18	19 52 14	23 ♋ 32 15	29 ♋ 35 35	08 33	01 48	05 D 31	18 11	12 08	08 28	23 19	07 52
12	07 24 14	20 49 29	05 ♌ 41 21	11 ♌ 49 44	10 36	03 02	05 32	18 24	12 03	08 31	23 20	07 51
13	07 28 11	21 46 43	18 ♌ 00 56	24 ♌ 15 06	12 40	04 16	05 34	18 37	11 59	08 34	23 21	07 50
14	07 32 08	22 43 58	00 ♍ 32 25	06 ♍ 53 02	14 45	05 30	05 36	18 49	11 55	08 37	23 23	07 49
15	07 36 04	23 41 13	13 ♍ 17 09	19 ♍ 44 55	16 52	06 43	05 40	19 02	11 50	08 40	23 24	07 49
16	07 40 01	24 38 27	26 ♍ 16 31	02 ♎ 52 07	18 59	07 57	05 44	19 15	11 46	08 43	23 25	07 48
17	07 43 57	25 35 42	09 ♎ 31 53	16 ♎ 15 57	21 08	09 11	05 49	19 27	11 42	08 45	23 27	07 47
18	07 47 54	26 32 58	23 ♎ 04 27	29 ♎ 57 28	23 16	10 24	05 55	19 40	11 37	08 48	23 28	07 46
19	07 51 50	27 30 13	06 ♏ 55 03	13 ♏ 57 11	25 25	11 38	06 01	19 52	11 33	08 51	23 29	07 45
20	07 55 47	28 27 28	21 ♏ 03 45	28 ♏ 14 35	27 33	12 51	06 09	20 05	11 29	08 55	23 30	07 44
21	07 59 43	29 ♋ 24 44	05 ♐ 29 22	12 ♐ 47 42	29 ♋ 41	14 05	06 17	20 17	11 25	08 58	23 32	07 43
22	08 03 40	00 ♌ 22 00	20 ♐ 09 04	27 ♐ 32 48	01 ♌ 48	15 19	06 26	20 29	11 21	09 01	23 33	07 42
23	08 07 37	01 19 16	04 ♑ 58 10	12 ♑ 24 16	03 54	16 32	06 36	20 41	11 17	09 04	23 34	07 41
24	08 11 33	02 16 33	19 ♑ 50 13	27 ♑ 15 00	05 59	17 46	06 46	20 53	11 13	09 07	23 35	07 40
25	08 15 30	03 13 50	04 ♒ 37 40	11 ♒ 57 16	08 02	19 00	06 57	21 05	11 09	09 10	23 36	07 39
26	08 19 26	04 11 08	19 ♒ 12 53	26 ♒ 23 44	10 05	20 13	07 09	21 17	11 05	09 14	23 37	07 38
27	08 23 23	05 08 27	03 ♓ 29 10	10 ♓ 28 39	12 06	21 27	07 22	21 29	11 01	09 17	23 38	07 37
28	08 27 19	06 05 46	17 ♓ 21 49	24 ♓ 08 30	14 05	22 40	07 35	21 41	10 57	09 20	23 39	07 36
29	08 31 16	07 03 06	00 ♈ 48 36	07 ♈ 22 15	16 03	23 54	07 49	21 53	10 54	09 23	23 40	07 35
30	08 35 12	08 00 28	13 ♈ 49 41	20 ♈ 11 12	18 00	25 08	08 03	22 05	10 50	09 27	23 41	07 34
31	08 39 09	08 ♌ 57 50	26 ♈ 27 16	02 ♉ 38 23	19 ♌ 55	26 ♌ 21	08 ♐ 19	22 ♊ 16	10 ♑ 46	09 ♍ 30	23 ♉ 42	07 ♓ 33

Date	Moon True ☊ ° '	Moon Mean ☊ ° '	Moon ☽ Latitude ° '
01	27 ♐ 33	27 ♐ 01	05 N 16
02	27 R 32	26 58	05 14
03	27 D 32	26 55	04 57
04	27 33	26 52	04 27
05	27 34	26 49	03 45
06	27 36	26 45	02 54
07	27 37	26 42	01 57
08	27 38	26 39	00 N 54
09	27 R 38	26 36	00 S 11
10	27 38	26 33	01 16
11	27 36	26 30	02 18
12	27 33	26 26	03 15
13	27 29	26 23	04 03
14	27 26	26 20	04 40
15	27 22	26 17	05 04
16	27 18	26 14	05 14
17	27 16	26 10	05 07
18	27 15	26 07	04 43
19	27 D 16	26 04	04 02
20	27 17	26 01	03 06
21	27 18	25 58	01 57
22	27 19	25 55	00 S 39
23	27 R 19	25 51	00 N 42
24	27 18	25 48	02 00
25	27 14	25 45	03 10
26	27 10	25 42	04 06
27	27 04	25 39	04 45
28	26 58	25 36	05 06
29	26 53	25 32	05 10
30	26 49	25 29	04 57
31	26 ♐ 46	25 ♐ 26	04 N 30

DECLINATIONS

Date	Sun ☉ ° '	Moon ☽ ° '	Mercury ☿ ° '	Venus ♀ ° '	Mars ♂ ° '	Jupiter ♃ ° '	Saturn ♄ ° '	Uranus ♅ ° '	Neptune ♆ ° '	Pluto ♇ ° '
01	23 N 02	01 N 51	21 N 23	23 N 01	24 S 41	22 N 09	22 S 16	09 N 15	16 N 53	20 S 15
02	22 57	06 58	21 41	22 52	24 41	22 11	22 16	09 14	16 53	20 16
03	22 52	11 39	21 59	22 42	24 41	22 12	22 17	09 13	16 54	20 16
04	22 47	15 44	22 15	22 32	24 40	22 13	22 17	09 12	16 54	20 17
05	22 41	19 07	22 31	22 21	24 41	22 15	22 18	09 11	16 54	20 17
06	22 35	21 39	22 45	22 09	24 41	22 16	22 18	09 10	16 55	20 18
07	22 28	23 14	22 57	21 56	24 41	22 17	22 19	09 09	16 55	20 18
08	22 21	23 46	23 08	21 43	24 41	22 19	22 19	09 08	16 55	20 19
09	22 14	23 15	23 16	21 29	24 41	22 20	22 20	09 07	16 56	20 19
10	22 06	21 40	23 23	21 15	24 42	22 21	22 20	09 06	16 56	20 20
11	21 58	19 07	23 27	21 00	24 42	22 22	22 21	09 05	16 56	20 20
12	21 49	15 42	23 29	20 44	24 43	22 23	22 21	09 04	16 57	20 21
13	21 40	11 34	23 28	20 28	24 44	22 25	22 22	09 03	16 57	20 21
14	21 31	06 54	23 25	20 11	24 45	22 26	22 22	09 02	16 57	20 22
15	21 21	01 N 53	23 19	19 54	24 46	22 27	22 23	09 01	16 58	20 23
16	21 11	03 S 19	23 10	19 36	24 47	22 28	22 23	08 59	16 58	20 23
17	21 01	08 28	22 59	19 18	24 48	22 29	22 24	08 58	16 58	20 24
18	20 50	13 20	22 45	18 59	24 49	22 30	22 24	08 57	16 58	20 24
19	20 39	17 37	22 28	18 39	24 50	22 31	22 25	08 56	16 59	20 25
20	20 28	21 00	22 08	18 19	24 52	22 32	22 25	08 55	16 59	20 25
21	20 16	23 08	21 47	17 59	24 53	22 33	22 26	08 54	16 59	20 26
22	20 04	23 43	21 22	17 38	24 55	22 34	22 26	08 52	16 59	20 27
23	19 52	22 39	20 56	17 16	24 56	22 35	22 26	08 51	17 00	20 27
24	19 39	19 59	20 28	16 54	24 58	22 35	22 27	08 50	17 00	20 28
25	19 26	16 02	19 58	16 32	25 00	22 36	22 27	08 49	17 00	20 28
26	19 12	11 10	19 26	16 09	25 02	22 37	22 28	08 48	17 00	20 29
27	18 59	05 48	18 53	15 46	25 03	22 38	22 28	08 46	17 00	20 29
28	18 45	00 S 17	18 18	15 22	25 05	22 39	22 29	08 45	17 01	20 30
29	18 30	05 N 04	17 42	14 58	25 07	22 39	22 29	08 44	17 01	20 31
30	18 16	10 01	17 05	14 33	25 10	22 40	22 29	08 43	17 01	20 31
31	18 N 01	14 N 24	16 N 26	14 N 09	25 S 12	22 N 41	22 S 30	08 N 41	17 N 01	20 S 32

ZODIAC SIGN ENTRIES

Date	h	m	Planets
02	01	49	☽ ♈
04	11	23	☽ ♉
06	23	40	☽ ♊
07	01	42	☿ ♋
09	12	43	☽ ♋
10	00	40	♀ ♌
12	00	48	☽ ♌
14	10	58	☽ ♍
16	18	48	☽ ♎
19	00	04	☽ ♏
21	02	55	☽ ♐
21	15	39	☿ ♌
22	02	47	☉ ♌
23	03	58	☽ ♑
25	04	28	☽ ♒
27	06	05	☽ ♓
29	10	32	☽ ♈
31	18	52	☽ ♉

LATITUDES

Date	Mercury ☿ ° '	Venus ♀ ° '	Mars ♂ ° '	Jupiter ♃ ° '	Saturn ♄ ° '	Uranus ♅ ° '	Neptune ♆ ° '	Pluto ♇ ° '
01	01 S 43	01 N 01	03 S 24	00 S 33	00 N 33	00 N 45	01 S 42	12 S 38
04	01 06	01 06	03 28	00 33	00 33	00 45	01 42	12 39
07	00 S 29	01 11	03 30	00 33	00 32	00 45	01 42	12 39
10	00 N 07	01 15	03 32	00 32	00 32	00 45	01 43	12 40
13	00 39	01 19	03 34	00 32	00 32	00 45	01 43	12 41
16	01 06	01 22	03 35	00 32	00 32	00 45	01 43	12 42
19	01 26	01 25	03 35	00 32	00 31	00 44	01 43	12 43
22	01 40	01 27	03 35	00 32	00 31	00 44	01 43	12 43
25	01 46	01 29	03 35	00 32	00 31	00 44	01 43	12 44
28	01 47	01 30	03 34	00 32	00 30	00 44	01 44	12 45
31	01 N 41	01 N 30	03 S 33	00 S 32	00 N 30	00 N 44	01 S 44	12 S 45

DATA

Julian Date	2469259
Delta T	+80 seconds
Ayanamsa	24° 32' 23"
Synetic vernal point	04° ♓ 34' 36"
True obliquity of ecliptic	23° 25' 58"

MOON'S PHASES, APSIDES AND POSITIONS ☽

Date	h	m	Phase	Longitude ° '	Eclipse Indicator
02	23	58	☾	11 ♈ 46	
11	04	04	●	19 ♋ 33	
18	18	32	◐	26 ♎ 49	
25	09	34	○	03 ♒ 08	

Day	h	m		
08	03	24	Apogee	
23	16	24	Perigee	
01	03	37	0N	
08	12	17	Max dec	23° N 46'
15	20	43	0S	
22	08	37	Max dec	23° S 44'
28	13	15	0N	

LONGITUDES

Date	Chiron ⚷ ° '	Ceres ⚳ ° '	Pallas ⚴ ° '	Juno ⚵ ° '	Vesta ⚶ ° '	Black Moon Lilith ⚸ ° '
01	05 ♏ 20	11 ♈ 23	20 ♒ 35	29 ♒ 55	06 ♉ 53	16 ♊ 48
11	05 ♏ 24	13 ♈ 08	18 ♒ 47	29 ♒ 21	10 ♉ 34	17 ♊ 54
21	05 ♏ 38	14 ♈ 27	16 ♒ 30	28 ♒ 08	13 ♉ 39	19 ♊ 01
31	06 ♏ 02	15 ♈ 17	13 ♒ 57	26 ♒ 19	16 ♉ 35	20 ♊ 07

All ephemeris data is given at 12.00 UT and the Moon's longitude is additionally given for 24.00 UT

ASPECTARIAN

h m Aspects

01 Wednesday
00 04 ☽ □ ♃
06 04 ☽ △ ♀
08 10 ☽ □ ☿
09 34 ☉ ∥ ♀
13 02 ☽ ✶ ♆
16 19 ☽ Q ♄

02 Thursday
09 39 ☽ Q ♃
13 04 ☽ △ ♂
16 43 ♀ ⚼ ♂
16 45 ☽ ⊻ ♇
16 59 ☽ ⊼ ♅
16 59 ☽ ∠ ♆
22 14 ☽ Q ☿
23 16 ☽ ∥ ♅
23 58 ☽ □ ☉

03 Friday
01 47 ☿ ⊻ ♆
01 50 ☽ □ ♄
04 09 ☽ ⊥ ♇
04 27 ☽ ± ♅
04 27 ☉ ± ♂
08 52 ☽ ✶ ♃
10 13 ☽ ⊥ ♆
17 31 ☽ ⚼ ♂
20 31 ☽ □ ♀
21 33 ☽ ∠ ♇
21 53 ☽ ⊻ ♆
21 54 ☽ ⚼ ♅
22 52 ♀ ⊥ ♃
23 04 ☉ ☍ ♄

04 Saturday
00 56 ☽ ✶ ☿
06 49 ♀ ⚼ ♇
08 41 ☿ ∥ ♃
10 30 ♀ ✶ ♆
10 59 ♀ ∠ ♅
11 01 ☽ ± ♂
13 53 ☽ Q ☉
14 43 ☽ ∠ ♃
15 08 ☿ # ♄
19 41 ☽ ∥ ♆
22 04 ☿ Q ♅
22 49 ☽ ⊼ ♂

05 Sunday
02 45 ☿ ∥ ♀
03 08 ☽ ✶ ♇
03 38 ☽ △ ♅
08 58 ☽ ⊥ ♃
11 35 ☽ ∠ ☿
12 10 ☽ Q ♀
12 25 ☽ △ ♄
15 54 ☽ ✶ ☉
17 16 ♀ # ♄
21 15 ☽ ⊻ ♃
22 05 ☽ # ♇
22 46 ♀ ∥ ♃

06 Monday
00 01 ☉ ∥ ☿
03 14 ☽ Q ♇
08 55 ☽ ⊥ ☿
09 50 ☽ ☌ ♆
15 02 ☿ ⊥ ♆
15 12 ☽ ✶ ♀
17 31 ☽ ∥ ♀
18 32 ☽ ⚼ ♄
19 52 ☽ ∥ ♃
20 21 ☽ # ♄
23 18 ☽ ⊻ ☿
23 29 ☽ ∥ ☉

07 Tuesday
00 55 ☽ ∠ ☉
05 30 ☽ ∥ ☿
11 05 ☽ ☍ ♂
12 44 ☽ ± ♄
15 46 ☽ □ ♇
16 34 ☽ □ ♅
20 52 ☽ ⊥ ☉

08 Wednesday
00 53 ☽ ⊼ ♄
01 17 ☽ ∠ ♀
10 10 ☽ ⊻ ☉
11 23 ☽ ☌ ♃
17 19 ☉ # ♄
19 04 ☉ ∥ ♃
21 44 ☽ ⊥ ♀
23 01 ☽ ⊻ ♆

09 Thursday
05 07 ☉ ⊻ ♃
05 21 ☽ Q ♅
11 13 ☽ ⊥ ♆
11 20 ☽ ⊻ ♀
11 29 ☽ ∥ ☿
13 21 ♃ # ♄
23 30 ☿ ⊼ ♂
23 54 ☽ ⊼ ♂
23 58 ☽ ☌ ☿

10 Friday
03 32 ☽ ∥ ♃
03 38 ☽ # ♄
04 39 ☽ △ ♇
05 27 ☽ ∠ ♆
05 43 ☽ ✶ ♅
06 24 ☽ ∥ ☉
11 58 ☽ ± ♂
13 20 ☽ ☍ ♄
17 13 ☽ ∥ ♀
22 38 ♂ St D

11 Saturday
01 08 ☽ ⊻ ♃
01 37 ☽ # ♇
03 48 ☿ △ ♇
04 04 ☽ ☌ ☉
05 59 ☽ ⚼ ♂
09 04 ☿ ∠ ♆
10 40 ☽ ⚼ ♇
10 57 ☿ ✶ ♅
11 33 ☽ ✶ ♆
11 52 ☽ ∠ ♅
13 03 ♀ ± ♇
13 18 ☽ ⊥ ♃

12 Sunday
01 27 ♀ ⊥ ♅
03 52 ☽ ∥ ♆
04 28 ☽ ± ♇
04 33 ☉ ⚼ ♂
05 44 ☽ ⊥ ♅
06 12 ☽ ☌ ♀
07 25 ☽ ∠ ♃
11 18 ☽ Q ♆
11 42 ☽ △ ♂
16 14 ☽ ⊼ ♇
17 33 ☽ ⊻ ♅
20 32 ♀ ∠ ♃
23 07 ☿ ± ♂
23 31 ☽ ⊻ ☿

13 Monday
00 22 ☽ ⊼ ♄
04 25 ☿ ☍ ♄
11 56 ☽ ± ♄
13 10 ☽ ✶ ♃
13 30 ☽ ⊥ ☿
19 51 ☽ ⊻ ☉
21 23 ♀ # ♇
22 18 ☽ □ ♆

14 Tuesday
01 20 ☽ ∥ ♅
05 07 ☽ ⚼ ♄
08 17 ☽ ⊥ ☉
09 47 ♀ Q ♆
10 12 ☽ ∠ ☿
12 33 ☽ Q ♃
14 13 ☉ ⚼ ♇
14 20 ♀ △ ♂
21 38 ☽ □ ♂
22 23 ☽ ⊻ ♀

15 Wednesday
01 45 ☽ ☍ ♇
02 42 ☽ ∠ ☉
03 18 ☽ ☌ ♅
04 41 ☉ ✶ ♆
09 19 ☽ △ ♄
10 50 ☽ ⊥ ♀
11 17 ☉ ∠ ♅
19 58 ☽ ✶ ☿
22 51 ☽ □ ♃

16 Thursday
05 16 ☽ ∠ ♀
06 46 ☽ △ ♆
07 19 ☽ Q ♂
08 46 ☽ ✶ ☉
09 02 ♀ ⊼ ♇
15 09 ☿ ⊻ ♃
17 15 ♂ ⊥ ♄
22 15 ☽ Q ☿

17 Friday
03 30 ♀ ⊻ ♅
05 17 ☽ ✶ ♂
07 27 ☉ ⊥ ♃
08 16 ☽ Q ☉
08 22 ☿ ⚼ ♂
08 52 ☽ ⊼ ♇
10 03 ☽ ⚼ ♆
10 36 ☽ ⊻ ♅
11 18 ☽ ✶ ♀
14 22 ☽ # ♅
15 51 ☽ □ ♄
19 35 ☽ ± ♇
21 22 ☽ ⊥ ♅

18 Saturday
02 07 ☽ ± ♆
05 55 ☽ △ ♃
06 23 ☿ ⚼ ♇
08 11 ☽ ∠ ♂
10 42 ☽ Q ♀
11 28 ☽ ⚼ ♇
12 24 ☽ □ ☿
12 41 ☽ ⊼ ♆
13 17 ☽ ∠ ♅
14 13 ☿ ✶ ♆
18 10 ☿ ∠ ♅
18 32 ☽ □ ☉
23 22 ☽ Q ♄

19 Sunday
00 01 ☽ ⊥ ♂
08 01 ☿ ∥ ♃
08 05 ☽ # ♆
08 26 ☽ ⚼ ♃
10 27 ☽ ⊻ ♂
10 36 ♀ ⊼ ♄
13 26 ☽ △ ♇
15 20 ☽ ✶ ♅
15 56 ☿ # ♄
17 42 ☿ ⊥ ♃
18 01 ☽ # ♀
19 53 ☽ ✶ ♄
20 49 ☽ □ ♀

20 Monday
00 02 ☽ ± ♃
07 14 ☽ ∥ ♇
07 51 ☽ # ☉
10 19 ☽ ⊼ ♃
11 44 ☽ Q ♅
16 06 ☽ ☍ ♆
16 50 ☉ # ♇
21 02 ☽ ∠ ♄
21 24 ☽ # ☿

21 Tuesday
00 43 ☽ △ ☿
01 14 ☽ △ ☉
02 06 ☽ ∥ ♄
03 32 ☽ # ♃
06 35 ☉ ☌ ☿
11 53 ☽ ⊥ ♄
13 19 ☽ ☌ ♂
15 40 ☽ □ ♇
17 44 ☽ □ ♅
21 42 ☽ ⊻ ♄

22 Wednesday
03 24 ☽ △ ♀
03 40 ☽ ⚼ ☉
05 37 ☽ ⚼ ☿
10 59 ☿ ± ♇
12 33 ☽ ☍ ♃
17 31 ☽ ⊼ ♆
19 19 ☽ ± ☉
21 00 ☽ Q ♇
22 41 ☽ ± ☿

23 Thursday
02 16 ☿ ⊥ ♅
03 15 ☽ ± ♆
05 42 ☽ ⊼ ☉
05 58 ☽ ⚼ ♀
09 59 ☽ ⊼ ☿
12 51 ☽ # ♃
14 28 ☽ ∥ ♄
14 39 ☽ ⊻ ♂
16 23 ☽ ✶ ♇
17 48 ☽ ⚼ ♆
18 38 ☽ △ ♅
21 03 ☉ ± ♇
21 48 ☽ ± ♀
22 09 ☽ ☌ ♄

24 Friday
00 27 ☽ ⊥ ♂
01 46 ♀ ± ♄
06 09 ♀ ∥ ♆
07 22 ☿ Q ♆
07 52 ☽ # ☿
08 21 ☽ ⊼ ♀
08 32 ☽ ∥ ♇
10 53 ☿ ∠ ♃
12 12 ☿ # ♇
13 44 ☽ ⊼ ♃
14 33 ☽ # ☉
15 10 ☽ ∠ ♂
16 35 ☽ ∠ ♇
18 04 ☽ △ ♆
18 57 ☽ ⚼ ♅
22 04 ☿ △ ♂
23 34 ☽ ± ♃

25 Saturday
06 42 ☽ # ♆
07 10 ☽ ⊥ ♇
07 32 ☿ ⊼ ♇
09 02 ☽ # ♀
09 34 ☽ ☍ ☉
09 37 ☽ ± ♅
10 26 ☉ ⊥ ♅
14 25 ☽ ⚼ ♃
15 51 ☽ ✶ ♂
16 56 ☽ ⊻ ♇
18 29 ☽ ☍ ☿
19 27 ☽ ⊼ ♅
22 38 ☽ ⊻ ♄

26 Sunday
01 38 ☿ ⊻ ♅
08 29 ☽ ⊥ ♄
11 54 ☽ Q ♂
13 50 ☽ ☍ ♀
15 30 ☽ △ ♃
19 21 ☽ □ ♆
22 48 ☽ # ♅
23 26 ☽ ∠ ♄
23 32 ☿ ⊼ ♄

27 Monday
04 48 ☉ ∥ ☿
12 57 ♀ ✶ ♃
15 02 ☽ ⊼ ☉
18 44 ☽ □ ♂
19 04 ☽ ☌ ♇
21 58 ☽ ☍ ♅

28 Tuesday
00 32 ☉ Q ♆
00 53 ☽ ✶ ♄
02 01 ☽ Q ♆
02 07 ☽ ± ☉
05 19 ☽ ⊼ ☿
14 08 ♂ □ ♇
17 37 ☽ ± ☿
19 05 ☽ ⚼ ☉
19 44 ☽ □ ♃
21 51 ☽ Q ♄
22 20 ☽ ⊼ ♀
23 08 ☽ ✶ ♆

29 Wednesday
06 37 ☉ ∠ ♃
07 17 ♀ □ ♆
10 11 ☽ ± ♀
12 32 ☽ ⚼ ☿
21 59 ☿ ± ♄

30 Thursday
00 19 ☽ △ ☉
00 23 ☽ ⊻ ♇
01 03 ☽ △ ♂
01 10 ☉ ⊼ ♇
02 24 ☽ ∠ ♆
03 48 ☽ ⊼ ♅
04 23 ☽ ⚼ ♀
04 54 ☽ Q ♃
05 28 ☽ ∥ ♅
06 26 ☽ □ ♄
11 31 ☽ ⊥ ♇
13 38 ☉ △ ♂
14 18 ☿ ∥ ♆
15 03 ☽ ± ♅
19 15 ☽ ⊥ ♆
21 15 ☽ △ ☿

31 Friday
01 09 ♀ ⚼ ♄
03 51 ☽ ✶ ♃
04 31 ☽ ∠ ♇
05 50 ☽ ⚼ ♂
06 41 ☽ ⊻ ♆
08 13 ☽ ⚼ ♅
10 38 ☽ ∥ ♀
11 47 ☽ △ ♀
22 49 ☽ ∥ ☿
23 36 ☽ ± ♂

APRIL 2048

LONGITUDES

Date	Sidereal time h m s	Sun ☉ ° ' "	Moon ☽ ° ' "	Moon ☽ 24.00 ° ' "	Mercury ☿ ° '	Venus ♀ ° '	Mars ♂ ° '	Jupiter ♃ ° '	Saturn ♄ ° '	Uranus ♅ ° '	Neptune ♆ ° '	Pluto ♇ ° '
01	00 42 06	12 ♈ 26 10	15 ♏ 29 33	22 ♏ 50 55	06 ♈ 12	27 ♓ [illegible]	15 ♐ 48	25 ♉ 24	15 ♑ 34	07 ♍ 47	19 ♉ 53	07 ♓ 14
02	00 46 02	13 25 20	00 ♐ 10 34	07 ♐ 27 49	08 09	28 42	16 06	25 36	15 36	07 R 45	19 55	07 15
03	00 49 59	14 24 29	14 ♐ 42 08	21 ♐ 53 04	10 08	29 56	16 22	25 48	15 38	07 43	19 57	07 17
04	00 53 55	15 23 36	29 ♐ 00 21	06 ♑ 03 50	12 08	[illegible] 11	16 38	26 00	15 40	07 41	19 59	07 18
05	00 57 52	16 22 41	13 ♑ 03 27	19 ♑ 59 13	14 10	[illegible] 25	16 54	26 13	15 42	07 39	20 01	07 20
06	01 01 49	17 21 44	26 ♑ 51 12	03 ♒ 39 31	16 [illegible]	03 39	17 09	26 25	15 44	07 37	20 03	07 21
07	01 05 45	18 20 46	10 ♒ 24 20	17 ♒ 05 45	18 [illegible]	04 53	17 24	26 38	15 45	07 35	20 05	07 22
08	01 09 42	19 19 46	23 ♒ 43 56	00 ♓ 18 59	20 [illegible]	06 07	17 38	26 50	15 47	07 34	20 07	07 24
09	01 13 38	20 18 44	06 ♓ 51 00	13 ♓ 20 02	[illegible]	07 22	17 52	27 03	15 48	07 32	20 09	07 25
10	01 17 35	21 17 40	19 ♓ 46 10	26 ♓ 09 24	[illegible] 30	08 36	18 05	27 16	15 50	07 30	20 11	07 26
11	01 21 31	22 16 34	02 ♈ 29 46	08 ♈ 47 [illegible]	[illegible] 36	09 50	18 18	27 28	15 51	07 29	20 13	07 27
12	01 25 28	23 15 27	15 ♈ 01 55	21 ♈ 13 [illegible]	[illegible] 42	11 04	18 30	27 41	15 52	07 27	20 15	07 29
13	01 29 24	24 14 17	27 ♈ 22 49	03 ♉ [illegible]	[illegible] ♉ 48	12 18	18 42	27 54	15 54	07 25	20 17	07 30
14	01 33 21	25 13 06	09 ♉ 32 59	15 [illegible]	02 53	13 32	18 53	28 07	15 55	07 24	20 19	07 31
15	01 37 18	26 11 52	21 ♉ 33 34	27 [illegible]	04 57	14 46	19 03	28 20	15 55	07 23	20 21	07 32
16	01 41 14	27 10 37	03 ♊ 26 29	[illegible]	07 01	16 01	19 13	28 33	15 56	07 21	20 23	07 33
17	01 45 11	28 09 19	15 ♊ 14 29	[illegible]	09 03	17 15	19 22	28 46	15 57	07 20	20 25	07 35
18	01 49 07	29 ♈ 07 59	27 ♊ 01 [illegible]	[illegible]	11 03	18 29	19 31	28 59	15 58	07 18	20 28	07 36
19	01 53 04	00 ♉ 06 37	08 ♋ [illegible]	[illegible]	13 01	19 43	19 38	29 13	15 58	07 17	20 30	07 37
20	01 57 00	01 05 13	20 ♋ [illegible]	[illegible] 52 01	14 56	20 57	19 46	29 26	15 59	07 16	20 32	07 38
21	02 00 57	02 03 47	02 [illegible]	[illegible] 11 31	16 49	22 11	19 53	29 39	15 59	07 15	20 34	07 39
22	02 04 53	03 02 18	1[illegible]	[illegible] ♌ 51 58	18 39	23 25	19 59	29 ♉ 52	15 59	07 14	20 36	07 40
23	02 08 50	04 00 48	[illegible] 01	[illegible] ♍ 57 39	20 26	24 39	20 04	00 ♊ 06	15 59	07 12	20 38	07 41
24	02 12 47	04 59 15	[illegible] 18	18 ♍ 31 20	22 09	25 53	20 09	00 19	15 R 59	07 11	20 41	07 42
25	02 16 43	05 57 [illegible]	[illegible] 24 36	02 ♎ 33 29	23 48	27 07	20 13	00 33	15 59	07 10	20 43	07 43
26	02 20 40	06 5[illegible]	[illegible] 27 26	17 ♎ 01 43	25 24	28 21	20 16	00 46	15 59	07 09	20 45	07 44
27	02 24 36	07 [illegible]	[illegible]	01 ♏ 50 34	26 55	29 ♈ 35	20 19	01 00	15 59	07 09	20 47	07 45
28	02 28 33	08 [illegible]	[illegible]	16 ♏ 52 03	28 22	00 ♉ 49	20 21	01 13	15 58	07 08	20 49	07 46
29	02 32 29	0[illegible]	[illegible]	01 ♐ 56 47	29 ♉ 45	02 03	20 22	01 27	15 58	07 07	20 52	07 47
30	02 36 26	[illegible]	[illegible]	16 ♐ 55 34	01 ♊ 04	03 ♉ 17	20 ♐ 22	01 ♊ 40	15 ♑ 57	07 ♍ 06	20 ♉ 54	07 ♓ 48

Date	[illegible] ° '	[illegible] ° '	…on ☽ Latitude ° '
01	[illegible]	[illegible]	03 S 35
02	[illegible]	[illegible]	02 34
03	[illegible]	[illegible]4	01 24
04	[illegible]	[illegible] 41	00 S 09
05	[illegible]	[illegible] 38	01 N 05
06	[illegible]	[illegible] 35	02 14
07	[illegible]	[illegible] 31	03 14
08	[illegible]	01 28	04 02
09	[illegible]	01 25	04 37
10	[illegible]	01 22	04 57
11	[illegible]	01 19	05 02
12	[illegible] ♐ 59	01 15	04 52
13	29 47	01 12	04 28
14	29 36	01 09	03 53
15	29 28	01 06	03 07
16	29 22	01 03	02 14
17	29 18	01 00	01 14
18	29 17	00 56	00 N 12
19	29 D 18	00 53	00 S 51
20	29 18	00 50	01 53
21	29 R 19	00 47	02 51
22	29 18	00 44	03 42
23	29 15	00 41	04 24
24	29 10	00 37	04 53
25	29 04	00 34	05 06
26	28 56	00 31	05 01
27	28 48	00 28	04 37
28	28 40	00 25	03 53
29	28 35	00 21	02 53
30	28 ♐ 31	00 ♑ 18	01 S 41

DECLINATIONS

Date	Sun ☉ ° '	Moon ☽ ° '	Mercury ☿ ° '	Venus ♀ ° '	Mars ♂ ° '	Jupiter ♃ ° '	Saturn ♄ ° '	Uranus ♅ ° '	Neptune ♆ ° '	Pluto ♇ ° '
01	04 N 55	19 S 54	00 N 56	02 S 22	22 S 10	18 N 25	21 S 55	09 N 24	16 N 04	20 S 06
02	05 18	22 42	01 48	01 52	22 13	18 28	21 55	09 25	16 05	20 05
03	05 41	23 57	02 41	01 23	22 16	18 32	21 55	09 25	16 05	20 05
04	06 04	23 35	03 35	00 54	22 19	18 35	21 55	09 26	16 06	20 05
05	06 26	21 43	04 29	00 S 24	22 22	18 38	21 54	09 27	16 06	20 05
06	06 49	18 36	05 24	00 N 05	22 25	18 41	21 54	09 27	16 07	20 04
07	07 11	14 31	06 19	00 35	22 28	18 44	21 54	09 28	16 07	20 04
08	07 34	09 47	07 15	01 04	22 31	18 47	21 54	09 29	16 08	20 04
09	07 56	04 S 42	08 10	01 34	22 34	18 50	21 54	09 29	16 09	20 03
10	08 18	00 N 30	09 06	02 03	22 37	18 53	21 53	09 30	16 09	20 03
11	08 40	05 37	10 01	02 33	22 40	18 56	21 53	09 30	16 10	20 03
12	09 02	10 24	10 56	03 02	22 43	18 59	21 53	09 31	16 10	20 03
13	09 24	14 43	11 50	03 31	22 46	19 02	21 53	09 32	16 11	20 02
14	09 45	18 21	12 44	04 01	22 49	19 05	21 53	09 32	16 11	20 02
15	10 07	21 09	13 37	04 30	22 51	19 08	21 53	09 33	16 12	20 02
16	10 28	23 01	14 28	04 59	22 54	19 11	21 53	09 33	16 13	20 02
17	10 49	23 51	15 18	05 28	22 57	19 14	21 53	09 34	16 13	20 01
18	11 10	23 36	16 06	05 57	23 00	19 17	21 53	09 34	16 14	20 01
19	11 30	22 17	16 53	06 26	23 03	19 20	21 53	09 34	16 14	20 01
20	11 51	19 58	17 37	06 54	23 05	19 23	21 53	09 35	16 15	20 01
21	12 11	16 43	18 19	07 23	23 08	19 26	21 53	09 35	16 15	20 01
22	12 31	12 39	18 59	07 51	23 11	19 29	21 53	09 36	16 16	20 01
23	12 51	07 55	19 37	08 19	23 14	19 32	21 53	09 36	16 17	20 01
24	13 11	02 N 40	20 13	08 47	23 16	19 35	21 53	09 36	16 17	20 00
25	13 30	02 S 53	20 45	09 15	23 19	19 38	21 53	09 37	16 18	20 00
26	13 50	08 28	21 16	09 42	23 22	19 41	21 53	09 37	16 18	20 00
27	14 09	13 45	21 43	10 10	23 25	19 44	21 53	09 37	16 19	20 00
28	14 27	18 17	22 09	10 37	23 27	19 46	21 53	09 38	16 20	20 00
29	14 46	21 40	22 31	11 04	23 30	19 49	21 53	09 38	16 20	20 00
30	15 N 04	23 S 32	22 N 52	11 N 30	23 S 33	19 N 52	21 S 53	09 N 38	16 N 21	20 S 00

ZODIAC SIGN ENTRIES

Date	h	m	Planets
02	11	43	☽ ♐
03	13	11	♀ ♈
04	13	41	☽ ♑
06	17	32	☽ ♒
08	23	25	☽ ♓
11	07	16	☽ ♈
13	02	55	☿ ♉
13	17	08	☽ ♉
16	05	02	☽ ♊
18	18	04	☽ ♋
19	09	17	☉ ♉
21	06	09	☽ ♌
23	01	43	♃ ♊
23	15	00	☽ ♍
25	19	41	☽ ♎
27	20	05	♀ ♉
27	21	02	☽ ♏
29	16	21	☿ ♊
29	20	54	☽ ♐

LATITUDES

Date	Mercury ☿ ° '	Venus ♀ ° '	Mars ♂ ° '	Jupiter ♃ ° '	Saturn ♄ ° '	Uranus ♅ ° '	Neptune ♆ ° '	Pluto ♇ ° '
01	01 S 40	01 S 29	00 N 31	00 S 42	00 N 37	00 N 48	01 S 42	12 S 09
04	01 19	01 29	00 27	00 42	00 37	00 48	01 42	12 10
07	00 54	01 29	00 22	00 41	00 36	00 48	01 42	12 11
10	00 S 26	01 28	00 17	00 41	00 36	00 48	01 42	12 11
13	00 N 06	01 27	00 11	00 40	00 36	00 48	01 42	12 12
16	00 39	01 25	00 N 05	00 40	00 36	00 48	01 42	12 13
19	01 12	01 23	00 S 01	00 39	00 36	00 48	01 41	12 14
22	01 41	01 20	00 08	00 39	00 36	00 48	01 41	12 14
25	02 06	01 17	00 15	00 39	00 36	00 48	01 41	12 15
28	02 25	01 13	00 22	00 38	00 36	00 48	01 41	12 16
31	02 N 36	01 S 09	00 S 30	00 S 38	00 N 36	00 N 47	01 S 41	12 S 17

DATA

Julian Date	2469168
Delta T	+80 seconds
Ayanamsa	24° 32' 09"
Synetic vernal point	04° ♓ 34' 50"
True obliquity of ecliptic	23° 26' 00"

MOON'S PHASES, APSIDES AND POSITIONS ☽

Date	h	m	Phase	Longitude ° '	Eclipse Indicator
05	18	11	☾	16 ♑ 38	
13	05	20	●	23 ♈ 58	
21	10	02	☽	01 ♌ 59	
28	11	13	○	08 ♏ 51	

Day	h	m		
01	10	21	Perigee	
17	07	28	Apogee	
29	14	39	Perigee	
03	18	31	Max dec	24° S 00'
10	09	40	0N	
17	18	29	Max dec	23° N 53'
24	23	37	0S	

LONGITUDES

Date	Chiron ⚷ ° '	Ceres ⚳ ° '	Pallas ⚴ ° '	Juno ⚵ ° '	Vesta ⚶ ° '	Black Moon Lilith ⚸ ° '
01	10 ♏ 28	13 ♓ 28	12 ♒ 09	12 ♒ 37	27 ♓ 27	06 ♊ 43
11	09 ♏ 48	17 ♓ 10	14 ♒ 47	15 ♒ 46	02 ♈ 12	07 ♊ 49
21	09 ♏ 04	20 ♓ 46	17 ♒ 07	18 ♒ 43	06 ♈ 54	08 ♊ 56
31	08 ♏ 18	24 ♓ 15	19 ♒ 06	21 ♒ 27	11 ♈ 30	10 ♊ 02

ASPECTARIAN

h m	Aspects
01 Wednesday	
02 23	☽ ⋕ ♃
02 34	☽ ⊥ ♂
05 49	☽ ± ☿
06 37	☽ ⚼ ♀
06 40	☽ ⚻ ☉
12 07	☽ ⚹ ♄
12 31	☽ ⚺ ♂
13 20	☽ ∥ ♇
13 56	♀ Q ♄
17 09	☽ ± ☉
18 59	☽ Q ♅
19 11	☽ ☍ ♆
22 43	☽ ⚼ ☿
02 Thursday	
00 55	☿ ⚺ ♇
03 54	☽ ∥ ♄
04 23	☽ ☍ ♃
06 40	☽ ∥ ♂
07 07	☿ ⚻ ♅
07 53	☉ ⊥ ♇
08 55	☽ ⚼ ☉
09 21	☽ △ ♀
12 42	☽ ∠ ♄
13 19	☿ ⋕ ♀
19 40	☉ ± ♅
23 41	☽ □ ♇
03 Friday	
00 26	☽ □ ♅
00 33	☉ ⊥ ♆
03 13	☽ △ ☿
03 34	☽ ⊥ ♄
11 28	☽ △ ☉
13 33	☽ ⚺ ♄
14 50	☽ ☌ ♂
20 47	☽ ⚻ ♆
20 53	☿ ∠ ♃
04 Saturday	
05 44	☽ Q ♇
06 52	☽ ⚻ ♃
06 53	☽ ± ♆
16 02	☽ □ ♀
17 10	☽ ± ♃
18 52	☉ □ ♄
22 11	☽ ⚼ ♆
05 Sunday	
01 59	☿ ⊥ ♇
02 09	☽ ⚹ ♇
02 44	☽ △ ♅
05 31	☽ ∥ ♂
06 03	☿ ± ♅
08 47	☽ ⚼ ♃
10 11	☽ ∥ ♄
10 15	☿ ⊥ ♆
14 14	☽ □ ☿
16 34	☽ ☌ ♄
18 11	☽ □ ☉
18 46	☽ ⚺ ♂
06 Monday	
00 05	☽ △ ♆
01 40	☽ ∥ ♇
02 00	☽ Q ♀
04 06	☽ ∠ ♇
04 36	☽ ⚼ ♅
05 15	☉ △ ♂
05 24	☽ ⊥ ♂
06 20	☿ □ ♄
11 14	☽ △ ♃
11 26	☽ ⋕ ♃
19 56	☽ ⊥ ♇
20 23	☽ ± ♅
21 31	☽ ∠ ♂
07 Tuesday	
00 40	☿ △ ♂
01 11	☽ ⚹ ♀
03 05	☽ ⋕ ♆
03 17	☽ Q ☿
04 12	☽ Q ☉
06 35	☽ ⚺ ♇
07 00	☽ ⚻ ♅
13 55	☉ ☌ ☿
15 55	♀ ∠ ♆
21 36	☽ ⚺ ♄
08 Wednesday	
00 47	☽ ⚹ ♂
03 23	☽ ⚹ ☉
04 42	☽ ⚹ ☿
05 26	☽ □ ♆
06 47	☽ ∠ ♀
08 27	☽ ⊥ ♄
09 29	☿ ⚺ ♆
13 31	☽ ⋕ ♅
17 45	☽ □ ♃
18 32	☿ ⊥ ♃
21 55	☽ ⋕ ☉
22 17	☽ ⋕ ☿
22 57	☽ Q ♂
09 Thursday	
00 51	☽ ⊥ ♀
00 53	☽ ∠ ♄
01 54	☉ ∥ ☿
07 52	☉ ⚺ ♆
08 56	☽ ∠ ☉
12 01	☿ ∠ ♇
13 02	☽ ⚺ ♀
13 03	☽ ☌ ♇
13 06	♀ ⚺ ♇
13 14	☽ ∠ ☿
13 15	☽ ☍ ♅
13 22	☿ ⚼ ♅
14 24	☽ Q ♆
15 18	♀ ⚻ ♅
10 Friday	
01 13	☽ ⋕ ♀
02 57	☽ ⊥ ☉
03 27	☽ Q ♃
04 38	☽ ⚹ ♄
08 48	☽ □ ♂
09 10	☽ ⊥ ☿
10 58	☉ ⊥ ♃
12 47	☽ ⚹ ♆
15 06	☽ ⚺ ☉
19 56	☽ ∥ ♀
22 32	☿ ∥ ♅
22 38	☽ ⚺ ☿
11 Saturday	
02 19	☽ ⚹ ♃
03 11	☽ Q ♄
16 31	☉ ∠ ♇
16 47	☉ ⚼ ♅
17 11	☽ ∠ ♆
21 28	☽ ⚺ ♇
21 28	☽ ⚻ ♅
22 16	♅ ☍ ♇
23 08	☿ ⚺ ♃
12 Sunday	
03 32	☽ ☌ ♀
04 20	☽ ∥ ☉
07 23	☽ ∥ ♅
07 24	☽ ∠ ♃
08 58	☽ ± ♅
09 00	☽ ⊥ ♇
10 29	☽ ⊥ ♆
13 38	☽ □ ♄
15 30	☽ ∥ ☿
18 49	☽ △ ♂
22 08	☽ ⚺ ♆
13 Monday	
01 07	☽ ⊥ ♃
02 21	☽ ⚼ ♅
02 27	☽ ∠ ♇
05 20	☽ ☌ ☉
13 02	☽ ⚺ ♃
20 05	☽ ☌ ☿
20 49	☉ ∥ ♅
21 09	☽ ∥ ♆
14 Tuesday	
00 36	☽ ⚼ ♂
02 06	♀ ∠ ♃
07 45	☽ △ ♅
07 58	☽ ⚹ ♇
09 21	♀ ± ♅
11 36	♀ ⊥ ♇
17 43	☽ ∥ ♃
18 43	☽ ± ♂
20 51	☽ ⚺ ♀
15 Wednesday	
00 34	☿ ⚼ ♂
00 41	☽ △ ♄
01 28	☽ ⋕ ♇
03 36	♀ ⊥ ♆
06 53	☽ ⚻ ♂
07 56	☽ Q ♇
09 34	☽ ☌ ♆
10 14	☽ ⊥ ♀
19 54	☽ ⋕ ♄
22 11	☽ ⚺ ☉
16 Thursday	
01 55	☽ ☌ ♃
06 30	☽ ∠ ♀
06 55	☽ ⚼ ♄
09 55	☽ ⋕ ♂
10 36	♀ □ ♄
11 25	☽ ⊥ ☉
15 58	☿ △ ♅
18 30	☿ ⚹ ♇
19 55	☽ □ ♅
20 22	☽ □ ♇
20 46	☽ ⚺ ☿
17 Friday	
01 13	☽ ± ♄
07 22	☽ ∠ ☉
11 31	☽ ⊥ ☿
13 27	☽ ⚻ ♄
16 34	☽ ⚹ ♀
20 31	☽ ☍ ♂
22 36	☽ ⚺ ♆
18 Saturday	
07 26	☉ ⚺ ♃
08 31	☽ Q ♅
09 37	☽ ∠ ☿
10 51	☽ ⊥ ♆
15 53	☿ ∥ ♆
16 05	☽ ⚺ ♃
16 41	☽ ⚹ ☉
19 52	☽ Q ♀
19 Sunday	
00 57	☽ ⋕ ♂
04 30	☽ ⊥ ♃
05 12	☽ ∠ ♆
08 51	☽ ⚹ ♅
09 30	☽ △ ♇
10 25	♀ △ ♂
17 05	☽ ⋕ ♄
19 10	☽ Q ☉
20 19	☿ ± ♂
22 01	☽ ⚹ ☿
23 00	☽ ∠ ♃
20 Monday	
02 20	☽ ☍ ♄
03 38	♀ ⚺ ♆
09 54	☽ ⚻ ♂
11 27	☽ ⚹ ♆
11 33	☽ ⋕ ♇
12 19	☽ □ ♀
14 53	☽ ∠ ♅
15 38	☽ ⚼ ♇
16 44	☽ ∥ ♃
21 55	☽ ± ♂
21 Tuesday	
01 12	☿ △ ♄
02 22	☽ Q ☿
02 50	☽ ∥ ☿
05 21	☽ ⚹ ♃
08 36	☽ ⊥ ♅
09 23	☽ ± ♇
10 02	☽ □ ☉
11 10	☽ Q ♆
13 11	♀ ⚼ ♅
14 56	☽ ∥ ♆
15 42	☽ ⚼ ♂
20 14	☽ ⚺ ♅
21 03	☽ ⚻ ♇
21 15	♀ ∠ ♇
22 Wednesday	
05 01	☽ Q ♃
12 40	☽ ∥ ☉
12 57	☽ ⚻ ♄
18 58	☽ □ ☿
20 32	☽ △ ♂
21 40	☽ □ ♆
22 49	♀ ⊥ ♃
23 Thursday	
00 14	☽ ± ♄
01 47	☿ Q ♇
03 49	☽ ∥ ♅
04 27	☽ △ ♀
06 47	☿ ⚻ ♂
08 12	☿ ∥ ♃
10 14	☽ ∥ ♀
14 57	☿ ☌ ♆
15 14	☽ □ ♃
16 48	☽ ⚼ ♄
23 06	☽ △ ☉
24 Friday	
03 36	☿ ⋕ ♇
04 01	☽ ☌ ♅
04 55	☽ ☍ ♇
07 54	♄ St R
10 27	☽ ⚼ ♀
16 11	☉ ⚼ ♂
19 35	☽ △ ♄
25 Saturday	
02 53	☽ □ ♂
03 39	☽ ⚼ ☉
03 46	☽ ± ♀
03 47	☽ △ ♆
08 45	☽ △ ☿
15 04	☽ ⚻ ♀
20 11	☽ ± ☉
20 44	☽ △ ♃
26 Sunday	
05 20	☽ ⚼ ♆
06 59	☽ ⚻ ☉
07 13	♀ ∥ ♅
07 42	☽ ⚺ ♅
08 40	☽ ⚻ ♇
09 32	☽ Q ♂
13 13	☽ ⚼ ☿
17 02	☽ ⋕ ♅
17 26	☉ △ ♅
17 38	☽ ⊥ ♅
17 55	☽ ⋕ ♀
18 36	☽ ± ♇
20 17	☽ ± ♆
22 05	☽ ⚼ ♃
22 17	☽ □ ♄
27 Monday	
05 21	☽ ⚹ ♂
05 43	☽ ± ☿
06 07	☽ ⚻ ♆
08 10	☉ ⚹ ♇
08 21	☽ ∠ ♅
09 20	☽ ⚼ ♇
10 42	♃ ⚼ ♄
12 58	☽ ± ♃
14 03	☽ ⋕ ☉
16 32	☽ ⚻ ☿
20 39	☿ ⋕ ♄
21 07	☽ ☍ ♀
22 48	☽ ⚻ ♃
28 Tuesday	
00 59	☽ ⋕ ♆
03 25	☽ Q ♄
05 36	☽ ∠ ♂
08 28	☽ ⚹ ♅
09 29	☽ △ ♇
11 13	☽ ☍ ☉
20 00	☽ ⊥ ♂
21 33	☽ ⋕ ♃
21 34	♀ ⚺ ♃
22 34	☽ ⚹ ♄
22 58	☽ ∥ ♇
29 Wednesday	
03 35	☽ Q ♅
05 33	☽ ⚺ ♂
06 20	☽ ☍ ♆
13 58	☽ ∥ ♄
21 20	☽ ☍ ☿
22 22	☽ ⋕ ☿
22 26	☽ ∠ ♄
23 22	☽ ☍ ♃
30 Thursday	
01 15	☽ ⚻ ♀
08 14	☽ □ ♅
09 21	☽ □ ♇
09 58	☿ ⚼ ♄
11 42	☽ ± ♀
12 25	☽ ∥ ♂
12 48	☽ ⊥ ♄
14 20	☽ ⚻ ☉
16 53	♂ St R
22 26	☽ ⚺ ♄

All ephemeris data is given at 12.00 UT and the Moon's longitude is additionally given for 24.00 UT

LONGITUDES

Date	Sidereal time h m s	Sun ☉ ° ' "	Moon ☽ ° ' "	Moon ☽ 24.00 ° ' "	Mercury ☿ ° '	Venus ♀ ° '	Mars ♂ ° '	Jupiter ♃ ° '	Saturn ♄ ° '	Uranus ♅ ° '	Neptune ♆ ° '	Pluto ♇ ° '
01	02 40 22	11 ♉ 47 30	24 ♐ 20 18	01 ♑ 40 52	02 ♊ 18	04 ♉ 31	20 ♐ 22	01 ♊ 54	15 ♑ 57	07 ♍ [illegible]	20 ♉ 56	07 ♓ 49
02	02 44 19	12 45 43	08 ♑ 56 45	16 ♑ 07 31	03 28	05 45	20 R 21	02 08	15 R 56	07 R [illegible]	20 58	07 50
03	02 48 16	13 43 54	23 ♑ 12 56	00 ♒ 12 54	04 32	06 59	20 20	02 21	15 55	07 0[illegible]	21 00	07 51
04	02 52 12	14 42 03	07 ♒ 07 26	13 ♒ 56 36	05 33	08 13	20 17	02 35	15 54	07 04	[illegible]	07 51
05	02 56 09	15 40 12	20 ♒ 40 36	27 ♒ 19 41	06 28	09 27	20 14	02 49	15 53	07 03	21 03	07 52
06	03 00 05	16 38 18	03 ♓ 54 04	10 ♓ 24 05	07 18	10 41	20 10	03 03	15 52	07 03	21 05	07 53
07	03 04 02	17 36 24	16 ♓ 49 59	23 ♓ 12 06	08 04	11 55	20 05	03 16	15 51	07 03	21 [illegible]	07 54
08	03 07 58	18 34 28	29 ♓ 30 40	05 ♈ 45 59	08 44	13 08	19 59	03 30	15 50	07 02	21 [illegible]	07 54
09	03 11 55	19 32 30	11 ♈ 58 17	18 ♈ 07 48	09 20	14 22	19 53	03 44	15 48	07 02	21 14	07 55
10	03 15 51	20 30 31	24 ♈ 14 46	00 ♉ 19 22	09 51	15 36	19 46	03 58	15 47	07 02	21 16	[illegible]
11	03 19 48	21 28 31	06 ♉ 21 49	12 ♉ 22 19	10 16	16 50	19 38	04 12	15 45	07 02	21 18	07 [illegible]
12	03 23 45	22 26 29	18 ♉ 21 03	24 ♉ 18 15	10 36	18 04	19 30	04 26	15 43	07 01	21 21	07 [illegible]
13	03 27 41	23 24 26	00 ♊ 14 07	06 ♊ 08 55	10 52	19 18	19 20	04 40	15 42	07 01	21 23	07 [illegible]
14	03 31 38	24 22 21	12 ♊ 02 54	17 ♊ 56 23	11 02	20 32	19 10	04 54	15 40	07 D 01	21 25	07 [illegible]
15	03 35 34	25 20 15	23 ♊ 49 41	29 ♊ 43 09	11 07	21 46	18 59	05 08	15 38	07 01	21 28	07 [illegible]
16	03 39 31	26 18 07	05 ♋ 37 12	11 ♋ 32 15	11 R 07	23 00	18 48	05 22	15 36	07 02	21 30	07 59
17	03 43 27	27 15 57	17 ♋ 28 47	23 ♋ 27 16	11 03	24 13	18 36	05 36	15 34	07 02	21 32	08 00
18	03 47 24	28 13 46	29 ♋ 28 16	05 ♌ 32 18	10 54	25 27	18 23	05 50	15 31	07 02	21 34	08 00
19	03 51 20	29 ♉ 11 33	11 ♌ 39 57	17 ♌ 51 47	10 40	26 41	18 09	06 04	15 29	07 02	21 37	08 01
20	03 55 17	00 ♊ 09 18	24 ♌ 08 21	00 ♍ 30 12	10 23	27 55	17 55	06 18	15 27	07 03	21 39	08 01
21	03 59 14	01 07 02	06 ♍ 57 50	13 ♍ 31 43	10 02	29 ♉ 09	17 40	06 32	15 24	07 03	21 41	08 02
22	04 03 10	02 04 44	20 ♍ 12 12	26 ♍ 59 34	09 37	00 ♊ 22	17 25	06 46	15 22	07 03	21 43	08 02
23	04 07 07	03 02 25	03 ♎ 53 59	10 ♎ 55 24	09 10	01 36	17 09	07 00	15 19	07 04	21 45	08 03
24	04 11 03	04 00 04	18 ♎ 03 41	25 ♎ 18 29	08 40	02 50	16 52	07 14	15 17	07 04	21 48	08 03
25	04 15 00	04 57 41	02 ♏ 39 15	10 ♏ 05 14	08 08	04 04	16 35	07 28	15 14	07 05	21 50	08 03
26	04 18 56	05 55 17	17 ♏ 35 35	25 ♏ 09 13	07 35	05 18	16 18	07 42	15 11	07 06	21 52	08 04
27	04 22 53	06 52 52	02 ♐ 44 59	10 ♐ 21 40	07 01	06 31	16 00	07 56	15 08	07 06	21 54	08 04
28	04 26 49	07 50 26	17 ♐ 58 02	25 ♐ 32 51	06 28	07 45	15 41	08 10	15 05	07 07	21 56	08 04
29	04 30 46	08 47 59	03 ♑ 05 00	10 ♑ 33 27	05 54	08 59	15 23	08 24	15 02	07 08	21 59	08 04
30	04 34 43	09 45 31	17 ♑ 57 20	25 ♑ 15 57	05 21	10 13	15 04	08 38	14 59	07 09	22 01	08 05
31	04 38 39	10 ♊ 43 01	02 ♒ 28 44	09 ♒ 35 21	04 ♊ 50	11 ♊ 26	14 ♐ 44	08 ♊ 52	14 ♑ 56	07 ♍ 10	22 ♉ 03	08 ♓ 05

Date	Moon True ☊ ° '	Moon Mean ☊ ° '	Moon ☽ Latitude ° '
01	28 ♐ 30	00 ♑ 15	00 S 22
02	28 D 30	00 12	00 N 56
03	28 31	00 09	02 10
04	28 32	00 06	03 14
05	28 R 32	00 ♑ 02	04 05
06	28 31	29 ♐ 59	04 42
07	28 28	29 56	05 04
08	28 23	29 53	05 10
09	28 17	29 50	05 01
10	28 10	29 47	04 39
11	28 03	29 43	04 04
12	27 57	29 40	03 19
13	27 53	29 37	02 25
14	27 50	29 34	01 25
15	27 49	29 31	00 N 22
16	27 D 49	29 27	00 S 43
17	27 50	29 24	01 46
18	27 52	29 21	02 45
19	27 53	29 18	03 38
20	27 54	29 15	04 22
21	27 R 54	29 12	04 54
22	27 53	29 08	05 12
23	27 51	29 05	05 13
24	27 48	29 02	04 55
25	27 45	28 59	04 19
26	27 42	28 56	03 23
27	27 40	28 53	02 13
28	27 39	28 49	00 S 53
29	27 D 39	28 46	00 N 30
30	27 39	28 43	01 50
31	27 ♐ 41	28 ♐ 40	03 N 01

DECLINATIONS

Date	Sun ☉ ° '	Moon ☽ ° '	Mercury ☿ ° '	Venus ♀ ° '	Mars ♂ ° '	Jupiter ♃ ° '	Saturn ♄ ° '	Uranus ♅ ° '	Neptune ♆ ° '	Pluto ♇ ° '
01	15 N 22	23 S 41	23 N 09	11 N 57	23 S 35	19 N 55	21 S 53	09 N 38	16 N 21	20 00
02	15 40	22 12	23 25	12 23	23 38	19 58	21 53	09 39	16 22	20 00
03	15 57	19 18	23 38	12 49	23 41	20 01	21 53	09 39	16 23	20 00
04	16 15	15 22	23 49	13 14	23 43	20 03	21 53	09 39	16 23	20 00
05	16 32	10 43	23 57	13 39	23 46	20 06	21 54	09 39	16 24	20 00
06	16 48	05 41	24 04	14 04	23 49	20 09	21 54	09 39	16 24	20 00
07	17 05	00 S 32	24 08	14 28	23 51	20 12	21 54	09 39	16 25	20 00
08	17 21	04 N 33	24 10	14 53	23 54	20 14	21 54	09 39	16 25	20 00
09	17 37	09 21	24 10	15 16	23 57	20 17	21 54	09 39	16 26	20 00
10	17 52	13 43	24 08	15 40	23 59	20 20	21 54	09 39	16 27	20 00
11	18 08	17 28	24 04	16 03	24 02	20 23	21 55	09 39	16 27	20 00
12	18 23	20 28	23 58	16 25	24 04	20 25	21 55	09 39	16 28	20 00
13	18 37	22 33	23 51	16 47	24 07	20 28	21 55	09 39	16 28	20 00
14	18 52	23 38	23 41	17 09	24 09	20 31	21 55	09 39	16 29	20 00
15	19 06	23 39	23 30	17 30	24 12	20 33	21 56	09 39	16 30	20 00
16	19 19	22 36	23 18	17 51	24 14	20 36	21 56	09 39	16 30	20 00
17	19 33	20 32	23 03	18 12	24 17	20 38	21 56	09 39	16 31	20 00
18	19 46	17 34	22 47	18 32	24 19	20 41	21 56	09 39	16 31	20 00
19	19 58	13 47	22 30	18 51	24 21	20 43	21 57	09 39	16 32	20 01
20	20 11	09 21	22 12	19 10	24 23	20 46	21 57	09 39	16 32	20 01
21	20 23	04 N 24	21 52	19 28	24 25	20 48	21 57	09 39	16 33	20 01
22	20 34	00 S 54	21 31	19 46	24 27	20 51	21 58	09 38	16 34	20 01
23	20 46	06 20	21 10	20 04	24 29	20 53	21 58	09 38	16 34	20 01
24	20 57	11 38	20 48	20 20	24 31	20 56	21 58	09 38	16 35	20 01
25	21 07	16 26	20 26	20 37	24 33	20 58	21 59	09 38	16 35	20 02
26	21 17	20 20	20 03	20 52	24 35	21 00	21 59	09 37	16 36	20 02
27	21 27	22 53	19 41	21 07	24 36	21 03	22 00	09 37	16 36	20 02
28	21 37	23 46	19 19	21 22	24 38	21 05	22 00	09 37	16 37	20 02
29	21 46	22 54	18 58	21 36	24 39	21 07	22 00	09 36	16 37	20 03
30	21 54	20 25	18 37	21 49	24 40	21 10	22 01	09 36	16 38	20 03
31	22 N 03	16 S 40	18 N 18	22 N 02	24 S 42	21 N 12	22 S 01	09 N 36	16 N 39	20 S 03

ZODIAC SIGN ENTRIES

Date	h	m	Planets
01	21	15	☽ ♑
03	23	38	☽ ♒
06	04	52	☽ ♓
08	12	56	☽ ♈
10	23	22	☽ ♉
13	11	31	☽ ♊
16	00	34	☽ ♋
18	13	03	☽ ♌
20	08	08	☉ ♊
20	23	03	☽ ♍
22	04	41	♀ ♊
23	05	15	☽ ♎
25	07	41	☽ ♏
27	07	40	☽ ♐
29	07	05	☽ ♑
31	07	51	☽ ♒

LATITUDES

Date	Mercury ☿ ° '	Venus ♀ ° '	Mars ♂ ° '	Jupiter ♃ ° '	Saturn ♄ ° '	Uranus ♅ ° '	Neptune ♆ ° '	Pluto ♇ ° '
01	02 N 36	01 S 09	00 S 30	00 S 38	00 N 36	00 N 47	01 S 41	12 S 17
04	02 38	01 04	00 39	00 38	00 36	00 47	01 41	12 18
07	02 31	00 59	00 48	00 37	00 36	00 47	01 41	12 19
10	02 14	00 53	00 57	00 37	00 36	00 47	01 41	12 20
13	01 48	00 47	01 07	00 37	00 36	00 47	01 41	12 21
16	01 12	00 41	01 17	00 36	00 36	00 47	01 41	12 22
19	00 N 28	00 35	01 27	00 36	00 36	00 47	01 41	12 23
22	00 S 22	00 28	01 38	00 36	00 35	00 47	01 41	12 24
25	01 15	00 21	01 48	00 35	00 35	00 46	01 41	12 25
28	02 06	00 14	01 59	00 35	00 35	00 46	01 41	12 26
31	02 S 51	00 S 07	02 S 09	00 S 35	00 N 35	00 N 46	01 S 41	12 S 27

DATA

Julian Date	2469198
Delta T	+80 seconds
Ayanamsa	24° 32' 12"
Synetic vernal point	04° ♓ 34' 46"
True obliquity of ecliptic	23° 25' 59"

MOON'S PHASES, APSIDES AND POSITIONS ☽

Date	h	m	Phase	Longitude ° '	Eclipse Indicator
05	02	22	☾	15 ♒ 17	
12	20	58	●	22 ♉ 48	
21	00	16	☽	00 ♍ 39	
27	18	57	○	07 ♐ 10	

Day	h	m		
14	17	30	Apogee	
28	00	06	Perigee	
01	02	11	Max dec	23° S 50'
07	14	29	ON	
15	00	20	Max dec	23° N 47'
22	07	59	OS	
28	12	02	Max dec	23° S 46'

LONGITUDES

Date	Chiron ⚷ ° '	Ceres ⚳ ° '	Pallas ⚴ ° '	Juno ⚵ ° '	Vesta ⚶ ° '	Black Moon Lilith ⚸ ° '
01	08 ♏ 18	24 ♓ 15	19 ♒ 06	21 ♒ 27	11 ♈ 30	10 ♊ 02
11	07 ♏ 33	27 ♓ 36	20 ♒ 41	23 ♒ 54	16 ♈ 00	11 ♊ 09
21	06 ♏ 51	00 ♈ 47	21 ♒ 49	26 ♒ 01	20 ♈ 24	12 ♊ 15
31	06 ♏ 15	03 ♈ 46	22 ♒ 26	27 ♒ 45	24 ♈ 40	13 ♊ 22

All ephemeris data is given at 12.00 UT and the Moon's longitude is additionally given for 24.00 UT

ASPECTARIAN

h m Aspects

01 Friday
00 39 ☽ ± ☉
02 12 ☿ ☌ ♃
03 28 ☽ ⚼ ♀
05 34 ☽ ☌ ♂
06 28 ☽ ⊼ ♆
14 24 ☽ Q ♇
14 47 ☽ ∥ ♂
16 14 ☽ ± ♆
16 17 ☽ ⚼ ☉
21 48 ☽ # ☿

02 Saturday
00 33 ☽ ⊼ ♃
02 10 ☽ ⊼ ☿
04 26 ♀ ⚼ ♂
06 13 ☽ △ ♀
07 03 ☽ ⚼ ♆
[illegible] 55 ☽ △ ♅
[illegible] ☽ ✶ ♇
[illegible] ☽ ± ♃
[illegible] ± ☿

03 [illegible]
01 55 [illegible]
04 08 [illegible]
05 16 [illegible]
07 02 ☽ [illegible]
07 04 ☽ [illegible]
07 06 ☽ ⊻ [illegible]
08 14 ☽ △ [illegible]
10 03 ☽ ⚼ ♅
11 22 ☽ ∠ ♇
13 46 ♀ △ ♅
17 18 ☽ ⊥ ♂
19 42 ☿ # ♂

04 Monday
01 28 ☽ ± ♅
02 08 ☉ ± ♂
02 50 ☽ ⊥ ♇
03 58 ☽ △ ♃
04 59 ♀ ✶ ♇
06 14 ☽ # ♆
07 21 ☽ # ☉
08 48 ☽ ∠ ♂
09 02 ☽ △ ☿
11 54 ☽ ⊼ ♅
13 17 ☽ ⊻ ♇
14 06 ☽ □ ♀
22 24 ☽ # ♀

05 Tuesday
00 23 ☉ ∥ ♆
02 22 ☽ □ ☉
03 28 ☽ ⊻ ♄
11 12 ☽ ✶ ♂
12 44 ☽ □ ♆
14 10 ☽ ⊥ ♄
17 13 ☽ # ♅
17 18 ☉ △ ♄

06 Wednesday
01 28 ☽ Q ♀
04 28 ☿ □ ♅
06 27 ☽ ∠ ♄
08 50 ☽ Q ♂
10 24 ☽ □ ♃
13 28 ☽ Q ☉
17 47 ☽ ☍ ♅
18 40 ☽ □ ☿
19 21 ☽ ☌ ♇
21 39 ☽ Q ♆

07 Thursday
01 50 ☽ ✶ ♀
06 19 ☿ □ ♇
10 10 ☽ ✶ ♄
13 34 ☽ ✶ ☉
18 04 ☽ □ ♂
20 10 ☽ ✶ ♆
20 31 ☽ Q ♃

08 Friday
06 26 ☽ Q ☿
08 48 ☽ Q ♄
09 06 ☽ ∠ ♀
19 48 ☽ ✶ ♃
20 26 ☽ ∠ ☉

09 Saturday
00 52 ☽ ∠ ♆
02 27 ☽ ⊼ ♅
03 17 ♀ ± ♂
04 09 ☽ ⊻ ♇
04 16 ☽ ⊥ ♀
06 39 ☽ ✶ ☿
13 35 ☽ ∥ ♅
14 04 ☽ ± ♅
15 19 ☽ ⊥ ☉
15 47 ☽ ⊥ ♇
17 11 ☽ ⊻ ♀
18 22 ☽ ⊥ ♆
19 26 ☽ □ ♄
19 39 ☉ ⊼ ♂
21 29 ☉ Q ♇

10 Sunday
01 26 ☽ ∠ ♃
03 18 ☽ △ ♂
04 02 ☽ ⊻ ☉
06 08 ☽ ⊻ ♆
07 38 ☽ ⚼ ♅
08 55 ☿ ± ♄
09 25 ☽ ∠ ♇
13 13 ☽ ∠ ☿
15 19 ♀ △ ♄
19 29 ☽ ⊥ ♃

11 Monday
01 14 ☽ ∥ ♀
05 00 ☽ ∥ ♆
07 37 ☽ ⊻ ♃
07 41 ☉ ☌ ♆
07 41 ☽ ⊥ ☿
08 36 ☽ ⚼ ♂
13 19 ☽ △ ♅
15 09 ☽ ✶ ♇
17 07 ☽ ∥ ☉
19 17 ☿ # ♂
20 02 ☽ ⊻ ☿

12 Tuesday
02 22 ☽ ± ♂
06 44 ☽ △ ♄
07 45 ☽ # ♇
11 22 ☽ ☌ ♀
11 34 ☽ ∥ ♃
14 16 ☽ ⊼ ♂
14 53 ♀ ∥ ♆
15 14 ☽ Q ♇
18 03 ☽ ☌ ♆
20 58 ☽ ☌ ☉

13 Wednesday
03 23 ☽ # ♄
12 40 ♀ ⊼ ♂
12 56 ☽ ⚼ ♄
21 10 ☽ ☌ ♃
22 38 ♅ St D

14 Thursday
01 03 ♀ Q ♇
[illegible] 47 ☽ □ ♅
[illegible] ☽ □ ♇
[illegible] ☽ ± ♄
[illegible] ☽ ☌ ☿
07 [illegible]
14 [illegible]
15 21 [illegible]
19 18 [illegible]
19 25 [illegible]
20 57 [illegible]

16 Saturday
01 32 ☿ St [illegible]
04 39 ☽ ⊥ [illegible]
11 28 ☽ ⊻ ♃
13 47 ☽ ∠ ♆
14 51 ☽ ✶ ♅
16 49 ☽ △ ♇
17 22 ☽ ∠ ♀
21 07 ☽ # ♄
23 06 ☽ ⊻ ☿
23 53 ☽ ⊥ ♃

17 Sunday
00 33 ☽ ∠ ☉
08 09 ☽ ☍ ♄
11 05 ☽ ∥ ♃
11 08 ☽ ⊥ ☿
14 12 ☽ ⊼ ♂
16 53 ☽ # ♇
18 24 ☽ ∠ ♃
20 11 ☽ ✶ ♆
20 13 ☽ ∥ ☉
21 09 ☽ ∠ ♅
23 06 ☽ ⚼ ♇

18 Monday
02 02 ☽ ± ♂
03 05 ☽ ✶ ♀
04 59 ☽ ∠ ☿
05 37 ☽ ∥ ♀
09 19 ☽ ✶ ☉
15 06 ☽ ⊥ ♅
17 02 ☽ ± ♇
19 05 ☽ ∥ ♆
19 36 ☽ ⚼ ♂
20 09 ☽ Q ♆

19 Tuesday
00 49 ☽ ✶ ♃
02 57 ☽ ⊻ ♅
04 52 ☽ ⊼ ♇
05 32 ☽ Q ♀
10 06 ☽ ✶ ☿
11 00 ☽ Q ☉
16 30 ☉ # ♇
19 24 ☽ ⊼ ♄

20 Wednesday
00 19 ☽ △ ♂
00 37 ☽ Q ♃
06 53 ☽ ± ♄
07 14 ☽ □ ♆
08 45 ☽ Q ☿
10 30 ☽ ∥ ♅
18 59 ☉ ⚼ ♄
19 54 ☽ □ ♀
23 51 ☽ ⚼ ♄

21 Thursday
00 16 ☽ □ ☉
05 35 ☿ # ♄
11 11 ☽ □ ♃
12 09 ☽ ☌ ♅
13 58 ☽ ☍ ♇
17 28 ☽ □ ☿

22 Friday
03 21 ☽ △ ♄
07 06 ☽ □ ♂
11 48 ♀ ⚼ ♄
14 42 ☽ △ ♆

23 Saturday
03 18 ☿ ± ♄
07 39 ☽ △ ♀
08 44 ♀ # ♇
10 24 ☽ △ ☉
14 06 ☽ Q ♂
16 55 ☽ ⚼ ♆
17 25 ☽ △ ♃
17 26 ☽ ⊻ ♅
18 56 ♃ □ ♅
19 06 ☽ ⊼ ♇
20 43 ☽ △ ☿

24 Sunday
02 47 ☽ # ♅
03 38 ☽ ⊥ ♅
04 36 ☿ ∥ ♃
05 17 ☽ ± ♇
05 50 ☉ ∥ ☿
07 21 ☽ □ ♄
08 12 ☽ ± ♆
09 25 ☉ ∥ ♃
10 03 ☽ ✶ ♂
11 35 ☽ ⚼ ♀
13 41 ☽ ⚼ ☉
18 13 ☽ ⊼ ♆
18 40 ☽ ∠ ♅
19 03 ☽ ⚼ ♃
20 17 ☽ ⚼ ♇
20 59 ☽ ⚼ ☿

25 Monday
03 50 ☽ ± ♀
05 13 ☿ ∥ ♀
05 34 ☽ ± ☉
10 02 ☽ ± ♃
10 18 ☽ ∠ ♂
11 12 ☽ ± ☿
12 51 ☽ # ♆
12 56 ☽ Q ♄
14 30 ☽ ⊼ ♀
15 42 ☿ □ ♇
16 00 ☽ ⊼ ☉
19 10 ☽ ✶ ♅
19 55 ☽ ⊼ ♃
20 33 ☽ ⊼ ☿
20 44 ☽ △ ♇

[illegible]6 Tuesday
[illegible]33 ☽ ⊥ ♂
[illegible] ☽ ✶ ♄
[illegible] ☽ [illegible] ☌ ♃
[illegible] ∥ ♇
[illegible] ♂
2[illegible]
01 [illegible]
03 0[illegible]
07 53 [illegible]
08 39 [illegible]
14 15 [illegible]
17 41 ☉ [illegible]
18 28 ☽ [illegible]
18 30 ☽ ☍ [illegible]
18 43 ☿ ☌ [illegible]
18 52 ☽ □ ♅
18 57 ☽ ☍ ☉
20 18 ☽ ☍ ♃
20 23 ☽ □ ♇
22 02 ☽ ⊥ ♄
23 29 ♀ □ ♅

28 Thursday
01 27 ♃ □ ♇
07 28 ☽ ⊻ ♄
08 29 ☽ ☌ ♂
17 44 ☉ □ ♇
18 12 ♀ □ ♇
18 18 ☽ ⊼ ♆
19 51 ☉ ☌ ♀
22 04 ♀ ☌ ♃
22 54 ☉ ☌ ♃

29 Friday
00 50 ☽ Q ♇
03 50 ☽ ± ♆
12 56 ♀ ± ♄
16 21 ☽ ⊼ ☿
17 31 ☉ ± ♄
18 15 ☽ ⚼ ♆
18 29 ☽ △ ♅
20 00 ☽ ✶ ♇
20 40 ☽ ⊼ ♃
21 48 ☽ ⊼ ☉
22 19 ☽ ⊼ ♀
22 19 ☽ ∥ ♄

30 Saturday
00 00 ☽ # ☉
01 05 ☽ # ♀
01 40 ☽ ± ☿
06 07 ☽ # ♃
06 31 ☽ ± ♃
07 10 ☽ ☌ ♄
07 23 ☽ ⊻ ♂
08 10 ☽ ± ☉
08 54 ☽ ± ♀
14 41 ☽ ∥ ♇
15 47 ☽ ⚼ ☿
16 58 ☽ ⊥ ♂
18 40 ☽ △ ♆
18 52 ☽ ⚼ ♅
19 15 ♂ ⊻ ♄
20 23 ☽ ∠ ♇
21 28 ☽ ⚼ ♃
23 57 ☽ ⚼ ☉

31 Sunday
01 01 ☽ ⚼ ♀
01 29 ☽ # ☿
06 36 ☉ # ♄
07 31 ☽ ∠ ♂
09 48 ☽ ± ♅
10 04 ♀ # ♄
11 20 ☽ ⊥ ♇
12 06 ☽ # ♆
15 50 ☽ △ ☿
16 26 ☉ ∥ ♀
16 31 ♃ ± ♄
19 53 ☽ ⊼ ♅
21 26 ☽ ⊻ ♇
22 58 ☽ △ ♃

AUGUST 2048

LONGITUDES

Date	Sidereal time h m s	Sun ☉ ° ' "	Moon ☽ ° ' "	Moon ☽ 24.00 ° ' "	Mercury ☿ ° '	Venus ♀ ° '	Mars ♂ ° '	Jupiter ♃ ° '	Saturn ♄ ° '	Uranus ♅ ° '	Neptune ♆ ° '	Pluto ♇ ° '
01	08 43 06	09 ♌ 55 14	08 ♉ 45 06	14 ♉ 48 03	21 ♌ 48	27 ♌ 35	08 ♐ 34	22 ♊ 28	10 ♑ 43	09 ♍ 33	23 ♉ 42	07 ♓ 32
02	08 47 02	10 52 38	20 ♉ 47 50	26 ♉ 45 07	23 40	28 ♌ 48	08 51	22 39	10 R 39	09 37	23 43	07 R 31
03	08 50 59	11 50 04	02 ♊ 40 33	08 ♊ 34 45	25 30	00 ♍ 02	09 08	22 50	10 36	09 40	23 44	07 30
04	08 54 55	12 47 31	14 ♊ 28 21	20 ♊ 21 57	27 18	01 16	09 26	23 02	10 32	09 44	23 45	07 29
05	08 58 52	13 44 59	26 ♊ 16 06	02 ♋ 11 20	29 ♌ 05	02 29	09 44	23 13	10 29	09 47	23 45	07 27
06	09 02 48	14 42 29	08 ♋ 08 08	14 ♋ 06 55	00 ♍ 50	03 43	10 03	23 24	10 26	09 51	23 46	07 26
07	09 06 45	15 39 59	20 ♋ 08 05	26 ♋ 11 56	02 34	04 56	10 22	23 35	10 22	09 54	23 47	07 25
08	09 10 41	16 37 31	02 ♌ 18 45	08 ♌ 28 45	04 16	06 10	10 42	23 46	10 19	09 58	23 47	07 24
09	09 14 38	17 35 04	14 ♌ 42 04	20 ♌ 58 49	05 56	07 23	11 03	23 57	10 16	10 01	23 48	07 23
10	09 18 35	18 32 37	27 ♌ 19 02	03 ♍ 42 43	07 35	08 37	11 24	24 07	10 13	10 05	23 48	07 22
11	09 22 31	19 30 12	10 ♍ 09 49	16 ♍ 40 18	09 12	09 50	11 46	24 18	10 10	10 08	23 49	07 20
12	09 26 28	20 27 48	23 ♍ 14 02	29 ♍ 50 57	10 48	11 04	12 08	24 29	10 07	10 12	23 49	07 19
13	09 30 24	21 25 25	06 ♎ 30 56	13 ♎ 13 53	12 22	12 18	12 31	24 39	10 05	10 16	23 50	07 18
14	09 34 21	22 23 03	19 ♎ 59 43	26 ♎ 48 21	13 55	13 31	12 54	24 49	10 02	10 19	23 50	07 17
15	09 38 17	23 20 42	03 ♏ 39 44	10 ♏ 33 49	15 26	14 45	13 18	25 00	09 59	10 23	23 51	07 16
16	09 42 14	24 18 22	17 ♏ 30 33	24 ♏ 29 54	16 56	15 58	13 42	25 10	09 57	10 26	23 51	07 14
17	09 46 10	25 16 03	01 ♐ 31 48	08 ♐ 36 11	18 24	17 11	14 07	25 20	09 54	10 30	23 51	07 13
18	09 50 07	26 13 45	15 ♐ 42 55	22 ♐ 51 49	19 50	18 25	14 32	25 30	09 52	10 34	23 52	07 12
19	09 54 04	27 11 28	00 ♑ 02 38	07 ♑ 15 01	21 15	19 38	14 58	25 39	09 50	10 37	23 52	07 11
20	09 58 00	28 09 12	14 ♑ 28 35	21 ♑ 42 47	22 38	20 52	15 24	25 49	09 48	10 41	23 52	07 09
21	10 01 57	29 ♌ 06 57	28 ♑ 57 03	06 ♒ 10 42	23 59	22 05	15 50	25 59	09 45	10 45	23 52	07 08
22	10 05 53	00 ♍ 04 43	13 ♒ 23 01	20 ♒ 33 15	25 19	23 19	16 17	26 08	09 43	10 49	23 52	07 07
23	10 09 50	01 02 31	27 ♒ 40 40	04 ♓ 44 32	26 37	24 32	16 44	26 17	09 41	10 52	23 53	07 06
24	10 13 46	02 00 20	11 ♓ 44 13	18 ♓ 39 08	27 53	25 45	17 12	26 27	09 40	10 56	23 53	07 04
25	10 17 43	02 58 10	25 ♓ 28 50	02 ♈ 12 59	29 ♍ 08	26 59	17 40	26 36	09 38	11 00	23 53	07 03
26	10 21 39	03 56 02	08 ♈ 51 24	15 ♈ 24 01	00 ♎ 20	28 12	18 09	26 45	09 36	11 04	23 53	07 02
27	10 25 36	04 53 56	21 ♈ 50 54	28 ♈ 12 15	01 31	29 ♍ 25	18 38	26 54	09 34	11 07	23 R 53	07 00
28	10 29 33	05 51 52	04 ♉ 28 22	10 ♉ 39 39	02 39	00 ♎ 39	19 07	27 02	09 33	11 11	23 53	06 59
29	10 33 29	06 49 49	16 ♉ 46 36	22 ♉ 49 45	03 46	01 52	19 36	27 11	09 32	11 15	23 53	06 58
30	10 37 26	07 47 48	28 ♉ 49 44	04 ♊ 47 10	04 50	03 05	20 06	27 19	09 30	11 19	23 53	06 57
31	10 41 22	08 ♍ 45 50	10 ♊ 42 45	16 ♊ 37 09	05 ♎ 51	04 ♎ 18	20 ♐ 37	27 ♊ 28	09 ♑ 29	11 ♍ 22	23 ♉ 52	06 ♓ 55

Date	Moon True ☊ ° '	Moon Mean ☊ ° '	Moon ☽ Latitude ° '
01	26 ♐ 45	25 ♐ 23	03 N 51
02	26 D 46	25 20	03 03
03	26 47	25 16	02 07
04	26 48	25 13	01 06
05	26 49	25 10	00 N 03
06	26 R 48	25 07	01 S 01
07	26 46	25 04	02 03
08	26 41	25 01	02 59
09	26 34	24 57	03 49
10	26 25	24 54	04 28
11	26 16	24 51	04 54
12	26 07	24 48	05 06
13	25 59	24 45	05 01
14	25 53	24 42	04 39
15	25 49	24 38	04 02
16	25 47	24 35	03 10
17	25 D 47	24 32	02 06
18	25 48	24 29	00 S 54
19	25 R 48	24 26	00 N 23
20	25 47	24 22	01 38
21	25 43	24 19	02 47
22	25 37	24 16	03 45
23	25 28	24 13	04 29
24	25 18	24 10	04 55
25	25 07	24 07	05 03
26	24 57	24 03	04 54
27	24 48	24 00	04 30
28	24 42	23 57	03 54
29	24 38	23 54	03 07
30	24 36	23 51	02 13
31	24 ♐ 36	23 ♐ 48	01 N 13

DECLINATIONS

Date	Sun ☉ ° '	Moon ☽ ° '	Mercury ☿ ° '	Venus ♀ ° '	Mars ♂ ° '	Jupiter ♃ ° '	Saturn ♄ ° '	Uranus ♅ ° '	Neptune ♆ ° '	Pluto ♇ ° '
01	17 N 45	18 N 04	15 N 47	13 N 43	25 S 14	22 N 42	22 S 30	08 N 40	17 N 01	20 S 32
02	17 30	20 53	15 07	13 18	25 16	22 42	22 31	08 39	17 01	20 33
03	17 14	22 46	14 27	12 52	25 18	22 43	22 31	08 37	17 01	20 34
04	16 58	23 38	13 46	12 25	25 21	22 44	22 31	08 36	17 02	20 34
05	16 42	23 26	13 04	11 59	25 23	22 44	22 32	08 35	17 02	20 35
06	16 25	22 10	12 22	11 32	25 25	22 45	22 32	08 33	17 02	20 35
07	16 08	19 54	11 39	11 04	25 28	22 45	22 32	08 32	17 02	20 36
08	15 51	16 43	10 57	10 37	25 30	22 46	22 33	08 31	17 02	20 37
09	15 34	12 46	10 14	10 09	25 33	22 46	22 33	08 29	17 02	20 37
10	15 16	08 12	09 31	09 41	25 35	22 47	22 34	08 28	17 02	20 38
11	14 58	03 N 13	08 48	09 12	25 37	22 47	22 34	08 27	17 02	20 38
12	14 40	01 S 59	08 05	08 44	25 40	22 48	22 34	08 25	17 02	20 39
13	14 21	07 11	07 23	08 15	25 42	22 48	22 35	08 24	17 02	20 40
14	14 03	12 07	06 40	07 46	25 45	22 49	22 35	08 23	17 02	20 40
15	13 44	16 31	05 57	07 17	25 47	22 49	22 35	08 21	17 02	20 41
16	13 25	20 05	05 15	06 47	25 49	22 50	22 35	08 20	17 03	20 41
17	13 06	22 31	04 33	06 18	25 52	22 50	22 36	08 18	17 03	20 42
18	12 46	23 33	03 51	05 48	25 54	22 50	22 36	08 17	17 03	20 42
19	12 27	23 03	03 10	05 18	25 56	22 51	22 36	08 16	17 03	20 43
20	12 07	21 01	02 29	04 48	25 58	22 51	22 37	08 14	17 03	20 44
21	11 47	17 38	01 48	04 18	26 00	22 51	22 37	08 13	17 03	20 44
22	11 26	13 12	01 08	03 47	26 02	22 52	22 37	08 11	17 03	20 45
23	11 06	08 04	00 N 29	03 17	26 04	22 52	22 37	08 10	17 03	20 45
24	10 45	02 S 37	00 S 10	02 46	26 06	22 52	22 38	08 08	17 02	20 46
25	10 25	02 N 50	00 48	02 16	26 08	22 52	22 38	08 07	17 02	20 46
26	10 04	08 01	01 25	01 45	26 10	22 53	22 38	08 06	17 02	20 47
27	09 43	12 41	02 02	01 14	26 12	22 53	22 38	08 04	17 02	20 47
28	09 22	16 40	02 38	00 43	26 13	22 53	22 39	08 03	17 02	20 48
29	09 00	19 50	03 13	00 N 12	26 15	22 53	22 39	08 01	17 02	20 49
30	08 39	22 03	03 47	00 S 19	26 16	22 53	22 39	08 00	17 02	20 49
31	08 N 17	23 N 15	04 S 20	00 S 50	26 S 18	22 N 54	22 S 39	07 N 58	17 N 02	20 S 50

ZODIAC SIGN ENTRIES

Date	h	m	Planets
03	06	34	☽ ♊
03	11	21	♀ ♍
05	19	34	☽ ♋
06	00	33	☿ ♍
08	07	28	☽ ♌
10	17	03	☽ ♍
13	00	16	☽ ♎
15	05	36	☽ ♏
17	09	24	☽ ♐
19	11	56	☽ ♑
21	13	44	☽ ♒
22	10	02	☉ ♍
23	15	56	☽ ♓
25	20	02	☽ ♈
26	05	12	☿ ♎
27	23	21	♀ ♎
28	03	25	☽ ♉
30	14	21	☽ ♊

LATITUDES

Date	Mercury ☿ ° '	Venus ♀ ° '	Mars ♂ ° '	Jupiter ♃ ° '	Saturn ♄ ° '	Uranus ♅ ° '	Neptune ♆ ° '	Pluto ♇ ° '
01	01 N 38	01 N 30	03 S 33	00 S 32	00 N 30	00 N 44	01 S 44	12 S 46
04	01 26	01 30	03 32	00 31	00 30	00 44	01 44	12 46
07	01 11	01 29	03 30	00 31	00 30	00 44	01 44	12 47
10	00 52	01 27	03 28	00 31	00 29	00 44	01 44	12 47
13	00 30	01 25	03 27	00 31	00 29	00 44	01 44	12 48
16	00 N 06	01 22	03 24	00 31	00 28	00 44	01 45	12 48
19	00 S 20	01 18	03 22	00 31	00 28	00 44	01 45	12 48
22	00 47	01 14	03 20	00 31	00 27	00 44	01 45	12 48
25	01 15	01 09	03 17	00 31	00 27	00 44	01 45	12 49
28	01 43	01 04	03 15	00 31	00 27	00 44	01 45	12 49
31	02 S 11	00 N 58	03 S 12	00 S 31	00 N 26	00 N 44	01 S 45	12 S 49

DATA

Julian Date	2469290
Delta T	+80 seconds
Ayanamsa	24° 32' 27"
Synetic vernal point	04° ♓ 34' 31"
True obliquity of ecliptic	23° 25' 58"

MOON'S PHASES, APSIDES AND POSITIONS ☽

Date	h	m	Phase	Longitude ° '	Eclipse Indicator
01	14	30	☾	10 ♉ 01	
09	17	59	●	17 ♌ 49	
17	00	32	☽	24 ♏ 48	
23	18	07	○	01 ♓ 17	
31	07	42	☾	08 ♊ 35	

Day	h	m		
04	17	12	Apogee	
20	12	36	Perigee	
04	19	31	Max dec	23° N 41'
12	02	53	0S	
18	16	18	Max dec	23° S 35'
24	23	26	0N	

LONGITUDES

Date	Chiron ⚷ ° '	Ceres ⚳ ° '	Pallas ⚴ ° '	Juno ⚵ ° '	Vesta ⚶ ° '	Black Moon Lilith ⚸ ° '
01	06 ♏ 05	15 ♈ 20	13 ♒ 41	26 ♒ 07	16 ♉ 52	20 ♊ 14
11	06 ♏ 41	15 ♈ 33	11 ♒ 04	23 ♒ 50	19 ♉ 24	21 ♊ 20
21	07 ♏ 26	15 ♈ 10	08 ♒ 40	21 ♒ 22	21 ♉ 30	22 ♊ 27
31	08 ♏ 18	14 ♈ 11	06 ♒ 38	19 ♒ 02	23 ♉ 05	23 ♊ 33

ASPECTARIAN

h m	Aspects
01 Saturday	
02 19	☉ ⚺ ♅
04 37	☽ ∥ ♆
09 25	☽ ∠ ♃
09 36	☽ ⚹ ♇
09 57	☽ ∥ ☉
11 38	☽ ⚻ ♂
13 36	☽ △ ♅
14 30	☽ □ ☉
15 51	☽ △ ♄
21 27	☿ ⚹ ♃
02 Sunday	
03 34	☽ ⊥ ♃
06 43	☉ ⚻ ♄
08 41	☽ # ♇
09 26	☽ Q ♇
12 46	☿ □ ♆
15 47	☽ ⚺ ♃
17 53	☽ ☌ ♆
18 49	☽ □ ☿
21 44	☽ bQ ♄
03 Monday	
05 44	☽ Q ☉
06 01	☽ □ ♀
07 53	☽ # ♄
11 05	☽ ∥ ♃
13 18	☿ bQ ♄
15 53	☽ ± ♄
21 47	☽ □ ♇
04 Tuesday	
01 28	☽ ☍ ♂
02 17	☽ □ ♅
04 02	☽ ⚻ ♄
06 44	☉ ∥ ♆
08 16	☽ ⚹ ☉
13 59	☽ Q ☿
22 53	☽ Q ♀
05 Wednesday	
05 42	☽ ☌ ♃
06 54	☽ ⚺ ♆
15 06	☽ Q ♅
17 03	♂ □ ♅
17 29	☽ ∠ ☉
18 42	☽ ⚹ ☿
19 05	☽ ⊥ ♆
06 Thursday	
02 03	☽ ⚹ ♀
03 24	☽ ∥ ♃
06 46	☽ # ♄
10 36	☽ △ ♇
13 15	☽ ⊥ ☉
13 16	☽ ∠ ♆
15 27	☽ ⚹ ♅
15 57	☽ ⚻ ♂
16 35	☽ ☍ ♄
07 Friday	
02 20	☽ ⚺ ☉
04 18	☽ ± ♂
05 38	☽ # ♇
06 02	☽ ∠ ☿
11 34	☽ ∠ ♀
12 10	♂ ⚺ ♄
16 31	☽ bQ ♇
18 56	☽ ⚺ ♃
19 14	☽ ⚹ ♆
21 29	☽ ∠ ♅
22 40	☽ bQ ♂
08 Saturday	
02 47	☽ ⊥ ☿
02 49	♀ Q ♃
04 49	☉ ± ♄
06 56	☽ ⊥ ♃
07 20	☽ ⊥ ♀
09 54	☽ ∥ ♆
10 13	☽ ± ♇
15 14	☽ ⊥ ♅
15 25	♃ ⚺ ♆
16 24	☽ ⚺ ☿
18 10	☽ ∥ ☉
18 47	☽ Q ♆
20 20	☽ ⚺ ♀
21 53	☽ ⚻ ♇
09 Sunday	
00 44	☽ ∠ ♃
02 56	☽ ⚺ ♅
03 30	☽ ⚻ ♄
04 46	☽ △ ♂
11 48	♀ ☍ ♇
12 11	☿ Q ♃
15 00	☽ ± ♄
17 59	☽ ☌ ☉
20 21	☿ ∥ ♀
10 Monday	
03 41	☽ ∥ ♀
04 09	☽ ∥ ☿
05 22	☽ □ ♆
05 53	☽ ⚹ ♃
08 03	☽ bQ ♄
08 47	☿ ☍ ♇
10 41	☽ ∥ ♅
11 Tuesday	
04 44	☽ Q ♃
06 46	☽ ☍ ♇
09 58	☽ ☌ ☿
11 20	☽ ☌ ♀
11 57	☽ ☌ ♅
12 01	☽ △ ♄
15 03	☽ □ ♂
18 08	♀ ☌ ♅
18 14	♀ △ ♄
19 20	♄ △ ♅
12 Wednesday	
00 30	☿ ∥ ♅
02 06	☿ △ ♄
02 34	☿ ☌ ♅
06 33	☽ ⚺ ☉
13 05	☽ △ ♆
14 17	☽ □ ♃
18 20	☽ ⊥ ☉
13 Thursday	
00 53	☽ Q ♂
04 15	♀ ∥ ♅
06 24	☿ ☌ ♀
11 49	☽ ∠ ☉
12 47	☽ # ☿
13 24	☽ ⚻ ♇
14 56	☿ □ ♂
16 09	☽ bQ ♆
16 35	☽ # ♀
17 42	☽ # ♅
18 19	♀ □ ♂
18 21	☽ □ ♄
18 44	☽ ⚺ ♅
23 02	☽ ⚹ ♂
23 22	☽ ⚺ ♀
23 50	☽ ⚺ ☿
14 Friday	
00 06	☽ ± ♇
05 28	☽ ⊥ ♅
08 11	☽ ± ♆
11 04	☽ ⊥ ♀
11 50	☽ ⊥ ☿
16 02	☽ bQ ♇
16 32	☽ ⚹ ☉
18 47	☽ ⚻ ♆
20 37	☽ △ ♃
21 23	☽ # ☉
21 26	☽ ∠ ♅
15 Saturday	
02 07	☽ Q ♄
02 20	☽ ∠ ♂
04 29	☽ ∠ ♀
05 40	☽ ∠ ☿
15 07	☽ # ♆
15 09	☽ Q ☉
18 15	☽ △ ♇
18 31	☽ ⊥ ♂
22 58	☽ ⚹ ♄
23 09	☽ bQ ♃
23 44	☽ ⚹ ♅
16 Sunday	
00 34	☉ □ ♆
05 14	☽ ⚺ ♂
09 05	☽ ⚹ ♀
10 53	☽ ⚹ ☿
14 53	☽ ± ♃
16 56	☽ ∥ ♇
20 31	☽ Q ♅
22 54	☽ ☍ ♆
17 Monday	
00 32	☽ □ ☉
00 44	☽ ∠ ♄
01 18	☽ ⚻ ♃
03 21	☉ bQ ♄
07 38	☽ Q ♀
09 50	☽ Q ☿
13 04	☽ ∥ ♄
13 49	☉ ⚹ ♃
16 02	☽ ⊥ ♄
16 36	☽ # ♃
21 38	☽ □ ♇
18 Tuesday	
02 10	☽ ⚺ ♄
03 17	☽ □ ♅
09 57	☽ ☌ ♂
16 58	☽ □ ♀
19 41	☽ □ ☿
19 Wednesday	
01 40	☽ ⚻ ♆
03 53	☽ Q ♇
04 35	☽ ☍ ♃
06 54	☽ △ ☉
11 42	☽ ± ♆
15 36	☽ # ♃
19 06	☽ ∥ ♄
23 52	☽ ⚹ ♇
20 Thursday	
02 41	☽ bQ ♆
04 15	☽ ☌ ♄
05 41	☽ △ ♅
09 39	☽ bQ ☉
13 34	☽ ⚺ ♂
14 30	☽ ∥ ♇
23 34	☽ △ ♀
23 50	☽ ⊥ ♂
21 Friday	
00 43	☽ ∠ ♇
01 38	☽ ± ☉
02 56	☽ △ ☿
03 35	☽ △ ♆
06 40	☽ bQ ♅
07 01	☽ ⚻ ♃
09 55	☿ △ ♆
12 18	☽ ⚻ ☉
15 14	☽ ∠ ♂
15 31	☽ # ♆
15 37	☽ ⊥ ♇
17 05	☽ ± ♃
21 40	☽ ± ♅
22 Saturday	
01 34	☽ ⚺ ♇
02 46	☽ bQ ♀
05 55	☽ ⚺ ♄
06 22	☽ bQ ☿
07 41	☽ ⚻ ♅
08 12	☽ bQ ♃
15 54	☽ ⊥ ♄
17 00	☽ ⚹ ♂
19 10	☽ ± ♀
21 07	☽ # ☉
22 56	☽ ± ☿
23 07	♀ △ ♆
23 Sunday	
05 02	☿ □ ♃
05 35	☽ □ ♆
06 12	☽ ⚻ ♀
06 58	☽ ∠ ♄
09 38	☽ △ ♃
10 02	☽ ⚻ ☿
11 34	☽ # ♅
13 51	☽ Q ♂
18 07	☽ ☍ ☉
24 Monday	
04 00	☽ ☌ ♇
08 26	☽ ⚹ ♄
10 37	☽ ☍ ♅
11 16	☽ # ♀
12 15	☽ Q ♆
21 35	☽ ∥ ☿
21 48	☽ □ ♂
25 Tuesday	
01 45	☽ # ☿
03 26	♀ □ ♃
05 14	☽ Q ♄
09 10	☽ ⚹ ♆
09 38	☽ ∥ ♀
14 00	☽ □ ♃
14 55	☽ ☍ ♀
19 08	☽ ☍ ☿
26 Wednesday	
02 24	☽ ⚻ ☉
08 41	☽ ⚺ ♇
12 03	☽ ∠ ♆
12 23	☽ ∥ ♅
13 21	☽ □ ♄
14 07	☽ ± ☉
15 05	♆ St R
16 02	☽ ⚻ ♅
18 51	☿ # ♀
19 37	☽ ⊥ ♇
21 27	☽ ∥ ☉
22 55	☽ Q ♃
27 Thursday	
03 09	☽ ± ♅
04 36	☽ ⊥ ♆
05 45	☽ △ ♂
08 03	☽ bQ ☉
12 18	☽ ∠ ♇
15 49	☽ ⚺ ♆
20 05	☽ bQ ♅
21 37	☽ ⚹ ♃
28 Friday	
03 52	☽ ⚻ ♀
08 09	☽ ⚻ ☿
11 17	☽ bQ ♂
14 29	☽ ∥ ♆
14 55	☽ △ ☉
16 39	☽ ± ♀
16 51	☽ ⚹ ♇
20 54	☽ ± ☿
21 49	☽ △ ♄
29 Saturday	
01 05	☽ △ ♅
02 52	☽ ∠ ♃
05 30	☽ ± ♂
12 12	☽ bQ ♀
15 16	☉ ☍ ♇
16 18	☽ bQ ☿
16 19	☽ Q ♇
17 50	☽ ⚻ ♂
20 50	☽ ⊥ ♃
21 22	☽ # ♇
30 Sunday	
02 05	☽ ☌ ♆
03 21	☽ bQ ♄
08 57	☽ ⚺ ♃
21 24	☽ ± ♄
21 33	☽ △ ♀
21 33	☽ # ♄
31 Monday	
01 14	☽ △ ☿
02 19	☽ ∥ ♃
04 20	☽ □ ♇
07 42	☽ □ ☉
09 31	☽ ⚻ ♄
13 21	☽ □ ♅

All ephemeris data is given at 12.00 UT and the Moon's longitude is additionally given for 24.00 UT

SEPTEMBER 2048

LONGITUDES

Date	Sidereal time h m s	Sun ☉	Moon ☽	Moon ☽ 24.00	Mercury ☿	Venus ♀	Mars ♂	Jupiter ♃	Saturn ♄	Uranus ♅	Neptune ♆	Pluto ♇
01	10 45 19	09 ♍ 43 53	22 ♊ 31 03	28 ♊ 25 09	06 ♎ 50	05 ♎ 32	21 ♐ 07	27 ♊ 36	09 ♑ 28	11 ♍ 26	23 ♉ 52	06 ♓ 54
02	10 49 15	10 41 58	04 ♋ 36 50	10 ♋ 16 34	07 47	06 45	21 38	27 44	09 R 27	11 30	23 R 52	06 R 53
03	10 53 12	11 40 05	16 ♋ 15 07	22 ♋ 16 18	08 40	07 58	22 10	27 52	09 26	11 34	23 52	06 52
04	10 57 08	12 38 13	28 ♋ 20 39	04 ♌ 28 35	09 31	09 11	22 41	28 00	09 25	11 37	23 51	06 50
05	11 01 05	13 36 24	10 ♌ 40 27	16 ♌ 56 32	10 18	10 25	23 13	28 08	09 25	11 41	23 51	06 49
06	11 05 02	14 34 36	23 ♌ 17 01	29 ♌ 42 01	11 02	11 38	23 46	28 15	09 24	11 45	23 51	06 48
07	11 08 58	15 32 51	06 ♍ 11 31	12 ♍ 45 26	11 42	12 51	24 18	28 22	09 23	11 49	23 50	06 46
08	11 12 55	16 31 07	19 ♍ 23 56	26 ♍ 05 42	12 19	14 04	24 51	28 30	09 23	11 53	23 50	06 45
09	11 16 51	17 29 24	02 ♎ 51 29	09 ♎ 40 33	12 51	15 17	25 24	28 37	09 22	11 56	23 50	06 44
10	11 20 48	18 27 44	16 ♎ 32 31	23 ♎ 26 58	13 19	16 30	25 58	28 44	09 22	12 00	23 49	06 43
11	11 24 44	19 26 05	00 ♏ 21 31	07 ♏ 21 45	13 41	17 44	26 32	28 51	09 22	12 04	23 49	06 41
12	11 28 41	20 24 28	14 ♏ 21 23	21 ♏ 22 04	13 59	18 57	27 06	28 57	09 22	12 08	23 48	06 40
13	11 32 37	21 22 52	28 ♏ 23 36	05 ♐ 25 46	14 11	20 10	27 40	29 04	09 D 22	12 11	23 48	06 39
14	11 36 34	22 21 19	12 ♐ 28 25	19 ♐ 31 26	14 18	21 23	28 15	29 10	09 22	12 15	23 47	06 38
15	11 40 31	23 19 46	26 ♐ 34 43	03 ♑ 38 10	14 R 18	22 36	28 50	29 16	09 22	12 19	23 46	06 37
16	11 44 27	24 18 16	10 ♑ 41 40	17 ♑ 45 05	14 12	23 49	29 ♐ 25	29 22	09 22	12 22	23 46	06 36
17	11 48 24	25 16 46	24 ♑ 48 15	01 ♒ 50 56	13 59	25 02	00 ♑ 00	29 28	09 23	12 26	23 45	06 34
18	11 52 20	26 15 19	08 ♒ 52 51	15 ♒ 53 41	13 39	26 15	00 36	29 34	09 23	12 30	23 44	06 33
19	11 56 17	27 13 53	22 ♒ 53 01	29 ♒ 50 27	13 13	27 28	01 12	29 39	09 24	12 34	23 44	06 32
20	12 00 13	28 12 28	06 ♓ 45 31	13 ♓ 37 45	12 39	28 41	01 48	29 45	09 25	12 37	23 43	06 31
21	12 04 10	29 ♍ 11 06	20 ♓ 26 43	27 ♓ 11 58	11 58	29 ♎ 54	02 24	29 50	09 26	12 41	23 42	06 29
22	12 08 06	00 ♎ 09 45	03 ♈ 53 10	10 ♈ 29 59	11 11	01 ♏ 06	03 01	29 ♊ 55	09 26	12 45	23 41	06 28
23	12 12 03	01 08 27	17 ♈ 02 15	23 ♈ 29 48	10 18	02 19	03 37	00 ♋ 00	09 27	12 48	23 40	06 27
24	12 16 00	02 07 10	29 ♈ 52 38	06 ♉ 10 50	09 19	03 32	04 14	00 05	09 29	12 52	23 39	06 26
25	12 19 56	03 05 56	12 ♉ 24 35	18 ♉ 34 08	08 17	04 45	04 51	00 09	09 30	12 55	23 38	06 25
26	12 23 53	04 04 44	24 ♉ 39 50	00 ♊ 42 09	07 12	05 58	05 29	00 13	09 31	12 59	23 38	06 24
27	12 27 49	05 03 34	06 ♊ 41 34	12 ♊ 38 38	06 05	07 10	06 06	00 18	09 32	13 03	23 37	06 23
28	12 31 46	06 02 26	18 ♊ 33 58	24 ♊ 28 13	04 58	08 23	06 44	00 22	09 34	13 06	23 36	06 22
29	12 35 42	07 01 21	00 ♋ 22 02	06 ♋ 16 08	03 53	09 36	07 22	00 25	09 36	13 10	23 35	06 21
30	12 39 39	08 ♎ 00 18	12 ♋ 11 10	18 ♋ 07 52	02 ♎ 52	10 ♏ 49	08 ♑ 00	00 ♋ 29	09 ♑ 37	13 ♍ 13	23 ♉ 33	06 ♓ 20

Moon Nodes, Latitude and DECLINATIONS

Date	Moon True ☊	Moon Mean ☊	Moon ☽ Latitude	Sun ☉	Moon ☽	Mercury ☿	Venus ♀	Mars ♂	Jupiter ♃	Saturn ♄	Uranus ♅	Neptune ♆	Pluto ♇
01	24 ♐ 36	23 ♐ 44	00 N 11	07 N 55	23 N 24	04 S 52	01 S 20	26 S 19	22 N 54	22 S 40	07 N 57	17 N 02	20 S 50
02	24 R 36	23 41	00 S 52	07 33	22 30	05 23	01 51	26 20	22 54	22 40	07 56	17 02	20 51
03	24 34	23 38	01 52	07 11	20 35	05 52	02 22	26 21	22 54	22 40	07 55	17 02	20 51
04	24 30	23 35	02 49	06 49	17 44	06 20	02 53	26 22	22 54	22 40	07 53	17 02	20 52
05	24 23	23 32	03 38	06 27	14 03	06 47	03 24	26 22	22 54	22 40	07 51	17 01	20 52
06	24 14	23 28	04 19	06 04	09 41	07 12	03 55	26 23	22 55	22 40	07 50	17 01	20 53
07	24 02	23 25	04 46	05 42	04 N 48	07 35	04 26	26 24	22 55	22 41	07 48	17 01	20 53
08	23 50	23 22	05 00	05 19	00 S 24	07 56	04 56	26 24	22 55	22 41	07 47	17 01	20 53
09	23 37	23 19	04 57	04 56	05 40	08 16	05 27	26 24	22 55	22 41	07 46	17 01	20 54
10	23 26	23 16	04 37	04 34	10 46	08 33	05 57	26 24	22 55	22 41	07 44	17 01	20 54
11	23 17	23 13	04 01	04 11	15 21	08 47	06 27	26 24	22 55	22 41	07 43	17 01	20 55
12	23 11	23 09	03 10	03 48	19 09	09 00	06 58	26 24	22 55	22 41	07 41	17 00	20 55
13	23 08	23 06	02 07	03 25	21 51	09 09	07 28	26 24	22 55	22 41	07 40	17 00	20 56
14	23 D 07	23 03	00 S 56	03 02	23 13	09 15	07 58	26 23	22 55	22 42	07 38	17 00	20 56
15	23 R 07	23 00	00 N 18	02 39	23 05	09 18	08 27	26 23	22 55	22 42	07 37	17 00	20 57
16	23 06	22 57	01 32	02 16	21 29	09 18	08 57	26 22	22 55	22 42	07 36	17 00	20 57
17	23 04	22 53	02 39	01 53	18 33	09 14	09 26	26 21	22 55	22 42	07 34	16 59	20 57
18	23 00	22 50	03 37	01 29	14 33	09 06	09 56	26 20	22 55	22 42	07 33	16 59	20 58
19	22 52	22 47	04 21	01 06	09 46	08 54	10 25	26 18	22 55	22 42	07 31	16 59	20 58
20	22 42	22 44	04 49	00 43	04 S 33	08 38	10 53	26 17	22 55	22 42	07 30	16 59	20 59
21	22 31	22 41	05 00	00 N 19	00 N 49	08 17	11 22	26 15	22 55	22 42	07 29	16 58	20 59
22	22 18	22 38	04 55	00 S 04	06 03	07 52	11 50	26 14	22 55	22 42	07 27	16 58	20 59
23	22 06	22 34	04 33	00 27	10 54	07 24	12 18	26 12	22 55	22 42	07 26	16 58	21 00
24	21 56	22 31	03 58	00 51	15 08	06 51	12 46	26 09	22 55	22 42	07 24	16 58	21 00
25	21 47	22 28	03 12	01 14	18 36	06 15	13 13	26 07	22 56	22 42	07 23	16 57	21 00
26	21 42	22 25	02 18	01 37	21 10	05 36	13 40	26 04	22 56	22 42	07 22	16 57	21 01
27	21 39	22 22	01 19	02 01	22 43	04 54	14 07	26 01	22 56	22 42	07 20	16 57	21 01
28	21 38	22 19	00 N 16	02 24	23 13	04 12	14 34	25 58	22 56	22 42	07 19	16 56	21 01
29	21 D 38	22 15	00 S 46	02 47	22 40	03 29	15 00	25 55	22 56	22 42	07 18	16 56	21 01
30	21 ♐ 39	22 ♐ 12	01 S 47	03 S 10	21 N 06	02 S 47	15 S 26	25 S 51	22 N 56	22 S 42	07 N 16	16 N 56	21 S 02

ZODIAC SIGN ENTRIES

Date	h	m	Planets
02	03	13	☽ ♋
04	15	15	☽ ♌
07	00	33	☽ ♍
09	06	56	☽ ♎
11	11	19	☽ ♏
13	14	44	☽ ♐
15	17	49	☽ ♑
17	11	50	♂ ♑
17	20	51	☽ ♒
20	00	17	☽ ♓
21	14	07	♀ ♏
22	05	01	☽ ♈
22	08	01	☉ ♎
23	12	56	♃ ♋
24	12	14	☽ ♉
26	22	36	☽ ♊
29	11	15	☽ ♋

LATITUDES

Date	Mercury ☿	Venus ♀	Mars ♂	Jupiter ♃	Saturn ♄	Uranus ♅	Neptune ♆	Pluto ♇
01	02 S 21	00 N 56	03 S 11	00 S 31	00 N 26	00 N 44	01 S 45	12 S 49
04	02 48	00 49	03 08	00 31	00 26	00 44	01 46	12 49
07	03 13	00 42	03 05	00 31	00 26	00 44	01 46	12 49
10	03 35	00 35	03 02	00 31	00 25	00 44	01 46	12 49
13	03 52	00 27	02 59	00 31	00 25	00 44	01 46	12 49
16	04 01	00 19	02 56	00 31	00 24	00 44	01 46	12 49
19	04 00	00 10	02 53	00 31	00 24	00 44	01 46	12 49
22	03 45	00 N 01	02 49	00 31	00 24	00 44	01 47	12 48
25	03 13	00 S 08	02 46	00 30	00 23	00 44	01 47	12 48
28	02 25	00 17	02 43	00 30	00 23	00 44	01 47	12 48
31	01 S 27	00 S 26	02 S 39	00 S 30	00 N 23	00 N 44	01 S 47	12 S 47

DATA

Julian Date	2469321
Delta T	+80 seconds
Ayanamsa	24° 32' 31"
Synetic vernal point	04° ♓ 34' 28"
True obliquity of ecliptic	23° 25' 58"

MOON'S PHASES, APSIDES AND POSITIONS ☽

Date	h	m	Phase	Longitude	Eclipse Indicator
08	06	24	●	16 ♍ 18	
15	06	04	☽	23 ♐ 05	
22	04	46	○	29 ♓ 52	
30	02	45	☾	07 ♋ 38	

Day	h	m		
01	10	51	Apogee	
15	15	06	Perigee	
29	06	19	Apogee	
01	03	26	Max dec	23° N 28'
08	10	11	0S	
14	21	58	Max dec	23° S 20'
21	08	19	0N	
28	11	21	Max dec	23° N 13'

LONGITUDES

Date	Chiron ⚷	Ceres ⚳	Pallas ⚴	Juno ⚵	Vesta ⚶	Black Moon Lilith ⚸
01	08 ♏ 24	14 ♈ 03	06 ♒ 28	18 ♒ 49	23 ♉ 12	23 ♊ 40
11	09 ♏ 25	12 ♈ 28	05 ♒ 01	16 ♒ 55	24 ♉ 08	24 ♊ 46
21	10 ♏ 33	10 ♈ 27	04 ♒ 11	15 ♒ 38	24 ♉ 22	25 ♊ 53
31	11 ♏ 45	08 ♈ 13	04 ♒ 56	15 ♒ 05	23 ♉ 53	26 ♊ 59

ASPECTARIAN

h m Aspects

01 Tuesday
05 30 ☉ △ ♄
08 12 ☉ Q ♃
09 02 ☽ ☍ ♂
09 45 ☉ ∥ ♅
13 33 ☿ ⊼ ♇
14 45 ☽ ⊻ ♆
22 27 ☽ ☌ ♃
23 33 ☽ Q ☉

02 Wednesday
02 08 ☽ Q ♅
02 57 ☽ ⊥ ♆
04 28 ☽ ∥ ♃
09 13 ☽ # ♄
14 31 ♀ ⊼ ♇
17 08 ☽ △ ♇
17 27 ☽ □ ♀
19 32 ☽ □ ☿
21 09 ☽ ∠ ♆
22 19 ☽ ☍ ♄

03 Thursday
01 59 ☽ ✶ ☉
02 32 ☽ ✶ ♅
09 10 ☉ ☌ ♅
09 18 ☽ # ♇
17 18 ☿ bQ ♆
23 10 ☽ bQ ♇

04 Friday
00 19 ☽ ⊼ ♂
03 09 ☽ ✶ ♆
05 29 ♀ bQ ♆
08 36 ☽ ∠ ♅
09 11 ☿ □ ♄
09 29 ☽ Q ♀
10 15 ☽ Q ☿
10 29 ☽ ∠ ☉
11 19 ☽ ⊻ ♃
12 43 ☽ ± ♂
16 24 ♀ □ ♄
16 53 ☽ ± ♇
17 00 ☽ ∥ ♆
23 11 ☽ ⊥ ♃

05 Saturday
02 03 ☉ # ☿
02 18 ☽ ⊥ ♅
02 41 ☽ Q ♆
04 34 ☽ ⊼ ♇
05 34 ☽ ⊥ ☉
06 18 ☿ ☌ ♀
07 03 ☽ bQ ♂
09 33 ☽ ⊼ ♄
11 14 ☽ ✶ ☿
11 26 ☽ ✶ ♀
13 57 ☽ ⊻ ♅
16 45 ☽ ∠ ♃
18 06 ☽ ⊻ ☉
21 04 ☽ ± ♄

06 Sunday
12 56 ☽ △ ♂
13 04 ☽ □ ♆
14 05 ☽ bQ ♄
14 28 ♀ ⊻ ♅
15 45 ♂ ⊼ ♆
16 38 ♀ Q ♂
17 28 ☽ ∠ ☿
18 56 ☽ ∠ ♀
21 23 ☽ ∥ ♅
21 24 ☽ ✶ ♃
23 32 ☽ # ☿

07 Monday
07 23 ☽ ∥ ☉
10 31 ♀ ± ♇
11 04 ☽ ⊥ ☿
13 04 ☽ ☍ ♇
13 20 ☽ ⊥ ♀
13 36 ☽ # ♀
16 25 ☿ ⊻ ♅
17 51 ☽ △ ♄
19 44 ☽ Q ♃
22 20 ☽ ☌ ♅
22 36 ☽ ⊻ ☿

08 Tuesday
01 24 ☽ ⊻ ♀
01 55 ☿ # ♅
06 24 ☽ ☌ ☉
07 46 ♂ Q ♇
19 58 ☽ △ ♆
22 12 ☽ □ ♂
22 23 ☉ # ♀

09 Wednesday
04 25 ☽ □ ♃
06 43 ☿ ± ♇
08 52 ☽ # ☉
10 50 ☽ ∥ ♀
18 49 ☽ ⊼ ♇
21 37 ☽ # ♅
22 30 ☽ bQ ♆
23 28 ☽ □ ♄

10 Thursday
00 45 ☽ ∥ ☿
04 02 ☽ ⊻ ♅
05 20 ☽ ± ♇
06 11 ☽ ☌ ☿
07 19 ☽ Q ♂
11 56 ☽ ☌ ♀
14 13 ☽ ± ♆
14 33 ☽ ⊥ ♅
15 36 ☽ ⊻ ☉
20 58 ☽ bQ ♇

11 Friday
00 38 ☽ ⊼ ♆
02 47 ☽ ⊥ ☉
05 03 ☽ ✶ ♂
06 14 ☽ ∠ ♅
06 46 ☽ Q ♄
09 18 ☽ △ ♃
13 39 ♀ ± ♆
18 59 ♀ ⊥ ♅
19 29 ☽ ∠ ☉
21 42 ☽ # ♆
22 50 ☽ △ ♇

12 Saturday
03 26 ☽ ✶ ♄
07 58 ☽ ∠ ♂
08 10 ☽ ✶ ♅
11 18 ☽ bQ ♃
11 21 ☽ ⊻ ☿
16 56 ♄ St D
20 36 ☽ ⊻ ♀
21 48 ☽ ⊥ ☿
23 08 ☽ ✶ ☉

13 Sunday
00 02 ☽ ⊥ ♂
02 24 ☽ ∥ ♇
02 49 ☽ ± ♃
04 09 ☽ ☍ ♆
04 47 ☽ Q ♅
05 07 ☽ ∠ ♄
07 50 ☽ ⊥ ♀
10 43 ☽ ⊻ ♂
13 09 ☽ ⊼ ♃
13 22 ☽ ∠ ☿
20 29 ☽ ⊥ ♄
21 08 ☽ Q ☉
21 15 ♀ # ♅
23 36 ☽ ∥ ♄

14 Monday
00 38 ☽ ∠ ♀
02 04 ☽ □ ♇
03 59 ☽ # ♃
06 42 ☽ ⊻ ♄
11 37 ☽ □ ♅
15 07 ☽ ✶ ☿
16 50 ♀ bQ ♇

15 Tuesday
01 08 ☿ St R
04 35 ☽ ✶ ♀
06 04 ☽ □ ☉
07 14 ☽ ⊼ ♆
08 39 ☽ Q ♇
11 32 ☽ Q ☿
15 54 ☽ # ♃
16 00 ☽ ☌ ♂
16 37 ☽ ☍ ♃
17 26 ☽ ± ♆
20 15 ☽ ∥ ♄
22 47 ☉ △ ♆

16 Wednesday
02 55 ☽ Q ♀
05 02 ☽ ✶ ♇
08 43 ☽ bQ ♆
09 45 ☽ ☌ ♄
09 51 ♂ ☍ ♃
10 58 ♀ ⊼ ♆
14 52 ☽ △ ♅
17 12 ☽ ∥ ♇
17 53 ☽ □ ☿

17 Thursday
01 53 ♂ ± ♆
03 25 ☿ ∥ ♀
06 30 ☽ ∠ ♇
10 12 ☽ △ ♆
12 25 ☽ □ ♀
12 52 ☽ △ ☉
16 30 ☽ bQ ♅
20 00 ☽ ⊼ ♃
21 15 ☽ ⊻ ♂
21 48 ☽ ⊥ ♇
22 03 ☽ # ♆

18 Friday
06 18 ☽ ± ♃
07 55 ☽ ± ♅
07 56 ☽ ⊥ ♂
08 01 ☽ ⊻ ♇
12 47 ☉ ⊻ ♀
12 52 ☽ ⊻ ♄
16 22 ☽ bQ ☉
18 13 ☽ ⊼ ♅
19 56 ☽ △ ☿
21 47 ☽ bQ ♃
23 09 ☽ ⊥ ♄

19 Saturday
00 00 ☽ ∠ ♂
08 57 ☽ ± ☉
09 13 ☽ ∥ ♀
10 44 ♀ Q ♄
13 27 ☽ □ ♆
14 00 ♀ ∠ ♅
14 37 ☽ ∠ ♄
16 18 ☽ ∥ ☿
20 03 ☽ ⊼ ☉
20 39 ☽ △ ♀
20 51 ☽ bQ ☿
22 33 ☽ # ♅
23 45 ☽ △ ♃

20 Sunday
03 00 ☽ ✶ ♂
04 25 ♃ ⊥ ♆
11 34 ☽ ☌ ♇
11 49 ☽ ± ☿
12 53 ☿ ⊻ ♅
16 38 ☽ ✶ ♄
17 17 ☿ ± ♇
20 38 ☽ Q ♆
21 49 ☽ ⊼ ☿
22 17 ☽ ☍ ♅

21 Monday
00 52 ☽ Q ♂
01 16 ☽ bQ ♀
06 28 ☽ # ☉
09 56 ☽ ∥ ☉
10 42 ♀ △ ♃
13 44 ☽ Q ♄
17 46 ☽ ✶ ♆
18 43 ☽ ± ♀

22 Tuesday
04 46 ☽ ☍ ☉
04 49 ☽ □ ♃
05 23 ☉ □ ♃
06 30 ☽ ⊼ ♀
10 21 ☽ □ ♂
16 40 ☽ ⊻ ♇
18 41 ☽ ∥ ♅
20 01 ☽ # ☿
20 41 ☽ ∠ ♆
22 05 ☽ □ ♄

23 Wednesday
00 26 ☽ ☍ ☿
03 35 ☽ ⊥ ♇
04 10 ☽ ⊼ ♅
10 07 ☿ # ♅
13 10 ☽ ⊥ ♆
13 47 ☽ Q ♃
15 17 ☽ ± ♅
20 10 ☽ ∠ ♇
20 28 ☽ # ♀

24 Thursday
00 19 ☽ ⊻ ♆
08 11 ☽ bQ ♅
08 26 ☿ □ ♄
12 23 ☽ ✶ ♃
16 37 ☽ ⊼ ☉
19 41 ☽ ☍ ♀
20 43 ☽ △ ♂
23 53 ☽ ∥ ♆

25 Friday
00 28 ☽ ✶ ♇
03 45 ☿ bQ ♆
04 39 ☽ ⊼ ☿
05 04 ☽ ± ☉
06 22 ☽ △ ♄
13 00 ☽ △ ♅
15 21 ☽ ± ☿
16 26 ♀ ✶ ♂
17 21 ☽ ∠ ♃
23 41 ☽ Q ♇

26 Saturday
00 03 ☽ bQ ☉
03 18 ☽ bQ ♂
07 32 ☽ bQ ☿
09 57 ☽ ☌ ♆
10 17 ☽ # ♇
11 07 ☽ ⊥ ♃
11 43 ☽ bQ ♄
20 29 ♀ △ ♇
22 05 ☽ ± ♂
23 06 ☽ ⊻ ♃

27 Sunday
00 43 ☿ ⊻ ♀
05 27 ☿ ⊼ ♇
05 40 ☽ ± ♄
08 26 ☽ △ ☉
10 45 ☽ ⊼ ♂
10 52 ☽ △ ☿
11 22 ☽ □ ♇
11 37 ☿ □ ♂
11 51 ☽ # ♄
13 05 ☽ ⊼ ♀
17 40 ☽ ∥ ♃
17 45 ☽ ⊼ ♄
22 06 ♂ ✶ ♇
23 41 ☉ ☌ ☿

28 Monday
00 52 ☽ □ ♅
02 34 ☽ ± ♀
19 41 ☉ ⊼ ♇
22 12 ☽ ⊻ ♆
22 55 ☽ bQ ♀

29 Tuesday
05 06 ☽ ∥ ♃
10 23 ☽ ⊥ ♆
10 59 ☽ # ♄
11 52 ♀ ✶ ♄
12 07 ☽ ☌ ♃
13 37 ☽ Q ♅
15 07 ☿ ⊥ ♀
18 35 ☽ □ ☿

30 Wednesday
00 08 ☽ △ ♇
02 45 ☽ □ ☉
03 02 ☽ ☍ ♂
03 14 ☉ ∥ ☿
04 40 ☽ ∠ ♆
06 47 ☽ ☍ ♄
08 54 ☽ △ ♀
11 54 ☉ □ ♂
12 47 ☽ # ♇
14 06 ☽ ✶ ♅

All ephemeris data is given at 12.00 UT and the Moon's longitude is additionally given for 24.00 UT

OCTOBER 2048

LONGITUDES

Date	Sidereal time h m s	Sun ☉ ° ' "	Moon ☽ ° ' "	Moon ☽ 24.00 ° ' "	Mercury ☿ ° '	Venus ♀ ° '	Mars ♂ ° '	Jupiter ♃ ° '	Saturn ♄ ° '	Uranus ♅ ° '	Neptune ♆ ° '	Pluto ♇ ° '
01	12 43 35	08 ♎ 59 17	24 ♋ 06 53	00 ♌ 08 53	01 ♎ 57	12 ♏ 01	08 ♑ 39	00 ♋ 32	09 ♑ 39	13 ♍ 17	23 ♉ 32	06 ♓ 18
02	12 47 32	09 58 18	06 ♌ 14 27	12 ♌ 24 10	01 R 08	13 14	09 17	00 36	09 41	13 20	23 R 31	06 R 17
03	12 51 29	10 57 22	18 ♌ 38 32	24 ♌ 57 56	00 ♎ 27	14 27	09 56	00 39	09 43	13 24	23 30	06 16
04	12 55 25	11 56 28	01 ♍ 22 42	07 ♍ 53 02	29 ♍ 56	15 39	10 35	00 42	09 45	13 27	23 29	06 15
05	12 59 22	12 55 36	14 ♍ 29 01	21 ♍ 10 38	29 35	16 52	11 14	00 44	09 47	13 31	23 28	06 14
06	13 03 18	13 54 46	27 ♍ 57 41	04 ♎ 49 53	29 24	18 04	11 53	00 47	09 49	13 34	23 27	06 13
07	13 07 15	14 53 58	11 ♎ 46 48	18 ♎ 47 54	29 D 24	19 17	12 32	00 49	09 52	13 37	23 25	06 13
08	13 11 11	15 53 13	25 ♎ 52 35	03 ♏ 00 09	29 34	20 29	13 12	00 51	09 54	13 41	23 24	06 12
09	13 15 08	16 52 29	10 ♏ 09 55	17 ♏ 21 10	29 ♍ 55	21 42	13 51	00 53	09 57	13 44	23 23	06 11
10	13 19 04	17 51 48	24 ♏ 33 12	01 ♐ 45 25	00 ♎ 25	22 54	14 31	00 55	09 59	13 47	23 22	06 10
11	13 23 01	18 51 08	08 ♐ 57 14	16 ♐ 08 11	01 05	24 07	15 11	00 56	10 02	13 51	23 20	06 09
12	13 26 58	19 50 30	23 ♐ 17 50	00 ♑ 25 54	01 54	25 19	15 51	00 58	10 05	13 54	23 19	06 08
13	13 30 54	20 49 54	07 ♑ 32 08	14 ♑ 36 21	02 50	26 31	16 32	00 59	10 08	13 57	23 18	06 07
14	13 34 51	21 49 20	21 ♑ 38 27	28 ♑ 38 20	03 53	27 44	17 12	01 00	10 11	14 00	23 16	06 06
15	13 38 47	22 48 47	05 ♒ 35 57	12 ♒ 31 14	05 03	28 ♏ 56	17 53	01 01	10 14	14 03	23 15	06 06
16	13 42 44	23 48 16	19 ♒ 24 07	26 ♒ 14 34	06 18	00 ♐ 08	18 33	01 01	10 17	14 06	23 13	06 05
17	13 46 40	24 47 47	03 ♓ 02 28	09 ♓ 47 43	07 37	01 21	19 14	01 01	10 20	14 10	23 12	06 04
18	13 50 37	25 47 19	16 ♓ 30 11	23 ♓ 09 44	09 01	02 33	19 55	01 02	10 24	14 13	23 11	06 03
19	13 54 33	26 46 53	29 ♓ 46 14	06 ♈ 19 30	10 29	03 45	20 36	01 R 01	10 27	14 16	23 09	06 03
20	13 58 30	27 46 30	12 ♈ 49 26	19 ♈ 15 55	11 59	04 57	21 18	01 01	10 31	14 19	23 08	06 02
21	14 02 27	28 46 08	25 ♈ 38 51	01 ♉ 58 12	13 32	06 09	21 59	01 01	10 34	14 22	23 06	06 01
22	14 06 23	29 ♎ 45 48	08 ♉ 13 59	14 ♉ 26 15	15 06	07 21	22 40	01 00	10 38	14 25	23 05	06 00
23	14 10 20	00 ♏ 45 30	20 ♉ 35 08	26 ♉ 40 48	16 42	08 33	23 22	00 59	10 42	14 27	23 03	06 00
24	14 14 16	01 45 14	02 ♊ 43 31	08 ♊ 43 36	18 20	09 45	24 03	00 58	10 45	14 30	23 02	05 59
25	14 18 13	02 45 01	14 ♊ 41 23	20 ♊ 37 20	19 58	10 57	24 45	00 57	10 49	14 33	23 00	05 59
26	14 22 09	03 44 49	26 ♊ 31 53	02 ♋ 25 35	21 37	12 09	25 27	00 55	10 53	14 36	22 58	05 58
27	14 26 06	04 44 40	08 ♋ 18 59	14 ♋ 12 41	23 16	13 20	26 09	00 54	10 57	14 39	22 57	05 57
28	14 30 02	05 44 33	20 ♋ 07 18	26 ♋ 03 30	24 56	14 32	26 51	00 52	11 02	14 41	22 55	05 57
29	14 33 59	06 44 28	02 ♌ 01 56	08 ♌ 03 15	26 36	15 44	27 33	00 50	11 06	14 44	22 54	05 56
30	14 37 56	07 44 26	14 ♌ 08 06	20 ♌ 17 06	28 16	16 56	28 16	00 48	11 10	14 47	22 52	05 56
31	14 41 52	08 ♏ 44 25	26 ♌ 30 52	02 ♍ 49 55	29 ♎ 56	18 ♐ 07	28 ♑ 58	00 ♋ 45	11 ♑ 15	14 ♍ 49	22 ♉ 50	05 ♓ 55

Date	Moon True ☊ ° '	Moon Mean ☊ ° '	Moon ☽ Latitude ° '
01	21 ♐ 37	22 ♐ 09	02 S 44
02	21 R 34	22 06	03 34
03	21 29	22 03	04 15
04	21 21	21 59	04 45
05	21 11	21 56	05 01
06	21 00	21 53	05 01
07	20 49	21 50	04 43
08	20 39	21 47	04 08
09	20 31	21 44	03 17
10	20 26	21 40	02 13
11	20 23	21 37	01 S 01
12	20 D 23	21 34	00 N 16
13	20 23	21 31	01 31
14	20 24	21 28	02 39
15	20 R 23	21 25	03 38
16	20 20	21 21	04 23
17	20 15	21 18	04 52
18	20 08	21 15	05 05
19	19 59	21 12	05 02
20	19 49	21 09	04 42
21	19 40	21 05	04 08
22	19 32	21 02	03 23
23	19 26	20 59	02 29
24	19 22	20 56	01 29
25	19 21	20 53	00 N 25
26	19 D 21	20 50	00 S 39
27	19 22	20 46	01 41
28	19 24	20 43	02 39
29	19 25	20 40	03 31
30	19 R 25	20 37	04 14
31	19 ♐ 23	20 ♐ 34	04 S 47

DECLINATIONS

Date	Sun ☉ ° '	Moon ☽ ° '	Mercury ☿ ° '	Venus ♀ ° '	Mars ♂ ° '	Jupiter ♃ ° '	Saturn ♄ ° '	Uranus ♅ ° '	Neptune ♆ ° '	Pluto ♇ ° '
01	03 S 34	18 N 36	02 S 06	15 S 51	25 S 48	22 N 56	22 S 42	07 N 15	16 N 56	21 S 02
02	03 57	15 15	01 28	16 17	25 44	22 56	22 42	07 14	16 55	21 02
03	04 20	11 11	00 53	16 41	25 40	22 56	22 42	07 12	16 55	21 03
04	04 43	06 32	00 S 22	17 06	25 36	22 56	22 42	07 11	16 55	21 03
05	05 06	01 N 29	00 N 04	17 29	25 31	22 56	22 42	07 10	16 54	21 03
06	05 29	03 S 47	00 24	17 53	25 27	22 56	22 42	07 08	16 54	21 03
07	05 52	09 00	00 40	18 16	25 22	22 56	22 42	07 07	16 54	21 04
08	06 15	13 51	00 50	18 39	25 17	22 56	22 42	07 06	16 53	21 04
09	06 38	17 59	00 55	19 01	25 12	22 56	22 42	07 05	16 53	21 04
10	07 00	21 03	00 54	19 23	25 06	22 56	22 42	07 03	16 53	21 04
11	07 23	22 47	00 49	19 44	25 00	22 56	22 42	07 02	16 52	21 04
12	07 45	23 00	00 39	20 04	24 54	22 56	22 42	07 01	16 52	21 04
13	08 08	21 43	00 24	20 25	24 48	22 56	22 42	07 00	16 51	21 05
14	08 30	19 04	00 N 05	20 44	24 42	22 56	22 41	06 59	16 51	21 05
15	08 52	15 21	00 S 18	21 03	24 35	22 56	22 41	06 57	16 51	21 05
16	09 14	10 50	00 43	21 22	24 29	22 56	22 41	06 56	16 50	21 05
17	09 36	05 50	01 12	21 40	24 22	22 56	22 41	06 55	16 50	21 05
18	09 58	00 S 38	01 43	21 57	24 14	22 56	22 41	06 54	16 50	21 05
19	10 19	04 N 31	02 17	22 14	24 07	22 56	22 41	06 53	16 49	21 05
20	10 41	09 23	02 52	22 31	23 59	22 56	22 41	06 52	16 49	21 05
21	11 02	13 46	03 28	22 46	23 51	22 56	22 40	06 50	16 48	21 05
22	11 23	17 27	04 06	23 01	23 43	22 56	22 40	06 49	16 48	21 05
23	11 44	20 17	04 45	23 16	23 35	22 56	22 40	06 48	16 48	21 06
24	12 05	22 09	05 25	23 30	23 27	22 56	22 40	06 47	16 47	21 06
25	12 25	22 58	06 05	23 43	23 18	22 56	22 40	06 46	16 47	21 06
26	12 46	22 44	06 46	23 56	23 09	22 56	22 39	06 45	16 46	21 06
27	13 06	21 29	07 27	24 07	23 00	22 56	22 39	06 44	16 46	21 06
28	13 26	19 18	08 08	24 19	22 51	22 56	22 39	06 43	16 46	21 06
29	13 46	16 16	08 48	24 29	22 41	22 56	22 39	06 42	16 45	21 06
30	14 05	12 31	09 29	24 39	22 31	22 56	22 38	06 41	16 45	21 05
31	14 S 25	08 N 10	10 S 10	24 S 48	22 S 21	22 N 56	22 S 38	06 N 40	16 N 44	21 S 05

ZODIAC SIGN ENTRIES

Date	h	m	Planets
01	23	42	☽ ♌
04	08	36	☿ ♍
04	09	26	☽ ♍
06	15	35	☽ ♎
08	18	57	☽ ♏
09	16	42	☿ ♎
10	21	04	☽ ♐
12	23	16	☽ ♑
15	02	20	☽ ♒
16	09	15	♀ ♐
17	06	37	☽ ♓
19	12	25	☽ ♈
21	20	15	☽ ♉
22	17	43	☉ ♏
24	06	35	☽ ♊
26	19	04	☽ ♋
29	07	56	☽ ♌
31	12	53	☿ ♏
31	18	39	☽ ♍

LATITUDES

Date	Mercury ☿ ° '	Venus ♀ ° '	Mars ♂ ° '	Jupiter ♃ ° '	Saturn ♄ ° '	Uranus ♅ ° '	Neptune ♆ ° '	Pluto ♇ ° '
01	01 S 27	00 S 26	02 S 39	00 S 30	00 N 23	00 N 44	01 S 47	12 S 47
04	00 S 26	00 36	02 36	00 30	00 22	00 44	01 47	12 47
07	00 N 28	00 45	02 32	00 30	00 22	00 44	01 47	12 46
10	01 10	00 55	02 28	00 30	00 22	00 44	01 47	12 46
13	01 39	01 04	02 25	00 30	00 21	00 44	01 47	12 45
16	01 56	01 13	02 21	00 30	00 21	00 44	01 47	12 45
19	02 02	01 22	02 17	00 30	00 21	00 45	01 48	12 44
22	02 00	01 31	02 14	00 30	00 20	00 45	01 48	12 44
25	01 51	01 39	02 10	00 30	00 20	00 45	01 48	12 43
28	01 38	01 47	02 06	00 30	00 20	00 45	01 48	12 42
31	01 N 22	01 S 55	02 S 02	00 S 29	00 N 19	00 N 45	01 S 48	12 S 42

DATA

Julian Date	2469351
Delta T	+80 seconds
Ayanamsa	24° 32' 34"
Synetic vernal point	04° ♓ 34' 25"
True obliquity of ecliptic	23° 25' 58"

MOON'S PHASES, APSIDES AND POSITIONS ☽

Date	h	m	Phase	Longitude ° '	Eclipse Indicator
07	17	45	●	15 ♎ 08	
14	12	20	☽	21 ♑ 50	
21	18	25	○	29 ♈ 02	
29	22	14	☾	07 ♌ 10	

Day	h	m		
11	04	44	Perigee	
27	01	53	Apogee	
05	18	47	OS	
12	03	22	Max dec	23° S 06'
18	14	55	ON	
25	18	39	Max dec	23° N 01'

LONGITUDES

Date	Chiron ⚷ ° '	Ceres ⚳ ° '	Pallas ⚴ ° '	Juno ⚵ ° '	Vesta ⚶ ° '	Black Moon Lilith ⚸ ° '
01	11 ♏ 45	08 ♈ 13	03 ♒ 56	15 ♒ 05	23 ♉ 53	26 ♊ 59
11	13 ♏ 02	06 ♈ 00	04 ♒ 15	15 ♒ 17	22 ♉ 39	28 ♊ 06
21	14 ♏ 21	04 ♈ 02	05 ♒ 04	16 ♒ 13	20 ♉ 46	29 ♊ 12
31	15 ♏ 43	02 ♈ 31	06 ♒ 21	17 ♒ 49	18 ♉ 23	00 ♋ 19

ASPECTARIAN

h m	Aspects
01 Thursday	
01 16	☉ bQ ♆
04 14	☽ Q ☿
06 23	☽ bQ ♇
08 17	♂ bQ ♆
10 51	☽ ✶ ♆
18 14	☽ Q ☉
20 20	☽ ∠ ♅
02 Friday	
00 18	☽ ± ♇
00 42	☽ ∥ ♆
00 50	☽ ⊻ ♃
02 33	☽ ✶ ☿
04 41	☉ □ ♄
05 56	☽ # ♀
10 36	☽ Q ♆
12 06	☽ ⊼ ♇
12 42	☽ ⊥ ♃
14 09	☽ ⊥ ♅
14 10	♀ ✶ ♅
18 16	☽ ⊼ ♂
18 44	☽ ⊼ ♄
19 55	☽ ✶ ☉
03 Saturday	
01 52	☽ ⊻ ♅
03 04	☽ □ ♀
03 37	♂ ☌ ♄
05 19	☿ □ ♃
06 11	☽ ∠ ☿
06 14	☽ ∠ ♃
06 22	☽ ± ♄
06 31	☽ ± ♂
21 14	☽ □ ♆
22 34	☽ ⊥ ☿
23 33	☽ bQ ♄
04 Sunday	
00 34	☽ bQ ♂
01 19	♀ # ♆
01 48	☿ ∠ ♀
03 01	☽ ∠ ☉
08 48	☽ ∥ ♅
09 24	☽ ⊻ ☿
10 43	☽ ✶ ♃
12 51	♀ bQ ♃
16 39	☽ Q ♀
19 34	☉ ± ♇
20 11	☽ # ☉
21 00	☽ ☍ ♇
21 07	☽ ⊥ ☉
05 Monday	
03 27	☽ △ ♄
05 48	☽ △ ♂
08 50	☽ Q ♃
08 57	☽ ⊻ ☉
10 14	☽ ☌ ♅
16 43	☽ ✶ ♀
18 04	☽ ∥ ☿
19 36	☽ # ☿
06 Tuesday	
03 01	☉ ⊻ ♅
04 02	☽ △ ♆
14 30	☽ ☌ ☿
16 57	☽ □ ♃
20 21	☽ ∥ ☉
21 48	☽ ∠ ♀
07 Wednesday	
00 30	☿ St D
02 24	☽ ⊼ ♇
03 16	☽ # ♅
06 14	☽ bQ ♆
08 41	☽ □ ♄
12 44	☽ ± ♇
13 22	☽ □ ♂
14 49	☽ ⊥ ♀
15 10	☽ ⊻ ♅
17 45	☽ ☌ ☉
21 39	☽ ± ♆
08 Thursday	
01 27	☽ ⊥ ♅
02 01	☽ ⊻ ♀
04 05	☽ bQ ♇
07 49	☽ ⊼ ♆
15 26	☽ Q ♄
16 45	☽ ∠ ♅
18 21	☽ ⊻ ☿
20 25	☽ △ ♃
21 24	☽ Q ♂
09 Friday	
04 41	☽ ⊥ ☿
05 05	☽ # ♆
05 20	☽ △ ♇
07 05	♂ △ ♅
11 38	☽ ✶ ♄
17 59	☽ ✶ ♅
18 28	☽ ✶ ♂
19 57	☽ ∥ ♀
20 11	☽ ∠ ☿
21 34	☽ bQ ♃
10 Saturday	
00 02	☽ ⊻ ☉
00 03	☉ ± ♆
09 00	☽ ☌ ♀
10 01	☽ ☍ ♆
10 46	☽ ⊥ ☉
12 06	☽ ∥ ♇
12 36	☽ ± ♃
12 44	☽ ∠ ♄
14 04	☽ Q ♅
15 02	☉ # ♅
20 41	☽ ∠ ♂
20 54	♀ ☍ ♆
22 13	☽ ✶ ☿
22 37	☽ ⊼ ♃
11 Sunday	
02 52	☽ ∠ ☉
03 46	☽ ⊥ ♄
06 58	☿ □ ♃
07 19	☽ □ ♇
09 56	☽ ∥ ♄
12 25	☽ ⊥ ♂
13 49	☽ ⊻ ♄
15 55	☽ # ♃
19 17	☽ Q ☿
20 12	☽ □ ♅
22 55	☽ ⊻ ♂
12 Monday	
04 48	♀ ± ♃
05 46	☽ ✶ ☉
07 06	♀ ∠ ♄
12 02	☽ ⊼ ♆
13 23	☉ ⊥ ♅
13 24	☽ Q ♇
14 50	☽ # ♃
15 42	☽ ⊻ ♀
20 56	☽ ∥ ♄
22 06	☽ ± ♆
13 Tuesday	
00 02	♀ Q ♅
00 55	☽ ☍ ♃
02 45	☽ ⊥ ♀
03 27	☽ Q ☉
03 28	☽ □ ☿
09 36	☽ ✶ ♇
13 17	☽ bQ ♆
16 25	☽ ☌ ♄
18 53	☉ bQ ♇
18 56	☽ ∥ ♇
19 24	☽ ∠ ♀
22 55	☽ △ ♅
23 46	☽ ∥ ♀
14 Wednesday	
04 02	☽ ☌ ♂
11 05	☽ ∠ ♇
12 20	☽ □ ☉
14 47	☽ △ ♆
22 25	☉ ⊥ ♀
23 25	☽ ✶ ♀
15 Thursday	
00 40	☽ bQ ♅
02 31	☽ ⊥ ♇
03 01	☽ # ♆
04 04	☽ ⊼ ♃
10 57	☽ △ ☿
12 51	☽ ⊻ ♇
13 53	♀ ∥ ♇
14 26	☽ ± ♃
16 16	☽ ± ♅
20 03	☽ ⊻ ♄
22 08	☽ Q ♀
22 17	☉ ⊼ ♆
16 Friday	
02 44	☽ ⊼ ♅
06 05	☽ bQ ♃
06 32	☽ ⊥ ♄
08 02	☿ ⊼ ♇
10 27	☽ ⊻ ♂
15 40	☽ bQ ☿
18 41	☽ □ ♆
19 19	☽ ∥ ☉
20 19	☽ △ ☉
21 31	☽ ⊥ ♂
22 21	☽ ∠ ♄
17 Saturday	
05 37	♀ ⊼ ♃
06 55	☽ # ♅
08 26	☽ △ ♃
08 42	☽ □ ♀
09 13	☽ ± ☿
14 14	☽ ∠ ♂
17 22	☽ ☌ ♇
21 04	☽ ⊼ ☿
21 51	☿ bQ ♆
18 Sunday	
00 57	☽ bQ ☉
01 01	☽ ✶ ♄
02 29	☽ Q ♆
07 28	☽ ∥ ☿
07 52	☽ ☍ ♅
15 57	♃ St R
18 23	☽ ± ☉
18 29	☽ ✶ ♂
22 39	☽ Q ♄
19 Monday	
00 00	☽ ✶ ♆
00 09	☽ # ☿
06 07	☽ ⊼ ☉
11 33	☿ □ ♄
14 17	☽ □ ♃
17 28	☽ Q ♂
20 00	☽ △ ♀
23 20	☽ ∥ ♅
23 28	☽ ⊻ ♇
20 Tuesday	
02 47	♀ ⊥ ♄
03 00	♂ ∠ ♇
03 20	☽ ∠ ♆
07 42	☽ □ ♄
10 14	☽ ☍ ☿
10 32	☽ ⊥ ♇
12 46	☿ ± ♇
14 46	☽ ⊼ ♅
19 20	☽ # ☉
19 59	☽ ⊥ ♆
23 32	☽ Q ♃
21 Wednesday	
02 01	☽ ± ♅
02 39	☽ bQ ♀
02 47	♀ ∥ ♄
03 18	☽ ∠ ♇
04 42	☽ □ ♂
06 53	☉ Q ♄
07 13	☽ ⊻ ♆
09 25	♀ □ ♇
18 25	☽ ☍ ☉
19 04	☽ bQ ♅
21 26	☽ ± ♀
22 10	☽ ✶ ♃
22 Thursday	
01 10	☿ ⊻ ♅
03 00	☉ ∠ ♅
03 04	♀ # ♃
07 23	☽ ∥ ♆
07 44	☽ ✶ ♇
10 07	☽ ⊼ ♀
16 39	☽ △ ♄
23 59	☽ △ ♅
23 Friday	
01 33	♂ △ ♆
03 01	☽ ∠ ♃
03 12	♀ ∠ ♂
03 16	☽ ⊼ ☿
06 56	☽ Q ♇
16 48	☽ ± ☿
16 50	☽ ☌ ♆
17 05	☿ ± ♆
17 25	☉ △ ♃
17 47	☽ △ ♂
20 39	☽ ⊥ ♃
20 52	☽ # ♇
22 06	☽ bQ ♄
24 Saturday	
08 31	☽ ⊻ ♃
08 36	♀ ∥ ♂
09 53	☽ ⊼ ☉
13 23	☽ bQ ☿
16 05	☽ ± ♄
18 30	☽ □ ♇
22 58	☽ ± ☉
23 06	☽ # ♄
25 Sunday	
01 27	☽ bQ ♂
03 37	☽ ☍ ♀
04 10	☽ ⊼ ♄
09 23	♀ ⊻ ♄
09 26	☽ ∥ ♃
11 43	☽ □ ♅
18 45	☽ bQ ☉
20 44	☽ ± ♂
20 48	☿ ⊥ ♅
26 Monday	
00 24	☽ △ ☿
02 39	☿ bQ ♇
03 51	☽ ∥ ♃
04 47	☽ ⊻ ♆
09 40	☽ ⊼ ♂
11 36	☿ # ♅
14 30	☽ # ♄
16 57	☽ ⊥ ♆
20 55	☽ ☌ ♃
27 Tuesday	
00 24	☽ Q ♅
04 03	☽ △ ☉
07 12	☽ △ ♇
07 21	☿ ⊼ ♆
11 15	☽ ∠ ♆
17 11	☽ # ♇
17 25	☽ ☍ ♄
21 38	♂ # ♃
23 23	☽ ⊼ ♀
28 Wednesday	
00 56	☽ ✶ ♅
12 56	☽ ± ♀
13 41	☽ bQ ♇
15 10	♀ □ ♅
16 56	☉ △ ♇
17 39	☽ ✶ ♆
23 20	☽ □ ☿
29 Thursday	
02 27	☽ ☍ ♂
07 23	☽ ∠ ♅
07 49	☽ ± ♇
08 33	☽ ∥ ♆
09 07	☽ bQ ♀
09 36	☽ ⊻ ♃
17 42	☽ Q ♆
18 09	♂ ∥ ♄
19 48	☽ ⊼ ♇
21 32	☽ ⊥ ♃
22 14	☽ □ ☉
30 Friday	
01 24	☽ ⊥ ♅
03 13	☽ # ☉
06 08	☽ ⊼ ♄
11 41	☿ □ ♂
13 16	☽ ⊻ ♅
15 14	☽ ∠ ♃
16 51	☽ Q ☿
17 59	☽ ± ♄
18 03	☽ △ ♀
31 Saturday	
01 32	☿ Q ♄
02 48	☽ # ☿
04 57	☽ □ ♆
10 16	☿ ∠ ♅
11 29	☽ bQ ♄
12 28	☽ Q ☉
16 57	☽ ⊼ ♂
19 31	☽ ✶ ☿
19 46	☽ ∥ ♅
20 02	☽ ✶ ♃
23 24	☿ △ ♃

All ephemeris data is given at 12.00 UT and the Moon's longitude is additionally given for 24.00 UT

NOVEMBER 2048

LONGITUDES

Date	Sidereal time h m s	Sun ☉ ° ' "	Moon ☽ ° ' "	Moon ☽ 24.00 ° ' "	Mercury ☿ ° '	Venus ♀ ° '	Mars ♂ ° '	Jupiter ♃ ° '	Saturn ♄ ° '	Uranus ♅ ° '	Neptune ♆ ° '	Pluto ♇ ° '
01	14 45 49	09 ♏ 44 26	09 ♍ 14 43	15 ♍ 45 39	01 ♏ 36	19 ♐ 19	29 ♑ 40	00 ♋ 42	11 ♑ 19	14 ♍ 52	22 ♉ 49	05 ♓ 55
02	14 49 45	10 44 30	22 ♍ 22 59	29 ♍ 06 52	03 16	20 30	00 ♒ 23	00 R 40	11 24	14 54	22 R 47	05 R 55
03	14 53 42	11 44 36	05 ♎ 57 16	12 ♎ 54 04	04 55	21 42	01 05	00 36	11 28	14 57	22 45	05 54
04	14 57 38	12 44 44	19 ♎ 56 57	27 ♎ 05 25	06 34	22 53	01 48	00 33	11 33	14 59	22 44	05 54
05	15 01 35	13 44 53	04 ♏ 18 51	11 ♏ 36 31	08 13	24 05	02 31	00 30	11 38	15 02	22 42	05 54
06	15 05 31	14 45 05	18 ♏ 57 32	26 ♏ 20 57	09 52	25 16	03 14	00 26	11 43	15 04	22 40	05 53
07	15 09 28	15 45 19	03 ♐ 45 49	11 ♐ 11 09	11 30	26 27	03 57	00 22	11 47	15 06	22 39	05 53
08	15 13 25	16 45 34	18 ♐ 36 00	25 ♐ 59 29	13 08	27 39	04 40	00 18	11 52	15 09	22 37	05 53
09	15 17 21	17 45 51	03 ♑ 20 51	10 ♑ 39 24	14 46	28 ♐ 50	05 23	00 14	11 57	15 11	22 35	05 52
10	15 21 18	18 46 09	17 ♑ 54 37	25 ♑ 06 02	16 23	00 ♑ 01	06 06	00 10	12 03	15 13	22 34	05 52
11	15 25 14	19 46 29	02 ♒ 13 22	09 ♒ 16 24	18 00	01 12	06 49	00 05	12 08	15 15	22 32	05 52
12	15 29 11	20 46 50	16 ♒ 15 01	23 ♒ 09 13	19 36	02 23	07 33	00 ♋ 00	12 13	15 17	22 30	05 52
13	15 33 07	21 47 12	29 ♒ 59 00	06 ♓ 44 28	21 12	03 34	08 16	29 ♊ 56	12 18	15 19	22 29	05 52
14	15 37 04	22 47 36	13 ♓ 25 44	20 ♓ 02 57	22 48	04 45	09 00	29 50	12 24	15 21	22 27	05 52
15	15 41 00	23 48 01	26 ♓ 36 15	03 ♈ 05 49	24 24	05 55	09 43	29 45	12 29	15 23	22 25	05 51
16	15 44 57	24 48 27	09 ♈ 31 48	15 ♈ 54 22	25 59	07 06	10 27	29 40	12 35	15 25	22 24	05 51
17	15 48 54	25 48 55	22 ♈ 13 39	28 ♈ 29 49	27 34	08 17	11 10	29 34	12 40	15 27	22 22	05 51
18	15 52 50	26 49 24	04 ♉ 43 01	10 ♉ 53 23	29 ♏ 09	09 27	11 54	29 29	12 46	15 29	22 20	05 51
19	15 56 47	27 49 55	17 ♉ 01 05	23 ♉ 06 17	00 ♐ 43	10 37	12 38	29 23	12 51	15 30	22 19	05 D 51
20	16 00 43	28 50 28	29 ♉ 09 09	05 ♊ 09 52	02 17	11 48	13 21	29 17	12 57	15 32	22 17	05 51
21	16 04 40	29 ♏ 51 01	11 ♊ 08 41	17 ♊ 05 49	03 51	12 58	14 05	29 10	13 03	15 34	22 15	05 51
22	16 08 36	00 ♐ 51 37	23 ♊ 01 32	28 ♊ 56 10	05 25	14 08	14 49	29 04	13 09	15 35	22 14	05 52
23	16 12 33	01 52 14	04 ♋ 50 01	10 ♋ 43 29	06 59	15 18	15 33	28 58	13 15	15 37	22 12	05 52
24	16 16 29	02 52 52	16 ♋ 36 59	22 ♋ 30 56	08 33	16 28	16 17	28 51	13 20	15 38	22 10	05 52
25	16 20 26	03 53 32	28 ♋ 25 49	04 ♌ 22 08	10 06	17 38	17 01	28 44	13 26	15 40	22 09	05 52
26	16 24 23	04 54 13	10 ♌ 20 27	16 ♌ 21 17	11 39	18 48	17 45	28 38	13 32	15 41	22 07	05 52
27	16 28 19	05 54 57	22 ♌ 25 14	28 ♌ 32 52	13 13	19 57	18 29	28 31	13 39	15 42	22 05	05 52
28	16 32 16	06 55 41	04 ♍ 44 44	11 ♍ 01 26	14 46	21 07	19 13	28 24	13 45	15 44	22 04	05 53
29	16 36 12	07 56 27	17 ♍ 23 29	23 ♍ 51 22	16 19	22 16	19 57	28 16	13 51	15 45	22 02	05 53
30	16 40 09	08 ♐ 57 15	00 ♎ 25 29	07 ♎ 06 12	17 ♐ 51	23 ♑ 25	20 ♒ 42	28 ♊ 09	13 ♑ 57	15 ♍ 46	22 ♉ 00	05 ♓ 53

Date	Moon True ☊ ° '	Moon Mean ☊ ° '	Moon ☽ Latitude ° '
01	19 ♐ 19	20 ♐ 31	05 S 06
02	19 R 15	20 27	05 11
03	19 09	20 24	04 58
04	19 03	20 21	04 27
05	18 57	20 18	03 39
06	18 53	20 15	02 36
07	18 50	20 11	01 22
08	18 49	20 08	00 S 01
09	18 D 50	20 05	01 N 19
10	18 51	20 02	02 33
11	18 53	19 59	03 36
12	18 54	19 56	04 25
13	18 R 54	19 52	04 58
14	18 52	19 49	05 13
15	18 50	19 46	05 12
16	18 46	19 43	04 54
17	18 42	19 40	04 23
18	18 38	19 37	03 39
19	18 35	19 33	02 45
20	18 33	19 30	01 45
21	18 32	19 27	00 N 41
22	18 D 32	19 24	00 S 25
23	18 33	19 21	01 29
24	18 34	19 17	02 30
25	18 35	19 14	03 24
26	18 37	19 11	04 10
27	18 38	19 08	04 45
28	18 38	19 05	05 08
29	18 R 38	19 02	05 18
30	18 ♐ 37	18 ♐ 58	05 S 11

DECLINATIONS

Date	Sun ☉ ° '	Moon ☽ ° '	Mercury ☿ ° '	Venus ♀ ° '	Mars ♂ ° '	Jupiter ♃ ° '	Saturn ♄ ° '	Uranus ♅ ° '	Neptune ♆ ° '	Pluto ♇ ° '
01	14 S 44	03 N 22	10 S 50	24 S 57	22 S 11	22 N 56	22 S 38	06 N 39	16 N 44	21 S 05
02	15 03	01 S 44	11 30	25 05	22 01	22 57	22 38	06 38	16 43	21 05
03	15 21	06 55	12 09	25 12	21 50	22 57	22 37	06 37	16 43	21 05
04	15 40	11 55	12 48	25 18	21 39	22 57	22 37	06 36	16 43	21 05
05	15 58	16 24	13 26	25 24	21 28	22 57	22 37	06 35	16 42	21 05
06	16 16	19 57	14 04	25 29	21 17	22 57	22 36	06 35	16 42	21 05
07	16 33	22 14	14 40	25 33	21 06	22 57	22 36	06 34	16 41	21 05
08	16 50	22 58	15 17	25 37	20 54	22 57	22 36	06 33	16 41	21 05
09	17 07	22 05	15 52	25 39	20 43	22 57	22 35	06 32	16 40	21 05
10	17 24	19 42	16 27	25 42	20 31	22 57	22 35	06 31	16 40	21 04
11	17 41	16 09	17 00	25 43	20 19	22 57	22 35	06 30	16 40	21 04
12	17 57	11 45	17 33	25 44	20 06	22 57	22 34	06 30	16 39	21 04
13	18 12	06 50	18 06	25 43	19 54	22 57	22 34	06 29	16 39	21 04
14	18 28	01 S 42	18 37	25 43	19 41	22 57	22 33	06 28	16 38	21 04
15	18 43	03 N 25	19 07	25 41	19 28	22 57	22 33	06 27	16 38	21 03
16	18 58	08 17	19 36	25 39	19 15	22 57	22 33	06 27	16 37	21 03
17	19 12	12 43	20 05	25 36	19 02	22 58	22 32	06 26	16 37	21 03
18	19 26	16 31	20 32	25 32	18 49	22 58	22 32	06 25	16 37	21 03
19	19 40	19 33	20 59	25 28	18 35	22 58	22 31	06 25	16 36	21 02
20	19 54	21 40	21 24	25 23	18 21	22 58	22 31	06 24	16 36	21 02
21	20 07	22 47	21 48	25 17	18 07	22 58	22 30	06 24	16 35	21 02
22	20 19	22 50	22 11	25 10	17 53	22 58	22 30	06 23	16 35	21 02
23	20 32	21 52	22 33	25 03	17 39	22 58	22 29	06 22	16 35	21 01
24	20 44	19 56	22 54	24 55	17 25	22 58	22 29	06 22	16 34	21 01
25	20 55	17 09	23 14	24 46	17 10	22 58	22 28	06 21	16 34	21 01
26	21 06	13 38	23 33	24 37	16 56	22 58	22 28	06 21	16 33	21 00
27	21 17	09 32	23 50	24 27	16 41	22 58	22 27	06 20	16 33	21 00
28	21 28	04 59	24 06	24 17	16 26	22 58	22 27	06 20	16 33	21 00
29	21 38	00 06	24 21	24 05	16 11	22 58	22 26	06 20	16 32	20 59
30	21 S 47	04 S 55	24 S 35	23 S 53	15 S 55	22 N 58	22 S 26	06 N 19	16 N 32	20 S 59

ZODIAC SIGN ENTRIES

Date	h	m	Planets
01	23	07	♂ ♒
03	01	34	☽ ♎
05	04	51	☽ ♏
07	05	55	☽ ♐
09	06	32	☽ ♑
10	11	41	♀ ♑
11	08	15	☽ ♒
12	14	06	♃ ♊
13	12	02	☽ ♓
15	18	16	☽ ♈
18	02	53	☽ ♉
19	01	03	☿ ♐
20	13	41	☽ ♊
21	15	33	☉ ♐
23	02	10	☽ ♋
25	15	11	☽ ♌
28	02	49	☽ ♍
30	11	14	☽ ♎

LATITUDES

Date	Mercury ☿ ° '	Venus ♀ ° '	Mars ♂ ° '	Jupiter ♃ ° '	Saturn ♄ ° '	Uranus ♅ ° '	Neptune ♆ ° '	Pluto ♇ ° '
01	01 N 16	01 S 57	02 S 01	00 S 29	00 N 19	00 N 45	01 S 48	12 S 41
04	00 58	02 04	01 57	00 29	00 19	00 45	01 48	12 41
07	00 38	02 10	01 53	00 29	00 19	00 45	01 48	12 40
10	00 N 18	02 16	01 50	00 29	00 18	00 45	01 48	12 39
13	00 S 02	02 20	01 46	00 29	00 18	00 45	01 48	12 38
16	00 22	02 24	01 42	00 28	00 18	00 46	01 48	12 37
19	00 42	02 28	01 38	00 28	00 18	00 46	01 48	12 37
22	01 00	02 30	01 34	00 28	00 17	00 46	01 48	12 36
25	01 18	02 31	01 31	00 28	00 17	00 46	01 48	12 35
28	01 33	02 32	01 27	00 27	00 17	00 46	01 48	12 34
31	01 S 47	02 S 31	01 S 23	00 S 27	00 N 16	00 N 46	01 S 48	12 S 33

DATA

Julian Date	2469382
Delta T	+80 seconds
Ayanamsa	24° 32' 37"
Synetic vernal point	04° ♓ 34' 21"
True obliquity of ecliptic	23° 25' 57"

MOON'S PHASES, APSIDES AND POSITIONS ☽

Date	h	m	Phase	Longitude ° '	Eclipse Indicator
06	04	38	●	14 ♏ 27	
12	20	29	☽	21 ♒ 08	
20	11	19	○	28 ♉ 49	
28	16	33	☾	07 ♍ 07	

Day	h	m		
07	23	11	Perigee	
23	18	11	Apogee	
02	03	56	0S	
08	10	46	Max dec	22° S 58'
14	19	53	0N	
22	01	15	Max dec	22° N 56'
29	12	30	0S	

LONGITUDES

Date	Chiron ⚷ ° '	Ceres ⚳ ° '	Pallas ⚴ ° '	Juno ⚵ ° '	Vesta ⚶ ° '	Black Moon Lilith ⚸ ° '
01	15 ♏ 51	02 ♈ 24	06 ♒ 30	18 ♒ 01	18 ♉ 08	00 ♋ 25
11	17 ♏ 13	01 ♈ 32	08 ♒ 13	20 ♒ 16	15 ♉ 31	01 ♋ 32
21	18 ♏ 35	01 ♈ 17	10 ♒ 15	23 ♒ 02	13 ♉ 03	02 ♋ 38
31	19 ♏ 56	01 ♈ 40	12 ♒ 34	26 ♒ 15	11 ♉ 01	03 ♋ 45

All ephemeris data is given at 12.00 UT and the Moon's longitude is additionally given for 24.00 UT

ASPECTARIAN

h m Aspects

01 Sunday
04 57 ☽ ± ♂
05 48 ☽ ☍ ♇
13 00 ☽ ⚹ ☉
15 52 ☽ △ ♄
18 22 ☽ Q ♃
18 56 ♂ ⊡ ♅
20 15 ♂ ⊥ ♇
22 24 ☽ ☌ ♅
22 35 ☽ ⊡ ♂

02 Monday
03 29 ☽ ∠ ☿
08 17 ☽ □ ♀
12 43 ☽ △ ♆
18 30 ☽ ∠ ☉
20 46 ♂ ⚻ ♃
21 56 ☽ ⊥ ☿

03 Tuesday
02 40 ☽ □ ♃
03 01 ☽ △ ♂
04 55 ☉ ⚹ ♄
09 57 ☽ ⊻ ☿
10 36 ☽ ⫲ ♅
11 36 ☽ ⊥ ☉
11 55 ☽ ⚻ ♇
15 08 ☽ ⊡ ♆
19 06 ☽ Q ♀
21 36 ☽ □ ♄
22 17 ☽ ± ♇
22 47 ☽ ⊻ ☉

04 Wednesday
02 12 ☿ △ ♇
03 33 ☽ ⊻ ♅
06 33 ☽ ± ♆
08 51 ♀ ⚻ ♆
13 36 ☽ ⊡ ♇
13 46 ☽ ⊥ ♅
16 41 ☽ ⚻ ♆
17 08 ☽ ∥ ☿
17 25 ☽ ⚹ ♀

05 Thursday
04 11 ☽ Q ♄
04 53 ☽ ∠ ♅
05 42 ☽ △ ♃
08 14 ♀ Q ♇
08 52 ☽ □ ♂
09 18 ☽ ∥ ☉
13 50 ☽ ⫲ ♆
14 36 ☽ △ ♇
19 16 ☽ ☌ ☿
20 33 ☽ ∠ ♀

06 Friday
00 06 ☽ ⚹ ♄
04 38 ☽ ☌ ☉
05 38 ☽ ⚹ ♅
06 17 ☽ ⊡ ♃
12 33 ☽ ⊥ ♀
15 53 ☽ Q ♂
18 02 ☽ ☍ ♆
19 51 ☉ ⚹ ♅
20 52 ☽ ± ♃
22 02 ☽ ∥ ♇
23 09 ☽ ⊻ ♀
23 14 ☽ ∥ ♂

07 Saturday
00 39 ☽ ∠ ♄
01 12 ☽ Q ♅
03 23 ☉ ⊡ ♃
06 32 ☽ ⚻ ♃
09 34 ☿ ∠ ♀
12 19 ☽ ⚹ ♂
14 22 ♂ ∥ ♇
15 18 ☽ ⊥ ♄
15 25 ☽ □ ♇
16 27 ☿ ⚹ ♄
18 40 ☽ ∥ ♄
23 07 ☉ ⫲ ♆

08 Sunday
01 03 ☽ ⊻ ♄
02 03 ☽ ⊻ ☿
06 23 ☽ □ ♅
07 13 ☽ ⫲ ♃
08 48 ☽ ⊻ ☉
12 58 ☽ ⊥ ☿
13 49 ☽ ∠ ♂
14 16 ☽ ⫲ ♃
18 30 ☽ ⚻ ♆
19 14 ☽ ⊥ ☉
20 34 ☽ Q ♇

09 Monday
03 06 ☽ ∥ ♄
03 59 ☽ ☌ ♀
04 15 ☽ ± ♆
05 11 ☽ ⊥ ♂
05 25 ☽ ∠ ☿
06 56 ☽ ☍ ♃
07 16 ♀ ± ♆
10 59 ☽ ∠ ☉
15 30 ☽ ⊻ ♂
16 08 ☽ ⚹ ♇
18 23 ☿ ⚹ ♅
18 45 ☿ ⊡ ♃
18 57 ☽ ⊡ ♆
23 53 ☽ ∥ ♇

10 Tuesday
02 14 ☽ ☌ ♄
04 16 ♂ ⊻ ♇
04 41 ☽ ∥ ♂
07 31 ☽ △ ♅
09 08 ☽ ⚹ ☿
13 32 ☽ ⚹ ☉
13 47 ♂ ± ♃
14 47 ♀ ☍ ♃
16 31 ♀ ⊥ ♂
16 55 ☽ ∠ ♇
19 44 ☽ △ ♆
21 19 ☿ ⫲ ♆

11 Wednesday
03 10 ☽ ∥ ☉
07 26 ☽ ∥ ☿
07 45 ☽ Q ☿
08 01 ☽ ⊥ ♇
08 24 ☽ ⚻ ♃
08 40 ☽ ⊡ ♅
08 55 ☽ ⫲ ♆
10 07 ☽ ⊻ ♀
11 11 ☽ Q ☉
18 11 ☽ ⊻ ♇
18 31 ☽ ± ♃
20 14 ☽ ☌ ♂
21 14 ☽ ⊥ ♀

12 Thursday
00 00 ☽ ± ♅
05 00 ☽ ⊻ ♄
09 52 ☽ ⊡ ♃
10 20 ☽ ⚻ ♅
14 08 ☽ ∠ ♀
15 26 ☽ ⊥ ♄
18 34 ☽ □ ☿
20 29 ☽ □ ☉
22 51 ☽ □ ♆

13 Friday
07 15 ☽ ∠ ♄
11 54 ☽ △ ♃
13 38 ☽ ⫲ ♅
18 57 ☽ ⚹ ♀
22 20 ☉ ∥ ☿
22 26 ☽ ☌ ♇

14 Saturday
03 34 ☽ ⊻ ♂
04 02 ☉ ☍ ♆
06 39 ☽ Q ♆
06 49 ☿ ☍ ♆
10 07 ☽ ⚹ ♄
11 42 ☉ ☌ ☿
14 59 ☽ ⊥ ♂
15 29 ☽ ☍ ♅
18 35 ☽ Q ♀

15 Sunday
00 27 ♂ ± ♅
02 51 ☿ ± ♃
04 21 ☽ ⚹ ♆
06 25 ☽ △ ☉
07 23 ☽ △ ☿
08 05 ☽ Q ♄
08 20 ☽ ∠ ♂
10 40 ♀ ⚹ ♇
10 59 ☉ ± ♃
17 46 ☽ □ ♃
23 53 ☿ ∥ ♂

16 Monday
02 46 ☽ ∥ ♅
05 08 ☽ ⊻ ♇
07 00 ☽ □ ♀
08 01 ☽ ∠ ♆
12 34 ☽ ⊡ ☉
13 49 ☽ ⚹ ♂
15 06 ☽ ⊡ ☿
16 22 ☽ ⊥ ♇
17 46 ☽ □ ♄
17 50 ♀ ⊡ ♆
23 06 ☽ ⚻ ♅

17 Tuesday
00 54 ☽ ⊥ ♆
02 56 ☉ ∥ ♂
03 13 ☽ Q ♃
07 01 ☽ ± ☉
09 23 ☽ ∠ ♇
10 11 ☿ Q ♅
10 31 ☽ ± ♅
10 33 ☽ ± ☿
12 16 ☽ ⊻ ♆
13 40 ☿ ∠ ♄
13 55 ☽ Q ♂
19 27 ☽ ⚻ ☉
23 41 ☽ ⚻ ☿

18 Wednesday
01 58 ☽ ⚹ ♃
03 47 ☽ ⊡ ♅
12 37 ☽ ∥ ♆
14 12 ☽ ⚹ ♇
16 46 ☿ ⚻ ♃
16 54 ♇ St D
22 10 ☽ △ ♀

19 Thursday
02 51 ☽ □ ♂
03 47 ☽ △ ♄
04 01 ☉ Q ♅
04 08 ☽ ⫲ ♂
06 52 ☽ ∠ ♃
09 01 ☽ △ ♅
09 27 ☿ Q ♂
12 37 ☉ ∠ ♄
13 11 ☽ ⫲ ☉
13 39 ☽ Q ♇
15 37 ☿ ∥ ♇
20 35 ♂ ⊻ ♄
22 24 ☽ ☌ ♆

20 Friday
00 26 ☽ ⊥ ♃
03 33 ☽ ⫲ ♇
06 49 ☽ ⊡ ♀
06 59 ☽ ⫲ ☿
09 35 ☽ ⊡ ♄
11 19 ☽ ☍ ☉
12 15 ☽ ⊻ ♃
19 12 ☽ ☍ ☿
21 25 ☉ ⚻ ♃

21 Saturday
01 23 ☽ □ ♇
02 42 ☽ ± ♀
03 31 ☽ ⫲ ♄
03 42 ☽ ± ♄
13 48 ♀ ☌ ♄
14 28 ♂ ⊡ ♃
15 52 ☽ ⚻ ♄
16 04 ☽ ⚻ ♀
18 19 ☽ △ ♂
20 55 ☽ □ ♅

22 Sunday
10 23 ☽ ⊻ ♆
18 42 ☿ □ ♇
22 32 ☽ ⊥ ♆
23 29 ☽ ⫲ ♄

23 Monday
00 10 ☽ ☌ ♃
01 42 ☽ ⫲ ☿
02 43 ☽ ⊡ ♂
05 24 ☽ ⚻ ☉
07 35 ☿ ∥ ♄
09 31 ☽ Q ♅
14 05 ♂ ⚻ ♅
14 05 ☽ △ ♇
16 13 ☿ ⊥ ♄
16 48 ☽ ∠ ♆
17 03 ☽ ⚻ ☿
18 33 ♀ △ ♅
18 46 ☽ ± ☉
22 15 ☽ ± ♂
23 50 ☽ ⫲ ♇

24 Tuesday
01 50 ♀ ⊻ ♂
04 05 ☽ ⫲ ☉
05 16 ☽ ☍ ♄
07 08 ☽ ± ☿
10 00 ☽ ⚹ ♅
11 17 ☽ ⚻ ♂
11 40 ☽ ☍ ♀
14 49 ☽ ⊡ ☉
16 07 ☿ ⫲ ♃
20 39 ☽ ⊡ ♇
23 16 ☽ ⚹ ♆

25 Wednesday
04 14 ☽ ⊡ ☿
11 47 ☽ ⫲ ♂
12 37 ☽ ⊻ ♃
14 54 ☽ ± ♇
16 19 ☽ ∥ ♆
16 31 ☽ ∠ ♅
23 15 ☉ ∥ ♇
23 31 ☽ Q ♆

26 Thursday
00 04 ☽ △ ☉
00 38 ☽ ⊥ ♃
03 01 ☽ ⚻ ♇
10 41 ☽ ⊥ ♅
15 01 ☽ △ ☿
18 27 ☽ ⚻ ♄
18 30 ☽ ∠ ♃
22 41 ☽ ⊻ ♅

27 Friday
03 44 ☽ ☍ ♂
06 28 ☽ ± ♄
06 37 ☽ ⚻ ♀
10 58 ☉ □ ♇
11 21 ☽ □ ♆
19 11 ☿ ⊻ ♄
19 39 ☽ ± ♀
23 49 ☽ ⚹ ♃

28 Saturday
00 17 ☽ ⊡ ♄
00 51 ♂ ⫲ ♆
05 04 ☽ ∥ ♅
07 08 ♀ ∠ ♇
14 10 ☽ ☍ ♇
14 53 ☽ ⊡ ♀
16 33 ☽ □ ☉
21 13 ☿ ∥ ♀
22 42 ☽ Q ♃

29 Sunday
03 12 ☿ □ ♅
05 17 ☽ △ ♄
07 16 ♀ △ ♆
07 50 ♂ ⊥ ♄
08 55 ☽ ☌ ♅
09 22 ☿ ⊥ ♀
09 32 ☉ ⊥ ♄
09 42 ☽ □ ☿
13 24 ☉ Q ♂
17 04 ☽ ⚻ ♂
20 37 ☽ △ ♆
21 57 ☽ △ ♀

30 Monday
04 48 ☽ ± ♂
05 09 ☽ Q ☉
07 55 ☽ □ ♃
18 40 ☽ ⫲ ♅
21 50 ☽ ⚻ ♇
22 02 ☽ ⊡ ♂
23 03 ☽ Q ☿
23 48 ☽ ⊡ ♆

DECEMBER 2048

LONGITUDES

Date	Sidereal time h m s	Sun ☉ ° ' "	Moon ☽ ° ' "	Moon ☽ 24.00 ° ' "	Mercury ☿ ° '	Venus ♀ ° '	Mars ♂ ° '	Jupiter ♃ ° '	Saturn ♄ ° '	Uranus ♅ ° '	Neptune ♆ ° '	Pluto ♇ ° '
01	16 44 05	09 ♐ 58 04	13 ♎ 53 44	20 ♎ 48 11	19 ♐ 24	24 ♑ 34	21 ♒ 26	28 ♊ 02	14 ♑ 03	15 ♍ 47	21 ♉ 59	05 ♓ 53
02	16 48 02	10 58 55	27 ♎ 49 31	04 ♏ 57 31	20 57	25 43	22 10	27 R 54	14 10	15 48	21 R 57	05 54
03	16 51 58	11 59 47	12 ♏ 11 48	19 ♏ 31 49	22 29	26 52	22 54	27 47	14 16	15 49	21 56	05 54
04	16 55 55	13 00 40	26 ♏ 56 49	04 ♐ 25 56	24 02	28 01	23 39	27 39	14 22	15 50	21 54	05 55
05	16 59 52	14 01 35	11 ♐ 58 08	19 ♐ 32 17	25 34	29 ♑ 09	24 23	27 31	14 29	15 51	21 53	05 55
06	17 03 48	15 02 31	27 ♐ 07 13	04 ♑ 41 44	27 06	00 ♒ 18	25 07	27 24	14 35	15 52	21 51	05 55
07	17 07 45	16 03 27	12 ♑ 14 40	19 ♑ 44 56	28 ♐ 39	01 26	25 52	27 16	14 42	15 53	21 49	05 56
08	17 11 41	17 04 25	27 ♑ 11 33	04 ♒ 33 42	00 ♑ 10	02 34	26 36	27 08	14 48	15 53	21 48	05 56
09	17 15 38	18 05 23	11 ♒ 50 43	19 ♒ 02 03	01 42	03 42	27 21	27 00	14 55	15 54	21 46	05 57
10	17 19 34	19 06 22	26 ♒ 07 24	03 ♓ 06 33	03 13	04 50	28 05	26 52	15 02	15 55	21 45	05 57
11	17 23 31	20 07 21	09 ♓ 59 29	16 ♓ 46 15	04 44	05 58	28 50	26 44	15 08	15 55	21 44	05 58
12	17 27 27	21 08 21	23 ♓ 27 03	00 ♈ 02 08	06 15	07 05	29 ♒ 34	26 36	15 15	15 56	21 42	05 59
13	17 31 24	22 09 21	06 ♈ 31 51	12 ♈ 56 33	07 45	08 12	00 ♓ 19	26 27	15 22	15 56	21 41	05 59
14	17 35 21	23 10 22	19 ♈ 16 39	25 ♈ 32 34	09 15	09 19	01 03	26 19	15 28	15 56	21 39	06 00
15	17 39 17	24 11 24	01 ♉ 44 42	07 ♉ 53 29	10 44	10 26	01 48	26 11	15 35	15 57	21 38	06 01
16	17 43 14	25 12 26	13 ♉ 59 18	20 ♉ 02 32	12 12	11 33	02 32	26 03	15 42	15 57	21 37	06 01
17	17 47 10	26 13 28	26 ♉ 03 33	02 ♊ 02 42	13 39	12 39	03 17	25 55	15 49	15 57	21 35	06 02
18	17 51 07	27 14 31	08 ♊ 00 17	13 ♊ 56 37	15 05	13 45	04 02	25 47	15 56	15 57	21 34	06 03
19	17 55 03	28 15 35	19 ♊ 51 59	25 ♊ 46 39	16 29	14 51	04 46	25 39	16 02	15 58	21 33	06 03
20	17 59 00	29 ♐ 16 39	01 ♋ 40 52	07 ♋ 34 54	17 52	15 57	05 31	25 30	16 09	15 58	21 31	06 04
21	18 02 56	00 ♑ 17 43	13 ♋ 29 01	19 ♋ 23 27	19 12	17 02	06 15	25 22	16 16	15 R 58	21 30	06 05
22	18 06 53	01 18 49	25 ♋ 18 30	01 ♌ 14 26	20 31	18 08	07 00	25 14	16 23	15 58	21 29	06 06
23	18 10 50	02 19 54	07 ♌ 11 34	13 ♌ 10 13	21 46	19 12	07 45	25 06	16 30	15 57	21 28	06 07
24	18 14 46	03 21 01	19 ♌ 10 43	25 ♌ 13 27	22 58	20 17	08 29	24 58	16 37	15 57	21 27	06 08
25	18 18 43	04 22 07	01 ♍ 18 48	07 ♍ 27 12	24 07	21 21	09 14	24 50	16 44	15 57	21 25	06 09
26	18 22 39	05 23 15	13 ♍ 39 04	19 ♍ 54 51	25 11	22 25	09 58	24 42	16 51	15 57	21 24	06 09
27	18 26 36	06 24 23	26 ♍ 15 01	02 ♎ 40 00	26 10	23 29	10 43	24 34	16 58	15 56	21 23	06 10
28	18 30 32	07 25 31	09 ♎ 10 15	15 ♎ 46 10	27 03	24 32	11 28	24 27	17 05	15 56	21 22	06 11
29	18 34 29	08 26 40	22 ♎ 28 06	29 ♎ 16 20	27 49	25 35	12 12	24 19	17 12	15 55	21 21	06 12
30	18 38 25	09 27 50	06 ♏ 11 03	13 ♏ 12 22	28 28	26 38	12 57	24 11	17 19	15 55	21 20	06 13
31	18 42 22	10 ♑ 29 00	20 ♏ 20 12	27 ♏ 34 21	28 ♑ 58	27 ♒ 41	13 ♓ 41	24 ♊ 04	17 ♑ 26	15 ♍ 54	21 ♉ 19	06 ♓ 14

Moon nodes and latitude

Date	Moon True ☊ ° '	Moon Mean ☊ ° '	Moon ☽ Latitude ° '
01	18 ♐ 36	18 ♐ 55	04 S 47
02	18 R 35	18 52	04 06
03	18 34	18 49	03 09
04	18 34	18 46	01 57
05	18 34	18 42	00 S 37
06	18 D 34	18 39	00 N 47
07	18 34	18 36	02 08
08	18 34	18 33	03 19
09	18 34	18 30	04 15
10	18 R 34	18 27	04 54
11	18 34	18 23	05 15
12	18 D 34	18 20	05 17
13	18 34	18 17	05 03
14	18 34	18 14	04 34
15	18 35	18 11	03 52
16	18 36	18 08	03 01
17	18 36	18 04	02 02
18	18 37	18 01	00 N 59
19	18 R 37	17 58	00 S 07
20	18 37	17 55	01 12
21	18 36	17 52	02 14
22	18 34	17 48	03 10
23	18 32	17 45	03 58
24	18 30	17 42	04 36
25	18 28	17 39	05 02
26	18 26	17 36	05 15
27	18 24	17 33	05 13
28	18 D 24	17 29	04 56
29	18 25	17 26	04 22
30	18 26	17 23	03 33
31	18 ♐ 27	17 ♐ 20	02 S 29

DECLINATIONS

Date	Sun ☉ ° '	Moon ☽ ° '	Mercury ☿ ° '	Venus ♀ ° '	Mars ♂ ° '	Jupiter ♃ ° '	Saturn ♄ ° '	Uranus ♅ ° '	Neptune ♆ ° '	Pluto ♇ ° '
01	21 S 56	09 S 53	24 S 48	23 S 41	15 S 40	22 N 58	22 S 25	06 N 19	16 N 31	20 S 59
02	22 05	14 32	24 59	23 28	15 24	22 58	22 24	06 18	16 31	20 58
03	22 13	18 29	25 09	23 14	15 09	22 58	22 24	06 18	16 31	20 58
04	22 21	21 22	25 17	23 00	14 53	22 58	22 23	06 18	16 30	20 58
05	22 29	22 49	25 24	22 45	14 37	22 58	22 23	06 17	16 30	20 57
06	22 36	22 37	25 30	22 29	14 21	22 58	22 22	06 17	16 30	20 57
07	22 42	20 45	25 34	22 13	14 05	22 58	22 21	06 17	16 29	20 56
08	22 48	17 28	25 37	21 56	13 49	22 58	22 21	06 17	16 29	20 56
09	22 54	13 08	25 38	21 39	13 32	22 58	22 20	06 16	16 28	20 55
10	22 59	08 11	25 38	21 21	13 16	22 58	22 20	06 16	16 28	20 55
11	23 04	02 S 58	25 37	21 03	12 59	22 58	22 19	06 16	16 28	20 54
12	23 08	02 N 15	25 34	20 44	12 42	22 58	22 18	06 16	16 27	20 54
13	23 12	07 13	25 29	20 25	12 26	22 57	22 18	06 16	16 27	20 54
14	23 15	11 46	25 24	20 05	12 09	22 57	22 17	06 16	16 27	20 53
15	23 18	15 42	25 16	19 45	11 52	22 57	22 16	06 16	16 26	20 53
16	23 21	18 55	25 07	19 25	11 34	22 57	22 15	06 16	16 26	20 52
17	23 23	21 15	24 57	19 04	11 17	22 57	22 15	06 15	16 26	20 52
18	23 24	22 36	24 45	18 42	11 00	22 57	22 14	06 15	16 26	20 51
19	23 25	22 56	24 32	18 20	10 43	22 57	22 13	06 15	16 25	20 51
20	23 26	22 13	24 18	17 58	10 25	22 57	22 13	06 15	16 25	20 50
21	23 26	20 32	24 02	17 35	10 08	22 57	22 12	06 15	16 25	20 49
22	23 26	17 57	23 45	17 12	09 50	22 56	22 11	06 15	16 24	20 49
23	23 25	14 38	23 27	16 49	09 32	22 56	22 10	06 16	16 24	20 48
24	23 23	10 41	23 09	16 25	09 14	22 56	22 09	06 16	16 24	20 48
25	23 22	06 17	22 49	16 01	08 57	22 56	22 09	06 16	16 24	20 47
26	23 19	01 N 34	22 28	15 36	08 39	22 56	22 08	06 16	16 24	20 47
27	23 17	03 S 18	22 07	15 12	08 21	22 56	22 07	06 16	16 23	20 46
28	23 13	08 10	21 46	14 47	08 03	22 56	22 06	06 16	16 23	20 46
29	23 10	12 48	21 25	14 22	07 45	22 55	22 05	06 17	16 23	20 45
30	23 06	16 55	21 03	13 56	07 27	22 55	22 05	06 17	16 23	20 44
31	23 S 01	20 S 13	20 S 43	13 S 30	07 S 08	22 N 55	22 S 04	06 N 17	16 N 22	20 S 44

ZODIAC SIGN ENTRIES

Date	h	m	Planets
02	15	41	☽ ♏
04	16	54	☽ ♐
06	05	42	♀ ♒
06	16	34	☽ ♑
08	09	17	☿ ♑
08	16	33	☽ ♒
10	18	38	☽ ♓
12	23	56	☽ ♈
13	01	52	♂ ♓
15	08	37	☽ ♉
17	19	54	☽ ♊
20	08	35	☽ ♋
21	05	02	☉ ♑
22	21	30	☽ ♌
25	09	25	☽ ♍
27	19	02	☽ ♎
30	01	16	☽ ♏

LATITUDES

Date	Mercury ☿ ° '	Venus ♀ ° '	Mars ♂ ° '	Jupiter ♃ ° '	Saturn ♄ ° '	Uranus ♅ ° '	Neptune ♆ ° '	Pluto ♇ ° '
01	01 S 47	02 S 31	01 S 23	00 S 27	00 N 16	00 N 46	01 S 48	12 S 33
04	01 59	02 30	01 19	00 27	00 16	00 46	01 48	12 32
07	02 09	02 27	01 16	00 26	00 16	00 46	01 47	12 32
10	02 15	02 23	01 12	00 26	00 16	00 47	01 47	12 31
13	02 17	02 17	01 08	00 26	00 15	00 47	01 47	12 30
16	02 16	02 11	01 05	00 25	00 15	00 47	01 47	12 29
19	02 08	02 03	01 01	00 25	00 15	00 47	01 47	12 29
22	01 55	01 54	00 58	00 24	00 15	00 47	01 47	12 28
25	01 33	01 44	00 54	00 24	00 15	00 47	01 47	12 27
28	01 00	01 32	00 51	00 23	00 15	00 47	01 47	12 27
31	00 S 22	01 S 18	00 S 47	00 S 23	00 N 14	00 N 47	01 S 47	12 S 26

DATA

Julian Date	2469412
Delta T	+80 seconds
Ayanamsa	24° 32' 42"
Synetic vernal point	04° ♓ 34' 17"
True obliquity of ecliptic	23° 25' 56"

MOON'S PHASES, APSIDES AND POSITIONS ☽

Date	h	m	Phase	Longitude ° '	Eclipse Indicator
05	15	30	●	14 ♐ 10	Total
12	07	29	☽	20 ♓ 57	
20	06	39	○	29 ♊ 03	
28	08	31	☾	07 ♎ 17	

Day	h	m		
06	07	55	Perigee	
21	00	14	Apogee	
05	21	00	Max dec	22° S 57'
12	01	33	0N	
19	07	33	Max dec	22° N 57'
26	19	48	0S	

LONGITUDES

Date	Chiron ⚷ ° '	Ceres ⚳ ° '	Pallas ⚴ ° '	Juno ⚵ ° '	Vesta ⚶ ° '	Black Moon Lilith ⚸ ° '
01	19 ♏ 56	01 ♈ 40	12 ♒ 34	26 ♒ 15	11 ♉ 01	03 ♋ 45
11	21 ♏ 13	02 ♈ 38	15 ♒ 07	29 ♒ 52	09 ♉ 36	04 ♋ 51
21	22 ♏ 26	04 ♈ 08	17 ♒ 52	03 ♓ 49	08 ♉ 55	05 ♋ 58
31	23 ♏ 34	06 ♈ 05	20 ♒ 46	08 ♓ 04	08 ♉ 59	07 ♋ 04

ASPECTARIAN

h m Aspects

01 Tuesday
04 32 ☽ ⚹ ☉
08 29 ☽ ± ♇
12 17 ☽ □ ♄
15 19 ☽ ⊻ ♅
15 38 ☽ ± ♆
22 47 ☽ ⚹ ☿

02 Wednesday
00 08 ♃ ⊥ ♆
00 09 ☽ bQ ♇
01 43 ☽ ⊥ ♅
01 49 ☽ △ ♂
02 00 ☽ ⊼ ♆
05 15 ♂ □ ♆
08 06 ☽ □ ♀
08 38 ☽ ∠ ☉
12 08 ☽ △ ♃
16 39 ☽ ∥ ♂
17 02 ☽ ∠ ♅
19 22 ☽ Q ♄
23 23 ☽ # ♆

03 Thursday
00 57 ☽ ⊥ ☉
01 34 ☽ △ ♇
03 17 ☽ ∠ ☿
03 23 ☿ ⊼ ♆
11 39 ☽ ⊻ ☉
12 57 ☽ bQ ♃
15 26 ☽ ⚹ ♄
16 46 ☽ Q ♀
17 57 ☽ ⚹ ♅
19 52 ☽ ⊥ ☿

04 Friday
00 27 ☿ ⚹ ♂
03 31 ☽ ± ♃
03 52 ☽ ☍ ♆
05 08 ♀ ⊼ ♃
06 24 ☽ □ ♂
06 45 ☽ ⊻ ☿
07 46 ☽ ∥ ♇
10 05 ☿ Q ♇
11 18 ☉ ∠ ♀
13 07 ☽ ⊼ ♃
13 26 ☽ Q ♅
13 52 ☽ ⚹ ♀
14 48 ♀ # ♃
15 56 ☽ ∠ ♄
18 14 ☉ ∥ ♄

05 Saturday
01 28 ☽ ∥ ♄
02 16 ☽ ∥ ☉
02 22 ☽ □ ♇
06 25 ☽ ⊥ ♄
10 03 ☽ ∥ ♀
12 42 ☽ Q ♂
15 30 ☽ ☌ ☉
15 45 ☽ ∠ ♀
16 01 ☽ ⊻ ♄
18 10 ☽ □ ♅

06 Sunday
00 00 ☉ ⊻ ♄
03 40 ☽ ⊼ ♆
04 02 ♀ ⊥ ♇
05 07 ☉ ∥ ♀
06 56 ☽ Q ♇
07 10 ☽ ⊥ ♀
08 41 ☽ ⚹ ♂
11 59 ☽ ☌ ☿
12 20 ☽ ∥ ☉
12 26 ☽ ☍ ♃
13 09 ☽ ± ♆
15 16 ☽ ∥ ♀
16 06 ☿ ☍ ♃
16 46 ☽ ∥ ♄
17 26 ☽ ⊻ ♀
23 01 ♀ ∥ ♄
23 25 ☿ ± ♆

07 Monday
00 04 ♀ bQ ♅
01 57 ☽ ⚹ ♇
03 24 ☽ bQ ♆
07 42 ☉ □ ♅
09 41 ☽ ∠ ♂
10 11 ☽ ∥ ♇
15 56 ☽ ☌ ♄
17 48 ☽ △ ♅
18 32 ☽ ⊻ ☉

08 Tuesday
00 49 ☽ ⊥ ♂
01 54 ☽ ∠ ♇
03 19 ☽ △ ♆
04 52 ☽ ⊥ ☉
11 00 ☽ ⊻ ♂
11 54 ☽ ⊼ ♃
16 28 ☽ ⊥ ♇
17 24 ☽ ⊻ ☿
17 54 ☽ # ♆
18 01 ☽ bQ ♅
20 31 ☽ ∠ ☉
21 29 ☽ ☌ ♀
21 34 ☽ ± ♃
22 33 ♀ ± ♃

09 Wednesday
02 16 ☽ ⊻ ♇
02 23 ♂ △ ♃
04 21 ☽ ⊥ ☿
08 47 ☽ ± ♅
09 51 ☽ ∥ ♂
12 15 ☽ bQ ♃
17 09 ☽ ⊻ ♄
18 45 ☽ ⊼ ♅
21 03 ☽ ∠ ☿
23 13 ☽ ⚹ ☉

10 Thursday
03 17 ☽ ⊥ ♄
04 35 ☽ □ ♆
05 07 ☉ # ♃
13 15 ☽ △ ♃
15 33 ☽ ☌ ♂
18 44 ☽ ∠ ♄
20 53 ☽ # ♅
21 13 ☽ Q ☉

11 Friday
01 42 ☽ ⚹ ☿
04 20 ☽ ⊻ ♀
04 57 ☽ ☌ ♇
11 32 ☽ Q ♆
12 06 ♀ ⊻ ♇
15 47 ☽ ⊥ ♀
21 10 ☽ ⚹ ♄
22 29 ☽ ☍ ♅
23 30 ♀ ∥ ♇

12 Saturday
01 28 ☽ Q ☿
07 29 ☽ □ ☉
07 36 ☿ ⚹ ♇
08 51 ☽ ⚹ ♆
09 18 ☽ ∠ ♀
17 39 ☽ □ ♃
18 58 ☽ Q ♄
19 04 ☿ bQ ♆
23 49 ☽ ⊻ ♂

13 Sunday
00 59 ☉ ⊼ ♆
07 13 ☽ ∥ ♅
10 59 ☽ ⊻ ♇
11 34 ☽ ⊥ ♂
12 16 ☽ ∠ ♆
13 46 ♂ ∠ ♄
14 35 ☽ □ ☿
15 25 ☽ ⚹ ♀
22 13 ☽ ⊥ ♇

14 Monday
02 42 ☽ Q ♃
04 43 ☽ □ ♄
05 08 ☽ ⊥ ♆
05 30 ☽ ∠ ♂
05 40 ☽ ⊼ ♅
14 01 ☽ # ♂
15 17 ☽ ∠ ♇
16 17 ☽ Q ♀
16 32 ☽ ⊻ ♆
16 55 ☿ ⊻ ♀
17 05 ☽ ± ♅
20 06 ☽ △ ☉

15 Tuesday
01 21 ☽ ⚹ ♃
01 23 ♀ ± ♅
07 41 ☉ Q ♇
10 27 ☽ bQ ♅
12 07 ☽ ⚹ ♂
17 00 ☽ ∥ ♆
20 19 ☽ ⚹ ♇

16 Wednesday
02 25 ♀ bQ ♃
03 52 ☽ bQ ☉
06 16 ☽ ∠ ♃
06 42 ☽ □ ♀
07 59 ☽ △ ☿
13 10 ☽ Q ♂
15 25 ☽ △ ♄
15 53 ☽ △ ♅
15 55 ☽ # ♀
19 59 ☽ Q ♇
23 17 ☽ ± ☉
23 53 ☽ ⊥ ♃

17 Thursday
03 06 ☽ ☌ ♆
05 33 ☉ ☍ ♃
07 19 ☽ # ♇
11 43 ☽ ⊻ ♃
12 22 ☽ ⊼ ☉
17 53 ☽ bQ ☿
21 37 ☽ bQ ♄

18 Friday
03 27 ☽ □ ♂
03 31 ☽ # ♄
08 03 ☽ □ ♇
14 27 ☽ ± ☿
15 55 ☽ ± ♄
18 38 ♄ △ ♅
19 29 ☉ ± ♆

19 Saturday
00 48 ☽ △ ♀
02 59 ☿ △ ♅
03 43 ☿ ☌ ♄
04 05 ☽ □ ♅
04 10 ☽ ⊼ ♄
04 14 ☽ ⊼ ☿
06 38 ☽ ∥ ♃
08 33 ☽ ∥ ♃
15 24 ☽ ⊻ ♆
23 36 ☽ ☌ ♃

20 Sunday
03 34 ☽ ⊥ ♆
06 39 ☽ ☍ ☉
10 22 ☽ bQ ♀
12 14 ♀ ⊼ ♅
12 17 ☽ # ♄
16 38 ☽ Q ♅
17 05 ♀ ⊻ ♄
17 42 ♅ St R
20 19 ☽ △ ♂
20 56 ☽ △ ♇
21 50 ☽ ∠ ♆

21 Monday
06 18 ♂ ☌ ♇
06 32 ☽ ± ♀
08 37 ☽ # ♇
17 02 ☽ ⚹ ♅
17 43 ☽ ☍ ♄
19 57 ☽ ⊼ ♀

22 Tuesday
01 05 ☽ ☍ ☿
03 27 ☽ bQ ♇
04 15 ☽ ⚹ ♆
04 51 ☽ bQ ♂
11 51 ☽ ⊻ ♃
18 45 ☽ # ♀
21 42 ☽ ± ♇
23 11 ☿ ∠ ♇
23 26 ☽ ∠ ♅
23 49 ☽ ∥ ♆
23 51 ☽ ⊥ ♃

23 Wednesday
00 17 ☽ ± ♂
01 17 ☽ ⊼ ☉
04 30 ☽ Q ♆
06 07 ☿ △ ♆
09 49 ☽ ⊼ ♇
13 11 ☽ ⊼ ♂
14 30 ☽ ± ☉
15 47 ☉ ∥ ☿
17 33 ☽ ⊥ ♅
17 47 ☽ ∠ ♃

24 Thursday
05 34 ☽ ⊻ ♅
06 51 ☽ ⊼ ♄
10 12 ☽ bQ ☉
13 04 ♀ # ♆
14 25 ☽ ☍ ♀
16 30 ☽ □ ♆
18 54 ☽ ± ♄
20 21 ☽ ⊼ ☿
20 45 ☽ # ♂
23 22 ☽ ⚹ ♃

25 Friday
03 14 ☿ # ♃
09 25 ☽ ± ☿
12 08 ☽ ∥ ♅
12 50 ☽ bQ ♄
13 30 ♀ □ ♆
18 03 ♂ Q ♆
18 32 ☽ △ ☉
19 57 ☿ ∠ ♂
21 27 ☽ ☍ ♇
22 41 ☽ Q ♃

26 Saturday
02 15 ☿ ⊼ ♃
04 27 ☽ ☍ ♂
04 41 ☽ bQ ☿
16 24 ☽ ☌ ♅
18 13 ☽ △ ♄
22 55 ♀ ⊥ ♄

27 Sunday
02 49 ☽ △ ♆
06 18 ☽ ⊼ ♀
06 25 ☉ ⚹ ♇
08 52 ☽ □ ♃
11 32 ☉ bQ ♆
11 49 ☽ △ ☿
12 14 ☿ ∥ ♄
18 37 ☽ ± ♀

28 Monday
02 36 ☽ # ♅
06 31 ☽ ⊼ ♇
06 51 ☽ bQ ♆
08 31 ☽ □ ☉
10 02 ♀ △ ♃
11 27 ☽ ∥ ♂
12 44 ☽ bQ ♀
16 26 ☽ ⊼ ♂
17 31 ☽ ± ♇
23 16 ☽ ± ♆

29 Tuesday
00 17 ☽ ⊻ ♅
02 30 ☽ □ ♄
03 56 ☽ ± ♂
09 45 ☽ bQ ♇
10 01 ☽ ⊼ ♆
11 02 ☽ ⊥ ♅
15 15 ☽ △ ♃
18 00 ☽ △ ♀
19 36 ☽ Q ☉
19 52 ☽ ∥ ♀
20 51 ☽ bQ ♂
21 57 ☽ □ ☿

30 Wednesday
02 53 ☽ ∠ ♅
08 33 ☽ # ♆
10 30 ☽ Q ♄
12 04 ☽ △ ♇
17 07 ☽ bQ ♃
18 04 ☽ ⚹ ☉

31 Thursday
00 12 ☽ △ ♂
04 34 ☽ ⚹ ♅
06 10 ☽ Q ☿
07 06 ☽ ⚹ ♄
08 13 ☽ ± ♃
10 46 ☿ ∥ ♇
13 38 ☽ ☍ ♆
15 56 ☽ ∥ ☿
16 36 ☽ ∥ ♇
18 09 ☽ ⊼ ♃
21 12 ☽ ∠ ☉

All ephemeris data is given at 12.00 UT and the Moon's longitude is additionally given for 24.00 UT

JANUARY 2049

LONGITUDES

Date	Sidereal time h m s	Sun ☉ ° ' "	Moon ☽ ° ' "	Moon ☽ 24.00 ° ' "	Mercury ☿ ° '	Venus ♀ ° '	Mars ♂ ° '	Jupiter ♃ ° '	Saturn ♄ ° '	Uranus ♅ ° '	Neptune ♆ ° '	Pluto ♇ ° '
01	18 46 19	11 ♑ 30 10	04 ♐ 54 25	12 ♐ 19 50	29 ♑ 19	28 ♒ 43	14 ♓ 26	23 ♊ 56	17 ♑ 33	15 ♍ 54	21 ♉ 18	06 ♓ 15
02	18 50 15	12 31 21	19 ♐ 49 52	27 ♐ 23 33	29 30	29 ♒ 44	15 11	23 R 49	17 41	15 R 53	21 R 17	06 16
03	18 54 12	13 32 32	04 ♑ 59 48	12 ♑ 37 26	29 R 29	00 ♓ 45	15 55	23 42	17 48	15 52	21 16	06 18
04	18 58 08	14 33 43	20 ♑ 15 08	27 ♑ 51 37	29 17	01 46	16 40	23 34	17 55	15 51	21 15	06 19
05	19 02 05	15 34 54	05 ♒ 25 35	12 ♒ 55 50	28 53	02 47	17 24	23 27	18 02	15 51	21 15	06 20
06	19 06 01	16 36 05	20 ♒ 21 20	27 ♒ 41 10	28 18	03 47	18 09	23 20	18 09	15 50	21 14	06 21
07	19 09 58	17 37 15	04 ♓ 54 37	12 ♓ 01 13	27 31	04 46	18 54	23 14	18 16	15 49	21 13	06 22
08	19 13 54	18 38 25	19 ♓ 00 38	25 ♓ 52 47	26 34	05 45	19 38	23 07	18 23	15 48	21 12	06 23
09	19 17 51	19 39 34	02 ♈ 37 43	09 ♈ 15 41	25 28	06 44	20 23	23 00	18 30	15 47	21 11	06 25
10	19 21 48	20 40 43	15 ♈ 46 58	22 ♈ 12 04	24 15	07 42	21 07	22 54	18 37	15 46	21 11	06 26
11	19 25 44	21 41 51	28 ♈ 31 26	04 ♉ 45 40	22 57	08 40	21 52	22 48	18 45	15 44	21 10	06 27
12	19 29 41	22 42 59	10 ♉ 55 20	17 ♉ 01 02	21 37	09 37	22 36	22 41	18 52	15 43	21 09	06 28
13	19 33 37	23 44 06	23 ♉ 03 22	29 ♉ 02 56	20 18	10 33	23 21	22 35	18 59	15 42	21 09	06 30
14	19 37 34	24 45 13	05 ♊ 00 16	10 ♊ 55 56	19 02	11 29	24 05	22 30	19 06	15 41	21 08	06 31
15	19 41 30	25 46 19	16 ♊ 50 25	22 ♊ 44 10	17 50	12 24	24 50	22 24	19 13	15 39	21 08	06 32
16	19 45 27	26 47 25	28 ♊ 37 37	04 ♋ 31 10	16 46	13 19	25 34	22 18	19 20	15 38	21 07	06 33
17	19 49 23	27 48 30	10 ♋ 25 07	16 ♋ 19 48	15 49	14 12	26 19	22 13	19 27	15 36	21 07	06 35
18	19 53 20	28 49 34	22 ♋ 15 29	28 ♋ 12 23	15 01	15 06	27 03	22 08	19 34	15 35	21 06	06 36
19	19 57 17	29 ♑ 50 37	04 ♌ 10 43	10 ♌ 10 40	14 23	15 58	27 47	22 03	19 41	15 33	21 06	06 38
20	20 01 13	00 ♒ 51 40	16 ♌ 12 25	22 ♌ 16 07	13 54	16 50	28 32	21 58	19 48	15 32	21 06	06 39
21	20 05 10	01 52 43	28 ♌ 21 56	04 ♍ 30 02	13 34	17 41	29 ♓ 16	21 53	19 55	15 30	21 05	06 40
22	20 09 06	02 53 45	10 ♍ 40 35	16 ♍ 53 48	13 23	18 31	00 ♈ 00	21 49	20 02	15 29	21 05	06 42
23	20 13 03	03 54 46	23 ♍ 09 52	29 ♍ 29 03	13 D 21	19 20	00 44	21 44	20 09	15 27	21 05	06 43
24	20 16 59	04 55 47	05 ♎ 51 35	12 ♎ 17 46	13 27	20 09	01 29	21 40	20 16	15 25	21 05	06 45
25	20 20 56	05 56 47	18 ♎ 47 52	25 ♎ 22 14	13 40	20 56	02 13	21 36	20 23	15 23	21 04	06 46
26	20 24 52	06 57 47	02 ♏ 01 09	08 ♏ 44 54	14 00	21 43	02 57	21 32	20 30	15 21	21 04	06 47
27	20 28 49	07 58 46	15 ♏ 33 45	22 ♏ 27 55	14 27	22 28	03 41	21 29	20 37	15 20	21 04	06 49
28	20 32 46	08 59 45	29 ♏ 27 33	06 ♐ 32 40	14 59	23 13	04 25	21 25	20 44	15 18	21 04	06 50
29	20 36 42	10 00 43	13 ♐ 43 11	20 ♐ 58 54	15 36	23 57	05 10	21 22	20 51	15 16	21 04	06 52
30	20 40 39	11 01 40	28 ♐ 19 25	05 ♑ 44 10	16 18	24 39	05 54	21 19	20 58	15 14	21 04	06 53
31	20 44 35	12 ♒ 02 37	13 ♑ 12 25	20 ♑ 43 15	17 ♑ 04	25 ♓ 20	06 ♈ 38	21 ♊ 16	21 ♑ 05	15 ♍ 12	21 ♉ 04	06 ♓ 55

Date	Moon True ☊ ° '	Moon Mean ☊ ° '	Moon ☽ Latitude ° '
01	18 ♐ 29	17 ♐ 17	01 S 14
02	18 R 29	17 14	00 N 07
03	18 28	17 10	01 30
04	18 26	17 07	02 46
05	18 23	17 04	03 50
06	18 19	17 01	04 37
07	18 14	16 58	05 05
08	18 11	16 54	05 14
09	18 08	16 51	05 04
10	18 07	16 48	04 38
11	18 D 07	16 45	03 59
12	18 08	16 42	03 10
13	18 10	16 39	02 13
14	18 11	16 35	01 12
15	18 12	16 32	00 N 07
16	18 R 12	16 29	00 S 57
17	18 09	16 26	01 58
18	18 04	16 23	02 54
19	17 58	16 20	03 44
20	17 50	16 16	04 23
21	17 41	16 13	04 51
22	17 33	16 10	05 06
23	17 25	16 07	05 07
24	17 19	16 04	04 52
25	17 15	16 00	04 23
26	17 14	15 57	03 39
27	17 D 14	15 54	02 42
28	17 15	15 51	01 34
29	17 16	15 48	00 S 19
30	17 R 15	15 45	00 N 59
31	17 ♐ 13	15 ♐ 41	02 N 15

DECLINATIONS

Date	Sun ☉ ° '	Moon ☽ ° '	Mercury ☿ ° '	Venus ♀ ° '	Mars ♂ ° '	Jupiter ♃ ° '	Saturn ♄ ° '	Uranus ♅ ° '	Neptune ♆ ° '	Pluto ♇ ° '
01	22 S 56	22 S 20	20 S 23	13 S 04	06 S 50	22 N 55	22 S 03	06 N 17	16 N 22	20 S 43
02	22 51	22 55	20 04	12 38	06 32	22 55	22 02	06 18	16 22	20 43
03	22 45	21 51	19 47	12 12	06 14	22 54	22 01	06 18	16 22	20 42
04	22 38	19 10	19 31	11 45	05 55	22 54	22 00	06 18	16 21	20 41
05	22 31	15 11	19 17	11 18	05 37	22 54	21 59	06 19	16 21	20 41
06	22 24	10 19	19 06	10 51	05 19	22 54	21 59	06 19	16 21	20 40
07	22 16	04 S 58	18 56	10 24	05 00	22 54	21 58	06 20	16 21	20 40
08	22 08	00 N 28	18 49	09 57	04 42	22 54	21 57	06 20	16 21	20 39
09	22 00	05 41	18 44	09 30	04 23	22 53	21 56	06 20	16 21	20 38
10	21 51	10 29	18 41	09 03	04 05	22 53	21 55	06 21	16 21	20 38
11	21 41	14 40	18 40	08 35	03 46	22 53	21 54	06 21	16 20	20 37
12	21 31	18 06	18 41	08 08	03 28	22 53	21 53	06 22	16 20	20 36
13	21 21	20 41	18 44	07 40	03 09	22 53	21 52	06 22	16 20	20 36
14	21 10	22 18	18 47	07 13	02 50	22 53	21 51	06 23	16 20	20 35
15	20 59	22 54	18 53	06 45	02 32	22 53	21 50	06 24	16 20	20 35
16	20 48	22 29	18 59	06 18	02 13	22 52	21 50	06 24	16 20	20 34
17	20 36	21 04	19 06	05 50	01 55	22 52	21 49	06 25	16 20	20 33
18	20 23	18 43	19 13	05 23	01 36	22 52	21 48	06 25	16 20	20 33
19	20 11	15 35	19 21	04 55	01 18	22 52	21 47	06 26	16 20	20 32
20	19 58	11 47	19 30	04 28	00 59	22 52	21 46	06 27	16 20	20 31
21	19 44	07 29	19 39	04 01	00 40	22 52	21 45	06 27	16 20	20 31
22	19 30	02 N 50	19 48	03 33	00 22	22 52	21 44	06 28	16 20	20 30
23	19 16	01 S 59	19 57	03 06	00 S 03	22 52	21 43	06 29	16 20	20 29
24	19 02	06 48	20 06	02 39	00 N 15	22 52	21 42	06 29	16 20	20 29
25	18 47	11 25	20 14	02 13	00 34	22 51	21 41	06 30	16 20	20 28
26	18 32	15 36	20 23	01 46	00 52	22 51	21 40	06 31	16 20	20 27
27	18 16	19 05	20 30	01 20	01 10	22 51	21 39	06 32	16 20	20 27
28	18 00	21 34	20 38	00 54	01 29	22 51	21 38	06 32	16 20	20 26
29	17 44	22 45	20 44	00 28	01 47	22 51	21 37	06 33	16 20	20 25
30	17 27	22 26	20 50	00 02	02 06	22 51	21 36	06 34	16 20	20 25
31	17 S 11	20 S 33	20 S 55	00 N 23	02 N 24	22 N 51	21 S 35	06 N 35	16 N 20	20 S 24

ZODIAC SIGN ENTRIES

Date	h	m	Planets
01	03	59	☽ ♐
02	18	10	♀ ♓
03	04	07	☽ ♑
05	03	23	☽ ♒
07	03	49	☽ ♓
09	07	18	☽ ♈
11	14	50	☽ ♉
14	01	55	☽ ♊
16	14	48	☽ ♋
19	03	37	☽ ♌
19	15	41	☉ ♒
21	15	12	☽ ♍
22	11	54	♂ ♈
24	00	58	☽ ♎
26	08	22	☽ ♏
28	12	55	☽ ♐
30	14	43	☽ ♑

LATITUDES

Date	Mercury ☿ ° '	Venus ♀ ° '	Mars ♂ ° '	Jupiter ♃ ° '	Saturn ♄ ° '	Uranus ♅ ° '	Neptune ♆ ° '	Pluto ♇ ° '
01	00 S 06	01 S 14	00 S 46	00 S 23	00 N 14	00 N 48	01 S 46	12 S 26
04	00 N 48	00 59	00 43	00 22	00 14	00 48	01 46	12 25
07	01 45	00 42	00 40	00 22	00 14	00 48	01 46	12 24
10	02 37	00 24	00 36	00 21	00 13	00 48	01 46	12 24
13	03 12	00 S 04	00 33	00 21	00 13	00 48	01 46	12 23
16	03 26	00 N 17	00 30	00 20	00 13	00 48	01 46	12 23
19	03 19	00 40	00 27	00 20	00 13	00 48	01 45	12 22
22	02 58	01 04	00 24	00 19	00 13	00 48	01 45	12 22
25	02 30	01 30	00 21	00 19	00 12	00 48	01 45	12 22
28	01 59	01 57	00 18	00 18	00 12	00 48	01 45	12 21
31	01 N 26	02 N 26	00 S 15	00 S 17	00 N 12	00 N 49	01 S 45	12 S 21

DATA

Julian Date	2469443
Delta T	+80 seconds
Ayanamsa	24° 32' 47"
Synetic vernal point	04° ♓ 34' 11"
True obliquity of ecliptic	23° 25' 56"

MOON'S PHASES, APSIDES AND POSITIONS ☽

Date	h	m	Phase	Longitude ° '	Eclipse Indicator
04	02	24	●	14 ♑ 09	
10	21	56	☽	21 ♈ 06	
19	02	29	○	29 ♋ 26	
26	21	33	☾	07 ♏ 22	

Day	h	m		
03	20	56	Perigee	
17	00	36	Apogee	
02	08	40	Max dec	22° S 56'
08	09	56	0N	
15	14	07	Max dec	22° N 55'
23	02	11	0S	
29	19	09	Max dec	22° S 49'

LONGITUDES

Date	Chiron ⚷ ° '	Ceres ⚳ ° '	Pallas ⚴ ° '	Juno ⚵ ° '	Vesta ⚶ ° '	Black Moon Lilith ⚸ ° '
01	23 ♏ 41	06 ♈ 18	21 ♒ 04	08 ♓ 30	09 ♉ 02	07 ♋ 11
11	24 ♏ 42	08 ♈ 41	24 ♒ 07	13 ♓ 02	09 ♉ 51	08 ♋ 17
21	25 ♏ 35	11 ♈ 24	27 ♒ 16	17 ♓ 46	11 ♉ 17	09 ♋ 24
31	26 ♏ 20	14 ♈ 24	00 ♓ 30	22 ♓ 43	13 ♉ 15	10 ♋ 31

All ephemeris data is given at 12.00 UT and the Moon's longitude is additionally given for 24.00 UT

ASPECTARIAN

h m	Aspects
01 Friday	
00 32	☽ Q ♅
01 06	☽ □ ♀
02 41	☽ ⚹ ☿
05 45	☿ ∠ ♂
07 42	☽ ∥ ♄
08 09	☽ ∠ ♄
13 02	☽ ⊥ ☉
14 12	☽ □ ♇
17 42	☉ # ♃
22 51	☽ ⊥ ♄
23 27	☽ ⊻ ☉
02 Saturday	
02 31	☽ ∥ ☉
03 24	☽ ∠ ☿
04 11	☽ □ ♂
04 41	☽ # ♃
05 35	☿ ⊻ ♀
05 42	☽ □ ♅
08 25	☽ Q ♀
08 32	☽ ⊻ ♄
12 45	☽ # ♃
14 19	☽ ⊼ ♆
17 32	☽ ∥ ☉
17 51	☽ ⊥ ☿
18 17	☽ ☍ ♃
19 04	☽ Q ♇
23 13	☿ St R
23 49	☽ ± ♆
03 Sunday	
03 21	☽ ⊻ ☿
04 50	☽ ⚹ ♀
06 19	♂ # ♅
09 39	☽ ∥ ♄
10 13	☽ Q ♂
10 24	♂ ☍ ♅
14 00	☽ ⚼ ♆
14 03	☽ ⚹ ♇
23 57	☽ ∥ ♇
04 Monday	
02 24	☽ ☌ ☉
05 06	☽ △ ♅
06 04	☽ ⚹ ♂
06 08	☽ ∠ ♀
08 17	☽ ☌ ♄
09 15	☽ ∥ ☿
13 35	☽ △ ♆
13 40	☽ ∠ ♇
17 11	☽ ⊼ ♃
21 19	☽ ⊥ ♀
05 Tuesday	
01 56	☽ ☌ ☿
02 36	☽ ± ♃
03 54	☽ ⊥ ♇
04 44	☽ ⚼ ♅
05 35	☽ # ♆
06 57	☽ ∠ ♂
07 30	☽ ⊻ ♀
13 27	☽ ⊻ ♇
16 48	☽ ⚼ ♃
18 05	☉ △ ♅
18 51	♀ ∠ ♄
19 03	☽ ± ♅
22 03	☽ ⊥ ♂
06 Wednesday	
04 41	☽ ⊼ ♅
05 28	☽ ⊻ ☉
08 14	☽ ⊻ ♂
08 24	☽ ⊻ ♄
09 12	☽ ∥ ♀
11 59	♂ ⚹ ♄
13 25	☽ □ ♆
15 56	☽ ⊥ ☉
16 50	☽ △ ♃
18 15	☽ ⊥ ♄
19 15	☿ ⊥ ♀
07 Thursday	
00 23	☽ ⊻ ☿
06 00	☽ # ♅
07 53	☽ ∠ ☉
09 14	☽ ∠ ♄
09 48	☽ ⊥ ☿
11 45	☽ ☌ ♀
11 51	☽ ∥ ♂
14 27	☽ ☌ ♇
19 14	☽ Q ♆
08 Friday	
00 04	☽ ∠ ☿
05 16	☉ ☌ ♄
06 28	☽ ☍ ♅
10 55	☽ ⚹ ♄
11 18	☽ ⚹ ☉
13 09	☽ ☌ ♂
15 48	☽ ⚹ ♆
19 05	☽ □ ♃
09 Saturday	
00 14	☽ ⚹ ☿
03 50	♀ ☌ ♇
06 13	☽ # ♂
08 10	☽ Q ♄
10 07	☽ Q ☉
15 08	☽ ∥ ♅
18 25	☽ ∠ ♆
18 49	☽ ⊻ ♇
19 59	☽ ⊻ ♀
20 00	☽ Q ☿
22 44	☉ ∥ ♄
10 Sunday	
03 04	☽ Q ♃
05 11	☽ # ♀
05 48	☽ ⊥ ♇
07 50	☽ ⊥ ♀
10 53	☽ ⊥ ♆
11 57	☽ ⊼ ♅
13 52	♂ ⚹ ♆
17 20	☽ □ ♄
21 56	☽ □ ☉
22 04	☽ ⊻ ♆
22 34	☽ ∠ ♇
22 35	☽ ⊻ ♂
23 09	☽ ± ♅
23 40	☉ △ ♆
11 Monday	
01 12	☽ ⚹ ♃
02 00	☽ ∠ ♀
02 24	☽ □ ☿
04 30	☿ ∠ ♀
06 03	☉ ∠ ♇
10 40	☽ ⊥ ♂
15 09	☿ ⊼ ♃
16 14	☽ ⚼ ♅
22 59	☽ ∥ ♆
12 Tuesday	
00 38	☿ ⚹ ♂
00 41	♀ Q ♆
00 50	☉ ☌ ☿
02 19	☉ ⚹ ♂
03 18	☽ ⚹ ♇
05 06	☽ ∠ ♂
05 44	☽ ∠ ♃
09 13	☽ ⚹ ♀
11 25	☉ ⊼ ♃
14 25	♂ □ ♃
14 43	☿ ∠ ♇
16 47	☽ # ☿
20 28	☿ △ ♆
21 25	☽ △ ♅
23 15	☽ ⊥ ♃
13 Wednesday	
02 54	☽ Q ♇
03 49	☽ △ ♄
07 03	☽ △ ☿
08 12	☽ ☌ ♆
10 54	☽ Q ♀
11 05	☽ ⊻ ♃
11 05	☽ # ♇
12 37	☽ ⚹ ♂
13 29	☽ △ ☉
19 31	☽ # ☉
14 Thursday	
03 53	☽ # ♄
10 09	☽ ⚼ ♄
10 13	☽ ⚼ ☿
10 48	☿ ☌ ♄
14 20	☽ Q ♂
15 03	☽ □ ♇
22 31	☽ ⚼ ☉
15 Friday	
02 13	☽ □ ♀
02 45	☽ ± ☿
04 34	☽ ± ♄
08 01	☽ ∥ ♃
09 36	☽ □ ♅
13 51	☽ ⊼ ☿
16 53	☽ ⊼ ♄
18 32	☽ ± ☉
20 19	☽ ∥ ♃
20 43	☽ ⊻ ♆
23 13	☽ ☌ ♃
16 Saturday	
05 21	☽ □ ♂
06 27	♀ # ♅
07 54	☽ ⊼ ☉
08 56	☽ ⊥ ♆
22 10	☽ Q ♅
17 Sunday	
01 16	☽ # ♄
03 15	☽ ∠ ♆
04 11	☽ △ ♇
17 04	☉ ∥ ♇
18 05	☿ △ ♅
18 06	☽ # ♇
18 12	☽ # ☉
20 19	☽ △ ♀
20 50	☉ ± ♃
22 14	☽ ☍ ☿
22 31	☽ ⚹ ♅
18 Monday	
06 31	☽ ☍ ♄
07 47	☽ # ☿
09 40	☽ ⚹ ♆
10 40	☽ ⚼ ♇
10 56	☿ ⚹ ♀
11 34	☿ Q ♂
11 44	☽ ⊻ ♃
22 18	☽ △ ♂
23 46	☽ ⊥ ♃
19 Tuesday	
01 02	♀ ☍ ♅
02 29	☽ ☍ ☉
04 45	☽ ∠ ♅
04 51	☽ ± ♇
05 03	☽ ⚼ ♀
06 46	☽ ∥ ♆
09 50	☽ Q ♆
16 55	☽ ⊼ ♇
17 42	☽ ∠ ♃
22 44	☽ ⊥ ♅
20 Wednesday	
00 29	☽ ± ♀
04 25	☉ ⚼ ♅
06 19	☽ ⚼ ♂
06 52	☉ ⊥ ♇
07 34	☽ ⊼ ☿
10 40	☽ ⊻ ♅
13 20	☽ ⊼ ♀
19 05	☽ ± ☿
19 12	☽ ⊼ ♄
21 41	☽ □ ♆
23 19	☽ ⚹ ♃
21 Thursday	
01 19	☽ ± ♂
07 09	☽ ± ♄
12 23	☽ ⚼ ☿
13 53	☽ ⊼ ♂
17 26	☽ ∥ ♅
17 34	☉ ∥ ☿
19 30	☽ ⊼ ☉
22 44	☽ Q ♃
22 Friday	
00 57	☽ ⚼ ♄
04 16	☽ ☍ ♇
07 59	☽ # ♀
08 14	☽ ± ☉
17 12	☽ △ ☿
21 15	☽ ☌ ♅
23 Saturday	
01 13	☽ # ♂
03 02	☽ ∥ ♂
03 09	☽ ⚼ ☉
04 10	☽ ☍ ♀
06 12	☽ △ ♄
06 20	☿ St D
08 01	☽ △ ♆
09 17	☽ □ ♃
17 05	☽ ∥ ♀
24 Sunday	
03 16	☽ ☍ ♂
10 06	☽ △ ☉
10 26	☽ # ♅
12 24	☽ ⚼ ♆
13 39	☽ ⊼ ♇
16 30	♀ ⚹ ♄
25 Monday	
00 51	☽ ± ♇
02 21	☽ □ ☿
05 08	☽ ± ♆
05 44	☽ ⊻ ♅
10 51	☉ ∠ ♀
14 56	☽ □ ♄
16 10	☽ ⊼ ♀
16 10	☽ ⊼ ♆
16 11	♀ ⚹ ♆
16 44	☽ ⊥ ♅
17 07	☽ △ ♃
17 27	☽ ⚼ ♇
18 39	♂ Q ♄
26 Tuesday	
02 32	☉ ⚼ ♃
03 46	☽ ± ♀
06 59	♀ □ ♃
07 51	☉ ⊻ ♇
09 01	☽ ∠ ♅
11 58	☽ Q ☿
13 46	☽ ⊼ ♂
16 36	☽ # ♆
20 02	☽ ⚼ ♃
20 32	☽ △ ♇
20 53	☽ ⚼ ♀
21 33	☽ □ ☉
23 40	☽ Q ♄
27 Wednesday	
01 04	☽ ± ♂
01 11	☿ ∥ ♇
06 16	☽ ∥ ☉
09 58	☽ ⚹ ☿
11 35	☽ ⚹ ♅
11 51	☽ ± ♃
17 01	♀ # ♂
17 46	☽ ⚼ ♂
20 53	☽ ⚹ ♄
21 35	☽ ☍ ♆
22 15	☽ ⊼ ♃
23 43	☽ ∥ ♇
28 Thursday	
00 42	☽ △ ♀
01 01	☽ ∥ ☿
07 28	☽ Q ☉
08 19	☽ Q ♅
12 53	☽ ∥ ♄
12 55	☽ ∠ ☿
18 50	☉ ± ♅
20 53	☽ △ ♂
22 43	☽ ∠ ♄
29 Friday	
00 04	☿ △ ♅
00 31	☽ □ ♇
04 48	☽ ⊥ ☿
05 21	☽ ⚹ ☉
13 53	☽ ⊥ ♄
14 33	☽ □ ♅
15 16	☽ ⊻ ☿
23 52	☽ ⊻ ♄
30 Saturday	
00 08	☽ ⊼ ♆
00 35	☽ ☍ ♃
05 43	☽ □ ♀
06 24	☽ Q ♇
07 59	☽ ∠ ☉
09 57	☽ ± ♆
13 23	♆ St D
17 34	♂ ∠ ♆
23 39	☽ ⊥ ☉
31 Sunday	
00 32	☽ ⚼ ♆
00 53	☽ □ ♂
01 07	☽ ∥ ♄
01 53	☽ ⚹ ♇
08 35	☽ ∥ ☿
09 49	♄ △ ♆
10 00	☽ ⊻ ☉
12 13	☽ Q ♀
13 17	☽ ∥ ♇
15 10	☽ △ ♅
18 31	☽ ☌ ☿
21 45	♂ ⊻ ♇

FEBRUARY 2049

LONGITUDES

Date	Sidereal time h m s	Sun ☉ ° ′ ″	Moon ☽ ° ′ ″	Moon ☽ 24.00 ° ′ ″	Mercury ☿ ° ′	Venus ♀ ° ′	Mars ♂ ° ′	Jupiter ♃ ° ′	Saturn ♄ ° ′	Uranus ♅ ° ′	Neptune ♆ ° ′	Pluto ⯓ ° ′
01	20 48 32	13 ♒ 03 33	28 ♑ 15 37	05 ♒ 48 21	17 ♑ 54	26 ♓ 01	07 ♈ 22	21 ♊ 14	21 ♑ 11	15 ♍ 10	21 ♉ 04	06 ♓ 57
02	20 52 28	14 04 28	13 ♒ 20 13	20 ♒ 49 58	18 48	26 40	08 06	21 R 11	21 18	15 R 08	21 D 04	06 58
03	20 56 25	15 05 21	28 ♒ 16 24	05 ♓ 38 24	19 44	27 17	08 50	21 09	21 25	15 05	21 04	07 00
04	21 00 21	16 06 14	12 ♓ 55 01	20 ♓ 05 27	20 44	27 54	09 34	21 07	21 32	15 03	21 04	07 01
05	21 04 18	17 07 05	27 ♓ 09 08	04 ♈ 05 40	21 47	28 29	10 18	21 05	21 38	15 01	21 04	07 03
06	21 08 15	18 07 55	10 ♈ 54 53	17 ♈ 36 49	22 52	29 02	11 02	21 03	21 45	14 59	21 05	07 04
07	21 12 11	19 08 43	24 ♈ 11 38	00 ♉ 39 40	23 59	29 ♓ 34	11 46	21 02	21 51	14 57	21 05	07 06
08	21 16 08	20 09 30	07 ♉ 01 22	13 ♉ 17 17	25 08	00 ♈ 05	12 29	21 01	21 58	14 54	21 05	07 08
09	21 20 04	21 10 15	19 ♉ 28 01	25 ♉ 34 13	26 19	00 34	13 13	21 00	22 05	14 52	21 06	07 09
10	21 24 01	22 10 59	01 ♊ 36 34	07 ♊ 35 46	27 32	01 01	13 57	20 59	22 11	14 50	21 06	07 11
11	21 27 57	23 11 42	13 ♊ 32 30	19 ♊ 27 26	28 ♑ 47	01 26	14 41	20 58	22 18	14 47	21 06	07 12
12	21 31 54	24 12 22	25 ♊ 21 13	01 ♋ 14 27	00 ♒ 03	01 50	15 24	20 58	22 24	14 45	21 07	07 14
13	21 35 50	25 13 01	07 ♋ 07 43	13 ♋ 01 32	01 21	02 12	16 08	20 58	22 30	14 42	21 07	07 16
14	21 39 47	26 13 39	18 ♋ 56 23	24 ♋ 52 41	02 40	02 31	16 52	20 D 58	22 37	14 40	21 08	07 17
15	21 43 44	27 14 15	00 ♌ 50 47	06 ♌ 50 59	04 01	02 49	17 35	20 58	22 43	14 38	21 08	07 19
16	21 47 40	28 14 49	12 ♌ 53 32	18 ♌ 58 38	05 23	03 05	18 19	20 58	22 49	14 35	21 09	07 20
17	21 51 37	29 ♒ 15 21	25 ♌ 06 25	01 ♍ 16 57	06 46	03 19	19 03	20 59	22 55	14 33	21 09	07 22
18	21 55 33	00 ♓ 15 53	07 ♍ 30 19	13 ♍ 46 30	08 10	03 30	19 46	21 00	23 02	14 30	21 10	07 24
19	21 59 30	01 16 22	20 ♍ 05 32	26 ♍ 27 22	09 36	03 39	20 30	21 01	23 08	14 28	21 11	07 25
20	22 03 26	02 16 50	02 ♎ 52 00	09 ♎ 19 24	11 02	03 46	21 13	21 02	23 14	14 25	21 11	07 27
21	22 07 23	03 17 17	15 ♎ 49 36	22 ♎ 22 37	12 30	03 50	21 56	21 03	23 20	14 22	21 12	07 29
22	22 11 19	04 17 42	28 ♎ 58 31	05 ♏ 37 22	13 58	03 53	22 40	21 05	23 26	14 20	21 13	07 30
23	22 15 16	05 18 06	12 ♏ 19 17	19 ♏ 04 25	15 28	03 R 52	23 23	21 07	23 32	14 17	21 14	07 32
24	22 19 13	06 18 29	25 ♏ 52 54	02 ♐ 44 52	16 59	03 49	24 06	21 09	23 38	14 15	21 15	07 34
25	22 23 09	07 18 50	09 ♐ 40 28	16 ♐ 39 46	18 31	03 44	24 50	21 11	23 43	14 12	21 15	07 35
26	22 27 06	08 19 10	23 ♐ 42 49	00 ♑ 49 32	20 03	03 36	25 33	21 13	23 49	14 10	21 16	07 37
27	22 31 02	09 19 28	07 ♑ 59 48	15 ♑ 13 20	21 37	03 26	26 16	21 16	23 55	14 07	21 17	07 39
28	22 34 59	10 ♓ 19 45	22 ♑ 29 44	29 ♑ 48 26	23 ♒ 12	03 ♈ 13	26 ♈ 59	21 ♊ 18	24 ♑ 00	14 ♍ 04	21 ♉ 18	07 ♓ 40

DECLINATIONS

Date	Moon True ☊ ° ′	Moon Mean ☊ ° ′	Moon ☽ Latitude ° ′	Sun ☉ ° ′	Moon ☽ ° ′	Mercury ☿ ° ′	Venus ♀ ° ′	Mars ♂ ° ′	Jupiter ♃ ° ′	Saturn ♄ ° ′	Uranus ♅ ° ′	Neptune ♆ ° ′	Pluto ⯓ ° ′
01	17 ♐ 08	15 ♐ 38	03 N 22	16 S 53	17 S 13	20 S 59	00 N 48	02 N 42	22 N 51	21 S 34	06 N 36	16 N 20	20 S 23
02	17 R 00	15 35	04 14	16 36	12 45	21 02	01 13	03 00	22 51	21 33	06 36	16 20	20 23
03	16 51	15 32	04 49	16 18	07 33	21 05	01 38	03 19	22 51	21 32	06 37	16 20	20 22
04	16 41	15 29	05 04	16 00	02 S 01	21 06	02 02	03 37	22 51	21 31	06 38	16 20	20 21
05	16 32	15 26	05 00	15 42	03 N 27	21 06	02 25	03 55	22 51	21 30	06 39	16 20	20 21
06	16 24	15 22	04 38	15 23	08 34	21 05	02 48	04 13	22 52	21 29	06 40	16 21	20 20
07	16 19	15 19	04 01	15 05	13 07	21 04	03 11	04 31	22 52	21 28	06 41	16 21	20 19
08	16 16	15 16	03 14	14 46	16 54	21 00	03 33	04 49	22 52	21 27	06 42	16 21	20 19
09	16 15	15 13	02 18	14 26	19 48	20 56	03 55	05 07	22 52	21 26	06 43	16 21	20 18
10	16 D 15	15 10	01 17	14 07	21 44	20 51	04 16	05 24	22 52	21 25	06 43	16 21	20 17
11	16 16	15 06	00 N 15	13 47	22 40	20 44	04 36	05 42	22 52	21 24	06 44	16 21	20 17
12	16 R 15	15 03	00 S 48	13 27	22 33	20 36	04 56	06 00	22 52	21 23	06 45	16 22	20 16
13	16 13	15 00	01 49	13 07	21 26	20 27	05 15	06 18	22 52	21 22	06 46	16 22	20 15
14	16 08	14 57	02 45	12 46	19 23	20 17	05 34	06 35	22 53	21 21	06 47	16 22	20 15
15	16 01	14 54	03 33	12 26	16 29	20 06	05 52	06 53	22 53	21 20	06 48	16 22	20 14
16	15 50	14 51	04 13	12 05	12 53	19 53	06 09	07 10	22 53	21 19	06 49	16 22	20 13
17	15 38	14 47	04 42	11 44	08 43	19 39	06 25	07 27	22 53	21 18	06 50	16 23	20 13
18	15 24	14 44	04 58	11 23	04 N 08	19 24	06 40	07 44	22 53	21 17	06 51	16 23	20 12
19	15 11	14 41	05 00	11 01	00 S 40	19 07	06 54	08 02	22 54	21 16	06 52	16 23	20 12
20	14 59	14 38	04 47	10 40	05 31	18 49	07 07	08 19	22 54	21 16	06 53	16 23	20 11
21	14 49	14 35	04 19	10 18	10 12	18 30	07 20	08 36	22 54	21 15	06 54	16 23	20 11
22	14 42	14 31	03 36	09 56	14 29	18 10	07 31	08 53	22 54	21 14	06 55	16 24	20 10
23	14 38	14 28	02 42	09 34	18 06	17 48	07 41	09 09	22 55	21 13	06 56	16 24	20 09
24	14 36	14 25	01 37	09 12	20 48	17 25	07 50	09 26	22 55	21 12	06 57	16 24	20 09
25	14 D 36	14 22	00 S 26	08 49	22 19	17 01	07 58	09 43	22 55	21 11	06 58	16 24	20 08
26	14 R 36	14 19	00 N 48	08 27	22 29	16 35	08 04	09 59	22 56	21 10	06 59	16 25	20 07
27	14 34	14 16	02 00	08 04	21 11	16 08	08 09	10 16	22 56	21 09	07 00	16 25	20 07
28	14 ♐ 31	14 ♐ 12	03 N 06	07 S 42	18 S 30	15 S 40	08 N 13	10 N 32	22 N 56	21 S 08	07 N 01	16 N 25	20 S 06

ZODIAC SIGN ENTRIES

Date	h	m	Planets
01	14	46	☽ ♒
03	14	48	☽ ♓
05	16	54	☽ ♈
07	22	46	☽ ♉
08	08	10	♀ ♈
10	08	47	☽ ♊
10	10	55	☿ ♒
12	21	28	☽ ♋
15	10	18	☽ ♌
17	21	31	☽ ♍
18	05	42	☉ ♓
20	06	39	☽ ♎
22	13	51	☽ ♏
24	19	13	☽ ♐
26	22	37	☽ ♑

LATITUDES

Date	Mercury ☿ ° ′	Venus ♀ ° ′	Mars ♂ ° ′	Jupiter ♃ ° ′	Saturn ♄ ° ′	Uranus ♅ ° ′	Neptune ♆ ° ′	Pluto ⯓ ° ′
01	01 N 16	02 N 36	00 S 14	00 S 17	00 N 12	00 N 49	01 S 45	12 S 21
04	00 44	03 07	00 11	00 17	00 12	00 49	01 44	12 21
07	00 N 15	03 39	00 09	00 16	00 11	00 49	01 44	12 21
10	00 S 12	04 13	00 06	00 16	00 11	00 49	01 44	12 20
13	00 37	04 47	00 03	00 15	00 11	00 49	01 44	12 20
16	01 00	05 22	00 S 01	00 15	00 11	00 49	01 44	12 20
19	01 19	05 57	00 N 02	00 14	00 11	00 49	01 44	12 20
22	01 36	06 31	00 04	00 14	00 11	00 49	01 43	12 20
25	01 50	07 04	00 06	00 13	00 10	00 49	01 43	12 20
28	02 00	07 34	00 09	00 13	00 10	00 49	01 43	12 20
31	02 S 07	08 N 01	00 N 11	00 S 12	00 N 10	00 N 49	01 S 43	12 S 21

DATA

Julian Date	2469474
Delta T	+80 seconds
Ayanamsa	24° 32′ 53″
Synetic vernal point	04° ♓ 34′ 06″
True obliquity of ecliptic	23° 25′ 56″

LONGITUDES

Date	Chiron ⚷ ° ′	Ceres ⚳ ° ′	Pallas ⚴ ° ′	Juno ⚵ ° ′	Vesta ⚶ ° ′	Black Moon Lilith ⚸ ° ′
01	26 ♏ 24	14 ♈ 43	00 ♓ 50	23 ♓ 13	13 ♉ 29	10 ♋ 37
11	26 ♏ 57	17 ♈ 58	04 ♓ 07	28 ♓ 21	15 ♉ 56	11 ♋ 44
21	27 ♏ 21	21 ♈ 24	07 ♓ 26	03 ♈ 37	18 ♉ 45	12 ♋ 50
31	27 ♏ 33	25 ♈ 00	10 ♓ 46	09 ♈ 52	21 ♉ 52	13 ♋ 57

MOON'S PHASES, APSIDES AND POSITIONS ☽

Date	h	m	Phase	Longitude ° ′	Eclipse Indicator
02	13	16	●	14 ♒ 08	
09	15	38	☽	21 ♉ 19	
17	20	47	○	29 ♌ 38	
25	07	36	☾	07 ♐ 08	

Day	h	m		
01	08	46	Perigee	
13	12	36	Apogee	
04	20	46	0N	
11	21	20	Max dec	22° N 44′
19	08	41	0S	
26	02	42	Max dec	22° S 36′

ASPECTARIAN

h m	Aspects
01 Monday	
00 33	☽ △ ♆
00 40	☽ ☌ ♄
00 50	☽ ⊼ ♃
01 56	☽ ∠ ⯓
07 09	☽ Q ♂
08 15	☽ ⚹ ♀
10 22	☽ ± ♃
14 04	☽ ∥ ☉
15 01	☽ ♇ ♅
16 16	☽ ⊥ ⯓
17 09	☽ # ♆
17 46	♃ ⊼ ♄
02 Tuesday	
00 38	☽ ♇ ♃
01 50	☽ ⊻ ⯓
03 13	☽ ⚹ ♂
05 18	☽ ± ♅
09 12	☽ ∠ ♀
13 16	☽ ☌ ☉
14 51	☽ ⊼ ♅
21 18	☽ ⊻ ☿
03 Wednesday	
00 14	☽ ⊥ ♀
00 23	☽ □ ♆
00 32	☽ △ ♃
00 51	☽ ⊻ ♄
04 27	☽ ∠ ♂
07 37	☽ ⊥ ☿
09 21	☉ # ♆
10 20	☽ ⊻ ♀
10 36	☽ ⊥ ♄
12 01	☉ ⊼ ♅
16 03	☽ # ♅
19 48	☽ ⊥ ♂
21 55	♂ Q ♃
23 17	☽ ∠ ☿
04 Thursday	
01 22	☽ ∠ ♄
02 15	☽ ☌ ⯓
05 29	☽ # ♂
05 38	☽ Q ♆
06 09	☽ ⊻ ♂
11 59	☽ # ♀
15 33	☽ ☍ ♅
17 43	☽ ⊻ ☉
19 49	☿ △ ♆
20 31	☿ ⊼ ♃
05 Friday	
01 39	☽ ⚹ ♆
01 42	☽ □ ♃
02 07	☽ ⚹ ☿
02 32	☽ ⚹ ♄
04 35	☽ ⊥ ☉
07 02	☽ ∥ ♀
08 23	☿ ☌ ♄
14 13	☽ ∥ ♂
14 22	☽ ☌ ♀
18 09	☿ ∠ ⯓
18 27	♃ ⊻ ♆
21 14	☽ ∠ ☉
23 18	☽ Q ♄
06 Saturday	
00 25	☽ Q ☿
02 45	☽ ∥ ♅
03 28	☽ ∠ ♆
05 12	☽ ⊻ ⯓
08 43	☽ Q ♃
12 13	☽ ☌ ♂
15 51	☽ ⊥ ⯓
19 14	☽ ⊼ ♅
19 26	☽ ⊥ ♆
07 Sunday	
02 01	☽ ⚹ ☉
06 04	☽ ± ♅
06 13	☽ ⚹ ♃
06 18	☽ ⊻ ♆
07 41	☽ □ ♄
08 09	☽ ∠ ⯓
11 34	☽ □ ☿
22 23	☽ ⊻ ♀
22 37	☽ ♇ ♅
22 53	☽ # ☉
08 Monday	
02 01	☽ Q ☉
08 08	☽ ∥ ♆
10 05	☽ ∠ ♃
10 08	☽ ⊥ ♀
12 12	☽ ⚹ ⯓
23 07	☽ ⊻ ♂
09 Tuesday	
03 05	☽ △ ♅
03 19	☽ ⊥ ♃
04 06	☽ ∠ ♀
07 53	☉ △ ♃
09 40	♂ ⊥ ⯓
10 08	☉ □ ♆
11 23	☽ Q ⯓
11 29	☽ ⊥ ♂
14 59	☽ ⊻ ♃
15 11	☽ ☌ ♆
15 38	☽ □ ☉
17 05	☽ # ⯓
17 09	☽ △ ♄
23 57	☽ # ☿
10 Wednesday	
01 09	☿ ± ♃
02 59	☽ △ ☿
06 22	☽ ∠ ♂
07 03	☽ # ♄
10 27	♄ ∠ ⯓
10 46	☽ ⚹ ♀
12 01	☉ ⊻ ♄
23 11	☽ □ ⯓
23 16	☽ ♇ ♄
11 Thursday	
11 47	☽ Q ♀
12 33	☽ ♇ ☿
14 27	☽ ⚹ ♂
14 31	☽ □ ♅
15 22	♂ ⊼ ♅
17 37	☽ ± ♄
12 Friday	
02 11	♂ ⊥ ♆
03 04	☽ ☌ ♃
03 22	☽ ⊻ ♆
05 56	☽ ⊼ ♄
06 22	☿ ♇ ♅
09 02	☽ ± ☿
09 27	☽ △ ☉
15 35	☽ ⊥ ♆
16 28	☽ Q ♂
22 46	☽ ⊼ ☿
13 Saturday	
01 38	☽ □ ♀
03 01	☽ Q ♅
09 57	☽ ∠ ♆
10 13	☿ ⊥ ⯓
12 16	☽ △ ⯓
12 53	☽ # ♄
18 53	☽ ♇ ☉
14 Sunday	
00 25	♃ St D
01 47	☽ # ☿
03 01	☽ # ⯓
03 22	☽ ⚹ ♅
07 31	☽ □ ♂
08 26	☿ ⚹ ♀
14 51	☽ ± ☉
16 05	☽ ⊻ ♃
16 26	☽ ⚹ ♆
17 27	☿ ∥ ⯓
18 47	☽ ♇ ⯓
19 29	☽ ☍ ♄
22 07	☉ ⊥ ♀
15 Monday	
04 05	☽ ⊼ ☉
04 12	☽ ⊥ ♃
05 41	♂ ∥ ♅
09 34	☽ ∠ ♅
12 52	☽ ∥ ♆
12 56	☽ ± ⯓
16 03	☽ △ ♀
16 36	☽ Q ♆
19 09	☽ ☍ ☿
22 14	☽ ∠ ♃
16 Tuesday	
00 57	☽ ⊼ ⯓
03 29	☽ ⊥ ♅
15 20	☽ ⊻ ♅
17 20	☽ # ☉
22 21	☿ ♇ ♃
22 27	☽ ♇ ♀
23 23	☽ △ ♂
17 Wednesday	
03 11	☉ ⊥ ♄
03 55	☽ ⚹ ♃
04 16	☽ □ ♆
07 42	☽ ⊼ ♄
16 22	☽ ± ♀
18 23	☽ ∥ ♂
19 29	☽ ± ♄
20 47	☽ ☍ ☉
21 54	☿ Q ♂
22 01	☽ ∥ ♅
22 34	☿ ⊻ ⯓
23 38	☽ ∥ ♀
18 Thursday	
03 18	☽ Q ♃
04 10	☽ ⊼ ♀
06 24	☽ ♇ ♂
11 47	☽ ☍ ⯓
13 01	☽ ♇ ♄
13 26	☽ ⊼ ☿
17 29	☿ ± ♅
19 Friday	
00 43	☽ ± ♂
01 20	☽ ☌ ♅
02 22	☽ ± ☿
08 12	♀ ∥ ♅
10 59	♂ ± ♅
12 48	☽ ⊼ ♂
13 44	☽ □ ♃
14 03	☽ △ ♆
17 47	☽ △ ♄
21 34	☽ ♇ ☿
20 Saturday	
05 40	♂ ⚹ ♃
10 49	☽ ⊼ ☉
11 09	♂ ⊻ ♆
13 41	☽ ☍ ♀
18 12	☽ ♇ ♆
18 53	☽ # ♅
20 27	☽ # ♀
20 33	☽ ⊼ ⯓
22 55	☽ ± ☉
21 Sunday	
03 03	☽ # ♂
05 05	☽ △ ☿
07 40	☽ ± ⯓
09 20	☽ ⊻ ♅
10 51	☽ ± ♆
12 29	☽ ∥ ☉
16 54	☽ ♇ ☉
20 19	☽ ⊥ ♅
21 36	☽ △ ♃
21 52	☽ ⊼ ♆
23 51	☽ ☍ ♂
22 Monday	
00 13	☽ ♇ ⯓
01 47	☉ ⊻ ♀
01 51	☽ □ ♄
06 38	♂ ∠ ⯓
12 39	☽ ∠ ♅
17 39	☿ ⊼ ♅
20 42	♀ St R
20 52	☽ ⊼ ♀
22 24	☽ △ ☉
23 Tuesday	
00 05	☽ # ♆
00 51	☽ ♇ ♃
03 25	☽ △ ⯓
07 38	☽ ± ♀
10 01	☽ ∥ ☿
10 34	☽ Q ♄
15 30	☽ ⚹ ♅
16 59	☽ ± ♃
17 33	♂ □ ♄
18 18	☽ □ ☿
23 37	☽ ♇ ♀
24 Wednesday	
03 07	☉ # ♂
03 38	☽ ⊼ ♃
03 50	☽ ☍ ♆
05 14	☽ ∥ ⯓
08 00	☽ ⚹ ♄
08 42	☽ ⊼ ♂
12 38	☽ Q ♅
16 40	☽ ∥ ♄
19 48	☽ ± ♂
25 Thursday	
01 48	☽ △ ♀
05 51	☽ Q ☿
07 36	☽ □ ☉
08 23	☽ □ ⯓
10 21	☽ ∠ ♄
12 17	☽ ♇ ♂
15 16	☿ ∠ ♀
18 45	☉ ☌ ⯓
19 46	☽ □ ♅
26 Friday	
01 54	☽ ⊥ ♄
05 01	☽ ⚹ ☿
07 45	☽ ☍ ♃
07 51	☽ ⊼ ♆
12 11	☽ ⊻ ♄
15 14	☽ Q ⯓
15 16	☽ △ ♂
16 45	☽ Q ☉
18 01	☽ ± ♆
21 13	☿ # ♆
27 Saturday	
01 10	☉ ∠ ♄
04 28	☽ □ ♀
06 20	☿ △ ♃
06 50	☿ □ ♆
07 25	☉ # ♀
09 09	☽ ♇ ♆
09 26	☽ ∠ ☿
11 07	☉ Q ♆
11 25	☽ ⚹ ⯓
12 28	☽ ∥ ♄
14 23	☽ ⚹ ☉
22 08	☽ △ ♅
23 07	☽ ∥ ⯓
28 Sunday	
02 12	☽ ⊥ ☿
09 27	♃ ⊻ ♆
09 56	☽ Q ♀
10 02	☽ △ ♆
10 02	☽ ⊼ ♃
12 17	☽ ∠ ⯓
13 18	☽ ⊻ ☿
14 30	☽ ☌ ♄
17 00	☽ ∠ ☉
19 45	☽ □ ♂
19 56	☽ ± ♃
22 46	☽ ♇ ♅

All ephemeris data is given at 12.00 UT and the Moon's longitude is additionally given for 24.00 UT

MARCH 2049

LONGITUDES

Date	Sidereal time h m s	Sun ☉ ° ' "	Moon ☽ ° ' "	Moon ☽ 24.00 ° ' "	Mercury ☿ ° '	Venus ♀ ° '	Mars ♂ ° '	Jupiter ♃ ° '	Saturn ♄ ° '	Uranus ♅ ° '	Neptune ♆ ° '	Pluto ♇ ° '
01	22 38 55	11 ♓ 20 01	07 ♒ 08 46	14 ♒ 29 57	24 ♒ 48	02 ♈ 58	27 ♈ 42	21 ♊ 21	24 ♑ 06	14 ♍ 02	21 ♉ 19	07 ♓ 42
02	22 42 52	12 20 15	21 ♒ 51 02	29 ♒ 11 06	26 25	02 R 40	28 25	21 25	24 12	13 R 59	21 20	07 44
03	22 46 48	13 20 27	06 ♓ 29 07	13 ♓ 44 08	28 03	02 20	29 08	21 28	24 17	13 56	21 21	07 45
04	22 50 45	14 20 37	20 ♓ 55 13	28 ♓ 01 33	29 ♒ 42	01 58	29 ♈ 51	21 31	24 22	13 54	21 22	07 47
05	22 54 42	15 20 46	05 ♈ 02 28	11 ♈ 57 26	01 ♓ 22	01 33	00 ♉ 34	21 35	24 28	13 51	21 24	07 49
06	22 58 38	16 20 52	18 ♈ 46 07	25 ♈ 28 19	03 04	01 07	01 17	21 39	24 33	13 49	21 25	07 50
07	23 02 35	17 20 57	02 ♉ 04 01	08 ♉ 33 22	04 46	00 38	02 00	21 43	24 38	13 46	21 26	07 52
08	23 06 31	18 20 59	14 ♉ 56 38	21 ♉ 14 14	06 29	00 ♈ 08	02 43	21 47	24 44	13 43	21 27	07 53
09	23 10 28	19 21 00	27 ♉ 26 37	03 ♊ 34 23	08 14	29 ♓ 35	03 26	21 52	24 49	13 41	21 28	07 55
10	23 14 24	20 20 58	09 ♊ 38 08	15 ♊ 38 32	09 59	29 02	04 09	21 56	24 54	13 38	21 30	07 56
11	23 18 21	21 20 54	21 ♊ 36 18	27 ♊ 32 06	11 46	28 27	04 51	22 01	24 59	13 35	21 31	07 58
12	23 22 17	22 20 48	03 ♋ 26 39	09 ♋ 20 38	13 34	27 51	05 34	22 06	25 04	13 33	21 32	08 00
13	23 26 14	23 20 40	15 ♋ 14 43	21 ♋ 09 31	15 23	27 14	06 17	22 11	25 08	13 30	21 34	08 02
14	23 30 11	24 20 29	27 ♋ 05 39	03 ♌ 03 38	17 13	26 37	06 59	22 16	25 13	13 28	21 35	08 03
15	23 34 07	25 20 17	09 ♌ 03 59	15 ♌ 07 07	19 05	25 59	07 42	22 22	25 18	13 25	21 36	08 05
16	23 38 04	26 20 02	21 ♌ 13 22	27 ♌ 23 03	20 57	25 21	08 24	22 27	25 22	13 23	21 38	08 06
17	23 42 00	27 19 45	03 ♍ 36 22	09 ♍ 53 27	22 51	24 43	09 07	22 33	25 27	13 20	21 39	08 08
18	23 45 57	28 19 26	16 ♍ 14 20	22 ♍ 39 02	24 46	24 06	09 49	22 39	25 31	13 18	21 41	08 10
19	23 49 53	29 ♓ 19 04	29 ♍ 07 27	05 ♎ 39 28	26 42	23 30	10 32	22 45	25 36	13 15	21 42	08 11
20	23 53 50	00 ♈ 18 41	12 ♎ 14 52	18 ♎ 53 27	28 ♓ 39	22 54	11 14	22 51	25 40	13 13	21 44	08 13
21	23 57 46	01 18 16	25 ♎ 35 01	02 ♏ 19 17	00 ♈ 37	22 20	11 56	22 58	25 44	13 10	21 45	08 14
22	00 01 43	02 17 49	09 ♏ 06 04	15 ♏ 55 08	02 36	21 47	12 39	23 04	25 48	13 08	21 47	08 16
23	00 05 40	03 17 20	22 ♏ 46 20	29 ♏ 39 30	04 36	21 16	13 21	23 11	25 53	13 05	21 48	08 17
24	00 09 36	04 16 50	06 ♐ 34 32	13 ♐ 31 21	06 36	20 46	14 03	23 18	25 57	13 03	21 50	08 19
25	00 13 33	05 16 18	20 ♐ 29 53	27 ♐ 30 05	08 38	20 18	14 45	23 25	26 00	13 01	21 52	08 20
26	00 17 29	06 15 44	04 ♑ 31 54	11 ♑ 35 14	10 40	19 53	15 27	23 32	26 04	12 58	21 53	08 22
27	00 21 26	07 15 08	18 ♑ 40 00	25 ♑ 46 00	12 42	19 29	16 09	23 39	26 08	12 56	21 55	08 23
28	00 25 22	08 14 31	02 ♒ 53 03	10 ♒ 00 48	14 44	19 08	16 52	23 47	26 12	12 54	21 57	08 25
29	00 29 19	09 13 51	17 ♒ 08 55	24 ♒ 16 55	16 46	18 49	17 34	23 54	26 15	12 51	21 59	08 26
30	00 33 15	10 13 10	01 ♓ 24 06	08 ♓ 30 30	18 48	18 32	18 16	24 02	26 19	12 49	22 00	08 28
31	00 37 12	11 ♈ 12 28	15 ♓ 34 48	22 ♓ 36 43	20 ♈ 48	18 ♓ 18	18 ♉ 57	24 ♊ 10	26 ♑ 22	12 ♍ 47	22 ♉ 02	08 ♓ 29

Moon Nodes and Latitude

Date	Moon True ☊ ° '	Moon Mean ☊ ° '	Moon ☽ Latitude ° '
01	14 ♐ 24	14 ♐ 09	04 N 00
02	14 R 15	14 06	04 38
03	14 03	14 03	04 58
04	13 51	14 00	04 58
05	13 39	13 57	04 40
06	13 28	13 53	04 06
07	13 21	13 50	03 19
08	13 16	13 47	02 24
09	13 13	13 44	01 23
10	13 D 12	13 41	00 N 19
11	13 R 12	13 37	00 S 44
12	13 12	13 34	01 45
13	13 10	13 31	02 41
14	13 06	13 28	03 30
15	12 59	13 25	04 11
16	12 50	13 22	04 40
17	12 38	13 18	04 57
18	12 25	13 15	05 00
19	12 13	13 12	04 48
20	12 01	13 09	04 21
21	11 51	13 06	03 39
22	11 45	13 03	02 44
23	11 41	12 59	01 39
24	11 39	12 56	00 S 27
25	11 D 39	12 53	00 N 47
26	11 40	12 50	01 59
27	11 R 39	12 47	03 04
28	11 37	12 43	03 58
29	11 32	12 40	04 37
30	11 25	12 37	05 00
31	11 ♐ 16	12 ♐ 34	05 N 03

DECLINATIONS

Date	Sun ☉	Moon ☽	Mercury ☿	Venus ♀	Mars ♂	Jupiter ♃	Saturn ♄	Uranus ♅	Neptune ♆	Pluto ♇
01	07 S 19	14 S 37	15 S 11	08 N 16	10 N 48	22 N 57	21 S 07	07 N 02	16 N 26	20 S 06
02	06 56	09 50	14 40	08 17	11 04	22 57	21 06	07 03	16 26	20 05
03	06 33	04 S 31	14 08	08 17	11 20	22 57	21 05	07 04	16 26	20 04
04	06 10	00 N 58	13 35	08 15	11 36	22 58	21 04	07 05	16 27	20 04
05	05 46	06 17	13 00	08 12	11 52	22 58	21 04	07 06	16 27	20 03
06	05 23	11 08	12 25	08 07	12 08	22 59	21 03	07 07	16 27	20 03
07	05 00	15 18	11 48	08 01	12 23	22 59	21 02	07 08	16 28	20 02
08	04 36	18 36	11 09	07 53	12 39	22 59	21 01	07 09	16 28	20 02
09	04 13	20 55	10 30	07 44	12 54	23 00	21 00	07 10	16 29	20 01
10	03 49	22 12	09 49	07 34	13 09	23 00	20 59	07 11	16 29	20 01
11	03 26	22 26	09 07	07 22	13 24	23 01	20 58	07 12	16 29	20 00
12	03 02	21 38	08 24	07 09	13 39	23 01	20 58	07 13	16 30	19 59
13	02 39	19 54	07 39	06 55	13 54	23 02	20 57	07 14	16 30	19 59
14	02 15	17 18	06 54	06 39	14 09	23 02	20 56	07 15	16 30	19 58
15	01 51	13 57	06 07	06 23	14 23	23 02	20 55	07 16	16 31	19 58
16	01 27	09 59	05 19	06 06	14 37	23 02	20 54	07 17	16 31	19 57
17	01 04	05 33	04 30	05 48	14 52	23 03	20 54	07 18	16 32	19 57
18	00 40	00 N 49	03 40	05 29	15 06	23 03	20 53	07 19	16 32	19 57
19	00 S 16	04 S 04	02 49	05 10	15 20	23 04	20 52	07 20	16 33	19 56
20	00 N 07	08 50	01 57	04 50	15 34	23 05	20 51	07 21	16 33	19 56
21	00 31	13 17	01 04	04 30	15 47	23 05	20 51	07 22	16 33	19 55
22	00 55	17 07	00 S 10	04 10	16 01	23 06	20 50	07 23	16 34	19 55
23	01 18	20 03	00 N 45	03 50	16 14	23 06	20 49	07 24	16 34	19 54
24	01 42	21 51	01 40	03 30	16 28	23 07	20 49	07 25	16 35	19 54
25	02 06	22 19	02 36	03 10	16 41	23 07	20 48	07 26	16 35	19 53
26	02 29	21 23	03 32	02 50	16 54	23 08	20 47	07 26	16 36	19 53
27	02 53	19 06	04 29	02 31	17 06	23 08	20 47	07 27	16 36	19 53
28	03 16	15 39	05 25	02 12	17 19	23 09	20 46	07 28	16 37	19 52
29	03 39	11 17	06 22	01 54	17 31	23 09	20 45	07 29	16 37	19 52
30	04 03	06 18	07 18	01 37	17 44	23 10	20 45	07 30	16 38	19 51
31	04 N 26	01 S 01	08 N 14	01 N 20	17 N 56	23 N 10	20 S 44	07 N 31	16 N 38	19 S 51

ZODIAC SIGN ENTRIES

Date	h m	Planets
01	00 19	☽ ♒
03	01 20	☽ ♓
04	16 17	☿ ♓
04	16 50	♂ ♉
05	03 22	☽ ♈
07	08 13	☽ ♉
08	17 48	♀ ♓
09	16 59	☽ ♊
12	05 00	☽ ♋
14	17 51	☽ ♌
17	05 04	☽ ♍
19	13 37	☽ ♎
20	04 29	☉ ♈
21	04 32	☿ ♈
21	19 52	☽ ♏
24	00 36	☽ ♐
26	04 16	☽ ♑
28	07 08	☽ ♒
30	09 38	☽ ♓

LATITUDES

Date	Mercury ☿	Venus ♀	Mars ♂	Jupiter ♃	Saturn ♄	Uranus ♅	Neptune ♆	Pluto ♇
01	02 S 03	07 N 44	00 N 10	00 S 12	00 N 10	00 N 49	01 S 43	12 S 21
04	02 09	08 08	00 12	00 12	00 10	00 49	01 43	12 21
07	02 11	08 28	00 14	00 12	00 10	00 49	01 43	12 21
10	02 09	08 40	00 16	00 11	00 10	00 49	01 42	12 21
13	02 03	08 44	00 18	00 11	00 09	00 49	01 42	12 22
16	01 53	08 39	00 20	00 11	00 09	00 49	01 42	12 22
19	01 38	08 26	00 22	00 10	00 09	00 49	01 42	12 22
22	01 18	08 05	00 24	00 10	00 09	00 49	01 42	12 23
25	00 54	07 37	00 26	00 09	00 09	00 49	01 42	12 23
28	00 S 25	07 04	00 28	00 09	00 08	00 49	01 42	12 24
31	00 N 08	06 N 28	00 N 30	00 S 08	00 N 08	00 N 49	01 S 41	12 S 24

DATA

Julian Date	2469502
Delta T	+80 seconds
Ayanamsa	24° 32' 56"
Synetic vernal point	04° ♓ 34' 03"
True obliquity of ecliptic	23° 25' 56"

MOON'S PHASES, APSIDES AND POSITIONS ☽

Date	h m	Phase	Longitude ° '	Eclipse Indicator
04	00 11	●	13 ♓ 51	
11	11 26	☽	21 ♊ 19	
19	12 23	○	29 ♍ 20	
26	15 10	☾	06 ♑ 24	

Day	h m		
01	12 41	Perigee	
13	07 22	Apogee	
28	12 52	Perigee	
04	07 45	ON	
11	05 13	Max dec	22° N 28'
18	16 01	OS	
25	08 03	Max dec	22° S 20'
31	16 36	ON	

LONGITUDES

Date	Chiron ⚷ ° '	Ceres ⚳ ° '	Pallas ⚴ ° '	Juno ⚵ ° '	Vesta ⚶ ° '	Black Moon Lilith ⚸ ° '
01	27 ♏ 32	24 ♈ 16	10 ♓ 06	07 ♈ 56	21 ♉ 13	13 ♋ 44
11	27 ♏ 35	27 ♈ 58	13 ♓ 26	13 ♈ 26	24 ♉ 33	14 ♋ 50
21	27 ♏ 28	01 ♉ 47	16 ♓ 45	19 ♈ 02	28 ♉ 05	15 ♋ 57
31	27 ♏ 11	05 ♉ 41	20 ♓ 02	24 ♈ 43	01 ♊ 48	17 ♋ 04

ASPECTARIAN

h m Aspects

01 Monday
00 51 ☿ ⊻ ♄
01 35 ☽ # ♆
03 05 ☽ ⊥ ♇
05 18 ☽ ⚹ ♀
08 26 ☽ ∥ ☿
08 49 ☽ ⊥ ☉
10 42 ☽ ⚼ ♃
12 54 ☽ ⊻ ♇
13 26 ☽ ± ♅
19 20 ☽ ⊻ ☉
23 12 ☽ ⊼ ♅

02 Tuesday
02 41 ☽ Q ♂
04 37 ☉ # ♅
05 19 ☽ ∠ ♀
06 23 ☽ # ♂
11 10 ☽ □ ♆
11 17 ☽ △ ♃
15 09 ☿ ⊥ ♀
15 51 ☽ ⊻ ♄
19 10 ☽ # ♀
19 43 ☽ ⊥ ♀
20 24 ☽ ☌ ☿
23 18 ☽ ⚹ ♂

03 Wednesday
00 42 ☽ # ♅
01 44 ☽ ⊥ ♄
02 19 ☽ ∥ ☉
05 20 ☽ ⊻ ♀
05 45 ♂ ⚼ ♅
14 06 ☽ ☌ ♇
16 39 ☽ ∠ ♄
16 45 ☽ Q ♆

04 Thursday
00 11 ☽ ☌ ☉
00 18 ☽ ☍ ♅
01 20 ☽ ∠ ♂
01 45 ☉ ☍ ♅
12 46 ☽ ⚹ ♆
13 01 ☽ □ ♃
15 52 ☿ ⚹ ♂
17 12 ☽ ⊥ ♂
17 51 ☽ ⚹ ♄
22 14 ☿ ⊥ ♄

05 Friday
03 56 ☽ ⊻ ♂
04 51 ☽ ⊻ ☿
06 12 ☽ ☌ ♀
09 47 ☽ # ☉
14 07 ☿ ⊻ ♀
14 20 ☽ ∠ ♆
14 28 ☽ Q ♄
15 53 ☽ ∥ ♅
16 35 ☽ ⊥ ☿
16 48 ☽ ⊻ ♇
19 54 ☽ Q ♃
20 58 ☽ ∥ ♀

06 Saturday
03 16 ☽ ⊼ ♅
03 17 ☽ ⊥ ♇
06 03 ☽ ⊥ ♆
06 57 ☉ ∠ ♂
07 22 ☽ ⊻ ☉
08 26 ♀ ⊻ ♂
10 34 ☽ ∠ ☿
13 51 ☽ ± ♅
16 43 ☽ ⊻ ♆
17 09 ☽ ⚹ ♃
17 39 ☽ ∥ ♂
18 01 ☽ # ☿
18 54 ☽ ⊥ ☉
19 16 ☽ ∠ ♇
19 49 ☿ # ♂
22 25 ☽ □ ♄

07 Sunday
05 59 ☽ ⚼ ♅
09 28 ☽ ⊻ ♀
11 52 ☽ ☌ ♂
12 34 ☽ ∠ ☉
17 43 ☽ ⚹ ☿
19 43 ☽ ∥ ♆
20 07 ☽ ⊥ ♀
20 37 ☽ ∠ ♃
22 44 ☽ ⚹ ♇

08 Monday
09 42 ☽ △ ♅
12 20 ☽ ∠ ♀
13 36 ☽ ⊥ ♃
19 02 ☽ ⚹ ☉
19 49 ☽ Q ☿
21 26 ☽ Q ♇

09 Tuesday
00 26 ☽ ☌ ♆
01 08 ☽ ⊻ ♃
01 23 ☽ # ♇
06 51 ☽ △ ♄
07 41 ☿ ☌ ♇
13 04 ☽ # ♄
16 00 ☽ ⚹ ♀
20 19 ☽ Q ☉

10 Wednesday
00 27 ☽ ⊻ ♂
05 11 ☿ Q ♆
08 38 ☽ □ ♇
10 39 ☿ ∠ ♄
12 31 ☽ ⚼ ♄
12 49 ☽ □ ☿
13 04 ☽ ⊥ ♂
14 39 ☽ Q ♀
19 57 ☽ □ ♅

11 Thursday
04 29 ♀ ⊥ ♂
06 40 ☽ ± ♄
08 15 ☽ ∠ ♂
11 26 ☽ □ ☉
11 49 ☽ ⊻ ♆
12 50 ☽ ☌ ♃
16 05 ☉ ⚹ ♆
18 52 ☽ ⊼ ♄
23 59 ☽ ⊥ ♆

12 Friday
01 11 ☽ □ ♀
04 41 ♀ ∥ ♅
05 29 ☉ □ ♃
08 10 ☽ Q ♅
11 46 ☿ ☍ ♅
16 36 ☽ ⚹ ♂
18 18 ☽ ∠ ♆
21 17 ☽ △ ♇
22 56 ☽ # ♄

13 Saturday
08 29 ☽ ⚹ ♅
11 00 ☽ # ♇
12 20 ☽ △ ☿
18 33 ☽ Q ♂

14 Sunday
00 50 ☽ ⚹ ♆
01 03 ☿ # ♅
02 11 ☽ ⊻ ♃
03 49 ☽ ⚼ ♇
05 56 ☽ △ ☉
08 11 ☽ ☍ ♄
11 04 ☽ △ ♀
14 23 ☽ ⊥ ♃
14 45 ☽ ∠ ♅
18 06 ☽ ∥ ♆
22 00 ☽ ± ♇
22 58 ♂ ∠ ♃
23 27 ☿ # ♀

15 Monday
00 12 ☽ ⚼ ☿
01 04 ☽ Q ♆
08 35 ☽ ∠ ♃
08 44 ☽ ⊥ ♅
09 06 ☽ □ ♂
09 19 ☽ ∥ ♂
10 02 ☽ ⊼ ♇
10 54 ☉ ⚹ ♄
14 45 ☽ ⚼ ☉
15 37 ☽ ⚼ ♀
20 37 ☽ ⊻ ♅
21 25 ☽ ± ☿
21 30 ☉ ☌ ♀

16 Tuesday
01 27 ♂ ⚹ ♇
08 31 ☽ ± ♀
10 06 ☽ ± ☉
11 16 ♀ ⚹ ♄
11 23 ☽ ⊼ ☿
12 48 ☽ □ ♆
14 26 ☽ ⚹ ♃
19 40 ☽ ⊼ ♀
20 09 ☽ ⊼ ♄
20 42 ☿ ⚹ ♆
22 50 ☽ ⊼ ☉

17 Wednesday
02 50 ☽ ∥ ♅
07 50 ☽ ± ♄
08 03 ☿ □ ♃
10 40 ☽ ∥ ♀
13 50 ☽ Q ♃
18 36 ☽ # ☿
20 40 ☽ ☍ ♇
23 00 ♀ ∠ ♂
23 09 ☽ △ ♂

18 Thursday
01 08 ☽ ⚼ ♄
05 47 ☿ ☌ ♀
06 28 ☽ ☌ ♅
12 47 ☽ # ☉
13 10 ☿ ∠ ♂
18 45 ☽ ∥ ☉
21 50 ☿ ⚹ ♄
22 12 ☽ △ ♆

19 Friday
00 06 ☽ □ ♃
02 02 ☽ ☍ ♀
04 58 ☽ ⚼ ♂
05 26 ☽ △ ♄
06 43 ☽ ☍ ☿
06 47 ☽ ∥ ☿
12 23 ☽ ☍ ☉
17 07 ☽ # ♀
22 30 ☽ ± ♂

20 Saturday
01 56 ☽ ⚼ ♆
04 22 ☽ # ♅
04 39 ☽ ⊼ ♇
10 03 ☽ ⊼ ♂
13 42 ♀ □ ♃
13 44 ☽ ⊻ ♅
15 34 ☽ ± ♇
18 19 ☽ ± ♆

21 Sunday
00 32 ☽ ⊥ ♅
05 08 ☽ ⊼ ♆
06 25 ☽ ⊼ ♀
07 16 ☽ △ ♃
07 48 ☽ ⚼ ♇
12 17 ☽ □ ♄
16 36 ☽ ∠ ♅
16 43 ☽ ± ♀
22 09 ☉ # ☿
22 30 ☽ ⊼ ☿
23 00 ☽ ⊼ ☉

22 Monday
04 08 ☽ # ♂
04 45 ☉ ☌ ☿
08 04 ☽ ⚼ ♀
08 15 ☽ # ♆
10 10 ☽ ⚼ ♃
10 28 ☽ ± ☉
10 31 ☽ △ ♇
10 58 ☽ ± ☿
12 03 ♀ ⚹ ♆
18 35 ☽ ☍ ♂
19 05 ☽ ⚹ ♅
20 20 ☽ Q ♄

23 Tuesday
02 08 ☽ ± ♃
03 33 ☽ ⚼ ☉
03 36 ♂ △ ♅
05 30 ☽ ⚼ ☿
09 27 ☽ △ ♀
10 19 ☽ ☍ ♆
10 34 ☽ ∥ ♇
12 43 ☽ ⊼ ♃
16 02 ☽ Q ♅
17 27 ☽ ⚹ ♄
20 23 ☽ ∥ ♄

24 Wednesday
07 43 ☽ △ ☉
12 04 ☽ △ ☿
13 34 ☉ ∥ ☿
14 44 ☿ ∠ ♆
15 01 ☽ □ ♇
19 35 ☽ ∠ ♄
23 09 ☽ □ ♅

25 Thursday
01 36 ☽ ⊼ ♂
01 56 ♂ ∥ ♆
04 23 ☿ Q ♄
08 30 ☿ ⊻ ♇
11 09 ☽ ⊥ ♄
11 41 ☽ □ ♀
12 28 ☽ ± ♂
14 15 ☿ ⊥ ♂
14 21 ☽ ⊼ ♆
17 03 ☽ ☍ ♃
21 29 ☽ ⊻ ♄
22 02 ☽ Q ♇
22 48 ☿ ∥ ♀

26 Friday
00 39 ☽ ± ♆
04 41 ☽ ⚼ ♂
15 10 ☽ □ ☉
16 02 ☽ ⚼ ♆
17 32 ☽ Q ♀
18 32 ☽ ⚹ ♇
19 43 ☽ ∥ ♄
22 57 ☿ Q ♃
23 52 ☉ ∥ ♀

27 Saturday
00 11 ☽ □ ☿
02 18 ☽ △ ♅
03 41 ☉ ∠ ♆
05 09 ☽ ∥ ♇
07 32 ☽ △ ♂
13 21 ☽ ⚹ ♀
14 42 ☿ ⊼ ♅
17 31 ☽ △ ♆
20 00 ☽ ∠ ♇
20 31 ☽ ⊼ ♃
23 58 ☽ Q ☉

28 Sunday
00 41 ☽ ☌ ♄
01 54 ☽ # ♂
03 37 ☽ ⚼ ♅
05 59 ☽ # ♆
06 43 ☽ ± ♃
08 11 ☿ ⊥ ♇
10 49 ☉ Q ♄
11 12 ☽ ⊥ ♇
11 42 ☽ Q ☿
14 03 ☽ ∠ ♀
16 17 ☉ ⊻ ♇
18 44 ☽ ± ♅
21 19 ☽ ⊻ ♇
21 42 ☽ ⚹ ☉
22 01 ☽ ⚼ ♃

29 Monday
02 32 ☿ ⊥ ♆
04 48 ☽ ⊼ ♅
04 52 ☽ ⊥ ♀
11 15 ☽ ⚹ ☿
12 44 ☽ □ ♂
14 45 ☽ ⊻ ♀
20 08 ☽ □ ♆
23 28 ☽ △ ♃

30 Tuesday
00 48 ☽ ∠ ☉
02 18 ☿ ⊻ ♂
02 33 ♂ ⊥ ♃
03 23 ☽ ⊻ ♄
06 26 ☽ # ♅
08 03 ☽ # ☿
09 21 ☿ ⊻ ♀
12 16 ☿ ± ♅
13 32 ☽ ⊥ ♄
16 42 ☽ ∠ ☿
17 00 ☿ ∥ ♅
17 06 ☽ ⊥ ☉
19 10 ♀ ⚹ ♂
20 37 ☽ Q ♂
21 38 ☽ # ☉
23 57 ☽ ☌ ♇

31 Wednesday
02 34 ☽ Q ♆
04 01 ☽ ⊻ ☉
04 49 ☽ ∠ ♄
07 15 ☽ ☍ ♅
10 28 ☽ ⊥ ☿
10 32 ☽ # ♀
16 35 ☽ ☌ ♀
18 03 ☽ ⚹ ♂
22 06 ☽ ∥ ♀
22 24 ☽ ⊻ ☿
23 02 ☽ ⚹ ♆

All ephemeris data is given at 12.00 UT and the Moon's longitude is additionally given for 24.00 UT

JUNE 2049

LONGITUDES

Date	Sidereal time h m s	Sun ☉ ° ' "	Moon ☽ ° ' "	Moon ☽ 24.00 ° ' "	Mercury ☿ ° '	Venus ♀ ° '	Mars ♂ ° '	Jupiter ♃ ° '	Saturn ♄ ° '	Uranus ♅ ° '	Neptune ♆ ° '	Pluto ♇ ° '
01	04 41 38	11 ♊ 26 38	21 ♊ 41 23	27 ♊ 42 27	17 ♉ 26	25 ♈ 44	01 ♋ 04	05 ♋ 31	26 ♑ 52	11 ♍ 53	24 ♉ 15	09 ♓ 23
02	04 45 35	12 24 09	03 ♋ 41 41	09 ♋ 39 21	18 17	26 45	01 44	05 44	26 R 50	11 54	24 17	09 23
03	04 49 32	13 21 40	15 ♋ 35 46	21 ♋ 31 17	19 13	27 46	02 23	05 57	26 47	11 55	24 19	09 23
04	04 53 28	14 19 09	27 ♋ 26 14	03 ♌ 21 03	20 11	28 47	03 03	06 10	26 45	11 56	24 21	09 24
05	04 57 25	15 16 37	09 ♌ 16 08	15 ♌ 11 58	21 13	29 ♈ 49	03 42	06 24	26 42	11 57	24 23	09 24
06	05 01 21	16 14 04	21 ♌ 09 01	27 ♌ 07 49	22 19	00 ♉ 50	04 22	06 37	26 39	11 58	24 26	09 24
07	05 05 18	17 11 29	03 ♍ 08 52	09 ♍ 12 43	23 27	01 52	05 02	06 50	26 36	11 59	24 28	09 24
08	05 09 14	18 08 54	15 ♍ 19 56	21 ♍ 31 02	24 39	02 55	05 41	07 03	26 33	12 00	24 30	09 24
09	05 13 11	19 06 17	27 ♍ 46 34	04 ♎ 07 02	25 54	03 57	06 21	07 16	26 30	12 01	24 32	09 R 24
10	05 17 07	20 03 40	10 ♎ 32 55	17 ♎ 04 36	27 12	05 00	07 00	07 30	26 27	12 02	24 34	09 24
11	05 21 04	21 01 01	23 ♎ 42 28	00 ♏ 26 43	28 33	06 03	07 39	07 43	26 24	12 03	24 36	09 24
12	05 25 01	21 58 21	07 ♏ 17 32	14 ♏ 14 55	29 ♉ 57	07 06	08 19	07 56	26 21	12 04	24 38	09 24
13	05 28 57	22 55 41	21 ♏ 18 44	28 ♏ 28 42	01 ♊ 24	08 10	08 58	08 10	26 18	12 06	24 40	09 24
14	05 32 54	23 52 59	05 ♐ 44 22	13 ♐ 05 06	02 54	09 13	09 38	08 23	26 14	12 07	24 42	09 23
15	05 36 50	24 50 17	20 ♐ 30 10	27 ♐ 58 37	04 26	10 17	10 17	08 37	26 11	12 09	24 44	09 23
16	05 40 47	25 47 35	05 ♑ 29 27	13 ♑ 01 34	06 02	11 22	10 56	08 50	26 07	12 10	24 46	09 23
17	05 44 43	26 44 51	20 ♑ 33 50	28 ♑ 05 06	07 41	12 26	11 35	09 03	26 04	12 11	24 48	09 23
18	05 48 40	27 42 08	05 ♒ 34 18	13 ♒ 00 26	09 22	13 30	12 15	09 17	26 00	12 13	24 50	09 23
19	05 52 36	28 39 24	20 ♒ 22 39	27 ♒ 40 12	11 06	14 35	12 54	09 30	25 57	12 15	24 52	09 22
20	05 56 33	29 ♊ 36 39	04 ♓ 52 33	11 ♓ 59 17	12 53	15 40	13 33	09 44	25 53	12 16	24 54	09 22
21	06 00 30	00 ♋ 33 54	19 ♓ 00 09	25 ♓ 55 04	14 43	16 45	14 12	09 57	25 49	12 18	24 56	09 22
22	06 04 26	01 31 10	02 ♈ 44 04	09 ♈ 27 16	16 36	17 50	14 51	10 11	25 45	12 20	24 58	09 22
23	06 08 23	02 28 25	16 ♈ 04 56	22 ♈ 37 19	18 31	18 56	15 31	10 25	25 42	12 22	25 00	09 21
24	06 12 19	03 25 40	29 ♈ 04 47	05 ♉ 27 42	20 28	20 02	16 10	10 38	25 38	12 23	25 02	09 21
25	06 16 16	04 22 55	11 ♉ 46 27	18 ♉ 01 27	22 28	21 07	16 49	10 52	25 34	12 25	25 03	09 20
26	06 20 12	05 20 10	24 ♉ 13 03	00 ♊ 21 39	24 30	22 13	17 28	11 05	25 30	12 27	25 05	09 20
27	06 24 09	06 17 24	06 ♊ 27 35	12 ♊ 31 12	26 34	23 19	18 07	11 19	25 26	12 29	25 07	09 20
28	06 28 05	07 14 39	18 ♊ 32 47	24 ♊ 32 40	28 ♊ 39	24 26	18 46	11 32	25 22	12 31	25 09	09 19
29	06 32 02	08 11 54	00 ♋ 31 06	06 ♋ 28 20	00 ♋ 46	25 32	19 25	11 46	25 18	12 33	25 11	09 19
30	06 35 59	09 ♋ 09 08	12 ♋ 24 38	18 ♋ 20 13	02 ♋ 55	26 ♉ 39	20 ♋ 04	12 ♋ 00	25 ♑ 13	12 ♍ 35	25 ♉ 12	09 ♓ 18

Date	Moon True ☊ ° '	Moon Mean ☊ ° '	Moon ☽ Latitude ° '
01	09 ♐ 23	09 ♐ 17	01 S 08
02	09 R 23	09 14	02 11
03	09 23	09 11	03 08
04	09 22	09 07	03 56
05	09 21	09 04	04 35
06	09 21	09 01	05 01
07	09 20	08 58	05 15
08	09 20	08 55	05 15
09	09 D 20	08 52	05 01
10	09 20	08 48	04 31
11	09 21	08 45	03 47
12	09 22	08 42	02 49
13	09 23	08 39	01 38
14	09 24	08 36	00 S 20
15	09 R 23	08 32	01 N 01
16	09 22	08 29	02 19
17	09 21	08 26	03 28
18	09 19	08 23	04 23
19	09 16	08 20	04 58
20	09 14	08 17	05 14
21	09 13	08 13	05 11
22	09 D 13	08 10	04 49
23	09 13	08 07	04 13
24	09 14	08 04	03 24
25	09 16	08 01	02 26
26	09 17	07 58	01 22
27	09 18	07 54	00 N 16
28	09 R 18	07 51	00 S 51
29	09 16	07 48	01 54
30	09 ♐ 13	07 ♐ 45	02 S 51

DECLINATIONS

Date	Sun ☉ ° '	Moon ☽ ° '	Mercury ☿ ° '	Venus ♀ ° '	Mars ♂ ° '	Jupiter ♃ ° '	Saturn ♄ ° '	Uranus ♅ ° '	Neptune ♆ ° '	Pluto ♇ ° '
01	22 N 09	22 N 03	13 N 30	07 N 53	24 N 22	23 N 18	20 S 43	07 N 49	17 N 12	19 S 47
02	22 16	21 12	13 47	08 13	24 22	23 18	20 43	07 49	17 13	19 48
03	22 24	19 25	14 05	08 32	24 22	23 17	20 44	07 48	17 13	19 48
04	22 31	16 48	14 24	08 52	24 21	23 17	20 45	07 48	17 14	19 48
05	22 37	13 31	14 45	09 12	24 21	23 16	20 45	07 48	17 14	19 48
06	22 43	09 41	15 07	09 31	24 20	23 15	20 46	07 47	17 15	19 49
07	22 49	05 27	15 30	09 51	24 18	23 15	20 46	07 47	17 15	19 49
08	22 54	00 N 56	15 53	10 11	24 17	23 14	20 47	07 46	17 16	19 49
09	22 59	03 S 43	16 18	10 30	24 16	23 14	20 48	07 46	17 16	19 50
10	23 04	08 20	16 43	10 50	24 14	23 13	20 48	07 45	17 16	19 50
11	23 08	12 43	17 09	11 10	24 12	23 12	20 49	07 45	17 17	19 50
12	23 11	16 36	17 36	11 29	24 10	23 12	20 50	07 44	17 18	19 51
13	23 15	19 40	18 02	11 49	24 07	23 11	20 50	07 44	17 18	19 51
14	23 17	21 35	18 29	12 08	24 05	23 10	20 51	07 43	17 18	19 51
15	23 20	22 05	18 57	12 28	24 02	23 09	20 52	07 43	17 19	19 52
16	23 22	21 00	19 24	12 47	23 59	23 09	20 53	07 42	17 19	19 52
17	23 24	18 26	19 51	13 06	23 56	23 08	20 53	07 41	17 20	19 52
18	23 25	14 38	20 17	13 25	23 53	23 07	20 54	07 41	17 20	19 53
19	23 26	09 58	20 43	13 44	23 49	23 06	20 55	07 40	17 21	19 53
20	23 26	04 S 50	21 08	14 02	23 46	23 05	20 56	07 39	17 21	19 54
21	23 26	00 N 25	21 33	14 21	23 42	23 04	20 56	07 39	17 22	19 54
22	23 25	05 31	21 56	14 39	23 38	23 03	20 57	07 38	17 22	19 55
23	23 25	10 12	22 19	14 57	23 34	23 03	20 58	07 37	17 22	19 55
24	23 23	14 19	22 39	15 15	23 29	23 02	20 59	07 37	17 23	19 55
25	23 22	17 40	22 58	15 33	23 25	23 01	21 00	07 36	17 23	19 56
26	23 19	20 09	23 16	15 51	23 20	23 00	21 00	07 35	17 24	19 56
27	23 17	21 38	23 31	16 08	23 15	22 58	21 01	07 34	17 24	19 57
28	23 14	22 06	23 44	16 25	23 10	22 57	21 02	07 34	17 24	19 57
29	23 11	21 32	23 55	16 41	23 04	22 56	21 03	07 33	17 25	19 58
30	23 N 07	20 N 00	24 N 03	16 N 58	22 N 59	22 N 55	21 S 04	07 N 32	17 N 25	19 S 58

ZODIAC SIGN ENTRIES

Date	h	m	Planets
02	04	35	☽ ♋
04	17	12	☽ ♌
05	16	28	♀ ♉
07	05	44	☽ ♍
09	16	14	☽ ♎
11	23	13	☽ ♏
12	12	53	☿ ♊
14	02	32	☽ ♐
16	03	14	☽ ♑
18	03	04	☽ ♒
20	03	52	☽ ♓
20	21	47	☉ ♋
22	07	10	☽ ♈
24	13	43	☽ ♉
26	23	18	☽ ♊
29	03	16	☿ ♋
29	10	57	☽ ♋

LATITUDES

Date	Mercury ☿ ° '	Venus ♀ ° '	Mars ♂ ° '	Jupiter ♃ ° '	Saturn ♄ ° '	Uranus ♅ ° '	Neptune ♆ ° '	Pluto ♇ ° '
01	03 S 40	02 S 12	00 N 57	00 S 01	00 N 04	00 N 46	01 S 41	12 S 43
04	03 30	02 19	00 57	00 S 01	00 04	00 46	01 41	12 44
07	03 14	02 25	00 58	00 00	00 03	00 46	01 41	12 45
10	02 53	02 30	00 59	00 00	00 03	00 46	01 41	12 46
13	02 27	02 33	01 00	00 00	00 03	00 46	01 41	12 47
16	01 57	02 35	01 01	00 00	00 03	00 46	01 41	12 48
19	01 24	02 36	01 01	00 N 01	00 02	00 46	01 41	12 49
22	00 49	02 35	01 02	00 01	00 02	00 46	01 41	12 50
25	00 S 15	02 34	01 03	00 01	00 02	00 45	01 41	12 51
28	00 N 19	02 32	01 03	00 02	00 02	00 45	01 41	12 52
31	00 N 48	02 S 29	01 N 04	00 N 02	00 N 01	00 N 45	01 S 41	12 S 53

DATA

Julian Date	2469594
Delta T	+80 seconds
Ayanamsa	24° 33' 06"
Synetic vernal point	04° ♓ 33' 52"
True obliquity of ecliptic	23° 25' 55"

MOON'S PHASES, APSIDES AND POSITIONS ☽

Date	h	m	Phase	Longitude ° '	Eclipse Indicator
08	17	56	☽	18 ♍ 23	
15	19	27	○	25 ♐ 08	
22	09	41	☾	01 ♈ 26	
30	04	50	●	08 ♋ 52	

Day	h	m		
04	14	46	Apogee	
16	22	48	Perigee	
01	04	09	Max dec	22° N 06'
08	16	50	OS	
15	07	35	Max dec	22° S 06'
21	10	05	ON	
28	10	41	Max dec	22° N 06'

LONGITUDES

Date	Chiron ⚷ ° '	Ceres ⚳ ° '	Pallas ⚴ ° '	Juno ⚵ ° '	Vesta ⚶ ° '	Black Moon Lilith ⚸ ° '
01	23 ♏ 09	00 ♊ 50	08 ♈ 35	01 ♊ 06	27 ♊ 13	23 ♋ 58
11	22 ♏ 31	04 ♊ 55	11 ♈ 03	07 ♊ 01	01 ♋ 32	25 ♋ 05
21	21 ♏ 59	08 ♊ 58	13 ♈ 16	12 ♊ 56	05 ♋ 52	26 ♋ 11
31	21 ♏ 35	13 ♊ 00	15 ♈ 13	18 ♊ 49	10 ♋ 14	27 ♋ 18

All ephemeris data is given at 12.00 UT and the Moon's longitude is additionally given for 24.00 UT

ASPECTARIAN

h m	Aspects
01 Tuesday	
02 21	♄ ⚼ ♅
02 55	☽ ⊻ ☿
03 17	☽ ∥ ☉
06 44	♀ ∥ ♅
10 22	☽ ± ♄
15 43	☽ ⊥ ☿
17 06	☽ ⊻ ♆
20 48	☽ ✶ ♀
22 12	☉ ⚼ ♄
22 17	☽ ⊼ ♄
23 17	☉ □ ♅
02 Wednesday	
04 23	☽ Q ♅
05 08	☽ ⊥ ♆
07 49	☽ ☌ ♂
11 07	☽ ∠ ☿
13 49	♀ □ ♄
15 41	♀ ⚼ ♅
16 11	☽ ☌ ♃
19 38	☽ # ♄
23 07	☽ Q ♀
23 17	☽ ∠ ♆
23 28	☽ △ ♇
03 Thursday	
04 33	☽ ✶ ♅
07 05	☽ ⊻ ☉
07 43	☽ # ♇
19 58	☽ ✶ ☿
20 18	☽ ⊥ ☉
04 Friday	
05 44	☽ ✶ ♆
05 49	☽ ⚼ ♇
08 32	☽ ∥ ♆
10 36	☽ ☍ ♄
10 58	☽ ∠ ♅
14 59	☽ □ ♀
16 09	☽ ∠ ☉
22 33	☽ Q ☿
05 Saturday	
00 03	☽ ⊻ ♂
00 05	☽ ± ♇
04 18	☽ ∥ ☿
05 15	☽ ⊥ ♅
06 04	☽ ⊻ ♃
06 09	☽ Q ♆
12 15	☽ ⊼ ♇
12 56	☽ ⊥ ♂
15 50	☿ Q ♇
16 45	☿ ∠ ♃
17 25	☽ ⊻ ♅
18 27	☽ ⊥ ♃
06 Sunday	
01 13	☽ ✶ ☉
08 12	☽ ∠ ♂
12 52	☽ ∥ ♀
12 57	☽ ∠ ♃
14 35	☽ □ ☿
18 36	☽ □ ♆
23 00	☽ ∥ ♅
23 00	☽ ⊼ ♄
07 Monday	
03 26	☽ Q ☉
09 14	☽ △ ♀
10 55	☽ ± ♄
15 56	☽ ✶ ♂
19 26	☽ ✶ ♃
08 Tuesday	
00 22	☽ ☍ ♇
04 38	☽ ⚼ ♄
05 28	☽ ☌ ♅
08 51	☿ ☌ ♆
16 50	☽ Q ♂
17 29	☽ ⚼ ♀
17 56	☽ □ ☉
19 12	♇ St R
19 22	☽ Q ♃
09 Wednesday	
05 47	☽ △ ♆
08 01	☽ △ ☿
09 35	☽ △ ♄
12 22	☽ ± ♀
22 52	☿ △ ♄
10 Thursday	
00 44	☽ ⊼ ♀
05 03	☽ □ ♂
06 14	☽ □ ♃
08 58	☽ # ♅
09 52	☽ ⊼ ♇
10 10	☽ ⚼ ♆
14 45	☽ ⊻ ♅
15 24	☽ ⚼ ☿
20 56	☽ ± ♇
21 21	☉ ± ♄
11 Friday	
01 46	☽ ⊥ ♅
02 34	☽ # ♀
02 45	☽ ± ♆
03 39	☉ ∠ ♀
06 47	☽ △ ☉
09 41	☽ ± ☿
13 14	☽ ⚼ ♇
13 36	☽ ⊼ ♆
15 22	♂ ☌ ♃
16 48	☽ □ ♄
17 59	☽ ∠ ♅
19 13	☿ ∥ ♆
21 38	☽ ⊼ ☿
12 Saturday	
11 24	☽ ⚼ ☉
11 39	☽ ☍ ♀
13 09	☽ △ ♃
13 52	☽ △ ♂
14 03	☉ ∥ ♃
15 39	☽ △ ♇
16 51	☽ # ♆
20 09	☽ # ☿
20 17	☽ ✶ ♅
13 Sunday	
00 07	☽ Q ♄
04 02	☽ ± ☉
12 01	♀ ✶ ♃
13 45	☽ ∥ ♇
14 55	☽ ⊼ ☉
15 10	☽ ⚼ ♃
16 41	☽ ⚼ ♂
16 41	☽ Q ♅
17 39	☽ ☍ ♆
20 20	☽ ✶ ♄
14 Monday	
00 42	☽ ∥ ♄
02 33	☿ ⊥ ♃
03 25	♂ △ ♇
06 23	☽ ± ♃
06 46	☽ ☍ ☿
08 21	☽ ± ♂
15 00	♂ ∠ ♆
15 44	♀ ✶ ♇
16 24	☽ ⊼ ♃
17 59	☽ □ ♇
18 09	☽ ⊼ ♀
18 40	☽ ⊼ ♂
20 58	☽ ∠ ♄
22 27	☽ □ ♅
15 Tuesday	
04 40	☽ ± ♀
07 47	☿ ⊥ ♂
09 22	☉ ⊻ ♆
11 29	♀ ✶ ♂
11 29	☽ ⊥ ♄
18 49	☽ ⊼ ♆
19 27	☽ ☍ ☉
20 17	☽ ⚼ ♀
21 05	☽ ⊻ ♄
23 03	☽ Q ♇
16 Wednesday	
04 27	☽ ± ♆
12 58	☽ ⊼ ☿
13 31	☽ ∥ ♄
17 24	☽ ☍ ♃
18 12	☽ ✶ ♇
18 50	☽ ⚼ ♆
19 50	☉ ⊼ ♄
21 04	☽ ☍ ♂
22 04	☽ △ ♀
22 39	☽ △ ♅
23 41	☽ ± ☿
17 Thursday	
00 08	☽ ∥ ♇
02 04	☽ # ☿
06 31	♀ △ ♅
13 46	☿ # ♇
15 47	☽ ⚼ ☿
18 05	☽ ∠ ♇
18 46	☽ △ ♆
19 38	☽ # ♆
20 44	☽ ☌ ♄
22 32	☽ ⊼ ☉
22 35	☽ ⚼ ♅
18 Friday	
08 29	☽ ⊥ ♇
08 48	☽ ± ☉
10 36	☿ ⊻ ♃
10 59	♂ ✶ ♅
12 07	☿ □ ♇
13 02	☽ ± ♅
18 04	☽ ⊼ ♃
18 08	☽ ⊻ ♇
18 09	☽ # ♀
18 55	☽ △ ☿
22 02	♃ △ ♇
22 44	☽ ⊼ ♅
23 16	☽ ⊼ ♂
19 Saturday	
00 18	☽ ⚼ ☉
01 49	☽ □ ♀
03 56	☽ ± ♃
09 28	☽ ± ♂
09 51	☿ ⚼ ♄
18 53	☽ ⚼ ♃
19 23	☽ □ ♆
21 07	☽ ⊻ ♄
22 57	☽ # ♅
23 23	☿ # ♄
20 Sunday	
00 58	☽ ⚼ ♂
02 36	☽ △ ☉
03 36	☿ □ ♅
07 01	☽ ⊥ ♄
09 49	☽ Q ♀
19 34	☽ ☌ ♇
20 18	☽ △ ♃
22 05	☽ ∠ ♄
21 Monday	
00 30	☽ ☍ ♅
01 34	☿ ⊻ ♂
01 35	☽ Q ♆
03 22	☽ △ ♂
03 33	☽ □ ☿
05 08	☉ Q ♅
07 48	☽ ✶ ♀
08 57	♃ ∠ ♆
21 34	☉ ⊥ ♆
22 18	☽ ✶ ♆
23 47	☽ ✶ ♄
22 Tuesday	
09 41	☽ □ ☉
12 12	☽ ∠ ♀
15 51	☽ Q ☿
20 55	☽ Q ♄
22 31	☽ ∥ ♅
23 49	☽ ⊻ ♇
23 Wednesday	
00 57	☽ ∠ ♆
01 32	☽ □ ♃
05 13	☽ ⊼ ♅
05 46	☽ ⊥ ♀
10 40	☽ ⊥ ♇
10 54	☽ □ ♂
16 10	☽ ± ♅
17 13	☽ ✶ ☿
17 20	☽ ⊥ ♆
17 41	☽ ⊻ ♀
20 41	☽ Q ☉
23 49	☿ ⊻ ♀
24 Thursday	
02 05	☿ ± ♄
03 12	☽ ∠ ♇
04 26	☽ ⊻ ♆
05 36	☽ □ ♄
08 50	☽ ⚼ ♅
11 09	☽ Q ♃
18 44	☽ ∥ ♀
20 49	☽ ✶ ☉
22 03	☽ Q ♂
25 Friday	
02 15	☽ ∠ ☿
07 22	☽ ✶ ♇
09 42	☽ ∥ ♆
10 13	☽ ✶ ♃
13 14	☽ △ ♅
14 35	☿ ∥ ♃
16 47	♀ Q ♇
22 12	☽ ✶ ♂
22 44	☽ ⊥ ☿
26 Saturday	
03 50	☽ ∠ ☉
06 24	☽ Q ♇
07 44	☽ ☌ ♀
09 34	☽ # ♇
12 39	☽ ⊻ ☿
13 42	☽ ☌ ♆
14 29	☽ △ ♄
14 55	☉ ∥ ♂
15 43	☽ ∠ ♃
16 33	☿ ∥ ♂
16 47	☉ ∥ ☿
19 02	☿ ⊻ ♆
22 50	☽ ⊥ ☉
23 18	☿ ⊼ ♄
23 56	☽ # ♄
27 Sunday	
05 02	☽ ∠ ♂
09 42	☽ ⊥ ♃
11 38	☽ ⊻ ☉
17 40	☽ □ ♇
19 48	☽ ⚼ ♄
21 47	☽ ⊻ ♃
23 50	☽ ⊥ ♂
23 58	☽ □ ♅
28 Monday	
12 28	☽ ⊻ ♂
13 37	☽ ± ♄
29 Tuesday	
00 58	☽ ⊻ ♀
01 15	☽ ⊻ ♆
01 34	☽ ⊼ ♄
04 05	♀ ☌ ♆
07 04	♀ △ ♄
09 30	☿ Q ♅
12 05	☽ Q ♅
12 38	☽ ☌ ☿
13 20	☽ ⊥ ♆
14 15	☽ ⊥ ♀
16 37	☿ ⊥ ♆
21 16	☽ # ♄
30 Wednesday	
04 50	☽ ☌ ☉
05 43	☽ △ ♇
05 44	☿ ⊥ ♀
07 32	☽ ∠ ♆
10 17	☽ ∠ ♀
11 08	☽ ☌ ♃
12 22	☽ ✶ ♅
12 27	☽ # ♇
15 46	♄ △ ♆
15 47	☉ △ ♇
21 30	♀ ∠ ♃

JULY 2049

LONGITUDES

Date	Sidereal time h m s	Sun ☉ ° ' "	Moon ☽ ° ' "	Moon ☽ 24.00 ° ' "	Mercury ☿ ° '	Venus ♀ ° '	Mars ♂ ° '	Jupiter ♃ ° '	Saturn ♄ ° '	Uranus ♅ ° '	Neptune ♆ ° '	Pluto ♇ ° '
01	06 39 55	10 ♋ 06 22	24 ♋ 15 21	00 ♌ 10 15	05 ♋ 04	27 ♉ 45	20 ♋ 43	12 ♋ 13	25 ♑ 09	12 ♍ 38	25 ♉ 14	09 ♓ 18
02	06 43 52	11 03 36	06 ♌ 05 12	12 ♌ 00 27	07 14	28 52	21 22	12 27	25 R 05	12 40	25 16	09 R 17
03	06 47 48	12 00 50	17 ♌ 56 18	23 ♌ 53 04	09 25	29 ♉ 59	22 01	12 40	25 01	12 42	25 18	09 17
04	06 51 45	12 58 04	29 ♌ 51 06	05 ♍ 50 44	11 35	01 ♊ 06	22 40	12 54	24 56	12 44	25 19	09 16
05	06 55 41	13 55 17	11 ♍ 52 25	17 ♍ 56 32	13 46	02 13	23 18	13 08	24 52	12 46	25 21	09 15
06	06 59 38	14 52 30	24 ♍ 03 33	00 ♎ 13 58	15 56	03 21	23 57	13 21	24 48	12 49	25 22	09 15
07	07 03 34	15 49 43	06 ♎ 28 15	12 ♎ 46 55	18 05	04 28	24 36	13 35	24 43	12 51	25 24	09 14
08	07 07 31	16 46 56	19 ♎ 10 28	25 ♎ 39 22	20 14	05 36	25 15	13 48	24 39	12 54	25 26	09 14
09	07 11 28	17 44 08	02 ♏ 14 05	08 ♏ 55 00	22 21	06 43	25 54	14 02	24 35	12 56	25 27	09 13
10	07 15 24	18 41 20	15 ♏ 42 27	22 ♏ 36 39	24 28	07 51	26 32	14 16	24 30	12 59	25 29	09 12
11	07 19 21	19 38 33	29 ♏ 37 42	06 ♐ 45 34	26 32	08 59	27 11	14 29	24 26	13 02	25 30	09 11
12	07 23 17	20 35 45	14 ♐ 00 01	21 ♐ 20 37	28 ♋ 36	10 07	27 50	14 43	24 21	13 04	25 32	09 11
13	07 27 14	21 32 58	28 ♐ 46 47	06 ♑ 17 42	00 ♌ 37	11 15	28 29	14 56	24 17	13 07	25 33	09 10
14	07 31 10	22 30 10	13 ♑ 52 20	21 ♑ 29 33	02 38	12 24	29 07	15 10	24 12	13 09	25 35	09 09
15	07 35 07	23 27 23	29 ♑ 08 02	06 ♒ 46 27	04 36	13 32	29 ♋ 46	15 23	24 08	13 12	25 36	09 08
16	07 39 03	24 24 36	14 ♒ 23 25	21 ♒ 57 38	06 33	14 40	00 ♌ 25	15 37	24 04	13 15	25 37	09 08
17	07 43 00	25 21 49	29 ♒ 27 54	06 ♓ 53 09	08 27	15 49	01 03	15 50	23 59	13 18	25 39	09 07
18	07 46 57	26 19 03	14 ♓ 12 33	21 ♓ 25 26	10 20	16 58	01 42	16 04	23 55	13 21	25 40	09 06
19	07 50 53	27 16 18	28 ♓ 31 24	05 ♈ 30 13	12 12	18 07	02 21	16 17	23 50	13 23	25 41	09 05
20	07 54 50	28 13 33	12 ♈ 21 50	19 ♈ 06 25	14 01	19 16	02 59	16 31	23 46	13 26	25 42	09 04
21	07 58 46	29 ♋ 10 49	25 ♈ 44 14	02 ♉ 15 39	15 49	20 25	03 38	16 44	23 42	13 29	25 44	09 03
22	08 02 43	00 ♌ 08 06	08 ♉ 41 09	15 ♉ 01 14	17 34	21 34	04 16	16 57	23 37	13 32	25 45	09 02
23	08 06 39	01 05 24	21 ♉ 16 28	27 ♉ 27 25	19 18	22 43	04 55	17 11	23 33	13 35	25 46	09 01
24	08 10 36	02 02 43	03 ♊ 34 39	09 ♊ 38 43	21 01	23 52	05 33	17 24	23 28	13 38	25 47	09 00
25	08 14 32	03 00 03	15 ♊ 40 09	21 ♊ 39 27	22 41	25 02	06 12	17 37	23 24	13 41	25 48	08 59
26	08 18 29	03 57 23	27 ♊ 37 03	03 ♋ 33 25	24 20	26 11	06 51	17 51	23 20	13 44	25 49	08 58
27	08 22 26	04 54 44	09 ♋ 28 54	15 ♋ 23 51	25 56	27 21	07 29	18 04	23 15	13 47	25 51	08 57
28	08 26 22	05 52 06	21 ♋ 18 34	27 ♋ 13 20	27 31	28 30	08 08	18 17	23 11	13 51	25 52	08 56
29	08 30 19	06 49 29	03 ♌ 08 24	09 ♌ 03 57	29 ♌ 05	29 ♊ 40	08 46	18 30	23 07	13 54	25 53	08 55
30	08 34 15	07 46 52	15 ♌ 00 12	20 ♌ 57 21	00 ♍ 36	00 ♋ 50	09 24	18 44	23 03	13 57	25 54	08 54
31	08 38 12	08 ♌ 44 17	26 ♌ 55 34	02 ♍ 55 04	02 ♍ 06	02 ♋ 00	10 ♌ 03	18 ♋ 57	22 ♑ 58	14 ♍ 00	25 ♉ 54	08 ♓ 53

DECLINATIONS

Date	Moon True ☊ ° '	Moon Mean ☊ ° '	Moon ☽ Latitude ° '	Sun ☉ ° '	Moon ☽ ° '	Mercury ☿ ° '	Venus ♀ ° '	Mars ♂ ° '	Jupiter ♃ ° '	Saturn ♄ ° '	Uranus ♅ ° '	Neptune ♆ ° '	Pluto ♇ ° '
01	09 ♐ 09	07 ♐ 42	03 S 41	23 N 03	17 N 37	24 N 09	17 N 14	22 N 53	22 N 54	21 S 05	07 N 31	17 N 26	19 S 59
02	09 R 03	07 38	04 22	22 58	14 31	24 11	17 30	22 47	22 53	21 05	07 30	17 26	19 59
03	08 57	07 35	04 51	22 53	10 50	24 11	17 45	22 41	22 52	21 06	07 29	17 26	20 00
04	08 52	07 32	05 07	22 48	06 43	24 09	18 00	22 35	22 51	21 07	07 28	17 27	20 00
05	08 47	07 29	05 11	22 42	02 N 19	24 03	18 15	22 29	22 49	21 08	07 27	17 27	20 01
06	08 44	07 26	05 00	22 36	02 S 14	23 55	18 29	22 22	22 48	21 09	07 26	17 27	20 01
07	08 42	07 23	04 36	22 30	06 47	23 44	18 43	22 15	22 47	21 10	07 25	17 28	20 02
08	08 D 41	07 19	03 57	22 23	11 10	23 30	18 57	22 09	22 45	21 11	07 24	17 28	20 02
09	08 42	07 16	03 05	22 15	15 09	23 14	19 10	22 02	22 44	21 12	07 24	17 28	20 03
10	08 43	07 13	02 02	22 08	18 29	22 56	19 23	21 54	22 43	21 12	07 23	17 29	20 03
11	08 44	07 10	00 S 49	22 00	20 52	22 35	19 36	21 47	22 41	21 13	07 22	17 29	20 04
12	08 R 45	07 07	00 N 28	21 51	22 00	22 12	19 48	21 40	22 40	21 14	07 20	17 29	20 04
13	08 43	07 04	01 46	21 42	21 39	21 48	19 59	21 32	22 39	21 15	07 19	17 30	20 05
14	08 40	07 00	02 59	21 33	19 45	21 21	20 10	21 24	22 37	21 16	07 18	17 30	20 06
15	08 35	06 57	03 59	21 24	16 26	20 53	20 21	21 16	22 36	21 17	07 17	17 30	20 06
16	08 28	06 54	04 42	21 14	12 01	20 23	20 31	21 08	22 34	21 18	07 16	17 30	20 07
17	08 22	06 51	05 04	21 04	06 54	19 52	20 41	21 00	22 33	21 18	07 15	17 31	20 07
18	08 15	06 48	05 06	20 53	01 S 30	19 19	20 50	20 51	22 31	21 19	07 14	17 31	20 08
19	08 10	06 44	04 49	20 42	03 N 50	18 46	20 59	20 43	22 30	21 20	07 13	17 31	20 08
20	08 07	06 41	04 15	20 31	08 48	18 11	21 07	20 34	22 28	21 21	07 12	17 32	20 09
21	08 06	06 38	03 28	20 19	13 10	17 36	21 15	20 25	22 27	21 22	07 11	17 32	20 10
22	08 D 06	06 35	02 32	20 07	16 47	17 00	21 22	20 16	22 25	21 22	07 09	17 32	20 10
23	08 07	06 32	01 30	19 55	19 31	16 23	21 29	20 07	22 24	21 23	07 08	17 32	20 11
24	08 08	06 29	00 N 25	19 42	21 16	15 45	21 35	19 58	22 22	21 24	07 07	17 32	20 11
25	08 R 08	06 25	00 S 40	19 29	22 00	15 07	21 41	19 48	22 20	21 25	07 06	17 33	20 12
26	08 06	06 22	01 43	19 16	21 42	14 28	21 46	19 39	22 19	21 26	07 05	17 33	20 12
27	08 02	06 19	02 40	19 02	20 26	13 50	21 51	19 29	22 17	21 27	07 03	17 33	20 13
28	07 55	06 16	03 30	18 48	18 17	13 11	21 54	19 19	22 15	21 28	07 02	17 33	20 14
29	07 46	06 13	04 10	18 34	15 23	12 31	21 58	19 09	22 14	21 28	07 01	17 33	20 14
30	07 35	06 09	04 40	18 19	11 51	11 52	22 01	18 59	22 12	21 29	07 00	17 34	20 15
31	07 ♐ 24	06 ♐ 06	04 S 58	18 N 04	07 N 51	11 N 12	22 N 03	18 N 49	22 N 10	21 S 30	06 N 58	17 N 34	20 S 15

ZODIAC SIGN ENTRIES

Date	h	m	Planets
01	23	39	☽ ♌
03	12	20	♀ ♊
04	12	18	☽ ♍
06	23	33	☽ ♎
09	07	57	☽ ♏
11	12	38	☽ ♐
13	04	35	☿ ♌
13	13	57	☽ ♑
15	13	23	☽ ♒
15	20	40	♂ ♌
17	12	52	☽ ♓
19	14	31	☽ ♈
21	19	49	☽ ♉
22	08	36	☉ ♌
24	04	58	☽ ♊
26	16	49	☽ ♋
29	05	38	☽ ♌
29	18	47	♀ ♋
30	02	29	☿ ♍
31	18	10	☽ ♍

LATITUDES

Date	Mercury ☿ ° '	Venus ♀ ° '	Mars ♂ ° '	Jupiter ♃ ° '	Saturn ♄ ° '	Uranus ♅ ° '	Neptune ♆ ° '	Pluto ♇ ° '
01	00 N 48	02 S 29	01 N 04	00 N 02	00 N 01	00 N 45	01 S 41	12 S 53
04	01 13	02 25	01 04	00 02	00 01	00 45	01 41	12 54
07	01 32	02 21	01 05	00 03	00 N 01	00 45	01 42	12 55
10	01 44	02 15	01 05	00 03	00 00	00 45	01 42	12 56
13	01 50	02 09	01 06	00 03	00 00	00 45	01 42	12 57
16	01 49	02 03	01 06	00 03	00 00	00 45	01 42	12 58
19	01 42	01 55	01 07	00 04	00 00	00 45	01 42	12 58
22	01 30	01 48	01 07	00 04	00 S 01	00 45	01 42	12 59
25	01 14	01 39	01 07	00 04	00 01	00 45	01 42	13 00
28	00 54	01 31	01 08	00 04	00 01	00 44	01 43	13 00
31	00 N 31	01 S 22	01 N 08	00 N 05	00 S 01	00 N 44	01 S 43	13 S 01

DATA

Julian Date	2469624
Delta T	+80 seconds
Ayanamsa	24° 33' 12"
Synetic vernal point	04° ♓ 33' 47"
True obliquity of ecliptic	23° 25' 54"

MOON'S PHASES, APSIDES AND POSITIONS ☽

Date	h	m	Phase	Longitude ° '	Eclipse Indicator
08	07	10	☽	16 ♎ 35	
15	02	30	○	23 ♑ 05	
21	18	49	☾	29 ♈ 27	
29	20	07	●	07 ♌ 09	

Day	h	m		
02	01	13	Apogee	
15	07	28	Perigee	
29	04	44	Apogee	
06	00	15	0S	
12	18	36	Max dec	22° S 04'
18	18	39	0N	
25	17	03	Max dec	22° N 01'

LONGITUDES

Date	Chiron ⚷ ° '	Ceres ⚳ ° '	Pallas ⚴ ° '	Juno ⚵ ° '	Vesta ⚶ ° '	Black Moon Lilith ⚸ ° '
01	21 ♏ 35	13 ♊ 00	15 ♈ 13	18 ♊ 49	10 ♋ 14	27 ♋ 18
11	21 ♏ 21	16 ♊ 58	16 ♈ 50	24 ♊ 38	14 ♋ 36	28 ♋ 25
21	21 ♏ 16	20 ♊ 53	18 ♈ 03	00 ♋ 06	18 ♋ 59	29 ♋ 32
31	21 ♏ 12	24 ♊ 43	18 ♈ 49	06 ♋ 06	23 ♋ 21	00 ♌ 39

ASPECTARIAN

h m	Aspects
01 Thursday	
04 24	☽ ☌ ♂
07 13	♂ ∥ ♃
12 05	☽ ⚼ ♇
13 41	☽ ∥ ♆
13 49	☽ ☍ ♄
14 00	☽ ⚹ ♆
15 03	☽ ∥ ♀
15 22	☉ ∠ ♆
18 52	☽ ∠ ♅
19 50	☽ ⚹ ♀
02 Friday	
06 17	♀ ∥ ♆
06 20	☽ ± ♇
13 10	☽ ⊥ ♅
14 24	☽ Q ♆
14 52	☽ ⊻ ☿
18 29	☽ ⊼ ♇
22 42	☽ Q ♀
22 58	☽ ⊻ ☉
03 Saturday	
01 08	☽ ⊻ ♃
01 22	☽ ⊻ ♅
05 45	☽ ⊥ ☿
10 30	☿ △ ♇
12 10	☽ ⊥ ☉
13 31	☽ ⊥ ♃
15 31	♃ ⚹ ♅
20 42	☽ ⊻ ♂
21 34	☉ ∥ ♃
21 48	☿ ∠ ♆
04 Sunday	
02 11	☽ ⊼ ♄
02 52	☽ □ ♆
03 59	☽ ∠ ☿
06 01	☉ ⚹ ♅
07 44	☽ ∥ ♅
07 53	☽ ∠ ☉
08 00	☽ ∠ ♃
09 28	☽ ⊥ ♂
09 46	☉ ☌ ♃
14 10	☽ ± ♄
14 46	☽ □ ♀
05 Monday	
00 53	☿ ⚹ ♅
04 07	☿ ☌ ♃
04 30	☽ ∠ ♂
06 48	☽ ☍ ♇
08 02	☽ ⚼ ♄
13 48	☽ ☌ ♅
14 32	☽ ⚹ ♃
15 04	☉ ☌ ☿
16 24	☽ ⚹ ☉
16 34	☽ ⚹ ☿
06 Tuesday	
11 47	☽ ⚹ ♂
13 26	☽ △ ♄
14 34	☽ △ ♆
14 34	☽ Q ♃
17 57	☽ Q ☉
21 08	☽ Q ☿
22 41	♂ ⚼ ♇
07 Wednesday	
07 47	☽ △ ♀
12 16	☽ Q ♂
15 24	☽ # ♅
16 04	♂ ☍ ♄
17 16	☽ ⊼ ♇
19 30	☽ ⚼ ♆
08 Thursday	
00 11	☽ ⊻ ♅
01 45	☽ □ ♃
04 37	☽ ± ♇
07 10	☽ □ ☉
11 29	☽ ⊥ ♅
12 28	☽ ± ♆
14 22	☽ □ ☿
14 54	☽ ⚼ ♀
18 54	♂ ⚹ ♆
20 42	☿ ∠ ♀
21 21	☽ ⚼ ♇
22 06	☽ □ ♄
23 36	☽ ⊼ ♆
23 50	☽ □ ♂
09 Friday	
04 10	☽ ∠ ♅
09 00	☽ ± ♀
20 50	☽ ⊼ ♀
10 Saturday	
00 31	☽ △ ♇
04 08	☽ # ♆
06 24	☽ Q ♄
07 12	☽ ⚹ ♅
09 04	☿ ⚼ ♇
09 25	☽ △ ♃
12 30	☿ ☍ ♄
17 36	☽ △ ☉
20 36	☽ # ♀
22 44	♀ ⊥ ♃
23 52	☿ ⚹ ♆
11 Sunday	
02 33	☽ ∥ ♇
03 11	☽ ⚹ ♄
04 08	☽ Q ♅
04 24	☿ ∥ ♃
04 57	☽ ☍ ♆
05 50	☽ △ ☿
07 39	☽ △ ♂
11 45	☽ ⚼ ♃
16 17	♀ □ ♇
17 06	☽ ∥ ♄
20 51	♀ ⚼ ♄
21 04	☽ ⚼ ☉
22 58	☿ ☌ ♂
12 Monday	
02 16	☽ # ♂
03 07	☽ ± ♃
04 02	☽ □ ♇
04 22	☽ ∠ ♄
05 03	☽ ☍ ♀
05 42	☿ ∠ ♅
07 03	☽ # ☉
09 59	☽ ⚼ ♂
10 28	☽ □ ♅
11 13	☽ ⚼ ☿
13 03	☽ ± ☉
13 11	☽ ⊼ ♃
19 06	☽ ⊥ ♄
21 12	☽ # ☿
21 26	♂ ∠ ♅
23 32	☽ ⊼ ☉
13 Tuesday	
01 23	☽ ± ♂
04 16	☽ ± ☿
04 48	☽ ⊻ ♄
04 51	☽ # ☿
06 48	☽ ⊼ ♆
09 25	☽ Q ♇
10 35	☽ # ☉
11 30	☽ ⊼ ♂
14 39	☽ # ♂
15 25	☽ ⊼ ☿
16 27	☽ ± ♆
19 02	☽ ∥ ♄
19 18	☉ ∥ ☿
14 Wednesday	
00 50	♀ # ♇
04 32	☽ ⚹ ♇
06 47	☽ ⚼ ♆
08 12	☿ ∥ ♂
08 15	☽ # ♀
08 47	☽ ∥ ♇
09 29	☽ ⊼ ♀
10 52	☽ △ ♅
14 04	☽ ☍ ♃
16 23	☿ # ♄
18 20	☿ ± ♇
19 42	☽ ± ♀
15 Thursday	
02 30	☽ ☍ ☉
04 10	☽ ∠ ♇
04 11	☽ ☌ ♄
04 46	♀ □ ♅
05 11	☽ # ♆
06 26	☽ △ ♆
10 28	♂ # ♄
10 31	♄ ∠ ♇
10 32	☽ ⚼ ♅
10 59	☽ ⚼ ♀
13 02	☽ ☍ ♂
18 17	☽ ⊥ ♇
21 50	☽ ☍ ☿
16 Friday	
00 43	☽ ± ♅
03 42	☽ ⊻ ♇
03 49	☉ # ♄
03 51	☉ ☍ ♄
04 56	☉ ⚼ ♇
07 01	☿ ∥ ♀
10 12	☽ ⊼ ♅
12 29	☽ △ ♀
13 58	☽ ⊼ ♃
21 02	☿ ⊥ ♅
23 37	☽ ± ♃
17 Saturday	
00 27	☿ # ♇
01 39	☿ Q ♆
03 16	☽ ⊻ ♄
04 59	☽ ⊼ ☉
05 52	☽ □ ♆
10 24	☽ # ♅
12 29	♀ ⊻ ♃
12 50	☽ ⊥ ♄
14 14	☽ ⚼ ♃
14 41	☽ ⊼ ♂
15 16	☽ ± ☉
19 13	☉ ⚹ ♆
20 15	☿ ⊼ ♇
23 24	♀ ∠ ♂
18 Sunday	
00 50	☽ ± ♂
03 21	☽ ∠ ♄
03 37	☽ ☌ ♇
04 42	☽ ⊼ ☿
06 55	☽ ⚼ ☉
10 34	☽ ☍ ♅
11 06	☽ Q ♆
13 11	♀ ∥ ♂
15 04	☉ ∥ ♀
15 07	☽ △ ♃
16 03	☽ ± ☿
16 19	☽ ⚼ ♂
16 57	☽ □ ♀
19 Monday	
04 06	☽ ⚹ ♄
04 55	☉ ∥ ♂
06 41	♀ ± ♄
07 10	☽ ⚹ ♆
09 24	☽ ⚼ ☿
09 43	☽ △ ☉
18 51	☽ △ ♂
20 Tuesday	
00 31	☽ Q ♄
02 14	☽ Q ♀
04 01	☽ ∥ ♅
04 08	☿ ⊻ ♅
06 13	☽ ⊻ ♇
09 05	☽ ∠ ♆
13 54	☽ ⊼ ♅
14 58	♂ ± ♇
15 22	☽ △ ☿
16 47	☽ ⊥ ♇
17 38	☉ ∠ ♅
19 29	☽ □ ♃
21 Wednesday	
00 38	☽ ± ♅
01 06	☽ ⊥ ♆
01 26	☽ ⚹ ♀
08 18	☽ □ ♄
08 56	☽ ∠ ♇
11 59	☽ ⊻ ♆
14 41	☿ ∥ ♆
17 03	☽ ⚼ ♅
18 49	☽ □ ☉
22 Thursday	
02 19	☿ ⊻ ♃
03 18	☽ □ ♂
04 53	☽ Q ♃
05 56	☉ # ♇
07 37	☽ ∠ ♀
12 40	☽ ⚹ ♇
12 50	♀ # ♄
13 18	☽ ∥ ☿
17 49	☽ ∥ ♆
21 13	☽ △ ♅
23 Friday	
02 21	☽ ⊥ ♀
02 34	♂ # ♇
03 59	☽ ⚹ ♃
07 27	☽ Q ☉
07 36	☽ □ ☿
11 31	☽ Q ♇
15 04	☽ ⊻ ♀
15 21	☽ Q ♂
16 00	☽ ∥ ☉
16 22	☽ △ ♄
18 21	☽ ∥ ♂
19 41	☽ # ♇
20 43	☽ ☌ ♆
24 Saturday	
04 13	♀ ⊼ ♄
08 44	☽ ⚹ ☉
09 39	☽ ∠ ♃
14 55	☽ # ♄
16 07	☽ ⚹ ♂
20 00	☽ ∥ ♀
21 37	☽ ⚼ ♄
22 43	☽ □ ♇
25 Sunday	
00 28	☽ Q ☿
03 47	☽ ⊥ ♃
08 02	☽ □ ♅
11 44	☉ ± ♇
15 26	☽ ± ♄
15 59	☽ ⊻ ♃
17 04	☽ ∠ ☉
21 59	☿ ⊼ ♄
23 43	☽ ∠ ♂
26 Monday	
03 25	☽ ⊼ ♄
03 50	☿ ⊥ ♃
04 19	☽ ⚹ ☿
04 24	♀ ⊻ ♆
08 23	☽ ⊻ ♆
08 48	☽ ☌ ♀
10 06	☽ ∥ ♀
12 45	☽ ⊥ ☉
18 50	☽ # ♄
18 53	☽ ⊥ ♂
20 22	☽ Q ♅
20 31	☽ ⊥ ♆
27 Tuesday	
01 56	☽ ⊻ ☉
07 43	☽ ⊻ ♂
10 31	☿ □ ♆
10 56	☽ △ ♇
14 46	☽ ∠ ♆
14 56	☽ # ♇
15 25	☽ ∠ ☿
20 47	☽ ⚹ ♅
28 Wednesday	
00 32	♂ ⊥ ♅
00 49	☽ ∥ ♂
01 47	♂ Q ♆
05 45	☽ ☌ ♃
06 32	☽ ∥ ☉
12 30	☽ ⊥ ☿
15 47	☽ ☍ ♄
17 20	☽ ⚼ ♇
18 40	☽ ∥ ♆
21 15	☽ ⚹ ♆
29 Thursday	
02 32	☽ ⊻ ☿
03 21	☽ ∠ ♅
04 12	☽ ⊻ ♀
11 34	☽ ± ♇
12 32	☿ ± ♄
17 41	☽ ⊥ ♀
17 42	♂ ⊼ ♇
20 07	☽ ☌ ☉
21 36	☽ Q ♆
21 41	☽ ⊥ ♅
23 42	☽ ⊼ ♇
30 Friday	
00 03	☽ ☌ ♂
09 52	☽ ⊻ ♅
11 57	☽ ∥ ☿
13 52	☽ ∠ ♀
14 50	☉ Q ♆
16 30	☉ ⊥ ♅
19 39	☽ ⊻ ♃
31 Saturday	
04 06	☽ ⊼ ♄
04 55	☿ ⚹ ♀
07 57	☽ ⊥ ♃
09 57	☽ □ ♆
10 03	♀ ⊥ ♆
12 04	♀ Q ♅
15 41	☉ ⊼ ♇
16 05	☽ ± ♄
17 03	☽ ∥ ♅
23 16	☽ ⚹ ♀
23 48	☽ ☌ ☿

All ephemeris data is given at 12.00 UT and the Moon's longitude is additionally given for 24.00 UT

AUGUST 2049

LONGITUDES

Date	Sidereal time h m s	Sun ☉ ° ' "	Moon ☽ ° ' "	Moon ☽ 24.00 ° ' "	Mercury ☿ ° '	Venus ♀ ° '	Mars ♂ ° '	Jupiter ♃ ° '	Saturn ♄ ° '	Uranus ♅ ° '	Neptune ♆ ° '	Pluto ♇ ° '
01	08 42 08	09 ♌ 41 42	08 ♍ 56 02	14 ♍ 58 42	03 ♍ 33	03 ♋ 10	10 ♌ 41	19 ♋ 10	22 ♑ 54	14 ♍ 03	25 ♉ 55	08 ♓ 52
02	08 46 05	10 39 07	21 ♍ 03 18	27 ♍ 10 07	04 59	04 20	11 20	19 23	22 R 50	14 07	25 56	08 R 51
03	08 50 01	11 36 34	03 ♎ 19 27	09 ♎ 31 39	06 24	05 30	11 58	19 36	22 46	14 10	25 57	08 50
04	08 53 58	12 34 01	15 ♎ 47 04	22 ♎ 06 08	07 46	06 41	12 37	19 49	22 42	14 13	25 58	08 49
05	08 57 55	13 31 28	28 ♎ 29 14	04 ♏ 56 50	09 06	07 51	13 15	20 02	22 38	14 17	25 59	08 48
06	09 01 51	14 28 57	11 ♏ 29 22	18 ♏ 07 14	10 24	09 02	13 53	20 15	22 34	14 20	25 59	08 47
07	09 05 48	15 26 26	24 ♏ 50 49	01 ♐ 40 29	11 41	10 12	14 32	20 28	22 30	14 24	26 00	08 46
08	09 09 44	16 23 56	08 ♐ 36 27	15 ♐ 38 51	12 55	11 23	15 10	20 41	22 26	14 27	26 01	08 44
09	09 13 41	17 21 27	22 ♐ 47 42	00 ♑ 02 47	14 07	12 33	15 48	20 54	22 22	14 30	26 02	08 43
10	09 17 37	18 18 59	07 ♑ 23 46	14 ♑ 50 03	15 17	13 44	16 27	21 06	22 18	14 34	26 02	08 42
11	09 21 34	19 16 31	22 ♑ 20 52	29 ♑ 55 12	16 24	14 55	17 05	21 19	22 15	14 37	26 03	08 41
12	09 25 30	20 14 05	07 ♒ 31 54	15 ♒ 09 40	17 29	16 06	17 43	21 32	22 11	14 41	26 03	08 40
13	09 29 27	21 11 39	22 ♒ 47 06	00 ♓ 22 49	18 32	17 17	18 22	21 44	22 08	14 44	26 04	08 39
14	09 33 24	22 09 15	07 ♓ 55 28	15 ♓ 23 48	19 32	18 28	19 00	21 57	22 04	14 48	26 04	08 37
15	09 37 20	23 06 51	22 ♓ 46 45	00 ♈ 03 27	20 29	19 39	19 38	22 10	22 00	14 52	26 05	08 36
16	09 41 17	24 04 30	07 ♈ 13 15	14 ♈ 15 44	21 23	20 50	20 17	22 22	21 57	14 55	26 05	08 35
17	09 45 13	25 02 09	21 ♈ 10 42	27 ♈ 58 10	22 15	22 02	20 55	22 34	21 54	14 59	26 06	08 34
18	09 49 10	25 59 51	04 ♉ 38 18	11 ♉ 11 27	23 03	23 13	21 33	22 47	21 50	15 02	26 06	08 32
19	09 53 06	26 57 34	17 ♉ 38 04	23 ♉ 58 40	23 48	24 25	22 11	22 59	21 47	15 06	26 06	08 31
20	09 57 03	27 55 18	00 ♊ 13 51	06 ♊ 24 16	24 29	25 36	22 50	23 11	21 44	15 10	26 07	08 30
21	10 00 59	28 53 05	12 ♊ 30 34	18 ♊ 33 25	25 07	26 48	23 28	23 24	21 41	15 13	26 07	08 29
22	10 04 56	29 ♌ 50 53	24 ♊ 33 26	00 ♋ 31 15	25 40	27 59	24 06	23 36	21 38	15 17	26 07	08 27
23	10 08 53	00 ♍ 48 42	06 ♋ 27 27	12 ♋ 22 35	26 10	29 ♋ 11	24 44	23 48	21 35	15 21	26 07	08 26
24	10 12 49	01 46 33	18 ♋ 17 09	24 ♋ 11 36	26 34	00 ♌ 23	25 22	24 00	21 32	15 24	26 08	08 25
25	10 16 46	02 44 26	00 ♌ 06 21	06 ♌ 01 44	26 55	01 35	26 01	24 12	21 29	15 28	26 08	08 24
26	10 20 42	03 42 21	11 ♌ 58 05	17 ♌ 55 38	27 10	02 47	26 39	24 24	21 27	15 32	26 08	08 22
27	10 24 39	04 40 17	23 ♌ 54 36	29 ♌ 55 09	27 20	03 59	27 17	24 36	21 24	15 35	26 08	08 21
28	10 28 35	05 38 14	05 ♍ 57 28	12 ♍ 01 38	27 25	05 11	27 55	24 47	21 21	15 39	26 08	08 20
29	10 32 32	06 36 13	18 ♍ 07 47	24 ♍ 16 00	27 R 23	06 23	28 33	24 59	21 19	15 43	26 R 08	08 19
30	10 36 28	07 34 14	00 ♎ 26 23	06 ♎ 39 04	27 16	07 36	29 12	25 11	21 17	15 47	26 08	08 17
31	10 40 25	08 ♍ 32 16	12 ♎ 54 11	19 ♎ 11 53	27 ♍ 03	08 ♌ 48	29 ♌ 50	25 ♋ 22	21 ♑ 14	15 ♍ 50	26 ♉ 08	08 ♓ 16

DECLINATIONS

Date	Moon True ☊ ° '	Moon Mean ☊ ° '	Moon ☽ Latitude ° '	Sun ☉ ° '	Moon ☽ ° '	Mercury ☿ ° '	Venus ♀ ° '	Mars ♂ ° '	Jupiter ♃ ° '	Saturn ♄ ° '	Uranus ♅ ° '	Neptune ♆ ° '	Pluto ♇ ° '
01	07 ♐ 12	06 ♐ 03	05 S 03	17 N 49	03 N 32	10 N 33	22 N 05	18 N 38	22 N 09	21 S 31	06 N 57	17 N 34	20 S 16
02	07 R 02	06 00	04 54	17 34	00 S 58	09 54	22 06	18 28	22 07	21 32	06 56	17 34	20 17
03	06 54	05 57	04 32	17 18	05 29	09 14	22 06	18 17	22 05	21 32	06 55	17 34	20 17
04	06 49	05 54	03 57	17 02	09 51	08 36	22 06	18 06	22 03	21 33	06 53	17 34	20 18
05	06 46	05 50	03 09	16 45	13 52	07 57	22 05	17 56	22 01	21 34	06 52	17 35	20 18
06	06 45	05 47	02 10	16 29	17 20	07 18	22 04	17 45	22 00	21 35	06 51	17 35	20 19
07	06 D 45	05 44	01 S 03	16 12	19 59	06 41	22 02	17 34	21 58	21 35	06 49	17 35	20 20
08	06 R 45	05 41	00 N 10	15 55	21 34	06 03	21 59	17 22	21 56	21 36	06 48	17 35	20 20
09	06 44	05 38	01 24	15 38	21 50	05 26	21 56	17 11	21 54	21 37	06 47	17 35	20 21
10	06 41	05 35	02 35	15 20	20 39	04 50	21 53	17 00	21 52	21 37	06 45	17 35	20 21
11	06 36	05 31	03 37	15 02	18 01	04 14	21 48	16 48	21 50	21 38	06 44	17 35	20 22
12	06 28	05 28	04 24	14 44	14 07	03 39	21 43	16 36	21 49	21 39	06 43	17 35	20 23
13	06 18	05 25	04 53	14 26	09 18	03 05	21 38	16 25	21 47	21 39	06 41	17 35	20 23
14	06 08	05 22	05 01	14 07	03 S 56	02 32	21 32	16 13	21 45	21 40	06 40	17 35	20 24
15	05 58	05 19	04 49	13 49	01 N 33	01 59	21 25	16 01	21 43	21 41	06 38	17 35	20 24
16	05 50	05 15	04 18	13 30	06 49	01 28	21 17	15 49	21 41	21 41	06 37	17 35	20 25
17	05 44	05 12	03 32	13 10	11 32	00 58	21 09	15 37	21 39	21 42	06 36	17 35	20 26
18	05 40	05 09	02 36	12 51	15 31	00 29	21 01	15 24	21 37	21 43	06 34	17 36	20 26
19	05 39	05 06	01 34	12 31	18 35	00 N 02	20 52	15 12	21 35	21 43	06 33	17 36	20 27
20	05 D 38	05 03	00 N 29	12 12	20 40	00 S 24	20 42	14 59	21 33	21 44	06 31	17 36	20 27
21	05 R 38	05 00	00 S 36	11 52	21 41	00 48	20 31	14 47	21 31	21 44	06 30	17 36	20 28
22	05 37	04 56	01 38	11 31	21 41	01 11	20 21	14 34	21 29	21 45	06 28	17 36	20 28
23	05 34	04 53	02 35	11 11	20 41	01 31	20 09	14 22	21 27	21 45	06 27	17 36	20 29
24	05 29	04 50	03 25	10 50	18 48	01 50	19 57	14 09	21 25	21 46	06 26	17 36	20 30
25	05 20	04 47	04 06	10 30	16 07	02 06	19 44	13 56	21 23	21 46	06 24	17 36	20 30
26	05 09	04 44	04 36	10 09	12 47	02 20	19 31	13 43	21 21	21 47	06 23	17 36	20 31
27	04 56	04 41	04 54	09 48	08 55	02 32	19 17	13 30	21 19	21 47	06 21	17 35	20 31
28	04 42	04 37	05 00	09 26	04 41	02 40	19 03	13 17	21 17	21 48	06 20	17 35	20 32
29	04 28	04 34	04 51	09 05	00 N 13	02 46	18 48	13 03	21 15	21 48	06 18	17 35	20 32
30	04 16	04 31	04 29	08 44	04 S 18	02 48	18 33	12 50	21 13	21 49	06 17	17 35	20 33
31	04 ♐ 06	04 ♐ 28	03 S 55	08 N 22	08 S 42	02 S 47	18 N 17	12 N 37	21 N 11	21 S 49	06 N 15	17 N 35	20 S 33

ZODIAC SIGN ENTRIES

Date	h	m	Planets
03	05	32	☽ ♎
05	14	49	☽ ♏
07	21	04	☽ ♐
09	23	55	☽ ♑
12	00	08	☽ ♒
13	23	24	☽ ♓
15	23	54	☽ ♈
18	03	38	☽ ♉
20	11	33	☽ ♊
22	15	47	☉ ♍
22	22	57	☽ ♋
24	04	18	♀ ♌
25	11	47	☽ ♌
28	00	10	☽ ♍
30	11	09	☽ ♎
31	18	23	♂ ♍

LATITUDES

Date	Mercury ☿ ° '	Venus ♀ ° '	Mars ♂ ° '	Jupiter ♃ ° '	Saturn ♄ ° '	Uranus ♅ ° '	Neptune ♆ ° '	Pluto ♇ ° '
01	00 N 23	01 S 19	01 N 08	00 N 05	00 S 01	00 N 44	01 S 43	13 S 01
04	00 S 04	01 10	01 08	00 05	00 02	00 44	01 43	13 02
07	00 33	01 01	01 08	00 05	00 02	00 44	01 43	13 02
10	01 03	00 51	01 09	00 06	00 02	00 44	01 43	13 03
13	01 35	00 42	01 09	00 06	00 02	00 44	01 43	13 03
16	02 07	00 32	01 09	00 06	00 03	00 44	01 43	13 04
19	02 39	00 23	01 09	00 07	00 03	00 44	01 44	13 04
22	03 10	00 13	01 09	00 07	00 03	00 44	01 44	13 04
25	03 38	00 S 04	01 09	00 07	00 03	00 44	01 44	13 05
28	04 02	00 N 05	01 09	00 08	00 04	00 44	01 44	13 05
31	04 S 19	00 N 14	01 N 09	00 N 08	00 S 04	00 N 44	01 S 44	13 S 05

DATA

Julian Date	2469655
Delta T	+80 seconds
Ayanamsa	24° 33' 16"
Synetic vernal point	04° ♓ 33' 42"
True obliquity of ecliptic	23° 25' 55"

MOON'S PHASES, APSIDES AND POSITIONS ☽

Date	h	m	Phase	Longitude ° '	Eclipse Indicator
06	17	52	☽	14 ♏ 43	
13	09	19	○	21 ♒ 05	
20	07	11	☾	27 ♉ 44	
28	11	19	●	05 ♍ 37	

Day	h	m		
12	17	10	Perigee	
25	10	41	Apogee	
02	06	53	0S	
09	04	34	Max dec	21° S 54'
15	05	09	0N	
21	23	48	Max dec	21° N 49'
29	13	10	0S	

LONGITUDES

Date	Chiron ⚷ ° '	Ceres ⚳ ° '	Pallas ⚴ ° '	Juno ⚵ ° '	Vesta ⚶ ° '	Black Moon Lilith ⚸ ° '
01	21 ♏ 22	25 ♊ 06	18 ♈ 51	06 ♋ 40	23 ♋ 47	00 ♌ 46
11	21 ♏ 39	28 ♊ 49	19 ♈ 01	12 ♋ 15	28 ♋ 08	01 ♌ 53
21	22 ♏ 06	02 ♋ 25	18 ♈ 35	17 ♋ 43	02 ♌ 27	03 ♌ 00
31	22 ♏ 42	05 ♋ 51	17 ♈ 29	23 ♋ 03	06 ♌ 44	04 ♌ 07

All ephemeris data is given at 12.00 UT and the Moon's longitude is additionally given for 24.00 UT

ASPECTARIAN

h m Aspects

01 Sunday
02 19 ☽ ∠ ♃
09 57 ☽ ⚼ ♄
11 52 ☽ ☍ ♇
13 39 ☽ ⊻ ☉
15 41 ☽ ⊻ ♂
22 13 ☽ ☌ ♅
23 56 ☿ ∠ ♃

02 Monday
01 42 ☽ Q ♀
02 34 ☽ ⊥ ☉
04 15 ☽ ⊥ ♂
08 39 ☽ ⚹ ♃
11 11 ☉ ∥ ♆
15 29 ☽ △ ♄
21 36 ☽ △ ♆
21 48 ☽ ∠ ☉
22 56 ☽ ∠ ♂

03 Tuesday
00 08 ♀ ∥ ♃
08 35 ☽ Q ♃
16 41 ☽ □ ♀
18 42 ☽ ⊻ ☿
19 41 ☽ # ♅
22 39 ☽ ⊼ ♇
23 27 ☉ ⊥ ♀

04 Wednesday
02 45 ☽ ⚼ ♆
05 20 ☽ ⚹ ☉
05 36 ☽ ⚹ ♂
05 53 ☽ # ☿
07 40 ☽ ⊥ ☿
08 54 ♀ ⊥ ♂
09 00 ☽ ⊻ ♅
10 09 ☽ ± ♇
10 55 ☿ ⚼ ♄
15 16 ☉ ☌ ♂
19 49 ☽ □ ♃
19 58 ☽ ± ♆
20 29 ☽ ⊥ ♅

05 Thursday
01 03 ☽ □ ♄
02 49 ☽ ∠ ☿
03 13 ☽ ⚼ ♇
05 37 ☽ Q ♂
06 00 ☽ Q ☉
06 35 ☿ ☍ ♇
07 18 ☽ ⊼ ♆
13 29 ☽ ∠ ♅

06 Friday
06 07 ☽ # ☉
07 01 ♀ △ ♇
07 03 ☽ △ ♇
07 03 ☽ △ ♀
08 06 ☉ ⊻ ♅
09 48 ☽ ⚹ ☿
10 19 ☽ Q ♄
13 53 ☽ # ♆
15 01 ☽ # ♂
16 35 ☽ □ ♂
17 12 ☽ ⚹ ♅
17 52 ☽ □ ☉

07 Saturday
04 05 ☽ △ ♃
06 13 ☿ ∥ ♅
06 23 ♂ ⊻ ♅
07 51 ☽ ⚹ ♄
09 25 ♂ ∥ ♆
09 43 ☽ Q ☿
12 41 ☽ ⚼ ♀
14 03 ☽ ☍ ♆
14 45 ☽ Q ♅
15 55 ☽ ∥ ♇

08 Sunday
04 32 ♀ ∠ ♆
05 55 ☽ ± ♀
06 53 ☽ ⚼ ♃
10 00 ☽ ∠ ♄
12 14 ☽ □ ♇
12 47 ☽ ∥ ♄
17 11 ☽ ⊼ ♀
20 03 ☽ □ ☿
22 01 ☽ □ ♅
22 31 ☽ ± ♃
23 43 ☽ △ ♂

09 Monday
01 16 ☽ ⊥ ♄
02 14 ☽ △ ☉
08 47 ☽ ⊼ ♃
11 18 ☽ ⊻ ♄
17 22 ☽ ⊼ ♆
18 30 ☽ Q ♇
19 43 ☽ ∥ ♄
20 29 ☿ ☌ ♅

10 Tuesday
01 51 ☽ ⚼ ♂
03 16 ☽ ± ♆
04 54 ☽ ⚼ ☉
14 07 ☽ ⚹ ♇
14 55 ♀ ∥ ♃
15 24 ☽ ∥ ♇
17 09 ☽ ± ♂
17 54 ☽ ⚼ ♆
20 30 ☽ ± ☉
23 07 ☽ ☍ ♀
23 37 ☽ △ ♅

11 Wednesday
01 45 ☽ △ ☿
03 14 ☽ ⊼ ♂
05 44 ♀ ⚹ ♅
06 46 ☽ ⊼ ☉
10 20 ☽ ☍ ♃
11 50 ☽ ☌ ♄
14 07 ☽ ∠ ♇
15 01 ☽ # ♆
17 53 ☽ △ ♆
20 43 ☽ # ♂
23 35 ☽ ⚼ ♅

12 Thursday
03 27 ☽ ⚼ ☿
04 20 ☽ ⊥ ♇
08 21 ☽ # ☉
13 47 ☽ ⊻ ♇
13 49 ☽ ± ♅
18 41 ☽ ± ☿
23 17 ☽ ⊼ ♅

13 Friday
01 48 ☿ ⊻ ♂
02 36 ☽ ⊼ ♀
04 44 ☽ ☍ ♂
04 49 ☽ ⊼ ☿
05 39 ♀ # ♄
09 19 ☽ ☍ ☉
10 20 ☽ ⊼ ♃
10 58 ☽ ⊻ ♄
12 51 ☽ ± ♀
17 11 ☽ □ ♆
19 56 ☽ ± ♃
20 24 ☽ ⊥ ♄
23 54 ☽ # ♅

14 Saturday
04 18 ☽ ⚼ ♀
05 28 ☉ ⊻ ♃
09 56 ☉ ⊼ ♄
10 25 ☽ ⚼ ♃
10 38 ☽ ∠ ♄
13 07 ☽ ☌ ♇
18 48 ☽ # ☿
21 52 ☽ Q ♆
22 25 ♃ ☍ ♄
23 05 ☽ ☍ ♅

15 Sunday
06 27 ☽ △ ♀
06 39 ☽ ⊼ ♂
08 00 ☽ ☍ ☿
10 45 ☽ ⚹ ♄
10 58 ☽ △ ♃
11 21 ♀ ⊻ ♂
12 35 ☽ ⊼ ☉
13 45 ☽ ∥ ☿
16 54 ☽ ± ♂
17 25 ☽ ⚹ ♆
23 11 ☽ ± ☉

16 Monday
06 31 ☽ Q ♄
08 16 ♃ # ♄
08 34 ☽ ⚼ ♂
11 05 ☽ ∥ ♅
14 18 ☽ ⊻ ♇
15 22 ☽ ⚼ ☉
18 34 ☽ ∠ ♆

17 Tuesday
00 32 ☽ ⊥ ♇
01 11 ☽ ⊼ ♅
02 37 ☿ △ ♄
09 26 ♀ ☍ ♄
10 06 ☽ ⊥ ♆
11 31 ☽ △ ♂
11 39 ☽ ± ♅
13 15 ☽ □ ♄
13 38 ☽ □ ♀
13 59 ☽ ⊼ ☿
14 29 ☽ □ ♃
16 11 ☽ ∠ ♇
19 18 ☽ △ ☉
20 35 ☽ ∥ ☉
20 40 ☽ ⊻ ♆

18 Wednesday
00 57 ☿ ⚹ ♃
01 18 ☽ ± ☿
01 20 ♀ ☌ ♃
01 55 ☿ ⚹ ♀
03 40 ☽ ⚼ ♅
11 18 ☽ ∥ ♂
14 37 ☉ □ ♆
18 23 ♀ ⚼ ♇
18 36 ☽ ⚼ ☿
19 07 ☽ ⚹ ♇
22 03 ♂ ⊼ ♄
23 25 ☽ Q ♃

19 Thursday
01 16 ☽ Q ♀
03 21 ☽ ∥ ♆
07 14 ☽ △ ♅
17 26 ☽ Q ♇
19 48 ☽ △ ♄
21 03 ☽ □ ♂
22 17 ☽ ⚹ ♃

20 Friday
00 21 ☽ △ ☿
02 10 ☽ ⚹ ♀
04 04 ☽ ☌ ♆
07 11 ☽ □ ☉
07 33 ☉ ± ♄
08 57 ☽ # ♇
12 32 ☽ ∥ ♀
22 19 ♀ ⚹ ♆

21 Saturday
00 35 ☽ ⚼ ♄
03 46 ☽ ∠ ♃
04 05 ☽ □ ♇
05 58 ☽ ∥ ♃
08 10 ♂ ⊻ ♃
09 49 ☽ Q ♂
10 26 ☽ ∠ ♀
14 43 ☽ # ♄
17 24 ☽ □ ♅
18 15 ☽ ± ♄
19 43 ♀ # ♇
21 06 ♃ ⚼ ♇
21 25 ☽ Q ☉
21 51 ☽ ⊥ ♃

22 Sunday
04 03 ☉ ⊥ ♃
06 10 ☽ ⊼ ♄
06 17 ☽ ⊥ ♀
08 29 ☽ # ♄
10 02 ☽ ⊻ ♃
11 02 ☽ ⚹ ♂
14 20 ☽ □ ☿
15 08 ☽ ⊻ ♆
19 25 ☽ ∥ ♃
19 40 ☽ ⊻ ♀
23 35 ☽ ⚹ ☉

23 Monday
03 14 ☽ ⊥ ♆
05 40 ☽ Q ♅
10 10 ☿ △ ♆
15 15 ☽ # ♇
16 00 ☽ △ ♇
19 01 ☽ ∠ ♂
21 04 ☽ ∥ ♀
21 28 ☽ ∠ ♆

24 Tuesday
04 12 ☽ Q ☿
06 07 ☽ ⚹ ♅
08 40 ☽ ∠ ☉
12 26 ♀ ∠ ♅
14 20 ☽ ⊥ ♂
18 34 ☽ ☍ ♄
22 24 ☽ ⚼ ♇
23 36 ☽ ∥ ♆
23 48 ☽ ☌ ♃

25 Wednesday
03 13 ☽ ⊻ ♂
03 56 ☽ ⚹ ♆
04 34 ☽ ⊥ ☉
05 21 ☽ ⚹ ☿
12 44 ☽ ∠ ♅
15 20 ☽ ☌ ♀
16 29 ♂ □ ♆
16 38 ☽ ± ♇
17 49 ☽ ⊻ ☉

26 Thursday
03 58 ♀ ± ♇
04 15 ☽ Q ♆
04 45 ☽ ⊼ ♇
05 15 ☽ ∥ ♂
07 03 ☽ ⊥ ♅
12 24 ☽ ∠ ☿
19 13 ☽ ⊻ ♅

27 Friday
06 15 ☽ ∥ ☉
06 47 ☽ ⊥ ☿
06 59 ☽ ⊼ ♄
13 23 ☽ ⊻ ♃
14 13 ☿ ⊻ ♂
16 02 ♂ ± ♄
16 27 ☽ □ ♆
18 54 ☽ ⊻ ☿
18 57 ☽ ± ♄
19 07 ☽ ☌ ♂

28 Saturday
01 34 ☽ ⊥ ♃
02 47 ☽ ∥ ♅
10 18 ☽ ⊻ ♀
11 19 ☽ ☌ ☉
12 47 ☽ ⚼ ♄
16 41 ☽ ☍ ♇
19 21 ☿ St R
19 42 ☽ ∠ ♃
22 40 ☽ # ☿
23 28 ☽ ⊥ ♀

29 Sunday
03 10 ♆ St R
05 08 ☉ ⚼ ♄
07 14 ☽ ☌ ♅
18 13 ☽ △ ♄
19 04 ☽ ∠ ♀

30 Monday
01 37 ☽ ⚹ ♃
03 38 ☽ △ ♆
04 00 ☽ ∥ ☿
05 56 ☽ ☌ ☿
09 27 ☽ ⊻ ♂
09 47 ☉ ⊻ ♀
21 41 ☽ ⊥ ♂
22 39 ☽ # ♅
22 45 ♀ Q ♆

31 Tuesday
01 13 ☽ Q ♃
01 38 ♀ ⊼ ♇
02 56 ☽ ⊻ ☉
03 07 ☽ ⊼ ♇
03 17 ☽ ⚹ ♀
05 26 ☉ ☍ ♇
08 37 ☽ ⚼ ♆
10 19 ☽ # ☉
14 36 ☽ ± ♇
15 23 ☽ ⊥ ☉
15 53 ☽ ∠ ♂
17 38 ☽ ⊻ ♅

SEPTEMBER 2049

Date	Sidereal time h m s	Sun ☉ ° ' "	Moon ☽ ° ' "	Moon ☽ 24.00 ° ' "	Mercury ☿ ° '	Venus ♀ ° '	Mars ♂ ° '	Jupiter ♃ ° '	Saturn ♄ ° '	Uranus ♅ ° '	Neptune ♆ ° '	Pluto ♇ ° '
01	10 44 22	09 ♍ 30 19	25 ♎ 32 22	01 ♏ 55 51	26 ♍ 44	10 ♌ 00	00 ♍ 28	25 ♋ 34	21 ♑ 12	15 ♍ 54	26 ♉ 08	08 ♓ 15
02	10 48 18	10 28 24	08 ♏ 22 34	14 ♏ 52 50	26 R 18	11 13	01 06	25 45	21 R 10	15 58	26 R 08	08 R 14
03	10 52 15	11 26 31	21 ♏ 26 56	28 ♏ 05 10	25 47	12 25	01 44	25 56	21 08	16 02	26 08	08 12
04	10 56 11	12 24 39	04 ♐ 47 51	11 ♐ 35 16	25 09	13 38	02 23	26 08	21 06	16 05	26 07	08 11
05	11 00 08	13 22 48	18 ♐ 27 40	25 ♐ 25 12	24 26	14 50	03 01	26 19	21 04	16 09	26 07	08 10
06	11 04 04	14 20 59	02 ♑ 27 58	09 ♑ 35 55	23 38	16 03	03 39	26 30	21 02	16 13	26 07	08 08
07	11 08 01	15 19 11	16 ♑ 48 53	24 ♑ 06 30	22 45	17 16	04 17	26 41	21 01	16 17	26 07	08 07
08	11 11 57	16 17 25	01 ♒ 28 16	08 ♒ 53 29	21 49	18 29	04 55	26 52	20 59	16 21	26 06	08 06
09	11 15 54	17 15 40	16 ♒ 21 16	23 ♒ 50 37	20 50	19 41	05 33	27 02	20 57	16 24	26 06	08 05
10	11 19 50	18 13 56	01 ♓ 20 24	08 ♓ 49 26	19 50	20 54	06 11	27 13	20 56	16 28	26 05	08 03
11	11 23 47	19 12 15	16 ♓ 16 30	23 ♓ 40 26	18 50	22 07	06 49	27 24	20 55	16 32	26 05	08 02
12	11 27 44	20 10 35	01 ♈ 00 12	08 ♈ 14 50	17 50	23 20	07 28	27 34	20 53	16 36	26 05	08 01
13	11 31 40	21 08 57	15 ♈ 23 37	22 ♈ 25 57	16 54	24 34	08 06	27 45	20 52	16 39	26 04	08 00
14	11 35 37	22 07 20	29 ♈ 21 29	06 ♉ 10 03	16 02	25 47	08 44	27 55	20 51	16 43	26 04	07 58
15	11 39 33	23 05 46	12 ♉ 51 39	19 ♉ 26 28	15 15	27 00	09 22	28 05	20 50	16 47	26 03	07 57
16	11 43 30	24 04 15	25 ♉ 54 47	02 ♊ 17 03	14 34	28 13	10 00	28 16	20 49	16 51	26 02	07 56
17	11 47 26	25 02 45	08 ♊ 33 46	14 ♊ 45 31	14 02	29 ♌ 27	10 38	28 26	20 49	16 54	26 02	07 55
18	11 51 23	26 01 18	20 ♊ 52 55	26 ♊ 56 37	13 37	00 ♍ 40	11 16	28 36	20 48	16 58	26 01	07 54
19	11 55 19	26 59 52	02 ♋ 57 17	08 ♋ 55 36	13 22	01 53	11 54	28 46	20 48	17 02	26 01	07 52
20	11 59 16	27 58 29	14 ♋ 52 12	20 ♋ 47 43	13 17	03 07	12 32	28 55	20 47	17 06	26 00	07 51
21	12 03 13	28 57 08	26 ♋ 42 44	02 ♌ 37 50	13 D 21	04 21	13 11	29 05	20 47	17 09	25 59	07 50
22	12 07 09	29 ♍ 55 49	08 ♌ 33 30	14 ♌ 30 14	13 34	05 34	13 49	29 15	20 46	17 13	25 58	07 49
23	12 11 06	00 ♎ 54 32	20 ♌ 28 26	26 ♌ 28 27	13 58	06 48	14 27	29 24	20 46	17 17	25 58	07 48
24	12 15 02	01 53 18	02 ♍ 30 36	08 ♍ 35 07	14 30	08 02	15 05	29 33	20 46	17 21	25 57	07 47
25	12 18 59	02 52 05	14 ♍ 42 12	20 ♍ 51 59	15 12	09 15	15 43	29 42	20 D 46	17 24	25 56	07 45
26	12 22 55	03 50 55	27 ♍ 04 33	03 ♎ 19 59	16 02	10 29	16 21	29 ♋ 52	20 46	17 28	25 55	07 44
27	12 26 52	04 49 46	09 ♎ 38 16	15 ♎ 59 26	17 00	11 43	16 59	00 ♌ 01	20 47	17 32	25 54	07 43
28	12 30 48	05 48 40	22 ♎ 23 27	28 ♎ 50 18	18 05	12 57	17 37	00 09	20 47	17 35	25 53	07 42
29	12 34 45	06 47 35	05 ♏ 19 58	11 ♏ 52 28	19 17	14 11	18 15	00 18	20 47	17 39	25 52	07 41
30	12 38 42	07 ♎ 46 33	18 ♏ 27 47	25 ♏ 05 59	20 ♍ 35	15 ♍ 25	18 ♍ 54	00 ♌ 27	20 ♑ 48	17 ♍ 42	25 ♉ 51	07 ♓ 40

Date	Moon True ☊ ° '	Moon Mean ☊ ° '	Moon ☽ Latitude ° '
01	03 ♐ 59	04 ♐ 25	03 S 08
02	03 R 54	04 21	02 10
03	03 52	04 18	01 S 05
04	03 D 52	04 15	00 N 05
05	03 R 52	04 12	01 16
06	03 51	04 09	02 25
07	03 48	04 06	03 26
08	03 43	04 02	04 15
09	03 35	03 59	04 48
10	03 25	03 56	05 01
11	03 14	03 53	04 54
12	03 04	03 50	04 27
13	02 55	03 47	03 43
14	02 49	03 43	02 47
15	02 45	03 40	01 43
16	02 43	03 37	00 N 36
17	02 D 43	03 34	00 S 31
18	02 43	03 31	01 35
19	02 R 43	03 27	02 34
20	02 42	03 24	03 25
21	02 38	03 21	04 07
22	02 32	03 18	04 38
23	02 23	03 15	04 57
24	02 13	03 12	05 04
25	02 02	03 08	04 56
26	01 51	03 05	04 35
27	01 41	03 02	04 00
28	01 33	02 59	03 13
29	01 28	02 56	02 15
30	01 ♐ 25	02 ♐ 53	01 S 09

DECLINATIONS

Date	Sun ☉ ° '	Moon ☽ ° '	Mercury ☿ ° '	Venus ♀ ° '	Mars ♂ ° '	Jupiter ♃ ° '	Saturn ♄ ° '	Uranus ♅ ° '	Neptune ♆ ° '	Pluto ♇ ° '
01	08 N 00	12 S 47	02 S 43	18 N 00	12 N 23	21 N 09	21 S 50	06 N 14	17 N 35	20 S 34
02	07 38	16 21	02 35	17 43	12 10	21 07	21 50	06 13	17 35	20 34
03	07 16	19 10	02 23	17 26	11 56	21 05	21 51	06 11	17 35	20 35
04	06 54	21 00	02 07	17 08	11 42	21 03	21 51	06 10	17 35	20 35
05	06 32	21 40	01 48	16 49	11 28	21 01	21 51	06 08	17 35	20 36
06	06 09	20 59	01 25	16 30	11 15	20 59	21 52	06 07	17 35	20 36
07	05 47	18 58	00 58	16 11	11 01	20 58	21 52	06 05	17 35	20 37
08	05 24	15 40	00 S 29	15 51	10 47	20 56	21 52	06 04	17 34	20 37
09	05 02	11 21	00 N 03	15 31	10 33	20 54	21 53	06 02	17 34	20 38
10	04 39	06 18	00 38	15 10	10 19	20 52	21 53	06 01	17 34	20 38
11	04 16	00 S 54	01 14	14 49	10 04	20 50	21 53	05 59	17 34	20 39
12	03 53	04 N 29	01 51	14 27	09 50	20 48	21 53	05 58	17 34	20 39
13	03 30	09 29	02 29	14 05	09 36	20 46	21 54	05 56	17 34	20 40
14	03 07	13 51	03 06	13 43	09 22	20 44	21 54	05 55	17 33	20 40
15	02 44	17 20	03 41	13 20	09 07	20 42	21 54	05 53	17 33	20 40
16	02 21	19 49	04 15	12 57	08 53	20 40	21 54	05 52	17 33	20 41
17	01 58	21 13	04 45	12 33	08 38	20 38	21 54	05 51	17 33	20 41
18	01 35	21 32	05 13	12 09	08 24	20 36	21 55	05 49	17 33	20 42
19	01 12	20 50	05 37	11 45	08 09	20 34	21 55	05 48	17 33	20 42
20	00 48	19 13	05 57	11 20	07 55	20 32	21 55	05 46	17 32	20 42
21	00 25	16 46	06 12	10 55	07 40	20 31	21 55	05 45	17 32	20 43
22	00 N 02	13 38	06 23	10 30	07 25	20 29	21 55	05 43	17 32	20 43
23	00 S 22	09 57	06 30	10 05	07 11	20 27	21 55	05 42	17 32	20 44
24	00 45	05 51	06 31	09 39	06 56	20 25	21 55	05 40	17 31	20 44
25	01 08	01 N 28	06 28	09 12	06 41	20 23	21 55	05 39	17 31	20 44
26	01 32	03 S 03	06 21	08 46	06 26	20 21	21 55	05 38	17 31	20 45
27	01 55	07 30	06 09	08 19	06 11	20 20	21 56	05 36	17 31	20 45
28	02 18	11 42	05 53	07 52	05 56	20 18	21 56	05 35	17 30	20 45
29	02 42	15 25	05 33	07 25	05 41	20 16	21 56	05 33	17 30	20 46
30	03 S 05	18 S 25	05 N 10	06 N 58	05 N 26	20 N 14	21 S 56	05 N 32	17 N 30	20 S 46

ZODIAC SIGN ENTRIES

Date	h m	Planets
01	20 23	☽ ♏
04	03 26	☽ ♐
06	07 49	☽ ♑
08	09 37	☽ ♒
10	09 51	☽ ♓
12	10 21	☽ ♈
14	13 07	☽ ♉
16	19 41	☽ ♊
17	22 56	♀ ♍
19	06 05	☽ ♋
21	18 40	☽ ♌
22	13 43	☉ ♎
24	07 01	☽ ♍
26	17 37	☽ ♎
27	10 28	♃ ♌
29	02 09	☽ ♏

LATITUDES

Date	Mercury ☿ ° '	Venus ♀ ° '	Mars ♂ ° '	Jupiter ♃ ° '	Saturn ♄ ° '	Uranus ♅ ° '	Neptune ♆ ° '	Pluto ♇ ° '
01	04 S 22	00 N 17	01 N 09	00 N 08	00 S 04	00 N 44	01 S 45	13 S 05
04	04 24	00 25	01 09	00 09	00 04	00 44	01 45	13 05
07	04 11	00 33	01 09	00 09	00 05	00 44	01 45	13 05
10	03 41	00 41	01 09	00 09	00 05	00 44	01 45	13 05
13	02 55	00 48	01 09	00 10	00 05	00 44	01 45	13 05
16	01 59	00 55	01 09	00 10	00 05	00 44	01 45	13 05
19	01 00	01 01	01 09	00 10	00 05	00 44	01 45	13 05
22	00 S 04	01 07	01 09	00 11	00 06	00 44	01 46	13 04
25	00 N 42	01 12	01 09	00 11	00 06	00 44	01 46	13 04
28	01 17	01 17	01 08	00 12	00 06	00 44	01 46	13 04
31	01 N 40	01 N 21	01 N 08	00 N 12	00 S 06	00 N 44	01 S 46	13 S 03

DATA

Julian Date	2469686
Delta T	+80 seconds
Ayanamsa	24° 33' 20"
Synetic vernal point	04° ♓ 33' 39"
True obliquity of ecliptic	23° 25' 55"

LONGITUDES

Date	Chiron ⚷ ° '	Ceres ⚳ ° '	Pallas ⚴ ° '	Juno ⚵ ° '	Vesta ⚶ ° '	Black Moon Lilith ⚸ ° '
01	22 ♏ 46	06 ♋ 12	17 ♈ 20	23 ♋ 34	07 ♌ 09	04 ♌ 14
11	23 ♏ 31	09 ♋ 26	15 ♈ 32	28 ♋ 43	11 ♌ 23	05 ♌ 21
21	24 ♏ 24	12 ♋ 26	13 ♈ 10	03 ♌ 41	15 ♌ 32	06 ♌ 28
31	25 ♏ 24	15 ♋ 10	10 ♈ 25	08 ♌ 26	19 ♌ 36	07 ♌ 35

MOON'S PHASES, APSIDES AND POSITIONS ☽

Date	h m	Phase	Longitude ° '	Eclipse Indicator
05	02 28	☽	13 ♐ 00	
11	17 04	○	19 ♓ 25	
18	23 04	☾	26 ♊ 28	
27	02 05	●	04 ♎ 25	

Day	h m		
10	00 18	Perigee	
22	00 13	Apogee	
05	12 06	Max dec	21° S 40'
11	15 58	ON	
18	07 21	Max dec	21° N 33'
25	19 48	OS	

ASPECTARIAN

h m	Aspects
01 Wednesday	
01 46	☽ ± ♆
03 49	☽ □ ♄
04 37	☽ Q ♀
05 06	☽ ⊥ ♅
07 41	☽ ℙ ♇
09 41	☽ # ♂
09 51	♀ ⊥ ♅
09 53	☽ ∠ ☉
12 02	☽ □ ♃
13 07	☽ ⊼ ♆
14 11	☽ ⊻ ☿
21 45	☽ ⚹ ♂
22 07	☽ ∠ ♅
02 Thursday	
01 05	☽ ⊥ ☿
11 43	☽ △ ♇
13 23	☿ ∠ ♀
13 27	☽ Q ♄
16 12	☽ ⚹ ☉
17 14	☽ ∠ ☿
17 47	☽ □ ♀
20 31	☉ ∠ ♃
20 48	☿ △ ♆
21 11	☽ Q ♂
21 38	☽ # ♆
21 48	☽ # ♀
23 25	♀ ∥ ♆
03 Friday	
02 03	☽ ⚹ ♅
07 00	☿ ⚹ ♃
11 25	☽ ⚹ ♄
15 54	☽ Q ☉
19 31	☽ ⚹ ☿
20 15	☽ △ ♃
20 28	☽ ☍ ♆
22 41	♂ ⊥ ♃
23 57	☽ Q ♅
04 Saturday	
05 02	☽ ∥ ♇
07 28	☽ □ ♂
11 30	♃ ⚹ ♆
12 56	☽ # ♃
14 18	☽ ∠ ♄
15 59	☽ Q ☿
17 59	☽ □ ♇
23 21	☽ ℙ ♃
05 Sunday	
02 28	☽ □ ☉
05 05	☽ △ ♀
06 06	☽ ⊥ ♄
07 58	☽ □ ♅
15 15	☽ ± ♃
16 30	☽ ⊻ ♄
21 46	☽ □ ☿
06 Monday	
01 12	☽ ⊼ ♆
01 15	☽ Q ♇
01 42	☽ ⊼ ♃
09 23	☽ ℙ ♀
11 24	☽ ± ♆
11 58	☽ # ♃
14 05	☽ △ ♂
15 14	☉ ∥ ♅
15 27	♀ ⊻ ♅
18 00	☽ ∥ ♇
21 33	☽ ⚹ ♇
07 Tuesday	
01 56	☽ ± ♀
02 32	☽ ℙ ♆
06 14	☿ ⊥ ♀
09 21	☽ △ ☉
11 07	☽ △ ♅
12 49	☽ ⊼ ♀
16 16	☽ ℙ ♂
18 54	☽ ☌ ♄
21 12	☽ △ ☿
22 22	☽ ∠ ♇
23 11	☽ # ♆
08 Wednesday	
03 16	☽ △ ♆
04 24	☽ ☍ ♃
07 40	☽ ± ♂
10 45	☽ # ♀
11 41	☽ ℙ ☉
11 47	☽ ℙ ♅
13 01	☽ ⊥ ♇
13 22	☉ ☌ ♅
17 50	☽ ⊼ ♂
20 07	☽ ℙ ☿
22 42	☽ ⊻ ♇
09 Thursday	
02 24	☽ ± ♅
03 15	☽ ± ☉
09 00	☿ △ ♄
09 43	☽ ± ☿
12 05	☽ ⊼ ♅
13 33	☽ ⊼ ☉
16 12	☽ # ♂
17 49	☽ ☍ ♀
18 44	☽ ⊼ ☿
19 22	☽ ⊻ ♄
10 Friday	
00 24	☿ ⊻ ♀
02 41	♂ ℙ ♄
03 36	☽ □ ♆
04 57	☽ ⊥ ♄
05 20	☽ ⊼ ♃
12 31	♀ ⊼ ♄
13 18	☽ # ♅
15 03	☽ ± ♃
19 21	☽ ∠ ♄
19 59	☽ # ☉
20 07	☽ ☍ ♂
22 45	☽ ☌ ♇
11 Saturday	
05 40	☽ ℙ ♃
07 24	☉ ☌ ☿
08 28	☽ Q ♆
10 40	☽ # ☿
12 25	☽ ☍ ♅
15 52	☽ ☍ ☿
17 04	☽ ☍ ☉
19 30	☽ ⚹ ♄
22 20	☽ ⊼ ♀
22 40	☽ ∥ ☿
12 Sunday	
03 55	☽ ⚹ ♆
06 18	☽ △ ♃
09 01	☽ ± ♀
09 31	☽ ∥ ☉
15 06	☽ Q ♄
18 51	☽ ∥ ♅
23 11	☽ ⊼ ♂
23 36	☽ ⊻ ♇
13 Monday	
01 17	☽ ℙ ♀
04 43	☽ ∠ ♆
05 17	☉ △ ♄
08 21	♂ ☍ ♇
09 38	☽ ⊥ ♇
09 42	☽ ± ♂
12 32	☽ ∥ ♂
14 09	☽ ⊼ ♅
14 24	☽ ⊼ ☿
18 05	☿ ☌ ♅
19 57	☽ ⊥ ♆
21 19	☽ □ ♄
22 32	☽ ⊼ ☉
14 Tuesday	
00 02	☽ ± ☿
00 26	☽ ± ♅
00 57	☽ ∠ ♇
01 46	☽ ℙ ♂
05 10	☽ △ ♀
06 16	☽ ⊻ ♆
09 28	☽ □ ♃
09 41	☽ ± ☉
11 15	☽ ∥ ♀
12 46	☉ ∥ ☿
14 46	☽ ℙ ☿
16 09	☽ ℙ ♅
17 30	♀ □ ♆
15 Wednesday	
02 46	☽ ℙ ☉
03 12	☽ ⚹ ♇
05 24	☽ △ ♂
08 53	♀ ± ♄
13 46	☽ ∥ ♆
16 06	☽ △ ☿
17 57	☽ Q ♃
19 10	☽ △ ♅
16 Thursday	
00 55	☽ Q ♇
02 30	♃ # ♇
02 34	☽ △ ♄
08 17	☽ △ ☉
12 14	☽ ☌ ♆
12 55	♀ ⊻ ♃
16 28	☽ ⚹ ♃
16 47	☽ □ ♀
17 Friday	
00 01	☽ ∥ ♃
00 37	☽ # ♇
06 44	☽ ℙ ♄
10 45	☽ □ ♇
16 13	☽ □ ♂
21 32	☽ ∠ ♃
22 13	☽ □ ☿
18 Saturday	
00 06	☽ ± ♄
04 17	☽ □ ♅
07 09	☽ Q ♀
11 50	☽ ⊼ ♄
11 59	☉ △ ♆
13 26	☿ ∠ ♃
15 26	☽ ⊥ ♃
22 09	☽ ⊻ ♆
23 04	☽ □ ☉
19 Sunday	
03 30	☽ ⊻ ♃
05 34	☽ Q ♂
08 53	☽ Q ☿
09 38	☽ ⚹ ♀
10 07	☽ ⊥ ♆
14 37	☽ # ♇
16 11	☽ Q ♅
17 06	☽ ∥ ♃
21 52	☽ △ ♇
23 21	☿ ∥ ♅
20 Monday	
04 11	☽ ∠ ♆
07 02	☽ ⚹ ♂
08 47	☽ ⚹ ☿
14 09	☿ St D
14 26	☽ Q ☉
16 32	☽ ⚹ ♅
19 20	☽ ∠ ♀
23 58	☽ ☍ ♄
21 Tuesday	
04 09	☽ ℙ ♇
05 12	☽ ∥ ♆
10 32	☽ ⚹ ♆
15 08	☽ ∠ ♂
15 21	☽ ∠ ☿
15 41	☽ ⊥ ♀
15 50	☉ ⚹ ♃
16 52	☽ ☌ ♃
16 57	☽ ⚹ ☉
20 42	☿ ☌ ♂
22 22	☽ ± ♇
23 06	☽ ∠ ♅
22 Wednesday	
04 40	♀ ⊥ ♃
05 15	☽ ⊻ ♀
09 57	☽ ⊥ ☿
10 24	☽ ⊥ ♂
10 30	☽ ⊼ ♇
10 49	☽ Q ♆
16 00	♀ ℙ ♄
17 24	☽ ⊥ ♅
22 25	☽ ⊻ ☿
23 12	☽ ⊻ ♂
23 Thursday	
02 01	☽ ∠ ☉
05 33	☽ ⊻ ♅
09 38	♂ ∠ ♃
11 11	☽ ∥ ♀
12 36	☽ ⊼ ♄
21 40	☽ ⊥ ☉
22 58	☽ □ ♆
24 Friday	
00 35	☽ ± ♄
05 29	☽ ∥ ♂
06 03	☽ ⊻ ♃
07 13	♀ ☍ ♇
08 12	☽ ∥ ☿
10 40	☽ ⊻ ☉
12 59	☽ ∥ ♅
14 23	☿ ∠ ♃
15 35	♄ St D
18 06	☽ ⊥ ♃
18 27	☽ ℙ ♄
22 23	☽ ☍ ♇
25 Saturday	
00 07	☽ ☌ ♀
12 01	☽ ∠ ♃
13 02	☽ ☌ ☿
13 35	☽ # ☉
14 05	☽ ☌ ♂
17 18	☽ ☌ ♅
23 49	☽ △ ♄
26 Sunday	
03 10	☽ ∥ ☉
09 46	☽ △ ♆
17 25	☽ ⚹ ♃
27 Monday	
01 43	☽ # ♅
02 05	☽ ☌ ☉
04 58	☽ # ☿
05 12	☽ # ♂
08 22	☽ ⊼ ♇
10 59	☿ ☌ ♂
14 24	☽ ℙ ♆
16 07	☽ # ♀
16 22	☽ ⊻ ♀
16 33	☽ Q ♃
19 43	☽ ± ♇
28 Tuesday	
00 32	☿ ☌ ♅
02 36	☽ ⊻ ♂
02 57	☽ ⊻ ♅
03 10	☽ ⊻ ☿
04 52	☽ ⊥ ♀
07 19	☽ ± ♆
08 59	☽ □ ♄
10 28	♂ ☌ ♅
12 35	☽ ℙ ♇
14 15	☽ ⊥ ♅
14 25	☽ ⊥ ♂
15 29	☽ ⊥ ☿
18 31	☽ ⊼ ♆
23 26	☽ ∠ ♀
29 Wednesday	
02 36	☽ □ ♃
07 01	☽ ∠ ♅
07 59	☽ ∠ ♂
09 52	☽ ∠ ☿
11 42	☿ ∥ ♅
14 54	☽ ⊻ ☉
16 19	☽ △ ♇
18 21	☽ Q ♄
30 Thursday	
02 05	♂ ∥ ♅
02 47	☽ ⊥ ☉
03 53	☽ # ♆
05 53	☽ ⚹ ♀
09 22	☉ ⊼ ♇
10 37	☽ ⚹ ♅
12 41	♀ ∠ ♃
12 49	☽ ⚹ ♂
15 59	☿ △ ♄
16 14	☽ ⚹ ♄
16 16	☽ ⚹ ☿
20 26	☽ ∠ ☉

All ephemeris data is given at 12.00 UT and the Moon's longitude is additionally given for 24.00 UT